P9-AGK-602

SITTIG'S HANDBOOK OF TOXIC AND HAZARDOUS CHEMICALS AND CARCINOGENS

Fourth Edition

VOLUME 1

ABOUT THE AUTHOR:

Developed and edited through the first three editions by Marshall Sittig, former Director of Government Relations, Princeton University, this revised edition has been expanded and edited by Richard P. Pohanish who has been active in the environmental health and safety field both as a publisher and author for more than 30 years. He is the former Executive Vice President of Van Nostrand Reinhold, president of VNR Information Systems, and cofounder of ChemTox, Inc. Currently, he is president of Chem-Data Systems and has coauthored more than a dozen professional books and CD-ROM products.

SITTIG'S HANDBOOK OF TOXIC AND HAZARDOUS CHEMICALS AND CARCINOGENS

Fourth Edition

Richard P. Pohanish
Editor

Volume 1
A – H

NOYES PUBLICATIONS
WILLIAM ANDREW PUBLISHING
Norwich, New York, U.S.A.

Library of Congress Catalog Card Number: 2001056289
ISBN 0-8155-1459-X
Printed in the United States of America

Published in the United States of America by
Noyes Publications / William Andrew Publishing
13 Eaton Avenue
Norwich, New York 13815
1-800-932-7045
www.williamandrew.com

10 9 8 7 6 5 4 3 2 1

Library of Congress Cataloging-in-Publication Data

Sittig's handbook of toxic and hazardous chemicals and carcinogens / edited by Richard Pohanish. -- 4th ed.
p. cm.
Rev. ed. of: Handbook of toxic and hazardous chemicals and carcinogens / by Marshall Sittig. 3rd ed. c1991.
Includes index.
Contents: v. 1. A-H -- v. 2. I-Z. Appendices 1.8.
ISBN 0-8155-1459-X (alk. paper)
1. Toxicology--Handbooks, manuals, etc. 2. Industrial toxicology--Handbooks, manuals, etc. 3. Carcinogens--Handbooks, manuals, etc. I. Pohanish, Richard P. II. Sittig, Marshall. Handbook of toxic and hazardous chemicals and carcinogens. III. Title: Handbook of toxic and hazardous chemicals and carcinogens.
RA1215.S58 2002
615.9'02--dc21

2001056289

DEDICATION

Henry A. Suchwarzewsky
He left the world a better place.
October 30, 1907 – December 28, 2000

Contents

C

D

E

F

G

M

N

O

P

T

Preface to the Fourth Edition

Sittig's Handbook of Toxic and Hazardous Chemicals and Carcinogens was first published more than 20 years ago. This work continues to provide occupational and environmental health and safety professionals with an accessible and portable reference source. The handbook contains chemical, health, and safety information on nearly 1,500 toxic and hazardous chemicals (up from nearly 600 in the first edition, nearly 800 in the second edition, and 1,300 in the third edition) so that responsible decisions can be made by all who may have contact with or interest in these chemicals.

The addition of more regulated chemicals and the expansion of data on each material has nearly doubled the size of this fourth edition over its predecessor. Some material from the previous edition has been eliminated or moved to the most appropriate section. All of this has been done to make the work more relevant, more inclusive, and easier to use. The utility of the work has been enhanced by the addition of eight appendices following the main chemical records section.

Appendix 1 lists oxidizing materials.

Appendix 2 contains a list of confirmed and suspected carcinogens.

Appendix 3 is a glossary of chemical, health, safety, medical, and environmental terms used in the handbook.

Appendix 4 is the RTECS number cross index.

Appendix 5 is the DOT ID cross index.

Appendix 6 is the synonym and trade name cross index.

Appendix 7 is the molecular formula cross index.

Appendix 8 is the CAS number cross index.

Following the Introduction is a key to the abbreviations and acronyms used in the handbook.

In keeping with the broad changes initiated with the third edition, the contents of the fourth edition are focused on the concept of "regulated chemicals." This implies recognition by some government agency, or rulemaking body. For example, the "Regulatory Authority" section has been expanded and many of the individual sections or "fields" now contain useful advice sought after by the regulated community. As a result the new volume should be even more useful to handlers and users of these chemicals and to those concerned with both adherence to, and enforcement of, regulations.

These are all then officially recognized substances, defined as carcinogens or as belonging to some designated category of hazardous or toxic materials or those with numerically-defined safe limits in air or water in the workplace, in waste effluents, or in the ambient air or water. For the most part these are materials of commerce that are heavily-used and possibly transported in bulk, and not laboratory curiosities or research chemicals encountered only in very small quantities.

Data is furnished, to the extent currently available, in a multi-section uniform format to make it easy for users who must find information quickly or compare data within various records in any or all of these important categories:

- Chemical Description
- Code Numbers
- Synonyms
- Potential Exposure
- Incompatibilities
- Permissible Exposure Limits in Air
- Determination in Air
- Permissible Concentration in Water
- Determination in Water
- Routes of Entry
- Short Term Exposure
- Long Term Exposure
- Points of Attack
- Medical Surveillance
- First Aid
- Personal Protective Methods
- Respirator Selection
- Storage
- Shipping
- Spill Handling
- Fire Extinguishing
- Disposal Method Suggested
- References

New and additional information has been included in the fourth edition in the following sections: Synonyms, EEC Number (new), EINECS Number (new), Regulatory Authority, Description, Incompatibilities, Short Term Exposure, Long Term Exposure, First Aid, Personal Protective Methods, Respirator Selection, Storage, Shipping, Spill Handling, Fire Extinguishing, Disposal. Specifically, additions include more regulatory information, identifiers, chemical and physical properties including flash points, explosive limits, water solubility, and odor thresholds, hazard ratings, DOT isolation and protective distances, and full text of NIOSH respirator recommendations. Many records now contain special warnings, notes to EMS personnel and other health care professionals, and occupational analytical methods from NIOSH and OSHA.

Although every effort has been made to produce an accurate and highly useful handbook, the author appreciates the need for constant improvement. Any comments, corrections, or advice from users of this book are welcomed by the author who asks that all correspondence be submitted in writing and mailed to the publisher who maintains a file for reprints and future editions.

Foreword to the Third Edition

In April 1980, an explosive fire broke out at an abandoned dump site in Elizabeth, New Jersey. When thousands of barrels containing unidentified toxic waste erupted in flames, an ominous black cloud filled the sky. No one knew for sure whether lethal fumes would jeopardize the surrounding area–one of America's most densely populated. Fortunately, nobody was hurt seriously, perhaps in part because prior to the fire, more than 500 pounds of such dangerous agents as nitroglycerine, picric acid, and mustard gas were removed by state environmental protection personnel.

That near-disaster gave the nation fresh evidence of the urgent need to control the disposal of toxic wastes. In New Jersey alone, there are 100 toxic waste sites on the Environmental Protection Agency's national list of the 951 worst sites. That is more than any other state.

Toxic wastes are a serious problem in our country. They threaten our water, our air, and our health. We must find better ways to handle toxic chemicals and dispose of our toxic wastes. Over the last decade, we have finally recognized the hazards posed by years of irresponsible disposal of wastes. Unfortunately, our belated response has often been to push pollution around from one place to another. When we recognized the risks of air pollution, we devised systems to prevent the discharge of particulates. The systems gathered soot and sludge, so we took that soot and sludge and dumped them in the ocean. When we became concerned about pollution in the ocean, we ordered the wastes buried in the ground. Now we have learned that what is buried in the ground can seep into our drinking water and poison it and us.

"Out of sight, out of mind" is no longer an acceptable way of handling toxic wastes. The public consciousness now has recognized that these wastes must be disposed of properly. Public officials, scientists, industry, and all interested parties now must consider the implications of eliminating and containing toxic waste rather than simply pushing it around.

In response to public concern, Congress has initiated numerous legislative measures. I was a strong supporter of the Superfund legislation. which created a fund to clean up existing hazardous waste disposal sites and deal with emergency situations in the future. This was a major step forward in providing the funding needed to attack the problem.

The intent of legislation such as Superfund is not to restrict unnecessarily the use of chemicals. Modern science has come to depend on them for scientific and technological advancement as well as for our health. But the increase in production has led to large-scale use of some chemicals known to have adverse health effects and others whose effects, especially in the long run, are uncertain. The recent development of a vast number of new chemicals has also added greatly to the difficulty of deciding which chemicals to regulate and of establishing safeguards for their manufacture, use, and disposal. Sometimes the effects of human exposure to chemicals are immediate, but more often years of research are required to determine whether a particular substance is hazardous. Appropriate management of the risks we face calls for a difficult but necessary balancing act.

In addition to environmental concerns, we must prevent the dangerous levels of exposure which can threaten the health of workers. Many people are harmed as a result of physical and chemical hazards at work, and the long-term effects of certain occupational conditions are still uncertain.

Protecting against potential public health hazards requires widespread knowledge about commercial chemicals-their mixtures, by-products, and uses. We must know more about their persistence and fate in the environment, the effects they will have, and most importantly, the ways in which we can minimize the risks they pose.

For these reasons, Marshall Sittig's book is an important addition to the literature on toxic chemicals. It brings together information contained in hundreds of government publications, with particular attention to the identification of carcinogenic materials. References at the end of many entries provide useful bibliographies listing thousands of original publications which describe the effect of toxic and hazardous chemicals on man and environment. This comprehensive information makes it a valuable desk reference.

In short, Marshall Sittig has made it easier for us to understand how to control the handling of toxic chemicals and the disposal of toxic wastes. I hope this understanding will produce more effective action.

June 1991

Bill Bradley
United States Senator, NJ

Preface to the Third Edition

This handbook presents concise chemical, health and safety information on some 1,300 toxic and hazardous chemicals (up from nearly 600 in the first edition and nearly 800 in the second edition) so that responsible decisions can be made by chemical manufactures, safety equipment producers, toxicologists, industrial safety engineers, waste disposal operators, health care professionals, and the many others who may have contact with or interest in these chemicals due to their own or third party exposure.

In this Third Edition, some less critical material from the Second Edition has been deleted and much new material has been added, all in an attempt to make the volume more relevant and more inclusive yet more concise.

The contents have been changed to incorporate the concept of "regulated chemicals." This implies recognition by some rulemaking body. It should make the new volume even more useful to handlers of these chemicals and to those concerned with both adherence to regulations and enforcement of regulations.

The volume now contains only important toxic materials whose inclusion is defined by official recognition by state or federal governments or by supra-governmental agencies such as the United Nations or ACGIH or the German Research Society (DFG). Such recognition usually implies that the material in question is a material of commerce and not just a laboratory curiosity encountered only in gram quantities.

These are all then officially recognized substances, defined as carcinogens or as belonging to some designated category of hazardous or toxic materials or those with numerically-defined safe limits in air or water in the workplace, in waste effluents, or in the ambient air or water.

Additional information, over that presented in the second edition, has been added in the third edition, where available in the following categories: Storage, Shipping, Spill Handling and Fire Extinguishing. In addition, information on flash points, explosive limits and odor thresholds has been added to round out inclusiveness in the spirit of the "material safety data sheets" now required by many states in the U.S.A. as well as in Canada. Substances included in this book are almost all of:

> The substances whose allowable concentrations in workplace air are adopted or proposed by ACGIH (1).

The substances whose allowable concentrations in air have been adopted by the Deutsche Forschungsgemeinschaft (German Research Society) (3).

The substances whose allowable concentrations in workplace air, in residential air and in water have been set forth by the USSR-UNEP/IRPTC Project (43).

The substances whose allowable concentrations in air have been set forth in the U.K. by the Health and Safety Executive (33).

The substances set forth by the Occupational Safety and Health Administration (OSHA) of the U.S. Department of Labor in a completely revised document (58) issued in final form in 1989.

The substances whose allowable concentrations in air and other safety considerations have been considered by NIOSH (2).

The chemicals reviewed in EPA "CHIPS" documents (see references in individual substance entries).

The chemicals reviewed in NIOSH Information Profiles (see references in individual substance entries).

The chemicals designated by EPA as "hazardous substances" (4).

All of the chemicals designated by EPA as "hazardous wastes" under RCRA (5).

All of the chemicals designated by EPA as "priority toxic pollutants" (6).

All of the chemicals designated by EPA as "extremely hazardous substances" under S.A.R.A. (7). Particular attention has been given to the "top 100" on that list (39).

All of the chemicals designated by EPA as "toxic chemicals" (8).

The carcinogens identified by the U.S. National Toxicology Program (10).

Most of the chemicals described in the ILO *Encyclopedia of Occupational Health and Safety* (1983) (30).

Most of the chemicals in the United Nations' *IRPTC Legal File* (35).

Most of the chemicals described in the journal publication: *Dangerous Properties of Industrial Materials Report.*

Many of the chemicals cited in "right-to-know" legislation of one or more of the 50 states in the U.S.A.

The necessity for informed handling and controlled disposal of hazardous and toxic materials has been spotlighted over and over in recent days as news of fires and explosions at factories and waste sites and groundwater contamination near dump sites has been widely publicized. In late 1980 the EPA imposed long-delayed regulations governing the handling of hazardous wastes — from creation to disposal. Prerequisite to control of hazardous substances, however, is knowledge of the extent of possible danger and toxic effects posed by any particular chemical. This book provides the prerequisites.

The 1984 tragedy at Bhopal, India involving methyl isocyanate (see entry on page 1122) may well stimulate the enactment of "right-to-know" legislation requiring that workers be furnished data akin to the entries in this volume on chemicals they encounter in the workplace.

The chemicals are presented alphabetically and each is classified as a "carcinogen," "hazardous substance," "hazardous waste," and/or a "priority toxic pollutant" — as defined by the various federal agencies, and explained in the comprehensive introduction to the book.

Particular attention is given in this third edition to delineation of the identity and properties of those chemicals now known to be carcinogens.

Data is furnished, to the extent currently available, on any or all of these important categories:

Chemical Description
Code Numbers
Synonyms
Potential Exposure
Incompatibilities
Permissible Exposure Limits in Air
Determination in Air
Permissible Concentration in Water
Determination in Water
Routes of Entry
Harmful Effects and Symptoms
Points of Attack
Medical Surveillance
First Aid
Personal Protective Methods
Respirator Selection
Storage
Shipping
Spill Handling
Fire Extinguishing
Disposal Method Suggested
References

Essentially the book attempts to answer seven questions about each compound (to the extent information is available):

(1) What is it?
(2) Where do you encounter it?
(3) How much can one tolerate?
(4) How does one measure it?
(5) What are its harmful effects?
(6) How does one protect against it?
(7) How does one handle it and protect against mishaps?
(8) Where can I learn more?

Under category (8), "Where can I learn more?" this third edition provides hundreds of new citations to secondary reference sources which in turn provide access to thousands of references on properties and toxicology and safe handling of all the compounds listed.

An outstanding and noteworthy feature of this book is the Index of Carcinogens. There were 92 listed in the first edition, 178 in the second edition and there are 265 in this edition.

This book will thus be a valuable addition to industrial and medical and legal libraries.

Advanced composition and production methods developed by Noyes Publications are employed to bring this durably bound book to you in a minimum of time. Special techniques are used to close the gap between "manuscript" and "completed book." Industrial technology is progressing so rapidly that time-honored conventional typesetting, binding and shipping methods are no longer suitable. We have bypassed the delays in the conventional book publishing cycle and provide the user with an effective and convenient means of reviewing up-to-date information in depth.

The alphabetical table of contents serves as a subject index and provides easy access to the information contained in the book.

Since the last edition of this volume in 1985, several categories of information covered in that volume have changed radically. These include:

DOT (U.S. Department of Transportation) Designation and accompanying shipping regulations (19). These have changed to conform to a system like that used by the United Nations (21).

Permissible Exposure Limits in Air — As of 1985, European limits had not been promulgated. Now the German Research Society (3) has independently set forth values and other country groups including the Health and Safety Executive (33) in the U.K. have adopted ACGIH standards for their country.

Respirator Selection — The detailed recommendations set forth by NIOSH have been completely revised.

In addition, a careful comparative study of other sources including materials safety data sheets has spurred the author and publisher on to broaden the content of individual entries to include the following: Storage, Shipping, Spill Handling, and Fire Extinguishing.

Acknowledgments

The author would like to thank some individuals and institutions, without whose generous help, the fourth edition would not have been possible. To each I am indebted and offer thanks. At the U.S. EPA: Daniel R. Bushman, Ph.D., Office of Pollution Prevention and Toxics and Maria Doa, Ph.D., Office of Toxic Substances. At the U.S. Coast Guard: Alan Schneider, D.Sc., Marine Technical and Hazardous Materials Division. At the Federal Emergency Management Agency: Brad Pabody, Fire Management Division. At the Agency for Toxic Substances and Disease Registry: Dr. Les Smith, Chief of Toxicology Information. To my friend and oft-time co-author, Stanley A. Greene, who helped gather some of material for the appendices to this handbook. I am grateful to my publisher, Bill Woishnis and his excellent staff including Sasha Gurke, Dudley Kay, Millicent Treloar, and Kathy Breed for their encouragement and help in the preparation of this work. And my wife, Dina, deserves special thanks for her patience and constant support.

Introduction

In the United States, over 60,000 chemicals are currently in commercial use. Each year, over 350 billion pounds of toxic chemicals are manufactured and more than 8 billion pounds of these hazardous materials are transported through populated areas. The average American household generates approximately 15 pounds of hazardous waste per year. Nearly five million poisonings occur in the United States annually, resulting in thousands of deaths. The toxic chemicals problem in the United States and indeed in all the world is frightening with news stories about Love Canal, the Valley of the Drums, the Valley of Death in Brazil, major chemical spills, and terrorist attacks in Tokyo and Matsumoto, Japan and the like. All of these incidents generate emotional responses, often from people uninformed about science or technology. On the other hand, one encounters some industrialists who tell us that toxic chemicals are present in nature and that industrial contributions are just the price we have to pay for progress. Somewhere in between lies the truth — or at least an area in which we can function. Information is vital in a world where virtually every aspect of our lives is touched by chemical hazards. It is the aim of this book to present data on specific industrial chemicals from an unemotional point of view so that decisions can be made by:

- Chemicals manufacturers
- Protective safety equipment producers
- Environmental management
- Toxicologists
- Industrial hygienists
- Industrial safety engineers
- Lawyers
- Doctors
- Analytical chemists
- Industrial waste disposal operators
- Legislators
- Enforcement officials
- First response personnel
- Technical librarians
- The informed public

This book gives the highlights of available data on nearly 1,450 important and/or regulated toxic and hazardous chemicals. Importance is defined by inclusion in official and semiofficial listings found below in the "Regulatory Authority" section. Most of this information found in U.S. government sources has been supplemented by a careful search of publications from the United Kingdom, European, Japanese, and other sources including United Nations and World Health Organization (WHO) publications. This is a handbook that is becoming more encyclopedic in nature. When one looks at most handbooks, one simply expects to find numerical data. Here, we have tried, wherever possible, to provide literature references to review documents which hopefully opens the door to a much broader field of published materials. It is recommended that this book be used as a guide. This book is not meant to be a substitute for workplace hazard communication programs require by regulatory bodies such as OSHA, and/or any other U.S., foreign, or international government agencies. If data are required for legal purposes, the original source documents and appropriate agencies, which are referenced, should be consulted.

In the pages which follow, these categories will be discussed with reference to scope, sources, nomenclature employed, and the like. Omission of a category indicates a lack of available information.

Chemical name: Each record is arranged alphabetically by a chemical name used by regulatory and advisory bodies. In most cases, this is not a product name or trade name.

Formula: Generally, this has been limited to a commonly used one-line empirical or atomic formula. In the *Molecular Formula* field, the Hill system has been used showing number of carbons (if present), number of hydrogen (if present), and then alphabetically by element.

Synonyms: This section contains scientific, product, trade, and other synonym names that are commonly used for each hazardous substance. Some of these names are registered trade names. Some are provided in other major languages other than English including Spanish, French, German, Dutch, Czech, Polish, and Italian. In some cases, "trivial" and nicknames (such as MEK for methyl ethyl ketone) have been included because they are commonly used in general communications and in the workplace. This section is important because the various "regulatory" lists published by federal, state, international, and advisory bodies and agencies do not always use the same name for a specific hazardous substance. Every attempt has been made to ensure the accuracy of the synonyms and trade names found in this volume, but errors are inevitable in compilations of this magnitude. Please note that this volume may not include the names of all products currently in commerce, particularly mixtures, that may contain regulated chemicals.

The synonym index contains all synonym names listed in alphabetical order. It should be noted that organic chemical prefixes and interpolations such as α, β, γ, δ etc.; (*o-*) *ortho-*, (*m-*) *meta-*; (*p-*) *para-*; *sec-* (*secondary-*), *trans-*, *cis-*, (*n-*) *normal-*, etc are not treated as part of the chemical name for the purposes of alphabetization.

CAS Number: The CAS number is a unique identifier assigned to each chemical registered with the Chemical Abstracts Service (CAS) of the American Chemical Society. This number is used to identify chemicals on the basis of their molecular structure. CAS numbers, in the format xxx-xx-x, can be used in conjunction with chemical names for positive identification. CAS numbers should always be used in conjunction with substance names to avoid confusion with like-sounding names, like benzene (71-43-2) and benzine (8032-32-4).

RTECS Number: The RTECS numbers (Registry of Toxic Effects of Chemical Substances) are assigned and published by NIOSH. The RTECS number in the format ABxxxxxxx may be useful for online searching for additional toxicologic information on specific substances. For example, it can be used to provide access to the MEDLARS® computerized literature retrieval services of the National Library of Medicine in Washington, DC.

DOT ID: The DOT hazard ID number is assigned to the substance by the U.S. Department of Transportation (DOT). The DOT ID number format is UN xxxx or NA xxxx. This ID number identifies substances regulated by DOT and must appear on shipping documents, the exterior of packages, and on specified containers. Identification numbers containing a UN prefix are also know as United Nations numbers and are authorized for use with all international shipments of hazardous materials. The "NA" prefix is used for shipments between Canada and the United States only, and may not be used for other international shipments.

EEC Number: The EEC and identification number is used by the European Economic Community.

EINECS Number: An identification number from *"European Inventory of Existing Commercial Chemical Substances,"* published by the European Community, Luxembourg, Brussels.

Use of these identification numbers for hazardous materials will (a) serve to verify descriptions of chemicals; (b) provide for rapid identification of materials when it might be inappropriate or confusing to require the display of lengthy chemical names on vehicles; (c) aid in speeding communication of information on materials from accident scenes and in the receipt of more accurate emergency response information; and (d) provide a means for quick access to immediate emergency response information in the *"North American Emergency Response Guidebook."*[31] In this latter volume, the various compounds have assigned "ID Numbers" or identification numbers which correspond closely (but not always precisely) to the UN listing.[20]

Regulatory Authority: Contains a listing of major regulatory and advisory lists containing the chemical of concern, including OSHA, EPA, DOT, ACGIH, IARC, NTP, WHMIS (Canada). Many law or regulatory references in this work have been abbreviated. For example, Title 40 of the Code of Federal Regulations, Part 261, subpart 32 has been abbreviated as 40CFR261.32. The symbol "§" may be used as well to designate a section or part.

Under the title of each substance, there are designations indicating whether the substance is:

- A carcinogen (the agency making such a determination, the nature of the carcinogenicity — whether human or animal and whether positive or suspected, are given in each case). These are frequently cited by IARC (International Agency for Research on Cancer),[12] DFG (Deutche Forschungsgemeinschaft),[3] NIOSH (U.S. National Institute for Occupational Safety and Health),[58] or the NTP (U.S. National Toxicology Programs).[10] It should be noted that the DFG have designated some substances as carcinogens not so classified by other agencies.
- A banned or severely restricted product as designated by the United Nations[13] or by the U.S. EPA Office of Pesticide Programs under FIFRA (The Federal Insecticide, Fungicide and Rodenticide Act).[14]
- A substance cited by the World Bank.[15]
- A substance with an air pollutant standard set or recommended by OSHA and/or NIOSH,[58] ACGIH,[1] DFG,[3] or HSE.[33] The OSHA limits are the enforceable pre-1989 PELs. The transitional limits that were vacated by court order have not been included. The NIOSH and ACGIH airborne limits are recommendations that do not carry the force of law.
- A substance whose allowable concentrations in workplace air are adopted or proposed by the American Conference of Government Industrial Hygienists,[1] Deutsche Forschungsgemeinschaft (German Research Society).[3] Substances whose allowable concentrations in air and other safety considerations have been considered by OSHA and NIOSH.[2] Substances which have limits set in workplace air, in residential air, in water for domestic purposes or in water for fishery purposes as set forth by the former USSR-UNEP/IRPTC Project.[43]
- Substances that are specifically regulated by OSHA under 29CFR1910.1001 to 29CFR1910.1050
- Highly hazardous chemicals, toxics, and reactives regulated by OSHA's "Process Safety Management of Highly Hazardous Chemicals" under 29CFR1910.119, Appendix A.
- Substances that are Hazardous Air Pollutants (Title I, Part A, Section 112) as amended under 42USC7412. This list provided for regulating at least 189 specific substances

using technology-based standards that employ Maximum Achievable Control Technology (MACT) standards; and, possibly health-based standards if required at a later time. Section 112 of the Clean Air Act (CAA) requires emission control by the EPA on a source-by-source basis. Therefore, the emission of substances on this list does not necessarily mean that a firm is subject to regulation.

- Regulated Toxic Substances and Threshold Quantities for Accidental Release Prevention. These appear as Accidental Release Prevention/Flammable Substances, Clean Air Act (CAA) §112(r), Table 3, TQ (threshold quantity) in pounds and kilograms under 40 CFR68.130. The accidental release prevention regulations applies to stationary sources that have present more than a threshold quantity of a CAA Section 112(r) regulated substance.
- Clean Air Act (CAA) Public Law 101–549, Title VI, "Protection of Stratospheric Ozone," Subpart A, Appendix A-Class I and Appendix B, Class II, Controlled Substances, (CFCs) Ozone depleting substances under 40CFR82.
- Clean Water Act (CWA) Priority toxic water pollutants defined by the U.S. Environmental Protection Agency for 65 pollutants and classes of pollutants which yielded 129 specific substances.[6]
- Chemicals designated by EPA as "Hazardous Substances"[4] under the Clean Water Act (CWA) 40CFR116.4, Table 116.4A.
- Clean Water Act (CWA) Section 311 Hazardous Materials Discharge Reportable Quantities (RQs). This regulation establishes reportable quantities for substances designated as hazardous (see §116.4, above) and sets forth requirements for notification in the event of discharges into navigable waters. Source: 40 CFR117.3, amended at 60FR30937.
- Clean Water Act (CWA) Section 307 List of Toxic Pollutants. Source: 40CFR401.15.
- Clean Water Act (CWA) Section 307 Priority Pollutant List. This list was developed from the List of Toxic Pollutants classes discussed above and includes substances with known toxic effects on human and aquatic life, and those known to be, or suspected of being, carcinogens, mutagens, or teratogens. Source: 40CFR423, Appendix A.
- Clean water Act, Section 313 Water Priority Chemicals. Source: 57FR41331.
- RCRA Maximum Concentration of Contaminants for the Toxicity Characteristic with Regulatory levels in mg/l. Source: 40CFR261.24.
- RCRA Hazardous Constituents. Source: 40CFR261, Appendix VIII. Substances listed in this list have been shown, in scientific studies, to have carcinogenic, mutagenic, teratogenic or toxic effects on humans and other life forms. This list also contains RCRA waste codes. The words, "waste number not listed" appears when a RCRA number is NOT provided in Appendix VIII.

Characteristic Hazardous Wastes	
Ignitability	• A nonaqueous solution containing less than 24% alcohol by volume and having a closed cup flashpoint below 60°C (140°F) using Pensky-Martens tester or equivalent. • An ignitible compressed gas. • A non-liquid capable of burning vigorously when ignited or causes fire by friction, moisture absorption, spontaneous chemical changes at standard pressure and temperature. • An oxidizer. See §261.21.
Corrosivity	• Liquids with a pH equal to or less than 2 or equal to or more than 12.5 or which corrode steel at a rate greater than 6.35 mm (0.25 in) per year @ 55°C (130°F). See §261.22.
Reactivity	• Unstable substances that undergo violent changes without detonating. • Reacts violently with water or other substances to create toxic gases. • Forms potentially explosive mixtures with air. See §261.23.
Toxicity	• A waste that leaches specified amounts of metals, pesticides, or organic chemicals using Toxicity Characteristic Leaching Procedure (TCLP). See §261, Appendix II, and §268, Appendix I.

Listed Hazardous Wastes	
"F" wastes	• Hazardous wastes from nonspecific sources §261.31.
"K" Wastes	• Hazardous wastes from specific sources §261.32.
"U" Wastes	• Hazardous wastes from discarded commercial products, off-specification species, container residues§261.34. Covers some 455 compounds and their salts and some isomers of these compounds.
"P" Wastes	• Acutely hazardous wastes from discarded commercial products, off-specification species, container residues§261.33. Covers some 203 compounds and their salts plus soluble cyanide salts.

Note: If a waste is not found on any of these lists, it may be found on a state hazardous waste list.

- RCRA Maximum Concentration of Contaminants for the Toxicity Characteristic. Source: 40CFR261.24, Table I. These are listed with regulatory level in mg/l and "D" waste numbers representing the broad waste classes of ignitability, corrosivity, and reactivity.

- EPA Hazardous Waste code(s), or RCRA number, appears in its own field. Acute hazardous wastes from commercial chemical products are identified with the prefix "P." Nonacutely hazardous wastes from commercial chemical products are identified with the prefix "U."
- RCRA Universal Treatment Standards. Lists hazardous wastes that are banned from land disposal unless treated to meet standards established by the regulations. Treatment standard levels for wastewater (reported in mg/l) and nonwastewater [reported in mg/kg or mg/l TCLP (Toxicity Characteristic Leachability Procedure)] have been provided. Source: 40CFR268.48 and revision, 61FR15654.
- RCRA Ground Water Monitoring List. Sets standards for owners and operators of hazardous waste treatment, storage, and disposal facilities, and contains test methods suggested by the EPA (see Report SW-846) followed by the Practical Quantitation Limit (PQL) shown in parentheses. The regulation applies only to the listed chemical; and, although both the test methods and PQL are provided, they are *advisory only*. Source: 40CFR264, Appendix IX.
- Safe Drinking Water Act (SDWA) Maximum Contaminant Level Goals (MCLG) for Organic Contaminants. Source: 40CFR141 and 40CFR141.50, amended 57FR31776.
- Maximum Contaminant Levels (MCL) for Organic Contaminants. Source: 40CFR141.61.
- Maximum Contaminant Level Goals (MCLG) for Inorganic Contaminants. Source: 40CFR141.51.
- Maximum Contaminant Levels (MCL) for Inorganic Contaminants. Source: 40CFR141.62.
- Maximum Contaminant Levels for Inorganic Chemicals. The maximum contaminant level for arsenic applies only to community water systems. Compliance with the MCL for arsenic is calculated pursuant to §141.23. Source: 40CFR141.11.
- Secondary Maximum Contaminant Levels (SMCL). Federal advisory standards for the States concerning substances that effect physical characteristics (i.e., smell, taste, color, etc.) of public drinking water systems. Source: 40CFR143.3.
- CERCLA Hazardous Substances ("RQ" Chemicals). From "*Consolidated List of Chemicals Subject to the Emergency Planning and Community Right-to-Know Act (EPCRA) and Section 112(r) of the Clean Air Act, as Amended.*" Source: EPA 550-B-98-017 "Title III List of Lists."
- Releases of CERCLA hazardous substances in quantities equal to or greater than their reportable quantity (RQ), are subject to reporting to the National response Center under CERCLA. Such releases are also subject to state and local reporting under §304 of SARA Title III (EPCRA). CERCLA hazardous substances, and their reportable quantities, are listed in 40CFR302, Table 302.4. RQs are shown in pounds and kilograms for chemicals that are CERCLA hazardous substances. For metals listed under CERCLA (antimony, arsenic, beryllium, cadmium, chromium, copper, lead, nickel, selenium, silver, thallium, and zinc), no reporting of releases of the solid for is required if the diameter of the pieces of solid metal released is 100 micrometers (0.004 inches) or greater. The RQs shown apply to smaller particles.
- EPCRA §302 Extremely Hazardous Substances (EHS). From "*Consolidated List of Chemicals Subject to the Emergency Planning and Community Right-to-Know Act (EPCRA) and Section 112(r) of the Clean Air Act, as Amended.*" Source: EPA document 550-B-98-017 "*Title III List of Lists*" The presence of Extremely Hazardous Substances in quantities in excess of the Threshold Planning Quantity (TPQ), requires certain emergency planning activities to be conducted. The Extremely Hazardous Substances and their TPQs are listed in 40CFR355, Appendices A & B. For chemicals that are solids, there may be two TPQs given (e.g., 500/10,000). In these cases, the lower quantity applies for solids in powder form with particle size less than 100 microns; or, if the substance is in solution or in molten form. Otherwise, the higher quantity (10,000 pounds in the example) TPQ applies.
- EPCRA §304 Reportable Quantities (RQ). In the event of a release or spill exceeding the reportable quantity, facilities are required to notify State emergency response commissions (SERCs) and Local Emergency Planning Committees (LEPCs). From "*Consolidated List of Chemicals Subject to the Emergency Planning and Community Right-to-Know Act (EPCRA) and Section 112(r) of the Clean Air Act, as Amended.*" Source: EPA document 550-B-98-017, "*Title III List of Lists.*"
- EPCRA Section 313 Toxic Chemicals. From "*Consolidated List of Chemicals Subject to the Emergency Planning and Community Right-to-Know Act (EPCRA) and Section 112(r) of the Clean Air Act, as Amended.*" Source: EPA document 550-B-98-017 "*Title III List of Lists.*" Chemicals on this list are reportable under §313 and §6607 of the Pollution Prevention Act. Some chemicals are reportable by category under §313. Category codes needed for reporting are provided for the EPCRA§313 categories. Information and Federal Register references have been provided where a chemical is subject to an administrative stay, and not reportable until further notice.
- From "*Toxic Chemical Release Inventory Reporting Form R and Instructions, Revised March 1996,*" EPA document 745-K-96-001 was used for *de minimis* concentrations, toxic chemical categories. Missing from the category listing was "chlorophenols" which has a reportable *de minimis* concentration of 1.0%.
- Chemicals which EPA has made the subject of Chemical Hazard Information Profiles or "CHIPS" review documents.

- Chemicals which NIOSH has made the subject of "Information Profile" review documents on "Current Intelligence Bulletins."
- Carcinogens identified by the National Toxicology Program of the U.S. Department of Health and Human Services at Research Triangle Park, NC.[10]
- Chemicals that were covered in the periodical "Dangerous Properties of Industrial Materials Report" formerly edited by N. Irving Sax, Richard Lewis, and Jan C. Prager, published by Van Nostrand Reinhold, New York.
- Chemicals described in the 2-volume "Encyclopedia of Occupational Health and Safety" published by the International Labor Office (ILO).[30]
- Most of the chemicals covered in the Legal File published by International Register of Potentially Toxic Chemicals Program (IRPTC) of the United Nations.[35] The reader who is particularly concerned with legal standards (allowable concentration in air, in water or in foods) is advised to check these most recent references because data may exist in this UN publication which has not been quoted *in toto* in this volume because of time and space limitations.
- A substances regulated by EPA[7] under the major environmental laws: Clean Air Act, Clean Water Act, Safe Drinking Water Act, RCRA, CERCLA, EPCRA, etc. A more detailed list appears above. And substance with a environmental standards set by some international bodies including Canada and the former USSR.[43]

If additional guidance or compliance assistance is needed, you are encouraged to use the information resources found in *Appendix 1*

Cited in U.S. State Regulations: A substance defined by one or more of the 50 states in the U.S.A. as being a material of concern either from the standpoint of discharges of materials believed to be carcinogenic, mutagenic, or causes of reproductive toxicity; or possible worker exposure and required availability of information to workers under "right-to-know" laws in the U.S. or Canada. The letter (G) indicates simply a reference to the chemical by name, (A) indicates an air pollutant numerical level cited and (W) a water pollutant numerical level cited.

Description: This section contains a quick summary of physical properties of the substance including state (solid, liquid or gas), color, odor description,* boiling point, freezing/melting point, flash point, autoignition temperature, explosion limits in air, Hazard Identification (based on NFPA-704 M Rating System) in the format, Health x, Flammability x, Reactivity x (see also below for a detailed explanation of the System and Fire Diamond), and solubility or miscibility in water. This section may also contain relevant comments about the substance.

Boiling Point at 1 atm: The value is the temperature of a liquid when its vapor pressure is 1 atm. For example, when water is heated to 100°C (212°F) its vapor pressure rises to 1 atm and the liquid boils. The boiling point at 1 atm indicates whether a liquid will boil and become a gas at any particular temperature and sea-level atmospheric pressure.

Melting/Freezing Point: The melting/freezing point is the temperature at which a solid changes to liquid or a liquid changes to a solid. For example, liquid water changes to solid ice at 0°C (32°F). Some liquids solidify very slowly even when cooled below their melting/freezing point. When liquids are not pure (for example, salt water) their melting/freezing points are lowered slightly.

Flash Point: This is defined as the lowest temperature at which vapors above a volatile combustible substance will ignite in air when exposed to a flame. Depending on the test method used, the values given are either Tag Closed Cup (cc.) (ASTM D56) or Cleveland Open Cup (oc) (ASTM D93). The values, along with those in *Flammable Limits in Air* and *Autoignition Temperature* below, give an indication of the relative flammability of the chemical. In general, the open cup value is slightly higher (perhaps 10 – 15°F higher) than the closed cup value. The flash points of flammable gases are often far below 0° (F or C) and these values are of little practical value, so the term "flammable gas" is often used instead of the flash point value.

Autoignition Temperature: This is the minimum temperature at which the material will ignite without a spark or flame being present. Values given are only approximate and may change substantially with changes in geometry, gas, or vapor concentrations, presence of catalysts, or other factors.

Flammable Limits in Air: The percent concentration in air (by volume) is given for the LEL (lower explosive-flammable-limit in air, % by volume) and UEL (upper explosive-flammable-limit in air, % by volume), at room temperature, unless other specified. The values, along with those in Flash Point and Autoignition Temperature give an indication of the relative flammability of the chemical.

NFPA Hazard Classifications: The NFPA 704 Hazard Ratings (Classifications) are reprinted with permission from "Fire Protection Guide to Hazardous Materials," 11th edition, National Fire Protection Association, Quincy, MA, ©1994. The classifications are defined in Table 1 below.

* *odor threshold:* This is the lowest concentration in air that most humans can detect by smell. Some values ranges are reported. The value cannot be relied on to prevent over-exposure, because human sensitivity to odors varies over wide limits, some chemicals cannot be smelled at toxic concentrations, odors can be masked by other odors, and some compounds rapidly deaden the sense of smell.

Table 1. Explanation of NFPA Hazard Classifications

HEALTH HAZARD (blue)

Classification	*Definition*
4	Materials which on very short exposure could cause death or major residual injury (even though prompt medical treatment were given), including those that are too dangerous to be approached without specialized protective equipment.
3	Materials which on short exposure could cause serious temporary or residual injury (even though prompt medical treatment were given), including those requiring protection from all bodily contact.
2	Materials that, on intense or continued (but not chronic) exposure, could cause temporary incapacitation or possible residual injury, including those requiring the use of protective clothing that has an independent air supply.
1	Materials which on exposure would cause irritation but only minor residual injury including those requiring the use of an approved air-purifying respirator.
0	Materials that, on exposure under fire conditions offer no hazard beyond that of ordinary combustible material.

FLAMMABILITY (red)

Classification	*Definition*
4	This degree includes flammable gases, pyrophoric liquids, and Class IA flammable liquids. Materials which will rapidly or completely vaporize at atmospheric pressure and normal ambient temperature, or which are readily dispersed in air and which will burn readily.
3	Includes Class IB and IC flammable liquids and materials that can be easily ignited under almost all normal temperature conditions.
2	Materials that must be moderately heated before ignition will occur and includes Class II and Class IIIA combustible liquids and solids and semi-solids that readily give off ignitable vapors.
1	Materials that must be preheated before ignition will occur, such as Class IIIB combustible liquids, and solids and semi-solids whose flash point exceeds 200°F (93.4°C), as well as most ordinary combustible materials.
0	Materials that will not burn.

REACTIVITY (yellow)

Classification	*Definition*
4	Materials that, in themselves, are readily capable of detonation, explosive decomposition or explosive reaction at normal temperatures and pressures.
3	Materials that, in themselves, are capable of detonation, explosive reaction or explosive reaction, but require a strong initiating source or heating under confinement. This includes materials that are sensitive to thermal and mechanical shock at elevated temperatures and pressures and materials that react explosively with water.
2	Materials that are normally unstable and readily undergo violent chemical change, but are not capable of detonation. This includes materials that can undergo chemical change with rapid release of energy at normal temperatures and pressures. This also includes materials that may react violently with water or that may form potentially explosive mixtures in water.
1	Materials that are normally stable, but that may become unstable at elevated temperatures and pressures and materials that will react with water with some release of energy, but not violently.
0	Materials that are normally stable, even under fire exposure conditions, and that do not reactive with water.

OTHER (white)

Classification	*Definition*
~~W~~	Materials which react so violently with water that a possible hazard results when they come in contact with water, as in a fire situation. Similar to Reactivity Classification 2.
Oxy	Oxidizing material; any solid or liquid that readily yields oxygen or other oxidizing gas, or that readily reacts to oxidize combustible materials.

It should be noted that OSHA and DOT have differing definitions for the term "flammable liquid" and "combustible liquid." DOT defines a flammable liquid as one which, under specified procedures, has a flashpoint of 140°F (60°C) or less. A combustible liquid is defined as "having a flashpoint above 140°F (60°C) and below 200°F (93°C)." OSHA defines a combustible liquid as having a flash point above 100°F (37.7°C).

Potential Exposure: A brief indication is given of the nature of exposure to each compound in the industrial environment. Where pertinent, some indications are given of background concentration and occurrence from other then industrial discharges such as water purification plants. Obviously in a volume of this size, this coverage must be very brief. It is of course recognized that non-occupational exposures may be important as well.

Incompatibilities: Important, potentially hazardous incompatibilities of each substance are listed where available. Where a hazard with water exists, it is described. Reactivity with other materials are described including structural materials such as metal, wood, plastics, cement, and glass. The nature of the hazard, such as severe corrosion formation of a flammable gas, is described. This list is by no means complete or all inclusive. In some cases a very small quantity of material can act as a catalyst and produce violent reactions such as polymerization, disassociation and condensation. Some chemicals can undergo rapid polymerization to form sticky, resinous materials, with the liberation of much heat. The containers may explode. For these chemicals the conditions under which the reaction can occur are given.

Permissible Exposure Limits in Air: The permissible exposure limit (PEL), has been cited as the Federal Standard where one exists. Inasmuch as OSHA has made the decision to enforce only pre-1989 PELs, we decided to use these values rather than the transitional limits that were vacated by court order. Except where otherwise noted, the PELs are 8-hour work-shift time-weighted average (TWA) levels. Ceiling limits, Short Term Exposure Limits (STEL), and TWAs that are averaged over other than full work-shifts are noted.

The short-term exposure limit (STEL) values are derived from NIOSH,[58] ACGIH,[1] and HSE[33] publications. This value is the maximal concentration to which workers can be exposed for a period up to 15 minutes continuously without suffering from: irritation; chronic or irreversible tissue change; or narcosis of sufficient degree to increase accident proneness, impair self-rescue, or materially reduce work efficiency, provided that no more than four excursions per day are permitted, with at least 60 minutes between exposure periods, and provided that the daily TWA also is not exceeded.

The "Immediately Dangerous to Life or Health" (IDLH) concentration represents a maximum level from which one could escape within 30 minutes without any escape-impairing symptoms or any irreversible health effects. However, the 30-minute period is meant to represent a MARGIN OF SAFETY and is NOT meant to imply that any person should stay in the work environments any longer than necessary. In fact, every effort should be made to exit immediately. The concentrations are reported in either parts per million (ppm) or milligrams per cubic meter (mg/m^3).

Most U.S. specifications on permissible exposure limits in air have come from ACGIH[1] or NIOSH.[2] In the U.K. the Health and Safety Executive has set forth Occupational Exposure Limits.[33] In Germany the DFG has established Maximum Concentrations in the workplace[3] and the former USSR-UNEP/IRPTC project has set maximum allowable concentrations and tentative safe exposure levels of harmful substance in workplace air and residential air for many substances.[43] This section also contains numerical values for allowable limits of various materials in ambient air[60] as assembled by the U.S. EPA.

Where available, this field contains legally enforceable airborne Permissible Exposure Limits (PELs) from OSHA. It also contains recommended airborne exposure limits from NIOSH, ACGIH, and international sources and special warnings when a chemical substance is a Special Health Hazard Substance. Each are described below.

TLVs have not been developed as legal standards and the ACGIH does not advocate their use as such. The TLV is defined as the time weighted average (TWA) concentration for a normal 8-hour workday and a 40-hour workweek, to which nearly all workers may be repeatedly exposed, day after day, without adverse effects. A ceiling value (TLV-C) is the concentration that should not be exceeded during any part of the working exposure. If instantaneous monitoring is not feasible, then the TLV-C can be assessed by sampling over a 15-minute periods except for those substances that may cause immediate irritation when exposures are short. As some people become ill after exposure to concentrations lower than the exposure limits, this value cannot be used to define exactly what is a "safe" or "dangerous" concentration. ACGIH threshold limit values (TLVs) are reprinted with permission of the American Conference of Governmental Industrial Hygienists, Inc., from the booklet entitled, "*Threshold Limit Values for Chemical Substances and Physical Agents and Biological Exposure Indices.*" This booklet is revise on an annual basis. No entry appears when the chemical is a mixture; it is possible to calculate the TLV for a mixture only when the TLV for each component of the mixture is known and the composition of the mixture by weight is also known. According to ACGIH, "Documentation of the Threshold Limit Values and Biological Exposure Indices, 6th Edition" is necessary to fully interpret and implement the TLVs.

OSHA permissible exposure limits (PELs), are found in Tables Z-1, Z-2, and Z-3 of OSHA General Industry Air Contaminants Standard (29CFR1910.1000) that were effective on July 1, 2001 and which are currently enforced by OSHA.

Unless otherwise noted, PELs are the time weighted average (TWA) concentrations that must not be exceeded during any 8-hour shift of a 40-hour workweek. An OSHA ceiling concentration must not be exceeded during any part of the workday; if instantaneous monitoring is not feasible, the ceiling must be assessed as a 15-minute TWA exposure. In addition there are a number of substances from Table Z-2 that have PEL ceiling values that must not be exceeded except for a maximum peak over a specified period (e.g., a 5-minute maximum peak in any 2 hours).

NIOSH Recommended Exposure Limits (RELs) are time weighted average (TWA) concentrations for up to a 10-hour work day during a 40-hour work week. A ceiling REL should not be exceeded at any time. Exposure limits are usually expressed in units of parts per million (ppm) — i.e., the parts of vapor (gas) per million parts of contaminated air by volume at 25°C (77°F) and one atmosphere pressure. For a chemical that forms a fine mist or dust, the concentration is given in milligrams per cubic meter (mg/m^3).

Short-Term Exposure Limits (15 minute TWA): This field contains Short Term Exposure Limits (STELs) from ACGIH, NIOSH and OSHA. The parts of vapor (gas per million parts of contaminated air by volume at 25°C (77°F) and one atmosphere pressure is given. The limits are given in milligrams per cubic meter (mg/m^3) for chemicals that can form a fine mist or dust. Unless otherwise specified, the STEL is a 15-minute TWA exposure that should not be exceeded at any time during the workday.

Determination in Air: The citations to analytical methods are drawn from various sources, such as the *NIOSH Manual of Analytical Methods.*[18] In addition, methods have been cited in the latest US Department of Health and Human Services publications including the *"NIOSH Pocket Guide to Chemical Hazards"* published June 1997.[2]

Permissible Concentrations in Water: The permissible concentrations in water are drawn from various sources also, including:

- The National Academy of Sciences/National Research Council publication. *Drinking Water and Health* published in 1980.[16]
- The priority toxic pollutant criteria published by U.S. EPA in draft form in 1979 and in final form in 1980.[6]
- The multimedia environmental goals for environmental assessment study conducted by EPA.[32] Values are cited from this source when not available from other sources.

The U.S. EPA has come forth with a variety of allowable concentration levels:

(1) For allowable concentrations in "California List" wastes.[38] The California list consists of liquid hazardous wastes containing certain metals, free cyanides, polychlorinated biphenyls (PCBs), corrosives with a pH of less than or equal to 2.0, and liquid and nonliquid hazardous wastes containing halogenated organic compounds (HOCs).

(2) For regulatory levels in leachates from landfills.[37]

(3) For concentrations of various materials in effluents from the organic chemicals and plastics and synthetic fiber industries.[51]

(4) For contaminants in drinking water.[36]

(5) For National Primary and Secondary Drinking Water Regulations.[62]

(6) In the form of health advisories for 16 pesticides,[47] 25 organics,[48] and 7 inorganics.[49]

(7) For primary drinking water standards starting with a priority list of 8 Volatile Organic Chemicals.[40]

(8) State drinking water standards and guidelines[61] as assembled by the US EPA.

Determination in Water: The sources of information in this field have been primarily US EPA publications including the test procedures for priority pollutant analysis[25] and later modifications.[42]

Routes of Entry: The toxicologically important routes of entry of each substance are listed. These are primarily taken from the *NIOSH Pocket Guide*[2] but are drawn from other sources as well.

Harmful Effects and Symptoms: These are primarily drawn from NIOSH, EPA publications, and New Jersey and New York State fact sheets on individual chemicals, and are supplemented from information from the draft criteria documents for priority toxic pollutants[26] and from other sources. The other sources include:

- EPA Chemical Hazard Information Profiles (CHIPS) cited under individual entries.
- NIOSH Information Profiles cited under individual entries.
- EPA Health and Environmental Effect Profiles cited under individual entries.

Particular attention has been paid to cancer as a "harmful effect" and special effort has been expended to include the latest data on carcinogenicity.

Short Term Exposure: These are brief descriptions of the effects observed in humans when the vapor (gas) is inhaled, when the liquid or solid is ingested (swallowed), and when the liquid or solid comes in contact with the eyes or skin. The term LD_{50} signifies that about 50% of the animals given the specified dose by mouth will die. Thus, for a Grade 4 chemical (below 50 mg/kg) the toxic dose for 50% of animals weighing 70 kg (150 lb) is 70 × 50 = 3500 mg = 3.5 g, or less than 1 teaspoonful; it might be as little as a few drops. For a Grade 1 chemical (5 – 15 g/kg), the LD_{50} would be between a pint and a quart for a 150-lb man. All LD_{50} values have been obtained using small laboratory animals such as rodents, cats, and dogs. The substantial risks taken in using these values for estimating human toxicity are the same as those taken when new drugs are administered to humans for the first time.

Long Term Exposure: Where there is evidence that the chemical can cause cancer, mutagenic effects, teratogenic effects, or a delayed injury to vital organs such as the liver or kidney, a description of the effect is given.

Points of Attack: This category is based in part on the "Target Organs" in the *NIOSH Pocket Guide*[2] but the title has been changed as many of the points of attack are not organs (blood, for example).

Medical Surveillance: This information is often drawn from a NIOSH publication[27] but also from New Jersey State Fact Sheets on individual chemicals. Where additional information is desired in areas of diagnosis, treatment and medical control, the reader is referred to a private publication[28] which is adapted from the products of the NIOSH Standards Completion Program.

First Aid: Simple first aid procedures are listed for response to eye contact, skin contact, inhalation, and ingestion of the toxic substance as drawn to a large extent from the *NIOSH Pocket Guide*[2] but supplemented by information from recent commercially available volumes in the U.S.,[29] in the U.K., and in Japan[24] as well as from state fact sheets. They deal with exposure to the vapor (gas), liquid, or solid and include inhalation, ingestion (swallowing) and contact with eyes or skin. The instruction "Do NOT induce vomiting" is given if an unusual hazard is associated with the chemical being sucked into the lungs (aspiration) while the patient is vomiting. "Seek medical attention" or "Call a doctor" is recommended in those cases where only competent medical personnel can treat the injury properly. In all cases of human exposure, seek medical assistance as soon as possible. In many cases, medical advice has been included for guidance only.

Personal Protective Methods: This information is drawn heavily from NIOSH publications[2] and supplemented by information from the U.S.,[29] the U.K. and Japan.[24] There are indeed other "personal protective methods" which space limitations prohibit describing here in full. One of these involves limiting the quantities of carcinogens to which a worker is exposed in the laboratory. The items listed are those recommended by (a) NIOSH and/or OSHA (b) manufacturers, either in technical bulletins or in material safety data sheets (MSDS), (c) the Chemical Manufacturers Association (CMA), or (d) the National Safety Council (NSC), for use by personnel while responding to fire or accidental discharge of the chemical. They are intended to protect the lungs, eyes, and skin.

Respirator Selection: The fourth edition has done away with the "hieroglyphics" to designate respirator selection. However, respirator codes found in the *NIOSH Pocket Guide* have been included to ease updating. For each line a maximum use concentration (in ppm, mg/m^3, μg/m^3, fibers/m^3, or mppcf) condition (e.g., escape) followed by the NIOSH code and full text related to respirator recommendations. All recommended respirators of a given class, can be utilized at any concentration equal to or less than the class's listed maximum use concentration. Respirator selection should follow recommendations that provide the greatest degree of protection. Respirator codes found in the *NIOSH Pocket Guide* have been included to ease updating.

In this section, single use respirators refer to those disposable-type respirators which have only been tested under provisions of 30CFR1111.140-5 for which a fivefold protection factor has been established. To determine if certain single-use respirators can be used in other applications, including a higher protection factor, consult the NIOSH (DHHS-NIOSH Publication No. 85–101) to identify the conditions under which they may be used. If a higher protection factor is utilized, the wearer must first be properly fitted with a respirator using either a quantitative fit test or an "improved" qualitative fit test as accepted in the OSHA lead standard. Respirator selection recommendations are based on the NIOSH REL's. For a chemical where no REL exists, recommendations are based on meeting the OSHA PEL's except where the ACGIH TLV for the chemical is lower.

All respirators selected should be approved by NIOSH and the Mine Safety and Health Administration (MSHA). Pesticides are not identified as such in the respirator selection tables. For those substances that are pesticides, the recommended air-purifying respirator must be specifically approved by NIOSH/MSHA.

Additionally, a complete respirator protection program should be implemented including all requirements in 29CFR1910.134. At a minimum, a respirator protection program should include regular training, fit-testing, periodic environmental monitoring, maintenance inspection, and cleaning. The selection of the actual respirator to be used within the classes of recommended respirators depends on the particular use situation, and should only be made by a knowledgeable person. Remember, air-purifying respirators will not protect from oxygen-deficient atmospheres. For firefighting, only self-contained breathing apparatuses with full facepieces operated in pressure-demand or other positive pressure modes are recommended for all chemicals in the *NIOSH Pocket Guide*.

Storage: This material safety data sheet information is drawn from information from the NFPA,[17] from Japanese sources[24] and from publications such as the *Hazardous Substance Fact Sheets* published by the New Jersey Department of Health and Senior Services.

Shipping: Label(s) required (if not excepted): The section refers to the type label or placard required by regulation on any container or packaging of the subject compound being shipped. In some cases a material may require more than one hazardous materials label. Hazard class or division: This number refers to the division number or hazard class that must appear on shipping papers. This information is drawn from DOT publications[19] as well as U.N. publications[20] and also

NFPA publications.[17] The U.S. Department of Transportation[19] has published listings of chemical substances which give a hazard classification and required labels. The U.S. DOT listing now corresponds with the U.N. listing[20] and specifies first a hazard class of chemicals as defined in the following table, and then a packing group (I, II or III) within each of the classes. These groups are variously defined depending on the hazard class but in general define materials presenting: I-a very severe risk (great danger); II-a serious risk (medium danger); and III-a relatively low risk (minor danger).

Class No.	Division No. (if any)	Name of Class or Division	49CFR Reference for Definitions
None	—	Forbidden materials	173.21
None	—	Forbidden explosives	173.53
1	1.1	Explosives (with a mass explosion hazard)	173.50
1	1.2	Explosives (with a projection hazard)	173.50
1	1.3	Explosives (with predominately a fire hazard)	173.50
1	1.4	Explosives (with no significant blast hazard)	173.50
1	1.5	Very insensitive explosives; blasting agents	173.50
2	2.1	Flammable gas	173.115
2	2.2	Non-flammable compressed gas	173.115
2	2.3	Poisonous gas	173.115
3	—	Flammable and combustible liquids	173.120
4	4.1	Flammable solids	173.124
4	4.2	Spontaneously combustible materials	173.124
4	4.3	Dangerous when wet materials	173.124
5	5.1	Oxidizers	173.128
5	5.2	Organic peroxides	173.128
6	6.1	Poisonous materials	173.132
6	6.1	Irritating materials	173.381
6	6.2	Etiologic or infectious substances	173.134
7	—	Radioactive materials	173.403
8	—	Corrosive materials	173.136
9	—	Miscellaneous hazardous materials	173.140
None	—	Other regulated materials: ORM-D and ORM-E	173.144

Spill Handling: Spill or leak information provided is intended to be used only as a guide.

Issue warning is used when the chemical is a poison, has a high flammability, is a water contaminant, is an air contaminant (so as to be hazardous to life), is an oxidizing material, or is corrosive. *Restrict access* is used for those chemicals that are unusually and immediately hazardous to personnel unless they are protected properly by appropriate protective clothing, eye protection, and respiratory protection equipment, etc. *Evacuate area* is used primarily for unusually poisonous chemicals or these that ignite easily.

Mechanical containment is used for water-insoluble chemicals that float and do not evaporate readily. *Should be removed* is used for chemicals that cannot be allowed to disperse because of potentially harmful effect on humans or on the ecological system in general. The term is not used unless there is a reasonable chance of preventing dispersal, after a discharge or leak, by chemical and physical treatment. *Chemical and physical treatment* is recommended for chemicals that can be removed by skimming, pumping, dredging, burning, neutralization, absorption, coagulation, or precipitation. The corrective response may also include the use of dispersing agents, sinking agents, and biological treatment. *Disperse and flush* is used for chemicals that can be made non-hazardous to humans by simple dilution with water. In a few cases the response is indicated even when the compound reacts with water because, when proper care is taken, dilution is still the most effective way of removing the primary hazard. This material safety data sheet information is drawn from a variety of sources including New Jersey Department of Health and Senior Services *Hazardous Substance Fact Sheets* and EPA *Profiles on Extremely Hazardous Substances.*

Fire Extinguishing: Fire information provided is intended to be used only as a guide. For fire-fighting, only self-contained breathing apparatuses with full facepieces operated in pressure-demand or other positive pressure modes are recommended for all chemicals in the *NIOSH Pocket Guide*. Certain extinguishing agents should not to be used because the listed agents react with the chemical and have the potential to create an additional hazard. In some cases they are listed because they are ineffective in putting out the fire. Some chemicals decompose or burn to give off toxic and irritating gases. Such gases may also be given off by chemicals that vaporize in the heat of a fire without either decomposing or burning. If no entry appears, the combustion products are thought to be similar to those formed by the burning of oil, gasoline, or alcohol; they include carbon monoxide (poisonous), carbon dioxide, and water vapor. The specific combustion products are usually not well known over the wide variety of conditions existing in fires; some may be hazardous. This information is drawn from NFPA publications[17] and other sources. Any characteristic behavior that might increase significantly the hazard involved in a fire is described. The formation of flammable vapor clouds or dense smoke, and the possibility of polymerization and explosions is stated in this section or the incompatibility section. Unusual difficulty in extinguishing the fire is noted.

Disposal Method Suggested: The disposal methods for various chemical substances have been drawn from a recent U.N. publication.[22]

References: The general bibliography for this volume follows immediately. It includes general reference sources and references dealing with analytical methods. The references at the end of individual product entries are generally restricted to: references dealing only with that particular compound; and references which in turn contain bibliographies giving references to the original literature on toxicological and other behavior of the substance in question.

Bibliography

1. American Conference of Governmental Industrial Hygienists, *Threshold Limit Values for Chemical Substances and Physical Agents in the Workroom Environment with Intended Changes*, ACGIH, Cincinnati, OH, (1998/99)
2. National Institute for Occupational Safety and Health, *NIOSH/OSHA Pocket Guide to Chemical Hazards*. DHHS (NIOSH) Publication No. 97–140, Washington, DC (June 1997)
3. Deutsche Forschungsgemeinschaft (DFG), "List of MAK and BAT Values 1999," Wiley-VCH Publishers, New York, (1999)
4. U.S. Environmental Protection Agency, "Water Programs: Hazardous Substances," *Federal Register*, **43**:49, pp. 10474–10508 (March 13, 1978)
5. U.S. Environmental Protection Agency, "Identification and Testing of Hazardous Waste," *Federal Register*, **45**:98, pp. 33084–33133 (May 19, 1980)
6. U.S. Environmental Protection Agency, *Federal Register*, **43**, p. 4109 (January 31, 1978); See also *Federal Register*, **44**, p. 44501 (July 30, 1979); and also *Federal Register*, **45**, pp. 79318–79379 (November 28, 1980)
7. U.S. Environmental Protection Agency, "Emergency Planning and Community Right-to-Know Programs," *Federal Register*, **51**:221, pp. 41570–41594 (November 17, 1986)
8. U.S. Environmental Protection Agency, "Toxic Chemical Release Reporting; Community Right-to-Know," *Federal Register*, **52**:107, pp. 21152–21208 (June 4, 1987)
9. National Institute for Occupational Safety and Health, *Registry of Toxic Effects of Chemical Substances*, DHEW (NIOSH) Publication No. 87–114 (1985–86)
10. U.S. Department of Health and Human Services, *Fourth Annual Report on Carcinogens*, Research Triangle Park, NC, National Toxicology Program (1985)
11. Verscheuren, K., "Handbook of Environmental Data on Organic Chemicals," 3rd ed., Van Nostrand Reinhold Co., New York (1996)
12. International Agency for Research on Cancer, "IARC Monographs on the Carcinogenic Risks of Chemicals to Humans," Lyon, France (Various years)
13. United Nations, "Consolidated List of Products Whose Consumption and/or Sale Have Been Banned, Withdrawn, Severely Restricted or Not Approved by Governments," Second Issue, U.N. Sales No. E. 87.IV.1, United Nations, Geneva (1987)
14. U.S. Environmental Protection Agency, "Report on the Status of Chemicals in the Special Review Program, Registration Program and Data Call-in Program," Office of Pesticide Programs, Washington, DC.
15. World Bank, "Manual of Industrial Hazard Assessment Techniques," Office of Environmental and Scientific Affairs, Washington, DC (1985)
16. National Research Council, "Drinking Water and Health," National Academy Press, Washington, DC, (1980). See also Ref. 46.
17. National Fire Protection Association, "Fire Protection Guide on Hazardous Materials," 11th ed., Quincy, MA, (1994)
18. National Institute of Occupational Safety and Health, "NIOSH Manual of Analytical Methods," 1985 Supplement to 3rd ed., NIOSH Publication No. 84–100, Cincinnati, OH (1985)
19. U.S. Department of Transportation, "Performance-Oriented Packaging Standards," 49CFR17179, *Federal Register*. **52**:215, pp. 42772–43000 (November 6, 1987)
20. United Nations, "Recommendations on the Transport of Dangerous Goods," Fourth Revised Edition, U.N. Sales No. E.85.VIII.3, United Nations, New York (1986)
21. United Nations, "Recommendations on the Transport of Dangerous Goods; Tests and Criteria," 1st ed., U.N. Sale No. E.85.VIII.2, United Nations, New York (1986)
22. International Register of Potentially Toxic Chemicals, "Treatment and Disposal Methods for Waste Chemicals," U.N. Sale No. E.85.111.2, United Nations Environment Programme, Geneva, Switzerland (1985)

23. Worthing, C. R., and Walker, S. B., Eds., "The Pesticide Manual," 8th ed., Thornton Heath, The British Crop Protection Council, U.K. (1987)

24. The International Technical Information Institute, "Toxic and Hazardous Industrial Chemicals Safety Manual for Handling and Disposal with Toxicity and Hazard Data," Tokyo (1986)

25. U.S. Environmental Protection Agency, "Guidelines Establishing Test Procedures for the Analysis of Pollutants; Proposed Regulations," *Federal Register,* **44**:233, pp. 69464–69575 (December 3, 1979); and also a corrected version in *Federal Register,* **44**:244, pp. 75028–75052 (December 18, 1979)

26. Sittig, M., Ed., *Priority Toxic Pollutants: Health Impacts and Allowable Limits,* Noyes Data Corp., Park Ridge, NJ (1980)

27. National Institute for Occupational Safety and Health, *Occupational Diseases: A Guide to Their Recognition,* DHEW (NIOSH) Publication No. 77–181, Washington, DC (June 1977)

28. Proctor, N. H., Hughes, J. P., and Fischman, M. L, *Chemical Hazards of the Workplace,* 3rd ed., Van Nostrand Reinhold, New York (1991)

29. Plunkett, E. R., *Handbook of Industrial Toxicology,* 3rd ed., Chemical Publishing Co., Inc., New York, NY (1987)

30. Parmeggiani, L., Ed., *Encyclopedia of Occupational Health and Safety,* 3rd ed., Geneva, International Labor Office (ILO) (1983)

31. Research and Special Programs Administration, U.S. Department of Transportation, *North American Emergency Response Guidebook,* Washington, DC (1996); See also Ref. 56.

32. U.S. Environmental Protection Agency, "Multimedia Environmental Goals for Environmental Assessment," Report EPA-600/7-77-136, Research Triangle Park, NC (November 1977)

33. Health and Safety Executive, "Occupational Exposure Limits 1988," Guidance Note EH 40/88, London (1988)

34. Health and Safety Executive, "Monitoring Strategies for Toxic Substances," Guidance Note EH 42, London (November 1984)

35. International Register of Potentially Toxic Chemicals (IRPTC), "IRPTC Legal File (1986)," U.N. Sales No. E.87.III.D5, United Nations Environment Programme, Geneva, Switzerland (1987)

36. U.S. Environmental Protection Agency, "Drinking Water; Proposed Substitution of Contaminants and Proposed List of Additional Substances Which May Require Regulation Under the Safe Drinking Water Act," *Federal Register,* **52**:130, pp. 25720–25734 (July 8, 1987)

37. U.S. Environmental Protection Agency, "Hazardous Waste Management System; Identification and Testing of Hazardous Waste; Notification Requirements; Reportable Quantity Adjustments; Proposed Rule," *Federal Register*, **51**:114, pp. 21648–21693 (June 13,1986)

38. U.S. Environmental Protection Agency, "Land Disposal Restrictions for Certain California List Hazardous Wastes and Modifications to the Framework; Final Rule," *Federal Register,* **52**:130, pp. 25760–25792 (July 8, 1987)

39. U.S. Environmental Protection Agency, "Notice of the First Priority List of Hazardous Substances That Will Be the Subject of Toxicological Profiles and Guidelines for the Development of Toxicological Profiles," *Federal Register,* **52**:74, 12868–12874 (April 17, 1987)

40. U.S. Environmental Protection Agency, "National Primary Drinking Water Regulations-Synthetic Organic Chemicals; Monitoring for Unregulated Contaminants; Final Rule," *Federal Register,* **52**:130, pp. 25690–25717 (July 8, 1987)

41. Weiss, G., Ed., "Hazardous Chemicals Data Book," 2nd ed., Noyes Data Corp., Park Ridge, NJ (1986)

42. U.S. Environmental Protection Agency, "Guidelines Establishing Test Procedures for the Analysis of Pollutants; Interim Final Rule and Request for Comments and Proposed Regulation," *Federal Register,* **52**:171, pp. 33542–33557 (September 3, 1987)

43. United Nations Environment Program, "Maximum Allowable Concentrations and Tentative Safe Exposure Levels of Harmful Substances in the Environmental Media (Hygiene Standards Officially Approved in the USSR)," Center of International Projects, Moscow (1984)

44. Lewis, R. J., Sr., "Hazardous Chemicals Desk Reference," 4th ed., Van Nostrand Reinhold Co., New York (1997)

45. National Institute for Occupational Safety and Health, "NIOSH Recommendations for Occupational Safety and Health Standards," Supplement to Morbidity and Mortality Weekly Report, Centers for Disease Control, Atlanta, GA (September 26, 1986)

46. National Research Council, "Drinking Water and Health," National Academy of Sciences, Washington, DC, (1977); See also Ref. 16.

47. U.S. Environmental Protection Agency, "Health Advisories for 16 Pesticides," Report PB-87-200176, Office of Drinking Water, Washington, DC (March 1987)

48. U.S. Environmental Protection Agency, "Health Advisories for 25 Organics," Report PB 87-235578, Office of Drinking Water, Washington, DC (March 1987)

49. U.S. Environmental Protection Agency, "Health Advisories for Legionella and Seven Inorganics," Report PB 87-235586, Office of Drinking Water, Washington, DC (March 1987)

50. Lewis, R. J., Sr., "Rapid Guide to Hazardous Chemicals in the Workplace," 3rd ed., Van Nostrand Reinhold Co., New York (1994)

51. U.S. Environmental Protection Agency, "Organic Chemicals and Plastics and Synthetic Fibers Category Effluent Limitations Guidelines, Pretreatment Standards and New Source Performance Standards," *Federal Register,* **52**:214, pp. 42522–42584 (November 5, 1987)

52. Keith, L. H., and Walters, D. B., Eds, "Compendium of Safety Data Sheets for Research and Industrial Chemicals," VCH Publishers, Inc., Vols. I–III, New York, NY (1985); and Vols. IV–VI, New York, NY (1987)

53. American Conference of Governmental Industrial Hygienists, "Documentation of the Threshold Limit Values and Biological Exposure Indices, 1998–1999," ACGIH, Cincinnati, OH (1999)

54. U.S. Congress, Office of Technology Assessment, "Identifying and Regulating Carcinogens," Report OTA-BP-H-42, U.S. Government Printing Office, Washington, DC (November 1987)

55. U.S. Environmental Protection Agency, "Pesticide Fact Handbook," Noyes Data Corp., Park Ridge, NJ (1988)

56. U.S. Department of Transportation, "1996 Emergency Response Guidebook," Washington, DC (1996); See also Ref. 31.

57. Dutch Association of Safety Experts, Dutch Chemical Industry Association and Dutch Safety Institute, "Handling Chemicals Safely," 2nd ed., Amsterdam (1980)

58. U.S. Department of Labor, "Air Contaminants-Final Rule," 29CFR1910, *Federal Register,* **54**:12, pp. 2332–2983, Occupational Safety and Health Administration (January 19, 1989)

59. New Jersey Drinking Water Institute, "Maximum Contaminant Level Recommendations for Hazardous Contaminants in Drinking Water," Appendix B: Health-Based Maximum Contaminant Level Support Documents, Trenton, NJ (March 26, 1987)

60. U.S. Environmental Protection Agency, "NATICH Data Base Report on State, Local and EPA Air Toxics Activities," Research Triangle Park, NC, Office of Air Quality Planning and Standards (July 1988)

61. U.S. Environmental Protection Agency, "Summary of State and Federal Drinking Water Standards and Guidelines," Federal-State Toxicology and Regulatory Alliance Committee (FSTRAC), Office of Drinking Water, Washington, DC (March 1988)

62. U.S. Environmental Protection Agency, "National Primary and Secondary Drinking Water Regulations," *Federal Register,* **54**:97, pp. 22062–22160 (May 22, 1989)

63. U.S. Department of Labor, "Occupational Safety and Health Standards," 29CFR1910, Washington, DC (July 1, 1988)

64. U.S. Department of Transportation, "Chemical Data Guide for Bulk Shipment by Water," United States Coast Guard, Washington, DC (1990)

65. New York State Department of Health, "Chemical Fact Sheets," Bureau of Toxic Substance Assessment, Albany, NY, various issues and dates.

66. U.S. Environmental Protection Agency, "Consolidated List of Chemicals Subject to the Emergency Planning and Community Right-to-Know Act (EPCRA) and Section 112 (r) of The Clean Air Act, As Amended," EPA 550-B-98-017, Washington, DC (1998)

67. Bomgardner, P. M., Ed., "Handling Hazardous Materials," American Trucking Association, Alexandria, VA (1997)

68. Lewis, R. J., Sr., "Hawley's Condensed Chemical Dictionary," 13th ed., Van Nostrand Reinhold, New York, NY (1998)

69. Pohanish, R. P., and Greene, S. A., "Hazardous Substance Resource Guide," 2nd ed., Gale Research, Inc., Detroit, MI (1997)

70. New Jersey Department of Health and Senior Services, Right-to-Know Project, "Hazardous Substance Fact Sheets," Trenton, NJ (various dates from 1985–2001)

71. Stricoff, S. R., and Partridge, L. J., Jr., Eds., "NIOSH/OSHA Occupational Health Guidelines for Chemical Hazards," A.D. Little, Inc., United States Department of Health and Human Services, United States Department of Labor, Occupational Safety and Health Administration, Cincinnati, OH (1997)

72. U.S. Environmental Protection Agency, "Pollution Prevention Fact Sheets: Chemical Production, FREG-1 (PPIC)," United States Environmental Protection Agency, Washington, DC (Various years)

73. U.S. Environmental Protection Agency, "Polychlorinated Biphenyl (PWB) Information Package," TSCA Information Service, Washington, DC (April, 1993)

74. Pohanish, R. P., and Stanley A. G., "Rapid Guide to Chemical Incompatibilities," Van Nostrand Reinhold, New York, NY (1997)

75. Lewis, R. J., Sr., "Sax's Dangerous Properties of Industrial Materials," 9th ed., Van Nostrand Reinhold, New York, NY (1996)

76. Agency for Toxic Substances and Disease Registry, "Toxicological Fact Sheets," U.S. Department of Health and Human Services, Public Health Service, Atlanta, GA (various dates)

77. U.S. Department of Health and Human Services/National Institute for Occupational Safety and Health, *CD-ROM NIOSH/OSHA Pocket Guide to Chemical Hazards and other Databases.* DHHS (NIOSH) Publication No. 99–115, Washington, DC (April 1999)

78. Federal Emergency Management Agency/United States Fire Administration, "Hazardous Materials for First Responders," Washington, DC (1999)

Key to Abbreviations and Acronyms

α	the Greek letter alpha; used as a prefix to denote the carbon atom in a straight chain compound to which the principal group is attached.
as-	prefix for asymmetric.
ACGIH	American Conference of Governmental Industrial Hygienists.
approx.	approximately.
asym-	prefix for asymmetric.
atm.	atmosphere.
β	the Greek letter beta.
BEI	Biological Exposure Indexes (ACGIH).
BP	boiling point.
C	centigrade.
CAA	Clean Air Act.
CAAA	Clean Air Act Amendments of 1990.
carc.	carcinogen.
CAS	Chemical Abstract Service.
cc	cubic centimeter.
c.c. or cc	closed cup.
CEPA	Canadian Environmental Protection Act.
CERCLA	Comprehensive Environmental Response, Compensation, And Liability Act.
CFCs	chlorofluorocarbons.
CFR	*Code of Federal Regulations.*
CHEMTREC	Chemical Manufacturers Association (CMA) Transportation Emergency Center.
cis-	(Latin, on this side). Indicating one of two geometrical isomers in which certain atoms or groups are on the same side of a plane.
CMA	Chemical Manufacturers Association.
comp.	compound.
CWA	Clean Water Act.
cyclo-	(Greek, circle). Cyclic, ring structure; as cyclohexane.
deriv.	derivative.
DFG	Deutsche Forschungsgemeinschaft.
DOT	U.S. Department of Transportation.
DOT ID	U.S. Department of Transportation Identification Numbers.
EEC	European Economic Community.
EHS	Extremely Hazardous Substances.
EINECS	European Inventory of Existing Commercial Chemical Substances.
EPA	U.S. Environmental Protection Agency.
EPCRA	Emergency Planning and Community Right-to-Know Act.
F	Fahrenheit.
FDA	Food and Drug Administration.
FEMA	Federal Emergency Management Agency.
FR	*Federal Register.*
h	hour(s).
HAPs	Hazardous Air Pollutants (CAA).
HCFC	hydrochlorofluorocarbons.
HCS	Hazard Communication Standard.
HOC	Halogenated Organic Compounds.
IARC	International Agency for Research on Cancer.
IDLH	Immediately Dangerous to Life or Health.
iso-	(Greek, equal, alike). Usually denoting an isomer of a compound.
kg	kilogram(s).
l	liter(s).
lb	pound(s).
LEL	Lower explosive (flammable) limit in air, % by volume at room temperature or other temperture as noted.
LEPC	Local Emergency Planning Committees.
m-	an abbreviation fort "meta-," a prefix used to distinguish between isomers or nearly related compounds.
m^3	cubic meter.
MACT	Maximum Achievable Control Technology (CAA).
MAK	airborne exposure limit used by the Deutsche Forschungsgemeinschaft (DFG).
MCLs	Maximum Contaminant Levels (SDWA).
MCLGs	Maximum Contaminant Level Goals (SDWA).
mg	milligram(s).
μ	micro.

µg	microgram(s).
min	minute(s).
mppcf	million particles per cubic foot.
MSDS	Material Safety Data Sheets.
n-	abbreviation for "normal," referring to the arrangement of carbon atoms in a chemical molecule prefix for normal.
N-	Symbol used in some chemical names, indicating that the next section of the name refers to a chemical group attached to a nitrogen atom. The bond to the nitrogen atom.
NCI	National Cancer Institute.
NO_X	Nitrogen Oxide.
NPRI	National Pollutant Release Inventory (Canada).
NTP	National Toxicology Program.
o-	ortho-, a prefix used to distinguish between isomers or nearly related compounds.
o.c. or oc	open cup.
OSHA	Occupational Safety and Health Administration.
Oxy	Oxidizer or oxidizing agent.
p-	an abbreviation for "para-," a prefix used to distinguish between isomers or nearly related compounds.
PCB	polychlorinated biphenyl.
PE	polyethylene.
PEL	Permissible Exposure Limit (OSHA).
pot. carc	potential carcinogen.
POTW	Publicly Owned Treatments Works.
PP	polypropylene.
ppb	parts per billion.
PPE	Personal Protective Equipment.
ppm	parts per million.
PQL	Practical Quantitation Limit (RCRA).
prim-	prefix for primary.
REL	Recommended Exposure Limits (NIOSH).
RQ	Reportable Quantity.
RTECS	Registry of Toxic Effects of Chemical Substances.
RTK	Right-to-Know.
SARA	Superfund Amendments and Reauthorization Act.
s. carc	Suspected Carcinogen.
SCBA	Self-Contained Breathing Apparatus.
SDWA	Safe Drinking Water Act.
sec-	prefix for secondary.
SERC	State emergency response commissions.
SMCL	Secondary Maximum Contaminant Levels (SDWA).
soln.	solution.
STEL	Short-Term Exposure Limit.
sus. carc.	Suspected Carcinogen.
sym-	abbreviation for "symmetrical," referring to a particular arrangement of elements within a chemical molecule.
t-	prefix for tertiary.
TRK	Technical Guiding Concentrations (DFG) for workplace control of carcinogens.
temp.	temperature.
tert-	abbreviation for "tertiary," referring to a particular arrangement of elements within a chemical molecule.
TLV	Threshold Limit Value (ACGIH).
TQ	Threshold Quantity.
trans-	(Latin, across). Indicating that one of two geometrical isomers in which certain atoms or groups are on opposite sides of a plane.
TSCA	Toxic Substances Control Act.
TWA	Time-Weighted Average.
UEL	Upper explosive (flammable) limit in air, % by volume at room temperature or other temperature as noted.
unsym-	prefix for asymmetric.
USDA	U.S. Department of Agriculture.
VOCs	Volatile Organic Compounds.
$>$	symbol for "greater than."
$<$	symbol for "less than."
$\leq$	symbol for "less than or equal to."
$\geq$	symbol for "greater than or equal to."

Acenaphthene

Molecular Formula: $C_{12}H_{10}$

Synonyms: Acenafeno (Spanish); Acenaphthylene, 1,2-dihydro; 1,8-Dihydroacenaphthalene; 1,2-Dihydroacenaphthylene; 1,8-Dihydroacenaphthylene; 1,8-Ethylenenaphthalene; Ethylenenaphthalene; Naphthyleneethylene; Periethylenenaphthalene

CAS Registry Number: 83-32-9

RTECS Number: AB0000000

DOT ID: UN 3077

Regulatory Authority

- OSHA, 29CFR1910 Specifically Regulated Chemicals (See CFR1910.1002) as coal tar pitch volatiles
- Clean Water Act: 40CFR401.15 Section 307 Toxic Pollutants, 40CFR413.02, Total Toxic Organics, 40CFR423, Priority Pollutants
- RCRA 40CFR258, Appendix 2
- RCRA 40CFR268.48; 61FR15654, Universal Treatment Standards: Wastewater (mg/l), 0.059; Nonwastewater (mg/kg), 3.4
- RCRA, 40CFR264, Appendix 9, Ground Water Monitoring List, Suggested Testing Methods (PQL μg/l): 8100 (200); 8270 (10)
- SUPERFUND/EPCRA 40CFR302.4, Appendix A, Reportable Quantity (RQ): 100 lb (45.4 kg)
- Canada, WHMIS, Ingredients Disclosure List
- Mexico, Drinking Water, Criteria (Ecological): 0.02 mg/l; wastewater: organic toxic pollutant.

Cited in U.S. State Regulations: California (G), Illinois (G), Massachusetts (G), New Hampshire (G), New Jersey (G), Pennsylvania (G)

Description: Acenaphthene, $C_{12}H_{10}$, is a white crystalline, combustible solid. Freezing/Melting point = 95 – 97°C. Insoluble in water. It is a polyaromatic hydrocarbon (PAH).

Potential Exposure: Acenaphthene occurs in coal tar produced during the high-temperature carbonization or coking of coal, petroleum processing, shale oil processing. It is used as a dye intermediate, in the manufacture of some plastics, as an insecticide and fungicide, and has been detected in cigarette smoke and gasoline exhaust condensates.

Incompatibilities: Ozone and oxidizing agents such as perchlorates, peroxides, permanganates, chlorates, nitrates, chlorine, bromine and fluorine.

Permissible Exposure Limits in Air: No standards have been established for acenaphthene. The ACGIH and OSHA recommended TLV as coal tar pitch volatiles as benzene solubles is 0.2 mg/m^3. NIOSH considers coal tar products to be occupational carcinogens; the NIOSH REL (10-hour TWA) for coal tar products is 0.1 mg/m^3.

Determination in Air: See NIOSH Method 5506 (HPLC) and 5515 (GC).[18]

Permissible Concentration in Water: To protect freshwater aquatic life:1,700 μg/l. To protect saltwater aquatic life – on an acute basis 970 μg/l and on a chronic basis 520 μg/l. To protect human health: 20.0 μg/l (based on organoleptic data).[6] See also Regulatory Authority for U.S. and Mexico regulatory levels.

Determination in Water: Gas chromatography or high performance liquid chromatograph (EPA Method 610) or gas chromatography and mass spectrometry (EPA Method 625).

Routes of Entry: Ingestion, inhalation, eye and/or skin contact. Absorbed through the skin.

Harmful Effects and Symptoms

Short Term Exposure: Acenaphthene is irritating to eyes, skin and respiratory tract causing coughing and wheezing. May cause vomiting if swallowed in large quantities.

Long Term Exposure: Although acenaphthene has not been identified as a carcinogen, it should be handled with care as several related polycyclic aromatic hydrocarbons (PAHs) are carcinogens. Repeated or high exposures may cause lung irritation, bronchitis with cough, phlegm, and/or shortness of breath. Acenaphthene may affect the liver and kidneys. The most thoroughly investigated effect of acenaphthene is its ability to produce nuclear and cytological changes in microbial and plant species. Most of these changes, such as an increase in cell size and DNA content, are associated with disruption of the spindle mechanism during mitosis and the biological impact of acenaphthene on mammalian cells, these effects are reported here because they are the only substantially investigated effects of acenaphthene. Reported to be a mutagen.[11]

Points of Attack: Liver, kidneys, skin.

Medical Surveillance: Preplacement and regular physical examinations are indicated for workers having contact with acenaphthene in the workplace. Liver and kidney function tests recommended.

First Aid: If this chemical gets into the eyes, remove any contact lenses at once and irrigate immediately for at least 15 minutes, occasionally lifting upper and lower lids. If this chemical contacts the skin, remove contaminated clothing and wash with soap immediately. When this chemical has been swallowed, get medical attention. Give large quantities of water and induce vomiting. Do not make an unconscious person vomit. If this chemical has been inhaled, remove from exposure and transfer promptly to a medical facility.

Personal Protective Methods: Wear protective gloves and clothing to prevent any reasonable probability of skin contact. Safety equipment suppliers/manufacturers can provide recommendations on the most protective glove/clothing material for your operation. All protective clothing (suits, gloves, footwear, headgear) should be clean, available each day, and put on before work. Contact lenses should not be worn when working with this chemical. Wear dust-proof chemical goggles and face shield unless full facepiece respiratory protection is worn. Employees should wash immediately with soap when skin is wet or contaminated. Provide emergency showers and eyewash.

Respirator Selection: *NIOSH* (as coal tar pitch volatiles): *At any concentrations above the NIOSH REL, or where there is no REL, at any detectable concentration:* SCBAF:PD,PP (any self-contained breathing apparatus that has a full facepiece and is operated in a pressure-demand or other positive-pressure mode); or SAF:PD,PP:ASCBA (any supplied-air respirator that has a full facepiece and is operated in a pressure-demand or other positive-pressure mode in combination with an auxiliary, self-contained breathing apparatus operated in a pressure-demand or other positive pressure mode). *Escape:* GMFOVHiE [any air-purifying, full-facepiece respirator (gas mask) with a chin-style, front- or back-mounted organic vapor canister having a high-efficiency particulate filter]; or SCBAE (any appropriate escape-type, self-contained breathing apparatus).

Storage: Prior to working with this chemical you should be trained on its proper handling and storage. Store in tightly closed containers in a cool, well ventilated area. Sources of ignition such as smoking and open flames, are prohibited where this chemical is used, handled, or stored in a manner that could create a potential fire or explosion hazard.

Shipping: Environmentally hazardous solid, n.o.s. Hazard Class: 9. Label: "Class 9." Packing Group: III.

Spill Handling: Evacuate persons not wearing protective equipment from area of spill or leak until clean-up is complete. Remove all ignition sources. Collect powdered material in the most convenient and safe manner and deposit in sealed containers. Ventilate area after clean-up is complete. It may be necessary to contain and dispose of this chemical as a hazardous waste. If material or contaminated runoff enters waterways, notify downstream users of potentially contaminated waters. Contact your Department of Environmental Protection or your regional office of the federal EPA for specific recommendations. If employees are required to clean-up spills, they must be properly trained and equipped. OSHA 1910.120(q) may be applicable. May be isolated using bentonite lined dam.

Fire Extinguishing: This chemical is a combustible solid. Use dry chemical, carbon dioxide, water spray, or alcohol foam extinguishers. Poisonous gases are produced in fire including carbon monoxide. If material or contaminated runoff enters waterways, notify downstream users of potentially contaminated waters. Notify local health and fire officials and pollution control agencies. From a secure, explosion-proof location, use water spray to cool exposed containers. If cooling streams are ineffective (venting sound increases in volume and pitch, tank discolors, or shows any signs of deforming), withdraw immediately to a secure position. If employees are expected to fight fires, they must be trained and equipped in OSHA 1910.156.

Disposal Method Suggested: Consult with environmental regulatory agencies for guidance on acceptable disposal practices. Generators of waste containing this contaminant (≥100 kg/mo) must conform with EPA regulations governing storage, transportation, treatment, and waste disposal. In accordance with 40CFR165 recommendations for the disposal of pesticides and pesticide containers. Must be disposed properly by following package label directions or by contacting your state pesticide or environmental control agency or by contacting your regional EPA office. Incineration or permanganate oxidation.

References

U.S. Environmental Protection Agency, Acenaphthene: Ambient Water Criteria, Report PB 296–782, Washington, DC (1980)

Lewis, R. J., Sax's Dangerous Properties of Industrial Materials, 9th edition, New York, (1998)

Sax, N. I., Ed., Dangerous Properties of Industrial Materials Report, 4, No. 1, 38–41 (1984)

New Jersey Department of Health and Senior Services, Hazardous Substance Fact Sheet, Trenton NJ (1998)

Acenaphthylene

Molecular Formula: $C_{12}H_8$

Synonyms: Cyclopenta(de)naphthalene (French)

CAS Registry Number: 208-96-8

RTECS Number: AB1254000

DOT ID: UN 3077

Regulatory Authority

- Clean Water Act 40CFR413.02, Total Toxic Organics; 40CFR423, Appendix A, Priority Pollutants
- RCRA 40CFR258, Appendix 2, List of Hazardous Constituents
- RCRA 40CFR268.48; 61FR15654, Universal Treatment Standards: Wastewater (mg/l), 0.059; Nonwastewater (mg/kg), 3.4
- RCRA, 40CFR264, Appendix 9, Ground Water Monitoring List, Suggested Testing Methods (PQL µg/l): 8100 (200); 8270 (10)
- SUPERFUND/EPCRA 40CFR302.4, Appendix A, Reportable Quantity (RQ): 5,000 lb (2,270 kg)
- Mexico: Wastewater, Organic Toxic Pollutant

Cited in U.S. State Regulations: California (G), Kansas (W), Massachusetts (G), New Jersey (G), Pennsylvania (G).

Description: Acenaphthylene, $C_{12}H_8$, is a solid. Freezing/Melting point = 80 – 83°C. Water soluble. It is a polynuclear aromatic hydrocarbon (PAH).

Potential Exposure: In coal tar processing.

Incompatibilities: Ozone and strong oxidizing agents such as perchlorates, peroxides, permanganates, chlorates, nitrates, chlorine, bromine and fluorine.

Permissible Exposure Limits in Air: The OSHA 8-hour TWA as coal tar pitch volatiles (benzene-soluble fraction) is 0.2 mg/m^3 OSHA defines "coal tar pitch volatiles" in 29 CFR1910.1002 as the fused polycyclic hydrocarbons that volatilize from the distillation residues of coal, petroleum (excluding asphalt), wood, and other organic matter and includes substances such as anthracene, benzo(a)pyrene (BaP), phenanthrene, acridine, chrysene, pyrene, etc. NIOSH considers coal tar products to be occupational carcinogens. The NIOSH REL (10-hour TWA) for coal tar products is 0.1 mg/m^3 (cyclohexane-extractable fraction).

Determination in Air: Filter; Benzene; Gravimetric; OSHA Method #58. See also NIOSH Methods[18] as follows: #5506 (HPLC) and #5515 (GC).

Permissible Concentration in Water: See the entry on "Polynuclear Aromatic Hydrocarbons." *Note:* Kansas[61] has set a guideline for drinking water of 0.03 µg/l.

Determination in Water: Extraction then HPLC/UV. EPA Method #610.

Routes of Entry: Ingestion, inhalation, skin and/or eye contact.

Harmful Effects and Symptoms

Long Term Exposure: May cause dermatitis. May cause lung irritation; bronchitis may develop. A potential occupational carcinogen. See also Polynuclear Aromatic Hydrocarbons entry.

Points of Attack: Respiratory system, skin, bladder, kidneys. Cancer Site: lung, kidney, and skin cancer.

Medical Surveillance: Preplacement and regular physical examination indicated. Kidney function tests. Urine cytology test (a test for abnormal cells in urine).

First Aid: If this chemical gets into the eyes, remove any contact lenses at once and irrigate immediately for at least 15 minutes, occasionally lifting upper and lower lids. If this chemical contacts the skin, remove contaminated clothing and wash with soap immediately. When this chemical has been swallowed, get medical attention. Give large quantities of water and induce vomiting. Do not make an unconscious person vomit. If this chemical has been inhaled, remove from exposure and transfer promptly to a medical facility.

Personal Protective Methods: Good particulate emission controls are the indicated engineering control scheme. Contact lenses should not be worn when working with coal tar pitch volatiles. Wear protective gloves and clothing to prevent any reasonable probability of skin contact. Safety equipment suppliers/manufacturers can provide recommendations on the most protective glove/clothing material for your operation. All protective clothing (suits, gloves, footwear, headgear) should be clean, available each day, and put on before work. Contact lenses should not be worn when working with this chemical. Wear dust-proof chemical goggles and face shield unless full facepiece respiratory protection is worn. Employees should wash immediately with soap when skin is wet or contaminated. Provide emergency showers and eyewash.

Respirator Selection: *NIOSH* (as coal tar pitch volatiles): *At any concentrations above the NIOSH REL, or where there is no REL, at any detectable concentration:* SCBAF:PD,PP (any self-contained breathing apparatus that has a full facepiece and is operated in a pressure-demand or other positive-pressure mode); or SAF:PD,PP:ASCBA (any supplied-air respirator that has a full facepiece and is operated in a pressure-demand or other positive-pressure mode in combination with an auxiliary, self-contained breathing apparatus operated in a pressure-demand or other positive pressure mode). *Escape:* GMFOVHiE [any air-purifying, full-facepiece respirator (gas mask) with a chin-style, front- or back-mounted organic vapor canister having a high-efficiency particulate filter]; or SCBAE (any appropriate escape-type, self-contained breathing apparatus).

Storage: Prior to working with this chemical you should be trained on its proper handling and storage. Store in tightly closed containers in a cool, well ventilated area.

Shipping: Environmentally hazardous solid, n.o.s. Hazard Class: 9. Label: "Class 9." Packing Group: III.

Spill Handling: Evacuate persons not wearing protective equipment from area of spill or leak until clean-up is complete. Remove all ignition sources. Collect powdered material in the most convenient and safe manner and deposit in sealed containers. Ventilate area after clean-up is complete. It may be necessary to contain and dispose of this chemical as a hazardous waste. If material or contaminated runoff enters waterways, notify downstream users of potentially contaminated waters. Contact your Department of Environmental Protection or your regional office of the federal EPA for specific recommendations. If employees are required to clean-up spills, they must be properly trained and equipped. OSHA 1910.120(q) may be applicable.

Fire Extinguishing: Irritating fumes are produced in fire. Use foam, dry chemical, and carbon dioxide. If material or contaminated runoff enters waterways, notify downstream users of potentially contaminated waters. Notify local health and fire officials and pollution control agencies. From a secure, explosion-proof location, use water spray to cool exposed containers. If cooling streams are ineffective (venting sound increases in volume and pitch, tank discolors, or shows any signs of deforming), withdraw immediately to a secure position. If employees are expected to fight fires, they must be trained and equipped in OSHA 1910.156.

Disposal Method Suggested: Consult with environmental regulatory agencies for guidance on acceptable disposal practices. Generators of waste containing this contaminant (≥100 kg/mo) must conform with EPA regulations governing storage, transportation, treatment, and waste disposal. Product residues and sorbent media may be packaged in epoxylined drums then destroyed by incineration, permanganate oxidation or microwave plasma treatment. The USEPA has investigated chemical precipitation for wastewater treatment.

References

Sax, N. I., Ed., Dangerous Properties of Industrial Materials Report 4, No. 2, 35–37, New York (1984)

USEPA, Management of Hazardous Waste Leachate, Washington, DC (1982)

Acetal

Molecular Formula: $C_4H_{14}O_2$

Common Formula: $CH_3CH(OC_2H_5)_2$

Synonyms: Acetaal (Dutch); Acetal (Spanish); Acetaldehyde diethyl acetal; Acetal diethylique (French); Acetale (Italian); Acetehyde; 1,1-Diaethoxy-aethan (German); Diethyl acetal; 1,1-Diethoxy acetal; 1,1-Diethoxy-ethaan (Dutch); 1,1-Diethoxyethane; Diethyl acetal; 1,1-Dietossietano (Italian); Ethane, 1,1-diethoxy-; Ethylidene diethyl ether

CAS Registry Number: 105-57-7

RTECS Number: AB2800000

DOT ID: UN 1088

EEC Number: 605-015-00-1

EINECS Number: 203-310-6

Regulatory Authority

- U.S. DOT Regulated Marine Pollutant (49CFR172.101, Appendix B)

Cited in U.S. State Regulations: California (G), Florida (G), Maine (G), Massachusetts (G), New Hampshire (G), New Jersey (G), Pennsylvania (G).

Description: Acetal, $C_4H_{14}O_2$, $CH_3CH(OC_2H_5)_2$, is a clear, volatile liquid with an agreeable odor. Boiling point = 103°C, Flash point = -21°C. Hazard Identification (based on NFPA-704 M Rating System): Health 2; Flammability, 3; Reactivity, 0. Explosive limits: LEL = 1.6%; UEL = 10.4%. Slightly water soluble.

Potential Exposure: Used as a solvent, in synthetic perfumes such as jasmine, cosmetics, flavors, in organic synthesis.

Incompatibilities: Oxidizing materials. Presumed to form explosive peroxides on contact with air and light. May accumulate static electrical charges, and may cause ignition of its vapors.

Permissible Exposure Limits in Air: No standards have been established.

Permissible Concentration in Water: No criteria established.

Routes of Entry: Ingestion, inhalation, eye and/or skin contact.

Harmful Effects and Symptoms

Short Term Exposure: Irritation of the eyes with redness and pain. Inhalation can cause coughing, headache, and dizziness. Skin contact can cause irritation with redness and pain. Ingestion can cause stomach pain, nausea, sleepiness and high exposure can cause unconsciousness. Affects the central nervous system; acts as a narcotic or hypnotic.

Points of Attack: Inhalation, ingestion, skin and/or eye contact.

First Aid: If this chemical gets into the eyes, remove any contact lenses at once and irrigate immediately for at least 15 minutes, occasionally lifting upper and lower lids. If this chemical contacts the skin, remove contaminated clothing and wash with soap immediately. When this chemical has been swallowed, get medical attention. Give large quantities of water and induce vomiting. Do not make an unconscious person vomit. If this chemical has been inhaled, remove from exposure and transfer promptly to a medical facility.

Personal Protective Methods: Wear protective gloves and clothing to prevent skin contact. Safety equipment suppliers/manufacturers can provide recommendations on the most protective glove/clothing material for your operation. All protective clothing (suits, gloves, footwear, headgear) should be clean, available each day, and put on before work. Contact lenses should not be worn when working with this chemical. Wear splash-proof chemical goggles and face shield unless full facepiece respiratory

protection is worn. Employees should wash immediately with soap when skin is wet or contaminated. Provide emergency showers and eyewash.

Respirator Selection: *At concentrations above the NIOSH REL, or where there is no REL, at any detectable concentration:* SCBAF:PD,PP (any self-contained breathing apparatus that has a full facepiece and is operated in a pressure-demand or other positive-pressure mode); or SAF:PD,PP:ASCBA (any supplied-air respirator that has a full facepiece and is operated in a pressure-demand or other positive-pressure mode in combination with an auxiliary self-contained breathing apparatus operated in a pressure-demand or other positive-pressure mode).

Storage: Prior to working with this chemical you should be trained on its proper handling and storage. Before entering confined space where this chemical may be present, check to make sure that an explosive concentration does not exist. Store in tightly closed containers in a cool, well ventilated area away from oxidizers and other incompatible materials. Where possible, automatically pump liquid from drums or other storage containers to process containers. Sources of ignition such as smoking and open flames are prohibited where this chemical is handled, used, or stored. Metal containers involving the transfer of 5 gallons or more of this chemical should be grounded and bonded. Drums must be equipped with self-closing valves, pressure vacuum bungs, and flame arresters. Use only non-sparking tools and equipment, especially when opening and closing containers of this chemical. Wherever this chemical is used, handled, manufactured, or stored, use explosion-proof electrical equipment and fittings.

Shipping: Label required is "Flammable Liquid." Is in DOT/UN Hazard Class 3, Packing Group II.[19][20]

Spill Handling: Evacuate persons not wearing protective equipment from area of spill or leak until clean-up is complete. Remove all ignition sources. Establish ventilation to keep levels below explosive limit. Collect powdered material in the most convenient and safe manner and deposit in sealed containers. Ventilate area after clean-up is complete. It may be necessary to contain and dispose of this chemical as a hazardous waste. If material or contaminated runoff enters waterways, notify downstream users of potentially contaminated waters. Contact your Department of Environmental Protection or your regional office of the federal EPA for specific recommendations. If employees are required to clean-up spills, they must be properly trained and equipped. OSHA 1910.120(q) may be applicable.

Fire Extinguishing: This chemical is a flammable liquid. Vapors are heavier than air and will collect in low areas. Vapors in confined areas may explode when exposed to fire. If material or contaminated runoff enters waterways, notify downstream users of potentially contaminated waters. Notify local health and fire officials and pollution control agencies. Use dry chemicals, alcohol-resistant foam, carbon dioxide. From a secure, explosion-proof location, use water spray to cool exposed containers. If cooling streams are ineffective (venting sound increases in volume and pitch, tank discolors, or shows any signs of deforming), withdraw immediately to a secure position. If employees are expected to fight fires, they must be trained and equipped in OSHA 1910.156.

Disposal Method Suggested: Incineration.

Acetaldehyde

Molecular Formula: C_2H_4O

Common Formula: CH_3CHO

Synonyms: Acetaldehido (Spanish); Acetaldehyd (German); Acetehyde; Acetic aldehyde; Aldehyde acetique (French); Aldeide acetica (Italian); Ethanal; Ethyl aldehyde; Octowy aldehyd (Polish)

CAS Registry Number: 75-07-0

RTECS Number: AB1925000

DOT ID: UN 1089

EEC Number: 605-003-00-6

EINECS Number: 200-836-8

Regulatory Authority

- Carcinogen (suspected, DFG)[3]
- Pregnancy Risk Group (DFG)[3]
- Air Pollutant standard Set (ACGIH),[1] (DFG)[3] HSE[33] (OSHA)[58] (Many States)[60] (Various Canadian Provinces) (Australia) (Israel) (Mexico)
- OSHA 29CFR1910.119, Appendix A, Process Safety List of Highly Hazardous Chemicals, TQ = 2,500 lb (1,135 kg)
- Clean Air Act 42USC7412; Title I, Part A, §112 hazardous pollutants; Accidental Release Prevention/Flammable Substances (Section 112[r], Table 3), TQ = 10,000 lb (4,540 kg)
- Clean Water Act 40CFR116.4A, hazardous substances, 40CFR413.02, Total Toxic Organics
- RCRA 40CFR266, Appendix 5, Air concentrations
- RCRA Land Ban Restrictions
- EPA Hazardous Waste Number (RCRA No.): U001
- SUPERFUND/EPCRA 40CFR302.4, Appendix A, Reportable Quantity (RQ): 1,000 lb (454 kg), SARA 313: Form R *de minimis* Concentration Reporting Level: 0.1%.
- Canada, WHMIS, Ingredients Disclosure List; National Pollutant Release Inventory (NPRI)
- TSCA: 40CFR716.120(d) (aldehydes)
- U.S. DOT Regulated Marine Pollutant (49CFR172.101, Appendix B)

Cited in U.S. State Regulations: Alaska (G), Connecticut (A), Florida (G), Illinois (G), Indiana (A), Kansas (G), Louisiana (G), Maine (G), Maryland (G), Massachusetts (G, A), Michigan (A), Minnesota (G), Nevada (A), New Hampshire (G), New

Jersey (G), New York (G, A), North Carolina (A), North Dakota (A), Oklahoma (G), Pennsylvania (G), Rhode Island (G), South Carolina (A), Vermont (G), Virginia (G, A), Washington (G), West Virginia (G), Wisconsin (G).

Description: CH_3CHO, acetaldehyde, is a flammable, volatile, colorless liquid or gas with a characteristic penetrating, fruit odor. Odor threshold = 0.067 ppm. Boiling point = 20 – 21°C. Hazard Identification (based on NFPA-704 M Rating System): Health 3; Flammability, 4; Reactivity, 2. Flash point = -38°C. Flammable limits: LEL = 4%, UEL = 60%. Autoignition Temperature: 175°C. Soluble in water.

Potential Exposure: Acetaldehyde can be reduced or oxidized to form acetic acid, acetic anhydride, acrolein, aldol, butanol, chloral, paraldehyde, and pentaerythritol. It is also used in the manufacture of disinfectants, drugs, dyes, explosives, flavorings, lacquers, mirrors (silvering), perfume, photographic chemicals, phenolic and urea resins, rubber accelerators and antioxidants, varnishes, vinegar, and yeast. It is also a pesticide intermediate. Acetaldehyde is the product of most hydrocarbon oxidations; it is a normal intermediate product in the respiration of higher plants; it occurs in all ripe fruits and may form in wine and other alcoholic beverages after exposure to air. Acetaldehyde is an intermediate product in the metabolism of sugars in the body and hence occurs in traces in blood. It has been reported in fresh leaf tobacco as well as in tobacco smoke and in automobile and diesel exhaust. It has been found in 5 of 10 water supplies surveyed by EPA with the highest concentrations in Philadelphia and Seattle at 0.1 μg/l.

Incompatibilities: Reacts with air to form unstable peroxides which can explode. Contact with air causes acetaldehyde to chemically degrade to acetic acid. Strong oxidizers, acids, bases, alcohols, ammonia, amines, halogens, phenols, acid anhydrides, ketones, hydrogen cyanide, H_2S. May dissolve rubber. Slightly corrosive to mild steel. May explode without warning when exposed to heat, dust, corrosives, or oxidizers.

Permissible Exposure Limits in Air: The Federal OSHA standard is 200 ppm (360 mg/m^3).[58] NIOSH Potential occupational carcinogen; reduce exposure to lowest feasible level. ACGIH recommended TLV recommends a ceiling of 25 ppm (45 mg/m^3). The NIOSH IDLH level is 2,000 ppm. The DFG (German) MAK value is 50 ppm (91 mg/m^3) and Peak Limitation (5 minute) of 2 times normal MAK; do not exceed more than 8 times during workshift. Australia, Mexico and Israel have set values of 100 ppm (180 mg/m^3) and Mexico and Israel STEL of 150 ppm (270 mg/m^3) and Israel has an Action Level of 50 ppm (90 mg/m^3). The Canadian Provincial Standards are: Alberta, British Columbia, Ontario, Quebec: 100 ppm (180 mg/m^3) TWA (TWAEV) and STEL (STEV) 150 ppm (270 mg/m^3); The state of California's standard is 100 ppm (180 mg/m^3) TWA and STEL of 150 (270 mg/m^3). The former USSR-UNEP/IRPTC MAC value in workplace air is 5 mg/m^3 and for ambient residential air is 0.01 mg/m^3.[43] In addition a number of states have set guidelines or standards for acetaldehyde in ambient air[60] ranging from 0.4 μg/m^3 (Michigan) to 4.6 μg/m^3 (Massachusetts) to 600 μg/m^3 (New York) to 1.800 μg/m^3 (North Dakota, South Carolina) to 2,700 μg/m^3 (Nevada) to 18,000 μg/m^3 (Indiana) to 27,000 μg/m^3 (North Carolina).

Determination in Air: Acetaldehyde may be collected by XAD® tube with a special coating added. See NIOSH Method 3507 (HPLC), 2538 (GC with flame ionization detection), OSHA Method 68.

Permissible Concentration in Water: Human exposure to acetaldehyde probably antedates recorded history, inasmuch as acetaldehyde is the major metabolite of ethyl alcohol. An additional source of widespread human exposure is tobacco smoke. The pharmacology and toxicology of acetaldehyde have been studied most extensively in its relationship to alcohol toxicity and human metabolism. Because or this background of human and laboratory experience, there appears to be no need to establish limits for acetaldehyde in drinking water.[32] However, EPA has set an ambient environmental goal of 2,480 μg/l for acetaldehyde on a health basis. This compares with a standard of 0.2 mg/l for domestic water supplies set by the former USSR.[43]

Routes of Entry: Inhalation, ingestion, eye and/or skin contact.

Harmful Effects and Symptoms

Short Term Exposure: This chemical can irritate and cause severe eye burns. Breathing can cause irritate the nose, throat and lungs causing coughing and/or shortness of breath. Higher exposure can cause sleepiness, dizziness, and cause unconsciousness and pulmonary edema, a medical emergency, with severe shortness of breath.

Long Term Exposure: Acetaldehyde may be a carcinogen in humans since it has been shown to cause cancer in animals and may be a teratogen in humans since it has been shown to be a teratogen in animals. Exposure to acetaldehyde has produced nasal tumors in rats and laryngeal tumors in hamsters, and exposure to malonaldehyde has produced thyroid gland and pancreatic islet cell tumors in rats. NIOSH therefore recommends that acetaldehyde and malonaldehyde be considered potential occupational carcinogens in conformance with the OSHA carcinogen policy. This chemical may cause skin allergy. If allergy develops, very low future exposure can cause itching and skin rash. Repeated exposure may cause chronic irritation of the eyes leading to permanent damage. See also "NIOSH Current Intelligence Bulletin 55: Carcinogenicity of Acetaldehyde and Malonaldehyde, and Mutagenicity of Related Low-Molecular-Weight Aldehydes" [DHHS (NIOSH) Publication No. 91–112].

Points of Attack: Eyes, skin, respiratory system, kidneys, central nervous system, reproductive system. Cancer site in animals: nasal cancer.

Medical Surveillance: Consideration should be given to skin, eyes, and respiratory tract in any preplacement or periodic examinations. Lung function tests are recommended. If symptoms develop or overexposure is suspected consider chest x-ray and evaluation by a qualified allergist, including careful exposure history and special testing may help diagnose skin allergy.

First Aid: If this chemical gets into the eyes remove any contact lenses at once and irrigate immediately for at least 30 minutes, ocassionally lifting upper and lower lids. Seek medical attention immediately. If this chemical contacts the skin, quickly remove contaminated clothing. Immediately wash area with large amounts of soap and water. If a person breathes in large amounts of this chemical, move the exposed person to fresh air at once and perform artificial respiration. Transfer promptly to medical facility. Medical observation is recommended for 24 – 48 hours after breathing overexposure, as pulmonary edema may be delayed. When this chemical has been swallowed, get medical attention. Give large quantities of water and induce vomiting. Do not make an unconscious person vomit.

Personal Protective Methods: Wear appropriate clothing to prevent repeated or prolonged skin contact. ACGIH recommends butyl rubber as a protective material. Wear eye protection to prevent any potential for eye contact. Employees should wash promptly when skin is wet. Remove clothing immediately if wet or contaminated to avoid flammability hazard. Provide eye-wash.

Respirator Selection: *At any concentrations above the NIOSH REL, or where there is no REL, at any detectable concentration:* SCBAF:PD,PP (Any self-contained breathing apparatus that has a full facepiece and is operated in a pressure-demand or other positive-pressure mode); or SAF:PD,PP: ASCBA (Any supplied-air respirator that has a full facepiece and is operated in a pressure-demand or other positive-pressure mode in combination with an auxiliary self-contained breathing apparatus operated in a pressure-demand or other positive pressure mode). *Escape:* GMFOV [Any air-purifying, full-facepiece respirator (gas mask) with a chin-style, front-or back-mounted organic vapor canister]; or SCBAE (Any appropriate escape-type, self-contained breathing apparatus).

Storage: Prior to working with this chemical you should be trained on its proper handling and storage. Before entering confined space where this chemical may be present, check to make sure that an explosive concentration does not exist. See incompatibilities. Acetaldehyde should be stored in tightly closed airtight containers in a cool, dark, well-ventilated area. Nitrogen or other inert gas should be used as an "inert gas blanket" over liquid acetaldehyde in storage containers. Sources of ignition such as smoking and open flames, are prohibited where this chemical is used, handled, or stored in a manner that could create a potential fire or explosion hazard. Metal containers involving the transfer of acetaldehyde should be grounded and bonded. Drums must be equipped with self-closing valves, pressure vacuum bungs, and flame arresters. Use only non-sparking tools and equipment, especially when opening and closing containers.

Shipping: Should be labeled "Flammable Liquid." Shipment by passenger aircraft or railcar forbidden. Is in Hazard Class 3 and packing group I.[19][20]

Spill Handling: *Liquid:* Evacuate persons not wearing protective equipment from area of spill or leak until clean-up is complete. Remove all ignition sources. Establish forced ventilation to keep levels below explosive limit. Absorb liquids in vermiculite, dry sand, earth, or a similar material and deposit in sealed containers. Keep liquid out of a confined space, such as a sewer, because of the potential for an explosion, unless the sewer is designed to prevent the build-up of explosive concentrations. It may be necessary to contain and dispose of this chemical as a hazardous waste. If material or contaminated runoff enters waterways, notify downstream users of potentially contaminated waters. Contact your Department of Environmental Protection or your regional office of the federal EPA for specific recommendations. If employees are required to clean-up spills, they must be properly trained and equipped. OSHA 1910.120(q) may be applicable.

Gas: Evacuate persons not wearing protective equipment from area of spill or leak until clean-up is complete. Remove all ignition sources. Establish forced ventilation to keep levels below explosive limit. Stop flow of gas. If source of leak is a cylinder and the leak cannot be stopped in place, remove the leaking cylinder to a safe place in the open air, and repair leak or allow cylinder to empty.

Fire Extinguishing: Acetaldehyde is a flammable and reactive liquid or gas. Poisonous gases including carbon monoxide is released in fire. Vapors are heavier than air and will collect in low areas. Vapors in confined areas may explode when exposed to fire. If material or contaminated runoff enters waterways, notify downstream users of potentially contaminated waters. Notify local health and fire officials and pollution control agencies. Use dry chemicals, alcohol-resistant foam, CO_2. Water may be ineffective on fire. From a secure, explosion-proof location, use water spray to cool exposed containers. If cooling streams are ineffective (venting sound increases in volume and pitch, tank discolors, or shows any signs of deforming), withdraw immediately to a secure position. If employees are expected to fight fires, they must be trained and equipped in OSHA 1910.156.

Disposal Method Suggested: Consult with environmental regulatory agencies for guidance on acceptable disposal practices. Generators of waste containing this contaminant ($\geq$100 kg/mo) must conform with EPA regulations governing storage, transportation, treatment, and waste disposal. Incineration.

References

U.S. Environmental Protection Agency, Chemical Hazard Information Profile: Acetaldehyde, (Preliminary), Washington, DC (1979)

U.S. Environmental Protection Agency, Acetaldehyde, Health and Environmental Effects Profile No. 1, Washington, DC, Office of Solid Waste (April 30, 1980)

Sax, N. I., Ed., Dangerous Properties of Industrial Materials Report, 1, No. 1, 25–26 (1980) and 3, No. 6, 23–27, (Nov./Dec. 1983)

New Jersey Dept. of Health and Senior Services, Hazardous Substance Fact Sheet: Acetaldehyde, Trenton, NJ (May, 1998)

U.S. Environmental Protection Agency, Chemical Hazard Information Profile Draft Report; Acetaldehyde, Washington, DC (April 29, 1983)

Acetaldehyde Ammonia

Molecular Formula: C_2H_7NO

Synonyms: Acetaldehidato amonico (Spanish); Acetaldehyde, amine salt; Aldehyde ammonia; 1-Aminoethanol; α-Aminoethyl Alcohol; Ethanol, 1-amino-

CAS Registry Number: 75-39-8

RTECS Number: AB1950000

DOT ID: UN 1841

Cited in U.S. State Regulations: New Jersey (G), Pennsylvania (G)

Description: Acetaldehyde ammonia is a combustible, colorless, white, yellow, or brown crystalline solid melting at 97°C. Highly soluble in water.

Potential Exposure: Acetaldehyde ammonia is used to make acetaldehyde and other chemicals, organic synthesis, and to vulcanize rubber.

Incompatibilities: Strong oxidizing agents such as chlorine, bromine, and fluorine and strong acids such as hydrochloric, sulfuric and nitric since violent reactions occur.

Permissible Exposure Limits in Air: No standards have been established.

Routes of Entry: Inhalation, eye and/or skin contact. Absorbed through the skin.

Harmful Effects and Symptoms

Short Term Exposure: Contact with acetaldehyde ammonia can irritate and may burn eyes and skin. Inhalation can irritate the nose, throat, and lungs causing coughing and/or shortness of breath.

Long Term Exposure: Repeated exposure may cause bronchitis to develop with cough, phlegm, and/or shortness of breath.

Points of Attack: Skin, lungs.

Medical Surveillance: If illness occurs or overexposure is suspected, medical attention is recommended.

First Aid: If this chemical gets into the eyes, remove any contact lenses at once and irrigate immediately for at least 15 minutes, occasionally lifting upper and lower lids. Seek medical attention immediately. If this chemical contacts the skin, quickly remove contaminated clothing and wash contaminated skin with large amounts of water immediately. When this chemical has been swallowed, get medical attention. Give large quantities of water and induce vomiting. Do not make an unconscious person vomit. If this chemical has been inhaled, remove from exposure and transfer promptly to a medical facility.

Personal Protective Methods: Prevent repeated or prolonged skin contact. Wear impact-resistant eye protection with side shields or goggles when eye exposure is reasonably probable. Wash skin and change clothing upon contamination. Eye wash fountains should be provided in the immediate work area for emergency use. If there is the potential for skin exposure, emergency shower facilities should be provided. Do not eat, smoke, or drink where this chemical is handled, processed or stored. Use vacuum or wet method to reduce dust during clean-up. Do not dry sweep.

Respirator Selection: MSHA/NIOSH approved supplied-air respirator with a full facepiece operated in a pressure-demand or other positive-pressure mode. For increased protection use in combination with an auxiliary self-contained breathing apparatus operated in a pressure-demand or other positive-pressure mode.

Storage: Prior to working with this chemical you should be trained on its proper handling and storage. Store in tightly closed containers in a cool, well ventilated area. Sources of ignition such as smoking and open flames, are prohibited where this chemical is used, handled, or stored in a manner that could create a potential fire or explosion hazard.

Shipping: Should be labeled "Class 9."

Spill Handling: Evacuate persons not wearing protective equipment from area of spill or leak until clean-up is complete. Remove all ignition sources. Collect powdered material in the most convenient and safe manner and deposit in sealed containers. Ventilate area after clean-up is complete. It may be necessary to contain and dispose of this chemical as a hazardous waste. If material or contaminated runoff enters waterways, notify downstream users of potentially contaminated waters. Contact your Department of Environmental Protection or your regional office of the federal EPA for specific recommendations. If employees are required to clean-up spills, they must be properly trained and equipped. OSHA 1910.120(q) may be applicable.

Fire Extinguishing: This chemical is a combustible solid. Use dry chemical, carbon dioxide, water spray, or alcohol foam extinguishers. Poisonous gases are produced in fire including acetaldehyde, ammonia, nitrogen odes, and carbon monoxide. If material or contaminated runoff enters waterways, notify downstream users of potentially contaminated waters. Notify local health and fire officials and pollution control agencies. From a secure, explosion-proof location, use water spray to cool exposed containers. If cooling streams are ineffective (venting sound increases in volume and pitch, tank discolors, or shows any signs of deforming), withdraw immediately to a secure position. If employees are expected to fight fires, they must be trained and equipped in OSHA 1910.156.

References

New Jersey Dept. of Health and Senior Services, Hazardous Substance Fact Sheet: Acetaldehyde, Trenton, NJ (May, 1998)

Acetaldehyde Oxime

Molecular Formula: C_2H_5NO

Common Formula: $CH_3CH{=}NOH$

Synonyms: α-Acetaldoxime; β-Acetaldoxime; Acetaldoxime; Aldoxime; Ethanal oxime; Ethylidenehydroxylamine

CAS Registry Number: 107-29-9

RTECS Number: AB2975000

DOT ID: UN 2332

Cited in U.S. State Regulations: New Hampshire (G), New Jersey (G), Pennsylvania (G).

Description: Acetaldehyde oxime, is an extremely flammable, colorless liquid or crystalline solid; low melting crystalline compound. Boiling point = 115°C. Freezing/Melting point = 12°C (beta-); 46.5°C (alpha-). Flash point ≤ -22°C. Soluble in water.

Potential Exposure: Used as a chemical intermediate.

Incompatibilities: Oxidizers (such as perchlorates, peroxides, permanganates, chlorates and nitrates) and strong acids (such as hydrochloric, sulfuric, and nitric). Vapor forms explosive mixture with air. The *beta-* form is able to form unstable peroxides.

Permissible Exposure Limits in Air: No standards have been established.

Permissible Concentration in Water: No criteria set.

Routes of Entry: Inhalation, skin and/or eye contact.

Harmful Effects and Symptoms

Short Term Exposure: Contact can irritate the eyes and skin. Inhalation can irritate nose, throat, and respiratory tract.

Long Term Exposure: Chronic health effects are unknown at this time.

Points of Attack: Eyes, skin, respiratory system.

Medical Surveillance: Pre-employment and regular physical exams are recommended. For those with frequent or high exposure, lung function tests are recommended.

First Aid: Immediately remove any contact lenses and flush eyes with water for 15 minutes, occasionally lifting upper and lower lids. Quickly remove contaminated clothing. Immediately wash contaminated skin with soap and water. Remove person from exposure area and begin rescue breathing if breathing has stopped and CPR if heart action has stopped. Transfer promptly to a medical facility.

Personal Protective Methods: Wear protective gloves and clothing to prevent any reasonable probability of skin contact. Safety equipment suppliers/manufacturers can provide recommendations on the most protective glove/clothing material for your operation. All protective clothing (suits, gloves, footwear, headgear) should be clean, available each day, and put on before work. Contact lenses should not be worn when working with this chemical. Wear impact- and splash-proof chemical goggles and face shield unless full facepiece respiratory protection is worn. Employees should wash immediately with soap when skin is wet or contaminated. Provide emergency showers and eyewash.

Respirator Selection: Where the potential for exposure to this chemical, use a MSHA/NIOSH approved supplied-air respirator with a full facepiece operated in the positive pressure mode or with a full facepiece, hood, or helmet in the continuous flow mode, or use a MSHA/NIOSH approved self-contained breathing apparatus with a full facepiece operated in pressure-demand or other positive pressure mode.

Storage: Prior to working with this chemical you should be trained on its proper handling and storage. Store in tightly closed containers in a cool, well ventilated area. Sources of ignition such as smoking and open flames, are prohibited where this chemical is used, handled, or stored in a manner that could create a potential fire or explosion hazard. Metal containers involving the transfer of acetaldehyde should be grounded and bonded. Drums must be equipped with self-closing valves, pressure vacuum bungs, and flame arresters. Use only non-sparking tools and equipment, especially when opening and closing containers.

Shipping: Label required is "Flammable Liquid." Is in DOT/UN Hazard Class 3 and Packing Group III.[19][20]

Spill Handling: Restrict persons not wearing protective equipment from area of spill or leak until cleanup is complete. Remove all ignition sources. Use foam spray to reduce vapors. Absorb liquids in vermiculite, dry sand, earth, or a similar material and deposit in sealed containers. Ventilate area of spill or leak after clean-up is complete. Collect powdered material in the most convenient and safe manner and deposit in sealed containers. Keep acetaldehyde oxime out of a confined space, such as a sewer, because of the potential for an explosion, unless the sewer is designed to prevent the build-up of explosive concentrations. It may be necessary to contain and dispose of this chemical as a hazardous waste. If material or contaminated runoff enters waterways, notify downstream users of potentially contaminated waters. Contact your Department of Environmental Protection or your regional office of the federal EPA for specific recommendations. If employees are required to clean-up spills, they must be properly trained and equipped. OSHA 1910.120(q) may be applicable.

Fire Extinguishing: This chemical is a flammable solid. Poisonous gases including nitrous oxides are produced in fire. Use dry chemical, carbon dioxide, water spray, or alcohol foam extinguishers. Poisonous gases are produced in fire including carbon monoxide. Vapors are heavier than air and will collect in low areas. Vapors in confined areas may explode when exposed to fire. If material or contaminated runoff enters waterways, notify downstream users of potentially contaminated waters. Notify local health and fire officials and pollution control agencies. From a secure, explosion-proof location, use water spray to cool exposed containers. If cooling streams are ineffective (venting sound increases in volume and pitch, tank discolors, or shows any signs of deforming), withdraw immediately to a secure position. If employees are expected to fight fires, they must be trained and equipped in OSHA 1910.156.

Disposal Method Suggested: Incineration.

References

New Jersey Department of Health and Senior Services, Hazardous Substance Fact Sheet: Acetaldehyde Oxime, Trenton, NJ (August 1998)

Acetamide

Molecular Formula: C_2H_5NO

Common Formula: CH_3CONH_2

Synonyms: Acetamido (Spanish); Acetic acid amide; Acetimidic acid; Amid kyseliny octove (Polish); Ethanamide; Methanecarboxamide

CAS Registry Number: 60-35-5

RTECS Number: AB4025000

DOT ID: NA 9188

EEC Number: 616-022-00-4

Regulatory Authority

- Carcinogen (suspected) (DFG),[3] (animal positive) (IARC)
- Clean Air Act 42USC7412; Title I, Part A, §112 hazardous pollutants
- SUPERFUND/EPCRA 40CFR302.4, Appendix A, Reportable Quantity (RQ): 1 lb (0.454 kg), SARA 313: Form R *de minimis* Concentration Reporting Level: 0.1%.
- TSCA 40CFR704.225(a), CAIR list reporting required

Cited in U.S. State Regulations: California (G), Maryland (G), New Jersey (G), New York (A), Pennsylvania (G).

Description: Acetamide, CH_3CONH_2, is a colorless to yellow, deliquescent, crystalline solid. Odorless if pure, "mousy" odor if impure. Boiling point = 222°C. Freezing/Melting point = 81°C. Hazard Identification (based on NFPA-704 M Rating System): Health 0, Flammability 1, Reactivity 0. Decomposes slowly in cold water.

Potential Exposure: Used as a stabilizer, plasticizer, and solvent in plastics, lacquers, explosives, soldering flux, and chemical manufacturing.

Incompatibilities: Reacts with strong acids such as hydrochloric, sulfuric, and nitric, strong oxidizers, strong bases, strong reducing agents, ammonia, isocyanates, phenols, cresols. Contact with water causes slow hydrolyzation to ammonia and acetate salts.

Permissible Exposure Limits in Air: New York State has set 0.03 μg/m³ for ambient air.[60]

Determination in Air: Odor Threshold: 140 – 160 mg/m³

Permissible Concentration in Water: No criteria set; not very toxic to fish but increases B.O.D.

Routes of Entry: Inhalation, skin and/or eye contact.

Harmful Effects and Symptoms

Short Term Exposure: Irritates eyes and respiratory tract.

Long Term Exposure: Acetamide may be a carcinogen in humans since it has been shown to cause cancers in animals. May cause liver damage.

Points of Attack: Cancer site in animals: liver and lymph cancer.

Medical Surveillance: Liver function tests.

First Aid: If this chemical gets into the eyes, remove any contact lenses at once and irrigate immediately for at least 15 minutes, occasionally lifting upper and lower lids. If this chemical contacts the skin, remove contaminated clothing and wash immediately with soap and water. When this chemical has been swallowed, get medical attention. Give large quantities of water and induce vomiting. Do not make an unconscious person vomit. If this chemical has been inhaled, remove from exposure and transfer promptly to a medical facility.

Personal Protective Methods: Wear protective gloves and clothing to prevent any reasonable probability of skin contact. Safety equipment suppliers/manufacturers can provide recommendations on the most protective glove/clothing material for your operation. All protective clothing (suits, gloves, footwear, headgear) should be clean, available each day, and put on before work. Contact lenses should not be worn when working with this chemical. Wear dust-proof chemical goggles and face shield unless full facepiece respiratory protection is worn.

Respirator Selection: *At any concentrations above the NIOSH REL, or where there is no REL, at any detectable concentration:* SCBAF:PD,PP (any self-contained breathing apparatus that has a full facepiece and is operated in a pressure-demand or other positive-pressure mode); or SAF:PD,PP: ASCBA (any supplied-air respirator that has a full facepiece and is operated in a pressure-demand or other positive-pressure mode in combination with an auxiliary, self-contained breathing apparatus operated in a pressure-demand or other positive pressure mode). *Escape:* GMFOVHiE [any air-purifying, full-facepiece respirator (gas mask) with a chin-style, front- or back-mounted organic vapor canister having a high-efficiency particulate filter]; or SCBAE (any appropriate escape-type, self-contained breathing apparatus).

Storage: Prior to working with this chemical you should be trained on its proper handling and storage. Should be stored in cool, well-ventilated area. Store in a dry area away from water because of deliquescent properties. Sources of ignition such as smoking and open flames, are prohibited where this chemical is used, handled, or stored in a manner that could create a potential fire or explosion hazard.

Shipping: Label required is "Hazard Class 9." Hazardous substances, solid, n.o.s. Falls into Hazard Class 9. Packing Group III.

Spill Handling: Evacuate persons not wearing protective equipment from area of spill or leak until clean-up is complete. Remove all ignition sources. Collect powdered material in the most convenient and safe manner and deposit in sealed containers. Ventilate area after clean-up is complete. Material is very water soluble and hydrolyzes slowly to ammonia and acetate salts. May be removed from alkaline solutions with adsorbent carbon. It may be necessary to contain and dispose of this chemical as a hazardous waste. If material or contaminated runoff enters waterways, notify downstream users of potentially contaminated waters. Contact your Department of Environmental Protection or your regional office of the federal EPA for specific recommendations. If employees are required to clean-up spills, they must be properly trained and equipped. OSHA 1910.120(q) may be applicable.

Fire Extinguishing: This substance is a combustible solid. Poisonous gases are produced in fire including nitrogen oxides and carbon monoxide. Use dry chemical, carbon dioxide, water spray, or alcohol foam extinguishers. Poisonous gases are produced in fire including carbon monoxide. If material or contaminated runoff enters waterways, notify downstream users of potentially contaminated waters. Notify local health and fire officials and pollution control agencies. From a secure, explosion-proof location, use water spray to cool exposed containers. If cooling streams are ineffective (venting sound increases in volume and pitch, tank discolors, or shows any signs of deforming), withdraw immediately to a secure position. If employees are expected to fight fires, they must be trained and equipped in OSHA 1910.156.

Disposal Method Suggested: Add to alcohol or benzene as a flammable solvent and incinerate; oxides of nitrogen produced may be scrubbed out with alkaline solution.

References

New Jersey Department of Health and Senior Services, Hazardous Substance Fact Sheet: Acetamide, Trenton, NJ (September 1998)

Sax, N. I., Ed., Dangerous Properties of Industrial Materials Report, 1, No. 4, 20–21 (1981), and 3, No. 6, 29–31 (Nov./Dec. 1983)

Acetanilide

Molecular Formula: C_8H_9NO

Synonyms: Acetamide, *N*-phenyl-; Acetamidobenzene; Acetanil; Acetanilid; Acetic acid anilide; Acetoanilide; Acetylaminobenzene; *N*-Acetylaniline; Acetylaniline; AN; Aniline, *N*-acetyl-; Antifebrin; Benzenamine, *N*-acetyl; *N*-Fenilacetamida (Spanish); Phenalgene; Phenalgin; *N*-Phenylacetamide

CAS Registry Number: 103-84-4

RTECS Number: AD7350000

DOT ID: UN 1693

Regulatory Authority

- Banned or Severely Restricted (In analgesic drugs) (Japan)[13]
- Water Pollution Standard Set (former USSR)[43]

Cited in U.S. State Regulations: Florida (G), Massachusetts (G), New Hampshire (G), Pennsylvania (G).

Description: Acetanilide, $CH_3CONHC_6H_5$, odorless, orthorhombic plates or scales or white, shining, crystalline solid or powder. Boiling point = 304°C. Freezing/Melting point = 114°C. Flash point = 169°C (oc). Autoignition temperature = 530°C. Hazard Identification (based on NFPA-704 M Rating System): Health 3, Flammability 1, Reactivity 0. Sinks in water; slightly soluble.

Potential Exposure: In rubber industry as accelerator, in plastics industry as cellulose ester stabilizer, in pharmaceutical manufacture, stabilizer for hydrogen peroxide, azo dye manufacture.

Incompatibilities: Alkyl nitrates, alkalis (liberate aniline), chloral hydrate, phenols, ferric salts.

Permissible Exposure Limits in Air: No standards set.

Permissible Concentration in Water: The former USSR-UNEP/IRPTC MAC is 0.004 mg/l in water bodies used for fishery purposes.[43]

Routes of Entry: Ingestion, eye and/or skin contact.

Harmful Effects and Symptoms

Short Term Exposure: Poisonous if ingested; may cause hallucinations, sleepiness, cyanosis, respiratory, kidney damage, cyanosis, methemoglobinemia, decreased body temperature.

Long Term Exposure: May cause kidney damage, skin allergy, eczema. An allergen. Causes contact dermatitis; inhalation or ingestion can cause eczema and cyanosis and methemoglobinemia. Animals tolerate doses of 200 – 400 mg/kg for many weeks.[11] Has been lethal to man at 59 mg/kg.

Points of Attack: Skin and blood stream.

First Aid: If this chemical gets into the eyes, remove any contact lenses at once and irrigate immediately for at least 15 minutes, occasionally lifting upper and lower lids. If this chemical contacts the skin, remove contaminated clothing and wash immediately with soap and water. When this chemical has been swallowed, get medical attention. Give large quantities of water and induce vomiting. Do not make an unconscious person vomit. If this chemical has been inhaled, remove from exposure and transfer promptly to a medical facility.

Note to Physician: Treat for methemoglobinemia. Spectrophotometry may be required for precise determination of levels of methemoglobinemia in urine.

Personal Protective Methods: Wear skin protection, avoid dust inhalation (see respirator selection below).

Respirator Selection: Wear filter mask unless high vapor concentrations are encountered; then use MSHA/NIOSH approved supplied-air respirator with a full facepiece operated in a pressure-demand or other positive-pressure mode. For increased protection use in combination with an auxiliary self-contained breathing apparatus operated in a pressure-demand or other positive-pressure mode.

Storage: Prior to working with this chemical you should be trained on its proper handling and storage. Store in tightly closed containers in a cool, well ventilated area. Metal containers involving the transfer of this chemical should be grounded and bonded. Drums must be equipped with self-closing valves, pressure vacuum bungs, and flame arresters. Use only non-sparking tools and equipment, especially when opening and closing containers of this chemical. Sources of ignition such as smoking and open flames, are prohibited where this chemical is used, handled, or stored in a manner that could create a potential fire or explosion hazard.

Shipping: DOT has not cited acetanilide with regard to label requirements or maximum shipping quantities. Classified as a Tear Gas Substance n.o.s. solid, shipment by passenger aircraft or railcar is forbidden; cargo aircraft shipment is limited. It falls in Hazard Class 6.1 and Packing Group II.[19][20]

Spill Handling: Evacuate persons not wearing protective equipment from area of spill or leak until clean-up is complete. Remove all ignition sources. Collect powdered material in the most convenient and safe manner and deposit in sealed containers. Ventilate area after clean-up is complete. Solids may be dredged. Carbon adsorbent may be used on dissolved portion. It may be necessary to contain and dispose of this chemical as a hazardous waste. If material or contaminated runoff enters waterways, notify downstream users of potentially contaminated waters. Contact your Department of Environmental Protection or your regional office of the federal EPA for specific recommendations. If employees are required to clean-up spills, they must be properly trained and equipped. OSHA 1910.120(q) may be applicable.

Fire Extinguishing: This chemical is a combustible solid. Use dry chemical, carbon dioxide, water spray, or alcohol foam extinguishers. Poisonous gases are produced in fire including nitrous oxides. If employees are expected to fight fires, they must be trained and equipped in OSHA 1910.156.

Disposal Method Suggested: Add to flammable solvents (alcohol or benzene) and incinerate. Oxides of nitrogen may be scrubbed from combustion gases with alkaline solution. See Ref. 22.

References

Sax, N. I., Ed., Dangerous Properties of Industrial Materials Report, 1, No. 4, 21–23 (1981) and 3, No. 6, 27–29 (Nov./Dec. 1983)

Acetic Acid

Molecular Formula: $C_2H_4O_2$

Common Formula: CH_3COOH

Synonyms: 777 etch; Acetic acid (aqueous solution); Acetic acid, glacial; Acide acetique (French); Acido acetico (Italian, Spanish); Aluminum etch 16-1-1-2; Aluminum etch 82-3-15-0; As-1; As-1400; As-18CZ10A; As-18CZ6E; As-1CE; As-5CE; As-CZ5E; Azijnzuur (Dutch); CEA-100 microchrome etchant; Copper, brass brite DIP 1127; Copper, brass brite DIP 127; Dazzlens cleaner; EPF B20 fixer; Essigsaeure (German); Ethanoic acid; Ethylic acid; Fema No. 2006; Freckle etch; Glacial acetic acid; Glass etch; Kodak 33 stop bath; Kovar bright DIP (412X); KTI aluminum etch I/II; Lens cleaner M6015; Mae etchants; Metal etch; Methane carboxylic acid; Mixed acid etch (5-2-2); Mixed acid etch (6-1-1); Octowy kwas (Polish); Pad etch; PFC; Poly etch 95%; Processor fixer concentrate; Rapid film fix; RDH lime solvent; Silicon etch solution; Stress relief etch; Vinegar (4- 6% solution in water); Vinegar acid; Wet K-etch; Wright etch

CAS Registry Number: 64-19-7

RTECS Number: AF1225000

DOT ID: UN 2789 (>80%); UN 2790 (10 – 80%)

EEC Number: 607-002-00-6 (85% in water); 607-002-01-3 (100%)

EINECS Number: 200-580-7

Regulatory Authority

- Air Pollutant Standard Set (ACGIH)[1] (DFG)[3] (HSE)[33] former USSR[43] (OSHA)[58] (Many U.S. States) (various Canadian Provinces)[60] (Australia) (Israel) (Mexico)
- Clean Water Act: 40CFR116.4 Hazardous Substances; RQ 40CFR117.3, (same as CERCLA)
- SUPERFUND/EPCRA 40CFR302.4, Appendix A, Reportable Quantity (RQ): 5,000 lb (2,270 kg)
- Canada WHMIS Ingredients Disclosure List

Cited in U.S. State Regulations: Alaska (G), California (A), Connecticut (A), Florida (G), Illinois (G), Maine (G), Massachusetts (G), Minnesota (G), Nevada (A), New Hampshire (G), New Jersey (G), New York (G), North Carolina (A), North Dakota (A), Oklahoma (G), Pennsylvania (G), Rhode Island (G), South Dakota (A), Virginia (A), West Virginia (G)

Description: Acetic acid, CH_3COOH, is a colorless liquid or crystals with a sour, vinegar-like odor. Pure compound is a solid below 62°F. Often used in an aqueous solution. Odor threshold = 0.016 ppm. Glacial acetic acid contains 99% acid. Boiling point = 117 – 118°C. Flash point = 39°C. Hazard Identification Ratings. Health 3, Flammability 2, Reactivity 0. Autoignition Temperature = 516°C. Explosive limits: LEL = 4.0%; UEL = 19.9%.

Potential Exposure: Acetic acid is widely used as a chemical feedstock for the production of vinyl plastics, acetic anhydride, acetone, acetanilide, acetyl chloride, ethyl alcohol, ketene, methyl ethyl ketone, acetate esters, and cellulose acetates. It is also used alone in the dye, rubber, pharmaceutical, food preserving, textile, and laundry industries. It is utilized, too, in the manufacture of Paris green, white lead, tint rinse, photographic chemicals, stain removers, insecticides and plastics.

Incompatibilities: Vapor forms explosive mixture with air. Violent reaction with oxidizers, organic amines, and bases such as hydroxides and carbonates. Incompatible with strong acids, aliphatic amines, alkanolamines, isocyanates, alkylene oxides, epichlorohydrin, acetaldehyde, 2-aminoethanol, ammonia, ammonium nitrate, chlorosulfonic acid, chromic acid, ethylene diamine, ethyleneimine, halides, peroxides, perchlorates, perchloric acid, permanganates, phosphorus isocyanate, phosphorus trichloride, potassium tert-butoxide, and xylene. Attacks cast iron, stainless steel, and other metals forming flammable hydrogen gas. Will attack many forms of rubber or plastic.

Permissible Exposure Limits in Air: The Federal OSHA standard (TWA),[50] ACGIH 1999, and German MAK[3] value is 10 ppm (25 mg/m^3) TWA for an 8-hour workshift. The NIOSH REL (10 hour) and ACGIH TLV (8 hour) is 10 ppm (25 mg/m^3) TWA and the STEL is 15 ppm (37 mg/m^3), not to be exceeded during any 15 minute work period. The NIOSH IDLH value is 50 ppm. The DFG (German) MAK is 10 ppm (25 mg/m^3) TWA and Peak Limitation (5 min.) of 2 times the normal MAK; do not exceed more than 8 times during an 8-hour workshift. The Australia, Mexico, and Israel standard is 10 ppm (25 mg/m^3) TWA and STEL of 15 ppm (37 mg/m^3) and Israel's Action Level is 5 ppm (12.5 mg/m^3). The Canadian Provincial Limits for Alberta, British Columbia, Ontario, Quebec: 10 ppm (25 mg/m^3) TWA and STEL of 15 ppm (37 mg/m^3). The former USSR-UNEP/IRPTC[43] MAC in workplace air is 5 mg/m^3 for acetic acid and the MAC for ambient air in residential areas is 0.2 mg/m^3 on a momentary basis and 0.06 mg/m^3 on a daily average basis. In addition, several states have set guidelines or standards for acetic acid in ambient air.[60] They range from 0.25 mg/m^3 (North Dakota) to 0.4 mg/m^3 (Virginia) to 0.5 mg/m^3 (Connecticut and South Dakota) to 0.595 mg/m^3 (North Carolina).

Determination in Air: Use OSHA Method 118 or Charcoal tube; formic acid; gas chromatography/flame ionization detection; NIOSH (IV) Method #1603.

Permissible Concentration in Water: No U.S. limit has been established. However, EPA,[32] has proposed an ambient environmental goal of 345 μg/l based on health effects.

Determination in Water: Acetic acid in water may be determined by titration.

Routes of Entry: Inhalation, skin and/or eye contact.

Harmful Effects and Symptoms

Short Term Exposure: Can cause severe irritation, burns, and permanent eye damage. Skin contact can cause severe irritation and burns. Breathing can cause irritation of the mouth, nose, and throat, coughing, and shortness of breath. Higher exposures can cause bronchopneumonia and pulmonary edema, a medical emergency.

Long Term Exposure: Repeated exposure may cause bronchitis to develop, with cough, phlegm, and/or shortness of breath. Repeated skin exposure can cause thickening and cracking of the skin, particularly the skin of the hands. Chronic exposure may result in pharyngitis and catarrhal bronchitis. Ingestion, though not likely to occur in industry, may result in penetration of the esophagus, bloody vomiting, diarrhea, shock, hemolysis, and hemoglobinuria which is followed by anuria. Repeated or prolonged exposure to acetic acid may cause darkening, irritation of the skin, erosion of the exposed front teeth, and chronic inflammation of the nose, throat and bronchi.

Points of Attack: Respiratory system, skin, eyes, teeth.

Medical Surveillance: Lung function tests. Consider chest x-ray following acute overexposure. Consideration should be given to the skin, eyes, teeth, and respiratory tract in placement or periodic examinations.

First Aid: If this chemical gets into the eyes, remove any contact lenses at once and irrigate immediately for at least 30 minutes, occasionally lifting upper and lower lids. Seek medical attention immediately. If this chemical contacts the skin, remove contaminated clothing and wash with soap immediately. When this chemical has been swallowed, get medical attention. If victim is conscious, administer water or milk. Do not induce vomiting. If this chemical has been inhaled, remove from exposure, begin rescue breathing if breathing has stopped and CPR if heart action has stopped. If swallowed, do not induce vomiting. Transfer promptly to a medical facility. Medical observation recommended for 24 – 48 hours following inhalation overexposure, as pulmonary edema may be delayed.

Personal Protective Methods: Contact lenses should not be worn when working with acetic acid. When working with glacial acetic acid, personal protective equipment, protective clothing, gloves, and splash-proof chemical goggles should be worn. Eye fountains and showers should be available in areas of potential exposure. Wear appropriate clothing to prevent any potential for skin contact with liquids of >50% content and promptly if liquids of 10 – 49% acetic acid are involved. ACGIH recommends neoprene, nitrile, polyethylene, and polyvinyl chloride as protective material. Remove clothing immediately if wet or contaminated with liquids containing 50% and promptly remove if liquid contains 10 – 49% acetic acid. Provide emergency eyewash if liquids containing >5% acetic acid are involved, drench if >50% acetic acid is involved.

Respirator Selection: NIOSH/OSHA *50 ppm:* SA:CF (any supplied-air respirator operated in a continuous-flow mode); or PAPROV [any powered, air-purifying respirator with organic vapor cartridge(s)]; CCRFOV [any air-purifying, full-facepiece respirator (gas mask) with a chin-style, front-or back-mounted acid gas canister]; or GMFOV [any air-purifying, full-facepiece respirator (gas mask) with a chin-style, front-or back-mounted organic vapor canister]; or SCBAF (any self-contained breathing apparatus with a full facepiece); or SAF (any supplied-air respirator with a full facepiece). *Emergency or planned entry into unknown concentrations or IDLH conditions:* SCBAF:PD,PP (any self-contained breathing apparatus that has a full facepiece and is operated in a pressure-demand or other positive-pressure mode); or SAF:PD,PP:ASCBA (any supplied-air respirator that has a full facepiece and is operated in a pressure-demand or other positive-pressure mode in combination with an auxiliary self-contained breathing apparatus operated in a pressure-demand or other positive-pressure mode). *Escape:* GMFOV [any air-purifying, full-facepiece respirator (gas mask) with a chin-style, front-or back-mounted organic vapor canister] or SCBAE (any appropriate escape-type, self-contained breathing apparatus).

Note: Substance causes eye irritation or damage; eye protection needed.

Storage: Prior to working with this chemical you should be trained on its proper handling and storage. Before entering confined space where this chemical may be present, check to make sure that an explosive concentration does not existShould be stored in cool dry place away from heat and incompatible substances listed above. Sources of ignition such as smoking and open flames, are prohibited where this chemical is used, handled, or stored in a manner that could create a potential fire or explosion hazard. Metal containers involving the transfer of this chemical should be grounded and bonded. Drums must be equipped with self-closing valves, pressure vacuum bungs, and flame arresters. Use only non-sparking tools and equipment, especially when opening and closing containers of this chemical.

Shipping: *Acetic acid >80%:* Should be labeled "Corrosive, Flammable Liquid." UN/DOT Hazard Class 8. Shipping Group II.[19][20] *Acetic acid 10–80%:* Should be labeled "Corrosive," shipped in glass and polyethylene carboys, metal drums aluminum tank cars and wooden barrels. UN/DOT Hazard Class 8. Shipping Group II.[19][20]

Spill Handling: Warn other workers of spill. Evacuate persons not wearing protective equipment from area of spill or leak until clean-up is complete. Remove all ignition sources. Establish forced ventilation to keep levels below explosive limit. Absorb liquids in vermiculite, dry sand, or similar material and deposit in sealed containers, and transport to outdoor location. Neutralize with lime or sodium bicarbonate. Alternatively cover with soda ash and then flush to sewer with water.[24] Ventilate area after clean-up is complete. It may be necessary to contain and dispose of this chemical as a hazardous waste. If material or contaminated runoff enters waterways, notify downstream users of potentially contaminated waters. Contact your Department of Environmental Protection or your regional office of the federal EPA for specific recommendations. If employees are required to clean-up spills, they must be properly trained and equipped. OSHA 1910.120(q) may be applicable.

Points of Attack: Use water spray, dry chemical, carbon dioxide, water spray, or alcohol foam extinguishers. Wear goggles and self-contained breathing apparatus when fighting fires. Poisonous gases are produced in fire including carbon monoxide. Vapors are heavier than air and will collect in low areas. Vapors in confined areas may explode when exposed to fire. If material or contaminated runoff enters waterways, notify downstream users of potentially contaminated waters. Notify local health and fire officials and pollution control agencies. From a secure, explosion-proof location, use water spray to cool exposed containers. If cooling streams are ineffective (venting sound increases in volume and pitch, tank discolors, or shows any signs of deforming), withdraw immediately to a secure position. If employees are expected to fight fires, they must be trained and equipped in OSHA 1910.156.

Disposal Method Suggested: Incineration after mixing with flammable solvent.[22]

References

Sax, N. I., Ed., Dangerous Properties of Industrial Materials Report, 1, No. 4, 23–25 (1981), and 3, No. 6, 31–35 (Nov./Dec. 1983)

New Jersey Dept. of Health and Senior Services, Hazardous Substance Fact Sheet: Acetic Acid, Trenton, NJ (June 1998)

New York State Dept. of Health, Bureau of Toxic Substances Assessment Chemical Fact Sheet: Acetic Acid, Albany, NY, (March 1986)

Acetic Anhydride

Molecular Formula: $C_4H_6O_3$

Common Formula: $(CH_3CO)_2O$

Synonyms: Acetic acid, anhydride; Acetic oxide; Acetyl anhydride; Acetyl ether; Acetyl oxide; Anhidrido acetico (Spanish); Anhydride acetique (French); Anidride acetica (Italian); Azijnzuuranhydride (Dutch); Essigsaeureanhydrid (German); Ethanoic anhydrate; Ethanoic anhydride; Octowy bezwodnik (Polish)

CAS Registry Number: 108-24-7

RTECS Number: AK1925000

DOT ID: UN 1715

EEC Number: 607-008-00-9

EINECS Number: 203-564-8

Regulatory Authority

- Banned or Severely Restricted (Singapore) (UN)[13]
- FDA, Controlled Substance Act, Essential Chemicals
- Air Pollutant Standard Set (ACGIH),[1] (DFG),[3] (HSE),[33] (OSHA),[58] (Many States),[60] (Various Canadian Provinces) (Australia) (Israel) (Mexico)

- Clean Water Act: 40CFR116.4 Hazardous Substances; RQ 40CFR117.3 (same as CERCLA)
- SUPERFUND/EPCRA 40CFR302.4, Appendix A, Reportable Quantity (RQ): 5,000 lb (2,270 kg)
- Controlled Subststance Act (FDA): TV (domestic and import/export) = 250 gallons (1,023 kg weight)
- Canada, WHMIS, Ingredient Disclosure

Cited in U.S. State Regulations: Alaska (G), California (G), Connecticut (A), Florida (G), Illinois (G), Maine (G), Massachusetts (G), Minnesota (G), Nevada (A), New Hampshire (G), New Jersey (G), New York (G, A), North Dakota (A), Pennsylvania (G), Rhode Island (G), South Carolina (A), Virginia (A), West Virginia (G)

Description: Acetic anhydride, $(CH_3CO)_2O$, is a combustible, colorless, strongly refractive, liquid which has a strongly irritating odor. Odor threshold = 0.14 ppm. Boiling point = 140°C. Hazard Identification (based on NFPA-704 M Rating System): Health 3, Flammability 2, Reactivity 1. Flash point = 49.4°C. Autoignition Temperature = 316°C. Explosive limits: LEL = 2.7%, UEL = 10.3%. Soluble in water.

Potential Exposure: Acetic anhydride is used as an acetylating agent or as a solvent in the manufacture of cellulose acetate, acetanilide, aspirin, synthetic fibers, plastics, explosives, resins, perfumes, and flavorings; and it is used in the textile dyeing industry. It is widely used as a pharmaceutical intermediate and as a pesticide intermediate.

Incompatibilities: Water, alcohols, strong acids, strong oxidizers, chromic acid (violent reaction), amines, strong caustics, finely divided metals. Contact with water forms acetic acid and liberates a large amount of heat. Corrosive to iron, steel and other metals.

Permissible Exposure Limits in Air: The Federal OSHA standard (TWA),[58] the German (DFG/MAK) standard[3] and the ACGIH 1999 value is 5 ppm (21 mg/m^3) TWA. NIOSH has a ceiling value of 5 ppm (21 mg/m^3). The German Peak limitation Value is 2 × normal MAK value (5 minute momentary value); not to be exceeded 8 times during work shift. There is no proposed STEL value in the U.S. The HSE[33] has only a STEL value of 5 ppm (20 mg/m^3). The Australia and Mexico limit is 5 ppm (21 mg/m^3) TWA. The Israel limit is Ceiling is 5 ppm (21 mg/m^3). The Canadian Provincial limits are: Alberta, British Columbia, Ontario, Quebec: Ceiling 5 ppm (21 mg/m^3). The NIOSH IDLH level is 200 ppm. The former USSR-UNEP/IRPTC project[43] has set a MAC value of 0.1 mg/m^3 on a momentary basis and 0.03 mg/m^3 on an average daily basis for air in residential areas. In addition, a number of states have set guidelines or standards for acetic anhydride in ambient air[60] ranging from 67 μg/m^3 (NY) to 160 μg/m^3 (Virginia) to 200 μg/m^3 (North Dakota) to 400 μg/m^3 (Connecticut) to 476 μg/m^3 (Nevada) to 500 μg/m^3 (South Carolina).

Determination in Air: The sample is bubbled through hydroxylamine, worked up with $FeCl_3$ and analyzed colorimetrically. See NIOSH Method #3506.[18]

Permissible Concentration in Water: No criteria set.

Routes of Entry: Inhalation, ingestion, and eye and/or skin contact.

Harmful Effects and Symptoms

Short Term Exposure: This chemical can cause severe skin and eye irritation. Permanent damage to the eyes may result from exposure to high concentrations. Breathing acetic anhydride can irritate the respiratory tract and high concentrations can cause severe lung damage and/or coughing and shortness of breath. In high concentrations, vapor may cause conjunctivitis, photophobia, lacrimation, and severe irritation of the nose and throat. Liquid acetic anhydride does not cause a severe burning sensation when it comes in contact with the skin. If it is not removed, the skin may become white and wrinkled, and delayed severe burns may occur. Both liquid and vapor may cause conjunctival edema and corneal burns, which may develop into temporary or permanent interstitial keratitis with corneal opacity due to progression of the infiltration. Contact and, occasionally, hypersensitivity dermatitis may develop. Immediate complaints following concentrated vapor exposure include conjunctival and nasopharyngeal irritation, cough, and dyspnea. Necrotic areas of mucous membranes may be present following acute exposure.

Long Term Exposure: This chemical may cause skin allergy. If allergy develops, very low future exposure can cause itching and a skin rash.

Points of Attack: Respiratory system, eyes, skin.

Medical Surveillance: Consideration should be given to the skin, eyes, and respiratory tract (lung function tests) in any placement or periodic examinations. Evaluation by a qualified allergist, including careful exposure history and special testing may help diagnose skin allergy.

First Aid: If this chemical gets into the eyes, remove any contact lenses at once and irrigate immediately with large amounts os water. If this chemical contacts the skin, quickly remove clothing and immediately wash area with large amounts of water. If a person breathes this chemical, move the exposed person to fresh air at once and perform rescue breathing if breathing has stopped and CPR if hearth action has stopped. When this chemical has been swallowed, get medical attention immediately. If victim is conscious, administer water or milk. Do not induce vomiting.

Personal Protective Methods: Wear appropriate clothing to prevent any reasonable probability of skin contact. Safety manufacturers recommend butyl rubber as a protective material. Contact lenses should not be worn when working with this chemical. Wear splash proof chemical goggles and face shield unless full facepiece respiratory protection is worn. Employees should wash immediately with soap when skin is wet or contaminated. Remove nonimpervious clothing immediately if wet or contaminated. Provide emergency showers and eyewash.

Respirator Selection: NIOSH/OSHA, *125 ppm:* SA:CF (any supplied-air respirator operated in a continuous-flow mode); or PAPROV [any powered, air-purifying respirator with organic vapor cartridge(s)]. *200 ppm:* CCRFOV [any air-purifying, full-facepiece respirator (gas mask) with a chin-style, front- or back-mounted acid gas canister]; or GMFOV [any air-purifying, full-facepiece respirator (gas mask) with a chin-style, front- or back-, mounted organic vapor canister]; or PAPRTOV [any powered, air-purifying respirator with a tight-fitting facepiece and organic vapor cartridge(s)]; or SCBAF (any self-contained breathing apparatus with a full facepiece); or SAF (any supplied-air respirator with a full facepiece). *Emergency or planned entry into unknown concentrations or IDLH conditions:* SCBAF:PD,PP (any self-contained breathing apparatus that has a full facepiece and is operated in a pressure-demand or other positive-pressure mode); or SAF:PD,PP:ASCBA (any supplied-air respirator that has a full facepiece and is operated in a pressure-demand or other positive-pressure mode in combination with an auxiliary self-contained breathing apparatus operated in a pressure-demand or other positive pressure mode). *Escape:* GMFOV [any air-purifying, full-facepiece respirator (gas mask) with a chin-style, front- or back-, mounted organic vapor canister]; or SCBAE (any appropriate escape-type, self-contained breathing apparatus).

Note: Substance causes eye irritation or damage; eye protection needed.

Storage: Prior to working with this chemical you should be trained on its proper handling and storage. Before entering confined space where this chemical may be present, check to make sure that an explosive concentration does not exist. Protect against physical damage. Outside or detached storage is preferred. Store in tightly closed containers in a cool, well-ventilated place, away from moisture, sources of ignition, and heat. Avoid pits, depressions and basements. Separate from other storage. Inside storage should be in a standard flammable liquids storage room or cabinet. Sources of ignition such as smoking and open flames, are prohibited where this chemical is used, handled, or stored in a manner that could create a potential fire or explosion hazard. Metal containers involving the transfer of this chemical should be grounded and bonded. Drums must be equipped with self-closing valves, pressure vacuum bungs, and flame arresters. Use only non-sparking tools and equipment, especially when opening and closing containers of this chemical.

Shipping: Label required is "Corrosive, Flammable Liquid." Is in UN/DOT Hazard Class 8, Packing Group II.[19][20]

Spill Handling: Evacuate and restrict persons not wearing protective equipment from area of spill or leak until cleanup is complete. Remove all ignition sources. Use foam spray to reduce vapors. Cover with vermiculite, dry sand, earth, or similar absorbent material and neutralize with lime or sodium bicarbonate. Deposit absorbent material in sealed containers. Alternatively cover with soda ash and then flush away with water.[24] Ventilate area of spill or leak after clean-up is complete. Collect powdered material in the most convenient and safe manner and deposit in sealed containers. Keep acetic anhydride out of a confined space, such as a sewer, because of the potential for an explosion, unless the sewer is designed to prevent the build-up of explosive concentrations. It may be necessary to contain and dispose of this chemical as a hazardous waste. If material or contaminated runoff enters waterways, notify downstream users of potentially contaminated waters. You may want to seek assistance from EPAs Environmental Response Team at (908)548-8730. Contact your Department of Environmental Protection or your regional office of the federal EPA for specific recommendations. If employees are required to clean-up spills, they must be properly trained and equipped. OSHA 1910.120(q) may be applicable.

Fire Extinguishing: This chemical is a flammable liquid. Firefighting gear (including SCBA) may not provide adequate protection. If exposure occurs, remove and isolate gear immediately and thoroughly decontaminate personnel. Poisonous gases including carbon monoxide and acetic acid are produced in fire. Use dry chemical, carbon dioxide, water spray, or alcohol foam extinguishers. Vapors are heavier than air and will collect in low areas. Vapors in confined areas may explode when exposed to fire. Vapors may travel long distances to ignition sources and flashback. If material or contaminated runoff enters waterways, notify downstream users of potentially contaminated waters. Notify local health and fire officials and pollution control agencies. Use water spray to disperse vapors only, as water contact will form acetic acid. From a secure, explosion-proof location, use water spray to cool exposed containers. If cooling streams are ineffective (venting sound increases in volume and pitch, tank discolors, or shows any signs of deforming), withdraw immediately to a secure position. If employees are expected to fight fires, they must be trained and equipped in OSHA 1910.156.

Disposal Method Suggested: Incineration of solution in flammable solvent.[24]

References

Sax, N. I., Ed., Dangerous Properties of Industrial Materials Report, 3, No. 3, 32–35 (1983)

New Jersey Dept. of Health and Senior Services, Hazardous Substance Fact Sheet: Acetic Anhydride, Trenton, NJ (May, 1998)

New York State Dept. of Health, Chemical Fact Sheet: Acetic Anhydride, Albany, NY, Bureau of Toxic Substances Assessment (Jan. 1986)

Acetone

Molecular Formula: C_3H_6O

Common Formula: CH_3COCH_3

Synonyms: Aceton (Dutch, German, Polish); Acetona (Spanish); Dimethylformaldehyde; Dimethylformehyde; Dimethylketal; Dimethyl ketone; Ketone; Ketone, dimethyl; Ketone propane; β-Ketopropane; Methyl ketone; 2-Propanone; Propanone; Pyroacetic acid; Pyroacetic ether

CAS Registry Number: 67-64-1

RTECS Number: AL3150000

DOT ID: UN 1090

EEC Number: 606-001-00-8

EINECS Number: 200-662-2

Regulatory Authority

- Air Pollutant Standard Set (ACGIH),[1] (DFG),[3] (HSE),[33] (former USSR),[43] (OSHA),[58] (MEXICO), (Many States)[60] (Various Canadian Provinces)
- FDA, Controlled Substance Act, Essential Chemicals
- Safe Drinking Water Act, 40CFR148.10, solvent waste prohibitions
- EPA Hazardous Waste Number (RCRA No.): U002
- RCRA 40CFR258, Appendix 1, constituents for detection monitoring
- RCRA 40CFR258, Appendix 2, list of inorganic and organic constituents
- RCRA, 40CFR264, Appendix 9, Ground Water Monitoring List, Suggested Testing Methods (PQL μg/l): 8240 (100)
- RCRA Land Ban Substance
- SUPERFUND/EPCRA 40CFR302.4, Appendix A, Reportable Quantity (RQ): 5,000 lb (2,270 kg)
- TSCA: 40CFR799.5000; 40CFR716.120(d)1 as aldehydes
- Canada WHMIS, Ingredients Disclosure List

Cited in U.S. State Regulations: Alaska (G), California (G, A), Connecticut (A), Florida (G, A), Illinois (G), Kansas (G), Louisiana (G), Maine (G), Maryland (G, W), Massachusetts (G, A, W), Minnesota (G), Nevada (A), New Hampshire (G, W), New Jersey (G), New York (G, A), North Dakota (A), Oklahoma (G), Pennsylvania (G), Rhode Island (G), South Dakota (A), Vermont (G), Virginia (G, A), Washington (G), West Virginia (G), Wisconsin (G)

Description: Acetone, CH_3COCH_3, is a highly flammable, colorless liquid with a sweet, mint-like odor. Odor threshold = 4.58 ppm; AIHA geometric mean air odor threshold, 62 ppm (detectable); 130 ppm (recognizable). Boiling point = 56.5°C. Flash point = -20°C. Hazard Identification (based on NFPA-704 M Rating System): Health 1, Flammability 3, Reactivity 0. Explosive limits: LEL = 2.5%, UEL = 12.8%. Water soluble.

Potential Exposure: It is used as a solvent in nail polish remover and many other chemicals. Used in the production of lubricating oils and as an intermediate in the manufacture of chloroform and of various pharmaceuticals and pesticides.

Incompatibilities: May explode when mixed with chloroform, chromic anhydride. Incompatible with acids, bases, and oxidizing materials such as peroxides, chlorates, perchlorates, nitrates, and permanganates. Unstable peroxides formed with strong oxidizers. May accumulate static electrical charges and may cause ignition of its vapors. Dissolves most rubber, resins, and plastics.

Permissible Exposure Limits in Air: The Federal OSHA standard[58] is 1,000 ppm (2,400 mg/m^3), the DFG/MAK value is 500 ppm (1,200 mg/m^3),[3] Peak Limitations are 2 × normal MAK (30 minute average value); not to exceed 4 times per shift. NIOSH has a TWA of 250 ppm (590 mg/m^3). The ACGIH has a TWA of 500 ppm (1,188 mg/m^3) and a STEL of 750 ppm (1,782 mg/m^3). The NIOSH IDLH level is 2,500 ppm [10% LEL]. The Australia limit is 500 ppm (1,185 mg/m^3) TWA and STEL of 1,000 ppm (2,375 mg/m^3). The Mexico limit is 1,000 ppm (2,400 mg/m^3) TWA and STEL of 1,260 ppm (3,000 mg/m^3). The Israel limit is 750 ppm (1,780 mg/m^3) TWA and STEL of 1,000 ppm (2,380 mg/m^3) and Action Level of 375 ppm (890 mg/m^3). The Canada Provincial levels are: Alberta, Ontario, Quebec: 750 ppm (1,782 mg/m^3) TWA and STEL of 1,000 ppm (2,375 mg/m^3); British Columbia: 250 ppm and STEL 500 ppm. The HSE[33] has set an 8-hour TWA of 750 ppm (1,780 mg/m^3) as well as a STEL of 1,500 ppm (3,560 mg/m^3). The former USSR-UNEP/IRPTC project[43] has set a MAC of 200 mg/m^3 for the workplace and 0.35 mg/m^3 for ambient air in residential areas (the latter both on a momentary and an average daily basis). California's workplace Permissible Exposure Limit is 750 ppm (1,780 mg/m^3) TWA, with STEL of 1,000 ppm (2,400 mg/m^3), Ceiling of 3,000 ppm. In addition, a number of states have set guidelines or standards for acetone in ambient air[60] ranging from 8 mg/m^3 (Massachusetts) to 11.8 mg/m^3 (Connecticut and South Dakota) to 17.8 – 23.75 mg/m^3 (North Dakota) to 30 mg/m^3 (Virginia) to 35.6 mg/m^3 (Florida and New York) to 42.4 mg/m^3 (Nevada).

Determination in Air: Charcoal adsorption followed by CS_2 treatment and gas chromatographic analysis. See NIOSH Method 1300 for ketones.[1][18]

Permissible Concentration in Water: Massachusetts has set a guideline of 250 μg/l and Maryland a guideline of 3,600 μg/l.[61]

Routes of Entry: Eyes, skin and/or eye contact.

Harmful Effects and Symptoms

Short Term Exposure: Contact can irritate the skin. Exposure can irritate the eyes and respiratory tract. Exposure to high concentrations can cause dizziness, lightheadedness, and unconsciousness.

Long Term Exposure: Repeated skin exposure can cause dryness and skin cracking. This chemical has not been adequately evaluated to determine whether brain or nerve damage could occur with repeated exposure. However, many solvents and other petroleum-based chemicals have been shown to cause such damage. Effects may include reduced memory and concentration, personality changes (withdrawal, irritability), and fatigue, sleep disturbances, reduced coordination, and/or effects on the nerves to the arms and legs (weakness, "pins and needles").

Points of Attack: Respiratory system, skin.

Medical Surveillance: Preplacement examinations should evaluate skin and respiratory conditions. Acetone can be detected in the blood, urine and expired air and can be used as an index of exposure. Evaluation for brain effects such as changes in memory, concentration, sleeping patterns and mood, as well as headaches and fatigue. Consider evaluations of the cerebellar, autonomic and peripheral nervous systems. Positive and borderline individuals should be referred for neuropsychological testing.

First Aid: If this chemical gets into the eyes, remove any contact lenses at once and irrigate immediately with large amounts of water for at least 15 minutes. If this chemical contacts the skin, quickly remove contaminated clothing and wash with large amounts of soap immediately. If a person breathes in large amounts of this chemical, move the exposed person to fresh air at once and perform rescue breathing if breathing has stopped and CPR if heart action has stopped. When this chemical has been swallowed, get medical attention. Give large quantities of water and induce vomiting. Do not make an unconscious person vomit.

Personal Protective Methods: Wear appropriate solvent-resistant gloves and clothing to prevent repeated or prolonged skin contact. ACGIH and safety equipment manufacturers recommend butyl rubber as a protective material. Wear splash-proof chemical goggles and face shield unless full facepiece respiratory protection is worn. Provide emergency showers and eyewash.

Respirator Selection: NIOSH, *2,500 ppm:* CCRFOV [any chemical cartridge respirator with a full facepiece and organic vapor cartridge(s)]; or PAPROV [any powered, air-purifying respirator with organic vapor cartridge(s)]; or GMFOV [any air-purifying, full-facepiece respirator (gas mask) with a chin-style, front-or back-mounted acid gas canister]; or SA (any supplied-air respirator); or SCBAF (any self-contained breathing apparatus with a full facepiece). *Emergency or planned entry into unknown concentrations or IDLH conditions:* SCBAF:PD,PP (any self-contained breathing apparatus that has a full facepiece and is operated in a pressure-demand or other positive-pressure mode); or SAF:PD,PP:ASCBA (any supplied-air respirator that has a full facepiece and is operated in a pressure-demand or other positive-pressure mode in combination with an auxiliary self-contained breathing apparatus operated in a pressure-demand or other positive-pressure mode). *Escape:* GMFOV [any air-purifying, full-facepiece respirator (gas mask) with a chin-style, front-or back-mounted organic vapor canister]; or SCBAE (any appropriate escape-type, self-contained breathing apparatus).

Note: Substance reported to cause eye irritation or damage.

Storage: Prior to working with this chemical you should be trained on its proper handling and storage. Before entering confined space where this chemical may be present, check to make sure that an explosive concentration does not exist. Store in tightly closed containers in a cool, well ventilated area. Metal containers involving the transfer of this chemical should be grounded and bonded. Drums must be equipped with self-closing valves, pressure vacuum bungs, and flame arresters. Use only non-sparking tools and equipment, especially when opening and closing containers of this chemical. Sources of ignition such as smoking and open flames, are prohibited where this chemical is used, handled, or stored in a manner that could create a potential fire or explosion hazard.

Shipping: The required DOT shipping designation is "Flammable Liquid." Is in UN/DOT Hazard Class 3, Packing Group II.[19][20]

Spill Handling: Restrict persons not wearing protective equipment from area of spill or leak until cleanup is complete. Remove all ignition sources. Establish forced ventilation to keep levels below explosive limit. Absorb liquids in vermiculite, dry sand, earth, or a similar material and deposit in sealed containers. Keep acetone out of a confined space, such as a sewer, because of the potential for an explosion, unless the sewer is designed to prevent the build-up of explosive concentrations. It may be necessary to contain and dispose of this chemical as a hazardous waste. If material or contaminated runoff enters waterways, notify downstream users of potentially contaminated waters. Contact your Department of Environmental Protection or your regional office of the federal EPA for specific recommendations. If employees are required to clean-up spills, they must be properly trained and equipped. OSHA 1910.120(q) may be applicable.

Fire Extinguishing: Acetone is a highly flammable liquid. Use dry chemical, carbon dioxide, or alcohol foam extinguishers. Poisonous gases are produced in fire. Vapors are heavier than air and will collect in low areas. Vapors in confined areas may explode when exposed to fire. Vapors may travel long distances to ignition sources and flashback. If material or contaminated runoff enters waterways, notify downstream users of potentially contaminated waters. Notify local health and fire officials and pollution control agencies. Containers may explode in fire. From a secure, explosion-proof location, use water spray to cool exposed containers. If cooling streams are ineffective (venting sound increases in volume and pitch, tank discolors, or shows any signs of deforming), withdraw immediately to a secure position. If employees are expected to fight fires, they must be trained and equipped in OSHA 1910.156.

Disposal Method Suggested: Consult with environmental regulatory agencies for guidance on acceptable disposal practices. Generators of waste containing this contaminant ($\geq$100 kg/mo) must conform with EPA regulations governing storage, transportation, treatment, and waste disposal. Incineration.[22]

References

Nat. Inst. for Occup. Safety and Health, Criteria for a Recommended Standard. Occupational Exposure to Ketones, NIOSH Pub. No. 78–173, Washington, DC (1978)

Sax, N. I., Ed., Dangerous Properties of Industrial Materials Report, 1, No. 4, 25–27 (1981)

New Jersey Department of Health and Senor Services, Hazardous Substance Fact Sheet: Acetone, Trenton, NJ (May 1998)

New York State Department of Health, "Chemical Fact Sheet: Acetone," Albany, NY, Bureau of Toxic Substance Assessment (March 1986)

Acetone Cyanohydrin (stabilized)

Molecular Formula: C_4H_7NO

Common Formula: $(CH_3)_2C(OH)CN$

Synonyms: Acetoncianhidrinei (Roumanian); Acetoncianidrina (Italian); Acetoncyaanhydrine (Dutch); Acetoncyanhydrin (German); Acetonecyanhydrine (French); Acetone cyanohydrin; Acetonkyanhydrin (Czech); Cianhidrina de acetona (Spanish); Cyanhydrine d'acetone (French); 2-Cyano-2-proponal; α-Hydroxyisobutyronitrile; 2-Hydroxyisobutyronitrile; Hydroxy isobutyronitrite; 2-Hydroxy-2-methylpropionitrile; 2-Methylactonitrile; 2-Propane cyanohydrin; Propanenitrile, 2-hydroxy-2-methyl-

CAS Registry Number: 75-86-5

RTECS Number: OD9275000

DOT ID: UN 1541 (stabilized)

EEC Number: 608-004-00-x

Regulatory Authority

- Air Pollutant Standard Set (NIOSH)[9] (former USSR)[43]
- Clean Water Act: 40CFR116.4 Hazardous Substances; RQ 40CFR117.3 (same as CERCLA); 40CFR423, Priority Pollutants
- EPA Hazardous Waste Number (RCRA No.): P069
- RCRA, 40CFR261, Appendix 8 Hazardous Constituents.
- RCRA Land Ban Chemical
- CERCLA/SARA 40CFR302, Extremely Hazardous Substances: TPQ = 1,000 lb (454 kg).
- SUPERFUND/EPCRA 40CFR302.4, Appendix A, Reportable Quantity (RQ): 10 lb (4.54 kg).
- EPCRA Section 313 Form R *de minimis* concentration reporting level: 1.0%.
- U.S. DOT Regulated Marine Pollutant (49CFR172.101, Appendix B)
- Toxic Substance (World Bank)[15]
- Canada, WHMIS, Ingredients Disclosure List

Cited in U.S. State Regulations: California (G), Florida (G), Illinois (G), Kansas (G), Louisiana (G), Maine (G), Massachusetts (G), Michigan (G), Minnesota (G), New Hampshire (G), New Jersey (G), Oklahoma (G), Pennsylvania (G), Rhode Island (G), Vermont (G), Washington (G), Wisconsin (G).

Description: Acetone cyanohydrin, C_4H_7NO, is a very flammable, colorless to light yellow liquid with an almond-like odor. Boiling point = 120°C (decomposes). Freezing/Melting point = -19°C. Hazard Identification (based on NFPA-704 M Rating System): Health 4, Flammability 2, Reactivity 2. Flash point = 74°C; Autoignition Temperature = 688°C. LEL = 2.2%, UEL = 12.0%. Extremely soluble in water; decomposes to form hydrogen cyanide.

Potential Exposure: Used in the manufacture of insecticides and making other chemicals such as methyl methacrylate.

Incompatibilities: Forms explosive mixture with air. Not compatible with strong reducers, strong bases, strong oxidizers and strong acids such as hydrochloric, sulfuric (explosive), and nitric. Contact with strong acid and strong bases may cause explosions. Slowly decomposes to acetone and HCN at room temperatures; rate is accelerated by and increase in pH, water content, or temperature.

Permissible Exposure Limits in Air: There is no OSHA PEL. A NIOSH-recommended standard is 1 ppm (4 mg/m^3) on a 15-minute exposure basis. The ACGIH TLV is 5 mg/m^3 as CN.[41] IDLH value = 25 ppm as CN (3/95 revision from NIOSH). The above are airborne exposure levels only. If skin contact occurs at the same time, you may be overexposed. Note: Forms CN in the body.

Determination in Air: Porapak® tube; Ethyl acetate; Gas chromatography/Nitrogen/phosphorus detection; NIOSH IV [#2506].[18]

Permissible Concentration in Water: The former USSR-UNEP/IRPTC project has set a MAC of 0.001 mg/l for water used for domestic purposes.

Determination in Water: The substance is very toxic to aquatic organisms. Avoid release to the environment in circumstances different to normal use.

Routes of Entry: Inhalation, ingestion, skin and/or eye contact.

Harmful Effects and Symptoms

Short Term Exposure: Contact can cause eye and skin irritation. Breathing this chemical can irritate the respiratory tract causing wheezing and shortness of breath. High exposure can cause sudden death without warning. Symptoms of exposure include weakness, headache, confusion, nausea, vomiting, and a pounding heart. High exposure can cause liver and kidney damage. Inhalation may cause pulmonary edema, which can be delayed for several hours; there is a risk of death in serious cases.

Long Term Exposure: Can cause thyroid gland to enlarge and interfere with normal thyroid function. May cause kidney and liver injury.

Points of Attack: Eyes, skin, respiratory system, central nervous system, cardiovascular system, liver, kidneys, GI tract.

Medical Surveillance: Test for urine thiocyanate and blood cyanide levels.

First Aid: Move victim to fresh air; call emergency medical care. If this chemical gets into the eyes, remove any contact lenses at once and irrigate immediately for at least 15 minutes, occasionally lifting upper and lower lids. Seek medical attention immediately. If this chemical contacts the skin, quickly remove contaminated clothing and wash with large amounts of water. Speed in removing material from skin is of extreme importance. Seek medical attention immediately. When this chemical has been swallowed, get medical attention immediately. If this chemical has been inhaled, remove from exposure and transfer promptly to a medical facility. If not breathing, give artificial respiration (avoid mouth to mouth resuscitation). If breathing is difficult, give oxygen. If heart has stopped begin CPR. Keep victim quiet and maintain normal body temperature. Effects may be delayed; keep victim under observation. Avoid contact with contaminated skin. A cyanide antidote kit should be kept in the immediate work area and must be rapidly available. Kit ingredients should be replaced every 1 – 2 years to ensure freshness. Persons trained in the use of this kit, oxygen use, and CPR must be available within 1 – 2 minutes.

Personal Protective Methods: Wear protective gloves and clothing to prevent any reasonable probability of skin contact. Safety equipment suppliers/manufacturers can provide recommendations on the most protective glove/clothing material for your operation. All protective clothing (suits, gloves, footwear, headgear) should be clean, available each day, and put on before work. Contact lenses should not be worn when working with this chemical. Wear splash-proof chemical goggles and face shield unless full facepiece respiratory protection is worn. Employees should wash immediately with soap when skin is wet or contaminated. Provide emergency showers and eyewash. See NIOSH Criteria Document 78–212 NITRILES.

Respirator Selection: NIOSH, *10 ppm:* SA (any supplied-air respirator); *25 ppm:* SA:CF (any supplied-air respirator operated in a continuous-flow mode); *50 ppm:* SCBAF (any self-contained breathing apparatus with a full facepiece); or SAF (any supplied-air respirator with a full facepiece); *250 ppm:* SAF:PD,PP (Any supplied-air respirator that has a full facepiece and is operated in a pressure-demand or other positive-pressure mode). *Emergency or planned entry into unknown concentrations or IDLH conditions:* SCBAF:PD,PP (any self-contained breathing apparatus that has a full facepiece and is operated in a pressure-demand or other positive-pressure mode); or SAF:PD,PP:ASCBA (Any supplied-air respirator that has a full facepiece and is operated in a pressure-demand or other positive-pressure mode in combination with an auxiliary self-contained breathing apparatus operated in a pressure-demand or other positive pressure mode. *Escape*: GMFOV [Any air-purifying, full-facepiece respirator (gas mask) with a chin-style, front-or back-, mounted organic vapor canister] or SCBAE (Any appropriate escape-type, self-contained breathing apparatus).

Storage: Prior to working with this chemical you should be trained on its proper handling and storage. Before entering confined space where this chemical may be present, check to make sure that an explosive concentration does not exist. Do not store for long periods of time; toxic fumes may form in closed container. Store in tightly closed containers in a cool, well ventilated area. Sources of ignition such as smoking and open flames, are prohibited where this chemical is used, handled, or stored in a manner that could create a potential fire or explosion hazard. Outside storage and separated storage is preferred.

Shipping: The DOT label required is "Poison." Shipment by passenger aircraft or railcar is forbidden. Is in UN/DOT Hazard Class 6.1, Packing Group I.[19][20]

Spill Handling: Issue poison warning. Evacuate and restrict persons not wearing protective equipment from area of spill or leak until cleanup is complete. Remove all ignition sources. Establish forced ventilation to keep levels below explosive limit. Absorb liquids in vermiculite, dry sand, earth, or a similar material and deposit in sealed containers. Keep this chemical out of a confined space, such as a sewer, because of the potential for an explosion, unless the sewer is designed to prevent the build-up of explosive concentrations. It may be necessary to contain and dispose of this chemical as a hazardous waste. If material or contaminated runoff enters waterways, notify downstream users of potentially contaminated waters. Contact your Department of Environmental Protection or your regional office of the federal EPA for specific recommendations. If employees are required to clean-up spills, they must be properly trained and equipped. OSHA 1910.120(q) may be applicable.

Initial Isolation and Protective Action Distances

Small Spills (From a small package or a small leak from a large package)

First: Isolate in all directions (feet)..... 300

Then: Protect persons downwind (miles)

Day 0.2

Night 0.8

Large Spills (From a large package or from many small packages)

First: Isolate in all directions (feet)..... 800

Then: Protect persons downwind (miles)

Day 0.7

Night 3.0

Fire Extinguishing: Acetone cyanohydrin is a combustible liquid. Poisonous gases including hydrogen cyanide and nitrogen oxides are produced in fire. Use dry chemical, carbon dioxide, water spray, or alcohol foam extinguishers. Vapors are heavier than air and will collect in low areas. Vapors in confined areas may explode when exposed to fire. If material or contaminated runoff enters waterways, notify downstream users of potentially contaminated waters. Notify local health and fire officials and pollution control agencies. From a secure, explosion-proof location, use water spray to cool exposed containers. If cooling streams are ineffective (venting sound increases in volume and pitch, tank discolors, or shows any signs of deforming), withdraw immediately to a secure position. If employees are expected to fight fires, they must be trained and equipped in OSHA 1910.156. *Note:* Water may cause frothing if it gets below surface of liquid and turns to steam. Water fog gently applied to surface will cause frothing which may extinguish fire.

Disposal Method Suggested: Consult with environmental regulatory agencies for guidance on acceptable disposal practices. Generators of waste containing this contaminant (≥100 kg/mo) must conform with EPA regulations governing storage, transportation, treatment, and waste disposal. Add with stirring to strong alkaline calcium hypochlorite solution. Alternatively dissolve in flammable solvent and burn in incinerator with afterburner and scrubber.

References

U.S. Environmental Protection Agency, "Chemical Profile: Acetone Cyanohydrin," Washington, DC, Chemical Emergency Preparedness Program (Nov. 30, 1987)

National Institute for Occupational Safety and Health, Criteria for a Recommended Standard: Occupational Exposure to Nitriles, Washington, DC

Sax, N. I., Ed., Dangerous Properties of Industrial Materials Report 4, No. 1, 41–43 (1984)

New Jersey Department of Health and Senor Services, Hazardous Substance Fact Sheet: Acetone cyanohydrin, Trenton, NJ (September 1996)

Acetone Thiosemicarbazide

Molecular Formula: $C_4H_9N_3S$

Synonyms: Acetone, thiosemicarbazone; Hydrazinecarbothioamide, 2-(1-methylethylidene); Thiosemicarbazone acetone; Tiosemicarbazida de la acetona (Spanish)

CAS Registry Number: 1752-30-3

RTECS Number: AL7350000

DOT ID: UN 3077

Regulatory Authority

- CERCLA/SARA 40CFR302, Extremely Hazardous Substances: TPQ = 1,000/10,000 lb (454/4,540 kg)[7] *Note:* The lower quantity applies for solids in powder form with particulate size less than 100 microns. Otherwise the 10,000 pound TPQ applies.

Cited in U.S. State Regulations: Florida (G), Massachusetts (G), New Jersey (G), Pennsylvania (G).

Description: Acetone thiosemicarbazide, $C_4H_9N_3S$, is a combustible, yellow crystalline solid. FEMA Hazard Identification (based on NFPA-704 M Rating System): Health 2, Flammability 1, Reactivity 0.

Potential Exposure: An agricultural chemical.

Incompatibilities: Contact with strong oxidizers may cause fore and explosions.

Permissible Exposure Limits in Air: No standard set.

Permissible Concentration in Water: No criteria set.

Routes of Entry: Ingestion.

Harmful Effects and Symptoms

High oral toxicity reported. The LD_{low}oral (rat) = 10 mg/kg (highly toxic).

Points of Attack: Eyes, skin, respiratory system, central nervous system, cardiovascular system, liver, kidneys, gastrointestinal tract.

First Aid: If this chemical gets into the eyes, remove any contact lenses at once and irrigate immediately for at least 15 minutes, occasionally lifting upper and lower lids. If this chemical contacts the skin, remove contaminated clothing and wash immediately with soap and water. When this chemical has been swallowed, get medical attention. Give large quantities of water and induce vomiting. Do not make an unconscious person vomit. If this chemical has been inhaled, remove from exposure and transfer promptly to a medical facility.

Personal Protective Methods: Contact lenses should not be worn when working with this chemical. Wear splash proof chemical goggles and face shield unless full facepiece respiratory protection is worn. Employees should wash immediately with soap when skin is wet or contaminated. Remove nonimpervious clothing immediately if wet or contaminated. Provide emergency showers and eyewash.

Respirator Selection: For emergency situations, wear a MSHA/NIOSH approved supplied-air respirator with a full facepiece operated in a pressure-demand or other positive-pressure mode. For increased protection use in combination with an auxiliary self-contained breathing apparatus operated in a pressure-demand or other positive-pressure mode.

Storage: Prior to working with this chemical you should be trained on its proper handling and storage. Store in tightly closed containers in a cool, well ventilated area. Sources of ignition such as smoking and open flames, are prohibited where this chemical is used, handled, or stored in a manner that could create a potential fire or explosion hazard.

Spill Handling: Restrict persons not wearing protective equipment from area of spill or leak until cleanup is complete. Remove all ignition sources. Ventilate area of spill or leak after clean-up is complete. It may be necessary to contain and dispose of this chemical as a hazardous waste. If material or contaminated runoff enters waterways, notify downstream users of potentially contaminated waters. Contact your Department of Environmental Protection or your regional office of the federal EPA for specific recommendations. If employees are required to clean-up spills, they must be properly trained and equipped. OSHA 1910.120(q) may be applicable.

Fire Extinguishing: When heated to decomposition, it emits very toxic fumes of nitrogen oxide and sulfur oxides. If material or contaminated runoff enters waterways, notify downstream users of potentially contaminated waters. Notify local health and fire officials and pollution control agencies. From a secure, explosion-proof location, use water spray to cool exposed containers. If cooling streams are ineffective (venting sound increases in volume and pitch, tank discolors, or shows any signs of deforming), withdraw immediately to a secure position.

References

U.S. Environmental Protection Agency, "Chemical Profile: Acetone Thiosemicarbazide," Washington, DC, Chemical Emergency Preparedness Program (Nov. 30, 1987)

Acetonitrile

Molecular Formula: C_2H_3N

Common Formula: CH_3CN

Synonyms: Acetonitril (German, Dutch); Acetonitrilo (Spanish); Cyanomethane; Cyanure de methyl (French); Ethanenitrile; Ethyl nitril; Ethylnitrile; Methanecarbonitril; Methanecarbonitrile; Methane, cyano-; Methyl cyanide

CAS Registry Number: 75-05-8

RTECS Number: AL7700000

DOT ID: UN 1648

EEC Number: 608-001-00-3

EINECS Number: 200-835-2

Regulatory Authority

- Air Pollutant Standard Set (ACGIH)[1] (DFG)[3] (HSE)[33] (former USSR)[43] (OSHA)[58] (Many States)[60] (Various Canadian Provinces) (Mexico) (Australia) (Israel)
- Clean Air Act 42USC7412; Title I, Part A, §112 hazardous pollutants
- Clean Water Act 40CFR116.4A, hazardous substances, 40CFR413.02, Total Toxic Organics
- RCRA 40CFR258, Appendix 2, list of inorganic and organic constituents.
- RCRA 40CFR261, Appendix 8; 40CFR261.11 Hazardous Constituents
- RCRA 40CFR268.48; 61FR15654, Universal Treatment Standards: Wastewater (mg/l), 5.6; Nonwastewater (mg/kg), 1.8
- RCRA 40CFR264, Appendix 9; Ground Water Monitoring List Suggested methods (PQL μg/l): 8015 (100)
- RCRA 40CFR266, Appendix 4, Air Concentrations List
- RCRA 40CFR266, Appendix 7, Basis for Listing Hazardous Waste
- RCRA Land Ban Waste Restrictions
- EPA Hazardous Waste Number (RCRA No.): U003
- SUPERFUND/EPCRA 40CFR302.4, Appendix A, Reportable Quantity (RQ): 5,000 lb (2,270 kg), SARA 313: Form R *de minimis* Concentration Reporting Level: 1.0%.
- TSCA: 40CFR716.120(d) (aldehydes)
- Canada WHMIS Ingredients Disclosure List

Cited in U.S. State Regulations: Alaska (G), California (A, G), Connecticut (A), Florida (G, A), Illinois (G), Kansas (G), Louisiana (G), Maine (G), Maryland (G), Massachusetts (G), Minnesota (G), Nevada (A), New Hampshire (G), New Jersey (G), New York (A), North Dakota (A), Oklahoma (G), Pennsylvania (G), Rhode Island (G), South Carolina (A), Vermont (G), Virginia (G, A), Washington (G), West Virginia (G), Wisconsin (G).

Description: CH_3CN, acetonitrile, is an extremely flammable, colorless liquid with an ether-like odor. Odor threshold = 40 ppm. Freezing/Melting point = -45°C. Boiling point = 81°C. Hazard Identification (based on NFPA-704 M Rating System): Health 2, Flammability 3, Reactivity 0. Flash point = 6°C (oc). Autoignition temperature = 524°C. Explosive limits LEL = 3.0%, UEL = 16.0%. Freely soluble in water.

Potential Exposure: Acetonitrile is used as an extractant for animal and vegetable oils, as a solvent, particularly in the pharmaceutical industry, and as a chemical intermediate in pesticide manufacture; making batteries and rubber products. It is present in cigarette smoke.

Incompatibilities: Strong oxidizers such as chlorine, bromine, and fluorine; chlorosulfonic acid; oleum or sulfuric acid. May accumulate static electrical charges, and may cause ignition of its vapors.

Permissible Exposure Limits in Air: The legal airborne permissible exposure limit (PEL) is 40 ppm (67 mg/m^3) averaged over an 8-hour workshift. (OSHA) (DFG) (ACGIH). ACGIH also recommends a STEL value of 60 ppm (101 mg/m^3). NIOSH REL is 20 ppm (34 mg/m^3). The NIOSH IDLH level is 500 ppm. The above exposure limits are for air levels only. When skin contact also occurs, you may be overexposed, even though air levels are less than the limits listed above.

The former USSR-UNEP/IRPTC program[43] has set a MAC of 10 mg/m^3 in workplace air. The HSE[33] has set an 8-hour TWA of 40 ppm (70 mg/m^3) as well as a STEL of 60 ppm (105 mg/m^3).

In addition, a number of states have set guidelines or standards for acetonitrile in ambient air:[60] California TWA PEL 40 ppm (70 mg/m^3) 0.68 mg/m^3 (Connecticut) to 0.70 – 1.05 mg/m^3 (North Dakota) to 1.1 mg/m^3 (Virginia) to 1.4 mg/m^3 (Florida and New York) to 1.67 mg/m^3 (South Carolina).

Determination in Air: Charcoal absorption followed by benzene workup and gas chromatographic analysis. See NIOSH Method 1606.[18]

Permissible Concentration in Water: Acetonitrile is infinitely soluble and stable in water. No criteria have been set, but EPA has proposed[32] an ambient environmental goal of 970 μg/l based on health effects. The former USSR-UNEP/IRPTC program[43] has set a MAC of 0.7 mg/ml for water bodies used for domestic purposes.

Routes of Entry: Inhalation, percutaneous absorption, ingestion, skin and/or eye contact.

Harmful Effects and Symptoms

Short Term Exposure: Irritates eyes, skin and respiratory tract. Exposure can cause fatal cyanide poisoning. Symptoms include flushing of the face, chest tightness, headache, nausea, and vomiting, weakness, and shortness of breath. These reactions may begin hours following overexposure.

Long Term Exposure: Repeated exposure may cause the thyroid gland to enlarge and cause permanent damage. Acetonitrile may cause damage to the developing fetus.

Points of Attack: Kidneys, liver, lungs, skin, eyes, central nervous system, cardiovascular system.

Medical Surveillance: Consider the lung, skin, respiratory tract, heart, central nervous system, renal and liver function in placement and periodic examinations. A history of fainting spells or convulsive disorders might present an added risk to persons working with toxic nitriles. Blood cyanide test and/or urine thiocyanate test. Blood cyanide over 0.1 mg/l or urine thiocyanate over 20 mg/l indicate overexposure. Maintain close medical monitoring. Slow release of cyanide from absorbed acetonitrile may cause delayed symptoms.

First Aid: Acetonitrile can cause fatal Cyanide poisoning. A cyanide antidote kit should be kept in the immediate work area and must be rapidly available. Kit ingredients should be replaced every 1 – 2 years to ensure freshness. Persons trained in the use of this kit, oxygen use, and CPR must be available within 1 – 2 minutes. In the event of overexposure, and/or symptoms: move victim to fresh air; call emergency medical care. Give Amyl Nitrate capsules (as directed, by trained personnel only). If this chemical gets into the eyes, remove any contact lenses at once and irrigate immediately for at least 15 minutes, occasionally lifting upper and lower lids. Seek medical attention immediately. If this chemical contacts the skin, quickly remove contaminated clothing and wash with large amounts of water. Speed in removing material from skin is of extreme importance. Seek medical attention immediately. When this chemical has been swallowed, get medical attention immediately. If this chemical has been inhaled, remove from exposure and transfer promptly to a medical facility. If not breathing, give artificial respiration (avoid mouth to mouth resuscitation). If breathing is difficult, give oxygen. If heart has stopped begin CPR. Keep victim quiet and maintain normal body temperature. Effects may be delayed; keep victim under observation. Avoid contact with contaminated skin. Observe victim for 24 – 48 hours.

Personal Protective Methods: Wear solvent-resistant protective gloves and clothing to prevent any reasonable probability of skin contact. Safety equipment suppliers/manufacturers can provide recommendations on the most protective glove/clothing material for your operation. Safety equipment manufacturers recommend butyl rubber and polyvinyl alcohol (PVA) as protective material. All protective clothing (suits, gloves, footwear, headgear) should be clean, available each day, and put on before work. Contact lenses should not be worn when working with this chemical. Wear splash-proof chemical goggles and face shield unless full facepiece respiratory protection is worn. Employees should wash immediately with soap when skin is wet or contaminated. Provide emergency showers and eyewash. See also NIOSH Criteria Document 78–212 NITRILES.

Respirator Selection: NIOSH: *200 ppm*: CCRFOV [Any air-purifying, full-facepiece respirator (gas mask) with a chin-style, front- or back-mounted acid gas canister]; SA (any supplied-air respirator); or SCBA (any self-contained breathing apparatus); *500 ppm*: SA:CF (any supplied-air respirator operated in a continuous-flow mode); or PAPROV [any powered, air-purifying respirator with organic vapor cartridge(s)]; or CCRFOV [any chemical cartridge respirator with a full facepiece and organic vapor cartridge(s)]; or GMFOV [Any air-purifying, full-facepiece respirator (gas mask) with a chin-style, front- or back-mounted acid gas canister]; or SCBAF (any self-contained breathing apparatus with a full facepiece); or SAF (any supplied-air respirator with a full facepiece). *Emergency or planned entry into unknown concentrations or IDLH conditions:* SCBAF:PD,PP (any self-contained breathing apparatus that has a full facepiece and is operated in a pressure-demand or other positive-pressure mode); or SAF:PD,PP:ASCBA (Any supplied-air respirator that has a full facepiece and is operated in a pressure-demand or other positive-pressure mode in combination with an auxiliary self-contained breathing apparatus operated in a pressure-demand or other positive pressure mode. *Escape:* GMFOV [Any air-purifying, full-facepiece respirator (gas mask) with a chin-style, front- or back-, mounted organic vapor canister]; or SCBAE (Any appropriate escape-type, self-contained breathing apparatus).

Storage: Prior to working with this chemical you should be trained on its proper handling and storage. Before entering confined space where acetonitrile may be present, check to make sure that an explosive concentration does not exist. Store in tightly closed containers in a cool, well ventilated area. Metal containers involving the transfer of this chemical should be grounded and bonded. Where possible, automatically pump liquid from drums or other storage containers to process containers. Drums must be equipped with self-closing valves,

pressure vacuum bungs, and flame arresters. Use only non-sparking tools and equipment, especially when opening and closing containers of this chemical. Sources of ignition such as smoking and open flames, are prohibited where this chemical is used, handled, or stored in a manner that could create a potential fire or explosion hazard.

Shipping: The label required is "Flammable Liquid" Is in UN/DOT Hazard Class 3, Packing Group II.[19][20]

Spill Handling: Restrict persons not wearing protective equipment from area of spill or leak until cleanup is complete. Remove all ignition sources. Establish forced ventilation to keep levels below explosive limit. Use foam spray to reduce vapors. Absorb liquids in vermiculite, dry sand, earth, or a similar material and deposit in sealed containers. Keep acetonitrile out of a confined space, such as a sewer, because of the potential for an explosion, unless the sewer is designed to prevent the build-up of explosive concentrations. It may be necessary to contain and dispose of this chemical as a hazardous waste. If material or contaminated runoff enters waterways, notify downstream users of potentially contaminated waters. Contact your Department of Environmental Protection or your regional office of the federal EPA for specific recommendations. If employees are required to clean-up spills, they must be properly trained and equipped. OSHA 1910.120(q) may be applicable.

Fire Extinguishing: This chemical is a flammable liquid. Poisonous gases including hydrogen cyanide and nitrogen oxides are produced in fire. Use dry chemical, carbon dioxide, or alcohol foam extinguishers. Water may be ineffective for fighting fires. Vapors are heavier than air and will collect in low areas. Vapors may travel long distances to ignition sources and flashback. Vapors in confined areas may explode when exposed to fire. If material or contaminated runoff enters waterways, notify downstream users of potentially contaminated waters. Notify local health and fire officials and pollution control agencies. From a secure, explosion-proof location, use water spray to cool exposed containers. If cooling streams are ineffective (venting sound increases in volume and pitch, tank discolors, or shows any signs of deforming), withdraw immediately to a secure position. If employees are expected to fight fires, they must be trained and equipped in OSHA 1910.156.

Disposal Method Suggested: Consult with environmental regulatory agencies for guidance on acceptable disposal practices. Generators of waste containing this contaminant (≥100 kg/mo) must conform with EPA regulations governing storage, transportation, treatment, and waste disposal. Incineration with nitrogen oxide removal from effluent gases by scrubbers or incinerators.[22]

References

U.S. Environmental Protection Agency, Chemical Hazard Information Profile: Acetonitrile, Washington, DC (March 9, 1979)

U.S. Environmental Protection Agency, Acetonitrile, Health and Environmental Effects Profile No. 2, Washington, DC, Office of Solid Waste (April 30, 1980)

Sax, N. I., Ed., Dangerous Properties of Industrial Materials Report, 4, No. 1, 44–46 (Jan./Feb. 1984)

New Jersey Department of Health and Seniro Services, Hazardous Substance Fact Sheet: Acetonitrile, Trenton, NJ (June 1998)

Acetophenetidin

Molecular Formula: $C_{10}H_{13}NO_2$

Common Formula: $C_2H_5OC_6H_4NHCOCH_3$

Synonyms: 1-Acetamido-4-ethoxybenzene; Acetofenetidna (Spanish); Aceto-*p*-phenalide; *p*-Acetophenetide; Aceto-*p*-phenetidide; Acetophenetidin; Aceto-4-phenetidine; Acetophenetidine; Acetophenetin; Acet-*p*-phenalide; *p*-Acetphenetidin; Acet-*p*-phenetidin; Acetphenetidin; Acetylphenetidin; *N*-Acetyl-*p*-phenetidine; Achrocidin; Anapac; APC; ASA compound; *p*-Ethoxyacetanilide; 4-Ethoxyacetanilide; *N*-(4-Ethoxyphenyl)acetamide; *N,p*-Ethoxyphenylacetamide; *p*-Phenacetin; Phenacetin

CAS Registry Number: 62-44-2

RTECS Number: AM4375000

DOT ID: UN 9188

Regulatory Authority

- Carcinogen (IARC)[9] (NTP)
- Banned or Severely Restricted (Many Countries) (UN)[13]
- Air Pollutant Standard Set (former USSR)[43]
- RCRA 40CFR258, Appendix 2, list of inorganic and organic constituents
- EPA Hazardous Waste Number (RCRA No.): U187, as phenacetin
- RCRA 40CFR261, Appendix 8; 40CFR261.11 Hazardous Constituents
- RCRA 40CFR264, Appendix 9; Ground Water Monitoring List Suggested methods (PQL µg/l): 8270 (10)
- RCRA 40CFR268.48; 61FR15654, Universal Treatment Standards: Wastewater (mg/l), 0.081; Nonwastewater (mg/kg), 16
- RCRA Land Ban Waste Restrictions
- SUPERFUND/EPCRA 40CFR302.4, Appendix A, Reportable Quantity (RQ): 100 lb (45.4 kg)

Cited in U.S. State Regulations: California (G), Florida (G), Kansas (G), Louisiana (G), Maine (G), Massachusetts (G), New Hampshire (G), New Jersey (G), Pennsylvania (G), Vermont (G), Virginia (G), Washington (G), West Virginia (G), Wisconsin (G).

Description: Acetophenetidin,$C_2H_5OC_6H_4NHCOCH_3$, is a white, crystalline powder or solid with a slightly bitter taste. Freezing/Melting point = 137 – 138°C. Slightly soluble in water.

Potential Exposure: Phenacetin is used as an analgesic and antipyretic drug. It is used alone or in combination with aspirin and caffeine for mild to moderate muscle pain relief.

Phenacetin has also been used as a stabilizer for hydrogen peroxide in hair bleaching preparations. In veterinary medicine, it is used as an analgesic and antipyretic.

Incompatibilities: Oxidizing agents, iodine and nitrating agents.[52]

Permissible Exposure Limits in Air: The former USSR-UNEP/IRPTC project[43] has set a MAC of 0.5 mg/m³ in working zones for phenacetin.

Permissible Concentration in Water: No criteria set.

Routes of Entry: Ingestion, inhalation, eye and/or skin contact.

Harmful Effects and Symptoms

Short Term Exposure: Exposure to high levels of this chemical can cause methemaglobinemia which lowers the ability of the blood to carry oxygen. This can result in a bluish color to skin and lips (cyanosis), headache, dizziness, collapse and possible death.

Long Term Exposure: This chemical is a probable cancer-causing agent in humans. It has been shown to cause bladder, urinary tract and nose cancer in animals. Mutation, reproductive, teratogenic data reported. There is limited evidence that this chemical may damage the developing fetus. High or repeated exposures can destroy red blood cells, causing low blood count, aplastic anemia, jaundice, kidney damage, and brownish color to urine. Can lead to a general allergic reaction with rash and itching.

Points of Attack: Bladder, kidneys, eyes, skin

Medical Surveillance: Before beginning employment and at regular times after that, for those with frequent or potentially high exposures, the following are recommended: Kidney function tests. If symptoms develop or overexposure is suspected, the following may be useful: Blood test for methemoglobin level, Complete blood count and reticulocyte count, Blood and urine bilirubin, Blood Phenacetin level.

First Aid: If this chemical gets into the eyes, remove any contact lenses at once and irrigate immediately for at least 15 minutes, occasionally lifting upper and lower lids. If this chemical contacts the skin, remove contaminated clothing and wash with soap immediately. Do not make an unconscious person vomit. If this chemical has been inhaled, remove from exposure and begin rescue breathing if breathing has stopped and CPR if heart action has stopped. Transfer promptly to a medical facility.

Note to Physician: Treat for methemoglobinemia. Spectrophotometry may be required for precise determination of levels of methemoglobinemia in urine.

Personal Protective Methods: Wear protective gloves and clothing to prevent any reasonable probability of skin contact. Safety equipment suppliers/manufacturers can provide recommendations on the most protective glove/clothing material for your operation. Wear chemical goggles and face shield unless full facepiece respiratory protection is worn. Employees should wash immediately with soap when skin is wet or contaminated. Remove clothing immediately if wet or contaminated. Provide emergency showers and eyewash.

Respirator Selection: *At any exposure level:* SCBAF:PD, PP (any MSHA/NIOSH approved self-contained breathing apparatus that has a full facepiece and is operated in a pressure-demand or other positive-pressure mode); or SAF:PD,PP: ASCBA (any supplied-air respirator that has a full facepiece and is operated in a pressure-demand or other positive-pressure mode in combination with an auxiliary, self-contained breathing apparatus operated in a pressure-demand or other positive pressure mode). *Escape:* GMFOVHiE [any air-purifying, full-facepiece respirator (gas mask) with a chin-style, front- or back-mounted organic vapor canister having a high-efficiency particulate filter]; or SCBAE (any appropriate escape-type, self-contained breathing apparatus).

Storage: Prior to working with this chemical you should be trained on its proper handling and storage. A regulated, marked area should be established where Phenacetin is handled, used, or stored. Store in tightly-closed containers in a cool, well-ventilated area away from heat.

Shipping: May be shipped in 100, 200 and 1000 pound drums. It is not on the DOT list of materials[19] for label and packaging standards, but may be classified[52] as a Hazardous Substance, solid n.o.s. which falls in Hazard Class 9 and Packing Group III. Label required is "Class 9."

Spill Handling: Restrict persons not wearing protective equipment from area of spill or leak until cleanup is complete. Remove all ignition sources. Ventilate area of spill or leak after clean-up is complete. Collect powdered material in the most convenient and safe manner and deposit in sealed containers. It may be necessary to contain and dispose of this chemical as a hazardous waste. If material or contaminated runoff enters waterways, notify downstream users of potentially contaminated waters. Contact your Department of Environmental Protection or your regional office of the federal EPA for specific recommendations. If employees are required to clean-up spills, they must be properly trained and equipped. OSHA 1910.120(q) may be applicable.

Fire Extinguishing: Poisonous gases including nitrogen oxides are produced in fire. Use any extinguishing agent suitable for surrounding fire. If material or contaminated runoff enters waterways, notify downstream users of potentially contaminated waters. Notify local health and fire officials and pollution control agencies. If employees are expected to fight fires, they must be trained and equipped in OSHA 1910.156.

Disposal Method Suggested: Consult with environmental regulatory agencies for guidance on acceptable disposal practices. Generators of waste containing this contaminant (≥100 kg/mo) must conform with EPA regulations governing storage, transportation, treatment, and waste disposal.

Permanganate oxidation, microwave plasma treatment, alkaline hydrolysis or incineration.

References

Sax, N. I., Ed., Dangerous Properties of Industrial Materials Report, 1, No. 1, 26–27, (as Phenacetin) (1980); and 6, No. 1, 107–110 (1986)

New Jersey Department of Health and Senior Services, Hazardous Substance Fact Sheet: Phenacetin, Trenton, NJ (March 1986)

Acetophenone

Molecular Formula: C_8H_8O

Common Formula: $CH_3COC_6H_5$

Synonyms: Acetofenona (Spanish); Acetylbenzene; Benzoyl methide hypnone; Dymex; Ethanone, 1-phenyl-; Hypnone; Ketone methyl phenyl; Methyl phenyl ketone; 1-Phenylethanone; Phenyl methyl ketone

CAS Registry Number: 98-86-2

RTECS Number: AM5250000

DOT ID: NA 3082

Regulatory Authority

- Clean Air Act 42USC7412; Title I, Part A, §112 hazardous pollutants
- RCRA 40CFR261, Appendix 8; 40CFR261.11 Hazardous Constituents
- RCRA 40CFR264, Appendix 9; Ground Water Monitoring List Suggested methods (PQL µg/l): 8270 (10)
- RCRA 40CFR268.48; 61FR15654, Universal Treatment Standards: Wastewater (mg/l), 0.010; Nonwastewater (mg/kg), 9.7
- RCRA 40CFR266, Appendix 7, Basis for Listing Hazardous Waste
- RCRA Land Ban Waste
- EPA Hazardous Waste Number (RCRA No.): U0034
- SUPERFUND/EPCRA 40CFR302.4, Appendix A, Reportable Quantity (RQ): 5,000 lb (2,270 kg), SARA 313: Form R *de minimis* Concentration Reporting Level: 1.0%.

Cited in U.S. State Regulations: California (G), Kansas (G), Louisiana (G), Massachusetts (G), Minnesota (G), New Hampshire (G), New Jersey (G), Pennsylvania (G), Rhode Island (G), Vermont (G), Virginia (G), Washington (G), West Virginia (G), Wisconsin (G).

Description: Acetophenone, $CH_3COC_6H_5$ is a combustible, colorless, oily liquid with a sweet, floral odor. Odor Threshold = 0.0363 ppm. Freezing/Melting point = 19°C. Boiling point = 202°C. Hazard Identification (based on NFPA-704 M Rating System): Health 1, Flammability 2, Reactivity 0. Flash point = 77°C. Autoignition temperature = 570°C. Not soluble in water.

Potential Exposure: Acetophenone is used in perfume manufacture to impact a pleasant jasmine or orange-blossom odor. It is used as a catalyst in olefin polymerization and as a flavorant in tobacco. It is used in the synthesis of pharmaceuticals, and as a solvent.

Incompatibilities: Forms explosive mixture with air. Strong oxidizers, strong bases, and strong reducing agents.

Permissible Exposure Limits in Air: ACGIH has recommended an airborne exposure limit of 10 ppm (49 mg/m^3) TWA over an 8-hour workshift. The former USSR-UNEP/IRPTC project[43] has set MAC of 5 mg/m^3 in working zones and 0.003 mg/m^3 in ambient air of residential areas, both on a momentary and on an average daily basis.

Permissible Concentration in Water: The former USSR-UNEP/IRPTC project[43] has set a MAC of 0.1 mg/l in water for domestic purposes and 0.04 mg/l in water used for fishery purposes.

Routes of Entry: Inhalation, skin and/or eye contact.

Harmful Effects and Symptoms

Short Term Exposure: Irritates eyes, skin, and respiratory tract. Skin contact can cause burning and rash. Exposure can result in headache, dizziness, nausea, and loss of coordination. A hypnotic, high levels of exposure may affect the nervous system.

Long Term Exposure: There is evidence that this chemical can cause genetic changes, mutations, and acne-like skin rash. Long term exposure may cause central nervous system damage.

Points of Attack: Skin, eyes, central nervous system.

Medical Surveillance: Hippuric acid levels in the urine is recommended.

First Aid: Remove any contact lenses at once, then flush eyes, wash contaminated areas of body with soap and water. If this chemical gets into the eyes, remove any contact lenses at once and irrigate immediately for at least 15 minutes, occasionally lifting upper and lower lids. If this chemical contacts the skin, remove contaminated clothing and wash immediately with soap and water. If this chemical has been inhaled, remove from exposure, begin rescue breathing (using universal precautions) if breathing has stopped and CPR if heart action has stopped. Transfer promptly to a medical facility. When this chemical has been swallowed, get medical attention. Give large quantities of water and induce vomiting. Do not make an unconscious person vomit.

Personal Protective Methods: Wear solvent-resistant gloves and clothing to prevent any reasonable probability of skin contact. ACGIH recommends Teflon® as a protective material. Contact lenses should not be worn when working with this chemical. Wear splash-proof chemical goggles and face shield unless full facepiece respiratory protection is worn. Employees should wash immediately with soap when skin is wet or contaminated. Remove nonimpervious clothing immediately if wet or contaminated. Provide emergency showers and eyewash.

Respirator Selection: *At any concentrations above the NIOSH REL, or where there is no REL, at any detectable concentration:* SCBAF:PD,PP (any MSHA/NIOSH approved self-contained breathing apparatus that has a full facepiece and is operated in a pressure-demand or other positive-pressure mode); or SAF:PD,PP:ASCBA (any supplied-air respirator that has a full facepiece and is operated in a pressure-demand or other positive-pressure mode in combination with an auxiliary, self-contained breathing apparatus operated in a pressure-demand or other positive pressure mode). *Escape:* GMFOVHiE [any air-purifying, full-facepiece respirator (gas mask) with a chin-style, front- or back-mounted organic vapor canister having a high-efficiency particulate filter]; or SCBAE (any appropriate escape-type, self-contained breathing apparatus).

Storage: Prior to working with this chemical you should be trained on its proper handling and storage. Before entering confined space where acetophenone may be present, check to make sure that an explosive concentration does not exist. Store in tightly closed containers in a cool, well ventilated area. Metal containers involving the transfer of this chemical should be grounded and bonded. Where possible, automatically pump liquid from drums or other storage containers to process containers. Drums must be equipped with self-closing valves, pressure vacuum bungs, and flame arresters. Use only non-sparking tools and equipment, especially when opening and closing containers of this chemical. Sources of ignition such as smoking and open flames, are prohibited where this chemical is used, handled, or stored in a manner that could create a potential fire or explosion hazard.

Shipping: No special conditions cited by DOT.

Spill Handling: Restrict persons not wearing protective equipment from area of spill or leak until cleanup is complete. Remove all ignition sources. Use foam spray to reduce vapors. Absorb liquids in vermiculite, dry sand, earth, or a similar material and deposit in sealed containers. Ventilate area of spill or leak after clean-up is complete. Collect powdered material in the most convenient and safe manner and deposit in sealed containers. Keep acetophenone out of a confined space, such as a sewer, because of the potential for an explosion, unless the sewer is designed to prevent the build-up of explosive concentrations. It may be necessary to contain and dispose of this chemical as a hazardous waste. If material or contaminated runoff enters waterways, notify downstream users of potentially contaminated waters. Contact your Department of Environmental Protection or your regional office of the federal EPA for specific recommendations. If employees are required to clean-up spills, they must be properly trained and equipped. OSHA 1910.120(q) may be applicable.

Fire Extinguishing: This chemical is a flammable liquid. Poisonous gases including carbon monoxide is produced in fire. Use dry chemical, carbon dioxide, water spray, or polymer foam extinguishers. Vapors are heavier than air and will collect in low areas. Vapors may travel long distances to ignition sources and flashback. Vapors in confined areas may explode when exposed to fire. If material or contaminated runoff enters waterways, notify downstream users of potentially contaminated waters. Notify local health and fire officials and pollution control agencies. From a secure, explosion-proof location, use water spray to cool exposed containers. If cooling streams are ineffective (venting sound increases in volume and pitch, tank discolors, or shows any signs of deforming), withdraw immediately to a secure position. If employees are expected to fight fires, they must be trained and equipped in OSHA 1910.156.

Disposal Method Suggested: Consult with environmental regulatory agencies for guidance on acceptable disposal practices. Generators of waste containing this contaminant (≥100 kg/mo) must conform with EPA regulations governing storage, transportation, treatment, and waste disposal. Incineration, preferably with a flammable solvent.

References

U.S. Environmental Protection Agency, Acetophenone, Health and Environmental Effects Profile No. 3, Washington, DC, Office of Solid Waste (April 30, 1980)

New Jersey Department of Health and Senior Services, Hazardous Substance Fact Sheet: Acetophenone, Trenton, NJ (November 1998)

Acetyl Acetone Peroxide

Molecular Formula: A C_{10}-ketone with no single definite structure, coexists with isomers. (Patnaik)

Synonyms: Lupersol 224; 2.4-Pentanedione, peroxide; Percure A; Peroxido de acetilacetona (Spanish); Trigonox 40

CAS Registry Number: 37187-22-7

RTECS Number: SA2400000

DOT ID: UN 3105 (solution); UN 3106 (paste)

Cited in U.S. State Regulations: New Jersey (G).

Description: Acetyl acetone peroxide is an extremely flammable and dangerously explosive colorless to light yellow liquid with a sharp smell. An organic peroxide. Often shipped as a solution or a paste.

Potential Exposure: A catalyst to make resins, vinyl, polyolefins and silicones; a curing agent for unsaturated thermoset resins.

Incompatibilities: An organic peroxide. May ignite combustibles such as wood, cloth, oil, etc. Contact with oxidizers, heat, sparks, flame, shock, or contamination can cause explosions.

Permissible Exposure Limits in Air: None established.

Routes of Entry: Inhalation, skin and/or eye contact. Absorbed through the skin.

Harmful Effects and Symptoms

Short Term Exposure: Acetyl acetone peroxide can be absorbed through the skin, thereby increasing exposure. Irritates eyes, skin and respiratory tract. Can cause wheezing and coughing, dizziness, nausea, headache, and loss of consciousness.

Long Term Exposure: Unknown at this time.

Points of Attack: Eyes, skin, central nervous system.

First Aid: Remove any contact lenses at once, then flush eyes, wash contaminated areas of body with soap and water. If this chemical gets into the eyes, remove any contact lenses at once and irrigate immediately for at least 15 minutes, occasionally lifting upper and lower lids. Seek medical attention immediately. If this chemical contacts the skin, remove contaminated clothing and wash immediately with soap and water. If this chemical has been inhaled, remove from exposure, begin rescue breathing (using universal precautions) if breathing has stopped and CPR if heart action has stopped. Transfer promptly to a medical facility. When this chemical has been swallowed, get medical attention.

Personal Protective Methods: Wear solvent-resistant gloves and clothing to prevent any reasonable probability of skin contact. Contact lenses should not be worn when working with this chemical. Wear splash-proof chemical goggles and face shield unless full facepiece respiratory protection is worn. Employees should wash immediately with soap when skin is wet or contaminated. Remove nonimpervious clothing immediately if wet or contaminated. Provide emergency showers and eyewash.

Respirator Selection: *At any concentrations above the NIOSH REL, or where there is no REL, at any detectable concentration:* SCBAF:PD,PP (any MSHA/NIOSH approved self-contained breathing apparatus that has a full facepiece and is operated in a pressure-demand or other positive-pressure mode); or SAF:PD,PP:ASCBA (any supplied-air respirator that has a full facepiece and is operated in a pressure-demand or other positive-pressure mode in combination with an auxiliary, self-contained breathing apparatus operated in a pressure-demand or other positive pressure mode). *Escape:* GMFOVHiE [any air-purifying, full-facepiece respirator (gas mask) with a chin-style, front- or back-mounted organic vapor canister having a high-efficiency particulate filter]; or SCBAE (any appropriate escape-type, self-contained breathing apparatus).

Storage: Prior to working with this chemical you should be trained on its proper handling and storage. Before entering confined space where acetyl acetone peroxide may be present, check to make sure that an explosive concentration does not exist. Store in tightly closed containers in a cool, well ventilated area. Metal containers involving the transfer of this chemical should be grounded and bonded. Drums must be equipped with self-closing valves, pressure vacuum bungs, and flame arresters. Use only non-sparking tools and equipment, especially when opening and closing containers of this chemical. Sources of ignition such as smoking and open flames, are prohibited where this chemical is used, handled, or stored in a manner that could create a potential fire or explosion hazard. See OSHA standard 1910.104 and NFPA 43A *Code for the Storage of Liquid and Solid Oxidizers* for detailed handling and storage regulations.

Shipping: DOT Label: "Organic Peroxide." Forbidden from Transport by the USDOT, when active oxygen content >9%.

Spill Handling: Evacuate and restrict persons not wearing protective equipment from area of spill or leak until cleanup is complete. Remove all ignition sources. Absorb liquids in vermiculite, dry sand, earth, or a similar material and deposit in sealed containers. Ventilate area of spill or leak after clean-up is complete. Keep this chemical out of a confined space, such as a sewer, because of the potential for an explosion, unless the sewer is designed to prevent the build-up of explosive concentrations. It may be necessary to contain and dispose of this chemical as a hazardous waste. If material or contaminated runoff enters waterways, notify downstream users of potentially contaminated waters. Contact your Department of Environmental Protection or your regional office of the federal EPA for specific recommendations. If employees are required to clean-up spills, they must be properly trained and equipped. OSHA 1910.120(q) may be applicable.

Fire Extinguishing: This chemical is a flammable liquid or paste. Poisonous gases including nitrogen oxides are produced in fire. Use dry chemical, carbon dioxide, or alcohol foam extinguishers. Vapors are heavier than air and will collect in low areas. Vapors may travel long distances to ignition sources and flashback. Vapors in confined areas may explode when exposed to fire. If material or contaminated runoff enters waterways, notify downstream users of potentially contaminated waters. Notify local health and fire officials and pollution control agencies. From a secure, explosion-proof location, use water spray to cool exposed containers. If cooling streams are ineffective (venting sound increases in volume and pitch, tank discolors, or shows any signs of deforming), withdraw immediately to a secure position. If employees are expected to fight fires, they must be trained and equipped in OSHA 1910.156.

References

New Jersey Department of Health and Senior Services, Hazardous Substance Fact Sheet, Acetyl Acetone Peroxide, Trenton, NJ (November 1998)

Acetylaminofluorene

Molecular Formula: $C_{15}H_{13}NO$

Synonyms: 2-AAF; AAF; Acetamide, *N*-9H-fluoren-2-yl; Acetamide, *N*-fluoren-2-yl-; 2-Acetamidofluorene; Acetominofluorine; 2-2-Acetylamidofluorene; 2-Acetylaminofluoren (German); *N*-Acetyl-2-aminofluorene; 2-Acetylaminofluorene; Azetylaminofluoren; 2-FAA; FAA; *N*-2-Fluoren-2-yl acetamide; 2-Fluorenylacetamide

CAS Registry Number: 53-96-3

RTECS Number: AB9450000

DOT ID: UN 3077

Regulatory Authority

- Carcinogen (OSHA, NTP, State of California)[9]
- OSHA, 29CFR1910 Specifically Regulated Chemicals (See CFR1910.1014)
- Clean Air Act 42USC7412; Title I, Part A, §112 hazardous pollutants
- RCRA 40CFR261, Appendix 8; 40CFR261.11 Hazardous Constituents
- RCRA 40CFR264, Appendix 9; Ground Water Monitoring List Suggested methods (PQL μg/l): 8270 (10)
- RCRA 40CFR268.48; 61FR15654, Universal Treatment Standards: Wastewater (mg/l), 0.059; Nonwastewater (mg/kg), 140
- RCRA 40CFR266, Appendix 7, Basis for Listing Hazardous Waste
- RCRA Land Ban Waste
- EPA Hazardous Waste Number (RCRA No.): U005
- SUPERFUND/EPCRA 40CFR302.4, Appendix A, Reportable Quantity (RQ): 1 lb (0.454 kg), SARA 313: Form R *de minimis* Concentration Reporting Level: 0.1%.
- Banned or Severely Restricted (In Industrial Chemicals) (Belgium, Finland, Sweden) (UN)[13]
- Air Pollutant Standard Set (New York)[60]
- Canada WHMIS Ingredients Disclosure List

Cited in U.S. State Regulations: Alaska (G), California (G), Florida (G), Illinois (G), Kansas (G), Louisiana (G), Maryland (G), Massachusetts (G), Minnesota (G), Michigan (G), New Hampshire (G), New Jersey (G), New York (W), Oklahoma (G), Pennsylvania (G), Vermont (G), Virginia (G), Washington (G), West Virginia (G), Wisconsin (G).

Description: 2-Acetylaminofluorene, $C_{15}H_{13}NO$, is a combustible, tan powder or crystalline solid. Freezing/Melting point = 194°C. Hazard Identification (based on NFPA-704 M Rating System): Health 1, Flammability 1, Reactivity 0.

Potential Exposure: 2-Acetylaminofluorene (AAF) was intended to be used as a pesticide, but it was never marketed because this chemical was found to be carcinogenic. AAF is used frequently by biochemists and technicians engaged in the study of liver enzymes and the carcinogenicity and mutagenicity of aromatic amines as a positive control. Therefore, these persons may be exposed to AAF.

Incompatibilities: Contact with strong oxidizers may cause fire and explosions. Not compatible with cyanides, acids, and acid anhydrides.

Permissible Exposure Limits in Air: NIOSH recommends that exposure to occupational carcinogens be limited to the lowest feasible concentration. 0.03 μg/m³ (New York) for ambient air.[60]

Permissible Concentration in Water: No criteria set.

Routes of Entry: Ingestion, inhalation, mucous membrane, skin absorption, skin and/or eye contact.

Harmful Effects and Symptoms

A carcinogen. Handle with extreme care.

Short Term Exposure: This chemical has limited use in industry, and contact is kept to a minimum to prevent cancer. Reduced function of liver, kidneys, bladder, pancreas (Potential occupational carcinogen).

Long Term Exposure: Incorporation of this compound in feed caused increased incidences of malignant tumors in a variety of organs in the rat. Long-term studies in which mice were given 2-acetylaminofluorene in their diet showed that this compound caused increased incidences of tumors and cancer of the liver, kidney, urinary bladder, lung, skin, and pancreas. There is limited evidence that this chemical is a teratogen in animals.

References

US DOL OSHA, Reduced Immunologic Competence, Code of Federal Regulations. 29CFR1910, Air Contaminants, July 1, 1996.

US DHHS NIOSH and US DOL OSHA, Urine (chemical/metabolite), NIOSH/OSHA Occupational Health Guidelines for Chemical Hazards. DHHS (NIOSH) Pub Nos. 81–123

Points of Attack: Liver, bladder, kidney, pancreas, skin, lungs.

Medical Surveillance: Urine cytology for abnormal cells in the urine.

First Aid: If this chemical gets into the eyes, remove any contact lenses at once and irrigate immediately for at least 15 minutes, occasionally lifting upper and lower lids. If this chemical contacts the skin, remove contaminated clothing and wash immediately with soap and water. If this chemical has been inhaled, remove from exposure. Transfer promptly to a medical facility. When this chemical has been swallowed, get medical attention.

Personal Protective Methods: Because AAF is carcinogen, on February 11, 1974, OSHA promulgated a standard for this chemical designating protective clothing, hygiene procedures for workers, and special engineering requirements for the manufacture or processing of AAF. Open vessel operations are prohibited. Wear protective gloves and clothing to prevent any reasonable probability of skin contact. Eye protection is included in the recommended respiratory protection. Employees should wash immediately with soap when skin is wet or contaminated. Remove nonimpervious clothing immediately if wet or contaminated. Provide emergency showers and eyewash.

Respirator Selection: Exposures of workers to these 13 chemicals are to be controlled through the required use of engineering controls, work practices, and personal protective equipment, including respirators. See 29CFR1910.1003–

1910.1016 for specific details of these requirements. NIOSH: *At any detectable concentration:* SCBAF:PD,PP (any MSHA/NIOSH approved self-contained breathing apparatus that has a full facepiece and is operated in a pressure-demand or other positive-pressure mode); or SAF:PD,PP: ASCBA (any supplied-air respirator that has a full facepiece and is operated in a pressure-demand or other positive-pressure mode in combination with an auxiliary, self-contained breathing apparatus operated in a pressure-demand or other positive pressure mode). *Escape:* HiEF (Any air-purifying, full-facepiece respirator with a high-efficiency particulate filter); or SCBAE (Any appropriate escape-type, self-contained breathing apparatus).

Storage: Prior to working with this chemical you should be trained on its proper handling and storage. A regulated, marked area should be established where 2-Acetylaminofluorene is handled, used, or stored. 2-Acetylaminofluorene must be stored to avoid contact with cyanides, since violent reactions occur. Store in tightly closed containers in a cool, well ventilated area. Sources of ignition such as smoking and open flames, are prohibited where this chemical is used, handled, or stored in a manner that could create a potential fire or explosion hazard.

Shipping: Usually shipped in small laboratory bottles as noted above.

Spill Handling: If 2-Acetylaminofluorene is spilled or leaked, only specifically trained personnel should be involved in the clean-up. Restrict persons not wearing protective equipment from area of spill or leak until cleanup is complete. Remove all ignition sources. Cover with lime or soda ash; collect material in the most convenient and safe manner and deposit in sealed containers. Ventilate area of spill or leak after clean-up is complete. It may be necessary to contain and dispose of this chemical as a hazardous waste. If material or contaminated runoff enters waterways, notify downstream users of potentially contaminated waters. Contact your Department of Environmental Protection or your regional office of the federal EPA for specific recommendations. If employees are required to clean-up spills, they must be properly trained and equipped. OSHA 1910.120(q) may be applicable.

Fire Extinguishing: This chemical is a combustible solid. Use dry chemical, carbon dioxide, water spray, or alcohol foam extinguishers. Poisonous gases are produced in fire including nitrogen oxides. If employees are expected to fight fires, they must be trained and equipped in OSHA 1910.156.

Disposal Method Suggested: Consult with environmental regulatory agencies for guidance on acceptable disposal practices. Generators of waste containing this contaminant (≥100 kg/mo) must conform with EPA regulations governing storage, transportation, treatment, and waste disposal. Presumably high temperature incineration with scrubber for No_x produced can be used.

References

New Jersey Department of Health and Senior Services, "Hazardous Substance Fact Sheet: 2-Acetylaminofluorene," Trenton, NJ (June 1998)

Acetyl Benzoyl Peroxide

Molecular Formula: $C_9H_8O_4$

Common Formula: $CH_3CO{\cdot}OO{\cdot}COC_6H_5$

Synonyms: Acetozone; Benzozone; Peroxide, acetyl benzoyl; Peroxido de acetil benzoilo (Spanish)

CAS Registry Number: 644-31-5

RTECS Number: SD7860000

DOT ID: UN 2081

Cited in U.S. State Regulations: Maine (G), New Hampshire (G), New Jersey (G), Oklahoma (G).

Description: Acetyl benzoyl peroxide, $CH_3CO{\cdot}OO{\cdot}COC_6H_5$, is a white crystalline solid, forming needles. Freezing/Melting point = 36 – 37°C. Boiling point = 130°C. Decomposes in water; violent reaction.

Potential Exposure: Used in disinfectants, to bleach flour, and in medications.

Incompatibilities: Acetyl benzoyl peroxide is an organic peroxide which can detonate if shocked, heated or on contact with contaminants. Avoid contact with moisture, water, steam, sources of ignition, combustible materials. Not compatible with strong bases, reducing materials, oxidizers.

Permissible Exposure Limits in Air: No standards set.

Permissible Concentration in Water: No criteria set. Reacts violently with water.

Routes of Entry: Inhalation, ingestion, skin and/or eye contact.

Harmful Effects and Symptoms

Short Term Exposure: This chemical can be absorbed through the skin, thereby increasing exposure. Contact can severely irritate and burn the eyes and skin. It may cause permanent damage. Breathing Acetyl benzoyl peroxide can irritate the nose, throat and lungs causing a cough, difficulty breathing and chest tightness. Higher levels of exposure can cause pulmonary edema, a medical emergency, which can be delayed for several hours. This can cause death.

Long Term Exposure: Repeated exposure may cause chronic irritation of the skin and eczema-like rash. Repeated lung exposure can cause bronchitis with cough, phlegm, and/or shortness of breath.

Points of Attack: Eyes, skin, nose, throat, lungs.

Medical Surveillance: Before beginning employment and at regular times after that, for those with frequent or potentially high exposures, the following are recommended: Lung function tests. If symptoms develop or overexposure is suspected, the following may be useful: Consider chest x-ray after acute overexposure.

First Aid: If this chemical gets into the eyes, remove any contact lenses at once and irrigate immediately for at least 15 minutes, occasionally lifting upper and lower lids. If this chemical contacts the skin, remove contaminated clothing and wash immediately with soap and water. If this chemical has been inhaled, remove from exposure, begin rescue breathing (using universal precautions) if breathing has stopped and CPR if heart action has stopped. Transfer promptly to a medical facility. When this chemical has been swallowed, get medical attention. Medical observation is recommended for 24 – 48 hours after breathing overexposure, as pulmonary edema may be delayed.

Personal Protective Methods: Avoid skin contact with Acetyl Benzoyl Peroxide. Wear protective gloves and clothing. Safety equipment suppliers/manufacturers can provide recommendations on the most protective glove/clothing material for your operation. All protective clothing (suits, gloves, footwear, headgear) should be clean, available each day and put on before work. Contact lenses should not be worn when working with this chemical. Wear impact-resistant goggles and face shield when working with powders or dust, unless full facepiece respiratory protection is worn. Wear splash-proof chemical goggles and face shield when working with liquid, unless full facepiece respiratory protection is worn.

Respirator Selection: Where the potential exists for exposures to solid Acetyl Benzoyl Peroxide, use a MSHA/NIOSH approved full facepiece respirator with a high efficiency particulate filter. Greater protection is provided by a powered-air purifying respirator. Where the potential for high exposures and to the liquid form of Acetyl Benzoyl Peroxide exists, use a MSHA/NIOSH approved supplied-air respirator with a full facepiece operated in the positive pressure mode or with a full facepiece, hood, or helmet in the continuous flow mode, or use a MSHA/NIOSH approved self-contained breathing apparatus with a full facepiece operated in pressure-demand or other positive pressure mode. See OSHA standard 1910.104 and NFPA 43A *Code for the Storage of Liquid and Solid Oxidizers* for detailed handling and storage regulations.

Storage: Prior to working with this chemical you should be trained on its proper handling and storage. This chemical is an organic peroxide which can detonate if shocked or heated. Before entering confined space where acetyl benzoyl peroxide may be present, check to make sure that an explosive concentration does not exist. Store in tightly closed containers in a cool, well ventilated area. Where possible, automatically pump liquid from drums or other storage containers to process containers. Drums must be equipped with self-closing valves, pressure vacuum bungs, and flame arresters. Use only non-sparking tools and equipment, especially when opening and closing containers of this chemical. Sources of ignition such as smoking and open flames, are prohibited where this chemical is used, handled, or stored in a manner that could create a potential fire or explosion hazard. See OSHA standard 1910.104 and NFPA 43A *Code for the Storage of Liquid and Solid Oxidizers* for detailed handling and storage regulations.

Shipping: Solid acetyl benzoyl peroxide falls in the forbidden category according to DOT.[19] Solutions of 40% or more are also forbidden. Solutions below 45% fall in Hazard Class 5.2 and Packing Group II. They must be labeled "Organic Peroxide."

Spill Handling: Restrict persons not wearing protective equipment from area of spill or leak until cleanup is complete. Remove all ignition sources. Collect powdered material in the most convenient and safe manner and deposit in sealed containers. Keep this substance out of a confined space, such as a sewer, because of the potential for an explosion, unless the sewer is designed to prevent the build-up of explosive concentrations. Ventilate area of spill or leak after clean-up is complete. It may be necessary to contain and dispose of this chemical as a hazardous waste. If material or contaminated runoff enters waterways, notify downstream users of potentially contaminated waters. Contact your Department of Environmental Protection or your regional office of the federal EPA for specific recommendations. If employees are required to clean-up spills, they must be properly trained and equipped. OSHA 1910.120(q) may be applicable.

Fire Extinguishing: This chemical is a combustible solid. Use dry chemical or carbon dioxide extinguishers. Irritating fumes are produced in fire. Vapors in confined areas may explode when exposed to fire. If material or contaminated runoff enters waterways, notify downstream users of potentially contaminated waters. Notify local health and fire officials and pollution control agencies. From a secure, explosion-proof location, use water spray to cool exposed containers. If cooling streams are ineffective (venting sound increases in volume and pitch, tank discolors, or shows any signs of deforming), withdraw immediately to a secure position. If employees are expected to fight fires, they must be trained and equipped in OSHA 1910.156.

References

New Jersey Department of Health and Senior Services, Hazardous Substance Fact Sheet: Acetyl Benzoyl Peroxide, Trenton, NJ (September 1998)

Acetyl Bromide

Molecular Formula: C_2H_3BrO

Common Formula: CH_3COBr

Synonyms: Acetic acid bromide; Acetic bromide; Acetilo de bromura (Spanish); Ethanoyl bromide

CAS Registry Number: 506-96-7

RTECS Number: AO5955000

DOT ID: UN 1716

Regulatory Authority

- Clean Water Act: 40CFR116.4 Hazardous Substances; RQ 40CFR117.3 (same as CERCLA)
- SUPERFUND/EPCRA 40CFR302.4, Appendix A, Reportable Quantity (RQ): 5,000 lb (2,270 kg)
- TSCA: 40CFR716.120(a), substance list
- Canada, WMIS Ingredients Disclosure List

Cited in U.S. State Regulations: California (G), Illinois (G), Maine (G), Massachusetts (G), New Hampshire (G), New Jersey (G), Pennsylvania (G).

Description: Acetyl bromide, CH_3COBr, is a colorless fuming liquid that turns yellow in air and has a sharp, unpleasant odor. Boiling point = 76.7°C. Freezing/Melting point = -96°C. Sinks in water; violent decomposition, forming toxic hydrogen bromide. Combustible but difficult to ignite. Flash point = 110°C. Odor threshold 5×10^{-4} ppm.

Potential Exposure: Those involved in using this chemical in dye manufacture and other organic syntheses.

Incompatibilities: Contact with moisture, water, steam, alcohols cause a violent reaction releasing corrosive carbonyl bromide, hydrogen bromide, and bromine gases. Incompatible with organic solvents, ethers, oxidizers, and strong bases. Corrodes or attacks most metals and wood in the presence of moisture. Contact with combustibles may cause ignition.

Permissible Exposure Limits in Air: No standard set. The NIOSH IDLH value is 3 ppm as Br.[41]

Permissible Concentration in Water: No criteria set.

Routes of Entry: Eye and/or skin contact, inhalation, and ingestion.

Harmful Effects and Symptoms

Short Term Exposure: This chemical can be absorbed through the skin, thereby increasing exposure. Skin and eye contact can cause severe irritation and possible permanent damage including blindness. Inhalation and swallowing are very toxic Exposure can irritate the nose, throat, air passages, and lungs with coughing and/or shortness of breath. Higher exposures may cause pulmonary edema, a medical emergency that can be delayed for several hours. This can cause death.

Long Term Exposure: Repeated respiratory exposure can cause bronchitis to develop with cough, phlegm, and/or shortness of breath. Repeated skin exposure can cause chronic skin irritation.

Points of Attack: Skin, lungs.

Medical Surveillance: Before beginning employment and at regular times after that, for those with frequent or potentially high exposures, lung function tests are recommended. If symptoms develop or overexposure is suspected, consider chest x-ray.

First Aid: If this chemical gets into the eyes, remove any contact lenses at once and irrigate immediately for at least 30 minutes, occasionally lifting upper and lower lids. Seek medical attention immediately. If this chemical contacts the skin, remove contaminated clothing and wash immediately with soap and water. Seek medical attention immediately. If this chemical has been inhaled, remove from exposure, begin rescue breathing (using universal precautions) if breathing has stopped and CPR if heart action has stopped. Transfer promptly to a medical facility. When this chemical has been swallowed, get medical attention. Do not induce vomiting. Medical observation is recommended for 24 – 48 hours after breathing overexposure, as pulmonary edema may be delayed.

Personal Protective Methods: Wear corrosive-resistant gloves and clothing to prevent any reasonable probability of skin contact. Safety equipment suppliers/manufacturers can provide recommendations on the most protective glove/clothing material for your operation. Contact lenses should not be worn when working with this chemical. Wear splash-proof chemical goggles and face shield unless full facepiece respiratory protection is worn. Employees should wash immediately with soap when skin is wet or contaminated. Remove nonimpervious clothing immediately if wet or contaminated. Provide emergency showers and eyewash.

Respirator Selection: Where the potential for exposures to Acetyl Bromide exists, use a MSHA/NIOSH approved supplied-air respirator with a full facepiece operated in the positive pressure mode or with a full facepiece, hood, or helmet in the continuous flow mode, or use a MSHA/NIOSH approved self-contained breathing apparatus with a full facepiece operated in pressure-demand or other positive pressure mode.

Storage: Prior to working with this chemical you should be trained on its proper handling and storage. Store in tightly closed containers in a cool, well ventilated area away from moisture and sunlight. Where possible, automatically pump liquid from drums or other storage containers to process containers. Drums must be equipped with self-closing valves, pressure vacuum bungs, and flame arresters. Use only non-sparking tools and equipment, especially when opening and closing containers of this chemical. Sources of ignition such as smoking and open flames, are prohibited where this chemical is used, handled, or stored in a manner that could create a potential fire or explosion hazard.

Shipping: Label required is "Corrosive." Is in UN/DOT Hazard Class 8, Packing Group II.[19][20]

Spill Handling: Prevent liquid from reaching water if possible. This material creates large amounts of toxic hydrogen bromide (HBr) vapor when spilled in water, and is considered by the North American Emergency Response Guide to be dangerous from 0.5 – 10 km (0.3 – 6.0 miles downwind). Enter area from upwind side. Evacuate area and issue warning of danger, including possible explosion. Evacuate and restrict persons not wearing protective equipment from area of spill or leak until cleanup is complete. Remove all ignition sources. Absorb liquids using dry lime or soda ash. *Do not use water* *or wet method.* Ventilate area of spill or leak after clean-up is complete. It may be necessary to contain and dispose of this chemical as a hazardous waste. If material or contaminated runoff enters waterways, notify downstream users of potentially contaminated waters. Contact your Department of Environmental Protection or your regional office of the federal EPA for specific recommendations. If employees are required to clean-up spills, they must be properly trained and equipped. OSHA 1910.120(q) may be applicable.

Initial isolation and protective action distances: See above. Distances shown are likely to be affected during the first 30 minutes after materials are spilled and could increase with time. If more than one tank car, cargo tank, portable tank, or large cylinder is involved in the incident is leaking, the protective action distance may need to be increased. You may need to seek emergency information from CHEMTREC at (800) 424-9300 and seek professional environmental engineering assistance from the U.S. EPA Environmental Response Team at (908) 548-8730 (24-hour response line).

Fire Extinguishing: This chemical is a combustible liquid, but does not ignite readily. Poisonous gases including hydrogen bromide, carbonyl bromide, and bromine are produced in fire. *Do not use water or foam extinguishers.* Use dry chemical or carbon dioxide extinguishers. Vapors are heavier than air and will collect in low areas. Vapors in confined areas may explode when exposed to fire. If material or contaminated runoff enters waterways, notify downstream users of potentially contaminated waters. Notify local health and fire officials and pollution control agencies. From a secure, explosion-proof location, use water spray to cool exposed containers. If cooling streams are ineffective (venting sound increases in volume and pitch, tank discolors, or shows any signs of deforming), withdraw immediately to a secure position. If employees are expected to fight fires, they must be trained and equipped in OSHA 1910.156.

Disposal Method Suggested: Slow addition to sodium bicarbonate solution.

References

Sax, I. N., Ed., Dangerous Properties of Industrial Materials Report, 1, No. 8, 29–30 (1981)

New Jersey Department of Health and Senior Services, Hazardous Substance Fact Sheet: Acetyl Bromide, Trenton, NJ (September 1998)

Acetyl Chloride

Molecular Formula: C_2H_3ClO

Common Formula: CH_3COCl

Synonyms: Acetic acid chloride; Acetic chloride; Cloruro de acetilo (Spanish); Ethanoyl chloride

CAS Registry Number: 75-36-5

RTECS Number: AO6390000

DOT ID: UN 1717

EEC Number: 607-011-00-5

EINECS Number: 200-865-6

Regulatory Authority

- Clean Air Act 42USC7412; Title I, Part A, §112 hazardous pollutants; RQ 40CFR117.3 (same as CERCLA)
- EPA Hazardous Waste Number (RCRA No.): U006
- RCRA 40CFR261, Appendix 8; 40CFR261.11 Hazardous Constituents
- RCRA Land Ban Waste Restrictions
- SUPERFUND/EPCRA 40CFR302.4, Appendix A, Reportable Quantity (RQ): 5,000 lb (2,270 kg)
- Banned or Severely Restricted (Singapore) (UN)[13]

Cited in U.S. State Regulations: Florida (G), Illinois (G), Kansas (G), Louisiana (G), Maine (G), Massachusetts (G), New Hampshire (G), New Jersey (G), North Dakota (G, W), Oklahoma (G), Pennsylvania (G), Rhode Island (G), Vermont (G), Virginia (G), Washington (G), Wisconsin (G).

Description: Acetyl chloride, CH_3COCl, is a highly flammable, colorless, fuming liquid with a pungent odor. The odor threshold is 1.0 ppm (as acetic acid or as HCl). Boiling point = 51 – 52°C. Flash point = 4°C. Autoignition temperature = 390°C. Hazard Identification (based on NFPA-704 M Rating System): Health 3, Flammability 3, Reactivity 2. Violent decomposition in water.

Potential Exposure: Acetyl chloride is used in organic synthesis as an acetylating agent, in the pharmaceutical industry and in pesticide manufacture, for example.

Incompatibilities: Avoid contact with moisture, steam, water, alcohols, dimethylsulfoxide, strong bases, phosphorus trichloride, oxidizers, and amines since violent reactions may occur.

Permissible Exposure Limits in Air: 2 $\mu g/m^3$ (North Dakota) in ambient air.[60] An IDLH value of 50 ppm (HCl) has been given.[41]

Permissible Concentration in Water: No criteria set. However, acetyl chloride reacts violently with water. Thus, its half-life in ambient water should be short and exposure from water should be nil. The degradation products should likewise pose no exposure problems if the pH of the water remains stable.

Routes of Entry: Inhalation, ingestion, eye and/or skin contact.

Harmful Effects and Symptoms

Acetyl chloride is an irritant and a corrosive. Cutaneous exposure results in skin burns, while vapor exposure causes extreme irritation of the eyes and mucous membranes. Inhalation of 2 ppm acetyl chloride has been found irritating to humans. Death or permanent injury may result after short exposures to small quantities of acetyl chloride. An aquatic toxicity rating has been estimated to range from 10 – 100 ppm.

Short Term Exposure: This chemical can be absorbed through the skin, thereby increasing exposure. Skin and eye contact can cause severe irritation and possible permanent damage including blindness. Inhalation and swallowing are very toxic Exposure can irritate the nose, throat, air passages and lungs with coughing and/or shortness of breath. Higher exposures may cause pulmonary edema, a medical emergency that can be delayed for several hours. This can cause death.

Long Term Exposure: Repeated respiratory exposure can cause bronchitis to develop with cough, phlegm, and/or shortness of breath. Repeated skin exposure can cause chronic skin irritation.

Points of Attack: Skin, lungs.

Medical Surveillance: Before beginning employment and at regular times after that, for those with frequent or potentially high exposures, lung function tests are recommended. If symptoms develop or overexposure is suspected, consider chest x-ray.

First Aid: If this chemical gets into the eyes, remove any contact lenses at once and irrigate immediately for at least 30 minutes, occasionally lifting upper and lower lids. Seek medical attention immediately. If this chemical contacts the skin, remove contaminated clothing and wash immediately with soap and water. Seek medical attention immediately. If this chemical has been inhaled, remove from exposure, begin rescue breathing (using universal precautions) if breathing has stopped and CPR if heart action has stopped. Transfer promptly to a medical facility. When this chemical has been swallowed, get medical attention. Do not induce vomiting. Medical observation is recommended for 24 – 48 hours after breathing overexposure, as pulmonary edema may be delayed.

Personal Protective Methods: Wear corrosive-resistant gloves and clothing to prevent any reasonable probability of skin contact. Safety equipment suppliers/manufacturers can provide recommendations on the most protective glove/clothing material for your operation. Contact lenses should not be worn when working with this chemical. Wear splash-proof chemical goggles and face shield unless full facepiece respiratory protection is worn. Employees should wash immediately with soap when skin is wet or contaminated. Remove nonimpervious clothing immediately if wet or contaminated. Provide emergency showers and eyewash.

Respirator Selection: Where the potential exists for exposures to Acetyl Chloride, use a MSHA/NIOSH approved supplied-air respirator with a full facepiece operated in the positive pressure mode or with a full facepiece, hood, or helmet in the continuous flow mode, or use a MSHA/NIOSH approved self-contained breathing apparatus with a full facepiece operated in pressure-demand or other positive pressure mode.

Storage: Prior to working with this chemical you should be trained on its proper handling and storage. Store in tightly closed containers in a cool, well ventilated area away from moisture and sunlight. Where possible, automatically pump liquid from drums or other storage containers to process containers. Drums must be equipped with self-closing valves, pressure vacuum bungs, and flame arresters. Use only nonsparking tools and equipment, especially when opening and closing containers of this chemical. Sources of ignition such as smoking and open flames, are prohibited where this chemical is used, handled, or stored in a manner that could create a potential fire or explosion hazard.

Shipping: Should bear "Flammable liquid, corrosive" label. Is in UN/DOT Hazard Class 3, Packing Group II.[19][20]

Spill Handling: Prevent liquid from reaching water if possible. This material creates large amounts of toxic hydrogen chloride (HCl) vapor when spilled in water, and is considered by the North American Emergency Response Guide to be dangerous from 0.5 – 10 km (0.3 – 6.0 miles downwind). Enter area from upwind side. Evacuate area and issue warning of danger, including possible explosion. Evacuate and restrict persons not wearing protective equipment from area of spill or leak until cleanup is complete. Remove all ignition sources. Absorb liquids using dry lime or soda ash. *Do not use water or wet method.* Ventilate area of spill or leak after clean-up is complete. It may be necessary to contain and dispose of this chemical as a hazardous waste. If material or contaminated runoff enters waterways, notify downstream users of potentially contaminated waters. Contact your Department of Environmental Protection or your regional office of the federal EPA for specific recommendations. If employees are required to clean-up spills, they must be properly trained and equipped. OSHA 1910.120(q) may be applicable.

Initial isolation and protective action distances: See above. Distances shown are likely to be affected during the first 30 minutes after materials are spilled and could increase with time. If more than one tank car, cargo tank, portable tank, or large cylinder is involved in the incident is leaking, the protective action distance may need to be increased. You may need to seek emergency information from CHEMTREC at (800) 424-9300 or seek professional environmental engineering assistance from the U.S. EPA Environmental Response Team at (908) 548-8730 (24-hour response line).

Fire Extinguishing: This chemical is a combustible liquid, but does not ignite readily. Poisonous gases including hydrogen bromide, carbonyl bromide and bromine are produced in fire. *Do not use water or foam extinguishers.* Use dry chemical or carbon dioxide extinguishers. Vapors are heavier than air and will collect in low areas. Vapors in confined areas may explode when exposed to fire. If material or contaminated runoff enters waterways, notify downstream users of potentially contaminated waters. Notify local health and fire officials and pollution control agencies. From a secure, explosion-proof location, use water spray to cool exposed containers. If cooling streams are ineffective (venting sound increases in volume and pitch, tank discolors, or shows any signs of deforming), withdraw immediately to a secure position. If employees are expected to fight fires, they must be trained and equipped in OSHA 1910.156.

Disposal Method Suggested: Consult with environmental regulatory agencies for guidance on acceptable disposal practices. Generators of waste containing this contaminant (≥100 kg/mo) must conform with EPA regulations governing storage, transportation, treatment, and waste disposal. May be mixed slowly with sodium bicarbonate solution and then flushed to sewer with large volumes of water. May also be incinerated.

References

U.S. Environmental Protection Agency, Acetyl Chloride, Health and Environmental Effects Profile No. 4, Washington, DC, Office of Solid Waste (April 30, 1980)

Sax, N. I., Ed., Dangerous Properties of Industrial Materials Report, 1, No. 8, 30–32 (1981) and 3, No. 3, 35–36 (1983)

New Jersey Department of Health and Senior Services, Hazardous Substance Fact Sheet: Acetyl Chloride, Trenton, NJ (September 1998)

Acetyl Cyclohexane Sulfonyl Peroxide

Molecular Formula: $C_8H_{14}O_5S$

Synonyms: Acetyl cyclohexylsulfonyl peroxide; Lupersol 228Z; Peroxide, acetyl cyclohexylsulfonyl

CAS Registry Number: 3179-56-4

RTECS Number: SD7864200 (>82%); SD7864000 (<82%); SD7864100 (<32% in solution)

DOT ID: UN 2082; UN 2083; UN3112 (organic peroxide type B, solid, temperature controlled); UN3115 (organic peroxide, type D, liquid, temperature controlled)

Cited in U.S. State Regulations: New Hampshire (G), New Jersey (G).

Description: Acetyl cyclohexane sulfonyl peroxide, $C_8H_{14}O_5S$ is a white solid which is often used in a liquid solution. An unstable organic peroxide. Flash point = 63°C.

Potential Exposure: Used in paint, rubber and plastics manufacture.

Incompatibilities: A very unstable organic peroxide that is heat, shock, and contamination sensitive. Store away from combustible materials, oxidizers.

Permissible Exposure Limits in Air: No standards have been established.

Permissible Concentration in Water: No criteria set.

Routes of Entry: Inhalation, ingestion, skin and or eye contact.

Harmful Effects and Symptoms

Short Term Exposure: This chemical can be absorbed through the skin, thereby increasing exposure. Eye and skin contact may cause irritation and burns. Inhalation can irritate the lungs causing coughing and/or shortness of breath. Higher exposures can cause pulmonary edema, a medical emergency that can be delayed for several hours. This can cause death.

Long Term Exposure: Can cause lung irritation with coughing and shortness of breath.

Points of Attack: Eyes, skin, respiratory system.

Medical Surveillance: For those with frequent or potentially high exposure, lung function tests are recommended before beginning work and at regular times after that. If symptoms develop or overexposure is suspected, consider chest x-ray.

First Aid: If this chemical gets into the eyes, remove any contact lenses at once and irrigate immediately for at least 15 minutes, occasionally lifting upper and lower lids. If this chemical contacts the skin, remove contaminated clothing and wash immediately with soap and water. If this chemical has been inhaled, remove from exposure, begin rescue breathing (using universal precautions) if breathing has stopped and CPR if heart action has stopped. Transfer promptly to a medical facility. When this chemical has been swallowed, get medical attention. Give large quantities of water and induce vomiting. Do not make an unconscious person vomit. Medical observation is recommended for 24 – 48 hours after breathing overexposure, as pulmonary edema may be delayed.

Personal Protective Methods: Wear solvent-resistant gloves and clothing to prevent any reasonable probability of skin contact. Safety equipment suppliers/manufacturers can provide recommendations on the most protective glove/clothing material for your operation. One manufacturer of Acetyl Cyclohexane Sulfonyl Peroxide recommends Rubber as a protective material. Contact lenses should not be worn when working with this chemical. Wear splash-proof chemical goggles and face shield unless full facepiece respiratory protection is worn. Employees should wash immediately with soap when skin is wet or contaminated. Remove nonimpervious clothing immediately if wet or contaminated. Provide emergency showers and eyewash.

Respirator Selection: Where the potential exists for exposures to *solid* Acetyl Cyclohexane Sulfonyl Peroxide, use a MSHA/NIOSH approved full facepiece respirator with a high efficiency particulate filter. Greater protection is provided by a powered-air purifying respirator. Where the potential for high exposures exists or to *liquid* Acetyl Cyclohexane Sulfonyl Peroxide, use a MSHA/NIOSH approved supplied-air respirator with a full facepiece operated in the positive pressure mode or with a full facepiece, hood, or helmet in the continuous flow mode, or use a MSHA/NIOSH approved self-contained breathing apparatus with a full facepiece operated in pressure-demand or other positive pressure mode.

Storage: Prior to working with this chemical you should be trained on its proper handling and storage. Before entering confined space where this chemical may be present, check to make sure that an explosive concentration does not exist. Acetyl Cyclohexane Sulfonyl Peroxide must be stored to avoid contact with combustible materials such as wood, paper and oil since violent reactions occur. Detached storage is recommended. Store in tightly closed containers in a cool, well ventilated area. Metal containers involving the transfer of this chemical should be grounded and bonded. Where possible, automatically pump liquid from drums or other storage containers to process containers. Drums must be equipped with self-closing valves, pressure vacuum bungs, and flame arresters. Use only non-sparking tools and equipment, especially when opening and closing containers of this chemical. Sources of ignition such as smoking and open flames, are prohibited where this chemical is used, handled, or stored in a manner that could create a potential fire or explosion hazard. See OSHA standard 1910.104 and NFPA 43A *Code for the Storage of Liquid and Solid Oxidizers* for detailed handling and storage regulations.

Shipping: Shipment of material >82% in concentration or containing <12% H_2O is forbidden. Shipment of material <82% wetted with >12% water falls in Hazard Class 5.2 and Packing Group I; such materials must be labeled "Organic Peroxide." Shipment of material <32% in solution falls in Hazard Class 5.2 and Packing Group II; such material must be labeled "Organic Peroxide." Shipment of any of these solutions by either passenger or cargo aircraft is forbidden.

Spill Handling: Evacuate and restrict persons not wearing protective equipment from area of spill or leak until cleanup is complete. Remove all ignition sources. Keep spilled material wet. Because this chemical is a severe explosion hazard, contact the manufacturer regarding the safest method for clean-up and disposal. Keep Acetyl Cyclohexane Sulfonyl Peroxide out of a confined space, such as a sewer, because of the potential for an explosion, unless the sewer is designed to prevent the build-up of explosive concentrations. Ventilate area of spill or leak after clean-up is complete. It may be necessary to contain and dispose of this chemical as a hazardous waste. If material or contaminated runoff enters waterways, notify downstream users of potentially contaminated waters. Contact your Department of Environmental Protection or your regional office of the federal EPA for specific recommendations. If employees are required to clean-up spills, they must be properly trained and equipped. OSHA 1910.120(q) may be applicable.

Fire Extinguishing: This chemical is a flammable solid or in solution. Poisonous gases including sulfur oxides are produced in fire. Use dry chemical, carbon dioxide, water spray, or foam extinguishers. Containers may explode in fire. *Evacuate area* if temperature control cannot be maintained. Cool containers with liquid nitrogen, dry ice, or ice. If material or contaminated runoff enters waterways, notify downstream users of potentially contaminated waters. Notify local health and fire officials and pollution control agencies. From a secure, explosion-proof location, use water spray to cool exposed containers. If cooling streams are ineffective (venting sound increases in volume and pitch, tank discolors, or shows any signs of deforming), withdraw immediately to a secure position. If employees are expected to fight fires, they must be trained and equipped in OSHA 1910.156.

References

New Jersey Department of Health and Senior Services, Hazardous Substance Fact Sheet: Acetyl Cyclohexane Sulfonyl Peroxide, Trenton, NJ (September, 1998)

Acetylene

Molecular Formula: C_2H_2

Common Formula: HC≡CH

Synonyms: Acetileno (Spanish); Acetylen; Acetylene, dissolved; Ethene; Ethine; Ethyne; Narcylen

CAS Registry Number: 74-86-2

RTECS Number: AO9600000

DOT ID: UN 1001

EEC Number: 601-015-00-0

Regulatory Authority

- Clean Air Act 42USC7412; Title I, Part A, §112 hazardous pollutants, Accidental Release Prevention/Flammable substances (Section 68.130) TQ = 10,000 lb (4,540 kg)
- EPA Hazardous Waste Number (RCRA No.): D001
- Highly Reactive Substance and Explosive (World Band)[15]

Cited in U.S. State Regulations: Alaska (G), California (G), Florida (G), Illinois (G), Maine (G), Massachusetts (G), Minnesota (G), New Hampshire (G), New Jersey (G), New York (G), Pennsylvania (G), Rhode Island (G), Virginia (A).

Description: Acetylene, HC≡CH, is a highly flammable, colorless, compressed gas with a faint ethereal odor when pure; a garlic-like odor when contaminated. Flash point = -18°C. Autoignition temperature = 305°C. Hazard Identification (based on NFPA-704 M Rating System): Health 0, Flammability 4, Reactivity 3. Explosive limits: LEL = 2.5%, UEL = 82%. Soluble in water.

Potential Exposure: Acetylene can be burned in air or oxygen and is used for brazing, welding, cutting, metallizing, hardening, flame scarfing, and local heating in metallurgy. The flame is also used in the glass industry. Chemically, acetylene is used in the manufacture of vinyl chloride, acrylinitrile, synthetic rubber, vinyl acetate, trichloroethylene, acrylate, butyrolactone, 1,4-butanediol, vinyl alkyl ethers, pyrrolidone, and other substances.

Incompatibilities: The substance may polymerize due to heating. The substance decomposes on heating and increasing pressure, causing fire and explosion hazard. The substance is a strong reducing agent and reacts violently with oxidants and with fluorine or chlorine under influence of light, causing fire and explosion hazard. Reacts with copper, silver, and mercury or their salts, forming shock-sensitive compounds (acetylides).

Forms explosive mixture with air. Forms shock-sensitive mixture with copper and copper salts, mercury and mercury salts, and silver and silver salts. Reacts with brass, bromine, cesium hydride, chlorine, cobalt, cuprous acetylise, fluorine, iodine, mercuric nitrate, nitric acid, potassium, rubidium hydride, trifluoromethyl hypofluorite, and sodium hydride.

Permissible Exposure Limits in Air: No federal standard has been established. NIOSH has recommended a ceiling limit of 2,500 ppm (2,662 mg/m^3). ACGIH classifies acetylene as a simple asphyxiant with no TLV value. Virginia has set a guideline or standard of 3.0 µg/m^3 for ambient air.[60]

Determination in Air: Gas collection bag; none; Gas chromatography/Flame ionization detection; NIOSH Acetylene Criteria Document.

Permissible Concentration in Water: No criteria set but EPA[32] suggests an ambient water limit of 73,000 µg/l based on health effects.

Routes of Entry: Inhalation, ingestion, eye and/or skin contact.

Harmful Effects and Symptoms

Short Term Exposure: Rapid evaporation of the liquid may cause frostbite. Initial signs and symptoms of exposure to harmful concentrations of impure acetylene are rapid respiration, air hunger, followed by impaired mental alertness and muscular incoordination. Other manifestations include cyanosis, weak and irregular pulse, nausea, vomiting, prostration, impairment of judgment and sensation, loss of consciousness, convulsions, and death. Low order sensitization of myocardium to epinephrine resulting in ventricular fibrillation may be possible. At high concentrations pure acetylene may act as a mild narcotic and asphyxiant. Most accounted cases of illness or death can be attributed to acetylene containing impurities of arsine, hydrogen sulfide, phosphine, carbon disulfide, or carbon monoxide.

Long Term Exposure: The substance may cause effects on the nervous system.

Points of Attack: Central nervous system, respiratory system.

Medical Surveillance: No specific considerations are needed.

First Aid: Move victim to fresh air. Call emergency medical care. Apply artificial respiration if victim is not breathing. If breathing is difficult, give oxygen. Remove and isolate contaminated clothing and shoes. In case of contact with liquefied gas, thaw frosted parts with lukewarm water. Keep victim warm and quiet. Ensure that medical personnel are aware of the material(s) involved and take precautions to protect themselves. See also NIOSH criteria document cited below. If frostbite has occurred, seek medical attention immediately; do *NOT* rub the affected areas or flush them with water. In order to prevent further tissue damage, do *NOT* attempt to remove frozen clothing from frostbitten areas. If frostbite has *NOT* occurred, immediately and thoroughly wash contaminated skin with soap and water.

Personal Protective Methods: Acetylene poisoning can quite easily be prevented if (1) there is adequate ventilation and (2) impurities are removed when acetylene is used in poorly ventilated areas. General industrial hygiene practices for welding, brazing, and other metallurgical processes should also be observed. Where exposure to the liquefied compressed gas may occur, employees should be provided with special clothing designed to prevent frostbite.

Respirator Selection: See NIOSH criteria document cited below.

Storage: Before entering confined space where this chemical may be present, check to make sure that an explosive concentration does not exist. After use for welding, turn valve off, regularly check tubing, etc., and test for leaks with soap and water. Acetylene in process may be stored in atmospheric gas holders. May be stored in conventional compressed gas cylinders. Storage of liquid acetylene should be avoided. Procedures for the handling, use and storage of cylinders should be in compliance with OSHA 1910.101 and 1910.169 as with the recommendations of the Compressed Gas Association.

Shipping: Label required: "Flammable Gas," usually shipped under pressure in acetone solution. Shipment of dissolved acetylene forbidden in passenger aircraft or railcar. Shipment of liquid acetylene forbidden under any conditions.

Spill Handling: Evacuate and restrict persons not wearing protective equipment from area of spill or leak until cleanup is complete. Remove all ignition sources. Establish forced ventilation to keep levels below explosive limit and to disperse the gas. Stop the flow of gas. If source of leak is a cylinder and the leak cannot be stopped in place, remove leaking cylinder to a safe place in the open air, and repair leak or allow cylinder to empty. Keep this chemical out of confined space, such as sewer because of the possibility of explosion, unless the sewer is designed to prevent the buildup of explosive concentrations. It may be necessary to contain and dispose of this chemical as a hazardous waste. Contact your Department of Environmental Protection or your regional office of the federal EPA for specific recommendations. If employees are required to clean-up spills, they must be properly trained and equipped. OSHA 1910.120(q) may be applicable.

Fire Extinguishing: Stop flow of gas. Gases in confined areas may explode when exposed to fire. Gases may travel long distances to ignition sources and flash back. Storage containers and parts of containers may rocket great distances, in many directions. If material or contaminated runoff enters waterways, notify downstream users of potentially contaminated waters. Do not extinguish the fire unless the flow of the gas can be stopped and any remaining gas is out of the line. Specially trained personnel may use fog lines to cool exposures and let the fire burn itself out. Use dry chemicals, carbon dioxide. From a secure explosion-proof location, use water spray to cool exposed containers. If cooling streams are ineffective (venting sound increases in volume and pitch, tank discolors or shows any signs of deforming), withdraw immediately to a secure position.

Disposal Method Suggested: Consult with environmental regulatory agencies for guidance on acceptable disposal practices. Generators of waste containing this contaminant (≥100 kg/mo) must conform with EPA regulations governing storage, transportation, treatment, and waste disposal. Incineration.

References

National Institute for Occupational Safety and Health, Criteria for a Recommended Standard: Occupational Exposure to Acetylene, NIOSH Doc. No. 76–195, Wash., DC (1976)

Sax, N. I., Ed., Dangerous Properties of Industrial Materials Report, 1, No. 2, 23–25 (1980)

New York StateDepartment of Health, "Chemical Fact Sheet: Acetylene." Bureau of Toxic Substance Assessment (March 1986)

New Jersey Department of Health and Senior Services, "Hazardous Substance Fact Sheet: Acetylene," Trenton, NJ (Feb. 22, 1988)

Acetylene Tetrabromide

Molecular Formula: $C_2H_2Br_4$

Common Formula: $CHBr_2CHBr_2$

Synonyms: Ethane, 1,1,2,2-tetrabromo-; Muthmanns liquid; TBE; 1,1,2,2-Tetrabromaethan (German); Tetrabromoacetylene; 1,1,2,2-Tetrabromoetano (Italian); *sym*-Tetrabromoethane; 1,1,2,2-Tetrabromoethane; 1,1,2,2-Tetrabromoethane, *sym*-; Tetrabromuro de acetileno (Spansih); 1,1,2,2-Tetrabroomethaan (Dutch)

CAS Registry Number: 79-27-6

RTECS Number: K18225000

DOT ID: UN 2504

EEC Number: 602-016-00-9

Regulatory Authority

- Air Pollutant Standard Set (ACGIH)[1] (DFG)[3] (HSE)[33] (former USSR)[43] OSHA[58] (Various States)[60] (Various Canadian Provinces) (Mexico) (Israel)
- Canada, WHMIS, Ingredient Disclosure List
- U.S. DOT Regulated Marine Pollutant (49CFR172.101, Appendix B)

Cited in U.S. State Regulations: Alaska (G), California (G), Connecticut (A), Florida (G), Illinois (G), Maine (G), Massachusetts (G), Minnesota (G), Nevada (A), New Hampshire (G), New Jersey (G), North Dakota (A), Pennsylvania (G), Rhode Island (G), Virginia (A), West Virginia (G).

Description: Acetylene tetrabromide, $CHBr_2CHBr_2$, is a combustible, colorless to yellow liquid with a strong odor. Freezing/Melting point = -1°C. Boiling point = 135°C. Autoignition temperature = 335°C. Hazard Identification (based on NFPA-704 M Rating System): Health 3, Flammability 0, Reactivity 1. Slightly soluble in water.

Potential Exposure: Acetylene tetrabromide is used as a gauge fluid, as a solvent, and as a refractive index liquid in microscopy.

Incompatibilities: Chemically active metals (sodium, potassium, magnesium, zinc), strong caustics, hot iron. Contact with strong oxidizers may cause fire and explosions.

Permissible Exposure Limits in Air: The Federal OSHA standard[58] and ACGIH 1999 value is 1 ppm (14 mg/m^3) TWA, as is the DFG MAK value[3] and DFG set Peak Limitation (30 min.) of 2 times normal MAK, do not exceed four times during a workshift. The NIOSH IDLH value is 8 ppm. The former USSR-UNEP/IRPTC project[43] has set a limit in workplace air of 0.1 mg/m^3 for tetrabromoethane. HSE (U.K.)[33] has set an 8-hour TWA of 0.5 ppm (7 mg/m^3) with the notation "skin" indicating the potential for cutaneous absorption. The Australia, Israel, and Mexico standard is 1 ppm (14 mg/m^3) TWA and Israel set an Action level of 0.5 ppm (7 mg/m^3) and Mexico's STEL is 1.5 ppm (20 mg/m^3). Canadian Provincial levels are: Alberta and British Columbia: 1 ppm (14 mg/m^3) TWA and STEL of 1.5 ppm (21 mg/m^3) TWA. Ontario and Quebec 1 ppm (14 mg/m^3) TWAEV. In addition, several states have set guidelines or standards for acetylene tetrabromide in ambient air[60] ranging from 0.15 mg/m^3 (North Dakota) to 0.25 mg/m^3 (Virginia) to 0.28 mg/m^3 (Connecticut) to 0.357 mg/m^3 (Nevada).

Determination in Air: Silica absorption followed by THF treatment and gas chromatographic analysis. See NIOSH Method #2003.

Permissible Concentration in Water: No criteria set.

Routes of Entry: Inhalation, ingestion, skin and/or eye contact.

Harmful Effects and Symptoms

Short Term Exposure: Irritates eyes, skin and respiratory tract. Can cause headaches, fatigue, dizziness, lightheadedness, and unconsciousness. High concentrations can cause death. Prolonged skin contact can cause burns.

Long Term Exposure: There is limited evidence that acetylene tetrabromide causes cancer; possibly stomach cancer. Repeated exposures can cause liver, kidney, and lung damage and drying and cracking of the skin.

Points of Attack: Eyes, upper respiratory system, liver, kidneys, lungs, central nervous system.

Medical Surveillance: Consider the points of attack in preplacement and periodic physical examinations.

First Aid: If this chemical gets into the eyes, remove any contact lenses at once and irrigate immediately for at least 15 minutes, occasionally lifting upper and lower lids. If this chemical contacts the skin, remove contaminated clothing and wash immediately with soap and water. If this chemical has been inhaled, remove from exposure, begin rescue breathing (using universal precautions) if breathing has stopped and CPR if heart action has stopped. Transfer promptly to a medical facility. When this chemical has been swallowed, get medical attention. Give large quantities of salt water and induce vomiting. Do not make an unconscious person vomit.

Personal Protective Methods: Wear solvent-resistant gloves and clothing to prevent any reasonable probability of skin contact. Contact lenses should not be worn when working with this chemical. Wear splash-proof chemical goggles and face shield unless full facepiece respiratory protection is worn. Employees should wash immediately with soap when skin is wet or contaminated. Remove nonimpervious clothing immediately if wet or contaminated. Provide emergency showers and eyewash.

Respirator Selection: OSHA: *8 ppm:* SA (any supplied-air respirator); or SCBAF (any self-contained breathing apparatus with a full facepiece); or SAF (any supplied-air respirator with a full facepiece). *Emergency or planned entry in unknown concentrations or IDLH conditions:* SCBAF:PD,PP (any self-contained breathing apparatus that has a full facepiece and is operated in a pressure-demand or other positive-pressure mode); or SAF:PD,PP:ASCBA (Any supplied-air respirator that has a full facepiece and is operated in a pressure-demand or other positive-pressure mode in combination with an auxiliary self-contained breathing apparatus operated in a pressure-demand or other positive pressure mode. *Escape:* GMFOV [Any air-purifying, full-facepiece respirator (gas mask) with a chin-style, front-or back-mounted organic vapor canister]; or SCBAE (Any appropriate escape-type, self-contained breathing apparatus).

Storage: Prior to working with this chemical you should be trained on its proper handling and storage. Acetylene Tetrabromide decomposes at 474°F producing flammable and highly toxic vapors of bromine and carbonyl bromide. Avoid contact with incompatible materials cited above. Sources of ignition such as smoking and open flames, are prohibited where this chemical is used, handled, or stored in a manner that could create a potential fire or explosion hazard.

Shipping: Label required: "Keep Away from Food." Is in UN/DOT Hazard Class 6.1 and Packing Group III.[19][20]

Spill Handling: Evacuate and restrict persons not wearing protective equipment from area of spill or leak until cleanup is complete. Remove all ignition sources. Absorb liquids in vermiculite, dry sand, earth, or a similar material and deposit in sealed containers. Ventilate area of spill or leak after clean-up is complete. Collect powdered material in the most convenient and safe manner and deposit in sealed containers. It may be necessary to contain and dispose of this chemical as a hazardous waste. If material or contaminated runoff enters waterways, notify downstream users of potentially contaminated waters. Contact your Department of Environmental Protection or your regional office of the federal EPA for specific recommendations. If employees are required to clean-up spills, they must be properly trained and equipped. OSHA 1910.120(q) may be applicable.

Fire Extinguishing: Extinguish fire using an agent suitable for type of surrounding fire. Poisonous gases are produced in fire including bromine and carbonyl bromide. If material or contaminated runoff enters waterways, notify downstream users of potentially contaminated waters. Notify local health and fire officials and pollution control agencies. From a secure, explosion-proof location, use water spray to cool exposed containers. If cooling streams are ineffective (venting sound increases in volume and pitch, tank discolors, or shows any signs of deforming), withdraw immediately to a secure position. If employees are expected to fight fires, they must be trained and equipped in OSHA 1910.156.

Disposal Method Suggested: Incineration in admixture with combustible fuel and with scrubber to remove halo acids produced.

References

U.S. Environmental Protection Agency, Chemical Hazard Information Profile Draft Report: Tetrabromoethane, Washington, DC (June 14, 1983)

New Jersey Department of Health and Senior Services, Hazardous Substance Fact Sheet: Acetylene Tetrabromide, Trenton, NJ (February, 19896)

NIOSH/OSHA Occupational Health Guidelines for Chemical Hazards. DHHS (NIOSH) Pub Nos. 81–123; 88–118, Suppls. I–IV. 1981–1995

Acetyl Iodide

Molecular Formula: C_2H_3IO

Common Formula: CH_3COI

Synonyms: Yoduro de acetilo (Spanish)

CAS Registry Number: 507-02-8

RTECS Number: AP4670000

DOT ID: UN 1898

Regulatory Authority

- U.S. DOT Regulated Marine Pollutant (49CFR172.101, Appendix B)
- Canada, WHMIS, Ingredients Disclosure List

Cited in U.S. State Regulations: Maine (G), New Hampshire (G), New Jersey (G)

Description: Acetyl iodide, CH_3COI, is a corrosive, colorless liquid that turns brown on contact with air or moisture. It has a suffocating odor. Boiling point = 108°C. Decomposes in water.

Potential Exposure: Used as an organic intermediate; to make other chemicals.

Incompatibilities: Moisture, water, steam, or alcohol. Corrosive to metals.

Permissible Exposure Limits in Air: No standards set.

Permissible Concentration in Water: No criteria set.

Routes of Entry: Inhalation, skin contact.

Harmful Effects and Symptoms

Short Term Exposure: Can cause severe skin and eye burns. Breathing this chemical can irritate the lungs causing coughing and/or shortness of breath. Higher exposures can cause pulmonary edema, a medical emergency. This can cause death.

Long Term Exposure: Repeated exposure can cause lung irritation, bronchitis with cough, phlegm, and/or shortness of breath.

Points of Attack: Eyes, skin, respiratory system.

Medical Surveillance: For those with frequent or potentially high exposure, lung function tests are recommended before beginning work and at regular times after that. If symptoms develop or overexposure is suspected, consider chest x-ray.

First Aid: If this chemical gets into the eyes, remove any contact lenses at once and irrigate immediately for at least 15 minutes, occasionally lifting upper and lower lids. If this chemical contacts the skin, remove contaminated clothing and wash immediately with soap and water. When this chemical has been swallowed, get medical attention. If victim is conscious, administer water or milk. Do not induce vomiting. If this chemical has been inhaled, remove from exposure and transfer promptly to a medical facility. Continue observation for 24 – 48 hours after breathing exposure as pulmonary edema may be delayed.

Personal Protective Methods: Wear solvent-resistant gloves and clothing to prevent any reasonable probability of skin contact. Safety equipment suppliers/manufacturers can provide recommendations on the most protective glove/clothing material for your operation. All protective clothing (suits, gloves, footwear, headgear) should be clean, available each day, and put on before work. Contact lenses should not be worn when working with this chemical. Wear splash-proof chemical goggles and face shield unless full facepiece respiratory protection is worn. Employees should wash immediately with soap when skin is wet or contaminated. Remove nonimpervious clothing immediately if wet or contaminated. Provide emergency showers and eyewash.

Respirator Selection: *At any concentrations above the NIOSH REL, or where there is no REL, at any detectable concentration:* SCBAF:PD,PP (any MSHA/NIOSH approved self-contained breathing apparatus that has a full facepiece and is operated in a pressure-demand or other positive-pressure mode); or SAF:PD,PP:ASCBA (any supplied-air respirator that has a full facepiece and is operated in a pressure-demand or other positive-pressure mode in combination with an auxiliary, self-contained breathing apparatus operated in a pressure-demand or other positive pressure mode). *Escape:* GMFOVHiE [any air-purifying, full-facepiece respirator (gas mask) with a chin-style, front- or back-mounted organic vapor canister having a high-efficiency particulate filter]; or SCBAE (any appropriate escape-type, self-contained breathing apparatus).

Storage: Prior to working with this chemical you should be trained on its proper handling and storage. Store in tightly closed containers in a cool, well ventilated area away from all forms of moisture and alcohols. Where possible, automatically pump liquid from drums or other storage containers to process containers.

Shipping: Shipping label required is "Corrosive." Is in DOT/UN Hazard Class 8, Packing Group II.[19][20]

Spill Handling: Evacuate and restrict persons not wearing protective equipment from area of spill or leak until cleanup is complete. Remove all ignition sources. Absorb liquids in vermiculite, dry sand, earth, or a similar material and deposit in sealed containers. Collect powdered material in the most convenient and safe manner and deposit in sealed containers. Ventilate area of spill or leak after clean-up is complete. It may be necessary to contain and dispose of this chemical as a hazardous waste. If material or contaminated runoff enters waterways, notify downstream users of potentially contaminated waters. Contact your Department of Environmental Protection or your regional office of the federal EPA for specific recommendations. If employees are required to clean-up spills, they must be properly trained and equipped. OSHA 1910.120(q) may be applicable.

Fire Extinguishing: Acetyl iodide may burn but does not readily ignite. Poisonous gases are produced in fire including iodine vapors and iodides. Do not use water as this chemical reacts forming corrosive and toxic fumes. Use dry chemical, carbon dioxide, or foam extinguishers. If employees are expected to fight fires, they must be trained and equipped in OSHA 1910.156.

Disposal Method Suggested: Incineration.

References

New Jersey Department of Health and Senior Services, "Hazardous Substance Fact Sheet: Acetyl Iodide," Trenton, NJ (June 1998)

Acetylsalicylic Acid

Molecular Formula: $C_9H_8O_4$

Common Formula: $HOOCC_6H_4OCOCH_3$

Synonyms: Acenterine®; Acesa®; Acetaal (Dutch); Acetaldehyde diethyl acetal; Acetal diethylique (French); Acetale (Italian); Aceticyl; Acetol®; Acetophen®; Acetosal®; Acetosalin®; Acetylin®; 2-(Acetyloxybenzoic) acid; Acetylsal®; Acisal®; Acylpyrin®; Asagran®; Aspirin; Aspro®; Asteric®; Benzoic acid, 2-(acetyloxy)-; Caprin; 1,1-Diaethoxy-aethan; Diaethylacetal (German); 1,1-Diethoxy-ethaan (Dutch); 1,1-Diethoxyethane; Diethyl acetal; 1,1-Dietossi-etano (Italian); Duramax®; Ecotrin®; Empirin®; Ethylidene diethyl ether; Neuronika®; Polopiryna®; Rhodine®; Salacetin®; Salicylic acid, acetate; XAXA®

CAS Registry Number: 50-78-2

RTECS Number: VO0700000

DOT ID: UN 9188

EEC Number: 605-015-00-1

Regulatory Authority

- Air Pollutant Standard Set (ACGIH)[1] (U.K.)[33] (OSHA)[58] (Israel) Several States[60] (Australia) (Israel) (Various Canadian Provinces)

Cited in U.S. State Regulations: Alaska (G), California (G), Connecticut (A), Florida (G), Illinois (G), Maine (G), Massachusetts (G), Minnesota (G), Nevada (A), New Hampshire (G), New Jersey (G), Pennsylvania (G), Rhode Island (G), Virginia (A), West Virginia (G).

Description: Acetylsalicylic acid, $HOOCC_6H_4OCOCH_3$, is a white crystalline solid with a slightly bitter taste. It is odorless but hydrolyzes in moist air to give an acetic acid odor. Freezing/Melting point = 135°C. Boiling point = 140°C. Soluble in water. Octanol/water partition coefficient as log Pow: 1.19

Potential Exposure: Those engaged in manufacture of aspirin or, more likely, in its consumption in widespread use as an analgesic, antipyretic and anti-inflammatory agent.

Incompatibilities: Strong oxidizers, strong acids, strong bases, carbonates, moisture. Dust dispersed in air is explosive.

Permissible Exposure Limits in Air: NIOSH has recommended 5 mg/m^3 as has the U.K.[33] ACGIH has set a TWA value of 5 mg/m^3 but no STEL value. In addition, some states have set guidelines or standards for aspirin in ambient air[60] ranging from 80 μg/m^3 (Virginia) to 100 μg/m^3 (Connecticut) to 119 μg/m^3 (Nevada).

Determination in Air: Filter; none; Gravimetric; NIOSH (IV) Method #0500, Particulates NOR (total).

Permissible Concentration in Water: No criteria set.

Routes of Entry: Primarily oral in medicinal use.

Harmful Effects and Symptoms

Short Term Exposure: Eye and skin contact can cause irritation. Burns to the eyes and scarring can occur. High exposures may cause headache, dizziness, depression, and irritability. This chemical can be absorbed through the skin, thereby increasing exposure. Adverse effects from the usual doses of aspirin are infrequent, most common are gastrointestinal disturbances. Prolonged administration of large doses results in occult bleeding and may result in anemia.

Long Term Exposure: Repeated exposures may cause headache, dizziness, depression, and irritability; allergy may develop causing hives. Animal tests show that this substance possibly causes toxic effects upon human reproduction; possible teratogen. This chemical can decrease the clotting ability of blood.

Points of Attack: Eyes, skin, respiratory system, blood, liver, kidneys.

Medical Surveillance: Tests of blood clotting ability. People taking anticoagulants may be at increased risk. Skin testing with dilute Acetylsalicylic Acid may help diagnose allergy, if done by a qualified allergist. If allergy is confirmed, all future exposure to this chemical should be avoided, since even small exposure can cause an allergic reaction. It is possible for severe allergic reaction to occur.

First Aid: If this chemical gets into the eyes, remove any contact lenses at once and irrigate immediately for at least 15 minutes, occasionally lifting upper and lower lids. If this chemical contacts the skin, remove contaminated clothing and wash immediately with soap and water. If this chemical has been inhaled, remove from exposure, begin rescue breathing (using universal precautions) if breathing has stopped and CPR if heart action has stopped. Transfer promptly to a medical facility.

Personal Protective Methods: Wear protective gloves and clothing to prevent skin contact. Safety equipment suppliers/manufacturers can provide recommendations on the most protective glove/clothing material for your operation. All protective clothing (suits, gloves, footwear, headgear) should be clean, available each day, and put on before work. Contact lenses should not be worn when working with this chemical. Wear dust-proof chemical goggles and face shield unless full facepiece respiratory protection is worn. Employees should wash immediately with soap when skin is wet or contaminated. Remove nonimpervious clothing immediately if wet or contaminated. Provide emergency showers and eyewash.

Respirator Selection: *Where potential for exposure exists:* SCBAF:PD,PP (any MSHA/NIOSH approved self-contained breathing apparatus that has a full facepiece and is operated in a pressure-demand or other positive-pressure mode); or SAF: PD,PP:ASCBA (any supplied-air respirator that has a full facepiece and is operated in a pressure-demand or other positive-pressure mode in combination with an auxiliary, self-contained breathing apparatus operated in a pressure-demand or other positive pressure mode). *Escape:* HiEF (Any air-purifying, full-facepiece respirator with a high-efficiency particulate filter); or SCBAE (Any appropriate escape-type, self-contained breathing apparatus).

Storage: Prior to working with this chemical you should be trained on its proper handling and storage. Store in tightly closed containers in a cool, well ventilated area away from all forms of moisture.

Shipping: Acetylsalicylic acid may be classified[52] as a Hazardous Substance, solid n.o.s. which puts it in Hazard Class ORM-E and Packing Group III but imposes no weight limits on passenger aircraft or railcar or cargo aircraft shipment.

Spill Handling: Evacuate and restrict persons not wearing protective equipment from area of spill or leak until cleanup is complete. Remove all ignition sources. Sweep spilled substance into containers; if appropriate, moisten first to prevent dusting. Carefully collect remainder, then remove to safe place (extra personal protection: P2 filter respirator for harmful particles). Use a vacuum or wet method to reduce dust during clean-up. *Do not dry sweep.* The spilled material may be dampened with 60 – 70% ethanol to avoid airborne dust and the material then scooped up for disposal. Collect powdered material in the most convenient and safe manner and deposit in sealed containers. Ventilate area of spill or leak after clean-up is complete. It may be necessary to contain and dispose of this chemical as a hazardous waste. If material or contaminated runoff enters waterways, notify downstream users of potentially contaminated waters. Contact your Department of Environmental Protection or your regional office of the federal EPA for specific recommendations. If employees are required to clean-up spills, they must be properly trained and equipped. OSHA 1910.120(q) may be applicable.

Fire Extinguishing: Acetylsalicylic acid itself may burn but does not readily ignite. Dust of this chemical dispersed in air is explosive. Use dry chemical, carbon dioxide, water spray, or alcohol foam extinguishers. Poisonous gases are produced in fire including carbon monoxide. If employees are expected to fight fires, they must be trained and equipped in OSHA 1910.156.

Disposal Method Suggested: May be flushed to sewer with large volumes of water.

References

Sax, N. I., Ed., Dangerous Properties of Industrial Materials Report, 1, No. 3, 20–22 (1981) (as Acetol)

New Jersey Department of Health and Senior Services, "Hazardous Substance Fact Sheet: Acetylsalicylic Acid," Trenton, NJ (June 1998)

1-Acetyl-2-Thiourea

Molecular Formula: $C_3H_6N_2OS$

Common Formula: $CH_3CONHCSNH_2$

Synonyms: Acetamide, *N*-(aminothioxomethyl)-; Acetyl thiourea

CAS Registry Number: 591-08-2

RTECS Number: YR7700000

Regulatory Authority

- EPA Hazardous Waste Number (RCRA No.): P002
- RCRA Land Ban Waste Restrictions
- SUPERFUND/EPCRA 40CFR302.4, Appendix A, Reportable Quantity (RQ): 1,000 lb (454 kg)
- TSCA: 40CFR716.120(a)

Cited in U.S. State Regulations: Kansas (G), Louisiana (G), Massachusetts (G), New Jersey (G), Pennsylvania (G), Vermont (G), Virginia (G), Washington (G), Wisconsin (G).

Description: 1-Acetyl-2-thiourea, $CH_3CONHCSNH2$, is a solid forming needles. Freezing/Melting point = 166 – 167°C. Soluble in water.

Potential Exposure: Studied as possible rodenticide.

Incompatibilities: Strong oxidizers, strong acids.

Permissible Exposure Limits in Air: No standards set.

Permissible Concentration in Water: No criteria set.

Routes of Entry: Ingestion.

Harmful Effects and Symptoms

An LD_{50} value (oral-rat) has been reported as 50 mg/kg and is the apparent basis for EPA classification as hazardous substance[4] and hazardous waste.[5] Acts as poison in humans by ingestion and intraperitoneal routes.

First Aid: If this chemical gets into the eyes, remove any contact lenses at once and irrigate immediately for at least 15 minutes, occasionally lifting upper and lower lids. If this chemical contacts the skin, remove contaminated clothing and wash immediately with soap and water. When this chemical has been swallowed, get medical attention. Give large quantities of water and induce vomiting. Do not make an unconscious person vomit. If this chemical has been inhaled, remove from exposure and transfer promptly to a medical facility.

Personal Protective Methods: Wear protective gloves and clothing to prevent any reasonable probability of skin contact. Safety equipment suppliers/manufacturers can provide recommendations on the most protective glove/clothing material for your operation. All protective clothing (suits, gloves, footwear, headgear) should be clean, available each day, and put on before work. Contact lenses should not be worn when working with this chemical. Wear dust-proof chemical goggles and face shield unless full facepiece respiratory protection is worn. Employees should wash immediately with soap when skin is wet or contaminated. Provide emergency showers and eyewash.

Respirator Selection: *Where potential for exposure exists:* SCBAF:PD,PP (any MSHA/NIOSH approved self-contained breathing apparatus that has a full facepiece and is operated in a pressure-demand or other positive-pressure mode); or SAF:PD,PP:ASCBA (any supplied-air respirator that has a full facepiece and is operated in a pressure-demand or other positive-pressure mode in combination with an auxiliary, self-contained breathing apparatus operated in a pressure-demand or other positive pressure mode). *Escape:* HiEF (Any air-purifying, full-facepiece respirator with a high-efficiency particulate filter); or SCBAE (Any appropriate escape-type, self-contained breathing apparatus).

Storage: Prior to working with this chemical you should be trained on its proper handling and storage. Store in tightly closed containers in a cool, well ventilated area away from moisture.

Spill Handling: Evacuate and restrict persons not wearing protective equipment from area of spill or leak until cleanup is complete. Remove all ignition sources. Collect powdered material in the most convenient and safe manner and deposit in sealed containers. Ventilate area of spill or leak after clean-up is complete. It may be necessary to contain and dispose of this chemical as a hazardous waste. If material or contaminated runoff enters waterways, notify downstream users of potentially contaminated waters. Contact your Department of Environmental Protection or your regional office of the federal EPA for specific recommendations. If employees are required to clean-up spills, they must be properly trained and equipped. OSHA 1910.120(q) may be applicable.

Fire Extinguishing: Use dry chemical, carbon dioxide, water spray, or alcohol foam extinguishers. Poisonous gases are produced in fire including nitrogen oxides and sulfur oxides. If employees are expected to fight fires, they must be trained and equipped in OSHA 1910.156.

Disposal Method Suggested: Consult with environmental regulatory agencies for guidance on acceptable disposal practices. Generators of waste containing this contaminant (≥100 kg/mo) must conform with EPA regulations governing storage, transportation, treatment, and waste disposal.

Acifluorfen

Molecular Formula: $C_{14}H_7ClF_3NO_5$

Common Formula: $F_3C\text{-}C_6H_3(Cl)\text{-}O\text{-}C_6H_3(NO_2)(COOH)$

Synonyms: Acifluorfen; Acifluorfene; Benzoic acid, 5-[2-chloro-4-(trifluoromethyl)phenoxy]-2-nitro-; Blazer®; Carbofluorfen; 5-[2-Chloro-4-(trifluoromethyl)phenoxy]-2-nitrobenzoic acid; 5-(2-Chloro-α,α,α-trifluoro-*p*-tolyloxy)-2-nitrobenzoic acid; Tackle®

CAS Registry Number: 50594-66-6; 62476-59-9 (sodium salt)

RTECS Number: DG5643200

DOT ID: UN 2588 (pesticide, solid, toxic, n.o.s.)

Regulatory Authority

- Carcinogen (possible human; animal positive) (EPA) (Reference below) (California Prop 65)
- SARA 313: Form R *de minimis* Concentration Reporting Level: 1.0% (Sodium salt)

Cited in U.S. State Regulations: California (G).

Description: Acifluorfen, $C_{14}H_7ClF_3NO_5$, $F_3C\text{-}C_6H_3(Cl)\text{-}O\text{-}C_6H_3(NO_2)(COOH)$ is a combustible, off-white, light tan to brown solid. Freezing/Melting point = 152 – 157°C;[23] 124 – 126°C (sodium salt). The sodium salt is soluble in water.

Potential Exposure: Those involved in the manufacture, formulation and application of this selective pre-emergence and post-emergence herbicide used to control weeds and grass in soybean and peanut crops.

Incompatibilities: Strong oxidizers. Avoid contact with all sources of ignition.

Permissible Exposure Limits in Air: Not established.

Permissible Concentration in Water: A no-adverse-effect level (NOAEL) has been determined to be 20 mg/kg. Body weight/day based on fetotoxicity. However a NOAEL of 5.6 was determined based on increase in liver size of male rats; further a NOAEL of 1.25 mg/kg/day was determined in a 2-generation rat reproduction study. On this last basis, a long term health advisory of acifluorfen has been set at 0.44 mg/l for a 70-kg adult. A lifetime health advisory for that same adult of 0.009 mg/l. The EPA has also determined a reference dose (acceptable daily intake) of 0.013 mg/kg/day.

Determination in Water: Analysis of acifluorfen is by a gas chromatographic (GC) method applicable to the determination of certain chlorinated acid pesticides in water samples. In this method, approximately 1 liter of sample is acidified. The compounds are extracted with ethyl ether using a separatory funnel. The derivatives are hydrolyzed with potassium hydroxide, and extraneous organic material is removed by a solvent wash. After acidification, the acids are extracted and converted to their methyl esters using diazomethane as the derivatizing agent. Excess reagent is removed, and the esters are determined by electron capture GC. The method detection limit has not been determined for this compound, but it is estimated that the detection limits for analytes included in this method are in the range of 0.5 – 2 μg/l.

Routes of Entry: Ingestion.

Harmful Effects and Symptoms

The acute oral LD_{50} for male rats is 2,025 mg/kg; for female rats is 1,370 mg/kg. Acifluorfen is a moderate dermal irritant.

Long Term Exposure: A known carcinogen. Similar chlorinated diphenyl ethers have caused liver damage in laboratory animals.

Points of Attack: Skin and liver.

Medical Surveillance: Liver function tests.

First Aid: If this chemical gets into the eyes, remove any contact lenses at once and irrigate immediately for at least 15 minutes, occasionally lifting upper and lower lids. If this chemical contacts the skin, remove contaminated clothing and wash immediately with soap and water. If this chemical has been inhaled, remove from exposure, begin rescue breathing (using universal precautions) if breathing has stopped and CPR if heart action has stopped. Transfer promptly to a medical facility. When this chemical has been swallowed, get medical attention.

Personal Protective Methods: Wear protective gloves and clothing to prevent any reasonable probability of skin contact. Safety equipment suppliers/manufacturers can provide recommendations on the most protective glove/clothing material for your operation. All protective clothing (suits, gloves, footwear, headgear) should be clean, available each day, and put on before work. Contact lenses should not be worn when working with this chemical. Wear dust-proof chemical goggles and face shield unless full facepiece respiratory protection is worn. Employees should wash immediately with soap when skin is wet or contaminated. Provide emergency showers and eyewash.

Respirator Selection: *At any detectable concentration:* SCBAF:PD,PP (any MSHA/NIOSH approved self-contained breathing apparatus that has a full facepiece and is operated in a pressure-demand or other positive-pressure mode); or SAF:PD,PP:ASCBA (any supplied-air respirator that has a full facepiece and is operated in a pressure-demand or other positive-pressure mode in combination with an auxiliary, self-contained breathing apparatus operated in a pressure-demand or other positive pressure mode). *Escape:* HiEF (Any air-purifying, full-facepiece respirator with a high-efficiency particulate filter); or SCBAE (Any appropriate escape-type, self-contained breathing apparatus).

Storage: Prior to working with this chemical you should be trained on its proper handling and storage. Store in tightly closed containers in a cool, well ventilated area. Sources of ignition such as smoking and open flames, are prohibited where this chemical is used, handled, or stored in a manner that could create a potential fire or explosion hazard.

Shipping: Acifluorfen may be classified as a pesticide, solid, toxic n.o.s. It would then fall in Hazard Class 6.1 and Packing Group III. This requires a "Keep Away from Food" label.

Spill Handling: Evacuate and restrict persons not wearing protective equipment from area of spill or leak until cleanup is complete. Remove all ignition sources. Collect powdered material in the most convenient and safe manner and deposit in sealed containers. Ventilate area of spill or leak after clean-up is complete. Do not flush spilled material into sewer. It may be necessary to contain and dispose of this chemical as a hazardous waste. If material or contaminated runoff enters waterways, notify downstream users of potentially contaminated waters. Contact your Department of Environmental Protection or your regional office of the federal EPA for specific recommendations. If employees are required to clean-up spills, they must be properly trained and equipped. OSHA 1910.120(q) may be applicable.

Reverse osmosis (RO) is a promising treatment method for pesticide-contaminated water. As a general rule, organic compounds with molecular weights greater than 100 are candidates for removal by RO, which yields 99% removal efficiency of chlorinated pesticides by a thin-film composite polyamide membrane operating at a maximum pressure of 1,000 psi and at a maximum temperature of 113°F. More operational data are required, however, to specifically determine the effectiveness and feasibility of applying RO for the removal of acifluorfen from water. Also, membrane adsorption must be considered when evaluating RO performance in the treatment of acifluorfen-contaminated drinking water supplies.

Fire Extinguishing: This chemical is a combustible solid but does not easily ignite. Use dry chemical, carbon dioxide, water spray, or foam extinguishers. Poisonous gases are produced in fire including nitrogen oxides, carbon monoxide, chlorides, and fluorides. If material or contaminated runoff enters waterways, notify downstream users of potentially contaminated waters. Notify local health and fire officials and pollution control agencies. From a secure, explosion-proof location, use water spray to cool exposed containers. If cooling streams are ineffective (venting sound increases in volume and pitch, tank discolors, or shows any signs of deforming), withdraw immediately to a secure position. If employees are expected to fight fires, they must be trained and equipped in OSHA 1910.156.

Disposal Method Suggested: In accordance with 40CFR165 recommendations for the disposal of pesticides and pesticide containers.

References

U.S. Environmental Protection Agency, "Health Advisory: Acifluorfen," Washington DC, Office of Drinking Water (August 1987)

Acridine

Molecular Formula: $C_{13}H_9N$

Synonyms: 10-Azaanthracene; 9-Azaanthracene; 2,3-Benzoquinoline; Benzo (b) quinoline; Dibenzo (b,e) pyridine; Dibenzopyridine

CAS Registry Number: 260-94-6

RTECS Number: AR7175000

DOT ID: UN 2713

Regulatory Authority

- See also Coal Tar Pitch Volatiles
- OSHA, 29CFR1910 Specifically Regulated Chemicals (See CFR1910.1002) as coal tar pitch volatiles
- U.S. DOT Regulated Marine Pollutant (49CFR172.101, Appendix B)

Cited in U.S. State Regulations: New Jersey (G), Oklahoma (G). See also Coal Tar Pitch Volatiles

Description: Acridine, $C_{13}H_9N$, is made up of small, colorless or light yellow crystalline needles. Boiling point = 346°C. Freezing/Melting point = 109 – 111°C. Hazard Identification (based on NFPA-704 M Rating System): Health 2, Flammability 2, Reactivity 0. Highly soluble in boiling water.

Potential Exposure: Acridine and its derivatives are widely used in the production of dyestuffs such as acriflavine, benzoflavine, and chrysaniline, and in the synthesis of pharmaceuticals such as aurinacrine, proflavine, and rivanol. A constituent of coal tar, coal tar creosote; found in wastes from gas and tar plants and coke oven emissions.

Incompatibilities: Strong acids, strong oxidizers.

Permissible Exposure Limits in Air: There is no Federal standard for acridine, nor are there ACGIH values. However, EPA[32] suggests an ambient air limit of 162 μg/m^3 based on health effects.

Determination in Air: By fluorometry.

Permissible Concentration in Water: No criteria set but EPA[32] suggests an ambient level goal of 800 μg/l based on health effects.

Determination in Water: Harmful to aquatic life in small quantities.

Routes of Entry: Inhalation, ingestion, eye and/or skin contact.

Harmful Effects and Symptoms

Short Term Exposure: Acridine is a severe irritant to the conjunctiva of the eyes, the mucous membranes of the respiratory tract, and the skin. It is a powerful photosensitizer of the skin. Acridine causes sneezing on inhalation. Poisonous by ingestion and subcutaneous routes.

Long Term Exposure: Yellowish discoloration of sclera and conjunctiva may occur. Mutational properties have been ascribed to acridine, but its effect on humans is not known.

Points of Attack: Eyes, skin and respiratory tract.

Medical Surveillance: Evaluate the skin, eyes, and respiratory tract in the course of any placement or periodic examinations.

First Aid: If this chemical gets into the eyes, remove any contact lenses at once and irrigate immediately for at least 15 minutes, occasionally lifting upper and lower lids. If this chemical contacts the skin, remove contaminated clothing and wash immediately with soap and water. When this chemical has been swallowed, get medical attention. Give large quantities of water and induce vomiting. Do not make an unconscious person vomit. If this chemical has been inhaled, remove from exposure and transfer promptly to a medical facility.

Personal Protective Methods: Wear protective gloves and clothing to prevent any reasonable probability of skin contact. A protective layer of petroleum jelly, lanolin or castor oil has bee recommended in ILO literature. Safety equipment suppliers/manufacturers can provide recommendations on the most protective glove/clothing material for your operation. All protective clothing (suits, gloves, footwear, headgear) should be clean, available each day, and put on before work. Contact lenses should not be worn when working with this chemical. Wear dust-proof chemical goggles and face shield unless full facepiece respiratory protection is worn. Employees should wash immediately with soap when skin is wet or contaminated. Provide emergency showers and eyewash.

Respirator Selection: *Where a potential for exposure exists:* SCBAF:PD,PP (any MSHA/NIOSH approved self-contained breathing apparatus that has a full facepiece and is operated in a pressure-demand or other positive-pressure mode); or SAF:PD,PP:ASCBA (any supplied-air respirator that has a full facepiece and is operated in a pressure-demand or other positive-pressure mode in combination with an auxiliary, self-contained breathing apparatus operated in a pressure-demand or other positive pressure mode). *Escape:* HiEF (Any air-purifying, full-facepiece respirator with a high-efficiency particulate filter); or SCBAE (Any appropriate escape-type, self-contained breathing apparatus).

Storage: Prior to working with this chemical you should be trained on its proper handling and storage. Store in tightly closed containers in a cool, well ventilated area. Use only non-sparking tools and equipment, especially when opening and closing containers of this chemical. Sources of ignition such as smoking and open flames, are prohibited where this chemical is used, handled, or stored in a manner that could create a potential fire or explosion hazard.

Shipping: Label required is "Flammable Solid." Falls in DOT/UN Hazard Class 4.1 and Packing Group III.[19][20]

Spill Handling: Evacuate and restrict persons not wearing protective equipment from area of spill or leak until cleanup is complete. Remove all ignition sources. Apply carbon or peat absorbants to dissolved material. Dredge up solid material for removal to disposal area. Collect powdered material in the most convenient and safe manner and deposit in sealed containers. Ventilate area of spill or leak after clean-up is complete. It may be necessary to contain and dispose of this chemical as a hazardous waste. If material or contaminated runoff enters waterways, notify downstream users of potentially contaminated waters. Contact your Department of Environmental Protection or your regional office of the federal EPA for specific recommendations. If employees are required to clean-up spills, they must be properly trained and equipped. OSHA 1910.120(q) may be applicable.

Fire Extinguishing: This chemical is a combustible solid. Use water spray, or foam extinguishers, CO_2 and dry chemicals may not be effective. Poisonous gases are produced in fire including nitrogen oxides. If employees are expected to fight fires, they must be trained and equipped in OSHA 1910.156.

Disposal Method Suggested: Incineration with nitrogen oxide removal from the effluent gas by scrubber, catalytic or thermal device.

References

Sax, N. I., Ed., Dangerous Properties of Industrial Materials Report, 1, No, 8, 32–33 (1981) and 8, No. 5, 49–55 (1988)

Acrolein

Molecular Formula: C_3H_4O

Common Formula: CH_2CHCHO

Synonyms: Acquinite; Acrehyde; Acroleina (Italian); Acroleine (Dutch, French); Acrylaldehyde; Acrylehyd (German); Acrylehyde; Acrylic aldehyde; Akrolein (Czech); Akroleina (Polish); Aldehyde acrylique (French); Aldeide acrilica (Italian); Allylaldehyde; Aqualin; Aqualine; Biocide; Ethylene aldehyde; Propenal (Czech); 2-Propenal; Prop-2-en-1-al; Propenal; 2-Propen-1-one; Propylene aldehyde; Slimicide

CAS Registry Number: 107-02-8

RTECS Number: AS1050000

DOT ID: UN 1092

EEC Number: 605-008-00-3

Regulatory Authority

- Toxic Substance (World Bank)[15]
- Air Pollutant Standard Set (ACGIH)[1] (DFG)[3] (HSE)[33] (former USSR)[43] (OSHA)[58] (Various States)[60] (Various Canadian Provinces), Mexico, Israel, Australia
- OSHA 29CFR1910.119, Appendix A, Process Safety List of Highly Hazardous Chemicals, TQ = 150 lb
- Clean Air Act 42USC7412; Title I, Part A, §112 hazardous pollutants; Part A, §112(r), Accidental Release Prevention/Flammable substances (Section 68.130) TQ = 5,000 lb (1,275 kg)
- Clean Water Act: 40CFR116.4 Hazardous Substances; RQ 40CFR117.3 (same as CERCLA); 40CFR423, Appendix A Priority Pollutants; 40CFR401.15 Toxic Pollutant
- EPA HAZARDOUS WASTE NUMBER (RCRA No.): P003
- RCRA 40CFR261, Appendix 8; 40CFR261.11 Hazardous Constituents
- RCRA Land Ban Waste.
- RCRA 40CFR268.48; 61FR15654, Universal Treatment Standards: Wastewater (mg/l), 0.29; Nonwastewater, N/A.
- RCRA 40CFR264, Appendix 9; Ground Water Monitoring List Suggested methods (PQL μg/l): 8030 (5); 8240 (5)
- CERCLA/SARA Section 302, Extremely Hazardous Substances: TPQ = 500 lb (228 kg).
- SUPERFUND/EPCRA 40CFR302.4, Appendix A, Reportable Quantity (RQ): 1 lb (0.454 kg), SARA 313: Form R *de minimis* Concentration Reporting Level: 1.0%.
- U.S. DOT Regulated Marine Pollutant (49CFR172.101, Appendix B)
- Canada, WHMIS, Ingredients Disclosure List
- Mexico: Drinking Water 0.3 mg/l (ecological criteria); Listed as an organic toxic pollutant in wastewater.

Cited in U.S. State Regulations: Alaska (G), Arizona (W), California (G, A), Connecticut (A), Florida (G, A), Illinois (G), Kansas (G, W), Louisiana (G), Maine (G), Maryland (G), Massachusetts (G), Michigan (G), Minnesota (G), Nevada (A), New Hampshire (G), New Jersey (G), New York (A), North Carolina (A), North Dakota (A), Oklahoma (G), Pennsylvania (G), Rhode Island (G), South Carolina (A), Vermont (G), Virginia (G, A) Washington (G), West Virginia (G), Wisconsin (G).

Description: Acrolein, CH_2CHCHO, is a highly flammable, clear to yellowish liquid. It has a piercing, disagreeable odor and causes tears. Boiling point = 53°C. Freezing/Melting point = -87.7°C. Flash point = -26°C (cc). Hazard Identification (based on NFPA-704 M Rating System): Health 4, Flammability 3, Reactivity 3. Explosive limits: LEL = 2.8%, UEL = 31.0%. Autoignition temperature (unstable) = 220°C. Odor threshold = 0.174 ppm. Soluble in water.

Potential Exposure: Acrolein is primarily used as an intermediate in the production of glycerine and in the production of methionine analogs (poultry feed protein supplements). It is also used in chemical synthesis (1,3,6-hexametriol and glutaraldehyde), as a liquid fuel, antimicrobial agent, in algae and aquatic weed control, and as a slimicide in paper manufacture. Used in making plastics, drugs, and tear gas.

Incompatibilities: Forms explosive mixture with air. Elevated temperatures or sunlight may cause explosive polymerization. A strong reducing agent; reacts violently with oxidizers. Reacts with acids, alkalis, ammonia, amines, oxygen, peroxides. Shock-sensitive peroxides or acids may be formed over time. Attacks zinc and cadmium.

Permissible Exposure Limits in Air: The Federal OSHA standard[58] the ACGIH and NIOSH recommendations for exposure to acrolein is 0.1 ppm (0.25 mg/m^3) TWA. The ACGIH, NIOSH, and HSE (U.K.) STEL value is 0.3 ppm (0.8 mg/m^3). The Australian, Mexican, Israeli and Canadian Provincial (Alberta, British Columbia, Ontario Quebec) TWA and STEL values are the same as ACGIH and NIOSH and Israel has a Action Level of 0.05 ppm (0.115 mg/m^3). The DFG (German) TWA value is also the same and Peak Limitation (5 min) is 2 times the normal MAK; do not exceed more than 8 times during a workshift. The NIOSH IDLH value is 2.0 ppm. The former USSR-UNEP/IRPTC project[43] has set a value in workplace air of 0.2 mg/m^3 (1993) and of 0.03 mg/m^3 for ambient air in residential areas on either a momentary or an average daily basis. In addition, a number of states have set guidelines or standards for acrolein in ambient air[60] ranging from 0.83 μg/m^3 (N.Y.) to 1.25 μg/m^3 (South Carolina) to 2.5 μg/m^3 (Florida, North Dakota) to 4 μg/m^3 (Virginia) to 5 μg/m^3 (Connecticut) to 6.9 μg/m^3 (Nevada) to 80.0 μg/m^3 (North Carolina).

Determination in Air: See NIOSH Method #2501 and OSHA Method #52. See website *http://www.osha-slc.gov/* [and applicable method (e.g., OSHA Method #21)].

Permissible Concentration in Water: To protect freshwater aquatic life-on an acute basis 68 μg/l and on a chronic basis

21 μg/l. To protect saltwater aquatic life-55 μg/l on an acute toxicity basis. To protect human health-320 μg/l.[6] In addition, two states have set guidelines for acrolein in drinking water.[61] These are both 320 μg/l as set by Arizona and Kansas.

Determination in Water: Gas chromatography (EPA Method #603) or gas chromatography and mass spectrometry (EPA Method #624).

Routes of Entry: Inhalation, ingestion, skin and/or eye contact. Absorbed through the skin.

Harmful Effects and Symptoms

Short Term Exposure: This chemical can be absorbed through the skin, thereby increasing exposure. Eye and skin contact may cause intense tearing, irritation, blisters, and burns. Inhalation can irritate the lungs causing irritation, coughing, wheezing, and/or shortness of breath. Higher exposures can cause pulmonary edema, a medical emergency that can be delayed for several hours. This can cause death. If swallowed, produces acute abdominal pains. Extremely toxic; probable oral human lethal dose is 5 – 50 mg/kg, between 7 drops and one teaspoon for a 70 kg (150 lb) person. Inhalation of air containing 10 ppm of acrolein may be fatal in a few minutes. Death from cardiac failure accompanied by hyperemia and hemorrhage of the lungs and degeneration of the bronchial epithelium is possible. Acrolein causes acute respiratory and eye irritation, severe gastrointestinal distress with slowly developing pulmonary edema (lung fill up with fluid), and skin irritation.

Long Term Exposure: This chemical is a metabolite of cyclophosphamide, a well-recognized animal teratogen. Acrolein may cause mutations. Such chemicals have a cancer risk. Long-term exposure can cause drying and cracking of the skin. High or repeated lower exposure may cause permanent lung damage. NIOSH testing has not been completed to determine the carcinogenicity of acrolein. However, the limited studies to date indicate that this substance has chemical reactivity and mutagenicity similar to acetaldehyde and malonaldehyde. Therefore, NIOSH recommends that careful consideration should be given to reducing exposures to this related aldehyde.

Points of Attack: Heart, lungs, eyes, skin, respiratory system.

Medical Surveillance: Preplacement and periodic medical examinations should consider respiratory, skin, and eye disease. For those with frequent or potentially high exposure, lung function tests are recommended before beginning work and at regular times after that. If symptoms develop or overexposure is suspected, consider chest x-ray.

First Aid: If this chemical gets into the eyes, remove any contact lenses at once and irrigate immediately for at least 15 minutes, occasionally lifting upper and lower lids. If this chemical contacts the skin, remove contaminated clothing and wash immediately with soap and water. If this chemical has been inhaled, remove from exposure, begin rescue breathing (using universal precautions) if breathing has stopped and CPR if heart action has stopped. Transfer promptly to a medical facility. When this chemical has been swallowed, get medical attention. Give large quantities of water and induce vomiting. Do not make an unconscious person vomit. Medical observation is recommended for 24 – 48 hours after breathing overexposure, as pulmonary edema may be delayed.

Personal Protective Methods: Wear protective gloves and clothing to prevent any reasonable probability of skin contact. Safety equipment manufacturers recommend butyl rubber and viton as a protective materials. Contact lenses should not be worn when working with this chemical. Wear splash-proof chemical goggles and face shield unless full facepiece respiratory protection is worn. Employees should wash immediately with soap when skin is wet or contaminated. Remove nonimpervious clothing immediately if wet or contaminated. Provide emergency showers and eyewash.

Respirator Selection: OSHA/NIOSH *2 ppm:* SA:CF (any supplied-air respirator operated in a continuous-flow mode); or PAPROV [any powered, air-purifying respirator with organic vapor cartridge(s)]; or CCRFOV [any air-purifying, full-facepiece respirator (gas mask) with a chin-style, front-or back-mounted acid gas canister]; or GMFOV [any air-purifying, full-facepiece respirator (gas mask) with a chin-style, front- or back-mounted organic vapor canister]; or SCBAF (any self-contained breathing apparatus with a full facepiece); or SAF (any supplied-air respirator with a full facepiece). *Emergency or planned entry into unknown concentrations or IDLH conditions:* SCBAF:PD,PP (any self-contained breathing apparatus that has a full facepiece and is operated in a pressure-demand or other positive-pressure mode); or SAF:PD, PP:ASCBA (any supplied-air respirator that has a full facepiece and is operated in a pressure-demand or other positive-pressure mode in combination with an auxiliary, self-contained breathing apparatus operated in a pressure-demand or other positive pressure mode. *Escape:* GMFOV [any air-purifying, full-facepiece respirator (gas mask) with a chin-style, front- or back-mounted organic vapor canister]; or SCBAE (any appropriate escape-type, self-contained breathing apparatus).

Storage: Prior to working with this chemical you should be trained on its proper handling and storage. *Do not* store uninhibited acrolein under any circumstances. Protect against physical damage. Outside or detached storage is preferable. Inside storage should be in a standard flammable liquids storage room or cabinet. Before entering confined space where acrolein may be present, check to make sure that an explosive concentration does not exist. Store in tightly closed containers in a cool, well ventilated area away from heat and light. Metal containers involving the transfer of this chemical should be grounded and bonded. Where possible, automatically pump liquid from drums or other storage containers to process containers. Drums must be equipped with self-closing valves,

pressure vacuum bungs, and flame arresters. Use only non-sparking tools and equipment, especially when opening and closing containers of this chemical. Sources of ignition such as smoking and open flames, are prohibited where this chemical is used, handled, or stored in a manner that could create a potential fire or explosion hazard.

Shipping: Ship in unbreakable packaging. Do not transport with human or animal food and feedstuffs. Label required is "Poison, Flammable liquid." Shipment by passenger railcar and passenger or cargo aircraft is forbidden in any amount whatsoever. Falls in Hazard Class 6.1 and Packing Group I.

Spill Handling: Evacuate and restrict persons not wearing protective equipment from area of spill or leak until cleanup is complete. Remove all ignition sources. Establish forced ventilation to keep levels below explosive limit. Take up very small spills for disposal by absorbing it in vermiculite, dry sand, or earth and disposing in a secured landfill or combustion chamber. Alternatively, cover with sodium bisulfite, add small amount of water and mix. Then, after 1 hour, flush with large amounts of water and wash site with soap solution. Liquid should not be allowed to enter confined space, such as sewer, because of potential for explosion. It may be necessary to contain and dispose of this chemical as a hazardous waste. If material or contaminated runoff enters waterways, notify downstream users of potentially contaminated waters. Contact your Department of Environmental Protection or your regional office of the federal EPA for specific recommendations. If employees are required to clean-up spills, they must be properly trained and equipped. OSHA 1910.120(q) may be applicable.

Initial isolation and protective action distances: Distances shown are likely to be affected during the first 30 minutes after materials are spilled and could increase with time. If more than one tank car, cargo tank, portable tank, or large cylinder is involved in the incident is leaking, the protective action distance may need to be increased. You may need to seek emergency information from CHEMTREC at (800) 424-9300 or seek professional environmental engineering assistance from the U.S. EPA Environmental Response Team at (908) 548-8730 (24-hour response line).

Small spills (From a small package or a small leak from a large package)

First: Isolate in all directions (feet)..... 400

Then: Protect persons downwind (miles)

Day 0.3

Night 1.4

Large spills (From a large package or from many small packages)

First: Isolate in all directions (feet)..... 1,000

Then: Protect persons downwind (miles)

Day 1.2

Night 5.2

Fire Extinguishing: This chemical is a flammable liquid and explosion hazard. Under fire conditions, polymerization may occur, blocking relief valves leading to tank explosion. Poisonous gases are produced in fire. Use dry chemical, carbon dioxide, or alcohol foam extinguishers. Vapors are heavier than air and will collect in low areas. Vapors may travel long distances to ignition sources and flashback. Vapors in confined areas may explode when exposed to fire. Storage containers and parts of containers may rocket great distances, in many directions. If material or contaminated runoff enters waterways, notify downstream users of potentially contaminated waters. Notify local health and fire officials and pollution control agencies. From a secure, explosion-proof location, use water spray to cool exposed containers. If cooling streams are ineffective (venting sound increases in volume and pitch, tank discolors, or shows any signs of deforming), withdraw immediately to a secure position. If employees are expected to fight fires, they must be trained and equipped in OSHA 1910.156.

Disposal Method Suggested: Consult with environmental regulatory agencies for guidance on acceptable disposal practices. Generators of waste containing this contaminant (≥100 kg/mo) must conform with EPA regulations governing storage, transportation, treatment, and waste disposal. Incineration. Conditions are 1,500°F, 0.5 second minimum for primary combustion; 2,000°F, 1.0 second for secondary combustion.

References

U.S. Environmental Protection Agency, Chemical Hazard Information Profile: Acrolein, Washington DC (March 10, 1978)

U.S. Environmental Protection Agency, Acrolein: Ambient Water Quality Criteria, Washington DC (1980)

Nat. Inst. for Occup. Safety and Health, Information Profiles on Potential Occupational Hazards-Single Chemicals: Acrolein, Report TR 79–607, Rockville, MD, pp 1–18 (December 1979)

U.S. Environmental Protection Agency, Acrolein, Health and Environmental Effects Profile No. 3, Washington, DC, Office of Solid Waste (April 30, 1980)

Sax, N. I., Ed., Dangerous Properties of Industrial Materials Report, 1, No. 4, 28–31 (1981) and 3, No. 3, 36–41 (1983)

U.S. Environmental Protection Agency, "Chemical Profile: Acrolein," Washington, DC, Chemical Emergency Preparedness Program (Nov. 30, 1987)

New Jersey Department of Health and Senior Services, Hazardous Substance Fact Sheet: Acrolein, Trenton, NJ (May 1998)

Acrylamide

Molecular Formula: C_3H_5NO

Common Formula: $CH_2CHCONH_2$

Synonyms: Acrilamida (Spanish); Acrylamide monomer; Acrylic acid amide (50%); Acrylic amide; Acrylic amide 50%; Akrylamid (Czech); Ethylenecarboxamide; Ethylene monoclinic tablets carboxamide; 2-Propenamide; Propenamide; Vinyl amide

CAS Registry Number: 79-06-1

RTECS Number: AS3325000

DOT ID: UN 2074

EEC Number: 616-003-00-0

EINECS Number: 201-173-7

Regulatory Authority

- Carcinogen (Human Suspected) (ACGIH)[1] (IARC) (DRG)[3]
- Air Pollutant Standard Set (ACGIH)[1] (HSE)[33] (OSHA)[58] (Various States)[60] (Australia) (Various Canadian Provinces) (Israel) (Mexico)
- Water Pollution Standard Proposed (EPA)[48] (Minnesota)[61]
- Clean Air Act 42USC7412; Title I, Part A, §112 hazardous pollutants
- EPA Hazardous Waste Number (RCRA No.): U007
- RCRA 40CFR261, Appendix 8; 40CFR261.11 Hazardous Constituents
- RCRA 40CFR268.48; 61FR15654, Universal Treatment Standards: Wastewater (mg/l), 19; Nonwastewater, 23
- Safe Drinking Water Act, MCL, treatment technique; MCLG = zero; Regulated Chemical (47FR9352)
- CERCLA/SARA 40CFR302, Extremely Hazardous Substances: TPQ = 1,000/10,000 lb (454/4,540 kg).
- SUPERFUND/EPCRA 40CFR302.4, Appendix A, Reportable Quantity (RQ): 5,000 lb (2,270 kg), SARA 313: Form R *de minimis* Concentration Reporting Level: 0.1%.
- TSCA: 716.120(a), listed chemical
- U.S. DOT Regulated Marine Pollutant (49CFR172.101, Appendix B)
- Canada, WHMIS, Ingredients Disclosure List

Cited in U.S. State Regulations: Alaska (G), California (G), Connecticut (A), Florida (G), Illinois (G, W), Kansas (G), Louisiana (G), Maine (G), Maryland (G), Massachusetts (G), Minnesota (W), Nevada (A), New Hampshire (G), New Jersey (G), New York (G, A), Pennsylvania (G), Rhode Island (G), South Carolina (A), South Dakota (A), Vermont (G), Virginia (G, A), Washington (G), West Virginia (G), Wisconsin (G).

Description: Acrylamide, $CH_2CHCONH_2$, in monomeric form is an odorless, flake-like crystals. Freezing/Melting point = 84.5°C. Flash point = 138°C. Hazard Identification (based on NFPA-704 M Rating System): Health 2, Flammability 2, Reactivity 2. Autoignition temperature = 240°C. Soluble in water.

Potential Exposure: The major application for monomeric acrylamide is in the production of polymers as polyacrylamides. Polyacrylamides are used for soil stabilization, gel chromatography, electrophoresis, papermaking strengtheners, clarifications and treatment of potable water, and foods.

Incompatibilities: May decompose with heat and polymerize at temperatures above 84°C, or exposure to light, releasing ammonia gas. Reacts violently with strong oxidizers. Reacts with reducing agents, peroxides, acids, bases, and vinyl polymerization initiators. Fine particles of dust form explosive mixture with air.

Permissible Exposure Limits in Air: The Federal OSHA standard is 0.3 mg/m^3 as a time-weighted average (TWA) concentration for up to a 10-hour workshift. NIOSH and ACGIH has a TWA of 0.03 mg/m^3.[58] The notation "skin" indicated possible cutaneous absorption. The NIOSH IDLH, (potential occupational carcinogen) is [60 mg/m^3]. HSE (U.K.),[33] Australia, Israel, and Mexico TWA is 0.3 mg/m^3 and Israel Action Limit is 0.015 mg/m^3 and Mexico STEL is 2.6 mg/m^3. Canadian Provincial TWAs are: Alberta and British Columbia: 0.3 mg/m^3 and STEL of 0.6 mg/m^3; Ontario and Quebec TWAEVs are 0.03 mg/m^3. California's PEL is 0.3 mg/m^3 TWA. In addition several states have set guidelines or standards for acrylamide ambient concentrations in air:[60] 0.3 μg/m^3 (South Carolina) to 1.0 μg/m^3 (New York) to 3.0 μg/m^3 (South Dakota) to 5.0 μg/m^3 (Virginia) to 6.0 μg/m^3 (Connecticut) to 7.0 μg/m^3 (Nevada).

Determination in Air: Filter/Si gel; Methanol; Gas chromatography/Nitrogen/phosphorus detection; OSHA (#21).

Permissible Concentration in Water: Health advisories have been developed by EPA[48] on a long term (7-year) basis as 0.02 mg/l for a 10 kg child and 0.07 mg/l for a 70 mg adult. A guideline for acrylamide in drinking water of 0.10 μg/l has been developed by the State of Minnesota.[61]

Determination in Water: There is no standardized method for the determination of acrylamide in drinking water. An analytical procedure for the determination of acrylamide has been reported in the literature. This procedure consists ob bromination, extraction of the brominated product from water with ethyl acetate and quantification using high performance liquid chromatography (HPLC) with an ultraviolet detector. The concentration of the ethyl acetate to dryness and dissolution in a small volume of distilled water prior to HPLC analysis allows the detection of acrylamide at concentrations of 0.2 μg/l.[48]

Routes of Entry: Eyes, skin, central and peripheral nervous systems, reproductive system. Acrylamide can be absorbed through unbroken skin.

Harmful Effects and Symptoms

Short Term Exposure: Irritates the eyes, skin, and respiratory tract. Symptoms of exposure include: complaints of drowsiness, fatigue, tingling of fingers, and a stumbling, propulsive type of walking with sense of unsteadiness have been reported. Motor and sensory impairment, numbness, tremor, abnormal feelings in the lower limbs accompanied by weakness, and speech disturbances were also reported. Classified as very toxic; probably oral lethal human dose is between 50 and 500 mg/kg or between 1 teaspoon and 1 ounce for a 150 lb person. Polymerized acrylamide may not be toxic, but the monomer can cause peripheral nerve damage.

Long Term Exposure: There is evidence that acrylamide causes cancer in animals. It may cause skin and lung cancer in humans. There is limited evidence that this chemical damages the male testes. Can cause damage to the central nervous system, causing numbness, and weakness of the hands and feet. Acrylamide is a cumulative neurotoxin and repeated exposure to small amounts may cause serious injury to the nervous system. The neurological effects may be delayed. Polymer inhibitors or stabilizers added to the monomer may also produce toxicity. The symptoms of acrylamide toxicity are consistent with mid-brain lesions and blocked transport along both motor and sensory axons.

Points of Attack: Central nervous system, peripheral nervous system, skin and eyes.

Medical Surveillance: Since skin contact with the substance may result in localized or systemic effects, NIOSH recommends that medical surveillance be made available to all employees working in an area where acrylamide is stored, produced, processed, or otherwise used, except as an unintentional contaminant in other materials at a concentration of less than 1% by weight. For those with frequent or potentially high exposure, nerve condition tests should be considered The use of alcoholic beverages may enhance the harmful effects. BOLD and listed first.

References: LaDou, J., Nerve Conduction Studies, Occupational Medicine. Appleton and Lange, 1990.

First Aid: If this chemical gets into the eyes, remove any contact lenses at once and irrigate immediately. If this chemical contacts the skin, flush with water immediately. If a person breathes in large amounts of this chemical, move the exposed person to fresh air at once and perform artificial respiration. When this chemical has been swallowed, get medical attention. Give large quantities of water and induce vomiting. Do not make an unconscious person vomit.

Personal Protective Methods: Engineering controls should be used wherever feasible to maintain airborne concentrations of this chemical below the prescribed exposure limit. Respirators and protective equipment less effective than engineering controls, and should be used only in non-routine or emergency situations which may result in exposure concentrations in excess of the TWA environmental limit. Wear protective gloves and clothing to prevent repeated or prolonged skin contact. Wear dust-proof eye protection to prevent any reasonable probability of eye contact. Employees should wash immediately when skin is wet or contaminated. Work clothing should be changed daily if clothing is contaminated. Remove nonimpervious clothing immediately if wet or contaminated. Provide emergency showers.

Respirator Selection: NIOSH: *At any concentrations above the NIOSH REL, or where there is no REL, at any detectable concentration:* SCBAF:PD,PP (any self-contained breathing apparatus that has a full facepiece and is operated in a pressure-demand or other positive-pressure mode); or SAF:PD,PP:ASCBA (any supplied-air respirator that has a full facepiece and is operated in a pressure-demand or other positive-pressure mode in combination with an auxiliary self-contained breathing apparatus operated in a pressure-demand or other positive-pressure mode). *Escape:* GMFOV [any air-purifying, full-facepiece respirator (gas mask) with a chin-style, front-or back-,mounted organic vapor canister]; or SCBAE (any appropriate escape-type, self-contained breathing apparatus).

Storage: Prior to working with this chemical you should be trained on its proper handling and storage. Store only if stabilized, under inert gas. Before entering confined space where acrylamide may be present, check to make sure that an explosive concentration does not exist. Store in tightly closed containers in a cool, well ventilated area. Metal containers involving the transfer of this chemical should be grounded and bonded. Where possible, automatically pump liquid from drums or other storage containers to process containers. Drums must be equipped with self-closing valves, pressure vacuum bungs, and flame arresters. Use only non-sparking tools and equipment, especially when opening and closing containers of this chemical. Sources of ignition such as smoking and open flames, are prohibited where this chemical is used, handled, or stored in a manner that could create a potential fire or explosion hazard.

Shipping: Label required is "Keep Away from Food." Falls in UN/DOT Hazard Class 6.1, Packing Group III.[20][21]

Spill Handling: Evacuate and restrict persons not wearing protective equipment from area of spill or leak until cleanup is complete. Stay upwind; keep out of low areas. Ventilate closed spaces before entering them. Use water spray to reduce vapors. Remove all ignition sources. *Small spills:* absorb with sand or other non-combustible absorbent material and place into containers for later disposal. Small dry spills: with clean shovel place material into clean, dry container and cover; move containers from spill area. *Large spills:* dike far ahead of spill for later disposal. Collect powdered material in the most convenient and safe manner and deposit in sealed containers. Ventilate area of spill or leak after clean-up is complete. It may be necessary to contain and dispose of this chemical as a hazardous waste. If material or contaminated runoff enters waterways, notify downstream users of potentially contaminated waters. Contact your Department of Environmental Protection or your regional office of the federal EPA for specific recommendations. If employees are required to clean-up spills, they must be properly trained and equipped. OSHA 1910.120(q) may be applicable. Should you need to seek emergency information, call CHEMTREC at (800) 424-9300, or seek professional environmental engineering assistance from the U.S. EPA Environmental Response Team at (908) 548-8730 (24-hour response line).

Fire Extinguishing: A combustible solid. Use dry chemical, carbon dioxide, water spray, or foam extinguishers. Poisonous gases are produced in fire including nitrogen oxides. If employees are expected to fight fires, they must be trained and equipped in OSHA 1910.156. If material or contaminated runoff enters waterways, notify downstream users of potentially contaminated waters. Notify local health and fire officials and pollution control agencies. From a secure, explosion-proof location, use water spray to cool exposed containers. If cooling streams are ineffective (venting sound increases in volume and pitch, tank discolors, or shows any signs of deforming), withdraw immediately to a secure position. If employees are expected to fight fires, they must be trained and equipped in OSHA 1910.156. Dike fire control water for later disposal; do not scatter the material.

Disposal Method Suggested: Consult with environmental regulatory agencies for guidance on acceptable disposal practices. Generators of waste containing this contaminant (≥100 kg/mo) must conform with EPA regulations governing storage, transportation, treatment, and waste disposal. Acrylamide residue and sorbent material may be packaged in epoxy-lined drums and taken to an EPA-approved disposal site. Incineration with provisions for scrubbing of nitrogen oxides from flue gases. Deep well injection.

References

National Institute for Occupational Safety and Health, Criteria for a Recommended Standard: Occupational Exposure to Acrylamide, NIOSH Doc. No. 77–112, Washington, DC (1977)

U.S. Environmental Protection Agency, Assessment of Testing Needs: Acrylamide, Report No. EPA-560/11-80-016, Washington, DC, Office of Toxic Substances (July 1980)

Sax, N. I., Ed., Dangerous Properties of Industrial Materials Report, 2, No. 4, 24–27 (1982)

U.S. Environmental Protection Agency, "Chemical Profile: Acrylamide," Washington, DC, Chemical Emergency Preparedness Program (Nov. 30, 1987)

New Jersey Department of Health and Senior Services, "Hazardous Substance Fact Sheet: Acrylamide," Trenton, NJ (February 1989)

New York State Department of Health, "Chemical Fact Sheet: Acrylamide," Albany, NY, Bureau of Toxic Substance Assessment (May 1986)

Acrylic Acid

Molecular Formula: $C_3H_4O_2$

Common Formula: $CH_2CHCOOH$

Synonyms: Acido acrilico (Spanish); Acroleic acid; Acrylic acid, glacial; Acrylic acid, inhibited; Aqueous acrylic acid (technical grade is 94%); Ethylenecarboxylic acid; Glacial acrylic acid; Kyselina akrylova; Propene acid; 2-Propenoic acid; Propenoic acid; Vinylformic acid

CAS Registry Number: 79-10-7

RTECS Number: AS4375000

DOT ID: UN 2218 (inhibited)

EEC Number: 607-061-00-8

EINECS Number: 201-177-9

Regulatory Authority

- Air Pollutant Standard Set (ACGIH)[1] (former USSR)[43] (DFG)[3] (HSE)[33] (OSHA)[58] (Several States)[60] (Australia) (Israel) (Canada)
- Water Pollution Standard Set (former USSR)[43]
- CLEAN AIR ACT: Hazardous Air Pollutants (Title I, Part A, Section 112)
- EPA HAZARDOUS WASTE NUMBER (RCRA No.): U008, D002
- RCRA, 40CFR261, Appendix 8 Hazardous Constituents.
- SUPERFUND/EPCRA 40CFR302.4 Reportable Quantity (RQ): CERCLA, 5,000 lb (2,270 kg).
- EPCRA Section 313 Form R *de minimis* concentration reporting level: 1.0%.
- Canada, WHMIS, Ingredients Disclosure List; National Pollutant Release Inventory (NPRI).

Cited in U.S. State Regulations: Alaska (G), Connecticut (A), Florida (G), Illinois (G), Kansas (G), Louisiana (G), Maine (G), Maryland (G), Massachusetts (G), Nevada (A), New Hampshire (G), New Jersey (G), New York (A), North Dakota (A), Pennsylvania (G), Rhode Island (G), Vermont (G), Virginia (G, A), Washington (G), West Virginia (G), Wisconsin (G).

Description: Acrylic acid, $CH_2CHCOOH$, is a colorless, flammable and corrosive liquid or solid (below 55 °F) with an irritating, rancid, odor. Sinks and mixes with water; irritating vapor is produced. Freezing/Melting point = 12° to 14°C. Boiling point = 141°C. Flash Point = 50°C (o.c.) glacial; 54°C(o.c.). Autoignition temperature = 820°F (438°C). Hazard Identification (based on NFPA-704 M Rating System): Health 3, Flammability 2, Reactivity 2. Odor threshold = 0.092 ppm. Soluble in water.

Potential Exposure: Acrylic acid is chiefly used as a monomer in the manufacture of acrylic resins and plastic products, leather treatment, and paper coatings. Also, it is used as a tackifier and flocculant.

Incompatibilities: Forms explosive mixture with air. Light, heat, and peroxides can cause polymerization. Violent reaction with strong oxidizers. Incompatible with sulfuric acid, caustics, ammonia, amines, isocyanates, alkylene oxides, epichlorohydrin, toluene diamine, oleum, pyridine, methyl pyridine, n-methyl pyrrolidone, 2-methyl-6-ethyl aniline, aniline, ethylene diamine, ethyleneimine, and 2-aminoethanol. Severely corrodes carbon steel and iron; attacks other metals. May accumulate static electrical charges and may cause ignition of its vapors.

Permissible Exposure Limits in Air: There is no OSHA PEL. NIOSH, ACGIH, Australia, Israel, level is 2 ppm (5.9 mg/m^3) TWA and Israel Action Level is 1 ppm (2.95 mg/m^3). HSE (U.K.) TWA is 10 ppm (30 mg/m^3) and STEL of 20 ppm (60 mg/m^3). Acrylic acid can be absorbed through the skin, thereby increasing exposure. Canadian Provincial values are: Alberta 10 ppm (30 mg/m^3) TWA and STEL (15 min.) Of 20 ppm (60 mg/m^3); Ontario and Quebec TWAEV 10 ppm (29 mg/m^3). The former USSR UNEP/IRPTC project has set a limit in workplace air of 5 mg/m^3.[43] In addition, several states have set guidelines or standards for acrylic acid in ambient air:[60] 0.003 mg/m^3 (North Dakota) to 0.1 mg/m^3 (New York) to 0.45 mg/m^3 (Virginia) to 0.6 mg/m^3 (Connecticut) to 0.714 mg/m^3 (Nevada).

Determination in Air: XAD®[2]; Methanol/Water; High-pressure liquid chromatography/Ultraviolet detection; OSHA Method #28.

Permissible Concentration in Water: No U.S. criteria set, but former USSR has set 0.5 mg/l as a limit in drinking water.[43]

Routes of Entry: Inhalation, skin and eye contact, ingestion.

Harmful Effects and Symptoms

Short Term Exposure: Skin and eye contact can cause burns and permanent damage. Inhaling this chemical can cause respiratory tract irritation.

Long Term Exposure: Acrylic acid may cause skin allergy and lung and kidney damage.

Points of Attack: Skin, eyes, respiratory system.

Medical Surveillance: Consider Lung function tests in preplacement and regular physical examinations. Kidney function tests if symptoms develop or overexposure is suspected. Skin testing with dilute acrylic acid may be used by a qualified allergist to diagnose allergy.

First Aid: If this chemical gets into the eyes, remove any contact lenses at once and irrigate immediately for at least 15 minutes, occasionally lifting upper and lower lids. Seek medical attention immediately. If this chemical contacts the skin, remove contaminated clothing and wash immediately with soap and water. Seek medical attention immediately. If this chemical has been inhaled, remove from exposure, begin rescue breathing (using universal precautions) if breathing has stopped and CPR if heart action has stopped. Transfer promptly to a medical facility. When this chemical has been swallowed, get medical attention. If victim is conscious, administer water or milk. Do not induce vomiting.

Personal Protective Methods: Wear protective gloves and clothing to prevent any reasonable probability of skin contact. Safety equipment suppliers/manufacturers recommend Teflon® or butyl rubber as a protective material. All protective clothing (suits, gloves, footwear, headgear) should be clean, available each day, and put on before work. Contact lenses should not be worn when working with this chemical. Wear gas-proof chemical goggles and face shield unless full facepiece respiratory protection is worn. Employees should wash immediately with soap when skin is wet or contaminated. Remove nonimpervious clothing immediately if wet or contaminated. Provide emergency showers and eyewash.

Respirator Selection: *At any concentrations above the NIOSH REL, or where there is no REL, at any detectable concentration:* SCBAF:PD,PP (any MSHA/NIOSH approved self-contained breathing apparatus that has a full facepiece and is operated in a pressure-demand or other positive-pressure mode); or SAF:PD,PP:ASCBA (any supplied-air respirator that has a full facepiece and is operated in a pressure-demand or other positive-pressure mode in combination with an auxiliary self-contained breathing apparatus operated in a pressure-demand or other positive-pressure mode). *Escape:* GMFOV [any air-purifying, full-facepiece respirator (gas mask) with a chin-style, front-or back-mounted organic vapor canister] or SCBAE (any appropriate escape-type, self-contained breathing apparatus).

Storage: Prior to working with this chemical you should be trained on its proper handling and storage. Do not allow to solidify. Can be stored only in glass, stainless steel, aluminum or polyethylene-lined containers. Before entering confined space where acrylic acid may be present, check to make sure that an explosive concentration does not exist. Acrylic acid is a dangerous explosion hazard unless it is stored with an inhibitor. Store in tightly closed containers in a cool, well ventilated area away from heat and sunlight. Do not freeze or refrigerate acrylic acid. Use only non-sparking tools and equipment, especially when opening and closing containers of this chemical. Sources of ignition such as smoking and open flames, are prohibited where this chemical is used, handled, or stored in a manner that could create a potential fire or explosion hazard.

Shipping: Shipped with an inhibitor (e.g., hydroquinone) since it readily polymerizes. Do not ship with human food or animal feedstuffs. Label required is "Corrosive." Falls in DOT/UN Hazard Class 8 and Packing Group II.

Spill Handling: Evacuate and restrict persons not wearing protective equipment from area of spill or leak until cleanup is complete. Remove all ignition sources. Cover spill with soda ash or sodium bicarbonate. Mix and add water. Neutralize and flush into sewer. Collect powdered material in the most convenient and safe manner and deposit in sealed containers. Ventilate area of spill or leak after clean-up is complete. Keep acrylic acid out of a confined space, such as a sewer, because of the potential for an explosion, unless the sewer is designed to prevent the build-up of explosive concentrations. It may be necessary to contain and dispose of this chemical as a hazardous waste. If material or contaminated runoff enters waterways, notify downstream users of potentially contaminated waters. Contact your Department of Environmental Protection or your regional office of the federal EPA for specific recommendations. If employees are required to clean-up spills, they must be properly trained and equipped. OSHA 1910.120(q) may be applicable.

Fire Extinguishing: This chemical is a combustible liquid. Poisonous gases including nitrogen oxides are produced in fire. Use dry chemical, carbon dioxide, or alcohol foam extinguishers. Vapors are heavier than air and will collect in low areas. Vapors may travel long distances to ignition sources and flashback. Vapors in confined areas may explode when exposed to fire. Storage containers and parts of containers may rocket great distances, in many directions. In advanced or massive fires, fire fighting should be done from a safe distance or from a protected location. If a leak or spill has not ignited, use water spray to disperse the vapors. Water spray may be used to flush spills away from exposures and to dilute spills to nonflammable mixtures. If material or contaminated runoff enters waterways, notify downstream users of potentially contaminated waters. Notify local health and fire officials and pollution control agencies. From a secure, explosion-proof location, use water spray to cool exposed containers. If cooling streams are ineffective (venting sound increases in volume and pitch, tank discolors, or shows any signs of deforming), withdraw immediately to a secure position. If employees are expected to fight fires, they must be trained and equipped in OSHA 1910.156.

Disposal Method Suggested: Consult with environmental regulatory agencies for guidance on acceptable disposal practices. Generators of waste containing this contaminant (≥100 kg/mo) must conform with EPA regulations governing storage, transportation, treatment, and waste disposal. Incineration. 100 – 500 ppm potassium permanganate will degrade acrylic acid to a hydroxy acid which can be disposed of at a sewage treatment.

References

Sax, N. I., Ed., Dangerous Properties of Industrial Materials Report, 1, No. 7, 26–28 (1981)

New Jersey Department of Health and Senior Services, "Hazardous Substance Fact Sheet: Acrylic Acid," Trenton, NJ (June, 1998)

Acrylonitrile

Molecular Formula: $C_3H_3N_4$

Common Formula: CH_2CHCN_4

Synonyms: Acrilonitrilo (Spanish); Acrinet®; Acrylnitril (Dutch); Acrylnitril (German); Acrylon®; Acrylonitrile monomer; Akrylonitryl (Polish); AN; Carbacryl; Cianuro di vinile (Italian); Cyanoethylene; Cyanure de vinyle (French); ENT 54; Fumigrain; Miller's fumigrain; Nitrile acrilico (Italian); Nitrile acrylique (French); 2-Propenenitrile; Propenenitrile; TL 314; VCN; Ventox; Vinyl cyanide; Vinyl cyanide, propenenitrile

CAS Registry Number: 107-13-1

RTECS Number: AT5250000

DOT ID: UN 1093 (inhibited)

EEC Number: 608-003-00-4

EINECS Number: 203-466-5

Regulatory Authority

- Carcinogen (IARC),[12] (NTP),[9] (DFG)[3]
- OSHA, 29CFR1910 Specifically Regulated Chemicals (See CFR1910.1045)
- Banned or Severely Restricted (Germany) (U.N.)[13]
- Toxic Substance (World Bank)[15]
- Air Pollutant Standard Set (ACGIH)[1] (HSE)[33] (UNEP)[43] (Several States and Canadian Provinces)[60] (Mexico) (Israel) (Australia)
- Clean Air Act, 42USC7412; Title I, Part A, §112 hazardous pollutants; Section 112[r], Accidental Release Prevention/Flammable Substances (Section 68.130), TQ = 20,000 lb (9,150 kg)
- Clean Water Act, 40CFR116.4 Hazardous Substances; RQ 40CFR117.3, (same as CERCLA)
- EPA HAZARDOUS WASTE NUMBER (RCRA No.): U009
- RCRA 40CFR261, Appendix 8; 40CFR261.11 Hazardous Constituents
- RCRA Land Ban Waste Restrictions
- RCRA 40CFR268.48; 61FR15654, Universal Treatment Standards: Wastewater (mg/l), 0.24; Nonwastewater, 84
- RCRA 40CFR264, Appendix 9; Ground Water Monitoring List Suggested methods (PQL μg/l): 8030 (5); 8240 (5).
- Safe Drinking Water Act, 55FR1470 Priority List
- CERCLA/SARA 40CFR302, Extremely Hazardous Substances: TPQ = 10,000 lb (4,540 kg).
- SUPERFUND/EPCRA 40CFR302.4, Appendix A, Reportable Quantity (RQ): 100 lb (45.5 kg), SARA 313: Form R *de minimis* Concentration Reporting Level: 0.1%.
- Canada, WHMIS, Ingredients Disclosure List
- Mexico, Wastewater, organic pollutants

Cited in U.S. State Regulations: Alaska (G), Arizona (W), California (G, A, W), Connecticut (A, W), Florida (G, A), Illinois (G), Indiana (A), Kansas (G, W), Louisiana (G), Maine (G), Maryland (G), Massachusetts (G, A), Michigan (G), Minnesota (G,W), Nevada (A), New Hampshire (G), New Jersey (G), New York (G, A, W), North Carolina (A), Oklahoma (G), Pennsylvania (A, G), Rhode Island (G, A), South Carolina (A), Vermont (G), Virginia (G, A), Washington (G), West Virginia (G), Wisconsin (G).

Description: Acrylonitrile, CH_2CHCN_4, is a highly flammable, clear, colorless or light yellowish liquid with an irritating, faint garlic- or onion-like odor. Its odor threshold is 17 ppm. Odor can only be detected above the PEL. Boiling point = 77°C. Flash point = 0°C. Hazard Identification (based on NFPA-704 M Rating System): Health 4, Flammability 3, Reactivity 2. Explosive limits: LEL= 3%, UEL = 17%. Floats on water and is moderately soluble.

Potential Exposure: Acrylonitrile is used in the manufacture of synthetic fibers, polymers, acrylostyrene plastics, acrylonitrile-butadiene-styrene plastics, nitrile rubbers, chemicals, and adhesives. It is also used as a pesticide. In the past, this chemical was used as a room fumigant and pediculicide (an agent used to destroy lice).

Incompatibilities: Forms explosive mixture with air. Reacts violently with strong acids, strong alkalis, bromine, and tetrahydrocarbazole. Copper, copper alloys, ammonia and amines may cause breakdown to poisonous products. Unless inhibited (usually with methylhydroquinone) acrylonitrile may polymerize spontaneously. It may also polymerize on contact with oxygen, heat, strong light, peroxides, and concentrated or heated alkalis. Reacts with oxidizers, acids, bromine, amines. Attacks copper and copper alloys. Attacks aluminum in high concentrations. Heat and flame may cause release of poisonous cyanide gas and nitrogen oxides.

Permissible Exposure Limits in Air: The Federal OSHA standard is 2 ppm TWA over am 8-hour workshift *and* 10 ppm not to be exceeded during any 15 minute work period. This chemical can be absorbed through the skin, thereby increasing exposure. NIOSH[2] has a recommended airborne exposure limit of 1 ppm TWA on an 10-hour workshift *and* a 10 ppm ceiling not to be exceeded during any 15 minute work period. ACGIH's recommendation is a TLV of 2 ppm averaged over an 8-hour workshift, with the notation that acrylonitrile is a human carcinogen. The NIOSH IDLH level = 85 ppm. This chemical is a probable carcinogen in humans; there may be no safe level and all contact should be reduced to lowest possible level. The DFG TRK is 3 ppm (7 mg/m^3), Animal Carcinogen, Suspected Human Carcinogen Australia's and Israel's limit is 2 ppm (4.3 mg/m^3) TWA. The HSE (U.K) limit is 2 ppm (4 mg/m^3) TWA.[33] Mexico's limit is 2 ppm (5.4 mg/m^3) TWA. The former USSR-UNEP/IRPTC project[43] has set a MAC of 0.5 mg/m^3 in workplace air and a limit of 0.03 mg/m^3 in ambient air in residential areas on a daily, average basis. In addition, several states have set guidelines or standards for acrylonitrile in ambient air:[60] 0.0147 μg/m^3 (Indiana) to 0.145 μg/m^3 (North Carolina) to 0.15 μg/m^3 (Massachusetts) to 11.3 μg/m^3 (Pennsylvania) to 15.0 μg/m^3 (New York) to 22.0 μg/m^3 (Connecticut and South Dakota) to 22.5 μg/m^3 (South Carolina) to 45.0 μg/m^3 (Florida and Virginia).

Determination in Air: Charcoal adsorption followed by acetone extraction and gas chromatographic analysis. See NIOSH Method 1604,[18] or OSHA Method 37.

Permissible Concentration in Water: The substance is toxic to aquatic organisms. Acrylonitrile usually breaks down in about 1 or 2 weeks, but this can vary depending on conditions. For example, high concentrations (such as might occur following a spill) tend to be broken down more slowly. In one case, measurable amounts of acrylonitrile were found in nearby wells 1 year after a spill (ATSDR public Health Statement, December 19990). See RCRA and Clean Water Act under Regulatory Authority. To protect freshwater aquatic life – on an acute basis, 7,550 μg/l and on a chronic basis, 2,600 μg/l over 30 days. To protect saltwater aquatic life-insufficient data to yield a value. To protect human health-preferably zero. Water concentration should be below 0.58 μg/l to keep lifetime cancer risk below 10^{-5}. The former USSR-UNEP/IRPTC project[43] has set a MAC of 2.0 mg/l for water bodies used for domestic purposes. The Mexico drinking water ecological criteria is 0.0006 mg/l, reduce human exposure to a minimum. In addition, several states have set guidelines for acrylonitrile in drinking water[61] ranging from 0.67 μg/l (Minnesota) to 3.8 μg/l (Kansas) to 10 μg/l (Arizona) to 35 μg/l (Connecticut).

Determination in Water: Charcoal tube; Acetone/CS2; Gas chromatography/Flame ionization detection; NIOSH (IV) Method #1604. Also, by gas chromatography (EPA Method #603) or gas chromatography plus mass spectrometry (EPA Method #624).

Routes of Entry: Inhalation and percutaneous absorption. It may be absorbed from contaminated rubber or leather. Routes include ingestion and eye and skin contact. Acrylonitrile vapor is absorbed readily from the lungs, and inhalation is an important route of exposure. This chemical's odor generally provides inadequate warning of hazardous concentrations and olfactory fatigue develops rapidly.

Harmful Effects and Symptoms

Short Term Exposure: Irritates eyes, skin and respiratory tract. Splashes in the eye may result in corneal damage. Skin contact can cause severe irritation and blistering. Breathing acrylonitrile can irritate the lungs causing coughing and shortness of breath. Higher exposures can cause pulmonary edema, a medical emergency that can result in death. Skin contact contributes significantly in overall exposure and can lead to systemic toxicity. Acrylonitrile reaction causes redness, blisters and some systemic signs. Symptoms derive from tissue anoxia in order of onset: limb weakness, dyspnea (difficult breathing), burning sensation in throat, dizziness, impaired judgment, cyanosis (turning blue), nausea, collapse, irregular breathing, convulsions and death. In later stages collapse, irregular breathing or convulsions and cardiac arrest may occur without warning. Some patients appear hysterical or may even be violent. Acrylonitrile is classified as very toxic. Probable oral lethal dose for human is 50 – 500 mg/kg (between 1 teaspoon and 1 oz.) for a 70 kg (150 lb) person. Toxic concentrations have been reported at 16 ppm/20 min. Acute toxicity is similar to that due to cyanide poisoning and the level of cyanide ion in blood is related to the level of poisoning. Inhalation or ingestion can results in fatal systemic poisoning, collapse and death due to tissue anoxia (lack of oxygen) and cardiac arrest (heart failure). At higher concentrations there may be damage to red blood cells and the liver. Jaundice may develop 24 hours following exposure and persist for several days. Because of continued metabolic release of cyanide, symptoms of severe poisoning may recur and the patient may relapse.

Long Term Exposure: This chemical is a probable carcinogen in humans. There is some evidence that it causes lung and large intestine cancer in humans and has been shown to cause brain and stomach cancer in animals. Exposure may cause the thyroid gland to enlarge and interfere with normal thyroid function. There is limited evidence that acrylonitrile may damage the developing fetus and the male reproductive system. Repeated exposure can irritate the nose, causing discharge, nose bleeds, and sores inside the nose. Acrylonitrile may affect the liver function.

Points of Attack: Eyes, skin, cardiovascular system, liver, kidneys, central nervous system. Cancer Site: brain tumors, lung and bowel cancer.

Medical Surveillance: For those with frequent or high exposure, consider urine thiocyanate levels, blood cyanide levels, liver function tests, fecal occult blood screening, pulmonary function tests. Consider chest x-ray following acute exposure. Consider the skin, respiratory tract, heart, central nervous system, renal and liver function in placement and periodic examinations. A history of fainting spells or convulsive disorders might present and added risk to persons working with toxic nitriles.

First Aid: If this chemical gets into the eyes, remove any contact lenses at once and irrigate immediately for at least 15 minutes, occasionally lifting upper and lower lids. Seek medical attention immediately. If this chemical contacts the skin, remove contaminated clothing and wash immediately with soap and water. Seek medical attention immediately. If this chemical has been inhaled, remove from exposure, begin rescue breathing (using universal precautions) if breathing has stopped and CPR if heart action has stopped. Transfer promptly to a medical facility. When this chemical has been swallowed, get medical attention. Give large quantities of water and induce vomiting. Do not make an unconscious person vomit. Medical observation is recommended for 24 – 48 hours after breathing overexposure, as pulmonary edema may be delayed. Use amyl nitrate capsules if symptoms develop. All area employees should be trained regularly in emergency measures for cyanide poisoning and in CPR. A cyanide antidote kit should be kept in the immediate work area and must be rapidly available. Kit ingredients should be replaced every 1 – 2 years to ensure freshness. Persons trained in the use of this kit, oxygen use, and CPR must be quickly available.

Personal Protective Methods: Wear protective gloves and clothing to prevent repeated or prolonged skin contact. Safety equipment suppliers and manufacturers recommend butyl rubber as a protective material. Leather should not be used in protective clothing since it is readily penetrated by acrylonitrile; contaminated leather shoes and gloves should be destroyed. Rubber clothing should be frequently washed and inspected because it will soften and swell. Wear splash-proof chemical goggles and face shield unless full facepiece respiratory protection is worn. Contact lenses should not be worn when working with this substance. Employees should wash immediately when skin is wet or contaminated. Remove clothing immediately if wet or contaminated to avoid flammability hazard. Provide emergency showers. See also NIOSH Criteria Document 78–212 NITRILES.

Respirator Selection: NIOSH: *At any concentrations above the NIOSH REL, or where there is no REL, at any detectable concentration:* SCBAF:PD,PP (any self-contained breathing apparatus that has a full facepiece and is operated in a pressure-demand or other positive-pressure mode); or SAF:PD, PP:ASCBA (any supplied-air respirator that has a full facepiece and is operated in a pressure-demand or other positive-pressure mode in combination with an auxiliary, self-contained breathing apparatus operated in a pressure-demand or other positive-pressure mode). *Escape:* GMFOV [any air-purifying, full-facepiece respirator (gas mask) with a chin-style, front-or back-mounted organic vapor canister]; or SCBAE (any appropriate escape-type, self-contained breathing apparatus).

Storage: Prior to working with this chemical you should be trained on its proper handling and storage. Before entering confined space where this chemical may be present, check to make sure that an explosive concentration does not exist. Protect against physical damage. See Incompatibilities for materials and physical conditions not permitted in storage room or cabinet. Sources of ignition such as smoking and open flames, are prohibited where this chemical is used, handled, or stored in a manner that could create a potential fire or explosion hazard. Store in tightly closed containers in a cool, well ventilated area. Do not store uninhibited acrylonitrile under any conditions. Store drums on end with bungs up, no more than two high. Outside tanks should be above ground and surrounded with dikes of sufficient capacity to hold entire tank contents. Metal containers involving the transfer of this chemical should be grounded and bonded. Drums must be equipped with self-closing valves, pressure vacuum bungs, and flame arresters. Use only non-sparking tools and equipment, especially when opening and closing containers of this chemical. A regulated, marked area should be established where this chemical is handled, used, or stored in compliance with OSHA standard 1910.1045.

Shipping: Ship in airtight, unbreakable packaging. Do not transport with human food and animal feedstuffs. UN/DOT label required is "Flammable Liquid, Poison." Shipment by air is forbidden (DOT). Shipped in lined pails, drums, trucks and tank cars. Shipment in passenger vessels forbidden. Falls in DOT/UN Hazard Class 3 and Packing Group I.[19][20]

Spill Handling: Evacuate and restrict persons not wearing protective equipment from area of spill or leak until cleanup is complete. Remove all ignition sources. Establish forced ventilation to keep levels below explosive limit. Absorb liquids in vermiculite, dry sand, earth, or a similar material and deposit in sealed containers. Collect powdered material in the most convenient and safe manner and deposit in sealed containers. Keep acrylonitrile out of a confined space, such as a sewer, because of the potential for an explosion, unless the sewer is designed to prevent the build-up of explosive concentrations. It may be necessary to contain and dispose of this chemical as a hazardous waste. If material or contaminated runoff enters waterways, notify downstream users of potentially contaminated waters. Contact your Department of Environmental Protection or your regional office of the federal EPA for specific recommendations. If employees are required to clean-up spills, they must be properly trained and equipped. OSHA 1910.120(q) may be applicable.

Initial isolation and protective action distances (as hydrogen cyanide)

Small spills (From a small package or a small leak from a large package)

First: Isolate in all directions (feet)..... 200

Then: Protect persons downwind (miles)

Day 0.1

Night 0.5

Large spills (From a large package or from many small packages)

First: Isolate in all directions (feet)..... 600

Then: Protect persons downwind (miles)

Day 0.4

Night 1.7

Fire Extinguishing: Acrylonitrile is a dangerously reactive and flammable liquid. Poisonous gases are produced in fire including hydrogen cyanide. A few "whiffs" of vapor could cause death. Vapor or liquid could be fatal on penetrating the fire fighter's normal full protective clothing. Firefighting gear (including SCBA) does not provide adequate protection. If exposure occurs, remove and isolate gear immediately and thoroughly decontaminate personnel. May react with itself without warning, blocking relief valves, and leading to container explosion. Vapors are heavier than air and will collect in low areas. Vapors in confined areas may explode when exposed to fire. Vapors may travel long distances to ignition sources and flash back. Storage containers and parts of containers may rocket great distances, in many directions. Small fires: dry chemical, carbon dioxide, water spray or foam. Large fires: water spray, fog or foam. Stay away from ends of tanks. Do not get water inside container. If material or contaminated runoff enters waterways, notify downstream users of potentially contaminated waters. Notify local health and fire officials and pollution control agencies. For massive fire in cargo area, use unmanned hose holder or monitor nozzles; if this is impossible, withdraw from area and let fire burn. From a secure explosion-proof location, use water spray to cool exposed containers. If cooling streams are ineffective (venting sound increases in volume and pitch, tank discolors or shows any signs of deforming), withdraw immediately to a secure position. If employees are expected to fight fires, they must be trained and equipped in OSHA 1910.156.

Disposal Method Suggested: Consult with environmental regulatory agencies for guidance on acceptable disposal practices. Generators of waste containing this contaminant (≥100 kg/mo) must conform with EPA regulations governing storage, transportation, treatment, and waste disposal. Incineration with provision for No_x removal from effluent gases by scrubbers or afterburners. A chemical disposal method has also been suggested involving treatment with alcoholic NaOH; the alcohol is evaporated and calcium hypochlorite added; after 24 hours the product is flushed to the sewer with large volumes of water. Recovery of acrylonitrile from acrylonitrile process effluents is an alternative to disposal.

References

U.S. Department of Health and Human Services, Public Health Service, "Managing Hazardous Materials Incidents, Volume III," Atlanta, GA (1990)

U.S. Department of Health and Human Services, Public Health Service, "Public Health Statements: Acrylonitrile," Atlanta, GA (December, 1990)

National Institute for Occupational Safety and Health, Criteria for a Recommended Standard: Occupational Exposure to Acrylonitrile, NIOSH Doc, No. 78–116, Washington DC (1978)

Department of Labor, Economic Impact Assessment for Acrylonitrile, Washington, DC, Occupational Safety and Health Administration (February 21, 1978)

U.S. Environmental Protection Agency, Status Assessment of Toxic Chemicals: Acrylonitrile, Report EPA-600/2-79-210A, Washington, DC (December 1979)

U.S. Environmental Protection Agency, Acrylonitrile: Ambient Water Quality Criteria, Washington, DC (1980)

U.S. Environmental Protection Agency, Investigation of Selected Potential Environmental Contaminants: Acrylonitrile, Report EPA-560/2-78-003, Washington, DC (May 1978)

U.S. Environmental Protection Agency, Acrylonitrile, Health and Environmental Effects Profile No. 7, Washington, DC, Office of Solid Waste (April 30, 1980)

Sax, N. I., Ed., Dangerous Properties of Industrial Materials Report, 1, No. 2, 25–27 (1980) and 3, No. 3, 41–46 (1988) and 5, No. 4, 31–33 (1985)

U.S. Environmental Protection Agency, "Chemical Profile: Acrylonitrile," Washington, DC, Chemical Emergency Preparedness Program (November 30, 1987)

New Jersey Department of Health and Senior Services, Hazardous Substance Fact Sheet: Acrylonitrile, Trenton, NJ (May, 1998)

New York State Department of Health, "Chemical Fact Sheet: Acrylonitrile," Albany, NY, Bureau of Toxic Substance Assessment (March 1986)

Acrylyl Chloride

Molecular Formula: C_3H_3ClO

Common Formula: CH_2=CHCOCl

Synonyms: Acrylic acid chloride; Acryloyl chloride; Cloruro de acriloilo (Spanish); 2-Propenoyl chloride

CAS Registry Number: 814-68-6

RTECS Number: AI7350000

Regulatory Authority

- Air Pollutant Standard Set (former USSR)[43]
- OSHA 29CFR1910.119, Appendix A. Process Safety List of Highly Hazardous Chemicals, TQ = 250 lb (114 kg)
- Clean Air Act 42USC7412; Title I, Part A, §112(r), Regulated Chemicals for Accidental Release Prevention/Flammable Substances (Section 68.130) TQ = 5,000 lb (2,270 kg)
- CERCLA/SARA 40CFR302, Extremely Hazardous Substances: TPQ = 100 lb (45.4 kg)[7]

Cited in U.S. State Regulations: Florida (G), Massachusetts (G), New Jersey (G), Pennsylvania (G).

Description: Acrylyl Chloride, CH_2=CHCOCl is a liquid. Boiling point = 75°C.

Potential Exposure: May be used as a monomer in preparation of specialty polymers or as a chemical intermediate.

Incompatibilities: Decomposes in water.

Permissible Exposure Limits in Air: Regulated by the Clean Air Act, OSHA, and CERCLA. The former USSR-UNEP/IRPTC project[43] has set a MAC of 0.3 mg/m^3 in workplace air.

Permissible Concentration in Water: No criteria set.

Routes of Entry: Inhalation.

Harmful Effects and Symptoms

The LC (low) for inhalation by rats is 0.093 mg/liter/4 hr; this was apparently the basis for inclusion in the EPA "Extremely Hazardous Substances" list[7] even though little is known of this compound otherwise.

First Aid: If this chemical gets into the eyes, remove any contact lenses at once and irrigate immediately for at least 15 minutes, occasionally lifting upper and lower lids. If this chemical contacts the skin, remove contaminated clothing and wash immediately with soap and water. When this chemical has been swallowed, immediately get medical attention. Give large quantities of water and induce vomiting. Do not make an unconscious person vomit. If this chemical has been inhaled, remove from exposure and transfer promptly to a medical facility.

Personal Protective Methods: Wear protective gloves and clothing to prevent any reasonable probability of skin contact. Safety equipment suppliers/manufacturers can provide recommendations on the most protective glove/clothing material for your operation. All protective clothing (suits, gloves, footwear, headgear) should be clean, available each day, and put on before work. Contact lenses should not be worn when working with this chemical. Wear splash-proof chemical goggles and face shield unless full facepiece respiratory protection is worn. Employees should wash immediately with soap when skin is wet or contaminated. Provide emergency showers and eyewash.

Storage: Prior to working with this chemical you should be trained on its proper handling and storage. Store in tightly closed containers in a cool, well ventilated area. Metal containers involving the transfer of this chemical should be grounded and bonded. Drums must be equipped with self-closing valves, pressure vacuum bungs, and flame arresters. Use only non-sparking tools and equipment, especially when opening and closing containers of this chemical. Sources of ignition such as smoking and open flames, are prohibited where this chemical is used, handled, or stored in a manner that could create a potential fire or explosion hazard.

Spill Handling: Evacuate and restrict persons not wearing protective equipment from area of spill or leak until cleanup is complete. Remove all ignition sources. Absorb liquids in vermiculite, dry sand, earth, or a similar material and deposit in sealed containers. Collect powdered material in the most convenient and safe manner and deposit in sealed containers. Ventilate area of spill or leak after clean-up is complete. It may be necessary to contain and dispose of this chemical as a hazardous waste. If material or contaminated runoff enters waterways, notify downstream users of potentially contaminated waters. Contact your Department of Environmental Protection or your regional office of the federal EPA for specific recommendations. If employees are required to clean-up spills, they must be properly trained and equipped. OSHA 1910.120(q) may be applicable.

Fire Extinguishing: Use dry chemical, carbon dioxide, water spray, or alcohol foam extinguishers. Poisonous gases are produced in fire including carbon monoxide. If employees are expected to fight fires, they must be trained and equipped in OSHA 1910.156.

References

U.S. Environmental Protection Agency, "Chemical Profile: Acrylyl Chloride," Washington, DC, Chemical Emergency Preparedness Program (November 30, 1987)

Actinomycin D

Molecular Formula: $C_{62}H_{86}N_{12}O_{16}$

Synonyms: Actinomycindioic d acid, dilactone; Actinomycin I; AD; Cosmegen; Dactinomycin; Dilactone actinomycindioic d acid; HBF 386; Lyovac cosmegen; Meractinomycin; Oncostatin K

CAS Registry Number: 50-76-0

RTECS Number: AU1575000

DOT ID: UN 2811 (medicines, toxic, solid, n.o.s.)

Regulatory Authority

- Carcinogen [NCI (animal positive), IARC (Group 3, not classifiable), California, Minnesota, New Jersey][9]

Cited in U.S. State Regulations: California (A,G), Florida (G), Illinois (G), Massachusetts (G), Michigan (G), Minnesota (G), New Hampshire (G), New Jersey (G), Pennsylvania (G), Rhode Island (G).

Description: Actinomycin D, $C_{62}H_{86}N_{12}O_{16}$, is a bright red crystalline solid. Freezing/Melting point = 241.5° to 243°C. Soluble in water.

Potential Exposure: An antibiotic product from streptomyces, used as anticancer drug.

Incompatibilities: Strong oxidizers, strong acids, and strong bases.

Permissible Exposure Limits in Air: No standards have been established.

Permissible Concentration in Water: No criteria set.

Routes of Entry: intravenous, skin, eyes.

Harmful Effects and Symptoms

Short Term Exposure: Highly Toxic. Irritates eyes and skin. A poison if ingested.

Long Term Exposure: Has been shown to cause peritoneal and local sarcomas in animals. May damage the developing fetus.

Medical Surveillance: Complete blood count (CBC), EKG

First Aid: If this chemical gets into the eyes, remove any contact lenses at once and irrigate immediately for at least 15 minutes, occasionally lifting upper and lower lids. If this chemical contacts the skin, remove contaminated clothing and wash immediately with soap and water. When this chemical has been swallowed, get medical attention. Give large quantities of water and induce vomiting. Do not make an unconscious person vomit. If this chemical has been inhaled, remove from exposure and transfer promptly to a medical facility.

Personal Protective Methods: Wear protective gloves and clothing to prevent any reasonable probability of skin contact. Safety equipment suppliers/manufacturers may be able to provide recommendations on the most protective glove/clothing material for your operation. All protective clothing (suits, gloves, footwear, headgear) should be clean, available each day, and put on before work. Contact lenses should not be worn when working with this chemical. Wear dust-proof chemical goggles and face shield unless full facepiece respiratory protection is worn. Employees should wash immediately with soap when skin is wet or contaminated. Provide emergency showers and eyewash.

Respirator Selection: *At any concentrations above the NIOSH REL, or where there is no REL, at any detectable concentration:* SCBAF:PD,PP (any MSHA/NIOSH approved self-contained breathing apparatus that has a full facepiece and is operated in a pressure-demand or other positive-pressure mode); or SAF:PD,PP:ASCBA (any supplied-air respirator that has a full facepiece and is operated in a pressure-demand or other positive-pressure mode in combination with an auxiliary, self-contained breathing apparatus operated in a pressure-demand or other positive pressure mode). *Escape:* GMFOVHiE [any air-purifying, full-facepiece respirator (gas mask) with a chin-style, front- or back-mounted organic vapor canister having a high-efficiency particulate filter]; or SCBAE (any appropriate escape-type, self-contained breathing apparatus).

Storage: Prior to working with this chemical you should be trained on its proper handling and storage. Store in tightly closed containers in a refrigerator or other cool, dry place.

Shipping: Classified as medicines, toxic solid, n.o.s. Actinomycin D falls in Shipping Class 6.1 and Packing Group II. The label required is "Poison."

Spill Handling: Evacuate and restrict persons not wearing protective equipment from area of spill or leak until cleanup is complete. Remove all ignition sources. Dampen with 60 – 70% ethanol to avoid airborne dust then transfer to a suitable sealed container in the most convenient and safe manner. Ventilate area of spill or leak after clean-up is complete. It may be necessary to contain and dispose of this chemical as a hazardous waste. If material or contaminated runoff enters waterways, notify downstream users of potentially contaminated waters. Contact your Department of Environmental Protection or your regional office of the federal EPA for specific recommendations. If employees are required to clean-up spills, they must be properly trained and equipped. OSHA 1910.120(q) may be applicable.

Fire Extinguishing: Use dry chemical, carbon dioxide, water spray, or foam extinguishers. Poisonous gases are produced in fire including nitrogen oxides. If employees are expected to fight fires, they must be trained and equipped in OSHA 1910.156.

References

Sax, N. I., Ed., Dangerous Properties of Industrial Materials Report, 1, No. 3, 23–24 (1981)

New Jersey Department of Health and Senior Services, Hazardous Substance Fact Sheet, Acinomycin D, Trenton, NJ (April, 1998)

Adipic Acid

Molecular Formula: $C_6H_{10}O_4$

Common Formula: $HOOC(CH_2)_4COOH$

Synonyms: Acido adipico (Spanish); Acifloctin; Acinetten; Adilac-tetten; Adipinic acid; 1,4-Butanedicarboxylic acid; Dicarboxylic acid C_6; 1,6-Hexanedioic acid; Hexanedioic acid; Kyselina adipova (Czech)

CAS Registry Number: 124-04-9

RTECS Number: AU8400000

DOT ID: NA 9077

EEC Number: 607-144-00-9

EINECS Number: 204-673-3

Regulatory Authority

- Clean Water Act, 40CFR116.4 Hazardous Substances; RQ 40CFR117.3 (same as CERCLA)

- SUPERFUND/EPCRA 40CFR302.4, Appendix A, Reportable Quantity (RQ): 5,000 lb (2,270 kg), SARA 313[4]
- Canada, WHMIS, Ingredients Disclosure List

Cited in U.S. State Regulations: California (G), Massachusetts (G), New Hampshire (G), New Jersey (G), New York (G), Pennsylvania (G), Rhode Island (G).

Description: Adipic Acid, $HOOC(CH_2)_4COOH$, is a combustible, white crystalline solid. Odorless. Freezing/Melting point = 152°C. Flash point = 196°C (cc). Hazard Identification (based on NFPA-704 M Rating System): Health –, Flammability 1, Reactivity 0. Autoignition temperature = 420°C. Slightly in hot water.

Potential Exposure: Workers in manufacture of nylon, plasticizers, urethanes, adhesives, and food additives.

Incompatibilities: Oxidizers, reducing agents, and strong bases. Dust forms an explosive mixture with air. Friction from stirring, pouring, or pneumatic transfer can form electrostatic charge on dry material.

Permissible Exposure Limits in Air: ACGIH recommends and airborne exposure limit of 5 mg/m^3 TWA, averaged over an 8-hour workshift.

Permissible Concentration in Water: Regulated as a hazardous substance by the Clean Water Act.

Routes of Entry: inhalation.

Harmful Effects and Symptoms

Short Term Exposure: Inhalation can cause burns to nose, throat and respiratory tract. Can cause eye irritation and burns.

Long Term Exposure: Due to the availability of insufficient data on long term effects, caution should be exercised.

Medical Surveillance: No special test for this chemical.

First Aid: If this chemical gets into the eyes, remove any contact lenses at once and irrigate immediately for at least 15 minutes, occasionally lifting upper and lower lids. If this chemical contacts the skin, remove contaminated clothing and wash immediately with soap and water. When this chemical has been swallowed, get medical attention. Give large quantities of water and induce vomiting. Do not make an unconscious person vomit. If this chemical has been inhaled, remove from exposure and transfer promptly to a medical facility.

Personal Protective Methods: Wear protective gloves and clothing to prevent any reasonable probability of skin contact. Safety equipment suppliers/manufacturers can provide recommendations on the most protective glove/clothing material for your operation. All protective clothing (suits, gloves, footwear, headgear) should be clean, available each day, and put on before work. Contact lenses should not be worn when working with this chemical. Wear dust-proof chemical goggles and face shield unless full facepiece respiratory protection is worn. Employees should wash immediately with soap when skin is wet or contaminated. Provide emergency showers and eyewash.

Respirator Selection: *At any concentrations above the NIOSH REL, or where there is no REL, at any detectable concentration:* SCBAF:PD,PP (any MSHA/NIOSH approved self-contained breathing apparatus that has a full facepiece and is operated in a pressure-demand or other positive-pressure mode); or SAF:PD,PP:ASCBA (any supplied-air respirator that has a full facepiece and is operated in a pressure-demand or other positive-pressure mode in combination with an auxiliary, self-contained breathing apparatus operated in a pressure-demand or other positive pressure mode). *Escape:* GMFOVHiE [any air-purifying, full-facepiece respirator (gas mask) with a chin-style, front- or back-mounted organic vapor canister having a high-efficiency particulate filter]; or SCBAE (any appropriate escape-type, self-contained breathing apparatus).

Storage: Prior to working with this chemical you should be trained on its proper handling and storage. Store in tightly-closed containers in a cool, well-ventilated area.

Shipping: There are no UN/DOT labeling requirements for adipic acid. Adipic acid falls in DOT Hazard Class ORM-E and UN Hazard Class 9.2 and Packing Group III.

Spill Handling: Evacuate and restrict persons not wearing protective equipment from area of spill or leak until cleanup is complete. Remove all ignition sources. Use lime or sodium bicarbonate to neutralize, and absorb with peat, vermiculite, or carbon. Collect powdered material in the most convenient and safe manner and deposit in sealed containers. Ventilate area of spill or leak after clean-up is complete. It may be necessary to contain and dispose of this chemical as a hazardous waste. If material or contaminated runoff enters waterways, notify downstream users of potentially contaminated waters. Contact your Department of Environmental Protection or your regional office of the federal EPA for specific recommendations. If employees are required to clean-up spills, they must be properly trained and equipped. OSHA 1910.120(q) may be applicable.

Fire Extinguishing: This chemical is a combustible solid. Use dry chemical, carbon dioxide, water spray, or foam extinguishers. Poisonous gases are produced in fire including acidic vapors and valeric acid. If employees are expected to fight fires, they must be trained and equipped in OSHA 1910.156.

Disposal Method Suggested: Incineration.

References

Sax, N. I., Ed., Dangerous Properties of Industrial Materials Report, 1, No. 7, 28–29 (1981) and 3, No. 3, 46–49 (1983)

New Jersey Department of Health and Senior Services, "Hazardous Substance Fact Sheet: Adipic Acid," Trenton, NJ (September 1998)

New York State Department of Health, "Chemical Fact Sheet: Adipic Acid," Albany, NY, Bureau of Toxic Substance Assessment (March 1986)

Adiponitrile

Molecular Formula: $C_6H_8N_2$

Common Formula: $NC(CH_2)_4CN$

Synonyms: Adipic acid dinitrile; Adipic acid nitrile; Adipodinitrile; Adiponitrilo (Spanish); 1,4-Dicyanobutane; Hexanedinitrile; Hexanedioic acid, dinitrile; Nitrile adipico (Italian); Tetramethylene cyanide

CAS Registry Number: 111-69-3

RTECS Number: AV2625000

DOT ID: UN 2205

Regulatory Authority

- Air Pollutant Standard Set (former USSR)[43] (Connecticut)[60]
- Water Pollution Standard Set (former USSR)[43]
- CERCLA/SARA 40CFR302, Extremely Hazardous Substances: TPQ = 1,000 lb (454 kg)[7]
- CERCLA/SARA 40CFR304, Appendix A, Reportable Quantity (EHS, RQ): 1 lb (0.454 kg), SARA 313: (as cyanide compounds) Form R *de minimis* Concentration Reporting Level: 1.0%
- U.S. DOT Regulated Marine Pollutant (49CFR172.101, Appendix B) as cyanide mixtures or solutions
- Canada, WHMIS, Ingredients Disclosure List

Cited in U.S. State Regulations: Connecticut (A), Florida (G), Massachusetts (G), New Hampshire (G), New Jersey (G), Oklahoma (G), Pennsylvania (G), Rhode Island (G).

Description: Adiponitrile is a combustible, water-white to yellow, oily liquid; a solid below 1.1°C. Practically odorless. Boiling point = 295°C. Freezing/Melting point = 2.3°C. Flash point = 93°C, technical grade (oc); 163°C (pure). Hazard Identification (based on NFPA-704 M Rating System): Health 2, Flammability 2, Reactivity 1. Explosive limits: LEL = 1%, UEL = undetermined. Autoignition temperature = 550°C Floats on water; slightly soluble.

Potential Exposure: Is used to manufacture corrosion inhibitors, rubber accelerators, and Nylon 66 and in organic synthesis.

Incompatibilities: Forms explosive mixture with air. Violent reaction with oxidizers. Also incompatible with strong acids, strong bases, and reducing agents.

Permissible Exposure Limits in Air: NIOSH recommended airborne exposure limit is 4 ppm (18 mg/m^3). TWA for a 10-hour workshift. ACGIH recommended airborne exposure limit is 2 ppm (8.8 mg/m^3). TWA for an 8-hour workshift. The former USSR-UNEP/IRPTC project[43] has set a MAC of 20 mg/m^3 in workplace air. In addition, a guideline or standard for adiponitrile in ambient air has been set in Connecticut[60] at 360 μg/m^3.

Determination in Air: Charcoal tube; Toluene; Gas chromatography/Flame ionization detection; NIOSH Nitriles Criteria Document.

Permissible Concentration in Water: The former USSR-UNEP/IRPTC project[43] has set a MAC of 0.1 mg/l in water bodies used for domestic purposes.

Routes of Entry: Inhalation, ingestion, skin contact.

Harmful Effects and Symptoms

Short Term Exposure: Irritates eyes, skin and respiratory tract. Skin and eye contact can cause burns. This chemical can be absorbed through the skin, thereby increasing exposure. Inhalation can cause coughing and shortness of breath. Exposure can cause fatal cyanide poisoning with symptoms of flushing of the face, chest tightness, nausea and vomiting, weakness, lightheadedness, confusion, headache, trouble breathing, and convulsions. Higher exposure can lead to convulsions, irregular heartbeat, coma, and death. Ingestion of a few milliliters may cause weakness, mental confusion, vomiting, rapid respiration, fast heartbeat, and convulsions.

Long Term Exposure: Adiponitrile produces disturbances of the respiration and circulation, irritation of the stomach and intestine, and loss of weight. Repeated exposure may cause personality changes, depression, anxiety, and irritability, thyroid gland damage and enlargement, and nervous system damage. Adiponitrile may damage the developing fetus. Adiponitrile may have effects on the blood and adrenal gland, resulting in anemia and tissue lesions.

Points of Attack: Eyes, skin, respiratory system, central nervous system, cardiovascular system.

Medical Surveillance: Consider urine thiocyanate test for preemployment and regular medical testing. If symptoms develop or overexposure is suspected, have blood cyanide levels, nervous system, and thyroid function tested.

First Aid: If this chemical gets into the eyes, remove any contact lenses at once and irrigate immediately for at least 15 minutes, occasionally lifting upper and lower lids. Seek medical attention immediately. If this chemical contacts the skin, remove contaminated clothing and wash immediately with soap and water. Seek medical attention immediately. If this chemical has been inhaled, remove from exposure, begin rescue breathing (using universal precautions) if breathing has stopped and CPR if heart action has stopped. Transfer promptly to a medical facility. When this chemical has been swallowed, get medical attention. Give large quantities of water and induce vomiting. Do not make an unconscious person vomit.

Use amyl nitrate capsules if symptoms develop. All area employees should be trained regularly in emergency measures for cyanide poisoning and in CPR. A cyanide antidote kit should be kept in the immediate work area and must be rapidly available. Kit ingredients should be replaced every 1 – 2 years to ensure freshness. Persons trained in the use of this kit, oxygen use, and CPR must be quickly available. Effects may be delayed; keep victim under observation.

Personal Protective Methods: Wear protective long rubber gloves and protective clothing to prevent any reasonable probability of skin contact. Safety equipment suppliers/manufacturers can provide recommendations on the most protective glove/clothing material for your operation. All protective clothing (suits, gloves, footwear, headgear) should be clean, available each day, and put on before work. Contact lenses should not be worn when working with this chemical. Wear splash-proof chemical goggles and face shield unless full facepiece respiratory protection is worn. Employees should wash immediately with soap when skin is wet or contaminated. Remove nonimpervious clothing immediately if wet or contaminated. Provide emergency showers and eyewash. See also NIOSH Criteria Document 78–212 NITRILES.

Respirator Selection: Engineering controls should be used wherever feasible to maintain airborne concentrations of this chemical below the prescribed exposure limit. Respirators and protective equipment less effective than engineering controls, and should be used only in non-routine or emergency situations which may result in exposure concentrations in excess of the TWA environmental limit.

Emergency or planned entry into unknown concentrations or IDLH conditions: SCBAF:PD,PP (any self-contained breathing apparatus that has a full facepiece and is operated in a pressure-demand or other positive-pressure mode); or SAF:PD,PP:ASCBA (Any supplied-air respirator that has a full facepiece and is operated in a pressure-demand or other positive-pressure mode in combination with an auxiliary self-contained breathing apparatus operated in a pressure-demand or other positive pressure mode. *Escape:* GMFOV [Any air-purifying, full-facepiece respirator (gas mask) with a chin-style, front-or back-mounted organic vapor canister]; or SCBAE (Any appropriate escape-type, self-contained breathing apparatus).

Storage: Prior to working with this chemical you should be trained on its proper handling and storage. Outside or detached storage is preferred. Before entering confined space where adiponitrile may be present, check to make sure that an explosive concentration does not exist. Store in tightly closed containers in a cool, well ventilated area. Metal containers involving the transfer of this chemical should be grounded and bonded. Where possible, automatically pump liquid from drums or other storage containers to process containers. Drums must be equipped with self-closing valves, pressure vacuum bungs, and flame arresters. Use only non-sparking tools and equipment, especially when opening and closing containers of this chemical. Sources of ignition such as smoking and open flames, are prohibited where this chemical is used, handled, or stored in a manner that could create a potential fire or explosion hazard.

Shipping: Label required is "Keep Away from Food." Falls in DOT/UN Hazard Class 6.1 and Packing Group III.[19][20]

Spill Handling: Evacuate and restrict persons not wearing protective equipment from area of spill or leak until cleanup is complete. Remove all ignition sources. Establish forced ventilation to keep levels below explosive limit. *Small spills:* absorb liquids in vermiculite, dry sand, earth, peat, activated carbon, or other non-combustible absorbent material and place into containers for later disposal. *Large spills:* dike far ahead of spill for later disposal. Seek professional environmental engineering assistance from the U.S. EPA Environmental Response Team at (908) 548-8730 (24-hour response line). Collect powdered material in the most convenient and safe manner and deposit in sealed containers. Keep adiponitrile out of a confined space, such as a sewer, because of the potential for an explosion It may be necessary to contain and dispose of this chemical as a hazardous waste. If material or contaminated runoff enters waterways, notify downstream users of potentially contaminated waters. Contact your Department of Environmental Protection or your regional office of the federal EPA for specific recommendations. If employees are required to clean-up spills, they must be properly trained and equipped. OSHA 1910.120(q) may be applicable.

Fire Extinguishing: This chemical is a combustible liquid. Poisonous gases including nitrogen oxides, hydrogen, and carbon monoxide are produced in fire. Use dry chemical, carbon dioxide, water spray, or alcohol or polymer foam extinguishers. Vapors are heavier than air and will collect in low areas. Vapors may travel long distances to ignition sources and flashback. Vapors in confined areas may explode when exposed to fire. Storage containers and parts of containers may rocket great distances, in many directions. If material or contaminated runoff enters waterways, notify downstream users of potentially contaminated waters. Notify local health and fire officials and pollution control agencies. From a secure, explosion-proof location, use water spray to cool exposed containers. If cooling streams are ineffective (venting sound increases in volume and pitch, tank discolors, or shows any signs of deforming), withdraw immediately to a secure position. If employees are expected to fight fires, they must be trained and equipped in OSHA 1910.156.

Disposal Method Suggested: Add excess alcoholic KOH. Than evaporate alcohol and add calcium hypochlorite. After 24 hours, flush to sewer with water.[24] Can also be incinerated with afterburner and scrubber to remove nitrogen oxides.

References

U.S. Environmental Protection Agency, "Chemical Profile: Adiponitrile," Washington, DC, Chemical Emergency Preparedness Program (November 30, 1987)

Sax, N. I., Ed., "Dangerous Properties of Industrial Materials Report," 1, No. 6, 22–24 (1981)

New Jersey Department of Health and Senior Services, Hazardous Substance Fact Sheet, Adiponitrile, Trenton, NJ (November, 1998)

Adriamycin

Molecular Formula: $C_{27}H_{29}O_{11}N$

Synonyms: ADM; Adriablastine; Adriamycin-HCl; Adriamycin semiquinone; Adriblastina; Doxorubicin; DX;

14'-Hydroxydaunomycin; 14-Hydroxydaunomycin; 14-Hydroxydaunorubicine

CAS Registry Number: 23214-92-8

RTECS Number: AV9800000

DOT ID: UN 2811 (as medicines, toxic solid, n.o.s.)

Regulatory Authority

- Carcinogen, NTP, IARC Group 2A (limited human data), OSHA (possible select carcinogen), California, Massachusetts, Minnesota, New Jersey[10]

Cited in U.S. State Regulations: California (G), Florida (G), Illinois (G), Massachusetts (G), Minnesota (G), New Jersey (G), Pennsylvania (G), Rhode Island (G).

Description: Adriamycin, $C_{27}H_{29}O_{11}N$, is orange to red cake-like or needle-like crystalline solid. Freezing/Melting point = 205°C. Slightly soluble in water.

Potential Exposure: An antibiotic product from streptomyces, used as anticancer drug.

Incompatibilities: Strong oxidizers.

Permissible Exposure Limits in Air: No standards set.

Permissible Concentration in Water: No criteria set.

Routes of Entry: Intravenous, skin, eyes.

Harmful Effects and Symptoms

Short Term Exposure: Highly Toxic. Irritates eyes and skin. A poison if ingested. This chemical can be absorbed through the skin, thereby increasing exposure.

Long Term Exposure: Has been shown to cause breast cancer animals. There is limited evidence that adriamycin is a teratogen in animals. Causes baldness, stomatitis and bone marrow aphasia in humans. Fatal human cardiac disturbances have been reported.

Points of Attack: Cardiovascular system.

Medical Surveillance: Complete blood count (CBC), EKG, examination of the nervous system.

First Aid: If this chemical gets into the eyes, remove any contact lenses at once and irrigate immediately for at least 15 minutes, occasionally lifting upper and lower lids. If this chemical contacts the skin, remove contaminated clothing and wash immediately with soap and water. When this chemical has been swallowed, get medical attention. Give large quantities of water and induce vomiting. Do not make an unconscious person vomit. If this chemical has been inhaled, remove from exposure and transfer promptly to a medical facility.

Personal Protective Methods: A regulated, marked area should be established where Adriamycin is handled, used, or stored. All operations should be enclosed and locally ventilated through HEPA filters. Wear protective work clothing. Wash thoroughly immediately after exposure to Adriamycin and at the end of the workshift. Medical personnel preparing solutions of Adriamycin should only do so under a biological safety hood with a vertical laminar flow. Workers whose clothing has been contaminated by Adriamycin should change into clean clothing promptly. Do not take contaminated work clothes home. Family members could be exposed. Contaminated work clothes should be laundered by individuals who have been informed off the hazards of exposure to Adriamycin. Do not eat, smoke, or drink where Adriamycin is handled, processed, or stored, since the chemical can be swallowed. Wash hands carefully before eating or smoking. When vacuuming, a high efficiency particulate absolute (HEPA) filter should be used, not a standard shop vacuum.

Respirator Selection: *At any concentrations above the NIOSH REL, or where there is no REL, at any detectable concentration:* SCBAF:PD,PP (any MSHA/NIOSH approved self-contained breathing apparatus that has a full facepiece and is operated in a pressure-demand or other positive-pressure mode); or SAF:PD,PP:ASCBA (any supplied-air respirator that has a full facepiece and is operated in a pressure-demand or other positive-pressure mode in combination with an auxiliary, self-contained breathing apparatus operated in a pressure-demand or other positive pressure mode). *Escape:* GMFOVHiE [any air-purifying, full-facepiece respirator (gas mask) with a chin-style, front- or back-mounted organic vapor canister having a high-efficiency particulate filter]; or SCBAE (any appropriate escape-type, self-contained breathing apparatus).

Storage: Prior to working with this chemical you should be trained on its proper handling and storage. Store in tightly closed containers in a cool, well-ventilated area. A regulated, marked area should be established where Adriamycin is handled, used, or stored.

Shipping: Classified as a medicines, toxic solid, n.o.s. Adriamycin falls in Shipping Class 6.1 and Packing Group II. The label required is "Poison."

Spill Handling: Evacuate and restrict persons not wearing protective equipment from area of spill or leak until cleanup is complete. Remove all ignition sources. Collect powdered material using a HEPA filter vacuum cleaner and deposit in sealed containers. Ventilate area of spill or leak after clean-up is complete. It may be necessary to contain and dispose of this chemical as a hazardous waste. If material or contaminated runoff enters waterways, notify downstream users of potentially contaminated waters. Contact your Department of Environmental Protection or your regional office of the federal EPA for specific recommendations. If employees are required to clean-up spills, they must be properly trained and equipped. OSHA 1910.120(q) may be applicable.

Fire Extinguishing: Use dry chemical, carbon dioxide, water spray, or alcohol or polymer foam extinguishers. Poisonous gases are produced in fire including nitrogen oxides and hydrogen chloride. If employees are expected to fight fires, they must be trained and equipped in OSHA 1910.156.

References

Sax, N. I., Ed., Dangerous Properties of Industrial Materials Report, 1, No. 3, 24–25 (1981)

New Jersey Department of Health and Senior Services, "Hazardous Substance Fact Sheet: Adriamycin," Trenton, NJ (October 1998)

Aflatoxins

Molecular Formula: $C_{16-17}H_{10-14}O_{6-7}$; $C_{17}H_{12}O_6$ (B; B1; B2); $C_{17}H_{12}O_7$ (G1; M1); $C_{17}H_{14}O_7$ (G2; M2); $C_{16}H_{10}O_6$ (P1)

Synonyms: AFL; Aflatoxicol; Aflatoxin B1 dichloride; Cyclopenta[c]furo(3',2':4,5)furo(2,3-H)(1)benzopyran-1,11-dione, 2,3,6a,8,9,9a-hexahydro-8,9-dichloro-4-methoxy-, [6aS-(6a-A-8-B,9-A-9aa-)]-; 2,3-Dichloroaflatoxin B1; Dihydroaflatoxin B1; Dihydroaflatoxin G1; 4-Dihydroxy-aflatoxin B1

CAS Registry Number: 1402-68-2 (aflatoxin); 1162-65-8 (B); 1162-65-8 (B1); 7220-81-7 (B2); 1165-39-5 (G1); 7241-98-7 (G2); 6795-23-9 (M1); 6885-57-0 (M2); 32215-02-4 (P1); 29611-03-8 (Ro)

RTECS Number: AW5950000 (aflatoxin); GY1925000 (B); GY1925000 (B1); GY1722000 (B2); LV1700000 (G1); LV1700000 (G2); GY1880000 (M1); GY1720000 (M2); GY1775000 (P1)

DOT ID: UN 2811 (toxic solid, organic, n.o.s.)

Regulatory Authority

- Carcinogen (Human Suspected) IARC, NTP, OSHA
- RCRA 40CFR261, Appendix 8; 40CFR261.11 Hazardous Constituents[5]

Cited in U.S. State Regulations: California (G), Illinois (G), Kansas (G), Louisiana (G), Massachusetts (G), Michigan (G), Minnesota (G), Pennsylvania (G), Rhode Island (G), Vermont (G), Virginia (G), Washington (G), West Virginia (G), Wisconsin (G).

Description: The aflatoxins are a group of molds produced by the fungus *Aspergillus flavus*. They are natural contaminants of fruits, vegetables, and grains. They are also described as a series of condensed ring heterocyclic compounds. Freezing/Melting point = 268 – 269°C (B1), 247 – 250°C (G1), and 237 – 240°C (G2), 290°C (M1). They form colorless to pale yellow crystals. Slightly soluble in water.

Potential Exposure: Aflatoxins are a group o toxic metabolites produced by certain types of fungi. Aflatoxins are not commercially manufactured; they are naturally occurring contaminants that are formed by fungi on food during conditions of high temperatures and high humidity. Most human exposure to aflatoxins occurs through ingestion of contaminated food. The estimated amount of aflatoxins that Americans consume daily is estimated to be 0.15 – 0.50 μg. Grains, peanuts, tree nuts, and cottonseed meal are among the more common foods on which these fungi grow. Meat, eggs, milk, and other edible products from animals that consume aflatoxin-contaminated feed may also contain aflatoxins. Aflatoxins can also be breathed in.

Permissible Exposure Limits in Air: No standards set.

Permissible Concentration in Water: No criteria set.

Routes of Entry: Ingestion and inhalation. As unavoidable contaminant in foods. The FDA limits the levels of aflatoxin contamination that are permitted in food. The complete elimination of aflatoxin contamination of food is probably not technically feasible. The FDA has lowered the maximum amount allowed in food products as methods of detection and methods of control have improved. Upper limits of 20 ppb (total B1, B2, G1, and G2) in foods and feeds and 0.5 ppm (M1) in milk are now in effect.

Harmful Effects and Symptoms

Aflatoxins are carcinogenic in mice, rats, fish, ducks, marmosets, tree shrews and monkeys by several routes of administration (including oral), producing mainly cancers of the liver, colon and kidney. Epidemiological studies have shown a positive correlation between the average dietary concentrations of aflatoxins in populations and the incidence of primary liver cancer. These studies were undertaken to test this specific hypothesis; however, no studies have been carried out which could link an increased risk of liver cancer to actual aflatoxin intake in individuals.

Short Term Exposure: No acute health effects are known at this time.

Long Term Exposure: Aflatoxins are carcinogens in humans. They may cause liver cancer. They have also been shown to be teratogens in animals. Aflatoxins may cause liver and kidney damage.

Points of Attack: Liver, kidneys.

Medical Surveillance: Liver and kidney function tests.

First Aid: If this chemical gets into the eyes, remove any contact lenses at once and irrigate immediately for at least 15 minutes, occasionally lifting upper and lower lids. If this chemical contacts the skin, remove contaminated clothing and wash immediately with soap and water. When this chemical has been swallowed, get medical attention. If this chemical has been inhaled, remove from exposure and transfer promptly to a medical facility.

Personal Protective Methods: A Class I, Type B, biological safety hood should be used when handling, mixing, or preparing aflatoxins. Wear protective gloves and clothing to prevent any reasonable probability of skin contact. Latex gloves are recommended as a protective material, *unless using aflatoxins in chloroform*. Contact lenses should not be worn when working with this chemical. Wear indirect-vent, impact and splash-proof chemical goggles and face shield unless full face-piece respiratory protection is worn. Employees should wash immediately with soap when skin is wet or contaminated. Remove nonimpervious clothing immediately if wet or contaminated. Provide emergency showers and eyewash.

Respirator Selection: *At any detectable concentration:* SCBAF:PD,PP (any MSHA/NIOSH approved self-contained breathing apparatus that has a full facepiece and is operated in a pressure-demand or other positive-pressure mode); or SAF:PD,PP:ASCBA (any supplied-air respirator that has a full facepiece and is operated in a pressure-demand or other positive-pressure mode in combination with an auxiliary, self-contained breathing apparatus operated in a pressure-demand or other positive pressure mode). *Escape:* HiEF (Any air-purifying, full-facepiece respirator with a high-efficiency particulate filter); or SCBAE (Any appropriate escape-type, self-contained breathing apparatus).

Storage: Prior to working with this chemical you should be trained on its proper handling and storage. A regulated, marked area should be established where aflatoxins are handled, used, or stored. Store in a refrigerator and protect from exposure to air and light.

Shipping: Falling in the DOT category of Toxic solid, organic, n.o.s. Aflatoxins require a "Poison" label. They fall in Hazard Class 6.1 and Packing Group I.

Spill Handling: Evacuate and restrict persons not wearing protective equipment from area of spill or leak until cleanup is complete. Remove all ignition sources. Dampen spilled material with 60 – 70% of acetone to avoid airborne dust then transfer to a suitable sealed container for disposal. Ventilate area of spill or leak after clean-up is complete. It may be necessary to contain and dispose of this chemical as a hazardous waste. If material or contaminated runoff enters waterways, notify downstream users of potentially contaminated waters. Contact your Department of Environmental Protection or your regional office of the federal EPA for specific recommendations. If employees are required to clean-up spills, they must be properly trained and equipped. OSHA 1910.120(q) may be applicable.

Fire Extinguishing: This material may burn, but will not easily ignite. Use dry chemical, carbon dioxide, water spray, or foam extinguishers. If employees are expected to fight fires, they must be trained and equipped in OSHA 1910.156.

Disposal Method Suggested: Consult with environmental regulatory agencies for guidance on acceptable disposal practices. Generators of waste containing this contaminant (≥100 kg/mo) must conform with EPA regulations governing storage, transportation, treatment, and waste disposal. Use of oxidizing agents such as hydrogen peroxide or 5% sodium hypochlorite bleach. Acids an bases may also be used.[22]

References

Sax, N. I., Ed., Dangerous Properties of Industrial Materials Report, 1, No. 4, 31–33 (1981), and 7, No. 2, 36–43 (1987)

New Jersey Department of Health and Senior Services, Hazardous Substance Fact Sheet, Aflatoxins, Trenton, NJ (August, 1998)

Alachlor

Molecular Formula: $C_{14}H_{20}ClNO_2$

Synonyms: Acetamide, 2-chloro-*N*-(2,6-diethylphenyl)-*N*-(methoxymethyl)-; Acetanilide, 2-chloro-2',6'-diethyl-*N*-methoxymethyl)-; A13-51506; Alachlore; Alanex; Alanox®; Alatox480®; α-Chloro-2',6'-diethyl-*N*-(methoxymethyl)acetanilide; 2-Chloro-*N*-(2,6-diethylphenyl)-*N*-(methoxymethyl) acetamide; EPA pesticide chemical code 090501; Glyphosate isopropylamine salt; Lassagrin; Lasso®; Lasso® Micro-tech; Lazo®; Metachlor; Methachlor; *N*-(Methoxymethyl)2,6-diethylchloroacetamide; Pillarzo®

CAS Registry Number: 15972-60-8

RTECS Number: AE1225000

DOT ID: UN 2588

Regulatory Authority

- Carcinogen (Probable human) EPA,[47] California
- Banned or Severely Restricted (EPA-FIFRA) (See reference below)
- Safe Drinking Water Act, MCL, 0.002 mg/l; MGLC, zero; Regulated chemical (47 FR 9352)
- SUPERFUND/EPCRA 40CFR302.4, Appendix A, Reportable Quantity (RQ): 1 lb (0.454 kg), SARA 313: Form R *de minimis* Concentration Reporting Level: 1.0%.
- Water Pollution Standard Proposed (Various States)[61]

Cited in U.S. State Regulations: Arizona (W), California (W), Illinois (W), Kansas (W), Maine (W), Massachusetts (W), Minnesota (W).

Description: Alachlor, $C_{14}H_{20}ClNO_2$, a poly-substituted single aromatic nucleus, is a cream-colored solid. Freezing/Melting point = 39.5 – 41.5°C. Hazard Identification (based on NFPA-704 M Rating System): Health 2, Flammability 0, Reactivity 0. Slightly soluble in water.

Potential Exposure: A chloracetanilide herbicide. In manufacture, formulation and application of this pre-emergence herbicide, personnel may be exposed. Its major use (99%) is as a pre-emergence herbicide for field crops (corn, soybeans, and peanuts, etc).

Incompatibilities: Strong oxidizers. Corrosive to iron and steel.

Permissible Exposure Limits in Air: No standards set.

Permissible Concentration in Water: No adverse effect level in drinking water has been calculated by NSA/NRC[46] as 0.7 mg/l. Allowable daily intake (ADI) has been calculated at 0.1 mg/kg/day. More recently, the EPA[47] has reviewed alachlor and determined a ten-day health advisory value of 0.1 mg/l for a 10 kg child. An acceptable daily intake was calculated as 0.01 mg/kg/day in the study. A maximum level in drinking water of 0.002 mg/l has been proposed by EPA.[62] In addition, a number of states[61] have set guidelines for alachlor in drinking water ranging from 0.15 μg/l (Arizona) to 0.2 μg/l (Illinois) to 2.0 μg/l (Massachusetts) to 10.0 μg/l (Minnesota) to 15 μg/l (Kansas) to 200 μg/l (Maine).

Determination in Water: May be accomplished by liquid-liquid extraction gas chromatographic procedure.[47]

Routes of Entry: Ingestion.

Harmful Effects and Symptoms

No effects found in human studies.[47] Exhibits relatively low acute oral toxicity; the LD_{50} value for rats is 0.93 g/kg. The technical product has only slight skin and eye irritation potential after an acute exposure.[47] However, alachlor feeding studies have demonstrated oncogenic effects including lung tumors in mice and stomach, thyroid and nasal turbinate tumors in rats.

Short Term Exposure: Toxic by skin contact, ingestion, and inhalation. Eye contact may cause severe irritation or injury. Skin contact may irritate and burn skin.

Long Term Exposure: Human mutation data reported. Suspected carcinogen; some experimental data reported.

First Aid: If this chemical gets into the eyes, remove any contact lenses at once and irrigate immediately for at least 15 minutes, occasionally lifting upper and lower lids. If this chemical contacts the skin, remove contaminated clothing and wash immediately with soap and water. When this chemical has been swallowed, get medical attention. Give large quantities of water and induce vomiting. Do not make an unconscious person vomit. If this chemical has been inhaled, remove from exposure and transfer promptly to a medical facility.

Personal Protective Methods: Wear protective gloves and clothing to prevent any reasonable probability of skin contact. Safety equipment suppliers/manufacturers can provide recommendations on the most protective glove/clothing material for your operation. All protective clothing (suits, gloves, footwear, headgear) should be clean, available each day, and put on before work. Contact lenses should not be worn when working with this chemical. Wear dust-proof chemical goggles and face shield unless full facepiece respiratory protection is worn. Employees should wash immediately with soap when skin is wet or contaminated. Provide emergency showers and eyewash.

Respirator Selection: *At any concentrations above the NIOSH REL, or where there is no REL, at any detectable concentration:* SCBAF:PD,PP (any MSHA/NIOSH approved self-contained breathing apparatus that has a full facepiece and is operated in a pressure-demand or other positive-pressure mode); or SAF:PD,PP:ASCBA (any supplied-air respirator that has a full facepiece and is operated in a pressure-demand or other positive-pressure mode in combination with an auxiliary, self-contained breathing apparatus operated in a pressure-demand or other positive pressure mode).

Storage: Prior to working with this chemical you should be trained on its proper handling and storage. Store in tightly closed containers in a cool, well ventilated area. Do not reuse container; see disposal methods, below.

Spill Handling: Evacuate persons not wearing protective equipment from area of spill or leak until clean-up is complete. Remove all ignition sources. Ventilate area of spill or leak. Use water spray to reduce vapors. Take up with diatomite, clay, expanded mineral, foamed glass, or synthetic treated absorbent material and deposit in sealed containers for later disposal. It may be necessary to contain and dispose of this chemical as a hazardous waste. If material or contaminated runoff enters waterways, notify downstream users of potentially contaminated waters. Contact your Department of Environmental Protection or your regional office of the federal EPA for specific recommendations. If employees are required to clean-up spills, they must be properly trained and equipped. OSHA 1910.120(q) may be applicable. For large spills seek professional environmental engineering assistance from the U.S. EPA Environmental Response Team at (908) 548-8730 (24-hour response line).

Fire Extinguishing: This chemical is not flammable, but may support combustion. Stay upwind of fire. Use dry chemical, carbon dioxide, water spray, or foam extinguishers. Poisonous gases are produced in fire including nitrogen oxides and chlorine. If material or contaminated runoff enters waterways, notify downstream users of potentially contaminated waters. Notify local health and fire officials and pollution control agencies. From a secure, explosion-proof location, use water spray to cool exposed containers. If cooling streams are ineffective (venting sound increases in volume and pitch, tank discolors, or shows any signs of deforming), withdraw immediately to a secure position. If employees are expected to fight fires, they must be trained and equipped in OSHA 1910.156. If employees are expected to fight fires, they must be trained and equipped in OSHA 1910.156.

Disposal Method Suggested: This compound is hydrolyzed under strongly acid or alkaline conditions, to chloroacetic acid, methanol, formaldehyde and 2,6-diethylanilne. Incineration is recommended as a disposal procedure. Techniques for alachlor removal from potable water have been reviewed by EPA[47] but the data revealed no superior method. Improper disposal of pesticides is a violation of federal law. Dispose in accordance with 40CFR165 recommendations for the disposal of pesticides and pesticide containers.

References

U.S. Environmental Protection Agency, "Alachlor: Notice of Intent to Cancel Registrations; Conclusion of Special Review," Federal Register 52, No. 251, pp. 49480–49504 incl. (December 31, 1987)

Waxman, Michael F and Kammel, David W, A Guidebook for the Safe Use of Hazardous Agricultural Farm Chemicals and Pesticides, North Central Regional Publication 402, Madison WI, (1991)

Aldicarb

Molecular Formula: $C_7H_{14}N_2O_2S$

Common Formula: $CH_3SC(CH_3)_2CH{=}NOCONHCH_3$

Synonyms: A13-27093; Aldecarb; Aldecarbe (French); Ambush®; Carbamic acid, methyl-,*O*-([2-methyl-2-(methylthio)propylidene]amino) deriv.; Carbanolate; Caswell No. 011A; EPA pesticide chemical code 098301; 2-Methyl-2-(methylthio)propanaldehyde, *O*-(methylcarbamoyl) oxime; 2-Methyl-2-(methylthio)propanal, *O*-[(methylamino)carbonyl] oxime; 2-Methyl-2-methylthio-propionaldehyd-*O*-(*n*-methyl-carbamoyl)-oxim (German); 2-Metil-2-tiometil-propionaldeid- *O*-(*n*-metil-carbamoil)-ossima (Italian); Permethrin; Propanal, 2-methyl-2-(methythio)-, *O*-[(methylamino)carbonyl] oxime; Propionaldehyde, 2-methyl-2-(methylthio)-, *O*-(methylcarbamoyl) oxime; Sulfone aldoxycarb; Temic®; Temik®; Temik 10 G®; Temik G10®; UC 21149® (Union Carbide); Union Carbide 21149; Union Carbide UC-21149

CAS Registry Number: 116-06-3

RTECS Number: UE2275000

DOT ID: UN 2757

Regulatory Authority

- Classified by the EPA as Restricted Use Pesticide (RUP)
- Banned or Severely Restricted (Austria, Belgium, Germany, Israel, Norway, Philippines) (UN)[13]
- Very Toxic Substance (World Bank)[15]
- EPA Hazardous waste number (RCRA No.): P070
- RCRA 40CFR261, Appendix 8; 40CFR261.11 Hazardous Constituents
- RCRA Land Ban Waste Restrictions
- Safe Drinking Water Act, MCL, 0.003 mg/l; MCLG 0.001 mg/l; Regulated chemical (47 FR 9352).
- CERCLA/SARA 40CFR302, Extremely Hazardous Substances: TPQ = 100 /10,000 lb (45.4/4,540 kg).
- SUPERFUND/EPCRA 40CFR302.4, Appendix A, Reportable Quantity (RQ): CERCLA 1 lb (0.454 kg), SARA 313: Form R *de minimis* Concentration Reporting Level: 1.0%.
- U.S. DOT Regulated Marine Pollutant (49CFR172.101, Appendix B)
- Air Pollutant Standard Set (New York, South Carolina)[60]
- Water Pollution Standard Proposed (EPA)[47] (Several States)[61]

Cited in U.S. State Regulations: Arizona (W), California (W), Florida (G), Illinois (G), Kansas (G), Louisiana (G), Maine (W), Massachusetts (G, W), Michigan (G), Minnesota (W), New Jersey (G), New York (W), Pennsylvania (G), Vermont (G), Virginia (G), Washington (G), Wisconsin (G, W).

Description: Aldicarb, $CH_3SC(CH_3)_2CH{=}NOCONHCH_3$, $C_7H_{14}N_2O_2S$ is a noncombustible, white crystalline solid with a slight sulfurous odor. Freezing/Melting point = 98 – 100°C. Slightly soluble in water.

Potential Exposure: A systemic, restricted use (RUP), carbamate pesticide. It is available for purchase and use by certified pesticide applicators or by those under their direct supervision. Personnel involved in manufacture, formulation or application of this insecticide to crops are at risk.

Incompatibilities: Strong alkalis.

Permissible Exposure Limits in Air: The American Industrial Hygiene Association (AIHA/WEEL) recommends a TWA level of 0.07 mg/m^3, and warns that aldicarb can be absorbed through the skin, thereby increasing exposure. Guidelines or standards have been set for aldicarb in ambient air[60] ranging from 2.0 μg/m^3 (New York) to 6.0 μg/m^3 (South Carolina).

Permissible Concentration in Water: EPA/Safe Drinking Water Act levels: MCL, 0.003 mg/l; MCLG 0.001 mg/l. Canada's Drinking Water Quality = 0.009 mg/l MAC.

Determination in Water: Aldicarb may be determined in water by gas-liquid chromatography with flame photometric detection after oxidation to the sulfone (aldoxycarb) by peracetic acid or 3-chloro-perbenzoic acid. Colorimetric methods have also been used based on hydrolysis to hydroxyl-amine which is oxidized to nitrous acid, the latter used to diazotize sulfanilic acid which is then coupled to give a dye.[23]

Routes of Entry: Ingestion, skin contact.

Harmful Effects and Symptoms

Aldicarb is a carbamate pesticide. This material is super toxic; the probable oral lethal dose for humans is less than 5 mg/kg, or a taste (less than 7 drops) for a 150-lb person; it is extremely toxic by both oral and dermal routes. Symptoms include headache, blurred vision, nausea, vomiting, diarrhea, and abdominal pain. In severe cases, unconsciousness and convulsions may occur.

Short Term Exposure: This chemical is one of the most highly toxic pesticides. It can be harmful or fatal if swallowed, inhaled or absorbed through the skin. Exposure can cause rapid severe poisoning with headache, sweating, nausea and vomiting, diarrhea, loss of coordination, and death.

Long Term Exposure: Human mutation data reported. Aldicarb is questionable carcinogen with no firm human evidence.

Points of Attack: skin, lungs.

Medical Surveillance: Plasma and red blood cell cholinesterase levels (test for enzyme poisoned by this chemical). For this substance, these tests are accurate only if done within about two hours of exposure.

First Aid: Move victim to fresh air; call emergency medical care. Speed in removing material from eyes and skin is of extreme importance. If this chemical gets into the eyes, remove any contact lenses at once and irrigate immediately for at least 15 minutes, occasionally lifting upper and lower lids. Seek medical attention immediately. If this chemical contacts the skin, remove contaminated clothing and wash immediately with soap and water. Seek medical attention immediately. If this chemical has been inhaled, remove from exposure, begin rescue breathing (using universal precautions) if breathing has stopped and CPR if heart action has stopped. Transfer promptly to a medical facility. When this chemical has been swallowed, get medical attention. Give large quantities of water and induce vomiting. Do not make an unconscious person vomit. Keep victim quiet and maintain normal body temperature. Effects may be delayed; keep victim under observation.

Personal Protective Methods: Wear protective gloves and clothing to prevent any reasonable probability of skin contact. Safety equipment suppliers/manufacturers can provide recommendations on the most protective glove/clothing material for your operation. All protective clothing (suits, gloves, footwear, headgear) should be clean, available each day, and put on before work. Contact lenses should not be worn when working with this chemical. Wear dust-proof chemical goggles and face shield unless full facepiece respiratory protection is worn. Employees should wash immediately with soap when skin is wet or contaminated. Provide emergency showers and eyewash.

Respirator Selection: *At any detectable concentration:* SCBAF:PD,PP (any MSHA/NIOSH approved self-contained breathing apparatus that has a full facepiece and is operated in a pressure-demand or other positive-pressure mode); or SAF:PD,PP:ASCBA (any supplied-air respirator that has a full facepiece and is operated in a pressure-demand or other positive-pressure mode in combination with an auxiliary, self-contained breathing apparatus operated in a pressure-demand or other positive pressure mode). *Escape:* HiEF (Any air-purifying, full-facepiece respirator with a high-efficiency particulate filter); or SCBAE (Any appropriate escape-type, self-contained breathing apparatus).

Storage: Prior to working with this chemical you should be trained on its proper handling and storage. A regulated, marked area should be established where Aldicarb is handled, used, or stored. Store in tightly closed containers in a cool, well-ventilated area away from strong alkalis (such as Sodium Hydroxide, Sodium Bicarbonate, etc.).

Shipping: Aldicarb is a solid toxic carbamate and should be labeled "Poison." It falls into DOT/UN Hazard Class 6.1.

Spill Handling: Evacuate and restrict persons not wearing protective equipment from area of spill or leak until cleanup is complete. Remove all ignition sources. Stay upwind; keep out of low areas. Ventilate closed spaces before entering them. Wear positive pressure breathing apparatus and special protective clothing. Remove and isolate contaminated clothing at the site. Do not touch spilled material; stop leak if you can do it without risk. Use water spray to reduce vapors. *Small spills:* take up with diatomite, clay, expanded mineral, vermiculite, sand or other noncombustible absorbent material and place into containers for later disposal. Small dry spills: with clean shovel place material into clean, dry container and cover; move containers from spill area. *Large spills:* dike far ahead of spill for later disposal. Seek professional environmental engineering assistance from the U.S. EPA Environmental Response Team at (908) 548-8730 (24-hour response line). It may be necessary to contain and dispose of this chemical as a hazardous waste. If material or contaminated runoff enters waterways, notify downstream users of potentially contaminated waters. Contact your Department of Environmental Protection or your regional office of the federal EPA for specific recommendations. If employees are required to clean-up spills, they must be properly trained and equipped. OSHA 1910.120(q) may be applicable.

Fire Extinguishing: May burn but does not ignite readily. Poisonous gases including nitrogen and sulfur oxides are produced in fire. Small fires: dry chemical, carbon dioxide, water spray, or alcohol foam. Large fires: water spray, fog or alcohol foam. Stay upwind; keep out of low areas. Ventilate closed spaces before entering them. Wear positive pressure breathing apparatus and special protective clothing. Move container from fire area if you can do it without risk. Fight fire from maximum distance. Dike fire control water for later disposal; do not scatter the material. If material or contaminated runoff enters waterways, notify downstream users of potentially contaminated waters. Notify local health and fire officials and pollution control agencies. From a secure, explosion-proof location, use water spray to cool exposed containers. If cooling streams are ineffective (venting sound increases in volume and pitch, tank discolors, or shows any signs of deforming), withdraw immediately to a secure position. If employees are expected to fight fires, they must be trained and equipped in OSHA 1910.156.

Disposal Method Suggested: Consult with environmental regulatory agencies for guidance on acceptable disposal practices. Generators of waste containing this contaminant ($\geq$100 kg/mo) must conform with EPA regulations governing storage, transportation, treatment, and waste disposal. Incineration with effluent gas scrubbing is recommended.[22] In accordance with 40CFR165 recommendations for the disposal of pesticides and pesticide containers. In accordance with 40CFR165 recommendations for the disposal of pesticides and pesticide containers. Must be disposed properly by following package label directions or by contacting your state pesticide or environmental control agency or by contacting your regional EPA office.

References

Sax, N. I., Ed., Dangerous Properties of Industrial Materials Report, 4, No. 2, 37–41 (1984)

U.S. Environmental Protection Agency, "Chemical Profile: Aldicarb," Washington, DC, Chemical Emergency Preparedness Program (November 30, 1987)

New Jersey Department of Health and Senior Services, "Hazardous Substance Fact Sheet: Aldicarb," Trenton, NJ (January 1986)

Aldol

Molecular Formula: $C_4H_8O_2$

Synonyms: Acetaldol; Aldehido, β-hidroxibutirico (Spanish); 3-Butanolal; Butyraldehyde, 3-hydroxy-; 3-Hydroxybutanal; β-Hydroxybutyraldehyde; 3-Hydroxybutyraldehyde; Oxybutanal; Oxybutyric aldehyde

CAS Registry Number: 107-89-1

RTECS Number: ES1500000

DOT ID: UN 2839

Cited in U.S. State Regulations: Florida (G), Massachusetts (G), New Jersey (G), Pennsylvania (G).

Description: Aldol is a flammable, colorless to pale yellow, syrupy liquid. Boiling point = 83°C @ 20 mm. Freezing/Melting point = 0°C. Flash point = 66°C. Autoignition temperature = 250°C. Hazard Identification (based on NFPA-704 M Rating System): Health 3, Flammability 2, Reactivity 2. Soluble in water.

Potential Exposure: Aldol is used as a solvent and to manufacture rubber accelerators, perfumes, and fungicides, and in engraving, cadmium plating.

Incompatibilities: Contact with oxidizers may cause fire and explosions. Contact with metals may cause the formation of flammable and explosive hydrogen gas. Heat caused the formation of crotonaldehyde and water.

Permissible Exposure Limits in Air: Not established.

Routes of Entry: Inhalation, skin and/or eye contact.

Harmful Effects and Symptoms

Short Term Exposure: Skin and eye contact causes irritation. This chemical can be absorbed through the skin, thereby increasing exposure. Exposure can cause dizziness, lightheadedness, and unconsciousness.

Long Term Exposure: Unknown at this time.

First Aid: If this chemical gets into the eyes, remove any contact lenses at once and irrigate immediately for at least 15 minutes, occasionally lifting upper and lower lids. Seek medical attention immediately. If this chemical contacts the skin, remove contaminated clothing and wash immediately with soap and water. Seek medical attention immediately. If this chemical has been inhaled, remove from exposure, begin rescue breathing (using universal precautions) if breathing has stopped and CPR if heart action has stopped. Transfer promptly to a medical facility. When this chemical has been swallowed, get medical attention. Give large quantities of water and induce vomiting. Do not make an unconscious person vomit.

Personal Protective Methods: Wear solvent-resistant gloves and clothing to prevent any reasonable probability of skin contact. Safety equipment suppliers/manufacturers can provide recommendations on the most protective glove/clothing material for your operation. All protective clothing (suits, gloves, footwear, headgear) should be clean, available each day, and put on before work. Contact lenses should not be worn when working with this chemical. Wear splash-proof chemical goggles and face shield unless full facepiece respiratory protection is worn. Employees should wash immediately with soap when skin is wet or contaminated. Remove nonimpervious clothing immediately if wet or contaminated. Provide emergency showers and eyewash.

Respirator Selection: *At any concentrations above the NIOSH REL, or where there is no REL, at any detectable concentration:* SCBAF:PD,PP (any MSHA/NIOSH approved self-contained breathing apparatus that has a full facepiece and is operated in a pressure-demand or other positive-pressure mode); or SAF:PD,PP:ASCBA (any supplied-air respirator that has a full facepiece and is operated in a pressure-demand or other positive-pressure mode in combination with an auxiliary, self-contained breathing apparatus operated in a pressure-demand or other positive pressure mode).

Storage: Prior to working with this chemical you should be trained on its proper handling and storage. Before entering confined space where Aldol may be present, check to make sure that an explosive concentration does not exist. Store in tightly closed containers in a cool, well ventilated area. Metal containers involving the transfer of this chemical should be grounded and bonded. Where possible, automatically pump liquid from drums or other storage containers to process containers. Drums must be equipped with self-closing valves, pressure vacuum bungs, and flame arresters. Use only non-sparking tools and equipment, especially when opening and closing containers of this chemical. Sources of ignition such as smoking and open flames, are prohibited where this chemical is used, handled, or stored in a manner that could create a potential fire or explosion hazard.

Shipping: Aldol is a liquid toxic pesticide and should be labeled "Poison." It falls into DOT/UN Hazard Class 6.1.

Spill Handling: Evacuate persons not wearing protective equipment from area of spill or leak until clean-up is complete. Remove all ignition sources. Ventilate area of spill or leak. Use water spray to reduce vapor. Absorb liquids in vermiculite, dry sand, earth, or a similar material and deposit in sealed containers. It may be necessary to contain and dispose of this chemical as a hazardous waste. If material or contaminated runoff enters waterways, notify downstream users of potentially contaminated waters. Contact your Department of Environmental Protection or your regional office of the federal EPA for specific recommendations. If employees are required to clean-up spills, they must be properly trained and equipped. OSHA 1910.120(q) may be applicable.

Fire Extinguishing: This chemical is a flammable liquid. Poisonous gases including nitrogen oxides are produced in fire. Use dry chemical, carbon dioxide, or alcohol foam extinguishers. Vapors are heavier than air and will collect in low areas. Vapors may travel long distances to ignition sources and flashback. Vapors in confined areas may explode when exposed to fire. Storage containers and parts of containers may rocket great distances, in many directions. If material or contaminated runoff enters waterways, notify downstream users of potentially contaminated waters. Notify local health and fire officials and pollution control agencies. From a secure, explosion-proof location, use water spray to cool exposed containers. If cooling streams are ineffective (venting sound increases in volume and pitch, tank discolors, or shows any signs of deforming), withdraw immediately to a secure position. If employees are expected to fight fires, they must be trained and equipped in OSHA 1910.156.

Disposal Method Suggested: Incineration.

References

New Jersey Department of Health and Senior Services, Hazardous Substance Fact Sheet, Aldol, Trenton NJ (September, 1998)

Aldrin

Molecular Formula: $C_{12}H_8Cl_6$

Synonyms: Aldocit®; Aldrec®; Aldrex®; Aldrex-30®; Aldrex-40®; Aldrina® (Spanish); Aldrin y dieldrin (Spanish); Aldrine (French); Aldrite®; Aldron®; Aldrosol®; Algran®; Altox®; 1,4,5,8-Dimethanonaphthalene,1,2,3,4,10,10-hexachloro-1,4,4a,5,8,8a-hexahydro-(1α,4α,4β,5α,8α,8β)-; 1,4:5,8-Dimethanonaphthalene, 1,2,3,4,10,10-hexachloro-1,4,4a,5,8,8a-hexahydro-, endoexo-; Drinox®; ENT15949; 1,2,3,4,10,10-Hexachloro-1,4,4a,5,8,8a-hexahydro-1,4,5,8-dimethanonaphthalene; 1,2,3,4,10,10-Hexachloro-1,4,4a,5,8,8a-hexahydro-1,4-endoexo-5,8-dimethanonaphthalene; 1,2,3,4,10-10-Hexachloro-1,4,4a,5,8,8a-hexahydro-1,4,5,8-endoexo-dimethanonaphthalene; Hexachlorohexahydro-endoexo-dimethanonaphthalene; 1,2,3,4,10,10-Hexachloro-1,4,4a,5,8,8a -hexahydro-exo-1,4-endo-5,8-dimethanonaphthalene; HHDN; Octalene; Octalene®; Seedrin; Seedrin®; Toxadrin®

CAS Registry Number: 309-00-2

RTECS Number: IO2100000

DOT ID: UN 2761

EEC Number: 602-048-00-3

Regulatory Authority

- Carcinogen (Probable Human) (USPHS) (See reference below), California, New Jersey, NIOSH (potential occupational carcinogen)
- Banned or Severely Restricted (Many Countries) (UN)[13]
- Air Pollutant Standard Set (ACGIH)[1] (NIOSH)[2] (DFG)[3] (HSE)[33] (former USSR)[43] (Several States)[60] (OSHA)[58] (Several Canadian Provinces) (Australia) (Israel) (Mexico)
- Water Pollution Standard Set (EPA) (Mexico)
- CLEAN WATER ACT: 40CFR116.4 Hazardous Substances; 40CFR117.3 (same as CERCLA); 40CFR423, Appendix A Priority Pollutants; 57FR41331 Priority Chemicals; 40CFR401.15 Toxic Pollutant
- EPA HAZARDOUS WASTE NUMBER (RCRA No.): P004
- RCRA 40CFR261, Appendix 8; 40CFR261.11 Hazardous Constituents
- RCRA Land Ban Waste Restrictions
- RCRA 40CFR268.48; 61FR15654, Universal Treatment Standards: Wastewater (mg/l), 0.021; Nonwastewater (mg/kg), 0.066
- RCRA 40CFR264, Appendix 9; Ground Water Monitoring List: Suggested methods (PQL μg/l): 8080 (0.05); 8270 (10)
- CERCLA/SARA 40CFR302, Extremely Hazardous Substances: Extremely Hazardous Substances: TPQ = 500/10,000 lb (227/4,540 kg).
- SUPERFUND/EPCRA 40CFR302.4, Appendix A, Reportable Quantity (RQ): 1 lb (0.454 kg); SARA 313: Form R *de minimis* Concentration Reporting Level: 1.0%.
- U.S. DOT Regulated Marine Pollutant (49CFR172.101, Appendix B)

Cited in U.S. State Regulations: Alaska (G), California (W), Florida (G), Illinois (G, W), Kansas (G, A, W), Louisiana (G), Maine (G, W), Massachusetts (G), Michigan (G), Minnesota (W), Nevada (A), New Hampshire (G), New Jersey (G), New York (G), North Dakota (A), Oklahoma (G), Pennsylvania (G, A), Vermont (G), Virginia (G, A), Washington (G), West Virginia (G), Wisconsin (G).

Description: Aldrin, $C_{12}H_8Cl_6$, is a colorless, crystalline solid. The technical grade is a tan to dark brown solid. It has a mild, chemical odor. Freezing/Melting point = 104°C (pure); 49 – 60°C (technical grade). Although noncombustible, Aldrin may be dissolved in a flammable liquid. Odor threshold = 0.3 mg/m^3. Hazard Identification (based on NFPA-704 M Rating System): Health 4, Flammability 0, Reactivity 0. Practically insoluble in water. Octanol/water partition coefficient as log Pow: 7.4.

Potential Exposure: Formerly used as an insecticide in the U.S. Pesticide manufacturers, formulators and applicators. Some people in the U.S. may be exposed to aldrin (and dieldrin) in air, water and food because of its persistence in the environment. See dieldrin for more details.

Incompatibilities: Avoid concentrated mineral acids, acid catalysts, acid oxidizing agents, phenol, or active metals.

Permissible Exposure Limits in Air: The OSHA, ACGIH, and NIOSH TWA is 0.25 mg/m^3 as is the DFG MAK (total dust), Australian, HSE (U.K.), Mexico and Israel value, with the notation "skin" indicating potential for cutaneous adsorption. HSE and Mexico suggests a STEL of 0.75 mg/m^3. NIOSH has recommends that Aldrin be held to the lowest feasible concentration (LFC). The NIOSH IDLH level is 25 mg/m^3, with the notation (potential occupational carcinogen).

The former USSR-UNEP/IRPTC project[43] has set a MAC in workplace air of 0.01 mg/m^3. In addition, several states have set guidelines or standards for Aldrin in ambient air:[60] 0.035 μg/m^3 (Pennsylvania) to 0.595 μg/m^3 (Kansas) to 1.5 μg/m^3 (Connecticut) to 2.5 μg/m^3 (North Dakota) to 4.0 μg/m^3 (Virginia) to 6.0 μg/m^3 (Nevada).

Determination in Air: A filter plus bubbler containing isooctane followed by workup with isooctane and analysis by gas chromatography. See NIOSH Method #5502.[18]

Permissible Concentration in Water: To protect freshwater aquatic life-not to exceed 3.0 μg/l at any time. To protect saltwater aquatic life-not to exceed 1.3 μg/l at any time. To protect human health-preferably zero. A limit of 0.00074 μg/l is believed to keep lifetime cancer risk below 10^{-5}.[6] The former USSR-UNEP/IRPTC project[43] has set a MAC of 0.01 mg/l in water used for domestic purposes. Mexico limit in drinking water 0.00003 mg/l. In addition, several states have set guidelines and standards for aldrin in drinking water.[61] Illinois has set a standard of 0.1 μg/l. Guidelines in other states range from 0.013 μg/l (Kansas) to 0.03 μg/l (Minnesota) to 0.05 μg/l (California). Aldrin is highly toxic to aquatic organisms and every care must be taken to avoid release to the environment. In the food chain important to humans, bioaccumulation takes place, specifically in aquatic organisms. It is strongly advised not to let the chemical enter into the environment because of its persistence.

Determination in Water: Gas chromatography (EPA Method 608) or gas chromatography plus mass spectrometry (EPA Method 625).

Routes of Entry: Inhalation, skin absorption, ingestion and eye and skin contact.

Harmful Effects and Symptoms

Short Term Exposure: Eye and skin contact can cause irritation and burns. Can be fatal if swallowed or absorbed through the skin. Aldrin tends to produce convulsions before other, less serious signs of illness have appeared. Victims have reported headache, nausea, vomiting, dizziness, and mild clonic jerking. Some victims have convulsions without warning. Poisoning by Aldrin usually involves convulsions due to its effects on the central nervous system. Probable oral lethal dose for humans is between 7 drops and one ounce for a 150 pound adult human.

Long Term Exposure: May cause tumors, cancer, mutations, reproductive effects. Reproductive effects and liver effects have also been reported. It is classified as an extremely toxic chemical. Conflicting reports of carcinogenicity of this compound remain an area of controversy. Similar chemically and toxicologically to dieldrin.

Points of Attack: Central nervous system, liver, kidneys, skin.

Medical Surveillance: Consider the points of attack in preplacement and periodic physical examinations.

First Aid: If this chemical gets into the eyes, remove any contact lenses at once and irrigate immediately for at least 15 minutes, occasionally lifting upper and lower lids. Seek medical attention immediately. If this chemical contacts the skin, remove contaminated clothing and wash immediately with soap and water. Seek medical attention immediately. If this chemical has been inhaled, remove from exposure, begin rescue breathing (using universal precautions) if breathing has stopped and CPR if heart action has stopped. Transfer promptly to a medical facility. When this chemical has been swallowed, get medical attention. Give a slurry of activated charcoal in water to drink. Do NOT induce vomiting.

Personal Protective Methods: Wear protective gloves and clothing to prevent any reasonable probability of skin contact. Safety equipment suppliers/manufacturers can provide recommendations on the most protective glove/clothing material for your operation. All protective clothing (suits, gloves, footwear, headgear) should be clean, available each day, and put on before work. Contact lenses should not be worn when working with this chemical. Wear dust-proof chemical goggles and face shield unless full facepiece respiratory protection is worn. Employees should wash immediately with soap when skin is wet or contaminated. Provide emergency showers and eyewash.

Respirator Selection: NIOSH: *At any detectable concentration:* SCBAF:PD,PP (any MSHA/NIOSH approved self-contained breathing apparatus that has a full facepiece and is operated in a pressure-demand or other positive-pressure mode); or SAF:PD,PP:ASCBA (any supplied-air respirator that has a full facepiece and is operated in a pressure-demand or other positive-pressure mode in combination with an auxiliary, self-contained breathing apparatus operated in a pressure-demand or other positive pressure mode). *Escape:* GMFOVHiE [any air-purifying, full-facepiece respirator (gas mask) with a chin-style, front- or back-mounted organic vapor canister having a high-efficiency particulate filter]; or SCBAE (any appropriate escape-type, self-contained breathing apparatus).

Storage: Prior to working with this chemical you should be trained on its proper handling and storage. Store in tightly closed containers in a cool, well ventilated area, preferably outdoors.

Shipping: Required shipping label is "Poison." Aldrin falls in DOT/UN Hazard Class 6.1 and Packing Group II.[19][20]

Spill Handling: Evacuate persons not wearing protective equipment from area of spill or leak until clean-up is complete. Remove all ignition sources. Ventilate area of spill or leak. Use water spray to reduce vapor. Absorb liquids in vermiculite, dry sand, earth, or a similar material and deposit in sealed containers. It may be necessary to contain and dispose of this chemical as a hazardous waste. If material or contaminated runoff enters waterways, notify downstream users of potentially contaminated waters. Contact your Department of Environmental Protection or your regional office of the federal EPA for specific recommendations. If employees are required to clean-up spills, they must be properly trained and equipped. OSHA 1910.120(q) may be applicable. For large spill seek professional environmental engineering assistance from the

U.S. EPA Environmental Response Team at (908) 548-8730 (24-hour response line).

Fire Extinguishing: Solid is not combustible, but Aldrin is often dissolved in a flammable liquid. Poisonous gases including hydrogen chloride and chlorine are produced in fire. Move container from fire area if you can do so without risk. Fight fire from maximum distance. Dike fire control water for later disposal; do not scatter the material. Use dry chemical, carbon dioxide, or alcohol foam extinguishers. If material or contaminated runoff enters waterways, notify downstream users of potentially contaminated waters. Notify local health and fire officials and pollution control agencies. From a secure, explosion-proof location, use water spray to cool exposed containers. If cooling streams are ineffective (venting sound increases in volume and pitch, tank discolors, or shows any signs of deforming), withdraw immediately to a secure position. If employees are expected to fight fires, they must be trained and equipped in OSHA 1910.156.

Disposal Method Suggested: Consult with environmental regulatory agencies for guidance on acceptable disposal practices. Generators of waste containing this contaminant (≥100 kg/mo) must conform with EPA regulations governing storage, transportation, treatment, and waste disposal. Aldrin is very stable thermally with no decomposition noted at 250°C. Aldrin (along with the structurally related compounds dieldrin and isodrin) is remarkably stable to alkali (in contrast to chlordane and heptachlor) and refluxing with aqueous or alcoholic caustic has no effect. Incineration methods for aldrin disposal involving 1,500°F, 0.5 second minimum for primary combustion, 3,200°F, 1.0 second for secondary combustion, with adequate scrubbing and ash disposal facilities have been recommended. The combustion of aldrin in polyethylene on a small scale gave more than 99% decomposition. Aldrin can be degraded by active metals such as sodium in alcohol (a reaction which forms the basis of the analytical method for total chlorine), but this method is not suitable for the layman. A disposal method suggested for materials contaminated with aldrin, dieldrin or endrin consists ob boring 8 – 12 feet underground in an isolated area away from water suppliers, with a layer of clay, a layer of lye and a second layer of clay beneath the wastes. In accordance with 40CFR165 recommendations for the disposal of pesticides and pesticide containers. Must be disposed properly by following package label directions or by contacting your state pesticide or environmental control agency or by contacting your regional EPA office.

References

U.S. Environmental Protection Agency, Aldrin/Dieldrin: Ambient Water Quality Criteria, Washington, DC (1979)

U.S. Environmental Protection Agency, Aldrin, Health and Environmental Effects Profile No. 8, Washington, DC, Office of Solid Waste (April 30, 1980)

Sax, N. I., Ed., Dangerous Properties of Industrial Materials Report, 1, No. 5, 31–32 (1981) and 3, No. 5, 25–29 (1983), and 5, No. 2, 23–39 (1988)

U.S. Environmental Protection Agency, "Chemical Profile: Aldrin," Washington, DC, Chemical Emergency Preparedness Program (November 30, 1987)

New York State Department of Health, "Chemical Fact Sheet: Aldrin," Albany, NY, Bureau of Toxic Substance Assessment (January 1986)

U.S. Public Health Service, "Toxicological Profile for Aldrin/Dieldrin," Atlanta, Georgia, Agency for Toxic Substances and Disease Registry (November 1987)

Allethrin

Molecular Formula: $C_{19}H_{26}O_3$

Synonyms: (+)-Allelrethonyl; d-Allethrin; d-*trans* Allethrin; Allethrin I; Alleviate; Allyl cinerin; Allyl homolog of cinerin I; d,l-2-Allyl-4-hydroxy-3-methyl-2-cyclopenten-1-one-d,l-chrysanthemum monocarboxylate; 3-Allyl-4-keto-2-methylcyclopentenyl chrysanthemummonocarboxylate 3; 3-Allyl-2-methyl-4-oxo-2-cyclopenten-1-yl chrysanthemate; dl-3-Allyl-2-methyl-4-oxocyclopent-2-enyl dl-*cis,trans*-chrysanthemate; Allylrethronyl dl-*cis,trans*-chrysanthemate; Bioallethrin; Bioaltrina; (+)-*cis*, *trans*-Chrysanthemate; Cinerin I allyl homolog; Depallethrin; ENT 17,510; Exthrin; FDA 1446; FMC 249; Necarboxylic acid; NIA 249; OMS 468; Pallethrine; Pynamin; Pynamin-Forte; Pyresin; Pyresyn; Synthetic pyrethrins

CAS Registry Number: 584-79-2

RTECS Number: GZ1925000

DOT ID: UN 2902

Cited in U.S. State Regulations: New Jersey (G), Pennsylvania (G).

Description: Allethrin, $C_{19}H_{26}O_3$, is a clear, yellow to amber, oily liquid which is also available as a wettable powder or granules. Boiling point = approx. 160°C. Hazard Identification (based on NFPA-704 M Rating System): Health 1, Flammability 1, Reactivity 0. Insoluble in water.

Potential Exposure: Allethrin is used to control insects in homes and animal shelters, and to treat lice in men.

Incompatibilities: Strong alkalis and oxidizers.

Permissible Exposure Limits in Air: Not established.

Routes of Entry: Skin, inhalation.

Harmful Effects and Symptoms

Short Term Exposure: Skin and eye contact causes irritation and burns. Inhalation can cause respiratory tract irritation with coughing and wheezing. High exposure may cause dizziness, shaking, irritability, seizures, and unconsciousness.

Long Term Exposure: May cause skin allergy. If the allergy develops, very low future exposure can cause itching and skin rash. Allethrin may cause an asthma-like allergy. Future exposure can cause asthma attacks with shortness of breath, wheezing, cough and/or chest tightness. Allethrin can cause bronchitis to develop with cough, phlegm, and/or shortness of breath. This chemical May cause liver and kidney damage. There is some evidence that allethrin may cause mutations.

Points of Attack: Skin, lungs, liver, and kidneys.

Medical Surveillance: Liver and kidney function tests. Lung function test. (These could be normal if the person is not having an attack at the time of the test). Evaluation by a qualified allergist may help to diagnose skin allergy.

First Aid: If this chemical gets into the eyes, remove any contact lenses at once and irrigate immediately for at least 15 minutes, occasionally lifting upper and lower lids. Seek medical attention immediately. If this chemical contacts the skin, remove contaminated clothing and wash immediately with soap and water. Seek medical attention immediately. If this chemical has been inhaled, remove from exposure, begin rescue breathing (using universal precautions) if breathing has stopped and CPR if heart action has stopped. Transfer promptly to a medical facility. When this chemical has been swallowed, get medical attention. Give large quantities of water and induce vomiting. Do not make an unconscious person vomit.

Personal Protective Methods: Wear protective gloves and clothing to prevent any reasonable probability of skin contact. Safety equipment suppliers/manufacturers can provide recommendations on the most protective glove/clothing material for your operation. All protective clothing (suits, gloves, footwear, headgear) should be clean, available each day, and put on before work. Contact lenses should not be worn when working with this chemical. Wear splash-proof chemical goggles and face shield unless full facepiece respiratory protection is worn. Employees should wash immediately with soap when skin is wet or contaminated. Remove nonimpervious clothing immediately if wet or contaminated. Provide emergency showers and eyewash.

Respirator Selection: *At any concentrations above the NIOSH REL, or where there is no REL, at any detectable concentration:* SCBAF:PD,PP (any MSHA/NIOSH approved self-contained breathing apparatus that has a full facepiece and is operated in a pressure-demand or other positive-pressure mode); or SAF:PD,PP:ASCBA (any supplied-air respirator that has a full facepiece and is operated in a pressure-demand or other positive-pressure mode in combination with an auxiliary, self-contained breathing apparatus operated in a pressure-demand or other positive pressure mode).

Storage: Prior to working with this chemical you should be trained on its proper handling and storage. Store in tightly closed containers in a cool, well ventilated area.

Shipping: Allethrin is a liquid toxic pesticide and should be labeled "Poison." It falls into DOT/UN Hazard Class 6.1.

Spill Handling: Evacuate persons not wearing protective equipment from area of spill or leak until clean-up is complete. Remove all ignition sources. Ventilate area of spill or leak. Use water spray to reduce vapor. Absorb liquids in vermiculite, dry sand, earth, or a similar material and deposit in sealed containers. It may be necessary to contain and dispose of this chemical as a hazardous waste. If material or contaminated runoff enters waterways, notify downstream users of potentially contaminated waters. Contact your Department of Environmental Protection or your regional office of the federal EPA for specific recommendations. If employees are required to clean-up spills, they must be properly trained and equipped. OSHA 1910.120(q) may be applicable.

Fire Extinguishing: Irritating gases are produced in fire. Use dry chemical, carbon dioxide, or alcohol foam extinguishers. Vapors are heavier than air and will collect in low areas. Vapors may travel long distances to ignition sources and flashback. Vapors in confined areas may explode when exposed to fire. Storage containers and parts of containers may rocket great distances, in many directions. If material or contaminated runoff enters waterways, notify downstream users of potentially contaminated waters. Notify local health and fire officials and pollution control agencies. From a secure, explosion-proof location, use water spray to cool exposed containers. If cooling streams are ineffective (venting sound increases in volume and pitch, tank discolors, or shows any signs of deforming), withdraw immediately to a secure position. If employees are expected to fight fires, they must be trained and equipped in OSHA 1910.156.

Disposal Method Suggested: Incineration. In accordance with 40CFR165 recommendations for the disposal of pesticides and pesticide containers.

References

New Jersey Department of Health and Senior Services, Hazardous Substance Fact Sheet, Allethrin, Trenton, NJ (November, 1998)

Allyl Acetate

Molecular Formula: $C_5H_8O_2$

Synonyms: Acetic acid, allyl acetate; Acetic acid, 2-propenyl ester; 3-Acetoxypropene; 2-Propenyl methanoate

CAS Registry Number: 591-87-7

RTECS Number: AF1750000

DOT ID: UN 2333

Regulatory Authority

- Canada, WHMIS, Ingredients Disclosure List

Cited in U.S. State Regulations: Florida (G), Massachusetts (G), New Jersey (G), Pennsylvania (G).

Description: Allyl acetate is a flammable, colorless liquid with an acrid odor. Boiling point = 103 – 104°C. Flash point = 22°C. Autoignition temperature = 374°C. Hazard Identification (based on NFPA-704 M Rating System): Health 1, Flammability 3, Reactivity 0. Insoluble in water.

Potential Exposure: Allyl Acetate is used to control insects in homes and animal shelters, and to treat lice in men.

Incompatibilities: Contact with oxidizers may cause fire and explosions.

Permissible Exposure Limits in Air: Not established.

Routes of Entry: Skin, inhalation.

Harmful Effects and Symptoms

Short Term Exposure: This chemical can be absorbed through the skin, thereby increasing exposure. Prolonged or repeated contact can cause rash, redness, and itching. Skin and eye contact causes irritation, severe burns, and permanent damage. Inhalation can cause respiratory tract irritation with coughing and wheezing. High exposure may cause pulmonary edema, a medical emergency, with severe shortness of breath. This can cause death.

Long Term Exposure: Allyl acetate can cause bronchitis to develop with cough, phlegm, and/or shortness of breath. This chemical May cause liver and kidney damage.

Points of Attack: Skin, lungs.

Medical Surveillance: If symptoms develop or overexposure is suspected, chest x-ray should be considered.

First Aid: If this chemical gets into the eyes, remove any contact lenses at once and irrigate immediately for at least 15 minutes, occasionally lifting upper and lower lids. Seek medical attention immediately. If this chemical contacts the skin, remove contaminated clothing and wash immediately with soap and water. Seek medical attention immediately. If this chemical has been inhaled, remove from exposure, begin rescue breathing (using universal precautions) if breathing has stopped and CPR if heart action has stopped. Transfer promptly to a medical facility. When this chemical has been swallowed, get medical attention. Give large quantities of water and induce vomiting. Do not make an unconscious person vomit. Medical observation is recommended for 24 – 48 hours after breathing overexposure, as pulmonary edema may be delayed.

Personal Protective Methods: Wear protective gloves and clothing to prevent any reasonable probability of skin contact. Safety equipment suppliers/manufacturers can provide recommendations on the most protective glove/clothing material for your operation. All protective clothing (suits, gloves, footwear, headgear) should be clean, available each day, and put on before work. Contact lenses should not be worn when working with this chemical. Wear splash-proof chemical goggles and face shield unless full facepiece respiratory protection is worn. Employees should wash immediately with soap when skin is wet or contaminated. Remove nonimpervious clothing immediately if wet or contaminated. Provide emergency showers and eyewash.

Respirator Selection: *At any concentrations above the NIOSH REL, or where there is no REL, at any detectable concentration:* SCBAF:PD,PP (any MSHA/NIOSH approved self-contained breathing apparatus that has a full facepiece and is operated in a pressure-demand or other positive-pressure mode); or SAF:PD,PP:ASCBA (any supplied-air respirator that has a full facepiece and is operated in a pressure-demand or other positive-pressure mode in combination with an auxiliary, self-contained breathing apparatus operated in a pressure-demand or other positive pressure mode).

Storage: Prior to working with this chemical you should be trained on its proper handling and storage. Before entering confined space where allyl acetate may be present, check to make sure that an explosive concentration does not exist. Store in tightly closed containers in a cool, well ventilated area. Metal containers involving the transfer of this chemical should be grounded and bonded. Where possible, automatically pump liquid from drums or other storage containers to process containers. Drums must be equipped with self-closing valves, pressure vacuum bungs, and flame arresters. Use only non-sparking tools and equipment, especially when opening and closing containers of this chemical. Sources of ignition such as smoking and open flames, are prohibited where this chemical is used, handled, or stored in a manner that could create a potential fire or explosion hazard.

Shipping: Allyl acetate labeled "Flammable liquid, Poison." It falls into DOT/UN Hazard Class 3.

Spill Handling: Evacuate persons not wearing protective equipment from area of spill or leak until clean-up is complete. Remove all ignition sources. Ventilate area of spill or leak. Use water spray to reduce vapor. Absorb liquids in vermiculite, dry sand, earth, or a similar material and deposit in sealed containers. It may be necessary to contain and dispose of this chemical as a hazardous waste. If material or contaminated runoff enters waterways, notify downstream users of potentially contaminated waters. Contact your Department of Environmental Protection or your regional office of the federal EPA for specific recommendations. If employees are required to clean-up spills, they must be properly trained and equipped. OSHA 1910.120(q) may be applicable.

Fire Extinguishing: Allyl acetate is a highly flammable liquid. Irritating gases are produced in fire. Use propylene foam extinguishers. Water may be ineffective. Vapors are heavier than air and will collect in low areas. Vapors may travel long distances to ignition sources and flashback. Vapors in confined areas may explode when exposed to fire. Storage containers and parts of containers may rocket great distances, in many directions. If material or contaminated runoff enters waterways, notify downstream users of potentially contaminated waters. Notify local health and fire officials and pollution control agencies. From a secure, explosion-proof location, use water spray to cool exposed containers. If cooling streams are ineffective (venting sound increases in volume and pitch, tank discolors, or shows any signs of deforming), withdraw immediately to a secure position. If employees are expected to fight fires, they must be trained and equipped in OSHA 1910.156.

Disposal Method Suggested: Incineration.

References

New Jersey Department of Health and Senior Services, Hazardous Substance Fact Sheet, Allyl Acetate, Trenton, NJ (September, 1998)

Allyl Alcohol

Molecular Formula: C_3H_6O

Common Formula: CH_2CHCH_2OH

Synonyms: AA; Alcool allilco (Italian); Alcool allylique (French); Alilico alcohol (Spanish); Allilowy alkohol (Polish); Allyl al; Allyl alcohol; Allylalkohol (German); Allylic alcohol; 3-Hydroxypropene; Orvinylcarbinol; 1-Propen-3-ol; 2-Propen-1-ol; 2-Propenol; Propenol; Propen-1-ol-3; 2-Propenyl alcohol; Propenyl alcohol; Shell Unkrautted A; Vinyl carbinol; Vinyl carbinol,2-propenol

CAS Registry Number: 107-18-6

RTECS Number: BA5075000

DOT ID: UN 1098

EEC Number: 603-015-00-6

Regulatory Authority

- Toxic Substance (World Bank)[15]
- Air Pollutant Standard Set (ACGIH)[1] (DFG)[3] (HSE)[33] (OSHA)[58] (Several States),[60] (Various Canadian Provinces) (Australia) (Israel) (Mexico)
- Clean Air Act 42USC7412; Title I, Part A, §112(r), Accidental Release Prevention/Flammable substances (Section 68.130) TQ = 15,000 lb (5,825 kg)
- Clean Water Act,40CFR116.4 Hazardous Substances; RQ 40CFR117.3 (same as CERCLA)
- EPA Hazardous Waste Number (RCRA No.): P005
- RCRA Land Ban Waste Restrictions
- CERCLA/SARA 40CFR302, Extremely Hazardous Substances: TPQ = 1,000 lb (454 kg)
- SUPERFUND/EPCRA 40CFR302.4, Appendix A, Reportable Quantity (RQ): 100 lb (45.5 kg), SARA 313: Form R *de minimis* Concentration Reporting Level: 1.0%.
- Canada, WHMIS, Ingredients Disclosure List; National pollutant Release Inventory (NPRI)

Cited in U.S. State Regulations: Alaska (G), California (G), Connecticut (A), Florida (G), Illinois (G), Kansas (G), Louisiana (G), Maine (G), Massachusetts (G), Minnesota (G), Nevada (A), New Hampshire (G), New Jersey (G), North Dakota (A), Oklahoma (G), Pennsylvania (G), Rhode Island (G), Vermont (G), Virginia (G, A), Washington (G), West Virginia (G), Wisconsin (G).

Description: Allyl alcohol, CH_2CHCH_2OH, is a severe health hazard that is a flammable, colorless liquid with a pungent mustard-like odor. Odor threshold = 1.4 – 2.1 ppm. The odor and irritant properties of allyl alcohol should be sufficient warning to prevent serous injury. Boiling point = 97°C. Flash point = 21°C (cc); 32°C (oc). Autoignition temperature = 378°C. Explosive limits: LEL = 2.5%, UEL = 18.0%. Hazard Identification (based on NFPA-704 M Rating System): Health 4, Flammability 3, Reactivity 1. Soluble in water.

Potential Exposure: Allyl alcohol is used in the production of allyl esters. These compounds are used as monomers and prepolymers in the manufacture of resins and plastics. Allyl alcohol is also used in the preparation of pharmaceuticals, in organic syntheses of glycerol and acrolein and as a fungicide and herbicide.

Incompatibilities: Forms explosive mixture with air. Reacts explosively with carbon tetrachloride, strong bases. Also incompatible with strong acids. Contact with oxidizers may cause fire and explosions. Polymerization may be caused by heat, peroxides, or oxidizers.

Permissible Exposure Limits in Air: The Federal OSHA standard (TWA),[58] ACGIH, NIOSH, DFG MAK, Australian, Mexico, Israel value[3] is 2 ppm (5 mg/m^3). ACGIH and NIOSH add the notation "skin" indicating potential for cutaneous absorption. The NIOSH, ACGIH, HSE, Israel, and Mexico STEL is 4 ppm (10 mg/m^3). The DFG Peak Limitation (30 min.) is 2 times normal MAK; do not exceed more than 4 times during a workshift. The Canadian Provincial TWA limits are: 2 ppm (4.7 – 5.0 mg/m^3) TWA or TWAEV and STEL is 4 ppm (9.5 – 10 mg/m^3) [Alberta, British Columbia, Ontario, Quebec]. The NIOSH IDLH value is 20 ppm. Guidelines or standards for allyl alcohol in ambient air have been set[60] by various states: 5 mg/m^3/STEL 10 mg/m^3 (California), 5 mg/m^3 (North Dakota); 8 mg/m^3 (Virginia); 10 mg/m^3 (Connecticut); 11.9 mg/m^3 (Nevada).

Determination in Air: Adsorption on charcoal, workup with CS_2 and gas chromatographic analysis. See NIOSH Method #1402.[18]

Permissible Concentration in Water: No criteria set.

Routes of Entry: Inhalation, ingestion, eye and/or skin contact. Absorbed through the skin.

Harmful Effects and Symptoms

Short Term Exposure: Allyl alcohol vapor can cause serious irritation and burns of eyes, nose and throat. Eye irritation may be accompanied by sensitivity to light, pain, blurred vision leading to permanent damage. The pain may not begin until 6 hours after exposure. Contact with the liquid may cause first and second degree burns of skin and blister formation. Areas of contact will become swollen and painful and local muscle spasms may occur. Allyl alcohol causes burns on contact, and may cause pulmonary edema, a medical emergency, if inhaled. It is poisonous in small quantities. The probable oral lethal dose is 50 – 500 mg/kg, or between 1 tea-spoonful and 1 ounce for a 150 lb person.

Long Term Exposure: Allyl alcohol may cause mutations; such chemicals may have a cancer or reproductive risk. This chemical may cause liver and kidney damage. Repeated exposure may cause bronchitis with cough, phlegm, and/or shortness of breath.

Points of Attack: Eyes, skin, respiratory system.

Medical Surveillance: Preplacement and periodic examinations should include lung function tests, liver and kidney function tests. Following acute exposure, chest x-ray should be considered.

First Aid: If this chemical gets into the eyes, remove any contact lenses at once and irrigate immediately for at least 15 minutes, occasionally lifting upper and lower lids. Seek medical attention immediately. If this chemical contacts the skin, remove contaminated clothing and wash immediately with soap and water. Seek medical attention immediately. If this chemical has been inhaled, remove from exposure, begin rescue breathing (using universal precautions) if breathing has stopped and CPR if heart action has stopped. Transfer promptly to a medical facility. When this chemical has been swallowed, get medical attention. Give large quantities of water and induce vomiting. Do not make an unconscious person vomit. Medical observation is recommended for 24 – 48 hours after breathing overexposure, as pulmonary edema may be delayed.

Personal Protective Methods: Wear solvent-resistant gloves and clothing to prevent any reasonable probability of skin contact. Safety equipment manufacturers recommend neoprene, teflon, and viton/neoprene as a protective material. Contact lenses should not be worn when working with this chemical. Wear splash-proof chemical goggles and face shield unless full facepiece respiratory protection is worn. Employees should wash immediately with soap when skin is wet or contaminated. Remove nonimpervious clothing immediately if wet or contaminated. Provide emergency showers and eyewash.

Respirator Selection: NIOSH: *20 ppm:* SA:CF* (any supplied-air respirator operated in a continuous-flow mode); or PAPROV* [any powered, air-purifying respirator with organic vapor cartridge(s)]; or CCRFOV [Any air-purifying, full-facepiece respirator (gas mask) with a chin-style, front-or back-mounted acid gas canister]; or GMFOV [Any air-purifying, full-facepiece respirator (gas mask) with a chin-style, front-or back-,mounted organic vapor canister]; or SCBAF (any self-contained breathing apparatus with a full facepiece); or SAF (any supplied-air respirator with a full facepiece); or SAF (any supplied-air respirator with a full facepiece). *Emergency or planned entry into unknown concentrations or IDLH conditions:* SCBAF: PD,PP (any self-contained breathing apparatus that has a full facepiece and is operated in a pressure-demand or other positive-pressure mode); or SAF:PD,PP:ASCBA (Any supplied-air respirator that has a full facepiece and is operated in a pressure-demand or other positive-pressure mode in combination with an auxiliary self-contained breathing apparatus operated in a pressure-demand or other positive-pressure mode. *Escape:* GMFOV [Any air-purifying, full-facepiece respirator (gas mask) with a chin-style, front-or back-,mounted organic vapor canister]; or SCBAE (Any appropriate escape-type, self-contained breathing apparatus).

* Substance reported to cause eye irritation or damage; may require eye protection.

Storage: Prior to working with this chemical you should be trained on its proper handling and storage. Before entering confined space where this chemical may be present, check to make sure that an explosive concentration does not exist. Store in tightly-closed containers in a cool, well-ventilated area away from heat. Store to avoid contact with strong oxidizers such as halogens. Avoid smoking, sparks or open flames in storage areas.

Shipping: Label required is "Flammable Liquid, Poison," Shipment by passenger aircraft, railcar, or cargo aircraft is forbidden. The DOT/UN Hazard Class is 6.1 and the Packing Group is I.[20][21]

Spill Handling: Evacuate and restrict persons not wearing protective equipment from area of spill or leak until cleanup is complete. Remove all ignition sources. Establish ventilation to keep levels below explosive limit. Absorb liquids in vermiculite, dry sand, earth, or a similar material and deposit in sealed containers. Keep allyl alcohol out of a confined space, such as a sewer, because of the potential for an explosion, unless the sewer is designed to prevent the build-up of explosive concentrations. It may be necessary to contain and dispose of this chemical as a hazardous waste. If material or contaminated runoff enters waterways, notify downstream users of potentially contaminated waters. Contact your Department of Environmental Protection or your regional office of the federal EPA for specific recommendations. If employees are required to clean-up spills, they must be properly trained and equipped. OSHA 1910.120(q) may be applicable.

Initial isolation and protective action distances

Small spills (From a small package or a small leak from a large package)

First: Isolate in all directions (feet)..... 200

Then: Protect persons downwind (miles)

Day 0.1

Night 0.3

Large spills (From a large package or from many small packages)

First: Isolate in all directions (feet)..... 500

Then: Protect persons downwind (miles)

Day 0.3

Night 1.2

Fire Extinguishing: This chemical is a flammable liquid. Firefighting gear (including SCBA) does not provide adequate protection. If exposure occurs, remove and isolate gear immediately and thoroughly decontaminate personnel. Poisonous gases including nitrogen oxides are produced in fire. Use dry chemical, carbon dioxide, or alcohol foam extinguishers. Vapors are heavier than air and will collect in low areas. Vapors may travel long distances to ignition sources and flashback. Vapors in confined areas may explode when exposed to fire. Storage containers and parts of containers may rocket great distances, in many directions. If material or contaminated runoff enters waterways, notify downstream users of potentially contaminated waters. Notify local health and fire officials and pollution control agencies. From a secure, explosion-proof location, use water spray to cool exposed containers. If cooling streams are ineffective (venting sound increases in volume and pitch, tank discolors, or shows any signs of deforming), withdraw immediately to a secure position. If employees are expected to fight fires, they must be trained and equipped in OSHA 1910.156.

Disposal Method Suggested: Consult with environmental regulatory agencies for guidance on acceptable disposal practices. Generators of waste containing this contaminant (≥100 kg/mo) must conform with EPA regulations governing storage, transportation, treatment, and waste disposal. Incineration after dilution with a flammable solvent.

References

U.S. Environmental Protection Agency, Allyl Alcohol, Health and Environmental Effects Profile No. 9, Washington, DC, Office of Solid Waste (April 30, 1980)

Sax, N. I., Ed., Dangerous Properties of Industrial Materials Report, 1, No. 7, 29–31 (1981)

U.S. Environmental Protection Agency, "Chemical Profile: Allyl Alcohol," Washington, DC, Chemical Emergency Preparedness Program (November 30, 1987)

New Jersey Department of Health and Senior Services, "Hazardous Substance Fact Sheet: Allyl Alcohol," Trenton, NJ (June 1998)

Allylamine

Molecular Formula: C_3H_7N

Common Formula: $CH_2=CHCH_2NH_2$

Synonyms: Alilamina (Spanish); 3-Aminopropene; 3-Aminopropylene; Monoallylamine; 2-Propen-1-amine; 2-Propenamine; 2-Propenylamine

CAS Registry Number: 107-11-9

RTECS Number: BA5425000

DOT ID: UN 2334

Regulatory Authority

- U.S. DOT 49CFR172.101, Inhalation Hazardous Chemical
- Toxic Substance (World Bank)[15]
- Air Pollutant Standard Set (former USSR)[43]
- OSHA 29CFR1910.119, Appendix A, Process Safety List of Highly Hazardous Chemicals, TQ = 10,000 lb (4,540 kg)
- Clean Air Act 42USC7412; Title I, Part A, §112(r), Accidental Release Prevention/Flammable substances (Section 68.130) TQ =10,000 lb (4,540 kg)
- CERCLA/SARA 40CFR302 Extremely Hazardous Substances: TPQ = 500 lb (227 kg)[7]
- SUPERFUND/EPCRA 40CFR302.4, Appendix A, Reportable Quantity (RQ): EHS lb (0.454 kg), SARA 313: Form R *de minimis* Concentration Reporting Level: 1.0%.
- Canada, WHMIS, Ingredients Disclosure List

Cited in U.S. State Regulations: Florida (G), Massachusetts (G), New Hampshire (G), New Jersey (G), Pennsylvania (G), Rhode Island (G).

Description: Allylamine, $CH_2=CHCH_2NH_2$, a highly flammable, colorless liquid with a strong ammonia odor. Freezing/Melting point = -88°C. Boiling point = 56° to 58°C. Flash point = -29°C. Autoignition temperature = 374°C. Hazard Identification (based on NFPA-704 M Rating System): Health 4, Flammability 3, Reactivity 1. Explosive limits: LEL = 2.2%, UEL = 22.0%. Odor threshold = 2.5 ppm. Soluble in water.

Potential Exposure: Used in manufacture of pharmaceuticals (mercurial diuretics, e.g.) and in organic synthesis. Used to improve dyeability of acrylic fibers.

Incompatibilities: Forms explosive mixture with air. Oxidizing materials and acids may cause a violent reaction. Attacks copper and corrodes active metals (i.e. aluminum, zinc, etc).

Permissible Exposure Limits in Air: The former USSR-UNEP/IRPTC project[43] has set a MAC in workplace air of 0.5 mg/m^3. See also Regulatory Authority, above.

Permissible Concentration in Water: No criteria set.

Routes of Entry: Inhalation, ingestion and skin contact.

Harmful Effects and Symptoms

Short Term Exposure: Corrosive to eyes, skin, and respiratory tract which is highly toxic if inhaled or ingested and moderately toxic if absorbed on skin. Ingestion or inhalation may cause death or permanent injury after very short exposure to small quantities. Inhalation may cause pulmonary edema, which can be delayed for several hours; there is a risk of death in serious cases. Swallowing liquid may cause pneumonia. Vapors are extremely unpleasant and may ensure voluntary avoidance of dangerous concentrations. Symptoms include irritation of nose, eyes, and mouth with tearing, runny nose, and sneezing. Can cause excitement, convulsions, and death. Skin absorption may cause irreversible and reversible changes. Toxic air concentration in humans is 5 ppm over 5 minutes.

Long Term Exposure: Mutation data reported.

Points of Attack: Skin, pulmonary, system, cardiovascular system.

Medical Surveillance: Lung function tests. Following acute exposure, consider chest x-ray.

First Aid: If this chemical gets into the eyes, remove any contact lenses at once and irrigate immediately for at least 15 minutes, occasionally lifting upper and lower lids. Seek medical attention immediately. If this chemical contacts the skin, remove contaminated clothing and wash immediately with soap and water. Seek medical attention immediately. If this chemical has been inhaled, remove from exposure, begin rescue breathing (using universal precautions) if breathing has stopped and CPR if heart action has stopped. Transfer promptly to a medical facility. When this chemical has been swallowed, get medical attention. Give large quantities of water and induce vomiting. Do not make an unconscious person vomit. Medical observation is recommended for 24 – 48 hours after breathing overexposure, as pulmonary edema may be delayed. As first aid for pulmonary edema, a doctor or authorized paramedic may consider administering a corticosteroid spray.

Personal Protective Methods: Wear solvent-resistant gloves and clothing to prevent any reasonable probability of skin contact. Safety equipment suppliers/manufacturers recommend butyl rubber as a protective material. All protective clothing (suits, gloves, footwear, headgear) should be clean, available each day, and put on before work. Contact lenses should not be worn when working with this chemical. Wear splash-proof chemical goggles and face shield unless full facepiece respiratory protection is worn. Employees should wash immediately with soap when skin is wet or contaminated. Remove nonimpervious clothing immediately if wet or contaminated. Provide emergency showers and eyewash.

Respirator Selection: *At any concentrations above the NIOSH REL, or where there is no REL, at any detectable concentration:* SCBAF:PD,PP (any MSHA/NIOSH approved self-contained breathing apparatus that has a full facepiece and is operated in a pressure-demand or other positive-pressure mode); or SAF:PD,PP:ASCBA (any supplied-air respirator that has a full facepiece and is operated in a pressure-demand or other positive-pressure mode in combination with an auxiliary, self-contained breathing apparatus operated in a pressure-demand or other positive pressure mode).

Storage: Prior to working with this chemical you should be trained on its proper handling and storage. Before entering confined space where allylamine may be present, check to make sure that an explosive concentration does not exist. Outside or detached storage preferred. Keep containers closed and away from oxidizers, acids, or combustible materials. If stored indoors, use standard flammable liquids storage room or cabinet. Store in tightly closed containers in a cool, well ventilated area. Metal containers involving the transfer of this chemical should be grounded and bonded. Where possible, automatically pump liquid from drums or other storage containers to process containers. Drums must be equipped with self-closing valves, pressure vacuum bungs, and flame arresters. Use only non-sparking tools and equipment, especially when opening and closing containers of this chemical. Sources of ignition such as smoking and open flames, are prohibited where this chemical is used, handled, or stored in a manner that could create a potential fire or explosion hazard.

Shipping: Label required is "Flammable Liquid, Poison." Shipment by passenger aircraft or railcar or even by cargo aircraft is forbidden. The DON/UN Hazard Class is 3 and the Packing Group is I.[20][21]

Spill Handling: Evacuate and restrict persons not wearing protective equipment from area of spill or leak until cleanup is complete. Remove all ignition sources. Establish ventilation to keep levels below explosive limit. Collect powdered material in the most convenient and safe manner and deposit in sealed containers. Ventilate area of spill or leak after cleanup is complete. Keep allylamine out of a confined space, such as a sewer, because of the potential for an explosion, unless the sewer is designed to prevent the build-up of explosive concentrations. It may be necessary to contain and dispose of this chemical as a hazardous waste. If material or contaminated runoff enters waterways, notify downstream users of potentially contaminated waters. Contact your Department of Environmental Protection or your regional office of the federal EPA for specific recommendations. If employees are required to clean-up spills, they must be properly trained and equipped. OSHA 1910.120(q) may be applicable.

Initial isolation and protective action distances

Distances shown are likely to be affected during the first 30 minutes after materials are spilled and could increase with time. If more than one tank car, cargo tank, portable tank, or large cylinder is involved in the incident is leaking, the protective action distance may need to be increased. You may need to seek emergency information from CHEMTREC at (800) 424-9300 or seek professional environmental engineering assistance from the U.S. EPA Environmental Response Team at (908) 548-8730 (24-hour response line).

Small spills (From a small package or a small leak from a large package)

First: Isolate in all directions (feet)..... 200

Then: Protect persons downwind (miles)

Day 0.1

Night 0.5

Large spills (From a large package or from many small packages)

First: Isolate in all directions (feet)..... 600

Then: Protect persons downwind (miles)

Day 0.4

Night 1.7

Fire Extinguishing: This chemical is a combustible liquid. Poisonous gases including nitrogen oxides are produced in fire. Use dry chemical, carbon dioxide, or alcohol foam extinguishers. Vapors are heavier than air and will collect in low areas. Vapors may travel long distances to ignition sources and flashback. Vapors in confined areas may explode when exposed to fire. Storage containers and parts of containers may rocket great distances, in many directions. If material or contaminated runoff enters waterways, notify downstream users of potentially contaminated waters. Notify local health and fire officials and pollution control agencies. From a secure, explosion-proof location, use water spray to cool exposed containers. If cooling streams are ineffective (venting sound increases in volume and pitch, tank discolors, or shows any signs of deforming), withdraw immediately to a secure position. If employees are expected to fight fires, they must be trained and equipped in OSHA 1910.156.

Disposal Method Suggested: High temperature incineration; encapsulation by resin or silicate fixation.

References

Sax, N. I., Ed., Dangerous Properties of Industrial materials Report, 2, No. 6, 28–30 (1982)

U.S. Environmental Protection Agency, "Chemical Profile: Allylamine," Washington DC, Chemical Emergency Preparedness Program (November 30, 1987)

Allyl Bromide

Molecular Formula: C_3H_5Br

Synonyms: Bromallylene; 1-Bromo-2-propene; 3-Bromopropeno (Spanish); 3-Bromopropylene; Bromuro de alilo (Spanish); 1-Propene, 3-bromo-

CAS Registry Number: 106-95-6

RTECS Number: UC7090000

DOT ID: UN 1099

Regulatory Authority

- Air and Water Pollutant Standard Set: see Bromine
- U.S. DOT Regulated Marine Pollutant (49CFR172.101, Appendix B)
- Canada, WHMIS, Ingredients Disclosure List

Cited in U.S. State Regulations: Florida (G), Massachusetts (G), New Jersey (G), Pennsylvania (G)

Description: Allyl bromide is a highly flammable, colorless to light yellow liquid with an unpleasant, pungent odor. Boiling point = 71.3°C. Flash point = -2°C. Autoignition temperature = 295°C. Explosive limits: LEL = 4.4%, UEL = 7.3%. Hazard Identification (based on NFPA-704 M Rating System): Health 3, Flammability 3, Reactivity 1. Slightly soluble in water.

Potential Exposure: Used as an insecticide, an in the manufacture of resins, fragrances, and other chemicals.

Incompatibilities: Forms explosive mixture with air. Contact with oxidizers may cause fire and explosions. Heat or light exposure may cause decomposition and corrosive vapors.

Permissible Exposure Limits in Air: The Federal OSHA standard for *bromine* is 0.1 ppm (TWA, 8-hour workshift). NIOSH recommends an airborne exposure limit of 0.1 ppm (TWA, 10-hour workshift), with an STEL of 0.3 ppm, not to be exceeded during any 15 minute work period. ACGIH recommends 0.1 ppm (TWA, 8-hour workshift), and STEL of 0.2 ppm.

Note: It should be recognized that allyl bromide can be absorbed through the skin, thereby increasing exposure. The NIOSH IDLH for *bromine* is 3 ppm.

Routes of Entry: Skin, inhalation.

Harmful Effects and Symptoms

Short Term Exposure: Poisonous. This chemical can be absorbed through the skin, thereby increasing exposure. Irritates eyes, skin, and respiratory tract, and can cause burns and permanent damage. Inhalation can cause respiratory tract irritation with coughing and wheezing. Exposure can cause headache, dizziness, nausea, and vomiting. High exposure may cause pulmonary edema, a medical emergency, with severe shortness of breath. This can cause death.

Long Term Exposure: Allyl bromide can cause bronchitis to develop with cough, phlegm, and/or shortness of breath. This chemical May cause liver and kidney damage, and mutations. May cause skin disorders.

Points of Attack: Skin, lungs.

Medical Surveillance: If symptoms develop or overexposure is suspected, chest x-ray should be considered.

First Aid: If this chemical gets into the eyes, remove any contact lenses at once and irrigate immediately for at least 15 minutes, occasionally lifting upper and lower lids. Seek medical attention immediately. If this chemical contacts the skin, remove contaminated clothing and wash immediately with soap and water. Seek medical attention immediately. If this chemical has been inhaled, remove from exposure, begin rescue breathing (using universal precautions) if breathing has stopped and CPR if heart action has stopped. Transfer promptly to a medical facility. When this chemical has been swallowed, get medical attention. Give large quantities of water and induce vomiting. Do not make an unconscious person vomit. Medical observation is recommended for 24 – 48 hours after breathing overexposure, as pulmonary edema may be delayed. As first aid for pulmonary edema, a doctor or authorized paramedic may consider administering a corticosteroid spray.

Personal Protective Methods: Wear protective gloves and clothing to prevent any reasonable probability of skin contact. Safety equipment suppliers/manufacturers can provide recommendations on the most protective glove/clothing material for your operation. All protective clothing (suits, gloves, footwear, headgear) should be clean, available each day, and put on before work. Contact lenses should not be worn when working with this chemical. Wear splash-proof chemical goggles and face shield unless full facepiece respiratory protection is worn. Employees should wash immediately with soap when skin is wet or contaminated. Remove nonimpervious clothing immediately if wet or contaminated. Provide emergency showers and eyewash.

Respirator Selection: For bromine, OSHA/NIOSH: *2.5 ppm:* SA:CF (any supplied-air respirator operated in a continuous-flow mode); or PAPRS (any powered, air-purifying respirator with cartridge(s) providing protection against the compound of concern. *3 ppm:* CCRFS (any chemical cartridge respirator with a full facepiece and cartridge(s) providing protection against the compound of concern) organic vapor and acid gas cartridge(s); or GMFS [any air-purifying, full-facepiece respirator (gas mask) with a chin-style, front- or back-mounted canister providing protection against the compound of concern]; or PAPRTS (any powered, air-purifying respirator with a tight-fitting facepiece and cartridge(s) providing protection against the compound of concern); or SCBAF (any self-contained breathing apparatus with a full facepiece); or SAF (any supplied-air respirator with a full facepiece). *Emergency or planned entry into unknown concentrations or IDLH conditions:* SCBAF:PD,PP (any self-contained breathing apparatus that has a full facepiece and is operated in a pressure-demand or other positive-pressure mode); or SAF:PD,PP:ASCBA (any supplied-air respirator that has a full facepiece and is operated in a pressure-demand or other positive-pressure mode in combination with an auxiliary, self-contained breathing apparatus operated in a pressure-demand or other positive-pressure mode. *Escape:* GMFS [any air-purifying, full-facepiece respirator (gas mask) with a chin-style, front- or back-mounted canister providing protection against the compound of concern]; or SCBAE (Any appropriate escape-type, self-contained breathing apparatus).

Note: Substance causes eye irritation or damage; eye protection needed. Only nonoxidizable sorbents are allowed (not charcoal).

Storage: Prior to working with this chemical you should be trained on its proper handling and storage. Before entering confined space where allyl bromide may be present, check to make sure that an explosive concentration does not exist. Store in tightly closed containers in a cool, well ventilated area. Metal containers involving the transfer of this chemical should be grounded and bonded. Where possible, automatically pump liquid from drums or other storage containers to process containers. Drums must be equipped with self-closing valves, pressure vacuum bungs, and flame arresters. Use only non-sparking tools and equipment, especially when opening and closing containers of this chemical. Sources of ignition such as smoking and open flames, are prohibited where this chemical is used, handled, or stored in a manner that could create a potential fire or explosion hazard.

Shipping: Allyl bromide should be labeled "Flammable liquid, Poison." It falls into DOT/UN Hazard Class 3.

Spill Handling: Evacuate persons not wearing protective equipment from area of spill or leak until clean-up is complete. Remove all ignition sources. Establish forced ventilation to keep levels below explosive limit. Cover with activated charcoal adsorbent and deposit in sealed containers. Keep allyl bromide out of a confined space, such as a sewer, because of the potential for an explosion, unless the sewer is designed to prevent the build-up of explosive concentrations. It may be necessary to contain and dispose of this chemical as a hazardous waste. If material or contaminated runoff enters waterways, notify downstream users of potentially contaminated waters. Contact your Department of Environmental Protection or your regional office of the federal EPA for specific recommendations. If employees are required to clean-up spills, they must be properly trained and equipped. OSHA 1910.120(q) may be applicable.

Fire Extinguishing: Allyl bromide a highly flammable liquid. Irritating gases, including carbon monoxide and hydrogen bromide gas are produced in fire. Use dry chemical, carbon dioxide, water spray, or alcohol foam extinguishers. Vapors are heavier than air and will collect in low areas. Vapors may travel long distances to ignition sources and flashback. Vapors in confined areas may explode when exposed to fire. Storage containers and parts of containers may rocket great distances, in many directions. If material or contaminated runoff enters waterways, notify downstream users of potentially contaminated waters. Notify local health and fire officials and pollution control agencies. From a secure, explosion-proof location, use water spray to cool exposed containers. If cooling streams are ineffective (venting sound increases in volume and pitch, tank discolors, or shows any signs of deforming), withdraw immediately to a secure position. If employees are expected to fight fires, they must be trained and equipped in OSHA 1910.156.

Disposal Method Suggested: In accordance with 40CFR 165 recommendations for the disposal of pesticides and pesticide containers. Must be disposed properly by following package label directions or by contacting your state pesticide or environmental control agency or by contacting your regional EPA office.

References

New Jersey Department of Health and Senior Services, Hazardous Substance Fact Sheet, Allyl Bromide, Trenton NJ (November 1998)

Allyl Chloride

Molecular Formula: C_3H_5Cl

Common Formula: CH_2CHCH_2Cl

Synonyms: Allile (cloruro di) (Italian); Allylchlorid (German); Allyl chloride; Allyle (chlorure d') (French); Chlorallylene; 3-Chloroprene; 1-Chloro propene-2; 1-Chloro-2-propene; 3-Chloro-1-propene; 3-Chloropropene; 3-Chloropropene-1; α-Chloropropylene; 3-Chloropropylene; 3-Chlorpropen (German); Cloruro de alilo (Spanish)

CAS Registry Number: 107-05-1

RTECS Number: UC7350000

DOT ID: UN 1100

EEC Number: 602-029-00-x

Regulatory Authority

- Carcinogen (Suspected) (DFG)[3]
- Air Pollutant Standard Set (ACGIH)[1] (DFG)[3] (HSE)[33] (OSHA)[58] (Several States)[60] (Various Canadian Provinces) (Australia) (Mexico) (Israel)

- OSHA 29CFR1910.119, Appendix A. Process Safety List of Highly Hazardous Chemicals, TQ = 1,000 lb (450 kg)
- Clean Air Act 42USC7412; Title I, Part A, §112 hazardous pollutants
- Clean Water Act: 40CFR116.4 Hazardous Substances; RQ 40CFR117.3 (same as CERCLA); Section 313 Water Priority Chemicals (57FR41331, 9/9/92)
- RCRA Hazardous Constituents 40 CFR261, Appendix 8, waste number not listed
- RCRA Land Ban Waste Restrictions
- RCRA 40CFR268.48; 61FR15654, Universal Treatment Standards: Wastewater (mg/l), 0.036; Nonwastewater (mg/kg), 30
- RCRA 40CFR264, Appendix 9; Ground Water Monitoring List: Suggested methods (PQL μg/l): 8010 (5); 8240 (5)
- SUPERFUND/EPCRA 40CFR302.4, Appendix A, Reportable Quantity (RQ): CERCLA 1,000 lb (454 kg)
- EPCRA Section 313 Form R *de minimis* Concentration Reporting Level: 1.0%.
- U.S. DOT Regulated Marine Pollutant (49CFR172.101, Appendix B)
- Canada, WHMIS, Ingredients Disclosure List; National Pollutant Release Inventory (NPRI)

Cited in U.S. State Regulations: Alaska (G), Connecticut (A), Florida (G), Illinois (G), Maine (G), Maryland (G), Massachusetts (G), Michigan (G), Nevada (A), New Hampshire (G), New Jersey (G), New York (A), North Carolina (A), North Dakota (A), Oklahoma (G), Pennsylvania (G), Rhode Island (G), South Carolina (A), Virginia (A), West Virginia (G).

Description: Allyl chloride, CH_2CHCH_2Cl, is a highly- reactive and -flammable, colorless, brown or purple liquid, with an unpleasant, pungent odor. Odor threshold = 0.48 – 5.9 ppm. Boiling point = 44 – 45°C. Freezing/Melting point = -135°C. Flash point = -32°C. Explosive limits: LEL = 2.9%, UEL = 11.1%. Hazard Identification (based on NFPA-704 M Rating System): Health 3, Flammability 3, Reactivity 1. Slightly soluble in water.

Potential Exposure: Allyl chloride is used in making allyl compounds, epichlorohydrin, and glycerol.

Incompatibilities: Strong oxidizers, acids, aluminum, amines, peroxides, chlorides of iron and aluminum; magnesium, zinc.

Permissible Exposure Limits in Air: The federal OSHA standard is 1 ppm (3 mg/m^3) (TWA) for an 8-hour workshift. NIOSH and ACGIH recommend an airborne exposure limit of 1 ppm (3 mg/m^3) (TWA) for a 10-hour-workshift, and 2 ppm (6 mg/m^3) STEL, not to be exceeded during any 15 minute work period.[1][33][58] The NIOSH IDLH level is 250 ppm. Australia, Canadian Provinces of Alberta, British Columbia, Ontario, Quebec, DFG, Israel, Mexico, the former USSR, and state of California have set standards similar to the above for allyl chloride: 1 ppm (3 mg/m^3) (TWA), and 2 ppm (6 mg/m^3) STEL Israel's Action Level is 0.5 ppm (1.5 mg/m^3). The DFG Peak Limitation is 2 × MAK; not to be exceeded 8 times during workshift.

Determination in Air: Charcoal tube; Benzene; Gas chromatography/Flame ionization detection; see NIOSH Methods (IV) #1000.

Permissible Concentration in Water: The former USSR-UNEP/IRPTC project[43] has set a MAC of 0.3 mg/l in water bodies used for domestic purposes. Minnesota[61] has set a guideline for allyl chloride in drinking water of 2.94 mg/l.

Routes of Entry: Employees may be exposed by dermal or eye contact, inhalation, or ingestion. This chemical can be absorbed through the skin.

Harmful Effects and Symptoms

Short Term Exposure: This chemical can be absorbed through the skin, thereby increasing exposure. Corrosive to eyes, skin, and respiratory tract; can cause burns and permanent damage. In addition to burns, skin contact can cause deep aching and "bone pain." Inhalation can cause respiratory tract irritation with coughing and wheezing. Exposure can cause headache, dizziness, nausea, and vomiting. High exposure may cause pulmonary edema, a medical emergency, with severe shortness of breath. This can cause death.

Long Term Exposure: May cause liver and kidney damage, lung irritation and bronchitis, drying and cracking skin. Allyl chloride has caused mutations. There is limited evidence that this chemical causes cancer of the stomach. The potential for damage to the respiratory tract, liver, and kidneys from inhalation was recognized through experimental evidence during the early commercial development of the industry involving this compound and is reflected in the precautions taken during its manufacture and use. Industrial experience in the United States has pointed to such problems as orbital pain and deep-seated aches after eye or skin contact. Both of these phenomena are believed to be transient when they occur and have been minimized through improved work practices.

Points of Attack: Respiratory system, lungs, skin, eyes, liver and kidneys.

Medical Surveillance: Preplacement and periodic physical examinations have been detailed by NIOSH. They give special attention to the respiratory system, liver, kidneys, skin and eyes. Urine, blood, and pulmonary function testing are required by NIOSH.[45] If symptoms of over exposure develop, chest x-ray should be considered.

First Aid: If this chemical gets into the eyes, remove any contact lenses at once and irrigate immediately for at least 15 minutes, occasionally lifting upper and lower lids. Seek medical attention immediately. If this chemical contacts the skin, remove contaminated clothing and wash immediately with soap and water. Seek medical attention immediately. If this chemical has been inhaled, remove from exposure, begin

rescue breathing (using universal precautions) if breathing has stopped and CPR if heart action has stopped. Transfer promptly to a medical facility. When this chemical has been swallowed, get medical attention. Give large quantities of water and induce vomiting. Do not make an unconscious person vomit. Medical observation is recommended for 24 – 48 hours after breathing overexposure, as pulmonary edema may be delayed. As first aid for pulmonary edema, a doctor or authorized paramedic may consider administering a corticosteroid spray.

Personal Protective Methods: Wear protective gloves and clothing to prevent any reasonable probability of skin contact. Safety equipment suppliers/manufacturers recommend rubber, neoprene, teflon, PV acetate, and chlorinated polyethylene as protective material. All protective clothing (suits, gloves, footwear, headgear) should be clean, available each day, and put on before work. Contact lenses should not be worn when working with this chemical. Wear splash-proof chemical goggles and face shield unless full facepiece respiratory protection is worn. Employees should wash immediately with soap when skin is wet or contaminated. Remove nonimpervious clothing immediately if wet or contaminated. Provide emergency showers and eyewash.

Respirator Selection: Engineering controls shall be used to maintain allyl chloride vapor concentrations below the permissible exposure limits. Compliance with the permissible exposure limits may be achieved by the use of respirators only in the following situations: (1) During the time necessary to install or test the required engineering controls. (2) For nonroutine operations, such as maintenance or repair activities, in which concentrations in excess of the permissible exposure limits may occur. (3) During emergencies when air concentrations of allyl chloride may exceed the permissible limits.

When a respirator is permitted, it shall be selected as follows: NIOSH/OSHA: *25 ppm:* SA:CF (any supplied-air respirator operated in a continuous-flow mode). *50 ppm:* SCBAF (any self-contained breathing apparatus with a full facepiece); or SAF (any supplied-air respirator with a full facepiece). *250 ppm:* SAF:PD,PP (any supplied-air respirator that has a full facepiece and is operated in a pressure-demand or other positive-pressure mode). *Emergency or planned entry into unknown concentrations or IDLH conditions:* SCBAF:PD, PP (any self-contained breathing apparatus that has a full facepiece and is operated in a pressure-demand or other positive-pressure mode); or SAF:PD,PP:ASCBA (any supplied-air respirator that has a full facepiece and is operated in a pressure-demand or other positive-pressure mode in combination with an auxiliary, self-contained breathing apparatus operated in a pressure-demand or other positive-pressure mode. *Escape:* GMFOV [any air-purifying, full-facepiece respirator (gas mask) with a chin-style, front-or back-mounted organic vapor canister]; or SCBAE (any appropriate escape-type, self-contained breathing apparatus).

Note: Substance reported to cause eye irritation or damage; may require eye protection.

Storage: Prior to working with this chemical you should be trained on its proper handling and storage. Before entering confined space where allyl chloride may be present, check to make sure that an explosive concentration does not exist. Store in tightly closed containers in a dark, cool, well ventilated and fireproof area. Metal containers involving the transfer of this chemical should be grounded and bonded. Where possible, automatically pump liquid from drums or other storage containers to process containers. Drums must be equipped with self-closing valves, pressure vacuum bungs, and flame arresters. Use only non-sparking tools and equipment, especially when opening and closing containers of this chemical. Sources of ignition such as smoking and open flames, are prohibited where this chemical is used, handled, or stored in a manner that could create a potential fire or explosion hazard.

Shipping: Label required is "Flammable Liquid, Poison." Shipment by passenger aircraft or railcar is forbidden. The DOT/UN Hazard Class is 6.1 and the Packing Group is I.[20][21]

Spill Handling: Evacuate persons not wearing protective equipment from area of spill or leak until clean-up is complete. Remove all ignition sources. Establish forced ventilation to keep levels below explosive limit. Cover with activated charcoal adsorbent and deposit in sealed containers. Do not wash into sewer. It may be necessary to contain and dispose of this chemical as a hazardous waste. If material or contaminated runoff enters waterways, notify downstream users of potentially contaminated waters. Contact your Department of Environmental Protection or your regional office of the federal EPA for specific recommendations. If employees are required to clean-up spills, they must be properly trained and equipped. OSHA 1910.120(q) may be applicable.

Fire Extinguishing: Allyl chloride is a highly flammable liquid. Poisonous gases including hydrogen chloride and phosgene are produced in fire. Use dry chemical, carbon dioxide, polymer or alcohol foam extinguishers. Vapors are heavier than air and will collect in low areas. Vapors may travel long distances to ignition sources and flashback. Vapors in confined areas may explode when exposed to fire. Storage containers and parts of containers may rocket great distances, in many directions. If material or contaminated runoff enters waterways, notify downstream users of potentially contaminated waters. Notify local health and fire officials and pollution control agencies. From a secure, explosion-proof location, use water spray to cool exposed containers. If cooling streams are ineffective (venting sound increases in volume and pitch, tank discolors, or shows any signs of deforming), withdraw immediately to a secure position. If employees are expected to fight fires, they must be trained and equipped in OSHA 1910.156.

Disposal Method Suggested: Consult with environmental regulatory agencies for guidance on acceptable disposal practices. Generators of waste containing this contaminant (≥100 kg/mo) must conform with EPA regulations governing storage, transportation, treatment, and waste disposal. Controlled incineration at a temperature of 1,800°F for 2 seconds minimum.

References

National Institute for Occupational Safety and Health, Criteria for a Recommended Standard: Occupational Exposure to Allyl Chloride, NIOSH, Doc. No. 76–204, Washington, DC (1976)

U.S. Environmental Protection Agency, Chemical Hazard Information Profile: Allyl Chloride, Washington, DC (July 1979)

Sax, N. I., Ed., Dangerous Properties of Industrial Materials Report, 1, No. 7, 32–34 (1981) and 8, No. 1, 20–28 (1988)

New Jersey Department of Health and Senior Services, "Hazardous Substance Fact Sheet: Allyl Chloride," Trenton, NJ (June 1998)

Allyl Ethyl Ether

Molecular Formula: $C_6H_{10}O$

Synonyms: 1-Propene, 3-ethoxy-

CAS Registry Number: 557-31-3

RTECS Number: KM9120000

DOT ID: UN 2335

Regulatory Authority

- DOT Appendix B, §172.101
- Canada, WHMIS, Ingredients Disclosure List

Cited in U.S. State Regulations: New Jersey (G), Pennsylvania (G).

Description: Allyl ethyl ether is a highly flammable liquid. Boiling point = 66°C. Flash point = -21°C. Hazard Identification (based on NFPA-704 M Rating System): Health 2, Flammability 4, Reactivity 4. Insoluble in water.

Potential Exposure: Used for making other chemicals.

Incompatibilities: Forms explosive mixture with air. Contact with oxidizers may cause fire and explosions. Incompatible with strong acids. May form explosive peroxides during storage.

Permissible Exposure Limits in Air: No occupational airborne exposure limits have been established.

Routes of Entry: Through the skin, inhalation.

Harmful Effects and Symptoms

Short Term Exposure: This chemical can be absorbed through the skin, thereby increasing exposure. Irritates eyes, skin, and respiratory tract. Exposure can cause headache, dizziness, nausea, and vomiting.

Long Term Exposure: May cause central nervous system damage. Removes the skin's natural oils leading to dryness, irritation, redness, possible cracking.

Points of Attack: Skin, lungs.

Medical Surveillance: If symptoms develop or overexposure is suspected, chest x-ray and nervous system tests should be considered.

First Aid: If this chemical gets into the eyes, remove any contact lenses at once and irrigate immediately for at least 15 minutes, occasionally lifting upper and lower lids. Seek medical attention immediately. If this chemical contacts the skin, remove contaminated clothing and wash immediately with soap and water. Seek medical attention immediately. If this chemical has been inhaled, remove from exposure, begin rescue breathing (using universal precautions) if breathing has stopped and CPR if heart action has stopped. Transfer promptly to a medical facility. When this chemical has been swallowed, get medical attention. Give large quantities of water and induce vomiting. Do not make an unconscious person vomit. Medical observation is recommended for 24 – 48 hours after breathing overexposure, as pulmonary edema may be delayed. As first aid for pulmonary edema, a doctor or authorized paramedic may consider administering a corticosteroid spray.

Personal Protective Methods: Wear protective gloves and clothing to prevent any reasonable probability of skin contact. Safety equipment suppliers/manufacturers can provide recommendations on the most protective glove/clothing material for your operation. All protective clothing (suits, gloves, footwear, headgear) should be clean, available each day, and put on before work. Contact lenses should not be worn when working with this chemical. Wear splash-proof chemical goggles and face shield unless full facepiece respiratory protection is worn. Employees should wash immediately with soap when skin is wet or contaminated. Remove nonimpervious clothing immediately if wet or contaminated. Provide emergency showers and eyewash.

Respirator Selection: *At any concentrations above the NIOSH REL, or where there is no REL, at any detectable concentration:* SCBAF:PD,PP (Any MSHA/NIOSH approved self-contained breathing apparatus that has a full facepiece and is operated in a pressure-demand or other positive-pressure mode) or SAF:PD,PP:ASCBA (Any supplied-air respirator that has a full facepiece and is operated in a pressure-demand or other positive-pressure mode in combination with an auxiliary self-contained breathing apparatus operated in a pressure-demand or other positive pressure mode). *Escape:* GMFOV [Any air-purifying, full-facepiece respirator (gas mask) with a chin-style, front- or back-mounted organic vapor canister]; or SCBAE (Any appropriate escape-type, self-contained breathing apparatus).

Storage: Prior to working with this chemical you should be trained on its proper handling and storage. Before entering confined space where allyl ethyl ether may be present, check to make sure that an explosive concentration does not exist. Check for peroxides and inhibit if necessary. Store in tightly closed containers in a cool, well ventilated area. Metal containers involving the transfer of this chemical should be grounded and bonded. Where possible, automatically pump liquid from drums or other storage containers to process containers. Drums must be equipped with self-closing valves,

pressure vacuum bungs, and flame arresters. Use only non-sparking tools and equipment, especially when opening and closing containers of this chemical. Sources of ignition such as smoking and open flames, are prohibited where this chemical is used, handled, or stored in a manner that could create a potential fire or explosion hazard.

Shipping: Allyl ethyl ether should be labeled "Flammable liquid, Poison." It falls into DOT/UN Hazard Class 3.

Spill Handling: Evacuate persons not wearing protective equipment from area of spill or leak until clean-up is complete. Remove all ignition sources. Establish forced ventilation to keep levels below explosive limit. Cover with activated charcoal adsorbent and deposit in sealed containers. Keep allyl chloride out of a confined space, such as a sewer, because of the potential for an explosion, unless the sewer is designed to prevent the build-up of explosive concentrations. It may be necessary to contain and dispose of this chemical as a hazardous waste. If material or contaminated runoff enters waterways, notify downstream users of potentially contaminated waters. Contact your Department of Environmental Protection or your regional office of the federal EPA for specific recommendations. If employees are required to clean-up spills, they must be properly trained and equipped. OSHA 1910.120(q) may be applicable.

Fire Extinguishing: Allyl ethyl ether is a highly flammable and reactive liquid. Poisonous gases are produced in fire. Use dry chemical, carbon dioxide, or polymer foam extinguishers. Vapors are heavier than air and will collect in low areas. Vapors may travel long distances to ignition sources and flashback. Vapors in confined areas may explode when exposed to fire. Storage containers and parts of containers may rocket great distances, in many directions. If material or contaminated runoff enters waterways, notify downstream users of potentially contaminated waters. Notify local health and fire officials and pollution control agencies. From a secure, explosion-proof location, use water spray to cool exposed containers. If cooling streams are ineffective (venting sound increases in volume and pitch, tank discolors, or shows any signs of deforming), withdraw immediately to a secure position. If employees are expected to fight fires, they must be trained and equipped in OSHA 1910.156.

References

New Jersey Department of Health and Senior Services, Hazardous Substance Fact Sheet, Allyl Ethyl Ether, Trenton, NJ (August 1998).

Allyl Glycidyl Ether

Molecular Formula: $C_6H_{10}O_2$

Synonyms: AGE; Alil glicidikico eter (Spanish); Allilglicidil etere (Italian); Allyl-2,3-epoxypropyl ether; Allylglycidaether (German); 1-Allyloxy-2,3-epoxy-propaan (Dutch); 1-Allyloxy-2,3-epoxy-propan (German); 1-(Allyloxy)-2,3-epoxypropane; 1,2-Epoxy-3-allyloxypropane; Oxirane, [(2-propenyloxy)methyl]; [(2-Propenyloxy) methyl]oxirane

CAS Registry Number: 106-92-3

RTECS Number: RR0875000

DOT ID: UN 2219

EEC Number: 603-038-00-1

Regulatory Authority

- Carcinogen (DFG) sufficient animal evidence.
- Air Pollutant Standard Set (OSHA)[58] (ACGIH)[1] (NIOSH)[2] (DFG)[3] (HSE)[33] (Several States)[60] (Australia) (Various Canadian Provinces) (Israel) (Mexico)
- Canada, WHMIS, Ingredients Disclosure List

Cited in U.S. State Regulations: Alaska (G), California (A, G), Connecticut (A), Florida (G), Illinois (G), Maine (G), Massachusetts (G), Minnesota (G), Nevada (A), New Hampshire (G), New Jersey (G), North Dakota (A), Pennsylvania (G), Rhode Island (G), Virginia (A), West Virginia (G).

Description: Allyl glycidyl ether, $C_6H_{10}O_2$, is a colorless liquid with a strong, sweet odor. Boiling point = 154°C. Freezing/Melting point = -100°C. Flash point = 57°C. Soluble in water.

Potential Exposure: Used in making epoxy resins, chlorinated compounds and rubber.

Incompatibilities: Forms explosive mixture with air. Contact with acids or bases may cause explosive polymerization. Contact with oxidizers or amines may cause fire and explosions.

Permissible Exposure Limits in Air: The Federal OSHA limit[58] is a 10 ppm (45 mg/m^3) ceiling, not to be exceeded during any 15 minute work period. NIOSH recommends 5 ppm (TWA) for a 10-hour workshift and a STEL of 10 ppm (44 mg/m^3). Australia, Israel, Mexico and the Canadian provinces of Albera, British Columbia, Ontario, and Quebec have airborne exposure limits limts similar to those recommended by NIOSH. The ACGIH-TWA as of 1999 is 1 ppm. IDLH = 50 ppm. NIOSH, Mexico, DFG, and other authorities add the notation "skin" indicating the potential for cutaneous absorption. In addition, several states have set guidelines or standards for AGE in ambient air:[60] 22 mg/m^3; STEL 44 22 mg/m^3 (California), 22 mg/m^3 (Connecticut), 44 mg/m^3 (North Dakota), 40 mg/m^3 (Virginia), 52.4 mg/m^3 (Nevada).

Determination in Air: Adsorption in a Tenax-tilled tube, workup with ethyl ether, analysis by gas chromatography. See NIOSH Method #2545.

Permissible Concentration in Water: No criteria set.

Routes of Entry: Inhalation, ingestion, eye and/or skin contact. Absorbed by the skin.

Harmful Effects and Symptoms

Short Term Exposure: Skin contact contributes significantly to overall exposure. Inhalation of vapors may irritate the eyes, skin and respiratory tract, causing shortness of breath and coughing. Higher exposure may cause dizziness, lightheadedness, unconsciousness, pulmonary edema, a medical emergency that can cause death. Skin or eye contact with the liquid may cause irritation, burns and skin rash. The DFG warns of the danger of skin sensitization. Use of alcoholic beverages enhances the harmful effect.

Long Term Exposure: This chemical may be a human carcinogen. It may cause mutations, and damage male reproductive glands. Repeated exposures may cause permanent lung damage.

Points of Attack: Eyes, skin, respiratory system, blood, reproductive system.

Medical Surveillance: Consider the points of attack in preplacement and periodic physical examinations. Liver and kidney function tests, lung function tests, examination of the eyes for corneal opacities. A qualified allergist may be consulted concerning potential skin allergy. Following acute overexposure, consider chest x-ray.

First Aid: If this chemical gets into the eyes, remove any contact lenses at once and irrigate immediately for at least 15 minutes, occasionally lifting upper and lower lids. Seek medical attention immediately. If this chemical contacts the skin, remove contaminated clothing and wash immediately with soap and water. Seek medical attention immediately. If this chemical has been inhaled, remove from exposure, begin rescue breathing (using universal precautions) if breathing has stopped and CPR if heart action has stopped. Transfer promptly to a medical facility. When this chemical has been swallowed, get medical attention. Give large quantities of water and induce vomiting. Do not make an unconscious person vomit. Medical observation is recommended for 24 – 48 hours after breathing overexposure, as pulmonary edema may be delayed. As first aid for pulmonary edema, a doctor or authorized paramedic may consider administering a corticosteroid spray.

Personal Protective Methods: Wear protective gloves and clothing to prevent any reasonable probability of skin contact. Safety equipment suppliers/manufacturers recommend natural rubber, butyl rubber, nitrile, PV alcohol, polyvinyl chloride, neoprene/natural rubber material. All protective clothing (suits, gloves, footwear, headgear) should be clean, available each day, and put on before work. Contact lenses should not be worn when working with this chemical. Wear splash-proof chemical goggles and face shield unless full facepiece respiratory protection is worn. Employees should wash immediately with soap when skin is wet or contaminated. Provide emergency showers and eyewash.

Respirator Selection: NIOSH: *50 ppm:* CCROV [any chemical cartridge respirator with organic vapor cartridge(s)]; or PAPROV [any powered, air-purifying respirator with organic vapor cartridge(s)]; or GMFOV [any air-purifying, full-facepiece respirator (gas mask) with a chin-style, front-or back-mounted acid gas canister]; or SA (any supplied-air respirator); or SCBAF (any self-contained breathing apparatus with a full facepiece). *At any concentrations above the NIOSH REL, or where there is no REL, at any detectable concentration:* SCBAF:PD,PP (any self-contained breathing apparatus that has a full facepiece and is operated in a pressure-demand or other positive-pressure mode); or SAF: PD,PP:ASCBA (any supplied-air respirator that has a full facepiece and is operated in a pressure-demand or other positive-pressure mode in combination with an auxiliary self-contained breathing apparatus operated in a pressure-demand or other positive-pressure mode). *Escape:* GMFOV [any air-purifying, full-facepiece respirator (gas mask) with a chin-style, front-or back-mounted organic vapor canister]; or SCBAE (any appropriate escape-type, self-contained breathing apparatus).

Storage: Prior to working with this chemical you should be trained on its proper handling and storage. Store in tightly closed containers in a dark, cool, well ventilated area away from heat. Metal containers involving the transfer of this chemical should be grounded and bonded. Where possible, automatically pump liquid from drums or other storage containers to process containers. Drums must be equipped with self-closing valves, pressure vacuum bungs, and flame arresters. Use only non-sparking tools and equipment, especially when opening and closing containers of this chemical. Sources of ignition such as smoking and open flames, are prohibited where this chemical is used, handled, or stored in a manner that could create a potential fire or explosion hazard.

Shipping: Label required "Flammable Liquid." Falls in DOT/UN Hazard Class 6.1 and Packing Group III.[19][20]

Spill Handling: Evacuate persons not wearing protective equipment from area of spill or leak until clean-up is complete. Remove all ignition sources. Ventilate area of spill or leak. Absorb liquid in vermiculite, dry sand, earth, or similar material and deposit in sealed containers. Wash away residue with plenty of water. It may be necessary to contain and dispose of this chemical as a hazardous waste. If material or contaminated runoff enters waterways, notify downstream users of potentially contaminated waters. Contact your Department of Environmental Protection or your regional office of the federal EPA for specific recommendations. If employees are required to clean-up spills, they must be properly trained and equipped. OSHA 1910.120(q) may be applicable.

Fire Extinguishing: This chemical is a combustible liquid. Poisonous gases are produced in fire. Use dry chemical, carbon dioxide, water spray, or foam extinguishers. Vapors are heavier than air and will collect in low areas. Vapors may travel long distances to ignition sources and flashback. Vapors in confined areas may explode when exposed to fire. If material or contaminated runoff enters waterways, notify downstream users of potentially contaminated waters. Notify local health and fire officials and pollution control agencies. From a secure, explosion-proof location, use water spray to cool exposed containers. If cooling streams are ineffective (venting sound increases in volume and pitch, tank discolors, or shows any signs of deforming), withdraw immediately to a secure position. If employees are expected to fight fires, they must be trained and equipped in OSHA 1910.156.

Disposal Method Suggested: Incineration.

References

National Institute for Occupational Safety and Health, Information Profiles on Potential Occupational Hazards: Glycidyl Ethers, Report No. PB 276–678, Rockville, MD, pp. 116–123 (October 1977)

New Jersey Department of Health and Senior Services, "Hazardous Substance Fact Sheet: Allyl Glycidyl Ether," Trenton, NJ (March 1989)

Allyl Iodide

Molecular Formula: C_3H_5I

Common Formula: $CH_2{=}CHCH_2I$

Synonyms: 3-Iodo-1-propene; 3-Iodopropene; 3-Iodopropylene; 1-Propene, 3-iodo-

CAS Registry Number: 556-56-9

RTECS Number: UD0450000

DOT ID: UN 1723

Cited in U.S. State Regulations: New Hampshire (G), New Jersey (G).

Description: Allyl Iodide, $CH_2{=}CHCH_2I$, is a highly flammable, corrosive, yellowish liquid that darkens on contact with air. It has an unpleasant, pungent odor. Boiling point = 103°C. Freezing/Melting point = -99°C. Flash point = 18°C. Hazard Identification (based on NFPA-704 M Rating System): Health 2, Flammability 3, Reactivity 2. Insoluble in water.

Potential Exposure: Allyl iodide may be used as an organic intermediate and in polymer manufacture.

Incompatibilities: Oxidizing materials. Do not expose to heat, light, or air.

Permissible Exposure Limits in Air: No occupation exposure limits have been established for allyl iodide. However, because this chemical can release Iodides and Iodine gas, the OSHA/NIOSH ceiling limit of 0.1 ppm for Iodine should not be exceeded during any 15 minute work period The above exposure limits are for air levels only. When skin contact also occurs, you may be overexposed, even though air levels are less than the limits listed above.

Permissible Concentration in Water: No criteria set.

Routes of Entry: Inhalation, ingestion, skin contact.

Harmful Effects and Symptoms

Is a powerful irritant. Acts as a poison by inhalation and ingestion.

Short Term Exposure: Irritates eyes, skin and respiratory tract. Eye and skin contact can cause burns and permanent damage. Inhalation can cause coughing and/or shortness of breath; may cause pulmonary edema, which can be delayed for several hours; there is a risk of death in serious cases.

Long Term Exposure: May cause mutations, liver, kidney damage, and lung damage with the development of bronchitis with cough, phlegm, and/or shortness of breath.

Points of Attack: Eyes, skin, nose, throat and lungs.

Medical Surveillance: Pre-employment and annual lung function tests are recommended. If symptoms develop or if overexposure is suspected, liver and kidney function tests may be useful. Consider chest x-rays after acute overexposure.

First Aid: If this chemical gets into the eyes, remove any contact lenses at once and irrigate immediately for at least 15 minutes, occasionally lifting upper and lower lids. Seek medical attention immediately. If this chemical contacts the skin, remove contaminated clothing and wash immediately with soap and water. Seek medical attention immediately. If this chemical has been inhaled, remove from exposure, begin rescue breathing (using universal precautions) if breathing has stopped and CPR if heart action has stopped. Transfer promptly to a medical facility. When this chemical has been swallowed, get medical attention. If victim is conscious, administer water or milk. Do not induce vomiting. Medical observation is recommended for 24 – 48 hours after breathing overexposure, as pulmonary edema may be delayed. As first aid for pulmonary edema, a doctor or authorized paramedic may consider administering a corticosteroid spray.

Personal Protective Methods: Wear protective gloves and clothing to prevent any reasonable probability of skin contact. Safety equipment suppliers/manufacturers can provide recommendations on the most protective glove/clothing material for your operation. All protective clothing (suits, gloves, footwear, headgear) should be clean, available each day, and put on before work. Contact lenses should not be worn when working with this chemical. Wear splash-proof chemical goggles and face shield unless full facepiece respiratory protection is worn. Employees should wash immediately with soap when skin is wet or contaminated. Remove nonimpervious clothing immediately if wet or contaminated. Provide emergency showers and eyewash.

Respirator Selection: *At any concentrations above the NIOSH REL, or where there is no REL, at any detectable concentration:* SCBAF:PD,PP (Any MSHA/NIOSH approved self-contained breathing apparatus that has a full facepiece and is operated in a pressure-demand or other positive-pressure mode) or SAF:PD,PP:ASCBA (Any supplied-air respirator that has a full facepiece and is operated in a pressure-demand or other positive-pressure mode in combination with an auxiliary self-contained breathing apparatus operated in a pressure-demand or other positive pressure mode). *Escape:* GMFOV [Any air-purifying, full-facepiece respirator (gas mask) with a chin-style, front-or back-mounted organic vapor canister]; or SCBAE (Any appropriate escape-type, self-contained breathing apparatus).

Storage: Prior to working with this chemical you should be trained on its proper handling and storage. Before entering confined space where allyl iodide may be present, check to make sure that an explosive concentration does not exist. Store in tightly closed containers in a cool, well ventilated area away from heat, light or air. Metal containers involving the transfer of this chemical should be grounded and bonded. Where possible, automatically pump liquid from drums or other storage containers to process containers. Drums must be equipped with self-closing valves, pressure vacuum bungs, and flame arresters. Use only non-sparking tools and equipment, especially when opening and closing containers of this chemical. Sources of ignition such as smoking and open flames, are prohibited where this chemical is used, handled, or stored in a manner that could create a potential fire or explosion hazard.

Shipping: Label required is "Flammable Liquid, Corrosive." The DOT/UN Hazard Class 3, the Packing Group is I.[19][20]

Spill Handling: Evacuate persons not wearing protective equipment from area of spill or leak until clean-up is complete. Remove all ignition sources. Establish forced ventilation to keep levels below explosive limit. Absorb liquids in vermiculite, dry sand, earth, or a similar material and deposit in sealed containers. Keep allyl iodide out of a confined space, such as a sewer, because of the potential for an explosion, unless the sewer is designed to prevent the build-up of explosive concentrations. It may be necessary to contain and dispose of this chemical as a hazardous waste. If material or contaminated runoff enters waterways, notify downstream users of potentially contaminated waters. Contact your Department of Environmental Protection or your regional office of the federal EPA for specific recommendations. If employees are required to clean-up spills, they must be properly trained and equipped. OSHA 1910.120(q) may be applicable.

Fire Extinguishing: This chemical is a flammable liquid. Poisonous gases including hydrogen isodide are produced in fire. Use dry chemical, carbon dioxide, water spray, or foam extinguishers. Vapors are heavier than air and will collect in low areas. Vapors may travel long distances to ignition sources and flashback. Vapors in confined areas may explode when exposed to fire. If material or contaminated runoff enters waterways, notify downstream users of potentially contaminated waters. Notify local health and fire officials and pollution control agencies. From a secure, explosion-proof location, use water spray to cool exposed containers. If cooling streams are ineffective (venting sound increases in volume and pitch, tank discolors, or shows any signs of deforming), withdraw immediately to a secure position. If employees are expected to fight fires, they must be trained and equipped in OSHA 1910.156.

References

New Jersey Department of Health and Senior Services, "Hazardous Substance Fact Sheet: Allyl Iodide," Trenton, NJ (June 1998)

Allyl Isothiocyanate

Molecular Formula: C_4H_5NS

Common Formula: $CH_2{=}CHCH_2NCS$

Synonyms: AITC; Allyl isorhodanide; Allyl isosulfocyanate; Allyl mustard oil; Allylsenfoel (German); Allyl sevenolum; Allyl thiocarbonimide; Artificial mustard oil; Carbospol; Isothiocyanate d'allyle (French); 3-Isothiocyanato-1-propene; Mustard oil; Oil of mustard, artificial; Oleum sinapis volatile; 2-Propenyl isothiocyanate; Redskin; Senf oel (German); Synthetic mustard oil; Volatile oil of mustard

CAS Registry Number: 57-06-7

RTECS Number: NX8225000

DOT ID: UN 1545 (stabilized)

Regulatory Authority

- Carcinogen (animal evidence) (NTP)[9]
- U.S. DOT 49CFR172.101, Inhalation Hazardous Chemical
- Canada, WHMIS, Ingredients Disclosure List

Cited in U.S. State Regulations: California (G): Florida (G), Massachusetts (G), New Hampshire (G), New Jersey (G), Pennsylvania (G).

Description: Allyl isothiocyanate, $CH_2{=}CHCH_2NCS$, is a highly flammable, colorless to pale yellow liquid with a pungent, irritating odor and acrid taste. Boiling point = 151°C. Flash point = 46°C. Hazard Identification (based on NFPA-704 M Rating System): Health 3, Flammability 2, Reactivity 0. Insoluble in water.

Potential Exposure: Used in fumigants, ointments and counter irritants, mustard plasters, and as a flavoring agent.

Incompatibilities: Alcohols, strong bases, and strong acids, amines.

Permissible Exposure Limits in Air: No standards set.

Permissible Concentration in Water: No criteria set.

Harmful Effects and Symptoms

Short Term Exposure: Irritates the eyes, skin and respiratory tract. Eye and skin contact can cause skin irritation. Prolonged contact can cause burns and blisters. This chemical can be absorbed through the skin, thereby increasing exposure.

Long Term Exposure: There is limited evidence that this chemical causes cancer in animals, It may cause bladder cancer in male rats. May damage the developing fetus. Exposure can cause an allergy-type reaction to develop with symptoms of asthma, watery eyes, sneezing, runny nose, couching, sneezing, chest tightness. One allergy develops, small exposures can cause symptoms to develop.

Points of Attack: Eyes, respiratory system.

Medical Surveillance: Pre-employment and regular lung function tests are recommended (these may be normal if the persons is not having an attack at the time of the test). If symptoms develop or if overexposure is suspected, evaluation by a qualified allergist may help diagnose skin allergy.

First Aid: If this chemical gets into the eyes, remove any contact lenses at once and irrigate immediately for at least 15 minutes, occasionally lifting upper and lower lids. Seek medical attention immediately. If this chemical contacts the skin, remove contaminated clothing and wash immediately with soap and water. Seek medical attention immediately. If this chemical has been inhaled, remove from exposure, begin rescue breathing (using universal precautions) if breathing has stopped and CPR if heart action has stopped. Transfer promptly to a medical facility. When this chemical has been swallowed, get medical attention. Give large quantities of water and induce vomiting. Do not make an unconscious person vomit. Medical observation is recommended for 24 – 48 hours after breathing overexposure, as pulmonary edema may be delayed. As first aid for pulmonary edema, a doctor or authorized paramedic may consider administering a corticosteroid spray.

Personal Protective Methods: Wear protective gloves and clothing to prevent any reasonable probability of skin contact. Safety equipment suppliers/manufacturers can provide recommendations on the most protective glove/clothing material for your operation. All protective clothing (suits, gloves, footwear, headgear) should be clean, available each day, and put on before work. Contact lenses should not be worn when working with this chemical. Wear splash-proof chemical goggles and face shield unless full facepiece respiratory protection is worn. Employees should wash immediately with soap when skin is wet or contaminated. Remove nonimpervious clothing immediately if wet or contaminated. Provide emergency showers and eyewash.

Respirator Selection: *At any concentrations above the NIOSH REL, or where there is no REL, at any detectable concentration:* SCBAF:PD,PP (Any MSHA/NIOSH approved self-contained breathing apparatus that has a full facepiece and is operated in a pressure-demand or other positive-pressure mode) or SAF:PD,PP:ASCBA (Any supplied-air respirator that has a full facepiece and is operated in a pressure-demand or other positive-pressure mode in combination with an auxiliary self-contained breathing apparatus operated in a pressure-demand or other positive pressure mode). *Escape:* GMFOV [Any air-purifying, full-facepiece respirator (gas mask) with a chin-style, front-or back-mounted organic vapor canister]; or SCBAE (Any appropriate escape-type, self-contained breathing apparatus).

Storage: Prior to working with this chemical you should be trained on its proper handling and storage. Before entering confined space where this chemical may be present, check to make sure that an explosive concentration does not exist. Store in tightly closed containers in a cool, well ventilated area. Metal containers involving the transfer of this chemical should be grounded and bonded. Where possible, automatically pump liquid from drums or other storage containers to process containers. Drums must be equipped with self-closing valves, pressure vacuum bungs, and flame arresters. Use only non-sparking tools and equipment, especially when opening and closing containers of this chemical. Sources of ignition such as smoking and open flames, are prohibited where this chemical is used, handled, or stored in a manner that could create a potential fire or explosion hazard.

Shipping: Label required is "Poison." Shipment by passenger aircraft or railcar is forbidden; shipment by cargo aircraft limited to 60 liter. Falls in DOT/UN Hazard Class 6.1 and Packing Group II.[19][20] This chemical must be inhibited.

Spill Handling: Evacuate persons not wearing protective equipment from area of spill or leak until clean-up is complete. Remove all ignition sources. Establish forced ventilation to keep levels below explosive limit. Cover with vermiculite, dry sand, soil, or similar adsorbent, and deposit in sealed containers. Keep this chemical out of a confined space, such as a sewer, because of the potential for an explosion, unless the sewer is designed to prevent the build-up of explosive concentrations. It may be necessary to contain and dispose of this chemical as a hazardous waste. If material or contaminated runoff enters waterways, notify downstream users of potentially contaminated waters. Contact your Department of Environmental Protection or your regional office of the federal EPA for specific recommendations. If employees are required to clean-up spills, they must be properly trained and equipped. OSHA 1910.120(q) may be applicable.

Fire Extinguishing: Allyl isothiocyanate is a highly flammable liquid. Poisonous gases, including nitrogen oxides, sulfur oxides, and hydrogen cyanide are produced in fire. Use dry chemical, carbon dioxide, or foam extinguishers. Vapors are heavier than air and will collect in low areas. Vapors may travel long distances to ignition sources and flashback. Vapors in confined areas may explode when exposed to fire. Storage containers and parts of containers may rocket great distances, in many directions. If material or contaminated runoff enters waterways, notify downstream users of potentially contaminated waters. Notify local health and fire officials and pollution control agencies. From a secure, explosion-proof location, use water spray to cool exposed containers. If cooling streams are ineffective (venting sound increases in volume and pitch, tank discolors, or shows any signs of deforming), withdraw immediately to a secure position. If employees are expected to fight fires, they must be trained and equipped in OSHA 1910.156. Use dry chemical, CO_2, or foam extinguishers. Poisonous gas is produced in fire. Containers may explode in fire. Water may be used to cool fire-exposed containers.

References

Sax, N. I., Ed., Dangerous Properties of Industrial Materials Report, 1, No. 1, 28–29 (1980)

New Jersey Department of Health and Senior Services, "Hazardous Substance Fact Sheet: Allyl Isothiocyanate," Trenton, NJ (June, 1998)

Allyl Propyl Disulfide

Molecular Formula: $C_6H_{12}S_2$

Common Formula: $CH_2=CHCH_2SSC_3H_7$

Synonyms: Disulfide, 2-propenyl propyl; Disulfuro de alil propilo (Spanish); 4,5-Dithia-1-octene; Onion oil; 2-Propenyl propyl disulfide; Propyl allyl disulfide

CAS Registry Number: 2179-59-1

RTECS Number: JO0350000

DOT ID: None assigned

Regulatory Authority

- Air Pollutant Standard Set (ACGIH)[1] (DFG)[3] (OSHA)[58] (Various States)[60] (Various Canadian Provinces), (Australia), (Israel)
- Canada, WHMIS, Ingredients Disclosure List

Cited in U.S. State Regulations: Alaska (G), California (G), Connecticut (A), Florida (G), Illinois (G), Maine (G),. Massachusetts (G), New Jersey (G), Pennsylvania (G), Rhode Island (G), Virginia (A), West Virginia (G).

Description: Allyl propyl disulfide, $CH_2=CHCH_2SSC_3H_7$, is a combustible, pale yellow liquid with a strong, irritating onion-like odor. Boiling point = 67 – 69°C. Insoluble in water.

Potential Exposure: The primary ingredient in onion oil. Workers in onion processing (slicing and dehydration) operations.

Incompatibilities: Strong oxidizers.

Permissible Exposure Limits in Air: The Federal OSHA airborne exposure limit is 2 ppm (12 mg/m^3) (TWA) for an 8-hour workshift. ACGIH, NIOSH, DFG, Australia, and the Canadian Provinces of Alberta, British Columbia, Ontario, Quebec, and Israel recommended limit is 2 ppm (12 mg/m^3) (TWA) and STEL value of 3 ppm (18 mg/m^3). Israel also lists an action level of 1 ppm. In addition, several states have set guidelines or standards for allyl propyl disulfide in ambient air:[60] 2 ppm (PEL); 3 ppm STEL (California), 0.12 – 0.18 mg/m^3 (North Dakota), 0.20 mg/m^3 (Virginia), 0.24 mg/m^3 (Connecticut), 0.286 mg/m^3 (Nevada).

Determination in Air: Sample collection by charcoal tube, analysis by gas liquid chromatography.

Permissible Concentration in Water: No criteria set.

Routes of Entry: Inhalation, ingestion, skin and eye contact.

Harmful Effects and Symptoms

Irritation of eyes, nose and throat. Is a poison by inhalation and ingestion.[44]

Short Term Exposure: Contact with the eyes produces tears. The vapor can irritate the eyes and respiratory tract, producing runny nose and tears.

Long Term Exposure: Unknown at this time.

Points of Attack: Eyes, nose and throat.

First Aid: If this chemical gets into the eyes, remove any contact lenses at once and irrigate immediately for at least 15 minutes, occasionally lifting upper and lower lids. Seek medical attention immediately. If this chemical contacts the skin, remove contaminated clothing and wash immediately with soap and water. Seek medical attention immediately. If this chemical has been inhaled, remove from exposure, begin rescue breathing (using universal precautions) if breathing has stopped and CPR if heart action has stopped. Transfer promptly to a medical facility. When this chemical has been swallowed, get medical attention. Give large quantities of water and induce vomiting. Do not make an unconscious person vomit.

Personal Protective Methods: Wear protective gloves and clothing to prevent any reasonable probability of skin contact. Safety equipment suppliers/manufacturers can provide recommendations on the most protective glove/clothing material for your operation. All protective clothing (suits, gloves, footwear, headgear) should be clean, available each day, and put on before work. Contact lenses should not be worn when working with this chemical. Wear splash- and gas-proof chemical goggles and face shield unless full facepiece respiratory protection is worn. Employees should wash immediately with soap when skin is wet or contaminated. Remove nonimpervious clothing immediately if wet or contaminated. Provide emergency showers and eyewash.

Respirator Selection: *At any concentrations above the NIOSH REL, or where there is no REL, at any detectable concentration:* SCBAF:PD,PP (Any MSHA/NIOSH approved self-contained breathing apparatus that has a full facepiece and is operated in a pressure-demand or other positive-pressure mode) or SAF:PD,PP:ASCBA (Any supplied-air respirator that has a full facepiece and is operated in a pressure-demand or other positive-pressure mode in combination with an auxiliary self-contained breathing apparatus operated in a pressure-demand or other positive pressure mode).

Storage: Prior to working with this chemical you should be trained on its proper handling and storage. Store in tightly closed containers in a cool, well ventilated area away from oxidizers. Where possible, automatically pump liquid from drums or other storage containers to process containers. Drums must be equipped with self-closing valves, pressure vacuum bungs, and flame arresters. Use only non-sparking tools and equipment, especially when opening and closing containers of this chemical. Sources of ignition such as smoking and open flames, are prohibited where this chemical is used, handled, or stored in a manner that could create a potential fire or explosion hazard.

Shipping: No specific labeling or shipping precaution specified by DON/UN.[19][20]

Spill Handling: Evacuate persons not wearing protective equipment from area of spill or leak until clean-up is complete. Remove all ignition sources. Ventilate area of spill or leak. Cover with vermiculite, dry sand, earth or similar material and deposit in sealed containers. It may be necessary to contain and dispose of this chemical as a hazardous waste. If material or contaminated runoff enters waterways, notify downstream users of potentially contaminated waters. Contact your Department of Environmental Protection or your regional office of the federal EPA for specific recommendations. If employees are required to clean-up spills, they must be properly trained and equipped. OSHA 1910.120(q) may be applicable.

Fire Extinguishing: This chemical is a combustible liquid. Poisonous gases, include sulfur oxides, are produced in fire. Use dry chemical, carbon dioxide, or foam extinguishers. Vapors are heavier than air and will collect in low areas. If material or contaminated runoff enters waterways, notify downstream users of potentially contaminated waters. Notify local health and fire officials and pollution control agencies. From a secure, explosion-proof location, use water spray to cool exposed containers. If cooling streams are ineffective (venting sound increases in volume and pitch, tank discolors, or shows any signs of deforming), withdraw immediately to a secure position. If employees are expected to fight fires, they must be trained and equipped in OSHA 1910.156.

Disposal Method Suggested: Incineration.

References

Sax, N. I., Ed., Dangerous Properties of Industrial Materials Report, 1, No. 5, 32–33 (1981)

New Jersey Department of Health and Senior Services, "Hazardous Substance Fact Sheet: Allyl Propyl Disulfide," Trenton, NJ (January 1986)

Allyl Trichlorosilane

Molecular Formula: $C_3H_5Cl_3Si$

Common Formula: $CH_2{=}CHCH_2SiCl_3$

Synonyms: Aliltriclorosilano (Spanish); Allylsilicone trichloride; Allyltrichlorosilane; Allyl trichorosilane, stabilized; Silane, allyltrichloro-; Silane, trichloroallyl-; Silane, trichloro-2-propenyl-; Trichloroallylsilane; Trichloro-2-propenylsilane

CAS Registry Number: 107-37-9

RTECS Number: VV1530000

DOT ID: UN 1724

Regulatory Authority

- Canada, WHMIS, Ingredients Disclosure List

Cited in U.S. State Regulations: Florida (G), Massachusetts (G), New Hampshire (G), New Jersey (G), Oklahoma (G), Pennsylvania (G).

Description: Allyl trichhlorosilane, $CH_2{=}CHCH_2SiCl_3$, is a volatile, corrosive, flammable, colorless liquid with an irritating odor. Boiling point = 118°C. Flash point = 35°C (oc). Hazard Identification (based on NFPA-704 M Rating System): Health 3, Flammability 3, Reactivity 2. Water reactive.

Potential Exposure: Used to make silicones and glass fiber finishes.

Incompatibilities: Water reaction is violent and forms toxic and corrosive hydrogen chloride gas. Contact with oxidizers may cause fire and explosions. Avoid all sources of ignition.

Permissible Exposure Limits in Air: No standards set.

Permissible Concentration in Water: No criteria set. Reacts vigorously with water.

Routes of Entry: Inhalation, ingestion.

Harmful Effects and Symptoms

Short Term Exposure: Allyl trichlorosilane is corrosive and contact can cause severe eye and skin burns. Exposure can irritate the eyes, nose and respiratory tract. Higher levels can irritate the lungs causing coughing and shortness of breath; still higher exposures can cause pulmonary edema, a medical emergency. This can cause death.

Long Term Exposure: Corrosives may cause long-term lung problems, although it is not known at this time that this chemical causes lung damage.

Points of Attack: Eyes, skin, respiratory tract.

Medical Surveillance: Lung function tests. Following acute overexposure, consider chest x-ray.

First Aid: If this chemical gets into the eyes, remove any contact lenses at once and irrigate immediately for at least 15 minutes, occasionally lifting upper and lower lids. Seek medical attention immediately. If this chemical contacts the skin, remove contaminated clothing and wash immediately with soap and water. Seek medical attention immediately. If this chemical has been inhaled, remove from exposure, begin rescue breathing (using universal precautions) if breathing has stopped and CPR if heart action has stopped. Transfer promptly to a medical facility. When this chemical has been swallowed, get medical attention. If victim is conscious, administer water or milk. Do not induce vomiting. Medical observation is recommended for 24 – 48 hours after breathing overexposure, as pulmonary edema may be delayed. As first aid for pulmonary edema, a doctor or authorized paramedic may consider administering a corticosteroid spray.

Personal Protective Methods: Wear protective gloves and clothing to prevent any reasonable probability of skin contact. Safety equipment suppliers/manufacturers can provide recommendations on the most protective glove/clothing material for your operation. All protective clothing (suits, gloves, footwear, headgear) should be clean, available each day, and put on before work. Contact lenses should not be worn when working with this chemical. Wear splash-proof chemical goggles and face shield unless full facepiece respiratory protection is worn. Employees should wash immediately with soap when skin is wet or contaminated. Provide emergency showers and eyewash.

Respirator Selection: NIOSH/OSHA for *Hydrogen Chloride*: *50 ppm:* CCRS [any chemical cartridge respirator with cartridge(s) providing protection against the compound of concern]; or GMFS [any air-purifying, full-facepiece respirator (gas mask) with a chin-style, front- or back-mounted canister providing protection against the compound of concern]; or PAPRS (any powered, air-purifying respirator with cartridge(s) providing protection against the compound of concern); or SA (any supplied-air respirator); or SCBAF (any self-contained breathing apparatus with a full facepiece).

Emergency or planned entry into unknown concentrations or IDLH conditions: SCBAF:PD,PP (any self-contained breathing apparatus that has a full facepiece and is operated in a pressure-demand or other positive-pressure mode); or SAF:PD,PP:ASCBA (any supplied-air respirator that has a full facepiece and is operated in a pressure-demand or other positive-pressure mode in combination with an auxiliary, self-contained breathing apparatus operated in a pressure-demand or other positive-pressure mode). *Escape:* GMFAG [any air-purifying, full-facepiece respirator (gas mask) with a chin-style, front- or back-mounted organic vapor canister]; or SCBAE (any appropriate escape-type, self-contained breathing apparatus).

Note: Substance reported to cause eye irritation or damage; may require eye protection.

Storage: Prior to working with this chemical you should be trained on its proper handling and storage. Before entering confined space where allyl trichlorosilane may be present, check to make sure that an explosive concentration does not exist. Store in tightly closed containers in a cool, well ventilated area away from contact with water and sources of ignition. Use and store allyl trichlorosilane under nitrogen. Metal containers involving the transfer of this chemical should be grounded and bonded. Where possible, automatically pump liquid from drums or other storage containers to process containers. Drums must be equipped with self-closing valves, pressure vacuum bungs, and flame arresters. Use only non-sparking tools and equipment, especially when opening and closing containers of this chemical. Sources of ignition such as smoking and open flames, are prohibited where this chemical is used, handled, or stored in a manner that could create a potential fire or explosion hazard. Where this chemical is used, handled, manufactured, or stored, use explosion-proof electrical equipment and fittings.

Shipping: Label required is "Corrosive." Shipment by passenger aircraft or railcar is forbidden. Falls in DOT/UN Hazard Class 8, Packing Group II.[19][20]

Spill Handling: Evacuate and restrict persons not wearing protective equipment from area of spill or leak until cleanup is complete. Remove all ignition sources. Establish forced ventilation to keep levels below explosive limit. Use foam spray to reduce vapors. Cover with dry lime, dry sand, soda ash, or a similar material and deposit in sealed containers. Collect powdered material in the most convenient and safe manner and deposit in sealed containers. Keep allyl trichlorosilane out of a confined space, such as a sewer, because of the potential for an explosion, unless the sewer is designed to prevent the build-up of explosive concentrations. It may be necessary to contain and dispose of this chemical as a hazardous waste. If material or contaminated runoff enters waterways, notify downstream users of potentially contaminated waters. Contact your Department of Environmental Protection or your regional office of the federal EPA for specific recommendations. If employees are required to clean-up spills, they must be properly trained and equipped. OSHA 1910.120(q) may be applicable.

Fire Extinguishing: Firefighting gear (including SCBA) does not provide adequate protection. If exposure occurs, remove and isolate gear immediately and thoroughly decontaminate personnel. This chemical is a highly flammable liquid. Poisonous gases, including hydrogen chloride and phosgene, are produced in fire. Do Not Use Water. Use dry chemical, CO_2 or foam extinguishers. Low or medium expansion AFFF foam or dry chemical if available in sufficient amounts (FEMA). Vapors are heavier than air and will collect in low areas. Vapors may travel long distances to ignition sources and flashback. Vapors in confined areas may explode when exposed to fire. If material or contaminated runoff enters waterways, notify downstream users of potentially contaminated waters. Notify local health and fire officials and pollution control agencies. From a secure, explosion-proof location, use water spray to cool exposed containers. If cooling streams are ineffective (venting sound increases in volume and pitch, tank discolors, or shows any signs of deforming), withdraw immediately to a secure position. If employees are expected to fight fires, they must be trained and equipped in OSHA 1910.156.

References

New Jersey Department of Health and Senior Services, "Hazardous Substance Fact Sheet: Allyl Trichlorosilane," Trenton, NJ (June, 1998)

Aluminum Alkyl Halides

Molecular Formula: $C_4H_{10}AlBr$ (Diethylaluminum bromide); $C_4H_{10}AlCl$ (Diethylaluminum chloride); $C_2H_5AlI_2$ (Ethylaluminum diiodide); $C_2H_5AlCl_2$ (Ethylaluminum dichlroride); $C_6H_{15}Al_2Cl_3$ (Ethylaluminum sesquichloride); $C_3H_9Al_2Cl_3$ (Methylaluminum sesquichloride)

Synonyms: *diethylaluminum chloride:* Chlorodiethylaluminum; Diethylaluminum monochloride; Diethylchloroaluminum

ethylaluminum dichloride: Aluminum, dichloroethyl-; Dichloroethylaluminum; Dichloromonoethylaluminum; Ethyldichloroaluminum

ethylaluminum sesquichloride: Riethyldialuminum trichloride; Sesquiethylaluminum chloride; Trichlorotriethyldialuminium; Trichlorotriethyldialuminum; Triethylaluminum sesquichloride; Triethyltrichlorodialuminum

methylaluminum sesquichloride: Aluminum, trichlorotrimethyldi-; Methyl aluminium sesquichloride; Trichlorotrimethyldialuminum

CAS Registry Number: 760-19-0 (diethylaluminum bromide); 96-10-6 (diethylaluminum chloride); 563-43-9 (ethylaluminum dichloride); 2938-73-0 (ethylalumium diiodide); 12075-68-2 (ethylaluminum sesquichloride); 12542-85-7 (methylaluminum sesquichloride)

RTECS Number: BD0558000 (diethylaluminum chloride); BD0705000 (ethylaluminum dichloride); BD1950000 (ethylaluminum sesquichloride); BD1970000 (methylaluminum sesquichloride)

DOT ID: UN 3052 (Aluminum alkyl halides)

Regulatory Authority

- OSHA 29CFR1910.119, Appendix A, Process Safety List of Highly Hazardous Chemicals, TQ = 5,000 lb (2,270 kg)
- Air Pollutant Standard Set (HSE)[33] (NIOSH)[58]
- Canada, WHMIS, Ingredients Disclosure List (aluminum alkyls)

Cited in U.S. State Regulations: California (A, G), Massachusetts (G), New Hampshire (G), New Jersey (G), Oklahoma (G), Pennsylvania (G).

Description: The aluminum alkyl halides are flammable, colorless to yellow liquids that are spontaneously flammable in air.

As examples, ethylaluminum sesquichloride: Boiling point = 114.5 – 116.5°C @ 50 mm. Flash point = –20°C Methylaluminum sesquichloride; Boiling point = 120 – 140°C. Flash point = –17°C.

Potential Exposure: These materials are used as components of olefin polymerization catalysts. The reader is referred to the entry on "Aluminum Alkyls" for additional information on this entry. The aluminum alkyl halides parallel very closely the aluminum alkyls.

Incompatibilities: Reacts violently with nitromethane. Ethylaluminum sesquichloride reacts explosively with carbon tetrachloride at room temperature. Diethylaluminum chloride may form an explosive product with chlorine azide.

Permissible Exposure Limits in Air: NIOSH and ACGIH recommend an airborne limit of 2 mg/m^3 (TWA) for aluminum soluble salts and alkyls, measured as aluminum.

Determination in Air: Use OSHA Method ID-25G, NIOSH Method #7013 Aluminum. See also Method #7300, Elements.

Routes of Entry: Inhalation, ingestion, skin and/or eye contact.

Points of Attack: Skin, respiratory system.

References

New Jersey Department of Health and Senior Services, "Hazardous Substance Fact Sheet: Diethyl Aluminum Chloride," Trenton, NJ (February 1986).

New Jersey Department of Health and Senior Services, "Hazardous Substance Fact Sheet: Ethyl Aluminum Dichloride," Trenton, NJ (January 1997)

New Jersey Department of Health and Senior Services, "Hazardous Substance Fact Sheet: Ethyl Aluminum Sesquichloride," Trenton, NJ (January 1997)

Aluminum Alkyls

Molecular Formula: C_3H_9Al (trimethyl-); $C_6H_{15}Al$ (triethyl-); $C_9H_{21}Al$ (tripropyl-); $C_{12}H_{27}Al$ (triisobutyl)

Synonyms: Trialkylaluminum (general)

tributyl-: Aluminum, Tributyl-; TNBA; Tributylalane; tri-*n*-Butyl aluminum

triethyl-: Aluminum, triethyl-; TEA; Triethylalane

triisobutyl-: Aluminum, triisobutyl-; Aluminum, tris(2-methylpropyl); Triisobutylalane; Tris(2-methylpropyl)aluminum

trimethyl-: Trimethylalane

tripropyl-: Aluminum, tripropyl-; Tripropylalane

CAS Registry Number: 75-24-1 (trimethyl-); 97-93-8 (triethyl-); 102-67-0 (tripropyl-); 1116-70-7 (tributyl-); 100-99-2 (triisobutyl-)

RTECS Number: BD2204000 (trimethyl-); BD2050000 (triethyl-); BD2208000 (tripropyl-); BD1820000 (tributyl-); BD2203500 (triisobutyl-)

DOT ID: UN 3051 (Aluminum alkyls group)

Regulatory Authority

- Air Pollutant Standard Set (ACGIH)[1] (HSE)[33] NIOSH[58] (California), (Various Canadian Provinces) (Mexico)
- OSHA 29CFR1910.119, Appendix A, Process Safety List of Highly Hazardous Chemicals, TQ = 5,000 lb (2,270 kg)
- Canada, WHMIS, Ingredients Disclosure List

Cited in U.S. State Regulations: Alaska (G), California (G), Illinois (G), Maine (G), Massachusetts (G), New Hampshire (G), New Jersey (G), Oklahoma (G), Pennsylvania (G), Rhode Island (G).

Description: The aluminum alkyls are highly flammable and reactive, colorless to yellow liquids at room temperature. The lighter trialkylaluminums ignite spontaneously in air. They are normally supplied and used in a 20% solution with a hydrocarbon solvent such as hexane, heptane, benzene, toluene. Properties may depend on solvent. Reacts violently with water. Boiling point = 212°C (triisobutyl-); 185°C (triethyl-). Freezing/Melting point = 12°C (triisobutyl); -46°C (triethyl). Flash point = ignites spontaneously in air. Hazard Identification (triethyl-, triisobutyl-): Health 3, Flammability 4, Reactivity 3. Water reactive.

Potential Exposure: Alkyl aluminum compounds are used as components of olefin polymerization catalysts. They are also used in the synthesis of higher primary alcohols and in pyrophoric fuels, as a catalyst in making ethylene gas, and in plating aluminum.

Incompatibilities: The lighter trialkylaluminums ignite spontaneously in air. These compounds are strong reducing agents. Violent reaction with oxidizers. Incompatible with water, oxygen (air), acids, alcohols, phenols, amines, carbon dioxide, sulfur oxides, halogenated compounds, and many other substances.

Permissible Exposure Limits in Air: NIOSH and ACGIH recommend an airborne limit of 2 mg/m^3 (TWA) for aluminum alkyls, measured as aluminum. The NIOSH limit is averaged over a 10-hour work period.

Determination in Air: Use OSHA Method ID-25G, NIOSH #7013 Aluminum; Also, #7300, Elements.

Permissible Concentration in Water: No criteria set. These compounds react violently with water.

Routes of Entry: Inhalation, skin and/or eye contact.

Harmful Effects and Symptoms

Short Term Exposure: Corrosive. Can cause severe eye and skin irritation and burns. Inhaling vapors or fumes can irritate the respiratory tract, causing coughing, wheezing, and/or shortness of breath. They can cause "metal fume fever" with symptoms of head-ache, nausea, vomiting, chills, cough and shortness of breath.

Points of Attack: Eyes, skin, respiratory system.

Medical Surveillance: Lung function tests. Consider chest x-ray following acute overexposure. See also NIOSH #8310, Metals in Urine.

First Aid: I If this chemical gets into the eyes, remove any contact lenses at once and irrigate immediately for at least 15 minutes, occasionally lifting upper and lower lids. Seek medical attention immediately. If this chemical contacts the skin, remove contaminated clothing and wash immediately with soap and water. Seek medical attention immediately. If this chemical has been inhaled, remove from exposure, begin rescue breathing (using universal precautions) if breathing has stopped and CPR if heart action has stopped. Transfer promptly to a medical facility. When this chemical has been swallowed, get medical attention. Give large quantities of water and induce vomiting. Do not make an unconscious person vomit.

Note to Physician: In case of fume inhalation, treat pulmonary edema. Consider administering prednisone or other corticosteroid orally to reduce tissue response to fume. Positive pressure ventilation may be necessary. Treat metal fume fever with bed rest, analgesics and antipyretics. The symptoms of metal fume fever may be delayed for 4 – 12 hours following exposure: it may last less than 36 hours.

Personal Protective Methods: Wear protective gloves and clothing to prevent any reasonable probability of skin contact. Safety equipment suppliers/manufacturers can provide recommendations on the most protective glove/clothing material for your operation Exposure to large quantities, as in plant transfers, require full body aluminized proximity suit. Gloves, used in plant transfers or operations should be aluminized leather. Preformed neoprene, aluminized vinyl, or other fire-resistant, nonreactive material. Preformed rubber gloves may be used in the laboratory. All gloves used when handling aluminum alkyls should be loose-fitting for instant removal if necessary. All protective clothing (suits, gloves, footwear, headgear) should be clean, available each day, and put on before work. Contact lenses should not be worn when working with this chemical. Wear splash-proof chemical goggles and face shield unless full facepiece respiratory protection is worn. Employees should wash immediately with soap when skin is wet or contaminated. Provide emergency showers and eyewash.

Respirator Selection: *At any concentrations above the NIOSH REL: 2 mg/m^3* SCBAF:PD,PP (any MSHA/NIOSH approved self-contained breathing apparatus that has a full facepiece and is operated in a pressure-demand or other positive-pressure mode); or SAF:PD,PP:ASCBA (any supplied-air respirator that has a full facepiece and is operated in a pressure-demand or other positive-pressure mode in combination with an auxiliary, self-contained breathing apparatus operated in a pressure-demand or other positive pressure mode).

Storage: Prior to working with this chemical you should be trained on its proper handling and storage. Protect against physical damage. Outside or detached storage is preferable. Before entering confined space where aluminum alkyls may be present, check to make sure that an explosive concentration does not exist. Store in tightly closed containers in a cool, well ventilated area. All transfer lines must be free of water, oxygen and other substances that react with aluminum alkyls. All vessels must be dry and oxygen free. Transfer lines should be blanked when not in use. Vessels should be top unloading when feasible with remote manual pressure relief. Metal containers involving the transfer of this chemical should be grounded and bonded. Use only non-sparking tools and equipment, especially when opening and closing containers of this chemical. Sources of ignition such as smoking and open flames, are prohibited where this chemical is used, handled, or stored in a manner that could create a potential fire or explosion hazard.

Shipping: Shipped in containers containing a blanket of nitrogen gas. Aluminum alkyls require a shipping label of "Spontaneously Combustible, Dangerous when Wet." Shipment by passenger aircraft or railcar or even by cargo aircraft is forbidden. The UN/DOT Hazard Class is 4.2 and the Packing Group is I.[19][20]

Spill Handling: Evacuate and restrict persons not wearing protective equipment from area of spill or leak until cleanup is complete. These chemicals ignite spontaneously in air. Remove all ignition sources. Absorb liquids in vermiculite, dry sand, earth, or a similar material and deposit in sealed containers. Collect powdered material in the most convenient and safe manner and deposit in sealed containers. It may be necessary to contain and dispose of this chemical as a hazardous waste. If material or contaminated runoff enters waterways, notify downstream users of potentially contaminated waters.

Contact your Department of Environmental Protection or your regional office of the federal EPA for specific recommendations. If employees are required to clean-up spills, they must be properly trained and equipped. OSHA 1910.120(q) may be applicable.

Fire Extinguishing: Stop flow of liquid, if possible, before extinguishing fire. These chemicals may ignite spontaneously in air, and are flammable liquids. Poisonous and flammable gases including aluminum oxides, and ethylene are produced in fire. *Do not use water, foam*, or a *halogenated* extinguishing agent. Use dry chemical, graphite powder, soda ash, lime extinguishers. On solvent-based material use carbon dioxide or dry chemical. Vapors are heavier than air and will collect in low areas. Vapors may travel long distances to ignition sources and flashback. Vapors in confined areas may explode when exposed to fire. Storage containers and parts of containers may rocket great distances, in many directions. If material or contaminated runoff enters waterways, notify downstream users of potentially contaminated waters. Notify local health and fire officials and pollution control agencies. From a secure, explosion-proof location, carefully use water spray to cool fire-exposed containers. If cooling streams are ineffective (venting sound increases in volume and pitch, tank discolors, or shows any signs of deforming), withdraw immediately to a secure position. If employees are expected to fight fires, they must be trained and equipped in OSHA 1910.156. Do not use water, foam or halogenated extinguishing agents. Do use dry chemical or CO_2 extinguishers on fires involving solvent solutions of aluminum alkyls.

Disposal Method Suggested: Careful incineration.

References

New Jersey Department of Health and Senior Services, "Hazardous Substance Fact Sheet: Triethyl Aluminum," Trenton, NJ (December 1996). There are very similar publications on: Trimethyl Aluminum, Tripropyl Aluminum, Tributyl Aluminum, Triisobutyl Aluminum.

Aluminum and Aluminum Oxide

Molecular Formula: Al

Synonyms: *aluminum:* A 00; A 95; A 99; A 995; A 999; AA 1099; AA 1199; AD 1; AD1M; ADO; AE; Alaun (German); Allbri aluminum paste and powder; Alumina fibre; Aluminio (Spanish); Aluminium; Aluminium flake; Aluminum 27; Aluminum dehydrated; Aluminum, metallic powder; Aluminum powder; AO A1; AR2; AV00; AV000; C.I. 77000; Emanay atomized aluminum powder; JISC 3108; JISC 3110; L16; Metana; Metana aluminum paste; Noral aluminum; Noral extra fine lining grade; Noral non-leafing grade; PAP-1

aluminum oxide: A-1 (sorbent); A-2; Alcan AA-100; Alcan C-70; Alcan C-71; Alcan C-72; Alcan C-73; Alcoa F1; Alexite; Almite; Alon; Alon C; Aloxite; Alufrit; α-Alumina; β-Alumina; γ-Alumina; Alumina; Aluminite 37; α-Aluminum oxide; β-Aluminum oxide; γ-Aluminum oxide; Aluminum oxide (2:3); Aluminum oxide C; Aluminum sesquioxide; Aluminum trioxide; Alumite; Alundum; Alundum 600; Backlap slurry; Bauxite; Bayerite; Boehmite; Brasivol; Brockmann, aluminum oxide; C-1; Cab-O-Grip; Catapal S; Compalox; Conopal; Corundum; D 201; Dialuminum trioxide; Diaspore dirubin; Dispal; Dispal alumina; Dispal M; Dotment 324; Dotment 358; Dural; Dycron; Exolon XW 60; F 360 (Alumina); Faserton; Fasertonerde; Fast cure 45 epoxy; Flame guard; G2 (Oxide); Gibbsite; GK (Oxide); GO (Oxide); Hypalox II; Itaclor; Jubenon R; KA 101; Ketjen B; KHP 2; LA 6; Lucalox; Ludox Cl; Maftecmartipol; Martisorb; Martoxin; Microgrit WCA; Micropolish alumina; Oxido aluminico (Spanish); PS-1; PS-1 (alumina); Purdox; Q-Loid A 30; RC 172DBM; Realox

CAS Registry Number: 7429-90-5 (aluminum powder); 1344-28-1 (aluminum oxide); 1302-74-5 (emery/corundum, natural form of aluminum oxide)

RTECS Number: BD0330000 (aluminum metal powder); BD1200000 (aluminum oxide)

DOT ID: UN 1309 (coated Al powder); UN 1396 (uncoated Al powder)

EINECS Number: 215-691-6

Regulatory Authority

- Air Pollutant Standard Set (ACGIH)[1] (DFG)[3] (HSE)[33] (OSHA)[58] (Several States)[60] (Various Canadian Provinces) (Israel) (Mexico) (Australia)
- Canada, WHMIS, Ingredients Disclosure List
- CLEAN WATER ACT: Section 313 (57FR41331) Water Priority Chemicals (Al, dust or fume)
- EPA HAZARDOUS WASTE NUMBER (RCRA No.): D003
- SAFE DRINKING WATER ACT: 47FR9352, regulated contaminants; 53FR1896, list of contaminants, 55FR1470, Priority List; 40 CFR143.3, SMLC, 0.05 – 0.2 mg/l
- SARA 313: Form R *de minimis* Concentration Reporting Level: 1.0% (Al, dust and fume)
- SARA 313: Form R *de minimis* Concentration Reporting Level: 0.1% (Aluminum oxide, fibrous form only)

Cited in U.S. State Regulations: Alaska (G), California (A, G), Connecticut (A), Florida (G), Illinois (G), Kansas (W), Maine (G, W), Maryland (G), Massachusetts (G), Minnesota (G), New Hampshire (G), New Jersey (G), New York (G), Oklahoma (G), Pennsylvania (G), Rhode Island (G), West Virginia (G).

Description: Aluminum, is a combustible, light, silvery-white, soft, ductile, malleable, amphoteric metal. Freezing/Melting point = 660°C, Boiling point = 2,470°C. Autoignition temperature = 590°C. Hazard Identification (dust): Health 0, Flammability 3, Reactivity 1. Insoluble in water. The primary sources are the ores cryolite, gibbsite and bauxite, found as boehmite; aluminum is never found in the elemental state. Aluminum oxide is a noncombustible, white, crystalline powder or granules. Boiling point = about 3,000°C. Freezing/Melting point = 2,050°C. Insoluble in water.

Potential Exposure: Most hazardous exposures to aluminum occur in smelting and refining processes. Aluminum is mostly produced by electrolysis of Al_2O_3 dissolved in molten cryolite (Na_3AlF_6). Aluminum is alloyed with copper, zinc, silicon, magnesium, manganese, and nickel; special additives may include chromium, lead, bismuth, titanium, zirconium, and vanadium. Aluminum and its alloys can be extruded or processed in rolling mills, wireworks, forges, or foundries, and are used in the shipbuilding, electrical, building, aircraft, automobile, light engineering, and jewelry industries. Aluminum foil is widely used in packaging. Powdered aluminum is used in the paints and pyrotechnic industries. Alumina, emery, corundum has been used for abrasives, refractories, and catalysts, and in the past in the first firing of china and pottery.

Incompatibilities: Aluminum powder forms an explosive mixture with air and is a strong reducing agent that reacts violently with oxidizers, strong bases, strong acids, some halogenated hydrocarbons, nitrates, sulfates, metal oxides and many other substances. Keep away from combustible materials.

Permissible Exposure Limits in Air: The Federal OSHA TWA standard is15 mg/m^3, total dust; 5 mg/m^3, respirable fraction. NIOSH recommends 10 mg/m^3 total dust; 5 mg/m^3 aluminum metal, respirable fraction. A TWA value of 10 mg/m^3 is recommended by ACGIH for aluminum oxide [for total dust containing no asbestos and less than 1% crystalline silica). Also ACGIH recommends a TWA value of 5 mg/m^3 was set for aluminum pyro powders and for aluminum welding fumes. Australia and Israel have set TWA values similar to ACGIH. Mexico's TWA is 10 mg/m^3 and STEL of 20 mg/m^3. The German DFG MAK value for dust is established as a concentration of the respirable fraction (previously "fine dust") of 1.5 mg/m^3 and a concentration of the inhalable fraction (previously "total dust") of 4 mg/m^3. In addition, some states and Canadian provinces have set guidelines or standards for aluminum in ambient air:[60] 10 mg/m^3, total dust; 5 mg/m^3, respirable fraction (California), 3 mg/m^3 (Virginia), 4 mg/m^3 (Connecticut), 10 mg/m^3 (North Dakota), 23.8 mg/m^3 (Nevada), 10 mg/m^3 and STEL of 20 mg/m^3 (Alberta).

Determination in Air: Filter collection and atomic absorption analysis. See OSHA Method #ID-125G, NIOSH Methods 7013, and 7300.[18] See also NIOSH Method 0500 and 0600 for alumina dusts (respirable = 0600; total = 0500).

Permissible Concentration in Water: Under the Safe Drinking water Act, secondary MCLs = 0.05 – 0.2 mg/l Mexican drinking water criteria is 0.02 mg/l. In addition, guidelines for aluminum in drinking water have been set[61] ranging from 1,430 μg/l (Maine) to 5,000 μg/l (Kansas).

Routes of Entry: Inhalation, eye contact.

Harmful Effects and Symptoms

Short Term Exposure: Aluminum dust can cause irritation, and particles can scratch the eyes. Aluminum oxide can irritate the eyes, nose and respiratory tract. Particles of aluminum deposited in the eye may cause necrosis of the cornea. Salts of aluminum may cause dermatoses, eczema, conjunctivitis, and irritation of the mucous membranes of the upper respiratory system by the acid liberated by hydrolysis. The effects on the human body caused by inhalation of aluminum dust and fumes are not known with certainty at this time. Present data suggest that pneumoconiosis might be a possible outcome. In the majority of causes investigated, however, it was found that exposure was not to aluminum dust alone, but to a mixture of aluminum, silica fume, iron dusts, and other materials.

Long Term Exposure: There is evidence of an increase in bladder, lung, and other cancers among aluminum smelter workers. The increase appears to be due to polycyclic aromatic compound exposure, not to aluminum compounds. Aluminum salts are toxic to the animal fetus and cause fetal damage. Exposure to fine dust from aluminum or aluminum oxide can cause lung damage, pneumonia, and pulmonary fibrosis, with symptoms of coughing, wheezing and shortness of breath. Very high levels of aluminum may cause brain damage.

Points of Attack: Lungs, eyes.

Medical Surveillance: Pre-employment and periodic physical examinations should give special consideration to the skin, eyes, and lungs. Lung function should be followed, including chest x-ray, pulmonary function tests.

First Aid: If this chemical gets into the eyes, remove any contact lenses at once and irrigate immediately for at least 15 minutes, occasionally lifting upper and lower lids. Seek medical attention immediately. If this chemical contacts the skin, remove contaminated clothing and wash immediately with soap and water. Seek medical attention immediately. If this chemical has been inhaled, remove from exposure, begin rescue breathing (using universal precautions) if breathing has stopped and CPR if heart action has stopped. Transfer promptly to a medical facility. When this chemical has been swallowed, get medical attention. Give large quantities of water and induce vomiting. Do not make an unconscious person vomit.

Personal Protective Methods: Workers in electrolysis manufacturing plants should be provided with respirators for protection from fluoride fumes. Dust masks are recommended in areas exceeding the nuisance levels. Aluminum workers generally should receive training in the proper use of personal protective equipment. Workers involved with salts of aluminum may require protective clothing, barrier creams, and where heavy concentrations exist, full face air supplied respirators may be indicated. Wear protective gloves and clothing to prevent any reasonable probability of skin contact. Safety equipment suppliers/manufacturers can provide recommendations on the most protective glove/clothing material for your operation. All protective clothing (suits, gloves, footwear, headgear) should be clean, available each day, and put on before work. Contact lenses should not be worn when working with this chemical. Wear dust-proof chemical goggles and face shield unless full facepiece respiratory protection is worn. Employees should wash immediately with soap when skin is wet or contaminated. Provide emergency showers and eyewash.

Respirator Selection: Engineering controls should be used wherever feasible to maintain airborne concentrations of this chemical below the prescribed exposure limit. Respirators and protective equipment less effective than engineering controls, and should be used only in non-routine or emergency situations which may result in exposure concentrations in excess of the TWA environmental limit. *At any concentrations above the NIOSH REL:* SCBAF:PD,PP (any MSHA/NIOSH approved self-contained breathing apparatus that has a full facepiece and is operated in a pressure-demand or other positive-pressure mode); or SAF:PD,PP:ASCBA (any supplied-air respirator that has a full facepiece and is operated in a pressure-demand or other positive-pressure mode in combination with an auxiliary, self-contained breathing apparatus operated in a pressure-demand or other positive pressure mode). *Escape:* GMFOVHiE [any air-purifying, full-facepiece respirator (gas mask) with a chin-style, front- or back-mounted organic vapor canister having a high-efficiency particulate filter]; or SCBAE (any appropriate escape-type, self-contained breathing apparatus).

Storage: Prior to working with this chemical you should be trained on its proper handling and storage. Keep aluminum powder dry and isolate from acids, caustics, chlorinated hydrocarbons, oxidizers, and combustible materials. Store both metal and oxide in tightly closed containers, in a cool, well-ventilated area, protected from physical damage.

Shipping: Coated aluminum powder requires a shipping label of "Flammable Solid." It falls in DOT Hazard Class 4.1 and Packing Group II.

Spill Handling: For aluminum metal spills, special care must be taken with aluminum powder which may be very reactive. Evacuate persons not wearing protective equipment from area of spill or leak until clean-up is complete. Remove all ignition sources. Do not use water to clean up spilled aluminum powder. Do not raise dust level by sweeping. Collect powdered material in the most convenient and safe manner and deposit in sealed containers. Ventilate area after clean-up is complete. It may be necessary to contain and dispose of this chemical as a hazardous waste. If material or contaminated runoff enters waterways, notify downstream users of potentially contaminated waters. Contact your Department of Environmental Protection or your regional office of the federal EPA for specific recommendations. If employees are required to clean-up spills, they must be properly trained and equipped. OSHA 1910.120(q) may be applicable. May be disposed of as an inert solid in a landfill.

Fire Extinguishing: With aluminum metal, do not use water or halogenated agents. Aluminum powder is a combustible solid. Aluminum oxide is not flammable. Use dry chemical, carbon dioxide, foam extinguishers. If employees are expected to fight fires, they must be trained and equipped in OSHA 1910.156.

Disposal Method Suggested: Consult with environmental regulatory agencies for guidance on acceptable disposal practices. Generators of waste containing this contaminant (≥100 kg/mo) must conform with EPA regulations governing storage, transportation, treatment, and waste disposal. Aluminum Oxide-Disposal in a sanitary landfill. Mixing of industrial process wastes and municipal wasters at such sites is not encouraged however. Aluminum powder may be recovered and sold as scrap. Recycle and recovery is a viable option to disposal for aluminum metal and aluminum fluoride (A-57).

References

U.S. Environmental Protection Agency, Chemical Hazard Information Profile: Aluminum and Aluminum Compounds, Washington, DC (September 1976).

National Institute for Occupational Safety and Health, Profiles on Occupational Hazards for Criteria Document Priorities: Aluminum and Its Compounds, Report PB 274–073, Washington, DC, pp 80–84 (1977)

U.S. Environmental Protection Agency, Toxicology of Metals, Vol. II: aluminum, Report EPA 600/1-77-022, Research Triangle Park, NC, pp 4–14 (May 1977)

Sax, N. I., Ed., Dangerous Properties of Industrial Materials Report, 1, No. 4, 34, (1981) (Aluminum)

Sax, N. I., Ed., Dangerous Properties of Industrial Materials Report, 1, No. 5, 33–34, (1981) (Aluminum Oxide and Aluminum Silicate)

New Jersey Department of Health and Senior Services, "Hazardous Substance Fact Sheet: Aluminum," Trenton, NJ (February 1989)

New Jersey Department of Health and Senior Services, "Hazardous Substance Fact Sheet: Aluminum Oxide," Trenton, NJ (January 1989)

New York State Department of Health, "Chemical Fact Sheet: Aluminum Oxide," Albany, NY, Bureau of Toxic Substance Assessment (March 1986)

Aluminum Chloride

Molecular Formula: $AlCl_3$

Synonyms: Alluminio (cloruro di); Aluminumchlorid (German); Aluminum chloride (1:3); Aluminum chloride, anhydrous; Aluminum chloride solution; Aluminum trichloride; Anhydrol forte; Anhydrous aluminum chloride; Chlorure d'aluminum (French); Clorato aluminico (Spanish); Driclor; PAC; PAC (Van); Pearsall; Praestol K2001; Trichloro aluminum

CAS Registry Number: 7446-70-0

RTECS Number: BD0525000

DOT ID: UN 1726 (anhydrous); UN 2581 (Solutions).

EEC Number: 013-003-00-7

EINECS Number: 231-208-1

Regulatory Authority

- Air Pollutant Standard Set (ACGIH)[1] (HSE)[33] (NIOSH)[58]
- Canada, WHMIS, Ingredients Disclosure List

Cited in U.S. State Regulations: Alaska (G), Florida (G), Maine (G), Massachusetts (G), New Hampshire (G), New Jersey (G), New York (G), Oklahoma (G), Pennsylvania (G), Rhode Island (G), West Virginia (G).

Description: Aluminum chloride, $AlCl_3$, is a noncombustible but highly reactive whitish-gray, yellow, or green powder or liquid. Strong, acidic, irritating odor like hydrochloric acid. Sinks in water and reacts, forming toxic and corrosive hydrogen chloride gas. Odor threshold = 6.31 ppm (as hydrogen chloride). Hazard Identification (based on NFPA-704 M Rating System): Health 3, Flammability 0, Reactivity 2. Water reactive.

Potential Exposure: It is used in making other chemicals and dyes, astringents, deodorants, and in the petroleum refining and the rubber industries.

Incompatibilities: Contact with air or water forms hydrochloric acid and hydrogen chloride gas. Reaction with water may be violent. Water, alcohol, and alkenes can cause polymerization. Incompatible with nitrobenzene, organic material, and bases. Attacks metal in presence of moisture, forming flammable hydrogen gas.

Permissible Exposure Limits in Air: NIOSH recommends an airborne exposure limit of 2 mg/m^3 TWA over a 10-hour workshift. ACGIH recommends an airborne exposure limits of 2 mg/m^3 TWA over an 8-hour workshift.

Permissible Concentration in Water: An ambient water limit of 73 μg/l for aluminum compounds has been suggested by EPA[32] based on health effects.

Routes of Entry: Inhalation, ingestion, skin contact.

Harmful Effects and Symptoms

Short Term Exposure: Corrosive to eyes, skin, and respiratory tract. Inhalation may cause pulmonary edema, which can be delayed for several hours; there is a risk of death in serious cases. Ingestion can cause severe burns of mouth, throat and stomach, vomiting, watery or bloody diarrhea, kidney damage, jaundice and liver damage, collapse and convulsions. Estimated lethal dose is about 8 ounces for an average 150 pound adult.

Long Term Exposure: May cause pulmonary fibrosis, and reduced lung function with symptoms of coughing, wheezing, and shortness of breath.

Points of Attack: Eyes, skin, respiratory system.

Medical Surveillance: Lung function tests. If symptoms develop or overexposure is suspected, chest x-ray should be considered.

First Aid: If this chemical gets into the eyes, remove any contact lenses at once and irrigate immediately for at least 30 minutes, occasionally lifting upper and lower lids. Seek medical attention immediately. If this chemical contacts the skin, remove contaminated clothing and wash immediately with soap and water. Seek medical attention immediately. If this chemical has been inhaled, remove from exposure, begin rescue breathing (using universal precautions) if breathing has stopped and CPR if heart action has stopped. Transfer promptly to a medical facility. When this chemical has been swallowed, get medical attention. If victim is conscious, administer water or milk. Do not induce vomiting. Medical observation is recommended for 24 – 48 hours after breathing overexposure, as pulmonary edema may be delayed.

Personal Protective Methods: Wear protective gloves and clothing to prevent any reasonable probability of skin contact. Safety equipment suppliers/manufacturers can provide recommendations on the most protective glove/clothing material for your operation. All protective clothing (suits, gloves, footwear, headgear) should be clean, available each day, and put on before work. Contact lenses should not be worn when working with this chemical. When working with liquid or solid, wear indirect vent, splash or dust-proof chemical goggles and face shield unless full facepiece respiratory protection is worn. Employees should wash immediately with soap when skin is wet or contaminated. Provide emergency showers and eyewash.

Respirator Selection: If dust levels are high, wear a dust mask. If hydrogen chloride is present, use *50 ppm:* CCRS (any MSHA/NIOSH approved chemical cartridge respirator with cartridge(s) providing protection against the compound of concern); or GMFS [any air-purifying, full-facepiece respirator (gas mask) with a chin-style, front-or back-mounted canister providing protection against the compound of concern]; or PAPRS [any powered, air-purifying respirator with cartridge(s) providing protection against the compound of concern]; or SA (any supplied-air respirator); or SCBAF (any self-contained breathing apparatus with a full facepiece). *Emergency or planned entry into unknown concentrations or IDLH conditions:* SCBAF:PD,PP (any self-contained breathing apparatus that has a full facepiece and is operated in a pressure-demand or other positive-pressure mode); or SAF:PD,PP:ASCBA (any supplied-air respirator that has a full facepiece and is operated in a pressure-demand or other positive-pressure mode in combination with an auxiliary, self-contained breathing apparatus operated in a pressure-demand or other positive-pressure mode). *Escape:* GMFAG [any air-purifying, full-facepiece respirator (gas mask) with a chin-style, front-or back-mounted organic vapor canister]; or SCBAE (any appropriate escape-type, self-contained breathing apparatus).

Storage: Prior to working with this chemical you should be trained on its proper handling and storage. Store in a cool, dry area in tightly closed container away from moisture, heat, and sunlight.

Shipping: Anhydrous $AlCl_3$ as well as the solution requires a "Corrosive" label according to DOT. Both salt and solution fall in DOT/UN Hazard Class 8.[19][20] The salt falls in Packing Group II and the solution in Packing Group III.

Spill Handling: Evacuate persons not wearing protective equipment from area of spill or leak until clean-up is complete. Remove all ignition sources. Sweep carefully, being careful not to raise dust. Do not get water inside containers; keep combustibles away. For liquid Aluminum Chloride, absorb liquids in vermiculite, dry sand, earth, or a similar material and deposit in sealed containers. Collect powdered material in the most convenient and safe manner and deposit in sealed containers. Ventilate area after clean-up is complete. It may be necessary to contain and dispose of this chemical as a hazardous waste. If material or contaminated runoff enters waterways, notify downstream users of potentially contaminated waters. Contact your Department of Environmental Protection or your regional office of the federal EPA for specific recommendations. If employees are required to clean-up spills, they must be properly trained and equipped. OSHA 1910.120(q) may be applicable.

Fire Extinguishing: Do not use water; will form poisonous fumes of hydrochloric acid. On solid material use any extinguishing medium suitable for surrounding fire. On solution, use dry chemical, carbon dioxide, or polymer foam extinguishers. Poisonous gases are produced in fire including hydrochloric acid, aluminum oxide, and nitrogen oxides. Storage containers and parts of containers may rocket great distances, in many directions. If material or contaminated runoff enters waterways, notify downstream users of potentially contaminated waters. Notify local health and fire officials and pollution control agencies. From a secure, explosion-proof location, use water spray to cool exposed containers. If cooling streams are ineffective (venting sound increases in volume and pitch, tank discolors, or shows any signs of deforming), withdraw immediately to a secure position. If employees are expected to fight fires, they must be trained and equipped in OSHA 1910.156.

Disposal Method Suggested: May be sprayed with aqueous ammonia in the presence of ice and, when reaction is complete, flushed down drain with running water.[22]

References

New Jersey Department of Health and Senior Services, "Hazardous Substance Fact Sheet: Aluminum Chloride," Trenton, NJ (January 1999)

New York State Department of Health, "Chemical Fact Sheet: Aluminum Chloride," Albany, NY, Bureau of Toxic Substance Assessment (March 1986)

Aluminum Fluoride

Molecular Formula: AlF_3

Synonyms: Aluminum fluoride, anhydrous; Aluminum fluorure (French); Aluminum trifluoride; Fluorid hlinity (Czech); Fluoruro aluminico hidratado (Spanish)

CAS Registry Number: 7784-18-1

RTECS Number: BD0725000

DOT ID: UN 3077

Regulatory Authority

- Air Pollutant Standard Set (ACGIH)[1] (DFG)[3] (HSE)[33] (former USSR)[43] (OSHA)[58] (Several States)[60]
- SAFE DRINKING WATER ACT: Regulated chemical (47 FR 9352); MCL, 4.0 mg/l; MCLG, 4.0 mg/l; SMCL, 2.0 mg/l.
- Canada, WHMIS, Ingredients Disclosure List

Cited in U.S. State Regulations: New Jersey (G), Pennsylvania (G).

Description: Aluminum fluoride, AlF_3, is a white, odorless powder or granule. Freezing/Melting point = 1,291°C. Boiling point = 1,537°C. Hazard Identification (based on NFPA-704 M Rating System): Health 3, Flammability 0, Reactivity 0. Slightly soluble in water.

Potential Exposure: Used in the manufacture of ceramics, enamels, aluminum, and aluminum silicate, as flux in metallurgy, fermentation inhibitor,

Incompatibilities: Reacts violently with potassium or sodium.

Permissible Exposure Limits in Air: Aluminum fluoride is an *aluminum soluble salt.* NIOSH recommends an exposure limit of 2 mg/m^3 TWA for a 10-hour workshift. ACGIH recommends the same level TWA for an 8-hour workshift. For *fluorides* (measured as fluoride), the Federal OSHA permissible exposure limit is 2.5 mg/m^3 TWA over an 8-hour workshift. ACGIH recommends the same limit for fluoride. Aluminum fluoride can be absorbed through the skin, thereby increasing exposure.

Permissible Concentration in Water: Fluoride is a safe drinking water act regulated chemical (47FR9352, 56FR 3594); MCL = 4.0 mg/l; MCLG = 4.0 mg/l; SMCL = 2.0 mg/l. The State of Maine has set 2.4 mg/l as a guideline for drinking water. Arizona[61] has set 1.8 mg/l as a standard for drinking water.

Routes of Entry: Inhalation, cutaneous absorption.

Harmful Effects and Symptoms

Short Term Exposure: This chemical irritates eyes, skin, and respiratory tract. Skin and eye contact can cause severe burns. Inhalation can cause nose and throat irritation, with possible nose bleeding.

Long Term Exposure: May cause lung irritation, the development of bronchitis, with coughing, shortness of breath, phlegm. This chemical may cause asthma-like allergy, and may lead to dense but brittle bones. It may also lead to stiffening of the joints.

Points of Attack: Lungs, bones, joints.

Medical Surveillance: Urine tests for fluoride. Lung function tests. If symptoms develop or overexposure is suspected, chest x-ray should be considered.

First Aid: If this chemical gets into the eyes, remove any contact lenses at once and irrigate immediately for at least 15 minutes, occasionally lifting upper and lower lids. Seek medical attention immediately. If this chemical contacts the skin, remove contaminated clothing and wash immediately with soap and water. Seek medical attention immediately. If this chemical has been inhaled, remove from exposure, begin rescue breathing (using universal precautions) if breathing has stopped and CPR if heart action has stopped. Transfer promptly to a medical facility. When this chemical has been swallowed, get medical attention. Give large quantities of water and induce vomiting. Do not make an unconscious person vomit.

Personal Protective Methods: Wear protective gloves and clothing to prevent any reasonable probability of skin contact. Safety equipment suppliers/manufacturers can provide recommendations on the most protective glove/clothing material for your operation. All protective clothing (suits, gloves, footwear, headgear) should be clean, available each day, and put on before work. Contact lenses should not be worn when working with this chemical. Wear dust-proof chemical goggles and face shield unless full facepiece respiratory protection is worn. Employees should wash immediately with soap when skin is wet or contaminated. Remove nonimpervious clothing immediately if wet or contaminated. Provide emergency showers and eyewash.

Respirator Selection: NIOSH/OSHA as fluorine:1 ppm: SA* (any supplied-air respirator); 2.5 ppm: SA:CF* (any supplied-air respirator operated in a continuous-flow mode); 5 ppm: SCBAF (any self-contained breathing apparatus with a full facepiece); or SAF (any supplied-air respirator with a full facepiece); 25 ppm: SAF:PD,PP (any supplied-air respirator that has a full facepiece and is operated in a pressure-demand or other positive-pressure mode). *Emergency or planned entry into unknown concentrations or IDLH conditions:* SCBAF:PD,PP (any self-contained breathing apparatus that has a full facepiece and is operated in a pressure-demand or other positive-pressure mode); or SAF:PD,PP: ASCBA (any supplied-air respirator that has a full facepiece and is operated in a pressure-demand or other positive-pressure mode in combination with an auxiliary self-contained breathing apparatus operated in a pressure-demand or other positive-pressure mode). *Escape:* GMFS [end of service life indicator (ESLI) required] (any air-purifying, full-facepiece respirator (gas mask) with a chin-style, front- or back-mounted canister providing protection against the compound of concern]; or SCBAE (any appropriate escape-type, self-contained breathing apparatus).

* Substance reported to cause eye irritation or damage; may require eye protection.

Storage: Prior to working with this chemical you should be trained on its proper handling and storage. Store in tightly closed containers in a cool, well ventilated area away from sodium, potassium, and glass.

Shipping: Environmentally hazardous solid, n.o.s. Hazard Class: 9. Label: "Class 9."

Spill Handling: Avoid contact with dust. Evacuate persons not wearing protective equipment from area of spill or leak until clean-up is complete. Ventilate area after clean-up is complete. It may be necessary to contain and dispose of this chemical as a hazardous waste. If material or contaminated runoff enters waterways, notify downstream users of potentially contaminated waters. Contact your Department of Environmental Protection or your regional office of the federal EPA for specific recommendations. If employees are required to clean-up spills, they must be properly trained and equipped. OSHA 1910.120(q) may be applicable.

Fire Extinguishing: Aluminum fluoride is not combustible. Poisonous gases including fluorines, are produced in fire. Do not use water as flammable hydrogen gas if formed. Use any extinguishing agent suitable for surrounding fires. If material or contaminated runoff enters waterways, notify downstream users of potentially contaminated waters. Notify local health and fire officials and pollution control agencies. If employees are expected to fight fires, they must be trained and equipped in OSHA 1910.156.

Disposal Method Suggested: Neutralize with soda ash; add slaked lime; let stand for 24 hours. Transfer sludge to sewage facility.

References

New Jersey Department of Health and Senior Services, Hazardous Substance Fact Sheet, Aluminum Hydride, Trenton NJ (November 1998)

Aluminum Nitrate

Molecular Formula: $AlH_{18}N_3O_{18}$

Common Formula: $AlN_3O_9{\cdot}9H_2O$

Synonyms: Aluminum nitrate, nonahydrate; Aluminum(III) nitrate, Nonahydrate (1:3:9); Aluminum trinitrate nonahydrate; Nitrato aluminico (Spanish); Nitric acid, aluminum salt; Nitric acid, aluminum(3+) salt; Nitric acid, aluminum(III) salt

CAS Registry Number: 13473-90-0

RTECS Number: BD1040000

DOT ID: UN 1438

Regulatory Authority

- Air Pollutant Standard Set (ACGIH)[1] (HSE)[33] (OSHA)[58]
- Canada, WHMIS, Ingredients Disclosure List

Cited in U.S. State Regulations: Alaska (G), Maine (G), New Hampshire (G), New Jersey (G), Oklahoma (G), West Virginia (G)

Description: Aluminum nitrate, is an odorless, white crystalline solid, often in liquid solution. Boiling point = 135 °C (decomposes). Freezing/Melting point = 70 – 74°C. Hazard

Identification (based on NFPA-704 M Rating System): Health 2, Flammability 1, Reactivity 0. Soluble in water.

Potential Exposure: Aluminum nitrate is used in tanning leather, as an antiperspirant, as a corrosion inhibitor, in the extraction of uranium and as a nitrating agent.

Incompatibilities: Aluminum nitrate is a strong oxidizer; avoid contact with flammable or combustible materials, and reducing agents. In solution this chemical is a strong acid; avoid contact with bases. Explosions may occur when aluminum nitrate is shocked or exposed to heat.

Permissible Exposure Limits in Air: The TWA for aluminum soluble salts, measured as aluminum. NIOSH recommends a TWA of 2.0 mg/m^3 averaged over a 10-hour workshift. ACGIH and HSE[33] recommend the same limit averaged over an 8-hour workshift.

Permissible Concentration in Water: An ambient water level of 73 μg/l for aluminum compounds has been suggested by EPA[32] based on health effects.

Routes of Entry: Inhalation

Harmful Effects and Symptoms

Short Term Exposure: Irritates the eyes, skin and respiratory tract. Eye contact may cause permanent damage. High exposures can cause unconsciousness. Prolonged contact can cause skin disorders. Ingestion can cause stomach cramps, nausea, blue skin, weakness, possible blood problems. The oral LD_{50} for rats is 264 mg/kg.

Long Term Exposure: Repeated contact can cause skin problems and eczema. This chemical is a corrosive, and it may cause lung problems.

Points of Attack: Respiratory system, eyes, skin, gastric system.

Medical Surveillance: Lung function tests. A qualified allergist should be consulted if skin problems occur.

First Aid: If this chemical gets into the eyes, remove any contact lenses at once and irrigate immediately for at least 15 minutes, occasionally lifting upper and lower lids. Seek medical attention immediately. If this chemical contacts the skin, remove contaminated clothing and wash immediately with soap and water. Seek medical attention immediately. If this chemical has been inhaled, remove from exposure, begin rescue breathing (using universal precautions) if breathing has stopped and CPR if heart action has stopped. Transfer promptly to a medical facility. When this chemical has been swallowed, get medical attention. Give large quantities of water and induce vomiting. Do not make an unconscious person vomit.

Personal Protective Methods: Wear acid-resistant gloves and clothing to prevent any reasonable probability of skin contact. Safety equipment suppliers/manufacturers can provide recommendations on the most protective glove/clothing material for your operation. All protective clothing (suits, gloves, footwear, headgear) should be clean, available each day, and put on before work. Contact lenses should not be worn when working with this chemical. Wear splash- or dust-proof chemical goggles and face shield unless full facepiece respiratory protection is worn. Employees should wash immediately with soap when skin is wet or contaminated. Provide emergency showers and eyewash.

Respirator Selection: Engineering controls should be used wherever feasible to maintain airborne concentrations of this chemical below the prescribed exposure limit. Respirators and protective equipment less effective than engineering controls, and should be used only in non-routine or emergency situations which may result in exposure concentrations in excess of the TWA environmental limit. *At any concentrations above the NIOSH REL:* SCBAF:PD,PP (any MSHA/NIOSH approved self-contained breathing apparatus that has a full facepiece and is operated in a pressure-demand or other positive-pressure mode); or SAF:PD,PP:ASCBA (any supplied-air respirator that has a full facepiece and is operated in a pressure-demand or other positive-pressure mode in combination with an auxiliary, self-contained breathing apparatus operated in a pressure-demand or other positive pressure mode). *Escape:* GMFOVHiE [any air-purifying, full-facepiece respirator (gas mask) with a chin-style, front- or back-mounted organic vapor canister having a high-efficiency particulate filter]; or SCBAE (any appropriate escape-type, self-contained breathing apparatus).

Storage: Prior to working with this chemical you should be trained on its proper handling and storage. Prior to working with this chemical you should be trained on its proper handling and storage. Store in tightly capped or sealed containers in a cool, well ventilated area away from combustible materials, heat or flame. Protect containers from physical shock. See OSHA standard 1910.104 and NFPA 43A *Code for the Storage of Liquid and Solid Oxidizers* for detailed handling and storage regulations.

Shipping: Label required is "Oxidizer." Falls in DOT/UN Hazard Class 5.1 and Packing Group III.[19][20]

Spill Handling: Evacuate persons not wearing protective equipment from area of spill or leak until clean-up is complete. Remove all ignition sources. Sweep into a beaker. Dilute by adding slowly to water. Add soda ash then neutralize with HCl. Flush to sewer with large volume of water.[24] Ventilate area after clean-up is complete. It may be necessary to contain and dispose of this chemical as a hazardous waste. If material or contaminated runoff enters waterways, notify downstream users of potentially contaminated waters. Contact your Department of Environmental Protection or your regional office of the federal EPA for specific recommendations. If employees are required to clean-up spills, they must be properly trained and equipped. OSHA 1910.120(q) may be applicable.

Fire Extinguishing: Not flammable per se. Poisonous gases including nitrogen oxides are produced in fire. Use dry chemical, carbon dioxide, or water spray. Vapors are heavier than air and will collect in low areas. Storage containers and parts of containers may rocket great distances, in many directions. If material or contaminated runoff enters waterways, notify downstream users of potentially contaminated waters. Notify local health and fire officials and pollution control agencies. From a secure, explosion-proof location, use water spray to cool exposed containers. If cooling streams are ineffective (venting sound increases in volume and pitch, tank discolors, or shows any signs of deforming), withdraw immediately to a secure position. If employees are expected to fight fires, they must be trained and equipped in OSHA 1910.156.

Disposal Method Suggested: See Spill Handling above.

References

New Jersey Department of Health and Senior Services, "Hazardous Substance Fact Sheet: Aluminum Nitrate," Trenton, NJ (February 1989)

Aluminum Phosphate

Molecular Formula: $AlPO_4$

Synonyms: Aluminophosphoric acid; Aluminum acid phosphate; Aluminum monophsophate; Aluminum orthophosphate; Aluphos; Fosfato aluminico (Spanish); Monoaluminum phosphate; Ortofosfato aluminico (Spanish); Phosphalugel; Phosphoric acid, aluminum salt

CAS Registry Number: 7784-30-7 (solution)

RTECS Number: TB6450000

DOT ID: NA 1760

Regulatory Authority

- Air Pollutant Standard Set (ACGIH)[1] (NIOSH)
- See also: Aluminum

Cited in U.S. State Regulations: Alaska (G), Maine (G), New Hampshire (G), New Jersey (G), West Virginia (G).

Description: Aluminum phosphate, $AlPO_4$, is a white crystal which is often used in liquid or gel form. Freezing/Melting point ≥1,450°C.

Potential Exposure: Used as a flux in ceramics, used in dental cements, in the manufacture of special glasses, paints and varnishes, cosmetics, making pulp and paper; as an antacid.

Incompatibilities: A strong oxidizer,; keep away from combustible materials. Violent reaction with reducing agents, strong bases.

Permissible Exposure Limits in Air: The NIOSH recommended airborne exposure limit for soluble aluminum salts (measured as aluminum) is 2 mg/m^3 TWA for a 10-hour workshift. ACGIH recommends the same criterion for an 8-hour workshift.

Permissible Concentration in Water: An ambient water level of 73 μg/l for aluminum compounds has been suggested by EPA[32] based on health effects.

Routes of Entry: Inhalation.

Harmful Effects and Symptoms

Short Term Exposure: Aluminum Phosphate can affect you when breathed in. Aluminum Phosphate is a corrosive substance. It can cause severe burns of the eyes and skin on contact. Exposure to the dust can irritate the eyes, nose, throat, and bronchial tubes. Fine powder can irritate the lungs.

Long Term Exposure: Can irritate the lungs, causing bronchitis with coughing, shortness of breath, and phlegm.

Points of Attack: Lungs, eyes, skin.

Medical Surveillance: Lung function tests.

First Aid: If this chemical gets into the eyes, remove any contact lenses at once and irrigate immediately for at least 15 minutes, occasionally lifting upper and lower lids. Seek medical attention immediately. If this chemical contacts the skin, remove contaminated clothing and wash immediately with soap and water. Seek medical attention immediately. If this chemical has been inhaled, remove from exposure, begin rescue breathing (using universal precautions) if breathing has stopped and CPR if heart action has stopped. Transfer promptly to a medical facility. When this chemical has been swallowed, get medical attention. If victim is *conscious*, administer water or milk. Do not induce vomiting.

Personal Protective Methods: Wear acid-resistant gloves and clothing to prevent any reasonable probability of skin contact. Safety equipment suppliers/manufacturers can provide recommendations on the most protective glove/clothing material for your operation. All protective clothing (suits, gloves, footwear, headgear) should be clean, available each day, and put on before work. Contact lenses should not be worn when working with this chemical. Wear splash-proof chemical goggles and face shield unless full facepiece respiratory protection is worn. Employees should wash immediately with soap when skin is wet or contaminated. Provide emergency showers and eyewash.

Respirator Selection: Where the potential exists for exposures over 2 mg/m^3 of aluminum, use a MSHA/NIOSH approved full facepiece respirator equipped with particulate (dust/fume/mist) filters. Particulate filters must be checked every day before work for physical damage, such as rips or tears, and replaced as needed. Where the potential for high exposures exists, use a MSHA/NIOSH approved supplied-air respirator with a full facepiece operated in the positive pressure mode or with a full facepiece hood, or helmet in the continuous flow mode, or use a MSHA/NIOSH approved self-contained breathing apparatus with a full facepiece operated in pressure-demand or other positive pressure mode.

Storage: Prior to working with this chemical you should be trained on its proper handling and storage. Store in tightly closed containers in a cool, well-ventilated area away from strong bases and combustible materials, such as wood, paper, and oil.

Shipping: Aluminum phosphate (solid) is not cited in DOT regulations. Aluminum phosphate solutions are covered under Corrosive Liquids n.o.s.[19] Hazard Class: 8. Label: "Corrosive."

Spill Handling: Restrict persons not wearing protective equipment from area of spill until clean-up is complete. Ventilate area of spill or leak. Collect spilled material in the most convenient and safe manner and deposit in sealed containers for reclamation or for disposal in an approved facility. Absorb liquid containing aluminum phosphate in vermiculite, dry sand, earth or similar material. It may be necessary to contain and dispose of this chemical as a hazardous waste. If material or contaminated runoff enters waterways, notify downstream users of potentially contaminated waters. Contact your Department of Environmental Protection or your regional office of the federal EPA for specific recommendations. If employees are required to clean-up spills, they must be properly trained and equipped. OSHA 1910.120(q) may be applicable.

Fire Extinguishing: Aluminum Phosphate is non-flammable. Poisonous gases, including phosphine are produced in fire. Use dry chemical powder extinguishers. If material or contaminated runoff enters waterways, notify downstream users of potentially contaminated waters. Notify local health and fire officials and pollution control agencies. If employees are expected to fight fires, they must be trained and equipped in OSHA 1910.156.

References

New Jersey Department of Health and Senior Services, "Hazardous Substance Fact Sheet: Aluminum Phosphate," Trenton, NJ (June 1998)

Aluminum Phosphide

Molecular Formula: AlP

Synonyms: AlP; Al-phos; Aluminum fosfide (Dutch); Aluminum monophosphide; Celphide; Celphos; Delicia; Fosfuri di alluminio (Italian); Fosfuro aluminico (Spanish); Phosphures d'aluminum (French); Phostoxin®; Quickphos

CAS Registry Number: 20859-73-8

RTECS Number: BD1400000

DOT ID: UN 1397; UN 3048 (insecticide)

Regulatory Authority

- Banned or Severely Restricted (Belgium)[13]
- U.S. DOT 49CFR172.101, Inhalation Hazardous Chemical
- Canada, WHMIS, Ingredients Disclosure List
- EPA HAZARDOUS WASTE NUMBER (RCRA No.): P006[5]
- RCRA, 40CFR261, Appendix 8 Hazardous Constituents
- SUPERFUND/EPCRA 40CFR355, Appendix B Extremely Hazardous Substances: TPQ = 500 lb (228 kg)
- SUPERFUND/EPCRA 40CFR302.4 Reportable Quantity (RQ): CERCLA, 100 lb (45.4 kg)
- EPCRA Section 313 Form R *de minimis* concentration reporting level: 1.0%

Cited in U.S. State Regulations: California (G), Illinois (G), Kansas (G), Louisiana (G), Maine (G), Massachusetts (G), New Hampshire (G), New Jersey (G), Oklahoma (G), Pennsylvania (G), Vermont (G), Virginia (G), Washington (G), Wisconsin (G).

Description: Aluminum phosphide is a pyrophoric, dark gray or dark yellow crystalline solid. Freezing/Melting point ≥ 1,000°C. Decomposes in water forming poisonous and flammable phosphine gas. NFPA-704 Hazard Identification (based on NFPA-704 M Rating System): Health 4, Flammability 4, Reactivity 2, Water reactive; dangerous when wet.

Potential Exposure: Used as an insecticidal fumigant for grain, peanuts, processed food, animal feed, leaf tobacco, cottonseed, and as space fumigant for flour mills, warehouses and railcars. Used as a source of phosphine; in semiconductor research.

Incompatibilities: Reacts violently with water, carbon dioxide, and foam fire extinguishers. Contact with water and bases slowly releases flammable phosphine gas. Contact with steam and acids may be violent. Can ignite spontaneously in air.

Permissible Exposure Limits in Air: The NIOSH recommended airborne exposure limit for soluble aluminum salts (measured as aluminum) is 2 mg/m^3 TWA for a 10-hour workshift. ACGIH recommends the same criterion for an 8-hour workshift.

Routes of Entry: Inhalation, ingestion.

Harmful Effects and Symptoms

Acute toxicity occurs primarily by the inhalation route when aluminum phosphide decomposes into the toxic gas, phosphine. The human median lethal dose for aluminum phosphide has been reported to be 20 mg/kg. Rated as super toxic: probable oral lethal dose is less than 5 mg/kg or less than 7 drops for a 70 kg (150 lb) person. Symptoms of phosphine gas poisoning include restlessness, headache, dizziness, fatigue, nausea, vomiting, coma, convulsions; lowered blood pressure, pulmonary edema, respiratory failure, and disorders of the kidney, liver, heart, and brain may be observed.

Short Term Exposure: A severe health hazard. Irritates the eye, skin and respiratory tract. Inhalation can cause lung irritation with coughing, wheezing, and shortness of breath. Affects metabolism and the central nervous system; exposure can lead to death. Higher exposures can cause pulmonary edema, a medical emergency that can be delayed for several hours. This can cause death.

Long Term Exposure: May cause lung, kidney, and liver damage. May be able to cause skin rash or eczema.

Points of Attack: Central nervous system, liver, kidney, lungs.

Medical Surveillance: Lung, liver, kidney, and nervous system function tests.

First Aid: If this chemical gets into the eyes, remove any contact lenses at once and irrigate immediately for at least 15 minutes, occasionally lifting upper and lower lids. Seek medical attention immediately. If this chemical contacts the skin, remove contaminated clothing and wash immediately with soap and water. Seek medical attention immediately. If this chemical has been inhaled, remove from exposure, begin rescue breathing (using universal precautions) if breathing has stopped and CPR if heart action has stopped. Transfer promptly to a medical facility. When this chemical has been swallowed, get medical attention. Give large quantities of water and induce vomiting. Do not make an unconscious person vomit. Medical observation is recommended for 24 – 48 hours after breathing overexposure, as pulmonary edema may be delayed. As first aid for pulmonary edema, a doctor or authorized paramedic may consider administering a corticosteroid spray.

Personal Protective Methods: Wear protective gloves and clothing to prevent any reasonable probability of skin contact. Safety equipment suppliers/manufacturers can provide recommendations on the most protective glove/clothing material for your operation. All protective clothing (suits, gloves, footwear, headgear) should be clean, available each day, and put on before work. Contact lenses should not be worn when working with this chemical. Wear dust-proof chemical goggles and face shield unless full facepiece respiratory protection is worn. Employees should wash immediately with soap when skin is wet or contaminated. Provide emergency showers and eyewash.

Respirator Selection: Where the potential exists for exposures over 2 mg/m^3 (aluminum), use a MSHA/NIOSH approved respirator equipped with particulate (dust/fume/mist) filters. Particulate filters must be checked every day before work for physical damage such as rips or tears, and replaced as needed, or where any potential for exposures to Phosphine gas exist, use a MSHA/NIOSH approved gas mask (Approval number TC-14-98) equipped with a canister offering protection against Phosphine, Chlorine, Hydrogen Sulfide, organic vapors, acid gases, and dusts and mists. If the potential for exposure to more than 15 ppm of Phosphine gas exists, use a MSHA/NIOSH approved supplied-air respirator with a full facepiece operated in the positive pressure mode or with a full facepiece, hood, or helmet in the continuous flow mode, or use a MSHA/NIOSH approved self-contained breathing apparatus with a full facepiece operated in pressure-demand or other positive pressure mode**.** See Phosphine entry.

Storage: Prior to working with this chemical you should be trained on its proper handling and storage. Prior to working with aluminum phosphide you should be trained on its proper handling and storage. Store in a noncombustible, nonsprinklered building or location, in tightly closed containers in a cool well-ventilated area away from all forms of moisture and strong acids. Aluminum phosphide decomposes in water forming phosphine gas. Consult the entry on Phosphine for more information.

Shipping: Label required is "Dangerous when Wet, Poison." Shipment by passenger aircraft or railcar is forbidden. Falls in UN/DOT Hazard Class 4.3 and Packing group I.[19][20]

Spill Handling: Shut off ignition; no flares, smoking, or flames in hazard area. Do not touch spilled material. Do not get water on spilled material or inside container. Dike spill for later disposal. Blanket release with dry sand, clay, or ground limestone. Shovel small spill into clean, dry container, and cover. Move containers from spill area. Avoid breathing dusts. Wear appropriate protective clothing and use appropriate respiratory protection. Cover large powder spill with plastic sheet or tarp to minimize spreading. Clean-up only under supervision of an expert. DOT warns that this chemical is spilled in water: Dangerous from 0.5 – 10 km (0.3 – 6.0 miles) downwind.

Fire Extinguishing: This chemical is a combustible solid. Poisonous gases are produced in fire including phosphorus oxides and aluminum fumes. Do not use water or foam. Small fires can be extinguished with dry chemical, soda ash, clay, or ground limestone, or use an approved Class D extinguisher. Large fires: withdraw from area and let fire burn. Move container from fire only if you can do it without risk. If employees are expected to fight fires, they must be trained and equipped in OSHA 1910.156. Wear self-contained breathing apparatus when fighting fires involving this material. If contact with the material is anticipated, wear full protective clothing.

Disposal Method Suggested: Consult with environmental regulatory agencies for guidance on acceptable disposal practices. Generators of waste containing this contaminant (≥100 kg/mo) must conform with EPA regulations governing storage, transportation, treatment, and waste disposal. Allow to react slowly with moisture in the open, being sure that phosphine gas evolved is dissipated. Alternatively, mix with dry diluent and incinerate at temperature above 1,000°C with effluent gas scrubbing.[22] In accordance with 40CFR165 recommendations for the disposal of pesticides and pesticide containers. Must be disposed properly by following package label directions or by contacting your state pesticide or environmental control agency or by contacting your regional EPA office.

References

U.S. Environmental Protection Agency, "Chemical Profile: Aluminum Phosphide," Washington, DC, Chemical Emergency Preparedness Program (November 30, 1987)

New Jersey Department of Health and Senior Services, "Hazardous Substance Fact Sheet: Aluminum Phosphide," Trenton, NJ (April 1998)

Aluminum Sulfate

Molecular Formula: $Al_2S_3O_{12}$

Common Formula: $Al_2(SO_4)_3$

Synonyms: Alum; Aluminum alum; Aluminum trisulfate; Cake alum; Diaaluminum trisulfate; Dialuminum sulfate;

Paper maker's alum; Sulfato aluminico (Spanish); Sulfuric acid, aluminum salt

CAS Registry Number: 10043-01-3

RTECS Number: BD1700000

DOT ID: NA 1760 (solution)

Regulatory Authority

- Air Pollutant Standard Set (ACGIH)[1] (HSE)[33] (OSHA)[58] (North Dakota)[60]
- CLEAN WATER ACT: Section 311 Hazardous Substances/ RQ 40CFR117.3 (same as CERCLA, see below)
- SUPERFUND/EPCRA 40CFR302.4 Reportable Quantity (RQ): CERCLA, 5,000 lb (2,270 kg)[4]

Cited in U.S. State Regulations: Alaska (G), New Hampshire (G), New Jersey (G), Massachusetts (G), New Jersey (G), New York (G), North Dakota (A), Pennsylvania (G).

Description: Aluminum sulfate is a white powder, often used in water solution. Freezing/Melting point = 770°C (decomposes). Soluble in water.

Potential Exposure: Widely used in tanning leather, sizing paper, mordant in dyeing, purifying water, fireproofing and waterproofing cloth, clarifying oils and fats, treating sewage, in antiperspirants, in agricultural pesticides, manufacturing aluminum salts and others.

Incompatibilities: In aqueous solution aluminum sulfate forms sulfuric acid; reacts with bases and many other substances.

Permissible Exposure Limits in Air: The Federal OSHA standard for soluble aluminum salt, including aluminum sulfate, is 2 mg/m^3 TWA averaged over an 8-hour workshift. This same standard is recommended by ACGIH,[1] and HSE.[33] In addition, North Dakota[60] has set a guideline for in ambient air of 0.02 mg/m^3.

Permissible Concentration in Water: An ambient water level of 73 μg/l for aluminum compounds has been suggested by EPA[32] based on health effects.

Routes of Entry: Inhalation, ingestion.

Harmful Effects and Symptoms

Short Term Exposure: Aluminum sulfate powder can irritate the eyes, skin and respiratory tract. It is capable of causing eye damage. Ingestion of large doses can cause stomach irritation, nausea and vomiting.

Long Term Exposure: Aluminum sulfate may cause skin disorders, and may cause lung problems.

Points of Attack: Lungs, skin.

Medical Surveillance: Lung function tests.

First Aid: If this chemical gets into the eyes, remove any contact lenses at once and irrigate immediately for at least 15 minutes, occasionally lifting upper and lower lids. Seek medical attention immediately. If this chemical contacts the skin, remove contaminated clothing and wash immediately with soap and water. Seek medical attention immediately. If this chemical has been inhaled, remove from exposure, begin rescue breathing (using universal precautions) if breathing has stopped and CPR if heart action has stopped. Transfer promptly to a medical facility. When this chemical has been swallowed, get medical attention. If victim is *conscious*, administer water or milk. Do not induce vomiting.

Personal Protective Methods: Wear protective gloves and clothing to prevent any reasonable probability of skin contact. Safety equipment suppliers/manufacturers can provide recommendations on the most protective glove/clothing material for your operation. All protective clothing (suits, gloves, footwear, headgear) should be clean, available each day, and put on before work. Contact lenses should not be worn when working with this chemical. When working with liquid wear splash -proof chemical goggles and face shield unless full facepiece respiratory protection is worn. When working with solid or powder wear dust-proof chemical goggles and face shield unless full facepiece respiratory protection is worn. Employees should wash immediately with soap when skin is wet or contaminated. Provide emergency showers and eyewash.

Respirator Selection: *At any concentrations above 2 mg/m^3:* SA:CF (any supplied-air respirator operated in a continuous-flow mode); or PAPROV [any powered, air-purifying respirator with organic vapor cartridge(s)]; CCRFOV [any air-purifying, full-facepiece respirator (gas mask) with a chin-style, front-or back-mounted acid gas canister]; or GMFOV [any air-purifying, full-facepiece respirator (gas mask) with a chin-style, front-or back-mounted organic vapor canister]; or SCBAF (any self-contained breathing apparatus with a full facepiece); or SAF (any supplied-air respirator with a full facepiece). *Emergency or planned entry into unknown concentrations or IDLH conditions:* SCBAF:PD,PP (any self-contained breathing apparatus that has a full facepiece and is operated in a pressure-demand or other positive-pressure mode); or SAF:PD,PP:ASCBA (any supplied-air respirator that has a full facepiece and is operated in a pressure-demand or other positive-pressure mode in combination with an auxiliary self-contained breathing apparatus operated in a pressure-demand or other positive-pressure mode). *Escape:* GMFOV [any air-purifying, full-facepiece respirator (gas mask) with a chin-style, front-or back-mounted organic vapor canister] or SCBAE (any appropriate escape-type, self-contained breathing apparatus). Wear a dust mask.

Storage: Prior to working with this chemical you should be trained on its proper handling and storage. Store in tightly closed containers in a cool, well-ventilated area. Aluminum sulfate powder should be kept dry since it forms sulfuric acid when wet.

Shipping: The solid requires no label; solutions must be labeled "Corrosive." Aluminum sulfate solutions fall into DOT/UN Hazard Class 8,[19][20] and Packing Group III.

Spill Handling: Evacuate persons not wearing protective equipment from area of spill or leak until clean-up is complete. Remove all ignition sources. Sweep up or vacuum powdered material being careful not to raise dust. Collect powdered material in the most convenient and safe manner and deposit in sealed containers. Absorb liquids in vermiculite, dry sand, earth, or a similar material and deposit in sealed containers. Ventilate area after clean-up is complete. It may be necessary to contain and dispose of this chemical as a hazardous waste. If material or contaminated runoff enters waterways, notify downstream users of potentially contaminated waters. Contact your Department of Environmental Protection or your regional office of the federal EPA for specific recommendations. If employees are required to clean-up spills, they must be properly trained and equipped. OSHA 1910.120(q) may be applicable.

Fire Extinguishing: Not flammable. Poisonous gases, including sulfur oxides, are produced in fire. Use extinguishing agents suitable for surrounding fire. If material or contaminated runoff enters waterways, notify downstream users of potentially contaminated waters. Notify local health and fire officials and pollution control agencies. If employees are expected to fight fires, they must be trained and equipped in OSHA 1910.156. When involved in fire, wear goggles and self-contained breathing apparatus and rubber clothing including gloves.[41]

Disposal Method Suggested: Pretreatment involves hydrolysis followed by neutralization with NaOH. The insoluble aluminum hydroxide formed is removed by filtration and can be heated to decomposition to yield alumina which has valuable industrial applications. The neutral solution of sodium sulfate can be discharged into sewers and waterways as long as its concentration is below the recommended provisional limit of 250 mg/l.

References

New Jersey Department of Health and Senior Services, "Hazardous Substance Fact Sheet: Aluminum Sulfate," Trenton, NJ (February 1989)

New York State Department of Health, "Chemical Fact Sheet: Aluminum Sulfate," Albany, NY, Bureau of Toxic Substance Assessment (March 1986)

Ametryn

Molecular Formula: $C_9H_{17}N_5S$

Synonyms: Ametrex; Amyphyt; Cemerim; Doruplant; EPA pesticide code 080801; 2-Ethylamino-4-isopropylamino-6-methylmercarpo-*s*-triazine; 2 Ethylamino-4-isopropylamino 6 methylthio triazine; 2-Ethylamino-4-isopropylamino-6-methylthio-1,3,5-triazine; Evik®; Gesapax®; 2-Methylmercapto-4-ethylamino-6-isopropylamino-*s*-triazine; 2-Methylmercapto-4-isopropylamino-6-ethylamino-*s*-triazine; 2-Methylthio-4-ethylamino-6-isopropylamino-*s*-triazine

CAS Registry Number: 834-12-8

RTECS Number: XY9100000

DOT ID: UN 2763 (triazine, pesticides, solid, toxic)

Regulatory Authority

- SUPERFUND/EPCRA 40CFR302.4, Appendix A, Reportable Quantity (RQ): 100 lb (45.5 kg), 40CFR372.65: Form R *de minimis* Concentration Reporting Level: 1.0%.

Description: Ametryn, $C_9H_{17}N_5S$, a triazine compound, is a colorless powder. Freezing/Melting point = 84 – 86°C.

Potential Exposure: Those involved in the manufacture, formulation and application of this selective herbicide.

Incompatibilities: Triazi9nes are incompatible with nitric acid.

Permissible Concentration in Water: The No-Adverse-Effect-Level (NOAEL) has been found to be 100 mg/kg/day and on that basis a ten-day health advisory of 8.6 mg/l was determined for a 10-kg child. If, however, on assumes a NOAEL of 10 mg/kg/day one arrives at a long term health advisory of 0.86 mg/l for a 70-kg adult. The lifetime health advisory for an adult is 0.06 mg/l using a NOAEL of 10.

Determination in Water: Extraction with methylene chloride may be followed by gas chromatography using a nitrogen phosphorus detector. The detection limits are in the range of 0.1 – 2.0 μg/l.

Routes of Entry: Ingestion, skin.

Harmful Effects and Symptoms

Short Term Exposure: Ametryn is an eye and skin irritant. It is mildly toxic by skin contact. Poisonous if swallowed or inhaled.

Long Term Exposure: It apparently causes liver degeneration. The LD_{50} value for male Charles River rats was 1,207 mg/kg and 1,543 mg/kg for female rats.

Points of Attack: Liver.

Medical Surveillance: Liver function tests.

First Aid: If this chemical gets into the eyes, remove any contact lenses at once and irrigate immediately for at least 15 minutes, occasionally lifting upper and lower lids. Seek medical attention immediately. If this chemical contacts the skin, remove contaminated clothing and wash immediately with soap and water. Seek medical attention immediately. If this chemical has been inhaled, remove from exposure, begin rescue breathing (using universal precautions) if breathing has stopped and CPR if heart action has stopped. Transfer promptly to a medical facility. When this chemical has been swallowed, get medical attention. Give large quantities of water and induce vomiting. Do not make an unconscious person vomit.

Personal Protective Methods: Wear protective gloves and clothing to prevent any reasonable probability of skin contact. Safety equipment suppliers/manufacturers can provide recommendations on the most protective glove/clothing material for

your operation. All protective clothing (suits, gloves, footwear, headgear) should be clean, available each day, and put on before work. Contact lenses should not be worn when working with this chemical. Wear dust-proof chemical goggles and face shield unless full facepiece respiratory protection is worn. Employees should wash immediately with soap when skin is wet or contaminated. Provide emergency showers and eyewash.

Respirator Selection: *Where there is no NIOSH REL, at any detectable concentration:* SCBAF:PD,PP (any MSHA/NIOSH approved self-contained breathing apparatus that has a full facepiece and is operated in a pressure-demand or other positive-pressure mode); or SAF:PD,PP:ASCBA (any supplied-air respirator that has a full facepiece and is operated in a pressure-demand or other positive-pressure mode in combination with an auxiliary, self-contained breathing apparatus operated in a pressure-demand or other positive pressure mode). *Escape:* GMFOVHiE [any air-purifying, full-facepiece respirator (gas mask) with a chin-style, front- or back-mounted organic vapor canister having a high-efficiency particulate filter]; or SCBAE (any appropriate escape-type, self-contained breathing apparatus).

Storage: Prior to working with this chemical you should be trained on its proper handling and storage. Store in tightly closed containers in a cool, well-ventilated area away from strong acids.

Shipping: Triazine pesticides, solid, toxic, n.o.s. fall in Hazard Class 6.1 and in Packing Group III which requires a "Keep away from Food" label.

Spill Handling: Evacuate and restrict persons not wearing protective equipment from area of spill or leak until cleanup is complete. Remove all ignition sources. Avoid inhalation of dust; wear respirator. Collect powdered material in the most convenient and safe manner and deposit in sealed containers. Ventilate area of spill or leak after clean-up is complete. It may be necessary to contain and dispose of this chemical as a hazardous waste. If material or contaminated runoff enters waterways, notify downstream users of potentially contaminated waters. Contact your Department of Environmental Protection or your regional office of the federal EPA for specific recommendations. If employees are required to clean-up spills, they must be properly trained and equipped. OSHA 1910.120(q) may be applicable. Granular activated carbon will remove ametryn from water.

Fire Extinguishing: Poisonous gases are produced in fire including carbon monoxide. Wear positive-pressure self-contained breathing apparatus when fighting fires involving this herbicide. Avoid breathing dusts and fumes from burning material. Keep upwind. If employees are expected to fight fires, they must be trained and equipped in OSHA 1910.156. Use extinguishing agents suitable for surrounding fire or dry chemical, carbondioxide, or alcohol foam. Use water in flooding quantities as fog.

Disposal Method Suggested: In accordance with 40CFR165 recommendations for the disposal of pesticides and pesticide containers.

References

U.S. Environmental Protection Agency, "Health Advisory: Ametryn," Washington, DC, Office of Drinking Water (August 1987)

2-Aminoanthraquinone

Molecular Formula: $C_{14}H_9NO_2$

Common Formula: $C_6H_4(CO)_2C_6H_3NH_2$

Synonyms: AAQ; 2-Amino-9,10-aminoanthraquinone; 2-Amino-9,10-anthracenedione; β-Aminoanthraquinone; Aminoantraquinona (Spanish); β-Anthraquinonylamine

CAS Registry Number: 117-79-3

RTECS Number: CB5120000

DOT ID: UN 9188

Regulatory Authority

- Carcinogen (Animal Positive) (NCI)[9] (NTP)
- Canada, WHMIS, Ingredients Disclosure List
- CERCLA/SARA 313: Form R *de minimis* Concentration Reporting Level: 0.1%.

Cited in U.S. State Regulations: California (G), Florida (G), Illinois (G), Maine (G), Maryland (G), Massachusetts (G), Michigan (G), Minnesota (G), New Jersey (G), Pennsylvania (G), West Virginia (G).

Description: 2-Aminoanthraquinone, $C_{14}H_9NO_2$, forms red or orange-brown needle-shaped crystalline solid. Freezing/Melting point = 302 – 306°C. Boiling point = sublimes. Insoluble in water.

Potential Exposure: AAQ is used as an intermediate in the industrial synthesis of anthraquinone dyes and pharmaceuticals. It is the precursor of five dyes and one pigment, including Color Index Vat Blues 4, 6, 12, and 24; Vat Yellow 1; and Pigment Blue 22. Because AAQ is used on a commercial scale solely by the dye industry, the potential for exposure to the compound is greatest for workers at dye manufacturing facilities. However, no additional data are available on the number of facilities using AAQ. The Consumer Product Safety Commission staff believes that trace amounts of unreacted AAQ may possibly be present in some dyes based on this chemical and in the final consumer product. Exposure even to trace amounts may be a cause for concern. This concern is based on experience with other dyes derived from aromatic amines.

Incompatibilities: Strong oxidizing and/or reducing agents.

Permissible Exposure Limits in Air: No standards set. However, the 1-amino analog has a MAC of 5 mg/m^3 set by the former USSR-UNEP/IRPTC project.[43]

Permissible Concentration in Water: No criteria set.

Routes of Entry: Inhalation and skin contact.

Harmful Effects and Symptoms

Technical grade 2-aminoanthraquinone (impurities unspecified), administered in the feed, was carcinogenic in male Fisher 344 rats, causing a combination of hepato-cellular carcinomas and neoplastic nodules of the liver. The compound was also carcinogenic in B6C3F1 mice, causing hepato-cellular carcinomas in both sexes and malignant gematopoietic lymphomas in females. An IARC working group considered that the evidence for the carcinogenicity in experimental animals of the material tested was limited. In view of another evaluation of NCl bioassay results, the evidence can be considered as sufficient.

Short Term Exposure: Irritates eyes and skin.

Long Term Exposure: A confirmed carcinogen in animals; causes liver and lymph system cancer. May cause mutations.

Medical Surveillance: This chemical may cause cancer. There is no specific tests. However, if illness occurs or overexposure is suspected, medical attention is recommended.

First Aid: If this chemical gets into the eyes, remove any contact lenses at once and irrigate immediately for at least 15 minutes, occasionally lifting upper and lower lids. Seek medical attention immediately. If this chemical contacts the skin, remove contaminated clothing and wash immediately with soap and water. Seek medical attention immediately. If this chemical has been inhaled, remove from exposure, begin rescue breathing (using universal precautions) if breathing has stopped and CPR if heart action has stopped. Transfer promptly to a medical facility. When this chemical has been swallowed, get medical attention. Give large quantities of water and induce vomiting. Do not make an unconscious person vomit.

Personal Protective Methods: Wear protective gloves and clothing to prevent any reasonable probability of skin contact. Safety equipment suppliers/manufacturers can provide recommendations on the most protective glove/clothing material for your operation. All protective clothing (suits, gloves, footwear, headgear) should be clean, available each day, and put on before work. Contact lenses should not be worn when working with this chemical. Wear dust-proof chemical goggles and face shield unless full facepiece respiratory protection is worn. Employees should wash immediately with soap when skin is wet or contaminated. Provide emergency showers and eyewash.

Respirator Selection: *At any detectable concentration:* SCBAF:PD,PP (any MSHA/NIOSH approved self-contained breathing apparatus that has a full facepiece and is operated in a pressure-demand or other positive-pressure mode); or SAF:PD,PP:ASCBA (any supplied-air respirator that has a full facepiece and is operated in a pressure-demand or other positive-pressure mode in combination with an auxiliary, self-contained breathing apparatus operated in a pressure-demand or other positive pressure mode). *Escape:* HiEF (Any air-purifying, full-facepiece respirator with a high-efficiency particulate filter); or SCBAE (Any appropriate escape-type, self-contained breathing apparatus).

Storage: Prior to working with this chemical you should be trained on its proper handling and storage. Store in tightly closed containers in a refrigerator or cool, well ventilated area away from strong oxidizers.

Shipping: This may be classified as a hazardous substance, solid, n.o.s., which has no label requirements or weight limits on shipments. The Hazard Class is ORM-E and the Packing Group is III.

Spill Handling: Evacuate persons not wearing protective equipment from area of spill or leak until clean-up is complete. Remove all ignition sources. Dampen spilled material with toluene to avoid airborne dust. Collect powdered material in the most convenient and safe manner and deposit in vapor-tight, sealed containers. Ventilate area after clean-up is complete. It may be necessary to contain and dispose of this chemical as a hazardous waste. If material or contaminated runoff enters waterways, notify downstream users of potentially contaminated waters. Contact your Department of Environmental Protection or your regional office of the federal EPA for specific recommendations. If employees are required to clean-up spills, they must be properly trained and equipped. OSHA 1910.120(q) may be applicable.

Fire Extinguishing: Use dry chemical, carbon dioxide, water spray, or alcohol foam extinguishers. Poisonous gases are produced in fire including nitrogen oxides. If material or contaminated runoff enters waterways, notify downstream users of potentially contaminated waters. Notify local health and fire officials and pollution control agencies. From a secure, explosion-proof location, use water spray to cool exposed containers. If cooling streams are ineffective (venting sound increases in volume and pitch, tank discolors, or shows any signs of deforming), withdraw immediately to a secure position. If employees are expected to fight fires, they must be trained and equipped in OSHA 1910.156.

References

Sax, N. I., Ed., Dangerous Properties of Industrial Materials Report 4, No. 6, 66–70 (1984)

New Jersey Department of Health and Senior Services, Hazardous Substance Fact Sheet, 2-Aminoanthroquinone, Trenton NJ (April, 1997)

Aminoazobenzene

Molecular Formula: $C_{12}H_{11}N_3$

Common Formula: $C_6H_5N{=}NC_6H_4NH_2$

Synonyms: AAB; *p*-Aminoazobenzene; 4-Amino-1,1'-azobenzene; 4-Aminoazobenzene; Aminoazobenzene; *p*-Aminoazobenzol; 4-Aminoazobenzol; *p*-Aminodiphenylimide; Aniline yellow; 4-Benzeneazoaniline; Brasilazina oil yellow G; Ceres yellow R; C.I. solvent blue 7; C.I. solvent yellow 1; Fast spirit yellow AAB; Oil soluble aniline yellow;

Oil yellow AAB; Organol yellow; Paraphenolazoaniline; *p*-Phenolazoaniline; *p*-(Phenylazo)aniline; 4-(Phenylazo)aniline; 4-(Phenylazo)benzenamine' *p*-phenylazophenylamine; Solvent yellow 1; Sudan yellow R

CAS Registry Number: 60-09-3

RTECS Number: BY8225000

Regulatory Authority

- Carcinogen (Animal Positive) (IARC)[9]
- Canada, WHMIS, Ingredients Disclosure List
- CERCLA/SARA 313: Form R *de minimis* Concentration Reporting Level: 0.1%.

Cited in U.S. State Regulations: California (G), Maryland (G), Massachusetts (G), Michigan (G), Minnesota (G), New Jersey (G), Pennsylvania (G).

Description: 4-Aminoazobenzene, $C_6H_5N{=}NC_6H_4NH_2$, $C_{12}H_{11}N_3$, forms yellow to tan crystals or orange needles. Freezing/Melting point = 126 – 128°C. Boiling point ≥ 360°C. Hazard Identification (based on NFPA-704 M Rating System): Health 0, Flammability 2, Reactivity 1. Slightly soluble in water.

Potential Exposure: Used in form of its salts in dyeing; used as intermediate in manufacture of acid yellow and diazo dyes; in insecticides, waxes, lacquers, varnishes, stains, styrene resins.

Permissible Exposure Limits in Air: No standards set.

Permissible Concentration in Water: No criteria set.

Harmful Effects and Symptoms

Carcinogen as noted above.

First Aid: If this chemical gets into the eyes, remove any contact lenses at once and irrigate immediately for at least 15 minutes, occasionally lifting upper and lower lids. Seek medical attention immediately. If this chemical contacts the skin, remove contaminated clothing and wash immediately with soap and water. Seek medical attention immediately. If this chemical has been inhaled, remove from exposure, begin rescue breathing (using universal precautions) if breathing has stopped and CPR if heart action has stopped. Transfer promptly to a medical facility. When this chemical has been swallowed, get medical attention. Give large quantities of water and induce vomiting. Do not make an unconscious person vomit.

Personal Protective Methods: Wear protective gloves and clothing to prevent any reasonable probability of skin contact. Safety equipment suppliers/manufacturers can provide recommendations on the most protective glove/clothing material for your operation. All protective clothing (suits, gloves, footwear, headgear) should be clean, available each day, and put on before work. Contact lenses should not be worn when working with this chemical. Wear dust-proof chemical goggles and face shield unless full facepiece respiratory protection is worn. Employees should wash immediately with soap when skin is wet or contaminated. Provide emergency showers and eyewash.

Respirator Selection: *At any detectable concentration:* SCBAF:PD,PP (any MSHA/NIOSH approved self-contained breathing apparatus that has a full facepiece and is operated in a pressure-demand or other positive-pressure mode); or SAF:PD,PP:ASCBA (any supplied-air respirator that has a full facepiece and is operated in a pressure-demand or other positive-pressure mode in combination with an auxiliary, self-contained breathing apparatus operated in a pressure-demand or other positive pressure mode). *Escape:* HiEF (Any air-purifying, full-facepiece respirator with a high-efficiency particulate filter); or SCBAE (Any appropriate escape-type, self-contained breathing apparatus).

Storage: Prior to working with this chemical you should be trained on its proper handling and storage. Store in tightly closed containers in a cool, well ventilated area

Spill Handling: Evacuate persons not wearing protective equipment from area of spill or leak until clean-up is complete. Remove all ignition sources. Collect powdered material in the most convenient and safe manner and deposit in sealed containers. Ventilate area after clean-up is complete. It may be necessary to contain and dispose of this chemical as a hazardous waste. If material or contaminated runoff enters waterways, notify downstream users of potentially contaminated waters. Contact your Department of Environmental Protection or your regional office of the federal EPA for specific recommendations. If employees are required to clean-up spills, they must be properly trained and equipped. OSHA 1910.120(q) may be applicable.

Fire Extinguishing: Use dry chemical, carbon dioxide, water spray, or alcohol foam extinguishers. Poisonous gases are produced in fire including nitrogen oxides. If material or contaminated runoff enters waterways, notify downstream users of potentially contaminated waters. Notify local health and fire officials and pollution control agencies. From a secure, explosion-proof location, use water spray to cool exposed containers. If cooling streams are ineffective (venting sound increases in volume and pitch, tank discolors, or shows any signs of deforming), withdraw immediately to a secure position. If employees are expected to fight fires, they must be trained and equipped in OSHA 1910.156.

References

Sax, N. I., Ed., Dangerous Properties of Industrial Materials Report 1, No. 3, 27–28 (1981)

Aminoazotoluene

Molecular Formula: $C_{14}H_{15}N_3$

Common Formula: $CH_3C_6H_4N{=}NC_6H_3(NH_2)CH_3$

Synonyms: o-AAT; o-Amidoazotoluol (German); Aminoazotoluene (Indicator); o-Aminoazotoluene; 4'-Amino-2,3'-Azotoluene; 4'-Amino-2:3'-Azotoluene; o-Aminoazotolueno (Spanish); o-Aminoazotoluol; 4-Amino-2',3-Dimethylazobenzene; 4'-Amino-2,3'-Dimethylazobenzene; o-AT;

Brasilazina Oil Yellow R; Butter Yellow; C.I. 11160; C.I. 11160B; C.I. Solvent Yellow 3; 2',3-Dimethyl-4-Aminoazobenzene; Fast Garnet GBC Base; Fast Oil Yellow; Fast Yellow AT; Fast Yellow B; Hidaco Oil Yellow; 2-Methyl-4-[(2-Methylphenyl)Azo]Benzenamine; OAAT; Oil Yellow; Oil Yellow 21; Oil Yellow 2681; Oil Yellow A; Oil Yellow AT; Oil Yellow C; Oil Yellow I; Oil Yellow 2R; Oil Yellow T; Organol Yellow 25; Somalia Yellow R; Sudan Yellow RRA; o-Tolueneazo-o-Toluidine; o-Toluol-Azo-o-Toluidin (German); 5-(o-Tolylazo)-2-Aminotoluene; 4-(o-Tolylazo)-o-Toluidine; Tulabase Fast Garnet GB; Tulabase Fast Garnet GBC; Waxakol Yellow NL

CAS Registry Number: 97-56-3

RTECS Number: XU8800000

Regulatory Authority

- Carcinogen (Animal Positive) (IARC)[9] (DFG)[3] possible select carcinogen (OSHA)
- Canada, WHMIS, Ingredients Disclosure List
- CERCLA/SARA 313: Form R *de minimis* Concentration Reporting Level: 0.1%.

Cited in U.S. State Regulations: California (A, G), Florida (G), Illinois (G), Massachusetts (G), Michigan (G), Minnesota (G), New Hampshire (G), New Jersey (G), Pennsylvania (G).

Description: Aminoazotoluene, $CH_3C_6H_4N{=}NC_6H_3(NH_2)CH_3$, forms golden yellow or reddish-brown crystals. Freezing/Melting point = 101 – 102°C. Slightly soluble in water.

Potential Exposure: Used in dyes, medicines, as a colorant in shoe polishes and other wax-based polishes.

Permissible Exposure Limits in Air: No standards set.

Permissible Concentration in Water: No criteria set.

Harmful Effects and Symptoms

Stated to be moderately toxic by several routes.[44]

Long Term Exposure: A known animal carcinogen.

First Aid: If this chemical gets into the eyes, remove any contact lenses at once and irrigate immediately for at least 15 minutes, occasionally lifting upper and lower lids. Seek medical attention immediately. If this chemical contacts the skin, remove contaminated clothing and wash immediately with soap and water. Seek medical attention immediately. If this chemical has been inhaled, remove from exposure, begin rescue breathing (using universal precautions) if breathing has stopped and CPR if heart action has stopped. Transfer promptly to a medical facility. When this chemical has been swallowed, get medical attention. Give large quantities of water and induce vomiting. Do not make an unconscious person vomit.

Personal Protective Methods: Wear protective gloves and clothing to prevent any reasonable probability of skin contact. Safety equipment suppliers/manufacturers can provide recommendations on the most protective glove/clothing material for your operation. All protective clothing (suits, gloves, footwear, headgear) should be clean, available each day, and put on before work. Contact lenses should not be worn when working with this chemical. Wear dust-proof chemical goggles and face shield unless full facepiece respiratory protection is worn. Employees should wash immediately with soap when skin is wet or contaminated. Provide emergency showers and eyewash.

Respirator Selection: *At any detectable concentration:* SCBAF:PD,PP (any MSHA/NIOSH approved self-contained breathing apparatus that has a full facepiece and is operated in a pressure-demand or other positive-pressure mode); or SAF: PD,PP:ASCBA (any supplied-air respirator that has a full facepiece and is operated in a pressure-demand or other positive-pressure mode in combination with an auxiliary, self-contained breathing apparatus operated in a pressure-demand or other positive pressure mode). *Escape:* HiEF (Any air-purifying, full-facepiece respirator with a high-efficiency particulate filter); or SCBAE (Any appropriate escape-type, self-contained breathing apparatus).

Storage: Prior to working with this chemical you should be trained on its proper handling and storage. Store in tightly closed containers in a cool, well ventilated area.

Shipping: No specific DON/UN requirements for labels or special shipping requirements.

Fire Extinguishing: Use dry chemical, carbon dioxide, water spray, or alcohol foam extinguishers. Poisonous gases are produced in fire including nitrogen oxides. If material or contaminated runoff enters waterways, notify downstream users of potentially contaminated waters. Notify local health and fire officials and pollution control agencies. From a secure, explosion-proof location, use water spray to cool exposed containers. If cooling streams are ineffective (venting sound increases in volume and pitch, tank discolors, or shows any signs of deforming), withdraw immediately to a secure position. If employees are expected to fight fires, they must be trained and equipped in OSHA 1910.156.

References

Sax, N. I., Ed., Dangerous Properties of Industrial Materials Report 6, No. 4, 54–63 (1986)

4-Aminobiphenyl

Molecular Formula: $C_{12}H_{11}N$

Common Formula: $C_6H_5C_6H_4NH_2$

Synonyms: *p*-Aminobifenilo (Spanish); 4-Aminobifenilo (Spanish); *p*-Aminobiphenyl; 4-Aminobiphenyl; 4-Aminodifenil (Spanish); *p*-Aminodiphenyl; 4-Aminodiphenyl; (1,1'-Biphenyl)-4-amine; *p*-Biphenylamine; 4-Biphenylamine; Biphenyline; *p*-Phenylaniline; 4-Phenylaniline; Xenylamin (Czech); *p*-Xenylamine; Xenylamine

CAS Registry Number: 92-67-1

RTECS Number: DU8925000

DOT ID: UN 2811 (toxic solid, organic, n.o.s.)

EEC Number: 612-072-00-6

Regulatory Authority

- Carcinogen (Animal Positive, Human Positive) (IARC)[9] (ACGIH)[1] (DFG)[3] (Varous states) (Various Canadian Provinces) (Mexico)
- Banned or Severely Restricted (Many Countries) (UN)[13][35]
- Very Toxic Substance (World Bank)[15]
- OSHA, 29CFR1910 Specifically Regulated Chemicals (See CFR1910.1011)
- Clean Air Act 42USC7412; Title I, Part A, §112 hazardous pollutants
- RCRA 40CFR261, Appendix 8 Hazardous Constituents
- RCRA Land Ban Waste.
- RCRA 40CFR268.48; 61FR15654, Universal Treatment Standards: Wastewater (mg/l), 0.13; Nonwastewater (mg/kg), N/A
- RCRA 40CFR264, Appendix 9; Ground Water Monitoring List Suggested methods (PQL μg/l): 8270 (10).
- SUPERFUND/EPCRA 40CFR302.4, Appendix A, Reportable Quantity (RQ): CERCLA, 1 lb (0.455 kg), 40CFR372.65: Form R *de minimis* Concentration Reporting Level: 0.1%
- Canada, WHMIS, Ingredients Disclosure List

Cited in U.S. State Regulations: Alaska (G), California (G, W), Florida (G), Illinois (G), Kansas (G), Louisiana (G), Maine (G), Maryland (G), Massachusetts (G), Michigan (G), New Hampshire (G), New Jersey (G), New York (A), North Dakota (A), Oklahoma (G), Pennsylvania (G, A), Rhode Island (G), South Carolina (A), Vermont (G), Virginia (G, A), Washington (G), West Virginia (G).

Description: 4-Aminobiphenyl, $C_6H_5C_6H_4NH_2$, is a combustible, colorless to tan crystalline solid that turns purple on exposure to air. May be used in a liquid solution. Floral odor. Freezing/Melting point = 49°C. Boiling point = 302°C. Hazard Identification (based on NFPA-704 M Rating System): Health 2, Flammability 1, Reactivity 0. Autoignition temperature = 450°C. Insoluble in water. Octanol/water partition coefficient as log Pow: 2.8.

Potential Exposure: It is no longer manufactured commercially and is only used for research purposes. 4-Aminobiphenyl was formerly used as a rubber anitoxidant and as a dye intermediate. Is a contaminant in 2-aminobiphenyl.

Incompatibilities: Strong oxidizers, strong acids, and acid anhydrides.

Permissible Exposure Limits in Air: 4-Aminobiphenyl is included in the Federal standards for carcinogens; all contact with it should be avoided. The compound also has the notation "skin" indicating the potential for cutaneous absorption. Several States have set guidelines or standards for 4-aminobiphenyl in ambient air:[60] zero for North Dakota, New York and South Carolina; 0.8 μg/m³ (Pennsylvania); 4.0 μg/m³ (Virginia).

Determination in Air: Collection on a filter and colorimetric analysis. Collection by filter, workup with 2-propanol and analysis by gas chromatography may also be employed.

Permissible Concentration in Water: No criteria set, but EPA[32] has suggested an ambient water limit of 200 μg/l based on health effects.

Routes of Entry: Inhalation and percutaneous absorption.

Harmful Effects and Symptoms

Short Term Exposure: This chemical can be absorbed through the skin, thereby increasing exposure. Irritates the skin. Exposure may cause methemoglobinemia, which interferes with the blood's ability to carry oxygen. This can cause headache, dizziness, fatigue, fast heart rate, blue color of the lips and skin. Higher levels can cause difficult breathing, collapse and death.

Long Term Exposure: 4-Aminobiphenyl is a known human bladder carcinogen. An exposure of only 133 days has been reported to have ultimately resulted in a bladder tumor, and blood in the urine. The latent period is generally 15 – 35 years. Acute exposure produces headaches,, lethargy, cyanosis, urinary burning, and hematuria. Cystoscopy reveals diffuse hyperemia, edema, and frank slough.

Points of Attack: Bladder, skin, blood.

Medical Surveillance: Blood test for hemoglobin levels. Placement and periodic examinations should include an evaluation of exposure to other carcinogens; use of alcohol, smoking, and medications; and family history. Special attention should be given on a regular basis to urine sediment and cytology. If red cells or positive smears are seen, cystoscopy should be done at once. The general health of exposed persons should also be evaluated in periodic examinations. Blood methemoglobin level. Complete blood count (CBC).

First Aid: If this chemical gets into the eyes, remove any contact lenses at once and irrigate immediately for at least 15 minutes, occasionally lifting upper and lower lids. Seek medical attention immediately. If this chemical contacts the skin, remove contaminated clothing and wash immediately with soap and water. Seek medical attention immediately. If this chemical has been inhaled, remove from exposure, begin rescue breathing (using universal precautions) if breathing has stopped and CPR if heart action has stopped. Transfer promptly to a medical facility. When this chemical has been swallowed, get medical attention. Give large quantities of water and induce vomiting. Do not make an unconscious person vomit.

Note to Physician: Treat for methemoglobinemia. Spectrophotometry may be required for precise determination of levels of methemoglobinemia in urine.

Personal Protective Methods: Wear protective gloves and clothing to prevent any reasonable probability of skin contact.

Safety equipment suppliers/manufacturers can provide recommendations on the most protective glove/clothing material for your operation. All protective clothing (suits, gloves, footwear, headgear) should be clean, available each day, and put on before work. Contact lenses should not be worn when working with this chemical. Wear splash- or dust-proof chemical goggles and face shield unless full facepiece respiratory protection is worn. Employees should wash immediately with soap when skin is wet or contaminated. Provide emergency showers and eyewash. These are designed to supplement engineering controls (such as the prohibition of open-vessel operations) and to prevent all skin or respiratory contact. Full body protective clothing and gloves should be used by those employed in handling operations. Fullface, supplied air respirators of continuous flow or pressure demand type should also be used. On exit from a regulated area, employees should shower and change into street clothes, leaving their clothing and equipment at the point of exit to be placed in impervious containers at the end of the work shift for decontamination or disposal. Effective methods should be used to clean and decontaminate gloves and clothing.

Respirator Selection: *At any detectable concentration:* SCBAF:PD,PP (any MSHA/NIOSH approved self-contained breathing apparatus that has a full facepiece and is operated in a pressure-demand or other positive-pressure mode); or SAF:PD,PP:ASCBA (any supplied-air respirator that has a full facepiece and is operated in a pressure-demand or other positive-pressure mode in combination with an auxiliary, self-contained breathing apparatus operated in a pressure-demand or other positive pressure mode). *Escape:* HiEF (Any air-purifying, full-facepiece respirator with a high-efficiency particulate filter); or SCBAE (Any appropriate escape-type, self-contained breathing apparatus).

Storage: Prior to working with this chemical you should be trained on its proper handling and storage. Store in accordance with OSHA standard 1910.1011. Storage area should be marked, regulated and maintained under negative pressure. Keep away from heat and sources of ignition, oxidizers.

Shipping: This chemical is a solid, toxic, organic, n.o.s. should be labeled "Poison" if in Packing Group I. This class of materials is in DOT/UN Hazard Class 6.1.[19][20]

Spill Handling: Evacuate persons not wearing protective equipment from area of spill or leak until clean-up is complete. Remove all ignition sources. Cover spills with dry lime or soda ash and collect powdered material, then collect powdered material in the most convenient and safe manner and deposit in sealed containers. Absorb liquids in vermiculite, dry sand, earth, or a similar material and deposit in sealed containers. Ventilate area after clean-up is complete. It may be necessary to contain and dispose of this chemical as a hazardous waste. If material or contaminated runoff enters waterways, notify downstream users of potentially contaminated waters. Contact your Department of Environmental Protection or your regional office of the federal EPA for specific recommendations. If employees are required to clean-up spills, they must be properly trained and equipped. OSHA 1910.120(q) may be applicable.

Fire Extinguishing: This chemical is a combustible solid. Use dry chemical, carbon dioxide, water spray, or alcohol foam extinguishers. Poisonous gases are produced in fire including nitrogen oxides. If material or contaminated runoff enters waterways, notify downstream users of potentially contaminated waters. Notify local health and fire officials and pollution control agencies. If employees are expected to fight fires, they must be trained and equipped in OSHA 1910.156.

Disposal Method Suggested: Consult with environmental regulatory agencies for guidance on acceptable disposal practices. Generators of waste containing this contaminant (≥100 kg/mo) must conform with EPA regulations governing storage, transportation, treatment, and waste disposal. Controlled incineration whereby oxides of nitrogen are removed from the effluent gas by scrubber, catalytic or thermal devices.[22]

References

New Jersey Department of Health and Senior Services, "Hazardous Substance Fact Sheet: 4-Aminodiphenyl," Trenton, NJ (June 1998)

2-Amino-4-Chlorophenol

Molecular Formula: C_6H_6ClNO

Common Formula: $HOC_6H_3(NH_2)Cl$

Synonyms: 2-Amino-4-clorofenol (Spanish); *p*-Chloro-*o*-aminophenol

CAS Registry Number: 95-85-2

RTECS Number: SJ5700000

DOT ID: UN 2673

Regulatory Authority

- Canada, WHMIS, Ingredients Disclosure List

Cited in U.S. State Regulations: New Hampshire (G), New Jersey (G).

Description: 2-Amino-4-chlorophenol, $HOC_6H_3(NH_2)Cl$, is a grayish to light brown crystalline solid or powder. Insoluble in water. Freezing/Melting point = 140°C.

Potential Exposure: Used as a chemical raw material, especially in dye manufacture.

Incompatibilities: Oxidizers. Keep away from iron, moisture and temperatures above 110°F (43°C).

Permissible Exposure Limits in Air: No standards set.

Permissible Concentration in Water: No criteria set.

Routes of Entry: Inhalation.

Harmful Effects and Symptoms

Short Term Exposure: Poisonous if inhaled or ingested. Exposure can lower the ability of the blood to carry oxygen.

This can result in a bluish color to the skin and lips, headaches, dizziness, collapse and even death.

Long Term Exposure: Unknown at this time.

Points of Attack: Blood.

Medical Surveillance: Test for blood hemoglobin level.

First Aid: If this chemical gets into the eyes, remove any contact lenses at once and irrigate immediately for at least 15 minutes, occasionally lifting upper and lower lids. Seek medical attention immediately. If this chemical contacts the skin, remove contaminated clothing and wash immediately with soap and water. Seek medical attention immediately. If this chemical has been inhaled, remove from exposure, begin rescue breathing (using universal precautions) if breathing has stopped and CPR if heart action has stopped. Transfer promptly to a medical facility. When this chemical has been swallowed, get medical attention. Give large quantities of water and induce vomiting. Do not make an unconscious person vomit.

Personal Protective Methods: Wear protective gloves and clothing to prevent any reasonable probability of skin contact. Safety equipment suppliers/manufacturers can provide recommendations on the most protective glove/clothing material for your operation. All protective clothing (suits, gloves, footwear, headgear) should be clean, available each day, and put on before work. Contact lenses should not be worn when working with this chemical. Wear dust-proof chemical goggles and face shield unless full facepiece respiratory protection is worn. Employees should wash immediately with soap when skin is wet or contaminated. Provide emergency showers and eyewash.

Respirator Selection: *At any detectable concentration:* SCBAF:PD,PP (any MSHA/NIOSH approved self-contained breathing apparatus that has a full facepiece and is operated in a pressure-demand or other positive-pressure mode); or SAF: PD,PP:ASCBA (any supplied-air respirator that has a full facepiece and is operated in a pressure-demand or other positive-pressure mode in combination with an auxiliary, self-contained breathing apparatus operated in a pressure-demand or other positive pressure mode). *Escape:* HiEF (Any air-purifying, full-facepiece respirator with a high-efficiency particulate filter); or SCBAE (Any appropriate escape-type, self-contained breathing apparatus).

Storage: Prior to working with this chemical you should be trained on its proper handling and storage. Store in tightly closed containers in a cool, well ventilated area way from Iron, moisture and temperatures above 110°C.

Shipping: DOT/UN label required is "Poison." Falls in DOT/UN Hazard Class 6.1 and Packing Group II.[19][20]

Spill Handling: Evacuate persons not wearing protective equipment from area of spill or leak until clean-up is complete. Remove all ignition sources. Collect powdered material in the most convenient and safe manner and deposit in sealed containers. Ventilate area after clean-up is complete. It may be necessary to contain and dispose of this chemical as a hazardous waste. If material or contaminated runoff enters waterways, notify downstream users of potentially contaminated waters. Contact your Department of Environmental Protection or your regional office of the federal EPA for specific recommendations. If employees are required to clean-up spills, they must be properly trained and equipped. OSHA 1910.120(q) may be applicable.

Fire Extinguishing: 2-Amino-4-chlorophenol may burn, but does not readily ignite. Poisonous gases are produced in fire, including Chlorides and Nitrogen Oxides. Use dry chemical, CO_2, water spray, or foam extinguishers. If material or contaminated runoff enters waterways, notify downstream users of potentially contaminated waters. Notify local health and fire officials and pollution control agencies. From a secure, explosion-proof location, use water spray to cool exposed containers. If cooling streams are ineffective (venting sound increases in volume and pitch, tank discolors, or shows any signs of deforming), withdraw immediately to a secure position. If employees are expected to fight fires, they must be trained and equipped in OSHA 1910.156.

References

New Jersey Department of Health and Senior Services, "Hazardous Substance Fact Sheet: 2-Amino-4-Chlorophenol," Trenton, NJ (August 1998)

2-Amino-5-Diethylaminopentane

Molecular Formula: $C_9H_{22}N_2$

Synonyms: *N,N*-Diethyl-1,4-pentanediamine; Novoldiamine; Tetramethylenediamine, *N,N*-Diethyl-4-methyl-

CAS Registry Number: 140-80-7

RTECS Number: SA0242000

DOT ID: UN 2946

Cited in U.S. State Regulations: New Jersey (G), Pennsylvania (G).

Description: 2-Amino-5-diethylaminopentane, $C_9H_{22}N_2$, is a flammable liquid with a fishy odor. Flash point = 68°C. Boiling point = 200°C. Soluble in water.

Potential Exposure: Used in the manufacture of quinacrine and other antimalarials having the same basic side chain.

Incompatibilities: Contact with oxidizers may cause fore and explosions.

Permissible Exposure Limits in Air: No standards set.

Permissible Concentration in Water: No criteria set.

Routes of Entry: Inhalation, ingestion, skin contact.

Harmful Effects and Symptoms

Short Term Exposure: May cause irritation of the nose and throat. Contact with the eyes or skin may cause severe irritation and burns. Passes through the skin; exposure may be increased.

Long Term Exposure: Allergy can develop, and future exposures, even if low, can cause rash and itching. Very irritating substance; may injure the lungs.

Points of Attack: Eyes, skin, respiratory system.

Medical Surveillance: Before beginning employment and at regular times after that, for those with frequent or potentially high exposures, the following are recommended: Lung function tests. If symptoms develop or overexposure is suspected, the following may be useful: Evaluation by a qualified allergist, including careful exposure history and special testing, may help diagnose skin allergy.

First Aid: If this chemical gets into the eyes, remove any contact lenses at once and irrigate immediately for at least 15 minutes, occasionally lifting upper and lower lids. Seek medical attention immediately. If this chemical contacts the skin, remove contaminated clothing and wash immediately with soap and water. Seek medical attention immediately. If this chemical has been inhaled, remove from exposure, begin rescue breathing (using universal precautions) if breathing has stopped and CPR if heart action has stopped. Transfer promptly to a medical facility. When this chemical has been swallowed, get medical attention. Give large quantities of water and induce vomiting. Do not make an unconscious person vomit.

Personal Protective Methods: Wear protective gloves and clothing to prevent any reasonable probability of skin contact. Safety equipment suppliers/manufacturers can provide recommendations on the most protective glove/clothing material for your operation. All protective clothing (suits, gloves, footwear, headgear) should be clean, available each day, and put on before work. Contact lenses should not be worn when working with this chemical. Wear splash-proof chemical goggles and face shield unless full facepiece respiratory protection is worn. Employees should wash immediately with soap when skin is wet or contaminated. Provide emergency showers and eyewash.

Respirator Selection: Where the potential for exposures to 2-Amino-5-Diethyl Aminopentane exists, use a MSHA/NIOSH approved supplied-air respirator with a full facepiece operated in the positive pressure mode or with a full facepiece, hood, or helmet in the continuous flow mode, or use a MSHA/NIOSH approved, self-contained breathing apparatus with a full facepiece operated in pressure-demand or other positive pressure mode.

Storage: Prior to working with this chemical you should be trained on its proper handling and storage. Store in tightly closed containers in a cool, well-ventilated area. Sources of ignition such as smoking and open flames, are prohibited where 2-Amino-5-Diethyl Aminopentane is used, handled, or stored in a manner that could create a potential fire or explosion hazard.

Shipping: This compound requires a "Keep away from Food" label. It falls in Hazard Class 6.1 and Packing Group III.

Spill Handling: Evacuate persons not wearing protective equipment from area of spill or leak until clean-up is complete. Remove all ignition sources. Ventilate area of spill or leak. Absorb liquids in vermiculite, dry sand, earth, or a similar material and deposit in sealed containers. It may be necessary to contain and dispose of this chemical as a hazardous waste. If material or contaminated runoff enters waterways, notify downstream users of potentially contaminated waters. Contact your Department of Environmental Protection or your regional office of the federal EPA for specific recommendations. If employees are required to clean-up spills, they must be properly trained and equipped. OSHA 1910.120(q) may be applicable.

Fire Extinguishing: 2-Amino-5-Diethyl Aminopentane is a flammable liquid. Poisonous gas, including nitrogen oxides is produced in fire. Use dry chemical, carbon dioxide, or alcohol foam extinguishers. Vapors are heavier than air and will collect in low areas. Vapors may travel long distances to ignition sources and flashback. Vapors in confined areas may explode when exposed to fire. Storage containers and parts of containers may rocket great distances, in many directions. If material or contaminated runoff enters waterways, notify downstream users of potentially contaminated waters. Notify local health and fire officials and pollution control agencies. From a secure, explosion-proof location, use water spray to cool exposed containers. If cooling streams are ineffective (venting sound increases in volume and pitch, tank discolors, or shows any signs of deforming), withdraw immediately to a secure position. If employees are expected to fight fires, they must be trained and equipped in OSHA 1910.156.

Disposal Method Suggested: Incinerator with scrubber for No_x absorption.

References

New Jersey Department of Health and Senior Services, "Hazardous Substance Fact Sheet: 2-Amino-5-Diethyl Aminopentane," Trenton, NJ (September, 1987)

2-(2-Aminoethoxy)Ethanol

Molecular Formula: $C_4H_{11}NO_2$

Synonyms: 2-Aminoethyoxyethanol; DGA; Diglycoamine; Ehhanol, 2-(2-aminoethioxy)-

CAS Registry Number: 929-06-6

RTECS Number: KJ6125000

DOT ID: UN 3055

Regulatory Authority

- Canada, WHMIS, Ingredients Disclosure List

Cited in U.S. State Regulations: Massachusetts (G), New Jersey (G), Pennsylvania (G).

Description: 2-(2-Aminoethoxy)ethanol Combustible Colorless, thick liquid with a fish-like odor. Flash point = 127°C. Hazard Identification (based on NFPA-704 M Rating System): Health 1, Flammability 1, Reactivity 0. Soluble in water.

Potential Exposure: Used to remove gases from natural gas, in coatings in plastics, textiles, fibers, and metals, and in making other chemicals.

Incompatibilities: Reacts with Oxidizers, strong acids, and chemically active metals such as potassium, sodium, magnesium and zinc.

Permissible Exposure Limits in Air: None established, but this chemical is absorbed through the skin and contributes significantly to overall exposure.

Routes of Entry: Inhalation, skin contact.

Harmful Effects and Symptoms

Short Term Exposure: This chemical is highly corrosive. Contact can severely irritate and burn the eyes and skin. Inhalation can irritate the respiratory tract and lungs causing shortness of breath, coughing and wheezing. Higher exposures can cause pulmonary edema, a medical emergency. This can cause death.

Long Term Exposure: May cause lung irritation, the development of bronchitis, with coughing, shortness of breath, phlegm.

Points of Attack: Lungs.

Medical Surveillance: Lung function tests. If symptoms develop or overexposure is suspected, chest x-ray should be considered.

First Aid: If this chemical gets into the eyes, remove any contact lenses at once and irrigate immediately for at least 15 minutes, occasionally lifting upper and lower lids. Seek medical attention immediately. If this chemical contacts the skin, remove contaminated clothing and wash immediately with soap and water. Seek medical attention immediately. If this chemical has been inhaled, remove from exposure, begin rescue breathing (using universal precautions) if breathing has stopped and CPR if heart action has stopped. Transfer promptly to a medical facility. When this chemical has been swallowed, get medical attention. If victim is *conscious,* administer water or milk. Do not induce vomiting. Medical observation is recommended for 24 – 48 hours after breathing overexposure, as pulmonary edema may be delayed.

Personal Protective Methods: Wear protective gloves and clothing to prevent any reasonable probability of skin contact. Safety equipment suppliers/manufacturers can provide recommendations on the most protective glove/clothing material for your operation. All protective clothing (suits, gloves, footwear, headgear) should be clean, available each day, and put on before work. Contact lenses should not be worn when working with this chemical. Wear splash-proof chemical goggles and face shield unless full facepiece respiratory protection is worn. Employees should wash immediately with soap when skin is wet or contaminated. Remove nonimpervious clothing immediately if wet or contaminated. Provide emergency showers and eyewash.

Respirator Selection: *At any concentrations above the NIOSH REL:* (Any MSHA/NIOSH approved self-contained breathing apparatus that has a full facepiece and is operated in a pressure-demand or other positive-pressure mode) or SAF:PD,PP:ASCBA (Any supplied-air respirator that has a full facepiece and is operated in a pressure-demand or other positive-pressure mode in combination with an auxiliary self-contained breathing apparatus operated in a pressure-demand or other positive pressure mode). *Escape:* GMFOV [Any air-purifying, full-facepiece respirator (gas mask) with a chin-style, front-or back-mounted organic vapor canister]; or SCBAE (Any appropriate escape-type, self-contained breathing apparatus).

Storage: Prior to working with this chemical you should be trained on its proper handling and storage. Store in tightly closed containers in a cool, well ventilated area. Sources of ignition such as smoking and open flames, are prohibited where this chemical is used, handled, or stored in a manner that could create a potential fire or explosion hazard.

Shipping: This chemical should be labeled "CORROSIVE." It falls into DOT/UN Hazard Class 8, and Packing Group III.

Spill Handling: Evacuate persons not wearing protective equipment from area of spill or leak until clean-up is complete. Remove all ignition sources. Ventilate area of spill or leak. Absorb liquids in vermiculite, dry sand, earth, or a similar material and deposit in sealed containers. It may be necessary to contain and dispose of this chemical as a hazardous waste. If material or contaminated runoff enters waterways, notify downstream users of potentially contaminated waters. Contact your Department of Environmental Protection or your regional office of the federal EPA for specific recommendations. If employees are required to clean-up spills, they must be properly trained and equipped. OSHA 1910.120(q) may be applicable.

Fire Extinguishing: 2-(2-aminoethoxy)ethanol may burn, but does not readily ignite. Poisonous gases including nitrogen oxides are produced in fire. Use dry chemical, carbon dioxide, water spray, or alcohol resistant foam to extinguish fire. Storage containers and parts of containers may rocket great distances, in many directions. If material or contaminated runoff enters waterways, notify downstream users of potentially contaminated waters. Notify local health and fire officials and pollution control agencies. From a secure, explosion-proof location, use water spray to cool exposed containers. If cooling streams are ineffective (venting sound increases in volume and pitch, tank discolors, or shows any signs of deforming), withdraw immediately to a secure position. If employees are expected to fight fires, they must be trained and equipped in OSHA 1910.156.

References

New Jersey Department of Health and Senior Services, Hazardous Substance Fact Sheet, 2-(2-Aminoethoxy)Ethanol, Trenton NJ (August 1998)

3-Amino-9-Ethylcarbazole

Molecular Formula: $C_{14}H_{14}N_2$; $C_{14}H_{14}N_2 \cdot ClH$ (hydrochloride)

Synonyms: 3-Amino-*N*-ethylcarbazole; 3-Amino-9-ethylcarbazole HCl (hydrochloride)

CAS Registry Number: 132-32-1; 6109-97-3 (hydrochloride)

RTECS Number: FE3590000; FE3675000 (hydrochloride)

DOT ID: UN 9188 (Hydrochloride)

Regulatory Authority

- Carcinogen (Hydrochloride – Animal Positive) (NCI)[9] (Suspected) (DFG)[3]

Cited in U.S. State Regulations: California (G), Massachusetts (G), Michigan (G).

Description: 3-Amino-9-ethylcarbazole, $C_{14}H_{14}N_2$, is a tan crystalline compound. Freezing/Melting point = 98 – 100°C. The hydrochloride forms blue-green crystals.

Potential Exposure: Plant workers engaged in the manufacture of this compound and its use in pigment manufacture. Laboratory workers using this material in colorimetric enzyme assays and as a biological stain.

Permissible Exposure Limits in Air: No standards set.

Permissible Concentration in Water: No criteria set.

Harmful Effects and Symptoms

The NCI carcinogenic assay proved positive.

Short Term Exposure: Poisonous if swallowed.

Long Term Exposure: Suspected carcinogen and mutagen.

First Aid: If this chemical gets into the eyes, remove any contact lenses at once and irrigate immediately for at least 15 minutes, occasionally lifting upper and lower lids. Seek medical attention immediately. If this chemical contacts the skin, remove contaminated clothing and wash immediately with soap and water. Seek medical attention immediately. If this chemical has been inhaled, remove from exposure, begin rescue breathing (using universal precautions) if breathing has stopped and CPR if heart action has stopped. Transfer promptly to a medical facility. When this chemical has been swallowed, get medical attention. Give large quantities of water and induce vomiting. Do not make an unconscious person vomit.

Personal Protective Methods: Wear protective gloves and clothing to prevent any reasonable probability of skin contact. Safety equipment suppliers/manufacturers can provide recommendations on the most protective glove/clothing material for your operation. All protective clothing (suits, gloves, footwear, headgear) should be clean, available each day, and put on before work. Contact lenses should not be worn when working with this chemical. Wear dust-proof chemical goggles and face shield unless full facepiece respiratory protection is worn. Employees should wash immediately with soap when skin is wet or contaminated. Provide emergency showers and eyewash.

Respirator Selection: *At any detectable concentration:* SCBAF:PD,PP (any MSHA/NIOSH approved self-contained breathing apparatus that has a full facepiece and is operated in a pressure-demand or other positive-pressure mode); or SAF: PD,PP:ASCBA (any supplied-air respirator that has a full facepiece and is operated in a pressure-demand or other positive-pressure mode in combination with an auxiliary, self-contained breathing apparatus operated in a pressure-demand or other positive pressure mode). *Escape:* HiEF (Any air-purifying, full-facepiece respirator with a high-efficiency particulate filter); or SCBAE (Any appropriate escape-type, self-contained breathing apparatus).

Storage: Prior to working with this chemical you should be trained on its proper handling and storage. Store in a cool, dry place. Protect from exposure to air.

Shipping: The hydrochloride is classified as a hazardous substance solid n.o.s. and falls in Hazard Class ORV-E and Packing Group III.

Spill Handling: Evacuate persons not wearing protective equipment from area of spill or leak until clean-up is complete. Remove all ignition sources. Collect powdered material in the most convenient and safe manner and deposit in sealed containers. Ventilate area after clean-up is complete. It may be necessary to contain and dispose of this chemical as a hazardous waste. If material or contaminated runoff enters waterways, notify downstream users of potentially contaminated waters. Contact your Department of Environmental Protection or your regional office of the federal EPA for specific recommendations. If employees are required to clean-up spills, they must be properly trained and equipped. OSHA 1910.120(q) may be applicable.

Fire Extinguishing: Use dry chemical, carbon dioxide, water spray, or alcohol foam extinguishers. Poisonous gases are produced in fire including nitrogen oxides and HCl (for the hydrochloride). If material or contaminated runoff enters waterways, notify downstream users of potentially contaminated waters. Notify local health and fire officials and pollution control agencies. From a secure, explosion-proof location, use water spray to cool exposed containers. If cooling streams are ineffective (venting sound increases in volume and pitch, tank discolors, or shows any signs of deforming), withdraw immediately to a secure position. If employees are expected to fight fires, they must be trained and equipped in OSHA 1910.156.

Disposal Method Suggested: Incinerator is equipped with a scrubber or thermal unit to reduce No_x emissions.

References

U.S. Environmental Protection Agency, Chemical Hazard Information Profile: 3-Amino-9-ethylcarbazole, Washington, DC (1979)

Sax, N. I., Ed., Dangerous Properties of Industrial Materials Report 4, No. 6, 70–72 (1984) and 6, No. 2, 41–43 (Hydrochloride) (1986)

Aminoethylethanolamine

Molecular Formula: $C_4H_{12}N_2O$

Synonyms: 2-(Aminoethyl)amino-; 2-[(2-Aminoethyl)amino]ethanol; *N*-(2-Aminoethyl)ethanolamine; Hidroxietiletilendiamina (Spanish); Hydroxyethylenediamine

CAS Registry Number: 111-41-1

RTECS Number: KJ6300000

DOT ID: UN 2735 (amines, liquid, corrosive, n.o.s.)

Regulatory Authority

- DOT Appendix B, §172.101
- Canada, WHMIS, Ingredients Disclosure List

Cited in U.S. State Regulations: Florida (G), Massachusetts (G), New Jersey (G), Pennsylvania (G)

Description: Aminoethylethanolamine is combustible, colorless, liquid with an ammonia-like odor. Flash point = 132°C. Autoignition temperature = 368°C. Hazard Identification (based on NFPA-704 M Rating System): Health 2, Flammability 1, Reactivity 0. Soluble in water.

Potential Exposure: Used to make textile finishing compounds, dyes, resins, rubber, insecticides, medicines, and other chemicals.

Incompatibilities: Contact with cellulose nitrate may cause fires upon contact. Reacts with Oxidizers, strong acids.

Permissible Exposure Limits in Air: None established. Due to the availability of insufficient data on long term effects, caution should be exercised.

Routes of Entry: Inhalation, skin contact.

Harmful Effects and Symptoms

Short Term Exposure: Contact can severely irritate and burn the eyes and skin; this can lead to permanent damage. Inhalation can irritate the respiratory tract and lungs causing shortness of breath, coughing and wheezing. High exposures can cause pulmonary edema, a medical emergency that can be delayed for several hours. This can cause death.

Long Term Exposure: May cause lung irritation, the development of bronchitis, with coughing, shortness of breath, phlegm. May cause skin allergy. If allergy develops, very low future exposure can cause itching and skin rash.

Points of Attack: Lungs, skin.

Medical Surveillance: Lung function tests. If symptoms develop or overexposure is suspected, chest x-ray should be considered. Evaluation by a qualified allergist may diagnose skin allergy.

First Aid: If this chemical gets into the eyes, remove any contact lenses at once and irrigate immediately for at least 15 minutes, occasionally lifting upper and lower lids. Seek medical attention immediately. If this chemical contacts the skin, remove contaminated clothing and wash immediately with soap and water. Seek medical attention immediately. If this chemical has been inhaled, remove from exposure, begin rescue breathing (using universal precautions) if breathing has stopped and CPR if heart action has stopped. Transfer promptly to a medical facility. When this chemical has been swallowed, get medical attention. If victim is conscious, administer water or milk. Do not induce vomiting. Medical observation is recommended for 24 – 48 hours after breathing overexposure, as pulmonary edema may be delayed.

Personal Protective Methods: Wear protective gloves and clothing to prevent any reasonable probability of skin contact. Safety equipment suppliers/manufacturers can provide recommendations on the most protective glove/clothing material for your operation. All protective clothing (suits, gloves, footwear, headgear) should be clean, available each day, and put on before work. Contact lenses should not be worn when working with this chemical. Wear splash-proof chemical goggles and face shield unless full facepiece respiratory protection is worn. Employees should wash immediately with soap when skin is wet or contaminated. Remove nonimpervious clothing immediately if wet or contaminated. Provide emergency showers and eyewash.

Respirator Selection: *At any concentrations above the NIOSH REL:* (Any MSHA/NIOSH approved self-contained breathing apparatus that has a full facepiece and is operated in a pressure-demand or other positive-pressure mode) or SAF:PD,PP:ASCBA (Any supplied-air respirator that has a full facepiece and is operated in a pressure-demand or other positive-pressure mode in combination with an auxiliary self-contained breathing apparatus operated in a pressure-demand or other positive pressure mode). *Escape:* GMFOV [Any air-purifying, full-facepiece respirator (gas mask) with a chin-style, front-or back-mounted organic vapor canister]; or SCBAE (Any appropriate escape-type, self-contained breathing apparatus).

Storage: Prior to working with this chemical you should be trained on its proper handling and storage. Store in tightly closed containers in a cool, well ventilated area. Sources of ignition such as smoking and open flames, are prohibited where this chemical is used, handled, or stored in a manner that could create a potential fire or explosion hazard.

Shipping: This chemical should be labeled "CORROSIVE." It falls into DOT/UN Hazard Class 8, and Packing Group I.

Spill Handling: Evacuate persons not wearing protective equipment from area of spill or leak until clean-up is complete. Remove all ignition sources. Ventilate area of spill or leak. Absorb liquids in vermiculite, dry sand, earth, or a similar material and deposit in sealed containers. It may be necessary to contain and dispose of this chemical as a hazardous waste. If material or contaminated runoff enters waterways, notify downstream users of potentially contaminated waters. Contact your Department of Environmental Protection or your regional office of the federal EPA for specific recommendations. If employees are required to clean-up spills, they must be properly trained and equipped. OSHA 1910.120(q) may be applicable.

Fire Extinguishing: Aminoethylethanolamine is a combustible liquid. Poisonous gases including nitrogen oxides are produced in fire. Use dry chemical, carbon dioxide, or alcohol resistant foam to extinguish fire. Storage containers and parts of containers may rocket great distances, in many directions. If material or contaminated runoff enters waterways, notify downstream users of potentially contaminated waters. Notify local health and fire officials and pollution control agencies. From a secure, explosion-proof location, use water spray to cool exposed containers. If cooling streams are ineffective (venting sound increases in volume and pitch, tank discolors, or shows any signs of deforming), withdraw immediately to a secure position. If employees are expected to fight fires, they must be trained and equipped in OSHA 1910.156.

References

New Jersey Department of Health and Senior Services, Hazardous Substance Fact Sheet, Aminoethylethanolamine, Trenton, NJ (December 1998).

N-Aminoethylpiperazine

Molecular Formula: $C_6H_{15}N_3$

Synonyms: *N*-(2-Aminoethyl) piperazine; 1-(2-Aminoethyl) piperazine; Piperazine, 1-(2-aminoethyl)-; 1-Piperazine ethanamine

CAS Registry Number: 140-31-8

RTECS Number: TK8050000

DOT ID: UN 2815

Regulatory Authority

- DOT Appendix B, §172.101
- Canada, WHMIS, Ingredients Disclosure List

Cited in U.S. State Regulations: Florida (G), Massachusetts (G), New Jersey (G), Pennsylvania (G).

Description: N-Aminoethypiperazine is a combustible aliphatic amine, corrosive, colorless, to light colored liquid. Boiling point = 222°C. Flash point = 93°C. Hazard Identification (based on NFPA-704 M Rating System): Health 2, Flammability 2, Reactivity 0. Soluble in water.

Potential Exposure: Used as an epoxy curing agent and making pharmaceuticals, synthetic fibers, and other chemicals.

Incompatibilities: Solution is a strong base. Incompatible with non-oxidizing mineral acids, strong acids, organic acids, acid chlorides, acid anhydrides, organic anhydrides, isocyanates, chloroformates, vinyl acetate, acrylates, substituted allyls, alkylene oxides, epichlorohydrin, ketones, aldehydes, alcohols, glycols, phenols, cresols, caprolactum solution, strong oxidizers. Contact with copper alloys, zinc or galvanized steel may cause violent reaction.

Permissible Exposure Limits in Air: None established. Due to the availability of insufficient data, caution should be exercised.

Routes of Entry: Inhalation, skin contact.

Harmful Effects and Symptoms

Short Term Exposure: Contact can severely irritate and burn the eyes and skin (can cause second and third degree burns); this can lead to permanent damage. Inhalation can irritate the respiratory tract and lungs causing shortness of breath, coughing and wheezing. Exposure can cause headache, nausea, and vomiting. Higher exposures can cause pulmonary edema, a medical emergency that can be delayed for several hours. This can cause death.

Long Term Exposure: May cause lung irritation, the development of bronchitis, with coughing, shortness of breath, phlegm. Mutation data reported.

Points of Attack: Lungs, skin.

Medical Surveillance: Lung function tests. If symptoms develop or overexposure is suspected, chest x-ray should be considered.

First Aid: If this chemical gets into the eyes, remove any contact lenses at once and irrigate immediately for at least 15 minutes, occasionally lifting upper and lower lids. Seek medical attention immediately. If this chemical contacts the skin, remove contaminated clothing and wash immediately with soap and water. Seek medical attention immediately. If this chemical has been inhaled, remove from exposure, begin rescue breathing (using universal precautions) if breathing has stopped and CPR if heart action has stopped. Transfer promptly to a medical facility. When this chemical has been swallowed, get medical attention. If victim is conscious, administer water or milk. Do not induce vomiting.

Personal Protective Methods: Wear protective gloves and clothing to prevent any reasonable probability of skin contact. Safety equipment suppliers/manufacturers can provide recommendations on the most protective glove/clothing material for your operation. All protective clothing (suits, gloves, footwear, headgear) should be clean, available each day, and put on before work. Contact lenses should not be worn when working with this chemical. Wear indirect-vent, impact and splash-proof chemical goggles and face shield unless full facepiece respiratory protection is worn. Employees should wash immediately with soap when skin is wet or contaminated. Remove nonimpervious clothing immediately if wet or contaminated. Provide emergency showers and eyewash.

Respirator Selection: *At any concentrations above the NIOSH REL:* (Any MSHA/NIOSH approved self-contained breathing apparatus that has a full facepiece and is operated in a pressure-demand or other positive-pressure mode) or SAF: PD,PP:ASCBA (Any supplied-air respirator that has a full facepiece and is operated in a pressure-demand or other positive-pressure mode in combination with an auxiliary self-contained breathing apparatus operated in a pressure-demand or other positive pressure mode). *Escape:* GMFOV [Any air-purifying, full-facepiece respirator (gas mask) with a chin-style, front-or back-mounted organic vapor canister]; or SCBAE (Any appropriate escape-type, self-contained breathing apparatus).

Storage: Prior to working with this chemical you should be trained on its proper handling and storage. Store in tightly closed containers in a cool, well ventilated area. Sources of ignition such as smoking and open flames, are prohibited where this chemical is used, handled, or stored in a manner that could create a potential fire or explosion hazard.

Shipping: This chemical should be labeled "CORROSIVE." It falls into DOT/UN Hazard Class 8, and Packing Group III.

Spill Handling: Evacuate persons not wearing protective equipment from area of spill or leak until clean-up is complete. Remove all ignition sources. Ventilate area of spill or leak. Absorb liquids in vermiculite, dry sand, earth, or a similar material and deposit in sealed containers. It may be necessary to contain and dispose of this chemical as a hazardous waste. If material or contaminated runoff enters waterways, notify downstream users of potentially contaminated waters. Contact your Department of Environmental Protection or your regional office of the federal EPA for specific recommendations. If employees are required to clean-up spills, they must be properly trained and equipped. OSHA 1910.120(q) may be applicable.

Fire Extinguishing: This chemical is a combustible liquid. Poisonous gases including nitrogen oxides and carbon monoxide are produced in fire. Use dry chemical, water, or alcohol or polymer foam to extinguish fire. Storage containers and parts of containers may rocket great distances, in many directions. If material or contaminated runoff enters waterways, notify downstream users of potentially contaminated waters. Notify local health and fire officials and pollution control agencies. From a secure, explosion-proof location, use water spray to cool exposed containers. If cooling streams are ineffective (venting sound increases in volume and pitch, tank discolors, or shows any signs of deforming), withdraw immediately to a secure position. If employees are expected to fight fires, they must be trained and equipped in OSHA 1910.156.

References

New Jersey Department of Health and Senior Services, Hazardous Substance Fact Sheet, N-Aminoethylpiperazine, Trenton NJ (December 1998)

1-Amino-2-Methylanthraquinone

Molecular Formula: $C_{15}H_{11}NO_2$

Synonyms: Acetate fast orange R; Acetoquinone light orange JL; 1-Amino-2-methyl-9,10-anthracenedione; 1-Amino-2-metilantraquinona (Spanish); 9,10-Anthracenedione, 1-amino-2-methyl-; Artisil orange 3RP; Celliton orange R; C.I. 60700; C.I. disperse orange 11; Cilla orange R; Disperse orange; Duranol orange G; 2-Methyl-1-anthraquinonylamine; Microsetile orange RA; Nyloquinone orange JR; Perliton orange 3R; Serisol orange yl; Supracet orange R

CAS Registry Number: 82-28-0

RTECS Number: CB5740000

DOT ID: UN 2811 (toxic solids, organic, n.o.s.)

Regulatory Authority

- Carcinogen (Animal Positive, NCI)[9] (Possible Select Carcinogen, OSHA) (California)
- CERCLA/SARA Section 313, Form R *de minimis* Concentration Reporting Level: 0.1%.
- Canada, WHMIS, Ingredients Disclosure List

Cited in U.S. State Regulations: California (G), Florida (G), Maine (G), Maryland (G), Massachusetts (G), Michigan (G), Minnesota (G), New Jersey (G), Pennsylvania (G), West Virginia (G).

Description: 1-Amino-2-methylanthraquinone, $C_{15}H_{11}NO_2$, is a crystalline substance. Freezing/Melting point = 205 – 206°C. Insoluble in water.

Potential Exposure: 1-Amino-2-methylanthraquinone is used almost exclusively as a dye intermediate for the production of a variety of anthraquinone dyes. The Society of Dyers and Colorists reported that it can be used as a dye for a variety of synthetic fibers, especially acetates, as well as wool, sheepskins, furs, and surface dyeing of thermoplastics. None of the dyes that can be prepared from it are presently produced in commercial quantities.1-Amino-2-methylanthraquinone had been produced commercially in the United States since 1948, but production was last reported by one company in 1970. The potential for exposure is greatest among workers engaged in the dyeing of textiles. 1-Amino-2-methylanthraquinone is not presently used in consumer products according to the CPSC.

Permissible Exposure Limits in Air: No standards set. However, this chemical is a known carcinogen in humans; contact should be lowered to the lowest possible levels.

Permissible Concentration in Water: No criteria set.

Routes of Entry: Inhalation, eye and/or skin contact.

Harmful Effects and Symptoms

Short Term Exposure: Contact may irritate the eyes and skin.

Long Term Exposure: There is evidence that this chemical causes liver and kidney cancer. A mutagen. Technical-grade 1-amino-2-methylanthraquinone (impurities unspecified), administered in the feed, was carcinogenic in Fischer 344 rats, inducing hepatocellular carcinomas in rats of both sexes, and kidney tumors (such as tubular-cell adenomas and adenocarcinomas) in males.

Points of Attack: Liver, kidneys.

First Aid: If this chemical gets into the eyes, remove any contact lenses at once and irrigate immediately for at least 15 minutes, occasionally lifting upper and lower lids. Seek medical attention immediately. If this chemical contacts the skin, remove contaminated clothing and wash immediately with

soap and water. Seek medical attention immediately. If this chemical has been inhaled, remove from exposure, begin rescue breathing (using universal precautions) if breathing has stopped and CPR if heart action has stopped. Transfer promptly to a medical facility. When this chemical has been swallowed, get medical attention. Give large quantities of water and induce vomiting. Do not make an unconscious person vomit.

Personal Protective Methods: Wear protective gloves and clothing to prevent any reasonable probability of skin contact. Safety equipment suppliers/manufacturers can provide recommendations on the most protective glove/clothing material for your operation. All protective clothing (suits, gloves, footwear, headgear) should be clean, available each day, and put on before work. Contact lenses should not be worn when working with this chemical. Wear splash-proof chemical goggles and face shield unless full facepiece respiratory protection is worn. Employees should wash immediately with soap when skin is wet or contaminated. Provide emergency showers and eyewash.

Respirator Selection: *At any detectable concentration:* SCBAF:PD,PP (any MSHA/NIOSH approved self-contained breathing apparatus that has a full facepiece and is operated in a pressure-demand or other positive-pressure mode); or SAF: PD,PP:ASCBA (any supplied-air respirator that has a full facepiece and is operated in a pressure-demand or other positive-pressure mode in combination with an auxiliary, self-contained breathing apparatus operated in a pressure-demand or other positive pressure mode). *Escape:* HiEF (Any air-purifying, full-facepiece respirator with a high-efficiency particulate filter); or SCBAE (Any appropriate escape-type, self-contained breathing apparatus).

Storage: Prior to working with this chemical you should be trained on its proper handling and storage. A regulated, marked area should be established where this chemical is handled, used, or stored. Should be stored in a refrigerator in a tightly closed container under an inert atmosphere.[52]

Shipping: The DOT/UN regulations does not set out any specific label requirements or maximum shipping quantities for this chemical. However in solid form this chemical is a toxic solid, organic, n.o.s and should be labeled "POISON," Hazard Class 6.1.

Spill Handling: Evacuate and restrict persons not wearing protective equipment from area of spill or leak until cleanup is complete. Remove all ignition sources. Dampen spilled material with toluene to avoid airborne dust. Collect powdered material in the most convenient and safe manner and deposit in sealed containers. Ventilate area of spill or leak after clean-up is complete. Keep this chemical out of a confined space, such as a sewer, because of the potential for an explosion, unless the sewer is designed to prevent the build-up of explosive concentrations. It may be necessary to contain and dispose of this chemical as a hazardous waste. If material or contaminated runoff enters waterways, notify downstream users of potentially contaminated waters. Contact your Department of Environmental Protection or your regional office of the federal EPA for specific recommendations. If employees are required to clean-up spills, they must be properly trained and equipped. OSHA 1910.120(q) may be applicable.

Fire Extinguishing: Poisonous gases including nitrogen oxides are produced in fire. Use dry chemical, carbon dioxide, or alcohol foam extinguishers. Vapors are heavier than air and will collect in low areas. Vapors may travel long distances to ignition sources and flashback. Vapors in confined areas may explode when exposed to fire. Storage containers and parts of containers may rocket great distances, in many directions. If material or contaminated runoff enters waterways, notify downstream users of potentially contaminated waters. Notify local health and fire officials and pollution control agencies. From a secure, explosion-proof location, use water spray to cool exposed containers. If cooling streams are ineffective (venting sound increases in volume and pitch, tank discolors, or shows any signs of deforming), withdraw immediately to a secure position. If employees are expected to fight fires, they must be trained and equipped in OSHA 1910.156.

References

New Jersey Department of Health and Senior Services, Hazardous Substance Fact Sheet, 1-Amino-2-Methylanthraquinone (June, 1988)

4-Amino-2-Nitrophenol

Molecular Formula: $C_6H_6N_2O_2$

Common Formula: $C_6H_3NO_2NH_2OH$

Synonyms: C.I.-76555; Fourrine 57; Fourrine brown PR; Fourrine brown propyl; 4-Hydroxy-3-nitroaniline; 2-Nitro-4-amenophenol; *o*-Nitro-*p*-aminophenol

CAS Registry Number: 119-34-6

RTECS Number: SJ6303000

Regulatory Authority

- Carcinogen (Rat Positive, Mouse Negative) (NCI)[9] (Suspected) (DFG)[3]
- TSCA 40CFR704.225

Cited in U.S. State Regulations: Massachusetts (G).

Description: 4-Amino-2-nitrophenol, $C_6H_6N_2O_3$, as dark red crystals. Freezing/Melting point = 131°C.

Potential Exposure: In dye formulation for furs and hair.

Incompatibilities: Strong oxidizers, mineral acids, strong bases.

Permissible Exposure Limits in Air: No standards set.

Permissible Concentration in Water: No criteria set.

Harmful Effects and Symptoms

Severe eye irritant in rabbits. Rat carcinogenicity positive but mouse carcinogenicity negative according to NCl.

Short Term Exposure: Poisonous if swallowed. Irritates eyes and may irritate skin and respiratory tract.

Long Term Exposure: May be carcinogenic in humans.

First Aid: If this chemical gets into the eyes, remove any contact lenses at once and irrigate immediately for at least 15 minutes, occasionally lifting upper and lower lids. Seek medical attention immediately. If this chemical contacts the skin, remove contaminated clothing and wash immediately with soap and water. Seek medical attention immediately. If this chemical has been inhaled, remove from exposure, begin rescue breathing (using universal precautions) if breathing has stopped and CPR if heart action has stopped. Transfer promptly to a medical facility. When this chemical has been swallowed, get medical attention. Give large quantities of water and induce vomiting. Do not make an unconscious person vomit.

Personal Protective Methods: Wear protective gloves and clothing to prevent any reasonable probability of skin contact. Safety equipment suppliers/manufacturers can provide recommendations on the most protective glove/clothing material for your operation. All protective clothing (suits, gloves, footwear, headgear) should be clean, available each day, and put on before work. Contact lenses should not be worn when working with this chemical. Wear dust-proof chemical goggles and face shield unless full facepiece respiratory protection is worn. Employees should wash immediately with soap when skin is wet or contaminated. Provide emergency showers and eyewash.

Respirator Selection: *At any detectable concentration:* SCBAF:PD,PP (any MSHA/NIOSH approved self-contained breathing apparatus that has a full facepiece and is operated in a pressure-demand or other positive-pressure mode); or SAF:PD,PP:ASCBA (any supplied-air respirator that has a full facepiece and is operated in a pressure-demand or other positive-pressure mode in combination with an auxiliary, self-contained breathing apparatus operated in a pressure-demand or other positive pressure mode). *Escape:* HiEF (any air-purifying, full-facepiece respirator with a high-efficiency particulate filter); or SCBAE (any appropriate escape-type, self-contained breathing apparatus).

Storage: Prior to working with this chemical you should be trained on its proper handling and storage. Store in cool dry place away from air and away from mineral acids and bases.[52]

Shipping: In solid form this chemical is a toxic solid, organic, n.o.s, and should be labeled "POISON," Hazard Class 6.1.

Spill Handling: Evacuate persons not wearing protective equipment from area of spill or leak until clean-up is complete. Remove all ignition sources. Dampen spilled material with 60 – 70% acetone to avoid airborne dust and remove to sealed containers for disposal.[52] Ventilate area after clean-up is complete. It may be necessary to contain and dispose of this chemical as a hazardous waste. If material or contaminated runoff enters waterways, notify downstream users of potentially contaminated waters. Contact your Department of Environmental Protection or your regional office of the federal EPA for specific recommendations. If employees are required to clean-up spills, they must be properly trained and equipped. OSHA 1910.120(q) may be applicable.

Fire Extinguishing: Use dry chemical, carbon dioxide, water spray, or alcohol foam extinguishers. Poisonous gases are produced in fire including nitrogen oxides. If material or contaminated runoff enters waterways, notify downstream users of potentially contaminated waters. Notify local health and fire officials and pollution control agencies. From a secure, explosion-proof location, use water spray to cool exposed containers. If cooling streams are ineffective (venting sound increases in volume and pitch, tank discolors, or shows any signs of deforming), withdraw immediately to a secure position. If employees are expected to fight fires, they must be trained and equipped in OSHA 1910.156.

References

Sax, N. I., Ed., Dangerous Properties of Industrial Materials Report, 1, No.7, 34–35 (1981)

Aminophenols

Molecular Formula: C_6H_7NO

Common Formula: $C_6H_4(OH)(NH_2)$

Synonyms: *ortho-:* 2-Amino-1-hydroxybenzene; *o*-Hydroxyaniline

meta-: 3-Amino-1-hydroxybenzene; 3-Hydroxyaniline

para-: 4-Amino-1-hydroxybenzene; *p*-Hydroxyaniline

CAS Registry Number: 95-55-6 (o-); 591-27-5 (m-); 123-30-8 (p-); 27598-85-2 (mixed isomers)

RTECS Number: SJ4950000 (o-); SJ4900000 (m-); SJ5075000 (p-)

DOT ID: UN 2512

EEC Number: 612-033-00-3 (o-)

Regulatory Authority

- Canada, WHMIS, Ingredients Disclosure List (m-, o-, p-)
- TSCA CFR721.5820

Cited in U.S. State Regulations: New Hampshire (G), New Jersey (G).

Description: $C_6H_7NO[C_6H_4(OH)(NH_2)]$; the amine group and the hydroxyl group may be in the ortho, meta, or para positions. o-Aminophenol is a colorless needles or white crystalline substance turning brown on exposure to air. Freezing/Melting point = 170 – 174°C. Soluble in water. p-Aminophenol is a white or reddish yellow crystalline substance. Freezing/Melting point = 190°C (decomposes). Boiling point = 284°C. Slightly soluble in water. m-Aminophenol is a white crystalline substance. Freezing/Melting point = 122 – 123°C. Soluble in water.

Potential Exposure: Workers may be exposed to o-Aminophenol during its use as a chemical intermediate, in the manufacture of azo and sulfur dyes, and in the photographic industry. There is potential for consumer exposure to o-Aminophenol because of its use in dyeing hair, fur, and leather. The compound is a constituent of 75 registered cosmetic products suggesting the potential for widespread consumer exposure. p-Aminophenol is used mainly as a dye and dye intermediate and as a photographic developer, and in small quantities in analgesic drug preparation. Consumer exposure to p-aminophenol may occur from use as a hair-dye or as a component in cosmetic preparations. m-Aminophenol is used mainly as a dye intermediate.

Incompatibilities: Strong oxidants.

Permissible Exposure Limits in Air: No limits set.

Permissible Concentration in Water: The former USSR-UNEP/IRPTC project[43] has set MAC values in water used for domestic purposes. The value for o-aminophenol is 0.01 mg/l and for p-aminophenol is 0.05 mg/l.

Routes of Entry: Inhalation, ingestion and skin absorption.

Harmful Effects and Symptoms

Short Term Exposure: Aminophenols can be absorbed through the skin, thereby increasing exposure. Can cause lung irritation. Poisonous if swallowed. These chemicals lower the blood's ability to carry oxygen (methemoglobinemia). This condition causes a bluish color to the skin and lips, headaches, dizziness; higher exposures can result in unconsciousness and death. Irritates eyes, skin and respiratory tract. Skin contact can cause burning sensation and rash. the o- isomer can affect the nervous system. o-Aminophenol has an oral LD_{50} for rats of 1,300 mg/kg and p-aminophenol has produced LD_{50} values in rats of 375, 671 and 1,270 mg/kg which are all of low acute toxicity. The oral LD_{50} in rats for m-aminophenol is 1,000 mg/kg. May produce dermatitits, methemoglobinemia, bronchial asthma and restlessness.

Long Term Exposure: Prolonged or repeated contact can cause blood damage, skin disorders, liver, kidney and brain damage. Aminophenols may cause mutations, and there is limited teratogenic evidence. Skin allergy or asthma may develop; future exposures, even in low doses can cause symptoms to occur.

Points of Attack: Blood, liver, kidneys, brain, skin and lungs.

Medical Surveillance: Blood methemoglobin level; lung function tests. Evaluation by a qualified allergist.

First Aid: If this chemical gets into the eyes, remove any contact lenses at once and irrigate immediately for at least 15 minutes, occasionally lifting upper and lower lids. Seek medical attention immediately. If this chemical contacts the skin, remove contaminated clothing and wash immediately with soap and water. Seek medical attention immediately. If this chemical has been inhaled, remove from exposure, begin rescue breathing (using universal precautions) if breathing has stopped and CPR if heart action has stopped. Transfer promptly to a medical facility. When this chemical has been swallowed, get medical attention. Give large quantities of water and induce vomiting. Do not make an unconscious person vomit.

Note to Physician: Treat for methemoglobinemia. Spectrophotometry may be required for precise determination of levels of methemoglobinemia in urine.

Personal Protective Methods: Wear protective gloves and clothing to prevent any reasonable probability of skin contact. Safety equipment suppliers/manufacturers can provide recommendations on the most protective glove/clothing material for your operation. All protective clothing (suits, gloves, footwear, headgear) should be clean, available each day, and put on before work. Contact lenses should not be worn when working with this chemical. Wear dust-proof chemical goggles and face shield unless full facepiece respiratory protection is worn. Employees should wash immediately with soap when skin is wet or contaminated. Provide emergency showers and eyewash.

Respirator Selection: *At any detectable concentration:* SCBAF:PD,PP (any MSHA/NIOSH approved self-contained breathing apparatus that has a full facepiece and is operated in a pressure-demand or other positive-pressure mode); or SAF: PD,PP:ASCBA (any supplied-air respirator that has a full facepiece and is operated in a pressure-demand or other positive-pressure mode in combination with an auxiliary, self-contained breathing apparatus operated in a pressure-demand or other positive pressure mode). *Escape:* GMFOVHiE [any air-purifying, full-facepiece respirator (gas mask) with a chin-style, front- or back-mounted organic vapor canister having a high-efficiency particulate filter]; or SCBAE (any appropriate escape-type, self-contained breathing apparatus).

Storage: Prior to working with this chemical you should be trained on its proper handling and storage. Store in tightly closed containers in a cool, well ventilated area. Aminophenols must be stored to avoid contact with strong oxidizers (such as chlorine, bromine, and fluorine) since violent reactions occur.

Shipping: The DOT/UN label requirements for o-, m-, and p-aminophenols is "Keep away from Food." The aminophenols fall in Hazard Class 6.1 and Packing Group III.[19][20]

Spill Handling: Keep dust under control. Use a vacuum or wet method to reduce dust during clean-up. Do not sweep. Evacuate persons not wearing protective equipment from area of spill or leak until clean-up is complete. Remove all ignition sources. Collect powdered material in the most convenient and safe manner and deposit in sealed containers. Ventilate area after clean-up is complete. It may be necessary to contain and dispose of this chemical as a hazardous waste. If material or contaminated runoff enters waterways, notify downstream users of potentially contaminated waters. Contact your Department of Environmental Protection or your regional office of the federal EPA for specific recommendations. If employees are required to clean-up spills, they must be properly trained and equipped. OSHA 1910.120(q) may be applicable.

Fire Extinguishing: Aminophenols may burn, but do not readily ignite. Use dry chemical, CO_2, water spray, or alcohol foam extinguishers. Poisonous gases are produced in fire, Including nitrogen oxides. If material or contaminated runoff enters waterways, notify downstream users of potentially contaminated waters. Notify local health and fire officials and pollution control agencies. From a secure, explosion-proof location, use water spray to cool exposed containers. If cooling streams are ineffective (venting sound increases in volume and pitch, tank discolors, or shows any signs of deforming), withdraw immediately to a secure position. If employees are expected to fight fires, they must be trained and equipped in OSHA 1910.156.

Disposal Method Suggested: Incineration.

References

U.S. Environmental Protection Agency, Chemical Hazard Information Profile Draft Report: o-Aminophenol, sulfate and hydrochloride, Washington, DC (March 29, 1984)

New Jersey Department of Health and Senior Services, "Hazardous Substance Fact Sheet: Aminophenols," Trenton, NJ (May 1986)

Aminopterin

Molecular Formula: $C_{19}H_{20}N_8O_5$

Synonyms: 4-Amino-4-deoxypteroyl glutamate; 4-Aminofolic acid; 4-Amino-PGA; Aminopteridine; 4-Aminopteroylglumatic acid; APGA; Folic acid, 4-amino-

CAS Registry Number: 54-62-6

RTECS Number: MA1050000

Regulatory Authority

- CERCLA/SARA 40CFR302 Extremely Hazardous Substances: TPQ = 500/10,000 lb (227/4,540 kg).
- SUPERFUND/EPCRA 40CFR302.4 Reportable Quantity (RQ): EHS, 1 lb (0.454 kg).

Cited in U.S. State Regulations: California (G), Florida (G), Massachusetts (G), New Jersey (G), Pennsylvania (G).

Description: Aminopterin, $C_{19}H_{20}N_8O_5$, is commonly used as the dihydrate which forms clusters of yellow needles.

Potential Exposure: Has found use as a medicine, rodenticide, and agricultural chemical.

Permissible Exposure Limits in Air: No standards set.

Permissible Concentration in Water: No criteria set.

Routes of Entry: Ingestion.

Harmful Effects and Symptoms

Short Term Exposure: Acts as antimetabolite; antagonizes the utilization of folic acid by the body. Highly toxic by ingestion. Has an LD_{low} oral (rat) of 2.5 mg/kg.

Long Term Exposure: A mutagen, and a questionable carcinogen. Listed as a reproductive toxicant by the State of California.

First Aid: If this chemical gets into the eyes, remove any contact lenses at once and irrigate immediately for at least 15 minutes, occasionally lifting upper and lower lids. Seek medical attention immediately. If this chemical contacts the skin, remove contaminated clothing and wash immediately with soap and water. Seek medical attention immediately. If this chemical has been inhaled, remove from exposure, begin rescue breathing (using universal precautions) if breathing has stopped and CPR if heart action has stopped. Transfer promptly to a medical facility. When this chemical has been swallowed, get medical attention. Give large quantities of water and induce vomiting. Do not make an unconscious person vomit.

Personal Protective Methods: Wear protective gloves and clothing to prevent any reasonable probability of skin contact. Safety equipment suppliers/manufacturers can provide recommendations on the most protective glove/clothing material for your operation. All protective clothing (suits, gloves, footwear, headgear) should be clean, available each day, and put on before work. Contact lenses should not be worn when working with this chemical. Wear dust-proof chemical goggles and face shield unless full facepiece respiratory protection is worn. Employees should wash immediately with soap when skin is wet or contaminated. Provide emergency showers and eyewash.

Storage: Prior to working with this chemical you should be trained on its proper handling and storage. Store in tightly closed containers in a cool, well ventilated area.

Shipping: The DOT/UN regulations do not set out any specific label requirements or maximum shipping quantities for this chemical. However this chemical is a toxic solid, organic, n.o.s and should be labeled "POISON," Hazard Class 6.1.

Spill Handling: Evacuate persons not wearing protective equipment from area of spill or leak until clean-up is complete. Remove all ignition sources. Do not touch spilled material; stop leak if you can do so without risk. Do not create dust. Use a vacuum and water spray to reduce vapors. *Small wet spills*: absorb with sand or other noncombustible absorbent material and place into containers for later disposal. *Small dry spills*: with clean shovel place material into clean, dry container and cover; move containers from spill area. *Large spills:* dike far ahead of spill for later disposal. Ventilate area after clean-up is complete. It may be necessary to contain and dispose of this chemical as a hazardous waste. If material or contaminated runoff enters waterways, notify downstream users of potentially contaminated waters. Contact your Department of Environmental Protection or your regional office of the federal EPA for specific recommendations. If employees are required to clean-up spills, they must be properly trained and equipped. OSHA 1910.120(q) may be applicable.

Fire Extinguishing: Use dry chemical, carbon dioxide, water spray, or alcohol foam extinguishers. Poisonous gases are produced in fire including nitrogen oxides. If material or contaminated runoff enters waterways, notify downstream users of potentially contaminated waters. Notify local health and fire officials and pollution control agencies. From a secure, explosion-proof location, use water spray to cool exposed containers. If cooling streams are ineffective (venting sound increases in volume and pitch, tank discolors, or shows any signs of deforming), withdraw immediately to a secure position. If employees are expected to fight fires, they must be trained and equipped in OSHA 1910.156.

References

U.S. Environmental Protection Agency, "Chemical Profile: Aminopterin," Washington, DC, Chemical Emergency Preparedness Program (November 30, 1987)

2-Aminopyridine

Molecular Formula: $C_5H_6N_2$

Synonyms: α-Aminopyridine; Amino-2-Pyridine; 1,2-Dihydro-2-Iminopyridine; α-Piridilamina (Spanish); α-Pyridinamine; 2-Pyridylamine

CAS Registry Number: 504-29-0

RTECS Number: US1575000

DOT ID: UN 2671

Regulatory Authority

- Air Pollutant Standard Set (ACGIH)[1] (DFG)[3] (HSE)[33] (OSHA)[58] (Several States)[60] (Various Canadian Provinces) (Australia) (Israel)
- Canada, WHMIS, Ingredients Disclosure List

Cited in U.S. State Regulations: Alaska (G), California (G), Connecticut (A), Florida (G), Illinois (G), Maine (G, W), Massachusetts (G), Minnesota (G), Nevada (A), New Hampshire (G), New Jersey (G), North Dakota (A), Oklahoma (G), Pennsylvania (G), Rhode Island (G), Virginia (A), West Virginia (G).

Description: 2-Aminopyridine is a flammable, colorless crystalline solid, or white leaflets or powder, or colorless liquid with a characteristic odor. Freezing/Melting point = 58°C. Boiling point = 210°C. Flash point = 68°C. Highly soluble in water.

Potential Exposure: 2-Aminopyridine is used in the manufacture of pharmaceuticals, especially antihistamines.

Incompatibilities: Strong oxidizers.

Permissible Exposure Limits in Air: The Federal OSHA standard[58] and ACGIH value is 0.5 ppm (2 mg/m^3) as is the DFG, HSE, Australia, and Israel values.[3] IDLH level is 5 ppm. In addition, several states have set guidelines or standards for 2-aminopyridine in ambient air:[61] 0.5 ppm (2 mg/m^3) TWA (California), 2 mg/m^3 (North Dakota), 3 mg/m^3 (Virginia), 4 mg/m^3 (Connecticut), 4.8 mg/m^3 (Nevada). Canadian standards are: 0.5 ppm (2 mg/m^3) TWA and 2 ppm STEL (Alberta, British Columbia), 0.5 ppm (2 mg/m^3) TWA (Ontario, Quebec).

Determination in Air: Adsorption on Tenax GC,[2] thermal desorption and gas chromatographic analysis. NIOSH Method #S158.

Permissible Concentration in Water: No criteria set.

Routes of Entry: Inhalation, ingestion, eye, and skin contact (absorption through the skin).

Harmful Effects and Symptoms

Headaches, dizziness, excited state, nausea, flushed appearance, high blood pressure, respiratory distress, weakness, convulsions, stupor.

Short Term Exposure: Skin contact contributes significantly to overall exposure. Irritates eyes, skin, and respiratory tract. Inhalation or skin exposure can cause a fatal reaction that begins with a headache, dizziness, heaviness and weakness of the limbs, and may progress to convulsions, stupor, and coma.

Long Term Exposure: May cause lung allergy.

Points of Attack: Central nervous system, respiratory system.

Medical Surveillance: Consider the points of attack in preplacement and periodic physical examinations. Interview for brain effects including recent memory loss, change in mood, concentration, headache, listless feeling, altered sleep patterns. Consider cerebellar, autonomic, and peripheral nervous system evaluation. Positive and borderline individuals should be referred for neuropsychological testing.

First Aid: If this chemical gets into the eyes, remove any contact lenses at once and irrigate immediately for at least 15 minutes, occasionally lifting upper and lower lids. Seek medical attention immediately. If this chemical contacts the skin, remove contaminated clothing and wash immediately with soap and water. Seek medical attention immediately. If this chemical has been inhaled, remove from exposure, begin rescue breathing (using universal precautions) if breathing has stopped and CPR if heart action has stopped. Transfer promptly to a medical facility. When this chemical has been swallowed, get medical attention. Give large quantities of water and induce vomiting. Do not make an unconscious person vomit.

Personal Protective Methods: Wear protective gloves and clothing to prevent any reasonable probability of skin contact. Safety equipment suppliers/manufacturers can provide recommendations on the most protective glove/clothing material for your operation. All protective clothing (suits, gloves, footwear, headgear) should be clean, available each day, and put on before work. Contact lenses should not be worn when working with this chemical. Wear splash or dust-proof chemical goggles and face shield unless full facepiece respiratory protection is worn. Employees should wash immediately with soap when skin is wet or contaminated. Provide emergency showers and eyewash.

Respirator Selection: NIOSH/OSHA: 5 ppm: SA (any supplied-air respirator); or SCBAF (any self-contained breathing apparatus with a full facepiece). *Emergency or planned entry* *into unknown concentrations or IDLH conditions:* SCBAF: PD,PP (any self-contained breathing apparatus that has a full facepiece and is operated in a pressure-demand or other positive-pressure mode); or SAF:PD,PP:ASCBA (any supplied-air respirator that has a full facepiece and is operated in a pressure-demand or other positive-pressure mode in combination with an auxiliary, self-contained breathing apparatus operated in a pressure-demand or other positive-pressure mode). *Escape:* GMFOVHiE [any air-purifying, full-facepiece respirator (gas mask) with a chin-style, front- or back-mounted organic vapor canister having a high-efficiency particulate filter]; or SCBAE (any appropriate escape-type, self-contained breathing apparatus).

Storage: Prior to working with this chemical you should be trained on its proper handling and storage. Store in a refrigerator under an inert atmosphere;[52] or store in tightly closed containers in a cool, well ventilated area away from oxidizers. Sources of ignition such as smoking and open flames, are prohibited where this chemical is used, handled, or stored in a manner that could create a potential fire or explosion hazard.

Shipping: UN/DOT label required is "Poison." Falls in UN/DOT Hazard Class 6.1 and Packing Group II.[19][20][21]

Spill Handling: Evacuate persons not wearing protective equipment from area of spill or leak until clean-up is complete. Remove all ignition sources. Establish forced ventilation to keep levels below explosive limit. Use water spray to reduce dust or vapors. Absorb liquids in vermiculite, dry sand, earth, or a similar material and deposit in sealed containers. Collect powdered material in the most convenient and safe manner and deposit in sealed containers. Dike large spills. It may be necessary to contain and dispose of this chemical as a hazardous waste. If material or contaminated runoff enters waterways, notify downstream users of potentially contaminated waters. Contact your Department of Environmental Protection or your regional office of the federal EPA for specific recommendations. If employees are required to clean-up spills, they must be properly trained and equipped. OSHA 1910.120(q) may be applicable.

Fire Extinguishing: This chemical is a combustible liquid. Poisonous gases including nitrogen oxides are produced in fire. Use dry chemical, carbon dioxide, water spray, or foam extinguishers. Vapors are heavier than air and will collect in low areas. Vapors in confined areas may explode when exposed to fire. Storage containers and parts of containers may rocket great distances, in many directions. If material or contaminated runoff enters waterways, notify downstream users of potentially contaminated waters. Notify local health and fire officials and pollution control agencies. From a secure, explosion-proof location, use water spray to cool exposed containers. If cooling streams are ineffective (venting sound increases in volume and pitch, tank discolors, or shows any signs of deforming), withdraw immediately to a secure position. If employees are expected to fight fires, they must be trained and equipped in OSHA 1910.156.

Disposal Method Suggested: Incineration with No_x removal from effluent gas.

References

New Jersey Department of Health and Senior Services, "Hazardous Substance Fact Sheet: 2-Aminopyridine," Trenton, NJ (October 1985)

4-Aminopyridine

Molecular Formula: $C_5H_6N_2$

Synonyms: 4-Aminopiridina (Spanish); γ-Aminopyridine; *p*-Aminopyridine; Amino-4-pyridine; Avitrol; 4-Pyridinamin; 4-Pyridylamine; VMI 10-3

CAS Registry Number: 504-24-5

RTECS Number: US1750000

DOT ID: UN 2671

Regulatory Authority

- EPA HAZARDOUS WASTE NUMBER (RCRA No.): P008
- RCRA 40CFR261, Appendix 8; 40CFR261.11 Hazardous Constituents
- CERCLA/SARA 40CFR302 Extremely Hazardous Substances: TPQ = 500/10,000 lb (227/4,540 kg).
- SUPERFUND/EPCRA 40CFR302.4 Reportable Quantity (RQ): CERCLA, 1,000 lb (454 kg)
- Canada, WHMIS, Ingredients Disclosure List

Cited in U.S. State Regulations: California (G), Florida (G), Illinois (G), Kansas (G), Louisiana (G), Massachusetts (G), Michigan (G), New Hampshire (G), New Jersey (G), Pennsylvania (G), Vermont (G), Virginia (G), Washington (G), Wisconsin (G).

Description: 4-Aminopyridine, $C_5H_6N_2$, is white to tan or brown crystalline material. Odorless. Boiling point = 274°C. Freezing/Melting point = 159°C. Flash point = 164°C. Moderately soluble in water.

Potential Exposure: Used as a chemical intermediate in pharmaceuticals, as an agricultural chemical for field crops, and as a bird repellent and poison.

Incompatibilities: Sodium nitrite, strong oxidizers. Avoid contact with acid anhydrides, acid chlorides, and strong acids.

Permissible Exposure Limits in Air: No standards set.

Permissible Concentration in Water: No criteria set. Harmful to aquatic life in very low concentrations.

Routes of Entry: Inhalation, ingestion, skin absorption.

Harmful Effects and Symptoms

Short Term Exposure: May be fatal if swallowed or absorbed through the skin. Symptoms of exposure include rapid onset of disagreeable taste, immediate burning of throat, and abdominal discomfort; in addition, weakness, dizziness, disorientation, and seizures may occur. Delayed symptoms of oral ingestion include elevated liver enzymes, and respiratory arrest. Material may be fatal if inhaled, swallowed or absorbed through skin. Contact may cause burns to skin and eyes. Material affects neural transmission. In sufficient concentration, material may cause metabolic acidosis, respiratory arrest, and cardiac arrhythmias. The fatal dose to a 70 kg person is about 5 grams.

Long Term Exposure: High exposure or repeated exposure may cause liver damage.

Points of Attack: Central nervous system, liver.

Medical Surveillance: Pre-employment and regular physical examinations with emphasis on central nervous system. Liver function tests. Persons exposed to *strychnine* or other chemicals capable of causing seizures are probably at increased risk.

First Aid: If this chemical gets into the eyes, remove any contact lenses at once and irrigate immediately for at least 15 minutes, occasionally lifting upper and lower lids. Seek medical attention immediately. If this chemical contacts the skin, remove contaminated clothing and wash immediately with soap and water. Seek medical attention immediately. If this chemical has been inhaled, remove from exposure, begin rescue breathing (using universal precautions) if breathing has stopped and CPR if heart action has stopped. Transfer promptly to a medical facility. When this chemical has been swallowed, get medical attention. Give large quantities of water and induce vomiting. Do not make an unconscious person vomit.

Personal Protective Methods: Wear protective gloves and clothing to prevent any reasonable probability of skin contact. Safety equipment suppliers/manufacturers can provide recommendations on the most protective glove/clothing material for your operation. All protective clothing (suits, gloves, footwear, headgear) should be clean, available each day, and put on before work. Contact lenses should not be worn when working with this chemical. Wear dust-proof chemical goggles and face shield unless full facepiece respiratory protection is worn. Employees should wash immediately with soap when skin is wet or contaminated. Provide emergency showers and eyewash.

Respirator Selection: *For 2-Aminopiridine:* NIOSH/OSHA: 5 ppm: SA (any supplied-air respirator); or SCBAF (any self-contained breathing apparatus with a full facepiece). *Emergency or planned entry into unknown concentrations or IDLH conditions:* SCBAF:PD,PP (any self-contained breathing apparatus that has a full facepiece and is operated in a pressure-demand or other positive-pressure mode); or SAF:PD,PP:ASCBA (any supplied-air respirator that has a full facepiece and is operated in a pressure-demand or other positive-pressure mode in combination with an auxiliary, self-contained breathing apparatus operated in a pressure-demand or other positive-pressure mode). *Escape:* GMFOVHiE [any air-purifying, full-facepiece respirator (gas mask) with a chin-style, front- or back-mounted organic vapor canister having a high-efficiency particulate filter]; or SCBAE (any appropriate escape-type, self-contained breathing apparatus).

Storage: Prior to working with this chemical you should be trained on its proper handling and storage. Store in tightly closed containers in a cool, well ventilated area. Avoid contact with acid anhydrides, acid chlorides, strong acids and strong oxidizers (such as halogens). Sources of ignition such as smoking and open flames, are prohibited where this chemical is used, handled, or stored in a manner that could create a potential fire or explosion hazard.

Shipping: UN/DOT label required is "Poison." Falls in UN/DOT Hazard Class 6.1 and Packing Group II.[19][20][21]

Spill Handling: Evacuate persons not wearing protective equipment from area of spill or leak until clean-up is complete. Remove all ignition sources. Ventilate area of spill or leak. Use water spray to reduce dust or vapors. Absorb liquids in vermiculite, dry sand, earth, or a similar material and deposit in sealed containers. Dike large spills. It may be necessary to contain and dispose of this chemical as a hazardous waste. If material or contaminated runoff enters waterways, notify downstream users of potentially contaminated waters. Contact your Department of Environmental Protection or your regional office of the federal EPA for specific recommendations. If employees are required to clean-up spills, they must be properly trained and equipped. OSHA 1910.120(q) may be applicable.

Fire Extinguishing: This chemical is a flammable liquid. Poisonous gases including nitrogen oxides are produced in fire. Use dry chemical, carbon dioxide, water spray, or foam extinguishers. Vapors are heavier than air and will collect in low areas. Vapors may travel long distances to ignition sources and flashback. Vapors in confined areas may explode when exposed to fire. Storage containers and parts of containers may rocket great distances, in many directions. If material or contaminated runoff enters waterways, notify downstream users of potentially contaminated waters. Notify local health and fire officials and pollution control agencies. From a secure, explosion-proof location, use water spray to cool exposed containers. If cooling streams are ineffective (venting sound increases in volume and pitch, tank discolors, or shows any signs of deforming), withdraw immediately to a secure position. If employees are expected to fight fires, they must be trained and equipped in OSHA 1910.156.

Disposal Method Suggested: Consult with environmental regulatory agencies for guidance on acceptable disposal practices. Generators of waste containing this contaminant (≥100 kg/mo) must conform with EPA regulations governing storage, transportation, treatment, and waste disposal. Incineration with No_x removal from effluent gas.

References

Sax, N. I., Ed., Dangerous Properties of Industrial Materials Report 5, No. 5, 39–41 (1985)

New Jersey Department of Health and Senior Services, "Hazardous Substance Fact Sheet: Avitrol," Trenton, NJ (April 1997)

U.S. Environmental Protection Agency, "Chemical Profile: 4-Aminopyridine," Washington, DC, Chemical Emergency Preparedness Program (November 30, 1987)

3-Amino-1,2,4-Triazole

Molecular Formula: $C_2H_4N_4$

Synonyms: Amerol; 2-Amino-1,3,4-triazole; 2-Aminotriazole; 3-Amino-*s*-triazole; 3-Amino-1,2,4-triazole; 3-Amino-1H-1,2,4-triazole; 3-Aminotriazole; Aminotriazole; Aminotriazole Bayer; Amino triazole weedkiller 90; Amitol; Amitril; Amitrol 90; Amitrole; Amitrol-T; Amizol; Amizol DP NAU; Amizol F; 3-AT; AT; AT (Liquid); AT-90; ATA; Atlazin; Atlazine flowable; Atraflow plus; Azaplant; Azaplant kombi; Azolan; Azole; Boroflow A/ATA; Boroflow S/ATA; Campaprim A 1544; Caswell No. 040; CDA Simflow plus; Chipman path weedkiller; Clearway; Cytrol; Cytrol Amitrole-T; Cytrole; Diurol; Diurol 5030; Domatol; Domatol 88; Elmasil; Emisol; Emisol F; ENT 25445; EPA pesticide chemical code 004401; Farmco; Fenamine; Fenavar; Herbazin plus SC; Herbicide total; Herbizole; Kleer-lot; Mascot highway; MSS aminotriazole; MSS simazine; Orga-414; Primatol AD 85 WP; Primatrol SE 500 FW; Radoxone TL; Ramizol; Rassapron; Simazol; Simflow plus; Solution Cncentree T271; Synchemicals total weed killer; Syntox total weed killer; Torapron; 1,2,4-Triazol-3-amine; 1H-1,2,4-Triazol-3-amine; Triazolamine; *s*-Triazole, 3-amino-; δ-2-1,2,2,4-Triazoline, 5-imino-; 1H-1,2,4-Triazol-3-ylamine; Vorox; Vorox AS; Weedar ADS; Weedar AT; Weedazin; Weedazin arginit; Weedazol; Weedazol GP2; Weedazol super; Weedazol T; Weedazol TL; Weedex granulat; Weedoclor; X-All (liquid)

CAS Registry Number: 61-82-5

RTECS Number: XZ3850000

DOT ID: UN 2588

Regulatory Authority

- Carcinogen (Animal Positive) (IARC) (suspected Carcinogen) (NTP)[9]
- Banned or Severely Restricted (UN) (Scandinavia)[13]
- Air Pollutant Standard Set (ACGIH)[1] (DFG)[3] (OSHA)[58] (Several States)[60] (Several Canadian Provinces) (Australia) (Israel)
- EPA HAZARDOUS WASTE NUMBER (RCRA No.): U011
- RCRA, 40CFR261, Appendix 8 Hazardous Constituents
- RCRA Land Ban Waste Restrictions
- SUPERFUND/EPCRA 40CFR302.4 Reportable Quantity (RQ): CERCLA, 10 lb (4.54 kg)
- EPCRA Section 313 Form R *de minimis* concentration reporting level: 0.1%
- Canada, WHMIS, Ingredients Disclosure List

Cited in U.S. State Regulations: Alaska (G), California (A, G), Florida (G), Illinois (G), Kansas (G, A), Louisiana (G), Maine (G), Massachusetts (G), Michigan (G), Minnesota (G), New Hampshire (G), New Jersey (G), North Dakota (A), Pennsylvania (G, A), Rhode Island (G), Vermont (G), Virginia (G, A), Washington (G), West Virginia (G), Wisconsin (G).

Description: Amitrol, $C_2H_4N_4$, is a colorless to off white crystalline solid. Odorless when pure. Freezing/Melting point = 158° to 159°C. Soluble in water.

Potential Exposure: Those involved in the manufacture, formulation and application of this herbicide, which is now limited to noncrop applications as a herbicide and plant growth regulator.

Incompatibilities: Strong oxidizers, strong acids, and light (decomposes). Corrosive to iron, aluminum, and copper.

Permissible Exposure Limits in Air: NIOSH recommends a limit of 0.2 mg/m^3 for a 10-hour workshift. ACGIH has set a TLV of 0.2 mg/m^3 TWA for an 8-hour workshift. Australia and Israel use the same level. DFG set the MAK (total dust) at 0.2 mg/m^3. In addition, several states have set guidelines or standards for amitrole in ambient air:[60] 0.2 mg/m^3 (California), 0.476 μg/m^3 (Kansas), 1.8 μg/m^3 (Pennsylvania), 2.0 μg/m^3 (North Dakota), 3,000 μg/m^3 (Virginia). Canadian province levels: 0.2 mg/m^3 TWA and 0.5 mg/m^3 (Alberta), 0.2 mg/m^3 TWA (Ontario, Quebec).

Permissible Concentration in Water: No criteria set.

Harmful Effects and Symptoms

Carcinogenicity is the primary observed effect. Amitrol is carcinogenic in mice and rats, producing thyroid and liver tumours following oral or subcutaneous administration. Railroad workers who were exposed to amitrole and other herbicides showed a slight (but statistically significant) excess of cancer when all sites were considered together. Because the workers were exposed to several different herbicides, however, no conclusions could be made regarding the carcinogenicity of amitrole alone.

Short Term Exposure: Amitrol can be absorbed through the skin, thereby increasing exposure.

Long Term Exposure: Causes liver, thyroid, and pituitary cancer in animals. May damage the developing fetus. May cause liver, thyroid gland (possible goiter or underactive thyroid), and pituitary gland damage.

Points of Attack: Liver, thyroid, and pituitary gland.

Medical Surveillance: Before beginning employment and at regular times after that, the following is recommended: Physical examination of the thyroid and thyroid function tests (T_4, TSH, and T_3). If symptoms develop or overexposure is suspected, the following may be useful: Liver function tests. Pituitary gland function tests.

First Aid: If this chemical gets into the eyes, remove any contact lenses at once and irrigate immediately for at least 15

minutes, occasionally lifting upper and lower lids. Seek medical attention immediately. If this chemical contacts the skin, remove contaminated clothing and wash immediately with soap and water. Seek medical attention immediately. If this chemical has been inhaled, remove from exposure, begin rescue breathing (using universal precautions) if breathing has stopped and CPR if heart action has stopped. Transfer promptly to a medical facility. When this chemical has been swallowed, get medical attention. Give large quantities of water and induce vomiting. Do not make an unconscious person vomit.

Personal Protective Methods: Wear protective gloves and clothing to prevent any reasonable probability of skin contact. Safety equipment suppliers/manufacturers can provide recommendations on the most protective glove/clothing material for your operation. All protective clothing (suits, gloves, footwear, headgear) should be clean, available each day, and put on before work. Contact lenses should not be worn when working with this chemical. Wear dust-proof chemical goggles and face shield unless full facepiece respiratory protection is worn. Employees should wash immediately with soap when skin is wet or contaminated. Provide emergency showers and eyewash.

Respirator Selection: *At any detectable concentration:* SCBAF:PD,PP (any MSHA/NIOSH approved self-contained breathing apparatus that has a full facepiece and is operated in a pressure-demand or other positive-pressure mode); or SAF:PD,PP:ASCBA (any supplied-air respirator that has a full facepiece and is operated in a pressure-demand or other positive-pressure mode in combination with an auxiliary, self-contained breathing apparatus operated in a pressure-demand or other positive pressure mode). *Escape:* HiEF (Any air-purifying, full-facepiece respirator with a high-efficiency particulate filter); or SCBAE (Any appropriate escape-type, self-contained breathing apparatus).

Storage: Prior to working with this chemical you should be trained on its proper handling and storage. Store in tightly closed containers in a cool, well ventilated area or refrigerator or in glass containers under an inert atmosphere. Protect from air and light.[52]

Shipping: The DOT/UN label requirement for pesticides, solid, toxic, n.o.s. is "Poison." Amitrole is a carcinogen. Packing Group I. The DOT/UN Hazard Class is 6.1.[19][20]

Spill Handling: Evacuate persons not wearing protective equipment from area of spill or leak until clean-up is complete. Remove all ignition sources. Contain and isolate spill. Product residues and sorbent media may be packaged in epoxy-lined drums. Collect powdered material in the most convenient and safe manner and deposit in sealed containers. Ventilate area after clean-up is complete. It may be necessary to contain and dispose of this chemical as a hazardous waste. If material or contaminated runoff enters waterways, notify downstream users of potentially contaminated waters. Contact your Department of Environmental Protection or your regional office of the federal EPA for specific recommendations. If employees are required to clean-up spills, they must be properly trained and equipped. OSHA 1910.120(q) may be applicable.

Fire Extinguishing: Use dry chemical, carbon dioxide, water spray, or alcohol foam extinguishers. Poisonous gases are produced in fire including. If material or contaminated runoff enters waterways, notify downstream users of potentially contaminated waters. Notify local health and fire officials and pollution control agencies. If employees are expected to fight fires, they must be trained and equipped in OSHA 1910.156.

Disposal Method Suggested: Consult with environmental regulatory agencies for guidance on acceptable disposal practices. Generators of waste containing this contaminant (≥100 kg/mo) must conform with EPA regulations governing storage, transportation, treatment, and waste disposal. Amitrol is resistant to hydrolysis and the action of oxidizing agents. Burning the compound with polyethylene is reported to result in >99% decomposition.

References

Sax, N. I., Ed., Dangerous Properties of Industrial Materials Report, 1, No. 4, 34–35 (1981) and 4, No. 2, 41–43 (1984)

New Jersey Department of Health and Senior Services, "Hazardous Substance Fact Sheet: Amitrol," Trenton, NJ (June 1998)

Amiton

Molecular Formula: $C_{10}H_{24}NO_3PS$

Common Formula: $(C_2H_5O)_2POSCH_2CH_2N(C_2H_5)_2$

Synonyms: Chipman 6200; Citram; *S*-(2-Diethylamino) ethyl]phosphorothioic acid, *O,O*-diethyl ester; *O,O*-Diethyl *S*-2-diethylaminoethyl phosphorothioate; Diethyl *S*-2-diethylaminoethyl phosphorothioate; *O,O*-Diethyl *S*-(β-diethylamino)ethyl phosphorothiolate; *O,O*-Diethyl *S*-2-diethylaminoethyl phosphorothiolate; *O,O*-Diethyl *S*-diethylamino ethyl phosphorothiolate; *O,O*-Diethyl *S*-(2-diethylaminoethyl) thiophosphate; DSDP; Inferno; Metramac; Metramak; R-5,158; Rhodia-6200; Tetram

CAS Registry Number: 78-53-5

RTECS Number: TF0525000

DOT ID: UN 3018

Regulatory Authority

- Very Toxic Substance (World Bank)[15]
- CERCLA/SARA 40CFR302 Extremely Hazardous Substances: TPQ = 500 lb (227 kg).
- CERCLA/SARA Section 304 Reportable Quantity (RQ): EHS, 1 lb (0.454 kg)
- Classified by EPA as a Restricted Use Pesticide (RUP)

Cited in U.S. State Regulations: Florida (G), Massachusetts (G), New Jersey (G), Pennsylvania (G)

Description: Amiton, $(C_2H_5O)_2POSCH_2CH_2N(C_2H_5)_2$, $C_{10}H_{24}NO_3PS$, is an organophosphate liquid. Boiling point = 110°C @ 0.2 mm pressure. Hazard Identification (based on

NFPA-704 M Rating System): Health 4, Flammability 2, Reactivity 1. Soluble in water.

Potential Exposure: Amiton is used as an acaricide and an insecticide; exposure may occur in manufacture and in application and use.

Permissible Exposure Limits in Air: No standards set.

Determination in Air: OSHA versatile sampler-2; Toluene/Acetone; Gas chromatography/Flame photometric detection for sulfur, nitrogen, or phosphorus; NIOSH Method IV Method #5600, Organophosphorus Pesticides.

Permissible Concentration in Water: No criteria set.

Routes of Entry: Inhalation, ingestion, skin and/or eye contact.

Harmful Effects and Symptoms

Short Term Exposure: Danger-poisonous; can be fatal if swallowed, inhaled, or absorbed through the skin or eyes. This material is highly toxic orally. It is a cholinesterase inhibitor. The LD_{50} oral (rat) is 3.3 mg/kg. The toxic effects are similar to parathion.

Organic phosphorus insecticides are absorbed by the skin, as well as by the respiratory and gastrointestinal tracts. They are cholinesterase inhibitors. Symptoms of exposure include headache, giddiness, blurred vision, nervousness, weakness, nausea, cramps, diarrhea, and discomfort in the chest. Signs include sweating, tearing, salivation, vomiting, cyanosis, convulsions, coma, loss of reflexes and loss of sphincter control.

Long Term Exposure: Cholinesterase inhibitor; cumulative effect is possible. this chemical may damage the nervous system with repeated exposure, resulting in convulsions, respiratory failure. May cause liver damage.

Points of Attack: Respiratory system, lungs, central nervous system, cardiovascular system, skin, eyes, plasma and red blood cell cholinesterase.

Medical Surveillance: Before employment and at regular times after that, the following are recommended: Plasma and red blood cell cholinesterase levels (tests for the enzyme poisoned by this chemical). If exposure stops, plasma levels return to normal in 1 – 2 weeks while red blood cell levels may be reduced for 1 – 3 months.

When cholinesterase enzyme levels are reduced by 25% or more below preemployment levels, risk of poisoning is increased, even if results are in lower ranges of "normal." Reassignment to work not involving organophosphate or carbamate pesticides is recommended until enzyme levels recover. If symptoms develop or overexposure occurs, repeat the above tests as soon as possible and get an exam of the nervous system. Also consider complete blood count. Consider chest x-ray following acute overexposure.

First Aid: If this chemical gets into the eyes, remove any contact lenses at once and irrigate immediately for at least 15 minutes, occasionally lifting upper and lower lids. Seek medical attention immediately. If this chemical contacts the skin, remove contaminated clothing and wash immediately with soap and water. Speed in removing material from skin is of extreme importance. Shampoo hair promptly if contaminated. Seek medical attention immediately. If this chemical has been inhaled, remove from exposure, begin rescue breathing (using universal precautions) if breathing has stopped and CPR if heart action has stopped. Transfer promptly to a medical facility. When this chemical has been swallowed, get medical attention. Give large quantities of water and induce vomiting. Do not make an unconscious person vomit.

Note to Physician: 1,1'-trimethylenebis(4-formylpyridinium bromide)dioxime (a.k.a TMB-4 DIBROMIDE and TMV-4) have been used as an antidote for organophosphate poisoning.

Personal Protective Methods: Wear protective gloves and clothing to prevent any reasonable probability of skin contact. Safety equipment suppliers/manufacturers can provide recommendations on the most protective glove/clothing material for your operation. All protective clothing (suits, gloves, footwear, headgear) should be clean, available each day, and put on before work. Contact lenses should not be worn when working with this chemical. Wear splash chemical goggles and face shield unless full facepiece respiratory protection is worn. Employees should wash immediately with soap when skin is wet or contaminated. Provide emergency showers and eyewash. See entry on Parathion.

Respirator Selection: NIOSH: *(for Parathion) 0.5 mg/m³*: CCROVDMFu (Any chemical cartridge respirator with organic vapor cartridge(s) in combination with a dust, mist, and fume filter; or SA (any supplied-air respirator). *1.25 mg/m³*: SA:CF (any supplied-air respirator operated in a continuous-flow mode); or PAPROVDMFu (any powered, air purifying respirator with organic vapor cartridge (s) in combination with a dust, mist and, fume filter). *2.5 mg/m³*: CCRFOVHiE (Any chemical cartridge respirator with a full facepiece and organic vapor cartridge(s) in combination with a high efficiency particulate filter); or SAT:CF (any supplied-air respirator that has a tight-fitting facepiece and is operated in a continuous-flow mode); or PAPRTOVHiE (any powered, air-purifying respirator with a tight-fitting facepiece and organic vapor cartridge(s) in combination with a high-efficiency particulate filter); or SCBAF (any self-contained breathing apparatus with full facepiece); or SAF (any supplied-air respirator with a full facepiece). *10 mg/m³*: SA:PD,PP (any supplied-air respirator operated in a pressure-demand or other positive-pressure mode). *Emergency or planned entry into unknown concentrations or IDLH conditions*: SCBAF:PD,PP (any self-contained breathing apparatus that has a full facepiece and is operated in a pressure-demand or other positive-pressure mode); or SAF: PD,PP:ASCBA (Any supplied-air respirator that has a full facepiece and is operated in a pressure-demand or other positive-pressure mode in combination with an auxiliary self-contained breathing apparatus operated in a pressure-demand or other positive pressure mode. *Escape:* GMFOVHiE (Any air-purifying, full-facepiece respirator (gas mask) with a chin-style, front-or back-mounted canister having a high efficiency particulate filter); or SCBAE (Any appropriate escape-type, self-contained breathing apparatus).

Storage: Prior to working with this chemical you should be trained on its proper handling and storage. Where possible, automatically transfer material from other storage containers to process containers

Shipping: Organophosphorus pesticides, liquid, toxic, n.o.s. must be labeled "Poison." This material falls in DOT/UN Hazard Class 6.1.[20][21]

Spill Handling: Evacuate persons not wearing protective equipment from area of spill or leak until clean-up is complete. Remove all ignition sources. Stay upwind; keep out of low areas. Ventilate area of spill or leak. Remove and isolate contaminated clothing at the site. Do not touch spilled material; stop leak if you can do so without risk. Use water spray to reduce vapors. *Small spills:* absorb with sand or other non-combustible absorbent material and place into containers for later disposal. *Large spills:* dike far ahead of spill for later disposal. Collect powdered material in the most convenient and safe manner and deposit in sealed containers. It may be necessary to contain and dispose of this chemical as a hazardous waste. If material or contaminated runoff enters waterways, notify downstream users of potentially contaminated waters. Contact your Department of Environmental Protection or your regional office of the federal EPA for specific recommendations. If employees are required to clean-up spills, they must be properly trained and equipped. OSHA 1910.120(q) may be applicable. Seek professional environmental engineering assistance from the U.S. EPA Environmental Response Team at (908) 548-8730 (24-hour response line).

Fire Extinguishing: Amiton is a combustible liquid. Poisonous gases including *nitrogen oxides, sulfur oxides, and phosphorus oxides* are produced in fire. Small fires: dry chemical, carbon dioxide, water spray, or foam. Large fires: water spray, fog, or foam. Stay upwind; keep out of low areas. Ventilate closed spaces before entering them. Firefighting gear (including SCBA) may not provide adequate protection. If exposure occurs, remove and isolate gear immediately and thoroughly decontaminate personnel. Wear positive pressure breathing apparatus and special protective clothing. Remove and isolate contaminated clothing at the site. Move container from fire area if you can do so without risk. Fight fire from maximum distance. Dike fire control water for later disposal; do not scatter the material. If material or contaminated runoff enters waterways, notify downstream users of potentially contaminated waters. Notify local health and fire officials and pollution control agencies. From a secure, explosion-proof location, use water spray to cool exposed containers. If cooling streams are ineffective (venting sound increases in volume and pitch, tank discolors, or shows any signs of deforming), withdraw immediately to a secure position. If employees are expected to fight fires, they must be trained and equipped in OSHA 1910.156.

Disposal Method Suggested: High-temperature incineration. Hydrolysis may also be used.[22] In accordance with 40CFR165 recommendations for the disposal of pesticides and pesticide containers. Must be disposed properly by following package label directions or by contacting your state pesticide or environmental control agency or by contacting your regional EPA office.

References

U.S. Environmental Protection Agency, "Chemical Profile: Amiton," Washington, DC, Chemical Emergency Preparedness Program (November 30, 19897)

Amiton Oxalate

Molecular Formula: $C_{12}H_{26}NO_7PS$

Common Formula: $(C_2H_5O)_2POSCH_2CH_2N(C_2H_5)_2 \cdot HOOCCOOH$

Synonyms: Acid oxalate; Chipman® 6199; Chipman® R-6, 199; Citram; 2-(2-diethylamino)ethyl] *O,O*-diethyl ester, oxalate (1:1); *S*-(2-Diethylaminoethyl) *O,O*-diethyl phosphorothioate hydrogen oxalate; *O,O*-Diethyl *S*-(β-diethylamino)ethyl phosphorothioate hydrogen oxalate; *O,O*-Diethyl *S*-(2-diethylamino)ethyl phosphorothioate hydrogen oxalate; *O,O*-Diethyl *S*-(2-ethyl-*N,N*-diethylamino)ethyl phosphorothioate hydrogen oxalate; Hydrogen oxalate of amiton; Phosphorothioc acid; Tetram; Tetram monooxalate, *S*-

CAS Registry Number: 3734-97-2

RTECS Number: TF1400000

DOT ID: UN 3018

Regulatory Authority

- CERCLA/SARA 40CFR302 Extremely Hazardous Substances: TPQ = 100/10,000 lb (45.4/4,540 kg)
- CERCLA/SARA Section 304 Reportable Quantity (RQ): EHS, 1 lb (0.454 kg)
- Classified by EPA as a Restricted Use Pesticide (RUP)

Cited in U.S. State Regulations: Florida (G), Massachusetts (G), New Jersey (G), Pennsylvania (G).

Description: $(C_2H_5O)_2POSCH_2CH_2N(C_2H_5)_2 \cdot HOOCCOOH$, is a crystalline solid. Freezing/Melting point = 98 – 99°C. Hazard Identification (based on NFPA-704 M Rating System): Health 4, Flammability 2, Reactivity 1. Soluble in water.

Potential Exposure: Those involved in manufacture and application of this insecticide.

Permissible Exposure Limits in Air: No standards set.

Permissible Concentration in Water: No criteria set.

Routes of Entry: Inhalation, ingestion, skin contact.

Harmful Effects and Symptoms

Short Term Exposure: Amiton oxalate is a cholinesterase inhibitor. Symptoms include headache, giddiness, nervousness, blurred vision, weakness, nausea, cramps, diarrhea, and discomfort in the chest. Signs include sweating, miosis, tearing, salivation and other excessive respiratory tract secretion, vomiting, cyanosis, uncontrollable muscle twitching followed by muscular weakness, convulsions, coma, loss of reflexes, and loss of muscular control. The LD_{50} oral rat is 3 mg/kg.

Points of Attack: See entry on Parathion as referred to under Amiton. Bear in mind that the oxalate is a solid whereas Amiton is a high-boiling liquid.

Medical Surveillance: See entry on Parathion as referred to under Amiton. Bear in mind that the oxalate is a solid whereas Amiton is a high-boiling liquid.

First Aid: If this chemical gets into the eyes, remove any contact lenses at once and irrigate immediately for at least 15 minutes, occasionally lifting upper and lower lids. Seek medical attention immediately. If this chemical contacts the skin, remove contaminated clothing and wash immediately with soap and water. Speed in removing material from skin is of extreme importance. Shampoo hair promptly if contaminated. Seek medical attention immediately. If this chemical has been inhaled, remove from exposure, begin rescue breathing (using universal precautions) if breathing has stopped and CPR if heart action has stopped. Transfer promptly to a medical facility. When this chemical has been swallowed, get medical attention. Give large quantities of water and induce vomiting. Do not make an unconscious person vomit.

Personal Protective Methods: Wear protective gloves and clothing to prevent any reasonable probability of skin contact. Safety equipment suppliers/manufacturers can provide recommendations on the most protective glove/clothing material for your operation. All protective clothing (suits, gloves, footwear, headgear) should be clean, available each day, and put on before work. Contact lenses should not be worn when working with this chemical. Wear dust-proof chemical goggles and face shield unless full facepiece respiratory protection is worn. Employees should wash immediately with soap when skin is wet or contaminated. Provide emergency showers and eyewash. See entry on Parathion. Bear in mind that the oxalate is a solid whereas Amiton is a high-boiling liquid.

Respirator Selection: NIOSH: *(for Parathion) 0.5 mg/m³*: CCROVDMFu (any chemical cartridge respirator with organic vapor cartridge(s) in combination with a dust, mist, and fume filter; or SA (any supplied-air respirator). *1.25 mg/m³*: SA:CF (any supplied-air respirator operated in a continuous-flow mode); or PAPROVDMFu (any powered, air purifying respirator with organic vapor cartridge (s) in combination with a dust, mist and, fume filter). *2.5 mg/m³*: CCRFOVHiE (any chemical cartridge respirator with a full facepiece and organic vapor cartridge(s) in combination with a high efficiency particulate filter); or SAT:CF (any supplied-air respirator that has a tight-fitting facepiece and is operated in a continuous-flow mode); or PAPRTOVHiE [any powered, air-purifying respirator with a tight-fitting facepiece and organic vapor cartridge(s) in combination with a high-efficiency particulate filter]; or SCBAF (any self-contained breathing apparatus with full facepiece); or SAF (any supplied-air respirator with a full facepiece). *10 mg/m³* SA:PD,PP (any supplied-air respirator operated in a pressure-demand or other positive-pressure mode). *Emergency or planned entry into unknown concentrations or IDLH conditions*: SCBAF:PD,PP (any self-contained breathing apparatus that has a full facepiece and is operated in a pressure-demand or other positive-pressure mode); or SAF:PD,PP:ASCBA (Any supplied-air respirator that has a full facepiece and is operated in a pressure-demand or other positive-pressure mode in combination with an auxiliary self-contained breathing apparatus operated in a pressure-demand or other positive pressure mode. *Escape:* GMFOVHiE (Any air-purifying, full-facepiece respirator (gas mask) with a chin-style, front-or back-mounted canister having a high efficiency particulate filter); or SCBAE (Any appropriate escape-type, self-contained breathing apparatus). Bear in mind that the oxalate is a solid whereas Amiton is a high-boiling liquid.

Storage: Prior to working with this chemical you should be trained on its proper handling and storage. Store in tightly closed containers in a cool, well ventilated area.

Shipping: Organophosphorus pesticides, solid, toxic, n.o.s. should be labeled "Poison" if in Packing Group I. This class of materials is in DOT/UN Hazard Class 6.1.[19][20]

Spill Handling: Evacuate persons not wearing protective equipment from area of spill or leak until clean-up is complete. Remove all ignition sources. Stay upwind; keep out of low areas. Ventilate closed spaces before entering them. Do not touch spilled material; stop leak if you can do so without risk. Use water spray to reduce vapors. *Small spills:* absorb with sand or other noncombustible absorbent material and place into containers for later disposal. *Large spills:* dike far ahead of spill for later disposal. Collect powdered material in the most convenient and safe manner and deposit in sealed containers. Ventilate area after clean-up is complete. It may be necessary to contain and dispose of this chemical as a hazardous waste. If material or contaminated runoff enters waterways, notify downstream users of potentially contaminated waters. Contact your Department of Environmental Protection or your regional office of the federal EPA for specific recommendations. If employees are required to clean-up spills, they must be properly trained and equipped. OSHA 1910.120(q) may be applicable. Seek professional environmental engineering assistance from the U.S. EPA Environmental Response Team at (908) 548-8730 (24-hour response line).

Fire Extinguishing: Poisonous gases including *nitrogen oxides, sulfur oxides, and phosphorus oxides* are produced in fire. Small fires: dry chemical, carbon dioxide, water spray, or foam. Large fires: water spray, fog, or foam. Stay upwind; keep out of low areas. Firefighting gear (including SCBA) may not provide adequate protection. If exposure occurs, remove and isolate gear immediately and thoroughly decontaminate personnel. Wear positive pressure breathing apparatus and special protective clothing. Remove and isolate contaminated clothing at the site. Move container from fire area if you can do so without risk. Fight fire from maximum distance. Dike fire control water for later disposal; do not scatter the material. If material or contaminated runoff enters waterways, notify downstream users of potentially contaminated waters. Notify local health and fire officials and pollution control agencies. From a secure, explosion-proof location, use water spray to cool exposed containers. If cooling streams are ineffective (venting sound increases in volume and pitch, tank discolors, or shows any signs of deforming), withdraw immediately to a secure position. If employees are expected to fight fires, they must be trained and equipped in OSHA 1910.156.

Disposal Method Suggested: In accordance with 40CFR165 recommendations for the disposal of pesticides and pesticide containers. Must be disposed properly by following package label directions or by contacting your state pesticide or environmental control agency or by contacting your regional EPA office.

References

U.S. Environmental Protection Agency, "Chemical Profile: Amiton Oxalate," Washington, DC, Chemical Emergency Preparedness Program (November 30, 1987)

Amitraz

Molecular Formula: $C_{19}H_{23}N_3$,N,N'

Synonyms: A13-27967; Acarac; Amitraz estrella; BAAM; *N,N*-Bis(2,4-xylyliminomethyl)methylamine; Boots® BTS 27419; BTS 27,419; 1,5-Di-(2,4-dimethylphenyl)-3-methyl-1,3,5-triazapenta-1,4-diene; *N'*-(2,4-Dimethylphenyl)-*N*-([(2,4-dimethylphenyl)imino]methyl)-*N*-methylmethanimidamide; *N'*-(2,4-Dimethylphenyl)-3-methyl-1,3,5-triazapenta-1,4-diene; *N,N*-Di-(2,4-xylyliminomethyl)methylamine; ENT 27967; EPA pesticide chemical code 106201; Formamidine, *N*-Methyl-*N'*-2,4-xylyl-*N*-(N-2,4-xylylformimidoyl)-; *N*-Methylbis(2,4-xylyliminomethyl)amine; 2Methyl-1,3-di(2,4-xylylimino)-2-azapropane; *N,N'*-[(Methylimino)dimethylidyne]bis(2,4-xylidine); *N,N'*-[(Methylimino)dimethylidyne] D-2,4-xylidine; Mitac®; NSC 324552; R.D. 27419; Taktic®; Triatox®; Upjohn® U-36059; 2,4-Xylidine, *N,N'*-(methyliminodimethylidyne)bis-

CAS Registry Number: 33089-61-1

RTECS Number: ZF0480000

DOT ID: UN 2763 (triazine pesticides, solid toxic)

Regulatory Authority

- Carcinogen (Animal Positive) (EPA)[13]
- Banned or Severely Restricted (Argentina, U.S.)[13]
- CERCLA/SARA 40CFR372.65: Form R *de minimis* Concentration Reporting Level: 1.0%.

Cited in U.S. State Regulations: Massachusetts (G).

Description: Amitraz, $C_{19}H_{23}N_3$,N,N'-(Methyliminodimethylidyne)bis-2,4-xylidine, forms colorless needle-like crystals. Freezing/Melting point = 86 – 87°C.

Potential Exposure: Those engaged in the manufacture, formulation and application of this insecticide and acaricide. A rebuttable presumption against registration for amitraz was issued on April 6, 1977 by US EPA on the basis of oncogenicity.

Permissible Exposure Limits in Air: No standards set.

Permissible Concentration in Water: No criteria set.

Harmful Effects and Symptoms

Amitraz metabolizes to 2,4-dimethylaniline which is a potential human carcinogen. A mouse oncogenic bioassay was conducted by Boots Chemical Company and reported by EPA; the results of that study have been disputed. Acute oral LD_{50} for rats is 800 mg/kg; for mice is greater than 1,600 mg/kg.

Short Term Exposure: Poisonous if ingested or absorbed through the skin. Eye or skin contact can cause irritation.

First Aid: If this chemical gets into the eyes, remove any contact lenses at once and irrigate immediately for at least 15 minutes, occasionally lifting upper and lower lids. Seek medical attention immediately. If this chemical contacts the skin, remove contaminated clothing and wash immediately with soap and water. Seek medical attention immediately. If this chemical has been inhaled, remove from exposure, begin rescue breathing (using universal precautions) if breathing has stopped and CPR if heart action has stopped. Transfer promptly to a medical facility. When this chemical has been swallowed, get medical attention. Give large quantities of water and induce vomiting. Do not make an unconscious person vomit.

Personal Protective Methods: Wear protective gloves and clothing to prevent any reasonable probability of skin contact. Safety equipment suppliers/manufacturers can provide recommendations on the most protective glove/clothing material for your operation. All protective clothing (suits, gloves, footwear, headgear) should be clean, available each day, and put on before work. Contact lenses should not be worn when working with this chemical. Wear dust-proof chemical goggles and face shield unless full facepiece respiratory protection is worn. Employees should wash immediately with soap when skin is wet or contaminated. Provide emergency showers and eyewash.

Respirator Selection: Use only MSHA/NIOSH approved air purifying respirators for pesticides.

Storage: Prior to working with this chemical you should be trained on its proper handling and storage. Store in tightly closed containers in a cool, well-ventilated area, away from fertilizers, seed, and other pesticides, flammable materials and sources of heat and flame. Do not reuse containers.

Shipping: Triazine pesticides, solid, toxic, n.o.s. require a shipping label of "Poison" if in Packing Groups I or II. These materials fall in DOT/UN Hazard Class 6.1.[19][20]

Spill Handling: Evacuate persons not wearing protective equipment from area of spill or leak until clean-up is complete. Remove all ignition sources. Stay upwind; keep out of low areas. Ventilate closed spaces before entering them. Remove and isolate contaminated clothing at the site. Do not touch spilled material; stop leak if you can do so without risk. Use water spray to reduce vapors. *Small spills:* absorb with sand or other noncombustible absorbent material and place into containers for later disposal. *Large spills:* dike far ahead of spill for later disposal. Collect powdered material in the most convenient and safe manner and deposit in sealed containers. It may be necessary to contain and dispose of this chemical as a hazardous waste. If material or contaminated runoff enters waterways, notify downstream users of potentially contaminated waters. Contact your Department of Environmental Protection or your regional office of the federal EPA for specific recommendations. If employees are required to clean-up spills, they must be properly trained and equipped. OSHA 1910.120(q) may be applicable. Seek professional environmental engineering assistance from the U.S. EPA Environmental Response Team at (908) 548-8730 (24-hour response line).

Fire Extinguishing: Poisonous gases including *nitrogen oxides* are produced in fire. Small fires: dry chemical, carbon dioxide, water spray, or foam. Large fires: water spray, fog, or foam. Stay upwind; keep out of low areas. Firefighting gear (including SCBA) may not provide adequate protection. If exposure occurs, remove and isolate gear immediately and thoroughly decontaminate personnel. Wear positive pressure breathing apparatus and special protective clothing. Remove and isolate contaminated clothing at the site. Move container from fire area if you can do so without risk. Fight fire from maximum distance. Dike fire control water for later disposal; do not scatter the material. If material or contaminated runoff enters waterways, notify downstream users of potentially contaminated waters. Notify local health and fire officials and pollution control agencies. From a secure, explosion-proof location, use water spray to cool exposed containers. If cooling streams are ineffective (venting sound increases in volume and pitch, tank discolors, or shows any signs of deforming), withdraw immediately to a secure position. If employees are expected to fight fires, they must be trained and equipped in OSHA 1910.156.

Disposal Method Suggested: In accordance with 40CFR 165 recommendations for the disposal of pesticides and pesticide containers. Must be disposed properly by following package label directions or by contacting your state pesticide or environmental control agency or by contacting your regional EPA office.

References

U.S. Environmental Protection Agency, Rebuttable Presumption Against Registration (RPAR) of Pesticide Products Containing Amitraz, Washington, DC (April 6, 1977)

Ammonia

Molecular Formula: NH_3

Synonyms: Am-Fol; Ammonia, anhydrous; Ammoniac (French); Ammoniaca (Italian); Ammonia gas; Ammoniale (German); Ammonium amide; Ammonium hydroxide; Amoniaco (Spanish); Amoniaco anhidro (Spanish); Amoniak (Polish); Anhydrous ammonia; Aqua ammonia; Daxad-32S; Liquid ammonia; Pro 330 clear thin spread; R717; Spirit of hartshorn; STCC 4904210

CAS Registry Number: 7664-41-7

RTECS Number: BO0875000

DOT ID: UN 1005 (anhydrous, liquefied, and >50% solutions); UN 2073 (35 – 50% solutions); UN 2672 (10 – 35% solutions).

EEC Number: 007-001-00-5 (anhydrous)

EINECS Number: 231-635-3

Regulatory Authority

- Toxic Substance (World Bank)[15]
- Air Pollutant Standard Set (NIOSH)[2] (ACGIH)[1] (DFG)[3] (HSE)[33] (former USSR)[43] (OSHA)[58] (Several States)[60] (Several Canadian Provinces) (Australia) (Israel) (Mexico)
- Water Pollution Standard Set (former USSR)[43]
- OSHA 29CFR1910.119, Appendix A, Process Safety List of Highly Hazardous Chemicals, TQ = 10,000 lb (4,540 kg) (anhydrous); TQ = 15,000 lb (6,815 kg) (solution >44% NH_3)
- Clean Air Act 42USC7412; Title I, Part A, §112(r), Accidental Release Prevention/Flammable substances (Section 68.130); (anhydrous) TQ = 10,000 lb (4,540 kg) (anhydrous); (concentrations ≥20% NH_3) TQ =20,000 lb (9,150 kg)
- Clean Water Act: 40CFR116.4 Hazardous Substances; RQ 40CFR117.3 (same as CERCLA); Section 313 Water Priority Chemicals (57FR41331, 9/9/92).
- CERCLA/SARA 40CFR302, Extremely Hazardous Substances: TPQ = 500 lb (228 kg).
- SUPERFUND/EPCRA 40CFR302.4, Appendix A, Reportable Quantity (RQ): 100 lb (45.4 kg), 40CFR372.65: Form R *de minimis* Concentration Reporting Level: 1.0%; includes anhydrous ammonia and aqueous ammonia from water dissociable ammonium salts and other sources; 10% of total aqueous ammonia, and 100% of anhydrous forms of ammonia is reportable under this listing. If a facility manufactures, processes, or otherwise uses anhydrous ammonia or aqueous ammonia, they must report under the ammonia listing. Solutions containing aqueous ammonia at a concentration in excess of 1% of the 10% reportable under this listing should be factored into threshold and release determinations.
- U.S. DOT 49CFR172.10; Poisonous by inhalation substances (anhydrous UN1005)
- Canada, WHMIS, Ingredients Disclosure List; National Pollutant Release Inventory

Cited in U.S. State Regulations: Alaska (G), California (A, G), Connecticut (A), Florida (G, A), Illinois (G), Kansas (A),

Maine (G), Maryland (G), Massachusetts (G, A), Minnesota (G), Nevada (A), New Hampshire (G), New Jersey (G), New York (G, A), North Carolina (A), North Dakota (A), Pennsylvania (G), Rhode Island (G), South Dakota (A), Virginia (A), West Virginia (G), Wyoming (A).

Description: Ammonia, NH_3, is a colorless, strongly alkaline, and extremely soluble gas with a pungent, suffocating odor. Odor threshold = 5.75 ppm. Boiling point = –33°C. Flash point = (flammable gas). Autoignition temperature = 630°C. Hazard Identification (based on NFPA-704 M Rating System): Health 3, Flammability 1, Reactivity 0. Explosive limits: LEL = 15%; UEL = 28%. Anhydrous ammonia is a colorless, highly irritating gas at room temperature with a pungent, suffocating odor. Ammonia gas is lighter than air; hugs the ground when cool; and flammable at high concentrations and temperatures. Easily compressed, it forms a clear, colorless liquid under pressure. Floats and "boils" on water. Readily dissolves in water, forming ammonium hydroxide, an alkaline, corrosive solution. Poisonous, visible vapor cloud is produced.

Potential Exposure: Ammonia is used as a nitrogen source for many nitrogen-containing compounds. It is used in the production of ammonium sulfate and ammonium nitrate for fertilizers and in the manufacture of nitric acid, soda, synthetic urea, synthetic fibers, dyes, and plastics. It is also utilized as a refrigerant and in the petroleum refining and chemical industries. It is used in the production of many drugs and pesticides. Other sources of occupational exposure include the silvering of mirrors, gluemaking, tanning of leather, and around nitriding furnaces. Ammonia is produced as a by-product in coal distillation and by the action of steam on calcium cyanamide, and from the decomposition of nitrogenous materials.

Incompatibilities: Violent reaction with strong oxidizers and acids. Shock-sensitive compounds may be formed with gold, halogens, mercury, mercury oxide, and silver oxide. Fire and explosions may be caused by trimethylammonium amide, 1-chloro-2,4-dinitrobenzene, o-chloronitrobenzene, platinum, trioxygen difluoride, selenium difluoride dioxide, boron halides, mercury, chlorine, iodine, bromine, hypochlorites, chlorine bleach, amides, organic anhydrides, isocyanates, vinyl acetate, alkylene oxides, epichlorohydrin, and aldehydes. Attacks some coatings, plastics, and rubber, copper, brass, bronze, aluminum, steel, tin, zinc, and their alloys.

Permissible Exposure Limits in Air: The Federal OSHA standard is 50 ppm (35 mg/m^3) TWA, averaged over an 8-hour workshift. NIOSH recommended limit is 25 ppm (17 mg/m^3) averaged over a 10-hour workshift and 35 ppm (27 mg/m^3) not to be exceeded during any 15 minute work period. Australia, HSE, Israel, and Mexico limits are similar to NIOSH. Israel has an action level of 12.5 ppm. ACGIH recommends values of 25 ppm (18 mg/m^3) TWA and STEL 35 ppm (27 mg/m^3). The NIOSH IDLH value is 300 ppm. The DFG MAC is 20 ppm (14 mg/m^3) TWA. In addition, several states have set airborne guidelines or standards for ammonia in ambient air:[60] 25 ppm (18 mg/m^3) TWA and STEL of 35 ppm (27 mg/m^3) (California), 0.024 mg/m^3 (Massachusetts), 0.042857 mg/m^3 (Kansas), 0.18 – 0.27 mg/m^3 (North Dakota), 0.25 mg/m^3 (Virginia), 0.36 mg/m^3 (Connecticut, Florida, New York, South Dakota), 0.429 mg/m^3 (Nevada, Wyoming), 2.7 mg/m^3 (North Carolina). Canadian Provinces of Alberta, British Columbia, Ontario, and Quebec have limits of 25 ppm TWA/TWAEV and STEL/STEV of 35 ppm. The former USSR-UNEP/IRPTC project[43] has set a MAC of 20 mg/m^3 in workplace air and a MAC of 0.2 mg/m^3 in ambient air in residential areas.

Determination in Air: Sampling by absorption in sulfuric acid followed by measurement by ion chromatography, conductivity. See NIOSH Method #6015, and #6016.

Permissible Concentration in Water: The former USSR-UNEP/IRPTC project[43] has set a MAC of 2.0 mg/ml in water bodies used for domestic purposes and 0.05 mg/ml in water bodies used for fishery purposes.

Routes of Entry: Inhalation, ingestion, skin and eye contact.

Harmful Effects and Symptoms

Short Term Exposure: Eye or skin contact with ammonia can cause irritation, burns, frostbite (anhydrous), and permanent damage. Irritates the respiratory tract causing coughing, wheezing and shortness of breath. Higher exposure can cause pulmonary edema, a medical emergency, that can be delayed for several hours and is life threatening. Exposure can cause headache, loss of sense of smell, nausea, and vomiting.

Inhalation: Nose and throat irritation have been reported at 72 ppm after 5 minutes exposure. Exposures of 500 ppm for 30 minutes have caused upper respiratory irritation, tearing, increased pulse rate and blood pressure. Death has been reported after an exposure to 10,000 ppm for an unknown duration.

Skin: Solutions of 2% ammonia can cause burns and blisters after 15 minutes of exposure. These burns may be slow to heal. Anhydrous ammonia may cause skin to freeze.

Eyes: Levels of 70 ppm (gas) have caused eye irritation. If not flushed with water immediately contact with eye may cause partial or complete blindness.

Ingestion: Ammonia will cause pain if swallowed and burning of the throat and stomach. May cause vomiting. One teaspoon of 28% aqua ammonia may cause death.

Long Term Exposure: Repeated exposure can cause chronic eye, nose, and throat irritation. Repeated lung irritation can result in bronchitis with coughing, shortness of breath, and phlegm. Levels of 170 ppm of ammonia vapor has caused mild changes in the spleens, kidneys and livers of guinea pigs.

Points of Attack: Skin, respiratory system, eyes.

Medical Surveillance: Pre-employment physical examinations for workers in ammonia exposure areas should be directed toward significant changes in the skin, eyes, and respiratory system. Persons with corneal disease, and glaucoma, or chronic respiratory diseases may suffer increased risk. Periodic examinations should include evaluation of skin, eyes, and respiratory system, and pulmonary function test to compare with baselines established at pre-employment examination. Consider chest x-ray following acute exposure.

First Aid: If this chemical gets into the eyes, remove any contact lenses at once and irrigate immediately for at least 30 minutes, occasionally lifting upper and lower lids. Seek medical attention immediately. If this chemical contacts the skin, remove contaminated clothing and wash immediately with soap and water. Seek medical attention immediately. If this chemical has been inhaled, remove from exposure, begin rescue breathing (using universal precautions) if breathing has stopped and CPR if heart action has stopped. Transfer promptly to a medical facility. When this chemical has been swallowed, get medical attention. If victim is conscious, administer water or milk. Do not induce vomiting. Medical observation is recommended for 24 – 48 hours after breathing overexposure, as pulmonary edema may be delayed. As first aid for pulmonary edema, a doctor or authorized paramedic may consider administering a corticosteroid spray.

Personal Protective Methods: Wear protective gloves and clothing to prevent any reasonable probability of skin contact. Safety equipment suppliers/manufacturers recommend butyl/neoprene or viton/neoprene as protective material. Appropriate clothing should be worn to prevent any possible skin contact with liquids of >10% content or reasonable probability of contact with liquids of <10% content. Wear eye protection to prevent any potential for eye contact with liquids of >10% NH_3 content. Employees should wash immediately when skin is wet or contaminated with liquids of >10% content. Remove nonimpervious clothing immediately if wet or contaminated with liquids containing >10% and promptly remove if liquid contains <10% NH_3 are involved. All protective clothing (suits, gloves, footwear, headgear) should be clean, available each day, and put on before work. Contact lenses should not be worn when working with this chemical. Wear gas- or splash-proof chemical goggles and face shield unless full facepiece respiratory protection is worn. Employees should wash immediately with soap when skin is wet or contaminated. Provide emergency showers and eyewash.

Respirator Selection: NIOSH: *250 ppm*: CCRS (any chemical cartridge respirator with cartridge(s) providing protection against the compound of concern); or SA (any supplied-air respirator). *300 ppm*: SA:CF (any supplied-air respirator operated in a continuous-flow mode); or PAPRS (any powered, air-purifying respirator with cartridge(s) providing protection against the compound of concern); or CCRFS (any chemical cartridge respirator with a full facepiece and cartridge(s) providing protection against the compound of concern); or GMFS [any air-purifying, full-facepiece respirator (gas mask) with a chin-style, front-or back-mounted canister providing protection against the compound of concern]; or SCBAF (any self-contained breathing apparatus with a full facepiece); or SAF (any supplied-air respirator with a full facepiece). *Emergency or planned entry into unknown concentrations or IDLH conditions*: SCBAF:PD,PP (any self-contained breathing apparatus that has a full facepiece and is operated in a pressure-demand or other positive-pressure mode); or SAF:PD,PP: ASCBA (any supplied-air respirator that has a full facepiece and is operated in a pressure-demand or other positive-pressure mode in combination with an auxiliary, self-contained breathing apparatus operated in a pressure-demand or other positive-pressure mode). *Escape:* GMFS [any air-purifying, full-facepiece respirator (gas mask) with a chin-style, front- or back-mounted canister providing protection against the compound of concern]; or SCBAE (any appropriate escape-type, self-contained breathing apparatus).

Storage: Prior to working with this chemical you should be trained on its proper handling and storage. Before entering confined space where this chemical may be present, check to make sure that an explosive concentration does not exist. Protect against physical damage. Outside or detached storage is preferred. Inside storage should be in a cool, well-ventilated, noncombustible location, preferably with automatic monitoring systems, away from all possible sources of ignition. Separate from other chemicals, particularly oxidizing gases, chlorine, bromine, iodine, and acids.

Shipping: Shipped in tank cars, tank trucks, barges and steel cylinders. Labeling and restrictions vary with concentration:[19][20]

Form	*Label Required*	*Hazard Class*	*Packing Group*
Anhydrous, liquefied, or solutions, rel. density <0.880 @15°C in H_2O, with >50% NH_3	Poison gas (Int'l) Nonflammable gas (Domestic)	2.3 (Int'l) 2.2 (Domestic)	—
Solution, rel. density between 0.880 – 0.957 @15°C in H_2O, with 10 – 35% NH_3	Corrosive	8	III
Solution, rel. density <0.880 @15°C in H_2O, with 35 – 50% NH_3	Nonflammable gas	2.2	—
Solution, <10% NH_3	None	ORM-E	III

Spill Handling: Evacuate and restrict persons not wearing protective equipment from area of spill or leak until clean-up is complete. Remove all ignition sources. Establish forced ventilation to keep levels below explosive limit. Ventilate area of spill or leak. Stop the flow of gas it can be done safely. Stay upwind; keep out of low areas. Ventilate closed spaces before entering them. Wear positive pressure breathing apparatus and full protective clothing. If the leak is a cylinder and the leak cannot be stopped in place, remove the leaking cylinder to a safe place in the open air, and repair the leak or allow the cylinder to empty. For small liquid spills, neutralize with hydrochloric acid. Wipe or mop or use water aspirator. Drain into sewer with sufficient water. It may be necessary to contain and dispose of this chemical as a hazardous waste. If material or contaminated runoff enters waterways, notify downstream users of potentially contaminated waters. Contact your Department of Environmental Protection or your regional office of the federal EPA for specific recommendations. If employees are required to clean-up spills, they must be properly trained and equipped. OSHA 1910.120(q) may be applicable.

Initial isolation and protective action distances

Distances shown are likely to be affected during the first 30 minutes after materials are spilled and could increase with time. If more than one tank car, cargo tank, portable tank, or large cylinder is involved in the incident is leaking, the protective action distance may need to be increased. You may need to seek emergency information from CHEMTREC at (800) 424-9300 or seek professional environmental engineering assistance from the U.S. EPA Environmental Response Team at (908) 548-8730 (24-hour response line).

Small spills (From a small package or a small leak from a large package)

First: Isolate in all directions (feet)..... 100

Then: Protect persons downwind (miles)

Day 0.1

Night 0.2

Large spills (From a large package or from many small packages)

First: Isolate in all directions (feet)..... 300

Then: Protect persons downwind (miles)

Day 0.2

Night 0.5

Fire Extinguishing: Firefighting gear (including SCBA) does not provide adequate protection. Poisonous gases, including nitrogen oxides, produced in fire. If exposure occurs, remove and isolate gear immediately and thoroughly decontaminate personnel. Move container from fire area if you can do it without risk. Vapors are heavier than air and will collect in low areas. Vapors in confined areas may explode when exposed to fire. Vapors may travel long distances to ignition sources and flash back. Storage containers and parts of containers may rocket great distances, in many directions. If material or contaminated runoff enters waterways, notify downstream users of potentially contaminated waters. Notify local health and fire officials and pollution control agencies. *Do not put water on liquid ammonia; will increase evaporation.* Small fires: dry chemical or carbon dioxide. Large fires: water spray, fog or foam. Apply water gently to the surface. Do not get water inside container. From a secure explosion-proof location, use water spray to cool exposed containers. If cooling streams are ineffective (venting sound increases in volume and pitch, tank discolors or shows any signs of deforming), withdraw immediately to a secure position. Isolate area until gas has dispersed.

Disposal Method Suggested: Dilute with water, neutralize with HCl and discharge to sewer. Recovery is an option to disposal which should be considered for paper manufacture, textile treating, fertilizer manufacture and chemical process wastes.

References

National Institute for Occupational Safety and Health, Criteria for a Recommended Standard: Occupational Exposure to Ammonia, NIOSH Doc. No. 74–136, Washington, DC (1974)

U.S. Environmental Protection Agency, "Toxic Pollutant List: Proposal to Add Ammonia," Federal Register, 45, No. 2, 803–806 (January 3, 1980) Rescinded by Federal Register, 45, No. 232, 79692–79693 (December 1, 1980)

National Research Council, Committee on Medical and Biologic Effects of Environmental Pollutants, Ammonia, Baltimore, MD, University Park Press (1979)

Sax, N. I., Ed., Dangerous Properties of Industrial Materials Report 2, No. 1, 65–68 (1982) and 3, No. 3, 49–53, (1983)

U.S. Environmental Properties Agency, "Chemical Profile: Ammonia," Washington, DC, Chemical Emergency Preparedness Program (November 30, 1987)

New Jersey Department of Health and Senior Services, "Hazardous Substance Fact Sheet: Ammonia," Trenton, NJ (June 1998)

New York State Department of Health, "Chemical Fact Sheet: Ammonia," Albany, NY, Bureau of Toxic Substance Assessment (January 1986)

Ammonium Acetate

Molecular Formula: $C_2H_7NO_2$

Common Formula: CH_3COONH_4

CAS Registry Number: 631-61-8

RTECS Number: AF3675000

DOT ID: NA 9079

EINECS Number: 211-162-9

Regulatory Authority

- Clean Water Act: Section 311 Hazardous Substances/RQ 40CFR117.3 (same as CERCLA, see below)
- SUPERFUND/EPCRA 40CFR302.4, Appendix A, Reportable Quantity (RQ): 5,000 lb (2,270 kg), 40CFR372.65: Form R *de minimis* Concentration Reporting Level: 1.0%.: Source of aqueous ammonia. Molecular weight 77.08. NH_3 equivalent weight 22.09

Cited in U.S. State Regulations: California (G), Massachusetts (G), New Hampshire (G), New Jersey (G), Pennsylvania (G).

Description: Ammonium acetate, CH_3COONH_4, is a white crystalline solid with an acetic odor. Freezing/Melting point = 114°C. Soluble in water.

Potential Exposure: It is used as a chemical reagent, to make drugs, foam rubber, vinyl plastics, explosives and to preserve meats.

Incompatibilities: Sodium hypochlorite, potassium chlorate, sodium nitrite, strong oxidizers, strong acids.

Permissible Exposure Limits in Air: No standards set.

Permissible Concentration in Water: No criteria set.

Harmful Effects and Symptoms

Short Term Exposure: Can irritate the eyes, skin and respiratory tract. Eye contact can cause burns and permanent damage. Inhalation can irritate the nose, throat, and lungs; high levels can cause pulmonary edema, a medical emergency that can be delayed for several hours. This can cause death.

Long Term Exposure: Repeated exposures can cause lung irritation and the development of bronchitis with coughing, shortness of breath, and phlegm.

Points of Attack: Respiratory system, eyes, skin.

Medical Surveillance: Lung function tests. Consider chest x-ray following acute exposure.

First Aid: If this chemical gets into the eyes, remove any contact lenses at once and irrigate immediately for at least 15 minutes, occasionally lifting upper and lower lids. Seek medical attention immediately. If this chemical contacts the skin, remove contaminated clothing and wash immediately with soap and water. Seek medical attention immediately. If this chemical has been inhaled, remove from exposure, begin rescue breathing (using universal precautions) if breathing has stopped and CPR if heart action has stopped. Transfer promptly to a medical facility. When this chemical has been swallowed, get medical attention. Give large quantities of water and induce vomiting. Do not make an unconscious person vomit. Medical observation is recommended for 24 – 48 hours after breathing overexposure, as pulmonary edema may be delayed. As first aid for pulmonary edema, a doctor or authorized paramedic may consider administering a corticosteroid spray.

Personal Protective Methods: Wear protective gloves and clothing to prevent any reasonable probability of skin contact. Safety equipment suppliers/manufacturers can provide recommendations on the most protective glove/clothing material for your operation. All protective clothing (suits, gloves, footwear, headgear) should be clean, available each day, and put on before work. Contact lenses should not be worn when working with this chemical. Wear dust-proof chemical goggles and face shield unless full facepiece respiratory protection is worn. Employees should wash immediately with soap when skin is wet or contaminated. Provide emergency showers and eyewash. Wash thoroughly immediately after exposure to Ammonium Acetate.

Respirator Selection: Where there is the potential for exposure to ammonium acetate: SCBAF:PD,PP (any self-contained breathing apparatus that has a full facepiece and is operated in a pressure-demand or other positive-pressure mode); or SAF:PD,PP:ASCBA (any supplied-air respirator that has a full facepiece and is operated in a pressure-demand or other positive-pressure mode in combination with an auxiliary self-contained breathing apparatus operated in a pressure-demand or other positive-pressure mode). *Escape:* GMFOV [any air-purifying, full-facepiece respirator (gas mask) with a chin-style, front-or back-mounted organic vapor canister]; or SCBAE (any appropriate escape-type, self-contained breathing apparatus).

Storage: Prior to working with this chemical you should be trained on its proper handling and storage. Store in tightly closed containers in a cool, well-ventilated area away from sodium hypochlorite, potassium chlorate and sodium nitrite since violent reactions occur.

Shipping: No special labels are required by DOT/UN.[19][20] Falls in DOT/UN Hazard Class ORM-E and Packing Group III.

Spill Handling: Evacuate persons not wearing protective equipment from area of spill or leak until clean-up is complete. Remove all ignition sources. Collect powdered material in the most convenient and safe manner and deposit in sealed containers. Ventilate area after clean-up is complete. It may be necessary to contain and dispose of this chemical as a hazardous waste. If material or contaminated runoff enters waterways, notify downstream users of potentially contaminated waters. Contact your Department of Environmental Protection or your regional office of the federal EPA for specific recommendations. If employees are required to clean-up spills, they must be properly trained and equipped. OSHA 1910.120(q) may be applicable.

Fire Extinguishing: Ammonium Acetate may burn, but does not readily ignite. Use dry chemical, carbon dioxide, water spray, or foam extinguishers. Poisonous gases are produced in fire including ammonia, oxides of nitrogen and carbon. If material or contaminated runoff enters waterways, notify downstream users of potentially contaminated waters. Notify local health and fire officials and pollution control agencies. From a secure, explosion-proof location, use water spray to cool exposed containers. If cooling streams are ineffective (venting sound increases in volume and pitch, tank discolors, or shows any signs of deforming), withdraw immediately to a secure position. If employees are expected to fight fires, they must be trained and equipped in OSHA 1910.156.

References

Sax, N. I., Ed., Dangerous Properties of Industrial Materials Report 2, No. 3, 30–31 (1982)

New Jersey Department of Health and Senior Services, Hazardous Substance Fact Sheet: Ammonium Acetate, Trenton, NJ (April, 1996)

Ammonium Arsenate

Molecular Formula: $AsH_3O_4{\cdot}2H_3N$

Synonyms: Ammonium acid arsenate; Arsenic acid, diammonium salt; Diammonium arsenate; Diammonium monohydrogen arsenate; Dibasic ammonium arsenate

CAS Registry Number: 7784-44-3

RTECS Number: CG0850000

DOT ID: UN 1546

Regulatory Authority

- See also Arsenic and Ammonia entries
- Carcinogen (arsenic compounds, n.o.s.) (IARC, Group I, carcinogenic to humans) (OSHA, select carcinogens)
- Clean Air Act, 42USC7412; Title I, Part A, §112 hazardous pollutants (arsenic compounds)
- OSHA 29CFR1910.119, Appendix A, Process Safety List of Highly Hazardous Chemicals, TQ = 100 lb (45 kg)
- Clean Water Act 40CFR401.15 Section 307 Toxic Pollutants, as arsenic and compounds
- RCRA, 40CFR261, Appendix 8 Hazardous Constituents., waste number not listed (arsenic compounds)
- SUPERFUND/EPCRA 40CFR302.4 Reportable Quantity (RQ): CERCLA, 1 lb (0.454 kg) (arsenic compounds)
- EPCRA Section 313: Form R *de minimis* concentration reporting level:1.0%. Source of aqueous ammonia. Molecular weight 176. NH_3 equivalent weight 19.35.
- EPCRA Section 313: Form R *de minimis* concentration reporting level: 0.1% (inorganic arsenic)
- U.S. DOT Regulated Marine Pollutant (49CFR172.101, Appendix B) arsenic compounds
- Canada: Priority Substance List & Restricted Substances/ Ocean Dumping Forbidden (CEPA), National Pollutant Release Inventory (NPRI) (arsenic compounds)

Cited in U.S. State Regulations: California (A,G), New Jersey (G), Pennsylvania.

Description: Ammonium Arsenate a combustible, white powder or plate-like colorless crystal. Freezing/Melting point = (decomposes). Soluble in water.

Incompatibilities: Hydrogen gas forms highly toxic arsine gas on contact with inorganic arsenic.

Permissible Exposure Limits in Air: The Federal OSHA PEL for inorganic arsenic (measured as arsenic) is 0.01 mg/m^3 TWA for an 8-hour workshift. NIOSH recommends a limit of 0.002 mg/m^3, which should not be exceeded at any time. ACGIH recommends a limit of 0.01 mg/m^3 TWA for an 8-hour workshift. The HSE limit for arsenic compounds (not including arsine) is 0.1 mg/m^3 TWA. The Canadian Provinces of British Columbia and Quebec airborne limits for arsenic compounds is 0.5 mg/m^3 TWA.

Determination in Air: NIOSH Method #7900 (Arsenic), also #7300 (Elements).

Routes of Entry: Inhalation, ingestion.

Harmful Effects and Symptoms

Short Term Exposure: Skin contact can contribute to overall exposure. Irritates eyes, skin and respiratory tract. Eye contact may cause burns and permanent damage. Inhalation may irritate the nose and throat causing coughing and wheezing. Repeated exposures may cause an ulcer or hole in the cartilage dividing the inner nose and poor appetite, a metallic or garlic taste, nausea, vomiting, stomach pain and diarrhea.

Long Term Exposure: May cause liver damage. Prolonged or repeated exposure may cause nerve damage, with a feeling of "pins and needles" and loss of coordination. While ammonium arsenate has not been identified as a carcinogen, arsenic, and certain arsenic compounds have been determined to be carcinogens. This chemical should be handled with extreme care.

Points of Attack: Liver, central nervous system, eyes, skin.

Medical Surveillance: Liver function, nervous system, and skin tests. Urine tests for arsenic.

First Aid: If this chemical gets into the eyes, remove any contact lenses at once and irrigate immediately for at least 15 minutes, occasionally lifting upper and lower lids. Seek medical attention immediately. If this chemical contacts the skin, remove contaminated clothing and wash immediately with soap and water. Seek medical attention immediately. If this chemical has been inhaled, remove from exposure, begin rescue breathing (using universal precautions) if breathing has stopped and CPR if heart action has stopped. Transfer promptly to a medical facility. When this chemical has been swallowed, get medical attention. Give large quantities of water and induce vomiting. Do not make an unconscious person vomit.

Personal Protective Methods: Wear protective gloves and clothing to prevent any reasonable probability of skin contact. Safety equipment suppliers/manufacturers can provide recommendations on the most protective glove/clothing material for your operation. All protective clothing (suits, gloves, footwear, headgear) should be clean, available each day, and put on before work. Contact lenses should not be worn when working with this chemical. Wear dust-proof chemical goggles and face shield unless full facepiece respiratory protection is worn. Employees should wash immediately with soap when skin is wet or contaminated. Provide emergency showers and eyewash.

Respirator Selection: SA (any supplied-air respirator); or SCBAF (any self-contained breathing apparatus with a full facepiece). *Emergency or planned entry into unknown concentrations or IDLH conditions:* SCBAF:PD,PP (any self-contained breathing apparatus that has a full facepiece and is operated in a pressure-demand or other positive-pressure mode); or SAF:PD,PP:ASCBA (any supplied-air respirator that has a full facepiece and is operated in a pressure-demand or other positive-pressure mode in combination with an auxiliary, self-contained breathing apparatus operated in a pressure-demand or other positive-pressure mode). *Escape:* GMFOVHiE [any air-purifying, full-facepiece respirator (gas mask) with a chin-style, front- or back-mounted organic vapor canister having a high-efficiency particulate filter]; or SCBAE (any appropriate escape-type, self-contained breathing apparatus).

Storage: Prior to working with this chemical you should be trained on its proper handling and storage. A regulated, marked area should be established where ammonium arsenate is handled, used, or stored. Store in tightly closed containers in a cool, well-ventilated area

Shipping: Label required is "POISON." Ammonium arsenate is in DOT/UN Hazard Class 6.1 and Packing Group II.

Spill Handling: Evacuate persons not wearing protective equipment from area of spill or leak until clean-up is complete. Remove all ignition sources. Stay upwind; keep out of low areas. Ventilate closed spaces before entering them. Remove and isolate contaminated clothing at the site. Do not touch spilled material; stop leak if you can do so without risk. Use water spray to reduce vapors. *Small spills:* absorb with sand or other noncombustible absorbent material and place into containers for later disposal. *Large spills:* dike far ahead of spill for later disposal. Collect powdered material in the most convenient and safe manner and deposit in sealed containers. Ventilate area after clean-up is complete. It may be necessary to contain and dispose of this chemical as a hazardous waste. If material or contaminated runoff enters waterways, notify downstream users of potentially contaminated waters. Contact your Department of Environmental Protection or your regional office of the federal EPA for specific recommendations. If employees are required to clean-up spills, they must be properly trained and equipped. OSHA 1910.120(q) may be applicable.

Fire Extinguishing: Ammonium arsenate is combustible solid. Poisonous gases including arsenic, ammonia, and nitrogen oxides are produced in fire. Use dry chemical, carbon dioxide, water spray, or foam. Firefighting gear (including SCBA) may not provide adequate protection. If exposure occurs, remove and isolate gear immediately and thoroughly decontaminate personnel. Wear positive pressure breathing apparatus and special protective clothing. Remove and isolate contaminated clothing at the site. Move container from fire area if you can do so without risk. Fight fire from maximum distance. Dike fire control water for later disposal; do not scatter the material. If material or contaminated runoff enters waterways, notify downstream users of potentially contaminated waters. Notify local health and fire officials and pollution control agencies. From a secure, explosion-proof location, use water spray to cool exposed containers. If cooling streams are ineffective (venting sound increases in volume and pitch, tank discolors, or shows any signs of deforming), withdraw immediately to a secure position. If employees are expected to fight fires, they must be trained and equipped in OSHA 1910.156.

Disposal Method Suggested: Consult with environmental regulatory agencies for guidance on acceptable disposal practices. Generators of waste containing this contaminant (≥100 kg/mo) must conform with EPA regulations governing storage, transportation, treatment, and waste disposal.

References

New Jersey Department of Health and Senior Services, Hazardous Substance Fact Sheet, Ammonium Arsenate, Trenton, NJ (December, 1998)

Ammonium Bicarbonate

Molecular Formula: CH_5NO_3

Common Formula: NH_4HCO_3

Synonyms: ABC-Trieb; Acid ammonium carbonate; Acid ammonium carbonate, monoammonium salt; Ammonium hydrogen carbonate; Bicarbonato amonico (Spanish); Carbonic acid, monoammonium salt

CAS Registry Number: 1066-33-7

RTECS Number: BO0860000

DOT ID: NA 9081

Regulatory Authority

- CLEAN WATER ACT: Section 311 Hazardous Substances/RQ 40CFR117.3 (same as CERCLA, see below)
- SUPERFUND/EPCRA 40CFR302.4 Reportable Quantity (RQ): CERCLA, 5,000 lb (2,270 kg)
- EPCRA Section 313: See ammonia

Cited in U.S. State Regulations: New Hampshire (G), New Jersey (G), Pennsylvania (G).

Description: Ammonium Bicarbonate, NH_4HCO_3, is a white crystalline solid with a faint ammonia odor. Freezing/Melting point = 107.5°C (if heated very rapidly; decomposition starts at about 60°C).

Potential Exposure: It is used in baking powders and fire extinguishers, to make dyes and pigments, in the manufacture of porous plastics and as an expectorant.

Incompatibilities: Contact with strong caustics such as potassium hydroxide or sodium hydroxide will cause the release of ammonia gas.

Permissible Exposure Limits in Air: No standards set.

Permissible Concentration in Water: No criteria set.

Harmful Effects and Symptoms

The dust can irritate skin, eyes and mucous membranes. Higher concentrations can cause temporary blindness, pulmonary edema and cyanosis. It can severely injure respiratory membranes with possible fatal results.

Short Term Exposure: Irritates the eyes, skin and respiratory tract.

Long Term Exposure: Prolonged or repeated exposure may cause lung damage.

Points of Attack: Lungs, skin, eyes.

Medical Surveillance: Lung function tests. Consider chest x-ray following acute exposure.

First Aid: If this chemical gets into the eyes, remove any contact lenses at once and irrigate immediately for at least 15 minutes, occasionally lifting upper and lower lids. Seek medical attention immediately. If this chemical contacts the skin, remove contaminated clothing and wash immediately with soap and water. Seek medical attention immediately. If this chemical has been inhaled, remove from exposure, begin rescue breathing (using universal precautions) if breathing has stopped and CPR if heart action has stopped. Transfer promptly to a medical facility. When this chemical has been swallowed, get medical attention. Give large quantities of water and induce vomiting. Do not make an unconscious person vomit.

Personal Protective Methods: Where possible, enclose operations and use local exhaust ventilation at the site of chemical release. Wear protective gloves and clothing to prevent any reasonable probability of skin contact. Safety equipment suppliers/manufacturers can provide recommendations on the most protective glove/clothing material for your operation. All protective clothing (suits, gloves, footwear, headgear) should be clean, available each day, and put on before work. Contact lenses should not be worn when working with this chemical. Wear dust-proof chemical goggles and face shield unless full facepiece respiratory protection is worn. Employees should wash immediately with soap when skin is wet or contaminated. Provide emergency showers and eyewash.

Respirator Selection: *At any concentrations above the NIOSH REL, or where there is no REL, at any detectable concentration:* SCBAF:PD,PP (Any MSHA/NIOSH approved self-contained breathing apparatus that has a full facepiece and is operated in a pressure-demand or other positive-pressure mode) or SAF:PD,PP:ASCBA (Any supplied-air respirator that has a full facepiece and is operated in a pressure-demand or other positive-pressure mode in combination with an auxiliary self-contained breathing apparatus operated in a pressure-demand or other positive pressure mode). *Escape:* GMFOV [Any air-purifying, full-facepiece respirator (gas mask) with a chin-style, front-or back-mounted organic vapor canister]; or SCBAE (Any appropriate escape-type, self-contained breathing apparatus).

Storage: Prior to working with this chemical you should be trained on its proper handling and storage. Store in tightly closed containers in a cool, well-ventilated area away from caustics (such as sodium hydroxide or potassium hydroxide), because ammonia gas is released.

Shipping: There are no DOT requirements for shipping labels. There are no quantity limitations on shipment by passenger or cargo aircraft. The DOT/UN Hazard Class is ORM-E and the Packing Group is III.[19][20]

Spill Handling: Evacuate persons not wearing protective equipment from area of spill or leak until clean-up is complete. Remove all ignition sources. Collect powdered material in the most convenient and safe manner and deposit in sealed containers. Ventilate area after clean-up is complete. It may be necessary to contain and dispose of this chemical as a hazardous waste. If material or contaminated runoff enters waterways, notify downstream users of potentially contaminated waters. Contact your Department of Environmental Protection or your regional office of the federal EPA for specific recommendations. If employees are required to clean-up spills, they must be properly trained and equipped. OSHA 1910.120(q) may be applicable.

Fire Extinguishing: Ammonium Bicarbonate may burn, but does not readily ignite. Use dry chemical, carbon dioxide, water spray, or alcohol foam extinguishers. Poisonous gases are produced in fire including ammonia gas and nitrogen oxides. If material or contaminated runoff enters waterways, notify downstream users of potentially contaminated waters. Notify local health and fire officials and pollution control agencies. From a secure, explosion-proof location, use water spray to cool exposed containers. If cooling streams are ineffective (venting sound increases in volume and pitch, tank discolors, or shows any signs of deforming), withdraw immediately to a secure position. If employees are expected to fight fires, they must be trained and equipped in OSHA 1910.156.

Disposal Method Suggested: May be buried in a chemical waste landfill. If neutralized, is amenable to treatment at a municipal sewage treatment plant.

References

Sax, N. I., Ed., "Dangerous Properties of Industrial Materials Report 4," No. 2, 43–45 (1984)

New Jersey Department of Health and Senior Services, "Hazardous Substance Fact Sheet: Ammonium Bicarbonate," Trenton, NJ (February 1987)

Ammonium Bifluoride

Molecular Formula: F_2H_5N

Common Formula: NH_4HF_2, $NH_3 \cdot H_2F_2$

Synonyms: Acid ammonium fluoride; Ammonium acid fluoride; Ammonium hydrogen difluoride; Ammonium hydrogen fluoride; Bifluoruro amonico (Spanish)

CAS Registry Number: 1341-49-7

RTECS Number: BQ9200000

DOT ID: UN 1727 (solid); UN 2817 (solution)

Regulatory Authority

- Air Pollutant Standard Set (NIOSH)[2] (ACGIH)[1] (HSE)[33] (DFG)[3] (former USSR)[43]
- Clean Water Act: 40CFR116.4 Hazardous Substances; RQ 40CFR117.3 (same as CERCLA)
- SUPERFUND/EPCRA 40CFR302.4, Appendix A, Reportable Quantity (RQ): CERCLA, 5,000 lb (2,270 kg); Section 313: Form R *de minimis* concentration reporting level: 1.0% (as ammonia) Molecular weight: 57.04; NH_3 Equivalent weight: 29.86

Cited in U.S. State Regulations: Massachusetts (G), New Hampshire (G), New Jersey (G), Oklahoma (G), Pennsylvania (G).

Description: Ammonium bifluoride, NH_4HF_2,$NH_3 \cdot H_2F_2$, is a white crystalline compound that is commonly found in solution. Freezing/Melting point = 125°C. Hazard Identification (based on NFPA-704 M Rating System): Health 3 Flammability 0, Reactivity 0. Soluble in water.

Potential Exposure: It is used as a sterilizer, in dairy and brewery operations; used in ceramic, glass, and electroplating industries; as a laundry sour.

Incompatibilities: Strong oxidizers, acids. In the presence of moisture corrodes, concrete, metals, glass.

Permissible Exposure Limits in Air: The exposure limits recommended for fluorides (measured as fluorine) are: OSHA,[2] the legal airborne permissible exposure limit (PEL) is 2.5 mg/m^3 averaged over an 8-hour workshift. This is also the HSE value in the U.K.[33] and the MAC value in Germany.[3] NIOSH,[2] the recommended airborne exposure limit is 2.5 mg/m^3 averaged over a 10-hour workshift. ACGIH,[1] the recommended airborne exposure limit is 2.5 mg/m^3 averaged over an 8-hour workshift. The former USSR-UNEP/IRPTC project[43] has not set a MAC in workplace air but readily soluble fluorides have assigned a MAC in residential air of 0.03 mg/m^3 on a momentary basis and 0.01/mg/m^3 on an average daily basis. An IDLH value has been given as 500 mg/m^3.[41]

Determination in Air: Fluorides (as F) may be collected on a filter and measured by an ion-specific electrode according to NIOSH Method 7902.[18]

Permissible Concentration in Water: No criteria set.

Routes of Entry: Inhalation, ingestion, skin absorption.

Harmful Effects and Symptoms

Short Term Exposure: The dust irritates and burns the eyes, skin, nose, throat and lungs. Eye contact may cause permanent damage. Inhalation can irritate the lungs causing coughing and/or shortness of breath. Higher exposures can cause pulmonary edema, a medical emergency that can be delayed for several hours. This can cause death. Between 1 teaspoonful and one ounce may be fatal to humans by ingestion. Ingestion will in any case cause irritation of the mouth and stomach and can cause vomiting, convulsions, collapse and acute toxic nephritis. If absorbed through the skin this chemical may cause added exposure.

Long Term Exposure: May cause chronic lung irritation and kidney and liver damage. Bronchitis may develop. Chronic exposure may cause weight loss, nausea, vomiting, weakness, shortness of breath.

Points of Attack: Lungs, skin.

Medical Surveillance: Pre-employment and periodic examinations should consider possible effects on the skin, eyes, teeth, respiratory tract, and kidneys. Chest x-ray and pulmonary function should be followed. Kidney function should be evaluated. If exposures have been heavy and skeletal fluorosis is suspected, pelvic x-rays may be helpful. Intake of fluoride from natural sources in food or water should be known. In the case of exposure to fluoride dusts, periodic urinary fluoride excretion levels have been very useful in evaluating industrial exposures and environmental dietary sources.

First Aid: If this chemical gets into the eyes, remove any contact lenses at once and irrigate immediately for at least 15 minutes, occasionally lifting upper and lower lids. Seek medical attention immediately. If this chemical contacts the skin, remove contaminated clothing and wash immediately with soap and water. Seek medical attention immediately. If this chemical has been inhaled, remove from exposure, begin rescue breathing (using universal precautions) if breathing has stopped and CPR if heart action has stopped. Transfer promptly to a medical facility. When this chemical has been swallowed, get medical attention. If victim is *conscious,* administer water or milk. Do not induce vomiting. Medical observation is recommended for 24 – 48 hours after breathing overexposure, as pulmonary edema may be delayed. As first aid for pulmonary edema, a doctor or authorized paramedic may consider administering a corticosteroid spray.

Personal Protective Methods: Where possible, enclose operations and use local exhaust ventilation at the site of chemical release. Wear protective gloves and clothing to prevent any reasonable probability of skin contact. Safety equipment suppliers/manufacturers can provide recommendations on the most protective glove/clothing material for your operation. All protective clothing (suits, gloves, footwear, headgear) should be clean, available each day, and put on before work. Contact lenses should not be worn when working with this chemical. Wear splash or dust-proof chemical goggles and face shield unless full facepiece respiratory protection is worn. Employees should wash immediately with soap when skin is wet or contaminated. Provide emergency showers and eyewash.

Respirator Selection: NIOSH: *(fluorides) 12.5* mg/m^3: DM (any dust and mist respirator). *25 mg/m^3*: DMXSQ* (any dust and mist respirator except single-use and quarter mask respirators); or SA* (any supplied-air respirator). *62.5 mg/m^3*: SA:CF* (any supplied-air respirator operated in a continuous-flow mode); or PAPRDM* ** *if not present as a fume* (any powered, air-purifying respirator with a dust and mist filter). *125 mg/m^3*: HiEF** (any air-purifying, full-facepiece respirator with a high-efficiency particulate filter); or SCBAF (any self-contained breathing apparatus with a full facepiece); or SAF (any supplied-air respirator with a full facepiece). 250 mg/m^3: SA:PD,PP (any supplied-air respirator operated in a pressure-demand or other positive-pressure mode). *Emergency or planned entry into unknown concentrations or IDLH conditions*: SCBAF: PD,PP (any self-contained breathing apparatus that has a full faceplate and is operated in a pressure-demand or other positive-pressure mode); or SAF:PD,PP: ASCBA (any supplied-air respirator that has a full facepiece and is operated in a pressure-demand or other positive-pressure mode in combination with an auxiliary, self-contained breathing apparatus operated in a pressure-demand or other positive-pressure mode). *Escape:* HiEF** (any air-purifying, full-facepiece respirator with a high-efficiency particulate filter); or SCBAE (any appropriate escape-type, self-contained breathing apparatus).

* Substance reported to cause eye irritation or damage; may require eye protection.

** May need acid gas sorbent. Exposure to 500 mg/m^3 (IDLH) is immediately dangerous to life and health.

Storage: Prior to working with this chemical you should be trained on its proper handling and storage. Store in tightly closed containers in a cool, well-ventilated area away from strong oxidizers (such as chlorine, bromine and fluorine) since highly poisonous hydrogen fluoride gas is released. Keep in plastic, rubber or paraffined containers (because it easily etches glass).

Shipping: The shipping label required for the solid is "Corrosive" and for solutions is "Corrosive, Poison." The DOT/UN Hazard Class in either case is 8 and the Packing Group is II.[19][20]

Spill Handling: Evacuate persons not wearing protective equipment from area of spill or leak until clean-up is complete. Remove all ignition sources. Absorb liquids in vermiculite, dry sand, earth, or a similar material and deposit in sealed containers. Collect powdered material in the most convenient and safe manner and deposit in sealed containers. Ventilate area after clean-up is complete. It may be necessary to contain and dispose of this chemical as a hazardous waste. If material or contaminated runoff enters waterways, notify downstream users of potentially contaminated waters. Contact your Department of Environmental Protection or your regional office of the federal EPA for specific recommendations. If employees are required to clean-up spills, they must be properly trained and equipped. OSHA 1910.120(q) may be applicable.

Fire Extinguishing: Ammonium bifluoride may burn, but does not readily ignite. Use dry chemical, carbon dioxide, water spray, or alcohol foam extinguishers. Poisonous gases are produced in fire including hydrogen fluoride gas. If material or contaminated runoff enters waterways, notify downstream users of potentially contaminated waters. Notify local health and fire officials and pollution control agencies. From a secure, explosion-proof location, use water spray to cool exposed containers. If cooling streams are ineffective (venting sound increases in volume and pitch, tank discolors, or shows any signs of deforming), withdraw immediately to a secure position. If employees are expected to fight fires, they must be trained and equipped in OSHA 1910.156.

Disposal Method Suggested: May be buried in a specially designated chemical landfill. Aqueous wastes may be reacted with an excess of lime followed by lagooning and either recovery or land disposal of the separated calcium fluoride.

References

National Institute for Occupational Safety and Health, "Criteria for a Recommended Standard: Occupational Exposure to Inorganic Fluoride," NIOSH Doc. No. 76–103 (1976)

Sax, N. I., Ed., Dangerous Properties of Industrial Materials Report 3, No. 5, 34–37 (1983)

New Jersey Department of Health and Senior Services, "Hazardous Substance Fact Sheet: Ammonium Bifluoride," Trenton, NJ (September 1986)

Ammonium Bisulfite

Molecular Formula: $H_3N{\cdot}H_2O_3S$

Synonyms: Ammonium hydrogen sulfite; Ammonium monosulfite; Monoammonium sulfite; Sulfurous acid, monoammonium salt

CAS Registry Number: 10192-30-0

RTECS Number: WT3595000

DOT ID: UN 2693 (solution)

Regulatory Authority

- Clean Water Act: 40CFR116.4 Hazardous Substances; RQ 40CFR117.3 (same as CERCLA)
- SUPERFUND/EPCRA 40CFR302.4, Appendix A, Reportable Quantity (RQ): CERCLA, 100 lb (45.4 kg); Section 313: Form R *de minimis* concentration reporting level: 1.0% (as ammonia) Molecular weight: 99.10; NH_3 Equivalent weight: 17.18
- Canada, WHMIS, Ingredients Disclosure List

Cited in U.S. State Regulations: California (G), Massachusetts (G), New Jersey (G), Pennsylvania (G)

Description: Ammonium bisulfite is a white crystalline solid or colorless to yellow solution. Hazard Identification (based on NFPA-704 M Rating System): Health 2 Flammability 0, Reactivity 0. Soluble in water.

Potential Exposure: Used as a preservative, in drilling fluids, making industrial explosives, hair waving and bleaching agent, for making other chemicals. A source of sulfur in liquid fertilizers

Incompatibilities: Sulfites react explosively on contact with oxidizers. Reacts with acids, acid fumes and water. Corrosive to metals.

Permissible Exposure Limits in Air: None established. However due to this chemical's corrosivity, care must be taken in its use.

Routes of Entry: Inhalation, ingestion.

Harmful Effects and Symptoms

Short Term Exposure: Ammonium bisulfite is highly corrosive. Eye and skin contact can cause severe irritation and burns with possible permanent damage. Inhalation can irritate the lungs, causing coughing and shortness of breath. Higher exposure can cause pulmonary edema, a medical emergency that can be delayed for several hours. This can cause death.

Long Term Exposure: May cause lung damage.

Points of Attack: Lungs, skin, eyes.

Medical Surveillance: Lung function tests. Consider chest x-ray following acute exposure.

First Aid: If this chemical gets into the eyes, remove any contact lenses at once and irrigate immediately for at least 15 minutes, occasionally lifting upper and lower lids. Seek medical attention immediately. If this chemical contacts the skin, remove contaminated clothing and wash immediately with soap and water. Seek medical attention immediately. If this chemical has been inhaled, remove from exposure, begin

rescue breathing (using universal precautions) if breathing has stopped and CPR if heart action has stopped. Transfer promptly to a medical facility. When this chemical has been swallowed, get medical attention. If victim is *conscious*, administer water or milk. Do not induce vomiting. Medical observation is recommended for 24 – 48 hours after breathing overexposure, as pulmonary edema may be delayed. As first aid for pulmonary edema, a doctor or authorized paramedic may consider administering a corticosteroid spray.

Personal Protective Methods: Wear protective gloves and clothing to prevent any reasonable probability of skin contact. Safety equipment suppliers/manufacturers can provide recommendations on the most protective glove/clothing material for your operation. All protective clothing (suits, gloves, footwear, headgear) should be clean, available each day, and put on before work. Contact lenses should not be worn when working with this chemical. Wear splash- or dust-proof chemical goggles and face shield unless full facepiece respiratory protection is worn. Employees should wash immediately with soap when skin is wet or contaminated. Provide emergency showers and eyewash.

Respirator Selection: SA (any supplied-air respirator); or SCBAF (any self-contained breathing apparatus with a full facepiece). *Emergency or planned entry into unknown concentrations or IDLH conditions:* SCBAF:PD,PP (any self-contained breathing apparatus that has a full facepiece and is operated in a pressure-demand or other positive-pressure mode); or SAF:PD,PP:ASCBA (any supplied-air respirator that has a full facepiece and is operated in a pressure-demand or other positive-pressure mode in combination with an auxiliary, self-contained breathing apparatus operated in a pressure-demand or other positive-pressure mode). *Escape:* GMFOVHiE [any air-purifying, full-facepiece respirator (gas mask) with a chin-style, front- or back-mounted organic vapor canister having a high-efficiency particulate filter]; or SCBAE (any appropriate escape-type, self-contained breathing apparatus).

Storage: Prior to working with this chemical you should be trained on its proper handling and storage. Store in tightly closed containers in a cool, well-ventilated area away from oxidizers, acids, water, combustible materials.

Shipping: Label required is "CORROSIVE." Ammonium bisulfite is in DOT/UN Hazard Class 8 and Packing Group II or III.

Spill Handling: Evacuate persons not wearing protective equipment from area of spill or leak until clean-up is complete. Remove all ignition sources. Ventilate closed spaces before entering them. Absorb liquid with sand, vermiculite, earth, or similar absorbent material and place into containers for later disposal. Collect powdered material in the most convenient and safe manner and deposit in sealed containers. Ventilate area after clean-up is complete. It may be necessary to contain and dispose of this chemical as a hazardous waste. If material or contaminated runoff enters waterways, notify downstream users of potentially contaminated waters. Contact your Department of Environmental Protection or your regional office of the federal EPA for specific recommendations. If employees are required to clean-up spills, they must be properly trained and equipped. OSHA 1910.120(q) may be applicable.

Fire Extinguishing: Ammonium bisulfite is noncombustible. Poisonous gases including *ammonia, sulfur oxides, and nitrogen oxides* are produced in fire. Use extinguishing agents suitable for surrounding fire. If material or contaminated runoff enters waterways, notify downstream users of potentially contaminated waters. Notify local health and fire officials and pollution control agencies. From a secure, explosion-proof location, use water spray to cool exposed containers. If cooling streams are ineffective (venting sound increases in volume and pitch, tank discolors, or shows any signs of deforming), withdraw immediately to a secure position. If employees are expected to fight fires, they must be trained and equipped in OSHA 1910.156.

Disposal Method Suggested: Incinerate. It may be possible to dispose of waste material at a municipal facility if treated, neutralized and oxidized.

References

New Jersey Department of Health and Senior Services, Hazardous Substance Fact Sheet, Ammonium Bisulfite, Trenton NJ (September, 1998)

Ammonium Carbamate

Molecular Formula: $CH_6N_2O_2$

Common Formula: NH_4COONH_2

Synonyms: Ammonium aminoformate; Anhydride of ammonium carbonate; Carbamato amonico (Spanish); Carbamic acid, ammonium salt; Carbamic acid, monoammonium salt

CAS Registry Number: 1111-78-0

RTECS Number: EY8575000

DOT ID: UN 2757 (carbamate pesticide, solid, toxic)

Regulatory Authority

- Clean Water Act: 40CFR116.4 Hazardous Substances; RQ 40CFR117.3, (same as CERCLA)
- SUPERFUND/EPCRA 40CFR302.4, Appendix A, Reportable Quantity (RQ): 5,000 lb (2,270 kg); Section 313: Form R *de minimis* concentration reporting level: 1.0% (as ammonia) Molecular weight: 99.10; NH_3 Equivalent weight: 21.81Canada, WHMIS, Ingredients Disclosure List

Cited in U.S. State Regulations: California (G), Illinois (G), Massachusetts (G), New Hampshire (G), New Jersey (G), Pennsylvania (G).

Description: Ammonium carbamate, NH_4COONH_2, is a colorless crystalline powder or white powder with an ammonia odor. Freezing/Melting point = about 60°C (sublimes). The odor threshold is 5 ppm as NH_3 (detection) and 46.8 ppm as NH_3 (recognition). Boiling point = 60°C. Highly soluble in water.

Potential Exposure: It is used as a fertilizer and ammoniating agent.

Incompatibilities: Strong bases, strong oxidizers. Keep away from heat (forms urea), moisture, and direct sunlight.

Permissible Exposure Limits in Air: No standards set. Loses ammonia in air, changing to ammonia carbonate.

Permissible Concentration in Water: No criteria set.

Harmful Effects and Symptoms

Irritates skin, respiratory tract and mucous membranes on contact.

Short Term Exposure: Exposure can irritate the eyes and skin causing redness and tearing. Inhalation can irritate the nose and lungs with coughing, and/or shortness of breath.

Long Term Exposure: Repeated or prolonged exposure can cause lung irritation and the development of bronchitis.

Points of Attack: Respiratory system, eyes, skin.

Medical Surveillance: Lung function testing.

First Aid: If this chemical gets into the eyes, remove any contact lenses at once and irrigate immediately for at least 15 minutes, occasionally lifting upper and lower lids. Seek medical attention immediately. If this chemical contacts the skin, remove contaminated clothing and wash immediately with soap and water. Seek medical attention immediately. If this chemical has been inhaled, remove from exposure, begin rescue breathing (using universal precautions) if breathing has stopped and CPR if heart action has stopped. Transfer promptly to a medical facility. When this chemical has been swallowed, get medical attention. Give large quantities of water and induce vomiting. Do not make an unconscious person vomit.

Personal Protective Methods: Where possible, enclose operations and use local exhaust ventilation at the site of chemical release. Wear protective gloves and clothing to prevent any reasonable probability of skin contact. Safety equipment suppliers/manufacturers can provide recommendations on the most protective glove/clothing material for your operation. All protective clothing (suits, gloves, footwear, headgear) should be clean, available each day, and put on before work. Contact lenses should not be worn when working with this chemical. Wear dust-proof chemical goggles and face shield unless full facepiece respiratory protection is worn. Employees should wash immediately with soap when skin is wet or contaminated. Provide emergency showers and eyewash.

Respirator Selection: *At any concentrations above the NIOSH REL, or where there is no REL, at any detectable concentration:* SCBAF:PD,PP (any MSHA/NIOSH approved self-contained breathing apparatus that has a full facepiece and is operated in a pressure-demand or other positive-pressure mode); or SAF:PD,PP:ASCBA (any supplied-air respirator that has a full facepiece and is operated in a pressure-demand or other positive-pressure mode in combination with an auxiliary, self-contained breathing apparatus operated in a pressure-demand or other positive pressure mode). *Escape:* GMFOVHiE [any air-purifying, full-facepiece respirator (gas mask) with a chin-style, front- or back-mounted organic vapor canister having a high-efficiency particulate filter]; or SCBAE (any appropriate escape-type, self-contained breathing apparatus).

Storage: Prior to working with this chemical you should be trained on its proper handling and storage. Store in tightly closed containers in a cool, well-ventilated area away from heat, moisture, and direct sunlight. See also Incompatibilities.

Shipping: Label required is "POISON" in Packing Groups I and II; "KEEP AWAY FROM FOOD" in Packing Group III. Ammonium carbamate is in DOT/UN Hazard Class 6.1.

Spill Handling: Evacuate persons not wearing protective equipment from area of spill or leak until clean-up is complete. Remove all ignition sources. Collect powdered material in the most convenient and safe manner and deposit in sealed containers. Ventilate area after clean-up is complete. It may be necessary to contain and dispose of this chemical as a hazardous waste. If material or contaminated runoff enters waterways, notify downstream users of potentially contaminated waters. Contact your Department of Environmental Protection or your regional office of the federal EPA for specific recommendations. If employees are required to clean-up spills, they must be properly trained and equipped. OSHA 1910.120(q) may be applicable.

Fire Extinguishing: Use dry chemical, carbon dioxide, water spray, or alcohol foam extinguishers. Poisonous gases are produced in fire including *ammonia, urea, and nitrogen oxides*. If material or contaminated runoff enters waterways, notify downstream users of potentially contaminated waters. Notify local health and fire officials and pollution control agencies. From a secure, explosion-proof location, use water spray to cool exposed containers. If cooling streams are ineffective (venting sound increases in volume and pitch, withdraw immediately to a secure position. If employees are expected to fight fires, they must be trained and equipped in OSHA 1910.156.

References

Sax, N. I., "Dangerous Properties of Industrial Materials Report," 2, No. 3, 31–33 (1982)

New Jersey Department of Health and Senior Services, "Hazardous Substance Fact Sheet: Ammonium Carbamate," Trenton, NJ (January, 1996)

Ammonium Carbonate

Molecular Formula: $CH_8N_2O_3$

Common Formula: $(NH_4)_2CO_3$

Synonyms: Ammoniumcarbonat (German); Carbonato amonico (Spanish); Carbonic acid, Ammonium salt; Carbonic acid, diammonium salt; Crystal ammonia; Diammonium carbonate; Hartshorn; Sal volatile

CAS Registry Number: 506-87-6

RTECS Number: BP1925000

DOT ID: NA 9084

Regulatory Authority

- Clean Water Act: 40CFR116.4 Hazardous Substances; RQ 40CFR117.3, (same as CERCLA)
- SUPERFUND/EPCRA 40CFR302.4, Appendix A, Reportable Quantity (RQ): CERCLA, 100 lb (45.4 kg); Section 313: Form R *de minimis* concentration reporting level: 1.0% (as ammonia) Molecular weight: 96.09; NH_3 Equivalent weight: 35.45

Cited in U.S. State Regulations: California (G), Massachusetts (G), New Hampshire (G), New Jersey (G), Pennsylvania (G).

Description: Ammonium carbonate, $(NH_4)_2CO_3$, is a colorless crystal or white lumpy powder with a strong ammonia odor. The odor threshold is <5 ppm (as ammonia gas). Freezing/Melting point = 58°C and volatilizes at about 60°C. Hazard Identification (based on NFPA-704 M Rating System): Health 1, Flammability 0, Reactivity 0. Slightly soluble in water.

Potential Exposure: It is used in dyeing, tanning, medicines, fire extinguishers; to make casein glue, ammonia salts, and baking powders. A laboratory reagent.

Incompatibilities: Acids, acid salts, salts of iron and zinc, alkaloids, calomel and tartar emetic. Keep cool, below 38°C.

Permissible Exposure Limits in Air: No standards set.

Permissible Concentration in Water: No criteria set.

Routes of Entry: Inhalation, ingestion, eye and/or skin contact.

Harmful Effects and Symptoms

Short Term Exposure: Contact can irritate eyes and nose. Breathing Ammonium Carbonate can irritate the nose, throat and lungs, causing a cough and difficulty in breathing.

Long Term Exposure: May cause lung problems.

Points of Attack: Lungs, eyes, skin.

Medical Surveillance: Lung function tests.

First Aid: If this chemical gets into the eyes, remove any contact lenses at once and irrigate immediately for at least 15 minutes, occasionally lifting upper and lower lids. Seek medical attention immediately. If this chemical contacts the skin, remove contaminated clothing and wash immediately with soap and water. Seek medical attention immediately. If this chemical has been inhaled, remove from exposure, begin rescue breathing (using universal precautions) if breathing has stopped and CPR if heart action has stopped. Transfer promptly to a medical facility. When this chemical has been swallowed, get medical attention. Give large quantities of water and induce vomiting. Do not make an unconscious person vomit.

Personal Protective Methods: Where possible, enclose operations and use local exhaust ventilation at the site of chemical release. Wear protective gloves and clothing to prevent any reasonable probability of skin contact. Safety equipment suppliers/manufacturers can provide recommendations on the most protective glove/clothing material for your operation. All protective clothing (suits, gloves, footwear, headgear) should be clean, available each day, and put on before work. Contact lenses should not be worn when working with this chemical. Wear dust-proof chemical goggles and face shield unless full facepiece respiratory protection is worn. Employees should wash immediately with soap when skin is wet or contaminated. Provide emergency showers and eyewash.

Respirator Selection: SA (any supplied-air respirator); or SCBAF (any self-contained breathing apparatus with a full facepiece). *Emergency or planned entry into unknown concentrations or IDLH conditions:* SCBAF:PD,PP (any self-contained breathing apparatus that has a full facepiece and is operated in a pressure-demand or other positive-pressure mode); or SAF:PD,PP:ASCBA (any supplied-air respirator that has a full facepiece and is operated in a pressure-demand or other positive-pressure mode in combination with an auxiliary, self-contained breathing apparatus operated in a pressure-demand or other positive-pressure mode). *Escape:* GMFOVHiE [any air-purifying, full-facepiece respirator (gas mask) with a chin-style, front- or back-mounted organic vapor canister having a high-efficiency particulate filter]; or SCBAE (any appropriate escape-type, self-contained breathing apparatus).

Storage: Prior to working with this chemical you should be trained on its proper handling and storage. Store in tightly closed containers in a cool, well-ventilated area away from temperatures above 38°C/100°F.

Shipping: There are no DOT/UN label requirements and no quantity limits on shipment by passenger or cargo aircraft. Falls in Hazard Class ORM-E and Packing Group III.[19][20]

Spill Handling: Evacuate persons not wearing protective equipment from area of spill or leak until clean-up is complete. Remove all ignition sources. Collect powdered material in the most convenient and safe manner and deposit in sealed containers. Ventilate area after clean-up is complete. It may be necessary to contain and dispose of this chemical as a hazardous waste. If material or contaminated runoff enters waterways, notify downstream users of potentially contaminated waters. Contact your Department of Environmental Protection or your regional office of the federal EPA for specific recommendations. If employees are required to clean-up spills, they must be properly trained and equipped. OSHA 1910.120(q) may be applicable.

Fire Extinguishing: This material may burn but does not readily ignite. Use dry chemical, carbon dioxide, water spray, or alcohol foam extinguishers. Poisonous gases are produced in fire including. If material or contaminated runoff enters waterways, notify downstream users of potentially contaminated waters. Notify local health and fire officials and pollution control agencies. From a secure, explosion-proof location, use water spray to cool exposed containers. If cooling

streams are ineffective (venting sound increases in volume and pitch, tank discolors, or shows any signs of deforming), withdraw immediately to a secure position. If employees are expected to fight fires, they must be trained and equipped in OSHA 1910.156.

Disposal Method Suggested: Slowly deposit in a large container of water. Add excess amounts of soda ash and let stand for 24 hours. Decant to another container, neutralize with hydrochloric acid, and drain with and excess of water. Ship to landfill.

References

Sax, N. I., Ed., "Dangerous Properties of Industrial Materials Report," 2, No. 3, 33–34 (1982)

New Jersey Department of Health and Senior Services, "Hazardous Substance Fact Sheet: Ammonium Carbonate," Trenton, NJ (September 1986)

Ammonium Chloride

Molecular Formula: NH_4Cl

Synonyms: Amchlor; Amchloride; Ammoneric; Ammonium muriate; AM solder flux; Cloruro amonico (Spanish); Salamac; Sal ammoniac; Salmiac; Slago

CAS Registry Number: 12125-02-9

RTECS Number: BP4550000

DOT ID: UN (NA 9085)

EEC Number: 017-014-00-8

EINECS Number: 235-186-4

Regulatory Authority

- Air Pollutant Standard Set (ACGIH)[1] (HSE)[33] (OSHA)[58] (Czechoslovakia)[35] (Several States)[60] (Seveal Canadian Provinces) (Australia) (Israel) (Mexico)
- Water Pollution Standard Proposed (former USSR)[43]
- Clean Water Act: 40CFR116.4 Hazardous Substances; RQ 40CFR117.3 (same as CERCLA)
- SUPERFUND/EPCRA 40CFR302.4, Appendix A, Reportable Quantity (RQ): CERCLA, 5,000 lb (2,270 kg); Section 313: Form R *de minimis* concentration reporting level: 1.0% (as ammonia) Molecular weight: 53.49; NH_3 Equivalent weight: 31.38
- Canada, WHMIS, Ingredients Disclosure List

Cited in U.S. State Regulations: Alaska (G), California (G), Connecticut (A), Florida (G), Illinois (G), Massachusetts (G), Minnesota (G), Nevada (A), New Hampshire (G), New Jersey (G), North Dakota (A), Oklahoma (G), Pennsylvania (G), Rhode Island (G, A, W), South Carolina (A), South Dakota (A), Virginia (A), West Virginia (G).

Description: Ammonium chloride, NH_4Cl, is a white crystalline odorless solid. Freezing/Melting point = 338°C with decomposition. Hazard Identification (based on NFPA-704 M Rating System): Health 1 Flammability 0, Reactivity 0. Soluble in water.

Potential Exposure: Ammonium chloride is used to make dry batteries, in galvanizing, and as a soldering flux.

Incompatibilities: Acids, alkalis, and silver salts.

Permissible Exposure Limits in Air: The Federal OSHA exposure limit (as fume) is 10 mg/m^3 TWA for an 8-hour workshift and 20 mg/m^3 15 minute STEL. ACGIH recommends the same levels. Australia, HSE, Israel, and Mexico have similar limits, and Israel has an action level of 5 mg/m^3. Czechoslovakia[35] has set values for ammonium chloride of 0.1 mg/m^3 on a daily average basis and 0.3 mg/m^3 on a half-hour exposure basis. Several states have set guidelines or standards for ammonium chloride in ambient air:[60] 0.1 mg/m^3 (North Dakota), 0.15 mg/m^3 (Virginia), 0.2 mg/m^3 (Connecticut, South Dakota), 0.238 mg/m^3 (Nevada) to 0.25 mg/m^3 (South Carolina). Several Canadian Provinces set levels similar to OSHA and ACGIH (Alberta, British Columbia, Ontario, and Quebec).

Determination in Air: Collection on a filter and colorimetric analysis.

Permissible Concentration in Water: The former USSR-UNEP/IRPTC project[43] has set a MAC of 1.2 mg/l in water for fishery purposes.

Routes of Entry: Inhalation, ingestion, skin and/or eye contact.

Harmful Effects and Symptoms

Short Term Exposure: Ammonium chloride is an eye, skin and respiratory system irritant with a low grade systemic toxicity by ingestion.

Points of Attack: Skin, respiratory system.

First Aid: If this chemical gets into the eyes, remove any contact lenses at once and irrigate immediately for at least 15 minutes, occasionally lifting upper and lower lids. Seek medical attention immediately. If this chemical contacts the skin, remove contaminated clothing and wash immediately with soap and water. Seek medical attention immediately. If this chemical has been inhaled, remove from exposure, begin rescue breathing (using universal precautions) if breathing has stopped and CPR if heart action has stopped. Transfer promptly to a medical facility. When this chemical has been swallowed, get medical attention. Give large quantities of water and induce vomiting. Do not make an unconscious person vomit.

Personal Protective Methods: Wear protective gloves and clothing to prevent any reasonable probability of skin contact. Safety equipment suppliers/manufacturers can provide recommendations on the most protective glove/clothing material for your operation. All protective clothing (suits, gloves, footwear, headgear) should be clean, available each day, and put on before work. Contact lenses should not be worn when working with this chemical. Wear dust-proof chemical goggles and face shield unless full facepiece respiratory protection is worn. Employees should wash immediately with soap when skin is wet or contaminated. Provide emergency showers and eyewash.

Respirator Selection: Where the potential exists for exposures over 10 mg/m^3, use a MSHA/NIOSH approved respirator equipped with particulate (dust/fume/mist) filters. Particulate filters must be checked every day before work for physical damage, such as rips or tears, and replaced as needed. Where the potential for high exposures exists, use a MSHA/NIOSH approved supplied-air respirator with a full facepiece operated in the positive pressure mode or with a full facepiece, hood, or helmet in the continuous flow mode, or use a MSHA/NIOSH approved self-contained breathing apparatus with a full facepiece operated in pressure-demand or other positive pressure mode.

Storage: Prior to working with this chemical you should be trained on its proper handling and storage. Store in tightly closed containers in a cool, well ventilated area away from acids, alkalis and silver salts. Protect against physical damage.[17]

Shipping: No label required by DOT/UN and no quantity limits set on passenger aircraft or railcar shipment or cargo aircraft shipment. Falls in Hazard Class ORV-E and packing Group III.[19][20]

Spill Handling: Evacuate persons not wearing protective equipment from area of spill or leak until clean-up is complete. Remove all ignition sources. Collect powdered material in the most convenient and safe manner and deposit in sealed containers. Ventilate area after clean-up is complete. Absorb liquids in vermiculite, dry sand, earth, or a similar material and deposit in sealed containers. It may be necessary to contain and dispose of this chemical as a hazardous waste. If material or contaminated runoff enters waterways, notify downstream users of potentially contaminated waters. Contact your Department of Environmental Protection or your regional office of the federal EPA for specific recommendations. If employees are required to clean-up spills, they must be properly trained and equipped. OSHA 1910.120(q) may be applicable.

Fire Extinguishing: Use dry chemical, carbon dioxide, water spray, or alcohol foam extinguishers. Poisonous gases are produced in fire including ammonia, chlorine gas, and nitrogen oxides. If material or contaminated runoff enters waterways, notify downstream users of potentially contaminated waters. Notify local health and fire officials and pollution control agencies. From a secure, explosion-proof location, use water spray to cool exposed containers. If cooling streams are ineffective (venting sound increases in volume and pitch, tank discolors, or shows any signs of deforming), withdraw immediately to a secure position. If employees are expected to fight fires, they must be trained and equipped in OSHA 1910.156.

Disposal Method Suggested: Pretreatment involves addition of sodium hydroxide to liberate ammonia and form the soluble sodium salt. The liberated ammonia can be recovered and sold. After dilution to the permitted provisional limit, the sodium salt can be discharged into a stream or sewer.[22]

References

Sax, N. I., Ed., Dangerous Properties of Industrial Materials Report, 2, No. 3, 34–36 (1982)

New Jersey Department of Health and Senior Services, "Hazardous Substance Fact Sheet: Ammonium Chloride," Trenton, NJ (February 1989)

Ammonium Chloroplatinate

Molecular Formula: $Cl_6H_8N_2Pt$

Common Formula: $(NH_4)_2PtCl_6$

Synonyms: Ammonium platinic chloride; Diammonium hexachloroplatinate(2-); Diammonium hexachloroplatinate(VI); Platinate(2-), platinic ammonium chloride

CAS Registry Number: 16919-58-7

RTECS Number: BP5425000

DOT ID: Not listed.

Regulatory Authority

- Air Pollutant Standard Set (ACGIH)[1] (HSE)[33] (DFG)[3]
- Canada, WHMIS, Ingredients Disclosure List

Note: Dropped from CERCLA/SARA EHS listing in 1988

Cited in U.S. State Regulations: Massachusetts (G), New Jersey (G).

Description: Ammonium chloroplatinate, $(NH_4)_2PtCl_6$, is an orange-red crystalline solid or yellow powder which decomposes when heated.

Potential Exposure: It is used in photography, platinum plating and in the manufacture of platinum sponge.

Incompatibilities: Acids and strong oxidizers.

Permissible Exposure Limits in Air: The Federal OSHA PEL and ACGIH recommended TLV for platinum soluble salts (measure as platinum) is 0.002 mg/m^3 TWA for an 8-hour workshift. The NIOSH REL is the same for a 10-hour workshift. HSE[33] and the DFG MAK[3] limits are the same as OSHA and ACGIH. The NIOSH IDLH is 4 mg/m^3.

Determination in Air: Platinum in air can be determined by particulate filter collection followed by graphic furnace atomic absorption spectrometry. NIOSH Method #S191 (NIOSH Manual of Analytical Methods, 2nd edition, Volume 7).

Permissible Concentration in Water: No criteria set.

Routes of Entry: Inhalation, ingestion, skin and /or eye contact.

Harmful Effects and Symptoms

Short Term Exposure: Eye and skin irritation can cause irritation and burns. Inhalation can cause respiratory tract irritation, causing coughing and shortness of breath. First symptoms are pronounced irritation of the nose and upper respiratory passages, with sneezing, running of eyes, and coughing. Later, "asthmatic syndrome," with cough, tightness of chest, wheezing, and shortness of breath, develops. It is toxic by inhalation or ingestion. Toxic concentrations for inhalation are 0.9 µg/minute. The oral LD_{50} for rats is 0.44 mg/kg.

Long Term Exposure: Platinum salts can cause asthma-like allergy and skin allergy. Symptoms begin during exposure and grow worse with chronic exposure.

Points of Attack: Lungs, skin.

Medical Surveillance: Lung function tests. Evaluation by a qualified allergist.

First Aid: If this chemical gets into the eyes, remove any contact lenses at once and irrigate immediately for at least 15 minutes, occasionally lifting upper and lower lids. Seek medical attention immediately. Like treatment for other platinum salts remove victim to fresh air and give oxygen. Nasal washing to remove salts is recommended. In case of skin or eye contact, irrigate with water immediately. If this chemical contacts the skin, remove contaminated clothing and wash immediately with soap and water. Seek medical attention immediately. If this chemical has been inhaled, remove from exposure, begin rescue breathing (using universal precautions) if breathing has stopped and CPR if heart action has stopped. Transfer promptly to a medical facility. When this chemical has been swallowed, get medical attention. Give large quantities of water and induce vomiting. Do not make an unconscious person vomit.

Personal Protective Methods: Where possible, enclose operations and use local exhaust ventilation at the site of chemical release. If local exhaust ventilation or enclosure is not used, respirators should be worn. Wear protective gloves and clothing to prevent any reasonable probability of skin contact. Safety equipment suppliers/manufacturers can provide recommendations on the most protective glove/clothing material for your operation. All protective clothing (suits, gloves, footwear, headgear) should be clean, available each day, and put on before work. Contact lenses should not be worn when working with this chemical. Wear dust-proof chemical goggles and face shield unless full facepiece respiratory protection is worn. Employees should wash immediately with soap when skin is wet or contaminated. Provide emergency showers and eyewash.

Respirator Selection: *NIOSH/ OSHA: At any detectable concentration above 0.05* mg/m^3: SA:CF* (any supplied-air respirator operated in a continuous-flow mode); or 0.1 mg/m^3: HiEF (Any air-purifying, full-facepiece respirator with a high-efficiency particulate filter); or SCBAE (Any appropriate escape-type, self-contained breathing apparatus); or SAF (any supplied-air respirator with a full facepiece). *4* mg/m^3*:* SCBAF:PD,PP (any self-contained breathing apparatus that has a full facepiece and is operated in a pressure-demand or other positive-pressure mode). *Emergency or planned entry into unknown concentrations or IDLH conditions:* SAF:PD, PP:ASCBA (any supplied-air respirator that has a full facepiece and is operated in a pressure-demand or other positive-pressure mode in combination with an auxiliary, self-contained breathing apparatus operated in a pressure-demand or other positive pressure mode). *Escape:* HiEF (Any air-purifying, full-facepiece respirator with a high-efficiency particulate filter); or SCBAE (Any appropriate escape-type, self-contained breathing apparatus).

* Causes eye irritation or damage; eye protection needed.

Storage: Prior to working with this chemical you should be trained on its proper handling and storage. Store in tightly closed containers in a cool, well-ventilated area away from acids and strong oxidizers.

Shipping: No UN/DOT specifications given.

Spill Handling: Evacuate persons not wearing protective equipment from area of spill or leak until clean-up is complete. Remove all ignition sources. Collect powdered material in the most convenient and safe manner and deposit in sealed containers. Ventilate area after clean-up is complete. It may be necessary to contain and dispose of this chemical as a hazardous waste. If material or contaminated runoff enters waterways, notify downstream users of potentially contaminated waters. Contact your Department of Environmental Protection or your regional office of the federal EPA for specific recommendations. If employees are required to clean-up spills, they must be properly trained and equipped. OSHA 1910.120(q) may be applicable.

Fire Extinguishing: Ammonium chloroplatinate may burn, but is hard to ignite. Use dry chemical, carbon dioxide, water spray, or foam extinguishers. Poisonous gases are produced in fire including ammonia, nitrogen oxides, chlorine. If material or contaminated runoff enters waterways, notify downstream users of potentially contaminated waters. Notify local health and fire officials and pollution control agencies. From a secure, explosion-proof location, use water spray to cool exposed containers. If cooling streams are ineffective (venting sound increases in volume and pitch, tank discolors, or shows any signs of deforming), withdraw immediately to a secure position. If employees are expected to fight fires, they must be trained and equipped in OSHA 1910.156.

Disposal Method Suggested: Because of the high value of the metal, recovery is the economically-indicated disposal method.

References

U.S. Environmental Protection Agency, "Chemical Profile: Ammonium Chloroplatinate," Washington, DC, Chemical Emergency Preparedness Program (October 31, 1985)

New Jersey Department of Health and Senior Services, "Hazardous Substance Fact Sheet: Ammonium Chloroplatinate," Trenton, NJ (February 1998)

Ammonium Chromate

Molecular Formula: $CrH_8N_2O_4$

Common Formula: $(NH_4)_2CrO_4$

Synonyms: Ammonium chromate(VI); Chromic acid, diammonium salt; Cromato amonico (Spanish); Diammonium chromate; Neutral ammonium chromate

CAS Registry Number: 7788-98-9

RTECS Number: GB2880000

DOT ID: NA 9086

Regulatory Authority

- Carcinogen: (California's Proposition 65) [as Chromium(VI)]
- Air Pollutant Standard Set (NIOSH)[2] (ACGIH)[1] (former USSR)[43] (Australia) (Various States), (Israel) (Various Canadian Provinces)
- Clean Air Act 42USC7412; Title I, Part A, §112 hazardous pollutants (as chromium compounds)
- Clean Water Act: 40CFR116.4 Hazardous Substances; RQ 40CFR117.3, (same as CERCLA); 40CFR423, Appendix A, Priority Pollutants
- RCRA, 40CFR261, Appendix 8 Hazardous Constituents., waste number not listed (chromium compounds)
- SUPERFUND/EPCRA 40CFR302.4, Appendix A, Reportable Quantity (RQ): CERCLA, 10 lb (4.54 kg)
- EPCRA Section 313 Form R *de minimis* concentration reporting level: 1.0% (as ammonia) Molecular weight: 152.07; NH_3 Equivalent weight: 22.04. *Also must be reported as a chromium compound:* "Includes any unique chemical substances that contains chromium as part of that chemical's infrastructure." Form R *de minimis* concentration reporting level: Chromium (VI) compounds: 0.1%.
- Canada, WHMIS, Ingredients Disclosure List

Cited in U.S. State Regulations: California (A,G), Massachusetts (G), Minnesota (G), New Hampshire (G), New Jersey (G), Oklahoma (G), Pennsylvania (G).

Description: $(NH_4)_2CrO_4$ is a yellow crystalline compound which can be used in solution which is yellow with an ammonia odor. Freezing/Melting point = 185°C (decomposes). Hazard Identification (based on NFPA-704 M Rating System): Health 2, Flammability 0, Reactivity 1 Soluble in water.

Potential Exposure: It is used to inhibit corrosion and in dyeing, photography and many chemical reactions. Used as a fungicide and fire retardant.

Incompatibilities: A strong oxidizer. Contact with combustible, organic and other readily oxidizable substances may cause fire and explosions. Hydrazine, other reducing agents. Corrosive to metals.

Permissible Exposure Limits in Air: OSHA – the legal airborne permissible exposure limit (PEL) is 0.1 mg/m^3 for chromic acid and chromates (as Cr), not to be exceeded at any time. NIOSH – the recommended airborne exposure limit is 0.001 mg/m^3 averaged over a 10-hour workshift. ACGIH – the recommended airborne exposure limit is 0.05 mg/m^3 for Chromium compounds (as Cr) averaged over an 8-hour workshift. These exposure limits are for air levels only. When skin contact also occurs, you may be overexposed, even though air levels are less than the limits listed. The NIOSH IDLH for chromates is 15 mg/m^3 as Cr(VI) (Carcinogen). The former USSR-UNEP/IRPTC project has set a MAK value of 0.01 mg/m^3 for chromates and bichromates in the workplace.[43] California Prop. 65 No significant risk level (inhalation) = 0.001 μg/day.

Determination in Air: Hexavalent chromium may be determined by filtration followed by visible absorption spectrophotometry according to NIOSH Method 7600.[18] Also Method #7604.

Permissible Concentration in Water: To protect human health, hexavalent chromium should be held below 0.05 mg/l according to EPA[6] in studies on priority toxic pollutants, This is also a WHO recommendation for total chromium in drinking water.

Determination in Water: Chromium (VI) may be determined by extraction and atomic absorption or colorimetry (using diphenylhydrazide).

Routes of Entry: Ingestion, skin and/or eye contact.

Harmful Effects and Symptoms

Short Term Exposure: Eye contact can cause severe damage with possible loss of vision. Breathing Ammonium Chromate can cause a sore or hole through the inner nose (septum), sometimes with bleeding, discharge or crusting. Irritation of nose, throat and bronchial tubes can also occur, with cough and/or wheezing. Skin contact can cause deep ulcers or an allergic skin rash.

Long Term Exposure: Some water-soluble chromium[16] compounds are inferred non-carcinogens; the water-insoluble compounds are generally deemed to be carcinogens but the border line is not precise nor universally agreed to. Ammonium chromate is a hexavalent chromium compound which may be carcinogenic and should be handled with extreme caution. Breathing ammonium chromate can cause sor or hole in the septum dividing the inner nose, sometimes with bleeding, discharge, and/or formation of a crust. May cause skin allergy and kidney damage.

Points of Attack: Blood, respiratory system, liver kidneys, eyes, skin.

Medical Surveillance: Skin and nose examination, kidney function tests, evaluation by a qualified allergist.

First Aid: If this chemical gets into the eyes, remove any contact lenses at once and irrigate immediately for at least 30 minutes, occasionally lifting upper and lower lids. Seek medical attention immediately. If this chemical contacts the skin, remove contaminated clothing and wash immediately with soap and water. Seek medical attention immediately. If this chemical has been inhaled, remove from exposure, begin rescue breathing (using universal precautions) if breathing has stopped and CPR if heart action has stopped. Transfer promptly to a medical facility. When this chemical has been swallowed, get medical attention. Give large quantities of water and induce vomiting. Do not make an unconscious person vomit.

Personal Protective Methods: Wear protective gloves and clothing to prevent any reasonable probability of skin contact. Safety equipment suppliers/manufacturers can provide recommendations on the most protective glove/clothing material for your operation. All protective clothing (suits, gloves, footwear, headgear) should be clean, available each day, and put on before work. Contact lenses should not be worn when working with this chemical. Wear splash or dust-proof chemical goggles and face shield unless full facepiece respiratory protection is worn. Employees should wash immediately with soap when skin is wet or contaminated. Provide emergency showers and eyewash.

Respirator Selection: NIOSH: *At any detectable concentration above 0.001* mg/m^3*:* SCBAF:PD,PP (any MSHA/NIOSH approved self-contained breathing apparatus that has a full facepiece and is operated in a pressure-demand or other positive-pressure mode); or SAF:PD,PP:ASCBA (any supplied-air respirator that has a full facepiece and is operated in a pressure-demand or other positive-pressure mode in combination with an auxiliary, self-contained breathing apparatus operated in a pressure-demand or other positive pressure mode). *Escape:* HiEF (Any air-purifying, full-facepiece respirator with a high-efficiency particulate filter); or SCBAE (Any appropriate escape-type, self-contained breathing apparatus).

Storage: Prior to working with this chemical you should be trained on its proper handling and storage. Store in tightly closed containers in a cool, well-ventilated area away from heat and away from contact with easily-oxidized or combustible materials, heat or any condition which could shock it. Sources of ignition such as smoking and open flames, are prohibited where this chemical is used, handled, or stored in a manner that could create a potential fire or explosion hazard.

Shipping: There are no DOT/UN label requirements or quantity limits on air shipment or on shipment in passenger railcar. Falls in Hazard Class ORM-E and packing Group III.[19][20]

Spill Handling: Evacuate persons not wearing protective equipment from area of spill or leak until clean-up is complete. Remove all ignition sources. Absorb liquids in vermiculite, dry sand, earth, or a similar material and deposit in sealed containers. Collect powdered material in the most convenient and safe manner and deposit in sealed containers. Ventilate area after clean-up is complete. It may be necessary to contain and dispose of this chemical as a hazardous waste. If material or contaminated runoff enters waterways, notify downstream users of potentially contaminated waters. Contact your Department of Environmental Protection or your regional office of the federal EPA for specific recommendations. If employees are required to clean-up spills, they must be properly trained and equipped. OSHA 1910.120(q) may be applicable.

Fire Extinguishing: Ammonium Chromate explodes when heated. Poisonous gases including ammonia and nitrogen oxides are produced in fire. Use water only. *Do not* use dry chemical, carbon dioxide, halon or foam extinguishers. Storage containers may explode and parts of containers may rocket great distances, in many directions. If material or contaminated runoff enters waterways, notify downstream users of potentially contaminated waters. Notify local health and fire officials and pollution control agencies. From a secure, explosion-proof location, use water spray to cool exposed containers. If cooling streams are ineffective (venting sound increases in volume and pitch, tank discolors, or shows any signs of deforming), withdraw immediately to a secure position. If employees are expected to fight fires, they must be trained and equipped in OSHA 1910.156.

Disposal Method Suggested: Consult with environmental regulatory agencies for guidance on acceptable disposal practices. Generators of waste containing this contaminant (≥100 kg/mo) must conform with EPA regulations governing storage, transportation, treatment, and waste disposal. Addition of a large volume of reducing agent solution (hypo, bisulfite or ferrous salt, and acidify with 3M sulfuric acid). When reduction is complete, flush to drain with large volumes of water.

References

Sax, N. I., Ed., Dangerous Properties of Industrial Materials Report 2, No. 3, 36–38 (1982)

New Jersey Department of Health and Senior Services, "Hazardous Substance Fact Sheet: Ammonium Chromate," Trenton, NJ (February 1998)

National Institute for Occupational Safety and Health, "Criteria for a Recommended Standard: Occupational Exposure to Chromium (VI), NIOSH Document No. 76–129 (1976)

U.S. Environmental Protection Agency: "Chromium: Ambient Water Quality Criteria," Washington, DC (1980)

Agency for Toxic Substances and Disease Registry, "Toxicological Profile for Chromium," Atlanta, Georgia (1988)

Ammonium Citrate

Molecular Formula: $C_6H_{14}N_2O_7$

Common Formula: $NH_4COOCH_2C(OH)(COOH)CH_2COONH_4$

Synonyms: Ammonium citrate, dibasic; Citrato amonico dibasico (Spanish); Citric acid, ammonium salt; Citric acid, diammonium salt; Diammonium citrate; Diammonium hysrogen citrate; Dibasic ammonium citrate; 1,2,3-Propane tricarboxylic acid, 2-hydroxy-, ammonium salt

CAS Registry Number: 3012-65-5

RTECS Number: GE7573000

DOT ID: NA 99087

Regulatory Authority

- Clean Water Act: 40CFR116.4 Hazardous Substances; RQ 40CFR117.3 (same as CERCLA)
- SUPERFUND/EPCRA 40CFR302.4, Appendix A, Reportable Quantity (RQ): CERCLA, 5,000 lb (2,270 kg); Section 313: Form R *de minimis* concentration reporting level:

1.0% (as ammonia) Molecular weight: 226.19; NH_3 Equivalent weight: 15.06

Cited in U.S. State Regulations: California (G), Massachusetts (G), New Hampshire (G), New Jersey (G), Pennsylvania (G).

Description: Ammonium Citrate, dibasic, $NH_4COOCH_2C(OH)(COOH)CH_2COONH_4$, is a white powdery material with a slight ammonia odor. Hazard Identification (based on NFPA-704 M Rating System): Health 0, Flammability 1, Reactivity 0. Soluble in water.

Potential Exposure: It is used to make pharmaceuticals, rust-proofing compounds and in plasticizers. A food additive.

Incompatibilities: Contact with strong caustics causes the release of ammonia gas.

Permissible Exposure Limits in Air: No standards set.

Routes of Entry: Inhalation, eye and/or skin contact.

Harmful Effects and Symptoms

Short Term Exposure: Exposure can irritate the eyes and may irritate the nose, throat and lungs. Contact may irritate the skin.

Points of Attack: Eyes, lungs, skin.

Medical Surveillance: Lung function tests.

First Aid: If this chemical gets into the eyes, remove any contact lenses at once and irrigate immediately for at least 15 minutes, occasionally lifting upper and lower lids. Seek medical attention immediately. If this chemical contacts the skin, remove contaminated clothing and wash immediately with soap and water. Seek medical attention immediately. If this chemical has been inhaled, remove from exposure, begin rescue breathing (using universal precautions) if breathing has stopped and CPR if heart action has stopped. Transfer promptly to a medical facility. When this chemical has been swallowed, get medical attention. Give large quantities of water and induce vomiting. Do not make an unconscious person vomit.

Personal Protective Methods: Wear protective gloves and clothing to prevent any reasonable probability of skin contact. Safety equipment suppliers/manufacturers can provide recommendations on the most protective glove/clothing material for your operation. All protective clothing (suits, gloves, footwear, headgear) should be clean, available each day, and put on before work. Contact lenses should not be worn when working with this chemical. Wear splash or dust-proof chemical goggles and face shield unless full facepiece respiratory protection is worn. Employees should wash immediately with soap when skin is wet or contaminated. Provide emergency showers and eyewash.

Where possible, enclose operations and use local exhaust ventilation at the site of chemical release. If local exhaust ventilation or enclosure is not used, respirators should be worn. Wear protective work clothing and goggles. Wash thoroughly immediately after exposure to Ammonium Citrate.

Respirator Selection: *At any detectable concentration:* PAPRTHiE (any MSHA/NIOSH approved air-purifying respirator with a tight-fitting full facepiece and a high-efficiency particulate filter); or SCBAF:PD,PP (any MSHA/NIOSH approved self-contained breathing apparatus that has a full facepiece and is operated in a pressure-demand or other positive-pressure mode); or SAF:PD,PP:ASCBA (any supplied-air respirator that has a full facepiece and is operated in a pressure-demand or other positive-pressure mode in combination with an auxiliary, self-contained breathing apparatus operated in a pressure-demand or other positive pressure mode). *Escape:* GMFOVHiE [any air-purifying, full-facepiece respirator (gas mask) with a chin-style, front- or back-mounted organic vapor canister having a high-efficiency particulate filter]; or SCBAE (any appropriate escape-type, self-contained breathing apparatus).

Storage: Prior to working with this chemical you should be trained on its proper handling and storage. Store in tightly closed containers in a cool, well-ventilated area away from caustics (such as sodium hydroxide or potassium hydroxide) because ammonia gas is released.

Shipping: There are no DOT/UN label requirements nor are there size limitations on aircraft shipment. The DOT/UN Hazard Class is ORM-E and the Packing Group is III.[20][21]

Spill Handling: Evacuate persons not wearing protective equipment from area of spill or leak until clean-up is complete. Remove all ignition sources. Collect powdered material in the most convenient and safe manner and deposit in sealed containers. Ventilate area after clean-up is complete. It may be necessary to contain and dispose of this chemical as a hazardous waste. If material or contaminated runoff enters waterways, notify downstream users of potentially contaminated waters. Contact your Department of Environmental Protection or your regional office of the federal EPA for specific recommendations. If employees are required to clean-up spills, they must be properly trained and equipped. OSHA 1910.120(q) may be applicable.

Fire Extinguishing: Ammonium Citrate may burn, but does not readily ignite. Use dry chemical, carbon dioxide, water spray, or alcohol foam extinguishers. Poisonous gases are produced in fire including ammonia gas and nitrogen oxides. If material or contaminated runoff enters waterways, notify downstream users of potentially contaminated waters. Notify local health and fire officials and pollution control agencies. From a secure, explosion-proof location, use water spray to cool exposed containers. If cooling streams are ineffective (venting sound increases in volume and pitch, tank discolors, or shows any signs of deforming), withdraw immediately to a secure position. If employees are expected to fight fires, they must be trained and equipped in OSHA 1910.156.

Disposal Method Suggested: May be flushed to sewer with large volumes of water.

References

New Jersey Department of Health and Senior Services, "Hazardous Substance Fact Sheet: Ammonium Citrate," Trenton, NJ (February 1987)

Ammonium Dichromate

Molecular Formula: $Cr_2H_8N_2O_7$

Common Formula: $(NH_4)_2Cr_2O_7$

Synonyms: Ammonium bichromate; Ammonium dichromate (VI); Bicromato amonico (Spanish); Chromic acid, diammonium salt; Diammonium dichromate

CAS Registry Number: 7789-09-5

RTECS Number: HX7650000

DOT ID: UN 1439

EEC Number: 024-003-00-1

Regulatory Authority

- Carcinogen: (California's Proposition 65) [as Chromium(VI)]
- Air Pollutant Standard Set (NIOSH)[2] (ACGIH)[1] (former USSR)[43] (Australia) (Various States), (Israel) (Various Canadian Provinces)
- Clean Air Act 42USC7412; Title I, Part A, §112 hazardous pollutants (as chromium compounds)
- Clean Water Act: 40CFR116.4 Hazardous Substances; RQ 40CFR117.3 (same as CERCLA); 40CFR423, Appendix A, Priority Pollutants
- RCRA, 40CFR261, Appendix 8 Hazardous Constituents., waste number not listed (chromium compounds)
- SUPERFUND/EPCRA 40CFR302.4, Appendix A, Reportable Quantity (RQ): CERCLA, 10 lb (4.54 kg)
- EPCRA Section 313 Form R *de minimis* concentration reporting level: 1.0% (as ammonia) Molecular weight: 252.06; NH_3. Equivalent weight: 13.51. *Also must be reported as a chromium compound:* "Includes any unique chemical substances that contains chromium as part of that chemical's infrastructure." Form R *de minimis* concentration reporting level: Chromium (VI) compounds: 0.1%. Form R Toxic Chemical Category Code: N090
- Canada, WHMIS, Ingredients Disclosure List

Cited in U.S. State Regulations: Florida (G), Massachusetts (G), New Hampshire (G), New Jersey (G), Oklahoma (G), Pennsylvania (G), Rhode Island (G).

Description: Ammonium dichromate, $(NH_4)_2Cr_2O_7$, is a combustible, orange-red crystalline solid which is used in solution. Freezing/Melting point = 170°C (see Incompatibilities). Autoignition temperature = 225°C. Hazard Identification (based on NFPA-704 M Rating System): Health 2, Flammability 1, Reactivity 1 (oxidizer). Soluble in water.

Potential Exposure: It is used in dyeing, leather tanning and to make fireworks and *chromic oxide*.

Incompatibilities: An unstable oxidizer. Decomposes at about 185°C; decomposition becomes self-sustaining and violent at about 225°C. Contact with combustible, organic or other easily oxidized materials, strong acids, hydrazine and other reducing agents, alcohols, sodium nitrite may cause fire and explosions.

Permissible Exposure Limits in Air: OSHA:[2] The legal airborne permissible exposure limit (PEL) is 0.1 mg/m^3, not to be exceeded at any time. NIOSH:[2] The recommended airborne exposure limit is 0.001 mg/m^3 averaged over a 10-hour workshift and 0.050 mg/m^3, not to be exceeded during any 15 minute work period. ACGIH:[1] The recommended airborne exposure limit is 0.05 mg/m^3 averaged over an 8-hour workshift. The above exposure limits are for air levels only. When skin contact also occurs, one may be overexposed, even though air levels are less than the limits listed above. The former USSR-UNEP/IRPTC project[43] has set a MAK value of 0.01 mg/m^3 for chromates and bichromates in workplace air. California Prop. 65 No significant risk level (inhalation) = 0.001 μg/day.

Determination in Air: Hexavalent chromium may be determined by filtration followed by visible absorption spectrophotometry according to NIOSH Method 7600.[18] Also Method #7604.

Permissible Concentration in Water: To protect human health, hexavalent chromium should be held below 50 μg/l according to EPA[6] in studies on priority toxic pollutants. This is also a WHO recommendation for total chromium in drinking water.

Determination in Water: Chromium (VI) may be determined by extraction and atomic absorption or colorimetry using diphenylhydrazole.

Routes of Entry: Ingestion, skin and/or eye contact.

Harmful Effects and Symptoms

Short Term Exposure: Should be handled as a carcinogen, with extreme caution. Ammonium dichromate can pass through the skin and may cause overexposure. Eye contact can cause severe damage with possible loss of vision. Irritation of nose, throat and bronchial tubes can also occur, with cough and/or wheezing. Skin contact can cause deep ulcers or an allergic skin rash.

Long Term Exposure: Ammonium dichromate is a hexavalent chromium compound which may be carcinogenic and should be handled with extreme caution. Breathing ammonium chromate can cause sores or hole in the septum dividing the inner nose, sometimes with bleeding, discharge, and/or formation of a crust. May cause skin allergy and kidney damage.

Points of Attack: Blood, respiratory system, liver kidneys, eyes, skin.

Medical Surveillance: Skin and nose examination, kidney function tests, evaluation by a qualified allergist.

First Aid: If this chemical gets into the eyes, remove any contact lenses at once and irrigate immediately for at least 15 minutes, occasionally lifting upper and lower lids. Seek medical attention immediately. If this chemical contacts the skin, remove contaminated clothing and wash immediately with

soap and water. Seek medical attention immediately. If this chemical has been inhaled, remove from exposure, begin rescue breathing (using universal precautions) if breathing has stopped and CPR if heart action has stopped. Transfer promptly to a medical facility. When this chemical has been swallowed, get medical attention. Give large quantities of water and induce vomiting. Do not make an unconscious person vomit.

Personal Protective Methods: Wear protective gloves and clothing to prevent any reasonable probability of skin contact. Safety equipment suppliers/manufacturers can provide recommendations on the most protective glove/clothing material for your operation. All protective clothing (suits, gloves, footwear, headgear) should be clean, available each day, and put on before work. Contact lenses should not be worn when working with this chemical. Wear splash or dust-proof chemical goggles and face shield unless full facepiece respiratory protection is worn. Employees should wash immediately with soap when skin is wet or contaminated. Provide emergency showers and eyewash.

Respirator Selection: NIOSH: *At any detectable concentration above 0.001* mg/m^3*:* SCBAF:PD,PP (any MSHA/NIOSH approved self-contained breathing apparatus that has a full facepiece and is operated in a pressure-demand or other positive-pressure mode); or SAF:PD,PP:ASCBA (any supplied-air respirator that has a full facepiece and is operated in a pressure-demand or other positive-pressure mode in combination with an auxiliary, self-contained breathing apparatus operated in a pressure-demand or other positive pressure mode). *Escape:* HiEF (Any air-purifying, full-facepiece respirator with a high-efficiency particulate filter); or SCBAE (Any appropriate escape-type, self-contained breathing apparatus).

Storage: Prior to working with this chemical you should be trained on its proper handling and storage. Store in tightly closed containers in a cool, well-ventilated area away from heat and away from contact with easily-oxidized or combustible materials, heat or any condition which could shock it. Sources of ignition such as smoking and open flames, are prohibited where this chemical is used, handled, or stored in a manner that could create a potential fire or explosion hazard. See OSHA standard 1910.104 and NFPA 43A *Code for the Storage of Liquid and Solid Oxidizers* for detailed handling and storage regulations.

Shipping: UN/DOT label required is "Oxidizer." The DOT/UN Hazard Class is 5.1 and the Packing Group is II.[20][21]

Spill Handling: Restrict persons not wearing protective equipment from area of spill or leak until clean-up is complete. Remove all ignition sources. Ventilate area of spill or leak. Absorb liquids in vermiculite, dry sand, earth, or a similar material and deposit in sealed containers. Collect powdered material in the most convenient and safe manner and deposit in sealed containers.

Fire Extinguishing: Ammonium Chromate is a combustible solid and explodes when heated. Poisonous gases including ammonia and nitrogen oxides are produced in fire. Use water only. *Do not* use dry chemical, carbon dioxide, halon or foam extinguishers. Storage containers may explode and parts of containers may rocket great distances, in many directions. If material or contaminated runoff enters waterways, notify downstream users of potentially contaminated waters. Notify local health and fire officials and pollution control agencies. From a secure, explosion-proof location, use water spray to cool exposed containers. If cooling streams are ineffective (venting sound increases in volume and pitch, tank discolors, or shows any signs of deforming), withdraw immediately to a secure position. If employees are expected to fight fires, they must be trained and equipped in OSHA 1910.156.

Disposal Method Suggested: Consult with environmental regulatory agencies for guidance on acceptable disposal practices. Generators of waste containing this contaminant (≥100 kg/mo) must conform with EPA regulations governing storage, transportation, treatment, and waste disposal. Add a large volume of reductant solution (hypo, bisulfite or ferrous salt and acidify with sulfuric acid). Neutralize when reduction is complete and flush to sewer with large volume of water.

References

New Jersey Department of Health and Senior Services, "Hazardous Substance Fact Sheet: Ammonium Dichromate," Trenton, NJ (April 1998)

National Institute for Occupational Safety and Health, "Criteria for a Recommended Standard: Occupational Exposure to Chromium(VI), NIOSH Document No. 76–129 (1976)

U.S. Environmental Protection Agency, "Chromium: Ambient Water Quality Criteria," Washington, DC (1980)

Sax, N. I., Ed., "Dangerous Properties of Industrial Materials Report," 2, No. 3, 38–40 (1982) (as Ammonium Dichromate) and 3, No. 5, 29–32 (as Ammonium Bichromate)

Agency for Toxic Substances and Disease Registry, "Toxicological Profile for Chromium," Atlanta, Georgia (1988)

Ammonium Fluoride

Molecular Formula: FH_4N

Common Formula: NH_4F

Synonyms: 777 etch; Ammonium fluorure (French); B-etch; BOE (buffered oxide etch); Fluouro amonico (Spanish); Glass etch; Imahe etch; KTI buffered oxide etch 50:1; KTI buffered oxide etch 6:1; Neutral ammonium fluoride; Pad etch; Poly silicon etch

CAS Registry Number: 12125-01-8

RTECS Number: BQ6300000

DOT ID: UN 2505

EEC Number: 009-006-00-8

Regulatory Authority

- Air Pollutant Standard Set (OSHA)[2] (NIOSH)[2] (ACGIH)[1] (HSE)[33] (DFG)[3] (former USSR)[43]

- Clean Water Act: 40CFR116.4 Hazardous Substances; RQ 40CFR117.3 (same as CERCLA); 40CFR423, Appendix A, Priority Pollutants (as inorganic fluorides)
- RCRA, 40CFR264, Appendix 9, Ground Water Monitoring List, Suggested Testing Methods (PQL μg/l) 8100 (200): 8270 (10) (as inorganic fluorides)
- RCRA Universal Treatment Standards: Wastewater (mg/l), 0.059; Nonwastewater (mg/kg), 3.4 (as inorganic fluorides)
- SUPERFUND/EPCRA 40CFR302.4, Appendix A, Reportable Quantity (RQ): CERCLA, 100 lb (45.4 kg); Section 313: Form R *de minimis* concentration reporting level: 1.0% (as ammonia) Molecular weight: 37.04; NH_3 Equivalent weight: 45.98
- Canada, WHMIS, Ingredients Disclosure List (as fluorides)
- Mexico, Wastewater (inorganic fluorides)

Cited in U.S. State Regulations: Florida (G), Maine (G), Massachusetts (G), New Hampshire (G), New Jersey (G), Oklahoma (G), Pennsylvania (G), Rhode Island (G).

Description: Ammonium fluoride, NH_4F, is a white crystalline solid. Freezing/Melting point = sublimes. Soluble in water.

Potential Exposure: Ammonium Fluoride is used in printing and dyeing textiles, glass etching, moth-proofing and wood preserving.

Incompatibilities: Acids, alkalis, chlorine trifluoride. Corrodes glass, cement, most metals.

Permissible Exposure Limits in Air: OSHA: The legal airborne permissible exposure limit (PEL) is 2.5 mg/m^3 for fluorides measured as fluorine averaged over an 8-hour workshift. This is also the HSE value in the U.K.[33] and the MAC value in Germany.[3] NIOSH: The recommended airborne exposure limit is 2.5 mg/m^3 for fluorides, inorganic measured as fluorine averaged over a 10-hour workshift. ACGIH: The recommended airborne exposure limit is 2.5 mg/m^3 for fluorides measured as fluorine averaged over an 8-hour workshift. The former USSR-UNEP/IRPTC project[43] has not set a MAC in workplace air but readily soluble fluorides have assigned a MAC in residential air of 0.03 mg/m^3 on a momentary basis and 0.01/mg/m^3 on an average daily basis. An IDLH value has been given as 500 mg/m^3.[41]

Determination in Air: Fluoride aerosols may be measured by collection on a filter followed by fluoride ion measurement using an ion-specific electrode. See NIOSH Method 7902.

Permissible Concentration in Water: No criteria set.

Routes of Entry: Inhalation, skin and /or eye contact, ingestion.

Harmful Effects and Symptoms

Short Term Exposure: Exposure can irritate the nose, throat and lungs. Contact can irritate or burn the skin and eyes. Higher exposures could cause pulmonary edema, a medical emergency that can be delayed for several hours. This can cause death.

Long Term Exposure: With repeated exposure, some persons may notice poor appetite, nausea, constipation or diarrhea. Repeated overexposure to fluoride can cause brittle bones, stiff muscles and joints and eventual crippling. This usually takes years to develop. High or repeated exposure may scar the lungs, causing shortness of breath and reduced lung function. These effects do not occur with prescribed fluoride levels in drinking water or dental use to prevent cavities.

Points of Attack: Skin, eyes, respiratory system, bones, muscles.

Medical Surveillance: For those with frequent or potentially high exposure (half the TLV or greater), the following are recommended before beginning work and at regular times after that. Lung function tests. If symptoms develop or overexposure is suspected, the following may be useful: Consider chest x-ray after acute overexposure. Urine fluoride level (should be less than 5 mg/l of urine).

First Aid: If this chemical gets into the eyes, remove any contact lenses at once and irrigate immediately for at least 15 minutes, occasionally lifting upper and lower lids. Seek medical attention immediately. If this chemical contacts the skin, remove contaminated clothing and wash immediately with soap and water. Seek medical attention immediately. If this chemical has been inhaled, remove from exposure, begin rescue breathing (using universal precautions) if breathing has stopped and CPR if heart action has stopped. Transfer promptly to a medical facility. When this chemical has been swallowed, get medical attention. Give large quantities of water and induce vomiting. Do not make an unconscious person vomit. Medical observation is recommended for 24 – 48 hours after breathing overexposure, as pulmonary edema may be delayed. As first aid for pulmonary edema, a doctor or authorized paramedic may consider administering a corticosteroid spray.

Personal Protective Methods: Wear protective gloves and clothing to prevent any reasonable probability of skin contact. Safety equipment suppliers/manufacturers can provide recommendations on the most protective glove/clothing material for your operation. All protective clothing (suits, gloves, footwear, headgear) should be clean, available each day, and put on before work. Contact lenses should not be worn when working with this chemical. Wear dust-proof chemical goggles and face shield unless full facepiece respiratory protection is worn. Employees should wash immediately with soap when skin is wet or contaminated. Provide emergency showers and eyewash.

Respirator Selection: NIOSH: *(fluorides) 12.5* mg/m^3: DM (any dust and mist respirator). *25 mg/m^3*: DMXSQ* (any dust and mist respirator except single-use and quarter mask respirators); or SA* (any supplied-air respirator). *62.5 mg/m^3*: SA:CF* (any supplied-air respirator operated in a continuous-flow mode); or PAPRDM* ** *if not present as a fume* (any powered, air-purifying respirator with a dust and mist filter).

125 mg/m³: HiEF** (any air-purifying, full-facepiece respirator with a high-efficiency particulate filter); or SCBAF (any self-contained breathing apparatus with a full facepiece); or SAF (any supplied-air respirator with a full facepiece). 250 mg/m³: SA:PD,PP (any supplied-air respirator operated in a pressure-demand or other positive-pressure mode). *Emergency or planned entry into unknown concentrations or IDLH conditions*: SCBAF:PD,PP (any self-contained breathing apparatus that has a full faceplate and is operated in a pressure-demand or other positive-pressure mode); or SAF:PD,PP: ASCBA (any supplied-air respirator that has a full facepiece and is operated in a pressure-demand or other positive-pressure mode in combination with an auxiliary, self-contained breathing apparatus operated in a pressure-demand or other positive-pressure mode). *Escape:* HiEF** (any air-purifying, full-facepiece respirator with a high-efficiency particulate filter); or SCBAE (any appropriate escape-type, self-contained breathing apparatus).

* Substance reported to cause eye irritation or damage; may require eye protection.

** May need acid gas sorbent. Exposure to 500 mg/m³ (IDLH) is immediately dangerous to life and health.

Storage: Prior to working with this chemical you should be trained on its proper handling and storage. Ammonium Fluoride must be stored to avoid contact with acids or alkalis since violent reactions occur. Store in tightly closed containers in a cool, well-ventilated area. Keep in plastic, rubber or paraffined containers (because it easily etches glass).

Shipping: Ammonium fluoride requires a "Keep away from Food" label. It falls in Hazard Class 6.1 and Packing Group III.

Spill Handling: Evacuate persons not wearing protective equipment from area of spill or leak until clean-up is complete. Remove all ignition sources. Absorb liquids in vermiculite, dry sand, earth, or a similar material and deposit in sealed containers. Collect powdered material in the most convenient and safe manner and deposit in sealed containers. Ventilate area after clean-up is complete. It may be necessary to contain and dispose of this chemical as a hazardous waste. If material or contaminated runoff enters waterways, notify downstream users of potentially contaminated waters. Contact your Department of Environmental Protection or your regional office of the federal EPA for specific recommendations. If employees are required to clean-up spills, they must be properly trained and equipped. OSHA 1910.120(q) may be applicable.

Fire Extinguishing: Ammonium fluoride may burn, but does not readily ignite. Use dry chemical, carbon dioxide, water spray, or alcohol foam extinguishers. Poisonous gases are produced in fire including hydrogen fluoride gas. If material or contaminated runoff enters waterways, notify downstream users of potentially contaminated waters. Notify local health and fire officials and pollution control agencies. From a secure, explosion-proof location, use water spray to cool exposed containers. If cooling streams are ineffective (venting sound increases in volume and pitch, tank discolors, or shows any signs of deforming), withdraw immediately to a secure position. If employees are expected to fight fires, they must be trained and equipped in OSHA 1910.156.

Disposal Method Suggested: Consult with environmental regulatory agencies for guidance on acceptable disposal practices. Generators of waste containing this contaminant (≥100 kg/mo) must conform with EPA regulations governing storage, transportation, treatment, and waste disposal.

References

New Jersey Department of Health and Senior Services, "Hazardous Substance Fact Sheet: Ammonium Fluoride," Trenton, NJ (April 1996)

Sax, N. I., Ed., "Dangerous Properties of Industrial Materials Report," 3, No. 5, 32–34 (1983)

Ammonium Hexafluorosilicate

Molecular Formula: $F_6H_8N_2Si$

Common Formula: $(NH_4)_2SiF_6$

Synonyms: Ammonium fluorosilicate; Ammonium hexafluorosilicate; Ammonium silicon fluoride; Diammonium fluosilicate; Diammonium silicon hexafluoride; Fluosilicate de ammonium (French); Fluosilicato amonico (Spanish); Picrato amonico (Spanish); Silicofluoruro amonico (Spanish)

CAS Registry Number: 16919-19-0; 1309-32-6

RTECS Number: GQ9450000

DOT ID: UN 2854

Regulatory Authority

- Air Pollutant Standard Set (NIOSH)[2] (ACGIH)[1] (HSE)[33] (DFG)[3] (former USSR)[43]
- Water Pollution Standard Proposed (former USSR)[43]
- Clean Water Act: 40CFR116.4 Hazardous Substances; RQ 40CFR117.3 (same as CERCLA); 40CFR423, Priority Pollutants (as inorganic fluorides)
- RCRA Universal Treatment Standards: Wastewater (mg/l), 0.059; Nonwastewater (mg/kg), 3.4 (as inorganic fluorides)
- RCRA, 40CFR264, Appendix 9, Ground Water Monitoring List, Suggested Testing Methods (PQL μg/l): 8100 (200); 8270 (10) (as inorganic fluorides)
- SUPERFUND/EPCRA 40CFR302.4, Appendix A, Reportable Quantity (RQ): CERCLA, 1,000 lb (454 kg); Section 313: Form R *de minimis* concentration reporting level: 1.0% (as ammonia) Molecular weight: 178.18; NH_3 Equivalent weight: 19.12
- Canada, WHMIS, Ingredients Disclosure List (as fluorides)
- Mexico, Wastewater (inorganic fluorides)

Cited in U.S. State Regulations: California (G), Massachusetts (G), New Hampshire (G), New Jersey (G), New York (G), Pennsylvania (G).

Description: Ammonium Hexafluorosilicate, $(NH_4)_2SiF_6$, is a white crystalline powder. Odorless. Boiling point = decomposes. Sinks and mixes with water.

Potential Exposure: This material is used as a pesticide and miticide, wood preservative, soldering flux, light metal casting, and in the etching of glass.

Incompatibilities: Liquid is corrosive. Contact with acids reacts to form hydrogen fluoride, which is a highly corrosive and toxic gas. Corrosive to aluminum. Keep away from strong oxidizers.

Permissible Exposure Limits in Air: OSHA: The legal airborne permissible exposure limit (PEL) is 2.5 mg/m^3 for fluorides (measured as fluorine) averaged over an 8-hour workshift. This is also the HSE value in the U.K.[33] and the MAC value in Germany.[3] NIOSH: The recommended airborne exposure limit is 2.5 mg/m^3 for fluorides, inorganic(measured as fluorine) averaged over a 10-hour workshift. ACGIH: The recommended airborne exposure limit is 2.5 mg/m^3 for fluorides (measured as fluorine) averaged over an 8-hour workshift. The former USSR-UNEP/IRPTC project[43] has not set a MAC in workplace air but readily soluble fluorides have assigned a MAC in residential air of 0.03 mg/m^3 on a momentary basis and 0.01/mg/m^3 on an average daily basis. An IDLH value has been given as 500 mg/m^3.[41]

Determination in Air: Fluorides may be measured by collection on a filter and measurement by ion-specific electrode according to NIOSH Method 7902.[18]

Permissible Concentration in Water: No criteria set for ammonium fluorosolicate as such. The former USSR-UNEP/IRPTC project[43] has set a limit of 1.5 mg/l for fluorine in water used for domestic purposes.

Routes of Entry: Inhalation, eyes and/or skin contact.

Harmful Effects and Symptoms

Short Term Exposure: Inhalation may cause difficult breathing and burning of the mouth, throat and nose which may result in bleeding. These may be felt at 7.5 mg/m^3. Nausea, vomiting, profuse sweating and excess thirst may occur at higher levels. May cause pulmonary edema, which can be delayed for several hours; there is a risk of death in serious cases. Skin contact may cause rash, itching and burning and ulceration of skin. Solutions of 1% strength may cause sores if not removed promptly. Eye contact may cause severe irritation. Most reported instances of fluoride toxicity are due to accidental ingestion and it is difficult to associate symptoms with dose. A dose of 5 – 40 mg may cause nausea, diarrhea and vomiting. More severe symptoms of burning and painful abdomen, sores in mouth, throat and digestive tract, tremors, convulsions and shock will occur around a dose of 1 gram. Death may result by ingestion of 2 – 5 grams. Also reported as 1 teaspoon to 1 ounce.

Long Term Exposure: May cause chronic lung irritation and kidney and liver damage. Bronchitis may develop. Chronic exposure may cause weight loss, nausea, vomiting, weakness, shortness of breath. Fluoride may increase bone density, stimulate new bone growth or cause calcium deposits in ligaments. This may become a problem at levels of 20 – 50 mg/m^3 or higher. May cause mottling of teeth at this level.

Points of Attack: Lungs, eyes.

Medical Surveillance: Pre-employment and periodic examinations should consider possible effects on the skin, eyes, teeth, respiratory tract, and kidneys. Chest x-ray and pulmonary function should be followed. Kidney function should be evaluated. If exposures have been heavy and skeletal fluorosis is suspected, pelvic x-rays may be helpful. Intake of fluoride from natural sources in food or water should be known. In the case of exposure to fluoride dusts, periodic urinary fluoride excretion levels have been very useful in evaluating industrial exposures and environmental dietary sources.

First Aid: If this chemical gets into the eyes, remove any contact lenses at once and irrigate immediately for at least 15 minutes, occasionally lifting upper and lower lids. Seek medical attention immediately. If this chemical contacts the skin, remove contaminated clothing and wash immediately with soap and water. Seek medical attention immediately. If this chemical has been inhaled, remove from exposure, begin rescue breathing (using universal precautions) if breathing has stopped and CPR if heart action has stopped. Transfer promptly to a medical facility. When this chemical has been swallowed, get medical attention. Give large quantities of water and induce vomiting. Do not make an unconscious person vomit. Medical observation is recommended for 24 – 48 hours after breathing overexposure, as pulmonary edema may be delayed. As first aid for pulmonary edema, a physician or authorized paramedic may consider administering a corticosteroid spray. *Note to Physician:* Ingestion: Give aluminum hydroxide gel, if conscious. Inject intravenously 10 ml of 10% calcium gluconate solution. Gastric lavage with lime water of 1% calcium chloride.

Personal Protective Methods: Use only with an effective and properly maintained exhaust ventilation or with a fully enclosed process. Wear protective gloves and clothing to prevent any reasonable probability of skin contact. Safety equipment suppliers/manufacturers can provide recommendations on the most protective glove/clothing material for your operation. All protective clothing (suits, gloves, footwear, headgear) should be clean, available each day, and put on before work. Contact lenses should not be worn when working with this chemical. Wear splash or dust-proof chemical goggles and face shield unless full facepiece respiratory protection is worn. Employees should wash immediately with soap when skin is wet or contaminated. Provide emergency showers and eyewash.

Respirator Selection: NIOSH: *(fluorides) 12.5 mg/m^3:* DM (any dust and mist respirator). *25 mg/m^3*: DMXSQ* (any dust and mist respirator except single-use and quarter mask respirators); or SA* (any supplied-air respirator). *62.5 mg/m^3:*

SA:CF* (any supplied-air respirator operated in a continuous-flow mode); or PAPRDM*** *if not present as a fume* (any powered, air-purifying respirator with a dust and mist filter). *125 mg/m³*: HiEF** (any air-purifying, full-facepiece respirator with a high-efficiency particulate filter); or SCBAF (any self-contained breathing apparatus with a full facepiece); or SAF (any supplied-air respirator with a full facepiece). *250 mg/m³:* SA:PD,PP (any supplied-air respirator operated in a pressure-demand or other positive-pressure mode). *Emergency or planned entry into unknown concentrations or IDLH conditions*: SCBAF:PD,PP (any self-contained breathing apparatus that has a full faceplate and is operated in a pressure-demand or other positive-pressure mode); or SAF:PD,PP:ASCBA (any supplied-air respirator that has a full facepiece and is operated in a pressure-demand or other positive-pressure mode in combination with an auxiliary, self-contained breathing apparatus operated in a pressure-demand or other positive-pressure mode). *Escape:* HiEF** (any air-purifying, full-facepiece respirator with a high-efficiency particulate filter); or SCBAE (any appropriate escape-type, self-contained breathing apparatus).

* Substance reported to cause eye irritation or damage; may require eye protection.

** May need acid gas sorbent. Exposure to 500 mg/m³ (IDLH) is immediately dangerous to life and health.

Storage: Prior to working with this chemical you should be trained on its proper handling and storage. Store in tightly closed containers in a cool, well-ventilated area away from strong oxidizers (such as chlorine, bromine and fluorine) since highly poisonous hydrogen fluoride gas is released. Keep in plastic, rubber or paraffined containers (because it easily etches glass).

Shipping: DOT/UN label requirement is "Keep Away From Food." The DOT/UN Hazard Class is 6.1 and the Packing Group is III.[19][20]

Spill Handling: Evacuate persons not wearing protective equipment from area of spill or leak until clean-up is complete. Remove all ignition sources. Treat with soda ash or slaked lime. Use an industrial vacuum cleaner to remove the spill. Absorb liquids in vermiculite, dry sand, earth, or a similar material and deposit in sealed containers. Collect powdered material in the most convenient and safe manner and deposit in sealed containers. Ventilate area after clean-up is complete. It may be necessary to contain and dispose of this chemical as a hazardous waste. If material or contaminated runoff enters waterways, notify downstream users of potentially contaminated waters. Contact your Department of Environmental Protection or your regional office of the federal EPA for specific recommendations. If employees are required to clean-up spills, they must be properly trained and equipped. OSHA 1910.120(q) may be applicable.

Fire Extinguishing: This chemical may burn, but does not readily ignite. Use dry chemical, carbon dioxide, water spray, or alcohol foam extinguishers. Poisonous gases are produced in fire including fluorine, ammonia and nitrogen oxides. If material or contaminated runoff enters waterways, notify downstream users of potentially contaminated waters. Notify local health and fire officials and pollution control agencies. From a secure, explosion-proof location, use water spray to cool exposed containers. If cooling streams are ineffective (venting sound increases in volume and pitch, tank discolors, or shows any signs of deforming), withdraw immediately to a secure position. If employees are expected to fight fires, they must be trained and equipped in OSHA 1910.156. Water may be used but should be contained to prevent fluoride run-off.

Disposal Method Suggested: Consult with environmental regulatory agencies for guidance on acceptable disposal practices. Generators of waste containing this contaminant (≥100 kg/mo) must conform with EPA regulations governing storage, transportation, treatment, and waste disposal. Incineration.

References

National Institute for Occupational Safety and Health, "Criteria for a Recommended Standard: Occupational Exposure to Inorganic Fluorides," NIOSH Doc. No. 76–103 (1976)

Sax, N. I., Ed., "Dangerous Properties of Industrial Materials Report," 4, No. 3, 36–38 (1984). (Al Ammonium Silicofluoride)

New York State Department of Health, "Chemical Fact Sheet: Ammonium Hexafluorosilicate," Albany, NY, Bureau of Toxic Substance Assessment (March 1986)

Ammonium Hydroxide

Molecular Formula: H_5NO

Common Formula: NH_4OH

Synonyms: Ammonia aqueous; Ammonia water; Aqua ammonia; Aqueous ammonia; Burmar lab clean; Enplate NI-418B; Enstrip NP-1; Hidroxido amonico (Spanish); Household ammonia; Poly silicon etch; PPD 5932 developer; Pre-metal etch; RCA clean (step 1); Scan kleen; Scrubber-Vapox

CAS Registry Number: 1336-31-6

RTECS Number: BQ9625000

DOT ID: NA 2672

EEC Number: 007-001-01-2

Regulatory Authority

- Clean Water Act: 40CFR116.4 Hazardous Substances; RQ 40CFR117.3 (same as CERCLA)
- SUPERFUND/EPCRA 40CFR302.4, Appendix A, Reportable Quantity (RQ): CERCLA, 1,000 lb (454 kg); Section 313: Form R *de minimis* concentration reporting level: 1.0% (as ammonia) Molecular weight: 35.05; NH_3 Equivalent weight: 48.59
- Canada, WHMIS, Ingredients Disclosure List

Cited in U.S. State Regulations: California (G), Illinois (G), Massachusetts (G), New Hampshire (G), New Jersey (G), Oklahoma (G), Pennsylvania (G).

Description: Ammonium Hydroxide, NH_4OH, is a colorless to milky-white solution of ammonia, with a strong, irritating odor. Freezing/Melting point = -77°C. Vapor pressure = 11.9 mm Hg @ 20°C. The odor threshold for ammonia is 0.043 – 47 ppm. Soluble in water.

Potential Exposure: It is used in detergents, stain removers, bleaches, dyes, fibers, and resins.

Incompatibilities: Solution is a strongly alkaline. Violent reaction with strong oxidizers, acids (exothermic reaction with strong mineral acids). Shock-sensitive compounds may be formed with halogens, mercury oxide, silver oxide. Fire and explosions may be caused by contact with β-propiolactone, silver nitrate, ethyl alcohol, silver permanganate, trimethylammonium amide, 1-chloro-2,4-dinitrobenzene, o-chloronitrobenzene, platinum, trioxygen difluoride, selenium difluoride dioxide, boron halides, mercury, chlorine, iodine, bromine, hypochlorites, chlorine bleach, amides, organic anhydrides, isocyanates, vinyl acetate, alkylene oxides, epichlorohydrin, aldehydes. Attacks some coatings, plastics and rubber. Attacks copper, brass, bronze, aluminum, steel, zinc, and their alloys.

Permissible Exposure Limits in Air: The pertinent limits are those for ammonia. See the entry for ammonia.

Determination in Air: Sampling by absorption in sulfuric acid followed by measurement by ion chromatography, conductivity. See NIOSH Method #6015, #6016.

Permissible Concentration in Water: The former USSR-UNEP/IRPTC project[43] has set a MAC of 2.0 mg/ml in water bodies used for domestic purposes and 0.05 mg/ml in water bodies used for fishery purposes.

Routes of Entry: Ingestion, skin and/or eye contact.

Harmful Effects and Symptoms

Short Term Exposure: Ammonium hydroxide is a corrosive chemical and can severely burn the skin and eyes causing permanent damage. Exposure can severely irritate the nose, throat and lungs. Inhalation may cause pulmonary edema, which can be delayed for several hours; there is a risk of death in serious cases.

Long Term Exposure: Long-term exposure at low levels may cause chronic bronchitis. Repeated skin contact can cause dermatitis, dryness, itching, and redness.

Points of Attack: Lungs, skin, eyes.

Medical Surveillance: Lung function tests. Consider chest x-ray following acute exposure.

First Aid: If this chemical gets into the eyes, remove any contact lenses at once and irrigate immediately for at least 15 minutes, occasionally lifting upper and lower lids. Seek medical attention immediately. If this chemical contacts the skin, remove contaminated clothing and wash immediately with soap and water. Seek medical attention immediately. If this chemical has been inhaled, remove from exposure, begin rescue breathing (using universal precautions) if breathing has stopped and CPR if heart action has stopped. Transfer promptly to a medical facility. When this chemical has been swallowed, get medical attention. If victim is conscious, administer water or milk. Do not induce vomiting. Medical observation is recommended for 24 – 48 hours after breathing overexposure, as pulmonary edema may be delayed. As first aid for pulmonary edema, a doctor or authorized paramedic may consider administering a corticosteroid spray.

Personal Protective Methods: Wear protective gloves and clothing to prevent any reasonable probability of skin contact. Safety equipment suppliers/manufacturers can provide recommendations on the most protective glove/clothing material for your operation. All protective clothing (suits, gloves, footwear, headgear) should be clean, available each day, and put on before work. Contact lenses should not be worn when working with this chemical. Wear splash-proof chemical goggles and face shield unless full facepiece respiratory protection is worn. Employees should wash immediately with soap when skin is wet or contaminated. Provide emergency showers and eyewash. Specific engineering controls are recommended for ammonia by NIOSH in criteria document #74-136 (Ammonia).

Respirator Selection: NIOSH (as ammonia): *250 ppm*: CCRS (any chemical cartridge respirator with cartridge(s) providing protection against the compound of concern); or SA (any supplied-air respirator). *300 ppm*: SA:CF (any supplied-air respirator operated in a continuous-flow mode); or PAPRS (any powered, air-purifying respirator with cartridge(s) providing protection against the compound of concern); or CCRFS (any chemical cartridge respirator with a full facepiece and cartridge(s) providing protection against the compound of concern); or GMFS [any air-purifying, full-facepiece respirator (gas mask) with a chin-style, front-or back-mounted canister providing protection against the compound of concern]; or SCBAF (any self-contained breathing apparatus with a full facepiece); or SAF (any supplied-air respirator with a full facepiece). *Emergency or planned entry into unknown concentrations or IDLH conditions*: SCBAF:PD,PP (any self-contained breathing apparatus that has a full facepiece and is operated in a pressure-demand or other positive-pressure mode); or SAF:PD,PP:ASCBA (any supplied-air respirator that has a full facepiece and is operated in a pressure-demand or other positive-pressure mode in combination with an auxiliary, self-contained breathing apparatus operated in a pressure-demand or other positive-pressure mode). *Escape:* GMFS [any air-purifying, full-facepiece respirator (gas mask) with a chin-style, front-or back-mounted canister providing protection against the compound of concern]; or SCBAE (any appropriate escape-type, self-contained breathing apparatus).

Storage: Prior to working with this chemical you should be trained on its proper handling and storage. Store in temperatures below 25°C (77°F). Do not fill bottles completely. Store in tightly-closed, strong glass, plastic or rubber-stoppered containers in a cool, well ventilated area. Where possible, automatically pump liquid from drums or other storage containers to process containers. Sources of ignition such as smoking and open flames, are prohibited where this chemical is used, handled, or stored in a manner that could create a potential fire or explosion hazard.

Shipping: See the entry on ammonia for more details.

Spill Handling: Evacuate persons not wearing protective equipment from area of spill or leak until clean-up is complete. Remove all ignition sources. Ventilate closed spaces before entering them. Ventilate area of spill or leak. Absorb liquids in vermiculite, dry sand, earth, or a similar material and deposit in sealed containers. It may be necessary to contain and dispose of this chemical as a hazardous waste. If material or contaminated runoff enters waterways, notify downstream users of potentially contaminated waters. Contact your Department of Environmental Protection or your regional office of the federal EPA for specific recommendations. If employees are required to clean-up spills, they must be properly trained and equipped. OSHA 1910.120(q) may be applicable. If professional environmental engineering assistance is required, contact the US EPA Environmental Response Team at (908) 548-8730 (24-hour response line).

Fire Extinguishing: Poisonous gases including ammonia and nitrogen oxides are produced in fire. Use dry chemical, carbon dioxide, water spray, or foam extinguishers. Vapors are heavier than air and will collect in low areas. Vapors in confined areas may explode when exposed to fire. Storage containers and parts of containers may rocket great distances, in many directions. If material or contaminated runoff enters waterways, notify downstream users of potentially contaminated waters. Notify local health and fire officials and pollution control agencies. From a secure, explosion-proof location, use water spray to cool exposed containers. If cooling streams are ineffective (venting sound increases in volume and pitch, tank discolors, or shows any signs of deforming), withdraw immediately to a secure position. If employees are expected to fight fires, they must be trained and equipped in OSHA 1910.156.

Disposal Method Suggested: Dilute with water, neutralize with HCl and discharge to sewer.[22]

References

Sax, N. I., Ed., "Dangerous Properties of Industrial Materials Report," 2, No. 3, 41–44 (1982)

New Jersey Department of Health and Senior Services, "Hazardous Substance Fact Sheet: Ammonium Hydroxide," Trenton, NJ (September 1986)

See also the references in the entry under "Ammonia" in this volume

Ammonium Metavanadate

Molecular Formula: H_4NO_3V

Common Formula: NH_4VO_3

Synonyms: Ammonium vanadate; Vanadate (V031-), ammonium; Vanadato amonico (Spanish); Vanadic acid, ammonium salt

CAS Registry Number: 7803-55-6

RTECS Number: YW0875000

DOT ID: UN 2859

Regulatory Authority

- Air Pollutant Standard Set (OSHA) (NIOSH) (ACGIH)
- EPA HAZARDOUS WASTE NUMBER (RCRA No.): P119
- RCRA 40CFR261, Appendix 8, Hazardous Constituents
- RCRA Land Ban Waste Restrictions
- SUPERFUND/EPCRA 40CFR302.4, Appendix A, Reportable Quantity (RQ): CERCLA, 1,000 lb (454 kg)
- Canada, WHMIS, Ingredients Disclosure List

Cited in U.S. State Regulations: Louisiana (G), Massachusetts (G), New Hampshire (G), New Jersey (G), Pennsylvania (G), Vermont (G), Virginia (G), Washington (G), Wisconsin (G).

Description: Ammonium metavanadate, NH_4VO_3, is a white or slightly yellow crystalline powder. Freezing/Melting point = 200°C (decomposes). Slightly soluble in water, with decomposition.

Potential Exposure: It is used in the metals industry to make alloys, chemical reactions, dyes, inks, varnishes, printing, medicines, and photography.

Incompatibilities: Moisture.

Permissible Exposure Limits in Air: The Federal OSHA exposure limit is 0.5 mg/m^3 for respirable vanadium dust and 0.1 mg/m^3 for vanadium fume, these levels are not to be exceeded at any time. NIOSH recommends a limit of 0.05 mg/m^3 for vanadium dust and fume, not to be exceeded at any time. ACGIH recommends a limit of 0.05 mg/m^3 TWA for vanadium dust and fume during an 8-hour workshift. These levels are for air levels only. When skin contact also occurs, overexposure is possible even though air levels are less than the airborne permissible limits.

Determination in Air: Vanadium in air can be determined by filter collection followed by determination by atomic emission spectroscopy according to NIOSH Method 7300[18]

Permissible Concentration in Water: The former USSR-UNEP/IRPTC project has set[43] a MAC of 0.1 mg/l for pentavalent vanadium in water for domestic purposes.

Routes of Entry: Inhalation, skin and/or eye contact.

Harmful Effects and Symptoms

Short Term Exposure: Irritates the eyes, nose, and respiratory tract. Inhalation can cause coughing, wheezing and phlegm. Exposure may cause headache and a green coating on the tongue. Higher exposures may cause pneumonia an/or pulmonary edema, which can be delayed for several hours;

there is a risk of death in serious cases. The oral LD_{50} for rats is 160 mg/kg.

Long Term Exposure: Ammonium metavanadate may be a reproductive hazard. Repeated exposure may cause lung irritation and bronchitis.

Points of Attack: Eyes, skin and lungs.

Medical Surveillance: Lung function tests. Consider x-ray following acute overexposure. Regulatory exams that include lung function, x-ray, and skin tests are proposed in criteria document NIOSH 77–222.

First Aid: If this chemical gets into the eyes, remove any contact lenses at once and irrigate immediately for at least 15 minutes, occasionally lifting upper and lower lids. Seek medical attention immediately. If this chemical contacts the skin, remove contaminated clothing and wash immediately with soap and water. Seek medical attention immediately. If this chemical has been inhaled, remove from exposure, begin rescue breathing (using universal precautions) if breathing has stopped and CPR if heart action has stopped. Transfer promptly to a medical facility. When this chemical has been swallowed, get medical attention. Give large quantities of water and induce vomiting. Do not make an unconscious person vomit. Medical observation is recommended for 24 – 48 hours after breathing overexposure, as pulmonary edema may be delayed. As first aid for pulmonary edema, a doctor or authorized paramedic may consider administering a corticosteroid spray.

Personal Protective Methods: Wear protective gloves and clothing to prevent any reasonable probability of skin contact. Safety equipment suppliers/manufacturers can provide recommendations on the most protective glove/clothing material for your operation. All protective clothing (suits, gloves, footwear, headgear) should be clean, available each day, and put on before work. Contact lenses should not be worn when working with this chemical. Wear dust-proof chemical goggles and face shield unless full facepiece respiratory protection is worn. Employees should wash immediately with soap when skin is wet or contaminated. Provide emergency showers and eyewash.

Storage: Prior to working with this chemical you should be trained on its proper handling and storage. Store in tightly closed containers in a cool, well-ventilated area away from heat.

Shipping: Ammonium metavanadate requires a "Poison" label. It falls in UN/DOT Hazard Class 6.1 and Packing Group II.[19][20]

Spill Handling: Evacuate persons not wearing protective equipment from area of spill or leak until clean-up is complete. Remove all ignition sources. Collect powdered material in the most convenient and safe manner and deposit in sealed containers. Ventilate area after clean-up is complete. It may be necessary to contain and dispose of this chemical as a hazardous waste. If material or contaminated runoff enters waterways, notify downstream users of potentially contaminated waters. Contact your Department of Environmental Protection or your regional office of the federal EPA for specific recommendations. If employees are required to clean-up spills, they must be properly trained and equipped. OSHA 1910.120(q) may be applicable.

Fire Extinguishing: Poisonous gases including ammonia, vanadium oxides, and nitrogen oxides are produced in fire. Extinguish fire using an agent suitable for type of surrounding fire. Ammonium Metavanadate itself does not burn. If material or contaminated runoff enters waterways, notify downstream users of potentially contaminated waters. Notify local health and fire officials and pollution control agencies. From a secure, explosion-proof location, use water spray to cool exposed containers. If cooling streams are ineffective (venting sound increases in volume and pitch, tank discolors, or shows any signs of deforming), withdraw immediately to a secure position. If employees are expected to fight fires, they must be trained and equipped in OSHA 1910.156.

References

National Institute for Occupational Safety and Health, "Criteria for a Recommended Standard: Occupational Exposure to Vanadium," NIOSH Doc. No. 77–222 (1977)

New Jersey Department of Health and Senior Services, "Hazardous Substance Fact Sheet: Ammonium Metavanadate," Trenton, NJ (September 1998)

Ammonium Molybdate

Molecular Formula: $H_8MoN_2O_4$

Common Formula: $(NH_4)_2MoO_4$

Synonyms: Ammonium paramolybdate; Diammonium molybdate; Molibdato amonico (Spanish); Molybdic acid, diammonium salt

CAS Registry Number: 13106-76-8

RTECS Number: QA4900000

Regulatory Authority

- Air Pollutant Standard Set (ACGIH)[1] (former USSR)[43] (DFG)[3] (HSE)[33] (OSHA)[58]
- CERCLA/SARA Section 313: Form R *de minimis* concentration reporting level: 1.0% (as ammonia) Molecular weight: 57.04; NH_3 Equivalent weight: 17.38
- Canada, WHMIS, Ingredients Disclosure List

Cited in U.S. State Regulations: New Jersey (G), Oklahoma (G).

Description: Ammonium molybdate, $(NH_4)_2MoO_4$, is a colorless, white or slightly greenish-yellow powder. Hazard Identification (based on NFPA-704 M Rating System): Health 1, Flammability 0, Reactivity 0. Soluble in water.

Potential Exposure: It is used as an analytical reagent, in pigments and in the production of molybdenum metal and ceramics.

Incompatibilities: Contact with strong oxidizers or chemically-active metals (such as potassium, sodium, magnesium and zinc) may cause a violent reaction.

Permissible Exposure Limits in Air: The Federal OSHA PEL is 15 mg/m^3 TWA for molybdenum soluble compounds for an 8-hour workshift. ACGIH recommends 10 mg/m^3 TWA for molybdenum soluble compounds for an 8-hour workshift.

The former USSR-UNEP/IRPTC project[43] has set a MAC for molybdenum in workplace air at 2.0 mg/m^3 for aerosol condensates and 4 mg/m^3 for soluble compounds in the form of dust.

Determination in Air: Molybdenum may be sampled by filter and measured by inductively coupled argon plasma atomic emission spectroscopy as described in NIOSH Method 7300 for Metals.[18]

Permissible Concentration in Water: No criteria set.

Routes of Entry: Inhalation, ingestion, skin and/or eye contact.

Harmful Effects and Symptoms

Short Term Exposure: This chemical can be absorbed through the skin, thereby increasing exposure. Symptoms of molybdenum poisoning includes stomach upset, diarrhea, coma; death from heart failure can occur. Exposure to high concentrations may cause irritation of the eyes, nose and throat. Very high exposure may cause kidney and liver damage.

Long Term Exposure: May cause kidney and liver damage. Inhalation of dust can cause lung disease.

Points of Attack: Skin, lungs.

Medical Surveillance: Kidney, liver and lung function tests.

First Aid: If this chemical gets into the eyes, remove any contact lenses at once and irrigate immediately for at least 15 minutes, occasionally lifting upper and lower lids. Seek medical attention immediately. If this chemical contacts the skin, remove contaminated clothing and wash immediately with soap and water. Seek medical attention immediately. If this chemical has been inhaled, remove from exposure, begin rescue breathing (using universal precautions) if breathing has stopped and CPR if heart action has stopped. Transfer promptly to a medical facility. When this chemical has been swallowed, get medical attention. Give large quantities of water and induce vomiting. Do not make an unconscious person vomit.

Personal Protective Methods: Where possible, enclose operations and use local exhaust ventilation at this site of chemical release. Wear protective gloves and clothing to prevent any reasonable probability of skin contact. Safety equipment suppliers/manufacturers can provide recommendations on the most protective glove/clothing material for your operation. All protective clothing (suits, gloves, footwear, headgear) should be clean, available each day, and put on before work. Contact lenses should not be worn when working with this chemical. Wear dust-proof chemical goggles and face shield unless full facepiece respiratory protection is worn. Employees should wash immediately with soap when skin is wet or contaminated. Provide emergency showers and eyewash.

Respirator Selection: OSHA (for soluble compounds of molybdenum) *25 mg/m^3*: DM (any dust and mist respirator). *50 mg/m^3*: DMXSQ (any dust and mist respirator except single-use and quarter mask respirators); or SA (any supplied-air respirator). *125 mg/m^3*: SA:CF (any supplied-air respirator operated in a continuous-flow mode); or PAPRDM, if not present as a fume (any powered, air-purifying respirator with a dust and mist filter). *250 mg/m^3*: HiEF (any air-purifying, full-facepiece respirator with a high-efficiency particulate filter); or SAT:CF (any supplied-air respirator that has a tight-fitting facepiece and is operated in a continuous-flow mode); or PAPRTHiE (any powered, air-purifying respirator with a tight-fitting facepiece and a high-efficiency particulate filter); or SCBAF (any self-contained breathing apparatus with a full facepiece); or SAF (any supplied-air respirator with a full facepiece). *1,000 mg/m^3*: SAF:PD,PP (any supplied-air respirator that has a full facepiece and is operated in a pressure-demand or other positive-pressure mode). *Emergency or planned entry into unknown concentrations or IDLH conditions:* SCBAF:PD,PP (any self-contained breathing apparatus that has a full facepiece and is operated in a pressure-demand or other positive-pressure mode); or SAF:PD,PP: ASCBA (any supplied-air respirator that has a full facepiece and is operated in a pressure-demand or other positive-pressure mode in combination with an auxiliary self-contained breathing apparatus operated in a pressure-demand or other positive-pressure mode). *Escape:* HiEF (any air-purifying, full-facepiece respirator with a high-efficiency particulate filter); or SCBAE (any appropriate escape-type, self-contained breathing apparatus). *Note:* Substance reported to cause eye irritation or damage; may require eye protection.

Storage: Prior to working with this chemical you should be trained on its proper handling and storage. Store to avoid possible contact with active metals.

Shipping: There are no UN/DOT restrictions.

Spill Handling: Evacuate persons not wearing protective equipment from area of spill or leak until clean-up is complete. Remove all ignition sources. Collect powdered material in the most convenient and safe manner and deposit in sealed containers. Ventilate area after clean-up is complete. It may be necessary to contain and dispose of this chemical as a hazardous waste. If material or contaminated runoff enters waterways, notify downstream users of potentially contaminated waters. Contact your Department of Environmental Protection or your regional office of the federal EPA for specific recommendations. If employees are required to clean-up spills, they must be properly trained and equipped. OSHA 1910.120(q) may be applicable.

Fire Extinguishing: Extinguish fire using an agent for type of surrounding fire. Ammonium molybdate itself does not burn. Poisonous gases including ammonia and nitrogen oxides are produced in fire. Use dry chemical appropriate for extinguishing metal fires. *Do not* use water. If material or contaminated runoff enters waterways, notify downstream users of potentially contaminated waters. Notify local health and fire officials and pollution control agencies. From a secure, explosion-proof location, use water spray to cool exposed containers. If cooling streams are ineffective (venting sound increases in volume and pitch, tank discolors, or shows any signs of deforming), withdraw immediately to a secure position. If employees are expected to fight fires, they must be trained and equipped in OSHA 1910.156.

References

New Jersey Department of Health and Senior Services, "Hazardous Substance Fact Sheet: Ammonium Molybdate," Trenton, NJ (May 1986)

Ammonium Nitrate

Molecular Formula: $H_4N_2O_3$

Common Formula: NH_4NO_3

Synonyms: Ammonium(I) nitrate(1:1); Ammonium saltpeter; Ansax; Herco Prills; Nitram; Nitrato amonico (Spanish); Nitric acid, ammonium salt; Norway saltpeter; Varioform I

CAS Registry Number: 6484-52-2

RTECS Number: BR9050000

DOT ID: UN 0222 (NH_4NO_3 fertilizer with <2% combustible material);[56] UN 0223 (NH_4NO_3 fertilizer more likely to explode than 0222); UN 1942 (NH_4NO_3 with <2% combustible material); UN 2067 (NH_4NO_3 fertilizer); UN 2068 (NH_4NO_3 fertilizer with $CaCO_3$); UN 2069 [NH_4NO_3 fertilizer with $(NH_4)_2SO_4$]; UN 2070 (NH_4NO_3 fertilizer with phosphate or potash); UN 2071 (NH_4NO_3 fertilizer with <45% NH_4NO_3 and <0.4% combustibles); UN 2072 (NH_4NO_3 fertilizer n.o.s.); UN 2426 (NH_4NO_3 solutions)[56]

Regulatory Authority

- Highly Reactive Substance and Explosive (World Bank)[15]
- Air Pollutant Standard Set (former USSR)[35]
- U.S. DOT Regulated Marine Pollutant (49CFR172.101, Appendix B) as nitrate compounds
- Clean Water Act: 40CFR116.4 Hazardous Substances; RQ 40CFR117.3 (same as CERCLA)
- CERCLA/SARA Section 313: Form R *de minimis* concentration reporting level: 1.0% (as ammonia) Molecular weight: 80.04; NH_3 Equivalent weight: 21.28 also reportable as a nitrate compound, water dissociable, (reportable only when in an aqueous solution), at the same reporting level (1.0%).
- Canada, National Pollutant Release Inventory (solution only)

Cited in U.S. State Regulations: California (G), Florida (G), Maryland (G), Massachusetts (G), Minnesota (G) (as inorganic nitrates), New Hampshire (G), New Jersey (G), New York (G), Oklahoma (G), Pennsylvania (G), Rhode Island (G).

Description: Ammonium nitrate, NH_4NO_3, is an odorless white to gray to brown, odorless beads, pellets or flakes. Freezing/Melting point = about 169°C with slow decomposition; the decomposition accelerates at about 210°C and may become explosive. Hazard Identification (based on NFPA-704 M Rating System): Health 0, Flammability 0, Reactivity 3, Oxidizer. Soluble in water.

Potential Exposure: Used in the manufacture of liquid and solid fertilizer compositions and industrial explosives and blasting agents from ammonium nitrate, matches, and antibiotics.

Incompatibilities: A strong oxidizer. Reducing agents, combustible materials, organic materials, finely divided (powdered) metals may form explosive mixtures or cause fire and explosions. When contaminated with oil, charcoal or flammable liquids, can be considered an explosive which can be detonated by combustion or shock.

Permissible Exposure Limits in Air: A MAC value of 0.3 mg/m^3 on a daily average basis in ambient air has been set in the former USSR.[35]

Permissible Concentration in Water: The former USSR-UNEP/IRPTC project[43] has set a MAC of 0.5 mg/l in water bodies used for fishery purposes.

Routes of Entry: Inhalation, skin and/or eye contact.

Harmful Effects and Symptoms

Short Term Exposure: The potential for ammonia poisoning in the course of NH_4NO_3 and fertilizer manufacture is the chief toxic effect associated with ammonium nitrate. Exposure may irritate the skin, eyes, nose, throat and lungs. Overexposure can cause nausea and vomiting, headaches, weakness, faintness and collapse. Severe overexposure may lower the ability of the blood to carry oxygen. This can result in a bluish color to skin and lips, headaches, dizziness, collapse and even death.

Long Term Exposure: Unknown at this time.

Points of Attack: Inhalation, skin.

Medical Surveillance: Consider the points of attack in preplacement and periodic physical examinations.

First Aid: If this chemical gets into the eyes, remove any contact lenses at once and irrigate immediately for at least 15 minutes, occasionally lifting upper and lower lids. Seek medical attention immediately. If this chemical contacts the skin, remove contaminated clothing and wash immediately with soap and water. Seek medical attention immediately. If this chemical has been inhaled, remove from exposure, begin rescue breathing (using universal precautions) if breathing has stopped and CPR if heart action has stopped. Transfer promptly to a medical facility. When this chemical has been swallowed, get medical attention. Give large quantities of water and induce vomiting. Do not make an unconscious person vomit.

Personal Protective Methods: Wear protective gloves and clothing to prevent any reasonable probability of skin contact. Safety equipment suppliers/manufacturers can provide recommendations on the most protective glove/clothing material for your operation. All protective clothing (suits, gloves, footwear, headgear) should be clean, available each day, and put on before work. Contact lenses should not be worn when working with this chemical. Wear dust-proof chemical goggles and face shield unless full facepiece respiratory protection is worn. Employees should wash immediately with soap when skin is wet or contaminated. Provide emergency showers and eyewash.

Respirator Selection: *At any concentrations above the NIOSH REL, or where there is no REL, at any detectable concentration:* SCBAF:PD,PP (any MSHA/NIOSH approved self-contained breathing apparatus that has a full facepiece and is operated in a pressure-demand or other positive-pressure mode); or SAF:PD,PP:ASCBA (any supplied-air respirator that has a full facepiece and is operated in a pressure-demand or other positive-pressure mode in combination with an auxiliary, self-contained breathing apparatus operated in a pressure-demand or other positive pressure mode). *Escape:* GMFOVHiE [any air-purifying, full-facepiece respirator (gas mask) with a chin-style, front- or back-mounted organic vapor canister having a high-efficiency particulate filter]; or SCBAE (any appropriate escape-type, self-contained breathing apparatus).

Storage: Prior to working with this chemical you should be trained on its proper handling and storage. Store in a well-ventilated area, away from sparks and flames preferably in a non-combustible location equipped with automatic sprinkler protection. Use only non-sparking tools and equipment, especially when opening and closing containers of this chemical. Sources of ignition such as smoking and open flames, are prohibited where this chemical is used, handled, or stored in a manner that could create a potential fire or explosion hazard. Keep well closed, dry, separated from materials and labeled "oxidizer." Storage of 500 tons or more should be avoided[35] or at least very carefully regulated. See NFPA 490, *Code for the Storage of Ammonium Nitrate.*

Shipping: Shipping label required is "Oxidizer." Falls in DOT/UN Hazard Class 5.1 and Packing Group II for liquid and III for solid.[19][20]

Spill Handling: Evacuate and isolate the area of the spill and restrict persons not wearing protective equipment from area of spill or leak until clean-up is complete. Remove all ignition sources. Cover the spill with dry lime or soda ash and collect powdered material in the most convenient and safe manner and deposit in sealed containers. Flush area with water. Ventilate area after clean-up is complete. Keep ammonium nitrate out of confined space such as a sewer, because of the potential for an explosion. It may be necessary to contain and dispose of this chemical as a hazardous waste. If material or contaminated runoff enters waterways, notify downstream users of potentially contaminated waters. Contact your Department of Environmental Protection or your regional office of the federal EPA for specific recommendations. If employees are required to clean-up spills, they must be properly trained and equipped. OSHA 1910.120(q) may be applicable.

Fire Extinguishing: Does not burn but supports combustion. May explode under confinement and high temperatures. Exercise extreme caution. Use flooding amounts of water. Poisonous gases are produced in fire including nitrogen oxides and ammonia. If material or contaminated runoff enters waterways, notify downstream users of potentially contaminated waters. Notify local health and fire officials and pollution control agencies. From a secure, explosion-proof location, use water spray to cool exposed containers. If cooling streams are ineffective (venting sound increases in volume and pitch, tank discolors, or shows any signs of deforming), withdraw immediately to a secure position. If employees are expected to fight fires, they must be trained and equipped in OSHA 1910.156.

Disposal Method Suggested: Pretreatment involves addition of sodium hydroxide to liberate ammonia and form the soluble sodium salt. The liberated ammonia can be recovered and sold. After dilution to the permitted provisional limit, the sodium salt can be discharged into a stream or sewer.[22]

References

National Institute for Occupational Safety and Health, Profiles on Occupational Hazard for Criteria Document Priorities: Ammonium Nitrate, pp 281–285, Report PB-274,073, Washington, DC (1977)

Sax, N. I., Ed., Dangerous Properties of Industrial Materials Report, 2, No. 3, 44–46 (1982)

New Jersey Department of Health and Senior Services, "Hazardous Substance Fact Sheet: Ammonium Nitrate," Trenton, NJ (June 1998)

New York State Department of Health, "Chemical Fact Sheet: Ammonium Nitrate," Albany, NY, Bureau of Toxic Substance Assessment (January 1986)

Ammonium Oxalate

Molecular Formula: $C_2H_8N_2O_4$

Common Formula: $NH_4OOCCOONH_4$

Synonyms: Ethanedioic acid, diammonium salt; Oxalato amonico (Spanish)

CAS Registry Number: 1113-38-8

RTECS Number: RO2750000

DOT ID: NA 2449

Regulatory Authority

- Clean Water Act: 40CFR116.4 Hazardous Substances; RQ 40CFR117.3 (same as CERCLA)
- SUPERFUND/EPCRA 40CFR302.4, Appendix A, Reportable Quantity (RQ): CERCLA, 2,000 lb (2,270 kg); Sec-

tion 313: Form R *de minimis* concentration reporting level: 1.0% (as ammonia). Molecular weight: 124.10; NH_3 Equivalent weight: 27.45

- Canada, WHMIS, Ingredients Disclosure List

Cited in U.S. State Regulations: Illinois (G), New Hampshire (G), New Jersey (G), Pennsylvania (G).

Description: Ammonium oxalate, $NH_4OOCCOONH_4$, is an odorless, colorless crystalline material. Freezing/Melting point = decomposes. Hazard Identification (based on NFPA-704 M Rating System): Health 1, Flammability 1, Reactivity 0. Slightly soluble in water.

Potential Exposure: It is used in chemical analysis and to make explosives and rust-removal chemicals.

Permissible Exposure Limits in Air: No standards set.

Permissible Concentration in Water: No criteria set.

Routes of Entry: Inhalation, ingestion, skin and/or eye contact.

Harmful Effects and Symptoms

Short Term Exposure: Breathing Ammonium Oxalate can irritate the nose, throat and lungs. Contact can irritate the skin. Overexposure may cause a kidney stone and kidney damage.

Long Term Exposure: Repeated exposures may cause bronchitis, kidney stones, and kidney damage. Repeated contact may cause cracking of the skin and slow healing ulcers.

Points of Attack: Lungs, skin, kidneys.

Medical Surveillance: Lung function tests, urinalysis, kidney function tests.

First Aid: If this chemical gets into the eyes, remove any contact lenses at once and irrigate immediately for at least 15 minutes, occasionally lifting upper and lower lids. Seek medical attention immediately. If this chemical contacts the skin, remove contaminated clothing and wash immediately with soap and water. Seek medical attention immediately. If this chemical has been inhaled, remove from exposure, begin rescue breathing (using universal precautions) if breathing has stopped and CPR if heart action has stopped. Transfer promptly to a medical facility. When this chemical has been swallowed, get medical attention. Give large quantities of water and induce vomiting. Do not make an unconscious person vomit.

Personal Protective Methods: Where possible, enclose operations and use local exhaust ventilation at the site of chemical release. Wear protective gloves and clothing to prevent any reasonable probability of skin contact. Safety equipment suppliers/manufacturers can provide recommendations on the most protective glove/clothing material for your operation. All protective clothing (suits, gloves, footwear, headgear) should be clean, available each day, and put on before work. Contact lenses should not be worn when working with this chemical. Wear dust-proof chemical goggles and face shield unless full facepiece respiratory protection is worn. Employees should wash immediately with soap when skin is wet or contaminated. Provide emergency showers and eyewash.

Respirator Selection: *Where there is no REL, at any detectable concentration:* SCBAF:PD,PP (any MSHA/NIOSH approved self-contained breathing apparatus that has a full facepiece and is operated in a pressure-demand or other positive-pressure mode); or SAF:PD,PP:ASCBA (any supplied-air respirator that has a full facepiece and is operated in a pressure-demand or other positive-pressure mode in combination with an auxiliary, self-contained breathing apparatus operated in a pressure-demand or other positive pressure mode). *Escape:* GMFOVHiE [any air-purifying, full-facepiece respirator (gas mask) with a chin-style, front- or back-mounted organic vapor canister having a high-efficiency particulate filter]; or SCBAE (any appropriate escape-type, self-contained breathing apparatus).

Storage: Prior to working with this chemical you should be trained on its proper handling and storage. Store in tightly closed containers in a cool, well-ventilated area.

Shipping: For oxalates, water soluble, the label required is "Keep away from Food." Falls in DOT/UN Hazard Class 6.1 and Packing Group III.[19][20]

Spill Handling: Evacuate persons not wearing protective equipment from area of spill or leak until clean-up is complete. Remove all ignition sources. Collect powdered material in the most convenient and safe manner and deposit in sealed containers. Ventilate area after clean-up is complete. It may be necessary to contain and dispose of this chemical as a hazardous waste. If material or contaminated runoff enters waterways, notify downstream users of potentially contaminated waters. Contact your Department of Environmental Protection or your regional office of the federal EPA for specific recommendations. If employees are required to clean-up spills, they must be properly trained and equipped. OSHA 1910.120(q) may be applicable.

Fire Extinguishing: Ammonium Oxalate may burn, but does not readily ignite. Use dry chemical, carbon dioxide, water spray, or alcohol foam extinguishers. Poisonous gases are produced in fire including nitrogen oxides and ammonia gas. If material or contaminated runoff enters waterways, notify downstream users of potentially contaminated waters. Notify local health and fire officials and pollution control agencies. From a secure, explosion-proof location, use water spray to cool exposed containers. If cooling streams are ineffective (venting sound increases in volume and pitch, tank discolors, or shows any signs of deforming), withdraw immediately to a secure position. If employees are expected to fight fires, they must be trained and equipped in OSHA 1910.156.

References

New Jersey Department of Health and Senior Services, "Hazardous Substance Fact Sheet: Ammonium Oxalate," Trenton, NJ (November 1986)

Ammonium Perfluorooctanoate

Molecular Formula: $C_8H_{19}NO_2$

Common Formula: $C_7H_{15}COONH_4$

Synonyms: Ammonium pentadecafluorooctanoate; Ammonium perfluorocaprilate; Ammonium perfluorocaprylate; APFO; Perfluoroammonium octonate

CAS Registry Number: 3825-26-1

RTECS Number: RH0782000

Regulatory Authority

- Carcinogen, (animal positive) (ACGIH)
- Air Pollutant Standard Set (ACGIH) (Australia) (Israel) (Ontario, Quebec, Canada)[1]

Cited in U.S. State Regulations: California (A, G), Florida (G), Massachusetts (G), Minnesota (G), Pennsylvania (G).

Description: Ammonium Perfluorooctanate, $C_7H_{15}COONH_4$, is a free-flowing powder. Boiling point = 125°C (sublines). Highly soluble in water.

Potential Exposure: This compound is used commercially in the polymerization of fluorinated monomers.

Permissible Exposure Limits in Air: ACGIH TLV is 0.1 mg/m^3 TWA for an 8-hour workshift,[1] with the notation that it may be absorbed through the skin.

Permissible Concentration in Water: No criteria set.

Routes of Entry: Inhalation, ingestion, eye and/or skin contact.

Harmful Effects and Symptoms

Short Term Exposure: Ammonium perfluorooctanoate has a moderate oral toxicity[53] with an LD_{50} in rats of 540 mg/kg. The material was non-irritating to the skin and produced moderate eye irritation characterized by iridal and conjunctival effects which persisted for at least 7 days. Lethality was produced dermally in rats and rabbits only following large doses, but repeated doses of 20 – 2,000 mg/kg to rats produced liver damage in a dose-related fashion and elevated blood organofluoride levels. Blood levels were reduced but still detectable 42 days after the last exposure.

Long Term Exposure: A possible human carcinogen and reproductive toxin.

First Aid: If this chemical gets into the eyes, remove any contact lenses at once and irrigate immediately for at least 15 minutes, occasionally lifting upper and lower lids. Seek medical attention immediately. If this chemical contacts the skin, remove contaminated clothing and wash immediately with soap and water. Seek medical attention immediately. If this chemical has been inhaled, remove from exposure, begin rescue breathing (using universal precautions) if breathing has stopped and CPR if heart action has stopped. Transfer promptly to a medical facility. When this chemical has been swallowed, get medical attention. Give large quantities of water and induce vomiting. Do not make an unconscious person vomit.

Personal Protective Methods: Wear protective gloves and clothing to prevent any reasonable probability of skin contact. Safety equipment suppliers/manufacturers can provide recommendations on the most protective glove/clothing material for your operation. All protective clothing (suits, gloves, footwear, headgear) should be clean, available each day, and put on before work. Contact lenses should not be worn when working with this chemical. Wear dust-proof chemical goggles and face shield unless full facepiece respiratory protection is worn. Employees should wash immediately with soap when skin is wet or contaminated. Provide emergency showers and eyewash.

Respirator Selection: *At any detectable concentration:* SCBAF:PD,PP (any MSHA/NIOSH approved self-contained breathing apparatus that has a full facepiece and is operated in a pressure-demand or other positive-pressure mode); or SAF:PD,PP:ASCBA (any supplied-air respirator that has a full facepiece and is operated in a pressure-demand or other positive-pressure mode in combination with an auxiliary, self-contained breathing apparatus operated in a pressure-demand or other positive pressure mode). *Escape:* HiEF (any air-purifying, full-facepiece respirator with a high-efficiency particulate filter); or SCBAE (any appropriate escape-type, self-contained breathing apparatus).

Storage: Prior to working with this chemical you should be trained on its proper handling and storage. Store in tightly closed containers in a cool, well ventilated area.

Spill Handling: The DOT/UN regulations does not set out any specific label requirements or maximum shipping quantities for this chemical. However in solid form this chemical is a toxic solid, organic, n.o.s and should be labeled "POISON," Hazard Class 6.1.

Fire Extinguishing: Use dry chemical, carbon dioxide, water spray, or alcohol foam extinguishers. Poisonous gases are produced in fire. If material or contaminated runoff enters waterways, notify downstream users of potentially contaminated waters. Notify local health and fire officials and pollution control agencies. From a secure, explosion-proof location, use water spray to cool exposed containers. If cooling streams are ineffective (venting sound increases in volume and pitch, tank discolors, or shows any signs of deforming), withdraw immediately to a secure position. If employees are expected to fight fires, they must be trained and equipped in OSHA 1910.156.

Ammonium Permanganate

Molecular Formula: H_4MnNO_4

Common Formula: NH_4MnO_4

Synonyms: Permanganato amonico (Spanish); Permanganic acid ammonium salt

CAS Registry Number: 13446-10-1

RTECS Number: SD6400000

DOT ID: NA 9190

Regulatory Authority

- CERCLA/SARA Section 313: Form R *de minimis* concentration reporting level: 1.0% (as manganese compounds)

Cited in U.S. State Regulations: Florida (G), Massachusetts (G), New Hampshire (G), New Jersey (G), Oklahoma (G), Pennsylvania (G), Rhode Island (G).

Description: Ammonium permanganate, NH_4MnO_4, is a violet-brown or dark purple crystalline (sugar-like or sand-like) solid. Freezing/Melting point = 110°C. It explodes when heated to 60°C. Hazard Identification (based on NFPA-704 M Rating System): Health 0, Flammability 0, Reactivity 3. Oxidizer. Soluble in water.

Potential Exposure: This material is used in bleaching and dyeing operations in the textile and leather industries.

Incompatibilities: A strong oxidizing agent. Contact with reducing agents, fuels and other combustible materials, heat, friction may cause a violent reaction.

Permissible Exposure Limits in Air: No standards set.

Permissible Concentration in Water: No criteria set.

Routes of Entry: Inhalation, skin and/or eye contact, ingestion.

Harmful Effects and Symptoms

Ammonium permanganate can affect you when breathed in. Exposure to Ammonium permanganate can cause irritation of the eyes, nose, throat and lungs. Contact can irritate the skin and eyes.

Short Term Exposure: Highly irritatint to eyes, skin and respiratory tract.

Long Term Exposure: Very irritating substances may cause lung effects and possible damage.

Points of Attack: Skin, eyes, nasal passages, throat and lungs.

Medical Surveillance: Lung function tests.

First Aid: If this chemical gets into the eyes, remove any contact lenses at once and irrigate immediately for at least 15 minutes, occasionally lifting upper and lower lids. Seek medical attention immediately. If this chemical contacts the skin, remove contaminated clothing and wash immediately with soap and water. Seek medical attention immediately. If this chemical has been inhaled, remove from exposure, begin rescue breathing (using universal precautions) if breathing has stopped and CPR if heart action has stopped. Transfer promptly to a medical facility. When this chemical has been swallowed, get medical attention. Give large quantities of water and induce vomiting. Do not make an unconscious person vomit.

Personal Protective Methods: Wear protective gloves and clothing to prevent any reasonable probability of skin contact. Safety equipment suppliers/manufacturers can provide recommendations on the most protective glove/clothing material for your operation. All protective clothing (suits, gloves, footwear, headgear) should be clean, available each day, and put on before work. Contact lenses should not be worn when working with this chemical. Wear dust-proof chemical goggles and face shield unless full facepiece respiratory protection is worn. Employees should wash immediately with soap when skin is wet or contaminated. Provide emergency showers and eyewash.

Respirator Selection: Where the potential exists for exposure to Ammonium Permanganate, use a MSHA/NIOAH approved respirator equipped with particulate (dust/fume/mist) filters. More protection is provided by a full facepiece respirator than by a half-mask respirator, and even greater protection is provided by a powered-air purifying respirator. Particulate filters must be checked every day before work for physical damage, such as rips or tears, and replaced as needed. Where the potential for high exposures exists, use a MSHA/NIOSH approved supplied-air respirator with a full facepiece operated in the positive pressure mode or with a full facepiece, hood, or helmet in the continuous flow mode, or use a MSHA/NIOSH approved self-contained breathing apparatus with a full facepiece operated in pressure-demand or other positive pressure mode.

Storage: Prior to working with this chemical you should be trained on its proper handling and storage. Ammonium Permanganate must be stored to avoid contact with heat, friction, organic and oxidizable material, fuels and combustibles since violent reactions occur. Always store ammonium permanganate at temperatures below 60°C/140°F. Protect containers from shock as Ammonium Permanganate may explode. See OSHA standard 1910.104 and NFPA 43A *Code for the Storage of Liquid and Solid Oxidizers* for detailed handling and storage regulations.

Shipping: Permanganate, inorganic, n.o.s. must carry the label "Oxidizer." They fall in Hazard Class 5.1 and Packing Group II.

Spill Handling: Evacuate and restrict persons not wearing protective equipment from area of spill or leak until cleanup is complete. Remove all ignition sources. Collect powdered material in the most convenient and safe manner and deposit in sealed containers. Ventilate area of spill or leak after clean-up is complete. Keep ammonium permanganate out of a confined space, such as a sewer, because of the potential for an explosion, unless the sewer is designed to prevent the build-up of explosive concentrations. It may be necessary to contain and dispose of this chemical as a hazardous waste. If material or contaminated runoff enters waterways, notify downstream users of potentially contaminated waters. Contact your Department of Environmental Protection or your regional office of the federal EPA for specific recommendations. If employees are required to clean-up spills, they must be properly trained and equipped. OSHA 1910.120(q) may be applicable.

Fire Extinguishing: This chemical is highly reactive and heat sensitive. Use dry chemical, carbon dioxide, water spray, or alcohol foam extinguishers. Poisonous gases are produced in fire including nitrogen oxides and ammonia. Storage containers and parts of containers may rocket great distances, in many directions. If material or contaminated runoff enters waterways, notify downstream users of potentially contaminated waters. Notify local health and fire officials and pollution control agencies. From a secure, explosion-proof location, use water spray to cool exposed containers. If cooling streams are ineffective (venting sound increases in volume and pitch, tank discolors, or shows any signs of deforming), withdraw immediately to a secure position. If employees are expected to fight fires, they must be trained and equipped in OSHA 1910.156.

References

New Jersey Department of Health and Senior Services, "Hazardous Substance Fact Sheet: Ammonium Permanganate," Trenton, NJ (February 9, 1988)

Ammonium Persulfate

Molecular Formula: $H_8N_2O_8S_2$

Common Formula: $(NH_4)_2S_2O_8$

Synonyms: Ammonium perosycisulfate; Ammonium perosysulfate; Ammonium peroxydisulfate; Ammonium peroxysulfate; Diammonium peroxydisulfate; Diammonium persulfate; Peroxydisulfuric acid diammonium salt; Persulfato amonico (Spanish)

CAS Registry Number: 7727-54-0

RTECS Number: SE0350000

DOT ID: UN 1444

Regulatory Authority

- Air Pollutant Standard Set (Ontario, Canada, as persulfates) (HSE)
- SUPERFUND/EPCRA 40CFR302.4, Appendix A, Reportable Quantity (RQ): CERCLA, 100 lb (45.4 kg)
- EPCRA Section 313: Form R *de minimis* concentration reporting level: 1.0% (as ammonia). Molecular weight: 228.19; NH_3 Equivalent weight: 14.93

Cited in U.S. State Regulations: Alaska (G), New Hampshire (G), New Jersey (G), Oklahoma (G), West Virginia (G).

Description: Ammonium persulfate, $(NH_4)_2S_2O_8$, is a colorless or white crystalline solid. Decomposes below metling point = 120°C. Hazard Identification (based on NFPA-704 M Rating System): Health 1, Flammability 0, Reactivity 1. Soluble in water.

Potential Exposure: It is used as a bleaching agent and in photographic chemicals and to make dyes. It is also used as an ingredient of polymerization catalysts.

Incompatibilities: Combustibles, sodium peroxide, aluminum, water.

Permissible Exposure Limits in Air: No standards set although values of 2 – 5 mg/m^3 variously reported in the literature.[9][50]

Permissible Concentration in Water: No criteria set.

Routes of Entry: Inhalation, ingestion, skin and/or eye contact.

Harmful Effects and Symptoms

Short Term Exposure: Exposure can irritate the skin, eyes, nose, throat and lungs. Exposure may cause an allergy-like reaction, with eye tearing, nose congestion, asthma-like wheezing and difficulty in breathing. Life-threatening shock may result. The oral LD_{50} for rats is 820 mg/kg.[9]

Points of Attack: Eye, lungs.

Medical Surveillance: Lung function tests. Examination by a qualified allergist.

First Aid: If this chemical gets into the eyes, remove any contact lenses at once and irrigate immediately for at least 15 minutes, occasionally lifting upper and lower lids. Seek medical attention immediately. If this chemical contacts the skin, remove contaminated clothing and wash immediately with soap and water. Seek medical attention immediately. If this chemical has been inhaled, remove from exposure, begin rescue breathing (using universal precautions) if breathing has stopped and CPR if heart action has stopped. Transfer promptly to a medical facility. When this chemical has been swallowed, get medical attention. Give large quantities of water and induce vomiting. Do not make an unconscious person vomit. Medical observation is recommended; effects may be delayed.

Personal Protective Methods: Wear protective gloves and clothing to prevent any reasonable probability of skin contact. Safety equipment suppliers/manufacturers can provide recommendations on the most protective glove/clothing material for your operation. All protective clothing (suits, gloves, footwear, headgear) should be clean, available each day, and put on before work. Contact lenses should not be worn when working with this chemical. Wear dust-proof chemical goggles and face shield unless full facepiece respiratory protection is worn. Employees should wash immediately with soap when skin is wet or contaminated. Provide emergency showers and eyewash.

Where possible, enclose operations and use local exhaust ventilation at the site of chemical release. If local exhaust ventilation or enclosure is not used, respirators should be worn. Wear protective work clothing and goggles. Wash thoroughly immediately after exposure to Ammonium Persulfate.

Storage: Prior to working with this chemical you should be trained on its proper handling and storage. Ammonium Persulfate must be stored to avoid contact with combustibles (such as wood, paper, and oil), sodium peroxide, aluminum and water since violent reactions occur. Store in tightly closed containers in a cool, dry and well-ventilated area. See OSHA standard 1910.104 and NFPA 43A *Code for the Storage of Liquid and Solid Oxidizers* for detailed handling and storage regulations.

Shipping: Label required by DOT is "Oxidizer."[19] Ammonium persulfate is in DOT/UN Hazard Class 5.1 and Packing Group III.[19][20]

Spill Handling: Restrict persons not wearing protective equipment from area of spill until clean-up is complete. Remove all ignition sources. Collect powdered material in the most convenient and safe manner and deposit in sealed containers. Wash area down with water.

Fire Extinguishing: Ammonium Persulfate is a strong oxidizing agent that can cause combustible materials such as wood, paper, and oil to ignite. Poisonous gases are produced in fire, including Ammonia, oxides of sulfur and nitrogen. Use dry chemical, CO_2 or water spray extinguishers.

Disposal Method Suggested: May be treated with large volumes of water, neutralized and flushed to sewer.[22] This applies to small quantities only.

References

New Jersey Department of Health and Senior Services, "Hazardous Substance Fact Sheet: Ammonium Persulfate," Trenton, NJ (May 1986)

Sax, N. I., Ed., Dangerous Properties of Industrial Materials Report, 2, No. 3, 48–49 (1982)

Ammonium Phosphate

Molecular Formula: $H_9N_2O_4P$

Common Formula: $(NH_4)_2HPO_4$

Synonyms: Ammonium orthophosphate, dibasic; Ammonium orthophosphate, monohydrogen; Ammonium phosphate, dibasic; Ammonium phosphate, hydrogen; Diammonium orthophosphate; Diammonium orthophosphate, hydrogen; Diammonium phosphate; Diammonium phosphate, hydrogen; Diammonium phosphate, monohydrogen; Dibasic ammonium phosphate; Secondary ammonium phosphate

CAS Registry Number: 7783-28-0

RTECS Number: TB9375000

Regulatory Authority

- CERCLA/SARA Section 313: Form R *de minimis* concentration reporting level: 1.0% (as ammonia) Molecular weight: 132.06; NH_3 Equivalent weight: 25.79

Cited in U.S. State Regulations: California (G), New Jersey (G), New York (G), Pennsylvania (G).

Description: Ammonium phosphate, $(NH_4)_2HPO_4$, is a white crystalline or powdery substance. Freezing/Melting point = 185°C (decomposes). Soluble in water.

Potential Exposure: Used in fireproofing or textiles, wood and paper; in soldering flux, as a fertilizer; a buffer; in baking powder and food additives.

Incompatibilities: Incompatible with strong oxidizers, strong bases. Contact with air causes this chemical to produce anhydrous ammonia fumes.

Permissible Exposure Limits in Air: No standards set.

Permissible Concentration in Water: No criteria set.

Routes of Entry: Inhalation, ingestion, skin and/or eye contact.

Harmful Effects and Symptoms

Short Term Exposure: On short term exposure, may cause skin and eye irritation; ammonia fumes can cause eye irritation above 70 ppm. In closed spaces, inhalation of ammonia fumes may cause nose and throat irritation (70 ppm, 5 minutes). Levels of 500 ppm for 30 minutes may cause irritation to throat and lungs. High levels may result in accumulation of fluid in the lung and suffocation. Ammonia poisoning upon ingestion is characterized by sagging of facial muscles, tremors, anxiety, difficulty in controlling muscles, stupor and coma. There is only a slight chance of this happening from ingestion of ammonium phosphates, except in persons with impaired liver function. Large doses may cause calcium imbalance and an increased flow of urine.

Points of Attack: Liver, skin, eyes.

Medical Surveillance: Liver function tests.

First Aid: If this chemical gets into the eyes, remove any contact lenses at once and irrigate immediately for at least 15 minutes, occasionally lifting upper and lower lids. Seek medical attention immediately. If this chemical contacts the skin, remove contaminated clothing and wash immediately with soap and water. Seek medical attention immediately. If this chemical has been inhaled, remove from exposure, begin rescue breathing (using universal precautions) if breathing has stopped and CPR if heart action has stopped. Transfer promptly to a medical facility. When this chemical has been swallowed, get medical attention. Seek medical attention, if necessary. Give large quantities of water or milk *Inhalation:* Move to fresh air. Give oxygen or artificial respiration if required. Seek medical attention, if necessary.

Personal Protective Methods: Wear protective gloves and clothing to prevent any reasonable probability of skin contact. Safety equipment suppliers/manufacturers can provide recommendations on the most protective glove/clothing material for your operation. All protective clothing (suits, gloves, footwear, headgear) should be clean, available each day, and put on before work. Contact lenses should not be worn when working with this chemical. Wear dust-proof chemical goggles and face shield unless full facepiece respiratory protection is worn. Employees should wash immediately with soap when skin is wet or contaminated. Provide emergency showers and eyewash.

Respirator Selection: Use a dust mask if necessary. A self-contained breathing apparatus may be necessary if ammonia fumes are present.

Storage: Prior to working with this chemical you should be trained on its proper handling and storage. Store in tightly sealed containers in a cool location away from oxidizers, bases.

Shipping: There are no DOT/UN requirements for labels or for maximum allowable shipping quantities.

Spill Handling: Evacuate persons not wearing protective equipment from area of spill or leak until clean-up is complete. Remove all ignition sources. Ventilate closed spaces before entering them. Absorb liquid with sand, vermiculite, earth, or similar absorbent material and place into containers for later disposal. Collect powdered material in the most convenient and safe manner and deposit in sealed containers. Ventilate area after clean-up is complete. It may be necessary to contain and dispose of this chemical as a hazardous waste. If material or contaminated runoff enters waterways, notify downstream users of potentially contaminated waters. Contact your Department of Environmental Protection or your regional office of the federal EPA for specific recommendations. If employees are required to clean-up spills, they must be properly trained and equipped. OSHA 1910.120(q) may be applicable.

Fire Extinguishing: This material is not flammable. Poisonous gases including *ammonia, phosphorus oxides, and nitrogen oxides* are produced in fire. Use extinguishing agents suitable for surrounding fire. If material or contaminated runoff enters waterways, notify downstream users of potentially contaminated waters. Notify local health and fire officials and pollution control agencies. From a secure, explosion-proof location, use water spray to cool exposed containers. If cooling streams are ineffective (venting sound increases in volume and pitch, tank discolors, or shows any signs of deforming), withdraw immediately to a secure position. If employees are expected to fight fires, they must be trained and equipped in OSHA 1910.156.

Disposal Method Suggested: May be flushed to sewer with large volumes of water.

References

New York State Department of Health, "Chemical Fact Sheet: Ammonium Phosphate," Albany, NY, Bureau of Toxic Substance Assessment (March 1986)

Ammonium Picrate

Molecular Formula: $C_6H_6NO_7$

Common Formula: $NH_4OC_6H_2(NO_2)_3$

Synonyms: Ammonium carbazoate; Ammonium picrate, dry; Ammonium picrate, wet; Ammonium picrate (yellow); Ammonium picronitrate; Explosive D; Obeline picrate; Phenol, 2,4,6-trinitro-, ammonium salt; Picratol; Picric acid, ammonium salt; Pictarol; 2,4,6-Trinitrophenol ammonium salt

CAS Registry Number: 131-74-8

RTECS Number: BS3855000

DOT ID: UN 0004 (dry or wetted with <10% water, by weight); UN 1310 (wetted with not less than 10% water, by weight)

Regulatory Authority

- EPA HAZARDOUS WASTE NUMBER (RCRA No.): P009
- RCRA 40CFR261, Appendix 8; 40CFR261.11 Hazardous Constituents
- RCRA Land Ban Waste Restrictions
- SUPERFUND/EPCRA 40CFR302.4, Appendix A, Reportable Quantity (RQ): 10 lb (4.54 kg)

Cited in U.S. State Regulations: California (G), Kansas (G), Louisiana (G), Massachusetts (G), New Hampshire (G), New Jersey (G), Oklahoma (G), Pennsylvania (G), Vermont (G), Virginia (G), Washington (G), Wisconsin (G).

Description: Ammonium Picrate, $NH_4OC_6H_2(NO_2)_3$ is bright yellow crystalline solid, that turns red if contaminated. Freezing/Melting point = decomposes on heating and explodes at 423°C. Slightly soluble in water. A high explosive when dry, and flammable when wet. Hazard Identification (Based on NFPA Rating System): Health 3, Flammability 3, Reactivity 3. Oxidizer.

Potential Exposure: Used in explosives, fireworks and rocket propellants.

Incompatibilities: A powerful oxidizer that reacts violently with reducing agents. Dangerous when heated or shocked. Keep away from metals, sodium nitrite, shock, perchlorates, peroxides, permanganates.

Permissible Exposure Limits in Air: No standards set, nevertheless, this chemical is harmful and can be absorbed through the skin, thereby increasing any exposure.

Permissible Concentration in Water: No criteria set.

Routes of Entry: Inhalation of dust, ingestion, skin contact.

Harmful Effects and Symptoms

Short Term Exposure: Ammonium Picrate can pass through the skin. This chemical can irritate the eyes and skin, and is an allergen. Ingestion can cause a bitter taste, nausea, diarrhea, vomiting, abdominal pain, skin eruptions, stupor and possible death. Breathing high levels can damage the kidneys, liver, and red blood cells. Urine may become reddish, scant or even stop, there may be drowsiness, coma, and even death.

Long Term Exposure: Repeated exposure can cause the skin and eyes to turn yellow, skin allergy, liver kidney and blood cell damage.

Points of Attack: Eyes, skin, respiratory system, kidneys, liver.

Medical Surveillance: If symptoms develop or overexposure is suspected, the following may be useful: Evaluation by a qualified allergist, including careful exposure history and special testing, may help diagnose skin allergy. Complete blood count. Liver function tests. Kidney function tests.

First Aid: If this chemical gets into the eyes, remove any contact lenses at once and irrigate immediately for at least 15 minutes, occasionally lifting upper and lower lids. Seek

medical attention immediately. If this chemical contacts the skin, remove contaminated clothing and wash immediately with soap and water. Seek medical attention immediately. If this chemical has been inhaled, remove from exposure, begin rescue breathing (using universal precautions) if breathing has stopped and CPR if heart action has stopped. Transfer promptly to a medical facility. When this chemical has been swallowed, get medical attention. Give large quantities of water and induce vomiting. Do not make an unconscious person vomit.

Personal Protective Methods: Wear protective gloves and clothing to prevent any reasonable probability of skin contact. Safety equipment suppliers/manufacturers can provide recommendations on the most protective glove/clothing material for your operation. All protective clothing (suits, gloves, footwear, headgear) should be clean, available each day, and put on before work. Contact lenses should not be worn when working with this chemical. Wear dust-proof chemical goggles and face shield unless full facepiece respiratory protection is worn. Employees should wash immediately with soap when skin is wet or contaminated. Provide emergency showers and eyewash.

Respirator Selection: Where the potential exists for exposures to Ammonium Picrate, use a MSHA/NIOSH approved full facepiece respirator with a high efficiency particulate filter. Greater protection is provided by a powered-air purifying respirator. Where the potential for high exposures exists, use a MSHA/NIOSH approved supplied-air respirator with a full facepiece operated in the positive pressure mode or with a full facepiece, hood, or helmet in the continuous flow mode, or use a MSHA/NIOSH approved self-contained breathing apparatus with a full facepiece operated in pressure-demand or other positive pressure mode.

Storage: Incompatible with strong oxidizers, strong bases. Contact with air causes substance to give off corrosive anhydrous ammonia fumes. Outside, detached storage is recommended. Store to avoid heat, shock or the presence off reducing materials. Use only non-sparking tools and equipment, especially when opening and closing containers of this chemical. Sources of ignition such as smoking and open flames, are prohibited where this chemical is used, handled, or stored in a manner that could create a potential fire or explosion hazard. See OSHA standard 1910.104 and NFPA 43A *Code for the Storage of Liquid and Solid Oxidizers* for detailed handling and storage regulations.

Shipping: The label requirement for wetted (>10% H_2O) material is "Flammable Solid." The dry material falls in DOT/NU Hazard Class 1.1D and wet material (containing at least 10% water by wt) in Hazard Class 4.1 The wet material falls in Packing Group I.[19][20]

Spill Handling: Evacuate persons not wearing protective equipment from area of spill or leak until clean-up is complete. Remove all ignition sources. Flood area with water. *Keep Material Wet. Do not dry sweep.* Keep Ammonium Picrate out of a confined space, such as a sewer, because of the potential for an explosion, unless the sewer is designed to prevent the build-up of explosive concentrations. Ventilate area after clean-up is complete. It may be necessary to contain and dispose of this chemical as a hazardous waste. If material or contaminated runoff enters waterways, notify downstream users of potentially contaminated waters. Contact your Department of Environmental Protection or your regional office of the federal EPA for specific recommendations. If employees are required to clean-up spills, they must be properly trained and equipped. OSHA 1910.120(q) may be applicable. If necessary, seek professional environmental engineering assistance from the U.S. EPA Environmental Response Team at (908) 548-8730 (24-hour response line).

Fire Extinguishing: Use extreme care as Ammonium Picrate will explode when heated or shocked, especially when dry. This chemical is a flammable solid. Use flooding quantities of water, applied from a distance. Poisonous gases are produced in fire including ammonia and nitrogen oxides. If material or contaminated runoff enters waterways, notify downstream users of potentially contaminated waters. Notify local health and fire officials and pollution control agencies. From a secure, explosion-proof location, use water spray to cool exposed containers. If cooling streams are ineffective (venting sound increases in volume and pitch, tank discolors, or shows any signs of deforming), withdraw immediately to a secure position. If employees are expected to fight fires, they must be trained and equipped in OSHA 1910.156.

Disposal Method Suggested: Consult with environmental regulatory agencies for guidance on acceptable disposal practices. Generators of waste containing this contaminant (≥100 kg/mo) must conform with EPA regulations governing storage, transportation, treatment, and waste disposal. May be poured onto soda ash, packaged in paper and burned. May also be mixed with flammable solvent and sprayed into an incinerator equipped with afterburner and scrubber.[24]

References

Sax, N. I., Ed., "Dangerous Properties of Industrial Materials Report," 2, No. 3, 49–51 (1982) and 8, No. 2, 42–44 (1988)

New Jersey Department of Health and Senior Services, "Hazardous Substance Fact Sheet: Ammonium Picrate," Trenton, NJ (August 3, 1987)

Ammonium Sulfamate

Molecular Formula: $H_6N_2O_3S$

Common Formula: $NH_2SO_3NH_4$

Synonyms: Amcide®; Amicide®; Ammat; Ammate®; Ammate herbicide; Ammonium amidosulfonate; Ammonium amidosulphate; Ammonium aminosulfonate; Ammonium salz der amidosulfonsaure (German); Ammonium sulphamate; AMS; Ikurin; Monoammonium salt of sulfamic acid; Monoammonium sulfamate; Sulfamate; Sulfamato amonico

(Spanish); Sulfamic acid, monoammonium salt; Sulfaminsaure (German)

CAS Registry Number: 7773-06-0

RTECS Number: WO6125000

DOT ID: NA 9089

Regulatory Authority

- Air Pollutant Standard Set (NIOSH)[2] (DFG)[3] (ACGIH)[1] (former USSR)[43] (HSE)[33] (OSHA)[58] (Several States)[60] (Several Canadian Provinces) (Australia) (Israel)
- RCRA 40CFR261, Appendix 8; 40CFR261.11 Hazardous Constituents
- Clean Water Act: 40CFR116.4 Hazardous Substances; 40CFR117.3, RQ (same as CERCLA)
- SUPERFUND/EPCRA 40CFR302.4, Appendix A, Reportable Quantity (RQ): CERCLA, 5,000 (2,270 kg). Section 313: Form R *de minimis* concentration reporting level: 1.0% (as ammonia) Molecular weight: 114.12; NH_3 Equivalent weight: 14.92
- Canada, WHMIS, Ingredients Disclosure List

Cited in U.S. State Regulations: Alaska (G), California (A, G), Connecticut (A), Florida (G), Illinois (G), Massachusetts (G), Minnesota (G), Nevada (A), New Hampshire (G), New Jersey (G), North Dakota (A), Pennsylvania (G), Rhode Island (G), Virginia (A), West Virginia (G).

Description: Ammonium Sulfamate, $NH_2SO_3NH_4$, is a white to yellow crystalline solid. Freezing/Melting point = 131°C (with decomposition), Boiling point = 160°C.

Potential Exposure: Ammonium Sulfamate is used as a herbicide and in fire retardant compositions.

Incompatibilities: Strong oxidizers, potassium, potassium chlorate, sodium nitrite, metal chlorates, and hot acid solutions. Elevated temperatures cause a highly exothermic reaction with water.

Permissible Exposure Limits in Air: The OSHA PEL is (total dust) 15 mg/m^3 TWA and (respirable fraction) 5 mg/m^3 TWA for an 8-hour workshift. The NIOSH IDLH level is 1,5,000 mg/m^3. NIOSH REL is (total dust) 10 mg/m^3 and (respirable fraction) 5 mg/m^3. The MAK set by DFG is 15 mg/m^3.[3] The ACGIH-TWA value is 10 mg/m^3. Australia limit is 10 mg/m^3. Israel limit is 10 mg/m^3 with a 5 mg/m^3 Action Level. HSE level is 10 mg/m^3 TWA and a STEL of 20 mg/m^3. The DFG MAK value for total dust is 15 mg/m^3. In addition, several states have set guidelines or standards for Ammonium Sulfamate in ambient air:[60] (total dust) 15 mg/m^3 TWA and (respirable fraction). The California PEL is 5 mg/m^3 TWA. In addition, several states have set guidelines or standards for Ammonium sulfamate in ambient air:[60] 0.1 mg/m^3 (North Dakota), 0.15 mg/m^3 (Virginia), 0.2 mg/m^3 (Connecticut), 0.238 mg/m^3 (Nevada). Canadian provincial guidelines follow: 10 mg/m^3 TWA (8-hour workshift) and STEL of 20 mg/m^3 (Alberta, British Columbia), 10 mg/m^3 TWA (Ontario) 10 mg/m^3 TWAEV (Quebec).

Determination in Air: Collection on a filter followed by gravimetric analysis. See NIOSH Method #S348.

Permissible Concentration in Water: The No-Observed-Adverse-Effect-Level (NOAEL) is 250 mg/kg/day according to the EPA Health Advisory cited below. From this a health advisory of 21.4 mg/l of water was derived for a 10 kg child on a one-day, ten-day or longer term basis. An acceptable daily intake has been determined to be 0.214 mg/kg/day and a lifetime health advisory for a 70 kg adult is 1.5 mg/l.

Determination in Water: There is no standard method for determining Ammonium Sulfamate in water. There is, however, a method for detection in foods which is a colorimetric method based on liberation of SO_4, reduction to H_2S which is measured after treatment with zinc, p-aminodimethylaniline and ferric chloride to give methylene blue.

Routes of Entry: Inhalation, ingestion, skin and/or eye contact.

Harmful Effects and Symptoms

Short Term Exposure: This material is moderately toxic by ingestion and may cause gastrointestinal disease. High levels may irritate the eyes, skin and respiratory tract, nausea and vomiting. The oral LD_{50} for rat is 3,900 mg/kg.

Long Term Exposure: Unknown at this time.

Medical Surveillance: Nothing special indicated.

First Aid: If this chemical gets into the eyes, remove any contact lenses at once and irrigate immediately for at least 15 minutes, occasionally lifting upper and lower lids. Seek medical attention immediately. If this chemical contacts the skin, remove contaminated clothing and wash immediately with soap and water. Seek medical attention immediately. If this chemical has been inhaled, remove from exposure, begin rescue breathing (using universal precautions) if breathing has stopped and CPR if heart action has stopped. Transfer promptly to a medical facility. When this chemical has been swallowed, get medical attention. Give large quantities of water and induce vomiting. Do not make an unconscious person vomit.

Personal Protective Methods: Wear protective gloves and clothing to prevent any reasonable probability of skin contact. Safety equipment suppliers/manufacturers can provide recommendations on the most protective glove/clothing material for your operation. All protective clothing (suits, gloves, footwear, headgear) should be clean, available each day, and put on before work. Contact lenses should not be worn when working with this chemical. Wear dust-proof chemical goggles and face shield unless full facepiece respiratory protection is worn. Employees should wash immediately with soap when skin is wet or contaminated. Provide emergency showers and eyewash.

Respirator Selection: NIOSH: *50 mg/m³*: DM (any dust and mist respirator). *100 mg/m³*: DMXSQ (any dust and mist respirator except single-use and quarter mask respirators); or SA (any supplied-air respirator). *250 mg/m³*: SA:CF (any supplied-air respirator operated in a continuous-flow mode); or PAPRDM, if not present as a fume (any powered, air-purifying respirator with a dust and mist filter). *500 mg/m³*: SAT:CF (any supplied-air respirator that has a tight-fitting facepiece and is operated in a continuous-flow mode); or PAPRTHiE (any powered, air-purifying respirator with a tight-fitting facepiece and a high-efficiency particulate filter); or HiEF (any air-purifying, full-facepiece respirator with a high-efficiency particulate filter); or SCBAF (any self-contained breathing apparatus with a full facepiece); or SAF (any supplied-air respirator with a full facepiece). *1,500 mg/m³*: SA:PD,PP (any supplied-air respirator operated in a pressure-demand or other positive-pressure mode). *Emergency or planned entry into unknown concentrations or IDLH conditions:* SCBAF:PD,PP (any self-contained breathing apparatus that has a full facepiece and is operated in a pressure-demand or other positive-pressure mode); or SAF:PD,PP:ASCBA (Any supplied-air respirator that has a full facepiece and is operated in a pressure-demand or other positive-pressure mode in combination with an auxiliary, self-contained breathing apparatus operated in a pressure-demand or other positive-pressure mode. *Escape:* HiEF (Any air-purifying, full-facepiece respirator with a high-efficiency particulate filter); or SCBAE (Any appropriate escape-type, self-contained breathing apparatus).

Storage: Store in tightly closed containers in a cool, well-ventilated area away from Oxidizers (such as Perchlorates, Peroxides, Permanganates, Chlorates and Nitrates), Water, Potassium, Potassium Chlorate, Sodium Nitrite, Metal Chlorates and Hot Acid Solutions.

Shipping: There are no label requirements and no maximum quantity limits imposed for air shipment.

Spill Handling: Evacuate persons not wearing protective equipment from area of spill or leak until clean-up is complete. Remove all ignition sources. Collect powdered material in the most convenient and safe manner and deposit in sealed containers. Ventilate area after clean-up is complete. It may be necessary to contain and dispose of this chemical as a hazardous waste. If material or contaminated runoff enters waterways, notify downstream users of potentially contaminated waters. Contact your Department of Environmental Protection or your regional office of the federal EPA for specific recommendations. If employees are required to clean-up spills, they must be properly trained and equipped. OSHA 1910.120(q) may be applicable.

Fire Extinguishing: Ammonium Sulfamate may burn, but does not readily ignite. Use dry chemical, CO_2, or foam extinguishers.

Disposal Method Suggested: Consult with environmental regulatory agencies for guidance on acceptable disposal practices. Generators of waste containing this contaminant (≥100 kg/mo) must conform with EPA regulations governing storage, transportation, treatment, and waste disposal. Dilute with water, make neutral with acid or base and flush into sewer with more water.

References

Sax, N. I., Ed., Dangerous Properties of Industrial Materials Report, 2, No. 3, 52–54 (1982)

New Jersey Department of Health and Senior Services, "Hazardous Substance Fact Sheet: Ammonium Sulfamate," Trenton, NJ (February 1987)

U.S. Environmental Protection Agency, "Health Advisory: Ammonium Sulfamate," Washington, DC, Office of Drinking Water (August 1987)

Ammonium Sulfide

Molecular Formula: H_8N_2S

Common Formula: $(NH_4)_2S$

Synonyms: Ammonium bisulfide; Ammonium sulfide, hydrogen; Diammonium sulfide; Sulfuro amonico (Spanish); True ammonium sulfide

CAS Registry Number: 12124-99-1 (solid); 12135-76-1 (solution)

RTECS Number: BS4900000 (solid); BS4920000 (solution)

DOT ID: NA 2683 (solid); UN 2683 (solution)

Regulatory Authority

- Clean Water Act: 40CFR116.4 Hazardous Substances; 40CFR117.3, RQ (same as CERCLA)
- CERCLA/SARA Section 313: 40CFR302.4, Appendix A, Reportable Quantity (RQ): CERCLA, 100 lb (45.4 kg); Form R *de minimis* concentration reporting level: 1.0% (as ammonia) Molecular weight: 51.11; NH_3 Equivalent weight: 33.32

Cited in U.S. State Regulations: California (G), Illinois (G), New Hampshire (G), New Jersey (G), Oklahoma (G), Pennsylvania (G).

Description: Ammonium sulfide, $(NH_4)_2$, is a yellow crystalline (sugar or sand-like) material, commonly found in liquid solution, which are flammable, and with an odor of rotten eggs. Flash point = 72°C. Freezing/Melting point = (decomposes). Soluble in water.

Potential Exposure: It is used in photographic developers, synthetic flavors, coloring metals, and to make textiles.

Incompatibilities: Strong oxidizers, acids, acid fumes (forms hydrogen sulfide). Keep away form moisture.

Permissible Exposure Limits in Air: No standards set.

Permissible Concentration in Water: No criteria set.

Routes of Entry: Ingestion, skin absorption, inhalation.

Harmful Effects and Symptoms

Short Term Exposure: Ammonium sulfide can be absorbed through the skin, thereby increasing exposure. This substance is a corrosive chemical and contact can irritate and burn the eyes and skin. Exposure can irritate the nose, throat and lungs, causing a cough and difficulty in breathing. Very high levels could cause pulmonary edema, a medical emergency that can be delayed for several hours. This can cause death. Very high levels could cause you to feel dizzy, lightheaded and to pass out.

Long Term Exposure: Repeated exposure can cause lung irritation and bronchitis.

Points of Attack: Lungs.

Medical Surveillance: Lung function tests.

First Aid: If this chemical gets into the eyes, remove any contact lenses at once and irrigate immediately for at least 15 minutes, occasionally lifting upper and lower lids. Seek medical attention immediately. If this chemical contacts the skin, remove contaminated clothing and wash immediately with soap and water. Seek medical attention immediately. If this chemical has been inhaled, remove from exposure, begin rescue breathing (using universal precautions) if breathing has stopped and CPR if heart action has stopped. Transfer promptly to a medical facility. When this chemical has been swallowed, get medical attention. If victim is conscious, administer water or milk. Do not induce vomiting. Medical observation is recommended for 24 - 48 hours after breathing overexposure, as pulmonary edema may be delayed. As first aid for pulmonary edema, a doctor or authorized paramedic may consider administering a corticosteroid spray.

Personal Protective Methods: Where possible, enclose operations and use local exhaust ventilation at the site of chemical release. If local exhaust ventilation or enclosure is not used, respirators should be worn. Wear protective work clothing and goggles. Wash thoroughly immediately after exposure to Ammonium Sulfide and at the end of the workshift.

Respirator Selection: *Where there is no REL, at any detectable concentration:* SCBAF:PD,PP (any MSHA/NIOSH approved self-contained breathing apparatus that has a full facepiece and is operated in a pressure-demand or other positive-pressure mode); or SAF:PD,PP:ASCBA (any supplied-air respirator that has a full facepiece and is operated in a pressure-demand or other positive-pressure mode in combination with an auxiliary, self-contained breathing apparatus operated in a pressure-demand or other positive pressure mode). *Escape:* GMFOVHiE [any air-purifying, full-facepiece respirator (gas mask) with a chin-style, front- or back-mounted organic vapor canister having a high-efficiency particulate filter]; or SCBAE (any appropriate escape-type, self-contained breathing apparatus).

Storage: Prior to working with this chemical you should be trained on its proper handling and storage. Store to avoid contact with acids and acid fumes since violent reactions can occur. Store in tightly-closed containers in cool, well-ventilated area away from moisture. Use only non-sparking tools and equipment, especially when opening and closing containers of this chemical. Sources of ignition such as smoking and open flames, are prohibited where this chemical is used, handled, or stored in a manner that could create a potential fire or explosion hazard.

Shipping: The solution has a DOT/UN label requirement of "Corrosive, Poison, Flammable Liquid." The DOT/UN Hazard Class is 8 and the Packing Group is II.[19][20]

Spill Handling: Evacuate persons not wearing protective equipment from area of spill or leak until clean-up is complete. Remove all ignition sources. Add Ferric Chloride solution to spilled material. Stir and add Soda Ash. Sweep solid material up and deposit in sealed containers. Keep Ammonium Sulfide out of a confined space, such as a sewer, because of the potential for an explosion, unless the sewer is designed to prevent the build-up of explosive concentrations. Ventilate area after clean-up is complete. It may be necessary to contain and dispose of this chemical as a hazardous waste. If material or contaminated runoff enters waterways, notify downstream users of potentially contaminated waters. Contact your Department of Environmental Protection or your regional office of the federal EPA for specific recommendations. If employees are required to clean-up spills, they must be properly trained and equipped. OSHA 1910.120(q) may be applicable.

Fire Extinguishing: Ammonium sulfide solution is a Flammable Liquid. Poisonous gases are produced in fire, including flammable hydrogen sulfide and ammonia. Use dry chemical, carbon dioxide, or alcohol foam extinguishers. Vapors are heavier than air and will collect in low areas. Vapors in confined areas may explode when exposed to fire. Storage containers and parts of containers may rocket great distances, in many directions. If material or contaminated runoff enters waterways, notify downstream users of potentially contaminated waters. Notify local health and fire officials and pollution control agencies. From a secure, explosion-proof location, use water spray to cool exposed containers. If cooling streams are ineffective (venting sound increases in volume and pitch, tank discolors, or shows any signs of deforming), withdraw immediately to a secure position. If employees are expected to fight fires, they must be trained and equipped in OSHA 1910.156.

Disposal Method Suggested: Add to a large volume of ferric chloride solution with stirring. Then neutralize with soda ash. Then flush to drain with water.

References

Sax, N. I., Ed., "Dangerous Properties of Industrial Materials Report," 2, No. 4, 27–29 (1982)

New Jersey Department of Health and Senior Services, "Hazardous Substance Fact Sheet: Ammonium Sulfide," Trenton, NJ (January 1996)

Ammonium Sulfite

Molecular Formula: $H_8N_2O_3S$

Common Formula: $(NH_4)_2SO_3$

Synonyms: *monoammonium salt:* Ammonium acid sulfite; Ammonium hydrogen sulfite; Ammonium hydrosulfite; Ammonium monosulfite; Ammonium sulfite, hydrogen; Monosodium sulfite; Sulfito amonico (Spanish); Sulfurous acid, Monoammonium salt

diammonium salt: Diammonium sulfite; Sulfito amonico (Spanish); Sulfurous acid, Diammonium salt

CAS Registry Number: 10196-04-0 (diammonium salt); 10192-30-0 (monoammonium salt)

RTECS Number: WT3505000 (diammonium); WT3595000 (monoammonium)

DOT ID: NA 9090 (ammonium sulfite)

Regulatory Authority

- Clean Water Act: 40CFR116.4 Hazardous Substances; 40CFR117.3, RQ (same as CERCLA)
- (diammonium salt) SUPERFUND/EPCRA 40CFR302.4, Appendix A, Reportable Quantity (RQ): CERCLA, 5,000 lb (2,270 kg); Section 313: Form R *de minimis* concentration reporting level: 1.0% (as ammonia) Molecular weight: 99.10; NH_3 Equivalent weight: 17.18
- (monoammonium salt) SUPERFUND/EPCRA 40CFR302.4, Appendix A, Reportable Quantity (RQ): CERCLA, 5,000 lb (2,270 kg); Section 313: Form R *de minimis* concentration reporting level: 1.0% (as ammonia) Molecular weight: 116.13; NH_3 Equivalent weight: 29.33

Cited in U.S. State Regulations: California (G), Illinois (G), Massachusetts (G), New Hampshire (G), New Jersey (G), Pennsylvania (G).

Description: Ammonium sulfite, $(NH_4)_2SO_3$, is a colorless to yellow crystalline (sand-like or sugar-like) solid, normally sold or used in a solution. Freezing/Melting point = 150°C (sublimes). Soluble in water.

Potential Exposure: Ammonium sulfite is used in medicines, metal lubricants, explosives, photography, hair wave solutions, and to make other chemicals. It is also used as a preservative, and treating agricultural grain.

Incompatibilities: A strong reducing agent. Reacts violently with strong oxidizers, acids.

Permissible Exposure Limits in Air: No standards set.

Permissible Concentration in Water: No criteria set.

Routes of Entry: Inhalation of dust, ingestion.

Harmful Effects and Symptoms

Short Term Exposure: Irritates the eyes, skin, and respiratory tract. Ammonium sulfite can affect you when breathed in; exposure can irritate the nose, throat, bronchial tubes, and lungs. Higher exposures can cause pulmonary edema, a medical emergency that can be delayed for several hours. This can cause death.

Long Term Exposure: Ammonium sulfite may cause an asthma-like allergy. Future exposures could then cause asthma attacks with cough, shortness of breath and wheezing. Very severe (anaphylactic) reactions could also occur, and could be fatal.

Points of Attack: Skin, eyes, respiratory system.

Medical Surveillance: Before beginning employment and at regular times after that, for those with frequent or potentially high exposure, the following are recommended: lung function tests; seek prompt medical attention if symptoms are suspected.

First Aid: If this chemical gets into the eyes, remove any contact lenses at once and irrigate immediately for at least 15 minutes, occasionally lifting upper and lower lids. Seek medical attention immediately. If this chemical contacts the skin, remove contaminated clothing and wash immediately with soap and water. Seek medical attention immediately. If this chemical has been inhaled, remove from exposure, begin rescue breathing (using universal precautions) if breathing has stopped and CPR if heart action has stopped. Transfer promptly to a medical facility. When this chemical has been swallowed, get medical attention. Give large quantities of water and induce vomiting. Do not make an unconscious person vomit. Medical observation is recommended for 24 - 48 hours after breathing overexposure, as pulmonary edema may be delayed. As first aid for pulmonary edema, a doctor or authorized paramedic may consider administering a corticosteroid spray.

Personal Protective Methods: Wear protective gloves and clothing to prevent any reasonable probability of skin contact. Safety equipment suppliers/manufacturers can provide recommendations on the most protective glove/clothing material for your operation. All protective clothing (suits, gloves, footwear, headgear) should be clean, available each day, and put on before work. Contact lenses should not be worn when working with this chemical. Wear splash or dust-proof chemical goggles and face shield unless full facepiece respiratory protection is worn. Employees should wash immediately with soap when skin is wet or contaminated. Provide emergency showers and eyewash.

Respirator Selection: Where the potential exists for exposures over Ammonium Sulfite, use a MSHA/NIOSH approved full facepiece respirator with a high efficiency particulate filter. Greater protection is provided by a powered-air purifying respirator. Where the potential exists for high exposures to Ammonium Sulfite, use a MSHA/NIOSH approved supplied-air respirator with a full facepiece operated in the positive pressure mode or with a full facepiece, hood, or helmet in the continuous flow mode, or use a MSHA/NIOSH approved self-contained breathing apparatus with a full facepiece operated in pressure-demand or other positive pressure mode.

Storage: Prior to working with this chemical you should be trained on its proper handling and storage. Ammonium Sulfite must be stored in tightly closed containers in a cool, well-ventilated place to avoid contact with oxidizers and acids as flammable hydrogen sulfide gas is produced.

Shipping: Ammonium sulfite is not specifically cited in DOT's Performance Oriented Packaging Standards.

Spill Handling: Evacuate and restrict persons not wearing protective equipment from area of spill or leak until cleanup is complete. Remove all ignition sources. Ventilate area of spill or leak. Absorb liquids in vermiculite, dry sand, earth, or a similar material and deposit in sealed containers. Collect powdered material in the most convenient and safe manner and deposit in sealed containers. It may be necessary to contain and dispose of this chemical as a hazardous waste. If material or contaminated runoff enters waterways, notify downstream users of potentially contaminated waters. Contact your Department of Environmental Protection or your regional office of the federal EPA for specific recommendations. If employees are required to clean-up spills, they must be properly trained and equipped. OSHA 1910.120(q) may be applicable.

Fire Extinguishing: Ammonium Sulfite may burn, but does not readily ignite. Use dry chemical, carbon dioxide, water spray, or alcohol foam extinguishers. Poisonous gases are produced in fire including ammonia, sulfur oxides, and nitrogen oxides. If material or contaminated runoff enters waterways, notify downstream users of potentially contaminated waters. Notify local health and fire officials and pollution control agencies. From a secure, explosion-proof location, use water spray to cool exposed containers. If cooling streams are ineffective (venting sound increases in volume and pitch, tank discolors, or shows any signs of deforming), withdraw immediately to a secure position. If employees are expected to fight fires, they must be trained and equipped in OSHA 1910.156.

Disposal Method Suggested: Incineration. May be buried in a chemical waste landfill in accordance with federal, state, and local statutes; or, if oxidized and neutralized, it may be sent to a municipal sewage treatment plant for biological treatment Incineration. Or, if oxidized and neutralized it can be sent to a municipal sewage treatment plant for biological treatment.

References

New Jersey Department of Health and Senior Services, "Hazardous Substance Fact Sheet: Ammonium Sulfite," Trenton, NJ (January 11, 1988)

Ammonium Tartrate

Molecular Formula: $C_4H_{12}N_2O_6$

Common Formula: $NH_4OOCCHOHCHOHCOONH_4$

Synonyms: Ammonium D-tartrate; Ammonium tartrate; Butanedioic acid, 2,3-dihydroxy-[R-(R*,R*)]-, diammonium salt; Diammonium tartrate; 2,3-Dihydroxy-butanedioic acid, diammonium salt; Tartaric acid, ammonium salt; 1-Tartaric acid, diammonium salt; Tartaric acid, diammonium salt; Tartrato amonico (Spanish)

CAS Registry Number: 14307-43-8; 3164-29-2 (diammonium salt)

RTECS Number: WW8050000

DOT ID: NA 9091

Regulatory Authority

- Clean Water Act: 40CFR116.4 Hazardous Substances; 40CFR117.3, RQ (same as CERCLA)
- SUPERFUND/EPCRA 40CFR302.4 Reportable Quantity (RQ): CERCLA, 5,000 lb (2,270 kg)

Cited in U.S. State Regulations: California (G), Massachusetts (G), New Hampshire (G), New Jersey (G), Pennsylvania (G).

Description: Ammonium tartrate, $NH_4OOCCHOHCHOH COONH_4$, is a colorless crystalline or white granular solid. Hazard Identification (based on NFPA-704 M Rating System): Health 1, Flammability 1, Reactivity 0. Soluble in water.

Potential Exposure: It is used in the textile industry and in medicine.

Incompatibilities: Strong oxidizers, especially potassium chlorate, sodium nitrite.

Permissible Exposure Limits in Air: No standards set.

Permissible Concentration in Water: No criteria set.

Routes of Entry: Inhalation, skin contact, ingestion.

Harmful Effects and Symptoms

Short Term Exposure: A corrosive. Contact can irritate the eyes and skin, and cause permanent damage. May irritate the nose, throat and lungs. Higher exposures can cause pulmonary edema, a medical emergency that can be delayed for several hours. This can cause death.

Long Term Exposure: Unknown at this time. However corrosive substances may cause lung effects and damage.

Points of Attack: Lungs, eyes, skin.

Medical Surveillance: Lung function tests. Consider x-ray following acute overexposure.

First Aid: If this chemical gets into the eyes, remove any contact lenses at once and irrigate immediately for at least 15 minutes, occasionally lifting upper and lower lids. Seek medical attention immediately. If this chemical contacts the skin, remove contaminated clothing and wash immediately with soap and water. Seek medical attention immediately. If this chemical has been inhaled, remove from exposure, begin rescue breathing (using universal precautions) if breathing has stopped and CPR if heart action has stopped. Transfer promptly to a medical facility. When this chemical has been swallowed, get medical attention. Give large quantities of water and induce vomiting. Do not make an unconscious person vomit. Medical observation is recommended for 24 - 48 hours after breathing overexposure, as pulmonary edema may be delayed. As first aid for pulmonary edema, a doctor or authorized paramedic may consider administering a corticosteroid spray.

Personal Protective Methods: Where possible, enclose operations and use local exhaust ventilation at the site of chemical release. Wear protective gloves and clothing to prevent any reasonable probability of skin contact. Safety equipment suppliers/manufacturers can provide recommendations on the most protective glove/clothing material for your operation. All protective clothing (suits, gloves, footwear, headgear) should be clean, available each day, and put on before work. Contact lenses should not be worn when working with this chemical. Wear dust-proof chemical goggles and face shield unless full facepiece respiratory protection is worn. Employees should wash immediately with soap when skin is wet or contaminated. Provide emergency showers and eyewash.

Respirator Selection: Engineering controls should be used wherever feasible to maintain airborne concentrations of this chemical below the prescribed exposure limit. Respirators and protective equipment less effective than engineering controls, and should be used only in non-routine or emergency situations which may result in exposure concentrations in excess of the TWA environmental limit.

At any concentrations above the NIOSH REL, or where there is no REL, at any detectable concentration: SCBAF:PD,PP (any MSHA/NIOSH approved self-contained breathing apparatus that has a full facepiece and is operated in a pressure-demand or other positive-pressure mode); or SAF:PD,PP:ASCBA (any supplied-air respirator that has a full facepiece and is operated in a pressure-demand or other positive-pressure mode in combination with an auxiliary, self-contained breathing apparatus operated in a pressure-demand or other positive pressure mode). *Escape:* GMFOVHiE [any air-purifying, full-facepiece respirator (gas mask) with a chin-style, front- or back-mounted organic vapor canister having a high-efficiency particulate filter]; or SCBAE (any appropriate escape-type, self-contained breathing apparatus).

Storage: Prior to working with this chemical you should be trained on its proper handling and storage. Store in tightly closed containers in a cool, well-ventilated area away from potassium chloride and sodium nitrite.

Shipping: There are no label requirements or maximum limits on quantities, which can be shipped by air. The DOT/UN Hazard Class is ORM-E and the Packing Group is III.[19][20]

Spill Handling: Evacuate persons not wearing protective equipment from area of spill or leak until clean-up is complete. Remove all ignition sources. Collect powdered material in the most convenient and safe manner and deposit in sealed containers. Ventilate area after clean-up is complete. It may be necessary to contain and dispose of this chemical as a hazardous waste. If material or contaminated runoff enters waterways, notify downstream users of potentially contaminated waters. Contact your Department of Environmental Protection or your regional office of the federal EPA for specific recommendations. If employees are required to clean-up spills, they must be properly trained and equipped. OSHA 1910.120(q) may be applicable.

Fire Extinguishing: Ammonium Tartrate may burn, but does not readily ignite. Poisonous gases are produced in fire, including Ammonia and Oxides of Nitrogen. Use dry chemical, carbon dioxide, water spray, or alcohol foam extinguishers. Poisonous gases are produced in fire. If material or contaminated runoff enters waterways, notify downstream users of potentially contaminated waters. Notify local health and fire officials and pollution control agencies. From a secure, explosion-proof location, use water spray to cool exposed containers. If cooling streams are ineffective (venting sound increases in volume and pitch, tank discolors, or shows any signs of deforming), withdraw immediately to a secure position. If employees are expected to fight fires, they must be trained and equipped in OSHA 1910.156.

Disposal Method Suggested: Incineration. May be buried in a chemical waste landfill in accordance with federal, state, and local statutes; or, if oxidized and neutralized, it may be sent to a municipal sewage treatment plant for biological treatment.

References

New Jersey Department of Health and Senior Services, "Hazardous Substance Fact Sheet: Ammonium Tartrate," Trenton, NJ (February 1987)

Ammonium Tetrachloroplatinate

Molecular Formula: $Cl_4H_8N_2Pt$

Common Formula: $(NH_4)_2PtCl_4$

Synonyms: Ammonium chloropalladate(2+); Ammonium chloropalladate(II); Ammonium palladium chloride; Ammonium tetrachloropalladate(2+); Ammonium tetrachloropalladate(II); Diammonium tetrachloropalladate; Tetramine platinum (II) chloride

CAS Registry Number: 13820-41-2

RTECS Number: TP1840000

Regulatory Authority

- Air Pollutant Standard Set (ACGIH)[1] (OSHA)[58]
- SUPERFUND/EPCRA 40CFR302.4, Appendix A, Reportable Quantity (RQ): CERCLA, 100 lb (45.4 kg); Section 313: Form R *de minimis* concentration reporting level: 1.0% (as ammonia) Molecular weight: 284.81; NH_3 Equivalent weight: 11.98

Cited in U.S. State Regulations: California (G), New Jersey (G), Pennsylvania (G).

Description: Ammonium tetrachloroplatinate, $(NH_4)_2PtCl_4$, is a dark ruby-red crystalline solid. Freezing/Melting point = 140 – 150°C. Soluble in water.

Potential Exposure: This material is used in photography.

Incompatibilities: Oxidizers, strong acids, strong bases.

Permissible Exposure Limits in Air: These exposure limits are recommended for platinum soluble salts as platinum. OSHA: The legal airborne permissible exposure limit (PEL) is 0.002 mg/m^3 averaged over an 8-hour workshift. ACGIH: The recommended airborne exposure limit is 0.002 mg/m^3 averaged over an 8-hour workshift. NIOSH: The recommended airborne exposure limit is 0.002 mg/m^3 averaged over an 10-hour workshift.

Permissible Concentration in Water: No criteria set.

Routes of Entry: Inhalation of vapor or dust, skin contact.

Harmful Effects and Symptoms

Short Term Exposure: Ammomium Tetrachloroplatinate can affect you when breathed in. Ammonium Tetrachloroplatinate may irritate the eyes, nose and throat, high exposures can cause irritability and even seizures ("fits").

Long Term Exposure: Severe allergy can develop to Ammonium Tetrachloroplatinate. Symptoms may include asthma (with cough, wheezing and/or shortness of breath), runny nose and/or skin rash, sometimes with hives. If allergy develops, even small future exposure can trigger significant symptoms. Some persons exposed to this type of chemical have developed lung scarring.

Points of Attack: Eyes, skin, nose, throat.

Medical Surveillance: Before beginning employment and at regular times after that, the following are recommended: Lung function tests. These may be normal if person is not having an attack at the time. If symptoms develop or overexposure is suspected, the following may be useful: Chest x-ray every 3 years should be considered if above tests are not normal. Evaluation by a qualified allergist, including careful exposure history and special testing, may help diagnose skin allergy.

First Aid: If this chemical gets into the eyes, remove any contact lenses at once and irrigate immediately for at least 15 minutes, occasionally lifting upper and lower lids. Seek medical attention immediately. If this chemical contacts the skin, remove contaminated clothing and wash immediately with soap and water. Seek medical attention immediately. If this chemical has been inhaled, remove from exposure, begin rescue breathing (using universal precautions) if breathing has stopped and CPR if heart action has stopped. Transfer promptly to a medical facility. When this chemical has been swallowed, get medical attention. Give large quantities of water and induce vomiting. Do not make an unconscious person vomit.

Personal Protective Methods: Wear protective gloves and clothing to prevent any reasonable probability of skin contact. Safety equipment suppliers/manufacturers can provide recommendations on the most protective glove/clothing material for your operation. All protective clothing (suits, gloves, footwear, headgear) should be clean, available each day, and put on before work. Contact lenses should not be worn when working with this chemical. Wear dust-proof chemical goggles and face shield unless full facepiece respiratory protection is worn. Employees should wash immediately with soap when skin is wet or contaminated. Provide emergency showers and eyewash.

Respirator Selection: Where the potential exists for exposures over 0.002 mg/m^3, use a MSHA/NIOSH approved supplied-air respirator with a full facepiece operated in the positive pressure mode or with a full facepiece, hood, or helmet in the continuous flow mode, or use a MSHA/NIOSH approved self-contained breathing apparatus with a full facepiece operated in pressure-demand or other positive pressure mode.

Storage: Prior to working with this chemical you should be trained on its proper handling and storage. Store in tightly closed containers in a cool, well-ventilated area away from oxidizers, strong acids, strong bases.

Shipping: This material is not specifically cited in the DOT performance-oriented packaging standards.

Spill Handling: Evacuate persons not wearing protective equipment from area of spill or leak until clean-up is complete. Remove all ignition sources. Collect powdered material in the most convenient and safe manner and deposit in sealed containers. Ventilate area after clean-up is complete. It may be necessary to contain and dispose of this chemical as a hazardous waste. If material or contaminated runoff enters waterways, notify downstream users of potentially contaminated waters. Contact your Department of Environmental Protection or your regional office of the federal EPA for specific recommendations. If employees are required to clean-up spills, they must be properly trained and equipped. OSHA 1910.120(q) may be applicable.

Fire Extinguishing: Use dry chemical, CO_2, water spray, or alcohol foam extinguishers. Poisonous gases are produced in fire including ammonia, chlorine, and platinum. If material or contaminated runoff enters waterways, notify downstream users of potentially contaminated waters. Notify local health and fire officials and pollution control agencies. From a secure, explosion-proof location, use water spray to cool exposed containers. If cooling streams are ineffective (venting sound increases in volume and pitch, tank discolors, or shows any signs of deforming), withdraw immediately to a secure position. If employees are expected to fight fires, they must be trained and equipped in OSHA 1910.156.

References

New Jersey Department of Health and Senior Services, "Hazardous Substance Fact Sheet: Ammonium Tetrachloroplatinate," Trenton, NJ (January 1996)

Ammonium Thiocyanate

Molecular Formula: CH_4N_2S

Common Formula: NH_4SCN

Synonyms: Ammonium isothiocyanate; Ammonium rhodanate; Ammonium rhodanide; Ammonium sulfocyanate; Ammonium sulfocyanide; Carbo-Tech ammonium thiocyanate; Degussa ammonium thiocyanate; Rhodanid; Thiocyanic acid, ammonium salt; Tiocianato amonico (Spanish)

CAS Registry Number: 1762-95-4

RTECS Number: XK7875000

DOT ID: UN 9092

EINECS Number: 217-175-6

Regulatory Authority

- Clean Air Act 42USC7412; Title I, Part A, §112 hazardous pollutants (as cyanide compounds)
- Clean Water Act: 40CFR116.4 Hazardous Substances; 40CFR117.3, RQ (same as CERCLA); 40CFR423, Appendix A, Priority Pollutants (as cyanide compounds)
- EPA HAZARDOUS WASTE NUMBER (RCRA No.): P030
- RCRA 40CFR261, Appendix 8; 40CFR261.11 Hazardous Constituents
- RCRA Land Ban Waste Restrictions
- SUPERFUND/EPCRA 40CFR302.4, Appendix A, Reportable Quantity (RQ): CERCLA, 5,000 lb (2,270 kg)
- EPCRA Section 313 Form R *de minimis* concentration reporting level: 1.0% (as ammonia) Molecular weight: 76.12; NH_3 Equivalent weight: 22.37 *Note:* May also be reportable as a cyanide compounds (X+CN- where X = H+ or any other group where a formal dissociation may occur. Form R *de minimis* concentration reporting level: 1.0%
- U.S. DOT Regulated Marine Pollutant (49CFR172.101, Appendix B) as cyanide compounds

Cited in U.S. State Regulations: California (A, G), Illinois (G), Massachusetts (G), New Hampshire (G), New Jersey (G), Pennsylvania (G).

Description: Ammonium Thiocyanate, NH_4SCN, is a colorless solid which absorbs moisture and becomes liquid. Freezing/Melting point = 150°C; decomposes at 170°C. Highly soluble in water.

Potential Exposure: It has many uses in making matches, fabric processing, metals processing, chemical manufacturing, electroplating, zinc coating, liquid rocket propellents, herbicides, weed killers, and defoliants, fabric dyeing, polymerization catalyst, in photography. Used as a laboratory chemical.

Incompatibilities: Lead nitrate, chlorates, nitric acid, acid, acid fumes. In the presence of moisture, corrosive to brass, copper, iron.

Permissible Exposure Limits in Air: No occupational exposure limits have been established for this chemical; however, inasmuch as it is a cyanide compound, the exposure limits are listed below:

OSHA and ACGIH: 5 mg/m^3 TWA; NIOSH: Ceiling limit, 4.7 ppm; 5 mg/m^3 per 10 minutes as cyanides. All have notations that skin contact contributes significantly in overall exposure. IDLH = 25 mg/m^3 as CN

Permissible Concentration in Water: No criteria set.

Routes of Entry: Skin and/or eyes.

Harmful Effects and Symptoms

Short Term Exposure: Contact may irritate and cause burns to the eyes and skin. This chemical can be absorbed through the skin, and significantly to overall exposure.

Long Term Exposure: Repeated exposure may cause nausea, loss of appetite, runny nose, abdominal problems, loss of weight, weakness, and skin rashes. Prolonged exposure may cause thyroid gland problems, blood cell damage, nervous system damage with personality and mood changes.

Points of Attack: Thyroid, nervous system, blood cells.

Medical Surveillance: Thyroid function tests, (CBC) complete blood count.

First Aid: If this chemical gets into the eyes, remove any contact lenses at once and irrigate immediately for at least 15 minutes, occasionally lifting upper and lower lids. Seek medical attention immediately. If this chemical contacts the skin, remove contaminated clothing and wash immediately with soap and water. Seek medical attention immediately. If this chemical has been inhaled, remove from exposure, begin rescue breathing (using universal precautions) if breathing has stopped and CPR if heart action has stopped. Transfer promptly to a medical facility. When this chemical has been swallowed, get medical attention. Give large quantities of water and induce vomiting. Do not make an unconscious person vomit.

Personal Protective Methods: Wear protective gloves and clothing to prevent any reasonable probability of skin contact. Safety equipment suppliers/manufacturers can provide recommendations on the most protective glove/clothing material for your operation. Non-absorbent material are recommended. All protective clothing (suits, gloves, footwear, headgear) should be clean, available each day, and put on before work. Contact lenses should not be worn when working with this chemical. Wear splash or dust-proof chemical goggles and face shield unless full facepiece respiratory protection is worn. Employees should wash immediately with soap when skin is wet or contaminated. Provide emergency showers and eyewash.

Respirator Selection: NIOSH/OSHA (as cyanides): *up to 25 mg/m^3:* SA (any supplied-air respirator); or SCBAF (any self-contained breathing apparatus with full facepiece). *Emergency or planned entry into unknown concentrations or IDLH conditions:* SCBAF:PD,PP (any self-contained breathing apparatus that has a full facepiece and is operated in a pressure-demand or other positive-pressure mode); or SAF:PD,PP:ASCBA (Any supplied-air respirator that has a full facepiece and is operated in a pressure-demand or other positive-pressure mode in combination with an auxiliary self-contained breathing apparatus operated in a pressure-demand or other positive pressure mode. *Escape:* GMFSHiE (Any air-purifying, full-facepiece respirator (gas mask) with a chin-style, front-or back-, mounted canister providing protection against the compound of concern and having a high efficiency particulate filter); or SCBAE (Any appropriate escape-type, self-contained breathing apparatus).

Storage: Prior to working with this chemical you should be trained on its proper handling and storage. Ammonium Thiocyanate must be stored to avoid contact with Potassium Chlorate and Lead Nitrate since violent reactions occur. Store in tightly closed containers in a cool, well-ventilated area away from moisture, acid, acid fumes, or chlorine because toxic fumes are released. Where possible, automatically pump liquid from drums or other storage containers to process containers.

Shipping: There are no special shipping labels required nor are there any quantity limitations n air shipments. The UN/DOT Hazard Class is ORM-E and the Packing Group III.[19][20]

Spill Handling: Evacuate persons not wearing protective equipment from area of spill or leak until clean-up is complete. Remove all ignition sources. Use vacuum to reduce dust during clean up. *Do not dry sweep.* Collect powdered material in the most convenient and safe manner and deposit in sealed containers. Ventilate area after clean-up is complete. It may be necessary to contain and dispose of this chemical as a hazardous waste. If material or contaminated runoff enters waterways, notify downstream users of potentially contaminated waters. Contact your Department of Environmental Protection or your regional office of the federal EPA for specific recommendations. If employees are required to clean-up spills, they must be properly trained and equipped. OSHA 1910.120(q) may be applicable.

Fire Extinguishing: Ammonium Thiocyanate may burn, but does not readily ignite. Use dry chemical, CO_2, water spray, or foam extinguishers. Poisonous gases are produced in fire, including ammonia, nitrogen oxides, sulfur oxides and cyanide. If material or contaminated runoff enters waterways, notify downstream users of potentially contaminated waters. Notify local health and fire officials and pollution control agencies. From a secure, explosion-proof location, use water spray to cool exposed containers. If cooling streams are ineffective (venting sound increases in volume and pitch, tank discolors, or shows any signs of deforming), withdraw immediately to a secure position. If employees are expected to fight fires, they must be trained and equipped in OSHA 1910.156.

Disposal Method Suggested: Consult with environmental regulatory agencies for guidance on acceptable disposal practices. Generators of waste containing this contaminant (≥100 kg/mo) must conform with EPA regulations governing storage, transportation, treatment, and waste disposal. Slowly add to large container of water. Stir in slight excess of soda ash. Decant or siphon liquid from sludge, neutralize with HCl and flush to sewer. Sludge may be landfilled.

References

Sax, N. I., Ed., "Dangerous Properties of Industrial Materials Report" 2, No. 3, 54–5, (1982)

New Jersey Department of Health and Senior Services, "Hazardous Substance Fact Sheet: Ammonium Thiocyanate," Trenton, NJ (January 1996)

Ammonium Thiosulfate

Molecular Formula: $H_8N_2O_3S_2$

Synonyms: Ammonium hyposulfite; Amthio; Diammonium thiosulfate; Thiosulfuric acid, diammonium salt; Tiosulfato amonico (Spanish)

CAS Registry Number: 7783-18-8

RTECS Number: XN6465000

DOT ID: UN 3077 (solid); UN 3082 (solution)

Regulatory Authority

- CERCLA/SARA Section 313: Form R *de minimis* concentration reporting level: 1.0% (as ammonia) Molecular weight: 148.20; NH_3 Equivalent weight: 22.98

Cited in U.S. State Regulations: California (G), Massachusetts (G), New Jersey (G), Pennsylvania (G).

Description: Ammonium Thiosulfate is a white crystalline solid with an ammonia odor. Freezing/Melting point = 150°C (decomposes below Freezing/Melting point. solution: <50°C, anhydrous crystals> 100°C). Hazard Identification (based on NFPA-704 M Rating System): Health 1, Flammability 0, Reactivity 0. Highly soluble in water.

Potential Exposure: Used as an agricultural chemical and fungicide, metal lubricant, cleaning metals, and in photographic chemicals, making other chemicals, a laboratory reagent.

Incompatibilities: Contact with sodium chlorate may cause a violent reaction. Corrodes brass, copper, and copper-based metals.

Permissible Exposure Limits in Air: None established. However, care must be taken in its use.

Routes of Entry: Skin contact.

Harmful Effects and Symptoms

Short Term Exposure: Exposure can irritate eyes, skin and respiratory tract.

First Aid: If this chemical gets into the eyes, remove any contact lenses at once and irrigate immediately for at least 15 minutes, occasionally lifting upper and lower lids. Seek medical attention immediately. If this chemical contacts the skin, remove contaminated clothing and wash immediately with soap and water. Seek medical attention immediately. If this chemical has been inhaled, remove from exposure, begin rescue breathing (using universal precautions) if breathing has stopped and CPR if heart action has stopped. Transfer promptly to a medical facility. When this chemical has been swallowed, get medical attention. Give large quantities of water and induce vomiting. Do not make an unconscious person vomit.

Personal Protective Methods: Wear protective gloves and clothing to prevent any reasonable probability of skin contact. Safety equipment suppliers/manufacturers can provide recommendations on the most protective glove/clothing material for

your operation. All protective clothing (suits, gloves, footwear, headgear) should be clean, available each day, and put on before work. Contact lenses should not be worn when working with this chemical. Wear splash- or dust-proof chemical goggles and face shield unless full facepiece respiratory protection is worn. Employees should wash immediately with soap when skin is wet or contaminated. Provide emergency showers and eyewash.

Respirator Selection: SA (any supplied-air respirator); or SCBAF (any self-contained breathing apparatus with a full facepiece). *Emergency or planned entry into unknown concentrations or IDLH conditions:* SCBAF:PD,PP (any self-contained breathing apparatus that has a full facepiece and is operated in a pressure-demand or other positive-pressure mode); or SAF:PD,PP:ASCBA (any supplied-air respirator that has a full facepiece and is operated in a pressure-demand or other positive-pressure mode in combination with an auxiliary, self-contained breathing apparatus operated in a pressure-demand or other positive-pressure mode). *Escape:* GMFOVHiE [any air-purifying, full-facepiece respirator (gas mask) with a chin-style, front- or back-mounted organic vapor canister having a high-efficiency particulate filter]; or SCBAE (any appropriate escape-type, self-contained breathing apparatus).

Storage: Prior to working with this chemical you should be trained on its proper handling and storage. Store in tightly closed containers in a cool, well-ventilated area away from oxidizers, acids, water, combustible materials.

Shipping: Environmental hazards, both solid and liquid, are in Packing Group III, Hazard Class 9, and must be labeled "Class 9."

Spill Handling: Evacuate persons not wearing protective equipment from area of spill or leak until clean-up is complete. Remove all ignition sources. Ventilate closed spaces before entering them. Absorb liquid with sand, vermiculite, earth, or similar absorbent material and place into containers for later disposal. Collect powdered material in the most convenient and safe manner and deposit in sealed containers. Ventilate area after clean-up is complete. It may be necessary to contain and dispose of this chemical as a hazardous waste. If material or contaminated runoff enters waterways, notify downstream users of potentially contaminated waters. Contact your Department of Environmental Protection or your regional office of the federal EPA for specific recommendations. If employees are required to clean-up spills, they must be properly trained and equipped. OSHA 1910.120(q) may be applicable.

Fire Extinguishing: Ammonium thiosulfate may burn, but does not readily ignite. Poisonous gases including ammonia, hydrogen sulfide, sulfur oxides, and nitrogen oxides are produced in fire. Use extinguishing agents suitable for surrounding fire. If material or contaminated runoff enters waterways, notify downstream users of potentially contaminated waters. Notify local health and fire officials and pollution control agencies. From a secure, explosion-proof location, use water spray to cool exposed containers. If cooling streams are ineffective (venting sound increases in volume and pitch, tank discolors, or shows any signs of deforming), withdraw immediately to a secure position. If employees are expected to fight fires, they must be trained and equipped in OSHA 1910.156.

Disposal Method Suggested: Incinerate. It may be possible to dispose of waste material at a municipal facility if treated, neutralized and oxidized.

References

New Jersey Department of Health and Senior Services, Hazardous Substance Fact Sheet, Ammonium Thiosulfate, Trenton, NJ (June 1988)

Amphetamine

Molecular Formula: $C_9H_{13}N$

Common Formula: $C_6H_5CH_2CH(NH_2)CH_3$

Synonyms: Actedron; Adipan; Allodene; DL-Amphetamine; Anfetamina (Spanish); Anorexide; Benzedrine; Deoxynorephedrine; racemic-Desoxynorephedrine; DL-Benzedrine; DL-α-Methylbenzeneethaneamine; DL-1-Phenyl-2-aminopropane; Elastonon; Isoamycin; Isomyn; Mecodrin; α-Methylbenzeneethaneamine; Norephedrane; Novydrine; Ortedrine; Phenedrine; 1-Phenyl isopropyl amine; Profamina; Propisamine; Psychedrine; Raphetamine; Simpatedrin; Sympamine; Sympatedrine; Weckamine

CAS Registry Number: 300-62-9

Note: There are various other "amphetamines" listed in the literature, however this CAS is specifically regulated by EPA. RTECS lists this CAS as benzedrine.

RTECS Number: SH9450000

DOT ID: UN 2811

Regulatory Authority

- Banned or Severely Restricted (UN) (U.S.)[13]
- SUPERFUND/EPCRA 40CFR355, Appendix A, Extremely Hazardous Substances: TPQ = 1,000 lb (454 kg)
- SUPERFUND/CERCLA 40CFR302.4 Reportable Quantity (RQ): EHS, 1 lb (0.454 kg)

Cited in U.S. State Regulations: California (G), Florida (G), Massachusetts (G), New Jersey (G), Pennsylvania (G).

Description: Amphetamine, $C_6H_5CH_2CH(NH_2)CH_3$, $C_9H_{13}N$, is a mobile liquid with an amine odor. Boiling point = 200 – 203°C. Flash point ≤100°C. Slightly soluble in water.

Potential Exposure: Amphetamine is used as a pharmaceutical. It is a central nervous system stimulant.

Incompatibilities: Oxidizing materials.

Permissible Exposure Limits in Air: No standard set.

Permissible Concentration in Water: No criteria set.

Routes of Entry: Inhalation, ingestion.

Harmful Effects and Symptoms

Short Term Exposure: Symptoms of exposure include dry mouth, metallic taste, loss of appetite, nausea, vomiting, diarrhea, abdominal cramps, headache, chilliness, flushing or pallor, palpitation, restlessness, dizziness, tremor, hyperactive reflexes, talkativeness, tenseness, irritability, weakness, insomnia, fever, confusion. With large doses, irregular heartbeat, pain and difficulty in urination. Convulsions, coma, circulatory collapse. This chemical is classified as extremely hazardous. Probable lethal dose in humans is 5 – 50 mg/kg or 7 drops to 1 teaspoon for a 70 kg (150 lb) person.

Long Term Exposure: Habit forming drug which affects the central nervous system.

Points of Attack: Central nervous system.

First Aid: If this chemical gets into the eyes, remove any contact lenses at once and irrigate immediately for at least 15 minutes, occasionally lifting upper and lower lids. Seek medical attention immediately. If this chemical contacts the skin, remove contaminated clothing and wash immediately with soap and water. Seek medical attention immediately. If this chemical has been inhaled, remove from exposure, begin rescue breathing (using universal precautions) if breathing has stopped and CPR if heart action has stopped. Transfer promptly to a medical facility. When this chemical has been swallowed, get medical attention. Give large quantities of water and induce vomiting. Do not make an unconscious person vomit.

Storage: Prior to working with this chemical you should be trained on its proper handling and storage. Store in a refrigerator or a cool, dry place.[52]

Shipping: Label required is "Poison." Poisonous solids n.o.s. fall in Hazard Class 6.1 and Packing Group II. The limit on passenger aircraft or railcar shipment is 25 kg; on cargo aircraft shipment is 100 kg.

Spill Handling: Evacuate persons not wearing protective equipment from area of spill or leak until clean-up is complete. Remove all ignition sources. Absorb liquid in dry sand, vermiculite or other absorbent material. Collect powdered material in the most convenient and safe manner and deposit in sealed containers. Ventilate area after clean-up is complete. It may be necessary to contain and dispose of this chemical as a hazardous waste. If material or contaminated runoff enters waterways, notify downstream users of potentially contaminated waters. Contact your Department of Environmental Protection or your regional office of the federal EPA for specific recommendations. If employees are required to clean-up spills, they must be properly trained and equipped. OSHA 1910.120(q) may be applicable.

Fire Extinguishing: Use dry chemical, carbon dioxide, water spray, or alcohol foam extinguishers. Poisonous gases are produced in fire including nitrogen oxides. If material or contaminated runoff enters waterways, notify downstream users of potentially contaminated waters. Notify local health and fire officials and pollution control agencies. From a secure, explosion-proof location, use water spray to cool exposed containers. If cooling streams are ineffective (venting sound increases in volume and pitch, tank discolors, or shows any signs of deforming), withdraw immediately to a secure position. If employees are expected to fight fires, they must be trained and equipped in OSHA 1910.156.

Disposal Method Suggested: Incineration.

References

U.S. Environmental Protection Agency, "Chemical Profile: Amphetamine," Washington, DC, Chemical Emergency Preparedness Program (November 30, 1987)

Ampicillin

Molecular Formula: $C_{16}H_{19}N_3O_4S$

Synonyms: Acillin; Adobacillin; Alpen; Amblosin; Amcill; Amfipen; Aminobenzyl penicillin; D-(-)-α-Aminobenzylpenicillin; D-(-)-α-Aminopenicillin; 6-[d(-)-α-Aminophenylacetamido]penicillanic acid 6; 6-(2-Amino-2-phenylacetamindo)-3,-3-dimethyl-7-oxo-4-thia-1-azabicyclo(3.2.0) heptane-2-carboxylic acid; (Aminophenylmethyl)-penicillin; Amipenix S; Amperil; Ampi-Bol; D-(-)-Ampicillin; D-Ampicillin; Ampicillin A; Ampicillin acid; Ampicillin anhydrate; Ampicillin (USDA); Ampicin; Ampikel; Ampimed; Ampipenin; Amplisom; Amplital; Ampy-penyl; Austrapen; AY-6108; Binotal; Bonapicillin; Britacil; BRL; BRL 1341; Copharcilin; Cymbi; Divercillin; Doktacillin; Grampenil; Guicitrina; Guicitrine; Lifeampil; Marisilan; NSC-528986; Nuvapen; Omnipen; P-50; Penbristol; Penbritin; Penbritin paediatric; Penbritin syrup; Penbrock penicline; Pentrex; Pfizerpen A; Polycillin; Ponecil; Principen; Qidamp; Ro-Ampen; Semicillin; SK-Ampicillin; Synpenin; Tokiocillin; Tolomol; Totacillin; Totalciclina; Totapen; Ultrabion; Ultrabron; Viccillin; Viccillin S; Vicillin; WY-5103

CAS Registry Number: 69-53-4; 7177-48-2 (trihydrate)

RTECS Number: XH8350000; HS8425000 (trihydrate)

DOT ID: NA 9188

Regulatory Authority

- Air Pollutant Standard Set (former USSR)[43]

Cited in U.S. State Regulations: New Jersey (G), Pennsylvania (G).

Description: Ampicillin, $C_{16}H_{19}N_3O_4S$, in anhydrous form occurs as crystals. Freezing/Melting point = 199 – 202°C (decomposes). Slightly soluble in water.

Potential Exposure: Used as an antibiotic.

Incompatibilities: Strong oxidizers.

Permissible Exposure Limits in Air: The former USSR-UNEP/IRPTC project[43] has set a MAC of 0.1 mg/m^3 in workplace air for ampicillin.

Permissible Concentration in Water: No criteria set.

Harmful Effects and Symptoms

Short Term Exposure: Ampicillin can affect you when breathed in. Exposure can cause skin rash. This may or may not be an allergic reaction, but if hives are present, allergy is likely. Exposure to high levels can cause upset stomach and diarrhea. The LD_{50} for rat is 10 gm/kg and for mouse is 28 mg/kg.[9] Ampicillin can cause hypersensitivity reactions in allergic persons.

Long Term Exposure: Exposure may cause allergy to develop, often accompanied with hives. Once allergy develops, even low future exposures can cause an allergic reaction. Persons having an allergy to penicillin may be more likely to develop an allergic reaction to ampicillin. Exposure can lead to a rare, but sometimes fatal reaction (aplastic anemia) in which blood cell count drops very low. Ampicillin can cause a liver-damaging reaction. Since ampicillin kills many normal germs, new resistant strains can grow after repeated exposure, resulting in a "yeast" or other types of infections.

Points of Attack: Liver, blood cells, skin.

Medical Surveillance: Liver function tests, (CBC) complete blood count. Examination by a qualified allergist.

First Aid: If this chemical gets into the eyes, remove any contact lenses at once and irrigate immediately for at least 15 minutes, occasionally lifting upper and lower lids. Seek medical attention immediately. If this chemical contacts the skin, remove contaminated clothing and wash immediately with soap and water. Seek medical attention immediately. If this chemical has been inhaled, remove from exposure, begin rescue breathing (using universal precautions) if breathing has stopped and CPR if heart action has stopped. Transfer promptly to a medical facility. When this chemical has been swallowed, get medical attention. Give large quantities of water and induce vomiting. Do not make an unconscious person vomit.

Personal Protective Methods: Where possible, enclose operations and use local exhaust ventilation at the site of chemical release. If local exhaust ventilation or enclosure is not used, respirators should be worn. Wear protective work clothing. Wash thoroughly at the end of the workshift. Post hazard and warning information in the work area. In addition, as part of an ongoing education and training effort, communicate all information on the health and safety hazards of Ampicillin to potentially exposed workers.

Respirator Selection: Engineering controls should be used wherever feasible to maintain airborne concentrations of this chemical below the prescribed exposure limit. Respirators and protective equipment less effective than engineering controls, and should be used only in non-routine or emergency situations which may result in exposure concentrations in excess of the TWA environmental limit. *Where there is no REL, at any detectable concentration:* SCBAF:PD,PP (any MSHA/NIOSH approved self-contained breathing apparatus that has a full facepiece and is operated in a pressure-demand or other positive-pressure mode); or SAF:PD,PP:ASCBA (any supplied-air respirator that has a full facepiece and is operated in a pressure-demand or other positive-pressure mode in combination with an auxiliary, self-contained breathing apparatus operated in a pressure-demand or other positive pressure mode). *Escape:* GMFOVHiE [any air-purifying, full-facepiece respirator (gas mask) with a chin-style, front- or back-mounted organic vapor canister having a high-efficiency particulate filter]; or SCBAE (any appropriate escape-type, self-contained breathing apparatus).

Storage: Prior to working with this chemical you should be trained on its proper handling and storage. Store in tightly closed containers in a cool, well-ventilated area away from strong oxidizers and moisture.

Shipping: Ampicillin trihydrate may be classified as a Hazardous Substance, solid, n.o.s.[52] As such, it falls in Hazard Class ORM-E and Packing Group III; there is no label requirement and no quantity limits on air or rail shipments.

Spill Handling: Evacuate persons not wearing protective equipment from area of spill or leak until clean-up is complete. Remove all ignition sources. Collect powdered material in the most convenient and safe manner and deposit in sealed containers. Ventilate area after clean-up is complete. It may be necessary to contain and dispose of this chemical as a hazardous waste. If material or contaminated runoff enters waterways, notify downstream users of potentially contaminated waters. Contact your Department of Environmental Protection or your regional office of the federal EPA for specific recommendations. If employees are required to clean-up spills, they must be properly trained and equipped. OSHA 1910.120(q) may be applicable.

Fire Extinguishing: Extinguish fire using an agent suitable for type of surrounding fire. Ampicillin itself does not burn. Poisonous gases are produced in fire including sulfur oxides and nitrogen oxides. If material or contaminated runoff enters waterways, notify downstream users of potentially contaminated waters. Notify local health and fire officials and pollution control agencies. From a secure, explosion-proof location, use water spray to cool exposed containers. If cooling streams are ineffective (venting sound increases in volume and pitch, tank discolors, or shows any signs of deforming), withdraw immediately to a secure position. If employees are expected to fight fires, they must be trained and equipped in OSHA 1910.156.

Disposal Method Suggested: Consult with environmental regulatory agencies for guidance on acceptable disposal practices. Generators of waste containing this contaminant (≥100 kg/mo) must conform with EPA regulations governing storage, transportation, treatment, and waste disposal.

References

New Jersey Department of Health and Senior Services, "Hazardous Substance Fact Sheet: Ampicillin," Trenton, NJ (February 1989)

Amyl Acetates

Molecular Formula: $C_6H_{12}O_2$ (n-); $C_7H_{14}O_2$ (sec-); $C_7H_{14}O_2$ (tert-); $C_8H_{14}O_2$ (iso-)

Common Formula: $CH_3COOCH_2CH_2CH_2CH_3$ (n-); $CH_3COOCH(CH_3)CH_2CH_2CH_3$ (sec-); $CH_3COOC(CH_3)_2C_2H_5$ (tert-); $CH_3COOCH_2CH_2CH(CH_3)CCH_3$ (iso-)

Synonyms: *n-:* Acetate d'amyle (French); Acetato de amilo (Spanish); Acetic acid *n*-amyl ester; Acetic acid pentyl ester; Amyazetat (German); *n*-Amyl acetate; Amyl acetate, mixed isomers; Amyl-acetate (*n*-); Amyl acetic acid; Amyl acetic ester; Amyl acetic ether; Banana oil; Birnenoel; Octan amylu (Polish); Pear oil; Pent acetate; 1-Pentanol acetate; Pentyl acetates; Pentyl ester of acetic acid; *n*-Pentyl ethanoate; Primary amyl acetate

sec-: Acetic acid, 2-pentyl ester; 2-Acetoxypentane; Banana oil; 1-Methylbutyl acetate; 2-Pentanol, acetate; 2-Pentyl acetate

tert-: Acetic acid, isopentyl ester; Amylacetic ester; Banana oil; Isoamyl ethanoate; Isopentyl acetate; Isopentyl alcohol acetate; 3-Methyl-1-butanol acetate; 3-Methyl-1-butyl acetate; 3-Methylbutyl acetate; 3-Methylbutyl ethanoate; Pear oil

iso-: Acetic acid, isopentyl ester; Amylacetic ester; Isoamyl ethanoate; Isopentyl acetate; Isopentyl alcohol acetate; 3-Methyl-1-butanol acetate; 3-Methyl-1-butyl acetate; 3-Methylbutyl acetate; 3-Methylbutyl ethanoate; Pear oil

CAS Registry Number: 625-16-1 (all isomers); 628-63-7 (n-); 626-38-0 (sec-); 675-16-1 (tert-); 123-92-2 (iso-)

RTECS Number: AJ1925000 (n-); AJ2100000 (sec-); NS9800000 (iso-)

DOT ID: UN 1104

EEC Number: 607-130-00-2

EINECS Number: 211-047-3 (n-)

Regulatory Authority

- Air Pollutant Standard Set (ACGIH)[1] (former USSR)[43] (DFG)[3] UK)[33] (OSHA)[58] (Several States)[60] (Several Canadian Provinces) (Australia) (Israel) (Mexico)
- Clean Water Act: 40CFR116.4 Hazardous Substances; 40CFR117.3, RQ (same as CERCLA)
- SUPERFUND/EPCRA 40CFR302.4 Reportable Quantity (RQ): CERCLA, 5,000 lb (2,270 kg)
- Canada, WHMIS, Ingredients Disclosure List

Cited in U.S. State Regulations: Alaska (G), California (A, G), Connecticut (A), Florida (G), Illinois (G), Maine (G), Massachusetts (G), Minnesota (G), Nevada (A), New Hampshire (G), New Jersey (G), New York (G), North Dakota (A), Oklahoma (G), Pennsylvania (G), Rhode Island (G), Virginia (A), West Virginia (G).

Description: There are four amyl acetate isomers: (n-) $CH_3COOCH_2CH_2CH_2CH_3$; (sec-) $CH_3COOCH(CH_3)CH_2CH_2CH_3$; (iso-) $CH_3COOCH_2CH_2CH(CH_3)CCH_3$, (tert-) $CH_3COOC(CH_3)_2C_2H_5$. All isomers are highly flammable, colorless to yellow, watery liquids. The n- and Iso- isomers have a persistent, fruity, banana-like or pear-like odor. The sec-isomer has a mild fruity odor. They floats on water with negligible to slight solubility. Water contact forms a flammable, irritating vapor. Hazard Identification (based on NFPA-704M Rating System): Health 1, Flammability 3, Reactivity 0. Autoignition temperature = 360°C (n-); 350°C (iso-).

Some physical properties and fire-related data are as follows:

	Boiling Point (°C)	Flash Point (°C)	Explosive Limits. LEL (%)	Explosive Limits. UEL (%)	Odor Threshold (ppm)
n-amyl	149	16 – 21	1.1	7.5	0.0075 – 7.3
sec-amyl	121	32	1.1	7.0	0.002
iso-amyl	142	25	1.2 @100°C	7.5	0.015*
tert-amyl	125	26	1.0	7.5	0.08

* 0.0006 in water.

Note: The range of accepted odor threshold values is quite broad and caution should be used in relying on odor alone as a warning of potentially hazardous exposure.

Potential Exposure: Amyl acetates are used as industrial solvents and in the manufacturing and dry-cleaning industry; making artificial fruit-flavoring agents, cements, coated papers, lacquers; in medications as an inflammatory agent; pet repellents, insecticides and miticide. Many other uses.

Incompatibilities: Nitrates, strong oxidizers, strong alkalis, strong acids. May soften certain plastics. Forms explosive mixture with air.

Permissible Exposure Limits in Air:

Isomer	TWA (1,2,58)	IDLH[2]
n-	100 ppm (525) mg/m^3)	1,000 ppm
sec-	125 ppm (650 mg/m^3)	1,000 ppm
iso-	100 ppm (525 mg/m^3)	1,000 ppm
tert-	100 ppm (525 mg/m^3)	1,000 ppm

The Australian standard is 100 ppm (532 mg/m^3) TWA for the n-isomer and 125 ppm (665 mg/m^3) fir the sec-isomer. The HSE (UK) standard is 100 ppm (530 mg/m^3) TWA and 150 ppm (800 mg/m^3) STEL for the n- and sec-isomers. The DFG (German) standard is 50 ppm (270 mg/m^3) TWA and 2 ppm Peak (5 min.) for all isomers. The Israel standard is 100 ppm (532 mg/m^3) TWA and 50 ppm (266 mg/m^3) Action Level for the n-isomer; 125 ppm (665 mg/m^3) TWA and 62.5 ppm (332.5 mg/m^3) Action Level for the iso-isomer. The Mexico standard is 100 ppm (530 mg/m^3) TWA and 150 ppm (800 mg/m^3) STEL for the n-isomer; 125 ppm (670 mg/m^3) TWA and 150 ppm (800 mg/m^3) STEL for the iso-isomer. The former USSR-IRPTC project[43] has set a MAC of 100 mg/m^3 in workplace air and 0.1 mg/m^3 in residential air, both on a momentary and on an average daily basis for n-amyl acetate. The MAC value set by the DFG[3] is 100 ppm (525 mg/m^3)

for all isomers. In addition, California and various Canadian provinces set standards for amyl acetates as follows (all values in mg/m^3):

State	*n-Amyl*	*sec-Amyl*
California	532 TWA	665 TWA
Alberta	530 TWA/800 STEL	665 TWA/800 STEL
British Columbia	539 TWA/800 STEL	670 TWA/800 STEL
Ontario	530 TWAEV	660 TWAEV
Quebec	532 TWAEV	665 TWAEV

Determination in Air: Charcoal adsorption, workup with CS_2, and gas chromatographic analysis. See NIOSH Method 1450 for esters, also 2549 for VOCs (screening).

Permissible Concentration in Water: No criteria set.

Routes of Entry: Inhalation, ingestion, eye, and skin contact. Passes through the skin.

Harmful Effects and Symptoms

Short Term Exposure: Amyl acetates can be absorbed through the skin, thereby increasing exposure. Irritates the eyes causing burning sensation. Inhalation can irritate the respiratory tract causing cough and wheezing. Higher exposure can cause headache, drowsiness, weakness, and loss of consciousness.

Long Term Exposure: (n-) May cause liver damage. (sec-) May cause slight changes in the nervous system (brain wave changes). Prolonged or repeated skin contact to amyl acetates can cause irritation, dryness and cracking. Although all of these chemicals have not been adequately tested, many similar petroleum-based chemicals can cause brain or other nerve damage. Effects may include reduced memory and concentration, personality changes such as withdrawl and irritability, fatigue, sleep disturbances, reduced coordination, and/or effects on autonomic nerves, and/or nerves to the arms and legs with weakness and sensation of "pins and needles."

Points of Attack: Eyes, skin and respiratory system, central nervous system.

Medical Surveillance: Liver function tests. Positive and borderline individuals showing brain effects, changes in memory, concentration, mood and sleeping patterns, as well as headaches and fatigue should be referred for neuropsychological testing.

First Aid: If this chemical gets into the eyes, remove any contact lenses at once and irrigate immediately for at least 15 minutes, occasionally lifting upper and lower lids. Seek medical attention immediately. If this chemical contacts the skin, remove contaminated clothing and wash immediately with soap and water. Seek medical attention immediately. If this chemical has been inhaled, remove from exposure, begin rescue breathing (using universal precautions) if breathing has stopped and CPR if heart action has stopped. Transfer promptly to a medical facility. When this chemical has been swallowed, get medical attention. Give large quantities of water and induce vomiting. Do not make an unconscious person vomit.

Personal Protective Methods: Employees should wash promptly when skin is wet or contaminated. Remove clothing immediately if wet or contaminated to avoid flammability hazard. Wear solvent-resistant protective gloves and clothing to prevent any reasonable probability of skin contact. ACGIH recommends butyl rubber or polyvinyl alcohol as a protective material. Neoprene + natural rubber is also recommended by safety equipment suppliers/manufacturers who can provide recommendations on the most protective glove/clothing material for your operation. All protective clothing (suits, gloves, footwear, headgear) should be clean, available each day, and put on before work. Contact lenses should not be worn when working with this chemical. Wear solvent-resistant, splash-proof chemical goggles and face shield unless full facepiece respiratory protection is worn. Employees should wash immediately with soap when skin is wet or contaminated. Provide emergency showers and eyewash.

Respirator Selection: NIOSH/OSHA (for n-, sec-, and iso-isomers): *1,000 ppm:* CCRFOV (Any air-purifying, full-facepiece respirator (gas mask) with a chin-style, front- or back-mounted acid gas canister); or GMFOV (Any air-purifying, full-facepiece respirator (gas mask) with a chin-style, front- or back-mounted organic vapor canister; PAPROV [any powered, air-purifying respirator with organic vapor cartridge(s)]; or SA (any supplied-air respirator); or SCBAF (any self-contained breathing apparatus with a full facepiece); or SAF (any supplied-air respirator with a full facepiece). *Emergency or planned entry into unknown concentrations or IDLH conditions:* SCBAF:PD,PP (any self-contained breathing apparatus that has a full facepiece and is operated in a pressure-demand or other positive-pressure mode); or SAF:PD,PP: ASCBA (any supplied-air respirator that has a full facepiece and is operated in a pressure-demand or other positive-pressure mode in combination with an auxiliary self-contained breathing apparatus operated in a pressure-demand or other positive pressure mode. *Escape:* GMFOV(any air-purifying, full-facepiece respirator (gas mask) with a chin-style, front- or back-mounted organic vapor canister; or SCBAE (Any appropriate escape-type, self-contained breathing apparatus).

Note: Substance reported to cause eye irritation or damage; may require eye protection. There are no specific respirator specs for tert-amyl acetate.

Storage: Prior to working with this chemical you should be trained on its proper handling and storage. Before entering confined space where amyl acetates may be present, check to make sure that an explosive concentration does not exist. Store in tightly closed containers in a cool, well ventilated area, preferably a detached shed. Metal containers involving the transfer of this chemical should be grounded and bonded. Where possible, automatically pump liquid from drums or other storage containers to process containers. Drums must be equipped with self-closing valves, pressure vacuum bungs, and flame arresters. Use only non-sparking tools and equipment, especially when opening and closing containers of this chemical. Sources of ignition such as smoking and open flames, are prohibited where this chemical is used, handled, or stored in a manner that could create a potential fire or explosion hazard.

Shipping: DOT label required is "Flammable Liquid." The DOT/UN Hazard Class is 3 and the Packing Group is III.[19][20]

Spill Handling: Avoid contact. Stay upwind. Evacuate and restrict persons not wearing protective equipment from area of spill or leak until cleanup is complete. Establish forced ventilation to keep levels below explosive limit. Remove all ignition sources. Absorb liquids in vermiculite, dry sand, earth, or a similar material and deposit in sealed containers. It may be necessary to contain and dispose of this chemical as a hazardous waste. If material or contaminated runoff enters waterways, notify downstream users of potentially contaminated waters. Contact your Department of Environmental Protection or your regional office of the federal EPA for specific recommendations. If employees are required to clean-up spills, they must be properly trained and equipped. OSHA 1910.120(q) may be applicable.

Fire Extinguishing: These chemicals are flammable liquids. Poisonous gases including nitrogen oxides are produced in fire. Use dry chemical, carbon dioxide, or alcohol foam extinguishers. Vapors are heavier than air and will collect in low areas. Vapors may travel long distances to ignition sources and flashback. Vapors in confined areas may explode when exposed to fire. Storage containers and parts of containers may rocket great distances, in many directions. If material or contaminated runoff enters waterways, notify downstream users of potentially contaminated waters. Notify local health and fire officials and pollution control agencies. From a secure, explosion-proof location, use water spray to cool exposed containers. If cooling streams are ineffective (venting sound increases in volume and pitch, tank discolors, or shows any signs of deforming), withdraw immediately to a secure position. If employees are expected to fight fires, they must be trained and equipped in OSHA 1910.156.

Disposal Method Suggested: Incineration. In accordance with 40CFR165 recommendations for the disposal of pesticides and pesticide containers. Must be disposed properly by following package label directions or by contacting your state pesticide or environmental control agency or by contacting your regional EPA office.

References

Sax, N. I., Ed., "Dangerous Properties of Industrial Materials Report" 2, No. 2, 39–40 (1982) (Isoamyl Acetate)

Sax, N. I., Ed., "Dangerous Properties of Industrial Materials Report" 3, No.6, 37–40 (November, December 1983) (t-Amyl Acetate)

New Jersey Department of Health and Senor Services, "Hazardous Substance Fact Sheet: n-Amyl Acetate," Trenton, NJ (February 1998)

New Jersey Department of Health and Senior Services, "Hazardous Substance Fact Sheet: sec-Amyl Acetate," Trenton, NJ (February 1998)

New Jersey Department of Health and Senior Services, "Hazardous Substance Fact Sheet: Isoamyl Acetate," Trenton, NJ (September 1986)

New York State Department of Health, "Chemical Fact Sheet: n-Amyl Acetate," Albany, NY, Bureau of Toxic Substance Assessment (March 1986)

Amyl Alcohols

Molecular Formula: $C_5H_{12}O$

Synonyms: *n-:* Alcohol *n*-amilico primario (Spanish); Alcohol C-5; Alcool *n*-amyl alcohol; 1-Amyl alcohol; Amyl alcohol; Amyl alcohol, *normal*; *n*-Amylalkohol (Czech); Amylique (French); Amylol; *n*-Butyl carbinol; *n*-Pentanol; 1-Pentanol; Pentan-1-ol; Pentanol; Pentanol-1; Pentasol; Pentyl alcohol; Prim-*n*-amyl alcohol; Primary amyl alcohol

sec-: Alcohol *n*-amilico secundario (Spanish); *sec*-Amyl alcohol; Methyl propyl carbinol; 2-Pentanol; Pentanol -2; *sec*-Pentyl alcohol

tert-: Alcohol amilico terciario (Spanish); *tert*-Amyl alcohol; Amylene hydrate; Dimethylethylcarbinol; 2-Methyl butanol-2; 2-Methyl-2-butanol; 3-Methylbutan-3-ol; *tert*-Pentanol; 3-Pentanol; *tert*-Pentyl alcohol

iso-: Alcohol isoamilico primario (Spanish); Alcool amilico (Italian); Alcool isoamylique (French); *iso*-Amylalkohol (German); Amylowy alkohol (Polish); Diethylcarbinol; Fermentation amyl alcohol; Isoamyl alcohol; Isoamyl alkohol (Czech); Isoamylol; Isobutylcarbinol; Isopentanol; Isopentyl alcohol; 2-Methyl-4-butanol; 3-Methyl butanol; 3-Methyl-1-butanol (Czech); 3-Methylbutan-1-ol; 3-Metil-butanolo (Italian); 3-Pentanol; Pentanol-3-ol

CAS Registry Number: 71-41-0 (n-); 6032-29-7 (sec-); 75-85-4 (tert-); 123-51-3 (iso-); 584-02-1[iso- (sec-)]

RTECS Number: SB9800000 (n-); SA4900000 (sec-); SC0175000 (tert-); EL5425000 (iso-); SA5075000 [iso-(sec-)]

DOT ID: UN 1105 (all isomers)

EEC Number: 603-006-00-7 (all isomers)

Cited in U.S. State Regulations: New Jersey [G (n-, iso-)], Pennsylvania (G).

Description: Amyl alcohols, $C_5H_{11}OH$, has eight isomers. All are flammable, colorless liquids, except the isomer 2,2-dimethyl-1-propanol, which is a crystalline solid. Odors are described as "acetone-like" and "similar to fuel oil." Hazard Identification (based on NFPA-704M Rating System): Health 1, Flammability 3, Reactivity 0.(n-, sec-); Hazard Identification (based on NFPA-704M Rating System): Health 1, Flammability 2, Reactivity 0 (iso-).

	Freezing/ Melting Point (°C)	Boiling Point (°C)	Flash Point (°C)	Explosive Limits. LEL (%)	UEL (%)	Odor Threshold (ppm)
n-	-78	138	33	1.2	10.0 @ 100°C	0.012 – 10
sec-	-50	119	34	1.2	9	
iso-	-117	131; 113 (sec-)	43; 30 (sec-)	1.2	9.0 @ 100°C	0.028 – 0.072
tert-	-8	102	19	1.2	9	

Note: The range of accepted odor threshold values is quite broad and caution should be used in relying on odor alone as a warning of potentially hazardous exposure.

Amyl alcohols are obtained from fused oil which forms during the fermentation of grain, potatoes, or beets for ethyl alcohol. The fusel oil is a mixture of amyl alcohol isomers, and the composition is determined somewhat by the sugar source. Slight to moderate solubility in water.

Potential Exposure: Used as a solvent in synthetic flavoring, pharmaceuticals, corrosion inhibitors, and making plastics and other chemicals.

Incompatibilities: Forms an explosive mixture with air. Contact with strong oxidizers and hydrogen trisulfide may cause fire and explosions. Incompatible with strong acids. Violent reaction with alkaline earth metals forming hydrogen, a flammable gas.

Permissible Exposure Limits in Air: There is no U.S. standard for the n-, sec, or tert- isomers; however, the Federal OSHA PEL for Isoamyl Alcohol (primary) is 100 ppm (360 mg/m^3) TWA. NIOSH and ACGIH recommend the same TWA *and* 125 ppm (450 mg/m^3) STEL. IDLH = 500 ppm. The former USSR-UNEP/IRPTC project[43] gives MAC values for "amyl alcohol" with no specific isomer indicated. Those values are quoted in the entry for "Isoamyl Alcohol."

Routes of Entry: Inhalation, skin contact, ingestion.

Harmful Effects and Symptoms

Short Term Exposure: Passes through the skin; contact contributes significantly to overall exposure. Skin contact can cause skin irritation. Inhalation can irritate the eyes and respiratory system causing headache, nausea, and vomiting. High exposure can cause dizziness, lightheadedness, confusion and unconsciousness. Very high exposure can cause death.

Long Term Exposure: Can cause liver damage and blood effects.

Points of Attack: Eyes, skin, respiratory system, central nervous system.

Medical Surveillance: Regular medical checkups are recommended, depending on degree of exposure.

First Aid: Skin contact contributes significantly to overall exposure. If this chemical gets into the eyes, remove any contact lenses at once and irrigate immediately for at least 15 minutes, occasionally lifting upper and lower lids. Seek medical attention immediately. If this chemical contacts the skin, remove contaminated clothing and wash immediately with soap and water. Seek medical attention immediately. If this chemical has been inhaled, remove from exposure, begin rescue breathing (using universal precautions) if breathing has stopped and CPR if heart action has stopped. Transfer promptly to a medical facility. When this chemical has been swallowed, get medical attention. Give large quantities of water and induce vomiting. Do not make an unconscious person vomit.

Personal Protective Methods: Wear solvent-resistant gloves and clothing to prevent any reasonable probability of skin contact. ACGIH and safety equipment suppliers/manufacturers recommend polyvinyl alcohol, polyvinyl chloride, neoprene, butyl rubber, and nitrile rubber, neoprene + styrene-butadiene rubber (SBR), polyurethane, SBR,, and SBR/neoprene as protective materials. All protective clothing (suits, gloves, footwear, headgear) should be clean, available each day, and put on before work. Contact lenses should not be worn when working with this chemical. Wear splash-proof chemical goggles and face shield unless full facepiece respiratory protection is worn. Employees should wash immediately with soap when skin is wet or contaminated. Remove nonimpervious clothing immediately if wet or contaminated. Provide emergency showers and eyewash.

Respirator Selection: NIOSH/OSHA: *500 ppm:* SA:CF* (any supplied-air respirator operated in a continuous-flow mode); CCRFOV [any chemical cartridge respirator with a full facepiece and organic vapor cartridge(s)]; or GMFOV [any air-purifying, full-facepiece respirator (gas mask) with a chin-style, front- or back-mounted acid gas canister]; PAPROV* [any powered, air-purifying respirator with organic vapor cartridge(s)]; or SCBAF (any self-contained breathing apparatus with a full facepiece); or SAF (any supplied-air respirator with a full facepiece). *Emergency or planned entry into unknown concentrations or IDLH conditions:* SCBAF:PD, PP (any self-contained breathing apparatus that has a full facepiece and is operated in a pressure-demand or other positive-pressure mode); or SAF:PD,PP:ASCBA (any supplied-air respirator that has a full facepiece and is operated in a pressure-demand or other positive-pressure mode in combination with an auxiliary, self-contained breathing apparatus operated in a pressure-demand or other positive pressure mode. *Escape:* GMFOV(any air-purifying, full-facepiece respirator (gas mask) with a chin-style, front- or back-mounted organic vapor canister); or SCBAE (any appropriate escape-type, self-contained breathing apparatus).

* Substance causes eye irritation or damage; eye protection needed.

Storage: Prior to working with this chemical you should be trained on its proper handling and storage. Before entering confined space where amyl alcohols may be present, check to make sure that an explosive concentration does not exist. Store in tightly closed containers in a cool, well ventilated area away from strong oxidizers, strong acids and hydrogen trifluoride since violent reactions occur. Metal containers involving the transfer of this chemical should be grounded and bonded. Where possible, automatically pump liquid from drums or other storage containers to process containers. Drums must be equipped with self-closing valves, pressure vacuum bungs, and flame arresters. Use only non-sparking tools and equipment, especially when opening and closing containers of this chemical. Sources of ignition such as smoking and open flames, are prohibited where this chemical is used, handled, or stored in a manner that could create a potential fire or explosion hazard.

Shipping: Label required is "Flammable." Is in DOT/UN Hazard Class 3 and Packing Group II or III.[19][20]

Spill Handling: Evacuate persons not wearing protective equipment from area of spill or leak until clean-up is complete. Remove all ignition sources. Establish forced ventilation to keep levels below explosive limit. Absorb liquids in vermiculite, dry sand, earth, or a similar material and deposit in sealed containers. It may be necessary to contain and dispose of this chemical as a hazardous waste. If material or contaminated runoff enters waterways, notify downstream users of potentially contaminated waters. Contact your Department of Environmental Protection or your regional office of the federal EPA for specific recommendations. If employees are required to clean-up spills, they must be properly trained and equipped. OSHA 1910.120(q) may be applicable.

Fire Extinguishing: This chemical is a flammable liquid. Poisonous gases including carbon monoxide are produced in fire. Use alcohol or polymer foam extinguishers. Vapors are heavier than air and will collect in low areas. Vapors may travel long distances to ignition sources and flashback. Vapors in confined areas may explode when exposed to fire. Storage containers and parts of containers may rocket great distances, in many directions. If material or contaminated runoff enters waterways, notify downstream users of potentially contaminated waters. Notify local health and fire officials and pollution control agencies. From a secure, explosion-proof location, use water spray to cool exposed containers. If cooling streams are ineffective (venting sound increases in volume and pitch, tank discolors, or shows any signs of deforming), withdraw immediately to a secure position. If employees are expected to fight fires, they must be trained and equipped in OSHA 1910.156.

Disposal Method Suggested: Incineration.

References

Sax, N. I., Ed., "Dangerous Properties of Industrial Materials Report," 2, No. 3, 55–56 (1982) (n-amyl Alcohol)

New Jersey Department of Health and Senior Services, "Hazardous Substance Fact Sheet: Amyl Alcohol," Trenton, NJ (June 1998)

Amyl Nitrate

Molecular Formula: $C_5H_{11}NO_3$

Common Formula: $C_5H_{11}ONO_2$

Synonyms: *n*-Amyl nitrate; Diesel ignition improver; Nitrate d'amyle (French); Nitric acid, pentyl ester

CAS Registry Number: 1002-16-0

RTECS Number: QV0600000

DOT ID: UN 1112

Cited in U.S. State Regulations: Florida (G), Maine (G), Massachusetts (G), New Hampshire (G), New Jersey (G), Pennsylvania (G), Rhode Island (G).

Description: Amyl nitrate, $C_5H_{11}ONO_2$, is a flammable, colorless liquid with an ether-like odor. Boiling point = 144 – 156°C. Flash point = 48°C (oc). Hazard Identification (based on NFPA-704M Rating System): Health 2, Flammability 2, Reactivity 0 Oxidizer. Floats on water; insoluble.

Potential Exposure: Amyl nitrate is used as an ignition additive in diesel fuels.

Incompatibilities: Contact with reducing agents or other easily oxidizable substances may cause fire and explosions.

Permissible Exposure Limits in Air: No standard set.

Permissible Concentration in Water: No criteria set.

Harmful Effects and Symptoms

Short Term Exposure: Contact can irritate the eyes and skin. Breathing amyl nitrate vapor can cause headaches, dizziness, weakness and nausea. Higher levels can cause convulsions and death. Exposure can interfere with the ability of the blood to carry oxygen, causing headaches, weakness and a blue color to the skin and lips.

Long Term Exposure: Repeated exposure may cause anemia. After repeated exposure tolerance develops. If exposure stops suddenly, chest pain and heart attack could occur.

Points of Attack: Blood cells.

Medical Surveillance: Blood tests for methemoglobin level.

First Aid: If this chemical gets into the eyes, remove any contact lenses at once and irrigate immediately for at least 15 minutes, occasionally lifting upper and lower lids. Seek medical attention immediately. If this chemical contacts the skin, remove contaminated clothing and wash immediately with soap and water. Seek medical attention immediately. If this chemical has been inhaled, remove from exposure, begin rescue breathing (using universal precautions) if breathing has stopped and CPR if heart action has stopped. Transfer promptly to a medical facility. When this chemical has been swallowed, get medical attention. Give large quantities of water and induce vomiting. Do not make an unconscious person vomit.

Note to Physician: Treat for methemoglobinemia. Spectrophotometry may be required for precise determination of levels of methemoglobinemia in urine.

Personal Protective Methods: Wear protective gloves and clothing to prevent any reasonable probability of skin contact. Safety equipment suppliers/manufacturers can provide

recommendations on the most protective glove/clothing material for your operation. All protective clothing (suits, gloves, footwear, headgear) should be clean, available each day, and put on before work. Contact lenses should not be worn when working with this chemical. Wear splash-proof chemical goggles and face shield unless full facepiece respiratory protection is worn. Employees should wash immediately with soap when skin is wet or contaminated. Provide emergency showers and eyewash.

Respirator Selection: Engineering controls should be used wherever feasible to maintain airborne concentrations of this chemical below the prescribed exposure limit. Respirators and protective equipment less effective than engineering controls, and should be used only in non-routine or emergency situations which may result in exposure concentrations in excess of the TWA environmental limit. *At any concentrations above the NIOSH REL, or where there is no REL, at any detectable concentration:* SCBAF:PD,PP (any MSHA/NIOSH approved self-contained breathing apparatus that has a full facepiece and is operated in a pressure-demand or other positive-pressure mode); or SAF:PD,PP:ASCBA (any supplied-air respirator that has a full facepiece and is operated in a pressure-demand or other positive-pressure mode in combination with an auxiliary, self-contained breathing apparatus operated in a pressure-demand or other positive pressure mode). *Escape:* GMFOVHiE [any air-purifying, full-facepiece respirator (gas mask) with a chin-style, front- or back-mounted organic vapor canister having a high-efficiency particulate filter]; or SCBAE (any appropriate escape-type, self-contained breathing apparatus).

Storage: Prior to working with this chemical you should be trained on its proper handling and storage. Store in tightly closed containers in a cool, well-ventilated area away from heat, sparks, flames and other combustible materials such as wood, paper or oil. Outside or detached storage is preferred. Before entering confined space where amyl nitrate may be present, check to make sure that an explosive concentration does not exist. Metal containers involving the transfer of this chemical should be grounded and bonded. Where possible, automatically pump liquid from drums or other storage containers to process containers. Drums must be equipped with self-closing valves, pressure vacuum bungs, and flame arresters. Use only non-sparking tools and equipment, especially when opening and closing containers of this chemical. Sources of ignition such as smoking and open flames, are prohibited where this chemical is used, handled, or stored in a manner that could create a potential fire or explosion hazard.

Shipping: The label required by DOT is "Flammable Liquid." The DOT/UN Hazard Class is 3 and the Packing Group is II.[19][20]

Spill Handling: Evacuate persons not wearing protective equipment from area of spill or leak until clean-up is complete. Remove all ignition sources. Ventilate site of spill or leak. Use water spray to reduce vapors. Absorb liquids in vermiculite, dry sand, earth, or a similar material and deposit in sealed containers. Keep Amyl Nitrate out of a confined space, such as a sewer, because of the potential for an explosion, unless the sewer is designed to prevent the build-up of explosive concentrations. It may be necessary to contain and dispose of this chemical as a hazardous waste. If material or contaminated runoff enters waterways, notify downstream users of potentially contaminated waters. Contact your Department of Environmental Protection or your regional office of the federal EPA for specific recommendations. If employees are required to clean-up spills, they must be properly trained and equipped. OSHA 1910.120(q) may be applicable.

Fire Extinguishing: This chemical is a flammable liquid. Poisonous gases including nitrogen oxides are produced in fire. Use dry chemical, carbon dioxide, or alcohol foam extinguishers. Vapors are heavier than air and will collect in low areas. Vapors may travel long distances to ignition sources and flashback. Vapors in confined areas may explode when exposed to fire. Storage containers and parts of containers may rocket great distances, in many directions. If material or contaminated runoff enters waterways, notify downstream users of potentially contaminated waters. Notify local health and fire officials and pollution control agencies. From a secure, explosion-proof location, use water spray to cool exposed containers. If cooling streams are ineffective (venting sound increases in volume and pitch, tank discolors, or shows any signs of deforming), withdraw immediately to a secure position. If employees are expected to fight fires, they must be trained and equipped in OSHA 1910.156.

Disposal Method Suggested: Incineration with scrubber to remove No_x in effluent gases.

References

New Jersey Department of Health and Senior Services, "Hazardous Substance Fact Sheet: Amyl Nitrate," Trenton, NJ (November 1986)

Amyl Nitrites

Molecular Formula: $C_5H_{11}NO_2$

Synonyms: *n-:* 1-Nitropentane

iso-: Nitrous Acid, 3-Methylbutyl Ester; Isoamyl Nitrite; Isopentyl Alcohol Nitrite; 3-Methylbutanol Nitrite; 3-Methylbutyl Nitrite; Nitropentane; Nitrous Acid, Pentyl Ester; Pentyl Nitrite

CAS Registry Number: 110-46-3 (iso-); 463-04-7 (n-)

RTECS Number: NT0187500 (iso-); RA1140000 (n-)

DOT ID: UN 1113

Cited in U.S. State Regulations: New Hampshire (G), New Jersey (G), Oklahoma (G).

Description: Amyl nitrite, $C_5H_{11}NO_2$, is a flammable, yellowish liquid with a penetrating, fruity odor. Boiling point = 97 – 99°C. Flash point = 10°C. Floats on water; very slightly soluble.

Potential Exposure: It is used to make pharmaceuticals, perfumes, diazonium compounds, and other chemicals.

Incompatibilities: Forms explosive mixture with air. A strong oxidizer. Contact with reducing agents and easily oxidizable materials may cause fire and explosions. Reported to be an explosion hazard when exposed to air and light. Keep away from alcohols, antipyrine, alkaline materials, alkaline carbonates, potassium iodide, bromides, and ferrous salts.

Permissible Exposure Limits in Air: No standards set.

Permissible Concentration in Water: No criteria set.

Routes of Entry: Inhalation, ingestion, skin absorption.

Harmful Effects and Symptoms

Short Term Exposure: Amyl nitrite can be absorbed through the skin, thereby increasing exposure. Contact can irritate the eyes and skin. A skin allergy can develop. Chronic overexposure can cause blood cell damage and anemia. Exposure can rapidly cause flushing, headaches, dizziness, sharp drop in blood pressure, fast pulse rate, confusion, a blue color to lips and fingernails and possible fainting and shock.

Long Term Exposure: Chronic overexposure can cause anemia and red blood cell damage. Repeated exposure causes tolerance to develop. If exposure stops suddenly, chest pain and heart attack could occur. May cause skin allergy to develop.

Points of Attack: Blood, skin.

Medical Surveillance: Blood methemoglobin level, complete blood count (CBC), evaluation by a qualified allergist.

First Aid: If this chemical gets into the eyes, remove any contact lenses at once and irrigate immediately for at least 15 minutes, occasionally lifting upper and lower lids. Seek medical attention immediately. If this chemical contacts the skin, remove contaminated clothing and wash immediately with soap and water. Seek medical attention immediately. If this chemical has been inhaled, remove from exposure, begin rescue breathing (using universal precautions) if breathing has stopped and CPR if heart action has stopped. Transfer promptly to a medical facility. When this chemical has been swallowed, get medical attention. Give large quantities of water and induce vomiting. Do not make an unconscious person vomit.

Note to Physician: Treat for methemoglobinemia. Spectrophotometry may be required for precise determination of levels of methemoglobinemia in urine.

Personal Protective Methods: Wear protective gloves and clothing to prevent any reasonable probability of skin contact. Safety equipment suppliers/manufacturers can provide recommendations on the most protective glove/clothing material for your operation. All protective clothing (suits, gloves, footwear, headgear) should be clean, available each day, and put on before work. Contact lenses should not be worn when working with this chemical. Wear splash-proof chemical goggles and face shield unless full facepiece respiratory protection is worn. Employees should wash immediately with soap when skin is wet or contaminated. Provide emergency showers and eyewash.

Respirator Selection: Engineering controls should be used wherever feasible to maintain airborne concentrations of this chemical below the prescribed exposure limit. Respirators and protective equipment less effective than engineering controls, and should be used only in non-routine or emergency situations which may result in exposure concentrations in excess of the TWA environmental limit. *Where there is no REL, at any detectable concentration:* SCBAF:PD,PP (any MSHA/NIOSH approved self-contained breathing apparatus that has a full facepiece and is operated in a pressure-demand or other positive-pressure mode); or SAF:PD,PP:ASCBA (any supplied-air respirator that has a full facepiece and is operated in a pressure-demand or other positive-pressure mode in combination with an auxiliary, self-contained breathing apparatus operated in a pressure-demand or other positive pressure mode). *Escape:* GMFOVHiE [any air-purifying, full-facepiece respirator (gas mask) with a chin-style, front- or back-mounted organic vapor canister having a high-efficiency particulate filter]; or SCBAE (any appropriate escape-type, self-contained breathing apparatus).

Storage: Prior to working with this chemical you should be trained on its proper handling and storage. Before entering confined space where amyl nitrite may be present, check to make sure that an explosive concentration does not exist. Store in an explosion-proof refrigerator.[52] Protect from light. Keep under an inert atmosphere. Metal containers involving the transfer of this chemical should be grounded and bonded. Where possible, automatically pump liquid from drums or other storage containers to process containers. Drums must be equipped with self-closing valves, pressure vacuum bungs, and flame arresters. Use only non-sparking tools and equipment, especially when opening and closing containers of this chemical. Sources of ignition such as smoking and open flames, are prohibited where this chemical is used, handled, or stored in a manner that could create a potential fire or explosion hazard.

Shipping: The DOT label required is "Flammable Liquid." The DOT/UN Hazard Class is 3 and the Packing Group is II.[19][20]

Spill Handling: Evacuate persons not wearing protective equipment from area of spill or leak until clean-up is complete. Remove all ignition sources. Ventilate area of spill or leak. Use water spray to reduce vapors. Absorb liquids in vermiculite, dry sand, earth, or a similar material and deposit in sealed containers. Keep amyl nitrite out of a confined space, such as a sewer, because of the potential for an explosion, unless the sewer is designed to prevent the build-up of explosive concentrations. It may be necessary to contain and dispose of this chemical as a hazardous waste. If material or contaminated runoff enters waterways, notify downstream users of potentially contaminated waters. Contact your Department of Environmental Protection or your regional office of the federal EPA for specific recommendations. If employees are required to clean-up spills, they must be properly trained and equipped. OSHA 1910.120(q) may be applicable.

Fire Extinguishing: This chemical is a flammable liquid. Poisonous gases including nitrogen oxides are produced in fire. Use dry chemical, carbon dioxide, or alcohol foam extinguishers. Vapors are heavier than air and will collect in low areas. Vapors may travel long distances to ignition sources and flashback. Vapors in confined areas may explode when exposed to fire. Storage containers and parts of containers may rocket great distances, in many directions. If material or contaminated runoff enters waterways, notify downstream users of potentially contaminated waters. Notify local health and fire officials and pollution control agencies. From a secure, explosion-proof location, use water spray to cool exposed containers. If cooling streams are ineffective (venting sound increases in volume and pitch, tank discolors, or shows any signs of deforming), withdraw immediately to a secure position. If employees are expected to fight fires, they must be trained and equipped in OSHA 1910.156.

Disposal Method Suggested: Incineration with scrubber to remove nitrogen oxides from the combustion gases.

References

New Jersey Department of Health and Senior Services, "Hazardous Substance Fact Sheet: Amyl Nitrite," Trenton, NJ (March 1987)

Amyltrichlorosilane

Molecular Formula: $C_5H_{11}SiCl_3$

Synonyms: Amiltriclorosilano (Spanish); Pentyltrichlorosilane; Pentylsilicon trichloride; Silane, trichloropentyl-; Trichloroamylsilane; Trichloropentylsilane

CAS Registry Number: 107-72-2

RTECS Number: VV4725000

DOT ID: UN 1728

Regulatory Authority

- Canada, WHMIS, Ingredients Disclosure List

Cited in U.S. State Regulations: Florida (G), Maine (G), Massachusetts (G), New Hampshire (G), New Jersey (G), Oklahoma (G), Pennsylvania (G).

Description: Amyl trichlorosilane, $C_5H_{11}SiCl_3$, is a colorless to yellowish liquid with a sharp acrid odor, like hydrochloric acid. Boiling point = 160°C. Flash point = 63°C. NFPA 704 M Hazard Identification (based on NFPA-704M Rating System): Health 3, Flammability 2, Reactivity 2, Water Reactive.

Potential Exposure: It is used to make silicones.

Incompatibilities: Water contact forms corrosive and toxic hydrogen chloride fumes. Contact with halogens may cause explosion. Corrodes metals.

Permissible Exposure Limits in Air: No standards set.

Permissible Concentration in Water: No criteria set.

Harmful Effects and Symptoms

Short Term Exposure: A corrosive chemical that can cause severe eye and skin burns. Exposure can irritate the eyes, nose and throat. Higher levels can irritate the lungs causing coughing and/or shortness of breath; still higher exposures can cause pulmonary edema, a medical emergency which can be delayed for several hours. This can cause death. The oral LD_{50} for rat is 2,340 mg/kg.

Long Term Exposure: Similar corrosive chemicals cause lung damage; bronchitis may develop.

Points of Attack: Lungs, skin.

Medical Surveillance: Lung function tests. Consider lung x-ray following acute overexposure.

First Aid: If this chemical gets into the eyes, remove any contact lenses at once and irrigate immediately for at least 15 minutes, occasionally lifting upper and lower lids. Seek medical attention immediately. If this chemical contacts the skin, remove contaminated clothing and wash immediately with soap and water. Seek medical attention immediately. If this chemical has been inhaled, remove from exposure, begin rescue breathing (using universal precautions) if breathing has stopped and CPR if heart action has stopped. Transfer promptly to a medical facility. When this chemical has been swallowed, get medical attention. If victim is conscious, administer water or milk. Do not induce vomiting. Medical observation is recommended for 24 - 48 hours after breathing overexposure, as pulmonary edema may be delayed. As first aid for pulmonary edema, a doctor or authorized paramedic may consider administering a corticosteroid spray.

Personal Protective Methods: Wear protective gloves and clothing to prevent any reasonable probability of skin contact. Safety equipment suppliers/manufacturers can provide recommendations on the most protective glove/clothing material for your operation. All protective clothing (suits, gloves, footwear, headgear) should be clean, available each day, and put on before work. Contact lenses should not be worn when working with this chemical. Wear splash-proof chemical goggles and face shield unless full facepiece respiratory protection is worn. Employees should wash immediately with soap when skin is wet or contaminated. Provide emergency showers and eyewash.

Respirator Selection: Acid gas: SA:CF (any supplied-air respirator operated in a continuous-flow mode); or PAPROV [any powered, air-purifying respirator with organic vapor cartridge(s)]; CCRFOV [any air-purifying, full-facepiece respirator (gas mask) with a chin-style, front-or back-mounted acid gas canister]; or GMFOV [any air-purifying, full-facepiece respirator (gas mask) with a chin-style, front-or back-mounted organic vapor canister]; or SCBAF (any self-contained breathing apparatus with a full facepiece); or SAF (any supplied-air respirator with a full facepiece). *Emergency or planned entry into unknown concentrations or IDLH conditions:* SCBAF:PD,PP (any self-contained breathing apparatus that has a full facepiece and is operated in a pressure-demand or other positive-pressure mode); or SAF:PD,PP: ASCBA (any supplied-air respirator that has a full facepiece and is operated in a pressure-demand or other positive-pressure mode in combination with an auxiliary self-contained breathing apparatus operated in a pressure-demand or other positive-pressure mode). *Escape*: GMFOV [any air-purifying, full-facepiece respirator (gas mask) with a chin-style, front-or back-mounted organic vapor canister] or SCBAE (any appropriate escape-type, self-contained breathing apparatus).

Storage: Prior to working with this chemical you should be trained on its proper handling and storage. Store in tightly closed containers in a cool, well ventilated area. Metal containers involving the transfer of this chemical should be grounded and bonded. Where possible, automatically pump liquid from drums or other storage containers to process containers. Drums must be equipped with self-closing valves, pressure vacuum bungs, and flame arresters. Use only non-sparking tools and equipment, especially when opening and closing containers of this chemical. Sources of ignition such as smoking and open flames, are prohibited where this chemical is used, handled, or stored in a manner that could create a potential fire or explosion hazard.

Shipping: The DOT label required is "Corrosive." Shipment by passenger aircraft or railcar is forbidden; cargo aircraft shipment is limited. The DOT/UN Hazard class is 8 and the Packing Group is II.[19][20]

Spill Handling: Evacuate persons not wearing protective equipment from area of spill or leak until clean-up is complete. Remove all ignition sources. Ventilate area of spill or leak. Absorb liquids in vermiculite, dry sand, earth, or a similar material and deposit in sealed containers. Keep amyl trichlorosilane out of a confined space, such as a sewer, because of the potential for an explosion, unless the sewer is designed to prevent the build-up of explosive concentrations. Collect powdered material in the most convenient and safe manner and deposit in sealed containers. It may be necessary to contain and dispose of this chemical as a hazardous waste. If material or contaminated runoff enters waterways, notify downstream users of potentially contaminated waters. Contact your Department of Environmental Protection or your regional office of the federal EPA for specific recommendations. If employees are required to clean-up spills, they must be properly trained and equipped. OSHA 1910.120(q) may be applicable.

Fire Extinguishing: This chemical is a combustible liquid. Poisonous gases including phosgene and hydrogen chloride are produced in fire. *Do not use water or foam.* Use dry chemical, carbon dioxide extinguishers. Vapors are heavier than air and will collect in low areas. Vapors may travel long distances to ignition sources and flashback. Vapors in confined areas may explode when exposed to fire. Storage containers and parts of containers may rocket great distances, in many directions. If material or contaminated runoff enters waterways, notify downstream users of potentially contaminated waters. Notify local health and fire officials and pollution control agencies. From a secure, explosion-proof location, use water spray to cool exposed containers. If cooling streams are ineffective (venting sound increases in volume and pitch, tank discolors, or shows any signs of deforming), withdraw immediately to a secure position. If employees are expected to fight fires, they must be trained and equipped in OSHA 1910.156.

References

New Jersey Department of Health and Senior Services, "Hazardous Substance Fact Sheet: Amyl Trichlorosilane," Trenton, NJ (May 1986)

Aniline

Molecular Formula: C_6H_5N

Common Formula: $C_6H_3NH_2$

Synonyms: Aminobenzene; Aminophen; Anilina (Spanish); Aniline oil; Anyvim; Arylamine; Benzeneamine; Benzene, amino-; Blue oil; C.I. 76000; Huile d'aniline (French); NCI-C03736; Phenylamine

CAS Registry Number: 62-53-3; 142-04-1 (hydrochloride)

RTECS Number: BW6650000; CY0875000 (hydrochloride)

DOT ID: UN 1547; UN 1548 (hydrochloride)

EEC Number: 612-008-00-7

Regulatory Authority

- Carcinogen (suspected) (DFG) (IARC)[3]
- Banned or Severely Restricted (Restricted in many countries) (UN)[35]
- Air Pollutant Standard Set (ACGIH)[1] (DFG)[3] (HSE)[33] (former USSR)[43] (OSHA)[58] (Several States)[60] (Several Canadian Provinces) (Australia) (Israel) (Mexico)
- Water Pollution Standard Proposed (former USSR)[43]
- Clean Air Act 42USC7412; Title I, Part A, §112 hazardous pollutants
- Clean Water Act: 40CFR116.4 Hazardous Substances; 40CFR117.3, RQ (same as CERCLA); Section 313 Water Priority Chemicals (57FR41331, 9/9/92)
- EPA Hazardous Waste Number (RCRA No.): U012
- RCRA 40CFR261, Appendix 8; 40CFR261.11 Hazardous Constituents
- RCRA Land Ban Waste
- RCRA 40CFR264, Appendix 9; Ground Water Monitoring List Suggested methods (PQL μg/l): 8270 (10)

- RCRA 40CFR268.48; 61FR15654, Universal Treatment Standards: Wastewater (mg/l), 0.081; Nonwastewater (mg/kg), 14
- CERCLA/SARA 40CFR302, Extremely Hazardous Substances: TPQ = 1,000 lb (454 kg)
- SUPERFUND/EPCRA 40CFR302.4, Appendix A, Reportable Quantity (RQ): CERCLA, 5,000 lb (2,270 kg), 40CFR372.65: Form R *de minimis* Concentration Reporting Level: 1.0%
- Canada, WHMIS, Ingredients Disclosure List; National Pollutant Release Inventory (NPRI); Priority Substance List

Cited in U.S. State Regulations: Alaska (G), California (A, G). Connecticut (A), Florida (G), Illinois (G), Kansas (G, A), Louisiana (G), Maine (G), Maryland (G), Massachusetts (G, A), Michigan (G), Minnesota (G), Nevada (A), New Hampshire (G), New Jersey (G), New York (G, A), North Carolina (A), North Dakota (A), Oklahoma (G), Pennsylvania (G), Rhode Island (G, A), South Carolina (A), Vermont (G), Virginia (G, A), Washington (G), West Virginia (G), Wisconsin (G).

Description: Aniline, $C_6H_3NH_2$, is a clear, colorless, oily liquid that darkens on exposure to light; with a characteristic amine-like odor. Boiling point = 184°C. Flash point = 70°C; 193°C (hydrochloride); NFPA 704 M Hazard Identification (based on NFPA-704M Rating System): Health 3, Flammability 2, Reactivity 0. Explosive Limits: LEL = 1.3%; UEL = 11%. Odor threshold = 0.58 – 10 ppm. Slightly soluble in water.

Potential Exposure: Aniline is widely used as an intermediate in the synthesis of dyestuffs. It is also used in the manufacture of rubber accelerators and antioxidants, pharmaceuticals, marking inks, tetryl, optical whitening agents, photographic developers, resins, varnishes, perfumes, shoe polishes, and many organic chemicals.

Incompatibilities: Forms explosive mixture with air. Unless inhibited (usually methanol), aniline is readily able to polymerize. Fires and explosions may result from contact with halogens, strong acids, oxidizers, strong bases, organic anhydrides, acetic anhydride, isocyanates, aldehydes, sodium peroxide. Strong reaction with toluene diisocyanate. Reacts with alkali metals and alkali earth metals. Attacks some plastics, rubber and coatings; copper and copper alloys.

Permissible Exposure Limits in Air: The Federal OSHA PEL is 5 ppm TWA for an 8- hour workshift. NIOSH recommends that exposure to occupational carcinogens be limited to the lowest feasible concentration. ACGIH recommends a value of 2 ppm (10 mg/m^3) TWA for an 8-hour workshift. OSHA and ACGIH add the notation "skin" indicating the potential for cutaneous absorption. The NIOSH IDLH level is 100 ppm. The Australian standard is 2 ppm (7.6 mg/m^3) TWA (skin). The Israel standard is 2 ppm (7.6 mg/m^3) TWA and an action level of 1 ppm (3.8 mg/m^3) TWA. The Mexico standard is 2 ppm (10 mg/m^3) TWA and 5 ppm (20 mg/m^3) STEL. The UK[33] has set a STEL of 5 ppm (20 mg/m^3). The DFG[3] has set a MAC of 2 ppm (7.7 mg/m^3) (skin) with the notation that aniline is a suspected carcinogen and peak limitation 5 times MAK (30 min. avg. value; don't exceed 2 times during workshift). The former USSR-UNEP/IRPTC project[43] has set a MAC of 0.1 mg/m^3 in workplace air and 0.05 m/mg^3 in residential air on a momentary basis and 0.03 mg/m^3 in residential air on an average daily basis. The California standard for workplace air is 2 ppm (7.6 mg/m^3) TWA.

In addition, several states have set guidelines or standards for aniline in ambient air:[60] 0.4 mg/m^3 (New York), 1.4 μg/m^3 (Massachusetts), 3.0 μg/m^3 (Rhode Island), 23.81 μg/m^3 (Kansas), 50 μg/m^3 (South Carolina), 100 μg/m^3 (North Dakota), 160 μg/m^3 (Virginia), 200 μg/m^3 (Connecticut), 238 μg/m^3 (Nevada), 1,000 μg/m^3 (North Carolina). Canadian Provincial levels follow: 2 ppm (7.6 mg/m^3) TWA and 5 ppm (19 mg/m^3) STEL (skin) Alberta, 5 ppm (19 mg/m^3) TWA (skin) British Columbia, 2 ppm (8 mg/m^3) TWA (skin) Ontario, 2 ppm (8 mg/m^3) TWAEV (skin) Quebec.

Determination in Air: Silica adsorption, workup in n-propanol, followed by gas chromatography. See NIOSH Method 2002 for aromatic amines.[18]

Permissible Concentration in Water: No criteria set but EPA[32] has suggested an ambient limit in water of 262 μg/ml based on health effects. The former USSR-UNEP/IRPTC project[43] has set a MAC of 0.1 mg/l in water for domestic purposes and 0.0001 mg/l in water for fishery purposes.

Routes of Entry: Inhalation of vapors, percutaneous absorption of liquid and vapor, ingestion, skin and/or eye contact.

Harmful Effects and Symptoms

Short Term Exposure: Skin contact contributes significantly to overall exposure. Direct contact with liquid aniline can cause skin burns, eye irritation and possible permanent damage. Inhalation can cause irritation of the respiratory tract with wheezing and coughing. High levels can interfere with the blood's ability to carry oxygen. Higher levels can cause difficult breathing, collapse and death. Symptoms of exposure include grayish blue skin, headache, nausea, sometimes vomiting, dryness of throat, confusion, vertigo, lack of muscle coordination, ringing in the ears, weakness, disorientation, lethargy, drowsiness and coma. Urinary signs include painful urinating, blood in the urine, the presence of hemoglobin in the urine, and diminished amounts of urine. Aniline is classified as very toxic. Probable oral lethal dose in humans is 50 – 500 mg/kg for a 150 lb person. The approximate minimum lethal dose for a 150 lb human is 10 grams. Serious poisoning may result from ingestion of 0.25 ml.

Long Term Exposure: Chronic exposure can cause anemia, anorexia, weight loss, and skin lesions. Chronic effects may be due to acute damage to the brain, heart, and kidneys. Loss of appetite, dizziness, insomnia, tremors, malignant bladder growths, liver damage, and jaundice. Anemia has been reported. Has been linked to bladder cancer according to a NIOSH study. Signs and symptoms include: blood in the urine, other changes in the appearance of the urine, changes in urinary habits, lumps in the groin and lower abdomen, and pain in the lower abdomen or back. Aniline can cross the placental barrier. Because fetal hemoglobin is more easily oxidized to methemoglobin than is adult hemoglobin and is less easily reduced back to normal hemoglobin, methemoglobin theoretically may be at higher levels in fetuses than in exposed mothers.

Points of Attack: Blood, cardiovascular system, liver, kidneys.

Medical Surveillance: Preplacement and periodic physical examinations should be performed on all employees working in aniline exposure areas. These should include a work history to elicit information on all past exposures to aniline, other aromatic amines, and nitro compounds known to cause chemical cyanosis, and the clinical history of any occurrence of chemical cyanosis; a personal history to elicit alcohol drinking habits; and general physical examination with particular reference to the cardiovascular system. Persons with impaired cardiovascular status may be at greater risk from the consequences of chemical cyanosis. A preplacement complete blood count and methemoglobin estimation should be performed as baseline levels, also follow-up studies including periodic blood counts and hematocrits. People at special risk include individuals with glucose-6-phosphate-dehydrogenase deficiency and those with liver and kidney disorders, blood diseases, or a history of alcoholism.

First Aid: If this chemical gets into the eyes, remove any contact lenses at once and irrigate immediately for at least 15 minutes, occasionally lifting upper and lower lids. Seek medical attention immediately. If this chemical contacts the skin, remove contaminated clothing and wash immediately with soap and water. Seek medical attention immediately. If this chemical has been inhaled, remove from exposure, begin rescue breathing (using universal precautions) if breathing has stopped and CPR if heart action has stopped. Transfer promptly to a medical facility. When this chemical has been swallowed, get medical attention. Give large quantities of water and induce vomiting. Do not make an unconscious person vomit.

Note to Physician: Treat for methemoglobinemia. Spectrophotometry may be required for precise determination of levels of methemoglobinemia in urine.

Personal Protective Methods: Aniline, the simplest aromatic amine, is a prototypical inducer of methemoglobinemia and impaired oxygen transport to tissues. Chemical protective clothing is recommended because aniline vapor and liquid can be dermally absorbed and may contribute to systemic toxicity. Wear protective gloves and clothing to prevent any reasonable probability of skin contact. In areas of vapor concentration, the use of respirators alone is not sufficient; skin protection by protective clothing should be provided even though there is no skin contact with liquid aniline. Safety equipment suppliers/manufacturers can provide recommendations on the most protective glove/clothing material for your operation Butyl rubber and polyvinyl alcohol protective clothing is reportedly superior to other materials. Chlorinated polyethylene, teflon, saranex, silvershield, and neoprene+ natural rubber also offer some protection. In severe exposure situations, complete body protection has been employed, consisting of air-conditioned suit with air supplied helmet and cape. Personal hygiene practices including prompt removal of clothing which has absorbed aniline, thorough showering after work and before changing to street clothes and clean working clothes daily are essential. Provide emergency showers and eyewash.

Respirator Selection: Engineering controls should be used wherever feasible to maintain airborne concentrations of this chemical below the prescribed exposure limit. Respirators and protective equipment less effective than engineering controls, and should be used only in non-routine or emergency situations which may result in exposure concentrations in excess of the TWA environmental limit.

NIOSH: *At any concentrations above the NIOSH REL, or where there is no REL, at any detectable concentration:* SCBAF:PD,PP (any MSHA/NIOSH approved self-contained breathing apparatus that has a full facepiece and is operated in a pressure-demand or other positive-pressure mode); or SAF:PD,PP:ASCBA (any supplied-air respirator that has a full facepiece and is operated in a pressure-demand or other positive-pressure mode in combination with an auxiliary, self-contained breathing apparatus operated in a pressure-demand or other positive pressure mode). *Escape:* GMFOVHiE [any air-purifying, full-facepiece respirator (gas mask) with a chin-style, front- or back-mounted organic vapor canister having a high-efficiency particulate filter]; or SCBAE (any appropriate escape-type, self-contained breathing apparatus).

Storage: Prior to working with this chemical you should be trained on its proper handling and storage. Before entering confined space where aniline may be present, check to make sure that an explosive concentration does not exist. Store in tightly closed containers in a cool, dry, dark, well ventilated area. Metal containers involving the transfer of this chemical should be grounded and bonded. Where possible, automatically pump liquid from drums or other storage containers to process containers. Drums must be equipped with self-closing valves, pressure vacuum bungs, and flame arresters. Use only non-sparking tools and equipment, especially when opening and closing containers of this chemical. Where this chemical is used, handled, manufactured, or stored, use explosion-proof electrical equipment and fittings. Sources of ignition such as smoking and open flames, are prohibited where this chemical is used, handled, or stored in a manner that could create a potential fire or explosion hazard. A regulated, marked area should be established where this chemical is handled, used, or stored in compliance with OSHA standard 1910.1045.

Shipping: The DOT label requirement is "Poison." The DOT/UN Hazard Class is 6.1 and the Packing Group is II.[19][20] The hydrochloride requires a "Keep away from Food" label and is in Hazard Class 6.1 but Packing Group III.

Spill Handling: Evacuate persons not wearing protective equipment from area of spill or leak until clean-up is complete. Stay upwind; keep out of low areas. Establish forced ventilation to keep levels below explosive limit. Wear positive pressure breathing apparatus and special protective clothing. Remove all ignition sources: no flares, smoking or flames in hazard area. Do not touch material; stop leak if you can do it without risk. Use water spray to reduce vapors. Small spills: take up with vermiculite, dry sand, earth, or other non-combustible absorbent material and place into containers for later disposal. Large spills: dike far ahead of spill for later disposal. It may be necessary to contain and dispose of this chemical as a hazardous waste. If material or contaminated runoff enters waterways, notify downstream users of potentially contaminated waters. Contact your Department of Environmental Protection or your regional office of the federal EPA for specific recommendations. If employees are required to clean-up spills, they must be properly trained and equipped. OSHA 1910.120(q) may be applicable.

Fire Extinguishing: This chemical is a combustible liquid. Poisonous gases including nitrogen oxides and carbon monoxide are produced in fire. Use alcohol foam extinguishers. Vapors are heavier than air and will collect in low areas. Vapors may travel long distances to ignition sources and flashback. Vapors in confined areas may explode when exposed to fire. Storage containers and parts of containers may rocket great distances, in many directions. If material or contaminated runoff enters waterways, notify downstream users of potentially contaminated waters. Notify local health and fire officials and pollution control agencies. Fight fire from maximum distance. Dike fire control water for later disposal and do not scatter material. If a leak or spill has not ignited, use water spray to control vapors. Wear self-contained breathing apparatus with a full facepiece operated in pressure-demand or other positive pressure mode and special protective clothing. From a secure, explosion-proof location, use water spray to cool exposed containers. If cooling streams are ineffective (venting sound increases in volume and pitch, tank discolors, or shows any signs of deforming), withdraw immediately to a secure position. If employees are expected to fight fires, they must be trained and equipped in OSHA 1910.156.

Disposal Method Suggested: Consult with environmental regulatory agencies for guidance on acceptable disposal practices. Generators of waste containing this contaminant (≥100 kg/mo) must conform with EPA regulations governing storage, transportation, treatment, and waste disposal. Incineration with provision for No_x removal from flue gases by scrubber, catalytic or thermal device.[22]

References

U.S. Environmental Protection Agency, "Chemical Hazard Information Profile: Aniline, Washington, DC (January 20, 1978)

Sax, N. I., Ed., "Dangerous Properties of Industrial Materials Report" 1, No. 3, 29–31 (1981) and 3, No. 5, 37–40 (1983) and 4, No. 4, 55–59 (1984) (Aniline hydrochloride)

U.S. Environmental Protection Agency, "Chemical Profile: Aniline," Washington, DC, Chemical Emergency Preparedness Program (November 30, 1987)

New Jersey Department of Health and Senior Services, "Hazardous Substance Fact Sheet: Aniline," Trenton, NJ (June 1998)

New York State Department of Health, "Chemical Fact Sheet: Aniline," NY, Bureau of Toxic Substance Assessment (April 1986)

Aniline, 2,4,6-Trimethyl

Molecular Formula: $C_9H_{13}N$

Common Formula: $C_6H_2(CH_3)_3NH_2$

Synonyms: Aminomesitylene; 2-Aminomesitylene 1-amino-2,4,6-trimethylbenzen (Czech); 2-Amino-1,3,5-trimethylbenzene; Benzenamine, 2,4,6-trimethyl-; Mesidin (Czech); Mesidine; Mesitylamine; Mezidine; 2,4,6-Trimethylaniline; 2,4,6-Trimethylbenzenamine

CAS Registry Number: 88-05-1

RTECS Number: BZ0700000

DOT ID: UN 3082

Regulatory Authority

- Carcinogen (suspected, Group 3) (IARC)
- CERCLA/SARA 40CFR302 Extremely Hazardous Substances: TPQ = 500 lb (227 kg); CERCLA/SARA Section 304 Reportable Quantity (RQ): EHS, 1 lb (0.454 kg)

Cited in U.S. State Regulations: California (G), Florida (G), Massachusetts (G), New Jersey (G), Pennsylvania (G).

Description: Aniline, 2,4,6-trimethyl-, $C_6H_2(CH_3)_3NH_2$, is a liquid. Boiling point = 233°C. Freezing/Melting point = -15°C.

Potential Exposure: Used on small scale in organic synthesis.

Permissible Exposure Limits in Air: No standards set.

Permissible Concentration in Water: No criteria set.

Routes of Entry: Ingestion.

Harmful Effects and Symptoms

Short Term Exposure: This material is moderately toxic orally. It is also considered highly toxic by unspecified routes. It is a skin and eye irritant. The danger of acute poisoning is represented by methemoglobinemia leading to adverse effects on the red cells. A number of the amines may act as skin sensitizers. Repeated exposure results in narrowing of peripheral vision, increase in size of blind spot and decrease in photosensitivity. The LC_{50} inhalation (mouse) is 0.29 mg/liter/2 hours.

Long Term Exposure: Suspect occupational carcinogen. This material is a suspect carcinogen on the basis of being an aromatic amine but unlike the 2,4,5-trimethyl isomer, is not animal positive.

Points of Attack: Blood, eyes, respiratory system.

First Aid: If this chemical gets into the eyes, remove any contact lenses at once and irrigate immediately for at least 15 minutes, occasionally lifting upper and lower lids. Seek medical attention immediately. If this chemical contacts the skin, remove contaminated clothing and wash immediately with soap and water. Seek medical attention immediately. If this chemical has been inhaled, remove from exposure, begin rescue breathing (using universal precautions) if breathing has stopped and CPR if heart action has stopped. Transfer promptly to a medical facility. When this chemical has been swallowed, get medical attention. Give large quantities of water and induce vomiting. Do not make an unconscious person vomit.

Note to Physician: Treat for methemoglobinemia. Spectrophotometry may be required for precise determination of levels of methemoglobinemia in urine.

Personal Protective Methods: Wear protective gloves and clothing to prevent any reasonable probability of skin contact. Safety equipment suppliers/manufacturers can provide recommendations on the most protective glove/clothing material for your operation. All protective clothing (suits, gloves, footwear, headgear) should be clean, available each day, and put on before work. Contact lenses should not be worn when working with this chemical. Wear splash-proof chemical goggles and face shield unless full facepiece respiratory protection is worn. Employees should wash immediately with soap when skin is wet or contaminated. Provide emergency showers and eyewash.

Respirator Selection: Where the potential for exposure to this chemical, use a MSHA/NIOSH approved supplied-air respirator with a full facepiece operated in the positive pressure mode or with a full facepiece, hood, or helmet in the continuous flow mode, or use a MSHA/NIOSH approved self-contained breathing apparatus with a full facepiece operated in pressure-demand or other positive pressure mode.

Storage: Prior to working with this chemical you should be trained on its proper handling and storage. Store in tightly closed containers in a cool, well ventilated area. Where possible, automatically pump liquid from drums or other storage containers to process containers Sources of ignition such as smoking and open flames are prohibited where this chemical is handled, used, or stored. Metal containers involving the transfer of 5 gallons or more of this chemical should be grounded and bonded. Drums must be equipped with self-closing valves, pressure vacuum bungs, and flame arresters. Use only non-sparking tools and equipment, especially when opening and closing containers of this chemical. Wherever this chemical is used, handled, manufactured, or stored, use explosion-proof electrical equipment and fittings. A regulated, marked area should be established where this chemical is handled, used, or stored in compliance with OSHA standard 1910.1045.

Shipping: Environmentally hazardous liquid, n.o.s. Hazard Class: 9. Label: "Class 9." Packing Group: III.

Spill Handling: Evacuate persons not wearing protective equipment from area of spill or leak until clean-up is complete. Remove all ignition sources. Do not touch spilled material; stop leak if you can do so without risk. Use water spray to reduce vapors. Small spills: absorb with sand or other noncombustible absorbent material and place into containers for later disposal. Large spills: dike far ahead of spill for later disposal. Wear positive pressure breathing apparatus and special protective clothing. It may be necessary to contain and dispose of this chemical as a hazardous waste. If material or contaminated runoff enters waterways, notify downstream users of potentially contaminated waters. Contact your Department of Environmental Protection or your regional office of the federal EPA for specific recommendations. If employees are required to clean-up spills, they must be properly trained and equipped. OSHA 1910.120(q) may be applicable.

Fire Extinguishing: This chemical is a combustible liquid. Poisonous gases including nitrogen oxides are produced in fire. *Small fires:* dry chemical, carbon dioxide, water spray or foam. *Large fires:* water spray, fog or foam. Move container from fire area if you can do so without risk. Fight fire from maximum distance. Dike fire control water for later disposal; do not scatter the material. Keep unnecessary people away; isolate hazard area and deny entry. Stay upwind; keep out of low areas. Ventilate closed spaces before entering them. Wear positive pressure breathing apparatus and special protective clothing. If water pollution occurs, notify appropriate authorities. Vapors are heavier than air and will collect in low areas. Vapors may travel long distances to ignition sources and flashback. Vapors in confined areas may explode when exposed to fire. Storage containers and parts of containers may rocket great distances, in many directions. If material or contaminated runoff enters waterways, notify downstream users of potentially contaminated waters. Notify local health and fire officials and pollution control agencies. From a secure, explosion-proof location, use water spray to cool exposed containers. If cooling streams are ineffective (venting sound increases in volume and pitch, tank discolors, or shows any signs of deforming), withdraw immediately to a secure position. If employees are expected to fight fires, they must be trained and equipped in OSHA 1910.156.

References

U.S. Environmental Protection Agency, "Chemical Profile: Aniline, 2,4,6-Trimethyl," Washington, DC, Chemical Emergency Preparedness Program (November 30, 1987)

Anisidines

Molecular Formula: C_7H_9NO

Common Formula: $H_2NC_6H_4OCH_3$

Synonyms: *o-:* Amine, *o*-methoxyphenylamine; *o*-Aminoanisole; 2-Aminoanisole; *o*-Anisidina (Spanish); 2-Anisidine; Anisidine-*o*; *o*-Anisylamine; *o*-Methoxyaniline;

2-Methoxyaniline; 2-Methoxybenzenamine; *o*-Methoxyphenylamine

p-: p-Aminoanisole; 4-Aminoanisole; 1-Amino-4-methoxybenzene; *p*-Anisidina (Spanish); 4-Anisidine; Anisidine-*p*; *p*-Methoxyaniline; 4-Methoxyaniline; 4-Methoxybenzenamine; 4-Methoxybenzeneamine

CAS Registry Number: 90-04-0 (o-); 104-94-9 (p-); 536-90-3 (m-); 29191-52-4 (mixed isomers)

RTECS Number: BZ5410000 (o-); BZ5450000 (p-)

DOT ID: UN 2431 (anisidines, solid or liquid)

Regulatory Authority

Note: The o- and p- isomers are the primary concern of federal, state, and local government regulators

- Carcinogen (human suspected) (IARC)[9]
- Air Pollutant Standard Set (ACGIH)[1] (DFG)[3] (HSE)[33] (Several States)[60]
- Clean Air Act 42USC7412; Title I, Part A, §112 hazardous pollutants (o-)
- SUPERFUND/EPCRA 40CFR302.4, Appendix A, Reportable Quantity (RQ): CERCLA 1 lb (0.454 kg) (o-); 40CFR372.65: Form R *de minimis* Concentration Reporting Level: 0.1%. (o-)
- CERCLA/SARA 40CFR372.65: Form R *de minimis* Concentration Reporting Level: 1.0% (p-)
- U.S. DOT Regulated Marine Pollutant (49CFR172.101, Appendix B) as o-anisidines
- Canada, WHMIS, Ingredients Disclosure List

Cited in U.S. State Regulations: Alaska (G), California (A, G), Connecticut (A), Florida (G), Illinois (G), Maine (G), Maryland (G), Massachusetts (G), Michigan (G), Minnesota (G), Nevada (A), New Hampshire (G), New Jersey (G), North Dakota (A), Pennsylvania (G), Rhode Island (G, A), Virginia (A), West Virginia (G).

Description: Anisidine, $H_2NC_6H_4OCH_3$, exists as *ortho-*, *meta-*, and *para-* isomers. *ortho-:* a colorless to pink liquid. Solid below 41°F (5°C). Boiling point = 286°C. Flash point = 118°C (oc). Hazard Identification (based on NFPA-704 M Rating System): for *o- isomer* Health 2, Flammability 1, Reactivity 0. *meta-:* a pale yellow liquid. Freezing/Melting point = 0°C. Boiling point = 251°C. *para-:* a reddish brown solid. Freezing/Melting point = 57°C They have characteristic amine odors (fishy). Insoluble in water.

Potential Exposure: Anisidines are used in the manufacture of azo dyes.

Incompatibilities: Incompatible with strong oxidizers, with risk of fire or explosions. Attacks some coatings and some forms of plastic and rubber.

Permissible Exposure Limits in Air: The Federal OSHA/NIOSH standard[58] and ACGIH recommended value of 0.1 ppm (0.5 mg/m^3) TWA has been set for *o-* and *p*-anisidine. The NIOSH IDLH level is 50 mg/m^3. The TWA value bears the notation "skin" indicating the potential for cutaneous absorption. *o*-anisidine is a probable carcinogen, and all contact should be reduced to the lowest possible level. The HSE (U.K.), Australia, Israel and Mexico standard is 0.1 ppm (0.5 mg/m^3) TWA (skin) and Israel ahas an action level of 0.05 ppm (0.25 mg/m^3). The DFG[3] has set MAK for both *o-* and *p*-anisidine at 0.1 ppm (0.5 mg/m^3) (skin) and a peak limitation of 2 times MAK during a 30 minute average value [don't exceed 4 times during workshift]. California's workplace PEL is 0.1 ppm (0.5 mg/m^3) TWA (skin). In addition, several states have set guidelines or standards for anisidines in ambient air:[60] 0.02 - 1.0 μg/m^3 (Rhode Island), 5.0 μg/m^3 (Connecticut and North Dakota), 8.0 μg/m^3 (Virginia), 12.0 μg/m^3 (Nevada). Canadian provincial workplace standards are 0.1 ppm (0.5 mg/m^3) TWA (skin) for Alberta, British Columbia, Ontario, and Quebec. Alberta also has a 15 minute STEL of 0.3 ppm (1.5 mg/m^3) (skin).

Determination in Air: Collection on XAD-2 sorbent, desorption by methanol and measurement by HPLC with UV detection according to NIOSH Method 2514.[18]

Permissible Concentration in Water: No criteria set.

Routes of Entry: Inhalation, ingestion, skin, eye, mucous membrane absorption.

Harmful Effects and Symptoms

Short Term Exposure: This chemical can be absorbed through the skin, eyes, or mucous membranes, thereby increasing exposure. Contact with anisidines can irritate the eyes, skin and respiratory tract; can cause a burning sensation and skin rash. Inhalation of p-anisidine can cause shortness of breath and coughing. Exposure can interfere with the blood's ability to carry hemoglobin (methemaglobinemia). This can cause headache, dizziness, cyanosis of the skin and lips. Higher levels can cause difficult breathing, collapse and death.

Long Term Exposure: o-anisidine is a probable carcinogen in humans (IARC Group 2B, limited human evidence). Related aromatic amines are carcinogens, and p-anisidine may be carcinogenic as well. Repeated exposure to these isomers may cause anemia, skin allergy, lung irritation and bronchitis, nerve damage, kidney damage.

Points of Attack: Blood, kidneys, liver, cardiovascular system, central nervous system.

Medical Surveillance: *o-anisidine*: evaluation by a qualified allergist, test for blood hemoglobin, complete blood count (CBC) and reticulocyte count. Consider nerve conduction studies. *p-anisidine:* Lung function tests, evaluation by a qualified allergist, kidney function tests, complete blood count (CBC), methemoglobin level.

First Aid: If this chemical gets into the eyes, remove any contact lenses at once and irrigate immediately for at least 15 minutes, occasionally lifting upper and lower lids. Seek medical attention immediately. If this chemical contacts the skin, remove contaminated clothing and wash immediately with

soap and water. Seek medical attention immediately. If this chemical has been inhaled, remove from exposure, begin rescue breathing (using universal precautions) if breathing has stopped and CPR if heart action has stopped. Transfer promptly to a medical facility. When this chemical has been swallowed, get medical attention. Give large quantities of water and induce vomiting. Do not make an unconscious person vomit.

Note to Physician: Treat for methemoglobinemia. Spectrophotometry may be required for precise determination of levels of methemoglobinemia in urine.

Personal Protective Methods: Wear protective gloves and clothing to prevent any reasonable probability of skin contact. Safety equipment suppliers/manufacturers can provide recommendations on the most protective glove/clothing material for your operation. All protective clothing (suits, gloves, footwear, headgear) should be clean, available each day, and put on before work. Contact lenses should not be worn when working with this chemical. Wear splash-proof (o-anisidine) or dust-proof (p-anisidine) chemical goggles and face shield unless full facepiece respiratory protection is worn. Employees should wash immediately with soap when skin is wet or contaminated. Provide emergency showers and eyewash.

Respirator Selection: NIOSH: *At any detectable concentration:* SCBAF:PD,PP (any MSHA/NIOSH approved self-contained breathing apparatus that has a full facepiece and is operated in a pressure-demand or other positive-pressure mode); or SAF:PD,PP:ASCBA (any supplied-air respirator that has a full facepiece and is operated in a pressure-demand or other positive-pressure mode in combination with an auxiliary, self-contained breathing apparatus operated in a pressure-demand or other positive pressure mode). *Escape:* HiEF (Any air-purifying, full-facepiece respirator with a high-efficiency particulate filter); or SCBAE (Any appropriate escape-type, self-contained breathing apparatus).

Storage: Prior to working with this chemical you should be trained on its proper handling and storage. Store in tightly closed containers in a cool, dark, well ventilated area. Protect against sunlight and strong oxidizers. Metal containers involving the transfer of this chemical should be grounded and bonded. Where possible, automatically pump liquid from drums or other storage containers to process containers. Drums must be equipped with self-closing valves, pressure vacuum bungs, and flame arresters. Use only non-sparking tools and equipment, especially when opening and closing containers of this chemical. Sources of ignition such as smoking and open flames, are prohibited where this chemical is used, handled, or stored in a manner that could create a potential fire or explosion hazard. A regulated, marked area should be established where this chemical is handled, used, or stored in compliance with OSHA standard 1910.1045.

Shipping: The DOT label required is "Keep away from Food." The DOT/UN Hazard Class is 6.1 and the Packing Group is III.[19][20]

Spill Handling: *Liquid:* Evacuate persons not wearing protective equipment from area of spill or leak until clean-up is complete. Remove all ignition sources. Ventilate area of spill or leak. Cover with sand and soda ash (9:1). After mixing, collect material in the most convenient and safe manner and deposit in sealed containers. Keep o-anisidine out of confined spaces, such as a sewer, because of the potential for an explosion, unless the sewer is designed to prevent the buildup of explosive concentrations. *Solid:* Evacuate persons not wearing protective equipment from area of spill or leak until clean-up is complete. Remove all ignition sources. Collect powdered material in the most convenient and safe manner and deposit in sealed containers. Ventilate area after clean-up is complete. If material or contaminated runoff enters waterways, notify downstream users of potentially contaminated waters. It may be necessary to contain and dispose of this chemical as a hazardous waste. Contact your Department of Environmental Protection or your regional office of the federal EPA for specific recommendations. If employees are required to clean-up spills, they must be properly trained and equipped. OSHA 1910.120(q) may be applicable.

Fire Extinguishing: This chemical is a combustible liquid. Poisonous gases including nitrogen oxides are produced in fire. Use dry chemical, carbon dioxide, or alcohol foam extinguishers. Vapors are heavier than air and will collect in low areas. Vapors may travel long distances to ignition sources and flashback. Vapors in confined areas may explode when exposed to fire. Storage containers and parts of containers may rocket great distances, in many directions. If material or contaminated runoff enters waterways, notify downstream users of potentially contaminated waters. Notify local health and fire officials and pollution control agencies. From a secure, explosion-proof location, use water spray to cool exposed containers. If cooling streams are ineffective (venting sound increases in volume and pitch, tank discolors, or shows any signs of deforming), withdraw immediately to a secure position. If employees are expected to fight fires, they must be trained and equipped in OSHA 1910.156.

Disposal Method Suggested: Dissolve in combustible solvent (alcohols, benzene, etc.) and spray solution into furnace equipped with afterburner and scrubber or burn spill residue on sand and soda ash absorbent in a furnace.

References

National Cancer Institute, Biossay of o-Ansidine Hydrochloride for Possible Carcinogenicity, Technical Report Series No. 89, Bethesda, MD (1978)

National Cancer Institute, Biossay of p-Anisidine Hydrochloride for Possible Carcinogenicity, Technical Report Series No. 116, Bethesda, MD (1978)

Sax, N. I., Ed., "Dangerous Properties of Industrial Materials Report" 1, No. 5, 34–36 (1981)

New Jersey Department of Health and Senior Services, "Hazardous Substance Fact Sheet: o-Anisidine," Trenton, NJ (April 1986)

New Jersey Department of Health and Senior Services, "Hazardous Substance Fact Sheet: p-Anisidine," Trenton, NJ (June 1998)

Anisole

Molecular Formula: C_7H_8O

Common Formula: $C_6H_5OCH_3$

Synonyms: Benzene, methoxy; Ether, methyl phenyl; Methoxybenzene; Methyl phenyl ether; Metil fenil eter (Spanish); Phenyl methyl ether

CAS Registry Number: 100-66-3

RTECS Number: BZ8050000

DOT ID: UN 2222

Regulatory Authority

- Water Pollution Standard Proposed (former USSR)[43]
- Canada, WHMIS, Ingredients Disclosure List

Cited in U.S. State Regulations: New Hampshire (G), New Jersey (G).

Description: Anisole, $C_6H_5OCH_3$, is a colorless to yellowish liquid with an agreeable, aromatic, spicy-sweet odor. Boiling point = 154°C. Freezing/Melting point = -37.3°C. Flash point = 51.6°C (oc). Autoignition temperature = 475°C. Hazard Identification (based on NFPA-704M Rating System): Health 1, Flammability 2, Reactivity 0. Insoluble in water.

Potential Exposure: Anisole is used as a solvent, a flavoring, vermicide, making perfumes, and in organic synthesis.

Incompatibilities: Oxidizers, strong acids.

Permissible Exposure Limits in Air: No standards set.

Permissible Concentration in Water: The former USSR-UNEP/IRTC Project[43] has set a MAC of 0.05 mg/l in water bodies used for domestic purposes.

Routes of Entry: Inhalation, absorbed through the skin, ingestion.

Harmful Effects and Symptoms

Short Term Exposure: A skin irritant since it degreases the skin; prolonged skin contact can cause drying and cracking. It irritates the eyes and respiratory tract if exposure occurs.[57] Exposure can cause dizziness, lightheadedness, and unconsciousness. It is moderately toxic by ingestion.[44] The oral LD_{50} for rat is 3,700 mg/kg and for mouse is 2,800 mg/kg.[9]

Long Term Exposure: Skin problems, dryness, cracking.

Points of Attack: Eyes, skin, respiratory system.

First Aid: If this chemical gets into the eyes, remove any contact lenses at once and irrigate immediately for at least 15 minutes, occasionally lifting upper and lower lids. Seek medical attention immediately. If this chemical contacts the skin, remove contaminated clothing and wash immediately with soap and water. Seek medical attention immediately. If this chemical has been inhaled, remove from exposure, begin rescue breathing (using universal precautions) if breathing has stopped and CPR if heart action has stopped. Transfer promptly to a medical facility. When this chemical has been swallowed, get medical attention. Give large quantities of water and induce vomiting. Do not make an unconscious person vomit.

Personal Protective Methods: Wear solvent-resistant gloves and clothing to prevent any reasonable probability of skin contact. Safety equipment suppliers/manufacturers can provide recommendations on the most protective glove/clothing material for your operation. All protective clothing (suits, gloves, footwear, headgear) should be clean, available each day, and put on before work. Contact lenses should not be worn when working with this chemical. Wear splash-proof chemical goggles and face shield unless full facepiece respiratory protection is worn. Employees should wash immediately with soap when skin is wet or contaminated. Remove nonimpervious clothing immediately if wet or contaminated. Provide emergency showers and eyewash.

Respirator Selection: Engineering controls should be used wherever feasible to maintain airborne concentrations of this chemical below the prescribed exposure limit. Respirators and protective equipment less effective than engineering controls, and should be used only in non-routine or emergency situations which may result in exposure concentrations in excess of the TWA environmental limit. *Where there is no REL, at any detectable concentration:* SCBAF:PD,PP (any MSHA/NIOSH approved self-contained breathing apparatus that has a full facepiece and is operated in a pressure-demand or other positive-pressure mode); or SAF:PD,PP:ASCBA (any supplied-air respirator that has a full facepiece and is operated in a pressure-demand or other positive-pressure mode in combination with an auxiliary, self-contained breathing apparatus operated in a pressure-demand or other positive pressure mode). *Escape:* GMFOVHiE [any air-purifying, full-facepiece respirator (gas mask) with a chin-style, front- or back-mounted organic vapor canister having a high-efficiency particulate filter]; or SCBAE (any appropriate escape-type, self-contained breathing apparatus).

Storage: Prior to working with this chemical you should be trained on its proper handling and storage. Before entering confined space where anisole may be present, check to make sure that an explosive concentration does not exist. Store in tightly closed containers in a cool, well ventilated, fireproof area. Metal containers involving the transfer of this chemical should be grounded and bonded. Where possible, automatically pump liquid from drums or other storage containers to process containers. Drums must be equipped with self-closing valves, pressure vacuum bungs, and flame arresters. Use only non-sparking tools and equipment, especially when opening and closing containers of this chemical. Sources of ignition such as smoking and open flames, are prohibited where this chemical is used, handled, or stored in a manner that could create a potential fire or explosion hazard.

Shipping: The DOT label requirement is "Flammable Liquid." Anisole falls in DOT/UN Hazard Class 3, Packing Group III.[19][20]

Spill Handling: Evacuate persons not wearing protective equipment from area of spill or leak until clean-up is complete. Remove all ignition sources. Ventilate spill or leak area. Absorb spilled liquid in sand or inert absorbent and put in sealed containers and remove for disposal. It may be necessary to contain and dispose of this chemical as a hazardous waste. If material or contaminated runoff enters waterways, notify downstream users of potentially contaminated waters. Contact your Department of Environmental Protection or your regional office of the federal EPA for specific recommendations. If employees are required to clean-up spills, they must be properly trained and equipped. OSHA 1910.120(q) may be applicable.

Fire Extinguishing: This chemical is a combustible liquid. Poisonous gases including carbon monoxide are produced in fire. Use dry chemical, carbon dioxide, or alcohol foam extinguishers. Vapors are heavier than air and will collect in low areas. Vapors may travel long distances to ignition sources and flashback. Vapors in confined areas may explode when exposed to fire. Storage containers and parts of containers may rocket great distances, in many directions. If material or contaminated runoff enters waterways, notify downstream users of potentially contaminated waters. Notify local health and fire officials and pollution control agencies. From a secure, explosion-proof location, use water spray to cool exposed containers. If cooling streams are ineffective (venting sound increases in volume and pitch, tank discolors, or shows any signs of deforming), withdraw immediately to a secure position. If employees are expected to fight fires, they must be trained and equipped in OSHA 1910.156.

Disposal Method Suggested: Incineration.

References

New Jersey Department of Health and Senior Services, Hazardous Substance Fact Sheet, Anisole, Trenton NJ (December, 1998)

Anthracene

Molecular Formula: $C_{14}H_{10}$

Synonyms: Anthracen (German); Anthracene oil; Anthracene polycyclic aromatic compound; Anthracin; Antraceno (Spanish); Green oil; Paranaphthalene; Sterilite Hop defoliant; Tetra olive N2G

CAS Registry Number: 120-12-7; 906-80-5 (anthracene oil)

RTECS Number: CA9350000

DOT ID: NA 9188

Regulatory Authority

- Banned or Severely Restricted (UN) (in cosmetic products in the EEC)[35]
- Air Pollutant Standard Set (ACGIH)[1] (HSE)[33]
- Water Pollution Standard Set (EPA)[6] (Kansas)[61] (Mexico)
- Clean Water Act: 40CFR423, Appendix A, Priority Pollutants; Section 313 Water Priority Chemicals (57FR41331, 9/9/92)
- RCRA, 40CFR261, Appendix 8 Hazardous Constituents., waste number not listed
- RCRA 40CFR268.48; 61FR15654, Universal Treatment Standards: Wastewater (mg/l), 0.059; Nonwastewater (mg/kg), 3.4
- RCRA 40CFR264, Appendix 9; Ground Water Monitoring List: Suggested methods (PQL μg/l): 8100 (200); 8270 (10)
- SUPERFUND/EPCRA 40CFR302.4, Appendix A, Reportable Quantity (RQ): 5,000 lb (2,270 kg), 40CFR 372.65: Form R *de minimis* Concentration Reporting Level: 1.0%
- TSCA: 716.120 (*a*), listed chemical
- Canada, WHMIS, Ingredients Disclosure List; National Pollutant Release Inventory (NPRI)

Cited in U.S. State Regulations: California (A, G), Kansas (W), Massachusetts (G), New Hampshire (G), New Jersey (G), Oklahoma (G), Pennsylvania (G).

Description: Anthracene, $C_{14}H_{10}$, is colorless, to pale yellow crystalline solid with a bluish fluorescence. Boiling point = 340°C. Freezing/Melting point = 216.5°C. Flash point = 121°C. Autoignition temperature = 540°C. NFPA 704 M Hazard Identification (based on NFPA-704M Rating System): Health 0, Flammability 1, Reactivity (not listed). Explosive limits: LEL = 0.6%.[17] NFPA 704 M Hazard Identification (based on NFPA-704M Rating System): Health 0, Flammability 1. Insoluble in water.

Potential Exposure: It is used in dye stuffs (alizarin), insecticides and wood preservatives, making synthetic fibers, anthraquinone and other chemicals. May be present in coke oven emissions, diesel fuel, and coal tar pitch volitiles.

Incompatibilities: Dust or fine powder forms an explosive mixture with air. Contact with strong oxidizers, chromic acid, calcium hypochlorite may cause violent reactions.

Permissible Exposure Limits in Air: No occupational limits have been established for anthracene. However this chemical may be present as *coke oven emissions* and *coal tar pitch volatiles*. See these entries for legal air standards.

Determination in Air: Use NIOSH Methods [for polynuclear aromatic hydrocarbons]: 5506 (HPLC), 5515 (GC), or OSHA Method #ID-58.

Permissible Concentration in Water: Anthracene falls in the "polynuclear aromatic hydrocarbon" category of priority toxic pollutants as defined by EPA.[6] The EPA has considered setting criteria in the range from 0.097 – 9.7 nanograms/liter for the protection of human health from polynuclear aromatic hydrocarbons. In addition, Kansas has set forth a guideline for anthracene in drinking water[61] of 0.029 μg/l.

Routes of Entry: Inhalation, skin and/or eye contact.

Harmful Effects and Symptoms

Short Term Exposure: Anthracene can affect you when breathed in. Skin contact can cause irritation or a skin allergy which is greatly aggravated by sunlight on contaminated skin. Breathing irritates the nose, throat and bronchial tubes. Eye contact or "fume" exposure can cause irritation and burns.

Long Term Exposure: Repeated skin contact can cause thickening, pigment changes and growths. Anthracene may cause mutations. Handle with extreme caution. The carcinogenic status of anthracene is a bit confusing: Animal negative[9] compares with ACGIH[1] and DFG[3] categorization of coal tar volatiles as proven carcinogens. The Lewis/Sax reference below states that it is a mutagen and questionable carcinogen.

Medical Surveillance: Evaluation by a qualified allergist.

First Aid: If this chemical gets into the eyes, remove any contact lenses at once and irrigate immediately for at least 15 minutes, occasionally lifting upper and lower lids. Seek medical attention immediately. If this chemical contacts the skin, remove contaminated clothing and wash immediately with soap and water. Seek medical attention immediately. If this chemical has been inhaled, remove from exposure, begin rescue breathing (using universal precautions) if breathing has stopped and CPR if heart action has stopped. Transfer promptly to a medical facility. When this chemical has been swallowed, get medical attention. Give large quantities of water and induce vomiting. Do not make an unconscious person vomit.

Personal Protective Methods: Where possible, enclose operations and use local exhaust ventilation at the site of chemical release. If local exhaust ventilation or enclosure is not used, respirators should be worn. Wear protective gloves and clothing to prevent any reasonable probability of skin contact. Safety equipment suppliers/manufacturers can provide recommendations on the most protective glove/clothing material for your operation. All protective clothing (suits, gloves, footwear, headgear) should be clean, available each day, and put on before work. Contact lenses should not be worn when working with this chemical. Wear dust-proof chemical goggles and face shield unless full facepiece respiratory protection is worn. Employees should wash immediately with soap when skin is wet or contaminated. Provide emergency showers and eyewash. *Antidotes and Special Procedures:* Sun screens for ultraviolet light may help prevent skin allergic reactions. These may need frequent re-applications if you are sweating. Ultraviolet-screening sunglasses can help with eye allergic reactions. Consult your doctor or pharmacist in selecting these.

Respirator Selection: Engineering controls should be used wherever feasible to maintain airborne concentrations of this chemical below the prescribed exposure limit. Respirators and protective equipment less effective than engineering controls, and should be used only in non-routine or emergency situations which may result in exposure concentrations in excess of the TWA environmental limit. *Where there is no REL, at any detectable concentration:* SCBAF:PD,PP (any MSHA/NIOSH approved self-contained breathing apparatus that has a full facepiece and is operated in a pressure-demand or other positive-pressure mode); or SAF:PD,PP:ASCBA (any supplied-air respirator that has a full facepiece and is operated in a pressure-demand or other positive-pressure mode in combination with an auxiliary, self-contained breathing apparatus operated in a pressure-demand or other positive pressure mode). *Escape:* GMFOVHiE [any air-purifying, full-facepiece respirator (gas mask) with a chin-style, front- or back-mounted organic vapor canister having a high-efficiency particulate filter]; or SCBAE (any appropriate escape-type, self-contained breathing apparatus).

Storage: Prior to working with this chemical you should be trained on its proper handling and storage. Before entering confined space where this chemical may be present, check to make sure that an explosive concentration does not exist. Anthracene must be stored to avoid contact with strong oxidizers (such as chlorine, bromine and fluorine), chromic acid and calcium hypochlorite since violent reactions occur. Store in tightly closed containers in a cool, well-ventilated area. Sources of ignition, such as smoking and open flames, are prohibited where Anthracene is used, handled, or stored in a manner that could create a potential fire or explosion hazard.

Shipping: There are no DOT label requirements. However, it could be classified[52] as a hazardous substance solid, n.o.s., which falls in Hazard Class ORM-E and Packing Group III but imposes no label requirements or shipping quantity restrictions.

Spill Handling: Evacuate persons not wearing protective equipment from area of spill or leak until clean-up is complete. Remove all ignition sources. Establish ventilation to keep levels below explosive limit. Collect powdered material in the most convenient and safe manner and deposit in sealed containers. Ventilate area after clean-up is complete. It may be necessary to contain and dispose of this chemical as a hazardous waste. If material or contaminated runoff enters waterways, notify downstream users of potentially contaminated waters. Contact your Department of Environmental Protection or your regional office of the federal EPA for specific recommendations. If employees are required to clean-up spills, they must be properly trained and equipped. OSHA 1910.120(q) may be applicable.

Fire Extinguishing: This chemical is a combustible solid. Use dry chemical, carbon dioxide, water spray, or alcohol foam extinguishers. Poisonous gases are produced in fire. If material or contaminated runoff enters waterways, notify downstream users of potentially contaminated waters. Notify local health and fire officials and pollution control agencies. From a secure, explosion-proof location, use water spray to cool exposed containers. If cooling streams are ineffective (venting sound increases in volume and pitch, tank discolors, or shows any signs of deforming), withdraw immediately to a secure position. If employees are expected to fight fires, they must be trained and equipped in OSHA 1910.156.

Disposal Method Suggested: Consult with environmental regulatory agencies for guidance on acceptable disposal practices. Generators of waste containing this contaminant (≥100 kg/mo) must conform with EPA regulations governing storage, transportation, treatment, and waste disposal. Incineration.[22]

References

Sax, N. I., Ed., "Dangerous Properties of Industrial Materials Report" 4, No. 6, 18–43 (1984)

New Jersey Department of Health and Senior Services, "Hazardous Substance Fact Sheet: Anthracene," Trenton, NJ (July 1996)

Anthraquinone

Molecular Formula: $C_{14}H_8O_2$

Synonyms: 9,10-Anthracenedione; Anthradione; 9,10-Anthraquinone; Antraquinona (Spanish); 9,10-Dioxoanthracene

CAS Registry Number: 84-65-1

RTECS Number: CB4725000

DOT ID: UN 3143

Regulatory Authority

- Air Pollutant Standard Set (former USSR)[43]
- TSCA 40CFR704.30; 40CFR716.120(a) List of substances; 40CFR712.30(m); 40CFR799.500 Testing Requirements. Export notification required by §12(b)

Cited in U.S. State Regulations: New Hampshire (G), New Jersey (G).

Description: Anthraquinone, $C_{14}H_8O_2$, is a combustible, light yellow to green crystalline solid. Freezing/Melting point= 286°C. Flash point = 185°C (CC). NFPA 704 M Hazard Identification (based on NFPA-704M Rating System): Health 0, Flammability 1, Reactivity not listed. Insoluble in water.

Potential Exposure: Anthraquinone is an important starting material for vat dye manufacture. Also used in making organics, and used as a bird repellent in seeds.

Incompatibilities: Contact with strong oxidizers may cause fire and explosions.

Permissible Exposure Limits in Air: The former USSR-UNEP/IRPTC project[43] has set a MAC value of 5 mg/m^3 in workplace air.

Permissible Concentration in Water: No criteria set.

Routes of Entry: Through the skin, inhalation.

Harmful Effects and Symptoms

Short Term Exposure: Can be absorbed through the skin, thereby increasing exposure. Eye or skin contact can cause irritation. An allergen, may cause skin irritation and sensitization. Severe poisoning may cause seizures and coma.

Long Term Exposure: May cause skin allergy, with itching and rash. It may be mutagenic.

Points of Attack: Skin, lungs.

Medical Surveillance: Evaluation by a qualified allergist.

First Aid: If this chemical gets into the eyes, remove any contact lenses at once and irrigate immediately for at least 15 minutes, occasionally lifting upper and lower lids. Seek medical attention immediately. If this chemical contacts the skin, remove contaminated clothing and wash immediately with soap and water. Seek medical attention immediately. If this chemical has been inhaled, remove from exposure, begin rescue breathing (using universal precautions) if breathing has stopped and CPR if heart action has stopped. Transfer promptly to a medical facility. When this chemical has been swallowed, get medical attention. Give large quantities of water and induce vomiting. Do not make an unconscious person vomit.

Personal Protective Methods: Wear protective gloves and clothing to prevent any reasonable probability of skin contact. Safety equipment suppliers/manufacturers can provide recommendations on the most protective glove/clothing material for your operation. All protective clothing (suits, gloves, footwear, headgear) should be clean, available each day, and put on before work. Contact lenses should not be worn when working with this chemical. Wear dust-proof chemical goggles and face shield unless full facepiece respiratory protection is worn. Employees should wash immediately with soap when skin is wet or contaminated. Provide emergency showers and eyewash.

Respirator Selection: Engineering controls should be used wherever feasible to maintain airborne concentrations of this chemical below the prescribed exposure limit. Respirators and protective equipment less effective than engineering controls, and should be used only in non-routine or emergency situations which may result in exposure concentrations in excess of the TWA environmental limit. *Where there is no REL, at any detectable concentration:* SCBAF:PD,PP (any MSHA/NIOSH approved self-contained breathing apparatus that has a full facepiece and is operated in a pressure-demand or other positive-pressure mode); or SAF:PD,PP:ASCBA (any supplied-air respirator that has a full facepiece and is operated in a pressure-demand or other positive-pressure mode in combination with an auxiliary, self-contained breathing apparatus operated in a pressure-demand or other positive pressure mode). *Escape:* GMFOVHiE [any air-purifying, full-facepiece respirator (gas mask) with a chin-style, front- or back-mounted organic vapor canister having a high-efficiency particulate filter]; or SCBAE (any appropriate escape-type, self-contained breathing apparatus).

Storage: Prior to working with this chemical you should be trained on its proper handling and storage. Store in tightly closed containers in a cool, well ventilated area away from oxidizers. Sources of ignition such as smoking and open flames, are prohibited where this chemical is used, handled, or stored in a manner that could create a potential fire or explosion hazard.

Shipping: This chemical can be classified by the hazardous materials description and proper shipping name as: "Dyes, solid, toxic, n.o.s. or dye intermediates, solid, toxic, n.o.s." Labeling is "Poison," in Packing Group I or II, Hazard Class 6.1.

Spill Handling: Evacuate persons not wearing protective equipment from area of spill or leak until clean-up is complete. Remove all ignition sources. Dampen spilled material with toluene to avoid airborne dust. Collect powdered material in the most convenient and safe manner and deposit in sealed containers. Ventilate area after clean-up is complete. It may be necessary to contain and dispose of this chemical as a hazardous waste. If material or contaminated runoff enters waterways, notify downstream users of potentially contaminated waters. Contact your Department of Environmental Protection or your regional office of the federal EPA for specific recommendations. If employees are required to clean-up spills, they must be properly trained and equipped. OSHA 1910.120(q) may be applicable.

Fire Extinguishing: This chemical is a combustible solid. Use dry chemical, water spray, or alcohol foam extinguishers. Poisonous gases are produced in fire including carbon monoxide. If material or contaminated runoff enters waterways, notify downstream users of potentially contaminated waters. Notify local health and fire officials and pollution control agencies. From a secure, explosion-proof location, use water spray to cool exposed containers. If cooling streams are ineffective (venting sound increases in volume and pitch, tank discolors, or shows any signs of deforming), withdraw immediately to a secure position. If employees are expected to fight fires, they must be trained and equipped in OSHA 1910.156.

Disposal Method Suggested: Incineration.

References

New Jersey Department of Health and Senior Services, Hazardous Substance Fact Sheet, Anthraquinone, Trenton NJ (January 1999)

Antimony

Molecular Formula: Sb

Synonyms: Amspec antimony; Antimonio (Spanish); Antimony black; Antimony powder; Antimony, regulus; Antymon (Polish); Atomergic antimony; C.I. 77050; Silver GLO 33BP; Silver GLO 3KBP; Silver GLO BP; Stibium; Thermoguard CPA

CAS Registry Number: 7440-36-0 (metal)

RTECS Number: CC4025000

DOT ID: UN 2871 (metal)

EINECS Number: 231-146-5

Regulatory Authority

- Banned or Severely Restricted (New Zealand)[13] (Many countries, especially in food) (UN)[35]
- Air Pollutant Standard Set (ACGIH)[1] (DFG)[3] (former USSR)[43] (HSE)[33] (OSHA)[58] (Several States)[60]
- Water Pollution Standards Set (EPA)[6][32] (former USSR)[43] (Kansas)[61] (Mexico)
- Clean Air Act, 42USC7412; Title I, Part A, §112 hazardous pollutants
- Clean Water Act 40CFR401.15 Section 307 Toxic Pollutants; 40CFR423, Priority Pollutants
- RCRA, 40CFR261, Appendix 8 Hazardous Constituents., waste number not listed
- RCRA Land Ban Waste
- Safe Drinking Water Act, MCL, treatment technique; MCL, 0.006 mg/l; MCLG, 0.006 mg/l.; Regulated Chemical (47FR9352)
- RCRA 40CFR264, Appendix 9; Ground Water Monitoring List: Suggested methods (PQL μg/l): 6010 (300); 7040 (2,000); 7041 (30)
- SUPERFUND/EPCRA 40CFR302.4, Appendix A, Reportable Quantity (RQ): 5,000 lb (2,270 kg) *only if the diameter of pieces of solid metal has a diameter equal to or greater than 0.004 in.*, 40CFR372.65: Form R *de minimis* Concentration Reporting Level: 1.0%
- Canada, WHMIS, Ingredients Disclosure List; National Pollutant Release Inventory

Cited in U.S. State Regulations: Alaska (G), California (A, G), Connecticut (A), Florida (G, A), Illinois (G), Kansas (G, A, W), Louisiana (G), Maine(G), Maryland (G), Massachusetts (G), Michigan (G), Minnesota (G), Nevada (A), New Hampshire (G), New Jersey (G), New York (A), North Dakota (A), Oklahoma (G), Pennsylvania (G, A), Rhode Island (G, A), Vermont (G), Virginia (G, A), Washington (G), West Virginia (G), Wisconsin (G).

Description: Antimony, Sb, is a noncombustible, silvery-white, lustrous, hard, brittle metal; scale-like crystals, or dark gray lustrous powder. Boiling point = 1,635°C. Freezing/Melting point = 630°C. Hazard Identification (based on NFPA-704 M Rating System): [powder] Health 2, Flammability 2, Reactivity 0. Insoluble in water and organic solvents.

Potential Exposure: Exposure to antimony may occur during mining, smelting or refining, alloy and abrasive manufacture, and typesetting in printing. Antimony is widely used in the production of alloys, imparting increased hardness, mechanical strength, corrosion resistance, and a low coefficient of friction. Some of the important alloys are Babbitt, pewter, white metal, Britannia metal and bearing metal (which are used in bearing shells), printing-type, metal, storage battery plates, cable sheathing, solder, ornamental castings, and ammunition. Pure antimony compounds are used as abrasives, pigments, flameproofing compounds, plasticizers, and catalysts in organic synthesis; they are also used in the manufacture of tartar emetic, paints, lacquers, glass, pottery, enamels, glazes, pharmaceuticals, pyrotechnics, matches, and explosives. In addition, they are used in dyeing, for blueing steel, and in coloring aluminum pewter, and zinc. A highly toxic gas, stibine, may be released from the metal under certain conditions.

Incompatibilities: Strong oxidizers, strong acids [especially halogenated acids], produce a violent reaction, and deadly stibine gas (antimony hydride). Heat forms stibine gas. Mixtures with nitrates or halogenated compounds may cause combustion. Forms an explosive mixture with chloric and perchloric acid.

Permissible Exposure Limits in Air: The Federal OSHA [8-hour workshift] standard and the NIOSH [10-hour workshift] and ACGIH [8-hour workshift] recommended values for antimony and its compounds is 0.5 mg/m^3 TWA. Australia, DFG/MAK, HSE (U.K.), Israel, and Mexico standard is also 0.5 mg/m^3 TWA *and* DFG has a peak limitation of 10 times the normal MAK, not to be exceeded during a workshift; Israel has an Action Level of 0.25 mg/m^3. The NIOSH IDLH value is 50 mg/m^3. The former USSR-UNEP/IRPTC project[43] has set a MAC value for metallic antimony dust at 0.5 mg/m^3 in the workplace with an added specification of 0.2 mg/m^3 as an average MAC value during the work shift. Canadian provincial standard for British Columbia, Ontario, Quebec, and Alberta is 0.5 mg/m^3 TWA *and* Alberta has a 15 minute STEL of 1.5 mg/m^3. In addition, several states have set guidelines or standards for antimony in ambient air[60] ranging from 0.67 μg/m^3 (New York) to 1.19 μg/m^3 (Kansas) to 1.2 μg/m^3 (Pennsylvania) to 5 μg/m^3 (North Dakota) to 8.0 μg/m^3 (Virginia) to 10 μg/m^3 (Connecticut) to 12 μg/m^3 (Nevada) to 40 μg/m^3 (Rhode Island).

Determination in Air: Collection by a particulate filter, workup with acid, analysis by atomic absorption. [P&CAM #261], or OSHA #ID-125G, or NIOSH 8005 (elements in blood and tissue).

Permissible Concentration in Water: To protect freshwater aquatic life – on an acute basis 9,000 μg/l and on a chronic basis, 1,600 μg/l. To protect saltwater aquatic life – no criterion due to insufficient data. To protect human health – 146 μg/l. EPA has also suggested an ambient limit of 7 μg/l[32] based on health effects. Mexico has set an Ecological Criteria guideline of 0.1 mg/l. The former USSR-UNEP/IRPTC project[43] has set a MAC of 0.05 mg/l of antimony dust in water used for domestic purposes. In addition, Kansas has set a guideline for antimony in drinking water[61] of 143 μg/l.

Determination in Water: Digestion followed by atomic absorption. See EPA Methods 204.1 and 204.2.

Routes of Entry: Inhalation of dust or fume, skin and/or eye contact.

Harmful Effects and Symptoms

Local: Antimony and its compounds are generally regarded as primary skin irritants. Lesions generally appear on exposed, moist areas of the body, but rarely on the face. The dust and fumes are also irritants to the eyes, nose, and throat, and may be associated with gingivitis, anemia, and ulceration of the nasal septum and larynx. Antimony trioxide causes a dermatitis known as "antimony spots." This form of dermatitis result in intense itching followed by skin eruptions. A diffuse erythema may occur, but usually the early lesions are small erythematous papules. They may enlarge, however, and become pustular. Lesions occur in hot weather and are due to dust accumulating on exposed areas that are moist due to sweating. No evidence of eczematous reaction is present, nor an allergic mechanism. *Systemic:* Systemic intoxication is uncommon from occupational exposure. However, miners of antimony may encounter dust containing free silica; cases of pneumoconiosis in miners have been termed "silico-antimoniosis." Antimony pneumoconiosis, per se, appears to be a benign process. Antimony metal dust and fumes are absorbed from the lungs into the blood stream. Principal organs attacked include certain enzyme systems (protein and carbohydrate metabolism), heart, lungs, and the mucous membrane of the respiratory tract. Symptoms of acute oral poisoning include violent irritation of the nose, mouth, stomach, and intestines, vomiting, bloody stools, slow shallow respiration, pulmonary congestion, coma, and sometimes death due to circulatory or respiratory failure. Chronic oral poisoning presents symptoms of dry throat, nausea, headache, sleeplessness, loss of appetite, and dizziness. Liver and kidney degenerative changes are late manifestations. Antimony compounds are generally less toxic than antimony. Antimony trisulfide, however, has been reported to cause myocardial changes in man and experimental animals. Antimony trichloride and pentachloride are highly toxic and can irritate and corrode the skin. Antimony fluoride is extremely toxic particularly to pulmonary tissue and skin.

Short Term Exposure: Passes through the skin, thereby increasing exposure. Eye and skin contact can cause irritation and itchy skin rash. Inhalation can cause respiratory tract irritation with wheezing, and shortness of breath. Exposure can cause headache, nausea, abdominal pain, and loss of sleep.

Long Term Exposure: Repeated exposure can cause ulcers and sores of the nose to develop, damage to the kidneys, liver, and heart, lung effects (abnormal chest x-ray).

Points of Attack: Respiratory system, cardiovascular system, skin, eyes and lungs.

Medical Surveillance: Preemployment and periodic examinations should give special attention to lung disease, skin disease, diseases of the nervous system, heart and gastrointestinal tract. Lung function, EKGs, blood, kidney, liver, and urine should be evaluated periodically. Consider chest x-ray following acute overexposure.

First Aid: If this chemical gets into the eyes, remove any contact lenses at once and irrigate immediately for at least 15 minutes, occasionally lifting upper and lower lids. Seek medical attention immediately. If this chemical contacts the

skin, remove contaminated clothing and wash immediately with soap and water. Seek medical attention immediately. If this chemical has been inhaled, remove from exposure, begin rescue breathing (using universal precautions) if breathing has stopped and CPR if heart action has stopped. Transfer promptly to a medical facility. When this chemical has been swallowed, get medical attention. Give large quantities of water and induce vomiting. Do not make an unconscious person vomit.

Personal Protective Methods: A combination of protective clothing, barrier creams, gloves, and personal hygiene will protect the skin. Wear protective gloves and clothing to prevent any reasonable probability of skin contact. Safety equipment suppliers/manufacturers can provide recommendations on the most protective glove/clothing material for your operation. All protective clothing (suits, gloves, footwear, headgear) should be clean, available each day, and put on before work. Contact lenses should not be worn when working with this chemical. Wear dust-proof chemical goggles and face shield unless full facepiece respiratory protection is worn. Employees should wash immediately with soap when skin is wet or contaminated. Provide washing facilities, emergency showers and eyewash. Eating should not be permitted in exposed areas.

Respirator Selection: NIOSH/OSHA: *5 mg/m³*: DMXSQ (if not present as a fume) (any dust and mist respirator except single-use and quarter mask respirators); or SA (any supplied-air respirator). *12.5 mg/m³*: SA:CF (any supplied-air respirator operated in a continuous-flow mode); PAPRDM (if not present as a fume) (any powered, air-purifying respirator with a dust and mist filter). *25 mg/m³*: HiEF (any air-purifying, full-facepiece respirator with a high-efficiency particulate filter); or SAT:CF (any supplied-air respirator that has a tight-fitting facepiece and is operated in a continuous-flow mode); or PAPRTHiE (any powered, air-purifying respirator with a tight-fitting facepiece and a high-efficiency particulate filter); or SCBAF (any self-contained breathing apparatus with a full facepiece); or SAF (any supplied-air respirator with a full facepiece). *50 mg/m³:* SA:PD,PP (any supplied-air respirator operated in a pressure-demand or other positive-pressure mode). *Emergency or planned entry into unknown concentrations or IDLH conditions:* SCBAF:PD,PP (any self-contained breathing apparatus that has a full faceplate and is operated in a pressure-demand or other positive-pressure mode); or SAF:PD,PP:ASCBA (any supplied-air respirator that has a full facepiece and is operated in a pressure-demand or other positive-pressure mode in combination with an auxiliary self-contained breathing apparatus operated in a pressure-demand or other positive-pressure mode). *Escape:* HiEF (any air-purifying, full-facepiece respirator with a high-efficiency particulate filter); or SCBAE (any appropriate escape-type, self-contained breathing apparatus).

Storage: Prior to working with this chemical you should be trained on its proper handling and storage. Store in tightly closed containers in a cool, well ventilated area away from oxidizers, halogens, strong acids, and heat. Sources of ignition such as smoking and open flames, are prohibited where this chemical is used, handled, or stored in a manner that could create a potential fire or explosion hazard. Contact with acids forms deadly stibine gas. Before entering confined space where this chemical may be present, check to make sure that an explosive concentration does not exist.

Shipping: Antimony powder must be labeled "Keep away from Food." The DOT/UN Hazard Class is 6.1 and Packing Group is III.[19][20]

Spill Handling: Evacuate persons not wearing protective equipment from area of spill or leak until clean-up is complete. Remove all ignition sources. Collect powdered material in the most convenient and safe manner and deposit in sealed containers. Ventilate area after clean-up is complete. It may be necessary to contain and dispose of this chemical as a hazardous waste. If material or contaminated runoff enters waterways, notify downstream users of potentially contaminated waters. Contact your Department of Environmental Protection or your regional office of the federal EPA for specific recommendations. If employees are required to clean-up spills, they must be properly trained and equipped. OSHA 1910.120(q) may be applicable.

Fire Extinguishing: Antimony dust can present a moderate hazard when exposed to flame. Self-contained breathing apparatus may be required when antimony is exposed to extreme temperatures in a fire. Use dry chemical appropriate for metal fires, water spray, fog or standard foam extinguishers. Toxic fumes are produced in fire including deadly *stibine.* If material or contaminated runoff enters waterways, notify downstream users of potentially contaminated waters. Notify local health and fire officials and pollution control agencies. From a secure, explosion-proof location, use water spray to cool exposed containers. If cooling streams are ineffective (venting sound increases in volume and pitch, tank discolors, or shows any signs of deforming), withdraw immediately to a secure position. If employees are expected to fight fires, they must be trained and equipped in OSHA 1910.156.

Disposal Method Suggested: Recovery and recycle is an option to disposal which should be considered for scrap antimony and spent catalysts containing antimony. Dissolve spilled material in minimum amount of concentrated HCl. Add water, until white precipitate appears. Then acidify to dissolve again. Saturate with H_2S. Filter, wash and dry the precipitate and return to supplier.[22] Consult with environmental regulatory agencies for guidance on acceptable disposal practices. Generators of waste containing this contaminant (≥100 kg/mo) must conform with EPA regulations governing storage, transportation, treatment, and waste disposal.

References

U.S. Environmental Protection Agency, Antimony: Ambient Water Quality criteria, Washington, DC (1980)

U.S. Environmental Protection Agency, Literature Study of Selected Potential Environmental Contaminants: Antimony and Its

Compounds, Report No. EPA-560/2-76-002, Washington, DC, Office of Toxic Substances (February 1976)

National Institute for Occupational Safety and Health, Environmental Exposure to Airborne Contaminants in the Antimony Industry, NIOSH Publ. No. 79–140, Cincinnati, OH (August 1979)

U.S. Environmental Protection Agency, Toxicology of Metals, Vol II: Antimony, Report EPA-600-/1-77-022, Research Triangle Park, NC, pp 15–29 (May 1977)

U.S. Environmental Protection Agency, Antimony, Health and Environmental Effects Profile No. 10, Washington, DC, Office of Solid Waste (April 30, 1980)

Sax, N. I., Ed., "Dangerous Properties of Industrial Materials Report" 2, No. 2, 68–69 (1982)

National Institute for Occupational Safety and Health, "Criteria for a Recommended Standard: Occupational Exposure to Antimony," Washington, DC (September 28, 1978)

New Jersey Department of Health and Senior Services, "Hazardous Substance Fact Sheet: Antimony," Trenton, NJ (February 1998)

Antimony Lactate

Molecular Formula: $C_9H_{15}O_9Sb$

Common Formula: $Sb(OCOCHOHCH_3)_3$

Synonyms: Antimony (3+) salt (3:1); 2-Hydroxypropanoic acid trainhydride with antimonic acid; 2-Hydroxy-, trianhydride with antimonic acid; Lactic acid, antimony salt; Propanic acid, 2-hydroxy-

CAS Registry Number: 58164-88-8

RTECS Number: CC5455000

DOT ID: UN 1550

Regulatory Authority

- Banned or Severely Restricted (New Zealand)[13] (Many countries, especially in food) (UN)[35]
- Air Pollutant Standard Set (ACGIH)[1] (OSHA)[2] (California) (HSE) (Ontario, Quebec)
- Clean Air Act: 42USC7412; Title I, Part A, §112 hazardous pollutants
- Clean Water Act: 40CFR401.15 Section 307 Toxic Pollutants
- RCRA, 40CFR261, Appendix 8 Hazardous Constituents, waste number not listed (as antimony compounds, n.o.s.)
- Safe Drinking Water Act, MCL, treatment technique; MCL, 0.006 mg/l; MCLG, 0.006 mg/l.; Regulated Chemical (47FR9352)
- EPCRA Section 313: Includes any unique chemical substance that contains antimony as part of that chemical's infrastructure. Form R *de minimis* concentration reporting level: 0.1%
- Canada, WHMIS, Ingredients Disclosure List; National Pollutant Release Inventory (as antimony compounds)

Cited in U.S. State Regulations: Alaska (G), California (A, G), Maryland (G), New Hampshire (G), New Jersey (G), Oklahoma (G), West Virginia (G).

Description: Antimony Lactate, $Sb(OCOCHOHCH_3)_3$, is a noncombustible, tan solid. Soluble in water. Antimony lactate may be contaminated with Arsenic or other toxic substances.

Potential Exposure: Antimony lactate is used in fabric dyeing.

Incompatibilities: Contact with strong oxidizers (chlorates, permanganates, peroxides and nitrates) may cause a violent reaction. Contact with acids can produce deadly stibine gas.

Permissible Exposure Limits in Air: The legal Federal OSHA[2] airborne permissible exposure limit (PEL) is 0.5 mg/m^3 averaged over an 8-hour workshift. The recommended NIOSH[2] airborne exposure limit is 0.5 mg/m^3 averaged over a 10-hour workshift. The recommended ACGIH[1] airborne exposure limit is 0.5 mg/m^3 averaged over a 8-hour workshift. The above exposure limits are for air levels only. When skin contact also occurs, you may be overexposed, even though air levels are lower than the limits listed above. The HSE (U.K.), California, Ontario, and Quebec airborne exposure limits for antimony compounds are the same as the OSHA levels show above. The former USSR-UNEP/IRPTC project[43] has set a MAC of 0.3 mg/m^3 in workplace air. The NIOSH IDLH for antimony compounds (as Sb) is 50 mg/m^3.

Permissible Concentration in Water: As part of the priority toxic pollutant program, EPA[6] has set a limit of 146 μg/l of antimony to protect human health. EPA has also suggested[32] an ambient limit of 7 μg/l of antimony based on health effects.

Routes of Entry: Inhalation, passing through the skin.

Harmful Effects and Symptoms

Short Term Exposure: Antimony Lactate can affect you when breathed and by passing through skin. Exposure can irritate the eyes, nose, throat and skin. Very high levels could cause Antimony poisoning, with symptoms of nausea, headaches, abdominal pain, trouble breathing and death. Antimony lactate may be contaminated with Arsenic or other toxic substances.

Long Term Exposure: Repeated exposure can cause abnormal chest x-ray and damage the heart and liver. Prolonged or repeated skin contact can cause sores and ulcers.

Points of Attack: Skin, lungs, heart, liver.

Medical Surveillance: EKG, Liver function tests, urine tests for antimony. Also test for arsenic if contamination is suspected. Consider chest x-ray following acute overexposure.

First Aid: If this chemical gets into the eyes, remove any contact lenses at once and irrigate immediately for at least 15 minutes, occasionally lifting upper and lower lids. Seek medical attention immediately. If this chemical contacts the skin, remove contaminated clothing and wash immediately with soap and water. Seek medical attention immediately. If this chemical has been inhaled, remove from exposure, begin rescue breathing (using universal precautions) if breathing has

stopped and CPR if heart action has stopped. Transfer promptly to a medical facility. When this chemical has been swallowed, get medical attention. Give large quantities of water and induce vomiting. Do not make an unconscious person vomit.

Personal Protective Methods: Specific engineering controls are recommended for this chemical in the NIOSH criteria document Antimony Number 72–216. Where possible, enclose operations and use local exhaust ventilation at the site of chemical release. If local exhaust ventilation or enclosure is not used, respirators should be worn. Wear protective work clothing. Wash thoroughly immediately after exposure and at the end of the workshift. Post hazard and warning information in the work area. In addition, as part of an ongoing education and training effort, communicate all information on the health and safety hazards of this antimony compound to potentially exposed workers.

Respirator Selection: NIOSH/OSHA: *5 mg/m³*: DMXSQ (if not present as a fume) (any dust and mist respirator except single-use and quarter mask respirators); or SA (any supplied-air respirator). *12.5 mg/m³*: SA:CF (any supplied-air respirator operated in a continuous-flow mode); PAPRDM (if not present as a fume) (any powered, air-purifying respirator with a dust and mist filter). *25 mg/m³*: HiEF (any air-purifying, full-facepiece respirator with a high-efficiency particulate filter); or SAT:CF (any supplied-air respirator that has a tight-fitting facepiece and is operated in a continuous-flow mode); or PAPRTHiE (any powered, air-purifying respirator with a tight-fitting facepiece and a high-efficiency particulate filter); or SCBAF (any self-contained breathing apparatus with a full facepiece); or SAF (any supplied-air respirator with a full facepiece). *50 mg/m³:* SA:PD,PP (any supplied-air respirator operated in a pressure-demand or other positive-pressure mode). *Emergency or planned entry into unknown concentrations or IDLH conditions:* SCBAF:PD,PP (any self-contained breathing apparatus that has a full faceplate and is operated in a pressure-demand or other positive-pressure mode); or SAF:PD,PP:ASCBA (any supplied-air respirator that has a full facepiece and is operated in a pressure-demand or other positive-pressure mode in combination with an auxiliary self-contained breathing apparatus operated in a pressure-demand or other positive-pressure mode). *Escape:* HiEF (any air-purifying, full-facepiece respirator with a high-efficiency particulate filter); or SCBAE (any appropriate escape-type, self-contained breathing apparatus).

Storage: Prior to working with this chemical you should be trained on its proper handling and storage. Store in tightly-closed containers in a cool, well-ventilated area away from heat, acids, and oxidizers.

Shipping: DOT label required is "Keep away from Food." The DOT/UN Hazard Class is 6.1 and the Packing Group is III.[19][20]

Spill Handling: Evacuate persons not wearing protective equipment from area of spill or leak until clean-up is complete. Remove all ignition sources. Collect powdered material in the most convenient and safe manner and deposit in sealed containers. Ventilate area after clean-up is complete. It may be necessary to contain and dispose of this chemical as a hazardous waste. If material or contaminated runoff enters waterways, notify downstream users of potentially contaminated waters. Contact your Department of Environmental Protection or your regional office of the federal EPA for specific recommendations. If employees are required to clean-up spills, they must be properly trained and equipped. OSHA 1910.120(q) may be applicable.

Fire Extinguishing: Extinguish fire using an agent suitable for type of surrounding fire. Antimony Lactate itself does not burn. Poisonous gases are produced in fire including deadly stibine gas. If material or contaminated runoff enters waterways, notify downstream users of potentially contaminated waters. Notify local health and fire officials and pollution control agencies. From a secure, explosion-proof location, use water spray to cool exposed containers. If cooling streams are ineffective (venting sound increases in volume and pitch, tank discolors, or shows any signs of deforming), withdraw immediately to a secure position. If employees are expected to fight fires, they must be trained and equipped in OSHA 1910.156.

Disposal Method Suggested: Consult with environmental regulatory agencies for guidance on acceptable disposal practices. Generators of waste containing this contaminant (≥100 kg/mo) must conform with EPA regulations governing storage, transportation, treatment, and waste disposal.

References

New Jersey Department of Health and Senior Services, "Hazardous Substance Fact Sheet: Antimony Lactate," Trenton, NJ (January 1987)

Antimony Pentachloride

Molecular Formula: Cl_5Sb

Common Formula: $SbCl_5$

Synonyms: Antimonic chloride; Antimonio (pentacloruro di) (Italian); Antimonpentachlorid (German); Antimony(V) chloride; Antimony perchloride; Antimoonpentachloride (Dutch); Atomergic antimony pentachloride; Butter of antimony; Pentachloroantimony; Pentacloruro de antimonio (Spanish); Perchlorure d'antimoine (French); Tentachlorure d'antimoine (French)

CAS Registry Number: 7647-18-9

RTECS Number: CC5075000

DOT ID: UN 1730 (liquid); UN 1731 (solution)

EEC Number: 051-022-00-3

EINECS Number: 231-601-8

Regulatory Authority

- Banned or Severely Restricted (New Zealand)[13] (Many countries, especially in food) (UN)[35]

- Air Pollutant Standard Set (OSHA) (ACGIH)[1] (former USSR)[43] (HSE) (California) (Canada: Ontario, Quebec)
- Clean Air Act, 42USC7412; Title I, Part A, §112 hazardous pollutants
- Clean Water Act: 40CFR401.15 Section 307Toxic Pollutants; Section 311 Hazardous Substances/RQ 40CFR117.3 (same as CERCLA, see below); Section 313 Water Priority Chemicals (57FR41331, 9/9/92)
- Safe Drinking Water Act, MCL, treatment technique; MCL, 0.006 mg/l; MCLG, 0.006 mg/l.; Regulated Chemical (47FR9352) (as antimony)
- RCRA, 40CFR261, Appendix 8 Hazardous Constituents., waste number not listed (as antimony compounds, n.o.s.)
- SUPERFUND/CERCLA 40CFR302.4, Appendix A, Reportable Quantity (RQ): 1,000 lb (454 kg); 40CFR372.65: Form R *de minimis* Concentration Reporting Level: 0.1%
- Canada, WHMIS, Ingredients Disclosure List; National Pollutant Release Inventory (as antimony compounds)

Cited in U.S. State Regulations: Alaska (G), California (A, G), Florida (G), Maine (G), Maryland (G), Massachusetts (G), New Hampshire (G), New Jersey (G), Oklahoma (G), Pennsylvania (G), Rhode Island (G), West Virginia (G).

Description: Antimony Pentachloride, $SbCl_5$, is a noncombustible, colorless to reddish-yellow oily liquid with an offensive odor. Boiling point = 77°C (decomposes). Freezing/Melting point = 3°C. Hazard Identification (based on NFPA-704M Rating System): Health 3, Flammability 0, Reactivity 1. Decomposes on contact with water.

Potential Exposure: It is used in dyeing, coloring metals and in many organic chemical reactions as a catalyst.

Incompatibilities: Decomposes on contact with heat, acids, alkalis, ammonia, water or other forms of moisture producing fumes of hydrogen chloride and antimony. Decomposes above 77°C forming chlorine and antimony trichloride. Attacks many metals in the presence of moisture forming explosive hydrogen gas. Reacts with air forming corrosive vapors.

Permissible Exposure Limits in Air: The legal Federal OSHA[2] airborne permissible exposure limit (PEL) is 0.5 mg/m^3 averaged over an 8-hour workshift. The recommended NIOSH[2] airborne exposure limit is 0.5 mg/m^3 averaged over a 10-hour workshift. The recommended ACGIH[1] airborne exposure limit is 0.5 mg/m^3 averaged over a 8-hour workshift. The above exposure limits are for air levels only. When skin contact also occurs, you may be overexposed, even though air levels are lower than the limits listed above. The HSE (U.K.), California, Ontario, and Quebec airborne exposure limits for antimony compounds are the same as the OSHA levels show above. The former USSR-UNEP/IRPTC project[43] has set a MAC of 0.3 mg/m^3 in workplace air. The NIOSH IDLH for antimony compounds (as Sb) is 50 mg/m^3.

Permissible Concentration in Water: As part of the priority toxic pollutant program, EPA[6] has set a limit of 146 μg/l of antimony to protect human health. EPA has also suggested[32] an ambient limit of 7 μg/l of antimony based on health effects.

Routes of Entry: Passing through the skin, inhalation.

Harmful Effects and Symptoms

Short Term Exposure: Antimony Pentachloride is a corrosive chemical; can affect you when breathed in and by passing through your skin. Exposure can cause sore throat, rash, poor appetite, abdominal pain, loss of sleep, and irritate the lungs. Higher levels can cause pulmonary edema, a medical emergency that can be delayed for several hours; irregular heartbeat which can cause death. High or repeated exposure may damage the liver, kidneys, and the heart muscle. If used near acid, a deadly gas (Stibine) can be formed. Antimony may contain Arsenic. It is a corrosive chemical. Contact can burn the skin or eyes. The oral LD_{50} for rat is 1,115 mg/kg.[9]

Long Term Exposure: Repeated contact can cause ulcers or sores in the nose, kidney, liver, and heart damage.

Points of Attack: Kidneys, liver, heart, respiratory system.

Medical Surveillance: EKG, liver and kidney function tests. Consider chest x-ray following acute overexposure.

First Aid: If this chemical gets into the eyes, remove any contact lenses at once and irrigate immediately for at least 15 minutes, occasionally lifting upper and lower lids. Seek medical attention immediately. If this chemical contacts the skin, remove contaminated clothing and wash immediately with soap and water. Seek medical attention immediately. If this chemical has been inhaled, remove from exposure, begin rescue breathing (using universal precautions) if breathing has stopped and CPR if heart action has stopped. Transfer promptly to a medical facility. When this chemical has been swallowed, get medical attention. If victim is conscious, administer water or milk. Do not induce vomiting. Medical observation is recommended for 24 - 48 hours after breathing overexposure, as pulmonary edema may be delayed. As first aid for pulmonary edema, a doctor or authorized paramedic may consider administering a corticosteroid spray.

Personal Protective Methods: Specific engineering controls are recommended for this chemical in the NIOSH criteria document Antimony Number 72–216. Where possible, enclose operations and use local exhaust ventilation at the site of chemical release. If local exhaust ventilation or enclosure is not used, respirators should be worn. Wear protective gloves and clothing to prevent any reasonable probability of skin contact. Safety equipment suppliers/manufacturers can provide recommendations on the most protective glove/clothing material for your operation. All protective clothing (suits, gloves, footwear, headgear) should be clean, available each day, and put on before work. Contact lenses should not be worn when working with this chemical. Wear splash-proof chemical goggles and face shield unless full facepiece respiratory protection is worn. Employees should wash immediately with

soap when skin is wet or contaminated. Provide emergency showers and eyewash.

Respirator Selection: NIOSH/OSHA: *5 mg/m³*: DMXSQ (if not present as a fume) (any dust and mist respirator except single-use and quarter mask respirators); or SA (any supplied-air respirator). *12.5 mg/m³*: SA:CF (any supplied-air respirator operated in a continuous-flow mode); PAPRDM (if not present as a fume) (any powered, air-purifying respirator with a dust and mist filter). *25 mg/m³*: HiEF (any air-purifying, full-facepiece respirator with a high-efficiency particulate filter); or SAT:CF (any supplied-air respirator that has a tight-fitting facepiece and is operated in a continuous-flow mode); or PAPRTHiE (any powered, air-purifying respirator with a tight-fitting facepiece and a high-efficiency particulate filter); or SCBAF (any self-contained breathing apparatus with a full facepiece); or SAF (any supplied-air respirator with a full facepiece). *50 mg/m³:* SA:PD,PP (any supplied-air respirator operated in a pressure-demand or other positive-pressure mode). *Emergency or planned entry into unknown concentrations or IDLH conditions:* SCBAF:PD,PP (any self-contained breathing apparatus that has a full faceplate and is operated in a pressure-demand or other positive-pressure mode); or SAF:PD,PP:ASCBA (any supplied-air respirator that has a full facepiece and is operated in a pressure-demand or other positive-pressure mode in combination with an auxiliary self-contained breathing apparatus operated in a pressure-demand or other positive-pressure mode). *Escape:* HiEF (any air-purifying, full-facepiece respirator with a high-efficiency particulate filter); or SCBAE (any appropriate escape-type, self-contained breathing apparatus).

Storage: Prior to working with this chemical you should be trained on its proper handling and storage. Antimony Pentachloride must be stored to avoid contact with organic or combustible materials (such as wood, paper and oil), since violent reactions occur. See incompatibilities. Store in tightly closed containers in a cool, well-ventilated area away from water or moisture and heat. Sources of ignition such as smoking and open flames, are prohibited where this chemical is used, handled, or stored in a manner that could create a potential fire or explosion hazard.

Shipping: The DOT label required is "Corrosive." The DOT/UN Hazard Class is 8, the Packing Group is II.[19][20]

Spill Handling: Evacuate persons not wearing protective equipment from area of spill or leak until clean-up is complete. Remove all ignition sources. Ventilate the area of spill or leak. Absorb liquids in vermiculite, dry sand, earth, or a similar material and deposit in sealed containers. It may be necessary to contain and dispose of this chemical as a hazardous waste. If material or contaminated runoff enters waterways, notify downstream users of potentially contaminated waters. Contact your Department of Environmental Protection or your regional office of the federal EPA for specific recommendations. If employees are required to clean-up spills, they must be properly trained and equipped. OSHA 1910.120(q) may be applicable.

Fire Extinguishing: Extinguish fire using an agent suitable for type of surrounding fire. Antimony Pentachloride itself does not burn. Poisonous gases are produced in fire including chlorine and hydrogen chloride. If material or contaminated runoff enters waterways, notify downstream users of potentially contaminated waters. Notify local health and fire officials and pollution control agencies. From a secure, explosion-proof location, use water spray to cool exposed containers. If cooling streams are ineffective (venting sound increases in volume and pitch, tank discolors, or shows any signs of deforming), withdraw immediately to a secure position. If employees are expected to fight fires, they must be trained and equipped in OSHA 1910.156.

Disposal Method Suggested: Encapsulate and transfer to an approve landfill. If chemically treated and naturalized, the chemical is amenable to biological treatment at municipal sewage treatment plant. Consult with environmental regulatory agencies for guidance on acceptable disposal practices. Generators of waste containing this contaminant (≥100 kg/mo) must conform with EPA regulations governing storage, transportation, treatment, and waste disposal.

References

New Jersey Department of Health and Senior Services, "Hazardous Substance Fact Sheet: Antimony Pentachloride," Trenton, NJ (February 1988)

Antimony Pentafluoride

Molecular Formula: F_5Sb

Common Formula: SbF_5

Synonyms: Antimony fluoride; Antimony(5+) fluoride; Antimony(V) fluoride; Antimony(5+) pentafluoride; Antimony(V) pentafluoride; Atochem antimony pentafluoride; Atomergic antimony pentafluoride; Pentafluoroantimony; Pentafluoruro de antimonio (Spanish)

CAS Registry Number: 7783-70-2

RTECS Number: CC5800000

DOT ID: UN 1732

Regulatory Authority

- Banned or Severely Restricted (New Zealand)[13] (Many countries, especially in food) (UN)[35]
- Air Pollutant Standard Set (ACGIH)[1] (OSHA)[2] (California) (HSE) (Ontario, Quebec)
- Clean Air Act, 42USC7412; Title I, Part A, §112 hazardous pollutants
- Clean Water Act: 40CFR401.15 Section 307Toxic Pollutants
- RCRA, 40CFR261, Appendix 8 Hazardous Constituents, waste number not listed (as antimony compounds, n.o.s.)

- Safe Drinking Water Act, MCL, treatment technique; MCL, 0.006 mg/l; MCLG, 0.006 mg/l.; Regulated Chemical (47FR9352)
- CERCLA/SARA 40CFR302 Extremely Hazardous Substances: TPQ = 500 lb (227 kg)
- CERCLA/SARA Section 304 Reportable Quantity (RQ): EHS, 1 lb (0.454 kg)
- EPCRA Section 313: Includes any unique chemical substance that contains antimony as part of that chemical's infrastructure. Form R *de minimis* concentration reporting level: 0.1%
- Canada, WHMIS, Ingredients Disclosure List; National Pollutant Release Inventory, as antimony compounds

Cited in U.S. State Regulations: Alaska (G), California (A, G), Florida (G), Maine (G), Maryland (G), Massachusetts (G), New Hampshire (G), New Jersey (G), Oklahoma (G), Pennsylvania (G), Rhode Island (G), West Virginia (G).

Description: Antimony Pentafluoride, SbF_5, is a noncombustible, oily, colorless liquid with a pungent odor. Boiling point = 150°C. Freezing/Melting point = 7°C. Hazard Identification (based on NFPA-704M Rating System): Health 4, Flammability 0, Reactivity 1. Soluble in water.

Potential Exposure: It is used as a catalyst in chemical reactions or as a source of fluorine in fluorination reactions.

Incompatibilities: Water and other forms of moisture (forms hydrofluoric acid); combustible organic and siliceous materials, phosphorus, and phosphate materials.

Permissible Exposure Limits in Air: The legal Federal OSHA[2] airborne permissible exposure limit (PEL) as Sb. The recommended NIOSH[2] airborne exposure limit (PEL) is 0.5 mg/m^3 averaged over a 10-hour workshift, *and* 2.5 mg/m^3 as F for an 8-hour workshift[9] is 0.5 mg/m^3 averaged over an 8-hour workshift. The recommended ACGIH[1] airborne exposure limit is 0.5 mg/m^3 averaged over a 8-hour workshift. The above exposure limits are for air levels only. When skin contact also occurs, you may be overexposed, even though air levels are lower than the limits listed above. The HSE (U.K.), California, Ontario, and Quebec airborne exposure limits for antimony compounds are the same as the OSHA levels show above. The former USSR-UNEP/IRPTC MAC value[43] is 0.3 mg/m^3. The NIOSH IDLH value for antimony and compounds is 50 mg/m^3.

Note: the OSHA PEL for fluorides (measured as fluorine) is 2.5 mg/m^3 TWA for an 8-hour workshift.

Permissible Concentration in Water: As part of the priority toxic pollutant program, EPA[6] has set a limit of 146 μg/l of antimony to protect human health. EPA has also suggested[32] an ambient limit of 7 μg/l of antimony based on health effects.

Routes of Entry: Inhalation, ingestion, skin contact.

Harmful Effects and Symptoms

Short Term Exposure: Severe health hazard. May be fatal if inhaled or absorbed through the skin. The compound corrosive and irritating to eyes, skin, and lungs. Contact with eyes or skin causes severe burns. Ingestion causes vomiting and sever burns of mouth and throat. Overexposure by any route can cause bloody stools, slow pulse, low blood pressure, coma, convulsions, and cardiac arrest. The probable oral lethal dose of 5 – 50 mg/kg or between 7 drops and a teaspoonful for a 150 pound person (antimony salts).

Long Term Exposure: Can damage the kidneys, liver, and heart. Repeated exposure may affect the lungs and cause an abnormal chest x-ray to develop.

Points of Attack: Eyes, skin, respiratory system, cardiovascular system.

Medical Surveillance: EKG, liver and kidney function tests. Consider lung function tests and chest x-ray.

First Aid: If this chemical gets into the eyes, remove any contact lenses at once and irrigate immediately for at least 15 minutes, occasionally lifting upper and lower lids. Seek medical attention immediately. If this chemical contacts the skin, remove contaminated clothing and wash immediately with soap and water. Seek medical attention immediately. If this chemical has been inhaled, remove from exposure, begin rescue breathing (using universal precautions) if breathing has stopped and CPR if heart action has stopped. Transfer promptly to a medical facility. When this chemical has been swallowed, get medical attention. If victim is conscious, administer water or milk. Do not induce vomiting. Medical observation is recommended for 24 – 48 hours after breathing overexposure, as pulmonary edema may be delayed. As first aid for pulmonary edema, a doctor or authorized paramedic may consider administering a corticosteroid spray.

Personal Protective Methods: Where possible, enclose operations and use local exhaust ventilation at the site of chemical release. If local exhaust ventilation or enclosure is not used, respirators should be worn. Wear protective gloves and clothing to prevent any reasonable probability of skin contact. Safety equipment suppliers/manufacturers can provide recommendations on the most protective glove/clothing material for your operation. All protective clothing (suits, gloves, footwear, headgear) should be clean, available each day, and put on before work. Contact lenses should not be worn when working with this chemical. Wear splash-proof chemical goggles and face shield unless full facepiece respiratory protection is worn. Employees should wash immediately with soap when skin is wet or contaminated. Provide emergency showers and eyewash.

Respirator Selection: NIOSH/OSHA: *(antimony) 5 mg/m^3:* DMXSQ (if not present as a fume) (any dust and mist respirator except single-use and quarter mask respirators); or SA (any supplied-air respirator). *12.5 mg/m^3:* SA:CF (any supplied-air respirator operated in a continuous-flow mode); PAPRDM (if not present as a fume) (any powered, air-purifying respirator with a dust and mist filter) *25 mg/m³:* HiEF (any air-purifying, full-facepiece respirator with a high-efficiency particulate filter); or SAT:CF (any supplied-air respirator that has a tight-fitting facepiece and is operated in a continuous-flow mode); or PAPRTHiE (any powered, air-purifying respirator with a tight-fitting facepiece and a high-efficiency particulate filter); or SCBAF (any self-contained breathing apparatus with a full facepiece); or SAF (any supplied-air respirator with a full facepiece). *50 mg/m³:* SA:PD,PP (any supplied-air respirator operated in a pressure-demand or other positive-pressure mode). *Emergency or planned entry into unknown concentrations or IDLH conditions:* SCBAF:PD, PP (any self-contained breathing apparatus that has a full faceplate and is operated in a pressure-demand or other positive-pressure mode); or SAF:PD,PP:ASCBA (any supplied-air respirator that has a full facepiece and is operated in a pressure-demand or other positive-pressure mode in combination with an auxiliary self-contained breathing apparatus operated in a pressure-demand or other positive-pressure mode). *Escape:* HiEF (any air-purifying, full-facepiece respirator with a high-efficiency particulate filter); or SCBAE (any appropriate escape-type, self-contained breathing apparatus).

NIOSH: *(fluorides) 12.5 mg/m³:* DM (any dust and mist respirator). *25 mg/m³:* DMXSQ* (any dust and mist respirator except single-use and quarter mask respirators); or SA* (any supplied-air respirator). *62.5 mg/m³:* SA:CF* (any supplied-air respirator operated in a continuous-flow mode); or PAPRDM*+ *if not present as a fume* (any powered, air-purifying respirator with a dust and mist filter). *125 mg/m³:* HiEF+ (any air-purifying, full-facepiece respirator with a high-efficiency particulate filter); or SCBAF (any self-contained breathing apparatus with a full facepiece); or SAF (any supplied-air respirator with a full facepiece). *250 mg/m³:* SA:PD,PP (any supplied-air respirator operated in a pressure-demand or other positive-pressure mode). *Emergency or planned entry into unknown concentrations or IDLH conditions:* SCBAF:PD,PP (any self-contained breathing apparatus that has a full faceplate and is operated in a pressure-demand or other positive-pressure mode); or SAF:PD, PP:ASCBA (any supplied-air respirator that has a full facepiece and is operated in a pressure-demand or other positive-pressure mode in combination with an auxiliary, self-contained breathing apparatus operated in a pressure-demand or other positive-pressure mode). *Escape:* HiEF+ (any air-purifying, full-facepiece respirator with a high-efficiency particulate filter); or SCBAE (any appropriate escape-type, self-contained breathing apparatus).

* Substance reported to cause eye irritation or damage; may require eye protection.

+ May need acid gas sorbent. Exposure to 500 mg/m³ (IDLH) is immediately dangerous to life and health.

Storage: Prior to working with this chemical you should be trained on its proper handling and storage. Antimony Pentafluoride must be stored to avoid contact with phosphorus, phosphates, siliceous, and combustible or organic materials since violent reactions occur. Store in tightly closed containers in a cool, well-ventilated area away from water or moisture and heat. Outside or detached storage is preferred.

Shipping: The DOT label requirement is "Corrosive, Poison." The DOT/UN Hazard Class is 8 and the Packing Group is II.[19][20]

Spill Handling: Evacuate persons not wearing protective equipment from area of spill or leak until clean-up is complete. Remove all ignition sources. Do not touch spilled material; stop leak if you can do so without risk. Use water spray to reduce vapors. For large spills dike far ahead. Absorb spills with noncombustible absorbent material such as vermiculite, dry sand, earth, etc., and deposit in sealed containers. Ventilate area after clean-up is complete. It may be necessary to contain and dispose of this chemical as a hazardous waste. If material or contaminated runoff enters waterways, notify downstream users of potentially contaminated waters. Contact your Department of Environmental Protection or your regional office of the federal EPA for specific recommendations. If employees are required to clean-up spills, they must be properly trained and equipped. OSHA 1910.120(q) may be applicable.

Fire Extinguishing: Approach fire from upwind; avoid hazardous vapors and toxic decomposition products. Do not use water or foam on fire or on adjacent fires; extinguish with dry chemicals or carbon dioxide. Water spray may be used to reduce vapors. Poisonous gases are produced in fire including hydrogen fluoride and antimony fumes. If material or contaminated runoff enters waterways, notify downstream users of potentially contaminated waters. Notify local health and fire officials and pollution control agencies. From a secure, explosion-proof location, use water spray to cool exposed containers. If cooling streams are ineffective (venting sound increases in volume and pitch, tank discolors, or shows any signs of deforming), withdraw immediately to a secure position. If employees are expected to fight fires, they must be trained and equipped in OSHA 1910.156.

Disposal Method Suggested: Consult with environmental regulatory agencies for guidance on acceptable disposal practices. Generators of waste containing this contaminant (≥100 kg/mo) must conform with EPA regulations governing storage, transportation, treatment, and waste disposal.

References

U.S. Environmental Protection Agency, "Chemical Profile: Antimony Pentafluoride," Washington, DC, Chemical Emergency Preparedness Program (November 30, 1987)

New Jersey Department of Health and Senior Services, "Hazardous Substance Fact Sheet: Antimony Pentafluoride," Trenton, NJ (February 1998)

Antimony Potassium Tartrate

Molecular Formula: $C_4H_4KO_7Sb$

Synonyms: Antimonate (2-), bis μ-2,3-dihydroxy-butanedioata (4-)-01,02:03,04di-, dipotassium, trihydrate, stereoisomer; Antimonyl potassium tartrate; Emetique (French); ENT 50,434; Potassium antimonyl-d-tartrate; Potassium antimonyl tartrate; Potassium antimony tartrate; Tartar emetic; Tartaric acid, antimony potassium salt; Tartarized antimony; Tartrated antimony; Tartrato de antimonio y potasio (Spanish); Tastox

CAS Registry Number: 28300-74-5

RTECS Number: CC6825000

DOT ID: UN 1551

Regulatory Authority

- Banned or Severely Restricted (New Zealand)[13] (Many countries, especially in food) (UN)[35]
- Air Pollutant Standard Set (ACGIH)[1] (OSHA)[2] (California) (HSE) (Ontario, Quebec)
- Clean Air Act, 42USC7412; Title I, Part A, §112 hazardous pollutants (as antimony compounds)
- Clean Water Act: 40CFR116.4 Hazardous Substances; 40CFR117.3, RQ (same as CERCLA); 40CFR423, Appendix A, Priority Pollutants; Section 313 Water Priority Chemicals (57FR41331, 9/9/92); 40CFR401.15 Section 307 Toxic Pollutants, as antimony compounds
- RCRA, 40CFR261, Appendix 8 Hazardous Constituents, waste number not listed (as antimony compounds, n.o.s.)
- Safe Drinking Water Act, MCL, treatment technique; MCL, 0.006 mg/l; MCLG, 0.006 mg/l.; Regulated Chemical (47FR9352)
- SUPERFUND/EPCRA 40CFR302.4 Reportable Quantity (RQ): CERCLA, 100 lb (45.4 kg)
- EPCRA Section 313: Includes any unique chemical substance that contains antimony as part of that chemical's infrastructure. Form R *de minimis* concentration reporting level: 0.1%
- Canada, WHMIS, Ingredients Disclosure List; National Pollutant Release Inventory (as antimony compounds)

Cited in U.S. State Regulations: Alaska (G), California (A, G), Maine (G), Maryland (G), Massachusetts (G), New Hampshire (G), New Jersey (G), Oklahoma (G), Pennsylvania (G), West Virginia (G).

Description: Antimony Potassium Tartrate, $C_4H_4KO_7Sb$, is an odorless, colorless, crystalline material or white powder with a sweetish, metallic taste. Hazard Identification (based on NFPA-704 M Rating System): Health 2, Flammability 0, Reactivity 0. Soluble in water; solution is slightly acidic.

Potential Exposure: It is used in medicine and textile and leather dyeing; and as an insecticide.

Incompatibilities: Solution will react with alkaline materials.

Permissible Exposure Limits in Air: The legal Federal OSHA[2] airborne permissible exposure limit (PEL) as Sb. The recommended NIOSH[2] airborne exposure limit (PEL) is 0.5 mg/m^3 averaged over a 10-hour workshift. *and* 2.5 mg/m^3 as F for an 8-hour workshift[9] is 0.5 mg/m^3 averaged over an 8-hour workshift. The recommended ACGIH[1] airborne exposure limit is 0.5 mg/m^3 averaged over a 8-hour workshift. The above exposure limits are for air levels only. When skin contact also occurs, you may be overexposed, even though air levels are lower than the limits listed above. The HSE (U.K.), California, Ontario, and Quebec airborne exposure limits for antimony compounds are the same as the OSHA levels show above. The former USSR-UNEP/IRPTC MAC value[43] is 0.3 mg/m^3. The NIOSH IDLH value for antimony and compounds is 50 mg/m^3.

Permissible Concentration in Water: As part of the priority toxic pollutant program, EPA[6] has set a limit of 146 μg/l of antimony to protect human health. EPA has also suggested[32] an ambient limit of 7 μg/l of antimony based on health effects.

Routes of Entry: Passing through the skin, inhalation.

Harmful Effects and Symptoms

Short Term Exposure: Antimony Potassium Tartrate can affect you when breathed in and by passing through your skin. Eye and skin contact can cause irritation and skin rash. Exposure can cause poor appetite, abdominal pain, nausea, headaches, sore throat and irritation of air passages, with cough. Higher levels can cause pulmonary edema, a medical emergency that can be delayed for several hours. This can cause death. Exposure may make the heart beat irregularly or stop. High or repeated exposure may damage the liver or heart muscle.

Long Term Exposure: Prolonged or repeated contact can cause ulcers or sores in the nose, kidney, liver, and heart damage.

Points of Attack: Skin, eyes, respiratory system, cardiovascular system, kidneys, liver.

Medical Surveillance: EKG. Liver and kidney function tests. Consider chest x-ray following acute overexposure.

First Aid: If this chemical gets into the eyes, remove any contact lenses at once and irrigate immediately for at least 15 minutes, occasionally lifting upper and lower lids. Seek medical attention immediately. If this chemical contacts the skin, remove contaminated clothing and wash immediately with soap and water. Seek medical attention immediately. If this chemical has been inhaled, remove from exposure, begin rescue breathing (using universal precautions) if breathing has stopped and CPR if heart action has stopped. Transfer promptly to a medical facility. When this chemical has been swallowed, get medical attention. Give large quantities of water and induce vomiting. Do not make an unconscious person vomit. Medical observation is recommended for 24 - 48 hours after breathing overexposure, as pulmonary edema may be delayed. As first aid for pulmonary edema, a doctor or authorized paramedic may consider administering a corticosteroid spray.

Personal Protective Methods: Specific engineering controls are recommended for this chemical in the NIOSH criteria document Antimony Number 72–216. Where possible, enclose operations and use local exhaust ventilation at the site of chemical release. If local exhaust ventilation or enclosure is not used, respirators should be worn. Wear protective gloves and clothing to prevent any reasonable probability of skin contact. Safety equipment suppliers/manufacturers can provide recommendations on the most protective glove/clothing material for your operation. All protective clothing (suits, gloves, footwear, headgear) should be clean, available each day, and put on before work. Contact lenses should not be worn when working with this chemical. Wear dust-proof chemical goggles and face shield unless full facepiece respiratory protection is worn. Employees should wash immediately with soap when skin is wet or contaminated. Provide emergency showers and eyewash.

Respirator Selection: NIOSH/OSHA: *5 mg/m³*: DMXSQ (if not present as a fume) (any dust and mist respirator except single-use and quarter mask respirators); or SA (any supplied-air respirator). *12.5 mg/m³*: SA:CF (any supplied-air respirator operated in a continuous-flow mode); PAPRDM (if not present as a fume) (any powered, air-purifying respirator with a dust and mist filter). *25 mg/m³*: HiEF (any air-purifying, full-facepiece respirator with a high-efficiency particulate filter); or SAT:CF (any supplied-air respirator that has a tight-fitting facepiece and is operated in a continuous-flow mode); or PAPRTHiE (any powered, air-purifying respirator with a tight-fitting facepiece and a high-efficiency particulate filter); or SCBAF (any self-contained breathing apparatus with a full facepiece); or SAF (any supplied-air respirator with a full facepiece). *50 mg/m³:* SA:PD,PP (any supplied-air respirator operated in a pressure-demand or other positive-pressure mode). *Emergency or planned entry into unknown concentrations or IDLH conditions:* SCBAF:PD,PP (any self-contained breathing apparatus that has a full faceplate and is operated in a pressure-demand or other positive-pressure mode); or SAF:PD,PP:ASCBA (any supplied-air respirator that has a full facepiece and is operated in a pressure-demand or other positive-pressure mode in combination with an auxiliary self-contained breathing apparatus operated in a pressure-demand or other positive-pressure mode). *Escape:* HiEF (any air-purifying, full-facepiece respirator with a high-efficiency particulate filter); or SCBAE (any appropriate escape-type, self-contained breathing apparatus).

Storage: Prior to working with this chemical you should be trained on its proper handling and storage. Store in tightly closed containers in a cool, well-ventilated area away from heat.

Shipping: The DOT/label requirement is "Keep away from Food." The UN/DOT Hazard Class is 6.1 and the Packing Group is III.[19][20]

Spill Handling: Evacuate persons not wearing protective equipment from area of spill or leak until clean-up is complete. Remove all ignition sources. Collect powdered material in the most convenient and safe manner and deposit in sealed containers. Ventilate area after clean-up is complete. It may be necessary to contain and dispose of this chemical as a hazardous waste. If material or contaminated runoff enters waterways, notify downstream users of potentially contaminated waters. Contact your Department of Environmental Protection or your regional office of the federal EPA for specific recommendations. If employees are required to clean-up spills, they must be properly trained and equipped. OSHA 1910.120(q) may be applicable.

Fire Extinguishing: Extinguish fire using an agent suitable for type of surrounding fire. Antimony Potassium Tartrate itself does not burn. Poisonous gases are produced in fire including potassium oxide and antimony. If material or contaminated runoff enters waterways, notify downstream users of potentially contaminated waters. Notify local health and fire officials and pollution control agencies. From a secure, explosion-proof location, use water spray to cool exposed containers. If cooling streams are ineffective (venting sound increases in volume and pitch, tank discolors, or shows any signs of deforming), withdraw immediately to a secure position. If employees are expected to fight fires, they must be trained and equipped in OSHA 1910.156.

Disposal Method Suggested: Consult with environmental regulatory agencies for guidance on acceptable disposal practices. Generators of waste containing this contaminant (≥100 kg/mo) must conform with EPA regulations governing storage, transportation, treatment, and waste disposal.

References

Sax, N. I., Ed., "Dangerous Properties of Industrial Materials Report" 1, No. 8, 33–35 (1981)

New Jersey Department of Health and Senior Services, "Hazardous Substance Fact Sheet: Antimony Potassium/Tartrate," Trenton, NJ (February 1998)

Antimony Tribromide

Molecular Formula: Br_3Sb

Common Formula: $SbBr_3$

Synonyms: Antimonous bromide; Antimony(3+) bromide; Antimony(III) bromide; Stibine, tribromo-; Tribromo stibine; Tribromuro de antimonio (Spanish)

CAS Registry Number: 7789-61-9

RTECS Number: CC4400000

DOT ID: UN 1549

Regulatory Authority

- Banned or Severely Restricted (New Zealand)[13] (Many countries, especially in food) (UN)[35]
- Air Pollutant Standard Set (ACGIH)[1] (OSHA)[2]
- Air Pollutant Standard Set (ACGIH)[1] (OSHA)[2] (California) (HSE) (Ontario, Quebec)
- Clean Air Act, 42USC7412; Title I, Part A, §112 hazardous pollutants (as antimony compounds)
- Clean Water Act: 40CFR116.4 Hazardous Substances; 40CFR117.3, RQ (same as CERCLA); 40CFR423, Appendix A, Priority Pollutants; Section 313 Water Priority Chemicals (57FR41331, 9/9/92); 40CFR401.15 Section 307 Toxic Pollutants, as antimony compounds
- RCRA, 40CFR261, Appendix 8 Hazardous Constituents, waste number not listed (as antimony compounds, n.o.s.)
- Safe Drinking Water Act, MCL, treatment technique; MCL, 0.006 mg/l; MCLG, 0.006 mg/l.; Regulated Chemical (47FR9352)
- SUPERFUND/EPCRA 40CFR302.4 Reportable Quantity (RQ): CERCLA, 1,000 lb (454 kg)
- EPCRA Section 313: Includes any unique chemical substance that contains antimony as part of that chemical's infrastructure. Form R *de minimis* concentration reporting level: 0.1%
- Canada, WHMIS, Ingredients Disclosure List; National Pollutant Release Inventory (as antimony compounds)

Cited in U.S. State Regulations: Alaska (G), California (A, G), Maryland (G), Massachusetts (G), New Hampshire (G), New Jersey (G), Pennsylvania (G), West Virginia (G).

Description: Antimony tribromide, $SbBr_3$, is a nonflammable, colorless to yellow crystalline solid. Boiling point = 288°C @ 749 mm Hg. Freezing/Melting point = 97°C. Hazard Identification (based on NFPA-704 M Rating System): Health 3, Flammability 0, Reactivity 1. Soluble in water; forms an acid.

Potential Exposure: It is used to make antimony salts, in dyeing, and analytical chemistry.

Incompatibilities: Potassium, sodium, and bases. Heat forms toxic bromides. Contact with water form an acid and liberates hydrogen bromide and antimony trioxide.

Permissible Exposure Limits in Air: The legal Federal OSHA[2] airborne permissible exposure limit (PEL) as Sb. The recommended NIOSH[2] airborne exposure limit (PEL) is 0.5 mg/m^3 averaged over a 10-hour workshift, *and* 2.5 mg/m^3 as F for an 8-hour workshift[9] is 0.5 mg/m^3 averaged over an 8-hour workshift. The recommended ACGIH[1] airborne exposure limit is 0.5 mg/m^3 averaged over a 8-hour workshift. The above exposure limits are for air levels only. When skin contact also occurs, you may be overexposed, even though air levels are lower than the limits listed above. The HSE (U.K.), California, Ontario, and Quebec airborne exposure limits for antimony compounds are the same as the OSHA levels show above. The former USSR-UNEP/IRPTC MAC value[43] is 0.3 mg/m^3. The NIOSH IDLH value for antimony and compounds is 50 mg/m^3.

Permissible Concentration in Water: As part of the priority toxic pollutant program, EPA[6] has set a limit of 146 μg/l of antimony to protect human health. EPA has also suggested[32] an ambient limit of 7 μg/l of antimony based on health effects.

Harmful Effects and Symptoms

Short Term Exposure: Antimony tribromide can affect you when breathed and by passing thought skin. Exposure can cause sore throat, skin rash, poor appetite and irritation of the air passages with cough. Higher exposures can cause pulmonary edema a medical emergency that can be delayed for several hours. This can cause death. This chemical can cause irregular heartbeat; this can cause death. High or repeated exposure may damage the liver and the heart muscle. Antimony Tribromide is a corrosive chemical and contact can burn the skin and eyes, with burns and possible permanent damage. If used near acid, a deadly gas (Stibine) can be released.

Long Term Exposure: Repeated exposure can cause liver, heart muscle damage, headaches, loss of appetite, dry throat and sleep disturbance. Corrosive substances such as antimony tribromide have the potential for causing lung damage.

Points of Attack: Skin, eyes, respiratory system, cardiovascular system.

Medical Surveillance: EKG, liver function tests, urine tests for antimony, lung function tests. Consider chest x-ray following acute overexposure.

First Aid: If this chemical gets into the eyes, remove any contact lenses at once and irrigate immediately for at least 15 minutes, occasionally lifting upper and lower lids. Seek medical attention immediately. If this chemical contacts the skin, remove contaminated clothing and wash immediately with soap and water. Seek medical attention immediately. If this chemical has been inhaled, remove from exposure, begin rescue breathing (using universal precautions) if breathing has stopped and CPR if heart action has stopped. Transfer promptly to a medical facility. When this chemical has been swallowed, get medical attention. If victim is conscious, administer water or milk. Do not induce vomiting. Medical observation is recommended for 24 - 48 hours after breathing overexposure, as pulmonary edema may be delayed. As first aid for pulmonary edema, a doctor or authorized paramedic may consider administering a corticosteroid spray.

Personal Protective Methods: Specific engineering controls are recommended for this chemical in the NIOSH criteria document Antimony Number 72–216. Where possible, enclose operations and use local exhaust ventilation at the site of chemical release. If local exhaust ventilation or enclosure is not used, respirators should be worn. Wear protective gloves and clothing to prevent any reasonable probability of skin contact. Safety equipment suppliers/manufacturers can provide recommendations on the most protective glove/clothing material for your operation. All protective clothing (suits, gloves, footwear, headgear) should be clean, available each day, and put on before work. Contact lenses should not be worn when working with this chemical. Wear dust-proof chemical goggles and face shield unless full facepiece respiratory protection is worn. Employees should wash immediately with soap when skin is wet or contaminated. Provide emergency showers and eyewash.

Respirator Selection: NIOSH/OSHA: *5 mg/m³*: DMXSQ (if not present as a fume) (any dust and mist respirator except single-use and quarter mask respirators); or SA (any supplied-air respirator). *12.5 mg/m³*: SA:CF (any supplied-air respirator operated in a continuous-flow mode); PAPRDM (if not present as a fume) (any powered, air-purifying respirator with a dust and mist filter). *25 mg/m³*: HiEF (any air-purifying, full-facepiece respirator with a high-efficiency particulate filter); or SAT:CF (any supplied-air respirator that has a tight-fitting facepiece and is operated in a continuous-flow mode); or PAPRTHiE (any powered, air-purifying respirator with a tight-fitting facepiece and a high-efficiency particulate filter); or SCBAF (any self-contained breathing apparatus with a full facepiece); or SAF (any supplied-air respirator with a full facepiece). *50 mg/m³:* SA:PD,PP (any supplied-air respirator operated in a pressure-demand or other positive-pressure mode). *Emergency or planned entry into unknown concentrations or IDLH conditions:* SCBAF:PD,PP (any self-contained breathing apparatus that has a full faceplate and is operated in a pressure-demand or other positive-pressure mode); or SAF:PD,PP:ASCBA (any supplied-air respirator that has a full facepiece and is operated in a pressure-demand or other positive-pressure mode in combination with an auxiliary self-contained breathing apparatus operated in a pressure-demand or other positive-pressure mode). *Escape:* HiEF (any air-purifying, full-facepiece respirator with a high-efficiency particulate filter); or SCBAE (any appropriate escape-type, self-contained breathing apparatus).

Storage: Prior to working with this chemical you should be trained on its proper handling and storage. Antimony Tribromide must be stored to avoid contact with potassium, sodium, and bases (such as sodium hydroxide, potassium hydroxide and ammonium hydroxide) since violent reactions occur. Store in tightly closed containers in a cool, well-ventilated area away from water or moisture and heat.

Shipping: The DOT shipping label requirement is "Corrosive." The UN/DOT Hazard Class is 8 and the Packing Group is II.[19][20]

Spill Handling: Evacuate persons not wearing protective equipment from area of spill or leak until clean-up is complete. Remove all ignition sources. Collect powdered material in the most convenient and safe manner and deposit in sealed containers. Ventilate area after clean-up is complete. It may be necessary to contain and dispose of this chemical as a hazardous waste. If material or contaminated runoff enters waterways, notify downstream users of potentially contaminated waters. Contact your Department of Environmental Protection or your regional office of the federal EPA for specific recommendations. If employees are required to clean-up spills, they must be properly trained and equipped. OSHA 1910.120(q) may be applicable.

Fire Extinguishing: Extinguish fire using an agent suitable for type of surrounding fire. Antimony Tribromide itself does not burn. Poisonous gases are produced in fire. If material or contaminated runoff enters waterways, notify downstream users of potentially contaminated waters. Notify local health and fire officials and pollution control agencies. From a secure, explosion-proof location, use water spray to cool exposed containers. If cooling streams are ineffective (venting sound increases in volume and pitch, tank discolors, or shows any signs of deforming), withdraw immediately to a secure position. If employees are expected to fight fires, they must be trained and equipped in OSHA 1910.156.

Disposal Method Suggested: Encapsulate and buried at an approved chemical landfill. Unacceptable for disposal at sewage treatment plants. Consult with environmental regulatory agencies for guidance on acceptable disposal practices. Generators of waste containing this contaminant (≥100 kg/mo) must conform with EPA regulations governing storage, transportation, treatment, and waste disposal.

References

Sax, N. I., Ed., "Dangerous Properties of Industrial Materials Report" 3, No. 5, 42–43 (1983) and 8, N0. 5, 56–59 (January 1987)

New Jersey Department of Health and Senior Services, "Hazardous Substance Fact Sheet: Antimony Tribromide," Trenton, NJ (January 1987)

Antimony Trichloride

Molecular Formula: Cl_3Sb

Common Formula: $SbCl_3$

Synonyms: Antimoine (trichlorure d') (French); Antimonio (trichloruro di) (Italian); Antimonius chloride; Antimony butter; Antimony(III) chloride; Antimoontrichlride (Dutch); Butter of antimony; Chlorid antimonity; C.I. 77056; Stibine, trichloro-; Trichloro stibine; Trichlorostibine; Trichlorure d' antimoine (French); Tricloruro de antimonio (Spanish)

CAS Registry Number: 10025-91-9

RTECS Number: CC4900000

DOT ID: UN 1733

Regulatory Authority

- Banned or Severely Restricted (New Zealand)[13] (Many countries, especially in food) (UN)[35]

- Air Pollutant Standard Set (ACGIH) (OSHA)[1] (former USSR)[43]
- Air Pollutant Standard Set (ACGIH)[1] (OSHA)[2] (California) (HSE) (Ontario, Quebec)
- Clean Air Act, 42USC7412; Title I, Part A, §112 hazardous pollutants (as antimony compounds)
- Clean Water Act: 40CFR116.4 Hazardous Substances; 40CFR117.3, RQ (same as CERCLA); 40CFR423, Appendix A, Priority Pollutants; Section 313 Water Priority Chemicals (57FR41331, 9/9/92); 40CFR401.15 Section 307 Toxic Pollutants, as antimony compounds
- RCRA, 40CFR261, Appendix 8 Hazardous Constituents, waste number not listed (as antimony compounds, n.o.s.)
- Safe Drinking Water Act, MCL, treatment technique; MCL, 0.006 mg/l; MCLG, 0.006 mg/l.; Regulated Chemical (47FR9352)
- SUPERFUND/EPCRA 40CFR302.4 Reportable Quantity (RQ): CERCLA, 1,000 lb (454 kg)
- EPCRA Section 313: Includes any unique chemical substance that contains antimony as part of that chemical's infrastructure. Form R *de minimis* concentration reporting level: 0.1%
- Canada, WHMIS, Ingredients Disclosure List; National Pollutant Release Inventory (as antimony compounds)

Cited in U.S. State Regulations: Alaska (G), California (A, G), Maine (G), Maryland (G), Massachusetts (G), New Hampshire (G), New Jersey (G), Oklahoma (G), Pennsylvania (G), West Virginia (G).

Description: Antimony trichloride, $SbCl_3$, is a noncombustible, clear, colorless, crystalline solid with an acrid, pungent odor. Boiling point = 220°C. Freezing/Melting point = 73°C. Hazard Identification (based on NFPA-704 M Rating System): Health 3, Flammability 0, Reactivity 2. Reacts with water.

Potential Exposure: It is used to make Antimony salts and drugs, to fireproof textiles, and as a catalyst in many organic reactions.

Incompatibilities: Decomposes in water forming hydrochloric acid and antimony oxychloride. Reacts violently with strong bases, ammonia, alkali metals, aluminum, potassium, sodium. Forms explosive mixture with perchloric acid when hot. Reacts with air forming hydrochloric acid. Attacks metals in the presence of moisture, forming explosive hydrogen gas.

Permissible Exposure Limits in Air: The legal Federal OSHA[2] airborne permissible exposure limit (PEL) as Sb. The recommended NIOSH[2] airborne exposure limit (PEL) is 0.5 mg/m^3 averaged over a 10-hour workshift. *and* 2.5 mg/m^3 as F for an 8-hour workshift[9] is 0.5 mg/m^3 averaged over an 8-hour workshift. The recommended ACGIH[1] airborne exposure limit is 0.5 mg/m^3 averaged over a 8-hour workshift. The above exposure limits are for air levels only. When skin contact also occurs, you may be overexposed, even though air levels are lower than the limits listed above. The HSE (U.K.), California, Ontario, and Quebec airborne exposure limits for antimony compounds are the same as the OSHA levels show above. The former USSR-UNEP/IRPTC MAC value[43] is 0.3 mg/m^3. The NIOSH IDLH value for antimony and compounds is 50 mg/m^3.

Permissible Concentration in Water: As part of the priority toxic pollutant program, EPA[6] has set a limit of 146 µg/l of antimony to protect human health. EPA has also suggested[32] an ambient limit of 7 µg/l of antimony based on health effects.

Routes of Entry: Inhalation.

Harmful Effects and Symptoms

Short Term Exposure: Antimony trichloride can affect you when inhaled. Exposure can cause sore throat, skin rash, poor appetite and irritation of the air passages, with cough. Higher exposures can cause pulmonary edema, a medical emergency that can be delayed for several hours. This can cause death. Higher exposures can also cause irregular heartbeat; This can cause death. High or repeated exposure may damage the liver and the heart muscle. Antimony Trichloride is a corrosive chemical and contact can burn the skin and eyes, with possible permanent damage. If used near acid, a deadly gas (Stibine) can be released. The oral LD_{50} for rat is 525 mg/kg.[9]

Long Term Exposure: May cause mutations. May damage the developing fetus. Repeated exposure can cause liver, heart muscle damage, headaches, loss of appetite, dry throat and sleep disturbance. Corrosive substances such as antimony trichloride have the potential for causing lung damage.

Points of Attack: Cardiovascular system, reproductive system, liver, heart, and lungs.

Medical Surveillance: EKG, liver function tests, urine test for antimony, lung function tests including chest x-ray following acute overexposure.

First Aid: If this chemical gets into the eyes, remove any contact lenses at once and irrigate immediately for at least 15 minutes, occasionally lifting upper and lower lids. Seek medical attention immediately. If this chemical contacts the skin, remove contaminated clothing and wash immediately with soap and water. Seek medical attention immediately. If this chemical has been inhaled, remove from exposure, begin rescue breathing (using universal precautions) if breathing has stopped and CPR if heart action has stopped. Transfer promptly to a medical facility. When this chemical has been swallowed, get medical attention. If victim is conscious, administer water or milk. Do not induce vomiting. Medical observation is recommended for 24 - 48 hours after breathing overexposure, as pulmonary edema may be delayed. As first aid for pulmonary edema, a doctor or authorized paramedic may consider administering a corticosteroid spray.

Personal Protective Methods: Specific engineering controls are recommended for this chemical in the NIOSH criteria document Antimony Number 72–216. Where possible, enclose operations and use local exhaust ventilation at the site of chemical release. If local exhaust ventilation or enclosure is not used, respirators should be worn. Wear protective gloves and clothing to prevent any reasonable probability of skin contact. Safety equipment suppliers/manufacturers can provide recommendations on the most protective glove/clothing material for your operation. All protective clothing (suits, gloves, footwear, headgear) should be clean, available each day, and put on before work. Contact lenses should not be worn when working with this chemical. Wear dust-proof chemical goggles and face shield unless full facepiece respiratory protection is worn. Employees should wash immediately with soap when skin is wet or contaminated. Provide emergency showers and eyewash.

Respirator Selection: NIOSH/OSHA: *5 mg/m³:* DMXSQ (if not present as a fume) (any dust and mist respirator except single-use and quarter mask respirators); or SA (any supplied-air respirator). *12.5 mg/m³:* SA:CF (any supplied-air respirator operated in a continuous-flow mode); PAPRDM (if not present as a fume) (any powered, air-purifying respirator with a dust and mist filter). *25 mg/m³:* HiEF (any air-purifying, full-facepiece respirator with a high-efficiency particulate filter); or SAT:CF (any supplied-air respirator that has a tight-fitting facepiece and is operated in a continuous-flow mode); or PAPRTHiE (any powered, air-purifying respirator with a tight-fitting facepiece and a high-efficiency particulate filter); or SCBAF (any self-contained breathing apparatus with a full facepiece); or SAF (any supplied-air respirator with a full facepiece). *50 mg/m³:* SA:PD,PP (any supplied-air respirator operated in a pressure-demand or other positive-pressure mode). *Emergency or planned entry into unknown concentrations or IDLH conditions:* SCBAF:PD,PP (any self-contained breathing apparatus that has a full faceplate and is operated in a pressure-demand or other positive-pressure mode); or SAF:PD,PP:ASCBA (any supplied-air respirator that has a full facepiece and is operated in a pressure-demand or other positive-pressure mode in combination with an auxiliary self-contained breathing apparatus operated in a pressure-demand or other positive-pressure mode). *Escape:* HiEF (any air-purifying, full-facepiece respirator with a high-efficiency particulate filter); or SCBAE (any appropriate escape-type, self-contained breathing apparatus).

Storage: Prior to working with this chemical you should be trained on its proper handling and storage. Store in tightly closed containers under nitrogen in a cool, well-ventilated area away from water or moisture, heat and incompatible substances such as strong bases, aluminum, potassium, and sodium.

Shipping: The DOT label requirement is "Corrosive." The UN/DOT Hazard Class is 8 and the Packing Group is II.[19][20]

Spill Handling: Evacuate persons not wearing protective equipment from area of spill or leak until clean-up is complete. Remove all ignition sources. Absorb liquids in vermiculite, dry sand, earth, or a similar material and deposit in sealed containers. Collect powdered material in the most convenient and safe manner and deposit in sealed containers. Ventilate area after clean-up is complete. It may be necessary to contain and dispose of this chemical as a hazardous waste. If material or contaminated runoff enters waterways, notify downstream users of potentially contaminated waters. Contact your Department of Environmental Protection or your regional office of the federal EPA for specific recommendations. If employees are required to clean-up spills, they must be properly trained and equipped. OSHA 1910.120(q) may be applicable.

Fire Extinguishing: Extinguish fire using an agent suitable for type of surrounding fire. Antimony Trichloride itself does not burn. Poisonous gases are produced in fire including chlorine and antimony. If material or contaminated runoff enters waterways, notify downstream users of potentially contaminated waters. Notify local health and fire officials and pollution control agencies. From a secure, explosion-proof location, use water spray to cool exposed containers. If cooling streams are ineffective (venting sound increases in volume and pitch, tank discolors, or shows any signs of deforming), withdraw immediately to a secure position. If employees are expected to fight fires, they must be trained and equipped in OSHA 1910.156.

Disposal Method Suggested: Consult with environmental regulatory agencies for guidance on acceptable disposal practices. Generators of waste containing this contaminant (≥100 kg/mo) must conform with EPA regulations governing storage, transportation, treatment, and waste disposal.

References

Sax, N. I., Ed., "Dangerous Properties of Industrial Materials Report" 2, No. 1, 73–74, New York, Van Nostrand Reinhold Co. (1982)

New Jersey Department of Health and Senior Services, "Hazardous Substance Fact Sheet: Antimony Trichloride," Trenton, NJ (September 1986)

Antimony Trifluoride

Molecular Formula: F_3Sb

Synonyms: Antimoine fluorure (French); Antimonous fluoride; Antimony(III) fluoride (1:3); Stibine, trifluoro-; Trifluoroantimony; Trifluoroantimony, stibine, trifluoro-; Trifluorostibine; Trifluoruro de antimonio (Spanish)

CAS Registry Number: 7783-56-4

RTECS Number: CC5150000

DOT ID: UN 1549

Regulatory Authority

- Banned or Severely Restricted (New Zealand)[13] (Many countries, especially in food) (UN)[35]
- Air Pollutant Standard Set (ACGIH)[1] (OSHA)[2] (California) (HSE) (Ontario, Quebec)

- Clean Air Act, 42USC7412; Title I, Part A, §112 hazardous pollutants (as antimony compounds)
- Clean Water Act: 40CFR116.4 Hazardous Substances; 40CFR117.3, RQ (same as CERCLA); 40CFR423, Appendix A, Priority Pollutants; Section 313 Water Priority Chemicals (57FR41331, 9/9/92); 40CFR401.15 Section 307 Toxic Pollutants, as antimony compounds
- RCRA, 40CFR261, Appendix 8 Hazardous Constituents, waste number not listed (as antimony compounds, n.o.s.)
- Safe Drinking Water Act, MCL, treatment technique; MCL, 0.006 mg/l; MCLG, 0.006 mg/l; Regulated Chemical (47FR9352)
- CERCLA/SARS Section 304 Reportable Quantity (RQ): CERCLA, 1,000 lb (454 kg)
- EPCRA Section 313: Includes any unique chemical substance that contains antimony as part of that chemical's infrastructure. Form R *de minimis* concentration reporting level: 0.1%
- Canada, WHMIS, Ingredients Disclosure List; National Pollutant Release Inventory (as antimony compounds)

Cited in U.S. State Regulations: Alaska (G), California (A, G), Maryland (G), Massachusetts (G), New Hampshire (G), New Jersey (G), Oklahoma (G), Pennsylvania (G), West Virginia (G).

Description: Antimony trifluoride, SbF_3, is a noncombustible, odorless, white to gray crystalline solid. Boiling point = 376°C. Freezing/Melting point = 292°C. Hazard Identification (based on NFPA-704 M Rating System): Health 3, Flammability 0, Reactivity 0. Soluble in water.

Potential Exposure: It is used in dyeing; to make porcelain and pottery; and as a fluorinating agent.

Incompatibilities: Hot perchloric acid.

Permissible Exposure Limits in Air: The legal Federal OSHA[2] airborne permissible exposure limit (PEL) as Sb. The recommended NIOSH[2] airborne exposure limit (PEL) is 0.5 mg/m^3 averaged over a 10-hour workshift. *and* 2.5 mg/m^3 as F for an 8-hour workshift[9] is 0.5 mg/m^3 averaged over an 8-hour workshift. The recommended ACGIH[1] airborne exposure limit is 0.5 mg/m^3 averaged over a 8-hour workshift. The above exposure limits are for air levels only. When skin contact also occurs, you may be overexposed, even though air levels are lower than the limits listed above. The HSE (U.K.), California, Ontario, and Quebec airborne exposure limits for antimony compounds are the same as the OSHA levels show above. The former USSR-UNEP/IRPTC MAC value[43] is 0.3 mg/m^3. The NIOSH IDLH value for antimony and compounds is 50 mg/m^3.

Permissible Concentration in Water: As part of the priority toxic pollutant program, EPA[6] has set a limit of 146 µg/l of antimony to protect human health. EPA has also suggested[32] an ambient limit of 7 µg/l of antimony based on health effects.

Harmful Effects and Symptoms

Short Term Exposure: Antimony trifluoride can affect you when breathed in and by passing through your skin. Exposure can cause sore throat, skin rash, poor appetite and irritate the lungs. Higher exposures can cause pulmonary edema, a medical emergency that can be delayed for several hours. This can cause death. Antimony trifluoride can cause irregular heartbeat; this can cause death. High or repeated exposure may damage the liver and the heart muscle. Antimony Trifluoride is a corrosive chemical and contact can burn the skin and eyes, causing damage. If used near acid, a deadly gas (Stibine) can be released. The oral LD_{50} for mouse is 804 mg/kg.[9]

Long Term Exposure: Repeated contact can cause ulcers and sores of the nose. Can damage the kidneys, liver, and heart. Repeated exposure may affect the lungs and cause an abnormal chest x-ray to develop.

Points of Attack: Eyes, skin, respiratory system, cardiovascular system.

Medical Surveillance: EKG, liver and kidney function tests. Consider lung function tests and chest x-ray.

First Aid: If this chemical gets into the eyes, remove any contact lenses at once and irrigate immediately for at least 15 minutes, occasionally lifting upper and lower lids. Seek medical attention immediately. If this chemical contacts the skin, remove contaminated clothing and wash immediately with soap and water. Seek medical attention immediately. If this chemical has been inhaled, remove from exposure, begin rescue breathing (using universal precautions) if breathing has stopped and CPR if heart action has stopped. Transfer promptly to a medical facility. When this chemical has been swallowed, get medical attention. If victim is conscious, administer water or milk. Do not induce vomiting. Medical observation is recommended for 24 - 48 hours after breathing overexposure, as pulmonary edema may be delayed. As first aid for pulmonary edema, a doctor or authorized paramedic may consider administering a corticosteroid spray.

Personal Protective Methods: Specific engineering controls are recommended for this chemical in the NIOSH criteria document Antimony Number 72–216. Where possible, enclose operations and use local exhaust ventilation at the site of chemical release. If local exhaust ventilation or enclosure is not used, respirators should be worn. Wear protective gloves and clothing to prevent any reasonable probability of skin contact. Safety equipment suppliers/manufacturers can provide recommendations on the most protective glove/clothing material for your operation. All protective clothing (suits, gloves, footwear, headgear) should be clean, available each day, and put on before work. Contact lenses should not be worn when working with this chemical. Wear dust-proof chemical goggles and face shield unless full facepiece respiratory protection is worn. Employees should wash immediately with soap when skin is wet or contaminated. Provide emergency showers and eyewash.

Respirator Selection: NIOSH/OSHA: *5 mg/m³*: DMXSQ (if not present as a fume) (any dust and mist respirator except single-use and quarter mask respirators); or SA (any supplied-air respirator). *12.5 mg/m³*: SA:CF (any supplied-air respirator operated in a continuous-flow mode); PAPRDM (if not present as a fume) (any powered, air-purifying respirator with a dust and mist filter). *25 mg/m³*: HiEF (any air-purifying, full-facepiece respirator with a high-efficiency particulate filter); or SAT:CF (any supplied-air respirator that has a tight-fitting facepiece and is operated in a continuous-flow mode); or PAPRTHiE (any powered, air-purifying respirator with a tight-fitting facepiece and a high-efficiency particulate filter); or SCBAF (any self-contained breathing apparatus with a full facepiece); or SAF (any supplied-air respirator with a full facepiece). *50 mg/m³:* SA:PD,PP (any supplied-air respirator operated in a pressure-demand or other positive-pressure mode). *Emergency or planned entry into unknown concentrations or IDLH conditions:* SCBAF:PD,PP (any self-contained breathing apparatus that has a full faceplate and is operated in a pressure-demand or other positive-pressure mode); or SAF:PD,PP:ASCBA (any supplied-air respirator that has a full facepiece and is operated in a pressure-demand or other positive-pressure mode in combination with an auxiliary self-contained breathing apparatus operated in a pressure-demand or other positive-pressure mode). *Escape:* HiEF (any air-purifying, full-facepiece respirator with a high-efficiency particulate filter); or SCBAE (any appropriate escape-type, self-contained breathing apparatus).

NIOSH: (fluorides) 12.5 mg/m³: DM (any dust and mist respirator). *25 mg/m³*: DMXSQ* (any dust and mist respirator except single-use and quarter mask respirators); or SA* (any supplied-air respirator). *62.5 mg/m³*: SA:CF* (any supplied-air respirator operated in a continuous-flow mode); or PAPRDM*+ *if not present as a fume* (any powered, air-purifying respirator with a dust and mist filter). *125 mg/m³*: HiEF+ (any air-purifying, full-facepiece respirator with a high-efficiency particulate filter); or SCBAF (any self-contained breathing apparatus with a full facepiece); or SAF (any supplied-air respirator with a full facepiece). 250 mg/m³: SA:PD,PP (any supplied-air respirator operated in a pressure-demand or other positive-pressure mode). *Emergency or planned entry into unknown concentrations or IDLH conditions:* SCBAF:PD,PP (any self-contained breathing apparatus that has a full faceplate and is operated in a pressure-demand or other positive-pressure mode); or SAF:PD,PP:ASCBA (any supplied-air respirator that has a full facepiece and is operated in a pressure-demand or other positive-pressure mode in combination with an auxiliary, self-contained breathing apparatus operated in a pressure-demand or other positive-pressure mode). *Escape:* HiEF+ (any air-purifying, full-facepiece respirator with a high-efficiency particulate filter); or SCBAE (any appropriate escape-type, self-contained breathing apparatus).

* Substance reported to cause eye irritation or damage; may require eye protection.

\+ May need acid gas sorbent. Exposure to 500 mg/m³ (IDLH) is immediately dangerous to life and health.

Storage: Prior to working with this chemical you should be trained on its proper handling and storage. Store in tightly closed containers in a cool, well-ventilated area away from heat.

Shipping: The DOT label required is "Corrosive." The UN/DOT Hazard Class is 8 and the Packing Group II.[19][20]

Spill Handling: Evacuate persons not wearing protective equipment from area of spill or leak until clean-up is complete. Remove all ignition sources. Absorb liquids in vermiculite, dry sand, earth, or a similar material and deposit in sealed containers. Collect powdered material in the most convenient and safe manner and deposit in sealed containers. Ventilate area after clean-up is complete. It may be necessary to contain and dispose of this chemical as a hazardous waste. If material or contaminated runoff enters waterways, notify downstream users of potentially contaminated waters. Contact your Department of Environmental Protection or your regional office of the federal EPA for specific recommendations. If employees are required to clean-up spills, they must be properly trained and equipped. OSHA 1910.120(q) may be applicable.

Fire Extinguishing: Extinguish fire using an agent suitable for type of surrounding fire. Antimony Trifluoride itself does not burn. Poisonous gases are produced in fire including fluorides and antimony. If material or contaminated runoff enters waterways, notify downstream users of potentially contaminated waters. Notify local health and fire officials and pollution control agencies. From a secure, explosion-proof location, use water spray to cool exposed containers. If cooling streams are ineffective (venting sound increases in volume and pitch, tank discolors, or shows any signs of deforming), withdraw immediately to a secure position. If employees are expected to fight fires, they must be trained and equipped in OSHA 1910.156.

Disposal Method Suggested: Consult with environmental regulatory agencies for guidance on acceptable disposal practices. Generators of waste containing this contaminant (≥100 kg/mo) must conform with EPA regulations governing storage, transportation, treatment, and waste disposal.

References

Sax, N. I., Ed., "Dangerous Properties of Industrial Materials Report" 2, No. 8, 34–36 (1981) and 3, No. 5, 40–42, New York, Van Nostrand Reinhold Co. (1983)

New Jersey Department of Health and Senior Services, "Hazardous Substance Fact Sheet: Antimony Trifluoride," Trenton, NJ (February 1988)

Antimony Trioxide

Molecular Formula: Sb_3O_3

Synonyms: Antimonous oxide; Antimony peroxide; Antimony sesquioxide; Antimony, white; Cystic prefil F; Diantimony trioxide; Exitelite; Fireshield H; Fireshield HPM; Fireshield L; Flowers of antimony; NCI-C55152; Nihon kagaku sangyo antimony trifluoride; Octoguard FR-10;

Octoguard FR-15; Petcat R-9; Senarmontite; STCC 4966905; Trioxido de antimonio (Spanish); Ultrafine II; UN 9201; Valentinite; Weisspiessglanz (German); White antimony

CAS Registry Number: 1309-64-4

RTECS Number: CC5650000

DOT ID: NA 9201

Regulatory Authority

- Carcinogen (animal Positive) (DFG)[3] (Suspected human) (ACGIH)[1] Banned or Severely Restricted (New Zealand)[13] (Many countries, especially in food) (UN)[35]
- Air Pollutant Standard Set (ACGIH)[1] (OSHA)[2] (California) (HSE) (Ontario, Quebec) (Australia) (Mexico)
- Clean Air Act, 42USC7412; Title I, Part A, §112 hazardous pollutants (as antimony compounds)
- Clean Water Act: 40CFR116.4 Hazardous Substances; 40CFR117.3, RQ (same as CERCLA); 40CFR423, Appendix A, Priority Pollutants; Section 313 Water Priority Chemicals (57FR41331, 9/9/92); 40CFR401.15 Section 307 Toxic Pollutants, as antimony compounds
- RCRA, 40CFR261, Appendix 8 Hazardous Constituents, waste number not listed, as antimony compounds, n.o.s
- Safe Drinking Water Act, MCL, treatment technique; MCL, 0.006 mg/l; MCLG, 0.006 mg/l; Regulated Chemical (47FR9352)
- CERCLA/SARA Section 304 Reportable Quantity (RQ): CERCLA, 1,000 lb (454 kg)
- EPCRA Section 313: Includes any unique chemical substance that contains antimony as part of that chemical's infrastructure. Form R *de minimis* concentration reporting level: 0.1%
- Canada, WHMIS, Ingredients Disclosure List; National Pollutant Release Inventory (as antimony compounds)

Cited in U.S. State Regulations: Alaska (G), Florida (G), Maine (G), Maryland (G), Massachusetts (G), New Hampshire (G), New Jersey (G), Oklahoma (G), Pennsylvania (G), Rhode Island (G), West Virginia (G).

Description: Antimony trioxide, Sb_3O_3, is a noncombustible, odorless, white crystalline powder. Freezing/Melting point = 655°C. Slightly soluble in water.

Potential Exposure: It is used in flame-proofing, pigments and ceramics, to stain iron and copper, and to decolorize glass.

Incompatibilities: Strong oxidizers, strong acids, halogenated acids or bases, chlorinated rubber, bromine trifluoride. Reduction with hydrogen forms toxic antimony hydride.

Permissible Exposure Limits in Air: The legal Federal OSHA[2] airborne permissible exposure limit (PEL) as Sb. The recommended NIOSH[2] airborne exposure limit (PEL) is 0.5 mg/m^3 averaged over a 10-hour workshift, *and* 2.5 mg/m^3 as F for an 8-hour workshift[9] is 0.5 mg/m^3 averaged over an 8-hour workshift. The above exposure limits are for air levels only. When skin contact also occurs, you may be overexposed, even though air levels are lower than the limits listed above. The HSE (U.K.), California, Ontario, and Quebec airborne exposure limits for antimony compounds are the same as the OSHA levels show above. The ACGIH[1] has set no TWA limits for Sb_2O_3 production on the grounds that it is a suspected human carcinogen. It has set a TWA for handling and use of 0.5 mg/m^3. The DFG has set no MAK value in any case on the grounds that it is an animal-positive carcinogen. The former USSR-UNEP/IRPTC project has set a MAC value of 1.0 mg/m^3 for Sb_2O_3 dust.

Permissible Concentration in Water: As part of the priority toxic pollutant program, EPA[6] has set a limit of 146 μg/l of antimony to protect human health. EPA has also suggested[32] an ambient limit of 7 μg/l of antimony based on health effects.

Harmful Effects and Symptoms

Short Term Exposure: Antimony Trioxide can affect you when breathed in. Exposure can cause sore throat, rash, poor appetite and irritation of the airways, with cough. High or repeated exposure may damage the liver and the heart muscle. If used near acid, a deadly gas (stribine) can be released.

Long Term Exposure: There is an association between Antimony Trioxide in smelting processes and increased lung cancer. There is some evidence that this chemical nay damage the developing fetus and cause miscarriage. Can damage the kidneys, liver, and heart. Repeated exposure may affect the lungs and cause an abnormal chest x-ray to develop.

Points of Attack: Eyes, skin, respiratory system, cardiovascular system, kidneys, liver, reproductive system.

Medical Surveillance: Complete blood count (CBC), urine test for antimony, EKG, liver and kidney function tests. Consider lung function tests and chest x-ray. Depending on the degree of exposure, periodic medical checkups are advisable.

First Aid: If this chemical gets into the eyes, remove any contact lenses at once and irrigate immediately for at least 15 minutes, occasionally lifting upper and lower lids. Seek medical attention immediately. If this chemical contacts the skin, remove contaminated clothing and wash immediately with soap and water. Seek medical attention immediately. If this chemical has been inhaled, remove from exposure, begin rescue breathing (using universal precautions) if breathing has stopped and CPR if heart action has stopped. Transfer promptly to a medical facility. When this chemical has been swallowed, get medical attention. Give large quantities of water and induce vomiting. Do not make an unconscious person vomit.

Personal Protective Methods: Specific engineering controls are recommended for this chemical in the NIOSH criteria document Antimony Number 72–216. Where possible, enclose operations and use local exhaust ventilation at the site of chemical release. If local exhaust ventilation or enclosure is not used, respirators should be worn. Wear protective gloves and clothing to prevent any reasonable probability of skin contact. Safety equipment suppliers/manufacturers can provide recommendations on the most protective glove/clothing material for your operation. All protective clothing (suits, gloves, footwear, headgear) should be clean, available each day, and put on before work. Contact lenses should not be worn when working with this chemical. Wear dust-proof chemical goggles and face shield unless full facepiece respiratory protection is worn. Employees should wash immediately with soap when skin is wet or contaminated. Provide emergency showers and eyewash.

Respirator Selection: NIOSH/OSHA: *5 mg/m³:* DMXSQ (if not present as a fume) (any dust and mist respirator except single-use and quarter mask respirators); or SA (any supplied-air respirator). *12.5 mg/m³:* SA:CF (any supplied-air respirator operated in a continuous-flow mode); PAPRDM (if not present as a fume) (any powered, air-purifying respirator with a dust and mist filter). *25 mg/m³:* HiEF (any air-purifying, full-facepiece respirator with a high-efficiency particulate filter); or SAT:CF (any supplied-air respirator that has a tight-fitting facepiece and is operated in a continuous-flow mode); or PAPRTHiE (any powered, air-purifying respirator with a tight-fitting facepiece and a high-efficiency particulate filter); or SCBAF (any self-contained breathing apparatus with a full facepiece); or SAF (any supplied-air respirator with a full facepiece). *50 mg/m³:* SA:PD,PP (any supplied-air respirator operated in a pressure-demand or other positive-pressure mode). *Emergency or planned entry into unknown concentrations or IDLH conditions:* SCBAF:PD,PP (any self-contained breathing apparatus that has a full faceplate and is operated in a pressure-demand or other positive-pressure mode); or SAF:PD,PP:ASCBA (any supplied-air respirator that has a full facepiece and is operated in a pressure-demand or other positive-pressure mode in combination with an auxiliary self-contained breathing apparatus operated in a pressure-demand or other positive-pressure mode). *Escape:* HiEF (any air-purifying, full-facepiece respirator with a high-efficiency particulate filter); or SCBAE (any appropriate escape-type, self-contained breathing apparatus).

Storage: Prior to working with this chemical you should be trained on its proper handling and storage. Store in tightly closed containers in a cool, well-ventilated area away from heat, strong oxidizers, acids. A regulated, marked area should be established where this chemical is handled, used, or stored in compliance with OSHA standard 1910.1045.

Shipping: There are no DOT label requirements nor are there maximum limits on air shipment.

Spill Handling: Evacuate persons not wearing protective equipment from area of spill or leak until clean-up is complete. Remove all ignition sources. Collect powdered material in the most convenient and safe manner and deposit in sealed containers. Ventilate area after clean-up is complete. It may be necessary to contain and dispose of this chemical as a hazardous waste. If material or contaminated runoff enters waterways, notify downstream users of potentially contaminated waters. Contact your Department of Environmental Protection or your regional office of the federal EPA for specific recommendations. If employees are required to clean-up spills, they must be properly trained and equipped. OSHA 1910.120(q) may be applicable.

Fire Extinguishing: Use any agent suitable for surrounding fires. Poisonous gases are produced in fire including toxic Sb fumes. If material or contaminated runoff enters waterways, notify downstream users of potentially contaminated waters. Notify local health and fire officials and pollution control agencies. From a secure, explosion-proof location, use water spray to cool exposed containers. If cooling streams are ineffective (venting sound increases in volume and pitch, tank discolors, or shows any signs of deforming), withdraw immediately to a secure position. If employees are expected to fight fires, they must be trained and equipped in OSHA 1910.156.

Disposal Method Suggested: Consult with environmental regulatory agencies for guidance on acceptable disposal practices. Generators of waste containing this contaminant (≥100 kg/mo) must conform with EPA regulations governing storage, transportation, treatment, and waste disposal.

References

Sax, N. I., Ed., "Dangerous Properties of Industrial Materials Report" 2, No. 1, 74–76

New Jersey Department of Health and Senior Services, "Hazardous Substance Fact Sheet: Antimony Trioxide," Trenton, NJ (February 1988)

Antimycin A

Molecular Formula: $C_{28}H_{40}N_2O_9$ (Antimycin A_1); $C_{26}H_{36}N_2O_9$ (Antimycin A_3); $C_{25}H_{34}N_2O_9$ (Antimycin A_4)

Synonyms: Antimicina A (Spanish); Antimycin A; Antipiricullin; Dihyrosamidin; Fintrol; Isovaleric acid 8-ester with 3-formamido-*N*-(7-hexyl-8-hydroxy-4,9-dimethyl-2,6-dioxo-1,5-dioxonan-3-yl)salicylamide isovaleric acid 8 ester; Virosin

CAS Registry Number: 1397-94-0 (A_1-); 642-15-9 (A_1-); 11118-72-2 (Antimycin)

Note: Both A_1 CAS Numbers are found in RTECS, with the same chemical formula, although EPA regulates only 1397-94-0 as Antimycin A

RTECS Number: CD0350000

Regulatory Authority

- CERCLA/SARA 40CFR302 Extremely Hazardous Substances: TPQ = 1000/10,000 lb (454/4,540 kg)[7]
- SUPERFUND/EPCRA 40CFR302.4 Reportable Quantity (RQ): CERCLA, 1 lb (0.454 kg)

Cited in U.S. State Regulations: California (G), Florida (G), Massachusetts (G), New Jersey (G), Michigan (G), Pennsylvania (G).

Description: $C_{26}H_{36}N_2O_9$ (Antimycin A_3) and $C_{28}H_{40}N_2O_9$ (Antimycin A_1) are crystalline solids. Freezing/Melting point = 170 – 175°C; (A_3); 149 – 150°C (A_1). They are complex 9-membered (2 oxygens and 7 carbons) ring derivatives with complex side chains. Practically insoluble in water.

Potential Exposure: Specific uses for Antimycin A were not found, however, Antimycin A_1, and Antimycin A_3 are reported to be antibiotic substances produced by Streptomyces for use as a fungicide, possible insecticide and miticide. Registered as a pesticide in the U.S.

Permissible Exposure Limits in Air: No standards set.

Permissible Concentration in Water: No criteria set.

Routes of Entry: Ingestion, intramuscular.

Harmful Effects and Symptoms

Short Term Exposure: Subcutaneous, intravenous, and intraperitoneal route poisons. Moderately toxic by ingestion and intramuscular routes. The oral LD_{50} for rat is 28 mg/kg.[9]

First Aid: If this chemical gets into the eyes, remove any contact lenses at once and irrigate immediately for at least 15 minutes, occasionally lifting upper and lower lids. Seek medical attention immediately. If this chemical contacts the skin, remove contaminated clothing and wash immediately with soap and water. Seek medical attention immediately. If this chemical has been inhaled, remove from exposure, begin rescue breathing (using universal precautions) if breathing has stopped and CPR if heart action has stopped. Transfer promptly to a medical facility. When this chemical has been swallowed, get medical attention. Give large quantities of water and induce vomiting. Do not make an unconscious person vomit.

Personal Protective Methods: Wear protective gloves and clothing to prevent any reasonable probability of skin contact. Safety equipment suppliers/manufacturers can provide recommendations on the most protective glove/clothing material for your operation. All protective clothing (suits, gloves, footwear, headgear) should be clean, available each day, and put on before work. Contact lenses should not be worn when working with this chemical. Wear dust-proof chemical goggles and face shield unless full facepiece respiratory protection is worn. Employees should wash immediately with soap when skin is wet or contaminated. Provide emergency showers and eyewash.

Storage: Prior to working with this chemical you should be trained on its proper handling and storage.

Shipping: This material is not listed in the DOT list of materials for performance-oriented packaging standards.[19]

Spill Handling: Evacuate persons not wearing protective equipment from area of spill or leak until clean-up is complete. Remove all ignition sources. (Non-specific – Pesticide, solid, n.o.s.). Keep unnecessary people away; isolate hazard area and deny entry. Stay upwind; keep out of low areas. Ventilate closed spaces before entering them. Wear positive pressure breathing apparatus and special protective clothing. Remove and isolate contaminated clothing at the site. Do not touch spilled material; stop leak if you can do so without risk. Use water spray to reduce vapors. Small spills: absorb with sand or other noncombustible absorbent material and place into containers for later disposal. Small dry spills: with clean shovel place material into clean, dry container and cover; move containers from spill area. Large spills: dike far ahead of spill for later disposal. It may be necessary to contain and dispose of this chemical as a hazardous waste. If material or contaminated runoff enters waterways, notify downstream users of potentially contaminated waters. Contact your Department of Environmental Protection or your regional office of the federal EPA for specific recommendations. If employees are required to clean-up spills, they must be properly trained and equipped. OSHA 1910.120(q) may be applicable.

Fire Extinguishing: (Non-specific, Pesticide, solid, n.o.s.). Small fires: dry chemical, carbon dioxide, water spray, or foam. Large fires: water spray, fog, or foam. Move container from fire area if you can do so without risk. Fight fire from maximum distance. Dike fire control water for later disposal; do not scatter the material. Poisonous gases are produced in fire including nitrogen oxides. If material or contaminated runoff enters waterways, notify downstream users of potentially contaminated waters. Notify local health and fire officials and pollution control agencies. If employees are expected to fight fires, they must be trained and equipped in OSHA 1910.156.

Disposal Method Suggested: In accordance with 40CFR165 recommendations for the disposal of pesticides and pesticide containers. Must be disposed properly by following package label directions or by contacting your state pesticide or environmental control agency or by contacting your regional EPA office.

References

U.S. Environmental Protection Agency, "Chemical Profile: Antimycin A," Washington, DC, Chemical Emergency Preparedness Program (November 30, 1987)

ANTU

Molecular Formula: $C_{11}H_{10}N_2S$

Synonyms: α-Naphthyl thiourea; α-Naphtyl thiouree (French); Alrato; Anturat; Bantu; Chemical 109; Dirax; Kill kantz; Krysid; Krysid PI; 1-Naftil-tiourea (Italian); 1-Naftylthioureum (Dutch); α-Naphthothiourea; α-Naphthyl-thiocarbamide; 1-Naphthyl-thioharnstoff (German); α-Naphthylthiourea; *N*-(1-Naphthyl)-2-thiourea; 1-(1-Naphthyl)-2-thiourea; 1-Naphthylthiourea; 1-Naphthyl-thiouree (French); Naphtox; Rattrack; Rat-TU; Smeesana; Thiourea, 1-Naphthalenyl-; Urea,1-(1-naphthyl)-2-thio-

CAS Registry Number: 86-88-4

RTECS Number: YT9275000

DOT ID: UN 1651

Regulatory Authority

- Air Pollutant Standard Set (ACGIH)[1] (DFG)[3] (OSHA)[58] (Several States)[60] (Australia) (Israel) (Mexico) (California) (Several Canadian Provinces)
- EPA HAZARDOUS WASTE NUMBER (RCRA No.): P072
- RCRA 40CFR261, Appendix 8; 40CFR261.11 Hazardous Constituents
- CERCLA/SARA 40CFR302 Extremely Hazardous Substances: TPQ = 500/10,000 lb (227/4,540 kg)
- CERCLA/SARA Section 304 Reportable Quantity (RQ): CERCLA, 100 lb (45.4 kg)

Cited in U.S. State Regulations: Alaska (G), California (A, G), Connecticut (A), Florida (G), Illinois (G), Kansas (G), Louisiana (G), Maine (G), Massachusetts (G), Minnesota (G), New Hampshire (G), New Jersey (G), Nevada (A), North Dakota (A), Pennsylvania (G), Rhode Island (G), Vermont (G), Virginia (G, A), Washington (G), West Virginia (G), Wisconsin (G).

Description: ANTU, $C_{11}H_{10}N_2S$, α-naphtylthiourea, is a noncombustible, white crystalline solid or gray powder. Odorless. Freezing/Melting point = 198°C. Hazard Identification (based on NFPA-704 M Rating System): Health 4, Flammability 1, Reactivity 0. Slightly soluble in water.

Potential Exposure: In production of ANTU or its formulations and use as a rodenticide.

Incompatibilities: Strong oxidizers, silver nitrate.

Permissible Exposure Limits in Air: The Federal OSHA/NIOSH standard[58] is 0.3 mg/m³ TWA. ACGIH TLV, DFG[3] MAK, Australia, Israel, Mexico have set the same value. The DFG peak limitation is 5 times the normal MAK (30 min), do not exceed 2 times during a workshift. Israel has an Action Level of 0.15 mg/m³. The Mexico STEL is 0.9 mg/m³. The NIOSH IDLH level is 100 mg/m³. Several states have set guidelines or standards for ANTU in ambient air[60] ranging from 3 μg/m³ (North Dakota) to 5 μg/m³ (Virginia) to 6 μg/m³ (Connecticut) to 7 μg/m³ (Nevada). Canadian Provincial level for Alberta, British Columbia, Ontario and Quebec are 0.2 mg/m³ TWA or TWAEV (Quebec). Alberta and British Columbia STEL is 0.9 mg/m³ (15 min).

Determination in Air: Collection on a filter and analysis by gas-liquid chromatography. See NIOSH Method 5276.

Permissible Concentration in Water: No criteria set.

Routes of Entry: Inhalation, ingestion and skin absorption.

Harmful Effects and Symptoms

Short Term Exposure: Poisonous. Symptoms include seizures, and dermal irritation. High exposures can cause pulmonary edema, a medical emergency that can be delayed for several hours. This can cause death. Ingestion may cause vomiting, shortness of breath, and bluish discoloration of the skin. ANTU is moderately toxic: probable oral lethal dose (human) 0.5 – 5 mg/kg, or between 1 ounce and 1 pint (or 1 lb) for 150 lb person. The LD_{50} for oral rat is 6 mg/kg.[9] Chronic sublethal exposure may cause antithyroid activity. Can produce hyperglycemia of three times normal in three hours.

Long Term Exposure: May cause chronic dermatitis, increased production of white blood cells. A questionable carcinogen (IARC, Group 3, inadequate human evidence) and a possible mutagen.

Group Points of Attack: Respiratory system.

Medical Surveillance: Consider the points of attack in preplacement and periodic physical examinations. People with chronic respiratory disease or liver disease may be especially at risk. Lung function tests. Consider chest x-ray following acute overexposure. Evaluation by a dermatologist.

First Aid: If this chemical gets into the eyes, remove any contact lenses at once and irrigate immediately for at least 15 minutes, occasionally lifting upper and lower lids. Seek medical attention immediately. If this chemical contacts the skin, remove contaminated clothing and wash immediately with soap and water. Seek medical attention immediately. If this chemical has been inhaled, remove from exposure, begin rescue breathing (using universal precautions) if breathing has stopped and CPR if heart action has stopped. Transfer promptly to a medical facility. When this chemical has been swallowed, get medical attention. Give large quantities of water and induce vomiting. Do not make an unconscious person vomit. Medical observation is recommended for 24 - 48 hours after breathing overexposure, as pulmonary edema may be delayed. As first aid for pulmonary edema, a doctor or authorized paramedic may consider administering a corticosteroid spray.

Personal Protective Methods: Wear protective gloves and clothing to prevent any reasonable probability of skin contact. Safety equipment suppliers/manufacturers can provide recommendations on the most protective glove/clothing material for your operation. All protective clothing (suits, gloves, footwear, headgear) should be clean, available each day, and put on before work. Contact lenses should not be worn when working with this chemical. Wear dust-proof chemical goggles and face shield unless full facepiece respiratory protection is worn. Employees should wash immediately with soap when skin is wet or contaminated. Provide emergency showers and eyewash.

Respirator Selection: NIOSH/OSHA: *3 mg/m³*: CCROV DMFu (any MSHA/NIOSH approved chemical cartridge respirator with a full facepiece and organic vapor cartridge(s) in combination with a dust, mist, and fume filter); or SA (any supplied-air respirator). *7.5 mg/m³*: SA:CF (any supplied-air respirator operated in a continuous-flow mode); or PAPROV DMFu (any powered, air-purifying respirator with organic vapor cartridge(s) in combination with a dust, mist, and fume filter). *15 mg/m³*: CCRFOVHiE (any chemical cartridge respirator with a full facepiece and organic vapor cartridge(s) in combination with a high efficiency particulate filter); or

PAPRTOVHiE [any powered, air-purifying respirator with a tight fitting facepiece and organic vapor cartridge(s) in combination with a high-efficiency particulate filter]; or GMFOVHiE [any air-purifying, full-facepiece respirator (gas mask) with a chin-style, front- or back-mounted organic vapor canister having a high-efficiency particulate filter]; or SAT:CF (any supplied-air respirator that has a tight-fitting facepiece and is operated in a continuous-flow mode); or SCBAF (any self-contained breathing apparatus with a full facepiece); or SAF (any supplied-air respirator with a full facepiece). *100 mg/m³*: SA: PD,PP (any supplied-air respirator operated in a pressure-demand or other positive-pressure mode). *Emergency or planned entry in unknown concentration or IDLH conditions:* SCBAF: PD,PP (any MSHA/NIOSH approved self-contained breathing apparatus that has a full facepiece and is operated in a pressure-demand or other positive-pressure mode); or SAF:PD,PP: ASCBA (any supplied-air respirator that has a full facepiece and is operated in a pressure-demand or other positive-pressure mode in combination with an auxiliary, self-contained breathing apparatus operated in a pressure-demand or other positive pressure mode). *Escape:* GMFOVHiE [any air-purifying, full-facepiece respirator (gas mask) with a chin-style, front- or back-mounted organic vapor canister having a high-efficiency particulate filter]; or SCBAE (any appropriate escape-type, self-contained breathing apparatus).

Storage: Prior to working with this chemical you should be trained on its proper handling and storage. Store in tightly closed containers in a cool, well ventilated area away from oxidizers and silver nitrate. Sources of ignition such as smoking and open flames, are prohibited where this chemical is used, handled, or stored in a manner that could create a potential fire or explosion hazard.

Shipping: The DOT label requirement is "Poison." The UN/DOT Hazard Class is 6.1 and the Packing Group is II.[19][20]

Spill Handling: Evacuate persons not wearing protective equipment from area of spill or leak until clean-up is complete. Remove all ignition sources. Avoid inhalation and skin contact; wear proper respiratory protection and protective clothing. Do not touch spilled material, stay upwind, keep out of low areas. Collect powdered material in the most convenient and safe manner and deposit in sealed containers. Ventilate area after clean-up is complete. It may be necessary to contain and dispose of this chemical as a hazardous waste. If material or contaminated runoff enters waterways, notify downstream users of potentially contaminated waters. Contact your Department of Environmental Protection or your regional office of the federal EPA for specific recommendations. If employees are required to clean-up spills, they must be properly trained and equipped. OSHA 1910.120(q) may be applicable.

Fire Extinguishing: ANTU may burn but will not ignite readily. Extinguish with dry chemical, carbon dioxide, water spray, fog, or foam. Poisonous gases are produced in fire including nitrogen and sulfur oxides. If material or contaminated runoff enters waterways, notify downstream users of potentially contaminated waters. Notify local health and fire officials and pollution control agencies. From a secure, explosion-proof location, use water spray to cool exposed containers. If cooling streams are ineffective (venting sound increases in volume and pitch, tank discolors, or shows any signs of deforming), withdraw immediately to a secure position. If employees are expected to fight fires, they must be trained and equipped in OSHA 1910.156.

Disposal Method Suggested: Incinerate in a furnace equipped with an alkaline scrubber.[22] Consult with environmental regulatory agencies for guidance on acceptable disposal practices. Generators of waste containing this contaminant (≥100 kg/mo) must conform with EPA regulations governing storage, transportation, treatment, and waste disposal.

References

Sax, N. I., Ed., "Dangerous Properties of Industrial Materials Report" 4, No. 2, 83–86 (1984)

U.S. Environmental Protection Agency, "Chemical Profile: ANTU," Washington, DC, Chemical Emergency Preparedness Program (November 30, 1987)

Argon

Molecular Formula: A

CAS Registry Number: 7440-37-1

RTECS Number: CF2300000

DOT ID: UN 1951 (liquid); UN 1006 (compressed gas)

Regulatory Authority

- Most regulatory authorities and advisory organizations (such as ACGIH) list argon as a "simple asphyxiant" or "asphyxiant."

Cited in U.S. State Regulations: Alaska (G), California (G), Florida (G), Illinois (G), Maine (G), Massachusetts (G), New Hampshire (G), New Jersey (G), Pennsylvania (G), Rhode Island (G).

Description: With the symbol A, argon is a nonflammable gas; one of the elements in the inert gas category. It is colorless. Boiling point = –186°C. Freezing/Melting point = 192°C. Slightly soluble in water.

Potential Exposure: Argon is used as an inert gas shield in arc wilding; it is used as an inert atmosphere in electric lamps. It is used as a blanketing agent in metals refining (especially titanium and zirconium).

Permissible Exposure Limits in Air: There is no Federal standard. ACGIH lists argon as a simple asphyxiant with no specified TLV. Australia has a TWA as "asphyxiant at less than 18% oxygen by volume."

Permissible Concentration in Water: No criteria set.

Routes of Entry: Inhalation and possibly skin contact with liquid argon.

Harmful Effects and Symptoms

Short Term Exposure: The gas is a simple asphyxiant as noted above. Contact with the liquid can cause frostbite.

First Aid: If contact with liquid argon occurs, seek medical attention immediately; do *NOT* rub the affected areas or flush them with water. In order to prevent further tissue damage, do *NOT* attempt to remove frozen clothing from frostbitten areas. If frostbite has *NOT* occurred, immediately and thoroughly wash contaminated skin with warm water. Seek medical attention immediately. If this chemical has been inhaled, remove from exposure, begin rescue breathing (using universal precautions) if breathing has stopped and CPR if heart action has stopped. Transfer promptly to a medical facility.

Personal Protective Methods: Where exposure to cold equipment, vapors, or liquids may occur, employees should be equipped with special clothing designed to prevent the freezing of body tissues. Avoid skin contact with liquid argon. All protective clothing (suits, gloves, footwear, headgear) should be clean, available each day, and put on before work. Wear splash-proof chemical goggles where exposure to liquid argon can occur.

Respirator Selection: Exposure to argon is dangerous because it can replace oxygen and lead to suffocation. Only MSHA/NIOSH approved self-contained breathing apparatus with a full facepiece operated in positive pressure mode should be used in oxygen deficient environments.

Storage: Prior to working with this chemical you should be trained on its proper handling and storage. Storage areas should be well-ventilated. Protect vessels which contain argon from physical damage.

Shipping: Argon must be labeled "Non-flammable Gas." It falls in DOT Hazard Class 2.2 and has no designated Packing Group.

Spill Handling: Evacuate persons not wearing protective equipment from area of spill or leak until clean-up is complete. Remove all ignition sources. Ventilate area of leak to disperse the gas. Stop flow of gas. If sources of leak is a cylinder and the leak cannot be stopped in place, remove the leaking cylinder to a safe place in the open air and allow cylinder to empty. If employees are required to clean-up spills, they must be properly trained and equipped. OSHA 1910.120(q) may be applicable.

Fire Extinguishing: Argon is non-flammable and indeed can act as an extinguishing agent itself. Therefore, use extinguishing agents suited for surrounding fires. If employees are expected to fight fires, they must be trained and equipped in OSHA 1910.156.

Disposal Method Suggested: Vent to atmosphere.

References

Sax, N. I., Ed., "Dangerous Properties of Industrial Materials Report" 1, No. 5, 36–37 (1981)

New Jersey Department of Health and Senior Services, "Hazardous Substance Fact Sheet: Argon," Trenton, NJ (August 1985)

Arsenic and Inorganic Arsenic Compounds

Molecular Formula: As

Synonyms: Accuspin ASX-10 Spin-On Dopant; Arsen (German, Polish); Arsenic-75; Arsenicals; Arsenic black; Arsenic, metallic; Arsenico (Spanish); Arsenic, solid; AS-120; AS-217; Butter of arsenic; Colloidal arsenic; Grey arsenic; Metallic arsenic; Realgar; Ruby arsenic

Note: The above synonyms are for metallic arsenic. Other inorganic synonyms vary depending on the specific As compound. The term "inorganic arsenic" does not include *arsine*.

CAS Registry Number: 7440-38-2 (metallic)

RTECS Number: CG0525000

DOT ID: UN 1558

Regulatory Authority

- Carcinogen (Human Positive) (IARC)[9] (DFG)[3] (NTP)
- OSHA, 29CFR1910 Specifically Regulated Chemicals (See CFR1910.1018) Inorganic compounds (except Arsine)
- Banned or Severely Restricted (In Agricultural, Pharmaceutical and Industrial Chemicals) (Many Countries)[13][35]
- Air Pollutant Standard Set (ACGIH)[1] (OSHA/NIOSH)[2] (HSE)[33] (former USSR)[43] (Several States)[60] (Australia) (Israel) (Mexico) (Several Canadian Provinces)
- Clean Air Act, 42USC7412; Title I, Part A, §112 hazardous pollutants
- Clean Water Act 40CFR401.15 Section 307Toxic Pollutants; 40CFR423, Appendix A Priority Pollutants; §313 Priority Chemicals
- RCRA 40CFR261.24 Toxicity Characteristics, Maximum Concentration of Contaminants (MCC), Regulatory level, 5.0 mg/l
- RCRA "D Series Waste" Number, D004, Chronic Toxicity Reference Level, 0.05 mg/l
- RCRA, 40CFR261, Appendix 8 Hazardous Constituents., waste number not listed
- RCRA 40CFR268.48; 61FR15654, Universal Treatment Standards: Wastewater (mg/l), 1.4; Nonwastewater (mg/l), 5.0 TCLP
- RCRA 40CFR264, Appendix 9; TSD Facilities Ground Water Monitoring List Suggested methods (PQL μg/l): (total) 6010 (500), 7060 (10), 7061 (20)
- Safe Drinking Water Act 47FR9352 Regulated chemical: MCL, 0.05 mg/l (Section 141.11) applies only to community water systems. *Note:* Effective January 2006 the MCL will be 0.01 mg/l
- SUPERFUND/EPCRA 40CFR302.4 Reportable Quantity (RQ): CERCLA, 1 lb (0.454 kg), no reporting required, if diameter of metal is equal to or exceeds 0.004 in
- EPCRA Section 313: Form R *de minimis* concentration reporting level: 0.1%

- U.S. DOT Regulated Marine Pollutant (49CFR172.101, Appendix B)
- Canada, WHMIS, Ingredients Disclosure List
- Canada: Priority Substance List & Restricted Substances/ Ocean Dumping Forbidden (CEPA), National Pollutant Release Inventory (NPRI) (arsenic compounds)

Cited in U.S. State Regulations: California (A, G), Connecticut (A), Florida (G), Illinois (G), Kansas (G), Louisiana (G), Maine (G, W), Maryland (G), Massachusetts (G), Michigan (G), Montana (A), Nevada (A), New Hampshire (G), New Jersey (G), New York (G, A), North Carolina (A), North Dakota (A), Oklahoma (G), Pennsylvania (G, A), Rhode Island (G, A), South Carolina (A), South Dakota (A, W), Tennessee (W), Utah (W), Vermont (G), Virginia (G, A), Washington (G), West Virginia (G), Wisconsin (G).

Description: Elemental arsenic, As, occurs to a limited extent in nature as a steel-gray, amorphous metalloid. Boiling point = 612°C (sublimes). Freezing/Melting point = 814°C @ 36 atm; 817°C @ 28 atm. Hazard Identification (based on NFPA-704 M Rating System): Health 3, Flammability 1, Reactivity 0. Insoluble in water. Arsenic in this entry includes the element and any of its inorganic compounds *excluding* arsine. Arsenic trioxide (As_2O_3), the principal form in which the element is used, is frequently designated as arsenic, white arsenic, or arsenous oxide. Arsenic is present as an impurity in many other metal ores and is generally produced as arsenic trioxide as a by-product in the smelting of these ores, particularly copper. Most other arsenic compounds are produced from the trioxide.

Potential Exposure: Arsenic compounds have a variety of uses. Arsenates and arsenites are used in agriculture as insecticides, herbicides, larvicides, and pesticides. Other arsenic compounds are used in pigment production, the manufacture of glass as a bronzing or decolorizing agent, the manufacture of opal glass and enamels, textile printing, tanning, taxidermy, and antifouling paints. They are also used to control sludge formation in lubricating oils. Metallic arsenic is used as an alloying agent for heavy metals, and in solders, medicines, herbicides. EPA has estimated that more than 6 million people living within twelve miles of major sources-copper, zinc, and lead smelters-may be exposed to 10 times the average U.S. atmospheric levels of arsenic. The agency says that 40,000 people living near some copper smelters may be exposed to 100 times the national atmospheric average.

Incompatibilities: Incompatible with strong acids, strong oxidizers, peroxides, bromine azide, bromine pentafluoride, bromine trifluoride, cesium acetylene carbide, chromium trioxide, nitrogen trichloride, silver nitrate. Can react vigorously with strong oxidizers (chlorine, dichromate, permanganate). Forms highly toxic fumes on contact with acids or active metals (iron, aluminum, zinc). Hydrogen gas can react with inorganic arsenic to form highly toxic arsine gas.

Permissible Exposure Limits in Air: The following exposure limits are for air levels only. When skin contact also occurs, overexposure is possible, even though air levels are less than the limits listed below. OSHA:[2] the legal airborne permissible exposure limit (PEL) is 0.010 mg/m^3 averaged over an 8-hour workshift. NIOSH:[2] the recommended airborne exposure limit is 0.002 mg/m^3 (ceiling), not to be exceeded during any 15 min. work period. ACGIH:[1] the recommended airborne exposure limit is 0.01 mg/m^3 averaged over an 8-hour workshift. The HSE (U.K.) Maximum Exposure Limit as As is 0.1 mg/m^3 TWA. California's workplace PEL is the same as ACGIH and an Action Level of 0.005 mg/m^3. The Australia limit is 0.05 mg/m^3 TWA (confirmed carcinogen); Israel 0.01 mg/m^3 TWA and Action Level 0.005 mg/m^3. Mexico level 0.2 mg/m^3 TWA. Canada: Alberta level 0.2 mg/m^3 TWA and STEL of 0.6 mg/m^3 (15 min); British Columbia level 0.5 mg/m^3 TWA; Ontario level 0.01 mg/m^3 TWAEV and STEV of 0.05; Quebec level 0.2 mg/m^3 TWAEV. The former USSR-UNEP/IRPTC project[43] has set a MAC of 0.003 mg/m^3 on an average daily basis for residential areas. In addition, several states have set guidelines or standards for arsenic in ambient air:[60] 0.06 mg/m^3 (California Prop. 65), 0.0002 μg/m^3 (Rhode Island), 0.00023 μg/m^3 (North Carolina), 0.024 μg/m^3 (Pennsylvania), 0.05 μg/m^3 (Connecticut), 0.07 - 0.39 μg/m^3 (Montana), 0.67 μg/m^3 (New York), 1.0 μg/m^3 (South Carolina), 2.0 μg/m^3 (North Dakota), 3.3 μg/m^3 (Virginia), 5 μg/m^3 (Nevada).

Determination in Air: Collection on a filter and analysis by atomic absorption spectrometry. See NIOSH Methods 7900 and 73000, Elements.[18] See also OSHA Method ID 105.

Permissible Concentration in Water: See Regulatory Authority for US EPA levels. To protect freshwater aquatic life-total recoverable trivalent inorganic arsenic never to exceed 440 μg/l. To protect saltwater aquatic life-508 μg/l on an acute basis. To protect human health-preferably zero. A value of 0.02 μg/l corresponds to a human health risk of 1 in 100,000. Allowable arsenic levels in drinking water have also been set by the former USSR-UNEP/IRPTC project[43] at 0.05 mg/l and in water for fishery purposes of 0.05 mg/l also. Maine (drinking water)[61] of 0.05 mg/l. Mexico 0.5 mg/l (reduce human exposure to a minimum).

Determination in Water: The atomic absorption graphite furnance technique is often used for measurement of total arsenic in water. It also has been standardized by EPA. Total arsenic may be determined by digestion followed by silver diethyldithiocarbamate; an alternative is atomic absorption; another is inductively coupled plasma (ICP) optical emission spectrometry.

Routes of Entry: Inhalation, through the skin, and ingestion of dust and fumes.

Harmful Effects and Symptoms

Local: Trivalent arsenic compounds are corrosive to the skin. Brief contact has no effect, but prolonged contact results in a local hyperemia and later vesicular or pustular eruption. The moist mucous membranes are most sensitive to the irritant action. Conjunctiva, moist and macerated areas of the skin, eyelids, the angles of the ears, nose, mouth, and respiratory mucosa are also vulnerable to the irritant effects. The wrists are common sites of dermatitis, as are the genitalia if personal hygiene is poor. Perforations of the nasal septum may occur. Arsenic trioxide and pentoxide are capable of producing skin sensitization and contact dermatitis. Arsenic is also capable of producing keratoses, especially of the palms and soles. Arsenic has been cited as a cause of skin cancer, but the incidence is low. *Systemic:* The acute toxic effects of arsenic are generally seen following ingestion of inorganic arsenical compounds. This rarely occurs in an industrial setting. Symptoms develop within ½ to 4 hours following ingestion and are usually characterized by constriction of the throat followed by dysphagia, epigastric pain, vomiting, and watery diarrhea. Blood may appear in vomitus and stools. If the amount ingested is sufficiently high, shock may develop due to sever fluid loss, and death may ensue in 24 hours. If the acute effects are survived, exfoliative dermatitis and peripheral neuritis may develop. Cases of acute arsenical poisoning due to inhalation are exceedingly rare in industry. When it does occur, respiratory tract symptoms-cough, chest pain, dyspnea-giddiness, headache, and extreme general weakness precede gastrointestinal symptoms. The acute toxic symptoms of trivalent arsenical poisoning are due to severe inflammation of the mucous membranes and greatly increased permeability of the blood capillaries. Chronic arsenical poisoning due to ingestion is rare and generally confined to patients taking prescribed medications. However, it can be a concomitant of inhaled inorganic arsenic from swallowed sputum and improper eating habits. Symptoms are weight loss, nausea and diarrhea alternating with constipation, pigmentation and eruption of the skin, loss of hair, and peripheral neuritis. Chronic hepatitis and cirrhosis have been described. Polyneuritis may be the salient feature, but more frequently there are numbness and paresthesias of "glove and sticking" distribution. The skin lesions are usually melanotic and keratotic and may occasionally take the form of an intradermal cancer of the squamous cell type, but without infiltrative properties. Horizontal white lines (striations) on the fingernails and toenails are commonly seen in chronic arsenical poisoning and are considered to be a diagnostic accompaniment of arsenical polyneuritis. Inhalation of inorganic arsenic compounds is the most common cause of chronic poisoning in the industrial situation. This condition is divided into three phases based on signs and symptoms. First Phase: The worker complains of weakness, loss of appetite, some nausea, occasional vomiting, a sense of heaviness in the stomach, and some diarrhea. Second Phase: The worker complains of conjunctivitis, and a catarrhal state of the mucous membranes of the nose, larynx, and respiratory passages. Coryza, hoarseness, and mild tracheobronchitis may occur. Perforation of the nasal septum is common, and is probably the most typical lesion of the upper respiratory tract in occupational exposure to arsenical dust. Skin lesions, eczematoid and allergic in type, are common. Third Phase: The worker complains of symptoms of peripheral neuritis, initially of hands and feet, which is essentially sensory. In more severe cases, motor paralyses occur; the first muscles affected are usually the toe extensors and the peronei. In only the most severe cases will paralysis of flexor muscles of the feet or of the extensor muscles of hands occur. Liver damage from chronic arsenical poisoning is still debated, and as yet the question is unanswered. In cases of chronic and acute arsenical poisoning, toxic effects to the myocardium have been reported based on EKG changes. These finding, however, are now largely discounted and the EKG changes are ascribed to electrolyte disturbances concomitant with arsenicalism. Inhalation of arsenic trioxide and other inorganic arsenical dusts does not give rise to radiological evidence of pneumoconiosis. Arsenic does have a depressant effect upon the bone marrow, with disturbances of both erythropoiesis and myelopoiesis. Evidence is now available incriminating arsenic compounds as a cause of lung cancer as well as skin cancer. Skin cancer in humans is causally associated with exposure to inorganic arsenic compounds in drugs, drinking water and the occupational environment. The risk of lung cancer was increased 4 - 12 times in certain smelter workers who inhaled high levels of arsenic trioxide. However, the influence of other constituents of the working environment cannot be excluded in these studies. Case reports have suggested an association between exposure to arsenic compounds and blood dyscrasias and liver tumors.

Short Term Exposure: Skin contact can cause irritation, itching, burning sensation, and rash. Eye contact can cause irritation and burns. Inhalation can cause irritation of the respiratory tract. High exposure can cause poor appetite, nausea, vomiting and muscle cramps. High exposure can cause nerve damage with numbness, "pins and needles" sensation, weakness of the arms and legs.

Long Term Exposure: Arsenic is a carcinogen; causes skin, lung, and lymphatic cancer, possible reproductive hazard (a teratogen in animals). Can cause an ulcer of the "bone" dividing the inner nose. It can cause hoarseness, sore eyes, nerve damage, thickening of the skin with patch areas of darkening and loss of pigment, liver damage and stomach problems. Small doses can accumulate in the body.

Points of Attack: Liver, kidneys, skin, lungs, lymphatic system.

Medical Surveillance: Before first exposure and every 6 - 12 months thereafter, OSHA 1910.1018 requires employers to provide (for persons exposed to 0.005 mg/m^3 of Arsenic) a medical history and exam which shall include: chest x-ray, exam of the nose, skin, and nails, sputum cytology examination, test for urine Arsenic (may not be accurate within 2 days of eating shellfish or fish; most accurate at the end of a workday. Levels should not be greater than 100 micrograms per gram creatinine in the urine. Exam of the nervous system. After suspected overexposure, repeat these tests and consider complete blood count (CBC) and liver function tests. Also examine skin periodically for abnormal growths. Skin cancer from arsenic can easily be cured when detected early. Employees have a legal right to testing information under OSHA 1910.20.

First Aid: If this chemical gets into the eyes, remove any contact lenses at once and irrigate immediately for at least 15 minutes, occasionally lifting upper and lower lids. Seek medical attention immediately. If this chemical contacts the skin, remove contaminated clothing and wash immediately with soap and water. Seek medical attention immediately. If this chemical has been inhaled, remove from exposure, begin rescue breathing (using universal precautions) if breathing has stopped and CPR if heart action has stopped. Transfer promptly to a medical facility. When this chemical has been swallowed, get medical attention. Give large quantities of water and induce vomiting. Do not make an unconscious person vomit.

Note: For severe poisoning BAL has been used. For milder poisoning *penicillamine (not penicillin)* has been used, both with mixed success. Side effects occur with such treatment and it is never a substitute for controlling exposure. It can only be done under strict medical care.

Personal Protective Methods: Workers should be trained in personal hygiene and sanitation, the use of personal protective equipment, and early recognition of symptoms of absorption, skin contact irritation, and sensitivity. With the exception of arsine and arsenic trichloride, the compounds of arsenic do not have odor or warning qualities. Wear protective gloves and clothing to prevent any reasonable probability of skin contact. Safety equipment suppliers/manufacturers can provide recommendations on the most protective glove/clothing material for your operation. All protective clothing (suits, gloves, footwear, headgear) should be clean, available each day, and put on before work. Wear dust-proof chemical goggles and face shield when working with powder or dust, unless full facepiece respiratory protection is worn. Employees should wash immediately with soap when skin is wet or contaminated. Provide emergency showers and eyewash. Specific engineering controls are required under OSHA 1910.1018, *Inorganic Arsenic.* See also NIOSH Criteria Document #75–149, *"Inorganic Arsenic."*

Respirator Selection: *At any concentrations above the NIOSH REL at any detectable concentration:* SCBAF:PD,PP (any self-contained breathing apparatus that has a full faceplate and is operated in a pressure-demand or other positive-pressure mode); or SAF:PD,PP:ASCBA (any supplied-air respirator that has a full facepiece and is operated in a pressure-demand or other positive-pressure mode in combination with an auxiliary self-contained breathing apparatus operated in a pressure-demand or other positive-pressure mode). *Escape:* GMFAGHiE [any air-purifying, full-facepiece respirator (gas mask) with a chin-style, front-or back-mounted acid gas canister having a high-efficiency particulate filter]; or SCBAE (any appropriate escape-type, self-contained breathing apparatus).

Note: Workers should be permitted to leave the work area every two hours to wash their faces and obtain clean respirators.

Storage: Prior to working with this chemical you should be trained on its proper handling and storage. Arsenic must be stored in a cool, dry place away from oxidizers (such as perchlorates, peroxides, permanganates, chlorates, and nitrates) and strong acids (such as hydrochloric, sulfuric, and nitric) since violent reactions occur. A regulated, marked area should be established where this chemical is handled, used, or stored in compliance with OSHA standard 1910.1045.

Shipping: The label required for solid arsenic and inorganic compounds is "Poison." These materials fall in UN/DOT Hazard Class 6.1 and Packing Group II for solid arsenic and I for solid arsenic compounds.[19][20]

Spill Handling: Evacuate persons not wearing protective equipment from area of spill or leak until clean-up is complete. Remove all ignition sources. Cover the spill with dry lime or soda ash and collect powdered material in a safe manner and deposit in sealed containers. Alternatively, if the spill is a solid, place in suitable container without raising dust. Use a high efficiency particulate absolute (HEPA) filter vacuum (*not* a standard shop vac), or wet method to reduce dust during cleanups. *Do not dry sweep.* If the spill is a liquid, cover with an absorbent and sweep into a suitable container. It may be necessary to contain and dispose of this chemical as a hazardous waste. If material or contaminated runoff enters waterways, notify downstream users of potentially contaminated waters. Contact your Department of Environmental Protection or your regional office of the federal EPA for specific recommendations. If employees are required to clean-up spills, they must be properly trained and equipped. OSHA 1910.120(q) may be applicable.

Fire Extinguishing: Arsenic metal is noncombustible, however arsenic dust can be flammable when exposed to heat or flame. Use dry chemical, CO_2, water spray, or foam extinguishers. Poisonous gases are produced in fire including arsine. If material or contaminated runoff enters waterways, notify downstream users of potentially contaminated waters. Notify local health and fire officials and pollution control agencies. From a secure, explosion-proof location, use water spray to cool exposed containers. If cooling streams are ineffective (venting sound increases in volume and pitch, tank discolors, or shows any signs of deforming), withdraw immediately to a secure position. If employees are expected to fight fires, they must be trained and equipped in OSHA 1910.156.

Disposal Method Suggested: Elemental arsenic wastes should be placed in long-term storage or returned to suppliers or manufacturers for reprocessing. Arsenic pentaselenide-wastes should be placed in long-term storage or returned to suppliers or manufacturers for reprocessing. Arsenic trichloride-hydrolyze to arsenic trioxide utilizing scrubbers for hydrogen chloride abatement. The trioxide may then be placed in long-term storage. Arsenic trioxide-long-term storage in large shiftproof and weatherproof silos. This compound may also be dissolved, precipitated as the sulfide and returned to the suppliers. Arsenic-containing sewage may be decontaminated by pyrolusite treatment.[22] Consult with environmental regulatory agencies for guidance on acceptable disposal practices. Generators of waste containing this contaminant (≥100 kg/mo) must conform with EPA regulations governing storage, transportation, treatment, and waste disposal. In accordance with 40CFR165 recommendations for the disposal of pesticides and pesticide containers. Must be disposed properly by following package label directions or by contacting your state pesticide or environmental control agency or by contacting your regional EPA office.

References

National Institute for Occupational Safety and Health, Criteria for a Recommended Standard: Occupational Exposure to Inorganic Arsenic, NIOSH Doc. No. 74–110, Washington, DC (1973)

National Institute for Occupational Safety and Health, Criteria for a Recommended Standard: Occupational Exposure to Inorganic Arsenic (Revised), NIOSH Doc. No. 75–149, Washington, DC (1975)

U.S. Environmental Protection Agency, Arsenic: Ambient Water Quality Criteria, Washington, DC (1979)

U.S. Environmental Protection Agency, Status Assessment of Toxic Chemicals: Arsenic, Report No. EPA-600/2-79-210B, Washington, DC (December 1979)

U.S. Environmental Protection Agency, Toxicology of Metals, Vol II: Arsenic, Report EPA-600/1-77-022, Research Triangle Park, NC, pp 30–70 (May 1977)

National Academy of Sciences, Medical and Biological Effects of Environmental Pollutants: Arsenic, Washington, DC (1977)

U.S. Environmental Protection Agency, Arsenic, Health and Environmental Effects Profile No. 11, Office of Solid Waste, Washington, DC (April 30, 1980)

Sax, N. I., Ed., "Dangerous Properties of Industrial Materials Report" 1, No. 3, 32–34 (1981)

Lederer, W. H., and Fensterheim, R. J., ARSENIC: Industrial, Biomedical and Environmental Perspectives, New York, Van Nostrand Reinhold Co. (1983)

New Jersey Department of Health and Senior Services, "Hazardous Substance Fact Sheet: Arsenic," Trenton, NJ (June 1998)

New York State Department of Health, "Chemical Fact Sheet: Arsenic," Albany, NY, Bureau of Toxic Substance Assessment (May 1986)

U.S. Public Health Service, "Toxicological Profile for Arsenic," Atlanta, GA, Agency for Toxic Substances and Disease Registry (November 1987)

Arsenic Acid

Molecular Formula: AsH_3O_4 (*ortho-*); $AsHO_3$ (*meta-*)

Synonyms: Acido arsenico (Spanish); Arsenate; *o*-Arsenic acid; Arsenic pentoxide; Orthoarsenic acid (*o*-); Scorch®; Zotox®

CAS Registry Number: 1327-52-2; 7778-39-4 (*ortho-*) (These two CAS number are regulated by the USEPA, New Jersey, California and others); 10102-53-1 (*meta-*)

RTECS Number: CG0700000

DOT ID: UN 1554 (solid); UN 1553 (liquid)

Regulatory Authority

- Very Toxic Substance (World Bank)[15]
- OSHA, 29CFR1910 Specifically Regulated Chemicals (See CFR1910.1018)
- Carcinogen (Human Positive) (IARC)[9] (DFG)[3] (NTP)
- Banned or Severely Restricted (In Agricultural, Pharmaceutical and Industrial Chemicals) (Many Countries)[13][35]
- Air Pollutant Standard Set (ACGIH)[1] (OSHA/NIOSH)[2] (HSE)[33] (former USSR)[43] (Several States)[60]
- Clean Air Act, 42USC7412; Title I, Part A, §112 hazardous pollutants
- Clean Water Act 40CFR401.15 Section 307 Toxic Pollutants; 40CFR423, Appendix A Priority Pollutants; §313 Priority Chemicals
- RCRA, 40CFR261, Appendix 8 Hazardous Constituents, waste number P010
- SUPERFUND/EPCRA 40CFR302.4 Reportable Quantity (RQ): CERCLA, 1 lb (0.454 kg)
- EPCRA Section 313: Form R *de minimis* concentration reporting level: 0.1%
- U.S. DOT Regulated Marine Pollutant (49CFR172.101, Appendix B)
- Canada, WHMIS, Ingredients Disclosure List
- Canada: Priority Substance List & Restricted Substances/ Ocean Dumping Forbidden (CEPA), National Pollutant Release Inventory (NPRI) (arsenic compounds)

Cited in U.S. State Regulations: California (A, G), Kansas (G), Louisiana (G), Maine (G), Maryland (G), Massachusetts (G), New Hampshire (G), New Jersey (G), Oklahoma (G), Pennsylvania (G), Vermont (G), Virginia (G), Washington (G), West Virginia (G), Wisconsin (G).

Description: Arsenic acid, H_3AsO_4, is an odorless, noncombustible, white semi-transparent crystalline material or in a commercial grade that is a pale yellow syrup-like liquid. Melting/Freezing/Melting point = 36°C. Hazard Identification (based on NFPA-704 M Rating System): Health 3, Flammability 0, Reactivity 0. It converts to As_2O_5 (arsenic pentoxide) when heated above 300°C. See also arsenic pentoxide.

Potential Exposure: It is used as a wood treatment, drying agent, soil sterilant and to make other arsenates. It has been used as a cotton defoliant.

Incompatibilities: Incompatible with sulfuric acid, caustics, ammonia, amines, isocyanates, alkylene oxides, oxidizers, epichlorohydrin, vinyl acetate, amides. Avoid contact with chemically active metals. Corrodes brass, mild steel and galvanized steel. Contact with acids or acid mists releases deadly arsine gas.

Permissible Exposure Limits in Air: OSHA: The legal airborne permissible exposure limit (PEL) is 0.01 mg/m^3 averaged over an 8-hour workshift for Arsenic and compounds as Arsenic, inorganic. NIOSH: The recommended airborne exposure limit is 0.002 mg/m^3, which should not be exceeded during any 15 minute work period for Arsenic, inorganic. ACGIH: The recommended airborne exposure limit is 0.2 mg/m^3 average over an 8-hour workshift for Arsenic and soluble compounds. The British HSE[33] has also adapted the ACGIH value of 0.2 mg/m^3 as an 8-hour TWA value. The DFG[3] has not set numerical limits for arsenic in air on the grounds that it is a proven human carcinogen. The former USSR-UNEP/IRPTC project[43] has set a MAC value for inorganic arsenic compounds (except arsine) of 0.003 mg/m^3 for ambient air in residential areas.

Determination in Air: Collection on a filter and analysis by atomic absorption spectrometry. See NIOSH Methods 7900 and 73000, Elements.[18] See also OSHA Method ID 105.

Permissible Concentration in Water: To protect freshwater aquatic life-total recoverable trivalent inorganic arsenic never to exceed 440 μg/l. To protect saltwater aquatic life-508 μg/l on an acute basis. To protect human health-preferably zero. A value of 0.02 μg/l corresponds to a human health risk of 1 in 100,000. EPA has established a maximum arsenic level of 0.05 mg/l. This does not address carcinogenicity and is under review. The former USSR-UNEP/IRPTC project[43] has set MAC values for inorganic arsenic compounds in water for domestic purposes at 0.05 mg/l and in water bodies for fishery purposes of 0.5 mg/l also.

Determination in Water: The atomic absorption graphite furnance technique is often used for measurement of total arsenic in water. It also has been standardized by EPA. Total arsenic may be determined by digestion followed by silver diethyldithiocarbamate; an alternative is atomic absorption; another is inductively coupled plasma (ICP) optical emission spectrometry.

Routes of Entry: Inhalation, ingestion, skin contact.

Harmful Effects and Symptoms

Short Term Exposure: Skin contact can cause irritation, itching, burning sensation, and rash. Eye contact can cause irritation and burns. Inhalation can cause irritation of the respiratory tract. High exposure can cause poor appetite, nausea, vomiting and muscle cramps. High exposure can cause nerve damage with numbness, "pins and needles" sensation, weakness of the arms and legs. Arsine, a very deadly gas is released in the presence of acid or acid mist. The oral LD_{50} for rat is 48 mg/kg.[9] Ingestion of 130 mg of arsenic may be fatal to humans. Smaller doses may become fatal since arsenic accumulates in the body.

Long Term Exposure: Arsenic acid is a mutagen that may cause changed to genetic material and an animal teratogen. Can cause an ulcer of the "bone" dividing the inner nose. It can cause nerve damage, thickening of the skin with patch areas of darkening and loss of pigment, or the development of white lines in the nails.

Points of Attack: Liver, kidneys, skin, lungs, nervous system, lymphatic system.

Medical Surveillance: Examination of the nose, skin, eyes, nails, and nervous system. Test for urine arsenic. At NIOSH recommended exposure limits, uring arsenic should not be greater than 50 – 100 micrograms per liter of urine. See also entry for Arsenic.

First Aid: If this chemical gets into the eyes, remove any contact lenses at once and irrigate immediately for at least 15 minutes, occasionally lifting upper and lower lids. Seek medical attention immediately. If this chemical contacts the skin, remove contaminated clothing and wash immediately with soap and water. Seek medical attention immediately. If this chemical has been inhaled, remove from exposure, begin rescue breathing (using universal precautions) if breathing has stopped and CPR if heart action has stopped. Transfer promptly to a medical facility. When this chemical has been swallowed, get medical attention. Give large quantities of water and induce vomiting. Do not make an unconscious person vomit.

Antidotes and Special Procedures: For severe poisoning BAL has been used. For milder poisoning *penicillamine (not penicillin)* has been used, both with mixed success. Side effects occur with such treatment and it is never a substitute for controlling exposure. It can only be done under strict medical care.

Personal Protective Methods: Where possible, enclose operations and use local exhaust ventilation at the site of chemical release. If local exhaust ventilation or enclosure is not used, respirators should be worn. A regulated, marked area should be established where Arsenic Acid is handled, used, or stored. Wear protective gloves and clothing to prevent any reasonable probability of skin contact. Safety equipment suppliers/manufacturers can provide recommendations on the most protective glove/clothing material for your operation. All protective clothing (suits, gloves, footwear, headgear) should be clean, available each day, and put on before work. Contact lenses should not be worn when working with this chemical. Wear full facepiece respiratory. Employees should wash immediately with soap when skin is wet or contaminated. Provide emergency showers and eyewash. Specific engineering controls are required under OSHA 1910.1018, *Inorganic Arsenic.* See also NIOSH Criteria Document #75–149, *"Inorganic Arsenic."*

Respirator Selection: *At any concentrations above the NIOSH REL at any detectable concentration:* SCBAF: PD,PP (any self-contained breathing apparatus that has a full faceplate and is operated in a pressure-demand or other positive-pressure mode); or SAF:PD,PP:ASCBA (any supplied-air respirator that has a full facepiece and is operated in a pressure-demand or other positive-pressure mode in combination with an auxiliary self-contained breathing apparatus operated in a pressure-demand or other positive-pressure mode). *Escape:* GMFAGHiE [any air-purifying, full-facepiece respirator (gas mask) with a chin-style, front-or back-mounted acid gas canister having a high-efficiency particulate filter]; or SCBAE (any appropriate escape-type, self-contained breathing apparatus).

Storage: Prior to working with this chemical you should be trained on its proper handling and storage. Arsenic Acid must be stored to avoid contact with heat and chemically active metals (such as Potassium, Sodium, Magnesium and Zinc) since violent reactions occur. Store in tightly closed containers in a cool, well-ventilated area away from heat. A regulated, marked area should be established where this chemical is handled, used, or stored in compliance with OSHA standard 1910.1045.

Shipping: The DOT label requirement is "Poison." The UN/DOT Hazard Class is 6.1 and the Shipping Group is II for solid and I for the liquid form.[19][20]

Spill Handling: Evacuate persons not wearing protective equipment from area of spill or leak until clean-up is complete. Remove all ignition sources. Neutralize spilled material with crushed limestone, soda ash or lime. Collect powdered material in the most convenient and safe manner and deposit in sealed containers. Ventilate area and wash spill site after clean-up is complete. It may be necessary to contain and dispose of this chemical as a hazardous waste. If material or contaminated runoff enters waterways, notify downstream users of potentially contaminated waters. Contact your Department of Environmental Protection or your regional office of the federal EPA for specific recommendations. If employees are required to clean-up spills, they must be properly trained and equipped. OSHA 1910.120(q) may be applicable.

Fire Extinguishing: Extinguish fire using an agent suitable for type of surrounding fire. Arsenic Acid itself does not burn. Poisonous gases are produced in fire including arsine and oxides of arsenic. If material or contaminated runoff enters waterways, notify downstream users of potentially contaminated waters. Notify local health and fire officials and pollution control agencies. From a secure, explosion-proof location, use water spray to cool exposed containers. If cooling streams are ineffective (venting sound increases in volume and pitch, tank discolors, or shows any signs of deforming), withdraw immediately to a secure position. If employees are expected to fight fires, they must be trained and equipped in OSHA 1910.156.

Disposal Method Suggested: Dissolve in a minimum of concentrated hydrochloric acid. Dilute with water until white precipitate forms. Add HCl to dissolve. Saturate with H_2S; filter and wash precipitate and return to supplier. Alternatively, precipitate with heavy metals such as lime or ferric hydroxide in lieu of H_2S.[22] Consult with environmental regulatory agencies for guidance on acceptable disposal practices. Generators of waste containing this contaminant (≥100 kg/mo) must conform with EPA regulations governing storage, transportation, treatment, and waste disposal.

References

Sax, N. I., Ed., "Dangerous Properties of Industrial Materials Report" 2, No. 3, 56–59 (1982)

New Jersey Department of Health and Senior Services, "Hazardous Substance Fact Sheet: Arsenic Acid," Trenton, NJ (April 1996)

Arsenic Pentoxide

Molecular Formula: As_2O_5

Synonyms: Anhydride arsenique (French); Arsenic acid anhydride; Arsenic anhydride; Arsenic oxide; Arsenic(V) oxide; Arsenic pentaoxide; Diarsenic pentoxide; Fotox; Peroxido de arsenico (Spanish)

CAS Registry Number: 1303-28-2

RTECS Number: CG2275000

DOT ID: UN 1559

Regulatory Authority

- See also Arsenic and Inorganic Compounds
- Banned or Severely Restricted (In Agricultural, Pharmaceutical and Industrial Chemicals) (Many Countries)[13][35]
- OSHA, 29CFR1910 Specifically Regulated Chemicals (See CFR1910.1018)
- Very Toxic Substance (World Bank)[15]
- Carcinogen (Human Positive) (IARC)[9] (DFG)[3] (NTP)
- Banned or Severely Restricted (In Agricultural, Pharmaceutical and Industrial Chemicals) (Many Countries)[13][35]
- Air Pollutant Standard Set (ACGIH)[1] (OSHA/NIOSH)[2] (HSE)[33] (former USSR)[43] (Several States)[60] (Australia) (Israel) (Mexico) (Several Canadian Provinces)
- Clean Air Act, 42USC7412; Title I, Part A, §112 hazardous pollutants
- Clean Water Act 40CFR401.15 Section 307Toxic Pollutants; 40CFR423, Appendix A Priority Pollutants; §313 Priority Chemicals
- RCRA 40CFR261.24 Toxicity Characteristics, Maximum Concentration of Contaminants (MCC), Regulatory level, 5.0 mg/l
- RCRA, 40CFR261, Appendix 8 Hazardous Constituents., waste number P011
- CERCLA/SARA 40CFR302 Extremely Hazardous Substances: TPQ = 100/10,000 lb (454/4,540 kg)
- SUPERFUND/EPCRA 40CFR302.4 Reportable Quantity (RQ): CERCLA, 1 lb (0.454 kg)
- EPCRA Section 313: Form R *de minimis* concentration reporting level: 0.1%

- U.S. DOT Regulated Marine Pollutant (49CFR172.101, Appendix B)
- Canada, WHMIS, Ingredients Disclosure List
- Canada: Priority Substance List & Restricted Substances/ Ocean Dumping Forbidden (CEPA), National Pollutant Release Inventory (NPRI) (arsenic compounds)

Cited in U.S. State Regulations: California (A, G), Connecticut (A), Florida (G), Illinois (G), Kansas (G), Louisiana (G), Maine (G, W), Maryland (G), Massachusetts (G), Michigan (G), Montana (A), Nevada (A), New Hampshire (G), New Jersey (G), New York (G, A), North Carolina (A), North Dakota (A), Oklahoma (G), Pennsylvania (G, A), Rhode Island (G, A), South Carolina (A), South Dakota (A, W), Tennessee (W), Utah (W), Vermont (G), Virginia (G, A), Washington (G), West Virginia (G), Wisconsin (G).

Description: Arsenic Pentoxide, As_2O_5, is an odorless white lumpy solid pr powder and non-flammable. Freezing/Melting point = 315°C (decomposes). Hazard Identification (based on NFPA-704 M Rating System): Health 3, Flammability 0, Reactivity 0. Highly soluble in water.

Potential Exposure: This material is used as a chemical intermediate, as an herbicide, and as an ingredient in wood preservatives and in glass. Other possible uses are as an insecticide and soil strerilant.

Incompatibilities: Chemically active metals such as aluminum and zinc. Incompatible with acids, strong alkalis, halogens, rubidium carbide, zinc. Corrosive to metals in the presence of moisture. Contact with acids or acid mists releases deadly arsine gas.

Permissible Exposure Limits in Air: OSHA: The legal airborne permissible exposure limit (PEL) is 0.01 mg/m^3 averaged over an 8-hour workshift for Arsenic and compounds as Arsenic, inorganic. NIOSH: The recommended airborne exposure limit is 0.002 mg/m^3, which should not be exceeded during any 15 minute work period for Arsenic, inorganic. ACGIH: The recommended airborne exposure limit is 0.2 mg/m^3 average over an 8-hour workshift for Arsenic and soluble compounds. The British HSE[33] has also adapted the ACGIH value of 0.2 mg/m^3 as an 8-hour TWA value. The DFG[3] has not set numerical limits for arsenic in air on the grounds that it is a proven human carcinogen. The former USSR-UNEP/IRPTC project[43] has set a MAC value for inorganic arsenic compounds (except arsine) of 0.003 mg/m^3 for ambient air in residential areas. In addition, several states have set specific guidelines or standards for arsenic pentoxide in ambient air[60] ranging from zero (New York) to 0.0002 μg/m^3 (North Carolina) to 1.0 μg/m^3 (South Carolina).

Determination in Air: Collection on a filter and analysis by atomic absorption spectrometry. See NIOSH Methods 7900 and 73000, Elements.[18] See also OSHA Method ID 105.

Permissible Concentration in Water: To protect freshwater aquatic life-total recoverable trivalent inorganic arsenic never to exceed 440 μg/l. To protect saltwater aquatic life-508 μg/l on an acute basis. To protect human health-preferably zero. A value of 0.02 μg/l corresponds to a human health risk of 1 in 100,000. EPA has established a maximum arsenic level of 0.05 mg/l. This does not address carcinogenicity and is under review. The former USSR-UNEP/IRPTC project[43] has set MAC values for inorganic arsenic compounds in water for domestic purposes at 0.05 mg/l and in water bodies for fishery purposes of 0.5 mg/l also.

Determination in Water: The atomic absorption graphite furnance technique is often used for measurement of total arsenic in water. It also has been standardized by EPA. Total arsenic may be determined by digestion followed by silver diethyldithiocarbamate; an alternative is atomic absorption; another is inductively coupled plasma (ICP) optical emission spectrometry.

Routes of Entry: Inhalation, ingestion, skin contact.

Harmful Effects and Symptoms

Short Term Exposure: It is irritating to eyes, nose, and respiratory system. This chemical can be absorbed through the skin, thereby increasing exposure. Skin contact can cause irritation, burning, itching, and a rash. Symptoms usually appear ½ to 1 hour after ingestion, but may be delayed. Symptoms include a sweetish, metallic taste and garlicky odor of breath; difficulty in swallowing; abdominal pain; vomiting and diarrhea; dehydration; feeble heart beat; dizziness and headache; and eventually coma, sometimes convulsions, general paralysis, and death. The oral LD_{50} for rat is 8 mg/kg.[9] This material is extremely toxic; the probable oral lethal dose for humans is 5 – 50 mg/kg, or between 7 drops and 1 teaspoonful for a 150-lb person.

Long Term Exposure: Arsenic pentoxide is a carcinogen in humans. It has been shown to cause skin cancer. May damage the male reproductive glands. Chronic exposure may cause nerve damage to the extremities, alter cellular composition of the blood, and cause structural changes in blood components. Repeated exposure can cause an ulcer in the "bone" dividing the inner nose. Long term skin contact can cause thickened skin and pigmentation changes. Some persons develop white lines in the finger nails.

Points of Attack: Liver, kidneys, skin, respiratory system, lymphatic system.

Medical Surveillance: See entry under Arsenic and Inorganic compounds.

First Aid: If this chemical gets into the eyes, remove any contact lenses at once and irrigate immediately for at least 15 minutes, occasionally lifting upper and lower lids. Seek medical attention immediately. If this chemical contacts the skin, remove contaminated clothing and wash immediately with soap and water. Seek medical attention immediately. If this chemical has been inhaled, remove from exposure, begin rescue breathing (using universal precautions) if breathing has

stopped and CPR if heart action has stopped. Transfer promptly to a medical facility. When this chemical has been swallowed, get medical attention. Give large quantities of water and induce vomiting. Do not make an unconscious person vomit.

Personal Protective Methods: Reduce contact to lowest possible level. Wear protective gloves and clothing to prevent any reasonable probability of skin contact. Safety equipment suppliers/manufacturers can provide recommendations on the most protective glove/clothing material for your operation. All protective clothing (suits, gloves, footwear, headgear) should be clean, available each day, and put on before work. Contact lenses should not be worn when working with this chemical. Wear full facepiece respiratory protection. Employees should wash immediately with soap when skin is wet or contaminated. Provide emergency showers and eyewash. Wash thoroughly immediately after exposure to Arsenic Pentoxide and at the end of the workshift. Specific engineering controls are required under OSHA 1910.1018, *Inorganic Arsenic.* See also NIOSH Criteria Document #75–149, *"Inorganic Arsenic."*

Respirator Selection: *At any concentrations above the NIOSH REL at any detectable concentration:* SCBAF:PD,PP (any self-contained breathing apparatus that has a full faceplate and is operated in a pressure-demand or other positive-pressure mode); or SAF:PD,PP:ASCBA (any supplied-air respirator that has a full facepiece and is operated in a pressure-demand or other positive-pressure mode in combination with an auxiliary self-contained breathing apparatus operated in a pressure-demand or other positive-pressure mode). *Escape:* GMFAGHiE [any air-purifying, full-facepiece respirator (gas mask) with a chin-style, front-or back-mounted acid gas canister having a high-efficiency particulate filter]; or SCBAE (any appropriate escape-type, self-contained breathing apparatus).

Storage: Prior to working with this chemical you should be trained on its proper handling and storage. Store in tightly closed containers in a cool, dry, well-ventilated area away from metals, acids and other incompatible materials. A regulated, marked area should be established where this chemical is handled, used, or stored in compliance with OSHA standard 1910.1045.

Shipping: The DOT label requirement is "Poison." The UN/DOT Hazard Class is 6.1, the Packing Group is II.[19][20]

Spill Handling: Evacuate persons not wearing protective equipment from area of spill or leak until clean-up is complete. Remove all ignition sources. Stay upwind; keep out of low areas. Wear self-contained (positive pressure if available) breathing apparatus and full protective clothing. Do not touch spilled material. Neutralize spilled material with crushed limestone, soda ash or lime. Absorb small liquid spills with sand or other noncombustible absorbent material and place into containers for later disposal. For large spills, dike far ahead of spill for later disposal. Ventilate area and wash spill site after clean-up is complete. It may be necessary to contain and dispose of this chemical as a hazardous waste. If material or contaminated runoff enters waterways, notify downstream users of potentially contaminated waters. Contact your Department of Environmental Protection or your regional office of the federal EPA for specific recommendations. If employees are required to clean-up spills, they must be properly trained and equipped. OSHA 1910.120(q) may be applicable.

Fire Extinguishing: As_2O_5 may burn but does not readily ignite. Poisonous gases including arsenic fumes are produced in fire. Use dry chemical, carbon dioxide, or alcohol foam extinguishers. If material or contaminated runoff enters waterways, notify downstream users of potentially contaminated waters. Notify local health and fire officials and pollution control agencies. From a secure, explosion-proof location, use water spray to cool exposed containers. If cooling streams are ineffective (venting sound increases in volume and pitch, tank discolors, or shows any signs of deforming), withdraw immediately to a secure position. If employees are expected to fight fires, they must be trained and equipped in OSHA 1910.156.

Disposal Method Suggested: Dissolve in a minimum of concentrated hydrochloric acid. Dilute with water until white precipitate forms. Add HCl to dissolve. Saturate with H_2S; filter and wash precipitate and return to supplier. Alternatively, precipitate with heavy metals such as lime or ferric hydroxide in lieu of H_2S.[22] If needed, seek professional environmental engineering assistance from the U.S. EPA Environmental Response Team at (908) 548-8730 (24-hour response line). Consult with environmental regulatory agencies for guidance on acceptable disposal practices. Generators of waste containing this contaminant (≥100 kg/mo) must conform with EPA regulations governing storage, transportation, treatment, and waste disposal. In accordance with 40CFR165 recommendations for the disposal of pesticides and pesticide containers. Must be disposed properly by following package label directions or by contacting your state pesticide or environmental control agency or by contacting your regional EPA office.

References

Sax, N. I., Ed., "Dangerous Properties of Industrial Materials Report" 2, No. 3, 59–61 (1982) and 8, No. 3, 45–55 (1988)

U.S. Environmental Protection Agency, "Chemical Profile: Arsenic Pentoxide," Washington, DC, Chemical Emergency Preparedness Program (November 30, 1987)

New Jersey Department of Health and Senior Services, "Hazardous Substance Fact Sheet: Arsenic Pentoxide," Trenton, NJ (January 1996)

Arsenic Trisulfide

Molecular Formula: As_2S_3

Synonyms: Arsenic sesquisulfide; Arsenic sulfide; Arsenic sulfide yellow; Arsenic tersulfide; Arsenic yellow; Arsenous sulfide; Auripigment; C.I. 77086; C.I. pigment yellow; Diarsenic trisulfide; King's gold; King's yellow; Orpiment; STCC 4923222; Trisulfuro de arsenico (Spanish); Yellow arsenic sulfide

CAS Registry Number: 1303-33-9

RTECS Number: CG2638000

DOT ID: NA 1557

Regulatory Authority

- Carcinogen (Human Positive) (IARC)[9] (DFG)[3] (NTP)
- OSHA, 29CFR1910 Specifically Regulated Chemicals (See CFR1910.1018)
- Banned or Severely Restricted (In Agricultural, Pharmaceutical and Industrial Chemicals) (Many Countries)[13][35]
- Air Pollutant Standard Set (ACGIH)[1] (OSHA/NIOSH)[2] (HSE)[33] (former USSR)[43] (Several States)[60] (Australia) (Israel) (Mexico) (Several Canadian Provinces)
- Clean Air Act, 42USC7412; Title I, Part A, §112 hazardous pollutants
- Clean Water Act 40CFR401.15 Section 307 Toxic Pollutants; 40CFR423, Appendix A Priority Pollutants; §313 Priority Chemicals
- RCRA, 40CFR261, Appendix 8 Hazardous Constituents., waste number not listed
- Safe Drinking Water Act 47FR9352 Regulated chemical: MCL, 0.05 mg/l (Section 141.11) applies only to community water systems
- SUPERFUND/EPCRA 40CFR302.4 Reportable Quantity (RQ): CERCLA, 1 lb (0.454 kg)
- EPCRA Section 313: Form R *de minimis* concentration reporting level: 0.1%
- U.S. DOT Regulated Marine Pollutant (49CFR172.101, Appendix B)
- Canada, WHMIS, Ingredients Disclosure List
- Canada: Priority Substance List & Restricted Substances/ Ocean Dumping Forbidden (CEPA), National Pollutant Release Inventory (NPRI) (arsenic compounds)

Cited in U.S. State Regulations: California (A, G), Connecticut (A), Florida (G), Illinois (G), Kansas (G), Louisiana (G), Maine (G, W), Maryland (G), Massachusetts (G), Michigan (G), Montana (A), Nevada (A), New Hampshire (G), New Jersey (G), New York (G, A), North Carolina (A), North Dakota (A), Oklahoma (G), Pennsylvania (G, A), Rhode Island (G, A), South Carolina (A), South Dakota (A, W), Tennessee (W), Utah (W), Vermont (G), Virginia (G, A), Washington (G), West Virginia (G), Wisconsin (G).

Description: Arsenic trisulfide, As_2S_3, is noncombustible, odorless, yellow or orange powder or red needles (changes at 170°C). Boiling point = 707°C. Freezing/Melting point = 300 – 327°C. Hazard Identification (based on NFPA-704 M Rating System): Health 3, Flammability 0, Reactivity 0. Insoluble in water.

Potential Exposure: Arsenic trisulfide is used in the manufacture of glass, oil cloth, linoleum, electrical semi-conductors, fireworks and used as a pigment.

Incompatibilities: Avoid contact with Oxidizers (such as Perchlorates, Peroxides, Permanganates, Chlorates, and Nitrates) and Potassium Nitrate mixed with Sulfur since violent reactions occur. Water contact forms hydrogen sulfide. Incompatible with acids, halogens. Contact with acids or acid mists releases deadly arsine gas.

Permissible Exposure Limits in Air: OSHA: The legal airborne permissible exposure limit (PEL) is 0.01 mg/m^3 averaged over an 8-hour workshift for Arsenic and compounds as Arsenic, inorganic. NIOSH: The recommended airborne exposure limit is 0.002 mg/m^3, which should not be exceeded during any 15 minute work period for Arsenic, inorganic. ACGIH: The recommended airborne exposure limit is 0.2 mg/m^3 average over an 8-hour workshift for Arsenic and soluble compounds. The British HSE[33] has also adapted the ACGIH value of 0.2 mg/m^3 as an 8-hour TWA value. The DFG[3] has not set numerical limits for arsenic in air on the grounds that it is a proven human carcinogen. The former USSR-UNEP/IRPTC project[43] has set a MAC value for inorganic arsenic compounds (except arsine) of 0.003 mg/m^3 for ambient air in residential areas. The above exposure limits are for air levels only. When skin contact also occurs, you may be overexposed, even though air levels are less than the limits listed above.

Determination in Air: The American Conference of Government Industrial Hygienists (ACGIH) Method 803 measures total particulate arsenic in air. The method involves filter collection of air samples, arsine generation, and silver diethyldithiocarbamate (SDDC) colorimetry. The most important interference is hydrogen sulfide removed by a lead acetate trap. High levels of antimony may result in stibine formation, which also interferes.

Permissible Concentration in Water: For the maximum protection of human health from the potential carcinogenic effects of exposure to arsenic through ingestion of water and contaminated aquatic organisms, the ambient water concentration is zero. Concentrations of arsenic estimated to result in additional lifetime cancer risks ranging from no additional risk to an additional risk of 1 - 100,000 are presented in the Criterion Formulation section of this document. The EPA is considering setting criteria at an interim target risk level in the range of 10^{-5}, 10^{-6}, or 10^{-7} with corresponding criteria of 0.02, 0.002, and 0.0002 μg/l, respectively.[6]

Determination in Water: The atomic absorption graphite furnance technique is often used for measurement of total arsenic in water. It also has been standardized by EPA. Total arsenic may be determined by digestion followed by silver diethyldithiocarbamate; an alternative is atomic absorption; another is inductively coupled plasma (ICP) optical emission spectrometry.

Routes of Entry: Inhalation of dust, skin contact, ingestion.

Harmful Effects and Symptoms

Short Term Exposure: Arsenic trisulfide can affect you when breathed in and may enter the body thought the skin. Arsenic Trisulfide is a carcinogen, handle with extreme caution. Contact can cause burning, itching, thickened skin, rash and color changes. Exposure can irritate the nose and throat, and cause an ulcer or hole in the inner nose. High or repeated exposures can cause nerve damage, "pins and needles," numbness and weakness of arms and legs. High or repeated exposure can cause poor appetite, nausea, vomiting, diarrhea and death.

Long Term Exposure: Arsenic trisulfide is a carcinogen in humans; it has been shown to cause skin, liver and lung cancer. It may be a teratogen, causing reproductive damage, such as reduced fertility and interference with the menstrual cycles. May cause liver damage and lower the red blood cell count. Repeated skin contact can cause thickened skin and pigmentation changes. Some persons develop white lines on the finger nails. High or repeated exposure can cause nerve damage with burning, "pins and needles" sensation and weakness of the extremities.

Points of Attack: Skin, respiratory system, nervous system.

Medical Surveillance: Before beginning employment and at regular times after that, the following are recommended: Exam of the nose, skin, eyes, nails and nervous system. Complete blood count (CBC). Examination of the nervous system. Test for urine Arsenic (may not be accurate within 2 days of eating shellfish or fish; most accurate at the end of a workday). At NIOSH recommended exposure levels, urine Arsenic should not be grater than 50 - 100 micrograms per liter of urine. After suspected overexposure, repeat these tests. Also examine your skin periodically for abnormal growths. Skin cancer from Arsenic is easily cured with early detection.

First Aid: If this chemical gets into the eyes, remove any contact lenses at once and irrigate immediately for at least 15 minutes, occasionally lifting upper and lower lids. Seek medical attention immediately. If this chemical contacts the skin, remove contaminated clothing and wash immediately with soap and water. Seek medical attention immediately. If this chemical has been inhaled, remove from exposure, begin rescue breathing (using universal precautions) if breathing has stopped and CPR if heart action has stopped. Transfer promptly to a medical facility. When this chemical has been swallowed, get medical attention. Give large quantities of water and induce vomiting. Do not make an unconscious person vomit.

Antidotes and Special Procedures: For severe poisoning BAL have been used. For milder poisoning, *penicillamine (not penicillin)* has been used, both with mixed success. Side effects occur with such treatment and it is never a substitute for controlling exposure. It can only be done under strict medical care.

Personal Protective Methods: Avoid skin contact with Arsenic Trisulfide. Wear protective gloves and clothing. Safety equipment suppliers/manufacturers can provide recommendations on the most protective glove/clothing material for your operations. All protective clothing (suits, gloves, footwear, headgear) should be clean, available each day, and put on before work. Eye protection is included in the recommended respiratory protection. Specific engineering controls are required under OSHA 1910.1018, *Inorganic Arsenic.* See also NIOSH Criteria Document #75–149, *"Inorganic Arsenic."*

Respirator Selection: *At any concentrations above the NIOSH REL at any detectable concentration:* SCBAF:PD,PP (any self-contained breathing apparatus that has a full faceplate and is operated in a pressure-demand or other positive-pressure mode); or SAF:PD,PP:ASCBA (any supplied-air respirator that has a full facepiece and is operated in a pressure-demand or other positive-pressure mode in combination with an auxiliary self-contained breathing apparatus operated in a pressure-demand or other positive-pressure mode). *Escape:* GMFAGHiE [any air-purifying, full-facepiece respirator (gas mask) with a chin-style, front-or back-mounted acid gas canister having a high-efficiency particulate filter]; or SCBAE (any appropriate escape-type, self-contained breathing apparatus).

Storage: Prior to working with this chemical you should be trained on its proper handling and storage. Avoid contact with the incompatible materials cited above. A regulated, marked area should be established where this chemical is handled, used, or stored in compliance with OSHA standard 1910.1045.

Shipping: Arsenic sulfide requires a "Poison" label. It falls in DOT Hazard Class 6.1 and Packing Group II.

Spill Handling: Evacuate persons not wearing protective equipment from area of spill or leak until clean-up is complete. Remove all ignition sources. Keep water and acids away from spilled Arsenic Trisulfide. Collect powdered materials using a vacuum equipped with a high efficiency particulate filter (do not use a standard shop vacuum) or wet cleaning methods and deposit in sealed containers. Ventilate area after clean-up is complete. It may be necessary to contain and dispose of this chemical as a hazardous waste. If material or contaminated runoff enters waterways, notify downstream users of potentially contaminated waters. Contact your Department of Environmental Protection or your regional office of the federal EPA for specific recommendations. If employees are required to clean-up spills, they must be properly trained and equipped. OSHA 1910.120(q) may be applicable.

Fire Extinguishing: Arsenic Trisulfide may burn, but does not readily ignite. Use dry chemical, CO_2, water spray, or foam extinguishers. Poisonous gases and fumes are produced in a fire, including hydrogen sulfide, arsine, sulfur oxides and arsenic fumes. If material or contaminated runoff enters waterways, notify downstream users of potentially contaminated waters. Notify local health and fire officials and pollution control agencies. From a secure, explosion-proof location, use water spray to cool exposed containers. If cooling streams are ineffective (venting sound increases in volume and pitch, tank discolors, or shows any signs of deforming), withdraw immediately to a secure position. If employees are expected to fight fires, they must be trained and equipped in OSHA 1910.156.

Disposal Method Suggested: Consult with environmental regulatory agencies for guidance on acceptable disposal practices. Generators of waste containing this contaminant (≥100 kg/mo) must conform with EPA regulations governing storage, transportation, treatment, and waste disposal.

References

New Jersey Department of Health and Senior Services, "Hazardous Substance Fact Sheet: Arsenic Trisulfide," Trenton, NJ (December 1998)

Arsenous Oxide

Molecular Formula: As_2O_3

Synonyms: Acide arsenieux (French); Anhydride arsenieux (French); Arsenic blanc (French); Arsenic(III) oxide; Arsenic sesquioxide; Arsenic trioxide; Arsenic trioxide, solid; Arsenicum album; Arsenigen saure (German); Arsenious acid; Arsenious oxide; Arsenious trioxide; Arsenite; Arsenolite; Arsenous acid; Arsenous acid anhydride; Arsenous anhydride; Arsenous oxide; Arsenous oxide anhydride; Arsodent; Claudelite; Claudetite; Crude arsenic; Diarsenic trioxide; Spinrite arsenic; Trioxido de arsenico (Spanish); White arsenic

CAS Registry Number: 1327-53-3

RTECS Number: CG3325000

DOT ID: UN 1561

Regulatory Authority

- See also Arsenic and Inorganic Arsenic Compounds entry
- Banned or Severely Restricted
- Carcinogen (Human Positive) (IARC)[9] (DFG)[3] (NTP)
- OSHA, 29CFR1910 Specifically Regulated Chemicals (See CFR1910.1018)
- Banned or Severely Restricted (In Agricultural, Pharmaceutical and Industrial Chemicals) (Many Countries)[13][35]
- Air Pollutant Standard Set (ACGIH)[1] (OSHA/NIOSH)[2] (HSE)[33] (former USSR)[43] (Several States)[60] (Australia) (Israel) (Mexico) (Several Canadian Provinces)
- Clean Air Act, 42USC7412; Title I, Part A, §112 hazardous pollutants
- Clean Water Act 40CFR401.15 Section 307 Toxic Pollutants; 40CFR423, Appendix A Priority Pollutants; §313 Priority Chemicals
- RCRA 40CFR261.24 Toxicity Characteristics, Maximum Concentration of Contaminants (MCC), Regulatory level, 5.0 mg/l
- RCRA, 40CFR261, Appendix 8, Appendix 8, Hazardous Constituents, waste number P 012
- RCRA Land Ban Waste
- Safe Drinking Water Act 47FR9352 Regulated chemical: MCL, 0.05 mg/l (Section 141.11) applies only to community water systems
- CERCLA/SARA 40CFR302 Extremely Hazardous Substances: TPQ = 100/10,000 lb (45.4/4,540 kg)
- SUPERFUND/EPCRA 40CFR302.4 Reportable Quantity (RQ): CERCLA, 1 lb (0.454 kg)
- EPCRA Section 313: Form R *de minimis* concentration reporting level: 0.1%
- U.S. DOT Regulated Marine Pollutant (49CFR172.101, Appendix B)
- Canada, WHMIS, Ingredients Disclosure List
- Canada: Priority Substance List & Restricted Substances/ Ocean Dumping Forbidden (CEPA), National Pollutant Release Inventory (NPRI) (arsenic compounds)

Cited in U.S. State Regulations: California (A, G), Connecticut (A), Florida (G), Illinois (G), Kansas (G), Louisiana (G), Maine (G, W), Maryland (G), Massachusetts (G), Michigan (G), Minnesota (G), Montana (A), Nevada (A), New Hampshire (G), New Jersey (G), New York (G, A), North Carolina (A), North Dakota (A), Oklahoma (G), Pennsylvania (G, A), Rhode Island (G, A), South Carolina (A), South Dakota (A, W), Tennessee (W), Utah (W), Vermont (G), Virginia (G, A), Washington (G), West Virginia (G), Wisconsin (G).

Description: Arsenic Trioxide, As_2O_3, is a noncombustible, odorless, white powder or colorless crystalline solid. Boiling point = 460°- 465°C. Freezing/Melting point = 312°C (sublimes at 193°C). Hazard Identification (based on NFPA-704 M Rating System): Health 3, Flammability 0, Reactivity 0. Slightly soluble in water.

Potential Exposure: This is a primary raw material for other arsenic compounds. It is an intermediate for insecticides, herbicides and fungicides. The material is used as a wood and tanning preservative and a decoloring and refining agent in glass manufacture. It is also used in pharmaceuticals and in the purification of synthetic gas.

Incompatibilities: Sodium chlorate; sodium hydroxide, sulfuric acid, fluorine; chlorine trifluoride; chromic oxide; aluminum chloride; phosphorus pentoxide; hydrogen fluoride; oxygen difluoride; tannic acid; infusion cinchona and other vegetable astringent infusions and decoctions; iron in solution. Contact with acids or acid mists releases deadly arsine gas.

Permissible Exposure Limits in Air: OSHA: The legal airborne permissible exposure limit (PEL) is 0.01 mg/m^3 averaged over an 8-hour workshift for Arsenic and compounds as Arsenic, inorganic. NIOSH: The recommended airborne exposure limit is 0.002 mg/m^3, which should not be exceeded during any 15 minute work period for Arsenic, inorganic. ACGIH: The recommended airborne exposure limit is 0.2 mg/m^3 average over an 8-hour workshift for Arsenic and soluble compounds. The British HSE[33] has also adapted the ACGIH value of 0.2 mg/m^3 as an 8-hour TWA value. The DFG[3] has not set numerical limits for arsenic in air on the grounds that it is a proven human carcinogen. The former USSR-UNEP/IRPTC project[43] has set a MAC value for inorganic arsenic compounds (except arsine) of 0.003 mg/m^3 for ambient air in residential areas. In addition, some states have set guidelines or standards for arsenic trioxide in ambient air:[60] zero (New York), 0.0002 μg/m^3 (North Carolina), 3.0 μg/m^3 (Virginia).

Determination in Air: Use NIOSH Method 7901, Arsenic trioxide.

Permissible Concentration in Water: To protect freshwater aquatic life-total recoverable trivalent inorganic arsenic never to exceed 440 μg/l. To protect saltwater aquatic life-508 μg/l on an acute basis. To protect human health-preferably zero. A value of 0.02 μg/l corresponds to a human health risk of 1 in 100,000. EPA has established a maximum arsenic level of 0.05 mg/l. This does not address carcinogenicity and is under review. The former USSR-UNEP/IRPTC project[43] has set MAC values for inorganic arsenic compounds in water for domestic purposes at 0.05 mg/l and in water bodies for fishery purposes of 0.5 mg/l also.

Determination in Water: The atomic absorption graphite furnance technique is often used for measurement of total arsenic in water. It also has been standardized by EPA. Total arsenic may be determined by digestion followed by silver diethyldithiocarbamate; an alternative is atomic absorption; another is inductively coupled plasma (ICP) optical emission spectrometry.

Routes of Entry: Inhalation, skin contact, ingestion.

Harmful Effects and Symptoms

Short Term Exposure: Skin contact can cause burning, itching, and rash. Inhalation can cause respiratory irritation. Eye contact can cause irritation and possible permanent damage. High exposures can cause an abnormal EKG. Symptoms of acute poisoning may take from ½ hour to several hours after ingestion to appear. They may include: sweetish metallic taste; garlicky odor of breath and feces; constriction in throat and difficulty in swallowing; burning and colicky pains in esophagus, stomach and bowel; vomiting and profuse painful diarrhea (stools are watery initially, later becoming bloody); dehydration with intense thirst and muscular cramps; bluing of skin; feeble pulse and cold extremities; vertigo, frontal headache, stupor, delirium and mania (these symptoms may occur without concurrent or preceding gastric symptoms); fainting, coma, convulsions, general paralysis and then death. This material is considered super toxic; probable oral lethal dose (human) is less than 5 mg/kg, i.e., a taste (less than 7 drops) for a 70 kg (150 lb) person. Material causes acute gastrointestinal and central nervous system symptoms.

Long Term Exposure: Arsenic trioxide is a human carcinogen; causes skin and liver cancer. Renal and hepatic damage have been observed. Chronic exposure to material has led to nasal septum perforation, dermatological symptoms (lesions, necrosis, etc.) and an increase in the incidence of lung and lymphatic cancers. Appreciable exposure to respiratory irritant promoters such as metal oxide fumes elicits a carcinogenic response from arsenic trioxide. Repeated or high exposure can cause nerve damage with "pins and needles" sensation, burning, numbness, and weakness in the extremities. Exposure can cause skin allergy to develop.

Points of Attack: Skin, liver, kidneys, lungs, lymphatic system.

Medical Surveillance: See entry under Arsenic and Inorganic Arsenic Compounds.

First Aid: If this chemical gets into the eyes, remove any contact lenses at once and irrigate immediately for at least 30 minutes, occasionally lifting upper and lower lids. Seek medical attention immediately. If this chemical contacts the skin, remove contaminated clothing and wash immediately with large amounts of soap and water. Seek medical attention immediately. If this chemical has been inhaled, remove from exposure, begin rescue breathing (using universal precautions) if breathing has stopped and CPR if heart action has stopped. Transfer promptly to a medical facility. When this chemical has been swallowed, get medical attention.

Antidotes and Special Procedures – For severe poisoning BAL have been used. For milder poisoning, *penicillamine (not penicillin)* has been used, both with mixed success. Side effects occur with such treatment and it is never a substitute for controlling exposure. It can only be done under strict medical care.

Personal Protective Methods: Where possible, enclose operations and use local exhaust ventilation at the site of chemical release. If local exhaust ventilation or enclosure is not used, respirators should be worn. A regulated, marked area should be established where Arsenic Trioxide is handled, used, or stored. Wear protective gloves and clothing to prevent any reasonable probability of skin contact. Safety equipment suppliers/manufacturers can provide recommendations on the most protective glove/clothing material for your operation. All protective clothing (suits, gloves, footwear, headgear) should be clean, available each day, and put on before work. Contact lenses should not be worn when working with this chemical. Wear full facepiece respiratory protection. Employees should wash immediately with soap when skin is wet or contaminated. Provide emergency showers and eyewash. Wash thoroughly immediately after exposure to Arsenic Trioxide and at the end of the workshift.

Respirator Selection: *At any concentrations above the NIOSH REL at any detectable concentration:* SCBAF:PD,PP (any self-contained breathing apparatus that has a full faceplate and is operated in a pressure-demand or other positive-pressure mode); or SAF:PD,PP:ASCBA (any supplied-air respirator that has a full facepiece and is operated in a pressure-demand or other positive-pressure mode in combination with an auxiliary self-contained breathing apparatus operated in a pressure-demand or other positive-pressure mode). *Escape:* GMFAGHiE [any air-purifying, full-facepiece respirator (gas mask) with a chin-style, front-or back-mounted acid gas canister having a high-efficiency particulate filter]; or SCBAE (any appropriate escape-type, self-contained breathing apparatus).

Storage: Prior to working with this chemical you should be trained on its proper handling and storage Store in tightly closed containers in a cool, dry well-ventilated area away from contact with incompatible materials. A regulated, marked area should be established where this chemical is handled, used, or stored in compliance with OSHA standard 1910.1045.

Shipping: The DOT label required is "Poison." The UN/DOT Hazard Class is 6.1 and the Packing Group is II.[19][20]

Spill Handling: Evacuate persons not wearing protective equipment from area of spill or leak until clean-up is complete. Remove all ignition sources. Avoid bodily contact with the material. Do not handle broken packages without protective equipment. Wash away any material which may have contacted the body with copious amounts of water or soap and water. Wear full protective clothing including gloves and eye protection. Collect powdered material in the most convenient and safe manner and deposit in sealed containers. Ventilate area after clean-up is complete. It may be necessary to contain and dispose of this chemical as a hazardous waste. If material or contaminated runoff enters waterways, notify downstream users of potentially contaminated waters. Contact your Department of Environmental Protection or your regional office of the federal EPA for specific recommendations. If employees are required to clean-up spills, they must be properly trained and equipped. OSHA 1910.120(q) may be applicable.

Fire Extinguishing: Avoid breathing dusts, and fumes from burning materials. Keep upwind. Wear self-contained breathing apparatus. Extinguish fire using agent suitable for type of surrounding fire (material itself does not burn or burns with difficulty). Use water in flooding quantities as fog. Poisonous gases are produced in fire. If material or contaminated runoff enters waterways, notify downstream users of potentially contaminated waters. Notify local health and fire officials and pollution control agencies. Containers may explode in fire. From a secure, explosion-proof location, use water spray to cool exposed containers. If cooling streams are ineffective (venting sound increases in volume and pitch, tank discolors, or shows any signs of deforming), withdraw immediately to a secure position. If employees are expected to fight fires, they must be trained and equipped in OSHA 1910.156.

Disposal Method Suggested: Consult with environmental regulatory agencies for guidance on acceptable disposal practices. Generators of waste containing this contaminant (≥100 kg/mo) must conform with EPA regulations governing storage, transportation, treatment, and waste disposal. Dissolve in a minimum of concentrated hydrochloric acid. Dilute with water until white precipitate forms. Add HCl to dissolve. Saturate with H_2S; filter and wash precipitate and return to supplier. Alternatively, precipitate with heavy metals such as lime or ferric hydroxide in lieu of H_2S.[22] If needed, seek professional environmental engineering assistance from the U.S. EPA Environmental Response Team at (908) 548-8730 (24-hour response line).

References

Sax, N. I., Ed., "Dangerous Properties of Industrial Materials Report" 3, No. 5, 50–58 (1983)

U.S. Environmental Protection Agency, "Chemical Profile: Arsenous Oxide," Washington, DC, Chemical Emergency Preparedness Program (November 30, 1987)

New Jersey Department of Health and Senior Services, "Hazardous Substance Fact Sheet: Arsenic Trioxide," Trenton, NJ (January 1987)

Arsenous Trichloride

Molecular Formula: $AsCl_3$

Synonyms: Arsenic butter; Arsenic chloride; Arsenic(III) chloride; Arsenic trichloride; Arsenous chloride; Arsenous trichloride; Butter of arsenic; Caustic arsenic chloride; Chlorure d'arsenic (French); Fuming liquid arsenic; Trichloroarsine; Trichlorure d'arsenic (French); Tricloruro de arsenico (Spanish)

CAS Registry Number: 7784-34-1

RTECS Number: CG1750000

DOT ID: UN 1560

Regulatory Authority

- See also Arsenic and Inorganic Arsenic Compounds entry
- Banned or Severely Restricted
- Carcinogen (Human Positive) (IARC)[9] (DFG)[3] (NTP)
- Banned or Severely Restricted (In Agricultural, Pharmaceutical and Industrial Chemicals) (Many Countries)[13][35]
- OSHA, 29CFR1910 Specifically Regulated Chemicals (See CFR1910.1018)
- Air Pollutant Standard Set (ACGIH)[1] (OSHA/NIOSH)[2] (HSE)[33] (former USSR)[43] (Several States)[60] (Australia) (Israel) (Mexico) (Several Canadian Provinces)
- Clean Air Act, 42USC7412; Title I, Part A, §112 hazardous pollutants; Section 112(r), Accidental Release Prevention/Flammable substances (Section 68.130), TQ = 15,000 lb (5,825 kg)
- Clean Water Act 40CFR401.15 Section 307 Toxic Pollutants; 40CFR423, Appendix A Priority Pollutants; §313 Priority Chemicals
- RCRA 40CFR261.24 Toxicity Characteristics, Maximum Concentration of Contaminants (MCC), Regulatory level, 5.0 mg/l
- RCRA, 40CFR261, Appendix 8, Appendix 8, Hazardous Constituents, waste number not listed
- CERCLA/SARA 40CFR302 Extremely Hazardous Substances: TPQ = 500 lb (228 kg)
- SUPERFUND/EPCRA 40CFR302.4 Reportable Quantity (RQ): EHS 1 lb (0.454 kg)

- EPCRA Section 313: Form R *de minimis* concentration reporting level: 0.1%
- U.S. DOT Regulated Marine Pollutant (49CFR172.101, Appendix B)
- Canada, WHMIS, Ingredients Disclosure List
- Canada: Priority Substance List & Restricted Substances/ Ocean Dumping Forbidden (CEPA), National Pollutant Release Inventory (NPRI) (arsenic compounds)

Cited in U.S. State Regulations: California (A, G), Connecticut (A), Florida (G), Illinois (G), Kansas (G), Louisiana (G), Maine (G, W), Maryland (G), Massachusetts (G), Michigan (G), Minnesota (G), Montana (A), Nevada (A), New Hampshire (G), New Jersey (G), New York (G, A), North Carolina (A), North Dakota (A), Oklahoma (G), Pennsylvania (G, A), Rhode Island (G, A), South Carolina (A), South Dakota (A, W), Tennessee (W), Utah (W), Vermont (G), Virginia (G, A), Washington (G), West Virginia (G), Wisconsin (G).

Description: Arsenous trichloride, $AsCl_3$, is a noncombustible, colorless or pale yellow oily liquid with an acrid odor. Freezing/Melting point = –16°C. Decomposes in water.

Potential Exposure: Used in the ceramics industry, in the synthesis of chlorine-containing arsenicals; as a chemical intermediate for arsenic insecticides, pharmaceuticals, and chemical warfare agents.

Incompatibilities: Contact with sodium, potassium, powdered aluminum may cause a violent reaction. it is decomposed by water to form arsenic hydroxide and hydrogen chloride. Exposure to light forms toxic gas. Violent reaction with anhydrous ammonia, strong acids, strong oxidizers and halogens. Corrodes metals in the presence of moisture. Incompatible with alkali metals, active metals such as arsenic, iron, aluminum, zinc.

Permissible Exposure Limits in Air: OSHA: The legal airborne permissible exposure limit (PEL) is 0.01 mg/m^3 averaged over an 8-hour workshift for Arsenic and compounds as Arsenic, inorganic. NIOSH: The recommended airborne exposure limit is 0.002 mg/m^3, which should not be exceeded during any 15 minute work period for Arsenic, inorganic. ACGIH: The recommended airborne exposure limit is 0.2 mg/m^3 average over an 8-hour workshift for Arsenic and soluble compounds. The British HSE[33] has also adapted the ACGIH value of 0.2 mg/m^3 as an 8-hour TWA value. The DFG[3] has not set numerical limits for arsenic in air on the grounds that it is a proven human carcinogen. The former USSR-UNEP/IRPTC project[43] has set a MAC value for inorganic arsenic compounds (except arsine) of 0.003 mg/m^3 for ambient air in residential areas. In addition, North Carolina has specifically set for arsenic chloride a guideline or standard in ambient air[60] of 0.0002 μg/m^3.

Permissible Concentration in Water: See the entry under Arsenic and Inorganic Compounds.

Determination in Water: The atomic absorption graphite furnance technique is often used for measurement of total arsenic in water. It also has been standardized by EPA. Total arsenic may be determined by digestion followed by silver diethyldithiocarbamate; an alternative is atomic absorption; another is inductively coupled plasma (ICP) optical emission spectrometry.

Routes of Entry: Inhalation, ingestion, skin contact.

Harmful Effects and Symptoms

Short Term Exposure: Irritates the eyes, skin and respiratory tract. Eye contact can cause burns, possibly causing permanent damage. Exposure to vapors causes spasm of eyelids, tearing, pain, and reddening. It can cause death. In acute exposures, it is extremely toxic and caustic, owing not only to the poisonous nature of arsenic, but also to the release of hydrochloric acid in the presence of water. Exposure to the skin causes local irritation and blisters. Inhalation or ingestion causes hemorrhagic gastroenteritis resulting in loss of fluids and electrolytes, collapse, shock and death. The fatal human dose is 70 – 180 mg depending on the weight of the victim. Symptoms usually appear one-half to one hour after ingestion. Symptoms include a sweetish metallic taste, garlicky odor on the breath and stools, constriction in throat, difficulty swallowing, abdominal pain, vomiting, diarrhea, bluing of the skin, weak pulse, dizziness, headaches, coma, and convulsions.

Long Term Exposure: Chronic poisoning can lead to peripheral nerve damage, skin conditions, liver damage. Arsenic and certain arsenic compounds have been implicated in the induction of skin and lung cancer. Based on several related arsenic compounds, arsenic trichloride may be a mutagen and teratogen.

Points of Attack: Liver, kidneys, skin, lungs, reproductive system.

Medical Surveillance: See entry under Arsenic and Inorganic Compounds.

First Aid: If this chemical gets into the eyes, remove any contact lenses at once and irrigate immediately for at least 15 minutes, occasionally lifting upper and lower lids. Seek medical attention immediately. If this chemical contacts the skin, remove contaminated clothing and wash immediately with soap and water. Seek medical attention immediately. If this chemical has been inhaled, remove from exposure, begin rescue breathing (using universal precautions) if breathing has stopped and CPR if heart action has stopped. Transfer promptly to a medical facility. When this chemical has been swallowed, get medical attention.

Antidotes and Special Procedures: For severe poisoning BAL have been used. For milder poisoning, *penicillamine (not penicillin)* has been used, both with mixed success. Side effects occur with such treatment and it is never a substitute for controlling exposure. It can only be done under strict medical care.

Personal Protective Methods: Where possible, enclose operations and use local exhaust ventilation at the site of chemical release. If local exhaust ventilation or enclosure is not used, respirators should be worn. A regulated, marked area should be established where Arsenic Trichloride is handled, used or stored. Wear protective gloves and clothing to prevent any reasonable probability of skin contact. Safety equipment suppliers/manufacturers can provide recommendations on the most protective glove/clothing material for your operation. All protective clothing (suits, gloves, footwear, headgear) should be clean, available each day, and put on before work. Contact lenses should not be worn when working with this chemical. Wear full facepiece respiratory protection. Employees should wash immediately with soap when skin is wet or contaminated. Provide emergency showers and eyewash.

Respirator Selection: *At any concentrations above the NIOSH REL at any detectable concentration:* SCBAF:PD,PP (any self-contained breathing apparatus that has a full faceplate and is operated in a pressure-demand or other positive-pressure mode); or SAF:PD,PP:ASCBA (any supplied-air respirator that has a full facepiece and is operated in a pressure-demand or other positive-pressure mode in combination with an auxiliary self-contained breathing apparatus operated in a pressure-demand or other positive-pressure mode). *Escape:* GMFAGHiE [any air-purifying, full-facepiece respirator (gas mask) with a chin-style, front-or back-mounted acid gas canister having a high-efficiency particulate filter]; or SCBAE (any appropriate escape-type, self-contained breathing apparatus).

Storage: Prior to working with this chemical you should be trained on its proper handling and storage. Arsenic Trichloride, $AsCl_3$, should be stored in tightly-closed containers in a cool, well-ventilated area away from heat, water and any possible contact with active metals or other incompatible materials. A regulated, marked area should be established where this chemical is handled, used, or stored in compliance with OSHA standard 1910.1045.

Shipping: $AsCl_3$ should be labeled "Poison" according to DOT.[19] Shipment by passenger aircraft or railcar and even by cargo aircraft is forbidden. It falls in UN/DOT Hazard Class 6.1 and Shipping Group I.[19][20]

Spill Handling: Evacuate persons not wearing protective equipment from area of spill or leak until clean-up is complete. Remove all ignition sources. Stay upwind; keep out of low areas. Ventilate closed spaces before entering them. Wear positive pressure breathing apparatus and special protective clothing. Remove and isolate contaminated clothing at the site. Do not touch-spilled material; stop leak if you can do so without risk. Use water spray to reduce vapors. Small spills: absorb with sand or other non-combustible absorbent material and place into sealed containers for later disposal. Large spills: dike far ahead of spill for later disposal. Ventilate area after clean-up is complete. It may be necessary to contain and dispose of this chemical as a hazardous waste. If material or contaminated runoff enters waterways, notify downstream users of potentially contaminated waters. Contact your Department of Environmental Protection or your regional office of the federal EPA for specific recommendations. If employees are required to clean-up spills, they must be properly trained and equipped. OSHA 1910.120(q) may be applicable.

Fire Extinguishing: Arsenic Trichloride is not flammable so agents must be used which are suitable for surrounding fire. Poisonous gases are produced in fire including arsine and chlorine. If material or contaminated runoff enters waterways, notify downstream users of potentially contaminated waters. Notify local health and fire officials and pollution control agencies. Containers may explode in fire. From a secure, explosion-proof location, use water spray to cool exposed containers. If cooling streams are ineffective (venting sound increases in volume and pitch, tank discolors, or shows any signs of deforming), withdraw immediately to a secure position. If employees are expected to fight fires, they must be trained and equipped in OSHA 1910.156.

Disposal Method Suggested: Consult with environmental regulatory agencies for guidance on acceptable disposal practices. Generators of waste containing this contaminant (≥100 kg/mo) must conform with EPA regulations governing storage, transportation, treatment, and waste disposal. Dissolve in a minimum of concentrated hydrochloric acid. Dilute with water until white precipitate forms. Add HCl to dissolve. Saturate with H_2S; filter and wash precipitate and return to supplier. Alternatively, precipitate with heavy metals such as lime or ferric hydroxide in lieu of H_2S.[22] If needed, seek professional environmental engineering assistance from the U.S. EPA Environmental Response Team at (908) 548-8730 (24-hour response line). In accordance with 40CFR165 recommendations for the disposal of pesticides and pesticide containers. Must be disposed properly by following package label directions or by contacting your state pesticide or environmental control agency or by contacting your regional EPA office.

References

U.S. Environmental Protection Agency, "Chemical Profile: Arsenous Trichloride," Washington, DC, Chemical Emergency Preparedness Program (November 30, 1987)

New Jersey Department of Health and Senior Services, "Hazardous Substance Fact Sheet: Arsenic Trichloride," Trenton, NJ (June 1996)

Arsine

Molecular Formula: AsH_3

Synonyms: Arsenic anhydride; Arsenic trihydride; Arseniuretted hydrogen; Arsenous hydride; Arsenowodor (Polish); Arsenwasserstoff (German); Arsina (Spanish); Hydrogen arsenide

CAS Registry Number: 7784-42-1

RTECS Number: CG6475000

DOT ID: UN 2188

EEC Number: 033-002-00-5

Regulatory Authority

- Carcinogen (Human Positive) (IARC)[9]
- Banned or Severely Restricted (UN)[35]
- Very Toxic Substance (World Bank)[15]
- Air Pollutant Standard Set (ACGIH)[1] (NIOSH)[2] (OSHA)[58] (Several States)[60] (Australia) (Israel) (Mexico) (Various Canadian Provinces)
- OSHA 29CFR1910.119, Appendix A, Process Safety List of Highly Hazardous Chemicals, TQ = 100 lb (45.4 kg)
- Clean Air Act, 42USC7412; Title I, Part A, §112 hazardous pollutants; 42USC7412; Title I, Part A, §112(r), Accidental Release Prevention/Flammable substances (Section 68.130) TQ = 1,000 lb (454 kg)
- CLEAN WATER ACT: 40CFR401.15 Toxic Pollutant
- RCRA, 40CFR261, Appendix 8 Hazardous Constituents, waste number not listed
- SUPERFUND/EPCRA 40CFR302.4, Appendix A, Reportable Quantity (RQ): EHS, 1 lb (0.454 kg)
- CERCLA/SARA 40CFR302, Extremely Hazardous Substances: TPQ = 100 lb (45.4 kg)
- EPCRA Section 313: Form R *de minimis* concentration reporting level: 0.1%
- U.S. DOT Regulated Marine Pollutant (49CFR172.101, Appendix B)
- U.S. DOT 49CFR172.101, Inhalation Hazardous Chemical

Cited in U.S. State Regulations: Alaska (G), California (A, G), Connecticut (A), Florida (G, A), Illinois (G), Maine (G), Massachusetts (G), Minnesota (G), Nevada (A), New Hampshire (G), New Jersey (G), New York (G, A), North Dakota (A), Oklahoma (G), Pennsylvania (G), Rhode Island (G), Virginia (A), West Virginia (G).

Description: Arsine, AsH_3, is an extremely flammable, colorless, liquefied compressed gas with a slight garlic-like odor. Boiling point = -62°C. Freezing/Melting point = -117°C. NFPA 704 M Hazard Identification (based on NFPA-704M Rating System): Health 4, Flammability 4, Reactivity 2. It has an odor threshold of <1.0 ppm, but the odor cannot be relied on. Decomposes at 300°C. Practically insoluble in water.

Potential Exposure: Used in making electronic components, in organic syntheses, and in making lead-acid storage batteries. Arsine may be generated by side reactions or unexpectedly; e.g., it may be generated in metal pickling operations, metal drossing operations, or when inorganic arsenic compounds contact sources of nascent hydrogen. It has been known to occur as an impurity in acetylene. Most occupational exposure occurs in chemical, smelting, and refining industries. Cases of exposure have come from workers dealing with zinc, tin, cadmium, galvanized coated aluminum, and silicon and steel metals. A regulated, marked area should be established where this chemical is handled, used, or stored in compliance with OSHA standard 1910.1045.

Incompatibilities: Thermally unstable and shock sensitive. Violent reaction with oxidizers, acids, halogens, mixtures of potassium and ammonia. Decomposes to arsenic on exposure to light.

Permissible Exposure Limits in Air: The legal OSHA PEL[58] and recommended ACGIH TLV for arsine is 0.05 ppm (0.2 mg/m^3) TWA for an 8-hour workshift. This is also the limit set by HSE (U.K.). NIOSH has recommended a limit of 0.0006 ppm (ceiling) not to be exceeded during an 15-minute period. Australia, Mexico, and California have the same standard as OSHA and ACGIH. Canadian Provinces, Alberta, British Columbia, Ontario, and Quebec also have the same basic standard as OSHA *and* Alberta has a 15 minute STEL of 0.15 ppm (0.48 mg/m^3). The NIOSH IDLH level is 3 ppm. In addition, several states have set guidelines or standards for arsine in ambient air[60] ranging from 0.67 μg/m^3 (New York) to 1.0 μg/m^3 (Connecticut) to 2.0 μg/m^3 (Florida and North Dakota) to 3.3. μg/m^3 (Virginia) to 5.0 μg/m^3 (Nevada).

Determination in Air: Charcoal adsorption followed by elution with nitric acid and flawless atomic absorption in a high temperature graphite analyzer. See NIOSH Method 6001.[18] Also OSHA Method ID-125G

Permissible Concentration in Water: No criteria set, but EPA[32] cites the same limits as earlier proposed for arsenic by EPA of 50 μg/l based on health effects and 10 μg/l based on ecological effects.

Routes of Entry: Inhalation of gas, ingestion, skin contact.

Harmful Effects and Symptoms

Arsine is an extremely toxic gas that can be fatal if inhaled in sufficient quantities. Acute poisoning is marked by a triad of main effects caused by massive intravascular homolysis of the circulating red cells. Early effects may occur within an hour or two and are commonly characterized by general malaise, apprehension, giddiness, headache, shivering, thirst, and abdominal pain with vomiting. In severe acute cases the vomitus may be blood stained and diarrhea ensues as with inorganic arsenical poisoning. Pulmonary edema has occurred in severe acute poisoning. Invariably, the first sign observed in arsine poisoning is hemoglobinuria, appearing with discoloration of the urine up to port wine hue (first to the triad). Jaundice (second of triad) sets in on the second or third day and may be intense, coloring the entire body surface a deep bronze hue. Coincident with these effects is a severe hemolytic-type anemia. Severe renal damage may occur with oliguria or complete suppression of urinary function (third of triad), leading to uremia and death. Severe hepatic damage may also occur, along with cardiac damage and EKG changes. Where death does not occur, recovery is prolonged. In cases where the amount of inhaled arsine is insufficient to produce acute effects, or where small quantities are inhaled over prolonged periods, the hemoglobin liberated by the destruction of red cells may be degraded by the reticuloendothelial system and the iron moiety taken up by the liver, without producing permanent damage. Some hemoglobin may be excreted unchanged by the kidneys. The only symptoms noted may be general tiredness, pallor, breathlessness on exertion, and palpitations as would be expected with severe secondary anemia.

Short Term Exposure: Inhaling arsine can irritate the lungs causing shortness of breath and coughing. Higher exposures can cause pulmonary edema, a medical emergency that can be delayed for several hours. This can cause death. High exposure can cause hemolysis (destruction of red blood cells), causing anemia with headache, weakness, nausea, vomiting and abdominal pain. Acute kidney failure may follow; this can cause death. High concentrations of arsine gas will cause damage to the eyes; however, many experts agree that before this occurs systemic effects can be expected. Skin or eye contact with compressed gas can cause frostbite.

Long Term Exposure: Arsenic may cause cancer and should be treated as a teratogen. Repeated exposure may damage the nerves causing weakness, "pins and needles" sensation, weakness in the limbs with loss of coordination.

Points of Attack: Blood, kidneys, liver, lungs.

Medical Surveillance: In preemployment physical examinations, special attention should be given to past or present kidney disease, liver disease, and anemia. Periodic physical examinations should include tests to determine arsenic levels in the blood and urine, complete blood count (CBC) with reticulocyte count, kidney and liver function tests, urine hemoglobin, examination of the nervous system, and chest x-ray (following acute exposure). Since arsine gas is a by-product of certain production processes, workers should be trained to recognize the symptoms of exposure and to use appropriate personal protective equipment.

First Aid: If a person breathes in large amounts of this chemical, move the exposed person to fresh air at once and perform artificial respiration. If this chemical has been inhaled, remove from exposure, begin rescue breathing (using universal precautions) if breathing has stopped and CPR if heart action has stopped. Transfer promptly to a medical facility. *Dimercaprol* treatment is indicated and blood transfusions may be necessary. If this chemical gets into the eyes, remove any contact lenses at once and irrigate immediately for at least 15 minutes, occasionally lifting upper and lower lids. Seek medical attention immediately. If frostbite has occurred, seek medical attention immediately; do *NOT* rub the affected areas or flush them with water. In order to prevent further tissue damage, do *NOT* attempt to remove frozen clothing from frostbitten areas. If frostbite has *NOT* occurred, immediately and thoroughly wash contaminated skin with soap and water. Seek medical attention immediately. Medical observation is recommended for 24 - 48 hours after breathing overexposure, as pulmonary edema may be delayed. As first aid for pulmonary edema, a doctor or authorized paramedic may consider administering a corticosteroid spray.

Personal Protective Methods: In most cases, arsine poisoning cannot be anticipated except through knowledge of the production processes. Where arsine is suspected in concentration above the acceptable standard, the worker should be supplied with a supplied air fullface respirator or a self-contained positive pressure-respirator with full facepiece. Safety equipment suppliers/manufacturers can provide recommendations on the most protective glove/clothing material for your operation. All protective clothing (suits, gloves, footwear, headgear) should be clean, available each day, and put on before work. Employees should wash immediately with soap when skin is wet or contaminated. Remove nonimpervious clothing immediately if wet or contaminated. Provide emergency showers and eyewash. Where exposure to the liquefied compressed gas may occur, employees should be provided with special clothing designed to prevent frostbite.

Respirator Selection: *At any concentrations above the NIOSH REL, or where there is no REL, at any detectable concentration:* SCBAF:PD,PP (any self-contained breathing apparatus that has a full facepiece and is operated in a pressure-demand or other positive-pressure mode); or SAF:PD,PP: ASCBA (any supplied-air respirator that has a full facepiece and is operated in a pressure-demand or other positive-pressure mode in combination with an auxiliary self-contained breathing apparatus operated in a pressure-demand or other positive-pressure mode). *Escape:* GMFS* [any air-purifying, full-facepiece respirator (gas mask) with a chin-style, front- or back-mounted canister providing protection against the compound of concern]; or SCBAE (any appropriate escape-type, self-contained breathing apparatus).

* Substance reported to cause eye irritation or damage; may require eye protection.

Storage: Prior to working with this chemical you should be trained on its proper handling and storage. Arsine must be stored to avoid contact with incompatible materials including oxidizers (such as Perchlorates, Peroxides, Permanganates, Chlorates and Nitrates) since violent reactions occur. Outside, detached storage is preferred. Arsine decomposes and deposits Arsenic on exposure to light and moisture. Sources of ignition, such as smoking and open flames, are prohibited where Arsine is handled, used, or stored. Metal containers involving the transfer of arsine should be grounded and bonded.

Shipping: The label required by DOT is "Poison Gas, Flammable Gas." Shipment by cargo aircraft or railcar or even by cargo aircraft is forbidden. The UN/DOT Hazard Class is 2.3 and the Packing Group is I.[19][20]

Spill Handling: No flares, smoking, or flames in area. Use water spray to reduce vapors. Isolate area until arsine gas has dispersed. Stay upwind; keep out of low areas. Ventilate area of leak to disperse the gas if it can be done without placing personnel at risk. Wear positive pressure breathing apparatus and full protective clothing. Evacuate persons not wearing protective equipment from area of spill or leak until clean-up is complete. Remove all ignition sources. Stop the flow of gas if it can be done safely, with undue risk. If the source of the leak is a cylinder and the leak cannot be stopped in place, remove the leaking cylinder to a safe place in the open air, and repair leak or allow cylinder to empty. Use large amounts of water to disperse vapors; contain runoff. It may be necessary to contain and dispose of this chemical as a hazardous waste. If material or contaminated runoff enters waterways, notify downstream users of potentially contaminated waters. Contact your Department of Environmental Protection or your regional office of the federal EPA for specific recommendations. If employees are required to clean-up spills, they must be properly trained and equipped. OSHA 1910.120(q) may be applicable.

Initial isolation and protective action distances

Distances shown are likely to be affected during the first 30 minutes after materials are spilled and could increase with time. If more than one tank car, cargo tank, portable tank, or large cylinder is involved in the incident is leaking, the protective action distance may need to be increased. You may need to seek emergency information from CHEMTREC at (800) 424-9300 or seek professional environmental engineering assistance from the U.S. EPA Environmental Response Team at (908) 548-8730 (24-hour response line).

Small spills (From a small package or a small leak from a large package)

First: Isolate in all directions (feet)..... 400

Then: Protect persons downwind (miles)

Day 0.4

Night 1.5

Large Spills (From a large package or from many small packages)

First: Isolate in all directions (feet)..... 1,100

Then: Protect persons downwind (miles)

Day 1.3

Night 5.7

Fire Extinguishing: This chemical is a highly flammable gas. Poisonous gases including nitrogen oxides are produced in fire. Firefighting gear (including SCBA) does not provide adequate protection. If exposure occurs, remove and isolate gear immediately and thoroughly decontaminate personnel. Approach fire from upwind. Fight fire from protected location or maximum possible distance. Let small fires burn out, if possible. Use water spray or foam extinguishers. For massive fire in cargo area use unmanned hose holder or monitor nozzles; if this is impossible, withdraw from area and let fire burn. Cool containers that are exposed to flames with water from the side until well after fire is out. See isolation distances above if tank car or truck is involved in fire. Vapors are heavier than air and will collect in low areas. Vapors may travel long distances to ignition sources and flashback. Vapors in confined areas may explode when exposed to fire. Storage containers and parts of containers may rocket great distances, in many directions. Notify local health and fire officials and pollution control agencies. From a secure, explosion-proof location, use water spray to cool exposed containers. If cooling streams are ineffective (venting sound increases in volume and pitch, tank discolors, or shows any signs of deforming), withdraw immediately to a secure position. If employees are expected to fight fires, they must be trained and equipped in OSHA 1910.156.

Disposal Method Suggested: Arsine may be disposed of by controlled burning. When possible, cylinders should be sealed and returned to suppliers.[22] Seek guidance from regulatory agencies as to proper disposal. Consult with environmental regulatory agencies for guidance on acceptable disposal practices. Generators of waste containing this contaminant (≥100 kg/mo) must conform with EPA regulations governing storage, transportation, treatment, and waste disposal.

References

U.S. Environmental Protection Agency, "Chemical Profile: Arsine," Washington, DC, Chemical Emergency Preparedness Program (November 30, 1987)

New Jersey Department of Health and Senior Services, "Hazardous Substance Fact Sheet: Arsine," Trenton, NJ (June 1998)

New York State Department of Health, "Chemical Fact Sheet: Arsine," Albany, NU, Bureau of Toxic Substance Assessment (March 1986)

Asbestos

Synonyms: Amianthus; Amosite; Amphibole; Asbest (German); Asbesto (Spanish); Asbestose (German); Asbestos fiber; Ascarite; Chrysotile (AKA white asbestos); Crocidolite (AKA brown asbestos, blue asbestos); Fiberous grunerite; Krokydolith (German); Mysorite; NCI-C08991; Serpentine; Tremolite

Note: asbestos is a generic name for various hydrated mineral silicates.

CAS Registry Number: 1332-21-4 (No Spec type); 77536-66-4 (Actinolite); 12172-73-5 (Amosite); 77536-67-5 (Anthophyllite); 17068-78-9 [Anthophyllite, (OSHA)]; 12001-29-5 (Chrysotile); 12001-28-4 (Crocidolite); 14567-73-8 (Tremolite)

RTECS Number: CI6475000 (No Spec type); CI6476000 (Actinolite); CI6477000 (Amosite); CI6478000 (Anthophyllite); CI6478500 (Chrysotile); CI6479000 (Crocidolite); CI6560000 (Tremolite)

DOT ID: UN 2212/2590 (No Spec type); NA 2212 (Actinolite); UN 2212 (Amosite); UN 2590 (Anthophyllite); UN 2590 (Chrysotile); UN 2212 (Crocidolite); UN 2590 (Tremolite)

Regulatory Authority

- Carcinogen (Animal and Human Positive) (IARC)[9] (DFG)[3] (ACGIH)[1] (NTP) (Australia) (California) (Massachusetts) (Minnesota) (New Jersey)
- Banned or Severely Restricted (Several Countries) (UN)[13][35]
- Air Pollutant Standard Set (ACGIH)[1] (NIOSH/OSHA)[2] (DFG)[3] (Several States)[60] (HSE) (Australia) (Israel) (Mexico) (Various Canadian Provinces)
- Water Pollution Standard Proposed (EPA)[6] (Kansas, Minnesota)[61] (Mexico)
- OSHA 29CFR1910 OSHA Specifically Regulated Substances. (See CFR1910.1001)
- Clean Air Act, 42USC7412; Title I, Part A, §112 hazardous pollutants; § 63.74 List of high risk pollutants, weighting factor: 100
- Clean Water Act 40CFR401.15 Section 307 Toxic Pollutants; §307 Priority Pollutants; §313 Priority Chemicals
- Safe Drinking Water Act: 40CFR141.62, MCL, 7 million fibers/l (longer than 10 microns); 40CFR141.51, MCLG, 7 million fibers/l (longer than 10 microns); 47FR9352 regulated chemical; 40CFR141.23, inorganic chemical sampling and analytical requirements; 40CFR141.32, public notification requirements
- SUPERFUND/EPCRA 40CFR302.4, Appendix A, Reportable Quantity (RQ): CERCLA, 1 lb (0.454 kg), 40CFR372.65: Form R *de minimis* Concentration Reporting Level: 0.1%
- Canada, WHMIS, Ingredients Disclosure List; National Pollutant Release Inventory (NPRI); CEPA Schedule I, Toxic Substances (atmospheric releases from mines and mills)
- TSCA 40CFR716.120.c6 asbestiform minerals

Cited in U.S. State Regulations: Alaska (G), California (A, G), Connecticut (A), Florida (G), Illinois (G), Kansas (W), Maine (G), (Maryland (G), Massachusetts (G), Michigan (G), Minnesota (W), Nevada (A), New Hampshire (G), New Jersey (G), New York (G, A), North Carolina (A), North Dakota (A), Oklahoma (G), Pennsylvania (G, A), Rhode Island (G), South Carolina (A), Virginia (A), West Virginia (G).

Description: White or greenish (chrysotile), blue (crocidolite), or gray-green (amosite) fibrous, odorless solids. Freezing/Melting point = 1,112°F (Decomposes). Hazard Identification (based on NFPA-704 M Rating System): Health 2, Flammability 0, Reactivity 0. Insoluble in water. Asbestos is a generic term that applies to a number of naturally occurring, hydrated mineral silicates incombustible in air and separable into filaments. The most widely used in industry in the United States is chrysotile, a fibrous form of serpentine. Other types include amosite, crocidolite, tremolite, anthophyllite, and actinolite.

Potential Exposure: Most asbestos is used in the construction industry. Much of it is firmly bonded, i.e., the asbestos is "locked in" in such products as floor tiles, asbestos cements, and roofing felts and shingles; while the remaining 8% is friable or in powder forms present in insulation materials, asbestos cement powders, and acoustical products. As expected, these latter materials generate more airborne fibers than the firmly bonded products. The asbestos used in non-construction industries is utilized in such products as textiles, friction material including brake linings, and clutch facings, paper, paints, plastics, roof coatings, floor tiles, and miscellaneous other products. Significant quantities of asbestos fibers appear in rivers and streams draining from areas where asbestos-rock outcroppings are found. Some of these outcroppings are being mined. Asbestos fibers have been found in a number of dinking water supplies, but the health implications of ingesting asbestos are not fully documented. Emissions of asbestos fibers into water and air are known to result from mining and processing of some minerals. Exposure to asbestos fibers may occur throughout urban environments perhaps resulting from asbestos from brake linings and the flaking of sprayed asbestos insulation material. In recent years, much effort has been put into removal of asbestos insulation, particularly from schools and other public buildings where worn or exposed asbestos causes public exposure.

Incompatibilities: None

Permissible Exposure Limits in Air: The OSHA PEL for asbestos fibers (i.e., actinolite asbestos, amosite, anthophyllite asbestos, chrysotile, crocidolite, and tremolite asbestos) is an 8-hour TWA airborne concentration of 0.1 fiber (longer than 5 micrometers and having a length-to-diameter ratio of at least 3:1) per cubic centimeter of air (0.1 fiber/cm^3), as determined by the membrane filter method at approximately 400X magnification with phase contrast illumination. No worker should be exposed in excess of 1 fiber/cm^3 (excursion limit)as averaged over a sampling period of 30 minutes.

NIOSH considers asbestos (i.e., actinolite, amosite, anthophyllite, chrysotile, crocidolite, and tremolite) to be a potential occupational carcinogen and recommends that exposures be reduced to the lowest possible concentration. For asbestos fibers >5 micrometers in length, NIOSH recommends a REL of 100,000 fibers per cubic meter of air (100,000 fibers/m^3), which is equal to 0.1 fiber per cubic centimeter of air (0.1 fiber/cm^3), as determined by NIOSH Analytical Method #7400, as found in 29 CFR1910.1001. ACGIH has categorized asbestos (all forms) as a human carcinogen and recommended TLV for all forms as : 0.1 fiber/c TWA. The DFG[3] MAK values have not been assigned because of the carcinogenic nature of asbestos. Australia has a limit of 0.1 fibers/ml of air TWA (confirmed carcinogen). Israel's TWA is 0.4 fibers/c [mean dust level (fibers longer than 5 microns and with an aspect ratio ≥3:1)] and Action Level of (mean dust level) of 0.1 fibers/c. The HSE (U.K.) level is 0.2 fibers/ml TWA

(4 hr) for croccidolite or amosite; 0.5 fibers/ml TWA (4 hr) for other forms and short term exposure limits of 0.6 fibers/ml STEL for crocidolite and amosite and 1.5 fibers/ml STEL for other forms. The California PEL is 0.2 fibers/c TWA [see also §5208 (requirements for respiratory protection)]. Canadian Provincial limits: British Columbia, 0.1 fibers/ml TWA (carcinogen); Ontario limits are 0.1 fibers/cm^3 TWAEV and short term exposure limit of 5.0 fibers/cm^3 STEV; Quebec limits are 0.1 mg/m^3 TWAEV (recirculation of respirable dusts). In addition, several states have set guidelines or standards for asbestos in ambient air[60] ranging from zero (Massachusetts, Nevada, North Carolina, North Dakota, South Carolina) to 0.001 μg/m^3 (Connecticut) to 0.005 μg/m^3 (Pennsylvania) to 0.005 μg/m^3 (Nevada) to 2.0 μg/m^3 (Virginia) to 5.0 μg/m^3 (New York).

Determination in Air: Sampling and analytical techniques should be performed as determined by a 400-liter air sample collected over 100 minutes as specified by NIOSH Method #7400. The actual determination involves a microscopic fiber count.

Permissible Concentration in Water: To protect freshwater and saltwater aquatic life – no criteria have been established due to insufficient data. To protect human health-preferably zero. A lifetime cancer risk of 1 in 100,000 corresponds to concentrations of 300,000 fibers/l.[6] A maximum level in drinking water of 7 million fibers/liter has been proposed by EPA.[62] In addition, Kansas and Minnesota have set guidelines for asbestos in drinking water.[61] Mexico drinking water ecological criteria is 3,000 mg/l.

Routes of Entry: Inhalation, ingestion, skin and/or eye contact.

Harmful Effects and Symptoms

Short Term Exposure: There is no known acute health effects. Persons who develop serious and fatal diseases later in life may feel fine at the time of exposure.

Long Term Exposure: Asbestosis (chronic exposure): dyspnea (breathing difficulty), interstitial fibrosis, restricted pulmonary function, finger clubbing; irritation eyes; (Potential occupational carcinogen). Available studies provided conclusive evidence that exposure to asbestos fibers causes cancer and asbestosis in man. Lung cancers and asbestosis have occurred following exposure to chrysotile, crocidolite, amosite, and anthophyllite. Mesotheliomas, lung and gastrointestinal cancers have been shown to be excessive in occupationally exposed persons, while mesotheliomas have developed also in individuals living in the neighborhood of asbestos factories and near crocidolite deposits, and in persons living with asbestos workers. Asbestosis has been identified among persons living near anthophyllite deposits. Likewise, all commercial forms of asbestos are carcinogenic in rats, producing lung carcinomas and mesotheliomas following their inhalation, and mesotheliomas after intrapleural or i.p. injection. Mesotheliomas and lung cancers were induced following even 1 day's exposure by inhalation. The size and shape of the fibers are important factors; fibers less than 0.5 μm in diameter are most active in producing tumors. Other fibers of a similar size including glass fibers can also produce mesotheliomas following intrapleural or i.p. injection. There are data that show that the lower the exposure, the lower the risk of developing cancer. Excessive cancer risks have been demonstrated at all fiber concentrations studied to date. Evaluation of all available human data provides no evidence for a threshold or for a "safe" level of asbestos exposure.

Points of Attack: Lungs [cancer].

Medical Surveillance: Medical surveillance is required, except where a variance from the medical requirements of this proposed standard have been granted, for all workers who are exposed to asbestos as part of their work environment. For the purposes of this requirement the term "exposed to Asbestos" will be interpreted as referring to time-weighted average exposures above 1 fiber/c or peak exposures above 5 fibers/c. The major objective off such surveillance will be to ensure proper medical management of individuals who show evidence of reaction to past dust exposures, either due to excessive exposures or unusual susceptibility. Medical management may range from recommendations as to job placement, improved work practices, cessation of smoking, to specific therapy for asbestos-related disease or its complications. Medical surveillance cannot be a guide to adequacy of current controls when environmental data and medical examinations only cover recent work experience because of the prolonged latent period required for the development of asbestosis and neoplasms. Required components of a medical surveillance program include periodic measurements of pulmonary function [forced vital capacity (FVC) and forced expiratory volume for one second (FEV_1)], and periodic chest roentgenograms (posteroanterior 14 × 17 inches). Additional medical requirement components include a history to describe smoking habits and details on past exposures to asbestos and other dusts and to determine presence or absence of pulmonary, cardiovascular, and gastrointestinal symptoms, and a physical examination, with special attention to pulmonary rales, clubbing of fingers and other signs related to cardiopulmonary systems. Chest roentgenograms and pulmonary function tests should be performed at least every 2 years on all employees exposed to asbestos. Such tests should be made annually on individuals: (a) who have a history of 10 or more years of employment involving exposure to asbestos or (b) who show roentgenographic findings (such as small opacities, pleural plaques, pleural thickening or pleural calcification) which suggest or indicate pneumoconiosis or other reactions to asbestos or (c) who have changes in pulmonary function which indicate restrictive or obstructive lung disease. Preplacement medical examinations and medical examinations on the termination of employment of asbestos-exposed workers are also required.

First Aid: If this chemical gets into the eyes, remove any contact lenses at once and irrigate immediately for at least 15 minutes, occasionally lifting upper and lower lids. Seek medical attention immediately. If this chemical contacts the skin, remove contaminated clothing and wash immediately with soap and water.

Personal Protective Methods: Use of respirators can be decided on the basis of time-weighted average or peak concentration. When the limits of exposure to asbestos dust cannot be met by limiting the concentration in the workplace, the employer must utilize a program of respiratory protection and furnishing of protective clothing to protect every worker exposed. Protective Clothing: (1) The employer shall provide each employee subject to exposure in a variance area with coveralls or similar full body protective clothing and hat, which shall be worn during the working hours in areas where there is exposure to asbestos dust. Non-disposable clothing should be placed in plastic bags for laundring or decontamination. (2) The employer shall provide for maintenance and laundering of the solid protective clothing, which shall be stored, transported and disposed of in sealed no reusable containers marked "Asbestos-Contaminated Clothing" in easy-to-read letters. (3) Protective clothing shall be vacuumed before removal. Clothes shall not be cleaned by blowing dust from the clothing or shaking. (4) If laundering is to be done by a private contractor, the employer shall inform the contractor of the potentially harmful effects of exposure to asbestos dust and of safe practices required in the laundering of the asbestos-soiled work clothes. (5) Resin-impregnated paper or similar protective clothing can be substituted for fabric-type of clothing. (6) It is recommended that in tightly contaminated operations (such as insulation and textiles) provisions be made for separate change rooms.

Respirator Selection: NIOSH *At any detectable concentration:* SCBAF:PD,PP (any MSHA/NIOSH approved self-contained breathing apparatus that has a full facepiece and is operated in a pressure-demand or other positive-pressure mode); or SAF:PD,PP:ASCBA (any supplied-air respirator that has a full facepiece and is operated in a pressure-demand or other positive-pressure mode in combination with an auxiliary, self-contained breathing apparatus operated in a pressure-demand or other positive pressure mode). *Escape:* HiEF (any air-purifying, full-facepiece respirator with a high-efficiency particulate filter); or SCBAE (any appropriate escape-type, self-contained breathing apparatus).

Storage: Prior to working with this chemical you should be trained on its proper handling and storage. Prior to working with asbestos workers *must* be, by law, trained in its proper handling and storage. Asbestos should be stored wet with special surfactants and water. Keep asbestos in closed, impermeable, sealed containers. Protect against physical damage. A regulated, marked area should be established where this chemical is handled, used, or stored in compliance with OSHA standard 1910.1045.

Shipping: There is no DOT label requirement for asbestos. Shipment of amosite or crocidolite (blue asbestos) is forbidden by passenger aircraft or railcar and even by cargo aircraft. The UN/DOT Hazard Class in any case is 9 and the Packing Group is II for blue and III for white asbestos.[19][20]

Spill Handling: Evacuate persons not wearing protective equipment from area of spill or leak until clean-up is complete. Proper procedures for repair or removal of the material must be followed by trained personnel. Strict hygiene is required; keep dust under control. Wearing protective equipment, use a wet mop or special high efficiency HEPA vacuum to clean area. Do not use common shop vacuum cleaner. *Note:* Avoid blowing, sweeping or dry brushing, and dry mopping, all of which may raise dust levels. Do not shovel. For storing asbestos wastes, use heavy-gauge impervious plastic bags. For final disposal contact your local environmental authority. It may be necessary to contain and dispose of this chemical as a hazardous waste. If material or contaminated runoff enters waterways, notify downstream users of potentially contaminated waters. Contact your Department of Environmental Protection or your regional office of the federal EPA for specific recommendations. If employees are required to clean-up spills, they must be properly trained and equipped. OSHA 1910.120(q) may be applicable.

Fire Extinguishing: Extinguish fire using an agent suitable for type of surrounding fire. Asbestos itself does not burn. Care should be taken to contain Asbestos materials disturbed in a fire. Becomes powder-like and loses its hazardous properties when heated above 1,200°C. If material or contaminated runoff enters waterways, notify downstream users of potentially contaminated waters. Notify local health and fire officials and pollution control agencies. If employees are expected to fight fires, they must be trained and equipped in OSHA 1910.156.

Disposal Method Suggested: Asbestos may be recovered from waste asbestos slurries as an alternative to disposal. Landfilling is an option[22] for disposal if carefully controlled.

References

National Institute for Occupational Safety and Health, Criteria for a Recommended Standard: Occupational Exposure to Asbestos, NIOSH Doc. No. HSV 72-1-267, Washington, DC (1972)

National Institute for Occupational Safety and Health, Revised Recommended Asbestos Standard: NIOSH Do. No. 77–169, Washington, DC (December 1976)

U.S. Environmental Protection Agency, Asbestos: Ambient Water Quality Criteria, Washington, DC (1980)

U.S. Environmental Protection Agency, Status Assessment of Toxic chemicals: Asbestos, Report No. EPA-600/2-79-210C, Washington, DC (December 1979)

National Academy of Sciences, Medical and Biologic Effects of Environmental Pollutants: Asbestos, Washington, DC (1971)

National Cancer Institute, Asbestos: An Information Resource, DHEW Publication No. (NIH)-79-1681, Bethesda, MD (May 1978)

U.S. Environmental Protection Agency, Asbestos, Health and Environmental Effects Profile No. 12, Washington, DC, Office of Solid Waste (April 30, 1980)

Sax, N. I., Ed., "Dangerous Properties of Industrial Materials Report" 1, No. 1, 29–31 (1981)

New Jersey Department of Health and Senior Services, "Hazardous Substance Fact Sheet: Asbestos," Trenton, NJ (February 25, 1987)

New York State Department of Health, "Chemical Fact Sheet: Asbestos," Albany, NY, Bureau of Toxic Substance Assessment (January 1986)

Asphalt Fumes

Synonyms: Asfalto (Spanish); Asphaltum; Bitumen (European term); Judean pitch; Mineral pitch; Petroleum asphalt; Petroleum bitumen; Pitch; Road asphalt; Road tar; Roofing asphalt

CAS Registry Number: 8052-42-4 (petroleum asphalt fumes)

RTECS Number: CI9900000

DOT ID: NA 1999

Regulatory Authority

- Air Pollutant Standard Set (ACGIH)[1] (HSE)[33] (OSHA)[58] (Several States)[60] (Australia) (Israel) (Mexico) (California)

Cited in U.S. State Regulations: Alaska (G), California (A, G), Connecticut (A), Florida (G), Illinois (G), Maine (G), Massachusetts (G), Nevada (A), New Hampshire (G), New Jersey (G), New York (G), North Dakota (A), Pennsylvania (G), Virginia (A), West Virginia (G).

Description: Asphalt fumes are flammable when hot and may contain hydrogen sulfide and human carcinogen such as benzo(a)pyrene and dibenz(a,h)anthracene. Fumes generated during the production or application of asphalt (a dark-brown to black cement-like substance manufactured by the vacuum distillation of crude petroleum oil). Flash point ≤10 – 225°C (as a general rule, the more liquid the type of asphalt, the lower the flashpoint; cutback <10°C; typical asphalt 225°C). Hazard Identification (based on NFPA-704 M Rating System) (typical asphalt): Health 0, Flammability 1, Reactivity 0. However, *asphalt cutback* has a flammability rating of 3. Insoluble in water. Hazard Identification (based on NFPA-704 M Rating System): *roofer's flux and straight run residue* Health 0, Flammability 1, Reactivity 0.

Asphalt fumes have been defined by NIOSH as the nimbose effusion of small, solid particles created by condensation from the vapor state after volatilization of asphalt. In additional to particles, a cloud of fume may contain materials still in the vapor state. The major constituent groups of asphalt are asphaltenes, resins, and oils made up of saturated and unsaturated hydrocarbons. The asphaltenes have molecular weights in the range of 1,000 - 2,600, those of the resins fall in the range of 370 - 500, and those of the oils is the range of 290 - 630. Asphalt has often been confused with tar because the two are similar in appearance and have sometimes been used interchangeably as construction materials. Tars are, however, produced by destructive distillation of coal, oil or wood whereas asphalt is a residue from fractional distillation or crude oil. The amounts of benzo(a)pyrene found in fumes collected from two different plants that prepared hot mix asphalt ranged from 3 - 22 ng/m^3; this is approximately 0.03% of the amount in coke oven emissions and 0.01% of that emitted from coal-burning home furnaces.

Potential Exposure: Occupational exposure to asphalt fumes can occur during the transport, storage, production, handling, or use of asphalt. The composition of the asphalt that is produced is dependent on the refining process applied to the crude oil, the source of the crude oil, and the penetration grade (viscosity) and other physical characteristics of the asphalt required by the consumer. The process for production of asphalt is essentially a closed-system distillation. Refinery workers are therefore potentially exposed to the fumes during loading of the asphalt for transport from the refinery during routine maintenance such as leaning of the asphalt storage tanks, or during accidental spills. Most asphalt is used out of doors, in paving and roofing, and the workers' exposure to the fumes is dependent on environmental conditions, work practices and other factors. These exposures are stated to be generally intermittent and at low concentrations. Workers are potentially exposed also to skin and eye contacts with hot, cut-back, or emulsified asphalts. Spray application of cut-back, or emulsified asphalts may involved respiratory exposure also.

Incompatibilities: None reported. *Note:* Asphalt becomes molten at about 200°F.

Permissible Exposure Limits in Air: Occupational exposure to asphalt fumes is defined as exposure in the workplace at a concentration of one-half or more of the recommended occupational exposure limit. If exposure to other chemicals also occurs, as is the case when asphalt is mixed with a solvent, emulsified, or used concurrently with other materials such as tar or pitch, provisions of any applicable standard for the other chemicals shall also be followed. Occupational exposure to asphalt fumes shall be controlled so that employees are not exposed to the airborne particulates at a concentration greater than 5 mg/m^3 (measured as total particulates) of air, determined during any 15-minute period (NIOSH ceiling). The recommended ACGIH airborne limit is 5 mg/m^3, not to be exceeded at any time. Australia, HSE (U.K.), Israel and Mexico limits are 5 mg/m^3 and HSE[33] and Mexico cite an STEL of 10 mg/m^3. Israel cites an Action Level of 2.5 mg/m^3. California's workplace PEL is 5 mg/m^3. Apart from workplace standards. Several states have set guidelines or standards for asphalt fumes in ambient air:[60] 50 μg/m^3 (North Dakota), 80 μg/m^3 (Virginia), 100 μg/m^3 (Connecticut), 119 μg/m^3 (Nevada).

Determination in Air: A gravimetric method is recommended for estimation of the air concentration of asphalt fumes. When large amounts of dust are present in the same atmosphere in which the asphalt fume is present, which may occur in road-building operations, the gravimetric method may lead to erroneously high estimates for asphalt fumes, and to possibly undeserved sanctions and citations for ostensibly exceeding the environmental limit for asphalt fumes or nuisance particulates. NIOSH recommends that where the resolution of such problems becomes necessary, a more specific procedure, which involves solvent extraction and gravimetric analysis, be employed for the determination of asphalt fumes. The best procedure now available seems to be ultrasonic agitation of the filter in benzene and weighing of the dried residue from an aliquot on the clear benzene extract. NIOSH is attempting to devise an even more specific method for asphalt fumes for use under such conditions.

Permissible Concentration in Water: No criteria set.

Routes of Entry: Inhalation of dusts and fumes. Skin exposure can cause thermal burns from hot asphalt.

Harmful Effects and Symptoms

Short Term Exposure: The principal adverse effects on health from exposure to asphalt fumes are irritation of the serous membranes of the conjunctivae and the mucous membranes of the respiratory tract. Hot asphalt can cause burns of the skin, and release vapors that irritate the eyes, throat, and possible bronchial tubes and lungs.

Long Term Exposure: In animals, there is evidence that asphalt left on the skin for long periods of time may result in local carcinomas, but there have been no reports of such effects of human skin that can be attributed to asphalt alone.

Points of Attack: Skin, respiratory system. In animals: skin tumors.

Medical Surveillance: Lung function tests, Details of recommended preplacement and periodic physical examinations and record keeping have been set forth by NIOSH in the Criteria Document cited below.

First Aid: If this chemical gets into the eyes, remove any contact lenses at once and irrigate immediately for at least 15 minutes, occasionally lifting upper and lower lids. Seek medical attention immediately. If this chemical contacts the skin, remove contaminated clothing and wash immediately with soap and water. Seek medical attention immediately. If this chemical has been inhaled, remove from exposure, begin rescue breathing (using universal precautions) if breathing has stopped and CPR if heart action has stopped. Transfer promptly to a medical facility. When this chemical has been swallowed, get medical attention. Give large quantities of water and induce vomiting. Do not make an unconscious person vomit.

Personal Protective Methods: Employees shall wear appropriate protective clothing, including gloves, suits, boots, face shields (8-inch minimum), or other clothing as needed, to prevent eye and/or skin contact with asphalt. NIOSH recommends thermally-insulated gloves if working with hot asphalt, long sleeve shirts, long cuffless trousers, and metal-toed shoes. All protective clothing should be clean and available each day, and put on prior to the workshift.

Respirator Selection: *At concentrations above the NIOSH REL, or where there is no REL, at any detectable concentration:* SCBAF:PD,PP (any MSHA/NIOSH approved self-contained breathing apparatus that has a full facepiece and is operated in a pressure-demand or other positive-pressure mode); or SAF:PD,PP:ASCBA (any supplied-air respirator that has a full facepiece and is operated in a pressure-demand or other positive-pressure mode in combination with an auxiliary, self-contained breathing apparatus operated in a pressure-demand or other positive pressure mode). *Escape:* HiEF (Any air-purifying, full-facepiece respirator (gas mask) with a chin-style, front- or back-mounted organic vapor canister having a high-efficiency particulate filter); or SCBAE (Any appropriate escape-type, self-contained breathing apparatus).

Engineering controls shall be used when needed to keep concentrations of asphalt fumes below the recommended exposure limit. The only conditions under which compliance with the recommended exposure limit may be achieved by the use of respirators are: (a) During the time required to install or test the necessary engineering controls. (b) For operations such as nonroutine maintenance or repair activities causing brief exposure at concentrations above the environmental limit. (c) During emergencies when concentrations of asphalt fumes may exceed the environmental limit. When a respirator is permitted above, it shall be selected from a list of respirators approved by NIOSH. See the Criteria Document cited below.

Storage: Prior to working with this chemical you should be trained on its proper handling and storage. Asphalt is normally shipped solid in steel drums. Fumes are encountered only in asphalt melting operations. Wherever lighter or liquid forms of asphalt are used, handled, manufactured, or stored, use explosion-proof electrical equipment and fittings. Sources of ignition such as smoking and open flames, are prohibited where this chemical is used, handled, or stored in a manner that could create a potential fire or explosion hazard.

Shipping: There is no DOT label requirement for asphalt. Shipment of asphalt at or above its flashpoint is forbidden in passenger railcars and in passenger or cargo aircraft. The Hazard Class is 3 and the Packing Group III[19][20] for asphalt at or above its flashpoint.

Spill Handling: Spilled asphalt is allowed to cool and solidify whereupon it is recovered, melted and used if clean. Ventilate area of spill or leak. Absorb liquid asphalt with sand or other non-combustible absorbent material and place in containers for disposal. It may be necessary to contain and dispose of this chemical as a hazardous waste. If material or contaminated runoff enters waterways, notify downstream users of potentially contaminated waters. Contact your Department of Environmental Protection or your regional office of the federal EPA for specific recommendations. If employees are required to clean-up spills, they must be properly trained and equipped. OSHA 1910.120(q) may be applicable.

Fire Extinguishing: The lighter, more liquid forms are flammable (especially cutback). The heavier, more solid forms may give off flammable gas when heated. Poisonous gases are produced in fire including hydrogen sulfide. For heavy forms of asphalt, use dry chemical, carbon dioxide, or foam extinguishers. For light forms, use dry chemical, fog, or water mist. If material or contaminated runoff enters waterways, notify downstream users of potentially contaminated waters. Notify local health and fire officials and pollution control agencies. From a secure, explosion-proof location, use water spray to cool exposed containers. If cooling streams are ineffective (venting sound increases in volume and pitch, tank discolors, or shows any signs of deforming), withdraw immediately to a secure position. If employees are expected to fight fires, they must be trained and equipped in OSHA 1910.156.

Disposal Method Suggested: Incineration. Asphalt solids may be landfilled.[22]

References

Sax, N. I., Ed., "Dangerous Properties of Industrial Materials Report" 2, No. 1, 76–77 (1982)

New Jersey Department of Health and Senior Services, "Hazardous Substance Fact Sheet: Asphalt Fumes," Trenton, NJ (August 1985)

National Institute for Occupational Safety and Health, Criteria for a Recommended Standard: Occupational Exposure to Asphalt Fumes, NIOSH Doc. No. 78–106, Washington, DC (September 1977)

New York State Department of Health, "Chemical Fact Sheet: Asphalt," Albany, NY, Bureau of Toxic Substance Assessment (March 1986)

Atrazine

Molecular Formula: $C_8H_{14}ClN_5$

Synonyms: A361; Aatram; Aatrex; Aatrex 4L; Aatrex 80W; Aatrex herbicide; Aatrex Nine-O; Actinite PK; 2-Aethylamino-4-chlor-6-isopropylamino-1,3,5-triazin (German); AI3-28244; Aktikon; Aktikon PK; Aktinit A; Aktinit PK; Argezin; Atazinax; Atranex; Atrasine; Atratol; Atratol A; Atrazin; Atrazin 80; Atrazina (Spanish); Atred; Atrex; Candex; Cekuzina-T; 2-Chloro-4-ethylamineisopropylamine-*s*-triazine; 1-Chloro-3-ethylamino-5-isopropylamino-*s*-triazine; 1-Chloro-3-ethylamino-5-isopropylamino-2,4,6-triazine; 2-Chloro-4-ethylamino-6-isopropylamino-*s*-triazine; 2-Chloro-4-ethylamino-6-isopropylamino-1,3,5-triazine; 2-Chloro-4-ethylamono-6-isopropylamino-; 6-Chloro-*N*-ethyl-*N*-isopropyl-1,3,5-triazinediyl-2,4-diamine; 6-Chloro-*N*-ethyl-*N*'-(1-methylethyl)-1,3,5-triazine-2,4-diamine; 2-Chloro-4-(2-propylamino)-6-ethylamino-*s*-triazine; Chromozin; Crisatrina; Crisazine; Cyazin; EPA pesticide chemical code 080803; 2-Ethylamino-4-isopropylamino-6-chloro-*s*-triazine; Farmco atrizine; Fenamin; Fenamine; Fenatrol; G30027; Geigy 30,027; Gesaprim; Gesaprim 50; Gesaprim 500L; Gesoprim; Griffex; Hungazin; Hungazin PK; Inakor; New chlorea; NSC 163046; Oleogesaprim; Penatrol; Plant extract, Corn grown in atrizine-treated soil; Primatol; Primatol A; Primaze; Radazin; Radizine; Residox; Shell Atrazine herbicide; Strazine; Triazine A 1294; *s*-Triazine, 2-chloro-4-(ethylamino)-6-(isopropylamino)-; 1,3,5-Triazine-2,4-diamine,6-chloro-*N*-ethyl-*N*'-(1-methylethyl)-; *s*-Triazine, zeazin; Tripart Atrazine 50 SC; Vectal; Vectal SC; Weedex; Weedex A; Wonuk; Zeapos; Zeazin; Zeazin 50; Zeazine

CAS Registry Number: 1912-24-9

RTECS Number: XY5600000

DOT ID: UN 2763

Regulatory Authority

- Air Pollutant Standard Set (ACGIH)[1] (DFG)[3] (HSE)[33] (former USSR)[43] (Several States)[60] (Australia) (Israel) (Mexico) (Various Canadian Provinces)
- Water Pollution Standard Set (EPA) (Canada)
- Safe Drinking Water Act: 40CFR141.61(c)5, MCL, 0.003 mg/l; 40CFR141.50(b)7, MCGL 0.003 mg/l; 40CFR142.62, Variances and Exceptions from the MCLs; 40CFR9352 Regulated Chemical; 40CFR141.24, Requirements for Sampling and Analytical Testing; 40CFR 141.32 Public Notification Requirements
- FIFRA 40CFR180.102-1147 et tolerances on raw agricultural commodities
- CERCLA/SARA 40CFR372.65: Form R *de minimis* Concentration Reporting Level: 0.1%

Cited in U.S. State Regulations: Alaska (G), California (A, G, W), Connecticut (A), Florida (G), Illinois (G), Kansas (W), Maine (G, W), Massachusetts (G, W), Minnesota (G), Nevada (A), New Hampshire (G), New Jersey (G), New York (W), North Dakota (A), Pennsylvania (G), Rhode Island (G), Virginia (A), West Virginia (G).

Description: Atrazine, $C_8H_{14}ClN_5$, is a white, odorless, crystalline solid or powder which is often mixed with a liquid. Freezing/Melting point = 175 – 177°C. Hazard Identification (based on NFPA-704 M Rating System): Health 1, Flammability 0, Reactivity 0. Slightly soluble in water.

Potential Exposure: Personnel involved in manufacture, formulation or application of this herbicide and plant growth regulator.

Incompatibilities: Strong oxidizers, acids.

Permissible Exposure Limits in Air: The NIOSH and ACGIH recommended airborne exposure limit is 5 mg/m^3 TWA for a 10-hour and 8-hour workshift, respectively. he DFG[3] has set a MAK limit value of 2 mg/m^3. The Australian and Israel limit is 5 mg/m^3 and Israel's Action Limit is 2.5 mg/m^3 The HSE (U.K.)[33] and Mexico limit is 10 mg/m^3 TWA. Canadian Provincial Limits are: Alberta: 5 mg/m^3 TWA and 15 min.

STEL of 10 mg/m^3; British Columbia: 10 mg/m^3; Ontario and Quebec: 5 mg/m^3 TWAEV. The former USSR-UNEP/IRPTC project[43] has set a MAC of 2 mg/m^3 in workplace air and of 0.02 mg/m^3 for both momentary and daily average exposure in ambient air in residential areas. In addition, several states have set guidelines or standards[60] for atrazine in ambient air ranging from 50 $\mu g/m^3$ (North Dakota) to 80 $\mu g/m^3$ (Virginia) to 100 $\mu g/m^3$ (Connecticut) to 119 $\mu g/m^3$ (Nevada).

Determination in Air: Use OSHA versatile sampler-2; Reagent; Gas chromatography/Electron capture detection; NIOSH #5602.

Permissible Concentration in Water: A maximum level (MCL and MCGL) in drinking water of 0.003 mg/l has been set by EPA.[62] The Canadian Drinking Water IMAC is 0.06 mg/l. A suggested no-adverse effect level in drinking water has been calculated by NAS/NRC as 0.15 mg/l. An acceptable daily intake (ADI) of 0.0215 mg/kg/day has been calculated for atrazine.[46] A limit of 0.5 mg/l of atrazine has been specified by the former USSR-UNEP/IRPTC program[43] in water bodies used for domestic purposes. Also, several states have set guidelines for atrazine in drinking water[61] ranging from 0.093 μg/l (Massachusetts) to 15 μg/l (California) to 25 μg/l (New York) to 43 μg/l (Maine) to 150 μg/l (Kansas).

Determination in Water: Analysis of atrazine is by a gas chromatographic (GC) method applicable to the determination of certain nitrogen-phosphorus containing pesticides in water samples. In this method, approximately 1 l of sample is extracted with methylene chloride. The extract is concentrated, and the compounds are separated using capillary column GC. Measurement is made using a nitrogen phosphorus detector. The method detection limit has not been determined for this compound, but it is estimated that the detection limits for the method analytes are in the range of 0.1 – 2 μg/l.

Routes of Entry: Inhalation, passing through the skin.

Harmful Effects and Symptoms

Atrazine is possibly carcinogenic to humans (IARC, 2B). There is inadequate evidence to confirm carcinogenicity of atrazine in humans. However, there is the increased risks for tumors known to be associated with hormonal factors. These have been observed in both animals and human beings, and are consistent with the known effects of atrazine on the hypothalamic pituitary gonadal axis.

Short Term Exposure: Contact may skin and severe eye irritation.

Long Term Exposure: Atrazine may be a carcinogen in humans since it has been show to cause mammary and uterine cancers in animals. There is limited evidence that atrazine may damage the developing fetus. Atrazine may cause skin allergy.

Points of Attack: Eyes, skin, respiratory system, central nervous system, liver.

Medical Surveillance: Evaluation by a qualified allergist. Examination of the nervous system.

First Aid: If this chemical gets into the eyes, remove any contact lenses at once and irrigate immediately for at least 15 minutes, occasionally lifting upper and lower lids. Seek medical attention immediately. If this chemical contacts the skin, remove contaminated clothing and wash immediately with soap and water. Seek medical attention immediately. If this chemical has been inhaled, remove from exposure, begin rescue breathing (using universal precautions) if breathing has stopped and CPR if heart action has stopped. Transfer promptly to a medical facility. When this chemical has been swallowed, get medical attention. Give large quantities of water or milk and induce vomiting. Do not make an unconscious person vomit.

Personal Protective Methods: Wear protective gloves and clothing to prevent any reasonable probability of skin contact. Safety equipment suppliers/manufacturers can provide recommendations on the most protective glove/clothing material for your operation. All protective clothing (suits, gloves, footwear, headgear) should be clean, available each day, and put on before work. Contact lenses should not be worn when working with this chemical. Wear splash-proof (for liquid) or dust-proof (for powders or dust) chemical goggles and face shield unless full facepiece respiratory protection is worn. Employees should wash immediately with soap when skin is wet or contaminated. Provide emergency showers and eyewash.

Respirator Selection: *At any concentrations above the NIOSH REL, or where there is no REL, at any detectable concentration:* SCBAF:PD,PP (any MSHA/NIOSH approved self-contained breathing apparatus that has a full facepiece and is operated in a pressure-demand or other positive-pressure mode); or SAF:PD,PP:ASCBA (any supplied-air respirator that has a full facepiece and is operated in a pressure-demand or other positive-pressure mode in combination with an auxiliary, self-contained breathing apparatus operated in a pressure-demand or other positive pressure mode).

Storage: Prior to working with this chemical you should be trained on its proper handling and storage. Should be stored in tightly-closed containers away from strong acids.

Shipping: There are no DOT specs for atrazine as such but Triazine pesticides, solid, toxic, n.o.s. should bear a "Poison" label. They fall in Hazard Class 6.1 and Packing Group I or II.[19][20]

Spill Handling: Evacuate persons not wearing protective equipment from area of spill or leak until clean-up is complete. Remove all ignition sources. Ventilate area of spill or leak. Collect spilled dry material in the most convenient and safe manner and deposit in sealed containers for reclamation or for disposal in an approved facility. Absorb liquid containing atrazine in vermiculite, dry sand, earth, or similar material. Treatment technologies which will remove atrazine from water include activated carbon adsorption, ion exchange, reverse osmosis, ozone oxidation and ultraviolet irradiation. Conventional treatment methods have been found to be ineffective for the removal of atrazine from drinking water.

Limited data suggest that aeration would not be effective in atrazine removal. It may be necessary to contain and dispose of this chemical as a hazardous waste. If material or contaminated runoff enters waterways, notify downstream users of potentially contaminated waters. Contact your Department of Environmental Protection or your regional office of the federal EPA for specific recommendations. If employees are required to clean-up spills, they must be properly trained and equipped. OSHA 1910.120(q) may be applicable.

Fire Extinguishing: Extinguish fire using an agent suitable for the type of surrounding fire; atrazine itself does not burn. Poisonous gases are produced in fire including hydrogen chloride and nitrogen oxide. If material or contaminated runoff enters waterways, notify downstream users of potentially contaminated waters. Notify local health and fire officials and pollution control agencies. Containers may explode in fire. From a secure, explosion-proof location, use water spray to cool exposed containers. If cooling streams are ineffective (venting sound increases in volume and pitch, tank discolors, or shows any signs of deforming), withdraw immediately to a secure position. If employees are expected to fight fires, they must be trained and equipped in OSHA 1910.156.

Disposal Method Suggested: Atrazine is hydrolyzed by either acid or base.[22] The hydroxy compounds are generally herbicidally inactive, but their complete environmental effects are uncertain. However, the method appears suitable for limited use and quantities of triazine. Atrazine underwent >99% decomposition when burned in a polyethylene bag, and combustion with a hydrocarbon fuel would appear to be a generally suitable method for small quantities. Combustion of larger quantities would probably require the use of a caustic wet scrubber to remove nitrogen oxides and HCl from the product gases.[22]

References

New Jersey Department of Health and Senior Services, "Hazardous Substance Fact Sheet: Atrazine," Trenton, NJ (June 1998)

U.S. Environmental Protection Agency, "Health Advisory: Atrazine," Washington, DC, Office of Drinking Water (September 1987)

Auramine

Molecular Formula: $C_{17}H_{21}N_3$

Common Formula: $(CH_3)_2N\text{-}C_6H_4CNH\text{-}C_6H_4\text{-}N(CH_3)_2$

Synonyms: Apyonine auramarine base; Auramina (Spanish); Auramine; Auramine base; Auramine N base; Auramine OAF; Auramine O base; Auramine SS; Basic yellow 2; Baso yellow 124; Benzeneamine, 4,4'-cabonimidoylbis(N-dimethyl-); Brilliant oil yellow; 4,4'-Carbonimidoylbis(*N,N*-dimethylbenzenamine); C.I. 41000B; C.I. basic yellow 2, free base; C.I. solvent yellow 34; 4,4'-Dimethylaminobenzophenonimide; Glauramine; 4,4-(Imidocarbonyl)bis (*N,N*-dimethylaniline); Tetramethyldiaminodiphenylacetimine; Waxoline yellow O; Yellow pyoctanine

CAS Registry Number: 492-80-8; 2465-27-2 (Hydrochloride salt)

RTECS Number: BY3500000; BY3675000 (Hydrochloride salt)

Regulatory Authority

- Carcinogenic (UN)[13] (Animal Positive) (DFG)[3] (IARC) (2B, Sufficient animal data)
- Banned or Severely Restricted (Italy, Sweden) (UN)[13]
- Air Pollutant Standard Set (North Dakota, New York)[60]
- EPA HAZARDOUS WASTE NUMBER (RCRA No.): U014 (as C.I. Solvent Yellow 34)
- RCRA, 40CFR261, Appendix 8 Hazardous Constituents
- SUPERFUND/EPCRA 40CFR302.4, Appendix A, Reportable Quantity (RQ): CERCLA, 100 lb (45.4 kg)
- CERCLA/SARA 40CFR372.65: Form R *de minimis* Concentration Reporting Level: 0.1%

Cited in U.S. State Regulations: California (G), Florida (G), Illinois (G), Kansas (G), Louisiana (G), Maine (G), Maryland (G), Massachusetts (G), Minnesota (G), New Hampshire (G), New Jersey (G), New York (A), North Dakota (A), Pennsylvania (G), Rhode Island (G), Vermont (G), Virginia (G), Washington (G), Wisconsin (G).

Description: Auramine, $C_{17}H_{21}N_3$, is a yellow, crystalline powder or flaky material. Freezing/Melting point = 136°C. Hazard Identification (based on NFPA-704 M Rating System): Health 2, Flammability 1, Reactivity 0. Soluble in water.

Potential Exposure: Auramine is used industrially as a dye or dye intermediate for coloring textiles, paper, and leather. Also used as an antiseptic (a powerful antiseptic in ear and nose surgery and in gonorrhea treatment) and fungicide. Human exposure to auramine occurs principally through skin absorption or inhalation of vapors. Low-level dermal exposure to the consumer may occur but would be limited to any migration of auramine from fabric, leather, or paper goods.

Incompatibilities: Strong oxidizers.

Permissible Exposure Limits in Air: No occupational exposure limits have been established. However auramine may be a carcinogen; there may be no safe level of exposure. This chemical can be absorbed through the skin, thereby increasing the potential for exposure. Zero in New York, North Dakota[60] in ambient air.

Permissible Concentration in Water: No criteria set.

Routes of Entry: Inhalation, ingestion, skin absorption.

Harmful Effects and Symptoms

Short Term Exposure: Contact can irritate the eyes, and may cause damage. Skin absorption may result in dermatitis and burns, nausea and vomiting.

Long Term Exposure: Commercial auramine is carcinogenic in mice and rats after oral administration, producing liver tumors, and after subcutaneous injection in rats, producing local sarcomas. The manufacture of auramine (which also involves exposure to other chemicals) has been shown in one study to be causally associated with an increase in bladder cancer. The actual carcinogenic compound(s) has not been specified precisely.

Points of Attack: Liver, bladder.

Medical Surveillance: Monthly urinalysis. Physical exam every 6 months focused on bladder.

First Aid: If this chemical gets into the eyes, remove any contact lenses at once and irrigate immediately for at least 15 minutes, occasionally lifting upper and lower lids. Seek medical attention immediately. If this chemical contacts the skin, remove contaminated clothing and wash immediately with soap and water. Seek medical attention immediately. If this chemical has been inhaled, remove from exposure, begin rescue breathing (using universal precautions) if breathing has stopped and CPR if heart action has stopped. Transfer promptly to a medical facility. When this chemical has been swallowed, get medical attention. Give large quantities of water and induce vomiting. Do not make an unconscious person vomit.

Personal Protective Methods: Wear protective gloves and clothing to prevent any reasonable probability of skin contact. Safety equipment suppliers/manufacturers can provide recommendations on the most protective glove/clothing material for your operation. All protective clothing (suits, gloves, footwear, headgear) should be clean, available each day, and put on before work. Contact lenses should not be worn when working with this chemical. Wear dust-proof chemical goggles and face shield unless full facepiece respiratory protection is worn. Employees should wash immediately with soap when skin is wet or contaminated. Provide emergency showers and eyewash.

Respirator Selection: Engineering controls should be used wherever feasible to maintain airborne concentrations of this chemical below the prescribed exposure limit. Respirators and protective equipment less effective than engineering controls, and should be used only in non-routine or emergency situations which may result in exposure concentrations in excess of the TWA environmental limit. *Where there is no REL, at any detectable concentration:* SCBAF:PD,PP (any MSHA/NIOSH approved self-contained breathing apparatus that has a full facepiece and is operated in a pressure-demand or other positive-pressure mode); or SAF:PD,PP:ASCBA (any supplied-air respirator that has a full facepiece and is operated in a pressure-demand or other positive-pressure mode in combination with an auxiliary, self-contained breathing apparatus operated in a pressure-demand or other positive pressure mode).

Storage: Prior to working with this chemical you should be trained on its proper handling and storage. Store in tightly closed containers in a cool, well ventilated area away from oxidizers. Sources of ignition such as smoking and open flames, are prohibited where this chemical is used, handled, or stored in a manner that could create a potential fire or explosion hazard. A regulated, marked area should be established where this chemical is handled, used, or stored in compliance with OSHA standard 1910.1045.

Shipping: Auramine per se is not on the DOT list[19] of hazardous wastes and chemical substances as no label requirement or limits on shipping amounts are given. Pesticides, solid, toxic, n.o.s. require a "Keep away from Food" label. They fall in Hazard Class 6.1 and Auramine falls in Packing Group III.

Spill Handling: Evacuate persons not wearing protective equipment from area of spill or leak until clean-up is complete. Remove all ignition sources. Cover with sand and soda ash. Collect powdered material in the most convenient and safe manner and deposit in sealed containers. Ventilate area after clean-up is complete. It may be necessary to contain and dispose of this chemical as a hazardous waste. If material or contaminated runoff enters waterways, notify downstream users of potentially contaminated waters. Contact your Department of Environmental Protection or your regional office of the federal EPA for specific recommendations. If employees are required to clean-up spills, they must be properly trained and equipped. OSHA 1910.120(q) may be applicable.

Fire Extinguishing: Use water, carbon dioxide, powdered agents or alcohol foam on auramine fires. Poisonous gases are produced in fire including nitrogen oxide. If material or contaminated runoff enters waterways, notify downstream users of potentially contaminated waters. Notify local health and fire officials and pollution control agencies. From a secure, explosion-proof location, use water spray to cool exposed containers. If cooling streams are ineffective (venting sound increases in volume and pitch, tank discolors, or shows any signs of deforming), withdraw immediately to a secure position. If employees are expected to fight fires, they must be trained and equipped in OSHA 1910.156.

Disposal Method Suggested: Consult with environmental regulatory agencies for guidance on acceptable disposal practices. Generators of waste containing this contaminant (≥100 kg/mo) must conform with EPA regulations governing storage, transportation, treatment, and waste disposal. Incinerate in furnace with afterburner and scrubber[24]

References

Sax, N. I., Ed., "Dangerous Properties of Industrial Materials Report" 1, No. 5, 37–38 (1981)

New Jersey Department of Health and Senior Services, Hazardous Substance Fact Sheet, Auramine, Trenton NJ (April 1997)

Azathioprine

Molecular Formula: $C_9H_7N_2O_2S$

Synonyms: Azanin azatioprin; Azothioprine; BW 57-322; Ccucol; Imuran; Imurek; Imurel; 6-(1'-Methyl-4'-nitro-5'-imidazolyl)-mercaptopurine; Methylnitroimidazolylmercaptopurine; 6-(1-Methyl-*p*-nitro-5-imidazolyl)-thiopurine; 6-(1-Methyl-4-nitroimidazol-5-ylthio)purine; 6-(Methyl-*p*-nitro-5-imidazolyl)-thiopurine; 6-[(1-Methyl-4-nitro-1H-imidazol-5-

yl)thio]-1H-purine; 6-[(1-Methyl-4-nitroimidazol-5-yl)thio]purine; NCI-C03474; NSC-39084; Rorasul

CAS Registry Number: 446-86-6

RTECS Number: UO8925000

DOT ID: UN 2811

Regulatory Authority

- Carcinogen (Human Positive) (NTP)[10] (IARC)[9]

Cited in U.S. State Regulations: California (A, G), Florida (G), Illinois (G), Maine (G), Massachusetts (G), Minnesota (G), New Jersey (G), Pennsylvania (G), Rhode Island (G).

Description: Azathioprine, $C_9H_7N_2O_2S$, is a complex heterocyclic compound which forms pale yellow crystals. Freezing/Melting point = 243 – 244°C (decomposes).

Potential Exposure: Azathioprine is an immunosuppressive agent, generally used in combination with a corticosteroid to prevent rejection following renal homotransplantations. It also is used following transplantation of other organs. Other uses of Azathioprine include the treatment of a variety of presumed autoimmune diseases, including rheumatoid arthritis, ankylosing spondylitis, systemic lupus erythematosus, dermatonyositis, periarteritis nodosa, scleroderma, refractory thombocytopenic purpura, autoimmune hemolytic anemia, chronic active liver disease, regional enteritis, ulcerative colitis, various autoimmune diseases of the eye, acute and chronic glomerulonephritis, the nephritic syndrome, Wegener's granulomatosis, and multiple sclerosis.

Permissible Exposure Limits in Air: No standards set.

Permissible Concentration in Water: No criteria set.

Routes of Entry: Human exposure to Azathioprine occurs because of its widespread use, since the 1970s, to prevent rejection following organ transplantation and to treat a variety of autoimmune diseases. Azathioprine is readily absorbed from the gut and is known to cross the human placenta.

Harmful Effects and Symptoms

There is sufficient evidence that Azathioprine is carcinogenic in humans. Two large prospective studies of kidney transplant patients (receiving Azathioprine and prednisone almost routinely) showed that these patients experienced increased incidences of non-Hodgkin's lymphomas, cancer of the skin, hepatobiliary carcinoma and other tumors. Patients treated with Azathioprine, but not transplant recipients, showed an increased incidence of the same cancers as patients with transplants, but to a lesser extent. Mice receiving Azathioprine by intraperitoneal, subcutaneous or intramuscular injection had suggestive evidence of induced lymph system neoplasms. Rats given azathioprine orally had suggestive evidence of induced ear cancers. Results of the animal studies provide limited evidence of carcinogenicity. Among reported symptoms of Azathioprine exposure are bone-marrow depression, especially leukopenia;[52] toxic hepatitis and jaundice; stomatitis; dermatitis; hair loss; fever; anorexia; vomiting and diarrhea.

First Aid: If this chemical gets into the eyes, remove any contact lenses at once and irrigate with water or normal saline immediately for at least 20 – 30 minutes, occasionally lifting upper and lower lids. Seek medical attention immediately. If this chemical contacts the skin, remove contaminated clothing and wash immediately with soap and water. Seek medical attention immediately. If this chemical has been inhaled, remove from exposure, begin rescue breathing (using universal precautions) if breathing has stopped and CPR if heart action has stopped. Transfer promptly to a medical facility. When this chemical has been swallowed, get medical attention. Give large quantities of water and induce vomiting. Do not make an unconscious person vomit.

Storage: Prior to working with this chemical you should be trained on its proper handling and storage. Store in a refrigerator under an inert atmosphere.[52] A regulated, marked area should be established where this chemical is handled, used, or stored in compliance with OSHA standard 1910.1045.

Shipping: Poisonous solids, n.o.s., in Hazard Class 6.1 and Packing Group III requires a "Keep away from Food" label.

Spill Handling: Evacuate persons not wearing protective equipment from area of spill or leak until clean-up is complete. Remove all ignition sources. Dampen spilled material with toluene to avoid airborne dust. Seal the accumulated wastes in vapor-tight plastic bags for eventual disposal. Ventilate area after clean-up is complete. It may be necessary to contain and dispose of this chemical as a hazardous waste. If material or contaminated runoff enters waterways, notify downstream users of potentially contaminated waters. Contact your Department of Environmental Protection or your regional office of the federal EPA for specific recommendations. If employees are required to clean-up spills, they must be properly trained and equipped. OSHA 1910.120(q) may be applicable.

Fire Extinguishing: Use dry chemical, carbon dioxide, water spray, or alcohol foam extinguishers. Poisonous gases are produced in fire including nitrogen and sulfur oxides. If material or contaminated runoff enters waterways, notify downstream users of potentially contaminated waters. Notify local health and fire officials and pollution control agencies. If employees are expected to fight fires, they must be trained and equipped in OSHA 1910.156.

References

Sax, N. I., Ed., "Dangerous Properties of Industrial Materials Report" 1, No. 4, 36–37 (1981)

Azinphos-Ethyl

Molecular Formula: $C_{12}H_{16}N_3O_3PS_2$

Synonyms: Athyl-gusathion; Azinfos-ethyl (Dutch); Azinos; Azinphos-aethyl (German); Azinphos etile (Italian); Bay 16225; Bayer 16259; Benzotriazine derivative of an ethyl dithiophosphate; Cotnion-ethyl; Crysthion; Crysthyon; *O,O*-Diethyl *S*-(4-oxo-3H-1,2,3-bezotriazine-3-yl)methyl] dithiophosphate; *O,O*-Diethyl-S-(4-oxobezotriazin-3-methyl)-dithiophosphat (German); *O,O*-Diethyl *S*-(4-oxobezotriazino-3-methyl) phosphorodithioate; *o,o*-Diethyl-S-[(4-oxo-3H-1,2,3-bezotriazin-3yl)methyl]-dithio fosfaat (Dutch); *O,O*-Diethyl-S-[(4-oxo-3H-1,2,3-bezotriazin-3-yl)-methyl]-dithiophosphat (German); *O,O*-Diethylphosphorodithioate ester with 3-(mercaptomethyl)-1,2,3-benzotriazin-4(3H)-one; *o,o*-Dietil-S-[(4-oxo-3H-1,2,3-bezotriazin-3il)metil]-ditiofosfato (Italian); *S*-(3,4-Dihydro-4-oxo-1,2,3-benzotriazin-3-ylmethyl) *O,O*-diethyl phosphorodithioate; 3,4-Dihydro-4-oxo-3-benzotriazinylmethyl *O,O*-diethyl phosphorodithioate; ENT 22,014; Ethyl azinphos; Ethyl guthion; Etil azinfos (Spanish); Etiltriazotion; Gusathion A; Gusathion A insecticide; Gusathion ethyl; Guthion ethyl; Guthion insecticide; R 1513; Triazotion (Russian)

CAS Registry Number: 2642-71-9

RTECS Number: TD8400000

DOT ID: UN 2783 (organophosphorus pesticides, solid, toxic)

Regulatory Authority

- Banned or Severely Restricted (UN)[35]
- Very Toxic Substance (World Bank)[15]
- CERCLA/SARA 40CFR302, Extremely Hazardous Substances: TPQ = 100/10,000 lb (45.4/4,540 kg)
- SUPERFUND/EPCRA 40CFR302.4, Appendix A, Reportable Quantity (RQ): EHS, 1 lb (0.454 kg)
- U.S. DOT Regulated Marine Pollutant (49CFR172.101, Appendix B)

Cited in U.S. State Regulations: Florida (G), Massachusetts (G), Michigan (G), New Jersey (G), Pennsylvania (G).

Description: Azinphos-ethyl, $C_{12}H_{16}N_3O_3PS_2$, is a colorless crystalline substance. Freezing/Melting point = 53°C. Boiling point = 111°C. Slightly soluble in water.

Potential Exposure: It is a non-systemic organophosphate insecticide and miticide with good ovicidal properties and long persistence. It is not registered for use in the U.S. Among other crops, it is used on cotton, citrus, vegetables, potatoes, tobacco, rice and cereals to control caterpillars, beetles, aphids, spiders and many other insects.

Incompatibilities: Strong oxidizers, strong acids.

Permissible Exposure Limits in Air: No standards set. However dusts or mists are poisonous.

Determination in Air: OSHA versatile sampler-2; Toluene/Acetone; Gas chromatography/Flame photometric detection for sulfur, nitrogen, or phosphorus; NIOSH Method IV Method #5600, Organophosphorus Pesticides.

Permissible Concentration in Water: No criteria set.

Routes of Entry: Inhalation, ingestion, skin absorption.

Harmful Effects and Symptoms

Short Term Exposure: The symptoms are similar to parathion. Nausea is often the first symptom followed by vomiting, abdominal cramps, diarrhea and excessive salivation. Also common in inhalation exposure are headache, giddiness, vertigo and weakness, nasal discharge and a sensation of tightness in the chest. Other symptoms include blurring or dimness of vision; tearing; eye muscle spasm and pain; pinpoint pupils; loss of muscle coordination; slurring of speech; muscle twitching (especially tongue and eyelids); difficulty in breathing; excessive secretions of mucous in mouth, nose, and respiratory tract; convulsions and coma. The systemic effects of this compound are similar to parathion. It is an extremely potent systemic toxicant via ingestion, inhalation and skin contact. It may cause death or permanent injury after very short exposure to small quantities. The oral LD_{50} for rat is 7 mg/kg.[9] A cholinesterase inhibitor. Like similar organic phosphorus poisons, guthion-ethyl may act as an irreversible inhibitor of the enzyme cholinesterase. This enzyme allows the accumulation of large amounts of acetylcholine. Death can be caused when a critical level of cholinesterase depletion is reached. Recovery may be complete when the poisoned victim has time to recover and regenerate cholinesterase.

Long Term Exposure: Cholinesterase inhibitor; cumulative effect is possible. This chemical may damage the nervous system with repeated exposure, resulting in convulsions, respiratory failure. May cause liver damage.

Points of Attack: Respiratory system, lungs, central nervous system, cardiovascular system, skin, eyes, plasma and red blood cell cholinesterase.

Medical Surveillance: Before employment and at regular times after that, the following are recommended: Plasma and red blood cell cholinesterase levels (tests for the enzyme poisoned by this chemical). If exposure stops, plasma levels return to normal in 1 – 2 weeks while red blood cell levels may be reduced for 1 – 3 months.

When cholinesterase enzyme levels are reduced by 25% or more below preemployment levels, risk of poisoning is increased, even if results are in lower ranges of "normal." Reassignment to work not involving organophosphate or carbamate pesticides is recommended until enzyme levels recover. If symptoms develop or overexposure occurs, repeat the above tests as soon as possible and get an exam of the nervous system. Also consider complete blood count. Consider chest x-ray following acute overexposure. Do not drink any alcoholic beverages before or during use. Alcohol promotes absorption of organic phosphates.

First Aid: If this chemical gets into the eyes, remove any contact lenses at once and irrigate immediately for at least 15 minutes, occasionally lifting upper and lower lids. Seek medical attention immediately. If this chemical contacts the skin, remove contaminated clothing and wash immediately with soap and water. Speed in removing material from skin is of extreme importance. Shampoo hair promptly if contaminated. Seek medical attention immediately. If this chemical has been inhaled, remove from exposure, begin rescue breathing (using universal precautions) if breathing has stopped and CPR if heart action has stopped. Transfer promptly to a medical facility. When this chemical has been swallowed, get medical attention. Give large quantities of water and induce vomiting. Do not make an unconscious person vomit.

Personal Protective Methods: Wear appropriate clothing to prevent any reasonable probability of skin contact. Wear eye protection to prevent any reasonable probability of eye contact. Employees should wash immediately when skin is wet or contaminated. Work clothing should be changed daily if it is at all possible that clothing is contaminated. Remove nonimpervious clothing immediately if wet or contaminated. Provide emergency showers.

Respirator Selection: See entry for Azinphos-Methyl. Use only MSHA/NIOSH approved air purifying respirators for pesticides.

Storage: Prior to working with this chemical you should be trained on its proper handling and storage. Store in tightly closed containers in a cool, well ventilated area away from oxidizers, acids, and sources of ignition. Although this compound is chemically stable in storage, it is decomposed at elevated temperatures with evolution of gas, and rapidly decomposed in cold alkali to form anthranilic acid and other decomposition products.[22]

Shipping: This chemical is an organophosphorus pesticide, solid, toxic. The required label is "Poison," it is in hazard class 6.1 and Packing Group II.

Spill Handling: Do not touch spilled material. Evacuate persons not wearing protective equipment from area of spill or leak until clean-up is complete. Use water spray to reduce vapors. Remove all ignition sources. If spill is wet take up with diatomite, clay, expanded mineral, foamed glass or synthetic absorbent material and deposit in sealed containers. For small dry spills use a high efficiency particulate absolute (HEPA) filter vacuum, *not* a standard shop vac, or wet method to reduce dust during cleanups. *Do not dry sweep or create airborne dust.* Wash spill area with common household detergent. Ventilate area after clean-up is complete. It may be necessary to contain and dispose of this chemical as a hazardous waste. If material or contaminated runoff enters waterways, notify downstream users of potentially contaminated waters. Contact your Department of Environmental Protection or your regional office of the federal EPA for specific recommendations. If employees are required to clean-up spills, they must be properly trained and equipped. OSHA 1910.120(q) may be applicable.

Fire Extinguishing: This material is noncombustible. Wear protective clothing and equipment. Use dry chemical, carbon dioxide, water spray, or standard foam extinguishers. Poisonous gases are produced in fire including nitrogen oxides, sulfur oxides, phosphorus oxides. If material or contaminated runoff enters waterways, notify downstream users of potentially contaminated waters. Notify local health and fire officials and pollution control agencies. From a secure, explosion-proof location, use water spray to cool exposed containers. If cooling streams are ineffective (venting sound increases in volume and pitch, tank discolors, or shows any signs of deforming), withdraw immediately to a secure position. If employees are expected to fight fires, they must be trained and equipped in OSHA 1910.156.

Disposal Method Suggested: In accordance with 40CFR 165 recommendations for the disposal of pesticides and pesticide containers. Must be disposed properly by following package label directions or by contacting your state pesticide or environmental control agency or by contacting your regional EPA office.

References

U.S. Environmental Protection Agency, "Chemical Profile: Azinphos-Ethyl," Washington, DC, Chemical Emergency Preparedness Program (November 30, 1987)

Azinphos-Methyl

Molecular Formula: $C_{10}H_{12}N_3O_3PS_2$

Synonyms: Azinfos-methyl (Dutch); Azinphos-methyl; Azinphos-methyl Guthion®; Azinphosmetile (Italian); Bay 9027; Bayer 17147; Benzotriazine derivative of a methyl dithiophosphate; Benzotriazinedithiophosphoric acid dimethoxy ester; Carfene; Cotnion methyl; Crysthion 2L; Crysthyon; DBD; *S*-(3,4-Dihydro-4-oxobenzo[d][1,2,3] triazin-3-ylmethyl) *O,O*-dimethyl phosphorodithioate; *S*-(3,4-Dihydro-4-oxo-1,2,3-benzotriazin-3-ylmethyl) *O,O*-dimethyl phosphorodithioate; *S*-(3,4-Dihydro-4-oxobenzo[a] [1,2,3]triazin-3-ylmethyl) *O,O*-dimethyl phosphorodithioate; *O,O*-Dimethyl *S*-(1,2,3-bezotriazinyl-4-keto)methyl phosphorodithioate; *O,O*-Dimethyl *S*-(3,4-dihydro-4-keto-1,2,3-bezotriazinyl-3-methyl) dithiophosphate; Dimethyldithiophosphoric acid *N*-methylbenzazimide ester; *O,O*-Dimethyl *S*-(4-oxo-3H-1,2,3-benzotriazine-3-methyl) phosphorodithioate; *O,O*-Dimethyl *S*-[4-oxobenzotriazino-3)-methyl] phosphorodithioate; *O,O*-Dimethyl *S*-[(4-oxo-1,2,3-benzotriazino-3)methyl] thiophosphorodithioate; *o,o*-Dimethyl-S-(4-oxo-3H-1,2,3-benzotriazin-3-yl)-methyl] dithiofosfaat (Dutch); *O,O*-Dimethyl-S-[(4-oxo-3H-1,2,3-benzotriazin-3-yl)-methyl]dithiophosphat (German); *O,O*-Dimethyl *S*-[oxo-1,2,3-benzotriazin-3-(4H)-yl-methyl] phosphodithioate; *O,O*-Dimethyl *S*-(4-oxo-1,2,3-bezotriazin-3(4H)-yl methyl) phosphorodithioate; *o,o*-Dimetil-S-[(4-oxo-3H-1,2,3-benzotriazin-3-il-metil)-ditiofosfato (Italian); ENT 23,233; Gothnion®; Gusathion®; Gusathion® M; Guthion®; 3-(Mercaptomethyl)-1,2,3-benzotriazin-4(3H)-one *O,O*-dimethyl phosphorodithioate; 3-(Mercaptomethyl)-1,2,3-benzotriazin-4(3H)-one *O,O*-dimethyl phosphorodithioate *S*-ester; Methyl azinphos; *N*-Methylbenzazimide, dimethyldithiophosphoric acid ester; Methyl guthion; Metil azinfos (Spanish); Metiltriazotion (Russian); NCI-C00066; R 1582 *o,o-*

CAS Registry Number: 86-50-0

RTECS Number: TE1925000

DOT ID: UN 2783

Regulatory Authority

- Banned or Severely Restricted (Various countries) (UN)[13][35]
- Very Toxic Substance (World Bank)[15]
- Air Pollutant Standard Set (ACGIH)[1] (DFG)[3] (HSE)[33] (OSHA)[58] (Australia) (Israel) (Mexico) (California)
- Clean Water Act: 40CFR116.4 Hazardous Substances; 40CFR117.3, RQ (same as CERCLA) as guthion
- CERCLA/SARA 40CFR302, Extremely Hazardous Substances: TPQ = 100 /10,000 lb (45.4/4,540 kg)
- SUPERFUND/EPCRA 40CFR302.4, Appendix A, Reportable Quantity (RQ): EHS, 1 lb (0.454 kg)
- U.S. DOT Regulated Marine Pollutant (49CFR172.101, Appendix B) severe pollutant
- Canada: Drinking Water MAC

Cited in U.S. State Regulations: Alaska (G), California (A, G), Florida (G), Illinois (G), Maine (G, W), Massachusetts (G), Michigan (G), New Hampshire (G), New Jersey (G), New York (G), Pennsylvania (G), Rhode Island (G), West Virginia (G).

Description: Azinphos-methyl, $C_{10}H_{12}N_3O_3PS_2$, is a brown, waxy solid or colorless, crystalline material. Its technical form is a brown waxy solid. Freezing/Melting point = 73° to 74°C. Practically insoluble in water.

Potential Exposure: Personnel engaged in the manufacture, formulation and application of this organophosphorus insecticide and acaricide.

Incompatibilities: Oxidizers, strong acids.

Permissible Exposure Limits in Air: The Federal NIOSH/OSHA value[58] is 0.2 mg/m^3 TWA with the notation "skin" indicating the potential for cutaneous absorption. The ACGIH value is the same, including "skin" notation. The DFG[3] has set a MAK of 0.2 mg/m^3 and Peak Limitation of 10 times normal MAK (30 min. average value; do not exceed during workshift). HSE level is 0.2 mg/m^3 *and* STEL value of 0.6 mg/m^3. California, Australia, Mexico, and Israel values are 0.2 mg/m^3 TWA. Israel set an Action Level of 0.1 mg/m^3 TWA. Mexico's STEL is 0.6 mg/m^3. Canadian Provincial: Alberta and British Columbia: 0.2 mg/m^3 TWA and STEL of 0.6 mg/m^3. Ontario and Quebec: 0.2 mg/m^3 TWAEV. All regulatory authorities have warnings about skin absorption.

The NIOSH IDLH level is 10 mg/m^3. Because no useful data on acute inhalation toxicity are available concerning the toxic effects produced by azinphos-methyl, the chosen IDLH has been based on an analogy with parathion, which has an IDLH of 20 mg/m^3 (NIOSH).

Determination in Air: Collection by impinger or fritter bubbler, analysis by gas liquid chromatography. OSHA versatile sampler-2; Toluene/Acetone; Gas chromatography/Flame photometric detection for sulfur, nitrogen, or phosphorus; NIOSH Method (IV) #5600, Organophosphorus Pesticides.

Permissible Concentration in Water: For the protection of freshwater and marine aquatic life, a criterion of 0.01 μg/l has been suggested by EPA. For the protection of human health, a no-adverse effect level in drinking water has been calculated by NAS/NRC[46] as 0.088 mg/l. An allowable daily intake of 0.0125 mg/kg/day was calculated. Canada's maximum allowable concentration (MAC) in drinking water is 0.02 mg/l. The State of Maine[61] has set a guideline of 25 μg/l for Azinphos-methyl in drinking water.

Determination in Water: Pesticide residue methods which should be applicable involve hydrolysis with KOH in isopropanol to give anthranilic acid which is diazotized and coupled to give a measurable color.

Routes of Entry: Inhalation, skin absorption, ingestion, skin and/or eye contact.

Harmful Effects and Symptoms

Symptoms include nausea, vomiting, diarrhea, excessive salivation, blurring of vision and other signs of cholinesterase inhibition, loss of muscle coordination, twitching of muscles, confusion, difficulty breathing, convulsions, and death are observed with this organophosphate poison. The oral LD_{50} for rat is 11 mg/kg. The acute toxicity rating is extremely toxic. Probable oral lethal dose in humans is 5 – 50 mg/kg or between 7 drops and 1 teaspoon for a 70 kg (150 lb) person. This is a potent cholinesterase inhibitor which can cause death.

Lethal concentration data:

Species	*Reference*	*LC50* (mg/m^3)	*LCLo*	*Time* (hr)	*0.5 hr LC (CF)* (mg/m^3)	*Derived Value* (mg/m^3)
Rat	Newell and Dilley 1978	69	—	1	86 (1.25)	8.6
Rat	Sanderson 1961	79	—	1	99 (1.25)	9.9

Short Term Exposure: Exposure can cause rapid, fatal organophosphorus poisoning. Inhalation can irritate the lungs causing coughing and/or shortness of breath. Higher exposures can cause pulmonary edema, a medical emergency that can be delayed for several hours. This can cause death. Organic phosphorus insecticides are absorbed by the skin, as well as by the respiratory and gastrointestinal tracts. They are cholinesterase inhibitors. Symptoms of exposure include headache, giddiness, blurred vision, nervousness, weakness, nausea, cramps, diarrhea, and discomfort in the chest. Signs include sweating, tearing, salivation, vomiting, cyanosis, convulsions, coma, loss of reflexes and loss of sphincter control.

Long Term Exposure: Cholinesterase inhibitor; cumulative effect is possible. This chemical may damage the nervous system causing weakness, "pins and needles," and poor coordination in arms and legs, with repeated exposure, resulting in convulsions, respiratory failure. May cause liver damage. Repeated exposure may cause personality changes of depression, anxiety, or irritability.

Points of Attack: Respiratory system, lungs, central nervous system, cardiovascular system, blood cholinesterase.

Medical Surveillance: Before employment and at regular times after that, the following are recommended: Plasma and red blood cell cholinesterase levels (tests for the enzyme poisoned by this chemical). If exposure stops, plasma levels return to normal in 1 – 2 weeks while red blood cell levels may be reduced for 1 – 3 months.

When cholinesterase enzyme levels are reduced by 25% or more below preemployment levels, risk of poisoning is increased, even if results are in lower ranges of "normal." Reassignment to work not involving organophosphate or carbamate pesticides is recommended until enzyme levels recover. If symptoms develop or overexposure occurs, repeat the above tests as soon as possible and get an exam of the nervous system. Also consider complete blood count. Consider chest x-ray following acute overexposure. Do not drink any alcoholic beverages before or during use. Alcohol promotes absorption of organic phosphates.

First Aid: If this chemical gets into the eyes, remove any contact lenses at once and irrigate immediately for at least 15 minutes, occasionally lifting upper and lower lids. Seek medical attention immediately. If this chemical contacts the skin, remove contaminated clothing and wash immediately with soap and water. Speed in removing material from skin is of extreme importance. Shampoo hair promptly if contaminated. Seek medical attention immediately. If this chemical has been inhaled, remove from exposure, begin rescue breathing (using universal precautions) if breathing has stopped and CPR if heart action has stopped. Transfer promptly to a medical facility. When this chemical has been swallowed, get medical attention. Give large quantities of water and induce vomiting. Do not make an unconscious person vomit. Medical observation is recommended for 24 – 48 hours after breathing overexposure, as pulmonary edema may be delayed. As first aid for pulmonary edema, a doctor or authorized paramedic may consider administering a corticosteroid spray.

Personal Protective Methods: Wear appropriate clothing to prevent any reasonable probability of skin contact. Wear eye protection to prevent any reasonable probability of eye contact. Employees should wash immediately when skin is wet or contaminated. Work clothing should be changed daily if it is at all possible that clothing is contaminated. Remove nonimpervious clothing immediately if wet or contaminated. Provide emergency showers.

Respirator Selection: NIOSH/OSHA: *Up to 2 mg/m³*: CCROVDMFu (any chemical cartridge respirator with organic vapor cartridge(s) in combination with a dust, mist, and fume filter); or SA (any supplied-air respirator). *Up to 5 mg/m³:* SA:CF (any supplied-air respirator operated in a continuous-flow mode); or PAPROVDMFu (any powered, air-purifying respirator with organic vapor cartridge(s) in combination with a dust, mist, and fume filter). *Up to 10 mg/m³*: CCRFOVHiE (any chemical cartridge respirator with a full facepiece and organic vapor cartridge(s) in combination with a high-efficiency particulate filter; or GMFOVHiE (any air-purifying, full-facepiece respirator (gas mask) with a chin-style, front- or back-mounted organic vapor canister having a high-efficiency particulate filter); or PAPRTOVHiE (any powered, air-purifying respirator with a tight-fitting facepiece and organic vapor cartridge(s) in combination with a high-efficiency particulate filter); or SAT:CF (any supplied-air respirator that has a tight-fitting facepiece and is operated in a continuous-flow mode; or SCBAF (any self-contained breathing apparatus with a full facepiece); or SAF (any supplied-air respirator with a full facepiece). *Emergency or planned entry into unknown concentrations or IDLH conditions:* SCBAF:PD,PP (any MSHA/NIOSH approved self-contained breathing apparatus that has a full facepiece and is operated in a pressure-demand or other positive-pressure mode); or SAF:PD,PP:ASCBA (any supplied-air respirator that has a full facepiece and is operated in a pressure-demand or other positive-pressure mode in combination with an auxiliary, self-contained breathing apparatus operated in a pressure-demand or other positive pressure mode). *Escape:* GMFOVHiE [any air-purifying, full-facepiece respirator (gas mask) with a chin-style, front- or back-mounted organic vapor canister having a high-efficiency particulate filter]; or SCBAE (any appropriate escape-type, self-contained breathing apparatus).

Storage: Prior to working with this chemical you should be trained on its proper handling and storage. Store in tightly closed containers in a cool, well ventilated area away from oxidizers, acids, and sources of ignition. Although this compound is chemically stable in storage, it is decomposed at elevated temperatures with evolution of gas, and rapidly decomposed in cold alkali to form anthranilic acid and other decomposition products.[22]

Shipping: The DOT label requirement is "Poison." The UN/DOT Hazard Class is 6.1 and the Packing Group is II.[19][20]

Spill Handling: Evacuate persons not wearing protective equipment from area of spill or leak until clean-up is complete. Remove all ignition sources. Stay upwind; keep out of low areas. Ventilate closed spaces before entering them. Wear positive pressure breathing apparatus and special protective clothing. Remove and isolate contaminated clothing at the site. Do not touch spilled material. Stop leak if you can do so without risk. Use water spray to reduce vapors. *Small spills:* absorb with sand or other noncombustible absorbent material and place into containers for later disposal. *Small dry spills:* with clean shovel place material into clean, dry container and cover; move containers from spill area. *Large spills:* dike far ahead of spill for later disposal. It may be necessary to contain and dispose of this chemical as a hazardous waste. If material or contaminated runoff enters waterways, notify downstream users of potentially contaminated waters. Contact your Department of Environmental Protection or your regional office of the federal EPA for specific recommendations. If employees are required to clean-up spills, they must be properly trained and equipped. OSHA 1910.120(q) may be applicable.

Fire Extinguishing: This product may burn but does not readily ignite. Poisonous gases including carbon monoxide, nitrogen oxides and sulfur are produced in fire. Fight fire from maximum distance. On small fires use dry chemical, carbon dioxide, water spray, or standard foam extinguishers. On large fires use water spray, fog or foam. Poisonous gases are produced in fire including nitrogen oxides, phosphorus oxides, sulfur oxides. Dike fire control water for later disposal. If material or contaminated runoff enters waterways, notify downstream users of potentially contaminated waters. Notify local health and fire officials and pollution control agencies. From a secure, explosion-proof location, use water spray to cool exposed containers. If cooling streams are ineffective (venting sound increases in volume and pitch, tank discolors, or shows any signs of deforming), withdraw immediately to a secure position. If employees are expected to fight fires, they must be trained and equipped in OSHA 1910.156.

Disposal Method Suggested: In accordance with 40CFR 165 recommendations for the disposal of pesticides and pesticide containers. Must be disposed properly by following package label directions or by contacting your state pesticide or environmental control agency or by contacting your regional EPA office.

References

Sax, N. I., Ed., "Dangerous Properties of Industrial Materials Report" 3, No. 4, 60–65 (1983)

U.S. Environmental Protection Agency, "Chemical Profile: Azinphos-Methyl," Washington, DC, Chemical Emergency Preparedness Program (November 30, 1987)

New York State Department of Health, "Chemical Fact Sheet: Guthion," Albany, NY, Bureau of Toxic Substance Assessment (March 1, 1986)

New Jersey Department of Health and Senior Services, "Hazardous Substance Fact Sheet: Guthion," Trenton, NJ (May 1999)

Azobenzene

Molecular Formula: $C_{12}H_{10}N_2$

Common Formula: $C_6H_5N{=}NC_6H_5$

Synonyms: Azobenzeen (Dutch); Azobenzide; Azobenzol; Azobisbenzene; Azodibenzene; Azodibenzeneazofume; Benzeneazobenzene; Diazobenzene; 1,2-Diphenyldiazene; Diphenyldiazene; Diphenyldiimide; NCI-C02926

CAS Registry Number: 103-33-3

RTECS Number: CN1400000

DOT ID: NA 9188

Regulatory Authority

- Carcinogen (Animal Positive) (IARC)[9] (NTP)
- Banned or Severely Restricted (Great Britain) (UN)[13]
- Canada, WHMIS, Ingredients Disclosure List

Cited in U.S. State Regulations: California (A, G), Massachusetts (G), Michigan (G), New Jersey (G), Pennsylvania (G).

Description: Azobenzene, $C_6H_5N{=}NC_6H_5$, is a combustible, orange-red crystalline solid. Freezing/Melting point = 68°C. Boiling point = 293 – 297°C (decomposes). Insoluble in water.

Potential Exposure: Those engaged in Azobenzene use in dye, rubber, chemical and pesticide manufacturing.

Incompatibilities: Strong oxidizers.

Permissible Exposure Limits in Air: No standards set.

Permissible Concentration in Water: No criteria set.

Routes of Entry: Inhalation, ingestion, skin absorption.

Harmful Effects and Symptoms

Short Term Exposure: Azobenzene irritates the eyes, skin and respiratory tract. In serious cases there is a risk of unconsciousness and death. The oral LD_{50} for rat is 1,000 mg/kg.

Long Term Exposure: Carcinogenesis bioassays by NCl were positive for rats and negative for mice. Can affect the blood and cause liver disorders.

Medical Surveillance: Liver function tests. Complete blood count (CBC).

First Aid: If this chemical gets into the eyes, remove any contact lenses at once and irrigate immediately for at least 15 minutes, occasionally lifting upper and lower lids. Seek medical attention immediately. If this chemical contacts the skin, remove contaminated clothing and wash immediately with soap and water. Seek medical attention immediately. If this chemical has been inhaled, remove from exposure, begin rescue breathing (using universal precautions) if breathing has stopped and CPR if heart action has stopped. Transfer promptly to a medical facility. When this chemical has been swallowed, get medical attention. Give large quantities of water and induce vomiting. Do not make an unconscious person vomit.

Personal Protective Methods: Wear protective gloves and clothing to prevent any reasonable probability of skin contact. Safety equipment suppliers/manufacturers can provide recommendations on the most protective glove/clothing material for your operation. All protective clothing (suits, gloves, footwear, headgear) should be clean, available each day, and put on before work. Contact lenses should not be worn when working with this chemical. Wear dust-proof chemical goggles and face shield unless full facepiece respiratory protection is worn. Employees should wash immediately with soap when skin is wet or contaminated. Provide emergency showers and eyewash.

Respirator Selection: *Where there is no REL, at any detectable concentration:* SCBAF:PD,PP (any MSHA/NIOSH approved self-contained breathing apparatus that has a full facepiece and is operated in a pressure-demand or other positive-pressure mode); or SAF:PD,PP:ASCBA (any supplied-air respirator that has a full facepiece and is operated in a pressure-demand or other positive-pressure mode in combination with an auxiliary, self-contained breathing apparatus operated in a pressure-demand or other positive pressure mode).

Storage: Prior to working with this chemical you should be trained on its proper handling and storage. Store under an inert atmosphere in a freezer or refrigerator. Protect from air and light.[52] Sources of ignition such as smoking and open flames, are prohibited where this chemical is used, handled, or stored in a manner that could create a potential fire or explosion hazard. A regulated, marked area should be established where this chemical is handled, used, or stored in compliance with OSHA standard 1910.1045.

Shipping: Azobenzene may be classified as a hazardous substance, solid, n.o.s. It falls in Hazard Class ORM-E and Packing Group III.

Spill Handling: Evacuate persons not wearing protective equipment from area of spill or leak until clean-up is complete. Remove all ignition sources. Dampen spilled material with alcohol to avoid dust. Transfer to vapor-tight plastic bags for eventual disposal. Collect powdered material in the most convenient and safe manner and deposit in sealed containers. Ventilate area after clean-up is complete. It may be necessary to contain and dispose of this chemical as a hazardous waste. If material or contaminated runoff enters waterways, notify downstream users of potentially contaminated waters. Contact your Department of Environmental Protection or your regional office of the federal EPA for specific recommendations. If employees are required to clean-up spills, they must be properly trained and equipped. OSHA 1910.120(q) may be applicable.

Fire Extinguishing: Use dry chemical, carbon dioxide, water spray, or alcohol foam extinguishers. Poisonous gases are produced in fire including Nitrogen oxides. If material or contaminated runoff enters waterways, notify downstream users of potentially contaminated waters. Notify local health and fire officials and pollution control agencies. From a secure, explosion-proof location, use water spray to cool exposed containers. If cooling streams are ineffective (venting sound increases in volume and pitch, tank discolors, or shows any signs of deforming), withdraw immediately to a secure position. If employees are expected to fight fires, they must be trained and equipped in OSHA 1910.156.

References

Sax, N. I., Ed., "Dangerous Properties of Industrial Materials Report" 1, No. 3, 35–36 (1981) and 7, No. 1, 38–47 (1987)

Azodiisobutyronitrile

Molecular Formula: $C_8H_{12}N_4$

Common Formula: $NCC(CH_3)_2N{=}NC(CH_3)_2CN$

Synonyms: Aceto azib; AIBN; α,α'-Azobisisobutylonitrile; 2,2'-Azobis(isobutyronitrile); Azobisisobutyronitrile; 2,2'-Azobis(2-methylpropionitrile); α,α'-Azodiisobutyronitrile; 2,2'-Azodiisobutyronitrile; Azodiisobutyronitrile; 2,2'-Dicyano-2,2'-azopropane; Poly-Zole AZDN; Porofor 57; Vazo 64

CAS Registry Number: 78-67-1

RTECS Number: UG0800000

DOT ID: UN 2952

Cited in U.S. State Regulations: Florida (G), Massachusetts (G), New Hampshire (G), New Jersey (G), Pennsylvania (G).

Description: $NCC(CH_3)_2N{=}NC(CH_3)_2CN$, $C_8H_{12}N_4$, is a white crystalline compound. Freezing/Melting point = 105°C (decomposes). Autoignition temperature = 64°C. NFPA 704 M Hazard Identification (based on NFPA-704M Rating System): Health 3, Flammability -, Reactivity 2. Insoluble in water.

Potential Exposure: A cyanide compound. Used as a catalyst in vinyl polymerizations. It is also used as a blowing agent for elastomers and plastics.

Incompatibilities: Acetone, lithium, aluminum, hydride, water.

Permissible Exposure Limits in Air: No occupational exposure limits have been established for this chemical; however, inasmuch as it is a cyanide compound, the exposure limits are listed below:

OSHA and ACGIH: 5 mg/m^3 TWA; NIOSH: Ceiling limit, 4.7 ppm; 5 mg/m^3 per 10 minutes as cyanides. All have notations that skin contact contributes significantly in overall exposure. IDLH = 25 mg/m^3 as CN.

Determination in Air: NIOSH Method #7904 (Cyanides). See also NIOSH Criteria Document 78–212 NITRILES

Permissible Concentration in Water: No criteria set.

Routes of Entry: Inhalation, skin contact, ingestion.

Harmful Effects and Symptoms

Short Term Exposure: Azodiisobutyronitrile can affect you by passing through your skin. Contact can cause irritation to the eyes and skin. Inhalation can irritate the nose and throat causing wheezing and coughing. High exposure can cause dizziness, vomiting, abdominal pain, flushing, headache, shortness of breath, and coma. Convulsions and death may follow.

Long Term Exposure: Azodiisobutyronitrile may cause liver and kidney damage.

Points of Attack: Respiratory system, liver.

Medical Surveillance: If symptoms develop or overexposure is suspected, the following may be useful: Blood

Cyanide level. Liver and kidney function tests. Consider thyroid evaluation and complete blood count.

First Aid: If this chemical gets into the eyes, remove any contact lenses at once and irrigate immediately for at least 15 minutes, occasionally lifting upper and lower lids. Seek medical attention immediately. If this chemical contacts the skin, remove contaminated clothing and wash immediately with soap and water. Seek medical attention immediately. If this chemical has been inhaled, remove from exposure, begin rescue breathing (using universal precautions) if breathing has stopped and CPR if heart action has stopped. Transfer promptly to a medical facility. When this chemical has been swallowed, get medical attention. Give large quantities of water and induce vomiting. Do not make an unconscious person vomit.

Note: Use amyl nitrate capsules if symptoms develop. All area employees should be trained regularly in emergency measures for cyanide poisoning and in CPR. A cyanide antidote kit should be kept in the immediate work area and must be rapidly available. Kit ingredients should be replaced every 1 – 2 years to ensure freshness. Persons trained in the use of this kit, oxygen use, and CPR must be quickly available.

Personal Protective Methods: Wear protective gloves and clothing to prevent any reasonable probability of skin contact. Safety equipment suppliers/manufacturers can provide recommendations on the most protective glove/clothing material for your operation. All protective clothing (suits, gloves, footwear, headgear) should be clean, available each day, and put on before work. Contact lenses should not be worn when working with this chemical. Wear dust-proof chemical goggles and face shield unless full facepiece respiratory protection is worn. Employees should wash immediately with soap when skin is wet or contaminated. Provide emergency showers and eyewash. See NIOSH Criteria Document 78–212 NITRILES.

Respirator Selection: Where the potential exists for exposures to Azodiisobutyronitrile, use a MSHA/NIOSH approved full facepiece respirator equipped with particulate (dust/fume/mist) filters. Particulate filters must be checked every day before work for physical damage, such as rips or tears, and replaced as needed. Where the potential for high exposures exists, use a MSHA/NIOSH approved supplied-air respirator with a full facepiece operated in the positive pressure mode or with a full facepiece, hood, or helmet in the continuous flow mode, or use a MSHA/NIOSH approved self-contained breathing apparatus with a full facepiece operated in pressure-demand or other positive pressure mode.

NIOSH (*as cyanides*): *25 mg/m³*: SA (any supplied-air respirator); or SCBAF (any self-contained breathing apparatus with full facepiece). *Emergency or planned entry into unknown concentrations or IDLH conditions:* SCBAF:PD,PP (any self-contained breathing apparatus that has a full facepiece and is operated in a pressure-demand or other positive-pressure mode); or SAF:PD,PP:ASCBA (any supplied-air respirator that has a full facepiece and is operated in a pressure-demand or other positive-pressure mode in combination with an auxiliary self-contained breathing apparatus operated in a pressure-demand or other positive-pressure mode). *Escape:* GMFSHiE [any air-purifying, full-facepiece respirator (gas mask) with a chin-style, front-or back-mounted canister providing protection against the compound of concern and having a high efficiency particulate filter]; or SCBAE (any appropriate escape-type, self-contained breathing apparatus).

Storage: Prior to working with this chemical you should be trained on its proper handling and storage. Azodiisobutyronitrile must be stored to avoid contact with acetone, lithium, aluminum hydride, and water since violent reactions occur. Azodiisobutyronitrile is self-reactive and will explode at elevated temperatures. It should be stored under nitrogen, dry ice or ice. Sources of ignition such as smoking and open flames, are prohibited where this chemical is used, handled, or stored in a manner that could create a potential fire or explosion hazard.

Shipping: This compound must be labeled "Flammable Solid." It falls in Hazard Class 4.1 and Packing Group II, according to DOT.[19] Shipment by passenger aircraft or railcar or even by cargo aircraft is forbidden.

Spill Handling: Evacuate persons not wearing protective equipment from area of spill or leak until clean-up is complete. Remove all ignition sources. Do not sweep or raise dust. Collect powdered material in the most convenient and safe manner and deposit in sealed containers. *Do not use water or wet method.* Ventilate area after clean-up is complete. It may be necessary to contain and dispose of this chemical as a hazardous waste. If material or contaminated runoff enters waterways, notify downstream users of potentially contaminated waters. Contact your Department of Environmental Protection or your regional office of the federal EPA for specific recommendations. If employees are required to clean-up spills, they must be properly trained and equipped. OSHA 1910.120(q) may be applicable.

Fire Extinguishing: Azodiisobutyronitrile may self-ignite at elevated temperatures. Use dry chemical, CO_2, water spray, or foam extinguishers. Use water spray to keep fire-exposed containers cool. Poisonous gases are produced in fire, including Oxides of Nitrogen and Cyanide Fumes. Containers may explode in fire. If material or contaminated runoff enters waterways, notify downstream users of potentially contaminated waters. Notify local health and fire officials and pollution control agencies. From a secure, explosion-proof location, use water spray to cool exposed containers. If cooling streams are ineffective (venting sound increases in volume and pitch, tank discolors, or shows any signs of deforming), withdraw immediately to a secure position. If employees are expected to fight fires, they must be trained and equipped in OSHA 1910.156.

References

New Jersey Department of Health and Senior Services, "Hazardous Substance Fact Sheet: Azodiisobutyronitrile," Trenton, NJ (November 1998)

B

Bacitracin

Molecular Formula: $C_{66}H_{103}N_{17}O_{16}S$

Synonyms: Ayfivin; Baciguent; Baci-Jel; Baciliquin; Bacitek Ointment; Fortracin; Parentracin; Penitracin; Topitracin; Zutracin

CAS Registry Number: 1405-87-4

RTECS Number: CP0175000

DOT ID: UN 1851

Regulatory Authority

- List of Acutely Toxic Chemicals, Chemical Emergency Preparedness Program (EPA) and formerly on CERCLA/SARA 40CFR302, Table 302.4 Extremely Hazardous Substances List. Dropped from listing in 1988

Cited in U.S. State Regulations: Massachusetts (G), New Jersey.

Description: Bacitracin is a white to light tan powder which is odorless or having a slight odor and very bitter taste. Highly soluble in water.

Potential Exposure: Bacitracin is used as an ingredient in antibiotic ointments to treat or prevent topical or eye infections. Commercial Bacitracin is a mixture of at least 9 bacitracins. Also used as a feed and drinking water additive in animals, and a food additive in food for human consumption.

Incompatibilities: Oxidizers such as peroxides, perchlorates, chlorates, nitrates, chlorine, bromine, and fluorine.

Permissible Exposure Limits in Air: No standards set.

Permissible Concentration in Water: No criteria set.

Routes of Entry: Through the skin, inhalation.

Harmful Effects and Symptoms

The oral LD_{50} mouse is 25 mg/kg which does put it in the "highly toxic" category. This has been questioned, however, and it has been stated that as a result of a mathematical miscalculation, bacitracin was wrongly included-on a list of hazardous chemicals drafted several years ago by the National Institute of Occupational Safety and Health. The mistake was remedied in 1988 when the substance was removed from the EHS list as note above.

Short Term Exposure: Bacitracin can be absorbed through the skin, thereby increasing exposure. May cause eye irritation. Hypersensitivity reactions may result from application of this compound to the skin, but this is uncommon. Exposure may cause nausea, vomiting, and diarrhea.

Long Term Exposure: May cause liver damage and skin allergy.

Points of Attack: Liver, skin.

Medical Surveillance: Evaluation by a qualified allergist. Kidney function tests.

First Aid: In case of large-scale exposure, the directions for medicines (non-specific, n.o.s.) would be applied as follows: Move victim to fresh air; call emergency medical care. If not breathing, give artificial respiration. If breathing is difficult, give oxygen. In case of contact with material, immediately flush skin or eyes with running water for at least 15 minutes. Speed in removing material from skin is of extreme importance. Remove and isolate contaminated clothing and shoes at the site. Keep victim quiet and maintain normal body temperature. Effects may be delayed; keep victim under observation.

Shipping: The DOT category of medicines, poisonous solid, n.o.s. calls for the label "Keep away from Food." Bacitracin would fall in Hazard Class 6.1 and in Packing Group III.

Spill Handling: Evacuate and restrict persons not wearing protective equipment from area of spill or leak until cleanup is complete. Remove all ignition sources. Collect powdered material in the most convenient and safe manner and deposit in sealed containers. Ventilate area of spill or leak after clean-up is complete. It may be necessary to contain and dispose of this chemical as a hazardous waste. If material or contaminated runoff enters waterways, notify downstream users of potentially contaminated waters. Contact your Department of Environmental Protection or your regional office of the federal EPA for specific recommendations. If employees are required to clean-up spills, they must be properly trained and equipped. OSHA 1910.120(q) may be applicable.

Fire Extinguishing: Use dry chemical, carbon dioxide, water spray, or polymer foam extinguishers. Poisonous gases are produced in fire including carbon monoxide, nitrogen oxides, and sulfur oxides. Small fires: dry chemical, carbon dioxide, water spray, or foam. Large fires: water spray, fog, or foam. Move container from fire area if you can do so without risk. Fight fire from maximum distance. Save fire control water for later disposal, do not scatter the material. If material or contaminated runoff enters waterways, notify downstream users of potentially contaminated waters. Notify local health and fire officials and pollution control agencies. From a secure, explosion-proof location, use water spray to cool exposed containers. If employees are expected to fight fires, they must be trained and equipped in OSHA 1910.156.

References

U.S. Environmental Protection Agency, "Chemical Profile: Bacitracin," Washington, DC, Chemical Emergency Preparedness Program (October 31, 1985)

New Jersey Department of Health and Senior Services, "Hazardous Substance Fact Sheet, Bacitracin," Trenton NJ (March, 1999)

Barium

Molecular Formula: Ba

Synonyms: Bario (Spanish); Barium, elemental; Barium metal

CAS Registry Number: 7440-39-3

RTECS Number: CQ8370000

DOT ID: UN 1400 (non-powder); UN 1854 (powder)

EINECS Number: 231-149-1

Regulatory Authority

- Air Pollutant Standard Set (ACGIH)[1] (HSE)[33] (DFG)[3] (OSHA)[58] (Several States)[60] (Australia) (Israel) (Mexico)
- Water Pollution Standards Set (EPA)[49] (former USSR)[43] (Several States)[61] (Canada) (Mexico)
- EPA HAZARDOUS WASTE NUMBER (RCRA No.): D005
- RCRA Toxicity Characteristic (Section 261.24), Maximum Concentration of Contaminants, regulatory level, 100.0 mg/l
- RCRA, 40CFR261, Appendix 8 Hazardous Constituents, waste number not listed
- RCRA Maximum Concentration Limit for Ground Water Protection (40CFR264.94), 1.0 mg/l
- RCRA 40CFR268.48; 61FR15654, Universal Treatment Standards: Wastewater (mg/l), 1.2; Nonwastewater (mg/l), 7.6 TCLP
- RCRA 40CFR264, Appendix 9; TSD Facilities Ground Water Monitoring List, Suggested methods (PQL μg/l): 6010 (20); 7080 (1,000)
- SAFE DRINKING WATER ACT: MCL, 2 mg/l; MCLG, 2 mg/l; Regulated chemical (47FR9352)
- EPCRA Section 313 Form R *de minimis* concentration reporting level: 1.0%
- Canada MAC for drinking water quality: 1.0 mg/l
- Mexico, Drinking Water 1.0 mg/l

Cited in U.S. State Regulations: Alaska (G), California (A, G), Connecticut (A), Florida (G, A), Illinois (G), Kansas (G), Louisiana (G), Maine (G, W), Maryland (G), Massachusetts (G, W), Minnesota (W), Nevada (A), New Hampshire (G), New Jersey (G), New York (A), North Dakota (A), Oklahoma (G), Pennsylvania (G), Vermont (G), Virginia (G, A), Washington (G), West Virginia (G), Wisconsin (G).

Description: Barium, Ba, a flammable, silver white or yellowish metal in various forms including powder. Barium may ignite spontaneously in air in the presence of moisture, evolving hydrogen. Freezing/Melting point = 725°C. Boiling point = 1,640°C. Hazard Identification (based on NFPA-704 M Rating System): Health 1, Flammability 4, Reactivity 3. Water reactive. It is produced by reduction of barium oxide. The primary sources are the minerals barite ($BaSO_4$) and witherite ($BaCO_3$).

Potential Exposure: Metallic barium is used for removal of residual gas in vacuum tubes and in alloys with nickel, lead, calcium, magnesium, sodium, and lithium. Barium compounds are used in the manufacture of lithopone (a white pigment in paints), chlorine, sodium hydroxide, valves, and green flares; in synthetic rubber vulcanization, x-ray diagnostic work, glassmaking, papermaking, beet-sugar purification, animal and vegetable oil refining. They are used in the brick and tile, pyrotechnics, and electronics industries. They are found in lubricants, pesticides, glazes, textile dyes and finishes, pharmaceuticals, and in cements which will be exposed to saltwater; and barium is used as a rodenticide, a flux for magnesium alloys, a stabilizer and mold lubricant in the rubber and plastics industries, an extender in paints, a loader for paper, soap, rubber, and linoleum, and as a fire extinguisher for uranium or plutonium fires.

Incompatibilities: Barium powder may spontaneously ignite on contact with air. It is a strong reducing agent and reacts violently with oxidizers and acids. Reacts with water, forming combustible hydrogen gas and barium hydroxide. Reacts violently with halogenated hydrocarbon solvents, causing a fire and explosion hazard.

Permissible Exposure Limits in Air: The Federal OSHA standard[58] and ACGIH recommended airborne limit for soluble barium compounds is 0.5 mg/m^3 TWA for an 8-hour workshift. The NIOSH level is the same for a 10-hour workshift. The DFG,[3] HSE,[33] Australia, Mexico, and Israel have adopted this same value and DFG has a Peak Limitation (5 min) of 2 times normal MAK; do not exceed more than 8 times during workshift. Israel's Action Level is one half the TWA. The NIOSH IDLH level is 50 mg/m^3 (Ba, soluble compounds). In addition, several states have set guidelines or standards for barium in ambient air[60] ranging from 0.67 μg/m^3 (New York) to 5.0 μg/m^3 (Florida and North Dakota) to 8.0 μg/m^3 (Virginia) to 10.0 μg/m^3 (Connecticut) to 12.0 μg/m^3 (Nevada).

Determination in Air: Filter; Water; Flame atomic absorption spectrometry; NIOSH Methods (IV) #7056, Barium, soluble compounds. Collection on a cellulose membrane filter, workup with hot water, analysis by atomic absorption. See NIOSH Method #8310.[18]

Permissible Concentration in Water: See Regulatory Authority for U.S. EPA, Canadian and Mexican levels. The former USSR-UNEP/IRPTC project[43] has set a MAC of 4.0 mg/l in water bodies used for domestic purposes. Also, these states have set standards for barium in drinking water:[61] a standard of 100 μg/l in Massachusetts and guidelines of 1,000 μg/l in Maine and 1,500 μg/l in Minnesota.

Determination in Water: Conventional flame atomization does not have sufficient sensitivity to determine barium in most water samples; however, a barium detection limit of 10 μg/l can be achieved, if a nitrous oxide flame is used. A concentration procedure for barium uses thenoyltrifluoro-acetone-methylisobutylketone extraction at a pH of 6.8. With a tantalum liner insert, the barium detection limit of the flameless atomic absorption procedure can be improved to 0.1 μg/l according to NAS/NRC.[46]

Routes of Entry: Ingestion or inhalation of dust or fume, skin and/or eye contact.

Harmful Effects and Symptoms

Short Term Exposure: Alkaline barium compounds, such as the hydroxide and carbonate, may cause local irritation to the eyes, nose, throat, and skin.

Long Term Exposure: Barium poisoning is virtually unknown in industry, although the potential exists when the soluble forms are used. When ingested or given orally, the soluble, ionized barium compounds exert a profound effect on all muscles and especially smooth muscles, markedly increasing their contractility. The heart rate is slowed and may stop in systole. Other effects are increased intestinal peristalsis, vascular constriction, bladder contraction, and increased voluntary muscle tension. The inhalation of the dust of barium sulfate may lead to deposition in the lungs in sufficient quantities to produce "baritosis" (a benign pneumoconiosis). This produces a radiologic picture in the absence of symptoms and abnormal physical signs. X-rays, however, will show disseminated nodular opacities throughout the lung fields, which are discrete, but sometimes overlap.

Points of Attack: Heart, lungs, central nervous system, skin, respiratory system, eyes.

Medical Surveillance: Consideration should be given to the skin, eye, heart, and lung in any placement or periodic examination.

First Aid: If a soluble barium compound gets into the eyes, remove any contact lenses at once and irrigate immediately. If a soluble barium compound contacts the skin, flush with water immediately. If a person breathes in large amounts of a soluble barium compound, move the exposed person to fresh air at once and perform artificial respiration. When a soluble barium compound has been swallowed, get medical attention. Give large quantities of water and induce vomiting. Do not make an unconscious person vomit.

Personal Protective Methods: Employees should receive instruction in personal hygiene and the importance of not eating in work areas. Good housekeeping and adequate ventilation are essential. Dust masks, respirators, or goggles may be needed where amounts of significant soluble or alkaline forms are encountered, as well as protective clothing.

Respirator Selection: NIOSH/OSHA: *5 mg/m³:* DMXSQ (any dust and mist respirator except single-use and quarter mask respirators); or SA (any supplied-air respirator). *12.5 mg/m³:* SA:CF (any supplied-air respirator operated in a continuous-flow mode); or PAPRDM (any powered, air-purifying respirator with a dust and mist filter). *25 mg/m³:* HiEF (any air-purifying, full-facepiece respirator with a high-efficiency particulate filter); or SAT:CF (any supplied-air respirator that has a tight-fitting facepiece and is operated in a continuous-flow mode); or PAPRTHiE (any powered, air-purifying respirator with a tight-fitting facepiece and a high-efficiency particulate filter); or SCBAF (any self-contained breathing apparatus with a full facepiece); or SAF (any supplied-air respirator with a full facepiece). *50 mg/m³:* SAF:PD,PP (any supplied-air respirator that has a full facepiece and is operated in a pressure-demand or other positive-pressure mode). *Emergency or planned entry into unknown concentrations or IDLH conditions:* SCBAF:PD,PP (any self-contained breathing apparatus that has a full facepiece and is operated in a pressure-demand or other positive-pressure mode); or SAF:PD,PP:ASCBA (any supplied-air respirator that has a full facepiece and is operated in a pressure-demand or other positive-pressure mode in combination with an auxiliary, self-contained breathing apparatus operated in a pressure-demand or other positive-pressure mode). *Escape:* HiEF (any air-purifying, full-facepiece respirator with a high-efficiency particulate filter); or SCBAE (any appropriate escape-type, self-contained breathing apparatus).

Storage: Barium metal should be stored in a dry area, separated from halogenated solvents, strong oxidants, acids, in tightly-closed containers under an inert gas blanket, petroleum or oxygen-free liquid. Rubber gloves, rubber protective clothing and apron, goggles and gas-filter mask should be worn when working in a barium storage area.

Shipping: The DOT label requirement for barium metal is "Dangerous when Wet." The metal falls in UN/DOT shipping class 4.3 and Packing Group II.[19][20] It should be noted that "Barium Alloys" have the same caveats but that (Barium Alloys, Pyrophoric) require a "Spontaneously Combustible" label, fall in Hazard Class 4.2 and that shipment by passenger aircraft or railcar or even by cargo aircraft is forbidden.

Spill Handling: Evacuate and restrict persons not wearing protective equipment from area of spill or leak until cleanup is complete. Remove all ignition sources. Small quantities of barium metal may be dissolved in large quantities of water. Soda ash is added and the solution then neutralized with HCl. Collect powdered material in the most convenient and safe manner and deposit in sealed containers. Ventilate area of spill or leak after clean-up is complete. It may be necessary to contain and dispose of this chemical as a hazardous waste. If material or contaminated runoff enters waterways, notify downstream users of potentially contaminated waters. Contact your Department of Environmental Protection or your regional office of the federal EPA for specific recommendations. If employees are required to clean-up spills, they must be properly trained and equipped. OSHA 1910.120(q) may be applicable.

Fire Extinguishing: Barium powder is a flammable solid. Reacts violently with fire extinguishing agents such as water, bicarbonate, powder, foam, and carbon dioxide. Use dry chemical, carbon dioxide, water spray, or alcohol foam extinguishers. Poisonous gases are produced in fire. If material or contaminated runoff enters waterways, notify downstream users of potentially contaminated waters. Notify local health and fire officials and pollution control agencies. From a secure, explosion-proof location, use water spray to cool exposed containers. If cooling streams are ineffective (venting sound increases in volume and pitch, tank discolors, or shows any signs of deforming), withdraw immediately to a secure position. If employees are expected to fight fires, they must be trained and equipped in OSHA 1910.156.

Disposal Method Suggested: Barium in solution (see spill handling) may be precipitated with soda ash and the sludge may be landfilled.

References

U.S. Environmental Protection Agency, Toxicology of Metals, Vol. 2: Barium, pp 71–84, Report EPA-600/1-77-022, Research Triangle Park, NC (May 1977)

U.S. Environmental Protection Agency, Barium, Health and Environmental Effects Profile No. 13, Washington, DC, Office of Solid Waste (April 30, 1980)

Sax, N. I., Ed., "Dangerous Properties of Industrial Materials Report" 1, No. 7, 35–40 (1981) and 3, No. 4, 29–30 (1983)

New Jersey Department of Health and Senior Services, "Hazardous Substance Fact Sheet: Barium," Trenton, NJ (January, 1996)

Barium Azide

Molecular Formula: BaN_6

Common Formula: $Ba(N_3)_2$

Synonyms: Azida de bario (Spanish)

CAS Registry Number: 18810-58-7

RTECS Number: CQ8500000 (dry); CQ8510000 (wet)

DOT ID: UN 0224 (dry); UN 1571 (wetted with not less than 50% water)

Regulatory Authority

- Explosive Substance (World Bank)[15]
- Air Pollutant Standard Set (ACGIH) (OSHA) (NIOSH) (barium, soluble compounds)
- RCRA 40CFR261, Appendix 8; 40CFR261.11 Hazardous Constituents
- EPCRA Section 313: Includes any unique chemical substance that contains barium as part of that chemical's infrastructure. This category does not include barium sulfate (7727-43-7). Form R *de minimis* concentration reporting level: 0.1%
- U.S. DOT 49CFR172.101, Appendix B, Regulated marine pollutant

Cited in U.S. State Regulations: California (G) (barium compounds), New Hampshire (G), New Jersey (G), Oklahoma (G).

Description: Barium Azide, $Ba(N_3)_2$, is a flammable, crystalline solid which can be used or transported in solution. Freezing/Melting point = 120°C (decomposes, losing nitrogen). Highly soluble in water.

Potential Exposure: Barium Azide is used in high explosives.

Incompatibilities: Carbon disulfide. It can explode when heated or shocked.

Permissible Exposure Limits in Air: For Ba soluble compounds, the Federal OSHA standard[58] and ACGIH recommended airborne limit is 0.5 mg/m^3 TWA for an 8-hour workshift. The NIOSH level is the same for a 10-hour workshift. The DFG,[3] HSE,[33] Australia, Mexico, and Israel have adopted this same value and DFG has a Peak Limitation (5 min) of 2 times normal MAK; do not exceed more than 8 times during workshift. Israel's Action Level is one half the TWA. The NIOSH IDLH level is 50 mg/m^3 (Ba, soluble compounds). In addition, several states have set guidelines or standards for barium in ambient air[60] ranging from 0.67 μg/m^3 (New York) to 5.0 μg/m^3 (Florida and North Dakota) to 8.0 μg/m^3 (Virginia) to 10.0 μg/m^3 (Connecticut) to 12.0 μg/m^3 (Nevada).

Determination in Air: No criteria set for Barium Azide. See entry under "Barium."

Permissible Concentration in Water: No criteria set for Barium Azide. See entry under "Barium."

Routes of Entry: Inhalation, skin and/or eyes.

Harmful Effects and Symptoms

Short Term Exposure: Barium Azide irritates the eyes, nose, and respiratory tract; with coughing. Overexposure can cause a drop in blood pressure, with dizziness, blurred vision, headaches and unconsciousness.

Long Term Exposure: Repeated exposure to the dust can cause spots on chest x-ray without lung scarring.

Points of Attack: Lungs.

Medical Surveillance: Lung function tests. Consider chest x-ray following acute overexposure.

First Aid: If this chemical gets into the eyes, remove any contact lenses at once and irrigate immediately for at least 15 minutes, occasionally lifting upper and lower lids. Seek medical attention immediately. If this chemical contacts the skin, remove contaminated clothing and wash immediately with soap and water. Seek medical attention immediately. If this chemical has been inhaled, remove from exposure, begin rescue breathing (using universal precautions) if breathing has stopped and CPR if heart action has stopped. Transfer promptly to a medical facility. When this chemical has been swallowed, get medical attention. Give large quantities of water and induce vomiting. Do not make an unconscious person vomit. *Eye Contact:* Immediately remove any contact lenses and flush with large amounts of water for at least 15 minutes, occasionally lifting upper and lower lids. If weakness or fainting are present, lay the person down flat with feet elevated.

Personal Protective Methods: Wear protective gloves and clothing to prevent any reasonable probability of skin contact. Safety equipment suppliers/manufacturers can provide recommendations on the most protective glove/clothing material for your operation. All protective clothing (suits, gloves, footwear, headgear) should be clean, available each day, and put on before work. Contact lenses should not be worn when working with this chemical. Wear splash- (for liquid) or dust-proof chemical goggles and face shield unless full facepiece respiratory protection is worn. Employees should wash immediately with soap when skin is wet or contaminated. Provide emergency showers and eyewash.

Respirator Selection: NIOSH/OSHA: (as soluble barium compounds) *5 mg/m³:* DMXSQ (any dust and mist respirator except single-use and quarter mask respirators); or SA (any supplied-air respirator). *12.5 mg/m³:* SA:CF (any supplied-air respirator operated in a continuous-flow mode); or PAPRDM (any powered, air-purifying respirator with a dust and mist filter). *25 mg/m³:* HiEF (any air-purifying, full-facepiece respirator with a high-efficiency particulate filter); or SAT:CF (any supplied-air respirator that has a tight-fitting facepiece and is operated in a continuous-flow mode); or PAPRTHiE (any powered, air-purifying respirator with a tight-fitting facepiece and a high-efficiency particulate filter); or SCBAF (any self-contained breathing apparatus with a full facepiece); or SAF (any supplied-air respirator with a full facepiece). *50 mg/m³:* SAF:PD,PP (any supplied-air respirator that has a full facepiece and is operated in a pressure-demand or other positive-pressure mode). *Emergency or planned entry into unknown concentrations or IDLH conditions:* SCBAF:PD,PP (any self-contained breathing apparatus that has a full facepiece and is operated in a pressure-demand or other positive-pressure mode); or SAF:PD, PP:ASCBA (any supplied-air respirator that has a full facepiece and is operated in a pressure-demand or other positive-pressure mode in combination with an auxiliary, self-contained breathing apparatus operated in a pressure-demand or other positive-pressure mode). *Escape:* HiEF (any air-purifying, full-facepiece respirator with a high-efficiency particulate filter); or SCBAE (any appropriate escape-type, self-contained breathing apparatus).

Storage: Barium Azide must be stored to avoid contact with Carbon Disulfide since violent reactions occur. Store in tightly closed containers in a cool, well-ventilated area from anything which could disturb or shock Barium Azide. Sources of ignition such as smoking and open flames are prohibited where Barium Azide is handled, used, or stored. Keeping Barium Azide wet greatly reduces its fire and explosion hazard. Wherever Barium Azide is used, handled, manufactured, or stored, use explosion-proof electrical equipment and fittings.

Shipping: Barium Azide must be labeled "Flammable Solid, Poison." Shipment by passenger aircraft or railcar is forbidden. Cargo aircraft shipments are limited to 0.5 kg. When dry or wet with less than 50% water, Barium Azide falls in Hazard Class 1.1A; when wet with 50% or more water, it falls in Hazard Class 4.1 and Packing Group I.[19][20]

Spill Handling: Evacuate and restrict persons not wearing protective equipment from area of spill or leak until cleanup is complete. Remove all ignition sources. Absorb liquids in vermiculite, dry sand, earth, or a similar material and deposit in sealed containers. Collect powdered material in the most convenient and safe manner and deposit in sealed containers. Ventilate area of spill or leak after clean-up is complete. It may be necessary to contain and dispose of this chemical as a hazardous waste. If material or contaminated runoff enters waterways, notify downstream users of potentially contaminated waters. Contact your Department of Environmental Protection or your regional office of the federal EPA for specific recommendations. If employees are required to clean-up spills, they must be properly trained and equipped. OSHA 1910.120(q) may be applicable.

Fire Extinguishing: Barium Azide will explode when heated or when shocked. If fire or explosion occurs, evacuate the area. Fight the fire from an explosion-resistant location as containers may explode in fire. Use dry chemical, carbon dioxide, water spray, or alcohol foam extinguishers. Poisonous gases are produced in fire including nitrogen oxides. If material or contaminated runoff enters waterways, notify downstream users of potentially contaminated waters. Notify local health and fire officials and pollution control agencies. From a secure, explosion-proof location, use water spray to cool exposed containers. If cooling streams are ineffective (venting sound increases in volume and pitch, tank discolors, or shows any signs of deforming), withdraw immediately to a secure position. If employees are expected to fight fires, they must be trained and equipped in OSHA 1910.156.

References

New Jersey Department of Health and Senior Services, "Hazardous Substance Fact Sheet: Barium Azide," Trenton, NJ (April 1998)

Barium Bromate

Molecular Formula: $BaBr_2O_6$

Common Formula: $Ba(BrO_3)_2$

Synonyms: Bromato barico (Spanish); Bromic acid, barium salt

CAS Registry Number: 13967-90-3

RTECS Number: EF8715000

DOT ID: UN 2719

Regulatory Authority

- Air Pollutant Standard Set (ACGIH) (OSHA) (NIOSH) (barium, soluble compounds)
- RCRA 40CFR261, Appendix 8; 40CFR261.11 Hazardous Constituents
- EPCRA Section 313: Includes any unique chemical substance that contains barium as part of that chemical's infrastructure. This category does not include barium sulfate (7727-43-7). Form R *de minimis* concentration reporting level: 0.1%
- U.S. DOT 49CFR172.101, Appendix B, Regulated marine pollutant (Ba compounds, soluble, n.o.s.)

Cited in U.S. State Regulations: New Hampshire (G), New Jersey (G).

Description: Barium bromate, $Ba(BrO_3)_2$, is a white crystalline powder. Freezing/Melting point = 260°C (decomposes). Slightly soluble in water.

Potential Exposure: This material is used as an analytical reagent, oxidizer and corrosion inhibitor.

Incompatibilities: A strong reducing agent. Keep away from oxidizers and oxidizable materials; aluminum, arsenic, carbon, copper, metal sulfides, phosphorus, sulfur, organic, and combustible materials (such as wood, paper, oil, fuels) since violent reactions occur.

Permissible Exposure Limits in Air: For Ba soluble compounds, the Federal OSHA standard[58] and ACGIH recommended airborne limit is 0.5 mg/m^3 TWA for an 8-hour workshift. The NIOSH level is the same for a 10-hour workshift. The DFG,[3] HSE,[33] Australia, Mexico, and Israel have adopted this same value and DFG has a Peak Limitation (5 min) of 2 times normal MAK; do not exceed more than 8 times during workshift. Israel's Action Level is one half the TWA. The NIOSH IDLH level is 50 mg/m^3 (Ba, soluble compounds). In addition, several states have set guidelines or standards for barium in ambient air[60] ranging from 0.67 μg/m^3 (New York) to 5.0 μg/m^3 (Florida and North Dakota) to 8.0 μg/m^3 (Virginia) to 10.0 μg/m^3 (Connecticut) to 12.0 μg/m^3 (Nevada).

Determination in Air: See entry under "Barium."

Permissible Concentration in Water: No criteria set for barium bromate per se. See entry under "Barium."

Routes of Entry: Inhalation, skin and eye contact.

Harmful Effects and Symptoms

Short Term Exposure: Barium Bromate can affect you when breathed in. Contact can irritate and even burn the eyes and skin. Breathing the dust or mist can irritate the nose, throat, and bronchial tubes, causing cough and phlegm.

Long Term Exposure: After repeated exposure, Barium may show up as spots on chest x-ray. Some Barium chemicals are contaminated with silica, which scars the lungs. Repeated exposure to Barium Bromate can cause Bromine to build up in the body. Consult the sheet on "bromine" entry. Repeated skin contact can cause chronic dryness and cracking.

Points of Attack: Lungs, skin.

Medical Surveillance: Serum Bromide levels. Lung function tests.

First Aid: If this chemical gets into the eyes, remove any contact lenses at once and irrigate immediately for at least 15 minutes, occasionally lifting upper and lower lids. Seek medical attention immediately. If this chemical contacts the skin, remove contaminated clothing and wash immediately with soap and water. Seek medical attention immediately. If this chemical has been inhaled, remove from exposure, begin rescue breathing (using universal precautions) if breathing has stopped and CPR if heart action has stopped. Transfer promptly to a medical facility. When this chemical has been swallowed, get medical attention. Give large quantities of water and induce vomiting. Do not make an unconscious person vomit.

Personal Protective Methods: Where possible, enclose operations and use local exhaust ventilation at the site of chemical release. If local exhaust ventilation or enclosure is not used, respirators should be worn. Wear protective gloves and clothing to prevent any reasonable probability of skin contact. Safety equipment suppliers/manufacturers can provide recommendations on the most protective glove/clothing material for your operation. All protective clothing (suits, gloves, footwear, headgear) should be clean, available each day, and put on before work. Contact lenses should not be worn when working with this chemical. Wear dust-proof chemical goggles and face shield unless full facepiece respiratory protection is worn. Employees should wash immediately with soap when skin is wet or contaminated. Provide emergency showers and eyewash.

Respirator Selection: NIOSH/OSHA: *5 mg/m^3:* DMXSQ (any dust and mist respirator except single-use and quarter mask respirators); or SA (any supplied-air respirator). *12.5 mg/m^3:* SA:CF (any supplied-air respirator operated in a continuous-flow mode); or PAPRDM (any powered, air-purifying respirator with a dust and mist filter). *25 mg/m^3:* HiEF (any air-purifying, full-facepiece respirator with a high-efficiency particulate filter); or SAT:CF (any supplied-air respirator that has a tight-fitting facepiece and is operated in a continuous-flow mode); or PAPRTHiE (any powered, air-purifying respirator with a tight-fitting facepiece and a high-efficiency particulate filter); or SCBAF (any self-contained breathing apparatus with a full facepiece); or SAF (any supplied-air respirator with a full facepiece). *50 mg/m³:* SAF:PD,PP (any supplied-air respirator that has a full facepiece and is operated in a pressure-demand or other positive-pressure mode). *Emergency or planned entry into unknown concentrations or IDLH conditions:* SCBAF:PD,PP (any self-contained breathing apparatus that has a full facepiece and is operated in a pressure-demand or other positive-pressure mode); or SAF:PD, PP:ASCBA (any supplied-air respirator that has a full facepiece and is operated in a pressure-demand or other positive-pressure mode in combination with an auxiliary, self-contained breathing apparatus operated in a pressure-demand or other positive-pressure mode). *Escape:* HiEF (any air-purifying, full-facepiece respirator with a high-efficiency particulate filter); or SCBAE (any appropriate escape-type, self-contained breathing apparatus).

Storage: Store in tightly closed containers in a cool, well-ventilated area. Sources of ignition, such as smoking and open flames, are prohibited where Barium Bromate is handled, used, or stored. Avoid any possible contact with incompatible materials. See OSHA standard 1910.104 and NFPA 43A *Code for the Storage of Liquid and Solid Oxidizers* for detailed handling and storage regulations.

Shipping: The DOT label requirement is "Oxidizer, Poison." This material falls in UN/DOT Hazard Class 5.1 and Packing Group II.[19][20]

Spill Handling: Evacuate and restrict persons not wearing protective equipment from area of spill or leak until cleanup is complete. Remove all ignition sources. Collect powdered material in the most convenient and safe manner and deposit in sealed containers. Ventilate area of spill or leak after clean-up is complete. It may be necessary to contain and dispose of this chemical as a hazardous waste. If material or contaminated runoff enters waterways, notify downstream users of potentially contaminated waters. Contact your Department of Environmental Protection or your regional office of the federal EPA for specific recommendations. If employees are required to clean-up spills, they must be properly trained and equipped. OSHA 1910.120(q) may be applicable.

Fire Extinguishing: Barium Bromate explodes at 275 – 300°C. Extinguish fire using an agent suitable for type of surrounding fire. Barium Bromate itself does not burn. Poisonous gases are produced in fire including bromine. If material or contaminated runoff enters waterways, notify downstream users of potentially contaminated waters. Notify local health and fire officials and pollution control agencies. From a secure, explosion-proof location, use water spray to cool exposed containers. If cooling streams are ineffective (venting sound increases in volume and pitch, tank discolors, or shows any signs of deforming), withdraw immediately to a secure position. If employees are expected to fight fires, they must be trained and equipped in OSHA 1910.156.

References

New Jersey Department of Health and Senior Services, "Hazardous Substance Fact Sheet: Barium Bromate," Trenton, NJ (June 1986)

Barium Chlorate

Molecular Formula: $BaCl_2O_6$

Common Formula: $Ba(ClO_3)_2$

Synonyms: Chloric acid, barium salt; Clorato barico (Spanish)

CAS Registry Number: 13477-00-4

RTECS Number: FN9770000

DOT ID: UN 1445

EEC Number: 017-003-00-8

Regulatory Authority

- Air Pollutant Standard Set (ACGIH) (OSHA) (NIOSH) (barium, soluble compounds)
- RCRA 40CFR261, Appendix 8; 40CFR261.11 Hazardous Constituents
- EPCRA Section 313: Includes any unique chemical substance that contains barium as part of that chemical's infrastructure. This category does not include barium sulfate (7727-43-7). Form R *de minimis* concentration reporting level: 0.1%
- U.S. DOT 49CFR172.101, Appendix B, Regulated marine pollutant (Ba compounds, soluble, n.o.s.)
- Canada, WHMIS, Ingredients Disclosure List

Cited in U.S. State Regulations: California (A, G), Florida (G), Maine (G), Massachusetts (G), New Hampshire (G), New Jersey (G), Oklahoma (G), Pennsylvania (G), Rhode Island (G).

Description: Barium chlorate, $Ba(ClO_3)_2$, is a combustible, colorless to white crystalline solid or powder. Freezing/Melting point = 250°C (loses oxygen); 414°C (melts). Hazard Identification (based on NFPA-704 M Rating System): Health 2, Flammability 0, Reactivity 1. Soluble in water.

Potential Exposure: It is used in fireworks and explosives manufacture, in textile dyeing and in the manufacture of other perchlorates.

Incompatibilities: A strong oxidizer. Barium Chlorate is a reactive chemical and is an explosion hazard. Violent reaction may occur with reducing materials, strong acids, powdered metals. Combustible materials will increase activity in fire.

Permissible Exposure Limits in Air: For Ba soluble compounds, the Federal OSHA standard[58] and ACGIH recommended airborne limit is 0.5 mg/m³ TWA for an 8-hour workshift. The NIOSH level is the same for a 10-hour workshift. The DFG,[3] HSE,[33] Australia, Mexico, and Israel have adopted this same value and DFG has a Peak Limitation (5 min) of 2 times normal MAK; do not exceed more than 8 times during workshift. Israel's Action Level is one half the TWA. The NIOSH IDLH level is 50 mg/m³ (Ba, soluble compounds). In addition, several states have set guidelines or standards for barium in ambient air[60] ranging from 0.67 μg/m³ (New York) to 5.0 μg/m³ (Florida and North Dakota) to 8.0 μg/m³ (Virginia) to 10.0 μg/m³ (Connecticut) to 12.0 μg/m³ (Nevada).

Determination in Air: See entry for "Barium."

Permissible Concentration in Water: No criteria set for barium chlorate per se. See entry for "Barium."

Routes of Entry: Inhalation, ingestion, eye and/or skin contact.

Harmful Effects and Symptoms

Short Term Exposure: Contact may burn the eyes and skin. Breathing the dust or mist can irrigate the nose, throat and bronchial tubes. Higher exposures can damage red blood cells. Symptoms include headache, weakness, abdominal pain, dark urine and jaundice. The symptoms of paralysis may be delayed for several hours.

Long Term Exposure: After repeated exposure, Barium may show up as spots in the lungs on chest x-ray. Some Barium chemicals are contaminated with silica, which scars the lungs. See entry for silica quartz. Chlorates can damage red blood cells, leading to kidney damage, or cause methemoglobin to form in the blood, reducing Oxygen supply to body organs. Repeated skin contact can cause chronic dryness and skin cracking.

Points of Attack: Lungs, blood cells, skin.

Medical Surveillance: Lung function tests, complete blood count (CBC), test for methemoglobin.

First Aid: If this chemical gets into the eyes, remove any contact lenses at once and irrigate immediately for at least 15 minutes, occasionally lifting upper and lower lids. Seek medical attention immediately. If this chemical contacts the skin, remove contaminated clothing and wash immediately with soap and water. Seek medical attention immediately. If this chemical has been inhaled, remove from exposure, begin rescue breathing (using universal precautions) if breathing has stopped and CPR if heart action has stopped. Transfer promptly to a medical facility. When this chemical has been swallowed, get medical attention. Give large quantities of water and induce vomiting. Do not make an unconscious person vomit. The symptoms of paralysis do not become obvious until some hours have passed. Keep under medical observation for 24 – 48 hours.

Note to Physician: Treat for methemoglobinemia. Spectrophotometry may be required for precise determination of levels of methemoglobinemia in urine.

Personal Protective Methods: Where possible, enclose operations and use local exhaust ventilation at the site of chemical release. If local exhaust ventilation or enclosure is not used, respirators should be worn. Wear protective work clothing. Wash thoroughly immediately after exposure to Barium Chlorate. Post hazard and warning information in the work area. In addition, as part of an ongoing education and training effort, communicate all information on the health and safety hazards of Barium Chlorate to potentially exposed workers.

Respirator Selection: NIOSH/OSHA: (Ba soluble compounds) *5 mg/m³:* DMXSQ (any dust and mist respirator except single-use and quarter mask respirators); or SA (any supplied-air respirator). *12.5 mg/m³:* SA:CF (any supplied-air respirator operated in a continuous-flow mode); or PAPRDM (any powered, air-purifying respirator with a dust and mist filter). *25 mg/m³:* HiEF (any air-purifying, full-facepiece respirator with a high-efficiency particulate filter); or SAT:CF (any supplied-air respirator that has a tight-fitting facepiece and is operated in a continuous-flow mode); or PAPRTHiE (any powered, air-purifying respirator with a tight-fitting facepiece and a high-efficiency particulate filter); or SCBAF (any self-contained breathing apparatus with a full facepiece); or SAF (any supplied-air respirator with a full facepiece). *50 mg/m³:* SAF:PD,PP (any supplied-air respirator that has a full facepiece and is operated in a pressure-demand or other positive-pressure mode). *Emergency or planned entry into unknown concentrations or IDLH conditions:* SCBAF:PD,PP (any self-contained breathing apparatus that has a full facepiece and is operated in a pressure-demand or other positive-pressure mode); or SAF:PD,PP:ASCBA (any supplied-air respirator that has a full facepiece and is operated in a pressure-demand or other positive-pressure mode in combination with an auxiliary, self-contained breathing apparatus operated in a pressure-demand or other positive-pressure mode). *Escape:* HiEF (any air-purifying, full-facepiece respirator with a high-efficiency particulate filter); or SCBAE (any appropriate escape-type, self-contained breathing apparatus).

Storage: Barium Chlorate must be stored to avoid contact with organic or combustible materials (such as wood, paper, oil, fuels and starch) and other easily oxidizable materials (such as sulfur, aluminum, cooper, metal sulfides, ammonium salts, etc.) since violent reactions occur. Store in tightly closed containers on non-wood floors in a cool, well-ventilated area. Wherever Barium Chlorate is used, handled, manufactured, or stored, use explosion-proof electrical equipment and fittings. See OSHA standard 1910.104 and NFPA 43A *Code for the Storage of Liquid and Solid Oxidizers* for detailed handling and storage regulations.

Shipping: The DOT-required shipping label is "Oxidizer, Poison." The UN/DOT Hazard Class is 5.1 and the Packing Group is II.[19][20]

Spill Handling: Evacuate and restrict persons not wearing protective equipment from area of spill or leak until cleanup is complete. Remove all ignition sources. Absorb liquids in vermiculite, dry sand, earth, or a similar non-organic material and deposit in sealed containers. May also be covered with weak reducing agents; resulting sludge neutralized and flushed to sewer. Collect powdered material in the most convenient and safe manner and deposit in sealed containers. Ventilate area of spill or leak after clean-up is complete. It may be necessary to contain and dispose of this chemical as a hazardous waste. If material or contaminated runoff enters waterways, notify downstream users of potentially contaminated waters. Contact your Department of Environmental Protection or your regional office of the federal EPA for specific recommendations. If employees are required to clean-up spills, they must be properly trained and equipped. OSHA 1910.120(q) may be applicable.

Fire Extinguishing: May explode when heated. Contact with combustible, organic, or other easily oxidizable materials, such as paper, oil, fuels or sawdust can cause fires. Rubbing of these mixtures can cause explosions. Use water to extinguish the fire. Poisonous gases are produced in fire. If material or contaminated runoff enters waterways, notify downstream users of potentially contaminated waters. Notify local health and fire officials and pollution control agencies. From a secure, explosion-proof location, use water spray to cool exposed containers. If cooling streams are ineffective (venting sound increases in volume and pitch, tank discolors, or shows any signs of deforming), withdraw immediately to a secure position. If employees are expected to fight fires, they must be trained and equipped in OSHA 1910.156.

Disposal Method Suggested: Use large volumes of reducing agent (bisulfite or ferrous salt) solutions. Neutralize and flush to sever with large volumes of water.[24]

References

New Jersey Department of Health and Senior Services, "Hazardous Substance Fact Sheet: Barium Chlorate," Trenton, NJ (June 1986)

Barium Cyanide

Molecular Formula: BaC_2N_2

Common Formula: $Ba(CN)_2$

Synonyms: Barium cyanide, solid; Barium dicyanide; Cianuro barico (Spanish)

CAS Registry Number: 542-62-1

RTECS Number: CQ8785000

DOT ID: UN 1565

Regulatory Authority

- CLEAN WATER ACT: Section 311 Hazardous Substances/ RQ 40CFR117.3 (same as CERCLA, see below); Section 313 Water Priority Chemicals (57FR41331, 9/9/92)
- EPA HAZARDOUS WASTE NUMBER (RCRA No.): P013
- RCRA, 40CFR261, Appendix 8 Hazardous Constituents
- SUPERFUND/EPCRA 40CFR302.4 Reportable Quantity (RQ): CERCLA, 10 lb (4.54 kg)
- RCRA Land Ban Waste Restrictions
- EPCRA Section 313: *as barium compounds*; Form R *de minimis* concentration reporting level: 0.1%
- U.S. DOT 49CFR172.101, Appendix B, Regulated marine pollutant as cyanide compounds
- CLEAN AIR ACT: Hazardous Air Pollutants (Title I, Part A, Section 112)
- CLEAN WATER ACT: 40CFR423, Appendix A, Priority Pollutants as cyanide, total
- EPA HAZARDOUS WASTE NUMBER (RCRA No.): P030 as cyanides soluble salts and complexes, n.o.s
- RCRA, 40CFR261, Appendix 8 Hazardous Constituents. as cyanides, soluble salts and complexes, n.o.s
- EPCRA (Section 313): X+CN- where X = H+ or any other group where a formal dissociation may occur. For example, KCN or $Ca(CN)_2$; Form R *de minimis* concentration reporting level: 1.0%
- U.S. DOT Regulated Marine Pollutant (49CFR172.101, Appendix B) as cyanide mixtures, cyanide solutions or cyanides, inorganic, n.o.s

Cited in U.S. State Regulations: California (G), Kansas (G), Louisiana (G), Maine (G), Massachusetts (G), New Hampshire (G), New Jersey (G), Oklahoma (G), Pennsylvania (G), Vermont (G), Virginia (G), Washington (G), Wisconsin (G).

Description: Barium cyanide, $Ba(CN)_2$, is a white crystalline powder. Often used in solution. Hazard Identification (based on NFPA-704 M Rating System): Health 3, Flammability 0, Reactivity 0. Soluble in water.

Potential Exposure: Barium cyanide is used in electroplating and in metallurgy.

Incompatibilities: Violent reactions may occur on contact with acids, acid salts, and strong oxidizers.

Permissible Exposure Limits in Air: For Ba soluble compounds, the Federal OSHA standard[58] and ACGIH recommended airborne limit is 0.5 mg/m^3 TWA for an 8-hour workshift. The NIOSH level is the same for a 10-hour workshift. The DFG,[3] HSE,[33] Australia, Mexico, and Israel have adopted this same value and DFG has a Peak Limitation (5 min) of 2 times normal MAK; do not exceed more than 8 times during workshift. Israel's Action Level is one half the TWA. The NIOSH IDLH level is 50 mg/m^3 (Ba, soluble compounds). The above exposure limits are for air levels only. When skin contact also occurs, overexposure can occur even though air levels are less than the limits listed above.

Permissible Concentration in Water: No criteria set for barium cyanide per se. See entry under "Barium." See also entry under "Cyanides."

Routes of Entry: Inhalation, ingestion, eye and/or skin contact. Passes through the skin.

Harmful Effects and Symptoms

Short Term Exposure: This chemical can be absorbed through the skin, thereby increasing exposure. Barium Cyanide is a deadly poison; can affect you when breathed and by passing through skin. Exposure can cause confusion, weakness, headaches, nausea, and vomiting, gasping for air, collapse and even death from cyanide poisoning. On contact with acids, acid mists, or acid salts, flammable hydrogen cyanide gas is formed which can cause rapid poisoning.

Long Term Exposure: Can interfere with the normal functioning of the thyroid gland, causing goiter (enlarged thyroid).

Points of Attack: Thyroid.

Medical Surveillance: Blood cyanide level. Thyroid function tests.

First Aid: If this chemical gets into the eyes, remove any contact lenses at once and irrigate immediately for at least 15 minutes, occasionally lifting upper and lower lids. Seek medical attention immediately. If this chemical contacts the skin, remove contaminated clothing and wash immediately with soap and water. Seek medical attention immediately. If this chemical has been inhaled, remove from exposure, begin rescue breathing (using universal precautions) if breathing has stopped and CPR if heart action has stopped. Transfer promptly to a medical facility. When this chemical has been swallowed, get medical attention. Give large quantities of water and induce vomiting. Do not make an unconscious person vomit. *If cyanide poisoning is confirmed*, use amyl nitrate capsules if symptoms develop. All area employees should be trained regularly in emergency measures for cyanide poisoning and in CPR. A cyanide antidote kit should be kept in the immediate work area and must be rapidly available. Kit ingredients should be replaced every 1 – 2 years to ensure freshness. Persons trained in the use of this kit, oxygen use, and CPR must be quickly available.

Personal Protective Methods: Where possible, enclose operations and use local exhaust ventilation at the site of chemical release. If local exhaust ventilation or enclosure is not used, respirators should be worn. Wear protective gloves and clothing to prevent any reasonable probability of skin contact. Safety equipment suppliers/manufacturers can provide recommendations on the most protective glove/clothing material for your operation. All protective clothing (suits, gloves, footwear, headgear) should be clean, available each day, and put on before work. Contact lenses should not be worn when working with this chemical. Wear splash- (for liquid) or dust-proof chemical goggles and face shield unless full facepiece respiratory protection is worn. Employees should wash immediately with soap when skin is wet or contaminated. Provide emergency showers and eyewash.

Respirator Selection: NIOSH/OSHA: (Ba soluble compounds) *5 mg/m³:* DMXSQ (any dust and mist respirator except single-use and quarter mask respirators); or SA (any supplied-air respirator). *12.5 mg/m³:* SA:CF (any supplied-air respirator operated in a continuous-flow mode); or PAPRDM (any powered, air-purifying respirator with a dust and mist filter). *25 mg/m³:* HiEF (any air-purifying, full-facepiece respirator with a high-efficiency particulate filter); or SAT:CF (any supplied-air respirator that has a tight-fitting facepiece and is operated in a continuous-flow mode); or PAPRTHiE (any powered, air-purifying respirator with a tight-fitting facepiece and a high-efficiency particulate filter); or SCBAF (any self-contained breathing apparatus with a full facepiece); or SAF (any supplied-air respirator with a full facepiece). *50 mg/m³:* SAF:PD,PP (any supplied-air respirator that has a full facepiece and is operated in a pressure-demand or other positive-pressure mode). *Emergency or planned entry into unknown concentrations or IDLH conditions:* SCBAF:PD, PP (any self-contained breathing apparatus that has a full facepiece and is operated in a pressure-demand or other positive-pressure mode); or SAF:PD,PP:ASCBA (any supplied-air respirator that has a full facepiece and is operated in a pressure-demand or other positive-pressure mode in combination with an auxiliary, self-contained breathing apparatus operated in a pressure-demand or other positive-pressure mode). *Escape:* HiEF (any air-purifying, full-facepiece respirator with a high-efficiency particulate filter); or SCBAE (any appropriate escape-type, self-contained breathing apparatus).

Storage: Barium Cyanide must be stored to avoid contact with acids; acid salt (such as potassium bisulfate, calcium biphosphate and calcium nitrate); carbon dioxide and strong oxidizers (such as nitrates, chlorates and chlorine) since violent reactions occur. Store in tightly closed containers in a cool, well-ventilated area.

Shipping: The DOT-required label is "Poison." The UN/DOT Hazard Class is 6.1 and the Packing Group is I.[19][20]

Spill Handling: Evacuate and restrict persons not wearing protective equipment from area of spill or leak until cleanup is complete. Remove all ignition sources. Collect powdered material in the most convenient and safe manner and deposit in sealed containers. Ventilate area of spill or leak after clean-up is complete. It may be necessary to contain and dispose of this chemical as a hazardous waste. If material or contaminated runoff enters waterways, notify downstream users of potentially contaminated waters. Contact your Department of Environmental Protection or your regional office of the federal EPA for specific recommendations. If employees are required to clean-up spills, they must be properly trained and equipped. OSHA 1910.120(q) may be applicable.

Restrict persons not wearing protective equipment from area of spill or leak until clean-up is complete. Ventilate the area of spill or leak. Absorb liquids in vermiculite, dry sand, earth, or a similar material and deposit in sealed containers. Collect powdered material in the most convenient and safe manner and deposit in sealed containers.

Fire Extinguishing: Barium Cyanide does not burn, but contact with acids, acid salts, or Carbon Dioxide in air may produce highly flammable Hydrogen Cyanide gas. Extinguish fire using an agent suitable for type of surrounding fire. Poisonous gases are produced in fire including cyanide. If material or contaminated runoff enters waterways, notify downstream users of potentially contaminated waters. Notify local health and fire officials and pollution control agencies. From a secure, explosion-proof location, use water spray to cool exposed containers. If cooling streams are ineffective (venting sound increases in volume and pitch, tank discolors, or shows any signs of deforming), withdraw immediately to a secure position. If employees are expected to fight fires, they must be trained and equipped in OSHA 1910.156.

Disposal Method Suggested: Precipitate barium with sulfate. Then add with stirring to alkaline calcium hypochlorite solution. Let stand 24 hours, then flush to sewer.

References

Sax, N. I., Ed., "Dangerous Properties of Industrial Materials Report" 1, No. 6, 33–35 (1981) and 3, No. 4, 31–32 (1983)

New Jersey Department of Health and Senior Services, "Hazardous Substance Fact Sheet: Barium Cyanide," Trenton, NJ (April 1998)

Barium Hypochlorite

Molecular Formula: $BaCl_2O_2$

Common Formula: $Ba(OCl)_2$

CAS Registry Number: 13477-10-6

RTECS Number: NH3480000

DOT ID: UN 2741

Regulatory Authority

- Air Pollutant Standard Set (ACGIH) (OSHA) (NIOSH) (barium, soluble compounds)
- RCRA 40CFR261, Appendix 8; 40CFR261.11 Hazardous Constituents
- EPCRA Section 313: Includes any unique chemical substance that contains barium as part of that chemical's infrastructure. This category does not include barium sulfate (7727-43-7). Form R *de minimis* concentration reporting level: 0.1%
- U.S. DOT 49CFR172.101, Appendix B, Regulated marine pollutant (Ba compounds, soluble, n.o.s.)

Cited in U.S. State Regulations: California (G), New Hampshire (G), New Jersey (G).

Description: Barium hypochlorite, $Ba(OCl)_2$, is a colorless crystalline solid which is often used in solution. Reacts with water.

Potential Exposure: This material is used as a bleaching agent and as an antiseptic.

Incompatibilities: Barium hypochlorite is a strong oxidizer. Avoid contact with organic and combustible materials (such as wood, oil, paper and fuels), acids and urea since violent reactions occur. Keep away from water or steam.

Permissible Exposure Limits in Air: For Ba soluble compounds, the Federal OSHA standard[58] and ACGIH recommended airborne limit is 0.5 mg/m^3 TWA for an 8-hour workshift. The NIOSH level is the same for a 10-hour workshift. The DFG,[3] HSE,[33] Australia, Mexico, and Israel have adopted this same value and DFG has a Peak Limitation (5 min) of 2 times normal MAK; do not exceed more than 8 times during workshift. Israel's Action Level is one half the TWA. The NIOSH IDLH level is 50 mg/m^3 (Ba, soluble compounds).

Determination in Air: See entry under "Barium."

Permissible Concentration in Water: No standards set for Barium Hypochlorite per se. See entry under "Barium."

Routes of Entry: Inhalation, ingestion, eye and/or skin contact. Absorbed through the skin.

Harmful Effects and Symptoms

Short Term Exposure: Barium Hypochlorite can affect you when breathed in and may enter the body through the skin. Contact can irritate and even burn the eyes and skin. Breathing the dust or mist can irritate the nose, throat and bronchial tubes, causing cough and phlegm.

Long Term Exposure: After repeated exposure, Barium may show up as spots in the lungs on chest x-ray. Some Barium chemicals are contaminated with Silica, which scars the lungs. Repeated skin contact can cause chronic dryness and cracking.

Points of Attack: Lungs, skin.

Medical Surveillance: Lung function tests.

First Aid: If this chemical gets into the eyes, remove any contact lenses at once and irrigate immediately for at least 15 minutes, occasionally lifting upper and lower lids. Seek medical attention immediately. If this chemical contacts the skin, remove contaminated clothing and wash immediately with soap and water. Seek medical attention immediately. If this chemical has been inhaled, remove from exposure, begin rescue breathing (using universal precautions) if breathing has stopped and CPR if heart action has stopped. Transfer promptly to a medical facility. When this chemical has been swallowed, get medical attention. Give large quantities of water and induce vomiting. Do not make an unconscious person vomit.

Personal Protective Methods: Where possible, enclose operations and use local exhaust ventilation at the site of chemical release. If local exhaust ventilation or enclosure is not used, respirators should be worn. Wear protective work clothing. Wash thoroughly immediately after exposure to Barium Hypochlorite and at the end of the workshift. Post hazard and warning information in the work area. In addition, as part of an ongoing education and training effort, communicate all information on the health and safety hazards of Barium hypochlorite to potentially exposed workers.

Respirator Selection: NIOSH/OSHA: (BA soluble compounds) *5 mg/m^3:* DMXSQ (any dust and mist respirator except single-use and quarter mask respirators); or SA (any supplied-air respirator). *12.5 mg/m³:* SA:CF (any supplied-air respirator operated in a continuous-flow mode); or PAPRDM (any powered, air-purifying respirator with a dust and mist filter). *25 mg/m³:* HiEF (any air-purifying, full-facepiece respirator with a high-efficiency particulate filter); or SAT:CF (any supplied-air respirator that has a tight-fitting facepiece and is operated in a continuous-flow mode); or PAPRTHiE (any powered, air-purifying respirator with a tight-fitting facepiece and a high-efficiency particulate filter); or SCBAF (any self-contained breathing apparatus with a full facepiece); or SAF (any supplied-air respirator with a full facepiece). *50 mg/m³:* SAF:PD,PP (any supplied-air respirator that has a full facepiece and is operated in a pressure-demand or other positive-pressure mode). *Emergency or planned entry into unknown concentrations or IDLH conditions:* SCBAF:PD,PP (any self-contained breathing apparatus that has a full facepiece and is operated in a pressure-demand or other positive-pressure mode); or SAF:PD, PP:ASCBA (any supplied-air respirator that has a full facepiece and is operated in a pressure-demand or other positive-pressure mode in combination with an auxiliary, self-contained breathing apparatus operated in a pressure-demand or other positive-pressure mode). *Escape:* HiEF (any air-purifying, full-facepiece respirator with a high-efficiency particulate filter); or SCBAE (any appropriate escape-type, self-contained breathing apparatus).

Storage: Store in tightly closed containers in a cool, well-ventilated area away from water and steam. Sources of ignition, such as smoking and open flames, are prohibited where Barium Hypochlorite is used, handled, or stored in a manner that could create a potential fire or explosion hazard. Avoid any possible contact with the incompatible materials cited above. See OSHA standard 1910.104 and NFPA 43A *Code for the Storage of Liquid and Solid Oxidizers* for detailed handling and storage regulations.

Shipping: Barium hypochlorite (with more than 22% available chlorine) requires an "Oxidizer" label. It falls in UN/DOT Hazard Class 5.1 and Packing Group II.

Spill Handling: Evacuate and restrict persons not wearing protective equipment from area of spill or leak until cleanup is complete. Remove all ignition sources. Collect powdered material in the most convenient and safe manner and deposit in sealed containers. Ventilate area of spill or leak after clean-up is complete. It may be necessary to contain and dispose of this chemical as a hazardous waste. If material or contaminated runoff enters waterways, notify downstream users of potentially contaminated waters. Contact your Department of Environmental Protection or your regional office of the federal EPA for specific recommendations. If employees are required to clean-up spills, they must be properly trained and equipped. OSHA 1910.120(q) may be applicable.

Fire Extinguishing: Do not use water. Extinguish fire using an agent suitable for type of surrounding fire. Barium Hypochlorite itself does not burn. Poisonous gases are produced in fire including chlorine and chlorides. If material or contaminated runoff enters waterways, notify downstream users of potentially contaminated waters. Notify local health and fire officials and pollution control agencies. From a secure, explosion-proof location, use water spray to cool exposed containers. If cooling streams are ineffective (venting sound increases in volume and pitch, tank discolors, or shows any signs of deforming), withdraw immediately to a secure position. If employees are expected to fight fires, they must be trained and equipped in OSHA 1910.156.

References

New Jersey Department of Health and Senior Services, "Hazardous Substance Fact Sheet: Barium Hypochlorite," Trenton, NJ (June 1986)

Barium Nitrate

Molecular Formula: BaN_2O_6

Common Formula: $Ba(NO_3)_2$

Synonyms: Barium dinitrate; Dusicnan barnaty (Czech); Nitrate de baryum (French); Nitrato barico (Spanish); Nitric acid, barium salt

CAS Registry Number: 10022-31-8

RTECS Number: CQ9625000

DOT ID: UN 1446

Regulatory Authority

- Air Pollutant Standard Set (ACGIH) (OSHA) (NIOSH) (barium, soluble compounds)
- RCRA 40CFR261, Appendix 8; 40CFR261.11 Hazardous Constituents
- EPCRA Section 313: Includes any unique chemical substance that contains barium as part of that chemical's infrastructure. This category does not include barium sulfate (7727-43-7). Form R *de minimis* concentration reporting level: 0.1%
- U.S. DOT 49CFR172.101, Appendix B, Regulated marine pollutant (Ba compounds, soluble, n.o.s.)

Cited in U.S. State Regulations: California (G), Florida (G), Maine (G), Massachusetts (G), New Hampshire (G), New Jersey (G), Oklahoma (G), Pennsylvania (G), Rhode Island (G).

Description: Barium Nitrate, $Ba(NO_3)_2$, is a shiny, white crystalline solid. Boiling point (decomposes). Freezing/Melting point = 592°C. Hazard Identification (based on NFPA-704 M Rating System): Health 2, Flammability 0, Reactivity 0.

Potential Exposure: Barium nitrate is used in fireworks, ceramics and in the electronics industry.

Incompatibilities: Acids, oxidizers. Contact with organic and combustible materials (such as wood, paper, oil and fuels); and aluminum-magnesium alloys since violent reactions occur. Contact with finely divided metals may form a shock-sensitive compounds.

Permissible Exposure Limits in Air: For Ba soluble compounds, the Federal OSHA standard[58] and ACGIH recommended airborne limit is 0.5 mg/m^3 TWA for an 8-hour workshift. The NIOSH level is the same for a 10-hour workshift. The DFG,[3] HSE,[33] Australia, Mexico, and Israel have adopted this same value and DFG has a Peak Limitation (5 min) of 2 times normal MAK; do not exceed more than 8 times during workshift. Israel's Action Level is one half the TWA. The NIOSH IDLH level is 50 mg/m^3 (Ba, soluble compounds).

Determination in Air: See entry for "Barium."

Permissible Concentration in Water: No criteria set for barium nitrate per se. See entry for "Barium."

Routes of Entry: Inhalation.

Harmful Effects and Symptoms

Short Term Exposure: Barium Nitrate can affect you when breathed in. Inhaling dust or mist can cause irritation of the respiratory system, causing cough and phlegm. Contact can irritate and even burn the eyes and skin. Exposure can irritate the eyes, nose, and throat. Very high exposure (such as swallowing or extremely high dust exposure) can cause barium poisoning with symptoms of vomiting and diarrhea, irregular heartbeat, paralysis, and death. The oral LD_{50} rat is 355 mg/kg.

Long Term Exposure: Repeated high exposure can irritate the lungs, causing cough and phlegm and may cause an abnormal chest x-ray.

Points of Attack: Lungs.

Medical Surveillance: Lung function tests.

First Aid: If this chemical gets into the eyes, remove any contact lenses at once and irrigate immediately for at least 15 minutes, occasionally lifting upper and lower lids. Seek medical attention immediately. If this chemical contacts the skin, remove contaminated clothing and wash immediately with soap and water. Seek medical attention immediately. If this chemical has been inhaled, remove from exposure, begin rescue breathing (using universal precautions) if breathing has stopped and CPR if heart action has stopped. Transfer promptly to a medical facility. When this chemical has been swallowed, get medical attention. Give large quantities of water and induce vomiting. Do not make an unconscious person vomit.

Personal Protective Methods: Where possible, enclose operations and use local exhaust ventilation at the site of chemical release. If local exhaust ventilation or enclosure is not used, respirators should be worn. Wear protective gloves and clothing to prevent any reasonable probability of skin contact. Safety equipment suppliers/manufacturers can provide recommendations on the most protective glove/clothing material for your operation. All protective clothing (suits, gloves, footwear, headgear) should be clean, available each day, and put on before work. Contact lenses should not be worn when working with this chemical. Wear splash or dust-proof chemical goggles and face shield unless full facepiece respiratory protection is worn. Employees should wash immediately with soap when skin is wet or contaminated. Provide emergency showers and eyewash.

Respirator Selection: NIOSH/OSHA: (Ba soluble compounds) *5 mg/m³:* DMXSQ (any dust and mist respirator except single-use and quarter mask respirators); or SA (any supplied-air respirator). *12.5 mg/m³:* SA:CF (any supplied-air respirator operated in a continuous-flow mode); or PAPRDM (any powered, air-purifying respirator with a dust and mist filter). *25 mg/m³:* HiEF (any air-purifying, full-facepiece respirator with a high-efficiency particulate filter); or SAT:CF (any supplied-air respirator that has a tight-fitting facepiece and is operated in a continuous-flow mode); or PAPRTHiE (any powered, air-purifying respirator with a tight-fitting facepiece and a high-efficiency particulate filter); or SCBAF (any self-contained breathing apparatus with a full facepiece); or SAF (any supplied-air respirator with a full facepiece). *50 mg/m³:* SAF:PD,PP (any supplied-air respirator that has a full facepiece and is operated in a pressure-demand or other positive-pressure mode). *Emergency or planned entry into unknown concentrations or IDLH conditions:* SCBAF:PD,PP (any self-contained breathing apparatus that has a full facepiece and is operated in a pressure-demand or other positive-pressure mode); or SAF:PD,PP:ASCBA (any supplied-air respirator that has a full facepiece and is operated in a pressure-demand or other positive-pressure mode in combination with an auxiliary, self-contained breathing apparatus operated in a pressure-demand or other positive-pressure mode). *Escape:* HiEF (any air-purifying, full-facepiece respirator with a high-efficiency particulate filter); or SCBAE (any appropriate escape-type, self-contained breathing apparatus).

Storage: Store in tightly closed containers in a cool, well-ventilated area. Avoid any possible contact with incompatible materials cited above. See OSHA standard 1910.104 and NFPA 43A *Code for the Storage of Liquid and Solid Oxidizers* for detailed handling and storage regulations.

Shipping: The label required by DOT is "Oxidizer, Poison." The limit on passenger aircraft or railcar shipment is 5 kg; on cargo aircraft shipment is 25 kg. The UN/DOT Hazard Class is 5.1 and the Packing Group is II.[19][20]

Spill Handling: Evacuate and restrict persons not wearing protective equipment from area of spill or leak until cleanup is complete. Remove all ignition sources. Collect powdered material in the most convenient and safe manner and deposit in sealed containers. Ventilate area of spill or leak after clean-up is complete. It may be necessary to contain and dispose of this chemical as a hazardous waste. If material or contaminated runoff enters waterways, notify downstream users of potentially contaminated waters. Contact your Department of Environmental Protection or your regional office of the federal EPA for specific recommendations. If employees are required to clean-up spills, they must be properly trained and equipped. OSHA 1910.120(q) may be applicable.

Fire Extinguishing: Extinguish fire using an agent suitable for type of surrounding fire. Barium Nitrate itself does not burn. Use flooding amounts of water in early stages of fire;[17] in large fires, the material may melt and water could scatter the molten material.

Poisonous gases are produced in fire including nitrogen oxides. If material or contaminated runoff enters waterways, notify downstream users of potentially contaminated waters. Notify local health and fire officials and pollution control agencies. From a secure, explosion-proof location, use water spray to cool exposed containers. If cooling streams are ineffective (venting sound increases in volume and pitch, tank discolors, or shows any signs of deforming), withdraw immediately to a secure position. If employees are expected to fight fires, they must be trained and equipped in OSHA 1910.156.

Disposal Method Suggested: Dissolve waste in 6-M HCl. Neutralize with NH_4OH. Precipitate with excess sodium carbonate. Filter, wash and dry precipitate and return to supplier.

References

Sax, N. I., Ed., "Dangerous Properties of Industrial Materials Report" 1, No. 6, 36–37 (1981)

New Jersey Department of Health and Senior Services, "Hazardous Substance Fact Sheet: Barium Nitrate," Trenton, NJ (April 1986)

Barium Oxide

Molecular Formula: BaO

Synonyms: Barium monoxide; Barium protoxide; Baryta; Calcined baryta; Monoxido barico (Spanish); Oxyde de baryum (French)

CAS Registry Number: 1304-28-5

RTECS Number: CQ9800000

DOT ID: UN 1884

EEC Number: 056-002-00-7

Regulatory Authority

- Air Pollutant Standard Set (ACGIH) (OSHA) (NIOSH) (Ontario) (barium, soluble compounds)
- RCRA 40CFR261, Appendix 8; 40CFR261.11 Hazardous Constituents
- EPCRA Section 313: Includes any unique chemical substance that contains barium as part of that chemical's infrastructure. This category does not include barium sulfate (7727-43-7). Form R *de minimis* concentration reporting level: 0.1%
- U.S. DOT 49CFR172.101, Appendix B, Regulated marine pollutant (Ba compounds, soluble, n.o.s.)
- Canada, WHMIS, Ingredients Disclosure List

Cited in U.S. State Regulations: California (G), Maine (G), New Hampshire (G), New Jersey (G), Oklahoma (G).

Description: Barium Oxide, BaO, is a white to yellowish-white, odorless powder. Freezing/Melting point = 1,923°C. Reacts violently with water forming barium hydroxide.

Potential Exposure: It is used to dry gases and solvents and in producing detergents for lubricating oils.

Incompatibilities: Reacts with water. Hydrogen sulfide, carbon dioxide, hydroxylamine, nitrogen tetroxide, sulfur trioxide since violent reactions occur. Reacts with triuranium.

Permissible Exposure Limits in Air: For Ba soluble compounds, the Federal OSHA standard[58] and ACGIH recommended airborne limit is 0.5 mg/m^3 TWA for an 8-hour workshift. The NIOSH level is the same for a 10-hour workshift. The DFG,[3] HSE,[33] Australia, Mexico, and Israel, and Ontario have adopted this same value and DFG has a Peak Limitation (5 min) of 2 times normal MAK; do not exceed more than 8 times during workshift. Israel's Action Level is one half the TWA. The NIOSH IDLH level is 50 mg/m^3 (Ba, soluble compounds).

Determination in Air: See entry for "Barium."

Permissible Concentration in Water: No criteria set for barium oxide per se. See entry for "Barium."

Determination in Water: Harmful to the environment.

Routes of Entry: Inhalation.

Harmful Effects and Symptoms

Short Term Exposure: Barium Oxide can affect you when breathed in. Contact can irritate the skin and burn the eyes, causing loss of vision. Breathing the dust or mist can irritate the nose, throat and bronchial tubes, causing cough and phlegm. High exposure may cause pulmonary edema, a medical emergency, that can be delayed for several hours. This can cause death.

Long Term Exposure: May cause lung irritation and bronchitis. After repeated exposure, Barium may show up as spots in the lungs on chest x-ray. Some Barium chemicals are contaminated with Silica, which scars the lungs.

Points of Attack: Lungs.

Medical Surveillance: Lung function tests.

First Aid: If this chemical gets into the eyes, remove any contact lenses at once and irrigate immediately for at least 15 minutes, occasionally lifting upper and lower lids. Seek medical attention immediately. If this chemical contacts the skin, remove contaminated clothing and wash immediately with soap and water. Seek medical attention immediately. If this chemical has been inhaled, remove from exposure, begin rescue breathing (using universal precautions) if breathing has stopped and CPR if heart action has stopped. Transfer promptly to a medical facility. When this chemical has been swallowed, get medical attention. Give large quantities of water and induce vomiting. Do not make an unconscious person vomit.

Medical observation is recommended for 24 – 48 hours after breathing overexposure, as pulmonary edema may be delayed. As first aid for pulmonary edema, a doctor or authorized paramedic may consider administering a corticosteroid spray.

Personal Protective Methods: Where possible, enclose operations and use local exhaust ventilation at the site of chemical release. If local exhaust ventilation or enclosure is not used, respirators should be worn. Wear protective gloves and clothing to prevent any reasonable probability of skin contact. Safety equipment suppliers/manufacturers can provide recommendations on the most protective glove/clothing material for your operation. All protective clothing (suits, gloves, footwear, headgear) should be clean, available each day, and put on before work. Contact lenses should not be worn when working with this chemical. Wear dust-proof chemical goggles and face shield unless full facepiece respiratory protection is worn. Employees should wash immediately with soap when skin is wet or contaminated. Provide emergency showers and eyewash.

Respirator Selection: NIOSH/OSHA: (Ba soluble compounds) *5 mg/m*^{3}*:* DMXSQ (any dust and mist respirator except single-use and quarter mask respirators); or SA (any supplied-air respirator). *12.5 mg/m*^{3}*:* SA:CF (any supplied-air respirator operated in a continuous-flow mode); or PAPRDM (any powered, air-purifying respirator with a dust and mist filter). *25 mg/m*^{3}*:* HiEF (any air-purifying, full-facepiece respirator with a high-efficiency particulate filter); or SAT:CF (any supplied-air respirator that has a tight-fitting facepiece and is operated in a continuous-flow mode); or PAPRTHiE (any powered, air-purifying respirator with a tight-fitting facepiece and a high-efficiency particulate filter); or SCBAF (any self-contained breathing apparatus with a full facepiece); or SAF (any supplied-air respirator with a full facepiece). *50 mg/m*^{3}*:* SAF:PD,PP (any supplied-air respirator that has a full facepiece and is operated in a pressure-demand or other positive-pressure mode). *Emergency or planned entry into unknown concentrations or IDLH conditions:* SCBAF:PD,PP (any self-contained breathing apparatus that has a full facepiece and is operated in a pressure-demand or other positive-pressure mode); or SAF:PD,PP: ASCBA (any supplied-air respirator that has a full facepiece and is operated in a pressure-demand or other positive-pressure mode in combination with an auxiliary, self-contained breathing apparatus operated in a pressure-demand or other positive-pressure mode). *Escape:* HiEF (any air-purifying, full-facepiece respirator with a high-efficiency particulate filter); or SCBAE (any appropriate escape-type, self-contained breathing apparatus).

Storage: Store in tightly closed containers in a dry, cool, well-ventilated area away from water and the incompatible substances cited above.

Shipping: The label required by DOT is "Keep away from Food." This material falls in UN/DOT Hazard Class 6.1 and Packing Group III.[19][20]

Spill Handling: Evacuate and restrict persons not wearing protective equipment from area of spill or leak until cleanup is complete. Remove all ignition sources. NEVER pour water into this substance; when dissolving or diluting always add it slowly to the water. Collect powdered material in the most convenient and safe manner and deposit in sealed containers. Ventilate area of spill or leak after clean-up is complete. It may be necessary to contain and dispose of this chemical as a hazardous waste. If material or contaminated runoff enters waterways, notify downstream users of potentially contaminated waters. Contact your Department of Environmental Protection or your regional office of the federal EPA for specific recommendations. If employees are required to clean-up spills, they must be properly trained and equipped. OSHA 1910.120(q) may be applicable.

Fire Extinguishing: Extinguish fire using an agent suitable for type of surrounding fire. Barium Oxide itself does not burn. Do not use water. Poisonous gases are produced in fire. If material or contaminated runoff enters waterways, notify downstream users of potentially contaminated waters. Notify local health and fire officials and pollution control agencies. From a secure, explosion-proof location, use water spray to cool exposed containers. If cooling streams are ineffective (venting sound increases in volume and pitch, tank discolors, or shows any signs of deforming), withdraw immediately to a secure position. If employees are expected to fight fires, they must be trained and equipped in OSHA 1910.156.

References

New Jersey Department of Health and Senior Services, "Hazardous Substance Fact Sheet: Barium Oxide," Trenton, NJ (June 1996)

Barium Perchlorate

Molecular Formula: $BaCl_2O_8$

Common Formula: $Ba(ClO_4)_2$

Synonyms: Barium perchlorate trihydrate; Perchloric acid, barium salt; Perclorato barico (Spanish)

CAS Registry Number: 13465-95-7

RTECS Number: SC7550000

DOT ID: UN 1447

EEC Number: 017-007-00-x

Regulatory Authority

- Air Pollutant Standard Set (ACGIH) (OSHA) (NIOSH) (barium, soluble compounds)
- RCRA 40CFR261, Appendix 8; 40CFR261.11 Hazardous Constituents
- EPCRA Section 313: Includes any unique chemical substance that contains barium as part of that chemical's infrastructure.

This category does not include barium sulfate (7727-43-7). Form R *de minimis* concentration reporting level: 0.1%

- U.S. DOT 49CFR172.101, Appendix B, Regulated marine pollutant (Ba compounds, soluble, n.o.s.)

Cited in U.S. State Regulations: California (G), Maine (G), New Hampshire (G), New Jersey (G), Oklahoma (G).

Description: Barium Perchlorate, $Ba(ClO_4)_2$, is a white crystalline solid. Freezing/Melting point = 505°C. Hazard Identification (based on NFPA-704 M Rating System): Health 2, Flammability 0, Reactivity 0. Highly soluble in water.

Potential Exposure: It is used to make explosives and in experimental rocket fuels.

Incompatibilities: An oxidizing agent. Contact with organic and combustible materials (such as paper, wood and oil), finely divided metals (specifically magnesium and aluminum), sulfur, calcium hydride and strontium hydride since violent reactions occur.

Permissible Exposure Limits in Air: OSHA: (Ba soluble compounds) the Federal OSHA standard[58] and ACGIH recommended airborne limit is 0.5 mg/m^3 TWA for an 8-hour workshift. The NIOSH level is the same for a 10-hour workshift. The DFG,[3] HSE,[33] Australia, Mexico, Israel, and Ontario have adopted this same value and DFG has a Peak Limitation (5 min) of 2 times normal MAK; do not exceed more than 8 times during workshift. Israel's Action Level is one half the TWA. The NIOSH IDLH level is 50 mg/m^3 (Ba, soluble compounds).

Determination in Air: See entry for "Barium."

Permissible Concentration in Water: No criteria set for Barium Perchlorate *per se*. See entry for "Barium."

Routes of Entry: Inhalation.

Harmful Effects and Symptoms

Short Term Exposure: Barium Perchlorate can affect you when breathed in. Contact can cause severe irritation and burn the eyes and skin. Breathing the dust or mist can irritate the nose, throat, and bronchial tubes, causing cough and phlegm. Overexposure can cause methemoglobinemia, causing dizziness, bluish color to the skin and lips. Higher levels can cause difficult breathing, collapse, and even death.

Long Term Exposure: After repeated exposure, Barium may show up as spots in the lungs on x-ray. Some Barium chemicals are contaminated with Silica, which scars the lungs. Perchlorates can interfere with thyroid function, affect the red blood cells (methemoglobinemia) or damage bone marrow (aplastic anemia).

Points of Attack: Lungs, red blood cells.

Medical Surveillance: Lung function tests, thyroid function tests, complete blood count (CBC), tests for methemoglobin.

First Aid: If this chemical gets into the eyes, remove any contact lenses at once and irrigate immediately for at least 15 minutes, occasionally lifting upper and lower lids. Seek medical attention immediately. If this chemical contacts the skin, remove contaminated clothing and wash immediately with soap and water. Seek medical attention immediately. If this chemical has been inhaled, remove from exposure, begin rescue breathing (using universal precautions) if breathing has stopped and CPR if heart action has stopped. Transfer promptly to a medical facility. When this chemical has been swallowed, get medical attention. Give large quantities of water and induce vomiting. Do not make an unconscious person vomit.

Note to Physician: Treat for methemoglobinemia. Spectrophotometry may be required for precise determination of levels of methemoglobinemia in urine.

Personal Protective Methods: Where possible, enclose operations and use local exhaust ventilation at the site of chemical release. If local exhaust ventilation or enclosure is not used, respirators should be worn. Wear protective gloves and clothing to prevent any reasonable probability of skin contact. Safety equipment suppliers/manufacturers can provide recommendations on the most protective glove/clothing material for your operation. All protective clothing (suits, gloves, footwear, headgear) should be clean, available each day, and put on before work. Contact lenses should not be worn when working with this chemical. Wear dust-proof chemical goggles and face shield unless full facepiece respiratory protection is worn. Employees should wash immediately with soap when skin is wet or contaminated. Provide emergency showers and eyewash.

Respirator Selection: NIOSH/OSHA: (Ba soluble compounds) *5 mg/m^3:* DMXSQ (any dust and mist respirator except single-use and quarter mask respirators); or SA (any supplied-air respirator). *12.5 mg/m^3:* SA:CF (any supplied-air respirator operated in a continuous-flow mode); or PAPRDM (any powered, air-purifying respirator with a dust and mist filter). *25 mg/m^3:* HiEF (any air-purifying, full-facepiece respirator with a high-efficiency particulate filter); or SAT:CF (any supplied-air respirator that has a tight-fitting facepiece and is operated in a continuous-flow mode); or PAPRTHiE (any powered, air-purifying respirator with a tight-fitting facepiece and a high-efficiency particulate filter); or SCBAF (any self-contained breathing apparatus with a full facepiece); or SAF (any supplied-air respirator with a full facepiece). *50 mg/m^3:* SAF:PD,PP (any supplied-air respirator that has a full facepiece and is operated in a pressure-demand or other positive-pressure mode). *Emergency or planned entry into unknown concentrations or IDLH conditions:* SCBAF:PD,PP (any self-contained breathing apparatus that has a full facepiece and is operated in a pressure-demand or other positive-pressure mode); or SAF:PD,PP:ASCBA (any supplied-air respirator that has a full facepiece and is operated in a pressure-demand or other positive-pressure mode in combination with an auxiliary, self-contained breathing apparatus operated in a pressure-demand or other positive-pressure mode). *Escape:* HiEF (any air-purifying, full-facepiece respirator with a high-efficiency particulate filter); or SCBAE (any appropriate escape-type, self-contained breathing apparatus).

Storage: Store in tightly closed containers in a cool, well-ventilated area away from heat sources, sources of shock, or the incompatible materials cited above. Sources of ignition, such as smoking and open flames, are prohibited where Barium Perchlorate is handled, used, or stored. See OSHA standard 1910.104 and NFPA 43A *Code for the Storage of Liquid and Solid Oxidizers* for detailed handling and storage regulations.

Shipping: The label required by DOT is "Oxidizer, Poison." This material falls in UN/DOT Hazard Class 5.1 and Packing Group II.[19][20]

Spill Handling: Evacuate and restrict persons not wearing protective equipment from area of spill or leak until cleanup is complete. Remove all ignition sources. Absorb liquids in vermiculite, dry sand, earth, or a similar material and deposit in sealed containers. Collect powdered material in the most convenient and safe manner and deposit in sealed containers. Ventilate area of spill or leak after clean-up is complete. It may be necessary to contain and dispose of this chemical as a hazardous waste. If material or contaminated runoff enters waterways, notify downstream users of potentially contaminated waters. Contact your Department of Environmental Protection or your regional office of the federal EPA for specific recommendations. If employees are required to clean-up spills, they must be properly trained and equipped. OSHA 1910.120(q) may be applicable.

Fire Extinguishing: Barium Perchlorate does not burn, but contact with organic and combustible materials or heat or shock may cause fires or explosions. In case of fire, evacuate the area and fight the fire from a safe, protected location. Poisonous gases are produced in fire including Chlorides. If material or contaminated runoff enters waterways, notify downstream users of potentially contaminated waters. Notify local health and fire officials and pollution control agencies. From a secure, explosion-proof location, use water spray to cool exposed containers. If cooling streams are ineffective (venting sound increases in volume and pitch, tank discolors, or shows any signs of deforming), withdraw immediately to a secure position. If employees are expected to fight fires, they must be trained and equipped in OSHA 1910.156.

References

New Jersey Department of Health and Senior Services, "Hazardous Substance Fact Sheet: Barium Perchlorate," Trenton, NJ (June 1986)

Barium Permanganate

Molecular Formula: $BaMn_2O_8$

Common Formula: $Ba(MnO_4)_2$

Synonyms: Barium manganate(VIII); Permanganato barico (Spanish); Permanganic acid, barium salt

CAS Registry Number: 7787-36-2

RTECS Number: SD6405000

DOT ID: UN 1448

Regulatory Authority

- Air Pollutant Standard Set (ACGIH) (OSHA) (NIOSH) (barium, soluble compounds)
- OSHA 29CFR1910.119, Appendix A, Process Safety List of Highly Hazardous Chemicals, TQ = 7,500 lb
- RCRA 40CFR261, Appendix 8; 40CFR261.11 Hazardous Constituents
- EPCRA Section 313: Includes any unique chemical substance that contains barium as part of that chemical's infrastructure. This category does not include barium sulfate (7727-43-7). Form R *de minimis* concentration reporting level: 0.1%
- U.S. DOT 49CFR172.101, Appendix B, Regulated marine pollutant (Ba compounds, soluble, n.o.s.)

Cited in U.S. State Regulations: California (G), New Hampshire (G), New Jersey (G), Oklahoma (G).

Description: Barium permananganate, $Ba(MnO_4)_2$, is a brownish-violet, dark purple to black crystalline solid. Hazard Identification (based on NFPA-704 M Rating System): Health 2, Flammability 0, Reactivity 0. Soluble in water.

Potential Exposure: It is used to make dry cells and other permanganates and as a disinfectant.

Incompatibilities: Acetic acid, acetic anhydride, and organic or combustible materials (such as wood, paper, oil and fuels) since violent reactions occur.

Permissible Exposure Limits in Air: For Ba soluble compounds, the Federal OSHA standard[58] and ACGIH recommended airborne limit is 0.5 mg/m^3 TWA for an 8-hour workshift. The NIOSH level is the same for a 10-hour workshift. The DFG,[3] HSE,[33] Australia, Mexico, and Israel have adopted this same value and DFG has a Peak Limitation (5 min) of 2 times normal MAK; do not exceed more than 8 times during workshift. Israel's Action Level is one half the TWA. The NIOSH IDLH level is 50 mg/m^3 (Ba, soluble compounds).

Determination in Air: See entry for "Barium."

Permissible Concentration in Water: No criteria set for barium permanganate per se. See entry for "Barium."

Routes of Entry: Inhalation.

Harmful Effects and Symptoms

Short Term Exposure: Skin and eye contact can cause severe irritation and burns. Barium Permanganate can affect you when breathed in. Breathing the dust or mist can irritate the nose, throat and bronchial tubes, causing cough and phlegm.

Long Term Exposure: After repeated exposure, Barium may show up as spots in the lungs on chest x-ray. Some barium chemicals are contaminated with Silica, which scars the lungs. Repeated contact may cause chronic drying and cracking skin.

Points of Attack: Lungs, skin.

Medical Surveillance: Lung function tests.

First Aid: If this chemical gets into the eyes, remove any contact lenses at once and irrigate immediately for at least 15 minutes, occasionally lifting upper and lower lids. Seek medical attention immediately. If this chemical contacts the skin, remove contaminated clothing and wash immediately with soap and water. Seek medical attention immediately. If this chemical has been inhaled, remove from exposure, begin rescue breathing (using universal precautions) if breathing has stopped and CPR if heart action has stopped. Transfer promptly to a medical facility. When this chemical has been swallowed, get medical attention. Give large quantities of water and induce vomiting. Do not make an unconscious person vomit.

Personal Protective Methods: Where possible, enclose operations and use local exhaust ventilation at the site of chemical release. If local exhaust ventilation or enclosure is not used, respirators should be worn. Wear protective gloves and clothing to prevent any reasonable probability of skin contact. Safety equipment suppliers/manufacturers can provide recommendations on the most protective glove/clothing material for your operation. All protective clothing (suits, gloves, footwear, headgear) should be clean, available each day, and put on before work. Contact lenses should not be worn when working with this chemical. Wear dust-proof chemical goggles and face shield unless full facepiece respiratory protection is worn. Employees should wash immediately with soap when skin is wet or contaminated. Provide emergency showers and eyewash. Post hazard and warning information in the work area. In addition, as part of an ongoing education and training effort, communicate all information on the health and safety hazards of Barium Permanganate to potentially exposed workers.

Respirator Selection: NIOSH/OSHA: (Ba soluble compounds) *5 mg/m³:* DMXSQ (any dust and mist respirator except single-use and quarter mask respirators); or SA (any supplied-air respirator). *12.5 mg/m³:* SA:CF (any supplied-air respirator operated in a continuous-flow mode); or PAPRDM (any powered, air-purifying respirator with a dust and mist filter). *25 mg/m³:* HiEF (any air-purifying, full-facepiece respirator with a high-efficiency particulate filter); or SAT:CF (any supplied-air respirator that has a tight-fitting facepiece and is operated in a continuous-flow mode); or PAPRTHiE (any powered, air-purifying respirator with a tight-fitting facepiece and a high-efficiency particulate filter); or SCBAF (any self-contained breathing apparatus with a full facepiece); or SAF (any supplied-air respirator with a full facepiece). *50 mg/m³:* SAF:PD,PP (any supplied-air respirator that has a full facepiece and is operated in a pressure-demand or other positive-pressure mode). *Emergency or planned entry into unknown concentrations or IDLH conditions:* SCBAF:PD, PP (any self-contained breathing apparatus that has a full facepiece and is operated in a pressure-demand or other positive-pressure mode); or SAF:PD,PP:ASCBA (any supplied-air respirator that has a full facepiece and is operated in a pressure-demand or other positive-pressure mode in combination with an auxiliary, self-contained breathing apparatus operated in a pressure-demand or other positive-pressure mode). *Escape:* HiEF (any air-purifying, full-facepiece respirator with a high-efficiency particulate filter); or SCBAE (any appropriate escape-type, self-contained breathing apparatus).

Storage: Store in tightly closed containers is a cool, well-ventilated area. Keep away from incompatible materials cited above. See OSHA standard 1910.104 and NFPA 43A *Code for the Storage of Liquid and Solid Oxidizers* for detailed handling and storage regulations.

Shipping: The DOT required shipping label is "Oxidizer, Poison." This material falls in UN/DOT Hazard Class 5.1 and Packing Group II.[19][20]

Spill Handling: Evacuate and restrict persons not wearing protective equipment from area of spill or leak until cleanup is complete. Remove all ignition sources. Collect powdered material in the most convenient and safe manner and deposit in sealed containers. Ventilate area of spill or leak after clean-up is complete. It may be necessary to contain and dispose of this chemical as a hazardous waste. If material or contaminated runoff enters waterways, notify downstream users of potentially contaminated waters. Contact your Department of Environmental Protection or your regional office of the federal EPA for specific recommendations. If employees are required to clean-up spills, they must be properly trained and equipped. OSHA 1910.120(q) may be applicable.

Fire Extinguishing: Barium Permanganate does not burn, but contact with organic and combustible materials may cause fires or explosions. In case of fire, evacuate the area and fight the fire from a safe, protected location. Poisonous gases are produced in fire. If material or contaminated run-off enters waterways, notify downstream users of potentially contaminated waters. Notify local health and fire officials and pollution control agencies. From a secure, explosion-proof location, use water spray to cool exposed containers. If cooling streams are ineffective (venting sound increases in volume and pitch, tank discolors, or shows any signs of deforming), withdraw immediately to a secure position. If employees are expected to fight fires, they must be trained and equipped in OSHA 1910.156.

References

New Jersey Department of Health and Senior Services, "Hazardous Substance Fact Sheet: Barium Permanganate," Trenton, NJ (June 1986)

Barium Peroxide

Molecular Formula: BaO_2

Synonyms: Bario (perossido di) (Italian); Barium binoxide; Barium dioxide; Bariumperoxid (German); Bariumperoxyde (Dutch); Barium superoxide; Dioxyde de baryum (French); Peroxido barico (Spanish); Peroxyde de baryum (French)

CAS Registry Number: 1304-29-6

RTECS Number: CR0175000

DOT ID: UN 1449

EEC Number: 056-001-00-1

Regulatory Authority

- Air Pollutant Standard Set (ACGIH) (OSHA) (NIOSH) (barium, soluble compounds)
- RCRA 40CFR261, Appendix 8; 40CFR261.11 Hazardous Constituents
- EPCRA Section 313: Includes any unique chemical substance that contains barium as part of that chemical's infrastructure. This category does not include barium sulfate (7727-43-7). Form R *de minimis* concentration reporting level: 0.1%
- U.S. DOT 49CFR172.101, Appendix B, Regulated marine pollutant (Ba compounds, soluble, n.o.s.)

Cited in U.S. State Regulations: California (G), Florida (G), Maine (G), Massachusetts (G), New Hampshire (G), New Jersey (G), Oklahoma (G), Pennsylvania (G), Rhode Island (G).

Description: Barium peroxide, BaO_2, is a grayish-white powder. Freezing/Melting point = 450°C. Boiling point = 800°C (decomposed below this point). Hazard Identification (based on NFPA-704 M Rating System): Health 3, Flammability 0, Reactivity 1. Very slightly soluble in water.

Potential Exposure: Is used as a bleaching agent, in making hydrogen peroxide, in aluminum welding and in textile dyeing.

Incompatibilities: A strong oxidizer. Keep away from organic and combustible materials (such as wood, paper, oil, fuels, and other easily oxidized materials) and peroxyformic acid, hydrogen sulfide and hydroxylamine solutions since violent reactions occur.

Permissible Exposure Limits in Air: For Ba soluble compounds, the Federal OSHA standard[58] and ACGIH recommended airborne limit is 0.5 mg/m^3 TWA for an 8-hour workshift. The NIOSH level is the same for a 10-hour workshift. The DFG,[3] HSE,[33] Australia, Mexico, and Israel have adopted this same value and DFG has a Peak Limitation (5 min) of 2 times normal MAK; do not exceed more than 8 times during workshift. Israel's Action Level is one half the TWA. The NIOSH IDLH level is 50 mg/m^3 (Ba, soluble compounds).

Determination in Air: See entry under "Barium."

Permissible Concentration in Water: No criteria set for Barium Peroxide per se. See entry under "Barium."

Determination in Water: Environmental hazard for aquatic organisms.

Routes of Entry: Inhalation, ingestion.

Harmful Effects and Symptoms

Short Term Exposure: Barium Peroxide can affect you when breathed in. Contact can irritate and burn the eyes and skin. Breathing the dust or mist can irritate the nose, throat and bronchial tubes, causing cough and phlegm.

Long Term Exposure: After repeated exposure, Barium may show up as spots in the lungs on chest x-ray. Some barium chemicals are contaminated with Silica, which scars the lungs. Repeated contact may cause chronic drying and cracking skin.

Points of Attack: Lungs, skin.

First Aid: If this chemical gets into the eyes, remove any contact lenses at once and irrigate immediately for at least 15 minutes, occasionally lifting upper and lower lids. Seek medical attention immediately. If this chemical contacts the skin, remove contaminated clothing and wash immediately with soap and water. Seek medical attention immediately. If this chemical has been inhaled, remove from exposure, begin rescue breathing (using universal precautions) if breathing has stopped and CPR if heart action has stopped. Transfer promptly to a medical facility. When this chemical has been swallowed, get medical attention. Give large quantities of water and induce vomiting. Do not make an unconscious person vomit.

Personal Protective Methods: Where possible, enclose operational and use local exhaust ventilation at the site of chemical release. If local exhaust ventilation or enclosure is not used, respirators should be worn. Wear protective gloves and clothing to prevent any reasonable probability of skin contact. Safety equipment suppliers/manufacturers can provide recommendations on the most protective glove/clothing material for your operation. All protective clothing (suits, gloves, footwear, headgear) should be clean, available each day, and put on before work. Contact lenses should not be worn when working with this chemical. Wear dust-proof chemical goggles and face shield unless full facepiece respiratory protection is worn. Employees should wash immediately with soap when skin is wet or contaminated. Provide emergency showers and eyewash.

Respirator Selection: NIOSH/OSHA: (Ba soluble compounds) *5 mg/m^3:* DMXSQ (any dust and mist respirator except single-use and quarter mask respirators); or SA (any supplied-air respirator). *12.5 mg/m^3:* SA:CF (any supplied-air respirator operated in a continuous-flow mode); or PAPRDM (any powered, air-purifying respirator with a dust and mist filter). *25 mg/m^3:* HiEF (any air-purifying, full-facepiece respirator with a high-efficiency particulate filter); or SAT:CF (any supplied-air respirator that has a tight-fitting facepiece and is operated in a continuous-flow mode); or PAPRTHiE (any powered, air-purifying respirator with a tight-fitting facepiece and a high-efficiency particulate filter); or SCBAF (any self-contained breathing apparatus with a full facepiece); or SAF (any supplied-air respirator with a full facepiece). *50 mg/m³:* SAF: PD,PP (any supplied-air respirator that has a full facepiece and is operated in a pressure-demand or other positive-pressure mode). *Emergency or planned entry into unknown concentrations or IDLH conditions:* SCBAF:PD, PP (any self-contained breathing apparatus that has a full facepiece and is operated in a pressure-demand or other positive-pressure mode); or SAF:PD,PP:ASCBA (any supplied-air respirator that has a full facepiece and is operated in a pressure-demand or other positive-pressure mode in combination with an auxiliary, self-contained breathing apparatus operated in a pressure-demand or other positive-pressure mode). *Escape:* HiEF (any air-purifying, full-facepiece respirator with a high-efficiency particulate filter); or SCBAE (any appropriate escape-type, self-contained breathing apparatus).

Storage: Store in tightly closed containers in a cool, well-ventilated area away from water or moisture, and away from contact with the incompatible materials cited above. See OSHA standard 1910.104 and NFPA 43A *Code for the Storage of Liquid and Solid Oxidizers* for detailed handling and storage regulations.

Shipping: The DOT required label is "Oxidizer, Poison." This material falls in UN/DOT Hazard Class 5.1 and Shipping Group II.[19][20]

Spill Handling: Evacuate and restrict persons not wearing protective equipment from area of spill or leak until cleanup is complete. Remove all ignition sources. Cover material with sand/soda ash 9:1 mixture. Mix thoroughly and while stirring add slowly to sodium bisulfite solution with plastic implements. Neutralize with dilute H_2SO_4. After setting, decant the solution with flushing water and transport the sand to a sanitary landfill.[24] Ventilate area of spill or leak after clean-up is complete. It may be necessary to contain and dispose of this chemical as a hazardous waste. If material or contaminated runoff enters waterways, notify downstream users of potentially contaminated waters. Contact your Department of Environmental Protection or your regional office of the federal EPA for specific recommendations. If employees are required to clean-up spills, they must be properly trained and equipped. OSHA 1910.120(q) may be applicable.

Fire Extinguishing: Barium Peroxide does not burn, but mixtures of Barium Peroxide and combustible, organic, or easily oxidized materials such as wood, fuels, paper and charcoal will burn or explode if rubbed or contact a small amount of water. Use large amounts of water to extinguish the fire. Poisonous gases are produced in fire. If material or contaminated runoff enters waterways, notify downstream users of potentially contaminated waters. Notify local health and fire officials and pollution control agencies. From a secure, explosion-proof location, use water spray to cool exposed containers. If cooling streams are ineffective (venting sound increases in volume and pitch, tank discolors, or shows any signs of deforming), withdraw immediately to a secure position. If employees are expected to fight fires, they must be trained and equipped in OSHA 1910.156.

Disposal Method Suggested: See Spill Handling.

References

New Jersey Department of Health and Senior Services, "Hazardous Substance Fact Sheet: Barium Peroxide," Trenton, NJ (May 1986)

Barium Sulfate

Molecular Formula: BaO_4S

Common Formula: $BaSO_4$

Synonyms: Actybaryte; Artificial brite; Artificial heavy spar; Bakontal; Baridol; Barite; Baritop; Barosperse; Barotrast; Baryta white; Barytes; Bayrites; Blanc fixe (French); C.I. 77120C.I.; Citobaryum; Colonatrast; Enamel white; Esophotrast; Eweisse-Z-paque; E-Z-paque; Finemeal; Lactobaryt; Liquibarine; Macropaque; Neobar; Oratrast; Permanent white; Pigment white 21; Polybar; Precipitated barium sulphate; Radiobaryt; Raybar; Redi-Flow; Solbar; Sulfato barico (Spanish); Sulfuric acid, barium salt (1:1); Supramike; Travad; Unibaryt

CAS Registry Number: 7727-43-7

RTECS Number: CR0600000

Regulatory Authority

- Air Pollutant Standard Set (ACGIH)[1] (HSE)[33] (OSHA)[58]
- RCRA, 40CFR261, Appendix 8 Hazardous Constituents, as barium compounds, n.o.s., waste number not listed
- EPCRA Section 313: This does *not* cover barium sulfate (7727-43-7)
- U.S. DOT Regulated Marine Pollutant (49CFR172.101, Appendix B)

Cited in U.S. State Regulations: New Hampshire (G).

Description: $BaSO_4$ is a white crystalline solid which melts at 1,580°C.

Potential Exposure: Barium sulfate is used as an opaque medium in radiography; as a mud weighting material in oil well drilling; in paper coating; as a paint pigment.

Incompatibilities: Aluminum powder, phosphorus.

Permissible Exposure Limits in Air: The ACGIH[1] has set a TWA of 10 mg/m³ with the notation that this value applies to a material free of asbestos and containing <1% of silica. OSHA[58] has proposed this same limit, with the provision that 10 mg/m³ is for total dust and that a limit of 5 mg/m³ be set for the respirable fraction. The DFG MAK is 4 mg/m³ (inhalable dust); 1.5 mg/m³ (respirable dust). The HSE[33] has set a limit of 2 mg/m³.

Determination in Air: Barium sulfate may be determined by filtration and gravimetric measurement.

Permissible Concentration in Water: No criteria set.

Routes of Entry: Inhalation of dust; ingestion.

Harmful Effects and Symptoms

Sax & Lewis[44] state that this material is an experimental carcinogen via the intrapleural route.

Long Term Exposure: Lungs may be affected by repeated or prolonged exposure to dust particles, resulting in baritosis (a form of benign pneumoconiosis) (WHO).

Points of Attack: Lungs.

Medical Surveillance: Lung function tests.

First Aid: If this chemical gets into the eyes, remove any contact lenses at once and irrigate immediately for at least 15 minutes, occasionally lifting upper and lower lids. Seek medical attention immediately. If this chemical contacts the skin, remove contaminated clothing and wash immediately with soap and water. Seek medical attention immediately. If this chemical has been inhaled, remove from exposure, begin rescue breathing (using universal precautions) if breathing has stopped and CPR if heart action has stopped. Transfer promptly to a medical facility. When this chemical has been swallowed, rinse mouth and get medical attention.

Personal Protective Methods: Wear protective gloves and clothing to prevent any reasonable probability of skin contact. Safety equipment suppliers/manufacturers can provide recommendations on the most protective glove/clothing material for your operation. All protective clothing (suits, gloves, footwear, headgear) should be clean, available each day, and put on before work. Contact lenses should not be worn when working with this chemical. Wear dust-proof chemical goggles and face shield unless full facepiece respiratory protection is worn. Employees should wash immediately with soap when skin is wet or contaminated. Provide emergency showers and eyewash.

Respirator Selection: NIOSH/OSHA: *5 mg/m³:* DMXSQ (any dust and mist respirator except single-use and quarter mask respirators); or SA (any supplied-air respirator). *12.5 mg/m³:* SA:CF (any supplied-air respirator operated in a continuous-flow mode); or PAPRDM (any powered, air-purifying respirator with a dust and mist filter). *25 mg/m³:* HiEF (any air-purifying, full-facepiece respirator with a high-efficiency particulate filter); or SAT:CF (any supplied-air respirator that has a tight-fitting facepiece and is operated in a continuous-flow mode); or PAPRTHiE (any powered, air-purifying respirator with a tight-fitting facepiece and a high-efficiency particulate filter); or SCBAF (any self-contained breathing apparatus with a full facepiece); or SAF (any supplied-air respirator with a full facepiece). *50 mg/m³:* SAF:PD,PP (any supplied-air respirator that has a full facepiece and is operated in a pressure-demand or other positive-pressure mode). *Emergency or planned entry into unknown concentrations or IDLH conditions:* SCBAF:PD,PP (any self-contained breathing apparatus that has a full facepiece and is operated in a pressure-demand or other positive-pressure mode); or SAF:PD,PP:ASCBA (any supplied-air respirator that has a full facepiece and is operated in a pressure-demand or other positive-pressure mode in combination with an auxiliary, self-contained breathing apparatus operated in a pressure-demand or other positive-pressure mode). *Escape:* HiEF (any air-purifying, full-facepiece respirator with a high-efficiency particulate filter); or SCBAE (any appropriate escape-type, self-contained breathing apparatus).

Storage: Store in tightly closed containers in a cool, well ventilated area away from incompatible materials listed above. A regulated, marked area should be established where this chemical is handled, used, or stored in compliance with OSHA standard 1910.1045.

Spill Handling: Evacuate and restrict persons not wearing protective equipment from area of spill or leak until cleanup is complete. Remove all ignition sources. Collect powdered material in the most convenient and safe manner and deposit in sealed containers. Ventilate area of spill or leak after cleanup is complete. It may be necessary to contain and dispose of this chemical as a hazardous waste. If material or contaminated runoff enters waterways, notify downstream users of potentially contaminated waters. Contact your Department of Environmental Protection or your regional office of the federal EPA for specific recommendations. If employees are required to clean-up spills, they must be properly trained and equipped. OSHA 1910.120(q) may be applicable.

Fire Extinguishing: Use dry chemical, carbon dioxide, water spray, or alcohol foam extinguishers. Poisonous gases are produced in fire. If material or contaminated runoff enters waterways, notify downstream users of potentially contaminated waters. Notify local health and fire officials and pollution control agencies. From a secure, explosion-proof location, use water spray to cool exposed containers. If cooling streams are ineffective (venting sound increases in volume and pitch, tank discolors, or shows any signs of deforming), withdraw immediately to a secure position. If employees are expected to fight fires, they must be trained and equipped in OSHA 1910.156.

References

Sax, N. I., Ed., "Dangerous Properties of Industrial Materials Report" 1, No. 1, 31 (1980)

Bendiocarb

Molecular Formula: $C_{11}H_{13}NO_4$

Synonyms: AI3-27695; Bencarbate; Bendiocarbe; 1,3-Benzodioxole, 2,2-dimethyl-1,3-benzodioxol-4-ol methylcarbamate; 1,3-Benzodioxole, 2,2-dimethyl-4-(N-methylcarbamato)-; 1,3-Benzodioxol-4-ol, 2,2-dimethyl-, methylcrbamate; Bicam ULV; Carbamic acid, methyl-, 2,3-(dimethylmethylenedioxy)phenyl ester; Carbamic acid, methyl-, 2,3-(isopropylidenedioxy)phenyl ester; 2,2-Dimethylbenzo-1,3-benzodioxol-

4-yl *N*-methylcarbamate; 2,2-Dimethyl-1,3-benzodioxol-4-yl *N*-methylcarbamate; 2,2-Dimethylbenzo-1,3-dioxol-4-yl methyl-carbamate; 2,2-Dimethyl-4-(N-methylaminocarboxylato)-; 2,2-Dimethyl-4-(N-methylaminocarboxylato)-1,3-benxodioxole; Dycarb; Ficam; Ficam 80W; Ficam D; Ficam ULV; Ficam W; Fuam; Garvox; Garvox3G; 2,3-Isopropylidene-dioxyphenyl methylcarbamate; MC 6897; Methylcarbamic acid 2,3-(isopropylidenedioxy)phenyl ester; Multamat; Multimet; NC 6897; Niomil; OMS-1394; Rotate; Seedox; Seedox SC; Tattoo; Turcam

CAS Registry Number: 22781-23-3

RTECS Number: FC1140000

DOT ID: UN 2757

Regulatory Authority

- EPA HAZARDOUS WASTE NUMBER (RCRA No.): U278
- RCRA, 40CFR261, Appendix 8 Hazardous Constituents
- RCRA 40CFR268.48; 61FR15654, Universal Treatment Standards: Wastewater (mg/l), 0.056; Nonwastewater (mg/kg), 1.4
- EPCRA Section 313 Form R *de minimis* concentration reporting level: 1.0%
- U.S. DOT 49CFR172.101, Appendix B, Regulated marine pollutant
- Canada: Drinking Water Quality Level Set

Cited in U.S. State Regulations: California (G), Michigan (G), New Jersey (G).

Description: Bendiocarb, $C_{11}H_{13}NO_4$, is a white odorless crystalline powder. Freezing/Melting point = 129 – 130°C. Slightly soluble in water.

Potential Exposure: Those involved in the manufacture, formulation and application of this insecticide which is used against household pests, in agriculture in seed treatment and as a foliar spray.

Incompatibilities: Keep away from flammable materials and sources of heat and flame.

Permissible Exposure Limits in Air: No standards set.

Permissible Concentration in Water: Canada's Drinking Water Quality is 0.04 mg/l MAC.

Routes of Entry: Inhalation, skin, contact, ingestion.

Harmful Effects and Symptoms

Short Term Exposure: Bendiocarb is a toxic carbamate chemical. Bendiocarb can affect you when inhaled. Exposure can cause rapid poisoning, with headaches, sweating, nausea and vomiting, diarrhea, loss of concentration and death. Eye contact can cause irritation and blurred vision.

Long Term Exposure: Similar carbamates can affect the central nervous system.

Medical Surveillance: Before starting work, at regular times after that, and if symptoms develop or over exposure occurs, the following is recommended: Serum and RBC cholinesterase levels (a test for the body substance affected by Bendiocarb). For this substance, these tests are accurate only if done within about two hours of exposure, and can return to normal before the person feels well.

First Aid: If this chemical gets into the eyes, remove any contact lenses at once and irrigate immediately for at least 15 minutes, occasionally lifting upper and lower lids. Seek medical attention immediately. If this chemical contacts the skin, remove contaminated clothing and wash immediately with soap and water. Shampoo hair. Seek medical attention immediately. If this chemical has been inhaled, remove from exposure, begin rescue breathing (using universal precautions) if breathing has stopped and CPR if heart action has stopped. Transfer promptly to a medical facility. When this chemical has been swallowed, get medical attention. Give large quantities of water and induce vomiting. Do not make an unconscious person vomit.

Personal Protective Methods: *Clothing:* Avoid skin contact with Bendiocarb. Wear protective gloves and clothing. Safety equipment suppliers/manufacturers can provide recommendations on the most protective glove/clothing material for your operation. All protective clothing (suits, gloves, footwear, headgear) should be clean, available each day, and put on before work. The Farm Chemicals Handbook recommends PVC or Rubber as a protective material. *Eye Protection:* Wear dust-proof goggles when working with powders or dust, unless full facepiece respiratory protection is worn.

Respirator Selection: Where the potential exists for exposures to Bendiocarb, use a MSHA/NIOSH approved full facepiece respirator with a pesticide cartridge. Increased protection is obtained from full facepiece air purifying respirators. Where the potential for high exposures exists, use a MSHA/NIOSH approved supplied-air respirator for pesticides with a full facepiece operated in the positive pressure mode or with a full facepiece hood, or helmet in the continuous flow mode, or use a MSHA/NIOSH approved self-contained breathing apparatus with a full facepiece operated in pressure-demand or other positive pressure mode.

Storage: Prior to working with Bendiocarb you should be trained on its proper handling and storage. Store in tightly closed containers in a cool, well-ventilated area away from food, fertilizers, other pesticides, flammable materials and sources of heat and flame.

Shipping: Carbamate pesticides, solid, toxic, n.o.s. It falls in Hazard Class 6.1 and Packing Group II and requires a "Poison" label.

Spill Handling: Evacuate and restrict persons not wearing protective equipment from area of spill or leak until cleanup is complete. Remove all ignition sources. Collect powdered material in the most convenient and safe manner and deposit in sealed containers. Ventilate area of spill or leak after clean-up is complete. It may be necessary to contain and dispose of this chemical as a hazardous waste. If material or contaminated runoff enters waterways, notify downstream users of potentially contaminated waters. Contact your Department of Environmental Protection or your regional office of the federal EPA for specific recommendations. If employees are required to clean-up spills, they must be properly trained and equipped. OSHA 1910.120(q) may be applicable.

Fire Extinguishing: Bendiocarb may burn, but does not readily ignite. Use dry chemical, carbon dioxide, halon, water spray, or standard foam extinguishers. Poisonous gases are produced in fire including nitrogen oxides and methyl isocyanate. If material or contaminated runoff enters waterways, notify downstream users of potentially contaminated waters. Notify local health and fire officials and pollution control agencies. From a secure, explosion-proof location, use water spray to cool exposed containers. If cooling streams are ineffective (venting sound increases in volume and pitch, tank discolors, or shows any signs of deforming), withdraw immediately to a secure position. If employees are expected to fight fires, they must be trained and equipped in OSHA 1910.156.

Disposal Method Suggested: Must be disposed properly by following package label directions or by contacting your state pesticide or environmental control agency or by contacting your regional EPA office. Dispose in accordance with 40CFR165 recommendations for the disposal of pesticides and pesticide containers.

References

New Jersey Department of Health and Senior Services, "Hazardous Substance Fact Sheet: Bendiocarb," Trenton, NJ (August 1997)

Benomyl

Molecular Formula: $C_{14}H_{18}N_4O_3$

Synonyms: Abortrine; Agrocite; Arilate; BBC; BBC6597; Benex; Benlat; Benlate®; Benlate® 40W; Benlate® 50W; Benlate® 50; Benomilo (Spanish); Benomyl 50W; 2-Benzimidazolecarbamic acid, 1-(butylcarbamoyl)-, methyl ester; BNM; 1-(Butylamino)carbonyl-1H-benzimidazol-2-yl-, methyl ester; 1-(Butylcarbamoyl)-2-benzimidazolec arbamic acid, methyl ester; 1-(N-Butylcarbamoyl)-2-(methoxy-carboxamido)-benzamidazol (German); 1-(N-Butylcarbamoyl)-2-(methoxy-carboxamido)-benzimidazol (German); Carbamic acid, 1-(butylamino)carbonyl- 1H-benzimidazol-2yl, methyl ester; D1991; Dupont 1991; F1991; Fundazol; Fungacide D-1991; Fungicide 1991; Fungochrom; MBC; Methyl 1-(butylcarbamoyl)-2-benzimidazolyl carbamate; Tarsan®; Tersan® 1991; UZGN

CAS Registry Number: 17804-35-2

RTECS Number: DD6475000

DOT ID: UN 2757

EEC Number: 613-049-00-3

Regulatory Authority

- Air Pollutant Standard Set (ACGIH)[1] (HSE)[33] (UNEP)[43] (former USSR)[35] (OSHA)[58] (Several States)[60] (Australia) (Israel) (Mexico) (Several Canadian Provinces)
- EPA HAZARDOUS WASTE NUMBER (RCRA No.): U271
- RCRA, 40CFR261, Appendix 8 Hazardous Constituents
- RCRA 40CFR268.48; 61FR15654, Universal Treatment Standards: Wastewater (mg/l), 0.056; Nonwastewater (mg/kg), 1.4
- EPCRA Section 313 Form R *de minimis* concentration reporting level: 1.0%
- CALIFORNIA'S PROPOSITION 65: Reproductive toxin (male)
- U.S. DOT 49CFR172.101, Appendix B, Regulated marine pollutant

Cited in U.S. State Regulations: Alaska (G), California (A, G), Connecticut (A), Florida (G), Illinois (G), Maine (G), Massachusetts (G), Michigan (G), Minnesota (G), Nevada (A), New Hampshire (G), New Jersey (G), North Dakota (A), Pennsylvania (G), Rhode Island (G), Virginia (A), West Virginia (G).

Description: Benomyl, $C_{14}H_{18}N_4O_3$, is a white crystalline solid with a faint acrid odor. Freezing/Melting point ≥ 572°F (Decomposes). Boiling point = decomposes. Slightly soluble in water.

Potential Exposure: Those involved in the manufacture, formulation and application of this fungicide.

Incompatibilities: Strong bases [forms toxic oxides of nitrogen], strong acids, peroxides and oxidizers.

Permissible Exposure Limits in Air: The Federal OSHA[58] standard is 10 mg/m^3 TWA for benomyl dust and 5 mg/m^3 TWA for the respirable fraction. California has set the same TWAs. ACGIH has set a TWA of 0.8 ppm (10 mg/m^3) but no STEL. The HSE[33] has set the same TWA plus a STEL of 1.3 ppm (15 mg/m^3). The former USSR-UNEP/IRPTC project[43] has set a tentative safe exposure limit in the workplace of 0.01 mg/m^3. In addition, The former USSR has set[35] a limit in ambient air of 0.35 mg/m^3 on a once-a-day basis and 0.05 mg/m^3 on an average daily basis. Several states have set guidelines or standards for benomyl in ambient air[60] ranging from 100 μg/m^3 (North Dakota and Virginia) to 200 μg/m^3 (Connecticut) to 238 μg/m^3 (Nevada).

Determination in Air: Filter; none; Gravimetric; NIOSH IV [Particulates NOR; #0500 (total), #0600 (respirable)]

Permissible Concentration in Water: The former USSR has set a MAC in surface water of 0.5 mg/l of benomyl.[35]

Routes of Entry: Inhalation.

Harmful Effects and Symptoms

Benomyl is generally felt to have a low order of acute and chronic toxicity.[53] However, a rebuttable presumption against registration for benomyl was issued on December 6, 1978 by U.S. EPA on the basis of reduction in nontarget species, mutagenicity, teratogenicity, reproductive effects, and hazard to wildlife. The ADI for man is 0.02 mg/kg.[23]

Short Term Exposure: The substance irritates the skin eyes and upper respiratory system. Exposure could cause depression of the central nervous system and lack of muscular co-ordination.

Long Term Exposure: Repeated or prolonged contact may cause skin sensitization and allergy. Human mutation data reported. Also experimental and reproductive effect. May damage the male reproductive system; cause heritable genetic damage in humans. Animal tests show that this substance possibly causes birth defects in human babies, miscarriage, or cancer.

Points of Attack: Eyes, skin, respiratory system, reproductive system.

Medical Surveillance: Evaluation by a qualified allergist.

First Aid: If this chemical gets into the eyes, remove any contact lenses at once and irrigate immediately for at least 15 minutes, occasionally lifting upper and lower lids. Seek medical attention immediately. If this chemical contacts the skin, remove contaminated clothing and wash immediately with soap and water. Seek medical attention immediately. If this chemical has been inhaled, remove from exposure, begin rescue breathing (using universal precautions) if breathing has stopped and CPR if heart action has stopped. Transfer promptly to a medical facility. When this chemical has been swallowed, get medical attention. Give large quantities of water and induce vomiting. Do not make an unconscious person vomit.

Personal Protective Methods: Wear protective gloves and clothing to prevent any reasonable probability of skin contact. Safety equipment suppliers/manufacturers can provide recommendations on the most protective glove/clothing material for your operation. All protective clothing (suits, gloves, footwear, headgear) should be clean, available each day, and put on before work. Contact lenses should not be worn when working with this chemical. Wear dust-proof chemical goggles and face shield unless full facepiece respiratory protection is worn. Employees should wash immediately with soap when skin is wet or contaminated. Provide emergency showers and eyewash.

Respirator Selection: *At any concentrations above the NIOSH REL, or where there is no REL, at any detectable concentration:* SCBAF:PD,PP (any MSHA/NIOSH approved self-contained breathing apparatus that has a full facepiece and is operated in a pressure-demand or other positive-pressure mode); or SAF:PD,PP:ASCBA (any supplied-air respirator that has a full facepiece and is operated in a pressure-demand or other positive-pressure mode in combination with an auxiliary, self-contained breathing apparatus operated in a pressure-demand or other positive pressure mode).

Storage: Store in a cool, dry place or in a refrigerator[52] away from strong bases, strong acids, heat.

Shipping: Label "Poison." Hazard Class 6.1

Spill Handling: Evacuate and restrict persons not wearing protective equipment from area of spill or leak until cleanup is complete. Remove all ignition sources. Dampen spilled material with toluene to avoid dust, then transfer material to a suitable container. Use absorbent paper dampened with toluene to pick up remaining material. Wash surfaces well with soap and water. Seal all wastes in vapor-thigh plastic bags for eventual disposal.[52] Ventilate area of spill or leak after clean-up is complete. It may be necessary to contain and dispose of this chemical as a hazardous waste. If material or contaminated runoff enters waterways, notify downstream users of potentially contaminated waters. Contact your Department of Environmental Protection or your regional office of the federal EPA for specific recommendations. If employees are required to clean-up spills, they must be properly trained and equipped. OSHA 1910.120(q) may be applicable.

Fire Extinguishing: Use dry chemical, carbon dioxide, water spray, or alcohol foam extinguishers. Poisonous gases are produced in fire including nitrogen oxides. If material or contaminated runoff enters waterways, notify downstream users of potentially contaminated waters. Notify local health and fire officials and pollution control agencies. From a secure, explosion-proof location, use water spray to cool exposed containers. If cooling streams are ineffective (venting sound increases in volume and pitch, tank discolors, or shows any signs of deforming), withdraw immediately to a secure position. If employees are expected to fight fires, they must be trained and equipped in OSHA 1910.156.

Extinguish fires using an agent suitable for the type of surrounding fire; Benomyl itself does not burn.

Disposal Method Suggested: In accordance with 40CFR165 recommendations for the disposal of pesticides and pesticide containers. Must be disposed properly by following package label directions or by contacting your state pesticide or environmental control agency or by contacting your regional EPA office.

References

New Jersey Department of Health and Senior Services, "Hazardous Substance Fact Sheet: Benomyl," Trenton, NJ (February 1989)

Sax, N. I., Ed., "Dangerous Properties of Industrial Materials Report" 4, No. 1, 20–21 (1984)

Bentazon

Molecular Formula: $C_{10}H_{12}N_2O_3S$

Synonyms: Asagio®; BAS351-H; Basagran®; Bendioxide; Bentazone; 1H-2,1,3-Benzothiadiazin-4(3H)-one, 3-(1-methylethyl)-, 2,2-dioxide; 3-Isopropyl-2,1,3-benzothiadiazinon-(4)-2,2-dioxid (German); 3-Isopropyl-1H-2,1,3-benzothiadiazin-4(3H)-one-2,2-dioxide; 3-(1-Methylethyl)-1H-2,1,3-benzothiazain-4(3H)-one-2,2-dioxide; Pledge®

CAS Registry Number: 25057-89-0

RTECS Number: DK9900000

DOT ID: UN 2588

EEC Number: 613-012-00-1

Regulatory Authority

Safe Drinking Water Act, 55FR1470 Priority List

Cited in U.S. State Regulations: California (W, G)

Description: Bentazon, $C_{10}H_{12}N_2O_3S$, is a colorless crystalline powder. Freezing/Melting point = 137 – 139°C. Boiling point = 200°C (decomposes). Hazard Identification (based on NFPA-704 M Rating System): Health 2, Flammability 2, Reactivity 0. Insoluble in water.

Potential Exposure: Those involved in the manufacture, formulation or application of this selective postemergent herbicide.

Incompatibilities: Keep away from flammable materials, heat and flame. Risk of fire and explosion if formulations contain flammable/explosive solvents.

Permissible Exposure Limits in Air: No standards set.

Permissible Concentration in Water: A no-observed adverse effect level (NOAEL) of 2.5 mg/kg/day has been determined by EPA based on the absence of prostatic effects in dogs. This led to the determination of a longer-term health advisory of 0.875 mg/l for a 70-kg adult. It also led to the establishment of a lifetime health advisory of 0.0175 mg/l. In addition, California[61] has set a guideline in drinking water of 8.0 μg/l.

Routes of Entry: Ingestion, inhalation.

Harmful Effects and Symptoms

Short Term Exposure: The oral LD_{50} rat is 1,100 mg/kg which puts Bentazon in the "slightly toxic" category. Avoid eye contact; may cause severe irritation or injury. May cause skin burns.

Long Term Exposure: May be a reproductive hazard.

First Aid: If this chemical gets into the eyes, remove any contact lenses at once and irrigate immediately for at least 15 minutes, occasionally lifting upper and lower lids. Seek medical attention immediately. If this chemical contacts the skin, remove contaminated clothing and wash immediately with soap and water. Seek medical attention immediately. If this chemical has been inhaled, remove from exposure, begin rescue breathing (using universal precautions) if breathing has stopped and CPR if heart action has stopped. Transfer promptly to a medical facility. When this chemical has been swallowed, get medical attention. Give large quantities of water and induce vomiting. Do not make an unconscious person vomit.

Personal Protective Methods: Wear protective gloves and clothing to prevent any reasonable probability of skin contact. Safety equipment suppliers/manufacturers can provide recommendations on the most protective glove/clothing material for your operation. All protective clothing (suits, gloves, footwear, headgear) should be clean, available each day, and put on before work. Contact lenses should not be worn when working with this chemical. Wear dust-proof chemical goggles and face shield unless full facepiece respiratory protection is worn. Employees should wash immediately with soap when skin is wet or contaminated. Provide emergency showers and eyewash.

Respirator Selection: Engineering controls should be used wherever feasible to maintain airborne concentrations of this chemical below the prescribed exposure limit. Respirators and protective equipment are less effective than engineering controls, and should be used only in non-routine or emergency situations which may result in exposure concentrations in excess of the TWA environmental limit. *Where there is no REL, at any detectable concentration:* SCBAF:PD,PP (any MSHA/NIOSH approved [for pesticides] self-contained breathing apparatus that has a full facepiece and is operated in a pressure-demand or other positive-pressure mode); or SAF:PD,PP: ASCBA (any supplied-air respirator that has a full facepiece and is operated in a pressure-demand or other positive-pressure mode in combination with an auxiliary, self-contained breathing apparatus operated in a pressure-demand or other positive pressure mode).

Storage: Store in tightly closed containers in a cool, well ventilated area away from flammable materials, sources of heat and fire.

Shipping: Pesticides, solid, toxic, n.o.s. require a "Keep away from Food" label in Packing Group III.

Spill Handling: Evacuate and restrict persons not wearing protective equipment from area of spill or leak until cleanup is complete. Remove all ignition sources. Collect powdered material in the most convenient and safe manner and deposit in sealed containers. Ventilate area of spill or leak after clean-up is complete. It may be necessary to contain and dispose of this chemical as a hazardous waste. If material or contaminated runoff enters waterways, notify downstream users of potentially contaminated waters. Contact your Department of Environmental Protection or your regional office of the federal EPA for specific recommendations. If employees are required to clean-up spills, they must be properly trained and equipped. OSHA 1910.120(q) may be applicable.

Fire Extinguishing: Bentazon is not combustible, but may support combustion under fire conditions. Risk of fire and explosion if formulations contain flammable/explosive solvents. Stay upwind of fire. Use dry chemical, carbon dioxide, water spray, or standard foam extinguishers. Poisonous gases are produced in fire including nitrogen oxides and sulfur oxides. If material or contaminated runoff enters waterways, notify downstream users of potentially contaminated waters. Notify local health and fire officials and pollution control agencies. From a secure, explosion-proof location, use water spray to cool exposed containers. If cooling streams are ineffective (venting sound increases in volume and pitch, tank discolors, or shows any signs of deforming), withdraw immediately to a secure position. If employees are expected to fight fires, they must be trained and equipped in OSHA 1910.156.

Disposal Method Suggested: In accordance with 40CFR165 recommendations for the disposal of pesticides and pesticide containers. Must be disposed properly by following package label directions or by contacting your state pesticide or environmental control agency or by contacting your regional EPA office.

References

U.S. Environmental Protection Agency, "Health Advisory: Bentazon," Washington, DC, Office of Drinking Water (August 1987)

Bentonite

Molecular Formula: $Al_2H_2O_{12}Si_4$

Common Formula: $Al_2O_3{\cdot}4SiO_2{\cdot}H_2O$

Synonyms: Albagel Premium USP 4444; Bentonite Magma; Entonite2073; Hi-Jel; Imvite I.G.B.A; Magbond; Montmorillonite; Panther Creek Bentonite; Southern Bentonite; Tixoton; Volclay; Volclay Bentonie BC; Wilkinite

CAS Registry Number: 1302-78-9

RTECS Number: CT9450000

Cited in U.S. State Regulations: New York (G).

Description: Bentonite, $Al_2O_3{\cdot}4SiO_2{\cdot}H_2O$, is a light yellow, cream, pale brown or gray to black powder or granules. Insoluble in water.

Potential Exposure: This material is used as Fuller's earth, as an emulsifier for oils, as a base for plasters, in cosmetics, in polishes and abrasives, as a food additive and others. Bentonites are aluminate silicate and can contain crystalline silica. The content varies widely from less than 1% to about 24%. (WHO)

Incompatibilities: Substance is a weak acid in water; avoid contact with strong alkaline material.

Permissible Exposure Limits in Air: No standards set. Under study by ACGIH.[1]

Permissible Concentration in Water: No criteria set.

Routes of Entry: Inhalation, ingestion.

Harmful Effects and Symptoms

Short Term Exposure: Dust may cause irritation to nose, throat and lungs. Dust may cause eye irritation. The intravenous LD_{50} rat is 35 mg/kg.[9]

Long Term Exposure: Repeated inhalation of dust can cause irritation and bronchial asthma. May cause silicosis due to the presence of crystalline silica

Points of Attack: Lungs.

Medical Surveillance: Lung function tests.

First Aid: *Inhalation:* Move person to fresh air. Seek medical attention if necessary. *Skin:* Wash with water. *Eyes:* Wash with water as needed. Seek medical attention if necessary. *Ingestion:* Seek medical attention if necessary.

Personal Protective Methods: Wear protective gloves and clothing to prevent any reasonable probability of skin contact. Safety equipment suppliers/manufacturers can provide recommendations on the most protective glove/clothing material for your operation. All protective clothing (suits, gloves, footwear, headgear) should be clean, available each day, and put on before work. Contact lenses should not be worn when working with this chemical. Wear dust-proof chemical goggles and face shield unless full facepiece respiratory protection is worn. Employees should wash immediately with soap when skin is wet or contaminated. Provide emergency showers and eyewash.

Respirator Selection: Wear dust mask.

Storage: Store in tightly closed containers in a cool, well ventilated area

Shipping: Bentonite is not cited in the DOT performance-oriented packaging standards.

Spill Handling: Evacuate and restrict persons not wearing protective equipment from area of spill or leak until cleanup is complete. Remove all ignition sources. Dampen dry material. Collect powdered material in the most convenient and safe manner and deposit in sealed containers. Ventilate area of spill or leak after clean-up is complete. Dispose with normal trash. If employees are required to clean-up spills, they must be properly trained and equipped. OSHA 1910.120(q) may be applicable.

Fire Extinguishing: Not combustible. Use extinguishers suitable for surrounding fires. Poisonous gases are produced in fire. If employees are expected to fight fires, they must be trained and equipped in OSHA 1910.156.

Disposal Method Suggested: Landfill disposal.

References

New York State Department of Health, "Chemical Fact Sheet: Bentonite," Albany, NY, Bureau of Toxic Substance Assessment (March 1986)

Benz[a]anthracene

Molecular Formula: $C_{18}H_{12}$

Synonyms: BA; B(a)A; BA.A13-50599; 1,2-Benzanthracene; Benzanthracene; 1,2-Benzanthrazen (German); 1,2-Benzanthrene; Benzanthrene; 1,2-Benz(a)antrhracene; 1,2-Benzo(a)anthracene; Benzo(a)anthracene; Benzo(a)anthrene; 2,3-Benzophenanthrene; Benzo(b)phenanthrene; 2,3-Benzphenanthrene; Naphthaanthracene; NSC 30970; Tetraphene

CAS Registry Number: 56-55-3

RTECS Number: CV9275000

DOT ID: UN 2811

Regulatory Authority

- Carcinogen (Animal Positive) (IARC) (NTP)[9] (DFG)
- OSHA, 29CFR1910 Specifically Regulated Chemicals (See CFR1910.1002) as coal tar pitch volatiles
- Air Pollutant Standard Set (ACGIH)[1] (NIOSH/OSHA)[2] (HSE)[33]
- Water Pollution Standard Proposed (EPA)[6] (Kansas)[61]
- Clean Water Act: 40CFR423, Appendix A, Priority Pollutants; 40CFR401.15 Section 307 Toxic Pollutants
- SUPERFUND/EPCRA 40CFR302.4 Reportable Quantity (RQ): CERCLA, 10 lb (4.54 kg)
- EPA HAZARDOUS WASTE NUMBER (RCRA No.): U018
- RCRA, 40CFR261, Appendix 8 Hazardous Constituents
- RCRA 40CFR268.48; 61FR15654, Universal Treatment Standards: Wastewater (mg/l), 0.059; Nonwastewater (mg/kg), 3.4
- RCRA 40CFR264, Appendix 9; TSD Facilities Ground Water Monitoring List. Suggested test method(s) (PQL μg/l): 8100 (200); 8270 (10)
- EPCRA (Section 313): as (PACs); Form R *de minimis* concentration reporting level: 0.1%
- Canada, WHMIS, Ingredients Disclosure List

Cited in U.S. State Regulations: California (G), Florida (G), Illinois (G), Kansas (G, W), Louisiana (G), Maine (G), Massachusetts (G), Michigan (G), Minnesota (G), Pennsylvania (G), Vermont (G), Virginia (G), Washington (G), West Virginia (G), Wisconsin (G).

Description: $C_{18}H_{12}$, a colorless plate-like material which is recrystallized from glacial acetic acid or a light yellow to tan powder. Boiling point = 400°C. Freezing/Melting point = 160°C; Hazard Identification (based on NFPA-704 M Rating System): Health 1, Flammability 1, Reactivity 0. Insoluble in water.

Potential Exposure: BA is a contaminant and does not have any reported commercial use or application, although one producer did report the substance for the Toxic Substances Control Act Inventory. BA has been reported present in cigarette smoke condensate, automobile exhaust gas, soot, and the emissions from coal and gas works and electric plants. BA also occurs in the aromatic fraction of mineral oil, commercial solvents, waxes, petrolatum, creosote, coal tar, petroleum asphalt, and coal tar pitch. Microgram quantities of BA can be found in various foods such as charcoal broiled, barbecued, or smoked meats and fish; certain vegetables and vegetable oils, and roasted coffee and coffee powders. Human subjects are exposed to BA through either inhalation or ingestion. Workers at facilities with likely exposure to fumes from burning or heating of organic materials have a potential for exposure to BA. Consumers can be exposed to this chemical through ingestion of various foods, with concentrations of 100 μg/kg in some instances. Cigarette smoke condensate has quantities of BA that range from 0.03 – 4.6 μg/g. BA is found in the atmosphere at levels that vary with geography and climatology. These values can range from up to 136 μg/1,000 m^3 in summer to 361 μg/1,000 m^3 in winter. Drinking water samples may contain up to 23 ng/l BA, and surface waters have been found to contain 4 – 185 ng/l. The soil near industrial centers has been shown to contain as much as 390 μg/kg of BA, whereas soil near highways can have levels of up to 1,500 μg/kg, and areas polluted with coal tar pitch can reach levels of 2,500 mg/kg.

Incompatibilities: Oxidizing agents.

Permissible Exposure Limits in Air: The current OSHA PEL for coal tar pitch volatiles (anthracene, B(a)P, phenanthrene, acridine, chrysene, pyrene) is 0.2 mg/m^3 Benzene-soluble fraction (8-hour TWA). NIOSH considers coal tar products to be carcinogenic (Ca); the NIOSH-recommended 10-hour TWA exposure limit for coal tar products is 0.1 mg/m^3 (cyclohexane-extractable fraction). ACGIH designates coal tar pitch volatiles as a human carcinogen (A1a) with an 8-hour TWA of 0.2 mg/m^3 (benzene-solubles). This is also the HSE value.[33]

Determination in Air: The EPA Toxicological Profile cited below summarizes methods of analysis. Generally they involve ultrasonic extraction or solvent extraction followed by chromatography and mass spectrometry. See NIOSH Methods #5506 and #5515.[18]

Permissible Concentration in Water: Water quality criteria document for polynuclear aromatic hydrocarbons published in final 11/28/80. Total PAH addressed. A concentration of 2.8 ng PAH/l is estimated to limit cancer risk to one in a million (EPA). Kansas[61] has set a guideline in drinking water of 0.029 μg/l.

Determination in Water: Extraction with methylene chloride followed by gas chromatography/mass spectrometry (EPA Methods 625 and 1625) are summarized in the EPA Toxicological Profile cited below.

Routes of Entry: Inhalation, skin contact.

Harmful Effects and Symptoms

BA is absorbed by the oral and dermal routes of exposure; no direct evidence is available for absorption of B(a)A via the lungs. Following oral absorption, B(a)A is distributed to several tissues and accumulates preferentially in the adipose and mammary tissues. It is metabolized to conjugated derivatives and eliminated. It is expected that absorbed B(a)A will be excreted predominantly in the feces, as is true for other PAHs. B(a)A is metabolized to reactive derivatives that are thought to be responsible for its mutagenic activity in experimental systems. B(a)A is a weak experimental carcinogen by the dermal route of exposure. There is some evidence that it is carcinogenic by the oral route as well. Its carcinogenicity by the inhalation route has not been studied. Mutation is thought to be a necessary (although insufficient) step for the carcinogenic activity of B(a)A.

Short Term Exposure: No acute health effects known at this time.

Long Term Exposure: May be a carcinogen in humans. Has shown to cause bladder and skin cancer in animals.

Medical Surveillance: There is no special test for this chemical. However if illness occurs or overexposure is suspected, medical attention is recommended.

First Aid: *Skin Contact:* Flood all areas of body that have contacted the substance with water. Don't wait to remove contaminated clothing; do it under the water stream. Use soap to help assure removal. Isolate contaminated clothing when removed to prevent contact by others. *Eye Contact:* Remove any contact lenses at once. Immediately flush eyes well with copious quantities of water or normal saline for at least 20 – 30 minutes. Seek medical attention. *Inhalation:* Leave contaminated area immediately; breathe fresh air. Proper respiratory protection must be supplied to any rescuers. If coughing, difficult breathing or any other symptoms develop; seek medical attention at once, even if symptoms develop many hours after exposure. Ingestion: Contact a physician, hospital or poison center at once. If the victim is unconscious or convulsing, do not induce vomiting or give anything by mouth. Assure that his airway is open and lay him on his side with his head lower than his body and transport immediately to a medical facility. If conscious and not convulsing, give a glass of water to dilute the substance. Vomiting should not be induced without a physician's advice.

Personal Protective Methods: Wear protective gloves and clothing to prevent any reasonable probability of skin contact. Safety equipment suppliers/manufacturers can provide recommendations on the most protective glove/clothing material for your operation. All protective clothing (suits, gloves, footwear, headgear) should be clean, available each day, and put on before work. Contact lenses should not be worn when working with this chemical. Wear dust-proof chemical goggles and face shield unless full facepiece respiratory protection is worn. Employees should wash immediately with soap when skin is wet or contaminated. Provide emergency showers and eyewash.

Respirator Selection: NIOSH: *At any detectable concentration:* SCBAF:PD,PP (any MSHA/NIOSH approved self-contained breathing apparatus that has a full facepiece and is operated in a pressure-demand or other positive-pressure mode); or SAF:PD,PP:ASCBA (any supplied-air respirator that has a full facepiece and is operated in a pressure-demand or other positive-pressure mode in combination with an auxiliary, self-contained breathing apparatus operated in a pressure-demand or other positive pressure mode). *Escape:* GMFOVHiE [any air-purifying, full-facepiece respirator (gas mask) with a chin-style, front- or back-mounted organic vapor canister having a high-efficiency particulate filter]; or SCBAE (any appropriate escape-type, self-contained breathing apparatus).

Storage: Store in a cool, dry place away from oxidizers. A regulated, marked area should be established where this chemical is handled, used, or stored in compliance with OSHA standard 1910.1045.

Shipping: Poisonous solids, n.o.s. require a "Poison" label. In Packing Group II, The UN/DOT Hazard Class is 6.1.[19][20]

Spill Handling: Evacuate and restrict persons not wearing protective equipment from area of spill or leak until cleanup is complete. Remove all ignition sources. Collect powdered material in the most convenient and safe manner and deposit in sealed containers. Ventilate area of spill or leak after clean-up is complete. It may be necessary to contain and dispose of this chemical as a hazardous waste. If material or contaminated runoff enters waterways, notify downstream users of potentially contaminated waters. Contact your Department of Environmental Protection or your regional office of the federal EPA for specific recommendations. If employees are required to clean-up spills, they must be properly trained and equipped. OSHA 1910.120(q) may be applicable.

Fire Extinguishing: Use dry chemical, carbon dioxide, water spray, or foam extinguishers. Poisonous gases are produced in fire. If material or contaminated runoff enters waterways, notify downstream users of potentially contaminated waters. Notify local health and fire officials and pollution control agencies. From a secure, explosion-proof location, use water spray to cool exposed containers. If cooling streams are ineffective (venting sound increases in volume and pitch, tank discolors, or shows any signs of deforming), withdraw immediately to a secure position. If employees are expected to fight fires, they must be trained and equipped in OSHA 1910.156.

Disposal Method Suggested: Atomize into incinerator with a flammable liquid.[22]

References

Agency for Toxic Substance and Disease Registry, U.S. Public Health Service, "Toxicological Profile for Benz[a]Anthracene," Atlanta, Georgia, ATSDR (October 1987)

Sax, N. I., Ed., "Dangerous Properties of Industrial Materials Report" 5, No. 1, 32–37 (1985)

New Jersey Department of Health and Senior Services, Hazardous Substance Fact Sheet, Benz[a]anthracene, Trenton NJ (September 1998)

Benzal Chloride

Molecular Formula: $C_7H_6Cl_2$

Common Formula: $C_6H_5CHCl_2$

Synonyms: Benzene, dichloro methyl-; Benzyl dichloride; Benzylene chloride; Benzylidene chloride; Chlorobenzal; Chlorure de benzylidene (French); Cloruro de benzal (Spanish); (Dichloromethyl)benzene; α,α-Dichlorotoluene; Toluene, α,α-dichloro-

CAS Registry Number: 98-87-3

RTECS Number: CZ5075000

DOT ID: UN 1886

EEC Number: 602-058-00-8

Regulatory Authority

- Carcinogen (animal suspected) (IARC)[9] (proven) (DFG as "alpha-chlorinated toluenes")[3]
- Banned or Severely Restricted (Sweden) (UN)[13]
- Air Pollutant Standard Set (former USSR/UNEP)[43]
- EPA HAZARDOUS WASTE NUMBER (RCRA No.): U017
- RCRA, 40CFR261, Appendix 8 Hazardous Constituents
- RCRA 40CFR268.48; 61FR15654, Universal Treatment Standards: Wastewater (mg/l), 0.055; Nonwastewater (mg/kg), 6.0
- SUPERFUND/EPCRA 40CFR355, Appendix B Extremely Hazardous Substances: TPQ = 500 lb (227 kg)
- SUPERFUND/EPCRA 40CFR302.4 Reportable Quantity (RQ): CERCLA, 5,000 lb (2,270 kg)
- EPCRA Section 313 Form R *de minimis* concentration reporting level: 1.0%
- Canada, WHMIS, Ingredients Disclosure List

Cited in U.S. State Regulations: California (A, G), Florida (G), Kansas (G), Louisiana (G), Massachusetts (G), New Jersey (G), Pennsylvania (G), Vermont (G), Virginia (G), Washington (G), Wisconsin (G)

Description: $C_6H_5CHCl_2$ is a combustible, fuming, colorless, oily liquid with a faint odor. Boiling point = 205 – 207°C. Freezing/Melting point = -16°C. Flash pt. = 65°C. Autoignition temperature = 585°C. Insoluble in water.

Potential Exposure: Benzal chloride is used almost exclusively for the manufacture of benzaldehyde. It can also be used to prepare cinnamic acid and benzoyl chloride.

Incompatibilities: Forms explosive mixture with air and reacts with air or heat, forming fumes of hydrochloric acid. Reacts (possibly violently) with acids, bases, strong oxidizers, many metals; potassium, sodium, aluminum. Attacks plastics and coatings.

Permissible Exposure Limits in Air: The former USSR-UNEP/IRPTC project[43] has set a MAC in workplace air of 0.5 mg/m^3.

Permissible Concentration in Water: No criteria set. Benzal chloride hydrolyzes to benzaldehyde and HCl on contact with water.

Routes of Entry: Inhalation, ingestion, passing through the skin.

Harmful Effects and Symptoms

Short Term Exposure: Benzal chloride can affect the nervous system, and may be fatal if inhaled, swallowed or absorbed through the skin. Benzal chloride is irritating to the skin and eyes, causing excessive tearing. Irritates the respiratory tract causing shortness of breath and cough. Higher exposures can cause pulmonary edema, a medical emergency that can be delayed for several hours. This can cause death.

Long Term Exposure: Benzal chloride may cause skin cancer. This chemical was found to induce carcinomas, leukemia, and papillomas in mice. Benzal chloride was shown to possess a longer latency period than benzotrichloride before the onset of harmful effects. May affect the central nervous system.

Points of Attack: Central nervous system, skin.

Medical Surveillance: Lung function and nervous system tests.

First Aid: If this chemical gets into the eyes, remove any contact lenses at once and irrigate immediately for at least 15 minutes, occasionally lifting upper and lower lids. Seek medical attention immediately. If this chemical contacts the skin, remove contaminated clothing and wash immediately with soap and water. Seek medical attention immediately. If this chemical has been inhaled, remove from exposure, begin rescue breathing (using universal precautions) if breathing has stopped and CPR if heart action has stopped. Transfer promptly to a medical facility. When this chemical has been swallowed, get medical attention. Give large quantities of water and induce vomiting. Do not make an unconscious person vomit. Medical observation is recommended for 24 – 48 hours after breathing overexposure, as pulmonary edema may be delayed. As first aid for pulmonary edema, a doctor or authorized paramedic may consider administering a corticosteroid spray.

Personal Protective Methods: Wear protective gloves and clothing to prevent any reasonable probability of skin contact. Safety equipment suppliers/manufacturers can provide recommendations on the most protective glove/clothing material for your operation. All protective clothing (suits, gloves, footwear, headgear) should be clean, available each day, and put on before work. Contact lenses should not be worn when working with this chemical. Wear splash-proof chemical goggles and face shield unless full facepiece respiratory protection is worn. Employees should wash immediately with soap when skin is wet or contaminated. Provide emergency showers and eyewash.

Respirator Selection: *Where there is no REL, at any detectable concentration:* SA:CF (any supplied-air respirator operated in a continuous-flow mode); or PAPROV [any powered, air-purifying respirator with organic vapor cartridge(s)]; CCRFOV [any air-purifying, full-facepiece respirator (gas mask) with a chin-style, front-or back-mounted acid gas canister]; or GMFOV [any air-purifying, full-facepiece respirator (gas mask) with a chin-style, front-or back-mounted organic vapor canister]; or SCBAF (any self-contained breathing apparatus with a full facepiece); or SAF (any supplied-air respirator with a full facepiece). *Emergency or planned entry into unknown concentrations or IDLH conditions:* SCBAF: PD,PP (any self-contained breathing apparatus that has a full facepiece and is operated in a pressure-demand or other positive-pressure mode); or SAF:PD,PP:ASCBA (any supplied-air respirator that has a full facepiece and is operated in a pressure-demand or other positive-pressure mode in combination with an auxiliary self-contained breathing apparatus operated in a pressure-demand or other positive-pressure mode). *Escape:* GMFOV [any air-purifying, full-facepiece respirator (gas mask) with a chin-style, front-or back-mounted organic vapor canister] or SCBAE (any appropriate escape-type, self-contained breathing apparatus).

Storage: Store in tightly-closed containers in a cool, well-ventilated area away from acids and acid fumes because of danger of phosgene formation. Store away from sources of ignition. Stabilize with additives such as propylene oxide. A regulated, marked area should be established where this chemical is handled, used, or stored in compliance with OSHA standard 1910.1045.

Shipping: Benzal chloride is on DOT list as requiring "Poison" label. It falls in Hazard Class 6.1 and in Packing Group II.[19][20]

Spill Handling: Evacuate and restrict persons not wearing protective equipment from area of spill or leak until cleanup is complete. Remove all ignition sources. Do not breathe vapors. Wear eye protection and proper respiratory protection. Wear full protective clothing. Do not touch material. Stop leak if possible. Use water spray to reduce vapors. For small spills, take up with sand or other noncombustible material and place in containers for later disposal. For small dry spills, place material in clean dry container with shovel and move containers from spill area. For large spills, dike far ahead of spills for later disposal.

Ventilate area of spill or leak after clean-up is complete. It may be necessary to contain and dispose of this chemical as a hazardous waste. If material or contaminated runoff enters waterways, notify downstream users of potentially contaminated waters. Contact your Department of Environmental Protection or your regional office of the federal EPA for specific recommendations. If employees are required to clean-up spills, they must be properly trained and equipped. OSHA 1910.120(q) may be applicable.

Fire Extinguishing: Use dry chemical, carbon dioxide, water spray, or alcohol foam extinguishers. Poisonous gases are produced in fire including hydrogen chloride. If material or contaminated runoff enters waterways, notify downstream users of potentially contaminated waters. Notify local health and fire officials and pollution control agencies. From a secure, explosion-proof location, use water spray to cool exposed containers. If cooling streams are ineffective (venting sound increases in volume and pitch, tank discolors, or shows any signs of deforming), withdraw immediately to a secure position. If employees are expected to fight fires, they must be trained and equipped in OSHA 1910.156.

Disposal Method Suggested: 1,500°F, 0.5 second minimum for primary combustion; 2,200°F, 1.0 second for secondary combustion; elemental chlorine formation may be alleviated through injection of steam or methane into the combustion process.

References

U.S. Environmental Protection Agency, Benzal Chloride, Health and Environmental Effects Profile No. 14, Washington, DC, Office of Solid Waste (April 30, 1980)

U.S. Environmental Protection Agency, "Chemical Profile: Benzal Chloride," Washington, DC, Chemical Emergency Preparedness Program (November 30, 1987)

New Jersey Department of Health and Senior Services, "Hazardous Substance Fact Sheet: Benzal Chloride," Trenton, NJ (September 1996)

Benzaldehyde

Molecular Formula: C_7H_6O

Common Formula: C_6H_5CHO

Synonyms: Almond artificial essential oil; Artificial almond oil; Benzene carbaldehyde; Benzenecarbonal; Benzene carcaboxaldehyde; Benzenemethtal; Benzoic aldehyde; NCI-C56133; Oil of bitter almond; Phenylmethanal

CAS Registry Number: 100-52-7

RTECS Number: CU4375000

DOT ID: UN 1990

EEC Number: 605-012-00-5

Regulatory Authority

- Air Pollutant Standard Set (UNEP)[43] (former USSR)[35]
- U.S. DOT 49CFR172.101, Appendix B, Regulated marine pollutant
- TSCA 40CFR712.30.(e)1 (aldehydes)

Cited in U.S. State Regulations: Florida (G), Massachusetts (G), Minnesota (G), New Hampshire (G), New Jersey (G), Pennsylvania (G), Rhode Island (G).

Description: C_6H_5CHO is a clear to yellowish liquid with an almond odor. The odor threshold is 0.042 ppm. Boiling point = 179°C. Flash point = 63°C (cc). Autoignition temperature = 192°C. Hazard Identification (based on NFPA-704 M Rating System): Health 2, Flammability 2, Reactivity 0. Insoluble in water. Octanol/water partition coefficient as Log P_{ow} = 1.5

Potential Exposure: In manufacture of perfumes, dyes and cinnamic acid; as solvent; in flavors.

Incompatibilities: The substance may be able to form explosive peroxides. Reacts violently with oxidants, aluminum, iron, bases and phenol causing fire and explosion hazard. May self-ignite if absorbed in combustible material with large surface area.

Permissible Exposure Limits in Air: The former USSR-UNEP/IRPTC project[43] has set a MAC of 5 mg/m^3 in workplace air. This is also The former USSR limit.[35] There is no DFG MAK established at present for this chemical.

Permissible Concentration in Water: No criteria set.

Routes of Entry: Inhalation, ingestion and skin absorption.

Harmful Effects and Symptoms

Short Term Exposure: Absorbed through the skin, thereby increasing exposure. The substance irritates the eyes, the skin and the respiratory tract, causing coughing and shortness of breath. May cause contact dermatitis. Acts as a narcotic in high concentrations; exposure can cause dizziness; and, at higher levels, unconsciousness. The oral LD_{50} rat is 1,300 mg/kg.

Long Term Exposure: Repeated or prolonged contact may cause skin sensitization and rashes. Causes mutations; may cause cancer or reproductive risks.

Points of Attack: Skin, central nervous system.

Medical Surveillance: Examination by a qualified allergist.

First Aid: If this chemical gets into the eyes, remove any contact lenses at once and irrigate immediately for at least 15 minutes, occasionally lifting upper and lower lids. Seek medical attention immediately. If this chemical contacts the skin, remove contaminated clothing and wash immediately with soap and water. Seek medical attention immediately. If this chemical has been inhaled, remove from exposure, begin rescue breathing (using universal precautions) if breathing has stopped and CPR if heart action has stopped. Transfer promptly to a medical facility. When this chemical has been swallowed, get medical attention. Give large quantities of water and induce vomiting. Do not make an unconscious person vomit.

Personal Protective Methods: Wear protective gloves and clothing to prevent any reasonable probability of skin contact. Safety equipment suppliers/manufacturers recommend polyvinyl alcohol (PVA), styrene-butadiene rubber (SBR), butyl rubber, and polethylene coated materials. All protective clothing (suits, gloves, footwear, headgear) should be clean, available each day, and put on before work. Contact lenses should not be worn when working with this chemical. Wear splash-proof chemical goggles and face shield unless full facepiece respiratory protection is worn. Employees should wash immediately with soap when skin is wet or contaminated. Provide emergency showers and eyewash.

Respirator Selection: Engineering controls should be used wherever feasible to maintain airborne concentrations of this chemical below the prescribed exposure limit. Respirators and protective equipment are less effective than engineering controls, and should be used only in non-routine or emergency situations which may result in exposure concentrations in excess of the TWA environmental limit. *Where there is no REL, at any detectable concentration*: SCBAF: PD,PP (any MSHA/NIOSH approved self-contained breathing apparatus that has a full facepiece and is operated in a pressure-demand or other positive-pressure mode); or SAF:PD,PP:ASCBA (any supplied-air respirator that has a full facepiece and is operated in a pressure-demand or other positive-pressure mode in combination with an auxiliary, self-contained breathing apparatus operated in a pressure-demand or other positive pressure mode). *Escape:* GMFOVHiE [any air-purifying, full-facepiece respirator (gas mask) with a chin-style, front- or back-mounted organic vapor canister having a high-efficiency particulate filter]; or SCBAE (any appropriate escape-type, self-contained breathing apparatus).

Storage: Store in tightly closed containers in a cool, well-ventilated area away from light. Benzaldehyde must be stored to avoid contact with oxidizing materials such as peroxyformic acid and combustible materials such as wood, paper and oil since violent reactions occur.

Shipping: Benzaldehyde label required is "Class 9." It is in Hazard Class 9 and Packaging Group III.

Spill Handling: Evacuate and restrict persons not wearing protective equipment from area of spill or leak until cleanup is complete. Remove all ignition sources. Ventilate area of spill or leak. Collect powdered material in the most convenient and safe manner and deposit in sealed containers. Ventilate area of spill or leak after clean-up is complete. It may be necessary to contain and dispose of this chemical as a hazardous waste. If material or contaminated runoff enters waterways, notify downstream users of potentially contaminated waters. Contact your Department of Environmental Protection or your regional office of the federal EPA for specific recommendations. If employees are required to clean-up spills, they must be properly trained and equipped. OSHA 1910.120(q) may be applicable.

Fire Extinguishing: Benzaldehyde is a combustible liquid. Poisonous gases are produced in fire. Use dry chemical, carbon dioxide, or alcohol foam extinguishers. Vapors are heavier than air and will collect in low areas. Vapors may travel long distances to ignition sources and flashback. Vapors in confined areas may explode when exposed to fire. Storage containers and parts of containers may rocket great distances, in many directions. If material or contaminated runoff enters waterways, notify downstream users of potentially contaminated waters. Notify local health and fire officials and pollution control agencies. From a secure, explosion-proof location, use water spray to cool exposed containers. If cooling streams are ineffective (venting sound increases in volume and pitch, tank discolors, or shows any signs of deforming), withdraw immediately to a secure position. If employees are expected to fight fires, they must be trained and equipped in OSHA 1910.156.

Disposal Method Suggested: Incineration; add combustible solvent and spray into incinerator with afterburner.[22]

References

Sax, N. I., Ed., "Dangerous Properties of Industrial Materials Report" 1, No. 8, 36–38 (1981)

New Jersey Department of Health and Senior Services, "Hazardous Substance Fact Sheet: Benzaldehyde," Trenton, NJ (July 1996)

Benzamide

Molecular Formula: C_7H_7NO

Common Formula: $C_6H_5CONH_2$

Synonyms: Benzamida (Spanish); Benzoic acid amide; Benzoylamide; Phenylcarboxyamide

CAS Registry Number: 55-21-0

RTECS Number: CU8700000

Regulatory Authority

- EPCRA Section 313 Form R *de minimis* concentration reporting level: 1.0%

Cited in U.S. State Regulations: California (A, G), Maryland (G), Massachusetts (G), New Jersey (G), Pennsylvania (G).

Description: Benzamide, $C_6H_5CONH_2$, is a combustible, colorless, crystalline solid. Freezing/Melting point = 132 – 133°C. Boiling point = 290°C. Soluble in hot water. The New Jersey Department of Health flammability rating is 2 (based on NFPA 704M)

Potential Exposure: Benzamide is used in organic synthesis.

Permissible Exposure Limits in Air: No standards set.

Permissible Concentration in Water: No criteria set.

Routes of Entry: Inhalation, passing through the skin.

Harmful Effects and Symptoms

Short Term Exposure: Irritates the eyes, nose, and throat. May produce gastric pain, nausea and vomiting.[52] The oral LD_{50} mouse is 1,160 mg/kg (slightly toxic).

Long Term Exposure: Similar chemicals cause methemoglobinemia and liver damage, but it is not known if benzamide has these effects.

Medical Surveillance: Completed blood count (CBC), liver function tests.

First Aid: If this chemical gets into the eyes, remove any contact lenses at once and irrigate immediately for at least 15 minutes, occasionally lifting upper and lower lids. Seek medical attention immediately. If this chemical contacts the skin, remove contaminated clothing and wash immediately with soap and water. Seek medical attention immediately. If this chemical has been inhaled, remove from exposure, begin rescue breathing (using universal precautions) if breathing has stopped and CPR if heart action has stopped. Transfer promptly to a medical facility. When this chemical has been swallowed, get medical attention. Give large quantities of water and induce vomiting. Do not make an unconscious person vomit.

Personal Protective Methods: Wear protective gloves and clothing to prevent any reasonable probability of skin contact. Safety equipment suppliers/manufacturers can provide recommendations on the most protective glove/clothing material for your operation. All protective clothing (suits, gloves, footwear, headgear) should be clean, available each day, and put on before work. Contact lenses should not be worn when working with this chemical. Wear dust-proof chemical goggles and face shield unless full facepiece respiratory protection is worn. Employees should wash immediately with soap when skin is wet or contaminated. Provide emergency showers and eyewash.

Respirator Selection: Engineering controls should be used wherever feasible to maintain airborne concentrations of this chemical below the prescribed exposure limit. Respirators and protective equipment less effective than engineering controls, and should be used only in non-routine or emergency situations which may result in exposure concentrations in excess of the TWA environmental limit. *Where there is no REL, at any detectable concentration*: SCBAF:PD,PP (any MSHA/NIOSH approved self-contained breathing apparatus that has a full facepiece and is operated in a pressure-demand or other positive-pressure mode); or SAF:PD, PP:ASCBA (any supplied-air respirator that has a full facepiece and is operated in a pressure-demand or other positive-pressure mode in combination with an auxiliary, self-contained breathing apparatus operated in a pressure-demand or other positive pressure mode).

Storage: Store in a refrigerator or a cool, dry place.

Shipping: There are no UN/DOT label or shipping quantity requirement.

Spill Handling: Evacuate and restrict persons not wearing protective equipment from area of spill or leak until cleanup is complete. Remove all ignition sources. Dampen spilled material with 60 – 70% ethanol to avoid dust and transfer to vapor-tight plastic bags for eventual disposal. Ventilate area of spill or leak after clean-up is complete. It may be necessary to contain and dispose of this chemical as a hazardous waste. If material or contaminated runoff enters waterways, notify downstream users of potentially contaminated waters. Contact your Department of Environmental Protection or your regional office of the federal EPA for specific recommendations. If employees are required to clean-up spills, they must be properly trained and equipped. OSHA 1910.120(q) may be applicable.

Fire Extinguishing: A combustible solid. Use dry chemical, carbon dioxide, water spray, or foam extinguishers. Poisonous gases are produced in fire including nitrogen oxides. If material or contaminated runoff enters waterways, notify downstream users of potentially contaminated waters. Notify local health and fire officials and pollution control agencies. From a secure, explosion-proof location, use water spray to cool exposed containers. If cooling streams are ineffective (venting sound increases in volume and pitch, tank discolors, or shows any signs of deforming), withdraw immediately to a secure position. If employees are expected to fight fires, they must be trained and equipped in OSHA 1910.156.

Benzenamine, 3-(Trifluoromethyl)-

Molecular Formula: $C_7H_6F_3N$

Common Formula: $H_2NC_6H_4CF_3$

Synonyms: *m*-Aminobenzal fluoride; *m*-Aminobenzal-trifluoride; *m*-Aminobenzotrifluoride; 3-Aminobenzot-rifluoride; 3-Amino-benzo-trifluoride; 1-Amino-3-(trifluoro-methyl)benzene; Toluene, 3-amino- α,α,α-trifluoro-; *m*-(Trifluoromethyl)aniline; 3-(Trifluoromethyl)aniline; *m*-(Trifluoromethyl)benzenamine; 3-(Trifluoromethyl)benzen-amine; α,α,α-Trifluoro-*m*-toluidine

CAS Registry Number: 98-16-8

RTECS Number: XU9180000

DOT ID: UN 2948

Regulatory Authority

- Water Pollution Standard Proposed (former USSR)[43]
- SUPERFUND/EPCRA 40CFR355, Appendix B Extremely Hazardous Substances: TPQ = 500 pounds
- SUPERFUND/EPCRA 40CFR302.4 Reportable Quantity (RQ): EHS, 500 lb
- Canada, WHMIS, Ingredients Disclosure List

Cited in U.S. State Regulations: Florida (G), Massachusetts (G), New Jersey (G), Pennsylvania (G).

Description: Benzenamine, 3-(trifluoromethyl)-, $H_2NC_6H_4CF_3$, is a combustible, colorless to yellow oily liquid with an aminelike odor. Freezing/Melting point = 3°C. Boiling point = 187.5°C. Flash point = 85°C.[52]

Potential Exposure: This material is used as a chemical intermediate for herbicides, antihypertensives, and diuretics.

Incompatibilities: Strong oxidizers.

Permissible Exposure Limits in Air: The exposure limits are recommended for inorganic Fluoride compounds and are measured as Fluorine. The OSHA PEL is 2.5 mg/m^3 TWA averaged over an 8-hour workshift. This is the ACGIH,[1] DFG,[3] HSE[33] and OSHA[58] value. The former USSR/UNEP joint project[43] has set a MAC in ambient air in residential areas of 0.2 mg/m^3 on a momentary basis or 0.03 mg/m^3 on an average daily basis.

Permissible Concentration in Water: The former USSR/UNEP/IRPTC project[43] has set a MAC in water used for domestic purposes of 0.02 mg/l.

Routes of Entry: Inhalation, ingestion, skin contact.

Harmful Effects and Symptoms

Short Term Exposure: Contact may cause burns to skin and eyes. May be poisonous if inhaled, swallowed or absorbed through the skin. The oral LD_{50} rat is 480 mg/kg; the oral LD_{50} mouse is 220 mg/kg.

Long Term Exposure: Repeated exposure to fluoride chemicals may cause stiffness in muscles or ligaments and even crippling. Fluoride may increase bone density, stimulate new bone growth or cause calcium deposits in ligaments. This may become a problem at levels of 20 – 50 mg/m^3 or higher. Mottling of the teeth may occur at this level.

First Aid: If this chemical gets into the eyes, remove any contact lenses at once and irrigate immediately for at least 15 minutes, occasionally lifting upper and lower lids. Seek medical attention immediately. If this chemical contacts the skin, remove contaminated clothing and wash immediately with soap and water. Seek medical attention immediately. If this chemical has been inhaled, remove from exposure, begin rescue breathing (using universal precautions) if breathing has stopped and CPR if heart action has stopped. Transfer promptly to a medical facility. When this chemical has been swallowed, get medical attention. Give large quantities of water and induce vomiting. Do not make an unconscious person vomit. Keep victim quiet and maintain normal body temperature. Effects may be delayed; keep victim under observation.

Personal Protective Methods: Wear protective gloves and clothing to prevent any reasonable probability of skin contact. Safety equipment suppliers/manufacturers can provide recommendations on the most protective glove/clothing material for your operation. All protective clothing (suits, gloves, footwear, headgear) should be clean, available each day, and put on before work. Contact lenses should not be worn when working with this chemical. Wear splash-proof chemical goggles and face shield unless full facepiece respiratory protection is worn. Employees should wash immediately with soap when skin is wet or contaminated. Provide emergency showers and eyewash.

Respirator Selection: Engineering controls should be used wherever feasible to maintain airborne concentrations of this chemical below the prescribed exposure limit. Respirators and protective equipment are less effective than engineering controls, and should be used only in non-routine or emergency situations which may result in exposure concentrations in excess of the TWA environmental limit. NIOSH: (fluorides) *12.5 mg/m^3:* DM (any dust and mist respirator). *25 mg/m^3:* DMXSQ* (any dust and mist respirator except single-use and quarter mask respirators); or SA* (any supplied-air respirator). *62.5 mg/m^3:* SA:CF* (any supplied-air respirator operated in a continuous-flow mode); or PAPRDM* ** if not present as a fume (any powered, air-purifying respirator with a dust and mist filter). *125 mg/m^3:* HiEF** (any air-purifying, full-facepiece respirator with a high-efficiency particulate filter); or SCBAF (any self-contained breathing apparatus with a full facepiece); or SAF (any supplied-air respirator with a full facepiece). *250 mg/m^3:* SA:PD,PP (any supplied-air respirator operated in a pressure-demand or other positive-pressure mode). *Emergency or planned entry into unknown concentrations or IDLH conditions:* SCBAF:PD,PP (any self-contained breathing apparatus that has a full faceplate and is operated in a pressure-demand or other positive-pressure mode); or SAF:PD,PP:ASCBA (any supplied-air respirator that has a full facepiece and is operated in a pressure-demand or other positive-pressure mode in combination with an auxiliary, self-contained breathing apparatus operated in a pressure-demand or other positive-pressure mode). *Escape:* HiEF** (any air-purifying, full-facepiece respirator with a high-efficiency particulate filter); or SCBAE (any appropriate escape-type, self-contained breathing apparatus).

* Substance reported to cause eye irritation or damage; may require eye protection.

** May need acid gas sorbent.

Storage: Store in tightly closed containers in a cool, well ventilated area away from oxidizers. Sources of ignition such as smoking and open flames, are prohibited where this chemical is used, handled, or stored in a manner that could create a potential fire or explosion hazard.

Shipping: 3-Trifluoromethylaniline requires a "Poison" label according to DOT.[19] This material falls in UN/DOT Hazard Class 6.1 and Packing Group II.[19][20]

Spill Handling: Evacuate and restrict persons not wearing protective equipment from area of spill or leak until cleanup is complete. Remove all ignition sources. Ventilate area of spill or leak. Do not touch spilled material; stop leak if you can do so without risk. Use water spray to reduce vapors. *Small spills:* absorb with sand or other non-combustible absorbent material and place into containers for later disposal. *Large spills:* dike far ahead of spill for later disposal. It may be necessary to contain and dispose of this chemical as a hazardous waste. If material or contaminated runoff enters waterways, notify downstream users of potentially contaminated waters. Contact your Department of Environmental Protection or your regional office of the federal EPA for specific recommendations. If employees are required to clean-up spills, they must be properly trained and equipped. OSHA 1910.120(q) may be applicable.

Fire Extinguishing: This chemical is a combustible liquid. Poisonous gases including nitrogen oxides are produced in fire. Use dry chemical, carbon dioxide, or alcohol foam extinguishers. Wear positive pressure breathing apparatus and special protective clothing. Move container from fire area if you can do so without risk. Fight fire from maximum distance. Dike fire control water for later disposal; do not scatter the material. Vapors are heavier than air and will collect in low areas. Vapors may travel long distances to ignition sources and flashback. Vapors in confined areas may explode when exposed to fire. Storage containers and parts of containers may rocket great distances, in many directions. If material or contaminated runoff enters waterways, notify downstream users of potentially contaminated waters. Notify local health and fire officials and pollution control agencies. From a secure, explosion-proof location, use water spray to cool exposed containers. If cooling streams are ineffective (venting sound increases in volume and pitch, tank discolors, or shows any signs of deforming), withdraw immediately to a secure position. If employees are expected to fight fires, they must be trained and equipped in OSHA 1910.156.

References

U.S. Environmental Protection Agency, "Chemical Profile: Benzenamine, 3-Trifluoromethyl)," Washington, DC, Chemical Emergency Preparedness Program (November 30, 1987)

Benzene

Molecular Formula: C_6H_6

Synonyms: (6) Annulene; Benceno (Spanish); Benzeen (Dutch); Benzelene; Benzen (Polish); Benzol; Benzole; Benzolo (Italian); Bicarburet of hydrogen; Carbon naphtha; Carbon oil; Coal naphtha; Coal naphtha, Phenyl hydride; Coal tar naphtha; Cyclohexatriene; Fenzen (Chech); Mineral naphtha; Motor benzol; NCI-C55276; Nitration benzene; Phene; Phenyl hydride; Pyrobenzol; Pyrobenzole

CAS Registry Number: 71-43-2

RTECS Number: CY1400000

DOT ID: UN 1114

EEC Number: 601-020-00-8

EINECS Number: 200-753-7

Regulatory Authority

- Carcinogen (Human Positive) (IARC)[9] (DFG)[3] (NTP)
- Banned or Severely Restricted (Several Countries) (UN)[13]
- OSHA, 29CFR1910 Specifically Regulated Chemicals (See CFR1910.1028)
- Air Pollutant Standard Set (ACGIH)[1] (HSE)[33] (OSHA)[58] (Several States)[60] (Australia) (Israel) (Mexico) (Several Canadian Provinces)
- Water Pollution Standard Proposed (EPA)[6][48] (former USSR)[43] (Several States)[61] (Canada) (Mexico)
- Clean Air Act: Hazardous Air Pollutants (Title I, Part A, Section 112). *Note:* Including benzene from gasoline
- Clean Water Act: Section 311 Hazardous Substances/RQ 40CFR117.3 (same as CERCLA, see below); 40CFR423, Appendix A, Priority Pollutants; Section 313 Water Priority Chemicals (57FR41331, 9/9/92); 40CFR401.15 Section 307 Toxic Pollutants
- EPA HAZARDOUS WASTE NUMBER (RCRA No.): U019
- RCRA Toxicity Characteristic (Section 261.24), Maximum Concentration of Contaminants, regulatory level, 0.5 mg/l
- RCRA, 40CFR261, Appendix 8 Hazardous Constituents
- RCRA 40CFR268.48; 61FR15654, Universal Treatment Standards: Wastewater (mg/l), 0.14; Nonwastewater (mg/kg), 10
- RCRA 40CFR264, Appendix 9; TSD Facilities Ground Water Monitoring List. Suggested test method(s) (PQL μg/l): 8020 (2); 8240 (5)
- Safe Drinking Water Act: MCL, 0.005 mg/l; MCLG, zero; Regulated chemical (47FR9352)
- SUPERFUND/EPCRA 40CFR302.4 Reportable Quantity (RQ): CERCLA, 10 lb (4.54 kg)

- EPCRA Section 313 Form R *de minimis* concentration reporting level: 0.1%
- U.S. DOT Regulated Marine Pollutant (49CFR172.101, Appendix B)
- Canada, WHMIS, Ingredients Disclosure List; National Pollutant Release Inventory (NPRI); Priority Substance List (CEPA)

Cited in U.S. State Regulations: Alaska (G), California (A, G, W), Connecticut (A, W), Florida (G, W), Illinois (G), Kansas (G), Louisiana (G), Maine (G, W), Maryland (G), Massachusetts (G, A), Michigan (G, A), Minnesota (G, W), Nevada (A), New Hampshire (G, W), New Jersey (G, W), New Mexico (W), New York (G, A, W), North Carolina (A), North Dakota (A), Oklahoma (G), Pennsylvania (G, A), Rhode Island (G, A), South Carolina (A), Vermont (G), Virginia (G, A), Washington (G), West Virginia (G), Wisconsin (G).

Description: Benzene, C_6H_6, is a clear, volatile, colorless, highly flammable liquid with a pleasant, characteristic odor. Boiling point = 80°C. Freezing/Melting point = 6°C. Flash point = –11°C. Autoignition temperature = 498°C. Explosive limits: LEL = 1.4%, UEL = 7.5%. Hazard Identification (based on NFPA-704 M Rating System): Health 2, Flammability 3, Reactivity 0. The odor threshold in air is 4.9 mg/m^3; the odor threshold in water is 2.0 mg/liter. Very slightly soluble in water.

Potential Exposure: Benzene is used as a constituent in motor fuels, as a solvent for fats, inks, oils, paints, plastics, and rubber, in the extraction of oils from seeds and nuts, and in photogravure printing. It is also used as a chemical intermediate. By alkylation, chlorination, nitration, and sulfonation, chemicals such as styrene, phenols, and malefic anhydride are produced. Benzene is also used in the manufacture of detergents, explosives, pharmaceuticals, and dye-stuffs. Increased concern for benzene as a significant environmental pollutant arises from public exposure to the presence of benzene in gasoline and the increased content in gasoline due to requirements for unleaded fuels for automobiles equipped with catalytic exhaust converters.

Incompatibilities: Strong oxidizers, many fluorides and perchlorates, nitric acid.

Permissible Exposure Limits in Air: OSHA[58] Federal standard is 1 ppm TWA and STEL of 5 ppm for an 8 hour workshift. NIOSH recommends an REL of 0.1 ppm TWA and STEL of 1 ppm for a 10-hour workshift. The ACGIH has designated benzene as an "Industrial Substance Suspect of Carcinogenic Potential for Man" with a TLV of 10 ppm (32 mg/m^3) but no STEL set. The British HSE has set the maximum Exposure Limit of 5 ppm (16 mg/m^3) TWA and no STEL. The German DFG has set no numerical limits, only the proven carcinogen notation. Other countries such as Sweden[35] have set a TWA of 5 ppm (16 mg/m^3) whereas Brazil has set 8 ppm (24 mg/m^3). The former USSR-UNEP/IRPTC project[43] has set a MAC of 5 mg/m^3 in workplace air, of 1.5 mg/m^3 in ambient air in residential areas on a momentary basis and 0.8 mg/m^3 in residential air on a daily average basis. The IDHL level is 2,000 ppm.[2] Several States have set guidelines or standards for benzene in ambient air[60] ranging from zero (North Dakota) to 0.12 μg/m^3 (North Carolina) to 0.14 μg/m^3 (Michigan) to 1.2 μg/m^3 (Massachusetts) to 72 μg/m^3 (Pennsylvania) to 100 μg/m^3 (New York and Rhode Island) to 150 μg/m^3 (Connecticut and South Carolina) to 300 μg/m^3 (Virginia).

The final OSHA Benzene standard in 1910.1028 applies to all occupational exposures to benzene except some subsegments of industry where exposures are consistently under the action level (i.e., distribution and sales of fuels, sealed containers and pipelines, coke production, oil and gas drilling and production, natural gas processing, and the percentage exclusion for liquid mixtures); *for the excepted subsegments, the benzene limits in Table Z-2 apply* (i.e., an 8-hour TWA of 10 ppm, an acceptable ceiling of 25 ppm, and 50 ppm for a maximum duration of 10 minutes as an acceptable maximum peak above the acceptable ceiling). IDLH = Carcinogen (500 ppm).

Determination in Air: Adsorption on charcoal, workups with CS_2, analysis by gas chromatography. See OSHA: #12 or NIOSH: Hydrocarbons, Aromatic, 1501; Hydrocarbons, BP 36 – 126°C, 1500.

Permissible Concentration in Water: To protect freshwater aquatic life: 5,300 μg/l on an acute basis. To protect saltwater aquatic life: 5,100 μg/l on an acute basis. To protect human health: preferably zero. An additional lifetime cancer risk of 1 in 100,000 results from a concentration of 6.6 μg/l.[6] In addition, several states have set standards and guidelines for benzene in drinking water.[61] The standards range from 1.0 μg/l (Florida and New Jersey) to 10.0 μg/l (New Mexico). A ten-day health advisory for benzene has been calculated[48] at 0.235 mg/l for a 10 kg child. A lifetime health advisory for humans cannot be calculated because of the carcinogenic potency of benzene. The former USSR-UNEP/IRPTC project[43] has set a MAC of 0.5 mg/l of benzene in water bodies used for domestic purposes and the same limit in water used for fishery purposes. The WHO has recommended a limit of 10 μg/l of benzene in drinking water.[35] Canada's drinking water quality MAC 0.005 mg/l. Mexico's drinking water ecological criteria is 0.01 mg/l.

Determination in Water: Gas chromatography (EPA Method 602) or gas chromatography plus mass spectrometry (EPA Method 624).

Routes of Entry: Inhalation, skin absorption, ingestion, skin and/or eye contact.

Harmful Effects and Symptoms

Short Term Exposure: Inhalation of Benzene may produce both nerve and blood effects. Irritation of the nose, throat and lungs may occur (3,000 ppm may be tolerated for only 30 – 60 minutes). Lung congestion may occur. Nerve effects may include an exaggerated feeling of well-being, excitement, headache, dizziness, and slurred speech. At high levels, slowed breathing and death may result. Death has occurred at 20,000 ppm for 5 – 10 minutes, or 7,500 ppm for 30 minutes. Skin contact: Irritation may occur, with redness and blistering if not promptly removed. Benzene is poorly absorbed. Whole body exposure for 30 minutes has been reported with no health effects. Eye contact may cause severe irritation. Ingestion may cause irritation of mouth, throat and stomach. Symptoms are similar to those listed under inhalation. One tablespoon may cause collapse, bronchitis, pneumonia and death. Use of alcoholic beverages enhances the harmful effect.

Long Term Exposure: Benzene is a known human carcinogen. Exposure has been linked to increased risk of several forms of leukemia. The liquid defats the skin. The substance may have effects on the blood forming organs, liver and immune system. May cause loss of appetite, nausea, weight loss, fatigue, muscle weakness, headache, dizziness, nervousness and irritability. Mild anemia has been reported from exposures of 25 ppm for several years and 100 ppm for 3 months. At levels between 100 and 200 ppm for periods of 6 months, or more, severe irreversible blood changes and damage to liver and heart may occur. Temporary partial paralysis has been reported.

Points of Attack: Eyes, skin, respiratory system, blood, central nervous system, bone marrow. Cancer site: leukemia.

Medical Surveillance: Preplacement and periodic examinations should be concerned especially with effects on the blood and bone marrow and with a history of exposure to other myelotoxic agents or drugs or of other diseases of the blood. Preplacement laboratory exams should include: (a) complete blood count (hematocrit, hemoglobin, mean corpuscular volume, white blood count, differential count, and platelet estimation); (b) reticulocyte count; (c) serum bilirubin; and (d) urinary phenol. The type and frequency of periodic hematologic studies should be related to the data obtained from biologic monitoring and industrial hygiene studies, as well as any symptoms or signs of hematologic effects. Recommendations for proposed examinations have been made in the criteria for a recommended standard. Examinations should also be concerned with other possible effects such as those on the skin, central nervous system, and liver and kidney functions. Biologic monitoring should be provided to all workers subject to benzene exposure. It consists of sampling and analysis of urine for total phenol content. The objective of such monitoring is to be certain that no worker absorbs an unacceptable amount of benzene. Unacceptable absorption of benzene, posing a risk of benzene poisoning, is considered to occur at levels of 75 mg phenol per liter of urine (with urine specific gravity, corrected to 1.024), when determined by methods specified in the NIOSH "Criteria for Recommended Standard-Benzene." Alternative methods shown to be equivalent in accuracy and precision may also be useful. Biological monitoring should be done at quarterly intervals. If environmental sampling and analysis are equal to or exceed accepted safe limits, the urinary phenol analysis should be conducted every two weeks. This increased monitoring frequency should continue for at least 2 months after the high environmental level has been demonstrated. Two follow-up urines should be obtained within one week after receipt of the original results, one at the beginning and the other at the end of the work week. If original elevated findings are confirmed, immediate steps should be taken to reduce the worker's absorption of benzene by improvement in environment control, personal protection, personal hygiene, and administrative control.

First Aid: If this chemical gets into the eyes, remove any contact lenses at once and irrigate immediately for at least 15 minutes, occasionally lifting upper and lower lids. Seek medical attention immediately. If this chemical contacts the skin, remove contaminated clothing and wash immediately with soap and water. Seek medical attention immediately. If this chemical has been inhaled, remove from exposure, begin rescue breathing (using universal precautions) if breathing has stopped and CPR if heart action has stopped. Transfer promptly to a medical facility. When this chemical has been swallowed, rinse mouth, get medical attention. Do not induce vomiting.

Personal Protective Methods: Wear solvent-resistant gloves and clothing to prevent any reasonable probability of skin contact. Viton/neoprene, butyl/neoprene, polyvinyl acetate and Silvershield® are recommended by safety equipment suppliers/manufacturers who can also provide recommendations on the most protective glove/clothing material for your operation. All protective clothing (suits, gloves, footwear, headgear) should be clean, available each day, and put on before work. Contact lenses should not be worn when working with this chemical. Wear splash-proof chemical goggles and face shield unless full facepiece respiratory protection is worn. Employees should wash immediately with soap when skin is wet or contaminated. Remove nonimpervious clothing immediately if wet or contaminated. Provide emergency showers and eyewash.

Respirator Selection: NIOSH: *At any detectable concentration:* SCBAF:PD,PP (any MSHA/NIOSH approved self-contained breathing apparatus that has a full facepiece and is operated in a pressure-demand or other positive-pressure mode); or SAF:PD,PP:ASCBA (any supplied-air respirator that has a full facepiece and is operated in a pressure-demand or other positive-pressure mode in combination with an auxiliary, self-contained breathing apparatus operated in a pressure-demand or other positive pressure mode). *Escape:* GMFOV [any air-purifying, full-facepiece respirator (gas mask) with a chin-style, front-or back-mounted organic vapor canister] or SCBAE (any appropriate escape-type, self-contained breathing apparatus).

Storage: Before entering confined space where benzene may be present, check to make sure that an explosive concentration does not exist. Store in tightly closed containers in a cool, well ventilated area. Protect containers against physical damage. Storage preferred in an outdoor or detached building. If storage is indoor, use a standard flammable liquid storage room. Metal containers involving the transfer of this chemical should be grounded and bonded. Where possible, automatically pump liquid from drums or other storage containers to process containers. Drums must be equipped with self-closing valves, pressure vacuum bungs, and flame arresters. Use only non-sparking tools and equipment, especially when opening and closing containers of this chemical. Sources of ignition such as smoking and open flames, are prohibited where this chemical is used, handled, or stored in a manner that could create a potential fire or explosion hazard. A regulated, marked area should be established where this chemical is handled, used, or stored in compliance with OSHA standard 1910.1045.

Shipping: The label required by DOT is "Flammable Liquid." Benzene falls in UN/DOT Hazard Class 3 and Packing Group II.[19][20]

Spill Handling: Evacuate and restrict persons not wearing protective equipment from area of spill or leak until cleanup is complete. Remove all ignition sources. Establish forced ventilation to keep levels below explosive limit. Absorb liquids in vermiculite, dry sand, earth, or a similar material and deposit in sealed containers. Keep benzene out of a confined space, such as a sewer, because of the possibility of an explosion, unless the sewer is designed to prevent the build-up of explosive concentrations. It may be necessary to contain and dispose of this chemical as a hazardous waste. If material or contaminated runoff enters waterways, notify downstream users of potentially contaminated waters. Contact your Department of Environmental Protection or your regional office of the federal EPA for specific recommendations. If employees are required to cleanup spills, they must be properly trained and equipped. OSHA 1910.120(q) may be applicable.

Fire Extinguishing: This chemical is highly flammable. Poisonous gases are produced in fire. Use dry chemical, carbon dioxide, or foam extinguishers. Vapors are heavier than air and will collect in low areas. Vapors may travel long distances to ignition sources and flashback. Vapors in confined areas may explode when exposed to fire. Storage containers and parts of containers may rocket great distances, in many directions. If material or contaminated runoff enters waterways, notify downstream users of potentially contaminated waters. Notify local health and fire officials and pollution control agencies. From a secure, explosion-proof location, use water spray to cool exposed containers. If cooling streams are ineffective (venting sound increases in volume and pitch, tank discolors, or shows any signs of deforming), withdraw immediately to a secure position. If employees are expected to fight fires, they must be trained and equipped in OSHA 1910.156.

Disposal Method Suggested: Incineration. Dilution with alcohol or acetone to minimize smoke is recommended.[22] Bacterial degradation is also possible.

References

National Academy of Sciences, A Review of Health Effects of Benzene, National Academy of Sciences, Washington, DC (June 1976)

National Institute for Occupational Safety and Health, Criteria for a Recommended Standard, Occupational Exposure to Benzene, NIOSH Doc. No. 74–137, Washington, DC (1974)

U.S. Environmental Protection Agency, Benzene: Ambient Water Quality Criteria, Washington, DC (1980)

U.S. Environmental Protection Agency, Status Assessment of Toxic Chemicals: Benzene, Report EPA-600/2-79-210D, Washington, DC (December 1979)

U.S. Environmental Protection Agency, Benzene, Health and Environmental Effects Profile No. 15, Washington, DC, Office of Solid Waste (April 30, 1980)

Sax, N. I., Ed., "Dangerous Properties of Industrial Materials Report" 1, No. 4, 38–41 (1981); 2, No. 4, 33–38 (1982); and 3, No. 3, 53–59 (1983) and 4, No. 1, 21–22 (1984); and 4, No. 6, 55 (1984)

Occupational Safety and Health Administration "Occupational Exposure to Benzene; Final Rule," Federal Register, 52, No. 176, 34460–34578 incl. (September 11, 1987)

New Jersey Department of Health and Senior Services, "Hazardous Substance Fact Sheet: Benzene," Trenton, NJ (April 1986; Revised January 31, 1988)

New York State Department of Health, "Chemical Fact Sheet: Benzene," Albany, NY, Bureau of Toxic Substance Assessment (Version 2)

Mehlman, M. A., Ed., Carcinogenicity and Toxicity of Benzene, Princeton, NJ, Princeton Scientific Publishers (1983)

Benzenearsonic Acid

Molecular Formula: $C_6H_7AsO_3$

Common Formula: $C_6H_5AsO(OH)_2$

Synonyms: Acido fenilarsonico (Spanish); Phenyl arsenic acid; Phenylarsonic acid

CAS Registry Number: 98-05-5

RTECS Number: CY3150000

Regulatory Authority

- Banned or Severely Restricted*
- Air Pollutant Standard Set*
- CLEAN AIR ACT: Hazardous Air Pollutants (Title I, Part A, Section 112); List of high risk pollutants (Section 63.74) as arsenic compounds
- CLEAN WATER ACT: 40CFR401.15 Section 307 Toxic Pollutants, as arsenic and compounds
- SUPERFUND/EPCRA 40CFR355, Appendix B Extremely Hazardous Substances: TPQ = 10/10,000 lb (4.54/4,540 kg)
- RCRA, 40CFR261, Appendix 8 Hazardous Constituents, waste number not listed

- SUPERFUND/EPCRA 40CFR302.4 Reportable Quantity (RQ): EHS, 1 lb (0.454 kg)
- EPCRA (Section 313): as an arsenic organic compound. Form R *de minimis* concentration reporting level: 1.0%
- U.S. DOT Regulated Marine Pollutant (49CFR172.101, Appendix B) as arsenates, liquid, n.o.s.; arsenates, solid, n.o.s.; arsenical pesticides liquid, toxic, flammable, n.o.s
- This compound is not specifically cited but falls in these categories since it is an arsenic compound. See entry on Arsenic.

Cited in U.S. State Regulations: Alaska (G), Florida (G), Kansas (G), Louisiana (G), Massachusetts (G), New Jersey (G), Pennsylvania (G), Vermont (G), Virginia (G), Washington (G), Wisconsin (G).

Description: Benzenearsonic Acid, $C_6H_5AsO(OH)_2$, is a colorless, crystalline powder. Freezing/Melting point = 160°C (decomposes). Hazard Identification (based on NFPA-704 M Rating System): Health 4, Flammability 1, Reactivity 0. Soluble in water.

Potential Exposure: This material is used as an analytical reagent for tin.

Permissible Exposure Limits in Air: The OSHA limit for organic arsenic compounds is 0.5 mg/m^3 TWA. For an 8-hour workshift. There is no NIOSH recommendation. The ACGIH TLV is 0.2 mg/m^3 TWA.

Determination in Air: See NIOSH Method #5022, Arsenic-organo.

Permissible Concentration in Water: See entry under arsenic.

Routes of Entry: Ingestion.

Harmful Effects and Symptoms

Short Term Exposure: Symptoms of arsenic poisoning usually appear one-half to one hour after ingestion, but may be delayed many hours. Symptoms include a sweetish metallic taste and garlicky odor; difficulty in swallowing; abdominal pain; vomiting and painful diarrhea; dehydration, thirst, and cramps; dizziness, stupor, and delirium, rapid heart beat, headache, skin disorders, and coma. Benzenearsonic acid is a deadly poison. The LD_{low} oral (rat) is 50 mg/kg; the oral LD_{50} mouse is 270 μg/kg.

Long Term Exposure: Chronic exposure to arsenic compounds can cause dermatitis and digestive disorders. Renal damage may develop.

Points of Attack: Skin, kidneys.

Medical Surveillance: Kidney function tests. Examination by a qualified allergist.

First Aid: If this chemical gets into the eyes, remove any contact lenses at once and irrigate immediately for at least 15 minutes, occasionally lifting upper and lower lids. Seek medical attention immediately. If this chemical contacts the skin, remove contaminated clothing and wash immediately with soap and water. Speed in removing material from skin is of extreme importance. Shampoo hair promptly if contaminated. Seek medical attention immediately. If this chemical has been inhaled, remove from exposure, begin rescue breathing (using universal precautions) if breathing has stopped and CPR if heart action has stopped. Transfer promptly to a medical facility. When this chemical has been swallowed, get medical attention. Give large quantities of water and induce vomiting. Do not make an unconscious person vomit.

Personal Protective Methods: Wear protective gloves and clothing to prevent any reasonable probability of skin contact. Safety equipment suppliers/manufacturers can provide recommendations on the most protective glove/clothing material for your operation. All protective clothing (suits, gloves, footwear, headgear) should be clean, available each day, and put on before work. Contact lenses should not be worn when working with this chemical. Wear dust-proof chemical goggles and face shield unless full facepiece respiratory protection is worn. Employees should wash immediately with soap when skin is wet or contaminated. Provide emergency showers and eyewash.

Respirator Selection: *Where there is no REL, at any detectable concentration:* SA:CF (any supplied-air respirator operated in a continuous-flow mode); or PAPROV [any powered, air-purifying respirator with organic vapor cartridge(s)]; CCRFOV [any air-purifying, full-facepiece respirator (gas mask) with a chin-style, front-or back-mounted acid gas canister]; or GMFOV [any air-purifying, full-facepiece respirator (gas mask) with a chin-style, front-or back-mounted organic vapor canister]; or SCBAF (any self-contained breathing apparatus with a full facepiece); or SAF (any supplied-air respirator with a full facepiece). *Emergency or planned entry into unknown concentrations or IDLH conditions:* SCBAF: PD,PP (any self-contained breathing apparatus that has a full facepiece and is operated in a pressure-demand or other positive-pressure mode); or SAF:PD,PP:ASCBA (any supplied-air respirator that has a full facepiece and is operated in a pressure-demand or other positive-pressure mode in combination with an auxiliary self-contained breathing apparatus operated in a pressure-demand or other positive-pressure mode). *Escape:* GMFOV [any air-purifying, full-facepiece respirator (gas mask) with a chin-style, front-or back-mounted organic vapor canister] or SCBAE (any appropriate escape-type, self-contained breathing apparatus).

Storage: Store in tightly closed containers in a cool, well ventilated area.

Shipping: Arsenic compounds, solid, n.o.s. including organic compounds of arsenic n.o.s. require a "Poison" label. They fall in Hazard Class 6.1 and Packing Group I.

Spill Handling: Evacuate and restrict persons not wearing protective equipment from area of spill or leak until cleanup is complete. Remove all ignition sources. Stay upwind; keep out of low areas. Wear self-contained (positive pressure if available) breathing apparatus and full protective clothing. Do not touch spilled material; stop leak if you can do it without risk. *Small liquid spills:* take up with sand or other noncombustible absorbent material into clean, dry container and cover; move containers from spill area. *Large spills:* dike far ahead of spill for later disposal. Collect powdered material in the most convenient and safe manner and deposit in sealed containers. Ventilate area of spill or leak after clean-up is complete. It may be necessary to contain and dispose of this chemical as a hazardous waste. If material or contaminated runoff enters waterways, notify downstream users of potentially contaminated waters. Contact your Department of Environmental Protection or your regional office of the federal EPA for specific recommendations. If employees are required to clean-up spills, they must be properly trained and equipped. OSHA 1910.120(q) may be applicable.

Fire Extinguishing: This material does not burn or burns with difficulty. Extinguish fire using agent suitable for surrounding fire. Use water in flooding quantities as fog. Avoid breathing dusts and fumes; keep upwind; wear self-contained breathing apparatus. Benzenearsonic acid emits poisonous fumes of arsenic when heated to decomposition. If material or contaminated runoff enters waterways, notify downstream users of potentially contaminated waters. Notify local health and fire officials and pollution control agencies. From a secure, explosion-proof location, use water spray to cool exposed containers. If cooling streams are ineffective (venting sound increases in volume and pitch, tank discolors, or shows any signs of deforming), withdraw immediately to a secure position. If employees are expected to fight fires, they must be trained and equipped in OSHA 1910.156.

References

U.S. Environmental Protection Agency, "Chemical Profile: Benzenearsonic Acid," Washington, DC, Chemical Emergency Preparedness Program (November 30, 1987)

Benzene, 1-(Chloromethyl)-4-Nitro-

Molecular Formula: $C_7H_6ClNO_2$

Common Formula: $ClCH_2C_6H_4NO_2$

Synonyms: Benzene, 1-(Chloromethyl)-4-Nitro-; p-(Chloromethyl)Nitrobenzene; 1-(Chloromethyl)-4-Nitrobenzene; 4-(Chloromethyl)Nitrobenzene; α-Chloro-p-Nitrotoluene; p-Nitrobenzyl Chloride

CAS Registry Number: 100-14-1

RTECS Number: XS9093000

DOT ID: UN 2433

Regulatory Authority

- SUPERFUND/EPCRA 40CFR355, Appendix B Extremely Hazardous Substances: TPQ = 500/10,000 lb (227/4,540 kg)
- SUPERFUND/EPCRA 40CFR302.4 Reportable Quantity (RQ): EHS, 500 lb (227 kg)

Cited in U.S. State Regulations: Florida (G), Massachusetts (G), New Jersey (G), Pennsylvania (G)

Description: Benzene, 1-(chloromethyl)-4-nitro-, $ClCH_2C_6H_4NO_2$, is a crystalline solid. Freezing/Melting point = 71°C.

Potential Exposure: Used in organic synthesis.

Incompatibilities: Can react with sulfuric acid.[52] Keep away from oxidizers.

Permissible Exposure Limits in Air: No standards set.

Permissible Concentration in Water: No criteria set.

Routes of Entry: Inhalation, ingestion, skin and/or eye contact.

Harmful Effects and Symptoms

Short Term Exposure: This chemical is a lachrymator.[52] Poisonous if swallowed or dust is inhaled. Can cause headaches, vomiting, cyanosis, difficulty in breathing. May cause skin irritation and sensitization.

Long Term Exposure: Prolonged chronic exposure to nitro compounds of aromatic hydrocarbons may cause liver and kidney damage.

Points of Attack: Liver, kidneys.

Medical Surveillance: Kidney and liver function tests.

First Aid: If this chemical gets into the eyes, remove any contact lenses at once and irrigate immediately for at least 15 minutes, occasionally lifting upper and lower lids. Seek medical attention immediately. If this chemical contacts the skin, remove contaminated clothing and wash immediately with soap and water. Seek medical attention immediately. If this chemical has been inhaled, remove from exposure, begin rescue breathing (using universal precautions) if breathing has stopped and CPR if heart action has stopped. Transfer promptly to a medical facility. When this chemical has been swallowed, get medical attention.

Personal Protective Methods: Wear protective gloves and clothing to prevent any reasonable probability of skin contact. Safety equipment suppliers/manufacturers can provide recommendations on the most protective glove/clothing material for your operation. All protective clothing (suits, gloves, footwear, headgear) should be clean, available each day, and put on before work. Contact lenses should not be worn when working with this chemical. Wear dust-proof chemical goggles and face shield unless full facepiece respiratory protection is worn. Employees should wash immediately with soap when skin is wet or contaminated. Provide emergency showers and eyewash.

Storage: Store a refrigerator or in a cool, dry place and protect from prolonged exposure to moisture.[52]

Shipping: Chloronitrotoluenes, solid require a "Keep away from Food" label. They fall in Hazard Class 6.1 and Packing Group III.

Spill Handling: Evacuate and restrict persons not wearing protective equipment from area of spill or leak until cleanup is complete. Remove all ignition sources. Do not touch spilled material; stop leak if you can do so without risk. *Small liquid spills:* absorb with sand or other non-combustible absorbent material and place into containers for later disposal. *Small dry spills:* with clean shovel, place material into clean, dry container and cover; move containers from spill area. *Large spills:* dike far ahead of spill for later disposal. Ventilate area of spill or leak after clean-up is complete. It may be necessary to contain and dispose of this chemical as a hazardous waste. If material or contaminated runoff enters waterways, notify downstream users of potentially contaminated waters. Contact your Department of Environmental Protection or your regional office of the federal EPA for specific recommendations. If employees are required to clean-up spills, they must be properly trained and equipped. OSHA 1910.120(q) may be applicable.

Fire Extinguishing: Use dry chemical, carbon dioxide, water spray, or foam extinguishers. Poisonous gases are produced in fire including nitrogen oxides. If material or contaminated runoff enters waterways, notify downstream users of potentially contaminated waters. Notify local health and fire officials and pollution control agencies. From a secure, explosion-proof location, use water spray to cool exposed containers. If cooling streams are ineffective (venting sound increases in volume and pitch, tank discolors, or shows any signs of deforming), withdraw immediately to a secure position. If employees are expected to fight fires, they must be trained and equipped in OSHA 1910.156.

References

U.S. Environmental Protection Agency, "Chemical Profile: Benzene, 1-(Chloromethyl)-4-Nitro," Washington, DC, Chemical Emergency Preparedness Program (November 30, 1987)

Benzene Sulfonyl Chloride

Molecular Formula: $C_6H_5ClO_2S$

Common Formula: $C_6H_5SO_2Cl$

Synonyms: Benzene sulfochloride; Benzene sulfonechloride; Benzene sulfone-chloride; Benzenesulfonic (acid) chloride; Benzenesulfonic acid chloride; Benzenesulfonyl chloride; Benzenosulfochlorek (Polish); Benzenosulphochloride; BSC-refined D

CAS Registry Number: 98-09-9

RTECS Number: DB8750000

DOT ID: UN 2225

Regulatory Authority

- Air Pollutant Standard Set (former USSR)[43]
- Water Pollution Standard Proposed (former USSR)[43]
- EPA HAZARDOUS WASTE NUMBER (RCRA No.): U020
- RCRA Land Ban Waste Restrictions
- SUPERFUND/EPCRA 40CFR302.4 Reportable Quantity (RQ): CERCLA, 100 lb (45.4 kg)
- TSCA 40CFR716.120(a)

Cited in U.S. State Regulations: Kansas (G), Louisiana (G), Massachusetts (G), New Hampshire (G), New Jersey (G), Pennsylvania (G), Vermont (G), Virginia (G), Washington (G), Wisconsin (G).

Description: Benzenesulfonyl chloride, $C_6H_5SO_2Cl$, is a colorless oily liquid with a pungent odor. Boiling point = 251 – 252°C (decomposes). Freezing/Melting point = 15°C. Flash point = 130°C. Hazard Identification (based on NFPA-704 M Rating System): Health 3, Flammability 1, Reactivity 1. Slightly soluble in water.

Potential Exposure: It is used as a chemical intermediate for benzenesulfonamides, thiophenol, glybuzole (hypoglycemic agent), N-2-chloroehtylamides, benzonitrile; for its esters – useful as insecticides, and miticides.

Incompatibilities: Violent reaction with strong oxidizers, dimethyl sulfoxide and methyl formamide. It is very reactive with bases and many organic compounds. Incompatible with ammonia, aliphatic amines. Water contact forms hydrochloric and chlorosulfonic acids. Aqueous solutions of this chemical are strong acids that react violently with bases. Attacks metals in presence of moisture.

Permissible Exposure Limits in Air: The former USSR-UNEP/IRPTC project[43] has set a MAC of 0.1 mg/m^3 in workplace air.

Permissible Concentration in Water: The former USSR-UNEP/PRPTC project[43] has set a Mac of 0.5 mg/l in water used for domestic purposes.

Routes of Entry: Inhalation, ingestion, skin and/or eye contact.

Harmful Effects and Symptoms

Short Term Exposure: Contact may cause severe irritation and burns to skin and eyes. Breathing this chemical may cause liver damage. Symptoms may include allergic reactions, and severe shock. Benzene sulfonyl chloride is poisonous; may be fatal if inhaled, swallowed or absorbed through the skin. Higher exposures can cause pulmonary edema, a medical emergency that can be delayed for several hours. This can cause death. Reversible toxic damage to the liver is possible after dermal exposure. The oral LD_{50} rat is 1,960 mg/kg.[9]

Long Term Exposure: May cause chronic irritation of the air passages and lungs, bronchitis with cough and phlegm. Repeated exposure may cause liver damage. Repeated skin contact may cause dry skin, redness, rash, and sores.

Points of Attack: Liver, lungs.

Medical Surveillance: Lung function tests, liver function tests. Consider x-ray following acute overexposure.

First Aid: If this chemical gets into the eyes, remove any contact lenses at once and irrigate immediately for at least 15

minutes, occasionally lifting upper and lower lids. Seek medical attention immediately. If this chemical contacts the skin, remove contaminated clothing and wash immediately with soap and water. Seek medical attention immediately. If this chemical has been inhaled, remove from exposure, begin rescue breathing (using universal precautions) if breathing has stopped and CPR if heart action has stopped. Transfer promptly to a medical facility. When this chemical has been swallowed, get medical attention. Do not induce vomiting. Medical observation is recommended for 24 – 48 hours after breathing overexposure, as pulmonary edema may be delayed. As first aid for pulmonary edema, a doctor or authorized paramedic may consider administering a corticosteroid spray.

Personal Protective Methods: Wear solvent-resistant gloves and clothing to prevent any reasonable probability of skin contact. Safety equipment suppliers/manufacturers can provide recommendations on the most protective glove/clothing material for your operation. All protective clothing (suits, gloves, footwear, headgear) should be clean, available each day, and put on before work. Contact lenses should not be worn when working with this chemical. Wear splash-proof chemical goggles and face shield unless full facepiece respiratory protection is worn. Employees should wash immediately with soap when skin is wet or contaminated. Remove nonimpervious clothing immediately if wet or contaminated. Provide emergency showers and eyewash.

Respirator Selection: *Where there is no REL, at any detectable concentration:* SA:CF (any supplied-air respirator operated in a continuous-flow mode); or PAPROV [any powered, air-purifying respirator with organic vapor cartridge(s)]; CCRFOV [any air-purifying, full-facepiece respirator (gas mask) with a chin-style, front-or back-mounted acid gas canister]; or GMFOV [any air-purifying, full-facepiece respirator (gas mask) with a chin-style, front-or back-mounted organic vapor canister]; or SCBAF (any self-contained breathing apparatus with a full facepiece); or SAF (any supplied-air respirator with a full facepiece). *Emergency or planned entry into unknown concentrations or IDLH conditions:* SCBAF:PD,PP (any self-contained breathing apparatus that has a full facepiece and is operated in a pressure-demand or other positive-pressure mode); or SAF:PD,PP:ASCBA (any supplied-air respirator that has a full facepiece and is operated in a pressure-demand or other positive-pressure mode in combination with an auxiliary self-contained breathing apparatus operated in a pressure-demand or other positive-pressure mode). *Escape:* GMFOV [any air-purifying, full-facepiece respirator (gas mask) with a chin-style, front-or back-mounted organic vapor canister] or SCBAE (any appropriate escape-type, self-contained breathing apparatus).

Storage: Store in tightly closed containers in a cool, well ventilated area. Metal containers involving the transfer of this chemical should be grounded and bonded. Drums must be equipped with self-closing valves, pressure vacuum bungs, and flame arresters. Use only non-sparking tools and equipment, especially when opening and closing containers of this chemical. Sources of ignition such as smoking and open flames, are prohibited where this chemical is used, handled, or stored in a manner that could create a potential fire or explosion hazard.

Shipping: This compound requires a "Corrosive" label.[19] It falls in Hazard Class 8 and Packing Group III.

Spill Handling: Evacuate and restrict persons not wearing protective equipment from area of spill or leak until cleanup is complete. Remove all ignition sources. Ventilate area of spill or leak. Do not breathe vapors. Wear proper respiratory protection, eye protection and full protective clothing. Do not touch spilled material; stop leak, use water spray to reduce vapors. *Small spills:* take up with sand or other noncombustible absorbent material and place into containers for later disposal. *Small dry spills:* with clean shovel place material into clean, dry container and cover; move containers from spill area. *Large spills:* dike far ahead of spill for later disposal. It may be necessary to contain and dispose of this chemical as a hazardous waste. If material or contaminated runoff enters waterways, notify downstream users of potentially contaminated waters. Contact your Department of Environmental Protection or your regional office of the federal EPA for specific recommendations. If employees are required to clean-up spills, they must be properly trained and equipped. OSHA 1910.120(q) may be applicable.

Fire Extinguishing: This chemical may burn but does not easily ignite. Use dry chemical, carbon dioxide, or foam extinguishers. *Do not use water.* Poisonous gases are produced in fire including hydrogen chloride and sulfur dioxide. If material or contaminated runoff enters waterways, notify downstream users of potentially contaminated waters. Notify local health and fire officials and pollution control agencies. Containers may explode in fire. From a secure, explosion-proof location, use water spray to cool exposed containers. If cooling streams are ineffective (venting sound increases in volume and pitch, tank discolors, or shows any signs of deforming), withdraw immediately to a secure position. If employees are expected to fight fires, they must be trained and equipped in OSHA 1910.156.

References

U.S. Environmental Protection Agency, "Chemical Profile: Benzene Sulfonyl Chloride," Washington, DC, Chemical Emergency Preparedness Program (October 31, 1985)

New Jersey Department of Health and Senior Services, "Hazardous Substance Fact Sheet: Benzene Sulfonyl Chloride," Trenton, NJ (February 1987)

Benzidine

Molecular Formula: $C_{12}H_{12}N_2$

Common Formula: $NH_2C_6H_4C_6H_4NH_2$

Synonyms: Bencidina (Spanish); Benzidin (Czech); Benzidina (Italian); Benzydyna (Polish); *p,p*-Bianiline; 4,4'-Bianiline; (1,1'-Bifenyl)-4,4'-diamine; (1,1'-Biphenyl)-4,4'diamine; (1,1'-Biphenyl)-4,4'-diamine (9CI); 4,4'-Biphenyldiamine; Biphenyl, 4,4'-diamino-; 4,4'-Biphenylenediamine; C.I. 37225; C.I. azoic diazo; C.I. azoic diazo component 112; Component 112; *p,p'*-Diaminobiphenyl; 4,4'-Diamino-1,1'-biphenyl; 4,4'-Diaminobiphenyl; *p*-Diaminodiphenyl; 4,4'-Diaminodiphenyl; *p,p'*-Dianiline; 4,4'-Diphenylenediamine; Fast Corinth base B; NCI-C03361

CAS Registry Number: 92-87-5

RTECS Number: DC9625000

DOT ID: UN 1885

EEC Number: 612-042-00-2

Regulatory Authority

- Carcinogen (Human Positive) (IARC)[9] (DFG)[3]
- Banned or Severely Restricted (Many Countries) (UN)[13][35]
- Very Toxic Substance (World Bank)[15]
- OSHA, 29CFR1910 Specifically Regulated Chemicals (See CFR1910.1010)
- Air Pollutant Standard Set (Several States)[60]
- CLEAN AIR ACT: Hazardous Air Pollutants (Title I, Part A, Section 112)
- CLEAN WATER ACT: 40CFR423, Appendix A, Priority Pollutants; 40CFR401.15 Section 307 Toxic Pollutants; Section 313 Water Priority Chemicals (57FR41331, 9/9/92)
- EPA HAZARDOUS WASTE NUMBER (RCRA No.): U021
- RCRA, 40CFR261, Appendix 8 Hazardous Constituents
- SUPERFUND/EPCRA 40CFR302.4 Reportable Quantity (RQ): CERCLA, 1 lb (0.45 kg)
- EPCRA Section 313 Form R *de minimis* concentration reporting level: 0.1%
- CALIFORNIA'S PROPOSITION 65: Carcinogen as benzidine and its salts; benzidine based dyes

Cited in U.S. State Regulations: Alaska (G), California (G), Florida (G), Illinois (G), Kansas (G, W), Louisiana (G), Maine (G), Maryland (G), Massachusetts (G), Michigan (G), New Hampshire (G), New Jersey (G), New York (G, A), North Carolina (A), North Dakota (A), Oklahoma (G), Pennsylvania (G, A), Rhode Island (G, A), South Carolina (A), Vermont (G), Virginia (G, A), Washington (G), West Virginia (G), Wisconsin (G).

Description: Benzidine, $NH_2C_6H_4C_6H_4NH_2$, is a white, grayish-yellow crystals or powder. Turns brownish-red on exposure to air and light. Freezing/Melting point = 128°C. Boiling point = about 400°C. Hazard Identification (based on NFPA-704 M Rating System): Health 2, Flammability 1, Reactivity 0. Soluble in water. Octanol/water partition coefficient as Log P_{ow} = 1.34.

Potential Exposure: Benzidine is used primarily in the manufacture of azo dyestuffs; there are over 250 of these produced. Other uses, including some which may have been discontinued, are in the rubber industry as a hardener, in the manufacture of plastic films, for detection of occult blood in feces, urine, and body fluids, in the detection of H_2O_2 in milk, in the production of security paper, and as a laboratory reagent in determining HCN, sulfate, nicotine, and certain sugars. No substitute has been found for its use in dyes. Free benzidine is present in the benzidine-derived azo dyes. According to industry, quality control specifications require that the level not exceed 20 ppm and in practice the level is usually below 10 ppm. Regulations in USA concerning this chemical define strict procedures to avoid worker contact: mixture containing 0.1% or more must be maintained in isolated or closed systems, employees must observe special personal hygiene rules, and certain procedures must be followed in case of emergencies (IARC).

Incompatibilities: Violent reaction with strong oxidizing materials. Contact with red fuming nitric acid may cause fire. Oxidizes in air.

Permissible Exposure Limits in Air: Benzidine and its salts are included in a federal standard for carcinogens; all contact with them should be avoided. Cutaneous absorption is possible. ACGIH lists benzidine as a human carcinogen with the notation that skin absorption may be significant. DFG[3] also lists benzidine as a carcinogen with no numerical limit. However, a number of states have set guidelines or standards for benzidine in ambient air[60] ranging from zero (New York, North Dakota, South Carolina, Virginia) to 0.014 nanograms/m^3 (North Carolina) to 0.02 nanograms/m^3 (Rhode Island) to 30 μg/m^3 (30,000 nanograms/m^3) (Pennsylvania).

Determination in Air: Collection on a filter followed by colorimetric analysis. See NIOSH Method #5013.

Permissible Concentration in Water: To protect freshwater aquatic life -2,500 μg/l on an acute basis; insufficient data to yield a value for saltwater aquatic life. To protect human health-preferably zero. An additional lifetime cancer risk of 1 in 100,000 results from a concentration of 0.0012 μg/l.[6] Kansas has set a guideline in drinking water of 0.0015 μg/l.[61]

Determination in Water: High performance liquid chromatography (EPA Method 605) or an oxidation/colorimetric method using Chloramine T (available from EPA) or a gas chromatography/mass spectrometric method (EPA Method 625).

Routes of Entry: Inhalation, percutaneous absorption, and ingestion of dust.

Harmful Effects and Symptoms

Short Term Exposure: Irritates the eyes. Corrosive to the skin and respiratory tract. Easily absorbed thought the skin. Inhalation can irritate the lungs causing coughing and/or shortness of breath. Higher exposures can cause pulmonary edema, a medical emergency that can be delayed for several hours. This can cause death. Ingestion: animal studies suggest 0.01 – 0.08% in food may cause a decrease in liver, kidney and body weight; an increase in spleen weight; swelling of the liver and blood in the urine.

Long Term Exposure: May cause skin allergy. Benzidine is a human carcinogen. Exposure may cause an increase in urination, blood in the urinary tract tumors. Can affect the blood and cause liver and kidney damage.

Points of Attack: Skin, bladder, kidney, liver.

Medical Surveillance: Placement and periodic examinations should include an evaluation of exposure to other carcinogens; use of alcohol, smoking, and medications: and family history. Special attention should be given on a regular basis to urine sediment and cytology. If red cells or positive smears are seen, cystoscopy should be done at once. The general health of exposed persons should also be evaluated in periodic examinations. Urine cytology. Evaluation by a certified allergist.

First Aid: If this chemical gets into the eyes, remove any contact lenses at once and irrigate immediately for at least 15 minutes, occasionally lifting upper and lower lids. Seek medical attention immediately. If this chemical contacts the skin, remove contaminated clothing and wash immediately with soap and water. Seek medical attention immediately. If this chemical has been inhaled, remove from exposure, begin rescue breathing (using universal precautions) if breathing has stopped and CPR if heart action has stopped. Transfer promptly to a medical facility. When this chemical has been swallowed, get medical attention. Use gastric lavage if ingested followed by saline catharsis. Medical observation is recommended for 24 – 48 hours after breathing overexposure, as pulmonary edema may be delayed. As first aid for pulmonary edema, a doctor or authorized paramedic may consider administering a corticosteroid spray.

Personal Protective Methods: These are designed to supplement engineering controls (such as a prohibition on open-vessel operation) and to prevent all skin or respiratory contact. Full body protective plastic clothing and butyl rubber gloves should also be used. On exit from a regulated area employees should shower and change into street clothes, leaving their protective clothing and equipment at the point of exit to be placed in impervious containers at the end of the work shift for decontamination or disposal. Effective methods should be used to clean and decontaminate gloves and clothing.

Respirator Selection: NIOSH: *At any detectable concentration:* SCBAF:PD,PP (any MSHA/NIOSH approved self-contained breathing apparatus that has a full facepiece and is operated in a pressure-demand or other positive-pressure mode); or SAF:PD,PP:ASCBA (any supplied-air respirator that has a full facepiece and is operated in a pressure-demand or other positive-pressure mode in combination with an auxiliary, self-contained breathing apparatus operated in a pressure-demand or other positive pressure mode). *Escape:* HiEF [any air-purifying, full-facepiece respirator (gas mask) with a chin-style, front- or back-mounted organic vapor canister having a high-efficiency particulate filter]; or SCBAE (any appropriate escape-type, self-contained breathing apparatus).

Storage: Store in a dark, cool, well-ventilated area in closed, sealed containers. Keep out of sunlight and away from heat. A regulated, marked area should be established where this chemical is handled, used, or stored in compliance with OSHA standard 1910.1045.

Shipping: The label required is "Poison." Benzidine falls in UN/DOT Hazard Class 6.1 and Packing Group II.[19][20]

Spill Handling: Evacuate and restrict persons not wearing protective equipment from area of spill or leak until cleanup is complete. Remove all ignition sources. Cover the spill with a mixture of 9 parts sand to 1 part soda ash. Collect powdered material in the most convenient and safe manner and deposit in sealed containers. Ventilate area of spill or leak after clean-up is complete. It may be necessary to contain and dispose of this chemical as a hazardous waste. If material or contaminated runoff enters waterways, notify downstream users of potentially contaminated waters. Contact your Department of Environmental Protection or your regional office of the federal EPA for specific recommendations. If employees are required to clean-up spills, they must be properly trained and equipped. OSHA 1910.120(q) may be applicable.

Fire Extinguishing: Benzidine may burn, but does not readily ignite. Use dry chemical, carbon dioxide, or foam extinguishers. Use water spray in flooding quantities as fog. Poisonous gases are produced in fire including nitrogen oxides. If material or contaminated runoff enters waterways, notify downstream users of potentially contaminated waters. Notify local health and fire officials and pollution control agencies. From a secure, explosion-proof location, use water spray to cool exposed containers. If cooling streams are ineffective (venting sound increases in volume and pitch, tank discolors, or shows any signs of deforming), withdraw immediately to a secure position. If employees are expected to fight fires, they must be trained and equipped in OSHA 1910.156.

Disposal Method Suggested: Incineration; oxides of nitrogen are removed from the effluent gas by scrubber, catalytic or thermal device.[22] Package spill residues and sorbent media in 17h epoxy-lined drums and move to an EPA-approved disposal site. Treatment may include destruction by potassium permanganate oxidation, high-temperature incineration, or microwave plasma methods. Encapsulation by organic polyester resin or silicate fixation. These disposal procedures should be confirmed with responsible environmental engineering and regulatory officials.

References

U.S. Environmental Protection Agency, Benzidine: Ambient Water Quality Criteria, Washington, DC (1980)

U.S. Environmental Protection Agency, Status Assessment of Toxic Chemicals: Benzidine, Report, EPA-600/2-79-210E, Washington, DC (December 1979)

U.S. Environmental Protection Agency, Reviews of the Environmental Effects of Pollutants: II, Benzidine, Cincinnati, Ohio, Health Effects Research Laboratory, Report No. EPA-600/1-78-024 (1978)

U.S. Environmental Protection Agency, Benzidine, Health and Environmental Effects Profile No. 16, Washington, DC, Office of Solid Waste (April 30, 1980)

U.S. Environmental Protection Agency, Benzidine, Its Congeners and Their Derivative Dyes and Pigments, TSCA Chemical Assessment Series; Preliminary Risk Assessment Phase I, Report EPA-560/11-80-019, Washington, DC (June 1980)

Sax, N. I., Ed., "Dangerous Properties of Industrial Materials Report" 1, No. 5, 38–39 (1981) and 2, No. 4, 38–43, and 3, No. 4, 32–37 (1983)

New Jersey Department of Health and Senior Services, "Hazardous Substance Fact Sheet: Benzidine," Trenton, NJ (July 1998)

New York State Department of Health, "Chemical Fact Sheet: Benzidine," Albany, NY, Bureau of Toxic Substance Assessment (January 1988)

U.S. Public Health Service, "Toxicological Profile for Benzidine," Atlanta, Georgia, Agency for Toxic Substances and Disease Registry (December 1988)

Benzo[b]fluoranthene

Molecular Formula: $C_{20}H_{12}$

Synonyms: B(b)F; 3,4-Benz[e]acephenanthrylene; Benz[e]acephenanthrylene; 2,3-Benzfluoranthene; 3,4-Benzfluoranthene; 2,3-Benzfluoranthrene; 3,4-Benzfluoranthrene; 2,3-Benzofluoranthene; 3,4-Benzofluoranthene; 4,5-Benzofluoranthene; Benzo[e]fluoranthene; NSC 89265

CAS Registry Number: 205-99-2

RTECS Number: CU1400000

DOT ID: UN 2811

Regulatory Authority

- Carcinogen (Animal Positive) (IARC)[9] (DFG)[3]
- OSHA, 29CFR1910 Specifically Regulated Chemicals (See CFR1910.1002) as coal tar pitch volatiles
- Air Pollutant Standard Set (ACGIH)[1] (HSE)[33] (NIOSH/OSHA)[2]
- Water Pollutant Standard Set (EPA) (Mexico)
- CLEAN WATER ACT: 40CFR423, Appendix A, Priority Pollutants; 40CFR401.15 Section 307 Toxic Pollutants
- RCRA, 40CFR261, Appendix 8 Hazardous Constituents, waste number not listed
- RCRA 40CFR268.48; 61FR15654, Universal Treatment Standards: Wastewater. *Note*: Difficult to distinguish from benzo(k) fluoranthene (mg/l), 0.11; Nonwastewater (mg/kg), 6.8
- RCRA 40CFR264, Appendix 9; TSD Facilities Ground Water Monitoring List. Suggested test method(s) (PQL μg/l): 8100 (200); 8270 (10)
- SUPERFUND/EPCRA 40CFR302.4 Reportable Quantity (RQ): CERCLA, 1 lb (0.454 kg)
- EPCRA (Section 313): as polycyclic aromatic compounds: Form R *de minimis* concentration reporting level: 0.1%
- Canada, WHMIS, Ingredients Disclosure List

Cited in U.S. State Regulations: California (A, G), Florida (G), Illinois (G), Kansas (G, W), Louisiana (G), Maine (G), Massachusetts (G), Minnesota (G), New Hampshire (G), New Jersey (G), Pennsylvania (G), Vermont (G), Virginia (G), Washington (G), West Virginia (G), Wisconsin (G).

Description: B(b)F, $C_{20}H_{12}$, is a colorless, needle-shaped solid. Freezing/Melting point = 167 – 168°C. Hazard Identification (based on NFPA-704 M Rating System): Health 3, Flammability 1, Reactivity 0. Insoluble in water. Octanol/water partition coefficient as Log P_{ow} = 6.0.

Potential Exposure: Benzo(b) fluoranthene [B(b)F] is a chemical substance formed during the incomplete burning of fossil fuel, garbage, or any organic matter and is found in smoke in general; it is carried into the air, where it condenses onto dust particles and is distributed into water and soil and on crops. B(b)F is a polycyclic aromatic hydrocarbon (PAH) and a component of coal tar pitch used in industry as a binder for electrodes. It is also a component of creosote, which is used to preserve wood. PAHs are also found in limited amounts in bitumens and asphalt used for paving, roofing, and insulation. B(b)F has some use as a research chemical. It is available from some specialty chemical firms in low quantities (25 – 100 mg).

Incompatibilities: Strong oxidizers.

Permissible Exposure Limits in Air: The National Institute for Occupational Safety and Health (NIOSH) concluded that inhalation of coal products can increase the risk of lung and skin cancer in workers and recommended an occupational exposure limit for coal tar products of 0.1 milligram of PAHs per cubic meter of air for a 10-hour workday, 40-hour workweek. Although the Secretary of Labor has taken the position that no safe occupational exposure can be established for a carcinogen, the Occupational Safety and Health Administration (OSHA) established a legally enforceable limit of 0.2 milligram of all PAHs per cubic meter of air. ACGIH[1] has set a similar limit for coal tar pitch volatiles as has HSE.[33]

Determination in Air: The most commonly used analytical methods for determining B(b)F in environmental samples are column gas chromatography (GC) and high-performance liquid chromatography (HPLC). The EPA report cited below summarizes common analytical methods, detection limits, and accuracy (percent recovery) for the determination of B(b)F in air, water, soil, and food. HPLC (Method 5506) or GC/MS (Method 5515) are the methods recommended by NIOSH[18] for analyzing PAHs in workplace air. The use of a high-volume sampler to sample a large quantity of air allows for the detection of small amounts of PAH.

Permissible Concentration in Water: The Environmental Protection Agency (EPA) developed guidelines for permissible levels of carcinogenic PAHs in ambient water based on data from a carcinogenicity study on benzo(a) pyrene. EPA recommended that for the protection of human health from potential carcinogenic effects of exposure to PAHs through the ingestion of contaminated water, fish, and shellfish, the ambient water concentrations of total carcinogenic PAHs may not be possible to achieve because of naturally occurring levels in the environment; EPA consequently estimated ambient water concentrations of total carcinogenic PAHs at 28, 2.8, and 0.28 nanograms per liter of water, corresponding to incremental lifetime cancer risk levels of one additional cancer case for every 100,000, 1 million, and 10 million people exposed, respectively, based on consumption of contaminated water, fish, and shellfish as an aid for developing water quality regulations. (One nanograms is one-billionth of a gram). Kansas has set a guideline for drinking water of 0.029 μg/l.

Determination in Water: The analytical methods required by EPA for the analysis of B(b)F in water are procedures 610 (HPLC/FS), 625 (GC/MS), and 1625 (GC/MS). These are required test procedures under the Clean Water Act for municipal and industrial wastewater-discharging sites. GC/MS is also the method recommended be the EPA Contract Laboratory Program for analysis of B(b)F and other PAHs in water and soil.

Routes of Entry: Inhalation, passing through the skin. B(b)F enters the body primarily through breathing polluted air containing the compound or from inhaling tobacco smoke. Drinking contaminated water or accidentally swallowing soil or dust particles containing B(b)F can also result in B(b)F entering the body. Smoking or charcoal-broiling food can cause B(b)F to be formed. B(b)F can then enter the body through consumption of the contaminated food. Under normal conditions of environmental exposure, B(b)F does not usually enter the body through the skin; however, small amounts could enter the body if contact occurs with soil that contains high levels of B(b)F (e.g., near a hazardous waste site) or if contact occurs with heavy oils containing B(b)F.

Harmful Effects and Symptoms

B(b)F is a toxic chemical that causes cancer in laboratory animals when it is applied to their skin. B(b)F has not been studied adequately, and, consequently, it is not known whether it causes cancer if it is breathed or ingested, or if it causes harmful effects other than cancer. Because B(b)F causes skin cancer in animals, it is possible that humans exposed in the same manner or by other routes could develop cancer as well.

Short Term Exposure: The substance can be absorbed into the body by inhalation of its aerosol and through the skin. Skin contact may cause irritation or skin allergy which is greatly aggravated by sunlight on contaminated skin. Eye contact or "fume" exposure may cause irritation and a reaction greatly aggravated by sunlight during or shortly following exposure. Direct contact or "fumes" can cause irritation of the nose, throat, and bronchial tubes, and skin irritation, redness, and possible swelling.

Long Term Exposure: Repeated skin contact may cause thickening, pigment changes and skin growths, including warts, pimples and skin cancer. B(b)F is a probable carcinogen in humans. May cause eye allergy. Repeated inhalation of fumes, especially with heating may cause chronic bronchitis with cough and phlegm.

Points of Attack: Skin, eyes.

Medical Surveillance: Periodic skin examination.

First Aid: If this chemical gets into the eyes, remove any contact lenses at once and irrigate immediately for at least 15 minutes, occasionally lifting upper and lower lids. Seek medical attention immediately. If this chemical contacts the skin, remove contaminated clothing and wash immediately with soap and water. Seek medical attention immediately. If this chemical has been inhaled, remove from exposure, begin rescue breathing (using universal precautions) if breathing has stopped and CPR if heart action has stopped. Transfer promptly to a medical facility. When this chemical has been swallowed, rinse mouth and get medical attention.

Personal Protective Methods: Sun screen lotion or creams with a high ability to screen out ultraviolet light can help prevent skin allergic reactions. These may need frequent reapplication if the user is sweating. Ultraviolet-screening sunglasses can help with eye allergic reactions. Consult your doctor or pharmacist in selecting these. Wear protective gloves and clothing to prevent any reasonable probability of skin contact. Safety equipment suppliers/manufacturers can provide recommendations on the most protective glove/clothing material for your operation. All protective clothing (suits, gloves, footwear, headgear) should be clean, available each day, and put on before work. Contact lenses should not be worn when working with this chemical. Wear dust-proof chemical goggles and face shield unless full facepiece respiratory protection is worn. Employees should wash immediately with soap when skin is wet or contaminated. Provide emergency showers and eyewash.

Respirator Selection: NIOSH: *At any detectable concentration:* SCBAF:PD,PP (any MSHA/NIOSH approved self-contained breathing apparatus that has a full facepiece and is operated in a pressure-demand or other positive-pressure mode); or SAF:PD,PP:ASCBA (any supplied-air respirator that has a full facepiece and is operated in a pressure-demand or other positive-pressure mode in combination with an auxiliary, self-contained breathing apparatus operated in a pressure-demand or other positive pressure mode). *Escape:* GMFOVHiE [any air-purifying, full-facepiece respirator (gas mask) with a chin-style, front- or back-mounted organic vapor canister having a high-efficiency particulate filter]; or SCBAE (any appropriate escape-type, self-contained breathing apparatus).

Storage: Store in a cool, dry place. A regulated, marked area should be established where this chemical is handled, used, or stored in compliance with OSHA standard 1910.1045.

Shipping: Poisonous solids, n.o.s. require a "Poison" label. In Packing Group II. The UN/DOT Hazard Class is 6.1.[19][20]

Spill Handling: Evacuate and restrict persons not wearing protective equipment from area of spill or leak until cleanup is complete. Remove all ignition sources. Collect powdered material in the most convenient and safe manner and deposit in sealed containers. Ventilate area of spill or leak after clean-up is complete. It may be necessary to contain and dispose of this chemical as a hazardous waste. If material or contaminated runoff enters waterways, notify downstream users of potentially contaminated waters. Contact your Department of Environmental Protection or your regional office of the federal EPA for specific recommendations. If employees are required to clean-up spills, they must be properly trained and equipped. OSHA 1910.120(q) may be applicable.

Fire Extinguishing: Use dry chemical, carbon dioxide, water spray, or alcohol foam extinguishers. Poisonous gases are produced in fire. If material or contaminated runoff enters waterways, notify downstream users of potentially contaminated waters. Notify local health and fire officials and pollution control agencies. From a secure, explosion-proof location, use water spray to cool exposed containers. If cooling streams are ineffective (venting sound increases in volume and pitch, tank discolors, or shows any signs of deforming), withdraw immediately to a secure position. If employees are expected to fight fires, they must be trained and equipped in OSHA 1910.156.

Disposal Method Suggested: Atomize into incinerator with a flammable liquid.

References

Agency for Toxic Substance and Disease Registry, U.S. Public Health Service, "Toxicological Profile for Benzo [b] fluoranthene," Atlanta, Georgia, ATSDR (November 1987)

Sax, N. I., Ed., "Dangerous Properties of Industrial Materials Report" 5, No. 1, 37–39 (1985)

New Jersey Department of Health and Senior Services, Hazardous Substance Fact Sheet, Benzo [f]fluororanthene, Trenton NJ (May 1988)

Benzoic Acid

Molecular Formula: $C_7H_6O_2$

Common Formula: C_6H_5COOH

Synonyms: Acide benzoique (French); Acido benzoico (Spanish); Benzenecarboxylic acid; Benzeneformic acid; Benzenemethanoic acid; Benzoate; Benzoesaeure (German); Carboxybenzene; Carboxylbenzene; Dracyclic acid; Kyselina benzoova (Czech); Phenyl carboxylic acid; Phenylformic acid; Retarder BA; Retarder Bax; Salvo; Tennplas

CAS Registry Number: 65-85-0

RTECS Number: DG0875000

DOT ID: UN 9094

EINECS Number: 200-618-2

Regulatory Authority

- Water Pollution Standard Proposed (former USSR)[35]
- CLEAN WATER ACT: Section 311 Hazardous Substances/ RQ 40CFR117.3 (same as CERCLA, see below)
- SUPERFUND/EPCRA 40CFR302.4 Reportable Quantity (RQ): CERCLA, 5,000 lb (2,270 kg)
- Canada, WHMIS, Ingredients Disclosure List

Cited in U.S. State Regulations: California (G), Florida (G), Illinois (G), Massachusetts (G), New Hampshire (G), New Jersey (G), New York (G), Pennsylvania (G).

Description: Benzoic acid, C_6H_5COOH, is a white crystalline or flaky solid with a faint, pleasant odor. Freezing/Melting point = 122.4°C. Boiling point = 249°C. Flash point of 121°C. Hazard Identification (based on NFPA-704 M Rating System): Health 2, Flammability 1, Reactivity 0. Highly soluble in water.

Potential Exposure: Those involved in the manufacture of benzoates; plasticizers, benzoyl chloride, alkyd resins, in the manufacture of food preservatives, in use as a dye binder in calico printing; in curing of tobacco, flavors, perfumes, dentifrices, standard in analytical chemistry, antifungal agent.

Incompatibilities: Incompatible with strong oxidizers, caustics, ammonia, amines, isocyanates.

Permissible Exposure Limits in Air: No standards set.

Permissible Concentration in Water: The former USSR has proposed a MAC of 0.6 mg/l in surface water.[35]

Routes of Entry: Inhalation and ingestion.

Harmful Effects and Symptoms

Short Term Exposure: Irritating to skin, eyes (possibly severe), and mucous membranes. Skin contact may cause irritation, skin rash, or burning feeling on contact. Ingestion causes nausea and G.I. troubles. For most people, ingestion of 1/10 – 2/10 ounce will have no effect although some sensitive people may experience allergic reactions. Larger amounts may cause stomach upset. Information from animal studies show that about 6 ounces may be lethal to a 150 pound person.

Long Term Exposure: Repeated or prolonged contact may cause skin sensitization. Mutation data reported.

Points of Attack: Skin, eyes, and mucous membranes.

First Aid: If this chemical gets into the eyes, remove any contact lenses at once and irrigate immediately for at least 15 minutes, occasionally lifting upper and lower lids. Seek medical attention immediately. If this chemical contacts the skin, remove contaminated clothing and wash immediately with soap and water. Seek medical attention immediately. If this chemical has been inhaled, remove from exposure, begin rescue breathing (using universal precautions) if breathing has stopped and CPR if heart action has stopped. Transfer promptly to a medical facility. When this chemical has been swallowed, get medical attention. Give large quantities of water and induce vomiting. Do not make an unconscious person vomit.

Personal Protective Methods: Wear protective gloves and clothing to prevent any reasonable probability of skin contact. Safety equipment suppliers/manufacturers can provide recommendations on the most protective glove/clothing material for your operation. All protective clothing (suits, gloves, footwear, headgear) should be clean, available each day, and put on before work. Contact lenses should not be worn when working with this chemical. Wear dust-proof chemical goggles and face shield unless full facepiece respiratory protection is worn. Employees should wash immediately with soap when skin is wet or contaminated. Provide emergency showers and eyewash.

Respirator Selection: Cannister mask recommended.

Storage: Store away from excessive heat. Benzoic Acid must be stored to avoid contact with strong oxidizers (such as chlorine, bromine and fluorine) since violent reactions occur.

Shipping: Benzoic acid is not specifically cited in DOT's performance-oriented packaging standards[19] as regards label requirements and maximum allowable shipping quantities.

Spill Handling: Evacuate and restrict persons not wearing protective equipment from area of spill or leak until cleanup is complete. Remove all ignition sources. Vapor may explode if ignited in an enclosed area. Solutions should be neutralized with soda ash. Collect powdered material in the most convenient and safe manner and deposit in sealed containers. Then flush the area with water. Ventilate area of spill or leak after clean-up is complete. It may be necessary to contain and dispose of this chemical as a hazardous waste. If material or contaminated runoff enters waterways, notify downstream users of potentially contaminated waters. Contact your Department of Environmental Protection or your regional office of the federal EPA for specific recommendations. If employees are required to clean-up spills, they must be properly trained and equipped. OSHA 1910.120(q) may be applicable.

Fire Extinguishing: Benzoic Acid is a combustible solid. Benzoic Acid may burn, but does not readily ignite. High levels of dust may form an explosive concentration in air. Use dry chemical, CO_2, or water spray extinguishers. Poisonous gases are produced in fire. If material or contaminated runoff enters waterways, notify downstream users of potentially contaminated waters. Notify local health and fire officials and pollution control agencies. Containers may explode in fire. From a secure, explosion-proof location, use water spray to cool exposed containers. If cooling streams are ineffective (venting sound increases in volume and pitch, tank discolors, or shows any signs of deforming), withdraw immediately to a secure position. If employees are expected to fight fires, they must be trained and equipped in OSHA 1910.156.

Disposal Method Suggested: Incineration.[22]

References

Sax, N. I., Ed., "Dangerous Properties of Industrial Materials Report" 1, No. 8, 38–40 (1981) and 3, No. 4, 37–40 (1983)

New Jersey Department of Health and Senior Services, "Hazardous Substance Fact Sheet: Benzoic Acid," Trenton, NJ (May 1986)

New York State Department of Health, "Chemical Fact Sheet: Benzoic Acid," Albany, NY, Bureau of Toxic Substance Assessment (January 1986)

Benzonitrile

Molecular Formula: C_7H_5N

Common Formula: C_6H_5CN

Synonyms: Benzene, cyano-; Benzenenitrile; Benzoic acid nitrile; Benzonitrilo (Spanish); Cyanobenzene; Fenylkyanid; Phenyl cyanide

CAS Registry Number: 100-47-0

RTECS Number: DI2450000

DOT ID: UN 2224

EEC Number: 608-012-00-3

EINECS Number: 202-855-7

Regulatory Authority

- Air Pollutant Standard Set (former USSR)[43]
- CLEAN WATER ACT: Section 311 Hazardous Substances/ RQ 40CFR117.3 (same as CERCLA, see below); Section 313 Water Priority Chemicals (57FR41331, 9/9/92)
- SUPERFUND/EPCRA 40CFR302.4 Reportable Quantity (RQ): CERCLA, 5,000 lb (2,270 kg)
- Canada, WHMIS, Ingredients Disclosure List

Cited in U.S. State Regulations: California (G), Illinois (G), Massachusetts (G), New Hampshire (G), New Jersey (G), Pennsylvania (G).

Description: Bensonitrile, C_6H_5CN, is a colorless, almond-smelling oily liquid. Boiling point = 191°C. Freezing/Melting point = -13°C. Explosive limits in air: Lower = 1.4%, Upper = 7.2%. Flash point = 71°C. Autoignition temperature = 550°C. Poor solubility in water. Octanol/water partition coefficient as Log P_{ow} = 1.6.

Potential Exposure: Workers in organic synthesis of pharmaceuticals, dyestuffs, rubber chemicals. Used as a solvent and chemical intermediate.

Incompatibilities: Strong acids which can liberate hydrogen cyanide. Forms explosive mixture with air.

Permissible Exposure Limits in Air: The former USSR-UNEP/IRPTC project[43] has set a MAC in workplace air of 1.0 mg/m^3.

Permissible Concentration in Water: No criteria set.

Routes of Entry: Inhalation, ingestion, skin absorption.

Harmful Effects and Symptoms

Short Term Exposure: Benzonitrile can affect you when breathed in by passing through your skin. Exposure to high levels, by breathing or skin contact, can cause trouble breathing, convulsions, coma and death. Lower exposures can cause you to feel dizzy and lightheaded. Contact can irritate the eyes and skin. Prolonged contact can cause skin burns. The oral LD_{50} cat and rabbit is 800 mg/kg.

Long Term Exposure: Repeated exposure may damage the liver and nervous system.

Points of Attack: Liver, nervous system.

Medical Surveillance: Liver function tests.

First Aid: If this chemical gets into the eyes, remove any contact lenses at once and irrigate immediately for at least 15 minutes, occasionally lifting upper and lower lids. Seek medical attention immediately. If this chemical contacts the skin, remove contaminated clothing and wash immediately with soap and water. Seek medical attention immediately. If this chemical has been inhaled, remove from exposure, begin rescue breathing (using universal precautions) if breathing has stopped and CPR if heart action has stopped. Transfer promptly to a medical facility. When this chemical has been swallowed, rinse mouth and get medical attention. If cyanide poisoning is suspected use amyl nitrate capsules if symptoms develop. All area employees should be trained regularly in emergency measures for cyanide poisoning and in CPR. A cyanide antidote kit should be kept in the immediate work area and must be rapidly available. Kit ingredients should be replaced every 1 – 2 years to ensure freshness. Persons trained in the use of this kit, oxygen use, and CPR must be quickly available.

Personal Protective Methods: Where possible, enclose operations and use local exhaust ventilation at the site of chemical release. If local exhaust ventilation or enclosure is not used, respirators should be worn. Wear protective work clothing. Butyl rubber and polyvinyl alcohol gloves have been tested and found to be resistant to permeation by benzonitrile. Wash thoroughly immediately after exposure to Benzonitrile and at the end off the workshift. Post hazard and warning information in the work area. In addition, as part of an ongoing education and training effort, communicate all information on the health and safety hazards of Benzonitrile to potentially exposed workers. See NIOSH Criteria Document 78–212 NITRILES.

Respirator Selection: *Where there is a potential for overexposure:* SCBAF:PD,PP (any MSHA/NIOSH approved self-contained breathing apparatus that has a full facepiece and is operated in a pressure-demand or other positive-pressure mode); or SAF:PD,PP:ASCBA (any supplied-air respirator that has a full facepiece and is operated in a pressure-demand or other positive-pressure mode in combination with an auxiliary, self-contained breathing apparatus operated in a pressure-demand or other positive pressure mode).

Storage: Store in tightly closed containers in a cool, well ventilated area. Away from strong acids. Metal containers involving the transfer of this chemical should be grounded and bonded. Drums must be equipped with self-closing valves, pressure vacuum bungs, and flame arresters. Use only non-sparking tools and equipment, especially when opening and closing containers of this chemical. Sources of ignition such as smoking and open flames, are prohibited where this chemical is used, handled, or stored in a manner that could create a potential fire or explosion hazard.

Shipping: The label required by DOT is "Poison." Benoznitrile falls in UN/DOT Hazard Class 6.1 and Packing Group II.[19][20]

Spill Handling: Before entering confined space where benzonitrile may be present, check to make sure that an explosive concentration does not exist. Evacuate and restrict persons not wearing protective equipment from area of spill or leak until cleanup is complete. Remove all ignition sources. Ventilate area of spill or leak. Absorb liquids in vermiculite, dry sand, earth, or a similar material and deposit in sealed containers. It may be necessary to contain and dispose of this chemical as a hazardous waste. If material or contaminated runoff enters waterways, notify downstream users of potentially contaminated waters. Contact your Department of Environmental Protection or your regional office of the federal EPA for specific recommendations. If employees are required to clean-up spills, they must be properly trained and equipped. OSHA 1910.120(q) may be applicable.

Fire Extinguishing: Benzonitrile is a combustible liquid. Use dry chemical, carbon dioxide, water spray, or foam extinguishers. Water may ineffective in controlling fire. Poisonous gases are produced in fire including nitrogen oxides and hydrogen cyanide. If material or contaminated runoff enters waterways, notify downstream users of potentially contaminated waters. Notify local health and fire officials and pollution control agencies. Containers may explode in fire. From a secure, explosion-proof location, use water spray to cool exposed containers. If cooling streams are ineffective (venting sound increases in volume and pitch, tank discolors, or shows any signs of deforming), withdraw immediately to a secure position. If employees are expected to fight fires, they must be trained and equipped in OSHA 1910.156.

Disposal Method Suggested: (1) Mix with calcium hypochlorite and flush to sewer with water or (2) incinerate.

References

Sax, N. I., Ed., "Dangerous Properties of Industrial Materials Report" 1, No. 8, 40–42 (1981); and 3, No. 4, 40–42 (1983)

New Jersey Department of Health and Senior Services, "Hazardous Substance Fact Sheet: Benzonitrile," Trenton, NJ (December 1986)

Benzophenone

Molecular Formula: $C_{13}H_{10}O$

Common Formula: $C_6H_5COC_6H_5$

Synonyms: Benzofenona (Spanish); Benzoyl Benzene; Diphenyl Ketone; Diphenyl Methanone; α-Oxodiphenylmethane; alpha-Oxoditane; Phenyl Ketone

CAS Registry Number: 119-61-9

RTECS Number: DI9950000

DOT ID: UN9188

Cited in U.S. State Regulations: Minnesota (G), New York (G).

Description: Benzophenone, $C_6H_5COC_6H_5$, is a white crystalline solid with a rose odor. Freezing/Melting point = 48.5°C. Boiling point = 305°C. Insoluble in water. Octanol/water partition coefficient as Log P_{ow}= 3.4

Potential Exposure: Benzophenone is used as an odor fixative, flavoring, soap fragrance, in the manufacture of pharmaceuticals and insecticides, in organic syntheses.

Incompatibilities: Oxidizing materials such as dichromates and permanganates.

Permissible Exposure Limits in Air: No standards set.

Permissible Concentration in Water: No criteria set.

Routes of Entry: Inhalation, through the skin and by ingestion.

Harmful Effects and Symptoms

Short Term Exposure: Skin and eye contact can cause irritation. Ingestion can cause nausea, vomiting and stomach pain. Estimated lethal dose is 8 oz.

First Aid: If this chemical gets into the eyes, remove any contact lenses at once and irrigate immediately for at least 15 minutes, occasionally lifting upper and lower lids. Seek medical attention immediately. If this chemical contacts the skin, remove contaminated clothing and wash immediately with soap and water. Seek medical attention immediately. If this chemical has been inhaled, remove from exposure, begin rescue breathing (using universal precautions) if breathing has stopped and CPR if heart action has stopped. Transfer promptly to a medical facility. When this chemical has been swallowed, rinse mouth and get medical attention.

Personal Protective Methods: Adequate ventilation; sinks, showers and eyewash stations should be available. Goggles or face shields and rubber gloves should be worn if contact is likely.

Respirator Selection: Local exhaust or breathing protection.

Storage: Store in a cool, well-ventilated area away from sources of ignition and incompatible materials.

Shipping: Hazardous substances, solid, n.o.s., require no special labels and have no quantity limits on shipment according to DOT.[19] They fall in Hazard Class ORM-E and Packing Group III.

Spill Handling: Evacuate and restrict persons not wearing protective equipment from area of spill or leak until cleanup is complete. Remove all ignition sources. Dampen spilled material with alcohol to avoid dust. Collect powdered material in the most convenient and safe manner and deposit in sealed containers. Ventilate area of spill or leak after clean-up is complete. It may be necessary to contain and dispose of this chemical as a hazardous waste. If material or contaminated runoff enters waterways, notify downstream users of potentially contaminated waters. Contact your Department of Environmental Protection or your regional office of the federal EPA for specific recommendations. If employees are required to clean-up spills, they must be properly trained and equipped. OSHA 1910.120(q) may be applicable.

Fire Extinguishing: Use dry chemical, carbon dioxide, water spray, or alcohol foam extinguishers. Poisonous gases are produced in fire. If material or contaminated runoff enters waterways, notify downstream users of potentially contaminated waters. Notify local health and fire officials and pollution control agencies. From a secure, explosion-proof location, use water spray to cool exposed containers. If cooling streams are ineffective (venting sound increases in volume and pitch, tank discolors, or shows any signs of deforming), withdraw immediately to a secure position. If employees are expected to fight fires, they must be trained and equipped in OSHA 1910.156.

Use dry chemical, alcohol foam, or CO_2.

References

New York State Department of Health, "Chemical Fact Sheet: Benzophenone," Albany, NY, Bureau of Toxic Substance Assessment (March 1986)

Sax, N. I., Ed., "Dangerous Properties of Industrial Materials Report" 1, No. 1, 77–78 (1982)

Benzo[a]pyrene

Molecular Formula: $C_{20}H_{12}$

Synonyms: B(a)P; BAP; Benzo(d,e,f)chrysene; 3,4-Benzopirene (Italian); 6,7-Benzopirene (Italian); Benzopireno (Spanish); 3,4-Benzopyrene; 6,7-Benzopyrene; Benzopyrene; 3,4-Benzpyren (German); 6,7-Benzpyren (German); 3,4-Benz(a)pyrene; 3,4-Benzypyrene; 3,4-BP; BP; NSC21914

CAS Registry Number: 50-32-8

RTECS Number: DJ3675000

DOT ID: UN9188

EEC Number: 601-032-00-3

Regulatory Authority

- Carcinogen (Animal Positive) (IARC)[9] (Suspected Human) (ACGIH)[1] (DFG)[3]

- OSHA, 29CFR1910 Specifically Regulated Chemicals (See CFR1910.1001) as coal tar pitch volatiles
- Air Pollutant Standard Set (OSHA/NIOSH)[2] (former USSR)[43] (Several States)[60] (Canada) (Australia)
- Water Pollution Standard Set (EPA) (Canada)
- CLEAN WATER ACT: 40CFR423, Appendix A, Priority Pollutants; 40CFR401.15 Section 307 Toxic Pollutants
- EPA HAZARDOUS WASTE NUMBER (RCRA No.): U022
- RCRA, 40CFR261, Appendix 8 Hazardous Constituents
- RCRA 40CFR268.48; 61FR15654, Universal Treatment Standards: Wastewater (mg/l), 0.061; Nonwastewater (mg/kg), 3.4
- RCRA 40CFR264, Appendix 9; TSD Facilities Ground Water Monitoring List. Suggested test method(s) (PQL µg/l): 8100 (200); 8270 (10)
- SAFE DRINKING WATER ACT: MCL, 0.0002 mg/l; MCLG, zero
- SUPERFUND/EPCRA 40CFR302.4 Reportable Quantity (RQ): CERCLA, 1 lb (0.454 kg)
- EPCRA (Section 313): as polycyclic aromatic compounds, Form R *de minimis* concentration reporting level: 0.1%
- Canada, WHMIS, Ingredients Disclosure List; Drinking water quality

Cited in U.S. State Regulations: Alaska (G), California (G), Connecticut (A), Florida (G), Illinois (G), Kansas (G, W), Louisiana (G), Maine (G), Massachusetts (G), Michigan (G), New Jersey (G), New Mexico (W), New York (G, A), Pennsylvania (G, A), Rhode Island (G), Vermont (G), Virginia (G, A), Washington (G), West Virginia (G), Wisconsin (G).

Description: B(a)P, $C_{20}H_{12}$, is an odorless, polynuclear aromatic hydrocarbon which forms yellowish crystals or powder. Freezing/Melting point = 179°C. Boiling point = 311°C. Hazard Identification (based on NFPA-704 M Rating System): Health 2, Flammability 1, Reactivity 0. Insoluble in water. Octanol/water partition coefficient as Log P_{ow} = 6.0.

Potential Exposure: Benzo[a]pyrene (B(a)P) is a polycyclic aromatic hydrocarbon (PAH) that has no commercial-scale production. B(a)P is produced in the United States by one chemical company and distributed by several specialty chemical companies in quantities from 100 mg to 5 g for research purposes. Although not manufactured in great quantity, B(a)P is a by-product of combustion. It is estimated that 1.8 million pounds per year are released from stationary sources, with 96% coming from: (1) coal refuse piles, outcrops, and abandoned coal mines; (2) residential external combustion of bituminous coal; (3) coke manufacture; and (4) residential external combustion of anthracite coal. Human exposure to B(a)P can occur from its presence as a by-product of chemical production. The number of persons exposed is not known. Persons working at airports in tarring operations refuse incinerator operations, power plants, and coke manufacturers may be exposed to higher B(a)P levels than the general population. Scientists involved in cancer research or in sampling toxic materials may also be occupationally exposed. The general population may be exposed to B(a)P from air pollution, cigarette smoke, and food sources. B(a)P has been detected in cigarette smoke at levels ranging from 0.2 to 12.2 µg per 100 cigarettes. B(a)P has been detected at low levels in foods ranging from 0.1 to 50 ppb.

Incompatibilities: Strong oxidizers, nitrogen dioxide and ozone.

Permissible Exposure Limits in Air: OSHA PEL for coal tar pitch volatiles is 0.2 mg/m^3 8-hr TWA for an 8-hour workshift. The NIOSH REL is and 0.1 mg/m^3 TWA. For a 10-hour workshift. ACGIH designates Benzo[a]pyrene as an industrial substance suspect of carcinogenic potential for man, and recommends that worker exposures, by all routes, be controlled to levels as low as can be resonably achieved. Sweden has set a TWA of 0.005 mg/m^3 and a STEL of 0.03 mg/m^3.[35] The former USSR-UNEP/IRPTC project[43] has set a MAC of 0.00015 mg/m^3 in workplace air and an average daily MAC of 0.001 µg/m^3 in ambient residential air. In addition. Several states have set guidelines or standards for benzo[a]pyrene in ambient air[60] ranging from zero (New York and Virginia) to 0.0007 µg/m^3 (Pennsylvania) to 0.10 µg/m^3 (Connecticut).

Determination in Air: For polynuclear aromatic hydrocarbons see NIOSH Methods #5506 (HPLC) and #5515 (GC).[18]

Permissible Concentration in Water: Water quality criteria document for PAH published in final 11/2/80. Total PAH addressed. A concentration of PAH 2.8 ng/l is estimated to limit a cancer risk to one in a million (EPA). Canada's MAC for drinking water quality is 0.00001 mg/l. The former USSR-UNEP/IRPTC project[43] has set a MAC of 0.005 µg/l water used for domestic purposes. The WHO[35] has set a maximum limit of 0.2 µg/l in drinking water and a guideline level of 0.01 µg/l. Kansas has set a guideline of 0.03 µg/l in drinking water[60] and New Mexico a standard of 10.0 µg/l.

Routes of Entry: Inhalation and through the skin.

Harmful Effects and Symptoms

B(a)P has produced tumours in all of the nine species for which data are reported following different administrations including oral, skin and intratracheal routes. It has both a local and a systemic carcinogenic effect. In sub-human primates, there is convincing evidence of the ability of B(a)P to produce local sarcomas following repeated subcutaneous injections and lung carcinomas following intratracheal instillation. It is also an initiator of skin carcinogenesis in mice, and it is carcinogenic in single-dose experiments and following prenatal exposure. In skin carcinogenesis studies in mice, B(a)P was consistently found to produce more tumours in a shorter period of time than did other polycyclic aromatic hydrocarbons, with the possible exception of DB(a, h)A. In a dose-response study involving subcutaneous injection in mice, the minimal dose at which carcinogenicity was detected was higher for B(a)P than for DB(a, h)A and for MC. However, the latent periods were shorter for B(a)P than for DB(a, h)A. In studies using intratracheal administration, B(a)P appeared to be less effective than 7H-dibenzo(c,g)carbazole in the hamster.[1] No epidemiological studies on the significance of B(a)P exposure to man are available, and studies are insufficient to prove that B(a)P is carcinogenic for man. However, coal-tar and other materials which are known to be carcinogenic to man may contain B(a)P. The substance has also been detected in other environmental situations. A 1% solution applied to the skin caused skin irritation, swelling, flaking, coloration of skin, and formation of warts.

Short Term Exposure: B(a)P can cause exposure by inhalation and passing through the unbroken skin. Can cause skin irritation with rash and/or burning sensations. Exposure to sunlight can increase these effects. Eye contact can cause irritations and burns.

Long Term Exposure: B(a)P is a probable carcinogenic in humans. May damage the developing fetus. May affect male reproductive glands and cause genetic damage. May be transferred to nursing infants through exposed mother's milk. Repeated exposure can cause skin changes such as thickening, darkening, and pimples. Later skin changes include loss of color, reddish areas, thinning of the skin, and warts. Sunlight may cause rash to develop, and increased risk of skin cancer. Skin cancer is very often easily cured when detected early.

Medical Surveillance: If symptoms develop or overexposure is suspected, the following may be useful: Lung function tests. Examination by a qualified dermatologist.

First Aid: If this chemical gets into the eyes, remove any contact lenses at once and irrigate immediately for at least 15 minutes, occasionally lifting upper and lower lids. Seek medical attention immediately. If this chemical contacts the skin, remove contaminated clothing and wash immediately with soap and water. Seek medical attention immediately. If this chemical has been inhaled, remove from exposure, begin rescue breathing (using universal precautions) if breathing has stopped and CPR if heart action has stopped. Transfer promptly to a medical facility. When this chemical has been swallowed, get medical attention. Give large quantities of water and induce vomiting. Do not make an unconscious person vomit.

Personal Protective Methods: *Avoid exposure of (pregnant) women*! Wear protective gloves and clothing to prevent any reasonable probability of skin contact. Safety equipment suppliers/manufacturers can provide recommendations on the most protective glove/clothing material for your operation. All protective clothing (suits, gloves, footwear, headgear) should be clean, available each day, and put on before work. Contact lenses should not be worn when working with this chemical. Wear dust-proof chemical goggles and face shield unless full facepiece respiratory protection is worn. Employees should wash immediately with soap when skin is wet or contaminated. Provide emergency showers and eyewash.

Respirator Selection: NIOSH: *At any detectable concentration:* SCBAF:PD,PP (any MSHA/NIOSH approved self-contained breathing apparatus that has a full facepiece and is operated in a pressure-demand or other positive-pressure mode); or SAF:PD,PP:ASCBA (any supplied-air respirator that has a full facepiece and is operated in a pressure-demand or other positive-pressure mode in combination with an auxiliary, self-contained breathing apparatus operated in a pressure-demand or other positive pressure mode). *Escape:* GMFOVHiE [any air-purifying, full-facepiece respirator (gas mask) with a chin-style, front- or back-mounted organic vapor canister having a high-efficiency particulate filter]; or SCBAE (any appropriate escape-type, self-contained breathing apparatus).

Storage: Store in tightly closed containers in a cool, well-ventilated area away from oxidizing chemicals (such as chlorates, perchlorates, permanganates, and nitrates). A regulated, marked area should be established where this chemical is handled, used, or stored in compliance with OSHA standard 1910.1045.

Shipping: Hazardous substances, solid, n.o.s. have no specific label requirements or quantity limitations on shipments. They fall in Hazard Class ORM-E and Packing Group III.

Spill Handling: Evacuate and restrict persons not wearing protective equipment from area of spill or leak until cleanup is complete. Remove all ignition sources. Collect powdered material in the most convenient and safe manner and deposit in sealed containers. Ventilate area of spill or leak after clean-up is complete. It may be necessary to contain and dispose of this chemical as a hazardous waste. If material or contaminated runoff enters waterways, notify downstream users of potentially contaminated waters. Contact your Department of Environmental Protection or your regional office of the federal EPA for specific recommendations. If employees are required to clean-up spills, they must be properly trained and equipped. OSHA 1910.120(q) may be applicable.

Fire Extinguishing: Flammable, but generally found in such low quantities it is not considered a fire hazard. Use dry chemical, carbon dioxide, water spray, or foam extinguishers. Poisonous gases are produced in fire including Carbon monoxide. If material or contaminated runoff enters waterways, notify downstream users of potentially contaminated waters. Notify local health and fire officials and pollution control agencies. From a secure, explosion-proof location, use water spray to cool exposed containers. If cooling streams are ineffective (venting sound increases in volume and pitch, tank discolors, or shows any signs of deforming), withdraw immediately to a secure position. If employees are expected to fight fires, they must be trained and equipped in OSHA 1910.156.

Disposal Method Suggested: Incineration in admixture with a flammable solvent.[22]

References

Sax, N. I., Ed., "Dangerous Properties of Industrial Materials Report" 5, No. 1, 42–49 (1985)

New York State Department of Health, "Chemical Fact Sheet: Benzso[a] Pyrene," Albany, NY, Bureau of Toxic Substance Assessment (January 1986)

Agency for Toxic Substance and Disease Registry, U.S. Public Health Service, "Toxicological Profile for Benzo[a] pyrene," Atlanta, Georgia, ATSDR (October 1987)

New Jersey Department of Health and Senior Services, "Hazardous Substance Fact Sheet: Benzo[a] Pyrene," Trenton, NJ (July 1998)

Benzotrichloride

Molecular Formula: $C_7H_5Cl_3$

Common Formula: $C_6H_5CCl_3$

Synonyms: Benzene, trichloromethyl-; Benzenyl chloride; Benzenyl trichloride; Benzotrichloride; Benzotricloruro (Spanish); Benzylidynechloride; Benzyl trichloride; Chlorure de benzenyle (French); Phenyl chloroform; Phenylchloroform; Phenyltrichloromethane; Toluene trichloride; Trichloormethylbenzeen (Dutch); Trichlormethylbenzol (German); 1-(Trichloromethyl)benzene; Trichloromethylbenzene; Trichlorophenylmethane; α,α,α-Trichlorotoluene; *o,o,o*-Trichlorotoluene; Triclorometilbenzene (Italian); Triclorotoluene

CAS Registry Number: 98-07-7

RTECS Number: XT9275000

DOT ID: UN 2226

EEC Number: 602-038-00-9

Regulatory Authority

- Carcinogen (Animal Positive, Human Suspected) (IARC)[9] (DFG)[3] (NTP)
- Banned or Severely Restricted (Sweden) (UN)[13]
- Air Pollutant Standard Set (former USSR)[43] (North Dakota, Rhode Island)[60]
- CLEAN AIR ACT: Hazardous Air Pollutants (Title I, Part A, Section 112)
- EPA HAZARDOUS WASTE NUMBER (RCRA No.): U023
- RCRA, 40CFR261, Appendix 8 Hazardous Constituents
- SUPERFUND/EPCRA 40CFR355, Appendix B Extremely Hazardous Substances: TPQ = 100 lb (45.4 kg)
- SUPERFUND/EPCRA 40CFR302.4 Reportable Quantity (RQ): CERCLA, 10 lb (4.54 kg)
- EPCRA Section 313 Form R *de minimis* concentration reporting level: 0.1%
- Canada, WHMIS, Ingredients Disclosure List

Cited in U.S. State Regulations: California (G), Florida (G), Illinois (G), Kansas (G), Louisiana (G), Maine (G), Maryland (G), Massachusetts (G), Minnesota (G), New Hampshire (G), New Jersey (G), New York (G), North Dakota (A), Pennsylvania (G), Rhode Island (A), Vermont (G), Virginia (G), Washington (G), Wisconsin (G).

Description: Benzotrichloride, $C_6H_5CCl_3$, is a combustible, colorless to yellow oily liquid which fumes on contact with air. It has a penetrating odor. Freezing/Melting point = -5°C. Boiling = 221°C. Flash point = 97°C (cc). Hazard Identification (based on NFPA-704 M Rating System): Health 1, Flammability 1, Reactivity 0. Autoignition temperature = 211°C. Explosive limits in air: Lower = 2.1%; Upper = 6.5% @ 160°C.

Potential Exposure: Benzotrichloride is used extensively in the dye industry for the production of Malachite green, Rosamine, Quinoline red, and Alizarin yellow A. It can also be used to produce ethyl benzoate. Commercial grades may contain hydrochloric acid, benzylidene chloride, or benzyl chloride.

Incompatibilities: Benzotrichloride decomposes on heating, on contact with acids and/or water, producing toxic and corrosive hydrogen chloride and benzoic acid. Reacts violently with strong oxidizers, iron and other metals, alkali and earth alkali metals, bases and organic substances, and may cause fire and explosions. On contact with air it emits toxic and corrosive hydrogen chloride. Attacks many metals in presence of water. Attacks many plastics.

Permissible Exposure Limits in Air: The former USSR-UNEP/IRPTC project has set a MAC in workplace air of 0.2 mg/m^3.[43] In the United States, the states have set guidelines or standards for benzotrichloride in ambient air ranging from zero in North Dakota to 0.0007 $\mu g/m^3$ in Rhode Island.

Permissible Concentration in Water: No criteria set. (Benzotrichloride decomposes in the presence of water to benzoic acid and hydrogen chloride).

Routes of Entry: Inhalation, ingestion and through the skin.

Harmful Effects and Symptoms

Short Term Exposure: Inhalation may cause irritation and chemical burns to the nose, throat and lungs. Higher exposures can cause pulmonary edema, a medical emergency that can be delayed for several hours. This can cause death. Based on animal study information, permanent injury or death may occur from exposure to 125 ppm for 4 hours. Skin contact may cause irritation and chemical burns. Eye contact can cause severe irritation and chemical burns. Ingestion may cause severe irritation and chemical burns to the mouth, throat and stomach. Additionally, it may be aspirated into the lungs with the risk of chemical pneumonitis. The damage may be permanent. The oral LD_{50} rat is 6 mg/kg.

Long Term Exposure: Benzotrichloride may affect the liver, kidneys, thyroid, central nervous system, and hematopoetic system causing impaired functions and anemia. Probable carcinogen in humans.

Points of Attack: Lungs, liver, kidneys, thyroid glands, blood.

Medical Surveillance: Liver function tests, examination of the nervous system, thyroid function, complete blood count (CBC).

First Aid: If this chemical gets into the eyes, remove any contact lenses at once and irrigate immediately for at least 15 minutes, occasionally lifting upper and lower lids. Seek medical attention immediately. If this chemical contacts the skin, remove contaminated clothing and wash immediately with soap and water. Seek medical attention immediately. If this chemical has been inhaled, remove from exposure, begin rescue breathing (using universal precautions) if breathing has stopped and CPR if heart action has stopped. Transfer promptly to a medical facility. When this chemical has been swallowed, get medical attention. If victim is conscious, administer water or milk. Do not induce vomiting. Medical observation is recommended for 24 – 48 hours after breathing overexposure, as pulmonary edema may be delayed. As first aid for pulmonary edema, a doctor or authorized paramedic may consider administering a corticosteroid spray.

Personal Protective Methods: Wear protective gloves and clothing to prevent any reasonable probability of skin contact. Safety equipment suppliers/manufacturers can provide recommendations on the most protective glove/clothing material for your operation. All protective clothing (suits, gloves, footwear, headgear) should be clean, available each day, and put on before work. Contact lenses should not be worn when working with this chemical. Wear splash-proof chemical goggles and face shield unless full facepiece respiratory protection is worn. Employees should wash immediately with soap when skin is wet or contaminated. Provide emergency showers and eyewash.

Respirator Selection: *At any detectable concentration:* SCBAF:PD,PP (any MSHA/NIOSH approved self-contained breathing apparatus that has a full facepiece and is operated in a pressure-demand or other positive-pressure mode); or SAF:PD,PP:ASCBA (any supplied-air respirator that has a full facepiece and is operated in a pressure-demand or other positive-pressure mode in combination with an auxiliary, self-contained breathing apparatus operated in a pressure-demand or other positive pressure mode). *Escape:* HiEF [any air-purifying, full-facepiece respirator (gas mask) with a chin-style, front- or back-mounted organic vapor canister having a high-efficiency particulate filter]; or SCBAE (any appropriate escape-type, self-contained breathing apparatus).

Storage: Before entering confined space where Benzotrichloride may be present, check to make sure that an explosive concentration does not exist. Store in tightly closed containers in a dark, cool, well ventilated area. Metal containers involving the transfer of this chemical should be grounded and bonded. Where possible, automatically pump liquid from drums or other storage containers to process containers. Drums must be equipped with self-closing valves, pressure vacuum bungs, and flame arresters. Use only non-sparking tools and equipment, especially when opening and closing containers of this chemical. Sources of ignition such as smoking and open flames, are prohibited where this chemical is used, handled, or stored in a manner that could create a potential fire or explosion hazard. A regulated, marked area should be established where this chemical is handled, used, or stored in compliance with OSHA standard 1910.1045.

Shipping: The label required by DOT[19] is "Corrosive." This material falls in UN/DOT Hazard Class 8 and Packing Group II.[19][20]

Spill Handling: Material is extremely hazardous to health but areas may be entered with extreme care. Full protective clothing including self-contained breathing apparatus should be provided. No skin surface should be exposed. Spilled material should not be touched. For large spills dike far ahead of spill for later disposal. Water should be used in copious amounts because of reaction with water and formation of toxic by-products. Evacuate and restrict persons not wearing protective equipment from area of spill or leak until cleanup is complete. Remove all ignition sources. Ventilate area of spill or leak. Absorb liquids in vermiculite, dry sand, earth, or a similar material and deposit in sealed containers. It may be necessary to contain and dispose of this chemical as a hazardous waste. If material or contaminated runoff enters waterways, notify downstream users of potentially contaminated waters. Contact your Department of Environmental Protection or your regional office of the federal EPA for specific recommendations. If employees are required to clean-up spills, they must be properly trained and equipped. OSHA 1910.120(q) may be applicable.

Fire Extinguishing: This material may react violently with water. Fire may produce irritating and/or poisonous gases. Small fires: dry chemical, carbon dioxide, water spray, or foam. Large fires: water spray, fog, or foam. If material or contaminated runoff enters waterways, notify downstream users of potentially contaminated waters. Notify local health and fire officials and pollution control agencies. From a secure, explosion-proof location, use water spray to cool exposed containers. If cooling streams are ineffective (venting sound increases in volume and pitch, tank discolors, or shows any signs of deforming), withdraw immediately to a secure position. If employees are expected to fight fires, they must be trained and equipped in OSHA 1910.156.

Disposal Method Suggested: Incineration with flammable solvent added in incinerator with afterburner and alkaline scrubber.

References

U.S. Environmental Protection Agency, "Chemical Profile: Benzotrichloride," Washington, DC, Chemical Emergency Preparedness Program (November 30, 1987)

U.S. Environmental Protection Agency, Benzotrichloride, Health and Environmental Effects Profile No. 20, Washington, DC, Office of Solid Waste (April 30, 1980)

New York State Department of Health, "Chemical Fact Sheet: Benzotrichloride," Albany, NY, Bureau of Toxic Substance Assessment (April 1997)

Benzoyl Chloride

Molecular Formula: C_7H_5ClO

Common Formula: C_6H_5COCl

Synonyms: Benzaldehyde, α-chloro-; Benzenecarbonyl chloride; Benzoic acid, chloride; α-Chlorobenzaldehyde; Cloruro de benzoilo (Spanish)

CAS Registry Number: 98-88-4

RTECS Number: DM6600000

DOT ID: UN 1736

EEC Number: 607-012-00-0

Regulatory Authority

- Air Pollutant Standard Set (UNEP)[43]
- CLEAN WATER ACT: Section 311 Hazardous Substances/RQ 40CFR117.3 (same as CERCLA, see below); Section 313 Water Priority Chemicals (57FR41331, 9/9/92)
- SUPERFUND/EPCRA 40CFR302.4 Reportable Quantity (RQ): CERCLA, 1,000 lb (454 kg)
- EPCRA Section 313 Form R *de minimis* concentration reporting level: 1.0%
- Canada, WHMIS, Ingredients Disclosure List

Cited in U.S. State Regulations: California (G), Florida (G), Illinois (G), Maine (G), Massachusetts (G), Minnesota (G), New Hampshire (G), New Jersey (G), New York (G), Oklahoma (G), Pennsylvania (G), Rhode Island (G).

Description: Benzoyl chloride, C_6H_5COCl, is a colorless to slight brown liquid with a strong, penetrating odor. Boiling point = 197.2°C @ 760 mm. Flash point = 72°C. Decomposes in water. Hazard Identification (based on NFPA-704 M Rating System): Health 3, Flammability 2, Reactivity 2. Water reactive. Autoignition temperature = 195°C. Explosive limits in air: Lower = 2.5%; Upper = 27%.

Potential Exposure: To workers in organic synthesis. Used in producing other chemicals, dyes, perfumes, herbicides, and medicines.

Incompatibilities: Forms explosive mixture with air. Contact with heat, hot surfaces, and flames form phosgene and hydrogen chloride. Water contact may be violent; forms hydrochloric acid. Reactions with amines, alcohols, alkali metals, dimethylsulfoxide, strong oxidizers may be violent. Attacks metals in the presence of moisture. Attacks some plastics, rubber or coatings.

Permissible Exposure Limits in Air: The recommended ACGIH TLV is 0.5 ppm, which should not be exceeded at any time. The AIHA WEEL ceiling is 1 ppm. The TLV set by the former USSR-UPEP/IRPTC project[43] is 5 mg/m^3 in workplace air. The NIOSH IDLH level is 10 ppm.[52]

Determination in Air: By photometry.[11]

Permissible Concentration in Water: EPA has suggested a limit of 69 μg/l.[52]

Routes of Entry: Inhalation, ingestion.

Harmful Effects and Symptoms

Short Term Exposure: Corrosive to the eyes, skin and respiratory tract. Higher exposures can cause pulmonary edema, a medical emergency that can be delayed for several hours. This can cause death.

Long Term Exposure: Can cause chronic rash, warts, sore throat and reduced sense of smell. Can cause bronchitis with cough and phlegm, and/or shortness of breath.

Points of Attack: Eyes, skin and mucous membranes.

Medical Surveillance: Lung function tests. Consider chest x-ray following acute exposure.

First Aid: If this chemical gets into the eyes, remove any contact lenses at once and irrigate immediately for at least 15 minutes, occasionally lifting upper and lower lids. Seek medical attention immediately. If this chemical contacts the skin, remove contaminated clothing and wash immediately with soap and water. Seek medical attention immediately. If this chemical has been inhaled, remove from exposure, begin rescue breathing (using universal precautions) if breathing has stopped and CPR if heart action has stopped. Transfer promptly to a medical facility. When this chemical has been swallowed, get medical attention. If victim is conscious, administer water or milk. Do not induce vomiting. Medical observation is recommended for 24 – 48 hours after breathing overexposure, as pulmonary edema may be delayed. As first aid for pulmonary edema, a doctor or authorized paramedic may consider administering a corticosteroid spray.

Personal Protective Methods: Use in a well-ventilated area. Showers, sinks, and eyewash stations should be readily available. Wear an apron, gauntlets, rubber gloves, boots and any additional protective clothing to prevent skin contact. Wear safety goggles.

Respirator Selection: For levels up to 50 ppm, use a chemical cartridge respirator with an acid gas cartridge, supplied-air respirator or a self-contained breathing apparatus. For up to 100 ppm use a chemical cartridge respirator with an acid gas cartridge and a full facepiece, gas mask with an acid gas canister, supplied-air respirator with a full facepiece, helmet or hood, or a self-contained breathing apparatus with a full facepiece. For escape from a contaminated area use a gas mask with an acid gas canister or a self-contained breathing apparatus.

Storage: Store in a cool, dry, well-ventilated area preferably in a detached warehouse. Keep away from sources of ignition. Avoid physical damage to the container.

Shipping: The label required is "Corrosive." The UN/DOT Hazard Class is 8 and the Packing Group is II.[19][20]

Spill Handling: Evacuate and restrict persons not wearing protective equipment from area of spill or leak until cleanup is complete. Remove all ignition sources. *Do not use water or wet method.* Absorb liquid in vermiculite, dry sand, earth, or similar material and deposit in sealed containers; or cover any spills with sodium bicarbonate. Remove the resulting mixture to a fiber container or plastic bag for disposal by incineration.[24] Ventilate area of spill or leak after clean-up is complete. It may be necessary to contain and dispose of this chemical as a hazardous waste. If material or contaminated runoff enters waterways, notify downstream users of potentially contaminated waters. Contact your Department of Environmental Protection or your regional office of the federal EPA for specific recommendations. If employees are required to clean-up spills, they must be properly trained and equipped. OSHA 1910.120(q) may be applicable.

Fire Extinguishing: Benzoyl Chloride is a combustible liquid. *Do not use water.* Reacts violently with water or steam releasing heat, phosgene and hydrogen chloride fumes. Use dry chemical, carbon dioxide, or foam extinguishers. Poisonous gases are produced in fire. If material or contaminated runoff enters waterways, notify downstream users of potentially contaminated waters. Notify local health and fire officials and pollution control agencies. From a secure, explosion-proof location, use water spray to cool exposed containers. If cooling streams are ineffective (venting sound increases in volume and pitch, tank discolors, or shows any signs of deforming), withdraw immediately to a secure position. If employees are expected to fight fires, they must be trained and equipped in OSHA 1910.156.

Disposal Method Suggested: Pour into sodium bicarbonate solution and flush to sewer.

References

Sax, N. I., Ed., "Dangerous Properties of Industrial Materials Report" 2, No. 1, 78–80 (1982)

New Jersey Department of Health and Senior Services, "Hazardous Substance Fact Sheet: Benzoyl Chloride," Trenton, NJ (November 1998)

New York State Department of Health, "Chemical Fact Sheet: Benzoyl Chloride," Albany, NY, Bureau of Toxic Substance Assessment (February 1986)

Benzoyl Peroxide

Molecular Formula: $C_{14}H_{10}O_4$

Common Formula: $C_6H_5CO\text{-}O\text{-}O\text{-}COC_6H_5$

Synonyms: Abcure S-40-25; Acetoxyl; Acnegel; Aztec benzoyl peroxide 70; Aztec benzoyl peroxide 77; Aztec BPO; Aztec BPO-Dry; Benox L-40V; Benoxyl; Benzac; Benzaknew; Benzoic acid; Benzoic acid benzoperoxide; Benzoic acid peroxide; Benzoperoxide; Benzoylperoxid (German); Benzoylperoxyde (Dutch); Benzoyl superoxide; BPO-W40; BPZ-250; BZF-60; Cadet; Cadet BPO-70W; Cadox; Cadox 40E; Cadox benzoyl peroxide-W40; Cadox BTW-50; Clearasil ACNE treatment cream; Clearasil antibacterial ACNE lotion; Clearasil benzoyl peroxide lotion; Clearasil super strength; Cuticura ACNE cream; Debroxide; Dermoxyl; Dibenzoylperoxid (German); Dibenzoyl peroxide; Diphenylglyoxal peroxide; Eloxyl; Epiclear; Florox; Fostex; Garox; Incidol; Loroxide-HC lotion; Lucidol; Lucidol 75-FP; Lucidol-78; Lucidol GS; Lucipal; LupercoAFR-250; Luperco A; Luperco AA; Luperco AC; Luperco AFR; Lupercol; Luperox FL; Nericur Gel 5; Norox; Novadelox; Oxy-10; Oxy-5 ACNE pimple medication; Oxylite; Oxy wash antibacterial skin wash; Pan oxyl; Panoxyl; Panoxyl aquagel; Panoxyl wash; Perlygel; Peroxide, dibenzoyl; Peroxido de benzoilo (Spanish); Peroxyde de benzoyle (French); Persadox; Persadox cream lotion; Persadox HP cream lotion; Persa-gel; Quinolor compound; Sulfoxyl lotion; Theraderm; Topex; Vanoxide-HC lotion; Xerac

CAS Registry Number: 94-36-0

RTECS Number: DM8575000; DM8576000 (≥30% <52% with inert solid); DM8576200 (>52% with inert solid); DM8577000 (≤72% as a paste); DM8578000 (≤77% with water); DM8579000 (>72% <95%); DM8579000 (>77% <95% with water)

DOT ID: UN 2085; UN 2087; UN 2088; UN 2089; UN 2090; UN 3102 (organic peroxide type B, solid)

EEC Number: 617-008-00-0

EINECS Number: 202-327-6

Regulatory Authority

- Banned or Severely Restricted (In Consumer Products) (Sweden)[13]
- Air Pollutant Standard Set (ACGIH)[1] (NIOSH/OSHA)[2] (OSHA)[58] (HSE)[33] (DFG)[3] (OSHA)[58] (Several States)[60] (Australia), (Israel), (Mexico) (Several Canadian Provinces)
- OSHA 29CFR1910.119, Appendix A, Process Safety List of Highly Hazardous Chemicals, TQ = 7,500 lb (3,504 kg)
- EPCRA Section 313 Form R *de minimis* concentration reporting level: 1.0%
- Canada, WHMIS, Ingredients Disclosure List

Cited in U.S. State Regulations: Alaska (G), California (G), Connecticut (A), Florida (G), Illinois (G), Maine (G), Massachusetts (G), Minnesota (G), Nevada (A), New Hampshire (G, W), New Jersey (G), New York (G), North Dakota (A), Oklahoma (G), Pennsylvania (G), Rhode Island (G), Virginia (A), West Virginia (G).

Description: Benzoyl peroxide, $C_6H_5CO\text{-}O\text{-}O\text{-}COC_6H_5$, is an odorless, white or colorless crystalline powder. Freezing/Melting point = 103 – 106°C (possible explosive decomposition heated). Flash point = 80°C. Poor solubility in water. Autoignition temperature = 80°C. Hazard Identification (based on NFPA-704 M Rating System): Health 1, Flammability 4, Reactivity 4 (Oxidizer).

Potential Exposure: Benzoyl peroxide, is used in the bleaching of fats, waxes, flour and edible oils, it is also used as a polymerization catalyst.

Incompatibilities: A strong oxidizer. Extremely explosion-sensitive to heat, shock, and friction. Combustible substances, wood, paper, lithium aluminum hydride.

Permissible Exposure Limits in Air: The OSHA legal airborned permissible exposure limit is 5 mg/m^3 TWA for an 8-hour workshift. ACGIH[1] HSE[33] Australia, Mexico, and DFG[3] have also adopted this value. DFG also has a Peak Limitation (5 min) of 2 times normal MAK, do not exceed 8 times during workshift. NIOSH has the same value averaged over a 10-hour workshift. IDLH: 1,500 mg/m^3. In addition, four states have set guidelines or standards for benzoyl peroxide in ambient air[60] ranging from 50 μg/m^3 (North Dakota) to 80 μg/m^3 (Virginia) to 100 μg/m^3 (Connecticut) to 119 μg/m^3 (Nevada).

Determination in Air: Collection on a filter, workup with ethyl ether, analysis by high performance liquid chromatography (HPLC). See NIOSH Method #5009.[18]

Permissible Concentration in Water: No criteria set.

Routes of Entry: Inhalation.

Harmful Effects and Symptoms

Short Term Exposure: Contact can irritate the eyes, skin and respiratory tract. *Inhalation:* 1.3 mg/m^3 may cause nose and throat irritation. 12 mg/m^3 may cause lung irritation, asthmatic wheezing, decreased pulse rate and temperature, difficult breathing and stupor. *Skin:* a 5% solution left on the skin for 12 hours has caused redness, swelling and burning. A 5% solution left on the skin for 48 hours may cause severe irritation. Solutions greater then 20% left on for more than a few minutes caused severe irritation. *Eyes:* 2.6 mg/m^3 has caused irritation. *Ingestion:* there is little or no information available on human ingestion. However, results from animal studies suggest that the lethal dose for humans is about ¾ pound. The oral LD_{50} rat is 7,710 mg/kg.[9]

Long Term Exposure: May cause skin sensitization; allergic reaction may develop. May cause lung irritation and bronchitis with cough, phlegm and/or shortness of breath.

Points of Attack: Skin, respiratory system and eyes.

Medical Surveillance: Preplacement and periodic medical examinations should be conducted with particular attention to skin conditions. Lung function tests. Examination by a qualified allergist.

First Aid: If this chemical gets into the eyes, remove any contact lenses at once and irrigate immediately for at least 15 minutes, occasionally lifting upper and lower lids. Seek medical attention immediately. If this chemical contacts the skin, remove contaminated clothing and wash immediately with soap and water. Seek medical attention immediately. If this chemical has been inhaled, remove from exposure, begin rescue breathing (using universal precautions) if breathing has stopped and CPR if heart action has stopped. Transfer promptly to a medical facility. When this chemical has been swallowed, get medical attention. Give large quantities of water and induce vomiting. Do not make an unconscious person vomit.

Medical observation is recommended for 24 – 48 hours after breathing overexposure, as pulmonary edema may be delayed. As first aid for pulmonary edema, a doctor or authorized paramedic may consider administering a corticosteroid spray.

Personal Protective Methods: Protective clothing and safety glasses with side shields or safety goggles should be worn by employees to reduce the possibility of skin contact and eye irritation. Such protection is especially important where benzoyl peroxide and other powder or granular benzoyl formulations may become airborne or where liquid or paste formulations of benzoyl peroxide might be spattered or spilled. Protective clothing should be fire resistant. Any fabric that generates static electricity is not recommended. To prevent the buildup of static electricity, appropriate conductive footwear should be worn. Gloves made of rubber; leather, or other appropriate material should be worn by employees purebenzoyl peroxide. Aprons made of rubber or another appropriate material are recommended for added protection when handling benzoyl peroxide and its formulations. Plastic aprons that may generate static electricity should not be used. Employees should wash promptly when skin is contaminated. Work clothing should be changed daily if it may be contaminated. Remove nonimpervious clothing promptly if contaminated.

Respirator Selection: Respiratory protection as follows must be used whenever airborne concentrations of benzoyl peroxide cannot be controlled to the recommended workplace environmental limit by either engineering or administrative controls. NIOSH: *50 mg/m^3:* DMXSQ (any dust and mist respirator except single-use and quarter mask respirators); or SA (any supplied-air respirator). *125 mg/m^3:* SA:CF (any supplied-air respirator operated in a continuous-flow mode); or PAPRDM, (any powered, air-purifying respirator with a dust and mist filter). *250 mg/m^3:* HiEF (any air-purifying, full-facepiece respirator with a high-efficiency particulate filter); or PAPRTHiE (any powered, air-purifying respirator with a tight-fitting facepiece and a high-efficiency particulate filter); or SCBAF (any self-contained breathing apparatus with a full facepiece); or SAF (any supplied-air respirator with a full facepiece). *1,500 mg/m^3:* SAF:PD,PP (any supplied-air respirator that has a full facepiece and is operated in a pressure-demand or other positive-pressure mode). *Emergency or planned entry into unknown concentrations or IDLH conditions:* SCBAF:PD, PP (any self-contained breathing apparatus that has a full facepiece and is operated in a pressure-demand or other positive-pressure mode); or SAF:PD,PP:ASCBA (any supplied-air respirator that has a full facepiece and is operated in a pressure-demand or other positive-pressure mode in combination with an auxiliary self-contained breathing apparatus operated in a pressure-demand or other positive-pressure mode). *Escape:* HiEF (any air-purifying, full-facepiece respirator with a high-efficiency particulate filter); or SCBAE (any appropriate escape-type, self-contained breathing apparatus).

Note: Substance reported to cause eye irritation or damage; may require eye protection.

Storage: Protect containers against physical damage. Isolate in well-detached, fire-resistive, cool and well-ventilated building with no other materials stored therein; provide explosion venting in a safe direction and prohibit any electrical installation or heating facilities. Dibenzoyl peroxide should be stored in and used from original containers. See OSHA standard 1910.104 and NFPA 43A *Code for the Storage of Liquid and Solid Oxidizers* for detailed handling and storage regulations.

Shipping: The material requires an "Organic Peroxide" label. Benzoyl peroxide falls in Hazard Class 5.2 but restriction vary with concentration and diluents as follows:

Concentration	Passenger Aircraft or Railcar	Cargo Aircraft	Packing Group
>77% <95% with water	Forbidden	Forbidden	I
≥30% <52% with inert solid	10 kg max.	25 kg max.	II
≤72% as a paste	10 kg max.	25 kg max.	II
≤77% with water	5 kg max.	10 kg max.	II
>52% with inert solid or technically pure material	Forbidden	Forbidden	I

Dry dibenzoyl peroxide is shipped in individual one-pound polyethylene-lined paper bags or fiber containers inside fiberboard or wooden boxes; wet dibenzoyl peroxide (30% water by weight) in individual one-pound polyethylene-lined paper bags or fiber containers, tightly sealed inside fiberboard container. Metal barrels or drums having liners of polyethylene or other suitable material may also be used.

Spill Handling: Evacuate and restrict persons not wearing protective equipment from area of spill or leak until cleanup is complete. Remove all ignition sources. *Mix spilled material with water-wetted vermiculite, and deposit in sealed containers. Do not use spark-generating materials or materials made of paper or wood for sweeping or handling spilled benzoyl peroxide.* Ventilate area of spill or leak after clean-up is complete. It may be necessary to contain and dispose of this chemical as a hazardous waste. If material or contaminated runoff enters waterways, notify downstream users of potentially contaminated waters. Contact your Department of Environmental Protection or your regional office of the federal EPA for specific recommendations. If employees are required to clean-up spills, they must be properly trained and equipped. OSHA 1910.120(q) may be applicable.

Fire Extinguishing: This chemical is a highly flammable liquid. At high temperatures it may explode. Poisonous gases are produced in fire. Use water only Do not use chemical or carbon dioxide extinguishers. Vapors are heavier than air and will collect in low areas. Vapors may travel long distances to ignition sources and flashback. Vapors in confined areas may explode when exposed to fire. Storage containers and parts of containers may rocket great distances, in many directions. If material or contaminated runoff enters waterways, notify downstream users of potentially contaminated waters. Notify local health and fire officials and pollution control agencies. From a secure, explosion-proof location, use water spray to cool exposed containers. If cooling streams are ineffective (venting sound increases in volume and pitch, tank discolors, or shows any signs of deforming), withdraw immediately to a secure position. If employees are expected to fight fires, they must be trained and equipped in OSHA 1910.156. Clean-up and salvage operations after a fire should not be attempted until all of the peroxide has cooled completely.

Disposal Method Suggested: Pretreatment involves decomposition with sodium hydroxide. The final solution of sodium benzoate, which is very biodegradable, may be flushed into the drain. Disposal of large quantities of solution may require pH adjustment before release into the sewer or controlled incineration after mixing with a noncombustible material.

References

National Institute for Occupational Safety and Health, Criteria for a Recommended Standard: Occupational Exposure to Benzoyl Peroxide, NIOSH Doc. No. 77–166, Washington, DC (1977)

Sax, N. I., Ed., "Dangerous Properties of Industrial Materials Report" 2, No. 1, 80–82 (1982)

New Jersey Department of Health and Senior Services, "Hazardous Substance Fact Sheet: Benzoyl Peroxide," Trenton, NJ (August 1985)

New York State Department of Health, "Chemical Fact Sheet: Benzoyl Peroxide," Albany, NY, Bureau of Toxic Substance Assessment (July 1998)

Benzyl Bromide

Molecular Formula: C_7H_7Br

Common Formula: $C_6H_5CH_2Br$

Synonyms: (Bromomethyl)benzene; *p*-(Bromomethyl)nitrobenzene; Bromophenylmethane; α-Bromotoluene; ω-Bromotoluene; Bromuro de bencilo (Spanish)

CAS Registry Number: 100-39-0

RTECS Number: X57965000

DOT ID: UN 1737

EEC Number: 602-057-00-2

Regulatory Authority

- Canada, WHMIS, Ingredients Disclosure List

Cited in U.S. State Regulations: Maine (G), Massachusetts, New Hampshire (G), New Jersey (G), Oklahoma (G).

Description: Benzyl bromide, $C_6H_5CH_2Br$, is a combustible, colorless to yellow liquid with a pleasant odor. Boiling point = 198°C. Flash point = 79°C (cc). Freezing/Melting point = -4.0°C. Insoluble in water (slowly decomposes). Octanol/water partition coefficient Log P_{ow} = 2.9.

Potential Exposure: It is used in organic syntheses and as a foaming and frothing agent.

Incompatibilities: Forms explosive mixture with air. Water forms hydrogen bromide and benzyl alcohol. Incompatible with strong oxidizers, bases. Attacks metals, except nickel and lead, in the presence of moisture.

Permissible Exposure Limits in Air: No standards set.

Permissible Concentration in Water: No criteria set.

Routes of Entry: Inhalation, ingestion, skin and/or eye contact

Harmful Effects and Symptoms

Short Term Exposure: Irritates eyes, skin and respiratory system. Eye contact can cause severe irritation and burns. Skin contact can cause severe irritation, redness, swelling and sores. High levels can cause dizziness because it attacks the nervous system

Long Term Exposure: Mutation data reported. Repeated skin contact can cause chronic irritation with dry skin, redness, rash, and sores. Similar very irritating substances can cause lung damage.

Points of Attack: Skin, lungs.

Medical Surveillance: Lung function tests.

First Aid: If this chemical gets into the eyes, remove any contact lenses at once and irrigate immediately for at least 15 minutes, occasionally lifting upper and lower lids. Seek medical attention immediately. If this chemical contacts the skin, remove contaminated clothing and wash immediately with soap and water. Seek medical attention immediately. If this chemical has been inhaled, remove from exposure, begin rescue breathing (using universal precautions) if breathing has stopped and CPR if heart action has stopped. Transfer promptly to a medical facility. When this chemical has been swallowed, get medical attention. If victim is conscious, administer water or milk. Do not induce vomiting. Medical observation is recommended for 24 – 48 hours after breathing overexposure, as pulmonary edema may be delayed. As first aid for pulmonary edema, a doctor or authorized paramedic may consider administering a corticosteroid spray.

Personal Protective Methods: Wear protective gloves and clothing to prevent any reasonable probability of skin contact. Safety equipment suppliers/manufacturers can provide recommendations on the most protective glove/clothing material for your operation. All protective clothing (suits, gloves, footwear, headgear) should be clean, available each day, and put on before work. Contact lenses should not be worn when working with this chemical. Wear splash-proof chemical goggles and face shield unless full facepiece respiratory protection is worn. Employees should wash immediately with soap when skin is wet or contaminated. Provide emergency showers and eyewash.

Storage: Store in tightly closed containers in a cool, well-ventilated area away from water and other incompatible materials listed above.

Shipping: The label required is "Poison, Corrosive." Benzyl bromide falls in UN/DOT Hazard Class 6.1 and Packing Group II.[19][20]

Spill Handling: Evacuate and restrict persons not wearing protective equipment from area of spill or leak until cleanup is complete. Remove all ignition sources. Ventilate area of spill or leak. Absorb liquid on vermiculite, dry sand, earth or similar noncombustible material. Collect material in the most convenient and safe manner and deposit in sealed containers. It may be necessary to contain and dispose of this chemical as a hazardous waste. If material or contaminated runoff enters waterways, notify downstream users of potentially contaminated waters. Contact your Department of Environmental Protection or your regional office of the federal EPA for specific recommendations. If employees are required to clean-up spills, they must be properly trained and equipped. OSHA 1910.120(q) may be applicable.

Fire Extinguishing: Use dry chemical, carbon dioxide, water spray, or foam extinguishers. Poisonous gases are produced in fire including bromine gas. If material or contaminated runoff enters waterways, notify downstream users of potentially contaminated waters. Notify local health and fire officials and pollution control agencies. From a secure, explosion-proof location, use water spray to cool exposed containers. If cooling streams are ineffective (venting sound increases in volume and pitch, tank discolors, or shows any signs of deforming), withdraw immediately to a secure position. If employees are expected to fight fires, they must be trained and equipped in OSHA 1910.156.

Disposal Method Suggested: Pour into vermiculite, sodium bicarbonate or a sand-soda ash mixture and transfer to paper boxes then to an open incinerator. Alternatively, mix with flammable solvent and spray into incinerator equipped with after burner and alkali scrubber.

References

New Jersey Department of Health and Senior Services, "Hazardous Substance Fact Sheet: Benzyl Bromide," Trenton, NJ (February 1987)

Sax, N. I., Ed., "Dangerous Properties of Industrial Materials Report" 2, No. 3, 66–68 (1982)

Benzyl Chloride

Molecular Formula: C_7H_4Cl

Common Formula: $C_6H_2CH_2Cl$

Synonyms: Benzene, (chloromethyl)-; Benzile (cloruro di) (Italian); Benzylchlorid (German); Benzyle (chlorure de) (French); Chloromethylbenzene; Chlorophenylmethane; α-Chlorotoluene; ω-Chlorotoluene; α-Chlortoluol (German); Chlorure de benzyle (French); Cloruro de bencilo (Spanish); NCI-C06360; Tolyl chloride

CAS Registry Number: 100-44-7

RTECS Number: XS8925000

DOT ID: UN 1738

EEC Number: 602-037-00-3

Regulatory Authority

- Carcinogen (Animal Positive) (IARC)[9] (proven) (DFG)[3]
- Air Pollutant Standard Set (ACGIH)[1] (HSE)[33] (former USSR)[43] (OSHA)[58] (Several States)[60] (Australia) (Israel) (Mexico) (Several Canadian Provinces)
- CLEAN AIR ACT: Hazardous Air Pollutants (Title I, Part A, Section 112)
- CLEAN WATER ACT: Section 311 Hazardous Substances/ RQ 40CFR117.3 (same as CERCLA, see below); Section 313 Water Priority Chemicals (57FR41331, 9/9/92)
- EPA HAZARDOUS WASTE NUMBER (RCRA No.): P028
- RCRA, 40CFR261, Appendix 8 Hazardous Constituents
- SUPERFUND/EPCRA 40CFR355, Appendix B Extremely Hazardous Substances: TPQ = 500 lb (227 kg)
- SUPERFUND/EPCRA 40CFR302.4 Reportable Quantity (RQ): CERCLA, 100 lb (45.4 kg)
- EPCRA Section 313 Form R *de minimis* concentration reporting level: 1.0%
- Canada, WHMIS, Ingredients Disclosure List; National Pollutant Release Inventory (NPRI)

Cited in U.S. State Regulations: Alaska (G), California (A, G), Connecticut (A), Florida (G), Illinois (G), Kansas (G), Louisiana (G), Maine (G), Maryland (G), Massachusetts (G, A), Michigan (G), Minnesota (G), Nevada (A), New Hampshire (G), New Jersey (G), New Mexico (W), New York (G, A, W), North Carolina (A), North Dakota (A), Oklahoma (G), Pennsylvania (G), Rhode Island (G, A, W), South Carolina (A), South Dakota (A, W), Tennessee (W), Utah (W), Vermont (G), Virginia (G, A), Washington (G), West Virginia (G), Wisconsin (G).

Description: Benzyl chloride, $C_6H_2CH_2Cl$, is a colorless to slightly yellow liquid with a strong, unpleasant, irritating odor. The odor threshold is 0.05 ppm.[41] Boiling point = 179.4°C. Freezing/Melting point = -43°C. Flash point = 67°C (cc). Flammable Limits in Air: LEL = 1.1%; UEL = 4.9%.[52] Hazard Identification (based on NFPA-704 M Rating System): Health 3, Flammability 2, Reactivity 1. Insoluble in water.

Potential Exposure: In contrast to phenyl halides, benzyl halides are very reactive. Benzyl chloride is used in production of benzal chloride, benzyl alcohol, and benzaldehyde. Industrial usage includes the manufacture of plastics, dyes, synthetic tannins, perfumes and resins. It is used in the manufacture of many pharmaceuticals. Suggested uses of benzyl chloride include: the vulcanization of fluororubbers and the benzylation of phenol and its derivatives for the production of possible disinfectants.

Incompatibilities: Forms explosive mixture with air. Contact with water forms hydrogen chloride fumes. Strong oxidizers may cause fire and explosions. Unstabilized benzyl chloride undergoes polymerization with copper, aluminum, iron, zinc, magnesium, tin, and other common metals except lead and nickel, with the liberation of heat and hydrogen chloride gas. May accumulate static electrical charges, and may cause ignition of its vapors. Attacks some plastics and rubber. Decomposition and polymerization reactions are inhibited, to a limited extent, by addition of triethylamine, propylene oxide, or sodium carbonate (NFPA).

Permissible Exposure Limits in Air: The Federal OSHA/ NIOSH standard[58] and ACGIH value is 1 ppm (5 mg/m^3). This is also the value set by Australia, Mexico, Israel, and HSE.[33] Israel's Action Level is set at one half the TWA. The DFG (Germany) lists this chemical as a proven carcinogen, listed under α-chlorinated toluenes. The NIOSH IDLH level is 10 ppm. The former USSR-UNEP/IRPTC project[43] has set a MAC value of 0.5 mg/m^3 in workplace air. Several States have set guidelines or standards for benzyl chloride in ambient air[60] ranging from 0.01 μg/m^3 (Rhode Island) to 0.7 μg/m^3 (Massachusetts) to 16.7 μg/m^3 (New York) to 20.0 μg/m^3 (Rhode Island) to 25 μg/m^3 (South Carolina) to 50 μg/m^3 (North Dakota) to 85 μg/m^3 (Virginia) to 100 μg/m^3 (Connecticut) to 119 μg/m^3 (Nevada) to 500 μg/m^3 (North Carolina).

Determination in Air: Adsorption on charcoal, workup with CS_2, analysis by gas chromatography. See NIOSH Methods 1003.[18]

Permissible Concentration in Water: No criteria set but EPA[32] has suggested an ambient limit of 69 μg/l based on health effects.

Routes of Entry: Inhalation of vapor, ingestion, eye and/or skin contact.

Harmful Effects and Symptoms

Short Term Exposure: Inhalation exposure may result in severe irritation of upper respiratory tract with coughing, burning of the throat, headache, dizziness, and weakness. Higher exposures can cause pulmonary edema, a medical emergency that can be delayed for several hours. This can cause death. Eye contact may result in immediate and severe eye irritation and prolonged exposure may cause permanent eye damage. Ingestion may cause severe burns of the mouth, throat, and gastrointestinal tract resulting in nausea, vomiting, cramps, and diarrhea. It is intensely irritating to skin, eyes, and mucous membranes. Highly toxic; may cause death or permanent injury after very short exposure to small quantities. Large doses cause central nervous system depression, with possible unconsciousness.

Long Term Exposure: Causes thyroid cancer in animals. Has been listed as a direct-acting or primary carcinogen. May damage the developing fetus. May cause liver damage and affect the nervous system.

Points of Attack: Eyes, skin and respiratory system.

Medical Surveillance: Preplacement and periodic examinations should include the skin, eyes, and an evaluation of the liver, kidney, respiratory tract, blood, and nervous system. Consider chest x-ray following acute overexposure.

First Aid: If this chemical gets into the eyes, remove any contact lenses at once and irrigate immediately for at least 15 minutes, occasionally lifting upper and lower lids. Seek medical attention immediately. If this chemical contacts the skin, remove contaminated clothing and wash immediately with soap and water. Seek medical attention immediately. If this chemical has been inhaled, remove from exposure, begin rescue breathing (using universal precautions) if breathing has stopped and CPR if heart action has stopped. Transfer promptly to a medical facility. When this chemical has been swallowed, get medical attention. If victim is conscious, administer water or milk. Do not induce vomiting. Medical observation is recommended for 24 – 48 hours after breathing overexposure, as pulmonary edema may be delayed. As first aid for pulmonary edema, a doctor or authorized paramedic may consider administering a corticosteroid spray.

Personal Protective Methods: Wear protective gloves and clothing to prevent any reasonable probability of skin contact. Safety equipment suppliers/manufacturers recommend teflon®, polyethylene, and ethylenevinyl alcohol as protective materials. All protective clothing (suits, gloves, footwear, headgear) should be clean, available each day, and put on before work. Contact lenses should not be worn when working with this chemical. Wear splash-proof chemical goggles and face shield unless full facepiece respiratory protection is worn. Employees should wash immediately with soap when skin is wet or contaminated. Provide emergency showers and eyewash.

Respirator Selection: NIOSH: *10 ppm:* CCROVAG [any chemical cartridge respirator with organic vapor and acid gas cartridge(s)]; or GMFOVAG [any air-purifying, full-facepiece respirator (gas mask) with a chin-style, front- or back-mounted organic vapor or acid gas canister]; or PAPROVAG [any powered, air-purifying respirator with organic vapor and acid gas cartridge(s)]; or SA (any supplied-air respirator); or SCBAF (any self-contained breathing apparatus with a full facepiece). *Emergency or planned entry into unknown concentrations or IDLH conditions:* SCBAF: PD,PP (any self-contained breathing apparatus that has a full facepiece and is operated in a pressure-demand or other positive-pressure mode); or SAF:PD,PP:ASCBA (any supplied-air respirator that has a full facepiece and is operated in a pressure-demand or other positive-pressure mode in combination with an auxiliary self-contained breathing apparatus operated in a pressure-demand or other positive pressure mode. *Escape:* GMFOVAG [any air-purifying, full-facepiece respirator (gas mask) with a chin-style, front- or back-mounted organic vapor or acid gas canister]; or SCBAE (any appropriate escape-type, self-contained breathing apparatus).

Note: Substance causes eye irritation or damage; eye protection needed.

Storage: Should be stored in tightly closed containers in a cool, well-ventilated area away from sunlight, heat, moisture, active metals, oxidizers. Metal containers involving the transfer of this chemical should be grounded and bonded. Where possible, automatically pump liquid from drums or other storage containers to process containers. Drums must be equipped with self-closing valves, pressure vacuum bungs, and flame arresters. Use only non-sparking tools and equipment, especially when opening and closing containers of this chemical. Sources of ignition such as smoking and open flames, are prohibited where this chemical is used, handled, or stored in a manner that could create a potential fire or explosion hazard. A regulated, marked area should be established where this chemical is handled, used, or stored in compliance with OSHA standard 1910.1045.

Shipping: The label required by DOT[19] is "Poison, Corrosive." The UN/DOT Hazard Class is 6.1 and the packing Group is II.[19][20]

Spill Handling: Evacuate and restrict persons not wearing protective equipment from area of spill or leak until cleanup is complete. Remove all ignition sources. Establish forced ventilation to keep levels below explosive limit. Ventilate area of spill or leak. If leak or spill has not ignited, use water spray to disperse vapors and to provide protection for persons attempting to stop leak. Use water spray to flush spills away from exposures. Take up small spills with sand or other noncombustible absorbent material and place into containers for later disposal. For larges spills, dike for later disposal. Always wear positive pressure breathing apparatus and special protective clothing. Absorb liquids in vermiculite, dry sand, earth, or a similar material and deposit in sealed containers. It may be necessary to contain and dispose of this chemical as a hazardous waste. If material or contaminated runoff enters waterways, notify downstream users of potentially contaminated waters. Contact your Department of Environmental Protection or your regional office of the federal EPA for specific recommendations. If employees are required to cleanup spills, they must be properly trained and equipped. OSHA 1910.120(q) may be applicable.

Fire Extinguishing: Benzyl chloride is a combustible liquid. Use dry chemical, carbon dioxide, or foam extinguishers. *Do not use water.* Poisonous gases are produced in fire. If material or contaminated runoff enters waterways, notify downstream users of potentially contaminated waters. Notify local health and fire officials and pollution control agencies. From a secure, explosion-proof location, use water spray to cool exposed containers. If cooling streams are ineffective (venting sound increases in volume and pitch, tank discolors, or shows any signs of deforming), withdraw immediately to a secure position. If employees are expected to fight fires, they must be trained and equipped in OSHA 1910.156.

Disposal Method Suggested: Incineration at 1,500°F for 0.5 second minimum for primary combustion and 2,200°F for 12.0 second for secondary combustion. Elemental chlorine formation may be alleviated by injection of steam or methane into the combustion process.[24]

References

U.S. Environmental Protection Agency, Chemical Hazard Information Profile: Benzyl Chloride, Washington, DC (December 9, 1977)

National Institute for Occupational Safety and Health, Information Profiles on Potential Occupational Hazards, Benzyl Chloride, Report PB 276,678, Rockville, Maryland, pp. 2–7 (October 1977)

U.S. Environmental Protection Agency, Benzyl Chloride, Health and Environmental Effects Profile No. 21, Washington, DC, Office of Solid Waste (April 30, 1980)

Sax, N. I., Ed., "Dangerous Properties of Industrial Materials Report" 2, No. 2, 9–11 (1982)

U.S. Environmental Protection Agency, "Chemical Profile: Benzyl Chloride," Washington, DC, Chemical Emergency Preparedness Program

New Jersey Department of Health and Senior Services, "Hazardous Substance Fact Sheet: Benzyl Chloride," Trenton, NJ (June 1996)

Benzyl Cyanide

Molecular Formula: C_8H_4N

Common Formula: $C_6H_2CH_2CN$

Synonyms: Benzeneacetonitrile; Benzylkyanid; Benzylnitrile; Cianuro de bencilo (Spanish); (Cyanomethyl) benzene; α-Cyanotoluene; 2-Phenylacetonitrile; Phenylacetonitrile; Phenyl acetyl nirtile; α-Tolunitrile

CAS Registry Number: 140-29-4

RTECS Number: AM1400000

DOT ID: UN 2470

Regulatory Authority

- Air Pollutant Standard Set (EPA) (former USSR)
- CLEAN AIR ACT: Hazardous Air Pollutants (Title I, Part A, Section 112) as cyanide compound
- SUPERFUND/EPCRA 40CFR355, Appendix B Extremely Hazardous Substances: TPQ = 500 lb (227 kg)
- SUPERFUND/EPCRA 40CFR302.4 Reportable Quantity (RQ): EHS, 1 lb (0.454 kg)
- EPCRA Section 313 Form R *de minimis* concentration reporting level: 1.0%
- U.S. DOT Regulated Marine Pollutant (49CFR172.101, Appendix B) as cyanide mixtures, cyanide solutions or cyanides, inorganic, n.o.s
- Canada, WHMIS, Ingredients Disclosure List (as cyanide compounds)

Cited in U.S. State Regulations: California (G), Florida (G), Massachusetts (G), New Hampshire (G), New Jersey (G), Pennsylvania (G).

Description: Benzyl cyanide, $C_6H_2CH_2CN$, is a colorless, oily liquid with an aromatic odor. Boiling point = 233.5°C. Flash point = 113°C (oc),[17] also cited as 101°C.[52] Hazard Identification (based on NFPA-704 M Rating System): Health 2, Flammability 1, Reactivity 0. Insoluble in water.

Potential Exposure: Benzyl cyanide is used in organic synthesis, especially of penicillin precursors. It is used as a chemical intermediate for amphetamines, phenobarbital, the stimulant, methyl phenidylacetate, esters as perfumes and flavors.

Incompatibilities: Violent reaction with strong oxidizers; sodium hypochlorite, lithium aluminum hydride.

Permissible Exposure Limits in Air: The former USSR-UNEP/IRPTC project[43] has set a MAC of 0.8 mg/m^3 in workplace air. No other occupational exposure limits have been established for this chemical; however, inasmuch as it is a cyanide compound, the exposure limits are listed here: OSHA and ACGIH: 5 mg/m^3 TWA; NIOSH: Ceiling limit, 4.7 ppm; 5 mg/m^3 per 10 minutes as cyanides. All have notations that skin contact contributes significantly in overall exposure. IDLH = 25 mg/m^3 as CN.

Permissible Concentration in Water: No criteria set.

Routes of Entry: Inhalation, ingestion, skin and/or eye contact.

Harmful Effects and Symptoms

Short Term Exposure: Poisonous. May be fatal if inhaled, swallowed, or absorbed through skin. Contact may cause burns to skin and eyes. The oral LD_{50} rat is 270 mg/kg.[9]

First Aid: If this chemical gets into the eyes, remove any contact lenses at once and irrigate immediately for at least 15 minutes, occasionally lifting upper and lower lids. Seek medical attention immediately. If this chemical contacts the skin, remove contaminated clothing and wash immediately with soap and water. Speed in removing material from skin is of extreme importance Seek medical attention immediately. If this chemical has been inhaled, remove from exposure, begin rescue breathing (using universal precautions) if breathing has stopped and CPR if heart action has stopped. Transfer promptly to a medical facility. When this chemical has been swallowed, get medical attention. Give large quantities of water and induce vomiting. Do not make an unconscious person vomit.

For cyanide poisoning, use amyl nitrate capsules if symptoms develop. All area employees should be trained regularly in emergency measures for cyanide poisoning and in CPR. A cyanide antidote kit should be kept in the immediate work area and must be rapidly available. Kit ingredients should be replaced every 1 – 2 years to ensure freshness. Persons trained in the use of this kit, oxygen use, and CPR must be quickly available.

Personal Protective Methods: Wear protective gloves and clothing to prevent any reasonable probability of skin contact. Safety equipment suppliers/manufacturers can provide recommendations on the most protective glove/clothing material for your operation. All protective clothing (suits, gloves, footwear, headgear) should be clean, available each day, and put on before work. Contact lenses should not be worn when working with this chemical. Wear splash-proof chemical goggles and face shield unless full facepiece respiratory protection is worn. Employees should wash immediately with soap when skin is wet or contaminated. Provide emergency showers and eyewash. See NIOSH Criteria Document 78–212 NITRILES.

Respirator Selection: NIOSH/OSHA (as cyanides): *up to 25 mg/m³:* SA (any supplied-air respirator); or SCBAF (any self-contained breathing apparatus with full facepiece). *Emergency or planned entry into unknown concentrations or IDLH conditions:* SCBAF:PD,PP (any self-contained breathing apparatus that has a full facepiece and is operated in a pressure-demand or other positive-pressure mode); or SAF:PD,PP:ASCBA (any supplied-air respirator that has a full facepiece and is operated in a pressure-demand or other positive-pressure mode in combination with an auxiliary self-contained breathing apparatus operated in a pressure-demand or other positive pressure mode. *Escape:* GMFSHiE [any air-purifying, full-facepiece respirator (gas mask) with a chin-style, front-or back-, mounted canister providing protection against the compound of concern and having a high efficiency particulate filter]; or SCBAE (any appropriate escape-type, self-contained breathing apparatus).

Storage: Store in tightly closed containers in a cool, well ventilated area away from oxidizers and other incompatible materials listed above. Sources of ignition such as smoking and open flames, are prohibited where this chemical is used, handled, or stored in a manner that could create a potential fire or explosion hazard.

Shipping: The label required for phenylaceto-nitrile is "Keep away from Food." The UN/DOT Hazard Class is 6.1 and the Packing Group is III.[19][20]

Spill Handling: Evacuate and restrict persons not wearing protective equipment from area of spill or leak until cleanup is complete. Remove all ignition sources. Ventilate area of spill or leak. Use water spray to reduce vapors. *Small spills:* take up with sand or other noncombustible absorbent material and place into containers for later disposal. *Large spills:* dike far ahead of spill for later disposal. It may be necessary to contain and dispose of this chemical as a hazardous waste. If material or contaminated runoff enters waterways, notify downstream users of potentially contaminated waters. Contact your Department of Environmental Protection or your regional office of the federal EPA for specific recommendations. If employees are required to clean-up spills, they must be properly trained and equipped. OSHA 1910.120(q) may be applicable.

Fire Extinguishing: This chemical is a combustible solid. *Small fires:* dry chemical, carbon dioxide, water spray, or foam. Large fires: water spray, fog, or foam. Move container from fire area if you can do it without risk. Fight fire from maximum distance. Dike fire control water for later disposal; do not scatter the material. Keep unnecessary people away; isolate hazard area and deny entry. Stay upwind; keep out of low areas. Ventilate closed spaces before entering them. Wear positive pressure breathing apparatus and special protective clothing. Remove and isolate contaminated clothing at the site. When heated to decomposition, it emits very toxic fumes of cyanide and nitrogen oxides. Container may explode in heat of fire. Runoff from fire control water may give off poisonous gases. If material or contaminated runoff enters waterways, notify downstream users of potentially contaminated waters. Notify local health and fire officials and pollution control agencies. From a secure, explosion-proof location, use water spray to cool exposed containers. If cooling streams are ineffective (venting sound increases in volume and pitch, tank discolors, or shows any signs of deforming), withdraw immediately to a secure position. If employees are expected to fight fires, they must be trained and equipped in OSHA 1910.156.

References

U.S. Environmental Protection Agency, "Chemical Profile: Benzyl Cyanide," Washington, DC, Chemical Emergency Preparedness Program (November 30, 1987)

Beryllium and Compounds

Molecular Formula: Be

Synonyms: Berilio (Spanish); Beryllium-9; Beryllium dust; Beryllium metal powder; Glucinium; Glucinum

CAS Registry Number: 7440-41-7

RTECS Number: DS1750000

DOT ID: UN 1566; UN 1567 (powder)

EEC Number: 004-001-00-7

EINECS Number: 231-150-7

Regulatory Authority

- Carcinogen (Animal Positive, Human Suspected) (IARC)[9] (Suspected) (NTP)
- Banned or Severely Restricted (UN)[35]
- Very Toxic Substance (World Bank)[15]

- Air Pollutant Standard Set (ACGIH)[1] (HSE)[33] (former USSR)[35][43] (OSHA)[58] (Several States)[60] (Australia) (Israel) (Mexico) (Various Canadian Provinces) (Japan)
- CLEAN WATER ACT: 40CFR423, Appendix A, Priority Pollutants; 40CFR401.15 Section 307 Toxic Pollutants
- EPA HAZARDOUS WASTE NUMBER (RCRA No.): P015
- RCRA, 40CFR261, Appendix 8 Hazardous Constituents
- RCRA 40CFR268.48; 61FR15654, Universal Treatment Standards: Wastewater (mg/l), 0.82; Nonwastewater (mg/l), 0.014 TCLP
- RCRA 40CFR264, Appendix 9; TSD Facilities Ground Water Monitoring List. Suggested test method(s) (PQL μg/l): (total) 6010 (3); 7090 (50); 7091 (2)
- SAFE DRINKING WATER ACT: MCL, 0.004 mg/l; MCLG, 0.004 mg/l
- SUPERFUND/EPCRA 40CFR302.4 Reportable Quantity (RQ): CERCLA, 10 lb (4.54 kg). *Note:* No report required if the diameter of the pieces of solid metal is equal to or exceeds 0.004 inches
- EPCRA Section 313 Form R *de minimis* concentration reporting level: 0.1%
- Canada, WHMIS, Ingredients Disclosure List; CEPA, Schedule 3, Part 2, Ocean Dumping Restriction

Cited in U.S. State Regulations: Alaska (G), California (G), Connecticut (A), Florida (G), Illinois (G), Kansas (G, W), Louisiana (G), Maine (G), Maryland (G), Massachusetts (G, A), Michigan (G), Minnesota (G), Nevada (A), New Hampshire (G), New Jersey (G), North Carolina (A), North Dakota (A), Oklahoma (G), Pennsylvania (G, A), Rhode Island (G, A, W), South Carolina (A), South Dakota (A), Tennessee (W), Utah (W), Vermont (G), Virginia (G, A), Washington (G), West Virginia (G), Wisconsin (G).

Description: Beryllium, Be, is a gray shiny metal or powder, or fine granules which resemble powdered aluminum. Beryllium is slightly soluble in water. All beryllium compounds are soluble to some degree in water. Beryl ore is the primary source of beryllium, although there are numerous other sources. Boiling point = 2,970°C. Freezing/Melting point = 1,278°C. Hazard Identification (based on NFPA-704 M Rating System): Health 3, Flammability 1, Reactivity 0.

Potential Exposure: Beryllium is used extensively in manufacturing electrical components, chemicals, ceramics, and x-ray tubes. A number of alloys are produced in which beryllium is added to yield greater tensile strength, electrical conductivity, and resistance to corrosion and fatigue. The metal is used as a neutron reflector in high-flux test reactors. Human exposure occurs mainly through inhalation of beryllium dust or fumes by beryllium ore miners, beryllium alloy makers and fabricators, phosphor manufacturers, ceramic workers, missile technicians, nuclear reactor workers, electric and electronic equipment workers, and jewelers. The major source of beryllium exposure of the general population is thought the burning of coal. Approximately 250.000 pounds of beryllium is released from coal and oil-fired burners. EPA estimates the total release of beryllium to the atmosphere from point sources is approximately 5,500 pounds per year. The principal emissions are from beryllium-copper alloy production. Approximately 721,000 persons living within 12,5 miles (20 km) of point sources are exposed to small amounts of beryllium (median concentration 0.005 μg/m^3). Levels of beryllium have been reported in drinking water supplies and in small amounts in food.

Incompatibilities: Beryllium metal reacts with strong acids, alkalis (forming combustible hydrogen gas), oxidizable materials. Forms shock sensitive mixtures with some chlorinated solvents, such as carbon tetrachloride and trichloroethylene.

Permissible Exposure Limits in Air: The present Federal OSHA PEL[58] for beryllium and beryllium compounds is 0.002 mg/m^3 as an 8-hour TWA and a ceiling concentration of 0.005 mg/m^3, not to be exceeded during any 15 minute work period, and 0.025 mg/m^3 as an acceptable maximum peak, permitted for any 30-minute period, above the ceiling limit. The recommended NIOSH Ceiling is 0.0005 mg/m^3, which should not be exceeded at any time. ACGIH recommended TLV is 0.002 mg/m^3 as an 8-hour TWA and STEL of 0.01 mg/m^3. Many other countries (Australia, Israel, Japan, Mexico, U.K.) also cite the 0.002 mg/m^3 limit in the workplace.[35] The NIOSH IDLH is 4 mg Be/m^3. The former USSR-UNEP/IRPTC project[43] sets a 0.001 μg/m^3 MAC in the workplace. The former USSR has set an allowable ambient MAC in residential air of 0.01 μg/m^3.[35] Several States have set guidelines or standards for beryllium in ambient air[60] ranging from zero (North Dakota) to 0.0042 μg/m^3 (Massachusetts) to 0.01 μg/m^3 (Connecticut, Pennsylvania, South Carolina) to 0.02 μg/m^3 (South Dakota and Virginia) to 0.1 μg/m^3 (Nevada).

Determination in Air: Collection on a filter, analysis by atomic absorption spectroscopy. See NIOSH Method #7102 and #7300.[18] See also the ATSDR profile cited below.

Permissible Concentration in Water: In 1980 the EPA set the following criteria: To protect freshwater aquatic life: 130 μg/l on an acute basis: 5.3 μg/l on a chronic basis. To protect saltwater aquatic life: insufficient data to set criteria. To protect human health: preferable zero. An additional lifetime cancer risk of 1 in 100,000 results from a concentration of 0.037 μg/l.[6] The former USSR-UNEP/IRPTC project[43] has set a MAC of 0.0002 mg/l (0.2 μg/l) in water bodies used for domestic purposes. Kansas and Rhode Island have set guidelines for beryllium in drinking water:[61] 0.13 μg/l in Kansas and 131 μg/l in Rhode Island.

Determination in Water: Total beryllium may be determined according to EPA, by digestion followed by atomic absorption or by a colorimetric method or by inductively coupled plasma (ZCP) optical emission spectrometry. Dissolved beryllium can be determined by 0.45 micron filtration prior to the above method for total beryllium. See also the ATSDR Profile cited below.

Routes of Entry: Inhalation of fume or dust.

Harmful Effects and Symptoms

Local: The soluble beryllium salts are Cutaneous sensitizes as well as primary irritants. Contact dermatitis of exposed parts of the body is caused by acid salts as well as primary irritants. Contact dermatitis of exposed parts of the body is caused by acid salts of beryllium. Onset is generally delayed about two weeks from the time of first exposure. Complete recovery occurs following cessation of exposure. Eye irritation and conjunctivitis can occur. Accidental implantation of beryllium metal or crystals of soluble beryllium compound in areas of broken or abraded skin may cause granulomatous lesions. There are hard lesions with a central nonhealing area. Surgical excision of the lesion is necessary. Exposure to soluble beryllium compounds may cause nasopharynigitis, a condition characterized by swollen and edematous mucous membranes, bleeding points, and ulceration. These symptoms are reversible when exposure is terminated. *Systemic:* Beryllium and its compounds are highly toxic substances. Entrance to the body is almost entirely by inhalation. The acute systemic effects of exposure to beryllium primarily involve the respiratory tract and are manifest by a nonproductive cough, substernal pain, moderate shortness of breath, and some weight loss. The character and speed of onset of these symptoms, as well as their severity, are dependent on the type and extent of exposure. An intense exposure, although brief, may result in severe chemical pneumonitis with pulmonary edema. Chronic beryllium disease can be classified by its clinical variants according to the disability the disease process produces. (1) Asympromatic nondisabling disease is usually diagnosed only by routine chest-x-ray changes and supported by urinary or tissue assay. (2) In its mildly disabling form, the disease results in some nonproductive cough and dyspnea following unusual levels of exertion. Joint pain and weakness are common complaints. Diagnosis is by x-ray changes. Renal calculi containing beryllium may be a complication. Usually, the patient remains stable for years, but eventually shows evidence of pulmonary or myocardial failure. (3) In its moderately severe disabling form, the disease produces symptoms of distressing cough and shortness of breath, with marked x-ray changes. The liver and spleen are frequently affected, and spontaneous pneumothorax may occur. There is generally weight loss, bone and joint pain, oxygen desaturation, increase in hematocrits, disturbed liver function, hypercalciuria, and spontaneous skin lesions similar to those of Boeck's sarcoid. Lung function studies show measurable decreases in diffusing capacity. Many people in this group survive for years with proper therapy. Bouts of chills and fever carry a bad prognosis. (4) The severely disabling disease will show all of the above mentioned signs and symptoms in addition to severe physical wasting and negative nitrogen balance. Right heart failure may appear causing a severe nonproductive cough which leads to vomiting after meals. Severe lack of oxygen is the predominant problem, and spontaneous pneumothorax can be a serious complication. Death is usually due to pulmonary insufficiency or right heart failure.

Short Term Exposure: Eye or skin contact can cause irritation, itching, and burning. Sometimes an allergic eye problem develops, breaking out with future exposure. Inhalation overexposure can severely irritate the airways and lungs, causing nasal discharge, tightness of the chest, cough, shortness of breath, and/or fever. Death can occur in severe cases. Seek prompt medical attention. Future exposures can cause further attacks. Symptoms may be delayed for days following exposure. Some persons later develop lung scarring after such exposures.

Long Term Exposure: Be is a probable cancer causing agent in humans. There is some evidence that it causes lung and bone cancer in humans and animals. High or repeated exposure can permanently scar the lungs or other body organs. If Be particles get under cuts in the skin, ulcers or lumps can develop; these must be surgically removed. Allergic skin rashes can occur. High or repeated exposure can cause kidney stones to develop.

Points of Attack: Skin, eyes, respiratory system, lungs, liver, spleen, heart.

Medical Surveillance: Preemployment history and physical examinations for worker applicants should include chest x-rays, baseline pulmonary function tests (FVC and FEV_1), and measurement of body weight. Beryllium workers should receive a periodic health evaluation that includes: spirometry (FVC and FEV_1), medical history questionnaire directed toward respiratory symptoms and a chest x-ray, blood/urine trace metals. General health, liver and kidney functions, and possible effects of the skin should be evaluated. See NIOSH Criteria Document 72-10268.

First Aid: If this chemical gets into the eyes, remove any contact lenses at once and irrigate immediately for at least 15 minutes, occasionally lifting upper and lower lids. Seek medical attention immediately. If this chemical contacts the skin, remove contaminated clothing and wash immediately with soap and water. Seek medical attention immediately. If this chemical has been inhaled, remove from exposure, begin rescue breathing (using universal precautions) if breathing has stopped and CPR if heart action has stopped. Transfer promptly to a medical facility. When this chemical has been swallowed, get medical attention. Give large quantities of water and induce vomiting. Do not make an unconscious person vomit.

Personal Protective Methods: Work areas should be monitored to limit and control levels of exposure. Personnel samplers are recommended. Good housekeeping, proper maintenance, and engineering control of processing equipment and technology are essential. The importance of safe work practices and personal hygiene should be stressed. When beryllium levels exceed the accepted standards, the workers should be provided with respiratory protective devices of the appropriate class, as determined on the basis of the actual or projected atmospheric concentration of airborne beryllium at the worksite. Protective clothing should be provided all workers who are subject to exposure in excess of the standard. This should include shoes or protective shoe covers as well as other clothing. The clothing should be reissued clean on a daily basis. Workers should shower following each shift prior to change to street clothes.

Respirator Selection: NIOSH: *At any detectable concentration:* SCBAF:PD,PP (any MSHA/NIOSH approved self-contained breathing apparatus that has a full facepiece and is operated in a pressure-demand or other positive-pressure mode); or SAF:PD,PP:ASCBA (any supplied-air respirator that has a full facepiece and is operated in a pressure-demand or other positive-pressure mode in combination with an auxiliary, self-contained breathing apparatus operated in a pressure-demand or other positive pressure mode). *Escape:* HiEF [any air-purifying, full-facepiece respirator (gas mask) with a chin-style, front- or back-mounted organic vapor canister having a high-efficiency particulate filter]; or SCBAE (any appropriate escape-type, self-contained breathing apparatus).

Storage: Beryllium must be stored to avoid contact with oxidizers (such as perchlorates, peroxides, permanganates, chlorates and nitrates), and strong acids (such as hydrochloric, sulfuric, and nitric) since violent reactions occur. Store in tightly closed containers in a cool, well-ventilated area away from heat. Protect storage containers from physical damage. Use only non-sparking tools, and equipment, especially when opening and closing containers of Beryllium. A regulated, marked area should be established where this chemical is handled, used, or stored in compliance with OSHA standard 1910.1045.

Shipping: Beryllium powder should bear a label of "Poison, Flammable Solid." It falls in UN/DOT Hazard Class 6.1 and Packing Group II.[19][20]

Spill Handling: Evacuate and restrict persons not wearing protective equipment from area of spill or leak until cleanup is complete. Remove all ignition sources. Collect powdered material in the most convenient and safe manner and deposit in sealed containers. Ventilate area of spill or leak after clean-up is complete. It may be necessary to contain and dispose of this chemical as a hazardous waste. If material or contaminated runoff enters waterways, notify downstream users of potentially contaminated waters. Contact your Department of Environmental Protection or your regional office of the federal EPA for specific recommendations. If employees are required to clean-up spills, they must be properly trained and equipped. OSHA 1910.120(q) may be applicable.

Fire Extinguishing: Be is a combustible solid. Smother fire with dry sand, dry clay, dry ground limestone, or use approved Class "D" extinguishers (NFPA). *Do not use carbon dioxide or hologenated extinguishing agents. Do not use water.* Poisonous gases including beryllium oxide fume are produced in fire. If material or contaminated runoff enters waterways, notify downstream users of potentially contaminated waters. Notify local health and fire officials and pollution control agencies. From a secure, explosion-proof location, use water spray to cool exposed containers. If cooling streams are ineffective (venting sound increases in volume and pitch, tank discolors, or shows any signs of deforming), withdraw immediately to a secure position. If employees are expected to fight fires, they must be trained and equipped in OSHA 1910.156.

Disposal Method Suggested: For beryllium (powder), waste should be converted into chemically inert oxides using incineration and particulate collection techniques. These oxides should be returned to suppliers if possible. Recovery and recycle is an alternative to disposal for beryllium scrap and pickle liquors containing beryllium.[22]

References

National Institute for Occupational Safety and Health, Criteria for a Recommended Standard: Occupational Exposure to Beryllium, NIOSH Doc. 72-10268, Washington, DC (1972)

U.S. Environmental Protection Agency, Beryllium: Ambient Water Quality Criteria, Washington, DC (1979)

U.S. Environmental Protection Agency, Toxicology of Metals, Vol. 2: Beryllium, Report No. EPA-600/1-77-022, Research Triangle Park, pp 85–109 (May 1977)

U.S. Environmental Protection Agency, Reviews of the Environmental Effects of Pollutants, VI, Beryllium, Report EPA-600/1-78-028, Cincinnati, Ohio, Health Effects Research Laboratory (1978)

U.S. Environmental Protection Agency, Beryllium, Health and Environmental Effects Profile No. 22, Washington, DC, Office of Solid Waste (April 30, 1980)

New Jersey Department of Health and Senior Services, "Hazardous Substance Fact Sheet: Beryllium (Dust and Powder)," Trenton, NJ (July 1998)

Agency for Toxic Substances and Disease Registry, U.S. Public Health Service, "Toxicological Profile for Beryllium," Atlanta, Georgia, ATSDR (October 1987)

Sax, N. I., Ed., "Dangerous Properties of Industrial Materials Report" 2, No. 2, 13–14 (1982)

Sax, N. I., Ed., "Dangerous Properties of Industrial Materials Report" 1, No. 1, 33–36 (1980). (Beryllium Fluoride, Oxide and Sulfate)

Sax, N. I., Ed., "Dangerous Properties of Industrial Materials Report" 2, No. 1, 84–88 (1982). (Beryllium Nitrate and Sulfate)

Sax, N. I., Ed., "Dangerous Properties of Industrial Materials Report" 3, No. 5, 59–61 (1983). (Beryllium Chloride)

Sax, N. I., Ed., "Dangerous Properties of Industrial Materials Report" 3, No. 5, 61–64 (1983). (Beryllium Fluoride)

Note: Beryllium Chloride, Fluoride, Nitrate and Oxide are the subject of fact sheets from the State of New Jersey, quite similar to one another. Also, the Chloride, Fluoride, Nitrate and Oxide are the topics of articles cited above. To avoid duplication, separate entries for those four compounds are therefore not included here.

Biphenyl

Molecular Formula: $C_{12}H_{10}$

Common Formula: $C_6H_5C_6H_5$

Synonyms: Bibenzene; 1,1'-Biphenyl; Dibenzene; 1,1'-Diphenyl; Diphenyl; Dowtherm A; Lemonene; Phenador-X; Phenylbenzene; PHPH; Xenene

CAS Registry Number: 92-52-4

RTECS Number: DU8050000

EINECS Number: 202-163-5

Regulatory Authority

- Air Pollutant Standard Set (ACGIH)[1] (DFG)[3] HSE)[33] (OSHA)[58] (Several States)[60] Australia) (Israel) (Mexico) (Several Canadian Provinces)
- CLEAN AIR ACT: Hazardous Air Pollutants (Title I, Part A, Section 112)
- SUPERFUND/EPCRA 40CFR302.4 Reportable Quantity (RQ): CERCLA, 1 lb (0.454 kg)
- EPCRA Section 313 Form R *de minimis* concentration reporting level: 1.0%
- Canada, WHMIS, Ingredients Disclosure List; National Pollutant Release Inventory (NPRI)

Cited in U.S. State Regulations: Alaska (G), California (G), Connecticut (A), Florida (G, A), Illinois (G), Maine (G), Maryland (G), Massachusetts (G, A), Minnesota (G), Nevada (A), New Hampshire (G), New Jersey (G), New York (A), North Dakota (A), Oklahoma (G), Pennsylvania (G), Rhode Island (G, A), Virginia (A), West Virginia (G).

Description: Biphenyl. $C_6H_5C_6H_5$, is a combustible, white flakes or crystalline solid with a pleasant, characteristic odor. Freezing/Melting point = 69 – 70°C. Boiling point = 256°C. Flash point = 113°C. Autoignition temperature = 540°C. Explosive limits in air: LEL = 0.6%, UEL = 5.8%.[17] Hazard Identification (based on NFPA-704 M Rating System): Health 2, Flammability 1, Reactivity 0. Insoluble in water.

Potential Exposure: Biphenyl is a fungicide. It is also used as a heat transfer agent and as an intermediate in organic synthesis.

Incompatibilities: Mist forms explosive mixture with air. Strong oxidizers may cause fire and explosions.

Permissible Exposure Limits in Air: The Federal OSHA PEL and NIOSH recommended standard is 0.2 ppm (1 mg/m^3)TWA. The ACGIH TLV is 0.2 ppm (1.3 mg/m^3). The HSE[33] and DFG[3] use this same value as do several other countries[35] including Australia, Mexico, Israel. The Canadian Provinces of Alberta, British Columbia, Ontario and Quebec use the same TWA/TWAEV. The ACGIH Has set no STEL value but the HSE,[33] Mexico, Alberta and British Columbia STEL value is 0.6 ppm (3.0 mg/m^3). The NIOSH IDLH level is 100 mg/m^3. Several states have set guidelines or standards for biphenyl in ambient air[60] ranging from 0.086 μg/m^3 (Massachusetts) to 0.4 μg/m^3 (Rhode Island) to 5.0 μg/m^3 (New York) to 15.0 μg/m^3 (Florida and North Dakota) to 20.0 μg/m^3 (Connecticut) to 25 μg/m^3 (Virginia) to 36 μg/m^3 (Nevada) to 40 μg/m^3 (North Dakota).

Determination in Air: Tenax Gas chromatography; CCl4; Gas chromatography/Flame ionization detection; NIOSH(IV) Method #2530.

Permissible Concentration in Water: No criteria set but EPA has suggested an ambient limit of 13.8 μg/l based on health effects.

Routes of Entry: Inhalation of vapor or dust; percutaneous absorption, ingestion, eye and/or skin contact.

Harmful Effects and Symptoms

Short Term Exposure: Skin contact contributes significantly to overall exposure. Repeated exposure to dust may result in irritation of skin and respiratory tract. The vapor may cause moderate eye irritation. Repeated skin contact may produce a sensitization dermatitis. In acute exposure, biphenyl exerts a toxic action on the central nervous system, on the peripheral nervous system, and on the liver. Symptoms of poisoning are headache, diffuse, gastrointestinal pain, nausea, indigestion, numbness and aching of limbs, and general fatigue. The oral LD_{50} rat is 3,280 mg/kg.[9]

Long Term Exposure: Chronic exposure is characterized mostly be central nervous system symptoms, fatigue, headache, tremor, insomnia, sensory impairment, and mood changes. However, such symptoms may be rare. May cause lung irritation and bronchitis. liver and kidney damage. May cause skin allergy with itching and rash.

Points of Attack: Liver, skin, central nervous system, upper respiratory system, eyes.

Medical Surveillance: Consider skin, eye, liver function, and respiratory tract irritation in any preplacement or periodic examination. Examination by a qualified allergist. Examination of the nervous system.

First Aid: If this chemical gets into the eyes, remove any contact lenses at once and irrigate immediately for at least 15 minutes, occasionally lifting upper and lower lids. Seek medical attention immediately. If this chemical contacts the skin, remove contaminated clothing and wash immediately with soap and water. Seek medical attention immediately. If this chemical has been inhaled, remove from exposure, begin rescue breathing (using universal precautions) if breathing has stopped and CPR if heart action has stopped. Transfer promptly to a medical facility. When this chemical has been swallowed, get medical attention. Do not induce vomiting.

Personal Protective Methods: Because of its low vapor pressure and low order of toxicity, it does not usually present a major problem in industry. Protective creams, gloves, and masks with organic vapor canisters for use in areas of elevated vapor concentrations should suffice. Elevated temperature may increase the requirement for protective methods or ventilation. Wear appropriate clothing to prevent repeated or prolonged skin contact. Wear eye protection to prevent any possibility of eye contact with molten biphenyl. Employees should wash promptly when skin is contaminated. Work clothing should be changed daily if it may be contaminated. Remove nonimpervious clothing immediately if wet or contaminated.

Respirator Selection: OSHA: *10 mg/m³*: CCROVDM [any chemical cartridge respirator with organic vapor cartridge(s) in combination with a dust and mist filter]; SA (any supplied-air respirator). *25 mg/m³:* SA:CF (any supplied-air respirator operated in a continuous-flow mode); or PAPROVDM [any powered, air-purifying respirator with organic vapor cartridge(s) in combination with a dust and mist filter]. *50 mg/m³:* CCRFOVHiE [any chemical cartridge respirator with a full facepiece and organic vapor cartridge(s) in combination with a high-efficiency particulate filter]; or GMFOVHiE [any air-purifying, full-facepiece respirator (gas mask) with a chin-style, front- or back-mounted organic vapor canister having a high-efficiency particulate filter]; or PAPRTOVHiE [any powered, air-purifying respirator with a tight-fitting facepiece and organic vapor cartridge(s) in combination with a high-efficiency particulate filter]; or SCBAF (any self-contained breathing apparatus with a full facepiece); or SAF (any supplied-air respirator with a full facepiece). *100 mg/m³*: SAF:PD,PP (any supplied-air respirator that has a full facepiece and is operated in a pressure-demand or other positive-pressure mode). *Emergency or planned entry into unknown concentrations or IDLH conditions:* SCBAF:PD,PP (any self-contained breathing apparatus that has a full facepiece and is operated in a pressure-demand or other positive-pressure mode); or SAF:PD,PP:ASCBA (any supplied-air respirator that has a full facepiece and is operated in a pressure-demand or other positive-pressure mode in combination with an auxiliary, self-contained breathing apparatus operated in a pressure-demand or other positive-pressure mode). *Escape:* GMFOVHiE [any air-purifying, full-facepiece respirator (gas mask) with a chin-style, front- or back-mounted organic vapor canister having a high-efficiency particulate filter]; or SCBAE [any appropriate escape-type, self-contained breathing apparat(s)].

Note: Substance reported to cause eye irritation or damage; may require eye protection.

Storage: Before entering confined space where biphenyl may be present, check to make sure that an explosive concentration does not exist. Store in tightly closed containers in a cool, well ventilated area. Metal containers involving the transfer of this chemical should be grounded and bonded. Where possible, automatically pump liquid from drums or other storage containers to process containers. Drums must be equipped with self-closing valves, pressure vacuum bungs, and flame arresters. Use only non-sparking tools and equipment, especially when opening and closing containers of this chemical. Sources of ignition such as smoking and open flames, are prohibited where this chemical is used, handled, or stored in a manner that could create a potential fire or explosion hazard.

Shipping: This material is not specifically cited in the DOT's Performance Oriented Packaging Standards[19] as regards label requirements or limitations on shipping quantities.

Spill Handling: Evacuate and restrict persons not wearing protective equipment from area of spill or leak until cleanup is complete. Remove all ignition sources. Establish ventilation to keep levels below explosive limit. Spill material should be dampened with alcohol to avoid dust. Collect powdered material in the most convenient and safe manner and deposit in sealed containers. Ventilate area of spill or leak following clean-up. It may be necessary to contain and dispose of this chemical as a hazardous waste. If material or contaminated runoff enters waterways, notify downstream users of potentially contaminated waters. Contact your Department of Environmental Protection or your regional office of the federal EPA for specific recommendations. If employees are required to clean-up spills, they must be properly trained and equipped. OSHA 1910.120(q) may be applicable.

Fire Extinguishing: Use dry chemical, carbon dioxide, water spray, or alcohol foam extinguishers. Poisonous gases are produced in fire including carbon monoxide and acrid smoke. If material or contaminated runoff enters waterways, notify downstream users of potentially contaminated waters. Notify local health and fire officials and pollution control agencies. From a secure, explosion-proof location, use water spray to cool exposed containers. If cooling streams are ineffective (venting sound increases in volume and pitch, tank discolors, or shows any signs of deforming), withdraw immediately to a secure position. If employees are expected to fight fires, they must be trained and equipped in OSHA 1910.156.

Use carbon dioxide, dry chemical, or water spray, mist or fog to fight fires.

Disposal Method Suggested: Incineration.[22]

References

National Institute for Occupational Safety and Health, Profiles on Occupational Hazards for Criteria Document Priorities: Diphenyl, Report PB 274,073, Cincinnati, OH pp 274–276 (1977)

Sax, N. I., Ed., "Dangerous Properties of Industrial Materials Report" 1, No. 5, 42–43 (1981)

New Jersey Department of Health and Senior Services, "Hazardous Substance Fact Sheet: Diphenyl," Trenton, NJ (December 1998)

Bis(2-Chloroethoxy)Methane

Molecular Formula: $C_5H_{10}C_{12}O_2$

Common Formula: $ClCH_2CH_2OCH_2OCH_2CH_2Cl$

Synonyms: A13-01455; Bis(β-chlorethyl)formal; Bis(chlorethyl)formal; Bis(2-cloroetoxi)metano (Spanish); β,β-Dichlorodiethyl formal; Dichlorodiethyl formal; Dichlorodiethyl methylal; 2,2-Dichloroethyl formal; Di-2-chloroethyl formal; Dichloroethyl formal; Dichloromethoxy ethane; Ethane,1,1'-

[methylenebis(oxy)]bis(2-chloro-); Formaldehyde bis(β-chloroethyl) acetal; Formaldehyde bis(2-chloroethyl) acetal; Methane, bis(2-chloroethoxy)-; 1,1-[Methylenebis(oxy)]bis(2-chloroethane)

CAS Registry Number: 111-91-1

RTECS Number: PA3675000

Regulatory Authority

- CLEAN WATER ACT: 40CFR423, Appendix A, Priority Pollutants
- EPA HAZARDOUS WASTE NUMBER (RCRA No.): U024
- RCRA, 40CFR261, Appendix 8 Hazardous Constituents
- RCRA 40CFR268.48; 61FR15654, Universal Treatment Standards: Wastewater (mg/l), 0.036; Nonwastewater (mg/kg), 7.2
- RCRA 40CFR264, Appendix 9; TSD Facilities Ground Water Monitoring List. Suggested test method(s) (PQL μg/l): 8270 (10)
- SUPERFUND/EPCRA 40CFR302.4 Reportable Quantity (RQ): CERCLA, 1,000 lb (454 kg)
- EPCRA Section 313 Form R *de minimis* concentration reporting level: 1.0%

Cited in U.S. State Regulations: Kansas (G), Louisiana (G), Massachusetts (G), Pennsylvania (G), Vermont (G), Virginia (G), Washington (G), Wisconsin (G).

Description: Bis(2-chloroethoxy)methane, $C_5H_{10}C_{12}O_2$, is a colorless liquid. Boiling point = 218°C. Flash point = 110°C (oc). Hazard Identification (based on NFPA-704 M Rating System): Health 2, Flammability 1, Reactivity 0.

Potential Exposure: The chloroalkyl ethers have a wide variety of industrial uses in organic synthesis, treatment of textiles, the manufacture of polymers and insecticides, as degreasing agents and solvents, and in the preparation of ion exchange resins.

Incompatibilities: Oxidizing materials.

Permissible Exposure Limits in Air: No standards set.

Permissible Concentration in Water: No criteria set because of inadequate data according to EPA.

Harmful Effects and Symptoms

Specific data on BCEXM are very sparse. The reader is referred to the sections on other chloroalkyl ethers: Chloromethyl methyl ether, CMME: Bis(chloromethyl) ether, BCME: Bis(2-chloroethyl) ether, BCEE - Bis (2-chlorosiopropyl) ether, BCIE The oral LD_{50} rat is 65 mg/kg.[9] This material is toxic by inhalation and ingestion and is a strong irritant.

First Aid: If this chemical gets into the eyes, remove any contact lenses at once and irrigate immediately for at least 15 minutes, occasionally lifting upper and lower lids. Seek medical attention immediately. If this chemical contacts the skin, remove contaminated clothing and wash immediately with soap and water. Seek medical attention immediately. If this chemical has been inhaled, remove from exposure, begin rescue breathing (using universal precautions) if breathing has stopped and CPR if heart action has stopped. Transfer promptly to a medical facility. When this chemical has been swallowed, get medical attention. Give large quantities of water and induce vomiting. Do not make an unconscious person vomit.

Personal Protective Methods: Wear protective gloves and clothing to prevent any reasonable probability of skin contact. Safety equipment suppliers/manufacturers can provide recommendations on the most protective glove/clothing material for your operation. All protective clothing (suits, gloves, footwear, headgear) should be clean, available each day, and put on before work. Contact lenses should not be worn when working with this chemical. Wear splash-proof chemical goggles and face shield unless full facepiece respiratory protection is worn. Employees should wash immediately with soap when skin is wet or contaminated. Provide emergency showers and eyewash.

Respirator Selection: SCBAF:PD,PP (any MSHA/NIOSH approved self-contained breathing apparatus that has a full facepiece and is operated in a pressure-demand or other positive-pressure mode); or SAF:PD,PP:ASCBA (any supplied-air respirator that has a full facepiece and is operated in a pressure-demand or other positive-pressure mode in combination with an auxiliary, self-contained breathing apparatus operated in a pressure-demand or other positive pressure mode)

Storage: Store in tightly closed containers in a cool, well ventilated area away from oxidizers. Sources of ignition such as smoking and open flames, are prohibited where this chemical is used, handled, or stored in a manner that could create a potential fire or explosion hazard.

Shipping: This material is not listed in the DOT performance-oriented packaging standards as regards label requirements or weight limitations on air or rail shipments.

Spill Handling: Evacuate and restrict persons not wearing protective equipment from area of spill or leak until cleanup is complete. Remove all ignition sources. Collect powdered material in the most convenient and safe manner and deposit in sealed containers. Ventilate area of spill or leak after clean-up is complete. It may be necessary to contain and dispose of this chemical as a hazardous waste. If material or contaminated runoff enters waterways, notify downstream users of potentially contaminated waters. Contact your Department of Environmental Protection or your regional office of the federal EPA for specific recommendations. If employees are required to clean-up spills, they must be properly trained and equipped. OSHA 1910.120(q) may be applicable.

Fire Extinguishing: This chemical is a combustible liquid. Use dry chemical, carbon dioxide, water spray, or alcohol foam extinguishers. Poisonous gases are produced in fire. If material or contaminated runoff enters waterways, notify downstream users of potentially contaminated waters. Notify local health and fire officials and pollution control agencies. From a secure, explosion-proof location, use water spray to cool exposed containers. If cooling streams are ineffective (venting sound increases in volume and pitch, tank discolors, or shows any signs of deforming), withdraw immediately to a secure position. If employees are expected to fight fires, they must be trained and equipped in OSHA 1910.156.

Disposal Method Suggested: Destroy by high-temperature incineration with HCl scrubber.

References

U.S. Environmental Protection Agency, Haloethers: Ambient Water Quality Criteria, Washington, DC (1980)

U.S. Environmental Protection Agency, Chloroalkyl Ethers: Ambient Water Quality Criteria, Washington, DC (1980)

U.S. Environmental Protection Agency, Bis(2-Chloroethoxy)Methane, Health and Environmental Effects Profile No. 23, Washington, DC, Office of solid Waste (April 30, 1980)

Sax, N. I., Ed., "Dangerous Properties of Industrial Materials Report" 7, No. 4, 39–42 (1987)

Bis(2-Chloroisopropyl)Ether

Molecular Formula: $C_6H_{12}Cl_2O$

Common Formula: $[ClCH_2CH(CH_3)]_2O$

Synonyms: BCIE; BCMEE; Bis(β-chloroisopropyl) ether; Bis(chloromethyl) ether; Bis(2-chloro-1-methylethyl) ether; Bis(1-chloro-2-propyl) ether; Bis(2-clorometil)eter (Spanish); (2-Chloro-1-methylethyl) ether; DCIP (nematocide); β,β'-Dichlorodiisopropyl ether; Dichlorodiisopropyl ether; 2,2'-Dichloroisopropyl ether; Dichloroisopropyl ether; Ether, bis(2-chloro-1-methylethyl); NCI-C50044; Nemamort; 2,2'-Oxybis(1-chloropropane); Propane, 2,2'-oxybis(1-chloro-); Propane,2,2'-oxybis(1-chloro)

CAS Registry Number: 108-60-1

RTECS Number: KN1750000

DOT ID: UN 2490

Regulatory Authority

- Air Pollutant Standard Set (former USSR)[35]
- CLEAN WATER ACT: 40CFR423, Appendix A, Priority Pollutants
- EPA HAZARDOUS WASTE NUMBER (RCRA No.): U027
- RCRA, 40CFR261, Appendix 8 Hazardous Constituents
- RCRA 40CFR268.48; 61FR15654, Universal Treatment Standards: Wastewater (mg/l), 0.055; Nonwastewater (mg/kg), 7.2
- RCRA 40CFR264, Appendix 9; TSD Facilities Ground Water Monitoring List. Suggested test method(s) (PQL μg/l): 8010 (100); 8270 (10)
- SUPERFUND/EPCRA 40CFR302.4 Reportable Quantity (RQ): CERCLA, 1,000 lb (454 kg)
- Canada, WHMIS, Ingredients Disclosure List

Cited in U.S. State Regulations: California (G), Florida (G), Kansas (G), Louisiana (G), Massachusetts (G), New Hampshire (G), New Jersey (G), Pennsylvania (G), Vermont (G), Virginia (G), Washington (G), Wisconsin (G).

Description: BCIE, $[ClCH_2CH(CH_3)]_2O$, is a colorless liquid. Boiling point = 187 – 189°C. Flash point = 85°C (oc).[17]

Potential Exposure: BCIE was previously used as a solvent and as an extractant. It may be formed as a by -product of propylene oxide production. It has been found in industrial waste water and in natural water. Flash point = 85°C. Hazard Identification (based on NFPA-704 M Rating System): Health 2, Flammability 2, Reactivity 0. Slightly soluble in water.

Incompatibilities: Strong oxidizers, strong acids and oxygen. It may form dangerous peroxides upon standing; may explode when heated.

Permissible Exposure Limits in Air: No standard set. A provisional occupational limit has been suggested of 15 ppm. The former USSR[35] has set a workplace ceiling value of 5 mg/m^3.

Permissible Concentration in Water: To protect freshwater aquatic life -238,000 μg/l on an acute basis for chloroalkyl ether in general. No criteria developed for saltwater aquatic life due to lack of data. For protection of human health, the ambient water criterion is 34.7 μg/l.[6] Kansas has set a guideline for this compound in drinking water of 34.7 μg/l.[61]

Determination in Water: Gas chromatography (EPA Method 611) or gas chromatography plus mass spectrometry (EPA Method 625).

Routes of Entry: Inhalation, passing through the skin.

Harmful Effects and Symptoms

There is no empirical evidence that BCIE is carcinogenic; however, some chronic toxic effects of the compound have been noted.

Short Term Exposure: Can cause irritation and burns on contact with eyes, nose and skin. Very high levels of BCIE may cause loss of appetite, fatigue, irritability and even death. The oral LD_{50} rat is 240 mg/kg (Moderately toxic).[9]

Long Term Exposure: There is limited evidence that this chemical may cause cancer in animals; lung adenomas. Many similar solvents can cause brain or nerve damage, this chemical has not been fully evaluated for these effects. May cause liver and kidney damage.

Medical Surveillance: Kidney and liver function tests. Evaluate for brain effects.

First Aid: If this chemical gets into the eyes, remove any contact lenses at once and irrigate immediately for at least 20 minutes, occasionally lifting upper and lower lids. Seek medical attention immediately. If this chemical contacts the skin, remove contaminated clothing and wash immediately with soap and water. Seek medical attention immediately. If this chemical has been inhaled, remove from exposure, begin

rescue breathing (using universal precautions) if breathing has stopped and CPR if heart action has stopped. Transfer promptly to a medical facility. When this chemical has been swallowed, get medical attention. Give large quantities of water and induce vomiting. Do not make an unconscious person vomit.

Personal Protective Methods: Wear protective gloves and clothing to prevent any reasonable probability of skin contact. Safety equipment suppliers/manufacturers can provide recommendations on the most protective glove/clothing material for your operation. All protective clothing (suits, gloves, footwear, headgear) should be clean, available each day, and put on before work. Contact lenses should not be worn when working with this chemical. Wear splash-proof chemical goggles and face shield unless full facepiece respiratory protection is worn. Employees should wash immediately with soap when skin is wet or contaminated. Provide emergency showers and eyewash.

Respirator Selection: SCBAF:PD,PP (any MSHA/NIOSH approved self-contained breathing apparatus that has a full facepiece and is operated in a pressure-demand or other positive-pressure mode); or SAF:PD,PP:ASCBA (any supplied-air respirator that has a full facepiece and is operated in a pressure-demand or other positive-pressure mode in combination with an auxiliary, self-contained breathing apparatus operated in a pressure-demand or other positive pressure mode).

Storage: Ethers tend to peroxidize forming unstable peroxides. Before entering confined space where BCIE may be present, check to make sure that an explosive concentration does not exist. Store in tightly closed containers in a cool, well ventilated area away from oxidizing materials. Metal containers involving the transfer of this chemical should be grounded and bonded. Where possible, automatically pump liquid from drums or other storage containers to process containers. Drums must be equipped with self-closing valves, pressure vacuum bungs, and flame arresters. Use only non-sparking tools and equipment, especially when opening and closing containers of this chemical. Sources of ignition such as smoking and open flames, are prohibited where this chemical is used, handled, or stored in a manner that could create a potential fire or explosion hazard.

Shipping: The label required is "Poison." This material falls in Hazard Class 6.1 and Packing Group II.

Spill Handling: Evacuate and restrict persons not wearing protective equipment from area of spill or leak until cleanup is complete. Remove all ignition sources. Collect powdered material in the most convenient and safe manner and deposit in sealed containers. Ventilate area of spill or leak after clean-up is complete. It may be necessary to contain and dispose of this chemical as a hazardous waste. If material or contaminated runoff enters waterways, notify downstream users of potentially contaminated waters. Contact your Department of Environmental Protection or your regional office of the federal EPA for specific recommendations. If employees are required to clean-up spills, they must be properly trained and equipped. OSHA 1910.120(q) may be applicable.

Fire Extinguishing: Use dry chemical, carbon dioxide, or foam extinguishers. Poisonous gases are produced in fire including hydrogen chloride. If material or contaminated runoff enters waterways, notify downstream users of potentially contaminated waters. Notify local health and fire officials and pollution control agencies. From a secure, explosion-proof location, use water spray to cool exposed containers. If cooling streams are ineffective (venting sound increases in volume and pitch, tank discolors, or shows any signs of deforming), withdraw immediately to a secure position. If employees are expected to fight fires, they must be trained and equipped in OSHA 1910.156.

Disposal Method Suggested: Use special incinerator due to high HCl content such as seagoing incinerator ships.[22]

References

U.S. Environmental Protection Agency, Chloroalkyl Ethers: Ambient Water Quality Criteria, Washington, DC (1980

U.S. Environmental Protection Agency, Bis(2-Chloroisopropyl) Ether, Health and Environmental Effects Profile No. 25, Washington, DC, Office of Solid Waste (April 30, 1980)

U.S. Environmental Protection Agency, Chemical Hazard Information Profile Draft Report: Bis(2-Chloro-1-Methylethyl) Ester (BCMEE); Washington, DC (July 29, 1983)

Sax, N. I., Ed., "Dangerous Properties of Industrial Materials Report" 6, No. 3, 47–49 (1986)

New Jersey Department of Health and Senior Services, Hazardous Substance Fact Sheet, Bis(2-chloro-1-methylethyl)ether, Trenton NJ (May, 1998)

Bis(Chloromethyl) Ether

Molecular Formula: $C_2H_4Cl_2O$

Common Formula: $ClCH_2OCH_2Cl$

Synonyms: BCME; Bis(2-chloromethyl) ether; Bis(clorometil)eter (Spanish); Bis-CME; Chloro(chloromethoxy)-methane; Chloromethyl ether; Dichlordimethylaether (German); α,α'-Dichlorodimethyl ether; *sym*-Dichlorodimethyl ether; Dichlorodimethyl ether; Dichlorodimethyl ether, *symmetrical*; *sym*-Dichloromethyl ether; Dichloromethyl ether; Dimethyl-1,1'-dichloroether; Ether, bis(chloromethyl); Methane oxybis(chloro-); Monochloromethyl ether; Oxybis(chloromethane)

CAS Registry Number: 542-88-1

RTECS Number: KN1575000

DOT ID: UN 2249

EEC Number: 603-046-00-5

Regulatory Authority

- Carcinogen (Human Positive) (IARC)[9] (DFG)[3] (OSHA)[58] (ACGIH) (Australia) (Mexico)

- Banned or Severely Restricted (Finland, Israel, Japan, Sweden) (UN)[13]
- Very Toxic Substance (World Bank)[15]
- OSHA, 29CFR1910 Specifically Regulated Chemicals (Scc CFR1910.1008)
- Air Pollutant Standard Set (ACGIH)[1] (HSE)[33] (Several Canadian Provinces)
- CLEAN AIR ACT: Hazardous Air Pollutants (Title I, Part A, Section 112); Accidental Release Prevention/Flammable substances, (Section 112[r], Table 3), TQ = 1,000 lb (454 kg)
- EPA HAZARDOUS WASTE NUMBER (RCRA No.): P016
- RCRA, 40CFR261, Appendix 8 Hazardous Constituents
- RCRA Land Ban Waste
- SUPERFUND/EPCRA 40CFR355, Appendix B Extremely Hazardous Substances: TPQ = 100 lb (45.4 kg)
- SUPERFUND/EPCRA 40CFR302.4 Reportable Quantity (RQ): EHS = 10 lb (4.54 kg)
- EPCRA Section 313 Form R *de minimis* concentration reporting level: 1.0%
- Canada, WHMIS, Ingredients Disclosure List; National Pollutant Release Inventory (NPRI)

Cited in U.S. State Regulations: Alaska (G), California (A, G), Florida (G), Illinois (G), Kansas (G, W), Louisiana (G), Maryland (G), Massachusetts (G), Minnesota (G), Michigan (G), New Jersey (G), New Hampshire (G), Oklahoma (G), Pennsylvania (G), Rhode Island (G), Vermont (G), Virginia (G), Washington (G), West Virginia (G), Wisconsin (G).

Description: Bis(chloromethyl) ether, $ClCH_2OCH_2Cl$, is a colorless, volatile liquid with a suffocating odor. Boiling point = 104°C. Freezing/Melting point = -42°C. Flash point ≤ 19°C. Hazard Identification (based on NFPA-704 M Rating System): Health 4, Flammability 3, Reactivity 1. Insoluble in water (decomposes). This substance may form spontaneously in warm moist air by the combination of formaldehyde and hydrogen chloride.

Potential Exposure: Exposure to bis(chloromethyl) ether may occur in industry and in the laboratory. This compound is used as an alkylating agent in the manufacture of polymers, as a solvent for polymerization reactions, in the preparation of ion exchange resins, and as an intermediate for organic synthesis. Haloethers, primarily α-chloromethyl ethers, represent a category of alkylating agents of increasing concern due to the establishment of a causal relationship between occupational exposure to two agents of this class and lung cancer in the United States and abroad. The cancers are mainly oat cell carcinomas. Potential sources of human exposure to BCME appear to exist primarily in areas including: (a) its use in chloromethylating (crosslinking) reaction mixtures in anion-exchange resin production; (b) segments of the textile industry using formaldehyde-containing reactants and resins in the finishing of fabric and as adhesive in the laminating and flocking of fabrics; and (c) the nonwoven industry which uses as binders, thermosetting acrylic emulsion polymers comprising methylol acrylamide, since a finite amount of formaldehyde is liberated on the drying and curing of these bonding agents. NIOSH has confirmed the spontaneous formation of BCME from the reaction of formaldehyde and hydrochloric acid in some textile plants and is now investigating the extent of possible worker exposure to the carcinogen. However, this finding has been disputed by industrial tests in which BCME was not formed in air by the reaction of textile systems employing hydrochloric acid and formaldehyde.

Incompatibilities: Forms explosive mixture with air. Incompatible with strong acids. Decomposes on contact with water, moist air, and heat forming corrosive hydrochloric acid and formaldehyde vapors. May form shock-sensitive compounds on contact with oxidizers, peroxides, and sunlight. Attacks many plastics.

Permissible Exposure Limits in Air: Bis(chloromethyl) ether is included in the Federal standard for carcinogens; all contact with it should be avoided. The ACGIH has set a TWA limit of 0.001 ppm (0.005 m/gm^3) with the notation that the material is a human carcinogen. DFG (Germany), Australia, Mexico, Quebec (Canada) have no numerical limits — only the human carcinogen notation. In Canada Alberta has a TWA of 0.005 ppm and STEL of 0.015 ppm. Czechoslovakia[35] has set a TWA of 0.00025 mg/m^3 and a ceiling value of 0.005 mg/m^3.

Determination in Air: Collection by charcoal tube, analysis by gas liquid chromatography.

Permissible Concentration in Water: For maximum protection of human health from potential carcinogenic effects of exposure to BCME through ingestion of water and contaminated aquatic organisms, the ambient water concentration is zero. Concentrations of BCME estimated to result in additional lifetime cancer risks of 1 in 100,000 are presented by a concentration of 0.038 ng/l (3.8×10^{-5} μg/l).[6] Kansas has set a guideline for drinking water also.[61]

Determination in Water: Gas chromatography (EPA Method 611) or gas chromatography plus mass spectrometry (EPA Method 625).

Routes of Entry: Inhalation of vapor, and percutaneous absorption.

Harmful Effects and Symptoms

Bis(chloromethyl) ether has an extremely suffocating odor even in minimal concentration so that experience with acute poisoning is not available. It is not considered a respiratory irritant at concentrations of 10 ppm Bis(chloromethyl) ether is a known human carcinogen. Animal experiments have shown increases in lung adenoma incidence; olfactory esthesioneuroepitheliomas which invaded the sinuses, cranial vault, and brain; skin papillomas and carcinomas; and subcutaneous fibrosarcomas. There have been several reports of increased incidence of human lung carcinomas (primarily small cell undifferentiated) among ether workers exposed to bis(chloromethyl) ether as an impurity. The latency period is relatively short — 10 – 15 years. Smokers as well as nonsmokers may be affected.

Short Term Exposure: This chemical is corrosive to the eyes, skin, and respiratory tract. Inhalation can cause pulmonary edema, a medical emergency that can be delayed for several hours. This can cause death. Affects the nervous system. Symptoms can be loss of appetite, nausea, and fatigue; higher exposures can cause irritability, anxiety, and weakness.

Long Term Exposure: Can cause liver and kidney damage. Lungs may be affected by repeated or prolonged exposure. This substance is carcinogenic to humans and has caused lung cancer in humans. May cause genetic damage in humans.

Points of Attack: Skin, respiratory tract, eyes, lungs.

Medical Surveillance: Preplacement and periodic medical examinations should include an examination of the skin and respiratory tract, including chest x-ray. Sputum cytology has been suggested as helpful in detecting early malignant changes, and in this connection a smoking history is of importance. Possible effects on the fetus should be considered. Consider chest x-ray following acute overexposure.

First Aid: If this chemical gets into the eyes, remove any contact lenses at once and irrigate immediately for at least 15 minutes, occasionally lifting upper and lower lids. Seek medical attention immediately. If this chemical contacts the skin, remove contaminated clothing and wash immediately with soap and water. Seek medical attention immediately. If this chemical has been inhaled, remove from exposure, begin rescue breathing (using universal precautions) if breathing has stopped and CPR if heart action has stopped. Transfer promptly to a medical facility. When this chemical has been swallowed, get medical attention. Give large quantities of water and induce vomiting. Do not make an unconscious person vomit.

Medical observation is recommended for 24 – 48 hours after breathing overexposure, as pulmonary edema may be delayed. As first aid for pulmonary edema, a doctor or authorized paramedic may consider administering a corticosteroid spray.

Personal Protective Methods: These are designed to supplement engineering controls and should be appropriate for protection of all skin or respiratory contact. Full body protective clothing and gloves should be used on entering areas of potential exposure. Those employed in handling operations should be provided with full face, supplied air respirators of continuous flow or pressure demand type. Wash thoroughly *immediately* following exposure to this chemical. On exit from a regulated area, employees should remove and leave protective clothing and equipment at the point of exit, to be placed in impervious containers at the end of the work shift for decontamination or disposal. Showers should be taken before dressing in street clothes.

Respirator Selection: NIOSH: *At any detectable concentration:* SCBAF:PD,PP (any MSHA/NIOSH approved self-contained breathing apparatus that has a full facepiece and is operated in a pressure-demand or other positive-pressure mode); or SAF:PD,PP:ASCBA (any supplied-air respirator that has a full facepiece and is operated in a pressure-demand or other positive-pressure mode in combination with an auxiliary, self-contained breathing apparatus operated in a pressure-demand or other positive pressure mode). *Escape:* GMFOV [any air-purifying, full-facepiece respirator (gas mask) with a chin-style, front-or back-mounted organic vapor canister]; or SCBAE (any appropriate escape-type, self-contained breathing apparatus).

Storage: Prior to working with this chemical you should be trained on its proper handling and storage. Before entering confined space where this chemical may be present, check to make sure that an explosive concentration does not exist. Store in airtight containers in a cool, dry, well ventilated area. Metal containers involving the transfer of this chemical should be grounded and bonded. Where possible, automatically pump liquid from drums or other storage containers to process containers. Drums must be equipped with self-closing valves, pressure vacuum bungs, and flame arresters. Use only nonsparking tools and equipment, especially when opening and closing containers of this chemical. Sources of ignition such as smoking and open flames, are prohibited where this chemical is used, handled, or stored in a manner that could create a potential fire or explosion hazard. A regulated, marked area should be established where this chemical is handled, used, or stored in compliance with OSHA standard 1910.1045.

Shipping: Dichlorodimethyl ether, symmetrical, has a DOT label requirement of "Poison." Shipment by passenger aircraft or railcar or even by cargo aircraft is forbidden. This material falls in UN/DOT Hazard Class 6.1 and Packing Group I.[19][20]

Spill Handling: Evacuate area. Seek professional environmental engineering assistance from the U.S. EPA Environmental Response Team at (908) 548-8730 (24-hour response line). Full body protective clothing and gloves should be used on entering areas of potential exposure. Stay upwind; keep out of low areas. Ventilate closed spaces before entering them. Wear positive pressure breathing apparatus and special protective clothing. Remove and isolate contaminated clothing at the site. *Spill or leak:* do not touch spilled material; stop leak if you can do so without risk. Use water spray to reduce vapors. *Small spills:* absorb with sand or other noncombustible material and place into containers for later disposal. *Large spills:* dike far ahead of spills for later disposal. Do NOT let this chemical enter the environment (extra personal protection: complete protective clothing including self-contained breathing apparatus). Evacuate and restrict persons not wearing protective equipment from area of spill or leak until cleanup is complete. Remove all ignition sources. It may be necessary to contain and dispose of this chemical as a hazardous waste. If material or contaminated runoff enters waterways, notify downstream users of potentially contaminated waters.

Contact your Department of Environmental Protection or your regional office of the federal EPA for specific recommendations. If employees are required to clean-up spills, they must be properly trained and equipped. OSHA 1910.120(q) may be applicable.

Fire Extinguishing: This chemical is a flammable liquid. Poisonous gases including hydrogen chloride are produced in fire. Use dry chemical, carbon dioxide, water spray, or foam extinguishers. Vapors are heavier than air and will collect in low areas. Vapors may travel long distances to ignition sources and flashback. Vapors in confined areas may explode when exposed to fire. Storage containers and parts of containers may rocket great distances, in many directions. If material or contaminated runoff enters waterways, notify downstream users of potentially contaminated waters. Notify local health and fire officials and pollution control agencies. From a secure, explosion-proof location, use water spray to cool exposed containers. If cooling streams are ineffective (venting sound increases in volume and pitch, tank discolors, or shows any signs of deforming), withdraw immediately to a secure position. If employees are expected to fight fires, they must be trained and equipped in OSHA 1910.156.

Disposal Method Suggested: Incineration, preferably after mixing with another combustible fuel. Care must be exercised to assure complete combustion to prevent the formation of phosgene. An acid scrubber is necessary to remove the halo acids produced.[22]

References

U.S. Environmental Protection Agency, Chloroalkyl Ethers: Ambient Water Quality Criteria, Washington, DC (1980)

U.S. Environmental Protection Agency, Haloethers: Ambient Water Quality Criteria, Washington, DC (1980)

U.S. Environmental Protection Agency, Bis(Chloromethyl) Ether, Health and Environmental Effects Profile No. 26, Washington, DC, Office of Solid Waste (April 30, 1980)

Sax, N. I., Ed., "Dangerous Properties of Industrial Materials Report" 6, No. 3, 49–52 (1986)

U.S. Environmental Protection Agency, "Chemical Profile: Chloromethyl Ether," Washington, DC, Chemical Emergency Preparedness Program (November 30, 1987)

U.S. Public Health Service, "Toxicological Profile for Bis(Chloromethyl) Ether," Atlanta, Georgia, Agency for Toxic Substances and Disease Registry (December 1988)

New Jersey Department of Health and Senior Services, Hazardous Substance Fact Sheet, Bis(2-chloromethyl)ether, Trenton NJ (July, 1998)

Bis(Chloromethyl) Ketone

Molecular Formula: $C_3H_4Cl_2O$

Common Formula: $ClCH_2COCH_2Cl$

Synonyms: Bis(chloromethyl) ketone; *sym*-Dichloroacetone; α,α'-Dichloroacetone; α,γ'-Dichloroacetone; 1,3-Dichloroacetone; 1,3-Dichloro-2-propanone

CAS Registry Number: 534-07-6

RTECS Number: UC1430000

DOT ID: UN 2649

Regulatory Authority

- SUPERFUND/EPCRA 40CFR355, Appendix B Extremely Hazardous Substances: TPQ = 10/10,000 lb (4.54/4,540 kg)
- Canada, WHMIS, Ingredients Disclosure List

Cited in U.S. State Regulations: Florida (G), Massachusetts (G), New Jersey (G), Pennsylvania (G).

Description: Bis(chhloromethyl)ketone, $ClCH_2COCH_2Cl$, is a crystalline solid. Freezing/Melting point = 45°C. Boiling point = 173°C. Soluble in water.

Potential Exposure: Formerly extensively used in textiles (especially polyester fabrics) and still employed in polyurethane foams, textile backcoating and adhesives.

Permissible Exposure Limits in Air: No standards set.

Permissible Concentration in Water: No criteria set.

Routes of Entry: Inhalation, ingestion, skin and/or eye contact.

Harmful Effects and Symptoms

Short Term Exposure: It causes tearing and blistering. It may be fatal if inhaled, swallowed or absorbed through skin. Contact may cause burns to skin and eyes.

Long Term Exposure: Due to the availability of insufficient data on short term effects, caution should be exercised.

Medical Surveillance: See: NIOSH Criteria Document: 78–173 KETONES.

First Aid: If this chemical gets into the eyes, remove any contact lenses at once and irrigate immediately for at least 15 minutes, occasionally lifting upper and lower lids. Seek medical attention immediately. If this chemical contacts the skin, remove contaminated clothing and wash immediately with soap and water. Seek medical attention immediately. If this chemical has been inhaled, remove from exposure, begin rescue breathing (using universal precautions) if breathing has stopped and CPR if heart action has stopped. Transfer promptly to a medical facility. When this chemical has been swallowed, get medical attention. Give large quantities of water and induce vomiting. Do not make an unconscious person vomit. Keep victim quiet and maintain normal body temperature. Effects may be delayed; keep victim under observation.

Shipping: The DOT label requirement[19] for 1,3-dichloroacetone is "Poison." The UN/DOT Hazard Class is 6.1 and the Packing Group is II.[19][20]

Spill Handling: Evacuate and restrict persons not wearing protective equipment from area of spill or leak until cleanup is complete. Remove all ignition sources. Ventilate area of spill or leak. Avoid inhalation; wear respiratory protection, eye protection and protective clothing. In case of contact, immediately flush skin or eyes with water. Do not touch spilled material; stop leak if you can do so without risk. Use water spray to reduce vapors. For small spills, absorb with sand or other noncombustible absorbent material and place into containers for later disposal. *Small dry spills:* with clean shovel place material into clean, dry container and cover; move containers from spill area. *For large spills:* dike far ahead of spill for later disposal. It may be necessary to contain and dispose of this chemical as a hazardous waste. If material or contaminated runoff enters waterways, notify downstream users of potentially contaminated waters. Contact your Department of Environmental Protection or your regional office of the federal EPA for specific recommendations. If employees are required to clean-up spills, they must be properly trained and equipped. OSHA 1910.120(q) may be applicable.

Fire Extinguishing: For small fires, use dry chemical, carbon dioxide, water spray, or foam. For large fires, use water spray, fog, or foam. Poisonous gases are produced in fire including chlorine. If material or contaminated runoff enters waterways, notify downstream users of potentially contaminated waters. Notify local health and fire officials and pollution control agencies. From a secure, explosion-proof location, use water spray to cool exposed containers. If cooling streams are ineffective (venting sound increases in volume and pitch, tank discolors, or shows any signs of deforming), withdraw immediately to a secure position. If employees are expected to fight fires, they must be trained and equipped in OSHA 1910.156.

References

U.S. Environmental Protection Agency, "Chemical Profile: Bis(chloromethyl)ketone," Washington, DC, Chemical Emergency Preparedness Program (November 30, 1987)

Bismuth and Compounds

Molecular Formula: Bi

Synonyms: Bismuth-209; Bismuto (Spanish)

CAS Registry Number: 7440-69-9

RTECS Number: EB2600000

Regulatory Authority

- Banned or Severely Restricted (In Medicine) (UN)[13]
- Mexico, wastewater as heavy metals

Cited in U.S. State Regulations: Oklahoma (G), Massachusetts (bismuth chromate) (G)

Description: Bismuth, Bi, is a pinkish-silver, hard, brittle metal. Freezing/Melting point = 271°C. Boiling point = 1,420 – 1,560°C. It is found as the free metal in ores such as bismutite and bismuthinite and in lead ores. Insoluble in water.

Potential Exposure: Bismuth is used as a constituent of tempering baths for steel alloys, in low Freezing/Melting point alloys which expand on cooling, in aluminum and steel alloys to increase machinability, and in printing type metal. Bismuth compounds are found primarily in pharmaceuticals as antiseptics, antacids, antiluetics, and as a medicament in the treatment of acute angina. They are also used as a contrast medium in roentgenoscopy and in cosmetics. For the general population the total intake from food is 5 – 20 μg with much smaller amounts contributed by air and water.

Incompatibilities: Reacts with strong acids and strong oxidants, chlorine, fused ammonium nitrates, iodine pentafluoride, and nitrosyl fluoride.

Permissible Exposure Limits in Air: There is no Federal standard for bismuth. ACGIH has set TWA values only for bismuth telluride (which see).

Permissible Concentration in Water: No criteria set but EPA[32] has suggested an ambient limit of 3.5 μg/l based on health effects. The former USSR-UNEP/IRPTC project[43] has set a MAC of 0.5 mg/l of trivalent bismuth and 0.1 mg/l of pentavalent bismuth.

Determination in Water: Atomic absorption spectrophotometry may be used.[1] Spark source mass spectrometry may also be used.

Routes of Entry: Ingestion of powder or inhalation of dust.

Harmful Effects and Symptoms

Most accounts of bismuth poisoning are from the soluble compounds used previously in therapeutics. Bismuth compounds have been withdrawn from pharmaceuticals because of reports of encephalopathy.[13] Fatalities and near fatalities have been reported chiefly as a result of intravenous or intramuscular injection of soluble salts.

Short Term Exposure: Bismuth and bismuth compounds have slight effect on intact skin and mucous membrane. Absorption occurs only minimally through broken skin.

Long Term Exposure: All bismuth compounds do not have equal toxicity. Although considered less hazardous than most heavy metals, can cause kidney and possible liver damage. Chronic intoxication from repeated oral or parenteral doses causes "bismuth line." This is a gum condition with black spots of buccal and colonic mucosa, superficial stomatitis, foul breath, and salivation.

Points of Attack: Kidneys, liver.

Medical Surveillance: No special considerations are necessary other than following good general health practices. Liver and kidney function should be followed if large amounts of soluble salts are ingested.

First Aid: If this chemical gets into the eyes, remove any contact lenses at once and irrigate immediately for at least 15 minutes, occasionally lifting upper and lower lids. Seek medical attention immediately. If this chemical contacts the skin, remove contaminated clothing and wash immediately with soap and water. Seek medical attention immediately. If this chemical has been inhaled, remove from exposure, begin rescue breathing (using universal precautions) if breathing has

stopped and CPR if heart action has stopped. Transfer promptly to a medical facility. When this chemical has been swallowed, get medical attention. Give large quantities of water and induce vomiting. Do not make an unconscious person vomit.

Note: Dimercaptol (BAL) brings good results in the treatment of bismuth poisoning if given early. Other measures include atropine and meperidine to relieve gastrointestinal discomfort.

Personal Protective Methods: Personal hygiene should be stressed, and eating should not be permitted in work areas. Dust masks should be worn in dusty areas to prevent inadvertent ingestion of the soluble bismuth compounds.

Shipping: Bismuth is not cited by DOT[19] in its performance-oriented packaging standards as regards label requirements or maximum permitted shipping quantities.

Spill Handling: Evacuate and restrict persons not wearing protective equipment from area of spill or leak until cleanup is complete. Remove all ignition sources. Collect powdered material in the most convenient and safe manner and deposit in sealed containers. Ventilate area of spill or leak after clean-up is complete. It may be necessary to contain and dispose of this chemical as a hazardous waste. If material or contaminated runoff enters waterways, notify downstream users of potentially contaminated waters. Contact your Department of Environmental Protection or your regional office of the federal EPA for specific recommendations. If employees are required to clean-up spills, they must be properly trained and equipped. OSHA 1910.120(q) may be applicable.

Fire Extinguishing: Use extinguishers suitable for surrounding fire. Poisonous gases are produced in fire. If material or contaminated runoff enters waterways, notify downstream users of potentially contaminated waters. Notify local health and fire officials and pollution control agencies. From a secure, explosion-proof location, use water spray to cool exposed containers. If cooling streams are ineffective (venting sound increases in volume and pitch, tank discolors, or shows any signs of deforming), withdraw immediately to a secure position. If employees are expected to fight fires, they must be trained and equipped in OSHA 1910.156.

Disposal Method Suggested: Dissolve in a minimum amount of concentrated HCl. Dilute with water until precipitate is formed. Redissolve in HCl. Then saturate with H_2S. Filter, wash, dry and return to supplier.

References

U.S. Environmental Protection Agency, Toxicology of Metals, Vol. II: Bismuth, Report EPA-600/1-77-022. Research Triangle Park, NC, pp 110–123 (May 1977)

Sax, N. I., Ed., "Dangerous Properties of Industrial Materials Report" 1, No. 5, 43–45 (1981) and 3, No. 5, 64–65 (1983)

Bismuth Telluride

Molecular Formula: Bi_2Te_3

Synonyms: Bismuth sesquitelluride; Dibismuth telluride

CAS Registry Number: 1304-82-1; 37293-14-4 (selenium-doped)

RTECS Number: EB3110000

Regulatory Authority

- Air Pollutant Standard Set (ACGIH)[1] (HSE)[33] (OSHA)[58] (Several States)[60] (Mexico) (Israel) (Australia) (Several Canadian Provinces)
- Canada, WHMIS, Ingredients Disclosure List

Cited in U.S. State Regulations: Alaska (G), California (G), Connecticut (A), Florida (G), Illinois (G), Maine (G), Massachusetts (G), Minnesota (G), (Nevada (A), New Hampshire (G, W), New Jersey (G), North Dakota (A), Pennsylvania (G), Rhode Island (G, A, W), South Carolina (A), South Dakota (A, W), Tennessee (W), Utah (W), Vermont (G), Virginia (A), West Virginia (G).

Description: Bismuth telluride, Bi_2Te_3, is a gray crystalline solid. Freezing/Melting point = 573°C.

Potential Exposure: Bismuth telluride is used thermoelectric cooling, power generation, and in semiconductor manufacture and exposure involves those working in "Silicon Valley" and similar areas around the world.

Incompatibilities: A violent reaction with strong oxidizers, and a toxic gas may evolve from contact with moisture.

Permissible Exposure Limits in Air: ACGIH[1] has set TWA values of 10 mg/m^3 for bismuth telluride and 5 mg/m^3 for Se-doped bismuth telluride but no STEL values. HSE[33] has set the same TWA values plus STEL values of 20 mg/m^3 for bismuth telluride and 10 mg/m^3 for Se-doped bismuth telluride. OSHA[58] has set a TWA of 15 mg/m^3 for the undoped material on a total dust basis and 5 mg/m^3 on a respirable fraction basis. Their limit for the doped material is 5 mg/m^3 — the same as ACGIH. Several states have set guidelines or standards for bismuth telluride in ambient air[60] ranging from 0.05 mg/m^3 (North Dakota) to 0.08 mg/m^3 (Virginia) to 0.2 mg/m^3 (Connecticut) to 0.238 mg/m^3 (Nevada).

Determination in Air: Bismuth telluride may be collected on a filter and analyzed by atomic absorption analysis.

Permissible Concentration in Water: No criteria set.

Routes of Entry: Inhalation of dust, ingestion.

Harmful Effects and Symptoms

Bismuth Telluride can affect you when breathed in. Exposure may irritate the eyes, nose and throat. Lung changes may occur. It is not known at this time whether these are permanent.

First Aid: If this chemical gets into the eyes, remove any contact lenses at once and irrigate immediately for at least 15 minutes, occasionally lifting upper and lower lids. Seek medical attention immediately. If this chemical contacts the skin, remove contaminated clothing and wash immediately with soap and water. Seek medical attention immediately. If

this chemical has been inhaled, remove from exposure, begin rescue breathing (using universal precautions) if breathing has stopped and CPR if heart action has stopped. Transfer promptly to a medical facility. When this chemical has been swallowed, get medical attention. Give large quantities of water and induce vomiting. Do not make an unconscious person vomit.

Personal Protective Methods: Clothing: Avoid skin contact with Bismuth Telluride. Wear protective gloves and clothing. Safety equipment suppliers/manufacturers can provide recommendations on the most protective glove/clothing material for your operation. All protective clothing (suits, gloves, footwear, headgear) should be clean, available each day, and put on before work. Eye Protection: Wear dust-proof goggles when working with powders or dust, unless full facepiece respiratory protection is worn.

Respirator Selection: Where the potential exists for exposures over 10 mg/m^3, use a MSHA/NIOSH approved respirator equipped with particulate (dust/fume/mist) filters. Particulate filters must be checked every day before work for physical damage, such as rips or tears, and replaced as needed. Where the potential for high exposures exists, use a MSHA/NIOSH approved self-contained breathing apparatus with a full facepiece operated in pressure demand or other positive pressure mode.

Storage: Bismuth Telluride must be stored to avoid contact with strong oxidizers (such as chlorine, bromine, and fluoride), since violent reactions occur. Store in tightly closed containers in a cool, well-ventilated area away from moisture.

Shipping: Bismuth Telluride is not specifically cited in DOT's Performance-Oriented Packaging Standards.[19]

Spill Handling: Evacuate and restrict persons not wearing protective equipment from area of spill or leak until cleanup is complete. Remove all ignition sources. Absorb liquid containing Bismuth Telluride in vermiculite, dry sand, earth, or similar material. Collect powdered material in the most convenient and safe manner and deposit in sealed containers. Ventilate area of spill or leak after clean-up is complete. It may be necessary to contain and dispose of this chemical as a hazardous waste. If material or contaminated runoff enters waterways, notify downstream users of potentially contaminated waters. Contact your Department of Environmental Protection or your regional office of the federal EPA for specific recommendations. If employees are required to clean-up spills, they must be properly trained and equipped. OSHA 1910.120(q) may be applicable.

Fire Extinguishing: Extinguish fire using an agent suitable for the type of surrounding fire; Bismuth Telluride itself does not burn. Poisonous gases are produced in fire including Te. If material or contaminated runoff enters waterways, notify downstream users of potentially contaminated waters. Notify local health and fire officials and pollution control agencies. From a secure, explosion-proof location, use water spray to cool exposed containers. If cooling streams are ineffective (venting sound increases in volume and pitch, tank discolors, or shows any signs of deforming), withdraw immediately to a secure position. If employees are expected to fight fires, they must be trained and equipped in OSHA 1910.156.

References

New Jersey Department of Health and Senior Services, "Hazardous Substance Fact Sheet: Bismuth Telluride," Trenton, NJ (Aug 1985)

Bisphenol A

Molecular Formula: $C_{15}H_{16}O_2$

Common Formula: $HOC_6H_4C(CH_3)_2$-C_6H_4OH

Synonyms: Bisfenol A (Spanish); Bisfenolo A (Italian); Bisferol A (German); 2,2-Bis-4'-hydroxyfenylpropan (Czech); Bis(4-hydroxyphenyl)dimethylmethane; β,β'-Bis(*p*-hydroxyphenyl)-propane; 2,2-Bis(*p*-hydroxyphenyl)propane; 2,2-Bis(4-hydroxyphenyl)propane; Bis(*p*-hydroxyphenyl)propane; *p,p*'-Bisphenol A; Dian; Diano; *p,p*'-Dihydroxydiphenyldimethylmetane; 4,4'-Dihydroxydiphenyldimethylmetane; *p,p*'-Dihydroxydiphenyl-propane; 2,2-(4,4'-Dihydroxydiphenyl)propane; 4,4'-Dihydroxy-2,2-diphenylpropane; 4,4'-Dihydroxydiphenyl-2,2-propane; 4,4'-Dihydroxydiphenylpropane; β-Di-*p*-hydroxyphenylpropane; 2,2-Di(4-hydroxyphenyl)propane; Dimethyl bis(*p*-hydroxyphenyl)-methane; Dimethylmethylene *p,p*'-diphenol; 2,2-Di(4-phenylol)propane; Diphenylolpropane; Ipognox 88; *p,p*'-Isopropilidendifenol (Spanish); Isopropylidenebis(4-hydroxybenzene); *p,p*'-Isopropylidenebisphenol; 4,4'-Isopropylidenebis(phenol); *p,p*'-Isopropylidenediphenol; 4,4'-Isopropylidenediphenol; 4,4'-(1-Methylethylidene)bisphenol; NCI-C50635; Parabis A; Phenol, 4,4'-isopropylidenedi-; Phenol, 4,4'-(1-methylethylidene)bis-; Pluracol 245; Rikabanol; UCAR Bisphenol HP

CAS Registry Number: 80-05-7

RTECS Number: SL6300000

EINECS Number: 201-245-8

Regulatory Authority

- Air Pollutant Standard Set (former USSR)
- EPCRA Section 313 Form R *de minimis* concentration reporting level: 1.0%
- TSCA 40CFR716.120(a)
- Canada, WHMIS, Ingredients Disclosure List; National Pollutant Release Inventory (NPRI)

Cited in U.S. State Regulations: California (G), Massachusetts (G), New Jersey (G), New York (G), Pennsylvania (G).

Description: Bisphenol-A, $HOC_6H_4C(CH_3)_2$-C_6H_4OH, is a white or tan crystals or flakes with a mild phenolic odor. Freezing/Melting point = 153°C. Boiling point = about 251°C. Flash point = 207°C. Autoignition temperature = 600°C. Hazard Identification (based on NFPA-704 M Rating System): Health 1, Flammability 1, Reactivity 0.

Potential Exposure: Workers engaged in the manufacture of epoxy, polysulfone, polycarbonate and certain polyester resins. It is also used in flame retardants and rubber chemicals, and as a fungicide.

Incompatibilities: Strong oxidizers, strong bases, acid chlorides and acid anhydrides.

Permissible Exposure Limits in Air: A standard of 5 mg/m^3 has been proposed in the former USSR according to the NIOSH Information Profile. It should be recognized that bisphenol-A can be absorbed through the skin, thereby increasing exposure.

Determination in Air: Collection by charcoal tube, analysis by gas liquid chromatography.

Permissible Concentration in Water: No criteria set.

Routes of Entry: Passes through the unbroken skin; inhalation; ingestion.

Harmful Effects and Symptoms

Bisphenol-A and its resins produce a typical contact dermatitis; redness and edema with weeping, followed by crusting and scaling, usually confined to the area of contact. Since the face is frequently affected, this may indicate that vapors are the cause, although contact with contaminated clothing can also be a factor. Seldom area areas other than the face and neck, back of hands, and forearms involved. The oral LD_{50} rat is 3,250 mg/kg[9] which is slightly toxic. Dusts may cause irritation to mouth, nose or throat and can cause eye irritation. As regards ingestion, comparison with phenol suggests probable symptoms would include nausea, burning of mouth, throat and stomach, severe stomach pain, stomach ulcers, vision disturbances, irregular breathing and pulse, dizziness, fainting, coma and death. Animal studies suggest that death may occur from ingestion of about 1/3 pound (6 ounces) for a 150 lb person. On long term exposure, susceptible individuals may become sensitized after repeated or prolonged contact and thereafter exhibit an allergic response. Allergy may include reaction to many epoxy resins containing Bisphenol A.

Short Term Exposure: Eye and skin contact can cause irritation and burns. May irritate the respiratory tract.

Long Term Exposure: May cause skin sensitization and allergy. There is limited evidence that this chemical may damage the developing fetus.

Points of Attack: Skin.

Medical Surveillance: Evaluation by a qualified allergist.

First Aid: If this chemical gets into the eyes, remove any contact lenses at once and irrigate immediately for at least 15 minutes, occasionally lifting upper and lower lids. Seek medical attention immediately. If this chemical contacts the skin, remove contaminated clothing and wash immediately with soap and water. Seek medical attention immediately. If this chemical has been inhaled, remove from exposure, begin rescue breathing (using universal precautions) if breathing has stopped and CPR if heart action has stopped. Transfer promptly to a medical facility. When this chemical has been swallowed, get medical attention. Give large quantities of water and induce vomiting. Do not make an unconscious person vomit.

Personal Protective Methods: Wear protective gloves and clothing to prevent any reasonable probability of skin contact. Safety equipment suppliers/manufacturers can provide recommendations on the most protective glove/clothing material for your operation. All protective clothing (suits, gloves, footwear, headgear) should be clean, available each day, and put on before work. Contact lenses should not be worn when working with this chemical. Wear dust-proof chemical goggles and face shield unless full facepiece respiratory protection is worn. Employees should wash immediately with soap when skin is wet or contaminated. Provide emergency showers and eyewash.

Respirator Selection: Dust mask should be worn to protect against inhaled dust. Fire fighters should wear self-contained breathing apparatus to protect against noxious fumes.

Storage: Store away from heat and strong oxidizers and the incompatible materials listed above.

Shipping: Bisphenol A is not cited in the DOT performance-oriented packaging standards as regards label requirements or maximum shipping quantities.

Spill Handling: Evacuate and restrict persons not wearing protective equipment from area of spill or leak until cleanup is complete. Remove all ignition sources. Vacuum cleaning is preferable to sweeping to keep dust levels down. Use special HEPA vacuum; not a shop vacuum. Ventilate area of spill or leak after clean-up is complete. It may be necessary to contain and dispose of this chemical as a hazardous waste. If material or contaminated runoff enters waterways, notify downstream users of potentially contaminated waters. Contact your Department of Environmental Protection or your regional office of the federal EPA for specific recommendations. If employees are required to clean-up spills, they must be properly trained and equipped. OSHA 1910.120(q) may be applicable.

Fire Extinguishing: Bisphenol A is a combustible solid. Use dry chemical, carbon dioxide, water spray, or foam extinguishers. Poisonous gases are produced in fire. If material or contaminated runoff enters waterways, notify downstream users of potentially contaminated waters. Notify local health and fire officials and pollution control agencies. Containers may explode in fire. From a secure, explosion-proof location, use water spray to cool exposed containers. If cooling streams are ineffective (venting sound increases in volume and pitch, tank discolors, or shows any signs of deforming), withdraw immediately to a secure position. If employees are expected to fight fires, they must be trained and equipped in OSHA 1910.156.

References

National Institute for Occupational Safety and Health, Information Profiles on Potential Occupational Hazards (Bisphenol-A), Rockville, Maryland (March 29, 1978)

New York State Department of Health, "Chemical Fact Sheet: Bisphenol-A," Albany, New York, Bureau of Toxic Substance Assessment (October 1984)

New Jersey Department of Health and Senior Services, Hazardous Substance Fact Sheet, Bisphenol A. Trenton NJ (May, 1998)

Bithionol

Molecular Formula: $C_{12}H_6Cl_4O_2S$

Synonyms: Actamer; Bidiphenbis(2-hydroxy-3,5-dichlorophenyl) sulfide; Bithinol sulfide; Bitin; CP3438; 2,2'-Dihydroxy-3,3',5,5'-tetrachlorodiphenyl sulfide; 2-Hydroxy-3,5-dichlorophenyl sulphide; Lorothidol; NCI-C60628; Neopellis; TBP; 2,2'-Thiobis(4,6-dichlorophenol); Vancide BL; XL 7

CAS Registry Number: 97-18-7

RTECS Number: SN0525000

Regulatory Authority

- Banned or Severely Restricted (USA, Japan) (UN)[13]
- TSCA 40CFR716.120(a)

Cited in U.S. State Regulations: New Jersey (G), Massachusetts (G), Pennsylvania (G)

Description: Bithionol, $C_{12}H_6Cl_4O_2S$ is a white or grayish powder with a slight phenolic odor. Freezing/Melting point = 188°C. Hazard Identification (based on NFPA-704 M Rating System): Health 3, Flammability 1, Reactivity 0.

Potential Exposure: It is used as a surfactant-formulated antimicrobial against bacteria, molds and yeast. It is proposed as an agricultural fungicide. Other uses include deodorant, germicide, fungistat and in the manufacture of pharmaceuticals. It is no longer allowed to be used in cosmetics. A food additive in feed and drinking water of animals. Also a food additive permitted in food for human consumption.

Incompatibilities: Strong oxidizers.

Permissible Exposure Limits in Air: No standards set.

Permissible Concentration in Water: No criteria set.

Routes of Entry: Ingestion, skin and/or eye contact

Harmful Effects and Symptoms

Probable oral lethal dose for humans is 5 – 15 g/kg for a 70 kg (150 lb) person. The toxicity of this compound is similar to that of phenol. Major hazard of phenol poisoning stems from its systemic effects which include central nervous system depression with coma, hypothermia, loss of vasoconstrictor tone, cardiac depression and respiratory arrest. Symptoms off exposure include burning pain in mouth and throat; white necrotic lesions in mouth, esophagus and stomach; abdominal pain; vomiting, bloody diarrhea; paleness; sweating; weakness; headache; dizziness; tinnitus; scanty, dark-colored urine; weak irregular pulse and shallow respiration.

First Aid: If this chemical gets into the eyes, remove any contact lenses at once and irrigate immediately for at least 15 minutes, occasionally lifting upper and lower lids. Seek medical attention immediately. If this chemical contacts the skin, remove contaminated clothing and wash immediately with soap and water. Seek medical attention immediately. If this chemical has been inhaled, remove from exposure, begin rescue breathing (using universal precautions) if breathing has stopped and CPR if heart action has stopped. Transfer promptly to a medical facility. When this chemical has been swallowed, get medical attention. Give large quantities of water and induce vomiting. Do not make an unconscious person vomit.

Personal Protective Methods: Wear protective gloves and clothing to prevent any reasonable probability of skin contact. Safety equipment suppliers/manufacturers can provide recommendations on the most protective glove/clothing material for your operation. All protective clothing (suits, gloves, footwear, headgear) should be clean, available each day, and put on before work. Contact lenses should not be worn when working with this chemical. Wear dust-proof chemical goggles and face shield unless full facepiece respiratory protection is worn. Employees should wash immediately with soap when skin is wet or contaminated. Provide emergency showers and eyewash.

Respirator Selection: For emergency situations, wear a positive pressure, pressure-demand, full facepiece self-contained breathing apparatus (SCBA) or pressure-demand supplied air respirator with escape SCBA.

Storage: Store in a refrigerator or a cool dry place.

Shipping: This material may be classified as a poisonous solid, n.o.s. This compound requires a shipping label of: "Poison." It falls in DOT Hazard Class 6.1 and Packing Group I. The limit on passenger aircraft or railcar shipment is 5 kg, and the limit on cargo aircraft shipment is 50 kg.

Spill Handling: Evacuate and restrict persons not wearing protective equipment from area of spill or leak until cleanup is complete. Remove all ignition sources. Dampen spilled material with 60 – 70% acetone to avoid airborne dust. Stay upwind. Do not touch spilled material. Use water spray to reduce vapors. Absorb spills with non-combustible absorbent material. For large spills dike far ahead for later disposal. Ventilate area of spill or leak after clean-up is complete. It may be necessary to contain and dispose of this chemical as a hazardous waste. If material or contaminated runoff enters waterways, notify downstream users of potentially contaminated waters. Contact your Department of Environmental Protection or your regional office of the federal EPA for specific recommendations. If employees are required to clean-up spills, they must be properly trained and equipped. OSHA 1910.120(q) may be applicable.

Fire Extinguishing: Use dry chemical, carbon dioxide, water spray, or alcohol foam extinguishers. Poisonous gases are produced in fire. If material or contaminated runoff enters waterways, notify downstream users of potentially contaminated waters. Notify local health and fire officials and pollution control agencies. From a secure, explosion-proof location, use water spray to cool exposed containers. If cooling streams are ineffective (venting sound increases in volume and pitch, tank discolors, or shows any signs of deforming), withdraw immediately to a secure position. If employees are expected to fight fires, they must be trained and equipped in OSHA 1910.156.

References

U.S. Environmental Protection Agency, "Chemical Profile: 2,2'-Thiobis (4,6-Dichlorophenol)," Washington, DC, Chemical Emergency Preparedness Program (November 30, 1987)

Bitoscanate

Molecular Formula: $C_8H_4N_2S_2$

Synonyms: Biscomate; 1,4-Diisothiocyanatobenzene; Isothiocyanic acid *p*-phenylene ester; Jonit; 1,4-Phenylene diisosthiocyanic acid; Phenylene 1,4-diisothiocyanate; Phenylene thiocyanate

CAS Registry Number: 4044-65-9

RTECS Number: NX9150000

Regulatory Authority

- SUPERFUND/EPCRA 40CFR355, Appendix B Extremely Hazardous Substances: TPQ = 500/10,000 lb (227/4,540 kg)
- SUPERFUND/EPCRA 40CFR302.4 Reportable Quantity (RQ): EHS, 1 lb (0.454 kg)
- EPCRA Section 313 Form R *de minimis* concentration reporting level: 1.0%
- As cyanide compound
- CLEAN AIR ACT: Hazardous Air Pollutants (Title I, Part A, Section 112) as cyanide compound
- CLEAN WATER ACT: 40CFR423, Appendix A, Priority Pollutants as cyanide, total
- EPA HAZARDOUS WASTE NUMBER (RCRA No.): P030 as cyanides soluble salts and complexes, n.o.s.
- RCRA, 40CFR261, Appendix 8 Hazardous Constituents. as cyanides, soluble salts and complexes, n.o.s.
- EPCRA (Section 313): X+CN- where X = H+ or any other group where a formal dissociation may occur. For example, KCN or $Ca(CN)_2$; Form R *de minimis* concentration reporting level: 1.0%
- U.S. DOT Regulated Marine Pollutant (49CFR172.101, Appendix B) as cyanide mixtures, cyanide solutions or cyanides, inorganic, n.o.s.
- CLEAN AIR ACT: Hazardous Air Pollutants (Title I, Part A, Section 112) as cyanide compound
- CLEAN WATER ACT: 40CFR423, Appendix A, Priority Pollutants as cyanide, total
- EPA HAZARDOUS WASTE NUMBER (RCRA No.): P030 as cyanides soluble salts and complexes, n.o.s.
- RCRA, 40CFR261, Appendix 8 Hazardous Constituents. as cyanides, soluble salts and complexes, n.o.s.
- EPCRA (Section 313): X+CN- where X = H+ or any other group where a formal dissociation may occur. For example, KCN or $Ca(CN)_2$; Form R *de minimis* concentration reporting level: 1.0%
- U.S. DOT Regulated Marine Pollutant (49CFR172.101, Appendix B) as cyanide mixtures, cyanide solutions or cyanides, inorganic, n.o.s.
- U.S. DOT Regulated Marine Pollutant (49CFR172.101, Appendix B) as cyanide mixtures, cyanide solutions or cyanides, inorganic, n.o.s.
- Canada, WHMIS, Ingredients Disclosure List (as cyanide compounds)

Cited in U.S. State Regulations: Florida (G), Massachusetts (G), New Jersey (G), Pennsylvania (G)

Description: Bitoscanate, $C_8H_4N_2S_2$, is a colorless, odorless, crystalline compound. Freezing/Melting point = 132°C. Hazard Identification (based on NFPA-704 M Rating System): Health 2, Flammability 1, Reactivity 0.

Potential Exposure: Those engaged in the manufacture, formulation and application of this anthelmintic compound.

Incompatibilities: This is a thiocyanate compound. Violent reactions may occur when upon contact with chlorates (potassium chlorate, sodium chlorate), nitrates, nitric acid, organic peroxides, peroxides.

Permissible Exposure Limits in Air: No standards set for this compound. However, inasmuch as it is a cyanide compound, the exposure limits are listed here: OSHA and ACGIH: 5 mg/m^3 TWA; NIOSH: Ceiling limit, 4.7 ppm; 5 mg/m^3 per 10 minutes as cyanides. All have notations that skin contact contributes significantly in overall exposure. IDLH = 25 mg/m^3 as CN.

Permissible Concentration in Water: No criteria set.

Routes of Entry: Ingestion.

Harmful Effects and Symptoms

This material is highly toxic if ingested. It is a central nervous system and gastrointestinal toxin in humans. The oral LD_{50} rat is 2 mg/kg (highly toxic).[9]

Medical Surveillance: Blood cyanide level.

First Aid: If this chemical gets into the eyes, remove any contact lenses at once and irrigate immediately for at least 15 minutes, occasionally lifting upper and lower lids. Seek medical attention immediately. If this chemical contacts the skin, remove contaminated clothing and wash immediately with soap and water. Seek medical attention immediately. If this chemical has been inhaled, remove from exposure, begin rescue breathing (using universal precautions) if breathing has stopped and CPR if heart action has stopped. Transfer promptly to a medical facility. When this chemical has been swallowed, get medical attention. Give large quantities of water and induce vomiting. Do not make an unconscious person vomit.

For cyanide poisoning, use amyl nitrate capsules if symptoms develop. All area employees should be trained regularly in emergency measures for cyanide poisoning and in CPR. A cyanide antidote kit should be kept in the immediate work area and must be rapidly available. Kit ingredients should be replaced every 1 – 2 years to ensure freshness. Persons trained in the use of this kit, oxygen use, and CPR must be quickly available.

Personal Protective Methods: Wear protective gloves and clothing to prevent any reasonable probability of skin contact. Safety equipment suppliers/manufacturers can provide recommendations on the most protective glove/clothing material for your operation. All protective clothing (suits, gloves, footwear, headgear) should be clean, available each day, and put on before work. Contact lenses should not be worn when working with this chemical. Wear dust-proof chemical goggles and face shield unless full facepiece respiratory protection is worn. Employees should wash immediately with soap when skin is wet or contaminated. Provide emergency showers and eyewash.

Respirator Selection: NIOSH/OSHA (as cyanides): *up to 25 mg/m³:* SA (any supplied-air respirator); or SCBAF (any self-contained breathing apparatus with full facepiece). *Emergency or planned entry into unknown concentrations or IDLH conditions:* SCBAF:PD,PP (any self-contained breathing apparatus that has a full facepiece and is operated in a pressure-demand or other positive-pressure mode); or SAF: PD,PP:ASCBA (any supplied-air respirator that has a full facepiece and is operated in a pressure-demand or other positive-pressure mode in combination with an auxiliary self-contained breathing apparatus operated in a pressure-demand or other positive pressure mode. *Escape:* GMFSHiE [any air-purifying, full-facepiece respirator (gas mask) with a chin-style, front- or back-, mounted canister providing protection against the compound of concern and having a high efficiency particulate filter]; or SCBAE (any appropriate escape-type, self-contained breathing apparatus).

Storage: Store in tightly closed containers in a cool, well ventilated area.

Spill Handling: Evacuate and restrict persons not wearing protective equipment from area of spill or leak until cleanup is complete. Remove all ignition sources. Collect powdered material in the most convenient and safe manner and deposit in sealed containers. Ventilate area of spill or leak after clean-up is complete. It may be necessary to contain and dispose of this chemical as a hazardous waste. If material or contaminated runoff enters waterways, notify downstream users of potentially contaminated waters. Contact your Department of Environmental Protection or your regional office of the federal EPA for specific recommendations. If employees are required to clean-up spills, they must be properly trained and equipped. OSHA 1910.120(q) may be applicable.

Fire Extinguishing: Use dry chemical, carbon dioxide, water spray, or alcohol foam extinguishers. Poisonous gases are produced in fire, including cyanide, nitrogen oxides, sulfur oxides. If material or contaminated runoff enters waterways, notify downstream users of potentially contaminated waters. Notify local health and fire officials and pollution control agencies. From a secure, explosion-proof location, use water spray to cool exposed containers. If cooling streams are ineffective (venting sound increases in volume and pitch, tank discolors, or shows any signs of deforming), withdraw immediately to a secure position. If employees are expected to fight fires, they must be trained and equipped in OSHA 1910.156.

References

U.S. Environmental Protection Agency, "Chemical Profile: Bitoscanate," Washington, DC, Chemical Emergency Preparedness Program (November 30, 1987)

Boron, Boric Acid and Borax

Molecular Formula: B (boron); BH_3O_3 (boric acid); $B_4H_2Na_2O_8$ (borax)

Common Formula: B (boron); H_3BO_3 (boric acid); $Na_2B_4O_7 \cdot H_2O$ (borax)

Synonyms: *elemental boron:* None

borax: Disodium tetraborate; Sodium borate; Sodium tetraborate

boric acid: Boracic acid; Othroboric acid

CAS Registry Number: 7440-42-8 (boron, elemental); 10043-35-3 (boric acid); 1303-96-4 (borax)

RTECS Number: ED7350000 (boron, elemental); ED4550000 (boric acid); VZ2275000 (borax)

DOT ID: UN 3077

Regulatory Authority

- Banned or Severely Restricted (In many products) (UN)[13]
- Air Pollutant Standard Set (ACGIH)[1] (HSE)[33] (former USSR)[43] (OSHA)[58] (Several States)[60]
- Safe Drinking Water Act, 55FR1470 Priority List
- Canada Drinking Water Quality: 5.0 mg/l IMAC
- Mexico Drinking Water Criteria: 1.0 mg/l

Cited in U.S. State Regulations: Alaska (G), California (G), Connecticut (A), Illinois (G), Maine (G), Nevada (A), New Hampshire (G), New Jersey (G), North Dakota (A), Pennsylvania (G), Rhode Island (G), Virginia (A), West Virginia (G).

Description: Boron, B, is a yellow or brownish-black powder and may be either crystalline or amorphous. It does not occur free in nature and is found in the minerals borax, colemanite, boronatrocalcite, and boracite. Freezing/Melting point = 2,190°C. Boiling point = 3,660°C. Practically insoluble in water, although boron is reported to be slightly soluble under certain conditions. Boric acid, H_3BO_3, is a white, amorphous powder or colorless, crystalline solid. Freezing/Melting point = 168 – 169°C (decomposes above 100°C). Saturated solutions: @ 0°C, 2.6% acid; @ 100°C, 28% acid. Boric acid is soluble in water (5 g/100 ml @ 20°C). Borax, $Na_2B_4O_7{\cdot}H_2O$, is a bluish-gray or green, odorless crystalline powder or granules. Boiling point = 320°C. Freezing/Melting point = 75°C (rapid heating) Borax is soluble in water (6g/100 ml @ 20°C).

Potential Exposure: Boron is used in metallurgy as a degasifying agent and is alloyed with aluminum, iron, and steel to increase hardness. It is also a neutron absorber in nuclear reactors. Boric acid is a fireproofing agent for wood, a preservative, and an antiseptic. It is used in the manufacture of glass, pottery, enamels, glazes, cosmetics, cements, porcelain, borates, leather, carpets, hats, soaps, and artificial gems, and in tanning, printing, dyeing, painting, and photography. It is a constituent in powders, ointments, nickeling baths, electric condensers and is used for impregnating wicks and hardening steel. Borax is used as a soldering flux, preservative against wood fungus, and as an antiseptic. It is used in the manufacture of enamels and glazes and in tanning, cleaning compounds, for fireproofing fabrics and wood, and in artificial aging of wood.

Incompatibilities: Contact with strong oxidizers may cause explosions. Boron dust is explosive on exposure with air. Boron is incompatible with ammonia, bromine tetrafluoride, cesium carbide, chlorine, fluorine, interhalogens, iodic acid, lead dioxide, nitric acid, nitric oxide, nitrosyl fluoride, nitrous oxide, potassium nitrite, rubidium carbide, silver fluoride. Boric acid decomposes in heat above 100 °C forming boric anhydride and water. Boric acid aqueous solution is a weak acid; incompatible with alkali carbonates and hydroxides.

Permissible Exposure Limits in Air: The TWA set by ACGIH[1] and by HSE[33] for anhydrous sodium tetraborate (borax) is 1.0 mg/m^3; for sodium tetraborate decahydrate it is 5.0 mg/m^3; for the pentahydrate it is 1.0 mg/m^3. OSHA[58] has set a TWA of 10 mg/m^3 for all sodium tetraborates. A MAC value in workplace air has been set by the former USSR-UNEP/IRPTC project[43] at 10 mg/m^3 for boric acid. Several states have set guidelines or standards for sodium tetraborates in ambient air,[60] ranging from 10 μg/m^3 (North Dakota) to 16 μg/m^3 (Virginia) to 20 μg/m^3 (Connecticut) to 24 μg/m^3 (Nevada) to 100 μg/m^3 (Connecticut).

Permissible Concentration in Water: EPA in July 1976 established a criterion for boron of 750 μg/l for long term irrigation on sensitive crops. More recently,[32] EPA has suggested an ambient water limit of 43 μg/l based on health effects. The former USSR-UNEP/IRPTC project[43] has set MAC values in mg/l, in water used for fishery purposes of 0.1 for boric acid and 0.05 for sodium tetraborate. See Regulatory Authority for Canadian and Mexican levels.

Routes of Entry: Inhalation of dust, fumes, and aerosols; ingestion.

Harmful Effects and Symptoms

Local: These boron compounds may produce irritation of the nasal mucous membranes, the respiratory tract, and eyes. *Systemic:* These effects vary greatly with the type of compound. Acute poisoning in man from boric acid or borax is usually the result of application of dressings, powders, or ointment to large areas of burned or abraded skin, or accidental ingestion. The sings are: nausea, abdominal pain, diarrhea and violent vomiting, sometimes bloody, which may be accompanied by headache and weakness. There is a characteristic erythematous rash followed by peeling. In severe cases, shock with fall in arterial pressure, tachycardia, and cyanosis occur. Marked CNS irritation, oliguria, and anuria may be present. The oral lethal dose in adults is over 30 g. Little information is available on chronic oral poisoning, although it is reported to be characterized by mild Gl irritation, loss of appetite, disturbed digestion, nausea, possibly vomiting, and erythematous rash. The rash may be "hard" with a tendency to become purpuric. Dryness off skin and mucous membranes, reddening of tongue, cracking of lips, loss of hair, conjunctivitis, palpebral edema, gastro-intestinal disturbances, and kidney injury have also been observed. Workers manufacturing boric acid had some atrophic changes in respiratory mucous membranes, weakness, joint pains, and other vague symptoms. The biochemical mechanism of boron toxicity is not clear but seems to involve action on the nervous system, enzyme activity, carbohydrate metabolism, hormone function, and oxidation processes coupled with allergic effects. Borates are excreted principally by the kidneys. No toxic effects have been attributed to elemental boron. The oral LD_{50} mouse for *boron* is 2,000 mg/kg. The oral LD_{50} rat for *boric acid* is 2,660 mg/kg (slightly toxic). The oral LD_{50} rat for *borax* is also 2,660 mg/kg (slightly toxic).

Short Term Exposure: Boric acid irritates the eyes, skin, and the respiratory tract. High exposure may cause effects on the gastrointestinal tract, liver and kidneys. Borax and Boric acid may affect the nervous system. Serious overexposure can cause seizures, unconsciousness and death.

Long Term Exposure: May cause brain, kidney and liver damage. Repeated or prolonged contact with skin may cause dermatitis. Animal tests show that this substance possibly causes toxic effects upon human reproduction. (WHO).

Medical Surveillance: No specific considerations are needed for boric acid or borates except for general health and liver and kidney function. In the case of boron trifluoride, the skin, eyes, and respiratory tract should receive special attention. In the case of the boranes, central nervous system and lung function will also be of special concern.

First Aid: If this chemical gets into the eyes, remove any contact lenses at once and irrigate immediately for at least 15 minutes, occasionally lifting upper and lower lids. Seek medical attention immediately. If this chemical contacts the skin, remove contaminated clothing and wash immediately with soap and water. Seek medical attention immediately. If this chemical has been inhaled, remove from exposure, begin rescue breathing (using universal precautions) if breathing has stopped and CPR if heart action has stopped. Transfer promptly to a medical facility. When this chemical has been swallowed, get medical attention. Give large quantities of water and induce vomiting. Do not make an unconscious person vomit.

Personal Protective Methods: Exposed workers should be educated in the proper use of protective equipment and there should be strict adherence to ventilating provisions in work areas. Workers involved with the manufacture of boric acid should be provided with masks to prevent inhalation of dust and fumes.

Storage: Store in a cool, dry place away from incompatible materials listed above.

Shipping: These materials are not specifically cited in the DOT performance-oriented packaging regulations.[19] They may be classified as Environmentally Hazardous Substance, solid, n.o.s. which require a label of Class 9.

Spill Handling: Evacuate and restrict persons not wearing protective equipment from area of spill or leak until cleanup is complete. Remove all ignition sources. The material may be dampened with water to avoid dust and than transferred to a sealed container for disposal. Ventilate area of spill or leak after clean-up is complete. It may be necessary to contain and dispose of this chemical as a hazardous waste. If material or contaminated runoff enters waterways, notify downstream users of potentially contaminated waters. Contact your Department of Environmental Protection or your regional office of the federal EPA for specific recommendations. If employees are required to clean-up spills, they must be properly trained and equipped. OSHA 1910.120(q) may be applicable.

Fire Extinguishing: Use dry chemical, carbon dioxide, water, or foam extinguishers. Irritation and toxic fumes are produced in fire. If material or contaminated runoff enters waterways, notify downstream users of potentially contaminated waters. Notify local health and fire officials and pollution control agencies. If employees are expected to fight fires, they must be trained and equipped in OSHA 1910.156.

Disposal Method Suggested: Borax, dehydrated: The material is diluted to the recommended provisional limit (0.10 mg/l) in water. The pH is adjusted to between 6.5 and 9.1 and then the material can be discharged into sewers or natural streams. Boric acids may be recovered from organic process wastes as an alternative to disposal.

References

Environmental Protection Agency, Preliminary Investigation of Effects on the Environment of Boron, Indium, Nickel, Selenium, Tin, Vanadium and Their Compounds, Volume 1: Boron, Report EPA-560/2-75-005A, Washington, DC, Office of Toxic Substances (August 1975)

National Institute for Occupational Safety and Health, Information Profiles on Potential Occupational Hazards: Boron and Its Compounds, Report PB 276,678, Rockville, Maryland, pp 63–75 (October 1977)

Sax, N. I., Ed., "Dangerous Properties of Industrial Materials Report" 1, No. 8, 42–45 (1981) (Boron and Boric Acid) and 3, No. 5, 65–67, New York, Van Nostrand Reinhold Co. (1983)

Sax, N. I., Ed., "Dangerous Properties of Industrial Materials Report" 2, No. 6, 76–78 (1982) (Sodium Borate)

New Jersey Department of Health and Senior Services, "Hazardous Substance Fact Sheet: Sodium Borates," Trenton, NJ (September 1985)

Boron Oxide

Molecular Formula: B_2O_3

Synonyms: Anhydrous boric acid; Boric anhydride; Boron sesquioxide; Boron trioxide; Diboron trioxide; Fused boric acid; Oxido de boro (Spanish)

CAS Registry Number: 1303-86-2

RTECS Number: ED7900000

DOT ID: UN 3077

Regulatory Authority

- Air Pollutant Standard Set (ACGIH)[1] (OSHA)[58] (HSE)[33] (former USSR)[43] (Several States)[60] (Australia) (Israel) (Mexico)
- Canada, WHMIS, Ingredients Disclosure List

Cited in U.S. State Regulations: Alaska (G), California (G), Connecticut (A), Florida (G), Illinois (G), Maine (G), Massachusetts (G), Minnesota (G), Nevada (A), New Hampshire (G), New Jersey (G), North Dakota (A), Pennsylvania (G), Virginia (A), West Virginia (G).

Description: boron oxide, B_2O_3, is a noncombustible, colorless, semitransparent lumps or hard, white, odorless crystals, with slightly bitter taste. Boiling point = about 1,860°C. Freezing/Melting point = about 450°C. Moderately soluble in water.

Potential Exposure: Boron oxide is used in glass manufacture and the production of other boron compounds. It is used in fluxes, enamels, drying agents and as a catalyst.

Incompatibilities: Bromine pentafluoride, calcium oxide. Water (Reacts slowly with water to form boric acid).

Permissible Exposure Limits in Air: The Federal OSHA standard is 15 mg/m³ TWA.[58] NIOSH and ACGIH has set a TWA of 10 mg/m³. The NIOSH IDLH is 2,000 mg/m³. The HSE[33] and Mexico have set a TWA of 10 mg/m³ and a STEL of 20 mg/m³. Australia and Israel set a TWA of 10 mg/m³ and Israel's Action Level is 5 mg/m³. The former USSR-UNEP/IRPTC project[43] has set a MAC of 5 mg/m³ in workplace air. The NIOSH IDLH is 2,000 mg/m³. In addition, several states have set guidelines or standards for boron oxide in ambient air[60] ranging from 10 μg/m³ (North Dakota) to 160 μg/m³ (Virginia) to 200 μg/m³ (Connecticut) to 238 μg/m³ (Nevada).

Determination in Air: Collection on a filter and gravimetric analysis. See NIOSH Method #0500 Particulates NOR (total).[18]

Permissible Concentration in Water: No criteria set but EPA has suggested[32] an ambient water limit of 138 μg/l based on health effects.

Routes of Entry: Inhalation, ingestion, skin and/or eye contact.

Harmful Effects and Symptoms

Short Term Exposure: May irritate the skin, causing a rash or burning feeling on contact. May cause nasal irritation, conjunctivitis, erythema. Ingestion causes abdominal pain, diarrhea, nausea, vomiting. Low toxicity. The oral LD_{50} mouse is 3,163 mg/kg.[9]

Long Term Exposure: Unknown at this time.

Points of Attack: Skin, eyes, respiratory system.

Medical Surveillance: Consider the points of attack in preplacement and periodic physical examinations.

First Aid: If this chemical gets into the eyes, remove any contact lenses at once and irrigate immediately for at least 15 minutes, occasionally lifting upper and lower lids. Seek medical attention immediately. If this chemical contacts the skin, remove contaminated clothing and wash immediately with soap and water. Seek medical attention immediately. If this chemical has been inhaled, remove from exposure, begin rescue breathing (using universal precautions) if breathing has stopped and CPR if heart action has stopped. Transfer promptly to a medical facility. When this chemical has been swallowed, get medical attention. Give large quantities of water and induce vomiting. Do not make an unconscious person vomit.

Personal Protective Methods: Wear protective gloves and clothing to prevent any reasonable probability of skin contact. Safety equipment suppliers/manufacturers can provide recommendations on the most protective glove/clothing material for your operation. All protective clothing (suits, gloves, footwear, headgear) should be clean, available each day, and put on before work. Contact lenses should not be worn when working with this chemical. Wear dust-proof chemical goggles and face shield unless full facepiece respiratory protection is worn. Employees should wash immediately with soap when skin is wet or contaminated. Provide emergency showers and eyewash.

Respirator Selection: NIOSH: *50 mg/m³*: DM (any dust and mist respirator). *100 mg/m³:* DMXSQ (any dust and mist respirator except single-use and quarter mask respirators); or SA (any supplied-air respirator). *250 mg/m³*: SA:CF (any supplied-air respirator operated in a continuous-flow mode); or PAPRDM (any powered, air-purifying respirator with a dust and mist filter). *500 mg/m³*: HiEF (any air-purifying, full-facepiece respirator with a high-efficiency particulate filter); or PAPRTHiE (any powered, air-purifying respirator with a tight-fitting facepiece and a high-efficiency particulate filter); or SCBAF (any self-contained breathing apparatus with a full facepiece); or SAF (any supplied-air respirator with a full facepiece). *2,000 mg/m³*: SAF:PD,PP (any supplied-air respirator that has a full facepiece and is operated in a pressure-demand or other positive-pressure mode). *Emergency or planned entry into unknown concentrations or IDLH conditions:* SCBAF:PD,PP (any self-contained breathing apparatus that has a full facepiece and is operated in a pressure-demand or other positive-pressure mode); or SAF:PD,PP:ASCBA (any supplied-air respirator that has a full facepiece and is operated in a pressure-demand or other positive-pressure mode in combination with an auxiliary self-contained breathing apparatus operated in a pressure-demand or other positive pressure mode. *Escape:* HiEF (any air-purifying, full-facepiece respirator with a high-efficiency particulate filter); or SCBAE (any appropriate escape-type, self-contained breathing apparatus).

Storage: Store in tightly closed containers in a dry, well-ventilated area away from incompatible materials listed above and water.

Shipping: These materials are not specifically cited in the DOT performance-oriented packaging regulations.[19] They may be classified as Environmentally Hazardous Substance, solid, n.o.s. which require a label of Class 9.

Spill Handling: Evacuate and restrict persons not wearing protective equipment from area of spill or leak until cleanup is complete. Remove all ignition sources. Moisten dry material to prevent dust. Collect powdered material in the most convenient and safe manner and deposit in sealed containers. Ventilate area of spill or leak after clean-up is complete. It may be necessary to contain and dispose of this chemical as a hazardous waste. If material or contaminated runoff enters waterways, notify downstream users of potentially contaminated waters. Contact your Department of Environmental Protection or your regional office of the federal EPA for specific recommendations. If employees are required to clean-up spills, they must be properly trained and equipped. OSHA 1910.120(q) may be applicable.

Fire Extinguishing: Boron Oxide is a noncombustible solid. Extinguish fire using an agent suitable for type of surrounding fire. Boron Oxide itself does not burn. Poisonous gases are produced in fire. If material or contaminated runoff enters waterways, notify downstream users of potentially contaminated waters. Notify local health and fire officials and pollution control agencies. From a secure, explosion-proof location, use water spray to cool exposed containers. If cooling streams are ineffective (venting sound increases in volume and pitch, tank discolors, or shows any signs of deforming), withdraw immediately to a secure position. If employees are expected to fight fires, they must be trained and equipped in OSHA 1910.156.

References

New Jersey Department of Health and Senior Services, "Hazardous Substance Fact Sheet: Boron Oxide," Trenton, NJ (January 1986)

Boron Tribromide

Molecular Formula: BBr_3

Synonyms: Borane, tribromo-; Boron bromide; Boron tribromide 6; Tribromoborand; Tribromuro de boro (Spanish); Trona

CAS Registry Number: 10294-33-4

RTECS Number: ED7400000

DOT ID: UN 2692

EEC Number: 005-003-00-0

Regulatory Authority

- Air Pollutant Standard Set (ACGIH)[1] (OSHA)[58] (HSE)[33] (Several States)[60] (Australia) (Israel) (Mexico) (Several Canadian Provinces)
- Canada, WHMIS, Ingredients Disclosure List

Cited in U.S. State Regulations: Alaska (G), Connecticut (A), Florida (G), Illinois (G), Maine (G), Massachusetts (G), Nevada (A), New Hampshire (G), New Jersey (G), North Dakota (A), Pennsylvania (G), Rhode Island (G), Virginia (A), West Virginia (G).

Description: Boron tribromide, BBr_3, is a colorless, fuming liquid. Boiling point = 90°C.

Potential Exposure: Boron tribromide is used as a catalyst in organic synthesis, making diborane, high purity boron, and semiconductors.

Incompatibilities: Reacts violently and explosively with water or steam forming hydrogen bromide gas. Mixtures with potassium or sodium can explode on impact. Incompatible with oxidizers, strong bases and alcohols. Attacks some metals, rubbers, and plastics.

Permissible Exposure Limits in Air: NIOSH and ACGIH recommends a ceiling value of 1 ppm (10 mg/m^3). The HSE[33] has set a TWA value of 1 ppm (10 mg/m^3) and a STEL value of 3 ppm (30 mg/m^3). Australia's TWA peak limitation, and Israel's and California's ceiling exposure limit is 1 ppm. In addition, several states have set guidelines or standards for boron tribromide in ambient air[60] ranging from 80 μg/m^3 (Virginia) to 100 μg/m^3 (Connecticut) to 238 μg/m^3 (Nevada).

Permissible Concentration in Water: No criteria set

Routes of Entry: Eyes, skin, respiratory system.

Harmful Effects and Symptoms

Short Term Exposure: Boron Tribromide can affect you when breathed in. Boron Tribromide is a corrosive liquid and exposure can cause severe burns of the eyes, nose, throat, lungs, and skin. Boron Tribromide may cause cough, headaches, nose bleeds, and shortness of breath. Higher exposures can cause pulmonary edema, a medical emergency that can be delayed for several hours. This can cause death.

Long Term Exposure: Repeated exposure may cause a brownish color of the tongue and/or runny nose. May cause irritation of the lungs and bronchitis to develop. May cause kidney damage and affect the nervous system.

Points of Attack: Kidneys, nervous system, lungs.

Medical Surveillance: Before beginning employment and at regular times after that, the following is recommended: lung function tests, kidney function tests, examination of the nervous system. If symptoms develop or overexposure is suspected the following may be useful: Consider chest x-ray following acute overexposure.

First Aid: If this chemical gets into the eyes, remove any contact lenses at once and irrigate immediately for at least 15 minutes, occasionally lifting upper and lower lids. Seek medical attention immediately. If this chemical contacts the skin, remove contaminated clothing and wash immediately with soap and water. Seek medical attention immediately. If this chemical has been inhaled, remove from exposure, begin rescue breathing (using universal precautions) if breathing has stopped and CPR if heart action has stopped. Transfer promptly to a medical facility. When this chemical has been swallowed, get medical attention. If victim is conscious, administer water or milk. Do not induce vomiting. Medical observation is recommended for 24 – 48 hours after breathing overexposure, as pulmonary edema may be delayed. As first aid for pulmonary edema, a doctor or authorized paramedic may consider administering a corticosteroid spray.

Personal Protective Methods: Wear acid resistant gloves and clothing to prevent any reasonable probability of skin contact. Safety equipment suppliers/manufacturers can provide recommendations on the most protective glove/clothing material for your operation. Gloves made of Chlorinated Polyethylene are considered fair to good protection for Boron Tribromide. All protective clothing (suits, gloves, footwear, headgear) should be clean, available each day, and put on before work. Contact lenses should not be worn when working with this chemical. Wear splash or dust-proof chemical goggles and face shield unless full facepiece respiratory protection is worn. Employees should wash immediately with soap when skin is wet or contaminated. Provide emergency showers and eyewash.

Respirator Selection: Where the potential exists for exposures over 1 ppm, use a MSHA/NIOSH approved full facepiece respirator with an acid gas canister. Increased protection is obtained from full facepiece powered-air purifying respirators. Where the potential for high exposures exists, use a MSHA/NIOSH approved supplied-air respirator with a full facepiece operated in the positive pressure mode or with a full facepiece, hood, or helmet in the continuous flow mode, or use a MSHA/NIOSH approved self-contained breathing apparatus with a full facepiece operated in pressure-demand or other positive pressure mode.

Storage: Before entering confined space where boron tribromide may be present, check to make sure that an explosive concentration does not exist. Store in airtight, unbreakable containers in a cool well-ventilated area away from water, steam, potassium, sodium, alcohol and other incompatible materials. Metal containers involving the transfer of this chemical should be grounded and bonded. Where possible, automatically pump liquid from drums or other storage containers to process containers. Drums must be equipped with self-closing valves, pressure vacuum bungs, and flame arresters. Use only non-sparking tools and equipment, especially when opening and closing containers of this chemical. Sources of ignition such as smoking and open flames, are prohibited where this chemical is used, handled, or stored in a manner that could create a potential fire or explosion hazard.

Shipping: Boron Tribromide requires a "Corrosive and Poison" label. It falls in UN/DOT Hazard Class 8 and Packing Group I. Shipment by passenger aircraft or railcar is forbidden.

Spill Handling: Evacuate and restrict persons not wearing protective equipment from area of spill or leak until cleanup is complete. Stop discharge if possible. Avoid contact with liquid or vapor.[41] Remove all ignition sources. Absorb liquids in vermiculite, dry sand, earth, or a similar material and deposit in sealed containers. It may be necessary to contain and dispose of this chemical as a hazardous waste. Ventilate area of leak or spill after clean-up is complete. If material or contaminated runoff enters waterways, notify downstream users of potentially contaminated waters. Contact your Department of Environmental Protection or your regional office of the federal EPA for specific recommendations. If employees are required to clean-up spills, they must be properly trained and equipped. OSHA 1910.120(q) may be applicable.

Fire Extinguishing: This chemical decomposes in heat and may explode. Use dry chemical or carbon dioxide. Do not use foam. Do not use water on material itself. Water can be used to cool intact containers and to absorb vapors. Poisonous gases are produced in fire including hydrogen bromide and boron oxides. Vapors are heavier than air and will collect in low areas. If material or contaminated runoff enters waterways, notify downstream users of potentially contaminated waters. Notify local health and fire officials and pollution control agencies. From a secure, explosion-proof location, use water spray to cool exposed containers. If cooling streams are ineffective (venting sound increases in volume and pitch, tank discolors, or shows any signs of deforming), withdraw immediately to a secure position. If employees are expected to fight fires, they must be trained and equipped in OSHA 1910.156.

References

New Jersey Department of Health and Senior Services, "Hazardous Substance Fact Sheet: Boron Tribromide," Trenton, NJ (July 1998)

Boron Trichloride

Molecular Formula: BCl_3

Synonyms: Borane, trichloro-; Boron chloride; Chlorure de bore (French); Trichloroborane; Trichloroboron; Tricloruro de boro (Spanish); Trona

CAS Registry Number: 10294-34-5

RTECS Number: ED1925000

DOT ID: UN 1741

EEC Number: 005-002-00-5

EINECS Number: 233-658-4

Regulatory Authority

- OSHA 29CFR1910.119, Appendix A, Process Safety List of Highly Hazardous Chemicals, TQ = 2,500 lb (1,135 kg)
- CLEAN AIR ACT: Accidental Release Prevention/Flammable substances, (Section 112[r], Table 3), TQ = 5,000 lb (2,270 kg)
- SUPERFUND/EPCRA 40CFR355, Appendix B Extremely Hazardous Substances: TPQ = 500 lb (227 kg)
- SUPERFUND/EPCRA 40CFR302.4 Reportable Quantity (RQ): EHS, 1 lb (0.454 kg)
- EPCRA Section 313 Form R *de minimis* concentration reporting level: 1.0%
- Canada, WHMIS, Ingredients Disclosure List

Cited in U.S. State Regulations: Maine (G), Massachusetts (G), New Hampshire (G), Oklahoma (G).

Description: Boron trichloride, BCl_3, is a colorless liquid with a pungent, irritating odor. Boiling point = 12.5°C. Freezing/Melting point = -107°C. Insoluble in water (reaction).

Potential Exposure: Manufacture and purification of boron; catalyst in organic reactions; semiconductors; bonding of iron or steel, purification of metal alloys to remove oxides, nitrides, and carbides; chemical intermediate for boron filaments; soldering flux; electrical resistors; and extinguishing magnesium fires in heat treating furnaces.

Incompatibilities: Incompatible with lead, graphite-impregnated asbestos, potassium, sodium. Vigorously attacks elastomers, packing materials, natural and synthetic rubber, viton, tygon, saran, silastic elastomers. Avoid aniline, hexafluorisopropylidene amino lithium, nitrogen dioxide, phosphine, grease, organic matter, and oxygen. Nitrogen peroxide, phosphine, fat or grease react vigorously with boron trichloride. It reacts with water or steam to produce heat, boric acid and hydrochloric acid fumes. Oxygen and boron trichloride react vigorously on sparking. Attacks most metals in the presence of moisture.

Permissible Exposure Limits in Air: No standards set. Since the hydrolysis product, HCl, tends to govern the effect in moist air, reference should be made to the limit for hydrogen chloride.

Permissible Concentration in Water: No criteria set.

Routes of Entry: Inhalation, ingestion, skin and/or eye contact

Harmful Effects and Symptoms

Short Term Exposure: Extremely corrosive to the eyes, skin and respiratory tract. Contact with eyes produces sever pain, swelling, corneal erosions and blindness. Viscid white or blood-stained foamy mucus and threads of tissue may appear in mouth. Inhalation can cause low blood oxygen, difficulty in breathing, chest pain and pulmonary edema, a medical emergency that can be delayed for several hours. This can cause death. Symptoms of overexposure include depression of circulation, persistent vomiting and diarrhea, profound shock and coma. Temperature becomes sub-normal and rash may cover entire body. Boron affects the central nervous system causing depression of circulation as well as shock and coma. May result in marked fluid and electrolyte loss and shock.

Long Term Exposure: Can cause liver, kidney and brain damage.

Medical Surveillance: Consider x-ray following acute overexposure. Liver, kidney, and lung function tests.

First Aid: If this chemical gets into the eyes, remove any contact lenses at once and irrigate immediately for at least 15 minutes, occasionally lifting upper and lower lids. Seek medical attention immediately. If this chemical contacts the skin, remove contaminated clothing and wash immediately with soap and water. Seek medical attention immediately. If this chemical has been inhaled, remove from exposure, begin rescue breathing (using universal precautions) if breathing has stopped and CPR if heart action has stopped. Transfer promptly to a medical facility. When this chemical has been swallowed, get medical attention immediately. Do not induce vomiting. Medical observation is recommended for 24 – 48 hours after breathing overexposure, as pulmonary edema may be delayed. As first aid for pulmonary edema, a doctor or authorized paramedic may consider administering a corticosteroid spray.

Personal Protective Methods: Wear acid-resistant gloves and clothing to prevent any reasonable probability of skin contact. Safety equipment suppliers/manufacturers can provide recommendations on the most protective glove/clothing material for your operation. All protective clothing (suits, gloves, footwear, headgear) should be clean, available each day, and put on before work. Contact lenses should not be worn when working with this chemical. Wear splash-proof chemical goggles and face shield unless full facepiece respiratory protection is worn. Employees should wash immediately with soap when skin is wet or contaminated. Remove nonimpervious clothing immediately if wet or contaminated. Provide emergency showers and eyewash.

Respirator Selection: *Where there is no REL, at any detectable concentration:* SA:CF (any supplied-air respirator operated in a continuous-flow mode); or PAPROV [any powered, air-purifying respirator with organic vapor cartridge(s)]; CCRFOV [any air-purifying, full-facepiece respirator (gas mask) with a chin-style, front- or back-mounted acid gas canister]; or GMFOV [any air-purifying, full-facepiece respirator (gas mask) with a chin-style, front- or back-mounted organic vapor canister]; or SCBAF (any self-contained breathing apparatus with a full facepiece); or SAF (any supplied-air respirator with a full facepiece). *Emergency or planned entry into unknown concentrations or IDLH conditions:* SCBAF:PD,PP (any self-contained breathing apparatus that has a full facepiece and is operated in a pressure-demand or other positive-pressure mode); or SAF:PD,PP:ASCBA (any supplied-air respirator that has a full facepiece and is operated in a pressure-demand or other positive-pressure mode in combination with an auxiliary self-contained breathing apparatus operated in a pressure-demand or other positive-pressure mode). *Escape:* GMFOV [any air-purifying, full-facepiece respirator (gas mask) with a chin-style, front- or back-mounted organic vapor canister] or SCBAE (any appropriate escape-type, self-contained breathing apparatus).

Storage: Before entering confined space where boron trichloride may be present, check to make sure that an explosive concentration does not exist. Store in tightly closed containers in a cool, well ventilated, fireproof place. Metal containers involving the transfer of this chemical should be grounded and bonded. Where possible, automatically pump liquid from drums or other storage containers to process containers. Drums must be equipped with self-closing valves, pressure vacuum bungs, and flame arresters. Use only non-sparking tools and equipment, especially when opening and closing containers of this chemical. Sources of ignition such as smoking and open flames, are prohibited where this chemical is used, handled, or stored in a manner that could create a potential fire or explosion hazard.

Shipping: The DOT label requirement is "Poison Gas, Corrosive." Shipment by passenger aircraft or even by cargo aircraft is forbidden. It falls in UN/DOT Hazard Class 2.3 and Packing Group II.[19][20]

Spill Handling: Seek expert help. Evacuate and restrict persons not wearing protective equipment from area of spill or leak until cleanup is complete. Avoid breathing vapors. Keep upwind. Remove all ignition sources. Ventilate area of leak or spill. Isolate area until gas has dispersed. Stop leak if you can do so without risk. Keep material out of water sources and sewers. Use water spray to knock down vapors. Do not use water on material itself. Neutralize spilled material with crushed limestone, soda ash or lime. Collect neutralized material in the most convenient and safe manner and deposit in sealed containers. It may be necessary to contain and dispose of this chemical as a hazardous waste. If material or contaminated runoff enters waterways, notify downstream users of potentially contaminated waters. Contact your Department of Environmental Protection or your regional office of the federal EPA for specific recommendations. If employees are required to clean-up spills, they must be properly trained and equipped. OSHA 1910.120(q) may be applicable.

Fire Extinguishing: Boron trichloride is not combustible but decomposes in heat forming chlorine. Reacts with water forming boric acid and hydrochloric acid fumes. Use dry chemical, carbon dioxide, or dry sand to extinguish. If large quantities of combustibles are involved, use water in flooding quantities as spray and fog. Use water spray to absorb vapors. For large fires use water spray, fog, or foam. *Do not get water on material itself.* Poisonous gases are produced in fire including hydrogen chloride. If material or contaminated runoff enters waterways, notify downstream users of potentially contaminated waters. Notify local health and fire officials and pollution control agencies. From a secure, explosion-proof location, use water spray to cool exposed, undamaged containers. If cooling streams are ineffective (venting sound increases in volume and pitch, tank discolors, or shows any signs of deforming), withdraw immediately to a secure position. If employees are expected to fight fires, they must be trained and equipped in OSHA 1910.156.

References

U.S. Environmental Protection Agency, "Chemical Profile: Boron Trichloride," DC, Chemical Emergency Preparedness Program (November 30, 1987)

Boron Trifluoride

Molecular Formula: BF_3

Synonyms: Borane, trifluoro-; Boron fluoride; Fluorure de bore (French); Leecure B; Leecure, B series; Trifluoroborane; Trifluoroboron; Trifluoruro de boro (Spanish)

CAS Registry Number: 7637-07-2

RTECS Number: ED2275000

DOT ID: UN 1008

EEC Number: 005-001-00-X

Regulatory Authority

- Air Pollutant Standard Set (ACGIH)[1] (DFG)[3] (HSE)[33] (former USSR)[43] (OSHA)[58] (Several States)[60] (Australia) (Israel) (Mexico)
- OSHA 29CFR1910.119, Appendix A. Process Safety List of Highly Hazardous Chemicals, TQ = 250 lb
- CLEAN AIR ACT: Accidental Release Prevention/Flammable substances, (Section 112[r], Table 3), TQ = 5,000 lb (2,270 kg)
- SAFE DRINKING WATER ACT as boron: Priority List (55FR1470)
- SUPERFUND/EPCRA 40CFR355, Appendix B Extremely Hazardous Substances: TPQ = 500 lb (227 kg)
- SUPERFUND/EPCRA 40CFR302.4 Reportable Quantity (RQ): EHS, 1 lb (0.454 kg)
- EPCRA Section 313 Form R *de minimis* concentration reporting level: 1.0%
- U.S. DOT 49CFR172.101, Inhalation Hazardous Chemical
- Canada, WHMIS, Ingredients Disclosure List

Cited in U.S. State Regulations: Alaska (G), California (A, G), Connecticut (A), Florida (G), Illinois (G), Maine (G), Massachusetts (G), Minnesota (G), Nevada (A), New Hampshire (G), New Jersey (G), North Dakota (A), Oklahoma (G), Pennsylvania (G), Rhode Island (G), Virginia (A), West Virginia (G).

Description: Boron trifluoride, BF_3, is a nonflammable, colorless gas with a pungent, suffocating odor. Forms dense white fumes in moist air. Shipped as a nonliquefied compressed gas. Boiling point = -100°C. Freezing/Melting point = -127°C. Reacts with water.

Potential Exposure: Boron trifluoride is a highly reactive chemical used primarily as a catalyst in chemical synthesis. It is stored and transported as a gas but can be reacted with a variety of materials to form both liquid and solid compounds. The magnesium industry utilizes the fire-retardant and antioxidant properties of boron trifluoride in casing and heat treating. Nuclear applications of boron trifluoride include neutron detector instruments, boron-10 enrichment and the production of neutroabsorbing salts for molten-salt breeder reactors.

Incompatibilities: Boron trifluoride reacts with polymerized unsaturated compounds. Decomposes on contact with water and moisture, forming toxic and corrosive hydrogen fluoride, fluoroboric acid and boric acid. Reacts violently with alkali and alkaline earth metals (except magnesium); metals such as sodium, potassium and calcium, and with alkyl nitrates. Attacks many metals in presence of water.

Permissible Exposure Limits in Air: The OSHA PEL [legal limit], NIOSH and ACGIH ceiling is 1 ppm, not to be exceeded at any time. The DFG has the same MAK[3] and the peal limitation (5 min) is 2 times normal MAK, not to be exceeded 8 times during a workshift. The Australian peak limitation is 1 ppm. Israel's and California's ceiling is the same, and Mexico's limit is 1 ppm TWA. The British HSE[33] has set the same value (1 ppm = 3 mg/m^3) as a STEL value. The former USSR-UNEP/IRPTC project[43] has set a MAC of 1 mg/m^3 in workplace air. The NIOSH IDLH is 25 ppm. Several states have set guidelines or standards for boron trifluoride in ambient air[60] ranging from zero (Connecticut) to 25 μg/m^3 (Virginia) to 30 μg/m^3 (North Dakota) to 71 μg/m^3 (Nevada).

Determination in Air: Collection by an impinger preceded by a filter followed by colorimetric analysis.

Permissible Concentration in Water: No criteria set. Reaction.

Routes of Entry: Inhalation, skin and/or eye contact.

Harmful Effects and Symptoms

Boron trifluoride gas, upon contact with air, immediately reacts with water vapor to form a mist which, if at a high enough concentration provides a visible warning of its presence. The gas or mist is irritating to the skin, eyes, and respiratory system. Boron trifluoride is highly toxic; may cause death or permanent injury after very short exposure to small quantities. Substance is irritating to the eyes, the skin, and the respiratory tract. The toxic action of the halogenated borons (boron trifluoride and trichloride) is considerably influenced by their halogenated decomposition products. They are primary irritants of the nasal passages, respiratory tract, and eyes in man. Animal experiments showed a fall in inorganic phosphorus level in blood and on auropsy, pneumonia, and degenerative changes in renal tubules. Long-term exposure leads to irrigation of the respiratory tract, dysporteinemia, reduction in cholinesterase activity, increased nervous system liability. High concentrations showed a reduction of acetyl carbonic acid and inorganic phosphorus in blood, and dental fluorosis.

Short Term Exposure: Corrosive to the eyes, skin and respiratory tract. Can cause burns to the skin and permanent eye damage. Inhalation can cause pulmonary edema, a medical emergency that can be delayed for several hours. This can cause death. Rapid evaporation of the liquid may cause frostbite.

Long Term Exposure: May cause kidney damage.

Points of Attack: Respiratory system, kidneys, eyes, skin.

Medical Surveillance: In the absence of a suitable monitoring method, NIOSH recommends that medical surveillance, including comprehensive preplacement and annual periodic examinations be made available to all workers employed in areas where boron trifluoride is manufactured, used, handled, or is evolved as a result of chemical processes.

First Aid: If contact with liquid, treat for frostbite. If this chemical gets into the eyes, remove any contact lenses at once and irrigate immediately for at least 30 minutes, occasionally lifting upper and lower lids. Seek medical attention immediately. If this chemical contacts the skin, remove contaminated clothing and wash immediately with soap and water. Seek medical attention immediately. If this chemical has been inhaled, remove from exposure, begin rescue breathing (using universal precautions) if breathing has stopped and CPR if heart action has stopped. Transfer promptly to a medical facility. When this chemical has been swallowed, get medical attention. Give large quantities of water and induce vomiting. Do not make an unconscious person vomit. Medical observation is recommended for 24 – 48 hours after breathing overexposure, as pulmonary edema may be delayed. As first aid for pulmonary edema, a doctor or authorized paramedic may consider administering a corticosteroid spray.

Personal Protective Methods: Engineering controls should be used to maintain boron trifluoride concentrations at the lowest feasible level. Wear corrosive-resistant gloves and clothing to prevent any reasonable probability of skin contact. Safety equipment suppliers/manufacturers can provide recommendations on the most protective glove/clothing material for your operation. All protective clothing (suits, gloves, footwear, headgear) should be clean, available each day, and put on before work. Contact lenses should not be worn when working with this chemical. Wear gas-proof goggles and face shield unless full facepiece respiratory protection is worn. Employees should wash immediately with soap when skin is wet or contaminated. Remove nonimpervious clothing immediately if wet or contaminated. Provide emergency showers and eyewash.

Respirator Selection: NIOSH/OSHA: *10 ppm:* SA* (any supplied-air respirator). *25 ppm*: SA:CF* (any supplied-air respirator operated in a continuous-flow mode); or SCBAF (any self-contained breathing apparatus with a full facepiece); or SAF (any supplied-air respirator with a full facepiece). *Emergency or planned entry into unknown concentrations or IDLH conditions:* SCBAF:PD,PP (any self-contained breathing apparatus that has a full facepiece and is operated in a pressure-demand or other positive-pressure mode); or SAF:PD,PP:ASCBA (any supplied-air respirator that has a full facepiece and is operated in a pressure-demand or other positive-pressure mode in combination with an auxiliary, self-contained breathing apparatus operated in a pressure-demand or other positive-pressure mode. *Escape:* GMFS [any air-purifying, full-facepiece respirator (gas mask) with a chin-style, front- or back-mounted canister providing protection against the compound of concern]; or SCBAE (any appropriate escape-type, self-contained breathing apparatus).

* Substance reported to cause eye irritation or damage; may require eye protection.

Storage: Protect cylinders against extreme temperature changes, dropping, falling, or physical damage. Store outdoors or in a cool, well-ventilated, dry area in a non-combustible structure away from incompatible materials listed above. Cylinders must not be exposed to temperatures below -28.9°C or above 54.4°C.

Shipping: The DOT label required is "Poison Gas," Shipment by passenger aircraft or railcar or even by cargo aircraft is forbidden. The UN/DOT Hazard Class is 2.3 and the Packing Group is II.[19][20]

Spill Handling: Evacuate and restrict persons not wearing protective equipment from area of spill or leak until cleanup is complete. Remove all ignition sources. Ventilate area of spill or leak to disperse gas. Stop leak if you can do it without risk. Use water spray to reduce vapor but do not put water on leak or spill area. *Small spills:* flush area with flooding amounts of water. *Large spills:* dike far ahead of spill for later disposal. Do not get water inside container. It may be necessary to contain and dispose of this chemical as a hazardous waste. If material or contaminated runoff enters waterways, notify downstream users of potentially contaminated waters. Contact your Department of Environmental Protection or your regional office of the federal EPA for specific recommendations. If employees are required to clean-up spills, they must be properly trained and equipped. OSHA 1910.120(q) may be applicable.

Initial isolation and protective action distances

Distances shown are likely to be affected during the first 30 minutes after materials are spilled and could increase with time. If more than one tank car, cargo tank, portable tank, or large cylinder is involved in the incident is leaking, the protective action distance may need to be increased. You may need to seek emergency information from CHEMTREC at (800) 424-9300 or seek professional environmental engineering assistance from the U.S. EPA Environmental Response Team at (908) 548-8730 (24-hour response line).

Small spills (From a small package or a small leak from a large package)

First: Isolate in all directions (feet) .. 200

Then: Protect persons downwind (miles)

Day 0.1

Night 0.4

Large spills (From a large package or from many small packages)

First: Isolate in all directions (feet) .. 600

Then: Protect persons downwind (miles)

Day 0.6

Night 2.4

Fire Extinguishing: Do not use water. For small fire use dry chemical or carbon dioxide. On larger fires use any agent suitable for surrounding fire. Poisonous gases are produced in fire. Vapors are heavier than air and will collect in low areas. If material or contaminated runoff enters waterways, notify downstream users of potentially contaminated waters. Notify local health and fire officials and pollution control agencies. From a secure, explosion-proof location, use water spray to cool exposed containers. If cooling streams are ineffective (venting sound increases in volume and pitch, tank discolors, or shows any signs of deforming), withdraw immediately to a secure position. If employees are expected to fight fires, they must be trained and equipped in OSHA 1910.156.

Disposal Method Suggested: Chemical reaction with water to form boric acid, and fluoroboric acid. The fluoroboric acid is reacted with limestone forming boric acid and calcium fluoride. The boric acid may be discharged into a sanitary sewer system while the calcium fluoride may be recovered or landfilled.

References

National Institute for Occupational Safety and Health, Criteria for a Recommended Standard: Occupational Exposure to Boron Trifluoride, NIOSH Doc. No. 77–122, Washington, DC (1977)

U.S. Environmental Protection Agency, "Chemical Profile: Boron Trifluoride," Washington, DC, Chemical Emergency Preparedness Program

New Jersey Department of Health and Senior Services, Hazardous Substance Fact Sheet: Boron Trifluoride, Trenton, NJ (July 1996)

Boron Trifluoride Etherates

Molecular Formula: $C_2H_6BF_3O$ (with Methyl Ether); $C_4H_{10}BF_3O$ (with Ethyl Ether)

Common Formula: $CH_3OCH_3{\cdot}BF_3$ (with Methyl Ether); $C_2H_5OC_2H_5{\cdot}BF_3$ (with Ethyl Ether)

Synonyms: Boron trifluoride diethyl etherate; Boron trifluoride-dimethyl ether; Boron trifluoride dimethyl etherate; Boron trifluoride etherate; Fluorid bority dimethyl ether

CAS Registry Number: 353-42-4 (with methyl ether); 109-63-7 (with ethyl ether)

RTECS Number: ED8400000 (with methyl ether); KX7375000 (with ethyl ether)

DOT ID: UN 2965 (dimethyl etherate); UN 2604 (diethyl etherate)

Regulatory Authority

The following is for 353-42-4 (with methyl ether), except as noted.

- CLEAN AIR ACT: Accidental Release Prevention/Flammable substances, (Section 112[r], Table 3), TQ = 15,000 lb (6,810 kg)
- SAFE DRINKING WATER ACT as boron: Priority List (55FR1470)
- SUPERFUND/EPCRA 40CFR302.4 Reportable Quantity (RQ): EHS, 1 lb (0.454 kg)
- SUPERFUND/EPCRA 40CFR355, Appendix B Extremely Hazardous Substances: TPQ = 1,000 lb (454 kg)
- Canada, WHMIS, Ingredients Disclosure List [109-63-7 (with ethyl ether)]

Cited in U.S. State Regulations: California (G), Florida (G), Massachusetts (G), New Jersey (G), Pennsylvania (G)

Description: $CH_3OCH_3{\cdot}BF_3$ (with methyl ether) is moisture-sensitive, corrosive, flammable liquid. Boiling point = 126–127°C. Freezing/Melting point = -14°C. $C_2H_5OC_2H_5{\cdot}BF_3$ (with ethyl ether) is a moisture-sensitive, corrosive, flammable liquid. Boiling point = 126°C. Freezing/Melting point = -58°C. Flash point = 64°C (oc). Decomposes in water. Hazard Identification (based on NFPA-704 M Rating System): Health 3, Flammability 2, Reactivity 1 Water reactive. Hazard Identification (based on NFPA-704 M Rating System): *(dimethyl etherate/UN 2965)* Health 4, Flammability 4, Reactivity 1.

Potential Exposure: Used as a catalyst.

Incompatibilities: Reacts with air forming corrosive hydrogen fluoride vapors. Incompatible with oxidizers (may cause fire and explosion), water, steam or heat, forming corrosive and flammable vapors. Peroxide containing etherate reacts explosively with aluminum lithium hydride, magnesium tetrahydroaluminate. Mixtures with phenol react explosively with 1,3-butadiene. Presumed to form explosive peroxides.

Permissible Exposure Limits in Air: There are no airborne exposure limits for the etherates. Reference can only be made to the parents boron trifluoride and methyl or ethyl ether.

Permissible Concentration in Water: No criteria set.

Routes of Entry: Inhalation, ingestion, skin and/or eye contact.

Harmful Effects and Symptoms

Short Term Exposure: These compounds are corrosive. Contact may cause severe burns to skin and eyes. The boron fluoride etherates are highly toxic by inhalation. Inhalation may cause pulmonary edema, a medical emergency that can be delayed for several hours. This can cause death.

Long Term Exposure: Can cause kidney damage.

Points of Attack: Kidneys.

Medical Surveillance: Consider chest x-ray following acute overexposure. Kidney and lung function tests.

First Aid: If this chemical gets into the eyes, remove any contact lenses at once and irrigate immediately for at least 15 minutes, occasionally lifting upper and lower lids. Seek medical attention immediately. If this chemical contacts the skin, remove contaminated clothing and wash immediately with soap and water. Seek medical attention immediately. If this chemical has been inhaled, remove from exposure, begin rescue breathing (using universal precautions) if breathing has stopped and CPR if heart action has stopped. Transfer promptly to a medical facility. When this chemical has been swallowed, get medical attention. If victim is conscious, administer water or milk. Do not induce vomiting. Medical observation is recommended for 24 – 48 hours after breathing overexposure, as pulmonary edema may be delayed. As first aid for pulmonary edema, a doctor or authorized paramedic may consider administering a corticosteroid spray.

Personal Protective Methods: Wear protective gloves and clothing to prevent any reasonable probability of skin contact. Safety equipment suppliers/manufacturers can provide recommendations on the most protective glove/clothing material for your operation. All protective clothing (suits, gloves, footwear, headgear) should be clean, available each day, and put on before work. Contact lenses should not be worn when working with this chemical. Wear splash-proof chemical goggles and face shield unless full facepiece respiratory protection is worn. Employees should wash immediately with soap when skin is wet or contaminated. Provide emergency showers and eyewash.

Respirator Selection: NIOSH/OSHA: (as boron trifluoride) *10 ppm:* SA* (any supplied-air respirator). *25 ppm:* SA:CF* (any supplied-air respirator operated in a continuous-flow mode); or SCBAF (any self-contained breathing apparatus with a full facepiece); or SAF (any supplied-air respirator with a full facepiece). *Emergency or planned entry into unknown concentrations or IDLH conditions:* SCBAF:PD,PP (any self-contained breathing apparatus that has a full facepiece and is operated in a pressure-demand or other positive-pressure mode); or SAF:PD,PP:ASCBA (any supplied-air respirator that has a full facepiece and is operated in a pressure-demand or other positive-pressure mode in combination with an auxiliary, self-contained breathing apparatus operated in a pressure-demand or other positive-pressure mode. *Escape:* GMFS [any air-purifying, full-facepiece respirator (gas mask) with a chin-style, front- or back-mounted canister providing protection against the compound of concern]; or SCBAE (any appropriate escape-type, self-contained breathing apparatus).

* Substance reported to cause eye irritation or damage; may require eye protection.

Storage: Before entering confined space where these chemical may be present, check to make sure that an explosive concentration does not exist. Store in tightly closed containers in a cool, well ventilated area. Metal containers involving the transfer of this chemical should be grounded and bonded. Where possible, automatically pump liquid from drums or other storage containers to process containers. Drums must be equipped with self-closing valves, pressure vacuum bungs, and flame arresters. Use only non-sparking tools and equipment, especially when opening and closing containers of this chemical. Sources of ignition such as smoking and open flames, are prohibited where this chemical is used, handled, or stored in a manner that could create a potential fire or explosion hazard.

Shipping: The DOT-required label for the dimethyl etherate is "Dangerous when wet, Corrosive, Flammable Liquid." The UN/DOT Hazard Class in 4.3 and the Packing Group is II.[19][20] The diethyl etherate requires a "Corrosive Flammable Liquid" label. It falls in Hazard Class 8 and Packing Group II.

Spill Handling: Evacuate and restrict persons not wearing protective equipment from area of spill or leak until cleanup is complete. Remove all ignition sources. Ventilate area of spill or leak. Do not touch spill material. Use water spray to reduce vapors, but do not get water inside containers. For small spills, absorb with sand or other noncombustible absorbent material and place into containers. For large spills, dike far ahead of spill for later disposal. It may be necessary to contain and dispose of this chemical as a hazardous waste. If material or contaminated runoff enters waterways, notify downstream users of potentially contaminated waters. Contact your Department of Environmental Protection or your regional office of the federal EPA for specific recommendations. If employees are required to clean-up spills, they must be properly trained and equipped. OSHA 1910.120(q) may be applicable.

Fire Extinguishing: For small fires use dry chemical, carbon dioxide, water spray, or foam extinguishers. For large fires, use water spray, fog, or foam. Poisonous gases are produced in fire. If material or contaminated runoff enters waterways, notify downstream users of potentially contaminated waters. Notify local health and fire officials and pollution control agencies. From a secure, explosion-proof location, use water spray to cool exposed containers. If cooling streams are ineffective (venting sound increases in volume and pitch, tank discolors, or shows any signs of deforming), withdraw immediately to a secure position. If employees are expected to fight fires, they must be trained and equipped in OSHA 1910.156.

References

U.S. Environmental Protection Agency, "Chemical Profile: Boron Trifluoride Compound with Methyl Ether (1:1)," Washington, DC, Chemical Emergency Preparedness Program (November 30, 1987)

Bromacil

Molecular Formula: $C_9H_{13}BrN_2O_2$

Synonyms: Borea; Borocil extra; Bromacil 1.5; alpha-Bromacil 80 WP; Bromax; Bromazil; 5-Bromo-3-*sec*-butyl-6-methyluracil; 5-Bromo-6-methyl-3-(1-methylpropyl)-2,4-(1H,3H)-pyrimidinedione; 5-Bromo-6-methyl-3-(1-methylpropyl)-2,4(1H,3H)-pyrimidinedione; 3-*sek*-Butyl-5-brom-6-methyluracil (German); Croptex onyx; Cynogan; Dupont herbicide 976; Eerex; Eerex granular weed killer; Eerex water soluble granular weed killer; Fenocil; Herbicide 976; Hydon; Hyvar; Hyvarex; Hyvar-EX; Hyvar X; Hyvar X-7; Hyvar X Bromacil; Hyvar X weed killer; Hyvar X-WS; Krovar II; Nalkil; 2,4(1H,3H)-Pyrimidinedione, 5-bromo-6-methyl-3-(1-methylpropyl)-; Uracil, 5-Bromo-3-*sec*-butyl-6-methyl; Uragan; Uragon; Urox; Urox B; Urox B water soluble concentrate weed killer; Urox-HX; Urox HX granular weed killer

CAS Registry Number: 314-40-9

RTECS Number: YQ9100000

DOT ID: UN 2588

Regulatory Authority

- Air Pollutant Standard Set (ACGIH)[1] (NIOSH) (HSE)[33] (Several States)[60] (Australia) (Israel) (Mexico) (Canada, Provincial)
- SAFE DRINKING WATER ACT: Priority List (55FR1470)
- EPCRA Section 313 Form R *de minimis* concentration reporting level: 1.0%
- Canada, WHMIS, Ingredients Disclosure List

Cited in U.S. State Regulations: Alaska (G), Connecticut (A), Florida (G), Illinois (G), Kansas (W), Maine (G, W), Massachusetts (G), Nevada (A), New Hampshire (G), New Jersey (G), North Dakota (A), Pennsylvania (G), Rhode Island (G), Virginia (A), West Virginia (G).

Description: Bromacil, $C_9H_{13}BrN_2O_2$, is a noncombustible colorless, crystalline solid, that may be dissolved in a flammable liquid. Boiling point = (sublimes). Freezing/Melting point = 158 – 159°C (sublimes). Hazard Identification (based on NFPA-704 M Rating System): Health 1, Flammability 0, Reactivity 0. Slightly soluble in water.

Potential Exposure: Bromacil is used primarily for the control of annual and perennial grasses and broadleaf weeds, both nonselectively on noncrop lands and selectively for weed-control in a few crops (citrus and pineapple). A limit of 0.1 mg/kg of agricultural products is set in several countries.[35] Those exposed will be those involved in manufacture, formulation and application.

Incompatibilities: Incompatible with strong acids, oxidizers, heat. Decomposes slowly in strong acids.

Permissible Exposure Limits in Air: A TWA value of 1 ppm (10 mg/m^3) has been recommended by NIOSH and ACGIH. This same TWA has been set by Australia, Israel, Mexico, and HSE[33] and HSE and Mexico set a STEL value set at 2 ppm (20 mg/m^3). The Canadian provinces of Alberta, Ontario and Quebec have the same TWAs and Alberta's STEL is 2 ppm (21 mg/m^3). For states have set guidelines or standards for Bromacil in ambient air[60] ranging from 100 μg/m^3 (North Dakota) to 160 μg/m^3 (Virginia) to 200 μg/m^3 (Connecticut) to 238 μg/m^3 (Nevada).

Determination in Air: Filter; none; Gravimetri; NIOSH Methods (IV) #0500, Particulates NOR (total).

Permissible Concentration in Water: A no-adverse effects level in drinking water has been calculated by NAS/NRC as 0.086 mg/l.[46] Some states have set guidelines for Bromacil in drinking water,[61] including Maine at 25 μg/l, and Kansas at 87.5 μg/l.

Routes of Entry: Inhalation, ingestion, skin and/or eye contact.

Harmful Effects and Symptoms

Short Term Exposure: Irritates the eyes, skin, upper respiratory system, lungs. Inhalation can cause irritation, coughing and wheezing.

Long Term Exposure: Has cause thyroid affects in animals.

Points of Attack: Eyes, skin, respiratory system, thyroid.

Medical Surveillance: Before beginning employment and at regular times after that, for those with frequent or potentially high exposures, the following is recommended: Lung function tests. Thyroid function tests. Consider x-ray following acute overexposure.

First Aid: If this chemical gets into the eyes, remove any contact lenses at once and irrigate immediately for at least 15 minutes, occasionally lifting upper and lower lids. Seek medical attention immediately. If this chemical contacts the skin, remove contaminated clothing and wash immediately with soap and water. Seek medical attention immediately. If this chemical has been inhaled, remove from exposure, begin rescue breathing (using universal precautions) if breathing has stopped and CPR if heart action has stopped. Transfer promptly to a medical facility. When this chemical has been swallowed, get medical attention. Give large quantities of water and induce vomiting. Do not make an unconscious person vomit.

Personal Protective Methods: Wear protective gloves and clothing to prevent any reasonable probability of skin contact. Safety equipment suppliers/manufacturers can provide recommendations on the most protective glove/clothing material for your operation. All protective clothing (suits, gloves, footwear, headgear) should be clean, available each day, and put on before work. Contact lenses should not be worn when working with this chemical. Wear splash- or dust-proof chemical goggles (depending on physical state of material) and face shield unless full facepiece respiratory protection is worn. Employees should wash immediately with soap when skin is wet or contaminated. Provide emergency showers and eyewash.

Respirator Selection: Where the potential exists for exposures over 1 ppm, use a MSHA/NIOSH approved respirator equipped with particulate (dust/fume/mist) filters. Particulate filters must be checked every day before work for physical damage, such as rips or tears, and replaced as needed. Where the potential for high exposures exists, use a MSHA/NIOSH approved supplied-air respirator with a full facepiece operated in the positive pressure mode or with a full facepiece, hood, or helmet in the continuous flow mode, or use a MSHA/NIOSH approved self-contained breathing apparatus with a full facepiece operated in pressure demand or other positive pressure mode.

Storage: Prior to working with Bromacil you should be trained on its proper handling and storage. Store in tightly closed containers in a cool, well-ventilated area away from strong acids, oxidizers, heat and open flame.

Shipping: Pesticides, solid, toxic, n.o.s. should bear a "Keep away from Food" label in Packing Group III.

Spill Handling: Evacuate and restrict persons not wearing protective equipment from area of spill or leak until cleanup is complete. Remove all ignition sources. Absorb liquid containing Bromacil in vermiculite, dry sand, earth, or similar material. Collect powdered material in the most convenient and safe manner and deposit in sealed containers. Ventilate area of spill or leak after clean-up is complete. It may be necessary to contain and dispose of this chemical as a hazardous waste. If material or contaminated runoff enters waterways, notify downstream users of potentially contaminated waters. Contact your Department of Environmental Protection or your regional office of the federal EPA for specific recommendations. If employees are required to cleanup spills, they must be properly trained and equipped. OSHA 1910.120(q) may be applicable.

Fire Extinguishing: Bromacil may be ignited by heat or open flame. Dust may cause and explosion. Use dry chemical, carbon dioxide, water spray, or foam extinguishers. Poisonous gases are produced in fire including bromine and nitrogen oxides. If material or contaminated runoff enters waterways, notify downstream users of potentially contaminated waters. Notify local health and fire officials and pollution control agencies. From a secure, explosion-proof location, use water spray to cool exposed containers. If cooling streams are ineffective (venting sound increases in volume and pitch, tank discolors, or shows any signs of deforming), withdraw immediately to a secure position. If employees are expected to fight fires, they must be trained and equipped in OSHA 1910.156.

Bromacil may burn, but does not readily ignite. Extinguish fire using an agent suitable for the type of surrounding fire.

Disposal Method Suggested: Bromacil should be incinerated in a unit operating at 850°C equipped with gas scrubbing equipment.[22]

References

New Jersey Department of Health and Senior Services, "Hazardous Substance Fact Sheet: Bromacil," Trenton, NJ (July 1998)

Bromadiolone

Molecular Formula: $C_{30}H_{23}BrO_4$

Synonyms: 2H-1-Benzopyran-2-one, 3-(3-[4'-Bromo(1,1'-biphenyl)-4-yl]-3-hydroxy-1-phenylpropyl)-4-hydroxy-; Bromadialone; 3-(3-[4'-Bromo(1,1'-biphenyl)-4-yl]3-hydroxy-1-phenylpropyl)-4-hydroxy-2H-1-benzopyran-2-one; 3-[3-(4'-Bromobiphenyl)-4-yl]3-hydroxy-1-phenylpropyl)-4-hydroxy-coumarin; 3-(α-[*p*-(*p*-Bromophenyl)-β-hydroxyphenethyl]benzyl)-4-hydroxy- coumarin; Canadien2000; Contrac; Coumarin, 3-(3-(4'-bromo-1,1'-biphenyl-4-yl)-3-hydroxy-1-phenylpropyl)-4-hydroxy-; Coumarin, 3-(α-[*p*-(*p*-bromophenyl)-β-hydroxyphenethyl)benzyl]-4-hydroxy-; (Hydroxy-4-coumarinyl 3)-3 phenyl-3(bromo-4 biphenyl-4)-1 propanol-1 (French); LM-637; MAKI; Ratimus; Rentokil Deadline; Slaymor; Supercaid; Super-Caid; Super-Rozol; Sup'orats; Temus

CAS Registry Number: 28772-56-7

RTECS Number: GN4934700

DOT ID: UN 3027

Regulatory Authority

- SUPERFUND/EPCRA 40CFR355, Appendix B Extremely Hazardous Substances: TPQ = 100/10,000 lb (45.4/4,540 kg)
- SUPERFUND/EPCRA 40CFR302.4 Reportable Quantity (RQ): EHS, 100 lb (45.4 kg)

Cited in U.S. State Regulations: California (G), Florida (G), Massachusetts (G), New Jersey (G), Pennsylvania (G).

Description: Bromadiolone, $C_{30}H_{23}BrO_4$, is a yellowish powder. Freezing/Melting point = 200 – 210°C. Hazard Identification (based on NFPA-704 M Rating System): Health 4, Flammability 1, Reactivity 0.

Potential Exposure: Bromadiolone is used as an anticoagulant rodenticide. It is bait for rodent control used against house mice, roof rats, warfarin-resistant Norway rats. It is also authorized by USDA for use in official establishments operating under the Federal meat, poultry, shell egg grading and egg products inspection program.

Permissible Exposure Limits in Air: No standards set.

Permissible Concentration in Water: No criteria set.

Routes of Entry: Ingestion, inhalation, skin and/or eye contact.

Harmful Effects and Symptoms

The compound is toxic by oral exposure, the oral LD_{50} rabbit is 1.0 mg/kg and the oral LD_{50} rat is 1.125 mg/kg (Extremely toxic).[9]

Short Term Exposure: May cause skin and eye irritation.

Long Term Exposure: Coumarin and its derivatives may be carcinogenic to humans.

First Aid: If this chemical gets into the eyes, remove any contact lenses at once and irrigate immediately for at least 15 minutes, occasionally lifting upper and lower lids. Seek medical attention immediately. If this chemical contacts the skin, remove contaminated clothing and wash immediately with soap and water. Seek medical attention immediately. If this chemical has been inhaled, remove from exposure, begin rescue breathing (using universal precautions) if breathing has stopped and CPR if heart action has stopped. Transfer promptly to a medical facility. When this chemical has been swallowed, get medical attention. Give large quantities of water and induce vomiting. Do not make an unconscious person vomit. Keep victim quiet and maintain normal body temperature. Effects may be delayed; keep victim under observation.

Personal Protective Methods: Wear protective gloves and clothing to prevent any reasonable probability of skin contact. Safety equipment suppliers/manufacturers can provide recommendations on the most protective glove/clothing material for your operation. All protective clothing (suits, gloves, footwear, headgear) should be clean, available each day, and put on before work. Contact lenses should not be worn when working with this chemical. Wear dust-proof chemical goggles and face shield unless full facepiece respiratory protection is worn. Employees should wash immediately with soap when skin is wet or contaminated. Provide emergency showers and eyewash.

Respirator Selection: SCBAF:PD,PP (any MSHA/NIOSH approved self-contained breathing apparatus that has a full facepiece and is operated in a pressure-demand or other positive-pressure mode); or SAF:PD,PP:ASCBA (any supplied-air respirator that has a full facepiece and is operated in a pressure-demand or other positive-pressure mode in combination with an auxiliary, self-contained breathing apparatus operated in a pressure-demand or other positive pressure mode).

Storage: Prior to working with Bromadiolone you should be trained on its proper handling and storage. Store in tightly closed containers in a cool, well ventilated area.

Shipping: Coumarin derivatives, solid, toxic, n.o.s. have a DOT label requirement of "Poison." The UN/DOT Hazard Class is 6.1 and the Packing Group is I.[19][20]

Spill Handling: Evacuate and restrict persons not wearing protective equipment from area of spill or leak until cleanup is complete. Remove all ignition sources. Collect powdered material in the most convenient and safe manner and deposit in sealed containers. Ventilate area of spill or leak after clean-up is complete. It may be necessary to contain and dispose of this chemical as a hazardous waste. If material or contaminated runoff enters waterways, notify downstream users of potentially contaminated waters. Contact your Department of Environmental Protection or your regional office of the federal EPA for specific recommendations. If employees are required to clean-up spills, they must be properly trained and equipped. OSHA 1910.120(q) may be applicable.

Do not touch spilled material; stop leak if you can do so without risk. Use water spray to reduce vapors. *Small spills:* absorb with sand or other non-combustible absorbent material and place into containers for later disposal. *Small dry spills:* with clean shovel place material into clean, dry container and cover; move containers from spill area. *Large spills:* dike far ahead of spill for later disposal.

Fire Extinguishing: This material may burn but does not ignite readily. *Small fires:* dry chemicals, carbon dioxide, water spray or foam. *Large fires:* water spray, fog or foam. Move container from fire area if you can do so without risk. Fight fire from maximum distance. Dike fire control water for later disposal; do not scatter the material. Poisonous gases are produced in fire. If material or contaminated runoff enters waterways, notify downstream users of potentially contaminated waters. Notify local health and fire officials and pollution control agencies. From a secure, explosion-proof location, use water spray to cool exposed containers. If cooling streams are ineffective (venting sound increases in volume and pitch, tank discolors, or shows any signs of deforming), withdraw immediately to a secure position. If employees are expected to fight fires, they must be trained and equipped in OSHA 1910.156.

References

U.S. Environmental Protection Agency, "Chemical Profile: Bromadiolone," Washington, DC, Chemical Emergency Preparedness Program (November 30, 1987)

Bromine

Molecular Formula: Br

Synonyms: Brom (German); Brome (French); Bromo (Italian, Spanish); Broom (Dutch)

CAS Registry Number: 7726-95-6

RTECS Number: EF9100000

DOT ID: UN 1744

EEC Number: 035-001-00-5

EINECS Number: 231-778-1

Regulatory Authority

- Toxic Chemical (World Bank)[15]
- Air Pollutant Standard Set (ACGIH)[1] (OSHA)[58] (HSE)[33] (DFG)[3] (former USSR)[43] (Several States)[60] (Australia) (Israel) (Mexico) (Canada, Provincial: Alberta, British Columbia, Ontario, Quebec)
- CLEAN AIR ACT: Accidental Release Prevention/Flammable substances, [Section 112(r)], TQ = 10,000 lb (4,540 kg)

- SUPERFUND/EPCRA 40CFR355, Appendix B Extremely Hazardous Substances: TPQ = 500 lb (227 kg)
- SUPERFUND/EPCRA 40CFR302.4 Reportable Quantity (RQ): EHS, 1 lb (0.454 kg)
- EPCRA Section 313 Form R *de minimis* concentration reporting level: 1.0%
- U.S. DOT 49CFR172.101, Inhalation Hazardous Chemical
- Canada, WHMIS, Ingredients Disclosure List

Cited in U.S. State Regulations: Alaska (G), California (G), Connecticut (A), Florida (G, A), Illinois (G), Maine (G, W), Massachusetts (G), Minnesota (G), Nevada (A), New Hampshire (G, W), New Jersey (G), New York (G, A), North Carolina (A), North Dakota (A), Oklahoma (G), Pennsylvania (G), Rhode Island (G), Virginia (A), West Virginia (G).

Description: Bromine, Br, is a fuming red to dark reddish-brown, non-flammable volatile liquid with a suffocating odor. Soluble in water and alcohol. Boiling point = about 59°C. Freezing/Melting point = -7.2°C. The odor threshold is 3.5 ppm.[41]

Potential Exposure: Bromine is primarily used in the manufacture of gasoline antiknock compounds (1,2-dibromoethane). Other uses are for gold extraction, in brominating hydrocarbons, in bleaching fibers and silk, in the manufacture of military gas, dyestuffs, and as an oxidizing agent. It is used in the manufacture of many pharmaceuticals and pesticides.

Incompatibilities: A powerful oxidizer. May cause fire and explosions in contact with organic or other readily oxidizable materials. Contact with aqueous ammonia, acetaldehyde, acetylene, acrylonitrile, or with metals may cause violent reactions. Anhydrous Br_2 reacts with aluminum, titanium, mercury, potassium; wet Br_2 with other metals. Also incompatible with alcohols, antimony, alkali hydroxides, arsenites, boron, calcium nitrite, cesium monoxide, carbonyls, dimethyl formamide, ethyl phosphine, fluorine, ferrous and mercurous salts, germanium, hypophosphites, iron carbide, isobutyronphenone, magnesium phosphide, methanol, nickel carbonyl, olefins, ozone, sodium and many other substances. Attacks some coatings, and some forms of plastic and rubber. Corrodes iron, steel, stainless steels and copper.

Permissible Exposure Limits in Air: The Federal standard[53] and the ACGIH TWA value for bromine is 0.1 ppm (0.66 mg/m^3). The Australian, British HSE, Israeli, Mexican, and the German DFG TWA values are the same. The NIOSH, Mexican, Israeli and Australian STEL is 0.3 ppm (2 mg/m^3). The STEL recommended by ACGIH is 0.2 ppm (1.3 mg/m^3). The DFG peak limitation (5 min) Is 2 times the normal MAK, not to be exceeded more than 8 times during a workshift. The NIOSH IDLH level is 3 ppm. The former USSR-UNEP/IRPTC project[43] has set a MAC value in workplace air of 0.5 mg/m^3. Several States have set guidelines or standards for bromine in ambient air[60] ranging from 2.33 μg/m^3 (New York) to 7.0 μg/m^3 (Florida and North Dakota) to 11.0 μg/m^3 (Connecticut) to 17 μg/m^3 (Nevada) to 200 μg/m^3 (North Carolina).

Determination in Air: Sample collection by impinger or fritted bubbler, colorimetric analysis. Filter; $Na_2S_2O_3$; Ion chromatography; NIOSH Method (IV) #6011.

Permissible Concentration in Water: The former USSR-UNEP/IRPTC project has set a MAC value of 0.2 mg/l in water bodies used for domestic purposes. Maine has set a guideline[61] for drinking water of 660 μg/l.

Routes of Entry: Inhalation, ingestion, eye and/or skin contact. Absorbed through the skin.

Harmful Effects and Symptoms

Inhalation: small amount will cause coughing, nose bleed, dizziness, and headache followed by abdominal pain and diarrhea and sometimes measles-like eruption on trunk and extremities. *Skin contact:* causes pustules and painful nodules in exposed areas of skin; if not removed will cause deep, painful ulcers. *Ingestion (of liquid):* causes burning pain in mouth and esophagus, lips and mucous membranes stained brown, severe gastroenteritis evidenced by abdominal pain and diarrhea, repaid headgear, cyanosis, and shock. Regular exposure to concentrations approaching the permissible exposure level causes irritability, loss of appetite, joint pains and dyspepsia. Other symptoms include loss of cornea reflexes, inflammation of the throat, thyroid dysfunction, cardiovascular disorders, disorders of digestive tract. Inhalation exposure to 11 – 23 mg/m^3 produces severe choking. 30 – 60 mg/m^3 is extremely dangerous. 200 mg/m^3 is fatal in a short time. Vapors can cause acute as well as chronic poisoning. It has cumulative properties. It is irritating to the eyes, and respiratory tract. Poisoning is due to the corrosive action on the gastrointestinal tract. Nervous, circulatory and renal disturbances occur after ingestion. Ingestion of liquid can cause death due to circulatory collapse and asphyxiation from swelling of the respiratory tract. The lowest oral lethal dose reported for humans is 14 mg/kg. The lowest lethal inhalation concentration reported for humans is 1,000 ppm.

Short Term Exposure: A corrosive liquid, bromine can cause severe eye and skin irritation and burns that may cause permanent damage and scarring. Inhalation can cause pulmonary edema, a medical emergency that can be delayed for several hours. This can cause death. Symptoms of exposure can include dizziness, headache; lacrimation (discharge of tears), epistaxis (nosebleed); cough, feeling of oppression; abdominal pain, diarrhea.

Long Term Exposure: May cause acnelike eruptions on the skin. Repeated exposure may cause headache, chest pain, joint pain, and indigestion. Bronchitis or pneumonia may develop with cough, shortness of breath, and phlegm.

Points of Attack: Respiratory system, eyes, lungs, central nervous system.

Medical Surveillance: The skin, eyes, and respiratory tract should be given special emphasis during preplacement and periodic examinations. Chest x-rays as well as general health, lung, blood, liver, and kidney function should be considered. Exposure to other irritants or bromine.

First Aid: If this chemical gets into the eyes, remove any contact lenses at once and irrigate immediately for at least 15 minutes, occasionally lifting upper and lower lids. Seek medical attention immediately. If this chemical contacts the skin, remove contaminated clothing and wash immediately with soap and water. Seek medical attention immediately. If this chemical has been inhaled, remove from exposure, begin rescue breathing (using universal precautions) if breathing has stopped and CPR if heart action has stopped. Transfer promptly to a medical facility. When this chemical has been swallowed, get medical attention. If victim is conscious, administer water or milk. Do not induce vomiting. Medical observation is recommended for 24 – 48 hours after breathing overexposure, as pulmonary edema may be delayed. As first aid for pulmonary edema, a doctor or authorized paramedic may consider administering a corticosteroid spray.

Personal Protective Methods: Wear protective gloves and clothing to prevent any reasonable probability of skin contact. Safety equipment suppliers/manufacturers can provide recommendations on the most protective glove/clothing material for your operation. All protective clothing (suits, gloves, footwear, headgear) should be clean, available each day, and put on before work. Contact lenses should not be worn when working with this chemical. Wear splash-proof chemical goggles and face shield unless full facepiece respiratory protection is worn. Employees should wash immediately with soap when skin is wet or contaminated. Provide emergency showers and eyewash.

Respirator Selection: OSHA: *2.5 ppm:* SA:CF (any supplied-air respirator operated in a continuous-flow mode); or PAPRS [any powered, air-purifying respirator with cartridge(s) providing protection against the compound of concern]. *3 ppm:* CCRFS [any chemical cartridge respirator with a full facepiece and cartridge(s) providing protection against the compound of concern] organic vapor and acid gas cartridge(s); or GMFS [any air-purifying, full-facepiece respirator (gas mask) with a chin-style, front- or back-mounted canister providing protection against the compound of concern]; or PAPRTS [any powered, air-purifying respirator with a tight-fitting facepiece and cartridge(s) providing protection against the compound of concern]; or SCBAF (any self-contained breathing apparatus with a full facepiece); or SAF (any supplied-air respirator with a full facepiece). *Emergency or planned entry into unknown concentrations or IDLH conditions:* SCBAF:PD,PP (any self-contained breathing apparatus that has a full facepiece and is operated in a pressure-demand or other positive-pressure mode); or SAF:PD,PP: ASCBA (any supplied-air respirator that has a full facepiece and is operated in a pressure-demand or other positive-pressure mode in combination with an auxiliary, self-contained breathing apparatus operated in a pressure-demand or other positive-pressure mode). *Escape:* GMFS [any air-purifying, full-facepiece respirator (gas mask) with a chin-style, front- or back-mounted canister providing protection against the compound of concern]; or SCBAE [any appropriate escape-type, self-contained breathing apparat(s)].

Note: Substance causes eye irritation or damage; eye protection needed. Only *nonoxidizable sorbents* are allowed (not charcoal).

Storage: Store in a cool, dry room with ventilation along the floor. Keep sealed or glass stoppered. Protect against physical damage. Keep out of direct sunlight. Separate from combustibles, organics or other readily oxidizable materials. Store above 20°F (-7°C) to prevent freezing but avoid heating above room temperature to prevent pressure increase which could rupture containers.

Shipping: The shipping label required by DOT is "Corrosive, Poison." Shipment by passenger aircraft or railcar or even by cargo aircraft is forbidden. The UN/DOT Hazard Class is 8 and the Packing Group is I.[19][20]

Spill Handling: Evacuate and restrict persons not wearing protective equipment from area of spill or leak until cleanup is complete. Remove all ignition sources. Wear eye protection. Wear positive pressure breathing apparatus and special protective clothing. Ventilate area of spill or leak. Collect for reclamation by absorbing it in vermiculite, dry sand, earth, or a similar material and deposit it in sealed containers in secured sanitary landfill. Potassium carbonate, sodium carbonate, sodium bicarbonate, lime, and sodium hydroxide solutions are neutralizing agents for liquid bromine spills. Do not touch material, stop leak if possible without risk. Use water spray to reduce vapors. *Do not* absorb in saw-dust or other combustible absorbents; avoid contact with metal implements. *Small spills:* absorb with sand or other noncombustible absorbent material and place in container. *Large spills:* dike spill for later disposal. It may be necessary to contain and dispose of this chemical as a hazardous waste. If material or contaminated runoff enters waterways, notify downstream users of potentially contaminated waters. Contact your Department of Environmental Protection or your regional office of the federal EPA for specific recommendations. If employees are required to clean-up spills, they must be properly trained and equipped. OSHA 1910.120(q) may be applicable.

Initial isolation and protective action distances

Distances shown are likely to be affected during the first 30 minutes after materials are spilled and could increase with time. If more than one tank car, cargo tank, portable tank, or large cylinder is involved in the incident is leaking, the protective action distance may need to be increased. You may need to seek emergency information from CHEMTREC at (800) 424-9300 or seek professional environmental engineering assistance from the U.S. EPA Environmental Response Team at (908) 548-8730 (24-hour response line).

Small spills (From a small package or a small leak from a large package)

First: Isolate in all directions (feet) 200

Then: Protect persons downwind (miles)

Day 0.2

Night 0.6

Large spills (From a large package or from many small packages)

First: Isolate in all directions (feet) 700

Then: Protect persons downwind (miles)

Day 0.5

Night 2.2

Fire Extinguishing: Bromine is not combustible but enhances combustion of other substances. Use dry chemical, carbon dioxide, water spray, or foam extinguishers. Poisonous gases are produced in fire. If material or contaminated runoff enters waterways, notify downstream users of potentially contaminated waters. Notify local health and fire officials and pollution control agencies. Containers may explode in fire. From a secure, explosion-proof location, use water spray to cool exposed containers. If cooling streams are ineffective (venting sound increases in volume and pitch, tank discolors, or shows any signs of deforming), withdraw immediately to a secure position. If employees are expected to fight fires, they must be trained and equipped in OSHA 1910.156.

Disposal Method Suggested: Large volumes of concentrated solutions of reducing agents (bisulfites or ferrous salts) may be added.[24] The mixture is neutralized with soda ash or dilute HCl and flushed to the sewer with large volumes of water.

References

U.S. Environmental Protection Agency, Chemical Hazard Information Profile: Bromine and Bromine Compounds, Washington, DC (November 1, 1976)

Sax, N. I., Ed., "Dangerous Properties of Industrial Materials Report" 1, No. 4, 41–43 (1981); and 3, No. 5, 67–69 (1983)

U.S. Environmental Protection Agency, "Chemical Profile: Bromine," Washington, DC, Chemical Emergency Preparedness Program (November 30, 1987)

New Jersey Department of Health and Senior Services, "Hazardous Substance Fact Sheet: Bromine," Trenton, NJ, (July 1998)

New York State Department of Health, "Chemical Fact Sheet: Bromine," Albany, NY, Bureau of Toxic Substance Assessment (May 1986)

Bromine Pentafluoride

Molecular Formula: BrF_5

Synonyms: Bromine fluoride; Pentafluoruro de bromo (Spanish)

CAS Registry Number: 7789-30-2

RTECS Number: EF9350000

DOT ID: UN 1745

Regulatory Authority

- Air Pollutant Standard Set (ACGIH)[1] (OSHA)[58] (HSE)[33] (Several States)[60] (Australia) (Israel) (Mexico) (Canada, Provincial: Alberta, British Columbia, Ontario, Quebec)
- OSHA 29CFR1910.119, Appendix A, Process Safety List of Highly Hazardous Chemicals, TQ = 2,500 lb (1,135 kg)
- U.S. DOT 49CFR172.101, Inhalation Hazardous Chemical
- Canada, WHMIS, Ingredients Disclosure List

Cited in U.S. State Regulations: Alaska (G), California (A, G), Connecticut (A), Florida (G), Illinois (G), Maine (G), Massachusetts (G), Minnesota (G), Nevada (A), New Hampshire (G), New Jersey (G), North Dakota (A), Oklahoma (G), Pennsylvania (G), Rhode Island (G), Virginia (A), West Virginia (G).

Description: Bromine pentafluoride, BrF_5, is colorless to pale yellow liquid, with pungent odor. At temperatures above boiling point this chemical is a colorless gas. Boiling point = 40.5°C (104°F). Freezing/Melting point = -61°C. Reacts explosively with water.

Potential Exposure: Bromine pentafluoride, BrF_5, is used as an oxidizer in liquid rocket propellant combinations; it may be used in chemical synthesis.

Incompatibilities: A powerful oxidizer. BrF_5 reacts with every known element except inert gases, nitrogen, and oxygen. It reacts violently with water, acids, acid fumes (releasing highly toxic fumes of bromine and fluorine). Incompatible with halogens, arsenic, selenium, alkaline halides, sulfur, iodine, glass, metallic halides, metal oxides, and metals (except copper, stainless steel, nickel and Monel®. Fire may result from contact with combustibles or organic matter at room temperature, and contact of this substance with water produces an explosion. Even under mild conditions this substance attacks organic compounds vigorously, often causing explosion. Decomposes in heat above 460°C.

Permissible Exposure Limits in Air: The Federal standards[58] and the ACGIH TWA value are 0.1 ppm (0.7 mg/m³). HSE[33] has set that same TWA value and a STEL of 0.3 ppm (2.0 mg/m³). Australia, Israel and Mexico have set the same TWA value and Mexico set the same STEL. In addition, several states have set guidelines or standards for bromine pentafluoride in ambient air[60] ranging from 7.0 μg/m³ (North Dakota) to 11.0 μg/m³ (Virginia) to 14.0 μg/m³ (Connecticut) to 17.0 μg/m³ (Nevada).

Determination in Air: Sample collection by impinger or fritted bubbler, analysis by ion-specific electrode.

Permissible Concentration in Water: No criteria set.

Routes of Entry: Inhalation, eye and/or skin contact.

Harmful Effects and Symptoms

Respiratory irritation is noted. See also fluorine and chlorine trifluoride.

Short Term Exposure: Can cause severe irritation and burns of the eyes and skin. Inhalation can irritate the nose and throat;

higher exposures can cause pulmonary edema, a medical emergency that can be delayed for several hours. This can cause death.

Long Term Exposure: Can cause kidney, liver and lung damage.

Points of Attack: Eyes, skin and mucous membranes.

Medical Surveillance: Kidney, liver, and lung function tests. Consider chest x-ray following acute overexposure.

First Aid: If this chemical gets into the eyes, remove any contact lenses at once and irrigate immediately for at least 15 minutes, occasionally lifting upper and lower lids. Seek medical attention immediately. If this chemical contacts the skin, remove contaminated clothing and wash immediately with soap and water. Seek medical attention immediately. If this chemical has been inhaled, remove from exposure, begin rescue breathing (using universal precautions) if breathing has stopped and CPR if heart action has stopped. Transfer promptly to a medical facility. When this chemical has been swallowed, get medical attention. If victim is conscious, administer water or milk. Do not induce vomiting. Medical observation is recommended for 24 – 48 hours after breathing overexposure, as pulmonary edema may be delayed. As first aid for pulmonary edema, a doctor or authorized paramedic may consider administering a corticosteroid spray.

Personal Protective Methods: Wear protective gloves and clothing to prevent any reasonable probability of skin contact. Safety equipment suppliers/manufacturers can provide recommendations on the most protective glove/clothing material for your operation. All protective clothing (suits, gloves, footwear, headgear) should be clean, available each day, and put on before work. Contact lenses should not be worn when working with this chemical. Wear splash-proof chemical goggles and face shield unless full facepiece respiratory protection is worn. Employees should wash immediately with soap when skin is wet or contaminated. Provide emergency showers and eyewash.

Respirator Selection: Use of self-contained breathing apparatus is recommended.[24] Where the potential exists for exposures over 0.1 ppm, use a MSHA/NIOSH approved supplied-air respirator with a full facepiece operated in the positive pressure mode or with a full facepiece, hood, or helmet in the continuous flow mode, or use a MSHA/NIOSH approved self-contained breathing apparatus with a full facepiece operated in pressure-demand or other positive pressure mode.

Storage: Prior to working with bromine pentafluoride you should be trained on its proper handling and storage. Store in tightly closed containers in a cool, well ventilated area away from water, steam, acids, acid fumes and combustibles. Metal containers involving the transfer of this chemical should be grounded and bonded. Drums must be equipped with self-closing valves, pressure vacuum bungs, and flame arresters. Use only non-sparking tools and equipment, especially when opening and closing containers of this chemical. Sources of ignition such as smoking and open flames, are prohibited where this chemical is used, handled, or stored in a manner that could create a potential fire or explosion hazard. See OSHA standard 1910.104 and NFPA 43A *Code for the Storage of Liquid and Solid Oxidizers* for detailed handling and storage regulations.

Shipping: BrF_5 requires a label reading "Oxidizer, Poison, Corrosive." Shipment by passenger aircraft or railcar or even by cargo aircraft is forbidden.[19] It falls in UN/DOT Hazard Class 5.1 and Packing Group I.

Spill Handling: Evacuate and restrict persons not wearing protective equipment from area of spill or leak until cleanup is complete. Remove all ignition sources. *Gas:* Ventilate area of spill or leak to disperse gas. If this is a gas leak, stop flow of gas. If the leak from a cylinder cannot be stopped, remove the container to a safe place in the open air, and repair leak, or allow cylinder to empty. *Liquid:* Cover spill with dry lime, sand, or soda ash and deposit in closed and sealed containers. Ventilate area of spill or leak after clean-up is complete. It may be necessary to contain and dispose of this chemical as a hazardous waste. If material or contaminated runoff enters waterways, notify downstream users of potentially contaminated waters. Contact your Department of Environmental Protection or your regional office of the federal EPA for specific recommendations. If employees are required to clean-up spills, they must be properly trained and equipped. OSHA 1910.120(q) may be applicable.

Initial isolation and protective action distances

Distances shown are likely to be affected during the first 30 minutes after materials are spilled and could increase with time. If more than one tank car, cargo tank, portable tank, or large cylinder is involved in the incident is leaking, the protective action distance may need to be increased. You may need to seek emergency information from CHEMTREC at (800) 424-9300 or seek professional environmental engineering assistance from the U.S. EPA Environmental Response Team at (908) 548-8730 (24-hour response line).

Small spills (From a small package or a small leak from a large package)

First: Isolate in all directions (feet) 300

Then: Protect persons downwind (miles)

Day 0.2

Night 0.8

Large spills (From a large package or from many small packages)

First: Isolate in all directions (feet) 800

Then: Protect persons downwind (miles)

Day 0.7

Night 3.0

Fire Extinguishing: Not combustible but enhances combustion of other substances. May spontaneously ignite combustible and organic materials. Do not use water to fight fire; reacts explosively with steam or water. An exception is when large amounts of combustible material are involved and firefighters can protect themselves by distance or barrier from the violent reaction with water. Use carbon dioxide or dry chemical on small fires.[17] Poisonous gases are produced in fire including bromine and fluorine. If material or contaminated runoff enters waterways, notify downstream users of potentially contaminated waters. Notify local health and fire officials and pollution control agencies. From a secure, explosion-proof location, use water spray to cool exposed containers. If cooling streams are ineffective (venting sound increases in volume and pitch, tank discolors, or shows any signs of deforming), withdraw immediately to a secure position. If employees are expected to fight fires, they must be trained and equipped in OSHA 1910.156.

Disposal Method Suggested: Allow gas to flow into mixed caustic soda and slaked lime solution.[24] Return unwanted cylinders to supplier if possible.

References

National Institute for Occupational Safety and Health, "Criteria for a Recommended Standard: Occupational Exposure to Inorganic Fluorides," NIOSH Document No. 76–103 (1976)

New Jersey Department of Health and Senior Services, "Hazardous Substance Fact Sheet: Bromine Pentafluoride," Trenton, NJ (July 1998)

Bromine Trifluoride

Molecular Formula: BrF_5

Synonyms: Boron fluoride; Bromine fluoride; Trifluoroborane

CAS Registry Number: 7787-71-5

RTECS Number: ED2275000

DOT ID: UN 1746

Regulatory Authority

- Air Pollutant Standard Set (ACGIH)[1] (OSHA)[58]
- OSHA 29CFR1910.119, Appendix A, Process Safety List of Highly Hazardous Chemicals, TQ = 15,000 lb (6,810 kg)
- U.S. DOT 49CFR172.101, Inhalation Hazardous Chemical
- Canada, WHMIS, Ingredients Disclosure List

Cited in U.S. State Regulations: California (G), Florida (G), Maine (G), Massachusetts (G), New Hampshire (G), New Jersey (G), Oklahoma (G), Pennsylvania (G), Rhode Island (G).

Description: Bromine trifluoride, BrF_5, is a noncombustible, colorless to gray-yellow fuming liquid with an extremely irritating odor. Boiling point = 135°C. Reacts with water forming corrosive gas.

Potential Exposure: Bromine Trifluoride is used as a fluorinating agent and an electrolytic solvent.

Incompatibilities: A powerful oxidizer; highly reactive and a dangerous explosion hazard. Contact with water or other hydrogen containing materials forms hydrogen fluoride gas. Reacts with almost all elements except for inert gases. Violent reaction with reducing agents, organic materials, strong acids, strong bases, halogens, salts (antimony salts), metal oxides, and many other materials. Attacks some plastics, rubber or coatings.

Permissible Exposure Limits in Air: OSHA: The legal airborne permissible exposure limit (PEL) is 0.1 ppm TWA as Bromine averaged over an 8-hour workshift. NIOSH: The recommended airborne exposure limit is 0.1 ppm TWA, averaged over a 10-hour workshift and STEL of 0.3 ppm, not to be exceeded during any 15 minute work period. ACGIH: The recommended airborne exposure limit is 0.1 ppm TWA averaged over an 8-hour workshift, and a STEL of 0.2 ppm. This chemical can be absorbed through the skin, thereby increasing the potential for exposure.

Permissible Concentration in Water: No criteria set.

Routes of Entry: Inhalation, skin contact, Ingestion.

Harmful Effects and Symptoms

Short Term Exposure: Bromine trifluoride can affect you when breathed in and passing through your skin. This substance is a corrosive chemical and contact can severely irritate and burn the skin and eyes (causing possible blindness). Exposure can severely irritate the nose, throat and lungs. Higher exposures can cause a build-up of fluid in the lungs which can cause death.

Long Term Exposure: Repeated exposure can cause skin rash, lung irritation and bronchitis, and may cause a build-up of bromine and fluorine in the body.

Points of Attack: Lungs, skin.

Medical Surveillance: Before beginning employment and at regular times after that, the following are recommended: Lung function tests. If symptoms develop or overexposure is suspected, the following may be useful: Consider chest x-ray after acute overexposure. Blood fluorine and bromine levels.

First Aid: If this chemical gets into the eyes, remove any contact lenses at once and irrigate immediately for at least 30 minutes without stopping, occasionally lifting upper and lower lids. Seek medical attention immediately. If this chemical contacts the skin, remove contaminated clothing and wash immediately with soap and water. Seek medical attention immediately. If this chemical has been inhaled, remove from exposure, begin rescue breathing (using universal precautions) if breathing has stopped and CPR if heart action has stopped. Transfer promptly to a medical facility. When this chemical has been swallowed, get medical attention. If victim is conscious, administer water or milk. Do not induce vomiting. Medical observation is recommended for 24 – 48 hours after breathing overexposure, as pulmonary edema may be delayed.

As first aid for pulmonary edema, a doctor or authorized paramedic may consider administering a corticosteroid spray.

Personal Protective Methods: *Clothing:* Avoid skin contact with Bromine Trifluoride. Wear protective gloves and clothing. Safety equipment suppliers/manufacturers can provide recommendations on the most protective glove/clothing material for your operation. All protective clothing (suits, gloves, footwear, headgear) should be clean, available each day and put on before work. *Eye Protection:* Wear splash-proof chemical goggles and face shield when working with liquid, unless full facepiece respiratory protection is worn.

Respirator Selection: Where the potential exists for exposures over 3 ppm (as bromine) use a MSHA/NIOSH approved supplied-air respirator with a full facepiece operated in the positive pressure mode or with a full facepiece, hood, or helmet in the continuous flow mode, or use a MSHA/NIOSH approved self-contained breathing apparatus with a full facepiece operated in pressure-demand or other positive pressure mode.

Storage: Bromine Trifluoride must be stored to avoid contact with Water (which releases Hydrogen Fluoride gas), Ammonium Halides, Antimony Trioxide, Antimony Chloride and Solvents (such as Ether, Acetone, Acetic Acid and Toluene), since violent reactions occur. Store in tightly closed containers in a cool, well-ventilated area away from Organic and/or Combustible Materials (such as Wood, Cotton and Straw), Chloride and Bromide salts and many metals. Whenever Bromine Trifluoride is used, handled, manufactured, or stored, use explosion-proof electrical equipment and fittings. See OSHA standard 1910.104 and NFPA 43A *Code for the Storage of Liquid and Solid Oxidizers* for detailed handling and storage regulations.

Shipping: BrF_5 must be labeled: "Oxidizer, Poison, Corrosive." It falls in Hazard Class 5.1 and Packing Group I. Shipment by passenger, aircraft or railcar or even by cargo aircraft is forbidden.

Spill Handling: Evacuate and restrict persons not wearing protective equipment from area of spill or leak until cleanup is complete. Remove all ignition sources. Collect powdered material in the most convenient and safe manner and deposit in sealed containers. Ventilate area of spill or leak after clean-up is complete. It may be necessary to contain and dispose of this chemical as a hazardous waste. If material or contaminated runoff enters waterways, notify downstream users of potentially contaminated waters. Contact your Department of Environmental Protection or your regional office of the federal EPA for specific recommendations. If employees are required to clean-up spills, they must be properly trained and equipped. OSHA 1910.120(q) may be applicable.

Initial isolation and protective action distances

Distances shown are likely to be affected during the first 30 minutes after materials are spilled and could increase with time. If more than one tank car, cargo tank, portable tank, or large cylinder is involved in the incident is leaking, the protective action distance may need to be increased. You may need to seek emergency information from CHEMTREC at (800) 424-9300 or seek professional environmental engineering assistance from the U.S. EPA Environmental Response Team at (908) 548-8730 (24-hour response line).

Small spills (From a small package or a small leak from a large package)

First: Isolate in all directions (feet) .. 200

Then: Protect persons downwind (miles)

Day 0.1

Night 0.3

Large spills (From a large package or from many small packages)

First: Isolate in all directions (feet) 500

Then: Protect persons downwind (miles)

Day 0.3

Night 1.2

Fire Extinguishing: Bromine Trifluoide does not burn but may ignite combustible materials. Do not use water or foam. Use dry chemical or CO_2 extinguishers. Poisonous gases are produced in fire including hydrogen fluoride and hydrogen bromide. If material or contaminated runoff enters waterways, notify downstream users of potentially contaminated waters. Notify local health and fire officials and pollution control agencies. Containers may explode in fire. From a secure, explosion-proof location, use water spray to cool exposed containers. If cooling streams are ineffective (venting sound increases in volume and pitch, tank discolors, or shows any signs of deforming), withdraw immediately to a secure position. If employees are expected to fight fires, they must be trained and equipped in OSHA 1910.156.

References

New Jersey Department of Health and Senior Services, "Hazardous Substance Fact Sheet: Bromine Trifluoride," Trenton, NJ (November, 1998)

Bromobenzene

Molecular Formula: C_6H_5Br

Synonyms: Bromobenceno (Spanish); Bromobenzol; Monobromobenzene; Phenyl bromide

CAS Registry Number: 108-86-1

RTECS Number: CY9000000

DOT ID: UN 2514

EEC Number: 602-060-00-9

Regulatory Authority

- Air Pollutant Standard Set (former USSR)[43]
- Safe Drinking Water Act, 55FR1470 Priority List; 40CFR 141.40(e), re: community and non-community water systems

- TSCA 40CFR766.38 precursor chemical substances reporting

Cited in U.S. State Regulations: California (G), Florida (G), Massachusetts (G), New Hampshire (G), New Jersey (G), Pennsylvania (G).

Description: Bromobenzene, C_6H_5Br, is a flammable, clear, colorless mobile liquid with a pleasant odor. Boiling point = 156°C. Freezing/Melting point = -31°C. Flash point = 51°C. Autoignition temperature = 565°C. Hazard Identification (based on NFPA-704 M Rating System): Health 2, Flammability 2, Reactivity 0. Explosive Limits in air: LEL = 0.5%; UEL = 2.5%. Very slightly soluble in water.

Potential Exposure: Bromobenzene is used as an intermediate in organic synthesis, and as an additive in motor oil and fuels. During chlorination water treatment, bromobenzene can be formed in small quantities.

Incompatibilities: Forms explosive mixture with air. Incompatible with strong oxidizers, alkaline earth metals (barium, calcium, magnesium, strontium, etc), metallic salts, with risk of violent reactions. May accumulate static electrical charges; may cause ignition of its vapors.

Permissible Exposure Limits in Air: No U.S. standard set, but the former USSR-UNEP/IRPTC project[43] has set a MAC of 3 mg/m^3 in workplace air and 0.03 mg/m^3 as an allowable daily average in ambient residential air. It should be recognized that this chemical can be absorbed through the skin, thereby increasing exposure.

Permissible Concentration in Water: No criteria set.

Routes of Entry: Inhalation, ingestion, skin absorption.

Harmful Effects and Symptoms

Observations in Man: Bromobenzene irritates the skin and is a central nervous system depressant in humans. In view of the relative paucity of data on the carcinogenicity, teratogenicity, and long-term oral toxicity of bromobenzene, estimates of the effects of chronic oral exposure at low levels cannot be made with any confidence. It is recommended that studies to produce such information be conducted before limits in drinking water can be established. Since bromobenzene was negative on the Salmonella/microsome mutagenicity test, there should be less concern than with those substances that are positive. The oral LD_{50} rat is 2,699 mg/kg (slightly toxic).[9]

Short Term Exposure: Irritates eyes, skin and respiratory tract. Exposure can cause dizziness, lightheadedness and unconsciousness.

Long Term Exposure: May cause liver and kidney damage.

Points of Attack: Skin, liver, kidneys.

Medical Surveillance: Liver and kidney function tests.

First Aid: If this chemical gets into the eyes, remove any contact lenses at once and irrigate immediately for at least 15 minutes, occasionally lifting upper and lower lids. Seek medical attention immediately. If this chemical contacts the skin, remove contaminated clothing and wash immediately with soap and water. Seek medical attention immediately. If this chemical has been inhaled, remove from exposure, begin rescue breathing (using universal precautions) if breathing has stopped and CPR if heart action has stopped. Transfer promptly to a medical facility. When this chemical has been swallowed, get medical attention. Give large quantities of water and induce vomiting. Do not make an unconscious person vomit.

Personal Protective Methods: Wear protective gloves and clothing to prevent any reasonable probability of skin contact. ACGIH recommends polyvinyl alcohol or Viton as protective materials. Also, safety equipment suppliers/manufacturers can provide recommendations on the most protective glove/clothing material for your operation. All protective clothing (suits, gloves, footwear, headgear) should be clean, available each day, and put on before work. Contact lenses should not be worn when working with this chemical. Wear splash-proof chemical goggles and face shield unless full facepiece respiratory protection is worn. Employees should wash immediately with soap when skin is wet or contaminated. Provide emergency showers and eyewash.

Respirator Selection: SCBAF:PD,PP (any MSHA/NIOSH approved self-contained breathing apparatus that has a full facepiece and is operated in a pressure-demand or other positive-pressure mode); or SAF:PD,PP:ASCBA (any supplied-air respirator that has a full facepiece and is operated in a pressure-demand or other positive-pressure mode in combination with an auxiliary, self-contained breathing apparatus operated in a pressure-demand or other positive pressure mode)

Storage: Prior to working with bromobenzene you should be trained on its proper handling and storage. Store in tightly closed containers in a refrigerated area away from incompatible materials listed above. Protect from light.[52] Metal containers involving the transfer of this chemical should be grounded and bonded. Drums must be equipped with self-closing valves, pressure vacuum bungs, and flame arresters. Use only non-sparking tools and equipment, especially when opening and closing containers of this chemical. Sources of ignition such as smoking and open flames, are prohibited where this chemical is used, handled, or stored in a manner that could create a potential fire or explosion hazard.

Shipping: Bromobenzene requires a "Flammable Liquid" label. It falls in UN/DOT Hazard Class 3 and Packing Group III.

Spill Handling: Evacuate and restrict persons not wearing protective equipment from area of spill or leak until cleanup is complete. Remove all ignition sources. Establish forced ventilation to keep levels below explosive limit. Absorb liquids in vermiculite, dry sand, earth, or a similar material and deposit in sealed containers. Follow by washing spill area with alcohol then with soap and water. Do not flush spilled material into sewer. It may be necessary to contain and dispose of this chemical as a hazardous waste. If material or contaminated runoff enters waterways, notify downstream users of potentially contaminated waters. Contact your Department of Environmental Protection or your regional office of the federal EPA for specific recommendations. If employees are required to clean-up spills, they must be properly trained and equipped. OSHA 1910.120(q) may be applicable.

Fire Extinguishing: This chemical is a flammable liquid. Poisonous gases including carbon monoxide and hydrogen bromide are produced in fire. Use dry chemical, carbon dioxide, or alcohol foam extinguishers. Vapors are heavier than air and will collect in low areas. Vapors may travel long distances to ignition sources and flashback. Vapors in confined areas may explode when exposed to fire. Storage containers and parts of containers may rocket great distances, in many directions. If material or contaminated runoff enters waterways, notify downstream users of potentially contaminated waters. Notify local health and fire officials and pollution control agencies. From a secure, explosion-proof location, use water spray to cool exposed containers. If cooling streams are ineffective (venting sound increases in volume and pitch, tank discolors, or shows any signs of deforming), withdraw immediately to a secure position. If employees are expected to fight fires, they must be trained and equipped in OSHA 1910.156.

Disposal Method Suggested: Incineration.[24]

References

National Institute for Occupational Safety and Health, Information Profiles on Potential Occupational Hazards: Brominated Aromatic Compounds, pp 76–85 incl, Report PB-276, 378, Rockville, MD (October 1977)

New Jersey Department of Health and Senior Services, Hazardous Substance Fact Sheet, Bromobenzene, Trenton NJ (January 1999)

Bromodichloromethane

Molecular Formula: $CHBrCl_2$

Common Formula: $BrCHCl_2$

Synonyms: BDCM; Dichlorobromomethane; Methane, bromodichloro; Monobromodichloromethane; NCI-C55243

CAS Registry Number: 75-27-4

RTECS Number: PA5310000

Regulatory Authority

- Carcinogen (sufficient animal data) IARC, NTP
- CLEAN WATER ACT: 40CFR423, Appendix A, Priority Pollutants; Section 313 Water Priority Chemicals (57FR41331, 9/9/92)
- RCRA 40CFR268.48; 61FR15654, Universal Treatment Standards: Wastewater (mg/l), 0.35; Nonwastewater (mg/kg), 15
- RCRA 40CFR264, Appendix 9; TSD Facilities Ground Water Monitoring List. Suggested test method(s) (PQL μg/l): 8010 (1); 8240 (5)
- SAFE DRINKING WATER ACT: Priority List (55FR1470) as bromodichloromethane
- SUPERFUND/EPCRA 40CFR302.4 Reportable Quantity (RQ): CERCLA, 5,000 lb (2,270 kg)
- EPCRA Section 313 Form R *de minimis* concentration reporting level: 1.0%
- TSCA 40CFR716.120(a); SNUR (PMN, P-90-299) Halomethanes
- Canada, WHMIS, Ingredients Disclosure List
- Mexico Wastewater Pollutant

Cited in U.S. State Regulations: California (G), Illinois (G, W), Massachusetts (G), New Hampshire (G), New Jersey (G), Pennsylvania (G), Vermont (W).

Description: Bromodichloromethane, $BrCHCl_2$, is a liquid. Boiling point = 90°C.

Potential Exposure: This compound may find application in organic synthesis.

Permissible Exposure Limits in Air: No standard set.

Permissible Concentration in Water: See Regulatory Authority above. Illinois and Vermont have set guidelines for Bromodichloromethane in drinking water.[61] Illinois at 1.0 μg/l and Vermont at 100 μg/l.

Determination in Water: Gas chromatography (EPA Method 601) or gas chromatography plus mass spectrometry (EPA Method 624).

Harmful Effects and Symptoms

Bromodichloromethane is acutely toxic to mice. It was mutagenic in the Salmonella typhimurium TA 100 bacterial test system and carcinogenic in mice with the same qualification for result significance as for dichloromethane noted. Positive correlations between cancer mortality rates and levels of brominated trihalomethanes in drinking water in epidemiological studies have been reported. The oral LD_{50} rat is 916 mg/kg (slightly toxic).[9]

Short Term Exposure: This material irritates the eyes, nose and mucous membranes.[52]

Long Term Exposure: May cause cancer in humans.

First Aid: If this chemical gets into the eyes, remove any contact lenses at once and irrigate immediately for at least 15 minutes, occasionally lifting upper and lower lids. Seek medical attention immediately. If this chemical contacts the skin, remove contaminated clothing and wash immediately with soap and water. Seek medical attention immediately. If this chemical has been inhaled, remove from exposure, begin rescue breathing (using universal precautions) if breathing has stopped and CPR if heart action has stopped. Transfer promptly to a medical facility. When this chemical has been swallowed, get medical attention. Give large quantities of water and induce vomiting. Do not make an unconscious person vomit.

Personal Protective Methods: Wear protective gloves and clothing to prevent any reasonable probability of skin contact.

Safety equipment suppliers/manufacturers can provide recommendations on the most protective glove/clothing material for your operation. All protective clothing (suits, gloves, footwear, headgear) should be clean, available each day, and put on before work. Contact lenses should not be worn when working with this chemical. Wear splash-proof chemical goggles and face shield unless full facepiece respiratory protection is worn. Employees should wash immediately with soap when skin is wet or contaminated. Provide emergency showers and eyewash.

Respirator Selection: *At any concentration:* SCBAF:PD,PP (any MSHA/NIOSH approved self-contained breathing apparatus that has a full facepiece and is operated in a pressure-demand or other positive-pressure mode); or SAF:PD,PP:ASCBA (any supplied-air respirator that has a full facepiece and is operated in a pressure-demand or other positive-pressure mode in combination with an auxiliary, self-contained breathing apparatus operated in a pressure-demand or other positive pressure mode).

Storage: Prior to working with this chemical you should be trained on its proper handling and storage. Store in a refrigerated space in a tightly-closed container. Protect from light.[52] A regulated, marked area should be established where this chemical is handled, used, or stored in compliance with OSHA standard 1910.1045.

Shipping: This material is not cited specifically in the DOT Performance-oriented packaging standards as regards label requirements or maximum limits on shipping quantities.

Spill Handling: Evacuate and restrict persons not wearing protective equipment from area of spill or leak until cleanup is complete. Remove all ignition sources. Ventilate area of spill or leak. Contain and isolate spill to limit spread. Construct clay-bentonite dams to isolate the spill. Treatment alternatives for contaminated waste include activated charcoal treatment. Absorb liquids in vermiculite, dry sand, earth, or a similar material and deposit in sealed containers. It may be necessary to contain and dispose of this chemical as a hazardous waste. If material or contaminated runoff enters waterways, notify downstream users of potentially contaminated waters. Contact your Department of Environmental Protection or your regional office of the federal EPA for specific recommendations. If employees are required to clean-up spills, they must be properly trained and equipped. OSHA 1910.120(q) may be applicable.

Fire Extinguishing: This material is not combustible. Use extinguishers suitable for surrounding fire. Poisonous gases are produced in fire. If material or contaminated runoff enters waterways, notify downstream users of potentially contaminated waters. Notify local health and fire officials and pollution control agencies. From a secure, explosion-proof location, use water spray to cool exposed containers. If cooling streams are ineffective (venting sound increases in volume and pitch, tank discolors, or shows any signs of deforming), withdraw immediately to a secure position. If employees are expected to fight fires, they must be trained and equipped in OSHA 1910.156.

Disposal Method Suggested: Package in epoxy-lined drums. Destroy by high-temperature incinerator equipped with an HCl scrubber.

References

U.S. Environmental Protection Agency, Halomethanes: Ambient Water Quality Criteria, Washington, DC (1980)

Sax, N. I., Ed., "Dangerous Properties of Industrial Materials Report" 6, No. 3, 39–41 (1986)

U.S. Public Health Service, "Toxicological Profile for Bromodichloromethane," Atlanta, Georgia, Agency for Toxic Substances and Disease Registry (December 1988)

Bromoform

Molecular Formula: $CHBr_3$

Synonyms: Bromoforme (French); Bromoformio (Italian); Bromoformo (Spanish); Methane, tribromo-; Methenyl tribromide; Methyl tribromide; NCI-C55130; Tribrommethaan (Dutch); Tribrommethan (German); Tribromometan (Italian); Tribromomethane

CAS Registry Number: 75-25-2

RTECS Number: PB5600000

DOT ID: UN 2515

EEC Number: 602-007-00-X

Regulatory Authority

- Air Pollutant Standard Set (ACGIH)[1] (OSHA)[58] (HSE)[33] (Several States)[60]
- Water Pollutant Standard Set (EPA), (Mexico)
- CLEAN AIR ACT: Hazardous Air Pollutants (Title I, Part A, Section 112)
- CLEAN WATER ACT: 40CFR423, Appendix A, Priority Pollutants; Section 313 Water Priority Chemicals (57FR41331, 9/9/92)
- EPA HAZARDOUS WASTE NUMBER (RCRA No.): U225
- RCRA, 40CFR261, Appendix 8 Hazardous Constituents
- RCRA 40CFR268.48; 61FR15654, Universal Treatment Standards: Wastewater (mg/l), 0.63; Nonwastewater (mg/kg), 15
- RCRA 40CFR264, Appendix 9; TSD Facilities Ground Water Monitoring List. Suggested test method(s) (PQL μg/l): 8010 (2); 8240 (5)
- SAFE DRINKING WATER ACT: Priority List (55FR1470)
- SUPERFUND/EPCRA 40CFR302.4 Reportable Quantity (RQ): CERCLA, 1,000 lb (454 kg)
- EPCRA Section 313 Form R *de minimis* concentration reporting level: 1.0%
- U.S. DOT 49CFR172.101, Appendix B, Regulated marine pollutant

- Canada, WHMIS, Ingredients Disclosure List
- Mexico Wastewater Pollutant, Drinking Water Pollutant
- TSCA 40CFR716.120(a); 40CFR712.30(e)10 dermal absorption testing; 40CFR799.5055(c), (d)2, hazardous waste constituents subject to testing; Section 12(b) export notification requirement

Cited in U.S. State Regulations: Alaska (G), California (A, G), Connecticut (A), Florida (G), Illinois (G, W), Kansas (G), Maine (G), Maryland (G, W), Massachusetts (G), Minnesota (G), Nevada (A), New Hampshire (G), New Jersey (G), New York (G), North Dakota (A), Pennsylvania (G), Rhode Island (G), Vermont (G), Virginia (G, A), Washington (G), West Virginia (G), Wisconsin (G).

Description: Bromoform, $CHBr_3$, a colorless (turns yellow on exposure to air) with a sweet-smelling, chloroformlike odor. Odor threshold = 0.447 ppm. Boiling point = 149.5°C. Freezing/Melting point = 8.3°C to hexagonal crystals. Hazard Identification (based on NFPA-704 M Rating System): Health 1, Flammability, Reactivity 0. Soluble in water.

Potential Exposure: Bromoform is used in pharmaceutical manufacturing, as an ingredient in fire-resistant chemicals and gauge fluid, and as a solvent for waxes, greases, and oils.

Incompatibilities: Heat causes bromoform to decompose forming toxic and corrosive hydrogen bromide and bromine. Bromoform is a weak acid. Reacts violently with oxidants, bases in powdered form. Reacts with chemically active metals (alkaline metals), powdered aluminium, potassium, sodium, zinc, and magnesium and acetone under basic conditions, causing fire and explosion hazard. Attacks some forms of plastic, rubber and coating. Corrosive to most metals.

Permissible Exposure Limits in Air: The Federal OSHA/NIOSH limit[58] is 0.5 ppm (5 mg/m^3) TWA with the notation "skin" indicating the possibility of cutaneous absorption. ACGIH, HSE,[33] Mexico, Australia, Israel all set the same TWA levels. In Canada, Alberta, British Columbia, Ontario, and Quebec use the same TWA and Alberta has a STEL of 1.5 ppm (16 mg/m^3) The NIOSH IDLH is 850 ppm. The former USSR has also set a TWA value of 5.0 mg/m^3 and Brazil has set 0.4 ppm (4.0 mg/m^3).[35] Several states have set guidelines or standards for bromoform in ambient air[60] ranging from 50 μg/m^3 (North Dakota) to 80 μg/m^3 (Virginia) to 100 μg/m^3 (Connecticut) to 110 μg/m^3 (Nevada).

Determination in Air: Adsorption on charcoal, workup with CS_2, analysis by gas chromatography. See NIOSH Method 1003.[18]

Permissible Concentration in Water: See Regulatory Authority above. Mexico's drinking water criteria is 0.002 mg/l. This substance is highly persistent; bioaccumulation or risk of cancer, reduce exposure by humans to minimum. Two states have set guidelines for bromoform in drinking water: 1.0 μg/l in Illinois and 40 μg/l in Maryland.[61]

Determination in Water: Gas chromatography (EPA Method 601) or gas chromatography plus mass spectrometry (EPA Method 624).

Routes of Entry: Inhalation, ingestion, skin absorption, eye and/or skin contact.

Harmful Effects and Symptoms

Short Term Exposure: *Inhalation:* Can cause irritation to the nose and throat, tearing, reddening of the face, dizziness, and death. Exposure of dogs to 7,000 ppm for 8 minutes causes death. *Skin:* Can be absorbed. Large quantities can lead to symptoms listed under ingestion. *Eyes:* Can cause irritation and tearing. *Ingestion:* Causes burning of mouth and throat. Can cause headache, dizziness, disorientation, slurred speech, difficulty breathing, tremors and unconsciousness. The estimated lethal dose is 1/3 ounce for a 150 pound adult. However, the oral LD_{50} rat is 1,147 mg/kg (slightly toxic).[9]

Long Term Exposure: Bromoform can cause liver damage. May have an effect on the nervous system. Repeated exposure can cause skin rash. Has caused cancer in laboratory animals; whether it does so in humans is unknown. Very irritating substances such as bromoform may affect the lungs.

Points of Attack: Skin, liver, kidneys, respiratory system, lungs, central nervous system.

Medical Surveillance: Consider the points of attack in preplacement and periodic physical examinations. Liver function tests. Lung function tests.

First Aid: If this chemical gets into the eyes, remove any contact lenses at once and irrigate immediately for at least 15 minutes, occasionally lifting upper and lower lids. Seek medical attention immediately. If this chemical contacts the skin, remove contaminated clothing and wash immediately with soap and water. Seek medical attention immediately. If this chemical has been inhaled, remove from exposure, begin rescue breathing (using universal precautions) if breathing has stopped and CPR if heart action has stopped. Transfer promptly to a medical facility. When this chemical has been swallowed, get medical attention. Give large quantities of water and induce vomiting. Do not make an unconscious person vomit.

Personal Protective Methods: Wear appropriate clothing to prevent repeated or prolonged skin contact. Wear eye protection to prevent any reasonable probability of eye contact. Employees should wash promptly when skin is wet or contaminated. Remove nonimpervious clothing promptly if wet or contaminated.

Respirator Selection: OSHA: *12.5 ppm:* SA:CF (any supplied-air respirator operated in a continuous-flow mode); or PAPROV [any powered, air-purifying respirator with organic vapor cartridge(s)]. *25 ppm:* CCRFOV [any chemical cartridge respirator with a full facepiece and organic vapor cartridge(s)]; or GMFOV [any air-purifying, full-facepiece respirator (gas mask) with a chin-style, front- or back-mounted acid gas canister]; or PAPRTOV [any powered, air-purifying respirator with a tight-fitting facepiece and organic vapor cartridge(s)]; or SCBAF (any self-contained breathing apparatus with a full facepiece); or SAF (any supplied-air respirator with a full facepiece). *850 ppm:* SAF:PD,PP (any supplied-air respirator that has a full facepiece and is operated in a pressure-demand or other positive-pressure mode). *Emergency or planned entry into unknown concentrations or IDLH conditions:* SCBAF:PD,PP (any self-contained breathing apparatus that has a full facepiece and is operated in a pressure-demand or other positive-pressure mode); or SAF: PD,PP:ASCBA (any supplied-air respirator that has a full facepiece and is operated in a pressure-demand or other positive-pressure mode in combination with an auxiliary, self-contained breathing apparatus operated in a pressure-demand or other positive-pressure mode. *Escape:* GMFOV [any air-purifying, full-facepiece respirator (gas mask) with a chin-style, front- or back-mounted organic vapor canister]; or SCBAE (any appropriate escape-type, self-contained breathing apparatus).

Note: Substance causes eye irritation or damage; eye protection needed.

Storage: Prior to working with this chemical you should be trained on its proper handling and storage. Bromoform must be stored to avoid contact with incompatible materials listed above. Store in tightly closed containers in a cool well-ventilated area away from heat and light.

Shipping: Bromoform requires a "Keep away from Food" label. It falls in Hazard Class 6.1 and Packing Group III.

Spill Handling: Evacuate and restrict persons not wearing protective equipment from area of spill or leak until cleanup is complete. Remove all ignition sources. Ventilate area of spill or leak. Absorb liquids in vermiculite, dry sand, earth, or a similar material and deposit in sealed containers. Collect powdered material in the most convenient and safe manner and deposit in sealed containers. It may be necessary to contain and dispose of this chemical as a hazardous waste. If material or contaminated runoff enters waterways, notify downstream users of potentially contaminated waters. Contact your Department of Environmental Protection or your regional office of the federal EPA for specific recommendations. If employees are required to clean-up spills, they must be properly trained and equipped. OSHA 1910.120(q) may be applicable.

Fire Extinguishing: Extinguish fire using an agent suitable for type of surrounding fire. Bromoform itself does not burn. Poisonous gases are produced in fire including bromine and hydrogen bromide. If material or contaminated runoff enters waterways, notify downstream users of potentially contaminated waters. Notify local health and fire officials and pollution control agencies. From a secure, explosion-proof location, use water spray to cool exposed containers. If cooling streams are ineffective (venting sound increases in volume and pitch, tank discolors, or shows any signs of deforming), withdraw immediately to a secure position. If employees are expected to fight fires, they must be trained and equipped in OSHA 1910.156.

Disposal Method Suggested: Purify by distillation and return to suppliers.[22] Alternatively. Incinerate with excess fuel.

References

U.S. Environmental Protection Agency, Halomethanes: Ambient Water Quality Criteria, Washington, DC (1980)

U.S. Environmental Protection Agency, Bromoform, Health and Environmental Effects Profile No, 28, Washington, DC, Office of Solid Waste (April 30, 1980)

Sax, N. I., Ed., "Dangerous Properties of Industrial Materials Report" 2, No. 6, 30–35 (1982)

New Jersey Department of Health and Senior Services, "Hazardous Substance Fact Sheet: Bromoform," Trenton, ND (January 1986)

New York State Department of Health, "Chemical Fact Sheet: Bromoform," Albany, NY, Bureau of Toxic Substance Assessment (March 1986)

4-Bromophenyl Phenyl Ether

Molecular Formula: $C_{12}H_9BrO$

Synonyms: Benzene, 1-bromo-4-phenoxy-; Benzene, 2-bromo-4-phenoxy-

CAS Registry Number: 101-55-3

Regulatory Authority

- CLEAN WATER ACT: 40CFR423, Appendix A, Priority Pollutants
- EPA HAZARDOUS WASTE NUMBER (RCRA No.): U030
- RCRA, 40CFR261, Appendix 8 Hazardous Constituents
- RCRA 40CFR268.48; 61FR15654, Universal Treatment Standards: Wastewater (mg/l), 0.055; Nonwastewater (mg/kg), 15
- RCRA 40CFR264, Appendix 9; TSD Facilities Ground Water Monitoring List. Suggested test method(s) (PQL μg/l): 8270 (10)
- SUPERFUND/EPCRA 40CFR302.4 Reportable Quantity (RQ): CERCLA, 100 lb (45.4 kg)

Cited in U.S. State Regulations: California (G), Kansas (G), Louisiana (G), Massachusetts (G), Pennsylvania (G), Vermont (G), Virginia (G), Washington (G), Wisconsin (G).

Description: 4-Bromophenyl phenyl ether, $BrC_6H_4OC_6H_5$, is a liquid haloether. Boiling point = 310°C. Hazard Identification (based on NFPA-704 M Rating System): Health 1, Flammability 1, Reactivity 0.

Potential Exposure: Very little information on 4-bromophenyl phenyl ether exists. 4-Bromophenyl phenyl ether has been identified in raw water, in drinking water and in river water.

Permissible Exposure Limits in Air: No standards set.

Permissible Concentration in Water: *Freshwater Aquatic Life:* For 4-bromophenyl phenyl ether the criterion to protect freshwater aquatic life as derived using the guidelines is 6.2 μg/l as a 24 hour average and the concentration should not exceed 14 µg/l at any time. *Saltwater Aquatic Life:* For saltwater aquatic life, no criterion for 4-bromophenyl phenyl ether can be derived using the guidelines, and there are insufficient data to estimate a criterion using other procedures. *Human Health:* Because of a lack of adequate toxicological data on nonhuman mammals and humans, protective criteria cannot be derived at this time for this compound.

Determination in Water: Methylene chloride extraction followed by gas chromatography with halogen-specific detector (EPA Method 611) or gas chromatography plus mass spectrometry (EPA Method 625).

Harmful Effects and Symptoms

4-Bromophenyl phenyl ether has been tested in the pulmonary adenoma assay, a short-termed carcinogenicity assay. Although the results were negative, several known carcinogens also gave negative results. No other health effects were available. Skin contact can cause tissue defatting and dehydration leading to dermatitis.

First Aid: If this chemical gets into the eyes, remove any contact lenses at once and irrigate immediately for at least 15 minutes, occasionally lifting upper and lower lids. Seek medical attention immediately. If this chemical contacts the skin, remove contaminated clothing and wash immediately with soap and water. Seek medical attention immediately. If this chemical has been inhaled, remove from exposure, begin rescue breathing (using universal precautions) if breathing has stopped and CPR if heart action has stopped. Transfer promptly to a medical facility. When this chemical has been swallowed, get medical attention. Give large quantities of water and induce vomiting. Do not make an unconscious person vomit.

Storage: Store away from heat, flames or oxidizing materials; ethers tend to peroxidize when exposed to air forming unstable peroxides that may detonate with extreme violence.

Shipping: This compound is not listed in the DOT Performance-Oriented Packaging Standards.[19] with regards to label requirements or maximum allowable shipping quantities.

Spill Handling: Evacuate and restrict persons not wearing protective equipment from area of spill or leak until cleanup is complete. Remove all ignition sources. Ventilate area of spill or leak. Absorb liquids in vermiculite, dry sand, earth, or a similar material and deposit in sealed containers. Collect powdered material in the most convenient and safe manner and deposit in sealed containers. It may be necessary to contain and dispose of this chemical as a hazardous waste. If material or contaminated runoff enters waterways, notify downstream users of potentially contaminated waters. Contact your Department of Environmental Protection or your regional office of the federal EPA for specific recommendations. If employees are required to clean-up spills, they must be properly trained and equipped. OSHA 1910.120(q) may be applicable.

Fire Extinguishing: Use dry chemical, carbon dioxide, water spray, or alcohol foam extinguishers. Poisonous gases are produced in fire. If material or contaminated runoff enters waterways, notify downstream users of potentially contaminated waters. Notify local health and fire officials and pollution control agencies. From a secure, explosion-proof location, use water spray to cool exposed containers. If cooling streams are ineffective (venting sound increases in volume and pitch, tank discolors, or shows any signs of deforming), withdraw immediately to a secure position. If employees are expected to fight fires, they must be trained and equipped in OSHA 1910.156.

Disposal Method Suggested: High-temperature incineration with alkaline flue gas scrubbing.

References

U.S. Environmental Protection Agency, 4-Bromophenyl Phenyl Ether, Health and Environmental Effects Profile No. 30, Washington, DC, Office of Solid Waste (April 30, 1980)

U.S. Environmental Protection Agency, Haloethers: Ambient Water Quality Criteria, Washington, DC (1980)

Sax, N. I., Ed., "Dangerous Properties of Industrial Materials Report" 6, No. 2, 43–45 (1986)

Bromopropane

Molecular Formula: C_4H_6

Synonyms: Propane, bromo-; *n*-Propylbromide; Propyl bromide

CAS Registry Number: 26446-77-5 (mixed isomers); 106-94-5 (1-bromopropane); 75-26-3 (2-bromopropane)

DOT ID: UN 2344

EEC Number: 602-019-00-5

EINECS Number: 203-445-0

Cited in U.S. State Regulations: New Jersey (G)

Description: Bromopropane is a flammable, colorless liquid. Flash point ≤22°C. Autoignition temperature = 490°C. Hazard Identification (based on NFPA-704 M Rating System): (as n-propyl bromide): Health 2, Flammability 3, Reactivity 0. Soluble in water.

Potential Exposure: Used for making other chemicals.

Incompatibilities: Forms explosive mixture with air. Contact with strong oxidizers may cause fire and explosion.

Routes of Entry: Inhalation, eye and/or skin contact. Absorbed through the skin.

Harmful Effects and Symptoms

Short Term Exposure: Contact can irritate the eyes and skin. Exposure can cause dizziness, lightheadedness and unconsciousness. Very high exposures can cause death.

Long Term Exposure: Repeated or high exposures may cause liver and lung damage.

Points of Attack: Liver, lungs.

Medical Surveillance: Lung and liver function tests.

First Aid: If this chemical gets into the eyes, remove any contact lenses at once and irrigate immediately for at least 15 minutes, occasionally lifting upper and lower lids. Seek medical attention immediately. If this chemical contacts the skin, remove contaminated clothing and wash immediately with soap and water. Seek medical attention immediately. If this chemical has been inhaled, remove from exposure, begin rescue breathing (using universal precautions) if breathing has stopped and CPR if heart action has stopped. Transfer promptly to a medical facility. When this chemical has been swallowed, get medical attention. Give large quantities of water and induce vomiting. Do not make an unconscious person vomit.

Personal Protective Methods: Wear solvent-resistant gloves and clothing to prevent any reasonable probability of skin contact. Safety equipment suppliers/manufacturers can provide recommendations on the most protective glove/clothing material for your operation. All protective clothing (suits, gloves, footwear, headgear) should be clean, available each day, and put on before work. Contact lenses should not be worn when working with this chemical. Wear splash-proof chemical goggles and face shield unless full facepiece respiratory protection is worn. Employees should wash immediately with soap when skin is wet or contaminated. Remove nonimpervious clothing immediately if wet or contaminated. Provide emergency showers and eyewash.

Respirator Selection: Engineering controls should be used wherever feasible to maintain airborne concentrations of this chemical below the prescribed exposure limit. Respirators and protective equipment are less effective than engineering controls, and should be used only in non-routine or emergency situations which may result in exposure concentrations in excess of the TWA environmental limit. *At any concentrations above the NIOSH REL, or where there is no REL, at any detectable concentration:* SCBAF:PD,PP (any MSHA/NIOSH approved self-contained breathing apparatus that has a full facepiece and is operated in a pressure-demand or other positive-pressure mode); or SAF:PD,PP:ASCBA (any supplied-air respirator that has a full facepiece and is operated in a pressure-demand or other positive-pressure mode in combination with an auxiliary, self-contained breathing apparatus operated in a pressure-demand or other positive pressure mode). *Escape:* GMFOVHiE [any air-purifying, full-facepiece respirator (gas mask) with a chin-style, front- or back-mounted organic vapor canister having a high-efficiency particulate filter]; or SCBAE (any appropriate escape-type, self-contained breathing apparatus).

Storage: Protect against physical damage. Outside or detached storage is preferred. Prior to working with Bromopropane you should be trained on its proper handling and storage. Store in tightly closed containers in a cool, well ventilated area away from incompatible materials listed above. Metal containers involving the transfer of this chemical should be grounded and bonded. Drums must be equipped with self-closing valves, pressure vacuum bungs, and flame arresters. Use only non-sparking tools and equipment, especially when opening and closing containers of this chemical. Sources of ignition such as smoking and open flames, are prohibited where this chemical is used, handled, or stored in a manner that could create a potential fire or explosion hazard.

Shipping: Bromopropane requires a "Flammable Liquid " label. It is in UN/DOT Hazard Class 3, Packing Group II.

Spill Handling: Evacuate and restrict persons not wearing protective equipment from area of spill or leak until cleanup is complete. Remove all ignition sources. Ventilate area of spill or leak. Absorb liquid in vermiculite, dry sand, earth or similar material and deposit in sealed containers. Ventilate area of spill or leak after clean-up is complete. It may be necessary to contain and dispose of this chemical as a hazardous waste. If material or contaminated runoff enters waterways, notify downstream users of potentially contaminated waters. Contact your Department of Environmental Protection or your regional office of the federal EPA for specific recommendations. If employees are required to clean-up spills, they must be properly trained and equipped. OSHA 1910.120(q) may be applicable.

Fire Extinguishing: This chemical is a combustible liquid. Poisonous gases, including hydrogen bromide, are produced in fire. Use dry chemical, carbon dioxide, or alcohol foam extinguishers. Vapors are heavier than air and will collect in low areas. Vapors may travel long distances to ignition sources and flashback. Vapors in confined areas may explode when exposed to fire. Containers may explode in fire. Storage containers and parts of containers may rocket great distances, in many directions. If material or contaminated runoff enters waterways, notify downstream users of potentially contaminated waters. Notify local health and fire officials and pollution control agencies. Containers may explode in fire. From a secure, explosion-proof location, use water spray to cool exposed containers. If cooling streams are ineffective (venting sound increases in volume and pitch, tank discolors, or shows any signs of deforming), withdraw immediately to a secure position. If employees are expected to fight fires, they must be trained and equipped in OSHA 1910.156.

Disposal Method Suggested: Incineration.

References

New Jersey Department of Health and Senior Services, "Hazardous Substance Fact Sheet: Bromopropane." Trenton, NJ (July, 1996)

Brucine

Molecular Formula: $C_{23}H_{26}O_4{\cdot}4H_2O$

Common Formula: $C_{23}H_{26}O_4{\cdot}4H_2O$

Synonyms: Brucina (Italian, Spanish); (-)Brucine; (-)Brucine dihydrate; Brucine hydrate; 2,3-Dimethoxystrichnidin-10-one;

10,11-Dimethoxystrychnine; 2,3-Dimethoxystrychnine; Dimethoxy strychnine; 10,11-Dimethylstrychnine; EEC No. 614-006-00-1; Strychnidin-10-one, 2,3-Dimethoxy-(9CI); Strychnine, 2,3-dimethoxy-

CAS Registry Number: 357-57-3

RTECS Number: EH8925000

DOT ID: UN 1570

EEC Number: 614-006-00-1

Regulatory Authority

- EPA HAZARDOUS WASTE NUMBER (RCRA No.): P018
- RCRA, 40CFR261, Appendix 8 Hazardous Constituents
- SUPERFUND/EPCRA 40CFR302.4 Reportable Quantity (RQ): CERCLA, 100 lbs. (45.4 kg)
- EPCRA Section 313 Form R *de minimis* concentration reporting level: 1.0%
- Canada, WHMIS, Ingredients Disclosure List

Cited in U.S. State Regulations: California (G), Massachusetts (G), New Jersey (G), Pennsylvania (G)

Description: Brucine is colorless to white, odorless, crystalline solid with a very bitter taste. Hazard Identification (based on NFPA-704 M Rating System): Health 2, Flammability 1, Reactivity 0.

Potential Exposure: Used in the manufacture of other chemicals, in perfumes, as a medication for animals, and as a poison for rodents.

Incompatibilities: Reacts with strong oxidizers. Finely dispersed material in air can cause dust explosions.

Permissible Exposure Limits in Air: None established.

Routes of Entry: Inhalation, ingestion, eye and/or skin contact. Absorbed through the skin.

Harmful Effects and Symptoms

Short Term Exposure: Irritates the eyes and respiratory tract. Exposure can cause headache, nausea, vomiting, ringing in the ears, disturbed vision, restlessness, excitement, twitching and convulsions, seizures, breathing difficulties. Severe poisoning can cause paralysis, unconsciousness and death.

First Aid: If this chemical gets into the eyes, remove any contact lenses at once and irrigate immediately for at least 15 minutes, occasionally lifting upper and lower lids. Seek medical attention immediately. If this chemical contacts the skin, remove contaminated clothing and wash immediately with soap and water. Seek medical attention immediately. If this chemical has been inhaled, remove from exposure, begin rescue breathing (using universal precautions) if breathing has stopped and CPR if heart action has stopped. Transfer promptly to a medical facility. When this chemical has been swallowed, get medical attention. Give large quantities of water and induce vomiting. Do not make an unconscious person vomit.

Personal Protective Methods: Wear protective gloves and clothing to prevent any reasonable probability of skin contact. Safety equipment suppliers/manufacturers can provide recommendations on the most protective glove/clothing material for your operation. All protective clothing (suits, gloves, footwear, headgear) should be clean, available each day, and put on before work. Contact lenses should not be worn when working with this chemical. Wear splash or dust-proof chemical goggles and face shield unless full facepiece respiratory protection is worn. Employees should wash immediately with soap when skin is wet or contaminated. Provide emergency showers and eyewash.

Respirator Selection: Engineering controls should be used wherever feasible to maintain airborne concentrations of this chemical below the prescribed exposure limit. Respirators and protective equipment less effective than engineering controls, and should be used only in non-routine or emergency situations which may result in exposure concentrations in excess of the TWA environmental limit. *At any concentrations above the NIOSH REL, or where there is no REL, at any detectable concentration:* SCBAF:PD,PP (any MSHA/NIOSH approved self-contained breathing apparatus that has a full facepiece and is operated in a pressure-demand or other positive-pressure mode); or SAF: PD,PP: ASCBA (any supplied-air respirator that has a full facepiece and is operated in a pressure-demand or other positive-pressure mode in combination with an auxiliary, self-contained breathing apparatus operated in a pressure-demand or other positive pressure mode). *Escape:* GMFOVHiE [any air-purifying, full-facepiece respirator (gas mask) with a chin-style, front- or back-mounted organic vapor canister having a high-efficiency particulate filter]; or SCBAE (any appropriate escape-type, self-contained breathing apparatus).

Storage: Protect against physical damage. Outside or detached storage is preferred. Prior to working with Brucine you should be trained on its proper handling and storage. Store in tightly closed containers in a cool, well ventilated area away from strong oxidizers. Shipping*:* Hazard Class: 6.1. Label: "Poison." Packing Group: I

Spill Handling: Evacuate and restrict persons not wearing protective equipment from area of spill or leak until cleanup is complete. Remove all ignition sources. Collect powdered material in the most convenient and safe manner and deposit in sealed containers. Ventilate area of spill or leak after clean-up is complete. It may be necessary to contain and dispose of this chemical as a hazardous waste. If material or contaminated runoff enters waterways, notify downstream users of potentially contaminated waters. Contact your Department of Environmental Protection or your regional office of the federal EPA for specific recommendations. If employees are required to clean-up spills, they must be properly trained and equipped. OSHA 1910.120(q) may be applicable.

Fire Extinguishing: Brucine may burn but does not readily ignite. Poisonous gases, including nitrogen oxides, are produced in fire. Use dry chemical, carbon dioxide, or foam extinguishers. If material or contaminated runoff enters waterways, notify downstream users of potentially contaminated waters. Notify local health and fire officials and pollution control agencies. Containers may explode in fire. From a secure, explosion-proof location, use water spray to cool exposed containers. If cooling streams are ineffective (venting sound increases in volume and pitch, tank discolors, or shows any signs of deforming), withdraw immediately to a secure position. If employees are expected to fight fires, they must be trained and equipped in OSHA 1910.156.

Disposal Method Suggested: Consult with environmental regulatory agencies for guidance on acceptable disposal practices. Generators of waste containing this contaminant (≥100 kg/mo) must conform with EPA regulations governing storage, transportation, treatment, and waste disposal.

References

New Jersey Department of Health and Senior Services, "Hazardous Substance Fact Sheet: Brucine," Trenton, NJ (January 1999)

Busulfan

Molecular Formula: $C_6H_{14}O_6S_2$

Common Formula: $CH_3SO_2O(CH_2)_4OSO_2CH_3$

Synonyms: 1,4-Bis(methanesulfonoxy)butane; [1,4-Bis(methanesulfonyloxy)butane]; Bisulfan; Bisulphane; 1,4-Butanediol dimethanesulphonate; 1,4-Butanediol dimethyl sulfonate; Buzulfan; C.B.2041; Citosulfan; 1,4-Dimesyloxybutane; 1,4-Dimethanesulfonoxbutane; 1,4-Di(methanesulfonyloxy)buane; 1,4-Dimethanesulphonyloxybutan; 1,4-Dimethylsulfonoxybutane; GT 2041; GT41; Leucosulfan; Mablin®; Methanesulfonic acid tetramethylene ester; Mielucin®; Misulban®; Mitostan®; Myeloleukon; Myleran®; NCI-C01592; NSC-750; Sulphabutin; Tetramethylene bis(methanesulfonate); Tetramethylene dimethane sulfonate; X 149

CAS Registry Number: 55-98-1

RTECS Number: EK1750000

Regulatory Authority

- Carcinogen (human carcinogen) (IARC) (NTP)[10]

Cited in U.S. State Regulations: California (G), Florida (G), Massachusetts (G), Minnesota (G), New Jersey (G), Pennsylvania (G).

Description: Busulfan, $C_6H_{14}O_6S_2$, is a white crystalline powder. Freezing/Melting point = 114 – 118°C. Decomposes in water.

Potential Exposure: Those involved in the manufacture, formulation or use of this compound which finds application as an insect sterilant and as a chemotherapeutic agent taken orally to treat some kinds of leukemia.

Incompatibilities: Oxidizers, moist air and water.

Permissible Exposure Limits in Air: No standards set.

Permissible Concentration in Water: No criteria set.

Routes of Entry: Inhalation.

Harmful Effects and Symptoms

Short Term Exposure: Irritates the skin causing rash. Exposure can cause nausea, vomiting, diarrhea, and seizures.

Long Term Exposure: A carcinogen in humans, causes leukemia and kidney and uterine cancer. A probable teratogen in humans. May damage the developing fetus. May cause testes damage in males (decrease sperm count, cause impotence), and decrease fertility in females. Long term exposure may cause cataracts, lung irritation, permanent lung scarring, bone marrow damage, liver damage. Symptoms of exposure include bleeding tendencies, decreased leucoyte count or depressed bone marrow activity.[52]

Points of Attack: See above.

Medical Surveillance: Completed blood count, chest x-ray, lung function tests, liver function tests.

First Aid: If this chemical gets into the eyes, remove any contact lenses at once and irrigate immediately with water or normal saline for at 20 – 30 minutes, occasionally lifting upper and lower lids. Seek medical attention immediately. If this chemical contacts the skin, remove contaminated clothing and wash immediately with soap and water. Seek medical attention immediately. If this chemical has been inhaled, remove from exposure, begin rescue breathing (using universal precautions) if breathing has stopped and CPR if heart action has stopped. Transfer promptly to a medical facility. When this chemical has been swallowed, get medical attention. Give large quantities of water and induce vomiting. Do not make an unconscious person vomit.

Storage: Store in a refrigerator and protect from exposure to oxidizers or moisture.[52] A regulated, marked area should be established where this chemical is handled, used, or stored in compliance with OSHA standard 1910.1045.

Shipping: Medicines, Poisonous, solid, n.o.s., fall in Hazard Class 6.1 and busulfan falls in Packing Group I. The label requirement is "Poison."

Spill Handling: Evacuate and restrict persons not wearing protective equipment from area of spill or leak until cleanup is complete. Remove all ignition sources. Dampen spilled material with alcohol to avoid dust. Collect powdered material in the most convenient and safe manner and deposit in sealed containers. Ventilate area of spill or leak after clean-up is complete. It may be necessary to contain and dispose of this chemical as a hazardous waste. If material or contaminated runoff enters waterways, notify downstream users of potentially contaminated waters. Contact your Department of Environmental Protection or your regional office of the federal EPA for specific recommendations. If employees are required to clean-up spills, they must be properly trained and equipped. OSHA 1910.120(q) may be applicable.

Fire Extinguishing: Use dry chemical, carbon dioxide, water spray, or alcohol foam extinguishers. Poisonous gases are produced in fire including sulfur oxides. If material or contaminated runoff enters waterways, notify downstream users of potentially contaminated waters. Notify local health and fire officials and pollution control agencies. From a secure, explosion-proof location, use water spray to cool exposed containers. If cooling streams are ineffective (venting sound increases in volume and pitch, tank discolors, or shows any signs of deforming), withdraw immediately to a secure position. If employees are expected to fight fires, they must be trained and equipped in OSHA 1910.156.

References

New Jersey Department of Health and Senior Services, Hazardous Substance Fact Sheet, Busulfan, Trenton NJ (December, 1998).

1,3-Butadiene

Molecular Formula: C_4H_6

Common Formula: H_2C=CH-CH=CH_2

Synonyms: Biethylene; Bivinyl; Buta-1,3-dieen (Dutch); Butadieen (Dutch); Buta-1,3-dien (German); Butadien (Polish); α-γ-Butadiene; Buta-1,3-diene; Butadiene; 1,3-Butadieno (Spanish); Divinyl; Erythrene; NCI-C50602; Pyrrolylene; Vinylethylene

CAS Registry Number: 106-99-0

RTECS Number: EI9275000

DOT ID: UN 1010

EEC Number: 601-013-00-X

Regulatory Authority

- Carcinogen (Human suspected) (ACGIH)[1] (animal positive) (DFG)[3]
- Air Pollutant Standard Set (ACGIH)[1] (HSE)[33] (former USSR)[43] (OSHA)[58] (Several States)[60] (Australia) (Israel) (Mexico) (Several Canadian Provinces)
- CLEAN AIR ACT: Hazardous Air Pollutants (Title I, Part A, Section 112); Accidental Release Prevention/Flammable substances, (Section 112[r], Table 3), TQ = 10,000 lb (4,540 kg)
- SUPERFUND/EPCRA 40CFR302.4 Reportable Quantity (RQ): CERCLA, 1 lb (0.454 kg)
- EPCRA Section 313 Form R *de minimis* concentration reporting level: 0.1%
- Canada, WHMIS, Ingredients Disclosure List

Cited in U.S. State Regulations: Alaska (G), California (G), Connecticut (A), Florida (G), Illinois (G), Maine (G), Maryland (G), Massachusetts (G, A), Michigan (G, A), Minnesota (G), Nevada (A), New Hampshire (G), New Jersey (G), New York (G), North Carolina (A), North Dakota (A), Pennsylvania (G), Rhode Island (G), Virginia (A), West Virginia (G).

Description: 1,3-butadiene, H_2C=CH-CH=CH_2, is a colorless, flammable, liquefied gas with a gasolinelike odor. Boiling point = –4°C to –5°C. Freezing/Melting point = –109°C. Flash point = -76°C. Autoignition temperature = 420°C. The explosive limits are: LEL = -2.0%; UEL = 11.5%.[17] Hazard Identification (based on NFPA-704 M Rating System): Health 2, Flammability 4, Reactivity 2. The odor threshold is 0.45 ppm. Floats and boils on water; slightly soluble.

Potential Exposure: 1,3-Butadiene is used chiefly as the principal monomer in the manufacture of many types of synthetic rubber and other chemicals. Butadiene is finding increasing usage in the formation of rocket fuels, plastics, and resins.

Incompatibilities: Self-reactive. May form explosive peroxides on exposure to air. High heat can cause a violent chemical reaction that will cause container rupture. Fires, explosions, or hazardous polymerization may result from contact with air, strong oxidizers, strong acids, ozone, rust, nitrogen dioxide, phenol, chlorine dioxide, crotonaldehyde, or a free radical polymerization initiator such as hydroquinone. Unsafe in contact with acetylide-forming materials such as monel, copper, and copper alloys (piping material for this gas must not contain more than 63% copper). Add inhibitor (such as tributylcatechol) to prevent self-polymerization and monitor to insure effective levels are maintained at all times. May accumulate static electrical charges, and may cause ignition of its vapors.

Permissible Exposure Limits in Air: The OSHA standard[58] for 1,3-butadiene is 1 ppm TWA for an 8-hour workshift, and STEL of 5 ppm not to be exceeded during any 15 minute work period. The recommended exposure limit from ACGIH is 2 ppm (4.4 mg/m^3) TWA. NIOSH recommends that exposure to occupational carcinogens be limited to the lowest feasible concentrations. The NIOSH IDLH level is 2,000 ppm. (Conversion factor: 1 ppm = 2.21 mg/m^3). The DFG[3] has replaced its TWA value with the notation that it is an "animal positive" carcinogen. The former USSR-UNEP/IRPTC project[43] has set a MAC of 100 mg/m^3 in workplace air and MAC values in ambient residential air of 3 mg/m^3 on a momentary basis and 1 mg/m^3 on a daily average basis. Czechoslovakia[35] has set a very stringent TWA of 20 mg/m^3 with a ceiling value of 40 mg/m^3. Several states have set guidelines or standards for butadiene in ambient air[60] ranging from zero (North Dakota) to 0.003 μg/m^3 (Michigan) to 0.035 μg/m^3 (Massachusetts and North Carolina) to 220 μg/m^3 (Virginia) to 22,000 μg/m^3 (Connecticut) to 52,400 μg/m^3 (Nevada).

Determination in Air: Adsorption by charcoal tube;[2] workup with CH_2C_{12}; analysis by gas chromatography/flame ionization detection See NIOSH Method #1024.[18]

Permissible Concentration in Water: The former USSR-UNEP/IRPTC project[43] has set a MAC of 0.05 mg/^{3} in water bodies used for domestic purposes.

Routes of Entry: Inhalation of gas or vapor, eye and/or skin contact.

Harmful Effects and Symptoms

Initial signs and symptoms of exposure include blurred vision, nausea, prickling and dryness of the mouth, throat, and nose, followed by fatigue, headache, vertigo, decreased blood pressure and pulse rate, unconsciousness, and respiratory paralysis. Concentrations above 8,000 ppm may cause narcotic effects, dizziness, headache, drowsiness and loss of consciousness. Death can result 23 minutes after inhaling air containing 25% butadiene. It is a central nervous system depressant in high concentrations. It may be irritating and cause burns to skin, mucous membranes and eyes. Contact with the liquid may cause frostbite. It can asphyxiate by the displacement of air.

Short Term Exposure: 1,3 butadiene irritates the eyes, skin, and respiratory tract. Rapid evaporation of the liquid may cause frostbite. Inhalation of the vapors may cause effects on the central nervous system, sleepiness and loss of consciousness. Very high exposures may cause death.

Long Term Exposure: This chemical is a probable carcinogen in humans; it may have effects on the bone marrow and liver. There is limited evidence that this chemical is a teratogen in animals, and that it may also damage the testes and ovaries. It may cause heritable genetic damage in humans. Animal tests show that this substance may cause toxic effects upon human reproduction.

Points of Attack: Eyes, respiratory system, central nervous system, reproductive system

Medical Surveillance: There is no special test for this chemical. However, if illness occurs or over-exposure is suspected, medical attention is recommended.

First Aid: If this chemical gets into the eyes, remove any contact lenses at once and irrigate immediately for at least 15 minutes, occasionally lifting upper and lower lids. Seek medical attention immediately. If this chemical contacts the skin, remove contaminated clothing and wash immediately with soap and water. Seek medical attention immediately. If this chemical has been inhaled, remove from exposure, begin rescue breathing (using universal precautions) if breathing has stopped and CPR if heart action has stopped. Transfer promptly to a medical facility. When this chemical has been swallowed, get medical attention. Give large quantities of water and induce vomiting. Do not make an unconscious person vomit.

If frostbite has occurred, seek medical attention immediately; do NOT rub the affected areas or flush them with water. In order to prevent further tissue damage, do NOT attempt to remove frozen clothing from frostbitten areas. If frostbite has NOT occurred, immediately and thoroughly wash contaminated skin with soap and water.

Personal Protective Methods: Wear protective gloves and clothing to prevent any reasonable probability of skin contact. Safety equipment suppliers/manufacturers can provide recommendations on the most protective glove/clothing material for your operation. All protective clothing (suits, gloves, footwear, headgear) should be clean, available each day, and put on before work. Contact lenses should not be worn when working with this chemical. Wear gas-proof chemical goggles and face shield unless full facepiece respiratory protection is worn. Employees should wash immediately with soap when skin is wet or contaminated. Provide emergency showers and eyewash. Where exposure to the liquefied compressed gas may occur, employees should be provided with special clothing designed to prevent frostbite.

Respirator Selection: NIOSH: *At any detectable concentration:* SCBAF:PD,PP (any MSHA/NIOSH approved self-contained breathing apparatus that has a full facepiece and is operated in a pressure-demand or other positive-pressure mode); or SAF:PD,PP:ASCBA (any supplied-air respirator that has a full facepiece and is operated in a pressure-demand or other positive-pressure mode in combination with an auxiliary, self-contained breathing apparatus operated in a pressure-demand or other positive pressure mode). *Escape:* GMFS [any air-purifying, full-facepiece respirator (gas mask) with a chin-style, front- or back-mounted canister providing protection against the compound of concern]; or SCBAE (any appropriate escape-type, self-contained breathing apparatus).

Storage: Outdoor or detached storage is preferred. Store cylinders upright. Prior to working with 1,3-butadiene, you should be trained on its proper handling and storage. Do not store uninhibited 1,3-butadiene. Before entering confined space where this chemical may be present, check to make sure that an explosive concentration does not exist. Store in tightly closed containers in a cool, well ventilated area away from incompatible materials listed above, heat and sunlight. Metal containers involving the transfer of this chemical should be grounded and bonded. Where possible, automatically pump liquid from drums or other storage containers to process containers. Drums must be equipped with self-closing valves, pressure vacuum bungs, and flame arresters. Use only non-sparking tools and equipment, especially when opening and closing containers of this chemical. Sources of ignition such as smoking and open flames, are prohibited where this chemical is used, handled, or stored in a manner that could create a potential fire or explosion hazard. Procedures for the handling, use and storage of cylinders should be in compliance with OSHA 1910.101 and 1910.169 as with the recommendations of the Compressed Gas Association. A regulated, marked area should be established where this chemical is handled, used, or stored in compliance with OSHA standard 1910.1045.

Shipping: The DOT-required shipping label[19] for inhibited butadiene is "Flammable Gas." Shipment by passenger aircraft or railcar is forbidden; shipment by cargo aircraft is limited. This material falls in UN/DOT Hazard Class 2.1 and does not have an assigned packing group.[19][20]

Spill Handling: Evacuate and restrict persons not wearing protective equipment from area of spill or leak until cleanup is complete. Remove all ignition sources. Establish forced ventilation to keep levels below explosive limit and to disperse gas. Attempt to stop leak if without hazard. Use water spray to knock down vapors. Avoid breathing vapors. Keep upwind. Do not handle broken packages without protective equipment. Absorb liquids in vermiculite, dry sand, earth, or a similar material and deposit in sealed containers It may be necessary to contain and dispose of this chemical as a hazardous waste. If material or contaminated runoff enters waterways, notify downstream users of potentially contaminated waters. Contact your Department of Environmental Protection or your regional office of the federal EPA for specific recommendations. If employees are required to clean-up spills, they must be properly trained and equipped. OSHA 1910.120(q) may be applicable.

Fire Extinguishing: This chemical is an extremely flammable liquid. Let tank car, tank truck or storage tank burn unless leak can be stopped; with smaller tanks or cylinders, extinguish/isolate from other flammables. Stop flow of gas before extinguishing fire. Small fires: dry chemical or carbon dioxide. Large fires: water spray, fog, or foam. Move container from fire area if you can do so without risk. Stay away from ends of tanks. For massive fire in cargo area, use unmanned hose holder or monitor nozzles; if this is impossible, withdraw from area and let fire burn. Withdraw immediately in case of rising sound from venting safety device or any discoloration of tank due to fire. Cool container with water using unmanned device until well after fire is out. Poisonous gases are produced in fire. Vapors are heavier than air and will collect in low areas. Vapors may travel long distances to ignition sources and flashback. Vapors in confined areas may explode when exposed to fire. Butadiene vapors are uninhibited and may form polymers in vents or flame arresters of storage tanks, resulting in stoppage of vents. Storage containers and parts of containers may rocket great distances, in many directions. If material or contaminated runoff enters waterways, notify downstream users of potentially contaminated waters. Notify local health and fire officials and pollution control agencies. From a secure, explosion-proof location, use water spray to cool exposed containers. If cooling streams are ineffective (venting sound increases in volume and pitch, tank discolors, or shows any signs of deforming), withdraw immediately to a secure position. If employees are expected to fight fires, they must be trained and equipped in OSHA 1910.156.

Disposal Method Suggested: Incineration.[22]

References

National Institute for Occupational Safety and Health, 1,3-Butadiene, Current Intelligence Bulletin 41, DHHS (NIOSH) Publication No 84–105, Cincinnati, Ohio (February 9, 1984)

U.S. Environmental Protection Agency, "Chemical Profile: Butadiene," Washington, DC, Chemical Emergency Preparedness Program (October 31, 1985)

New Jersey Department of Health and Senior Services, "Hazardous Substance Fact Sheet: 1,3-Butadiene," Trenton, NJ (July 1998)

New York State Department of Health, "Chemical Fact Sheet: Butadiene," Albany, NY, Bureau of Toxic Substance Assessment (March 1986)

Butanes

Molecular Formula: C_4H_{10}

Common Formula: $CH_3CH_2CH_2CH_3$

Synonyms: A-17; Bu-gas; *n*-Butane; Butanen (Dutch); Butani (Italian); Butano (Spanish); Butyl hydride; Diethyl; Diethyl, liquified petroleum gas; Methyl ethyl methane; Methylethylmethane; Twinkle stainless steel cleaner

iso-: 1,1-Dimethylethane; Isobutane; Isobutano (Spanish); 2-Methylpropane; Propane, 2-methyl; Trimethylmethane

CAS Registry Number: 106-97-8; 75-28-5 (*iso-*)

RTECS Number: EJ4200000; TZ4300000 (iso-)

DOT ID: UN 1011; UN 1969 (iso-)

EEC Number: 601-004-00-0

EINECS Number: 203-448-7; 200-857-2 (iso-)

Regulatory Authority

- Air Pollutant Standard Set (ACGIH)[1] (OSHA)[58] (HSE)[33] (DFG)[3] (former USSR)[43] (Several States)[60] (Australia) (Israel) (Mexico) (Several Canadian Provinces)
- CLEAN AIR ACT: Accidental Release Prevention/Flammable substances, (Section 112[r], Table 3), TQ = 10,000 lb (4,540 kg)
- Canada, WHMIS, Ingredients Disclosure List

iso-: CLEAN AIR ACT: Accidental Release Prevention/Flammable substances, [Section 112(r), Table 3], TQ = 10,000 lb (4,540 kg)

Cited in U.S. State Regulations: Alaska (G), California (G), Connecticut (A), Illinois (G), Maine (G), Massachusetts (G), Minnesota (G), Nevada (A), New Hampshire (G), New Jersey (G), North Dakota (A), Pennsylvania (G), Rhode Island (G), Virginia (A), West Virginia (G).

Description: Butane is a colorless, extremely flammable, liquefied, compressed gas. A liquid below 30°F (-1.1°C). Natural gas-like odor. Odor threshold 204 ppm. Boiling point = –1°C. Freezing/Melting point = –138°C. The explosive limits are: LEL = 1.9%; UEL = 8.5%. Flash point = –60°C (flammable gas). Autoignition temperature= 287°C. Hazard Identification (based on NFPA-704 M Rating System): Health 1, Flammability 4, Reactivity 0. Insoluble in water.

Isobutane is a colorless, extremely flammable, liquefied, compressed gas. A liquid below 11°F (-11°C). Gasoline-like odor. Boiling point = –12°C. Freezing/Melting point = –145°C. The explosive limits are: LEL = 1.8%; UEL = 8.4%. Flash point = flammable gas. Autoignition temperature= 460°C. Hazard Identification (based on NFPA-704 M Rating System): Health 1, Flammability 4, Reactivity 0. Insoluble in water.

Potential Exposure: It is used as a raw material for butadiene, as a fuel for household or industrial purposes (alone or in admixture with propane). It is also used as an extractant, solvent, and aerosol propellant. It is used in plastic foam production as a replacement for fluorocarbons.

Incompatibilities: Strong bases, strong oxidizers (e.g., nitrates & perchlorates), chlorine, fluorine, (nickel carbonyl + oxygen).

Permissible Exposure Limits in Air: For both isomers, the OSHA PEL[58] and ACGIH TWA value is 800 ppm (1,900 mg/m^3). The HSE[33] has set a TWA of 600 ppm (1,430 mg/m^3) and a STEL of 750 ppm (1,780 mg/m^3). The DFG[3] has set a MAK of 1,000 ppm (2,400 mg/m^3) and Peak limitation of 2 × normal MAK, 60 min momentary value. Australia, Israel and Mexico all have TWA values of 800 ppm (1,900 mg/m^3). Canadian TWA values range form 600 ppm in British Columbia to 800 ppm in Alberta, Ontario and Quebec. The former USSR-UPEN/IRPTC project[43] has set a MAC in workplace air of 300 mg/m^3 and a momentary MAC of 200 mg/m^3 for ambient air in residential areas. Several states have set forth guidelines or standards for butane in ambient air[60] ranging from 19 mg/m^3 (North Dakota) to 32 mg/m^3 (Virginia) to 38 mg/m^3 (Connecticut) to 45.2 mg/m^3 (Nevada).

Determination in Air: No measurement methods available.

Permissible Concentration in Water: No criteria set but EPA[32] has suggested an ambient limit of 19,000 μg/l based on health effects.

Routes of Entry: Inhalation, skin and/or eye contact (liquid).

Harmful Effects and Symptoms

Butane is not characterized by its toxicity but rather by its narcosis-producing potential at high exposure levels.

Short Term Exposure: Can cause headache, lightheadedness, drowsiness, and unconsciousness from lack of oxygen. Contact with the liquid can cause frostbite.

First Aid: Move victim to fresh air. Call emergency medical care. Apply artificial respiration if victim is not breathing. Administer oxygen if breathing is difficult. Remove and isolate contaminated clothing and shoes. Clothing frozen to the skin should be thawed before being removed. In case of contact with liquefied gas, thaw frosted parts with lukewarm water. Keep victim warm and quiet. Keep victim under observation. Ensure that medical personnel are aware of the material(s) involved and take precautions to protect themselves.

Personal Protective Methods: Where exposure to the liquefied compressed gas may occur, employees should be provided with special clothing designed to prevent frostbite. Safety equipment suppliers/manufacturers can provide recommendations on the most protective glove/clothing material for your operation. All protective clothing (suits, gloves, footwear, headgear) should be clean, available each day, and put on before work. Wear gas-proof chemical goggles and face shield unless full facepiece respiratory protection is worn. Employees should wash immediately with soap when skin is wet or contaminated. Provide emergency showers and eyewash.

Respirator Selection: Large amounts of butane (800 ppm and above) will replace the amount of available oxygen and lead to suffocation. Oxygen content should never be below 19%. Use MSHA/NIOSH approved self-contained breathing apparatus that has a full facepiece and is operated in a pressure-demand or other positive-pressure mode); or SAF: PD,PP:ASCBA (any supplied-air respirator that has a full facepiece and is operated in a pressure-demand or other positive-pressure mode in combination with an auxiliary, self-contained breathing apparatus operated in a pressure-demand or other positive-pressure mode).

Storage: Prior to working with butane you should be trained on its proper handling and storage. All appropriate section of the OSHA Standard 1910.111, Storage, Handling of Liquefied Petroleum Gases must be followed. Store in tightly closed containers in a cool, well ventilated area away from incompatible materials listed above and heat. Metal containers involving the transfer of this chemical should be grounded and bonded. Drums must be equipped with self-closing valves, pressure vacuum bungs, and flame arresters. Use only non-sparking tools and equipment, especially when opening and closing containers of this chemical. Sources of ignition such as smoking and open flames, are prohibited where this chemical is used, handled, or stored in a manner that could create a potential fire or explosion hazard. Procedures for the handling, use and storage of cylinders should be in compliance with OSHA 1910.101 and 1910.169 as with the recommendations of the Compressed Gas Association.

Shipping: The label requirement for butane is "Flammable Gas." Passenger aircraft or railcar shipment is forbidden. Butane falls in Hazard Class 2.1 with no Packing Group specified.

Spill Handling: Evacuate and restrict persons not wearing protective equipment from area of spill or leak until cleanup is complete. Remove all ignition sources. Keep the gas concentration below the explosive limit range by forced ventilation.[24] stop flow of gas. If source of leak is a cylinder and the leak cannot be stopped in place, remove the leaking cylinder to a safe place in the open air, and repair leak or allow cylinder to dissipate to the atmosphere. If employees are required to clean-up spills, they must be properly trained and equipped. OSHA 1910.120(q) may be applicable.

Fire Extinguishing: Butane is a flammable gas. In case of fire Stop the flow of gas if it can be done safely. Use dry chemical, carbon dioxide, or halon extinguishers. Use water to keep fire-exposed containers cool and to protect personnel doing the shut-off. If a leak or spill has caught fire, use water spray to disperse gas and to protect personnel shutting off leak. If cooling streams are ineffective (venting sound increases in volume and pitch, tank discolors, or shows any signs of deforming), withdraw immediately to a secure position. If material or contaminated runoff enters waterways, notify downstream users of potentially contaminated waters. Notify local health and fire officials and pollution control agencies. If employees are expected to fight fires, they must be trained and equipped in OSHA 1910.156.

Disposal Method Suggested: Controlled incineration.

References

New Jersey Department of Health and Senior Services, "Hazardous Substance Fact Sheet: Butane," Trenton, NJ (August 1998)

Butanedione

Molecular Formula: $C_4H_6O_2$

Common Formula: $CH_3COCOCH_3$

Synonyms: Biacetyl; Butadione; 2,3-Butanedione; Butanodiona (Spanish); Diacetyl; 2,3-Diketobutane; Dimethyl diketone; Dimethylglyoxal; Glyoxal dimethyl; Glyoxal, dimethyl-

CAS Registry Number: 431-03-8

RTECS Number: EK2625000

DOT ID: UN 2346

Regulatory Authority

- Canada, WHMIS, Ingredients Disclosure List

Cited in U.S. State Regulations: Florida (G), Maine (G), Massachusetts (G), New Hampshire (G), New Jersey (G), Pennsylvania (G).

Description: Butanedione, $CH_3COCOCH_3$, is a yellow-green, mobile liquid with a chlorine-like odor. Boiling point = 88°C. Freezing/Melting point = -2.4°C. Flash point = 27°C. Autoignition temperature = 365°C. Hazard Identification (based on NFPA-704 M Rating System): Health 1, Flammability 3, Reactivity 0. Highly soluble in water.

Potential Exposure: It is used as an aroma carrier food additive in butter, vinegar, coffee and other foods.

Incompatibilities: Contact with oxidizers may cause fire and explosions. High heat may cause violent combustion or explosion.

Permissible Exposure Limits in Air: No standards set.

Permissible Concentration in Water: No criteria set.

Routes of Entry: Inhalation, skin contact, ingestion.

Harmful Effects and Symptoms

Short Term Exposure: Butanedione can affect you when breathed in. Exposure can irritate the eyes, nose and throat. Contact can irritate the skin. May have a narcotic effect on the nervous system.

Long Term Exposure: Repeated exposure may affect the blood count and nervous system. Repeated or prolonged contact may cause skin sensitization.

Points of Attack: Skin, eyes, respiratory system, blood, nervous system.

Medical Surveillance: If symptoms develop or overexposure is suspected, the following may be useful: Complete blood count. Consider nerve conductions studies. Examination by a qualified allergist.

First Aid: If this chemical gets into the eyes, remove any contact lenses at once and irrigate immediately for at least 15 minutes, occasionally lifting upper and lower lids. Seek medical attention immediately. If this chemical contacts the skin, remove contaminated clothing and wash immediately with soap and water. Seek medical attention immediately. If this chemical has been inhaled, remove from exposure, begin rescue breathing (using universal precautions) if breathing has stopped and CPR if heart action has stopped. Transfer promptly to a medical facility. When this chemical has been swallowed, get medical attention. Give large quantities of water and induce vomiting. Do not make an unconscious person vomit.

Personal Protective Methods: *Clothing:* Avoid skin contact with Butanedione. Wear solvent-resistant gloves and clothing. Safety equipment suppliers/manufacturers can provide recommendations on the most protective glove/clothing material for your operation. All protective clothing (suits, gloves, footwear, headgear) should be clean, available each day and put on before work. *Eye Protection:* Wear splash-proof chemical goggles and face shield when working with liquid, unless full facepiece respiratory protection is worn.

Respirator Selection: Where the potential for exposures to Butanedione exists, use a MSHA/NIOSH approved supplied-air respirator with a full facepiece operated in the positive pressure mode or with a full facepiece, hood, or helmet in the continuous flow mode, or use a MSHA/NIOSH approved self-contained breathing apparatus with a full facepiece operated in pressure-demand or other positive pressure mode.

Storage: Store in tightly closed containers in a cool, well-ventilated area. Sources of ignition such as smoking and open flames, are prohibited where Butanedione is used, handled, or stored in a manner that could create a potential fire or explosion hazard. Metal containers involving the transfer of 5 gallons or more of Butanedione should be grounded and bonded. Drums must be equipped with self-closing valves, pressure vacuum bungs and flame arresters. Use only non-speaking tools and equipment, especially when opening and closing containers of Butanedione. Wherever Butanedione is used, handled, manufactured, or stored, use explosion-proof electrical equipment and fittings.

Shipping: Butanedione must be labeled "Flammable Liquid." It falls in Hazard Class 3 and Packing Group II.

Spill Handling: Evacuate and restrict persons not wearing protective equipment from area of spill or leak until cleanup is complete. Remove all ignition sources. Ventilate area of spill or leak. Absorb liquids in vermiculite, dry sand, earth, or a similar material and deposit in sealed containers. Keep Butanedione out of a confined space, such as a sewer, because of the possibility of an explosion, unless the sewer is designed to prevent the build-up of explosive concentrations. It may be necessary to contain and dispose of this chemical as a hazardous waste. If material or contaminated runoff enters waterways, notify downstream users of potentially contaminated waters. Contact your Department of Environmental Protection or your regional office of the federal EPA for specific recommendations. If employees are required to clean-up spills, they must be properly trained and equipped. OSHA 1910.120(q) may be applicable.

Fire Extinguishing: Butanedione is a flammable liquid. Poisonous gases are produced in fire. Use dry chemical, carbon dioxide, or alcohol foam extinguishers. Vapors are heavier than air and will collect in low areas. Vapors may travel long distances to ignition sources and flashback. Vapors in confined areas may explode when exposed to fire. Storage containers and parts of containers may rocket great distances, in many directions. If material or contaminated runoff enters waterways, notify downstream users of potentially contaminated waters. Notify local health and fire officials and pollution control agencies. From a secure, explosion-proof location, use water spray to cool exposed containers. If cooling streams are ineffective (venting sound increases in volume and pitch, tank discolors, or shows any signs of deforming), withdraw immediately to a secure position. If employees are expected to fight fires, they must be trained and equipped in OSHA 1910.156.

Disposal Method Suggested: Incineration.

References

New Jersey Department of Health and Senior Services, "Hazardous Substance Fact Sheet: Butanedione," Trenton, NJ (March 1987)

n-Butoxyethanol

Molecular Formula: $C_6H_{14}O_2$

Common Formula: $C_4H_9OCH_2CH_2OH$

Synonyms: BUCS; Butoksyetylowy alkohol (Polish); 2-Butossi-etanolo (Italian); 2-Butoxy-aethanol (German); 2-Butoxyethanol; Butyl cellosolve; Butyl oxitol; Dowanol EB; EGBE; Ektasolve EB solvent; Ethyleneglycol monobutyl ether; Glycol butyl ether; Jeffersol EB; Poly-Solv EB

CAS Registry Number: 111-76-2

RTECS Number: KJ8575000

DOT ID: UN 2369

EEC Number: 603-014-00-0

Regulatory Authority

- Air Pollutant Standard Set (ACGIH)[1] (OSHA)[58] (HSE)[33] (DFG)[3] (former USSR)[43] (Australia) (Israel) (Mexico)

As glycol ethers:

- CLEAN AIR ACT: Hazardous Air Pollutants (Title I, Part A, Section 112) includes mono- and di- ethers of ethylene glycol, diethyl glycol, and triethylene glycol $R\text{-}(OCH_2CH_2)_n$ -OR' where n = 1,2, or 3; R = alkyl or aryl groups; R' = R, H, or groups which when removed, yield glycol ethers with the structure: $R\text{-}(OCH_2CH)_n\text{-}OH$. Polymers are excluded from the glycol category
- EPCRA Section 313: Certain glycol ethers are covered. R-(OCH_2CH_2)n-OR'; Where n = 1,2 or 3; R = alkyl C7 or less; or R = phenyl or alkyl substituted phenyl; R' + H, or alkyl C7 or less; or OR' consisting of carboxylic ester, sulfate, phosphate, nitrate or sulfonate. Form R *de minimis* concentration reporting level: 1.0%
- TSCA 40CFR716.120(a)
- Canada, WHMIS, Ingredients Disclosure List

Cited in U.S. State Regulations: Alaska (G), California (A, G), Florida (G), Illinois (G), Maine (G), Massachusetts (G), Minnesota (G), New Hampshire (G), New Jersey (G), Pennsylvania (G), Rhode Island (G), West Virginia (G).

Description: 2-Butoxy ethanol, $C_4H_9OCH_2CH_2OH$, is a colorless liquid with a mild, etherlike odor. Boiling point = 171°C. Freezing/Melting point = -75°C. Autoignition temperature= 238°C. Hazard Identification (based on NFPA-704 M Rating System): Health 1, Flammability 2, Reactivity 0. The explosive limits are: LEL 1.1%; UEL 12.7%. The flash point is 61°C. Soluble in water.

Potential Exposure: This material is used as a solvent for resins in lacquers, varnishes and enamels. It is also used in varnish removers and in dry cleaning compounds.

Incompatibilities: Forms explosive mixture with air. Can form unstable and explosive peroxides; check for peroxides prior to distillation; render harmless if positive. Decomposes, producing toxic fumes. Violent reaction with strong caustics and strong oxidizers. Attacks some coatings, plastics and rubber. Attacks metallic aluminum at high temperatures.

Permissible Exposure Limits in Air: The Federal OSHA standard[58] 50 ppm (240 mg/m^3) TWA averaged over an 8-hour workshift. The NIOSH recommended airborne limit is 5 ppm (24 mg/m^3)TWA averaged over a 10-hour workshift. The ACGIH limit is 25 ppm (121 mg/m^3)TWA averaged over an 8-hour workshift. They add the notation "skin" indicating the possibility of cutaneous absorption. The NIOSH IDLH level is 700 ppm. The DFG[3] has set a MAK of 20 ppm (100 mg/m^3) and peak limitation (30 min) of 2 times normal MAK which cannot be exceeded more than 4 times during a normal workshift. Argentina[35] has set a TWA of 50 ppm (240 mg/m^3), and a STEL of 150 ppm (720 mg/m^3). The HSE[33] has also set a TWA of 25 ppm (120 mg/m^3). Australia, Israel,

Mexico and the state of California have also set a TWA of 25 ppm (120 mg/m^3)TWA and Mexico has a STEL of 75 ppm (360 mg/m^3). The former USSR-UNEP/IRPTC program[43] has set limits in ambient air of residential areas of 1.0 mg/m^3 on a momentary basis and 0.3 mg/m^3 on a daily average basis.

Determination in Air: Adsorption on charcoal, workup with methanol and methylene chloride, analysis by gas chromatography. See NIOSH Method 1403.[18]

Permissible Concentration in Water: No criteria set.

Routes of Entry: Inhalation, skin absorption, ingestion, skin and/or eye contact.

Harmful Effects and Symptoms

Irritation of eyes, nose and throat; hemolysis, hemoglobinuria. The oral LD_{50} rat is 1,480 mg/kg (slightly toxic).

Short Term Exposure: This chemical irritates the eyes, skin, and respiratory tract. High exposure caused dizziness, lightheadedness, and unconsciousness. breath. Higher exposures can cause pulmonary edema, a medical emergency that can be delayed for several hours. This can cause death. Exposure could cause central nervous system depression and liver and kidney damage

Long Term Exposure: The liquid defats the skin. This chemical can break down red blood cells, and cause anemia; effects the haematopoietic system, resulting in blood disorders. It can also damage the liver and kidneys.

Points of Attack: Eyes, skin, respiratory system, central nervous system, hematopoietic system, blood, kidneys, liver, lymphoid system

Medical Surveillance: Consider the points of attack in preplacement and periodic physical examinations. Lung function tests, Urinalysis and kidney function tests. Complete blood count (CBC) with reticulocyte count. Liver function tests. Consider chest x-ray following acute overexposure.

First Aid: If this chemical gets into the eyes, remove any contact lenses at once and irrigate immediately for at least 15 minutes, occasionally lifting upper and lower lids. Seek medical attention immediately. If this chemical contacts the skin, remove contaminated clothing and wash immediately with soap and water. Seek medical attention immediately. If this chemical has been inhaled, remove from exposure, begin rescue breathing (using universal precautions) if breathing has stopped and CPR if heart action has stopped. Transfer promptly to a medical facility. When this chemical has been swallowed, get medical attention. Give large quantities of water and induce vomiting. Do not make an unconscious person vomit. Medical observation is recommended for 24 – 48 hours after breathing overexposure, as pulmonary edema may be delayed. As first aid for pulmonary edema, a doctor or authorized paramedic may consider administering a corticosteroid spray.

Personal Protective Methods: Wear protective gloves and clothing to prevent any reasonable probability of skin contact. Safety equipment suppliers/manufacturers can provide recommendations on the most protective glove/clothing material for your operation. All protective clothing (suits, gloves, footwear, headgear) should be clean, available each day, and put on before work. Contact lenses should not be worn when working with this chemical. Wear splash-proof chemical goggles and face shield unless full facepiece respiratory protection is worn. Employees should wash immediately with soap when skin is wet or contaminated. Provide emergency showers and eyewash.

Respirator Selection: NIOSH/OSHA: *50 ppm:* CCROV [any chemical cartridge respirator with organic vapor cartridge(s)]; or SA (any supplied-air respirator). *125 ppm:* SA:CF (any supplied-air respirator operated in a continuous-flow mode); or PAPROV [any powered, air-purifying respirator with organic vapor cartridge(s)]. *250 ppm:* CCRFOV [any chemical cartridge respirator with a full facepiece and organic vapor cartridge(s)]; or GMFOV [any air-purifying, full-facepiece respirator (gas mask) with a chin-style, front- or back-mounted acid gas canister]; or PAPRTOV [any powered, air-purifying respirator with a tight-fitting facepiece and organic vapor cartridge(s)]; or SCBAF (any self-contained breathing apparatus with a full facepiece); or SAF (any supplied-air respirator with a full facepiece). *Emergency or planned entry into unknown concentrations or IDLH conditions:* SCBAF:PD,PP (any self-contained breathing apparatus that has a full facepiece and is operated in a pressure-demand or other positive-pressure mode); or SAF:PD, PP:ASCBA (any supplied-air respirator that has a full facepiece and is operated in a pressure-demand or other positive-pressure mode in combination with an auxiliary, self-contained breathing apparatus operated in a pressure-demand or other positive-pressure mode). *Escape:* GMFOV [any air-purifying, full-facepiece respirator (gas mask) with a chin-style, front- or back-mounted organic vapor canister]; or SCBAE (any appropriate escape-type, self-contained breathing apparatus).

Note: Substance reported to cause eye irritation or damage; may require eye protection.

Storage: Prior to working with this chemical you should be trained on its proper handling and storage. Store in tightly closed containers in a dark, cool, well ventilated area. Keep in dark due to possible formation of explosive peroxides. Metal containers involving the transfer of this chemical should be grounded and bonded. Drums must be equipped with self-closing valves, pressure vacuum bungs, and flame arresters. Use only non-sparking tools and equipment, especially when opening and closing containers of this chemical. Sources of ignition such as smoking and open flames, are prohibited where this chemical is used, handled, or stored in a manner that could create a potential fire or explosion hazard.

Shipping: Ethylene glycol monobutyl ether requires a "Keep away from Food" label. It falls in Hazard Class 6.1 and Packing Group III.

Spill Handling: Evacuate and restrict persons not wearing protective equipment from area of spill or leak until cleanup is complete. Remove all ignition sources. Enter spill area from upwind side. Establish forced ventilation to keep levels below explosive limit. Use absorbent material to permit spill removal to vapor-tight plastic bags for eventual disposal. It may be necessary to contain and dispose of this chemical as a hazardous waste. If material or contaminated runoff enters waterways, notify downstream users of potentially contaminated waters. Contact your Department of Environmental Protection or your regional office of the federal EPA for specific recommendations. If employees are required to clean-up spills, they must be properly trained and equipped. OSHA 1910.120(q) may be applicable.

Fire Extinguishing: Use dry chemical, carbon dioxide, water spray, or alcohol-resistant foam extinguishers. Poisonous gases are produced in fire. If material or contaminated runoff enters waterways, notify downstream users of potentially contaminated waters. Notify local health and fire officials and pollution control agencies. From a secure, explosion-proof location, use water spray to cool exposed containers. If cooling streams are ineffective (venting sound increases in volume and pitch, tank discolors, or shows any signs of deforming), withdraw immediately to a secure position. If employees are expected to fight fires, they must be trained and equipped in OSHA 1910.156.

Disposal Method Suggested: Incineration.

References

New Jersey Department of Health and Senior Services, "Hazardous Substance Fact Sheet: 2-Butoxyethanol," Trenton, NJ (February 1989)

Sax, N. I., Ed., "Dangerous Properties of Industrial Materials Report" 4, No. 2, 58–61 (1984)

Butoxyl

Molecular Formula: $C_7H_{14}O_3$

Common Formula: $CH_3COOCH_2CH_2CH(OCH_3)CH_3$

Synonyms: Acetic acid 3-methoxybutyl ester; 3-Methoxybutyl acetate; 3-Methoxybutylester kyseliny octove (Polish); Methyl-1,3-butylene glycol acetate

CAS Registry Number: 4435-53-4

RTECS Number: EL4725000

DOT ID: UN 2708

Cited in U.S. State Regulations: New Hampshire (G), New Jersey (G).

Description: Butoxyl, $CH_3COOCH_2CH_2CH(OCH_3)CH_3$, is a colorless liquid with a sharp odor. Boiling point = 135 – 173°C. Flash point = 77°C.[17] Hazard Identification (based on NFPA-704 M Rating System): Health 1, Flammability 2, Reactivity 0. Slightly soluble in water.

Potential Exposure: Those involved in the use of this material as a cleaning solvent and as a component in varnishes and casting molds.

Incompatibilities: Strong oxidizers.

Permissible Exposure Limits in Air: No standards set.

Permissible Concentration in Water: No criteria set.

Routes of Entry: Inhalation, skin absorption. Ingestion.

Harmful Effects and Symptoms

Short Term Exposure: Butoxyl can be an eye irritant. High vapor levels can cause dizziness. The oral LD_{50} rat is 4,210 mg/kg (slightly toxic).[9]

Long Term Exposure: Unknown at this time.

First Aid: If this chemical gets into the eyes, remove any contact lenses at once and irrigate immediately for at least 15 minutes, occasionally lifting upper and lower lids. Seek medical attention immediately. If this chemical contacts the skin, remove contaminated clothing and wash immediately with soap and water. Seek medical attention immediately. If this chemical has been inhaled, remove from exposure, begin rescue breathing (using universal precautions) if breathing has stopped and CPR if heart action has stopped. Transfer promptly to a medical facility. When this chemical has been swallowed, get medical attention. Give large quantities of water and induce vomiting. Do not make an unconscious person vomit.

Personal Protective Methods: Where possible, enclose operations and use local exhaust ventilation at the site of chemical release. If local exhaust ventilation or enclosure is not used, respirators should be worn. Wear protective work clothing. Wash thoroughly immediately after exposure to Butoxyl and at the end of the workshift. Post hazard and warning information in the work area. In addition, as part of an ongoing education and training effort, communicate all information on the health and safety hazards of Butoxyl to potentially exposed workers.

Respirator Selection: SCBAF:PD,PP (any MSHA/NIOSH approved self-contained breathing apparatus that has a full facepiece and is operated in a pressure-demand or other positive-pressure mode); or SAF:PD,PP:ASCBA (any supplied-air respirator that has a full facepiece and is operated in a pressure-demand or other positive-pressure mode in combination with an auxiliary, self-contained breathing apparatus operated in a pressure-demand or other positive pressure mode).

Storage: Butoxyl must be stored to avoid contact with Oxidizers (such as Perchlorates, Peroxides, Permanganates, Chlorates and Nitrates) since violent reactions occur. Store in tightly closed containers in a cool, well-ventilated area. Sources of ignition, such as smoking and open flames, are prohibited where Butoxyl is used, handled, or stored in a manner that could create a potential fire or explosion hazard.

Shipping: The label requirement for Butoxyl is "Flammable Liquid." The Hazard Class is 3 and the Packing Group is III.

Spill Handling: Evacuate and restrict persons not wearing protective equipment from area of spill or leak until cleanup is complete. Remove all ignition sources. Ventilate area of spill or leak. Use water spray to reduce vapors. Absorb liquids in vermiculite, dry sand, earth, or a similar material and deposit in sealed containers. Keep this chemical out of a confined space, such as a sewer, because of the possibility of an explosion, unless the sewer is designed to prevent the build-up of explosive concentrations. It may be necessary to contain and dispose of this chemical as a hazardous waste. If material or contaminated runoff enters waterways, notify downstream users of potentially contaminated waters. Contact your Department of Environmental Protection or your regional office of the federal EPA for specific recommendations. If employees are required to cleanup spills, they must be properly trained and equipped. OSHA 1910.120(q) may be applicable.

Fire Extinguishing: This chemical is a combustible liquid. Poisonous gases including nitrogen oxides are produced in fire. Use dry chemical, carbon dioxide, or alcohol foam extinguishers. Vapors are heavier than air and will collect in low areas. Vapors may travel long distances to ignition sources and flashback. Vapors in confined areas may explode when exposed to fire. Storage containers and parts of containers may rocket great distances, in many directions. If material or contaminated runoff enters waterways, notify downstream users of potentially contaminated waters. Notify local health and fire officials and pollution control agencies. From a secure, explosion-proof location, use water spray to cool exposed containers. If cooling streams are ineffective (venting sound increases in volume and pitch, tank discolors, or shows any signs of deforming), withdraw immediately to a secure position. If employees are expected to fight fires, they must be trained and equipped in OSHA 1910.156.

Disposal Method Suggested: Incineration.

References

New Jersey Department of Health and Senior Services, "Hazardous Substance Fact Sheet: Butoxyl," Trenton, NJ (November 1986)

Butyl Acetates

Molecular Formula: $C_6H_{12}O_2$

Common Formula: $C_4H_9OCOCH_3$

Synonyms: *n-:* Acetate de butyle (French); *n*-Acetato de butilo (Spanish); Acetato de butilo (Spanish); Acetic acid, *n*-butyl ester; Acetic acid, butyl ester; Aristoline (+); AZ1470 (+); AZ4140 (+); AZ4210 (+); AZ4330 (+); AZ4620 (+); AZ 1310-SF (+); AZ 1312-SFD (+); AZ 1350J (+); AZ 1370 (+); AZ 1370-SF (+); AZ 1375 (+); AZ thinner; Butile (acetati di) (Italian); Butylacetat (German); *n*-Butyl acetate; *normal* Butyl acetate; 1-Butyl acetate; Butylacetaten (Dutch); Butyle (acetate de) (French); *n*-Butyl ester of acetic acid; Butyl ethanoate; EINECS No. 204-658-1; 6-6 Epoxy Chem resin finish, clear curing agent; FEMA No. 2174; Goodrite Nr-R; KTI1470 (+); KTI 1300 thinner; KTI 1350 J (+); KTI 1370/1375 (+); KTI II (+); Microposit 111S (+); Microposit 119S (+); Microposit 119 thinner; Microposit 1375 (+); Microposit 1400-33 (+); Microposit 1400S (+); Microposit 1470 (+); Microposit 6009 (+); Microposit Sal 601-ER7 (+); Microposit XP-6012 (+); Octan *n*-butylu (Polish); TSMR 8800 (+); TSMR 8800 BE; Ultramac PR-1024 MB-628 resin; Ultramac solvent EPA; Waycoat 204 (+); Waycoat HPR 205/207 (+); Waycoat RX 507 (+); Xanthochrome (+); XIR-3000-T resin

iso-: Acetate d'isobutyle (French); Acetato de isobutilo (Spanish); Acetic acid, isobutyl ester; Acetic acid, 2-methylpropyl ester; Isobutyl acetate (DOT); Isobutylester kyseliny octove (Czech); 2-Methyl-1-propyl acetate; 2-Methylpropyl acetate; β-Methylpropyl ethanoate

sec-: Acetate de butyle secondaire (French); Acetato de butilo-*sec* (Spanish); Acetic acid, 2-butoxy ester; Acetic acid, 1-methylpropyl ester (9CI); *s*-Butyl acetate; *sec*-Butyl acetate; *secondary* Butyl acetate; 2-Butyl acetate; *sec*-Butyl alcohol acetate; 1-Methyl propyl acetate

tert-: Acetato de *terc*-butilo (Spanish); Acetic acid *t*-butyl ester; Acetic acid *tert*-butyl ester; Acetic acid, 1,1-dimethylethyl ester (9CI); Acetic acid, *tert*-butyl ester; *t*-Butyl acetate; Texaco lead appreciator; TLA

CAS Registry Number: 123-86-4 (n-); 105-46-4 (sec-); 10-19-0 (iso-); 540-88-5 (tert-)

RTECS Number: AF7350000 (n-); AF380000 (sec-); AI4025000 (iso-); AF7400000 (tert-)

DOT ID: UN 1123 (butyl acetates); UN 1213 (iso-)

EEC Number: 607-025-00-1 (n-); 607-026-00-7 (sec-); 607-026-00-7 (iso-)

EINECS Number: 204-658-1; 208-760-7 (tert-); 203-745-1 (iso-)

Regulatory Authority

- Air Pollutant Standard Set (ACGIH)[1] (OSHA)[58] (DFG)[3] (HSE)[33] (former USSR)[43] (Several States)[60] (Australia) (Israel) (Mexico) (Several Canadian Provinces)
- Water Pollution Standard Proposed (former USSR)[43]
- CLEAN WATER ACT: Section 311 Hazardous Substances/ RQ 40CFR117.3 (same as CERCLA, see below)
- SUPERFUND/EPCRA 40CFR302.4 Reportable Quantity (RQ): CERCLA, 5,000 lb (2,270 kg)
- Canada, WHMIS, Ingredients Disclosure List

Cited in U.S. State Regulations: Alaska (G), California (A, G), Connecticut (A), Florida (G, A), Illinois (G), Maine (G), Massachusetts (G, A), Minnesota (G), Nevada (A), New Hampshire (G), New Jersey (G), New York (A), North Dakota (A), Oklahoma (G), Pennsylvania (G), Rhode Island (G), South Dakota (G), Virginia (A), West Virginia (G).

Description: Butyl acetates, $C_4H_9OCOCH_3$, are colorless or yellowish liquids with pleasant, fruity odors. There are 4 isomers:

Isomer	Formula	Boiling Point (°C)	Flash Point (%)	Flammable Limits (LEL)	(UEL)
n-	$CH_3CH_2CH_2CH_2OCOCH_3$	126 – 127	22	1.7	7.6
sec-	$CH_3CH_2CH(CH_3)OCOCH_3$	112 – 113	31	1.7	9.8
iso-	$(CH_3)_2CHCH_2OCOCH_3$	117 – 118	18	1.3	10.5
tert-	$(CH_3)_3COCOCH_3$	97 – 98	—	—	—

Potential Exposure: n-Butyl acetate is an important solvent in the production of lacquers, leather and airplane dopes, and perfumes. It is used as a solvent and gasoline additive. sec-Butyl acetate is used as a widely used solvent for nitrocellulose, nail enamels and many different purposes. tert-Butyl acetate is common industrial solvent used in the making of lacquers, artificial leather, airplane dope, perfume, and as a food additive. Isobutyl acetate is used as a solvent and in perfumes and artificial flavoring materials.

Incompatibilities: All butyl acetates are incompatible with nitrates, strong oxidizers, strong alkalies, strong acids. Butyl acetates form explosive mixtures with air; react with water on standing to form acetic acid and n-butyl alcohol. Violent reaction with strong oxidizers and potassium-tert-butoxide. Dissolves rubber, many plastics, resins and some coatings. May accumulate static electrical charges, and may cause ignition of its vapors.

Permissible Exposure Limits in Air: n-Butyl and isobutyl acetates have a Federal and ACGIH limit of 150 ppm (710 mg/m^3) TWA. sec-butyl and tert-butyl have a Federal[58] and ACGIH[1] limit of 200 ppm (950 mg/m^3) TWA. The STEL values are: (n-) 200 ppm (950 mg/m^3);[1][33][58] (sec-) 250 ppm (1,190 mg/m^3);[33] (iso-) 187 ppm (875 mg/m^3);[33] (tert-) 250 ppm (1,190 mg/m^3).[33] The DFG[3] MAK for the n-isomer is 100 ppm (480 mg/m^3) as does the former USSR/UNEP-IRPTC.[43] The DFG MAK (*tert*-isomer) is 20 ppm (96 mg/m^3) with a Peak limitation level of 2x the MAK, 5-minute momentary value, maximum frequency of 8 times per shift. The DFG not set a MAK for the *sec*-isomer at this time. The NIOSH IDLH levels[2] are: (n-) 1,700 ppm; (sec-) 1,700 ppm; (iso-) 1,300 ppm; tert-Butyl 1,500 ppm In addition, The former USSR-UNEP/IRPTC project[43] cites a MAC of 0.1 mg/m^3 in ambient air in residential areas, either on a momentary basis or on a daily average basis. Further, several states have set guidelines or standards for butyl acetates in ambient air[60] as follows (all values in mg/m^3):

State	n-	sec-	tetr-	iso-
Connecticut	14.2	19.0	19.0	14.0
Florida	14.2	—	—	14.0
Massachusetts	—	—	—	0.97
North Dakota	7.1 – 9.5	9.5 – 11.9	9.5 – 11.9	7.0 – 8.75
Nevada	16.9	22.6	22.6	16.67
New York	14.2	—	—	14.0
South Dakota	14.2	—	—	—
Virginia	12.0	15.0	15.0	12.0

Determination in Air: Adsorption of charcoal, workup with CS_2, analysis by gas chromatography. See NIOSH Method #1450.[18]

Permissible Concentration in Water: The former USSR-UNEP/IRPTC project[43] cites a MAC in water used for domestic purposes of 0.1 mg/l (for any isomer).

Routes of Entry: Inhalation, ingestion, skin and/or eye contact. Passes through the unbroken skin.

Harmful Effects and Symptoms

Headaches, drowsiness, eye irritation, irritation of skin and upper respiratory system. Humans and animals that inhale comparatively low doses of n-butyl acetate experience irritation of the nasal and respiratory passages and of the eyes. At higher concentrations narcosis takes place, and repeated exposures have resulted in renal and blood changes in experimental animals.

Short Term Exposure: The substance irritates the eyes, skin, and respiratory tract. High exposures, above the occupational exposure levels, can cause weakness, headache, and drowsiness and may cause unconsciousness.

Long Term Exposure: n-Butyl acetate may cause skin allergy. n-Butyl acetate has been shown to damage the developing fetus in animals. Prolonged and repeated exposure to butyl acetates can cause defatting, drying and cracking of the skin. Although many solvents and petroleum based products cause lung, brain and nerve damage, these chemicals have not been adequately evaluated to determine these effects.

Points of Attack: Eyes, skin, respiratory system, central nervous system.

Medical Surveillance: Consider initial effects on skin and respiratory tract in any preplacement or periodical examinations, as well as liver, lung, and kidney function.

First Aid: If this chemical gets into the eyes, remove any contact lenses at once and irrigate immediately for at least 15 minutes, occasionally lifting upper and lower lids. Seek medical attention immediately. If this chemical contacts the skin, remove contaminated clothing and wash immediately with soap and water. Seek medical attention immediately. If this chemical has been inhaled, remove from exposure, begin rescue breathing (using universal precautions) if breathing has

stopped and CPR if heart action has stopped. Transfer promptly to a medical facility. When this chemical has been swallowed, get medical attention. Give large quantities of salt water and induce vomiting. Do not make an unconscious person vomit.

Personal Protective Methods: Wear solvent-resistant gloves and clothing to prevent any reasonable probability of skin contact. Safety equipment suppliers/manufacturers can provide recommendations on the most protective glove/clothing material for your operation. All protective clothing (suits, gloves, footwear, headgear) should be clean, available each day, and put on before work. Contact lenses should not be worn when working with this chemical. Wear splash-proof chemical goggles and face shield unless full facepiece respiratory protection is worn. Employees should wash immediately with soap when skin is wet or contaminated. Remove nonimpervious clothing immediately if wet or contaminated. Provide emergency showers and eyewash.

Respirator Selection: *n-:* OSHA: *1,500 ppm:* CCROV [any chemical cartridge respirator with a full facepiece and organic vapor cartridge(s)]; or SA (any supplied-air respirator). *1,700 ppm:* SA:CF (any supplied-air respirator operated in a continuous-flow mode); or PAPROV [any powered, air-purifying respirator with organic vapor cartridge(s)]; or CCRFOV [any air-purifying, full-facepiece respirator (gas mask) with a chin-style, front- or back-mounted acid gas canister]; or GMFOV [any air-purifying, full-facepiece respirator (gas mask) with a chin-style, front- or back-mounted acid gas canister]; or SCBAF (any self-contained breathing apparatus with a full facepiece); or SAF (any supplied-air respirator with a full facepiece). *Emergency or planned entry into unknown concentrations or IDLH conditions:* SCBAF:PD,PP (any self-contained breathing apparatus that has a full facepiece and is operated in a pressure-demand or other positive-pressure mode); or SAF:PD,PP:ASCBA (any supplied-air respirator that has a full facepiece and is operated in a pressure-demand or other positive-pressure mode in combination with an auxiliary, self-contained breathing apparatus operated in a pressure-demand or other positive-pressure mode). *Escape:* GMFOV [any air-purifying, full-facepiece respirator (gas mask) with a chin-style, front- or back-mounted organic vapor canister]; or SCBAE (any appropriate escape-type, self-contained breathing apparatus).

Note: Substance reported to cause eye irritation or damage; may require eye protection.

sec-: 1,700 ppm: SA:CF (any supplied-air respirator operated in a continuous-flow mode); or PAPROV [any powered, air-purifying respirator with organic vapor cartridge(s)]; or CCRFOV [any air-purifying, full-facepiece respirator (gas mask) with a chin-style, front- or back-mounted acid gas canister]; or GMFOV [any air-purifying, full-facepiece respirator (gas mask) with a chin-style, front- or back-mounted acid gas canister]; or SCBAF (any self-contained breathing apparatus with a full facepiece); or SAF (any supplied-air respirator with a full facepiece). *Emergency or planned entry into unknown concentrations or IDLH conditions:* SCBAF:PD,PP (any self-contained breathing apparatus that has a full facepiece and is operated in a pressure-demand or other positive-pressure mode); or SAF:PD,PP:ASCBA (any supplied-air respirator that has a full facepiece and is operated in a pressure-demand or other positive-pressure mode in combination with an auxiliary, self-contained breathing apparatus operated in a pressure-demand or other positive-pressure mode). *Escape:* GMFOV [any air-purifying, full-facepiece respirator (gas mask) with a chin-style, front- or back-mounted organic vapor canister]; or SCBAE (any appropriate escape-type, self-contained breathing apparatus).

Note: Substance reported to cause eye irritation or damage; may require eye protection.

tert-: 1,500 ppm: SA:CF (any supplied-air respirator operated in a continuous-flow mode); or PAPROV [any powered, air-purifying respirator with organic vapor cartridge(s)]; or CCRFOV [any air-purifying, full-facepiece respirator (gas mask) with a chin-style, front- or back-mounted acid gas canister]; or GMFOV [any air-purifying, full-facepiece respirator (gas mask) with a chin-style, front- or back-mounted acid gas canister]; or SCBAF (any self-contained breathing apparatus with a full facepiece); or SAF (any supplied-air respirator with a full facepiece). *Emergency or planned entry into unknown concentrations or IDLH conditions:* SCBAF:PD,PP (any self-contained breathing apparatus that has a full facepiece and is operated in a pressure-demand or other positive-pressure mode); or SAF:PD,PP:ASCBA (any supplied-air respirator that has a full facepiece and is operated in a pressure-demand or other positive-pressure mode in combination with an auxiliary, self-contained breathing apparatus operated in a pressure-demand or other positive-pressure mode). *Escape:* GMFOV [any air-purifying, full-facepiece respirator (gas mask) with a chin-style, front- or back-mounted organic vapor canister]; or SCBAE (any appropriate escape-type, self-contained breathing apparatus).

Note: Substance reported to cause eye irritation or damage; may require eye protection.

iso-: 1,300 ppm: SA:CF (any supplied-air respirator operated in a continuous-flow mode); or CCROV [any chemical cartridge respirator with a full facepiece and organic vapor cartridge(s)]; or PAPROV [any powered, air-purifying respirator with organic vapor cartridge(s)]; or GMFOV [any air-purifying, full-facepiece respirator (gas mask) with a chin-style, front- or back-mounted acid gas canister]; or SCBAF (any self-contained breathing apparatus with a full facepiece); or SAF (any supplied-air respirator with a full facepiece). *Emergency or planned entry into unknown concentrations or IDLH conditions:* SCBAF:PD,PP (any self-contained breathing apparatus that has a full facepiece and is operated in a pressure-demand or other positive-pressure mode); or SAF:PD,PP:ASCBA (any supplied-air respirator that has a full facepiece and is operated in a pressure-demand or other positive-pressure mode in combination with an auxiliary, self-contained breathing apparatus operated in a pressure-demand or other positive-pressure mode). *Escape:* GMFOV [any air-purifying, full-facepiece respirator (gas mask) with a chin-style, front- or back-mounted organic vapor canister]; or SCBAE (any appropriate escape-type, self-contained breathing apparatus).

Note: Substance reported to cause eye irritation or damage; may require eye protection.

Storage: Prior to working with butyl acetates you should be trained on its proper handling and storage. Before entering confined space where these chemical may be present, check to make sure that an explosive concentration does not exist. Store in tightly closed containers in a cool, well ventilated area. Metal containers involving the transfer of this chemical should be grounded and bonded. Where possible, automatically pump liquid from drums or other storage containers to process containers. Drums must be equipped with self-closing valves, pressure vacuum bungs, and flame arresters. Use only non-sparking tools and equipment, especially when opening and closing containers of this chemical. Sources of ignition such as smoking and open flames, are prohibited where this chemical is used, handled, or stored in a manner that could create a potential fire or explosion hazard.

Shipping: According to DOT,[19] butyl acetates require a "Flammable Liquid" label. They fall in Hazard Group 3 and Packing Group II.

Spill Handling: Evacuate and restrict persons not wearing protective equipment from area of spill or leak until cleanup is complete. Remove all ignition sources. Ventilate area of spill or leak. Vapor build-up may cause suffocation. For small quantities absorb on paper towels. Evaporate in a safe place (such as a fume hood). Allow sufficient time for the evaporated vapors to completely clear the hood duct work. Burn the paper in a suitable location away from combustible materials. Or, absorb liquids in activated carbon, vermiculite, dry sand, earth, or a similar material and deposit in sealed containers. Collect powdered material in the most convenient and safe manner and deposit in sealed containers. It may be necessary to contain and dispose of this chemical as a hazardous waste. If material or contaminated runoff enters waterways, notify downstream users of potentially contaminated waters. Contact your Department of Environmental Protection or your regional office of the federal EPA for specific recommendations. If employees are required to clean-up spills, they must be properly trained and equipped. OSHA 1910.120(q) may be applicable.

Fire Extinguishing: These chemicals are flammable liquids. Poisonous gases are produced in fire. Use dry chemical, carbon dioxide, or alcohol foam extinguishers. Water may be used as a fog to control heat and to dilute vapors and wash them from the air. Use water fog in conjunction with alcohol foam, dry chemical or carbon dioxide as extinguishing agents. Vapors are heavier than air and will collect in low areas. Vapors may travel long distances to ignition sources and flashback. Vapors in confined areas may explode when exposed to fire. Container may explode in fire. Storage containers and parts of containers may rocket great distances, in many directions. If material or contaminated runoff enters waterways, notify downstream users of potentially contaminated waters. Notify local health and fire officials and pollution control agencies. From a secure, explosion-proof location, use water spray to cool exposed containers. If cooling streams are ineffective (venting sound increases in volume and pitch, tank discolors, or shows any signs of deforming), withdraw immediately to a secure position. If employees are expected to fight fires, they must be trained and equipped in OSHA 1910.156.

Disposal Method Suggested: Incineration. See also Spill Handling above.

References

National Institute for Occupational Safety and Health, Information Profiles on Potential Occupational Hazards-Single Chemicals: n-Butyl Acetate, Report TR 79-607, Rockville, MD pp 19–27 (December 1979)

Sax, N. I., Ed., "Dangerous Properties of Industrial Materials Report" 2, No. 2, 41–43 (1982). (Isobutyl Acetate)

Sax, N. I., Ed., "Dangerous Properties of Industrial Materials Report" 3, No. 6, 35–37 (1983). (t-Butyl Acetate)

Sax, N. I., Ed., "Dangerous Properties of Industrial Materials Report" 4, No. 3, 38–41 (1984). (n-Butyl Acetate)

Sax, N. I., Ed., "Dangerous Properties of Industrial Materials Report" 4, No. 6, 82–83 (1984). (sec-Butyl Acetate)

New Jersey Department of Health and Senior Services, "Hazardous Substance Fact Sheet: n-Butyl Acetate," Trenton, NJ (February 1989)

New Jersey Department of Health and Senior Services, "Hazardous Substance Fact Sheet: sec-Butyl Acetate," Trenton, NJ (February 1989)

New Jersey Department of Health and Senior Services, "Hazardous Substance Fact Sheet: tert-Butyl Acetate," Trenton, NJ (September 1987)

New Jersey Department of Health and Senior Services, "Hazardous Substance Fact Sheet: Isobutyl Acetate," Trenton, NJ (April 1997)

Butyl Acid Phosphate

Molecular Formula: $C_8H_{21}O_4P$

Common Formula: $(C_4H_9O)_2PH(OH)_2$

Synonyms: Acid butyl phosphate; *n*-Butyl acid phosphate; Butyl phosphoric acid; Phosphoric acid, dibutyl ester

CAS Registry Number: 12788-93-1

RTECS Number: TB8490000

Regulatory Authority

- Canada, WHMIS, Ingredients Disclosure List

Cited in U.S. State Regulations: Maine (G), New Hampshire (G), New Jersey (G).

Description: Butyl acid phosphate, $(C_4H_9O)_2PH(OH)_2$, is a clear white liquid. Flash point = 110°C. Hazard Identification (based on NFPA-704 M Rating System): Health 2, Flammability 1, Reactivity 1. Insoluble in water.

Potential Exposure: It is used in industrial chemicals manufacture.

Incompatibilities: Strong oxidizers, strong acids.

Routes of Entry: Inhalation, skin contact, ingestion.

Harmful Effects and Symptoms

Short Term Exposure: Butyl Acid Phosphate can affect you when breathed in. Contact can severely irritate the eyes, skin and respiratory tract. Can cause permanent eye damage. Inhalation can irritate the nose, throat, and lungs causing difficult breathing and shortness of breath.

Long Term Exposure: Repeated or prolonged contact can cause skin rash. Very irritating substances such as butyl acid phosphate may affect the lungs.

Points of Attack: Eyes, skin, respiratory system.

Medical Surveillance: Before beginning employment and at regular times after that, for those with frequent or potentially high exposures, the following are recommended: Lung function tests.

First Aid: If this chemical gets into the eyes, remove any contact lenses at once and irrigate immediately for at least 15 minutes, occasionally lifting upper and lower lids. Seek medical attention immediately. If this chemical contacts the skin, remove contaminated clothing and wash immediately with soap and water. Seek medical attention immediately. If this chemical has been inhaled, remove from exposure, begin rescue breathing (using universal precautions) if breathing has stopped and CPR if heart action has stopped. Transfer promptly to a medical facility. When this chemical has been swallowed, get medical attention. If victim is conscious, administer water or milk. Do not induce vomiting. If pulmonary edema develops from high exposure, medical observation is recommended for 24 – 48 hours.

Personal Protective Methods: Wear solvent-resistant gloves and clothing to prevent any reasonable probability of skin contact. Safety equipment suppliers/manufacturers can provide recommendations on the most protective glove/clothing material for your operation. All protective clothing (suits, gloves, footwear, headgear) should be clean, available each day, and put on before work. Contact lenses should not be worn when working with this chemical. Wear splash-proof chemical goggles and face shield unless full facepiece respiratory protection is worn. Employees should wash immediately with soap when skin is wet or contaminated. Remove nonimpervious clothing immediately if wet or contaminated. Provide emergency showers and eyewash.

Respirator Selection: Where the potential for exposures to Butyl Acid Phosphate exists, use a MSHA/NIOSH approved supplied-air respirator with a full facepiece operated in the positive pressure mode or with a full facepiece, hood, or helmet in the continuous flow mode, or use a MSHA/NIOSH approved self-contained breathing apparatus with a full facepiece operated in pressure-demand or other positive pressure mode.

Storage: Store in tightly closed containers is a cool, well-ventilated area away from potentially high heat sources.

Shipping: Butyl Acid Phosphate requires a "Corrosive" label. It falls in Hazard Class 8 and Packing Group III.

Spill Handling: Evacuate and restrict persons not wearing protective equipment from area of spill or leak until cleanup is complete. Remove all ignition sources. Ventilate area of spill or leak. Absorb liquids in vermiculite, dry sand, earth, or a similar material and deposit in sealed containers. Collect powdered material in the most convenient and safe manner and deposit in sealed containers. It may be necessary to contain and dispose of this chemical as a hazardous waste. If material or contaminated runoff enters waterways, notify downstream users of potentially contaminated waters. Contact your Department of Environmental Protection or your regional office of the federal EPA for specific recommendations. If employees are required to clean-up spills, they must be properly trained and equipped. OSHA 1910.120(q) may be applicable.

Fire Extinguishing: Use dry chemical, carbon dioxide, water spray, or alcohol foam extinguishers. Poisonous gases are produced in fire. If material or contaminated runoff enters waterways, notify downstream users of potentially contaminated waters. Notify local health and fire officials and pollution control agencies. From a secure, explosion-proof location, use water spray to cool exposed containers. If cooling streams are ineffective (venting sound increases in volume and pitch, tank discolors, or shows any signs of deforming), withdraw immediately to a secure position. If employees are expected to fight fires, they must be trained and equipped in OSHA 1910.156

References

New Jersey Department of Health and Senior Services, "Hazardous Substance Fact Sheet: Butyl Acid Phosphate," Trenton, NJ (August 6, 1987)

n-Butyl Acrylate

Molecular Formula: $C_7H_{12}O_2$

Common Formula: $CH_2{=}CHCOOC_4H_9$

Synonyms: Acrilato de *n*-butilo (Spanish); Acrylic acid *n*-butyl ester; Acrylic acid, butyl ester; *n*-Butyl acrylate; *normal* Butyl acrylate; Butylacrylate, inhibited; Butyl 2-propenoate; 2-Propenoic acid, butyl ester

CAS Registry Number: 141-32-2

RTECS Number: UD3150000

DOT ID: UN 2348

EEC Number: 607-062-00-3

EINECS Number: 205-480-7

Regulatory Authority

- Air Pollutant Standard Set (ACGIH)[1] (OSHA)[58] (DFG)[3] HSE)[33] (former USSR)[43] (Several States)[60] (Australia) (Israel) (Mexico) (Several Canadian Provinces)
- EPCRA Section 313 Form R *de minimis* concentration reporting level: 1.0%
- Canada, WHMIS, Ingredients Disclosure List; National Pollutant Release Inventory (NPRI)

Cited in U.S. State Regulations: Alaska (G), California (G), Connecticut (A), Florida (G), Illinois (G), Maine (G), Maryland (G), Massachusetts (G), Nevada (A), New Hampshire (G), New Jersey (G), North Dakota (A), Pennsylvania (G), Rhode Island (G), Virginia (A), West Virginia (G).

Description: Butyl Acrylate, CH_2=$CHCOOC_4H_9$, is a colorless liquid. Boiling point = 146 – 148°C. Freezing/Melting point = -64°C. Flash point = 29°C (also reported to be 39.4°C). Autoignition temperature = 292°C. The explosive limits are: LEL = 1.3%; UEL = 9.9%.[17] Hazard Identification (based on NFPA-704 M Rating System): Health 2, Flammability 2, Reactivity 2. Slightly soluble in water.

Potential Exposure: This material is used as a monomer in the production of polymers, copolymers, and resins; for solvent coatings, adhesives, paints and binders.

Incompatibilities: Forms explosive mixture with air. Heat, sparks, open flame, light, reducing agents, or peroxides may cause explosive polymerization. Incompatible with strong acids, amines, halogens, hydrogen compounds, oxidizers, sunlight, or other catalysts.

Permissible Exposure Limits in Air: The NIOSH, ACGIH, Australia, Israel, Mexico, HSE[33] and DFG[3] TWA value is 10 ppm (55 mg/m^3). The former USSR-UNEP/IRPTC project[43] has set a MAC of 10 mg/m^3 in workplace air. In Canada, Alberta, British Columbia, Ontario, and Quebec have a TWA value of 10 ppm and Alberta has a STEL of 20 ppm. Three states have set guidelines or standards for Butyl Acrylate in ambient air[60] ranging from 900 μg/m^3 (Virginia) to 1,100 μg/m^3 (Connecticut) to 1,310 μg/m^3 (Nevada).

Permissible Concentration in Water: The former USSR-UNEP/IRPTC[43] has set a MAC of 0.01 mg/l in water bodies used for domestic purposes.

Routes of Entry: Ingestion, skin and eye contact.

Harmful Effects and Symptoms

n-Butyl Acrylate was found to be but moderately irritating to the skin. As an eye irritant it produced corneal necrosis in an unwashed rabbit eye, similar to that produced by ethyl alcohol. Exposure of rats at 1,000 ppm for 4 hours proved lethal to 5 of 6 rats exposed; however, rats survived a 30-minute exposure to 7,000 ppm. There is a close similarity in toxic response by inhalation, skin and eye to methyl acrylate. The oral LD_{50} rat is 900 mg/kg (slightly toxic).

Short Term Exposure: This chemical can pass through the skin. The substance severely irritates the eyes, skin and respiratory tract. Inhalation can cause pulmonary edema, a medical emergency that can be delayed for several hours. This can cause death. High exposure may cause liver damage.

Long Term Exposure: May cause liver and lung damage. May cause skin sensitization and allergy. Similar solvents and petroleum-based chemicals have been shown to cause brain and nerve damage.

Points of Attack: Skin, eyes

Medical Surveillance: Liver an lung function tests. Examination by a qualified allergist. Interview for brain effects.

First Aid: If this chemical gets into the eyes, remove any contact lenses at once and irrigate immediately for at least 30 minutes, occasionally lifting upper and lower lids. Seek medical attention immediately. If this chemical contacts the skin, remove contaminated clothing and wash immediately with soap and water. Seek medical attention immediately. If this chemical has been inhaled, remove from exposure, begin rescue breathing (using universal precautions) if breathing has stopped and CPR if heart action has stopped. Transfer promptly to a medical facility. When this chemical has been swallowed, get medical attention. Give large quantities of water and induce vomiting. Do not make an unconscious person vomit. Medical observation is recommended for 24 – 48 hours after breathing overexposure, as pulmonary edema may be delayed. As first aid for pulmonary edema, a doctor or authorized paramedic may consider administering a corticosteroid spray.

Personal Protective Methods: Wear solvent-resistant gloves and clothing to prevent any reasonable probability of skin contact. Safety equipment suppliers/manufacturers can provide recommendations on the most protective glove/clothing material for your operation. All protective clothing (suits, gloves, footwear, headgear) should be clean, available each day, and put on before work. Contact lenses should not be worn when working with this chemical. Wear splash-proof chemical goggles and face shield unless full facepiece respiratory protection is worn. Employees should wash immediately with soap when skin is wet or contaminated. Remove nonimpervious clothing immediately if wet or contaminated. Provide emergency showers and eyewash.

Respirator Selection: *At any detectable concentration*: SCBAF:PD,PP (any MSHA/NIOSH approved self-contained breathing apparatus that has a full facepiece and is operated in a pressure-demand or other positive-pressure mode); or SAF:PD,PP:ASCBA (any supplied-air respirator that has a full facepiece and is operated in a pressure-demand or other positive-pressure mode in combination with an auxiliary, self-contained breathing apparatus operated in a pressure-demand or other positive pressure mode).

Storage: Prior to working with Butyl Acrylate you should be trained on its proper handling and storage. Do not store unless stabilized. Before entering confined space where Butyl Acrylate may be present, check to make sure that an explosive concentration does not exist. Store in tightly closed containers in a cool, well ventilated, fireproof area. Metal containers involving the transfer of this chemical should be grounded and bonded. Where possible, automatically pump liquid from drums or other storage containers to process containers. Drums must be equipped with self-closing valves, pressure vacuum bungs, and flame arresters. Use only non-sparking tools and equipment, especially when opening and closing containers of this chemical. Sources of ignition such as smoking and open flames, are prohibited where this chemical is used, handled, or stored in a manner that could create a potential fire or explosion hazard.

Shipping: The DOT label requirement is "Flammable Liquid." In Packing Group III

Spill Handling: Evacuate and restrict persons not wearing protective equipment from area of spill or leak until cleanup is complete. Remove all ignition sources. Establish forced ventilation to keep levels below explosive limit. Cover liquids with dry lime or soda ash or absorb liquids in vermiculite, dry sand, earth, peat, carbon, or a similar material and deposit in sealed containers. It may be necessary to contain and dispose of this chemical as a hazardous waste. If material or contaminated runoff enters waterways, notify downstream users of potentially contaminated waters. Contact your Department of Environmental Protection or your regional office of the federal EPA for specific recommendations. If employees are required to clean-up spills, they must be properly trained and equipped. OSHA 1910.120(q) may be applicable.

Fire Extinguishing: This chemical is a combustible and highly reactive liquid. Poisonous gases are produced in fire. Use dry chemical, carbon dioxide, or foam extinguishers. Vapors are heavier than air and will collect in low areas. Vapors may travel long distances to ignition sources and flashback. Vapors in confined areas may explode when exposed to fire. Containers may explode in fire. Storage containers and parts of containers may rocket great distances, in many directions. If material or contaminated runoff enters waterways, notify downstream users of potentially contaminated waters. Notify local health and fire officials and pollution control agencies. From a secure, explosion-proof location, use water spray to cool exposed containers. If cooling streams are ineffective (venting sound increases in volume and pitch, tank discolors, or shows any signs of deforming), withdraw immediately to a secure position. If employees are expected to fight fires, they must be trained and equipped in OSHA 1910.156.

Disposal Method Suggested: Incineration.

References

Sax, N. I., Ed., "Dangerous Properties of Industrial Materials Report" 7, No. 3, 61–75 (1987)

New Jersey Department of Health and Senior Services, "Hazardous Substance Fact Sheet: Butyl Acrylate," Trenton, NJ (February 1986)

Butyl Alcohols

Molecular Formula: $C_4H_{10}O$

Common Formula: C_4H_9OH

Synonyms: *iso-:* 197 rosin flux; 4282 flux; Alcohol isobutilico (Spanish); Alcool isobutylique (French); Alcowipe; Alpha 100 flux; Alpha 850-33 flux; Aqua-Sol flux; Avantine; Boron B-30; Boron B-40; Boron B-50; Boron B-60; Burmar Lab Clean; C-589; Chemtranic flux stripper; Copper 2 reagent; CP290B activator; DAG 154; Dazzlens cleaner; EINECS No. 201-148-0; ENTAC 349 biocide; ENTEC 327 surfactant; Epoxy cure agent; FC-95; FEMA No. 2179; Fermentation butyl alcohol; Film remover; Glid-Guard epoxy safety blue; Hardness 2 test solution; High Grade 1086; 1-Hydroxymethylpropane; IBA; Isobutanol; Isobutyl-alkohol (Czech); Isopropylcarbinol; Kester 103 thinner; Kester 108 thinner; Kester 145 rosin flux; Kester 1585 rosin flux; Kester 185 rosin flux; KTI Cop Rinse I/II; KTI mask protective coating; KTI NMD-25(+); KTI PBS rinse; KTI PMMA rinse; Lens cleaner M6015; Magic glass cleaner and antifogging fluid; Markem 320 cleaner; 2-Methyl-1-propanol; 2-Methylpropyl alcohol; Microposit NPE-210 solution; Omega meter solution; Opti Skan scan cleaner; Organo flux 3355-11; PBS developer; PBS rinse; PC-96 solvent soluble resist; Primer 910-S; 1-Propanol, 2-methyl-; RCRA No. U140; RN-10 E-Beam negative resist rinse; RN-11 developer; RN-11 E-beam negative resist rinse; Rosin flux; Rosin flux Kester 135/1544 Mil; RP-10 E-beam positive resist rinse; Scan Kleen; Solder flux; Solder flux 2163 Organic; Solder flux thinner; Sterets pre-injection Swabs; Surfynol 104PA surfactant; True blue glass cleaner; Uvex primer 910S; Vandalex 124; Vandalex 20; VWR glass cleaner; Whirlwind glass cleaner; WRS200S solution; Xerox cleaner, Formulka A; Xerox film remover, Tip Wipes

n-: Alcohol butilico-*n* (Spanish); Alcool butylique (French); *n*-Butanol; 1-Butanol; Butan-1-ol; Butanol; Butanolen (Dutch); Butanolo (Italian); *normal* Butyl alcohol; Butyl alcohol (DOT); Butyl hydroxide; Butylowy alkohol (Polish); Butyric alcohol; CCS 203; CEM420; DAG 154; 6-6 Epoxy Chem resin finish, clear curing agent; Epoxy solvent cure agent; 1-Hydroxybutane; Isanol; Kester 5612 protecto; Methylolpropane; *normal* Primary butyl alcohol; Propyl carbinol; Propyl methanol; Protecto 5612; Tebol-88; Tebol-99

sec-: Alcohol *sec*-butilico (Spanish); Alcool butylique secondaire (French); *sec*-Butanol; 2-Butanol; Butan-2-ol; Butanol-2; *secondary* Butyl alcohol; 2-Butyl alcohol; Butylene hydrate; CCS301; Ethylmethyl carbinol; 2-Hydroxybutane; Methyl ethyl carbinol; 1-Methypropyl alcohol; RTECS No. EO1750000; S.B.A.; Tanol secondaire (French) *tert-:* Alcohol *terc*-butilico (Spanish); Alcool butylique tertiaire (French); *tert*-Butanol; 1-Butanol; Butanol tertiaire (French); *tert*-Butyl hydroxide; 1,1-Dimethylethanol; Methanol, trimethyl-; 2-Methyl-2-propanol; NCI-C55367; 2-Propanol, 2-methyl-; TBA; Tertiary butyl alcohol; Trimethyl carbinol; Trimethyl methanol

CAS Registry Number: 71-36-3 (n-); 78-92-2 (sec-); 78-83-1 (iso-); 75-65-0 (tert-)

RTECS Number: EO1400000 (n-); EO1750000 (sec-); NP9625000 (iso-); EO1925000 (tert-)

DOT ID: UN 1120 (butyl alcohol); UN 1212 (iso-)

EEC Number: 603-004-00-6 (n-), (sec-), (iso-); 603-005-00-1 (tert-)

EINECS Number: 200-751-6 (n-)

Regulatory Authority

- Air Pollutant Standard Set (ACGIH)[1] (OSHA)[58] (HSE)[33] (DFG)[3] (former USSR)[43] (Several States)[60] (Australia) (Israel) (Mexico) (Several Canadian Provinces)
- EPA HAZARDOUS WASTE NUMBER (RCRA No.): U031(n-); U140(iso-)
- RCRA, 40CFR261, Appendix 8 Hazardous Constituents. (n-) (iso-)

n-, iso-:

- RCRA 40CFR268.48; 61FR15654, Universal Treatment Standards: Wastewater (mg/l), 5.6; Nonwastewater (mg/kg), 2.6
- RCRA 40CFR268.48; 61FR15654, Universal Treatment Standards: Wastewater (mg/l), 5.6; Nonwastewater (mg/kg), 170 (iso-)
- SUPERFUND/EPCRA 40CFR302.4 Reportable Quantity (RQ): CERCLA, 5,000 lb (2,270 kg). (n-) (iso-)
- RCRA 40CFR264, Appendix 9; TSD Facilities Ground Water Monitoring List. Suggested test method(s) (PQL μg/l): 8015 (50) (iso-)
- EPCRA Section 313 Form R *de minimis* concentration reporting level: 1.0% (n-), (sec-), (tert-)

iso-: Canada, WHMIS, Ingredients Disclosure List; National Pollutant Release Inventory (NPRI)

Cited in U.S. State Regulations: Alaska (G), California (A, G), Connecticut (A), Florida (G), Illinois (G), Kansas (G), Louisiana (G), Maine (G), Maryland (G), Massachusetts (G, A), Minnesota (G), Nevada (A), New Hampshire (G), New Jersey (G), New York (A), North Dakota (A), Oklahoma (G), Pennsylvania (G), Rhode Island (G), South Dakota (A), Vermont (G), Virginia (G, A), Washington (G), West Virginia (G), Wisconsin (G).

Description: There are four isomers. n- and sec-Butyl alcohols, are colorless liquids with strong, sweet, alcoholic odor. tert-Butyl alcohol is a colorless crystalline powder or liquid (above 26°C) with a camphorlike odor. It is often used in aqueous solution. Iso-butyl alcohol is a colorless liquid with a mild, sweet, and musty odor.

Isomer	Chemical Formula	Boiling Point (°C)	Flash Point* (°C)	Expl. Limits (LEL) (%)	(UEL) (%)
n-	$CH_3CH_2CH_2CH_2OH$	117 – 118	37	1.4	11.2
sec-	$CH_3CH_2CH(CH_3)OH$	99 – 100	24	1.7 @ 100°C	9.8 @ 100°C
iso-	$(CH_3)_2CHCH_2OH$	108	28	1.7	10.6 @ 94°C
tert-	$(CH_3)_3COH_3$	82 – 83	11	2.4	8.0

* closed cup

The n- isomer is practically insoluble in water; the sec-isomer is slightly soluble in water; the tert- and iso- isomers are soluble in water. Hazard Identification (based on NFPA-704 M Rating System): Health 1, Flammability 3, Reactivity 0 (all isomers).

Potential Exposure: Butyl alcohols are used as solvents for paints, lacquers, varnishes, natural and synthetic resins, gums, vegetable oils, dyes, camphor, and alkaloids. They are also used as an intermediate in the manufacture of pharmaceuticals and chemicals and in the manufacture of artificial leather, safety glass, rubber and plastic cements, shellac, raincoats, photographic films, perfumes, and in plastic fabrication.

Incompatibilities: Butyl alcohols forms explosive mixtures with air. In all cases they are incompatible with strong oxidizers and attack some plastics, rubber and coatings. n-Butanol is incompatible with strong acids, halogens, caustics, alkali metals, aliphatic amines, isocyanates. sec-Butanol forms an explosive peroxide in air. Ignites with chromium trioxide. Incompatible with strong oxidizers, strong acids, aliphatic amines, isocyanates, organic peroxides. tert-Butanol is incompatible with strong acids (including mineral acid), including mineral acids, strong oxidizers or caustics, aliphatic amines, isocyanates, alkali metals (i.e., lithium, sodium, potassium, rubidium, cesium, francium). iso-Butanol is incompatible with strong acids, strong oxidizers, caustics, aliphatic amines, isocyanates, alkali metals and alkali earth. May react with aluminum at high temperatures.

Permissible Exposure Limits in Air:

Isomers	OSHA TWA	NIOSH TWA	NIOSH STEL	ASGIH TWA
n-	100 ppm (300 mg/m³)	—	50 ppm* (150 mg/m³)	25 ppm (75 mg/m³)
sec-	150 ppm (450 mg/m³)	100 ppm (305 mg/m³)	150 ppm (455 mg/m³)	100 ppm (305 mg/m³)
iso-	100 ppm (300 mg/m³)	50 ppm (150 mg/m³)	None	100 ppm (300 mg/m³)
tetr-	100 ppm (300 mg/m³)	100 ppm (300 mg/m³)	150 ppm (450 mg/m³)	100 ppm (300 mg/m³)

* Ceiling value (skin notation)

The ILDH levels are: n- 1,400 ppm sec- 2,000 ppm iso- 1,600 ppm tert- 1,600 ppm The former USSR-UNEP/IRPTC project has set a MAC of only 10 mg/m^3 in workplace air and 0.1 mg/m^3 on either a momentary or a daily average basis for ambient air in residential areas.[43] For the n-isomer, the DFG MAK is 100 ppm (310 mg/m^3) and Peak limitation of 2 × MAK, 5 minute momentary value not to be exceeded 8 times per workshift.; for the tert-isomer is 20 ppm (62 mg/m^3) and Peak limitation of 2 × MAK, 30 minute momentary value not to be exceeded 4 times per workshift.. There is no DFG MAK for the sec-isomer at present. In addition, several states have set guidelines or standards for butyl alcohols in ambient air[60] as follows (all values in mg/m^3):

State	n-	sec-	tetr-	iso-
Connecticut	6.0	6.1	—	—
Massachusetts	0.021	—	—	0.021
North Dakota	1.5	3.05 – 4.55	3.05 – 4.5	1.5 – 2.25
Nevada	3.57	7.26	—	—
New York	3.0	—	—	—
South Dakota	3.0	—	—	—
Virginia	1.25	3.05	—	2.5

Determination in Air: Adsorption on charcoal, workup with 2-propanol in CS_2, for all 4 isomers [except 2-butanone used instead of 2-propanol in tert-butanol measurement] analysis by gas chromatography. See NIOSH Method (IV) #1401.[18]

Permissible Concentration in Water: No criteria set, but EPA has suggested[32] ambient limits as follows based on health effects: n- 2,070 μg/l sec- 6,200 μg/l iso- 2,070 μg/l tert- 4,140 μg/l The former USSR-UNEP/IRPTC project[43] has set a MAC for butyl alcohol of 1.0 mg/ml in water bodies used for domestic purposes and 0.03 mg/l in water bodies used for fishery purposes. These are also the former USSR values.[35]

Routes of Entry: Ingestion, inhalation, skin and/or eye contact. Passes through the unbroken skin (n-, iso-).

Harmful Effects and Symptoms

Short Term Exposure: The vapors of butyl alcohols irritates the eyes and respiratory tract. They can irritate the skin and cause rash or burning feeling on contact. May affect the central nervous system. Exposure to high concentrations could cause headache, nausea, vomiting, and dizziness. Exposure to high levels of the n- isomer may cause unconsciousness and may lead to irregular heartbeat. The oral LD_{50} value for rats for the various isomers are as follows: (n-) 790 mg/kg; (sec-) 6,480 mg/kg; (iso-) 2,460 mg/kg; (tert-) 3,500 mg/kg.

Long Term Exposure: Repeated or prolonged contact with skin may cause dermatitis, drying and cracking of the skin. Exposure to the n- isomer can damage the liver, heart, and kidneys, cause hearing loss and affect sense of balance.

Points of Attack: Eyes, skin, respiratory system, central nervous system.

Medical Surveillance: *n-:* Liver and kidney function tests, hearing test (audiogram) and test for balance, EKG test. *tert-:* Liver and kidney function tests. *iso-:* Evaluate for brain effects and possible neuropsychological testing.

First Aid: If this chemical gets into the eyes, remove any contact lenses at once and irrigate immediately for at least 15 minutes, occasionally lifting upper and lower lids. Seek medical attention immediately. If this chemical contacts the skin, remove contaminated clothing and wash immediately with soap and water. Seek medical attention immediately. If this chemical has been inhaled, remove from exposure, begin rescue breathing (using universal precautions) if breathing has stopped and CPR if heart action has stopped. Transfer promptly to a medical facility. When this chemical has been swallowed, get medical attention. Give large quantities of water and induce vomiting. Do not make an unconscious person vomit.

Personal Protective Methods: Wear solvent-resistant gloves and clothing to prevent any reasonable probability of skin contact. Safety equipment suppliers/manufacturers and NIOSH recommend Neoprene, Nitrile, Polyethylene, butyl rubber, and Teflon as the most protective glove/clothing material for butylalcohol. All protective clothing (suits, gloves, footwear, headgear) should be clean, available each day, and put on before work. Contact lenses should not be worn when working with this chemical. Wear splash-proof chemical goggles and face shield unless full facepiece respiratory protection is worn. Employees should wash immediately with soap when skin is wet or contaminated. Remove nonimpervious clothing immediately if wet or contaminated. Provide emergency showers and eyewash.

Respirator Selection: *n-: NIOSH/OSHA: 1,250 ppm:* SA:CF* (any supplied-air respirator operated in a continuous-flow mode); or PAPROV* [any powered, air-purifying respirator with organic vapor cartridge(s)]. *1,400 ppm:* CCRFOV [any chemical cartridge respirator with a full facepiece and organic vapor cartridge(s)]; or GMFOV [any air-purifying, full-facepiece respirator (gas mask) with a chin-style, front- or back-mounted acid gas canister]; or PAPRTOV* [any powered, air-purifying respirator with a tight-fitting facepiece and organic vapor cartridge(s)]; SCBAF (any self-contained breathing apparatus with a full facepiece); or SAF (any supplied-air respirator with a full facepiece). *Emergency or planned entry into unknown concentrations or IDLH conditions:* SCBAF:PD,PP (any self-contained breathing apparatus that has a full facepiece and is operated in a pressure-demand or other positive-pressure mode); or SAF:PD,PP:ASCBA (any supplied-air respirator that has a full facepiece and is operated in a pressure-demand or other positive-pressure mode in combination with an auxiliary, self-contained breathing apparatus operated in a pressure-demand or other positive-pressure mode). *Escape:* GMFOV [any air-purifying, full-facepiece respirator (gas mask) with a chin-style, front- or back-mounted organic vapor canister]; or SCBAE (any appropriate escape-type, self-contained breathing apparatus).

* Substance causes eye irritation or damage; eye protection needed.

sec-: NIOSH: *1,000 ppm:* CCROV* [any chemical cartridge respirator with organic vapor cartridge(s)]; or SA* (any supplied-air respirator). *2,000 ppm:* SA:CF* (any supplied-air respirator operated in a continuous-flow mode); or PAPROV [any powered, air-purifying respirator with organic vapor cartridge(s)]; or CCRFOV [any chemical cartridge respirator with a full facepiece and organic vapor cartridge(s)]; or GMFOV [any air-purifying, full-facepiece respirator (gas mask) with a chin-style, front- or back-mounted acid gas canister]; or SCBAF (any self-contained breathing apparatus with a full facepiece); or SAF (any supplied-air respirator with a full facepiece). *Emergency or planned entry into unknown concentrations or IDLH conditions:* SCBAF:PD,PP (any self-contained breathing apparatus that has a full facepiece and is operated in a pressure-demand or other positive-pressure mode); or SAF:PD,PP:ASCBA (any supplied-air respirator that has a full facepiece and is operated in a pressure-demand or other positive-pressure mode in combination with an auxiliary, self-contained breathing apparatus operated in a pressure-demand or other positive-pressure mode). *Escape:* GMFOV [any air-purifying, full-facepiece respirator (gas mask) with a chin-style, front- or back-mounted organic vapor canister]; or SCBAE (any appropriate escape-type, self-contained breathing apparatus).

* Substance reported to cause eye irritation or damage; may require eye protection.

iso-: NIOSH/OSHA: *500 ppm:* CCROV [any chemical cartridge respirator with organic vapor cartridge(s)]; or SA (any supplied-air respirator). *1,250 ppm:* SA:CF (any supplied-air respirator operated in a continuous-flow mode); or PAPROV [any powered, air-purifying respirator with organic vapor cartridge(s)]. *1,600 ppm:* CCRFOV [any chemical cartridge respirator with a full facepiece and organic vapor cartridge(s)]; or GMFOV [any air-purifying, full-facepiece respirator (gas mask) with a chin-style, front- or back-mounted acid gas canister]; or PAPRTOV [any powered, air-purifying respirator with a tight-fitting facepiece and organic vapor cartridge(s)]; or SCBAF (any self-contained breathing apparatus with a full facepiece); or SAF (any supplied-air respirator with a full facepiece). *Emergency or planned entry into unknown concentrations or IDLH conditions:* SCBAF:PD,PP (any self-contained breathing apparatus that has a full facepiece and is operated in a pressure-demand or other positive-pressure mode); or SAF:PD,PP:ASCBA (any supplied-air respirator that has a full facepiece and is operated in a pressure-demand or other positive-pressure mode in combination with an auxiliary, self-contained breathing apparatus operated in a pressure-demand or other positive-pressure mode). *Escape:* GMFOV [any air-purifying, full-facepiece respirator (gas mask) with a chin-style, front- or back-mounted organic vapor canister]; or SCBAE (any appropriate escape-type, self-contained breathing apparatus).

Note: Substance reported to cause eye irritation or damage; may require eye protection.

tert-: NIOSH/OSHA: *1,600 ppm:* SA:CF* (any supplied-air respirator operated in a continuous-flow mode); or PAPROV* [any powered, air-purifying respirator with organic vapor cartridge(s)]; or CCRFOV [any chemical cartridge respirator with a full facepiece and organic vapor cartridge(s)]; or GMFOV [any air-purifying, full-facepiece respirator (gas mask) with a chin-style, front- or back-mounted acid gas canister]; or SCBAF (any self-contained breathing apparatus with a full facepiece); or SAF (any supplied-air respirator with a full facepiece). *Emergency or planned entry into unknown concentrations or IDLH conditions:* SCBAF:PD,PP (any self-contained breathing apparatus that has a full facepiece and is operated in a pressure-demand or other positive-pressure mode); or SAF:PD,PP:ASCBA (any supplied-air respirator that has a full facepiece and is operated in a pressure-demand or other positive-pressure mode in combination with an auxiliary, self-contained breathing apparatus operated in a pressure-demand or other positive-pressure mode). *Escape:* GMFOV [any air-purifying, full-facepiece respirator (gas mask) with a chin-style, front- or back-mounted organic vapor canister]; or SCBAE (any appropriate escape-type, self-contained breathing apparatus).

* Substance causes eye irritation or damage; eye protection needed.

Storage: Prior to working with Butyl Alcohols you should be trained on its proper handling and storage. Before entering confined space where these may be present, check to make sure that an explosive concentration does not exist. Store in tightly closed containers in a cool, well ventilated area. Metal containers involving the transfer of this chemical should be grounded and bonded. Where possible, automatically pump liquid from drums or other storage containers to process containers. Drums must be equipped with self-closing valves, pressure vacuum bungs, and flame arresters. Use only non-sparking tools and equipment, especially when opening and closing containers of this chemical. Sources of ignition such as smoking and open flames, are prohibited where this chemical is used, handled, or stored in a manner that could create a potential fire or explosion hazard.

Shipping: Butanols require "Flammable Liquid" labels. They all fall in Hazard Class 3. Tertiary Butanol falls in Packing Group II, all the others in Packing Group III based on flash points.

Spill Handling: Evacuate and restrict persons not wearing protective equipment from area of spill or leak until cleanup is complete. Remove all ignition sources. Establish forced ventilation to keep levels below explosive limit. Absorb liquids in activated charcoal, vermiculite, dry sand, earth, peat, carbon, or a similar material and deposit in sealed containers. The *tert-* isomer may be in powdered form, collect powdered material in the most convenient and safe manner and deposit in sealed containers. It may be necessary to contain and dispose of this chemical as a hazardous waste. If material or contaminated runoff enters waterways, notify downstream users of potentially contaminated waters. Contact your Department of Environmental Protection or your regional office of the federal EPA for specific recommendations. If employees are required to clean-up spills, they must be properly trained and equipped. OSHA 1910.120(q) may be applicable.

Fire Extinguishing: Butyl Alcohols are flammable liquids (*tert-* isomer may also be a flammable solid). Use dry chemical, carbon dioxide, water spray, or alcohol foam extinguishers (recommended). Water may be ineffective because of low flash point. Poisonous gases are produced in fire. If material or contaminated runoff enters waterways, notify downstream users of potentially contaminated waters. Notify local health and fire officials and pollution control agencies. From a secure, explosion-proof location, use water spray to cool exposed containers. If cooling streams are ineffective (venting sound increases in volume and pitch, tank discolors, or shows any signs of deforming), withdraw immediately to a secure position. If employees are expected to fight fires, they must be trained and equipped in OSHA 1910.156.

Disposal Method Suggested: Incineration, or bury absorbed waste in an approved landfill.[22]

References

New Jersey Department of Health and Senior Services, "Hazardous Substance Fact Sheet: n-Butyl Alcohol," Trenton, NJ (November 1998)

New Jersey Department of Health and Senior Services, "Hazardous Substance Fact Sheet: sec-Butyl Alcohol," Trenton, NJ (September 1998)

New Jersey Department of Health and Senior Services, "Hazardous Substance Fact Sheet: tert-Butyl Alcohol," Trenton, NJ (October 1998)

New Jersey Department of Health and Senior Services, "Hazardous Substance Fact Sheet: Isobutyl Alcohol," Trenton, NJ (April 1997)

Sax, N. I., Ed., "Dangerous Properties of Industrial Materials Report" 2, 44–46 (1982)

U.S. Environmental Protection Agency, Isobutyl Alcohol, Health and Environmental Effects Profile No. 120, Washington, DC, Office of Solid Waste (April 30, 1980)

U.S. Environmental Protection Agency, Chemical Hazard Information Profile: Isobutyl Alcohol, Washington, DC (March 31, 1983)

Butyl Amines

Molecular Formula: $C_4H_{11}N$

Common Formula: $C_4H_9NH_2$

Synonyms: *iso-:* 1-Amino-2-methylpropane; Isobutilamina (Spanish); Isobutylamine; 2-Methylpropylamine; Monoisobutylamine; 1-Propanamine, 2-methyl-; Valamine

n-: 1-Amino-butaan (Dutch); 1-Aminobutan (German); 1-Aminobutane; 1-Butanamine; *n*-Butilamina (Italian, Spanish); *n*-Butylamin (German); *n*-Butylamine; *normal* Butylamine; Mono-*n*-butylamine; Monobutylamine; Norvalamine

sec-: 2-AB; 2-Aminobutane; 2-Aminobutane base; Butafume; 2-Butanamine; *sec*-Butilamina (Spanish); Butilamina-*sec* (Spanish); *sec*-Butylamine, (*s*)-; *secondary* Butyl amine; Butyl 2-aminobutane; CSC 2-aminobutane; Deccotane; Decotane; Frucote; 1-Methylpropylamine; Propylamine, 1-methyl; Tutane

tert-: 2-Aminoisobutane; 2-Amino-2-methylpropane; Butilamina-*terc* (Spanish); *tert*-Butylamine; Butylamine, *tert*-; 1,1-Dimethylethylamine; 2-Methyl-2-propanamine; Trimethylaminomethane; Trimethylcarbinylamine

CAS Registry Number: 109-73-9 (n-); 13952-84-6 (sec-); 513-49-5 (sec-); 75-64-9 (tert-); 78-81-9 (iso-)

RTECS Number: EO2975000 (n-); EO3325000 (sec-); EO3327000 (sec-); EO3330000 (tert-); NP9900000 (iso-)

DOT ID: UN 1125 (n-); UN 1214 (iso-)

EEC Number: 612-005-00-0 (n-)

EINECS Number: 203-699-2 (n-)

Regulatory Authority

- Air Pollutant Standard Set (ACGIH)[1] (OSHA)[58] (HSE)[33] (DFG)[3] (Several States)[60] (Australia) (Israel) (Mexico) (Several Canadian Provinces)
- CLEAN WATER ACT: Section 311 Hazardous Substances/ RQ 40CFR117.3 (same as CERCLA, see below)
- SUPERFUND/EPCRA 40CFR302.4 Reportable Quantity (RQ): CERCLA, 1,000 lb (454 kg)
- Canada, WHMIS, Ingredients Disclosure List

Cited in U.S. State Regulations: Alaska (G), California (G), Florida (G), Illinois (G), Maine (G), Massachusetts (G), Minnesota (G), Nevada (A), New Jersey (G), New York (A), North Dakota (A), Oklahoma (G), Pennsylvania (G), Rhode Island (G), South Carolina (A), Virginia (A), West Virginia(G).

Description: Butylamines, $C_4H_9NH_2$, are highly flammable, colorless liquids (n- turns yellow on standing) with ammoniacal or fishlike odors. Odor threshold for n-butylamine = 0.24 – 13.9 ppm.

Isomer	Chemical Formula	Boiling Point (°C)	Flash Point* (°C)	Expl. Limits (LEL) (%)	(UEL) (%)
n-	$CH_3(CH_2)_3NH_2$	66	-12	1.7	9.8
sec-	$CH_3CH(NH_2)CH_2CH_3$	63	-9	1.7	10.0
tert-	$(CH_3)_3CNH_2$	45	+10.0	1.7	8.9
iso-	$CH_3CH(CH_3)CH_2NH_2$	68 – 69	-9	3.4	9.0

n-, sec-: Hazard Identification (based on NFPA-704 M Rating System): Health 0, Flammability 0, Reactivity 0. Health 1, Flammability 3, Reactivity 0.

iso-: Hazard Identification (based on NFPA-704 M Rating System): Health 2, Flammability 3, Reactivity 0.

Potential Exposure: n-Butylamine is used in pharmaceuticals, dyestuffs, rubber, chemicals, emulsifying agents, photography, desizing agents for textiles, pesticides, and synthetic agents. tert-Butylamine is used as a chemical intermediate in the production of tert-Butylaminoethyl methacrylate (a lube oil additive), as an intermediate in the production of rubber and in rust preventatives and emulsion deterrents in petroleum products. It is used in the manufacture of several drugs.

Incompatibilities: Forms explosive mixture with air. May accumulate static electrical charges, and may cause ignition of its vapors. n-Butylamine is a weak base; reacts with strong oxidizers and acids causing fire and explosion hazard. Incompatible with organic anhydrides, isocyanates, vinyl acetate, acrylates, substituted allyls, alkylene oxides, epichlorohydrin, ketones, aldehydes, alcohols, glycols, phenols, cresols, caprolactum solution. Attacks some metals in presence of moisture.

Permissible Exposure Limits in Air: The OSHA PEL, NIOSH REL and ACHIH TLV for n-Butylamine is 5 ppm (15 mg/m^3) ceiling value, not to be exceeded at any time. The HSE[33] has set 5 ppm as a STEL value with the notation "skin" indicating the possibility of skin absorption. The DFG[3] has set a MAK of 5 ppm for all four isomers. The NIOSH IDLH is 300 ppm for n-butylamine. The above exposure limits are for air levels only. When skin contact also occurs, you may be overexposed, even though air levels are less than the limits listed above. Several states have set guidelines or standards for n-butylamine in ambient air[60] ranging from 50.0 μg/m^3 (New York) to 75 μg/m^3 (South Carolina) to 80 μg/m^3 (Virginia) to 150 μg/m^3 (North Dakota) to 357 μg/m^3 (Nevada).

Determination in Air: Adsorption on H_2SO_4 treated silica gel, desorption with 50% methanol, analysis by gas chromatography/flame ionization detection, see NIOSH (IV) #2012.

Permissible Concentration in Water: No criteria set, but EPA[32] has suggested ambient water limits for Butylamines (n-, iso- or tert-) as 207 μg/l based on health effects.

Routes of Entry: Inhalation and percutaneous absorption, ingestion, and eye and/or skin contact.

Harmful Effects and Symptoms

Short Term Exposure: Contact with Butylamine can irritate and burn the eyes with possible permanent damage. Skin contact can cause irritation, burns and blisters. Inhalation can cause nose and throat irritation, "flushed" feeling, headache, and dizziness, coughing, shortness of breath. Higher exposures can cause pulmonary edema, a medical emergency that can be delayed for several hours. This can cause death. The LD_{50} rat value range from 78 mg/kg (tert) to 366 mg/kg (n-butyl) and all 4 are designated moderately toxic.

Long Term Exposure: Repeated exposure may cause itching and skin rash. May cause bronchitis to develop with phlegm, and/or shortness of breath. Exposure to high levels of isobutylamine can affect the heart.

Points of Attack: Skin, eyes, respiratory system.

Medical Surveillance: Consider the points of attack in preplacement and periodic physical examinations. Lung function tests.

First Aid: If this chemical gets into the eyes, remove any contact lenses at once and irrigate immediately for at least 15 minutes, occasionally lifting upper and lower lids. Seek medical attention immediately. If this chemical contacts the skin, remove contaminated clothing and wash immediately with soap and water. Seek medical attention immediately. If this chemical has been inhaled, remove from exposure, begin rescue breathing (using universal precautions) if breathing has stopped and CPR if heart action has stopped. Transfer promptly to a medical facility. When this chemical has been swallowed, get medical attention. Rinse out mouth and *do not* induce vomiting. Medical observation is recommended for 24 – 48 hours after breathing overexposure, as pulmonary edema may be delayed. As first aid for pulmonary edema, a doctor or authorized paramedic may consider administering a corticosteroid spray.

Personal Protective Methods: Wear protective gloves and clothing to prevent any reasonable probability of skin contact. Teflon is recommended by safety equipment suppliers/manufacturers for n-Butylamine and Chlorinated polyethylene (CPE) and VITON/chlorobutylene for Isobutylamine. All protective clothing (suits, gloves, footwear, headgear) should be clean, available each day, and put on before work. Contact lenses should not be worn when working with this chemical. Wear splash-proof chemical goggles and face shield unless full facepiece respiratory protection is worn. Employees should wash immediately with soap when skin is wet or contaminated. Provide emergency showers and eyewash.

Respirator Selection: *n-:* NIOSH/OSHA: *50 ppm:* CCRS [any chemical cartridge respirator with cartridge(s) providing protection against the compound of concern]; or SA (any supplied-air respirator). *125 ppm:* SA:CF (any supplied-air respirator operated in a continuous-flow mode); or PAPRS [any powered, air-purifying respirator with cartridge(s) providing protection against the compound of concern]. *250 ppm:* CCRFS [any chemical cartridge respirator with a full facepiece and cartridge(s) providing protection against the compound of concern]; or GMFS [any air-purifying, full-facepiece respirator (gas mask) with a chin-style, front- or back-mounted canister providing protection against the compound of concern]; or PAPRTS [any powered, air-purifying respirator with a tight-fitting facepiece and cartridge(s) providing protection against the compound of concern]; or SCBAF (any self-contained breathing apparatus with a full facepiece); or SAF (any supplied-air respirator with a full facepiece). *300 ppm:* SAF:PD,PP (any supplied-air respirator that has a full facepiece and is operated in a pressure-demand or other positive-pressure mode). *Emergency or planned entry into unknown concentrations or IDLH conditions:* SCBAF:PD,PP (any self-contained breathing apparatus that has a full facepiece and is operated in a pressure-demand or other positive-pressure mode); or SAF:PD, PP:ASCBA (any supplied-air respirator that has a full facepiece and is operated in a pressure-demand or other positive-pressure mode in combination with an auxiliary self-contained breathing apparatus operated in a pressure-demand or other positive pressure mode). *Escape:* GMFS [any air-purifying, full-facepiece respirator (gas mask) with a chin-style, front- or back-mounted canister providing protection against the compound of concern]; or SCBAE (any appropriate escape-type, self-contained breathing apparatus).

Note: Substance reported to cause eye irritation or damage; may require eye protection.

Storage: Store in a refrigerator. Keep under an inert atmosphere for long-term storage.[52]

Shipping: n-Butylamine and Isobutylamine are specifically cited by DOT[19] as requiring "Flammable Liquid" labels. They fall in Hazard Class 3 and Packing Group II.

Spill Handling: Evacuate and restrict persons not wearing protective equipment from area of spill or leak until cleanup is complete. Remove all ignition sources. Establish forced ventilation to keep levels below explosive limit. Absorb liquids in vermiculite, dry sand, earth, peat, carbon, or a similar material and deposit in sealed containers. It may be necessary to contain and dispose of this chemical as a hazardous waste. If material or contaminated runoff enters waterways, notify downstream users of potentially contaminated waters. Contact your Department of Environmental Protection or your regional office of the federal EPA for specific recommendations. If employees are required to clean-up spills, they must be properly trained and equipped. OSHA 1910.120(q) may be applicable.

Fire Extinguishing: This chemical is a combustible liquid. Poisonous gases including nitrogen oxides are produced in fire. Use dry chemical, carbon dioxide, or alcohol foam extinguishers. Vapors are heavier than air and will collect in low areas. Vapors may travel long distances to ignition sources and flashback. Vapors in confined areas may explode when exposed to fire. Containers may explode in fire. Storage containers and parts of containers may rocket great distances, in many directions. If material or contaminated runoff enters waterways, notify downstream users of potentially contaminated waters. Notify local health and fire officials and pollution control agencies. From a secure, explosion-proof location, use water spray to cool exposed containers. If cooling streams are ineffective (venting sound increases in volume and pitch, tank discolors, or shows any signs of deforming), withdraw immediately to a secure position. If employees are expected to fight fires, they must be trained and equipped in OSHA 1910.156.

Disposal Method Suggested: Incineration; incinerator is equipped with a scrubber or thermal unit to reduce No_x emissions.[22]

References

New Jersey Department of Health and Senior Services, "Hazardous Substance Fact Sheet: Butylamine," Trenton, NJ (August 1998)

New Jersey Department of Health and Senior Services, Hazardous Substance Fact Sheet, Isobutylamine, Trenton NJ (May 1999)

Sax, N. I., Ed., "Dangerous Properties of Industrial Materials Report" 2, No. 3, 68–70 (1982) and 6, No. 2, 45–48 (1986) (n-Butylamine)

Sax, N. I., Ed., "Dangerous Properties of Industrial Materials Report" 3, No. 6, 40–42 (1983) (sec-Butylamine)

Sax, N. I., Ed., "Dangerous Properties of Industrial Materials Report" 5, No. 6, 40–43 (1985) (t-Butylamine)

National Institute for Occupational Safety and Health, Information Profiles on Potential Occupational Exposures-Single Chemicals: tert-Butylamine, Report TR79-607, Rockville, MD pp 28–33 (December 1979)

Butylate

Molecular Formula: $C_{11}H_{23}NOS$

Common Formula: $CH_3CH_2SCON[CH_2CH(CH_3)_2]_2$

Synonyms: Bis(2-methylpropyl)carbamothioic acid *S*-ethyl ester; Butilate; Diisobutylthiocarbamic acid *S*-ethyl ester; Diisocarb; *S*-Ethyl bis(2-methylpropyl)carbamothioate; *S*-Ethyl *N,N*-diisobutylthiocarbamate; *S*-Ethyldiisobutyl thiocarbamate; Ethyl *N,N*-diisobutylthiocarbamate; Ethyl-*N,N*-diisobutyl thiolcarbamate; R-1910; Stauffer R-1910; Sutan

CAS Registry Number: 2008-41-5

RTECS Number: EZ7525000

Regulatory Authority

- Water Pollution Standard Proposed (EPA) (See reference below) (Wisconsin)[61]
- RCRA 40CFR268.48; 61FR15654, Universal Treatment Standards: Wastewater (mg/l), 0.003; Nonwastewater (mg/kg), 1.4

Cited in U.S. State Regulations: Wisconsin (W).

Description: $CH_3CH_2SCON[CH_2CH(CH_3)_2]_2$ is a clear liquid with an aromatic odor. Boiling point = 130°C under 10 mm mercury pressure.

Potential Exposure: Those involved in the manufacture, formulation or application of this herbicide which is used to control weed seeds in the soil prior to sowing a crop.

Permissible Exposure Limits in Air: No standards set.

Permissible Concentration in Water: A No-Adverse Effects Level (NOAEL) in the range of 24 – 40 mg/kg body weight/day/has been determined by USEPA. This leads to derivation of a 10-day health advisory for Butylate of 2.4 mg/l and a lifetime health advisory of 0.05 mg/l for a 70-kg man. Wisconsin has set a guideline for Butylate in drinking water of 200 μg/l.[61]

Determination in Water: Analysis of Butylate is by a gas chromatographic (GC) method applicable to the determination of certain nitrogen — and phosphorus-containing pesticides in water samples. In this method, approximately 1l of sample is extracted with methylene chloride. The extract is concentrated and the compounds are separated using capillary column GC. Measurement is made using a nitrogen-phosphorus detector. The method detection limit has not been determined for Butylate, but it is estimated that the detection limits for analytes included in this method are in the range of 0.1 – 2 μg/l.

Harmful Effects and Symptoms

The oral LD_{50} rat is 4,000 mg/kg (slightly toxic). Applying the criteria described in EPA's guidelines for assessment of carcinogenic risk, Butylate may be placed in Group C: a possible human carcinogen. This category is for substances that show limited evidence of carcinogenity in animals and inadequate evidence in humans.

First Aid: If this chemical gets into the eyes, remove any contact lenses at once and irrigate immediately for at least 15 minutes, occasionally lifting upper and lower lids. Seek medical attention immediately. If this chemical contacts the skin, remove contaminated clothing and wash immediately with soap and water. Seek medical attention immediately. If this chemical has been inhaled, remove from exposure, begin rescue breathing (using universal precautions) if breathing has stopped and CPR if heart action has stopped. Transfer promptly to a medical facility. When this chemical has been swallowed, get medical attention. Give large quantities of water and induce vomiting. Do not make an unconscious person vomit.

Spill Handling: Evacuate and restrict persons not wearing protective equipment from area of spill or leak until cleanup is complete. Remove all ignition sources. Collect powdered material in the most convenient and safe manner and deposit in sealed containers. Ventilate area of spill or leak after clean-up is complete. It may be necessary to contain and dispose of this chemical as a hazardous waste. If material or contaminated runoff enters waterways, notify downstream users of potentially contaminated waters. Contact your Department of Environmental Protection or your regional office of the federal EPA for specific recommendations. If employees are required to clean-up spills, they must be properly trained and equipped. OSHA 1910.120(q) may be applicable.

Fire Extinguishing: Use dry chemical, carbon dioxide, water spray, or alcohol foam extinguishers. Poisonous gases are produced in fire. If material or contaminated runoff enters waterways, notify downstream users of potentially contaminated waters. Notify local health and fire officials and pollution control agencies. From a secure, explosion-proof location, use water spray to cool exposed containers. If cooling streams are ineffective (venting sound increases in volume and pitch, tank discolors, or shows any signs of deforming), withdraw immediately to a secure position. If employees are expected to fight fires, they must be trained and equipped in OSHA 1910.156.

References

U.S. Environmental Protection Agency, "Health Advisory: Butylate," Washington, DC, Office of Drinking Water (August 1987)

Butyl Benzyl Phthalate

Molecular Formula: $C_{19}H_{20}O_4$

Common Formula: $C_6H_4(OCOC_4H_9)(OCOCH_2C_6H_5)$

Synonyms: Ashland butyl benzyl phthalate; BBP; 1,2-Benzenedicarboxylic acid, butyl phenylmethyl ester; *n*-Benzyl butyl phthalate; Benzyl butyl phthalate; *normal* Butyl benzyl phthalate; Ftalato de butilbencilo (Spanish); Monsanto butyl benzyl phthalate; NCI-C54375; Palatinol BB; Santicizer 160; Sicol; Unimoll BB

CAS Registry Number: 85-68-7

RTECS Number: TH9990000

Regulatory Authority

- Carcinogen (NCI/NTP)[54]
- Air Pollutant Standard Set (HSE)[33] (former USSR)[35] (Florida, New York)[60] (Mexico)
- CLEAN WATER ACT: 40CFR423, Appendix A, Priority Pollutants; Section 313 Water Priority Chemicals (57FR 41331, 9/9/92)
- RCRA, 40CFR261, Appendix 8 Hazardous Constituents, waste number not listed
- RCRA 40CFR268.48; 61FR15654, Universal Treatment Standards: Wastewater (mg/l), 0.017; Nonwastewater (mg/kg), 28
- RCRA 40CFR264, Appendix 9; TSD Facilities Ground Water Monitoring List. Suggested test method(s) (PQL μg/l): 8060 (5); 8270 (10)
- SUPERFUND/EPCRA 40CFR302.4 Reportable Quantity (RQ): CERCLA, 100 lb (45.4 kg)
- U.S. DOT Regulated Marine Pollutant (49CFR172.101, Appendix B)
- Canada, WHMIS, Ingredients Disclosure List; National Pollution Release Inventory (NPRI)

Cited in U.S. State Regulations: California (G), Florida (A), Kansas (G), Louisiana (G), Maryland (G), Massachusetts (G), Michigan (G), New Hampshire (G), New Jersey (G), New York (A), Pennsylvania (G), Vermont (G), Virginia (G), Washington (G), Wisconsin (G).

Description: Butyl Benzyl Phthalate, $C_6H_4(OCOC_4H_9)(OCOCH_2C_6H_5)$, is a clear, oily liquid with a slight odor. Boiling point = 370°. Freezing/Melting point = -35°C. Flash point = 199°C. Hazard Identification (based on NFPA-704 M Rating System): Health 1, Flammability 1, Reactivity 0.

Potential Exposure: This material is used as a plasticizer for polyvinyl and cellulosic resins. It is also used as an organic intermediate.

Incompatibilities: Incompatible with strong acids, nitrates, oxidizers. Destructive to rubber and paint.

Permissible Exposure Limits in Air: The HSE[33] has set an allowable TWA of 5 mg/m^3. The former USSR[35] has set a limit of 1.0 mg/m^3 in workplace air. Two states have set quidelines or standards for butyl benzyl phthalate in ambient air.[60] They are 100 μg/m^3 for both New York and Florida.

Permissible Concentration in Water: No criteria set butyl phthalate esters in general are classified as priority toxic pollutants by EPA.[6] Listed by Mexico for wastewater as phthalate esters.

Routes of Entry: Inhalation, can be absorbed through the skin.

Harmful Effects and Symptoms

Short Term Exposure: Irritates the eyes, the skin and the respiratory tract. Skin contact may cause a burning sensation. High levels of this chemical may cause dizziness and lightheadedness. The oral LD_{50} rat is 2,330 mg/kg (slightly toxic).

Long Term Exposure: Listed by NTP as an animal carcinogen. No data for humans. May affect liver and kidney function. Repeated exposure may damage the nervous system, causing weakness, "pins and needles," and poor coordination in arms and legs.

Medical Surveillance: Liver and kidney function tests. Examination of the nervous system, including nerve conduction tests.

First Aid: If this chemical gets into the eyes, remove any contact lenses at once and irrigate immediately for 20 – 30 minutes, occasionally lifting upper and lower lids. Seek medical attention immediately. If this chemical contacts the skin, remove contaminated clothing and wash immediately with soap and water. Seek medical attention immediately. If this chemical has been inhaled, remove from exposure, begin rescue breathing (using universal precautions) if breathing has stopped and CPR if heart action has stopped. Transfer promptly to a medical facility. When this chemical has been swallowed, get medical attention. Give large quantities of water and induce vomiting. Do not make an unconscious person vomit.

Personal Protective Methods: Wear protective gloves and clothing to prevent any reasonable probability of skin contact. Safety equipment suppliers/manufacturers can provide recommendations on the most protective glove/clothing material for your operation. All protective clothing (suits, gloves, footwear, headgear) should be clean, available each day, and put on before work. Contact lenses should not be worn when working with this chemical. Wear splash-proof chemical goggles and face shield unless full facepiece respiratory protection is worn. Employees should wash immediately with soap when skin is wet or contaminated. Provide emergency showers and eyewash.

Storage: Prior to working with this chemical you should be trained on its proper handling and storage. Store in tightly closed containers in a cool, dry place or refrigerator away from incompatible materials listed above. Metal containers involving the transfer of this chemical should be grounded and bonded. Drums must be equipped with self-closing valves, pressure vacuum bungs, and flame arresters. Use only non-sparking tools and equipment, especially when opening and closing containers of this chemical. Sources of ignition such as smoking and open flames, are prohibited where this chemical is used, handled, or stored in a manner that could create a potential fire or explosion hazard. A regulated, marked area should be established where this chemical is handled, used, or stored in compliance with OSHA standard 1910.1045.

Shipping: This material is not specifically cited in the DOT performance-oriented packaging standards.[19] It could be considered as a Hazardous Substance. Liquid, n.o.s., which has no label requirements or weight limits on shipments.

Spill Handling: Evacuate and restrict persons not wearing protective equipment from area of spill or leak until cleanup is complete. Remove all ignition sources. Ventilate area of spill or leak. Absorb liquids in vermiculite, dry sand, earth, peat, carbon, or a similar material and deposit in sealed containers. It may be necessary to contain and dispose of this chemical as a hazardous waste. If material or contaminated runoff enters waterways, notify downstream users of potentially contaminated waters. Contact your Department of Environmental Protection or your regional office of the federal EPA for specific recommendations. If employees are required to clean-up spills, they must be properly trained and equipped. OSHA 1910.120(q) may be applicable.

Fire Extinguishing: Use dry chemical, carbon dioxide, or foam extinguishers. Water may cause frothing. Poisonous gases are produced in fire. If material or contaminated runoff enters waterways, notify downstream users of potentially contaminated waters. Notify local health and fire officials and pollution control agencies. From a secure, explosion-proof location, use water spray to cool exposed containers. If cooling streams are ineffective (venting sound increases in volume and pitch, tank discolors, or shows any signs of deforming), withdraw immediately to a secure position. If employees are expected to fight fires, they must be trained and equipped in OSHA 1910.156.

Disposal Method Suggested: Atomize into an incinerator together with a flammable solvent.[22]

References

Sax, N. I., Ed., "Dangerous Properties of Industrial Materials Report" 2, No. 2, 15–17 (1982)

New Jersey Department of Health and Senior Services, Hazardous Substance Fact Sheet, "Butyl Benzyl Phthalate," Trenton NJ (May, 1998)

n-Butyl Bromide

Molecular Formula: $C_4H_{12}Br$

Common Formula: $CH_6(CH_2)_3Br$

Synonyms: 1-Bromobutane; Bromuro de *n*-butilo (Spanish); Butane, 1-bromo-; Butyl bromide; Methyl ethyl bromomethane

CAS Registry Number: 109-65-9. *Note:* Much of the same information in this record may apply to sec-Butyl Bromide (78-76-2)

RTECS Number: EJ6225000

DOT ID: UN 1126 (n-); UN 2339 (sec-)

Regulatory Authority

- Air Pollutant Standard Set (former-USSR/UNEP)[43]
- Canada, WHMIS, Ingredients Disclosure List

Cited in U.S. State Regulations: Florida (G), Maine (G), Massachusetts (G), New Hampshire (G), New Jersey (G), Pennsylvania (G).

Description: Butyl bromide, $CH_6(CH_2)_3Br$, is a highly flammable, colorless liquid with a pleasant odor. Boiling point = 101.6°C. Flash point = 18°C (21°C for sec- isomer). Autoigniton temperature = 265°C. Explosive limits: LEL = 2.6% @ 100°C; UEL = 6.6% @ 100°C. Hazard Identification (based on NFPA-704 M Rating System): Health 2, Flammability 3, Reactivity 0. Insoluble in water.

Potential Exposure: Butyl Bromide is used to make other chemicals and in making pharmaceuticals.

Incompatibilities: Contact with strong oxidizers may cause fire and explosion. Forms explosive mixture with air. Incompatible with strong acids. May accumulate static electrical charges and cause ignition of its vapors.

Permissible Exposure Limits in Air: The former-USSR/UNEP Joint Project[43] has set limits for Butyl Bromide in the ambient air of residential areas. The momentary MAC is 0.7 mg/m^3. The average daily MAC is 0.03 mg/m^3.

Permissible Concentration in Water: No criteria set.

Routes of Entry: Inhalation, ingestion (sec-butyl bromide passes through the skin)

Harmful Effects and Symptoms

Short Term Exposure: Causes eye, skin and respiratory tract irritation. Inhalation can cause coughing, wheezing and/or shortness of breath. High levels can cause you to feel dizzy, lightheaded, and to pass out. Very high levels can cause death.

Long Term Exposure: Repeated exposure may cause liver and kidney damage. There is limited evidence that Butyl Bromide may damage the developing fetus.

Points of Attack: Liver, kidney.

Medical Surveillance: For those with frequent or potentially high exposure, the following are recommended before beginning work and at regular times after that: Liver function tests. Kidney function tests.

First Aid: If this chemical gets into the eyes, remove any contact lenses at once and irrigate immediately for at least 15 minutes, occasionally lifting upper and lower lids. Seek medical attention immediately. If this chemical contacts the skin, remove contaminated clothing and wash immediately with soap and water. Seek medical attention immediately. If this chemical has been inhaled, remove from exposure, begin rescue breathing (using universal precautions) if breathing has stopped and CPR if heart action has stopped. Transfer promptly to a medical facility. When this chemical has been swallowed, get medical attention. Give large quantities of water and induce vomiting. Do not make an unconscious person vomit.

Personal Protective Methods: Wear protective gloves and clothing to prevent any reasonable probability of skin contact. Safety equipment suppliers/manufacturers can provide recommendations on the most protective glove/clothing material for your operation. All protective clothing (suits, gloves, footwear, headgear) should be clean, available each day, and put on before work. Contact lenses should not be worn when working with this chemical. Wear splash-proof chemical goggles and face shield unless full facepiece respiratory protection is worn. Employees should wash immediately with soap when skin is wet or contaminated. Provide emergency showers and eyewash.

Respirator Selection: Where the potential exists for exposures to Butyl Bromide, use a MSHA/NIOSH approved supplied-air respirator with a full facepiece operated in the positive pressure mode or with a full facepiece, hood, or helmet in the continuous flow mode, or use a MSHA/NIOSH approve self-contained breathing apparatus with a full facepiece operated in pressure-demand or other positive pressure mode.

Storage: Prior to working with Butyl Bromide you should be trained on its proper handling and storage. Store in tightly closed containers in a cool, well-ventilated area. Sources of ignition, such as smoking and open flames, are prohibited where Butyl Bromide is handled, used, or stored. Metal containers involving the transfer of 5 gallons or more of Butyl Bromide should be grounded and bonded. Drums must be equipped with self-closing valves, pressure vacuum bungs, and flame arresters. Use only non-sparking tools and equipment, especially when opening and closing containers of Butyl Bromide.

Shipping: n-Butyl Bromide must be labeled "Flammable Liquid." It falls in Hazard Class 3 and Packing Group II.

Spill Handling: Evacuate and restrict persons not wearing protective equipment from area of spill or leak until cleanup is complete. Remove all ignition sources. Establish forced ventilation to keep levels below explosive limit. Absorb liquids in vermiculite, dry sand, earth, or a similar material and deposit in sealed containers. Collect powdered material in the most convenient and safe manner and deposit in sealed containers. It may be necessary to contain and dispose of this chemical as a hazardous waste. If material or contaminated runoff enters waterways, notify downstream users of potentially contaminated waters. Contact your Department of Environmental Protection or your regional office of the federal EPA for specific recommendations. If employees are required to clean-up spills, they must be properly trained and equipped. OSHA 1910.120(q) may be applicable.

Fire Extinguishing: This chemical is a flammable liquid. Poisonous gases including bromine are produced in fire. Use dry chemical, carbon dioxide, or alcohol foam extinguishers. Vapors are heavier than air and will collect in low areas. Vapors may travel long distances to ignition sources and flashback. Vapors in confined areas may explode when exposed to fire. Storage containers and parts of containers may rocket great distances, in many directions. If material or contaminated runoff enters waterways, notify downstream users of potentially contaminated waters. Notify local health and fire officials and pollution control agencies. From a secure, explosion-proof location, use water spray to cool exposed containers. If cooling streams are ineffective (venting sound increases in volume and pitch, tank discolors, or shows any signs of deforming), withdraw immediately to a secure position. If employees are expected to fight fires, they must be trained and equipped in OSHA 1910.156.

Disposal Method Suggested: Incineration.

References

New Jersey Department of Health and Senior Services, "Hazardous Substance Fact Sheet: Butyl Bromide," Trenton, NJ (July 1996)

n-Butyl Chloride

Molecular Formula: C_4H_9Cl

Common Formula: $CH_3(CH_2)_3Cl$

Synonyms: Butane, 1-chloro-; *n*-Butyl chloride; 1-Chlorobutane; Chlorure de butyle (French); Cloruro de *n*-butilo (Spanish); NCI-C06155; *n*-Propylcarbinyl chloride

CAS Registry Number: 109-69-3

RTECS Number: EJ6300000

DOT ID: UN 1127

Cited in U.S. State Regulations: Florida (G), Maine (G), Massachusetts (G), New Hampshire (G), New Jersey (G), Pennsylvania (G).

Description: Butyl chloride, $CH_3(CH_2)_3Cl$, is a highly flammable, clear, colorless liquid. Boiling point = 78°C. Flash point = –9°C. Autoignition temperature = 240°C. Explosive limits: LEL = 1.8%; UEL = 10.1%. Insoluble in water. Hazard Identification (based on NFPA-704 M Rating System): Health 2, Flammability 3, Reactivity 0.

Potential Exposure: Butyl Chloride is used as a solvent, as a medicine to control worms, and to make other chemicals.

Incompatibilities: Forms explosive mixture with air. May accumulate static electrical charges, and may cause ignition of its vapors. Water contact slowly forms hydrochloric acid. Incompatible with strong oxidizers, alkaline earth and alkali metals, finely divided metal. Attacks metals in presence of moisture. Attacks some plastics, rubber or coatings.

Permissible Exposure Limits in Air: No standards set.

Permissible Concentration in Water: No criteria set.

Routes of Entry: Inhalation, ingestion.

Harmful Effects and Symptoms

Short Term Exposure: Irritates the eyes, skin and respiratory tract. Butyl Chloride can affect you when breathed in. High levels can cause you to feel dizzy, lightheaded, and to pass out. Very high levels can affect the nervous system and cause death. Sec-butyl chloride and tert-butyl chloride may have similar effects.

Long Term Exposure: Unknown at this time.

Points of Attack: Eyes, respiratory system.

First Aid: If this chemical gets into the eyes, remove any contact lenses at once and irrigate immediately for at least 15 minutes, occasionally lifting upper and lower lids. Seek medical attention immediately. If this chemical contacts the skin, remove contaminated clothing and wash immediately with soap and water. Seek medical attention immediately. If this chemical has been inhaled, remove from exposure, begin rescue breathing (using universal precautions) if breathing has stopped and CPR if heart action has stopped. Transfer promptly to a medical facility. When this chemical has been swallowed, get medical attention. Give large quantities of water and induce vomiting. Do not make an unconscious person vomit.

Personal Protective Methods: Wear protective gloves and clothing to prevent any reasonable probability of skin contact. Safety equipment suppliers/manufacturers can provide recommendations on the most protective glove/clothing material for your operation. All protective clothing (suits, gloves, footwear, headgear) should be clean, available each day, and put on before work. Contact lenses should not be worn when working with this chemical. Wear splash-proof chemical goggles and face shield unless full facepiece respiratory protection is worn. Employees should wash immediately with soap when skin is wet or contaminated. Provide emergency showers and eyewash.

Respirator Selection: Where the potential exists for exposures to Butyl Chloride, use a MSHA/NIOSH approved supplied-air respirator with a full facepiece operated in the positive pressure mode or with a full facepiece, hood, or helmet in the continuous flow mode, or use a MSHA/NIOSH approved self-contained breathing apparatus with a full facepiece operated in pressure-demand or other positive pressure mode.

Storage: Butyl Chloride is incompatible with oxidizers (such as perchlorates, peroxides, permanganates, chlorates and nitrates). Store in tightly closed containers in a cool, well-ventilated area. Sources of ignition, such as smoking and open flames, are prohibited where Butyl Chloride is handled, used, or stored. Metal containers involving the transfer of 5 gallons or more of Butyl Chloride should be grounded and bonded. Drums must be equipped with self-closing valves, pressure vacuum bungs, and flame arresters. Use only non-sparking tools and equipment, especially when opening and closing containers of Butyl Chloride.

Shipping: Chlorobutanes require a "Flammable Liquid" label. They fall in Hazard Class 3 and Packing Group II.

Spill Handling: Evacuate and restrict persons not wearing protective equipment from area of spill or leak until cleanup is complete. Remove all ignition sources. Establish forced ventilation to keep levels below explosive limit. Absorb liquids in vermiculite, dry sand, earth, or a similar material and deposit in sealed containers. It may be necessary to contain and dispose of this chemical as a hazardous waste. If material or contaminated runoff enters waterways, notify downstream users of potentially contaminated waters. Contact your Department of Environmental Protection or your regional office of the federal EPA for specific recommendations. If employees are required to clean-up spills, they must be properly trained and equipped. OSHA 1910.120(q) may be applicable.

Fire Extinguishing: This chemical is a combustible liquid. Poisonous gases including phosgene are produced in fire. Use dry chemical, carbon dioxide, or alcohol foam extinguishers. Vapors are heavier than air and will collect in low areas. Vapors may travel long distances to ignition sources and flashback. Vapors in confined areas may explode when exposed to fire. Storage containers and parts of containers may rocket great distances, in many directions. If material or contaminated runoff enters waterways, notify downstream users of potentially contaminated waters. Notify local health and fire officials and pollution control agencies. From a secure, explosion-proof location, use water spray to cool exposed containers. If cooling streams are ineffective (venting sound increases in volume and pitch, tank discolors, or shows any signs of deforming), withdraw immediately to a secure position. If employees are expected to fight fires, they must be trained and equipped in OSHA 1910.156.

Disposal Method Suggested: Incineration.

References

New Jersey Department of Health and Senior Services, "Hazardous Substance Fact Sheet: Butyl Chloride," Trenton, NJ (July 1996)

tert-Butyl Chromate

Molecular Formula: $C_8H_{18}CrO_4$

Common Formula: $[(CH_3)_3CO]_2CrO_2$

Synonyms: Bis(*tert*-butyl) chromate; Chromato *terc*-butilico (Spanish); Chromic acid, di-*tert*-butyl ester of chromic acid

CAS Registry Number: 1189-85-1

RTECS Number: GB2900000

Regulatory Authority

- Air Pollutant Standard Set (ACGIH)[1] (OSHA)[58] (Several States)[60] (Australia) (Israel) (Mexico) (Several Canadian Provinces)

as chromium compounds:

- Carcinogen: Chromium(VI) IARC, carcinogenic to humans. *Note:* this evaluation applies to the group of Chromium(VI) compounds, and not necessarily to all individual chemicals
- CLEAN AIR ACT: Hazardous Air Pollutants (Title I, Part A, Section 112)
- CLEAN WATER ACT: 40CFR401.15 Section 307 Toxic Pollutants as chromium and compounds
- RCRA, 40CFR261, Appendix 8 Hazardous Constituents, waste number not listed
- EPCRA (Section 313): Includes any unique chemical substances that contains chromium as part of that chemical's infrastructure. Form R *de minimis* concentration reporting level: Chromium(VI) compounds: 0.1%
- Canada: National Pollutant Release Inventory (NPRI); CEPA Priority Substance List

Cited in U.S. State Regulations: Alaska (G), California (G), Connecticut (A), Florida (G), Illinois (G), Maine (G), Massachusetts (G), Minnesota (G), Nevada (A), New Hampshire (G), New Jersey (G), North Dakota (A), Pennsylvania (G), Rhode Island (G), Virginia (A), West Virginia (G).

Description: tert-Butyl chromate, $[(CH_3)_3CO]_2CrO_2$, is a clear, colorless, flammable liquid or red crystals from petroleum ether. Freezing/Melting point = of -5 – 0°C. Soluble in water.

Potential Exposure: Butyl chromate is used in specialty reactions as an organic source of chromium, in making catalysts, and as a curing agent for urethane foams.

Incompatibilities: This chemical is a powerful oxidizer and a fire hazard. Incompatible with reducing agents, moisture, strong bases, alcohols, hydrazine and easily oxidized materials such as paper, wood, sulfur, aluminum and plastics.

Permissible Exposure Limits in Air: The Federal standard[58] and ACGIH recommended value is 0.1 mg/m^3 (as chromic acid) as a ceiling value, not to be exceeded at any time. The NIOSH recommended limit is 0.001 mg/m^3 (as hexavalent chromium). Australia, Israel, Alberta, British Columbia, Ontario and Quebec ceiling values are the same as OSHA. Mexico's TWA is 0.1 mg/m^3. Most agencies and advisory bodies add the notation "skin" indicating the possibility of cutaneous absorption. Several states have set guidelines or standards for t-butyl chromate in ambient air[60] ranging from 0.5 μg/m^3 (Virginia) to 1.0 μg/m^3 (North Dakota) to 2.0 μg/m^3 (Connecticut and Nevada).

Determination in Air: Use NIOSH: Chromium hexavalent #7600 (Also #1704).

Permissible Concentration in Water: No criteria set.

Routes of Entry: Inhalation, skin absorption, ingestion and skin and/or eye contact.

Harmful Effects and Symptoms

Short Term Exposure: Contact can cause severe eye and skin irritation and acidlike burns with possible permanent damage to the eyes. Inhalation can cause irritation of the respiratory tract with coughing and wheezing. Exposure can cause headache, nausea, vomiting, diarrhea, and wheezing. Higher exposures may cause pulmonary edema, a medical emergency that can be delayed for several hours. This can cause death.

Long Term Exposure: While tert-butyl chromate has not been identified as a carcinogen, certain hexavalent chromium compounds have been determined to be human carcinogens. Breathing this chemical can cause a hole in the "bone" dividing the inner nose (septum), sometime with discharge, bleeding and/or formation of a crust. Tert-butyl chromate can cause allergies to the skin and lung. Prolonged skin contact can cause, burns, blisters and deep necrotic ulcers. Can cause liver and kidney damage.

Points of Attack: Respiratory system, lungs, skin, eyes, central nervous system.

Medical Surveillance: Consider the points of attack in preplacement and periodic physical examinations. Examination of the skin and nose, Evaluation by a qualified allergist. Lung function tests. Liver and kidney function tests. Urine test for chromates. This test is most accurate shortly after exposure.

First Aid: If this chemical gets into the eyes, remove any contact lenses at once and irrigate immediately for at least 30 minutes, occasionally lifting upper and lower lids. Seek medical attention immediately. If this chemical contacts the skin, remove contaminated clothing and wash immediately with soap and water. Seek medical attention immediately. If this chemical has been inhaled, remove from exposure, begin rescue breathing (using universal precautions) if breathing has stopped and CPR if heart action has stopped. Transfer promptly to a medical facility. When this chemical has been swallowed, get medical attention. Give large quantities of water and induce vomiting. Do not make an unconscious person vomit. Medical observation is recommended for 24 – 48 hours after breathing overexposure, as pulmonary edema may be delayed. As first aid for pulmonary edema, a doctor or authorized paramedic may consider administering a corticosteroid spray.

Note to Physician: In case of fume inhalation, treat pulmonary edema. Consider administering prednisone or other corticosteroid orally to reduce tissue response to fume. Positive pressure ventilation may be necessary. Treat metal fume fever with bed rest, analgesics and antipyretics.

Personal Protective Methods: Wear appropriate clothing to prevent any possibility of skin contact. Wear eye protection to prevent any possibility of eye contact. Employees should wash immediately when skin is wet or contaminated and daily at the end of each work shift. Remove nonimpervious clothing immediately if wet or contaminated. Provide emergency showers and eyewash. Specific engineering controls are recommended in NIOSH Criteria Document #76–129 [Chromium (VI)].

Respirator Selection: NIOSH (as chromates): *At any concentrations above the NIOSH REL, or where there is no REL, at any detectable concentration:* SCBAF:PD,PP (any self-contained breathing apparatus that has a full facepiece and is operated in a pressure-demand or other positive-pressure mode); or SAF:PD,PP:ASCBA (any supplied-air respirator that has a full facepiece and is operated in a pressure-demand or other positive-pressure mode in combination with an auxiliary, self-contained breathing apparatus operated in a pressure-demand or other positive-pressure mode). *Escape:* GMFOV [any air-purifying, full-facepiece respirator (gas mask) with a chin-style, front- or back-mounted organic vapor canister]; or SCBAE (any appropriate escape-type, self-contained breathing apparatus).

Storage: Prior to working with tert-butyl chromate you should be trained on its proper handling and storage. Before entering confined space where this chemical may be present, check to make sure that an explosive concentration does not exist. Store in tightly closed containers in an explosion-proof refrigerator away from incompatible materials listed above. Metal containers involving the transfer of this chemical should be grounded and bonded. Where possible, automatically pump liquid from drums or other storage containers to process containers. Drums must be equipped with self-closing valves, pressure vacuum bungs, and flame arresters. Use only non-sparking tools and equipment, especially when opening and closing containers of this chemical. Sources of ignition such as smoking and open flames, are prohibited where this chemical is used, handled, or stored in a manner that could create a potential fire or explosion hazard. A regulated, marked area should be established where this chemical is handled, used, or stored in compliance with OSHA standard 1910.1045.

Shipping: t-Butyl Chromate is shipped in solution as a flammable liquid. It can therefore be classified as a flammable liquid, n.o.s. which requires a "Flammable Liquid" label. Depending on the flash point of the solvent used, the Packing Group is set in Hazard Class 3. In Class I.

Spill Handling: Evacuate and restrict persons not wearing protective equipment from area of spill or leak until cleanup is complete. Remove all ignition sources. Ventilate area of spill or leak. Absorb liquids in vermiculite, dry sand, earth, or a similar material and deposit in sealed containers. Collect powdered material in the most convenient and safe manner and deposit in sealed containers. It may be necessary to contain and dispose of this chemical as a hazardous waste. If material or contaminated runoff enters waterways, notify downstream users of potentially contaminated waters. Contact your Department of Environmental Protection or your regional office of the federal EPA for specific recommendations. If employees are required to clean-up spills, they must be properly trained and equipped. OSHA 1910.120(q) may be applicable.

Fire Extinguishing: Use dry chemical, sand, water spray, or foam extinguishers. Poisonous gases are produced in fire. If material or contaminated runoff enters waterways, notify downstream users of potentially contaminated waters. Notify local health and fire officials and pollution control agencies. From a secure, explosion-proof location, use water spray to cool exposed containers. If cooling streams are ineffective (venting sound increases in volume and pitch, tank discolors, or shows any signs of deforming), withdraw immediately to a secure position. If employees are expected to fight fires, they must be trained and equipped in OSHA 1910.156.

References

National Institute for Occupational Safety and Health, Criteria for a Recommended Standard: Occupational Exposure to Chromium, NIOSH Doc. No. 76–129, Washington DC (1979)

New Jersey Department of Health and Senior Services, "Hazardous Substance Fact Sheet: tert-Butyl Chromate," Trenton, NJ (November 1986)

1,2-Butylene Oxide

Molecular Formula: C_4H_8O

Synonyms: 1,2-Butene oxide; α-Butylene oxide; 1,2-Epoxybutane; 2-Ethyloxirane; Oxirane, ethyl-; Propyl oxirane

CAS Registry Number: 106-88-7

RTECS Number: EK3675000

DOT ID: UN 3022

Regulatory Authority

- Carcinogen (animal positive) (NTP) (DFG)
- CLEAN AIR ACT: Hazardous Air Pollutants (Title I, Part A, Section 112)
- SUPERFUND/EPCRA 40CFR302.4 Reportable Quantity (RQ): CERCLA, 1 lb (0.454 kg)
- EPCRA Section 313 Form R *de minimis* concentration reporting level: 1.0%
- TSCA 40CFR712.30(d); 40CFR716.120 (a)
- Canada, WHMIS, Ingredients Disclosure List; National Pollutant Release Inventory (NPRI)

Cited in U.S. State Regulations: California (A, G), Florida (G), Massachusetts (G), New Jersey (G), Pennsylvania (G).

Description: Butylene Oxide is a watery-white liquid with and etherial odor. Boiling point = 63°C. Freezing/Melting point = -130°C. Flash point = -22°C. Autoignition temperature = 439°C. Explosive limits in air: LEL: 1.7%; UEL: 19%. Hazard Identification (based on NFPA-704 M Rating System) Health 2, Flammability 3, Reactivity 2. Soluble in water. Log P octanol/water partition coefficient = 0.42

Potential Exposure: It is used as a stabilizer and to make certain other chemicals such as gasoline additives.

Incompatibilities: Forms explosive mixture with air. Unless inhibited, can form unstable and explosive peroxides. Before entering confined space where this chemical may be present, check to make sure that an explosive concentration does not exist. Polymerization will occur in the presence of acids, strong bases and chlorides of tin, iron and aluminum. Storage tanks and other equipment should be absolutely dry and free from air, ammonia, acetylene, hydrogen sulfide, rust and other contaminants. Reacts with strong oxidizers. Attacks some plastics. May accumulate static electric charges that can result in ignition of its vapors. A regulated, marked area should be established where this chemical is handled, used, or stored in compliance with OSHA standard 1910.1045.

Permissible Exposure Limits in Air: The American Industrial Health Association's WEEL recommendation is 2 ppm (5.9 mg/m^3) TWA. The German DFG has set no numerical limits, only the proven carcinogen notation.

Routes of Entry: Inhalation, ingestion, eye and/or skin contact. Absorbed through the skin.

Harmful Effects and Symptoms

Short Term Exposure: Butylene oxide can cause severe irritation of the eyes, skin, and respiratory tract, with coughing and/or shortness of breath. High exposures can cause dizziness, lightheadedness and unconsciousness. Higher exposures can cause pulmonary edema, a medical emergency that can be delayed for several hours. This can cause death.

Long Term Exposure: 1,2-Butylene oxide is possibly carcinogenic to humans. It may cause mutations and damage to the developing fetus. Prolonged or repeated skin contact may cause blisters or other disorders. 2,2-Butylene oxide may affect the nervous system.

Points of Attack: Skin, lungs, central nervous system, reproductive system.

Medical Surveillance: Lung function tests. Consider chest x-ray following acute overexposure.

First Aid: If this chemical gets into the eyes, remove any contact lenses at once and irrigate immediately for at least 15 minutes, occasionally lifting upper and lower lids. Seek medical attention immediately. If this chemical contacts the skin, remove contaminated clothing and wash immediately with soap and water. Seek medical attention immediately. If this chemical has been inhaled, remove from exposure, begin rescue breathing (using universal precautions) if breathing has stopped and CPR if heart action has stopped. Transfer promptly to a medical facility. When this chemical has been swallowed, get medical attention. Give large quantities of water and induce vomiting. Do not make an unconscious person vomit. Medical observation is recommended for 24 –

48 hours after breathing overexposure, as pulmonary edema may be delayed. As first aid for pulmonary edema, a doctor or authorized paramedic may consider administering a corticosteroid spray.

Personal Protective Methods: Wear solvent-resistant gloves and clothing to prevent any reasonable probability of skin contact. Safety equipment suppliers/manufacturers can provide recommendations on the most protective glove/clothing material for your operation. All protective clothing (suits, gloves, footwear, headgear) should be clean, available each day, and put on before work. Contact lenses should not be worn when working with this chemical. Wear splash-proof chemical goggles and face shield unless full facepiece respiratory protection is worn. Employees should wash immediately with soap when skin is wet or contaminated. Remove nonimpervious clothing immediately if wet or contaminated. Provide emergency showers and eyewash.

Respirator Selection: Engineering controls should be used wherever feasible to maintain airborne concentrations of this chemical below the prescribed exposure limit. Respirators and protective equipment less effective than engineering controls, and should be used only in non-routine or emergency situations which may result in exposure concentrations in excess of the TWA environmental limit. *At any concentrations above the NIOSH REL, or where there is no REL, at any detectable concentration:* SCBAF:PD,PP (any MSHA/NIOSH approved self-contained breathing apparatus that has a full facepiece and is operated in a pressure-demand or other positive-pressure mode); or SAF:PD,PP:ASCBA (any supplied-air respirator that has a full facepiece and is operated in a pressure-demand or other positive-pressure mode in combination with an auxiliary, self-contained breathing apparatus operated in a pressure-demand or other positive pressure mode). *Escape:* GMFOVHiE [any air-purifying, full-facepiece respirator (gas mask) with a chin-style, front- or back-mounted organic vapor canister having a high-efficiency particulate filter]; or SCBAE (any appropriate escape-type, self-contained breathing apparatus).

Storage: Prior to working with this chemical acid you should be trained on its proper handling and storage. Protect against physical damage. Store only if inhibited. Outside or detached storage is preferred. Store in tightly closed containers in a cool, well ventilated area away from incompatible materials listed above. Metal containers involving the transfer of this chemical should be grounded and bonded. Drums must be equipped with self-closing valves, pressure vacuum bungs, and flame arresters. Use only non-sparking tools and equipment, especially when opening and closing containers of this chemical. Sources of ignition such as smoking and open flames, are prohibited where this chemical is used, handled, or stored in a manner that could create a potential fire or explosion hazard. A regulated, marked area should be established where this chemical is handled, used, or stored in compliance with OSHA standard 1910.1045.

Shipping: 1,2-Butylene oxide requires a "Flammable Liquid" label. It is in UN/DOT Hazard Class 3, Packing Group II

Spill Handling: Evacuate and restrict persons not wearing protective equipment from area of spill or leak until cleanup is complete. Remove all ignition sources. Establish forced ventilation to keep levels below explosive limit. Absorb liquid in vermiculite, dry sand, earth or similar material and deposit in sealed containers. It may be necessary to contain and dispose of this chemical as a hazardous waste. If material or contaminated runoff enters waterways, notify downstream users of potentially contaminated waters. Contact your Department of Environmental Protection or your regional office of the federal EPA for specific recommendations. If employees are required to clean-up spills, they must be properly trained and equipped. OSHA 1910.120(q) may be applicable.

Fire Extinguishing: This chemical is a highly flammable liquid. Poisonous gases are produced in fire. Use dry chemical, carbon dioxide, or foam extinguishers. Vapors are heavier than air and will collect in low areas. Vapors may travel long distances to ignition sources and flashback. Vapors in confined areas may explode when exposed to fire. Containers may explode in fire. Storage containers and parts of containers may rocket great distances, in many directions. If material or contaminated runoff enters waterways, notify downstream users of potentially contaminated waters. Notify local health and fire officials and pollution control agencies. Containers may explode in fire. From a secure, explosion-proof location, use water spray to cool exposed containers. If cooling streams are ineffective (venting sound increases in volume and pitch, tank discolors, or shows any signs of deforming), withdraw immediately to a secure position. If employees are expected to fight fires, they must be trained and equipped in OSHA 1910.156.

References

New Jersey Department of Health and Senior Services, "Hazardous Substance Fact Sheet: "Butylene oxide," Trenton, NJ (September, 1996)

Butyl Ether

Molecular Formula: $C_8H_{18}O$

Common Formula: $C_4H_9OC_4H_9$

Synonyms: 1-Butoxybutane; Di-*n*-butyl ether; Dibutyl ether; Dibutyl oxide; Ether butylique (French); 1,1'-Oxybis(butane); 1,1-Oxybis-butane

CAS Registry Number: 142-96-1

RTECS Number: EK5425000

DOT ID: UN 1149

Cited in U.S. State Regulations: Florida (G), Massachusetts (G), New Hampshire (G), New Jersey (G), Oklahoma (G), Pennsylvania (G), Rhode Island (G).

Description: Butyl Ether, $C_4H_9OC_4H_9$, is a flammable, colorless liquid with a mild, ethereal odor. Boiling point = 142°C. Flash point = 25°C. Autoignition temperature = 194°C. Explosive limits: LEL = 1.5%; UEL = 7.6%. Hazard Identification (based on NFPA-704 M Rating System): Health 2, Flammability 3, Reactivity 1. Insoluble in water.

Potential Exposure: It is used as a solvent for hydrocarbons, fatty materials; extracting agent used metals separation; solvent purification, making other chemicals.

Incompatibilities: Forms explosive mixture with air. May accumulate static electrical charges, and may cause ignition of its vapors. Incompatible with strong acids, oxidizers. Contact with air or light may form unstable and explosive peroxides, especially anhydrous form.

Permissible Exposure Limits in Air: No standards set.

Permissible Concentration in Water: No criteria set.

Routes of Entry: Inhalation of vapor, skin contact, ingestion.

Harmful Effects and Symptoms

Short Term Exposure: May be poisonous if inhaled or absorbed through skin. Inhalation of vapors may cause dizziness or suffocation. Skin or eye contact may cause irritation. Repeated or prolonged skin contact may cause rash. The vapor irritates the nose, throat and bronchial tubes and may cause nose bleeds, hoarseness, cough, phlegm and/or tightness in the chest. Overexposure can also cause headache and make you feel dizzy and lightheaded. Higher levels can cause unconsciousness and even death.

Points of Attack: Skin, eyes, respiratory system.

First Aid: If this chemical gets into the eyes, remove any contact lenses at once and irrigate immediately for at least 15 minutes, occasionally lifting upper and lower lids. Seek medical attention immediately. If this chemical contacts the skin, remove contaminated clothing and wash immediately with soap and water. Seek medical attention immediately. If this chemical has been inhaled, remove from exposure, begin rescue breathing (using universal precautions) if breathing has stopped and CPR if heart action has stopped. Transfer promptly to a medical facility. When this chemical has been swallowed, get medical attention. Give large quantities of water and induce vomiting. Do not make an unconscious person vomit.

Personal Protective Methods: Wear protective gloves and clothing to prevent any reasonable probability of skin contact. Safety equipment suppliers/manufacturers can provide recommendations on the most protective glove/clothing material for your operation. All protective clothing (suits, gloves, footwear, headgear) should be clean, available each day, and put on before work. Contact lenses should not be worn when working with this chemical. Wear splash-proof chemical goggles and face shield unless full facepiece respiratory protection is worn. Employees should wash immediately with soap when skin is wet or contaminated. Provide emergency showers and eyewash.

Respirator Selection: Where the potential exists for exposures to Butyl Ether, use a MSHA/NIOSH approved supplied-air respirator with a full facepiece operated in the positive pressure mode or with a full facepiece, hood, or helmet in the continuous flow mode, or use a MSHA/NIOSH approved self-contained breathing apparatus with a full facepiece operated in pressure-demand or other positive pressure mode.

Storage: Prior to working with butyl ether you should be trained on its proper handling and storage. Before entering confined space where this chemical may be present, check to make sure that an explosive concentration does not exist. Butyl Ether must be stored in a cool, dark place, separated from oxidizers (such as perchlorates, peroxides, permanganates, chlorates, and nitrates) since violent reactions occur. Store in tightly closed containers in a cool, well-ventilated area. Protect storage containers from physical damage. Sources of ignition, such as smoking and open flames, are prohibited where Butyl Ether is handled, used or stored. Metal containers involving the transfer of 5 gallons or more of butyl ether should be grounded and bonded. Drums must be equipped with self-closing valves, pressure vacuum bungs, and flame arresters. Use only non-sparking tools and equipment, especially when opening and closing containers of Butyl Ether.

Shipping: Dibutyl Ethers require a "Flammable Liquid" label. They fall in Hazard Class 3 and Packing Group III.

Spill Handling: Evacuate and restrict persons not wearing protective equipment from area of spill or leak until cleanup is complete. Remove all ignition sources. Establish forced ventilation to keep levels below explosive limit. Absorb liquids in vermiculite, dry sand, earth, or a similar material and deposit in sealed containers. It may be necessary to contain and dispose of this chemical as a hazardous waste. If material or contaminated runoff enters waterways, notify downstream users of potentially contaminated waters. Contact your Department of Environmental Protection or your regional office of the federal EPA for specific recommendations. If employees are required to clean-up spills, they must be properly trained and equipped. OSHA 1910.120(q) may be applicable.

Fire Extinguishing: This chemical is a combustible liquid. Poisonous gases are produced in fire. Use dry chemical, carbon dioxide, or alcohol foam extinguishers. Vapors are heavier than air and will collect in low areas. Vapors may travel long distances to ignition sources and flashback. Vapors in confined areas may explode when exposed to fire. Storage containers and parts of containers may rocket great distances, in many directions. If material or contaminated runoff enters waterways, notify downstream users of potentially contaminated waters. Notify local health and fire officials and pollution control agencies. From a secure, explosion-proof location, use water spray to cool exposed containers. If cooling streams are ineffective (venting sound increases in volume and pitch, tank discolors, or shows any signs of deforming), withdraw immediately to a secure position. If employees are

expected to fight fires, they must be trained and equipped in OSHA 1910.156.

Disposal Method Suggested: Incineration.

References

New Jersey Department of Health and Senior Services, "Hazardous Substance Fact Sheet: Butyl Ether," Trenton, NJ (April 12, 1988)

n-Butyl Glycidyl Ether

Molecular Formula: $C_7H_{14}O_2$

Common Formula: $C_4H_9OCH_2CH\text{-}CH_2$

Synonyms: *n*-BGE; BGE; 1-Butoxy-2,3-epoxypropane; (Butoxymethyl) oxiraine; 1,2-Epoxy-3-butoxy propane; 2,3-Epoxypropyl butyl ether; Gylcidy butyl ether

CAS Registry Number: 2426-08-6

RTECS Number: TX4200000

DOT ID: UN 1993

EEC Number: 603-039-00-7

Regulatory Authority

- Air Pollutant Standard Set (ACGIH)[1] (OSHA)[58] (HSE)[33] (Several States)[60] (Australia) (Israel) (Mexico) (Several Canadian Provinces)
- Carcinogen: DFG (suspected)
- TSCA: 716.120(c); 40CFR712.30(d)
- Canada, WHMIS, Ingredients Disclosure List

Cited in U.S. State Regulations: Alaska (G), California (A, G), Connecticut (A), Florida (G), Illinois (G), Maine (G), Massachusetts (G), Minnesota (G), Nevada (A), New Hampshire (G), New Jersey (G), North Dakota (A), Pennsylvania (G), Rhode Island (G, A, W), South Carolina (A), South Dakota (A, W), Tennessee (W), Utah (W), Virginia (A), West Virginia (G).

Description: n-butyl glycidyl ether, $C_4H_9OCH_2CH\text{-}CH_2$ is a colorless liquid with slight irritating odor. Boiling point = 164°C. Flash point = 58°C. Soluble in water.

Potential Exposure: NIOSH has estimated human exposures at 18,000. It is used as in the manufacture of epoxy resins, and as a stabilizer, viscosity-reducing agent, as acid acceptor for solvents, and as a chemical intermediate.

Incompatibilities: Forms explosive mixture with air. Air and light form unstable and explosive peroxides. Contact with strong oxidizers may cause fire and explosions. Contact with strong caustics may cause polymerization. Attacks some plastics and rubber.

Permissible Exposure Limits in Air: The Federal OSHA PEL[58] is 50 ppm (270 mg/m^3) TWA, averaged over an 8-hour workshift. The HSE, Australian, Israeli, Mexican, and ACGIH TWA value is 25 ppm (135 mg/m^3). NIOSH recommends a 15-minute ceiling level of 5.6 ppm (30 mg/m^3). The NIOSH IDLH level is 250 ppm. DFG lists this chemical as a suspected carcinogen with danger of skin absorption. In Canada Alberta's TWA is 25 ppm and STEL of 38 ppm. British Columbia's level is 50 ppm TWA. Ontario and Quebec levels are 25 ppm TWAEV. Several states have set guidelines or standards for butyl glycidyl ether in ambient air[60] ranging from 1.35 mg/m^3 (Connecticut and North Dakota) to 2.25 mg/m^3 (Virginia) to 3.21 mg/m^3 (Nevada).

Determination in Air: Adsorption on charcoal, workup with CS_2, analysis by gas chromatography. See NIOSH Method S-81.

Permissible Concentration in Water: No criteria set.

Routes of Entry: Inhalation, ingestion, skin and/or eye contact.

Harmful Effects and Symptoms

Short Term Exposure: Irritation of eyes, skin, and respiratory tract with wheezing and coughing. Exposure can cause headache, lightheadedness, dizziness, lack of coordination, and fainting. High levels can cause unconsciousness and even death. The oral LD_{50} rat is 2,050 mg/kg (slightly toxic).

Long Term Exposure: There is limited evidence that this chemical can cause mutations. N-butyl glycidyl ether may cause skin allergy. DFG lists danger of skin and respiratory sensitization. May cause lung and liver disorders.

Points of Attack: Eyes, skin, respiratory system, central nervous system.

Medical Surveillance: Consider the points of attack in preplacement and periodic physical examinations. Evaluation by a qualified allergist. Lung function tests. Liver function tests.

First Aid: If this chemical gets into the eyes, remove any contact lenses at once and irrigate immediately for at least 15 minutes, occasionally lifting upper and lower lids. Seek medical attention immediately. If this chemical contacts the skin, remove contaminated clothing and wash immediately with soap and water. Seek medical attention immediately. If this chemical has been inhaled, remove from exposure, begin rescue breathing (using universal precautions) if breathing has stopped and CPR if heart action has stopped. Transfer promptly to a medical facility. When this chemical has been swallowed, get medical attention. Give large quantities of water and induce vomiting. Do not make an unconscious person vomit.

Personal Protective Methods: Wear protective gloves and clothing to prevent any reasonable probability of skin contact. Safety equipment suppliers/manufacturers can provide recommendations on the most protective glove/clothing material for your operation. All protective clothing (suits, gloves, footwear, headgear) should be clean, available each day, and put on before work. Contact lenses should not be worn when working with this chemical. Wear splash-proof chemical goggles and face shield unless full facepiece respiratory protection is worn. Employees should wash immediately with soap when skin is wet or contaminated. Provide emergency showers and eyewash.

Respirator Selection: NIOSH: *56 ppm:* CCROV [any chemical cartridge respirator with organic vapor cartridge(s)]; or SA (any supplied-air respirator). *140 ppm:* SA:CF (any supplied-air respirator operated in a continuous-flow mode); or PAPROV [any powered, air-purifying respirator with organic vapor cartridge(s)]. *250 ppm:* CCRFOV [any chemical cartridge respirator with a full facepiece and organic vapor cartridge(s)] or GMFOV [any air-purifying, full-facepiece respirator (gas mask) with a chin-style, front- or back-mounted acid gas canister]; or PAPRTOV [any powered, air-purifying respirator with a tight-fitting facepiece and organic vapor cartridge(s)]; or SCBAF (any self-contained breathing apparatus with a full facepiece); or SAF (any supplied-air respirator with a full facepiece). *Emergency or planned entry into unknown concentrations or IDLH conditions:* SCBAF:PD,PP (any self-contained breathing apparatus that has a full facepiece and is operated in a pressure-demand or other positive-pressure mode); or SAF:PD,PP:ASCBA (any supplied-air respirator that has a full facepiece and is operated in a pressure-demand or other positive-pressure mode in combination with an auxiliary self-contained breathing apparatus operated in a pressure-demand or other positive pressure mode). *Escape:* GMFOV [any air-purifying, full-facepiece respirator (gas mask) with a chin-style, front-or back-mounted organic vapor canister]; or SCBAE (any appropriate escape-type, self-contained breathing apparatus).

Note: Substance reported to cause eye irritation or damage; may require eye protection.

Storage: Prior to working with this chemical you should be trained on its proper handling and storage. Before entering confined space where n-butyl glycidyl ether may be present, check to make sure that an explosive concentration does not exist. Store in a fireproof refrigerator in tightly closed containers under an inert atmosphere,[52] separated from strong oxidants, strong bases, strong acids. Metal containers involving the transfer of this chemical should be grounded and bonded. Where possible, automatically pump liquid from drums or other storage containers to process containers. Drums must be equipped with self-closing valves, pressure vacuum bungs, and flame arresters. Use only non-sparking tools and equipment, especially when opening and closing containers of this chemical. Sources of ignition such as smoking and open flames, are prohibited where this chemical is used, handled, or stored in a manner that could create a potential fire or explosion hazard. A regulated, marked area should be established where this chemical is handled, used, or stored in compliance with OSHA standard 1910.1045.

Shipping: Butyl glycidyl ether can be classified as a combustible liquid n.o.s.

Spill Handling: Evacuate and restrict persons not wearing protective equipment from area of spill or leak until cleanup is complete. Remove all ignition sources. Absorb spills with paper or other absorbent material. Seal in vapor-tight plastic bags or sealed containers. Ventilate area of spill or leak after clean-up is complete. It may be necessary to contain and dispose of this chemical as a hazardous waste. If material or contaminated runoff enters waterways, notify downstream users of potentially contaminated waters. Contact your Department of Environmental Protection or your regional office of the federal EPA for specific recommendations. If employees are required to clean-up spills, they must be properly trained and equipped. OSHA 1910.120(q) may be applicable.

Fire Extinguishing: Use dry chemical, carbon dioxide, water spray, or alcohol or polymer foam extinguishers. Poisonous gases are produced in fire including carbon monoxide. If material or contaminated runoff enters waterways, notify downstream users of potentially contaminated waters. Notify local health and fire officials and pollution control agencies. From a secure, explosion-proof location, use water spray to cool exposed containers. If cooling streams are ineffective (venting sound increases in volume and pitch, tank discolors, or shows any signs of deforming), withdraw immediately to a secure position. If employees are expected to fight fires, they must be trained and equipped in OSHA 1910.156.

Disposal Method Suggested: Incineration.[24]

References

National Institute for Occupational Safety and Health, Criteria for a Recommended Standard: Occupational Exposure to Glycidyl Ethers, NIOSH Doc. No. 78–166, Washington, DC (1978)

National Institute for Occupational Safety and Health, Information Profiles on Potential Occupational Hazards: Clycidyl Ethers, Report PB 276–678, Rockville, MD, pp 116–123 (October 1977)

New Jersey Department of Health and Senior Services, Hazardous Substance Fact Sheet, n-butyl Glycidyl Ether, Trenton NJ (December, 1998)

Butyl Isovalerate

Molecular Formula: $C_9H_{18}O_2$

Common Formula: $CH_3CH_2CH(CH_3)COOC_4H_9$

Synonyms: Butanoic acid, 3-methyl-, butyl ester; *n*-Butyl isopentanoate; *n*-Butyl isovalerate; Butyl isovalerianate; Butyl 3-methyl-butyrate; Isovaleric acid, butyl ester

CAS Registry Number: 109-19-3

RTECS Number: NY1502000

Cited in U.S. State Regulations: Massachusetts (G), New Hampshire (G).

Description: Butyl Isovalerate, $CH_3CH_2CH(CH_3)COOC_4H_9$, $C_9H_{18}O_2$, is a clear liquid. Boiling point = 150°C. Flash point = 53°C. Hazard Identification (based on NFPA 704 M Rating System): Health 0, Flammability 2, Reactivity 0. Insoluble in water.

Potential Exposure: May be used as a specialty solvent.

Permissible Exposure Limits in Air: No standards set.

Permissible Concentration in Water: No criteria set.

Routes of Entry: Ingestion.

Harmful Effects and Symptoms

Short Term Exposure: Symptoms include headache, muscle weakness, giddiness nausea, vomiting, confusion, delirium, coughing, labored and difficult breathing, coma and even death. This ester is a skin irritant, and has a high oral toxicity; the oral LD_{50} for rabbit is 8.2 mg/kg. Toxicity information of this chemical is grouped with N-butyl acetate. It is classified as moderately toxic. Probable oral lethal dose for humans is 0.5 – 5 g/kg (between 1 ounce and a pint) for a 150 lb person. It is a mild irritant and central nervous depressant. Also, it is less toxic than the parent alcohol.

First Aid: If this chemical gets into the eyes, remove any contact lenses at once and irrigate immediately for at least 15 minutes, occasionally lifting upper and lower lids. Seek medical attention immediately. If this chemical contacts the skin, remove contaminated clothing and wash immediately with soap and water. Seek medical attention immediately. If this chemical has been inhaled, remove from exposure, begin rescue breathing (using universal precautions) if breathing has stopped and CPR if heart action has stopped. Transfer promptly to a medical facility. When this chemical has been swallowed, get medical attention. Give large quantities of water and induce vomiting. Do not make an unconscious person vomit.

Personal Protective Methods: Wear protective gloves and clothing to prevent any reasonable probability of skin contact. Safety equipment suppliers/manufacturers can provide recommendations on the most protective glove/clothing material for your operation. All protective clothing (suits, gloves, footwear, headgear) should be clean, available each day, and put on before work. Contact lenses should not be worn when working with this chemical. Wear splash-proof chemical goggles and face shield unless full facepiece respiratory protection is worn. Employees should wash immediately with soap when skin is wet or contaminated. Provide emergency showers and eyewash.

Storage: Prior to working with this chemical you should be trained on its proper handling and storage. Before entering confined space where Butyl Isovalerate may be present, check to make sure that an explosive concentration does not exist. Store in tightly closed containers in a cool, well ventilated area. Metal containers involving the transfer of this chemical should be grounded and bonded. Where possible, automatically pump liquid from drums or other storage containers to process containers. Drums must be equipped with self-closing valves, pressure vacuum bungs, and flame arresters. Use only non-sparking tools and equipment, especially when opening and closing containers of this chemical. Sources of ignition such as smoking and open flames, are prohibited where this chemical is used, handled, or stored in a manner that could create a potential fire or explosion hazard.

Shipping: This compound is not listed in the DOT performance-oriented packaging standards[17] with regard to label requirements or quantity limitations on shipments.

Spill Handling: Evacuate and restrict persons not wearing protective equipment from area of spill or leak until cleanup is complete. Remove all ignition sources. Ventilate area of spill or leak. Absorb liquids in vermiculite, dry sand, earth, peat, carbon, or a similar material and deposit in sealed containers. It may be necessary to contain and dispose of this chemical as a hazardous waste. If material or contaminated runoff enters waterways, notify downstream users of potentially contaminated waters. Contact your Department of Environmental Protection or your regional office of the federal EPA for specific recommendations. If employees are required to clean-up spills, they must be properly trained and equipped. OSHA 1910.120(q) may be applicable.

Fire Extinguishing: Use dry chemical, carbon dioxide, water spray, or alcohol foam extinguishers. Poisonous gases are produced in fire. If material or contaminated runoff enters waterways, notify downstream users of potentially contaminated waters. Notify local health and fire officials and pollution control agencies. From a secure, explosion-proof location, use water spray to cool exposed containers. If cooling streams are ineffective (venting sound increases in volume and pitch, tank discolors, or shows any signs of deforming), withdraw immediately to a secure position. If employees are expected to fight fires, they must be trained and equipped in OSHA 1910.156.

Disposal Method Suggested: Incineration.

References

U.S. Environmental Protection Agency, "Chemical Profile: Butyl Isovalerate," Washington, DC, Chemical Emergency Preparedness Program (October 31, 1985)

Butyl Lactate

Molecular Formula: $C_6H_{11}O_3$

Common Formula: $CH_3CHOHCOO(CH_2)_3$

Synonyms: Butyl α-hydroxypropionate; Butyl lactate; 2-Hydroxypropanoic acid, butyl ester; Lactato de *n*-butilo (Spanish); Lactic acid, butyl ester

CAS Registry Number: 138-22-7

RTECS Number: OD4025000

DOT ID: UN 1993 (combustible liquid, n.o.s.)

Regulatory Authority

- Air Pollutant Standard Set (ACGIH)[1] (OSHA)[58] (HSE)[33] (Australia) (Israel) (Mexico) (Several Canadian Provinces)
- Canada, WHMIS, Ingredients Disclosure List

Cited in U.S. State Regulations: Alaska (G), California (A, G), Florida (G), Illinois (G), Maine (G), Massachusetts (G), Minnesota (G), New Hampshire (G), New Jersey (G), Pennsylvania (G), Rhode Island (G), West Virginia (G).

Description: Butyl lactate, $CH_3CHOHCOO(CH_2)_3$, is a liquid. Boiling point = 160°C. Flash point = 71°C (oc).[17] Autoignition temperature = 382°C. Hazard Identification (based on NFPA-704 M Rating System): Health 1, Flammability 2, Reactivity 0. Slightly soluble in water.

Potential Exposure: Butyl lactate is used in making paints, inks, perfumes, dry cleaning fluids, as a resin solvent in varnishes and lacquers.

Incompatibilities: Forms explosive mixture with air. Incompatible with strong oxidizers, strong bases.

Permissible Exposure Limits in Air: There is no OSHA PEL. The NIOSH and ACGIH recommended level,[58] and the HSE,[33] Australian, Mexican, and Israeli airborne limit is 5 ppm (25 mg/m^3) TWA and Israel has an Action Level of 2.5 ppm. In Canada, the Alberta, British Columbia, Ontario and Quebec limit is 5 ppm and Alberta has a STEL of 10 ppm. There is no IDLH listed.

Determination in Air: Collection by filter and analysis by gas liquid chromatography. See OSHA Method 7.

Permissible Concentration in Water: No criteria set.

Routes of Entry: Inhalation, ingestion, skin and/or eye contact.

Harmful Effects and Symptoms

Short Term Exposure: Irritation eyes, skin, nose, throat, and may cause headaches and cough. Symptoms may also include drowsiness, central nervous system depression; nausea, vomiting. At concentrations of 7 ppm with short peaks of 11 ppm, workers experienced headaches, upper respiratory system irritation and coughing. Some complained of sleepiness and headache in the evening after work and occasional nausea and vomiting was experienced. When exposures were below 1.4 ppm, however, no symptoms were manifested.

Long Term Exposure: Headaches, feeling sleepy and nausea may develop in the evening after exposure during the day.

Points of Attack: Eyes, skin, respiratory system, central nervous system.

First Aid: If this chemical gets into the eyes, remove any contact lenses at once and irrigate immediately for at least 15 minutes, occasionally lifting upper and lower lids. Seek medical attention immediately. If this chemical contacts the skin, remove contaminated clothing and wash immediately with soap and water. Seek medical attention immediately. If this chemical has been inhaled, remove from exposure, begin rescue breathing (using universal precautions) if breathing has stopped and CPR if heart action has stopped. Transfer promptly to a medical facility. When this chemical has been swallowed, get medical attention. Give large quantities of water and induce vomiting. Do not make an unconscious person vomit.

Personal Protective Methods: Wear protective gloves and clothing to prevent any reasonable probability of skin contact. Safety equipment suppliers/manufacturers can provide recommendations on the most protective glove/clothing material for your operation. All protective clothing (suits, gloves, footwear, headgear) should be clean, available each day, and put on before work. Contact lenses should not be worn when working with this chemical. Wear splash-proof chemical goggles and face shield unless full facepiece respiratory protection is worn. Employees should wash immediately with soap when skin is wet or contaminated. Provide emergency showers and eyewash.

Storage: Prior to working with this chemical you should be trained on its proper handling and storage. Before entering confined space where butyl lactate may be present, check to make sure that an explosive concentration does not exist. Store in tightly closed containers in a cool, well ventilated area. Metal containers involving the transfer of this chemical should be grounded and bonded. Where possible, automatically pump liquid from drums or other storage containers to process containers. Drums must be equipped with self-closing valves, pressure vacuum bungs, and flame arresters. Use only non-sparking tools and equipment, especially when opening and closing containers of this chemical. Sources of ignition such as smoking and open flames, are prohibited where this chemical is used, handled, or stored in a manner that could create a potential fire or explosion hazard.

Shipping: Butyl Lactate is a combustible liquid, n.o.s.

Spill Handling: Evacuate and restrict persons not wearing protective equipment from area of spill or leak until cleanup is complete. Remove all ignition sources. Ventilate area of spill or leak. Absorb liquids in vermiculite, dry sand, earth, peat, carbon, or a similar material and deposit in sealed containers. It may be necessary to contain and dispose of this chemical as a hazardous waste. If material or contaminated runoff enters waterways, notify downstream users of potentially contaminated waters. Contact your Department of Environmental Protection or your regional office of the federal EPA for specific recommendations. If employees are required to clean-up spills, they must be properly trained and equipped. OSHA 1910.120(q) may be applicable.

Fire Extinguishing: This chemical is a combustible liquid. Poisonous gases are produced in fire. Use dry chemical, carbon dioxide, or foam extinguishers. Vapors are heavier than air and will collect in low areas. Vapors may travel long distances to ignition sources and flashback. Vapors in confined areas may explode when exposed to fire. Containers may explode in fire. Storage containers and parts of containers may rocket great distances, in many directions. If material or contaminated runoff enters waterways, notify downstream users of potentially contaminated waters. Notify local health and fire officials and pollution control agencies. From a secure, explosion-proof location, use water spray to cool exposed containers. If cooling streams are ineffective (venting sound increases in volume and pitch, tank discolors, or shows any signs of deforming), withdraw immediately to a secure position. If employees are expected to fight fires, they must be trained and equipped in OSHA 1910.156.

Disposal Method Suggested: Incineration.

References

New Jersey Department of Health and Senior Services, "Hazardous Substance Fact Sheet: n-Butyl Lactate," Trenton, NJ (April 1986)

Butyl Mercaptan

Molecular Formula: $C_4H_{10}S$

Common Formula: $CH_3CH_2CH_2CH_2SH$

Synonyms: *n*-Butanethiol; 1-Butanethiol; Butanethiol; Butane-thiol; *n*-Butyl mercaptan; *n*-Butyl thioalcohol; 1-Mercaptobutane; NCI-C60866; Thiobutyl alcohol

CAS Registry Number: 109-79-5

RTECS Number: EK6300000

DOT ID: UN 2347

Regulatory Authority

- Air Pollutant Standard Set (ACGIH)[1] (OSHA)[58] (DFG)[3] (Several States)[60] (Australia) (Israel) (Mexico) (Several Canadian Provinces)
- Canada, WHMIS, Ingredients Disclosure List

Cited in U.S. State Regulations: Alaska (G), California (A, G), Connecticut (A), Florida (G), Illinois (G), Maine (G), Massachusetts (G), Minnesota (G), Nevada (A), New Hampshire (G), New Jersey (G), New York (A), North Dakota (A), Oklahoma (G), Pennsylvania (G), Rhode Island (G), South Carolina (A), Virginia (A), West Virginia (G).

Description: Butyl Mercaptan, $CH_3CH_2CH_2CH_2SH$, is a flammable, colorless liquid with a strong, skunklike odor. Boiling point = 98°C. Freezing/Melting point = -116°C. Autoignition temperature ≤225°C. Flash point = 2°C.[17] Hazard Identification (based on NFPA-704 M Rating System): Health 2, Flammability 3, Reactivity 0. The odor threshold in air is 0.00097 ppm. Slightly soluble in water.

Potential Exposure: The major use is in the production of organophosphorus compounds, thiolcarbamates; more specifically insecticides, herbicides, acaricides and defoliants.

Incompatibilities: Forms explosive mixture with air. Incompatible with strong oxidizers such as dry bleaches and nitric acid. Attacks some plastics and rubber.

Permissible Exposure Limits in Air: The OSHA PEL is 10 ppm TWA for an 8-hour workshift. The NIOSH ceiling limit is 0.5 ppm. (1.8 mg/m^3). The DFG,[3] ACGIH, Australia, Israel, and Mexico have adopted 0.5 ppm as a TWA value. The NIOSH IDLH level is 500 ppm. Canadian Provincial limits are: British Columbia: Ceiling 3 ppm; Alberta 0.5 ppm TWA *and* STEL of 1.5 ppm, Ontario and Quebec have adopted a TWAEV of 0.5 ppm. Several states have set guidelines or standards for Butyl Mercaptan in ambient air[60] ranging from 5 μg/m^3 (New York) to 15 μg/m^3 (North Dakota and South Carolina) to 25 μg/m^3 (Virginia) to 30 μg/m^3 (Connecticut) to 36 μg/m^3 (Nevada).

Determination in Air: Adsorption on Chemisorb 104, desorption with acetone, analysis by gas chromatography. See NIOSH Method S-350.

Permissible Concentration in Water: No criteria set, but EPA[32] has suggested an ambient water limit of 21 μg/l based on health effects.

Routes of Entry: Inhalation, ingestion, skin and/or eye contact.

Harmful Effects and Symptoms

In animals-narcosis, in coordination, weakness; cyanosis, pulmonary irritation, eye irritation, paralysis. The oral LD_{50} rat is 1,500 mg/kg (slightly toxic).

Short Term Exposure: Irritates the eyes, skin, and respiratory tract. Skin contact can cause a skin rash. The substance may affect the thyroid gland. High concentrations can cause weakness, nausea, dizziness, headache and confusion. Very high concentrations (above the occupational exposure limit) exposure may affect the central nervous system and cause unconsciousness.

Long Term Exposure: Repeated exposure can cause skin rash.

Points of Attack: Eyes, skin, respiratory system, central nervous system, liver, kidneys.

Medical Surveillance: Consider the points of attack in preplacement and periodic physical examinations.

First Aid: If this chemical gets into the eyes, remove any contact lenses at once and irrigate immediately for at least 15 minutes, occasionally lifting upper and lower lids. Seek medical attention immediately. If this chemical contacts the skin, remove contaminated clothing and wash immediately with soap and water. Speed in removing material from skin is of extreme importance. Shampoo hair promptly if contaminated. Seek medical attention immediately. If this chemical has been inhaled, remove from exposure, begin rescue breathing (using universal precautions) if breathing has stopped and CPR if heart action has stopped. Transfer promptly to a medical facility. When this chemical has been swallowed, get medical attention. Give large quantities of water and induce vomiting. Do not make an unconscious person vomit.

Personal Protective Methods: Wear protective gloves and clothing to prevent any reasonable probability of skin contact. Safety equipment suppliers/manufacturers can provide recommendations on the most protective glove/clothing material for your operation. All protective clothing (suits, gloves, footwear, headgear) should be clean, available each day, and put on before work. Remove clothing immediately if wet or contaminated to avoid flammability hazard. Wear splash-proof chemical goggles and face shield unless full facepiece respiratory protection is worn. Employees should wash immediately with soap when skin is wet or contaminated. Provide emergency showers and eyewash.

Respirator Selection: NIOSH: *5 ppm:* CCRFOV [any chemical cartridge respirator with a full facepiece and organic vapor cartridge(s)]; or SA (any supplied-air respirator). *12.5 ppm:* SA:CF (any supplied-air respirator operated in a continuous-flow mode); or PAPROV [any powered, air-purifying respirator with organic vapor cartridge(s)]. *25 ppm:* CCRFOV [any chemical cartridge respirator with a full facepiece and organic vapor cartridge(s)]; or GMFOV [any air-purifying, full-facepiece respirator (gas mask) with a chin-style, front- or back-mounted acid gas canister]; or PAPRTOV [any powered, air-purifying respirator with a tight-fitting facepiece and organic vapor cartridge(s)]; SCBAF (any self-contained breathing apparatus with a full facepiece); or SAF (any supplied-air respirator with a full facepiece). *500 ppm:* SA:PD,PP (any supplied-air respirator operated in a pressure-demand or other positive-pressure mode). *Emergency or planned entry into unknown concentrations or IDLH conditions:* SCBAF:PD,PP (any self-contained breathing apparatus that has a full facepiece and is operated in a pressure-demand or other positive-pressure mode); or SAF:PD,PP:ASCBA (any supplied-air respirator that has a full facepiece and is operated in a pressure-demand or other positive-pressure mode in combination with an auxiliary self-contained breathing apparatus operated in a pressure-demand or other positive pressure mode. *Escape:* GMFOV [any air-purifying, full-facepiece respirator (gas mask) with a chin-style, front-or back-mounted organic vapor canister]; or SCBAE (any appropriate escape-type, self-contained breathing apparatus).

Note: Substance reported to cause eye irritation or damage; may require eye protection.

Storage: Store in tightly closed containers in a cool, well-ventilated area away from heat, oxidizers or acids.

Shipping: Butyl Mercaptan requires a "Flammable Liquid" label. It falls in UN/DOT Hazard Class 3 and Packing Group II.[19][20]

Spill Handling: Evacuate and restrict persons not wearing protective equipment from area of spill or leak until cleanup is complete. Remove all ignition sources. Stop discharge if possible. Evacuate area in case of large spill. Ventilate area of spill or leak. Absorb liquids in vermiculite, dry sand, earth, peat, carbon, or a similar material and deposit in sealed containers. It may be necessary to contain and dispose of this chemical as a hazardous waste. If material or contaminated runoff enters waterways, notify downstream users of potentially contaminated waters. Contact your Department of Environmental Protection or your regional office of the federal EPA for specific recommendations. If employees are required to clean-up spills, they must be properly trained and equipped. OSHA 1910.120(q) may be applicable.

Fire Extinguishing: Use dry chemical, carbon dioxide, water spray, or alcohol foam (preferred) extinguishers. Water may be ineffective in fire fighting. Poisonous gases are produced in fire. If material or contaminated runoff enters waterways, notify downstream users of potentially contaminated waters. Notify local health and fire officials and pollution control agencies. From a secure, explosion-proof location, use water spray to cool exposed containers. If cooling streams are ineffective (venting sound increases in volume and pitch, tank discolors, or shows any signs of deforming), withdraw immediately to a secure position. If employees are expected to fight fires, they must be trained and equipped in OSHA 1910.156.

Disposal Method Suggested: Incineration (2,000°F) followed by scrubbing with a caustic solution.

References

National Institute for Occupational Safety and Health, Information Profiles on Potential Occupational Hazards: Report PB 276,678, Rockville, MD pp 169–176 (October 1977)

New Jersey Department of Health and Senior Services, "Hazardous Substance Fact Sheet: Butyl Mercaptan," Trenton, NJ (September, 1996)

Sax, N. I., Ed., "Dangerous Properties of Industrial Materials Report" 1, No. 6, 39–40 (1981)

Butyl Methacrylate

Molecular Formula: $C_8H_{14}O_2$

Synonyms: Butilmetacrilato (Italian); Butyl methacrylaat (Dutch); *n*-Butyl methacrylate; Butyl 2-methacrylate; *n*-Butyl α-methylacrylate; Butyl 2-methyl-2-propenoate; EEC No. 607-033-00-5; EINECS No. 202-615-1; Metacrilato de *n*-butilo (Spanish); Methacrylate de butyle (French); Methacrylic acid, butyl ester; Methacrylsaeure butyl ester (German); 2-Methyl butylacrylate; 2-Propenic acid, 2-methyl-, butyl ester

CAS Registry Number: 97-88-1

RTECS Number: OZ3675000

DOT ID: UN 2227

EEC Number: 607-033-00-5

EINECS Number: 202-615-1

Regulatory Authority

TSCA 40CFR716.120(a)

Cited in U.S. State Regulations: California (G), Florida (G), Massachusetts (G), Minnesota (G), New Jersey (G), Pennsylvania (G).

Description: Butyl methacrylate is a flammable, colorless liquid with a mild odor. Flash point = 52°C. Autoignition temperature = 294°C. Hazard Identification (based on NFPA-704 M Rating System): Health 2, Flammability 2, Reactivity 0. Insoluble in water.

Potential Exposure: Butyl methacrylate is used in resins, solvents, coatings, adhesives, dental materials and textile emulsions.

Incompatibilities: Forms and explosive mixture with air. Unless inhibitor is maintained at the proper level, oxidizers, heat, ultraviolet light, contamination, or moisture may cause polymerization. May accumulate static electrical charges and cause ignition of its vapors.

Permissible Exposure Limits in Air: None established.

Routes of Entry: Inhalation, ingestion, eye and/or skin contact. Absorbed through the skin.

Harmful Effects and Symptoms

Short Term Exposure: Contact can irritate the eyes and skin. Inhalation can irritate the respiratory tract with coughing, wheezing and/or shortness of breath. Higher exposures can cause pulmonary edema, a medical emergency that can be delayed for several hours. This can cause death.

Long Term Exposure: Butyl Methacrylate may cause skin allergy. There is limited evidence that this chemical is teratogen in animals.

Points of Attack: Skin, reproductive system.

Medical Surveillance: Evaluation by a qualified allergist. Consider chest x-ray following acute overexposure.

First Aid: If this chemical gets into the eyes, remove any contact lenses at once and irrigate immediately for at least 15 minutes, occasionally lifting upper and lower lids. Seek medical attention immediately. If this chemical contacts the skin, remove contaminated clothing and wash immediately with soap and water. Seek medical attention immediately. If this chemical has been inhaled, remove from exposure, begin rescue breathing (using universal precautions) if breathing has stopped and CPR if heart action has stopped. Transfer promptly to a medical facility. When this chemical has been swallowed, get medical attention. Give large quantities of water and induce vomiting. Do not make an unconscious person vomit. Medical observation is recommended for 24 – 48 hours after breathing overexposure, as pulmonary edema may be delayed. As first aid for pulmonary edema, a doctor or authorized paramedic may consider administering a corticosteroid spray.

Personal Protective Methods: Wear solvent-resistant gloves and clothing to prevent any reasonable probability of skin contact. Safety equipment suppliers/manufacturers can provide recommendations on the most protective glove/ clothing material for your operation. All protective clothing (suits, gloves, footwear, headgear) should be clean, available each day, and put on before work. Contact lenses should not be worn when working with this chemical. Wear splash-proof chemical goggles and face shield unless full facepiece respiratory protection is worn. Employees should wash immediately with soap when skin is wet or contaminated. Remove nonimpervious clothing immediately if wet or contaminated. Provide emergency showers and eyewash.

Respirator Selection: Engineering controls should be used wherever feasible to maintain airborne concentrations of this chemical below the prescribed exposure limit. Respirators and protective equipment are less effective than engineering controls, and should be used only in non-routine or emergency situations which may result in exposure concentrations in excess of the TWA environmental limit. *At any concentrations above the NIOSH REL, or where there is no REL, at any detectable concentration:* SCBAF:PD,PP (any MSHA/NIOSH approved self-contained breathing apparatus that has a full facepiece and is operated in a pressure-demand or other positive-pressure mode); or SAF:PD,PP:ASCBA (any supplied-air respirator that has a full facepiece and is operated in a pressure-demand or other positive-pressure mode in combination with an auxiliary, self-contained breathing apparatus operated in a pressure-demand or other positive pressure mode). *Escape:* GMFOVHiE [any air-purifying, full-facepiece respirator (gas mask) with a chin-style, front- or back-mounted organic vapor canister having a high-efficiency particulate filter]; or SCBAE (any appropriate escape-type, self-contained breathing apparatus).

Storage: Protect against physical damage. Before entering confined space where this chemical may be present, check to make sure that an explosive concentration does not exist. Outside or detached storage is preferred. Prior to working with Butyl Methacrylate you should be trained on its proper handling and storage. Store in tightly closed containers in a cool, well ventilated area away from incompatible materials listed above, light and heat. Butyl Methacrylate should be kept refrigerated and inhibited with 10 ppm hydroquinone monomethylether. Metal containers involving the transfer of this chemical should be grounded and bonded. Drums must be equipped with self-closing valves, pressure vacuum bungs, and flame arresters. Use only non-sparking tools and equipment, especially when opening and closing containers of this chemical. Sources of ignition such as smoking and open flames, are prohibited where this chemical is used, handled, or stored in a manner that could create a potential fire or explosion hazard.

Shipping: Butyl Methacrylate requires a "Flammable Liquid " label. It is in UN/DOT Hazard Class 3, Packing Group II

Spill Handling: Evacuate and restrict persons not wearing protective equipment from area of spill or leak until cleanup is complete. Remove all ignition sources. Establish forced ventilation to keep levels below explosive limit. Absorb liquid in vermiculite, dry sand, earth or similar material and deposit in sealed containers. Ventilate area of spill or leak after cleanup is complete. It may be necessary to contain and dispose of this chemical as a hazardous waste. If material or contaminated runoff enters waterways, notify downstream users of potentially contaminated waters. Contact your Department of Environmental Protection or your regional office of the federal EPA for specific recommendations. If employees are required to clean-up spills, they must be properly trained and equipped. OSHA 1910.120(q) may be applicable.

Fire Extinguishing: This chemical is a flammable liquid. Poisonous gases are produced in fire. Use dry chemical, carbon dioxide, or foam extinguishers. Vapors are heavier than air and will collect in low areas. Vapors may travel long distances to ignition sources and flashback. Vapors in confined areas may explode when exposed to fire. Containers may explode in fire.

Storage containers and parts of containers may rocket great distances, in many directions. If material or contaminated run-off enters waterways, notify downstream users of potentially contaminated waters. Notify local health and fire officials and pollution control agencies. Containers may explode in fire. From a secure, explosion-proof location, use water spray to cool exposed containers. If cooling streams are ineffective (venting sound increases in volume and pitch, tank discolors, or shows any signs of deforming), withdraw immediately to a secure position. If employees are expected to fight fires, they must be trained and equipped in OSHA 1910.156.

Disposal Method Suggested: Incineration.

References

New Jersey Department of Health and Senior Services, "Hazardous Substance Fact Sheet: Butyl Methacrylate," Trenton, NJ (June, 1996)

Butylphenols

Molecular Formula: $C_{10}H_{14}O$

Common Formula: $C_4H_9C_6H_4OH$

Synonyms: *o-n-:* 2-*n*-Butylphenol

o-sec-: 2-*sec*-Butylfenol (Czech); *o,sec* Butylphenol; 2-*sec*-Butylphenol

o-tert-: 2-*t*-Butylphenol; Phenol, *o*-(*tert*-butyl)-

p-sec-: *p,sec*-Butylphenol; 4-*sec*-Butylphenol

p-tert-: *p,tert*-Butylfenol (Czech); Butylphen; *p,ter*-Butylphenol; 4-*t*-Butylphenol; 4-*tert*-Butylphenol; 4-(1,1-Demethylethyl)phenol; 1-Hydroxy-4-*tert*-butylbenzene; UCAR butylphenol 4-*t*

CAS Registry Number: 3180-09-4 (o-); 89-72-5 (o-sec-); 99-71-8 (p-sec-); 4074-43-5 (m-); 88-18-6 (o-tert-); 98-54-4 (p-tert-); 1638-22-8 (p-); 28805-86-9 (mixed isomers)

RTECS Number: SJ8850000 (o-n-); SJ8920000 (o-sec-); SJ8810000 (m-); SJ8924000 (p-sec-); SJ8925000 (p-tert-); SJ8922500 (p-n-)

DOT ID: UN 2228 (butyphenols, liquid); UN 2229 (butyphenols, solid)

Regulatory Authority

- *o-sec-:* Air Pollutant Standard Set (ACGIH)[1] (OSHA)[58] (DFG)[3] (HSE)[33] (Several States)[60] (Australia) (Israel) (Several Canadian Provinces)
- U.S. DOT 49CFR172.101, Appendix B, Regulated marine pollutant (as butyl phenol)
- Canada, WHMIS, Ingredients Disclosure List (o-sec-; p-tert-; p-; o-; m-)

Cited in U.S. State Regulations: Alaska (G), California (A, G), Connecticut (A), Florida (G), Maine (G), Massachusetts (G), Minnesota (G), Nevada (A), New Hampshire (G), New Jersey (G), New York (G), North Dakota (A), Pennsylvania (G), Rhode Island (G), Virginia (A), West Virginia (G). All refer to o-sec-butylphenol except NY which is p-tert-butylphenol. NJ also refers to butylphenol (total) SN 0249 and SN 1440.

Description: The butylphenols, $C_4H_9C_6H_4OH$ include a number of isomers, the two most highly regulated are o-sec-butylphenol and p-tert-butylphenol. Their properties are as follows: *o-sec-:* Colorless liquid or solid (below 61°F). Boiling point = 108°C. Flash point = 108°C. Insoluble in water. *p-tert-:* White crystalline solid. Freezing/Melting point = 97°C. Hazard Identification (based on NFPA-704 M Rating System): Health 1, Flammability 1, Reactivity 0. Insoluble in water.

Potential Exposure: Butylphenols may be used as intermediates in manufacturing varnish and lacquer resins; as a germicidal agent in detergent disinfectants; as a pour point depressant, motor-oil additive, de-emulsifier for oil, soap-antioxidant, plasticizer, fumigant and insecticide.

Incompatibilities: Incompatible with strong acids, caustics, aliphatic amines, amides, oxidizers.

Permissible Exposure Limits in Air: *o-sec- isomer:* There is no OSHA PEL for the o-sec- isomer. The NIOSH recommended REL is 5 ppm (30 mg/m^3) TWA with the notation that skin absorption is of concern. The ACGIH, Australia, HSE, and Israel TWA is the same as NIOSH. In Canada Alberta, Ontario, and Quebec TWA is 5 ppm and Alberta's STEL is 10 ppm. (p-tert- isomer) The DFG[3] has set a MAK of 0.08 ppm (0.5 mg/m^3) and Peak Limitation of 5 times the normal MAK (30 min), not to be exceeded 2 times during a workshift. Mexico's TWA is 10 ppm and STEL of 20 ppm. Several states have set guidelines or standards for o-sec-butylphenol in ambient air[60] ranging from 300 μg/m^3 (North Dakota) to 500 μg/m^3 (Virginia) to 600 μg/m^3 (Connecticut) to 714 μg/m^3 (Nevada).

Permissible Concentration in Water: No criteria set.

Routes of Entry: Inhalation, skin absorption, ingestion, skin and/or eye contact.

Harmful Effects and Symptoms

Short Term Exposure: Inhalation may cause irritation to nose, throat and lungs. Sensitization may occur. Skin contact studies with animals suggest that severe irritation at concentrations above 10% may occur. May cause rash, redness and irritation, especially when skin is wet. Absorption is significant and contact may lead to allergic reaction. Eye studies with animals suggest that severe irritation may occur. Ingestion studies on animals suggest that 8 oz. may be lethal to a 150 lb person.

Long Term Exposure: May cause skin color changes by contact or inhalation of levels between 10 and 100 ppm. Allergy may develop after repeated exposure. Liver damage may also occur. There is limited evidence that butylphenol causes skin cancer in animals. Repeated or prolonged skin contact can cause skin ulcers and lead to permanent loss of skin pigment in affected areas.

Points of Attack: Eyes, skin, respiratory system.

Medical Surveillance: Lung function tests. Liver function tests.

First Aid: If this chemical gets into the eyes, remove any contact lenses at once and irrigate immediately for at least 15 minutes, occasionally lifting upper and lower lids. Seek medical attention immediately. If this chemical contacts the skin, remove contaminated clothing and wash immediately with soap and water. Seek medical attention immediately. If this chemical has been inhaled, remove from exposure, begin rescue breathing (using universal precautions) if breathing has stopped and CPR if heart action has stopped. Transfer promptly to a medical facility. When this chemical has been swallowed, get medical attention. Give large quantities of water and induce vomiting. Do not make an unconscious person vomit.

Personal Protective Methods: Wear protective gloves and clothing to prevent any reasonable probability of skin contact. Safety equipment suppliers/manufacturers can provide recommendations on the most protective glove/clothing material for your operation. All protective clothing (suits, gloves, footwear, headgear) should be clean, available each day, and put on before work. Contact lenses should not be worn when working with this chemical. Wear splash- or dust-proof chemical goggles and face shield unless full facepiece respiratory protection is worn. Employees should wash immediately with soap when skin is wet or contaminated. Provide emergency showers and eyewash.

Respirator Selection: SCBAF:PD,PP (any MSHA/NIOSH approved self-contained breathing apparatus that has a full facepiece and is operated in a pressure-demand or other positive-pressure mode); or SAF:PD,PP:ASCBA (any supplied-air respirator that has a full facepiece and is operated in a pressure-demand or other positive-pressure mode in combination with an auxiliary, self-contained breathing apparatus operated in a pressure-demand or other positive pressure mode). *Escape:* HiEF [any air-purifying, full-facepiece respirator (gas mask) with a chin-style, front- or back-mounted organic vapor canister having a high-efficiency particulate filter]; or SCBAE (any appropriate escape-type, self-contained breathing apparatus).

Storage: Prior to working with butyl phenol you should be trained on its proper handling and storage. Store in tightly closed containers in a cool, well ventilated area. Metal containers involving the transfer of this chemical should be grounded and bonded. Drums must be equipped with self-closing valves, pressure vacuum bungs, and flame arresters. Use only non-sparking tools and equipment, especially when opening and closing containers of this chemical. Sources of ignition such as smoking and open flames, are prohibited where this chemical is used, handled, or stored in a manner that could create a potential fire or explosion hazard.

Shipping: Liquid butylphenols (such as o-sec-butylphenol) should have a "Keep away from Food" label as should solid butylphenols (such as p-tert-butylphenol). Both classes of material fall in UN/DOT Hazard Class 6.1 and Packing Group III.

Spill Handling: Evacuate and restrict persons not wearing protective equipment from area of spill or leak until cleanup is complete. Remove all ignition sources. Ventilate area of spill or leak. Absorb liquids in vermiculite, dry sand, earth, peat, carbon, or a similar material and deposit in sealed containers. Collect powdered material in the most convenient and safe manner and deposit in sealed containers. It may be necessary to contain and dispose of this chemical as a hazardous waste. If material or contaminated runoff enters waterways, notify downstream users of potentially contaminated waters. Contact your Department of Environmental Protection or your regional office of the federal EPA for specific recommendations. If employees are required to clean-up spills, they must be properly trained and equipped. OSHA 1910.120(q) may be applicable.

Fire Extinguishing: Butyl phenol is combustible. Use dry chemical, carbon dioxide, or foam extinguishers. Poisonous gases are produced in fire. If material or contaminated runoff enters waterways, notify downstream users of potentially contaminated waters. Notify local health and fire officials and pollution control agencies. From a secure, explosion-proof location, use water spray to cool exposed containers. If cooling streams are ineffective (venting sound increases in volume and pitch, tank discolors, or shows any signs of deforming), withdraw immediately to a secure position. If employees are expected to fight fires, they must be trained and equipped in OSHA 1910.156.

References

New Jersey Department of Health and Senior Services, "Hazardous Substance Fact Sheet: o-sec-Butylphenol," Trenton, NJ (April 1986)

New Jersey Department of Health and Senior Services, "Hazardous Substance Fact Sheet: Butylphenol," Trenton, NJ (May 1986)

New Jersey Department of Health and Senior Services, "Hazardous Substance Fact Sheet: o-sec-Butylphenol," Trenton, NJ (April 1986)

New York State Department of Health, "Chemical Fact Sheet: p-tert-Butylphenol," Albany, NY, Bureau of Toxic Substance Assessment (March 1986)

Butyl Propionate

Molecular Formula: $C_7H_{14}O_2$

Common Formula

$C_2H_5COOC_4H_9$

Synonyms: Butyl propanoate; Propanoic acid butyl ester; Propanoic acid butyl ester (9CI); Propionic acid butyl ester

CAS Registry Number: 590-01-2

RTECS Number: UE8245000

DOT ID: UN 1914

EEC Number: 607-029-00-3

Regulatory Authority

Canada, WHMIS, Ingredients Disclosure List

Cited in U.S. State Regulations: Florida (G), Massachusetts (G), New Jersey (G), Pennsylvania (G).

Description: Butyl Propionate, $C_2H_5COOC_4H_9$, is a flammable, colorless to straw-yellow liquid with an apple-like odor. Boiling point = 146°C. Freezing/Melting point= -90°C. Flash point = 32°C. Autoignition temperature = 426°C. Hazard Identification (based on NFPA-704 M Rating System): Health 2, Flammability 3, Reactivity 0. Insoluble in water.

Potential Exposure: It is used as a solvent or lacquer thinner, and in perfumes and flavorings.

Incompatibilities: Forms explosive mixture with air. Incompatible with strong oxidizers, strong acids.

Permissible Exposure Limits in Air: None established.

Routes of Entry: Inhalation, ingestion, eye and/or skin contact.

Harmful Effects and Symptoms

Short Term Exposure: The substance irritates the eyes, the skin and the respiratory tract.

Long Term Exposure: Unknown at this time.

First Aid: If this chemical gets into the eyes, remove any contact lenses at once and irrigate immediately for at least 15 minutes, occasionally lifting upper and lower lids. Seek medical attention immediately. If this chemical contacts the skin, remove contaminated clothing and wash immediately with soap and water. Seek medical attention immediately. If this chemical has been inhaled, remove from exposure, begin rescue breathing (using universal precautions) if breathing has stopped and CPR if heart action has stopped. Transfer promptly to a medical facility. When this chemical has been swallowed, get medical attention. Give large quantities of water and induce vomiting. Do not make an unconscious person vomit.

Personal Protective Methods: Wear solvent-resistant gloves and clothing to prevent any reasonable probability of skin contact. Safety equipment suppliers/manufacturers can provide recommendations on the most protective glove/clothing material for your operation. All protective clothing (suits, gloves, footwear, headgear) should be clean, available each day, and put on before work. Contact lenses should not be worn when working with this chemical. Wear splash-proof chemical goggles and face shield unless full facepiece respiratory protection is worn. Employees should wash immediately with soap when skin is wet or contaminated. Remove nonimpervious clothing immediately if wet or contaminated. Provide emergency showers and eyewash.

Respirator Selection: Engineering controls should be used wherever feasible to maintain airborne concentrations of this chemical below the prescribed exposure limit. Respirators and protective equipment are less effective than engineering controls, and should be used only in non-routine or emergency situations which may result in exposure concentrations in excess of the TWA environmental limit. *At any concentrations above the NIOSH REL, or where there is no REL, at any detectable concentration:* SCBAF:PD,PP (any MSHA/NIOSH approved self-contained breathing apparatus that has a full facepiece and is operated in a pressure-demand or other positive-pressure mode); or SAF:PD,PP:ASCBA (any supplied-air respirator that has a full facepiece and is operated in a pressure-demand or other positive-pressure mode in combination with an auxiliary, self-contained breathing apparatus operated in a pressure-demand or other positive pressure mode). *Escape:* GMFOVHiE [any air-purifying, full-facepiece respirator (gas mask) with a chin-style, front- or back-mounted organic vapor canister having a high-efficiency particulate filter]; or SCBAE (any appropriate escape-type, self-contained breathing apparatus).

Storage: Protect against physical damage. Outside or detached storage is preferred. Prior to working with Butyl Propionate you should be trained on its proper handling and storage. Store in tightly closed containers in a cool, well ventilated area away from incompatible materials listed above. Metal containers involving the transfer of this chemical should be grounded and bonded. Drums must be equipped with self-closing valves, pressure vacuum bungs, and flame arresters. Use only non-sparking tools and equipment, especially when opening and closing containers of this chemical. Sources of ignition such as smoking and open flames, are prohibited where this chemical is used, handled, or stored in a manner that could create a potential fire or explosion hazard.

Shipping: Butyl Propionate requires a "Flammable Liquid " label. It is in UN/DOT Hazard Class 3, Packing Group II

Spill Handling: Evacuate and restrict persons not wearing protective equipment from area of spill or leak until cleanup is complete. Remove all ignition sources. Ventilate area of spill or leak. Absorb liquid in vermiculite, dry sand, earth or similar material and deposit in sealed containers. Ventilate area of spill or leak after clean-up is complete. It may be necessary to contain and dispose of this chemical as a hazardous waste. If material or contaminated runoff enters waterways, notify downstream users of potentially contaminated waters. Contact your Department of Environmental Protection or your regional office of the federal EPA for specific recommendations. If employees are required to clean-up spills, they must be properly trained and equipped. OSHA 1910.120(q) may be applicable.

Fire Extinguishing: This chemical is a combustible liquid. Poisonous gases, including carbon monoxide, are produced in fire. Use dry chemical, carbon dioxide, or alcohol foam extinguishers. Vapors are heavier than air and will collect in low areas. Vapors may travel long distances to ignition sources and flashback. Vapors in confined areas may explode when exposed to fire. Containers may explode in fire. Storage containers and parts of containers may rocket great distances, in many directions. If material or contaminated runoff enters waterways, notify downstream users of potentially contaminated waters. Notify local health and fire officials and pollution control agencies. Containers may explode in fire. From a secure, explosion-proof location, use water spray to cool exposed containers. If cooling streams are ineffective (venting sound increases in volume and pitch, tank discolors, or shows any signs of deforming), withdraw immediately to a secure position. If employees are expected to fight fires, they must be trained and equipped in OSHA 1910.156.

Disposal Method Suggested: Incineration.

References

New Jersey Department of Health and Senior Services, "Hazardous Substance Fact Sheet: Butyl Propionate," Trenton, NJ (November, 1998)

p-tert-Butyltoluene

Molecular Formula: $C_{11}H_{16}$

Common Formula: p-$CH_3C_6H_4C_4H_9$

Synonyms: *p*-Methyl-*tert*-Butylbenzene; 1-Methyl-4-*tert*-butylbenzene; TBT

CAS Registry Number: 98-51-1

RTECS Number: XS8400000

DOT ID: UN 2667

Regulatory Authority

- Air Pollutant Standard Set (ACGIH)[1] (OSHA)[58] (DFG)[3] (Several States)[60] (Australia) (Israel) (Several Canadian Provinces)
- Canada, WHMIS, Ingredients Disclosure List

Cited in U.S. State Regulations: Alaska (G), California (A, G), Connecticut (A), Florida (G), Illinois (G), Maine (G), Massachusetts (G), Minnesota (G), Nevada (A), New Jersey (G), New Hampshire (G), North Dakota (A), Oklahoma (G), Pennsylvania (G), Rhode Island (G), Virginia (A), West Virginia (G).

Description: p-tert-Butyltoluene, $CH_3C_6H_4C_4H_9$, is a colorless liquid with an aromatic gasoline like odor. Boiling point = 193 – 194°C. Boiling point = 193°C. Freezing/Melting point = -52°C. Flash point = 24°C. Insoluble in water.

Potential Exposure: This material is used as a solvent for resins and as an intermediate in organic synthesis.

Incompatibilities: Forms explosive mixture with air. Reacts with strong oxidizers. May accumulate static electrical charges, and may cause ignition of its vapors.

Permissible Exposure Limits in Air: The Federal OSHA and NIOSH[58] and ACGIH[1] and DFG[3] value is 10 ppm (60 mg/m^3). The NIOSH recommended STEL (1,58) is 20 ppm (120 mg/m^3). The NIOSH IDLH level is 100 ppm. Virginia has set a limit in ambient air[60] of 1.0 mg/m^3. Several states have set guidelines or standards for this compound in ambient air[60] ranging from 600 to 1,200 μg/m^3 (North Dakota) to 1,000 μg/m^3 (Virginia) to 1,200 μg/m^3 (Connecticut) to 1,429 μg/m^3 (Nevada).

Determination in Air: Adsorption on charcoal, workup with CS_2, analysis by gas chromatography. See NIOSH Method 1501 Aromatic hydrocarbons.[18]

Permissible Concentration in Water: No criteria set.

Routes of Entry: Inhalation, ingestion, eye, and skin contact.

Harmful Effects and Symptoms

Short Term Exposure: Irritates the eyes, skin, and respiratory tract; dry nose and throat and headaches; low blood pressure; tachycardia; abnormal cardiovascular system behavior; central nervous system depression; hematopoietic depression. The oral LD_{50} rat is 1,500 mg/kg (slightly toxic). This chemical may cause effects on the central nervous system.

Long Term Exposure: The liquid defats the skin, and may have effects on the liver and kidneys.

Points of Attack: Eyes, skin, respiratory system, cardiovascular system, central nervous system, bone marrow, liver, kidneys.

Medical Surveillance: Consider the points of attack in preplacement and periodic physical examinations.

First Aid: If this chemical gets into the eyes, remove any contact lenses at once and irrigate immediately for at least 15 minutes, occasionally lifting upper and lower lids. Seek medical attention immediately. If this chemical contacts the skin, remove contaminated clothing and wash immediately with soap and water. Seek medical attention immediately. If this chemical has been inhaled, remove from exposure, begin rescue breathing (using universal precautions) if breathing has stopped and CPR if heart action has stopped. Transfer promptly to a medical facility. When this chemical has been swallowed, get medical attention. Give large quantities of water and induce vomiting. Do not make an unconscious person vomit.

Personal Protective Methods: Wear appropriate clothing to prevent repeated or prolonged skin contact. Wear eye protection to prevent any reasonable probability of eye contact. Employees should wash promptly when skin is wet or contaminated. Remove nonimpervious clothing promptly if wet or contaminated.

Respirator Selection: NIOSH/OSHA: *100 ppm:* SA:CF (any supplied-air respirator operated in a continuous-flow mode); PAPROV [any powered, air-purifying respirator with organic vapor cartridge(s)]; or CCRFOV [any chemical cartridge respirator with a full facepiece and organic vapor cartridge(s)]; GMFOV [any air-purifying, full-facepiece respirator (gas mask) with a chin-style, front- or back-mounted acid gas canister]; or SCBAF (any self-contained breathing apparatus with a full facepiece); or SAF (any supplied-air respirator with a full facepiece). *Emergency or planned entry into unknown concentrations or IDLH conditions:* SCBAF:PD,PP (any self-contained breathing apparatus that has a full facepiece and is operated in a pressure-demand or other positive-pressure mode); or SAF:PD,PP:ASCBA (any supplied-air respirator that has a full facepiece and is operated in a pressure-demand or other positive-pressure mode in combination with an auxiliary self-contained breathing apparatus operated in a pressure-demand or other positive pressure mode. *Escape:* GMFOV [any air-purifying, full-facepiece respirator (gas mask) with a chin-style, front-or back-mounted organic vapor canister]; or SCBAE (any appropriate escape-type, self-contained breathing apparatus).

Note: Substance causes eye irritation or damage; eye protection needed.

Storage: Prior to working with this chemical you should be trained on its proper handling and storage. Store in tightly closed containers in a cool, well ventilated area. Sources of ignition such as smoking and open flames, are prohibited where this chemical is used, handled, or stored in a manner that could create a potential fire or explosion hazard.

Shipping: Butyltoluenes fall in UN/DOT Hazard Class 6.1 and Packing Group III.[19][20] They require a "Keep away from Food" label.

Spill Handling: Evacuate and restrict persons not wearing protective equipment from area of spill or leak until cleanup is complete. Remove all ignition sources. Ventilate area of spill or leak. Absorb liquids in vermiculite, dry sand, earth, peat, carbon, or a similar material and deposit in sealed containers. It may be necessary to contain and dispose of this chemical as a hazardous waste. If material or contaminated runoff enters waterways, notify downstream users of potentially contaminated waters. Contact your Department of Environmental Protection or your regional office of the federal EPA for specific recommendations. If employees are required to clean-up spills, they must be properly trained and equipped. OSHA 1910.120(q) may be applicable.

Fire Extinguishing: Use dry chemical, carbon dioxide, water spray, or alcohol foam extinguishers. Poisonous gases are produced in fire. If material or contaminated runoff enters waterways, notify downstream users of potentially contaminated waters. Notify local health and fire officials and pollution control agencies. From a secure, explosion-proof location, use water spray to cool exposed containers. If cooling streams are ineffective (venting sound increases in volume and pitch, tank discolors, or shows any signs of deforming), withdraw immediately to a secure position. If employees are expected to fight fires, they must be trained and equipped in OSHA 1910.156.

Disposal Method Suggested: Incineration, preferably in admixture with a more flammable solvent.[24]

Butyl Trichlorosilane

Molecular Formula: $C_4H_9Cl_3S$

Common Formula: $CH_3(CH_2)_3SiCl_3$

Synonyms: *n*-Butiltriclorosilano (Spanish); Butylsilicon trichloride; *n*-Butyltrichlorosilane

CAS Registry Number: 7521-80-4

RTECS Number: VV2080000

DOT ID: UN 1747

Regulatory Authority

- Canada, WHMIS, Ingredients Disclosure List

Cited in U.S. State Regulations: Florida (G), Maine (G), Massachusetts (G), New Jersey (G), Oklahoma (G), Pennsylvania (G).

Description: Butyl trichlorosilane, $CH_3(CH_2)_3SiCl_3$, is a colorless liquid. Boiling point = 149°C. Flash point = 54°C (oc).[17] Hazard Identification (based on NFPA-704 M Rating System): Health 2, Flammability 2, Reactivity 0. Insoluble in water; reacts violently.

Potential Exposure: This is a raw material for silicone resin production.

Incompatibilities: Forms explosive mixture with air. Violent reaction with water forming hydrochloric acid and fumes. Contact with strong oxidizers may cause fire and explosions. Attacks metals in a moist environment.

Permissible Exposure Limits in Air: No standards set. Reference can be made to the entry on Hydrogen Chloride since HCl is the hydrolysis product.

Permissible Concentration in Water: No criteria set.

Routes of Entry: Inhalation, skin, eyes.

Harmful Effects and Symptoms

Short Term Exposure: Butyl Trichlorosilane is a corrosive chemical and can cause severe eye burns leading to permanent damage. Contact can cause severe skin burns. Exposure to vapors can irritate the eyes, nose and throat. Butyl Trichlorosilane can affect you when breathed in. Exposure can irritate the lungs causing coughing and/or shortness of breath. Higher exposure can cause a build-up of the fluid in the lungs (pulmonary edema). This can cause death.

Long Term Exposure: Repeated exposure may cause bronchitis to develop with cough, phlegm, and/or shortness of breath.

Points of Attack: Lungs.

Medical Surveillance: Lung function tests. Consider chest x-ray following acute overexposure.

First Aid: If this chemical gets into the eyes, remove any contact lenses at once and irrigate immediately for at least 30 minutes, occasionally lifting upper and lower lids. Seek medical attention immediately. If this chemical contacts the skin, remove contaminated clothing and wash immediately with soap and water. Seek medical attention immediately. If this chemical has been inhaled, remove from exposure, begin rescue breathing (using universal precautions) if breathing has stopped and CPR if heart action has stopped.

Transfer promptly to a medical facility. When this chemical has been swallowed, get medical attention. If victim is conscious, administer water or milk. Do not induce vomiting. Medical observation is recommended for 24 – 48 hours after breathing overexposure, as pulmonary edema may be delayed. As first aid for pulmonary edema, a doctor or authorized paramedic may consider administering a corticosteroid spray.

Personal Protective Methods: Where possible, enclose operations and use local exhaust ventilation at the site of chemical release. Wear protective gloves and clothing to prevent any reasonable probability of skin contact. Safety equipment suppliers/manufacturers can provide recommendations on the most protective glove/clothing material for your operation. All protective clothing (suits, gloves, footwear, headgear) should be clean, available each day, and put on before work. Contact lenses should not be worn when working with this chemical. Wear splash-proof chemical goggles and face shield unless full facepiece respiratory protection is worn. Employees should wash immediately with soap when skin is wet or contaminated. Provide emergency showers and eyewash. Post hazard and warning information in the work area. In addition, as part of an ongoing education and training effort, communicate all information on the health and safety hazards of Butyl Trichlorosilane to potentially exposed workers.

Respirator Selection: Engineering controls should be used wherever feasible to maintain airborne concentrations of this chemical below the prescribed exposure limit. Respirators and protective equipment less effective than engineering controls, and should be used only in non-routine or emergency situations which may result in exposure concentrations in excess of the TWA environmental limit. *At any concentrations above the NIOSH REL, or where there is no REL, at any detectable concentration:* SCBAF:PD,PP (any MSHA/NIOSH approved self-contained breathing apparatus that has a full facepiece and is operated in a pressure-demand or other positive-pressure mode); or SAF:PD,PP:ASCBA (any supplied-air respirator that has a full facepiece and is operated in a pressure-demand or other positive-pressure mode in combination with an auxiliary, self-contained breathing apparatus operated in a pressure-demand or other positive pressure mode). *Escape:* GMFAG [any air-purifying, full-facepiece respirator (gas mask) with a chin-style, front-or back-mounted organic vapor canister]; or SCBAE (any appropriate escape-type, self-contained breathing apparatus).

Note: Substance reported to cause eye irritation or damage; may require eye protection.

Storage: Prior to working with butyl trichlorosilane you should be trained on its proper handling and storage. Store in tightly closed containers in a cool, well ventilated area away from moisture and incompatible materials listed above. Metal containers involving the transfer of this chemical should be grounded and bonded. Drums must be equipped with self-closing valves, pressure vacuum bungs, and flame arresters. Use only non-sparking tools and equipment, especially when opening and closing containers of this chemical. Sources of ignition such as smoking and open flames, are prohibited where this chemical is used, handled, or stored in a manner that could create a potential fire or explosion hazard.

Shipping: Butyl Trichlorosilane falls in UN/DOT Hazard Class 8 and Packing Group II.[19][20] It requires a "Corrosive" label. Shipment by passenger aircraft or railcar is forbidden.

Spill Handling: Evacuate and restrict persons not wearing protective equipment from area of spill or leak until cleanup is complete. Remove all ignition sources. Ventilate area of spill or leak. Absorb liquids in vermiculite, dry sand, earth, peat, carbon, or a similar material and deposit in sealed containers. It may be necessary to contain and dispose of this chemical as a hazardous waste. If material or contaminated runoff enters waterways, notify downstream users of potentially contaminated waters. Contact your Department of Environmental Protection or your regional office of the federal EPA for specific recommendations. If employees are required to clean-up spills, they must be properly trained and equipped. OSHA 1910.120(q) may be applicable.

Fire Extinguishing: This chemical is a combustible liquid. Poisonous gases including hydrogen chloride, chlorine, and phosgene are produced in fire. Use dry chemical or carbon dioxide. Foam extinguishers are not recommended. Vapors are heavier than air and will collect in low areas. Vapors may travel long distances to ignition sources and flashback. Vapors in confined areas may explode when exposed to fire. Containers may explode in fire. Storage containers and parts of containers may rocket great distances, in many directions. If material or contaminated runoff enters waterways, notify downstream users of potentially contaminated waters. Notify local health and fire officials and pollution control agencies. Containers may explode in fire. From a secure, explosion-proof location, use water spray to cool exposed containers. If cooling streams are ineffective (venting sound increases in volume and pitch, tank discolors, or shows any signs of deforming), withdraw immediately to a secure position. If employees are expected to fight fires, they must be trained and equipped in OSHA 1910.156.

References

New Jersey Department of Health and Senior Services, "Hazardous Substance Fact Sheet: Butyl Trichlorosilane," Trenton, NJ (September 1998)

Butyl Vinyl Ether

Molecular Formula: $C_6H_{12}O$

Common Formula: $CH_2{=}CHOC_4H_9$

Synonyms: Butane, 1-(ethenyloxy)-; Butoxyethene; 1-(Ethenyloxy)butane; Ether, butyl vinyl; Vinyl *n*-butyl ether; Vinyl butyl ether

CAS Registry Number: 111-34-2

RTECS Number: KN5900000

DOT ID: UN 2352

Regulatory Authority

- Canada, WHMIS, Ingredients Disclosure List

Cited in U.S. State Regulations: Florida (G), Massachusetts (G), New Jersey (G), Pennsylvania (G).

Description: Butyl vinyl ether, $CH_2{=}CHOC_4H_9$, $C_6H_{12}O$, is an extremely flammable, colorless liquid with an etherlike odor. Boiling point = 94°C. Flash point= –9.4°C (oc).[17] Autoignition temperature = 255°C. Hazard Identification (based on NFPA-704 M Rating System): Health 2, Flammability 3, Reactivity 2. Slightly soluble in water.

Potential Exposure: This material may be used in organic synthesis and in copolymer manufacture.

Incompatibilities: Moderately explosive by spontaneous chemical reaction. Contact with oxidizers and strong acids may cause fire and explosions. Able to form unstable peroxides, which can cause polymerization.

Permissible Exposure Limits in Air: No standards set.

Permissible Concentration in Water: No criteria set.

Routes of Entry: Inhalation, skin and/or eye contact.

Harmful Effects and Symptoms

Short Term Exposure: May be poisonous if inhaled or absorbed through skin. Vapors may cause dizziness or suffocation. Contact may irritate or burn skin and eyes. The oral LD_{50} rat is 15,000 mg/kg (insignificantly toxic).

First Aid: If this chemical gets into the eyes, remove any contact lenses at once and irrigate immediately for at least 15 minutes, occasionally lifting upper and lower lids. Seek medical attention immediately. If this chemical contacts the skin, remove contaminated clothing and wash immediately with soap and water. Seek medical attention immediately. If this chemical has been inhaled, remove from exposure, begin rescue breathing (using universal precautions) if breathing has stopped and CPR if heart action has stopped. Transfer promptly to a medical facility. When this chemical has been swallowed, get medical attention. Give large quantities of water and induce vomiting. Do not make an unconscious person vomit.

Personal Protective Methods: Wear solvent-resistant gloves and clothing to prevent any reasonable probability of skin contact. Safety equipment suppliers/manufacturers can provide recommendations on the most protective glove/clothing material for your operation. All protective clothing (suits, gloves, footwear, headgear) should be clean, available each day, and put on before work. Contact lenses should not be worn when working with this chemical. Wear splash-proof chemical goggles and face shield unless full facepiece respiratory protection is worn. Employees should wash immediately with soap when skin is wet or contaminated. Remove nonimpervious clothing immediately if wet or contaminated. Provide emergency showers and eyewash.

Respirator Selection: Engineering controls should be used wherever feasible to maintain airborne concentrations of this chemical below the prescribed exposure limit. Respirators and protective equipment are less effective than engineering controls, and should be used only in non-routine or emergency situations which may result in exposure concentrations in excess of the TWA environmental limit. *At any concentrations above the NIOSH REL, or where there is no REL, at any detectable concentration:* SCBAF:PD,PP (any MSHA/NIOSH approved self-contained breathing apparatus that has a full facepiece and is operated in a pressure-demand or other positive-pressure mode); or SAF:PD,PP:ASCBA (any supplied-air respirator that has a full facepiece and is operated in a pressure-demand or other positive-pressure mode in combination with an auxiliary, self-contained breathing apparatus operated in a pressure-demand or other positive pressure mode). *Escape:* GMFOVHiE [any air-purifying, full-facepiece respirator (gas mask) with a chin-style, front- or back-mounted organic vapor canister having a high-efficiency particulate filter]; or SCBAE (any appropriate escape-type, self-contained breathing apparatus).

Storage: Prior to working with butyl vinyl ether you should be trained on its proper handling and storage. Store in tightly closed containers in a cool, well ventilated area. Metal containers involving the transfer of this chemical should be grounded and bonded. Drums must be equipped with self-closing valves, pressure vacuum bungs, and flame arresters. Use only non-sparking tools and equipment, especially when opening and closing containers of this chemical. Sources of ignition such as smoking and open flames, are prohibited where this chemical is used, handled, or stored in a manner that could create a potential fire or explosion hazard.

Shipping: The DOT label requirement for inhibited butyl vinyl ether is "Flammable Liquid."[19]

Spill Handling: Evacuate and restrict persons not wearing protective equipment from area of spill or leak until cleanup is complete. Stay upwind. Remove all ignition sources. Stop leak if you can do so without risk. Use water spray to reduce vapors. Collect powdered material in the most convenient and safe manner and deposit in sealed containers. Ventilate area of spill or leak after clean-up is complete. It may be necessary to contain and dispose of this chemical as a hazardous waste. If material or contaminated runoff enters waterways, notify downstream users of potentially contaminated waters. Contact your Department of Environmental Protection or your regional office of the federal EPA for specific recommendations. If employees are required to clean-up spills, they must be properly trained and equipped. OSHA 1910.120(q) may be applicable.

Fire Extinguishing: Dangerous, fire risk. For small fires use dry chemical, carbon dioxide, water spray, or alcohol foam extinguishers. For large fires use water spray, fog, or alcohol foam. Move containers from fire area if you can do so without risk. Poisonous gases are produced in fire. If material or contaminated runoff enters waterways, notify downstream users of potentially contaminated waters. Notify local health and fire officials and pollution control agencies. From a secure, explosion-proof location, use water spray to cool exposed containers. If cooling streams are ineffective (venting sound increases in volume and pitch, tank discolors, or shows any signs of deforming), withdraw immediately to a secure position. If employees are expected to fight fires, they must be trained and equipped in OSHA 1910.156.

Disposal Method Suggested: Controlled incineration.

References

U.S. Environmental Protection Agency, "Chemical Profile: Butyl Vinyl Ether," Washington, DC, Chemical Emergency Preparedness Program (October 31, 1985)

Butyraldehyde

Molecular Formula: C_4H_8O

Synonyms: Aldehyde butyrique (French); Aldeide butirrica (Italian); Butal; Butaldehyde; Butalyde; *n*-Butanal (Czech); Butanal; Butirraldehido (Spanish); *n*-Butyl aldehyde; Butyl aldehyde; Butyral; Butyraldehyd (German); *n*-Butyraldehyde; Butyric acid; Butyric aldehydenci-C56291

CAS Registry Number: 123-72-8

RTECS Number: ES2275000

DOT ID: UN 1129

EEC Number: 605-006-00-2

EINECS Number: 204-646-6

Regulatory Authority

- EPCRA Section 313 Form R *de minimis* concentration reporting level: 1.0%
- Canada, WHMIS, Ingredients Disclosure List; National Pollutant Release Inventory (NPRI)

Cited in U.S. State Regulations: California (A, G), Florida (G), Massachusetts (G), Minnesota (G), New Jersey (G), Pennsylvania (G)

Description: Butyraldehyde is a flammable, colorless liquid with a pungent odor. Odor threshold = 0.009 ppm. Boiling point = 75°C. Freezing/Melting point = -99°C. Flash point = -22°C. Autoignition temperature = 230°C. Explosive limits in air: LEL= 2.5%; UEL = 12.5%. Hazard Identification (based on NFPA-704 M Rating System): Health 3, Flammability 3, Reactivity 0. Slowly mixes with water. Octanol/water Log P = 1.2.

Potential Exposure: Used in making synthetic resins, solvents, and plasticizers.

Incompatibilities: Butyraldehyde can presumably form explosive peroxides, and may polymerize due to heat or contact with acids or alkalis. May accumulate static electrical charges, and may cause ignition of its vapors. Forms explosive mixture with air. Possible self-reaction in air; undergoes rapid oxidation to butyric acid in air. Incompatible with strong oxidizers (possible violent reaction), strong acids, caustics, ammonia, aliphatic amines, alkanolamines, aromatic amines. May corrode steel due to corrosive action of butyric acid.

Permissible Exposure Limits in Air: There is no OSHA PEL. However, the National Industrial Hygiene Association recommended WEEL is 25 ppm TWA.

Routes of Entry: Inhalation, ingestion, eye and/or skin contact.

Harmful Effects and Symptoms

Short Term Exposure: Butyraldehyde is corrosive. Irritates the eyes, skin, and respiratory tract. Eye or skin contact may cause burns and possible permanent damage. High exposure can cause dizziness and lightheadedness. Higher exposures can cause pulmonary edema, a medical emergency that can be delayed for several hours. This can cause death.

Long Term Exposure: Prolonged or repeated skin exposure may cause skin disorders.

Points of Attack: Lungs, skin

Medical Surveillance: Consider chest x-ray following acute overexposure.

First Aid: If this chemical gets into the eyes, remove any contact lenses at once and irrigate immediately for at least 15 minutes, occasionally lifting upper and lower lids. Seek medical attention immediately. If this chemical contacts the skin, remove contaminated clothing and wash immediately with soap and water. Seek medical attention immediately. If this chemical has been inhaled, remove from exposure, begin rescue breathing (using universal precautions) if breathing has stopped and CPR if heart action has stopped. Transfer promptly to a medical facility. When this chemical has been swallowed, get medical attention. Give large quantities of water and induce vomiting. Do not make an unconscious person vomit. Medical observation is recommended for 24 – 48 hours after breathing overexposure, as pulmonary edema may be delayed. As first aid for pulmonary edema, a doctor or authorized paramedic may consider administering a corticosteroid spray.

Personal Protective Methods: Wear solvent-resistant gloves and clothing to prevent any reasonable probability of skin contact. Safety equipment suppliers/manufacturers can provide recommendations on the most protective glove/clothing material for your operation. All protective clothing (suits, gloves, footwear, headgear) should be clean, available each day, and put on before work. Contact lenses should not be worn when working with this chemical. Wear splash-proof chemical goggles and face shield unless full facepiece respiratory protection is worn. Employees should wash immediately with soap when skin is wet or contaminated. Remove nonimpervious clothing immediately if wet or contaminated. Provide emergency showers and eyewash.

Respirator Selection: Engineering controls should be used wherever feasible to maintain airborne concentrations of this chemical below the prescribed exposure limit. Respirators and protective equipment are less effective than engineering controls, and should be used only in non-routine or emergency situations which may result in exposure concentrations in excess of the TWA environmental limit. *At any concentrations above the NIOSH REL, or where there is no REL, at any detectable concentration:* SCBAF:PD,PP (any MSHA/NIOSH approved self-contained breathing apparatus that has a full facepiece and is operated in a pressure-demand or other positive-pressure mode); or SAF:PD,PP:ASCBA (any supplied-air respirator that has a full facepiece and is operated in a pressure-demand or other positive-pressure mode in combination with an auxiliary, self-contained breathing apparatus operated in a pressure-demand or other positive pressure mode). *Escape:* GMFOVHiE [any air-purifying, full-facepiece respirator (gas mask) with a chin-style, front- or back-mounted organic vapor canister having a high-efficiency particulate filter]; or SCBAE (any appropriate escape-type, self-contained breathing apparatus).

Storage: Protect against physical damage. Before entering confined space where this chemical may be present, check to make sure that an explosive concentration does not exist. Outside or detached storage is preferred. Prior to working with Butyraldehyde you should be trained on its proper handling and storage. Store in tightly closed containers in a cool, well ventilated area away from incompatible materials listed above. Metal containers involving the transfer of this chemical should be grounded and bonded. Drums must be equipped with self-closing valves, pressure vacuum bungs, and flame arresters. Use only non-sparking tools and equipment, especially when opening and closing containers of this chemical. Sources of ignition such as smoking and open flames, are prohibited where this chemical is used, handled, or stored in a manner that could create a potential fire or explosion hazard.

Shipping: Butyl propionate requires a "Flammable Liquid" label. It is in UN/DOT Hazard Class 3, Packing Group II.

Spill Handling: Evacuate and restrict persons not wearing protective equipment from area of spill or leak until cleanup is complete. Remove all ignition sources. Establish forced ventilation to keep levels below explosive limit. Absorb liquid in vermiculite, dry sand, earth or similar noncombustible absorbent material and deposit in sealed containers. *Do not* use sawdust or other combustible absorbent. Ventilate area of spill or leak after clean-up is complete. It may be necessary to contain and dispose of this chemical as a hazardous waste. If material or contaminated runoff enters waterways, notify downstream users of potentially contaminated waters. Contact your Department of Environmental Protection or your regional office of the federal EPA for specific recommendations. If employees are required to clean-up spills, they must be properly trained and equipped. OSHA 1910.120(q) may be applicable.

Fire Extinguishing: This chemical is a flammable liquid. Poisonous gases, including carbon monoxide, are produced in fire. Use dry chemical, carbon dioxide, or alcohol foam extinguishers. Vapors are heavier than air and will collect in low areas. Vapors may travel long distances to ignition sources and flashback. Vapors in confined areas may explode when exposed to fire. Containers may explode in fire. Storage containers and parts of containers may rocket great distances, in many directions. If material or contaminated runoff enters waterways, notify downstream users of potentially contaminated waters. Notify local health and fire officials and pollution control agencies. Containers may explode in fire. From a secure, explosion-proof location, use water spray to cool exposed containers. If cooling streams are ineffective (venting sound increases in volume and pitch, tank discolors, or shows any signs of deforming), withdraw immediately to a secure position. If employees are expected to fight fires, they must be trained and equipped in OSHA 1910.156.

Disposal Method Suggested: Incineration.

References

New Jersey Department of Health and Senior Services, "Hazardous Substance Fact Sheet: Butyraldehyde," Trenton, NJ (September 1996)

Butyric Acid

Molecular Formula: $C_4H_8O_2$

Common Formula: $CH_3(CH_2)_2COOH$

Synonyms: Acido butirico (Spanish); Butanic acid; *n*-Butanoic acid; Butanoic acid; Buttersaeure (German); *n*-Butyric acid; *normal* Butyric acid; Butyric acid; Ethylacetic acid; 1-Propanecarboxylic acid; Propylformic acid

CAS Registry Number: 107-92-6

RTECS Number: ES5425000

DOT ID: UN 2820

Regulatory Authority

- Air Pollutant Standard Set (former USSR)[43]
- CLEAN WATER ACT: Section 311 Hazardous Substances/ RQ 40CFR117.3 (same as CERCLA, see below)
- SUPERFUND/EPCRA 40CFR302.4 Reportable Quantity (RQ): CERCLA, 5,000 lb (2,270 kg)
- Canada, WHMIS, Ingredients Disclosure List

Cited in U.S. State Regulations: California (G), Florida (G), Illinois (G), Massachusetts (G), New Hampshire (G), New Jersey (G), Pennsylvania (G), Rhode Island (G).

Description: Butyric acid, $CH_3(CH_2)_2COOH$, is a combustible, oily liquid with an unpleasant odor. The odor threshold is 0.0001 ppm.[41] Boiling point = 163.5°C. Flash point =

72°C. Autoignition temperature = 440°C. Hazard Identification (based on NFPA-704 M Rating System): Health 3, Flammability 2, Reactivity 0. The flammable limits in air are LEL = 2%; UEL = 10%. Highly soluble in water.

Potential Exposure: In manufacture of butyrate esters, some of which go into artificial flavoring.

Incompatibilities: Forms explosive mixture with air. Incompatible with sulfuric acid, caustics, ammonia, aliphatic amines, isocyanates, strong oxidizers, alkylene oxides, epichlorohydrin.

Permissible Exposure Limits in Air: The former-USSR-UNEP/IRPTC MAC value is 2.5 ppm (10 mg/m^3)[43] for workplace air. They also cite a momentary MAC value of 0.015 mg/m^3 and an allowable average daily MAC of 0.01 mg/m^3 in ambient air of residential areas.

Permissible Concentration in Water: No criteria set.

Routes of Entry: Inhalation, absorbed through the skin.

Harmful Effects and Symptoms

Note: The following information may also apply to isobutyric acid (79-31-2).

Short Term Exposure: Butyric acid is a medium-strong corrosive acid. Can cause sever eye and skin irritation and burns leading to permanent damage. Inhalation can cause respiratory tract irritation, coughing, wheezing and/or shortness of breath. The oral LD_{50} rat is 2,940 mg/kg (slightly toxic).[9]

Long Term Exposure: Can affect the blood. Repeated exposures may cause bronchitis to develop with coughing, phlegm, and/or shortness of breath. May cause kidney damage.

Points of Attack: Skin, eyes and respiratory system.

Medical Surveillance: Lung function tests. Kidney function tests. Complete blood count (CBC).

First Aid: If this chemical gets into the eyes, remove any contact lenses at once and irrigate immediately for at least 30 minutes, occasionally lifting upper and lower lids. Seek medical attention immediately. If this chemical contacts the skin, remove contaminated clothing and wash immediately with soap and water. Seek medical attention immediately. If this chemical has been inhaled, remove from exposure, begin rescue breathing (using universal precautions) if breathing has stopped and CPR if heart action has stopped. Transfer promptly to a medical facility. When this chemical has been swallowed, get medical attention. If victim is conscious, administer water or milk. Do not induce vomiting.

Personal Protective Methods: Wear solvent-resistant gloves and clothing to prevent any reasonable probability of skin contact. Safety equipment suppliers/manufacturers can provide recommendations on the most protective glove/clothing material for your operation. All protective clothing (suits, gloves, footwear, headgear) should be clean, available each day, and put on before work. Contact lenses should not be worn when working with this chemical. Wear splash-proof chemical goggles and face shield unless full facepiece respiratory protection is worn. Employees should wash immediately with soap when skin is wet or contaminated. Remove non-impervious clothing immediately if wet or contaminated. Provide emergency showers and eyewash.

Respirator Selection: Engineering controls should be used wherever feasible to maintain airborne concentrations of this chemical below the prescribed exposure limit. Respirators and protective equipment are less effective than engineering controls, and should be used only in non-routine or emergency situations which may result in exposure concentrations in excess of the TWA environmental limit. *At any concentrations above the NIOSH REL, or where there is no REL, at any detectable concentration:* SCBAF:PD,PP (any MSHA/NIOSH approved self-contained breathing apparatus that has a full facepiece and is operated in a pressure-demand or other positive-pressure mode); or SAF:PD,PP:ASCBA (any supplied-air respirator that has a full facepiece and is operated in a pressure-demand or other positive-pressure mode in combination with an auxiliary, self-contained breathing apparatus operated in a pressure-demand or other positive pressure mode). *Escape:* GMFOVHiE [any air-purifying, full-facepiece respirator (gas mask) with a chin-style, front- or back-mounted organic vapor canister having a high-efficiency particulate filter]; or SCBAE (any appropriate escape-type, self-contained breathing apparatus).

Storage: Protect against physical damage. Outside or detached storage is preferred. Prior to working with butyric acid you should be trained on its proper handling and storage. Store in tightly closed containers in a cool, well ventilated area away from incompatible materials listed above. Metal containers involving the transfer of this chemical should be grounded and bonded. Drums must be equipped with self-closing valves, pressure vacuum bungs, and flame arresters. Use only non-sparking tools and equipment, especially when opening and closing containers of this chemical. Sources of ignition such as smoking and open flames, are prohibited where this chemical is used, handled, or stored in a manner that could create a potential fire or explosion hazard.

Shipping: Butyric acid falls in UN/DOT Hazard Class 8 and Packing Group III.[19][20]

Spill Handling: Evacuate and restrict persons not wearing protective equipment from area of spill or leak until cleanup is complete. Remove all ignition sources. Establish forced ventilation to keep levels below explosive limit. Absorb liquid in vermiculite, dry sand, earth or similar material and deposit in sealed containers. Ventilate area of spill or leak after cleanup is complete. It may be necessary to contain and dispose of this chemical as a hazardous waste. If material or contaminated runoff enters waterways, notify downstream users of potentially contaminated waters. Contact your Department of Environmental Protection or your regional office of the federal EPA for specific recommendations. If employees are required to clean-up spills, they must be properly trained and equipped. OSHA 1910.120(q) may be applicable.

Fire Extinguishing: This chemical is a combustible liquid. Poisonous gases are produced in fire. Use dry chemical, carbon dioxide, or alcohol foam extinguishers. Vapors are heavier than air and will collect in low areas. Vapors may travel long distances to ignition sources and flashback. Vapors in confined areas may explode when exposed to fire. Containers may explode in fire. Storage containers and parts of containers may rocket great distances, in many directions. If material or contaminated runoff enters waterways, notify downstream users of potentially contaminated waters. Notify local health and fire officials and pollution control agencies. Containers may explode in fire. From a secure, explosion-proof location, use water spray to cool exposed containers. If cooling streams are ineffective (venting sound increases in volume and pitch, tank discolors, or shows any signs of deforming), withdraw immediately to a secure position. If employees are expected to fight fires, they must be trained and equipped in OSHA 1910.156.

Disposal Method Suggested: Incineration.

References

Sax, N. I., Ed., "Dangerous Properties of Industrial Materials Report" 2, No. 3, 71–73 (1982)

New Jersey Department of Health and Senior Services, "Hazardous Substance Fact Sheet: Butyric Acid," Trenton, NJ (April 1986)

C

Cacodylic Acid

Molecular Formula: $C_2H_7AsO_2$

Synonyms: Acide cacodylique (French); Acide dimethylarsinique (French); Acido cacodilico (Spanish); Agent blue; Ansan; Ansar; Arsinic acid, Dimethyl-(9CI); Bolls-Eye; Chexmate; Cottonaide HC; Dilic; Dimethylarsenic acid; Dimethylarsinic acid; Dimethylarsinic arsinic acid; DMAA; Erase; Hydroxydimethylarsine oxide; Kyselina kakodylova (Czech); Monocide; Montar; Phylar; Phytar 138; Phytar 560; Phytar 600; Rad-E-Cate 25; Salvo; Silvisar 510

CAS Registry Number: 75-60-5

RTECS Number: CH7525000

DOT ID: UN 1572

Regulatory Authority

- SUPERFUND/EPCRA 40CFR302.4 Reportable Quantity (RQ): CERCLA, 1 lb (0.454 kg)
- EPA HAZARDOUS WASTE NUMBER (RCRA No.): U136
- RCRA, 40CFR261, Appendix 8 Hazardous Constituents
- EPCRA Section 313 Form R (as organic arsenic compound) *de minimis* concentration reporting level: 1.0%
- U.S. DOT Regulated Marine Pollutant (49CFR172.101, Appendix B) as dimethylarsinic acid
- Canada, WHMIS, Ingredients Disclosure List
- Canada: Priority Substance List & Restricted Substances/Ocean Dumping Forbidden (CEPA), National Pollutant Release Inventory (NPRI) (arsenic compounds)

Cited in U.S. State Regulations: Massachusetts (G), New Jersey (G), Pennsylvania (G). See also Arsenic compounds.

Description: Cacodylic acid is a colorless, odorless, crystalline solid arsenic compound. Freezing/Melting point = 192°C. Hazard Identification (based on NFPA-704 M Rating System): Health 1, Flammability 0, Reactivity 0. Highly soluble in water.

Potential Exposure: Used as an herbicide, soil sterilant and in timber thinning. Has been used as a chemical warfare agent.

Incompatibilities: Aqueous solution react violently with chemically active metals releasing toxic arsenic fumes. Incompatible with oxidizers, sulfuric acid, caustics (strong bases), reducing agents, ammonia, amines, isocyanates, alkylene oxides, epichlorohydrin.

Permissible Exposure Limits in Air: The following exposure limits are for air levels only. When skin contact also occurs, overexposure is possible, even thought air levels are less than the limits listed below. The OSHA PEL 8-hour TWA is 0.5 mg(As)/m^3. ACGIH recommends a TWA of 0.2 mg(As)/m^3.

Determination in Air: Filter; Reagent; Ion chromatography/Hydride generation atomic absorption spectrometry; NIOSH IV Method #5022, Arsenic, Organo-.

Determination in Water: The atomic absorption graphite furnace technique is often used for measurement of total arsenic in water. It also has been standardized by EPA. Total arsenic may be determined by digestion followed by silver diethyldithiocarbamate; an alternative is atomic absorption; another is inductively coupled plasma (ICP) optical emission spectrometry.

Routes of Entry: Inhalation, ingestion, skin and/or eye contact This chemical can be absorbed through the skin, thereby increasing exposure.

Harmful Effects and Symptoms

Short Term Exposure: Irritates the eyes, skin, and respiratory tract. Skin contact can also cause burning, itching, thickening and color changes.

Long Term Exposure: Certain other arsenic compounds have been identified as carcinogen. Although this chemical has not been identified as a carcinogen should be handled with extreme caution. May cause an ulcer of the "bone" dividing the inner nose. It can cause nerve damage, thickening of the skin with patch areas of darkening and loss of pigment, or the development of white lines in the nails. May cause liver and kidney damage. Repeated exposure can lead to a metallic or garlic taste in the mouth, loss of appetite, nausea, vomiting, difficulty swallowing, stomach pain, diarrhea, and death.

Points of Attack: Skin, respiratory system, kidneys, central nervous system, liver, gastrointestinal tract, reproductive system.

Medical Surveillance: Test for urine arsenic. Levels should not be greater than 100 micrograms per gram of creatinin in the urine. Examine the skin for abnormal growths. Liver and kidney function tests.

First Aid: If this chemical gets into the eyes, remove any contact lenses at once and irrigate immediately for at least 15 minutes, occasionally lifting upper and lower lids. Seek medical attention immediately. If this chemical contacts the skin, remove contaminated clothing and wash immediately with soap and water. Seek medical attention immediately. If this chemical has been inhaled, remove from exposure, begin rescue breathing (using universal precautions) if breathing has stopped and CPR if heart action has stopped. Transfer promptly to a medical facility. When this chemical has been swallowed, get medical attention. Give large quantities of water and induce vomiting. Do not make an unconscious person vomit.

Personal Protective Methods: Wear protective gloves and clothing to prevent any reasonable probability of skin contact. Safety equipment suppliers/manufacturers can provide recommendations on the most protective glove/clothing material for your operation. All protective clothing (suits, gloves, footwear, headgear) should be clean, available each day, and put on before work. Contact lenses should not be worn when working with this chemical. Wear dust-proof chemical goggles and face shield unless full facepiece respiratory protection is worn. Employees should wash immediately with soap when skin is wet or contaminated. Provide emergency showers and eyewash.

Respirator Selection: *At any concentrations above the NIOSH REL at any detectable concentration:* SCBAF: PD,PP (any self-contained breathing apparatus that has a full faceplate and is operated in a pressure-demand or other positive-pressure mode); or SAF:PD,PP:ASCBA (any supplied-air respirator that has a full facepiece and is operated in a pressure-demand or other positive-pressure mode in combination with an auxiliary self-contained breathing apparatus operated in a pressure-demand or other positive-pressure mode). *Escape:* GMFAGHiE [any air-purifying, full-facepiece respirator (gas mask) with a chin-style, front-or back-mounted acid gas canister having a high-efficiency particulate filter]; or SCBAE (any appropriate escape-type, self-contained breathing apparatus).

Storage: Prior to working with cacodylic acid you should be trained on its proper handling and storage. A regulated, marked area should be established where this chemical is handled, used, or stored in compliance with OSHA standard 1910.1045. Store in tightly closed containers in a cool, well-ventilated area away from oxidizing agents, chemically active metals, strong bases, moisture, fertilizers, seeds, insecticides, and fungicides.

Shipping: The DOT label requirement is "Poison." The UN/DOT Hazard Class is 6.1 and the Shipping Group is II.

Spill Handling: Evacuate and restrict persons not wearing protective equipment from area of spill or leak until cleanup is complete. Remove all ignition sources. Ventilate area of spill or leak. Absorb liquids in vermiculite, dry sand, earth, peat, carbon, or a similar material and deposit in sealed containers. It may be necessary to contain and dispose of this chemical as a hazardous waste. If material or contaminated runoff enters waterways, notify downstream users of potentially contaminated waters. Contact your Department of Environmental Protection or your regional office of the federal EPA for specific recommendations. If employees are required to clean-up spills, they must be properly trained and equipped. OSHA 1910.120(q) may be applicable.

Fire Extinguishing: This chemical may burn, but does not readily ignite. Use dry chemical, carbon dioxide, water spray, or alcohol or polymer foam extinguishers. Poisonous gases are produced in fire including carbon monoxide and arsenic oxides. If material or contaminated runoff enters waterways, notify downstream users of potentially contaminated waters. Notify local health and fire officials and pollution control agencies. From a secure, explosion-proof location, use water spray to cool exposed containers. If cooling streams are ineffective (venting sound increases in volume and pitch, tank discolors, or shows any signs of deforming), withdraw immediately to a secure position. If employees are expected to fight fires, they must be trained and equipped in OSHA 1910.156.

References

New Jersey Department of Health and Senior Services, "Hazardous Substance Fact Sheet, Cacodylic Acid," Trenton, NJ (January 1999)

Cadmium

Molecular Formula: Cd

Synonyms: Cadmio (Spanish); C.I. 77180; Colloidal cadmium; Elemental cadmium; Kadmium (German); Kadmu (Polish)

CAS Registry Number: 7440-43-9 (metal)

RTECS Number: EU9800000 (metal)

DOT ID: UN 2570 (for cadmium compounds)

EINECS Number: 231-152-8

Regulatory Authority

- Carcinogen (Animal Positive) (IARC) (DFG)[9] (Suspected) (ACGIH) (NTP) (DFG)[3]
- Banned or Severely Restricted (Many Countries) (UN)[13][35]
- Air Pollutant Standard Set (OSHA) (ACGIH) (Australia) (Israel) (Mexico) (Several States) (Several Canadian Provinces)
- CLEAN AIR ACT: Hazardous Air Pollutants (Title I, Part A, Section 112). Includes any unique chemical substance that contains cadmium as part of that chemical's infrastructure

- CLEAN WATER ACT: 40CFR423, Appendix A, Priority Pollutants; Section 313 Water Priority Chemicals (57FR41331, 9/9/92); Toxic Pollutant (Section 401.15)
- EPA HAZARDOUS WASTE NUMBER (RCRA No.): D006
- RCRA, 40CFR261, Appendix 8 Hazardous Constituents, waste number not listed
- RCRA Toxicity Characteristic (Section 261.24), Maximum Concentration of Contaminants, regulatory level, 1.0 mg/l
- RCRA 40CFR268.48; 61FR15654, Universal Treatment Standards: Wastewater (mg/l), 0.69; Nonwastewater (mg/l), 0.19 TCLP
- RCRA 40CFR264, Appendix 9; TSD Facilities Ground Water Monitoring List. Suggested test method(s) (PQL μg/l): 601 (40); 7130 (50); 7131 (1)
- SAFE DRINKING WATER ACT: MCL, 0.005 mg/l; MCLG, 0.005 mg/l; Regulated chemical (47 FR 9352); Priority List (55 FR 1470)
- SUPERFUND/EPCRA 40CFR302.4 Reportable Quantity (RQ): CERCLA, 10 lb (4.54 kg). *Note*: No release report required if diameter of pieces is equal to or exceeds 0.004 in
- EPCRA Section 313 Form R *de minimis* concentration reporting level: 0.1%
- U.S. DOT Regulated Marine Pollutant (49CFR172.101, Appendix B): Severe pollutant, as cadmium compounds
- Canada, WHMIS, Ingredients Disclosure List; National Pollutant Release Inventory (NPRI); CEPA Priority Substance List, Ocean dumping prohibited.; Drinking Water Quality: 0.005 mg/l MAC
- Mexico Drinking Water Criteria: 0.01 mg/l

Cited in U.S. State Regulations: Alaska (G), California (A, G), Connecticut (A), Florida (G), Illinois (G), Kansas (G, W), Louisiana (G), Maine (G, W), Maryland (G), Massachusetts (G, A), Michigan (G), Minnesota (W), Montana (A), Nevada (A), New Hampshire (G), New Jersey (G), New York (G, A), North Carolina (A), North Dakota (A), Oklahoma (G), Pennsylvania (G, A), Rhode Island (G, A), South Carolina (A), South Dakota (A), Vermont (G), Virginia (G, A), Washington (G), West Virginia (G), Wisconsin (G).

Description: Cadmium, Cd, is a bluish-white metal. Boiling point = 765°C. Freezing/Melting point = 321°C. Cadmium metal dust has an Autoignition temperature = 250°C. Hazard Identification (based on NFPA-704 M Rating System): (powder) Health 2, Flammability 2, Reactivity 0. The only cadmium mineral, greenockite (CdS), is rare; however, small amounts of cadmium are found in zinc, copper, and lead ores. It is generally produced as a by-product of these metals, particularly zinc. Cadmium is insoluble in water but is soluble in acids. "Cadmium dust" includes dust of various cadmium compounds such as $CdCl_2$. "Cadmium fume" has the composition Cd/CdO.

Potential Exposure: Cadmium is highly corrosion resistant and is used as a protective coating for iron, steel, and copper; it is generally applied by electroplating, but hot dipping and spraying are possible. Cadmium may be alloyed with copper, nickel, gold, silver, bismuth, and aluminum to form easily fusible compounds. These alloys may be used as coatings for other materials, welding electrodes, solders, etc. It is also utilized in electrodes of alkaline storage batteries, as a neutron absorber in nuclear reactors, a stabilizer for polyvinyl chloride plastics, a deoxidizer in nickel plating, an amalgam in dentistry, in the manufacture of fluorescent lamps, semiconductors, photocells, and jewelry, in process engraving, in the automobile and aircraft industries, and to charge Jones reductors. Various cadmium compounds find use as fungicides, insecticides, nematocides, polymerization catalysts, pigments, paints, and glass; they are used in the photographic industry and in glazes. Cadmium is also a contaminant of superphosphate fertilizers. Human exposure to cadmium and certain cadmium compounds occurs through inhalation and ingestion. The entire population is exposed to low levels of cadmium in the diet because of the entry of cadmium into the food chain as a result of its natural occurrence. Tobacco smokers are exposed to an estimated 17 μg/cigarette. Cadmium is present in relatively low amounts in the earth's crust; as a component of zinc ores, cadmium may be released into the environment around smelters.

Incompatibilities: Air exposure with cadmium powder may cause self ignition. Moist air slowly oxidizes cadmium forming cadmium oxide. Cadmium dust is incompatible with strong oxidizers, ammonium nitrate, elemental sulfur, hydrazoic acid, selenium, zinc, tellurium. Contact with acids cause a violent reaction producing flammable hydrogen gas.

Permissible Exposure Limits in Air: The Federal standard[58] for cadmium fume or dust is 0.005 mg/m^3 (as Cd) as an 8-hour TWA. ACGIH has recommended a TWA value of 0.01 mg/m^3 for Cd dusts/salts and 0.002 mg/m^3 for respirable dust. NIOSH recommends that exposure to cadmium be at the lowest feasible level. The NIOSH IDLH: Carcinogen [9 mg/m^3 (as Cd)]. Several states have set guidelines or standards for cadmium in ambient air[60] ranging from zero (North Dakota) to 0.0006 μg/m^3 (Rhode Island) to 0.0055 μg/m^3 (North Carolina) to 0.0056 μg/m^3 (Massachusetts) to 0.07 (Montana) to 0.12 μg/m^3 (Pennsylvania) to 0.25 μg/m^3 (South Carolina) to 0.4 μg/m^3 (Connecticut and South Dakota) to 0.8 μg/m^3 (Virginia) to 1.0 μg/m^3 (Nevada) to 2.0 μg/m^3 (New York).

Determination in Air: Collection of particles on a filter, workup with acid and measurement by atomic absorption has been specified by NIOSH. See NIOSH Methods #7048 (Cd) and #7300 (Elements), #7200 (Welding and Brazing Fume) or OSHA: #ID-125G.

Permissible Concentration in Water: To protect freshwater aquatic life: $e^{[1.05 \ln (\text{hardness}) - 8.53]}$ μg/l as a 24-hour average, never to exceed: $e^{[1.05 \ln (\text{hardness}) - 3.73]}$ μg/l at any time. To protect saltwater aquatic life: 4.5 μg/l as a 24-hour average, never to exceed 59.0 μg/l at any time. EPA[62] and Canada set a limit of 0.005 mg/l in drinking water. Mexico has set a limit of 0.01 mg/l in drinking water. Effluent standards for cadmium in water have been set by Argentina, 0.1 mg/l; Japan, 0.1 mg/l. Drinking water standards have been set[35] by Czechoslovakia, 0.010 mg/l; EEC, 5.0 μg/l (0.005 mg/l); Japan, <0.01 mg/l; USSR/UNEP, 0.01 mg/l; WHO, 0.005 mg/l. Further, guidelines for cadmium in drinking water have been set[61] ranging from 5 μg/l (Kansas and Minnesota) to 10 μg/l (Maine).

Determination in Water: Total cadmium may be determined by digestion followed by atomic absorption of colorimetric (Dithizone) analysis or by Inductively Coupled Plasma (ICP) Optical Emission Spectrometry. Dissolved cadmium is determined by 0.45 μ filtration followed by the previously cited methods. See also reverence.[49]

Routes of Entry: Inhalation or ingestion of fumes or dust.

Harmful Effects and Symptoms

Short Term Exposure: *Inhalation:* Dust may cause irritation of the nose and throat. Non-fatal lung inflammation has been reported from concentrations of 0.5 – 2.5 mg/m^3. In 4 – 10 hours after exposure severe chest pain, with persistent cough and difficult breathing, headache, chills, muscle aches, nausea, vomiting, diarrhea. Fluid in the lungs (pulmonary edema) and dark-purple coloration of the skin may occur. Pulmonary edema is a medical emergency that can be delayed for several hours. This can cause death. Breathing becomes more difficult and is accompanied by wheezing or coughing of blood. Other symptoms which may occur 12 – 36 hours after exposure in addition to those above include dizziness, irritability, gastrointestinal disturbances, shortness of breath, fever, profuse sweating, exhaustion and inflammation of the lungs. Death may result within 7 – 10 days after exposure. The average concentrations of fume responsible for fatalities have been 40 – 50 mg/m^3 for 1 hour, 9 mg/m^3 for 5 hours, or 5 mg/m^3 for 8 hours. *Skin:* Absorption is negligible. *Eyes:* Cadmium compound dust may cause irritation. *Ingestion:* A dose of 15 – 30 mg (1/1000 oz) of metal or soluble compounds may cause increased salivation, choking, vomiting, abdominal pain, anemia, kidney malfunction, diarrhea, and persistent desire to empty the bladder. Symptoms may occur within 15 – 30 minutes after ingestion. May cause heart and lung failure.

Long Term Exposure: Continued exposure to low levels of cadmium in air may cause irreversible lung injury, abnormal lung function and kidney disease. Other consequences of cadmium exposures are inflammation of the nose and throat, open sores in the nose, soreness, bleeding and reduced nose size, loss of sense of smell, damage to the olfactory nerve, yellow cadmium stains on teeth, sleeplessness, nausea, lack of appetite, weight loss, anemia, lung distention with scar formation, and liver damage. May cause bone disease characterized by softening, bending and reduction in bone size. Difficulty walking, pain in back and extremities, and spontaneous fractures may result. Inhalation of 0.06 mg/m^3 – 0.68 mg/m^3 for 4 – 8 years may cause throat irritation, cough, chest pain, upset stomach and fatigue. Exposure to levels of 3.0 – 15.0 mg/m^3 of fumes or dust over a period of 20 years has caused lung distention, anemia, protein in urine and kidney dysfunction. Studies indicate that there is an increased incidence of prostatic cancer and possible kidney and respiratory cancer in cadmium workers. Cadmium causes birth defects in rats, mice and hamsters; whether it does so in humans is not known.

Points of Attack: Respiratory system, lungs, kidneys, prostate, blood.

Medical Surveillance: In preemployment physical examinations emphasis should be given to a history of, or actual presence of, significant kidney disease, smoking history, and respiratory disease. A chest x-ray and baseline pulmonary function study is recommended. Periodic examinations should emphasize the respiratory system, including pulmonary function tests, kidneys and blood. A low molecular weight proteinuria may be the earliest indication of renal toxicity. The trichloroacetic acid test may pick this up, but more specific quantitative studies would be preferable. If renal disease due to cadmium is present, there may also be increased excretion of calcium, amino acids, glucose, and phosphates.

First Aid: If this chemical gets into the eyes, remove any contact lenses at once and irrigate immediately for at least 15 minutes, occasionally lifting upper and lower lids. Seek medical attention immediately. If this chemical contacts the skin, remove contaminated clothing and wash immediately with soap and water. Seek medical attention immediately. If this chemical has been inhaled, remove from exposure, begin rescue breathing (using universal precautions) if breathing has stopped and CPR if heart action has stopped. Transfer promptly to a medical facility. When this chemical has been swallowed, get medical attention. Give large quantities of water and induce vomiting. Do not make an unconscious person vomit. Medical observation is recommended for 24 – 48 hours after breathing overexposure, as pulmonary edema may be delayed. As first aid for pulmonary edema, a doctor or authorized paramedic may consider administering a corticosteroid spray.

Personal Protective Methods: Wear protective gloves and clothing to prevent skin contact. Safety equipment suppliers/manufacturers can provide recommendations on the most protective glove/clothing material for your operation. All protective clothing (suits, gloves, footwear, headgear) should be clean, available each day, and put on before work. Contact lenses should not be worn when working with this chemical. Wear dust-proof chemical goggles and face shield to prevent any possibility of eye contact, unless full facepiece respiratory protection is worn. Employees should wash immediately with soap when skin is contaminated. Provide emergency showers and eyewash.

Respirator Selection: NIOSH: *at any detectable concentration:* SCBAF:PD,PP (any MSHA/NIOSH approved self-contained breathing apparatus that has a full facepiece and is operated in a pressure-demand or other positive-pressure mode); or SAF:PD,PP:ASCBA (any supplied-air respirator that has a full facepiece and is operated in a pressure-demand or other positive-pressure mode in combination with an auxiliary, self-contained breathing apparatus operated in a pressure-demand or other positive pressure mode). *Escape:* HiEF [any air-purifying, full-facepiece respirator (gas mask) with a chin-style, front- or back-mounted organic vapor canister having a high-efficiency particulate filter]; or SCBAE (any appropriate escape-type, self-contained breathing apparatus).

Storage: Prior to working with Cadmium you should be trained on its proper handling and storage. Cadmium must be stored to avoid contact with Sulfur, Selenium, Tellurium, Ammonium Nitrate, and Hydrazoic Acid since violent reactions occur. Store in tightly closed containers in a cool, well-ventilated area away from Oxidizers (such as Perchlorates, Peroxides, Permanganates, Chlorates, and Nitrates). Sources of ignition such as smoking and open flames are prohibited where Cadmium is used, handled, or stored in a manner that could create a potential fire or explosion hazard. A regulated, marked area should be established where this chemical is handled, used, or stored in compliance with OSHA standard 1910.1045.

Shipping: Cadmium metal and dust are not specifically cited in DOT's Performance Oriented Packaging Standards.[19]

Spill Handling: Evacuate and restrict persons not wearing protective equipment from area of spill or leak until cleanup is complete. Remove all ignition sources. Collect powdered material in the most convenient and safe manner and deposit in sealed containers. Ventilate area after clean-up is complete. It may be necessary to contain and dispose of this chemical as a hazardous waste. If material or contaminated runoff enters waterways, notify downstream users of potentially contaminated waters. Contact your Department of Environmental Protection or your regional office of the federal EPA for specific recommendations. If employees are required to clean-up spills, they must be properly trained and equipped. OSHA 1910.120(q) may be applicable.

Fire Extinguishing: This chemical is a flammable powder. Use dry chemicals appropriate for metal fires, or dry sand. Use no other extinguishing agents. *Do not use water.* Toxic gases are produced in fire. If material or contaminated runoff enters waterways, notify downstream users of potentially contaminated waters. Notify local health and fire officials and pollution control agencies. From a secure, explosion-proof location, use water spray to cool exposed containers. If cooling streams are ineffective (venting sound increases in volume and pitch, tank discolors, or shows any signs of deforming), withdraw immediately to a secure position. If employees are expected to fight fires, they must be trained and equipped in OSHA 1910.156.

Disposal Method Suggested: With cadmium compounds in general, precipitation from solution as sulfides, drying and return of the material to suppliers for recovery is recommended. Cadmium may be recovered from battery scrap as an alternative to disposal.[22] In accordance with 40CFR165 recommendations for the disposal of pesticides and pesticide containers. Must be disposed properly by following package label directions or by contacting your state pesticide or environmental control agency or by contacting your regional EPA office.

References

National Institute for Occupational Safety and Health, Criteria for a Recommended Standard: Occupational Exposure to Cadmium, NIOSH Doc. No. 76–192, Washington, DC (1976)

U.S. Environmental Protection Agency, Cadmium: Ambient Water Quality Criteria, Washington, DC (1980)

U.S. Environmental Protection Agency, Status Assessment of Toxic Chemicals: Cadmium, Report EPA-600/2-79-210F, Washington, DC (December 1979)

U.S. Environmental Protection Agency, Toxicology of Metals, Vol II: Cadmium, Report EPA-600/1-77-022, Research Triangle Park, NC, pp 124–163 (May 1977)

U.S. Environmental Protection Agency, Reviews of the Environmental Effects of Pollutants, IY: Cadmium, Report EPA-600/1-78-026, Health Effects Research Laboratory, Cincinnati, OH (1978)

U.S. Environmental Protection Agency, Health Assessment Document for Cadmium, Report EPA-600/8-79-003, Research Triangle Park, NC (1979)

U.S. Environmental Protection Agency, Cadmium, Health and Environmental Effects Profile No. 31, Office of Solid Wastes, Washington, DC (April 30, 1980)

Sax, N. I., Ed., "Dangerous Properties of Industrial Materials Report" 1, No. 1, 36–38 (1980) and 3, No. 5, 72–76 (1983)

New Jersey Department of Health and Senior Services, "Hazardous Substance Fact Sheet: Cadmium," Trenton, NJ (April 1986)

New York State Department of Health, "Chemical Fact Sheet: Cadmium Compounds," Albany, NY, Bureau of Toxic Substance Assessment (January 1986)

U.S. Environmental Protection Agency, "Health Advisory: Cadmium," Washington, DC, Office of Drinking Water (March 31, 1987)

U.S. Public Health Service, "Toxicological Profile for Cadmium," Agency for Toxic Substances and Disease Registry (November 1987)

Cadmium Acetate

Molecular Formula: $C_4H_6CdO_4$

Common Formula: $Cd(C_2H_3O_2)_2$

Synonyms: Acetic acid, cadmium salt; Aceto cadmio (Spanish); Bis(acetoxy)cadmium; Cadmium(II) acetate; Cadmium diacetate; C.I. 77185

CAS Registry Number: 543-90-8

RTECS Number: EU9810000

DOT ID: NA 2570

Regulatory Authority

- Carcinogen (NTP)[10]
- Air Pollutant Standard Set (See cadmium)
- CLEAN AIR ACT: Toxic Pollutant (Section 401.15), Hazardous Air Pollutants (Title I, Part A, Section 112). *Note:* Includes any unique chemical substance that contains cadmium as part of that chemical's infrastructure
- CLEAN WATER ACT: Section 311 Hazardous Substances/RQ 40CFR117.3 (same as CERCLA, see below); Section 313 Water Priority Chemicals (57FR41331, 9/9/92)
- SUPERFUND/EPCRA 40CFR302.4 Reportable Quantity (RQ): CERCLA, 10 lb (4.54 kg)
- EPCRA Section 313 Form R *de minimis* concentration reporting level: 1.0%
- U.S. DOT Regulated Marine Pollutant (49CFR172.101, Appendix B).: Severe pollutant, as cadmium compounds
- Canada, WHMIS, Ingredients Disclosure List; National Pollutant Release Inventory (NPRI); CEPA Priority Substance List, Ocean dumping prohibited, as cadmium compounds

Cited in U.S. State Regulations: Alaska (G), California (A, G), Maine (G), Maryland (G), Massachusetts (G), New Hampshire (G), New Jersey (G), New York (G), North Carolina (A), Oklahoma (G), Pennsylvania (G), West Virginia (G).

Description: Cadmium Acetate, $Cd(C_2H_3O_2)_2$, is a colorless crystalline solid. Freezing/Melting point = 130°C. Hazard Identification (based on NFPA-704 M Rating System): Health 2, Flammability 0, Reactivity 0. Soluble in water.

Potential Exposure: Cadmium Acetate is used in ceramics, textile dyeing and printing and electroplating and to make other acetate compounds.

Incompatibilities: Sulfides, strong acids.

Permissible Exposure Limits in Air: The Federal standard[58] for cadmium compounds is 0.005 mg/m^3 (as Cd) as an 8-hour TWA. ACGIH has recommended a TWA value of 0.01 mg/m^3 for Cd dusts/salts and 0.002 mg/m^3 for respirable dust. NIOSH recommends that exposure to cadmium be at the lowest feasible level. See also this section in the entry on Cadmium. North Carolina has set a guideline for cadmium acetate in ambient air[60] of 0.0055 μg/m^3.

Determination in Air: Collection of particles on a filter, workup with acid and measurement by atomic absorption has been specified by NIOSH. See NIOSH Methods #7048 (Cd).

Permissible Concentration in Water: See the entry on cadmium.

Determination in Water: Total cadmium may be determined by digestion followed by atomic absorption of colorimetric (Dithizone) analysis or by Inductively Coupled Plasma (ICP) Optical Emission Spectrometry. Dissolved cadmium is determined by 0.45 μ filtration followed by the previously cited methods. See also reverence.[49]

Routes of Entry: Inhalation of dust, ingestion.

Harmful Effects and Symptoms

Short Term Exposure: Cadmium Acetate can affect our when breathed in. Cadmium Acetate is a carcinogen; handle with extreme caution. High exposures can cause pulmonary edema, a medical emergency that can be delayed for several hours. This can cause death. Risk is greatest near dust or fume from heating or grinding cadmium acetate.

Long Term Exposure: Repeated lower exposure can cause permanent kidney damage, emphysema, low blood sugar and/or lowered sense of smell. Serious damage can occur at levels below the PEL. See also the preceding entry on Cadmium.

Medical Surveillance: Before beginning employment and at regular times after that, the following is recommended: Urine test for Cadmium (levels should be less than 10 micrograms per liter of urine). Urine test for "low molecular weight proteins" (electrophoresis method best). Urinalysis (UA). Complete blood count (CBC). Lung function test. These should be repeated after suspected overexposure. For those with frequent or potentially high exposure (half the TLV or greater), the following are recommended before beginning work and at regular times after that: Consider chest x-ray after acute overexposure.

First Aid: If this chemical gets into the eyes, remove any contact lenses at once and irrigate immediately for at least 15 minutes, occasionally lifting upper and lower lids. Seek medical attention immediately. If this chemical contacts the skin, remove contaminated clothing and wash immediately with soap and water. Seek medical attention immediately. If this chemical has been inhaled, remove from exposure, begin rescue breathing (using universal precautions) if breathing has stopped and CPR if heart action has stopped. Transfer promptly to a medical facility. When this chemical has been swallowed, get medical attention. Give large quantities of water and induce vomiting. Do not make an unconscious person vomit. Medical observation is recommended for 24 – 48 hours after breathing overexposure, as pulmonary edema may be delayed. As first aid for pulmonary edema, a doctor or authorized paramedic may consider administering a corticosteroid spray.

Personal Protective Methods: *Clothing:* Avoid skin contact with Cadmium Acetate. Wear protective gloves and clothing. Safety equipment suppliers/manufacturers can provide recommendations on the most protective glove/clothing material for your operation. All protective clothing (suits, gloves, footwear, headgear) should be clean, available each day and put on before work. *Eye Protection:* Eye protection is included in the recommended respiratory protection.

Respirator Selection: NIOSH (as Cd compounds): *at any detectable concentration:* SCBAF:PD,PP (any MSHA/NIOSH approved self-contained breathing apparatus that has a full facepiece and is operated in a pressure-demand or other positive-pressure mode); or SAF:PD,PP: ASCBA (any supplied-air respirator that has a full facepiece and is operated in a pressure-demand or other positive-pressure mode in combination with an auxiliary, self-contained breathing apparatus operated in a pressure-demand or other positive pressure mode). *Escape:* HiEF [any air-purifying, full-facepiece respirator (gas mask) with a chin-style, front- or back-mounted organic vapor canister having a high-efficiency particulate filter]; or SCBAE (any appropriate escape-type, self-contained breathing apparatus).

Storage: Prior to working with Cadmium Acetate you should be trained on its proper handling and storage. Store in tightly closed containers in a cool, well-ventilated area away from heat and incompatible materials listed above. A regulated, marked area should be established where this chemical is handled, used, or stored in compliance with OSHA standard 1910.1045.

Shipping: Cadmium acetate requires a "Poison" label. It falls in Hazard Class 6.1 and Packing Group II.

Spill Handling: Evacuate persons not wearing protective equipment from area of spill or leak until clean-up is complete. Remove all ignition sources. Collect powdered material in the most convenient and safe manner and deposit in sealed containers. Ventilate area after clean-up is complete. It may be necessary to contain and dispose of this chemical as a hazardous waste. If material or contaminated runoff enters waterways, notify downstream users of potentially contaminated waters. Contact your Department of Environmental Protection or your regional office of the federal EPA for specific recommendations. If employees are required to clean-up spills, they must be properly trained and equipped. OSHA 1910.120(q) may be applicable.

Fire Extinguishing: This chemical is a combustible solid. Use dry chemical, carbon dioxide, water spray, or alcohol foam extinguishers. Poisonous gases are produced in fire. If material or contaminated runoff enters waterways, notify downstream users of potentially contaminated waters. Notify local health and fire officials and pollution control agencies. From a secure, explosion-proof location, use water spray to cool exposed containers. If cooling streams are ineffective (venting sound increases in volume and pitch, tank discolors, or shows any signs of deforming), withdraw immediately to a secure position. If employees are expected to fight fires, they must be trained and equipped in OSHA 1910.156.

Cadmium Acetate itself does not burn. Poisonous gases are produced in fire, including Cadmium fumes. Extinguish fire using an agent suitable for type of surrounding fire. If material or contaminated runoff enters waterways, notify downstream users of potentially contaminated waters. Notify local health and fire officials and pollution control agencies. From a secure, explosion-proof location, use water spray to cool exposed containers. If cooling streams are ineffective (venting sound increases in volume and pitch, tank discolors, or shows any signs of deforming), withdraw immediately to a secure position. If employees are expected to fight fires, they must be trained and equipped in OSHA 1910.156.

Disposal Method Suggested: Precipitation as sulfide, drying and return to supplier.[22] Incineration is not recommended.

References

Sax, N. I., Ed., "Dangerous Properties of Industrial Materials Report" No. 4, 59–70 (1984)

New Jersey Department of Health and Senior Services, "Hazardous Substance Fact Sheet: Cadmium Acetate," Trenton, NJ (September 1986)

Cadmium Bromide

Molecular Formula: Br_2Cd

Common Formula: $CdBr_2$

Synonyms: Bromuro de cadmio (Spanish); Cadmium dibromide

CAS Registry Number: 7789-42-6

RTECS Number: EU9935000

DOT ID: UN 2570

Regulatory Authority

- Carcinogen (Sax, DPIMR)
- Air Pollutant Standard Set (See Cadmium) (North Carolina)[60]
- CLEAN AIR ACT: Hazardous Air Pollutants (Title I, Part A, Section 112). Includes any unique chemical substance that contains cadmium as part of that chemical's infrastructure
- CLEAN WATER ACT: Toxic Pollutant (Section 401.15); Section 311 Hazardous Substances/RQ 40CFR117.3 (same as CERCLA, see below); Section 313 Water Priority Chemicals (57FR41331, 9/9/92)
- SUPERFUND/EPCRA 40CFR302.4 Reportable Quantity (RQ): CERCLA, 10 lb (4.54 kg)
- EPCRA Section 313 Form R *de minimis* concentration reporting level: 0.1%
- U.S. DOT Regulated Marine Pollutant (49CFR172.101, Appendix B).: Severe pollutant, as cadmium compounds
- Canada, WHMIS, Ingredients Disclosure List; National Pollutant Release Inventory (NPRI); CEPA Priority Substance List, Ocean dumping prohibited, as cadmium compounds

Cited in U.S. State Regulations: Alaska (G), California (G), Maine (G), Maryland (G), Massachusetts (G), New Hampshire (G), New Jersey (G), New York (G), North Carolina (A), Oklahoma (G), Pennsylvania (G), West Virginia (G).

Description: $CdBr_2$ is a white to yellowish crystalline powder. Melting point = 567°C. Hazard Identification (based on NFPA-704 M Rating System): Health 2, Flammability 0, Reactivity 0. Soluble in water.

Potential Exposure: Cadmium Bromide is used in photography, engraving and lithography.

Permissible Exposure Limits in Air: The Federal standard[58] for cadmium compounds is 0.005 mg/m^3 (as Cd) as an 8-hour TWA. ACGIH has recommended a TWA value of 0.01 mg/m^3 for Cd dusts/salts and 0.002 mg/m^3 for respirable dust. NIOSH recommends that exposure to cadmium be at the lowest feasible level. See also this section in the entry on Cadmium.

Determination in Air: Collection of particles on a filter, workup with acid and measurement by atomic absorption has been specified by NIOSH. See NIOSH Methods #7048 (Cd).

Permissible Concentration in Water: See the entry on Cadmium.

Determination in Water: Total cadmium may be determined by digestion followed by atomic absorption of colorimetric (Dithizone) analysis or by Inductively Coupled Plasma (ICP) Optical Emission Spectrometry. Dissolved cadmium is determined by 0.45 μ filtration followed by the previously cited methods. See also reverence.[49]

Routes of Entry: Inhalation of dust, ingestion.

Harmful Effects and Symptoms

Short Term Exposure: Cadmium Bromide can affect you when breathed in. Cadmium Bromide is a carcinogen; handle with extreme caution. High exposures can cause pulmonary edema, a medical emergency that can be delayed for several hours. This can cause death. Risk is greatest near dust or fume from heating or grinding cadmium acetate.

Long Term Exposure: Repeated exposures can cause anemia, permanent kidney damage, emphysema, low blood count and/or loss of sense of smell, fatigue, yellow staining of teeth. Cadmium bromide should be handled as a potential teratogenic agent and reproductive hazard since several related cadmium compounds are know teratogens and decreases fertility in males and females. Repeated low exposures (below the OSHA PEL) can cause permanent kidney damage which can go unnoticed without testing until severe. Kidney stones can also occur. Emphysema and/or lung scarring can occur from a single high exposure or repeated lower exposures. Sax (see reference below) states that $CdBr_2$ is a recognized carcinogen of the connective tissue, lungs and liver.

Points of Attack: Respiratory system, kidneys, liver.

Medical Surveillance: Before beginning employment and at regular times after that, the following are recommended: Urine test for Cadmium (levels should be less than 10 micrograms of Cadmium per liter of urine). Urine test for low molecular weight proteins (electrophoresis method best). Urine analysis (UA). Complete blood count (CBC). Lung function tests. These should be repeated after suspected overexposure. If symptoms develop or overexposure is suspected, the following may be useful: Consider chest x-ray after acute overexposure.

First Aid: If this chemical gets into the eyes, remove any contact lenses at once and irrigate immediately for at least 15 minutes, occasionally lifting upper and lower lids. Seek medical attention immediately. If this chemical contacts the skin, remove contaminated clothing and wash immediately with soap and water. Seek medical attention immediately. If this chemical has been inhaled, remove from exposure, begin rescue breathing (using universal precautions) if breathing has stopped and CPR if heart action has stopped. Transfer promptly to a medical facility. When this chemical has been swallowed, get medical attention. Give large quantities of water and induce vomiting. Do not make an unconscious person vomit. Medical observation is recommended for 24 – 48 hours after breathing overexposure, as pulmonary edema may be delayed. As first aid for pulmonary edema, a doctor or authorized paramedic may consider administering a corticosteroid spray.

Personal Protective Methods: *Clothing:* Avoid skin contact with Cadmium Bromide. Wear protective gloves and clothing. Safety equipment suppliers/manufacturers can provide recommendations on the most protective glove/clothing material for your operation. All protective clothing (suits, gloves, footwear, headgear) should be clean, available each day and put on before work. *Eye Protection:* Eye protection is included in the recommended respiratory protection.

Respirator Selection: NIOSH (as Cd compounds): *at any detectable concentration:* SCBAF:PD,PP (any MSHA/NIOSH approved self-contained breathing apparatus that has a full facepiece and is operated in a pressure-demand or other positive-pressure mode); or SAF:PD,PP:ASCBA (any supplied-air respirator that has a full facepiece and is operated in a pressure-demand or other positive-pressure mode in combination with an auxiliary, self-contained breathing apparatus operated in a pressure-demand or other positive pressure mode). *Escape:* HiEF [any air-purifying, full-facepiece respirator (gas mask) with a chin-style, front- or back-mounted organic vapor canister having a high-efficiency particulate filter]; or SCBAE (any appropriate escape-type, self-contained breathing apparatus).

Storage: Prior to working with Cadmium Bromide you should be trained on its proper handling and storage. Cadmium Bromide must be stored to avoid contact with Potassium since violent reactions occur. A regulated, marked area should be established where this chemical is handled, used, or stored in compliance with OSHA standard 1910.1045.

Shipping: Cadmium bromide requires a "Poison" label. It falls in Hazard Class 6.1 and Packing Group II.

Spill Handling: Evacuate persons not wearing protective equipment from area of spill or leak until clean-up is complete. Remove all ignition sources. Collect powdered material in the most convenient and safe manner and deposit in sealed containers. Ventilate area after clean-up is complete. It may be necessary to contain and dispose of this chemical as a hazardous waste. If material or contaminated runoff enters waterways, notify downstream users of potentially contaminated waters. Contact your Department of Environmental Protection or your regional office of the federal EPA for specific recommendations. If employees are required to clean-up spills, they must be properly trained and equipped. OSHA 1910.120(q) may be applicable.

Fire Extinguishing: Extinguish fire using an agent suitable for type of surrounding fire. Cadmium Bromide itself does not burn. Poisonous gases are produced in fire including cadmium and bromine. If material or contaminated runoff enters waterways, notify downstream users of potentially contaminated waters. Notify local health and fire officials and pollution control agencies. From a secure, explosion-proof location, use water spray to cool exposed containers. If cooling streams are ineffective (venting sound increases in volume and pitch, tank discolors, or shows any signs of deforming), withdraw immediately to a secure position. If employees are expected to fight fires, they must be trained and equipped in OSHA 1910.156.

References

New Jersey Department of Health and Senior Services, "Hazardous Substance Fact Sheet: Cadmium Bromide," Trenton, NJ (November 1986)

Sax, N. I.., Ed., "Dangerous Properties of Industrial Materials Report" 3, No. 5, 76–79 (1983)

Cadmium Chloride

Molecular Formula: $CdCl_2$

Synonyms: Caddy; Cadmium dichloride; Cloruro de cadmio (Spanish); Dichlorocadmium; Kadmiumchlorid (Germany); VI-CAD

CAS Registry Number: 10108-64-2

RTECS Number: EV0175000

EEC Number: 048-008-00-3

Regulatory Authority

- Carcinogen (UK) (DFG)[3] (suspected) (NTP)
- Air Pollutant Standard Set (see Cadmium)
- Banned or Severely Restricted (In pesticides in UK)[13] (other) (UN)[35]
- Air Pollutant Standard Set (ACGIH)[1] (Several States)
- CLEAN AIR ACT: Hazardous Air Pollutants (Title I, Part A, Section 112). Includes any unique chemical substance that contains cadmium as part of that chemical's infrastructure
- CLEAN WATER ACT: Toxic Pollutant (Section 401.15); Section 311 Hazardous Substances/RQ 40CFR117.3 (same as CERCLA, see below); Section 313 Water Priority Chemicals (57FR41331, 9/9/92)
- SUPERFUND/EPCRA 40CFR302.4 Reportable Quantity (RQ): CERCLA, 10 lb (4.54 kg)
- EPCRA Section 313 Form R *de minimis* concentration reporting level: 0.1%
- U.S. DOT Regulated Marine Pollutant (49CFR172.101, Appendix B): Severe pollutant, as cadmium compounds
- Canada, WHMIS, Ingredients Disclosure List; National Pollutant Release Inventory (NPRI); CEPA Priority Substance List, Ocean dumping prohibited, as cadmium compounds

Cited in U.S. State Regulations: Alaska (G), California (G), Florida (G, A), Maine (G), Maryland (G), New Hampshire (G), New York (G, A), Oklahoma (G), Pennsylvania (G), West Virginia (G).

Description: $CdCl_2$ is a colorless, odorless, crystalline solid or powder. Boiling point = 960°C. Freezing/Melting point = 568°C. Hazard Identification (based on NFPA-704 M Rating System): Health 2, Flammability 0, Reactivity 0. Soluble in water.

Potential Exposure: Cadmium chloride is used in dyeing and printing of fabrics; in electronic component manufacture; in photography.

Incompatibilities: Sulfur, selenium, potassium, and strong oxidizers.

Permissible Exposure Limits in Air: The Federal standard[58] for cadmium compounds is 0.005 mg/m^3 (as Cd) as an 8-hour TWA. ACGIH has recommended a TWA value of 0.01 mg/m^3 for Cd dusts/salts and 0.002 mg/m^3 for respirable dust. NIOSH recommends that exposure to cadmium be at the lowest feasible level. See also this section in the entry on Cadmium. Guidelines for cadmium chloride in ambient air have been set[60] ranging from 1.67 μg/m^3 (New York) to 5.0 μg/m^3 (Florida).

Determination in Air: Collection of particles on a filter, workup with acid and measurement by atomic absorption has been specified by NIOSH. See NIOSH Methods #7048 (Cd).

Permissible Concentration in Water: See the entry on Cadmium.

Determination in Water: Total cadmium may be determined by digestion followed by atomic absorption of colorimetric (Dithizone) analysis or by Inductively Coupled Plasma (ICP) Optical Emission Spectrometry. Dissolved cadmium is determined by 0.45 μ filtration followed by the previously cited methods. See also reverence.[49]

Routes of Entry: Inhalation of dust, ingestion.

Harmful Effects and Symptoms

Short Term Exposure: Eye contact can cause irritation. Cadmium chloride can cause severe irritation of the gastrointestinal tract and the respiratory tract. Inhalation can cause nose, throat, and lung irritation. Fumes can cause flu-like illness with chills, headache, aching muscles and/or fever. Higher exposures can cause nausea, salivation, vomiting, cramps, diarrhea, and pulmonary edema, a medical emergency that can be delayed for several hours. This can cause death. Cadmium chloride is highly toxic. As little as 14.5 mg of Cd orally causes nausea and vomiting; 8.9 grams has caused death. Cadmium salts cause cramps, nausea, vomiting and diarrhea. Acute poisoning causes lung damage.

Long Term Exposure: This chemical is a probable carcinogen in humans, with some evidence that it causes prostate and kidney cancer in humans and it has been shown to cause lung and testes cancer in animals. It may also be a reproductive hazard in humans. Repeated low exposures may cause permanent kidney and liver damage, anemia, and/or loss of the sensed of smell. Chronic poisoning damages kidneys, lungs, bones and causes blood changes (anemia).

Points of Attack: See above on harmful effects and symptoms.

Medical Surveillance: Urine test for cadmium. Urine test for "low molecular weight proteins" to detect kidney damage. Urinalysis. Complete Blood Count (CBC). Lung function tests. Consider chest x-ray following acute overexposure.

First Aid: If this chemical gets into the eyes, remove any contact lenses at once and irrigate immediately for at least 15 minutes, occasionally lifting upper and lower lids. Seek medical attention immediately. If this chemical contacts the skin, remove contaminated clothing and wash immediately with soap and water. Seek medical attention immediately. If this chemical has been inhaled, remove from exposure, begin rescue breathing (using universal precautions) if breathing has stopped and CPR if heart action has stopped. Transfer promptly to a medical facility. When this chemical has been swallowed, get medical attention. Give large quantities of water and induce vomiting. Do not make an unconscious person vomit. Medical observation is recommended for 24 – 48 hours after breathing overexposure, as pulmonary edema may be delayed. As first aid for pulmonary edema, a doctor or authorized paramedic may consider administering a corticosteroid spray.

Personal Protective Methods: Wear protective gloves and clothing to prevent any reasonable probability of skin contact. Safety equipment suppliers/manufacturers can provide recommendations on the most protective glove/clothing material for your operation. All protective clothing (suits, gloves, footwear, headgear) should be clean, available each day, and put on before work. Contact lenses should not be worn when working with this chemical. Wear dust-proof chemical goggles and face shield unless full facepiece respiratory protection is worn. Employees should wash immediately with soap when skin is wet or contaminated. Provide emergency showers and eyewash.

Respirator Selection: NIOSH (as Cd compounds): *at any detectable concentration:* SCBAF:PD,PP (any MSHA/NIOSH approved self-contained breathing apparatus that has a full facepiece and is operated in a pressure-demand or other positive-pressure mode); or SAF:PD,PP:ASCBA (any supplied-air respirator that has a full facepiece and is operated in a pressure-demand or other positive-pressure mode in combination with an auxiliary, self-contained breathing apparatus operated in a pressure-demand or other positive pressure mode). *Escape:* HiEF [any air-purifying, full-facepiece respirator (gas mask) with a chin-style, front- or back-mounted organic vapor canister having a high-efficiency particulate filter]; or SCBAE (any appropriate escape-type, self-contained breathing apparatus).

Storage: Store in tightly closed containers. Avoid contact with strong acids and oxidizers or moisture.

Shipping: Cadmium chloride requires a "Keep Away from Food" label. It falls in Hazard Class 6.1 and Packing Group III.

Spill Handling: Evacuate persons not wearing protective equipment from area of spill or leak until clean-up is complete. Remove all ignition sources. Collect powdered material in the most convenient and safe manner and deposit in sealed containers. Ventilate area after clean-up is complete. It may be necessary to contain and dispose of this chemical as a hazardous waste. If material or contaminated runoff enters waterways, notify downstream users of potentially contaminated waters. Contact your Department of Environmental Protection or your regional office of the federal EPA for specific recommendations. If employees are required to clean-up spills, they must be properly trained and equipped. OSHA 1910.120(q) may be applicable.

Fire Extinguishing: Use agent suitable for surrounding fire. Cadmium chloride itself does not burn. Poisonous gases are produced in fire including cadmium and chlorine. If material or contaminated runoff enters waterways, notify downstream users of potentially contaminated waters. Notify local health and fire officials and pollution control agencies. From a secure, explosion-proof location, use water spray to cool exposed containers. If cooling streams are ineffective (venting sound increases in volume and pitch, tank discolors, or shows any signs of deforming), withdraw immediately to a secure position. If employees are expected to fight fires, they must be trained and equipped in OSHA 1910.156.

Disposal Method Suggested: It is preferred to convert the salt to the nitrate, precipitate it with H_2S, filter, wash and dry the precipitate and return it to the supplier.

References

New Jersey Department of Health and Senior Services and Senior Services, "Hazardous Substance Fact Sheet, Cadmium Chloride," Trenton, NJ (January, 1996)

Sax, N. I.., Ed., "Dangerous Properties of Industrial Materials Report" 2, No. 3, 73–76 (1982)

Cadmium Oxide

Molecular Formula: CdO

Synonyms: Cadmium oxide brown; Kadmu tlenek (Polish); Oxido de cadmio (Spanish)

CAS Registry Number: 1306-19-0

RTECS Number: EV1925000

DOT ID: UN 2570 (cadmium compounds)

EEC Number: 048-002-00-0

EINECS Number: 215-146-2

Regulatory Authority

- Carcinogen (Animal Positive) (IARC)[9] (Suspected Human) (ACGIH)[1] (DFG)[3]
- Air Pollutant Standard Set (See also cadmium) (Australia) (Israel) (Mexico) (Several States) (Several Canadian Provinces) (HSE)
- Banned or Severely Restricted (Czechoslovakia, Germany) (UN)[35]
- Air Pollutant Standard Set (ACGIH)[1] (OSHA)[2] (Several States)[6]
- CLEAN AIR ACT: Hazardous Air Pollutants (Title I, Part A, Section 112). Includes any unique chemical substance that contains cadmium as part of that chemical's infrastructure
- CLEAN WATER ACT: Toxic Pollutant (Section 401.15), as cadmium compounds
- SUPERFUND/EPCRA 40CFR302.4 Reportable Quantity (RQ): EHS, 1 lb (0.454 kg)
- EPCRA Section 313 Form R *de minimis* concentration reporting level: 0.1%
- SUPERFUND/EPCRA 40CFR355, Appendix B Extremely Hazardous Substances: TPQ = 10/10,000 lb (4.54/4,540 kg)
- U.S. DOT Regulated Marine Pollutant (49CFR172.101, Appendix B).: Severe pollutant, as cadmium compounds
- Canada, WHMIS, Ingredients Disclosure List; National Pollutant Release Inventory (NPRI); CEPA Priority Substance List, Ocean dumping prohibited, as cadmium compounds

Cited in U.S. State Regulations: Alaska (G), California (G), Connecticut (A), Florida (G, A), Illinois (G), Maine (G), Maryland (G), Massachusetts (G), Nevada (A), New Hampshire (G), New York (G, A), Oklahoma (G), Pennsylvania (G), South Carolina (A), Virginia (A), West Virginia (G).

Description: Cadmium oxide, CdO, forms brownish-red crystals or a yellow to dark brown amorphous powder. Freezing/Melting point = 900°C (slow decomposition begins at 700°C). Sublimation point = 1,559°C. Hazard Identification (based on NFPA-704 M Rating System): Health 2, Flammability 0, Reactivity 0. Very slightly soluble in water.

Potential Exposure: Cadmium oxide is used as an electroplating chemical and in the manufacture of semiconductors and cadmium electrodes. It is a component of silver alloys, phosphorus, glass and ceramic glazes, semiconductors, and batteries. Used as a vermicide.

Incompatibilities: Oxides of cadmium react explosively with magnesium, especially when heated. Heat above 700°C causes slow decomposition. Not compatible with oxidizers. May ignite combustibles such as wood, paper, oil, etc.

Permissible Exposure Limits in Air: The Federal standard[58] for cadmium compounds is 0.005 mg/m^3 (as Cd) as an 8-hour TWA. ACGIH has recommended a TWA value of 0.01 mg/m^3 TWA for Cd dusts/salts and 0.002 mg/m^3 TWA for respirable dust. NIOSH recommends that exposure to cadmium be at the lowest feasible level. See also this section in the entry on Cadmium. The British HSE[33] has set an 8-hour TWA of 50 μg/m^3 and a STEL value of 50 mg/m^3 also. The German DFG[3] has set no numerical limits, indicating simply that the substance is a suspect carcinogen. The former USSR-UNEP/IRPTC project[43] has set a MAC in the workplace of 100 μg/m^3 but an 8-hour average MAC of 30 μg/m^3. Several states have set guidelines or standards for cadmium oxide in ambient air[60] ranging from 0.167 μg/m^3 (New York) to 0.25 μg/m^3 (South Carolina) to 0.40 μg/m^3 (Virginia) to 0.5 μg/m^3 (Florida) to 1.0 μg/m^3 (Nevada). The NIOSH IDLH: Carcinogen [9 mg/m^3 (as Cd)].

Determination in Air: Collection of particles on a filter, workup with acid and measurement by atomic absorption has been specified by NIOSH. See NIOSH Methods #7048 (Cd).

Permissible Concentration in Water: See entry under "Cadmium and Compounds."

Determination in Water: Total cadmium may be determined by digestion followed by atomic absorption of colorimetric (Dithizone) analysis or by Inductively Coupled Plasma (ICP) Optical Emission Spectrometry. Dissolved cadmium is determined by 0.45 μ filtration followed by the previously cited methods. See also reverence.[49]

Routes of Entry: Inhalation and ingestion.

Harmful Effects and Symptoms

Short Term Exposure: Eye contact causes irritation. Inhalation can cause irritation of the nose and throat; irritation of the lungs with coughing and shortness of breath. Higher exposures can cause pulmonary edema, a medical emergency that can be delayed for several hours. This can cause death. Cadmium oxide can cause metal fume fever with chills, headache, aching muscles, metallic taste, and/or fever. Exposure can cause nausea, salivation, vomiting, cramps, and diarrhea. Symptoms for cadmium poisoning include metallic taste in the mouth, headache, shortness of breath, chest pain, cough with foamy or bloody sputum, pulmonary rales, weakness, leg pains and pulmonary edema. The lethal inhalation dose of cadmium oxide in humans is 2,500 mg/m^3 for a minute exposure. Lethal exposure has been established at 50 mg (cadmium)/m^3 for 1 hour for cadmium oxide dust and ½ hour for the fume. These concentrations may be inhaled without sufficient discomfort to warn worker of exposure. Acute exposure by inhalation may cause death by anoxia. The lowest human toxic inhalation concentration is 8,630 μg/m^3/5 hours for the fume.

Long Term Exposure: Lungs (tracheobronchitis, pneumonitis), kidneys (possible kidney stones), and liver may be affected or damaged by repeated or prolonged exposure. Cadmium oxide is probably carcinogenic to humans;

there is some evidence of lung and prostate cancer. There is limited evidence that this chemical is a teratogen in animals. Long term exposure can cause anemia, brittle and painful bones, loss of sense of smell, fatigue, and/or yellow staining of teeth.

Medical Surveillance: Urine test for cadmium. Urine test for "low molecular weight proteins" to detect kidney damage. Urinalysis. Complete Blood Count (CBC). Lung function tests. Liver function tests. Consider chest x-ray following acute overexposure.

First Aid: If this chemical gets into the eyes, remove any contact lenses at once and irrigate immediately for at least 15 minutes, occasionally lifting upper and lower lids. Seek medical attention immediately. If this chemical contacts the skin, remove contaminated clothing and wash immediately with soap and water. Seek medical attention immediately. If this chemical has been inhaled, remove from exposure, begin rescue breathing (using universal precautions) if breathing has stopped and CPR if heart action has stopped. Transfer promptly to a medical facility. When this chemical has been swallowed, get medical attention. Give large quantities of water and induce vomiting. Do not make an unconscious person vomit. Medical observation is recommended for 24 – 48 hours after breathing overexposure, as pulmonary edema may be delayed. As first aid for pulmonary edema, a doctor or authorized paramedic may consider administering a corticosteroid spray.

Note to Physician: In case of fume inhalation, treat pulmonary edema. Give prednisone or other corticosteroid orally to reduce tissue response to fume. Positive pressure ventilation may be necessary. Treat metal fume fever with bed rest, analgesics and antipyretics. The symptoms of metal fume fever may be delayed for 4 – 12 hours following exposure: it may last less than 36 hours.

Personal Protective Methods: Persons with respiratory disorders should be excluded from contact with this material. A Class I, Type B, biological safety hood should be used when mixing, handling, or preparing cadmium oxide. Wear protective gloves and clothing to prevent any skin contact. Safety equipment suppliers/manufacturers can provide recommendations on the most protective glove/clothing material for your operation. All protective clothing (suits, gloves, footwear, headgear) should be clean, available each day, and put on before work. Contact lenses should not be worn when working with this chemical. Wear dust-proof chemical goggles and face shield unless full facepiece respiratory protection is worn. Employees should wash immediately with soap when skin is wet or contaminated. Provide emergency showers and eyewash.

Respirator Selection: NIOSH (as Cd compounds): *at any detectable concentration:* SCBAF:PD,PP (any MSHA/NIOSH approved self-contained breathing apparatus that has a full facepiece and is operated in a pressure-demand or other positive-pressure mode); or SAF:PD,PP:ASCBA (any supplied-air respirator that has a full facepiece and is operated in a pressure-demand or other positive-pressure mode in combination with an auxiliary, self-contained breathing apparatus operated in a pressure-demand or other positive pressure mode). *Escape:* HiEF [any air-purifying, full-facepiece respirator (gas mask) with a chin-style, front- or back-mounted organic vapor canister having a high-efficiency particulate filter]; or SCBAE (any appropriate escape-type, self-contained breathing apparatus).

Storage: Prior to working with cadmium oxide you should be trained on its proper handling and storage. Store in tightly closed containers in a cool, well-ventilated area, away from magnesium, oxidizers, combustible materials, heat, moisture, and acids. Where this chemical is used, handled, manufactured, or stored, use explosion-proof electrical equipment and fittings. A regulated, marked area should be established where this chemical is handled, used, or stored in compliance with OSHA standard 1910.1045.

Shipping: The DOT label requirement is "Poison." The UN/DOT Hazard Class is 6.1 and Packing Group I.[19][20]

Spill Handling: Evacuate persons not wearing protective equipment from area of spill or leak until clean-up is complete. Remove all ignition sources. Stay upwind and out of low areas. Collect powdered material in the most convenient and safe manner and deposit in sealed containers. Ventilate area after clean-up is complete. It may be necessary to contain and dispose of this chemical as a hazardous waste. If material or contaminated runoff enters waterways, notify downstream users of potentially contaminated waters. Contact your Department of Environmental Protection or your regional office of the federal EPA for specific recommendations. If employees are required to clean-up spills, they must be properly trained and equipped. OSHA 1910.120(q) may be applicable.

Fire Extinguishing: This chemical is a noncombustible solid, but may increase fire activity. Use any extinguishing agent. Poisonous gases are produced in fire including toxic fumes of cadmium. If material or contaminated runoff enters waterways, notify downstream users of potentially contaminated waters. Notify local health and fire officials and pollution control agencies. Containers may explode in fire. From a secure, explosion-proof location, use water spray to cool exposed containers. If cooling streams are ineffective (venting sound increases in volume and pitch, tank discolors, or shows any signs of deforming), withdraw immediately to a secure position. If employees are expected to fight fires, they must be trained and equipped in OSHA 1910.156.

Disposal Method Suggested: Form nitrate with HNO_3, precipitate with H_2S, filter, package and return to supplier.[22]

References

Sax, N. I., Ed., "Dangerous Properties of Industrial Materials Report" 4, No. 4, 77–83 (1984) (Cadmium Oxide Fumes)

U.S. Environmental Protection Agency, "Chemical Profile: Cadmium Oxide," Washington, DC, Chemical Emergency Preparedness Program (November 30, 1987)

New Jersey Department of Health and Senior Services and Senior Services, "Hazardous Substance Fact Sheet, Cadmium Oxide," Trenton, NJ (September, 1998)

Cadmium Stearate

Molecular Formula: $C_{36}H_{72}CdO_4$

Common Formula: $Cd(C_{17}H_{36}COO)_2$

Synonyms: Alaixol II; Cadmium octadeconoate; Estearato de cadmio (Spanish); Kadmiumstearat (German); Octadecanoic acid, cadmium salt; Stearic acid, cadmium salt

CAS Registry Number: 2223-93-0

RTECS Number: RG1050000

DOT ID: UN 2570 (Cadmium compound)

Regulatory Authority

- Air Pollutant Standard Set (See cadmium)
- CLEAN AIR ACT: Hazardous Air Pollutants (Title I, Part A, Section 112). Includes any unique chemical substance that contains cadmium as part of that chemical's infrastructure
- CLEAN WATER ACT: Toxic Pollutant (Section 401.15), as cadmium compounds
- SUPERFUND/EPCRA 40CFR302.4 Reportable Quantity (RQ): EHS, 1 lb (0.454 kg)
- EPCRA Section 313 Form R *de minimis* concentration reporting level: 1.0%
- SUPERFUND/EPCRA 40CFR355, Appendix B Extremely Hazardous Substances: TPQ = 1,000/10,000 lb (454/4,540 kg)
- U.S. DOT Regulated Marine Pollutant (49CFR172.101, Appendix B).: Severe pollutant, as cadmium compounds
- Canada, WHMIS, Ingredients Disclosure List; National Pollutant Release Inventory (NPRI); CEPA Priority Substance List, Ocean dumping prohibited, as cadmium compounds

Cited in U.S. State Regulations: Alaska (G), California (G), Florida (G), Maine (G), Maryland (G), Massachusetts (G), New Jersey (G), New York (G), Oklahoma (G), Pennsylvania (G), West Virginia (G).

Description: Cadmium stearate, $Cd(C_{17}H_{36}COO)_2$, is a crystalline solid. Hazard Identification (based on NFPA-704 M Rating System): Health 2, Flammability 1, Reactivity 0.

Potential Exposure: Used as a lubricant and stabilizer in polyvinyl chloride plastics.

Permissible Exposure Limits in Air: The Federal standard[58] for cadmium compounds is 0.005 mg/m^3 (as Cd) as an 8-hour TWA. ACGIH has recommended a TWA value of 0.01 mg/m^3 for Cd dusts/salts and 0.002 mg/m^3 for respirable dust. NIOSH recommends that exposure to cadmium be at the lowest feasible level. See also this section in the entry on Cadmium. The NIOSH IDLH: Carcinogen [9 mg/m^3 (as Cd)]. The former USSR-UNEP/IRPTC project sets a MAC in workplace air of 0.1 mg/m^3.[43]

Determination in Air: Collection of particles on a filter, workup with acid and measurement by atomic absorption has been specified by NIOSH. See NIOSH Methods #7048 (Cd).

Permissible Concentration in Water: See entry on Cadmium and compounds.

Determination in Water: Total cadmium may be determined by digestion followed by atomic absorption of colorimetric (Dithizone) analysis or by Inductively Coupled Plasma (ICP) Optical Emission Spectrometry. Dissolved cadmium is determined by 0.45 μ filtration followed by the previously cited methods. See also reverence.[49]

Routes of Entry: Inhalation and ingestion.

Harmful Effects and Symptoms

Short Term Exposure: Acute poisoning produces severe nausea, vomiting, diarrhea, and abdominal and chest pains. Dry mouth, salivation, and metallic taste have been reported. If ingested, may result in exhaustion, collapse, shock and death within a period of 24 hours. Acute toxicity most notably occurs secondary to cadmium ingestion or inhalation of cadmium fumes. Poisoning from inhalation is relatively rare but dangerous, having a mortality rate of about 15 percent. Toxic inhaled concentrations in humans have been reported at 147 mg/m^3/35 minutes and at 1,800 μg/m^3/2 years. The oral LD_{50} rat is 1,125 mg/kg; the oral LD_{50} mouse is 590 mg/kg.

Long Term Exposure: Yellow rings may be seen in teeth when chronically exposed. A probable human carcinogen and teratogen.

First Aid: If this chemical gets into the eyes, remove any contact lenses at once and irrigate immediately for at least 15 minutes, occasionally lifting upper and lower lids. Seek medical attention immediately. If this chemical contacts the skin, remove contaminated clothing and wash immediately with soap and water. Seek medical attention immediately. If this chemical has been inhaled, remove from exposure, begin rescue breathing (using universal precautions) if breathing has stopped and CPR if heart action has stopped. Transfer promptly to a medical facility. When this chemical has been swallowed, get medical attention. Give large quantities of water and induce vomiting. Do not make an unconscious person vomit.

Personal Protective Methods: Wear protective gloves and clothing to prevent any reasonable probability of skin contact. Safety equipment suppliers/manufacturers can provide recommendations on the most protective glove/clothing material for your operation. All protective clothing (suits, gloves, footwear, headgear) should be clean, available each day, and put on before work. Contact lenses should not be worn when working with this chemical. Wear dust-proof chemical goggles and face shield unless full facepiece respiratory

protection is worn. Employees should wash immediately with soap when skin is wet or contaminated. Provide emergency showers and eyewash.

Respirator Selection: NIOSH (as Cd compounds): *at any detectable concentration:* SCBAF:PD,PP (any MSHA/NIOSH approved self-contained breathing apparatus that has a full facepiece and is operated in a pressure-demand or other positive-pressure mode); or SAF:PD,PP:ASCBA (any supplied-air respirator that has a full facepiece and is operated in a pressure-demand or other positive-pressure mode in combination with an auxiliary, self-contained breathing apparatus operated in a pressure-demand or other positive pressure mode). *Escape:* HiEF [any air-purifying, full-facepiece respirator (gas mask) with a chin-style, front- or back-mounted organic vapor canister having a high-efficiency particulate filter]; or SCBAE (any appropriate escape-type, self-contained breathing apparatus).

Storage: Prior to working with Cadmium stearate you should be trained on its proper handling and storage. A regulated, marked area should be established where this chemical is handled, used, or stored. Store in tightly closed containers in a cool, well-ventilated area, away from incompatible materials. Where this chemical is used, handled, manufactured, or stored, use explosion-proof electrical equipment and fittings.

Shipping: Cadmium stearate falls under Cadmium Compounds is DOT regulations. It is in Hazard Class 6.1 and Packing Group III.

Spill Handling: Do not touch spilled material. Stay upwind, keep out of low areas. Wear self-contained (positive pressure if available) breathing apparatus and full protective clothing. Evacuate persons not wearing protective equipment from area of spill or leak until clean-up is complete. Remove all ignition sources. Collect powdered material in the most convenient and safe manner and deposit in sealed containers. Ventilate area after clean-up is complete. It may be necessary to contain and dispose of this chemical as a hazardous waste. If material or contaminated runoff enters waterways, notify downstream users of potentially contaminated waters. Contact your Department of Environmental Protection or your regional office of the federal EPA for specific recommendations. If employees are required to clean-up spills, they must be properly trained and equipped. OSHA 1910.120(q) may be applicable.

Fire Extinguishing: Use dry chemical, carbon dioxide, water spray, or alcohol foam extinguishers. Poisonous gases are produced in fire. If material or contaminated runoff enters waterways, notify downstream users of potentially contaminated waters. Notify local health and fire officials and pollution control agencies. From a secure, explosion-proof location, use water spray to cool exposed containers. If cooling streams are ineffective (venting sound increases in volume and pitch, tank discolors, or shows any signs of deforming), withdraw immediately to a secure position. If employees are expected to fight fires, they must be trained and equipped in OSHA 1910.156.

References

U.S. Environmental Protection Agency, "Chemical Profile: Cadmium Stearate," Washington, DC, Chemical Emergency Preparedness Program (November 30, 1987)

Cadmium Sulfate

Molecular Formula: $O_4S{\cdot}Cd$

Synonyms: Cadmium monosulfate; Cadmium sulphate; Sulfuric acid, cadmium (2+) salt; Sulfuric acid, cadmium(II) salt; Sulphuric acid, cadmium salt

CAS Registry Number: 10124-36-4

DOT ID: UN 2570

Regulatory Authority

- Carcinogen (suspect) (NTP) (IARC) (animal positive) (DFG)
- CLEAN AIR ACT: Hazardous Air Pollutants (Title I, Part A, Section 112). Includes any unique chemical substance that contains cadmium as part of that chemical's infrastructure
- CLEAN WATER ACT: 40CFR401.15 Section 307 Toxic Pollutants
- RCRA, 40CFR261, Appendix 8 Hazardous Constituents, waste number not listed
- EPCRA (Section 313): Includes any unique chemical substance that contains cadmium as part of that chemical's infrastructure. Form R *de minimis* concentration reporting level: (inorganic compounds: 0.1%; organic compounds: 1.0%)
- Canada, WHMIS, Ingredients Disclosure List; National Pollutant Release Inventory (NPRI); CEPA Priority Substance List, Ocean dumping prohibited
- U.S. DOT Regulated Marine Pollutant (49CFR172.101, Appendix B).: Severe pollutant, as cadmium compounds
- Canada, WHMIS, Ingredients Disclosure List

Cited in U.S. State Regulations: California (A, G), Florida (G), Massachusetts (G), New Jersey (G), Pennsylvania (G)

Description: Cadmium sulfate is a white to colorless, odorless, crystalline substance. Freezing/Melting point = 1,000°C. Hazard Identification (based on NFPA-704 M Rating System): Health 2, Flammability 0, Reactivity 0. Soluble in water.

Potential Exposure: It is used in pigments, electroplating, as a fungicide, and in synthetic and analytical chemistry. Also used in fluorescent screens, as an electrolyte.

Incompatibilities: Incompatible with strong oxidizers, sulfur, selenium, tellurium, zinc. May ignite combustible materials.

Permissible Exposure Limits in Air: The Federal standard[58] for cadmium compounds is 0.005 mg/m^3 (as Cd) as an 8-hour TWA. ACGIH has recommended a TWA value of

0.01 mg/m^3 for Cd dusts/salts and 0.002 mg/m^3 for respirable dust. NIOSH recommends that exposure to cadmium be at the lowest feasible level.

Determination in Air: Collection of particles on a filter, workup with acid and measurement by atomic absorption has been specified by NIOSH. See NIOSH Methods #7048 (Cd).

Permissible Concentration in Water: See the entry on cadmium.

Determination in Water: Total Cadmium may be determined by digestion followed by atomic absorption of colorimetric (Dithizone) analysis or by Inductively Coupled Plasma (ICP) Optical Emission Spectrometry. Dissolved Cadmium is determined by 0.45 μ filtration followed by the previously cited methods. See also reverence.[49]

Routes of Entry: Inhalation, ingestion.

Harmful Effects and Symptoms

Short Term Exposure: Irritates the eyes on contact. Inhalation irritates the nose, throat and lungs with coughing, and/or shortness of breath. Higher exposures can cause pulmonary edema, a medical emergency that can be delayed for several hours. This can cause death. Cadmium sulfate can cause nausea, salivation, vomiting, cramps and diarrhea, metal fume fever with flu-like symptoms, chills, headache, weakness, metallic tasted in the mouth.

Long Term Exposure: Repeated exposure can cause anemia, brittle and painful bones, diminished or loss of the sense of smell, fatigue, and/or yellow staining of the teeth. May cause lung and prostate cancer, kidney damage with kidney stones, liver damage, lung damage with bronchitis, cough, phlegm, and/or shortness of breath. There is some evidence that Cadmium sulfate is a teratogen in humans.

Points of Attack: Lungs, liver, kidneys, blood.

Medical Surveillance: Urine test for Cd (levels should be less than 10 μg/l of urine). Urine test for low molecular weight proteins (β-2-microglobulin) to detect kidney damage. Complete blood Count (CBC). Lung function tests. Liver function tests. Consider chest x-ray following acute overexposure.

First Aid: If this chemical gets into the eyes, remove any contact lenses at once and irrigate immediately for at least 15 minutes, occasionally lifting upper and lower lids. Seek medical attention immediately. If this chemical contacts the skin, remove contaminated clothing and wash immediately with soap and water. Seek medical attention immediately. If this chemical has been inhaled, remove from exposure, begin rescue breathing (using universal precautions) if breathing has stopped and CPR if heart action has stopped. Transfer promptly to a medical facility. When this chemical has been swallowed, get medical attention. Give large quantities of water and induce vomiting. Do not make an unconscious person vomit. Medical observation is recommended for 24 – 48 hours after breathing overexposure, as pulmonary edema may be delayed. As first aid for pulmonary edema, a doctor or authorized paramedic may consider administering a corticosteroid spray.

Note to Physician: In case of fume inhalation, treat pulmonary edema. Give prednisone or other corticosteroid orally to reduce tissue response to fume. Positive pressure ventilation may be necessary. Treat metal fume fever with bed rest, analgesics and antipyretics. The symptoms of metal fume fever may be delayed for 4 – 12 hours following exposure: it may last less than 36 hours.

Personal Protective Methods: Wear protective gloves and clothing to prevent any reasonable probability of skin contact. Safety equipment suppliers/manufacturers can provide recommendations on the most protective glove/clothing material for your operation. All protective clothing (suits, gloves, footwear, headgear) should be clean, available each day, and put on before work. Contact lenses should not be worn when working with this chemical. Wear dust-proof chemical goggles and face shield unless full facepiece respiratory protection is worn. Employees should wash immediately with soap when skin is wet or contaminated. Provide emergency showers and eyewash.

Respirator Selection: NIOSH (as Cd compounds): *At any detectable concentration:* SCBAF:PD,PP (any MSHA/NIOSH approved self-contained breathing apparatus that has a full facepiece and is operated in a pressure-demand or other positive-pressure mode); or SAF:PD,PP:ASCBA (any supplied-air respirator that has a full facepiece and is operated in a pressure-demand or other positive-pressure mode in combination with an auxiliary, self-contained breathing apparatus operated in a pressure-demand or other positive pressure mode). *Escape:* HiEF [any air-purifying, full-facepiece respirator (gas mask) with a chin-style, front- or back-mounted organic vapor canister having a high-efficiency particulate filter/Any appropriate escape-type, self-contained breathing apparatus]; or SCBAE (any appropriate escape-type, self-contained breathing apparatus).

Storage: Prior to working with Cadmium sulfate, you should be trained on its proper handling and storage. A regulated, marked area should be established where this chemical is handled, used, or stored in compliance with OSHA standard 1910.1045. Store in tightly closed containers in a cool, well-ventilated area away from oxidizers and metals.

Shipping: DOT label requirement of "Poison." Cadmium compounds falls in UN/DOT Hazard Class 6.1 and Packing Group I and II.[19][20]

Spill Handling: Evacuate persons not wearing protective equipment from area of spill or leak until clean-up is complete. Remove all ignition sources. Collect powdered material in the most convenient and safe manner and deposit in sealed containers. Ventilate area after clean-up is complete. It may be necessary to contain and dispose of this chemical as a hazardous waste. If material or contaminated runoff enters waterways, notify downstream users of potentially contaminated waters. Contact your Department of Environmental Protection or your regional office of the federal EPA for specific recommendations. If employees are required to clean-up spills, they must be properly trained and equipped. OSHA 1910.120(q) may be applicable.

Fire Extinguishing: This chemical is a noncombustible solid that may ignite combustible materials. Use dry chemical, carbon dioxide, water spray, or alcohol-resistant foam extinguishers. Poisonous gases are produced in fire. If material or contaminated runoff enters waterways, notify downstream users of potentially contaminated waters. Notify local health and fire officials and pollution control agencies. From a secure, explosion-proof location, use water spray to cool exposed containers. If cooling streams are ineffective (venting sound increases in volume and pitch, tank discolors, or shows any signs of deforming), withdraw immediately to a secure position. If employees are expected to fight fires, they must be trained and equipped in OSHA 1910.156.

References

New Jersey Department of Health and Senior Services, "Hazardous Substance Fact Sheet, Cadmium Sulfate," Trenton, NJ (September 1998)

Cadmium Sulfide

Molecular Formula: CdS

Synonyms: Aurora yellow; Cadmium golden 366; Cadmium lemon yellow 527; Cadmium monosulfide; Cadmium orange; Cadmium primrose 819; Cadmium sulphide; Cadmium yellow; Cadmium yellow 000; Cadmium yellow 10G conc.; Cadmium yellow 892; Cadmium yellow conc. golden; Cadmium yellow conc. lemon; Cadmium yellow conc. primrose; Cadmium yellow oz dark; Cadmium yellow primrose 47-4100; Cadmopur golden yellow N; Cadmopur yellow; Capsebon capsebon; C.I. 77199; C.I. pigment orange 20; C.I. pigment yellow 37; Ferro lemon yellow; Ferro orange yellow; Ferro yellow; Greenockite; NCI-C02711

CAS Registry Number: 1306-23-6

DOT ID: UN 2570

Regulatory Authority

- CLEAN AIR ACT: Hazardous Air Pollutants (Title I, Part A, Section 112). Includes any unique chemical substance that contains cadmium as part of that chemical's infrastructure
- CLEAN WATER ACT: 40CFR401.15 Section 307 Toxic Pollutants
- RCRA, 40CFR261, Appendix 8 Hazardous Constituents, waste number not listed
- EPCRA (Section 313): Includes any unique chemical substance that contains cadmium as part of that chemical's infrastructure. Form R *de minimis* concentration reporting level: (inorganic compounds: 0.1%.; organic compounds: 1.0%)
- U.S. DOT Regulated Marine Pollutant (49CFR172.101, Appendix B).: Severe pollutant, as cadmium compounds

Description: Cadmium sulfide is an odorless, crystalline, lemon yellow to orange solid.

Potential Exposure: Used in pigments, dandruff shampoos, photoconductors, solar cells, and other electronic components.

Incompatibilities: Contact with water or moisture releases poisonous hydrogen sulfide gas. Incompatible with oxidizers, hydrogen azide, zinc, selenium, tellurium, and other metals, iodine monochloride, strong acids.

Permissible Exposure Limits in Air: The Federal standard[58] for cadmium compounds is 0.005 mg/m^3 (as Cd) as an 8-hour TWA. ACGIH has recommended a TWA value of 0.01 mg/m^3 for Cd dusts/salts and 0.002 mg/m^3 for respirable dust. NIOSH recommends that exposure to cadmium be at the lowest feasible level. See also this section in the entry on Cadmium.

Determination in Air: Collection of particles on a filter, workup with acid and measurement by atomic absorption has been specified by NIOSH. See NIOSH Methods #7048 (Cd).

Permissible Concentration in Water: See entry for Cadmium.

Determination in Water: Total cadmium may be determined by digestion followed by atomic absorption of colorimetric (Dithizone) analysis or by Inductively Coupled Plasma (ICP) Optical Emission Spectrometry. Dissolved cadmium is determined by 0.45 μ filtration followed by the previously cited methods. See also reverence.[49]

Harmful Effects and Symptoms

Short Term Exposure: Irritates the eyes on contact. Inhalation irritates the nose, throat and lungs with coughing, and/or shortness of breath. Higher exposures can cause pulmonary edema, a medical emergency that can be delayed for several hours. This can cause death. Cadmium sulfate can cause nausea, salivation, vomiting, cramps and diarrhea, metal fume fever with flu-like symptoms, chills, headache, weakness, metallic tasted in the mouth.

Long Term Exposure: Repeated exposure can cause anemia, brittle and painful bones, diminished or loss of the sense of smell, fatigue, and/or yellow staining of the teeth. May cause lung cancer, kidney damage with kidney stones, liver damage, lung damage with bronchitis, cough, phlegm, and/or shortness of breath, damage to the testes, and may damage the developing fetus. There is some evidence that cadmium sulfate is a teratogen in humans.

Points of Attack: Lungs, liver, kidneys, blood.

Medical Surveillance: Urine test for Cd (levels should be less than 10 μg/l of urine). Urine test for low molecular weight proteins (β-2-microglobulin) to detect kidney damage. Complete blood Count (CBC). Lung function tests. Liver function tests. Consider chest x-ray following acute overexposure.

First Aid: If this chemical gets into the eyes, remove any contact lenses at once and irrigate immediately for at least 15

minutes, occasionally lifting upper and lower lids. Seek medical attention immediately. If this chemical contacts the skin, remove contaminated clothing and wash immediately with soap and water. Seek medical attention immediately. If this chemical has been inhaled, remove from exposure, begin rescue breathing (using universal precautions) if breathing has stopped and CPR if heart action has stopped. Transfer promptly to a medical facility. When this chemical has been swallowed, get medical attention. Give large quantities of water and induce vomiting. Do not make an unconscious person vomit. Medical observation is recommended for 24 – 48 hours after breathing overexposure, as pulmonary edema may be delayed. As first aid for pulmonary edema, a doctor or authorized paramedic may consider administering a corticosteroid spray.

Note to Physician: In case of fume inhalation, treat pulmonary edema. Give prednisone or other corticosteroid orally to reduce tissue response to fume. Positive pressure ventilation may be necessary. Treat metal fume fever with bed rest, analgesics and antipyretics. The symptoms of metal fume fever may be delayed for 4 – 12 hours following exposure: it may last less than 36 hours.

Personal Protective Methods: Wear protective gloves and clothing to prevent any reasonable probability of skin contact. Safety equipment suppliers/manufacturers can provide recommendations on the most protective glove/clothing material for your operation. All protective clothing (suits, gloves, footwear, headgear) should be clean, available each day, and put on before work. Contact lenses should not be worn when working with this chemical. Wear-proof chemical goggles and face shield unless full facepiece respiratory protection is worn. Employees should wash immediately with soap when skin is wet or contaminated. Provide emergency showers and eyewash.

Respirator Selection: NIOSH (as Cd compounds): *At any detectable concentration:* SCBAF:PD,PP (any MSHA/NIOSH approved self-contained breathing apparatus that has a full facepiece and is operated in a pressure-demand or other positive-pressure mode); or SAF:PD,PP:ASCBA (any supplied-air respirator that has a full facepiece and is operated in a pressure-demand or other positive-pressure mode in combination with an auxiliary, self-contained breathing apparatus operated in a pressure-demand or other positive pressure mode). *Escape:* HiEF [any air-purifying, full-facepiece respirator (gas mask) with a chin-style, front- or back-mounted organic vapor canister having a high-efficiency particulate filter/Any appropriate escape-type, self-contained breathing apparatus]; or SCBAE (any appropriate escape-type, self-contained breathing apparatus).

Storage: Prior to working with Cadmium sulfide, you should be trained on its proper handling and storage. A regulated, marked area should be established where this chemical is handled, used, or stored in compliance with OSHA standard 1910.1045. Store in tightly closed containers in a cool, dark, well-ventilated area away from oxidizers and metals, strong acids, water or moisture, and other incompatible materials listed above.

Shipping: DOT label requirement of "Poison." Cadmium compounds falls in UN/DOT Hazard Class 6.1 and Packing Group I and II.[19][20]

Spill Handling: Evacuate persons not wearing protective equipment from area of spill or leak until clean-up is complete. Remove all ignition sources. Collect powdered material in the most convenient and safe manner and deposit in sealed containers. Ventilate area after clean-up is complete. It may be necessary to contain and dispose of this chemical as a hazardous waste. If material or contaminated runoff enters waterways, notify downstream users of potentially contaminated waters. Contact your Department of Environmental Protection or your regional office of the federal EPA for specific recommendations. If employees are required to clean-up spills, they must be properly trained and equipped. OSHA 1910.120(q) may be applicable.

Fire Extinguishing: DO NOT use water. Use dry chemical extinguishers appropriate for metal fires. Poisonous gases are produced in fire including sulfur oxides and hydrogen sulfide. If material or contaminated runoff enters waterways, notify downstream users of potentially contaminated waters. Notify local health and fire officials and pollution control agencies. From a secure, explosion-proof location, use water spray to cool exposed containers. If cooling streams are ineffective (venting sound increases in volume and pitch, tank discolors, or shows any signs of deforming), withdraw immediately to a secure position. If employees are expected to fight fires, they must be trained and equipped in OSHA 1910.156.

Disposal Method Suggested: References

New Jersey Department of Health and Senior Services, "Hazardous Substance Fact Sheet, Cadmium Sulfide," Trenton, NJ (September 1998)

Calcium

Molecular Formula: Ca

Synonyms: Calcicat; Calcium metal; Calcium metal, crystaline; Elemental calcium

CAS Registry Number: 7440-70-2

RTECS Number: EV8040000

DOT ID: UN 1401; UN 1855 (pyrophoric)

EEC Number: 020-001-00-X

Cited in U.S. State Regulations: California (G), Florida (G), Massachusetts (G), New Hampshire (G), New Jersey (G), Oklahoma (G), Pennsylvania (G), Rhode Island (G).

Description: Calcium, Ca, is a silvery-white metal when freshly cut which tarnishes to a blue-gray color in air. It can also be found as a powder. Boiling point = 1,485°C. Freezing/Melting point = 839°C. Hazard Identification (based on NFPA-704 M Rating System): Health 3, Flammability 1, Reactivity 2. Reacts with water.

Potential Exposure: Calcium is used as a raw material for aluminum, copper and lead alloys.

Incompatibilities: Forms hydrogen gas on contact with air; dust may ignite spontaneously. A strong reducing agent; reacts violently with water, acids, strong oxidizers (such as chlorine, bromine and fluorine), alkaline carbonates, dinitrogen tetroxide, halogenated hydrocarbons, lead chloride, halogens, alkaline hydroxides, oxygen, silicon, sulfur, chlorine, fluorine, chlorine trifluoride and many other substances. Reacts with water to produce flammable hydrogen gas.

Permissible Exposure Limits in Air: No standards set.

Determination in Air: Filter; Acid; Flame atomic absorption spectrometry; IV NIOSH Method #7020 (Calcium).

Permissible Concentration in Water: No criteria set.

Routes of Entry: Inhalation of dust.

Harmful Effects and Symptoms

Contact with the dust can severely irritate and burn the eyes and skin. Exposure to the dust can irritate the air passages and lungs. Calcium is a reactive chemical and is an explosion hazard.

Short Term Exposure: Eye contact can cause irritation and possible permanent damage. Skin contact can cause irritation and burns. Inhalation can irritate air passages and lungs, causing coughing and difficult breathing.

Points of Attack: Eyes, skin and respiratory system.

Medical Surveillance: Lung function tests are recommended on a pre-employment and regular post-employment basis.

First Aid: If this chemical gets into the eyes, remove any contact lenses at once and irrigate immediately for at least 15 minutes, occasionally lifting upper and lower lids. Seek medical attention immediately. If this chemical contacts the skin, remove contaminated clothing and wash immediately with soap and water. Seek medical attention immediately. If this chemical has been inhaled, remove from exposure, begin rescue breathing (using universal precautions) if breathing has stopped and CPR if heart action has stopped. Transfer promptly to a medical facility. When this chemical has been swallowed, get medical attention. Give large quantities of water and induce vomiting. Do not make an unconscious person vomit.

Personal Protective Methods: *Clothing:* Avoid skin contact with Calcium. Wear protective gloves and clothing. Safety equipment suppliers/manufacturers can provide recommendations on the most protective glove/clothing material for your operation. All protective clothing (suits, gloves, footwear, headgear) should be clean, available each day, and put on before work. *Eye Protection:* Wear chemical goggles and face shield when working with Calcium, unless full facepiece respiratory protection is worn.

Respirator Selection: Where the potential exists for exposure to Calcium, use a MSHA/NIOSH approved full facepiece respirator with a high efficiency particulate filter. Greater protection is provided by a powered-air purifying respirator. Where the potential for high exposures exists, use a MSHA/NIOSH approved supplied-air respirator with a full facepiece operated in the positive pressure mode or with a full facepiece, hood, or helmet in the continuous flow mode, or use a MSHA/NIOSH approved self-contained breathing apparatus with a full facepiece operated in pressure-demand or other positive pressure mode.

Storage: Store in tightly closed containers in a cool, well-ventilated area away from water, moisture, oxidizers, and acids. Wherever Calcium is used, handled, manufactured, or stored, use explosion-proof electrical equipment and fittings. Store in Kerosene or other neutral oil. Do not store large quantities of Calcium in rooms with sprinkler systems. A detached fire resistant building is recommended for large storage.

Shipping: Calcium or calcium alloys must be labeled: "Dangerous when Wet." They fall in Hazard Class 4.3 and Packing Group II.

Spill Handling: Evacuate persons not wearing protective equipment from area of spill or leak until clean-up is complete. Remove all ignition sources. Collect powdered material in the most convenient and safe manner and deposit in sealed containers. Ventilate area after clean-up is complete. It may be necessary to contain and dispose of this chemical as a hazardous waste. If material or contaminated runoff enters waterways, notify downstream users of potentially contaminated waters. Contact your Department of Environmental Protection or your regional office of the federal EPA for specific recommendations. If employees are required to clean-up spills, they must be properly trained and equipped. OSHA 1910.120(q) may be applicable.

Fire Extinguishing: This chemical is a combustible solid. Calcium dust may ignite spontaneously in air. In contact with water or moisture, Calcium releases Hydrogen gas which can be explosive. Containers may explode in fire. Fire may restart after in has been extinguished. Use dry graphite, soda ash, powdered salt or appropriate metal fire extinguisher. DO NOT USE WATER, CO_2, or dry chemical extinguishers since they are ineffective. Notify local health and fire officials and pollution control agencies. From a secure, explosion-proof location, use water spray to cool exposed containers. If cooling streams are ineffective (venting sound increases in volume and pitch, tank discolors, or shows any signs of deforming), withdraw immediately to a secure position. If employees are expected to fight fires, they must be trained and equipped in OSHA 1910.156.

Disposal Method Suggested: Calcium metal may be burned in an open furnace.[24] When burning calcium waste in a steel pan, dry steam may be directed to the waste with due care to avoid splashing.

References

New Jersey Department of Health and Senior Services, "Hazardous Substance Fact Sheet: Calcium," Trenton, NJ (June 1986).

Calcium Arsenate

Molecular Formula: $As_2Ca_3O_8$

Common Formula: $Ca_3(AsO_4)_2$

Synonyms: Arsenate de calcium (French); Arseniato calcico (Spanish); Arsenic acid, calcium salt (2:3); Calciumarsenat (German); Calcium orthoarsenate; Chip-Cal®; Cucumber dust; Fencal®; Flac®; Kalo®; Kalziumarseniat (German); Kilmag®; Pencal®; Protars®; Security; Spracal; Tricalciumarsenat (German); Tricalcium arsenate; Tricalcium orthoarsenate

CAS Registry Number: 7778-44-1

RTECS Number: CG0830000

DOT ID: UN 1573

EEC Number: 033-005-00-1

Regulatory Authority

- Carcinogen (Human) (IARC)[9] (DFG)[3] (NTP) (OSHA)
- Banned or Severely Restricted (UK, India) (UN)[13] (Various) (UN)[35]
- Air Pollutant Standard Set (USSR/UNEP)[43][60] (HSE) (Several Canadian Provinces) as As compounds
- CLEAN AIR ACT: List of high risk pollutants (Section 63.74) as arsenic compounds
- CLEAN WATER ACT: Section 311 Hazardous Substances/RQ 40CFR117.3 (same as CERCLA, see below); Section 313 Water Priority Chemicals (57FR41331, 9/9/92); Toxic Pollutant (Section 401.15) as arsenic and compounds
- RCRA, 40CFR261, Appendix 8 Hazardous Constituents, waste number not listed
- SUPERFUND/EPCRA 40CFR302.4 Reportable Quantity (RQ): CERCLA, 1 lb(0.454 kg)
- EPCRA Section 313 Form R *de minimis* concentration reporting level: 0.1%
- SUPERFUND/EPCRA 40CFR355, Appendix B Extremely Hazardous Substances: TPQ = 500/10,000 lb (227/4,540 kg)
- U.S. DOT Regulated Marine Pollutant (49CFR172.101, Appendix B), listed by name; also listed as calcium arsenate and calcium arsenite, mixtures, solid
- Canada, WHMIS, Ingredients Disclosure List; National Pollutant Release Inventory (NPRI); CEPA Priority Substance List, Ocean dumping prohibited

Cited in U.S. State Regulations: California (A, G), Maine (G), Massachusetts (G), New Hampshire (G), New Jersey (G), North Carolina (A), Oklahoma (G), Pennsylvania (G).

Description: $Ca_3(AsO_4)_2$, a white flocculent powder. It is not combustible. Boiling point = (decomposes). Hazard Identification (based on NFPA-704 M Rating System): Health 4, Flammability 0, Reactivity 0. Slightly soluble in water.

Potential Exposure: Workers engaged in manufacture, formulation, and application of pesticides containing calcium arsenate.

Incompatibilities: None reported, according to NIOSH. When heated produces As fumes.

Permissible Exposure Limits in Air: The Federal limit is 0.010 mg/m^3.[2] NIOSH has recommended a limit of 0.002 g/m^3 with a 15-minute ceiling. The NIOSH IDLH level is 5 mg/m^3 (as As). The German DFG[3] has set no MAK value because of the notation that this is a proven human carcinogen. The former USSR-UNEP/IPRTC project[43] has set a MAC for calcium arsenate in ambient air of residential areas as 0.009 mg/m^3 on a momentary basis and 0.004 mg/m^3 on an average daily basis. North Carolina has set a limit in ambient air[60] of 0.0002 μg/m^3.

Determination in Air: Filter collection followed by atomic absorption analysis.

Permissible Concentration in Water: No criteria set for Calcium arsenate. See "Arsenic and Arsenic Compounds."

Routes of Entry: Inhalation, skin absorption, ingestion, eye and skin contact.

Harmful Effects and Symptoms

Short Term Exposure: Irritates eyes, skin and respiratory tract. Can cause poor appetite, a metallic or garlic taste, nausea, vomiting, stomach pain, diarrhea, abnormal heart rhythm, seizures, pain in extremities and muscles, weakness, flushing of skin, numbness and tingling in extremities, intense thirst, and muscular cramps, delerium and even death. Kidney failure may occur. Jaundice may appear within an hour. In severe poisoning, death can occur within an hour, but the usual interval is 24 hours. This material is extremely toxic; the probable oral lethal dose for humans is 5 – 50 mg/kg; or between 7 drops and 1 teaspoonful for a 150 lb person. It is an irritant to eyes, respiratory tract, mouth and stomach.

Long Term Exposure: Calcium arsenate is a carcinogen in humans and may be a reproductive hazard. Damage to kidneys, liver and the nervous system have been reported. Chronic exposure can cause bone marrow damage, often leading to aplastic anemia. Long term exposure can cause an ulcer in the septum dividing the inner nose. High or repeated exposure may damage the nerves causing weakness, and poor coordination in the limbs. Repeated skin contact can cause thickened skin and/or patch areas of darkening and loss of pigment. Some may develop white lines on the nails. There is epidemiological evidence that chronic ingestion or arsenic compounds causes a predisposition to skin cancers. A rebuttable presumption against pesticide registration was issued on October 18, 1978 by U.S. EPA on the basis of oncogenicity, teratogenicity and mutagenicity.

Points of Attack: Eyes, respiratory system, liver, skin, lymphatics, lungs, central nervous system.

Medical Surveillance: Consider the points of attack in preplacement and periodic physical examinations. Examination of the nose, eyes, nails, and nervous system. Liver function tests. Tests for Urine arsenic. NIOSH recommends urine arsenic should not exceed 100 micrograms per gram of creatine in the urine.

First Aid: If this chemical gets into the eyes, remove any contact lenses at once and irrigate immediately for at least 15 minutes, occasionally lifting upper and lower lids. Seek medical attention immediately. If this chemical contacts the skin, remove contaminated clothing and wash immediately with soap and water. Seek medical attention immediately. If this chemical has been inhaled, remove from exposure, begin rescue breathing (using universal precautions) if breathing has stopped and CPR if heart action has stopped. Transfer promptly to a medical facility. When this chemical has been swallowed, get medical attention. Give large quantities of water and induce vomiting. Do not make an unconscious person vomit. Medical observation is recommended for 24 – 48 hours after breathing overexposure.

Personal Protective Methods: Wear protective gloves and clothing to prevent any reasonable probability of skin contact. Safety equipment suppliers/manufacturers can provide recommendations on the most protective glove/clothing material for your operation. All protective clothing (suits, gloves, footwear, headgear) should be clean, available each day, and put on before work. Contact lenses should not be worn when working with this chemical. Wear dust-proof chemical goggles and face shield unless full facepiece respiratory protection is worn. Employees should wash immediately with soap when skin is wet or contaminated. Provide emergency showers and eyewash.

Respirator Selection: NIOSH: *At any detectable concentration:* SCBAF:PD,PP (any MSHA/NIOSH approved self-contained breathing apparatus that has a full facepiece and is operated in a pressure-demand or other positive-pressure mode); or SAF:PD,PP:ASCBA (any supplied-air respirator that has a full facepiece and is operated in a pressure-demand or other positive-pressure mode in combination with an auxiliary, self-contained breathing apparatus operated in a pressure-demand or other positive pressure mode). *Escape:* HiEF (any air-purifying, full-facepiece respirator (gas mask) with a chin-style, front- or back-mounted organic vapor canister having a high-efficiency particulate filter/Any appropriate escape-type.

Storage: Prior to working with Calcium arsenate you should be trained on its proper handling and storage. Store in tightly closed containers in a cool, well-ventilated area. A regulated, marked area should be established where this chemical is handled, used, or stored in compliance with OSHA standard 1910.1045.

Shipping: The DOT label requirement[19] is "Poison." The UN/DOT Hazard Class is 6.1 and the Packing Group is II.[19][20]

Spill Handling: Evacuate persons not wearing protective equipment from area of spill or leak until clean-up is complete. Remove all ignition sources. Collect powdered material in the most convenient and safe manner and deposit in sealed containers. Ventilate area after clean-up is complete. It may be necessary to contain and dispose of this chemical as a hazardous waste. If material or contaminated runoff enters waterways, notify downstream users of potentially contaminated waters. Contact your Department of Environmental Protection or your regional office of the federal EPA for specific recommendations. If employees are required to clean-up spills, they must be properly trained and equipped. OSHA 1910.120(q) may be applicable.

Fire Extinguishing: This chemical is a noncombustible solid. Use any extinguishing agent suitable for surrounding fires. Poisonous gases are produced in fire including arsenic. If material or contaminated runoff enters waterways, notify downstream users of potentially contaminated waters. Notify local health and fire officials and pollution control agencies. Containers may explode in fire. From a secure, explosion-proof location, use water spray to cool exposed containers. If cooling streams are ineffective (venting sound increases in volume and pitch, tank discolors, or shows any signs of deforming), withdraw immediately to a secure position. If employees are expected to fight fires, they must be trained and equipped in OSHA 1910.156.

Disposal Method Suggested: Long-term storage in large, weatherproof, and sift-proof storage bins or silos; small amounts may be disposed in a chemical waste landfill. Alternatively, dissolve in HCl, precipitate as sulfide, with H_2S, dry and return to supplier.[23]

References

National Institute for Occupational Safety and Health, Criteria for a Recommended Standard: Occupational Exposure to Inorganic Arsenic, NIOSH Doc. No. 74–110, Washington, DC (1974)

Sax, N. I.., Ed., "Dangerous Properties of Industrial Materials Report" 2, No. 1, 89–91 (1982)

U.S. Environmental Protection Agency, "Chemical Profile: Calcium Arsenate," Washington, DC, Chemical Emergency Preparedness Program (November 30, 1987)

Calcium Carbide

Molecular Formula: C_2Ca

Common Formula: CaC_2

Synonyms: Acetylongen; Calcium acetylide; Calcium dicarbide; Carbide, acetylenogen

CAS Registry Number: 75-20-7

RTECS Number: EV9400000

DOT ID: UN 1402

Regulatory Authority

- CLEAN WATER ACT: Section 311 Hazardous Substances/RQ 40CFR117.3 (same as CERCLA, see below)
- SUPERFUND/EPCRA 40CFR302.4 Reportable Quantity (RQ): CERCLA, 10 lb (4.54 kg)

Cited in U.S. State Regulations: California (G), Florida (G), Illinois (G), Maine (G), Massachusetts (G), New Hampshire (G), New Jersey (G), Oklahoma (G), Pennsylvania (G).

Description: Calcium carbide, CaC_2, is a grayish-black granules or powder. Slight garlic odor. Boiling point ≥ 447°C. Freezing/Melting point = about 2,300°C. Hazard Identification (based on NFPA-704 M Rating System): Health 3, Flammability 3, Reactivity 2. Water reactive.

Potential Exposure: Those involved in the manufacture and handling of carbide and the generation of acetylene.

Incompatibilities: Water contact or moist air forms calcium hydroxide and explosive acetylene gas with risk of fire and explosion. Acids, oxidizers, hydrogen chloride, methanol, copper salt solutions, lead fluoride, magnesium, selenium, silver nitrate, iron trichloride, tin dichloride, sodium peroxide, stannous chloride, sulfur.

Permissible Exposure Limits in Air: No standards set.

Permissible Concentration in Water: No criteria set.

Routes of Entry: Inhalation and ingestion.

Harmful Effects and Symptoms

Irritation of skin, eyes and respiratory tract. Inhalation of dust may cause lung edema.

Short Term Exposure: Corrosive. Contact with eyes or skin causes severe irritation and burns with possible permanent eye damage and ulcers to the skin. Irritates the lungs with coughing and/or shortness of breath. Higher exposures can cause pulmonary edema, a medical emergency that can be delayed for several hours. This can cause death.

Long Term Exposure: Can irritate the lungs. Exposure may cause bronchitis with coughing, phlegm, and/or shortness of breath.

Points of Attack: Eyes, skin and respiratory tract.

Medical Surveillance: Lung function tests. Consider chest x-ray following acute overexposure.

First Aid: If this chemical gets into the eyes, remove any contact lenses at once and irrigate immediately for at least 15 minutes, occasionally lifting upper and lower lids. Seek medical attention immediately. If this chemical contacts the skin, remove contaminated clothing and wash immediately with soap and water. Seek medical attention immediately. If this chemical has been inhaled, remove from exposure, begin rescue breathing (using universal precautions) if breathing has stopped and CPR if heart action has stopped. Transfer promptly to a medical facility. When this chemical has been swallowed, get medical attention. Give large quantities of water and induce vomiting. Do not make an unconscious person vomit. Medical observation is recommended for 24 – 48 hours after breathing overexposure, as pulmonary edema may be delayed. As first aid for pulmonary edema, a doctor or authorized paramedic may consider administering a corticosteroid spray.

Personal Protective Methods: Wear protective gloves and clothing to prevent any reasonable probability of skin contact. Safety equipment suppliers/manufacturers can provide recommendations on the most protective glove/clothing material for your operation. All protective clothing (suits, gloves, footwear, headgear) should be clean, available each day, and put on before work. Contact lenses should not be worn when working with this chemical. Wear splash or dust-proof chemical goggles and face shield unless full facepiece respiratory protection is worn. Employees should wash immediately with soap when skin is wet or contaminated. Provide emergency showers and eyewash.

Respirator Selection: Engineering controls should be used wherever feasible to maintain airborne concentrations of this chemical below the prescribed exposure limit. Respirators and protective equipment are less effective than engineering controls, and should be used only in non-routine or emergency situations which may result in exposure concentrations in excess of the TWA environmental limit. *At any concentrations above the NIOSH REL, or where there is no REL, at any detectable concentration:* SCBAF:PD,PP (any MSHA/NIOSH approved self-contained breathing apparatus that has a full facepiece and is operated in a pressure-demand or other positive-pressure mode); or SAF:PD,PP:ASCBA (any supplied-air respirator that has a full facepiece and is operated in a pressure-demand or other positive-pressure mode in combination with an auxiliary, self-contained breathing apparatus operated in a pressure-demand or other positive pressure mode).

Storage: Prior to working with Calcium carbide you should be trained on its proper handling and storage. Store in tightly closed containers in a cool, well ventilated area away from moisture and without sprinkler protection and avoid contact with incompatible materials. Use only non-sparking tools and equipment especially when opening and closing containers of Calcium carbide. Metal containers involving the transfer of this chemical should be grounded and bonded. Use explosion-proof electrical equipment in the carbide-handling area.

Sources of ignition such as smoking and open flames, are prohibited where this chemical is used, handled, or stored in a manner that could create a potential fire or explosion hazard.

Shipping: Calcium carbide requires a "Dangerous when Wet" label. It falls in Hazard Class 4.3 and Packing Group II.

Spill Handling: Evacuate persons not wearing protective equipment from area of spill or leak until clean-up is complete. Remove all ignition sources. Ventilate area of spill. Collect powdered material in the most convenient and safe manner and deposit in sealed containers. It may be necessary to contain and dispose of this chemical as a hazardous waste. If material or contaminated runoff enters waterways, notify downstream users of potentially contaminated waters. Contact your Department of Environmental Protection or your regional office of the federal EPA for specific recommendations. If employees are required to clean-up spills, they must be properly trained and equipped. OSHA 1910.120(q) may be applicable.

Fire Extinguishing: This chemical is a combustible solid. Wet Calcium Carbide produces highly flammable *Acetylene gas*. Dry calcium carbide itself is not flammable. Do not use water, foam, carbon dioxide or halogen extinguishers on fire. Use dry chemical, sand, soda ash, or lime extinguishers. Poisonous gases are produced in fire. If material or contaminated runoff enters waterways, notify downstream users of potentially contaminated waters. Notify local health and fire officials and pollution control agencies. From a secure, explosion-proof location, use water spray to cool exposed containers. If cooling streams are ineffective (venting sound increases in volume and pitch, tank discolors, or shows any signs of deforming), withdraw immediately to a secure position. If employees are expected to fight fires, they must be trained and equipped in OSHA 1910.156.

Disposal Method Suggested: Mixing with large quantity of water using pilot flame to ignite evolved acetylene. Lime residue sent to landfill.

References

Sax, N. I.., Ed., "Dangerous Properties of Industrial Materials Report" 2, No. 1, 91–93 (1982)

New Jersey Department of Health and Senior Services, "Hazardous Substance Fact Sheet: Calcium Carbide," Trenton, NJ (September 1996)

Calcium Carbonate

Molecular Formula: $CCaO_3$

Common Formula: $CaCO_3$

Synonyms: Agricultural limestone; Agstone; Aragonite; Atomit; Bell mine pulverized limestone; Calcite; Calcium(II)carbonate (1:1); Carbonic acid, calcium salt (1:1); Chalk; Domolite; Franklin; Limestone; Lithographic stone; Marble; Portland stone; Sohnhofen stone; Vaterite

CAS Registry Number: 1317-65-3; 471-34-1 (monocarbonate)

RTECS Number: EV9580000; FF9335000 (monocarbonate)

Regulatory Authority

- Air Pollutant Standard Set (ACGIH)[1] (HSE)[33] (OSHA)[58] (Australia) (Israel) (Mexico) (Several Canadian Provinces)

Cited in U.S. State Regulations: Alaska (G), Maine (G), Massachusetts (G), Minnesota (G), New Hampshire (G), New York (G), Pennsylvania (G).

Description: Calcium carbonate, $CaCO_3$ is a white, odorless powder or crystalline solid. Freezing/Melting point = 1,517 – 2,442°F (decomposes). Very slightly soluble in water. The literature lists for CAS 471-34-1: Boiling point = 825°C (decomposes).

Potential Exposure: Calcium carbonate is used as a source of lime; neutralizing agent; manufacturing or rubber, plastics, paint, paper, dentifrices, ceramics, putty, polishes, insecticides, inks and cosmetics; whitewash; Portland cement; antacid; analytical chemistry and others.

Incompatibilities: Calcium carbonate decomposes in high temperature forming carbon dioxide and corrosive materials. Reacts with acids producing carbon dioxide gas release. Incompatible with acids, ammonium salts, fluorine. NIOSH also lists alum as incompatible, but this is questionable.

Permissible Exposure Limits in Air: ACGIH[1] has set a TWA of 10 mg/m^3 (for dust containing no asbestos and <1% free silica). The HSE[33] has set a TWA of 10 mg/m^3 for total inhalable dust and 5 mg/m^3 for respirable dust. OSHA[58] has set a TWA of 15 mg/m^3 on a total dust basis and 5 mg/m^3 on a respirable fraction basis. NIOSH[58] has set a TWA of 10 mg/m^3 on a total dust basis and 5 mg/m^3 on a respirable fraction basis.

Determination in Air: Filter; Acid; Flame atomic absorption spectrometry; IV NIOSH Method #7020 (Calcium). Calcium Carbonate may be measured by gravimetric means after filter collection. See NIOSH analytical methods #0500 for nuisance dust, total and 0600 for nuisance dust, respirable.

Permissible Concentration in Water: No criteria set.

Routes of Entry: Inhalation of dust, ingestion.

Harmful Effects and Symptoms

Short Term Exposure: Inhalation can cause irritation to nose. Eyes contact can cause irritation. *Ingestion:* Large amounts can cause irritability, nausea, dehydration and constipation. Estimated lethal dose is over 2 lb.

Long Term Exposure: Ingestion of more than 8 grams (1/3 ounce) a day can cause blood and kidney disorders.

Points of Attack: Eyes, respiratory system, digestive system.

First Aid: If this chemical gets into the eyes, remove any contact lenses at once and irrigate immediately for at least 15 minutes, occasionally lifting upper and lower lids. Seek medical attention immediately. If this chemical contacts the skin, remove contaminated clothing and wash immediately with

soap and water. Seek medical attention immediately. If this chemical has been inhaled, remove from exposure, begin rescue breathing (using universal precautions) if breathing has stopped and CPR if heart action has stopped. Transfer promptly to a medical facility. When this chemical has been swallowed, get medical attention. Give large quantities of water and induce vomiting. Do not make an unconscious person vomit.

Personal Protective Methods: Wear safety glasses.

Respirator Selection: Wear dust mask.

Storage: Store to avoid contact with acids.

Shipping: The DOT Performance-Oriented Packaging Standards[19] do not cite Calcium carbonate specifically.

Spill Handling: Scoop up and place in suitable container. Discard with regular trash.

Fire Extinguishing: This chemical is a noncombustible solid. Use extinguishing agents suitable for surrounding materials.

Disposal Method Suggested: Landfills.

References

New York State Department of Health, "Chemical Fact Sheet: Calcium Carbonate," Albany, NY, Bureau of Toxic Substance Assessment (March 1986)

Calcium Chlorate

Molecular Formula: $CaCl_2O_6$

Common Formula: $Ca(ClO_3)_2$

Synonyms: Calcium chlorate aqueous solution; Chlorate de calcium (French); Chloric acid, calcium salt

CAS Registry Number: 10137-74-3 (monohydrate)

RTECS Number: FN9800000

DOT ID: UN 1452 (solid); UN 2429 (solution)

Cited in U.S. State Regulations: Maine (G), Massachusetts (G), New Hampshire (G), New Jersey (G), Oklahoma (G), Pennsylvania (G).

Description: Calcium chlorate, $Ca(ClO_3)_2$, forms white to yellow deliquescent crystals. Freezing/Melting point = 340° (loses H_2O of crystallization @ > 100°). Hazard Identification (based on NFPA-704 M Rating System): Health 1, Flammability 0, Reactivity 0. Highly soluble in water.

Potential Exposure: Calcium Chlorate is used in making fireworks, herbicides (weed killers) and in photography.

Incompatibilities: A strong reducing agent. Reacts, possibly with risk of fire and explosion, with acids (especially organic), reducing agents, aluminum, arsenic, chemically active metals, combustible materials, ammonium compounds, charcoal, copper, cyanides, manganese dioxide, metal sulfides, phosphorus, sulfur.

Permissible Exposure Limits in Air: No standards set.

Permissible Concentration in Water: No criteria set.

Routes of Entry: Inhalation of dust, ingestion.

Harmful Effects and Symptoms

Short Term Exposure: Calcium Chlorate can affect you when breathed in. Contact can irritate the skin and eyes. Inhalation can cause irritation of the respiratory tract. Calcium Chlorate may damage the kidneys. Very high exposures can interfere with the ability of the blood to carry oxygen, causing headaches, dizziness, weakness, a bluish skin color and even death. High exposures can cause death. The oral LD_{50} rat is 4,500 mg/kg.[41]

Long Term Exposure: Can affect the kidneys, liver, heart, and blood.

Points of Attack: Eyes, skin, respiratory system, blood.

Medical Surveillance: If symptoms develop or overexposure is suspected, the following may be useful: Methemoglobin level. EKG. Kidney and liver function tests.

First Aid: If this chemical gets into the eyes, remove any contact lenses at once and irrigate immediately for at least 15 minutes, occasionally lifting upper and lower lids. Seek medical attention immediately. If this chemical contacts the skin, remove contaminated clothing and wash immediately with soap and water. Seek medical attention immediately. If this chemical has been inhaled, remove from exposure, begin rescue breathing (using universal precautions) if breathing has stopped and CPR if heart action has stopped. Transfer promptly to a medical facility. When this chemical has been swallowed, get medical attention. Give large quantities of water and induce vomiting. Do not make an unconscious person vomit.

Note to Physician: Treat for methemoglobinemia. Spectrophotometry may be required for precise determination of levels of methemoglobinemia in urine.

Personal Protective Methods: Avoid skin contact with Calcium Chlorate. Wear protective gloves and clothing. Safety equipment suppliers/manufacturers can provide recommendations on the most protective glove/clothing material for your operation. All protective clothing (suits, gloves, footwear, headgear) should be clean, available each day, and put on before work. Wear dust-proof goggles when working with powders or dust, unless full facepiece respiratory protection is worn.

Respirator Selection: Where the potential exists for exposure to Calcium Chlorate, use a MSHA/NIOSH approved respirator equipped with particulate (dust/fume/mist) filters. Particulate filters must be checked every day before work for physical damage, such as rips or tears, and replaced as needed. Where the potential for high exposures exists, use a MSHA/NIOSH approved supplied-air respirator with a full facepiece operated in the positive pressure mode or with a full facepiece, hood, or helmet in the continuous flow mode, or use a MSHA/NIOSH approved self-contained breathing apparatus with a full facepiece operated in pressure-demand or other positive pressure mode.

Storage: Store in tightly closed containers in a cool, well-ventilated area away from strong acids (such as Hydrochloric, Sulfuric and Nitric), chemically active metals (such as Potassium, Sodium, Magnesium and Zinc). Calcium Chlorate must be stored to avoid contact with organic matter, Ammonium Compounds, Aluminum, Copper, Cyanides, Flammable Vapors and other Oxidizable Materials since violent reactions occur. Avoid storage on wood floors. Friction, heat or physical shocks may cause Calcium Chlorate to ignite and explode. Wherever Calcium Chlorate is used, handled, manufactured, or stored, use explosion-proof electrical equipment and fittings. See OSHA standard 1910.104 and NFPA 43A *Code for the Storage of Liquid and Solid Oxidizers* for detailed handling and storage regulations.

Shipping: Calcium Chlorate requires an "Oxidizer" label. It falls in Hazard Class 5.1 and Packing Group II.

Spill Handling: *For dry material:* Evacuate persons not wearing protective equipment from area of spill or leak until clean-up is complete. Remove all ignition sources. Collect powdered material in the most convenient and safe manner and deposit in sealed containers. Ventilate area after clean-up is complete. It may be necessary to contain and dispose of this chemical as a hazardous waste. If material or contaminated runoff enters waterways, notify downstream users of potentially contaminated waters. Contact your Department of Environmental Protection or your regional office of the federal EPA for specific recommendations. If employees are required to clean-up spills, they must be properly trained and equipped. OSHA 1910.120(q) may be applicable.

For solution: Evacuate and restrict persons not wearing protective equipment from area of spill or leak until cleanup is complete. Remove all ignition sources. Ventilate area of spill or leak. Absorb liquids in vermiculite, dry sand, earth, peat, carbon, or a similar material and deposit in sealed containers. It may be necessary to contain and dispose of this chemical as a hazardous waste. If material or contaminated runoff enters waterways, notify downstream users of potentially contaminated waters. Contact your Department of Environmental Protection or your regional office of the federal EPA for specific recommendations. If employees are required to clean-up spills, they must be properly trained and equipped. OSHA 1910.120(q) may be applicable.

Fire Extinguishing: This chemical is a noncombustible solid that increases the combustion of other substances. Use dry chemical, carbon dioxide, water spray, or alcohol foam extinguishers. Poisonous gases are produced in fire including chlorine. If material or contaminated runoff enters waterways, notify downstream users of potentially contaminated waters. Notify local health and fire officials and pollution control agencies. Containers may explode in fire. From a secure, explosion-proof location, use water spray to cool exposed containers. If cooling streams are ineffective (venting sound increases in volume and pitch, tank discolors, or shows any signs of deforming), withdraw immediately to a secure position. If employees are expected to fight fires, they must be trained and equipped in OSHA 1910.156.

Disposal Method Suggested: For barium chlorate, the UN[22] recommends using a vast volume of reducing agent (bisulfites, ferrous salts or hypo) followed by neutralization and flushing to the sewer with abundant water. This should be applicable here as well.

References

New Jersey Department of Health and Senior Services, "Hazardous Substance Fact Sheet: Calcium Chlorate," Trenton, NJ (May 1986).

Calcium Chloride

Molecular Formula: $CaCl_2$

Synonyms: Calcium chloride, anhydrous; Calplus; Caltac

CAS Registry Number: 10043-52-4

RTECS Number: EV9800000

EEC Number: 017-013-00-2

Cited in U.S. State Regulations: New York (G).

Description: Calcium chloride, $CaCl_2$, is a colorless to off-white crystalline solid which is deliquescent (absorb water). When heated, crystals lose water at 100°C. Boiling point = 1,935°C (decomposes). Freezing/Melting point = 772°C. Hazard Identification (based on NFPA-704 M Rating System): Health 1, Flammability 0, Reactivity 0. Soluble in water.

Potential Exposure: Calcium chloride is used as road salt for melting snow, a drying agent in desiccators, for dehydrating organic liquids and gases, in refrigeration brines and antifreeze, as a dust-proofing agent, food additives, concrete hardening accelerator and others.

Incompatibilities: The solution in water is a weak base. Reacts with zinc in presence of water forming highly flammable hydrogen gas. Dissolves violently in water with generation of much heat. Incompatible with water, bromine trifluoride, 2-furan, percarboxylic acid. May attack some building materials and metals in the presence of moisture.

Permissible Exposure Limits in Air: No standards set.

Permissible Concentration in Water: No criteria set.

Routes of Entry: Inhalation of dust; ingestion.

Harmful Effects and Symptoms

Short Term Exposure: Inhalation of dust may cause burning, irritation of the nose, mouth and throat, nose bleeds and breakdown of nasal tissue. Skin contact with dry skin solid may cause severe irritation. Contact with wet skin or concentrated solutions can cause more severe irritation and burns. Ingestion may cause irritation of the mouth, throat and stomach, nausea and vomiting. Eye contact may cause irritation, burning and some damage to the surface of the eye. The oral LD_{50} rat is 1,000 mg/kg (slightly toxic).

Long Term Exposure: Repeated or prolonged contact with skin may cause dermatitis. Prolonged or repeated inhalation may cause ulcerations of the nasal mucous membrane.

First Aid: If this chemical gets into the eyes, remove any contact lenses at once and irrigate immediately for at least 15 minutes, occasionally lifting upper and lower lids. Seek medical attention immediately. If this chemical contacts the skin, remove contaminated clothing and wash immediately with soap and water. Seek medical attention immediately. If this chemical has been inhaled, remove from exposure, begin rescue breathing (using universal precautions) if breathing has stopped and CPR if heart action has stopped. Transfer promptly to a medical facility. When this chemical has been swallowed, get medical attention. Give large quantities of water and induce vomiting. Do not make an unconscious person vomit.

Personal Protective Methods: Wear goggles or face shield if eye hazard exists, coveralls and rubber gloves. Natural Rubber, Neoprene, and Polyvinyl Chloride are among the recommended protective materials.

Respirator Selection: Dust mask or dust respirator may be helpful in preventing inhalation exposures.

Storage: Keep tightly sealed in a cool, dry place away from incompatible materials.

Shipping: There are no label or maximum shipping quantity requirements set by DOT.[19]

Spill Handling: Evacuate and restrict persons not wearing protective equipment from area of spill or leak until cleanup is complete. Remove all ignition sources. Ventilate area of spill or leak. Absorb liquids in vermiculite, dry sand, earth, peat, carbon, or a similar material and deposit in sealed containers. It may be necessary to contain and dispose of this chemical as a hazardous waste. If material or contaminated runoff enters waterways, notify downstream users of potentially contaminated waters. Contact your Department of Environmental Protection or your regional office of the federal EPA for specific recommendations. If employees are required to clean-up spills, they must be properly trained and equipped. OSHA 1910.120(q) may be applicable.

Fire Extinguishing: Use extinguishing agent suitable for surrounding fire. Poisonous gases are produced in fire including chlorine. If material or contaminated runoff enters waterways, notify downstream users of potentially contaminated waters. Notify local health and fire officials and pollution control agencies. If employees are expected to fight fires, they must be trained and equipped in OSHA 1910.156.

Disposal Method Suggested: Add large volumes of water. Add excess soda ash then neutralize with HCl. Route to sewage plant or use as landfill sludge.

References

New York State Department of Health, "Chemical Fact Sheet: Calcium Chloride," Albany, NY, Bureau of Toxic Substance Assessment (March 1986)

Sax, N. I.., Ed., "Dangerous Properties of Industrial Materials Report" 2, No. 1, 93–94 (1982)

Calcium Chromate

Molecular Formula: $CaCrO_4$

Synonyms: Calcium chromate(IV); Calcium chrome yellow; Calcium chromium oxide; Calcium monochromate; Chromato calcico (Spanish); Chromic acid, calcium salt (1:1); C.I. 77223; C.I. pigment yellow 33; Gelbin; Yellow ultramarine

CAS Registry Number: 13765-19-0

RTECS Number: GB2750000

DOT ID: NA 9096

Regulatory Authority

- Carcinogen (Human) (ACGIH)[1] (IARC)[9] (UN)[35] (DFG)[3]
- Air Pollutant Standard Set (ACGIH)[1] (HSE)[33] (Massachusetts, North Carolina)[60]
- CLEAN AIR ACT: Hazardous Air Pollutants (Title I, Part A, Section 112)
- CLEAN WATER ACT: Section 311 Hazardous Substances/RQ 40CFR117.3 (same as CERCLA, see below); Section 313 Water Priority Chemicals (57FR41331, 9/9/92); Toxic Pollutant (Section 401.15)
- EPA HAZARDOUS WASTE NUMBER (RCRA No.): U032
- RCRA, 40CFR261, Appendix 8 Hazardous Constituents
- SUPERFUND/EPCRA 40CFR302.4 Reportable Quantity (RQ): CERCLA, 10 lb (4.54 kg)
- EPCRA Section 313 Form R *de minimis* concentration reporting level: 0.1%
- Canada, WHMIS, Ingredients Disclosure List

Cited in U.S. State Regulations: California (G), Florida (G), Kansas (G), Louisiana (G), Massachusetts (G, A), New Hampshire (G), New Jersey (G), New York (G), North Carolina (A), Pennsylvania (G), Vermont (G), Virginia (G), Washington (G), Wisconsin (G).

Description: Calcium chromate, $CaCrO_4$, is a yellow crystalline solid, often used in solution. Odorless. It normally occurs as the dehydrate and loses water at 200°C. Hazard Identification (based on NFPA-704 M Rating System): Health 2, Flammability 0, Reactivity 0. Insoluble in water.

Potential Exposure: Calcium chromate is used as a pigment, corrosion inhibitor, in the manufacture of chromium, in oxidizing reactions, in battery depolarization.

Incompatibilities: A strong oxidizer. Incompatible with boron (violent reaction), ethanol, combustible, organic or other easily oxidized materials.

Permissible Exposure Limits in Air: The legal OSHA airborne permissible exposure limit (PEL) is 0.1 mg/m^3 for Chromic Acid and Chromates (as Cr), (Ceiling) not to be exceeded at any time. NIOSH recommends an airborne exposure limit of 0.001 mg/m^3 for carcinogenic Chromium(VI) compounds (as Cr), averaged over an 8-hour workshift. ACGIH recommends an airborne exposure limit of 0.05 mg/m^3 for Chromium(IV) compounds (as Cr) averaged over an 8-hour workshift. This is also the HSE value.[33] The above exposure limits are for air levels only. When skin contact also occurs, you may be overexposed, even though air levels are less than the limits listed above. Two states have set guidelines or standards for calcium chromate in ambient air[60] ranging from zero for North Carolina to 0.0008 μg/m^3 for Massachusetts.

Determination in Air: Filter; Acid; Flame atomic absorption spectrometry; NIOSH IV Method #7024, Chromium.

Permissible Concentration in Water: See entry in Chromium for priority toxic pollutant limits in water.

Determination in Water: Be atomic absorption (AA) using either direct aspiration into a flame or a furnace technique.[49]

Routes of Entry: Skin contact, inhalation of dust, ingestion. This chemical can be absorbed through the skin, thereby increasing exposure.

Harmful Effects and Symptoms

Short Term Exposure: *Inhalation:* Exposure to 0.18 – 1.4 mg/m^3 can cause irritation of nose and throat within 2 weeks and disintegration of nasal tissue, coughing, wheezing, headache, painful breathing and fever within 8 weeks. Skin contact can cause severe irritation. Contact with damaged skin can cause deep sores known as "chrome holes." Eye contact can cause severe chemical burns and possible loss of vision. Ingestion can cause severe sore throat and irritation of the throat, stomach and intestine which can develop into tissue damage.

Long Term Exposure: Calcium chromate is a carcinogen in humans, and has been shown to cause lung, liver, bladder, etc., cancer. Inhalation can cause breakdown of nasal tissue and a hole in the septum dividing the inner nose. Exposed persons may develop skin allergy, bronchitis, lung allergy, and kidney damage.

Points of Attack: Lungs, kidneys, skin, respiratory system, eyes, gastrointestinal system.

Medical Surveillance: For those with frequent or potentially high exposure (half the TLV or greater), the following are recommended before beginning work and at regular times after that: Lung function tests. Urine test for chromates. This test is most accurate shortly after exposure. Exam of the skin and nose. If symptoms develop or overexposure is suspected, the following may be useful: Kidney function tests. Evaluation by a qualified allergist, including careful exposure history and special testing, may help diagnose skin allergy.

First Aid: If this chemical gets into the eyes, remove any contact lenses at once and irrigate immediately for at least 30 minutes, occasionally lifting upper and lower lids. Seek medical attention immediately. If this chemical contacts the skin, remove contaminated clothing and wash immediately with soap and water. Seek medical attention immediately. If this chemical has been inhaled, remove from exposure, begin rescue breathing (using universal precautions) if breathing has stopped and CPR if heart action has stopped. Transfer promptly to a medical facility. When this chemical has been swallowed, get medical attention. Give large quantities of water and induce vomiting. Do not make an unconscious person vomit.

Personal Protective Methods: Wear protective gloves and clothing to prevent any reasonable probability of skin contact. Safety equipment suppliers/manufacturers can provide recommendations on the most protective glove/clothing material for your operation. All protective clothing (suits, gloves, footwear, headgear) should be clean, available each day, and put on before work. Contact lenses should not be worn when working with this chemical. Wear splash- (if using solution) or dust-proof chemical goggles and face shield unless full facepiece respiratory protection is worn. Employees should wash immediately with soap when skin is wet or contaminated. Provide emergency showers and eyewash. Specific engineering controls are recommended in NIOSH Criteria Document #76-129 [Chromium (VI)].

Respirator Selection: NIOSH, as chromates: *at any concentrations above the NIOSH REL, or where there is no REL, at any detectable concentration:* SCBAF:PD,PP (any self-contained breathing apparatus that has a full facepiece and is operated in a pressure-demand or other positive-pressure mode); or SAF:PD,PP:ASCBA (any supplied-air respirator that has a full facepiece and is operated in a pressure-demand or other positive-pressure mode in combination with an auxiliary, self-contained breathing apparatus operated in a pressure-demand or other positive-pressure mode). *Escape:* HiEF (any air-purifying, full-facepiece respirator with a high-efficiency particulate filter); or SCBAE (any appropriate escape-type, self-contained breathing apparatus).

Storage: Calcium chromate must be stored to avoid contact with combustible, organic or other easily oxidized materials (such as Paper, Wood, Sulfur, Aluminum, Hydrazine and Plastics) since violent reactions occur. A regulated, marked area should be established where this chemical is handled, used, or stored in compliance with OSHA standard 1910.1045.

Shipping: Calcium chromate is not specifically cited in the DOT regulations.[19]

Spill Handling: *Dry material:* Evacuate persons not wearing protective equipment from area of spill or leak until clean-up is complete. Remove all ignition sources. Collect powdered material in the most convenient and safe manner and deposit in sealed containers. Ventilate area after clean-up is complete. It may be necessary to contain and dispose of this chemical as a hazardous waste. If material or contaminated runoff enters waterways, notify downstream users of potentially contaminated waters. Contact your Department of Environmental Protection or your regional office of the federal EPA for specific recommendations. If employees are required to clean-up spills, they must be properly trained and equipped. OSHA 1910.120(q) may be applicable.

Solution: Evacuate and restrict persons not wearing protective equipment from area of spill or leak until cleanup is complete. Remove all ignition sources. Ventilate area of spill or leak. Absorb liquids in vermiculite, dry sand, earth, peat, carbon, or a similar material and deposit in sealed containers. It may be necessary to contain and dispose of this chemical as a hazardous waste. If material or contaminated runoff enters waterways, notify downstream users of potentially contaminated waters. Contact your Department of Environmental Protection or your regional office of the federal EPA for specific recommendations. If employees are required to clean-up spills, they must be properly trained and equipped. OSHA 1910.120(q) may be applicable.

Fire Extinguishing: This chemical may burn but does not readily ignite. Use dry chemical, carbon dioxide, water spray, or alcohol foam extinguishers. Poisonous gases are produced in fire including chromium fumes. If material or contaminated runoff enters waterways, notify downstream users of potentially contaminated waters. Notify local health and fire officials and pollution control agencies. Container may explode in fire. From a secure, explosion-proof location, use water spray to cool exposed containers. If cooling streams are ineffective (venting sound increases in volume and pitch, tank discolors, or shows any signs of deforming), withdraw immediately to a secure position. If employees are expected to fight fires, they must be trained and equipped in OSHA 1910.156.

Disposal Method Suggested: Reduce to trivalent chromium and precipitate as chromium (III) hydroxide. Compact the sludge and dispose in single purpose special waste dumps.[22]

References

New Jersey Department of Health and Senior Services, "Hazardous Substance Fact Sheet: Calcium Chromate," Trenton, NJ (September 1998)

New York State Department of Health, "Chemical Fact Sheet: Calcium Chromate (YI)," Albany, NY, Bureau of Toxic Substance Assessment (January 1986)

U.S. Environmental Protection Agency, "Health Advisory: Chromium," Washington, DC, Office of Drinking Water (March 31, 1987)

U.S. Public Health Service, "Toxicological Profile for Chromium," Washington, DC, Agency for Toxic Substance and Disease Registry (October 1987)

Calcium Cyanamide

Molecular Formula: $CCaN_2$

Common Formula: $CaCN_2$

Synonyms: Aero-cyanamid; Aero-cyanamid, Special grade; Alzodef; Calcium carbimide; Calcium cyanamid; CCC; Cianamida calcica (Spanish); Cyanamid; Cyanamide; Cyanamide calcique (French); Cyanamide, calcium salt (1:1); Cyanamid granular; Cyanamid special grade; Cy-L 500; Lime nitrogen; NCI-C02937; Nitrogen lime; Nitrolime; USAF Cy-2

CAS Registry Number: 156-62-7

RTECS Number: GS6000000

DOT ID: UN 1403

Regulatory Authority

- Air Pollutant Standard Set (ACGIH)[1] (HSE)[33] (UNEP)[43] (OSHA)[58] (Several States)[60] (Australia) (DFG) (Israel) (Mexico) (Seveal Canadian Provinces)
- CLEAN AIR ACT: Hazardous Air Pollutants (Title I, Part A, Section 112)
- SUPERFUND/EPCRA 40CFR302.4 Reportable Quantity (RQ): CERCLA, 1 lb (0.454 kg)
- EPCRA Section 313 Form R *de minimis* concentration reporting level: 1.0%
- U.S. DOT Regulated Marine Pollutant (49CFR172.101, Appendix B) as cyanide mixtures, cyanide solutions or cyanides, inorganic, n.o.s.
- Canada, WHMIS, Ingredients Disclosure List; National Pollutant Release Inventory (NPRI)

Cited in U.S. State Regulations: Alaska (G), California (G), Connecticut (A), Florida (G), Illinois (G), Maine (G), Maryland (G), Massachusetts (G), Minnesota (G), Nevada (A), New Hampshire (G), New Jersey (G), North Dakota (A), Pennsylvania (G), Rhode Island (G), Virginia (A), West Virginia (G).

Description: Calcium cyanamide, $CaCN_2$, is a blackish-grey, shiny crystalline material or powder. Freezing/Melting point = 1,300°C (sublimes > 1,500°C). Insoluble in water.

Potential Exposure: Calcium cyanamide is used in agriculture as a fertilizer, herbicide, defoliant for cotton plants, and pesticide. It is also used in the manufacture of dicyandiamide and calcium cyanide as a desulfurizer in the iron and steel industry, and in steel hardening.

Incompatibilities: Commercial grades of calcium cyanamide may contain calcium carbide; contact with any form of moisture solutions may cause decomposition, liberating explosive acetylene gas and ammonia. May polymerize in water or alkaline solutions to dicyanamide. Contact with all solvents tested also cause decomposition.

Permissible Exposure Limits in Air: NIOSH and ACGIH TLV recommends 0.5 mg/m^3. The MAK set by DFG[3] is

1.0 mg/m^3. The HSE[33] TLV is 0.5 mg/m^3 and they have set a STEL of 1.0 mg/m^3. The former USSR/UNEP joint project[43] has set a MAC in ambient residential air of 0.02 mg/m^3 either on a momentary or a daily average basis. Several states have set guidelines or standards for calcium cyanamide in ambient air[60] ranging from 5.0 μg/m^3 (North Dakota) to 8.0 μg/m^3 (Virginia) to 10.0 μg/m^3 (Connecticut) to 12.0 μg/m^3 (Nevada).

Determination in Air: Filter; none; Gravimetric; IV NIOSH Method #0500, Particulates NOR (total).

Permissible Concentration in Water: The former USSR/UNEP joint project[43] has set a MAC of 1.0 mg/l in water for domestic purposes.

Routes of Entry: Inhalation of dust, ingestion.

Harmful Effects and Symptoms

Short Term Exposure: Calcium cyanamide can cause nausea, headache, dizziness, and flushing of the skin. It is a primary irritant of the mucous membranes of the respiratory tract, eyes, and skin. Drinking alcohol shortly before or within 1-2 days after exposure can cause a severe reaction. Inhalation may result in rhinitis, pharyngitis, laryngitis, and bronchitis. Conjunctivitis, keratitis, and corneal ulceration may occur. An itchy erythematous dermatitis has been reported and continued skin contact leads to the formation of slowly healing ulcerations on the palms and between the fingers. Sensitization occasionally develops. Chronic rhinitis and perforation on the nasal septum have been reported after long exposures. All local effects appear to be due to the caustic nature of cyanamide.

Long Term Exposure: Calcium cyanamide may damage the developing fetus. This chemical may damage the nervous system, causing numbness, and weakness in the hands and feet. Prolonged contact can cause skin ulcers. It causes a characteristic vasomotor reaction. There is erythema of the upper portions of the body, face and arms accompanied by nausea, fatigue, headache, dyspnea, vomiting, oppression in the chest and shivering. Circulatory collapse may follow in the more serious cases. The vasomotor response may be triggered or intensified by alcohol ingestion. Pneumonia or lung edema may develop. Cyanide ion is not released in the body, and the mechanism of toxic action is unknown.

Points of Attack: Eyes, skin, respiratory system, vasomotor system.

Medical Surveillance: Examination of the nervous system. Evaluation by a qualified allergist. Evaluate skin, respiratory tract and history of alcohol intake in placement or periodic examinations.

First Aid: If this chemical gets into the eyes, remove any contact lenses at once and irrigate immediately for at least 15 minutes, occasionally lifting upper and lower lids. Seek medical attention immediately. If this chemical contacts the skin, remove contaminated clothing and wash immediately with soap and water. Seek medical attention immediately. If this chemical has been inhaled, remove from exposure, begin rescue breathing (using universal precautions) if breathing has stopped and CPR if heart action has stopped. Transfer promptly to a medical facility. When this chemical has been swallowed, get medical attention. Give large quantities of water and induce vomiting. Do not make an unconscious person vomit.

Personal Protective Methods: Wear protective gloves and clothing to prevent any reasonable probability of skin contact. Safety equipment suppliers/manufacturers can provide recommendations on the most protective glove/clothing material for your operation. All protective clothing (suits, gloves, footwear, headgear) should be clean, available each day, and put on before work. Contact lenses should not be worn when working with this chemical. Wear dust-proof chemical goggles and face shield unless full facepiece respiratory protection is worn. Employees should wash immediately with soap when skin is wet or contaminated. Provide emergency showers and eyewash. In addition to personal protective equipment, waterproof barrier creams may be used to provide additional face and skin protection.

Respirator Selection: NIOSH/OSHA: *up to 25 mg/m^3:* SA (any supplied-air respirator); or SCBAF (any self-contained breathing apparatus with full facepiece). *Emergency or planned entry into unknown concentrations or IDLH conditions:* SCBAF:PD,PP (any self-contained breathing apparatus that has a full facepiece and is operated in a pressure-demand or other positive-pressure mode); or SAF:PD,PP: ASCBA (any supplied-air respirator that has a full facepiece and is operated in a pressure-demand or other positive-pressure mode in combination with an auxiliary self-contained breathing apparatus operated in a pressure-demand or other positive pressure mode). *Escape:* GMFSHiE (any air-purifying, full-facepiece respirator (gas mask) with a chin-style, front-or back-, mounted canister providing protection against the compound of concern and having a high efficiency particulate filter); or SCBAE (any appropriate escape-type, self-contained breathing apparatus).

Storage: Prior to working with Calcium cyanamide you should be trained on its proper handling and storage. Store in tightly closed containers in a cool, well-ventilated area away from moisture. A regulated, marked area should be established where this chemical is handled, used, or stored in compliance with OSHA standard 1910.1045.

Shipping: Calcium cyanamide with more than 1% calcium carbide requires a "Dangerous when Wet" label. It falls in Hazard Class 4.3 and Packing Group III. The limit on passenger aircraft or railcar shipment is 25 kg on cargo aircraft shipment is 100 kg.[19]

Spill Handling: Restrict persons not wearing protective equipment from area of spill until clean-up is complete.

Remove all ignition sources. Ventilate area of spill. Collect spilled material in the most convenient and safe manner and deposit in sealed containers for reclamation or for disposal in an approved facility. Absorb liquid containing Calcium Cyanamide in vermiculite, dry sand, earth, or similar material. If material or contaminated runoff enters waterways, notify downstream users of potentially contaminated waters. Contact your Department of Environmental Protection or your regional office of the federal EPA for specific recommendations. If employees are required to clean-up spills, they must be properly trained and equipped. OSHA 1910.120(q) may be applicable.

Fire Extinguishing: This chemical is a combustible solid. Do not use foam extinguishers or water. Use dry chemical, soda ash, or lime. Poisonous gases are produced in fire. If material or contaminated runoff enters waterways, notify downstream users of potentially contaminated waters. Notify local health and fire officials and pollution control agencies. From a secure, explosion-proof location, use water spray to cool exposed containers. If cooling streams are ineffective (venting sound increases in volume and pitch, tank discolors, or shows any signs of deforming), withdraw immediately to a secure position. If employees are expected to fight fires, they must be trained and equipped in OSHA 1910.156.

References

Sax, N. I.., Ed., "Dangerous Properties of Industrial Materials Report" 2, No. 6, 38–41 (1982)

New Jersey Department of Health and Senior Services, "Hazardous Substance Fact Sheet: Calcium Cyanamide," Trenton, NJ (April 1998)

Calcium Cyanide

Molecular Formula: C_2CaN_2

Common Formula: $Ca(CN)_2$

Synonyms: Calcid; Calcyan; Calcyanide; Cianuro calcico (Spanish); Cyanogas; Cyanure de calcium (French)

CAS Registry Number: 592-01-8

RTECS Number: EW0700000

DOT ID: UN 1575

EEC Number: 020-002-00-5

Regulatory Authority

- Air Pollutant Standard Set (ACGIH)[1] (DFG)[3] (HSE)[33] (UNEP)[43] (OSHA)[58]
- CLEAN AIR ACT: Hazardous Air Pollutants (Title I, Part A, Section 112)
- CLEAN WATER ACT: Section 311 Hazardous Substances/RQ 40CFR117.3 (same as CERCLA, see below); Section 313 Water Priority Chemicals (57FR41331, 9/9/92)
- EPA HAZARDOUS WASTE NUMBER (RCRA No.): P021
- RCRA, 40CFR261, Appendix 8 Hazardous Constituents
- SUPERFUND/EPCRA 40CFR302.4 Reportable Quantity (RQ): CERCLA, 10 lb (4.54 kg)
- U.S. DOT Regulated Marine Pollutant (49CFR172.101, Appendix B) as cyanide mixtures, cyanide solutions or cyanides, inorganic, n.o.s.

Cited in U.S. State Regulations: California (G), Florida (G), Kansas (G), Louisiana (G), Maine (G), Massachusetts (G), New Jersey (G), New Hampshire (G), Pennsylvania (G), Rhode Island (G), Vermont (G), Virginia (G), Washington (G), Wisconsin (G).

Description: Calcium cyanide, $Ca(CN)_2$, is a white crystalline solid or powder. Almond odor. Freezing/Melting point ≥ 350°C (dangerous decomposition below m.p.). Hazard Identification (based on NFPA-704 M Rating System): Health 3, Flammability 0, Reactivity 1. Soluble in water; violent reaction.

Potential Exposure: Calcium cyanide is used as a fumigant, as a rodenticide, in leaching precious metal ores; in the manufacture of stainless steel and as a stabilizer for cement.

Incompatibilities: Contact with water, acids, acidic salts, moist air, or carbon dioxide, forms highly toxic and flammable hydrogen cyanide. Incompatible with fluorine, magnesium. Reacts violently when heated with nitrites, nitrates, chlorates and perchlorates. Calcium cyanide decomposes in high heat forming hydrogen cyanide and nitrous oxides fumes.

Permissible Exposure Limits in Air: OSHA and ACGIH: 5 mg/m^3 TWA; NIOSH: Ceiling limit, 4.7 ppm; 5 mg/m^3 per 10 minutes as cyanides. All have notations that skin contact contributes significantly in overall exposure. IDLH = 25 mg/m^3 as CN The former USSR/UNEP Project[43] has set a MAC of 0.3 mg/m^3.

Determination in Air: Filter/Bubbler; Potassium hydroxide; Ion-specific electrode; IV NIOSH Method #7904, Cyanides.

Permissible Concentration in Water: A limit of $0.2CN^-$ per liter has been set by U.S.E.P.A.[6] as part of the priority toxic pollutant program.

Routes of Entry: Can be absorbed through the skin, inhalation, ingestion.

Harmful Effects and Symptoms

Calcium cyanide is highly toxic. The lethal human dose is 18 mg/kg. The hazard is that of hydrogen cyanide. The dust is irritating to the eyes, nose and throat.[41] Inhalation or ingestion causes headache, nausea, vomiting and weakness; high concentrations are rapidly fatal.

Short Term Exposure: The substance is corrosive to the eyes, skin, and respiratory tract. Higher exposures can cause pulmonary edema, a medical emergency that can be delayed for several hours. This can cause death. My affect the central nervous system, blood, heart and respiratory tract.

Long Term Exposure: Repeated or prolonged contact with skin may cause dermatitis. May be a reproductive toxin in humans.

Points of Attack: Skin, lungs.

Medical Surveillance: Lung function tests. Examination by a qualified allergist. Consider chest x-ray following acute overexposure.

First Aid: If this chemical gets into the eyes, remove any contact lenses at once and irrigate immediately for at least 15 minutes, occasionally lifting upper and lower lids. Do not allow water to enter nose or mouth. Seek medical attention immediately. If this chemical contacts the skin, remove contaminated clothing and wash immediately with soap and water. Seek medical attention immediately. If this chemical has been inhaled, remove from exposure, begin rescue breathing (using universal precautions) if breathing has stopped and CPR if heart action has stopped. Transfer promptly to a medical facility. When this chemical has been swallowed, get medical attention. Give large quantities of water and induce vomiting. Do not make an unconscious person vomit. Medical observation is recommended for 24 – 48 hours after breathing overexposure, as pulmonary edema may be delayed. As first aid for pulmonary edema, a doctor or authorized paramedic may consider administering a corticosteroid spray.

Use amyl nitrate capsules if symptoms of cyanide poisoning develop. All area employees should be trained regularly in emergency measures for cyanide poisoning and in CPR. A cyanide antidote kit should be kept in the immediate work area and must be rapidly available. Kit ingredients should be replaced every 1 – 2 years to ensure freshness. Persons trained in the use of this kit, oxygen use, and CPR must be quickly available.

Personal Protective Methods: Wear protective gloves and clothing to prevent any reasonable probability of skin contact. Safety equipment suppliers/manufacturers can provide recommendations on the most protective glove/clothing material for your operation. All protective clothing (suits, gloves, footwear, headgear) should be clean, available each day, and put on before work. Contact lenses should not be worn when working with this chemical. Wear dust-proof chemical goggles and face shield unless full facepiece respiratory protection is worn. Employees should wash immediately with soap when skin is wet or contaminated. Provide emergency showers and eyewash.

Respirator Selection: NIOSH/OSHA: *up to 25 mg/m³:* SA (any supplied-air respirator); or SCBAF (any self-contained breathing apparatus with full facepiece). *Emergency or planned entry into unknown concentrations or IDLH conditions:* SCBAF:PD,PP (any self-contained breathing apparatus that has a full facepiece and is operated in a pressure-demand or other positive-pressure mode); or SAF: PD,PP:ASCBA (any supplied-air respirator that has a full facepiece and is operated in a pressure-demand or other positive-pressure mode in combination with an auxiliary self-contained breathing apparatus operated in a pressure-demand or other positive pressure mode). *Escape:* GMFSHiE (any air-purifying, full-facepiece respirator (gas mask) with a chin-style, front-or back-, mounted canister providing protection against the compound of concern and having a high efficiency particulate filter); or SCBAE (any appropriate escape-type, self-contained breathing apparatus).

Storage: Prior to working with Calcium cyanide you should be trained on its proper handling and storage. A regulated, marked area should be established where this chemical is handled, used, or stored. Store in tightly closed containers in a cool, well-ventilated area away from moisture, water, acids, and oxidizers. Protect against physical damage to containers.

Shipping: Calcium cyanide requires a "Poison" label. It falls in Hazard Class 6.1 and Packing Group I.

Spill Handling: Evacuate persons not wearing protective equipment from area of spill or leak until clean-up is complete. Remove all ignition sources. Collect powdered material in the most convenient and safe manner and deposit in sealed containers. Ventilate area after clean-up is complete. Keep material out of drains, sewers, streams. It may be necessary to contain and dispose of this chemical as a hazardous waste. If material or contaminated runoff enters waterways, notify downstream users of potentially contaminated waters. Contact your Department of Environmental Protection or your regional office of the federal EPA for specific recommendations. If employees are required to clean-up spills, they must be properly trained and equipped. OSHA 1910.120(q) may be applicable.

Fire Extinguishing: Use dry chemical or dry sand. DO NOT USE foam, water, carbon dioxide. Poisonous gases are produced in fire. If material or contaminated runoff enters waterways, notify downstream users of potentially contaminated waters. Notify local health and fire officials and pollution control agencies. From a secure, explosion-proof location, use water spray to cool exposed containers. If cooling streams are ineffective (venting sound increases in volume and pitch, tank discolors, or shows any signs of deforming), withdraw immediately to a secure position. If employees are expected to fight fires, they must be trained and equipped in OSHA 1910.156.

Disposal Method Suggested: Add cyanide waste to strong alkaline sodium hypochlorite. Let stand 24 hours then flush to sewage plant.[22]

References

Sax, N. I.., Ed., "Dangerous Properties of Industrial Materials Report" 2, No. 1, 95–96 (1982)

Calcium Fluoride

Molecular Formula: CaF_2

Synonyms: Calcium difluoride; Fluorite; Fluorspar; Fluospar; Met-Spar

CAS Registry Number: 7789-75-5

RTECS Number: EW1760000

Regulatory Authority

- Air Pollutant Standard Set (ACGIH)[1] (DFG)[3] (HSE)[33] (OSHA)[58] (former USSR)[43]

Cited in U.S. State Regulations: New Jersey (G), New York (G), Oklahoma (G).

Description: Calcium fluoride, CaF_2, is colorless crystalline or white, powdery substance. Freezing/Melting point = 1,423°C. Hazard Identification (based on NFPA-704 M Rating System): Health 2, Flammability 0, Reactivity 0. Practically insoluble in water.

Potential Exposure: Calcium Fluoride is used in smelting, electric are welding, making steel, glass and ceramics and to fluoridate drinking water.

Permissible Exposure Limits in Air: The exposure limits are recommended for inorganic Fluoride compounds and are measured as Fluorine. The OSHA PEL is 2.5 mg/m^3 TWA averaged over an 8-hour workshift. This is the ACGIH,[1] DFG,[3] HSE[33] and OSHA[58] value. The former USSR/UNEP joint project[43] has set a MAC in ambient air in residential areas of 0.2 mg/m^3 on a momentary basis or 0.03 mg/m^3 on an average daily basis.

Permissible Concentration in Water: No criteria set.

Routes of Entry: Inhalation of dust, ingestion.

Harmful Effects and Symptoms

Short Term Exposure: *Inhalation:* May cause difficult breathing, burning of mouth, throat and nose which may result in bleeding. These may be felt at 7.5 mg/m^3. Nausea, vomiting, profuse sweating and excess thirst may occur at higher levels. *Skin:* May cause rash, itching and burning of skin. Solutions of 1% strength may cause sores if not removed promptly. *Eyes:* May cause severe irritation. *Ingestion:* Most reported instances of fluoride toxicity are due to accidental ingestion and it is difficult to associate symptoms with dose. 5 – 40 mg may cause nausea, diarrhea and vomiting. More severe symptoms of burning and painful abdomen, sores in mouth, throat and digestive tract, tremors, convulsions and shock will occur from about 1 gm. Death may result by ingestion of 2 – 5 grams (1/6 ounce).

Long Term Exposure: Repeated exposure may cause poor appetite, nausea, constipation, or diarrhea. Repeated exposure to fluoride chemicals may cause stiffness in muscles or ligaments and even crippling; this could take years to develop. Fluoride may increase bone density, stimulate new bone growth or cause calcium deposits in ligaments. This may become a problem at levels of 20 – 50 mg/m^3 or higher. Mottling of the teeth may occur at this level.

Points of Attack: Eyes, skin, respiratory system, bones.

Medical Surveillance: For those with frequent or potentially high exposure (half the TLV or greater), the following are recommended before beginning work and at regular times after that: Lung function tests. If symptoms develop or overexposure is suspected, the following may be useful: Consider chest x-ray after acute overexposure. Urine Fluoride level (normal is less than 4 mg/l).

First Aid: If this chemical gets into the eyes, remove any contact lenses at once and irrigate immediately for at least 15 minutes, occasionally lifting upper and lower lids. Seek medical attention immediately. If this chemical contacts the skin, remove contaminated clothing and wash immediately with soap and water. Seek medical attention immediately. If this chemical has been inhaled, remove from exposure, begin rescue breathing (using universal precautions) if breathing has stopped and CPR if heart action has stopped. Transfer promptly to a medical facility. When this chemical has been swallowed, get medical attention. A doctor or authorized paramedic may consider administering aluminum hydroxide gel, if conscious.

Personal Protective Methods: Wear protective gloves and clothing to prevent any reasonable probability of skin contact. Safety equipment suppliers/manufacturers can provide recommendations on the most protective glove/clothing material for your operation. All protective clothing (suits, gloves, footwear, headgear) should be clean, available each day, and put on before work. Contact lenses should not be worn when working with this chemical. Wear dust-proof chemical goggles and face shield unless full facepiece respiratory protection is worn. Employees should wash immediately with soap when skin is wet or contaminated. Provide emergency showers and eyewash.

Respirator Selection: Engineering controls should be used wherever feasible to maintain airborne concentrations of this chemical below the prescribed exposure limit. Respirators and protective equipment are less effective than engineering controls, and should be used only in non-routine or emergency situations which may result in exposure concentrations in excess of the TWA environmental limit. NIOSH: *fluorides: 12.5 mg/m^3:* DM (any dust and mist respirator). *25 mg/m^3:* DMXSQ* (any dust and mist respirator except single-use and quarter mask respirators); or SA* (any supplied-air respirator). *62.5 mg/m^3:* SA:CF* (any supplied-air respirator operated in a continuous-flow mode); or PAPRDM*+ *if not present as a fume* (any powered, air-purifying respirator with a dust and mist filter). *125 mg/m^3:* HiEF+ (any air-purifying, full-facepiece respirator with a high-efficiency particulate filter); or SCBAF (any self-contained breathing apparatus with a full facepiece); or SAF (any supplied-air respirator with a full facepiece). *250 mg/m^3:* SA:PD,PP (any supplied-air respirator operated in a pressure-demand or other positive-pressure mode). *Emergency or planned entry into unknown concentrations or IDLH conditions:* SCBAF:PD, PP (any self-contained breathing apparatus that has a full faceplate and is operated in a pressure-demand or other positive-pressure mode); or SAF:PD,PP:ASCBA (any supplied-air respirator that has a full facepiece and is operated in a pressure-demand or other positive-pressure mode in combination with an auxiliary, self-contained breathing apparatus operated in a pressure-demand or other positive-pressure mode). *Escape:* HiEF+ (any air-purifying, full-facepiece respirator with a high-efficiency particulate filter); or SCBAE (any appropriate escape-type, self-contained breathing apparatus).

* Substance reported to cause eye irritation or damage; may require eye protection.

\+ May need acid gas sorbent.

Storage: Store in tightly closed containers in a cool, well-ventilated area away from acids, and chemically active metals (such as potassium, sodium, magnesium, and zinc) because Corrosive Hydrogen Fluoride will be produced.

Shipping: Calcium Fluoride is not specifically covered by DOT[19] in its Performance-Oriented Packaging Standards.

Spill Handling: Evacuate persons not wearing protective equipment from area of spill or leak until clean-up is complete. Remove all ignition sources. Collect powdered material in the most convenient and safe manner and deposit in sealed containers. Ventilate area after clean-up is complete. It may be necessary to contain and dispose of this chemical as a hazardous waste. If material or contaminated runoff enters waterways, notify downstream users of potentially contaminated waters. Contact your Department of Environmental Protection or your regional office of the federal EPA for specific recommendations. If employees are required to clean-up spills, they must be properly trained and equipped. OSHA 1910.120(q) may be applicable.

Fire Extinguishing: This chemical is a noncombustible solid. Use any extinguishing agent suitable for surrounding fire. Poisonous gases are produced in fire. If material or contaminated runoff enters waterways, notify downstream users of potentially contaminated waters. Notify local health and fire officials and pollution control agencies. From a secure, explosion-proof location, use water spray to cool exposed containers. If cooling streams are ineffective (venting sound increases in volume and pitch, tank discolors, or shows any signs of deforming), withdraw immediately to a secure position. If employees are expected to fight fires, they must be trained and equipped in OSHA 1910.156.

References

Sax, N. I.., Ed., "Dangerous Properties of Industrial Materials Report" 1, No. 8, 47–48 (1981)

New Jersey Department of Health and Senior Services and Senior Services, "Hazardous Substance Fact Sheet: Calcium Fluoride," Trenton, NJ (May 1986)

New York State Department of Health, "Chemical Fact Sheet: Calcium Fluoride," Albany, NY, Bureau of Toxic Substance Assessment (March 1986)

Calcium Hydride

Molecular Formula: CaH_2

CAS Registry Number: 57308-10-8

RTECS Number: EW2440000

DOT ID: UN 1404

Cited in U.S. State Regulations: Maine (G), New Hampshire (G), New Jersey (G), Oklahoma (G).

Description: Calcium hydride, CaH_2, is a grayish-white crystalline solid. Melting point = 816°C (It decomposes at about 600°C). Hazard Identification (based on NFPA-704 M Rating System): Health 3, Flammability 4, Reactivity 2. Reacts with water.

Potential Exposure: Calcium Hydride is used as a drying and reducing agent and a cleaner for blocked up oil wells.

Incompatibilities: Reacts with water, moist air, and steam, releasing flammable hydrogen gas, and may self-ignite in air. Incompatible with metal halogenates, silver fluoride, and tetrahydrofuran.

Permissible Exposure Limits in Air: No standards set.

Permissible Concentration in Water: No criteria set (Calcium hydride reacts with water in any event).

Routes of Entry: Inhalation of dust, ingestion.

Harmful Effects and Symptoms

Short Term Exposure: Calcium Hydride can affect you when breathed in. Contact with skin or eyes can cause severe burns. Exposure can irritate the eyes, nose and throat. Breathing Calcium Hydride can irritate the lungs causing coughing and/or shortness of breath. Higher exposures can cause a build-up of fluid in the lungs (pulmonary edema). This can cause death.

Long Term Exposure: Although it is unknown whether calcium hydride causes lung damage, similar very irritation substances are capable of causing lung damage.

Points of Attack: Eyes, skin, respiratory system.

Medical Surveillance: Before beginning employment and at regular times after that, for those with frequent or potentially high exposures, the following are recommended: Lung function tests. IF symptoms develop or overexposure is suspected, the following may be useful: Consider chest x-ray after acute overexposure.

First Aid: If this chemical gets into the eyes, remove any contact lenses at once and irrigate immediately for at least 30 minutes, occasionally lifting upper and lower lids. Seek medical attention immediately. If this chemical contacts the skin, remove contaminated clothing and wash immediately with soap and water. Seek medical attention immediately. If this chemical has been inhaled, remove from exposure, begin rescue breathing (using universal precautions) if breathing has stopped and CPR if heart action has stopped. Transfer promptly to a medical facility. When this chemical has been swallowed, get medical attention. Give large quantities of water and induce vomiting. Do not make an unconscious person vomit. Medical observation is recommended for 24 – 48 hours after breathing overexposure, as pulmonary edema may be delayed. As first aid for pulmonary edema, a doctor or authorized paramedic may consider administering a corticosteroid spray.

Personal Protective Methods: *Clothing:* Avoid skin contact with Calcium Hydride. Wear protective gloves and clothing. Safety equipment suppliers/manufacturers can provide recommendations on the most protective glove/clothing material for your operation. All protective clothing (suits, gloves, footwear, headgear) should be clean, available each day, and put on before work. *Eye Protection:* Wear dust-proof goggles and face shield when working with powders or dust, unless full facepiece respiratory protection is worn.

Respirator Selection: Where the potential exists for exposure to Calcium Hydride use a MSHA/NIOSH approved full facepiece respirator equipped with particulate (dust/fume/mist) filters. Particulate filters must be checked every day before work for physical damage, such as rips or tears, and replaced as needed. Where the potential for high exposures exists, use a MSHA/NIOSH approved supplied-air respirator with a full facepiece operated in the positive pressure mode or with a full facepiece hood, or helmet in the continuous flow mode, or use a MSHA/NIOSH approved self-contained breathing apparatus with a full facepiece operated in pressure-demand or other positive pressure mode.

Storage: Calcium Hydride must be stored to avoid contact with water or steam since violent reactions occur and flammable Hydrogen gas is produced. Store in tightly closed containers in a cool. Well-ventilated area away from metal halogenates, silver fluoride, and tetrahydrofuran.

Shipping: Calcium Hydride should carry a "Dangerous when Wet" label. It falls in Hazard Class 4.3 and Packing Group I Shipment by passenger aircraft or railcar is forbidden.

Spill Handling: Evacuate persons not wearing protective equipment from area of spill or leak until clean-up is complete. Remove all ignition sources. Collect powdered material in the most convenient and safe manner and deposit in sealed containers. Ventilate area after clean-up is complete. It may be necessary to contain and dispose of this chemical as a hazardous waste. If material or contaminated runoff enters waterways, notify downstream users of potentially contaminated waters. Contact your Department of Environmental Protection or your regional office of the federal EPA for specific recommendations. If employees are required to clean-up spills, they must be properly trained and equipped. OSHA 1910.120(q) may be applicable.

Fire Extinguishing: Do not any hydrous (water, foam, etc.) extinguishing agents. Fire may restart after it has been extinguished. Use dry chemical, soda ash, or lime extinguishers. Poisonous gases are produced in fire. If material or contaminated runoff enters waterways, notify downstream users of potentially contaminated waters. Notify local health and fire officials and pollution control agencies. From a secure, explosion-proof location, use water spray to cool exposed containers. If cooling streams are ineffective (venting sound increases in volume and pitch, tank discolors, or shows any signs of deforming), withdraw immediately to a secure position. If employees are expected to fight fires, they must be trained and equipped in OSHA 1910.156.

References

New Jersey Department of Health and Senior Services and Senior Services, "Hazardous Substance Fact Sheet: Calcium Hydride," Trenton, NJ (March 1987).

Calcium Hydroxide

Molecular Formula: CaH_2O_2

Common Formula: $Ca(OH)_2$

Synonyms: Bell mine; Calcium hydrate; Carboxide; Hydrated kemikal; Hydrated lime; Lime water; Slaked lime

CAS Registry Number: 1305-62-0

RTECS Number: EW2800000

Regulatory Authority

- Air Pollutant Standard Set (ACGIH)[1] (HSE)[33] (OSHA)[58] (Several States)[60] (Australia) (Israel) (Mexico) (Several Canadian Provinces)

Cited in U.S. State Regulations: Alaska (G), California (G), Connecticut (A), Florida (G), Illinois (G), Maine (G), Massachusetts (G), Minnesota (G), Nevada (A), New Hampshire (G), New Jersey (G), New York (G), North Dakota (A), Oklahoma (G), Pennsylvania (G), Rhode Island (G), Virginia (A), West Virginia (G).

Description: Calcium hydroxide, $Ca(OH)_2$, is a soft white crystalline, odorless powder with an alkaline, bitter taste. Freezing/Melting point (decomposes; dehydrates to calcium oxide) = 580°C. Hazard Identification (based on NFPA-704 M Rating System): Health 1, Flammability 0, Reactivity 0. Insoluble in water.

Note: Readily absorbs CO_2 from the air to form calcium carbonate.

Potential Exposure: Calcium hydroxide is used in agriculture and in fertilizer manufacture; it is used in the formulation of mortar, plasters and cements; it is used as a scrubbing and neutralizing agent in the chemical industry.

Incompatibilities: May react violently with acids, maleic anhydride, nitromethane, nitroethane, nitropropane, nitroparaffins and phosphorus.

Permissible Exposure Limits in Air: NIOSH REL: TWA 5 mg/m^3. OSHA PEL: TWA 15 mg/m^3 (total); 5 mg/m^3 (respirable fraction). ACGIH recommends a TWA value of 5 mg/m^3. Several states have set guidelines or standards for calcium hydroxide in ambient air[60] ranging from 50 μg/m^3 (North Dakota) to 80 μg/m^3 (Virginia) to 100 μg/m^3 (Connecticut) to 119 μg/m^3 (Nevada).

Determination in Air: Filter collection followed by atomic absorption analysis. See NIOSH Method #7020 for Calcium. See also OSHA Method ID 121.

Permissible Concentration in Water: No criteria set.

Routes of Entry: Inhalation of dust, ingestion.

Harmful Effects and Symptoms

Short Term Exposure: *Inhalation:* May cause severe irritation to mouth, throat and lungs if dust is inhaled. Higher exposures can cause pulmonary edema, a medical emergency that can be delayed for several hours. This can cause death. *Skin:* May cause painful irritation and chemical burns on contact with open cuts or sores or on prolonged contact with intact skin. *Eyes:* Powders and slurries may cause severe chemical burns. Blindness can result. *Ingestion:* Powders, crystals or slurries may give rise to irritation, soreness and chemical burns. The estimated lethal dose is about one pound.

Long Term Exposure: Repeated or prolonged contact with skin may cause dermatitis. Lungs may be affected by repeated or prolonged exposure to dust particles.

Points of Attack: Eyes, skin, respiratory system.

Medical Surveillance: Lung function tests. Consider chest x-ray following acute overexposure.

First Aid: If this chemical gets into the eyes, remove any contact lenses at once and irrigate immediately for at least 30 minutes, occasionally lifting upper and lower lids. Seek medical attention immediately. If this chemical contacts the skin, remove contaminated clothing and wash immediately with soap and water. Seek medical attention immediately. If this chemical has been inhaled, remove from exposure, begin rescue breathing (using universal precautions) if breathing has stopped and CPR if heart action has stopped. Transfer promptly to a medical facility. When this chemical has been swallowed, get medical attention. Give large quantities of water and induce vomiting. Do not make an unconscious person vomit. Medical observation is recommended for 24 – 48 hours after breathing overexposure, as pulmonary edema may be delayed. As first aid for pulmonary edema, a doctor or authorized paramedic may consider administering a corticosteroid spray.

Personal Protective Methods: Gloves, eye protection, and coveralls should be worn if contact with Calcium hydroxide is likely.

Respirator Selection: A dust mask or respirator with dust cartridges.

Storage: Keep containers tightly closed. Store away from incompatible materials listed above.

Shipping: Calcium hydroxide is not specifically cited by DOT[19] as regards labels or maximum shipping quantities allowed.

Spill Handling: Evacuate persons not wearing protective equipment from area of spill or leak until clean-up is complete. Remove all ignition sources. Collect powdered material in the most convenient and safe manner and deposit in sealed containers. Ventilate area after clean-up is complete. It may be necessary to contain and dispose of this chemical as a hazardous waste. If material or contaminated runoff enters waterways, notify downstream users of potentially contaminated waters. Contact your Department of Environmental Protection or your regional office of the federal EPA for specific recommendations. If employees are required to clean-up spills, they must be properly trained and equipped. OSHA 1910.120(q) may be applicable.

Fire Extinguishing: This chemical is a non-combustible solid. Use any extinguishing agent suitable for surrounding fire. Poisonous gases are produced in fire. If material or contaminated runoff enters waterways, notify downstream users of potentially contaminated waters. Notify local health and fire officials and pollution control agencies. From a secure, explosion-proof location, use water spray to cool exposed containers. If cooling streams are ineffective (venting sound increases in volume and pitch, tank discolors, or shows any signs of deforming), withdraw immediately to a secure position. If employees are expected to fight fires, they must be trained and equipped in OSHA 1910.156.

Disposal Method Suggested: Landfill oradmixutre with acid industrial wastes prior to lagooning.

References

Sax, N. I.., Ed., "Dangerous Properties of Industrial Materials Report" 1, No. 8, 48–50 (1981)

New York State Department of Health, "Chemical Fact Sheet: Calcium Hydroxide," Albany, NY, Bureau of Toxic Substance Assessment (January 1986)

New Jersey Department of Health and Senior Services and Senior Services, "Hazardous Substance Fact Sheet: Calcium Hydroxide," Trenton, NJ (February 1989)

Calcium Hypochlorite

Molecular Formula: $CaCl_2O_2$

Common Formula: $Ca(OCl)_2$

Synonyms: B-K powder; Bleaching powder; Calcium chlorohydrochlorite; Calcium hypochloride; Calcium oxychloride; Caporit; CCH; Chloride of lime; Chlorinated lime; Hipoclorito calcico (Spanish); HTH; Hy-Chlor; Hypochlorous acid, calcium; Hyporit; Induclor; Lime chloride; Lo-Bax; Losantin; Perchloron; Pittabs; Pittchlor; Pittcide; Prestochlor; Pulsar; Stellos

CAS Registry Number: 7778-54-3

RTECS Number: NH3485000

DOT ID: UN 1748 (dry); UN 2208 (mixture); UN 2880 (hydrated)

EINECS Number: 231-908-7

Regulatory Authority

- CLEAN WATER ACT: Section 311 Hazardous Substances/RQ 40CFR117.3 (same as CERCLA, see below)
- SUPERFUND/EPCRA 40CFR302.4 Reportable Quantity (RQ): CERCLA, 10 lb (4.54 kg)

Cited in U.S. State Regulations: California (G), Florida (G), Illinois (G), Maine (G), Massachusetts (G), New Hampshire (G), New Jersey (G), New York (G), Oklahoma (G), Pennsylvania (G), Rhode Island (G).

Description: Calcium hypochlorite, $Ca(OCl)_2$, is a white powder, granule or pellets with a strong chlorine-like odor. It decomposes (possibly explosively) at about 100°C. Hazard Identification (based on NFPA-704 M Rating System): Health 3, Flammability 0, Reactivity 1. Soluble in water; reacts slowly releasing chlorine gas.

Potential Exposure: Calcium Hypochlorite is used to kill algae and bacteria, in bleach and in pool chemical products.

Incompatibilities: A strong oxidizer. Decomposes in heat or sunlight; becomes explosive above 100°C (212°F). Incompatible with strong acids, water and other forms of moisture, reducing agents, combustible materials, all other chemicals, especially acetylene, aniline and all other amines, anthracene, carbon tetrachloride, iron oxide, manganese oxide, mercaptans, diethylene glycol monomethyl ether, nitromethane, organic matter, organic sulfides, phenol, 1-propanethiol, propyl mercaptan, sulfur, turpentine, organic sulfur compounds

Permissible Exposure Limits in Air: No standards set.

Permissible Concentration in Water: No criteria set.

Routes of Entry: Inhalation of dust, ingestion.

Harmful Effects and Symptoms

Short Term Exposure: Calcium Hypochlorite can affect you when breathed in. Calcium Hypochlorite may cause mutations. Handle with extreme caution. Exposure can severely irritate the eyes, nose and throat. Contact can severely irritate the skin. Exposure can severely irritate the "voice box" (larynx), bronchial tubes and lungs. Higher levels can cause a build-up of fluid in the lungs (pulmonary edema). This can cause death.

Long Term Exposure: Repeated or prolonged contact can irritate the lungs; may cause bronchitis.

Points of Attack: Eyes, skin, respiratory system.

Medical Surveillance: Before beginning employment and at regular times after that, for those with frequent or potentially high exposures, the following are recommended: Lung function tests. If symptoms develop or overexposure is suspected, the following may be useful: Consider chest x-ray following acute overexposure.

First Aid: If this chemical gets into the eyes, remove any contact lenses at once and irrigate immediately for at least 15 minutes, occasionally lifting upper and lower lids. Seek medical attention immediately. If this chemical contacts the skin, remove contaminated clothing and wash immediately with soap and water. Seek medical attention immediately. If this chemical has been inhaled, remove from exposure, begin rescue breathing (using universal precautions) if breathing has stopped and CPR if heart action has stopped. Transfer promptly to a medical facility. When this chemical has been swallowed, get medical attention. Give large quantities of water and induce vomiting. Do not make an unconscious person vomit. Medical observation is recommended for 24 – 48 hours after breathing overexposure, as pulmonary edema may be delayed. As first aid for pulmonary edema, a doctor or authorized paramedic may consider administering a corticosteroid spray.

Personal Protective Methods: Avoid skin contact with Calcium Hypochlorite. Wear protective gloves and clothing. Safety equipment suppliers/manufacturers can provide recommendations on the most protective glove/clothing material for your operation. Natural Rubber, Neoprene and Polyvinyl Chloride are among the recommended protective materials. All protective clothing (suits, gloves, footwear, headgear) should be clean, available each day, and put on before work. Wear dust-proof goggles and face shield when working with powders or dust unless fullfacepiece respiratory protection is worn.

Respirator Selection: Where the potential exists for exposure to Calcium Hypochlorite, use a MSHA/NIOSH approved full facepiece respirator with a high efficiency particulate filter. Greater protection is provided by a powered-air purifying respirator. Where the potential for high exposures exists, use a MSHA/NIOSH approved supplied-air respirator with a full facepiece operated in the positive pressure mode or with a full facepiece, hood, or helmet, in the continuous flow mode, or use a MSHA/NIOSH approved self-contained breathing apparatus with a full facepiece operated in pressure-demand or other positive pressure mode.

Storage: Prior to working with Calcium Hypochlorite you should be trained on its proper handling and storage. Calcium Hypochlorite must be stored to avoid contact with strong acids (such as hydrochloric, sulfuric, and nitric); ammonium compounds (such as ammonia and ammonium hydroxide) and amines (such as aniline) since violent reactions occur. Store in tightly closed containers in a cool, well-ventilated area away from water or moisture and combustibles (such as wood, paper or oil). When heated above 100°C, Calcium Hypochlorite becomes explosive. Protect containers against physical damage. Avoid storage for long periods, particularly at summer temperatures. See OSHA standard 1910.104 and NFPA 43A *Code for the Storage of Liquid and Solid Oxidizers* for detailed handling and storage regulations.

Shipping: Calcium hypochlorite must be labeled "Oxidizer." It falls in DOT Hazard Class 5.1 and Packing Group II.[19]

Spill Handling: Evacuate persons not wearing protective equipment from area of spill or leak until clean-up is complete. Remove all ignition sources. Collect powdered material in the most convenient and safe manner and deposit in sealed containers. Ventilate area after clean-up is complete. It may be necessary to contain and dispose of this chemical as a hazardous waste. If material or contaminated runoff enters waterways, notify downstream users of potentially contaminated waters. Contact your Department of Environmental Protection or your regional office of the federal EPA for specific recommendations. If employees are required to clean-up spills, they must be properly trained and equipped. OSHA 1910.120(q) may be applicable.

Initial isolation and protective action distances:

Distances shown are likely to be affected during the first 30 minutes after materials are spilled and could increase with time. If more than one tank car, cargo tank, portable tank, or large cylinder is involved in the incident is leaking, the protective action distance may need to be increased. You may need to seek emergency information from CHEMTREC at (800) 424-9300 or seek professional environmental engineering assistance from the U.S. EPA Environmental Response Team at (908)548-8730 (24-hour response line).

UN 1748 (calcium hypochlorite, dry) is on the DOT's list of dangerous water-reactive materials which create large amounts of toxic vapor when *spilled in water:* Dangerous from 0.5 – 10 km (0.3 – 6.0 miles) downwind.

Fire Extinguishing: This chemical is a strong oxidizer and will increase the intensity of any fire. USE WATER ONLY. DO NOT USE chemical or carbon dioxide extinguishers. Poisonous gases are produced in fire. If material or contaminated runoff enters waterways, notify downstream users of potentially contaminated waters. Notify local health and fire officials and pollution control agencies. From a secure, explosion-proof location, use water spray to cool exposed containers. If cooling streams are ineffective (venting sound increases in volume and pitch, tank discolors, or shows any signs of deforming), withdraw immediately to a secure position. If employees are expected to fight fires, they must be trained and equipped in OSHA 1910.156.

Disposal Method Suggested: Dissolve the material in water and add to a large volume of concentrated reducing agent solution, then acidify the mixture with H_2SO_4. When reduction is complete, soda ash is added to make the solution alkaline. The alkaline liquid is decanted from any sludge produced, neutralized, and diluted before discharge to a sewer or stream. The sludge is landfilled.

References

Sax, N. I.., Ed., "Dangerous Properties of Industrial Materials Report" 1, No. 8, 50–52 (1981) and 4, No. 3, 76–79 (1984)

New Jersey Department of Health and Senior Services and Senior Services, "Hazardous Substance Fact Sheet: Calcium Hypochlorite," Trenton, NJ (September 1986)

New York State Department of Health, "Chemical Fact Sheet: Calcium Hypochlorite," Albany, NY, Bureau of Toxic Substance Assessment (March 1986)

Calcium Nitrate

Molecular Formula: CaN_2O_6

Common Formula: $Ca(NO_3)_2$

Synonyms: Calcium nitrate; Calcium(II) nitrate (1:2); Lime saltpeter; Nitric acid, calcium salt; Nitrocalcite; Norwegian saltpeter

CAS Registry Number: 10124-37-5

RTECS Number: EW2985000

DOT ID: UN 1454

Cited in U.S. State Regulations: Maine(G), New Hampshire (G), New Jersey (G), Oklahoma (G).

Description: Calcium Nitrate, $Ca(NO_3)_2$, is a colorless, moisture absorbing crystalline material. Freezing/Melting point = 561°C. Hazard Identification (based on NFPA-704 M Rating System): Health 2, Flammability 0, Reactivity 1. Oxidizer. Soluble in water.

Potential Exposure: It is used to make explosives, fertilizers, matches, fireworks and other industrial products.

Incompatibilities: A strong oxidizer. Incompatible with combustible materials, reducing agents, organics and other oxidizable materials, chemically active metals, aluminum nitrate, ammonium nitrate.

Permissible Exposure Limits in Air: No standards set.

Permissible Concentration in Water: No criteria set.

Routes of Entry: Eyes, skin, respiratory system.

Harmful Effects and Symptoms

Short Term Exposure: Calcium Nitrate can irritate the skin, eyes, nose, throat, bronchial tubes and lungs. Overexposure may cause nausea and vomiting, headaches, flushing, weakness, faintness and collapse. Severe overexposure can cause nausea, vomiting, flusing of the head and neck, headache, weakness faintness and collapse; the lips and fingernails may become bluish. There may be shortness of breath. Coma, convulsions and death are possible.

Points of Attack: Eyes, skin, respiratory system.

Medical Surveillance: For those with frequent or potentially high exposure, the following are recommended before beginning work and at regular times after that: Lung function tests. If overexposure is suspected, also consider: Complete blood count (CBC) and test for methemoglobin.

First Aid: If this chemical gets into the eyes, remove any contact lenses at once and irrigate immediately for at least 15 minutes, occasionally lifting upper and lower lids. Seek medical attention immediately. If this chemical contacts the skin, remove contaminated clothing and wash immediately with soap and water. Seek medical attention immediately. If this chemical has been inhaled, remove from exposure, begin rescue breathing (using universal precautions) if breathing has stopped and CPR if heart action has stopped. Transfer promptly to a medical facility. When this chemical has been swallowed, get medical attention. Give large quantities of water and induce vomiting. Do not make an unconscious person vomit.

Note to Physician: Treat for methemoglobinemia. Spectrophotometry may be required for precise determination of levels of methemoglobinemia in urine.

Personal Protective Methods: *Clothing:* Avoid skin contact with Calcium Nitrate. Wear protective gloves and clothing. Safety equipment suppliers/manufacturers can provide recommendations on the most protective glove/clothing material for your operation. All protective clothing (suits, gloves, footwear, headgear) should be clean, available each day, and put on before work. *Eye Protection:* Wear dust-proof goggles when working with powders or dust, unless full facepiece respiratory protection is worn.

Respirator Selection: Where the potential exists for exposure to Calcium Nitrate, use a MSHA/NIOSH approved full facepiece respirator with a high efficiency particulate filter. Greater protection is provided by a powered-air purifying respirator. Where the potential for high exposures exists, use a MSHA/NIOSH approved supplied-air respirator with a full facepiece operated in the positive pressure mode or with a full facepiece, hood, or helmet in the continuous flow mode, or use a MSHA/NIOSH approved self-contained breathing apparatus with a full facepiece operated in pressure-demand or other positive pressure mode.

Storage: Prior to working with Calcium Nitrate you should be trained on its proper handling and storage. Store in tightly closed containers in a cool, well-ventilated area away from flammables (such as fuel) or combustibles (such as wood, paper and oil). Calcium Nitrate may explode if shocked or heated. See OSHA standard 1910.104 and NFPA 43A *Code for the Storage of Liquid and Solid Oxidizers* for detailed handling and storage regulations.

Shipping: Calcium nitrate must be labeled "Oxidizer."[19] It falls in Hazard Class 5.1 and Packing Group III. Passenger aircraft and railcar shipment is limited to 25 kg; cargo aircraft shipment is limited to 100 kg.

Spill Handling: Evacuate persons not wearing protective equipment from area of spill or leak until clean-up is complete. Remove all ignition sources. Collect powdered material in the most convenient and safe manner and deposit in sealed containers. Ventilate area after clean-up is complete. It may be necessary to contain and dispose of this chemical as a hazardous waste. If material or contaminated runoff enters waterways, notify downstream users of potentially contaminated waters. Contact your Department of Environmental Protection or your regional office of the federal EPA for specific recommendations. If employees are required to clean-up spills, they must be properly trained and equipped. OSHA 1910.120(q) may be applicable.

Fire Extinguishing: This chemical is a strong oxidizer that will increase the intensity of any fire. Use dry chemical, carbon dioxide, or water spray extinguishers. Poisonous gases are produced in fire. If material or contaminated runoff enters waterways, notify downstream users of potentially contaminated waters. Notify local health and fire officials and pollution control agencies. From a secure, explosion-proof location, use water spray to cool exposed containers. If cooling streams are ineffective (venting sound increases in volume and pitch, tank discolors, or shows any signs of deforming), withdraw immediately to a secure position. If employees are expected to fight fires, they must be trained and equipped in OSHA 1910.156.

References

New Jersey Department of Health and Senior Services and Senior Services, "Hazardous Substance Fact Sheet: Calcium Nitrate," Trenton, NJ (September 1986)

Sax, N. I., Ed., "Dangerous Properties of Industrial Materials Report" 2, No. 1, 96–98 (1982) (Calcium Nitrate Tetrahydrate)

Calcium Oxide

Molecular Formula: CaO

Synonyms: Burnt lime; Calcia; Calx; Fluxing lime; Lime; Lime, burned; Lime, unslaked; Oxyde de calcium (French); Pebble lime; Quicklime; Wapniowy tlenek (Polish)

CAS Registry Number: 1305-78-8

RTECS Number: EW3100000

DOT ID: UN 1910

Regulatory Authority

- Air Pollutant Standard Set (ACGIH)[1] (DFG)[3] (HSE)[33] (OSHA)[58] (Several States)[60] (Australia) (Israel) (Mexico) (Several Canadian Provinces)

Cited in U.S. State Regulations: Alaska (G), California (G), Connecticut (A), Florida (G), Illinois (G), Maine (G), Massachusetts (G), Minnesota (G), Nevada (A), New Hampshire (G), New Jersey (G), New York (G), North Dakota (A), Oklahoma (G), Pennsylvania (G), Rhode Island (G), Virginia (A), West Virginia (G).

Description: Calcium oxide, CaO, occurs as white or grayish-white lumps or granular powder. The presence of iron gives it a yellowish or brownish tint. Hazard Identification (based on NFPA-704 M Rating System): Health 1, Flammability 0, Reactivity 0. Soluble in water (reactive).

Potential Exposure: Calcium oxide is used as a refractory material, a binding agent in bricks, plaster, mortar, stucco and other building materials, a dehydrating agent, a flux in steel manufacturing, and a laboratory agent to absorb CO_2; in the manufacture of aluminum, magnesium, glass, pulp and paper, sodium carbonate, calcium hydroxide, chlorinated lime, calcium salts, and other chemicals; in the flotation of nonferrous ores, water and sewage treatment, soil treatment in agriculture, dehairing hides, the clarification of cane and beet sugar juice, and in fungicides, insecticides, drilling fluids and lubricants.

Incompatibilities: The water solution is a medium strong base. Reacts with water generating calcium hydroxide and sufficient heat to ignite nearby combustible materials. Reacts violently with acids, halogens, metals.

Permissible Exposure Limits in Air: The OSHA TWA calcium oxide[58] is 5 mg/m^3. NIOSH, ACGIH, and HSE recommend a TWA value of 2 mg/m^3. The NIOSH IDLH level is 25 mg/m^3. There is no DFG MAK at present. Other countries regulations[35] include: Argentina – 10 mg/m^3, Czechoslovakia – 4 mg/m^3, USSR – 0.3 mg/m^3. In addition, several states have set guidelines or standards for calcium oxide in ambient air[60] ranging from 20 μg/m^3 (North Dakota) to 35 μg/m^3 (Virginia) to 40 μg/m^3 (Connecticut) to 48 μg/m^3 (Nevada).

Determination in Air: Filtration, workup with acid and analysis by atomic absorption are specified in NIOSH Method 7020 for calcium. See also OSHA Method ID121.

Permissible Concentration in Water: No criteria set.

Routes of Entry: Inhalation of dust.

Harmful Effects and Symptoms

Short Term Exposure: The corrosive action of calcium oxide is due primarily to its alkalinity and exothermic reaction with water. It is irritating and may be caustic to the skin, conjunctiva, cornea and mucous membranes of upper respiratory tract; may produce burns or dermatitis with desquamation and vesicular rash, lacrimation, spasmodic blinking, ulceration, and ocular perforation, ulceration and inflammation of the respiratory passages, ulceration of nasal and buccal mucosa, and perforation of nasal septum. Bronchitis and pneumonia have been reported from inhalation of dust. Higher exposures can cause pulmonary edema, a medical emergency that can be delayed for several hours. This can cause death. The lower respiratory tract may not be affected because irritation of upper respiratory passages is so severe that workers may be forced to leave the area.

Long Term Exposure: Repeated or prolonged contact with skin may cause brittle nails and thickening and cracking of the skin. Repeated or prolonged exposure to dust particles may cause lung problems. Calcium oxide may cause ulceration and perforation of the cartilage separating the nose (septum).

Points of Attack: Respiratory system, skin and eyes.

Medical Surveillance: Preemployment physical examinations should be directed to significant problems of the eyes, skin and the upper respiratory tract. Periodic examinations should evaluate the skin, changes in the eyes, especially the cornea and conjunctiva, mucosal ulcerations of the nose, mouth and nasal septum, and any pulmonary symptoms. Smoking history should be known.

First Aid: If this chemical gets into the eyes, remove any contact lenses at once and irrigate immediately for at least 15 minutes, occasionally lifting upper and lower lids. Seek medical attention immediately. If this chemical contacts the skin, remove contaminated clothing and wash immediately with soap and water. Seek medical attention immediately. If this chemical has been inhaled, remove from exposure, begin rescue breathing (using universal precautions) if breathing has stopped and CPR if heart action has stopped. Transfer promptly to a medical facility. When this chemical has been swallowed, get medical attention. If victim is conscious, administer water or milk. Do not induce vomiting. Medical observation is recommended for 24 – 48 hours after breathing overexposure, as pulmonary edema may be delayed. As first aid for pulmonary edema, a doctor or authorized paramedic may consider administering a corticosteroid spray.

Personal Protective Methods: Wear appropriate clothing to prevent any reasonable probability of skin contact. Wear eye protection to prevent any possibility of eye contact. Employees should wash promptly when skin is contaminated and daily at the end off each work shift. Work clothing should be changed daily if it is possible that clothing is contaminated. Remove nonimpervious clothing if contaminated. Provide emergency showers and eyewash.

Respirator Selection: NIOSH: *10 mg/m^3*: DM (any dust and mist respirator); *20 mg/m^3:* DMXSQ (any dust and mist respirator except single-use and quarter-mask respirators; or SA (any supplied-air respirator). *25 mg/m^3:* SA:CF (any supplied-air respirator operated in a continuous-flow mode); or PAPRHiE (any powered, air-purifying respirator with a high-efficiency particulate filter); or HiEF (any air-purifying, full-facepiece respirator with a high-efficiency particulate filter); or SCBAF (any self-contained breathing apparatus with a full facepiece); or SAF (any supplied-air respirator with a full facepiece). *at any concentrations above the NIOSH REL, or where there is no REL, at any detectable concentration:* SCBAF:PD,PP (any MSHA/NIOSH approved self-contained breathing apparatus that has a full facepiece and is operated in a pressure-demand or other positive-pressure mode); or SAF:PD,PP:ASCBA (any supplied-air respirator that has a full facepiece and is operated in a pressure-demand or other positive-pressure mode in combination with an auxiliary self-contained breathing apparatus operated in a pressure-demand or other positive pressure mode). *Escape:* HiEF [any air-purifying, full-facepiece respirator (gas mask) with a chin-style, front- or back-mounted organic vapor canister having a high-efficiency particulate filter]; or SCBAE (any appropriate escape-type, self-contained breathing apparatus).

Storage: Prior to working with Calcium oxide you should be trained on its proper handling and storage. Should be stored on dry flooring in a fire resistant room well protected from the weather. The area should be cool and adequately ventilated. Store in containers protected from physical damage, acids and oxidizing materials such as permanganate, dichromate or chlorine.

Shipping: Calcium oxide must bear a "Corrosive" label. It falls in DOT Hazard Class 8 and Packing Group III.

Spill Handling: Evacuate persons not wearing protective equipment from area of spill or leak until clean-up is complete. Remove all ignition sources. Collect powdered material in the most convenient and safe manner and deposit in sealed containers. Ventilate area after clean-up is complete. It may be necessary to contain and dispose of this chemical as a hazardous waste. If material or contaminated runoff enters waterways, notify downstream users of potentially contaminated waters. Contact your Department of Environmental Protection or your regional office of the federal EPA for specific recommendations. If employees are required to clean-up spills, they must be properly trained and equipped. OSHA 1910.120(q) may be applicable.

Fire Extinguishing: This chemical is a noncombustible solid. Contact with water or moisture may generate enough heat to ignite nearby combustible materials. Avoid the use of water. Do not use carbon dioxide, foam, or halogenated fire extinguishers. Poisonous gases are produced in fire. If material or contaminated runoff enters waterways, notify downstream users of potentially contaminated waters. Notify local health and fire officials and pollution control agencies. From a secure, explosion-proof location, use water spray to cool exposed containers. If cooling streams are ineffective (venting sound increases in volume and pitch, tank discolors, or shows any signs of deforming), withdraw immediately to a secure position. If employees are expected to fight fires, they must be trained and equipped in OSHA 1910.156.

Disposal Method Suggested: Pretreatment involves neutralization with hydrochloric acid to yield calcium chloride. The calcium chloride formed is treated with soda ash to yield the insoluble calcium carbonate. The remaining brine solution may be discharged into sewers and waterways.[22]

References

Sax, N. I.., Ed., "Dangerous Properties of Industrial Materials Report" 2, No. 1, 98–99 (1982)

New Jersey Department of Health and Senior Services and Senior Services, "Hazardous Substance Fact Sheet: Calcium Oxide," Trenton, NJ (October 1985)

New York State Department of Health, "Chemical Fact Sheet: Calcium Oxide," Albany, NY, Bureau of Toxic Substance Assessment (January 1996)

Calcium Peroxide

Molecular Formula: CaO_2

Synonyms: Calcium dioxide; Calcium superoxide

CAS Registry Number: 1305-79-9

RTECS Number: EW3865000

DOT ID: UN 1457

Cited in U.S. State Regulations: Maine (G), New Hampshire (G), New Jersey (G), Oklahoma (G).

Description: Calcium peroxide, CaO_2, is a grayish-white or yellowish odorless crystalline solid. It decomposes at 275°C. Hazard Identification (based on NFPA-704 M Rating System): Health 1, Flammability 0, Reactivity 1. Insoluble in water.

Potential Exposure: Calcium Peroxide is used as a seed disinfectant, an antiseptic and a rubber stabilizer.

Incompatibilities: A strong alkali, and strong oxidizer Incompatible with reducing agents, combustible materials, polysulfide polymers.

Permissible Exposure Limits in Air: No standards set.

Permissible Concentration in Water: No criteria set.

Routes of Entry: Inhalation of dust, ingestion.

Harmful Effects and Symptoms

Short Term Exposure: Calcium Peroxide can affect you when breathed in. Contact can severely irritate and may burn the skin and eyes. Exposure can irritate the eyes, nose and throat. Higher levels can irritate the lungs causing coughing and/or shortness of breath. Still higher exposures may cause a build-up of fluid in the lungs (pulmonary edema). This can cause death.

Long Term Exposure: Prolonged exposure can damage the skin. Very irritating substances may cause problems.

Points of Attack: Eyes, skin, respiratory system.

Medical Surveillance: Before beginning employment and at regular times after that, for those with frequent or potentially high exposures, the following are recommended: Lung function tests. If symptoms develop or overexposure is suspected, the following may be useful: Consider chest x-ray after acute overexposure.

First Aid: If this chemical gets into the eyes, remove any contact lenses at once and irrigate immediately for at least 30 minutes, occasionally lifting upper and lower lids. Seek medical attention immediately. If this chemical contacts the skin, remove contaminated clothing and wash immediately with soap and water. Seek medical attention immediately. If this chemical has been inhaled, remove from exposure, begin rescue breathing (using universal precautions) if breathing has stopped and CPR if heart action has stopped. Transfer promptly to a medical facility. When this chemical has been swallowed, get medical attention. Give large quantities of water and induce vomiting. Do not make an unconscious person vomit. Medical observation is recommended for 24 – 48 hours after breathing overexposure, as pulmonary edema may be delayed. As first aid for pulmonary edema, a doctor or authorized paramedic may consider administering a corticosteroid spray.

Personal Protective Methods: *Clothing:* Avoid skin contact with Calcium Peroxide. Wear protective gloves and clothing. Safety equipment suppliers/manufacturers can provide recommendations on the most protective glove/clothing material for your operation. All protective clothing (suits, gloves, footwear, headgear) should be clean, available each day and put on before work. *Eye Protection:* Wear dust-proof goggles with face shield when working with powders or dust, unless full facepiece respiratory protection is worn.

Respirator Selection: Where the potential exists for exposures to Calcium Peroxide, use a MSHA/NIOSH approved full facepiece respirator equipped with particulate (dust/fume/mist) filters. Greater protection is provided by a powered-air purifying respirator. Particulate filters must be checked every day before work for physical damage, such as rips or tears and replaced as needed. Where the potential for high exposures exists, use a MSHA/NIOSH approved supplied-air respirator with a full facepiece operated in the positive pressure mode or with a full facepiece, hood, or helmet in the continuous flow mode, or use a MSHA/NIOSH approved self-contained breathing apparatus with a full facepiece operated in pressure-demand or other positive pressure mode.

Storage: Prior to working with Calcium Peroxide you should be trained on its proper handling and storage. Calcium Peroxide must be stored to avoid contact with Combustible Materials (such as wood, paper, oil, fuels, etc.) since violent reactions occur. Store in tightly closed containers. See OSHA standard 1910.104 and NFPA 43A *Code for the Storage of Liquid and Solid Oxidizers* for detailed handling and storage regulations.

Shipping: Calcium peroxide must be labeled "Oxidizer."[19] It falls in Hazard Class 5.1 and Packing Group II.

Spill Handling: Evacuate persons not wearing protective equipment from area of spill or leak until clean-up is complete. Remove all ignition sources. Collect powdered material in the most convenient and safe manner and deposit in sealed containers. Ventilate area after clean-up is complete. It may be necessary to contain and dispose of this chemical as a hazardous waste. If material or contaminated runoff enters waterways, notify downstream users of potentially contaminated waters. Contact your Department of Environmental Protection or your regional office of the federal EPA for specific recommendations. If employees are required to clean-up spills, they must be properly trained and equipped. OSHA 1910.120(q) may be applicable.

Fire Extinguishing: Calcium Peroxide is an oxidizer and will greatly increase the intensity of a fire. Extinguish fire using an agent suitable for type of surrounding fire. Poisonous gases are produced in fire. If material or contaminated runoff enters waterways, notify downstream users of potentially contaminated waters. Notify local health and fire officials and pollution control agencies. From a secure, explosion-proof location, use water spray to cool exposed containers. If cooling streams are ineffective (venting sound increases in volume and pitch, tank discolors, or shows any signs of deforming), withdraw immediately to a secure position. If employees are expected to fight fires, they must be trained and equipped in OSHA 1910.156.

References

New Jersey Department of Health and Senior Services and Senior Services, "Hazardous Substance Fact Sheet: Calcium Peroxide," Trenton, NJ (August 6, 1987)

Calcium Phosphide

Molecular Formula: Ca_3P_2

Synonyms: Calcium phosphide; Photophor; Tricalcium diphosphide

CAS Registry Number: 1305-99-3

RTECS Number: EW3860000

DOT ID: UN 1360

Cited in U.S. State Regulations: Maine (G), New Hampshire (G), New Jersey (G), Oklahoma (G).

Description: Calcium phosphide, Ca_3P_2, is a gray granular solid or reddish-brown crystalline solid with a musty odor, somewhat like acetylene. Freezing/Melting point = about 1,600°C. Hazard Identification (based on NFPA-704 M Rating System): Health 4, Flammability 0, Reactivity 2. Water reactive.

Potential Exposure: Calcium Phosphide is used to kill rodents and in explosives and fireworks.

Incompatibilities: A strong reducing agent. Forms highly flammable phosphine gas in moist air. Contact with water or acids release phosphine gas, and can cause explosions. Incompatible with oxidizers, acids, chlorine, chlorine monoxide, halogens, halogen acids, oxygen, sulfur.

Permissible Exposure Limits in Air: No standards set.

Permissible Concentration in Water: No criteria set. (Reacts violently with water as noted above).

Routes of Entry: Inhalation of dust.

Harmful Effects and Symptoms

Short Term Exposure: Calcium phosphide can affect you when breathed in. Phosphine gas is a highly toxic gas released when Calcium phosphide is wet or has contacted moisture. Consult the entry on phosphine. As noted by Sax (reference below), phosphine is an acute local irritant, is toxic upon inhalation causing restlessness, tremors, fatigue, gastric pain, diarrhea, coma and convulsions. Also, Calcium phosphide is a dangerous fire and explosion hazard.

Long Term Exposure: Unknown at this time.

First Aid: If this chemical gets into the eyes, remove any contact lenses at once and irrigate immediately for at least 15 minutes, occasionally lifting upper and lower lids. Seek medical attention immediately. If this chemical contacts the skin, remove contaminated clothing and wash immediately with soap and water. Seek medical attention immediately. If this chemical has been inhaled, remove from exposure, begin rescue breathing (using universal precautions) if breathing has stopped and CPR if heart action has stopped. Transfer promptly to a medical facility. When this chemical has been swallowed, get medical attention. Give large quantities of water and induce vomiting. Do not make an unconscious person vomit.

Personal Protective Methods: *Clothing:* Avoid skin contact with Calcium Phosphide. Wear protective gloves and clothing. Safety equipment suppliers/manufacturers can provide recommendations on the most protective glove/clothing material for your operation. All protective clothing (suits, gloves, footwear, headgear) should be clean, available each day and put on before work. *Eye Protection:* Wear dust-proof goggles when working with powders or dust, unless full facepiece respiratory protection is worn.

Respirator Selection: Where the potential exists for exposure to Calcium Phosphide, use a MSHA/NIOSH approved supplied-air respirator with a full facepiece operated in the positive pressure mode or with a full facepiece, hood, or helmet in the continuous flow mode, or use a MSHA/NIOSH approved self-contained breathing apparatus with a full facepiece operated in pressure-demand or other positive pressure mode.

Storage: Store in tightly closed containers in a cool, well-ventilated area away from strong oxidizers (such as chlorine, bromine and fluorine), strong acids (such as hydrochloric, sulfuric, and nitric), oxygen, sulfur or moisture, since violent reactions occur. Sources of ignition, such as smoking, and open flames, are prohibited where Calcium Phosphide is handled, used, or stored. Use only nonsparking tools and equipment, especially when opening and closing containers of Calcium Phosphide. Wherever Calcium Phosphide is used, handled, manufactured, or stored, used explosion-proof electrical equipment and fittings. Do not store large amounts of this material in a room protected by water sprinkler systems. Protect containers against physical damage.

Shipping: Calcium Phosphide must carry a "Dangerous when Wet" label.[19] It falls in Hazard Class 4.3 and Packing Group I. The shipment of Calcium Phosphide by passenger aircraft or railcar is forbidden.

Spill Handling: Evacuate persons not wearing protective equipment from area of spill or leak until clean-up is complete. Remove all ignition sources. Collect powdered material in the most convenient and safe manner and deposit in sealed containers. Ventilate area after clean-up is complete. It may be necessary to contain and dispose of this chemical as a hazardous waste. If material or contaminated runoff enters waterways, notify downstream users of potentially contaminated waters. Contact your Department of Environmental Protection or your regional office of the federal EPA for specific recommendations. If employees are required to clean-up spills, they must be properly trained and equipped. OSHA 1910.120(q) may be applicable.

Initial isolation and protective action distances

Distances shown are likely to be affected during the first 30 minutes after materials are spilled and could increase with time. If more than one tank car, cargo tank, portable tank, or large cylinder is involved in the incident is leaking, the protective action distance may need to be increased. You may need to seek emergency information from CHEMTREC at (800) 424-9300 or seek professional environmental engineering assistance from the U.S. EPA Environmental Response Team at (908)548-8730 (24-hour response line).

UN 1360 (calcium phosphide) is on the DOT's list of dangerous water-reactive materials which create large amounts of toxic vapor when *spilled in water:* Dangerous from 0.5 – 10 km (0.3 – 6.0 miles) downwind.

Fire Extinguishing: Do not use water or foam extinguishers. Contact with water forms highly toxic and flammable phosphine gas. Use dry chemical, dry sand, soda ash, or lime extinguishers. Poisonous gases produced in fire. If material or contaminated runoff enters waterways, notify downstream users of potentially contaminated waters. Notify local health and fire officials and pollution control agencies. From a secure, explosion-proof location, use water spray to cool exposed containers. If cooling streams are ineffective (venting sound increases in volume and pitch, tank discolors, or shows any signs of deforming), withdraw immediately to a secure position. If employees are expected to fight fires, they must be trained and equipped in OSHA 1910.156.

References

New Jersey Department of Health and Senior Services and Senior Services, "Hazardous Substance Fact Sheet: Calcium Phosphide," Trenton, NJ (February 1987)

Sax, N. I., Ed., "Dangerous Properties of Industrial Materials Report" 2, No. 1, 102–103 (1982)

Calcium Sulfate

Molecular Formula: CaO_4S

Common Formula: $CaSO_4$

Synonyms: *anhydrous:* Anhydrite; Anhydrous calcium sulfate; Anhydrous gypsum; Anhydrous sulfate of lime; Calcium salt of sulfuric acid

dihydrate: Gypsum

hemihydrate: Plaster of Paris

CAS Registry Number: 7778-18-9 (anhydrous); 10101-41-4 (dihydrate)

RTECS Number: WS6920000 (anhydrous); EW4150000 (dihydrate)

Regulatory Authority

- Air Pollutant Standard Set (ACGIH)[1] (OSHA)[58] (Australia) (Israel) (Mexico) (DFG) (Quebec, Canada)

Cited in U.S. State Regulations: New York (G), Minnesota (G), Pennsylvania (G)

Description: Calcium sulfate, $CaSO_4$, forms white to clear crystals which melt at 1,450°C. It is commonly encountered in the anhydrous form or as the dihydrate. Freezing/Melting point = 2,840°F (decomposes).

Potential Exposure: Calcium sulfate is used as a pigment, in Portland cement, in tiles and plaster, in polishing powders, a filler in paints and paper coatings, in the drying of gases and liquids, a soil conditioner, in molds and surgical casts, in wall-board and many others.

Incompatibilities: Contact with diazomethane, aluminum, phosphorus, water may cause explosions. *Note:* Hygroscopic (i.e., absorbs moisture from the air). Reacts with water to form Gypsum and Plaster of Paris.

Permissible Exposure Limits in Air: OSHA[58] has set a TWA of 15 mg/m^3 on a total dust basis and 5 mg/m^3 on a respirable fraction basis. NIOSH recommends a TWA of 10 mg/m^3 on a total dust basis and 5 mg/m^3 on a respirable fraction basis. The ACGIH has set a TWA of 10 mg/m^3 (total dust containing no asbestos and < 1% fine silica). Australia, Israel, and the Province of Quebec have set the same TWA as ACGIH. The German DFG MAK is 6 mg/m^3.

Determination in Air: By filter collection and gravimetric means. Particulates NOR: NIOSH #0500 (total), #0600 (respirable)]

Permissible Concentration in Water: No criteria set.

Routes of Entry: Inhalation of dust, ingestion.

Harmful Effects and Symptoms

Short Term Exposure: *Inhalation:* May cause irritation of mouth, throat, nose and lungs. Senses of smell and taste may be lessened. Nose irritation may lead to bleeding. *Skin:* May cause irritation in open sores. The harsh washing and abrasive action necessary to remove this material may also lead to irritation. *Eyes:* Dust may irritate eyes. *Ingestion:* May cause blockage of digestive system if material hardens.

Long Term Exposure: May cause nose irritation accompanied by sneezing, tear formation, and excessive fluid secretion. Animal studies suggest that pneumonia and other more serious lung disorders may occur.

Points of Attack: Eyes, skin, respiratory system.

First Aid: If this chemical gets into the eyes, remove any contact lenses at once and irrigate immediately for at least 15 minutes, occasionally lifting upper and lower lids. Seek medical attention immediately. If this chemical contacts the skin, remove contaminated clothing and wash immediately with soap and water. Seek medical attention immediately. If this chemical has been inhaled, remove from exposure, begin rescue breathing (using universal precautions) if breathing has stopped and CPR if heart action has stopped. Transfer promptly to a medical facility. When this chemical has been swallowed, get medical attention. Give large quantities of water and induce vomiting. Do not make an unconscious person vomit.

Personal Protective Methods: Wear loose fitting clothing of dust-tight material and safety goggles.

Respirator Selection: A dust mask should be worn if irritation effects become apparent.

Storage: The hemihydrate and anhydrous forms should be stored in tightly sealed containers.

Shipping: DOT[19] does not cite Calcium sulfate specifically in its Performance-Oriented Packaging Standards.

Spill Handling: Evacuate persons not wearing protective equipment from area of spill or leak until clean-up is complete. Remove all ignition sources. Clean up using methods that do not raise dust, such as vacuum and deposit in sealed containers. Ventilate area after clean-up is complete. It may be necessary to contain and dispose of this chemical as a hazardous waste. If material or contaminated runoff enters waterways, notify downstream users of potentially contaminated waters. Contact your Department of Environmental Protection or your regional office of the federal EPA for specific recommendations. If employees are required to clean-up spills, they must be properly trained and equipped. OSHA 1910.120(q) may be applicable.

Fire Extinguishing: This chemical is a non-combustible solid. Use any extinguisher suitable for surrounding fire. Poisonous gases are produced in fire. If material or contaminated runoff enters waterways, notify downstream users of potentially contaminated waters. Notify local health and fire officials and pollution control agencies. From a secure, explosion-proof location, use water spray to cool exposed containers. If cooling streams are ineffective (venting sound increases in volume and pitch, tank discolors, or shows any signs of deforming), withdraw immediately to a secure position. If employees are expected to fight fires, they must be trained and equipped in OSHA 1910.156.

This chemical is a combustible liquid. Poisonous gases including sulfuroxides are produced in fire. Use dry chemical, carbon dioxide, or alcohol foam extinguishers. Vapors are heavier than air and will collect in low areas. Vapors may travel long distances to ignition sources and flashback. Vapors in confined areas may explode when exposed to fire. Containers may explode in fire. Storage containers and parts of containers may rocket great distances, in many directions. If material or contaminated runoff enters waterways, notify downstream users of potentially contaminated waters. Notify local health and fire officials and pollution control agencies. From a secure, explosion-proof location, use water spray to cool exposed containers. If cooling streams are ineffective (venting sound increases in volume and pitch, tank discolors, or shows any signs of deforming), withdraw immediately to a secure position. If employees are expected to fight fires, they must be trained and equipped in OSHA 1910.156.

Non-flammable so imposes no particular extinguisher requirements.

Disposal Method Suggested: Landfilling.

References

New York State Department of Health, "Chemical Fact Sheet: Calcium Sulfate," Albany, NY, Bureau of Toxic Substance Assessment (March 1986)

Camphene

Molecular Formula: $C_{10}H_{10}$

Synonyms: Bicyclo-(2.2.1)heptane; 3,3-Dimethylenenorcamphene; 2,2-Dimethyl-3-methylene-; 2-2-Dimethyl-3-methylene norborane; 3,3-Dimethyl-2-methylene norcamphone

CAS Registry Number: 79-92-5

RTECS Number: EX1055000

DOT ID: NA 9011

Cited in U.S. State Regulations: Maine (G), New Hampshire (G), New Jersey (G).

Description: Camphene, $C_{10}H_{10}$, is colorless to white crystalline substance with camphor like odor. Freezing/Melting point = 50°C. Boiling point = 154°C. Flash point = 42°C (oc) and 33°C (cc). Hazard Identification (based on NFPA-704 M Rating System): Health 2, Flammability 2, Reactivity 0. Insoluble in water.

Potential Exposure: Camphene is used for mothproofing and in the cosmetics, perfume and food flavoring industries.

Incompatibilities: Contact with strong oxidizers may cause fire and explosions. Emulsions in xylene may violently decompose on contact with iron or aluminum above 70°C.

Permissible Exposure Limits in Air: No standards set.

Permissible Concentration in Water: No criteria set.

Routes of Entry: Inhalation of vapors, ingestion.

Harmful Effects and Symptoms

Short Term Exposure: Camphene can affect you when breathed in and by passing through your skin. Contact can irritate the eyes and skin. Exposure can irritate the eyes, nose and throat. Higher levels can cause you to feel dizzy, excited, sweaty and have a headache. At very high levels confusion, nausea, drowsiness, coma and kidney damage can occur.

Long Term Exposure: Similar chemicals also can cause skin allergy. It is not know if camphene can cause the same problem.

Points of Attack: Eyes, skin, respiratory system, kidneys.

Medical Surveillance: If symptoms develop or overexposure is suspected, the following may be useful: Kidney function tests. Evaluation by a qualified allergist, including careful exposure history and special testing, may help diagnose skin allergy.

First Aid: If this chemical gets into the eyes, remove any contact lenses at once and irrigate immediately for at least 15 minutes, occasionally lifting upper and lower lids. Seek medical attention immediately. If this chemical contacts the skin, remove contaminated clothing and wash immediately with soap and water. Seek medical attention immediately. If this chemical has been inhaled, remove from exposure, begin rescue breathing (using universal precautions) if breathing has stopped and CPR if heart action has stopped. Transfer promptly to a medical facility. When this chemical has been swallowed, get medical attention. Give large quantities of water and induce vomiting. Do not make an unconscious person vomit.

Personal Protective Methods: *Clothing:* Avoid skin contact with Camphene. Wear protective gloves and clothing. Safety equipment suppliers/manufacturers can provide recommendations on the most protective glove/clothing material for your operation. All protective clothing (suits, gloves, footwear, headgear) should be clean, available each day and put on before work. *Eye Protection:* Eye protection is included in the recommended respiratory protection.

Respirator Selection: Where the potential exists for exposure to Camphene, use a MSHA/NIOSH approved full facepiece respirator with a high efficiency particulate filter. Greater protection is provided by a powered-air purifying respirator. Where the potential for high exposures exists, use a MSHA/NIOSH approved supplied-air respirator with a full facepiece operated in the positive pressure mode or with a full facepiece, hood, or helmet in the continuous flow mode, or use a MSHA/NIOSH approved self-contained breathing apparatus with a full facepiece operated in pressure-demand or other positive pressure mode.

Storage: Prior to working with Camphene you should be trained on its proper handling and storage. Camphene must be stored to avoid contact with strong oxidizers (such as chlorine, bromine, and fluorine) since violent reactions occur. Store in tightly closed containers in a cool, well-ventilated area. Sources of ignition, such as smoking and open flames, are prohibited where Camphene is used, handled, or stored in a manner that could create a potential fire or explosion hazard. Use only nonsparking tools and equipment, especially when opening and closing containers of Camphene.

Shipping: Camphene is not specifically cited in DOT's Performance-oriented packaging standards.[19]

Spill Handling: Evacuate persons not wearing protective equipment from area of spill or leak until clean-up is complete. Remove all ignition sources. Collect powdered material in the most convenient and safe manner, using non-sparking tools, and deposit in sealed containers. Ventilate area after clean-up is complete. It may be necessary to contain and dispose of this chemical as a hazardous waste. If material or contaminated runoff enters waterways, notify downstream users of potentially contaminated waters. Contact your Department of Environmental Protection or your regional office of the federal EPA for specific recommendations. If employees are required to clean-up spills, they must be properly trained and equipped. OSHA 1910.120(q) may be applicable.

Fire Extinguishing: This chemical is a flammable solid. The crystals do not easily ignite, but they release flammable vapor at room temperature. Heating greatly increases the release of these flammable vapors. Use dry chemical, carbon dioxide, water spray, or foam extinguishers. Poisonous gases are produced in fire. If material or contaminated runoff enters waterways, notify downstream users of potentially contaminated waters. Notify local health and fire officials and pollution control agencies. From a secure, explosion-proof location, use water spray to cool exposed containers. If cooling streams are ineffective (venting sound increases in volume and pitch, tank discolors, or shows any signs of deforming), withdraw immediately to a secure position. If employees are expected to fight fires, they must be trained and equipped in OSHA 1910.156.

Disposal Method Suggested: Incineration.

References

New Jersey Department of Health and Senior Services and Senior Services, "Hazardous Substance Fact Sheet: Camphene," Trenton, NJ (July 30, 1987)

Camphor

Molecular Formula: $C_{10}H_{16}O$

Synonyms: Bicyclo-(2.2.1.)-heptanone; Bicyclo 2.2.1 heptan-2-one,1,7,7-trimethyl-; Bornane, 2-oxo-; 2-Bornanone; 2-Camphanone; Camphor, natural; 2-Camphorone; Formosa camphor; Gum camphor; Huile de camphre (French); Japan camphor; Kampfer (German); 2-Keto-1,7,7-trimethylnorcamphane; Laurel camphor; Matricaria camphor; Norcamphor, synthetic camphor; 1,7,7-Trimethyl-; 1,7,7-Trimethylbicyclo(2.2.1)-2-heptanone; 1,7,7-Trimethylnorcamphor

CAS Registry Number: 76-22-2; 8008-51-3 (camphor oil)

RTECS Number: EX1225000

DOT ID: UN 2717; UN 1130 (camphor oil)

Regulatory Authority

- Banned or Severely Restricted (In juvenile drugs) (UN)[13]
- Air Pollutant Standard Set (ACGIH)[1] (DFG)[3] (HSE)[33] (UNEP)[43] (OSHA)[58] (Several States)[60] (Australia) (Israel) (Mexico) (Several Canadian Provinces)

Cited in U.S. State Regulations: Alaska (G), California (G), Connecticut (A), Florida (G), Illinois (G), Maine (G), Massachusetts (G), Minnesota (G), Nevada (A), New Hampshire (G), New Jersey (G), North Dakota (A), Pennsylvania (G), Rhode Island (G), Virginia (A), West Virginia (G).

Description: Camphor, $C_{10}H_{16}O$, is a colorless glassy solid with a penetrating, characteristic odor. Odor threshold = 0.079 mg/m^3. Freezing/Melting point = 180°C; 165°C (synthetic). Flash point = 66°C (solid); 47°C (oil). Autoignition temperature = 460°C. Explosive limits: LEL = 0.6%, UEL = 3.5%. *Solid:* Hazard Identification (based on NFPA-704 M Rating System): Health 0, Flammability 2, Reactivity 0. Insoluble in water. *Oil:* Hazard Identification (based on NFPA-704 M Rating System): Health 2, Flammability 2, Reactivity 0.

Potential Exposure: Camphor is used as a plasticizer for cellulose esters and ethers; it is used in lacquers and varnishes and in explosives and pyrotechnics formulations. It is used as a moth repellent and as a medicinal.

Incompatibilities: Forms explosive mixture with air. Violent, possibly explosive, reaction with strong oxidizers, especially chromic anhydride, potassium permanganate. May accumulate static electrical charges, and may cause ignition of its vapors.

Permissible Exposure Limits in Air: The OSHA PEL is 2 mg/m^3 TWA. NIOSH recommends the same level. ACGIH recommends a TWA of 2 ppm and STEL of 3 ppm. The NIOSH IDLH level is 200 mg/m^3. The DFG,[3] HSE,[33] Australian, Israeli, and Mexican values are also 2 ppm for a TWA and HSE, Australia, and Israel have set a STEL of 3 ppm. The former USSR/UNEP joint project[43] has set a MAC in workplace air of 0.5 ppm (3 mg/m^3). In Canada, Alberta, British Columbia, Ontario, and Quebec provinces set level of 2 ppm and STEL of 3 ppm. In addition, several states have set guidelines or standards for camphor in ambient air[60] ranging from 80 μg/m^3 (Connecticut) to 120 – 180 μg/m^3 (Virginia) to 286 μg/m^3 (Nevada).

Determination in Air: Charcoal absorption is followed by CS_2, workup and analysis by gas chromatography. See NIOSH Method #1301, Ketones II.

Permissible Concentration in Water: No criteria set.

Routes of Entry: Inhalation, ingestion, skin and eye contact.

Harmful Effects and Symptoms

Short Term Exposure: Contact can irritate the eyes and skin. Inhalation can cause respiratory tract irritation and coughing. Exposure can cause nausea, vomiting, diarrhea, headaches, dizziness, excitement, irrational behavior, mental confusion, epileptiform convulsions. Higher exposures can cause unconsciousness and death.

Long Term Exposure: Camphor may cause kidney damage.

Points of Attack: Central nervous system, eyes, skin, respiratory system.

Medical Surveillance: Consider the points of attack in preplacement and periodic physical examinations. Kidney function tests.

First Aid: If this chemical gets into the eyes, remove any contact lenses at once and irrigate immediately for at least 15 minutes, occasionally lifting upper and lower lids. Seek medical attention immediately. If this chemical contacts the skin, remove contaminated clothing and wash immediately with soap and water. Seek medical attention immediately. If this chemical has been inhaled, remove from exposure, begin rescue breathing (using universal precautions) if breathing has stopped and CPR if heart action has stopped. Transfer promptly to a medical facility. When this chemical has been swallowed, get medical attention. Give large quantities of water and induce vomiting. Do not make an unconscious person vomit.

Personal Protective Methods: Wear appropriate clothing to prevent repeated or prolonged skin contact. Wear eye protection to prevent any reasonable probability of eye contact. Employees should wash promptly when skin is wet or contaminated. Work clothing should be changed daily if it is possible that clothing is contaminated. Remove nonimpervious clothing promptly if wet or contaminated.

Respirator Selection: OSHA: *50 mg/m³:* SA:CF (any supplied-air respirator operated in a continuous-flow mode); PAPROVDM (any powered, air-purifying respirator with organic vapor cartridge(s) in combination with a dust and mist filter). *100 mg/m³:* CCRFOVHiE (any chemical cartridge respirator with a full facepiece and organic vapor cartridge(s) in combination with a high efficiency particulate filter); or GMFOVHiE [any air-purifying, full-facepiece respirator (gas mask) with a chin-style, front- or back-mounted organic vapor canister having a high-efficiency particulate filter]; or PAPRTOVHiE (any powered, air-purifying respirator with a tight-fitting facepiece and organic vapor cartridge(s) in combination with a high-efficiency particulate filter); or SCBAF (any self-contained breathing apparatus with a full facepiece); or SAF (any supplied-air respirator with a full facepiece). *200 mg/m³:* SAF:PD,PP (any supplied-air respirator that has a full facepiece and is operated in a pressure-demand or other positive-pressure mode). *Emergency or planned entry into unknown concentrations or IDLH conditions:* SCBAF:PD,PP (any self-contained breathing apparatus that has a full facepiece and is operated in a pressure-demand or other positive-pressure mode); or SAF:PD, PP:ASCBA (any supplied-air respirator that has a full facepiece and is operated in a pressure-demand or other positive-pressure mode in combination with an auxiliary self-contained breathing apparatus operated in a pressure-demand or other positive- pressure mode). *Escape:* GMFOVHiE [any air-purifying, full-facepiece respirator (gas mask) with a chin-style, front- or back-mounted organic vapor canister having a high-efficiency particulate filter]; or SCBAE (any appropriate escape-type, self-contained breathing apparatus).

Note: Substance causes eye irritation or damage; eye protection needed.

Storage: Prior to working with camphor you should be trained on its proper handling and storage. Before entering confined space where this chemical may be present, check to make sure that an explosive concentration does not exist. Camphor must be stored to avoid contact with oxidizers such as permanganates, nitrates, peroxides, chlorates, and perchlorates, and especially chromic anhydride, since violent reactions occur. Store in tightly closed containers in a cool, well-ventilated area away from heat, sparks, or flame. Sources of ignition such as smoking and open flames are prohibited where Camphor is used, handled, or stored in a manner that could create a potential fire or explosion hazard.

Shipping: Synthetic camphor should carry a "Flammable Solid" label. It falls in Hazard Class 4.1 and Packing Group III.

Spill Handling: Evacuate persons not wearing protective equipment from area of spill or leak until clean-up is complete. Remove all ignition sources. Collect powdered material in the most convenient and safe manner and deposit in sealed containers. Establish forced ventilation to keep levels below explosive limit. It may be necessary to contain and dispose of this chemical as a hazardous waste. If material or contaminated runoff enters waterways, notify downstream users of potentially contaminated waters. Contact your Department of Environmental Protection or your regional office of the federal EPA for specific recommendations. If employees are required to clean-up spills, they must be properly trained and equipped. OSHA 1910.120(q) may be applicable.

Fire Extinguishing: This chemical is a combustible solid. Combustion produces lots of soot. Use dry chemical, carbon dioxide, water spray, or alcohol foam extinguishers. Poisonous gases are produced in fire. If material or contaminated runoff enters waterways, notify downstream users of potentially contaminated waters. Notify local health and fire officials and pollution control agencies. From a secure, explosion-proof location, use water spray to cool exposed containers. If cooling streams are ineffective (venting sound increases in volume and pitch, tank discolors, or shows any signs of deforming), withdraw immediately to a secure position. If employees are expected to fight fires, they must be trained and equipped in OSHA 1910.156.

Disposal Method Suggested: Incineration of a solution in a flammable solvent.

References

Sax, N. I., Ed., "Dangerous Properties of Industrial Materials Report" 1, No. 8, 52–54 (1981)

New Jersey Department of Health and Senior Services and Senior Services, "Hazardous Substance Fact Sheet: Camphor," Trenton, NJ (April 1998)

Cantharidin

Molecular Formula: $C_{10}H_{12}O_4$

Synonyms: Can; Cantharides camphor; 1,2-Dimethyl-3,6-epoxyperhydrophthalic anhydride; 2,3-Dimethyl-7-oxabicyclo [2.2.1] heptane-2,3-dicarboxylic anhydride; 4,7-Epoxyisobenzofuran-1,3-dione, hexahydro-3a, 7a-dimethyl-, (3a α, 4 β, 7 β, 7a α)-; 7-Oxabicyclo[2.2.1]heptane-2,3-dicarboxylic anhydride, 2,3-dimethyl-

CAS Registry Number: 56-25-7

RTECS Number: RN8575000

Regulatory Authority

- Carcinogen (Animal Suspected) (IARC)[9]
- Extremely Hazardous Substance (EPA-SARA) (TPQ = 100)[7]

Cited in U.S. State Regulations: California (G), Florida (G), Massachusetts (G), New Jersey (G), Pennsylvania (G).

Description: Cantharidin, $C_{10}H_{12}O_4$, is a brown to black powder. Freezing/Melting point = 218°C (begins to sublime at 110°C). Hazard Identification (based on NFPA-704 M Rating System): Health 4, Flammability 1, Reactivity 0.

Potential Exposure: Formerly used as a counter-irritant and vesicant. Also used for the removal of benign epithelial growth, e.g., warts. Used as an experimental antitumor agent. Active ingredient in "spanish fly," a reputed aphrodisiac.

Permissible Exposure Limits in Air: No standards set.

Permissible Concentration in Water: No criteria set.

Routes of Entry: Inhalation, ingestion, skin contact.

Harmful Effects and Symptoms

Short Term Exposure: Symptoms from ingestion include vomiting, abdominal pain, shock, bloody diarrhea, pain in throat and stomach, swelling and blistering of tongue, difficulty swallowing, salivation, slow and painful urination, and thirst. There may be delirium, fainting, and titanic convulsions. Eye contact results in irritation with much swelling of the lids. Initial tissue reaction upon contact with the skin is swelling followed by blister formation within 24 hours. It is classified as super toxic. Probable oral lethal dose in humans is less than 5 mg/kg or a taste of less than 7 drops for a 70 kg (150 lb) person. It is very toxic by absorption through skin.

First Aid: If this chemical gets into the eyes, remove any contact lenses at once and irrigate immediately for at least 15 minutes, occasionally lifting upper and lower lids. Seek medical attention immediately. If this chemical contacts the skin, remove contaminated clothing and wash immediately with soap and water. Seek medical attention immediately. If this chemical has been inhaled, remove from exposure, begin rescue breathing (using universal precautions) if breathing has stopped and CPR if heart action has stopped. Transfer promptly to a medical facility. When this chemical has been swallowed, get medical attention. For ingestion, induce vomiting with syrup of ipecac.

Personal Protective Methods: Wear protective gloves and clothing to prevent any reasonable probability of skin contact. Safety equipment suppliers/manufacturers can provide recommendations on the most protective glove/clothing material for your operation. All protective clothing (suits, gloves, footwear, headgear) should be clean, available each day, and put on before work. Contact lenses should not be worn when working with this chemical. Wear dust-proof chemical goggles and face shield unless full facepiece respiratory protection is worn. Employees should wash immediately with soap when skin is wet or contaminated. Provide emergency showers and eyewash.

Storage: Prior to working with this chemical you should be trained on its proper handling and storage. Store in tightly closed containers in a cool, well ventilated area. A regulated, marked area should be established where this chemical is handled, used, or stored in compliance with OSHA standard 1910.1045.

Shipping: Cantharidin is not specifically listed in the DOT performance-oriented packaging standards[19] with respect to labeling requirements or restrictions on shipping quantities.

Spill Handling: Evacuate and restrict persons not wearing protective equipment from area of spill or leak until cleanup is complete. Remove all ignition sources. Ventilate area of spill or leak. Absorb liquids in vermiculite, dry sand, earth, peat, carbon, or a similar material and deposit in sealed containers. It may be necessary to contain and dispose of this chemical as a hazardous waste. If material or contaminated runoff enters waterways, notify downstream users of potentially contaminated waters. Contact your Department of Environmental Protection or your regional office of the federal EPA for specific recommendations. If employees are required to clean-up spills, they must be properly trained and equipped. OSHA 1910.120(q) may be applicable.

Fire Extinguishing: This chemical is a combustible solid. Use dry chemical, carbon dioxide, water spray, or alcohol foam extinguishers. Poisonous gases are produced in fire. If material or contaminated runoff enters waterways, notify downstream users of potentially contaminated waters. Notify local health and fire officials and pollution control agencies. From a secure, explosion-proof location, use water spray to cool exposed containers. If cooling streams are ineffective (venting sound increases in volume and pitch, tank discolors, or shows any signs of deforming), withdraw immediately to a secure position. If employees are expected to fight fires, they must be trained and equipped in OSHA 1910.156.

References

U.S. Environmental Protection Agency, "Chemical Profile: Cantharidin," Washington, DC, Chemical Emergency Preparedness Program (November 30, 1987)

Sax, N. I., Ed., "Dangerous Properties of Industrial Materials Report" 1, No. 2, 27–28 (1980)

Caprolactam

Molecular Formula: $C_6H_{11}NO$

Synonyms: Aminocaproic lactam; 6-Aminohexanoic acid cyclic lactam; 2-Azacycloheptanone; ε-Caprolactam; Caprolactama (Spanish); 6-Caprolactum; Caprolattame (French); Cyclohexanone isooxime; Epsylon kaprolaktam (Czech, Polish); Hexahydro-2H-azepine-2-one; Hexahydro-2-azepinone; Hexahydro-2H-azepin-2-one; 6-Hexanelactum; Hexanone isoxime; Hexanonisoxim (German); 1,6-Hexolactam; ε-Kaprolaktam (Czech, Polish); 2-Keto-hexamethyleneimine; 2-Ketohexamethylenimine; NCI-C50646; 2-Oxohexamethyleneimine; 2-Oxohexamethylenimine; 2-Perhydroazepinone

CAS Registry Number: 105-60-2

RTECS Number: CM3675000

DOT ID: NA 1693

EEC Number: 613-069-00-2

Regulatory Authority

- Air Pollutant Standard Set (ACGIH)[1] (DFG)[1] (HSE)[33] (UNEP)[43] (OSHA)[58] (Several States)[60] (Australia) (Israel) (Mexico) (Several Canadian Provinces)
- CLEAN AIR ACT: Hazardous Air Pollutants (Title I, Part A, Section 112)
- SUPERFUND/EPCRA 40CFR302.4 Reportable Quantity (RQ): CERCLA, 1 lb (0.454 kg)
- Canada, WHMIS, Ingredients Disclosure List

Cited in U.S. State Regulations: Alaska (G), California (G), Connecticut (A), Florida (G), Illinois (G), Maine (G), Massachusetts (G), Minnesota (G), Nevada (A), New Hampshire (G), New Jersey (G), North Dakota (A), Pennsylvania (G), Rhode Island (G), Virginia (A), West Virginia (G).

Description: Caprolactum, $C_6H_{11}NO$, is a white crystalline solid with an unpleasant odor. The odor threshold is 0.3 mg/m^3. Boiling point = 267°C. Freezing/Melting point = 69°C. Flash point = 125°C (oc); 139°C (cc). Autoignition temperature = 375°C. Explosive limits: LEL = 1.84%; UEL = 8.0%. Hazard Identification (based on NFPA-704 M Rating System): Health 1, Flammability 1, Reactivity 0. Highly soluble in water.

Potential Exposure: Caprolactam is used in the manufacture of nylon, plastics, bristles, film, coatings, synthetic leather, plasticizers, and paint vehicles; as a crossslinking agent for curing polyurethanes; and in the synthesis of lysine.

Incompatibilities: Caprolactum decomposes on heating and on burning producing toxic fumes including nitrogen oxides, ammonia. Reacts violently with strong oxidizers, producing toxic fumes. Toxic decompositon above 400°C.

Permissible Exposure Limits in Air: The ACGIH and NIOSH recommended airborne exposure limit for caprolactum *dust* is 1 mg/m^3 TWA and STEL of 3 mg/m^3. For *vapor* the NIOSH recommended airborne exposure limit is 0.22 ppm TLV (10-hour workshift) and STEL of 0.66 ppm; the ACGIH recommends a 5 ppm TWA (8-hour workshift) and STEL of 10 ppm. The DFG[3] has set a MAK of 5 mg/m^3 for vapor and dust. The former USSR/UNEP joint project[43] has set a MAC in workplace air of 10 mg/m^3 and in ambient air in residential areas of 0.06 mg/m^3 on either a momentary or a daily average basis. In addition, several states have set guidelines or standards for caprolactam in ambient air ranging from 10 μg/m^3 (North Dakota) to 24 μg/m^3 (Nevada) to 160 μg/m^3 (Virginia) to 400 μg/m^3 (Connecticut).

Determination in Air: Collection on a filter (for the dust) or by impinger (for the vapor) and analysis of gas liquid chromatography.

Permissible Concentration in Water: The former USSR/UNEP joint project[43] has adopted The former USSR value[35] of a MAC of 1 mg/l in water used for domestic purposes.

Routes of Entry: Inhalation, ingestion, skin and eye contact.

Harmful Effects and Symptoms

Short Term Exposure: The vapor irritates the eyes, skin, and respiratory tract. Inhalation may affect the central nervous system. Skin contact can cause irritation and serious burns if contact is prolonged and confined. Exposure in airborne dust at 5 mg/m^3 causes skin irritation in some people but not at 1 mg/m^3. Sensitivity has not been related to race, skin pigmentation, or other common indices of sensitivity. The prevalence of dermatoses among workers in a caprolactam manufacturing plant showed that contact dermatitis and eczema of the hands were most prevalent. Dry erythematous squamous foci on smooth skin was a typical manifestation. Light sensitivity of the eyes was produced by inhalation of caprolactam at 0.11 mg/m^3 and higher. The olfactory threshold was 0.30 mg/m^3. An oral dose of 3 – 6 g was given daily for 3 – 5 years for the treatment of obesity in 90 subjects. No toxic effects were observed. There was no effect on appetite, and only one person developed an allergy to Caprolactam. High exposures may cause irritability, confusion, and convulsions (seizures).

Long Term Exposure: Exposure may damage the developing fetus and may affect the reproductive ability of males. Caprolactam may damage the liver. Repeated or prolonged contact may cause skin sensitization and dermatitis. The substance may have effects on the nervous system, liver. Exposure to high concentrations over many years may cause irritability and confusion.

Points of Attack: Eyes, skin, respiratory system, liver.

Medical Surveillance: Before beginning employment and at regular times after that, for those with frequent or potentially high exposures (half the TLV or greater), the following are recommended: Liver function tests. Lung function tests.

If symptoms develop or overexposure has occurred, the following may also be useful: EEG (brain wave test), Skin testing with dilute Caprolactam may help diagnose allergy, if done by a qualified allergist.

First Aid: If this chemical gets into the eyes, remove any contact lenses at once and irrigate immediately for at least 15 minutes, occasionally lifting upper and lower lids. Seek medical attention immediately. If this chemical contacts the skin, remove contaminated clothing and wash immediately with soap and water. Seek medical attention immediately. If this chemical has been inhaled, remove from exposure, begin rescue breathing (using universal precautions) if breathing has stopped and CPR if heart action has stopped. Transfer promptly to a medical facility. When this chemical has been swallowed, get medical attention. Give large quantities of water and induce vomiting. Do not make an unconscious person vomit.

Personal Protective Methods: Wear protective gloves and clothing to prevent any reasonable probability of skin contact. Safety equipment suppliers/manufacturers can provide recommendations on the most protective glove/clothing material for your operation. All protective clothing (suits, gloves, footwear, headgear) should be clean, available each day, and put on before work. Contact lenses should not be worn when working with this chemical. Wear dust-proof chemical goggles and face shield unless full facepiece respiratory protection is worn. Employees should wash immediately with soap when skin is wet or contaminated. Provide emergency showers and eyewash.

Respirator Selection: Where the potential exists for exposures over 5 ppm (vapor) or 1 mg/m^3 (dust), use an MSHA/NIOSH approved respirator equipped with organic vapor cartridges and a particulate prefilter. More protection is provided by a full facepiece respirator than by a half-mask respirator, and even greater protection is provided by a powered-air purifying respirator. Where the potential for high exposures exists, use a MSHA/NIOSH approved supplied-air respirator with a full facepiece operated in the positive pressure mode or with a full facepiece, hood, or helmet in the continuous flow mode, or use a MSHA/NIOSH approved self-contained breathing apparatus with a full facepiece operated in pressure-demand or other positive pressure mode.

Storage: Prior to working with Caprolactum you should be trained on its proper handling and storage. Before entering confined space where this chemical may be present, check to make sure that an explosive concentration does not exist. Store in tightly closed containers in a cool, well ventilated area away from oxidizers and heat. Metal containers involving the transfer of this chemical should be grounded and bonded. Drums must be equipped with self-closing valves, pressure vacuum bungs, and flame arresters. Use only non-sparking tools and equipment, especially when opening and closing containers of this chemical. Sources of ignition such as smoking and open flames, are prohibited where this chemical is used, handled, or stored in a manner that could create a potential fire or explosion hazard.

Shipping: The DOT[19] has set no specific requirements for caprolactam in their performance oriented packaging standards.

Spill Handling: Restrict persons not wearing protective equipment from area of spill until clean-up is complete. Collect spilled material in the most convenient and safe manner and deposit in sealed containers for reclamation or for disposal in an approved facility. Establish forced ventilation to keep levels below explosive limit. Absorb liquids in vermiculite, dry sand, earth, or a similar material and deposit in sealed containers. It may be necessary to contain and dispose of this chemical as a hazardous waste. If material or contaminated runoff enters waterways, notify downstream users of potentially contaminated waters. Contact your Department of Environmental Protection or your regional office of the federal EPA for specific recommendations. If employees are required to clean-up spills, they must be properly trained and equipped. OSHA 1910.120(q) may be applicable.

Fire Extinguishing: This chemical will burn but does not easily ignite. Extinguish fire using any agent suitable for the type of surrounding fire. If heated to more than 100°C, Caprolactam boils, giving off poisonous gases (including oxides of Nitrogen). If material or contaminated runoff enters waterways, notify downstream users of potentially contaminated waters. Notify local health and fire officials and pollution control agencies. From a secure, explosion-proof location, use water spray to cool exposed containers. If cooling streams are ineffective (venting sound increases in volume and pitch, tank discolors, or shows any signs of deforming), withdraw immediately to a secure position. If employees are expected to fight fires, they must be trained and equipped in OSHA 1910.156.

Disposal Method Suggested: Controlled incineration (oxides of nitrogen are removed from the effluent gas by scrubbers and/or thermal devices). Also, Caprolactam may be recovered from Caprolactam still bottoms or nylon waste.[22]

References

New Jersey Department of Health and Senior Services and Senior Services, "Hazardous Substance Fact Sheet: Caprlactam," Trenton, NJ (September 1985)

Captafol

Molecular Formula: $C_{10}H_9Cl_4NO_2S$

Synonyms: Captafol; Captatol; Captofol; 4-Cyclohexene-1,2-dicarboximide, *N*-(1,1,2,2-tetrachloroethyl)thiol-; Difolatan®; Difosan; Folcid; 1H-Isoindole-1,3(2H)-dione,3a,4,7,7a-tetrahydro-2-(1,1,2,2-tetrachloroethyl)thio-; Ortho 5865; Sanspor; Sulfonimide; Sulpheimide; *N*-(1,1,2,2-Tetrachloraethylthio)-cyclohex-4-en-1,4-diacarboximid (German); *N*-[(1,1,2,2-Tetrachloroethyl)-thio]-4-cyclohexene-1,2-dicarboximide; *N*-1,1,2,2-Tetrachloroethylmercapto- 4-

cyclohexene-1,2-carboximide; *N*-[(1,1,2,2-Tetrachloroethyl)sulfenyl]-*cis*-4-cyclohexene-1,2-dicarboximide; *N*-(1,1,2,2-Tetrachloroethylthio)-4- cyclohexene-1,2-dicarboximide

CAS Registry Number: 2425-06-1

RTECS Number: GW4900000

DOT ID: UN 2773

EEC Number: 613-046-00-7

Regulatory Authority

- Carcinogen (RTECS)[9] (limited human data) (IARC)
- Banned or Severely Restricted (Germany, Norway) (UN)[13]
- Air Pollutant Standard Set (ACGIH)[1] (HSE)[33] (OSHA)[58] (Several States)[60] (Australia) (Israel) (Mexico) (Several Canadian Provinces)
- Canada, WHMIS, Ingredients Disclosure List

Cited in U.S. State Regulations: Alaska (G), California (A, G), Connecticut (A), Florida (G), Illinois (G), Maine (G), Massachusetts (G), Michigan (G), Minnesota (G), Nevada (A), New Hampshire (G), New Jersey (G), North Dakota (A), Pennsylvania (G), Rhode Island (G), Virginia (A), West Virginia (G).

Description: Captafol, $C_{10}H_9Cl_4NO_2S$, is a white crystalline solid. Freezing/Melting point = 160 – 161°C. Insoluble in water.

Potential Exposure: Those engaged in the manufacture, formulation and application of this fungicide.

Incompatibilities: Reacts violently with bases causing fire and explosion hazard. Not compatible with strong acids or acid vapor, oxidizers. Strong alkaline conditions contribute to instability. Attacks some metals.

Permissible Exposure Limits in Air: NIOSH and ACGIH recommended airborne exposure limit is 0.1 mg/m^3 TWA with the notation "skin" indicating the possibility of cutaneous absorption. HSE,[33] Mexico, Israel, and Australia have set this same value as has NIOSH. In addition, several states have set guidelines or standards for captafol in air[60] ranging from 1.0 μg/m^3 (North Dakota) to 1.5 μg/m^3 (Virginia) to 2.0 μg/m^3 (Connecticut and Nevada).

Permissible Concentration in Water: No criteria set.

Routes of Entry: Inhalation, ingestion, skin.

Harmful Effects and Symptoms

Short Term Exposure: Irritates eyes, skin and respiratory tract. Captafol can affect you when breathed in and by passing through your skin. Caprafol may cause an asthma-like allergy. Future exposures can cause asthma attacks with shortness of breath, wheezing, cough, and/or chest tightness. Exposure can irritate the skin. It can also cause a skin allergy to develop. Exposure to the sun (or other ultraviolet light) after exposure to Captafol may cause severe rash with itching, swelling, and blistering.

Long Term Exposure: Repeated or prolonged contact cause skin sensitization, dermatitis, allergic conjunctivitis. Repeated or prolonged inhalation exposure may cause asthma. The substance may have damaging effects on the liver and kidneys. Captafol is a probable carcinogen in humans. There is some evidence that it causes liver cancer in humans and it has caused kidney cancer in animals. Captafol may cause mutations. Handle with extreme caution.

Points of Attack: Skin, respiratory system, liver, kidneys.

Medical Surveillance: If symptoms develop or overexposure is suspected, the following may be useful: Liver and kidney function test; lung function test. Skin testing with dilute Captafol may help diagnose allergy, if done by a qualified allergist.

First Aid: If this chemical gets into the eyes, remove any contact lenses at once and irrigate immediately for at least 15 minutes, occasionally lifting upper and lower lids. Seek medical attention immediately. If this chemical contacts the skin, remove contaminated clothing and wash immediately with soap and water. Seek medical attention immediately. If this chemical has been inhaled, remove from exposure, begin rescue breathing (using universal precautions) if breathing has stopped and CPR if heart action has stopped. Transfer promptly to a medical facility. When this chemical has been swallowed, get medical attention. Give large quantities of water and induce vomiting. Do not make an unconscious person vomit.

Personal Protective Methods: *Clothing:* Avoid skin contact with Captafol. Wear protective gloves and clothing. Safety equipment suppliers/manufacturers can provide recommendations on the most protective glove/clothing material for your operation. All protective clothing (suits, gloves, footwear, headgear) should be clean, available each day, and put on before work. *Eye Protection:* Wear dust-proof goggles when working with powders or dust, unless full facepiece respiratory protection is worn. Use splash-proof chemical goggles and face shield when working with liquids containing Captafol.

Respirator Selection: Where the potential exists for exposures over 0.1 mg/m^3, use a MSHA/NIOSH approved full facepiece respirator with a pesticide cartridge. Increased protection is obtained from full facepiece air purifying respirators. Where the potential for high exposures exists, use a MSHA/NIOSH approved supplied-air respirator with a full facepiece operated in the positive pressure mode or with a full facepiece, hood, or helmet in the continuous flow mode, or use a MSHA/NIOAH approved self-contained breathing apparatus with a full facepiece operated in pressure-demand or other positive pressure mode.

Storage: Prior to working with Captafol you should be trained on its proper handling and storage. Store in tightly closed containers in a cool, well-ventilated area away from heat, acids, acid fumes or strong oxidizers (such as peroxides, chlorates, perchlorates, nitrates and permanganates) since violent reactions occur. A regulated, marked area should be established where this chemical is handled, used, or stored in compliance with OSHA standard 1910.1045.

Shipping: Captafol is not specifically cited by DOT[19] in its Performance-Oriented Packaging Standards.

Spill Handling: Evacuate persons not wearing protective equipment from area of spill or leak until clean-up is complete. Remove all ignition sources. Collect powdered material in the most convenient and safe manner and deposit in sealed containers. Ventilate area after clean-up is complete. If Captafol is in liquid or slurry form, absorb it with vermiculite, dry sand, earth or a similar material. Dispose of the abosrbing material in an approved facility. It may be necessary to contain and dispose of this chemical as a hazardous waste. If material or contaminated runoff enters waterways, notify downstream users of potentially contaminated waters. Contact your Department of Environmental Protection or your regional office of the federal EPA for specific recommendations. If employees are required to clean-up spills, they must be properly trained and equipped. OSHA 1910.120(q) may be applicable.

Fire Extinguishing: This chemical is a noncombustible solid but it may be dissolved in a flammable liquid. Use dry chemical, carbon dioxide, water spray, or alcohol foam extinguishers. The substance decomposes on heating or on burning producing toxic and corrosive fumes including hydrogen chloride, nitrogen oxides, sulfur oxides. If material or contaminated runoff enters waterways, notify downstream users of potentially contaminated waters. Notify local health and fire officials and pollution control agencies. From a secure, explosion-proof location, use water spray to cool exposed containers. If cooling streams are ineffective (venting sound increases in volume and pitch, tank discolors, or shows any signs of deforming), withdraw immediately to a secure position. If employees are expected to fight fires, they must be trained and equipped in OSHA 1910.156.

Disposal Method Suggested: Hydrolysis.[22]

References

New Jersey Department of Health and Senior Services and Senior Services, "Hazardous Substance Fact Sheet: Captafol," Trenton, NJ (April 1998)

Captan

Molecular Formula: $C_9H_8Cl_3NO_2S$

Synonyms: Aacaptan; Agrosol S; Agrox 2-Way and 3-Way; Amercide; Bangton; Bean seed protectant; Captaf; Captaf 85W; Captan 50W; Captancapteneet 26,538; Captane; Captex; 4-Cyclohexene-1,2-dicarboximide, *N*-[(Trichloromethyl)mercapto]; ENT 26538; ESSO Fungicide 406; Flit 406; Fungicide 406; Fungus Ban type II; Glyodex 37-22; Hexacap; 1H-Isoindole-1,3(2H)-dione,3a,4,7,7a-tetrahydro-2-[(trichloromethyl)thiol]-; Isopto carbachol; Isotox seed treater "D" and "F"; Kaptan; le Captane (French); Malipur; Merpan; Micro-Check 12; Miostat; NCI-0077; Neracid; Orthocide®; Orthocide® 406; Orthocide® 50; Orthocide® 7.5; Orthocide® 75; Orthocide® 83; Osocide; *N*-Trichloromethylmercapto-4-cyclohexene-1,2-dicarboximide; *N*-(Trichloromethylmercapto)-δ(sup 4)-tetrahydrophthalimide; *N*-[(Trichloromethyl)thio]-4-cyclohexene-1,2-dicarboximide; *N*-Trichloromethylthio-cyclohex-4-ene-1,2-dicarboximide; *N*-Trichloromethylthio-*cis*-δ(sup4)-cyclohexene-1,2-dicarboximide; *N*-[(Trichloromethyl)thio]-δ-4-tetrahydrophthalimide; *N*-[(Trichloromethyl)thio] tetrahydrophthalimide; *N*-Trichloromethylthio-3a,4,7,7a-tetrahydrophthalimide; Trimegol; Vancide 89; Vancide 89RE; Vancide P-75; Vangard K; Vanicide; Vondcaptan

CAS Registry Number: 133-06-2

RTECS Number: GW5075000

DOT ID: UN 2773

EEC Number: 613-044-00-6

Regulatory Authority

- Carcinogen (Animal Positive-mice) (NTP)
- Banned or Severely Restricted (Finland, Sweden) (UN)[13]
- Air Pollutant Standard Set (ACGIH)[1] (Australia) (HSE)[33] (Israel) (Mexico) (OSHA)[58] (Several States)[60] (Several Canadian Provinces)
- CLEAN AIR ACT: Hazardous Air Pollutants (Title I, Part A, Section 112)
- CLEAN WATER ACT: Section 311 Hazardous Substances/RQ 40CFR117.3 (same as CERCLA, see below); Section 313 Water Priority Chemicals (57FR41331, 9/9/92)
- SUPERFUND/EPCRA 40CFR302.4 Reportable Quantity (RQ): CERCLA, 10 lb (4.54 kg)
- EPCRA Section 313 Form R *de minimis* concentration reporting level: 1.0%

Cited in U.S. State Regulations: Alaska (G), California (A, W), Connecticut (A), Florida (G), Illinois (G), Kansas (A), Maine (G, W), Massachusetts (G), Michigan (G), Minnesota (G), Nevada (A), New Hampshire (G), New Jersey (G), Pennsylvania (G, A), Rhode Island (G), West Virginia (G).

Description: Captan, $C_9H_8Cl_3NO_2S$, when pure, is a colorless crystalline solid. The technical grade is a cream to yellow powder with a strong odor. It is commonly dissolved in a "carrier" which may be combustible or flammable. Freezing/Melting point = 178°C (decomposes). Hazard Identification (based on NFPA-704 M Rating System): Health 3, Flammability 2, Reactivity 0. Very slightly soluble in water. Log P_{ow} (octanol/water partition coefficient) = 2.35.

Incompatibilities: Incompatible with tetraethyl pyrophosphate, parathion. Keep away from strong alkaline materials (e.g., hydrated lime) as captan may become unstable. May react with water releasing hydrogen chloride gas. Corrosive to metals in the presence of moisture.

Permissible Exposure Limits in Air: ACGIH and NIOSH have recommended a TWA for captan of 5 mg/m^3. In addition, several states have set guidelines or standards for captan in ambient air[60] ranging from 11.9 μg/m^3 (Kansas) to 35 μg/m^3 (Pennsylvania) to 50 μg/m^3 (North Dakota) to 100 μg/m^3 (Connecticut) to 119 μg/m^3 (Nevada).

Determination in Air: OSHA versatile sampler-2; Reagent; High-pressure liquid chromatography/Ultraviolet detection; IV NIOSH Method #5601.

Permissible Concentration in Water: A no-adverse-effect level of drinking water has been calculated by NAS/NRC as 0.35 mg/l.

The former USSR/UPEN joint project[43] has set a MAC of 2.0 mg/l in water bodies used for domestic purposes.

Guidelines have been set in two states for Captan in drinking water ranging from 100 μg/l in Maine to 350 μg/l in California.

Routes of Entry: Skin contact, inhalation of dust, ingestion.

Harmful Effects and Symptoms

Short Term Exposure: The substance irritates the eyes and the skin. The acute oral LD_{50} value for rats in 9,000 mg/kg (insignificantly toxic). Most of the chronic-oral-toxicity data on captan suggest that the no-adverse-effect or toxicologically safe dosage is about 1,000 ppm (50 mg/kg/day). However, on the basis of fetal mortality observed in monkeys exposed to captan (12.5 mg/kg/day), the acceptable daily intake of Captan has been established at 0.1 mg/kg of body weight by the FAO/WHO. Based of long-term feeding studies results in rats and dogs, ADIs were calculated at 0.05 mg/kg/day for Captan. A rebuttal presumption against registration for captan was issued on August 19, 1980 by EPA on the basis of possible oncogenicity, mutagenitcity and teratogenicity.

Long Term Exposure: Repeated or prolonged contact with skin may cause skin allergy to develop. Once this occurs, even very small future exposures can cause itching and a skin rash. Exposure may cause mutations or damage the developing fetus; however, this needs further study. Animal studies have found the development of cancer in animals. Whether Captan is a human cancer hazard requires further study.

Points of Attack: Eyes, skin, respiratory system, gastrointestinal tract, liver, kidneys. Cancer site in animals: duodenal tumors.

Medical Surveillance: If symptoms develop or overexposure is suspected, the following may be useful: Skin testing with dilute captan may help diagnose allergy, if done by a qualified allergist.

First Aid: If this chemical gets into the eyes, remove any contact lenses at once and irrigate immediately for at least 15 minutes, occasionally lifting upper and lower lids. Seek medical attention immediately. If this chemical contacts the skin, remove contaminated clothing and wash immediately with soap and water. Seek medical attention immediately. If this chemical has been inhaled, remove from exposure, begin rescue breathing (using universal precautions) if breathing has stopped and CPR if heart action has stopped. Transfer promptly to a medical facility. When this chemical has been swallowed, get medical attention. Give large quantities of water and induce vomiting. Do not make an unconscious person vomit.

Personal Protective Methods: Wear protective gloves and clothing to prevent any reasonable probability of skin contact. Safety equipment suppliers/manufacturers can provide recommendations on the most protective glove/clothing material for your operation. All protective clothing (suits, gloves, footwear, headgear) should be clean, available each day, and put on before work. Contact lenses should not be worn when working with this chemical. Wear dust-proof chemical goggles and face shield unless full facepiece respiratory protection is worn. Employees should wash immediately with soap when skin is wet or contaminated. Provide emergency showers and eyewash.

Respirator Selection: NIOSH: *At any detectable concentration:* SCBAF:PD,PP (any MSHA/NIOSH approved self-contained breathing apparatus that has a full facepiece and is operated in a pressure-demand or other positive-pressure mode); or SAF:PD,PP:ASCBA (any supplied-air respirator that has a full facepiece and is operated in a pressure-demand or other positive-pressure mode in combination with an auxiliary, self-contained breathing apparatus operated in a pressure-demand or other positive pressure mode). *Escape:* GMFOV [any air-purifying, full-facepiece respirator (gas mask) with a chin-style, front-or back-mounted organic vapor canister]; or SCBAE (any appropriate escape-type, self-contained breathing apparatus).

Storage: Prior to working with Captan you should be trained on its proper handling and storage. Store in tightly closed containers in a cool, well ventilated area away from water, heat and incompatible materials. Metal containers involving the transfer of this chemical should be grounded and bonded. Where possible, automatically pump liquid from drums or other storage containers to process containers. Drums must be equipped with self-closing valves, pressure vacuum bungs, and flame arresters. Use only non-sparking tools and equipment, especially when opening and closing containers of this chemical. Sources of ignition such as smoking and open flames, are prohibited where this chemical is used, handled, or stored in a manner that could create a potential fire or explosion hazard. A regulated, marked area should be established where this chemical is handled, used, or stored in compliance with OSHA standard 1910.1045.

Shipping: DOT[19] does not specifically cite Captan in its Performance-Oriented Packaging Standards.

Spill Handling: Evacuate persons not wearing protective equipment from area of spill or leak until clean-up is complete. Remove all ignition sources. Collect powdered material in the most convenient and safe manner and deposit in sealed containers. Absorb liquid containing captan in vermiculite, dry sand, earth, or similar material. Ventilate area after clean-up is complete. It may be necessary to contain and dispose of this chemical as a hazardous waste. If material or contaminated runoff enters waterways, notify downstream users of potentially contaminated waters. Contact your Department of Environmental Protection or your regional office of the federal EPA for specific recommendations. If employees are required to clean-up spills, they must be properly trained and equipped. OSHA 1910.120(q) may be applicable.

Fire Extinguishing: Captan may burn, but does not ignite readily. Use dry chemical, CO_2, or foam extinguishers. Do not use water. At high temperatures, Captan decomposes and produces poisonous gases including oxides of sulfur and nitrogen, hydrogen chloride and phosgene). If material or contaminated runoff enters waterways, notify downstream users of potentially contaminated waters. Notify local health and fire officials and pollution control agencies. From a secure, explosion-proof location, use water spray to cool exposed containers. If cooling streams are ineffective (venting sound increases in volume and pitch, tank discolors, or shows any signs of deforming), withdraw immediately to a secure position. If employees are expected to fight fires, they must be trained and equipped in OSHA 1910.156.

Disposal Method Suggested: Captan decomposes fairly readily in alkaline media (pH>8). It is hydrolytically stable at neutral or acid pH but decomposes when heated alone at its Freezing/Melting point. Alkaline hydrolysis is recommended.[22]

Carbachol Chloride

Molecular Formula: $C_6H_{15}ClN_2O_2$

Synonyms: 2-[(Aminocarbonyl)oxy]-*N,N,N*-trimethylethanaminium chloride; Cabacolina; Carbachol; Carbacholin; Carbacholine chloride; Carbamic acid, Ester with choline chloride; Carbamiotin; Carbamoylcholine chloride; Carbamylcholine chloride; Carbochol; Carbocholin; Carbyl; Carcholin; Choline carbamate chloride; Choline chlorine carbamate; Choline, chlorine carbamate (ester); Coletyl; Doryl (pharmaceutical); (2-Hydroxyethyl)trimethylammonium chloride carbamate; Isopto carbachol; Jestryl; Lentin; Lentine (French); Miostat; Mistura C; Moryl; P.V. carbachol; RTECS No. GA0875000; TL 457; Vasoperif

CAS Registry Number: 51-83-2

RTECS Number: GA0875000

Regulatory Authority

- SUPERFUND/EPCRA 40CFR355, Appendix B Extremely Hazardous Substances: TPQ = 500/10,000 lb (227/4,540 kg)
- SUPERFUND/EPCRA 40CFR302.4 Reportable Quantity (RQ): EHS, 1 lb (0.454 kg)

Cited in U.S. State Regulations: Florida (G), Massachusetts (G), New Jersey (G), Pennsylvania (G).

Description: Carbachol Chloride, $C_6H_{15}N_2O_2Cl$, is a crystalline odorless powder which, on standing in an open container, develops a faint odor resembling that of an aliphatic amine. Freezing/Melting point = 200 – 205°C.

Potential Exposure: Used in veterinary medicine as a cholinergic; parasympathomimetic, used chiefly in large animals, especially for colic in the horse.

Incompatibilities: Strong oxidizers.

Permissible Exposure Limits in Air: No standards set.

Permissible Concentration in Water: No criteria set.

Routes of Entry: Ingestion, skin contact.

Harmful Effects and Symptoms

Highly toxic by mouth. The LD_{50} oral-rat is 40 mg/kg.

First Aid: If this chemical gets into the eyes, remove any contact lenses at once and irrigate immediately for at least 15 minutes, occasionally lifting upper and lower lids. Seek medical attention immediately. If this chemical contacts the skin, remove contaminated clothing and wash immediately with soap and water. Seek medical attention immediately. If this chemical has been inhaled, remove from exposure, begin rescue breathing (using universal precautions) if breathing has stopped and CPR if heart action has stopped. Transfer promptly to a medical facility. When this chemical has been swallowed, get medical attention. Give large quantities of water and induce vomiting. Do not make an unconscious person vomit.

Storage: Prior to working with Carbachol chloride you should be trained on its proper handling and storage. Store in tightly closed containers in a cool, well ventilated area away from incompatible materials.

Shipping: Carbachol chloride is not specifically listed in the DOT Performance-Oriented Packaging Standards[19] as regards to label requirements or shipping weight restrictions.

Spill Handling: Evacuate persons not wearing protective equipment from area of spill or leak until clean-up is complete. Remove all ignition sources. Collect powdered material in the most convenient and safe manner and deposit in sealed containers. Ventilate area after clean-up is complete. It may be necessary to contain and dispose of this chemical as a hazardous waste. If material or contaminated runoff enters waterways, notify downstream users of potentially contaminated waters. Contact your Department of Environmental Protection or your regional office of the federal EPA for specific recommendations. If employees are required to clean-up spills, they must be properly trained and equipped. OSHA 1910.120(q) may be applicable.

Fire Extinguishing: Use dry chemical, carbon dioxide, water spray, or alcohol foam extinguishers. Poisonous gases are produced in fire. If material or contaminated runoff enters waterways, notify downstream users of potentially contaminated waters. Notify local health and fire officials and pollution control agencies. If employees are expected to fight fires, they must be trained and equipped in OSHA 1910.156.

Disposal Method Suggested: High temperature incineration with scrubber for chloride and nitrogen oxide removal.

References

U.S. Environmental Protection Agency, "Chemical Profile: Carbachol Chloride," Washington, DC, Chemical Emergency Preparedness Program (November 30, 1987).

Sax, N. I., Ed., "Dangerous Properties of Industrial Materials Report," 1, No. 7, 40–41 (1981).

Carbaryl

Molecular Formula: $C_{12}H_{11}NO_2$

Common Formula: $C_{10}H_7OOCNHCH_3$

Synonyms: Arilat; Arilate; Arylam; Bercema NMC50; Caprolin; Carbamic acid, methyl-, 1-naphthyl ester; Carbamine; Carbaril (Italian); Carbaryl, NAC; Carbatox; Carbatox 60; Carbatox 75; Carbavur; Carbomate; Carpolin; Carylderm; Compound 7744; Crag Sevin; Denapon; Dicarbam; Dyna-Carbyl; ENT 23969; Experimental insecticide 7744; Gamonil; Germain's; Hexavin; Karbaryl (Polish); Karbaspray; Karbatox; Karbosep; Menapham; Methylcarbamate 1-naphthalenol; *N*-Methylcarbamate de 1-naphtyle (French); Methylcarbamic acid, 1-naphthyl ester; *N*-Methyl-1-naftyl-carbamaat (Dutch); *N*-Methyl-1-naphthyl-carbamat (German); *N*-Methyl α-naphthylcarbamate; *N*-Methyl-1-naphthyl carbamate; *N*-Methyl-α-naphthylurethan; *N*-Metil-1-naftil-carbammato (Italian); Microcarb; Mugan; Murvin; Murvin 85; NAC; α-Naftyl-*N*-methylkarbamat (Czech); 1-Naphthol; α-Naphthyl *N*-methylcarbamate; 1-Naphthyl *N*-methylcarbamate; 1-Naphthyl *N*-methyl-carbamate; 1-Naphthyl methylcarbamate; NMC 50; Oltitox; OMS-29; OMS 629; Panam; Pomex; Prosevor 85; Ravyon; Seffein; Septene; Sevimol; Sevin®; Sevin® 4; Sewin; Sok; Tercyl; Thinsec; Tornado; Tricarnam; UC 7744 (Union Carbide); Union Carbide 7,744; Vioxan

CAS Registry Number: 63-25-2

RTECS Number: FC5950000

DOT ID: UN 2757

EEC Number: 006-011-00-7

Regulatory Authority

- Air Pollutant Standard Set (ACGIH)[1] (Australia) (DFG)[3] (HSE)[33] (Israel) (Mexico) (former USSR)[43] (OSA)[58] (Several States)[60] (Several Canadian Provinces)
- CLEAN AIR ACT: Hazardous Air Pollutants (Title I, Part A, Section 112)
- CLEAN WATER ACT: Section 311 Hazardous Substances/RQ 40CFR117.3 (same as CERCLA, see below); Section 313 Water Priority Chemicals (57FR41331, 9/9/92)
- RCRA 40CFR268.48; 61FR15654, Universal Treatment Standards: Wastewater (mg/l), 0.006; Nonwastewater (mg/kg), 0.14
- SUPERFUND/EPCRA 40CFR302.4 Reportable Quantity (RQ): CERCLA, 100 lb (45.4 kg)
- EPCRA Section 313 Form R *de minimis* concentration reporting level: 1.0%
- U.S. DOT Regulated Marine Pollutant (49CFR172.101, Appendix B)
- Canada: Drinking water MAC = 0.09 mg/l

Cited in U.S. State Regulations: Alaska (G), California (W), Connecticut (A), Florida (G), Illinois (G), Kansas (A, W), Maine (G, W), Massachusetts (G), Michigan (G), Minnesota (G), Nevada (A), New Hampshire (G), New Jersey (G), New York (G), North Dakota (A), Pennsylvania (G, A), Rhode Island (G), Virginia (A), West Virginia (G), Wisconsin (W).

Description: Carbaryl, $C_{10}H_7OOCNHCH_3$, is a white or grayish, odorless, crystalline solid, or various other forms including liquid and paste. Boiling point = (decomposes below BP). Freezing/Melting point = 142°C. Hazard Identification (based on NFPA-704 M Rating System): Health 2, Flammability 0, Reactivity 0. Insoluble in water.

Potential Exposure: Workers engaged in production formulation and application of Carbaryl as a contact insecticide for fruits, vegetables, cotton and other crops.

Incompatibilities: Contact with strong oxidizers can cause fire and explosions.

Permissible Exposure Limits in Air: The OSHA PEL is 5 mg/m^3 TWA for an 8-hour workshift. NIOSH and ACGIH recommended limit is also 5 mg/m^3. Australia, Israel, Mexico, DFG[3] and HSE[33] have all set this same value. The STEL set by HSE[33] is 10 mg/m^3. The NIOSH IDLH level is 100 mg/m^3. The former USSR/UNEP joint project[43] sets a MAC in workplace air of 1 mg/m^3 and limits in the ambient air in residential areas of 0.02 mg/m^3 on a momentary basis and 0.01 mg/m^3 on an average daily basis. In addition, several states have set guidelines or standards for carbaryl in ambient air[60] ranging from 3.5 μg/m^3 (Pennsylvania) to 11.9050 μg/m^3 (Kansas) to 50 μg/m^3 (North Dakota) to 80 μg/m^3 (Virginia) to 100 μg/m^3 (Connecticut) to 119 μg/m^3 (Nevada).

Determination in Air: OSHA versatile sampler-2; Reagent; High-pressure liquid chromatography/Ultraviolet detection; IV NIOSH Method #5601; also NIOSH Method #5006.

Permissible Concentration in Water: A no-adverse effect level in drinking water has been calculated as 0.574 mg/l by NAS/NRC. The UNEP/USSR joint project[43] has set a MAC of 0.1 mg/l in water used for domestic purposes and 0.0005 mg/l in water bodies used for fishery purposes.

Further, several states have set guidelines for carbaryl in drinking water[61] ranging from 10 μg/l (Wisconsin) to 60 μg/l (California) to 164 μg/l (Maine) to 574 μg/l (Kansas). See Regulatory section for Canada drinking water level.

Routes of Entry: Inhalation, skin contact or eye contact, skin absorption.

Harmful Effects and Symptoms

Short Term Exposure: Carbaryl irritates the eyes, skin, and respiratory tract. The hot liquid may cause severe skin burns. The substance may affect the nervous system, resulting in convulsions and respiratory failure. The effects may be delayed. The oral LD_{50} rat is 250 mg/kg (moderately toxic). Single doses of up to about 140 mg (0.005 oz) have been reported to cause no effect. However, a single dose of about 200 mg has caused stomach pain and excessive sweating. Individual responses may vary. Several milliliters (0.1 fluid oz) of an 80% solution of carbaryl has caused nausea, salivation, headache, tremors, and excessive tearing. 500 ml (1 pint) of a 80% solution has resulted in death.

Long Term Exposure: The major health problem associated with occupational exposure to Carbaryl is related to its inhibition of the enzyme cholinesterase in the central, autonomic and peripheral nervous systems. The inhibition of cholinesterase allows acetylcholine to accumulate at these sites and thereby leads to over stimulation of innervated organs. The signs and symptoms observed as a consequence of exposure to carbaryl in the workplace environment are manifestations of excessive cholinergic stimulation, e.g., nausea, vomiting, mild abdominal cramping, dimness of vision, dizziness, headache, difficulty in breathing, and weakness. Carbaryl may affect the kidneys and nervous system. It may cause mutations and be a teratogen in humans. There is limited evidence that it reduces fertility in both males and females.

Points of Attack: Respiratory system, skin, central nervous system, cardiovascular system.

Medical Surveillance: NIOSH recommends that workers subject to Carbaryl exposure have comprehensive preplacement medical examinations, with subsequent annual medical surveillance. If symptoms develop or overexposure has occurred, the following may be useful: Kidney function tests. Exam of the nervous system. If done within 2-3 hours after exposure, serum and RBC cholinesterase levels may be helpful. Levels can return to normal before the exposed person feels well.

First Aid: If this chemical gets into the eyes, remove any contact lenses at once and irrigate immediately for at least 15 minutes, occasionally lifting upper and lower lids. Seek medical attention immediately. If this chemical contacts the skin, remove contaminated clothing and wash immediately with soap and water. Seek medical attention immediately. If this chemical has been inhaled, remove from exposure, begin rescue breathing (using universal precautions) if breathing has stopped and CPR if heart action has stopped. Transfer promptly to a medical facility. When this chemical has been swallowed, get medical attention. Give large quantities of water and induce vomiting. Do not make an unconscious person vomit.

Personal Protective Methods: Wear appropriate clothing to prevent repeated or prolonged skin contact. Wear eye protection to prevent any reasonable probability of eye contact. Employees should wash promptly when skin is contaminated. Work clothing should be changed daily if it is possible that clothing is contaminated. Remove nonimpervious clothing promptly if contaminated.

Any employee whose work involves likely exposure of the skin to Carbaryl or Carbaryl formulations, e.g., mixing of formulating, shall wear full-body coveralls or the equivalent, impervious gloves, i.e., highly resistant to the penetration of Carbaryl, impervious footwear and when there is danger of Carbaryl coming in contact with the eyes, goggles or a face shield.

Any employee engaged in field application of carbaryl shall be provided with, and required to wear, the following protective clothing and equipment: goggles, full-body coveralls, impervious footwear, and a protective head covering.

Employees working as flaggers in the aerial application of Carbaryl shall be provided with, and required to wear, full-body coveralls or waterproof rainsuits, protective head coverings, impervious gloves and impervious footwear. Significant engineering controls are recommended for this chemical in NIOSH Criteria Document #77-107.

Respirator Selection: Engineering controls should be used, wherever feasible to maintain Carbaryl concentrations below the prescribed limits, and respirators should only be used in certain nonroutine or emergency situations. During certain agricultural applications, however, respirators must be used. NIOSH/OSHA: *Up to 50 mg/m³:* SA (any supplied-air respirator).* *Up to 100 mg/m³:* SA:CF (any supplied-air respirator operated in a continuous-flow mode); or SCBAF (any self-contained breathing apparatus with a full facepiece); or SAF (any supplied-air respirator with a full facepiece). *Emergency or planned entry into unknown concentrations or IDLH conditions:* SCBAF: PD,PP (any self-contained breathing apparatus that has a full facepiece and is operated in a pressure-demand or other positive-pressure mode); or SAF:PD,PP:ASCBA (any supplied-air respirator that has a full facepiece and is operated in a pressure-demand or other positive-pressure mode in combination. *Escape:* GMFOVHiE [any air-purifying, full-facepiece respirator (gas mask) with a chin-style, front- or back-mounted organic vapor canister having a high-efficiency particulate filter]; or SCBAE (any appropriate escape-type, self-contained breathing apparatus).

* Substance reported to cause eye irritation or damage; may require eye protection.

Storage: Prior to working with Carbaryl you should be trained on its proper handling and storage. Store in tightly closed containers in a cool, well-ventilated area. Carbaryl must be stored to avoid contact with strong oxidizers (such as chlorine, bromine, and fluorine) since violent reactions occur. Sources of ignition such as smoking and open flames are prohibited where Carbaryl is used, handled, or stored in a manner that could create a potential fire or explosion hazard.

Shipping: Carbamate pesticides, solid, toxic, n.o.s., fall in Hazard Class 6.1, Carbaryl falls in Packing Group III and requires a "Keep away from Food" label.

Spill Handling: Evacuate persons not wearing protective equipment from area of spill or leak until clean-up is complete. Remove all ignition sources. If spill involves a liquid containing Carbaryl, absorb liquids in vermiculite, dry sand, earth, peat, carbon, or a similar material and deposit in sealed containers. Collect powdered material in the most convenient and safe manner and deposit in sealed containers. Ventilate area after clean-up is complete. It may be necessary to contain and dispose of this chemical as a hazardous waste. If material or contaminated runoff enters waterways, notify downstream users of potentially contaminated waters. Contact your Department of Environmental Protection or your regional office of the federal EPA for specific recommendations. If employees are required to clean-up spills, they must be properly trained and equipped. OSHA 1910.120(q) may be applicable.

Fire Extinguishing: Carbaryl is a non-combustible solid but may be dissolved in flammable liquids. Poisonous gases including nitrogen oxides are produced in fire. Use dry chemical, carbon dioxide, or foam extinguishers. Containers may explode in fire. Storage containers and parts of containers may rocket great distances, in many directions. If material or contaminated runoff enters waterways, notify downstream users of potentially contaminated waters. Notify local health and fire officials and pollution control agencies. From a secure, explosion-proof location, use water spray to cool exposed containers. If cooling streams are ineffective (venting sound increases in volume and pitch, tank discolors, or shows any signs of deforming), withdraw immediately to a secure position. If employees are expected to fight fires, they must be trained and equipped in OSHA 1910.156.

Disposal Method Suggested: Incineration. Submit to alkaline hydrolysis before disposal.[22] In accordance with 40CFR165 recommendations for the disposal of pesticides and pesticide containers. Must be disposed properly by following package label directions or by contacting your state pesticide or environmental control agency or by contacting your regional EPA office.

References

National Institute for Occupational Safety and Health, Criteria for a Recommended Standard: Occupational Exposure to Carbaryl, NIOSH Document No. 77–107 (1977)

Sax, N. I., Ed., "Dangerous Properties of Industrial Materials Report," 1, No. 5, 45–46 (1981) and 3, No. 6, 42–48 (1983)

U.S. Environmental Protection Agency, "Health Advisory: Carbaryl," Washington, DC, Office of Drinking Water (August 1987)

New Jersey Department of Health and Senior Services and Senior Services, "Hazardous Substance Fact Sheet: Carbaryl," Trenton, NJ (September 1996)

New Jersey Department of Health and Senior Services and Senior Services, "Chemical Fact Sheet: Carbaryl," Albany, NY, Bureau of Toxic Substance Assessment (March 1986)

Carbofuran

Molecular Formula: $C_{12}H_{15}NO_3$

Synonyms: A13-27164; Bay 70143; Bay 704143; Bay 78537; 7-Benzofuranol, 2,3-dihydro-2,2-dimethyl-, methylcarbamate; Brifur; Carbamic acid, methyl-, 2,2-dimethyl-2,3-dihydrobenzofuran-7-yl ester; Carbofurano (Spanish); Carbosip 5G; Chinufur; Crisfuran; Curaterr; D 1221; 2,3-Dihydro-2,2-dimethyl-7-benzofuranol *N*-methylcarbamate; 2,3-Dihydro-2,2-dimethyl-7-benzofuranol methylcarbamate; 2,3-Dihydro-2,2-dimethylbenzofuran-7-yl methylcarbamate; 2,3-Dihydro-2,2-dimethylbenzofuranyl-7 *N*-methylcarbamate; 2,2-Dimethyl-7-coumaranyl *N*-methylcarbamate; 2,2-Dimethyl-2,2-dihydrobenzofuranyl-7 *N*-methylcarbamate; 2,2-Dimethyl-2,3-dihydro-7-benzofuranyl *N*-methylcarbamate; ENT 27,164; FMC 10242; Furadan®; Furadan® 10G; Furadan® 3G; Furadan® 4F; Furadan® G; Furodan®; Kenofuran; Methyl carbamic acid 2,3-dihydro-2,2-dimethyl-7-benzofuranyl ester; Nex; NIA-10242; Niagara 10242; Niagra 10242; Niagra NIA-10242; NSC 167822; Pillarfuran; Yaltox

CAS Registry Number: 1563-66-2

RTECS Number: FB9450000

DOT ID: UN 2757

EEC Number: 006-026-00-9

Regulatory Authority

- Very Toxic Substance (World Bank)[15]
- Air Pollutant Standard Set (ACGIH)[1] (HSE)[33] (OSHA)[58] (Several States)[60]
- CLEAN WATER ACT: Section 311 Hazardous Substances/RQ 40CFR117.3 (same as CERCLA, see below)
- RCRA 40CFR268.48; 61FR15654, Universal Treatment Standards: Wastewater (mg/l), 0.006; Nonwastewater (mg/kg), 0.14
- SAFE DRINKING WATER ACT: MCL, 0.04 mg/l; MCLG, 0.04 mg/l; Regulated chemical (47 FR 9352)
- SUPERFUND/EPCRA 40CFR355, Appendix B Extremely Hazardous Substances: TPQ = 10/10,000 lb (4.54/4,540 kg)
- SUPERFUND/EPCRA 40CFR302.4 Reportable Quantity (RQ): CERCLA, 10 lb (4.54 kg)

- EPCRA Section 313 Form R *de minimis* concentration reporting level: 1.0%
- U.S. DOT Regulated Marine Pollutant (49CFR172.101, Appendix B)
- Canada: Drinking water quality, 0.09 mg/l MAC

Cited in U.S. State Regulations: Alaska (G), Arizona (W), California (G), Connecticut (A), Florida (G), Illinois (G), Kansas (W), Maine (G), Massachusetts (G, W), Michigan (G), Minnesota (W), Nevada (A), New Hampshire (G), New Jersey (G), New York (W), North Dakota (A), Pennsylvania (G), Rhode Island (G), Virginia (A),. West Virginia (G), Wisconsin (W).

Description: Carbofuran, $C_{12}H_{15}NO_3$, is white, odorless crystalline solid. Freezing/Melting point = 150 – 152°C. Soluble in water. Log P_{ow} (octanol/water partition coefficient) = 2.32.

Potential Exposure: Those involved in the manufacture, formulation and application of this insecticide, acaricide and nematocide.

Incompatibilities: Alkaline substances, acid, strong oxidizers such as perchlorates, peroxides, chlorates, nitrates, permanganates.

Permissible Exposure Limits in Air: NIOSH and ACGIH recommend an airborne exposure limit of 0.1 mg/m^3 TWA. This is also the British HSE, Israeli, Australian, Mexican, and Canadian Provincial (Alberta, British Columbia, Ontario, and Quebec) limit, *and* Alberta's STEL 0.3 mg/m^3. Several States have set guidelines or standards for carbofuran in ambient air[60] ranging from 1.0 μg/m^3 (North Dakota) to 1.6 μg/m^3 (Virginia) to 2.0 μg/m^3 (Connecticut and Nevada).

Permissible Concentration in Water: EPA[47] has determined one-day, ten-day and longer-term health advisories for a 10 kg. child of 50 μg/l of Carbofuran. The longer term (1 year) value for a 70 kg. adult is 0.18 mg/l or 180 μg/l. A lifetime health advisory for a 70 kg. adult has been determined to be 36 μg/l of Carbofuran. Most recently, EPA has proposed a limit of 40 μg/l in drinking water.[62]

Further, several states have set guidelines for carbofuran in drinking water[61] ranging from 10 μg/l (Massachusetts) to 15 μg/l (New York) to 36 μg/l (Arizona and Minnesota) to 50 μg/l (Kansas and Wisconsin).

Determination in Water: Analysis of Carbofuran is by a high performance liquid chromatographic procedure used for the determination of N-methyl carbamoyloximes and N-methylcarbamates in drinking water (U.S. EPA 1984). In this method, the water sample is filtered and a 400 μl aliquot is injected into a reverse phase HPLC column. Separation of compounds is achieved using gradient elution chromatography. After elution from the HPLC column, the compounds are hydrolyzed with sodium hydroxide. The methylamine formed during hydrolysis is reacted with o-phthalaldehyde (OPA) to form a fluorescent derivative which is detected using a fluorescence detector. The method detection limit has been estimated to be approximately 0.9 μg/l for carbofuran.

Routes of Entry: Inhalation, ingestion, skin contact.

Harmful Effects and Symptoms

Short Term Exposure: Carbofuran may affect the nervous system, resulting in convulsions and respiratory failure. Cholinesterase inhibitor. Exposure may result in death. The effects may be delayed and exposed personnel should be kept under medical observation. Symptoms include headache, giddiness, blurred vision weakness; nausea, cramps, diarrhea, chest discomfort, sweating, contraction of pupils, tearing; salivation, blue lips, lungs and abdomen fill with fluid, convulsions, coma, loss of reflexes and sphincter control. This material is extremely poisonous. The LD_{50} rat is 5.3 mg/kg. May be fatal if swallowed, inhaled, or absorbed through skin. Contact may burn skin or eyes. Probable lethal oral dose to humans 5 – 50 mg/kg or 7 drops to 1 teaspoon for 150 lb person.

Long Term Exposure: The major health problem associated with occupational exposure to Carbofuran is related to its inhibition of the enzyme cholinesterase in the central, autonomic and peripheral nervous systems. The inhibition of cholinesterase allows acetylcholine to accumulate at these sites and thereby leads to over stimulation of innervated organs. The signs and symptoms observed as a consequence of exposure to Carbofuran in the workplace environment are manifestations of excessive cholinergic stimulation, e.g., nausea, vomiting, mild abdominal cramping, dimness of vision, dizziness, headache, difficulty in breathing, and weakness. Carbofuran may affect the immune system.

Points of Attack: Central nervous system, peripheral nervous system, blood cholinesterase.

Medical Surveillance: Before starting work, at regular times after that, and if any symptoms develop, or overexposure occurs, the following is recommended: serum and red blood cell cholinesterase levels (a special test for the substance in the body that Carbofuran affects). For this substance these tests are accurate only if done within about two hours of exposure.

First Aid: If this chemical gets into the eyes, remove any contact lenses at once and irrigate immediately for at least 15 minutes, occasionally lifting upper and lower lids. Seek medical attention immediately. If this chemical contacts the skin, remove contaminated clothing and wash immediately with soap and water. Seek medical attention immediately. If this chemical has been inhaled, remove from exposure, begin rescue breathing (using universal precautions) if breathing has stopped and CPR if heart action has stopped. Transfer promptly to a medical facility. When this chemical has been swallowed, get medical attention. Give large quantities of water and induce vomiting. Do not make an unconscious person vomit. Effects may be delayed; keep victim under observation.

Personal Protective Methods: Wear protective gloves and clothing to prevent any reasonable probability of skin contact. Safety equipment suppliers/manufacturers can provide recommendations on the most protective glove/clothing material for your operation. All protective clothing (suits, gloves, footwear, headgear) should be clean, available each day, and put on before work. Contact lenses should not be worn when working with this chemical. Wear splash or dust-proof chemical goggles and face shield unless full facepiece respiratory protection is worn. Employees should wash immediately with soap when skin is wet or contaminated. Provide emergency showers and eyewash.

Respirator Selection: Where the potential exists for exposures over 0.1 mg/m^3, use a MSHA/NIOSH approved respirator with a pesticide cartridge. More protection is provided by a full facepiece respirator than by a half-mask respirator, and even greater protection is provided by a powered-air purifying respirator. Where the potential for high exposures exists, use a MSHA/NIOSH approved supplied-air respirator with a full facepiece operated in the positive pressure mode or with a full facepiece, hood, or helmet in the continuous flow mode, or use a MSHA/NIOSH approved self-contained breathing apparatus with a full facepiece operated in pressure-demand or other positive pressure mode.

Storage: Prior to working with Carbofuran you should be trained on its proper handling and storage. Carbofuran must be stored to avoid contact with acids and strong oxidizers (such as perchlorates, peroxides, chlorates, nitrates, and permanganates). Store in tightly closed containers in a cool, well-ventilated area.

Shipping: Carbamate insecticides, solid, toxic, n.o.s., have a DOT label requirement of "Poison." The UN/DOT Hazard Class is 6.1, the Packing Group is II.[19][20]

Spill Handling: Evacuate persons not wearing protective equipment from area of spill or leak until clean-up is complete. Remove all ignition sources. Collect powdered material in the most convenient and safe manner and deposit in sealed containers for reclamation or for disposal in an approved facility. Absorb liquid containing carbofuran in vermiculite, dry sand, earth, or similar material. Ventilate area after clean-up is complete. It may be necessary to contain and dispose of this chemical as a hazardous waste. If material or contaminated runoff enters waterways, notify downstream users of potentially contaminated waters. Contact your Department of Environmental Protection or your regional office of the federal EPA for specific recommendations. If employees are required to clean-up spills, they must be properly trained and equipped. OSHA 1910.120(q) may be applicable.

Fire Extinguishing: Carbofuran itself does not burn. The substance decomposes on heating producing toxic fumes including nitrogen oxides. Dike fire control water for later disposal, do not scatter the material. Stay at maximum distance from fire. Extinguish fire using an agent suitable for the type of surrounding fire; use dry chemical, carbon dioxide, water spray, or foam extinguishers. If material or contaminated runoff enters waterways, notify downstream users of potentially contaminated waters. Notify local health and fire officials and pollution control agencies. From a secure, explosion-proof location, use water spray to cool exposed containers. If cooling streams are ineffective (venting sound increases in volume and pitch, tank discolors, or shows any signs of deforming), withdraw immediately.

Disposal Method Suggested: Alkaline hydrolysis is the recommended mode of disposal.[22] In accordance with 40CFR165 recommendations for the disposal of pesticides and pesticide containers. Must be disposed properly by following package label directions or by contacting your state pesticide or environmental control agency or by contacting your regional EPA office.

References

U.S. Environmental Protection Agency, "Chemical Profile: Carbofuran," Washington, DC, Chemical Emergency Preparedness Program (November 30, 1987)

New Jersey Department of Health and Senior Services and Senior Services, "Hazardous Substance Fact Sheet: Carbofuran," Trenton, NJ (April 1998)

U.S. Environmental Protection Agency, "Preliminary Determination to Cancel Registrations of Carbofuran Products," Federal Register 54, No. 15, 3744-3754 (January 25, 1989)

Carbon Black

Molecular Formula: C

Synonyms: Acetylene black; Channel black; C.I. pigment black 7; Elemental carbon; Furnace black; Lamp black; Thermal black

CAS Registry Number: 1333-86-4

RTECS Number: FF5800000

DOT ID: UN 1361

Regulatory Authority

- Carcinogen (ACGIH)[1]
- Air Pollutant Standard Set (Australia) (ACGIH)[1] (HSE)[33] (Israel) (Japan)[35](Mexico) (UNEP)[43] (OSHA)[58] (Several States)[60] (Several Canadian Provinces)
- Water Pollution Standard Proposed (UNEP)[43]
- Canada, WHMIS, Ingredients Disclosure List

Cited in U.S. State Regulations: Alaska (G), California (G), Connecticut (A), Illinois (G), Maine (G), Massachusetts (G), Minnesota (G), Nevada (A), New Hampshire (G), New Jersey (G), New York (G, A), North Dakota (A), Pennsylvania (G), Rhode Island (G), Virginia (A), West Virginia (G).

Description: Carbon black (substantially elemental carbon), C, it is a black, odorless solid Freezing/Melting point = about

3,550°C. Insoluble in water. Combustible solid that may contain flammable hydrocarbons.

Potential Exposure: Workers in carbon black production or in its use in rubber compounding, ink manufacture or paint manufacture, plastics compounding, dry-cell battery manufacture.

Incompatibilities: Carbon blacks containing over 8% volatiles may pose an explosion hazard. Dust can form an explosive mixture in air. Strong oxidizers such as chlorates, bromates, nitrates.

Permissible Exposure Limits in Air: The OSHA legal limit[58] and ACGIH value[1] is 3.5 mg/m^3 TWA. The limit is the same in Australia, Israel, Mexico, United Kingdom (HSE) and the Canadian provinces of Alberta, British Columbia, Ontario, and Quebec, and the STEL value is 7 mg/m^3 in the UK and the forementioned Canadian provinces. NIOSH recommends that exposure to carbon black (as an occupational carcinogen) be limited to the lowest feasible concentrations. Also, NIOSH recommended airborne exposure limit is 0.1 mg (PHA)/m^3 [Carbon black in the presence of polycyclic aromatic hydrocarbons (PAH)]. The NIOSH IDLH is 1,750 mg/m^3. The former USSR/UNEP joint project[43] has set a MAC in ambient air in residential areas of 0.15 mg/m^3 on a momentary basis and 0.005 mg/m^3 on an average daily basis. Japan[35] has set a workplace TWA of 1.0 mg/m^3 (inhalable dust) and 4.0 mg/m^3 (total dust). In addition, several states have set guidelines or standards for carbon black in ambient air[60] ranging from 117 μg/m^3 (New York) to 35 μg/m^3 (North Dakota) to 50 μg/m^3 (Virginia) to 70 μg/m^3 (Connecticut) to 83 μg/m^3 (Nevada).

Determination in Air: Filtration from air is followed by gravimetric analysis as described in NIOSH Method #5000.[18]

Permissible Concentration in Water: The former USSR/UNEP joint project[43] has set a MAC in water bodies used for domestic purposes of 1.0 mg/l.

Routes of Entry: Inhalation, skin and/or eye contact.

Harmful Effects and Symptoms

Short Term Exposure: Inhalation may cause irritation to respiratory tract. Skin contact may cause irritation. Eye contact may cause irritation.

Ingestion: Animal studies show that toxic effects are unlikely, although Carbon black contains several substances that are toxic and known carcinogens.

Long Term Exposure: Exposure to levels well above 3.5 mg/m^3 for several months may result in damage to the skin and nails, temporary or permanent damage to the lungs and breathing passages, and adversely affect the heart. Carbon Black containing PAH greater than 0.1% should be considered a suspect carcinogen. Lungs may be affected by repeated or prolonged exposure at very high concentrations: Some Carbon blacks may contain compounds which are carcinogenic and as organic extracts of these have been classified as possibly carcinogenic to humans, special care should be taken to avoid exposure to such extracts. Lung effects remain controversial and may be due to contaminants. It is probable that minor effects reported are non-specific effects associated with exposure to nuisance dusts in general. Polyaromatic hydrocarbons (PAH) are reportedly present in some carbon blacks. Depending on the process of manufacture, there are variations in their chemical compositions.

Points of Attack: Eyes, skin, respiratory system.

Medical Surveillance: For those with frequent or potentially high exposure (half the TLV or greater) the following are recommended before beginning work and at regular times after that: Chest x-ray (to be read by a special NIOSH "B reader" radiologist); Lung function tests.

First Aid: If this chemical gets into the eyes, remove any contact lenses at once and irrigate immediately for at least 15 minutes, occasionally lifting upper and lower lids. Seek medical attention immediately. If this chemical contacts the skin [and PAH contamination is present], remove contaminated clothing and wash immediately with soap and water. Seek medical attention immediately. If this chemical has been inhaled, remove from exposure, begin rescue breathing (using universal precautions) if breathing has stopped and CPR if heart action has stopped. Transfer promptly to a medical facility. When this chemical has been swallowed, get medical attention. Give large quantities of water and induce vomiting. Do not make an unconscious person vomit.

Personal Protective Methods: Wear protective gloves and clothing to prevent any reasonable probability of skin contact. Safety equipment suppliers/manufacturers can provide recommendations on the most protective glove/clothing material for your operation. All protective clothing (suits, gloves, footwear, headgear) should be clean, available each day, and put on before work. Contact lenses should not be worn when working with this chemical. Wear dust-proof chemical goggles and face shield unless full facepiece respiratory protection is worn. Employees should wash immediately with soap when skin is wet or contaminated. Provide emergency showers and eyewash.

Respirator Selection: NIOSH/OSHA: *17.5 mg/m^3:* DM (any dust and mist respirator). *35 mg/m^3:* DMXSQ (any dust and mist respirator except single-use and quarter mask respirators); or SA (any supplied-air respirator). *87.5 mg/m^3:* SA:CF (any supplied-air respirator operated in a continuous-flow mode); or PAPRDM (any powered, air-purifying respirator with a dust and mist filter); or HiEF [any air-purifying, full-facepiece respirator with a high-efficiency particulate filter)]; or PAPRTHiE (any powered, air-purifying respirator with a tight-fitting facepiece and a high-efficiency particulate filter); or SCBAF (any self-contained breathing apparatus with a full facepiece); or SAF (any supplied-air respirator with a full facepiece); or SA:PD,PP (any supplied-air respirator operated in a pressure-demand or other positive-pressure mode).

Emergency or planned entry into unknown concentrations or IDLH conditions: SCBAF:PD,PP (any self-contained breathing apparatus that has a full facepiece and is operated in a pressure-demand or other positive-pressure mode); or SAF:PD,PP:ASCBA (any supplied-air respirator that has a full facepiece and is operated in a pressure-demand or other positive-pressure mode in combination with an auxiliary, self-contained breathing apparatus operated in a pressure-demand or other positive pressure mode. *Escape:* HiEF (any air-purifying, full-facepiece respirator with a high-efficiency particulate filter); or SCBAE (any appropriate escape-type, self-contained breathing apparatus).

In presence of polycyclic aromatic hydrocarbons NIOSH: *At any detectable concentration:* SCBAF:PD,PP (any MSHA/NIOSH approved self-contained breathing apparatus that has a full facepiece and is operated in a pressure-demand or other positive-pressure mode); or SAF:PD,PP:ASCBA (any supplied-air respirator that has a full facepiece and is operated in a pressure-demand or other positive-pressure mode in combination with an auxiliary, self-contained breathing apparatus operated in a pressure-demand or other positive pressure mode). *Escape:* HiEF [any air-purifying, full-facepiece respirator (gas mask) with a chin-style, front- or back-mounted organic vapor canister having a high-efficiency particulate filter]; or SCBAE (any appropriate escape-type, self-contained breathing apparatus).

Storage: Prior to working with carbon black you should be trained on its proper handling and storage. Carbon black must be stored to avoid contact with chlorates, bromates, and nitrates since violent reactions occur. Sources of ignition such as smoking and open flames are prohibited where Carbon black is used, handled, or stored in a manner that could create a potential fire or explosion hazard. If Carbon black contains more than 0.1% PAHs it should be used, handled and stored in a regulated area as a carcinogen. A regulated, marked area should be established where this chemical is handled, used, or stored in compliance with OSHA standard 1910.1045.

Shipping: Carbon in various forms must carry a "Spontaneously Combustible" label. It falls in DOT Hazard Class 4.2 and Packing Group III. Shipment by passenger aircraft or railcar or even by cargo aircraft is forbidden.

Spill Handling: Evacuate persons not wearing protective equipment from area of spill or leak until clean-up is complete. Remove all ignition sources. Collect powdered material in the most convenient and safe manner and deposit in sealed containers. Ventilate area after clean-up is complete. It may be necessary to contain and dispose of this chemical as a hazardous waste. If material or contaminated runoff enters waterways, notify downstream users of potentially contaminated waters. Contact your Department of Environmental Protection or your regional office of the federal EPA for specific recommendations. If employees are required to clean-up spills, they must be properly trained and equipped. OSHA 1910.120(q) may be applicable.

Fire Extinguishing: Carbon Black will ignite and burn slowly. Use dry chemical, sand, water spray, or foam extinguishers. If employees are expected to fight fires, they must be trained and equipped in OSHA 1910.156.

Disposal Method Suggested:

Dump into a landfill or incinerate as a slurry.[22]

References

National Institute for Occupational Safety and Health, Information Profiles on Potential Occupational Hazards: Carbon Black, Report PB-276,678, Rockville, MD (October 1977)

U.S. Environmental Protection Agency, "Chemical Hazard Information Profile: Carbon Black," Washington, DC (August 1, 1976)

National Institute for Occupational Safety and Health, Criteria for a Recommended Standard: Occupational Exposure to Carbon Black, NIOSH Doc. No. 78–204, Washington, DC (1978)

New Jersey Department of Health and Senior Services and Senior Services, "Hazardous Substance Fact Sheet: Carbon Black," Trenton, NJ (July 1998)

New York State Department of Health, "Chemical Fact Sheet: Carbon Black," Albany, NY, Bureau of Toxic Substance Assessment (January 1986 and Version 2)

Carbon Dioxide

Molecular Formula: CO_2

Synonyms: Acetylene black; Channel black; C.I. pigment black 7; Elemental carbon; Furnace black; Lamp black; Thermal black

CAS Registry Number: 124-38-9

RTECS Number: FF6400000

DOT ID: UN 1013 (gas), 2187 (liquid), 1845 (solid, dry ice).

Regulatory Authority

- Air Pollutant Standard Set (ACGIH)[1] (Australia) (DFG) (HSE)[33] (Israel) (OSHA)[58] (Mexico) (Several Canadian Provinces)

Cited in U.S. State Regulations: Alaska (G), California (A, G), Florida (G), Illinois (G), Maine (G), Massachusetts (G), Minnesota (G), Nevada (A), New Hampshire (G), New Jersey (G), New York (G), Pennsylvania (G), Rhode Island (G), Virginia (A), West Virginia (G).

Description: Carbon dioxide, CO_2, is a colorless, odorless, noncombustible gas. Boiling point = (sublimes). Freezing/Melting point = -78°C (sublimes). It is commonly shipped in the compressed liquid form, and the solid form (dry ice). Soluble in water.

Potential Exposure: Gaseous Carbon dioxide is used to carbonate beverages, as a weak acid in the textile, leather and chemical industries, in water treatment, and in the manufacture of aspirin and white lead, for hardening molds in foundries, in food preservation, in purging tanks and pipelines, as a fire extinguisher, in foams, and in welding. Because it is relatively inert, it is utilized as a pressure medium. It is also used as a propellant in aerosols, to promote plant growth in green houses; it used medically as a respiratory stimulant, in the manufacture of carbonates, and to produce an inert atmosphere when an explosive or flammable hazard exists. The liquid is used in fire extinguishing equipment, in cylinders for inflating life rafts, in the manufacturing of dry ice, and as a refrigerant. Dry ice is used primarily as a refrigerant.

Occupational exposure to Carbon dioxide may also occur in any place where fermentation processes may deplete oxygen with the formation of Carbon dioxide, e.g., in mines, silos, wells, vats, ships' holds, etc.

Incompatibilities: The substance decomposes on heating above 2,000°C producing toxic carbon monoxide. Reacts violently with strong bases and alkali metals. Various metal dusts from chemically active metals such as magnesium, zirconium, titanium, aluminum, chromium and manganese are ignitable and explosive when suspended and heated in carbon dioxide.

Permissible Exposure Limits in Air: The Federal standard[58] and ACGIH TWA value is 5,000 ppm (9,000 mg/m^3). The STEL set by ACGIH and OSHA is 30,000 ppm (54,000 mg/m^3). HSE, DFG, Australian, Israeli, and Mexican TWA is 5,000 ppm and the HSE STEL is 15,000 ppm (27,000 mg/m^3). The NIOSH IDLH level is 40,000 ppm. The Canadian Provincial TWA for Alberta, British Columbia, Ontario, and Quebec are also 5,000 ppm and the STEL for Alberta and B.C. are 15,000, and for Ontario and Quebec are 5,000 ppm. In addition, two states have set guidelines or standards for CO_2 in ambient air[60] which are 150 mg/m^3 (Virginia) and 214 mg/m^3 (Nevada).

Determination in Air: Collection in a bag followed by gas chromatography. See OSHA Method ID 172. Gas collection bag; none; Gas chromatography/Thermal conductivity detector; IV NIOSH Method #6603.

Permissible Concentration in Water: No criteria set.

Routes of Entry: Inhalation of gas.

Harmful Effects and Symptoms

Short Term Exposure: Inhalation of high concentrations of this gas may cause headache, shortness of breath, nausea, vomiting, dizziness, hyperventilation and unconciousness. Rapid evaporation of the liquid or skin contact with "dry ice" may cause frostbite. On loss of containment this liquid evaporates very quickly causing supersaturation of the air with serious risk of suffocation when in confined areas. Carbon dioxide is a simple asphyxiant. Concentrations of 10% (100,000 ppm) can produce unconsciousness and death from oxygen deficiency. A concentration of 5% may produce shortness of breath and headache. Continuous exposure to 1.5% CO_2 may cause changes in some physiological processes. The concentration of Carbon dioxide in the blood affects the rate of breathing.

Long Term Exposure: Long term exposure at levels between 5,000 and 20,000 ppm of Carbon dioxide can affect the acid-base balance, causing acidosis, and can affect calcium metabolism.

Points of Attack: Lungs, skin, cardiovascular system.

Medical Surveillance: Consider evaluation of body calcium and acid-base balance.

First Aid: If dry ice gets into the eyes, get medical attention. If this chemical contacts the skin, get medical attention for frostbite. If a person breathes in large amounts of this chemical, move the exposed person to fresh air at once and perform rescue breathing and CPR if hearth action has stopped. Transfer promptly to a medical facility.

If frostbite has occurred, seek medical attention immediately; do *NOT* rub the affected areas or flush them with water. In order to prevent further tissue damage, do *NOT* attempt to remove frozen clothing from frostbitten areas. If frostbite has *NOT* occurred, immediately and thoroughly wash contaminated skin with soap and water.

Personal Protective Methods: Carbon dioxide is a heavy gas and accumulates at low levels in depressions and along the floor. Generally, adequate ventilation will provide sufficient protection for the worker. Where concentrations are of a high order, supplied air respirators are recommended. Where exposure to the liquefied compressed gas may occur, employees should be provided with special clothing designed to prevent frostbite.

Respirator Selection: NIOSH/OSHA: *40,000 ppm:* SA (any supplied-air respirator); or SCBAF (any self-contained breathing apparatus with a full facepiece). *Emergency or planned entry into unknown concentrations or IDLH conditions:* SCBAF:PD,PP (any self-contained breathing apparatus that has a full facepiece and is operated in a pressure-demand or other positive-pressure mode); or SAF:PD,PP:ASCBA (any supplied-air respirator that has a full facepiece and is operated in a pressure-demand or other positive-pressure mode in combination with an auxiliary, self-contained breathing apparatus operated in a pressure-demand or other positive-pressure mode). *Escape:* SCBAE (any appropriate escape-type, self-contained breathing apparatus).

Storage: Prior to working with carbon dioxide you should be trained on its proper handling and storage. Carbon dioxide must be stored to avoid contact with chemically active metals (such as potassium, sodium, magnesium, and zinc) especially in combination with peroxides since violent reactions occur. Protect containers from physical damage. Procedures for the handling, use and storage of cylinders should be in compliance with OSHA 1910.101 and 1910.169 as with the recommendations of the Compressed Gas Association.

Shipping: Carbon dioxide must carry a "Nonflammable Gas" label. It falls in Hazard Class 2.2.

Spill Handling: If Carbon dioxide gas is leaked, take the following steps. Restrict persons not wearing protective equipment from area of leak until clean-up is complete. Ventilate area of leak to disperse the gas. Stop flow of gas. If source of leak is a cylinder and the leak cannot be stopped in place, remove the leaking cylinder to a safe place in the open air, and repair leak or allow cylinder to empty. If Carbon dioxide liquid or solid is spilled of leaked, take the following steps: Restrict persons not wearing protective equipment from area of spill or leak until clean-up is complete. Ventilate the area of spill or leak.

Fire Extinguishing: Containers may explode in fire. Extinguish fire using an agent suitable for type of surrounding fire. Carbon dioxide itself does not burn. If employees are expected to fight fires, they must be trained and equipped in OSHA 1910.156.

Disposal Method Suggested: Vent to atmosphere.[22]

References

National Institute for Occupational Safety and Health, Criteria for a Recommended Standard: Occupational Exposure to Carbon Dioxide, NIOSH Doc. No. 76–194 (1976)

New Jersey Department of Health and Senior Services and Senior Services, "Hazardous Substance Fact Sheet: Carbon Dioxide," Trenton, NJ (January 1986)

New York State Department of Health, "Chemical Fact Sheet: Carbon Dioxide," Albany, NY, Bureau of Toxic Substance Assessment (May 1986)

Carbon Disulfide

Molecular Formula: CS_2

Synonyms: Carbon bisulfide; Carbon bisulphide; Carbon disulphide; Carbone (sufure de) (French); Carbonio (solfuro di) (Italian); Carbon sulfide; Dithiocarbonic anhydride; Kohlendisulfid (schwefelkohlenstoff) (German); Koolstofdisulfide (zwavelkoolstof) (Dutch); NCI-C04591; Schwefelkohlenstoff (German); Solfuro di carbonio (Italian); Sulphocarbonic anhydride; Weeviltox; Wegla dwusiarczek (Polish)

CAS Registry Number: 75-15-0

RTECS Number: FF6650000

DOT ID: UN 1131

EEC Number: 006-003-00-3

EINECS Number: 200-843-6

Regulatory Authority

- Banned or Severely Restricted (In Agriculture) (Several Countries) (UN)[13]
- Toxic Substance (World Bank)[15]
- Air Pollutant Standard Set (ACGIH)[1] (DFG)[3] (HSE)[33] (OSHA)[58] (Several States)[60] (Australia) (Israel) (Mexico) (Several Canadian Provinces)
- CLEAN AIR ACT: Hazardous Air Pollutants (Title I, Part A, Section 112); Accidental Release Prevention/Flammable substances, (Section 112[r], Table 3), TQ = 20,000 lb (9,080 kg)
- CLEAN WATER ACT: Section 311 Hazardous Substances/RQ 40CFR117.3 (same as CERCLA, see below); Section 313 Water Priority Chemicals (57FR41331, 9/9/92)
- EPA HAZARDOUS WASTE NUMBER (RCRA No.): P022
- RCRA, 40CFR261, Appendix 8 Hazardous Constituents
- RCRA 40CFR268.48; 61FR15654, Universal Treatment Standards: Wastewater (mg/l), 3.8; Nonwastewater (mg/l), 4.8 TCLP
- RCRA 40CFR264, Appendix 9; TSD Facilities Ground Water Monitoring List. Suggested test method(s) (PQL μg/l): 8240 (5)
- SUPERFUND/EPCRA 40CFR355, Appendix B Extremely Hazardous Substances: TPQ = 10,000 lb (4,540 kg)
- SUPERFUND/EPCRA 40CFR302.4 Reportable Quantity (RQ): CERCLA, 100 lb (45.4 kg)
- EPCRA Section 313 Form R *de minimis* concentration reporting level: 1.0%
- U.S. DOT Regulated Marine Pollutant (49CFR172.101, Appendix B) as carbon bisulphide
- Canada, WHMIS, Ingredients Disclosure List; National Pollutant Release Inventory (NPRI)

Cited in U.S. State Regulations: Alaska (G), Arizona (W), California (A, G), Connecticut (A), Florida (G, A), Illinois (G), Kansas (G), Louisiana (G), Maine (G), Maryland (G), Massachusetts (G), Minnesota (G), Nevada (A), New Hampshire (G), New Jersey (G), New York (G, A), North Carolina (A), North Dakota (A), Oklahoma (G), Pennsylvania (G), Rhode Island (G), South Carolina (A), Vermont (G), Virginia (G, A), Washington (G), West Virginia (G), Wisconsin (G).

Description: Carbon disulfide, CS_2, is a highly refractive, flammable liquid boiling at 46°C which in pure form has a sweet odor and in commercial and reagent grades has a foul smell. It can be detected by odor at about 1 ppm but the sense of smell fatigues rapidly and, therefore, odor does not serve as a good warning property. Boiling point = 46.3°C. Freezing/Melting point = -110.8°C. Flash point = -30°C. Autoignition temperature = 90°C. The explosive limits are LEL = 1.3%; UEL = 50%. Hazard Identification (based on NFPA-704 M Rating System): Health 3, Flammability 4, Reactivity 0. Slightly soluble in water.

Potential Exposure: Carbon disulfide is used in the manufacture of viscose rayon, ammonium salts, carbon tetrachloride, carbanilide, xanthogenates, flotation agents, soil disinfectants, dyes, electronic vacuum tubes, optical glass, paints, enamels, paint removers, varnishes, varnish removers, tallow, textiles, explosives, rocket fuel, putty, preservatives, and rubber cement; as a solvent for phosphorus, sulfur, selenium, bromine, iodine, alkali cellulose, fats, waxes, lacquers, camphor, resins, and cold vulcanized rubber. It is also used in degreasing, chemical analysis, electroplating, grain fumigation, oil extraction, and dry-cleaning. It is widely used as a pesticide intermediate.

Incompatibilities: Strong oxidizers, chemically active metals (such as sodium, potassium, zinc), azides, organic amines, halogens. May explosively decompose on shock, friction, or concussion. May explode on heating. The substance may spontaneously ignite on contact with air and on contact with hot surfaces producing toxic fumes of sulphur dioxide. Reacts violently with oxidants to produce oxides of sulfur and carbon monoxide and causing fire and explosion hazard. Attacks some forms of plastic, rubber and coating.

Permissible Exposure Limits in Air: The OSHA legal PEL is 20 ppm TWA for an 8-hour workshift and 30 ppm as an acceptable ceiling, *and* 100 ppm as a maximum peak above the acceptance ceiling concentration not to be exceeded during any 30-minute work period. NIOSH recommended exposure limit is 1 ppm TWA for a 10-hour workshift, and 10 ppm not to be exceeded during any 15-minute work period. The ACGIH recommended TWA value is 10 ppm (30 mg/m^3) for an 8-hour workshift. They add the notation "skin" indicating the possibility of cutaneous absorption. The NIOSH IDLH level is 500 ppm. Australia, DFG, HSE, Israel, and Mexico airborne limit is 10 ppm. The former USSR-UNEP/IRPTC project[43] has set a MAC of 1.0 mg/m^3 in workplace air, of 0.03 mg/m^3 in ambient residential air on a momentary basis, and 0.003 mg/m^3 in residential ambient air on a daily average basis. Further, several states have set guidelines or standards for carbon disulfide in ambient air[60] ranging from 60 μg/m^3 (Connecticut) to 100 μg/m^3 (New York) to 150 μg/m^3 (South Carolina) to 186 μg/m^3 (North Carolina) to 300 μg/m^3 (Florida and North Dakota) to 714 μg/m^3 (Nevada). The WHO[35] has recommended a TWA of 10 mg/m^3 for male workers but a TWA of 3 mg/m^3 for women of fertile age.

Determination in Air: Adsorption on charcoal, workup with benzene, gas chromatographic analysis per NIOSH Method #1600.[18]

Permissible Concentration in Water: In view of the relative paucity of data on the mutagenicity, carcinogenicity, and long-term oral toxicity of carbon disulfide, it was stated that estimates of the effects of chronic oral exposure at low levels cannot be made with any confidence. It was recommended by NAS/NRC that studies to produce such information be conducted before limits in drinking water are established. Now, however, EPA[32] has suggested a permissible ambient goal of 830 μg/l. The former USSR-UNEP/IRPTC project[43] has suggested that limits in drinking water be set on an organoleptic basis and that a MAC of 1 mg/l be set in water bodies used for fishery purposes. Arizona has set a guideline for CS_2 in drinking water of 830 μg/l.[61]

Routes of Entry: Inhalation of vapor which may be compounded by percutaneous absorption of liquid or vapor, ingestion and skin and eye contact.

Harmful Effects and Symptoms

Short Term Exposure: Carbon disulfide irritates the eyes, skin, and respiratory tract. Swallowing the liquid may cause aspiration into the lungs with the risk of chemical pneumonitis. This chemical may affect the central nervous system. In acute poisoning, early excitation of the central nervous system occurs, followed by depression with stupor, restlessness, and unconsciousness. If recovery occurs, the patient usually passes through the after-stage of narcosis, with nausea, vomiting, headache, etc. Also possible are motor disturbances of the bowel, anemia, disturbances of cardiac rhythm, loss of weight, polyuria and menstrual disorders. Severe chronic poisoning may also result in liver degeneration and jaundice. Exposure can cause a loss of consciousness. Exposure far above the PEL may result in death. The probable oral lethal dose for a human is between 0.5 and 5 g/kg or between 1 ounce and 1 pint (or 1 pound) for a 70 kg (150 lb) person. In chronic exposures, the central nervous system is damaged and results in the disturbance of vision and sensory changes at the most common early symptoms. Lowest lethal dose for humans has been reported at 14 mg/kg or 0.98 grams for a 70 kg person. Alcoholics and those suffering from neuropsychic trouble are at special risk.

Long Term Exposure: Repeated or prolonged contact with skin may cause skin allergy, dermatitis, increased cholesterol, artherosclerosis, high blood pressure, heart disease, and damage to the eyes, and other organs from its effects on arteries. Carbon disulfide may affect the central nervous system, resulting in severe neurobehavioural effects, polyneuritis, psychoses. Animal tests show that this substance possibly causes toxic effects upon human reproduction.

Points of Attack: The material affects the central nervous system, cardiovascular system, eyes, kidneys, liver, and skin.

Medical Surveillance: Preplacement and periodic medical examinations should be concerned especially with skin, eyes, central and peripheral nervous system, cardiovascular disease, as well as liver and kidney function. Electrocardiograms should be taken, CS_2, can be determined in expired air, blood, and urine. The iodine-azide test detects carbon disulfide metabolites in the urine, and it may indicate other sulfur compounds. Examination of the nervous system, including metal status.

First Aid: If this chemical gets into the eyes, remove any contact lenses at once and irrigate immediately for at least 15 minutes, occasionally lifting upper and lower lids. Seek medical attention immediately. If this chemical contacts the skin, remove contaminated clothing and wash immediately with soap and water. Seek medical attention immediately. If this chemical has been inhaled, remove from exposure, begin rescue breathing (using universal precautions) if breathing has stopped and CPR if heart action has stopped. Transfer promptly to a medical facility. When this chemical has been swallowed, get medical attention. Give large quantities of water and induce vomiting. Do not make an unconscious person vomit.

Personal Protective Methods: Wear protective gloves and clothing to prevent any reasonable probability of skin contact. Safety equipment suppliers/manufacturers can provide recommendations on the most protective glove/clothing material for your operation. NIOSH recommends the use of VITON and Polyvinyl Alcohol as protective material. All protective clothing (suits, gloves, footwear, headgear) should be clean, available each day, and put on before work. Contact lenses should not be worn when working with this chemical. Wear splash-proof chemical goggles and face shield unless full facepiece respiratory protection is worn. Employees should wash immediately with soap when skin is wet or contaminated. Provide emergency showers and eyewash.

Respirator Selection: NIOSH: *10 ppm:* CCROV [any chemical cartridge respirator with organic vapor cartridge(s)]; or SA (any supplied-air respirator). *25 ppm:* SA:CF (any supplied-air respirator operated in a continuous-flow mode); or PAPROV [any powered, air-purifying respirator with organic vapor cartridge(s)]. *50 ppm:* CCRFOV [any chemical cartridge respirator with a full facepiece and organic vapor cartridge(s)]; or GMFOV [any air-purifying, full-facepiece respirator (gas mask) with a chin-style, front- or back-mounted acid gas canister]; or PAPRTOV [any powered, air-purifying respirator with a tight-fitting facepiece and organic vapor cartridge(s)]; or SCBAF (any self-contained breathing apparatus with a full facepiece); or SAF (any supplied-air respirator with a full facepiece). *500 ppm:* SA:PD,PP (any supplied-air respirator operated in a pressure-demand or other positive-pressure mode). *Emergency or planned entry into unknown concentrations or IDLH conditions:* SCBAF:PD,PP (any self-contained breathing apparatus that has a full facepiece and is operated in a pressure-demand or other positive-pressure mode); or SAF:PD,PP:ASCBA (any supplied-air respirator that has a full facepiece and is operated in a pressure-demand or other positive-pressure mode in combination with an auxiliary, self-contained breathing apparatus operated in a pressure-demand or other positive-pressure mode). *Escape:* GMFOV [any air-purifying, full-facepiece respirator (gas mask) with a chin-style, front- or back-mounted acid gas canister]; or SCBAE (any appropriate escape-type, self-contained breathing apparatus).

Storage: Prior to working with Carbon disulfide you should be trained on its proper handling and storage. Before entering confined space where this chemical may be present, check to make sure that an explosive concentration does not exist. Store in tightly closed containers in a cool, well ventilated area. Metal containers involving the transfer of this chemical should be grounded and bonded. Where possible, automatically pump liquid from drums or other storage containers to process containers. Drums must be equipped with self-closing valves, pressure vacuum bungs, and flame arresters. Use only non-sparking tools and equipment, especially when opening and closing containers of this chemical. Sources of ignition such as smoking and open flames, are prohibited where this chemical is used, handled, or stored in a manner that could create a potential fire or explosion hazard.

Shipping: The DOT-required shipping label if "Flammable Liquid, Poison." Shipment by passenger aircraft or railcar or even by cargo aircraft is forbidden. The UN/DOT Hazard Class is 3 and the Packing Group is I.[19][20]

Spill Handling: Evacuate and restrict persons not wearing protective equipment from area of spill or leak until cleanup is complete. Remove all ignition sources. Establish forced ventilation to keep levels below explosive limit. For small leaks, absorb on paper towels. Evaporate the spills in a safe place, such as a fume hood. Large quantities can be reclaimed or collected and atomized in a suitable combustion chamber equipped with an appropriate effluent gas-cleaning device. If Carbon disulfide is spilled in water, neutralize with agricultural lime, crushed limestone, or sodium bicarbonate. If dissolved, apply activated carbon at ten times the spilled amount. Use mechanical dredges or lifts to remove immobilized masses of pollutants and precipitates. In case of a spill or leak from a drum or smaller container or a small leak from a tank, isolate 50 feet in all directions. In case of a large spill, first isolate 100 feet in all directions, then evacuate in a downwind direction an area 0.2 miles wide and 0.3 miles long. Do not touch spilled material; stop leak if you can do it without risk. Use water spray to reduce vapors. Wear positive pressure breathing apparatus and special protective clothing. It may be necessary to contain and dispose of this chemical as a hazardous waste. If material or contaminated runoff enters waterways, notify downstream users of potentially contaminated waters. Contact your Department of Environmental Protection or your regional office of the federal EPA for specific recommendations. If employees are required to clean-up spills, they must be properly trained and equipped. OSHA 1910.120(q) may be applicable.

Fire Extinguishing: This chemical is a flammable liquid. Note that the ignition temperature is dangerously low: 100°C. Vapors may be ignited by contact with ordinary light bulb; when heated to decomposition, it emits highly toxic fumes of oxides of sulfur. If the vapor concentration excess 2 percent by volume or is unknown, self-contained breathing mask with full face should be used by all persons entering contaminated area to fight fires. Wear special protective clothing. Isolate for ½ mile in all directions if tank car or truck is involved in fire. Use dry chemical, carbon dioxide or other inert gas extinguishers. Foam may be ineffective. Vapors are heavier than air and will collect in low areas. Vapors may travel long distances to ignition sources and flashback. Vapors in confined areas may explode when exposed to fire. Containers may explode in fire. Storage containers and parts of containers may rocket great distances, in many directions. If material or contaminated runoff enters waterways, notify downstream users of potentially contaminated waters. Notify local health and fire officials and pollution control agencies. From a secure, explosion-proof location, use water spray to cool exposed containers. If cooling streams are ineffective (venting sound increases in volume and pitch, tank discolors, or shows any signs of deforming), withdraw immediately to a secure position. If employees are expected to fight fires, they must be trained and equipped in OSHA 1910.156.

Disposal Method Suggested: This compound is a very flammable liquid which evaporates rapidly. It burns with a blue flame to carbon dioxide (harmless) and sulfur dioxide. Sulfur dioxide has a strong suffocating odor; 1,000 ppm in air is lethal to rats. The pure liquid presents an acute fire and explosion hazard. The following disposal procedure is suggested:[22]

All equipment or contact surfaces should be grounded to avoid ignition by static charges. Absorb on vermiculite, sand, or ashes and cover with water. Transfer underwater in buckets to an open area. Ignite from a distance with an excelsior train. If quantity is large, Carbon disulfide may be recovered by distillation and repackaged for use.

References

National Institute for Occupational Safety and Health, Criteria for a Recommended Standard: Occupational Exposure to Carbon Disulfide, NIOSH Doc. No. 77–156 (1977)

World Health Organization, Carbon Disulfide, Environmental Health Criteria No. 10, Geneva (1979)

U.S. Environmental Protection Agency, Carbon Disulfide, Health and Environmental Effects Profile No. 32, Office of Solid Waste, Washington, DC (April 30, 1980)

Sax, N. I., Ed., "Dangerous Properties of Industrial Materials Report," 1, N0. 2, 28–30 (1980); and 3, No. 5, 84–87, New York, Van Nostrand Reinhold Co. (1983)

U.S. Environmental Protection Agency, "Chemical Profile: Carbon Disulfide," Washington, DC, Chemical Emergency Preparedness Program (October 31, 1985)

New Jersey Department of Health and Senior Services and Senior Services, "Hazardous Substance Fact Sheet: Carbon Disulfide," Trenton, NJ (November 1986)

New York State Department of Health, "Chemical Fact Sheet: Carbon Disulfide," Albany, NY, Bureau of Toxic Substance Assessment (May 1986)

Carbon Monoxide

Molecular Formula: CO

Synonyms: Carbone (oxyde de) (French); Carbonic oxide; Carbonio (ossido di) (Italian); Carbon oxide (CO); Exhaust gas; Flue gas; Kohlenmonoxid (German); Koolmonoxyde (Dutch); Oxyde de carbone (French); Wegla tlenek (Polish)

CAS Registry Number: 630-08-0

RTECS Number: FG3500000

DOT ID: UN 1016 (gas); UN 9202 (cryogenic liquid)

EEC Number: 006-001-00-2

Regulatory Authority

- Air Pollutant Standard Set (ACGIH)[1] (DFG)[3] (HSE)[33] (OSHA)[58] (Several States)[60] (Australia) (Israel) (Mexico) (Several Canadian Provinces)
- Canada, WHMIS, Ingredients Disclosure List

Cited in U.S. State Regulations: Alaska (G), Arizona (A), California (A, G). Connecticut (A), Florida (G), Illinois (G), Maine (G), Massachusetts (G), Minnesota (G), Nevada (A), New Hampshire (G, W), New Jersey (G), New York (G), Pennsylvania (G), Rhode Island (G), West Virginia (G).

Description: Carbon monoxide, CO, is a flammable, colorless, odorless, tasteless gas, partially soluble in water. Boiling point = -192°C. Freezing/Melting point = -205°C. Autoignition temperature = 605°C. The explosive limits are: LEL = 12.5%; UEL = 74.0%. Hazard Identification (based on NFPA-704 M Rating System): Health 3, Flammability 4, Reactivity 0. Soluble in water.

Potential Exposure: Carbon monoxide is used in metallurgy as a reducing agent, particularly in the Mond process for nickel; in organic synthesis, especially in the Fischer-Tropsch process for petroleum products and in the oxo reaction; and in the manufacture of metal carbonyls.

It is usually encountered in industry as a waste product of incomplete combustion of carbonaceous material (complete combustion produces CO_2). The major source of CO emission in the atmosphere is the gasoline-powered internal combustion engine.

Special industrial processes which contribute significantly to CO emission are iron foundries, particularly the cupola; fluid catalytic crackers, fluid coking, and moving-bed catalytic crackers in thermal operations in carbon black plants; beehive coke ovens, basic oxygen furnaces, sintering of blast furnace feed in steel mills; and formaldehyde manufacture. There are numerous other operations in which a flame touches a surface that is cooler than the ignition temperature of the gaseous part of the flame where exposure to CO may occur, e.g., arc welding, automobile repair, traffic control, tunnel construction, fire fighting, mines, use of explosives, etc.

Incompatibilities: Forms extremely explosive mixture with air. Strong oxidizers. In the presence of finely dispersed metal powders the substance forms toxic and flammable carbonyls. May react vigorously with oxygen, acetylene, chlorine, fluorine, nitrous oxide.

Permissible Exposure Limits in Air: OSHA,[58] the legal airborne PEL is 50 ppm (55 mg/m^3) TWA for an 8-hour workshift. NIOSH recommends an REL of 35 ppm (40 mg/m^3) TWA for a 10-hour workshift and a ceiling limit of 200 ppm not to be exceeded at any time. The ACGIH recommended TWA value is 25 ppm. The HSE[33] value is 50 ppm and STEL of 300 ppm. The DFG[3] has set a MAK of 30 ppm (33 mg/m^3). Mexico's, Israel's, and Australia's limit value is 50 ppm and STEL of 400 ppm. The former USSR/UNEP joint project[43] has set a MAC in workplace air of 20 mg/m^3 and MAC values for ambient air in residential areas of 3 mg/m^3 on a momentary basis and 1 mg/m^3 on an average daily basis. In addition, some states have set guidelines or standards for CO in ambient air[60] ranging from 10-40 μg/m^3 (Arizona) to 1,310 μg/m^3 (Nevada) to 10,000 μg/m^3 (10 mg/m^3) (Connecticut).

Determination in Air: Gas collection bag; none; Sensor; IV NIOSH Method #6604.

Permissible Concentration in Water: No criteria set, but EPA[32] has suggested a permissible ambient level of 552 μg/l based on health effects.

Routes of Entry: Inhalation of gas.

Harmful Effects and Symptoms

Short Term Exposure: Carbon monoxide may affect the blood, cardiovascular system and central nervous system. Exposure at high levels may result in a loss of consciousness and death. Carbon monoxide combines with hemoglobin to form carboxyhemoglobin which interferes with the oxygen-carrying capacity of blood, resulting in a state of tissue hypoxia. The typical signs and symptoms of acute CO poisoning are headache, dizziness, drowsiness, vomiting, collapse, coma, and death. Initially the victim is pale; late the skin and mucous membranes may be cherry-red in color. Loss of consciousness occurs at about the 50% carboxyhemoglobin level. The amount of carboxyhemoglobin formed is dependent on concentration and duration of CO exposure, ambient temperature, health, and metabolism of the individual. The formation of carboxyhemoglobin is a reversible process. Recovery from acute poisoning usually occurs without sequelae unless tissue hypoxia was severe enough to result in brain cell degeneration. Carbon monoxide at low levels may initiate or enhance deleterious myocardial alterations in individuals with restricted coronary artery blood flow and decreased myocardial lactate production. Severe carbon monoxide poisoning has been reported to permanently damage the extrapyramidal system, including the basal ganglia.

Long Term Exposure: Carbon monoxide may affect the nervous system and the cardiovascular system, causing neurological and cardiac disorders. Suspected to cause reproductive effects such as neurological problems, low birth weight, increased still births, and congenital heart problems. The DFG lists pregnancy risk/fetus probable.

Points of Attack: Central nervous system, lungs, blood, cardiovascular system.

Medical Surveillance: Preplacement and periodic medical examinations should give special attention to significant cardiovascular disease and any medical conditions which could be exacerbated by exposure to CO. Heavy smokers may be at greater risk. Methylene chloride exposure may also cause an increase of carboxyhemoglobin. Smokers usually have higher levels of carboxyhemoglobin than nonsmokers (often 5 – 10% or more). Carboxyhemoglobin levels are reliable indicators of exposure and hazard. Carboxyhemoglobin should be tested within a few hours following exposure to the gas. EKG. Examination of the nervous system. Persons with heart disease should not be exposed to levels of CO above 35 ppm.

First Aid: *Gas:* Move victim to fresh air. Call emergency medical care. Apply artificial respiration if victim is not breathing. Do not use mouth-to-mouth method if victim ingested or inhaled the substance; induce artificial respiration with the aid of a pocket mask equipped with a one-way valve or other proper respiratory medical device. Administer oxygen if breathing is difficult. Remove and isolate contaminated clothing and shoes. In case of contact with substance, immediately flush skin or eyes with running water for at least 20 minutes. In case of contact with liquefied gas, thaw frosted parts with lukewarm water. Keep victim warm and quiet. Keep victim under observation for 24 – 48 hours. Effects of contact or inhalation may be delayed. Ensure that medical personnel are aware of the material(s) involved and take precautions to protect themselves.

Refrigerated liquid: Move victims to fresh air. Call emergency medical care. Apply artificial respiration if victim is not breathing. Administer oxygen if breathing is difficult. Remove and isolate contaminated clothing and shoes. In case of contact with substance, immediately flush skin or eyes with running water for at least 20 minutes. In case of contact with liquefied gas, thaw frosted parts with lukewarm water. Keep victim warm and quiet. Keep victim under observation. Effects of contact or inhalation may be delayed. Ensure that medical personnel are aware of the material(s) involved and take precautions to protect themselves.

Personal Protective Methods: Under certain circumstances where Carbon monoxide levels are not exceedingly high, gas masks with proper canisters can be used for short periods, but are not recommended. In areas with high concentrations, self-contained air apparatus is recommended.

Respirator Selection: NIOSH: *350 ppm:* SA (any supplied-air respirator). *875 ppm:* SA:CF (any supplied-air respirator operated in a continuous-flow mode). *1,500 ppm:* GMFS *end of service life indicator (ESLI) required.* [any air-purifying, full-facepiece respirator (gas mask) with a chin-style, front- or back-mounted canister providing protection against the compound of concern]; or SCBAF (any self-contained breathing apparatus with a full facepiece); or SAF (any supplied-air respirator with a full facepiece). *Emergency or planned entry into unknown concentrations or IDLH conditions:* SCBAF:PD,PP (any self-contained breathing apparatus that has a full facepiece and is operated in a pressure-demand or other positive-pressure mode); or SAF:PD,PP:ASCBA (any supplied-air respirator that has a full facepiece and is operated in a pressure-demand or other positive-pressure mode in combination with an auxiliary, self-contained breathing apparatus operated in a pressure-demand or other positive-pressure mode). *Escape:* GMFS *end of service life indicator (ESLI) required.* [any air-purifying, full-facepiece respirator (gas mask) with a chin-style, front- or back-mounted canister providing protection against the compound of concern]; or SCBAE (any appropriate escape-type, self-contained breathing apparatus).

Storage: Prior to working with Carbon monoxide you should be trained on its proper handling and storage. Before entering confined space where this chemical may be present, check to make sure that an explosive concentration does not exist. Carbon monoxide must be stored to avoid contact with strong oxidizers, such as chlorine or chlorine dioxide, since violent reactions occur. Keep containers in a cool, well ventilated area away from heat, flame, and sunlight. Metal containers involving the transfer of 5 gallons or more of liquid Carbon monoxide should be grounded and bonded. Drums must be equipped with self-closing valves, pressure vacuum bungs and flame arresters. Use only non-sparking tools and equipment, especially when opening and closing containers of Carbon monoxide. Sources of ignition such as smoking and open flames are prohibited where Carbon monoxide is used, handled, or stored. Procedures for the handling, use and storage of cylinders should be in compliance with OSHA 1910.101 and 1910.169 as with the recommendations of the Compressed Gas Association.

Shipping: Carbon Monoxide must carry a "Poison Gas, Flammable Gas" label. It falls in DOT Hazard Class 2.3 and Packing Group III. Shipment by passenger aircraft or railcar or even by cargo aircraft is forbidden.

Spill Handling: Evacuate and restrict persons not wearing protective equipment from area of spill or leak until cleanup is complete. Remove all ignition sources. Establish forced ventilation to keep levels below explosive limit and to disperse the gas. Stop the flow of gas if it can be done safely. If source is a cylinder and the leak cannot be stopped in place, remove the leaking cylinder to a safe place, and repair leak or allow cylinder to empty. Keep this chemical out of confined spaces, such as a sewer, because of the possibility of explosion, unless the sewer is designed to prevent the buildup of explosive concentrations. If employees are required to cleanup spills, they must be properly trained and equipped. OSHA 1910.120(q) may be applicable.

Initial isolation and protective action distances

Distances shown are likely to be affected during the first 30 minutes after materials are spilled and could increase with time. If more than one tank car, cargo tank, portable tank, or large cylinder is involved in the incident is leaking, the protective action distance may need to be increased. You may need to seek emergency information from CHEMTREC at (800) 424-9300 or seek professional environmental engineering assistance from the U.S. EPA Environmental Response Team at (908)548-8730 (24-hour response line).

Small spills (From a small package or a small leak from a large package) UN 1016/UN 9202

First: Isolate in all directions (feet) 100

Then: Protect persons downwind (miles)

Day 0.1

Night 0.1

Large spills (From a large package or from many small packages)

First: Isolate in all directions (feet) 300

Then: Protect persons downwind (miles)

Day 0.1

Night 0.4

Fire Extinguishing: This chemical is a flammable gas that can cause explosion. Use dry chemical extinguishers. Vapors are heavier than air and will collect in low areas. Vapors may travel long distances to ignition sources and flashback. Vapors in confined areas may explode when exposed to fire. Containers may explode in fire. Storage containers and parts of containers may rocket great distances, in many directions. If material or contaminated runoff enters waterways, notify downstream users of potentially contaminated waters. Notify local health and fire officials and pollution control agencies. From a secure, explosion-proof location, use water spray to cool exposed containers. If cooling streams are ineffective (venting sound increases in volume and pitch, tank discolors, or shows any signs of deforming), withdraw immediately to a secure position. If employees are expected to fight fires, they must be trained and equipped in OSHA 1910.156.

Disposal Method Suggested: Incineration.[22] Carbon monoxide can also be recovered from gas mixtures as an alternative to disposal.

References

National Institute for Occupational Safety and Health, Criteria for a Recommended Standard: Occupational Exposure to Carbon Monoxide, NIOSH Doc. No. 73–11,000 (1973)

National Academy of Sciences, Medical and Biologic Effects of Environmental Pollutants: Carbon Monoxide, Washington, DC (1977)

U.S. Environmental Protection Agency, Air Quality Criteria for Carbon Monoxide, Report EPA 600/8-79-022, Environmental Criteria and Assessment Office, Research Triangle Park, NC (1979)

U.S. Environmental Protection Agency, "Carbon Monoxide: Proposed Revisions to the National Ambient Air Quality Standards," Federal Register, 45, No. 161, 55066-84 (August 18, 1980)

World Health Organization, Carbon Monoxide, Environmental Health Criteria No. 13, Geneva, Switzerland (1979)

Sax, N. I., Ed., "Dangerous Properties of Industrial Materials Report," 1, No. 7, 43–45 (1981); and 3, No. 5, 87–89, New York, Van Nostrand Reinhold Co. (1983)

New Jersey Department of Health and Senior Services and Senior Services, "Hazardous Substance Fact Sheet: Carbon Monoxide," Trenton, NJ (May 1998)

New York State Department of Health, "Chemical Fact Sheet: Carbon Monoxide," Albany, NY, Bureau of Toxic Substance Assessment (March 1986 and Version 2)

Carbon Oxysulfide

Molecular Formula: COS

Synonyms: Carbon monoxide monosulfide; Carbon oxide sulfide; Carbon oxygen sulfide; Carbon oxygen sulphide;

Carbon oxysulphide; Carbonyl sulfide-(32)S; Carbonyl sulphide; Oxycarbon sulfide; Oxycarbon sulphide; SCO; Sulfuro de carbonilo (Spanish)

CAS Registry Number: 463-58-1

RTECS Number: FG6475000; FG6400000

DOT ID: UN 2204

Regulatory Authority

- Air Pollutant Standard Set (UNEP)[43] (South Carolina)[60]
- CLEAN AIR ACT: Hazardous Air Pollutants (Title I, Part A, Section 112); Accidental Release Prevention/Flammable substances, (Section 112[r], Table 3), TQ = 10,000 lb (4,540 kg)
- SUPERFUND/EPCRA 40CFR355, Appendix B Extremely Hazardous Substances: TPQ = 10,000 lb (4,540 kg)
- SUPERFUND/EPCRA 40CFR302.4 Reportable Quantity (RQ): CERCLA, 1 lb (0.454 kg)
- EPCRA Section 313 Form R *de minimis* concentration reporting level: 1.0%
- U.S. DOT 49CFR172.101, Inhalation Hazardous Chemical
- Canada, WHMIS, Ingredients Disclosure List

Cited in U.S. State Regulations: California (A, G), Florida (G), Maryland (G), Massachusetts (G), New Hampshire (G), New Jersey (G), Pennsylvania (G), South Carolina (A).

Description: Carbonyl sulfide, COS, is a colorless gas or cold liquid. Boiling point = -50°C. Hazard Identification (based on NFPA-704 M Rating System): Health 3, Flammability 4, Reactivity 1. Flammable limits: LEL = 12%; UEL = 29%.

Potential Exposure: Carbon oxysulfide is an excellent source of usable atomic sulfur, therefore, it can be used in various chemical syntheses, such as the production of episulfides, alkenylthiols, and vinylicthiols. It is also used to make viscose rayon.

It is probable that the largest source of Carbon oxysulfide is as a by-product from various organic syntheses and petrochemical processes.

Carbon oxysulfide is always formed when carbon, oxygen, and sulfur, or their compounds, such as carbon monoxide, carbon disulfide, and sulfur dioxide, are brought together at high temperatures. Hence, carbon, oxysulfide is formed as an impurity in various types of manufactured gases and as a by-product in the manufacture of carbon disulfide. Carbon oxysulfide is also often present in refinery gases.

Incompatibilities: COS can form explosive mixtures with air. Incompatible with strong bases. Contact with strong oxidizers may cause fire and explosions.

Permissible Exposure Limits in Air: No occupational exposure limits have been established in the United States. It should be recognized that COS can be absorbed through the skin, thereby increasing the potential for exposure. The former-USSR/UNEP joint project[43] has set a MAC in workplace air of 10 mg/m^3. South Carolina[60] has set a guideline for COS in ambient air of 12.25 mg/m^3.

Permissible Concentration in Water: No criteria set.

Routes of Entry: Inhalation, absorbed through the skin.

Harmful Effects and Symptoms

Short Term Exposure: Can cause irritation of the eyes, skin and respiratory tract. Contact with the liquefied gas can cause frostbite. Inhalation can cause irritation, coughing and sneezing. High exposure can cause salivation, nausea, vomiting, diarrhea, sweating, weakness, and muscle cramps. It may cause tachycardia or arrhythmia. Higher exposures can cause pulmonary edema, a medical emergency that can be delayed for several hours. This can cause death. COS is an irritant to the lungs and trachea. It depresses the central nervous system. It can be fatal by paralysis of the respiratory system.[24] The acute toxicity of carbon oxysulfide was examined by Japanese workers. Exposure of laboratory animals to this contaminant of coal gas and petroleum gas was associated with pathological changes in the brain, medulla oblongata, liver, kidney, and lung. When rats were placed in chambers containing 0.05 and 0.2 percent Carbon oxysulfide, death occurred in 10 and 0.5 – 1.0 hours, respectively.

Long Term Exposure: High or repeated exposure may affect the nervous system causing headache, dizziness, and confusion with memory problems. May cause brain damage, reduced memory, inability to concentrate, and/or personality changes. COS can cause bronchitis with coughing, phlegm and/or shortness of breath.

Points of Attack: Lungs, brain, central nervous system.

Medical Surveillance: Evaluate the cerebellar, autonomic, and peripheral nervous systems. Brain functions. Test include: EKG, complete nervous system evaluation, chest x-ray following acute overexposure.

First Aid: *Gas:* Move victim to fresh air. Call emergency medical care. Apply artificial respiration if victim is not breathing. Do not use mouth-to-mouth method if victim ingested or inhaled the substance; induce artificial respiration with the aid of a pocket mask equipped with a one-way valve or other proper respiratory medical device. Administer oxygen if breathing is difficult. Remove and isolate contaminated clothing and shoes. In case of contact with substance, immediately flush skin or eyes with running water for at least 20 minutes. In case of contact with liquefied gas, thaw frosted parts with lukewarm water. Keep victim warm and quiet. Keep victim under observation. Effects of contact or inhalation may be delayed. Ensure that medical personnel are aware of the material(s) involved and take precautions to protect themselves.

Refrigerated liquid: Move victims to fresh air. Call emergency medical care. Apply artificial respiration if victim is not breathing. Administer oxygen if breathing is difficult. Remove and isolate contaminated clothing and shoes. In case of contact with substance, immediately flush skin or eyes with running water for at least 20 minutes. In case of contact with liquefied gas, thaw frosted parts with lukewarm water. Keep victim warm and quiet. Keep victim under observation. Effects of contact or inhalation may be delayed. Ensure that medical personnel are aware of the material(s) involved and take precautions to protect themselves. If frostbite has occurred, seek medical attention immediately; do NOT rub the affected areas or flush them with water. In order to prevent further tissue damage, do NOT attempt to remove frozen clothing from frostbitten areas. If frostbite has NOT occurred, immediately and thoroughly wash contaminated skin with soap and water.

Personal Protective Methods: Wear protective gloves and clothing to prevent any reasonable probability of skin contact. Safety equipment suppliers/manufacturers can provide recommendations on the most protective glove/clothing material for your operation. All protective clothing (suits, gloves, footwear, headgear) should be clean, available each day, and put on before work. Contact lenses should not be worn when working with this chemical. Wear non-vented impact resistant goggles when working with gasses. When working with liquid, wear splash-proof chemical goggles and face shield unless full facepiece respiratory protection is worn. Employees should wash immediately with soap when skin is wet or contaminated. Provide emergency showers and eyewash. Wear rubber gloves and coveralls.[24] Where exposure to the liquefied compressed gas may occur, employees should be provided with special clothing designed to prevent frostbite.

Respirator Selection: *At any concentrations above the NIOSH REL, or where there is no REL, at any detectable concentration:* SCBAF:PD,PP (any MSHA/NIOSH approved self-contained breathing apparatus that has a full facepiece and is operated in a pressure-demand or other positive-pressure mode); or SAF:PD,PP:ASCBA (any supplied-air respirator that has a full facepiece and is operated in a pressure-demand or other positive-pressure mode in combination with an auxiliary, self-contained breathing apparatus operated in a pressure-demand or other positive pressure mode).

Storage: Prior to working with Carbon oxysulfide you should be trained on its proper handling and storage. Before entering confined space where this chemical may be present, check to make sure that an explosive concentration does not exist. Carbon oxysulfide must be stored to avoid contact with bases and strong oxidizers since violent reactions occur. Keep containers in a cool, well ventilated area away from heat, flame, and sunlight. Metal containers involving the transfer of 5 gallons or more of liquid carbon oxysulfide should be grounded and bonded. Drums must be equipped with self-closing valves, pressure vacuum bungs and flame arresters. Use only non-sparking tools and equipment, especially when opening and closing containers of COS. Sources of ignition such as smoking and open flames are prohibited where carbon oxysulfide is used, handled, or stored. Procedures for the handling, use and storage of cylinders should be in compliance with OSHA 1910.101 and 1910.169 as with the recommendations of the Compressed Gas Association.

Shipping: Carbonyl sulfide must carry a "Poison Gas, Flammable Gas" label. It falls in Hazard Class 2.3 and Packing Group II. Shipment by passenger aircraft or railcar or even by cargo aircraft is forbidden.[19]

Spill Handling: *Liquid:* Evacuate and restrict persons not wearing protective equipment from area of spill or leak until cleanup is complete. Remove all ignition sources. Establish forced ventilation to keep levels below explosive limit and allow to vaporize. Or, cover the spill with weak hypochlorite solution (up to 15%). After 12 hours, the produce may be neutralized and flushed to a sewer with abundant water. It may be necessary to contain and dispose of this chemical as a hazardous waste. If material or contaminated runoff enters waterways, notify downstream users of potentially contaminated waters. Contact your Department of Environmental Protection or your regional office of the federal EPA for specific recommendations. If employees are required to clean-up spills, they must be properly trained and equipped. OSHA 1910.120(q) may be applicable.

Gas: Evacuate and restrict persons not wearing protective equipment from area of spill or leak until cleanup is complete. Remove all ignition sources. Establish forced ventilation to keep levels below explosive limit and to disperse the gas. Stop the flow of gas if it can be done safely. If source is a cylinder and the leak cannot be stopped in place, remove the leaking cylinder to a safe place, and repair leak or allow cylinder to empty. Keep this chemical out of confined spaces, such as a sewer, because of the possibility of explosion, unless the sewer is designed to prevent the buildup of explosive concentrations. If employees are required to clean-up spills, they must be properly trained and equipped. OSHA 1910.120(q) may be applicable.

Initial isolation and protective action distances

Distances shown are likely to be affected during the first 30 minutes after materials are spilled and could increase with time. If more than one tank car, cargo tank, portable tank, or large cylinder is involved in the incident is leaking, the protective action distance may need to be increased. You may need to seek emergency information from CHEMTREC at (800) 424-9300 or seek professional environmental engineering assistance from the U.S. EPA Environmental Response Team at (908)548-8730 (24-hour response line).

Small spills (From a small package or a small leak from a large package)

First: Isolate in all directions (feet)..... 200

Then: Protect persons downwind (miles)

Day 0.1

Night 0.3

Large spills (From a large package or from many small packages)

First: Isolate in all directions (feet)..... 500

Then: Protect persons downwind (miles)

Day 0.2

Night 1.0

Fire Extinguishing: COS is a flammable gas or liquid. Fire may restart after it has been extinguished. Poisonous gases including hydrogen sulfide and sulfur oxides are produced in fire. Use dry chemical, carbon dioxide, or alcohol foam extinguishers. Vapors are heavier than air and will collect in low areas. Vapors may travel long distances to ignition sources and flashback. Vapors in confined areas may explode when exposed to fire. Containers may explode in fire. Storage containers and parts of containers may rocket great distances, in many directions. If material or contaminated runoff enters waterways, notify downstream users of potentially contaminated waters. Notify local health and fire officials and pollution control agencies. From a secure, explosion-proof location, use water spray to cool exposed containers. If cooling streams are ineffective (venting sound increases in volume and pitch, tank discolors, or shows any signs of deforming), withdraw immediately to a secure position. If employees are expected to fight fires, they must be trained and equipped in OSHA 1910.156.

Disposal Method Suggested: Dissolve in a combustible solvent such as alcohol, benzene, etc. Burn in a furnace with afterburner and scrubber to remove SO_2.[22]

References

National Institute for Occupational Safety and Health, Information Profiles on Potential Occupational Hazards — Ingle Chemicals: Carbon Oxysulfide, pp 34–38, Publ. No. TR79-607, Rockville, MD (December 1979).

New Jersey Department of Health and Senior Services, Hazardous Substance Fact Sheet, Carbonyl Sulfide, Trenton NJ (September 1998).

Carbon Tetrabromide

Molecular Formula: CBr_4

Synonyms: Carbon bromide; Methane, tetrabromide; Methane, tetrabromo-; Tetrabromide methane; Tetrabromomethane

CAS Registry Number: 558-13-4

RTECS Number: FG4725000

DOT ID: UN 2516

Regulatory Authority

- Air Pollutant Standard Set (ACGIH)[1] (HSE)[33] (OSHA)[58] (Several States)[60] (Australia) (Israel) (Mexico) (Several Canadian Provinces)
- Canada, WHMIS, Ingredients Disclosure List

Cited in U.S. State Regulations: Alaska (G), California (A, G), Connecticut (A), Florida (G), Illinois (G), Maine (G), Massachusetts (G), Minnesota (G), Nevada (A), New Hampshire (G), New Jersey (G), North Dakota (A), Pennsylvania (G), Rhode Island (G), Virginia (A), West Virginia (G).

Description: Carbon tetrabromide, CBr_4, is a colorless powder or yellow-brown crystals with a slight odor. Boiling point = 190°C. Freezing/Melting point = 90°C. Very slightly soluble in water.

Potential Exposure: CBr_4 is used in organic synthesis.

Incompatibilities: Incompatible with strong oxidizers, lithium and hexacyclohexyldiilead since violent reactions occur.

Permissible Exposure Limits in Air: The NIOSH recommended airborne exposure limit is 0.1 ppm (1.4 mg/m^3) TWA for a 10-hour workshift and STEL of 0.3 ppm (4.0 mg/m^3). ACGIH has set this same value for an 8-hour workshift. Australia, HSE, Israel, Mexico and Canadian provinces of Alberta, British Columbia, Ontario and Quebec have set this same limit. In addition, several states have set guidelines or standards for CBr_4 in ambient air[60] ranging from 14 – 40 μg/m^3 (North Dakota) to 20 μg/m^3 (Virginia) to 28 μg/m^3 (Connecticut) to 33 μg/m^3 (Nevada).

Permissible Concentration in Water: No criteria set.

Routes of Entry: Inhalation, ingestion.

Harmful Effects and Symptoms

Short Term Exposure: The material is a potent lachrymator even at low concentrations. Carbon tetrabromide is corrosive to the eyes and skin and may cause permanent damage. Inhalation can cause severe irritation of the respiratory tract. Higher exposures can cause pulmonary edema, a medical emergency that can be delayed for several hours. This can cause death. It can affect the nervous system, liver, and kidneys. Exposure to high concentrations may result in unconsciousness.

Long Term Exposure: The substance may damage the liver and kidneys.

Points of Attack: Eyes, skin, respiratory system, liver, kidneys.

Medical Surveillance: For those with frequent or potentially high exposure (half the TLV or greater) the following are recommended before beginning work and at regular times after that: Liver, kidney, and lung function tests. If symptoms develop or overexposure is suspected, the following may be useful: kidney, liver and lung function tests. Consider chest x-ray after acute overexposure.

First Aid: If this chemical gets into the eyes, remove any contact lenses at once and irrigate immediately for at least 15 minutes, occasionally lifting upper and lower lids. Seek medical attention immediately. If this chemical contacts the skin, remove contaminated clothing and wash immediately with soap and water. Seek medical attention immediately. If this chemical has been inhaled, remove from exposure, begin rescue breathing (using universal precautions) if breathing has stopped and CPR if heart action has stopped. Transfer promptly to a medical facility. When this chemical has been swallowed, get medical attention. Give large quantities of water and induce vomiting. Do not make an unconscious person vomit. Medical observation is recommended for 24 – 48 hours after breathing overexposure, as pulmonary edema may be delayed. As first aid for pulmonary edema, a doctor or authorized paramedic may consider administering a corticosteroid spray.

Personal Protective Methods: Wear protective gloves and clothing to prevent any reasonable probability of skin contact. Safety equipment suppliers/manufacturers can provide recommendations on the most protective glove/clothing material for your operation. All protective clothing (suits, gloves, footwear, headgear) should be clean, available each day, and put on before work. Contact lenses should not be worn when working with this chemical. Wear dust-proof chemical goggles and face shield unless full facepiece respiratory protection is worn. Employees should wash immediately with soap when skin is wet or contaminated. Provide emergency showers and eyewash.

Respirator Selection: Where the potential exists for exposures over 0.1 ppm, use a MSHA/NIOSH approved supplied-air respirator with a full facepiece operated in the positive pressure mode or with a full facepiece, hood, or helmet in the continuous flow mode, or use a MSHA/NIOSH approved self-contained breathing apparatus with a full facepiece operated in pressure-demand or other positive pressure mode.

Storage: Prior to working with Carbon tetrabromide you should be trained on its proper handling and storage. Store in tightly closed containers in a cool, well-ventilated area away from oxidizers and other incompatible materials listed above.

Shipping: Carbon tetrabromide must carry a "Keep Away from Food" label. It falls in Hazard Class 6.1 and Packing Group III.

Spill Handling: Evacuate persons not wearing protective equipment from area of spill or leak until clean-up is complete. Remove all ignition sources. Collect powdered material in the most convenient and safe manner and deposit in sealed containers. Ventilate area after clean-up is complete. Absorb liquid containing Carbon Tetrabromide in vermiculite, dry sand, earth, or similar material. It may be necessary to contain and dispose of this chemical as a hazardous waste. If material or contaminated runoff enters waterways, notify downstream users of potentially contaminated waters. Contact your Department of Environmental Protection or your regional office of the federal EPA for specific recommendations. If employees are required to clean-up spills, they must be properly trained and equipped. OSHA 1910.120(q) may be applicable.

Fire Extinguishing: This chemical is a noncombustible liquid. Poisonous gases produced in fire. Use any agent suitable for surrounding fire. If material or contaminated runoff enters waterways, notify downstream users of potentially contaminated waters. Notify local health and fire officials and pollution control agencies. If employees are expected to fight fires, they must be trained and equipped in OSHA 1910.156.

Disposal Method Suggested: Purify by distillation and return to suppliers.

References

New Jersey Department of Health and Senior Services and Senior Services, "Hazardous Substance Fact Sheet: Carbon Tetrabromide," Trenton, NJ (April 1998)

Carbon Tetrachloride

Molecular Formula: CCl_4

Synonyms: Benzinoform; Carbona; Carbon chloride; Carbon tet; Czterochlorek wegla (Polish); ENT 4705; Fasciolin; Flukoids; Freon 10; Halon 104; Katharin; Methane tetrachloride; Methane, tetrachloro-; Necatorina; Necatorine; Perchloromethane; R 10; Tetrachloorkoolstof (Dutch); Tetrachloormetan; Tetrachlorkohlenstoff, tetra (German); Tetrachlormethan (German); Tetrachlorocarbon; Tetrachloromethane; Tetrachlorure de carbone (French); Tetraclorometano (Italian); Tetracloruro de carbono (Spanish); Tetracloruro di carbonio (Italian); Tetrafinol; Tetraform; Tetrasol; Twawpit; UN 1846; Univerm; Vermoestricid

CAS Registry Number: 56-23-5

RTECS Number: FG4900000

DOT ID: UN 1846

EEC Number: 602-008-00-5

EINECS Number: 200-262-8

Regulatory Authority

- Carcinogen (Suspected Human) (NTP/7) (ACGIH)[1] (Animal Positive) (IARC)[9]
- Banned or Severely Restricted (Several Countries) (UN)[13]
- Air Pollutant Standard Set (ACGIH)[1] (DFG)[3] (HSE)[33] (UNEP)[43] (OSHA)[58] (Several States)[60] (Australia) (Israel) (Mexico) (Several Canadian Provinces)
- CLEAN AIR ACT: Hazardous Air Pollutants (Title I, Part A, Section 112); Stratospheric ozone protection (Title VI, Subpart A, Appendix A), Class I, Ozone Depletion Potential = 1.1
- CLEAN WATER ACT: Section 311 Hazardous Substances/RQ 40CFR117.3 (same as CERCLA, see below); 40CFR423, Appendix A, Priority Pollutants; Section 313 Water Priority Chemicals (57FR41331, 9/9/92); Toxic Pollutant (Section 401.15)
- EPA HAZARDOUS WASTE NUMBER (RCRA No.): U211, D019
- RCRA Toxicity Characteristic (Section 261.24), Maximum
- Concentration of Contaminants, regulatory level, 0.5 mg/l
- RCRA, 40CFR261, Appendix 8 Hazardous Constituents
- RCRA 40CFR268.48; 61FR15654, Universal Treatment Standards: Wastewater (mg/l), 0.057; Nonwastewater (mg/kg), 6.0
- RCRA Maximum Concentration Limit for Ground Water Protection (Section 264.94): 8010 (1); 8240 (5)

- SAFE DRINKING WATER ACT: MCL, 0.005 mg/l; MCLG, zero; Regulated chemical (47 FR 9352)
- SUPERFUND/EPCRA 40CFR302.4 Reportable Quantity (RQ): CERCLA, 10 lb (4.54 kg)
- EPCRA Section 313 Form R *de minimis* concentration reporting level: 0.1%
- CALIFORNIA'S PROPOSITION 65: Carcinogen
- U.S. DOT Regulated Marine Pollutant (49CFR172.101, Appendix B)
- Canada, WHMIS, Ingredients Disclosure List; National Pollutant Release Inventory (NPRI); CEPA Toxic Substance List
- Mexico, Drinking water criteria, 0.004 mg/l

Cited in U.S. State Regulations: Alaska (G), California (A, G, W), Colorado (W), Connecticut (A), Florida (G, W), Illinois (G), Indiana (A), Kansas (G, W), Louisiana (G), Maine (G, W), Maryland (G), Massachusetts (G, A), Michigan (G), Minnesota (W), Nevada (A), New Hampshire (G), New Jersey (G, W), New Mexico (W), New York (G, A), North Carolina (A), North Dakota (A), Oklahoma (G), Pennsylvania (G, A), Rhode Island (G, A), South Carolina), Vermont (G), Virginia (G, A), Washington (G), West Virginia (G), Wisconsin (G).

Description: Carbon tetrachloride, CCl_4, is a colorless, nonflammable liquid with a characteristic ether-like odor. The odor threshold is 0.52 mg/l in water and 140 – 548 ppm in air. Boiling point = 76.5°C. Freezing/Melting point = -23°C. Very slightly soluble in water. Log P_{ow} (octanol/water partition coefficient) = 2.6.

Potential Exposure: Carbon tetrachloride is used as a solvent for oils, fats, lacquers, varnishes, rubber, waxes, and resins. Fluorocarbons are chemically synthesized from it. It is also used as an azeotropic drying agent for spark plugs, a drycleaning agent, a fire extinguishing agent, a fumigant, and an anthelmintic agent. The use of this solvent is widespread, and substitution of less toxic solvents when technically possible is recommended.

Incompatibilities: Oxidative decomposition on contact with hot surfaces, flames or welding arcs carbon tetrachloride decomposes forming toxic phosgene fumes and hydrogen chloride. Decomposes violently (producing heat) on contact chemically active metals such as aluminium, barium, magnesium, potassium, sodium, with fluorine gas, allyl alcohol, and other substances, causing fire and explosion hazard. Attacks copper, lead and zinc. Attacks some coatings, plastics and rubber. Becomes corrosive when in contact with water; corrosive to metals in the presence of moisture.

Permissible Exposure Limits in Air: The OSHA[58] PEL is 10 ppm TWA for an 8-hour workshift.; 25 ppm not to exceeded during any 15-minute workperiod, and 200 ppm as an acceptable maximum peak level for 5-minutes in any 4-hours. NIOSH recommends an airborne exposure limit of 2 ppm which should not be exceeded during any 1-hour period. ACGIH recommends a limit of 5 ppm TWA for an 8-hour workshift and STEL of 10 ppm. These limits contain notations that CCL_4 is a substance suspect of carcinogenic potential for man and with the further notation "skin" indicating the possibility of cutaneous absorption. The NIOSH IDLH level is 200 ppm with the notation "occupational carcinogen." HSE[33] has set a TWA of 10 pm (65 mg/m^3) and a STEL of 20 ppm (130 mg/m^3). DFG[3] has set a TWA of 10 ppm (65 mg/m^3). Australia's, Israel's and Mexico's TWA is 5 ppm (30 mg/m^3) and Mexico's STEL is 20 ppm (120 mg/m^3). The former-USSR/UNEP joint project[43] has set a MAC in workplace air of 20 mg/m^3 and a MAC in ambient air of residential areas of 4 mg/m^3 on a momentary basis and 2 mg/m^3 on an average daily basis. In addition, several states have set guidelines or standards for CCl_4 in ambient air[60] ranging from zero (North Dakota) to 0.03 μg/m^3 (Rhode Island) to 0.0667 μg/m^3 (Indiana) to 0.67 μg/m^3 (Massachusetts and North Carolina) to 72 μg/m^3 (Pennsylvania) to 100 μg/m^3 (New York) to 150 μg/m^3 (South Carolina) to 300 μg/m^3 (Connecticut and Virginia) to 714 μg/m^3 (Nevada).

Determination in Air: Charcoal adsorption followed by workup with CS_2 and analysis by gas chromatography; see NIOSH Method #1003 for Hydrocarbons, Chlorinated.

Permissible Concentration in Water: To protect freshwater aquatic life: 35,2000 μg/l on an acute toxicity basis. To protect saltwater aquatic life: 50,000 μg/l on an acute toxicity basis. To protect human health: preferably zero. An additional lifetime cancer risk of 1 in 100,000 is presented by a concentration of 4.0 μg/l. Mexico's drinking water criteria is 0.004 mg/l.

The former USSR/UNEP joint project[43] has set a MAC in water bodies used for domestic purposes of 0.3 mg/l. The USEPA[48] has set a lifetime health advisory of 0.0007 mg/kg/day and a drinking water equivalent level of 25 μg/l.

In addition, several states have set standards and guidelines for CCl_4 in drinking water[61] ran ranging from standards of 2 μg/l (New Jersey) to 3 μg/l (Florida) to 10 μg/l (New Mexico) and guidelines of 2.7 μg/l (Minnesota) to 5 μg/l (California, Kansas and Maine, Colorado).

Determination in Water: Gas chromatography (EPA Method 601) or gas chromatography plus mass spectrometry (EPA Method 624).

Routes of Entry: Inhalation of vapor, percutaneous absorption, ingestion and skin and eye contact.

Harmful Effects and Symptoms

Short Term Exposure: Carbon tetrachloride irritates the eyes, causing redness.

Inhalation: Levels of 20 ppm may cause dizziness, headache, vomiting, visual disturbances, extreme fatigue and nose and throat irritation. Other symptoms may include restlessness, loss of balance, twitching and tremors. Severe exposure can lead to liver, kidney, eye and nerve damage that may be delayed after exposure; can cause breathing stoppage, coma and death. 1,000 ppm for an unspecified time has caused death.

Skin: May cause irritation and redness; Carbon tetrachloride is readily absorbed through the skin. Symptoms as listed above may occur through skin absorption even when vapor concentrations are below OSHA standards.

Ingestion: May cause severe abdominal pain with diarrhea, followed by symptoms described under inhalation. Death may occur by ingestion of as little as ½ teaspoon.

Between 45 and 100 ppm, Carbon tetrachloride may cause headache, drowsiness, fatigue, nausea and vomiting. 100 – 300 ppm may cause additional effects of mental confusion, weight loss and sluggishness. Liver, kidney, eye and nerve damage can result from more severe exposures. Coma and death may occur.

Long Term Exposure: Repeated or prolonged skin contact may cause dermatitis. Carbon tetrachloride is a possible human carcinogen. Cancer site in animals: liver cancer.

Points of Attack: Central nervous system, eyes, lungs, liver, kidneys, skin.

Medical Surveillance: Preplacement and periodic examinations should include an evaluation of alcohol intake and appropriate tests for liver and kidney functions. Special attention should be given to the central and peripheral nervous system, the skin, and blood. Expired air and blood levels may be useful as indicators of exposure.

First Aid: If this chemical gets into the eyes, remove any contact lenses at once and irrigate immediately for at least 15 minutes, occasionally lifting upper and lower lids. Seek medical attention immediately. If this chemical contacts the skin, remove contaminated clothing and wash immediately with soap and water. Seek medical attention immediately. If this chemical has been inhaled, remove from exposure, begin rescue breathing (using universal precautions) if breathing has stopped and CPR if heart action has stopped. Transfer promptly to a medical facility. When this chemical has been swallowed, get medical attention. Give large quantities of water and induce vomiting. Do not make an unconscious person vomit.

Personal Protective Methods: Wear protective gloves and clothing to prevent any reasonable probability of skin contact. Safety equipment suppliers/manufacturers can provide recommendations on the most protective glove/clothing material for your operation. Polyvinyl Alcohol, VITON, Teflon, Polyvinyl Acetate and SILVERSHIELD are among the recommended protective materials. All protective clothing (suits, gloves, footwear, headgear) should be clean, available each day, and put on before work. Contact lenses should not be worn when working with this chemical. Wear splash-proof chemical goggles and face shield unless full facepiece respiratory protection is worn. Employees should wash immediately with soap when skin is wet or contaminated. Provide emergency showers and eyewash.

Respirator Selection: NIOSH: *At any detectable concentration:* SCBAF:PD,PP (any MSHA/NIOSH approved self-contained breathing apparatus that has a full facepiece and is operated in a pressure-demand or other positive-pressure mode); or SAF:PD,PP:ASCBA (any supplied-air respirator that has a full facepiece and is operated in a pressure-demand or other positive-pressure mode in combination with an auxiliary, self-contained breathing apparatus operated in a pressure-demand or other positive pressure mode). *Escape:* HiEF [any air-purifying, full-facepiece respirator (gas mask) with a chin-style, front- or back-mounted organic vapor canister having a high-efficiency particulate filter]; or SCBAE (any appropriate escape-type, self-contained breathing apparatus).

Storage: Prior to working with Carbon tetrachloride you should be trained on its proper handling and storage. Store in tightly closed containers in a cool, well ventilated area. Carbon tetrachloride must be stored to avoid contact with chemically active metals such as sodium, potassium and magnesium, since violent reactions occur. Store in tightly closed containers in a cool, well-ventilated area. A regulated, marked area should be established where this chemical is handled, used, or stored in compliance with OSHA standard 1910.1045.

Shipping: Carbon tetrachloride requires a "Poison" label. It falls in DOT Hazard Class 6.1 and Packing Group II.

Spill Handling: Evacuate and restrict persons not wearing protective equipment from area of spill or leak until cleanup is complete. Remove all ignition sources. Ventilate area of spill or leak. Absorb liquids in vermiculite, dry sand, earth, peat, carbon, or a similar material and deposit in sealed containers. It may be necessary to contain and dispose of this chemical as a hazardous waste. If material or contaminated runoff enters waterways, notify downstream users of potentially contaminated waters. Contact your Department of Environmental Protection or your regional office of the federal EPA for specific recommendations. If employees are required to clean-up spills, they must be properly trained and equipped. OSHA 1910.120(q) may be applicable.

Fire Extinguishing: This chemical is a noncombustible liquid. Use any extinguishing agent suitable for surrounding fire. Poisonous gases are produced in fire including phosgene and hydrogen chloride. If material or contaminated runoff enters waterways, notify downstream users of potentially contaminated waters. Notify local health and fire officials and pollution control agencies. From a secure, explosion-proof location, use water spray to cool exposed containers. If cooling streams are ineffective (venting sound increases in volume and pitch, tank discolors, or shows any signs of deforming), withdraw immediately to a secure position. If employees are expected to fight fires, they must be trained and equipped in OSHA 1910.156.

Disposal Method Suggested: Incineration, preferably after mixing with another combustible fuel; care must be exercised to assure complete combustion to prevent the formation of phosgene; an acid scrubber in necessary to remove the halo acids produced.[22] Recover and purify by distillation where possible.

References

National Institute for Occupational Safety and Health, Criteria for a Recommended Standard: Occupational Exposure to Carbon Tetrachloride, NIOSH Doc. No. 76–133 (1976)

U.S. Environmental Protection Agency, Carbon Tetrachloride: Ambient Water Quality Criteria, Washington, DC (1980)

National Academy of Sciences, Chloroform, Carbon Tetrachloride and Other Halomethanes: An Environmental Assessment, Washington, DC (1978)

Sax, N. I., Ed., "Dangerous Properties of Industrial Materials Report," 1, No. 2, 30–32 (1980) and 3, No. 5, 88–94 (1983)

U.S. Public Health Service, "Toxicological Profile for Carbon Tetrachloride," Atlanta, Georgia, Agency for Toxic Substance and Disease Registry (December 1988)

New Jersey Department of Health and Senior Services and Senior Services, "Hazardous Substance Fact Sheet: Carbon Tetrachloride," Trenton, NJ (August 1998)

New York State Department of Health, "Chemical Fact Sheet: Carbon Tetrachloride," Albany, NY, Bureau of Toxic Substance Assessment (January 1986 and Version 2)

Carbonyl Fluoride

Molecular Formula: CF_2O

Common Formula: COF_2

Synonyms: Carbon difluoride oxide; Carbon fluoride oxide; Carbonic difluoride; Carbon oxyfluoride; Carbonyl difluoride; Difluoroformaldehyde; Fluophosgene; Fluoroformyl fluoride; Fluorophosgene; Fluoruro de carbonilo (Spanish)

CAS Registry Number: 353-50-4

RTECS Number: FG6125000

DOT ID: UN 2417

Regulatory Authority

- Air Pollutant Standard Set (ACGIH)[1] (HSE)[33] (OSHA)[58] (Several States)[60] (Australia) (Israel)
- EPA HAZARDOUS WASTE NUMBER (RCRA No.): U033
- RCRA, 40CFR261, Appendix 8 Hazardous Constituents
- SUPERFUND/EPCRA 40CFR302.4 Reportable Quantity (RQ): CERCLA, 1,000 lb (454 kg)
- Canada, WHMIS, Ingredients Disclosure List

Cited in U.S. State Regulations: Alaska (G), California (A, G), Connecticut (A, G), Florida (G), Illinois (G), Kansas (G), Louisiana (G), Maine (G), Massachusetts (G), Minnesota (G), Nevada (A), New Hampshire (G), New Jersey (G), North Dakota (A), Pennsylvania (G), Rhode Island (G), Vermont (G), Virginia (G), Washington (G), West Virginia (G), Wisconsin (G).

Description: Carbonyl Fluoride, COF_2, is a colorless or light yellow, hygroscopic, compressed liquefied gas, with a pungent, highly irritating odor. Boiling point= -83°C. Freezing/Melting point = -114°C. Reacts with water.

Potential Exposure: The major source of exposure to COF_2 results from the thermal decomposition of fluorocarbon plastics such as PTFE in air. Carbonyl fluoride may also find use in organic synthesis. It has been suggested for use as a military poison gas.

Incompatibilities: Reacts with water to form hydrogen fluoride gas. The substance decomposes on heating at 450 – 490°C producing toxic gases including hydrogen fluoride. Not compatible with hexafluoroisopropylidene-amino lithium.

Permissible Exposure Limits in Air: ACGIH recommended airborne exposure limit is 2 ppm (5 mg/m^3) TWA for an 8-hour workshift and STEL of 5 ppm (15 mg/m^3). The NIOSH recommended limit is the same for a 10-hour workshift. HSE,[33] Australia, Israel, and the Canadian provinces of Alberta, Britihs Columbia, Ontario and Quebec have adopted these same values.

In addition, some states have set guidelines or standards for COF_2 in ambient air[60] ranging from 50 – 150 μg/m^3 (North Dakota) to 100 μg/m^3 (Connecticut) to 119 μg/m^3 (Nevada).

Determination in Air: None available.

Permissible Concentration in Water: No criteria set.

Routes of Entry: Inhalation, skin and/or eye contact. May be absorbed through the skin.

Harmful Effects and Symptoms

Short Term Exposure: The substance irritates the eyes, skin, and respiratory tract. Higher exposures can cause pulmonary edema, a medical emergency that can be delayed for several hours. This can cause death. Rapid evaporation of the liquid may cause frostbite. On an acute basis, COF_2 is about as toxic as HF as a respiratory irritant gas.

Long Term Exposure: The long-term effects are due to the fluoride ion generated by hydrolysis; this inhibits succinic dehydrogenase activity since this is a fluoride-sensitive enzyme. May cause liver and kidney damage. Repeated exposure may cause bronchitis. Chronic exposure: gastrointestinal pain, muscle fibrosis, skeletal fluorosis (NIOSH)

Points of Attack: Eyes, skin, respiratory system, bone, liver, kidneys.

Medical Surveillance: Liver and kidney function tests. Consider chest x-ray following acute overexposure.

First Aid: If this chemical gets into the eyes, remove any contact lenses at once and irrigate immediately for at least 15 minutes, occasionally lifting upper and lower lids. Seek medical attention immediately. If this chemical contacts the skin, remove contaminated clothing and wash immediately with soap and water. Seek medical attention immediately. If this chemical has been inhaled, remove from exposure, begin rescue breathing (using universal precautions) if breathing has stopped and CPR if heart action has stopped. Transfer promptly to a medical facility. When this chemical has been swallowed, get medical attention. Give large quantities of water and induce vomiting. Do not make an unconscious person vomit. Medical observation is recommended for 24 – 48 hours after breathing overexposure, as pulmonary edema may be delayed. As first aid for pulmonary edema, a doctor or authorized paramedic may consider administering a corticosteroid spray. If frostbite has occurred, seek medical attention immediately; do NOT rub the affected areas or flush them with water. In order to prevent further tissue damage, do NOT attempt to remove frozen clothing from frostbitten areas. If frostbite has NOT occurred, immediately and thoroughly wash contaminated skin with soap and water.

Personal Protective Methods: Wear protective gloves and clothing to prevent any reasonable probability of skin contact. Where exposure to the liquefied compressed gas may occur, employees should be provided with special clothing designed to prevent frostbite. Safety equipment suppliers/ manufacturers can provide recommendations on the most protective glove/clothing material for your operation. All protective clothing (suits, gloves, footwear, headgear) should be clean, available each day, and put on before work. Contact lenses should not be worn when working with this chemical. Wear non-vented, impact resistant goggles when working with gas. Wear indirect-vent and splash-proof chemical goggles and face shield when working with liquid unless full facepiece respiratory protection is worn. Employees should wash immediately with soap when skin is wet or contaminated. Provide emergency showers and eyewash.

Respirator Selection: Engineering controls should be used wherever feasible to maintain airborne concentrations of this chemical below the prescribed exposure limit. Respirators and protective equipment are less effective than engineering controls, and should be used only in non-routine or emergency situations which may result in exposure concentrations in excess of the TWA environmental limit. NIOSH: *(fluorides) 12.5 mg/m³:* DM (any dust and mist respirator). *25 mg/m³:* DMXSQ* (any dust and mist respirator except single-use and quarter mask respirators); or SA* (any supplied-air respirator). *62.5 mg/m³:* SA:CF* (any supplied-air respirator operated in a continuous-flow mode); or PAPRDM*+ *if not present as a fume* (any powered, air-purifying respirator with a dust and mist filter). *125 mg/m³*: HiEF+ (any air-purifying, full-facepiece respirator with a high-efficiency particulate filter); or SCBAF (any self-contained breathing apparatus with a full facepiece); or SAF (any supplied-air respirator with a full facepiece). *250 mg/m³:* SA:PD,PP (any supplied-air respirator operated in a pressure-demand or other positive-pressure mode). *Emergency or planned entry into unknown concentrations or IDLH conditions:* SCBAF:PD,PP (any self-contained breathing apparatus that has a full faceplate and is operated in a pressure-demand or other positive-pressure mode); or SAF:PD,PP:ASCBA (any supplied-air respirator that has a full facepiece and is operated in a pressure-demand or other positive-pressure mode in combination with an auxiliary, self-contained breathing apparatus operated in a pressure-demand or other positive-pressure mode). *Escape:* HiEF+ (any air-purifying, full-facepiece respirator with a high-efficiency particulate filter); or SCBAE (any appropriate escape-type, self-contained breathing apparatus).

* Substance reported to cause eye irritation or damage; may require eye protection.

\+ May need acid gas sorbent.

Shipping: Carbonyl Fluoride must be labeled "Poison Gas." It falls in Hazard Class 2.3 and Packing Group I. Shipment by passenger aircraft or railcar of even by cargo aircraft is forbidden.

Spill Handling: Evacuate and restrict persons not wearing protective equipment from area of spill or leak until cleanup is complete. Remove all ignition sources. Ventilate area of spill or leak to disperse the gas. Stop the flow of gas if it can be done safely. If source is a cylinder and the leak cannot be stopped in place, remove the leaking cylinder to a safe place, and repair leak or allow cylinder to empty. DO NOT USE WATER OR WET METHOD. Keep this chemical out of confined spaces, such as a sewer, because of the possibility of explosion, unless the sewer is designed to prevent the buildup of explosive concentrations. If employees are required to cleanup spills, they must be properly trained and equipped. OSHA 1910.120(q) may be applicable.

Initial isolation and protective action distances

Distances shown are likely to be affected during the first 30 minutes after materials are spilled and could increase with time. If more than one tank car, cargo tank, portable tank, or large cylinder is involved in the incident is leaking, the protective action distance may need to be increased. You may need to seek emergency information from CHEMTREC at (800) 424-9300 or seek professional environmental engineering assistance from the U.S. EPA Environmental Response Team at (908)548-8730 (24-hour response line).

Small spills (From a small package or a small leak from a large package)

First: Isolate in all directions (feet)..... 200

Then: Protect persons downwind (miles)

Day 0.2

Night 0.6

Large spills (From a large package or from many small packages)

First: Isolate in all directions (feet)..... 600

Then: Protect persons downwind (miles)

Day 0.5

Night 2.0

Fire Extinguishing: This chemical may burn but does not readily ignite. Use dry chemical, carbon dioxide, water spray, or foam extinguishers. Poisonous gases are produced in fire including hydrogen fluoride. Notify local health and fire officials and pollution control agencies. From a secure, explosion-proof location, use water spray to cool exposed containers. If cooling streams are ineffective (venting sound increases in volume and pitch, tank discolors, or shows any signs of deforming), withdraw immediately to a secure position. If employees are expected to fight fires, they must be trained and equipped in OSHA 1910.156.

References

New Jersey Department of Health and Senior Services, Hazardous Substance Fact Sheet, Carbonyl Fluoride, Trenton NJ (Mardh 1999)

Carbophenothion

Molecular Formula: $C_{11}H_{16}ClO_2PS_3$

Common Formula: $(C_2H_5O)_2PSSCH_2SC_6H_4Cl$

Synonyms: Acarithion®; Akarithion; Carbofenothion (Dutch); *S*-[(*p*-Chlorophenylthio)methyl] *O,O*-diethyl phosphorodithioate; *S*-(4-Chlorophenylthiomethyl)diethyl phosphorothiolothionate; Dagadip®; *O,O*-Diaethy-S-[(4-chlor-phenyl-thio)-methyl]dithiophosphat (German); *o,o*-Diethy-S-[(4-chloor-fenyl-thio)-methyl]dithiofosfaat (Dutch); *O,O*-Diethyl *S*-*p*-chlorophenylthiomethyl dithiophosphate; *O,O*-Diethyl *S*-(*p*-chlorophenylthiomethyl) phosphorodithioate; *O,O*-Diethyldithiophosphoric acid, *p*-chlorophenylthiomethyl ester; *O,O*-Diethyl-S-*p*-chlorfenylthiomethylester kyseliny dithiofosforecne (Czech); *O,O*-Diethyl *p*-chlorophenylmercaptomethyl dithiophosphate; *O,O*-Diethyl 4-chlorophenylmercaptomethyl dithiophosphate; *o,o*-Dietil-S-[(*p*-clorofenil-tio)-metile]-ditiofosfato (Italian); *o*,-Dietil-S-[(4-clorofenil-tio)-metile]-ditiofosfato (Italian); Dithiophosphate de *O,O*-diethyle et de (4-chlorophenyl) thiomethyle (French); Endyl; ENT 23,708; Garrathion®; Lethox; Nephocarb®; Oleoakarithion; R-1303; Stauffer R-1,303; Trithion® miticide

CAS Registry Number: 786-19-6

RTECS Number: ID5250000

DOT ID: UN 1615

EEC Number: 015-044-00-6

Regulatory Authority

- Banned or Severely Restricted (In Agriculture) (East Germany and India) (UN)[13]
- Very Toxic Substance (World Bank)[15]
- SUPERFUND/EPCRA 40CFR355, Appendix B Extremely Hazardous Substances: TPQ = 500 lb (227 kg)
- SUPERFUND/EPCRA 40CFR302.4 Reportable Quantity (RQ): EHS, 1 lb (0.454 kg)
- U.S. DOT Regulated Marine Pollutant (49CFR172.101, Appendix B), severe pollutant

Cited in U.S. State Regulations: California (G), Florida (G), Massachusetts (G), New Jersey (G), Michigan (G), New Hampshire (G), Pennsylvania (G).

Description: Carbophenothion, $C_{11}H_{16}ClO_2PS_3$, is a colorless to light amber liquid with a characteristic odor. Boiling point = 82°C at 0.01 mm Hg. Log P_{ow} (octanol/water partition coefficient) = 5.1

Potential Exposure: Those engaged in the manufacture or application of this material, which is an insecticide and acaricide, primarily for citrus crops and deciduous fruits and nuts.

Incompatibilities: The substance decomposes on heating or on burning producing toxic fumes including phosphorus oxides, sulfur oxides, hydrogen chloride.

Permissible Exposure Limits in Air: No standards set.

Permissible Concentration in Water: No criteria set.

Routes of Entry: Inhalation, ingestion and skin contact.

Harmful Effects and Symptoms

Short Term Exposure: The substance may affect the nervous system, resulting in convulsions and respiratory failure. Cholinesterase inhibitor. Exposure may result in death. Produces headaches, nausea, weakness and dizziness. Symptoms may include nausea, vomiting, abdominal cramps, diarrhea, excessive salivation, headache, giddiness, weakness, muscle twitching, difficult breathing, blurring or dimness of vision, and loss of muscle coordination. Death may occur from failure of the respiratory center, paralysis of the respiratory muscles, intense bronchoconstriction, or all three. This material is highly toxic; the estimated fatal oral dose is 0.6 g for a 150 lb (70 kg) person. Oral LD_{50} for rats is 6.8 mg/kg.

Long Term Exposure: Cholinesterase inhibitor; cumulative effect is possible: see acute hazards/symptoms. The state of Massachusetts lists this chemical as a neurotoxin.

Points of Attack: Respiratory system, lungs, central nervous system, cardiovascular system, skin, eyes, plasma and red blood cell cholinesterase.

Medical Surveillance: Before employment and at regular times after that, the following are recommended: Plasma and red blood cell cholinesterase levels (tests for the enzyme poisoned by this chemical). If exposure stops, plasma levels return to normal in 1 – 2 weeks while red blood cell levels may be reduced for 1 – 3 months.

When cholinesterase enzyme levels are reduced by 25% or more below preemployment levels, risk of poisoning is increased, even if results are in lower ranges of "normal." Reassignment to work not involving organophosphate or carbamate pesticides is recommended until enzyme levels recover. If symptoms develop or overexposure occurs, repeat the above tests as soon as possible and get an exam of the nervous system. Also consider complete blood count. Consider chest x-ray following acute overexposure. Do not drink any alcoholic beverages before or during use. Alcohol promotes absorption of organic phosphates.

First Aid: If this chemical gets into the eyes, remove any contact lenses at once and irrigate immediately for at least 15 minutes, occasionally lifting upper and lower lids. Seek

medical attention immediately. If this chemical contacts the skin, remove contaminated clothing and wash immediately with soap and water. Speed in removing material from skin is of extreme importance. Shampoo hair promptly if contaminated. Seek medical attention immediately. If this chemical has been inhaled, remove from exposure, begin rescue breathing (using universal precautions) if breathing has stopped and CPR if heart action has stopped. Transfer promptly to a medical facility. When this chemical has been swallowed, get medical attention. Give large quantities of water and induce vomiting. Do not make an unconscious person vomit.

Personal Protective Methods: Wear protective gloves and clothing to prevent any reasonable probability of skin contact. Safety equipment suppliers/manufacturers can provide recommendations on the most protective glove/clothing material for your operation. All protective clothing (suits, gloves, footwear, headgear) should be clean, available each day, and put on before work. Contact lenses should not be worn when working with this chemical. Wear splash-proof chemical goggles and face shield unless full facepiece respiratory protection is worn. Employees should wash immediately with soap when skin is wet or contaminated. Provide emergency showers and eyewash.

Respirator Selection: Engineering controls should be used wherever feasible to maintain airborne concentrations of this chemical below the prescribed exposure limit. Respirators and protective equipment are less effective than engineering controls, and should be used only in non-routine or emergency situations which may result in exposure concentrations in excess of the TWA environmental limit. *At any concentrations above the NIOSH REL, or where there is no REL, at any detectable concentration:* SCBAF:PD,PP (any MSHA/NIOSH approved self-contained breathing apparatus that has a full facepiece and is operated in a pressure-demand or other positive-pressure mode); or SAF:PD,PP:ASCBA (any supplied-air respirator that has a full facepiece and is operated in a pressure-demand or other positive-pressure mode in combination with an auxiliary, self-contained breathing apparatus operated in a pressure-demand or other positive pressure mode).

Storage: Prior to working with this chemical you should be trained on its proper handling and storage. Store in tightly closed containers in a cool, well-ventilated area.

Shipping: Organophosphorus pesticides, solid, toxic, n.o.s. have a DOT label requirement of "Poison." The limit on passenger aircraft or railcar shipment is 25 kg; on cargo aircraft shipment is 100 kg. The UN/DOT Hazard Class is 6.1 and the Shipping Group is II.[19][20]

Spill Handling: Stay upwind; keep out of low areas. Evacuate and restrict persons not wearing protective equipment from area of spill or leak until cleanup is complete. Remove all ignition sources. Ventilate area of spill or leak. Absorb liquids in vermiculite, dry sand, earth, peat, carbon, or a similar material and deposit in sealed containers. It may be necessary to contain and dispose of this chemical as a hazardous waste. If material or contaminated runoff enters waterways, notify downstream users of potentially contaminated waters. Contact your Department of Environmental Protection or your regional office of the federal EPA for specific recommendations. If employees are required to clean-up spills, they must be properly trained and equipped. OSHA 1910.120(q) may be applicable.

Fire Extinguishing: The substance decomposes on heating or on burning producing toxic fumes including phosphorus oxides, sulfur oxides, hydrogen chloride. This material may burn, but does not ignite readily. For small fires: use dry chemicals, carbon dioxide, water spray, or foam. For large fires: use water spray, fog or foam. Stay upwind; keep out of low areas. Move container from fire area if you can do it without risk. Fight fire from maximum distance. Dike fire control water for later disposal; do not scatter the material. Wear positive pressure breathing apparatus and special protective clothing. If material or contaminated runoff enters waterways, notify downstream users of potentially contaminated waters. Notify local health and fire officials and pollution control agencies. Container may explode in heat of fire. Fire and runoff from fire control water may produce irritating or poisonous gases. From a secure, explosion-proof location, use water spray to cool exposed containers. If cooling streams are ineffective (venting sound increases in volume and pitch, tank discolors, or shows any signs of deforming), withdraw immediately to a secure position. If employees are expected to fight fires, they must be trained and equipped in OSHA 1910.156.

Disposal Method Suggested: Hydrolysis by hypochlorites may be used as may incineration. In accordance with 40CFR165 recommendations for the disposal of pesticides and pesticide containers. Must be disposed properly by following package label directions or by contacting your state pesticide or environmental control agency or by contacting your regional EPA office.

References

Sax, N. I., Ed., "Dangerous Properties of Industrial Materials Report," 2, No. 4, 55–59, New York, Van Nostrand Reinhold Co. (1982)

U.S. Environmental Protection Agency, "Chemical Profile: Carbophenothion," Washington, DC, Chemical Emergency Preparedness Program (October 31, 1985)

Carboxin

Molecular Formula: $C_{12}H_{13}O_2NS$

Common Formula: $C_6H_2NHCO\text{-}C_5OSH_7$

Synonyms: Carbathiin; 5-Carboxanilido-2,3-dihydro-6-methyl-1,4-oxathiin; Carboxine; Carboxin oxathion pesticide; Caswell No. 165 A; D-735; DCMO; 2,3-Dihydro-5-carboxanilido-6-methyl-1,4-oxathiin; 2,3-Dihydro-6-methyl-1,4-oxathiin-5-carboxanilide; 5,6-Dihydro-2-methyl-1,4-oxathiin-3-carboxanilide; 2,3-Dihydro-6-methyl-5-phenylcarbamoyl-1,4-oxathiin; 5,6-Dihydro-2-methyl-*N*-phenyl-1,4-oxathiin-3-carboxamide; DMOC; EPA pesticide chemical code 090201; F-735; Flo Pro V seed protectant; NSC 263492; 1,4-Oxathiin-3-carboxamide,5,6-dihydro-2-methyl-*N*-phenyl; 1,4-Oxathiin-3-carboxanilide,5,6-dihydro-2-methyl; 1,4-Oxathiin-3-carboxanilide,5,6-dihydro-2-methyl-; 1,4-Oxathiin-2,3-dihydro-5-carboxanilido-6-methyl-; V 4X; Vitaflo; Vitavax; Vitavax 100; Vitavax 735d; Vitavax 75 PM; Vitavax 75W

CAS Registry Number: 5234-68-4

RTECS Number: RP4550000

Regulatory Authority

- EPCRA Section 313 Form R *de minimis* concentration reporting level: 1.0%

Description: Carboxin, $C_{12}H_{13}O_2NS$, is a crystalline solid. Freezing/Melting point = 93 – 95°C.

Potential Exposure: Those involved in the production, formulation and application of this systemic fungicide, seed protectant and wood preservative.

Permissible Exposure Limits in Air: No standard set.

Permissible Concentration in Water: The no-observed-adverse-effect level has been determined by EPA to be 10 mg/kg body weight/day. This results in a long-term health advisory of 3.5 mg/l and a lifetime health advisory of 0.7 mg/l.

Determination in Water: Analysis of Carboxin is by a gas chromatographic (GC) method applicable to the determination of certain nitrogen-phosphorus containing pesticides in water samples. In this method, approximately 1 liter of sample is extracted with Methylene chloride. The extract is concentrated and the compounds are separated using capillary column GC. Measurement is made using a nitrogen phosphorus detector. The method detection limit has not been determined for carboxin but it is estimated that detection limits for analyses included in this method are in the range of 0.1 – 2 µg/l.

Harmful Effects and Symptoms

The LD_{50} oral for mice has been reported to be 3,550 mg/kg (slightly toxic). A value for LD_{50} rat of 430 mg/kg puts carboxin in the moderately toxic category.

First Aid: If this chemical gets into the eyes, remove any contact lenses at once and irrigate immediately for at least 15 minutes, occasionally lifting upper and lower lids. Seek medical attention immediately. If this chemical contacts the skin, remove contaminated clothing and wash immediately with soap and water. Seek medical attention immediately. If this chemical has been inhaled, remove from exposure, begin rescue breathing (using universal precautions) if breathing has stopped and CPR if heart action has stopped. Transfer promptly to a medical facility. When this chemical has been swallowed, get medical attention. Give large quantities of water and induce vomiting. Do not make an unconscious person vomit.

Personal Protective Methods: Wear protective gloves and clothing to prevent any reasonable probability of skin contact. Safety equipment suppliers/manufacturers can provide recommendations on the most protective glove/clothing material for your operation. All protective clothing (suits, gloves, footwear, headgear) should be clean, available each day, and put on before work. Contact lenses should not be worn when working with this chemical. Wear dust-proof chemical goggles and face shield unless full facepiece respiratory protection is worn. Employees should wash immediately with soap when skin is wet or contaminated. Provide emergency showers and eyewash.

Respirator Selection: Engineering controls should be used wherever feasible to maintain airborne concentrations of this chemical below the prescribed exposure limit. Respirators and protective equipment are less effective than engineering controls, and should be used only in non-routine or emergency situations which may result in exposure concentrations in excess of the TWA environmental limit. *At any concentrations above the NIOSH REL, or where there is no REL, at any detectable concentration:* SCBAF:PD,PP (any MSHA/NIOSH approved self-contained breathing apparatus that has a full facepiece and is operated in a pressure-demand or other positive-pressure mode); or SAF:PD,PP:ASCBA (any supplied-air respirator that has a full facepiece and is operated in a pressure-demand or other positive-pressure mode in combination with an auxiliary, self-contained breathing apparatus operated in a pressure-demand or other positive pressure mode).

Storage: Prior to working with Carboxin you should be trained on its proper handling and storage. Store in tightly closed containers in a cool, well-ventilated area.

Spill Handling: Evacuate persons not wearing protective equipment from area of spill or leak until clean-up is complete. Remove all ignition sources. Collect powdered material in the most convenient and safe manner and deposit in sealed containers. Ventilate area after clean-up is complete. It may be necessary to contain and dispose of this chemical as a hazardous waste. If material or contaminated runoff enters waterways, notify downstream users of potentially contaminated waters. Contact your Department of Environmental Protection or your regional office of the federal EPA for specific recommendations. If employees are required to clean-up spills, they must be properly trained and equipped. OSHA 1910.120(q) may be applicable.

Fire Extinguishing: Use dry chemical, carbon dioxide, water spray, or alcohol foam extinguishers. Poisonous gases are produced in fire. If material or contaminated runoff enters waterways, notify downstream users of potentially contaminated waters. Notify local health and fire officials and pollution control agencies. From a secure, explosion-proof location, use water spray to cool exposed containers. If cooling streams are ineffective (venting sound increases in volume and pitch, tank discolors, or shows any signs of deforming), withdraw immediately to a secure position. If employees are

expected to fight fires, they must be trained and equipped in OSHA 1910.156.

Disposal Method Suggested: Incineration.[22]

References

U.S. Environmental Protection Agency, "Health Advisory: Carboxin," Washington, DC, Office of Drinking Water (August 1987)

Carmustine

Synonyms: BCNU; Bischloroethyl nitrosourea

CAS Registry Number: 154-93-8

RTECS Number: YS2625000

Regulatory Authority

- Carcinogen (Animal Positive, Human Suspected) (IARC)[9] (NTP)

Cited in U.S. State Regulations: California (G), Florida (G), Illinois (G), Massachusetts (G), Minnesota (G), New Jersey (G), Pennsylvania (G)

Description: Carmustine is a light yellow crystalline solid. Freezing/Melting point = 30 – 32°C.

Potential Exposure: BCNU has been used since 1971 as an antineoplastic agent in the treatment of Hodgkin's lymphoma, multiple meyloma and primary or metastatic brain tumors. It also has been reported to have antiviral, antibacterial, and antifungal activity, but no evidence was found that it is used in these ways.

BCNU is not known to be naturally occurring. Health professionals who handle this drug (for example, pharmacists, nurses, and physicians) may possibly be exposed to BCNU during drug preparation, administration, or cleanup; however, the risks can be avoided through use of containment equipment and proper work practices.

Incompatibilities: Acids.

Permissible Exposure Limits in Air: No standards set.

Permissible Concentration in Water: No criteria set.

Harmful Effects and Symptoms

Short Term Exposure: Symptoms include nausea, vomiting, and diarrhea, dyspnea; flushing of the skin; esophagitis; cytotoxic effects on the liver, kidneys and central nervous system; delayed bone-marrow suppression (e.g., leukopenia and thrombocytopenia). The oral LD_{50} rat is 120 mg/kg (highly toxic).

Long Term Exposure: May cause liver, kidney and nervous system damage.

Points of Attack: Liver, kidney, central nervous system, bone-marrow.

Medical Surveillance: Liver and kidney function tests. Examination of the nervous system.

First Aid: *Skin Contact:* Flood all area of body that have contacted the substance with water. Don't wait to remove contaminated clothing; do it under the water stream. Use soap to help assure removal. Isolate contaminated clothing when removed to prevent contact by others.

Eye Contact: Remove any contact lenses at once. Immediately flush eyes well with copious quantities of water or normal saline for at least 20 – 30 minutes. Seek medical attention.

Inhalation: Leave contaminated area immediately; breathe fresh air. Proper respiratory protection must be supplied to any rescuers. If coughing, difficult breathing or any other symptoms develop, seek medical attention at once, even if symptoms develop many hours after exposure.

Ingestion: Contact a physician, hospital or poison center at once. If the victim is unconscious or convulsing, do not induce vomiting or give anything by mouth. Assure that his airway is open and lay him on his side with his head lower than his body and transport immediately to a medical facility. If conscious and not convulsing, give a glass of water to dilute the substance. Vomiting should not be induced without a physician's advice.

Personal Protective Methods: Wear protective gloves and clothing to prevent any reasonable probability of skin contact. Safety equipment suppliers/manufacturers can provide recommendations on the most protective glove/clothing material for your operation. All protective clothing (suits, gloves, footwear, headgear) should be clean, available each day, and put on before work. Contact lenses should not be worn when working with this chemical. Wear dust-proof chemical goggles and face shield unless full facepiece respiratory protection is worn. Employees should wash immediately with soap when skin is wet or contaminated. Provide emergency showers and eyewash.

Respirator Selection: Engineering controls should be used wherever feasible to maintain airborne concentrations of this chemical below the prescribed exposure limit. Respirators and protective equipment are less effective than engineering controls, and should be used only in non-routine or emergency situations which may result in exposure concentrations in excess of the TWA environmental limit. *At any concentrations above the NIOSH REL, or where there is no REL, at any detectable concentration:* SCBAF:PD,PP (any MSHA/NIOSH approved self-contained breathing apparatus that has a full facepiece and is operated in a pressure-demand or other positive-pressure mode); or SAF:PD,PP:ASCBA (any supplied-air respirator that has a full facepiece and is operated in a pressure-demand or other positive-pressure mode in combination with an auxiliary, self-contained breathing apparatus operated in a pressure-demand or other positive pressure mode).

Storage: Prior to working with this chemical you should be trained on its proper handling and storage. Store in tightly closed containers in a cool, well-ventilated area away from acids.[52] A regulated, marked area should be established where this chemical is handled, used, or stored in compliance with OSHA standard 1910.1045.

Shipping: Poisonous solids n.o.s., fall in DOT Hazard Class 6.1 and Packing Group II.

Spill Handling: Evacuate persons not wearing protective equipment from area of spill or leak until clean-up is complete. Remove all ignition sources. Dampen spilled material with 60-70% ethanol to avoid airborne dust. Collect powdered material in the most convenient and safe manner and deposit in sealed containers. Ventilate area after clean-up is complete. It may be necessary to contain and dispose of this chemical as a hazardous waste. If material or contaminated runoff enters waterways, notify downstream users of potentially contaminated waters. Contact your Department of Environmental Protection or your regional office of the federal EPA for specific recommendations. If employees are required to clean-up spills, they must be properly trained and equipped. OSHA 1910.120(q) may be applicable.

Fire Extinguishing: Use dry chemical, carbon dioxide, water spray, or foam extinguishers. Poisonous gases are produced in fire. If material or contaminated runoff enters waterways, notify downstream users of potentially contaminated waters. Notify local health and fire officials and pollution control agencies. From a secure, explosion-proof location, use water spray to cool exposed containers. If cooling streams are ineffective (venting sound increases in volume and pitch, tank discolors, or shows any signs of deforming), withdraw immediately to a secure position. If employees are expected to fight fires, they must be trained and equipped in OSHA 1910.156.

Carvone

Molecular Formula: $C_{10}H_{14}O$

Synonyms: δ-Carvone; 1-6, 8(9)-*p*-Menthadien-2-one; δ-1-Methyl-4-isopropenyl-6-cyclohexen-2-one

CAS Registry Number: 2244-16-8

RTECS Number: OS8670000

Cited in U.S. State Regulations: Massachusetts (G), New Jersey (G)

Description: Carvone, $C_{10}H_{14}O$, is a pale yellow to white clear liquid. Boiling point = 230°C. Flash point = 93°C.

Potential Exposure: Carvone is found in various natural oils including caraway and dill seed and mandarin peel and spearmint oils. It is used in flavoring liqueurs and in perfumes and soaps.

Incompatibilities: Strong oxidizers.

Permissible Exposure Limits in Air: No standards set.

Permissible Concentration in Water: No criteria set.

Routes of Entry: Ingestion and skin contact.

Harmful Effects and Symptoms

Short Term Exposure: Irritates the eyes, skin, and respiratory tract. This material is highly toxic by ingestion and through the skin. The LD_{50} oral-rat is 3.71 mg/kg.

First Aid: *Skin Contact:* Flood all areas of body that have contacted the substance with water. Don't wait to remove contaminated clothing; do it under the water stream. Use soap to help assure removal. Isolate contaminated clothing when removed to prevent contact by others.

Eye contact: Remove any contact lenses at once. Flush eyes well with copious quantities of water or normal saline for at least 20 – 30 minutes. Seek medical attention.

Inhalation: Leave contaminated area immediately; breathe fresh air. Proper respiratory protection must be supplied to nay rescuers. If coughing, difficult breathing or any other symptoms develop, seek medical attention at once, even if symptoms develop many hours after exposure.

Ingestion: If convulsions are not present, give a glass or two of water or mild to dilute the substance. Assure that the person's airway is unobstructed and contact a hospital or poison center immediately for advice on whether or not to induce vomiting.

Storage: Store in a cool, dry place or in a refrigerator.

Shipping: Combustible liquids, n.o.s. have no DOT label requirements. The UN/DOT Hazard Class is 3 and the Packing Group is III.[19][20]

Spill Handling: Evacuate and restrict persons not wearing protective equipment from area of spill or leak until cleanup is complete. Remove all ignition sources. Ventilate area of spill or leak. Absorb liquids in vermiculite, dry sand, earth, peat, carbon, or a similar material and deposit in sealed containers. Follow by washing surfaces well first with alcohol, then with soap and water. It may be necessary to contain and dispose of this chemical as a hazardous waste. If material or contaminated runoff enters waterways, notify downstream users of potentially contaminated waters. Contact your Department of Environmental Protection or your regional office of the federal EPA for specific recommendations. If employees are required to clean-up spills, they must be properly trained and equipped. OSHA 1910.120(q) may be applicable.

Fire Extinguishing: This chemical is a combustible liquid. Poisonous gases are produced in fire. Use dry chemical, carbon dioxide, or alcohol foam extinguishers. Vapors are heavier than air and will collect in low areas. Vapors may travel long distances to ignition sources and flashback. Vapors in confined areas may explode when exposed to fire. Containers may explode in fire. Storage containers and parts of containers may rocket great distances, in many directions. If material or contaminated runoff enters waterways, notify downstream users of potentially contaminated waters. Notify local health and fire officials and pollution control agencies. From a secure, explosion-proof location, use water spray to cool exposed containers. If cooling streams are ineffective (venting sound increases in volume and pitch, tank discolors, or shows any signs of deforming), withdraw immediately to a secure position. If employees are

expected to fight fires, they must be trained and equipped in OSHA 1910.156.

References

U.S. Environmental Protection Agency, "Chemical Profile: Carvone," Washington, DC, Chemical Emergency Preparedness Program (October 31, 1985)

Catechol

Molecular Formula: $C_6H_6O_2$

Common Formula: $C_6H_4(OH)_2$

Synonyms: Benzene, *o*-dihydroxy-; *o*-Benzenediol; 1,2-Benzenediol; Burmar Nophenol-922 HB; Catacol (Spanish); Catechin; C.I. 76500; C.I. oxidation base 26; *o*-Dihydroxybenzene; 1,2-Dihydroxybenzene; *o*-Dioxybenzene; *o*-Diphenol; Durafur developer C; Fouramine PCH; Fourrine 68; *o*-Hydroquinone; *o*-Hydroxyphenol; 2-Hydroxyphenol; NCI-C55856; Oxyphenic acid; P-370; Pelagol grey C; *o*-Phenylenediol; Pyrocatechin; Pyrocatechine; Pyrocatechinic acid; Pyrocatechol; Pyrocatechuic acid

CAS Registry Number: 120-80-9

RTECS Number: UX1050000

DOT ID: UN 2811

EEC Number: 604-016-00-4

EINECS Number: 204-427-5

Regulatory Authority

- Banned or Severely Restricted (Czechoslovakia) (In Cosmetics)[35]
- Air Pollutant Standard Set (ACGIH)[1] (OSHA)[58] (Several States)[60] (HSE) (Australia) (Israel) (Mexico) (Several Canadian Provinces)
- CLEAN AIR ACT: Hazardous Air Pollutants (Title I, Part A, Section 112)
- SUPERFUND/EPCRA 40CFR302.4 Reportable Quantity (RQ): CERCLA, 1 lb (0.454 kg)
- EPCRA Section 313 Form R *de minimis* concentration reporting level: 1.0%
- Canada, WHMIS, Ingredients Disclosure List

Cited in U.S. State Regulations: Alaska (G), California (G), Connecticut (A), Florida (G), Illinois (G), Maine (G), Massachusetts (G), Nevada (A), New Hampshire (G), New Jersey (G), North Dakota (A), Pennsylvania (G), Rhode Island (G), Virginia (A), West Virginia (G).

Description: Catechol, $C_6H_4(OH)_2$, is a white crystalline solid. Turns brown on contact with light and air. Boiling point = 245.5°C. Freezing/Melting point = 104°C. It sublimes readily. Flash point = 165°C. Autoignition temperature = 515°C. Explosive Limits in air: LEL = 1.4%; UEL – unknown. Hazard Identification (based on NFPA-704 M Rating System): Health 2, Flammability 1, Reactivity 0. Log P_{ow} (octanol/water partition coefficient) = 0.88.

Potential Exposure: It is used as an antiseptic, in photography, in dyestuff manufacture and application. It is also used in electroplating, in the formulation of specialty inks and in antioxidants and light stabilizers.

Incompatibilities: Strong oxidizers, nitric acid.

Permissible Exposure Limits in Air: NIOSH recommended REL is 5 ppm (20 mg/m^3) TWA for a 10-hour workshift. The ACGIH recommended level is the same for an 8-hor workshift. In additions, several states have set guidelines or standards for catechol in ambient air[60] ranging from 200 μg/m^3 (North Dakota) to 350 μg/m^3 (Virginia) to 400 μg/m^3 (Connecticut) to 476 μg/m^3 (Nevada).

Determination in Air: None available.

Permissible Concentration in Water: EPA[32] has suggested a permissible ambient goal of 280 μg/l on a health basis. The former USSR[35] has set a MAC in surface water of 0.1 mg/l.

Routes of Entry: Skin absorption, skin and eye contact, inhalation of vapors, ingestion.

Harmful Effects and Symptoms

Short Term Exposure: Catechol can affect you when breathed in. It can also rapidly enter the body through the skin. Death can occur from extensive skin contact. Lower exposures can cause skin burns, headaches, nausea, muscle twitching and convulsions. Skin allergy with rash can also occur. Catechol is a lacramator. It irritates the respiratory and digestive tracts. It is corrosive to the eyes and can cause severe burns. The substance may affect the central nervous system, causing depression, convulsions and respiratory failure. Because this is a mutagen, handle it as a possible cancer-causing substance, with extreme caution. Exposure lowers the ability of the blood to carry oxygen, causing a bluish color of the skin. Absorption through the skin results in illness akin to that which phenol produces except convulsions are more pronounced. Catechol increases blood pressure, apparently from peripheral vasoconstriction. Catechol can cause death, apparently initiated by respiratory failure. The oral LD-rat is 260 mg/kg (moderately toxic).

Long Term Exposure: Repeated or prolonged contact may cause skin sensitization and allergy. A mutagen and may have a cancer or reproductive risk. High or repeated damage may cause kidney and liver damage. Repeated lower exposures can cause methemaglobinemia, with blue color to the skin, rapid breathing and dizziness.

Points of Attack: Eyes, skin, respiratory system, central nervous system, kidneys.

Medical Surveillance: If symptoms develop or overexposure is suspected, the following may be useful: Tests for liver and kidney function. Blood methemoglobin level. Evaluation by a qualified allergist, including careful exposure history and special testing, may help diagnose skin allergy.

First Aid: If this chemical gets into the eyes, remove any contact lenses at once and irrigate immediately for at least

15 minutes, occasionally lifting upper and lower lids. Seek medical attention immediately. If this chemical contacts the skin, remove contaminated clothing and wash immediately with soap and water. Seek medical attention immediately. If this chemical has been inhaled, remove from exposure, begin rescue breathing (using universal precautions) if breathing has stopped and CPR if heart action has stopped. Transfer promptly to a medical facility. When this chemical has been swallowed, get medical attention. Give large quantities of water and induce vomiting. Do not make an unconscious person vomit.

Note to Physician: Treat for methemoglobinemia. Spectrophotometry may be required for precise determination of levels of methemoglobinemia in urine.

Personal Protective Methods: Wear protective gloves and clothing to prevent any reasonable probability of skin contact. Safety equipment suppliers/manufacturers can provide recommendations on the most protective glove/clothing material for your operation. All protective clothing (suits, gloves, footwear, headgear) should be clean, available each day, and put on before work. Contact lenses should not be worn when working with this chemical. Wear dust-proof chemical goggles and face shield unless full facepiece respiratory protection is worn. Employees should wash immediately with soap when skin is wet or contaminated. Provide emergency showers and eyewash.

Respirator Selection: Where the potential exists for exposures over 5 ppm, use a MSHA/NIOSH approved full facepiece respirator with a high efficiency particulate filter. Greater protection is provided by a powered-air purifying respirator. If while wearing a filter, cartridge or canister respirator, you can smell, taste, or otherwise detect 1,2-Dihydroxybenznene, or in the case of a full facepiece respirator you experience eye irritation, leave the area immediately. Check to make sure the respirator-to-face seal is still good. If it is, replace the filter, cartridge, or canister. If the seal is no longer good, you may need a new respirator.

Where the potential for high exposures exists, use a MSHA/NIOSH approved supplied-air respirator with a full facepiece operated in the positive pressure mode or with a full facepiece, hood, or helmet in the continuous flow mode, or use a MSHA/NIOSH approved self-contained breathing apparatus with a full facepiece operated in pressure-demand or other positive pressure mode.

Storage: Prior to working with Catechol you should be trained on its proper handling and storage. Before entering confined space where this chemical may be present, check to make sure that an explosive concentration does not exist. Store in tightly closed containers in a cool, well ventilated area away from strong oxidizers and acids. Use only non-sparking tools and equipment, especially when opening and closing containers of this chemical. Sources of ignition such as smoking and open flames, are prohibited where this chemical is used, handled, or stored in a manner that could create a potential fire or explosion hazard.

Shipping: Irritating agents, n.o.s., solid, require a "Poison" label. They fall in DOT Hazard Class 6.1 and Packing Group III. Shipment by passenger aircraft or railcar is forbidden; cargo aircraft shipments are limited.

Spill Handling: Evacuate persons not wearing protective equipment from area of spill or leak until clean-up is complete. Remove all ignition sources. Dampen spilled material with water to avoid dust or use a vacuum. Do not dry sweep. Collect powdered material in the most convenient and safe manner and deposit in sealed containers. Ventilate area after clean-up is complete. It may be necessary to contain and dispose of this chemical as a hazardous waste. If material or contaminated runoff enters waterways, notify downstream users of potentially contaminated waters. Contact your Department of Environmental Protection or your regional office of the federal EPA for specific recommendations. If employees are required to clean-up spills, they must be properly trained and equipped. OSHA 1910.120(q) may be applicable.

Fire Extinguishing: This chemical is a combustible solid. Use dry chemical, alcohol foam, or carbon dioxide. Water and conventional foam may be ineffective.[41] Poisonous gases are produced in fire. If material or contaminated runoff enters waterways, notify downstream users of potentially contaminated waters. Notify local health and fire officials and pollution control agencies. Containers may explode in fire. From a secure, explosion-proof location, use water spray to cool exposed containers. If cooling streams are ineffective (venting sound increases in volume and pitch, tank discolors, or shows any signs of deforming), withdraw immediately to a secure position. If employees are expected to fight fires, they must be trained and equipped in OSHA 1910.156.

References

New Jersey Department of Health and Senior Services and Senior Services, "Hazardous Substance Fact Sheet: 1,2-Dihydroxy Benzene," Trenton, NJ (April 1986)

Cesium Hydroxide

Molecular Formula: CsHO

Common Formula: CsOH

Synonyms: Caesium hydroxide; Cesium hydrate; Cesium hydroxide dimer

CAS Registry Number: 21351-79-1

RTECS Number: FK9800000

DOT ID: UN 2682 (solid); UN 2681 (solution)

Regulatory Authority

- Air Pollutant Standard Set (ACGIH)[1] (UNEP)[43] (OSHA)[59] (Several States)[60] (Australia) (HSE) (Israel) (Mexico) (Several Canadian Provinces)

Cited in U.S. State Regulations: Alaska (G), California (G), Connecticut (A), Florida (G), Illinois (G), Maine (G), Massachusetts (G), Minnesota (G), Nevada (A), New Hampshire (G), New Jersey (G), North Dakota (A), Pennsylvania (G), Rhode Island (G), Virginia (A), West Virginia (G).

Description: Cesium Hydroxide, CsOH, is a colorless-to-yellow crystalline compound. It is often used in a water solution. Freezing/Melting point = 272°C. Highly soluble in water.

Potential Exposure: Cesium hydroxide may be used as a raw material for other cesium salts such as the chloride which in turn may be used to produce cesium metal. Cesium metal is used in electronic devices.

Incompatibilities: Cesium hydroxide is the strongest base known and must be stored in silver or platinum out of contact with air because of its reactivity with glass of CO_2. CsOH causes the generation of considerable heat in contact with water or moisture. Contact with many organic compounds, many metals (i.e., aluminum, lead, tin, zinc), glass, oxygen or carbon dioxide causes a violent reaction.

Permissible Exposure Limits in Air: The ACGIH recommended TLV is 2 mg/m^3 for an 8-hour workshift. The former USSR/UNEP joint project[43] has set a MAC in workplace air of 0.3 mg/m^3.

In addition, several states have set guidelines or standards for CsOH in ambient air[60] ranging from 20 μg/m^3 (North Dakota) to 35 μg/m^3 (Virginia) to 40 μg/m^3 (Connecticut) to 48 μg/m^3 (Nevada).

Determination in Air: No method available.

Permissible Concentration in Water: No criteria set.

Routes of Entry: Inhalation, ingestion, skin and/or eye contact.

Harmful Effects and Symptoms

Short Term Exposure: Irritates eyes, skin, and respiratory tract. This chemical is corrosive to the eyes and can cause permanent damage. The oral LD_{50} rat is 570 mg/kg (slightly toxic).

Long Term Exposure: May cause lung irritation with the development of bronchitis, shortness of breath, coughing, phlegm.

Points of Attack: Eyes, skin, respiratory system.

Medical Surveillance: For those with frequent or potentially high exposure (half the TLV or greater) the following are recommended before beginning work and at regular times after that: Lung function tests.

First Aid: If this chemical gets into the eyes, remove any contact lenses at once and irrigate immediately for at least 30 minutes, occasionally lifting upper and lower lids. Seek medical attention immediately. If this chemical contacts the skin, remove contaminated clothing and wash immediately with soap and water. Seek medical attention immediately. If this chemical has been inhaled, remove from exposure, begin rescue breathing (using universal precautions) if breathing has stopped and CPR if heart action has stopped. Transfer promptly to a medical facility. When this chemical has been swallowed, get medical attention. If victim is conscious, administer water or milk. Do not induce vomiting.

Personal Protective Methods: *Clothing:* Avoid skin contact with Cesium Hydroxide. Wear protective gloves and clothing. Safety equipment suppliers/manufacturers can provide recommendations on the most protective glove/clothing material for your operation. All protective clothing (suits, gloves, footwear, headgear) should be clean, available each day, and put on before work. ACGIH recommended protective gloves be made of Butyl Rubber, Natural Rubber, Nitrile Rubber, Neoprene or Polyvinyl Chloride.

Eye Protection: Wear dust-proof goggles and faceshield when working with powders or dusts unless full facepiece respiratory protection is worn. Wear gas-proof goggles and faceshield where Cesium Hydroxide is in solution, unless full facepiece respiratory protection is worn.

Respirator Selection: Where the potential exists for exposures over 2 mg/m^3, use a MSHA/NKIOSH approved full facepiece respirator equipped with a particulate (dust/fume/mist) filter. Where the potential for high exposures exists, use a MSHA/NIOSH approved supplied-air respirator with a full facepiece operated in the positive pressure mode or with a full facepiece, hood, or helmet in the continuous flow mode, or us a MSHA/NIOSH approved self-contained breathing apparatus with a full facepiece operated in pressure-demand or other positive pressure mode.

Storage: Prior to working with Cesium hydroxide you should be trained on its proper handling and storage. Cesium Hydroxide should be stored in silver or platinum away from air because it reacts violently with oxygen. Store in tightly closed containers in a cool, well-ventilated area away from moisture and incompatible materials listed above.

Shipping: Cesium hydroxide must bear a "Corrosive" label. It falls in DOT Hazard Class 8 and Packing Group II.

Spill Handling: Evacuate and restrict persons not wearing protective equipment from area of spill or leak until cleanup is complete. Remove all ignition sources. Do not use water in clean-up. Absorb liquids in vermiculite, dry sand, earth, peat, carbon, or a similar material and deposit in sealed containers. Collect powdered material in the most convenient and safe manner and deposit in sealed containers. Ventilate area of spill or leak after cleanup is complete. It may be necessary to contain and dispose of this chemical as a hazardous waste. If material or contaminated runoff enters waterways, notify downstream users of potentially contaminated waters. Contact your Department of Environmental Protection or your regional office of the federal EPA for specific recommendations. If employees are required to clean-up spills, they must be properly trained and equipped. OSHA 1910.120(q) may be applicable.

Fire Extinguishing: This chemical is a combustible solid. Do not use water. Use dry chemical appropriate for extinguishing metal fires. Poisonous gases are produced in fire.

If material or contaminated runoff enters waterways, notify downstream users of potentially contaminated waters. Notify local health and fire officials and pollution control agencies. Containers may explode in fire. From a secure, explosion-proof location, use water spray to cool exposed containers. If cooling streams are ineffective (venting sound increases in volume and pitch, tank discolors, or shows any signs of deforming), withdraw immediately to a secure position. If employees are expected to fight fires, they must be trained and equipped in OSHA 1910.156.

References

New Jersey Department of Health and Senior Services and Senior Services, "Hazardous Substance Fact Sheet: Cesium Hydroxide," Trenton, NJ (August 1998)

Chloral

Molecular Formula: C_2HCl_3O

Common Formula

CCl_3CHO

Synonyms: Acetaldehyde, trichloro-; Anhydrous chloral; Chloral, anhydrous, inhibited; Cloralio; Ethanal, trichloro-; Grasex; 2,2,2-Trichloroacetaldehyde; Trichloroacetaldehyde; Trichloroethanal; Tricloroacetaldehido (Spanish)

CAS Registry Number: 75-87-6 (chloral); 302-17-0 (chloral hydrate)

RTECS Number: FM7870000 (chloral); FM8750000 (chloral hydrate)

DOT ID: UN 2075 (chloral); UN 2811 (chloral hydrate)

EEC Number: 605-014-00-6 (chloral hydrate)

Regulatory Authority

- Air Pollutant Standard Set (former USSR) (See EPA reference below)
- EPA HAZARDOUS WASTE NUMBER (RCRA No.): U034, as chloral
- RCRA, 40CFR261, Appendix 8 Hazardous Constituents, as chloral
- SUPERFUND/EPCRA 40CFR302.4 Reportable Quantity (RQ): CERCLA, 5,000 lb (2,270 kg), as chloral
- Canada, WHMIS, Ingredients Disclosure List

Cited in U.S. State Regulations: Kansas (G), Louisiana (G), Massachusetts (G), New Hampshire (G), New Jersey (G), Oklahoma (G), Pennsylvania (G), Vermont (G), Virginia (G), Washington (G), Wisconsin (G).

Description: Chloral, CCl_3CHO, is a combustible, oily liquid with a pungent irritating odor. Boiling point = 97 – 98°C. Flash point = 75°C. Chloral hydrate is colorless crystals, with characteristic odor. Boiling point = 97°C (decomposes). Freezing/Melting point = 57 – 60°C.

Potential Exposure: Chloral is used as an intermediate in the manufacture of such pesticides as DDT, methoxychlor, DDVP, naled, trichlorfon, and TCA. Chloral is also used in the production of chloral hydrate, a therapeutic agent with hypnotic and sedative effects used prior to the introduction of barbiturates.

Incompatibilities: Chloral hydrate reacts with strong bases forming chloroform. Contact with acids, or exposure to light may cause polymerization. Reacts with water forming chloral hydrate. Reacts with oxidizers, with a risk of fire or explosions.

Permissible Exposure Limits in Air: There are no U.S. Standards. The former USSR has recommended a maximum concentration in workroom air of 220 mg/m^3.

Permissible Concentration in Water: There are no U.S. criteria but the former USSR has set 0.2 mg/l as the MAC for water bodies used for domestic purposes.[43]

Harmful Effects and Symptoms

Short Term Exposure: Irritates the eyes, skin, respiratory tract. Skin and eye contact may cause burns. Chloral may affect the central nervous system, kidneys, liver and the cardiovascular system, causing impaired functions or damage. Exposure to high levels may cause tiredness, dizziness, lightheadedness and loss of consciousness. Specific information on the pharmacokinetic behavior, carcinogenicity, mutagenitcity, teratogenicity, and other reproductive effects of chloral was not found in the available literature. However, the pharmacokinetic behavior of chloral may be similar to chloral hydrate where metabolism to ritchloroethanol and trichloroacetic acid and excretion via the urine (and possibly bile) have been observed. Chloral hydrate produced skin tumors in 4 of 20 mice dermally exposed. Alcohol synergistically increases the depressant effect of the compound, creating a potent depressant commonly referred to as "Mickey Finn" or "knock-out drops." Addiction to chloral hydrate through intentional abuse of the compound has been reported.

Long Term Exposure: Repeated skin contact may cause acne-like rash. Repeated contact may cause sedation. Chronic effects from respiratory exposure to Chloral as indicated in laboratory animals include reduction of kidney function and serum transaminase activity, change in central nervous system function (unspecified), decrease in antitoxic and enzyme-synthesizing function of the liver, and alteration of morphological characteristics of peripheral blood. Slowed growth rate, leukocytosis and changes in the arterial blood pressure were also observed.

Points of Attack: Inhalation, ingestion.

Medical Surveillance: Lung function tests. Serum trichloroethanol level.

First Aid: If this chemical gets into the eyes, remove any contact lenses at once and irrigate immediately for at least 15 minutes, occasionally lifting upper and lower lids. Seek medical attention immediately. If this chemical contacts the skin, remove contaminated clothing and wash immediately with soap and water. Seek medical attention immediately. If this chemical has been inhaled, remove from exposure, begin rescue breathing (using universal precautions) if breathing has stopped and CPR if heart action has stopped. Transfer promptly to a medical facility. When this chemical has been swallowed, get medical attention. Give large quantities of water and induce vomiting. Do not make an unconscious person vomit.

Personal Protective Methods: Wear protective gloves and clothing to prevent any reasonable probability of skin contact. Safety equipment suppliers/manufacturers can provide recommendations on the most protective glove/clothing material for your operation. All protective clothing (suits, gloves, footwear, headgear) should be clean, available each day, and put on before work. Contact lenses should not be worn when working with this chemical. Wear splash-proof chemical goggles and face shield unless full facepiece respiratory protection is worn. Employees should wash immediately with soap when skin is wet or contaminated. Provide emergency showers and eyewash.

Respirator Selection: *Where there is potential for exposure to chloral:* SCBAF:PD,PP (any MSHA/NIOSH approved self-contained breathing apparatus that has a full facepiece and is operated in a pressure-demand or other positive-pressure mode); or SAF:PD,PP:ASCBA (any supplied-air respirator that has a full facepiece and is operated in a pressure-demand or other positive-pressure mode in combination with an auxiliary, self-contained breathing apparatus operated in a pressure-demand or other positive pressure mode).

Storage: Prior to working with Chloral you should be trained on its proper handling and storage. Protect from light, moisture, air and acids. DEA regulations require storage in a locked storage area. Store in tightly closed containers in a cool, well ventilated area. Metal containers involving the transfer of this chemical should be grounded and bonded. Drums must be equipped with self-closing valves, pressure vacuum bungs, and flame arresters. Use only non-sparking tools and equipment, especially when opening and closing containers of this chemical. Sources of ignition such as smoking and open flames, are prohibited where this chemical is used, handled, or stored in a manner that could create a potential fire or explosion hazard.

Shipping: Chloral, anhydrous, inhibited requires a "Poison" label. It falls in Hazard Class 6.1 and Packing Group II.

Spill Handling: Evacuate and restrict persons not wearing protective equipment from area of spill or leak until cleanup is complete. Remove all ignition sources. Ventilate area of spill or leak. Absorb liquids in vermiculite, dry sand, earth, peat, carbon, or a similar material and deposit in sealed containers. It may be necessary to contain and dispose of this chemical as a hazardous waste. If material or contaminated runoff enters waterways, notify downstream users of potentially contaminated waters. Contact your Department of Environmental Protection or your regional office of the federal EPA for specific recommendations. If employees are required to clean-up spills, they must be properly trained and equipped. OSHA 1910.120(q) may be applicable.

Fire Extinguishing: This chemical is a combustible liquid. Poisonous gases are produced in fire, including hydrogen chloride. Use dry chemical, carbon dioxide, or foam extinguishers. Vapors are heavier than air and will collect in low areas. Vapors may travel long distances to ignition sources and flashback. Vapors in confined areas may explode when exposed to fire. Containers may explode in fire. Storage containers and parts of containers may rocket great distances, in many directions. If material or contaminated runoff enters waterways, notify downstream users of potentially contaminated waters. Notify local health and fire officials and pollution control agencies. From a secure, explosion-proof location, use water spray to cool exposed containers. If cooling streams are ineffective (venting sound increases in volume and pitch, tank discolors, or shows any signs of deforming), withdraw immediately to a secure position. If employees are expected to fight fires, they must be trained and equipped in OSHA 1910.156.

Disposal Method Suggested: Incineration after mixing with another combustible fuel; care must be taken to assure complete combustion to prevent phosgene formation; an acid scrubber is necessary to remove the halo acids produced.

References

U.S. Environmental Protection Agency, Chloral, Health and Environmental Effects Profile No. 34, Office of Solid Waste, Washington, DC (April 30, 1980)

New Jersey Department of Health and Senior Services, "Hazardous Substance Fact Sheet, Chloral," Trenton, NJ (April 1997)

Chloramben

Molecular Formula: $C_7H_5Cl_2NO_2$

Synonyms: ACP-M-728; Amben®; Ambiben®; Amiben®; Amibin®; 3-Amino-2,5-dichlorobenzoic acid; 3-Amino-2,6-dichlorobenzoic acid; Amoben; Benzoic acid, 3-amino-2,5-dichloro-; Chlorambed; Chloramben, aromatic carboxylic acid; Chloramben benzoic acid herbicide; Chlorambene; 2,5-Dichloro-3-aminobenzoic acid; NCI-C00055; Ornamental weeder; Vegaben®; Vegiben®; Weedone garden weeder

CAS Registry Number: 133-90-4

RTECS Number: DG1925000

DOT ID: UN 2588

Regulatory Authority

- Carcinogen, (animal positive) (NTP)
- Air Pollutant Standard Set (former USSR)[35] (Pennsylvania)[60]
- CLEAN AIR ACT: Hazardous Air Pollutants (Title I, Part A, Section 112)
- SUPERFUND/EPCRA 40CFR302.4 Reportable Quantity (RQ): CERCLA, 1 lb (0.454 kg)

- EPCRA Section 313 Form R *de minimis* concentration reporting level: 1.0%

Cited in U.S. State Regulations: California (A), Maine (W), Massachusetts (G), New Jersey (G), Pennsylvania (A), Wisconsin (W).

Description: Chloramben, is a colorless, odorless, crystalline solid. Freezing/Melting point = 200 – 201°C. Soluble in water.

Potential Exposure: It is used as a herbicide for grasses, and broadlasf weeds, on soybeans, beans, and some vegetables. Workers involved in the manufacture, formulation or application of this reemergence herbicide.

Incompatibilities: Rapidly decomposed by light. Strong acids and acid fumes.

Permissible Exposure Limits in Air: Although no occupational exposure limits have been established, this chemical can be absorbed through the skin. The former USSR[35] has set a MAC in ambient air in residential areas of 0.01 mg/m^3 on a momentary basis and 0.006 mg/m^3 on an average daily basis.

Pennsylvania[60] has set a guideline for chloramben in ambient air of 1.3333 mg/m^3.

Permissible Concentration in Water: The USR[35] has set a MAC of 0.5 mg/l in surface water. A lifetime health advisory of 0.105 mg/l has been determined by EPA (see reference below).

Routes of Entry: Inhalation, passes through the skin.

Harmful Effects and Symptoms

The available data on Chloramben are very sparse. Much additional information is needed regarding its chronic toxicity, teratogenicity, and carcinogenicity before limits can be confidently set.

No-observed-adverse-effect doses for chloramben were 15 mg/kg/day. Based on these data an ADI was calculated at 0.015 mg/kg/day.

The LD_{50} rat is 3,500 mg/kg (slightly toxic).

Short Term Exposure: Skin or eye contact may cause irritation.

Long Term Exposure: There is evidence that this chemical causes cancer in animals. It may cause cancer of the liver. Repeated exosure may cause skin rash with itching.

Points of Attack: Skin, liver.

Medical Surveillance: Liver function tests. Examination by a qualified allergist.

First Aid: If this chemical gets into the eyes, remove any contact lenses at once and irrigate immediately for at least 15 minutes, occasionally lifting upper and lower lids. Seek medical attention immediately. If this chemical contacts the skin, remove contaminated clothing and wash immediately with soap and water. Seek medical attention immediately. If this chemical has been inhaled, remove from exposure, begin rescue breathing (using universal precautions) if breathing has stopped and CPR if heart action has stopped. Transfer promptly to a medical facility. When this chemical has been swallowed, get medical attention. Give large quantities of water and induce vomiting. Do not make an unconscious person vomit.

Personal Protective Methods: Wear protective gloves and clothing to prevent any reasonable probability of skin contact. Safety equipment suppliers/manufacturers can provide recommendations on the most protective glove/clothing material for your operation. All protective clothing (suits, gloves, footwear, headgear) should be clean, available each day, and put on before work. Contact lenses should not be worn when working with this chemical. Wear dust-proof chemical goggles and face shield unless full facepiece respiratory protection is worn. Employees should wash immediately with soap when skin is wet or contaminated. Provide emergency showers and eyewash.

Respirator Selection: *Where there is a potential for overexposure:* SCBAF:PD,PP (any MSHA/NIOSH approved self-contained breathing apparatus that has a full facepiece and is operated in a pressure-demand or other positive-pressure mode); or SAF:PD,PP:ASCBA (any supplied-air respirator that has a full facepiece and is operated in a pressure-demand or other positive-pressure mode in combination with an auxiliary, self-contained breathing apparatus operated in a pressure-demand or other positive pressure mode).

Storage: Prior to working with Chloramben you should be trained on its proper handling and storage. Store in a cool, dry place or a refrigerator and avoid contact with strong acids, acid fumes and light. A regulated, marked area should be established where this chemical is handled, used, or stored in compliance with OSHA standard 1910.1045.

Shipping: Pesticides, solid, toxic, n.o.s., in Hazard Class 6.1 and Packing Group III require a "Keep Away from Food" label.

Spill Handling: Evacuate persons not wearing protective equipment from area of spill or leak until clean-up is complete. Remove all ignition sources. Collect powdered material in the most convenient and safe manner and deposit in sealed containers. Ventilate area after clean-up is complete. It may be necessary to contain and dispose of this chemical as a hazardous waste. If material or contaminated runoff enters waterways, notify downstream users of potentially contaminated waters. Contact your Department of Environmental Protection or your regional office of the federal EPA for specific recommendations. If employees are required to clean-up spills, they must be properly trained and equipped. OSHA 1910.120(q) may be applicable.

Fire Extinguishing: Use dry chemical, carbon dioxide, water spray, or foam extinguishers. Poisonous gases are produced in fire including toxic chloride fumes and nitrous oxides. If material or contaminated runoff enters waterways, notify downstream users of potentially contaminated waters.

Notify local health and fire officials and pollution control agencies. Containers may explode in fire. From a secure, explosion-proof location, use water spray to cool exposed containers. If cooling streams are ineffective (venting sound increases in volume and pitch, tank discolors, or shows any signs of deforming), withdraw immediately to a secure position. If employees are expected to fight fires, they must be trained and equipped in OSHA 1910.156.

Disposal Method Suggested: Chloramben is stable to heat, oxidation, and hydrolysis in acidic or basic media. The stability is comparable to that of benzoic acid. Wet oxidation or incineration are recommended disposal methods.[22]

References

National Cancer Institute, Bioassay of Chloramben for Possible Carcinogenicity, Technical Report Series No. 25, Bethesda, Maryland (1977)

Sax, N. I., Ed., "Dangerous Properties of Industrial Materials Report," 1, No. 3, 28–29, New York, Van Nostrand Reinhold Co. (1981). (As 3-Amino-2,5-Dichlorobenzoic Acid)

U.S. Environmental Protection Agency, "Health Advisory: Chloramben." Washington, DC, Office of Drinking Water (August 1987)

New Jersey Department of Health and Senior Services, Hazardous Substance Fact Sheet, Chloramben. Trenton NJ (September, 1998)

Chlorambucil

Molecular Formula: $C_{14}H_{19}Cl_2NO_2$

Common Formula: $(ClCH_2CH_2)_2N\text{-}C_6H_4\text{-}(CH_2)_3COOH$

Synonyms: Ambochlorin; Amboclorin; Benzenebutanoic acid, 4-[Bis(2-chloroethyl)amino]-; 4-[Bis(2-Chloroethyl) amino]benzenebutanoic acid; γ-[*p*-Bis(2-chloroethyl)aminophenyl]butyric acid; 4-(*p*-[Bis(2-chloroethyl)amino]phenyl)butyric acid; 4-[*p*-Bis(β-chloroethyl)aminophenyl]butyric acid; CB 1348; Chloraminophene; Chloroambucil; Chloroaminophen; Chlorobutin; Chlorobutine; Clorambucil (Spanish); γ(*p*-Di-(2-chloroethyl)aminophenyl)butyric acid; *N,N*-Di-2-chloroethyl-γ-*p*-aminophenylbutyric acid; *p*-(*N,N*-Di-2-chloroethyl)aminophenyl butyric acid; *p*-*N,N*-Di-(β-chloroethyl)aminophenylbutyric acid; Ecoril; Elcoril; Leukeran; Leukersan®; Leukoran®; Linfolizin; Linfolysin; NCI-CO3485; NSC-3088; Phenylbuyyric acid nitrogen mustard

CAS Registry Number: 305-03-3

RTECS Number: ES7525000

Regulatory Authority

- Carcinogen (Animal, Human) (IARC)[9] (NTP/7)
- EPA HAZARDOUS WASTE NUMBER (RCRA No.): U035
- RCRA, 40CFR261, Appendix 8 Hazardous Constituents
- SUPERFUND/EPCRA 40CFR302.4 Reportable Quantity (RQ): CERCLA, 10 lb (4.54 kg)

Cited in U.S. State Regulations: California (A, G), Florida (G), Illinois (G), Kansas (G), Louisiana (G), Maine (G), Massachusetts (G), New Hampshire (G), New Jersey (G), Pennsylvania (G), Rhode Island (G), Vermont (G), Virginia (G), Washington (G), West Virginia (G), Wisconsin (G).

Description: Chlorambucil, $(ClCH_2CH_2)_2N\text{-}C_6H_4\text{-}(CH_2)_3COOH$, is a crystalline solid. Freezing/Melting point = 64 – 66°C. Insoluble in water.

Potential Exposure: Chlorambucil, an anti-cancer drug, is a derivative of nitrogen mustard. This drug is primarily used as an antineoplastic agent for the treatment of lymphocytic leukemia, malignant lymphomas, follicular lymphoma, and Hodgkin's disease. The treatments are not curative but do produce some marked remissions. Chlorambucil has also been tested for treatment of chronic hepatitis, rheumatoid arthritis, and as an insect chemosterilant.

All of the chemical used in this country is imported from the United Kingdom. Work exposure in the United States would be limited to workers formulating the tablets, or to patients receiving the drug.

Incompatibilities: Moisture.

Permissible Exposure Limits in Air: No standards set.

Permissible Concentration in Water: No criteria set.

Harmful Effects and Symptoms

Chlorambucil is carcinogenic in rats and mice following intraperitoneal injection, producing lymphomas in rats, and lymphosarcomas, ovarian tumors in mice. Excesses of acute leukemia were reported in a number of epidemiological studies of people treated with Chlorambucil, either alone or in combination with other therapies, for both nonmalignant, and malignant diseases. Other cancers have also been associated with the use of Chlorambucil and other agents. An excess of acute leukemia in association with Chlorambucil was seen in a further study in which 431 previously untreated patients with polycythemia vera were given phlebotomy alone or Chlorambucil with phlebotomy, and followed for a mean of 6.5 years. Of the 26 cases of acute leukemia that occurred, 16 were in the group receiving Chlorambucil. The risk increased with increasing dose and time of treatment. The oral LD_{50} rat is 76 mg/kg (moderately toxic). Causes nausea and vomiting and CNS excitation in humans.

Short Term Exposure: Irritates the eyes, and respiratory tract. Exposure can cause dizziness, loss of coordination, numbness, weakness, muscle twitching, convulsions and unconsciousness.

Long Term Exposure: This chemical is a carcinogen in humans. It has been shown to cause lung cancer and leukemia. It is a probable teratogen in humans and may damage the testes in males and decrease fertility in females.

Points of Attack: Lung, kidney, liver, blood.

Medical Surveillance: Liver and kidney function tests, lung function tests, comblete blood count (CBC).

First Aid: *Skin Contact:* Flood all areas of body that have contacted the substance with water. Don't wait to remove contaminated clothing; do it under the water stream. Use soap to help assure removal. Isolate contaminated clothing when removed to prevent contact by others.

Eye Contact: Remove any contact lenses at once. Flush eyes well with copious quantities of water or normal saline for at least 20-30 minutes. Seek medical attention.

Inhalation: Leave contaminated area immediately; breathe fresh air. Proper respiratory protection must be supplied to any rescuers. It coughing, difficult breathing or any other symptoms develop, seek medical attention at once, even if symptoms develop many hours after exposure.

Ingestion: If convulsions are not present, give a glass or two of water or milk to dilute the substance. Assure that the person's airway is unobstructed and contact a hospital or poison center immediately for advice on whether or not to induce vomiting.

Personal Protective Methods: Wear protective gloves and clothing to prevent any reasonable probability of skin contact. Safety equipment suppliers/manufacturers can provide recommendations on the most protective glove/clothing material for your operation. All protective clothing (suits, gloves, footwear, headgear) should be clean, available each day, and put on before work. Contact lenses should not be worn when working with this chemical. Wear or dust-proof chemical goggles and face shield unless full facepiece respiratory protection is worn. Employees should wash immediately with soap when skin is wet or contaminated. Provide emergency showers and eyewash.

Respirator Selection: *Where there is a potential for overexposure:* SCBAF:PD,PP (any MSHA/NIOSH approved self-contained breathing apparatus that has a full facepiece and is operated in a pressure-demand or other positive-pressure mode); or SAF:PD,PP:ASCBA (any supplied-air respirator that has a full facepiece and is operated in a pressure-demand or other positive-pressure mode in combination with an auxiliary, self-contained breathing apparatus operated in a pressure-demand or other positive pressure mode).

Storage: Prior to working with Chlorambucil you should be trained on its proper handling and storage. Store in cool, dry place. Store in sealed ampules or in amber screw-capped bottles or vials with Teflon cap liners. Solutions may be stored in bottles or vials with a silicone system having a Teflon liner and sampled with needle and syringe. Prevent exposure to light. A regulated, marked area should be established where this chemical is handled, used, or stored in compliance with OSHA standard 1910.1045.

Shipping: Chlorambucil is not specifically cited by DOT but may be considered as a hazardous substance, solid, n.o.s., in UN 9188. This imposes no label requirement and there are no maximum quantity limits on aircraft or railcar shipment.

Spill Handling: Evacuate persons not wearing protective equipment from area of spill or leak until clean-up is complete. Remove all ignition sources. Collect powdered material in the most convenient and safe manner and deposit in sealed containers. Ventilate area after clean-up is complete. It may be necessary to contain and dispose of this chemical as a hazardous waste. If material or contaminated runoff enters waterways, notify downstream users of potentially contaminated waters. Contact your Department of Environmental Protection or your regional office of the federal EPA for specific recommendations. If employees are required to clean-up spills, they must be properly trained and equipped. OSHA 1910.120(q) may be applicable.

Fire Extinguishing: This chemical is a combustible solid. Use dry chemical, carbon dioxide, water spray, or alcohol foam extinguishers. Poisonous gases are produced in fire including carbon monoxide, nitrogen oxides, and hydrogen chloride. If material or contaminated runoff enters waterways, notify downstream users of potentially contaminated waters. Notify local health and fire officials and pollution control agencies. From a secure, explosion-proof location, use water spray to cool exposed containers. If cooling streams are ineffective (venting sound increases in volume and pitch, tank discolors, or shows any signs of deforming), withdraw immediately to a secure position. If employees are expected to fight fires, they must be trained and equipped in OSHA 1910.156.

Disposal Method Suggested: Permanganate oxidation, high temperature incineration with scrubbing equipment, or microwave plasma treatment.

References

New Jersey Department of Health and Senior Services, "Hazardous Substance Fact Sheet, Chlorambucil," Trenton, NJ (April 1999).

Sax, N. I., Ed., "Dangerous Properties of Industrial Materials Report," 1, No. 4, 43–44 (1981) and 5, No. 1, 49–53 (1985).

Chloramphenicol

Molecular Formula: $C_{11}H_{12}Cl_2N_2O_5$

Common Formula: $O_2NC_6H_4CHOHCH(NHCOCHCl_2)CH_2OH$

Synonyms: Acetamide, 2,2-dichloro-*N*-(β-hydroxy-α-(hydroxymethyl)-*p*-nitrophenethyl)-, d-(-)-Threo-; Acetamide, 2,2-dichloro-*N*-[2-hydroxy-1-(hydroxymethyl)-2-(4-nitrophenyl)ethyl]-; Acetamide, 2,2-dichloro-*N*-2-hydroxy-1-(hydroxymethyl)-2-(4-nitrophenyl)ethyl-, R-(R*,R*)-; Alficetyn; Ambofen; Amphenicol; Amphicol; Amseclor; Aquamycetin; Austracil; Austracol; Biocetin; Biophenicol; CAF; CAM; CAP; Catilan; Chemicetin; Chemicetina; Chlomin; Chlomycol; Chloramex; D-Chloramphenicol; Chloramsaar; Chlorasol; Chlora-tabs; Chloricol; Chloro-25 Vetag; Chlorocaps; Chlorocid; Chlorocide; Chlorocidin C; Chlorocidin C Tetran; Chlorocol; Chloromycetin; Chloromycetin R; Chloronitrin; Chloroptic; Cidocetine; Ciplamycetin; Cloramficin; Cloramicol; Cloramidina; Cloroamfenicolo (Italian); Clorocyn; Cloromisan; Clorosintex;

Comycetin; CPH; Cylphenicol; Desphen; Detreomycine; Dextromycetin; Doctamicina; D-Threochloramphenicol; D-Threo-*N*-dichloroacetyl-1-*p*-nitrophenyl-2-amino-1,3-propanediol; D-Threo-*N*-(1,1'-dihydroxy-1-*p*-nitrophenyl-isopropyl)dichloroacetamide; D-Threo-1-(*p*-nitrophenyl)-2-(dichloroacetylamino)-1,3-propanediol; Econochlor; Embacetin; Emetren; Enicol; Enteromycetin; Erbaplast; Ertilen; Farmicetina; Fenicol; Globenicol; Glorous; Halomycetin; Hortfenicol; I 337A; Intramycetin; Isicetin; Ismicetina; Isophenicol; Isopto Fenicol; Juvamycetin; Kamaver; Kemicetina; Kemicetine; Klorita; Klorocid S; Leukomyan; Leukomycin; Levomicetina; Levomycetin; Loromisan; Loromisin; Mastiphen; Mediamycetine; Micloretin; Micochlorine; Micoclorina; Microcetina; Mychel; Mycinol; NCI-C55709; Normimycin V; Novochlorocap; Novomycetin; Novophenicol; NSC 3069; Oftalent; Oleomycetin; Opclor; Opelor; Ophthochlor; Ophtochlor; Otachron; Otophen; Pantovernil; Paraxin; Pentamycetin; Quemicetina; Rivomycin; Romphenil; Septicol; Sificetina; Sintomicetina; Sintomicetine R; Stanomycetin; Synthomycetin; Synthomycetine; Synthomycine; Tevcocin; Tevcosin; D-(-)-Threochloramphenicol; D-(-)-Threo-2-dichloroacetamido-1-*p*-nitrophenyl-1,3-propanediol; D-(-)-Threo-2,2-dichloro-*N*-[β-hydroxy-α-(hydroxymethyl)]-*p*-nitrophenethylacetamide; D-(-)-Threo-1-*p*-nitrophenyl-2-dichloracetamido-1,3-propanediol; Tifomycin; Tifomycine; Treomicetina; U-6062; Unimycetin; Veticol

CAS Registry Number: 56-75-7

RTECS Number: AB6825000

DOT ID: UN 1851

Regulatory Authority

- Carcinogen (Human Suspected) (IARC)[9]
- Banned or Severely Restricted (Medical Uses Restricted) (UN)[13]

Cited in U.S. State Regulations: California (A, G), Florida (G), Massachusetts (G), Minnesota (G), New Jersey (G), Pennsylvania (G), Rhode Island (G).

Description: Chloramphenicol, $O_2NC_6H_4CHOHCH(NHCOCHCl_2)CH_2OH$, is a white to grayish-white or yellowish-white crystalline solid. Freezing/Melting point = 151°C. Slightly soluble in water.

Potential Exposure: Those involved in the manufacture, formulation and application of this antibiotic and antifungal agent.

Permissible Exposure Limits in Air: No standards set but the FDA (Food and Drug Administration) has set standards for Good Manufacturing Practices for Drugs and Pharmaceuticals. These should be followed for personal protection as well as product quality. See the FDA regulation 21 CFR 210. Also, there may be no safe level of exposure to a carcinogen, so all contact should be reduced to the lowest possible level. It should be recognized that this chemical can be absorbed through the skin, thereby increasing exposure.

Permissible Concentration in Water: No criteria set.

Routes of Entry: Inhalation, ingestion.

Harmful Effects and Symptoms

Short Term Exposure: Chloramphenicol can affect you when breathed in and by passing through your skin. Skin or eye contact can cause irritation. Exposure can damage the bone marrow's ability to make blood cells and/or platelets (for blood clotting). This can lead to severe illness or death.

Long Term Exposure: High or repeated exposure can damage the liver. Effects on the nervous system may also occur, such as numbness and tingling in the fingers or toes and blurred vision. Chloramphenicol is a carcinogen, mutagen and teratogen. The LD_{50} oral-rat is 2,500 mg/kg (slightly toxic).

Points of Attack: Liver, nervous system, blood.

Medical Surveillance: Before beginning employment and monthly after that, the following is recommended: Complete blood count (CBC) with platelet count. If symptoms develop or overexposure is suspected, the following may be useful: Liver function tests; exam of the nervous system.

First Aid: If this chemical gets into the eyes, remove any contact lenses at once and irrigate immediately for at least 15 minutes, occasionally lifting upper and lower lids. Seek medical attention immediately. If this chemical contacts the skin, remove contaminated clothing and wash immediately with soap and water. Seek medical attention immediately. If this chemical has been inhaled, remove from exposure, begin rescue breathing (using universal precautions) if breathing has stopped and CPR if heart action has stopped. Transfer promptly to a medical facility. When this chemical has been swallowed, get medical attention. Give large quantities of water and induce vomiting. Do not make an unconscious person vomit.

Personal Protective Methods: Wear protective gloves and clothing to prevent any reasonable probability of skin contact. Safety equipment suppliers/manufacturers can provide recommendations on the most protective glove/clothing material for your operation. All protective clothing (suits, gloves, footwear, headgear) should be clean, available each day, and put on before work. Contact lenses should not be worn when working with this chemical. Wear dust-proof chemical goggles and face shield unless full facepiece respiratory protection is worn. Employees should wash immediately with soap when skin is wet or contaminated. Provide emergency showers and eyewash.

Respirator Selection: Where the potential exists for exposures to Chloramphenicol, use a MSHA/NIOSH approved supplies-air respirator with a full facepiece operated in the positive pressure mode or with a full facepiece, hood, or helmet in the continuous flow mode, or use a MSHA/NIOSH approved self-contained breathing apparatus with a full facepiece operated in pressure-demand or other positive pressure mode.

Storage: Prior to working with Chloramphenicol you should be trained on its proper handling and storage. A regulated, marked area should be established where this chemical is handled, used, or stored in compliance with OSHA standard 1910.1045. Store in tightly closed containers in a cool, well-ventilated area.

Shipping: Chloramphenicol is not specifically cited by DOT but reference may be made to the category Medicines, poisonous, solid, n.o.s. where Category III requires a "Keep Away from Food" label.

Spill Handling: Evacuate persons not wearing protective equipment from area of spill or leak until clean-up is complete. Remove all ignition sources. Collect powdered material in the most convenient and safe manner and deposit in sealed containers. Ventilate area after clean-up is complete. It may be necessary to contain and dispose of this chemical as a hazardous waste. If material or contaminated runoff enters waterways, notify downstream users of potentially contaminated waters. Contact your Department of Environmental Protection or your regional office of the federal EPA for specific recommendations. If employees are required to clean-up spills, they must be properly trained and equipped. OSHA 1910.120(q) may be applicable.

Fire Extinguishing: Use dry chemical, carbon dioxide, water spray, or foam extinguishers. Poisonous gases are produced in fire including chlorine and nitrogen oxides. If material or contaminated runoff enters waterways, notify downstream users of potentially contaminated waters. Notify local health and fire officials and pollution control agencies. From a secure, explosion-proof location, use water spray to cool exposed containers. If cooling streams are ineffective (venting sound increases in volume and pitch, tank discolors, or shows any signs of deforming), withdraw immediately to a secure position. If employees are expected to fight fires, they must be trained and equipped in OSHA 1910.156.

References

New Jersey Department of Health and Senior Services and Senior Services, "Hazardous Substance Fact Sheet: Chloramphenicol," Trenton, NJ (April, 1998)

Chlordane

Molecular Formula: $C_{10}H_6Cl_8$

Synonyms: Aspon-Chlordane; Belt; CD 68; Chloordaan (Dutch); Chlordan; Chlorindan; Chlor kil; Chlorodane; Clordan (Italian); Clordano (Spanish); Corodane; Cortilan-Neu; Dichlorochlordene; Dowchlor; ENT 25,552-X; ENT 9,932; γ-Chlordan; HCS 3260; Kypchlor; M 140; M 410; 4,7-Methanoindan, 1,2,3,4,5,6,7,8,8-octachloro-2,3,3a, 4,7,7a-hexahydro-; 4,7-Methanoindan, 1,2,4,5,6,8,8-octachloro 3a,4,7,7a-tetrahydro; 4,7-Methano-1H-indene,1,2,4,5,6,7,8,8-octachloro-2,3,3a,4,7,7a-hexahydro-; NCI-C00099; Niran; 1,2,4,5,6,7,8,8-Octachloor-3a,4,7,7a-tetrahydro-4,7-endo-methano-indaan (Dutch); Octachlor; Octachlorodihydrodicyclopentadiene; 1,2,4,5,6,7,8,8-Octachloro-2,3,3a,4,7,7a-hexahydro-4,7-methano-1H-indene; 1,2,4,5,6,7,8,8-Octachloro-2,3,3a,4,7,7a-hexahydro-4,7-methanoindene; 1,2,4,5,6,7,8,8-Octachloro-3a,4,7,7a-hexahydro-4,7-methylene indane; Octachloro-4,7-methanohydroindane; 1,2,4,5,6,7,8,8-Octachloro-4,7-methano-3a,4,7,7a-tetrahydroindane; Octachloro-4,7-methanotetrahydroindane; 1,2,4,5,6,7,8,8-Octachloro-3a,4,7,7a-tetrahydro-4,7-methanoindan; 1,2,4,5,6,7,8,8-Octachloro-3a,4,7,7a-tetrahydro-4,7-methanoindane; 1,2,4,5,6,7,10,10-Octachloro-4,7,8,9-tetrahydro-4,7-methyleneindane; 1,2,4,5,6,7,8,8-Octachlor-3a,4,7,7a-tetrahydro-4,7-endo-methano-indan (German); Octa-klor; Oktaterr; OMS 1437; Ortho-klor; 1,2,4,5,6,7,8,8-Ottochloro-3a,4,7,7a-tetraidro-4,7-endo-metano-indano (Italian); SD 5532; Shell SD-5532; Synklor; TAT; TAT Chlor 4; Topichlor 20; Topiclor; Topiclor 20; Toxichlor; Velsicol 1068

CAS Registry Number: 57-74-9

RTECS Number: PB9800000

DOT ID: UN 2762 (organochlorine pesticide, liquid, flammable, poisonous)

EEC Number: 602-047-00-8

Regulatory Authority

- Carcinogen (Animal Positive) (IARC) (NCI)[9] (DFG)[3]
- Banned or Severely Restricted (In Agriculture) (Many Countries) (UN)[13][35]
- Air Pollutant Standard Set (ACGIH)[1] (DFG)[3] (HSE)[33] (UNEP)[43] (OSHA)[58] (Several States)[60] (Australia) (Israel) (Mexico) (Several Canadian Provinces)
- CLEAN AIR ACT: Hazardous Air Pollutants (Title I, Part A, Section 112)
- CLEAN WATER ACT: Section 311 Hazardous Substances/RQ 40CFR117.3 (same as CERCLA, see below); 40CFR423, Appendix A, Priority Pollutants; Section 313 Water Priority Chemicals (57FR41331, 9/9/92); Toxic Pollutant (Section 401.15) as technical mixture and metabolites
- EPA HAZARDOUS WASTE NUMBER (RCRA No.): U036
- RCRA Toxicity Characteristic (Section 261.24), Maximum Concentration of Contaminants, regulatory level, 0.03 mg/l
- RCRA, 40CFR261, Appendix 8 Hazardous Constituents
- RCRA 40CFR268.48; 61FR15654, Universal Treatment Standards: Wastewater (mg/l), (alpha- and gamma- isomers) 0.0033; Nonwastewater (mg/kg), 0.26
- RCRA 40CFR264, Appendix 9; TSD Facilities Ground Water Monitoring List. Suggested test method(s) (PQL μg/l): 8080 (0.1); 8250 (10)
- SAFE DRINKING WATER ACT: MCL, 0.002 mg/l; MCLG, zero; Regulated chemical (47 FR 9352)
- SUPERFUND/EPCRA 40CFR355, Appendix B Extremely Hazardous Substances: TPQ = 1,000 lb (454 kg)

- SUPERFUND/EPCRA 40CFR302.4 Reportable Quantity (RQ): CERCLA, 1 lb (0.454 kg)
- EPCRA Section 313 Form R *de minimis* concentration reporting level: 1.0%
- U.S. DOT Regulated Marine Pollutant (49CFR172.101, Appendix B), severe pollutant
- Canada, Drinking water quality: 0.007 mg/l MAC
- Mexico, Drinking water quality: 0.003 mg/l

Cited in U.S. State Regulations: Alaska (G), Arizona (W), California (A, W), Connecticut (A), Florida (G, A), Illinois (G, W), Kansas (G, A, W), Louisiana (G), Maine (G, W), Massachusetts (G, A), Michigan (G), Minnesota (W), Nevada (A), New Hampshire (G), New Jersey (G, W), New York (G, A), North Dakota (A), Oklahoma (G), Pennsylvania (G, A), Rhode Island (G), South Carolina (A), Vermont (G), Virginia (G, A), Washington (G), West Virginia (G), Wisconsin (G).

Description: Chlordane, $C_{10}H_6Cl_8$, is a colorless, or light-yellow or amber, thick liquid with a pungent, chlorine-like odor. It may occur as a crystalline solid. Boiling point = 175°C. Freezing/Melting point = 104 – 107°C, the commercial grade Noncombustible Liquid, but may be utilized in flammable solutions: The Flash point = 56°C has been found in the literature, but his may vary depending on carrier. Hazard Identification (based on NFPA-704 M Rating System): (in a flammable solution)Health 3, Flammability 3, Reactivity 0. Practically insoluble in water.

Potential Exposure: Chlordane is a broad spectrum insecticide of the group of polycyclic chlorinated hydrocarbons called cyclodiene insecticides. Chlordane has been used extensively since the 1950's for termite control, as an insecticide for homes and gardens, and as a control for soil insects during the production of crops such as corn. Both the uses and the production volume of chlordane have decreased extensively since the issuance of a registration suspension notice for all food crops and home and garden uses of chlordane by the U.S. Environmental Protection Agency, However, significant commercial use of Chlordane for termite control continues.

Special groups at risk include children as a result of milk consumed; fishermen and their families because of the high consumption of fish and shellfish, especially freshwater fish; persons living downwind from treated fields; and persons living in houses treated with Chlordane pesticide control agents.

Incompatibilities: Contact with strong oxidizers may cause fire and explosions. High heat and contact with alkaline solutions cause decomposition with the production of toxic fumes including chlorine, phosgene, hydrogen chloride. Attacks iron, zinc, plastics, rubber and coatings.

Permissible Exposure Limits in Air: The Federal (OSHA PEL) limit[58] is 0.5 mg/m^3 TWA for an 8-hour workshift. NIOSH and ACGIH recommend the same limit. Australia, DFG (MAK), Israel and Mexico TWA value is 0.5 mg/m^3 and Mexico's STEL value is 2.0 mg/m^3. Many of these regulatory and advisories add the notation "skin," indicating the possibility of cutaneous absorption. The NIOSH IDLH level is 100 mg/m^3. The Canadian provinces of Alberta, British Columbia, Ontario, and Quebec set a limit of 0.5 mg/m^3 TWA and STEL value of 2.0 mg/m^3.

The former USSR-UNEP/IRPTC project has set a MAC of 0.01 mg/m^3 in workplace air.[43]

Several states have set guidelines or standards for chlordane in ambient air[60] ranging from 0.068 μg/m^3 (Massachusetts) to 0.36 μg/m^3 (Pennsylvania) to 1.19 μg/m^3 (Kansas) to 1.7 μg/m^3 (New York) to 2.5 μg/m^3 (Connecticut and South Carolina) to 5.0 μg/m^3 (Florida) to 5 – 20 μg/m^3 (North Dakota) to 8.0 μg/m^3 (Virginia) to 12.0 μg/m^3 (Nevada).

Determination in Air: Filter/Chromosorb tube-102; Toluene; Gas chromatography/Electrochemical detection; NIOSH IV, Method #5510.

Permissible Concentration in Water: To protect freshwater aquatic life: 0.0043 μg/l as a 24-hour average, not to exceed 2.4 μg/l at any time. To protect saltwater aquatic life: 0.0040 μg/l as a 24-hour average, never to exceed 0.09 μg/l. To protect human health: preferably zero. An additional lifetime cancer risk of 1 in 100,000 is presented by a concentration of 0.0046 μg/l.[6]

The EPA[47] has found a lowest-observed-adverse-effect-level (LOAEL) of 0.045 mg/kg body weight/day which results in a lifetime health advisory of 2 μg/l.

Several states have set standards and guidelines for chlordane in drinking water.[61] Standards range from 0.5 μg/l (New Jersey) to 3.0 μg/l (Illinois) and guidelines range from 0.055 μg/l (California) to 0.22 μg/l (Kansas and Minnesota) to 0.50 μg/l (Arizona) to 0.55 μ/l (Maine).

See values listed under Regulatory Authority for Canada and Mexico.

It is strongly advised not to let the chemical enter into the environment because it persists in the environment. The substance may cause long-term effects in the aquatic environment.

Determination in Water: Filter/Chromosorb tube-102; Toluene; Gas chromatography/Electrochemical detection; NIOSH IV, Method #5510. Gas chromatography (EPA Method 608) or gas chromatography plus mass spectrometry (EPA Method 625).

Routes of Entry: Inhalation, skin absorption, ingestion and skin and eye contact.

Harmful Effects and Symptoms

Short Term Exposure: Chlordane can irritate the eyes and skin and can cause burns on contact. Skin rash or acne may develop. The vapor can irritate the respiratory tract. Exposure can cause blurred vision, nausea, headache, abdominal pain and vomiting, Exposure at high levels may result in disorientation, tremors, convulsions, respiratory failure and death. Medical observation is indicated. Symptoms include increased sensitivity to stimuli, tremors, muscular incoordination, and convulsions with or without coma. Fatal oral dose to adult humans is between 6 and 60 g with onset of symptoms within 45 minutes to several hours after ingestion, although symptoms have occurred following very small doses either orally or by skin exposure. Some reports of delayed development of liver disease, blood disorders and upset stomach. Chlordane is considered to be borderline between a moderately and highly toxic substance. The oral LD_{50} for rats is 283 mg/kg.

Long Term Exposure: This chemical has been shown to cause liver cancer in animals and may be a human carcinogen. It may damage the developing fetus. Chlordane may damage the kidneys, liver and affect the immune system. May cause an acne-like rash following skin contact.

Points of Attack: Central nervous system, eyes, lungs, liver, kidneys, skin.

Medical Surveillance: Consider the points of attack in preplacement and periodic physical examinations. Liver and kidney function tests. Examination by a qualified allergist. Complete blood count (CBC).

First Aid: If this chemical gets into the eyes, remove any contact lenses at once and irrigate immediately for at least 15 minutes, occasionally lifting upper and lower lids. Seek medical attention immediately. If this chemical contacts the skin, remove contaminated clothing and wash immediately with soap and water. Speed in removing material from skin is of extreme importance. Shampoo hair promptly if contaminated. Seek medical attention immediately. If this chemical has been inhaled, remove from exposure, begin rescue breathing (using universal precautions) if breathing has stopped and CPR if heart action has stopped. Transfer promptly to a medical facility. When this chemical has been swallowed, get medical attention. Give large quantities of water and induce vomiting. Do not make an unconscious person vomit.

Personal Protective Methods: Wear protective gloves and clothing to prevent any reasonable probability of skin contact. Safety equipment suppliers/manufacturers can provide recommendations on the most protective glove/clothing material for your operation. All protective clothing (suits, gloves, footwear, headgear) should be clean, available each day, and put on before work. Contact lenses should not be worn when working with this chemical. When working with liquid, wear splash-proof chemical goggles and face shield unless full facepiece respiratory protection is worn. Employees should wash immediately with soap when skin is wet or contaminated. Provide emergency showers and eyewash.

Respirator Selection: NIOSH: *At any concentrations above the NIOSH REL, or where there is no REL, at any detectable concentration:* SCBAF:PD,PP (any MSHA/NIOSH approved self-contained breathing apparatus that has a full facepiece and is operated in a pressure-demand or other positive-pressure mode); or SAF:PD,PP:ASCBA (any supplied-air respirator that has a full facepiece and is operated in a pressure-demand or other positive-pressure mode in combination with an auxiliary, self-contained breathing apparatus operated in a pressure-demand or other positive pressure mode). *Escape:* GMFOVHiE [any air-purifying, full-facepiece respirator (gas mask) with a chin-style, front- or back-mounted organic vapor canister having a high-efficiency particulate filter]; or SCBAE (any appropriate escape-type, self-contained breathing apparatus).

Storage: Prior to working with Chlordane you should be trained on its proper handling and storage. Chlordane must be stored to avoid contact with strong oxidizers (such as perchlorates, peroxides, permanganates, chlorates and nitrates) since violent reactions occur. Store in tightly closed containers in a cool, well ventilated area away from heat. A regulated, marked area should be established where this chemical is handled, used, or stored in compliance with OSHA standard 1910.1045.

Shipping: There are no shipping regulations specific to chlordane, perhaps because its use is so widely banned. However, Pesticides, liquid, toxic, flammable, n.o.s., or Organochlorine pesticide, liquid, flammable, poisonous, require a "Keep Away from Food" label. Chlordane falls in UN/DOT Hazard Class 6.1 and Packing Group III.[19][20] It is a DOT regulated severe marine pollutant.

Spill Handling: Do not touch spilled material. Use water spray to reduce vapors. Stay upwind. Avoid breathing vapors. Wear positive-pressure breathing apparatus and full protective clothing. Evacuate and restrict persons not wearing protective equipment from area of spill or leak until cleanup is complete. Remove all ignition sources. Ventilate area of spill or leak. Small spills: absorb liquids in vermiculite, dry sand, earth, peat, carbon, or a similar material and deposit in sealed containers. Large spills: dike far ahead of spill for later disposal. It may be necessary to contain and dispose of this chemical as a hazardous waste. If material or contaminated runoff enters waterways, notify downstream users of potentially contaminated waters. Contact your Department of Environmental Protection or your regional office of the federal EPA for specific recommendations. If employees are required to clean-up spills, they must be properly trained and equipped. OSHA 1910.120(q) may be applicable.

Fire Extinguishing: This chemical is a noncombustible liquid but it may be dissolved in flammable or combustible liquids for commercial application. Poisonous gases including chlorine, phosgene, hydrogen chloride are produced in fire. If the flammable or combustible commercial material catches fire, use dry chemical, carbon dioxide, or alcohol foam extinguishers. Vapors are heavier than air and will collect in low areas. Vapors may travel long distances to ignition sources and flashback. Vapors in confined areas may explode when exposed to fire. Containers may explode in fire. Storage containers and parts of containers may rocket great distances, in many directions. If material or contaminated runoff enters waterways, notify downstream users of potentially contaminated waters. Notify local health and fire officials and pollution control agencies. From a secure, explosion-proof location, use water spray to cool exposed containers. If cooling streams are ineffective (venting sound increases in volume and pitch, tank discolors, or shows any signs of deforming), withdraw immediately to a secure position. If employees are expected to fight fires, they must be trained and equipped in OSHA 1910.156.

Disposal Method Suggested: Chlordane is dehydrochlorinated in alkali to form "nontoxic" products, a reaction catalyzed by traces of iron, but the reaction is slow. The environmental hazards of the products are uncertain. Chlordane is completely dechlorinated by sodium in isopropyl alcohol. The UN Recommends incineration methods for disposal of chlordane.[22] In accordance with 40CFR165 recommendations for the disposal of pesticides and pesticide containers. Must be disposed properly by following package label directions or by contacting your state pesticide or environmental control agency or by contacting your regional EPA office.

References

U.S. Environmental Protection Agency, Chlordane: Ambient Water Quality Criteria, Washington, DC (1980)

U.S. Environmental Protection Agency, Chlordane, Health and Environmental Effects Profile No. 35, Office of Solid Waste, Washington, DC (April 30, 1980)

Sax, N. I., Ed., "Dangerous Properties of Industrial Materials Report," 1, No. 2, 33–35 (1980) and 3, No. 5, 94–99 (1983) and 7, No. 6, 46–55 (1987)

U.S. Environmental Protection Agency, "Chemical Profile: Chlordane," Washington, DC, Chemical Emergency Preparedness Program (October 31, 1985)

New Jersey Department of Health and Senior Services and Senior Services, "Hazardous Substance Fact Sheet: Chlordane," Trenton, NJ (April 1998)

New York State Department of Health, "Chemical Fact Sheet: Chlordane," Albany, NY, Bureau of Toxic Substance Assessment (March 1986)

U.S. Public Health Service, "Toxicological Profile for Chlordane," Atlanta, Georgia, Agency of Toxic Substances and Disease Registry (December 1988)

Chlordecone (Kepone)

Molecular Formula: $C_{10}Cl_{10}O$

Synonyms: Chlordecone; Ciba 8514; Compound 1189; 1,2,3,5,6,7,8,9,10,10-Decachloro(5.2.2.0$^{2.6}$.0$^{3.9}$.0$^{5.8}$)decano-4-one; Decachloroketone; Decachlorooctahydrokepone-2-one; 1,1a, 3, 3a, 4,5,5,5a, 5b, 6-Decachlorooctahydro-1,3,4-metheno-2H-cyclobuta(cd)pentalen-2-one; 1,1a,3,3a, 4,5,5,5a,5b,6-Decachlorooctahydro-1,3,4-metheno-2H-cyclobuta(c,d)pentalen-2-one; Decachlorooctahydro-1,3,4-metheno-2H-cyclobuta(cd)-pentalen-2-one; Decachlorooctahydro-1,3,4-metheno-2H-cyclobuta(cd)pentalen-2-one; Decachlorotetracyclodecanone; Decachlorotetrahydro-4,7-methanoindeneone; ENT 16,391; GC-1189; General chemicals 1189; Kepone; Merex®; 1,3,4- Metheno-2H-cyclobuta(cd)pentalen-2-one,1,1a,3,3a,4,5,5a,5b,6-decachlorooctahydro-, Kepone®; NCI-C00191

CAS Registry Number: 143-50-0

RTECS Number: PC8575000

DOT ID: UN 2761

Regulatory Authority

- Carcinogen (Animal Positive, Human Suspected) (IARC)[9] (NPT)[10] (DFG)[3]
- Banned or Severely Restricted (Many Countries) (UN)[13][35]
- Air Pollutant Standard Set (NIOSH) (See Reference below) (Several States)[60]
- CLEAN WATER ACT: Section 311 Hazardous Substances/RQ 40CFR117.3 (same as CERCLA, see below)
- EPA HAZARDOUS WASTE NUMBER (RCRA No.): U142
- RCRA, 40CFR261, Appendix 8 Hazardous Constituents
- RCRA 40CFR268.48; 61FR15654, Universal Treatment Standards: Wastewater (mg/l), 0.0011; Nonwastewater (mg/kg), 0.13
- RCRA 40CFR264, Appendix 9; TSD Facilities Ground Water Monitoring List. Suggested test method(s) (PQL μg/l): 8270 (10)
- SUPERFUND/EPCRA 40CFR302.4 Reportable Quantity (RQ): CERCLA, 1 lb (0.454 kg)

Cited in U.S. State Regulations: California (G), Florida (G), Illinois (G), Kansas (G), Louisiana (G), Maine (G), Massachusetts (G), Michigan (G), New Hampshire (G), New Jersey (G), New York (A), Pennsylvania (G, A), South Carolina (A), Vermont (G), Virginia (G), Washington (G), West Virginia (G), Wisconsin (G).

Description: Kepone, $C_{10}Cl_{10}O$, is a tan to white, odorless crystalline solid. Freezing/Melting point = 360°C (sublimes). Hazard Identification (based on NFPA-704 M Rating System): Health 3, Flammability 1, Reactivity 0. Soluble in water.

Potential Exposure: Kepone was registered for the control or rootborers on bananas with a residue tolerance of 0.01 ppm. This constituted the only food or feed use of Kepone. Nonfood uses included wireworm control in tobacco fields and bait to control ants and other insects in indoor and outdoor areas.

A rebuttable presumption against registration of chlordecone was issued by the U.S. EPA on March 25, 1976 on the basis of oncogenicity. The trademarked Kepone and products of six formulations were the subject of voluntary cancellation according to a U.S. EPA notice dated July 27, 1977. In a series of decisions, the first of which was issued on June 17, 1976, the EPA effectively cancelled all registered products containing Kepone as of May 1, 1978.

Incompatibilities: Acids, acid fumes.

Permissible Exposure Limits in Air: NIOSH recommends that the workplace environmental level be limited to 0.001 g/m^3 TWA for up to a 10-hour workday, 40-hour workweek, as an emergency standard.

Guidelines or standards for Kepone in ambient air have been set[60] ranging from zero (South Carolina) to 0.03 μg/m^3 (New York) to 0.88 μg/l (Pennsylvania).

Determination in Air: Collection by membrane filter and backup impinger containing NaOH solution, workup with benzene, analysis by gas chromatography with electron capture detector. See NIOSH Method 5508.

Permissible Concentration in Water: No criteria set.

Routes of Entry: Inhalation of dust, ingestion, skin absorption.

Harmful Effects and Symptoms

In July 1975, a private physician submitted a blood sample to the Center for Disease Control (CDC) to be analyzed for Kepone, a chlorinated hydrocarbon pesticide. The sample had been obtained from a Kepone production worker who suffered from weight loss, nystagmus, and tremors. CDC notified the State epidemiologist that high levels of Kepone were present in the blood sample, and he initiated an epidemiologic investigation which revealed other employees suffering with similar symptoms. It was evident to the State official after visiting the plant that the employees had been exposed to Kepone at extremely high concentrations through inhalation, ingestion, and skin absorption. He recommended that the plant be closed, and company management complied.

Of the 113 current and former employees of this Kepone-manufacturing plant examined, more than half exhibited clinical symptoms of Kepone poisoning. Medical histories of tremors (called "Kepone shakes" by employees), visual disturbances, loss of weight, nervousness, insomnia, pain in the chest and abdomen and, in some cases, infertility and loss of libido were reported. The employees also complained of vertigo and lack of muscular coordination. The intervals between exposure and onset of the signs and symptoms varied between patients but appeared to be dose related.

NIOSH has received a report on a carcinogenesis bioassay of technical grade Kepone which was conducted by the National Cancer Institute using Osborne-Mendel rats and B6C3F1 mice, Kepone was administered in the diet at two tolerated dosages. In addition to the clinical signs of toxicity, which were seen in both species, a significant increase ($P<0.05$) of hepatocellular carcinoma in rats given large dosages of Kepone and in mice at both dosages was found. Rats and mice also had extensive hyperplasia of the liver.

In view of these findings, NIOSH must assume that Kepone is a potential human carcinogen. The oral LD_{50} rat is 95 mg/kg (moderately toxic).

Short Term Exposure: May be poisonous if absorbed through the skin. Skin or eye contact may cause irritation and rash. Poisonous if swallowed. Exposure can cause headache, nervousness, tremor; liver, kidney damage; visual disturbance; ataxia, chest pain, skin erythema (skin redness).

Long Term Exposure: Has been shown to cause liver cancer in animals; potential human carcinogen. May cause testicular atrophy, low sperm count, damage to the developing fetus, reproductive damage; sterility, breast enlargement; skin changes; liver and kidney damage; brain and nervous system damage with hyperactivity, hyperexcitability, muscle spasms, tremors.

Points of Attack: Eyes, skin, respiratory system, central nervous system, liver, kidneys, reproductive system.

Medical Surveillance: Employers shall make medical surveillance available to all workers occupationally exposed to Kepone, including personnel periodically exposed during routine maintenance or emergency operations. Periodic examinations shall be made available at least on an annual basis.

First Aid: If this chemical gets into the eyes, remove any contact lenses at once and irrigate immediately for at least 15 minutes, occasionally lifting upper and lower lids. Seek medical attention immediately. If this chemical contacts the skin, remove contaminated clothing and wash immediately with soap and water. Speed in removing material from skin is of extreme importance. Shampoo hair promptly if contaminated. Seek medical attention immediately. If this chemical has been inhaled, remove from exposure, begin rescue breathing (using universal precautions) if breathing has stopped and CPR if heart action has stopped. Transfer promptly to a medical facility. When this chemical has been swallowed, get medical attention. Give large quantities of water and induce vomiting. Do not make an unconscious person vomit. Qualified medical personnes may consider the admistration of cholestyramine resin (QUESTRAN). Medical personnel should wear neoprene gloves as protection against contamination (Dreisbach).

Personal Protective Methods: *Protective Clothing:*

a: Coveralls or other full-body protective clothing shall be worn in areas where there is occupational exposure to Kepone. Protective clothing shall be changed at least daily at the end of the shift and more frequently if it should become grossly contaminated.

b: Impervious gloves, aprons and footwear shall be worn at operations where solutions of Kepone may contact the skin. Protective gloves shall be worn at operations where dry Kepone or materials containing Kepone are handled and may contact the skin.

c: Eye protective devices shall be provided by the employer and used by the employees where contact of Kepone with eyes is likely. Selection, use, and maintenance of eye protective equipment shall be in accordance with the provisions of the American National Standard Practice for Occupational and

Educational Eye and Face Protection, ANSI Z87.1-1968. Unless eye protection is afforded by a respirator hood or facepiece, protective goggles or a face shield shall be worn at operations where there is danger of contact of the eyes with dry or wet materials containing Kepone because of spills, splashes, or excessive dust or mists in the air.

d: The employer shall ensure that all personal protective devices are inspected regularly and maintained in clean and satisfactory working condition.

e: Work clothing may not be taken home be employees. The employer shall provide for maintenance and laundering of protective clothing.

f: The employer shall ensure that precautions necessary to protect laundry personnel are taken while soiled protective clothing is being laundered.

g: The employer shall ensure that Kepone is not discharged into municipal waste treatment systems or the community air.

Respiratory Protection from Kepone: Engineering controls shall be used wherever feasible to maintain airborne Kepone concentrations at or below that recommended. Compliance with the environmental exposure limit by the use of respirators is allowed only when airborne Kepone concentrations are in excess of the workplace environmental limit because required engineering controls are being installed or tested when nonroutine maintenance or repair is being accomplished, or during emergencies. When a respirator is thus permitted, it shall be selected and used in accordance with NIOSH requirements.

Respirator Selection: *At any concentrations above the NIOSH REL, or where there is no REL, at any detectable concentration:* SCBAF:PD,PP (any MSHA/NIOSH approved self-contained breathing apparatus that has a full facepiece and is operated in a pressure-demand or other positive-pressure mode); or SAF:PD,PP:ASCBA (any supplied-air respirator that has a full facepiece and is operated in a pressure-demand or other positive-pressure mode in combination with an auxiliary, self-contained breathing apparatus operated in a pressure-demand or other positive pressure mode). *Escape:* GMFOVHiE [any air-purifying, full-facepiece respirator (gas mask) with a chin-style, front- or back-mounted organic vapor canister having a high-efficiency particulate filter]; or SCBAE (any appropriate escape-type, self-contained breathing apparatus).

Storage: Prior to working with kepone you should be trained on its proper handling and storage. A regulated, marked area should be established where this chemical is handled, used, or stored in compliance with OSHA standard 1910.1045. Store in a cool, dry place in a refrigerator under inert atmosphere.[52] Keep away from acids and acid fumes.

Shipping: While not specifically cited, chlordecone may be classified under Organochlorine pesticides, solid, toxic n.o.s., which fall in Hazard Class 6.1. Chlordecone falls in Packing Group III and requires a "Keep Away from Food" label.

Spill Handling: Do not touch spilled material. Use water spray to reduce vapors. Evacuate persons not wearing protective equipment from area of spill or leak until clean-up is complete. Remove all ignition sources. Collect powdered material in the most convenient and safe manner and deposit in sealed containers. For larger spills, dike far ahead of spill for later disposal. Ventilate area after clean-up is complete. It may be necessary to contain and dispose of this chemical as a hazardous waste. If material or contaminated runoff enters waterways, notify downstream users of potentially contaminated waters. Contact your Department of Environmental Protection or your regional office of the federal EPA for specific recommendations. If employees are required to clean-up spills, they must be properly trained and equipped. OSHA 1910.120(q) may be applicable.

Fire Extinguishing: This chemical is a noncombustible solid. Use dry chemical, carbon dioxide, halon, water spray or standard foam. Poisonous gases are produced in fire including toxic chlorides. If material or contaminated runoff enters waterways, notify downstream users of potentially contaminated waters. Notify local health and fire officials and pollution control agencies. From a secure, explosion-proof location, use water spray to cool exposed containers. If cooling streams are ineffective (venting sound increases in volume and pitch, tank discolors, or shows any signs of deforming), withdraw immediately to a secure position. If employees are expected to fight fires, they must be trained and equipped in OSHA 1910.156.

Disposal Method Suggested: A process has been developed which effects Chlordecone degradation by treatment of aqueous wastes with UV radiation in the presence of hydrogen in aqueous sodium hydroxide solution. Up to 95% decomposition was effected by this process.

Chlordecone previously presented serious disposal problems because of its great resistance to bio- and photo degradation in the environment. It is highly toxic to normally-occur in degrading microorganisms. Although it can undergo some photodecomposition when exposed to sunlight to the dihydro compound (leaving a compound with 8 chloro substituents) that degradation product does not significantly reduce toxicity. Disposal by incineration with HCl scrubbing is recommended.[22]

References

National Institute for Occupational Safety and Health, Recommended Standard for Occupational Exposure to Kepone, Washington, DC (January 27, 1976)

U.S. Environmental Protection Agency, Reviews of the environmental Effects of Pollutants: I. Mirex and Kepone, Report EPA 600/1-78-013, Cincinnati, OH (1978)

National Academy of Sciences, Kepone, Mirex, Hexachlorocyclopentadiene: An Environmental Assessment, Washington, DC (1978)

Sax, N. I., Ed., "Dangerous Properties of Industrial Materials Report," 1, No. 4, 77–79 (1981) and 4, No. 4, 10–44 (1984)

Chlorfenvinphos

Molecular Formula: $C_{12}H_{14}Cl_3O_4P$

Synonyms: Apachlor; Benzyl alcohol,2,4-dichloro-α-(chloromethylene)-, diethyl phosphate; Birlane; Birlane liquid; C-10015; C8949; CFV; CGA 26351; *o*-2-Chloor-1-(2,4-dichloor-fenyl)-vinyl-o,o-diethylfosfaat (Dutch); *O*-2-Chlor-1-(2,4-dichlor-phenyl)-vinyl-*O,O*-diaethylphosphat (German); Chlorfenvinphos; β-2-Chloro-1-(2',4'-dichlorophenyl) vinyl diethylphosphate; 2-Chloro-1-(2,4-dichlorophenyl)vinyl diethyl phosphate; Chlorofenvinphos; Chlorphenvinfos; Chlorphenvinphos; Clofenvinfos; Clorfenvinfos (Spanish); *o*-2-Cloro-1-(2,4-dicloro-fenil)-vinil-*o,o*-di etilfosfato (Italian); Compound 4072; CVP; Diethyl1-(2,4-dichlorophenyl)-2-chlorovinyl phosphate; *O,O*-Diethyl *O*-[2-chloro-1-(2',4'-dichlorophenyl)vinyl] phosphate; ENT 24969; GC 4072; OMS 1328; Phosphate de *O,O*-diethyle etdeo-2-chloro-1-(2,4-dichlorophenyl) vinyle (French); Phosphoric acid, 2-chloro-1-(2,4-dichlorophenyl)ethenyldiethyl ester; Sapecron; Sapecron 10FGEC; Sapecron 240; Saprecon C; SD 4072; SD 7859; Shell 4072; Supona; Supone; Unitox; Vinylphare; Vinylphate

CAS Registry Number: 470-90-6

RTECS Number: TB8750000

DOT ID: UN 2783; UN 3018 (organophorus pesticide, liquid, poisonous)

Regulatory Authority

- Very Toxic Substance (World Bank)[15]
- SUPERFUND/EPCRA 40CFR355, Appendix B Extremely Hazardous Substances: TPQ = 500 lb (227 kg)
- SUPERFUND/EPCRA 40CFR302.4 Reportable Quantity (RQ): EHS, 1 lb (0.454 kg)
- U.S. DOT Regulated Marine Pollutant (49CFR172.101, Appendix B)

Cited in U.S. State Regulations: California (G), Florida (G), Illinois (G), Massachusetts (G), Michigan (G), New Hampshire (G), New Jersey (G), Oklahoma (G), Pennsylvania (G).

Description: Chlorfenvinphos, $C_{12}H_{14}Cl_3O_4P$, is a nonflammable, yellow or amber liquid with a mild odor. Boiling point = 110°C @ 0.001 mm Hg; 168 – 170°C @ 0.5 mm Hg. Freezing/Melting point = -23 – -19°C. Hazard Identification (based on NFPA-704 M Rating System): Health 4, Flammability 1, Reactivity 0. Slightly soluble in water.

Potential Exposure: Those engaged in the production, formulation and application of this insecticide.

Incompatibilities: Strong oxidizers, strong bases, moisture. May be corrosive to metals in the presence of moisture.

Permissible Exposure Limits in Air: No standards set. However, it should be recognized that this chemical can be absorbed through the skin, thereby increasing exposure.

Determination in Air: OSHA versatile sampler-2; Toluene/Acetone; Gas chromatography/Flame photometric detection for sulfur, nitrogen, or phosphorus; NIOSH Method IV #5600, Organophosphorus Pesticides.

Permissible Concentration in Water: 0.1 mg/l in drinking water is a recommended drinking water limit.

Routes of Entry: Inhalation, passes through the unbroken skin.

Harmful Effects and Symptoms

Short Term Exposure: Highly toxic (LD_{50} for rats is 10 mg/kg). Symptoms exhibited on chlorfenvinphos exposure are typical of cholinesterase poisoning. Nausea is often first symptom, with vomiting, abdominal cramps, diarrhea, and excessive salivation. Headache, giddiness, weakness, tightness in chest, blurring of vision, pinpoint pupils, loss of muscle coordination, and difficulty breathing. Convulsions and coma precede death. Higher exposures can cause pulmonary edema, a medical emergency that can be delayed for several hours. This can cause death. Chlorfenvinphos can cause the heart to beat slower (bradycardia) or irregularly (arrhythmia).

Long Term Exposure: There is limited evidence that this chemical may damage the developing fetus. Symptoms resembling influenza with headache, nausea, weakness have been reported. Cholinesterase inhibitor; cumulative effect is possible. Chlorfenvinphos may damage the nervous system with repeated exposure, resulting in impaired memory, depression, anxiety, or irritability, convulsions, respiratory failure. May cause liver damage.

Points of Attack: Respiratory system, lungs, central nervous system, cardiovascular system, skin, eyes, plasma and red blood cell cholinesterase.

Medical Surveillance: Before employment and at regular times after that, the following are recommended: Plasma and red blood cell cholinesterase levels (tests for the enzyme poisoned by this chemical). If exposure stops, plasma levels return to normal in 1 – 2 weeks while red blood cell levels may be reduced for 1 – 3 months.

When cholinesterase enzyme levels are reduced by 25% or more below preemployment levels, risk of poisoning is increased, even if results are in lower ranges of "normal." Reassignment to work not involving organophosphate or carbamate pesticides is recommended until enzyme levels recover. If symptoms develop or overexposure occurs, repeat the above tests as soon as possible and get an exam of the nervous system. Also consider complete blood count. Consider chest x-ray following acute overexposure. Do not drink any alcoholic beverages before or during use. Alcohol promotes absorption of organic phosphates.

First Aid: If this chemical gets into the eyes, remove any contact lenses at once and irrigate immediately for at least 15 minutes, occasionally lifting upper and lower lids. Seek medical attention immediately. If this chemical contacts the skin, remove contaminated clothing and wash immediately with soap and water. Speed in removing material from skin is of extreme importance. Shampoo hair promptly if contaminated. Seek medical attention immediately. If this chemical has been inhaled, remove from exposure, begin rescue breathing (using universal precautions) if breathing has stopped and CPR if heart action has stopped. Transfer promptly to a medical facility. When this chemical has been swallowed, get medical attention. Give large quantities of water and induce vomiting. Do not make an unconscious person vomit. Medical observation is recommended for 24 – 48 hours after breathing overexposure, as pulmonary edema may be delayed. As first aid for pulmonary edema, a doctor or authorized paramedic may consider administering a corticosteroid spray.

Personal Protective Methods: Wear protective gloves and clothing to prevent any reasonable probability of skin contact. Safety equipment suppliers/manufacturers can provide recommendations on the most protective glove/clothing material for your operation. All protective clothing (suits, gloves, footwear, headgear) should be clean, available each day, and put on before work. Contact lenses should not be worn when working with this chemical. Wear splash-proof chemical goggles and face shield unless full facepiece respiratory protection is worn. Employees should wash immediately with soap when skin is wet or contaminated. Provide emergency showers and eyewash.

Respirator Selection: *At any concentrations above the NIOSH REL, or where there is no REL, at any detectable concentration:* SCBAF:PD,PP (any MSHA/NIOSH approved self-contained breathing apparatus that has a full facepiece and is operated in a pressure-demand or other positive-pressure mode); or SAF:PD,PP:ASCBA (any supplied-air respirator that has a full facepiece and is operated in a pressure-demand or other positive-pressure mode in combination with an auxiliary, self-contained breathing apparatus operated in a pressure-demand or other positive pressure mode). *Escape:* GMFOVHiE [any air-purifying, full-facepiece respirator (gas mask) with a chin-style, front- or back-mounted organic vapor canister having a high-efficiency particulate filter]; or SCBAE (any appropriate escape-type, self-contained breathing apparatus).

Storage: Prior to working with Chlorfenvinphos you should be trained on its proper handling and storage. Should be protected from moisture and stored in glass-lined or polyethylene-lined containers. Keep away from strong bases. Sources of ignition such as smoking and open flames, are prohibited where this chemical is used, handled, or stored in a manner that could create a potential fire or explosion hazard.

Shipping: Organophosphorus pesticides, liquid, toxic, n.o.s. require a "Poison" label (in Packing Group II). The UN/DOT Hazard Class[19][20] is 6.1.

Spill Handling: Stay upwind; keep out of low areas. Ventilate closed spaces before entering them. Evacuate and restrict persons not wearing protective equipment from area of spill or leak until cleanup is complete. Remove all ignition sources. Ventilate area of spill or leak. Absorb liquids in vermiculite, dry sand, earth, peat, carbon, or a similar material and deposit in sealed containers. Large spills: dike far ahead of spill for later disposal. It may be necessary to contain and dispose of this chemical as a hazardous waste. If material or contaminated runoff enters waterways, notify downstream users of potentially contaminated waters. Contact your Department of Environmental Protection or your regional office of the federal EPA for specific recommendations. If employees are required to clean-up spills, they must be properly trained and equipped. OSHA 1910.120(q) may be applicable.

Fire Extinguishing: The state of New Jersey has assigned a flammability rating of "2" to Chlorfenvinphos. This chemical may burn, but does not readily ignite. Move container from fire area if you can do it without risk. Fight fire from maximum distance. Dike fire control water for later disposal; do not scatter the material. Wear positive pressure breathing apparatus and special protective clothing. Use dry chemical, carbon dioxide, water spray, or alcohol resistant foam extinguishers. Poisonous gases are produced in fire including hydrogen chloride, phosphorous oxides, and sulfur oxides. If material or contaminated runoff enters waterways, notify downstream users of potentially contaminated waters. Notify local health and fire officials and pollution control agencies. Containers may explode in fire. From a secure, explosion-proof location, use water spray to cool exposed containers. If cooling streams are ineffective (venting sound increases in volume and pitch, tank discolors, or shows any signs of deforming), withdraw immediately to a secure position. If employees are expected to fight fires, they must be trained and equipped in OSHA 1910.156.

Disposal Method Suggested: Destruction by alkali hydrolysis or incineration.[22] In accordance with 40CFR165 recommendations for the disposal of pesticides and pesticide containers. Must be disposed properly by following package label directions or by contacting your state pesticide or environmental control agency or by contacting your regional EPA office.

References

Sax, N. I., Ed., "Dangerous Properties of Industrial Materials Report," 2, No. 4, 63–67, New York, Van Nostrand Reinhold Co. (1982)

U.S. Environmental Protection Agency, "Chemical Profile: Chlorfenvinphos," Washington, DC, Chemical Emergency Preparedness Program (October 31, 1985)

Chlorinated Diphenyl Oxide

Molecular Formula: $C_{12}H_4Cl_6O$

Common Formula: $C_{12}H_{10-n}Cl_nO$ (general, for chlorinated diphenyl oxides)

Synonyms: Benzene, 1,1'-oxybis, hexachloro derivatives; Ether, hexachlorophenyl; Hexachlorodiphenyl ether;

Hexachlorodiphenyl oxide; Phenyl ether, hexachloro derivative; Trichloro diphenyl ether; Trichloro diphenyl oxide

hexachloro-: Benzene,1,1'-oxy bis chloro; Bis(trichlorophenyl) ether; Ether, hexachlorophenyl; Hexachlorodiphenyl oxide; Hexachlorophenyl ether; Phenyl ether, hexachloro-

CAS Registry Number: 31242-93-0; 55720-99-5 (hexachloro-); 57321-63-8 (trichloro-)

RTECS Number: KO0875000 (hexachloro-); KO4200000 (trichloro-)

Regulatory Authority

- Air Pollutant Standard Set (ACGIH)[1] (DFG, total dust)[3] (OSHA)[58] (Several States)[60] (Several Canadian Provinces) (Australia) (Israel)
- Canada, WHMIS, Ingredients Disclosure List

Cited in U.S. State Regulations: Alaska (G), California (A, G), Connecticut (A), Florida (A), Illinois (G), Maine (G), Massachusetts (G), Nevada (A), New Hampshire (G), New Jersey (G), North Dakota (A), Pennsylvania (G), Rhode Island (G), Virginia (A), West Virginia (G).

Description: Chlorinated Diphenyl Oxide, $C_{12}H_4Cl_6O$, is a white or yellowish waxy solid or very viscous liquid. Boiling point = 230 – 260°C @ 8 mm Hg. *Hexachloro Diphenyl Oxide:* Autoignition temperature = 620°C. Hazard Identification (based on NFPA-704 M Rating System): Health 2, Flammability 1, Reactivity 0.

Potential Exposure: These materials are used as dielectric fluids in the electrical industry; they may be used as organic intermediates to make other chemicals, and in dry cleaning detergents.

Incompatibilities: Strong oxidizers or heat may cause fire and explosion. May be able to form unstable peroxides.

Permissible Exposure Limits in Air: The OSHA TWA,[58] the DFG value[3] and the ACGIH TWA value is 0.5 mg/m^3 and the ACGIH STEL is 2.0 mg/m^3. The NIOSH IDLH level is 5.0 mg/m^3. Australia, Israel, and the Canadian provinces of Alberta, British Columbia, Ontario, and Quebec use the same TWA values.

Several states have set guidelines or standards for Chlorinated Diphenyl Oxide in ambient air[60] ranging from 5-20 μg/m^3 (North Dakota) to 8 μg/m^3 (Virginia) to 10 μg/m^3 (Connecticut) to 12 μg/m^3 (Nevada).

Determination in Air: Collect on filter, work up with isooctane, analyze by gas chromatography. See NIOSH Method IV #5025.[18]

Permissible Concentration in Water: Because of the lack of data on both toxicologic effects and environmental contamination, the hazard posed by these compounds cannot be estimated according to the Environmental Protection Agency.[6]

Routes of Entry: Inhalation, ingestion, eye and skin contact.

Harmful Effects and Symptoms

Acne-form dermatitis and liver damage.

Short Term Exposure: Contact can cause skin irritation, rash, burning sensation; chloracne.

Long Term Exposure: May cause acne-form dermatitis and liver damage. More than light alcohol consumption may increase the liver damage.

Points of Attack: Skin, liver.

Medical Surveillance: Consider the points of attack in preplacement and periodic physical examinations. Liver function tests. Examination by a qualified allergist.

First Aid: If this chemical gets into the eyes, remove any contact lenses at once and irrigate immediately for at least 15 minutes, occasionally lifting upper and lower lids. Seek medical attention immediately. If this chemical contacts the skin, remove contaminated clothing and wash immediately with soap and water. Seek medical attention immediately. If this chemical has been inhaled, remove from exposure, begin rescue breathing (using universal precautions) if breathing has stopped and CPR if heart action has stopped. Transfer promptly to a medical facility. When this chemical has been swallowed, get medical attention. Give large quantities of water and induce vomiting. Do not make an unconscious person vomit.

Personal Protective Methods: Wear protective gloves and clothing to prevent any reasonable probability of skin contact. Safety equipment suppliers/manufacturers can provide recommendations on the most protective glove/clothing material for your operation. All protective clothing (suits, gloves, footwear, headgear) should be clean, available each day, and put on before work. Contact lenses should not be worn when working with this chemical. Wear splash-proof chemical goggles and face shield unless full facepiece respiratory protection is worn. Employees should wash immediately with soap when skin is wet or contaminated. Provide emergency showers and eyewash.

Respirator Selection: OSHA: *5 mg/m^3:* SA (any self-contained breathing apparatus that has a full facepiece and is operated in a pressure-demand or other positive-pressure mode; SCBA (any supplied-air respirator that has a full facepiece and is operated in a pressure-demand or other positive-pressure mode in combination with an auxiliary self-contained positive-pressure breathing apparatus. *Emergency or planned entry into unknown concentrations or IDLH conditions:* SCBAF:PD,PP (any self-contained breathing apparatus that has a full facepiece and is operated in a pressure-demand or other positive-pressure mode); or SAF:PD, PP:ASCBA (any supplied-air respirator that has a full facepiece and is operated in a pressure-demand or other positive-pressure mode in combination with an auxiliary self-contained breathing apparatus operated in a pressure-demand or other positive-pressure mode). *Escape:* GMFHiOVAG [any air-purifying, full-facepiece respirator (gas mask) with a chin-style, front- or back-mounted organic vapor canister having a high-efficiency particulate filter]; or SCBAE (any appropriate escape-type, self-contained breathing apparatus).

Storage: Prior to working with Chlorinated diphenyl oxides you should be trained on its proper handling and storage. Store in tightly closed containers in a cool, well ventilated area. Metal containers involving the transfer of this chemical should be grounded and bonded. Where possible, automatically pump liquid from drums or other storage containers to process containers. Drums must be equipped with self-closing valves, pressure vacuum bungs, and flame arresters. Use only non-sparking tools and equipment, especially when opening and closing containers of this chemical. Sources of ignition such as smoking and open flames, are prohibited where this chemical is used, handled, or stored in a manner that could create a potential fire or explosion hazard.

Shipping: Chlorinated diphenyl oxides are not specifically sited by DOT in their Performance Oriented Packaging Standards.[19]

Spill Handling: Evacuate and restrict persons not wearing protective equipment from area of spill or leak until cleanup is complete. Remove all ignition sources. Ventilate area of spill or leak. Absorb liquids in vermiculite, dry sand, earth, peat, carbon, or a similar material and deposit in sealed containers. It may be necessary to contain and dispose of this chemical as a hazardous waste. If material or contaminated runoff enters waterways, notify downstream users of potentially contaminated waters. Contact your Department of Environmental Protection or your regional office of the federal EPA for specific recommendations. If employees are required to clean-up spills, they must be properly trained and equipped. OSHA 1910.120(q) may be applicable.

Fire Extinguishing: Although Chlorinated diphenyl oxide is listed as a "noncombustible solid or liquid" by the state of New Jersey, and in some of the literature; however, hexachloro diphenyl oxide has a NFPA Flammability rating of "1." Poisonous gases including chlorine are produced in fire. Use dry chemical or carbon dioxide extinguishers. Vapors in confined areas may explode when exposed to fire. Containers may explode in fire. Storage containers and parts of containers may rocket great distances, in many directions. If material or contaminated runoff enters waterways, notify downstream users of potentially contaminated waters. Notify local health and fire officials and pollution control agencies. From a secure, explosion-proof location, use water spray to cool exposed containers. If cooling streams are ineffective (venting sound increases in volume and pitch, tank discolors, or shows any signs of deforming), withdraw immediately to a secure position. If employees are expected to fight fires, they must be trained and equipped in OSHA 1910.156.

Disposal Method Suggested: For trichlorophenyl ether, solution in a flammable solvent and incineration in a furnace with afterburner and scrubber is recommended.[22]

References

U.S. Environmental Protection Agency, Haloethers: Ambient Water Quality Criteria, Washington, DC (1980)

New Jersey Department of Health and Senior Services and Senior Services, "Hazardous Substance Fact Sheet: Chlorinated Diphenyl Oxide," Trenton, NJ (April 1986)

Chlorinated Naphthalenes

Molecular Formula: $C_{10}H_{8-x}Cl_x$

Synonyms: *1-chloro-:* α-Chloronaphthalene; 1-Cloronaftaleno (Spanish)

2-chloro-: β-Chloronaphthalene; 2-Cloronaftaleno (Spanish)

hexachloro-: Halowax 1014; Hexachlornaftalen (Czech); Hexacloronaftaleno (Spanish); Naphthalene, hexachloro-

octachloro-: Halowax 1051

pentachloro-: Halowax 1013

tetrachloro-: Halowax; Nibren wax; Seekay wax

trichloro-: Halowax; Nibren wax; Seekay wax

CAS Registry Number: 90-13-1 (1-Chloro-); 91-58-7 (2-Chloro-); 1321-65-9 (Trichloro-); 1335-88-2 (Tetrachloro-); 1321-64-8 (Pentachloro-); 1335-87-1 (Hexachloro-); 2234-13-1 (Octachloro-)

RTECS Number: QJ2100000 (1-Chloro-); QJ2275000 (2-Chloro-); QK4025000 (Trichloro-); QK3700000 (Tetrachloro-); QK0300000 (Pentachloro-); QJ7350000 (Hexachloro-); QK0250000 (Octachloro-)

Regulatory Authority

- Air Pollutant Standard Set (ACGIH)[1] (DFG)[3] (HSE)[33] (Other countries)[35] (Several States)[60]

Chlorinated naphthalenes:

- CLEAN WATER ACT: Toxic Pollutant (Section 401.15)
- RCRA, 40CFR261, Appendix 8 Hazardous Constituents, waste number not listed

2-chloronaphthalene:

- CLEAN WATER ACT: 40CFR423, Appendix A, Priority Pollutants
- EPA HAZARDOUS WASTE NUMBER (RCRA No.): U047
- RCRA, 40CFR261, Appendix 8 Hazardous Constituents
- RCRA 40CFR268.48; 61FR15654, Universal Treatment Standards: Wastewater (mg/l), 0.055; Nonwastewater (mg/kg), 5.6
- RCRA 40CFR264, Appendix 9; TSD Facilities Ground Water Monitoring List. Suggested test method(s) (PQL μg/l): 8120 (10); 8270 (10)
- SUPERFUND/EPCRA 40CFR302.4 Reportable Quantity (RQ): CERCLA, 5,000 lb (2,270 kg)

Hexachloronaphthalene:

- EPCRA Section 313 Form R *de minimis* concentration reporting level: 1.0%

Cited in U.S. State Regulations: Alaska (G), California (A, G), Connecticut (A), Florida (G, A), Illinois (G), Kansas (G), Louisiana (G), Maine (G), Massachusetts (G), Nevada (A), New Hampshire (G), New York (A), North Dakota (A), Pennsylvania (G), Rhode Island (G), South Carolina (A), Vermont (G), Virginia (G, A), Washington (G), West Virginia (G), Wisconsin (G).

Description: The Chlorinated naphthalenes, $C_{10}H_{8-x}Cl_x$, in which one or more hydrogen atoms have been replaced by chlorine to form wax-like substances, beginning with monochloronaphthalene and going on to the octachlor derivatives. Their physical states vary from mobile liquids to waxy-solids depending on the degree of chlorination. Freezing/Melting points of the pure compounds range from 17°C for 1-chloronaphthalene to 198°C for 1,2,3,4-tetrachloronaphthalene.

2-chloronaphtalene: Freezing/Melting point = 61°. Boiling point = 256°. Insoluble in water. *Hexa-:* White to light-yellow solid with an aromatic odor. Boiling point = 343 – 388°C. Freezing/Melting point = 137°C. Insoluble in water.

Octa-: Waxy, pale yellow solid with an aromatic odor. Boiling point = 410 – 440°C. Freezing/Melting point = 185 – 192°C. Insoluble in water.

Penta-: Colorless to white crystalline solid with a benzene-like odor. Boiling point = 309°C. Freezing/Melting point = 190°C. Slightly soluble in water.

Tetra-: Colorless to pale yellow solid with an aromatic odor. Boiling point = 315 – 360°C. Freezing/Melting point = 182°C. Insoluble in water.

Tri-: Colorless to pale-yellow solid with an aromatic odor. Boiling point = 304 – 354°C. Freezing/Melting point = 93°C. Insoluble in water.

Potential Exposure: Industrial exposure from individual chlorinated naphthalenes is rarely encountered; rather it usually occurs from mixtures of two or more Chlorinated naphthalenes. Due to their stability, thermoplasticity, and non-flammability, these compounds enjoy wide industrial application. These compounds are used in the production of electric condensers, in the insulation of electric cables and wires, as additives to extreme pressure lubricants, as supports for storage batteries, and as a coating in foundry use.

Because of the possible potentiation of the toxicity of higher Chlorinated naphthalenes by ethanol and carbon tetrachloride, individuals who ingest enough alcohol to result in liver dysfunction would be a special group at risk. Individuals, e.g., analytical and synthetic chemists, mechanics and cleaners, who are routinely exposed to carbon tetrachloride or other hepatotoxic chemicals would also be at a greater risk than a population without such exposure. Individuals involved in the manufacture, utilization, or disposal of polychlorinated naphthalenes would be expected to have higher levels of exposure than the general population.

Incompatibilities: All are incompatible with strong oxidizers. Keep away from heat. Penta- is also incompatible with acids, alkalis.

Permissible Exposure Limits in Air: TWA or MAC values set by various countries are as follows:

Isomer	U.S. (OSHA),[58] (mg/m³)	Sweden,[35] (mg/m³)	U.K.[35] (mg/m³)	DFG[3] (mg/m³)	USSR[35] (mg/m³)
1-Chloro-	—	0.2	—	—	—
2-Chloro-	—	0.2	—	—	—
Trichloro-	5.0 (skin)	—	—	5.0	—
Tetrachloro-	2.0 (skin)	—	—	—	—
Pentachloro-	0.5 (skin)	0.2	—	0.5	0.5
Hexchloro-	0.2 (skin)	0.2	—	—	0.5
Octachloro-	0.1 (skin)	0.2	0.1	—	0.5

The only country that has set a STEL is the U.K.[33] which has set 0.3 mg/m³. The NIOSH IDLH levels[2] are 50 for trichloro, 2 mg/m³ for hexachloro and 200 for octachloro.

All the U.S. TWA values carry the notation "skin" indicating the possibility of cutaneous absorption.

In addition, a number of states have set guidelines or standards for Chlorinated naphthalenes in ambient air[60] as follows (all values in μg/m³):

	Tri-chloro-	Tetra-chloro-	Penta-chloro-	Hexa-chloro-	Octa-chloro-
Connecticut	100	40	10	4	2
Florida	—	—	—	2	1
Nevada	119	48	12	5	2
New York	—	—	—	0.67	0.33
North Dakota	50	20	5	2	1 – 3
South Carolina	—	—	—	1.0	0.5
Virginia	80	35	—	3.2	1.6

Determination in Air: *Hexa-:* Filter; Hexane; Gas chromatography/Electrochemical detection; NIOSH II(2) Method #S100.

Octa-: Filter; Hexane; Gas chromatography/Flame ionization detection; NIOSH II(2) Method #S97.

Penta-: Filter/Bubbler; Isooctane; Gas chromatography/Electrochemical detection; NIOSH II(2) Method #S96.

Tetra-: Filter/Bubbler; none; Gas chromatography/Flame ionization detection; NIOSH II(2) Method #S130.

Tri-: Filter/Bubbler; none; Gas chromatography/Flame ionization detection; NIOSH II(2) Method #S128.

Permissible Concentration in Water: To protect freshwater aquatic life: 1,600 μg/l on an acute toxicity basis. To protect saltwater aquatic life: 7.5 μg/l on an acute toxicity basis.

For the protection of human health from the toxic properties of chlorinated naphthalenes ingested through water and through contaminated aquatic organisms, there are insufficient data to permit establishment of criteria.[6]

The former USSR[35] has set MAC limits in surface water of 0.01 mg/l for 1-Chloronaphthalene and for 2-Chloronaphthalene.

Determination in Water: 2-Chloronaphthalene may be determined by gas chromatography (EPA Method 612) or by gas chromatography plus mass spectrometry (EPA Method 625).

Routes of Entry: Inhalation of fumes and percutaneous absorption of liquid, ingestion, and eye and skin contact.

Harmful Effects and Symptoms

Short Term Exposure: Contact can irritate eyes and skin. Cases of systemic poisoning are few in number and they may occur without the development of chloracne. It is believed that chloracne develops from skin contact and inhalation of fumes, while systemic effects result primarily from inhalation of fumes. Symptoms of poisoning may include headaches, fatigue, vertigo, and anorexia. Jaundice may occur from liver damage. Highly Chlorinated naphthalenes seem to be more toxic than those Chlorinated naphthalenes with a lower degree of substitution.

Long Term Exposure: May cause acne-like dermatitis. May affect the liver, resulting in jaundice. Can affect the nervous system. Chronic exposure to Chlorinated naphthalenes can cause chloracne, which consists of simple erythematous eruptions with pustules, papules, and comedones. Cysts may develop due to plugging of the sebaceous gland orifices.

Points of Attack: Skin, liver, nervous system.

Medical Surveillance: Preplacement and periodic examinations should be concerned particularly with skin lesions such as chloracne and with liver function. Liver function tests. Examination by a qualified allergist. Examination of the nervous system. More than light consumption of alcohol may increase liver damage.

First Aid: If this chemical gets into the eyes, remove any contact lenses at once and irrigate immediately for at least 15 minutes, occasionally lifting upper and lower lids. Seek medical attention immediately. If this chemical contacts the skin, remove contaminated clothing and wash immediately with soap and water. Seek medical attention immediately. If this chemical has been inhaled, remove from exposure, begin rescue breathing (using universal precautions) if breathing has stopped and CPR if heart action has stopped. Transfer promptly to a medical facility. When this chemical has been swallowed, get medical attention. Give large quantities of water and induce vomiting. Do not make an unconscious person vomit.

Personal Protective Methods: Wear protective gloves and clothing to prevent any reasonable probability of skin contact. Safety equipment suppliers/manufacturers can provide recommendations on the most protective glove/clothing material for your operation. All protective clothing (suits, gloves, footwear, headgear) should be clean, available each day, and put on before work. Contact lenses should not be worn when working with this chemical. Wear dust-proof chemical goggles and face shield unless full facepiece respiratory protection is worn. Employees should wash immediately with soap when skin is wet or contaminated. Provide emergency showers and eyewash.

Respirator Selection: *hexa-:* NIOSH/OSHA: *Up to 2 mg/m³:* SA: (any supplied-air respirator);* or SCBAF: (any self-contained breathing apparatus with a full facepiece). *Emergency or planned entry into unknown concentrations or IDLH conditions:* SCBAF:PD,PP (any self-contained breathing apparatus that has a full facepiece and is operated in a pressure-demand or other positive-pressure mode); or SAF: PD,PP:ASCBA (any supplied-air respirator that has a full facepiece and is operated in a pressure-demand or other positive-pressure mode in combination with an auxiliary self-contained breathing apparatus operated in a pressure-demand or other positive-pressure mode). *Escape:* GMFOVHiE [any air-purifying, full-facepiece respirator (gas mask) with a chin-style, front- or back-mounted organic vapor canister having a high-efficiency particulate filter]; or SCBAE (any appropriate escape-type, self-contained breathing apparatus).

* May require eye protection.

octa-: NIOSH/OSHA: *Up to 1 mg/m³:* SA: (any supplied-air respirator); or SCBAF: (any self-contained breathing apparatus with a full facepiece). *Emergency or planned entry into unknown concentrations or IDLH conditions:* SCBAF: PD, PP (any self-contained breathing apparatus that has a full facepiece and is operated in a pressure-demand or other positive-pressure mode); or SAF:PD,PP:ASCBA (any supplied-air respirator that has a full facepiece and is operated in a pressure-demand or other positive-pressure mode in combination with an auxiliary self-contained breathing apparatus operated in a pressure-demand or other positive-pressure mode). *Escape:* GMFOVHiE [any air-purifying, full-facepiece respirator (gas mask) with a chin-style, front- or back-mounted organic vapor canister having a high-efficiency particulate filter]; or SCBAE (any appropriate escape-type, self-contained breathing apparatus).

penta-: NIOSH/OSHA: *Up to 5 mg/m³:* SA: (any supplied-air respirator);* or SCBAF: (any self-contained breathing apparatus with a full facepiece). *Emergency or planned entry into unknown concentrations or IDLH conditions:* SCBAF:PD,PP (any self-contained breathing apparatus that has a full facepiece and is operated in a pressure-demand or other positive-pressure mode); or SAF:PD,PP:ASCBA (any supplied-air respirator that has a full facepiece and is operated in a pressure-demand or other positive-pressure mode in combination with an auxiliary self-contained breathing apparatus operated in a pressure-demand or other positive-pressure mode). *Escape:* GMFOVHiE [any air-purifying, full-facepiece respirator (gas mask) with a chin-style, front- or back-mounted organic vapor canister having a high-efficiency particulate filter]; or SCBAE (any appropriate escape-type, self-contained breathing apparatus).

* May require eye protection.

tetra-: NIOSH/OSHA: *Up to 20 mg/m³:* SCBAF: (any self-contained breathing apparatus with a full facepiece); or SAF: (any supplied-air respirator with a full facepiece). *Emergency or planned entry into unknown concentrations or IDLH conditions:* SCBAF:PD,PP (any self-contained breathing apparatus that has a full facepiece and is operated in a pressure-demand or other positive-pressure mode); or SAF:PD,PP:ASCBA (any supplied-air respirator that has a full facepiece and is operated in a pressure-demand or other positive-pressure mode in combination with an auxiliary self-contained breathing apparatus operated in a pressure-demand or other positive-pressure mode). *Escape:* GMFOVHiE [any air-purifying, full-facepiece respirator (gas mask) with a chin-style, front- or back-mounted organic vapor canister having a high-efficiency particulate filter]; or SCBAE (any appropriate escape-type, self-contained breathing apparatus).

tri-: NIOSH/OSHA: *Up to 50 mg/m³:* SCBAF: (any self-contained breathing apparatus with a full facepiece); or SAF: (any supplied-air respirator with a full facepiece). *Emergency or planned entry into unknown concentrations or IDLH conditions:* SCBAF:PD,PP (any self-contained breathing apparatus that has a full facepiece and is operated in a pressure-demand or other positive-pressure mode); or SAF:PD,PP: ASCBA (any supplied-air respirator that has a full facepiece and is operated in a pressure-demand or other positive-pressure mode in combination with an auxiliary self-contained breathing apparatus operated in a pressure-demand or other positive-pressure mode). *Escape:* GMFOVHiE [any air-purifying, full-facepiece respirator (gas mask) with a chin-style, front- or back-mounted organic vapor canister having a high-efficiency particulate filter]; or SCBAE (any appropriate escape-type, self-contained breathing apparatus).

Storage: Store in a refrigerator or a cool, dry place.[52]

Shipping: The chlorinated naphthalenes are not specifically cited in the DOT Performance-Oriented Packaging Standards.[19]

Spill Handling: Evacuate persons not wearing protective equipment from area of spill or leak until clean-up is complete. Remove all ignition sources. Dampen material with toluene. Collect material in the most convenient and safe manner and deposit in sealed containers. Ventilate area after clean-up is complete. It may be necessary to contain and dispose of this chemical as a hazardous waste. If material or contaminated runoff enters waterways, notify downstream users of potentially contaminated waters. Contact your Department of Environmental Protection or your regional office of the federal EPA for specific recommendations. If employees are required to clean-up spills, they must be properly trained and equipped. OSHA 1910.120(q) may be applicable.

Fire Extinguishing: These chemicals are noncombustible solids. Use dry chemical, carbon dioxide, water spray, or foam extinguishers. Poisonous gases are produced in fire including hydrogen chloride and phosgene. If material or contaminated runoff enters waterways, notify downstream users of potentially contaminated waters. Notify local health and fire officials and pollution control agencies. From a secure, explosion-proof location, use water spray to cool exposed containers. If cooling streams are ineffective (venting sound increases in volume and pitch, tank discolors, or shows any signs of deforming), withdraw immediately to a secure position. If employees are expected to fight fires, they must be trained and equipped in OSHA 1910.156.

Disposal Method Suggested: High-temperature incineration with flue gas scrubbing. Incineration, preferably after mixing with another combustible fuel. Care must be exercised to assure complete combustion to prevent the formation of phosgene. An acid scrubber is necessary to remove the halo acids produced.[22]

References

U.S. Environmental Protection Agency, Chlorinated Naphthalenes: Ambient Water Quality Criteria, Washington, DC (1980)

U.S. Environmental Protection Agency, Chlorinated Naphthalenes, Health and Environmental Effects Profile No. 38, Office of Solid Waste, Washington, DC (April 30, 1980)

U.S. Environmental Protection Agency, 2-Chloronaphthalene, Health and Environmental Effects Profile No. 49, Office of Solid Waste, Washington, DC (April 30, 1980)

New Jersey Department of Health and Senior Services and Senior Services, "Hazardous Substance Fact Sheet: Hexachloronaphthalene," Trenton, NJ (April 1999)

New Jersey Department of Health and Senior Services and Senior Services, "Hazardous Substance Fact Sheet: Octachloronaphthalene," Trenton, NJ (March 1986)

New Jersey Department of Health and Senior Services and Senior Services, "Hazardous Substance Fact Sheet: Pentachloronaphthalene," Trenton, NJ (April 1986)

New Jersey Department of Health and Senior Services and Senior Services, "Hazardous Substance Fact Sheet: Tetrachloronaphthalene," Trenton, NJ (November 1986)

The references to Sax, N. I., Ed., "Dangerous Properties of Industrial Materials Report," may be tabulated as follows:

Isomer	*Volume*	*Number*	*Pages*	*Year*
1-Chloro-	3	2	77–78	(1983)
2-Chloro-	4	6	85–88	(1984)
Trichloro-	6	6	78–80	(1986)
Tetrachloro-	6	6	76–78	(1986)
Pentachloro-	5	1	84–87	(1985)
Hexachloro-	5	1	81–84	(1985)
Octachloro-	4	5	40–45	(1984)

Chlorine

Molecular Formula: Cl_2

Synonyms: Bertholite; Chloor (Dutch); Chlor (German); Chlore (French); Chlorine molecular (C12); Cloro (Italian, Spanish); Diatomic chlorine; Dichlorine; Molecular chlorine; Poly I gas

CAS Registry Number: 7782-50-5

RTECS Number: FO2100000

DOT ID: UN 1017

EEC Number: 017-001-00-7

Regulatory Authority

- Toxic Substance (World Bank)[15]
- Air Pollutant Standard Set (ACGIH)[1] (Australia) (DFG)[3] (HSE)[33] (Israel) (Mexico) (OSHA)[58] (Several States)[60] (Several Canadian Provinces)
- CLEAN AIR ACT: Hazardous Air Pollutants (Title I, Part A, Section 112); Accidental Release Prevention/Flammable substances, (Section 112[r], Table 3), TQ = 2,500 lb (1,135 kg)
- CLEAN WATER ACT: Section 311 Hazardous Substances/RQ 40CFR117.3 (same as CERCLA, see below); Section 313 Water Priority Chemicals (57FR41331, 9/9/92)
- SAFE DRINKING WATER ACT: SMCL, 250 mg/l; Priority List (55 FR 1470)
- In 1998 EPA set an MCL for TTHM (total trihalomethane) at MCLs to 0.80 mg/l (down from 0.100 mg/l set in 1976), and Maximum Residual Disinfectant level Goals (MRDG) for chlorine was set at 4 mg/l
- SUPERFUND/EPCRA 40CFR355, Appendix B Extremely Hazardous Substances: TPQ = 100 lb (45.4 kg)
- SUPERFUND/EPCRA 40CFR302.4 Reportable Quantity (RQ): CERCLA, 10 lb (4.54 kg)
- EPCRA Section 313 Form R *de minimis* concentration reporting level: 1.0%
- U.S. DOT Regulated Marine Pollutant (49CFR172.101, Appendix B)
- Canada, WHMIS, Ingredients Disclosure List; National Pollutant Release Inventory (NPRI)

Cited in U.S. State Regulations: Alaska (G), California (A, G). Connecticut (A), Florida (G, A), Illinois (G), Kansas (A), Maine (G), Maryland (G), Massachusetts (G, A), Michigan (G), Minnesota (G), Nevada (A), New Hampshire (G), New Jersey (G), New York (G, A), North Carolina (A), North Dakota (A), Oklahoma (G), Pennsylvania (G), Rhode Island (G), South Carolina (A), Virginia (A), West Virginia (G).

Description: Chlorine, Cl_2, is a greenish-yellow gas with a pungent, irritating odor. Shipped as a liquefied compressed gas. The odor threshold is 0.01 ppm in air. Boiling point = -34.6°C. Freezing/Melting point = -101°C. Soluble in water. It is slightly soluble in water and is soluble in alkalis. It is the commonest of the four halogens which are among the most chemically reactive of all the elements. It is not flammable; but, it is a strong oxidizer, and contact with other materials may cause fire.

Potential Exposure: Gaseous chlorine is a bleaching agent in the paper and pulp and textile industries for bleaching cellulose for artificial fibers. It is used in the manufacture of chlorinated lime, inorganic and organic compounds such as metallic chlorides, chlorinated solvents, refrigerants, pesticides, and polymers, e.g., synthetic rubber and plastics; it is used as a disinfectant, particularly for water and refuse, and in detinning and dezincing iron.

Incompatibilities: A powerful oxidizer. Reacts explosively or forms explosive compounds with many organic compounds and common substances such as acetylene, ether, turpentine, ammonia, fuel gas, hydrogen, and finely divided metals. Keep away from combustible substances and reducing agents. Corrosive to some plastic, rubber, and coating materials. Reacts with water to form hypochlorous acid. Corrosive to many metals in presence of water.

Permissible Exposure Limits in Air: The Federal standard (OSHA PEL)[58] is a ceiling limit of 1 ppm (0.3 mg/m^3) not to be exceeded at any time. The NIOSH recommended ceiling limit is 0.5 ppm which should not be exceeded during any 15 minute work period. The basis for the NIOSH-recommended environmental limit is the prevention of irritation of the skin, eyes, and respiratory tract. The ACGIH recommended value is 0.5 ppm TWA and STEL is 1 ppm. The former USSR-UNEP/IRPTC project has set a MAC in workplace air of 1.0 mg/m^3 and values for ambient air in residential areas of 0.1 mg/m^3 on a momentary basis and 0.03 mg/m^3 on an average daily basis. In addition, a number of states have set guidelines or standards for chlorine in ambient air[60] ranging from zero (North Carolina) to 7.143 μg/m^3 (Kansas) to 10.0 μg/m^3 (New York) to 30.0 μg/m^3 (Florida) to 30 – 90 μg/m^3 (North Dakota) to 39 μg/m^3 (Massachusetts) to 50 μg/m^3 (Virginia) to 60 μg/m^3 (Connecticut) to 71 μg/m^3 (Nevada) to 75 μg/m^3 (South Carolina).

Determination in Air: Collection by fritted bubbler, colorimetric analysis using methyl orange which is bleached by free chlorine. See OSHA Method ID101. Filter; $Na_2S_2O_3$; Ion chromatography; NIOSH IV Method #6011.

Permissible Concentration in Water: EPA has suggested the following limits: Total residual chlorine: 2.0 μg/l for salmonid fish; 10.0 μg/l for other freshwater and marine organisms.

The former USSR-UNEP/IRPTC project[43] recommends an absence of active chlorine (taking into account the absorbing capacity of water) in domestic water supplies and a MAC of zero in water for fishery purposes.

Routes of Entry: Inhalation, eye and skin contact.

Harmful Effects and Symptoms

Chlorine reacts with body moisture to form acids. It is itself extremely irritating to skin, eyes, and mucous membranes, and it may cause corrosion of teeth. Prolonged exposure to low concentrations may produce chloracne.

Chlorine in high concentrations acts as an asphyxiant by causing cramps in the muscles of the larynx (choking), swelling of the mucous membranes, nausea, vomiting, anxiety, and syncope. Acute respiratory distress including cough, hemoptysis, chest pain, dyspnea, and cyanosis develop, and later tracheobronchitis, pulmonary edema, and pneumonia may supervene.

1.0 ppm may produce irritation of the nose, mouth and throat; at 1.3 ppm and above, irritation may be more pronounced with coughing and labored breathing; high concentrations may cause throat muscle spasm leading to suffocation and death; delayed effects may include accumulation of fluid in the lungs, bronchitis and pneumonia. Death may occur after a few breaths at 1,000 ppm.

Short Term Exposure: A lacramator. Chlorine is corrosive to the eyes, skin, and respiratory tract. Eye contact can cause permanent damage. Inhalation of the gase can cause pulmonary edema, a medical emergency that can be delayed for several hours. This can cause death. Rapid evaporation of the liquid may cause frostbite.

Long Term Exposure: Repeated exposure may permanently damage the lungs, or cause chronic bronchitis. Chlorine may affect the teeth, resulting in erosion, and cause skin rash. A single high exposure may cause similar health effects.

Points of Attack: Lungs, respiratory system.

Medical Surveillance: Special emphasis should be given to the skin, eyes, teeth, cardiovascular status in placement and periodic examinations. Chest x-rays should be taken and pulmonary function followed.

First Aid: If this chemical gets into the eyes, remove any contact lenses at once and irrigate immediately for at least 30 minutes, occasionally lifting upper and lower lids. Seek medical attention immediately. If this chemical contacts the skin, remove contaminated clothing and wash immediately with soap and water. Seek medical attention immediately. If this chemical has been inhaled, remove from exposure, begin rescue breathing (using universal precautions) if breathing has stopped and CPR if heart action has stopped. Transfer promptly to a medical facility. When this chemical has been swallowed, get medical attention. Give large quantities of water and induce vomiting. Do not make an unconscious person vomit. Medical observation is recommended for 24 – 48 hours after breathing overexposure, as pulmonary edema may be delayed. As first aid for pulmonary edema, a doctor or authorized paramedic may consider administering a corticosteroid spray.

If frostbite has occurred, seek medical attention immediately; do *NOT* rub the affected areas or flush them with water. In order to prevent further tissue damage, do *NOT* attempt to remove frozen clothing from frostbitten areas. If frostbite has *NOT* occurred, immediately and thoroughly wash contaminated skin with soap and water.

Personal Protective Methods: Whenever there is likelihood of excessive gas levels, workers should use respiratory protection in the form of fullface gas masks with proper canisters or supplied air respirators. The skin effects of chlorine can generally be controlled by good personal hygiene practices. Where very high gas concentrations or liquid chlorine may be present, full protective clothing, gloves, and eye protection should be used. Saranex, Butyl Rubber/Neoprene, Viton, Neoprene, Butyl Rubber, and Viton/Neoprene are among the recommended protective materials. Changing work clothes daily and showering following each shift where exposures exist are recommended. Where exposure to the liquefied compressed gas may occur, employees should be provided with special clothing designed to prevent frostbite.

Respirator Selection: NIOSH/OSHA: *5 ppm:* CCRS [any chemical cartridge respirator with cartridge(s) providing protection against the compound of concern]; or SA (any supplied-air respirator). *12.5 ppm:* SA:CF (any supplied-air respirator operated in a continuous-flow mode); or PAPRS (any powered, air-purifying respirator with cartridge(s) providing protection against the compound of concern); or CCRFS (any chemical cartridge respirator with a full facepiece and cartridge(s) providing protection against the compound of concern); or GMFS (any air-purifying, full-facepiece respirator (gas mask) with a chin-style, front- or back-mounted canister providing protection against the compound of concern); or SCBAF (any self-contained breathing apparatus with a full facepiece); or SAF (any supplied-air respirator with a full facepiece). *Emergency or planned entry into unknown concentrations or IDLH conditions:* SCBAF:PD,PP (any self-contained breathing apparatus that has a full facepiece and is operated in a pressure-demand or other positive-pressure mode); or SAF:PD,PP:ASCBA (any supplied-air respirator that has a full facepiece and is operated in a pressure-demand or other positive-pressure mode in combination with an auxiliary self-contained breathing apparatus operated in a pressure-demand or other positive pressure mode). *Escape:* GMFS (any air-purifying, full-facepiece respirator (gas mask) with a chin-style, front- or back-mounted canister providing protection against the compound of concern); or SCBAE (any appropriate escape-type, self-contained breathing apparatus).

Note: Substance reported to cause eye irritation or damage; may require eye protection.

Storage: Prior to working with Chlorine you should be trained on its proper handling and storage. Protect containers against physical damage. Store cylinders and containers in a cool, dry, relatively isolated area, protected from weather and extreme temperature changes.

Shipping: The DOT label required is "Poison Gas." Shipment by passenger aircraft or railcar or even by cargo aircraft is forbidden. The UN/DOT Hazard Class is 2.3 and the Packing Group is I.[19][20] Chlorine is a DOT regulated marine pollutant.

Spill Handling: Evacuate and restrict persons not wearing protective equipment from area of spill or leak until cleanup is complete. If the gas is leaked, stop the flow of gas if it can be done safely. If the source of the leak is a cylinder and the leak cannot be stopped in place, remove the leaking cylinder to a safe place in the open air, and, repair the leak or allow the cylinder to empty. If the leak can be stopped in place, bubble chlorine through a sodium sulfide and excess sodium bicarbonate solution including a trap in the line. For liquid spills, ventilate area and wash down spill with water. It may be necessary to contain and dispose of this chemical as a hazardous waste. If material or contaminated runoff enters waterways, notify downstream users of potentially contaminated waters. Contact your Department of Environmental Protection or your regional office of the federal EPA for specific recommendations. If employees are required to clean-up spills, they must be properly trained and equipped. OSHA 1910.120(q) may be applicable.

Initial isolation and protective action distances

Distances shown are likely to be affected during the first 30 minutes after materials are spilled and could increase with time. If more than one tank car, cargo tank, portable tank, or large cylinder is involved in the incident is leaking, the protective action distance may need to be increased. You may need to seek emergency information from CHEMTREC at (800) 424-9300 or seek professional environmental engineering assistance from the U.S. EPA Environmental Response Team at (908)548-8730 (24-hour response line).

Small spills (From a small package or a small leak from a large package)

First: Isolate in all directions (feet)..... 200

Then: Protect persons downwind (miles)

Day 0.2

Night 0.5

Large spills (From a large package or from many small packages)

First: Isolate in all directions (feet)..... 600

Then: Protect persons downwind (miles)

Day 0.5

Night 1.9

Fire Extinguishing: Chlorine is a noncombustible solid, but it will increase the intensity of a fire and cause fire upon contact with combustible materials. Firefighting gear (including SCBA) does not provide adequate protection. If exposure occurs, remove and isolate gear immediately and thoroughly decontaminate personnel. Vapors are heavier than air and will collect in low areas. Hydrogen and chlorine mixtures (5 – 95%) are exploded by almost any form of energy (heat, sunlight, sparks, etc.). May combine with water or steam to produce toxic and corrosive fumes of hydrochloric acid. Use any extinguishing agent suitable for surrounding fire. Poisonous gases are produced in fire. If material or contaminated runoff enters waterways, notify downstream users of potentially contaminated waters. Notify local health and fire officials and pollution control agencies. Containers may explode in fire. From a secure, explosion-proof location, use water spray to cool exposed containers. If cooling streams are ineffective (venting sound increases in volume and pitch, tank discolors, or shows any signs of deforming), withdraw immediately to a secure position. If employees are expected to fight fires, they must be trained and equipped in OSHA 1910.156.

Disposal Method Suggested: Introduce into large volume and solution of reducing agent (bisulfite, ferrous salts or hypo), neutralize and flush to sewer with water. Recovery is an option to disposal for Chlorine in the case of gases from aluminum chloride electrolysis and chlorine in wastewaters. See also Spill Handling.

References

National Institute for Occupational Safety and Health, Criteria for a Recommended Standard: Occupational Exposure to Chlorine, NIOSH Document No. 76–170 (1976)

National Academy of Sciences, Medical and Biological Effects of Environmental Pollutants: Chlorine and Hydrogen Chloride, Washington, DC (1976)

Sax, N. I., Ed., "Dangerous Properties of Industrial Materials Report," 1, No. 3, 41–43 (1981) and 2, No. 4, 67–70, New York, Van Nostrand Reinhold Co. (1982). (Chlorine-36)

U.S. Environmental Protection Agency, "Chemical Profile: Chlorine," Washington, DC, Chemical Emergency Preparedness Program (October 31, 1985)

New Jersey Department of Health and Senior Services and Senior Services, "Hazardous Substance Fact Sheet: Chlorine," Trenton, NJ (August 1998)

New York State Department of Health, "Chemical Fact Sheet: Chlorine," Albany, NY, Bureau of Toxic Substance Assessment (January 1986 and Version 2)

Chlorine Dioxide

Molecular Formula: ClO_2

Synonyms: Alcide; Anthium dioxcide; Chlorine oxide; Chlorine(IV) oxide; Chlorine peroxide; Chloroperoxyl; Chloryl radical: "ClO_2"; Dioxido de cloro (Spanish); Doxcide 50; Ez flow; Purogene

CAS Registry Number: 10049-04-4

RTECS Number: FO3000000

DOT ID: UN/NA 9191 (for frozen hydrate). Transport of gas is *FORBIDDEN*.

Regulatory Authority

- Air Pollutant Standard Set (ACGIH)[1] (DFG)[3] (HSE)[33] (OSHA)[58] (Several States)[60] (Australia) (Israel) (Mexico) (Several Canadian Provinces)

- CLEAN AIR ACT: Accidental Release Prevention/Flammable substances, (Section 112[r], Table 3), TQ = 1,000 lb (454 kg)
- SAFE DRINKING WATER ACT: Priority List (55 FR 1470)
- In 1998 EPA set an MCL for TTHM (total trihalomethane) at MCLs to 0.80 mg/l (down from 0.100 mg/l set in 1976), and Maximum Residual Disinfectant level Goals (MRDG) for chlorine dioxide was set at 0.8 mg/l
- EPCRA Section 313 Form R *de minimis* concentration reporting level: 1.0%
- Canada, WHMIS, Ingredients Disclosure List; National Pollutant Release Inventory (NPRI)

Cited in U.S. State Regulations: Alaska (G), California (A, G), Connecticut (A), Florida (G, A), Illinois (G), Maine (G, W), Maryland (G), Massachusetts (G), Minnesota (G), Nevada (A), New Hampshire (G, W), New Jersey (G), New York (G, A), North Dakota (A), Oklahoma (G), Pennsylvania (G), Rhode Island (G), Virginia (A), West Virginia (G).

Description: Chlorine dioxide, ClO_2, is a flammable, reddish-yellow gas or reddish-brown liquid (below 11°C/52°F) with an irritating odor like chlorine or nitric acid. Boiling point = 11°C. Freezing/Melting point = -59°C. Soluble in water (reactive).

Potential Exposure: Chlorine dioxide is used in bleaching cellulose pulp, bleaching flour, water purification. It is used as an oxidizing agent.

Incompatibilities: Unstable in light. A powerful oxidizer. Chlorine dioxide gas is explosive at concentrations over 10% and can be ignited by almost any form of energy, including sunlight, heat (explosions can occur in air in temperature above 130°C), or sparks, shock, friction, or concussion. This chemical reacts violently with dust, combustible materials, and reducing agents. Reacts violently with mercury, phosphorus, sulfur and many compounds, causing fire and explosion hazard. Contact with water forms perchloric and hydrochloric acid. Corrosive to metals.

Permissible Exposure Limits in Air: The OSHA PEL is 0.1 ppm TWA for an 8-hour workshift. NIOSH recommends a limit of 0,1 ppm TWA for a 10-hour workshift and STEL (15 min) of 0.3 ppm. ACGIH recommends the same TWA (8-hour workshift) and STEL as NIOSH. The NIOSH IDLH = 5 ppm. Australia, Israel, Mexico and the Canadian provinces of Alberta, British Columbia, Ontario and Quebec have set the airborne exposure limits and STELs as NIOSH. The former USSR/UNEP joint project[43] has set a MAC in workplace air of 0.1 mg/m^3. In addition, several states have set guidelines or standards for ClO_2 in ambient air[60] ranging from 1.0 μg/m^3 (New York) to 3.0 μg/m^3 (Florida) to 3.0 – 9.0 μg/m^3 (North Dakota) to 5.0 μg/m^3 (Virginia) to 6.0 μg/m^3 (Connecticut) to 7.0 μg/m^3 (Nevada).

Determination in Air: Collection by bubbler; Potassium iodide; Ion chromatography; OSHA Method #ID202.

Permissible Concentration in Water: Values in guidelines or standards for chlorine in drinking water[61] have been set by Maine at 110 μg/l and by the USEPA at 1,000 μg/l.

Routes of Entry: Inhalation, ingestion, eye and skin contact.

Harmful Effects and Symptoms

Short Term Exposure: A lacramator. Chlorine dioxide is corrosive to the eyes, skin, and respiratory tract. Inhalation can cause pulmonary edema, a medical emergency that can be delayed for several hours. This can cause death. Inhalation at levels above 0.25 ppm may cause slight irritation to the nose, throat and mouth. Levels above 5 ppm may cause severe irritation to the nose, throat, and mouth. 19 ppm for an unspecified time has caused death.

Long Term Exposure: Chlorine dioxide may affect the lungs, causing chronic bronchitis to develop with cough, phlegm, and/or shortness of breath. This chemical may affect the teeth, causing erosion. There is limited evidence that Chlorine dioxide may damage the developing fetus.

Points of Attack: Respiratory system, lungs, eyes.

Medical Surveillance: Consider the points of attack in preplacement and periodic physical examinations. Lung function tests. Consider x-ray following acute overexposure.

First Aid: If this chemical gets into the eyes, remove any contact lenses at once and irrigate immediately for at least 15 minutes, occasionally lifting upper and lower lids. Seek medical attention immediately. If this chemical contacts the skin, remove contaminated clothing and wash immediately with soap and water. Seek medical attention immediately. If this chemical has been inhaled, remove from exposure, begin rescue breathing (using universal precautions) if breathing has stopped and CPR if heart action has stopped. Transfer promptly to a medical facility. When this chemical has been swallowed, get medical attention. Give large quantities of water and induce vomiting. Do not make an unconscious person vomit. Medical observation is recommended for 24 – 48 hours after breathing overexposure, as pulmonary edema may be delayed. As first aid for pulmonary edema, a doctor or authorized paramedic may consider administering a corticosteroid spray.

Personal Protective Methods: Wear protective gloves and clothing to prevent any reasonable probability of skin contact. Safety equipment suppliers/manufacturers can provide recommendations on the most protective glove/clothing material for your operation. All protective clothing (suits, gloves, footwear, headgear) should be clean, available each day, and put on before work. Contact lenses should not be worn when working with this chemical. When working with liquid, wear splash-proof chemical goggles and face shield when there is a potential for exposure to gas, unless full facepiece respiratory protection is worn. Employees should wash immediately with soap when skin is wet or contaminated. Provide emergency showers and eyewash.

Respirator Selection: NIOSH/OSHA: *1 ppm:* CCRS (any chemical cartridge respirator with cartridge(s) providing protection against the compound of concern); or SA (any supplied-air respirator). *2.5 ppm:* SA:CF (any supplied-air respirator operated in a continuous-flow mode); or PAPRS [any powered, air-purifying respirator with cartridge(s) providing protection against the compound of concern]. *5 ppm:* CCRFS [any chemical cartridge respirator with a full facepiece and cartridge(s) providing protection against the compound of concern]; or GMFS [any air-purifying, full-facepiece respirator (gas mask) with a chin-style, front- or back-mounted canister providing protection against the compound of concern]; or SCBAF (any self-contained breathing apparatus with a full facepiece); or SAF (any supplied-air respirator with a full facepiece). *Emergency or planned entry into unknown concentrations or IDLH conditions:* SCBAF:PD,PP (any self-contained breathing apparatus that has a full facepiece and is operated in a pressure-demand or other positive-pressure mode); or SAF:PD,PP:ASCBA (any supplied-air respirator that has a full facepiece and is operated in a pressure-demand or other positive-pressure mode in combination with an auxiliary, self-contained breathing apparatus operated in a pressure-demand or other positive-pressure mode). *Escape:* GMFS [any air-purifying, full-facepiece respirator (gas mask) with a chin-style, front- or back-mounted canister providing protection against the compound of concern; or SCBAE (any appropriate escape-type, self-contained breathing apparatus).

Note: Substance reported to cause eye irritation or damage; may require eye protection. Only non-oxidizable sorbents allowed (not charcoal).

Storage: Prior to working with Chlorine dioxide you should be trained on its proper handling and storage. This chemical is a powerful oxidizer, and is shock-, light- and heat-sensitive. It is violently explosive in air at concentrations over 10%. Keep frozen when not in use. Store in tightly closed containers in a cool, dark, well-ventilated area at temperatures well below 130°C. Gas explosions may occur above 130°C. Use only non-sparking tools and equipment, especially when opening and closing containers of this chemical. Sources of ignition such as smoking and open flames, are prohibited where this chemical is used, handled, or stored in a manner that could create a potential fire or explosion hazard. Use explosion-proof electrical equipment and fittings in storage area. See OSHA standard 1910.104 and NFPA 43A *Code for the Storage of Liquid and Solid Oxidizers* for detailed handling and storage regulations.

Shipping: Shipment of Chlorine dioxide (not hydrated) is forbidden by any means according to DOT.[19] The frozen hydrate can be shipped but must be labeled "Oxidizer, Poison." It falls in Hazard Class 5.1 and Packing Group II. Shipment by passenger aircraft or railcar or even by cargo aircraft is forbidden.

Spill Handling: Evacuate and restrict persons not wearing protective equipment from area of spill or leak until cleanup is complete. Remove all ignition sources. Ventilate area of leak or spill. If the gas is leaked, stop the flow of gas if it can be done safely. If the source of the leak is a cylinder and the leak cannot be stopped in place, remove the leaking cylinder to a safe place in the open air, and, repair the leak or allow the cylinder to empty. If the leak can be stopped in place, bubble Chlorine dioxide through a solution made up of reducing agent sodium bisulfide and sodium bicarbonate with a trap in the line. For liquid spills, allow Chlorine dioxide to evaporate with all available ventilation. Keep Chlorine dioxide out of a confined space, such as a sewer, because of the possibility of an explosion, unless the sewer is designed to prevent the build-up of explosive concentrations. It may be necessary to contain and dispose of this chemical as a hazardous waste. If material or contaminated runoff enters waterways, notify downstream users of potentially contaminated waters. Contact your Department of Environmental Protection or your regional office of the federal EPA for specific recommendations. If employees are required to clean-up spills, they must be properly trained and equipped. OSHA 1910.120(q) may be applicable.

Initial isolation and protective action distances

Distances shown are likely to be affected during the first 30 minutes after materials are spilled and could increase with time. If more than one tank car, cargo tank, portable tank, or large cylinder is involved in the incident is leaking, the protective action distance may need to be increased. You may need to seek emergency information from CHEMTREC at (800) 424-9300 or seek professional environmental engineering assistance from the U.S. EPA Environmental Response Team at (908)548-8730 (24-hour response line).

NA 9191 (chlorine dioxide, hydrate, frozen) is on the DOT's list of dangerous water-reactive materials which create large amounts of toxic vapor when *spilled in water*: Dangerous from 0.5 – 10 km (0.3 – 6.0 miles) downwind.

Fire Extinguishing: A powerful oxidizer, this chemical will increase the intensity of a fire, and can cause fire upon contact with combustibles. This chemical is an explosive at concentrations over 10% and can be ignited by almost any form of energy. Firefighting gear (including SCBA) may not provide adequate protection. If exposure occurs, remove and isolate gear immediately and thoroughly decontaminate personnel. Poisonous gases including chlorine are produced in fire. Use water only. *Do not* use dry chemical or carbon dioxide extinguishers. Use water with caution as chlorine dioxide reacts with water forming hydrogen chloride gas. Vapors are heavier than air and will collect in low areas. Containers may explode in fire. Storage containers and parts of containers may rocket great distances, in many directions. If material or contaminated runoff enters waterways, notify downstream users of potentially contaminated waters. Notify local health and fire officials and pollution control agencies. From a secure, explosion-proof location, use water spray to cool exposed containers. If cooling streams are ineffective (venting sound increases in volume and pitch, tank discolors, or shows any signs of deforming), withdraw immediately to a secure position. If employees are expected to fight fires, they must be trained and equipped in OSHA 1910.156.

Disposal Method Suggested: Use large volume of concentrated solution of ferrous salt or bisulfite solution as reducing agent. Then neutralize and flush to sewer with abundant water.[24]

References

New York State Department of Health, "Chemical Fact Sheet: Chlorine Dioxide," Albany, NY, Bureau of Toxic Substance Assessment (March 1986 and revision)

New Jersey Department of Health and Senior Services and Senior Services, "Hazardous Substance Fact Sheet: Chlorine Dioxide," Trenton, NJ (June 1998)

Chlorine Trifluoride

Molecular Formula: ClF_3

Synonyms: Chlorine fluoride; Chlorine trifluoride; Chlorotrifluoride; Trifluorure de chlore (French)

CAS Registry Number: 7790-91-2

RTECS Number: FO2800000

DOT ID: UN 1749

Regulatory Authority

- Air Pollutant Standard Set (ACGIH)[1] (Australia) (DFG)[3] (HSE)[33] (Israel) (Mexico) (Several States)[60] (Several Canadian Provinces)

Cited in U.S. State Regulations: Alaska (G), California (A, G), Florida (G), Illinois (G), Maine (G), Massachusetts (G), Minnesota (G), Nevada (A), New Hampshire (G), New Jersey (G), North Dakota (A), Oklahoma (G), Pennsylvania (G), Rhode Island (G), Virginia (A), West Virginia (G).

Description: Chlorine trifluoride, ClF_3, is a greenish yellow, almost colorless, liquid or colorless gas with a sweet, irritating odor. Shipped as a liquefied compressed gas. Boiling point = 11°C. Freezing/Melting point = -76°C. Reacts with water.

Potential Exposure: Chlorine trifluoride, ClF_3, is used as a fluorinating agent. It may be used as an igniter and propellant in rockets. It is used in nuclear fuel processing.

Incompatibilities: Most combustible materials ignite spontaneously on contact with chlorine trifluoride. Explodes on contact with organic materials. The liquid can explode if mixed with halocarbons or hydrocarbons. It reacts violently with oxidizable materials, finely divided metals and metal oxides, sand, glass, asbestos, silicon-containing compounds. Emits highly toxic fumes on contact with acids. Chlorine trifluoride decomposes above 220°C forming poisonous gases including hydrogen chloride and hydrogen fluoride. Reacts violently with water, forming chlorine and hydrofluoric acid. Reacts with most forms of plastics, rubber, coatings, and resins, except the highly fluorinated polymers such as "Teflon" and "Kel-F."

Permissible Exposure Limits in Air: The Federal limit (OSHA PEL)[2] is 0.1 ppm (0.4 mg/m^3) as a ceiling value, not to be exceeded at any time. NIOSH and ACGIH recommends the same level as OSHA. The NIOSH IDLH level is 20 ppm. The German DFG MAK is the same as OSHA and a peak limitation of 2 times the normal MAK (5 minute momentary value) not to be exceeded 8 times during a workshift. Australia's peak limitation and Israel's ceiling limit is 0.1 ppm. Mexico's limit is 0.1 ppm. TWA. The Canadian provinces of Alberta, British Columbia, Ontario, and Quebec have a ceiling value of 0.1 ppm.

Values for guidelines or standards for ClF_3 in ambient air have been set[60] ranging from 3.0 μg/m^3 (Virginia) to 4.0 μg/m^3 (North Dakota) to 10.0 μg/m^3 (Nevada).

Determination in Air: No test available.

Permissible Concentration in Water: No criteria set.

Routes of Entry: Inhalation, ingestion, eye and skin contact.

Harmful Effects and Symptoms

Short Term Exposure: Chlorine trifluoride is corrosive to the eyes, skin, and respiratory tract. Inhalation can cause pulmonary edema, a medical emergency that can be delayed for several hours. This can cause death. Contact with the liquefied gas can cause frostbite.

Long Term Exposure: Can cause lung irritation, bronchitis may develop with cough, phlegm, and shortness of breath.

Points of Attack: Skin, eyes, respiratory tract.

Medical Surveillance: Consider the points of attack in preplacement and periodic physical examinations. Lung function tests. Consider chest x-ray following acute overexposure.

First Aid: If this chemical gets into the eyes, remove any contact lenses at once and irrigate immediately for at least 15 minutes, occasionally lifting upper and lower lids. Seek medical attention immediately. If this chemical contacts the skin, remove contaminated clothing and wash immediately with soap and water. Seek medical attention immediately. If this chemical has been inhaled, remove from exposure, begin rescue breathing (using universal precautions) if breathing has stopped and CPR if heart action has stopped. Transfer promptly to a medical facility. When this chemical has been swallowed, get medical attention. If victim is conscious, administer water or milk. Do not induce vomiting. Medical observation is recommended for 24 – 48 hours after breathing overexposure, as pulmonary edema may be delayed. As first aid for pulmonary edema, a doctor or authorized paramedic may consider administering a corticosteroid spray.

If frostbite has occurred, seek medical attention immediately; do *NOT* rub the affected areas or flush them with water. In order to prevent further tissue damage, do *NOT* attempt to remove frozen clothing from frostbitten areas. If frostbite has *NOT* occurred, immediately and thoroughly wash contaminated skin with soap and water.

Personal Protective Methods: Wear protective gloves and clothing to prevent any reasonable probability of skin contact. Safety equipment suppliers/manufacturers can provide recommendations on the most protective glove/clothing material for your operation. All protective clothing (suits, gloves, footwear, headgear) should be clean, available each day, and put on before work. Contact lenses should not be worn when working with this chemical. Wear splash-proof chemical goggles if working with the liquid and face shield when working with gas, unless full facepiece respiratory protection is worn. Employees should wash immediately with soap when skin is wet or contaminated. Provide emergency showers and eyewash. Where exposure to the liquefied compressed gas may occur, employees should be provided with special clothing designed to prevent frostbite.

Respirator Selection: NIOSH/OSHA: *2.5 ppm:* SA:CF (any supplied-air respirator operated in a continuous-flow mode). *5 ppm:* SCBAF (any self-contained breathing apparatus with a full facepiece); or SAF (any supplied-air respirator with a full facepiece). *20 ppm:* SAF:PD,PP (any supplied-air respirator that has a full facepiece and is operated in a pressure-demand or other positive-pressure mode). *Emergency or planned entry into unknown concentrations or IDLH conditions:* SCBAF:PD,PP (any self-contained breathing apparatus that has a full facepiece and is operated in a pressure-demand or other positive-pressure mode); or SAF:PD,PP: ASCBA (any supplied-air respirator that has a full facepiece and is operated in a pressure-demand or other positive-pressure mode in combination with an auxiliary self-contained breathing apparatus operated in a pressure-demand or other positive pressure mode). *Escape:* GMFS (any air-purifying, full-facepiece respirator (gas mask) with a chin-style, front- or back-mounted canister providing protection against the compound of concern); or SCBAE (any appropriate escape-type, self-contained breathing apparatus).

Note: Substance causes eye irritation or damage; eye protection needed.

Storage: Prior to working with Chlorine trifluoride you should be trained on its proper handling and storage. Chlorine trifluoride must be stored to avoid contact with water, sand, glass, silicon-containing compounds, asbestos, and combustible materials, since violent reactions occur. See Incompatibilities. Store in tightly closed containers in a cool, well-ventilated area away from heat.

Shipping: Chlorine trifluoride must be labeled: "Poison Gas, Oxidizer, Corrosive." It falls in Hazard Class 2.3 and Packing Group I. Shipment by passenger aircraft or railcar or even by cargo aircraft is forbidden.[19]

Spill Handling: Evacuate and restrict persons not wearing protective equipment from area of spill or leak until cleanup is complete. If in a building, shut down HVAC systems. Remove all ignition sources. Collect solid chlorine trifluoride in the most convenient and safe manner and deposit in sealed containers. If the gas has leaked, Stop the flow of gas if it can be done safely. Ventilate area of leak or spill. If the source of the leak is a cylinder and the leak cannot be stopped in place, remove the leaking cylinder to a safe place in the open air, and, repair the leak or allow the cylinder to empty. Keep Chlorine trifluoride out of a confined space, such as a sewer, because of the possibility of an explosion, unless the sewer is designed to prevent the build-up of explosive concentrations. It may be necessary to contain and dispose of this chemical as a hazardous waste. If material or contaminated runoff enters waterways, notify downstream users of potentially contaminated waters. Contact your Department of Environmental Protection or your regional office of the federal EPA for specific recommendations. If employees are required to clean-up spills, they must be properly trained and equipped. OSHA 1910.120(q) may be applicable.

Initial isolation and protective action distances

Distances shown are likely to be affected during the first 30 minutes after materials are spilled and could increase with time. If more than one tank car, cargo tank, portable tank, or large cylinder is involved in the incident is leaking, the protective action distance may need to be increased. You may need to seek emergency information from CHEMTREC at (800) 424-9300 or seek professional environmental engineering assistance from the U.S. EPA Environmental Response Team at (908)548-8730 (24-hour response line).

Small spills (From a small package or a small leak from a large package)

First: Isolate in all directions (feet) 300

Then: Protect persons downwind (miles)

Day 0.2

Night 0.8

Large spills (From a large package or from many small packages)

First: Isolate in all directions (feet)..... 800

Then: Protect persons downwind (miles)

Day 0.7

Night 2.9

Fire Extinguishing: This chemical does not burn but it will increase the activity of fire and will cause combustible to ignite. *Do not use water* or foam. Use dry chemical, carbon dioxide. Poisonous gases are produced in fire including hydrogen fluoride and hydrogen chloride. Firefighting gear (including SCBA) does not provide adequate protection. If exposure occurs, remove and isolate gear immediately and thoroughly decontaminate personnel. Vapors are heavier than air and will collect in low areas. If material or contaminated runoff enters waterways, notify downstream users of potentially contaminated waters. Notify local health and fire officials and pollution control agencies. Containers may explode in fire. Storage containers and parts of containers may rocket great distances, in many directions. From a secure, explosion-proof location, use water spray to cool exposed containers. If cooling streams are ineffective (venting sound increases in volume and pitch, tank discolors, or shows any signs of deforming), withdraw immediately to a secure position. If cylinders are exposed to excessive heat from fire or flame contact, withdraw immediately to a secure position. If employees are expected to fight fires, they must be trained and equipped in OSHA 1910.156.

References

New Jersey Department of Health and Senior Services and Senior Services, "Hazardous Substance Fact Sheet: Chlorine Trifluoride," Trenton, NJ (November 1998)

Chlormephos

Molecular Formula: $C_5H_{12}ClO_2PS_2$

Synonyms: *S*-(Chloromethyl) *O,O*-diethyl ester phosphorodithioic acid; *S*-(Chloromethyl) *O,O*-diethyl phosphorodithioate; *S*-Chloromethyl *O,O*-diethyl phosphorodithioate; *S*-(Chloromethyl) *O,O*-diethyl phosphorodithioic acid; *S*-Chloromethyl *O,O*-diethyl phosphorodithiothiolothionate; Dotan®; MC2188; Phosphorodithioic acid, *S*-(chloromethyl) *O,O*-diethyl ester

CAS Registry Number: 24934-91-6

RTECS Number: TD5170000

Regulatory Authority

- SUPERFUND/EPCRA 40CFR355, Appendix B Extremely Hazardous Substances: TPQ = 500 lb (227 kg)
- SUPERFUND/EPCRA 40CFR302.4 Reportable Quantity (RQ): EHS, 1 lb (0.454 kg)
- U.S. DOT Regulated Marine Pollutant (49CFR172.101, Appendix B)

Cited in U.S. State Regulations: California (G), Massachusetts (G), Florida (G), Massachusetts (G), New Jersey (G), Pennsylvania (G).

Description: Chlormephos is an organophosphate, colorless liquid. Boiling point = 81 – 85°C @t 0.1 mm Hg. Hazard Identification (based on NFPA-704 M Rating System): Health 3, Flammability 1, Reactivity 0. Soluble in water.

Potential Exposure: This material is a soil insecticide. Not registered as a pesticide in the U.S.

Incompatibilities: Contact with strong oxidizers may cause fire and explosion. May corrode metals.

Permissible Exposure Limits in Air: No standards set. Passes through the skin.

Determination in Air: OSHA versatile sampler-2; Toluene/Acetone; Gas chromatography/Flame photometric detection for sulfur, nitrogen, or phosphorus; NIOSH Method IV #5600, Organophosphorus Pesticides.

Routes of Entry: Inhalation, ingestion, skin contact, passes through the skin.

Harmful Effects and Symptoms

Short Term Exposure: This material is poisonous; it may be fatal if inhaled, swallowed, or absorbed through the skin. The acute oral LD_{50} for rats is 7 mg/kg (highly toxic).

Symptoms exhibited on Chlormephos exposure are typical of cholinesterase poisoning. Nausea is often first symptom, with vomiting, abdominal cramps, diarrhea, and excessive salivation. Headache, giddiness, weakness, tightness in chest, blurring of vision, pinpoint pupils, loss of muscle coordination, and difficulty breathing. Death may occur from failure of the respiratory center, paralysis of the respiratory muscles, intense bronchoconstriction, or all three. Convulsions and coma precede death. Higher exposures can cause pulmonary edema, a medical emergency that can be delayed for several hours. This can cause death. Chlormephos can cause the heart to beat slower (bradycardia) or irregularly (arrhythmia).

Long Term Exposure: There is limited evidence that this chemical may damage the developing fetus. Can affect the nervous system and cause impaired memory, depression, anxiety, or irritability. Symptoms resembling influenza with headache, nausea, weakness have been reported.

Points of Attack: Respiratory system, lungs, central nervous system, cardiovascular system, skin, eyes, plasma and red blood cell cholinesterase.

Medical Surveillance: Before employment and at regular times after that, the following are recommended: Plasma and red blood cell cholinesterase levels (tests for the enzyme poisoned by this chemical). If exposure stops, plasma levels return to normal in 1 – 2 weeks while red blood cell levels may be reduced for 1 – 3 months.

When cholinesterase enzyme levels are reduced by 25% or more below preemployment levels, risk of poisoning is increased, even if results are in lower ranges of "normal." Reassignment to work not involving organophosphate or carbamate pesticides is recommended until enzyme levels recover. If symptoms develop or overexposure occurs, repeat the above tests as soon as possible and get an exam of the nervous system. Also consider complete blood count. Consider chest x-ray following acute overexposure. Do not drink any alcoholic beverages before or during use. Alcohol promotes absorption of organic phosphates.

First Aid: If this chemical gets into the eyes, remove any contact lenses at once and irrigate immediately for at least 15 minutes, occasionally lifting upper and lower lids. Seek medical attention immediately. If this chemical contacts the skin, remove contaminated clothing and wash immediately with soap and water. Speed in removing material from skin is of extreme importance. Shampoo hair promptly if contaminated. Seek medical attention immediately. If this chemical has been inhaled, remove from exposure, begin rescue breathing (using universal precautions) if breathing has stopped and CPR if heart action has stopped. Transfer promptly to a medical facility. When this chemical has been swallowed, get medical attention. Give large quantities of water and induce vomiting. Do not make an unconscious person vomit. Medical observation is recommended for 24 – 48 hours after breathing overexposure, as pulmonary edema may be delayed. As first aid for pulmonary edema, a doctor or authorized paramedic may consider administering a corticosteroid spray.

Personal Protective Methods: Wear protective gloves and clothing to prevent any reasonable probability of skin contact. Safety equipment suppliers/manufacturers can provide recommendations on the most protective glove/clothing material for your operation. All protective clothing (suits, gloves, footwear, headgear) should be clean, available each day, and put on before work. Contact lenses should not be worn when working with this chemical. Wear splash-proof chemical goggles and face shield unless full facepiece respiratory protection is worn. Employees should wash immediately with soap when skin is wet or contaminated. Provide emergency showers and eyewash.

Respirator Selection: *At any concentrations above the NIOSH REL, or where there is no REL, at any detectable concentration:* SCBAF:PD,PP (any MSHA/NIOSH approved self-contained breathing apparatus that has a full facepiece and is operated in a pressure-demand or other positive-pressure mode); or SAF:PD,PP:ASCBA (any supplied-air respirator that has a full facepiece and is operated in a pressure-demand or other positive-pressure mode in combination with an auxiliary, self-contained breathing apparatus operated in a pressure-demand or other positive pressure mode). *Escape:* GMFOVHiE [any air-purifying, full-facepiece respirator (gas mask) with a chin-style, front- or back-mounted organic vapor canister having a high-efficiency particulate filter]; or SCBAE (any appropriate escape-type, self-contained breathing apparatus).

Storage: Prior to working with Chlormephos you should be trained on its proper handling and storage. Should be protected from moisture and stored in glass-lined or polyethylene-lined containers. Keep away from strong oxidizers. Sources of ignition such as smoking and open flames, are prohibited where this chemical is used, handled, or stored in a manner that could create a potential fire or explosion hazard.

Shipping: The DOT label required is "Poison" for organophosphorus pesticides, liquid, toxic, n.o.s. Packing Group II. The UN/DOT Hazard Class is 6.1.[19][20]

Spill Handling: Stay upwind; keep out of low areas. Ventilate closed spaces before entering them. Evacuate and restrict persons not wearing protective equipment from area of spill or leak until cleanup is complete. Remove all ignition sources. Ventilate area of spill or leak. Absorb liquids in vermiculite, dry sand, earth, peat, carbon, or a similar material and deposit in sealed containers. Large spills: dike far ahead of spill for later disposal. It may be necessary to contain and dispose of this chemical as a hazardous waste. If material or contaminated runoff enters waterways, notify downstream users of potentially contaminated waters. Contact your Department of Environmental Protection or your regional office of the federal EPA for specific recommendations. If employees are required to clean-up spills, they must be properly trained and equipped. OSHA 1910.120(q) may be applicable.

Fire Extinguishing: This chemical may burn, but does not readily ignite. Move container from fire area if you can do it without risk. Fight fire from maximum distance. Dike fire control water for later disposal; do not scatter the material. Wear positive pressure breathing apparatus and special protective clothing. Use dry chemical, carbon dioxide, water spray, or alcohol resistant foam extinguishers. Poisonous gases are produced in fire including hydrogen chloride, phosphorous oxides, and sulfur oxides. If material or contaminated runoff enters waterways, notify downstream users of potentially contaminated waters. Notify local health and fire officials and pollution control agencies. Containers may explode in fire. From a secure, explosion-proof location, use water spray to cool exposed containers. If cooling streams are ineffective (venting sound increases in volume and pitch, tank discolors, or shows any signs of deforming), withdraw immediately to a secure position. If employees are expected to fight fires, they must be trained and equipped in OSHA 1910.156.

References

U.S. Environmental Protection Agency, "Chemical Profile: Chlormephos," Washington, DC, Chemical Emergency Preparedness Program (October 31, 1985)

Chlormequat Chloride

Molecular Formula: $C_5H_{13}Cl_2N$

Common Formula: $ClCH_2CH_2N(CH_3)_3Cl$

Synonyms: 60-CS-16; AC 38555; Ammonium, (2-chloroethyl)trimethyl-, chloride 2-chloro-*N,N,N*-trimethylethanaminium chloride; Antywylegacz; CCC plant growth regulant; 2-Chloraethyl-trimethylammoniumchlorid (German); Chlorcholinchlorid; Chlorcholine chloride; Chlormequat; Chlorocholine chloride; (β-Chloroethyl) trimethylammonium chloride; (2-Chloroethyl)trimethylammonium chloride; 2-Chloroethyl trimethylammonium chloride; 2-Chloro-*N,N,N*-ethyl)trimethylethanaminium chloride; 2-Chloro-*N,N,N*-trimethylammonium chloride; Choline dichloride; Clormecuato de cloroacetilo (Spanish); Cyclocel; Cycocel; Cycocel-extra; Cycogan; Cycogan extra; Cyocel; EI 38,555; Ethanaminium, 2-chloro-*N,N,N*-trimethyl-, chloride (9CI); Hico CCC; Hormocel-2CCC; Increcel; Lihocin; NCI-C02960; Retacel; Stabilan; Trimethyl-β-chlorethylammoniumchlorid; Trimethyl-β-chloroethyl ammonium chloride; Tur

CAS Registry Number: 999-81-5

RTECS Number: BP5250000

EEC Number: 007-003-00-6

Regulatory Authority

- Air Pollutant Standard Set (former USSR)[35]
- SUPERFUND/EPCRA 40CFR355, Appendix B Extremely Hazardous Substances: TPQ = 100/10,000 lb (45.4/4,540 kg)
- SUPERFUND/EPCRA 40CFR302.4 Reportable Quantity (RQ): EHS, 1 lb (0.454 kg)

Cited in U.S. State Regulations: Massachusetts (G).

Description: Chlormequat chloride, $ClCH_2CH_2N(CH_3)_3Cl$, is a white to yellowish crystalline solid with a fish-like odor. Freezing/Melting point = 245°C (decomposes). Hazard Identification (based on NFPA-704 M Rating System): Health 3, Flammability 0, Reactivity 0. Highly soluble in water. Carrier solvents used in commercial products may alter physical and toxicological properties.

Potential Exposure: People engaged in the manufacture, formulation, and application of this plant growth regulator said to be effective for cereal grains, tomatoes, and peppers.

Incompatibilities: Chlormequat chloride decomposes on heating or in fire forming nitrogen oxides, carbon monoxide, and hydrogen chloride fumes. This chemical decomposes on heating with strong aqueous alkali solutions forming trimethylamine and other gaseous products. Contact with strong oxidizers may cause fire and explosions. Attacks many metals in presence of water.

Permissible Exposure Limits in Air: There is no U.S. airborne exposure limit. This chemical can be absorbed through the skin, thereby increasing exposure. The former USSR[35] has set a ceiling TLV in workplace air of 0.3 mg/m^3.

Permissible Concentration in Water: The former USSR[35] has set a MAC in surface water of 0.2 mg/l.

Routes of Entry: Inhalation, passing through the skin.

Harmful Effects and Symptoms

Short Term Exposure: The LD_{low} oral (human) is 10 mg/kg. It is an irritant and can be absorbed through the skin. Irritates the eyes and the respiratory tract. Higher exposures can cause pulmonary edema, a medical emergency that can be delayed for several hours. This can cause death. Exposure can cause nausea and vomiting. Higher levels can cause slow or irregular heartbeat, tremors, seizures, and coma. This can be fatal. Chlormequat chloride may affect the nervous system.

Long Term Exposure: May cause liver damage.

Points of Attack: Lungs, liver, nervous system.

Medical Surveillance: Liver function tests. Consider chest x-ray following acute overexposure. EKG examination of the nervous system.

First Aid: If this chemical gets into the eyes, remove any contact lenses at once and irrigate immediately for at least 15 minutes, occasionally lifting upper and lower lids. Seek medical attention immediately. If this chemical contacts the skin, remove contaminated clothing and wash immediately with soap and water. Seek medical attention immediately. If this chemical has been inhaled, remove from exposure, begin rescue breathing (using universal precautions) if breathing has stopped and CPR if heart action has stopped. Transfer promptly to a medical facility. When this chemical has been swallowed, get medical attention. Give large quantities of water and induce vomiting. Do not make an unconscious person vomit. Medical observation is recommended for 24 – 48 hours after breathing overexposure, as pulmonary edema may be delayed. As first aid for pulmonary edema, a doctor or authorized paramedic may consider administering a corticosteroid spray.

Personal Protective Methods: Wear protective gloves and clothing to prevent any reasonable probability of skin contact. Permeation data indicate that Neoprene gloves may provide protection from exposure to this compound. Safety equipment suppliers/manufacturers can provide recommendations on the most protective glove/clothing material for your operation. All protective clothing (suits, gloves, footwear, headgear) should be clean, available each day, and put on before work. Contact lenses should not be worn when working with this chemical. Wear dust-proof chemical goggles and face shield unless full facepiece respiratory protection is worn. Employees should wash immediately with soap when skin is wet or contaminated. Provide emergency showers and eyewash.

Respirator Selection: *Where there is a potential for overexposure:* SCBAF:PD,PP (any MSHA/NIOSH approved self-contained breathing apparatus that has a full facepiece and is operated in a pressure-demand or other positive-pressure mode); or SAF:PD,PP:ASCBA (any supplied-air respirator that has a full facepiece and is operated in a pressure-demand or other positive-pressure mode in combination with an auxiliary, self-contained breathing apparatus operated in a pressure-demand or other positive pressure mode).

Storage: Prior to working with Chlormequat chloride you should be trained on its proper handling and storage. Keep away from strong oxidizers. Store in a cool, dry place and protect from heat and moisture. Sources of ignition such as smoking and open flames, are prohibited where this chemical is used, handled, or stored in a manner that could create a potential fire or explosion hazard.

Shipping: The label required for Poisonous Solids, n.o.s., is "Poison" (for Packing Group II). This material falls in UN/DOT Hazard Class 6.1.[19][20]

Spill Handling: Evacuate persons not wearing protective equipment from area of spill or leak until clean-up is complete. Remove all ignition sources. Collect powdered material in the most convenient and safe manner and deposit in sealed containers. Ventilate area after clean-up is complete. It may be necessary to contain and dispose of this chemical as a hazardous waste. If material or contaminated runoff enters waterways, notify downstream users of potentially contaminated waters. Contact your Department of Environmental Protection or your regional office of the federal EPA for specific recommendations. If employees are required to clean-up spills, they must be properly trained and equipped. OSHA 1910.120(q) may be applicable.

Fire Extinguishing: This chemical is a noncombustible solid. Use dry chemical, carbon dioxide, or water extinguishers. Poisonous gases are produced in fire including carbon monoxide, nitrogen oxide, and hydrogen chloride. If material or contaminated runoff enters waterways, notify downstream users of potentially contaminated waters. Notify local health and fire officials and pollution control agencies. From a secure, explosion-proof location, use water spray to cool exposed containers. If cooling streams are ineffective (venting sound increases in volume and pitch, tank discolors, or shows any signs of deforming), withdraw immediately to a secure position. If employees are expected to fight fires, they must be trained and equipped in OSHA 1910.156.

Disposal Method Suggested: Incinerate in a unit with effluent gas scrubbing.[22]

References

New Jersey Department of Health and Senior Services, "Hazardous Substance Fact Sheet, Chlormequat Chloride," Trenton, NJ (April 1999)

U.S. Environmental Protection Agency, "Chemical Profile: Chlormequat Chloride," Washington, DC, Chemical Emergency Preparedness Program (October 31, 1985)

Chlornaphazine

Molecular Formula: $C_{14}H_{15}Cl_2N$

Common Formula: $(ClCH_2CH_2)_2NC_{10}H_7$

Synonyms: 2-Bis(2-chloroethyl)aminonaphthalene; *N,N*-Bis(2-chloroethyl)-2-naphthylamine; Bis(2-chloroethyl)-β-naphthylamine; Chlornaftina; Chlornaphazin; Chlornaphthin; Chloronaftina; Chloronaphthine; *N,N*-Di(2-chloroethyl)-β-naphthlamine; 2-*N,N*-Di(2-chloroethyl)naphthlamine; Di(2-chloroethyl)-β-naphthlamine; Dichloroethyl-β-naphthylamine; Erysan; 2-*N*aphthalenamine, *N,N*-bis(2-chloroethyl)-; Naphthlamine mustard; β-Naphthl-bis(β-chloroethyl)amine; 2-Naphthl-bis(β-chloroethyl)amine; β-Naphthl-di(2-chloroethyl)amine; NSC-62209; R48

CAS Registry Number: 494-03-1

RTECS Number: QM2450000

DOT ID: UN 9188

Regulatory Authority

- Carcinogen (IARC) (RTECS)[9]
- Banned or Severely Restricted (Israel) (UN)[13]
- EPA HAZARDOUS WASTE NUMBER (RCRA No.): U026
- RCRA, 40CFR261, Appendix 8 Hazardous Constituents
- SUPERFUND/EPCRA 40CFR302.4 Reportable Quantity (RQ): CERCLA, 100 lb (45.4 kg)

Cited in U.S. State Regulations: California (A, G), Florida (G), Kansas (G), Louisiana (G), Massachusetts (G), Minnesota (G), New Jersey (G), Pennsylvania (G), Rhode Island (G), Vermont (G), Virginia (G), Washington (G), Wisconsin (G).

Description: Clornaphazine, $C_{14}H_{15}Cl_2N$, forms crystals. Freezing/Melting point = 54 – 56°C.

Potential Exposure: Not produced or used commercially in the United States, Chlornaphazine has been used in other countries in the treatment of leukemia and related cancers. Currently, this drug does not have wide therapeutic usage.

Permissible Exposure Limits in Air: No standards set.

Permissible Concentration in Water: No criteria set.

First Aid: *Eye Contact:* Remove any contact lenses at once. Immediately flush eyes well with copious quantities of water or normal saline for at least 20 – 30 minutes. Seek medical attention.

Inhalation: Leave contaminated area immediately; breathe fresh air. Proper respiratory protection must be supplied to any rescuers. If coughing, difficult breathing or any other symptoms develop, seek medical attention at once, even if symptoms develop many hours after exposure.

Ingestion: Contact a physician, hospital or poison center at once. If the victim is unconscious or convulsing, do not induce vomiting or give anything by mouth. Assure that his airway is open and lay him on his side with his head lower than his body and transport immediately to a medical facility. If conscious and not convulsing, give a glass of water to dilute the substance. Vomiting should not be induced without a physician's advice.

Storage: Prior to working with this chemical you should be trained on its proper handling and storage. Store in a refrigerator or a cool, dry place. A regulated, marked area should be established where this chemical is handled, used, or stored in compliance with OSHA standard 1910.1045.

Shipping: Hazardous substances, n.o.s., fall in Class ORM-E and Packing Group III. They have no label requirements and there is a maximum allowable limit on shipment by passenger aircraft or railcar or by cargo aircraft.

Spill Handling: Evacuate persons not wearing protective equipment from area of spill or leak until clean-up is complete. Remove all ignition sources. Dampen spilled material with 60 – 70% ethanol to avoid airborne dust, then transfer material to a suitable container. Ventilate the spill area and use absorbent paper dampened with 60 – 70% ethanol to pick up remaining material. Wash surfaces well with soap and water. It may be necessary to contain and dispose of this chemical as a hazardous waste. If material or contaminated runoff enters waterways, notify downstream users of potentially contaminated waters. Contact your Department of Environmental Protection or your regional office of the federal EPA for specific recommendations. If employees are required to clean-up spills, they must be properly trained and equipped. OSHA 1910.120(q) may be applicable.

Fire Extinguishing: Use dry chemical, carbon dioxide, water spray, or alcohol foam extinguishers. Poisonous gases are produced in fire. If material or contaminated runoff enters waterways, notify downstream users of potentially contaminated waters. Notify local health and fire officials and pollution control agencies. From a secure, explosion-proof location, use water spray to cool exposed containers. If cooling streams are ineffective (venting sound increases in volume and pitch, tank discolors, or shows any signs of deforming), withdraw immediately to a secure position. If employees are expected to fight fires, they must be trained and equipped in OSHA 1910.156.

Chloroacetaldehyde

Molecular Formula: C_2H_3ClO

Common Formula: $ClCH_2CHO$

Synonyms: Acetaldehyde, chloro-; 2-Chloroacetaldehyde; Chloroacetaldehyde monomer; 2-Chloro-1-ethanal; 2-Chloroethanal; Cloroacetaldehido (Spanish); Monochloroacetaldehyde

CAS Registry Number: 107-20-0

RTECS Number: AB2450000

DOT ID: UN 2232

Regulatory Authority

- Air Pollutant Standard Set (ACGIH)[1] (Australia) (DFG)[3] (HSE)[33] (Israel) (Mexico) (Several States)[60] (Several Canadian Provinces)
- EPA HAZARDOUS WASTE NUMBER (RCRA No.): P023
- RCRA, 40CFR261, Appendix 8 Hazardous Constituents
- SUPERFUND/EPCRA 40CFR302.4 Reportable Quantity (RQ): CERCLA, 1,000 lb (454 kg)
- U.S. DOT 49CFR172.101, Inhalation Hazardous Chemical
- Canada, WHMIS, Ingredients Disclosure List

Cited in U.S. State Regulations: Alaska (G), California (A, G), Connecticut (A), Florida (G), Illinois (G), Kansas (G), Louisiana (G), Maine (G), Massachusetts (G), Minnesota (G), Nevada (A), New Hampshire (G), New Jersey (G), New York (A), North Dakota (A), Oklahoma (G), Pennsylvania (G), Rhode Island (G), Vermont (G), Virginia (G, A), Washington (G), West Virginia (G), Wisconsin (G).

Description: Chloroacetaldehyde, $ClCH_2CHO$, is a combustible, colorless liquid with a very sharp, irritating odor. Boiling point = 85 – 100°C. Freezing/Melting point = 16°C (40% solution). Soluble in water. Flash point = 88°C (40% solutions).

Potential Exposure: Chloroacetaldehyde is used as a fungicide, as an intermediate in 2-aminothiazole manufacture and in bark removal from tree trunks.

Incompatibilities: Reacts with oxidizers, acids. On heating, chloroacetaldehyde forms chlorine fumes.

Permissible Exposure Limits in Air: The Federal (OSHA PEL) ceiling limit is 1 ppm (3 mg/m^3), not to be exceeded at any time. NIOSH and ACGIH also recommend the same ceiling limit. The British (HSE), Australian, Israeli, Mexican and Canadian Provinces of Alberta, British Columbia, Ontario, and Quebec have set a limit of 1 ppm (3 mg/m^3) as a ceiling value and the British HSE's[33] STEL is 1 ppm (3 mg/m^3). The German DFG MAK is also 1 ppm and Peak Limitation is 2 times normal MAK (5 minute) not to be exceeded 8 times during a workshift. The NIOSH IDLH level is 45 ppm.

In addition, several states have set guidelines or standards for chloroacetaldehyde in ambient air[60] ranging from 1.0 μg/m^3 (New York) to 25 μg/m^3 (Virginia) to 30 μg/m^3 (North Dakota) to 60 μg/m^3 (Connecticut) to 71 μg/m^3 (Nevada).

Determination in Air: Si gel; Methanol; Gas chromatography/Electrochemical detection; NIOSH IV Method #2015.

Permissible Concentration in Water: No criteria set.

Routes of Entry: Inhalation, ingestion, eye and skin contact.

Harmful Effects and Symptoms

Irritation of skin, eyes and mucous membrane; skin burns; eye damage; pulmonary edema; sensitization of skin and respiratory system. Does have a mutagenic effect. The oral LD_{50} for mouse is 21 mg/kg.

Short Term Exposure: Corrosive to the eyes, skin, and respiratory tract Contact can cause burns and permanent damage. Higher exposures can cause pulmonary edema, a medical emergency that can be delayed for several hours. This can cause death.

Long Term Exposure: This chemical may cause mutations. It can cause skin allergy and an asthma-like lung allergy.

Points of Attack: Eyes, skin, respiratory system, lungs.

Medical Surveillance: Consider the points of attack in preplacement and periodic physical examinations. Lung function tests, Evaluation by a qualified allergist. Consider chest x-ray following acute overexposure.

First Aid: If this chemical gets into the eyes, remove any contact lenses at once and irrigate immediately for at least 30 minutes, occasionally lifting upper and lower lids. Seek medical attention immediately. If this chemical contacts the skin, remove contaminated clothing and wash immediately with soap and water. Seek medical attention immediately. If this chemical has been inhaled, remove from exposure, begin rescue breathing (using universal precautions) if breathing has stopped and CPR if heart action has stopped. Transfer promptly to a medical facility. When this chemical has been swallowed, get medical attention. Give large quantities of water and induce vomiting. Do not make an unconscious person vomit. Medical observation is recommended for 24 – 48 hours after breathing overexposure, as pulmonary edema may be delayed. As first aid for pulmonary edema, a doctor or authorized paramedic may consider administering a corticosteroid spray.

Personal Protective Methods: Wear protective gloves and clothing to prevent any reasonable probability of skin contact with liquids of >0.1% content or repeated or prolonged contact with liquids of <0.1% content. Safety equipment suppliers/manufacturers can provide recommendations on the most protective glove/clothing material for your operation. All protective clothing (suits, gloves, footwear, headgear) should be clean, available each day, and put on before work. Contact lenses should not be worn when working with this chemical. Wear splash-proof chemical goggles and face shield unless full facepiece respiratory protection is worn. Employees should wash immediately with soap when skin is wet or contaminated. Provide emergency showers and eyewash.

Respirator Selection: NIOSH/OSHA: *10 ppm:* CCROV [any chemical cartridge respirator with organic vapor cartridge(s)]; or SA (any supplied-air respirator). *25 ppm:* SA:CF (any supplied-air respirator operated in a continuous-flow mode); or PAPROV [any powered, air-purifying respirator with organic vapor cartridge(s)]. *45 ppm:* CCRFOV [any chemical cartridge respirator with a full facepiece and organic vapor cartridge(s)]; or GMFOV(any air-purifying, full-facepiece respirator (gas mask) with a chin-style, front- or back-mounted acid gas canister); or PAPRTOV [any powered, air-purifying respirator with a tight-fitting facepiece and organic vapor cartridge(s)]; or SCBAF (any self-contained breathing apparatus with a full facepiece); or SAF (any supplied-air respirator with a full facepiece). *Emergency or planned entry into unknown concentrations or IDLH conditions:* SCBAF:PD,PP (any self-contained breathing apparatus that has a full facepiece and is operated in a pressure-demand or other positive-pressure mode); or SAF:PD, PP:ASCBA (any supplied-air respirator that has a full facepiece and is operated in a pressure-demand or other positive-pressure mode in combination with an auxiliary self-contained breathing apparatus operated in a pressure-demand or other positive-pressure mode). *Escape:* GMFOV [any air-purifying, full-facepiece respirator (gas mask) with a chin-style, front-or back-mounted organic vapor canister] or SCBAE (any appropriate escape-type, self-contained breathing apparatus).

Storage: Prior to working with chloracetaldehyde you should be trained on its proper handling and storage. Store in tightly closed containers in a cool, well ventilated area. Metal containers involving the transfer of this chemical should be grounded and bonded. Where possible, automatically pump liquid from drums or other storage containers to process containers. Drums must be equipped with self-closing valves, pressure vacuum bungs, and flame arresters. Use only non-sparking tools and equipment, especially when opening and closing containers of this chemical. Sources of ignition such as smoking and open flames, are prohibited where this chemical is used, handled, or stored in a manner that could create a potential fire or explosion hazard.

Shipping: The DOT label required is "Poison." The UN/DOT Hazard Class is 6.1 and the Packing Group is II.[19][20]

Spill Handling: Avoid inhalation of vapors. Do not touch spilled material; stop leak; use water spray to reduce vapors. Remove all ignition sources. Ventilate area of spill or leak. Absorb liquids in vermiculite, dry sand, earth, peat, carbon, or a similar material and deposit in sealed containers. Large spills: dike far ahead of spill for later disposal Evacuate and restrict persons not wearing protective equipment from area of spill or leak until cleanup is complete. It may be necessary to contain and dispose of this chemical as a hazardous waste. If material or contaminated runoff enters waterways, notify downstream users of potentially contaminated waters. Contact your Department of Environmental Protection or your regional office of the federal EPA for specific recommendations. If employees are required to clean-up spills, they must be properly trained and equipped. OSHA 1910.120(q) may be applicable.

Initial isolation and protective action distances

Distances shown are likely to be affected during the first 30 minutes after materials are spilled and could increase with time. If more than one tank car, cargo tank, portable tank, or large cylinder is involved in the incident is leaking, the protective action distance may need to be increased. You may need to seek emergency information from CHEMTREC at (800) 424-9300 or seek professional environmental engineering assistance from the U.S. EPA Environmental Response Team at (908)548-8730 (24-hour response line).

Small spills (From a small package or a small leak from a large package)

First: Isolate in all directions (feet)..... 200

Then: Protect persons downwind (miles)

Day 0.2

Night 0.6

Large spills (From a large package or from many small packages)

First: Isolate in all directions (feet)..... 700

Then: Protect persons downwind (miles)

Day 0.5

Night 2.2

Fire Extinguishing: This chemical is a combustible liquid. Poisonous gases are produced in fire. Use dry chemical, carbon dioxide, or alcohol foam extinguishers. Vapors are heavier than air and will collect in low areas. Vapors may travel long distances to ignition sources and flashback. Vapors in confined areas may explode when exposed to fire. Containers may explode in fire. Storage containers and parts of containers may rocket great distances, in many directions. If material or contaminated runoff enters waterways, notify downstream users of potentially contaminated waters. Notify local health and fire officials and pollution control agencies. From a secure, explosion-proof location, use water spray to cool exposed containers. If cooling streams are ineffective (venting sound increases in volume and pitch, tank discolors, or shows any signs of deforming), withdraw immediately to a secure position. If employees are expected to fight fires, they must be trained and equipped in OSHA 1910.156.

Disposal Method Suggested: Incineration, preferably after mixing with another combustible fuel; care must be exercised to assure complete combustion to prevent the formation of phosgene; an acid scrubber is necessary to remove the halo acids produced.

References

U.S. Environmental Protection Agency, Chloroacetaldehyde, Health and Environmental Effects Profile No. 40, Office of Solid Waste, Washington, DC (April 30, 1980)

Sax, N. I., Ed., "Dangerous Properties of Industrial Materials Report," 2, No. 4, 70–72, New York, Van Nostrand Reinhold Co. (1982)

U.S. Environmental Protection Agency, "Chemical Profile: Chloroacetaldehyde," Washington, DC, Chemical Emergency Preparedness Program (October 31, 1985)

New Jersey Department of Health and Senior Services and Senior Services, "Hazardous Substance Fact Sheet: Chloroacetaldehyde," Trenton, NJ (April 1998)

Chloroacetic Acid

Molecular Formula: $C_2H_3ClO_2$

Common Formula: $ClCH_2COOH$

Synonyms: Acetic acid, chloro-; Acide chloracetique (French); Acide monochloracetique (French); Acido cloroacetico (Spanish); Acidomonocloroacetico (Italian); Chloracetic acid; Chloroethanoic acid; MCA; Monochloorazijnzuur (Dutch); Monochloracetic acid; Monochloressigsaeure (German); Monochloroacetic acid; Monochloroethanoic acid; NCI-C60231

CAS Registry Number: 79-11-8

RTECS Number: AF8575000

DOT ID: UN 1750 (liquid); UN 1751 (solid)

EEC Number: 607-003-00-1

Regulatory Authority

- Air Pollutant Standard Set (former USSR)[43] (HSE)
- CLEAN AIR ACT: Hazardous Air Pollutants (Title I, Part A, Section 112)
- SUPERFUND/EPCRA 40CFR355, Appendix B Extremely Hazardous Substances: TPQ = 100/10,000 lb (45.4/4,540 kg)
- SUPERFUND/EPCRA 40CFR302.4 Reportable Quantity (RQ): CERCLA, 1 lb (0.454 kg)
- EPCRA Section 313 Form R *de minimis* concentration reporting level: 1.0%
- Canada, WHMIS, Ingredients Disclosure List; National Pollutant Release Inventory (NPRI)

Cited in U.S. State Regulations: California (A, G), Florida (G), Maine (G), Maryland (G), Massachusetts (G), Minnesota (G), New Hampshire (G), New Jersey (G), New York (G), Pennsylvania (G).

Description: $ClCH_2COOH$ is a colorless to white crystalline solid. It has a strong vinegar-like odor and an odor threshold of 0.15 mg/m^3. Boiling point = 188°C. Freezing/Melting point = 63°C. Flash point = 126°C. Autoignition temperature ≥ 500°C. Hazard Identification (based on NFPA-704 M Rating System): Health 3, Flammability 1, Reactivity 0. Soluble in water.

Potential Exposure: Monochloracetic acid is used primarily as a chemical intermediate in the synthesis of sodium carboxymethyl cellulose, and such other diverse substances as ethyl chloroacetate, glycine, synthetic caffeine, sarcosine, thioglycolic acid, and various dyes. Hence, workers in these areas are affected. It is also used as an herbicide. Therefore, formulators and applicators of such herbicides are affected.

Incompatibilities: The solution in water is a strong acid. Contact with strong oxidizers, strong bases, and strong reducing agents can cause violent reactions. Chloracetic acid decomposes on heating producing toxic and corrosive hydrogen chloride, phosgene gases. Attacks metals in the presence of moisture.

Permissible Exposure Limits in Air: No US exposure limits have been set by OSHA, NIOSH, or ACGIH, however, the HSE in Great Britain sets, and the American Industrial Hygiene Association recommends an airborne exposure limit of 0.3 ppm (1 mg/m^3) TWA and the AIHA recommends a STEL of 1 ppm (4 mg/m^3). Skin absorbtion is indicated. This chemical can be absorbed through the skin, thereby increasing exposure. The former USSR-UNEP/IRPTC project[43] has set a MAC in workplace air of 1.0 mg/m^3.

Permissible Concentration in Water: No criteria set.

Routes of Entry: Inhalation, ingestion and skin contact. This chemical can be absorbed through the skin, thereby increasing exposure.

Harmful Effects and Symptoms

Short Term Exposure: Corrosive to the eyes, skin, and respiratory tract. Contact can cause severe irritation and burns. Inhaling this chemical can cause pulmonary edema, a medical emergency that can be delayed for several hours. This can cause death. Chloracetic acid can cause a feeling of anxiety, restlessness, blurred vision, a feeling of "pons and needles" in the limbs, muscle twitching, and/or hallucinations, may affects the cardiovascular system, central nervous system and kidneys, resulting in heart problems, convulsions, and kidney damage. These effects may be delayed. Symptoms of exposure include irritation and pain in skin. If chloroacetic acid is inhaled the patient may exhibit difficulty in breathing. Vomiting may occur if the material is ingested. It can burn the skin, cornea and respiratory tract.

This material is very toxic. The probable lethal oral dose is 50-500 mg/kg of body weight, between one teaspoon and one ounce, for a 150 lb person. Chloroacetic acid is irritating to the skin, cornea, and respiratory tract and causes burns. It may severely damage skin and mucous membranes. Ingestion may interfere with essential enzyme systems and cause perforation and peritonitis. Burns to skin result in marked fluid and electrolyte loss. Death may follow if more than 3% of the skin is exposed to this material. Other health hazards include central nervous system depression, and respiratory system depression.

Long Term Exposure: Repeated exposure may cause kidney damage and affect the lungs.

Points of Attack: Lungs, kidneys, central nervous system.

Medical Surveillance: Lung function tests, kidney function tests. Examination of the nervous system.

First Aid: If this chemical gets into the eyes, remove any contact lenses at once and irrigate immediately for at least 15 minutes, occasionally lifting upper and lower lids. Seek medical attention immediately. If this chemical contacts the skin, remove contaminated clothing and wash immediately with soap and water. Seek medical attention immediately. If this chemical has been inhaled, remove from exposure, begin rescue breathing (using universal precautions) if breathing has stopped and CPR if heart action has stopped. Transfer promptly to a medical facility. When this chemical has been swallowed, get medical attention. If victim is conscious, administer water or milk. Do not induce vomiting. Medical observation is recommended for 24 – 48 hours after breathing overexposure, as pulmonary edema may be delayed. As first aid for pulmonary edema, a doctor or authorized paramedic may consider administering a corticosteroid spray.

Personal Protective Methods: Wear protective gloves and clothing to prevent any reasonable probability of skin contact. Safety equipment suppliers/manufacturers can provide recommendations on the most protective glove/clothing material for your operation Saranex is among the recommended protective materials. All protective clothing (suits, gloves, footwear, headgear) should be clean, available each day, and put on before work. Contact lenses should not be worn when working with this chemical. If you are working with dry material wear dust-proof chemical goggles and face shield if you are working with the liquid, unless full facepiece respiratory protection is worn. Employees should wash immediately with soap when skin is wet or contaminated. Provide emergency showers and eyewash.

Respirator Selection: *Where there is a potential for overexposure:* SCBAF:PD,PP (any MSHA/NIOSH approved self-contained breathing apparatus that has a full facepiece and is operated in a pressure-demand or other positive-pressure mode); or SAF:PD,PP:ASCBA (any supplied-air respirator that has a full facepiece and is operated in a pressure-demand or other positive-pressure mode in combination with an auxiliary, self-contained breathing apparatus operated in a pressure-demand or other positive pressure mode).

Storage: Prior to working with chloroacetic acid you should be trained on its proper handling and storage. Store in tightly closed containers in a cool, well ventilated area away from metal, combustibles, strong oxidizers, strong bases, and reducing agents. Where possible, automatically pump liquid from drums or other storage containers to process containers. Drums must be equipped with self-closing valves, pressure vacuum bungs, and flame arresters. Use only non-sparking tools and equipment, especially when opening and closing containers of this chemical. Sources of ignition such as smoking and open flames, are prohibited where this chemical is used, handled, or stored in a manner that could create a potential fire or explosion hazard.

Shipping: The DOT label required for solid chloroacetic acid is "Corrosive." The UN/DOT Hazard Class is 8 and the Packing Group is II.

For liquid chloroacetic acid, the label is "Corrosive, Poison." Shipment by passenger aircraft or railcar or even by cargo aircraft is forbidden. The UN/DOT Hazard Class is 8 and the Packing Group is I.[19][20]

Spill Handling: Stay upwind; keep out of low areas. Evacuate persons not wearing protective equipment from area of spill or leak until clean-up is complete. Remove all ignition sources. Neutralize spilled materials with crushed limestone, soda ash, or lime. Waste water containing chloroacetic acid can be treated with ammonia, ammonium salts, or amines followed by separation of suspended solids. Collect spilled material powdered material in the most convenient and safe manner and deposit in sealed containers. Dike large spills far ahead of spill for later disposal. Ventilate area after clean-up is complete. It may be necessary to contain and dispose of this chemical as a hazardous waste. If material or contaminated runoff enters waterways, notify downstream users of potentially contaminated waters. Contact your Department of Environmental Protection or your regional office of the federal EPA for specific recommendations. If employees are required to clean-up spills, they must be properly trained and equipped. OSHA 1910.120(q) may be applicable.

Fire Extinguishing: This material is extremely hazardous to health, but fire fighters may enter areas with extreme care. Full protective clothing including a self-contained breathing apparatus, coat, pants, gloves, boots and bands around legs, arms and waist should be provided. No skin surface should be exposed. Move container from fire area if you can do so without risk. Spray cooling water on containers that are exposed to flames until well after fire is out. This chemical may burn but does not readily ignite. Use dry chemical, carbon dioxide, water spray, or foam extinguishers. Poisonous gases are produced in fire including chlorine and phosgene. If material or contaminated runoff enters waterways, notify downstream users of potentially contaminated waters. Notify local health and fire officials and pollution control agencies. From a secure, explosion-proof location, use water spray to cool exposed containers. If cooling streams are ineffective (venting sound increases in volume and pitch, tank discolors, or shows any signs of deforming), withdraw immediately to a secure position. If employees are expected to fight fires, they must be trained and equipped in OSHA 1910.156.

Disposal Method Suggested: Incineration, preferably after mixing with another combustible fuel; care must be exercised to assure complete combustion to prevent the formation of phosgene; an acid scrubber is necessary to remove the halo acids produced.

References

National Institute for Occupational Safety and Health, Profiles on Occupational Hazards for Criteria Document Priorities: Monochloroacetic Acid, pp 309–311, Report PB-274,073, Rockville, MD (1977)

Sax, N. I., Ed., "Dangerous Properties of Industrial Materials Report," 3, No. 5, 99–101, New York, Van Nostrand Reinhold Co. (1983)

U.S. Environmental Protection Agency, "Chemical Profile: Chloroacetic Acid," Washington, DC, Chemical Emergency Preparedness Program (October 31, 1985)

New York State Department of Health, "Chemical Fact Sheet: Chloroacetic Acid," Albany, NY, Bureau of Toxic Substance Assessment (January 1986)

New Jersey Department of Health and Senior Services and Senior Services, "Hazardous Substance Fact Sheet: Chloroacetic Acid," Trenton, NJ (July 1996)

2-Chloroacetophenone

Molecular Formula: C_8H_7ClO

Common Formula: $C_6H_5COCH_2Cl$

Synonyms: Acetophenone, 2-chloro-; CAF; CAP; Chemical mace; α-Chloroacetophenone; ε-Chloroacetophenone; 1-Chloroacetophenone; Chloroacetophenone (DOT); Chloromethyl phenyl ketone; 2-Chloro-1-phenylethanone; α-Cloroacetofenona (Spanish); CN; Ethanone, 2-Chloro-1-phenyl-; MACE (lacramator); NCI-C55107; Phenacyl chloride; Phenyl chloromethyl ketone; Tear gas

CAS Registry Number: 532-27-4

RTECS Number: AM6300000

DOT ID: UN 1697

Regulatory Authority

- Air Pollutant Standard Set (ACGIH)[1] (Australia) (HSE)[33] (Israel) (Mexico) (OSHA)[58] (Several States)[60] (Several Canadian Provinces)
- CLEAN AIR ACT: Hazardous Air Pollutants (Title I, Part A, Section 112)
- SUPERFUND/EPCRA 40CFR302.4 Reportable Quantity (RQ): CERCLA, 1 lb (0.454 kg)
- EPCRA Section 313 Form R *de minimis* concentration reporting level: 1.0%
- Canada, WHMIS, Ingredients Disclosure List

Cited in U.S. State Regulations: Alaska (G), California (A, G), Connecticut (A), Florida (G, A), Illinois (G), Maine (G), Maryland (G), Massachusetts (G), Minnesota (G), Nevada (A), New Hampshire (G, W), New Jersey (G), New York (A), North Dakota (A), Oklahoma (G), Pennsylvania (G), Rhode Island (G), South Carolina (A), West Virginia (G).

Description: 2-Chloroacetophenone, $C_6H_5COCH_2Cl$, is a combustible, colorless-to-gray solid with a sharp, irritating odor. Odor threshold = 0.015 ppm. Freezing/Melting point = 56.5°C. Boiling point = 247°C. Flash point = 118°C. Insoluble in water.

Potential Exposure: Chloroacetophenone is used as a chemical warfare agent (Agent CN) and as a principal ingredient in the riot control agent Mace. It is also used as a pharmaceutical intermediate.

Incompatibilities: Water, steam, strong oxidizers. Slowly corrodes metals.

Permissible Exposure Limits in Air: The Federal legal limit (OSHA PEL),[58] and the recommended NIOSH and ACGIH TWA value is 0.05 ppm (0.3 mg/m^3) as is the Mexican, Israeli, Australian, Great Britain (HSE), and Canadian provincial (Alberta, British Columbia, Ontario) value[33] and Alberta set a STEL of 0.15 ppm. The NIOSH IDLH level is 15 mg/m^3.

A number of states have set guidelines or standards[60] for chloroacetophenone in ambient air ranging from 1.0 $\mu g/m^3$ (New York) to 3.0 $\mu g/m^3$ (Florida and North Dakota) to 6.0 $\mu g/m^3$ (Connecticut) to 7.0 $\mu g/m^3$ (Nevada) to 7.5 $\mu g/m^3$ (South Carolina).

Determination in Air: Tenax Gas chromatography;[2] Thermal desorption; Gas chromatography/Flame ionization detection; NIOSH II(5) P&CAM Method #291.

Permissible Concentration in Water: No criteria set.

Routes of Entry: Inhalation, skin absorption, ingestion, skin and eye contact.

Harmful Effects and Symptoms

Short Term Exposure: A lacramator (a "tear gas"). This chemical can be absorbed through the skin, thereby increasing exposure. This chemical irritates the eyes, skin, and respiratory tract. Eye contact can cause severe irritation, burns and permanent damage. Breathing the vapor can cause lung irritation, coughing and shortness of breath. Higher exposures can cause pulmonary edema, a medical emergency that can be delayed for several hours. This can cause death.

Long Term Exposure: Repeated or prolonged contact with skin may cause skin sensitization and skin allergy with itching and rash.

Points of Attack: Eyes, skin, respiratory system, lungs.

Medical Surveillance: Consider the points of attack in preplacement and periodic physical examinations. Lung function tests. Examination by a qualified allergist. Consider x-ray following acute overexposure.

First Aid: If this chemical gets into the eyes, remove any contact lenses at once and irrigate immediately for at least 15 minutes, occasionally lifting upper and lower lids. Seek medical attention immediately. If this chemical contacts the skin, remove contaminated clothing and wash immediately with soap and water. Seek medical attention immediately. If this chemical has been inhaled, remove from exposure, begin rescue breathing (using universal precautions) if breathing has stopped and CPR if heart action has stopped. Transfer promptly to a medical facility. When this chemical has been swallowed, get medical attention. Give large quantities of water and induce vomiting. Do not make an unconscious person vomit. Medical observation is recommended for 24 – 48 hours after breathing overexposure, as pulmonary edema may be delayed. As first aid for pulmonary edema, a doctor or authorized paramedic may consider administering a corticosteroid spray.

Personal Protective Methods: Wear protective gloves and clothing to prevent any reasonable probability of skin contact. Safety equipment suppliers/manufacturers can provide recommendations on the most protective glove/clothing material for your operation. All protective clothing (suits, gloves, footwear, headgear) should be clean, available each day, and put on before work. Contact lenses should not be worn when working with this chemical. Wear indirect vent, impact and splash-proof chemical goggles and face shield unless full facepiece respiratory protection is worn. Employees should wash immediately with soap when skin is wet or contaminated. Provide emergency showers and eyewash.

Respirator Selection: NIOSH/OSHA: *3 mg/m³:* CCROVDM [any chemical cartridge respirator with organic vapor cartridge(s) in combination with a dust and mist filter]; SA (any supplied-air respirator). *7.5 mg/m³*: SA:CF (any supplied-air respirator operated in a continuous-flow mode); or PAPROVDM [any powered, air-purifying respirator with organic vapor cartridge(s) in combination with a dust and mist filter]. *15 mg/m³:* CCRFOVHiE [any chemical cartridge respirator with a full facepiece and organic vapor cartridge(s) in combination with a high efficiency particulate filter]; or GMFSHiE (any air-purifying, full-facepiece respirator (gas mask) with a chin-style, front- or back-mounted canister providing protection against the compound of concern and having a high-efficiency particulate filter); or SCBAF (any self-contained breathing apparatus with a full facepiece); or SAF (any supplied-air respirator with a full facepiece). *Emergency or planned entry into unknown concentrations or IDLH conditions:* SCBAF:PD,PP (any self-contained breathing apparatus that has a full facepiece and is operated in a pressure-demand or other positive-pressure mode); or SAF:PD,PP:ASCBA (any supplied-air respirator that has a full facepiece and is operated in a pressure-demand or other positive-pressure mode in combination with an auxiliary self-contained breathing apparatus operated in a pressure-demand or other positive pressure mode). *Escape:* GMFSHiE [any air-purifying, full-facepiece respirator (gas mask) with a chin-style, front- or back-mounted canister providing protection against the compound of concern and having a high-efficiency particulate filter]; or SCBAE (any appropriate escape-type, self-contained breathing apparatus).

Note: Substance causes eye irritation or damage; eye protection needed.

Storage: Prior to working with this chemical you should be trained on its proper handling and storage. Store in tightly closed containers in a refrigerator or cool, well ventilated area away from oxidizers, heat, water and steam. Metal containers involving the transfer of this chemical should be grounded and bonded. Drums must be equipped with self-closing valves, pressure vacuum bungs, and flame arresters. Use only non-sparking tools and equipment, especially when opening and closing containers of this chemical. Sources of ignition such as smoking and open flames, are prohibited where this chemical is used, handled, or stored in a manner that could create a potential fire or explosion hazard.

Shipping: Chloroacetophenone must bear a "Poison" label. It falls in Hazard Class 6.1 and Packing Group I. Shipment by passenger aircraft or railcar or even by cargo aircraft is forbidden.[19]

Spill Handling: Evacuate persons not wearing protective equipment from area of spill or leak until clean-up is complete. Remove all ignition sources and dampen spilled material with toluene to avoid airborne dust. Collect powdered material in the most convenient and safe manner and deposit in sealed containers. Ventilate area after clean-up is complete. It may be necessary to contain and dispose of this chemical as a hazardous waste. If material or contaminated runoff enters waterways, notify downstream users of potentially contaminated waters. Contact your Department of Environmental Protection or your regional office of the federal EPA for specific recommendations. If employees are required to clean-up spills, they must be properly trained and equipped. OSHA 1910.120(q) may be applicable.

Fire Extinguishing: This chemical is a combustible solid but does not readily ignite. Use dry chemical or carbon dioxide extinguishers. Poisonous gases are produced in fire including carbon monoxide and hydrogen chloride. If material or contaminated runoff enters waterways, notify downstream users of potentially contaminated waters. Notify local health and fire officials and pollution control agencies. From a secure, explosion-proof location, use water spray to cool exposed containers. If cooling streams are ineffective (venting sound increases in volume and pitch, tank discolors, or shows any signs of deforming), withdraw immediately to a secure position. If employees are expected to fight fires, they must be trained and equipped in OSHA 1910.156.

Disposal Method Suggested: Tear gas-containing waste is dissolved in an organic solvent and sprayed into an incinerator equipped with an afterburner and alkaline scrubber utilizing reaction with sodium sulfide in an alcohol-water solution. Hydrogen sulfide is liberated and collected by an alkaline scrubber.

References

Sax, N. I., Ed., "Dangerous Properties of Industrial Materials Report," 4, No. 1, 48–49 (1984)

New Jersey Department of Health and Senior Services, Hazardous Substance Fact Sheet, α-Chloroacetophenone, Trenton NJ (November 1998)

Chloroacetyl Chloride

Molecular Formula: $C_2H_2Cl_2O$

Common Formula: $ClCH_2COCl$

Synonyms: Acetyl chloride, chloro-; Chloroacetic acid chloride; Chloroacetic chloride; Chlorure de chloracetyle (French); Monochloroacetyl chloride

CAS Registry Number: 79-04-9

RTECS Number: AO6475000

DOT ID: UN 1752

EEC Number: 607-080-00-1

Regulatory Authority

- Air Pollutant Standard Set (ACGIH)[1] (Australia) (HSE)[33] (Israel) (Several States)[60] (Several Canadian Provinces)
- U.S. DOT 49CFR172.101, Inhalation Hazardous Chemical
- Canada, WHMIS, Ingredients Disclosure List

Cited in U.S. State Regulations: Alaska (G), California (A, G), Connecticut (A), Florida (G), Illinois (G), Maine (G), Massachusetts (G), Minnesota (G), Nevada (A), New Hampshire (G), New Jersey (G), Oklahoma (G), Pennsylvania (G), Rhode Island (G), Virginia (A), West Virginia (G).

Description: Chloroacetyl chloride, $ClCH_2COCl$, is a colorless to yellowish liquid with a pungent odor. Boiling point = 105 – 110°C. Freezing/Melting point = -22°C. Reacts violently with water.

Potential Exposure: Chloroacetyl chloride is used in the manufacture of acetophenone. It is used in the manufacture of a number of pesticides including: alachlor, allidochlor, butachlor, dimethachlor, formothion, mecarbam, metolachlor, propachlor. It is also used in the manufacture of pharmaceuticals such as chlordiazepoxide hydrochloride, diazepam, lidocaine, mianserin.

Incompatibilities: Reacts violently with water, steam forming chloroacetic acid and hydrogen chloride gas. Reacts with alcohols, powdered metals, sodium amide, combustibles, and many organics causing toxic fumes, fire and explosion hazard. On contact with air it emits corrosive gas. Decomposes when heated forming phosgene gas.

Permissible Exposure Limits in Air: The legal OSHA PEL is 0.05 ppm TWA for an 8-hour workshift. ACGIH recommends the same airborne exposure limit. The Australian, Israeli, and Canadian provinces of Alberta, Ontario and Quebec have set the same TWA as OSHA and Alberta and Quebec have set STELs of 0.15 ppm. This chemical can be absorbed through the skin, thereby increasing exposure.

In addition, some states have set guidelines or standards for Chloroacetyl chloride in ambient air[60] ranging from 3.0 μg/m³ (Virginia) to 4.0 μg/m³ (Connecticut) to 5.0 μg/m³ (Nevada).

Determination in Air: No test available.

Permissible Concentration in Water: No criteria set. (Chloroacetyl chloride decomposes in water).

Routes of Entry: Skin absorption, skin and eye contact, inhalation, ingestion.

Harmful Effects and Symptoms

Short Term Exposure: A lacramator (causes discharge of tears). Chloroacetyl chloride can severely irritate the eyes and cause permanent damage. It is corrosive to the skin, and respiratory tract. Corrosive on ingestion. Inhalation of vapor or aerosol can cause pulmonary edema, a medical emergency that can be delayed for several hours. This can cause death. The substance may affects the cardiovascular system. Exposure far above the PEL may cause death. Medical reports of the effects of acute exposures include: mild-to-moderate skin burns and erythema; lachrymation and mild eye burns; mild-to-moderate respiratory effects with cough, dyspnea and cyanosis; and mild gastrointestinal effects.

Long Term Exposure: Repeated or prolonged contact with skin may cause dermatitis. Lungs may be affected by repeated or prolonged exposure.

Points of Attack: Skin, eyes, respiratory system.

Medical Surveillance: Should include attention to skin, eyes and respiratory system in preplacement and regular physical examinations. Lung function tests. Consider chest x-ray following acute overexposure.

First Aid: If this chemical gets into the eyes, remove any contact lenses at once and irrigate immediately for at least 15 minutes, occasionally lifting upper and lower lids. Seek medical attention immediately. If this chemical contacts the skin, remove contaminated clothing and wash immediately with soap and water. Seek medical attention immediately. If this chemical has been inhaled, remove from exposure, begin rescue breathing (using universal precautions) if breathing has stopped and CPR if heart action has stopped. Transfer promptly to a medical facility. When this chemical has been swallowed, get medical attention. If victim is conscious, administer water or milk. Do not induce vomiting. Medical observation is recommended for 24 – 48 hours after breathing overexposure, as pulmonary edema may be delayed. As first aid for pulmonary edema, a doctor or authorized paramedic may consider administering a corticosteroid spray.

Personal Protective Methods: Wear protective gloves and clothing to prevent any reasonable probability of skin contact. Safety equipment suppliers/manufacturers can provide

recommendations on the most protective glove/clothing material for your operation. All protective clothing (suits, gloves, footwear, headgear) should be clean, available each day, and put on before work. Contact lenses should not be worn when working with this chemical. Wear splash-proof chemical goggles and face shield unless full facepiece respiratory protection is worn. Employees should wash immediately with soap when skin is wet or contaminated. Provide emergency showers and eyewash.

Respirator Selection: *At any concentrations above the NIOSH REL, or where there is no REL, at any detectable concentration:* SCBAF:PD,PP (any MSHA/NIOSH approved self-contained breathing apparatus that has a full facepiece and is operated in a pressure-demand or other positive-pressure mode); or SAF:PD,PP:ASCBA (any supplied-air respirator that has a full facepiece and is operated in a pressure-demand or other positive-pressure mode in combination with an auxiliary, self-contained breathing apparatus operated in a pressure-demand or other positive pressure mode).

Storage: Prior to working with this chemical you should be trained on its proper handling and storage. Store in tightly closed containers in a cool, well ventilated area away from water, steam, heat and combustibles. Sources of ignition such as smoking and open flames, are prohibited where this chemical is used, handled, or stored in a manner that could create a potential fire or explosion hazard.

Shipping: Chloroacetyl Chloride must carry a "Corrosive" label. It falls in DOT Hazard Class 8 and Packing Group II. Shipment by passenger aircraft or railcar or even by cargo aircraft is forbidden.

Spill Handling: Evacuate and restrict persons not wearing protective equipment from area of spill or leak until cleanup is complete. Remove all ignition sources. Ventilate area of spill or leak. Soak up spill with quicklime, sodium/potassium chloride or absorb liquids in vermiculite, dry sand, earth, peat, carbon, or a similar material and deposit in sealed containers. It may be necessary to contain and dispose of this chemical as a hazardous waste. If material or contaminated runoff enters waterways, notify downstream users of potentially contaminated waters. Contact your Department of Environmental Protection or your regional office of the federal EPA for specific recommendations. If employees are required to clean-up spills, they must be properly trained and equipped. OSHA 1910.120(q) may be applicable.

Initial isolation and protective action distances

Distances shown are likely to be affected during the first 30 minutes after materials are spilled and could increase with time. If more than one tank car, cargo tank, portable tank, or large cylinder is involved in the incident is leaking, the protective action distance may need to be increased. You may need to seek emergency information from CHEMTREC at (800) 424-9300 or seek professional environmental engineering assistance from the U.S. EPA Environmental Response Team at (908)548-8730 (24-hour response line).

Small spills (From a small package or a small leak from a large package)

First: Isolate in all directions (feet)..... 300

Then: Protect persons downwind (miles)

Day 0.2

Night 0.7

Large spills (From a large package or from many small packages)

First: Isolate in all directions (feet)..... 700

Then: Protect persons downwind (miles)

Day 0.6

Night 2.6

Fire Extinguishing: This chemical is a nonflammable liquid. Do not use water or water-based extiguishers. Use extinguishing agent suitable for surrounding fire. This chemical decomposes in heat producing phosgene, chlorine and hydrogen chloride. If material or contaminated runoff enters waterways, notify downstream users of potentially contaminated waters. Notify local health and fire officials and pollution control agencies. From a secure, explosion-proof location, use water spray to cool exposed containers. If cooling streams are ineffective (venting sound increases in volume and pitch, tank discolors, or shows any signs of deforming), withdraw immediately to a secure position. If employees are expected to fight fires, they must be trained and equipped in OSHA 1910.156.

Disposal Method Suggested: It may be discharged into sodium bicarbonate solution, then flushed to the sewer with water.

References

New Jersey Department of Health and Senior Services and Senior Services, "Hazardous Substance Fact Sheet: Chloroacetyl Chloride," Trenton, NJ (February 1989)

Chloroanilines

Molecular Formula: $C_6H_4ClNH_2$

Synonyms: *m-: m-*Aminochlorobenzene; 1-Amino-3-chlorobenzene; 3-Chloroaniline; 3-Chlorobenzeneamine; Orange GC base

mixed isomers: Aminochlorbenzene; Benzeneamine, chloro-; Benzene chloride; Chloorbenzeen (Dutch); Chlorbenzen; Chlorobenzen (Polish); Chlorobenzol; MCB; Monochloorbenzeen (Dutch); Monochlorbenzene; Monochlorbenzol (German); Monochlorobenzene; NCI-C54886; Phenyl chloride; Phenylchloride

*o-: o-*Aminochlorobenzene; 1-Amino-2-chlorobenzene; 2-Chloroaminobenzene; 2-Chloroaniline; Fast yellow GC base

*p-: p-*Aminochlorobenzene; 1-Amino-4-chlorobenzene; Benzeneamine, 4-chloro-; 4-Chloraniline; 4-Chloraniline

(Czech); *p*-Chloroaminobenzene; 4-Chloro-1-aminobenzene; 4-Chloroaniline; 4-Chlorobenzenamine; 4-Chlorobenzeneamine; 4-Chlorophenylamine; *p*-Cloroanilina (Spanish)

CAS Registry Number: 95-51-2 (o-); 108-42-9 (m-); 106-47-8 (p-); 27134-26-5 (mixed isomers)

RTECS Number: BX0525000 (o-); BX0350000 (m-); BX0700000 (m-)

DOT ID: UN 2018 (solid); UN 2019 (liquid)

EEC Number: 612-010-00-8

Regulatory Authority

- Carcinogen (sufficient animal data) (IARC)
- Air Pollutant Standard Set (former USSR/UNEP)[43] (Several States)[60]

para- isomer:

- EPA HAZARDOUS WASTE NUMBER (RCRA No.): P024 (p-)
- RCRA, 40CFR261, Appendix 8 Hazardous Constituents (p-)
- RCRA 40CFR268.48; 61FR15654, Universal Treatment Standards: Wastewater (mg/l), 0.46; Nonwastewater (mg/kg), 16 (p-)
- RCRA 40CFR264, Appendix 9; TSD Facilities Ground Water Monitoring List. Suggested test method(s) (PQL μg/l): 8270 (20) (p-)
- SUPERFUND/EPCRA 40CFR302.4 Reportable Quantity (RQ): CERCLA, 1,000 lb (454 kg) (p-)
- EPCRA Section 313 Form R *de minimis* concentration reporting level: 1.0% (p-)
- Canada, WHMIS, Ingredients Disclosure List as chloroanilines

Cited in U.S. State Regulations: Connecticut (A), Kansas (G), Louisiana (G), Massachusetts (G), New Hampshire (G), New Jersey (G), New York (A), Vermont (G), Virginia (G), Washington (G), Wisconsin (G).

Description: The chloroanilines, $C_6H_4ClNH_2$, have the following properties: the ortho- and meta- isomers are colorless to yellow liquids that may turn brown on exposure to air; the para- isomer is a white or pale-yellow crystalline solid.

Isomer	Melting Point, °C	Boiling Point, °C	Flash Point, °C	Autoignition Temperature, °C	Water Solubility
o-	-14	209	108	>500	slight
m-	-11	230	183 (cc)	>540	very slight
p-	70	232	120 – 123	685	soluble in hot only

Hazard Identification (based on NFPA-704 M Rating System): *p-isomer* Health 2, Flammability 1, Reactivity 0.

Potential Exposure: Chloroanilines are used to make dyes, other chemicals, insecticides and many other industrial products.

Incompatibilities: Contact with strong oxidizers may cause fire and explosions. The aqueous solution of the m-isomer is a weak base. Incompatible with strong acids, organic anhydrides, isocyanates, aldehydes.

Permissible Exposure Limits in Air: The former USSR/UNEP joint project[43] has set MAC limits in workplace air of 0.05 mg/m^3 for the meta isomer and 0.3 mg/m^3 for the para isomer, the MAC is 0.04 mg/m^3 on a temporary basis; 0.01 on an average daily basis. No standards were set for the ortho-isomer. This chemical can be absorbed through the skin, thereby increasing exposure.

States which have set guidelines or standards for p-chloroaniline in ambient air[60] include Connecticut at 0.06 μg/m^3 and New York at 6.0 μg/m^3.

Permissible Concentration in Water: Limits in water bodies used for domestic purposes have been set by the former USSR/UNEP joint project[43] of 0.2 mg/l for the meta- and para- isomers.

Routes of Entry: Inhalation of vapor, passing through the skin, ingestion.

Harmful Effects and Symptoms

Short Term Exposure: Chloroanilines can affect you when breathed in and by passing through your skin. Exposure can lower the ability of the blood to carry oxygen (a condition called methemoglobinemia). This can cause headaches, trouble breathing, weakness, a bluish color to the nose and lips, collapse and death. Contact can severely irritate and may burn the eyes. *o-:* Irritates the eyes, skin, and respiratory tract. *m-:* May affect the liver and kidneys. Exposure may result in death. The effects may be delayed. *p-:* Irritates the eyes, skin, and respiratory tract. May affect the red blood cells, resulting in formation of methemoglobin and hemolysis. Exposure could cause headaches, trouble breathing, weakness, a bluish color to the nose and lips, loss of consciousness, and possible death.

Long Term Exposure: *o-:* Repeated or prolonged contact with skin may cause dermatitis. The substance may have effects on the liver and kidneys and also the blood system, resulting in forming of methemoglobin. *m-:* The substance may have effects on the blood system, resulting in forming of methemoglobin. *p-:* Repeated or prolonged contact may cause skin sensitization. The substance may have effects on the spleen, liver and kidneys, resulting in organ damage. Tumors have been detected in experimental animals but may not be relevant to humans. The methemoglobinemia described above can occur gradually over weeks instead of all at once.

Points of Attack: Blood, skin.

Medical Surveillance: If symptoms develop or overexposure is suspected, the following may be useful: Methemoglobin level.

First Aid: If this chemical gets into the eyes, remove any contact lenses at once and irrigate immediately for at least 15

minutes, occasionally lifting upper and lower lids. Seek medical attention immediately. If this chemical contacts the skin, remove contaminated clothing and wash immediately with soap and water. Seek medical attention immediately. If this chemical has been inhaled, remove from exposure, begin rescue breathing (using universal precautions) if breathing has stopped and CPR if heart action has stopped. Transfer promptly to a medical facility. When this chemical has been swallowed, get medical attention. Give large quantities of water and induce vomiting. Do not make an unconscious person vomit. Medical observation is recommended.

Note to Physician: Treat for methemoglobinemia. Spectrophotometry may be required for precise determination of levels of methemoglobinemia in urine.

Personal Protective Methods: *Clothing:* Avoid skin contact with Chloroaniline. Wear solvent-resistant gloves and clothing. Safety equipment suppliers/manufacturers can provide recommendations on the most protective glove, clothing material for your operation. All protective clothing (suits, gloves, footwear, headgear) should be clean, available each day and put on before work.

Eye Protection: Wear splash-proof chemical goggles and face shield when working with liquid, unless full facepiece respiratory protection is worn. Wear dust-proof goggles and face shield when working with powders or dust, unless full facepiece respiratory protection is worn.

Respirator Selection: Where the potential exists for exposures to Chloroaniline, use a MSHA/NIOSH approved supplied-air respirator with a full facepiece operated in the positive pressure mode or with a full facepiece, hood, or helmet in the continuous flow mode, or use a MSHA/ NIOSH approved self-contained breathing apparatus with a full facepiece operated in pressure-demand or other positive pressure mode.

Storage: Prior to working with Chloroaniline you should be trained on its proper handling and storage. Store in tightly closed containers in a cool, well-ventilated area away from Oxidizers (such as Perchlorates, Peroxides, Permanganates, Chlorates and Nitrates) since violent reactions occur. A regulated, marked area should be established where this chemical is handled, used, or stored in compliance with OSHA standard 1910.1045.

Shipping: Chloroanilines are required by DOT[19] to carry a "Poison" label. They fall in Hazard Class 6.1 and Packing Group II.

Spill Handling: *ortho- and meta- iomers:* Evacuate and restrict persons not wearing protective equipment from area of spill or leak until cleanup is complete. Remove all ignition sources. Ventilate area of spill or leak. Absorb liquids in vermiculite, dry sand, earth, peat, carbon, or a similar material and deposit in sealed containers. It may be necessary to contain and dispose of this chemical as a hazardous waste. If material or contaminated runoff enters waterways, notify downstream users of potentially contaminated waters. Contact your Department of Environmental Protection or your regional office of the federal EPA for specific recommendations. If employees are required to clean-up spills, they must be properly trained and equipped. OSHA 1910.120(q) may be applicable.

para- isomer: Evacuate persons not wearing protective equipment from area of spill or leak until clean-up is complete. Remove all ignition sources. Collect powdered material in the most convenient and safe manner and deposit in sealed containers. Ventilate area after clean-up is complete. It may be necessary to contain and dispose of this chemical as a hazardous waste. If material or contaminated runoff enters waterways, notify downstream users of potentially contaminated waters. Contact your Department of Environmental Protection or your regional office of the federal EPA for specific recommendations. If employees are required to clean-up spills, they must be properly trained and equipped. OSHA 1910.120(q) may be applicable

Fire Extinguishing: *para isomer:* This chemical is a combustible solid. Use dry chemical, carbon dioxide, water spray, or alcohol foam extinguishers. Poisonous gases are produced in fire including nitrous oxides and hydrogen chloride. If material or contaminated runoff enters waterways, notify downstream users of potentially contaminated waters. Notify local health and fire officials and pollution control agencies. Containers may explode in fire. From a secure, explosion-proof location, use water spray to cool exposed containers. If cooling streams are ineffective (venting sound increases in volume and pitch, tank discolors, or shows any signs of deforming), withdraw immediately to a secure position. If employees are expected to fight fires, they must be trained and equipped in OSHA 1910.156.

ortho- and meta- isomers: These chemicals are combustible liquids. Poisonous gases are produced in fire including nitrous oxides and hydrogen chloride. Use dry chemical, carbon dioxide, or foam extinguishers. Vapors are heavier than air and will collect in low areas. Vapors may travel long distances to ignition sources and flashback. Vapors in confined areas may explode when exposed to fire. Containers may explode in fire. Storage containers and parts of containers may rocket great distances, in many directions. If material or contaminated runoff enters waterways, notify downstream users of potentially contaminated waters. Notify local health and fire officials and pollution control agencies. From a secure, explosion-proof location, use water spray to cool exposed containers. If cooling streams are ineffective (venting sound increases in volume and pitch, tank discolors, or shows any signs of deforming), withdraw immediately to a secure position. If employees are expected to fight fires, they must be trained and equipped in OSHA 1910.156.

Disposal Method Suggested: Dissolve in a combustible solvent such as alcohol or benzene and spray into a furnace equipped with afterburner and scrubber.[24] Alternatively, pour into a mixture of sand and soda ash and burn in a furnace with paper as a fuel. In accordance with 40CFR165 recommendations for the disposal of pesticides and pesticide containers. Must be disposed properly by following package label directions or by contacting your state pesticide or environmental control agency or by contacting your regional EPA office.

References

New Jersey Department of Health and Senior Services and Senior Services, "Hazardous Substance Fact Sheet: Chloroaniline," Trenton, NJ (August 3, 1987).

Sax, N. I., Ed., "Dangerous Properties of Industrial Materials Report," 6, No. 5, 64–70 (1986) (2-Chloroaniline).

Chlorobenzene

Molecular Formula: C_6H_5Cl

Synonyms: Abluton T-30; Benzene chloride; Benzene, chloro-; Chloorbenzeen (Dutch); Chlorbenzen; Chlorobenzen (Polish); Chlorobenzol; Clorobanceno (Spanish); Clorobenceno (Spanish); KTI PMMA-standard 496K/950K; MCB; Monochloorbenzeen (Dutch); Monochlorbenzol (German); Monochlorobenzene; NCI-C54886; Phenyl chloride

CAS Registry Number: 108-90-7

RTECS Number: CZ0175000

DOT ID: UN 1134

EEC Number: 602-033-00-1

EINECS Number: 203-628-5

Regulatory Authority

- Air Pollutant Standard Set (ACGIH)[1] (DFG)[3] (HSE)[33] (UNEP)[43] (Several States)[60] (Australia) (Israel) (Mexico) (Several Canadian Provinces)
- CLEAN AIR ACT: Hazardous Air Pollutants (Title I, Part A, Section 112)
- CLEAN WATER ACT: Section 311 Hazardous Substances/RQ 40CFR117.3 (same as CERCLA, see below); 40CFR423, Appendix A, Priority Pollutants; Section 313 Water Priority Chemicals (57FR41331, 9/9/92)
- EPA HAZARDOUS WASTE NUMBER (RCRA No.): U037
- RCRA Toxicity Characteristic (Section 261.24), Maximum
- Concentration of Contaminants, regulatory level, 100 mg/l
- RCRA, 40CFR261, Appendix 8 Hazardous Constituents
- RCRA 40CFR268.48; 61FR15654, Universal Treatment Standards: Wastewater (mg/l), 0.057; Nonwastewater (mg/kg), 6.0
- RCRA 40CFR264, Appendix 9; TSD Facilities Ground Water Monitoring List. Suggested test method(s) (PQL μg/l): 8010 (2); 8020 (2); 8240 (5)
- SAFE DRINKING WATER ACT: MCL, 0.1 mg/l; MCLG, 0.1 mg/l; Regulated chemical (47 FR 9352)
- SUPERFUND/EPCRA 40CFR302.4 Reportable Quantity (RQ): CERCLA, 100 lb (45.4 kg)
- EPCRA Section 313 Form R *de minimis* concentration reporting level: 1.0%
- Canada, WHMIS, Ingredients Disclosure List; National Pollutant Release Inventory (NPRI); CEPA Priority Substance List; Drinking Water Quality: 0.08 mg/l MAC

Cited in U.S. State Regulations: Alaska (G), Arizona (W), California (A, W), Connecticut (A), Florida (G, A), Illinois (G), Kansas (G, W), Louisiana (G), Maine (G, W), Maryland (G), Massachusetts (G, A), Michigan (G), Minnesota (W), Nevada (A), New Hampshire (G), New Jersey (G, W), New York (G, A), North Carolina (A), North Dakota (A), Oklahoma (G), Pennsylvania (G), Rhode Island (G), Vermont (G, W), Virginia (G, A), Washington (G), West Virginia (G), Wisconsin (G, W).

Description: Chlorobenzene, C_6H_5Cl, a colorless liquid with an almond-like odor. The odor threshold is between 0.1 and 3.0 μg/l.[35] Also reported in the literature at 0.68 ppm and 0.741 ppm. Boiling point = 131 – 132°C. Freezing/Melting point = -45°C. Flash point = 28°C. Autoignition temperature = 593°C. Flammable limits are: LEL = 1.3%; UEL = 9.6%. Hazard Identification (based on NFPA-704 M Rating System): Health 2, Flammability 3, Reactivity 0. Insoluble in water.

Potential Exposure: Chlorobenzene is used in the manufacture of aniline, phenol and chloronitrobenzene and as an intermediate in the manufacture off dyestuffs and many pesticides.

Incompatibilities: Reacts violently with strong oxidizers, dimethyl sulfoxide, sodium powder, silver perchlorate, causing fire and explosion hazard. Attacks some plastics, rubber and coatings. Decomposes on heating, producing phosgene and hydrogen chloride fumes.

Permissible Exposure Limits in Air: The Federal limit (OSHA PEL) is 75 ppm (350 mg/m³) TWA for an 8-hour workshift. This is also the limit recommended or set by ACGIH, Australia, Israel, Mexico, and Canadian provinces of Alberta, British Columbia, Ontario and Quebec and Alberta's STEL is 115 ppm. The NIOSH IDLH level is 1,000 ppm. This chemical can be absorbed through the skin, thereby increasing exposure.

Various countries have set other limits[35] including Argentina, 75 pm (350 mg/m³) and same value for STEL; Brazil, 59 ppm (275 mg/m³); HSE and Germany, 50 ppm (230 mg/m³); See reference[3] also; USSR, 50 mg/m³ in workplace air; USR, 0.1 mg/m³ in ambient air in residential areas.

In addition, several states have set guidelines or standards for Chlorobenzene in ambient air[60] ranging from 6.3 μg/m³ (Massachusetts) to 1,167 μg/m³ (New York) to 2,200 μg/m³ (New York) to 3,500 μg/m³ (Florida and North Dakota) to 6,000 μg/m³ (Virginia) to 7,000 μg/m³ (Connecticut) to 8,333 μg/m³ (Nevada).

Determination in Air: Charcoal absorption followed by workup with CS_2 and analysis by gas chromatography. See NIOSH Method #1003 for hydrocarbons, halogenated.

Permissible Concentration in Water: To protect freshwater aquatic life: 250 μg/l on an acute basis for chlorobenzenes as a class. To protect saltwater aquatic life: 160 μg/l on an acute basis and 129 μg/l on a chronic basis for chlorinated benzenes as a class. To protect human health: for the prevention of adverse toxicological effects, 488 μg/l; but to prevent adverse organoleptic effects, 20 μg/l.

The former USSR/UNEP joint project[43] has set a MAC in water bodies used for domestic purposes of 0.02 mg/l.

The USEPA has set a lifetime health advisory of 0.3 mg/l (300 μg/l).

In addition, several states have set guidelines or standards for Chlorobenzene in drinking water[61] ranging from 2 μg/l (New Jersey) to 30 μg/l (California) to 47 μg/l (Maine) to 60 μg/l (Arizona, Kansas and Minnesota) to 600 μg/l (Vermont and Wisconsin).

Determination in Water: Gas chromatography (EPA Methods 601 and 602) or gas chromatography plus mass spectrometry (EPA Method 624).

Routes of Entry: Inhalation, ingestion, eye and skin contact. This chemical can be absorbed through the skin, thereby increasing exposure.

Harmful Effects and Symptoms

Short Term Exposure: The liquid can irritate and burn the skin. The vapor can irritate the eyes, nose and throat. Chlorobenzene can affect you when breathed in and by passing through your skin. Exposure to high concentrations can cause you to become dizzy, lightheaded, and to pass out. Swallowing the liquid may cause aspiration into the lungs with the risk of chemical pneumonitis. The effects may be delayed. Medical observation is indicated.

Long Term Exposure: May cause damage to the lungs, blood, nervous system, liver, and kidneys. Repeated exposure to the liquid may cause skin burns. Similar petroleum-based solvents cause brain damage, with reduced memory and concentration, peronality changes, fatigue, sleep disturbances, reduced coordination.

Points of Attack: Respiratory system, eyes, skin, central nervous system, liver.

Medical Surveillance: For those with frequent or potentially high exposure (half the TLV or greater, or significant skin contact) the following are recommended before beginning work and at regular times after that: Liver function tests. If symptoms develop or overexposure has occurred, the following may be useful: Lung and kidney function tests. Interview for brain effects.

First Aid: If this chemical gets into the eyes, remove any contact lenses at once and irrigate immediately for at least 15 minutes, occasionally lifting upper and lower lids. Seek medical attention immediately. If this chemical contacts the skin, remove contaminated clothing and wash immediately with soap and water. Seek medical attention immediately. If this chemical has been inhaled, remove from exposure, begin rescue breathing (using universal precautions) if breathing has stopped and CPR if heart action has stopped. Transfer promptly to a medical facility. When this chemical has been swallowed, get medical attention. Give large quantities of water and induce vomiting. Do not make an unconscious person vomit. Medical observation is recommended.

Personal Protective Methods: Wear solvent-resistant gloves and clothing to prevent any reasonable probability of skin contact. Safety equipment suppliers/manufacturers can provide recommendations on the most protective glove/clothing material for your operation. VITON and Teflon are among the recommended protective materials. All protective clothing (suits, gloves, footwear, headgear) should be clean, available each day, and put on before work. Contact lenses should not be worn when working with this chemical. Wear splash-proof chemical goggles and face shield unless full facepiece respiratory protection is worn. Employees should wash immediately with soap when skin is wet or contaminated. Remove nonimpervious clothing immediately if wet or contaminated. Provide emergency showers and eyewash.

Respirator Selection: OSHA: *1,000 ppm:* SA:CF (any supplied-air respirator operated in a continuous-flow mode); or PAPROV [any powered, air-purifying respirator with organic vapor cartridge(s)]; or CCRFOV [any air-purifying, full-facepiece respirator (gas mask) with a chin-style, front- or back-mounted acid gas canister]; or GMFOV [any air-purifying, full-facepiece respirator (gas mask) with a chin-style, front-or back-, mounted organic vapor canister]; or SCBAF (any self-contained breathing apparatus with a full facepiece); or SAF (any supplied-air respirator with a full facepiece). *Emergency or planned entry into unknown concentrations or IDLH conditions:* PD,PP (any self-contained breathing apparatus that has a full facepiece and is operated in a pressure-demand or other positive-pressure mode); or SAF:PD,PP:ASCBA (any supplied-air respirator that has a full facepiece and is operated in a pressure-demand or other positive-pressure mode in combination with an auxiliary self-contained breathing apparatus operated in a pressure-demand or other positive pressure mode). *Escape:* GMFOV [any air-purifying, full-facepiece respirator (gas mask) with a chin-style, front-or back-, mounted organic vapor canister]; or SCBAE (any appropriate escape-type, self-contained breathing apparatus).

Note: Substance causes eye irritation or damage; eye protection needed.

Storage: Prior to working with Chlorobenzene you should be trained on its proper handling and storage. Before entering confined space where this chemical may be present, check to make sure that an explosive concentration does not exist. Chlorobenzene must be stored to avoid contact with strong oxidizers (such as chlorine, bromine, and fluorine), since violent reactions occur. Store in tightly closed containers in a cool, well-ventilated area away from heat, sparks or flames. Sources of ignition such as smoking and open flames are prohibited where Chlorobenzene is used, handled, or stored in a manner that could create a potential fire or explosion hazard. Metal containers involving the transfer of 5 gallons or more of Chlorobenzene should be grounded and bonded. Drums must be equipped with self-closing valves, pressure vacuum bungs, and flame arresters. Use only non-sparking tools and equipment, especially when opening and closing containers of Chlorobenzene.

Shipping: Chlorobenzene must be labeled "Flammable Liquid." It falls in Hazard Class 3 and Packing Group III.[19]

Spill Handling: Evacuate and restrict persons not wearing protective equipment from area of spill or leak until cleanup is complete. Remove all ignition sources. Establish forced ventilation to keep levels below explosive limit. Absorb liquids in vermiculite, dry sand, earth, peat, carbon, or a similar material and deposit in sealed containers. It may be necessary to contain and dispose of this chemical as a hazardous waste. Keep Chlorobenzene out of a confined space, such as a sewer, because of the possibility of an explosion, unless the sewer is designed to prevent the build-up of explosive concentrations. If material or contaminated runoff enters waterways, notify downstream users of potentially contaminated waters. Contact your Department of Environmental Protection or your regional office of the federal EPA for specific recommendations. If employees are required to clean up spills, they must be properly trained and equipped. OSHA 1910.120(q) may be applicable.

Fire Extinguishing: This chemical is a combustible liquid. Poisonous gases including phosgene and hydrogen chloride are produced in fire. Use dry chemical, carbon dioxide, or foam extinguishers. Vapors are heavier than air and will collect in low areas. Vapors may travel long distances to ignition sources and flashback. Vapors in confined areas may explode when exposed to fire. Containers may explode in fire. Storage containers and parts of containers may rocket great distances, in many directions. If material or contaminated runoff enters waterways, notify downstream users of potentially contaminated waters. Notify local health and fire officials and pollution control agencies. From a secure, explosion-proof location, use water spray to cool exposed containers. If cooling streams are ineffective (venting sound increases in volume and pitch, tank discolors, or shows any signs of deforming), withdraw immediately to a secure position. If employees are expected to fight fires, they must be trained and equipped in OSHA 1910.156.

Disposal Method Suggested: Incineration, preferably after mixing with another combustible fuel; care must be exercised to assure complete combustion to prevent the formation of phosgene; an acid scrubber is necessary to remove the halo acids produced.[22]

References

U.S. Environmental Protection Agency, Chlorinated Benzenes: Ambient Water Quality Criteria, Washington, DC (1980)

U.S. Environmental Protection Agency, Chlorobenzene, Health and Environmental Effects Profile No. 42, Office of Solid Waste, Washington, DC (April 30, 1980)

Sax, N. I., Ed., "Dangerous Properties of Industrial Materials Report," 2, No. 4, 72–75, New York, Van Nostrand Reinhold Co. (1982)

New Jersey Department of Health and Senior Services and Senior Services, "Hazardous Substance Fact Sheet: Chlorobenzene," Trenton, NJ (January 1986)

New York State Department of Health, "Chemical Fact Sheet: Chlorobenzene," Albany, NY, Bureau of Toxic Substance Assessment (May 1986)

U.S. Environmental Protection Agency, "Health Advisory: Chlorobenzene," Washington, DC, Office of Drinking Water (March 31, 1987)

p-Chlorobenzotrichloride

Molecular Formula: $C_7H_4Cl_4$

Common Formula: Cl_3C-C_6H_4-Cl

Synonyms: Benzene, 1-chloro-4-(trichloromethyl)-; 4-Chlorobenzotrichloride; *p*-Chlorophenyltrichloromethane; 1-Chloro-4-(trichloromethyl)benzene; *p*-α, α, α-Tetrachlorotoluene

CAS Registry Number: 5216-25-1

RTECS Number: XT8580000

Regulatory Authority

- Carcinogen (animal evidence) (DFG) (California)

Cited in U.S. State Regulations: California (A, G), New York (G).

Description: p-Chlorobenzotrichloride, Cl_3C-C_6H_4-Cl, is a flammable, water-white liquid. Boiling point = 245 – 257°C.

Potential Exposure: Used in pesticide manufacture as an intermediate; reaction with HF yields chlorobenzotrifluoride as a major intermediate for several pesticides.

Incompatibilities: Strong oxidizers.

Permissible Exposure Limits in Air: No limits set. However, this chemical can be absorbed through the skin, thereby increasing exposure. A carcinogen: exposure should be kept to lowest feasible level.

Determination in Air: No test listed.

Permissible Concentration in Water: No criteria set.

Routes of Entry: Inhalation, ingestion; absorbed through the skin.

Harmful Effects and Symptoms

Short Term Exposure: May cause irritation by any route of exposure. Poisonous by inhalation.

Long Term Exposure: A skin sensitizer; may cause rash. A carcinogen.

Points of Attack: Skin.

Medical Surveillance: Examination by a qualified allergist.

First Aid: If this chemical gets into the eyes, remove any contact lenses at once and irrigate immediately for at least 15 minutes, occasionally lifting upper and lower lids. Seek medical attention immediately. If this chemical contacts the skin, remove contaminated clothing and wash immediately with soap and water. Seek medical attention immediately. If this chemical has been inhaled, remove from exposure, begin rescue breathing (using universal precautions) if breathing has stopped and CPR if heart action has stopped. Transfer promptly to a medical facility. When this chemical has been swallowed, get medical attention. Give large quantities of water and induce vomiting. Do not make an unconscious person vomit.

Personal Protective Methods: Wear protective gloves and clothing to prevent any reasonable probability of skin contact. Safety equipment suppliers/manufacturers can provide recommendations on the most protective glove/clothing material for your operation. All protective clothing (suits, gloves, footwear, headgear) should be clean, available each day, and put on before work. Contact lenses should not be worn when working with this chemical. Wear splash-proof chemical goggles and face shield unless full facepiece respiratory protection is worn. Employees should wash immediately with soap when skin is wet or contaminated. Provide emergency showers and eyewash.

Respirator Selection: *At any detectable concentration:* SCBAF:PD,PP (any MSHA/NIOSH approved self-contained breathing apparatus that has a full facepiece and is operated in a pressure-demand or other positive-pressure mode); or SAF:PD,PP:ASCBA (any supplied-air respirator that has a full facepiece and is operated in a pressure-demand or other positive-pressure mode in combination with an auxiliary, self-contained breathing apparatus operated in a pressure-demand or other positive pressure mode). *Escape:* HiEF [any air-purifying, full-facepiece respirator (gas mask) with a chin-style, front- or back-mounted organic vapor canister having a high-efficiency particulate filter]; or SCBAE (any appropriate escape-type, self-contained breathing apparatus).

Storage: Prior to working with this chemical you should be trained on its proper handling and storage. A regulated, marked area should be established where this chemical is handled, used, or stored in compliance with OSHA standard 1910.1045. Store in tightly closed containers in a cool, well-ventilated area.

Shipping: Chlorobenzotrichloride is not specifically cited in DOT's Performance-Oriented Packaging Standards.[19] However, although no flash point can be found in available literature, it may have to be labeled "Flammable Liquid," and it would fall in Hazard Class 3 and Packing Group III.[19]

Spill Handling: Evacuate and restrict persons not wearing protective equipment from area of spill or leak until cleanup is complete. Remove all ignition sources. Ventilate area of spill or leak. Absorb liquids in vermiculite, dry sand, earth, peat, carbon, or a similar material and deposit in sealed containers. It may be necessary to contain and dispose of this chemical as a hazardous waste. If material or contaminated runoff enters waterways, notify downstream users of potentially contaminated waters. Contact your Department of Environmental Protection or your regional office of the federal EPA for specific recommendations. If employees are required to clean-up spills, they must be properly trained and equipped. OSHA 1910.120(q) may be applicable.

Fire Extinguishing: This chemical is a flammable liquid. Poisonous gases including chlorine are produced in fire. Use dry chemical, carbon dioxide, or foam extinguishers. Vapors are heavier than air and will collect in low areas. Vapors may travel long distances to ignition sources and flashback. Vapors in confined areas may explode when exposed to fire. Containers may explode in fire. Storage containers and parts of containers may rocket great distances, in many directions. If material or contaminated runoff enters waterways, notify downstream users of potentially contaminated waters. Notify local health and fire officials and pollution control agencies. From a secure, explosion-proof location, use water spray to cool exposed containers. If cooling streams are ineffective (venting sound increases in volume and pitch, tank discolors, or shows any signs of deforming), withdraw immediately to a secure position. If employees are expected to fight fires, they must be trained and equipped in OSHA 1910.156.

References

U.S. Environmental Protection Agency, "Chemical Hazard Information Profile Draft Report: p-Chlorobenzotrichloride," Washington, DC (February 24, 1983)

New York State Department of Health, "Chemical Fact Sheet: para-Chlorobenzotrichloride," Albany, NY, Bureau of Toxic Substance Assessment (January 1986)

Chlorobenzotrifluoride

Molecular Formula: $C_7H_4ClF_3$

Common Formula: $ClC_6H_4CF_3$

Synonyms: (*p*-Chlorophenyl)trifluoromethane; *p*-Chlorotrifluoromethylbenzene; 4-Chlorotrifluoromethylbenzene; 1-Chloro-4-(trimethyl)-benzene; a,a,a-Trifluoro-4-chlorotoluene; *p*-(Trifluoromethyl)chlorobenzene; *p*-Trifluoromethylphenyl chloride; Trifluoromethylphenyl chloride

CAS Registry Number: 98-56-6; 52181-51-8

RTECS Number: XS9145000

DOT ID: UN 2234

Cited in U.S. State Regulations: Florida (G), New Hampshire (G), New Jersey (G), New York (G).

Description: $ClC_6H_4CF_3$ is a colorless liquid. Boiling point = 139°C. Flash point = 47°C. Hazard Identification (based on NFPA-704 M Rating System): Health (Unknown), Flammability 2, Reactivity 0. The pure ortho- isomer: Boiling point = 152°C. Flash point = 59°C. Hazard Identification (based on NFPA-704 M Rating System): Health 2, Flammability 2, Reactivity 1.

Potential Exposure: This material is used in the manufacture of pharmaceuticals, dyes, dielectrics and insecticides.

Incompatibilities: Strong oxidizers such as permanganates and dichromates.

Permissible Exposure Limits in Air: No standards set.

Determination in Air: No tests available.

Permissible Concentration in Water: No criteria set.

Harmful Effects and Symptoms

Short Term Exposure: Causes local irritation to skin, eyes and mucous membranes. May cause irritation by any route of exposure. The LD_{50} rat is 13 gm/kg (13,000 mg/kg) (insignificantly toxic).

Long Term Exposure: There is evidence that this chemical is a mutagen.

First Aid: If this chemical gets into the eyes, remove any contact lenses at once and irrigate immediately for at least 15 minutes, occasionally lifting upper and lower lids. Seek medical attention immediately. If this chemical contacts the skin, remove contaminated clothing and wash immediately with soap and water. Seek medical attention immediately. If this chemical has been inhaled, remove from exposure, begin rescue breathing (using universal precautions) if breathing has stopped and CPR if heart action has stopped. Transfer promptly to a medical facility. When this chemical has been swallowed, get medical attention. Give large quantities of water and induce vomiting. Do not make an unconscious person vomit.

Personal Protective Methods: Wear solvent-resistant gloves and clothing to prevent any reasonable probability of skin contact. Safety equipment suppliers/manufacturers can provide recommendations on the most protective glove/clothing material for your operation. All protective clothing (suits, gloves, footwear, headgear) should be clean, available each day, and put on before work. Contact lenses should not be worn when working with this chemical. Wear splash-proof chemical goggles and face shield unless full facepiece respiratory protection is worn. Employees should wash immediately with soap when skin is wet or contaminated. Remove nonimpervious clothing immediately if wet or contaminated. Provide emergency showers and eyewash.

Respirator Selection: Engineering controls should be used wherever feasible to maintain airborne concentrations of this chemical below the prescribed exposure limit. Respirators and protective equipment less effective than engineering controls, and should be used only in non-routine or emergency situations which may result in exposure concentrations in excess of the TWA environmental limit. NIOSH: *(fluorides) 12.5 mg/m³:* DM (any dust and mist respirator). *25 mg/m³:* DMXSQ* (any dust and mist respirator except single-use and quarter mask respirators); or SA* (any supplied-air respirator). *62.5 mg/m³:* SA:CF* (any supplied-air respirator operated in a continuous-flow mode); or PAPRDM*+ *if not present as a fume* (any powered, air-purifying respirator with a dust and mist filter). *125 mg/m³:* HiEF+ (any air-purifying, full-facepiece respirator with a high-efficiency particulate filter); or SCBAF (any self-contained breathing apparatus with a full facepiece); or SAF (any supplied-air respirator with a full facepiece). *250 mg/m³:* SA:PD,PP (any supplied-air respirator operated in a pressure-demand or other positive-pressure mode). *Emergency or planned entry into unknown concentrations or IDLH conditions:* SCBAF:PD,PP (any self-contained breathing apparatus that has a full faceplate and is operated in a pressure-demand or other positive-pressure mode); or SAF:PD,PP:ASCBA (any supplied-air respirator that has a full facepiece and is operated in a pressure-demand or other positive-pressure mode in combination with an auxiliary, self-contained breathing apparatus operated in a pressure-demand or other positive-pressure mode). *Escape:* HiEF+ (any air-purifying, full-facepiece respirator with a high-efficiency particulate filter); or SCBAE (any appropriate escape-type, self-contained breathing apparatus).

* Substance reported to cause eye irritation or damage; may require eye protection.

\+ May need acid gas sorbent.

Storage: Prior to working with chlorobenzotrifluoride you should be trained on its proper handling and storage. Store in tightly closed containers in a cool, well ventilated area away from oxidizers. Metal containers involving the transfer of this chemical should be grounded and bonded. Drums must be equipped with self-closing valves, pressure vacuum bungs, and flame arresters. Use only non-sparking tools and equipment, especially when opening and closing containers of this chemical. Sources of ignition such as smoking and open flames, are prohibited where this chemical is used, handled, or stored in a manner that could create a potential fire or explosion hazard.

Shipping: Chlorobenzotrifluoride must be labeled "Flammable Liquid." It falls in Hazard Class 3 and Packing Group III.[19]

Spill Handling: Evacuate and restrict persons not wearing protective equipment from area of spill or leak until cleanup is complete. Remove all ignition sources. Ventilate area of spill or leak. Absorb liquids in vermiculite, dry sand, earth, peat, carbon, or a similar material and deposit in sealed containers. It may be necessary to contain and dispose of this chemical as a hazardous waste. If material or contaminated runoff enters waterways, notify downstream users of potentially contaminated waters. Contact your Department of Environmental Protection or your regional office of the federal EPA for specific recommendations. If employees are required to clean-up spills, they must be properly trained and equipped. OSHA 1910.120(q) may be applicable.

Fire Extinguishing: This chemical is a flammable liquid. Poisonous gases including fluorine and chlorine are produced in fire. Use dry chemical, carbon dioxide, or foam extinguishers. Vapors are heavier than air and will collect in low areas. Vapors may travel long distances to ignition sources and flashback. Vapors in confined areas may explode when exposed to fire. Containers may explode in fire. Storage containers and parts of containers may rocket great distances, in many directions. If material or contaminated runoff enters waterways, notify downstream users of potentially contaminated waters. Notify local health and fire officials and pollution control agencies. From a secure, explosion-proof location, use water spray to cool exposed containers. If cooling streams are ineffective (venting sound increases in volume and pitch, tank discolors, or shows any signs of deforming), withdraw immediately to a secure position. If employees are expected to fight fires, they must be trained and equipped in OSHA 1910.156.

Disposal Method Suggested: In accordance with 40CFR 165 recommendations for the disposal of pesticides and pesticide containers. Must be disposed properly by following package label directions or by contacting your state pesticide or environmental control agency or by contacting your regional EPA office.

References

New York State Department of Health, "Chemical Fact Sheet: para-Chlorobenzotrifluoride," Albany, NY, Bureau of Toxic Substance Assessment (January 1986)

o-Chlorobenzylidene Malonitrile

Molecular Formula: $C_{10}H_5ClN_2$

Common Formula: $ClC_6H_4CH{=}C(CN)_2$

Synonyms: *o*-Chlorobenzalmalononitrile; 2-Chlorobenzalmalononitrile; 2-Chlorobenzylidene malononitrile; CS; β,β-Dicyano-*o*-chlorostyrene; Propanedinitrile[(2-chlorophenyl)Methylene]

CAS Registry Number: 2698-41-1

RTECS Number: OO3675000

DOT ID: UN 1693; UN 1065

Regulatory Authority

- Air Pollutant Standard Set (ACGIH)[1] (Australia) (Israel) (Mexico) (OSHA)[58] (Several States)[60] (Several Canadian Provinces)

As a cyanide compound:

- CLEAN AIR ACT: Hazardous Air Pollutants (Title I, Part A, Section 112)
- CLEAN WATER ACT: 40CFR423, Appendix A, Priority Pollutants as cyanide, total
- U.S. DOT Regulated Marine Pollutant (49CFR172.101, Appendix B) as cyanide mixtures

Cited in U.S. State Regulations: Alaska (G), California (A, G), Connecticut (A), Florida (G), Illinois (G), Maine (G), Massachusetts (G), Minnesota (G), Nevada (A), New Hampshire (G), North Dakota (A), Oklahoma (G), Pennsylvania (G), Rhode Island (G), Virginia (A), West Virginia (G).

Description: o-Chlorobenzylidene malonitrile, $ClC_6H_4CH{=}C(CN)_2$, is a combustible, white crystalline solid. Boiling point = 310 – 315°C. Freezing/Melting point = 95°C. Insoluble in water.

Potential Exposure: OCBM is used as a riot control agent.

Incompatibilities: Contact with strong oxidizers may cause fire and explosion. May be explosive if dust mixes with air.

Permissible Exposure Limits in Air: The OSHA/NIOSH standard is 0.05 ppm (0.4 mg/m^3) as a ceiling value, not to be exceeded at any time. ACGIH recommends the same ceiling limit. The notation "skin" is added to indicate the possibility of cutaneous absorption. Mexico, Israel, Australia and the Canadian provinces of Alberta, British Columbia, Ontario, and Quebec hve all set the same ceiling limit as OSHA and B.C.'s STEL is 0.15 ppm. The NIOSH IDLH level is 2.0 mg/m^3. In addition, several states have set guidelines or standards for OCBM in ambient air[60] ranging from 3.0 μg/m^3 (Virginia) to 4.0 μg/m^3 (North Dakota) to 8.0 μg/m^3 (Connecticut) to 10.0 μg/m^3 (Nevada).

Determination in Air: Collection by charcoal tube, analysis by gas liquid chromatography. See NIOSH Method #304.

Permissible Concentration in Water: No criteria set.

Routes of Entry: Inhalation, ingestion, eye and skin contact.

Harmful Effects and Symptoms

Short Term Exposure: A lacramator. Irritates the eyes, skin, and respiratory tract. OCBM is extremely irritating and acts on exposed sensory nerve endings (primarily in the eyes and upper respiratory tract). The signs and symptoms from exposure to the vapor are conjunctivitis and pain in the eyes, lacrimation, erythema of the eyelids, blepharospasms, irritation and running of the nose, burning in the throat, coughing and constricted feeling in the chest, and excessive salivation. Vomiting may occur if saliva is swallowed. Most of the symptoms subside after exposure ceases. Burning on the exposed skin is increased by moisture. With heavy exposure, vesiculation and erythema occur. Photophobia has been reported.

Long Term Exposure: Repeated or prolonged contact may cause skin sensitization. Animal experiments indicate that OCBM has a relatively low toxicity. The systemic changes observed in human experiments are nonspecific reactions to stress. OCBM is capable of sensitizing guinea pigs; there also appears to be a cross-reaction in guinea pigs previously sensitized to 1-chloroacetophenone.

Points of Attack: Respiratory system, skin and eyes.

Medical Surveillance: Consideration should be given to the eyes, skin, and respiratory tract in any placement or periodic evaluations.

First Aid: If this chemical gets into the eyes, remove any contact lenses at once and irrigate immediately for at least 15 minutes, occasionally lifting upper and lower lids. Seek medical attention immediately. If this chemical contacts the skin, remove contaminated clothing and wash immediately with soap and water. Seek medical attention immediately. If this chemical has been inhaled, remove from exposure, begin rescue breathing (using universal precautions) if breathing has stopped and CPR if heart action has stopped. Transfer promptly to a medical facility. When this chemical has been swallowed, get medical attention. Give large quantities of water and induce vomiting. Do not make an unconscious person vomit.

Personal Protective Methods: Wear protective gloves and clothing to prevent any reasonable probability of skin contact. Safety equipment suppliers/manufacturers can provide recommendations on the most protective glove/clothing material for your operation. All protective clothing (suits, gloves, footwear, headgear) should be clean, available each day, and put on before work. Contact lenses should not be worn when working with this chemical. Wear splash-proof chemical goggles and face shield unless full facepiece respiratory protection is worn. Employees should wash immediately with soap when skin is wet or contaminated. Provide emergency showers and eyewash. See NIOSH Criteria Document 78–212 NITRILES.

Respirator Selection: OSHA: *2 mg/m³:* SA:CF (any supplied-air respirator operated in a continuous-flow mode); or GMFSHiE (any air-purifying, full-facepiece respirator (gas mask) with a chin-style, front- or back-mounted canister providing protection against the compound of concern and having a high efficiency particulate filter); or SCBAF (any self-contained breathing apparatus with a full facepiece); or SAF (any supplied-air respirator with a full facepiece). *Emergency or planned entry into unknown concentrations or IDLH conditions:* SCBAF:PD,PP (any self-contained breathing apparatus that has a full facepiece and is operated in a pressure-demand or other positive-pressure mode); or SAF:PD,PP:ASCBA (any supplied-air respirator that has a full facepiece and is operated in a pressure-demand or other positive-pressure mode in combination with an auxiliary self-contained breathing apparatus operated in a pressure-demand or other positive-pressure mode). *Escape:* GMFOVHiE [any air-purifying, full-facepiece respirator (gas mask) with a chin-style, front- or back-mounted organic vapor canister having a high-efficiency particulate filter]; or SCBAE (any appropriate escape-type, self-contained breathing apparatus).

Storage: Prior to working with this chemical you should be trained on its proper handling and storage. Store in tightly closed containers in a refrigerator or cool, well ventilated area. Use only non-sparking tools and equipment, especially when opening and closing containers of this chemical. Sources of ignition such as smoking and open flames, are prohibited where this chemical is used, handled, or stored in a manner that could create a potential fire or explosion hazard.

Shipping: Tear gas substances, n.o.s., solid must be labeled "Poison." They fall in Hazard Class 6.1 and packing Group II.

Spill Handling: Evacuate persons not wearing protective equipment from area of spill or leak until clean-up is complete. Remove all ignition sources. Dampen spilled material with 60 – 70% acetone to avoid airborne dust. Collect powdered material in the most convenient and safe manner and deposit in sealed containers. Ventilate area after clean-up is complete. It may be necessary to contain and dispose of this chemical as a hazardous waste. If material or contaminated runoff enters waterways, notify downstream users of potentially contaminated waters. Contact your Department of Environmental Protection or your regional office of the federal EPA for specific recommendations. If employees are required to clean-up spills, they must be properly trained and equipped. OSHA 1910.120(q) may be applicable.

Fire Extinguishing: This chemical is a combustible solid. Use dry chemical, carbon dioxide, water spray, or alcohol foam extinguishers. Poisonous gases are produced in fire including cyanide, chlorine and nitrogen oxides. If material or contaminated runoff enters waterways, notify downstream users of potentially contaminated waters. Notify local health and fire officials and pollution control agencies. From a secure, explosion-proof location, use water spray to cool exposed containers. If cooling streams are ineffective (venting sound increases in volume and pitch, tank discolors, or shows any signs of deforming), withdraw immediately to a secure position. If employees are expected to fight fires, they must be trained and equipped in OSHA 1910.156.

Chlorobromomethane

Molecular Formula: CH_2BrCl

Synonyms: Bromochloromethane; CB; CBM; Halon 1011; Metane, bromochloro-; Methylene chlorobromide; Mil-B-4394-B; Mono-chloro-mono-bromo-methane

CAS Registry Number: 74-97-5

RTECS Number: PA5250000

DOT ID: UN 1887

Regulatory Authority

- Air Pollutant Standard Set (ACIGH)[1] (Australia) (DFG)[3] (HSE)[33] (Israel) (Mexico) (OSHA)[58] (Several States)[60] (Several Canadian Provinces)
- Canada, WHMIS, Ingredients Disclosure List

Cited in U.S. State Regulations: California (A, G), Connecticut (A), Florida (G), Illinois (G), Maine (G), Massachusetts (G), Minnesota (G), Nevada (A), New Hampshire (G), New Jersey (G), North Dakota (A), Pennsylvania (G), Rhode Island (G), Virginia (A), West Virginia (G).

Description: Chlorobromomethane, CH_2BrCl, is a clear, colorless, to pale-yellow liquid with a chloroform-like odor. Odor threshold = 400 ppm. Boiling point = 68°C. Freezing/Melting point = -88°C.

Potential Exposure: This compound is used as a fire-fighting agent and in organic synthesis.

Incompatibilities: Chemical active metals such as calcium, powdered aluminum, zinc, magnesium. Liquid attacks some plastics, rubber and coatings.

Permissible Exposure Limits in Air: The Federal limit (OSHA PEL)[58] and ACGIH TWA value is 200 ppm (1,050 mg/m^3). The DFG[3] set the same TWA value as OSHA and Peak limitation of 2 times the MAK (30 min) not to be exceeded 4 times during a workshift. HSE,[33] Australia, Israel, Mexico, and the Canadian provinces of Alberta, British Columbia, Ontario and Quebec set a limit of 200 ppm TWA and STEL value of 250 ppm (1,300 mg/m^3). Quebec and Israel have no STEL value but Israel's Action level is 100 ppm. The NIOSH IDLH level is 2,000 ppm. In addition, several states have set guidelines or standards for CBM in ambient air[60] ranging from 10.5 – 13 mg/m^3 (North Dakota) to 17.5 mg/m^3 (Virginia) to 21.0 mg/m^3 (Connecticut) to 25.0 mg/m^3 (Nevada).

Determination in Air: Charcoal adsorption, workup with CS_2, followed by gas chromatography. See NIOSH Method #1003 for hydrocarbons, halogenated.

Permissible Concentration in Water: No criteria set.

Routes of Entry: Inhalation, ingestion, eye and skin contact.

Harmful Effects and Symptoms

Short Term Exposure: Contact can irritate and burn the skin and eyes. This chemical can irritate the lungs. Higher exposures can cause pulmonary edema, a medical emergency that can be delayed for several hours. This can cause death. Symptoms include disorientation, dizziness; irritation of eyes, throat and skin; headaches, anorexia, nausea, vomiting, abdominal pain, weakness, tremors and convulsions, narcosis. The LD_{50} oral-rat is 5,000 mg/kg (slightly toxic).

Long Term Exposure: May cause liver and kidney damage, which may be progressive; skin irritation and cracking. May affect the lungs and cause bronchitis to develop. May cause weight loss, memory impairment, paralysis. Skin contact can cause drying and cracking.

Points of Attack: Skin, liver, kidneys, respiratory system, lungs, central nervous system.

Medical Surveillance: Consider the points of attack in preplacement and periodic physical examinations. Lung function tests. Serum bromine level. Consider chest x-ray following acute overexposure.

First Aid: If this chemical gets into the eyes, remove any contact lenses at once and irrigate immediately for at least 15 minutes, occasionally lifting upper and lower lids. Seek medical attention immediately. If this chemical contacts the skin, remove contaminated clothing and wash immediately with soap and water. Seek medical attention immediately. If this chemical has been inhaled, remove from exposure, begin rescue breathing (using universal precautions) if breathing has stopped and CPR if heart action has stopped. Transfer promptly to a medical facility. When this chemical has been swallowed, get medical attention. Give large quantities of water and induce vomiting. Do not make an unconscious person vomit. Medical observation is recommended for 24 – 48 hours after breathing overexposure, as pulmonary edema may be delayed. As first aid for pulmonary edema, a doctor or authorized paramedic may consider administering a corticosteroid spray.

Personal Protective Methods: Wear protective gloves and clothing to prevent any reasonable probability of skin contact. Safety equipment suppliers/manufacturers can provide recommendations on the most protective glove/clothing material for your operation. All protective clothing (suits, gloves, footwear, headgear) should be clean, available each day, and put on before work. Contact lenses should not be worn when working with this chemical. Wear splash-proof chemical goggles and face shield unless full facepiece respiratory protection is worn. Employees should wash immediately with soap when skin is wet or contaminated. Provide emergency showers and eyewash.

Respirator Selection: NIOSH/OSHA: *2,000 ppm:* SA:CF* (any supplied-air respirator operated in a continuous-flow mode); or PAPROV [any powered, air-purifying respirator with organic vapor cartridge(s)]; CCRFOV* [any air-purifying, full-facepiece respirator (gas mask) with a chin-style, front-or back-mounted acid gas canister]; or GMFOV [any air-purifying, full-facepiece respirator (gas mask) with a chin-style, front-or back-mounted organic vapor canister]; or SCBAF (any self-contained breathing apparatus with a full facepiece); or SAF (any supplied-air respirator with a full facepiece). *Emergency or planned entry into unknown concentrations or IDLH conditions:* SCBAF:PD,PP (any self-contained breathing apparatus that has a full facepiece and is operated in a pressure-demand or other positive-pressure mode); or SAF:PD,PP:ASCBA (any supplied-air respirator that has a full facepiece and is operated in a pressure-demand or other positive-pressure mode in combination with an auxiliary self-contained breathing apparatus operated in a pressure-demand or other positive-pressure mode). *Escape:* GMFOV [any air-purifying, full-facepiece respirator (gas mask) with a chin-style, front-or back-mounted organic vapor canister] or SCBAE (any appropriate escape-type, self-contained breathing apparatus).

* Substance can cause eye irritation or damage; eye protection needed.

Storage: Prior to working with this chemical you should be trained on its proper handling and storage. Chlorobromomethane must be stored to avoid contact with chemically active metals since violent reactions occur. Store in tightly closed containers in a cool, well-ventilated area away from heat.

Shipping: Bromochloromethane must be labeled "Keep Away from Food." It falls in Hazard Class 6.1 and Packing Group III.[19]

Spill Handling: Evacuate and restrict persons not wearing protective equipment from area of spill or leak until cleanup is complete. Remove all ignition sources. Ventilate area of spill or leak. Absorb liquids in vermiculite, dry sand, earth, peat, carbon, or a similar material and deposit in sealed containers. It may be necessary to contain and dispose of this chemical as a hazardous waste. If material or contaminated runoff enters waterways, notify downstream users of potentially contaminated waters. Contact your Department of Environmental Protection or your regional office of the federal EPA for specific recommendations. If employees are required to clean-up spills, they must be properly trained and equipped. OSHA 1910.120(q) may be applicable.

Restrict persons not wearing protective equipment from area of spill or leak until clean-up is complete. Ventilate the area of spill or leak. Absorb liquids in vermiculite, dry sand, earth, or a similar material and deposit in sealed containers.

Fire Extinguishing: This chemical is a noncombustible liquid. Use any extinguishers suitable for surrounding fire. Poisonous gases are produced in fire including hydrogen chloride and hydrogen bromide. If material or contaminated runoff enters waterways, notify downstream users of potentially contaminated waters. Notify local health and fire officials and pollution control agencies. Containers may explode in fire. From a secure, explosion-proof location, use water spray to cool exposed containers. If cooling streams are ineffective (venting sound increases in volume and pitch, tank discolors, or shows any signs of deforming), withdraw immediately to a secure position. If employees are expected to fight fires, they must be trained and equipped in OSHA 1910.156.

Disposal Method Suggested: Incinerate together with flammable solvent in furnace equipped with afterburner and alkali scrubber.

References

New Jersey Department of Health and Senior Services and Senior Services, "Hazardous Substance Fact Sheet: Chlorobromomethane," Trenton, NJ (November 1998)

Chlorodifluorobromomethane

Molecular Formula: $CBrClF_2$

Synonyms: Bromochlorodifluoromethane; Chlorodifluorobromomethane; Flugex 12B1; Fluorocarbon 1211; Freon 12B1; Halon 1211; R12B1

CAS Registry Number: 353-59-3

RTECS Number: PA5270000

DOT ID: UN 1974

Regulatory Authority

- CLEAN AIR ACT: Stratospheric ozone protection (Title VI, Subpart A, Appendix A), Class I, Ozone Depletion Potential = 3.0
- EPCRA Section 313 Form R *de minimis* concentration reporting level: 1.0%
- Canada, CEPA Schedule I Toxic Substances (import/export, manufacturing, and processing restrictions)

Cited in U.S. State Regulations: California (G), New Jersey (G), Pennsylvania (G).

Description: Chlorodifluorobromomethane is a colorless gas or liquid under pressure. Boiling point = -4°C. Freezing/Melting point = -160.5°C.

Potential Exposure: Used as a refrigerant.

Permissible Exposure Limits in Air: No OELs have been established.

Routes of Entry: Inhalation.

Harmful Effects and Symptoms

Short Term Exposure: Irritates the eyes and respiratory tract. Contact with the liquid can cause frostbite. High exposure can cause dizziness, lightheadedness, and unconsciousness. Inhalation can cause irregular heartbeat.

Long Term Exposure: Can affect the heartbeat causing irregular rhythms and skipped beats.

Points of Attack: Heart.

Medical Surveillance: Special 24-hour EKG (Holter monitor) for irregular heartbeat.

First Aid: If this chemical gets into the eyes, remove any contact lenses at once and irrigate immediately for at least 15 minutes, occasionally lifting upper and lower lids. Seek medical attention immediately. If this chemical contacts the skin, remove contaminated clothing and wash immediately with soap and water. Seek medical attention immediately. If this chemical has been inhaled, remove from exposure, begin rescue breathing (using universal precautions) if breathing has stopped and CPR if heart action has stopped. Transfer promptly to a medical facility. When this chemical has been swallowed, get medical attention. Give large quantities of water and induce vomiting. Do not make an unconscious person vomit. If frostbite has occurred, seek medical attention immediately; do *NOT* rub the affected areas or flush them with water. In order to prevent further tissue damage, do *NOT* attempt to remove frozen clothing from frostbitten areas. If frostbite has *NOT* occurred, immediately and thoroughly wash contaminated skin with soap and water.

Personal Protective Methods: Wear protective gloves and clothing to prevent any reasonable probability of skin contact. Safety equipment suppliers/manufacturers can provide recommendations on the most protective glove/clothing material for your operation. All protective clothing (suits, gloves, footwear, headgear) should be clean, available each day, and put on before work. Contact lenses should not be worn when working with this chemical. Wear splash-proof chemical goggles and face shield unless full facepiece respiratory protection is worn. Employees should wash immediately with soap when skin is wet or contaminated. Provide emergency showers and eyewash.

Respirator Selection: Where there is a potential for overexposure*:* SCBAF:PD,PP (any MSHA/NIOSH approved self-contained breathing apparatus that has a full facepiece and is operated in a pressure-demand or other positive-pressure mode); or SAF:PD,PP:ASCBA (any supplied-air respirator that has a full facepiece and is operated in a pressure-demand or other positive-pressure mode in combination with an auxiliary, self-contained breathing apparatus operated in a pressure-demand or other positive pressure mode).

Storage: Prior to working with this chemical you should be trained on its proper handling and storage. Before entering confined space where this chemical may be present, check to make sure sufficient oxygen (19%) exists. Store in tightly closed containers in a cool, well-ventilated area away from heat and sparks. Procedures for the handling, use and storage of cylinders should be in compliance with OSHA 1910.101 and 1910.169 as with the recommendations of the Compressed Gas Association.

Shipping: Label required is "Non flammable Gas." Is in DOT/UN Hazard Class 2.2.

Spill Handling: *Gas:* If in a building, evacuate building and confine vapors by closing doors and shutting down HVAC systems. Restrict persons not wearing protective equipment from area of spill or leak until cleanup is complete. Remove all ignition sources. Ventilate area of spill or leak to disperse the gas. Stop the flow of gas, if it can be done safely from a distance. If source is a cylinder and the leak cannot be stopped in place, remove the leaking cylinder to a safe place, and repair leak or allow cylinder to empty. If employees are required to clean-up spills, they must be properly trained and equipped. OSHA 1910.120(q) may be applicable.

Liquid: Evacuate and restrict persons not wearing protective equipment from area of spill or leak until cleanup is complete. Remove all ignition sources. Ventilate area of spill or leak. Absorb liquids in vermiculite, dry sand, earth, peat, carbon, or a similar material and deposit in sealed containers. It may be necessary to contain and dispose of this chemical as a hazardous waste. If material or contaminated runoff enters waterways, notify downstream users of potentially contaminated waters. Contact your Department of Environmental Protection or your regional office of the federal EPA for specific recommendations. If employees are required to clean-up spills, they must be properly trained and equipped. OSHA 1910.120(q) may be applicable.

Fire Extinguishing: This chemical may burn, but does not readily ignite. Use dry chemical, carbon dioxide, water spray, or alcohol foam extinguishers. Poisonous gases are produced in fire including bromine, chlorine and fluoride compounds. If material or contaminated runoff enters waterways, notify downstream users of potentially contaminated waters. Notify local health and fire officials and pollution control agencies. Container may explode in fire. From a secure, explosion-proof location, use water spray to cool exposed containers. If cooling streams are ineffective (venting sound increases in volume and pitch, tank discolors, or shows any signs of deforming), withdraw immediately to a secure position. If employees are expected to fight fires, they must be trained and equipped in OSHA 1910.156.

References

New Jersey Department of Health and Senior Services, "Hazardous Substance Fact Sheet, CHLORODIFLUORO-MONO-BROMOMETHANE," Trenton, NJ (June 1998)

Chlorodifluoroethane

Molecular Formula: C_2H_5Cl

Common Formula: $ClCH_2CH_3$

Synonyms: CFC 142B; Chlofluorocarbon 142B; 1,1,1-Chlorodifluoroethane; 1-Chloro-1,1-difluoroethane; α-Chloroethylidene fluoride; Chloroethylidene fluoride; 1,1,1-Difluorochloroethane; 1,1-Difluoro-1-chloroethane; Difluoro-1-chloroethane; Difluoromonochloroethane; Ethane, 1-chloro-1,1-difluoro-; FC 142B; Fluorocarbon 142B; Fluorocarbon FC 142B; Freon 142; Freon 142B; Genetron 101; Genetron 142B; Gentron 142B; HCFC-142B; Hydrochlorofluorocarbon 142B; Propellant 142B; R 142B

CAS Registry Number: 75-68-3; 2547-29-4; 27497-51-4 (chlorodifluoroethanes)

RTECS Number: KH7650000; KH7630000 (chlorodifluoroethanes)

DOT ID: UN 2517 (chlorodifluoroethanes)

Regulatory Authority

- Banned or Severely Restricted (In Aerosol Sprays) (UN)[13]
- Air Pollutant Standard Set (UNEP)[43] (DFG)
- CLEAN AIR ACT: Stratospheric ozone protection (Title VI, Subpart A, Appendix B), Class II, Ozone Depletion Potential = 0.06
- EPCRA Section 313 Form R *de minimis* concentration reporting level: 1.0%
- Canada, WHMIS, Ingredients Disclosure List

Cited in U.S. State Regulations: Florida (G), Massachusetts (G), New Hampshire (G), New Jersey (G), Pennsylvania (G).

Description: Chlorodifluoroethane, $ClCH_2CH_3$, is a flammable, colorless, nearly odorless gas. Boiling point = -9.5°C. Freezing/Melting point = -131°C. The explosive limits are: LEL = 6.2%; UEL = 17.9%. Insoluble in water.

Potential Exposure: Chlorodifluoroethane is used in refrigerants, solvents, as a propellent in aerosol sprays, and to make other chemicals.

Incompatibilities: Oxidizers such as perchlorates, peroxides, permanganates, chlorates, and nitrates since vigorous reactions occur. Decomposes in heat to form phosgene, and hydrofluoric and hydrochloric acids.

Permissible Exposure Limits in Air: The former USSR/UNEP joint project[43] has set a MAC in workplace air of 3,000 mg/m^3.

Determination in Air: No test available.

Permissible Concentration in Water: No criteria set.

Routes of Entry: Inhalation.

Harmful Effects and Symptoms

Chlorodifluoroethane can affect you when breathed in. High levels can cause you to feel dizzy, lightheaded, and to pass out. Very high levels could cause death. High exposure could cause irregular heartbeat, which could lead to death.

Short Term Exposure: Chlorodifluoroethane can affect you when breathed in. High levels can cause you to feel dizzy, lightheaded, and to pass out. Very high levels could cause death. High exposure could cause irregular heartbeat, which could lead to death. Contact with the liquid may cause frostbite.

Medical Surveillance: If symptoms develop or overexposure is suspected, the following may be useful: Special 24 hour EKG (halter monitor) to look for irregular heartbeat. Evaluation by a qualified allergist, including careful exposure history and special testing, may help diagnose skin allergy.

First Aid: If this chemical gets into the eyes, remove any contact lenses at once and irrigate immediately for at least 15 minutes, occasionally lifting upper and lower lids. Seek medical attention immediately. If this chemical contacts the skin, remove contaminated clothing and wash affected parts in warm water. Seek medical attention immediately. If this chemical has been inhaled, remove from exposure, begin rescue breathing (using universal precautions) if breathing has stopped and CPR if heart action has stopped. Transfer promptly to a medical facility. When this chemical has been swallowed, get medical attention. Give large quantities of water and induce vomiting. Do not make an unconscious person vomit.

If frostbite has occurred, seek medical attention immediately; do *NOT* rub the affected areas or flush them with water. In order to prevent further tissue damage, do *NOT* attempt to remove frozen clothing from frostbitten areas. If frostbite has *NOT* occurred, immediately and thoroughly wash contaminated skin with soap and water.

Personal Protective Methods: Wear protective gloves and clothing to prevent any reasonable probability of skin contact. Safety equipment suppliers/manufacturers can provide recommendations on the most protective glove/clothing material for your operation. All protective clothing (suits, gloves, footwear, headgear) should be clean, available each day, and put on before work. Contact lenses should not be worn when working with this chemical. Wear non-vented, impact resistant chemical goggles when working with gas Employees should wash immediately with soap when skin is wet or contaminated. Provide emergency showers and eyewash. Where exposure to the liquefied compressed gas may occur, employees should be provided with special clothing designed to prevent frostbite.

Respirator Selection: Engineering controls should be used wherever feasible to maintain airborne concentrations of this chemical below the prescribed exposure limit. Respirators and protective equipment are less effective than engineering controls, and should be used only in non-routine or emergency situations which may result in exposure concentrations in excess of the TWA environmental limit. NIOSH: *(fluorides) 12.5 mg/m^3:* DM (any dust and mist respirator). *25 mg/m^3:* DMXSQ* (any dust and mist respirator except single-use and quarter mask respirators); or SA* (any supplied-air respirator). *62.5 mg/m^3:* SA:CF* (any supplied-air respirator operated in a continuous-flow mode); or PAPRDM*+ *if not present as a fume* (any powered, air-purifying respirator with a dust and mist filter). *125 mg/m^3:* HiEF+ (any air-purifying, full-facepiece respirator with a high-efficiency particulate filter); or SCBAF (any self-contained breathing apparatus with a full facepiece); or SAF (any supplied-air respirator with a full facepiece). *250 mg/m^3:* SA:PD,PP (any supplied-air respirator operated in a pressure-demand or other positive-pressure mode). *Emergency or planned entry into unknown concentrations or IDLH conditions:* SCBAF:PD,PP (any self-contained breathing apparatus that has a full faceplate and is operated in a pressure-demand or other positive-pressure mode); or SAF:PD,PP:ASCBA (any supplied-air respirator that has a full facepiece and is operated in a pressure-demand or other positive-pressure mode in combination with an auxiliary, self-contained breathing apparatus operated in a pressure-demand or other positive-pressure mode). *Escape:* HiEF+ (any air-purifying, full-facepiece respirator with a high-efficiency particulate filter); or SCBAE (any appropriate escape-type, self-contained breathing apparatus).

* Substance reported to cause eye irritation or damage; may require eye protection.

\+ May need acid gas sorbent.

Storage: Prior to working with Chlorodifluoroethane you should be trained on its proper handling and storage. Before entering confined space where this chemical may be present, check to make sure that an explosive concentration does not exist. Chlorodifluoroethane must be stored to avoid contact with Oxidizers (such as Perchlorates, Peroxides, Permanganates, Chlorates and Nitrates) since violent reactions occur.

Detached or outside storage is preferred. Metal containers involving the transfer of this chemical should be grounded and bonded. Where possible, automatically pump liquid from drums or other storage containers to process containers. Drums must be equipped with self-closing valves, pressure vacuum bungs, and flame arresters. Use only non-sparking tools and equipment, especially when opening and closing containers of this chemical. Sources of ignition such as smoking and open flames, are prohibited where this chemical is used, handled, or stored in a manner that could create a potential fire or explosion hazard. Procedures for the handling, use and storage of cylinders should be in compliance with OSHA 1910.101 and 1910.169 as with the recommendations of the Compressed Gas Association.

Shipping: Chlorodifluoroethanes must be labeled "Flammable Gas." They fall in Hazard Class 2.1. Shipment by passenger aircraft or railcar is forbidden. Shipment by cargo aircraft is limited.

Spill Handling: Evacuate and restrict persons not wearing protective equipment from area for leak until clean-up is complete. Remove all ignition sources. Establish forced ventilation to keep levels below explosive limit. Stop flow of gas. If source of leak is a cylinder and the leak cannot be stopped in place, remove the leaking cylinder to a safe place in the open air, and repair leak or allow cylinder to empty. Keep Chlorodifluoroethane out of a confined space, such as a sewer, because of the possibility of an explosion, unless the sewer is designed to prevent the build-up of explosive concentrations. It may be necessary to contain and dispose of this chemical as a hazardous waste. If material or contaminated runoff enters waterways, notify downstream users of potentially contaminated waters. Contact your Department of Environmental Protection or your regional office of the federal EPA for specific recommendations. If employees are required to clean-up spills, they must be properly trained and equipped. OSHA 1910.120(q) may be applicable.

Fire Extinguishing: This chemical is a flammable gas. Poisonous gases including phosgene, hydrogen fluoride, hydrogen chloride, and chlorine are produced in fire. Use dry chemical, carbon dioxide, or foam extinguishers. Vapors are heavier than air and will collect in low areas. Vapors may travel long distances to ignition sources and flashback. Vapors in confined areas may explode when exposed to fire. Containers may explode in fire. Storage containers and parts of containers may rocket great distances, in many directions. If material or contaminated runoff enters waterways, notify downstream users of potentially contaminated waters. Notify local health and fire officials and pollution control agencies. From a secure, explosion-proof location, use water spray to cool exposed containers. If cooling streams are ineffective (venting sound increases in volume and pitch, tank discolors, or shows any signs of deforming), withdraw immediately to a secure position. If employees are expected to fight fires, they must be trained and equipped in OSHA 1910.156.

References

New Jersey Department of Health and Senior Services and Senior Services, "Hazardous Substance Fact Sheet: Chlorodifluoroethane," Trenton, NJ (December 1998)

Chlorodifluoromethane

Molecular Formula: $CHClF_2$

Synonyms: Algeon 22; Algofrene 22; Algofrene type 6; Arcton 22; Arcton 4; CFC 22; Chlorofluorocarbon 22; Diaflon 22; Difluorochloromethane; Difluoromonochloromethane; Dymel 22; Electro-CF 22; Eskimon 22; F 22; FC 22; Flugene 22; Fluorocarbon 22; Forane 22; Forane 22 B; Freon; Freon 22; Frigen; Frigen 22; Genetron 22; HCFC-22; Hydrochlorofluorocarbon 22; Isceon 22; Isotron 22; Khaladon 22; Khladon 22; Methane, chlorodifluoro-; Monochlorodifluoromethane; Propellant 22; R-22; Refrigerant 22; Ucon 22; Ucon 22/Halocarbon 22

CAS Registry Number: 75-45-6

RTECS Number: PA6390000

DOT ID: UN 1018

Regulatory Authority

- Carcinogen (Animal Suspected) (IARC)[9]
- Banned or Severely Restricted (In Aerosol Sprays) (UN)[13]
- Air Pollutant Standard Set (ACGIH)[1] (Australia) (DFG)[3] (HSE)[33] (Israel) (Mexico) (USSR/UNEP)[43] (OSHA)[58] (Several States)[60] (Several Canadian Provinces)
- CLEAN AIR ACT: Stratospheric ozone protection (Title VI, Subpart A, Appendix B), Class II, Ozone Depletion Potential = 0.05
- EPCRA Section 313 Form R *de minimis* concentration reporting level: 1.0%
- Canada, WHMIS, Ingredients Disclosure List

Cited in U.S. State Regulations: Alaska (G), California (A, G), Connecticut (A), Florida (G), Maine (G), Massachusetts (G), Minnesota (G), Nevada (A), New Hampshire (G), New Jersey (G), North Dakota (A), Pennsylvania (G), Rhode Island (G), Virginia (A), West Virginia (G).

Description: Chlorodifluoromethane, $CHClF_2$, is a nonflammable, colorless, nearly odorless gas. Boiling point = -41°C. Freezing/Melting point = -146°C. Soluble in water.

Potential Exposure: Chlorodifluoromethane is used as an aerosol propellant, refrigerant and low-temperature solvent. It is used in the synthesis of polytetrafluoroethylene (PTFE).

Incompatibilities: Reacts violently with alkalies and alkaline earth metals, powdered aluminum, sodium, potassium, and zinc, causing fire and explosion hazard. Moisture and rust cause slow decomposition, forming toxic gases. Attacks some plastics, rubber and coatings. Decomposes in heat forming fumes of chlorine, hydrogen chloride, hydrogen fluoride, and phosgene. Attacks magnesium and its alloys.

Permissible Exposure Limits in Air: The NIOSH recommended limit is 1,000 ppm (3,500 mg/m^3) TWA for a 10-hour workshift and a STEL of 1,250 ppm (4,375 mg/m^3) not to be exceeded during any 15 minute work period. HSE, Mexico and the Canadian provinces of Alberta, British Columbia, and Ontario have set the same TWA and STEL as NIOSH. ACGIH recommends, Australia, Israel and the Canadian province of Quebec have set the same TWA with no STEL value. The DFG[3] has set a MAK of 500 ppm (1,800 mg/m^3) as has Sweden[35] and Sweden have, in addition, set a STEL of 750 ppm (2,500 mg/m^3). The DFG peak limitation is 2 times the MAK (1 hr. momentary value) not to be exceeded 3 times during a workshift. The former USSR/UNEP joint project[43] has set a MAC in workplace air of 3,000 mg/m^3 and has set limits for ambient air in residential areas of 100 mg/m^3 on a momentary basis and 10 mg/m^3 on a daily average basis. In addition, several states have set guidelines or standards for FC-22 in ambient air[60] ranging from 35.0 – 43.75 mg/m^3 (North Dakota) to 58 mg/m^3 (Virginia) to 70.0 mg/m^3 (Connecticut) to 83.3 mg/m^3 (Nevada).

Determination in Air: Charcoal tube;[2] Methylene chloride; Gas chromatography/Flame ionization detection; NIOSH IV Method #1018.

Permissible Concentration in Water: The former USSR/UNEP joint project[43] has set a MAC of 10 mg/l water bodies used for domestic purposes.

Harmful Effects and Symptoms

Short Term Exposure: Chlorodifluoromethane can affect you when breathed in. Inhalation can irritate the respiratory tract causing tightness in the chest and trouble breathing. Exposure can cause headache, nausea, dizziness and weakness, sleepiness, tremors, loss of coordination, cardiac arrythmia (irregular heartbeat), coma, and asphyxiation, which could lead to death. Chlorodifluoromethane can irritate the nose, throat and skin. It can also cause tightening in the chest and trouble breathing. Skin contact with the liquid can cause frostbite.

Long Term Exposure: Liver, kidney, spleen injury.

Points of Attack: Respiratory system, cardiovascular system, central nervous system, liver, kidneys, spleen.

Medical Surveillance: If symptoms develop or overexposure is suspected, the following may be useful: Special 24 hour EKG (Holter monitor) to look for irregular heart beat. Lung, liver and kidney function tests.

First Aid: If this chemical gets into the eyes, remove any contact lenses at once and irrigate immediately for at least 15 minutes, occasionally lifting upper and lower lids. Seek medical attention immediately. If this chemical contacts the skin, remove contaminated clothing and wash immediately with soap and water. Seek medical attention immediately. If this chemical has been inhaled, remove from exposure, begin rescue breathing (using universal precautions) if breathing has stopped and CPR if heart action has stopped. Transfer promptly to a medical facility. When this chemical has been swallowed, get medical attention. Give large quantities of water and induce vomiting. Do not make an unconscious person vomit.

If frostbite has occurred, seek medical attention immediately; do *NOT* rub the affected areas or flush them with water. In order to prevent further tissue damage, do *NOT* attempt to remove frozen clothing from frostbitten areas. If frostbite has *NOT* occurred, immediately and thoroughly wash contaminated skin with soap and water.

Note to physicians: Adrenergic agents are contraindicated.

Personal Protective Methods: Wear protective gloves and clothing to prevent any reasonable probability of skin contact. Safety equipment suppliers/manufacturers can provide recommendations on the most protective glove/clothing material for your operation. All protective clothing (suits, gloves, footwear, headgear) should be clean, available each day, and put on before work. Contact lenses should not be worn when working with this chemical. Wear splash-proof chemical goggles and face shield unless full facepiece respiratory protection is worn. Employees should wash immediately with soap when skin is wet or contaminated. Provide emergency showers and eyewash. Where exposure to the liquefied compressed gas may occur, employees should be provided with special clothing designed to prevent frostbite.

Respirator Selection: Where the potential exists for exposures over 1,000 ppm, use a MSHA/NIOSH approved supplied-air respirator with a full facepiece operated in the positive pressure mode or with a full facepiece, hood, or helmet in the continuous flow mode, or use a MSHA/NIOSH approved self-contained breathing apparatus with a full facepiece operated in pressure-demand or other positive pressure mode.

Storage: Prior to working with Chlorodifluoromethane you should be trained on its proper handling and storage. Store in tightly closed containers in a cool, well ventilated area. Sources of ignition such as smoking and open flames, are prohibited where this chemical is used, handled, or stored in a manner that could create a potential fire or explosion hazard. Procedures for the handling, use and storage of cylinders should be in compliance with OSHA 1910.101 and 1910.169 as with the recommendations of the Compressed Gas Association. A regulated, marked area should be established where this chemical is handled, used, or stored in compliance with OSHA standard 1910.1045.

Shipping: Chlorodifluoromethane must carry a "Non-Flammable Gas" label. It falls in DOT Hazard Class 2.2.[19]

Spill Handling: Evacuate and restrict persons not wearing protective equipment from area of spill or leak until cleanup is complete. Remove all ignition sources. Ventilate area of spill or leak. *Liquid:* allow to evaporate. *Gas:* Stop flow of gas. If source of leak is a cylinder and the leak cannot be stopped in place, remove the leaking cylinder to a safe place in the open air, and repair leak or allow cylinder to empty. If liquid or contaminated runoff enters waterways, notify downstream users of potentially contaminated waters. Contact your Department of Environmental Protection or your regional office of the federal EPA for specific recommendations. If employees are required to clean-up spills, they must be properly trained and equipped. OSHA 1910.120(q) may be applicable.

Fire Extinguishing: This chemical is nonflammable. Use dry chemical, carbon dioxide, water spray, or foam extinguishers. On contact with fire, this chemical decomposes forming poisonous gases including hydrogen fluoride, hydrogen chloride, phosgene, carbonyl fluoride, chloride fumes and fluoride. If material or contaminated runoff enters waterways, notify downstream users of potentially contaminated waters. Notify local health and fire officials and pollution control agencies. Containers may explode in fire. From a secure, explosion-proof location, use water spray to cool exposed containers. If cooling streams are ineffective (venting sound increases in volume and pitch, tank discolors, or shows any signs of deforming), withdraw immediately to a secure position. If employees are expected to fight fires, they must be trained and equipped in OSHA 1910.156.

Disposal Method Suggested: Return to vendor or send to licensed waste disposal company.[22]

References

New Jersey Department of Health and Senior Services and Senior Services, "Hazardous Substance Fact Sheet: Chlorodifluoromethane," Trenton, NJ (January 1999)

2-Chloroethyl Vinyl Ether

Molecular Formula: C_4H_7ClO

Common Formula: $ClCH_2CH_2OCH{=}CH_2$

Synonyms: 2-Chlorethyl vinyl ether; (2-Chloroethoxy)ethene; 2-Cloroetilo vinil eter (Spanish); Vinyl-β-chloroethyl ether; Vinyl-2-chloroethyl ether

CAS Registry Number: 110-75-8

RTECS Number: KN6300000

DOT ID: UN 1993

Regulatory Authority

- CLEAN WATER ACT: 40CFR423, Appendix A, Priority Pollutants as 2-chloroethyl vinyl ether (mixed)
- EPA HAZARDOUS WASTE NUMBER (RCRA No.): U042
- RCRA, 40CFR261, Appendix 8 Hazardous Constituents
- RCRA 40CFR268.48; 61FR15654, Universal Treatment Standards: Wastewater (mg/l), 0.062; Nonwastewater (mg/kg), N/A
- SUPERFUND/EPCRA 40CFR302.4 Reportable Quantity (RQ): CERCLA, 1,000 lb (454 kg)
- Canada, WHMIS, Ingredients Disclosure List

Cited in U.S. State Regulations: California (A, G), Florida (G), Kansas (G), Louisiana (G), Massachusetts (G), New Hampshire (G), New Jersey (G), Pennsylvania (G), Vermont (G), Virginia (G), Washington (G), Wisconsin (G).

Description: 2-Chloroethyl vinyl ether, $ClCH_2CH_2OCH{=}CH_2$, is a flammable, colorless liquid. Boiling point = 109°C. Flash point = 27°C. Hazard Identification (based on NFPA-704 M Rating System): Health 2, Flammability 3, Reactivity 2.

Potential Exposure: The compound finds use in the manufacture of anesthetics, sedatives, and cellulose ethers. The number of potentially exposed individuals is greatest for the following areas: fabricated metal products; wholesale trade; leather, rubber and plastic, and chemical products.

Incompatibilities: Presumed to form unstable peroxides that can cause polymerization. Forms explosive mixture with air. May accumulate static electrical charges, and may cause ignition of its vapors. Contact with oxidizing materials may cause fire or explosion hazard.

Permissible Exposure Limits in Air: No standards set.

Permissible Concentration in Water: No criteria have been developed for aquatic life or for the protection of human health.

Determination in Water: Inert gas purge followed by gas chromatography with halide specific detection (EPA Method 601) or gas chromatography plus mass spectrometry (EPA Method 624).

Routes of Entry: Eye, skin.

Harmful Effects and Symptoms

Short Term Exposure: Contact with skin or eyes may cause severe irritation. The oral LD_{50} for 2-Chloroethyl vinyl ether in rats is 250 mg/kg (moderately toxic). Primary skin irritation and eye irritation studies have also been conducted for 2-Chloroethyl vinyl ether. Dermal exposure to undiluted 2-chloroethyl vinyl ether did not cause even slight erythema. Application of undiluted 2-Chloroethyl vinyl ether to the eyes of rabbits resulted in severe eye injury.

First Aid: If this chemical gets into the eyes, remove any contact lenses at once and irrigate immediately for at least 15 minutes, occasionally lifting upper and lower lids. Seek medical attention immediately. If this chemical contacts the skin, remove contaminated clothing and wash immediately with soap and water. Seek medical attention immediately. If this chemical has been inhaled, remove from exposure, begin rescue breathing (using universal precautions) if breathing has stopped and CPR if heart action has stopped. Transfer promptly to a medical facility. When this chemical has been swallowed, get medical attention. Give large quantities of water and induce vomiting. Do not make an unconscious person vomit.

Personal Protective Methods: Wear protective gloves and clothing to prevent any reasonable probability of skin contact. Safety equipment suppliers/manufacturers can provide

recommendations on the most protective glove/clothing material for your operation. All protective clothing (suits, gloves, footwear, headgear) should be clean, available each day, and put on before work. Contact lenses should not be worn when working with this chemical. Wear splash-proof chemical goggles and face shield unless full facepiece respiratory protection is worn. Employees should wash immediately with soap when skin is wet or contaminated. Provide emergency showers and eyewash.

Respirator Selection: *Where there is a potential for overexposure:* SCBAF:PD,PP (any MSHA/NIOSH approved self-contained breathing apparatus that has a full facepiece and is operated in a pressure-demand or other positive-pressure mode); or SAF:PD,PP:ASCBA (any supplied-air respirator that has a full facepiece and is operated in a pressure-demand or other positive-pressure mode in combination with an auxiliary, self-contained breathing apparatus operated in a pressure-demand or other positive pressure mode).

Storage: Prior to working with this chemical you should be trained on its proper handling and storage. Before entering confined space where 2-Chloroethyl vinyl ether may be present, check to make sure that an explosive concentration does not exist. Store in tightly closed containers in a cool, well ventilated area. Metal containers involving the transfer of this chemical should be grounded and bonded. Where possible, automatically pump liquid from drums or other storage containers to process containers. Drums must be equipped with self-closing valves, pressure vacuum bungs, and flame arresters. Use only non-sparking tools and equipment, especially when opening and closing containers of this chemical. Sources of ignition such as smoking and open flames, are prohibited where this chemical is used, handled, or stored in a manner that could create a potential fire or explosion hazard.

Shipping: Combustible liquids n.o.s. fall in DOT Hazard Class 3 and Packing Group III.[19] There is no label requirement.

Spill Handling: Evacuate and restrict persons not wearing protective equipment from area of spill or leak until cleanup is complete. Remove all ignition sources. Ventilate area of spill or leak. Absorb liquids in vermiculite, dry sand, earth, peat, carbon, or a similar material and deposit in sealed containers. It may be necessary to contain and dispose of this chemical as a hazardous waste. If material or contaminated runoff enters waterways, notify downstream users of potentially contaminated waters. Contact your Department of Environmental Protection or your regional office of the federal EPA for specific recommendations. If employees are required to clean-up spills, they must be properly trained and equipped. OSHA 1910.120(q) may be applicable.

Fire Extinguishing: This chemical is a flammable liquid. Poisonous gases produced in fire including toxic chlorides. Use dry chemical or alcohol foam extinguishers. Vapors are heavier than air and will collect in low areas. Vapors may travel long distances to ignition sources and flashback. Vapors in confined areas may explode when exposed to fire. Containers may explode in fire. Storage containers and parts of containers may rocket great distances, in many directions. If material or contaminated runoff enters waterways, notify downstream users of potentially contaminated waters. Notify local health and fire officials and pollution control agencies. From a secure, explosion-proof location, use water spray to cool exposed containers. If cooling streams are ineffective (venting sound increases in volume and pitch, tank discolors, or shows any signs of deforming), withdraw immediately to a secure position. If employees are expected to fight fires, they must be trained and equipped in OSHA 1910.156.

Disposal Method Suggested: Residues may be packaged in epoxy-lined drums and disposed of by high temperature incineration with HCl scrubbing of effluent gases.

References

U.S. Environmental Protection Agency, 2-Chloroethyl Vinyl Ether, Health and Environmental Effects Profile No. 46, Office of Solid Waste, Washington, DC (April 30, 1980).

U.S. Environmental Protection Agency, Chloralkyl Ethers: Ambient Water Quality Criteria, Washington, DC (1980).

Sax, N. I., Ed., "Dangerous Properties of Industrial Materials Report," 7, No. 4, 46–50 (1987).

Chloroform

Molecular Formula: $CHCl_3$

Synonyms: Chloroforme (French); Cloroformio (Italian); Cloroformo (Spanish); Formyl trichloride; Freon 20; Methane trichloride; Methane, trichloro-; Methenyl trichloride; Methyl trichloride; NCI-C02686; R 20 refrigerant; Refrigerant 20; TCM; Trichloormethaan (Dutch); Trichlormethan (Czech); Trichloroform; Trichloromethane; Triclorometano (Italian)

CAS Registry Number: 67-66-3

RTECS Number: FS9100000

DOT ID: UN 1888

EEC Number: 602-006-00-4

EINECS Number: 200-663-8

Regulatory Authority

- Carcinogen (Animal Positive) (Human Suspected) (IARC)[9] (ACGIH)[1] (DFG)[3] (Australia) (Mexico)
- Banned or Severely Restricted (In Pharmaceuticals) (Many Countries) (UN)[13]
- Air Pollutant Standard Set (ACGIH)[1] (Australia) (DFG)[3] (HSE)[33] (Israel) (Mexico) (OSH)[58] (Several States)[60] (Several Canadian Provinces)
- CLEAN AIR ACT: Hazardous Air Pollutants (Title I, Part A, Section 112); Accidental Release Prevention/Flammable substances, (Section 112[r], Table 3), TQ = 20,000 lb (9,080 kg)

- CLEAN WATER ACT: Section 311 Hazardous Substances/ RQ 40CFR117.3 (same as CERCLA, see below); 40CFR 423, Appendix A, Priority Pollutants; Section 313 Water Priority Chemicals (57FR41331, 9/9/92); Toxic Pollutant (Section 401.15)
- EPA HAZARDOUS WASTE NUMBER (RCRA No.): U044; D022
- RCRA Toxicity Characteristic (Section 261.24), Maximum Concentration of Contaminants, regulatory level, 6.0 mg/l
- RCRA, 40CFR261, Appendix 8 Hazardous Constituents
- RCRA 40CFR268.48; 61FR15654, Universal Treatment Standards: Wastewater (mg/l), 0.046; Nonwastewater (mg/kg), 6.0
- RCRA 40CFR264, Appendix 9; TSD Facilities Ground Water Monitoring List. Suggested test method(s) (PQL μg/l): 8010 (0.5); 8240 (5)
- SAFE DRINKING WATER ACT: Priority List (55 FR 1470)
- SUPERFUND/EPCRA 40CFR355, Appendix B Extremely Hazardous Substances: TPQ = 10,000 lb (4,540 kg)
- SUPERFUND/EPCRA 40CFR302.4 Reportable Quantity (RQ): CERCLA, 10 lb (4.54 kg)
- EPCRA Section 313 Form R *de minimis* concentration reporting level: 0.1%
- Canada, WHMIS, Ingredients Disclosure List; National Pollutant Release Inventory (NPRI)
- Mexico, Drinking Water Criteria, 0.03 mg/l

Cited in U.S. State Regulations: Alaska (G), Arizona (W), California (G), Connecticut (A), Florida (G), Illinois (G, W), Indiana (A), Kansas (G), Louisiana (G), Maine (G), Maryland (G), Massachusetts (G, A), Michigan (G, A), Minnesota (W), Nevada (A), New Hampshire (G), New Jersey (G), New York (G, A), North Carolina (A), North Dakota (A), Oklahoma (G), Pennsylvania (G, A), Rhode Island (G, A), South Carolina (A), Vermont (G), Virginia (G, A), Washington (G), West Virginia (G), Wisconsin (G).

Description: Chloroform, $CHCl_3$, is a noncombustible, clear, colorless liquid with a pleasant, sweet odor. The odor threshold is 12 ppm. Boiling point = 62°C. Freezing/Melting point = -63°C. Soluble in water.

Potential Exposure: Chloroform was one of the earliest general anesthetics, but its use for this purpose has been abandoned because of toxic effects. Chloroform is widely used as a solvent (especially in the lacquer industry); in the extraction and purification of penicillin and other pharmaceuticals; in the manufacture of artificial silk, plastics, floor polishes, and fluorocarbons; and in sterilization of catgut. Chemists and support workers as well as hospital workers are believed to be at a higher risk than the general population.

Chloroform is widely distributed in the atmosphere and water (including municipal drinking water primarily as a consequence of chlorination). A survey of 80 American cities by EPA found chloroform in every water system in levels ranging from < 0.3 – 311 ppb.

Incompatibilities: Though nonflammable, chloroform decomposes to form hydrogen chloride, phosgene, and chlorine upon contact with a flame. Chloroform decomposes slowly in air and light. Reacts violently with strong caustics (bases), strong oxidants, chemically active metals (especially powders), such as aluminum, lithium, magnesium, potassium and sodium, causing fire and explosion hazard. Attacks plastic, rubber and coatings. Corrodes iron and other metals in the presence of moisture.

Permissible Exposure Limits in Air: The Federal standard (OSHA PEL)[58] Ceiling is 50 ppm (240 mg/m^3), not to be exceeded at any time. The NIOSH REL is 2 ppm (9.78 mg/m^3) 60-minute STEL; NIOSH considers chloroform to be a potential occupational carcinogen as defined by the OSHA carcinogen policy [29 CFR 1990]. ACGIH: 10 ppm (49 mg/m^3) TWA, A2, chloroform is an "Industrial Substance Suspect of Carcinogenic Potential for Man." The NIOSH IDLH value is 500 ppm. (carcinogen). The DFG MAK is 0.5 ppm (2.5 mg/m^3) and Peak limitation of 2 × MAK, 30 minute momentary value not to be exceeded 4 times per workshift. HSE[33] set a limit of 10 ppm TWA and STEL of 50 ppm (225 mg/m^3). The former USSR[35] has set a MAC for ambient air in residential areas of 0.3 mg/m^3. In addition, several states have set guidelines or standards for chloroform in ambient air[60] ranging from zero (North Dakota) to 0.04 μg/m^3 (Michigan and Rhode Island) to 0.43 μg/m^3 (Massachusetts and North Carolina) to 120 μg/m^3 (Pennsylvania) to 167 μg/m^3 (New York) to 250 μg/m^3 (Connecticut and South Carolina) to 500 μg/m^3 (Virginia) to 1,190 μg/m^3 (Nevada) to 1,200 μg/m^3 (Indiana).

Determination in Air: Charcoal adsorption, workup with CS_2, analysis by gas chromatography. See OSHA Method 5 and NIOSH Method #1003 for Hydrocarbons, halogenated.

Permissible Concentration in Water: To protect freshwater aquatic life: 28,900 μg/l on an acute basis and 1,240 μg/l on a chronic basis. To protect saltwater aquatic life: no value set due to insufficient data. To protect human health: preferably zero. An additional lifetime cancer risk of 1 in 100,000 results at a level of 1.9 μg/l.[6]

The former USSR has set[35] a limit in surface water of 0.06 mg/l.

Some states have set standards and guidelines for chloroform in drinking water[61] ranging from 1 μg/l (Illinois) to 3 μg/l (Arizona) to 5 μg/l (Minnesota).

Determination in Water: Gas chromatography (EPA Method 601) or gas chromatography plus mass spectrometry (EPA Method 624).

Routes of Entry: Inhalation of vapors, ingestion, skin and eye contact. This chemical can be absorbed through the skin, thereby increasing exposure.

Harmful Effects and Symptoms

Short Term Exposure: Chloroform irritates the eyes and contact can cause, tearing, conjunctivitis and permanent eye damage. Symptoms of acute chloroform exposure include fainting sensation, vomiting, dizziness, salivation, nausea, fatigue, and headache. Other symptoms are respiratory depression, coma, kidney damage, and liver damage. Chloroform is classified as moderately toxic. Probable oral lethal dose for humans is 0.5 – 5 g/kg (between 1 ounce and 1 pint) for a 150 lb person. The mean lethal dose is probably near 1 fluid ounce (44 g). It is a human suspected carcinogen. Also, it is a central nervous system depressant and a gastrointestinal irritant. It has caused rapid death attributable to cardiac arrest and delayed death from liver and kidney damage.

Long Term Exposure: Symptoms of chronic exposure include loss of appetite, hallucinations, moodiness and physical and mental sluggishness. Repeated or prolonged contact with skin may cause skin drying, cracking and dermatitis. May cause heart, thyroid, liver and kidney damage. This substance is possibly carcinogenic to humans. It has been shown to cause liver, kidney, and thyroid cancer in animals. There is limited evidence that chloroform is a teratogen in animals.

Points of Attack: Liver, kidneys, heart, eyes, skin, central nervous system.

Medical Surveillance: Preplacement and periodic examinations should include appropriate tests for thyroid, liver and kidney functions, and special attention should be given to the nervous system, the skin, and to any history of alcoholism. Expired air and blood levels may be useful in estimating levels of acute exposure. Special 24 hour (holster monitor) to detect irregular heart beat.

First Aid: If this chemical gets into the eyes, remove any contact lenses at once and irrigate immediately for at least 15 minutes, occasionally lifting upper and lower lids. Seek medical attention immediately. If this chemical contacts the skin, remove contaminated clothing and wash immediately with soap and water. Seek medical attention immediately. If this chemical has been inhaled, remove from exposure, begin rescue breathing (using universal precautions) if breathing has stopped and CPR if heart action has stopped. Transfer promptly to a medical facility. When this chemical has been swallowed, get medical attention. Give large quantities of water and induce vomiting. Do not make an unconscious person vomit. Medical observation is recommended.

Personal Protective Methods: Wear solvent-resistant gloves and clothing to prevent any reasonable probability of skin contact. Safety equipment suppliers/manufacturers can provide recommendations on the most protective glove/clothing material for your operation. Polyvinyl chloride, Teflon, polyurethane and VITON/Chlorobutyl have been recommended as protective materials in the literature. All protective clothing (suits, gloves, footwear, headgear) should be clean, available each day, and put on before work. Contact lenses should not be worn when working with this chemical. Wear splash-proof chemical goggles and face shield unless full facepiece respiratory protection is worn. Employees should wash immediately with soap when skin is wet or contaminated. Remove nonimpervious clothing immediately if wet or contaminated. Provide emergency showers and eyewash.

Respirator Selection: NIOSH: *At any detectable concentration:* SCBAF:PD,PP (any MSHA/NIOSH approved self-contained breathing apparatus that has a full facepiece and is operated in a pressure-demand or other positive-pressure mode); or SAF:PD,PP:ASCBA (any supplied-air respirator that has a full facepiece and is operated in a pressure-demand or other positive-pressure mode in combination with an auxiliary, self-contained breathing apparatus operated in a pressure-demand or other positive pressure mode). *Escape:* GMFOV [any air-purifying, full-facepiece respirator (gas mask) with a chin-style, front-or back-mounted organic vapor canister] or SCBAE (any appropriate escape-type, self-contained breathing apparatus).

Storage: Prior to working with Chloroform you should be trained on its proper handling and storage. A regulated, marked area should be established where this chemical is handled, used, or stored in compliance with OSHA standard 1910.1045. Store in tightly closed dark bottles or cans in a cool, well-ventilated area.

Shipping: The DOT label requirement is "Poison." The UN/DOT Hazard Class is 6.1 and the Packing Group is II.[19][20]

Spill Handling: Stay upwind; keep out of low areas. Do not touch spilled material; stop leak if you can do it without risk. Use water spray to reduce vapors. Evacuate and restrict persons not wearing protective equipment from area of spill or leak until cleanup is complete. Remove all ignition sources. Ventilate area of spill or leak. Small spills: absorb liquids in vermiculite, dry sand, earth, peat, carbon, or a similar material and deposit in sealed containers. Large spills: dike far ahead of spill for later disposal. It may be necessary to contain and dispose of this chemical as a hazardous waste. If material or contaminated runoff enters waterways, notify downstream users of potentially contaminated waters. Contact your Department of Environmental Protection or your regional office of the federal EPA for specific recommendations. If employees are required to clean-up spills, they must be properly trained and equipped. OSHA 1910.120(q) may be applicable.

Fire Extinguishing: This chemical is a noncombustible liquid. Use any extinguishing agents. Poisonous gases are produced in fire including hydrogen chloride, phosgene, and chlorine. If material or contaminated runoff enters waterways, notify downstream users of potentially contaminated waters. Notify local health and fire officials and pollution control agencies. Containers may explode in fire. Fight fire from maximum distance. Dike fire control water for later disposal; do not scatter the material. From a secure, explosion-proof location, use water spray to cool exposed containers. If cooling streams are ineffective (venting sound increases in volume and pitch, tank discolors, or shows any signs of deforming), withdraw immediately to a secure position. If employees are expected to fight fires, they must be trained and equipped in OSHA 1910.156.

Disposal Method Suggested: Incineration, preferably after mixing with another combustible fuel. Care must be exercised to assure complete combustion to prevent the formation of phosgene. An acid scrubber is necessary to remove the halo acids produced.[22]

Where possible it should be recovered, purified by distillation, and returned to the supplier.

References

National Institute for Occupational Safety and Health, Criteria for a Recommended Standard: Occupational Exposure to Chloroform, NIOSH Document No. 75–114, Washington, DC (1975)

National Institute for Occupational Safety and Health, Current Intelligence Bulletin No. 9-Chloroform, Washington, DC (1976)

U.S. Environmental Protection Agency, Chloroform: Ambient Water Quality Criteria, Washington, DC (1980)

National Academy of Sciences, Chloroform, Carbon Tetrachloride and Other Halomethanes: An Environmental Assessment, Washington, DC (1978)

U.S. Environmental Protection Agency, Chloroform, Health and Environmental Effects Profile No. 47, Office of Solid Waste, Washington, DC (April 30, 1980)

Sax, N. I., Ed., "Dangerous Properties of Industrial Materials Report," 1, No. 4, 44–47 (1981) and 3, No. 5, 101–106, New York, Van Nostrand Reinhold Co. (1983)

U.S. Public Health Service, "Toxicological Profile for Chloroform," Atlanta, Georgia, Agency for Toxic Substances and Disease Registry (October 1987)

U.S. Environmental Protection Agency, "Chemical Profile: Chloroform," Washington, DC, Chemical Emergency Preparedness Program (October 31, 1985)

New Jersey Department of Health and Senior Services and Senior Services, "Hazardous Substance Fact Sheet: Chloroform," Trenton, NJ (February 1989)

New York State Department of Health, "Chemical Fact Sheet: Chloroform," Albany, NY, Bureau of Toxic Substance Assessment (January 1986)

Chloromethyl Anilines

Molecular Formula: C_7H_8ClN

Common Formula: $C_6H_3Cl(CH_3)NH_2$

Synonyms: *3165-93-3:* Amarthol fast red TR base; Amarthol fast red TR salt; 2-Amino-5-chlorotoluene hydrochloride; Azanil red salt TRD; Azoene fast red TR salt; Azogene fast red TR; Azoic diazo component 11 base; Benzeneamine, 4-chloro-2-methyl-, hydrochloride; Brentamine fast red TR salt; Chlorhydrate de 4-chloroortho-toluidine (French); 5-Chloro-2-aminotoluene hydrochloride; 4-Chloro-2-methylaniline hydrochloride; 4-Chloro-6-methylaniline hydrochloride; 4-Chloro-2-methylbenzenamine hydrochloride; 4-Chloro-2-toluidine hydrochloride; C.I. 37085; C.I. azoic diazo component 11; Clorhidrato de 4-cloro-*o*-toluidina (Spanish); Daito red salt TR; Devol red K; Devol red TA salt; Devol red TR; Diazo fast red TR; Diazo fast red TRA; Fast red 5CT salt; Fast red salt TR; Fast red salt TRA; Fast red salt TRN; Fast red TR salt; Hindasol red TR salt; Kromon green B; 2-Methyl-4-chloroaniline hydrochloride; Natasol fast red TR salt; NCI-C02368; Neutrosel red TRVA; Ofna-Perl salt RRA; Red base ciba IX; Red base IRGA IX; Red salt Ciba IX; Red salt IRGA IX; Red TRS salt; Sanyo fast red salt TR; UN 1579

95-69-2: Amarthol fast red TR base; 2-Amino-5-chlorotoluene; *asym-m*-Chloro-*o*-toluidine; Asymmetric *m*-chloro-*o*-toluidine; Azoene fast red TR base; Azogene fast red TR; Azoic diazo component 11, base; Benzenamine, 4-chloro-2-methyl; Brentamine fast red TR base; 3-Chloro-6-aminotoluene; 5-Chloro-2-aminotoluene; 4-Chloro-2-methylaniline; 4-Chloro-6-methylaniline; 4-Chloro-2-methylbenzenamine; *p*-Chloro-*o*-toluidine; 4-Chloro-*o*-toluidine; 4-Chloro-2-toluidine; 4-Cloro-*o*-toluidina (Spanish); Daito red base TR; Deval red K; Deval red TR; Diazo fast red TRA; Fast red 5CT base; Fast red base TR; Fast red TR; Fast red TR11; Fast red TR base; Fast red TRO base; Kako red TR base; Kambamine red TR; 2-Methyl-4-chloroaniline; Mitsui red TR base; Red base Ciba IX; Red base IRGA IX; Red base NTR; Red TR base; Sanyo fast red TR base; *o*-Toluidine, 4-chloro-; Tulabase fast red TR

CAS Registry Number: 87-60-5 (3-chloro-o-); 95-69-2 (4-chloro-o-); 95-79-4 (5-chloro-o-); 95-81-8 (2-chloro-5-); 87-63-8 (6-chloro-o-); 615-65-6 (2-chloro-p-); 95-74-9 (3-chloro-para); 3165-93-3 (4-chloro-o-, hydrochloride); 7149-75-9 (4 chloro-m-); 29027-17-6 (2-chloro-m-)

RTECS Number: XU4760000 (3-chloro-o-); XU5000000 (4-chloro-o-); XU5075000 (5-chloro-o-); XU5100000 (6-chloro-o-); XU5110000 (2-chloro-para); XU5111000 (3-chloro-para); XU5250000 (4-chloro-o-, hydrochloride)

DOT ID: UN 2239 (chlorotoluidines)

Regulatory Authority

- Carcinogen (See Asterisked Isomers Below) (IARC) (NCI)[9] (DFG)[3]
- EPCRA Section 313 Form R *de minimis* concentration reporting level: 0.1%. (4-chloro-o-) (95-69-2)
- EPA HAZARDOUS WASTE NUMBER (RCRA No.): U049 (4-Chloro-o- HCl only) (3165-93-3)[4]
- RCRA, 40CFR261, Appendix 8 Hazardous Constituents. (4-Chloro-o- HCl only) (3165-93-3)[4]
- SUPERFUND/EPCRA 40CFR302.4 Reportable Quantity (RQ): CERCLA, 100 lb (45.4 kg). (4-chloro-o- HCl only) (3165-93-3)[5]

• Canada, WHMIS, Ingredients Disclosure List

Cited in U.S. State Regulations: Kansas (G), Louisiana (G), Massachusetts (G), Michigan (G), New Hampshire (G), New Jersey (Total) (G), Oklahoma (G), Pennsylvania (G), Vermont (G), Virginia (G), Washington (G), Wisconsin (G).

Description: The chloromethylanilines, $C_6H_3Cl(CH_3)NH_2$, are colorless or white crystalline solids or liquids, some have a mild fishy odor. All are soluble in water, and include:

Isomer	Melting Point, °C	Boiling Point, °C	Flash Point, °C
3-chloro-o-	0	245	>113
4-chloro-o-	29	241	100
5-chloro-o-	26	237	160
6-chloro-o-	—	—	99
2-chloro-para	7	257	99
3-chloro-para	26	238	100
2-chloro-5-	—	—	107

Potential Exposure: Most of the isomers are used in dyestuff manufacture. The 3-chloro-para isomer is used to kill birds. It is marketed as pelleted bait for control of bird populations.

Incompatibilities: Incompatible with oxidizers, strong acids, chloroformates, and acid anhydrides, isocyanates, aldehydes forming fire and explosive hazards.

Permissible Exposure Limits in Air: No standards set. However these chemicals can be absorbed through the skin, thereby increasing exposure.

Permissible Concentration in Water: No criteria set.

Routes of Entry: Inhalation, skin contact, ingestion. These chemicals can be absorbed through the skin, thereby increasing exposure.

Harmful Effects and Symptoms

Short Term Exposure: Chloromethyl Anilines can affect you when breathed in and by passing through your skin. Eye contact causes irritation and can lead to permanent damage. Skin contact can cause a rash and produce a burning feeling. Exposure can lower the ability of the blood to carry oxygen (methemoglobinemia) causing a bluish color of the skin, headaches, dizziness, nausea, and even death. They can damage the kidneys and bladder, causing painful, bloody urine.

Long Term Exposure: The methemoglobinemia condition described above may occur gradually from repeated exposure or all at once. Some of these chemicals like 4-chloro-2- and 5-chloro-2-) are probable carcinogens in humans; they cause bladder and liver cancer in animals.

Points of Attack: Eyes, skin, kidneys, bladder, liver.

Medical Surveillance: If symptoms develop or overexposure has occurred, the following may be useful: Blood tests for methemoglobin levels. Kidney function tests. Urine tests for blood, and for N-acetyl p-aminophenol.

First Aid: If this chemical gets into the eyes, remove any contact lenses at once and irrigate immediately for at least 15 minutes, occasionally lifting upper and lower lids. Seek medical attention immediately. If this chemical contacts the skin, remove contaminated clothing and wash immediately with soap and water. Seek medical attention immediately. If this chemical has been inhaled, remove from exposure, begin rescue breathing (using universal precautions) if breathing has stopped and CPR if heart action has stopped. Transfer promptly to a medical facility. When this chemical has been swallowed, get medical attention. Give large quantities of water and induce vomiting. Do not make an unconscious person vomit.

Note to Physician: Treat for methemoglobinemia. Spectrophotometry may be required for precise determination of levels of methemoglobinemia in urine.

Personal Protective Methods: Wear protective gloves and clothing to prevent any reasonable probability of skin contact. Safety equipment suppliers/manufacturers can provide recommendations on the most protective glove/clothing material for your operation. All protective clothing (suits, gloves, footwear, headgear) should be clean, available each day, and put on before work. Contact lenses should not be worn when working with this chemical. When working with liquids or solids wear indirect-vent, impact-resistant chemical goggles and face shield unless full facepiece respiratory protection is worn. Employees should wash immediately with soap when skin is wet or contaminated. Provide emergency showers and eyewash.

Respirator Selection: Where the potential exists for exposure to Chloromethyl Anilines, use a MSHA/NIOSH approved full facepiece respirator with a high efficiency particulate filter. Greater protection is provided by a powered-air purifying respirator.

Where the potential for high exposure to 3-Chloro-2-Methyl Aniline exists, or to liquid form use a MSHA/NIOSH approved supplied-air respirator with a full facepiece operated in the positive pressure mode or with a full facepiece, hood, or helmet in the continuous flow mode, or use a MSHA/NIOSH approved self-contained breathing apparatus with a full facepiece operated in pressure-demand or other positive pressure mode.

Storage: Prior to working with Chloromethyl Anilines you should be trained on its proper handling and storage. Store in tightly closed containers in a cool, well-ventilated area. Sources of ignition, such as smoking and open flames, are prohibited where Choromethyl Anilines are handled, or stored in a manner that could create a potential fire or explosion hazard. A regulated, marked area should be established where this chemical is handled, used, or stored in compliance with OSHA standard 1910.1045.

Shipping: Chlorotoluidines require a "Keep Away from Food" label. They fall in DOT Hazard Class 6.1 and Packing Group III.[19]

Spill Handling: *Liquid:* Evacuate and restrict persons not wearing protective equipment from area of spill or leak until cleanup is complete. Remove all ignition sources. Ventilate area of spill or leak. Absorb liquids in vermiculite, dry sand, earth, peat, carbon, or a similar material and deposit in sealed containers. It may be necessary to contain and dispose of this chemical as a hazardous waste. If material or contaminated runoff enters waterways, notify downstream users of potentially contaminated waters. Contact your Department of Environmental Protection or your regional office of the federal EPA for specific recommendations. If employees are required to clean up spills, they must be properly trained and equipped. OSHA 1910.120(q) may be applicable.

Solid: Evacuate persons not wearing protective equipment from area of spill or leak until clean up is complete. Remove all ignition sources. Collect powdered material in the most convenient and safe manner and deposit in sealed containers. Ventilate area after cleanup is complete. It may be necessary to contain and dispose of this chemical as a hazardous waste. If material or contaminated runoff enters waterways, notify downstream users of potentially contaminated waters. Contact your Department of Environmental Protection or your regional office of the federal EPA for specific recommendations. If employees are required to clean-up spills, they must be properly trained and equipped. OSHA 1910.120(q) may be applicable.

Fire Extinguishing: These chemicals may burn, but according the state of New Jersey, do not readily ignite. Use dry chemical or carbon dioxide extinguishers. Poisonous gases are produced in fire including chlorine, hydrogen chloride, nitrogen oxides, and carbon monoxides. If material or contaminated runoff enters waterways, notify downstream users of potentially contaminated waters. Notify local health and fire officials and pollution control agencies. From a secure, explosion-proof location, use water spray to cool exposed containers. If cooling streams are ineffective (venting sound increases in volume and pitch, tank discolors, or shows any signs of deforming), withdraw immediately to a secure position. If employees are expected to fight fires, they must be trained and equipped in OSHA 1910.156.

References

New Jersey Department of Health and Senior Services and Senior Services, "Hazardous Substance Fact Sheet: 3-Chloro-2-Methyl Aniline," Trenton, NJ (April 1986)

New Jersey Department of Health and Senior Services and Senior Services, "Hazardous Substance Fact Sheet: 4-Chloro-o-Methyl Aniline," Trenton, NJ (November 1998)

New Jersey Department of Health and Senior Services and Senior Services, "Hazardous Substance Fact Sheet: 5-Chloro-2-Methyl Aniline," Trenton, NJ (November 1998)

New Jersey Department of Health and Senior Services and Senior Services, "Hazardous Substance Fact Sheet: 6-Chloro-2-Methyl Aniline," Trenton, NJ (May 1986)

New Jersey Department of Health and Senior Services and Senior Services, "Hazardous Substance Fact Sheet: 2-Chloro-4-Methyl Aniline," Trenton, NJ (May 1986)

New Jersey Department of Health and Senior Services and Senior Services, "Hazardous Substance Fact Sheet: 3-Chloro-4-Methyl Aniline," Trenton, NJ (October 1998)

New Jersey Department of Health and Senior Services and Senior Services, "Hazardous Substance Fact Sheet: 2-Chloro-5-Methyl Aniline," Trenton, NJ (May 1986)

New Jersey Department of Health and Senior Services and Senior Services, "Hazardous Substance Fact Sheet: 4-Chloro-3-Methyl Aniline," Trenton, NJ (May 1986)

New Jersey Department of Health and Senior Services and Senior Services, "Hazardous Substance Fact Sheet: 2-Chloro-3-Methyl Aniline," Trenton, NJ (May 1986)

Chloromethyl Methyl Ether

Molecular Formula: C_2H_5ClO

Common Formula: $ClCH_2OCH_3$

Synonyms: Chlordimethylether (Czech); Chlorodimethyl ether; Chloromethoxymethane; CMME; α,α-Dichlorodimethyl ether; Dimethylchloroether; Ether, chloromethyl methyl; Ether, dimethyl chloro; Ether methylique monochlore (French); Methane, chloromethoxy-; Methoxychloromethane; Methoxymethyl chloride; Monochlorodimethyl ether; Monochloromethyl methyl ether

CAS Registry Number: 107-30-2

RTECS Number: KN6650000

DOT ID: UN 1239

EEC Number: 603-075-00-3

Regulatory Authority

- Carcinogen (Human Positive) (IARC)[9] (Known) (NTP/7th) (Human Suspected) (ACGIH)[1] (DFG)[3]
- Banned or Severely Restricted (Several Countries) (UN)[13][35]
- Very Toxic Substance (World Bank)[15]
- OSHA, 29CFR1910 Specifically Regulated Chemicals (See CFR 1910.1006)
- Air Pollutant Standard Set (Alberta, Canada)
- CLEAN AIR ACT: Hazardous Air Pollutants (Title I, Part A, Section 112); Accidental Release Prevention/Flammable substances, (Section 112[r], Table 3), TQ = 5,000 lb (2,270 kg)

- EPA HAZARDOUS WASTE NUMBER (RCRA No.): U046
- RCRA, 40CFR261, Appendix 8 Hazardous Constituents
- SUPERFUND/EPCRA 40CFR355, Appendix B Extremely Hazardous Substances: TPQ = 100 lb (45.4 kg)
- SUPERFUND/EPCRA 40CFR302.4 Reportable Quantity (RQ): CERCLA, 10 lb (4.54 kg)
- EPCRA Section 313 Form R *de minimis* concentration reporting level: 0.1%
- Canada, WHMIS, Ingredients Disclosure List; National Pollutant Release Inventory (NPRI)

Cited in U.S. State Regulations: Alaska (G), California (G), Florida (G), Illinois (G), Kansas (G), Louisiana (G), Maine (G), Maryland (G), Massachusetts (G), Minnesota (G), New Hampshire (G), New Jersey (G), New York (A), North Dakota (A), Pennsylvania (G, A), Rhode Island (G), Vermont (G), Virginia (G, A), Washington (G), Wisconsin (G).

Description: Chloromethyl methyl ether, $ClCH_2OCH_3$, is a volatile, corrosive liquid with an ether-like odor. Boiling point = 59°C. Freezing/Melting point = -104°C. Flash point of = -8°C. Hazard Identification (based on NFPA-704 M Rating System): Health 2, Flammability 3, Reactivity 1. Decomposes in water. Commercial chloromethyl methyl ether contains from 1 – 7% *bis(chloromethyl)ether*, a known carcinogen.

Potential Exposure: Chloromethyl methyl ether is a highly reactive methylating agent and is used in the chemical industry for synthesis of organic chemicals. Most industrial operations are carried out in closed process vessels so that exposure is minimized.

Incompatibilities: Forms explosive mixture with air. May be able to form unstable and explosive peroxides. Contact with oxidizers may cause fire and explosion. Decomposes on contact with water forming hydrochloric acid and formaldehyde. Attacks various metals in presence of water.

Permissible Exposure Limits in Air: Chloromethyl methyl ether is included in the Federal standard for carcinogens; all contact with it should be avoided. A full OSHA standard[1910.1006] has been adopted for this substance. ACGIH has designated it A2, suspected human carcinogen.

The HSE[33] has set no specific limits for this compound. The DFG[3] simply notes that it is a human carcinogen with no numerical limits; it notes that this is the technical guide material which can contain up to 7% of bis-chloromethyl ether.

The former USSR-UNEP/IRPTC project[43] on the other hand does give a MAC of 0.5 mg/m^3 in workplace air. Czechoslovakia has set a TWA of 0.003 mg/m^3.[35] Alberta, Canada set an airborne exposure limit of 0.005 ppm TWA and STEL of 0.015 ppm.

Several states have set guidelines or standards for CMME in ambient air[60] ranging from zero (North Dakota) to 0.02 ppb (Pennsylvania) to 0.03 μg/m^3 (New York) to 3.0 μg/m^3 (Virginia).

Determination in Air: Impinger; Hexane; Gas chromatography/Electrochemical detection; NIOSH II[1] P&CAM Method #220.

Permissible Concentration in Water: No criteria have been set for the protection of freshwater or saltwater aquatic life due to lack of data. For the protection of human health: preferably zero.[6]

Routes of Entry: Inhalation of vapor and possibly percutaneous absorption, and ingestion.

Harmful Effects and Symptoms

Short Term Exposure: This chemical can be absorbed through the skin, thereby increasing exposure. Corrosive to the eyes, skin, and respiratory tract. Inhalation can cause pulmonary edema, a medical emergency that can be delayed for several hours. This can cause death. Symptoms of exposure include sore throat, coughing, shortness of breath, fever, chills, difficulty in breathing, bronchial secretions from pulmonary edema. The liquid causes severe irritation of eyes and skin; and vapor exposure of 100 ppm is severely irritating to eyes and nose. This level is dangerous to life in 4 hours. Due to an increased death rate from respiratory cancer among exposed victims, it is a regulated carcinogen.

Long Term Exposure: There is evidence that this chemical caused lung cancer in humans, and has caused skin and lung cancer in animals. Lungs may be affected by repeated or prolonged exposure causing bronchitis with cough, phlegm, and/or shortness of breath.

Points of Attack: Eyes, skin, respiratory system.

Medical Surveillance: Preplacement and periodic medical examinations should include an examination of the skin and respiratory tract, including lung function tests and chest x-ray. Sputum cytology has been suggested as helpful in detecting early malignant changes, and in this connection a detailed smoking history is of importance. Possible effects on the fetus should be considered.

First Aid: If this chemical gets into the eyes, remove any contact lenses at once and irrigate immediately for at least 15 minutes, occasionally lifting upper and lower lids. Seek medical attention immediately. If this chemical contacts the skin, remove contaminated clothing and wash immediately with soap and water. Seek medical attention immediately. If this chemical has been inhaled, remove from exposure, begin rescue breathing (using universal precautions) if breathing has stopped and CPR if heart action has stopped. Transfer promptly to a medical facility. When this chemical has been swallowed, get medical attention. Give large quantities of water and induce vomiting. Do not make an unconscious person vomit. Medical observation is recommended for 24 – 48 hours after breathing overexposure, as pulmonary edema may be delayed. As first aid for pulmonary edema, a doctor or authorized paramedic may consider administering a corticosteroid spray.

Personal Protective Methods: Those designed to supplement engineering controls and to prevent all skin or respiratory contact. Full body protective clothing and gloves should be used on entering areas of partial exposure. Those employed in handling operations should be provided with fullface, supplied-air respirators of continuous-flow or pressure-demand type. On exit from a regulated area, employees should be required to remove and leave protective clothing and equipment at the point of exit, to be placed in impervious containers as the end of the workshift for decontamination or disposal. Showers should be taken prior to dressing in street clothes.

Respirator Selection: NIOSH: *At any detectable concentration:* SCBAF:PD,PP (any MSHA/NIOSH approved self-contained breathing apparatus that has a full facepiece and is operated in a pressure-demand or other positive-pressure mode); or SAF:PD,PP:ASCBA (any supplied-air respirator that has a full facepiece and is operated in a pressure-demand or other positive-pressure mode in combination with an auxiliary, self-contained breathing apparatus operated in a pressure-demand or other positive pressure mode). *Escape:* GMFOV [any air-purifying, full-facepiece respirator (gas mask) with a chin-style, front-or back-mounted organic vapor canister] or SCBAE (any appropriate escape-type, self-contained breathing apparatus).

Storage: Prior to working with this chemical you should be trained on its proper handling and storage. Before entering confined space where Chloromethyl methyl ether may be present, check to make sure that an explosive concentration does not exist. A regulated, marked area should be established where this chemical is handled, used, or stored in compliance with OSHA standard 1910.1045. Store in tightly closed containers in a cool, well ventilated area. Metal containers involving the transfer of this chemical should be grounded and bonded. Where possible, automatically pump liquid from drums or other storage containers to process containers. Drums must be equipped with self-closing valves, pressure vacuum bungs, and flame arresters. Use only non-sparking tools and equipment, especially when opening and closing containers of this chemical. Sources of ignition such as smoking and open flames, are prohibited where this chemical is used, handled, or stored in a manner that could create a potential fire or explosion hazard.

Shipping: This compound requires a shipping label of: "Flammable Liquid, Poison." It falls in DOT Hazard Class 6.1 and Packing Group I.

Spill Handling: Evacuate and restrict persons not wearing protective equipment from area of spill or leak until cleanup is complete. Remove all ignition sources. Ventilate area of spill or leak. Small spill: absorb liquids in vermiculite, dry sand, earth, peat, carbon, or a similar material and deposit in sealed containers. Large spill: dike far ahead of large spills for later disposal. Flood with water. Rinse with sodium bicarbonate or lime solution. It may be necessary to contain and dispose of this chemical as a hazardous waste. If material or contaminated runoff enters waterways, notify downstream users of potentially contaminated waters. Contact your Department of Environmental Protection or your regional office of the federal EPA for specific recommendations. If employees are required to clean-up spills, they must be properly trained and equipped. OSHA 1910.120(q) may be applicable.

Fire Extinguishing: This chemical is a highly flammable liquid. Poisonous gases including hydrogen chloride are produced in fire. Do not use water or water based extinguishers. Use dry chemical, carbon dioxide, AFFF. Vapors are heavier than air and will collect in low areas. Vapors may travel long distances to ignition sources and flashback. Vapors in confined areas may explode when exposed to fire. Containers may explode in fire. Storage containers and parts of containers may rocket great distances, in many directions. If material or contaminated runoff enters waterways, notify downstream users of potentially contaminated waters. Notify local health and fire officials and pollution control agencies. From a secure, explosion-proof location, use water spray to cool exposed containers. If cooling streams are ineffective (venting sound increases in volume and pitch, tank discolors, or shows any signs of deforming), withdraw immediately to a secure position. If employees are expected to fight fires, they must be trained and equipped in OSHA 1910.156.

Disposal Method Suggested: Incineration, preferably after mixing with another combustible fuel. Care must be exercised to assure complete combustion to prevent the formation of phosgene. An acid scrubber is necessary to remove the halo acids produced.[22]

References

U.S. Environmental Protection Agency, Chloroalkyl Ethers: Ambient Water Quality Criteria, Washington, DC (1980)

U.S. Environmental Protection Agency, Chloroalkyl Ethers, Health and Environmental Effects Profile No. 41, Office of Solid Waste, Washington, DC (April 30, 1980)

U.S. Environmental Protection Agency, "Chemical Profile: Chloromethyl Methyl Ether," Washington, DC, Chemical Emergency Preparedness Program (October 31, 1985)

New Jersey Department of Health and Senior Services and Senior Services, "Hazardous Substance Fact Sheet: ChloromethylMethyl Ether," Trenton, NJ (October 1985)

Sax, N. I., Ed., "Dangerous Properties of Industrial Materials Report," 7, No. 4, 51–54 (1987)

2-(4-Chloro-2-Methylphenoxy) Propionic Acid

Molecular Formula: $C_{10}H_{11}ClO_3$

Synonyms: 2M-4CP; 2M4KHP; Acide 2-(4-chloro-2-methyl-phenoxy)propionique (French); Acido 2-(4-cloro-2-metilfenossi)-propionico (Italian); Assassin; Banvel BP; Banvel P; BH Mecoprop; Ceridor; Chipco; Chipco Turf herbicide MCPP; 2-(4-Chlor-2-methyl-phenoxy)-propionsaeure (German); 2-(4-Chloro-2-methylphenoxy)propanoic acid; α-(4-Chloro-2-methylphenoxy)propionic acid; (+)-α-(4-Chloro-2-methylphenoxy)propionic acid; (4-Chloro-2-methylphenoxy)propionic acid; 2-(4-Chloro-2-methylphenoxy)propionic acid; 4-Chloro-2-methylphenoxy-α-propionic acid; 2-(4-Chlorophenoxy-2-methyl)propionic acid; 2-(4-Chloro-*o*-tolyl)oxylpropionic acid; 2-(*p*-Chloro-*o*-tolyloxy)propionic acid; Cleaval; Clenecorn; 2-(4-Cloor-2-methyl-fenoxy)-propionzuur (Dutch); Clovotox; CMPP; Compitox extra; *iso*-Cornox; *iso*-Cornox 64; Cornox plus; CR 205; Crusader; Docklene; EXP 419; Graslam; Harness; Harrier; Hedonal MCPP; Herrisol; Hymec; Hytane extra; Iotox; Kilprop; Liranox; 2-MCPP; MCPP; MCPP 2,4-D; MCPP-D-4; MCPP K-4; Mechlorprop; Mecobrom; Mecomec; Mecopeop; Mecoper; Mecopex; Mecoprop; Mecoturf; Mepro; Methoxone; 2-(2-Methyl-4-chlorophenoxy)propanoic acid; α-(2-Methyl-4-chlorophenoxy)propionic acid; 2-(2'-Methyl-4'-chlorophenoxy)propionic acid; 2-Methyl-4-chlorophenoxy-α-propionic acid; Musketeet; Mylone; N.B. Mecoprop; NSC 60282; Post-Kite; Propal; Propanoic acid, 2-(4-chloro-2-methylphenoxy)-; Propionic acid, 2-(4-chloro-2-methylphenoxy); Propionic acid, 2-[(4-chloro-*o*-tolyl)oxy)-; Propionic acid, 2-(2-methyl-4-chlorophenoxy)-; Proponex-plus; Rankotex; RD 4593; Runcatex; Scotlene; Seloxone; Sel-oxone; Super green and weed; Supoertox; Swipe 560 EC; Terset; Tetralen-plus; U 46; U 46 KV-ester; U 46 KV-fluid; Verdone; VI-Par; Vipex; VI-Pex

CAS Registry Number: 93-65-2

RTECS Number: UE9750000

DOT ID: UN 2765 (phenoxy pesticides, solid, toxic)

EEC Number: 607-049-00-2

Regulatory Authority

- Carcinogen (sufficient animal data) IARC
- EPCRA Section 313 Form R *de minimis* concentration reporting level: 0.1%

Cited in U.S. State Regulations: California (A, G), New Jersey (G), Minnesota (G)

Description: Mecoprop is a colorless, odorless, crystalline solid. Freezing/Melting point = 93 – 95°C. Slightly soluble in water.

Potential Exposure: It is a chlorophenoxy-herbicide, used as a herbicide and weed killer.

Incompatibilities: A weak acid. Incompatible with strong bases and oxidizers.

Permissible Exposure Limits in Air: This chemical can be absorbed through the skin, thereby increasing exposure.

Routes of Entry: Inhalation, absorbed through the skin.

Harmful Effects and Symptoms

Short Term Exposure: This chemical can be absorbed through the skin, thereby increasing exposure. Contact irritates the eyes and skin. Irritates the respiratory tract. Exposure can cause headache, weakness, convulsions, muscle cramps, loss of coordination, unconsciousness, and death.

Long Term Exposure: There is limited evidence that the chemical affects human reproduction. Exposure may damage blood cells, causing anemia, and damage the kidneys. Although this chemical has not been identified as a carcinogen, several related compounds have shown limited evidence of cancer.

Points of Attack: Blood, kidney, nervous system.

Medical Surveillance: Examination of the nervous system. Complete blood count. Kidney function tests.

First Aid: If this chemical gets into the eyes, remove any contact lenses at once and irrigate immediately for at least 15 minutes, occasionally lifting upper and lower lids. Seek medical attention immediately. If this chemical contacts the skin, remove contaminated clothing and wash immediately with soap and water. Seek medical attention immediately. If this chemical has been inhaled, remove from exposure, begin rescue breathing (using universal precautions) if breathing has stopped and CPR if heart action has stopped. Transfer promptly to a medical facility. When this chemical has been swallowed, get medical attention. Give large quantities of water and induce vomiting. Do not make an unconscious person vomit.

Personal Protective Methods: Wear protective gloves and clothing to prevent any reasonable probability of skin contact. Safety equipment suppliers/manufacturers can provide recommendations on the most protective glove/clothing material for your operation. All protective clothing (suits, gloves, footwear, headgear) should be clean, available each day, and put on before work. Contact lenses should not be worn when working with this chemical. Wear dust-proof chemical goggles and face shield unless full facepiece respiratory protection is worn. Employees should wash immediately with soap when skin is wet or contaminated. Provide emergency showers and eyewash.

Respirator Selection: Where there is a potential for over-exposure: SCBAF:PD,PP (any MSHA/NIOSH approved self-contained breathing apparatus that has a full facepiece and is operated in a pressure-demand or other positive-pressure mode); or SAF:PD,PP:ASCBA (any supplied-air respirator that has a full facepiece and is operated in a pressure-demand or other positive-pressure mode in combination with an auxiliary, self-contained breathing apparatus operated in a pressure-demand or other positive pressure mode).

Storage: Prior to working with this chemical you should be trained on its proper handling and storage. Store in tightly closed containers in a cool, well-ventilated area away from oxidizers and strong bases.

Shipping: Phenoxy pesticides, solid, toxic, n.o.s. have a DOT label requirement of "Poison." This chemical falls in UN/DOT Hazard Class 6.1 and Packing Group II.[19][20]

Spill Handling: Evacuate persons not wearing protective equipment from area of spill or leak until clean-up is complete. Remove all ignition sources. Collect powdered material in the most convenient and safe manner and deposit in sealed containers. Ventilate area after clean-up is complete. It may be necessary to contain and dispose of this chemical as a hazardous waste. If material or contaminated runoff enters waterways, notify downstream users of potentially contaminated waters. Contact your Department of Environmental Protection or your regional office of the federal EPA for specific recommendations. If employees are required to clean-up spills, they must be properly trained and equipped. OSHA 1910.120(q) may be applicable.

Fire Extinguishing: This chemical is a noncombustible solid. Use dry chemical, carbon dioxide, or alcohol or polymer foam extinguishers. Poisonous gases are produced in fire including carbon monoxide and hydrogen chloride. If material or contaminated runoff enters waterways, notify downstream users of potentially contaminated waters. Notify local health and fire officials and pollution control agencies. From a secure, explosion-proof location, use water spray to cool exposed containers. If cooling streams are ineffective (venting sound increases in volume and pitch, tank discolors, or shows any signs of deforming), withdraw immediately to a secure position. If employees are expected to fight fires, they must be trained and equipped in OSHA 1910.156.

Disposal Method Suggested: In accordance with 40CFR165 recommendations for the disposal of pesticides and pesticide containers. Must be disposed properly by following package label directions or by contacting your state pesticide or environmental control agency or by contacting your regional EPA office.

References

New Jersey Department of Health and Senior Services, "Hazardous Substance Fact Sheet, 2-(4-CHLORO-2-METHYLPHENOXY) PROPIONIC ACID," Trenton, NJ (April 1999)

Chloromethyl Phenyl Isocyanate

Molecular Formula: C_8H_6ClNO

Common Formula: $OCN\text{-}C_6H_3{\cdot}Cl{\cdot}CH_3$

Synonyms: 3-Chloro-4-methylphenyl isocyanate; Isocyanic acid, 3-chloro-*p*-tolyl ester

CAS Registry Number: 28479-22-3

RTECS Number: NQ8585000

DOT ID: UN 2236

Regulatory Authority

- Canada, WHMIS, Ingredients Disclosure List

Cited in U.S. State Regulations: New Hampshire (G), New Jersey (G).

Description: Chloromethyl Phenyl Isocyanate, $OCN\text{-}C_6H_3{\cdot}Cl{\cdot}CH_3$, is a colorless to yellow liquid or low-melting solid. Flash point ≥ 93°C. Insoluble in water (decomposes)

Potential Exposure: This material is used in organic synthesis.

Incompatibilities: Acids, alkalis, and amines. Keep away from heat, light, and moisture. Do not store in temperatures above 30°C (86°F).

Permissible Exposure Limits in Air: No standard set.

Permissible Concentration in Water: No criteria set.

Routes of Entry: Inhalation, ingestion.

Harmful Effects and Symptoms

Short Term Exposure: 3-Chloro-4-Methyl Phenyl Isocyanate can affect you when breathed in. Little is known about the health effects of 3-Chloro-4-Methyl Phenyl Isocyanate. However, organic isocyanates can irritate the eyes, skin, and respiratory tract and cause lung allergies.

Long Term Exposure: Although the long-term effects of this chemical are unknown, many isocyanates cause allergic reactions and asthama-like allergy.

Points of Attack: Respiratory system.

Medical Surveillance: For those with frequent or potentially high exposure, the following are recommended before beginning work and at regular times after that: Lung function tests. These may be normal if the person is not having an attack at the time of the test.

First Aid: If this chemical gets into the eyes, remove any contact lenses at once and irrigate immediately for at least 15 minutes, occasionally lifting upper and lower lids. Seek medical attention immediately. If this chemical contacts the skin, remove contaminated clothing and wash immediately with soap and water. Seek medical attention immediately. If this chemical has been inhaled, remove from exposure, begin rescue breathing (using universal precautions) if breathing has stopped and CPR if heart action has stopped. Transfer promptly to a medical facility. When this chemical has been swallowed, get medical attention. Give large quantities of water and induce vomiting. Do not make an unconscious person vomit.

Personal Protective Methods: Wear protective gloves and clothing to prevent any reasonable probability of skin contact. Safety equipment suppliers/manufacturers can provide recommendations on the most protective glove/clothing material for your operation. All protective clothing (suits, gloves, footwear, headgear) should be clean, available each day, and put on before work. Contact lenses should not be worn when working with this chemical. When working with liquids, wear splash-proof chemical goggles and face shield unless full facepiece respiratory protection is worn. When working with powders or dusts, wear dust-proof chemical goggles and face shield unless full facepiece respiratory protection is worn. Employees should wash immediately with soap when skin is wet or contaminated. Provide emergency showers and eyewash.

Respirator Selection: Where the potential for exposure to 3-Chloro-4-Methyl Phenyl Isocyanate exists, use a MSHA/NIOSH approved supplied-air respirator with a full facepiece operated in the positive pressure mode or with a full facepiece, hood, or helmet in the continuous flow mode, or use a MSHA/NIOSH approved self-contained breathing apparatus with a full facepiece operated in pressure demand or other positive pressure mode.

Storage: Prior to working with 3-Chloro-4-Methyl Phenyl Isocyanate you should be trained on its proper handling and storage. 3-Chloro-4-Methyl Phenyl Isocyanate must be stored to avoid contact with Acids (such as hydrochloric and nitric); alkalis (such as sodium hydroxide and potassium hydroxide); and amines (like ammonia) since violent reactions occur. Store in tightly closed containers in a cool, well-ventilated area away from heat, light and moisture. Do not store at temperatures above 86°F. Sources of ignition such as smoking and open flames are prohibited where 3-Chloro-4-Methyl Phenyl Isocyanate is used, handled, or stored in a manner that could create a potential fire or explosion hazard.

Shipping: This material must carry a "Poison" label. It falls in Hazard Class 6.1 and Packing Group II.[17]

Spill Handling: Evacuate persons not wearing protective equipment from area of spill or leak until clean-up is complete. Remove all ignition sources. Ventilate area of spill or leak. Absorb liquids in vermiculite, dry sand, earth, peat, carbon, or a similar material and deposit in sealed containers. Collect powdered material in the most convenient and safe manner and deposit in sealed containers. Ventilate area after clean-up is complete. It may be necessary to contain and dispose of this chemical as a hazardous waste. If material or contaminated runoff enters waterways, notify downstream users of potentially contaminated waters. Contact your Department of Environmental Protection or your regional office of the federal EPA for specific recommendations. If employees are required to clean-up spills, they must be properly trained and equipped. OSHA 1910.120(q) may be applicable.

Fire Extinguishing: 3-Chloro-4-Methyl Phenyl Isocyanate may burn, but does not readily ignite. Containers may explode in fire. Poisonous gases are produced in fire, including Hydrogen Cyanide, Nitrogen Oxides and Hydrogen Chloride. Use dry chemical, CO_2, water spray, or foam extinguishers. If material or contaminated runoff enters waterways, notify downstream users of potentially contaminated waters. Notify local health and fire officials and pollution control agencies. From a secure, explosion-proof location, use water spray to cool exposed containers. If cooling streams are ineffective (venting sound increases in volume and pitch, tank discolors, or shows any signs of deforming), withdraw immediately to a secure position. If employees are expected to fight fires, they must be trained and equipped in OSHA 1910.156.

References

New Jersey Department of Health and Senior Services and Senior Services, "Hazardous Substance Fact Sheet: 3-Chloro-4-Methyl Phenyl Isocyanate," Trenton, NJ (April 1997)

1-Chloro-1-Nitropropane

Molecular Formula: $C_3H_6ClNO_2$

Common Formula: $C_2H_5CHClNO_2$

Synonyms: 1,1-Chloronitropropane; Chloronitropropane; 1-Chloro-1-nitropropano (Spanish); Korax®; Korax 6®; Lanstan®; Propane, 1-chloro-1-nitro-

CAS Registry Number: 600-25-9

RTECS Number: TX5075000

DOT ID: No citation.

Regulatory Authority

- Air Pollutant Standard Set (ACGIH)[1] (Australia) (DFG)[3] (Israel) (Mexico) (OSHA)[58] (Several Canadian Provinces) (Several States)[60]
- Canada, WHMIS, Ingredients Disclosure List

Cited in U.S. State Regulations: Alaska (G), California (A, G), Connecticut (A), Florida (G), Illinois (G), Maine (G), Massachusetts (G), Minnesota (G), Nevada (A), New Hampshire (G), New Jersey (G), North Dakota (A), Pennsylvania (G), Rhode Island (G), Virginia (A), West Virginia (G).

Description: Chloronitroproane, $C_2H_5CHClNO_2$, is a flammable, colorless liquid with an unpleasant odor that causes tears (lachrymator). Boiling point = 141°C. Flash point = 62°C. Hazard Identification (based on NFPA-704 M Rating System): Health U (unknown), Flammability 3, Reactivity 2. Slightly soluble in water.

Potential Exposure: This compound is used in the synthetic rubber industry and as a component in rubber cements and as a fungicide.

Incompatibilities: Forms explosive mixture with air. Strong oxidizers may cause a fire and explosion hazard. May explode when exposed to heat.

Permissible Exposure Limits in Air: The Federal limit (OSHA PEL)[58] is 20 ppm (100 mg/m^3) as is the Mexico and DFG MAK value.[3] The TWA recommended by NIOSH and ACGIH is 2 ppm (10 mg/m^3) as is the Australian and Israeli value. The NIOSH IDLH level is 100 ppm. Canadian provincial limits are: British Columbia TWA 20 ppm and STEL of 30 ppm; Alberta, Ontario and Quebec TWA is 2 ppm and Alberta's STEL is 4 ppm.

In addition, several states have set guidelines or standards for chloronitropropane in ambient air[60] ranging from 10 μg/m^3

(North Dakota) to 150 μg/m^3 (Virginia) to 200 μg/m^3 (Connecticut) to 238 μg/m^3 (Nevada).

Determination in Air: Chromosorb tube-108; Ethyl acetate; Gas chromatography/Flame ionization detection; See NIOSH II(5) Method #S-211.

Permissible Concentration in Water: No criteria set.

Routes of Entry: Inhalation, ingestion, skin and eye contact.

Harmful Effects and Symptoms

Short Term Exposure: 1-Chloro-1-Nitropropane can affect you when breathed in. Exposure can irritate and burn the eyes, skin and cause respiratory tract irritation with coughing and /or shortness of breath. Higher exposures can cause pulmonary edema, a medical emergency that can be delayed for several hours. This can cause death.

Long Term Exposure: May cause damage to the heart, liver, and kidneys. Similar irritating substances can cause lung injury and bronchitis.

Points of Attack: In animals: respiratory system, lungs, liver, kidneys, cardiovascular system.

Medical Surveillance: Consider the points of attack in preplacement and periodic physical examinations. EKG. Lung function tests. Liver and kidney function tests. Consider chest x-ray following acute overexposure.

First Aid: If this chemical gets into the eyes, remove any contact lenses at once and irrigate immediately for at least 15 minutes, occasionally lifting upper and lower lids. Seek medical attention immediately. If this chemical contacts the skin, remove contaminated clothing and wash immediately with soap and water. Seek medical attention immediately. If this chemical has been inhaled, remove from exposure, begin rescue breathing (using universal precautions) if breathing has stopped and CPR if heart action has stopped. Transfer promptly to a medical facility. When this chemical has been swallowed, get medical attention. Give large quantities of water and induce vomiting. Do not make an unconscious person vomit. Medical observation is recommended for 24 – 48 hours after breathing overexposure, as pulmonary edema may be delayed. As first aid for pulmonary edema, a doctor or authorized paramedic may consider administering a corticosteroid spray.

Personal Protective Methods: Wear protective gloves and clothing to prevent any reasonable probability of skin contact. Safety equipment suppliers/manufacturers can provide recommendations on the most protective glove/clothing material for your operation. All protective clothing (suits, gloves, footwear, headgear) should be clean, available each day, and put on before work. Contact lenses should not be worn when working with this chemical. Wear splash-proof chemical goggles and face shield unless full facepiece respiratory protection is worn. Employees should wash immediately with soap when skin is wet or contaminated. Provide emergency showers and eyewash.

Respirator Selection: NIOSH: *20 ppm:* SA (any supplied-air respirator);* *50 ppm:* SA:CF (any supplied-air respirator operated in a continuous-flow mode;* or PAPROV [any powered, air-purifying respirator with organic vapor cartridge(s)].* *100 ppm:* CCRFOV [any chemical cartridge respirator with a full facepiece and organic vapor cartridge(s)]; or GMFOV [any air-purifying, full-facepiece respirator (gas mask) with a chin-style, front- or back-mounted organic vapor canister]; or PAPRTOV [any powered, air-purifying respirator with a tight-fitting facepiece and organic vapor cartridge(s)];* or SCBAF (any self-contained breathing apparatus with a full facepiece); or SAF (any supplied-air respirator with a full facepiece). *Emergency or planned entry into unknown concentrations or IDLH conditions:* SCBAF:PD,PP (any self-contained breathing apparatus that has a full facepiece and is operated in a pressure-demand or other positive-pressure mode); or SAF:PD,PP:ASCBA (any supplied-air respirator that has a full facepiece and is operated in a pressure-demand or other positive-pressure mode in combination with an auxiliary self-contained breathing apparatus operated in a pressure-demand or other positive-pressure mode). *Escape:* GMFOV [any air-purifying, full-facepiece respirator (gas mask) with a chin-style, front-or back-mounted organic vapor canister] or SCBAE (any appropriate escape-type, self-contained breathing apparatus).

* Substance reported to cause eye irritation or damage; requires eye protection.

Storage: Prior to working with 1-Chloro-1-Nitropropane you should be trained on its proper handling and storage. Before entering confined space where 1-Chloro-1-Nitropropane may be present, check to make sure that an explosive concentration does not exist. Store to avoid contact with strong oxidizers, (such as chlorine, bromine, and fluorine) since violent reactions occur. Store in tightly closed containers in a cool, well-ventilated area away from heat. Metal containers involving the transfer of this chemical should be grounded and bonded. Where possible, automatically pump liquid from drums or other storage containers to process containers. Drums must be equipped with self-closing valves, pressure vacuum bungs, and flame arresters. Use only non-sparking tools and equipment, especially when opening and closing containers of this chemical. Sources of ignition such as smoking and open flames, are prohibited where this chemical is used, handled, or stored in a manner that could create a potential fire or explosion hazard.

Shipping: 1-Chloro-1nitropropane is not specifically cited by DOT[19] in its performance-oriented packaging standards.

Spill Handling: Evacuate and restrict persons not wearing protective equipment from area of spill or leak until cleanup is complete. Remove all ignition sources. Ventilate area of spill or leak. Absorb liquids in vermiculite, dry sand, earth, peat, carbon, or a similar material and deposit in sealed containers. It may be necessary to contain and dispose of this chemical as a hazardous waste. If material or contaminated runoff enters waterways, notify downstream users of potentially contaminated waters. Contact your Department of Environmental Protection or your regional office of the federal EPA for specific recommendations. If employees are required to clean-up spills, they must be properly trained and equipped. OSHA 1910.120(q) may be applicable.

Fire Extinguishing: This chemical is a flammable. Poisonous gases including nitrogen oxides and chlorine are produced in fire. Use dry chemical, carbon dioxide, or foam extinguishers. Vapors are heavier than air and will collect in low areas. Vapors may travel long distances to ignition sources and flashback. Vapors in confined areas may explode when exposed to fire. Containers may explode in fire. Storage containers and parts of containers may rocket great distances, in many directions. If material or contaminated runoff enters waterways, notify downstream users of potentially contaminated waters. Notify local health and fire officials and pollution control agencies. From a secure, explosion-proof location, use water spray to cool exposed containers. If cooling streams are ineffective (venting sound increases in volume and pitch, tank discolors, or shows any signs of deforming), withdraw immediately to a secure position. If employees are expected to fight fires, they must be trained and equipped in OSHA 1910.156.

Disposal Method Suggested: Incineration (1,500°F, 0.5-second minimum for primary combustion; 2,200°F, 1.0 second for secondary combustion) after mixing with other fuel. The formation of elemental chlorine may be prevented by injection of steam or using methane as a fuel in the process. Alternatively it may be poured over soda ash, neutralized and flushed into the sewer with large volumes of water.

References

New Jersey Department of Health and Senior Services and Senior Services, "Hazardous Substance Fact Sheet: 1-Chloro-1-Nitropropane," Trenton, NJ (February 1989)

Chloropentafluoroethane

Molecular Formula: C_2ClF_5

Common Formula: ClF_2CCF_3

Synonyms: CFC-115; Chloropentafluoroethane; 1-Chloro-1,1,2,2,2-pentafluoromethane; Ethane, Chloropentafluoro-; F-115; FC 115; Flurocarbon 115; Freon 115; Genetron 115; Halocarbon 115; HCFC-115; Monocloropentafluoetano (Spanish); Pentafluoromonochloroethane; Propellent 115; R 115; Refrigerant 115

CAS Registry Number: 76-15-3

RTECS Number: KH7877500

DOT ID: UN 1020

Regulatory Authority

- Air Pollutant Standard Set (ACGIH)[1] (Australia) (HSE) (Israel) (UNEP)[43] (OSHA)[58] (Several States)[60] (Several Canadian Provinces)
- CLEAN AIR ACT: Stratospheric ozone protection (Title VI, Subpart A, Appendix A), Class I, Ozone Depletion Potential = 0.6
- EPCRA Section 313 Form R *de minimis* concentration reporting level: 1.0%
- Canada, WHMIS, Ingredients Disclosure List

Cited in U.S. State Regulations: Alaska (G), California (A, G), Connecticut (A), Florida (G), Illinois (G), Massachusetts (G), Minnesota (G), Nevada (A), New Hampshire (G), New Jersey (G), North Dakota (A), Pennsylvania (G), Rhode Island (G), Virginia (A), West Virginia (G).

Description: Chloropentafluoroethane, ClF_2CCF_3, is a colorless, odorless gas with and ethereal odor. Shipped as a liquefied compressed gas. Boiling point = -39°C. Freezing/Melting point = -142°C. Practically insoluble in water.

Potential Exposure: This material is used as a refrigerant, as a dielectric gas, and as a propellant in aerosol food preparations.

Incompatibilities: Keep away from strong oxidizers, strong bases (alkalis), alkaline earth metals (e.g., aluminum powder, sodium, potassium, zinc), and beryllium. Keep away from open flames or temperatures above 52°C (125°F); decomposes forming toxic fumes including hydrogen chloride and hydrogen fluoride.

Permissible Exposure Limits in Air: There is no OSHA PEL. The NIOSH and ACGIH recommended limit is1,000 ppm (6,320 mg/m^3) TWA. Australia, HSE (UK) Israel and the Canadian Provinces of Alberta, Ontario, Quebec have all set a limit of 1,000 ppm TWA, and Alberta's STEL (15 min) is 1,250 ppm. The former USSR/UNEP joint project[43] has set a MAC in workplace air of 3,000 mg/m^3. Several states have set guidelines or standards for FC-115 in ambient air[60] ranging from 0.1264 μg/m^3 (Connecticut) to 9,999 μg/m^3 (Virginia) to 63,200 μg/m^3 (North Dakota) to 151,000 μg/m^3 (Nevada).

Determination in Air: No Method available.

Permissible Concentration in Water: No criteria set.

Routes of Entry: Inhalation, ingestion.

Harmful Effects and Symptoms

Short Term Exposure: Chloropentafluoroethane can affect you when breathed in. Irritates the nose, throat and lungs causing coughing, chest tightness, wheezing, and shortness of breath. High levels can cause you to feel dizzy, lightheaded and to pass out. Very high levels can cause death. Chloropentafluoroethane may irritate the skin causing a rash or burning feeling on contact. Exposure may affect the heart, causing irregular heartbeat, which could lead to death. Rapid evaporation of liquid chloropentafluoroethane may cause frostbite of the eyes and skin.

Long Term Exposure: Similar very irritating substances can cause lung damage.

Points of Attack: Skin, central nervous system, cardiovascular system.

Medical Surveillance: If symptoms develop or overexposure is suspected, the following may be useful: Special 24 hour EKG (Holter monitor) to look for irregular heartbeat. Lung function tests.

First Aid: If this chemical gets into the eyes, remove any contact lenses at once and irrigate immediately for at least 15 minutes, occasionally lifting upper and lower lids. Seek medical attention immediately. If this chemical contacts the skin, remove contaminated clothing and wash immediately with soap and water. Seek medical attention immediately. If this chemical has been inhaled, remove from exposure, begin rescue breathing (using universal precautions) if breathing has stopped and CPR if heart action has stopped. Transfer promptly to a medical facility. When this chemical has been swallowed, get medical attention. Give large quantities of water and induce vomiting. Do not make an unconscious person vomit.

If frostbite has occurred, seek medical attention immediately; do *NOT* rub the affected areas or flush them with water. In order to prevent further tissue damage, do *NOT* attempt to remove frozen clothing from frostbitten areas. If frostbite has *NOT* occurred, immediately and thoroughly wash contaminated skin with soap and water.

Personal Protective Methods: Wear protective gloves and clothing to prevent any reasonable probability of skin contact. Safety equipment suppliers/manufacturers can provide recommendations on the most protective glove/clothing material for your operation. All protective clothing (suits, gloves, footwear, headgear) should be clean, available each day, and put on before work. Contact lenses should not be worn when working with this chemical. Wear splash-proof chemical goggles and face shield unless full facepiece respiratory protection is worn. Employees should wash immediately with soap when skin is wet or contaminated. Provide emergency showers and eyewash. Where exposure to the liquefied compressed gas may occur, employees should be provided with special clothing designed to prevent frostbite.

Respirator Selection: Where the potential exists for exposures over 1,000 ppm, use a MSHA/NIOSH approved supplied-air respirator with a full facepiece operated in the positive pressure mode or with a full facepiece, hood, or helmet in the continuous flow mode, or use a MSHA/NIOSH approved self-contained breathing apparatus with a full facepiece operated in pressure-demand or other positive pressure mode.

Storage: Prior to working with Chloropentafluoroethane you should be trained on its proper handling and storage. Before entering confined space where this chemical may be present, check to make sure that sufficient oxygen (19%) exists. Store in tightly closed containers in a cool, well-ventilated area away from metals, including aluminum, zinc, and beryllium, and from open flames or temperatures above 52°C (125°F). Procedures for the handling, use and storage of cylinders should be in compliance with OSHA 1910.101 and 1910.169 as with the recommendations of the Compressed Gas Association.

Shipping: Chloropentafluoroethane must be labeled "Nonflammable Gas." It falls in Hazard Class 2.2.[19]

Spill Handling: Evacuate and restrict persons not wearing protective equipment from area of spill or leak until cleanup is complete. Remove all ignition sources. Ventilate area of spill or leak to disperse the gas. Stop the flow of gas if it can be done safely. If source is a cylinder and the leak cannot be stopped in place, remove the leaking cylinder to a safe place, and repair leak or allow cylinder to empty. Absorb liquids in vermiculite, dry sand, earth, or a similar material and deposit in sealed containers. If employees are required to clean up spills, they must be properly trained and equipped. OSHA 1910.120(q) may be applicable.

Fire Extinguishing: Chloropentafluoroethane may burn, but does not readily ignite. Poisonous gases are produced in fire, including fluorides, chlorides, phosgene and acid gases. Use dry chemical or carbon dioxide extinguishers. If material or contaminated runoff enters waterways, notify downstream users of potentially contaminated waters. Notify local health and fire officials and pollution control agencies. Containers may explode in fire. From a secure, explosion-proof location, use water spray to cool exposed containers. If cooling streams are ineffective (venting sound increases in volume and pitch, tank discolors, or shows any signs of deforming), withdraw immediately to a secure position. If employees are expected to fight fires, they must be trained and equipped in OSHA 1910.156.

References

New Jersey Department of Health and Senior Services and Senior Services, "Hazardous Substance Fact Sheet: Chloropentafluoroethane," Trenton, NJ (May 1998)

Chlorophacinone

Molecular Formula: $C_{23}H_{15}ClO_3$

Synonyms: Afnor; Caid; Chlorfacinon (German); 2-(α-*p*-Chlorophenylacetyl)indane-1,3-dione; [(4-Chlorophenyl)-1-phenyl]-acetyl-1,3-indandion (German); 2-[(*p*-Chlorophenyl)phenylacetyl]-1,3-indandione; 2[2-(4-Chlorophenyl)-2-phenylacetyl]indan-1,3-dione; 2[(4-Chlorophenyl)phenylacetyl)-1H-indene-1,3(2H)-dione; Chlorphacinon (Italian); 1-(4-Chlorphenyl)-1-phenyl-acetyl indan-1,3-dion (German); Cloorfacinon (Dutch); 2[2-(4-Cloor-fenyl-2-fenyl)-acetyl]-indaan-1,3-dion (Dutch); Clorofacinona (Spanish); Clorphacinon (Italian); Delta; Drat; 1,3-Indandione, 2-[(*p*-chlorophenyl)phenylacetyl]-; 1H-Indene-1,3(2H)-dione, 2-[(4-Chlorophenyl)phenylacetyl]-; Liphadione; LM 91; Microzul; Muriol; 2-[2-Phenyl-2-(4-chlorophenyl)acetyl]-1,3-indandione; Quick; Ramucide; Ranac; Ratomet; Raviac; Rozol; Topitox

CAS Registry Number: 3691-35-8

RTECS Number: NK5335000

DOT ID: No citation.

Regulatory Authority

- SUPERFUND/EPCRA 40CFR355, Appendix B Extremely Hazardous Substances: TPQ = 100/10,000 lb (45.4/4,540 kg)
- SUPERFUND/EPCRA 40CFR302.4 Reportable Quantity (RQ): EHS, 1 lb (0.454 kg)

Cited in U.S. State Regulations: Florida (G), Massachusetts (G), New Jersey (G), Pennsylvania (G).

Description: 2-[(p-chlorphenyl)phenyl-actyl]-1,3-indandione, $C_{23}H_{15}ClO_3$, is a crystalline solid. Freezing/Melting point = 140°C.

Potential Exposure: This material is an anticoagulant rodenticide. Those involved in its manufacture, formulation and application are at risk.

Permissible Exposure Limits in Air: No standards set.

Permissible Concentration in Water: No criteria set.

Routes of Entry: Ingestion and skin contact.

Harmful Effects and Symptoms

Short Term Exposure: Contact may cause burns to skin and eyes. Symptoms of exposure are similar to those of warfarin. Symptoms develop after a few days or a few weeks or repeated ingestion and include nosebleed and bleeding gums; pallor and sometimes a rash; massive bruises, especially of the elbow, knees, and buttocks; blood in urine and feces; occasionally paralysis from cerebral hemorrhage; and hemorrhagic shock and death.

Chlorophacinone is highly toxic orally and by skin adsorption. The probable oral lethal dose for humans is less than 5 mg/kg to 50 mg/kg, or between a taste (less than 7 drops) and 1 teaspoonful for a 150 lb (70 kg) person. The LD_{50} oral (mouse) is 1.06 mg/kg.

Long Term Exposure: See above.

Points of Attack: Blood, cardiovascular system.

First Aid: If this chemical gets into the eyes, remove any contact lenses at once and irrigate immediately for at least 15 minutes, occasionally lifting upper and lower lids. Seek medical attention immediately. If this chemical contacts the skin, remove contaminated clothing and wash immediately with soap and water. Seek medical attention immediately. If this chemical has been inhaled, remove from exposure, begin rescue breathing (using universal precautions) if breathing has stopped and CPR if heart action has stopped. Transfer promptly to a medical facility. When this chemical has been swallowed, get medical attention. Give large quantities of water and induce vomiting. Do not make an unconscious person vomit.

Personal Protective Methods: Wear protective gloves and clothing to prevent any reasonable probability of skin contact. Safety equipment suppliers/manufacturers can provide recommendations on the most protective glove/clothing material for your operation. All protective clothing (suits, gloves, footwear, headgear) should be clean, available each day, and put on before work. Contact lenses should not be worn when working with this chemical. Wear dust-proof chemical goggles and face shield unless full facepiece respiratory protection is worn. Employees should wash immediately with soap when skin is wet or contaminated. Provide emergency showers and eyewash.

Respirator Selection: SCBAF:PD,PP (any MSHA/NIOSH approved self-contained breathing apparatus that has a full facepiece and is operated in a pressure-demand or other positive-pressure mode); or SAF:PD,PP:ASCBA (any supplied-air respirator that has a full facepiece and is operated in a pressure-demand or other positive-pressure mode in combination with an auxiliary, self-contained breathing apparatus operated in a pressure-demand or other positive pressure mode).

Storage: Prior to working with this chemical you should be trained on its proper handling and storage. Store in tightly closed containers in a cool, well-ventilated area.

Shipping: Coumarin derivative pesticides, solid, toxic, n.o.s., require a "Poison" label in packing Group I. The UN/DOT Hazard Class is 6.1.[19][20]

Spill Handling: Evacuate persons not wearing protective equipment from area of spill or leak until clean up is complete. Remove all ignition sources. Collect powdered material in the most convenient and safe manner and deposit in sealed containers. Ventilate area after clean up is complete. It may be necessary to contain and dispose of this chemical as a hazardous waste. If material or contaminated runoff enters waterways, notify downstream users of potentially contaminated waters. Contact your Department of Environmental Protection or your regional office of the federal EPA for specific recommendations. If employees are required to clean-up spills, they must be properly trained and equipped. OSHA 1910.120(q) may be applicable.

Fire Extinguishing: Small fires: dry chemicals, carbon dioxide, water spray or foam. Large fires: water spray, fog or foam. Move container form fire area if you can do so without risk. Fight fire from maximum distance. Dike fire control water for later disposal; do not scatter the material. Poisonous gases are produced in fire. If material or contaminated runoff enters waterways, notify downstream users of potentially contaminated waters. Notify local health and fire officials and pollution control agencies. Containers may explode in fire. From a secure, explosion-proof location, use water spray to cool exposed containers. If cooling streams are ineffective (venting sound increases in volume and pitch, tank discolors, or shows any signs of deforming), withdraw immediately to a secure position. If employees are expected to fight fires, they must be trained and equipped in OSHA 1910.156.

Disposal Method Suggested: Incineration at high temperature with effluent gas scrubbing.[22]

References

U.S. Environmental Protection Agency, "Chemical Profile: Chlorophacinone," Washington, DC, Chemical Emergency Preparedness Program (October 31, 1985)

Chlorophenols, Mono

Molecular Formula: C_6H_5ClO

Common Formula: C_6H_4ClOH

Synonyms: *m-: m*-Chlorophenate; *m*-Chlorophenol; *m*-Clorofenol (Czech, Spanish)

o-: o-Chlorophenol; *o*-Chlorphenol (German); *o*-Clorofenol (Czech, Spanish); Phenol, *o*-chloro; Phenol, *o*-chloro-; Phenol, 2-chloro; Phenol, 2-chloro-

p-: p-Chlorfenol (Czech, Spanish); *p*-Chlorophenate; *p*-Chlorophenol; Parachlorophenol

CAS Registry Number: 95-57-8 (o-); 108-43-0 (m-); 106-48-9 (p-); 25167-80-0 (mixed isomers)

RTECS Number: SK2625000 (o-); SK2450000 (m-); SK2800000 (p-)

DOT ID: UN 2020 (solid) (m- and p- isomers); UN 2021 (liquid) (o-isomer)

EEC Number: 604-008-00-0

Regulatory Authority

- Air Pollutant Standard Set (former USSR)[35] (Sweden)[35] (UNEP)[43]
- CLEAN WATER ACT: 40CFR423, Appendix A, Priority Pollutants; Section 313 Water Priority Chemicals (57FR41331, 9/9/92); Toxic Pollutant (Section 401.15)

o-isomer:

- SUPERFUND/EPCRA 40CFR302.4 Reportable Quantity (RQ): CERCLA, 100 lb (45.4 kg) (o-)
- EPA HAZARDOUS WASTE NUMBER (RCRA No.): U048 (o-)
- RCRA, 40CFR261, Appendix 8 Hazardous Constituents (o-)
- RCRA 40CFR268.48; 61FR15654, Universal Treatment Standards: Wastewater (mg/l), 0.44; Nonwastewater (mg/kg), 5.7 (o-)
- RCRA 40CFR264, Appendix 9; TSD Facilities Ground Water Monitoring List. Suggested test method(s) (PQL µg/l): 8040 (5); 8270 (10) (o-)
- EPCRA Section 313 Form R *de minimis* concentration reporting level: 0.1%. (chlorophenols)
- U.S. DOT Regulated Marine Pollutant (49CFR172.101, Appendix B) (chlorophenols)
- Canada, WHMIS, Ingredients Disclosure List (all isomers)
- Mexico, Drinking Water Criteria: 0.03 mg/l (o- isomer)

Cited in U.S. State Regulations: California (A, G), Florida (G), Illinois (G), Kansas (G, W), Louisiana (G), Maryland (G), Massachusetts (G), Minnesota (G), New Hampshire (G), New Jersey (G), New York (G), Pennsylvania (G), Rhode Island (G), Vermont (G), Virginia (G), Washington (G), Wisconsin (G).

Description: Monochlorophenols, C_6H_4ClOH: 2-chloro- is a colorless (in pure state) to pink or amber (technical grade, due to impurities) crystalline solid or liquid. 3-chloro- is a white crystalline (needle-like) solid. 4-chloro- is a white to straw-colored, needle-like crystalline solid. All have a characteristic odor. Odor threshold = 1.24 ppm (pure).

Isomer	Melting Point, °C	Boiling Point, °C	Flash Point, °C	Odor threshold, ppm	Solubility in H_2O
o-	9	175	64	—	slight
m-	33	214	>112	—	moderate
p-	43	220	121	30	soluble

o-isomer: Hazard Identification (based on NFPA-704 M Rating System): Health 3, Flammability 2, Reactivity 0.

p-isomer: Hazard Identification (based on NFPA-704 M Rating System): Health 3, Flammability 1, Reactivity 0.

Potential Exposure: Monochlorophenols are used in the manufacture of fungicides, slimicides, bactericides, pesticides, herbicides, disinfectants, wood and glue preservatives; in the production of phenolic resins; in the extraction of certain minerals from coal; as a denaturant for ethanol; as an antiseptic; as a disinfectant, and others.

Incompatibilities: Forms explosive mixture with air. Contact with oxidizing agents can cause fire and explosion hazard. Heat produces hydrogen chloride and chlorine. Corrosive to aluminum, copper and other chemically active metals.

Permissible Exposure Limits in Air: These chemicals can be absorbed through the skin, thereby increasing exposure. The former USSR/UNEP joint project[43] has set a MAC in workplace air of 1.0 mg/m^3. Sweden[35] has set a TWA in workplace air of 0.5 mg/m^3 for all 3 isomers in workplace air with a STEL of 1.5 mg/m^3. The former Soviet Union[35] has set limits in ambient air in residential areas of 0.01 mg/m^3 for the *m-* and *p-*chlorphenols and 0.02 mg/m^3 for 2-chlorophenol.

Determination in Air: Use NIOSH: (*o*-chlorophenol) P&CAM Method #337.

Permissible Concentration in Water: To protect freshwater aquatic life: 4,380 µg/l on an acute toxicity basis. To protect saltwater aquatic life: no criteria set due to insufficient data. To protect human health: 0.1 µg/l for the prevention of adverse effects due to organoleptic properties.[6]

The State of Kansas has set guidelines for chlorophenols in drinking water[61] of 0.1 µg/l for the *o*-isomer and 0.3 µg/l for the para-isomer.

Determination in Water: Gas chromatography (EPA Method 604) or gas chromatography plus mass spectrometry (EPA Method 625).

Routes of Entry: Skin absorption, inhalation, ingestion, skin and/or eye contact.

Harmful Effects and Symptoms

Short Term Exposure: Inhalation can cause severe irritation, burns to the nose and throat, headache, dizziness, vomiting, lung damage, muscle twitchings, spasms, tremors, weakness, staggering and collapse. Skin contact can cause severe irritation and burns. Can be absorbed through the skin to cause or increase the severity of symptoms listed above. Eye contact causes severe irritation. May cause burns. Ingestion can cause irritation, burns to the mouth and throat, low blood pressure, profuse sweating, intense thirst, nausea, abdominal pain, stupor, vomiting, red blood cell damage and accumulation of fluid in the lungs followed by pneumonia. May also cause restlessness and increased breathing rate followed by rapidly developing muscle weakness. Tremors, convulsions and coma can promptly set in and will continue until death. Based on animal studies, the estimated lethal dose is between one teaspoon and one ounce for a 150 pound adult. *p*-isomer: The substance irritates the eyes, the skin and the respiratory tract. The substance may cause effects on the central nervous system and bladder.

Long Term Exposure: Skin sensitivity may develop. May have effects on the blood, heart, liver, lung, kidney. The state of New Jersey lists 2-chloro- a probable carcinogen in humans, and that it causes leukemia and soft tissue cancers in humans.

Points of Attack: Eyes, skin, respiratory system.

Medical Surveillance: If symptoms develop or overexposure is suspected, the following may be useful: Liver function tests. Kidney function tests. Examination by a qualified allergist. EKG.

First Aid: If this chemical gets into the eyes, remove any contact lenses at once and irrigate immediately for at least 15 minutes, occasionally lifting upper and lower lids. Seek medical attention immediately. If this chemical contacts the skin, remove contaminated clothing and wash immediately with soap and water. Seek medical attention immediately. If this chemical has been inhaled, remove from exposure, begin rescue breathing (using universal precautions) if breathing has stopped and CPR if heart action has stopped. Transfer promptly to a medical facility. When this chemical has been swallowed, get medical attention. Give large quantities of water and induce vomiting. Do not make an unconscious person vomit.

Personal Protective Methods: Wear protective gloves and clothing to prevent any reasonable probability of skin contact. Safety equipment suppliers/manufacturers can provide recommendations on the most protective glove/clothing material for your operation. All protective clothing (suits, gloves, footwear, headgear) should be clean, available each day, and put on before work. Contact lenses should not be worn when working with this chemical. Wear splash-proof chemical goggles and face shield when working with liquid, unless full facepiece respiratory protection is worn. Wear dust-proof goggles when working with powders or dust, unless full facepiece respiratory protection is worn. Employees should wash immediately with soap when skin is wet or contaminated. Provide emergency showers and eyewash.

Respirator Selection: Where the potential exists for exposure to chlorophenols, use a MSHA/NIOSH approved full facepiece respirator with a high efficiency particulate filter. Greater protection is provided by a powered-air purifying respirator. Where the potential for high exposure to Chlorophenols exists, or to liquid form, use a MSHA/NIOSH approved supplied-air respirator with a full facepiece operated in the positive pressure mode or with a full facepiece, hood, or helmet in the continuous flow mode, or use a MSHA/NIOSH approved self-contained breathing apparatus with a full facepiece operated in pressure-demand or other positive pressure mode.

Storage: Prior to working with chlorophenols you should be trained on its proper handling and storage. Before entering confined space where chlorophenols may be present, check to make sure that an explosive concentration does not exist. Store in tightly closed containers in a cool, well ventilated area. Metal containers involving the transfer of this chemical should be grounded and bonded. Where possible, automatically pump liquid from drums or other storage containers to process containers. Drums must be equipped with self-closing valves, pressure vacuum bungs, and flame arresters. Use only non-sparking tools and equipment, especially when opening and closing containers of this chemical. Sources of ignition such as smoking and open flames, are prohibited where this chemical is used, handled, or stored in a manner that could create a potential fire or explosion hazard. A regulated, marked area should be established where this chemical is handled, used, or stored in compliance with OSHA standard 1910.1045.

Shipping: Chlorophenols must carry a "Keep Away from Food" label.[19] They fall in Hazard Class 6.1 and Packing Group III.

Spill Handling: Evacuate and restrict persons not wearing protective equipment from area of spill or leak until cleanup is complete. Remove all ignition sources. Ventilate area of spill or leak. Absorb liquids in vermiculite, dry sand, earth, peat, carbon, or a similar material and deposit in sealed containers. Collect powdered material in the most convenient and safe manner and deposit in sealed containers. It may be necessary to contain and dispose of this chemical as a hazardous waste. If material or contaminated runoff enters waterways, notify downstream users of potentially contaminated waters. Contact your Department of Environmental Protection or your regional office of the federal EPA for specific recommendations. If employees are required to clean-up spills, they must be properly trained and equipped. OSHA 1910.120(q) may be applicable.

Fire Extinguishing: 2-Chlorophenol is a combustible liquid/solid. 3- and 4- Chlorophenols may burn, but do not readily ignite. Poisonous gases including phenols and chlorides are produced in fire. Use dry chemical, carbon dioxide, or foam extinguishers. Vapors are heavier than air and will collect in low areas. Vapors from 2-Chloro-may travel long distances to ignition sources and flashback. Vapors in confined areas may explode when exposed to fire. Containers may explode in fire. Storage containers and parts of containers may rocket great distances, in many directions. If material or contaminated runoff enters waterways, notify downstream users of potentially contaminated waters. Notify local health and fire officials and pollution control agencies. From a secure, explosion-proof location, use water spray to cool exposed containers. If cooling streams are ineffective (venting sound increases in volume and pitch, tank discolors, or shows any signs of deforming), withdraw immediately to a secure position. If employees are expected to fight fires, they must be trained and equipped in OSHA 1910.156.

Disposal Method Suggested: Incinerate in admixture with flammable solvent in furnace equipped with afterburner and scrubber.[22]

References

U.S. Environmental Protection Agency, 2-Chlorophenol: Ambient Water Quality Criteria, Washington, DC (1980)

U.S. Environmental Protection Agency, 2-Chloropehnol, Health and Environmental Effects Profile No. 50, Office of Solid Waste, Washington, DC (April 30, 1980)

Sax, N. I.., Ed., "Dangerous Properties of Industrial Materials Report," 2, No. 6, 48–51 (1982) and 4, No. 6, 88–94 (1986) (2-Chlorophenol); 2, No. 6, 46–48 (1982) and 6, No. 5, 70–74 (1986) (3-Chlorophenol); 2, No. 6, 52–55 (1982) and 6, No. 5, 74–81 (1986) (4-Chlorophenol)

New Jersey Department of Health and Senior Services and Senior Services, "Hazardous Substance Fact Sheet: 2-Chlorophenol," Trenton, NJ (May 1986)

New Jersey Department of Health and Senior Services and Senior Services, "Hazardous Substance Fact Sheet: 3-Chlorophenol," Trenton, NJ (April 1986)

New Jersey Department of Health and Senior Services and Senior Services, "Hazardous Substance Fact Sheet: 4-Chloophenol," Trenton, NJ (April 1986)

New York State Department of Health, "Chemical Fact Sheet: Chlorophenols," Albany, NY, Bureau of Toxic Substance Assessment (May 1986)

New York State Department of Health, "Chemical Fact Sheet: 2-Chlorophenol," Albany, NY, Bureau of Toxic Substance Assessment (March 1986)

4-Chloro-o-Phenylenediamine

Molecular Formula: $C_6H_7ClN_2$

Common Formula: $ClC_6H_3(NH_2)_2$

Synonyms: 2-Amino-4-chloroaniline; 4-Chloro-1,2-benzenediamine; 4-Chloro-1,2-diaminobenzene; *p*-Chloro-*o*-phenylenediamine; 4-Chloro-1,2-phenylenediamine; 4-Cl-*o*-Pd; 1,2-Diamino-4-chlorobenzene; 3,4-Diamino-1-chlorobenzene; 3,4-Diaminochlorobenzene; NCI-C03292; Ursol olive 6G

CAS Registry Number: 95-83-0

RTECS Number: SS8850000

DOT ID: UN 9188

Regulatory Authority

- Carcinogen (animal sufficient data) (IARC) (animal positive) (NTP) (NCI) (EPA)[9]
- Air Pollutant Standard Set (North Dakota)[60]

Cited in U.S. State Regulations: California (A, G), Florida (G), Illinois (G), Massachusetts (G), Michigan (G), Minnesota (G), New Jersey (G), North Dakota (A), Pennsylvania (G).

Description: 4-Chloro-o-phenylenediamine, $ClC_6H_3(NH_2)_2$, is a brown crystalline powder or leaflets from water. Freezing/Melting point = 67 – 70°C. Also reported at 76°C.

Potential Exposure: This material has been patented as a hair dye component. It is believed to be used in production of photographic chemicals.

Incompatibilities: Strong oxidizers.

Permissible Exposure Limits in Air: North Dakota has set a guideline for ambient air of zero concentration.[60]

Permissible Concentration in Water: No criteria set.

Harmful Effects and Symptoms

Short Term Exposure: Can cause irritation of eyes, nose, skin and mucous membranes.

Long Term Exposure: There is sufficient evidence that 4-chloro-o-phenylenediamine is carcinogenic in experimental animals. In long-term feeding bioassays with technical grade 4-chloro-o-phenylenediamine, rats developed tumors of the urinary bladder and fore stomach. In mice, the compound induced hepatocellular carcinomas.

First Aid: If this chemical gets into the eyes, remove any contact lenses at once and irrigate immediately for at least 20 minutes, occasionally lifting upper and lower lids. Seek medical attention immediately. If this chemical contacts the skin, remove contaminated clothing and wash immediately with soap and water. Seek medical attention immediately. If this chemical has been inhaled, remove from exposure, begin rescue breathing (using universal precautions) if breathing has stopped and CPR if heart action has stopped. Transfer promptly to a medical facility. When this chemical has been swallowed, get medical attention. Give large quantities of water and induce vomiting. Do not make an unconscious person vomit.

Personal Protective Methods: Wear protective gloves and clothing to prevent any reasonable probability of skin contact. Safety equipment suppliers/manufacturers can provide recommendations on the most protective glove/clothing material for your operation. All protective clothing (suits, gloves, footwear, headgear) should be clean, available each day, and put on before work. Contact lenses should not be worn when working with this chemical. Wear dust-proof chemical goggles and face shield unless full facepiece respiratory protection is worn. Employees should wash immediately with soap when skin is wet or contaminated. Provide emergency showers and eyewash.

Respirator Selection: *At any concentrations above the NIOSH REL, or where there is no REL, at any detectable concentration:* SCBAF:PD,PP (any MSHA/NIOSH approved self-contained breathing apparatus that has a full facepiece and is operated in a pressure-demand or other positive-pressure mode); or SAF:PD,PP:ASCBA (any supplied-air respirator that has a full facepiece and is operated in a pressure-demand or other positive-pressure mode in combination with an auxiliary, self-contained breathing apparatus operated in a pressure-demand or other positive pressure mode). *Escape:* GMFOVHiE [any air-purifying, full-facepiece respirator (gas mask) with a chin-style, front- or back-mounted organic vapor canister having a high-efficiency particulate filter]; or SCBAE (any appropriate escape-type, self-contained breathing apparatus).

Storage: Prior to working with this chemical you should be trained on its proper handling and storage. A regulated, marked area should be established where this chemical is handled, used, or stored in compliance with OSHA standard 1910.1045. Store in tightly closed containers in a cool, well-ventilated place or a refrigerator.

Shipping: While not specifically cited by DOT, this may be classed as a hazardous substance, solid, n.o.s. Such materials fall in Hazard Class ORM-E and Packing Group III. There are no label requirements nor are there maximum limits on shipping quantities.

Spill Handling: Evacuate persons not wearing protective equipment from area of spill or leak until clean-up is complete. Remove all ignition sources. Ventilate area of spill. Dampen spilled material with alcohol to avoid dust, then transfer material to a suitable container. Use absorbent paper dampened with alcohol to pick up remaining material. Wash surfaces well with soap and water. It may be necessary to contain and dispose of this chemical as a hazardous waste. If material or contaminated runoff enters waterways, notify downstream users of potentially contaminated waters. Contact your Department of Environmental Protection or your regional office of the federal EPA for specific recommendations. If employees are required to clean-up spills, they must be properly trained and equipped. OSHA 1910.120(q) may be applicable.

Fire Extinguishing: Use dry chemical, carbon dioxide, water spray, or foam extinguishers. Poisonous gases are produced in fire. If material or contaminated runoff enters waterways, notify downstream users of potentially contaminated waters. Notify local health and fire officials and pollution control agencies. From a secure, explosion-proof location, use water spray to cool exposed containers. If cooling streams are ineffective (venting sound increases in volume and pitch, tank discolors, or shows any signs of deforming), withdraw immediately to a secure position. If employees are expected to fight fires, they must be trained and equipped in OSHA 1910.156.

Chlorophenyltrichlorosilane

Molecular Formula: $C_6H_4Cl_4Si$

Common Formula: $ClC_6H_4SiCl_3$

Synonyms: Chlorofeniltriclorosilano (Spanish); Chlorophenyl trichlorosilane; Trichloro (chlorophenyl)silane

CAS Registry Number: 26571-79-9

RTECS Number: VV2650000

DOT ID: UN 1753

Regulatory Authority

- U.S. DOT Regulated Marine Pollutant (49CFR172.101, Appendix B)
- Canada, WHMIS, Ingredients Disclosure List

Cited in U.S. State Regulations: Maine (G), New Hampshire (G), New Jersey (G).

Description: Chlorophenyltrichlorosilane, $ClC_6H_4SiCl_3$, is a combustible, colorless to pale yellow liquid. Boiling point = 230°C. Flash point = 125°C. Decomposes in water.

Potential Exposure: Chlorophenyltrichlorosilane is used as an intermediate for silicones manufacture.

Incompatibilities: Decomposes in water or on contact with steam or other forms of moisture producing hydrogen chloride. Attacks metals in the presence of moisture. Some chlorosilanes can self-ignite in air. Contact with ammonia, forms a self-igniting product. Keep away from combustible materials such as wood, paper and oil.

Permissible Exposure Limits in Air: No standards set.

Permissible Concentration in Water: No criteria set.

Routes of Entry: Skin contact, inhalation.

Harmful Effects and Symptoms

Short Term Exposure: Chlorophenyltrichlorosilane can affect you when breathed in. Chlorophenyltrichlorosilane is a corrosive chemical. Contact can cause severe eye and skin burns. Inhalation can cause irritation of the lungs causing coughing and/or shortness of breath. High exposures can cause pulmonary edema, a medical emergency which can be delayed for several hours. This can cause death.

Long Term Exposure: Can cause lung irritation; bronchitis may develop.

Points of Attack: Eyes, skin, respiratory system.

Medical Surveillance: Before beginning employment and at regular times after that, for those with frequent or potentially high exposures, the following is recommended: Lung function tests. Consider chest x-ray following acute overexposure.

First Aid: If this chemical gets into the eyes, remove any contact lenses at once and irrigate immediately for at least 30 minutes, occasionally lifting upper and lower lids. Seek medical attention immediately. If this chemical contacts the skin, remove contaminated clothing and wash immediately with soap and water. Seek medical attention immediately. If this chemical has been inhaled, remove from exposure, begin rescue breathing (using universal precautions) if breathing has stopped and CPR if heart action has stopped. Transfer promptly to a medical facility. When this chemical has been swallowed, get medical attention. If victim is conscious, administer water or milk. Do not induce vomiting. Medical observation is recommended for 24 – 48 hours after breathing overexposure, as pulmonary edema may be delayed. As first aid for pulmonary edema, a doctor or authorized paramedic may consider administering a corticosteroid spray.

Personal Protective Methods: Wear protective gloves and clothing to prevent any reasonable probability of skin contact. Safety equipment suppliers/manufacturers can provide recommendations on the most protective glove/clothing material for your operation. All protective clothing (suits, gloves, footwear, headgear) should be clean, available each day, and put on before work. Contact lenses should not be worn when working with this chemical. Wear splash-proof chemical goggles and face shield unless full facepiece respiratory protection is worn. Employees should wash immediately with soap when skin is wet or contaminated. Provide emergency showers and eyewash.

Respirator Selection: Where the potential for exposure exists, use a MSHA/NIOSH approved supplied-air respirator with a full facepiece operated in the positive pressure mode or with a full facepiece, hood, or helmet in the continuous flow mode, or use a MSHA/NIOSH approved self-contained breathing apparatus with a full facepiece operated in pressure-demand or other positive pressure mode.

Storage: Prior to working with Chlorophenyltrichlorosilane you should be trained on its proper handling and storage. Chlorophenyltrichlorosilane should be stored to avoid contact with combustible materials such as wood, paper and oil. Store in tightly closed containers in a cool, well-ventilated area away from water, steam, and moisture because toxic and corrosive gases including Hydrogen Chloride can be produced.

Shipping: This material must carry a "Corrosive" label.[19] It falls in hazard Class 8 and Packing Group II. Shipment by passenger aircraft or railcar is forbidden; cargo aircraft shipment is limited.

Spill Handling: Evacuate and restrict persons not wearing protective equipment from area of spill or leak until cleanup is complete. Remove all ignition sources. Ventilate area of spill or leak. Absorb liquids in vermiculite, dry sand, earth, peat, carbon, or a similar material and deposit in sealed containers. It may be necessary to contain and dispose of this chemical as a hazardous waste. If material or contaminated runoff enters waterways, notify downstream users of potentially contaminated waters. Contact your Department of Environmental Protection or your regional office of the federal EPA for specific recommendations. If employees are required to clean-up spills, they must be properly trained and equipped. OSHA 1910.120(q) may be applicable.

Fire Extinguishing: Chlorophenyltrichlorosilane may burn, but does not readily ignite. Use dry chemical, CO_2 or foam extinguishers. *Do not* use water. Poisonous gases are produced in fire including chlorides. If material or contaminated runoff enters waterways, notify downstream users of potentially contaminated waters. Notify local health and fire officials and pollution control agencies. From a secure, explosion-proof location, use water spray to cool exposed containers. If cooling streams are ineffective (venting sound increases in volume and pitch, tank discolors, or shows any signs of deforming), withdraw immediately to a secure position. If employees are expected to fight fires, they must be trained and equipped in OSHA 1910.156.

References

New Jersey Department of Health and Senior Services and Senior Services, "Hazardous Substance Fact Sheet: Chlorophenyltrichlorosilane," Trenton, NJ (April 1997)

Chloropicrin

Molecular Formula: CCl_3NO_2

Synonyms: Acquinite; Chloorpikrine (Dutch); Chlor-o-Pic; Chloropicrine (French); Chlorpikrin (German); Clorpicrina (Italian, Spanish); Larvacide 100; Methane, trichloronitro-; Mycrolysin; Nitrochloroform; Nitrotrichloromethane; Pic-Chlor; Picfume; Picride; Profume A; PS; Trichlor; Trichloronitromethane

CAS Registry Number: 76-06-2

RTECS Number: PB6300000

DOT ID: UN 1580

Regulatory Authority

- Banned or Severely Restricted (Various Countries) (UN)[13][35]
- OSHA 29CFR1910.119, Appendix A. Process Safety List of Highly Hazardous Chemicals, TQ = 500 lb (227 kg)
- Air Pollutant Standard Set (ACGIH)[1] (Australia) (DFG)[3] (HSE) (Israel) (Mexico)[33] (USSR, Japan)[35] (Several States)[60] (Several Canadian Provinces)

- SAFE DRINKING WATER ACT: Priority List (55 FR 1470)
- EPCRA Section 313 Form R *de minimis* concentration reporting level: 1.0%
- Canada, WHMIS, Ingredients Disclosure List

Cited in U.S. State Regulations: Alaska (G), California (A, G, W), Connecticut (A), Florida (G), Illinois (G), Maine (G), Massachusetts (G), Minnesota (G), Nevada (A), New Hampshire (G), New Jersey (G), North Dakota (A), Oklahoma (G), Pennsylvania (G), Rhode Island (G), Virginia (A), West Virginia (G).

Description: Chloropicrin, CCl_3NO_2, is a highly reactive, colorless, oily liquid with a sharp, penetrating odor that causes tears. Odor threshold = 1.1 ppm. Boiling point = 112°C. Freezing/Melting point = -64°. Slightly soluble in water.

Potential Exposure: Chloropicrin is used in the manufacture of the dye-stuff methyl violet and in other organic syntheses. It has been used as a military poison gas. It is used as a fumigant insecticide.

Incompatibilities: Can be self reactive. Chloropicrin may explode when heated under confinement. Quickly elevated temperatures, shock, contact with alkali metals or alkaline earth may cause explosions. A strong oxidizer; violent reaction with reducing agents, aniline (especially in presence of heat), alcoholic sodium hydroxide, combustible substances, sodium methoxide, propargyl bromide, metallic powders. Liquid attacks some plastics, rubber and coatings.

Permissible Exposure Limits in Air: The Federal limit (OSHA PEL),[58] the DFG,[3] the HSE,[33] the Japanese,[35] Australian, Israeli, Mexican, and the ACGIH TWA value is 0.1 ppm (0.7 mg/m^3) and the ACGIH,[1] Mexican, and HSE[33] STEL value is 0.3 ppm. The NIOSH IDLH level is 2 ppm. The former USSR[35] has set a MAC for ambient air in residential areas of 0.01 mg/m^3 on a momentary basis and 0.007 mg/m^3 on a daily average basis. Several states have set guidelines or standards for chloropicrin in ambient air[60] ranging from 7-20 μg/m^3 (North Dakota) to 11.7 μg/m^3 (Virginia) to 14 μg/m^3 (Connecticut) to 17 μg/m^3 (Nevada).

Determination in Air: No method available.

Permissible Concentration in Water: California[61] has set guidelines for Chloropicrin in drinking water of 50 μg/l on a taste basis and 37 μg/l on an odor basis.

Routes of Entry: Inhalation, ingestion, eye and skin contact.

Harmful Effects and Symptoms

Short Term Exposure: Chlorpicrin was used as poison gas in WW 1. Exposure causes intense tearing of the eyes, headache, nausea and vomiting, diarrhea, and cough. Contact can severely irritate the skin causing rash or burning sensation. Higher exposures can irritate and burn the lungs, causing a build-up of fluid (pulmonary edema), a medical emergency that can be delayed for several hours. This can cause death. The oral LD_{50} rat is 250 mg/kg (moderately toxic).

Long Term Exposure: Repeated exposure can damage the lungs, causing bronchitis. It may also damage the liver and kidneys.

Points of Attack: Eyes, skin, respiratory system, liver, kidneys.

Medical Surveillance: Before beginning employment and at regular times after that, the following is recommended: Lung function test. If symptoms develop or overexposure has occurred, the following may be useful: Liver and kidney function tests. Consider chest x-ray following acute overexposure.

First Aid: If this chemical gets into the eyes, remove any contact lenses at once and irrigate immediately for at least 15 minutes, occasionally lifting upper and lower lids. Seek medical attention immediately. If this chemical contacts the skin, remove contaminated clothing and wash immediately with soap and water. Seek medical attention immediately. If this chemical has been inhaled, remove from exposure, begin rescue breathing (using universal precautions) if breathing has stopped and CPR if heart action has stopped. Transfer promptly to a medical facility. When this chemical has been swallowed, get medical attention. Give large quantities of water and induce vomiting. Do not make an unconscious person vomit. Medical observation is recommended for 24 – 48 hours after breathing overexposure, as pulmonary edema may be delayed. As first aid for pulmonary edema, a doctor or authorized paramedic may consider administering a corticosteroid spray.

Personal Protective Methods: Wear protective gloves and clothing to prevent any reasonable probability of skin contact. Safety equipment suppliers/manufacturers can provide recommendations on the most protective glove/clothing material for your operation. The International Technical Information Institute recommends wearing Neoprene gloves. All protective clothing (suits, gloves, footwear, headgear) should be clean, available each day, and put on before work. Contact lenses should not be worn when working with this chemical. Wear splashproof chemical goggles and face shield unless full facepiece respiratory protection is worn. Employees should wash immediately with soap when skin is wet or contaminated. Provide emergency showers and eyewash.

Respirator Selection: NIOSH/OSHA: *2 ppm:* SA:CF (any supplied-air respirator operated in a continuous-flow mode); or PAPROV [any powered, air-purifying respirator with organic vapor cartridge(s)]; or CCRFOV [any chemical cartridge respirator with a full facepiece and organic vapor cartridge(s)]; or GMFOV (any air-purifying, full-facepiece respirator (gas mask) with a chin-style, front- or back-mounted acid gas canister);or SCBAF (any self-contained breathing apparatus with a full facepiece); or SAF (any supplied-air respirator with a full facepiece). *Emergency or planned entry into unknown concentrations or IDLH conditions:* SCBAF:PD,PP (any self-contained breathing apparatus that has a full facepiece and is operated in a pressure-demand or other positive-pressure mode);

or SAF:PD,PP:ASCBA (any supplied-air respirator that has a full facepiece and is operated in a pressure-demand or other positive-pressure mode in combination with an auxiliary self-contained breathing apparatus operated in a pressure-demand or other positive pressure mode). *Escape:* GMFOV [any air-purifying, full-facepiece respirator (gas mask) with a chin-style, front-or back-, mounted organic vapor canister] or SCBAE (any appropriate escape-type, self-contained breathing apparatus).

Note: Substance causes eye irritation or damage; eye protection needed.

Storage: Prior to working with chloropicrin you should be trained on its proper handling and storage. Before entering confined space where chloropicrin may be present, check to make sure that an explosive concentration does not exist. Chloropicrin must be stored to avoid contact with strong oxidizers, such as chlorine or chlorine dioxide, since violent reactions occur. Store in tightly closed containers in a cool, well-ventilated area away from heat. High temperatures or severe shock may cause an explosion, particularly with containers having capacities of greater than 30 gallons. Where possible, automatically pump liquid from drums or other storage containers to process containers. Sources of ignition such as smoking and open flames, are prohibited where this chemical is used, handled, or stored in a manner that could create a potential fire or explosion hazard.

Shipping: Chloropicrin must be labeled "Poison." It falls in DOT Hazard Class 6.1 and Packing Group I.[19] Shipment by passenger aircraft or railcar or even by cargo aircraft is forbidden.

Spill Handling: Evacuate and restrict persons not wearing protective equipment from area of spill or leak until cleanup is complete. Remove all ignition sources. Ventilate area of spill or leak. Absorb liquids in vermiculite, dry sand, earth, peat, carbon, or a similar material and deposit in sealed containers. It may be necessary to contain and dispose of this chemical as a hazardous waste. If material or contaminated runoff enters waterways, notify downstream users of potentially contaminated waters. Contact your Department of Environmental Protection or your regional office of the federal EPA for specific recommendations. If employees are required to clean-up spills, they must be properly trained and equipped. OSHA 1910.120(q) may be applicable.

Initial isolation and protective action distances

Distances shown are likely to be affected during the first 30 minutes after materials are spilled and could increase with time. If more than one tank car, cargo tank, portable tank, or large cylinder is involved in the incident is leaking, the protective action distance may need to be increased. You may need to seek emergency information from CHEMTREC at (800) 424-9300 or seek professional environmental engineering assistance from the U.S. EPA Environmental Response Team at (908)548-8730 (24-hour response line).

Small spills (From a small package or a small leak from a large package)

First: Isolate in all directions (feet) 300

Then: Protect persons downwind (miles)

Day 0.2

Night 0.7

Large spills (From a large package or from many small packages)

First: Isolate in all directions (feet) 700

Then: Protect persons downwind (miles)

Day 0.6

Night 2.4

Restrict persons not wearing protective equipment from area of spill or leak until clean-up is complete. Ventilate the area of spill or leak.

Fire Extinguishing: This chemical is a noncombustible liquid. Use any extinguishing agent suitable for surrounding fire. Heat can cause closed containers to explode. Fight fire from an explosion-resistant location. Poisonous gases are produced in fire including hydrogen chloride and nitrous vapors. If material or contaminated runoff enters waterways, notify downstream users of potentially contaminated waters. Notify local health and fire officials and pollution control agencies. From a secure, explosion-proof location, use water spray to cool exposed containers. If cooling streams are ineffective (venting sound increases in volume and pitch, tank discolors, or shows any signs of deforming), withdraw immediately to a secure position. If employees are expected to fight fires, they must be trained and equipped in OSHA 1910.156.

Disposal Method Suggested: Incineration (1,500°F, 0.5 second minimum for primary combustion; 2,200°F, 1.0 second for secondary combustion) after mixing with other fuel. The formation of elemental chlorine may be prevented by injection of steam or using methane as a fuel in the process.

Chloropicrin reacts readily with alcoholic sodium sulfite solutions to produce methanetrisulfonic acid (which is relatively nonvolatile and less harmful). This reaction has been recommended for treating spills and cleaning equipment. Although not specifically suggested as a decontamination procedure, the rapid reaction of chloropicrin with ammonia to produce guanidine (LD_{50} = 500 mg/kg) could be used for detoxification.

The Chemical Manufacturers' Association has suggested two procedures for disposal of Chloropicrin: (a) Pour or sift over soda ash. Mix and wash slowly into large tank. Neutralize and pass to sewer with excess water. (b) Absorb on vermiculite. Mix and shovel into paper boxes. Drop into incinerator with afterburner and scrubber.[22] In accordance with

40CFR165 recommendations for the disposal of pesticides and pesticide containers. Must be disposed properly by following package label directions or by contacting your state pesticide or environmental control agency or by contacting your regional EPA office.

References

Sax, N. I., Ed., "Dangerous Properties of Industrial Materials Report," 2, No. 2, 17–19, New York, Van Nostrand Reinhold Co. (1982)

New Jersey Department of Health and Senior Services and Senior Services, "Hazardous Substance Fact Sheet: Chloropicrin," Trenton, NJ (April 1998)

Chloroplatinic Acid

Molecular Formula: Cl_6H_2Pt

Common Formula: H_2PtCl_6

Synonyms: Dihydrogen hexachloroplatinate; Dihydrogenhexachloroplatinate (2-); Hexachloro dyhydrogen platinate; Hexachloroplatinic acid; Hexachloroplatinic(IV) acid; Hexachloroplatinic(4+) acid, hydrogen-; Hydrogen hexachloroplatinate(4+); Platinate, hexachloro-; Platinic chloride

CAS Registry Number: 16491-12-1

RTECS Number: TP1500000

DOT ID: UN 2507

Regulatory Authority

- Air Pollutant Standard Set (ACGIH)[1] (DFG)[3] (HSE)[33] (OSHA)[58]
- Canada, WHMIS, Ingredients Disclosure List

Cited in U.S. State Regulations: New Hampshire (G), New Jersey (G).

Description: Chloroplatinic acid, H_2PtCl_6, is a reddish-brown deliquescent solid. Freezing/Melting point = 60°C. Soluble in water.

Potential Exposure: Chloroplatinic Acid has many uses, among them are platinum plating, photography and catalysis.

Permissible Exposure Limits in Air: OSHA[58] - The legal airborne permissible exposure limit (PEL) for Platinum and its soluble salts is 0.002 mg/m^3 averaged over an 8-hour workshift. NIOSH recommends the same airborne limit for a 10-hour workshift. ACGIH[1] – The recommended airborne exposure limit is 0.002 mg/m^3, which should not be exceeded at any time. The NIOSH IDLH = 4 mg/m^3 (as Pt). The DFG[3] and the HSE[33] have adapted the same value.

Determination in Air: Filter; Acid/Reagent; Graphite furnace atomic absorption spectrometry; NIOSH II(7) Method #S191. Also NIOSH IV #7300, Elements.

Permissible Concentration in Water: No criteria set.

Routes of Entry: Skin contact, inhalation of vapors, ingestion.

Harmful Effects and Symptoms

Short Term Exposure: Chloroplatinic Acid can affect you when breathed in. It is a highly corrosive chemical; contact can severely irritate and burn the eyes. Inhalation can irritate the respiratory tract. Exposure can cause severe allergies affecting the nose, skin and lungs. Symptoms include sneezing, coughing, nose and throat irritation and nasal discharge. Irritation and even ulcers can develop in the nose. Once lung allergy develops, even very small future exposures cause cough, wheezing, chest tightness and shortness of breath. Skin allergy with a rash and itching can also develop.

Long Term Exposure: Repeated exposure may lead to permanent lung damage (pulmonary fibrosis), skin allergy, asthma-like allergy.

Points of Attack: Skin, respiratory system.

Medical Surveillance: Before beginning employment and at regular times after that, the following are recommended: Lung function tests. These may be normal if the person is not having an attack at the time of the test. If symptoms develop or overexposure is suspected, the following may be useful: Evaluation by a qualified allergist, including careful exposure history and special testing, may help diagnose skin allergy.

First Aid: If this chemical gets into the eyes, remove any contact lenses at once and irrigate immediately for at least 15 minutes, occasionally lifting upper and lower lids. Seek medical attention immediately. If this chemical contacts the skin, remove contaminated clothing and wash immediately with soap and water. Seek medical attention immediately. If this chemical has been inhaled, remove from exposure, begin rescue breathing (using universal precautions) if breathing has stopped and CPR if heart action has stopped. Transfer promptly to a medical facility. When this chemical has been swallowed, get medical attention. If victim is conscious, administer water or milk. Do not induce vomiting.

Personal Protective Methods: Wear protective gloves and clothing to prevent any reasonable probability of skin contact. Safety equipment suppliers/manufacturers can provide recommendations on the most protective glove/clothing material for your operation. All protective clothing (suits, gloves, footwear, headgear) should be clean, available each day, and put on before work. Contact lenses should not be worn when working with this chemical. Wear dust-proof chemical goggles and face shield unless full facepiece respiratory protection is worn. Employees should wash immediately with soap when skin is wet or contaminated. Provide emergency showers and eyewash.

Respirator Selection: NIOSH/OSHA: *up to 0.05 mg/m^3:* SA:CF (any supplied-air respirator operated in a continuous-flow mode), *Up to 0.1 mg/m^3:* HiEF (any air-purifying, full-facepiece respirator with a high-efficiency particulate filter); or SCBAF (any self-contained breathing apparatus with a full facepiece); or SAF (any supplied-air respirator with a full facepiece). *Up to 4 mg/m^3:* SAF: PD,PP (any supplied-air respirator that has a full facepiece and is operated in a pressure-demand or other positive-pressure mode). *Emergency or planned entry into unknown concentrations or IDLH conditions:* SCBAF: PD,PP (any self-contained breathing apparatus that has a full facepiece and is operated in a pressure-demand or other positive-pressure mode); or SAF:PD,PP: ASCBA (any supplied-air respirator that has a full facepiece and is operated in a pressure-demand or other positive-pressure mode in combination with an auxiliary self-contained breathing apparatus operated in a pressure-demand or other positive-pressure mode). *Escape:* HiEF [any air-purifying, full-facepiece respirator (gas mask) with a chin-style, front- or back-mounted organic vapor canister having a high-efficiency particulate filter]; or SCBAE (any appropriate escape-type, self-contained breathing apparatus).

Storage: Store in tightly closed containers in a cool, well-ventilated area.

Shipping: Solid chloroplatinic acid must carry a "Corrosive" label.[19] It falls in Hazard Class 8 and Packing Group III.

Spill Handling: Evacuate persons not wearing protective equipment from area of spill or leak until clean-up is complete. Remove all ignition sources. Collect powdered material in the most convenient and safe manner and deposit in sealed containers. Ventilate area after clean-up is complete. It may be necessary to contain and dispose of this chemical as a hazardous waste. If material or contaminated runoff enters waterways, notify downstream users of potentially contaminated waters. Contact your Department of Environmental Protection or your regional office of the federal EPA for specific recommendations. If employees are required to clean-up spills, they must be properly trained and equipped. OSHA 1910.120(q) may be applicable.

Fire Extinguishing: Use dry chemical, CO_2, water spray, or foam extinguishers. Chloroplatinic Acid may burn, but does not readily ignite. Poisonous gases are produced in fire. If material or contaminated runoff enters waterways, notify downstream users of potentially contaminated waters. Notify local health and fire officials and pollution control agencies. From a secure, explosion-proof location, use water spray to cool exposed containers. If cooling streams are ineffective (venting sound increases in volume and pitch, tank discolors, or shows any signs of deforming), withdraw immediately to a secure position. If employees are expected to fight fires, they must be trained and equipped in OSHA 1910.156.

References

New Jersey Department of Health and Senior Services and Senior Services, "Hazardous Substance Fact Sheet: Chloroplatinic Acid," Trenton, NJ (April 1998)

Chloroprene

Molecular Formula: C_4H_5Cl

Common Formula: $H_2C=CCl-CH=CH_2$

Synonyms: 1,3-Butadiene, 2-chloro-; 2-Chloor-1,3-butadieen (Dutch); 2-Chlor-1,3-butadien (German); 1,3-Chlor-2-butadiene; 2-Chloro-1,3-butadiene; 2-Chlorobuta-1,3-diene; 2-Chlorobutadiene; Chloropreen (Dutch); Chloropren (German); Chloropren (Polish); β-Chloroprene; 2-Cloro-1,3-butadiene (Italian); Cloroprene (Italian); β-Cloropreno (Spanish); Neoprene (polymerized product)

CAS Registry Number: 126-99-8

RTECS Number: EI9625000

DOT ID: UN 1991

EEC Number: 602-036-00-8

Regulatory Authority

- Carcinogen (ACGIH)[1] (DFG)
- Air Pollutant Standard Set (ACGIH)[1] (DFG)[3] (HSE)[33] (OSHA)[58] (former USSR)[35] (Several States)[60]
- CLEAN AIR ACT: Hazardous Air Pollutants (Title I, Part A, Section 112)
- RCRA, 40CFR261, Appendix 8 Hazardous Constituents, waste number not listed
- RCRA 40CFR268.48; 61FR15654, Universal Treatment Standards: Wastewater (mg/l), 0.057; Nonwastewater (mg/kg), 0.28
- RCRA 40CFR264, Appendix 9; TSD Facilities Ground Water Monitoring List. Suggested test method(s) (PQL μg/l): 8010 (50); 8240 (5)
- SUPERFUND/EPCRA 40CFR302.4 Reportable Quantity (RQ): CERCLA, 1 lb (0.454 kg)
- EPCRA Section 313 Form R *de minimis* concentration reporting level: 1.0%
- Canada, WHMIS, Ingredients Disclosure List

Cited in U.S. State Regulations: Alaska (G), California (A, G), Connecticut (A), Florida (G), Illinois (G), Maine (G), Maryland (G), Massachusetts (G, A), Michigan (G), Nevada (A), New Hampshire (G, W), New Jersey (G), North Carolina (A), North Dakota (A), Pennsylvania (G), Rhode Island (G), South Carolina (A), Virginia (A), West Virginia (G).

Description: Chloroprene, $H_2C=CCl-CH=CH_2$, is a colorless, flammable liquid possessing a pungent odor. The odor threshold is 0.4 mg/m^3.[41] Boiling point = 59°C. Freezing/Melting point = -130°C. Flash point = -20°C. The explosive limits are: LEL = 4.0%; UEL = 20.0%.[17] Hazard Identification (based on NFPA-704 M Rating System): Health 2, Flammability 3, Reactivity 0. Slightly soluble in water.

Potential Exposure: The only major use of chloroprene is in the production of artificial rubber (neoprene, duprene). Chloroprene is extremely reactive, e.g., it can polymerize spontaneously at room temperatures, the process being catalyzed by light, peroxides and other free radical initiators. It can also react with oxygen to form polymeric peroxides and because of its instability, flammability and toxicity, Chloroprene has no end product uses as such.

Incompatibilities: Can form unstable peroxides; chloroprene may polymerize on standing with fire or explosion hazard. Forms explosive mixture with air. Reacts with liquid or gaseous fluorine, alkali metals, metal powders, oxidizers, creating a fire or explosion hazard. Attacks some plastics, rubber and coatings. May accumulate static electrical charges, and may cause ignition of its vapors.

Permissible Exposure Limits in Air: The Federal standard (OSHA PEL) is 25 ppm (90 mg/m^3) TWA. NIOSH recommends a ceiling of 1 ppm (3.6 mg/m^3) (15-minute). NIOSH, in 1977, specified that the employer shall control exposure to chloroprene so that no employee is ever exposed at a concentration greater than 3.6 mg/m^3 of air (1 ppm) determined as a ceiling concentration for any 15-minute sampling period during a 40-hour workweek. ACGIH TWA values of 10 ppm (36 mg/m^3), with the notation "skin" indicating the possibility of cutaneous absorption. The NIOSH IDLH level is 300 ppm with [Ca] Potential Occupational Carcinogen notation. The DFG[3] simply notes that it is a potential human carcinogen with sufficient animal evidence and no numerical limits. The former Soviet Union has set an occupational ceiling value of 0.05 mg/m^3 and a MAC in ambient air in residential areas of 0.02 mg/m^3 on a momentary basis and 0.002 mg/m^3 on a daily average basis.[35] In addition, several states have set guidelines or standards for chloroprene in ambient air[60] ranging from 2.5 μg/m^3 (Massachusetts) to 175 μg/m^3 (South Carolina) to 350 μg/m^3 (North Dakota) to 420 – 3,500 μg/m^3 (North Carolina) to 800 μg/m^3 (Virginia) to 900 μg/m^3 (Connecticut) to 1,070 μg/m^3 (Nevada).

Determination in Air: Charcoal adsorption, workup with CS_2 and analysis by gas chromatography. See NIOSH Method #1002.[18]

Permissible Concentration in Water: A MAC in water bodies used for domestic purposes of 0.01 mg/l has been set by the former USSR.[35]

Routes of Entry: Inhalation of vapor and skin absorption, ingestion and skin and eye contact.

Harmful Effects and Symptoms

Short Term Exposure: Chloroprene irritates the eyes, skin, and respiratory tract. Chloroprene acts as a primary irritant on contact with skin, conjunctiva, and mucous membranes and may result in dermatitis, conjunctivitis, and circumscribed necrosis of the cornea. Inhalation of high concentrations may result in dizziness, lightheadedness and unconciousness; anesthesia and respiratory paralysis. Chloroprene may affects the central nervous system, kidneys and liver. The LD_{50} oral rat is only 900 mg/kg (slightly toxic).

Long Term Exposure: Chronic exposure may produce damage to the lungs, nervous system, liver, kidneys, spleen, and myocardium. Because this is a mutagen, handle it as a possible cancer-causing substance — with extreme caution. It may also damage the developing fetus, cause spontaneous abortions, and interfere with sperm production. Repeated or prolonged contact with skin may cause dermatitis. Chronic exposure may cause alopecia. Chloroprene is a potential occupational carcinogen. Temporary hair loss has been reported during the manufacture of polymers.

Points of Attack: Eyes, skin, respiratory system, liver, kidneys.

Medical Surveillance: Preplacement and periodic examinations should include an evaluation of the skin, eyes, respiratory tract, and central nervous system. Liver and kidney function should be evaluated.

First Aid: If this chemical gets into the eyes, remove any contact lenses at once and irrigate immediately for at least 15 minutes, occasionally lifting upper and lower lids. Seek medical attention immediately. If this chemical contacts the skin, remove contaminated clothing and wash immediately with soap and water. Seek medical attention immediately. If this chemical has been inhaled, remove from exposure, begin rescue breathing (using universal precautions) if breathing has stopped and CPR if heart action has stopped. Transfer promptly to a medical facility. When this chemical has been swallowed, get medical attention. Give large quantities of water and induce vomiting. Do not make an unconscious person vomit.

Personal Protective Methods: Wear protective gloves and clothing to prevent any reasonable probability of skin contact. Safety equipment suppliers/manufacturers can provide recommendations on the most protective glove/clothing material for your operation. All protective clothing (suits, gloves, footwear, headgear) should be clean, available each day, and put on before work. Contact lenses should not be worn when working with this chemical. Wear splash-proof chemical goggles and face shield unless full facepiece respiratory protection is worn. Employees should wash immediately with soap when skin is wet or contaminated. Provide emergency showers and eyewash. Engineering controls are recommended in NIOSH Criteria Document: 77-1210.

Respirator Selection: NIOSH *At any detectable concentration:* SCBAF:PD,PP (any self-contained breathing apparatus that has a full facepiece and is operated in a pressure-demand or other positive-pressure mode); or SAF:PD,PP:ASCBA (any supplied-air respirator that has a full facepiece and is operated in a pressure-demand or other positive-pressure mode in combination with an auxiliary self-contained breathing apparatus operated in a pressure-demand or other positive-pressure mode). *Escape:* GMFOV [any air-purifying, full-facepiece respirator (gas mask) with a chin-style, front-or back-mounted organic vapor canister] or SCBAE (any appropriate escape-type, self-contained breathing apparatus).

Storage: Prior to working with Chloroprene you should be trained on its proper handling and storage. Before entering confined space where this chemical may be present, check to make sure that an explosive concentration does not exist. Chloroprene must be stored to avoid contact with peroxides and other oxidizers, such as permanganates, nitrates, chlorates, and perchlorates, since violent reactions occur. Store in tightly closed containers in a cool, well-ventilated area at temperatures below 10°C (50°F). Sources of ignition such as smoking and open flames are prohibited where Chloroprene is handled, used, or stored. Metal containers involving the transfer of 5 gallons or more of Chloroprene should be grounded and bonded. Drums must be equipped with self-closing valves, pressure vacuum bungs, and flame arresters. Use only non-sparking tools and equipment, especially when opening and closing containers of Chloroprene. A regulated, marked area should be established where this chemical is handled, used, or stored in compliance with OSHA standard 1910.1045.

Shipping: Shipment of uninhibited Chloroprene is forbidden. Inhibited chloroprene must bear the label "Flammable Liquid, Poison." It falls in Hazard Class 3 and Packing Group I. Shipment by passenger aircraft or railcar is forbidden.

Spill Handling: Evacuate and restrict persons not wearing protective equipment from area of spill or leak until cleanup is complete. Remove all ignition sources. Establish forced ventilation to keep levels below explosive limit. Absorb liquids in vermiculite, dry sand, earth, peat, carbon, or a similar material and deposit in sealed containers. It may be necessary to contain and dispose of this chemical as a hazardous waste. If material or contaminated runoff enters waterways, notify downstream users of potentially contaminated waters. Contact your Department of Environmental Protection or your regional office of the federal EPA for specific recommendations. If employees are required to clean-up spills, they must be properly trained and equipped. OSHA 1910.120(q) may be applicable.

Fire Extinguishing: Chloroprene is a flammable liquid. Poisonous gases including hydrogen chloride and phosgene are produced in fire. Use dry chemical, carbon dioxide, or alcohol foam extinguishers. May react with itself without warning, blocking relief valves and leading to container explosions. Vapors are heavier than air and will collect in low areas. Vapors may travel long distances to ignition sources and flashback. Vapors in confined areas may explode when exposed to fire. Containers may explode in fire. Storage containers and parts of containers may rocket great distances, in many directions. If material or contaminated runoff enters waterways, notify downstream users of potentially contaminated waters. Notify local health and fire officials and pollution control agencies. From a secure, explosion-proof location, use water spray to cool exposed containers. If cooling streams are ineffective (venting sound increases in volume and pitch, tank discolors, or shows any signs of deforming), withdraw immediately to a secure position. If employees are expected to fight fires, they must be trained and equipped in OSHA 1910.156.

Disposal Method Suggested: Incineration, preferably after mixing with another combustible fuel. Care must be exercised to assure complete combustion to prevent the formation of phosgene. An acid scrubber is necessary to remove the halo acids produced.[22]

References

National Institute for Occupational Safety and Health, Criteria for a Recommended Standard: Occupational Exposure to Chloroprene, NIOSH Document No. 77–210 (1977)

Sax, N. I., Ed., "Dangerous Properties of Industrial Materials Report," 1, No. 4, 47–49, New York, Van Nostrand Reinhold Co. (1981)

New Jersey Department of Health and Senior Services and Senior Services, "Hazardous Substance Fact Sheet: Chloroprene," Trenton, NJ (February 1989)

3-Chloropropionitrile

Molecular Formula: C_3H_4ClN

Common Formula: $ClCH_2CH_2CN$

Synonyms: A13-28526; 1-Chloro-2-cyanoethane; 3-Chloropropanenitrile; 3-Chloropropanonitrile; β-Chloropropionitrile; 3-Chloropropionitrile; Propanenitrile, 3-chloro-; Propionitrile, 3-chloro-

CAS Registry Number: 542-76-7

RTECS Number: UG1400000

DOT ID: UN 3276

Regulatory Authority

- Extremely Hazardous Substance (EPA-SARA) (TPQ = 1000)[7]
- EPA HAZARDOUS WASTE NUMBER (RCRA No.): P027
- RCRA, 40CFR261, Appendix 8 Hazardous Constituents
- SUPERFUND/EPCRA 40CFR355, Appendix B Extremely Hazardous Substances: TPQ = 1,000 lb (454 kg)
- SUPERFUND/EPCRA 40CFR302.4 Reportable Quantity (RQ): CERCLA, 1,000 lb (454 kg)
- EPCRA Section 313 Form R *de minimis* concentration reporting level: 1.0%

Cited in U.S. State Regulations: Florida (G), Kansas (G), Louisiana (G), Massachusetts (G), New Hampshire (G), New Jersey (G), Pennsylvania (G), Vermont (G), Virginia (G), Washington (G), Wisconsin (G).

Description: 3-Chloropropionitrile, $ClCH_2CH_2CN$, is a colorless liquid with an acrid odor. Boiling point = 175 – 176°C (decomposes). Flash point = 76°C. Hazard Identification (based on NFPA-704 M Rating System): Flammability 2, Reactivity 1.

Potential Exposure: This material is used in pharmaceutical manufacture and in polymer synthesis.

Incompatibilities: Contact with strong oxidizers may cause a fire and explosion hazard.

Permissible Exposure Limits in Air: The NIOSH REL for nitriles is a ceiling limit of 6 mg/m^3, not to be exceeded in any 15-minute work period.

Permissible Concentration in Water: No criteria set.

Routes of Entry: Inhalation, ingestion, skin contact. This chemical can be absorbed through the skin, thereby increasing exposure.

Harmful Effects and Symptoms

Short Term Exposure: Symptoms of exposure include rapid and irregular breathing, anxiety, confusion, odor of bitter almonds (on breath or vomitus), nausea, vomiting (if oral exposure), irregular heart beat, a feeling of tightness in the chest, bright pink coloration of the skin, sweating, protruding eyeballs, dilated pupils, unconsciousness followed by convulsions, involuntary urination and defecation, paralysis and respiratory arrest (heart will beat after breathing stops).

Toxic effects are a result of systemic cyanide poisoning. Few poisons are more rapidly lethal. Average oral lethal dose for hydrogen cyanide is approximately 60 – 90 mg (corresponds to 200 mg of potassium cyanide). Cause of death is lack of oxygen to the body's cells (especially the brain and heart) as a result of the chemical inhibiting cell enzymes.

Medical Surveillance: Blood cyanide level.

First Aid: If this chemical gets into the eyes, remove any contact lenses at once and irrigate immediately for at least 15 minutes, occasionally lifting upper and lower lids. Seek medical attention immediately. If this chemical contacts the skin, remove contaminated clothing and wash immediately with soap and water. Seek medical attention immediately. If this chemical has been inhaled, remove from exposure, begin rescue breathing (using universal precautions) if breathing has stopped and CPR if heart action has stopped. Transfer promptly to a medical facility. When this chemical has been swallowed, get medical attention. Give large quantities of water and induce vomiting. Do not make an unconscious person vomit.

Use amyl nitrate capsules if symptoms develop. All area employees should be trained regularly in emergency measures for cyanide poisoning and in CPR. A cyanide antidote kit should be kept in the immediate work area and must be rapidly available. Kit ingredients should be replaced every 1-2 years to ensure freshness. Persons trained in the use of this kit, oxygen use, and CPR must be quickly available.

Personal Protective Methods: Wear protective gloves and clothing to prevent any reasonable probability of skin contact. Safety equipment suppliers/manufacturers can provide recommendations on the most protective glove/clothing material for your operation. All protective clothing (suits, gloves, footwear, headgear) should be clean, available each day, and put on before work. Contact lenses should not be worn when working with this chemical. Wear splash-proof chemical goggles and face shield unless full facepiece respiratory protection is worn. Employees should wash immediately with soap when skin is wet or contaminated. Provide emergency showers and eyewash. See NIOSH Criteria Document 78–212 NITRILES.

Respirator Selection: NIOSH/OSHA: *up to 25 mg/m^3:* SA (any supplied-air respirator); or SCBAF (any self-contained breathing apparatus with full facepiece). *Emergency or planned entry into unknown concentrations or IDLH conditions:* SCBAF:PD,PP (any self-contained breathing apparatus that has a full facepiece and is operated in a pressure-demand or other positive-pressure mode); or SAF:PD,PP:ASCBA (any supplied-air respirator that has a full facepiece and is operated in a pressure-demand or other positive-pressure mode in combination with an auxiliary self-contained breathing apparatus operated in a pressure-demand or other positive pressure mode). *Escape:* GMFSHiE (any air-purifying, full-facepiece respirator (gas mask) with a chin-style, front-or back-, mounted canister providing protection against the compound of concern and having a high efficiency particulate filter); or SCBAE (any appropriate escape-type, self-contained breathing apparatus).

Storage: Prior to working with this chemical you should be trained on its proper handling and storage. Before entering confined space where 3-Chloropropionitrile may be present, check to make sure that an explosive concentration does not exist. Store in tightly closed containers in a cool, well ventilated area. Metal containers involving the transfer of this chemical should be grounded and bonded. Where possible, automatically pump liquid from drums or other storage containers to process containers. Drums must be equipped with self-closing valves, pressure vacuum bungs, and flame arresters. Use only non-sparking tools and equipment, especially when opening and closing containers of this chemical. Sources of ignition such as smoking and open flames, are prohibited where this chemical is used, handled, or stored in a manner that could create a potential fire or explosion hazard.

Shipping: This compound requires a shipping label of: "Poison." It falls in DOT Hazard Class 6.1.

Spill Handling: Stay upwind. Evacuate and restrict persons not wearing protective equipment from area of spill or leak until cleanup is complete. Remove all ignition sources. Do not touch spilled material; stop leak if you can do so without risk. Use water spray to reduce vapors. Ventilate area of spill or leak. Absorb liquids in vermiculite, dry sand, earth, peat, carbon, or a similar material and deposit in sealed containers. Dike far ahead of spill for later disposal. It may be necessary to contain and dispose of this chemical as a hazardous waste. If material or contaminated runoff enters waterways, notify downstream users of potentially contaminated waters. Contact your Department of Environmental Protection or your regional office of the federal EPA for specific recommendations. If employees are required to clean-up spills, they must be properly trained and equipped. OSHA 1910.120(q) may be applicable.

Fire Extinguishing: This chemical is a combustible liquid. Poisonous gases including cyanides are produced in fire. Use alcohol foam extinguishers and water spray. Vapors are heavier than air and will collect in low areas. Vapors may travel long distances to ignition sources and flashback. Vapors in confined areas may explode when exposed to fire. Containers may explode in fire. Storage containers and parts of containers may rocket great distances, in many directions. If material or contaminated runoff enters waterways, notify downstream users of potentially contaminated waters. Notify local health and fire officials and pollution control agencies. From a secure, explosion-proof location, use water spray to cool exposed containers. If cooling streams are ineffective (venting sound increases in volume and pitch, tank discolors, or shows any signs of deforming), withdraw immediately to a secure position. If employees are expected to fight fires, they must be trained and equipped in OSHA 1910.156.

References

U.S. Environmental Protection Agency, "Chemical Profile: 3-Chloropropionitrile," Washington, DC, Chemical Emergency Preparedness Program (November 30, 1987)

o-Chlorostyrene

Molecular Formula: C_8H_7Cl

Common Formula: $ClC_6H_4CH{=}CH_2$

Synonyms: Benzene, 1-chloro-2-ethenyl-; 1-Chloro-2-ethenylbenzene; *o*-Chlorostyrene; 2-Chlorostyrene; Chlorostyrene

CAS Registry Number: 2039-87-4 (o-); 1331-28-8 (chlorostyrene)

RTECS Number: WL4150000

Regulatory Authority

- Air Pollutant Standard Set (ACGIH)[1] (Australia) (Israel) (NIOSH) (Several States)[60] (Several Canadian Provinces)
- Canada, WHMIS, Ingredients Disclosure List

Cited in U.S. State Regulations: Alaska (G), Connecticut (A), Florida (G), Illinois (G), Maine (G), Massachusetts (G), Minnesota (G), Nevada (A), New Hampshire (G), New Jersey (G), North Dakota (A), Pennsylvania (G), Rhode Island (G), Virginia (A), West Virginia (G).

Description: o-Chlorostyrene, $ClC_6H_4CH{=}CH_2$, is a flammable, colorless liquid. Boiling point = 189°C. Freezing/Melting point = -62°C. Flash point = 59°C. Insoluble in water.

Potential Exposure: In organic synthesis; in the preparation of specialty polymers.

Incompatibilities: Contact with strong oxidizers may cause a fire or explosion hazard.

Permissible Exposure Limits in Air: There is no OSHA PEL. The recommended NIOSH REL is 50 ppm (285 mg/m^3) TWA for a 10-hour workshift and STEL of 75 ppm (428 mg/m^3). Australia, Israel, and ACGIH also set or recommend a TWA of 50 ppm (285 mg/m^3) for an 8-hour workshift and STEL of 75 ppm (430 mg/m^3). The Canadian provinces of British Columbia and Ontario recommend the same TWA and STEL. In addition, several states have set guidelines or standards for chlorostyrene in ambient air[60] ranging from 2.85 – 4.30 mg/m^3 (North Dakota) to 4.8 mg/m^3 (Virginia) to 5.7 mg/m^3 (Connecticut) to 6.79 mg/m^3 (Nevada).

Determination in Air: No Method listed by NIOSH.

Permissible Concentration in Water: No criteria set.

Routes of Entry: Inhalation, ingestion, skin and/or eye contact.

Harmful Effects and Symptoms

Short Term Exposure: o-Chlorostyrene can affect you when breathed in and by passing through your skin. Exposure can irritate the eyes and skin.

Long Term Exposure: Repeated exposures may damage the liver and kidneys. Animal studies show hematuria (blood in the urine), proteinuria, acidosis; enlarged liver, jaundice

Points of Attack: Eyes, skin, liver, kidneys, central nervous system, peripheral nervous system.

Medical Surveillance: If symptoms develop or overexposure is suspected, the following may be useful: Liver and kidney function tests. Nervous system tests.

First Aid: If this chemical gets into the eyes, remove any contact lenses at once and irrigate immediately for at least 15 minutes, occasionally lifting upper and lower lids. Seek medical attention immediately. If this chemical contacts the skin, remove contaminated clothing and wash immediately with soap and water. Seek medical attention immediately. If this chemical has been inhaled, remove from exposure, begin rescue breathing (using universal precautions) if breathing has stopped and CPR if heart action has stopped. Transfer promptly to a medical facility. When this chemical has been swallowed, get medical attention. Give large quantities of water and induce vomiting. Do not make an unconscious person vomit.

Personal Protective Methods: Wear protective gloves and clothing to prevent any reasonable probability of skin contact. Safety equipment suppliers/manufacturers can provide recommendations on the most protective glove/clothing material for your operation. All protective clothing (suits, gloves, footwear, headgear) should be clean, available each day, and put on before work. Contact lenses should not be worn when working with this chemical. Wear splash-proof chemical goggles and face shield unless full facepiece respiratory protection is worn. Employees should wash immediately with

soap when skin is wet or contaminated. Provide emergency showers and eyewash.

Respirator Selection: Where the potential exists for exposures over 50 ppm, use a MSHA/NIOSH approved supplied-air respirator with a full facepiece operated in the positive pressure mode or with a full facepiece, hood, or helmet in the continuous flow mode, or use a MSHA/NIOSH approved self-contained breathing apparatus with a full facepiece operated in pressure-demand or other positive pressure mode.

Storage: Prior to working with o-Chlorostyrene you should be trained on its proper handling and storage. Before entering confined space where o-Chlorostyrene may be present, check to make sure that an explosive concentration does not exist. Store in tightly closed containers in a cool, well ventilated area away from strong oxidizers. Metal containers involving the transfer of this chemical should be grounded and bonded. Where possible, automatically pump liquid from drums or other storage containers to process containers. Drums must be equipped with self-closing valves, pressure vacuum bungs, and flame arresters. Use only non-sparking tools and equipment, especially when opening and closing containers of this chemical. Sources of ignition such as smoking and open flames, are prohibited where this chemical is used, handled, or stored in a manner that could create a potential fire or explosion hazard.

Shipping: Chlorostyrene is not specifically cited by DOT[19] in its performance-oriented packaging standards. It can be classified as a combustible liquid n.o.s. (UN 1993) in which case in falls in Hazard Class 3, Packing Group III where there are no label requirements.

Spill Handling: Evacuate and restrict persons not wearing protective equipment from area of spill or leak until cleanup is complete. Remove all ignition sources. Ventilate area of spill or leak. Absorb liquids in vermiculite, dry sand, earth, peat, carbon, or a similar material and deposit in sealed containers. It may be necessary to contain and dispose of this chemical as a hazardous waste. If material or contaminated runoff enters waterways, notify downstream users of potentially contaminated waters. Contact your Department of Environmental Protection or your regional office of the federal EPA for specific recommendations. If employees are required to clean-up spills, they must be properly trained and equipped. OSHA 1910.120(q) may be applicable.

Fire Extinguishing: This chemical is a combustible liquid. Poisonous gases including chlorine are produced in fire. Use dry chemical, carbon dioxide, or alcohol foam extinguishers. Vapors are heavier than air and will collect in low areas. Vapors may travel long distances to ignition sources and flashback. Vapors in confined areas may explode when exposed to fire. Containers may explode in fire. Storage containers and parts of containers may rocket great distances, in many directions. If material or contaminated runoff enters waterways, notify downstream users of potentially contaminated waters. Notify local health and fire officials and pollution control agencies. From a secure, explosion-proof location, use water spray to cool exposed containers. If cooling streams are ineffective (venting sound increases in volume and pitch, tank discolors, or shows any signs of deforming), withdraw immediately to a secure position. If employees are expected to fight fires, they must be trained and equipped in OSHA 1910.156.

References

New Jersey Department of Health and Senior Services and Senior Services, "Hazardous Substance Fact Sheet: o-Chlorostyrene," Trenton, NJ (April 1986)

Chlorosulphonic Acid

Molecular Formula: $ClHO_3S$

Synonyms: Acido clorosulfonico (Spanish); Chlorosulfuric acid; *p*-Chloro-*o*-toluidine hydrochloride; 4-Chloro-*o*-toluidine, hydrochloride; Monochlorosulfuric acid; Sulfonic acid, monochloride; Sulfuric chlorohydrin

CAS Registry Number: 7790-94-5

RTECS Number: FX5730000

DOT ID: UN 1754; UN 2240

Regulatory Authority

- CLEAN WATER ACT: Section 311 Hazardous Substances/RQ 40CFR117.3 (same as CERCLA, see below)
- SUPERFUND/EPCRA 40CFR302.4 Reportable Quantity (RQ): CERCLA, 1,000 lb (454 kg)
- U.S. DOT 49CFR172.101, Inhalation Hazardous Chemical
- Canada, WHMIS, Ingredients Disclosure List

Cited in U.S. State Regulations: California (A, G), Florida (G), Massachusetts (G), Minnesota (G), New Jersey (G), Pennsylvania (G).

Description: Chlorosulphonic acid is a highly corrosive, colorless to yellow, slightly cloudy, fuming liquid with a sharp odor. Reactive with water. Boiling point = 155°C. Freezing/Melting point = -80°C. Reactive with water.

Potential Exposure: Used to make pesticides, detergents, pharmaceuticals, dyes, and resins.

Incompatibilities: Explosively reacts with water forming sulfuric and hydrochloric acid and dense fumes. Dangerously reactive, avoid contact with all other material. Strong oxidizer and strong acid; violent reaction with bases, reducing agents, combustibles, acids (especially sulfuric acid), alcohols, diphenyl ether, finely divided metals, silver nitrate. Contact with phosphorous may cause fire and explosions. Forms explosive material with ethyl alcohol. Attacks many metals; reaction with steel drums, forms explosive hydrogen gas which must be periodically relieved.

Permissible Exposure Limits in Air: The American Hygiene Association (AIHA) recommends a TWA of 0.3 ppm (1.3 mg/m^3). Due to its highly corrosive nature, all contact with this material should be reduced to the lowest possible level.

Routes of Entry: Inhalation.

Harmful Effects and Symptoms

Short Term Exposure: Skin or eye contact can cause severe irritation, burns, and permanent eye damage. Irritates the respiratory tract causing coughing, wheezing and/or shortness of breath. Higher exposures can cause pulmonary edema, a medical emergency that can be delayed for several hours. This can cause death.

Long Term Exposure: Can cause bronchitis with cough, phlegm, and/or shortness of breath.

Points of Attack: Lungs.

Medical Surveillance: Lung function tests. Consider x-ray following acute overexposure.

First Aid: If this chemical gets into the eyes, remove any contact lenses at once and irrigate immediately for at least 15 minutes, occasionally lifting upper and lower lids. Seek medical attention immediately. If this chemical contacts the skin, remove contaminated clothing and wash immediately with soap and water. Seek medical attention immediately. If this chemical has been inhaled, remove from exposure, begin rescue breathing (using universal precautions) if breathing has stopped and CPR if heart action has stopped. Transfer promptly to a medical facility. When this chemical has been swallowed, get medical attention. If victim is conscious, administer water or milk. Do not induce vomiting.

Personal Protective Methods: Wear protective gloves and clothing to prevent any reasonable probability of skin contact. Safety equipment suppliers/manufacturers can provide recommendations on the most protective glove/clothing material for your operation. Saranex and Polyethylene are among the recommended protective materials. All protective clothing (suits, gloves, footwear, headgear) should be clean, available each day, and put on before work. Contact lenses should not be worn when working with this chemical. Wear splash-proof chemical goggles and face shield unless full facepiece respiratory protection is worn. Employees should wash immediately with soap when skin is wet or contaminated. Provide emergency showers and eyewash.

Respirator Selection: Where there is a potential for overexposure: SCBAF:PD,PP (any MSHA/NIOSH approved self-contained breathing apparatus that has a full facepiece and is operated in a pressure-demand or other positive-pressure mode); or SAF:PD,PP:ASCBA (any supplied-air respirator that has a full facepiece and is operated in a pressure-demand or other positive-pressure mode in combination with an auxiliary, self-contained breathing apparatus operated in a pressure-demand or other positive pressure mode).

Storage: Prior to working with chlorosulphonic acid you should be trained on its proper handling and storage. Store in tightly closed containers in a cool, well ventilated area away from water, acids, bases, alcohols, metal powders, and organic combustible materials. It is preferable to store this chemical under NITROGEN. Where possible, automatically pump liquid from drums or other storage containers to process containers.

Shipping: Label required is "Corrosive, Poison." Chlorosulphonic acid is in DOT/UN Hazard Class 8 and Packing Group I.[19][20]

Spill Handling: Evacuate and restrict persons not wearing protective equipment from area of spill or leak until cleanup is complete. Remove all ignition sources. Ventilate area of spill or leak. Absorb liquids in vermiculite, dry sand, earth, peat, carbon, or a similar material and deposit in sealed containers. It may be necessary to contain and dispose of this chemical as a hazardous waste. If material or contaminated runoff enters waterways, notify downstream users of potentially contaminated waters. Contact your Department of Environmental Protection or your regional office of the federal EPA for specific recommendations. If employees are required to clean-up spills, they must be properly trained and equipped. OSHA 1910.120(q) may be applicable.

Initial isolation and protective action distances

Distances shown are likely to be affected during the first 30 minutes after materials are spilled and could increase with time. If more than one tank car, cargo tank, portable tank, or large cylinder is involved in the incident is leaking, the protective action distance may need to be increased. You may need to seek emergency information from CHEMTREC at (800) 424-9300 or seek professional environmental engineering assistance from the U.S. EPA Environmental Response Team at (908)548-8730 (24-hour response line).

Small spills (From a small package or a small leak from a large package)

First: Isolate in all directions (feet)..... 200

Then: Protect persons downwind (miles)

Day 0.1

Night 0.5

Large spills (From a large package or from many small packages)

First: Isolate in all directions (feet)..... 600

Then: Protect persons downwind (miles)

Day 0.4

Night 1.8

Fire Extinguishing: Chlorosulphonic acid does not burn but can readily ignite combustible materials on contact and will increase fire activity. Poisonous gases including hydrogen chloride and sulfur oxides are produced in fire. Decomposes explosively on contact with water. Use dry chemical, carbon dioxide, or foam extinguishers. DO NOT use water. Vapors are heavier than air and will collect in low areas. Containers may explode in fire. Storage containers and parts of containers may rocket great distances, in many directions. If material or contaminated runoff enters waterways, notify downstream users of potentially contaminated waters. Notify local health and fire officials and pollution control agencies. From a secure, explosion-proof location, use water spray to cool exposed containers. If cooling streams are ineffective (venting sound increases in volume and pitch, tank discolors, or shows any signs of deforming), withdraw immediately to a secure position. If employees are expected to fight fires, they must be trained and equipped in OSHA 1910.156.

References

New Jersey Department of Health and Senior Services, Hazardous Substance Fact Sheet, Trenton, NJ (1998)

Chlorothalonil

Molecular Formula: $C_8Cl_4N_2$

Common Formula: $C_6Cl_4(CN)_2$

Synonyms: BB Chlorothalonil; 1,3-Benzenedicarbonitrile,2,4,6,6-tetrachloro-; Bombardier; Bravo; Bravo 500; Bravo 6F; Bravo-W-75; Chiltern Ole; Chlorothalonil; Chlorthalonil (German); Contact 75; DAC 2787; Daconil; Daconil 2787 fungicide; Daconil 2787 W; Daconil F; Daconil M; Daconil Turf; Dacosoil; 1,3-Dicyanotetrachlorobenzene; Exotherm; Exotherm termil; Forturf; Grouticide 75; Impact Excel; Isophthalonitrile, tetrachloro; Jupital; Meta-tetrachlorophthalodinitrile; NCI-C00102; Nopcocide; Nopcocide 54DB; Nopcocide N-96; Nopocide N-40-D; Nopocide N-96-S; Nuocide; Power chlorothalonil 50; Repulse; RTECS No. NT2600000; Siclor; Sipcam UK Rover 5000; Sweep; Ter-Mil; 2,4,5,6-Tetrachloro-1,3-benzenedicarbonitrile; 2,4,5,6-Tetrachloro-1,3-dicyanobenzene; Tetrachloroisophthalonitrile; *m*-Tetrachlorophthalodinitrile; Tetrachlorophthalodinitrile, *m*-; Thalonil; TPN; TPN (Pesticide); Tripart Faber; Tripart Ultrafaber

CAS Registry Number: 1897-45-6

RTECS Number: NT2600000

DOT ID: UN 2588

EEC Number: 608-014-00-4

Regulatory Authority

- Carcinogen (Animal Positive) (NCI) (NTP) (California)
- EPCRA Section 313 Form R *de minimis* concentration reporting level: 1.0%

As a cyanide compound:

- CLEAN AIR ACT: Hazardous Air Pollutants (Title I, Part A, Section 112)
- CLEAN WATER ACT: 40CFR423, Appendix A, Priority Pollutants as cyanide, total
- U.S. DOT Regulated Marine Pollutant (49CFR172.101, Appendix B) as cyanide mixtures

Cited in U.S. State Regulations: California (A, G), Massachusetts (G), New Jersey (G), Pennsylvania (G).

Description: Chlorothalonil, $C_6Cl_4(CN)_2$, is a combustible, white, odorless, crystalline solid. Boiling point = 350°C. Freezing/Melting point = 260°C. Insoluble in water. Based on the NFPA rating system. (FEMA) Hazard Identification: Health 3, Flammability 1, Reactivity 0.

Potential Exposure: Chlorothalonil is a broad spectrum fungicide. Therefore people involved in its manufacture, formulation and application can be exposed.

Incompatibilities: Contact with strong oxidizers may cause a fire and explosion hazard. Thermal decomposition may include fumes of hydrogen cyanide.

Permissible Exposure Limits in Air: No standards set. However, inasmuch as it is a cyanide compound, the exposure limits are listed here: OSHA and ACGIH: 5 mg/m^3 TWA; NIOSH: Ceiling limit, 4.7 ppm; 5 mg/m^3 per 10 minutes as cyanides. All have notations that skin contact contributes significantly in overall exposure. IDLH = 25 mg/m^3 as CN

Determination in Air: See NIOSH Criteria Document 78–212 NITRILES.

Permissible Concentration in Water: A ten-day health advisory for a 10 kg child has been calculated by EPA to be 0.25 mg/l.

A longer term health advisory for a 10-kg child was calculated to be 0.15 mg/l and for a 70-kg adult was calculated to be 0.525 mg/l.

The estimated excess cancer risk associated with lifetime exposure to drinking water containing 0.525 mg/l of Chlorothalonil is 3.5×10^{-4}.

Determination in Water: Analysis of Chlorothalonil is by a gas chromatographic (GC) method applicable to the determination of certain chlorinated pesticides in water samples. In this method, approximately 1 liter of sample is extracted with Methylene chloride. The extract is concentrated and the compounds are separated using capillary column GC. Measurement is made using an electron capture detector. The method detection limit has not been determined for Chlorothalonil, but it is estimated that the detection limits for analytes included in this method are in the range of 0.01 – 0.1 μg/l.

Routes of Entry: Inhalation, skin contact.

Harmful Effects and Symptoms

Johnson et al. (1983) reported that Chlorothalonil exposure resulted in contact dermatitis in 14 of 20 workers involved in woodenware preservation. The wood preservative used by the workers consisted mainly of "white spirit," with 0.5% chlorothalonil as a fungicide. Workers exhibited erythema and edema of the eyelids, especially the upper eyelids, and eruptions on the wrist and forearms. Results of patch test conducted with 0.1% chlorothalonil in acetone were positive in 7 of 14 subjects. Reactions ranged from a few erythematous papules to marked popular erythema with a brownish hue without infiltration.

Wilson et al. (1985) gave Chlorothalonil (98.1% pure with less than 0.03% hexachlorobenzene) to Fischer 344 rats (60/sex/dose) in their diet at dose levels of 0, 40, 80 or 175 mg/kg/day. Males were treated for 116 weeks, while females received the chemical for 129 weeks. Survival among the various groups was comparable. In both sexes, at the high dose level, there were significant decreases in body weights. In addition, there were also significant increases in blood urea nitrogen and creatinine, while there were decreases in serum glucose and albumin levels. In both sexes, there were dose-dependent increases in kidney carcinomas and adenomas at doses above 40 mg/kg/day. In the high-dose females, there was also a significant increase in stomach papillomas. The data show that, in the Fischer 344 rat, chlorothalonil is a carcinogen.

The oral LD_{50} rat is 10 g/kg (insignificantly toxic).

Short Term Exposure: Irritates the eyes, skin, respiratory tract. Inhalation can cause coughing, phlegm, and/or tightness in the chest.

Long Term Exposure: Repeated or prolonged contact with skin may cause nose bleeding, skin sensitization and dermatitis with skin rash. May affect the kidneys and gastrointestinal tract. This chemical causes cancer of the kidneys in animals.

Points of Attack: Skin, lungs, kidneys.

Medical Surveillance: Complete blood count (CBC). Lung function tests. Kidney function tests.

First Aid: If this chemical gets into the eyes, remove any contact lenses at once and irrigate immediately for at least 20-30 minutes, occasionally lifting upper and lower lids. Seek medical attention immediately. If this chemical contacts the skin, remove contaminated clothing and wash immediately with soap and water. Seek medical attention immediately. If this chemical has been inhaled, remove from exposure, begin rescue breathing (using universal precautions) if breathing has stopped and CPR if heart action has stopped. Transfer promptly to a medical facility. When this chemical has been swallowed, get medical attention. Give large quantities of water and induce vomiting. Do not make an unconscious person vomit.

Personal Protective Methods: Wear protective gloves and clothing to prevent any reasonable probability of skin contact. Safety equipment suppliers/manufacturers can provide recommendations on the most protective glove/clothing material for your operation. All protective clothing (suits, gloves, footwear, headgear) should be clean, available each day, and put on before work. Contact lenses should not be worn when working with this chemical. Wear dust-proof chemical goggles and face shield unless full facepiece respiratory protection is worn. Employees should wash immediately with soap when skin is wet or contaminated. Provide emergency showers and eyewash. See NIOSH Criteria Document 78-212 NITRILES.

Respirator Selection: *Where there is no REL, at any detectable concentration:* SCBAF:PD,PP (any MSHA/NIOSH approved self-contained breathing apparatus that has a full facepiece and is operated in a pressure-demand or other positive-pressure mode); or SAF:PD,PP:ASCBA (any supplied-air respirator that has a full facepiece and is operated in a pressure-demand or other positive-pressure mode in combination with an auxiliary, self-contained breathing apparatus operated in a pressure-demand or other positive pressure mode).

Storage: Prior to working with chlorothalonil you should be trained on its proper handling and storage. Store in tightly closed containers in a cool, well ventilated area. Metal containers involving the transfer of this chemical should be grounded and bonded. Drums must be equipped with self-closing valves, pressure vacuum bungs, and flame arresters. Use only non-sparking tools and equipment, especially when opening and closing containers of this chemical. Sources of ignition such as smoking and open flames, are prohibited where this chemical is used, handled, or stored in a manner that could create a potential fire or explosion hazard. A regulated, marked area should be established where this chemical is handled, used, or stored in compliance with OSHA standard 1910.1045.

Shipping: Chlorothalonil is not cited specifically in DOT's performance-oriented packaging standards.[19] However, hazardous substances, solid, n.o.s., may apply in which case no label is needed, and there is no limit on aircraft or railcar shipments. This puts one in UN category 9188, Hazard Class ORV-E and Packing Group III.

Spill Handling: Evacuate persons not wearing protective equipment from area of spill or leak until clean-up is complete. Remove all ignition sources. Dampen spilled material with toluene to avoid dust Collect powdered material in the most convenient and safe manner and deposit in sealed containers. Ventilate area after clean-up is complete. It may be necessary to contain and dispose of this chemical as a hazardous waste. If material or contaminated runoff enters waterways, notify downstream users of potentially contaminated waters. Contact your Department of Environmental Protection or your regional office of the federal EPA for specific recommendations. If employees are required to clean-up spills, they must be properly trained and equipped. OSHA 1910.120(q) may be applicable.

Fire Extinguishing: This chemical is a combustible solid. Use dry chemical, carbon dioxide, water spray, or alcohol foam extinguishers. Poisonous gases are produced in fire including hydrogen cyanide, hydrogen chloride and nitrogen oxides. If material or contaminated runoff enters waterways, notify downstream users of potentially contaminated waters. Notify local health and fire officials and pollution control agencies. From a secure, explosion-proof location, use water spray to cool exposed containers. If cooling streams are ineffective (venting sound increases in volume and pitch, tank discolors, or shows any signs of deforming), withdraw immediately to a secure position. If employees are expected to fight fires, they must be trained and equipped in OSHA 1910.156.

Disposal Method Suggested: Incineration in a unit operating at 850°C equipped with off-gas scrubbing equipment.

References

New Jersey Department of Health and Senior Services, Hazardous Substance Fact Sheet, Chlorothalonil, Trenton, NJ (April 1998)

U.S. Environmental Protection Agency, "Chemical Profile: Chlorothalonil," Washington, DC, Office of Drinking Water (August 1987)

o-Chlorotoluene

Molecular Formula: C_7H_7Cl

Common Formula: $CH_3C_6H_4Cl$

Synonyms: Benzene, 1-chloro-2-methyl-; 2-Chloro-1-methylbenzene; 2-Chlorotoluene; 1-Methyl-2-chlorobenzene; 2-Methylchlorobenzene; Toluene, *o*-chloro-; *o*-Tolylchloride

CAS Registry Number: 95-49-8

RTECS Number: XS9000000

DOT ID: UN 2238

Regulatory Authority

- Air Pollutant Standard Set (ACGIH)[1] (Australia) (HSE) (Israel) (OSHA)[58] (Several States)[60] (Several Canadian Provinces)
- Canada, WHMIS, Ingredients Disclosure List

Cited in U.S. State Regulations: Alaska (G), California (A, G), Connecticut (A), Florida (A), Illinois (G), Maine (G), Massachusetts (G), Minneosta (G), Nevada (A), New Hampshire (G), New Jersey (G), North Dakota (A), Pennsylvania (G), Rhode Island (G), Virginia (A), West Virginia (G).

Description: o-Chlorotoluene, $CH_3C_6H_4Cl$, is a flammable, colorless liquid with an aromatic odor. The odor threshold = 0.32 ppm. Boiling point = 159°C. Flash point = 52°C. Hazard Identification (based on NFPA-704 M Rating System): Health 2, Flammability 2, Reactivity 0. Slightly soluble in water.

Potential Exposure: o-Chlorotoluene is widely used as a solvent and intermediate in the synthesis of dyes, synthetic rubber, pharmaceuticals, and other organic chemicals. Used as an insecticide, bactericide.

Incompatibilities: Incompatible with acids, alkalis, oxidizers, reducing materials, water.

Permissible Exposure Limits in Air: There is no OSHA PEL. NIOSH recommends an airborne exposure limit of 50 ppm (250 mg/m^3) TWA for a 10-hour workshift and STEL of 75 ppm (375 mg/m^3) not to be exceeded during any 15 minute work period, with the notation "skin" indicating the possibility of cutaneous absorption. The ACGIH recommends the same TWA averaged over an 8-hour workshift and does not have an STEL value. This chemical can be absorbed through the skin, thereby increasing exposure. There is no NIOSH IDLH. The UK HSE, Australia, Israel, and Canadian provinces of Alberta, BC, Ontario, and Quebec have adopted the same TWA as NIOSH and Alberta, BC, Ontario use the same STEL value. Several states have set guidelines or standards for chlorotoluene in ambient air[60] ranging from 2.5 – 3.75 mg/m^3 (North Dakota) to 4.0 mg/m^3 (Virginia) to 5.0 mg/m^3 (Connecticut) to 5.95 mg/m^3 (Nevada).

Determination in Air: No Methods listed by NIOSH or OSHA.

Permissible Concentration in Water: No criteria set, but EPA[32] has suggested a permissible ambient goal of 3,450 μg/l based on health effects.

Routes of Entry: Inhalation, skin absorption, ingestion, skin and/or eye contact.

Harmful Effects and Symptoms

Short Term Exposure: Contact can irritate and burn the eyes and skin. Inhalation can irritate the respiratory tract, causing coughing, and/or shortness of breath. High exposure can cause dizziness, loss of coordination, convulsions and coma. Vasodilatation, labored respiration, and narcosis have been observed in test animals.

Long Term Exposure: May affect the liver and kidneys.

Points of Attack: Eyes, skin, respiratory system, central nervous system, liver, kidneys. Prolonged or repeated contact may cause dermatitis.

Medical Surveillance: If symptoms develop or overexposure is suspected, the following may be useful: liver function tests. Kidney function tests. Examination by a dermatologist.

First Aid: If this chemical gets into the eyes, remove any contact lenses at once and irrigate immediately for at least 15 minutes, occasionally lifting upper and lower lids. Seek medical attention immediately. If this chemical contacts the skin, remove contaminated clothing and wash immediately with soap and water. Seek medical attention immediately. If this chemical has been inhaled, remove from exposure, begin rescue breathing (using universal precautions) if breathing has stopped and CPR if heart action has stopped. Transfer promptly to a medical facility. When this chemical has been swallowed, get medical attention. Give large quantities of water and induce vomiting. Do not make an unconscious person vomit.

Personal Protective Methods: Wear solvent-resistant gloves and clothing to prevent any reasonable probability of skin contact. Safety equipment suppliers/manufacturers can provide recommendations on the most protective glove/clothing material for your operation. Viton is recommended in the literature. All protective clothing (suits, gloves, footwear, headgear) should be clean, available each day, and put on before work. Contact lenses should not be worn when working with this chemical. Wear splash-proof chemical goggles and face shield unless full facepiece respiratory protection is worn. Employees should wash immediately with soap when skin is wet or contaminated. Remove nonimpervious clothing immediately if wet or contaminated. Provide emergency showers and eyewash.

Respirator Selection: Where the potential exists for exposures over 50 ppm, use a MSHA/NIOSH approved full facepiece respirator with an organic vapor cartridge/canister. Increased protection is obtained from full facepiece powered-air purifying respirators.

Where the potential for high exposures exists, use a MSHA/NIOSH approved supplied-air respirator with a full facepiece operated in the positive pressure mode or with a full facepiece, hood, or helmet in the continuous flow mode, or use a MSHA/NIOSH approved self-contained breathing apparatus with a full facepiece operated in pressure-demand or other positive pressure mode.

Storage: Prior to working with o-Chlorotoluene you should be trained on its proper handling and storage. Store in tightly closed containers in a cool, well ventilated area. Metal containers involving the transfer of this chemical should be grounded and bonded. Drums must be equipped with self-closing valves, pressure vacuum bungs, and flame arresters. Use only non-sparking tools and equipment, especially when opening and closing containers of this chemical. Sources of ignition such as smoking and open flames, are prohibited where this chemical is used, handled, or stored in a manner that could create a potential fire or explosion hazard.

Shipping: Chlorotoluenes must carry a "Flammable Liquid" label. They fall in Hazard Class 3 and Packing Group III.[19]

Spill Handling: Evacuate and restrict persons not wearing protective equipment from area of spill or leak until cleanup is complete. Remove all ignition sources. Ventilate area of spill or leak. Absorb liquids in vermiculite, dry sand, earth, peat, carbon, or a similar material and deposit in sealed containers. It may be necessary to contain and dispose of this chemical as a hazardous waste. If material or contaminated runoff enters waterways, notify downstream users of potentially contaminated waters. Contact your Department of Environmental Protection or your regional office of the federal EPA for specific recommendations. If employees are required to clean-up spills, they must be properly trained and equipped. OSHA 1910.120(q) may be applicable.

Fire Extinguishing: This chemical is a combustible liquid. Poisonous gases including chlorine are produced in fire. Use dry chemical, carbon dioxide, or alcohol or polymer foam extinguishers. Vapors are heavier than air and will collect in low areas. Vapors may travel long distances to ignition sources and flashback. Vapors in confined areas may explode when exposed to fire. Containers may explode in fire. Storage containers and parts of containers may rocket great distances, in many directions. If material or contaminated runoff enters waterways, notify downstream users of potentially contaminated waters. Notify local health and fire officials and pollution control agencies. From a secure, explosion-proof location, use water spray to cool exposed containers. If cooling streams are ineffective (venting sound increases in volume and pitch, tank discolors, or shows any signs of deforming), withdraw immediately to a secure position. If employees are expected to fight fires, they must be trained and equipped in OSHA 1910.156.

Disposal Method Suggested: In accordance with 40CFR 165 recommendations for the disposal of pesticides and pesticide containers. Must be disposed properly by following package label directions or by contacting your state pesticide or environmental control agency or by contacting your regional EPA office.

References

New Jersey Department of Health and Senior Services and Senior Services, "Hazardous Substance Fact Sheet: o-Chlorotoluene," Trenton, NJ (November 1998)

Chloroxuron

Molecular Formula: $C_{15}H_{15}ClN_2O_2$

Common Formula: $(CH_3)_2NCONHC_6H_4OC_6H_4Cl$

Synonyms: C 1983; *N*'-[4-(4-Chlorophenoxy)phenyl]-*N,N*-dimethylurea; 3-[*p*-(*p*-Chlorophenoxy)phenyl-1,1]-dimethylurea; 3-[4-(4-Chlorophenoxy)phenyl]-1,1-dimethylurea; Chloroxifenidum; Ciba 1983; Cloroxuron (Spanish); Norex; Tenoran®; Urea, 3-[*p*-(*p*-chlorophenoxy)phenyl]-1,1-dimethyl-; Urea, N'-[4-(4-chlorophenoxy)phenyl]-*N,N*-dimethyl-

CAS Registry Number: 1982-47-4

RTECS Number: YS6125000

Regulatory Authority

- SUPERFUND/EPCRA 40CFR355, Appendix B Extremely Hazardous Substances: TPQ = 500/10,000 lb (227/4,540 kg)
- SUPERFUND/EPCRA 40CFR302.4 Reportable Quantity (RQ): EHS, 1 lb (0.454 kg)

Cited in U.S. State Regulations: Florida (G), Massachusetts (G), New Jersey (G), Pennsylvania (G).

Description: Chloroxuron, $(CH_3)_2NCONHC_6H_4OC_6H_4Cl$, is a combustible, colorless crystalline solid Freezing/Melting point = 151 – 152°C. Practically insoluble in water.

Potential Exposure: Those involved in the manufacture, formulation and application of chloroxuron for use as a selective pre- and early post-emergency herbicide in soybeans, strawberries, various vegetable croups and ornamentals. It is a root- and foliage-absorbed herbicide selective in leek, celery, onion, carrot and strawberry.

Incompatibilities: Contact with strong oxidizers may cause fire and explosion hazard.

Permissible Exposure Limits in Air: No standards set.

Permissible Concentration in Water: No criteria set.

Routes of Entry: Ingestion.

Harmful Effects and Symptoms

Short Term Exposure: Slightly irritating to eyes and skin. The LD_{50} oral (dog) is 10 mg/kg (highly toxic). The LD_{50} oral (rat) is 3,700 mg/kg. Chloroxuron is stated to be highly toxic to humans by ingestion and under certain conditions, it can form carcinogenic dimethylnitrosamine. A regulated, marked area should be established where this chemical is handled, used, or stored in compliance with OSHA standard 1910.1045.

Long Term Exposure: No data available.

First Aid: If this chemical gets into the eyes, remove any contact lenses at once and irrigate immediately for at least 15 minutes, occasionally lifting upper and lower lids. Seek medical attention immediately. If this chemical contacts the skin, remove contaminated clothing and wash immediately with soap and water. Seek medical attention immediately. If this chemical has been inhaled, remove from exposure, begin rescue breathing (using universal precautions) if breathing has stopped and CPR if heart action has stopped. Transfer promptly to a medical facility. When this chemical has been swallowed, get medical attention. Give large quantities of water and induce vomiting. Do not make an unconscious person vomit.

Personal Protective Methods: Wear protective gloves and clothing to prevent any reasonable probability of skin contact. Safety equipment suppliers/manufacturers can provide recommendations on the most protective glove/clothing material for your operation. All protective clothing (suits, gloves, footwear, headgear) should be clean, available each day, and put on before work. Contact lenses should not be worn when working with this chemical. Wear dust-proof chemical goggles and face shield unless full facepiece respiratory protection is worn. Employees should wash immediately with soap when skin is wet or contaminated. Provide emergency showers and eyewash.

Storage: Prior to working with Chloroxuron you should be trained on its proper handling and storage. Store in tightly closed containers in a cool, well-ventilated area.

Shipping: Pesticides, solid, toxic, n.o.s. require a label of "Poison" in Packing Group II. This compound falls in UN/DOT Hazard Class 6.1.[19][20]

Spill Handling: Evacuate persons not wearing protective equipment from area of spill or leak until clean up is complete. Remove all ignition sources. Do not touch spilled material; stop leak if you can do it without risk. Use water spray to reduce vapors. Collect powdered material in the most convenient and safe manner and deposit in sealed containers. *Large spills:* dike far ahead of spill for later disposal. Ventilate area after clean-up is complete. It may be necessary to contain and dispose of this chemical as a hazardous waste. If material or contaminated runoff enters waterways, notify downstream users of potentially contaminated waters. Contact your Department of Environmental Protection or your regional office of the federal EPA for specific recommendations. If employees are required to clean-up spills, they must be properly trained and equipped. OSHA 1910.120(q) may be applicable.

Fire Extinguishing: This chemical is a combustible solid. Use dry chemical, carbon dioxide, water spray, or standard foam extinguishers. Poisonous gases are produced in fire including oxides of nitrogen and carbon, and corrosive fumes of chlorides. If material or contaminated runoff enters waterways, notify downstream users of potentially contaminated waters. Notify local health and fire officials and pollution control agencies. From a secure, explosion-proof location, use water spray to cool exposed containers. If cooling streams are ineffective (venting sound increases in volume and pitch, tank discolors, or shows any signs of deforming), withdraw immediately to a secure position. If employees are expected to fight fires, they must be trained and equipped in OSHA 1910.156.

Disposal Method Suggested: Incinerate in a unit with effluent gas scrubbing.[22] In accordance with 40CFR165 recommendations for the disposal of pesticides and pesticide containers. Must be disposed properly by following package label directions or by contacting your state pesticide or environmental control agency or by contacting your regional EPA office.

References

U.S. Environmental Protection Agency, "Chemical Profile: Chloroxuron," Washington, DC, Chemical Emergency Preparedness Program (October 31, 1985)

Chlorpyrifos

Molecular Formula: $C_9H_{11}Cl_3NO_3PS$

Synonyms: Brodan; α- Chlorpyrifos 48EC (α); Chlorpyrifos-ethyl; Clorpirifos (Spanish); Detmol U.A.; *O,O*-Diaethyl-*O*-3,5,6-trichlor-2-pyridylmonothiophosphat (German); *o,o*-Diethyl; *O,O*-Dimethyl *O*-(3,5,6-trichloro-2-pyridinyl) phosphorothioate; Dowco® 179; Dursban®; Dursban® 4; Dursban® 5G; Dursban® F; EF 121; ENT 27311; Eradex; Global Crawling insect bait; Lorsban®; Murphy Super Root Guard; Phosphorothioic acid, *O,O*-diethyl *O*-(3,5,6-trichloro-2-pyridinyl) ester; 2-Pyridinol, 3,5,6-trichloro-, *O*-ester with *O,O*-diethyl phosphorothioate; Pyrinex; Spannit®; Talon®; *O*-3,5,6-Trichloro-2-pyridyl phosphorothioate; Twinspan®

CAS Registry Number: 2921-88-2

RTECS Number: TF6300000

DOT ID: UN 2783

EEC Number: 015-084-00-4

Regulatory Authority

- Air Pollutant Standard Set (ACGIH)[1] (Australia) (HSE)[33] (Israel) (Mexico) (OSHA)[58] (former USSR)[35] (Several States)[60] (Several Canadian Provinces)
- CLEAN WATER ACT: Section 311 Hazardous Substances/RQ 40CFR117.3 (same as CERCLA, see below)
- SUPERFUND/EPCRA 40CFR302.4 Reportable Quantity (RQ): CERCLA, 1 lb (0.454 kg)
- U.S. DOT Regulated Marine Pollutant (49CFR172.101, Appendix B), severe pollutant
- Canada, Drinking Water Quality MAC = 0.09 mg/l

Cited in U.S. State Regulations: Alaska (G), California (G), Connecticut (A), Florida (G), Illinois (G), Maine (G), Massachusetts (G), Michigan (G), Minnesota (G), Nevada (A), New Hampshire (G), New Jersey (G), North Dakota (A), Pennsylvania (G), Rhode Island (G), Virginia (A), West Virginia (G)

Description: Chlorpyrifos is a colorless crystalline compound with a mild mercaptan odor. The odor is also described as like natural gas. Freezing/Melting point = 41 – 43°C. Insoluble in water.

Potential Exposure: Those involved in the manufacture, formulation and application of this insecticide.

Incompatibilities: Above 130°C this chemical may undergo violent exothermic decomposition. The substance decomposes on heating at approximately 160°C and on burning producing toxic and corrosive fumes including hydrogen chloride, nitrogen oxides, phosphorous oxides, sulfur oxides. Reacts with strong acids, strong bases, causing hydrolysis. Attacks copper and brass.

Permissible Exposure Limits in Air: There is no OSHA PEL. NIOSH recommends a REL of 0.2 mg/m^3 TWA for a 10-hour workshift. and STEL of 0.6 mg/m^3 (skin). ACGIH, has recommended the same TWA (for an 8-hour workshift). Both limits bear the notation "skin" indicating the cutaneous absorption should be prevented so the threshold limit value is not invalidated. UK's HSE,[33] Australia, Mexico, and the Canadian provinces of Alberta, BC, Ontario, and Quebec have set these same TWA values as ACGIH. The former Soviet Union[35] has set a MAC in workplace air of 0.3 mg/m^3. Several states have set guidelines or standards for Chlorpyrifos in ambient air[60] ranging from 2-6 μg/m^3 (North Dakota) to 3.0 μg/m^3 (Virginia) to 4.0 μg/m^3 (Connecticut) to 5 μg/m^3 (Nevada).

Determination in Air: OSHA versatile sampler-2; Toluene/Acetone; Gas chromatography/Flame photometric detection for sulfur, nitrogen, or phosphorus; NIOSH IV Method #5600, Organophosphorus Pesticides.

Permissible Concentration in Water: Mexico[35] has set a limit of 3.0 μg/l in coastal waters and 0.03 mg/l in estuaries. The former USSR[35] has set a MAC in water bodies used for fishery purposes of 5.0 μg/l.

Routes of Entry: Skin absorption, inhalation of dust, ingestion.

Harmful Effects and Symptoms

Short Term Exposure: May cause eye and skin irritation. Cholinesterase inhibitor. Exposure at high levels may result in death. The effects may be delayed. The LD_{50} rat is 82 mg/kg (moderately toxic). Chlorpyrifos can affect you when breathed in and quickly enters the body be passing through the skin. Severe poisoning can occur from skin contact. It is a moderately toxic organophosphate chemical. Exposure can cause rapid severe poisoning with headache, sweating, nausea and vomiting, diarrhea, loss of coordination, and death.

Long Term Exposure: Cholinesterase inhibitor; cumulative effect is possible. Chlorpyrifos may damage the nervous system with repeated exposure, resulting in convulsions, respiratory failure. May cause liver damage.

Points of Attack: Respiratory system, central nervous system, peripheral nervous system, plasma cholinesterase.

Medical Surveillance: Before employment and at regular times after that, the following are recommended: Plasma and red blood cell cholinesterase levels (tests for the enzyme poisoned by this chemical). If exposure stops, plasma levels return to normal in 1-2 weeks while red blood cell levels may be reduced for 1-3 months.

When cholinesterase enzyme levels are reduced by 25% or more below preemployment levels, risk of poisoning is increased, even if results are in lower ranges of "normal." Reassignment to work not involving organophosphate or carbamate pesticides is recommended until enzyme levels recover. If symptoms develop or overexposure occurs, repeat the above tests as soon as possible and get an exam of the nervous system. Also consider complete blood count. Consider chest x-ray following acute overexposure. Do not drink any alcoholic beverages before or during use. Alcohol promotes absorption of organic phosphates.

First Aid: If this chemical gets into the eyes, remove any contact lenses at once and irrigate immediately for at least 15 minutes, occasionally lifting upper and lower lids. Seek medical attention immediately. If this chemical contacts the skin, remove contaminated clothing and wash immediately with soap and water. Speed in removing material from skin is of extreme importance. Shampoo hair promptly if contaminated. Seek medical attention immediately. If this chemical has been inhaled, remove from exposure, begin rescue breathing (using universal precautions) if breathing has stopped and CPR if heart action has stopped. Transfer promptly to a medical facility. When this chemical has been swallowed, get medical attention. Give large quantities of water and induce vomiting.

Do not make an unconscious person vomit. Medical observation is recommended.

Personal Protective Methods: *Clothing:* Avoid skin contact with Chlorpyrifos. Wear protective gloves and clothing. Safety equipment suppliers/manufacturers can provide recommendations on the most protective glove/clothing material for your operation. All protective clothing (suits, gloves, footwear, headgear) should be clean, available each day, and put on before work.

Eye Protection: Wear splash-proof chemical goggles and face shield when working with liquid, unless full facepiece respiratory protection is worn. Wear dust-proof goggles and face shield when working with powders or dust, unless full facepiece respiratory protection is worn.

Respirator Selection: Where the potential exists for exposures over 0.2 mg/m^3, use a MSHA/NIOSH approved full facepiece respirator with a pesticide cartridge. Greater protection is provided by a powered-air-purifying respirator. Where the potential for high exposures exists, use a MSHA/NIOSH approved supplied-air respirator with a full facepiece operated in the positive pressure mode or with a full facepiece, hood, or helmet in the continuous flow mode, or use a MSHA/NIOSH approved self-contained breathing apparatus with a full facepiece, hood, or helmet in the continuous flow mode, or use a MSHA/NIOSH approved self-contained breathing apparatus with a full facepiece operated in pressure-demand or other positive pressure mode.

Storage: Prior to working with Chlorpyrifos you should be trained on its proper handling and storage. Chlorpyrifos must be stored to avoid contact with strong bases, or acids, or acid fumes since violent reactions occur. Store in tightly closed containers in a cool, well-ventilated area away from sources of heat.

Shipping: Organophosphorus pesticides, solid, toxic n.o.s. require a "Keep Away from Food" label for Hazard Class 6.1 and Packing Group III which applies to Chlorpyrifos.

Spill Handling: Evacuate persons not wearing protective equipment from area of spill or leak until clean-up is complete. Remove all ignition sources. Collect powdered material in the most convenient and safe manner and deposit in sealed containers. Ventilate area after clean-up is complete. Absorb liquid containing Chlorpyrifos in vermiculite, dry sand, earth, or similar material. It may be necessary to contain and dispose of this chemical as a hazardous waste. If material or contaminated runoff enters waterways, notify downstream users of potentially contaminated waters. Contact your Department of Environmental Protection or your regional office of the federal EPA for specific recommendations. If employees are required to clean-up spills, they must be properly trained and equipped. OSHA 1910.120(q) may be applicable.

Fire Extinguishing: Chlorpyrifos may burn, but does not readily ignite. Poisonous gases are produced in fire, including hydrogen chloride, nitrogen oxides, phosphorous oxides, sulfur oxides, and organic sulfides. Use dry chemical, carbon dioxide, water spray, or standard foam extinguishers. If material or contaminated runoff enters waterways, notify downstream users of potentially contaminated waters. Notify local health and fire officials and pollution control agencies. Heat above 130°C may cause violent exothermic reaction. Containers may explode in fire. From a secure, explosion-proof location, use water spray to cool exposed containers. If cooling streams are ineffective (venting sound increases in volume and pitch, tank discolors, or shows any signs of deforming), withdraw immediately to a secure position. If employees are expected to fight fires, they must be trained and equipped in OSHA 1910.156.

Disposal Method Suggested: This compound is 50% hydrolyzed in aqueous MeOH solution at pH 6 in 1,930 days, and in 7.2 days at pH 9.96. Spray mixtures of <1% concentration are destroyed with an excess of 5.25% sodium hypochlorite in <30 minutes at 100°C, and in 24 hours at 30°C. Concentrated (61.5%) mixtures are essentially destroyed by treatment with 100:1 volumes of the above sodium hypochlorite solution and steam in 10 minutes.[22] In accordance with 40CFR165 recommendations for the disposal of pesticides and pesticide containers. Must be disposed properly by following package label directions or by contacting your state pesticide or environmental control agency or by contacting your regional EPA office.

References

New Jersey Department of Health and Senior Services and Senior Services, "Hazardous Substance Fact Sheet: Chlorpyrifos," Trenton, NJ (July 1996)

Chlorthiophos

Molecular Formula: $C_{11}H_{15}Cl_2O_3PS_2$

Synonyms: Celamerck S-2957; Cela S-2957; Celathion; CM S 2957; *O*-[Dichloro(methylthio)phenyl] *O,O*-diethyl phosphorothioate (3 isomers); *O,O*-[Diethyl-*O*-2,4,5-dichloro(methylthio)phenyl] thionophosphate; ENT 27,635; NSC 195164; OMS 1342

CAS Registry Number: 21923-23-9

RTECS Number: TF1590000

Regulatory Authority

- Banned or Severely Restricted (In Agriculture) (Germany, Malaysia) (UN)[13]
- SUPERFUND/EPCRA 40CFR355, Appendix B Extremely Hazardous Substances: TPQ = 500 lb (227 kg)
- U.S. DOT Regulated Marine Pollutant (49CFR172.101, Appendix B)

Cited in U.S. State Regulations: California (G), Florida (G), Massachusetts (G), New Jersey (G), Pennsylvania (G).

Description: Chlorthiophos is a yellowish-brown liquid. Boiling point = 153 – 158°C @ 13 mm Hg pressure and crystallizes at less than 25°C. Hazard Identification (based on NFPA-704 M Rating System): Health 3, Flammability 1, Reactivity 0.

Potential Exposure: Those involved in the manufacture, formulation and application of this insecticide and acaricide (now discontinued in some cases).

Incompatibilities: Strong acids, strong bases, strong oxidizers.

Permissible Exposure Limits in Air: No standards set. However, this chemical can be absorbed through the skin, thereby increasing exposure.

Determination in Air: OSHA versatile sampler-2; Toluene/Acetone; Gas chromatography/Flame photometric detection for sulfur, nitrogen, or phosphorus; NIOSH Method IV Method #5600, Organophosphorus Pesticides.

Routes of Entry: Inhalation, ingestion, skin contact. This chemical can be absorbed through the skin, thereby increasing exposure.

Harmful Effects and Symptoms

Short Term Exposure: The LD_{50} oral (rabbit) is 20 mg/kg which is in the highly toxic class.

Symptoms of Chlorthiophos exposure include headache, giddiness, blurred vision, nervousness, weakness, nausea, cramps, diarrhea, and discomfort in the chest. Signs include sweating, tearing, salivation, vomiting, cyanosis, convulsions, coma, loss of reflexes and loss of sphincter control.

Organic phosphorus insecticides are absorbed by the skin, as well as by the respiratory and gastrointestinal tracts. They are cholinesterase inhibitors.

Long Term Exposure: Cholinesterase inhibitor; cumulative effect is possible. Chlorthiophos may damage the nervous system with repeated exposure, resulting in convulsions, respiratory failure. May cause liver damage.

Points of Attack: Respiratory system, central nervous system, peripheral nervous system, plasma cholinesterase.

Medical Surveillance: Before employment and at regular times after that, the following are recommended: Plasma and red blood cell cholinesterase levels (tests for the enzyme poisoned by this chemical). If exposure stops, plasma levels return to normal in 1-2 weeks while red blood cell levels may be reduced for 1-3 months.

When cholinesterase enzyme levels are reduced by 25% or more below preemployment levels, risk of poisoning is increased, even if results are in lower ranges of "normal." Reassignment to work not involving organophosphate or carbamate pesticides is recommended until enzyme levels recover. If symptoms develop or overexposure occurs, repeat the above tests as soon as possible and get an exam of the nervous system. Also consider complete blood count. Consider chest x-ray following acute overexposure. Do not drink any alcoholic beverages before or during use. Alcohol promotes absorption of organic phosphates.

First Aid: If this chemical gets into the eyes, remove any contact lenses at once and irrigate immediately for at least 15 minutes, occasionally lifting upper and lower lids. Seek medical attention immediately. If this chemical contacts the skin, remove contaminated clothing and wash immediately with soap and water. Speed in removing material from skin is of extreme importance. Shampoo hair promptly if contaminated. Seek medical attention immediately. If this chemical has been inhaled, remove from exposure, begin rescue breathing (using universal precautions) if breathing has stopped and CPR if heart action has stopped. Transfer promptly to a medical facility. When this chemical has been swallowed, get medical attention. Give large quantities of water and induce vomiting. Do not make an unconscious person vomit. Effects may be delayed; keep victim under observation.

Personal Protective Methods: *Clothing:* Avoid skin contact with this chemical. Wear protective gloves and clothing. Safety equipment suppliers/manufacturers can provide recommendations on the most protective glove/clothing material for your operation. All protective clothing (suits, gloves, footwear, headgear) should be clean, available each day, and put on before work.

Eye Protection: Wear splash-proof chemical goggles and face shield when working with liquid, unless full facepiece respiratory protection is worn. Wear dust-proof goggles and face shield when working with powders or dust, unless full facepiece respiratory protection is worn.

Respirator Selection: Where the potential exists for exposures, use a MSHA/NIOSH approved full facepiece respirator with a pesticide cartridge. Greater protection is provided by a powered-air-purifying respirator. Where the potential for high exposures exists, use a MSHA/NIOSH approved supplied-air respirator with a full facepiece operated in the positive pressure mode or with a full facepiece, hood, or helmet in the continuous flow mode, or use a MSHA/NIOSH approved self-contained breathing apparatus with a full facepiece, hood, or helmet in the continuous flow mode, or use a MSHA/NIOSH approved self-contained breathing apparatus with a full facepiece operated in pressure-demand or other positive pressure mode.

Storage: Prior to working with Chlorthiophos you should be trained on its proper handling and storage. Store in tightly closed containers in a cool, well-ventilated area away from incompatible materials.

Shipping: Organophosphorus pesticides, liquid, toxic, n.o.s. have a DOT label requirement of "Poison." Chlorthiophos falls in UN/DOT Hazard Class 6.1 and Packing Group II.[19][20]

Spill Handling: Evacuate and restrict persons not wearing protective equipment from area of spill or leak until cleanup

is complete. Remove all ignition sources. Stay upwind; keep out of low areas. Ventilate closed spaces before entering them. Wear positive pressure breathing apparatus and special protective clothing. Do not touch spilled material; stop leak if you can do so without risk. Use water spray to reduce vapors. Small spills: absorb with sand or other non-combustible absorbent material and place into containers for later disposal. Large spills: dike far ahead of spill for later disposal It may be necessary to contain and dispose of this chemical as a hazardous waste. If material or contaminated runoff enters waterways, notify downstream users of potentially contaminated waters. Contact your Department of Environmental Protection or your regional office of the federal EPA for specific recommendations. If employees are required to clean-up spills, they must be properly trained and equipped. OSHA 1910.120(q) may be applicable.

Fire Extinguishing: This material may burn, but does not ignite readily. For small fires, use dry chemical, carbon dioxide, water spray, or foam. For large fires, use water spray, fog, or foam. Stay upwind; keep out of low areas. Move containers from fire area if you can do so without risk. Fight fire from maximum distance. Dike fire control water for later disposal; do not scatter the material. Poisonous gases are produced in fire. If material or contaminated runoff enters waterways, notify downstream users of potentially contaminated waters. Notify local health and fire officials and pollution control agencies. From a secure, explosion-proof location, use water spray to cool exposed containers. If cooling streams are ineffective (venting sound increases in volume and pitch, tank discolors, or shows any signs of deforming), withdraw immediately to a secure position. If employees are expected to fight fires, they must be trained and equipped in OSHA 1910.156.

Disposal Method Suggested: In accordance with 40CFR 165 recommendations for the disposal of pesticides and pesticide containers. Must be disposed properly by following package label directions or by contacting your state pesticide or environmental control agency or by contacting your regional EPA office.

References

U.S. Environmental Protection Agency, "Chemical Profile: Chlorthiophos," Washington, DC, Chemical Emergency Preparedness Program (November 30, 1987)

Chromic Acetate

Molecular Formula: $C_6H_9CrO_6$

Common Formula: $Cr(C_2H_3O_2)_3$

Synonyms: Acetato cromico (Spanish); Acetic acid, chromium(3+) salt; Chromic acetate(III); Chromium acetate; Chromium(III) acetate; Chromium triacetate

CAS Registry Number: 1066-30-4

RTECS Number: AG2975000

DOT ID: UN 9101

Regulatory Authority

- Carcinogen (NJ) (DFG)[3]
- Air Pollutant Standard Set (ACGIH)[1] (HSE)[33]
- CLEAN AIR ACT: Hazardous Air Pollutants (Title I, Part A, Section 112). as chromium compounds
- CLEAN WATER ACT: Section 311 Hazardous Substances/RQ 40CFR117.3 (same as CERCLA, see below); Section 313 Water Priority Chemicals (57FR41331, 9/9/92); 40CFR 401.15 Section 307 Toxic Pollutants
- SUPERFUND/EPCRA 40CFR302.4 Reportable Quantity (RQ): CERCLA, 1,000 lb (454 kg)
- EPCRA Section 313 Form R *de minimis* concentration reporting level: Chromium III compounds: 1.0%
- Canada, WHMIS, Ingredients Disclosure List

Cited in U.S. State Regulations: Alaska (G), California (G), Florida (G), Massachusetts (G), New Hampshire (G), New Jersey (G), Oklahoma (G), Pennsylvania (G), Rhode Island (G), West Virginia (G).

Description: Chromic acetate, $Cr(C_2H_3O_2)_3$ is a gray-green powder or blue-green pasty mass. Hazard Identification (based on NFPA-704 M Rating System): Health 1, Flammability 0, Reactivity 0. Soluble in water.

Potential Exposure: Chromic Acetate is used to fix certain textile dyes, to harden photographic emulsions in tanning and as a catalyst.

Incompatibilities: Contact with strong oxidizers may cause fire and explosion hazard.

Permissible Exposure Limits in Air: The OSHA PEL for Chromium(III) compounds (as Cr) is 0.5 mg/m^3 TWA for an 8-hour workshift. ACGIH[1] and HSE[33] recommend or have set the same TWA as OSHA. The NIOSH IDLH is 25 mg/m^3.

Determination in Air: Filter; Acid; Flame atomic absorption spectrometry; NIOSH IV Method #7024, Chromium.

Permissible Concentration in Water: For the protection of freshwater aquatic life: Trivalent chromium: not to exceed $e^{[1.08 \ln(\text{hardness}) + 3.48]}$ μg/l. For the protection of saltwater aquatic life: Trivalent chromium: 10,300 μg/l on an acute toxicity basis. To protect human health: Trivalent chromium: 170 μg/l; hexavalent chromium 50 μg/l according to EPA.[6]

EPA[49] has set a long-term health advisory for adults of 0.84 mg/l and a lifetime health advisory of 0.12 mg/l (120 μg/l) for chromium. EPA's maximum drinking water level (MCL) is 0.1 mg/l.[62]

Germany, Canada, EEC, and WHO[35] have set a limit of 0.05 mg/l in drinking water.

The states of Maine and Minnesota have set guidelines for chromium in drinking water[61] of 50 μg/l for Maine and 120 μg/l for Minnesota.

Determination in Water: Total chromium may be determined by digestion followed by atomic absorption or by colorimetry (diphenylcarbazide) or by inductively coupled plasma (CP) optical emission spectrometry. Chromium (VI) may be determined by extraction and atomic absorption or colorimetry (using diphenylhydrazide). Dissolved total Cr or Cr(VI) may be determined by 0.45 μ filtration followed by the above-cited methods.[49]

Routes of Entry: Inhalation, ingestion, skin and/or eye contact

Harmful Effects and Symptoms

Short Term Exposure: Inhalation may cause irritation of the eyes, nose and throat. Eye contact may cause irritation, redness and tearing. Skin contact may cause irritation, redness and tearing. Skin allergy sometimes occurs, with itching, redness and/or an eczema-like rash. If this happens, future contact can trigger symptoms

Long Term Exposure: Chromic acetate is a carconigen — handle with extreme caution. May cause lung or throat cancer; birth defects, miscarriage, skin allergy with redness, itching and rash. Repeated or prolonged skin contact may cause skin sensitization, irritation and dermatitis. Some chromium compounds can cause a sore or hole in the septum dividing the nose.

Points of Attack: Eyes, skin.

Medical Surveillance: If illness occurs or overexposure is suspected, medical attention is recommended. Evaluation by a qualified allergist, including careful exposure history and special testing, may help diagnose skin allergy.

First Aid: If this chemical gets into the eyes, remove any contact lenses at once and irrigate immediately for at least 15 minutes, occasionally lifting upper and lower lids. Seek medical attention immediately. If this chemical contacts the skin, remove contaminated clothing and wash immediately with soap and water. Seek medical attention immediately. If this chemical has been inhaled, remove from exposure, begin rescue breathing (using universal precautions) if breathing has stopped and CPR if heart action has stopped. Transfer promptly to a medical facility. When this chemical has been swallowed, get medical attention.

Personal Protective Methods: Wear protective gloves and clothing to prevent any reasonable probability of skin contact. Safety equipment suppliers/manufacturers can provide recommendations on the most protective glove/clothing material for your operation. All protective clothing (suits, gloves, footwear, headgear) should be clean, available each day, and put on before work. Contact lenses should not be worn when working with this chemical. Wear dust-proof chemical goggles and face shield unless full facepiece respiratory protection is worn. Employees should wash immediately with soap when skin is wet or contaminated. Provide emergency showers and eyewash.

Respirator Selection: NIOSH/OSHA as chromium(III) compounds: *2.5 mg/m³:* DM (any dust and mist respirator). *5 mg/m³:* DMXSQ (any dust and mist respirator except single-use and quarter mask respirators); or SA (any supplied-air respirator); *12.5 mg/m³:* SA:CF (any supplied-air respirator operated in a continuous-flow mode); or PAPRDM (any powered, air-purifying respirator with a dust and mist filter). *25 mg/m³:* HiEF (any air-purifying, full-facepiece respirator with a high-efficiency particulate filter); or PAPRTHiE (any powered, air-purifying respirator with a tight-fitting facepiece and a high-efficiency particulate filter); or SCBAF (any self-contained breathing apparatus with a full facepiece); or SAF (any supplied-air respirator with a full facepiece). *Emergency or planned entry into unknown concentrations or IDLH conditions:* SCBAF:PD,PP (any self-contained breathing apparatus that has a full facepiece and is operated in a pressure-demand or other positive-pressure mode); or SAF:PD,PP:ASCBA (any supplied-air respirator that has a full facepiece and is operated in a pressure-demand or other positive-pressure mode in combination with an auxiliary self-contained breathing apparatus operated in a pressure-demand or other positive pressure mode). *Escape:* HiEF (any air-purifying, full-facepiece respirator with a high-efficiency particulate filter); or SCBAE (any appropriate escape-type, self-contained breathing apparatus).

* Substance reported to cause eye irritation or damage; may require eye protection.

Storage: Prior to working with Chromic Acetate you should be trained on its proper handling and storage. Store in tightly closed containers in a cool, well-ventilated area. A regulated, marked area should be established where Chromic Acetate is handled, used, or stored. A regulated, marked area should be established where this chemical is handled, used, or stored in compliance with OSHA standard 1910.1045.

Shipping: Chromium acetate is not specifically identified in DOT's Performance-Oriented Packaging Standards.[19]

Spill Handling: Evacuate persons not wearing protective equipment from area of spill or leak until clean-up is complete. Remove all ignition sources. Collect powdered material in the most convenient and safe manner and deposit in sealed containers. Ventilate area after clean-up is complete. It may be necessary to contain and dispose of this chemical as a hazardous waste. If material or contaminated runoff enters waterways, notify downstream users of potentially contaminated waters. Contact your Department of Environmental Protection or your regional office of the federal EPA for specific recommendations. If employees are required to clean-up spills, they must be properly trained and equipped. OSHA 1910.120(q) may be applicable.

Fire Extinguishing: Chromic Acetate may burn, but does not readily ignite. Use dry chemical, CO_2, water spray, or foam extinguishers. Poisonous gases are produced in fire. If material or contaminated runoff enters waterways, notify downstream users of potentially contaminated waters. Notify local health and fire officials and pollution control agencies. From a secure, explosion-proof location, use water spray to cool exposed containers. If cooling streams are ineffective (venting sound increases in volume and pitch, tank discolors, or shows any signs of deforming), withdraw immediately to a secure position. If employees are expected to fight fires, they must be trained and equipped in OSHA 1910.156.

Disposal Method Suggested:

Dilute and stir in excess soda ash. Let stand, neutralize liquid and flush to sewer. Dispose of sludge in landfill.

References

New Jersey Department of Health and Senior Services and Senior Services, "Hazardous Substance Fact Sheet: Chromic Acetate," Trenton, NJ (September 1986)

Sax, N. I., Ed., "Dangerous Properties of Industrial Materials Report," 1, No. 3, 43–45 (1981) and 5, No. 6, 43–45 (1985)

Chromic Acid

Molecular Formula: $H_2Cr_2O_7$

Common Formula: H_2CrO_4 ($CrO_3 \cdot H_2O$)

Synonyms: Acide chromique (French); Acido cromico (Spanish); Chromic(6+) acid; Chromic(VI) acid; Chromic anhydride; Chromium anhydride; Chromium trioxide

ester: Acido cromico (Spanish); Chromic(IV) acid; Chromic acid ester; Chromic acid, solid; Chromic acid, solution, chromic anhydride; Chromic trioxide; Chromium oxide; Chromium(IV) oxidemonochromium oxide; Chromium trioxide; Chromium(4+) trioxide; Chromium trioxide, anhydrous; Monochromium trioxide; Puratronic chromium trioxide

CAS Registry Number: 7738-94-5; 11115-74-5 (ester)

RTECS Number: GB2450000; GB6650000 (ester)

DOT ID: UN 1463 (solid); UN 1755 (solution)

Regulatory Authority

- Carcinogen (DFG)[3] (NJ)
- Air Pollutant Standard Set (ACGIH)[1] (OSHA)[58] (Several States)[60]
- CLEAN AIR ACT: Hazardous Air Pollutants (Title I, Part A, Section 112). as chromium compounds
- CLEAN WATER ACT: 40CFR401.15 Section 307 Toxic Pollutants
- SUPERFUND/EPCRA 40CFR302.4 Reportable Quantity (RQ): CERCLA, 10 lb (4.54 kg)
- EPCRA Section 313 Form R *de minimis* concentration reporting level: Chromium VI compounds: 0.1%
- CLEAN AIR ACT: Hazardous Air Pollutants (Title I, Part A, Section 112). As chromium compounds. (ester)
- CLEAN WATER ACT: Section 311 Hazardous Substances/RQ 40CFR117.3 (same as CERCLA, see below); Section 313 Water Priority Chemicals (57FR41331, 9/9/92); 40CFR401.15 Section 307 Toxic Pollutants (ester)
- SUPERFUND/EPCRA 40CFR302.4 Reportable Quantity (RQ): CERCLA, 10 lb (4.54 kg) (ester)
- Canada, WHMIS, Ingredients Disclosure List

Cited in U.S. State Regulations: Alaska (G), California (G), Florida (G), Illinois (G, W), Maine (G), Massachusetts (G, A, W), Nevada (A), New Hampshire (G), New Jersey (G), North Carolina (A), Oklahoma (G), Pennsylvania (G), Rhode Island (G), West Virginia (G).

Description: Chromic acid, $H_2Cr_2O_7$, is a dark purplish-red odorless flakes or crystalline powder. Boiling point = 250°C (decomposes).

Freezing/Melting point = 197°C (decomposes). Often used in aqueous solution. Hazard Identification (based on NFPA-704 M Rating System): Health 3, Flammability 0, Reactivity 0, Oxidizer. Highly soluble in water. It may also be formulated as H_2CrO_4 ($CrO_3 \cdot H_2O$).

Potential Exposure: Chromic Acid is used in chromium plating, medicine, ceramic glazers, and paints.

Incompatibilities: A strong oxidizer. Aqueous solution is strongly acidic. Reacts with acetic acid, acetic anhydride, acetone, anthracene, chromous sulfide, diethyl ether, dimethyl formamide, ethanol, hydrogen sulfide, methanol, naphthalene, camphor, glycerol, potassium ferricyanide, pyridine, turpentine, combustibles, organics and other easily oxidized materials (such as paper, wood, sulfur, aluminum, and plastics). Attacks metals in presence of moisture.

Permissible Exposure Limits in Air: OSHA:[58] The legal airborne permissible exposure limit (PEL) is 0.1 mg/m^3, not be exceeded at any time. NIOSH recommends an airborne exposure limit of 0.001 mg/m^3 TWA averaged over a 10-hour workshift and 0.05 mg/m^3 not to be exceeded during any 15-minute work period. ACGIH:[1] The recommended airborne exposure limit is 0.05 mg/m^3 averaged over an 8-hour workshift. The NIOSH IDLH = 15 mg/m^3 [as Cr(VI)]. This chemical can be absorbed through the skin, thereby increasing exposure. In addition, some states have set guidelines or standards for chromic acid in ambient air[60] ranging from zero (North Carolina) to 0.001 μg/m^3 (Massachusetts) to 1.0 μg/m^3 (Nevada).

Determination in Air: Filter; Acid; Flame atomic absorption spectrometry; NIOSH IV Method #7024, Chromium.

Permissible Concentration in Water: For the protection of freshwater aquatic life: Hexavalent chromium: 0.29 μg/l as a 24 hour average, never to exceed 21.0 μg/l. For the protection of saltwater aquatic life: Hexavalent chromium: 18 μg/l as a 24 hour average, never to exceed 1,260 μg/l. To protect human health: hexavalent chromium 50 μg/l according to EPA.[6]

EPA[49] has set a long-term health advisory for adults of 0.84 mg/l and a lifetime health advisory of 0.12 mg/l (120 μg/l) for chromium. EPA's maximum drinking water level (MCL) is 0.1 mg/l.[62]

Germany, Canada, EEC, and WHO[35] have set a limit of 0.05 mg/l in drinking water.

The states of Maine and Minnesota have set guidelines for chromium in drinking water[61] of 50 μg/l for Maine and 120 μg/l for Minnesota.

Determination in Water: Total chromium may be determined by digestion followed by atomic absorption or by colorimetry (diphenylcarbazide) or by inductively coupled plasma (CP) optical emission spectrometry. Chromium (VI) may be determined by extraction and atomic absorption or colorimetry (using diphenylhydrazide). Dissolved total Cr or Cr(VI) may be determined by 0.45 μ filtration followed by the above-cited methods.[49]

Routes of Entry: Inhalation, ingestion, skin and/or eye contact.

Harmful Effects and Symptoms

Short Term Exposure: Chromic Acid can affect you when breathed in. It can also pass into inner layers of the skin. Chromic Acid should be handled as a Carcinogen with extreme caution. *Inhalation:* May be poisonous. Dust may cause severe irritation to the nose, throat and lungs, causing coughing, shortness of breath. May cause flu-like symptoms including chills, muscle ache, headache, fever. High exposure may cause nausea, salivation, vomiting, cramps, diarrhea, chest pains, cough, a build-up of fluids in the lungs (pulmonary edema) and possible death. Pulmonary edema is a medical emergency and may be delayed from 1 – 2 days following exposure. *Skin:* May cause severe irritation and thermal and acid burns, especially if skin is wet. *Eyes:* May cause severe irritation, burns, pain and possible blindness. *Swallowed:* May be poisonous. May cause severe burns of the mouth, throat and stomach, vomiting, watery or bloody diarrhea. Damage to kidneys and liver, collapse and convulsions can result.

Long Term Exposure: May cause lung cancer, birth defects, miscarriage, kidney and liver damage, skin allergy and ulcers; injury to the nasal septum (may cause a hole in the nose); discoloration of teeth; bronchitis; lung allergy.

Points of Attack: Blood, respiratory system, liver, kidneys, eyes, skin.

Medical Surveillance: Before beginning employment and at regular times after that, the following are recommended: Lung function tests. Exam of the mouth and larynx. If symptoms develop or overexposure is suspected, the following may also be useful: Kidney function tests. Evaluation by a qualified allergist, including careful exposure history and special testing, may help diagnose skin allergy. Specific engineering controls are recommended in NIOSH criteria documents: #73-11021 (chromic acid) and #76-129 [chromium (VI)].

First Aid: If this chemical gets into the eyes, remove any contact lenses at once and irrigate immediately for at least 15 minutes, occasionally lifting upper and lower lids. Seek medical attention immediately. If this chemical contacts the skin, remove contaminated clothing and wash immediately with soap and water. Seek medical attention immediately. If this chemical has been inhaled, remove from exposure, begin rescue breathing (using universal precautions) if breathing has stopped and CPR if heart action has stopped. Transfer promptly to a medical facility. When this chemical has been swallowed, get medical attention. If victim is conscious, administer water or milk. Do not induce vomiting. Medical observation is recommended for 24 – 48 hours after breathing overexposure, as pulmonary edema may be delayed. As first aid for pulmonary edema, a doctor or authorized paramedic may consider administering a corticosteroid spray.

Personal Protective Methods: Wear protective gloves and clothing to prevent any reasonable probability of skin contact. Safety equipment suppliers/manufacturers can provide recommendations on the most protective glove/clothing material for your operation. Viton and Nitrile+PVC are among the recommended protective materials. All protective clothing (suits, gloves, footwear, headgear) should be clean, available each day, and put on before work. Contact lenses should not be worn when working with this chemical. When working with liquids, wear splash-proof chemical goggles and face shield unless full facepiece respiratory protection is worn. When working with powders or dusts, wear dust-proof chemical goggles and face shield unless full facepiece respiratory protection is worn. Employees should wash immediately with soap when skin is wet or contaminated. Provide emergency showers and eyewash.

Respirator Selection: NIOSH: *at any concentrations above the NIOSH REL, or where there is no REL, at any detectable concentration:* SCBAF:PD,PP (any self-contained breathing apparatus that has a full facepiece and is operated in a pressure-demand or other positive-pressure mode); or SAF:PD,PP:ASCBA (any supplied-air respirator that has a full facepiece and is operated in a pressure-demand or other positive-pressure mode in combination with an auxiliary, self-contained breathing apparatus operated in a pressure-demand or other positive-pressure mode). *Escape:* HiEF (any air-purifying, full-facepiece respirator with a high-efficiency particulate filter); or SCBAE (any appropriate escape-type, self-contained breathing apparatus).

Storage: Prior to working with Chromic Acid you should be trained on its proper handling and storage. Store in tightly closed containers in a cool, well-ventilated area away from acetone, combustible, organic or other readily oxidizable material (such as paper, wood, sulfur, aluminum and plastics). Sources of ignition such as smoking and open flames, are prohibited where Chromic Acid is used, handled or stored in a manner that could create a potential fire or explosion hazard. A storage hazard; sealed containers may burst from CARBON DIOXIDE release. Store in tightly closed containers in a dry, cool, well-ventilated place with non-wood floors. Keep away from combustible materials, alcohols and acetone. Where possible, automatically transfer chromic acid from drums or other storage containers to process containers. Containers may explode in fire. See OSHA standard 1910.104 and NFPA 43A *Code for the Storage of Liquid and Solid Oxidizers* for detailed handling and storage regulations. A regulated, marked area should be established where this chemical is handled, used, or stored in compliance with OSHA standard 1910.1045.

Shipping: Chromic Acid, solid must be labeled "Oxidizer, Corrosive." It falls in Hazard Class 5.1 and Packing Group II.

Spill Handling: Evacuate persons not wearing protective equipment from area of spill or leak until clean-up is complete. Remove all ignition sources. Collect powdered material in the most convenient and safe manner and deposit in sealed containers. Ventilate area after clean-up is complete. It may be necessary to contain and dispose of this chemical as a hazardous waste. If material or contaminated runoff enters waterways, notify downstream users of potentially contaminated waters. Contact your Department of Environmental Protection or your regional office of the federal EPA for specific recommendations. If employees are required to clean-up spills, they must be properly trained and equipped. OSHA 1910.120(q) may be applicable.

Fire Extinguishing: Use dry chemical, soda ash, CO_2, water spray, or foam extinguishers. Chromic Acid may ignite other combustible materials such as paper and wood. Poisonous gases are produced in fire. If material or contaminated runoff enters waterways, notify downstream users of potentially contaminated waters. Notify local health and fire officials and pollution control agencies. From a secure, explosion-proof location, use water spray to cool exposed containers. If cooling streams are ineffective (venting sound increases in volume and pitch, tank discolors, or shows any signs of deforming), withdraw immediately to a secure position. If employees are expected to fight fires, they must be trained and equipped in OSHA 1910.156.

Disposal Method Suggested: Chemical reduction to Chromium (III) can be followed by landfill disposal of the sludge.[22]

References

New Jersey Department of Health and Senior Services and Senior Services, "Hazardous Substance Fact Sheet: Chromic Acid," Trenton, NJ (September 1996)

Sax, N. I., Ed., "Dangerous Properties of Industrial Materials Report," 2, 21–22 (1982) and 3, No. 3, 60–62 (1983)

Chromic Chloride

Molecular Formula: Cl_3Cr

Common Formula: $CrCl_3$

Synonyms: Chromium chloride; Chromium(III) chloride (1:3); Chromium chloride (III) anhydrous; Chromium chloride, anhydrous; Chromium sesquichloride; Chromium trichloride; C.I. 77295; Cloruro cromico (Spanish); Puratronic chromium chloride; Trichlorochromium

CAS Registry Number: 10025-73-7

RTECS Number: GB5425000

DOT ID: None

EINECS Number: 233-038-3

Regulatory Authority

- Air Pollutant Standard Set (ACGIH)[1] (HSE)[33] (UNEP)[43]
- CLEAN AIR ACT: Hazardous Air Pollutants (Title I, Part A, Section 112) as chromium compounds
- CLEAN WATER ACT: 40CFR401.15 Section 307 Toxic Pollutants
- SUPERFUND/EPCRA 40CFR355, Appendix B Extremely Hazardous Substances: TPQ = 1/10,000 lb (0.454/4,540 kg)
- SUPERFUND/EPCRA 40CFR302.4 Reportable Quantity (RQ): EHS, 1 lb (0.454 kg)
- EPCRA Section 313 Form R *de minimis* concentration reporting level: Chromium III compounds: 1.0%
- Canada, WHMIS, Ingredients Disclosure List

Cited in U.S. State Regulations: Alaska (G), Florida (G), Illinois (G), Maine (G), Massachusetts (G), New Hampshire (G), New Jersey (G). Oklahoma (G), Pennsylvania (G), Rhode Island (G).

Description: Chromic chloride, $CrCl_3$, is a red-violet or greenish to black crystalline solid. Freezing/Melting point = 1,152°C (dissociates above 1,300°C). Reacts with water.

Potential Exposure: Chromic chloride is used in chromizing; in the manufacture of chromium metal and compounds; as a catalyst for polymerization of olefins and other organic reactions; as a textile mordant; in tanning; in corrosion inhibitors; and as a waterproofing agent. A nutritional supplement.

Incompatibilities: Oxidizers. Reacts with water.

Permissible Exposure Limits in Air: The OSHA PEL for trivalent chromium compounds is 0.5 mg/m^3 TWA for an 8-hour workshift. The ACGIH[1] and HSE[33] TWA values are the same. The NIOSH IDLH level is 25 mg/m^3 [as Cr(III)].

Determination in Air: Filter; Acid; Flame atomic absorption spectrometry; NIOSH IV Method #7024, Chromium.[18]

Permissible Concentration in Water: For the protection of freshwater aquatic life: Trivalent chromium: not to exceed $e^{[1.08 \ln(\text{hardness}) + 3.48]}$µg/l. For the protection of saltwater aquatic life: Trivalent chromium: 10,300 µg/l on an acute toxicity basis. To protect human health: Trivalent chromium: 170 μg/l; hexavalent chromium 50 μg/l according to EPA.[6]

EPA[49] has set a long-term health advisory for adults of 0.84 mg/l and a lifetime health advisory of 0.12 mg/l (120 μg/l) for chromium. EPA's maximum drinking water level (MCL) is 0.1 mg/l.[62]

Germany, Canada, EEC, and WHO[35] have set a limit of 0.05 mg/l in drinking water.

The states of Maine and Minnesota have set guidelines for chromium in drinking water[61] of 50 μg/l for Maine and 120 μg/l for Minnesota.

Determination in Water: Total chromium may be determined by digestion followed by atomic absorption or by colorimetry (diphenylcarbazide) or by inductively coupled plasma (CP) optical emission spectrometry. Chromium (VI) may be determined by extraction and atomic absorption or colorimetry (using diphenylhydrazide). Dissolved total Cr or Cr(VI) may be determined by 0.45 μ filtration followed by the above-cited methods.[49]

Routes of Entry: Inhalation, ingestion, skin and/or eye contact.

Harmful Effects and Symptoms

Short Term Exposure: Chromic chloride is highly corrosive. Irritates and burns the eyes and skin. Inhalation can irritate the respiratory tract causing coughing and wheezing. It displays high dermal toxicity, and moderate oral toxicity. The oral toxicity-rat is given as 1,870 mg/kg.

Long Term Exposure: It causes histologic fibrosis off the lungs. Signs for exposure to Chromic chloride as for other chromium compounds include dermatitis, ulcers of the upper respiratory tract and inflammation of the larynx, lungs, gastrointestinal tract, and nasal passages. Repeated skin contact can cause sensitization and dermatitis.

Points of Attack: Eyes, skin, lungs.

Medical Surveillance: Examination by a qualified allergist. Lung function tests.

First Aid: If this chemical gets into the eyes, remove any contact lenses at once and irrigate immediately for at least 15 minutes, occasionally lifting upper and lower lids. Seek medical attention immediately. If this chemical contacts the skin, remove contaminated clothing and wash immediately with soap and water. Seek medical attention immediately. If this chemical has been inhaled, remove from exposure, begin rescue breathing (using universal precautions) if breathing has stopped and CPR if heart action has stopped. Transfer promptly to a medical facility. When this chemical has been swallowed, get medical attention. Give large quantities of water and induce vomiting. Do not make an unconscious person vomit.

Personal Protective Methods: Wear protective gloves and clothing to prevent any reasonable probability of skin contact. Safety equipment suppliers/manufacturers can provide recommendations on the most protective glove/clothing material for your operation. All protective clothing (suits, gloves, footwear, headgear) should be clean, available each day, and put on before work. Contact lenses should not be worn when working with this chemical. Wear dust-proof chemical goggles and face shield unless full facepiece respiratory protection is worn. Employees should wash immediately with soap when skin is wet or contaminated. Provide emergency showers and eyewash.

Respirator Selection: NIOSH/OSHA as chromium (III) compounds: *2.5 mg/m³:* DM (any dust and mist respirator). *5 mg/m³:* DMXSQ (any dust and mist respirator except single-use and quarter mask respirators); or SA (any supplied-air respirator); *12.5 mg/m³:* SA:CF (any supplied-air respirator operated in a continuous-flow mode); or PAPRDM (any powered, air-purifying respirator with a dust and mist filter). *25 mg/m³:* HiEF (any air-purifying, full-facepiece respirator with a high-efficiency particulate filter); or PAPRTHiE (any powered, air-purifying respirator with a tight-fitting facepiece and a high-efficiency particulate filter); or SCBAF (any self-contained breathing apparatus with a full facepiece); or SAF (any supplied-air respirator with a full facepiece). *Emergency or planned entry into unknown concentrations or IDLH conditions:* SCBAF:PD,PP (any self-contained breathing apparatus that has a full facepiece and is operated in a pressure-demand or other positive-pressure mode); or SAF:PD,PP:ASCBA (any supplied-air respirator that has a full facepiece and is operated in a pressure-demand or other positive-pressure mode in combination with an auxiliary self-contained breathing apparatus operated in a pressure-demand or other positive pressure mode). *Escape:* HiEF (any air-purifying, full-facepiece respirator with a high-efficiency particulate filter); or SCBAE (any appropriate escape-type, self-contained breathing apparatus).

* Substance reported to cause eye irritation or damage; may require eye protection.

Storage: Prior to working with this chemical you should be trained on its proper handling and storage. Store in tightly closed containers in a cool, well-ventilated area away from oxidizing agents and water.

Shipping: Chromic chloride is not a specifically cited compound in DOT's Performance-Oriented Packaging Standard[19] as regarding labels or quantity limitations on shipments.

Spill Handling: Do not touch spilled material. Avoid inhalation. Wear full protective clothing and proper respiratory protection. Evacuate persons not wearing protective equipment from area of spill or leak until clean-up is complete. Remove all ignition sources. Collect powdered material in the most convenient and safe manner and deposit in sealed containers. Ventilate area after clean-up is complete. It may be necessary to contain and dispose of this chemical as a hazardous waste. If material or contaminated runoff enters waterways, notify downstream users of potentially contaminated waters. Contact your Department of Environmental Protection or your regional office of the federal EPA for specific recommendations. If employees are required to clean-up spills, they must be properly trained and equipped. OSHA 1910.120(q) may be applicable.

Fire Extinguishing: Chromic chloride itself does not burn. It reacts with water, so use in flooding quantities only. Use extinguishing agents suitable for surrounding fire. Poisonous gases are produced in fire. If material or contaminated runoff enters waterways, notify downstream users of potentially contaminated waters. Notify local health and fire officials and pollution control agencies. From a secure, explosion-proof location, use water spray to cool exposed containers. If cooling streams are ineffective (venting sound increases in volume and pitch, tank discolors, or shows any signs of deforming), withdraw immediately to a secure position. If employees are expected to fight fires, they must be trained and equipped in OSHA 1910.156.

Disposal Method Suggested: Precipitate as chromium hydroxide. Dewater the sludge and dispose of the compacted sludge in single-purpose dumps.[22]

References

U.S. Environmental Protection Agency, "Chemical Profile: Chromic Chloride," Washington, DC, Chemical Emergency Preparedness Program (September 1998)

Chromic Sulfate

Molecular Formula: $Cr_2O_{12}S_3$

Common Formula: $Cr_2(SO_4)_3$

Synonyms: Chromic sulphate; Chromium sulfate; Chromium(3+) sulfate; Chromium(III) sulfate; Chromium sulphate; C.I.77305; Dichromium sulfate; Dichromium sulphate; Dichromium trisulfate; Dichromium trisulphate; Sulfato cromico (Spanish); Sulfuric acid, chromium(3+) salt

CAS Registry Number: 10101-53-8

RTECS Number: GB7200000

DOT ID: NA 9100

Regulatory Authority

- Air Pollutant Standard Set (ACGIH)[1]
- CLEAN AIR ACT: Hazardous Air Pollutants (Title I, Part A, Section 112). as chromium compounds
- CLEAN WATER ACT: Section 311 Hazardous Substances/RQ 40CFR117.3 (same as CERCLA, see below); Section 313 Water Priority Chemicals (57FR41331, 9/9/92); 40CFR401.15 Section 307 Toxic Pollutants
- SUPERFUND/EPCRA 40CFR302.4 Reportable Quantity (RQ): CERCLA, 1,000 lb (454 kg)
- EPCRA Section 313 Form R *de minimis* concentration reporting level: Chromium III compounds: 1.0%
- Canada, WHMIS, Ingredients Disclosure List

Cited in U.S. State Regulations: Alaska (G), California (G), Illinois (G), Maine (G), Massachusetts (G), New Hampshire (G), New Jersey (G), Oklahoma (G), Pennsylvania (G), Rhode Island (G), West Virginia (G).

Description: Chromic sulfate, $Cr_2(SO_4)_3$, is a peach colored, odorless powder. Hazard Identification (based on NFPA-704 M Rating System): Health 1, Flammability 0, Reactivity 0. Insoluble in water.

Potential Exposure: This compound is used in green paints, inks, dyes, and ceramics.

Incompatibilities: Strong oxidizers (perchlorates, peroxides, permanganates, chlorates, nitrates, chlorine, bromine, and fluorine). When heated this chemical decomposes to chromic acid.

Permissible Exposure Limits in Air: OSHA: The legal airborne permissible exposure limit (PEL) is 0.5 mg/m^3 as Chromium averaged over an 8-hour workshift.

ACGIH: The recommended airborne exposure limit is 0.5 mg/m^3 averaged over an 8-hour workshift for Chromium III compounds as Chromium. The NIOSH IDLH level is 25 mg/m^3 [as Cr(III)].

Determination in Air: Filter; Acid; Flame atomic absorption spectrometry; NIOSH IV Method #7024, Chromium.

Permissible Concentration in Water: For the protection of freshwater aquatic life: Trivalent chromium: not to exceed $e^{[1.08 \ln(\text{hardness}) + 3.48]}$µg/l. For the protection of saltwater aquatic life: Trivalent chromium: 10,300 µg/l on an acute toxicity basis. To protect human health: Trivalent chromium: 170 µg/l; hexavalent chromium 50 µg/l according to EPA.[6]

EPA[49] has set a long-term health advisory for adults of 0.84 mg/l and a lifetime health advisory of 0.12 mg/l (120 µg/l) for chromium. EPA's maximum drinking water level (MCL) is 0.1 mg/l.[62]

Germany, Canada, EEC, and WHO[35] have set a limit of 0.05 mg/l in drinking water.

The states of Maine and Minnesota have set guidelines for chromium in drinking water[61] of 50 µg/l for Maine and 120 µg/l for Minnesota.

Determination in Water: Total chromium may be determined by digestion followed by atomic absorption or by colorimetry (diphenylcarbazide) or by inductively coupled plasma (CP) optical emission spectrometry. Chromium (VI) may be determined by extraction and atomic absorption or colorimetry (using diphenylhydrazide). Dissolved total Cr or Cr(VI) may be determined by 0.45 µ filtration followed by the above-cited methods.[49]

Routes of Entry: Inhalation of dust or mist, skin, contact, ingestion.

Harmful Effects and Symptoms

Short Term Exposure: Chromic Sulfate can affect you when breathed in. Skin contact may cause irritation, especially if repeated or prolonged. Skin allergy sometimes occurs with itching, redness, and/or an eczema-like rash. If this happens future contact can trigger symptoms. Eye contact may cause irritation and burns.

Long Term Exposure: May cause skin allergy. Some chromium compounds can cause an ulcer in the septum separating the nose. It is uncertain whether this chemical has this effect.

Points of Attack: Eyes, skin.

Medical Surveillance: If symptoms develop or overexposure is suspected, the following may be useful: Evaluation by a qualified allergist, including careful exposure history and special testing, may help diagnose skin allergy.

First Aid: If this chemical gets into the eyes, remove any contact lenses at once and irrigate immediately for at least 15 minutes, occasionally lifting upper and lower lids. Seek medical attention immediately. If this chemical contacts the skin, remove contaminated clothing and wash immediately with soap and water. Seek medical attention immediately. If this chemical has been inhaled, remove from exposure, begin rescue breathing (using universal precautions) if breathing has stopped and CPR if heart action has stopped. Transfer promptly to a medical facility. When this chemical has been swallowed, get medical attention. Give large quantities of water and induce vomiting. Do not make an unconscious person vomit.

Personal Protective Methods: Wear protective gloves and clothing to prevent any reasonable probability of skin contact. Safety equipment suppliers/manufacturers can provide recommendations on the most protective glove/clothing material for your operation. All protective clothing (suits, gloves, footwear, headgear) should be clean, available each day, and put on before work. Contact lenses should not be worn when working with this chemical. Wear dust-proof chemical goggles and face shield unless full facepiece respiratory protection is worn. Employees should wash immediately with soap when skin is wet or contaminated. Provide emergency showers and eyewash.

Respirator Selection: NIOSH/OSHA as chromium (III) compounds: *2.5 mg/m³:* DM (any dust and mist respirator). *5 mg/m³:* DMXSQ (any dust and mist respirator except single-use and quarter mask respirators); or SA (any supplied-air respirator); *12.5 mg/m³:* SA:CF (any supplied-air respirator operated in a continuous-flow mode); or PAPRDM (any powered, air-purifying respirator with a dust and mist filter). *25 mg/m³:* HiEF (any air-purifying, full-facepiece respirator with a high-efficiency particulate filter); or PAPRTHiE (any powered, air-purifying respirator with a tight-fitting facepiece and a high-efficiency particulate filter); or SCBAF (any self-contained breathing apparatus with a full facepiece); or SAF (any supplied-air respirator with a full facepiece). *Emergency or planned entry into unknown concentrations or IDLH conditions:* SCBAF:PD,PP (any self-contained breathing apparatus that has a full facepiece and is operated in a pressure-demand or other positive-pressure mode); or SAF:PD,PP:ASCBA (any supplied-air respirator that has a full facepiece and is operated in a pressure-demand or other positive-pressure mode in combination with an auxiliary self-contained breathing apparatus operated in a pressure-demand or other positive pressure mode). *Escape:* HiEF (any air-purifying, full-facepiece respirator with a high-efficiency particulate filter); or SCBAE (any appropriate escape-type, self-contained breathing apparatus).

Note: Substance reported to cause eye irritation or damage; may require eye protection.

Storage: Prior to working with Chromic Sulfate you should be trained on its proper handling and storage. Store in tightly closed containers in a cool, well-ventilated area away from strong oxidizers and sources of heat.

Shipping: Chromic sulfate is not specifically cited in DOT's Performance-Oriented Packaging Standards.[19]

Spill Handling: Evacuate persons not wearing protective equipment from area of spill or leak until clean-up is complete. Remove all ignition sources. Collect powdered material in the most convenient and safe manner and deposit in sealed containers. Ventilate area after clean-up is complete. It may be necessary to contain and dispose of this chemical as a hazardous waste. If material or contaminated runoff enters waterways, notify downstream users of potentially contaminated waters. Contact your Department of Environmental Protection or your regional office of the federal EPA for specific recommendations. If employees are required to clean-up spills, they must be properly trained and equipped. OSHA 1910.120(q) may be applicable.

Fire Extinguishing: Chromic Sulfate may burn, but does not readily ignite. Use dry chemical, CO_2, water spray, or foam extinguishers. Poisonous gases are produced in fire. If material or contaminated runoff enters waterways, notify downstream users of potentially contaminated waters. Notify local health and fire officials and pollution control agencies. From a secure, explosion-proof location, use water spray to cool exposed containers. If cooling streams are ineffective (venting sound increases in volume and pitch, tank discolors, or shows any signs of deforming), withdraw immediately to a secure position. If employees are expected to fight fires, they must be trained and equipped in OSHA 1910.156.

Disposal Method Suggested: Return to supplier where possible. Where this is not practical, the material should be encapsulated and buried in a specially-designated chemical landfill.

References

New Jersey Department of Health and Senior Services and Senior Services, "Hazardous Substance Fact Sheet: Chromic Sulfate," Trenton, NJ (January 1996)

Sax, N. I., Ed., "Dangerous Properties of Industrial Materials Report," 3, No. 3, 62–65 (1983)

Chromium and Chromium Compounds

Molecular Formula: Cr

Synonyms: Chrome; Chromium, elemental; Chromium metal; Cromo (Spanish); Elemental chromium

CAS Registry Number: 7440-47-3 (elemental)

RTECS Number: GB4200000 (elemental)

EINECS Number: 231-157-5

Regulatory Authority

- Carcinogen (NTP)[10]
- Banned or Severely Restricted (Many countries, many categories) (UN)[35]
- Air Pollutant Standard Set (ACGIH) (Australia)[1] (HSE)[33] (Israel) (Mexico) (OSHA)[58] (Others)[35] (UNEP)[43] (Several States)[60]
- CLEAN AIR ACT: Hazardous Air Pollutants (Title I, Part A, Section 112). as chromium compounds
- CLEAN WATER ACT: 40CFR423, Appendix A, Priority Pollutants; Section 313 Water Priority Chemicals (57FR41331, 9/9/92); Toxic Pollutant (Section 401.15)
- EPA HAZARDOUS WASTE NUMBER (RCRA No.): D007
- RCRA, 40CFR261, Appendix 8 Hazardous Constituents, waste number not listed
- RCRA Toxicity Characteristic (Section 261.24), Maximum Concentration of Contaminants, regulatory level, 5.0 mg/l
- RCRA 40CFR268.48; 61FR15654, Universal Treatment Standards: Wastewater (mg/l), 2.77; Nonwastewater (mg/kg), 0.86 as chromium (total)
- RCRA 40CFR264, Appendix 9; TSD Facilities Ground Water Monitoring List. Suggested test method(s) (PQL μg/l): (total) 6010 (70); 7190 (500); 7191 (10)
- SAFE DRINKING WATER ACT: MCL, 0.1 mg/l; MCLG, 0.1 mg/l; Regulated chemical (47 FR 9352)
- SUPERFUND/EPCRA 40CFR302.4 Reportable Quantity (RQ): CERCLA, 5,000 lb (2,270 kg)
- EPCRA Section 313 Form R *de minimis* concentration reporting level: 0.1%
- Canada, WHMIS, Ingredients Disclosure List; National Pollutant Release Inventory (NPRI); CEPA Priority Substance List; Drinking Water Quality, 0.05 mg/l MAC

Cited in U.S. State Regulations: Alaska (G), California (A, G), Connecticut (A), Florida (G), Illinois (G), Maine (G, W), Maryland (G), Massachusetts (G, A), Michigan (G), Minnesota (W), Montana (A), Nevada (A), New Hampshire (G), New Jersey (G), New York (G, A), North Carolina (A), North Dakota (A), Pennsylvania (G, A), Rhode Island (G, A), Virginia (A), West Virginia (G).

Description: Chromium may exist in one of three valence states in compounds, +2, +3, and +6. The most stable oxidation state is trivalent chromium; hexavalent chromium is a less stable state. Chromium (element) blue-white to steel-gray, lustrous, brittle, hard, odorless solid. Elemental: Boiling point = 2,642°C. Freezing/Melting point = 1,900°C. Hazard Identification (based on NFPA-704 M Rating System): [dust] Health 0, Flammability 1, Reactivity 0. Insoluble in water. Noncombustible Solid in bulk form, but finely divided dust burns rapidly if heated in a flame.

Potential Exposure: Chromium metal is used for greatly increasing resistance and durability of metals and for chrome plating of other metals.

Incompatibilities: Chromium metal (especially in finely divided or powder form) and insoluble salts reacts violently with strong oxidants such as hydrogen peroxide, causing fire and explosion hazard. Reacts with diluted hydrochloric and sulfuric acids. Incompatible with alkalis and alkali carbonates.

Permissible Exposure Limits in Air: The legal airborne limit (OSHA PEL) is 1 mg/m^3 TWA for an 8-hour workshift. The PEL also applies to insoluble chromium salts. The ACGIH and NIOSH recommends a TWA value of 0.05 mg/m^3. Australia, Israel, Mexico, and the Canadian provinces of Alberta, BC, Ontario, and Quebec all set the same limits as NIOSH and ACGIH and Alberta has a STEL of 1.5 mg/m^3.

Sweden[35] has this same limit. However, Czechoslovakia has a TWA of 0.05 mg/m^3. Germany and Japan have set no numerical values, simply classifying chromium as a carcinogen.

The former USSR/UNEP joint project[43] has set a limit (both momentary and daily average) of 0.0015 mg/m^3 for ambient air in residential areas.

In addition, a number of states have set guidelines or standards for chromium in ambient air[60] ranging from zero (North Carolina) to 0.00009 μg/m^3 (Rhode Island) to 0.068 μg/m^3 (Massachusetts) to 0.07 – 0.39 μg/m^3 (Montana) to 0.12 μg/m^3 (Pennsylvania) to 0.167 μg/m^3 (Virginia) to 2.5 μg/m^3 (Connecticut) to 5.0 μg/m^3 (North Dakota) to 12.0 μg/m^3 (Nevada).

Determination in Air: For chromium metal and both insoluble and soluble salts: collection on a filter followed by acid workup and analysis by atomic absorption; see NIOSH Method #7024 for Chromium and OSHA Methods ID121 and ID125.[58]

Permissible Concentration in Water: For the protection of freshwater aquatic life: Trivalent chromium: not to exceed $e^{[1.08 \ln(\text{hardness}) + 3.48]}$μg/l. Hexavalent chromium: 0.29 μg/l as a 24 hour average, never to exceed 21.0 μg/l. For the protection of saltwater aquatic life: Trivalent chromium: 10,300 μg/l on an acute toxicity basis. Hexavalent chromium: 18 μg/l as a 24 hour average, never to exceed 1,260 μg/l. To protect human health: Trivalent chromium: 170 μg/l; hexavalent chromium 50 μg/l according to EPA.[6]

EPA[49] has set a long-term health advisory for adults of 0.84 mg/l and a lifetime health advisory of 0.12 mg/l (120 μg/l) for chromium. EPA's maximum drinking water level (MCL) is 0.1 mg/l.[62]

Germany, Canada, EEC, and WHO[35] have set a limit of 0.05 mg/l in drinking water.

The states of Maine and Minnesota have set guidelines for chromium in drinking water[61] of 50 μg/l for Maine and 120 μg/l for Minnesota.

Determination in Water: Total chromium may be determined by digestion followed by atomic absorption or by colorimetry (diphenylcarbazide) or by inductively coupled plasma (CP) optical emission spectrometry. Chromium (VI) may be determined by extraction and atomic absorption or colorimetry (using diphenylhydrazide). Dissolved total Cr or Cr(VI) may be determined by 0.45 μ filtration followed by the above-cited methods.[49]

Routes of Entry: Inhalation, ingestion, and eye and skin contact.

Harmful Effects and Symptoms

Chromium can affect you when breathed in. Chromium metal ore has been reported to cause lung allergy. Chromium fumes can cause "metal fume fever," a flu-like illness lasting about 24 hours with chills, aches, cough and fever. Chromium particles can irritate the eyes.

The above cautions apply to chromium metal. Since chromium is a reasonably reactive metal, thought must be given to the actions of combined chromium and particularly to the presence or absence of carcinogenic effects in various chromium compounds.

A table of differentiation between noncarciongenic and carcinogenic chromium (VI) compounds has been presented by NIOSH as shown below.

NIOSH has not conducted an in-depth study of the toxicity of chromium metal or compounds containing chromium in an oxidation state other than chromium (VI) compounds be reduced to 0.001 mg/m^3 and that these compounds be regulated as occupational carcinogens. NIOSH also recommends that the permissible exposure limit for noncarcinogenic chromium (VI) be reduced to 0.025 mg/m^3 averaged over a workshift of up to 10 hours per day, 40 hours per week, with a ceiling level of 0.05 mg/m^3 averaged over a 15-minute period. It is recommended further that chromium (VI) in the workplace be considered carcinogenic, unless it has been demonstrated that only the noncarcinogenic chromium (VI) compounds mentioned below are present. The NIOSH Criteria Documents for Chromic Acid and Chromium (VI) should be consulted for more detailed information.

Noncarcinogens:	
Evident	***Inferred***
Sodium bichromate	Lithium bichromate
Sodium chromate	Lithium chromate
Chromium(VI) oxide	Potassium bichromate
	Potassium chromate
	Rubidium bichromate
	Rubidium chromate
	Cesium bichromate
	Cesium chromate
	Ammonium bichromate
	Ammonium chromate
Carcinogens:	
Evident	***Inferred***
Calcium chromate	Alkaline earth chromates and bichromates
Sintered calcium chromate	Chromyl chloride
Alkaline lime-roasting	tert-Butyl chromate process residue
Zinc potassium chromate	Other chromium(VI) materials not listed in this table
Lead chromate	

Short Term Exposure: Chromium particles can irritate the eyes. Chromium fumes can cause "metal fume fever" a flu-like illness, lasting about 24 – 36 hours with chills, aches, cough, and fever.

Long Term Exposure: Repeated or prolonged contact may cause skin sensitization. Chromium metal ore has been reported to cause lung allergy.

Points of Attack: Chromium metal and insoluble salts: respiratory system and lungs.

Medical Surveillance: Preemployment physical examinations should include: a work history to determine past exposure to chromic acid and hexavalent chromium compounds, exposure to other carcinogens, smoking history, history of skin or pulmonary sensitization to chromium, history or presence of dermatitis, skin ulcers, or lesions of the nasal mucosa and/or perforation of the septum, and a chest x-ray. On periodic examinations and evaluation should be made of skin and respiratory complaints especially in workers who demonstrate allergic reactions. Chest x-rays should be taken yearly for workers over age 40, and every five years for younger workers. Blood, liver, lungs, and kidney function should be evaluated periodically. Urine test for chromates. This test is most accurate shortly after exposure. Regulatory exams are required for carcinogens by OSHA 1910.1002-1016.

First Aid: If this chemical gets into the eyes, remove any contact lenses at once and irrigate immediately for at least 15 minutes, occasionally lifting upper and lower lids. Seek

medical attention immediately. If this chemical contacts the skin, remove contaminated clothing and wash immediately with soap and water. Seek medical attention immediately. If this chemical has been inhaled, remove from exposure, begin rescue breathing (using universal precautions) if breathing has stopped and CPR if heart action has stopped. Transfer promptly to a medical facility. When this chemical has been swallowed, get medical attention. Give large quantities of water and induce vomiting. Do not make an unconscious person vomit.

Note to Physician: In case of fume inhalation, treat pulmonary edema. Give prednisone or other corticosteroid orally to reduce tissue response to fume. Positive pressure ventilation may be necessary. Treat metal fume fever with bed rest, analgesics and antipyretics. The symptoms of metal fume fever may be delayed for 4 – 12 hours following exposure: it may last less than 36 hours.

Personal Protective Methods: Wear protective gloves and clothing to prevent any reasonable probability of skin contact. Safety equipment suppliers/manufacturers can provide recommendations on the most protective glove/clothing material for your operation. All protective clothing (suits, gloves, footwear, headgear) should be clean, available each day, and put on before work. Contact lenses should not be worn when working with this chemical. Wear dust-proof chemical goggles and face shield unless full facepiece respiratory protection is worn. Employees should wash immediately with soap when skin is wet or contaminated. Provide emergency showers and eyewash. Respirators should be used in areas where dust, fumes or mist exposure exceeds Federal standards of where brief concentrations exceed the TWA, and for emergencies. Dust, fumes and mist filter-type respirators or supplied air respirators should be supplied all workers exposed, depending on concentration of exposure. Specific engineering controls are recommended in NIOSH Criteria Document #76-129.

Respirator Selection: Since Chromium exists in so many compounds in different oxidation states, the respirator requirements for various chromium compounds (in addition to chromium metal) will be reviewed here:

Chromic Acid and Chromates: NIOSH: *at any concentrations above the NIOSH REL, or where there is no REL, at any detectable concentration:* SCBAF:PD,PP (any self-contained breathing apparatus that has a full facepiece and is operated in a pressure-demand or other positive-pressure mode); or SAF:PD,PP:ASCBA (any supplied-air respirator that has a full facepiece and is operated in a pressure-demand or other positive-pressure mode in combination with an auxiliary, self-contained breathing apparatus operated in a pressure-demand or other positive-pressure mode). *Escape:* HiEF (any air-purifying, full-facepiece respirator with a high-efficiency particulate filter); or SCBAE (any appropriate escape-type, self-contained breathing apparatus).

Chromium Metal, Chromium (II) compounds and soluble chromous salts (as Cr): NIOSH/OSHA as chromium (III) compounds: *2.5 mg/m³:* DM (any dust and mist respirator). *5 mg/m³:* DMXSQ (any dust and mist respirator except single-use and quarter mask respirators); or SA (any supplied-air respirator); *12.5 mg/m³:* SA:CF (any supplied-air respirator operated in a continuous-flow mode); or PAPRDM (any powered, air-purifying respirator with a dust and mist filter). *25 mg/m³* HiEF (any air-purifying, full-facepiece respirator with a high-efficiency particulate filter); or PAPRTHiE (any powered, air-purifying respirator with a tight-fitting facepiece and a high-efficiency particulate filter); or SCBAF (any self-contained breathing apparatus with a full facepiece); or SAF (any supplied-air respirator with a full facepiece). *250 mg/m³* (any supplied-air respirator that has a full facepiece and is operated in a pressure-demand or other positive-pressure mode *Emergency or planned entry into unknown concentrations or IDLH conditions:* SCBAF:PD,PP (any self-contained breathing apparatus that has a full facepiece and is operated in a pressure-demand or other positive-pressure mode); or SAF:PD,PP:ASCBA (any supplied-air respirator that has a full facepiece and is operated in a pressure-demand or other positive-pressure mode in combination with an auxiliary self-contained breathing apparatus operated in a pressure-demand or other positive pressure mode). *Escape:* HiEF (any air-purifying, full-facepiece respirator with a high-efficiency particulate filter); or SCBAE (any appropriate escape-type, self-contained breathing apparatus).

Note: Substance reported to cause eye irritation or damage; may require eye protection.

Chromium(III) Compounds and soluble chromic salts (as Cr): NIOSH/OSHA as chromium (III) compounds: *2.5 mg/m³:* DM (any dust and mist respirator). *5 mg/m³:* DMXSQ (any dust and mist respirator except single-use and quarter mask respirators); or SA (any supplied-air respirator); *12.5 mg/m³:* SA:CF (any supplied-air respirator operated in a continuous-flow mode); or PAPRDM (any powered, air-purifying respirator with a dust and mist filter). *25 mg/m³* HiEF (any air-purifying, full-facepiece respirator with a high-efficiency particulate filter); or PAPRTHiE (any powered, air-purifying respirator with a tight-fitting facepiece and a high-efficiency particulate filter); or SCBAF (any self-contained breathing apparatus with a full facepiece); or SAF (any supplied-air respirator with a full facepiece). *Emergency or planned entry into unknown concentrations or IDLH conditions:* SCBAF: PD,PP (any self-contained breathing apparatus that has a full facepiece and is operated in a pressure-demand or other positive-pressure mode); or SAF:PD,PP:ASCBA (any supplied-air respirator that has a full facepiece and is operated in a pressure-demand or other positive-pressure mode in combination with an auxiliary self-contained breathing apparatus operated in a pressure-demand or other positive pressure mode). *Escape:* HiEF (any air-purifying, full-facepiece respirator with a high-efficiency particulate filter); or SCBAE (any appropriate escape-type, self-contained breathing apparatus).

Note: Substance reported to cause eye irritation or damage; may require eye protection.

Storage: Prior to working with chromium you should be trained on its proper handling and storage. A regulated, marked area should be established where this chemical is handled, used, or stored in compliance with OSHA standard 1910.1045. Store in tightly closed containers in a cool, well-ventilated area Chromium must be stored to avoid contact with strong oxidizers (such as chlorine, bromine, and fluorine) since violent reactions occur. Sources of ignition such as smoking and open flames are prohibited where Chromium is used, handled, or stored in a manner that could create a potential fire or explosion hazard.

Shipping: Chromium metal is not specifically cited in DOT's Performance-Oriented Packaging Standards.[19]

Spill Handling: Evacuate persons not wearing protective equipment from area of spill or leak until clean-up is complete. Remove all ignition sources. Collect powdered material in the most convenient and safe manner and deposit in sealed containers. Ventilate area after clean-up is complete. It may be necessary to contain and dispose of this chemical as a hazardous waste. If material or contaminated runoff enters waterways, notify downstream users of potentially contaminated waters. Contact your Department of Environmental Protection or your regional office of the federal EPA for specific recommendations. If employees are required to clean-up spills, they must be properly trained and equipped. OSHA 1910.120(q) may be applicable.

Fire Extinguishing: Chromium metal as dust/powder is combustible. The powder may explode in the air. In case of fire in the surroundings: all extinguishing agents allowed. Poisonous gases are produced in fire. If material or contaminated runoff enters waterways, notify downstream users of potentially contaminated waters. Notify local health and fire officials and pollution control agencies. From a secure, explosion-proof location, use water spray to cool exposed containers. If cooling streams are ineffective (venting sound increases in volume and pitch, tank discolors, or shows any signs of deforming), withdraw immediately to a secure position. If employees are expected to fight fires, they must be trained and equipped in OSHA 1910.156.

Disposal Method Suggested:

Recovery and recycle is a viable alternative to disposal for chromium in plating wastes, tannery wastes, cooling tower blowdown water and chemical plant wastes.

References

National Institute for Occupational Safety and Health, Criteria for a Recommended Standard: Occupational Exposure to Chromic Acid, NIOSH Publication No. 73–11021 (1973)

National Institute for Occupational Safety and Health, Criteria for a Recommended Standard: Occupational Exposure to Chromium (VI), NIOSH Document No. 76–129 (1976)

National Institute for Occupational Safety and Health, Information Profiles on Potential Occupational Hazards: Inorganic Chromium Compounds, Report PB-276,678, pp 136–142, Rockville, MD (October 1977)

U.S. Environmental Protection Agency, Chromium: Ambient Water Quality Criteria, Washington, DC (1980)

U.S. Environmental Protection Agency, Chromium, Health and Environmental Effects Profile No. 51, Office of Solid Waste, Washington, DC (April 30, 1980)

Sax, N. I., Ed., "Dangerous Properties of Industrial Materials Report," 1, No. 1, 40–41 (1980) and 3, No. 3, 65–68 (1983)

New Jersey Department of Health and Senior Services and Senior Services, "Hazardous Substance Fact Sheet: Chromium," Trenton, NJ (January 1986)

New York State Department of Health, "Chemical Fact Sheet: Chromium Metal," Albany, NY, Bureau of Toxic Substance Assessment (January 1986)

U.S. Public Health Service, "Toxicological Profile for Chromium," Atlanta, Georgia, Agency for Toxic Substances and Disease Registry (October 1987)

Chromium Carbonyl

Molecular Formula: C_6CrO_6

Common Formula: $Cr(CO)_6$

Synonyms: Chromium hexacarbonyl; Hexacarbonyl chromium.

CAS Registry Number: 13007-92-6

RTECS Number: GB5075000

Regulatory Authority

- Carcinogen (DFG)[3]
- Air Pollutant Standard Set (ACGIH)[1]
- CLEAN AIR ACT: Hazardous Air Pollutants (Title I, Part A, Section 112)
- CLEAN WATER ACT: Toxic Pollutant (Section 401.15); 40CFR401.15 Section 307 Toxic Pollutants as chromium and compounds
- RCRA, 40CFR261, Appendix 8 Hazardous Constituents, waste number not listed
- EPCRA (Section 313): Includes any unique chemical substances that contains chromium as part of that chemical's infrastructure. Form R *de minimis* concentration reporting level: Chromium VI compounds: 0.1%.; Chromium(III) compounds: 1.0%
- Canada, WHMIS, Ingredients Disclosure List

Cited in U.S. State Regulations: Alaska (G), California (G), Maine (G), West Virginia (G).

Description: Chromium carbonyl, $Cr(CO)_6$, is a colorless crystalline substance which sinters at 90°C. Freezing/Melting point = 110°C (decomposes). Explodes (in lieu of boiling) at 210°C.

Potential Exposure: Chromium Carbonyl is used as a catalyst for hydrogenation, isomerization, water gar shift reaction and alkylation of aromatic hydrocarbons; gasoline additive to increase octane number; preparation of chromous oxide, CrO.

Incompatibilities: Decomposed by chlorine and fuming nitric acid; sensitive to heat and light (undergoes photochemical decomposition).

Permissible Exposure Limits in Air: Under the category of "certain water insoluble chromium compounds." The TWA is 0.05 mg/m^3 with the notation that the material is a confirmed human carcinogen.

Determination in Air: Filter; Acid; Flame atomic absorption spectrometry; NIOSH IV Method #7024, Chromium.

Permissible Concentration in Water: For the protection of freshwater aquatic life: Trivalent chromium: not to exceed $e^{[1.08 \ln(\text{hardness}) + 3.48]}$ μg/l. Hexavalent chromium: 0.29 μg/l as a 24 hour average, never to exceed 21.0 μg/l. For the protection of saltwater aquatic life: Trivalent chromium: 10,300 μg/l on an acute toxicity basis. Hexavalent chromium: 18 μg/l as a 24 hour average, never to exceed 1,260 μg/l. To protect human health: Trivalent chromium: 170 μg/l; hexavalent chromium 50 μg/l according to EPA.[6]

EPA[49] has set a long-term health advisory for adults of 0.84 mg/l and a lifetime health advisory of 0.12 mg/l (120 μg/l) for chromium. EPA has recently proposed a maximum drinking water level of 0.1 mg/l.[62]

Germany, Canada, EEC, and WHO[35] have set a limit of 0.05 mg/l in drinking water.

The states of Maine and Minnesota have set guidelines for chromium in drinking water[61] of 50 μg/l for Maine and 120 μg/l for Minnesota.

Determination in Water: Total chromium may be determined by digestion followed by atomic absorption or by colorimetry (diphenylcarbazide) or by inductively coupled plasma (CP) optical emission spectrometry. Chromium (VI) may be determined by extraction and atomic absorption or colorimetry (using diphenylhydrazide). Dissolved total Cr or Cr(VI) may be determined by 0.45 μ filtration followed by the above-cited methods.[49]

Routes of Entry: Inhalation, ingestion.

Harmful Effects and Symptoms

Highly toxic. Also emits toxic fumes of CO when heated. Carcinogenic.

First Aid: *Skin Contact:* Flood all areas of body that have contacted the substance with water. Don't wait to remove contaminated clothing; do it under the water stream. Use soap to help assure removal. Isolate contaminated clothing when removed to prevent contact by others.

Eye Contact: Remove any contact lenses at once. Immediately flush eyes well with copious quantities of water or normal saline for at least 20 – 30 minutes. Seek medical attention.

Inhalation: Leave contaminated area immediately; breathe fresh air. Proper respiratory protection must be supplied to any rescuers. If coughing, difficult breathing or any other symptoms develop, seek medical attention at once, even if symptoms develop many hours after exposure.

Ingestion: Contact a physician, hospital or poison center at once. If the victim is unconscious or convulsing, do not induce vomiting or give anything by mouth. Assure that his airway is open and lay him on his side with his head lower than his body and transport immediately to a medical facility. If conscious and not convulsing, give a glass of water to dilute the substance. Vomiting should not be induced without a physician's advice.

Respirator Selection: NIOSH/OSHA as chromium (III) compounds: *2.5 mg/m³:* DM (any dust and mist respirator). *5 mg/m³:* DMXSQ (any dust and mist respirator except single-use and quarter mask respirators); or SA (any supplied-air respirator); *12.5 mg/m³:* SA:CF (any supplied-air respirator operated in a continuous-flow mode); or PAPRDM (any powered, air-purifying respirator with a dust and mist filter). *25 mg/m³:* HiEF (any air-purifying, full-facepiece respirator with a high-efficiency particulate filter); or PAPRTHiE (any powered, air-purifying respirator with a tight-fitting facepiece and a high-efficiency particulate filter); or SCBAF (any self-contained breathing apparatus with a full facepiece); or SAF (any supplied-air respirator with a full facepiece). *Emergency or planned entry into unknown concentrations or IDLH conditions:* SCBAF:PD,PP (any self-contained breathing apparatus that has a full facepiece and is operated in a pressure-demand or other positive-pressure mode); or SAF:PD,PP:ASCBA (any supplied-air respirator that has a full facepiece and is operated in a pressure-demand or other positive-pressure mode in combination with an auxiliary self-contained breathing apparatus operated in a pressure-demand or other positive pressure mode). *Escape:* HiEF (any air-purifying, full-facepiece respirator with a high-efficiency particulate filter); or SCBAE (any appropriate escape-type, self-contained breathing apparatus).

Note: Substance reported to cause eye irritation or damage; may require eye protection.

Storage: Prior to working with this chemical you should be trained on its proper handling and storage. Store in tightly closed containers in a cool, well-ventilated area or in a refrigerator, and protect from light. A regulated, marked area should be established where this chemical is handled, used, or stored in compliance with OSHA standard 1910.1045.

Shipping: Chromium carbonyl is not specifically cited in DOT's Performance-Oriented Packaging Standards.[19]

Spill Handling: Evacuate persons not wearing protective equipment from area of spill or leak until clean-up is complete. Remove all sources of ignition and dampen spilled material with 60 – 70% acetone to avoid dust, then transfer material to a suitable container. Ventilate the spill area and use absorbent paper dampened with 60 – 70% acetone to pick up remaining material. Wash surfaces well with soap and water. Ventilate area after clean-up is complete. It may be necessary to contain and dispose of this chemical as a hazardous waste. If material or contaminated runoff enters waterways, notify downstream users of potentially contaminated waters. Contact your Department of Environmental Protection or your regional office of the federal EPA for specific recommendations. If employees are required to clean-up spills, they must be properly trained and equipped. OSHA 1910.120(q) may be applicable.

Fire Extinguishing: Chromium carbonyl is a flammable compound and extreme care should be used in fire fighting both due to the carcinogenic nature of the material and the emission of CO as a decomposition product. Use dry chemical, carbon dioxide, water spray, or alcohol foam extinguishers. Poisonous gases are produced in fire. If material or contaminated runoff enters waterways, notify downstream users of potentially contaminated waters. Notify local health and fire officials and pollution control agencies. Containers may explode in fire. From a secure, explosion-proof location, use water spray to cool exposed containers. If cooling streams are ineffective (venting sound increases in volume and pitch, tank discolors, or shows any signs of deforming), withdraw immediately to a secure position. If employees are expected to fight fires, they must be trained and equipped in OSHA 1910.156.

Chromium Nitrate

Molecular Formula: CrN_3O_9

Common Formula: $Cr(NO_3)_3$

Synonyms: Chromic nitrate; Chromium(III) nitrate; Chromium+ (3+) nitrate; Chromium trinitrate; Nitric acid, chromium(3+) salt

CAS Registry Number: 13548-38-4

RTECS Number: GB6280000

DOT ID: UN 2720

Regulatory Authority

- Air Pollutant Standard Set (ACGIH)[1] (Australia) (HSE)[33] (Israel) (OSHA)
- CLEAN AIR ACT: Hazardous Air Pollutants (Title I, Part A, Section 112)
- CLEAN WATER ACT: Toxic Pollutant (Section 401.15); 40CFR401.15 Section 307 Toxic Pollutants as chromium and compounds
- RCRA, 40CFR261, Appendix 8 Hazardous Constituents, waste number not listed
- EPCRA (Section 313): Includes any unique chemical substances that contains chromium as part of that chemical's infrastructure. Form R *de minimis* concentration reporting level: Chromium(III) compounds: 1.0%
- Canada, WHMIS, Ingredients Disclosure List; National Pollutant Release Inventory (NPRI); CEPA Priority Substance List (as chromium compounds)

Cited in U.S. State Regulations: Alaska (G), California (A, G), Illinois (G), Maine (G), Minnesota (g), New Hampshire (G), New Jersey (G), Pennsylvania (G), Rhode Island (G), West Virginia (G).

Description: Chromium Nitrate, $Cr(NO_3)_3$, is a crystalline substance, variously stated to be green brown or purple and existing in various hydrated forms. Freezing/Melting point = 60 – 100°C, depending on the degree of hydration.

Potential Exposure: Chromium Nitrate is used in the preparation of chrome catalysts, in textile printing operations, and as a corrosion inhibitor.

Incompatibilities: This chemical is a strong oxidizer. Contact with reducing agents, fuels, ethers and other flammable and combustible materials cause a fire and explosion hazard.

Permissible Exposure Limits in Air: The OSHA PEL for trivalent chromium compounds is 0.5 mg/m^3 TWA for an 8-hour workshift. The ACGIH[1] and HSE[33] TWA values are the same. The NIOSH IDLH level is 25 mg/m^3 [as Cr(III)].

Determination in Air: Filter; Acid; Flame atomic absorption spectrometry; NIOSH IV Method #7024, Chromium.

Permissible Concentration in Water: For the protection of freshwater aquatic life: Trivalent chromium: not to exceed $e^{[1.08 \ln(\text{hardness}) + 3.48]}$µg/l. Trivalent chromium: 10,300 µg/l on an acute toxicity basis. To protect human health: Trivalent chromium: 170 µg/l; hexavalent chromium 50 µg/l according to EPA.[6]

EPA[49] has set a long-term health advisory for adults of 0.84 mg/l and a lifetime health advisory of 0.12 mg/l (120 µg/l) for chromium. EPA has recently proposed a maximum drinking water level of 0.1 mg/l.[62]

Germany, Canada, EEC, and WHO[35] have set a limit of 0.05 mg/l in drinking water.

The states of Maine and Minnesota have set guidelines for chromium in drinking water[61] of 50 µg/l for Maine and 120 µg/l for Minnesota.

Determination in Water: Total chromium may be determined by digestion followed by atomic absorption or by colorimetry (diphenylcarbazide) or by inductively coupled plasma (CP) optical emission spectrometry. Chromium (VI) may be determined by extraction and atomic absorption or colorimetry (using diphenylhydrazide). Dissolved total Cr or Cr(VI) may be determined by 0.45 µ filtration followed by the above-cited methods.[49]

Routes of Entry: Skin contact, inhalation, ingestion.

Harmful Effects and Symptoms

Short Term Exposure: Chromium Nitrate can affect you when breathed in. Chromium Nitrate may cause mutations. Handle with extreme caution. Skin contact may cause irritation, especially if repeated or prolonged. Skin allergy sometimes occurs, with itching, redness, and/or an eczema-like rash.

If this happens, future contact can trigger symptoms. Eye contact may cause irritation.

Long Term Exposure: May cause genetic changes, birth defects; skin allergy with rash and itching.

Points of Attack: Irritation eyes; sensitization dermatitis.

Medical Surveillance: If symptoms develop or overexposure is suspected, the following may be useful: Evaluation by a qualified allergist, including careful exposure history and special testing, may help diagnose skin allergy.

First Aid: If this chemical gets into the eyes, remove any contact lenses at once and irrigate immediately for at least 15 minutes, occasionally lifting upper and lower lids. Seek medical attention immediately. If this chemical contacts the skin, remove contaminated clothing and wash immediately with soap and water. Seek medical attention immediately. If this chemical has been inhaled, remove from exposure, begin rescue breathing (using universal precautions) if breathing has stopped and CPR if heart action has stopped. Transfer promptly to a medical facility. When this chemical has been swallowed, get medical attention. Give large quantities of water and induce vomiting. Do not make an unconscious person vomit.

Personal Protective Methods: Wear protective gloves and clothing to prevent any reasonable probability of skin contact. Safety equipment suppliers/manufacturers can provide recommendations on the most protective glove/clothing material for your operation. All protective clothing (suits, gloves, footwear, headgear) should be clean, available each day, and put on before work. Contact lenses should not be worn when working with this chemical. Wear -splash or dust-proof chemical goggles and face shield unless full facepiece respiratory protection is worn. Employees should wash immediately with soap when skin is wet or contaminated. Provide emergency showers and eyewash.

Respirator Selection: NIOSH/OSHA as chromium (III) compounds: *2.5 mg/m³:* DM (any dust and mist respirator). *5 mg/m³:* DMXSQ (any dust and mist respirator except single-use and quarter mask respirators); or SA (any supplied-air respirator); *12.5 mg/m³:* SA:CF (any supplied-air respirator operated in a continuous-flow mode); or PAPRDM (any powered, air-purifying respirator with a dust and mist filter). *25 mg/m³:* HiEF (any air-purifying, full-facepiece respirator with a high-efficiency particulate filter); or PAPRTHiE (any powered, air-purifying respirator with a tight-fitting facepiece and a high-efficiency particulate filter); or SCBAF (any self-contained breathing apparatus with a full facepiece); or SAF (any supplied-air respirator with a full facepiece). *Emergency or planned entry into unknown concentrations or IDLH conditions:* SCBAF:PD,PP (any self-contained breathing apparatus that has a full facepiece and is operated in a pressure-demand or other positive-pressure mode); or SAF:PD,PP:ASCBA (any supplied-air respirator that has a full facepiece and is operated in a pressure-demand or other positive-pressure mode in combination with an auxiliary self-contained breathing apparatus operated in a pressure-demand or other positive pressure mode). *Escape:* HiEF (any air-purifying, full-facepiece respirator with a high-efficiency particulate filter); or SCBAE (any appropriate escape-type, self-contained breathing apparatus).

Note: Substance reported to cause eye irritation or damage; may require eye protection.

Storage: Prior to working with this chemical you should be trained on its proper handling and storage. Chromium Nitrate must be stored to avoid contact with strong reducing agents, fuels, and ether since violent reactions occur. Store in tightly closed containers in a cool, well-ventilated area away from flammable and combustible materials. A regulated, marked area should be established where this chemical is handled, used, or stored in compliance with OSHA standard 1910.1045. See OSHA standard 1910.104 and NFPA 43A *Code for the Storage of Liquid and Solid Oxidizers* for detailed handling and storage regulations.

Shipping: Chromium nitrate must be labeled "Oxidizer." It falls in DOT Hazard Class 5.1 and Packing Group III. The limit on passenger aircraft or railcar shipment is 25 kg; on cargo aircraft shipment is 100 kg.[19]

Spill Handling: Evacuate persons not wearing protective equipment from area of spill or leak until clean-up is complete. Remove all ignition sources. Collect powdered material in the most convenient and safe manner and deposit in sealed containers. Ventilate area after clean-up is complete. Keep Chromium Nitrate out of a confined space, such as a sewer, because of the possibility of an explosion, unless the sewer is designed to prevent the build-up of explosive concentrations. It may be necessary to contain and dispose of this chemical as a hazardous waste. If material or contaminated runoff enters waterways, notify downstream users of potentially contaminated waters. Contact your Department of Environmental Protection or your regional office of the federal EPA for specific recommendations. If employees are required to clean-up spills, they must be properly trained and equipped. OSHA 1910.120(q) may be applicable.

Fire Extinguishing: Extinguish fire using an agent suitable for type of surrounding fire. Chromium Nitrate itself does not burn but it will increase the intensity of a fire since it is an oxidizer. Poisonous gases are produced in fire including nitrogen oxides and chromium oxide fumes. If material or contaminated runoff enters waterways, notify downstream users of potentially contaminated waters. Notify local health and fire officials and pollution control agencies. From a secure, explosion-proof location, use water spray to cool exposed containers. If cooling streams are ineffective (venting sound increases in volume and pitch, tank discolors, or shows any signs of deforming), withdraw immediately to a secure position. If employees are expected to fight fires, they must be trained and equipped in OSHA 1910.156.

Disposal Method Suggested: Precipitate chromium as the hydroxide. Dewater the sludge and dispose of in single-purpose dumps.[22]

References

New Jersey Department of Health and Senior Services and Senior Services, "Hazardous Substance Fact Sheet: Chromium Nitrate," Trenton, NJ (September 15, 1987).

Chromium(III) Oxide

Molecular Formula: Cr_2O_3

Synonyms: Anadomis green; Anidride cromique (French); Casalis green; Chrome green; Chrome oxide; Chromia; Chromic Acid; Chromic oxide; Chromium(3+) oxide; Chromium(III) oxide; Chromium sesquioxide; Chromium(3+) trioxide; C.I. 77288; C.I. pigment green; Dichromium trioxide; Green chromic oxide; Green cinnabar; Green GA; Green rouge; Leaf green; Levanox; Oil green; Ultramarine green

CAS Registry Number: 1308-38-9

RTECS Number: GB6475000

DOT ID: No citation.

Regulatory Authority

- Air Pollutant Standard Set (Australia) (HSE) (Israel) (Several Canadian Provinces)
- CLEAN AIR ACT: Hazardous Air Pollutants (Title I, Part A, Section 112)
- CLEAN WATER ACT: Toxic Pollutant (Section 401.15); 40CFR401.15 Section 307 Toxic Pollutants as chromium and compounds
- RCRA, 40CFR261, Appendix 8 Hazardous Constituents, waste number not listed
- EPCRA (Section 313): Includes any unique chemical substances that contains chromium as part of that chemical's infrastructure. Form R *de minimis* concentration reporting level: Chromium(III) compounds: 1.0%
- Canada, WHMIS, Ingredients Disclosure List; National Pollutant Release Inventory (NPRI); CEPA Priority Substance List

Cited in U.S. State regulations: Alaska (G), California (A, G), Illinois (G), Maine (G), Massachusetts (G), New Hampshire (G), New Jersey (G), New York (G), Pennsylvania (G), Rhode Island (G), West Virginia (G).

Description: Chromium oxide, Cr_2O_3, is a bright green, odorless powder. Freezing/Melting point = 2,266°C (±25°C). Insoluble in water.

Potential Exposure: Chromium(III) Oxide is used as a paint pigment, a fixative for certain textile dyes and a catalyst.

Incompatibilities: Contact with glycerol, oxygen difluoride, lithium, or strong oxidizers may cause fire and explosion hazard.

Permissible Exposure Limits in Air: The OSHA PEL for trivalent chromium compounds is 0.5 mg/m^3 TWA for an 8-hour workshift. The ACGIH[1] and HSE[33] TWA values are the same. The NIOSH IDLH level is 25 mg/m^3 [as Cr(III)].

Determination in Air: Filter; Acid; Flame atomic absorption spectrometry; NIOSH IV Method #7024, Chromium.

Permissible Concentration in Water: For the protection of freshwater aquatic life: Trivalent chromium: not to exceed $e^{[1.08 \ln(\text{hardness}) + 3.48]}$μg/l. For the protection of saltwater aquatic life: Trivalent chromium: 10,300 μg/l on an acute toxicity basis. To protect human health: Trivalent chromium: 170 μg/l; hexavalent chromium 50 μg/l according to EPA.[6]

EPA[49] has set a long-term health advisory for adults of 0.84 mg/l and a lifetime health advisory of 0.12 mg/l (120 μg/l) for chromium. EPA's maximum drinking water level (MCL) is 0.1 mg/l.[62]

Germany, Canada, EEC, and WHO[35] have set a limit of 0.05 mg/l in drinking water.

The states of Maine and Minnesota have set guidelines for chromium in drinking water[61] of 50 μg/l for Maine and 120 μg/l for Minnesota.

Determination in Water: Total chromium may be determined by digestion followed by atomic absorption or by colorimetry (diphenylcarbazide) or by inductively coupled plasma (CP) optical emission spectrometry. Chromium(VI) may be determined by extraction and atomic absorption or colorimetry (using diphenylhydrazide). Dissolved total Cr or Cr(VI) may be determined by 0.45 μ filtration followed by the above-cited methods.[49]

Routes of Entry: Inhalation of dust or mist, skin contact, ingestion.

Harmful Effects and Symptoms

Short Term Exposure: Eye contact can cause irritation. Skin allergy may occur with itching, redness, and/or an eczema-like rash. Persons allergic to other chromium compounds may be more likely to develop skin allergy to this chemical.

Long Term Exposure: There is limited evidence that chromium(III) oxide is a teratogen in animals, and until further testing is done, it should be treated as a possible teratogen in humans. Chromium(III) oxide has been tested and has *not* been shown to cause cancer in animals (NJ Department of Health, 5/1988). Skin allergy may develop from repeated exposure to this chemical.

Points of Attack: Irritation eyes; sensitization dermatitis.

Medical Surveillance: If symptoms develop or overexposure is suspected, the following may be useful: Evaluation by a qualified allergist, including careful exposure history and special testing, may help diagnose skin allergy.

First Aid: If this chemical gets into the eyes, remove any contact lenses at once and irrigate immediately for at least 15 minutes, occasionally lifting upper and lower lids. Seek

medical attention immediately. If this chemical contacts the skin, remove contaminated clothing and wash immediately with soap and water. Seek medical attention immediately. If this chemical has been inhaled, remove from exposure, begin rescue breathing (using universal precautions) if breathing has stopped and CPR if heart action has stopped. Transfer promptly to a medical facility. When this chemical has been swallowed, get medical attention. Give large quantities of water and induce vomiting. Do not make an unconscious person vomit.

Personal Protective Methods: *Clothing:* Avoid skin contact with Chromium(III) Oxide. Wear protective gloves and clothing. Safety equipment suppliers/manufacturers can provide recommendations on the most protective glove/clothing material for your operation. All protective clothing (suits, gloves, footwear, headgear) should be clean, available each day and put on before work.

Eye Protection: Eye protection is included in the recommended respiratory protection.

Respirator Selection: NIOSH/OSHA as chromium (III) compounds: *2.5 mg/m³:* DM (any dust and mist respirator). *5 mg/m³:* DMXSQ (any dust and mist respirator except single-use and quarter mask respirators); or SA (any supplied-air respirator); *12.5 mg/m³:* SA:CF (any supplied-air respirator operated in a continuous-flow mode); or PAPRDM (any powered, air-purifying respirator with a dust and mist filter). *25 mg/m³:* HiEF (any air-purifying, full-facepiece respirator with a high-efficiency particulate filter); or PAPRTHiE (any powered, air-purifying respirator with a tight-fitting facepiece and a high-efficiency particulate filter); or SCBAF (any self-contained breathing apparatus with a full facepiece); or SAF (any supplied-air respirator with a full facepiece). *Emergency or planned entry into unknown concentrations or IDLH conditions:* SCBAF:PD,PP (any self-contained breathing apparatus that has a full facepiece and is operated in a pressure-demand or other positive-pressure mode); or SAF:PD,PP: ASCBA (any supplied-air respirator that has a full facepiece and is operated in a pressure-demand or other positive-pressure mode in combination with an auxiliary self-contained breathing apparatus operated in a pressure-demand or other positive pressure mode). *Escape:* HiEF (any air-purifying, full-facepiece respirator with a high-efficiency particulate filter); or SCBAE (any appropriate escape-type, self-contained breathing apparatus).

Note: Substance reported to cause eye irritation or damage; may require eye protection.

Storage: Chromium(III) Oxide must be stored to avoid contact with strong oxidizers (such as Chlorine, Bromine and Fluorine), Glycerol and Oxygen Difluoride since violent reactions occur. A regulated, marked area should be established where Chromium(III) Oxide is handled, used, or stored. Store in tightly closed containers in a cool, well-ventilated area.

Shipping: Chromium Oxide is not specifically cited in DOT's Performance-Oriented Packaging Standards.[19]

Spill Handling: Evacuate persons not wearing protective equipment from area of spill or leak until clean-up is complete. Remove all ignition sources. Collect powdered material in the most convenient and safe manner and deposit in sealed containers. Ventilate area after clean-up is complete. It may be necessary to contain and dispose of this chemical as a hazardous waste. If material or contaminated runoff enters waterways, notify downstream users of potentially contaminated waters. Contact your Department of Environmental Protection or your regional office of the federal EPA for specific recommendations. If employees are required to clean-up spills, they must be properly trained and equipped. OSHA 1910.120(q) may be applicable.

Fire Extinguishing: Extinguish fire using an agent suitable for type of surrounding fire. Chromium(III) Oxide itself does not burn. Poisonous gases are produced in fire. If material or contaminated runoff enters waterways, notify downstream users of potentially contaminated waters. Notify local health and fire officials and pollution control agencies. From a secure, explosion-proof location, use water spray to cool exposed containers. If cooling streams are ineffective (venting sound increases in volume and pitch, tank discolors, or shows any signs of deforming), withdraw immediately to a secure position. If employees are expected to fight fires, they must be trained and equipped in OSHA 1910.156. OSHA 1910.156.

References

New Jersey Department of Health and Senior Services and Senior Services, "Hazardous Substance Fact Sheet: Chromium (III) Oxide," Trenton, NJ (May 1998)

New York State Department of Health, "Chemical Fact Sheet: Chromium (III) Oxide," Albany, NY, Bureau of Toxic Substance Assessment (March 1986 and Revised Version #2)

Chromium Potassium Sulfate

Molecular Formula: $CrKO_8S_2$

Common Formula: $CrK(SO_4)_2$

Synonyms: Chrome alum; Chrome potash alum; Chromic potassium sulfate; Chromic potassium sulphate; Chromium potassium sulfate; Chromium potassium sulphate; Crystal chrome alum; Potash alum; Potassium chromic sulphate; Potassium chromium alum; Potassium disulphatochromate(III); Sulfuric acid, chromium (3+) potassium salt

CAS Registry Number: 10141-00-1) (anhydrous); 7788-99-0 (dodecahydrate)

RTECS Number: GB6845000 (anhydrous); GB6850000 (dodecahydrate)

Regulatory Authority

- Air Pollutant Standard Set (ACGIH)[1] (Australia) (HSE)[33] (Israel) (OSHA) (Several Canadian Provinces)
- CLEAN AIR ACT: Hazardous Air Pollutants (Title I, Part A, Section 112)

- CLEAN WATER ACT: Toxic Pollutant (Section 401.15); 40CFR401.15 Section 307 Toxic Pollutants as chromium and compounds
- RCRA, 40CFR261, Appendix 8 Hazardous Constituents, waste number not listed
- EPCRA (Section 313): Includes any unique chemical substances that contains chromium as part of that chemical's infrastructure. Form R *de minimis* concentration reporting level: Chromium(III) compounds: 1.0%
- Canada, WHMIS, Ingredients Disclosure List; National Pollutant Release Inventory (NPRI); CEPA Priority Substance List

Cited in U.S. State Regulations: Alaska (G), California (A, G), Illinois (G), Maine (G), Massachusetts (G), Minnesota (G), New Hampshire (G), New Jersey (G), New York (G), Pennsylvania (G), Rhode Island (G), West Virginia (G).

Description: Chromium Potassium Sulfate, $CrK(SO_4)_2$, is a crystalline compound whose color may range from violet-red to black. Freezing/Melting point = 89°C. Soluble in water.

Potential Exposure: This material is used in tanning of leather, dyeing of fabrics, manufacture of glues and gums, chromium salts, ink, photographic emulsions, and ceramics.

Incompatibilities: Contact with magnesium, aluminum may cause fire and explosion hazard.

Permissible Exposure Limits in Air: The OSHA PEL for trivalent chromium compounds is 0.5 mg/m^3 TWA for an 8-hour workshift. The ACGIH[1] and HSE[33] TWA values are the same. The NIOSH IDLH level is 25 mg/m^3 [as Cr(III)].

Determination in Air: Filter; Acid; Flame atomic absorption spectrometry; NIOSH IV Method #7024, Chromium.

Permissible Concentration in Water: For the protection of freshwater aquatic life: Trivalent chromium: not to exceed $e^{[1.08 \ln(\text{hardness}) + 3.48]}$μg/l. For the protection of saltwater aquatic life: Trivalent chromium: 10,300 μg/l on an acute toxicity basis. To protect human health: Trivalent chromium: 170 μg/l; hexavalent chromium 50 μg/l according to EPA.[6]

EPA[49] has set a long-term health advisory for adults of 0.84 mg/l and a lifetime health advisory of 0.12 mg/l (120 μg/l) for chromium. EPA's maximum drinking water level (MCL) is 0.1 mg/l.[62]

Germany, Canada, EEC, and WHO[35] have set a limit of 0.05 mg/l in drinking water.

The states of Maine and Minnesota have set guidelines for chromium in drinking water[61] of 50 μg/l for Maine and 120 μg/l for Minnesota.

Determination in Water: Total chromium may be determined by digestion followed by atomic absorption or by colorimetry (diphenylcarbazide) or by inductively coupled plasma (CP) optical emission spectrometry. Chromium (VI) may be determined by extraction and atomic absorption or colorimetry (using diphenylhydrazide). Dissolved total Cr or Cr(VI) may be determined by 0.45 μ filtration followed by the above-cited methods.[49]

Routes of Entry: Inhalation, eyes, skin.

Harmful Effects and Symptoms

Short Term Exposure: *Inhalation:* May cause irritation to mouth, nose, throat and lungs. *Skin:* May cause irritation, redness and sores. Allergic reaction may occur in sensitive individuals. *Eyes:* May cause irritation. *Ingestion:* Large quantities may cause stomach irritation and nausea.

Long Term Exposure: Repeated skin contact may give rise to allergic sensitization.

Points of Attack: Skin.

Medical Surveillance: Examination by a qualified allergist.

First Aid: If this chemical gets into the eyes, remove any contact lenses at once and irrigate immediately for at least 15 minutes, occasionally lifting upper and lower lids. Seek medical attention immediately. If this chemical contacts the skin, remove contaminated clothing and wash immediately with soap and water. Seek medical attention immediately. If this chemical has been inhaled, remove from exposure, begin rescue breathing (using universal precautions) if breathing has stopped and CPR if heart action has stopped. Transfer promptly to a medical facility. When this chemical has been swallowed, get medical attention. Give large quantities of water and induce vomiting. Do not make an unconscious person vomit.

Personal Protective Methods: Wear protective gloves and clothing to prevent any reasonable probability of skin contact. Safety equipment suppliers/manufacturers can provide recommendations on the most protective glove/clothing material for your operation. All protective clothing (suits, gloves, footwear, headgear) should be clean, available each day, and put on before work. Contact lenses should not be worn when working with this chemical. Wear splash or dust-proof chemical goggles and face shield unless full facepiece respiratory protection is worn. Employees should wash immediately with soap when skin is wet or contaminated. Provide emergency showers and eyewash.

Respirator Selection: NIOSH/OSHA as chromium (III) compounds: *2.5 mg/m³:* DM (any dust and mist respirator). *5 mg/m³:* DMXSQ (any dust and mist respirator except single-use and quarter mask respirators); or SA (any supplied-air respirator); *12.5 mg/m³:* SA:CF (any supplied-air respirator operated in a continuous-flow mode); or PAPRDM (any powered, air-purifying respirator with a dust and mist filter). *25 mg/m³:* HiEF (any air-purifying, full-facepiece respirator with a high-efficiency particulate filter); or PAPRTHiE (any powered, air-purifying respirator with a tight-fitting facepiece and a high-efficiency particulate filter); or SCBAF (any self-contained breathing apparatus with a full facepiece); or SAF (any supplied-air respirator with a full facepiece). *250 mg/m³:* (any supplied-air respirator that has a full facepiece and is operated in a pressure-demand or other positive-pressure mode). *Emergency or planned entry into unknown concentrations or IDLH conditions:* SCBAF:PD,PP (any self-contained breathing apparatus that has a full facepiece and is operated in a pressure-demand or other positive-pressure mode); or SAF:PD,PP:ASCBA (any supplied-air respirator that has a full facepiece and is operated in a pressure-demand or other positive-pressure mode in combination with an auxiliary self-contained breathing apparatus operated in a pressure-demand or other positive pressure mode). *Escape:* HiEF (any air-purifying, full-facepiece respirator with a high-efficiency particulate filter); or SCBAE (any appropriate escape-type, self-contained breathing apparatus).

Note: Substance reported to cause eye irritation or damage; may require eye protection.

Storage: Prior to working with this chemical you should be trained on its proper handling and storage. Store in tightly closed containers in a cool, well-ventilated area. Keep away from heat.

Shipping: This material is not specifically cited in DOT's Performance-Oriented Packaging Standards.[19]

Spill Handling: Evacuate and restrict persons not wearing protective equipment from area of spill or leak until cleanup is complete. Remove all ignition sources. Ventilate area of spill or leak. Absorb liquids in vermiculite, dry sand, earth, peat, carbon, or a similar material and deposit in sealed containers. It may be necessary to contain and dispose of this chemical as a hazardous waste. If material or contaminated runoff enters waterways, notify downstream users of potentially contaminated waters. Contact your Department of Environmental Protection or your regional office of the federal EPA for specific recommendations. If employees are required to clean-up spills, they must be properly trained and equipped. OSHA 1910.120(q) may be applicable.

Fire Extinguishing: This material is not flammable. Use extinguisher appropriate for surrounding fire. Poisonous gases are produced in fire. If material or contaminated runoff enters waterways, notify downstream users of potentially contaminated waters. Notify local health and fire officials and pollution control agencies. From a secure, explosion-proof location, use water spray to cool exposed containers. If cooling streams are ineffective (venting sound increases in volume and pitch, tank discolors, or shows any signs of deforming), withdraw immediately to a secure position. If employees are expected to fight fires, they must be trained and equipped in OSHA 1910.156.

References

New York State Department of Health, "Chemical Fact Sheet: Chromium Potassium Sulfate," Albany, NY, Bureau of Toxic Substance Assessment (January 1986)

Chromium Trioxide

Molecular Formula: CrO_3

Synonyms: Anhydride chromique (French); Anidride cromica (Italian); Chrome (trioxyde de) (French); Chromic acid; Chromic(VI) acid; Chromic anhydride; Chromic trioxide; Chromium oxide; Chromium(VI) oxide; Chromium trioxide; Chromium(6+) trioxide; Chromo (triossido di) (Italian); Chromsaeureanhydrid (German); Chromtrioxid (German); Chroomtrioxyde (Dutch); Chroomzuuranhydride (Dutch); Monochromium oxide; Monochromium trioxide; Puratronic chromium trioxide

CAS Registry Number: 1333-82-0

RTECS Number: GB6650000

DOT ID: UN 1463 (anhydrous)

EEC Number: 024-001-00-0

Regulatory Authority

- Carcinogen (Human) (IARC)[9] (NTP)[10] (ACGIH)[1]
- Banned or Severely Restricted (Some countries) (UN)[35]
- Air Pollutant Standard Set (ACGIH)[1] (Australia) (DFG)[3] (HSE)[33] (Israel) (OSHA) (former USSR)[35][43] (Connecticut)[60] (Several Canadian Provinces)
- CLEAN AIR ACT: Hazardous Air Pollutants (Title I, Part A, Section 112)
- CLEAN WATER ACT: Toxic Pollutant (Section 401.15); 40CFR401.15 Section 307 Toxic Pollutants as chromium and compounds
- RCRA, 40CFR261, Appendix 8 Hazardous Constituents, waste number not listed
- EPCRA (Section 313): Includes any unique chemical substances that contains chromium as part of that chemical's infrastructure. Form R *de minimis* concentration reporting level: Chromium(VI) compounds: 0.1%
- Canada, WHMIS, Ingredients Disclosure List; National Pollutant Release Inventory (NPRI); CEPA Priority Substance List

Cited in U.S. State Regulations: Alaska (G), California (A, G), Connecticut (A), Florida (G), Illinois (G), Maine (G), Massachusetts (G), Minnesota (G), New Hampshire (G), New Jersey (G), New York (G), Oklahoma (G), Pennsylvania (G), Rhode Island (G), West Virginia (G).

Description: Cromium trioxide, CrO_3 is a dark-red crystalline substance. It is odorless. Freezing/Melting point = 196°C. It is deliquescent. Decomposes below boiling point at 250°C. Soluble in water.

Potential Exposure: Chromium trioxide is used in chrome plating, aluminum anodizing, dye, ink and paint manufacturing, tanning, engraving and photography.

Incompatibilities: Chromium trioxide is a strong oxidizer. The solution in water, is a strong acid. Reacts violently with bases and is corrosive. Contact with reducing agents, fuels, organic chemicals, flammable and combustible materials causing fire and explosion hazard. This chemical decomposes above 250°C to chromic oxide and oxygen with increased fire hazard. Attacks metals in the presence of moisture.

Permissible Exposure Limits in Air: The legal OSHA PEL is 0.1 mg/m^3 not to be exceeded at any time. The recommended NIOSH limit is 0.001 mg/m^3 TWA for a 10-hour workshift. The ACGIH recommended limit is 0.05 mg/m^3. The NIOSH IDLH = 15 mg/m^3 [as Cr(VI)]. The Soviet Union[35] and also The former USSR/UNEP joint project[43] give a MAC in workplace air of 0.01 mg/m^3. Connecticut[60] has set a guideline for chromium trioxide in ambient air of 0.25 $\mu g/m^3$.

Determination in Air: Filter; Acid; Flame atomic absorption spectrometry; NIOSH IV Method #7024, Chromium. See also OSHA Method ID103.

Permissible Concentration in Water: For the protection of freshwater aquatic life: Hexavalent chromium: 0.29 μg/l as a 24 hour average, never to exceed 21.0 μg/l. For the protection of saltwater aquatic life: Hexavalent chromium: 18 μg/l as a 24 hour average, never to exceed 1,260 μg/l. To protect human health: hexavalent chromium 50 μg/l according to EPA.[6]

EPA[49] has set a long-term health advisory for adults of 0.84 mg/l and a lifetime health advisory of 0.12 mg/l (120 μg/l) for chromium. EPA's maximum drinking water level (MCL) is 0.1 mg/l.[62]

Germany, Canada, EEC, and WHO[35] have set a limit of 0.05 mg/l in drinking water.

The states of Maine and Minnesota have set guidelines for chromium in drinking water[61] of 50 μg/l for Maine and 120 μg/l for Minnesota.

Determination in Water: Total chromium may be determined by digestion followed by atomic absorption or by colorimetry (diphenylcarbazide) or by inductively coupled plasma (CP) optical emission spectrometry. Chromium (VI) may be determined by extraction and atomic absorption or colorimetry (using diphenylhydrazide). Dissolved total Cr or Cr(VI) may be determined by 0.45 μ filtration followed by the above-cited methods.[49]

Routes of Entry: Inhalation, skin contact, ingestion.

Harmful Effects and Symptoms

Short Term Exposure: *Inhalation:* Human exposure to concentrations between 0.18 – 1.4 mg/m^3 of acid mist for two weeks produced irritation of the nose; after 4 weeks, ulcers of the nose developed; and after 8 weeks, holes formed in the tissue separating the nostrils with bleeding, discharge or formation of a crust in the inner nose. Additional effects noted in humans exposed to unknown concentrations are irritation of the throat, voice-box, lungs, asthmatic attacks, headache, wheezing, coughing, shortness of breath, and painful breathing. Chromic trioxide can cause severe allergic lung reaction.

Skin: Direct contact will produce severe irritation of the skin. Sensitization from such contact can occur and result in severe dermatitis from very small exposures at a later time.

Eyes: Can cause severe irritation, burns, and possible loss of vision.

Ingestion: Swallowing of Chromic acid solutions can result in severe irritation and damage to the mouth, throat and stomach.

Long Term Exposure: Prolonged exposure to chromic acid mist can result in perorations (holes) of the nasal septum (tissue separating the nostrils), lung irritation with symptoms similar to asthma and liver damage. Repeated or prolonged contact with skin may cause dermatitis and chrome ulcers, or "chrome holes." May cause skin sensitization, allergy, irritation, and rashes. Repeated or prolonged inhalation exposure may cause asthma-like reactions. Wearing away of the surfaces of teeth has been noted in workers exposed to chromic acid mist for a prolonged time. This substance is carcinogenic to humans. It has been shown to cause lung and throat cancers. There is limited evidence that this chemical is a teratogen in animals. May cause kidney and liver damage.

Points of Attack: Skin, lungs, kidney, liver.

Medical Surveillance: Before first exposure and every 6 – 12 months, a medical history and exam is recommended, with very careful attention to the nose, skin, lungs, and voice box. Tests for kidney, liver, or lung function should be considered. IF you notice skin, nose or lung effects, seek prompt medical attention. Also check your skin daily for little bumps or blisters, the first sign of "chrome ulcers." If not treated early, these can last for years after exposure.

If symptoms develop or overexposure is suspected, the following may be useful: Evaluation by a qualified allergist, including careful exposure history and special testing, may help diagnose skin allergy.

First Aid: If this chemical gets into the eyes, remove any contact lenses at once and irrigate immediately for at least 30 minutes, occasionally lifting upper and lower lids. Seek medical attention immediately. If this chemical contacts the skin, remove contaminated clothing and wash immediately with soap and water. Seek medical attention immediately. If this chemical has been inhaled, remove from exposure, begin rescue breathing (using universal precautions) if breathing has stopped and CPR if heart action has stopped. Transfer promptly to a medical facility. When this chemical has been swallowed, get medical attention. If victim is conscious, administer water or milk. Do not induce vomiting.

Personal Protective Methods: Wear acid resistant gloves and clothing to prevent any reasonable probability of skin contact. Safety equipment suppliers/manufacturers can provide recommendations on the most protective glove/clothing material for your operation. All protective clothing (suits, gloves, footwear, headgear) should be clean, available each day, and put on before work. AIHA recommends Polyvinyl Chloride for solutions of Chromium (VI) Oxide in water as a protective material. Contact lenses should not be worn when working with this chemical. Wear dust-proof chemical goggles and face shield unless full facepiece respiratory protection is worn. Employees should wash immediately with soap when skin is wet or contaminated. Provide emergency showers and eyewash. Specific engineering controls are recommended in NIOSH criteria document #76-129, chromium (VI).

Respirator Selection: NIOSH, as chromates: *at any concentrations above the NIOSH REL, or where there is no REL, at any detectable concentration:* SCBAF:PD,PP (any self-contained breathing apparatus that has a full facepiece and is operated in a pressure-demand or other positive-pressure mode); or SAF:PD,PP:ASCBA (any supplied-air respirator that has a full facepiece and is operated in a pressure-demand or other positive-pressure mode in combination with an auxiliary, self-contained breathing apparatus operated in a pressure-demand or other positive-pressure mode). *Escape:* GMFOV [any air-purifying, full-facepiece respirator (gas mask) with a chin-style, front- or back-mounted organic vapor canister]; or SCBAE (any appropriate escape-type, self-contained breathing apparatus).

Storage: Prior to working with Chromium (VI) Oxide you should be trained on its proper handling and storage. Chromium (VI) Oxide must be stored to avoid contact with reducing agents and organic chemicals since violent reactions occur. Store in tightly closed containers in a cool, well-ventilated area away from fuels and other flammable and combustible materials. Do not store Chromium (VI) Oxide on wood floors, because prolonged contact with wood can produce a fire hazard. Protect from excess moisture to minimize rusting of containers. A regulated, marked area should be established where Chromium (VI) Oxide is handled, used, or stored. See OSHA standard 1910.104 and NFPA 43A *Code for the Storage of Liquid and Solid Oxidizers* for detailed handling and storage regulations. A regulated, marked area should be established where this chemical is handled, used, or stored in compliance with OSHA standard 1910.1045.

Shipping: Chromic acid, solid, must be labeled "Oxidizer, Corrosive." It falls in Hazard Class 5.1 and Packing Group II.[19]

Spill Handling: Evacuate persons not wearing protective equipment from area of spill or leak until clean-up is complete. Remove all ignition sources. Collect powdered material in the most convenient and safe manner and deposit in sealed containers. Ventilate area after clean-up is complete. Keep Chromium (VI) Oxide out of a confined space, such as a sewer, because of the possibility of an explosion, unless the sewer is designed to prevent the build-up of explosive concentrations. It may be necessary to contain and dispose of this chemical as a hazardous waste. If material or contaminated runoff enters waterways, notify downstream users of potentially contaminated waters. Contact your Department of Environmental Protection or your regional office of the federal EPA for specific recommendations. If employees are required to clean-up spills, they must be properly trained and equipped. OSHA 1910.120(q) may be applicable.

Fire Extinguishing: Extinguish fire using an agent suitable for type of surrounding fire. Chromium (VI) oxide itself does not burn but it will increase the intensity of a fire since it is an oxidizer. The substance decomposes above 250°C to chromic oxide and oxygen with increased fire hazard. Poisonous gases are produced in fire. If material or contaminated runoff enters waterways, notify downstream users of potentially contaminated waters. Notify local health and fire officials and pollution control agencies. From a secure, explosion-proof location, use water spray to cool exposed containers. If cooling streams are ineffective (venting sound increases in volume and pitch, tank discolors, or shows any signs of deforming), withdraw immediately to a secure position. If employees are expected to fight fires, they must be trained and equipped in OSHA 1910.156.

Disposal Method Suggested: Reduce to Cr(III). If material cannot be recovered and recycled, dispose of sludge in a chemical waste landfill.[22]

References

Sax, N. I., Ed., "Dangerous Properties of Industrial Materials Report," 1, No. 7, 47–49 (1981)

New Jersey Department of Health and Senior Services and Senior Services, "Hazardous Substance Fact Sheet: Chromium (VI) Oxide," Trenton, NJ (April 1998)

New York State Department of Health, "Chemical Fact Sheet: Chromic Acid," Albany, NY, Bureau of Toxic Substance Assessment (January 1986)

Chromosulfuric Acid

Molecular Formula: $CrO_{4x}S_x$

Common Formula: $Cr(SO_4)_x$

Synonyms: Basic chromic sulfate; Basic chromic sulphate; Basic chromium sulfate; Basic chromium sulphate; Chromium hydroxide sulfate; Chromium sulfate; Chromium sulfate, basic; Chromium(III) sulfate, hexahydrate; Chromium sulphate; Chronisulfat (German); Koreon; Monobasic chromium sulfate; Monobasic chromium sulphate; Neochromium; Sulfuric acid, chromium salt; Sulfuric acid, chromium salt, basic

CAS Registry Number: 14489-25-9; 15005-90-0 (hexahydrate); 64093-79-4 (basic)

RTECS Number: WS6985000

DOT ID: UN 2240

Regulatory Authority

- Air Pollutant Standard Set (ACGIH)[1]
- CLEAN AIR ACT: Hazardous Air Pollutants (Title I, Part A, Section 112). as chromium compounds

- CLEAN WATER ACT: 40CFR423, Appendix A, Priority Pollutants; Section 313 Water Priority Chemicals (57FR41331, 9/9/92); Toxic Pollutant (Section 401.15)
- EPA HAZARDOUS WASTE NUMBER (RCRA No.): D007
- RCRA, 40CFR261, Appendix 8 Hazardous Constituents, waste number not listed
- RCRA Toxicity Characteristic (Section 261.24), Maximum
- Concentration of Contaminants, regulatory level, 5.0 mg/l
- RCRA 40CFR268.48; 61FR15654, Universal Treatment Standards: Wastewater (mg/l), 2.77; Nonwastewater (mg/kg), 0.86 as chromium (total)
- RCRA 40CFR264, Appendix 9; TSD Facilities Ground Water Monitoring List. Suggested test method(s) (PQL μg/l): (total) 6010 (70); 7190 (500); 7191 (10)
- SAFE DRINKING WATER ACT: MCL, 0.1 mg/l; MCLG, 0.1 mg/l; Regulated chemical (47 FR 9352)
- SUPERFUND/EPCRA 40CFR302.4 Reportable Quantity (RQ): CERCLA, 5,000 lb (2,270 kg)
- EPCRA Section 313 Form R *de minimis* concentration reporting level: 0.1%
- Canada, WHMIS, Ingredients Disclosure List; National Pollutant Release Inventory (NPRI); CEPA Priority Substance List as chromium compounds

Cited in U.S. State Regulations: Alaska (G), California (A, G), Illinois (G), Maine (G), New Hampshire (G), New Jersey (G), Rhode Island (G), West Virginia (G).

Description: Chromosulfuric acid, $Cr(SO_4)_x$ is a dark green powder. Highly soluble in water (reactive).

Potential Exposure: Chromosulfuric Acid is used in chrome plating, paint, ink and glaze manufacturing, tanning, catalyst preparation and as a fixative in textile dyeing.

Incompatibilities: Violent reaction with water. Incompatible with caustic materials, ammonia, aliphatic amines, alkanolmines, isocyanates, epichlorohydrin.

Permissible Exposure Limits in Air: OSHA PEL: the legal airborne permissible exposure limit is 0.5 mg/m^3 TWA, averaged over an 8-hour workshift for Chromium as soluble Chromic and Chromous salts measured as Chromium. The ACGIH[1] and HSE[33] TWA values for trivalent chromium are the same.

Determination in Air: Filter; Acid; Flame atomic absorption spectrometry; NIOSH IV Method #7024, Chromium.

Permissible Concentration in Water: For the protection of freshwater aquatic life: Trivalent chromium: not to exceed $e^{[1.08 \ln(\text{hardness}) + 3.48]}$μg/l. For the protection of saltwater aquatic life: Trivalent chromium: 10,300 μg/l on an acute toxicity basis. To protect human health: Trivalent chromium: 170 μg/l; hexavalent chromium 50 μg/l according to EPA.[6]

EPA[49] has set a long-term health advisory for adults of 0.84 mg/l and a lifetime health advisory of 0.12 mg/l (120 μg/l) for chromium. EPA's maximum drinking water level (MCL) is 0.1 mg/l.[62]

Germany, Canada, EEC, and WHO[35] have set a limit of 0.05 mg/l in drinking water.

The states of Maine and Minnesota have set guidelines for chromium in drinking water[61] of 50 μg/l for Maine and 120 μg/l for Minnesota

Determination in Water: Total chromium may be determined by digestion followed by atomic absorption or by colorimetry (diphenylcarbazide) or by inductively coupled plasma (CP) optical emission spectrometry. Chromium (VI) may be determined by extraction and atomic absorption or colorimetry (using diphenylhydrazide). Dissolved total Cr or Cr(VI) may be determined by 0.45 μ filtration followed by the above-cited methods.[49]

Routes of Entry: Inhalation, skin contact, ingestion.

Harmful Effects and Symptoms

Short Term Exposure: Chromosulfuric Acid is a corrosive chemical and skin or eye contact can cause severe irritation and burns. Inhalation can irritate the respiratory tract causing coughing and wheezing. Skin allergy sometimes occurs with itching, redness and/or an eczema-like rash. If this happens, future contact can trigger symptoms.

Long Term Exposure: may cause skin allergy with itching and rash. Some related chromium compounds are known carcinogens. Although this chemical has not been identified as a carcinogen, it should be handled with extreme caution.

Points of Attack: Skin, respiratory system.

Medical Surveillance: Lung function tests. If symptoms develop or overexposure is suspected, the following may be useful: Consider chest x-ray after acute overexposure. Evaluation by a qualified allergist, including careful exposure history and special testing, may help diagnose skin allergy.

First Aid: If this chemical gets into the eyes, remove any contact lenses at once and irrigate immediately for at least 15 minutes, occasionally lifting upper and lower lids. Seek medical attention immediately. If this chemical contacts the skin, remove contaminated clothing and wash immediately with soap and water. Seek medical attention immediately. If this chemical has been inhaled, remove from exposure, begin rescue breathing (using universal precautions) if breathing has stopped and CPR if heart action has stopped. Transfer promptly to a medical facility. When this chemical has been swallowed, get medical attention. If victim is conscious, administer water or milk. Do not induce vomiting.

Personal Protective Methods: Wear protective gloves and clothing to prevent any reasonable probability of skin contact. Safety equipment suppliers/manufacturers can provide recommendations on the most protective glove/clothing material for your operation. All protective clothing (suits, gloves, footwear, headgear) should be clean, available each day, and put on before work. Contact lenses should not be worn when working with this chemical. Wear dust-proof chemical goggles and face shield unless full facepiece respiratory protection is worn. Employees should wash immediately with soap when skin is wet or contaminated. Provide emergency showers and eyewash.

Respirator Selection: NIOSH/OSHA: *2.5 mg/m³:* DM (any dust and mist respirator). *5 mg/m³:* DMXSQ (any dust and mist respirator except single-use and quarter mask respirators); or SA (any supplied-air respirator); *12.5 mg/m³:* SA:CF (any supplied-air respirator operated in a continuous-flow mode); or PAPRDM (any powered, air-purifying respirator with a dust and mist filter). *25 mg/m³:* HiEF (any air-purifying, full-facepiece respirator with a high-efficiency particulate filter); or PAPRTHiE (any powered, air-purifying respirator with a tight-fitting facepiece and a high-efficiency particulate filter); or SCBAF (any self-contained breathing apparatus with a full facepiece); or SAF (any supplied-air respirator with a full facepiece). *250 mg/m³:* (any supplied-air respirator that has a full facepiece and is operated in a pressure-demand or other positive-pressure mode). *Emergency or planned entry into unknown concentrations or IDLH conditions:* SCBAF:PD,PP (any self-contained breathing apparatus that has a full facepiece and is operated in a pressure-demand or other positive-pressure mode); or SAF:PD,PP:ASCBA (any supplied-air respirator that has a full facepiece and is operated in a pressure-demand or other positive-pressure mode in combination with an auxiliary self-contained breathing apparatus operated in a pressure-demand or other positive pressure mode). *Escape:* HiEF (any air-purifying, full-facepiece respirator with a high-efficiency particulate filter); or SCBAE (any appropriate escape-type, self-contained breathing apparatus).

Note: Substance reported to cause eye irritation or damage; may require eye protection.

Storage: Prior to working with Chromosulfuric Acid you should be trained on its proper handling and storage. Chromosulfuric Acid must be stored to avoid contact with water and other incompatible materials listed above, since violent reactions occur. Store in tightly closed containers in a cool, well-ventilated area away from flammable and combustible materials.

Shipping: Chromosulfuric Acid requires a "Corrosive" label. It falls in DOT Hazard Class 8 and Packing Group I.[19]

Spill Handling: Evacuate persons not wearing protective equipment from area of spill or leak until clean-up is complete. Remove all ignition sources. Collect powdered material in the most convenient and safe manner and deposit in sealed containers. Ventilate area after clean-up is complete. Keep Chromosulfuric Acid out of confined spaces, such as a sewer, because of the possibility of an explosion, unless the sewer is designed to prevent the build-up of explosive concentrations. It may be necessary to contain and dispose of this chemical as a hazardous waste. If material or contaminated runoff enters waterways, notify downstream users of potentially contaminated waters. Contact your Department of Environmental Protection or your regional office of the federal EPA for specific recommendations. If employees are required to clean-up spills, they must be properly trained and equipped. OSHA 1910.120(q) may be applicable.

Fire Extinguishing: Chromosulfuric Acid may burn, but does not readily ignite. It will increase the intensity of a fire since it is an oxidizer. Use dry chemical or CO_2 extinguishers. Use water spray to keep fire-exposed containers cool, but do not get water inside containers. Poisonous gases are produced in fire including sulfur oxides and chromium. If material or contaminated runoff enters waterways, notify downstream users of potentially contaminated waters. Notify local health and fire officials and pollution control agencies. From a secure, explosion-proof location, use water spray to cool exposed containers. If cooling streams are ineffective (venting sound increases in volume and pitch, tank discolors, or shows any signs of deforming), withdraw immediately to a secure position. If employees are expected to fight fires, they must be trained and equipped in OSHA 1910.156.

References

New Jersey Department of Health and Senior Services and Senior Services, "Hazardous Substance Fact Sheet: Chromosulfuric Acid," Trenton, NJ (September 1996)

Chromous Chloride

Molecular Formula: Cl_2Cr

Synonyms: Chromium chloride; Chromium(2+) chloride; Chromium(II) chloride; Chromium dichloride; Cloruro cromoso (Spanish)

CAS Registry Number: 10049-05-5

RTECS Number: GB5250000

DOT ID: UN 9102

Regulatory Authority

- Air Pollutant Standard Set (OSHA) (ACGIH) (Australia) (Israel) (Mexico) (Several States) (Several Canadian Provinces)
- CLEAN AIR ACT: Hazardous Air Pollutants (Title I, Part A, Section 112). as chromium compounds
- CLEAN WATER ACT: Section 311 Hazardous Substances/RQ 40CFR117.3 (same as CERCLA, see below); Section 313 Water Priority Chemicals (57FR41331, 9/9/92); 40CFR401.15 Section 307 Toxic Pollutants
- SUPERFUND/EPCRA 40CFR302.4 Reportable Quantity (RQ): CERCLA, 1,000 lb (454 kg)
- Canada, WHMIS, Ingredients Disclosure List

Description: Chromous Chloride is a white-to-blue solid or lustrous needles which turn blue in water. Boiling point = 1,300°C. Melting point = 824°C. Hazard Identification (based on NFPA-704 M Rating System): Health 1, Flammability 0, Reactivity 0. Soluble in water.

Potential Exposure: It is used in metal alloys and metal finishing, textile treatment including mothproofing, waterproofing, printing, and dying, leather tanning, making photographic chemicals, and green pigments for various uses.

Incompatibilities: Very hygroscopic. Water solution forms flammable hydrogen gas.

Permissible Exposure Limits in Air: The legal airborne limit (OSHA PEL) measured as Cr is 0.5 mg/m^3 TWA for an 8-hour workshift. ACGIH and NIOSH recommend a TWA value of 0.05 mg/m^3. For Chromium(II) compounds, Australia, Israel, Mexico, and the Canadian provinces of Alberta, BC, Ontario, and Quebec all set the same limits as NIOSH and ACGIH and Alberta set a STEL of 1.5 mg/m^3.

Determination in Air: Filter; Acid; Flame atomic absorption spectrometry; NIOSH IV Method #7024, Chromium.

Permissible Concentration in Water: See Chromium entry.

Routes of Entry: Inhalation, eyes and/or skin.

Harmful Effects and Symptoms

Short Term Exposure: Irritates the eyes, skin, and respiratory tract. Eye contact may cause damage. Inhalation can cause coughing and/or shortness of breath. May cause pulmonary irritation. Moderately toxic to humans. Between 1 ounce, and 1 pound may be fatal. The LD_{50} oral- rat is 1,870 mg/kg.

Long Term Exposure: May cause skin allergy. Chrome ulcers or sore of the skin or nasal septum; hole in the nasal septum, sometimes with bleeding may result. May cause lung damage.

Points of Attack: Skin, lungs.

First Aid: If this chemical gets into the eyes, remove any contact lenses at once and irrigate immediately for at least 15 minutes, occasionally lifting upper and lower lids. Seek medical attention immediately. If this chemical contacts the skin, remove contaminated clothing and wash immediately with soap and water. Seek medical attention immediately. If this chemical has been inhaled, remove from exposure, begin rescue breathing (using universal precautions) if breathing has stopped and CPR if heart action has stopped. Transfer promptly to a medical facility. When this chemical has been swallowed, get medical attention. Give large quantities of water and induce vomiting. Do not make an unconscious person vomit.

Note to Physician: In case of fume inhalation, treat pulmonary edema. Consider administering prednisone or other corticosteroid to reduce tissue response to fume. Positive pressure ventilation may be necessary. Treat metal fume fever with bed rest, analgesics and antipyretics.

Personal Protective Methods: Wear protective gloves and clothing to prevent any reasonable probability of skin contact. Safety equipment suppliers/manufacturers can provide recommendations on the most protective glove/clothing material for your operation. All protective clothing (suits, gloves, footwear, headgear) should be clean, available each day, and put on before work. Contact lenses should not be worn when working with this chemical. Wear dust-proof chemical goggles and face shield unless full facepiece respiratory protection is worn. Employees should wash immediately with soap when skin is wet or contaminated. Provide emergency showers and eyewash. Specific engineering controls are recommended in NIOSH Criteria Document #76-129.

Respirator Selection: *NIOSH/OSHA: 2.5 mg/m^3:* DM (any dust and mist respirator). *5 mg/m^3:* DMXSQ (any dust and mist respirator except single-use and quarter mask respirators); or SA (any supplied-air respirator); *12.5 mg/m^3:* SA:CF (any supplied-air respirator operated in a continuous-flow mode); or PAPRDM (any powered, air-purifying respirator with a dust and mist filter). *25 mg/m^3:* HiEF (any air-purifying, full-facepiece respirator with a high-efficiency particulate filter); or PAPRTHiE (any powered, air-purifying respirator with a tight-fitting facepiece and a high-efficiency particulate filter); or SCBAF (any self-contained breathing apparatus with a full facepiece); or SAF (any supplied-air respirator with a full facepiece). *250 mg/m^3:* (any supplied-air respirator that has a full facepiece and is operated in a pressure-demand or other positive-pressure mode). *Emergency or planned entry into unknown concentrations or IDLH conditions:* SCBAF:PD,PP (any self-contained breathing apparatus that has a full facepiece and is operated in a pressure-demand or other positive-pressure mode); or SAF: PD,PP:ASCBA (any supplied-air respirator that has a full facepiece and is operated in a pressure-demand or other positive-pressure mode in combination with an auxiliary self-contained breathing apparatus operated in a pressure-demand or other positive pressure mode). *Escape:* HiEF (any air-purifying, full-facepiece respirator with a high-efficiency particulate filter); or SCBAE (any appropriate escape-type, self-contained breathing apparatus).

Note: Substance reported to cause eye irritation or damage; may require eye protection.

Storage: Prior to working with Chromous chloride you should be trained on its proper handling and storage. Store in tightly closed containers in a cool, well ventilated area away from water and moisture. Where possible, automatically pump liquid from drums or other storage containers to process containers.

Shipping: The DOT does not set out any specific label requirements or maximum shipping quantities for Chromous chloride.

Spill Handling: *Solid:* Evacuate persons not wearing protective equipment from area of spill or leak until clean-up is complete. Remove all ignition sources. Collect powdered material in the most convenient and safe manner and deposit in sealed containers. Ventilate area after clean-up is complete. It may be necessary to contain and dispose of this chemical as a hazardous waste. If material or contaminated runoff enters waterways, notify downstream users of potentially contaminated waters. Contact your Department of Environmental Protection or your regional office of the federal EPA for specific recommendations. If employees are required to clean-up spills, they must be properly trained and equipped. OSHA 1910.120(q) may be applicable.

Liquid solution: Evacuate and restrict persons not wearing protective equipment from area of spill or leak until cleanup is complete. Remove all ignition sources. Ventilate area of spill or leak. Absorb liquids in vermiculite, dry sand, earth, peat, carbon, or a similar material and deposit in sealed containers. It may be necessary to contain and dispose of this chemical as a hazardous waste. If material or contaminated runoff enters waterways, notify downstream users of potentially contaminated waters. Contact your Department of Environmental Protection or your regional office of the federal EPA for specific recommendations. If employees are required to clean-up spills, they must be properly trained and equipped. OSHA 1910.120(q) may be applicable.

Fire Extinguishing: Chromous chloride may burn, but does not readily ignite. Use dry chemical, carbon dioxide, water spray, or foam extinguishers. Poisonous gases are produced in fire including chromous and chromic salts and fumes of chromium. If material or contaminated runoff enters waterways, notify downstream users of potentially contaminated waters. Notify local health and fire officials and pollution control agencies. Containers may explode in fire. From a secure, explosion-proof location, use water spray to cool exposed containers. If cooling streams are ineffective (venting sound increases in volume and pitch, tank discolors, or shows any signs of deforming), withdraw immediately to a secure position. If employees are expected to fight fires, they must be trained and equipped in OSHA 1910.156.

References

New Jersey Department of Health and Senior Services, "Hazardous Substance Fact Sheet, Chromous Chloride." Trenton, NJ (September 1998)

Chromyl Chloride

Molecular Formula: Cl_2CrO_2

Common Formula: CrO_2Cl_2

Synonyms: Chlorure de chromyle (French); Chromic oxychloride; Chromium chloride oxide; Chromium, dichlorodioxo-; Chromium dioxychloride; Chromium(VI) dioxychloride; Chromium dioxychloride dioxide; Chromium oxychloride; Chromylchlorid (German); Chromyl chloride; Cromile, cloruro di (Italian); Croomoxylchloride (Dutch); Dioxodichlorochromium; Oxychlorure chromique (French)

CAS Registry Number: 14977-61-8

RTECS Number: GB5775000

DOT ID: UN 1758

Regulatory Authority

- Carcinogen (Suspected) (DFG)[3]
- Air Pollutant Standard Set (ACGIH)[1] (Israel) (OSHA)[58] (Several States)[60] (Several Canadian Provinces)
- CLEAN AIR ACT: Hazardous Air Pollutants (Title I, Part A, Section 112)
- CLEAN WATER ACT: Toxic Pollutant (Section 401.15); 40CFR401.15 Section 307 Toxic Pollutants as chromium and compounds
- RCRA, 40CFR261, Appendix 8 Hazardous Constituents, waste number not listed
- EPCRA (Section 313): Includes any unique chemical substances that contains chromium as part of that chemical's infrastructure. Form R *de minimis* concentration reporting level: Chromium(VI) compounds: 0.1%
- U.S. DOT 49CFR172.101, Appendix B, Regulated marine pollutant
- Canada, WHMIS, Ingredients Disclosure List

Cited in U.S. State Regulations: Alaska (G), California (A, G), Connecticut (A), Florida (G), Illinois (G), Maine (G), Massachusetts (G), Minnesota (G), Nevada (A), New Hampshire (G), New Jersey (G), North Dakota (A), Oklahoma (G), Pennsylvania (G), Rhode Island (G), West Virginia (G).

Description: Chromyl chloride, CrO_2Cl_2, is a dark red fuming liquid with a musty, burning odor. Freezing/Melting point = -96.5°C. Boiling point = 117°C. Soluble in water (decomposes).

Potential Exposure: Chromium Oxychloride is used in making Chromium complexes and dyes and in various organic oxidation and chlorination reactions.

Incompatibilities: Contact with water is violent and forms hydrochloric, chromic acids, and chlorine gas. A powerful oxidizer. Reacts violently with acetone, alcohol, ammonia, ether, fuels, organic solvents, moist phosphorus, phosphorus trichloride, sodium azide, sulfur, reducing agents, turpentine. Contact with nonmetal halides such as disulfur dichloride, phosphorus trichloride, and phosphorus tribromide, nonmetal hydrides such as hydrogen sulfide and hydrogen phosphide, and urea causes a danger fire and explosion hazard.

Permissible Exposure Limits in Air: The NIOSH recommended airborne exposure limit is 0.001 mg/m^3 TWA averaged over a 10-hour workshift. The ACGIH TLV is 0.025 ppm TWA. DFG lists this chemical as a carcinogen with no numerical value. NIOSH has not determined The NIOSH IDLH value for this specific chemical, but the NIOSH list of synonyms for this chemical suggests this chemical may be hexavalent; therefore, the following has been included for reference: NIOSH IDLH = 15 mg/m^3 [as Cr(VI)]. This chemical can be absorbed through the skin, thereby increasing exposure. Some states have set guidelines or standards for chromyl chloride in ambient air[60] ranging from 1.5 μg/m^3 (North Dakota) to 4.0 μg/m^3 (Nevada) to 15.0 μg/m^3 (Connecticut).

Determination in Air: Filter; Acid; Flame atomic absorption spectrometry; NIOSH IV Method #7024, Chromium.

Permissible Concentration in Water: This material reacts with water but to the extent that it affects the chromium content of water see the following. For the protection of freshwater aquatic life: Hexavalent chromium: 0.29 μg/l as a 24 hour average, never to exceed 21.0 μg/l. For the protection of saltwater aquatic life: Hexavalent chromium: 18 μg/l as a 24 hour average, never to exceed 1,260 μg/l. To protect human health: hexavalent chromium 50 μg/l according to EPA.[6]

EPA[49] has set a long-term health advisory for adults of 0.84 mg/l and a lifetime health advisory of 0.12 mg/l (120 μg/l) for chromium. EPA's maximum drinking water level (MCL) is 0.1 mg/l.[62]

Germany, Canada, EEC, and WHO[35] have set a limit of 0.05 mg/l in drinking water.

Determination in Water: Total chromium may be determined by digestion followed by atomic absorption or by colorimetry (diphenylcarbazide) or by inductively coupled plasma (CP) optical emission spectrometry. Chromium(VI) may be determined by extraction and atomic absorption or colorimetry (using diphenylhydrazide). Dissolved total Cr or Cr(VI) may be determined by 0.45 μ filtration followed by the above-cited methods.[49]

Routes of Entry: Inhalation, skin and/or eye contact.

Harmful Effects and Symptoms

Short Term Exposure: Chromium Oxychloride can affect you when breathed in and by passing through your skin. Chromium Oxychloride should be handled as a carcinogen, with extreme caution. Eye contact can cause severe damage with this corrosive chemical. Skin contact can cause deep ulcers or an allergic rash or severe irritation.

Long Term Exposure: May cause cancer of the lungs and throat; birth defects, fetus damage, possible miscarriage; skin allergy, with itching and rash; lung allergy with cough, wheezing and difficult breathing; kidney damage; damage to the bone (septum) in the nose.

Points of Attack: Eyes, skin, respiratory system.

Medical Surveillance: Before beginning employment and at regular times after that, the following are recommended: Urine test for chromates. This test is most accurate shortly after exposure. Exam of the nose and skin. Lung function tests. These may be normal if the person is not having an attack at the time of the test. Kidney function tests. If symptoms develop or overexposure is suspected, the following may be useful: Evaluation by a qualified allergist, including careful exposure history and special testing, may help diagnose skin allergy. If any skin bumps or blisters develop, seek medical attention promptly. If not treated early, "chrome ulcers" can develop which can last for years.

First Aid: If this chemical gets into the eyes, remove any contact lenses at once and irrigate immediately for at least 30 minutes, occasionally lifting upper and lower lids. Seek medical attention immediately. If this chemical contacts the skin, remove contaminated clothing and wash immediately with soap and water. Seek medical attention immediately. If this chemical has been inhaled, remove from exposure, begin rescue breathing (using universal precautions) if breathing has stopped and CPR if heart action has stopped. Transfer promptly to a medical facility. When this chemical has been swallowed, get medical attention. If victim is conscious, administer water or milk. Do not induce vomiting.

Personal Protective Methods: Wear protective gloves and clothing to prevent any reasonable probability of skin contact. Safety equipment suppliers/manufacturers can provide recommendations on the most protective glove/clothing material for your operation. All protective clothing (suits, gloves, footwear, headgear) should be clean, available each day, and put on before work. Contact lenses should not be worn when working with this chemical. Wear splash-proof chemical goggles and face shield unless full facepiece respiratory protection is worn. Employees should wash immediately with soap when skin is wet or contaminated. Provide emergency showers and eyewash. Specific engineering controls are recommended in NIOSH Criteria Document #76-129.

Respirator Selection: NIOSH, as chromates: *at any concentrations above the NIOSH REL, or where there is no REL, at any detectable concentration:* SCBAF:PD,PP (any self-contained breathing apparatus that has a full facepiece and is operated in a pressure-demand or other positive-pressure mode); or SAF:PD,PP:ASCBA (any supplied-air respirator that has a full facepiece and is operated in a pressure-demand or other positive-pressure mode in combination with an auxiliary, self-contained breathing apparatus operated in a pressure-demand or other positive-pressure mode). *Escape:* GMFOV [any air-purifying, full-facepiece respirator (gas mask) with a chin-style, front- or back-mounted organic vapor canister]; or SCBAE (any appropriate escape-type, self-contained breathing apparatus).

Storage: Prior to working with Chromium Oxychloride you should be trained on its proper handling and storage. A regulated, marked area should be established where this chemical is handled, used, or stored in compliance with OSHA standard 1910.1045. Chromium Oxychloride must be stored to avoid contact with water since violent reactions occur, releasing poisonous materials including chromic acid, hydrogen chloride, chromic chloride and chlorine. Store in tightly closed containers in a cool, well-ventilated area away from flammable and combustible materials, ammonia, alcohol and turpentine, and other incompatible materials listed above.

Shipping: Chromium Oxychloride must carry a "Corrosive" label. It falls in DOT Hazard Class 8 and Packing Group I.

Spill Handling: Evacuate and restrict persons not wearing protective equipment from area of spill or leak until cleanup is complete. Remove all ignition sources. Ventilate area of spill or leak. Absorb liquids in vermiculite, dry sand, earth, peat, carbon, or a similar material and deposit in sealed containers. Keep this chemical out of confined spaces, such as a sewer, because of the potential for an explosion, unless the sewer is designed to prevent the buildup of explosive concentrations. It may be necessary to contain and dispose of this chemical as a hazardous waste. If material or contaminated runoff enters waterways, notify downstream users of potentially contaminated waters. Contact your Department of Environmental Protection or your regional office of the federal EPA for specific recommendations. If employees are required to clean-up spills, they must be properly trained and equipped. OSHA 1910.120(q) may be applicable.

Initial isolation and protective action distances: distances shown are likely to be affected during the first 30 minutes after materials are spilled and could increase with time. If more than one tank car, cargo tank, portable tank, or large cylinder is involved in the incident is leaking, the protective action distance may need to be increased. You may need to seek emergency information from CHEMTREC at (800) 424-9300 or seek professional environmental engineering assistance from the U.S. EPA Environmental Response Team at (908)548-8730 (24-hour response line).

UN 1758 (Chromium oxychloride) is on the DOT's list of dangerous water-reactive materials which create large amounts of toxic vapor when *spilled in water*: Dangerous from 0.5 – 10 km (0.3 – 6.0 miles) downwind.

Fire Extinguishing: Chromium Oxychloride does not burn, but it will increase the intensity of a fire since it is an oxidizer. Extinguish surrounding fire with dry chemicals or CO_2. Do not use water to control a small fire. Poisonous gases are produced in fire including chromic acid, hydrogen chloride, chromic chloride and chlorine. Vapors may travel to a source of ignition and flash back. If material or contaminated runoff enters waterways, notify downstream users of potentially contaminated waters. Notify local health and fire officials and pollution control agencies. Containers may explode in fire especially if water gets in them. From a secure, explosion-proof location, use water spray to cool exposed containers. If cooling streams are ineffective (venting sound increases in volume and pitch, tank discolors, or shows any signs of deforming), withdraw immediately to a secure position. If employees are expected to fight fires, they must be trained and equipped in OSHA 1910.156.

References

New Jersey Department of Health and Senior Services and Senior Services, "Hazardous Substance Fact Sheet: Chromium Oxychloride," Trenton, NJ (August 6, 1987)

Chrysene

Molecular Formula: $C_{18}H_{12}$

Synonyms: A13-00867; 1,2-Benzofenantreno (Spanish); 1,2-Benzophenanthrene; Benzo(a)phenanthrene; 1,2-Benzphenanthrene; Benz(a)phenanthrene; Criseno (Spanish); 1,2,5,6-Dibenzonaphthalene

CAS Registry Number: 218-01-9

RTECS Number: GC0700000

Regulatory Authority

- Carcinogen (Animal Positive) (IARC)[9] (Suspected Human) (ACGIH)[1] (DFG)[3]
- OSHA, 29CFR1910 Specifically Regulated Chemicals (See CFR 1910.1002) as a coal tar pitch volatile
- Air Pollutant Standard Set (ACGIH)[1] (DFG)[3] (North Dakota, Virginia)[60]
- CLEAN WATER ACT: 40CFR423, Appendix A, Priority Pollutants; 40CFR401.15 Section 307 Toxic Pollutants, as polynuclear aromatic hydrocarbons
- EPA HAZARDOUS WASTE NUMBER (RCRA No.): U050
- RCRA, 40CFR261, Appendix 8 Hazardous Constituents
- RCRA 40CFR268.48; 61FR15654, Universal Treatment Standards: Wastewater (mg/l), 0.059; Nonwastewater (mg/kg), 3.4
- RCRA 40CFR264, Appendix 9; TSD Facilities Ground Water Monitoring List. Suggested test method(s) (PQL μg/l): 8100 (200); 8270 (10)
- SUPERFUND/EPCRA 40CFR302.4 Reportable Quantity (RQ): CERCLA, 100 lb (45.4 kg)
- EPCRA (Section 313): as polycyclic aromatic compounds; Form R *de minimis* concentration reporting level: 0.1%
- Canada, WHMIS, Ingredients Disclosure List

Cited in U.S. State Regulations: Alaska (G), California (A, G), Florida (G), Kansas (G, W), Louisiana (G), Maine (G), Massachusetts (G), Minnesota (G), New Hampshire (G), New Jersey (G), North Dakota (A), Pennsylvania (G), Rhode Island (G), Vermont (G), Virginia (G, A), Washington (G), Wisconsin (G).

Description: Chrysene, $C_{18}H_{12}$, is a combustible, red, blue, odorless, fluorescent crystals. Pure chrysene is a colorless crystal. Freezing/Melting point = 254°C. Hazard Identification (based on NFPA-704 M Rating System): Health 1, Flammability 1, Reactivity 0. Insoluble in water.

Potential Exposure: Almost never found by itself, Chrysene is found in gasoline and diesel exhaust as well as in cigarette smoke and in coal tar, coal tar pitch, creosote. It is used in organic synthesis.

Incompatibilities: Contact with strong oxidizers may cause fire and explosion hazard.

Permissible Exposure Limits in Air: NIOSH lists Chrysene as a coal tar pitch volatile substance. OSHA defines "coal tar pitch volatiles" in 29 CFR 1910.1002 as the fused polycyclic hydrocarbons that volatilize from the distillation residues of coal, petroleum (excluding asphalt), wood, and other organic matter and includes substances such as anthracene, benzo(a) pyrene (BaP), phenanthrene, acridine, chrysene, pyrene, etc. The legal airborne permissible exposure limit (OSHA PEL) is 0.2 mg/m^3 (benzene soluble fraction) TWA averaged over an 8-hour workshift. NIOSH lists this chemical as a potential occupational carcinogen and sets an REL of 0.1 mg/m^3 (cyclohexane-extractable fraction). The NIOSH IDLH = (carcinogen) 80 mg/m^3. ACGIH sets no numerical limits but simply lists chrysene as a suspected human carcinogen. DFG[3] also sets no limit but simply notes chrysene to be a proven animal carcinogen and a possible human carcinogen. Guidelines for chrysene concentration in ambient air[60] have been set at zero by North Dakota and at 3.0 μg/m^3 by Virginia.

Determination in Air: See NIOSH analytical methods for polynuclear hydrocarbons #5506 (GLC) and # 5515 (G.C). See also OSHA Method #58.

Permissible Concentration in Water: In view of the carcinogenicity of polynuclear aromatic, the concentration in water is preferably zero as noted by EPA.[6] Kansas[61] gives a guideline for chrysene in drinking water of 0.029 μg/l.

Determination in Water: Extraction with Methylene chloride may be followed by measurement by ultraviolet or by HPLC/flame spectrometry or by gas chromatography and mass spectrometry as reviewed in the ATSDR document referenced below.

Routes of Entry: Inhalation, skin contact, ingestion. This chemical can be absorbed through the skin, thereby increasing exposure.

Harmful Effects and Symptoms

Short Term Exposure: Chrysene can affect you when breathed in and by passing through your skin. Chrysene should be handled as a carcinogen — with extreme caution. Skin contact may cause a rash. If skin in exposed to sunlight, a "sunburn" can occur. Sunlight exposure on skin contaminated with coal tar chemicals such as Chrysene can cause rash and later, pigment changes. Persons who smoke cigarettes may be at increased lung cancer risk with this chemical. This can be significantly reduced by stopping smoking as well as by reducing exposures.

Long Term Exposure: May cause cancer of the skin and kidneys; birth defects, miscarriage; dermatitis, changes in skin pigment. May cause bronchitis with phlegm and/or shortness of breath.

Points of Attack: Respiratory system, skin, bladder, kidneys.

Medical Surveillance: Monthly, carefully look at any skin areas that are exposed. Any growth (like a mole) that increases in size or shows changes in color should be examined by a physician. Skin cancer is curable when detected early. Examination by a qualified allergist.

First Aid: If this chemical gets into the eyes, remove any contact lenses at once and irrigate immediately for at least 15 minutes, occasionally lifting upper and lower lids. Seek medical attention immediately. If this chemical contacts the skin, remove contaminated clothing and wash immediately with soap and water. Seek medical attention immediately. If this chemical has been inhaled, remove from exposure, begin rescue breathing (using universal precautions) if breathing has stopped and CPR if heart action has stopped. Transfer promptly to a medical facility. When this chemical has been swallowed, get medical attention. Give large quantities of water and induce vomiting. Do not make an unconscious person vomit.

Personal Protective Methods: Wear protective gloves and clothing to prevent any reasonable probability of skin contact. Safety equipment suppliers/manufacturers can provide recommendations on the most protective glove/clothing material for your operation. All protective clothing (suits, gloves, footwear, headgear) should be clean, available each day, and put on before work. Contact lenses should not be worn when working with this chemical. Wear dust-proof chemical goggles and face shield unless full facepiece respiratory protection is worn. Employees should wash immediately with soap when skin is wet or contaminated. Provide emergency showers and eyewash.

Respirator Selection: *At any detectable concentration:* SCBAF:PD,PP (any MSHA/NIOSH approved self-contained breathing apparatus that has a full facepiece and is operated in a pressure-demand or other positive-pressure mode); or SAF:PD,PP:ASCBA (any supplied-air respirator that has a full facepiece and is operated in a pressure-demand or other positive-pressure mode in combination with an auxiliary, self-contained breathing apparatus operated in a pressure-demand or other positive pressure mode). *Escape:* GMFOV [any air-purifying, full-facepiece respirator (gas mask) with a chin-style, front-or back-mounted organic vapor canister] or SCBAE (any appropriate escape-type, self-contained breathing apparatus).

Storage: Prior to working with Chrysene you should be trained on its proper handling and storage. A regulated, marked area should be established where this chemical is handled, used, or stored in compliance with OSHA standard 1910.1045. Store in tightly closed containers in a cool, well-ventilated area away from oxidizers. Sources of ignition such as smoking and open flames, are prohibited where this chemical is used, handled, or stored in a manner that could create a potential fire or explosion hazard.

Shipping: Chrysene is not specifically cited in DOT's Performance-Oriented Packaging Standards.[19]

Spill Handling: Evacuate and restrict persons not wearing protective equipment from area of spill or leak until cleanup is complete. Remove all ignition sources. Ventilate area of spill or leak. Absorb liquids in vermiculite, dry sand, earth, peat, carbon, or a similar material and deposit in sealed containers. It may be necessary to contain and dispose of this chemical as a hazardous waste. If material or contaminated runoff enters waterways, notify downstream users of potentially contaminated waters. Contact your Department of Environmental Protection or your regional office of the federal EPA for specific recommendations. If employees are required to clean-up spills, they must be properly trained and equipped. OSHA 1910.120(q) may be applicable.

Fire Extinguishing: Coal tar pitch volatiles are combustible, but may not readily ignite. Use dry chemical, carbon dioxide, water spray, or alcohol foam extinguishers. Poisonous gases are produced in fire. If material or contaminated runoff enters waterways, notify downstream users of potentially contaminated waters. Notify local health and fire officials and pollution control agencies. From a secure, explosion-proof location, use water spray to cool exposed containers. If cooling streams are ineffective (venting sound increases in volume and pitch, tank discolors, or shows any signs of deforming), withdraw immediately to a secure position. If employees are expected to fight fires, they must be trained and equipped in OSHA 1910.156.

Disposal Method Suggested: Chrysene may be destroyed by permanganate oxidation, by high-temperature incinerator with scrubbing equipment or by microwave plasma treatment.

References

New Jersey Department of Health and Senior Services and Senior Services, "Hazardous Substance Fact Sheet: Chrysene," Trenton, NJ (September 1985)

U.S. Public Health Service, "Toxicological Profile for Chrysene," Atlanta, Georgia, Agency for Toxic Substances and Disease Registry (October 1987)

Sax, N. I., Ed., "Dangerous Properties of Industrial Materials Report," 4, No. 4, 83–101 (1984)

C.I. Basic Green 1

Molecular Formula: $C_{27}H_{34}N_2O_4S$

Common Formula: $C_{27}H_{33}N_2 \cdot HSO_4$

Synonyms: Brilliant Green; Ethyl Green; Emerald Green; Malachite Green G

CAS Registry Number: 633-03-4

RTECS Number: BP6825000

DOT ID: No citation.

Cited in U.S. State Regulations: Massachusetts (G).

Description: C.I. Basic Green 1, $C_{27}H_{33}N_2 \cdot HSO_4$, is a metallic green, odorless crystal or powder. Soluble in water.

Potential Exposure: C.I. Basic Green 1 is used in dyeing silk, wool, leather, jute and cotton yellowish-green; manufacturing green ink; as staining constituent of bacteriological media; indicator, an intestinal anthelmintic; a wound antiseptic; treatment of mycotic infections; agricultural fungicide (Not registered as a pesticide in the U.S.).

Incompatibilities: Oxidizing agents, reducing agents, anionics, and aqueous solutions of bentonite. Keep away from moisture.

Permissible Exposure Limits in Air: No occupational exposure limits have been established.

Permissible Concentration in Water: No criteria set.

Routes of Entry: Ingestion.

Harmful Effects and Symptoms

It is classified as very toxic; probable lethal dose is 50 – 500 mg/kg in humans (between 1 teaspoon and 1 ounce for a 150-lb person). It is a skin irritant. Ingestion causes diarrhea and abdominal pain.

Short Term Exposure: C.I. Basic Green can irritate and burn the skin and eyes. Ingestion causes nausea, vomitine, diarrhea and abdominal pain. It is classified as very toxic; probable lethal dose is 50 – 500 mg/kg in humans (between 1 teaspoon and 1 ounce for a 150-lb person).

Long Term Exposure: Skin contact can cause drying and cracking.

Points of Attack: Skin.

Medical Surveillance: There is no special tests for this substance.

First Aid: If this chemical gets into the eyes, remove any contact lenses at once and irrigate immediately for at least 15 minutes, occasionally lifting upper and lower lids. Seek medical attention immediately. If this chemical contacts the skin, remove contaminated clothing and wash immediately with soap and water. Seek medical attention immediately. If this chemical has been inhaled, remove from exposure, begin rescue breathing (using universal precautions) if breathing has stopped and CPR if heart action has stopped. Transfer promptly to a medical facility. When this chemical has been swallowed, get medical attention. Give large quantities of water and induce vomiting. Do not make an unconscious person vomit.

Personal Protective Methods: Wear protective gloves and clothing to prevent any reasonable probability of skin contact. Safety equipment suppliers/manufacturers can provide recommendations on the most protective glove/clothing material for your operation. All protective clothing (suits, gloves, footwear, headgear) should be clean, available each day, and put on before work. Contact lenses should not be worn when working with this chemical. Wear dust-proof chemical goggles and face shield unless full facepiece respiratory protection is worn. Employees should wash immediately with soap when skin is wet or contaminated. Provide emergency showers and eyewash.

Respirator Selection: Where there is a potential for overexposure: SCBAF:PD,PP (any MSHA/NIOSH approved self-contained breathing apparatus that has a full facepiece and is operated in a pressure-demand or other positive-pressure mode); or SAF:PD,PP:ASCBA (any supplied-air respirator that has a full facepiece and is operated in a pressure-demand or other positive-pressure mode in combination with an auxiliary, self-contained breathing apparatus operated in a pressure-demand or other positive pressure mode).

Shipping: Poisonous solids, n.o.s., require a DOT shipping label reading "Poison" for Packing Group II. The UN/DOT Hazard Class is 6.1.[19][20]

Spill Handling: Evacuate persons not wearing protective equipment from area of spill or leak until clean-up is complete. Remove all ignition sources. Collect powdered material in the most convenient and safe manner and deposit in sealed containers. Ventilate area after clean-up is complete. It may be necessary to contain and dispose of this chemical as a hazardous waste. If material or contaminated runoff enters waterways, notify downstream users of potentially contaminated waters. Contact your Department of Environmental Protection or your regional office of the federal EPA for specific recommendations. If employees are required to clean-up spills, they must be properly trained and equipped. OSHA 1910.120(q) may be applicable.

Fire Extinguishing: This chemical may burn but does not easily ignite. For small fires use dry chemical, carbon dioxide, water spray, or foam. For large fires use water spray, fog, or foam. Keep unnecessary people away; isolate hazard area and deny entry. Stay upwind; keep out of low areas. Wear self-contained (positive pressure if available) breathing apparatus and full protective clothing. Poisonous gases are produced in fire. If material or contaminated runoff enters waterways, notify downstream users of potentially contaminated waters. Notify local health and fire officials and pollution control agencies. From a secure, explosion-proof location, use water spray to cool exposed containers. If cooling streams are ineffective (venting sound increases in volume and pitch, tank discolors, or shows any signs of deforming), withdraw immediately to a secure position. If employees are expected to fight fires, they must be trained and equipped in OSHA 1910.156.

References

U.S. Environmental Protection Agency, "Chemical Profile: C.I. Basic Green 1," Washington, DC, Chemical Emergency Preparedness Program (October 31, 1985)

C.I. Direct Red 28

Molecular Formula: $C_{32}H_{24}N_6O_6S_2{\cdot}2Na$

Synonyms: Atlantic Congo red; Atul Congo red; Azocard red Congo; Benzo Congo red; Brasilamina Congo 4B; C.I. 22120; C.I. direct red 28; C.I. direct red 28, disodium salt; Congo red; Cotton red L; Diacotton Congo red; Direct red 28; Erie Congo 4B; Hispamin Congo 4B; Kayaku Congo red; Mitsui Congo red; Peeramine Congo red; Sugai Congo red; Tertrodirect red C; Trisulfon Congo red; Vondacel red Cl

CAS Registry Number: 573-58-0

DOT ID: UN 3147

Cited in U.S. State Regulations: New Jersey (G).

Description: C.I. Direct Red 28 is an odorless, brownish-red powder.

Potential Exposure: It is used as an indicator dye, a biological stain, a diagnostic aid in medicine, and a dye for fabric and paper.

Incompatibilities: Contact with oxidizers may cause fire and explosion hazard. Incompatible with strong acids, reducing agents.

Permissible Exposure Limits in Air: No OELs established.

Determination in Air: See NIOSH Method # 5013, Dyes.

Routes of Entry: Inhalation.

Harmful Effects and Symptoms

Short Term Exposure: Skin and eye contact causes irritation. Exposure can cause nausea, vomiting, and diarrhea, and increase the formation of platelets, and increasing the ability of the blood to clot.

Long Term Exposure: While this chemical has not been designated a carcinogen, the parent compound, benzidine, causes bladder cancer. May decrease the fertility in males. May cause skin allergy with itching and rash.

Points of Attack: Skin, blood.

Medical Surveillance: Complete blood count (CBC). Evaluation by a qualified allergist.

First Aid: If this chemical gets into the eyes, remove any contact lenses at once and irrigate immediately for at least 15 minutes, occasionally lifting upper and lower lids. Seek medical attention immediately. If this chemical contacts the skin, remove contaminated clothing and wash immediately with soap and water. Seek medical attention immediately. If this chemical has been inhaled, remove from exposure, begin rescue breathing (using universal precautions) if breathing has stopped and CPR if heart action has stopped. Transfer promptly to a medical facility. When this chemical has been swallowed, get medical attention. If victim is conscious, administer water or milk. Do not induce vomiting.

Personal Protective Methods: Wear protective gloves and clothing to prevent any reasonable probability of skin contact. Safety equipment suppliers/manufacturers can provide recommendations on the most protective glove/clothing material for your operation. All protective clothing (suits, gloves, footwear, headgear) should be clean, available each day, and put on before work. Contact lenses should not be worn when working with this chemical. Wear dust-proof chemical goggles and face shield unless full facepiece respiratory protection is worn. Employees should wash immediately with soap when skin is wet or contaminated. Provide emergency showers and eyewash.

Respirator Selection: Where there is a potential for overexposure: SCBAF:PD,PP (any MSHA/NIOSH approved self-contained breathing apparatus that has a full facepiece and is operated in a pressure-demand or other positive-pressure mode); or SAF:PD,PP:ASCBA (any supplied-air respirator that has a full facepiece and is operated in a pressure-demand or other positive-pressure mode in combination with an auxiliary, self-contained breathing apparatus operated in a pressure-demand or other positive pressure mode).

Storage: Prior to working with this chemical you should be trained on its proper handling and storage. Store in tightly closed containers in a cool, well-ventilated area away from oxidizers, strong acids, reducing agents and heat. Where possible, automatically pump liquid from drums or other storage containers to process containers.

Shipping: Label required is "Corrosive." This chemical is in DOT/UN Hazard Class 8 and Packing Group II or III.

Spill Handling: Evacuate persons not wearing protective equipment from area of spill or leak until clean-up is complete. Remove all ignition sources. Collect powdered material in the most convenient and safe manner and deposit in sealed containers. Ventilate area after clean-up is complete. It may be necessary to contain and dispose of this chemical as a hazardous waste. If material or contaminated runoff enters waterways, notify downstream users of potentially contaminated waters. Contact your Department of Environmental Protection or your regional office of the federal EPA for specific recommendations. If employees are required to clean-up spills, they must be properly trained and equipped. OSHA 1910.120(q) may be applicable.

Fire Extinguishing: This chemical may burn, but does not easily ignite. Use dry chemical, carbon dioxide, water spray, or alcohol or polymer foam extinguishers. Poisonous gases are produced in fire including ammonia, carbon monoxide, nitrogen and sulfur oxides. If material or contaminated runoff enters waterways, notify downstream users of potentially contaminated waters. Notify local health and fire officials and pollution control agencies. Containers may explode in fire. From a secure, explosion-proof location, use water spray to cool exposed containers. If cooling streams are ineffective (venting sound increases in volume and pitch, tank discolors, or shows any signs of deforming), withdraw immediately to a secure position. If employees are expected to fight fires, they must be trained and equipped in OSHA 1910.156.

References

New Jersey Department of Health and Senior Services, "Hazardous Substance Fact Sheet, C.I Direct Red 28," Trenton, NJ (May 1999)

C.I. Food Red 15

Molecular Formula: $C_{28}H_{31}N_2O_3 \cdot Cl$

Synonyms: 11411 Red; Acid brilliant pink B; ADC Rhodamine B; Aizen Rhodamine BH; Aizen Rhodamine BHC; Akiriku Rhodamine B; Ammonium, [9(*o*-Carboxyphenyl)-6-(diethylamino)-3H-xanthen-3-ylidene]diethyl-, chloride; Basic Violet 10; Calcozine Red BX; Calcozine Rhodamine BXP; 9-*o*-Carboxyphenyl-6-diethylamino-3-ethylimino-3-isoxanthrene, 3-ethochloride; [9-(*o*-Carboxyphenyl)-6-(diethylamino)-3-xanthen-3-ylidene]diethylammonium chloride; Cerise toner X 1127; C.I. 45170; C.I. basic violet 10; Cosmetic brilliant pink bluish D conc.; D and C red No. 19; Diabasic Rhodamine B; Edicol Supra Rose B; Edicol Supra Rose BS; Eriosin Rhodamine B; Ethanaminium *N*-[9-(2-carboxyphenyl)-6-(diethylamino)-3H-xanthern-3-ylidene]-*N*-ethyl-, chloride; FD and C red No. 19; Flexco red 540; Hexacol Rhodamine B extra; Ikada Rhodamine B; Japan red 213; Japan red No. 213; Mitsui Rhodamine BX; Red No. 213; Rheonine B; Rhodamine B; Rhodamine B 500; Rhodamine B 500 hydrochloride; Rhodamine BA; Rhodamine BA export; Rhodamine B extra; Rhodamine B extra M 310; Rhodamine B extra S; Rhodamine BN; Rhodamine BS; Rhodamine BX; Rhodamine BXL; Rhodamine BXP; Rhodamine FB; Rhodamine lake red B; Rhodamine O; Rhodamine S; Rhodamine S (Russian); Rhodamine, tetraethyl-; Sicilian Cerise toner A 7127; Symulex magenta F; Symulex Rhodamine B toner F; Takaoka Rhodamine B; Tetraethyldiamino-*o*-carboxyphenyl-xanthenyl chloride; Tetraethylrhodamine; Xanthylium, 9-(2-carboxyphenyl)-3,6-bis(diethylamino)-, chloride

CAS Registry Number: 81-88-9

RTECS Number: BP3675000

DOT ID: No citation.

Regulatory Authority

- EPCRA Section 313 Form R *de minimis* concentration reporting level: 1.0%

Cited in U.S. State Regulations: California (G), New Jersey (G), Pennsylvania (G)

Description: C.I. Food Red 15 is a green crystalline or red-violet powdered solid. Hazard Identification (based on NFPA-704 M Rating System): Health 1, Flammability 1, Reactivity 0. Highly soluble in water.

Potential Exposure: It is used as a color additive in drugs, foods, cosmetics, and fabric dyes. It is also used as a tracing agent in water pollution studies.

Incompatibilities: Reducing agents and oxidizers.

Permissible Exposure Limits in Air: No OELs established.

Determination in Air: See NIOSH Method #5013, Dyes.

Routes of Entry: Inhalation.

Harmful Effects and Symptoms

Short Term Exposure: Irritates the skin, eyes, and respiratory tract. Eye or skin contact can cause burns and permanent damage. Can cause headaches, difficult breathing and chest tightness.

Long Term Exposure: May cause liver damage.

Points of Attack: Liver.

Medical Surveillance: Liver function tests.

First Aid: If this chemical gets into the eyes, remove any contact lenses at once and irrigate immediately for at least 15 minutes, occasionally lifting upper and lower lids. Seek medical attention immediately. If this chemical contacts the skin, remove contaminated clothing and wash immediately with soap and water. Seek medical attention immediately. If this chemical has been inhaled, remove from exposure, begin rescue breathing (using universal precautions) if breathing has stopped and CPR if heart action has stopped. Transfer promptly to a medical facility. When this chemical has been swallowed, get medical attention. Give large quantities of water and induce vomiting. Do not make an unconscious person vomit.

Personal Protective Methods: Wear protective gloves and clothing to prevent any reasonable probability of skin contact. Safety equipment suppliers/manufacturers can provide recommendations on the most protective glove/clothing material for your operation. All protective clothing (suits, gloves, footwear, headgear) should be clean, available each day, and put on before work. Contact lenses should not be worn when working with this chemical. Wear dust-proof chemical goggles and face shield unless full facepiece respiratory protection is worn. Employees should wash immediately with soap when skin is wet or contaminated. Provide emergency showers and eyewash.

Respirator Selection: Where there is a potential for overexposure: SCBAF:PD,PP (any MSHA/NIOSH approved self-contained breathing apparatus that has a full facepiece and is operated in a pressure-demand or other positive-pressure mode); or SAF:PD,PP:ASCBA (any supplied-air respirator that has a full facepiece and is operated in a pressure-demand or other positive-pressure mode in combination with an auxiliary, self-contained breathing apparatus operated in a pressure-demand or other positive pressure mode).

Storage: Prior to working with C.I. Food dye 15 you should be trained on its proper handling and storage. Store in tightly closed containers in a cool, well ventilated area away from oxidizers and reducing agents. Where possible, automatically pump liquid from drums or other storage containers to process containers.

Shipping: The DOT does not set out any specific label requirements or maximum shipping quantities for this chemical.

Spill Handling: Evacuate persons not wearing protective equipment from area of spill or leak until clean-up is complete. Remove all ignition sources. Collect powdered material in the most convenient and safe manner and deposit in sealed containers. Ventilate area after clean-up is complete. It may be necessary to contain and dispose of this chemical as a hazardous waste. If material or contaminated runoff enters waterways, notify downstream users of potentially contaminated waters. Contact your Department of Environmental Protection or your regional office of the federal EPA for specific recommendations. If employees are required to clean-up spills, they must be properly trained and equipped. OSHA 1910.120(q) may be applicable.

Fire Extinguishing: This chemical is a noncombustible solid. Use dry chemical, carbon dioxide, water spray, or alcohol or polymer foam extinguishers. Poisonous gases are produced in fire including carbon monoxide, nitrogen oxides, hydrogen chloride, and ammonia. If material or contaminated runoff enters waterways, notify downstream users of potentially contaminated waters. Notify local health and fire officials and pollution control agencies. From a secure, explosion-proof location, use water spray to cool exposed containers. If cooling streams are ineffective (venting sound increases in volume and pitch, tank discolors, or shows any signs of deforming), withdraw immediately to a secure position. If employees are expected to fight fires, they must be trained and equipped in OSHA 1910.156.

References

New Jersey Department of Health and Senior Services, "Hazardous Substance Fact Sheet, C.I. Food Red," Trenton, NJ (April 1999)

Cisplatin

Molecular Formula: $Cl_2H_6N_2Pt$

Common Formula: $Pt(NH_3)_2Cl_2$

Synonyms: CDDP; DDP; *cis*-Diamminedichloroplatinum; *cis*-Platinous diaminodichloride; *cis*-Platinum

CAS Registry Number: 15663-27-1

RTECS Number: TP2450000

DOT ID: UN 3249

Regulatory Authority

- Carcinogen[9] (limited human data) (IARC) (suspect) NTP
- Air Pollutant Standard Set (ACGIH)[1] (DFG)[3] (HSE)[33] (OSHA)[58]
- Canada, WHMIS, Ingredients Disclosure List

Cited in U.S. State Regulations: California (A, G), Florida (G), Maine (G), Massachusetts (G), Minnesota (G), New Jersey (G), Pennsylvania (G).

Description: Cisplatin, $Pt(NH_3)_2Cl_2$, is a white powder or yellow crystalline solid. Freezing/Melting point = 270°C (decomposes). Soluble in water.

Potential Exposure: Those involved in the manufacture, formulation and administration of this anticancer chemotherapy agent. Contact with water causes decomposition.

Incompatibilities: Aluminum reacts with Cisplatin and decreases the drug's effectiveness. Do not use any aluminum equipment to prepare or administer Cisplatin.

Permissible Exposure Limits in Air: These exposure limits are recommended for Soluble Platinum salts and measured as Platinum. OSHA PEL: The legal airborne permissible exposure limit (PEL) is 0.002 mg/m^3 averaged over an 8-hour workshift. ACGIH: The recommended airborne exposure limit is 0.002 mg/m^3 averaged over an 8-hour workshift. The NIOSH IDLH is 4 mg/m^3 (as Pt).

Determination in Air: Filter; Acid/Reagent; Graphite furnace atomic absorption spectrometry; NIOSH II[7] Method #S191. Also NIOSH IV Method #7300, Elements.

Permissible Concentration in Water: No criteria set.

Routes of Entry: Inhalation, ingestion, skin and/or eye contact.

Harmful Effects and Symptoms

Short Term Exposure: Contact with the skin or eyes can cause irritation with possible loss of vision. Inhalation can irritate the nose and throat. Exposure to high levels can cause tintinus (ringing in the ears) and possible hearing loss. The oral LD_{50} rat is 25.8 mg/kg (highly toxic).

Long Term Exposure: Cisplatin is a probable carcinogen in humans handle with extreme caution. It has been shown to cause lung and skin cancer in animals. It may damage the developing fetus and may damage the testes (male reproductive glands). Repeated exposure to high levels can cause the same side effects seen in patients. These include kidney damage, hearing loss, low blood cell count and nausea and committing. Exposure may cause anemia. May damage the nervous system causing numbness and weakness in the hands and feet. A regulated, marked area should be established where this chemical is handled, used, or stored in compliance with OSHA standard 1910.1045.

Points of Attack: Skin, lungs, nervous system, bone marrow, kidneys.

Medical Surveillance: Before beginning employment and at regular times after that, for those with frequent or potentially high exposures, the following are recommended: Examination of the nervous system. Audiogram (hearing test). Complete blood count (CBC). Kidney function tests. If symptoms develop or overexposure is suspected, the following may be useful: Blood levels of Cisplatin.

First Aid: If this chemical gets into the eyes, remove any contact lenses at once and irrigate immediately for at least 15 minutes, occasionally lifting upper and lower lids. Seek medical attention immediately. If this chemical contacts the skin, remove contaminated clothing and wash immediately with soap and water. Seek medical attention immediately. If this chemical has been inhaled, remove from exposure, begin rescue breathing (using universal precautions) if breathing has stopped and CPR if heart action has stopped. Transfer promptly to a medical facility. When this chemical has been swallowed, get medical attention. Give large quantities of water and induce vomiting. Do not make an unconscious person vomit.

Personal Protective Methods: Wear protective gloves and clothing to prevent any reasonable probability of skin contact. Safety equipment suppliers/manufacturers can provide recommendations on the most protective glove/clothing material for your operation. All protective clothing (suits, gloves, footwear, headgear) should be clean, available each day, and put on before work. Contact lenses should not be worn when working with this chemical. Wear dust-proof chemical goggles and face shield unless full facepiece respiratory protection is worn. Employees should wash immediately with soap when skin is wet or contaminated. Provide emergency showers and eyewash.

Respirator Selection: NIOSH/OSHA platinum as soluble salts as Pt: *Up to 0.05 mg/m^3:* SA:CF (any supplied-air respirator operated in a continuous-flow mode). *Up to 0.1 mg/m^3:* HiEF (any air-purifying, full-facepiece respirator with a high-efficiency particulate filter); or SCBAF (any self-contained breathing apparatus with a full facepiece). SAF (any supplied-air respirator with a full facepiece). *Up to 4 mg/m^3:* SAF:PD,PP (any supplied-air respirator that has a full facepiece and is operated in a pressure-demand or other positive-pressure mode). *Emergency or planned entry into unknown concentrations or IDLH conditions:* SCBAF:PD,PP (any self-contained breathing apparatus that has a full facepiece and is operated in a pressure-demand or other positive-pressure mode); or SAF:PD, PP:ASCBA (any supplied-air respirator that has a full facepiece and is operated in a pressure-demand or other positive-pressure mode in combination with an auxiliary self-contained breathing apparatus operated in a pressure-demand or other positive pressure mode). *Escape:* HiEF (any air-purifying, full-facepiece respirator with a high-efficiency particulate filter); or SCBAE (any appropriate escape-type, self-contained breathing apparatus).

Note: Substance reported to cause eye irritation or damage; may require eye protection.

Storage: Prior to working with Cisplatin you should be trained on its proper handling and storage. A regulated, marked area should be established where Cisplatin is handled, used, or stored in manufacturing and packaging operations. Store Cisplatin in sealed vials or tightly closed containers in a cool, well-ventilated area away from aluminum.

Shipping: Cisplatin is not specifically cited in DOT's Performance-Oriented Packaging Standards[19] but medicines, poisonous solid n.o.s. fall in Hazard Class 6.1 and Cisplatin in Packing Group II which requires a "Poison" label.

Spill Handling: Evacuate persons not wearing protective equipment from area of spill or leak until clean-up is complete. Remove all ignition sources. Collect powdered material in the most convenient and safe manner and deposit in sealed containers. Ventilate area after clean-up is complete. It may be necessary to contain and dispose of this chemical as a hazardous waste. If material or contaminated runoff enters waterways, notify downstream users of potentially contaminated waters. Contact your Department of Environmental Protection or your regional office of the federal EPA for specific recommendations. If employees are required to clean-up spills, they must be properly trained and equipped. OSHA 1910.120(q) may be applicable.

References

New Jersey Department of Health and Senior Services and Senior Services, "Hazardous Substance Fact Sheet: Cisplatin," Trenton, NJ (October 1998)

Clopidol

Synonyms: Clopindol; Coccidiostat C; Coyden®; 3,5-Dichloro-2,6-dimethyl-4-pyridinol; 3,5-Dichloro-4-pyridnol; Farmcoccid; Lerbek®; Methylchloropindol; Methylchlorpindol; Metilchlorpindol

CAS Registry Number: 2971-90-6

RTECS Number: UU7711500

DOT ID: No citation.

Regulatory Authority

- Air Pollutant Standard Set (ACGIH)[1] (Australia) (Israel) (Mexico) (OSHA)[58] (Several States)[60] (Several Canadian Provinces)

Cited in U.S. State Regulations: Alaska (G), California (A, W), Connecticut (A), Florida (G), Illinois (G), Maine (G), Massachusetts (G), Minnesota (G), Nevada (A), New Hampshire (G), New Jersey (G), North Dakota (A), Pennsylvania (G), Rhode Island (G), Virginia (A), West Virginia (G).

Description: Clopidol, is a white to light brown powder. Freezing/Melting point ≥ 320°C. Insoluble in water.

Potential Exposure: Those engaged in formulation, application or manufacture of this veterinary antibiotic.

Incompatibilities: Noncombustible solid, but dust may explode in cloud form. Contact with strong oxidizers may cause a fire or explosion hazard.

Permissible Exposure Limits in Air: The OSHA PEL 8-hour TWA is 15 mg/m^3 (total) TWA 5 mg/m^3 (respirable fraction). NIOSH recommended REL 10-hour TWA is 10 mg/m^3 (total) and STEL of 20 mg/m^3 (total); TWA 5 mg/m^3 (respirable fraction). ACGIH has adopted an 8-hour TWA value of 10 mg/m^3. Australia, Israel, Mexico and the Canadian provinces of Alberta, BC, Ontario, and Quebec set the same TWA level as ACGIH and Mexico, Alberta, BC, and Ontario set a STEL of 20 mg/m^3. Several States have set guidelines for clopidol in ambient air[60] ranging from 100 μg/m^3 (North Dakota) to 160 μg/m^3 (Virginia) to 200 μg/m^3 (Connecticut) to 238 μg/m^3 (Nevada).

Determination in Air: Filter; none; Gravimetric; NIOSH IV, Particulates NOR: Method #0500 (total), Method #0600 (respirable).

Permissible Concentration in Water: No criteria set.

Harmful Effects and Symptoms

Short Term Exposure: Clopidol may cause irritation of eyes, skin, nose, and throat. Clopidol has a low order off toxicity. Rats feed 15 mg/kg/day for 2 years showed no ill effects. The oral LD_{50} rat is 18 gm/kg (slightly toxic).

Long Term Exposure: Unknown at this time.

First Aid: If this chemical gets into the eyes, remove any contact lenses at once and irrigate immediately for at least 15 minutes, occasionally lifting upper and lower lids. Seek medical attention immediately. If this chemical contacts the skin, remove contaminated clothing and wash immediately with soap and water. Seek medical attention immediately. If this chemical has been inhaled, remove from exposure, begin rescue breathing (using universal precautions) if breathing has stopped and CPR if heart action has stopped. Transfer promptly to a medical facility. When this chemical has been swallowed, get medical attention. Give large quantities of water and induce vomiting. Do not make an unconscious person vomit.

Personal Protective Methods: Wear protective gloves and clothing to prevent any reasonable probability of skin contact. Safety equipment suppliers/manufacturers can provide recommendations on the most protective glove/clothing material for your operation. All protective clothing (suits, gloves, footwear, headgear) should be clean, available each day, and put on before work. Contact lenses should not be worn when working with this chemical. Wear dust-proof chemical goggles and face shield unless full facepiece respiratory protection is worn. Employees should wash immediately with soap when skin is wet or contaminated. Provide emergency showers and eyewash.

Respirator Selection: Where the potential exists for exposures over 10 mg/m^3, use a MSHA/NIOSH approved respirator equipped with particulate (dust/fume/mist) filters. Particulate filters must be checked every day before work for physical damage, such as rips or tears, and replaced as needed.

Where the potential for high exposures exists, use a MSHA/NIOSH approved supplied-air respirator with a full facepiece operated in the positive pressure mode or with a full facepiece, hood, or helmet in the continuous flow mode, or use a MSHA/NIOSH approved self-contained breathing apparatus with a full facepiece operated in pressure-demand or other positive pressure mode.

Storage: Prior to working with Clopidol you should be trained on its proper handling and storage. Store in tightly closed containers in a cool, well-ventilated area.

Shipping: There are no DOT regulations specific to Clopidol.

Spill Handling: Evacuate persons not wearing protective equipment from area of spill or leak until clean-up is complete. Remove all ignition sources. Collect powdered material in the most convenient and safe manner and deposit in sealed containers. Ventilate area after clean-up is complete. It may be necessary to contain and dispose of this chemical as a hazardous waste. If material or contaminated runoff enters waterways, notify downstream users of potentially contaminated waters. Contact your Department of Environmental Protection or your regional office of the federal EPA for specific recommendations. If employees are required to clean-up spills, they must be properly trained and equipped. OSHA 1910.120(q) may be applicable.

Fire Extinguishing: This chemical is a non combustible solid. Extinguish fire using and agent suitable for the type of surrounding fire; Clopidol itself does not burn. Poisonous gases are produced in fire including hydrogen chloride and organic nitrogens. If material or contaminated runoff enters waterways, notify downstream users of potentially contaminated waters. Notify local health and fire officials and pollution control agencies. From a secure, explosion-proof location, use water spray to cool exposed containers. If cooling streams are ineffective (venting sound increases in volume and pitch, tank discolors, or shows any signs of deforming), withdraw immediately to a secure position. If employees are expected to fight fires, they must be trained and equipped in OSHA 1910.156.

References

New Jersey Department of Health and Senior Services and Senior Services, "Hazardous Substance Fact Sheet: Clopidol," Trenton, NJ (January 1986)

Coal Dust

Molecular Formula: $C_{3n}H_{4n}$

Common Formula: $(C_3H_4)_n$

Synonyms: Anthracite coal dust; Bituminous coal dust; Coal facings; Lignite coal dust; Sea coal

RTECS Number: GF8281000

DOT ID: UN 1361

Regulatory Authority

- Air Pollutant Standard Set (ACGIH)[1] (Australia) (HSE)[33] (Israel) (OSHA)[58] (Several Canadian Provinces)

Cited in U.S. State Regulations: California (A), Florida (G), Illinois (G), Massachusetts (G), Minnesota (G), New Hampshire (G), Pennsylvania (G), Rhode Island (G), West Virginia (G).

Description: Coal dust, $(C_3H_4)_n$ is a combustible dark-brown black solid dispersed in the air. Properties vary depending on type of coal.

Potential Exposure: Those involved in the mining, preparation, delivery or use of powdered coal.

Incompatibilities: Strong oxidizers. Slightly explosive when exposed to flame.

Permissible Exposure Limits in Air: The OSHA PEL (8-hour TWA) for coal dust (as the respirable fraction) containing less than 5% SiO2 is 2.4 mg/m^3. The OSHA PEL (8-hour TWA) for coal dust (as the respirable fraction) containing greater than 5% SiO_2 is 10 mg/m^3 divided by the value "%SiO_{2+2}." ACGIH set a TWA of 0.9 mg/m^3 for Bituminous and 0.4 mg/m^3 for Anthracite.

Determination in Air: Coal dust may be determined gravimetrically: Filter; none; Gravimetric; NIOSH IV Method #0600, Particulates NOR (respirable). See also Method #7500. See also OSHA Method ID142.

Permissible Concentration in Water: No criteria set.

Routes of Entry: Inhalation.

Harmful Effects and Symptoms

Long Term Exposure: The inhalation of coal dust may cause coal workers' pneumoconiosis (CWP), chronic bronchitis, decreased pulmonary function, emphysema. This can result in reduction in ventilatory capacity, pulmonary hypertension and premature death.

Points of Attack: Respiratory system.

Medical Surveillance: Preplacement and annual physical examinations should be performed with emphasis on the respiratory system including chest x-rays.

First Aid: If this chemical gets into the eyes, remove any contact lenses at once and irrigate immediately for at least 15 minutes, occasionally lifting upper and lower lids. Seek medical attention immediately. If this chemical contacts the skin, remove contaminated clothing and wash immediately with soap and water. Seek medical attention immediately. If this chemical has been inhaled, remove from exposure, begin rescue breathing (using universal precautions) if breathing has stopped and CPR if heart action has stopped. Transfer promptly to a medical facility. When this chemical has been swallowed, get medical attention. Give large quantities of water and induce vomiting. Do not make an unconscious person vomit.

Personal Protective Methods: Wear thick working gloves and safety glasses.[24]

Respirator Selection: Use MSHA/NIOSH -approved dust respirator.

Storage: Prior to working with this material you should be trained on its proper handling and storage. Store in a cool, well-ventilated area away from strong oxidizers and open flame. Sources of ignition such as smoking and open flames, are prohibited where this chemical is used, handled, or stored in a manner that could create a potential fire or explosion hazard.

Shipping: Carbon, animal or vegetable origin falls in Hazard Class 4.2 and Packing Group III.[19] The label required is "Spontaneously Combustible" and shipment by passenger aircraft or railcar or even by cargo aircraft is forbidden.

Spill Handling: Evacuate persons not wearing protective equipment from area of spill or leak until clean-up is complete. Remove all ignition sources. Remove to containers which are readily disposable to land reclamation or dumps. Ventilate area after clean-up is complete. It may be necessary to contain and dispose of this chemical as a hazardous waste.

Fire Extinguishing: This chemical is a combustible solid. Coal dust is explosive when exposed to flame. Use dry chemical, carbon dioxide, water spray, or alcohol foam extinguishers. Poisonous gases can be produced in fire including carbon monoxide. If employees are expected to fight fires, they must be trained and equipped in OSHA 1910.156.

Disposal Method Suggested: Use land reclamation or dumps.[24]

Coal Tar Pitch Volatiles

Synonyms: *8001-58-9:* AWPA No. 1; Brick oil; Coal tar creosote; Coal tar distillate; Coal tar oil; Creosota de alquitran de hulla (Spanish); Creosote, coal tar; Creosote, from coal tar; Creosote oil; Creosote P1; Creosotum; Cresylic creosote; Dead oil; Heavy oil; Liquid pitch oil; Naphthalene oil; Preserv-o-Sote; Tar oil; Wash oil

CAS Registry Number: 8007-45-2 (coal tar); 65996-93-2 (coal tar pitch); 65996-92-1 (coal tar distillate); 8001-58-9 (coal tar creosote)

RTECS Number: GF8600000 (coal tar); GF8655000 (coal tar pitch); GF8617500 (coal tar distillate); GF8615000 (coal tar creosote)

DOT ID: UN 1999 (coal tar); UN 1136 and UN 1137 (coal tar distillate); UN 1993 (coal tar creosote)

Regulatory Authority

- Carcinogen (Animal and Human Positive) (IARC)[9] (ACGIH)[1] (DFG)[3]
- Banned or Severely Restricted (Coal Tar Oils) (UN)[35]
- OSHA, 29CFR1910 Specifically Regulated Chemicals (See CFR 1910.1002)
- Air Pollutant Standard Set (ACGIH)[1] (SFG)[3] (HSE)[33] (OSHA)[58] (Several States)[60]
- EPA HAZARDOUS WASTE NUMBER (RCRA No.): U051
- RCRA, 40CFR261, Appendix 8 Hazardous Constituents
- SUPERFUND/EPCRA 40CFR302.4 Reportable Quantity (RQ): CERCLA, 1 lb (0.454 kg)
- EPCRA Section 313 Form R *de minimis* concentration reporting level: 0.1%
- U.S. DOT Regulated Marine Pollutant (49CFR172.101, Appendix B)

Cited in U.S. State Regulations: Alaska (G), Connecticut (A), Illinois (G), Kansas (G, A), Louisiana (G), Maine (G), Massachusetts (G), Nevada (A), New Hampshire (G), New Jersey (G), North Carolina (A), Pennsylvania (G, A), Rhode Island (G), Vermont (G), Virginia (G, A), Washington (G), West Virginia (G), Wisconsin (G).

Description: The term "coal tar products," as used by NIOSH, includes coal tar and two of the fractionation products off coal tar, creosote and coal tar pitch, derived from the carbonization of bituminous coal. Coal tar, coal tar pitch, and creosote derived from bituminous coal often contain identifiable components which by themselves are carcinogenic, such as benzo(a)pyrene, benzanthracene, chrysene, and phenanthrene. Other chemicals from coal tar products, such as anthracene, carbazole, fluoranthene, and pyrene may also cause cancer, but these causal relationships have not been adequately documented. Coal tar pitch has a flash point = 207°C. Creosote oil has a Flash point = 74°C.

Potential Exposure: The coke-oven plant is the principal source of coal tar. The hot gases and vapors produced during the conversion of coal to coke are collected by means of a scrubber, which condenses the effluent into ammonia, water, crude tar, and other by-products. Crude tar is separated from the remainder of the condensate for refining and may undergo further processing.

Employees may be exposed to pitch and creosote in metal and foundry operations, when installing electrical equipment, and in construction, railway, utility, and briquette manufacturing.

Incompatibilities: Strong oxidizers.

Permissible Exposure Limits in Air: The OSHA TWA value is 0.2 mg/m^3 for coal tar pitch volatiles. ACGIH recommends the same level. NIOSH recommends a 10-hour TWA of 0.1 mg/m^3. The HSE[33] has set the same TWA value but the DFG[3] has set no MAK value, simply indicating that coal tar, coal tar pitch and coal tar oils are proved human carcinogens. Several states have set guidelines or standards for coal tar pitch volatiles in ambient air[60] ranging from zero (North Carolina) to 0.0161 μg/m^3 (Kansas) to 0.48 μg/m^3 (Pennsylvania) to 2.0 μg/m^3 (Connecticut and Virginia) to 5.0 μg/m^3 (Nevada).

Determination in Air: Collection on a filter, extraction, column chromatography, spectrophotometric measurement. Benzene soluble may be determined by collection of particulates on a filter, ultrasonic extraction with benzene, evaporation and gravimetric determination. (See NIOSH Method 5023). See also OSHA Method #ID-58; NIOSH Method 5506, Polynuclear Aromatic Hydrocarbons (HPLC); Method #5506 (GC), Method #5515.

Routes of Entry: Inhalation and skin and eye contact.

Harmful Effects and Symptoms

Based on a review of the toxicologic and epidemiologic evidence presented, it has been concluded that some materials contained in coal tar pitch, and therefore, in coal tar, can cause lung and skin cancer, and perhaps cancer at other sites. Based on a review of experimental toxicologic evidence, it is also concluded that creosote can cause skin and lung cancer. While the evidence on creosote is not so strong as that on pitch (in part because of difficulties in chemical characterization of such mixtures), the conclusion on the carcinogenic potential of creosote is supported by information on the presence of polynuclear aromatic hydrocarbons and imputations and evidence of the carcinogenicity of such hydrocarbons.

The overwhelming scientific evidence in the record supports the finding that coke oven emissions are carcinogenic. This finding rests on epidemiological surveys as well as animal studies and chemical analyses of coke oven emissions. Coke oven workers have an increased risk off developing cancer of the lung and urinary tract. In addition, observations of animals and of human populations have shown that skin tumors can be induced by the products of coal combustion and distillation. Chemical analyses of coke oven emissions reveal the presence of a large number of scientifically recognized carcinogens as well as several agents known to enhance the effect of chemical carcinogens especially on the respiratory tract.

Points of Attack: Respiratory system, lungs, bladder, kidneys, skin.

Medical Surveillance: Medical surveillance shall be made available, as specified below, to all employees occupationally exposed to coal tar products.

Preplacement Medical Examinations: These examinations shall include:

- Comprehensive initial medical and work histories, with special emphasis directed toward identifying preexisting disorders of the skin, respiratory tract, liver, and kidneys.
- A physical examination giving particular attention to the oral cavity, skin, and respiratory system. This shall include posteroanterior and lateral chest x-rays (35 × 42 cm). Pulmonary function tests, including forced vital capacity (FVC) and forced expiratory volume at 1 second (FEV 1.0), and a sputum cytology examination shall be offered as part of the medical examination of exposed employees. Other tests, such as liver function and urinalysis, should be performed as considered appropriate by the responsible physician. In addition, the mucous membranes of the oral cavity should be examined.
- A judgment of the employee's ability to use positive pressure respirators.

Periodic Examinations: These examinations shall be made available at least annually and shall include:

- Interim medical and work histories.
- A physical examination as outlined above.

Initial Medical Examinations: These examinations shall be made available to all workers as soon as practicable after the promulgation of a standard based on these recommendations.

Pertinent Medical Records: These records shall be maintained for at least 30 years after termination of employment. They shall be made available to medical representatives of the government, the employer or the employee.

First Aid: If this chemical gets into the eyes, remove any contact lenses at once and irrigate immediately for at least 15 minutes, occasionally lifting upper and lower lids. Seek medical attention immediately. If this chemical contacts the skin, remove contaminated clothing and wash immediately with soap and water. Seek medical attention immediately. If this chemical has been inhaled, remove from exposure, begin rescue breathing (using universal precautions) if breathing has stopped and CPR if heart action has stopped. Transfer promptly to a medical facility. When this chemical has been swallowed, get medical attention. Give large quantities of water and induce vomiting. Do not make an unconscious person vomit.

Personal Protective Methods: Employers shall use engineering controls when needed to keep the concentration of airborne coal tar products at, or below, the specified limit. Employers shall provide protective clothing and equipment impervious to coal tar products to employees whenever liquid coal tar products may contact the skin or eyes. Emergency equipment shall be located at well-marked and clearly identified stations and shall be adequate to permit all personnel to escape from the area or to cope safely with the emergency on reentry. Protective equipment shall include: eye and face protection; protective clothing; and respiratory protection as spelled out in detail by NIOSH.

Respirator Selection: NIOSH: *At any detectable concentration:* SCBAF:PD,PP (any MSHA/NIOSH approved self-contained breathing apparatus that has a full facepiece and is operated in a pressure-demand or other positive-pressure mode); or SAF:PD,PP:ASCBA (any supplied-air respirator that has a full facepiece and is operated in a pressure-demand or other positive-pressure mode in combination with an auxiliary, self-contained breathing apparatus operated in a pressure-demand or other positive pressure mode). *Escape:* GMFHiE [any air-purifying, full-facepiece respirator (gas mask) having a high-efficiency particulate filter]; or SCBAE (any appropriate escape-type, self-contained breathing apparatus).

Storage: Prior to working with this chemical you should be trained on its proper handling and storage. Before entering confined space where this chemical may be present, check to make sure that an explosive concentration does not exist. Store in tightly closed containers in a cool, well-ventilated area away from strong oxidizers (such as chlorine, bromine, and fluorine). Sources of ignition such as smoking and open flames are prohibited where Coal Tar Creosote is used, handled, or stored in a manner that could create a potential fire or explosion hazard. Metal containers involving the transfer of 5 gallons or more of Coal Tar Creosote should be grounded and bonded. Drums must be equipped with self-closing valves, pressure vacuum bungs, and flame arresters. A regulated, marked area should be established where this chemical is handled, used, or stored in compliance with OSHA standard 1910.1045. Entry into areas containing Coal Tar Creosote is to be controlled by permit only.

Shipping: Coal tar distillates, flammable fall in Hazard Class 3 and Packaging Group III. Such materials require a "Flammable Liquid" label.

Spill Handling: Evacuate and restrict persons not wearing protective equipment from area of spill or leak until cleanup is complete. Remove all ignition sources. Ventilate area of spill or leak. Absorb liquids in vermiculite, dry sand, earth, peat, carbon, or a similar material and deposit in sealed containers. It may be necessary to contain and dispose of this chemical as a hazardous waste. If material or contaminated runoff enters waterways, notify downstream users of potentially contaminated waters. Contact your Department of Environmental Protection or your regional office of the federal EPA for specific recommendations. If employees are required to clean-up spills, they must be properly trained and equipped. OSHA 1910.120(q) may be applicable. Restrict persons not wearing protective equipment from area of spill or leak until clean-up is complete. Remove all ignition sources. Keep Coal Tar Creosote out of a confined space, such as a sewer because of the possibility of an explosion, unless the sewer is designed to prevent the build-up of explosive concentrations.

Fire Extinguishing: This chemical is a combustible solid. Use dry chemical, carbon dioxide, water spray, or alcohol foam extinguishers. Poisonous gases are produced in fire. If material or contaminated runoff enters waterways, notify downstream users of potentially contaminated waters. Notify local health and fire officials and pollution control agencies. From a secure, explosion-proof location, use water spray to cool exposed containers. If cooling streams are ineffective (venting sound increases in volume and pitch, tank discolors, or shows any signs of deforming), withdraw immediately to a secure position. If employees are expected to fight fires, they must be trained and equipped in OSHA 1910.156.

This chemical is a combustible liquid. Poisonous gases are produced in fire. Use dry chemical, carbon dioxide, or alcohol foam extinguishers. Vapors are heavier than air and will collect in low areas. Vapors may travel long distances to ignition sources and flashback. Vapors in confined areas may explode when exposed to fire. Containers may explode in fire. Storage containers and parts of containers may rocket great distances, in many directions. If material or contaminated runoff enters waterways, notify downstream users of potentially contaminated waters. Notify local health and fire officials and pollution control agencies. From a secure, explosion-proof location, use water spray to cool exposed containers. If cooling streams are ineffective (venting sound increases in volume and pitch, tank discolors, or shows any signs of deforming), withdraw immediately to a secure position. If employees are expected to fight fires, they must be trained and equipped in OSHA 1910.156. Coal Tar Creosote is a combustible liquid. Use dry chemical, CO_2, water spray, or foam extinguishers.

Disposal Method Suggested: Incineration.[22]

References

National Institute for Occupational Safety and Health, Criteria for a Recommended Standard: Occupational Exposure to Coal Tar Products, NIOSH Document No. 78–107, Washington, DC (September 1977)

U.S. Environmental Protection Agency, Creosote, Health and Environmental Effects Profile No. 53, Office of Solid Waste, Washington, DC (April 30, 1980)

New Jersey Department of Health and Senior Services and Senior Services, "Hazardous Substance Fact Sheet: Coal Tar Creosote," Trenton, NJ (January 1986)

Cobalt and Oxides

Molecular Formula: Co

Synonyms: *cobalt metal:* Aquacat; C.I. 77320; Cobalt-59; Cobalto (Spanish); Kobalt (German, Polish); NCI-C60311; Super cobalt

CAS Registry Number: 7440-48-4 (cobalt metal); 1307-96-6 (cobaltous oxide); 1308-04-9 (cobaltic oxide)

RTECS Number: GF8750000 (cobalt metal); GG2800000 (cobaltous oxide); GG2900000 (cobaltic oxide)

EEC Number: 027-001-00-9 (cobalt metal)

EINECS Number: 231-158-0 (cobalt metal)

Regulatory Authority

- Carcinogen (Respirable Dust) (DFG)[3]
- Banned or Severely Restricted (In Pharmaceuticals) (USA)[13]
- Very Toxic Substance (World Bank)[15]
- Air Pollutant Standard Set (ACGIH)[1] (Australia) (HSE)[33] (Israel) (Mexico) (OSHA)[58] (UNEP)[43] (Several States)[60] (Several Canadian Provinces) (cobalt metal)
- CLEAN AIR ACT: Hazardous Air Pollutants (Title I, Part A, Section 112) as cobalt compounds
- RCRA 40CFR264, Appendix 9; TSD Facilities Ground Water Monitoring List. Suggested test method(s) (PQL μg/l): (total) 6010 (70); 7200 (500); 7201 (10) (cobalt metal)
- EPCRA (Section 313): Form R *de minimis* concentration reporting level: 1.0% (cobalt metal)
- CLEAN AIR ACT: Hazardous Air Pollutants (Title I, Part A, Section 112) Note: Includes any unique chemical substance that contains cobalt as part of that chemical's infrastructure (cobalt compounds)
- EPCRA (Section 313): Includes any unique chemical substance that contains cobalt as part of that chemical's infrastructure. Form R *de minimis* concentration reporting level: 1.0% (cobalt compounds)
- Canada, WHMIS, Ingredients Disclosure List; National Pollutant Release Inventory (NPRI)
- Mexico, Wastewater, Toxic Pollutant

Cited in U.S. State Regulations: Alaska (G), Connecticut (A), Florida (G, A), Illinois (G), Maine (G), Maryland (G), Massachusetts (G), Michigan (G), Nevada (A), New Hampshire (G), New Jersey (G), New York (A), North Dakota (A), Pennsylvania (G), Rhode Island (G), Virginia (A), West Virginia (G).

Description: Cobalt, Co, is a silver-gray to black, hard, brittle, magnetic metal. It is relatively rare; the important mineral sources are the arsenides, sulfides, and oxidized forms. It is generally obtained as a by-product of other metals, particularly copper. Boiling point = 2,870°C. Freezing/Melting point = 1,495°C. Hazard Identification (based on NFPA-704 M Rating System): Health 1, Flammability 3, Reactivity 0. Cobalt is insoluble in water, but soluble in acids. Cobalt fume and dust have the composition $Co/CoO/Co_2O_2/Co_2O_4$. Cobaltic oxide: Freezing/Melting point = (decomposes) 895°C. Insoluble in water.

Potential Exposure: Nickel-aluminum-cobalt alloys are used for permanent magnets. Alloys with nickel, aluminum, copper, beryllium, chromium, and molybdenum are used in the electrical, automobile, and aircraft industries. Cobalt is added to tool steels to improve their cutting qualities and is used as a binder in the manufacture of tungsten carbide tools.

Various cobalt compounds are used as pigments in enamels, glazes, and paints, as catalysts in afterburners, and in the glass, pottery, photographic, electroplating industries. Radioactive cobalt (^{60}Co) is used in the treatment of cancer.

Cobalt has been added to beer to promote formation of foam but cobalt acts with alcohol to produce severe cardiac effects at concentrations as low as 1.2 – 1.5 mg/l of beer.

Cobalt is part of the vitamin B_{12} molecule and as such is an essential nutrient. The requirement of humans for cobalt in the form of vitamin B_{12} is about 0.13 μg/day.

Incompatibilities: Cobalt metal dust/powder may spontaneously ignite on contact with air, when finely divided. Reacts with acids, strong oxidizers, ammonium nitrate causing fire and explosion hazard. Can promote decomposition of various organic substances. Cobaltic oxide reacts with reducing agents, and violently with hydrogen peroxide.

Permissible Exposure Limits in Air: The OSHA PEL 8-hour TWA is 0.1 mg/m^3. The NIOSH recommended REL 10-hour TWA is 0.05 mg/m^3. The ACGIH recommended TLV 8-hour TWA is 0.02 mg/m^3. Mexico, 0.1 mg/m^3. The NIOSH IDLH = 20 mg/m^3. Australia, Israel; Canada: Alberta, Ontario, Quebec TWA is 0.05 mg/m^3 and Alberta's and Ontario's STEL is 0.1 mg/m^3. British Columbia's TWA is 0.02 mg/m^3 DFG notes animal evidence of animal carcingenicity and does not set a numerical value for respirable dusts and aerosols. The former USSR-UNEP/IRPTC project has set a MAC of 0.5 mg/m^3 for cobalt metal and oxides in workplace air.[43] For ambient air in residential areas, 0.001 mg/m^3 is the MAC on an average daily basis for cobalt metal.

Several states have set guidelines or standards for cobalt in ambient air[60] ranging from 0.33 μg/m^3 (New York) to 0.8 μg/m^3 (Virginia) to 1.0 μg/m^3 (Florida and North Dakota) to 2.0 μg/m^3 (Connecticut and Nevada).

Determination in Air: Cobalt metal, dust and fume may be determined by filter collection, acid dissolution, digestion and measurement by atomic absorption spectrophotometry. See NIOSH IV Method #7027. See also Method #7300, Elements, and OSHA Methods ID121 and ID125.

Permissible Concentration in Water: The EPA[32] has suggested a permissible ambient goal of 0.7 μg/l based on health effects. The former USSR-UNEP/IRPTC project[43] has set a MAC of 1 mg/l in water bodies used for domestic purposes. Limits in water bodies for fishery purposes have been set at 0.01 mg/l for fresh water and 0.05 mg/l for sea water.

Determination in Water: Atomic absorption spectroscopy gives a detection limit of 0.05 mg/l in water. Neutron activation can detect cobalt in urine below 0.5 μg/l.

Routes of Entry: Inhalation of dust or fume, ingestion, skin or eye contact.

Harmful Effects and Symptoms

Short Term Exposure: Cobalt dust is mildly irritating to the eyes and to a lesser extent to the skin. It is an allergen and has caused allergic sensitivity type dermatitis in some industries where only minute quantities of cobalt are used. The eruptions appear in the flexure creases of the elbow, knee ankle and neck. Cross sensitization occurs between cobalt and nickel, and to chromium when cobalt and chromium are combined. Inhalation of dust and fume may cause irritation of the lungs with coughing and/or shortness of breath. Higher exposures can cause pulmonary edema, a medical emergency that can be delayed for several hours. This can cause death. Ingestion of cobalt or cobalt compounds is rare in industry. Vomiting, diarrhea, and a sensation of hotness may occur after ingestion or after the inhalation of excessive amounts of cobalt dust.

Long Term Exposure: Repeated or prolonged inhalation exposure may affect lungs and cause an asthma-like disease with cough and dyspnea. This situation may progress to interstitial pneumonia with marked fibrosis. Pneumoconiosis may develop which is believed to be reversible. Since cobalt dust is usually combined with other dusts, the role cobalt plays in causing the pneumoconiosis is not entirely clear. Cobalt may decrease fertility in males. Repeated or prolonged contact may cause skin sensitization. The substance may have effects on the heart, resulting in cardiomyopathy. Cobalt may affect the thyroid and kidneys. Cobalt is possibly carcinogenic and mutagenic to humans. Some isotopes of cobalt emit ionizing radiation; such exposure is associated with an increased risk of developing cancer.

Points of Attack: Respiratory system, skin.

Medical Surveillance: In preemployment examinations, special attention should be given to a history of skin diseases, allergic dermatitis, baseline allergic respiratory diseases, and smoking history. A baseline chest x-ray should be taken and chest x-ray for scarring should be done every 2 – 3 years following 5 or more years of exposure. Periodic examinations should be directed toward skin and respiratory symptoms and lung function. Evaluation for heart failure. Lidney and thyroid function tests. Evaluation buy a qualified allergist. Consider chest x-ray following acute overexposure.

First Aid: If this chemical gets into the eyes, remove any contact lenses at once and irrigate immediately for at least 15 minutes, occasionally lifting upper and lower lids. Seek medical attention immediately. If this chemical contacts the

skin, remove contaminated clothing and wash immediately with soap and water. Seek medical attention immediately. If this chemical has been inhaled, remove from exposure, begin rescue breathing (using universal precautions) if breathing has stopped and CPR if heart action has stopped. Transfer promptly to a medical facility. When this chemical has been swallowed, get medical attention. Give large quantities of water and induce vomiting. Do not make an unconscious person vomit. Medical observation is recommended for 24 – 48 hours after breathing overexposure, as pulmonary edema may be delayed. As first aid for pulmonary edema, a doctor or authorized paramedic may consider administering a corticosteroid spray.

Personal Protective Methods: Wear protective gloves and clothing to prevent any reasonable probability of skin contact. Safety equipment suppliers/manufacturers can provide recommendations on the most protective glove/clothing material for your operation. All protective clothing (suits, gloves, footwear, headgear) should be clean, available each day, and put on before work. Contact lenses should not be worn when working with this chemical. Wear dust-proof chemical goggles and face shield unless full facepiece respiratory protection is worn. Where dust levels are excessive, dust respirators should be used by all workers. Protective clothing should be issued to all workers and changed on a daily basis. Showering after each shift is encouraged prior to change to street clothes. Gloves and barrier creams may be helpful in preventing dermatitis. Employees should wash immediately with soap when skin is wet or contaminated. Provide emergency showers and eyewash.

Respirator Selection: NIOSH/OSHA, for cobalt metal dust and fume:: *0.25 mg/m³:* DM, *if not present as a fume* (any dust and mist respirator). *0.5 mg/m³:* DMXSQ, *if not present as a fume* (any dust and mist respirator except single-use and quarter mask respirators); or DMFu (any dust, mist, and fume respirator); or SA (any supplied-air respirator). *1.25 mg/m³:* SA:CF (any supplied-air respirator operated in a continuous-flow mode); or PAPRDM, *if not present as a fume* (any powered, air-purifying respirator with a dust and mist filter); or PAPRDMFu (any powered, air-purifying respirator with a dust, mist, and fume filter). *2.5 mg/m³:* HiEF (any air-purifying, full-facepiece respirator with a high-efficiency particulate filter); or SCBAF (any self-contained breathing apparatus with a full facepiece); or SAF (any supplied-air respirator with a full facepiece). *20 mg/m³:* SAF:PD,PP (any supplied-air respirator that has a full facepiece and is operated in a pressure-demand or other positive-pressure mode *Emergency or planned entry into unknown concentrations or IDLH conditions:* SCBAF:PD,PP (any self-contained breathing apparatus that has a full facepiece and is operated in a pressure-demand or other positive-pressure mode); or SAF:PD,PP: ASCBA (any supplied-air respirator that has a full facepiece and is operated in a pressure-demand or other positive-pressure mode in combination with an auxiliary, self-contained breathing apparatus operated in a pressure-demand or other positive-pressure mode). *Escape:* HiEF (any air-purifying, full-facepiece respirator with a high-efficiency particulate filter); or SCBAE (any appropriate escape-type, self-contained breathing apparatus).

Note: Substance reported to cause eye irritation or damage; may require eye protection.

Storage: Prior to working with this chemical you should be trained on its proper handling and storage. Store in tightly closed containers in a cool, well-ventilated area Cobalt must be stored to avoid contact with strong oxidizers (such as chlorine, bromine, and fluorine), acids and ammonium nitrate since violent reactions occur. It should be stored in a cool place under an inert atmosphere. A regulated, marked area should be established where this chemical is handled, used, or stored in compliance with OSHA standard 1910.1045.

Shipping: Cobalt metal and oxides are not among the specific compounds listed in the DOT's Performance-Oriented Packaging Standards[19] as regards labeling or quantity limits on shipments.

Spill Handling: Evacuate persons not wearing protective equipment from area of spill or leak until clean-up is complete. Remove all ignition sources. Collect powdered material in the most convenient and safe manner and deposit in sealed containers. If spill involves radioactive cobalt, evacuate area and delay clean up until properly instructed by qualified radiation authorities. Ventilate area after clean-up is complete. It may be necessary to contain and dispose of this chemical as a hazardous waste. If material or contaminated runoff enters waterways, notify downstream users of potentially contaminated waters. Contact your Department of Environmental Protection or your regional office of the federal EPA for specific recommendations. If employees are required to clean up spills, they must be properly trained and equipped. OSHA 1910.120(q) may be applicable.

Fire Extinguishing: This chemical is a noncombustible solid, however powdered cobalt with ignite. Use dry chemical such as sand, dolomite, and graphite powder for extinguishing powdered metal fires. Do not use water. Poisonous gases are produced in fire. If material or contaminated runoff enters waterways, notify downstream users of potentially contaminated waters. Notify local health and fire officials and pollution control agencies. From a secure, explosion-proof location, use water spray to cool exposed containers. If cooling streams are ineffective (venting sound increases in volume and pitch, tank discolors, or shows any signs of deforming), withdraw immediately to a secure position. If employees are expected to fight fires, they must be trained and equipped in OSHA 1910.156.

Disposal Method Suggested: For cobalt chloride: chemical reaction with water, caustic soda, and slaked lime, resulting in precipitation of the metal sludge, which may be landfilled. Cobalt metal may be recovered from scrap and cobalt compounds from spent catalysts as alternatives to disposal.[22]

References

U.S. Environmental Protection Agency, Toxicology of Metals, Vol. II: Cobalt, pp 188–205, Report EPA-600/1-77-022, Research Triangle Park, NC (May 1977)

Sax, N. I., Ed., "Dangerous Properties of Industrial Materials Report," 1, No. 3, 47–48, New York, Van Nostrand Reinhold Co. (1981). (Cobalt)

U.S. Environmental Protection Agency, "Chemical Profile: Cobalt," Washington, DC, Chemical Emergency Preparedness Program (October 31, 1985)

New Jersey Department of Health and Senior Services and Senior Services, "Hazardous Substance Fact Sheet: Cobalt," Trenton, NJ (May 1998)

Cobalt Carbonyl

Molecular Formula: $C_8Co_2O_8$

Common Formula: $Co_2(CO)_8$

Synonyms: Cobalt octacarbonyl; Cobalto tetracarbonilo (Spanish); Cobalt tetracarbonyl; Cobalt tetracarbonyl dimer; Di-μ-carbonylhexacarbonyldicobalt; Dicobalt carbonyl; Dicobalt octacarbonyl; Octacarbonyldicobalt

CAS Registry Number: 10210-68-1

RTECS Number: GG0300000

DOT ID: No citation.

Regulatory Authority

- Air Pollutant Standard Set (ACGIH)[1] (Australia) (Israel) (former USSR)[43] (OSHA)[58] (Several States)[60] (Several Canadian Provinces)
- CLEAN AIR ACT: Hazardous Air Pollutants (Title I, Part A, Section 112) as cobalt compounds
- SUPERFUND/EPCRA 40CFR302.4 Reportable Quantity (RQ): EHS, 1 lb (0.454 kg)
- EPCRA Section 313 Form R *de minimis* concentration reporting level: 1.0%
- SUPERFUND/EPCRA 40CFR355, Appendix B Extremely Hazardous Substances: TPQ = 10/10,000 lb (4.54/4,540 kg)
- Canada, WHMIS, Ingredients Disclosure List

Cited in U.S. State Regulations: Alaska (G), California (A, G), Connecticut (A), Florida (G), Illinois (G, W), Maine (G), Massachusetts (G), Minnesota (G), Nevada (A), New Hampshire (G, W), New Jersey (G), North Dakota (A), Pennsylvania (G), Rhode Island (G), Virginia (A), West Virginia (G).

Description: Cobalt carbonyl, $Co_2(CO)_8$, is an orange to dark-brown solid. The pure substance is white. Boiling point = 52°C (decomposes). Freezing/Melting point = 50°C (decomposes). Hazard Identification (based on NFPA-704 M Rating System): Health 2, Flammability 0, Reactivity 0. Insoluble in water.

Potential Exposure: This material is used as a catalyst for a number of reactions. It is also used in anti-knock gasoline and for high-purity cobalt salts.

Incompatibilities: Reacts with strong acids and strong oxidizers. Decomposes on exposure to air or heat (below 52°C) producing toxic fumes of carbon monoxide and cobalt; stable in atmosphere of hydrogen and carbon monoxide.

Permissible Exposure Limits in Air: There is no OSHA PEL. The recommended NIOSH REL 10-hour TWA is 0.1 mg/m³. The ACGIH TLV 8-hour TWA is the same. Australia, Israel, the Canadian provinces of Alberta, BC, Ontario and Quebec use the same TWA limit and Alberta's STEL is 0.3 mg/m³ The former USSR-UNEP/IRPTC project[43] has set a MAC of 0.01 mg/m³ in workplace air. Some states have set guidelines or standards for cobalt in ambient air[60] ranging from 1.0 μg/m³ (North Dakota) to 1.6 μg/m³ (Virginia) to 2.0 μg/m³ (Connecticut and Nevada).

Determination in Air: No method available.

Permissible Concentration in Water: The former USSR-UNEP/IRPTC project[43] has set a MAC in water for fishery purposes of 0.5 mg/l.

Routes of Entry: Inhalation, ingestion, skin contact.

Harmful Effects and Symptoms

Short Term Exposure: The carbonyls are direct irritants. Carbon monoxide causes breathlessness, headache, weakness, and fatigue, nausea and vomiting, dimness of vision, collapse and coma. Cobalt carbonyl is corrosive to the eyes, skin, and severely irritates the respiratory tract. Inhalation of the aerosols can cause pulmonary edema, a medical emergency that can be delayed for several hours. This can cause death. Cobalt carbonyls share the general high toxicity of carbonyls because of the direct irritant and systemic action of the compound coupled with the effects of carbon monoxide which is released from their decomposition. The oral LD_{50} rat is 754 mg/kg (slightly toxic).

Long Term Exposure: May cause lung irritation and decreased pulmonary function, wheezing, dyspnea (breathing difficulty). Animal tests produce liver, and kidney injury.

Points of Attack: Eyes, skin, respiratory system, blood, central nervous system.

Medical Surveillance: Lung function tests. Liver and kidney function tests. Complete Blood Count (CBC).

First Aid: If this chemical gets into the eyes, remove any contact lenses at once and irrigate immediately for at least 15 minutes, occasionally lifting upper and lower lids. Seek medical attention immediately. If this chemical contacts the skin, remove contaminated clothing and wash immediately with soap and water. Seek medical attention immediately. If this chemical has been inhaled, remove from exposure, begin rescue breathing (using universal precautions) if breathing has stopped and CPR if heart action has stopped. Transfer promptly to a medical facility. When this chemical has

been swallowed, get medical attention. Give large quantities of water and induce vomiting. Do not make an unconscious person vomit. Medical observation is recommended for 24 – 48 hours after breathing overexposure, as pulmonary edema may be delayed. As first aid for pulmonary edema, a doctor or authorized paramedic may consider administering a corticosteroid spray.

Personal Protective Methods: Wear protective gloves and clothing to prevent any reasonable probability of skin contact. Safety equipment suppliers/manufacturers can provide recommendations on the most protective glove/clothing material for your operation. All protective clothing (suits, gloves, footwear, headgear) should be clean, available each day, and put on before work. Contact lenses should not be worn when working with this chemical. Wear dust-proof chemical goggles and face shield unless full facepiece respiratory protection is worn. Employees should wash immediately with soap when skin is wet or contaminated. Provide emergency showers and eyewash.

Respirator Selection: Where there is a potential for overexposure: SCBAF:PD,PP (any MSHA/NIOSH approved self-contained breathing apparatus that has a full facepiece and is operated in a pressure-demand or other positive-pressure mode); or SAF:PD,PP:ASCBA (any supplied-air respirator that has a full facepiece and is operated in a pressure-demand or other positive-pressure mode in combination with an auxiliary, self-contained breathing apparatus operated in a pressure-demand or other positive pressure mode).

Storage: Prior to working with this chemical you should be trained on its proper handling and storage. Decomposes on exposure to air or heat; stable in atmosphere of hydrogen and carbon monoxide. Store in airtight, unbreakable containers in a cool, well-ventilated area away from strong oxidizers and acids.

Shipping: The DOT Performance-Oriented Packaging Standards[19] do not list Cobalt carbonyl.

Spill Handling: Evacuate persons not wearing protective equipment from area of spill or leak until clean-up is complete. Remove all ignition sources. If appropriate, moisten spilled material to prevent dust. Collect powdered material in the most convenient and safe manner and deposit in airtight, sealed containers. Ventilate area after clean-up is complete. It may be necessary to contain and dispose of this chemical as a hazardous waste. If material or contaminated runoff enters waterways, notify downstream users of potentially contaminated waters. Contact your Department of Environmental Protection or your regional office of the federal EPA for specific recommendations. If employees are required to clean up spills, they must be properly trained and equipped. OSHA 1910.120(q) may be applicable.

Fire Extinguishing: This chemical is a combustible solid. Use dry chemical, carbon dioxide, water spray, or alcohol foam extinguishers. Poisonous gases are produced in fire including cobalt and carbon monoxide. If material or contaminated runoff enters waterways, notify downstream users of potentially contaminated waters. Notify local health and fire officials and pollution control agencies. From a secure, explosion-proof location, use water spray to cool exposed containers. If cooling streams are ineffective (venting sound increases in volume and pitch, tank discolors, or shows any signs of deforming), withdraw immediately to a secure position. If employees are expected to fight fires, they must be trained and equipped in OSHA 1910.156.

References

U.S. Environmental Protection Agency, "Chemical Profile: Cobalt Carbonyl," Washington, DC, Chemical Emergency Preparedness Program (November 30, 1987)

Cobalt Hydrocarbonyl

Molecular Formula: C_5HO_5

Common Formula: $HCO(CO)_4$

Synonyms: Hydrocobalt tetracarbonyl; Tetracarbonylhydridocobalt; Tetracarbonylhydrocobalt

CAS Registry Number: 16842-03-8

RTECS Number: GG0900000

DOT ID: No citation.

Regulatory Authority

- Air Pollutant Standard Set (ACGIH)[1] (Australia) (former USSR)[43] (Israel) (OSHA)[58] (Several States)[60] (Several Canadian Provinces)
- CLEAN AIR ACT: Hazardous Air Pollutants (Title I, Part A, Section 112) *Note*: Includes any unique chemical substance that contains cobalt as part of that chemical's infrastructure
- EPCRA (Section 313): Includes any unique chemical substance that contains cobalt as part of that chemical's infrastructure. Form R *de minimis* concentration reporting level: 1.0%
- Canada, WHMIS, Ingredients Disclosure List

Cited in U.S. State Regulations: Alaska (G), California (G), Connecticut (A), Florida (G), Illinois (G), Maine (G), Massachusetts (G), Minnesota (G), Nevada (A), New Hampshire (G), New Jersey (G), North Dakota (A), Pennsylvania (G), Rhode Island (G), Virginia (A), West Virginia (G).

Description: Cobalt Hydrocarbonyl, $HCO(CO)_4$, is a flammable and toxic gas with an offensive odor, which decomposes rapidly at room temperature. Boiling point = 10°C. Freezing/Melting point = -26°C. Very slightly soluble in water.

Potential Exposure: Those involved in manufacture and use of this material as a catalyst for organic reactions.

Incompatibilities: Unstable gas that decomposes rapidly in air at room temperature to toxic cobalt carbonyl and explosive hydrogen.

Permissible Exposure Limits in Air: There is no OSHA PEL. The recommended NIOSH REL 10-hour TWA is 0.1 mg/m^3. The ACGIH TLV 8-hour TWA is the same. Australia, Israel, the Canadian provinces of Alberta, BC, Ontario and Quebec use the same TWA limit and Alberta's STEL is 0.3 mg/m^3 The former USSR-UNEP/IRPTC project[43] has set a MAC of 0.01 mg/m^3 in workplace air. Some states have set guidelines or standards for cobalt in ambient air[60] ranging from 1.0 $\mu g/m^3$ (North Dakota) to 1.6 $\mu g/m^3$ (Virginia) to 2.0 $\mu g/m^3$ (Connecticut and Nevada).

Determination in Air: Use OSHA: Method #ID (125 g).

Permissible Concentration in Water: No criteria set.

Routes of Entry: Inhalation, skin and/or eye contact.

Harmful Effects and Symptoms

Short Term Exposure: Irritates eyes, skin, and respiratory tract. Higher exposures can cause pulmonary edema, a medical emergency that can be delayed for several hours. This can cause death. The 30-minute LD_{50} in rats is 165 mg/m^3. The clinical effects are similar to nickel carbonyl and iron pentacarbonyl but it has about one-half the toxicity of nickel carbonyl. In animals: irritation respiratory system; dyspnea (breathing difficulty), cough.

Points of Attack: Eyes, skin, respiratory system.

Medical Surveillance: Preemployment physical examinations should give particular attention to the respiratory tract and skin. Periodic examinations should include the respiratory tract and nasal sinuses, smoking history as well as general health. A baseline chest x-ray should be available and pulmonary function followed.

First Aid: If this chemical gets into the eyes, remove any contact lenses at once and irrigate immediately for at least 15 minutes, occasionally lifting upper and lower lids. Seek medical attention immediately. If this chemical contacts the skin, remove contaminated clothing and wash immediately with soap and water. Seek medical attention immediately. If this chemical has been inhaled, remove from exposure, begin rescue breathing (using universal precautions) if breathing has stopped and CPR if heart action has stopped. Transfer promptly to a medical facility. When this chemical has been swallowed, get medical attention. Give large quantities of water and induce vomiting. Do not make an unconscious person vomit. Medical observation is recommended for 24 – 48 hours after breathing overexposure, as pulmonary edema may be delayed. As first aid for pulmonary edema, a doctor or authorized paramedic may consider administering a corticosteroid spray.

Personal Protective Methods: Wear protective gloves and clothing to prevent any reasonable probability of skin contact. Where the danger of splash or spill of liquids exists, impervious protective clothing should be used. Safety equipment suppliers/manufacturers can provide recommendations on the most protective glove/clothing material for your operation. All protective clothing (suits, gloves, footwear, headgear) should be clean, available each day, and put on before work. Contact lenses should not be worn when working with this chemical. Wear splash-proof chemical goggles and face shield unless full facepiece respiratory protection is worn. Employees should wash immediately with soap when skin is wet or contaminated. Provide emergency showers and eyewash.

Respirator Selection: Where there is a potential for overexposure: SCBAF:PD,PP (any MSHA/NIOSH approved self-contained breathing apparatus that has a full facepiece and is operated in a pressure-demand or other positive-pressure mode); or SAF:PD,PP:ASCBA (any supplied-air respirator that has a full facepiece and is operated in a pressure-demand or other positive-pressure mode in combination with an auxiliary, self-contained breathing apparatus operated in a pressure-demand or other positive pressure mode).

Storage: Prior to working with this chemical you should be trained on its proper handling and storage. Store in a cool, well-ventilated area. Procedures for the handling, use and storage of cylinders should be in compliance with OSHA 1910.101 and 1910.169 as with the recommendations of the Compressed Gas Association.

Shipping: Cobalt hydrocarbonyl is not specifically cited in the DOT Performance-Oriented Packaging Standards.[19]

Spill Handling: If in a building, evacuate building and confine vapors by closing doors and shutting down HVAC systems. Restrict persons not wearing protective equipment from area of spill or leak until cleanup is complete. Remove all ignition sources.

Establish forced ventilation to keep levels below explosive limit and to disperse the gas. Stop the flow of gas, if it can be done safely from a distance. If source is a cylinder and the leak cannot be stopped in place, remove the leaking cylinder to a safe place, and repair leak or allow cylinder to empty. Keep this chemical out of confined spaces, such as a sewer, because of the possibility of explosion, unless the sewer is designed to prevent the buildup of explosive concentrations. If employees are required to clean-up spills, they must be properly trained and equipped. OSHA 1910.120(q) may be applicable.

Fire Extinguishing: This chemical is a flammable gas. Poisonous gases are produced in fire. Do not extinguish the fire unless the flow of gas can be stopped and any remaining gas is out of the line. Specially trained personnel may use fog lines to cool exposures and let the fire burn itself out. Vapors are heavier than air and will collect in low areas. Vapors may travel long distances to ignition sources and flashback. Vapors in confined areas may explode when exposed to fire. Containers may explode in fire. Storage containers and parts of containers may rocket great distances, in many directions. If material or contaminated runoff enters waterways, notify downstream users of potentially contaminated waters. Notify local health and fire officials and pollution control agencies. From a secure, explosion-proof location, use water spray to cool exposed containers. If cooling streams are ineffective (venting sound increases in volume and pitch, tank discolors, or shows any signs of deforming), withdraw immediately to a secure position. If cylinders are exposed to excessive heat from fire or flame contact, withdraw immediately to a secure location. If employees are expected to fight fires, they must be trained and equipped in OSHA 1910.156.

References

American Council of Governmental Industrial Hygienists, Inc., Documentation of the Threshold Limit Values: Supplemental Documentation: Fifth Edition, Cincinnati, OH (1986)

Cobalt Naphthenate

Molecular Formula: $C_{12}H_{18}CoO_4$

Common Formula: $(C_5H_9COO)_2Co$

Synonyms: Cobalt naphtha; Cobalt naphthenate powder; Cobaltous naphthenate; Naphtenate de cobalt (French); Naphthenate de cobalt (French); Naphthenic acid, cobalt salt

CAS Registry Number: 61789-51-3

RTECS Number: QK8925000

DOT ID: UN 2001

Regulatory Authority

- CLEAN AIR ACT: Hazardous Air Pollutants (Title I, Part A, Section 112) *Note:* Includes any unique chemical substance that contains cobalt as part of that chemical's infrastructure
- EPCRA (Section 313): Includes any unique chemical substance that contains cobalt as part of that chemical's infrastructure. Form R *de minimis* concentration reporting level: 1.0%

Cited in U.S. State Regulations: California (G), New Hampshire (G), New Jersey (G), Pennsylvania (G).

Description: Cobalt Naphthenate, $(C_5H_9COO)_2Co$, is a brown powder or bluish-red solid. Freezing/Melting point = 140°C. Flash point = 49°C. Autoignition temperature = 276°C. Hazard Identification (based on NFPA-704 M Rating System): Health 1, Flammability 2, Reactivity 0. Insoluble in water.

Potential Exposure: Cobalt Naphthenate is used in paint, varnish and ink driers.

Incompatibilities: Contact with strong oxidizers cause a fire and explosion hazard. Powder or dust can form an explosive mixture with air.

Permissible Exposure Limits in Air: The OSHA PEL 8-hour TWA is 0.1 mg/m^3. The NIOSH recommended REL 10-hour TWA is 0.05 mg/m^3. The ACGIH recommended TLV 8-hour TWA is 0.02 mg/m^3. Mexico, 0.1 mg/m^3. The NIOSH IDLH = 20 mg/m^3. Australia, Israel; Canada: Alberta, Ontario, Quebec TWA is 0.05 mg/m^3 and Alberta's and Ontario's STEL is 0.1 mg/m^3. British Columbia's TWA is 0.02 mg/m^3 DFG notes animal evidence of animal carcingenicity and does not set a numerical value for respirable dusts and aerosols. The former USSR-UNEP/IRPTC project has set a MAC of 0.5 mg/m^3 for cobalt metal and oxides in workplace air.[43] For ambient air in residential areas, 0.001 mg/m^3 is the MAC on an average daily basis for cobalt metal.

Determination in Air: Cobalt metal, dust and fume may be determined by filter collection, acid dissolution, digestion and measurement by atomic absorption spectrophotometry. See NIOSH IV Method #7027. See also Method #7300, Elements, and OSHA Methods ID121 and ID125.

Permissible Concentration in Water: No criteria set. See Cobalt Metal.

Routes of Entry: Inhalation, ingestion.

Harmful Effects and Symptoms

Short Term Exposure: Cobalt Naphthenate can affect you when breathed in. Breathing the dust or fumes can cause lung allergy (asthma) to develop. Repeated exposures can cause lung scarring. High exposure can damage the heart and/or cause a large thyroid (goiter). Exposure can irritate the nose, throat, and lungs. Higher levels can cause a build-up of fluid (pulmonary edema). This can cause death. Repeated exposure can cause a loss of the sense of smell. Contact can cause a skin allergy to develop.

The oral LD_{50} rat is 3,900 mg/kg (slightly toxic).

Long Term Exposure: Repeated or prolonged contact may cause skin sensitization and allergy. Repeated exposure to Cobalt Naphthenate can cause lung irritation, bronchitis, lung scarring with shortness of breath, coughing. This chemical can cause kidney damage.

Points of Attack: Skin, lungs, kidneys.

Medical Surveillance: For those with frequent or potentially high exposure (half the TLV or greater) the following are recommended before beginning work and at regular times after that: Lung function tests. Chest x-ray (every 5 years) beginning 10 years after exposure. If symptoms develop or overexposure is suspected, the following may be useful: Evaluation by a qualified allergist, including careful exposure history and special testing, may help diagnose skin allergy. GTT and serum lipid studies (to check for blood sugar and fat changes). CBC exam of the cardiovascular system. Kidney function tests. Consider chest x-ray following acute overexposure.

First Aid: If this chemical gets into the eyes, remove any contact lenses at once and irrigate immediately for at least 15 minutes, occasionally lifting upper and lower lids. Seek medical attention immediately. If this chemical contacts the skin, remove contaminated clothing and wash immediately with soap and water. Seek medical attention immediately. If this chemical has been inhaled, remove from exposure, begin rescue breathing (using universal precautions) if breathing has stopped and CPR if heart action has stopped. Transfer promptly to a medical facility. When this chemical has been swallowed, get medical attention. Give large quantities of water and induce vomiting. Do not make an unconscious person vomit. Medical observation is recommended for 24 – 48 hours after breathing overexposure, as pulmonary edema may be delayed. As first aid for pulmonary edema, a doctor or authorized paramedic may consider administering a corticosteroid spray.

Personal Protective Methods: Wear protective gloves and clothing to prevent any reasonable probability of skin contact. Safety equipment suppliers/manufacturers can provide recommendations on the most protective glove/clothing material for your operation. All protective clothing (suits, gloves, footwear, headgear) should be clean, available each day, and put on before work. Contact lenses should not be worn when working with this chemical. Wear dust-proof chemical goggles and face shield unless full facepiece respiratory protection is worn. Employees should wash immediately with soap when skin is wet or contaminated. Provide emergency showers and eyewash.

Respirator Selection: NIOSH/OSHA, for cobalt metal dust and fume:: *0.25 mg/m³:* DM, *if not present as a fume* (any dust and mist respirator). *0.5 mg/m³:* DMXSQ, *if not present as a fume* (any dust and mist respirator except single-use and quarter mask respirators); or DMFu (any dust, mist, and fume respirator); or SA (any supplied-air respirator). *1.25 mg/m³:* SA:CF (any supplied-air respirator operated in a continuous-flow mode); or PAPRDM, *if not present as a fume* (any powered, air-purifying respirator with a dust and mist filter); or PAPRDMFu (any powered, air-purifying respirator with a dust, mist, and fume filter). *2.5 mg/m³:* HiEF (any air-purifying, full-facepiece respirator with a high-efficiency particulate filter); or SCBAF (any self-contained breathing apparatus with a full facepiece); or SAF (any supplied-air respirator with a full facepiece). *20 mg/m³:* SAF:PD,PP (any supplied-air respirator that has a full facepiece and is operated in a pressure-demand or other positive-pressure mode). *Emergency or planned entry into unknown concentrations or IDLH conditions:* SCBAF:PD,PP (any self-contained breathing apparatus that has a full facepiece and is operated in a pressure-demand or other positive-pressure mode); or SAF:PD,PP: ASCBA (any supplied-air respirator that has a full facepiece and is operated in a pressure-demand or other positive-pressure mode in combination with an auxiliary, self-contained breathing apparatus operated in a pressure-demand or other positive-pressure mode). *Escape:* HiEF (any air-purifying, full-facepiece respirator with a high-efficiency particulate filter); or SCBAE (any appropriate escape-type, self-contained breathing apparatus).

Note: Substance reported to cause eye irritation or damage; may require eye protection.

Storage: Prior to working with this chemical you should be trained on its proper handling and storage. Store in tightly closed containers in a cool, well-ventilated area away from strong oxidizers (such as chlorine, bromine, and fluorine). Sources of ignition such as smoking and open flames are prohibited where Cobalt Naphthenate is used, handled, or stored in a manner that could create a potential fire or explosion hazard.

Shipping: Cobalt Naphthenate powder falls in UN/DOT Hazard Class 4.1 and Packing Group III. They require a "Flammable Solid" label.

Spill Handling: Evacuate persons not wearing protective equipment from area of spill or leak until clean-up is complete. Remove all ignition sources. Collect powdered material in the most convenient and safe manner and deposit in sealed containers. Ventilate area after clean-up is complete. It may be necessary to contain and dispose of this chemical as a hazardous waste. If material or contaminated runoff enters waterways, notify downstream users of potentially contaminated waters. Contact your Department of Environmental Protection or your regional office of the federal EPA for specific recommendations. If employees are required to clean-up spills, they must be properly trained and equipped. OSHA 1910.120(q) may be applicable.

Fire Extinguishing: Cobalt Naphthenate is a combustible solid. Use dry chemical, sand, water spray, or alcohol foam extinguishers. Poisonous gases are produced in fire. If material or contaminated runoff enters waterways, notify downstream users of potentially contaminated waters. Notify local health and fire officials and pollution control agencies. Containers may explode in fire. Storage containers and parts of containers may rocket great distances, in many directions. From a secure, explosion-proof location, use water spray to cool exposed containers. If cooling streams are ineffective (venting sound increases in volume and pitch, tank discolors, or shows any signs of deforming), withdraw immediately to a secure position. If employees are expected to fight fires, they must be trained and equipped in OSHA 1910.156.

References

New Jersey Department of Health and Senior Services and Senior Services, "Hazardous Substance Fact Sheet: Cobalt Naphthenate," Trenton, NJ (September 1998)

Colchicine

Molecular Formula: $C_{22}H_{25}NO_6$

Synonyms: Acetamide, *N*-(5,6,7,9-Tetrahydro-1,2,3,10-tetramethoxy-9-oxobenzo[α]heptalen-7-yl); 7-Acetamido-6,7-dihydro-1,2,3,10-tetramethoxybenzo(a)heptalen-9(5H)-one; *N*-Acetyltrimethylcolchicinic acid methyl ether; Benzo[α]heptalen-9(5H)-one; 7-α-H-Colchicine; Colchineos; Colchisol; Colcin; Colquicina (Spanish); Colsaloid; Condylon; NSC 757; *N*-(5,6,7,9)-Tetrahydro-1,2,3,10-tetramethoxy-9-oxobenzo(α)heptalen-7-yl)-acetamide

CAS Registry Number: 64-86-8

RTECS Number: GH0700000

DOT ID: UN 3249

Regulatory Authority

- CLEAN AIR ACT: Hazardous Air Pollutants (Title I, Part A, Section 112)
- SUPERFUND/EPCRA 40CFR355, Appendix B Extremely Hazardous Substances: TPQ = 10/10,000 lb (4.54/4,540 kg)
- SUPERFUND/EPCRA 40CFR302.4 Reportable Quantity (RQ): EHS, 1 lb (0.454 kg)

Cited in U.S. State Regulations: California (A, G), Florida (G), Massachusetts (G), New Jersey (G), Pennsylvania (G).

Description: Colchicine, $C_{22}H_{25}NO_6$, is a pale yellow powder. Freezing/Melting point = 142 – 50°C. It has little or no odor and darkens on exposure to light. Hazard Identification (based on NFPA-704 M Rating System): Health 3, Flammability 1, Reactivity 0. Slightly soluble in water.

Potential Exposure: Colchicine is a drug used to treat gouty arthritis, pseudogout, sarcoidal arthritis and calcific tendonitis.

Incompatibilities: Oxidizers, mineral acids. Keep away from light.

Permissible Exposure Limits in Air: No standards set.

Permissible Concentration in Water: No criteria set.

Routes of Entry: Inhalation, ingestion.

Harmful Effects and Symptoms

Short Term Exposure: Colchine can irritate and burn the eyes, skin nose, and throat. Exposure can cause nausea, vomiting, diarrhea, loss of appetite, and abdominal pain may occur several hours after exposure. Inhalation can cause lung irritation with coughing and shortness of breath. Higher exposures can cause pulmonary edema, a medical emergency that can be delayed for several hours. This can cause death. Colchine can affect the heart causing arrhythmia. Shock occurs because of extensive vascular damage. Kidney damage resulting in bloody urine and diminished urine output may occur. It is classified as super toxic. The LD_{low} oral (dog, cat) is 0.125 mg/kg. Probable oral lethal dose in humans is less than 5 mg/kg. i.e., less than 7 drops for a 70 kg (150 lb) person. Death results from respiratory arrest. The fatal dose varies considerably; as little as 7 mg of colchicines has proved fatal.

Long Term Exposure: May cause genetic changes, liver and kidney damage. High exposure can cause headache, confusion, muscle weakness, coma and death.

Points of Attack: Heart, liver, kidneys, lungs.

Medical Surveillance: Complete blood count. Liver and kidney function tests. EKG. Lung function tests. Consider chest x-ray following acute overexposure.

First Aid: This material is an alkaloid. If this chemical gets into the eyes, remove any contact lenses at once and irrigate immediately for at least 15 minutes, occasionally lifting upper and lower lids. Seek medical attention immediately. If this chemical contacts the skin, remove contaminated clothing and wash immediately with soap and water. Seek medical attention immediately. If this chemical has been inhaled, remove from exposure, begin rescue breathing (using universal precautions) if breathing has stopped and CPR if heart action has stopped. Transfer promptly to a medical facility. When this chemical has been swallowed, get medical attention. Give large quantities of water and induce vomiting. Do not make an unconscious person vomit. Medical observation is recommended for 24 – 48 hours after breathing overexposure, as pulmonary edema may be delayed. As first aid for pulmonary edema, a doctor or authorized paramedic may consider administering a corticosteroid spray.

Personal Protective Methods: Wear protective gloves and clothing to prevent any reasonable probability of skin contact. Safety equipment suppliers/manufacturers can provide recommendations on the most protective glove/clothing material for your operation. All protective clothing (suits, gloves, footwear, headgear) should be clean, available each day, and put on before work. Contact lenses should not be worn when working with this chemical. Wear dust-proof chemical goggles and face shield unless full facepiece respiratory protection is worn. Employees should wash immediately with soap when skin is wet or contaminated. Provide emergency showers and eyewash.

Respirator Selection: Where there is a potential for overexposure: SCBAF:PD,PP (any MSHA/NIOSH approved self-contained breathing apparatus that has a full facepiece and is operated in a pressure-demand or other positive-pressure mode); or SAF:PD,PP:ASCBA (any supplied-air respirator that has a full facepiece and is operated in a pressure-demand or other positive-pressure mode in combination with an auxiliary, self-contained breathing apparatus operated in a pressure-demand or other positive pressure mode).

Storage: Prior to working with this chemical you should be trained on its proper handling and storage. Store in tightly closed containers in a cool, dry place or a refrigerator. Protect from exposure to light, and keep away from mineral acids and oxidizers.

Shipping: Colchine is not specifically cited in DOT's Performance-Oriented Packaging Standards[19] but medicines, poisonous solid n.o.s. fall in Hazard Class 6.1 and Cisplatin in Packing Group II which requires a "Poison" label.

Spill Handling: Evacuate persons not wearing protective equipment from area of spill or leak until clean-up is complete.

Remove all ignition sources. Collect powdered material in the most convenient and safe manner and deposit in sealed containers. Ventilate area after clean-up is complete. It may be necessary to contain and dispose of this chemical as a hazardous waste. If material or contaminated runoff enters waterways, notify downstream users of potentially contaminated waters. Contact your Department of Environmental Protection or your regional office of the federal EPA for specific recommendations. If employees are required to clean-up spills, they must be properly trained and equipped. OSHA 1910.120(q) may be applicable.

Fire Extinguishing: Extinguish fire using dry chemical, carbon dioxide, or water spray extinguishers. Poisonous gases are produce in fire, including nitrogen oxides and carbon monoxide. Avoid breathing dusts and fumes from burning material. Keep upwind. Wear full protective clothing. Wear self-contained breathing apparatus. Avoid bodily contact with the material. Wash away any material which may have contacted the body with copious amounts of water or soap and water. If material or contaminated runoff enters waterways, notify downstream users of potentially contaminated waters. Notify local health and fire officials and pollution control agencies. Containers may explode in fire. From a secure, explosion-proof location, use water spray to cool exposed containers. If cooling streams are ineffective (venting sound increases in volume and pitch, tank discolors, or shows any signs of deforming), withdraw immediately to a secure position. If employees are expected to fight fires, they must be trained and equipped in OSHA 1910.156.

References

U.S. Environmental Protection Agency, "Chemical Profile: Colchicine," Washington, DC, Chemical Emergency Preparedness Program (March 1999)

Conjugated Estrogens

Synonyms: Conjugated estrogenic hormones.; Equigyne (sodium estrone sulfate and sodium equilin sulfate, or synthetic estrogen piperazine estrone sulfate); Premarin

CAS Registry Number: 12126-59-9

RTECS Number: GL1224000

Regulatory Authority

- Carcinogen (NTP)[10] [some animal evidence (Equigyne)] (IARC)

Cited in U.S. State Regulations: California (A, G), Minnesota (G), Pennsylvania (G).

Description: Conjugated estrogens generally occur as butter-colored powders that are soluble in water. The sodium equilin sulfate component is unstable to light and air. Piperazine estrone sulfate occurs as a white to yellowish white crystalline powder that is slightly soluble in water. Piperazine estrone sulfate melts at 245°C with decomposition. Conjugated estrogens are naturally occurring substances excreted in the urine of pregnant mares; piperazine estrone sulfate is not known to occur naturally.

Potential Exposure: Conjugated estrogens are used to treat symptoms of the climacteric, vulvae dystrophies, female hypogonadism, and dysfunctional uterine bleeding. They also are used for treatment following: ovariectomy, for chemotherapy of mammary cancer and prostate carcinogema, and for prevention of postpartum breast engorgement. Additionally, conjugated estrogens have been found in cosmetic preparations.

Permissible Exposure Limits in Air: No standards set.

Permissible Concentration in Water: No criteria set. Because conjugated estrogens are used as pharmaceuticals and in low quantities relative to other chemicals, they are not regulated by EPA. There may be a small pollution problem relative to hospital wastes.

Harmful Effects and Symptoms

Long Term Exposure: There is sufficient evidence that conjugated estrogens are carcinogenic in humans. Liver tumors, endometrial cancer, ovarian cancer, breast cancer, vascular system, and testicular cancer are associated with the use of conjugated estrogens in humans. A large number of studies of cancer of the endometrium suggest that use of conjugated estrogens causes the disease. Several studies reported on the relative risk of breast cancer from use of conjugated estrogens; the evidence is conflicting, both overall and within subgroups.

First Aid: If this chemical gets into the eyes, remove any contact lenses at once and irrigate immediately for at least 15 minutes, occasionally lifting upper and lower lids. Seek medical attention immediately. If this chemical contacts the skin, remove contaminated clothing and wash immediately with soap and water. Seek medical attention immediately. If this chemical has been inhaled, remove from exposure, begin rescue breathing (using universal precautions) if breathing has stopped and CPR if heart action has stopped. Transfer promptly to a medical facility. When this chemical has been swallowed, get medical attention. Give large quantities of water and induce vomiting. Do not make an unconscious person vomit.

Storage: Prior to working with this chemical you should be trained on its proper handling and storage. A regulated, marked area should be established where this chemical is handled, used, or stored in compliance with OSHA standard 1910.1045.

Shipping: As USP XX pharmaceuticals, conjugated estrogens are covered regarding requirements for labeling directed to the physician and to the patient. See 21 CFR 310.515. Published 7/22/77. FD&CA: Labeling requirements for estrogens directed to the patient.

Copper

Molecular Formula: Cu

Synonyms: 1721 gold; Allbri natural copper; Anac 110; Arwood copper; Bronze powder; CDA 101; CDA 102; CDA 110; CDA 122; C.I. 77400; C.I. pigment metal 2; Cobre (Spanish); Copper bronze; Elemental copper; Gold bronze; Kafar copper; M2 copper; MI (copper); OFHC Cu; Raney copper

CAS Registry Number: 7440-50-8; 1317-38-0 (CuO, copper fume)

RTECS Number: GL5325000; GL7900000 (CuO, copper fume)

Regulatory Authority

- Air Pollutant Standard Set (ACGIH)[1] (Australia) (DFG)[3] (HSE)[33] (Israel) (Mexico) (OSHA) (former USSR)[43] (Several States)[60] (Several Canadian Provinces)
- CLEAN WATER ACT: 40CFR423, Appendix A, Priority Pollutants; Section 313 Water Priority Chemicals (57FR41331, 9/9/92); Toxic Pollutant (Section 401.15)
- RCRA 40CFR264, Appendix 9; TSD Facilities Ground Water Monitoring List. Suggested test method(s) (PQL μg/l): (total) 6010 (60); 7210 (200)
- SAFE DRINKING WATER ACT: MCL, 1.3 mg/l; MCLG, 1 mg/l; SMLC, 1.0 mg/l; Regulated chemical (47 FR 9352)
- SUPERFUND/EPCRA 40CFR302.4 Reportable Quantity (RQ): CERCLA, 5,000 lb (2,270 kg) (no reporting of releases of this hazardous substance is required if the diameter of the pieces of solid metal released is equal to 0.004 in)
- EPCRA Section 313 Form R *de minimis* concentration reporting level: 1.0%
- Canada, WHMIS, Ingredients Disclosure List; National Pollutant Release Inventory (NPRI); CEPA Priority Substance List, Ocean dumping prohibited; Drinking Water Quality ≤ 1.0 mg/l
- Mexico, Drinking Water = 1.0 mg/l

Cited in U.S. State Regulations: Alaska (G), California (A, G, W), Connecticut (A), Florida (G, A), Illinois (G), Maine (G), Maryland (G), Massachusetts (G), Michigan (G), Minnesota (W), Montana (A), Nevada (A), New Hampshire (G), New Jersey (G), New York (G, A), North Dakota (A), Pennsylvania (G), Rhode Island (G), Virginia (A), West Virginia (G).

Description: Copper, Cu, is a reddish-brown metal which occurs free or in ores such as malachite, cuprite, and chalcopyrite. Copper fume is finely divided black particulate dispersed in air. Copper dusts and mists have been assigned the formula $CuSO_4 \cdot 5H_2O/CuCl$ by NIOSH. Copper fume has been designated as $Cu/CuO/Cu_2O$ by NIOSH. Boiling point = 2,595°C. Freezing/Melting point = 1,083°C It may form both mono- and divalent compounds. Copper is insoluble in water, but soluble in nitric acid and hot sulfuric acid.

Potential Exposure: Metallic copper is an excellent conductor of electricity and is widely used in the electrical industry in all gauges of wire for circuitry, coil, and armature windings, high conductivity tubes, commutator bars, etc. It is made into castings, sheets, rods, tubing and wire, and is used in water and gas piping, roofing materials, cooking utensils, chemical and pharmaceutical equipment and coinage. Copper forms many important alloys: Be-Cu alloy, brass, bronze, gunmetal, bell metal, German silver, aluminum bronze, silicon bronze, phosphor bronze, and manganese bronze.

Copper compounds are used as insecticides, algicides, molluscicides, plant fungicides, mordants, pigments, catalysts, and as a copper supplement for pastures, and in the manufacture of powdered bronze paint and percussion caps. They are also utilized in analytical reagents, in paints for ships' bottoms, in electroplating, and in the solvent for cellulose in rayon manufacture.

Incompatibilities: Copper dust, fume, and mists form shock-sensitive compounds with acetylene gas, acetylenic compounds, azides, and ethylene oxides. Incompatible with acids, chemically active metals such as potassium, sodium, magnesium, and zinc, zirconium, strong bases. Violent reaction, possibly explosive, if finely-divided material come in contact with strong oxidizers.

Permissible Exposure Limits in Air: The Federal standard (OSHA PEL 8-hour TWA)[58] for *copper fume* is 0.1 mg/m^3, and 1 mg/m^3 for *copper dusts and mists.* NIOSH recommends the same level for a 10-hour workshift. ACGIH recommends a TWA of 0.2 mg/m^3 for copper fume and 1 mg/m^3 for dusts and mists, as has HSE,[33] Australia, Israel, and the Canadian provinces of Alberta, BC, Ontario, and Quebec and HSE, Alberta, BC set a STEL for dusts and mists of 2.0 mg/m^3. The NIOSH IDLH is 100 mg/m^3 (as Cu). The DFG MAK for total dust is 1 mg/m^3; 0.1 mg/m^3 for fine dust and Peak Limitation is 2 times MAK (30 min), not to be exceeded 4 times during a workshift. Mexico set a limit of 0.2 mg/m^3 TWA and STEL of 2 mg/m^3. The former USSR-UNEP/IRPTC project[43] has set a MAC value for copper in workplace air of 1.0 mg/m^3 and 0.5 mg/m^3 on an average value per workshift basis.

Several states have set guidelines or standards for copper in ambient air[60] ranging from 0.26 – 1.57 μg/m^3 (Montana) to 2.0 μg/m^3 (North Dakota) to 2.0 – 20.0 μg/m^3 (Connecticut) to 4.0-20.0 μg/m^3 (Florida) to 5.0 μg/m^3 (Nevada) to 16.0 μg/m^3 (Virginia) to 20.0 μg/m^3 (New York).

Determination in Air: Copper dusts and mists are collected on a filter, worked up with acid, measured by atomic absorption. See NIOSH Method #7029 for copper. For copper fume: filter collection, acid digestion, measurement by atomic absorption. See NIOSH Method #7200 for welding and brazing fume.

Permissible Concentration in Water: To protect freshwater aquatic life: 5.6 μg/l as a 24-hour average, never to exceed $e^{[0.94 \ln(\text{hardness}) - 1.23]}$ μg/l. To protect human health:

1,000 μg/l.[6] Canada: Drinking Water Quality (AO) ≤ 1.0 mg/l. Mexico, Drinking Water = 1.0 mg/l. Czechoslovakia[35] has set a MAC in surface water of 0.1 mg/l and in drinking water of 0.05 mg/l. The former USSR[35] and The former USSR-UNEP/IRPTC joint project have set a MAC in water used for domestic purposes of 1.0 mg/l and 0.001 mg/l in fresh water and 0.005 mg/l in seawater used for fishery purposes. Two states have set guidelines for copper in drinking water;[61] they are Kansas at 1,000 μg/l and Minnesota at 1,300 μg/l.

Determination in Water: Total copper may be determined by digestion followed by atomic absorption or by colorimetry (using neocuproine) or by Inductively Coupled Plasma (ICP) Optical Emission Spectrometry. Dissolved Copper may be determined by 0.45 μ filtration followed by the preceding methods.

Routes of Entry: Inhalation of dust or fume, ingestion, or skin or eye contact.

Harmful Effects and Symptoms

Short Term Exposure: Copper salts act as irritants to the intact skin causing itching, erythema, and dermatitis. In the eyes, copper salts may cause conjunctivitis and even ulceration and turbidity of the cornea. Metallic copper may cause keratinization of the hands and soles of the feet, but it is not commonly associated with industrial dermatitis. The fumes and dust cause irritation of the upper respiratory tract, metallic taste in the mouth, nausea, metal fume fever. Inhalation of dusts, fumes, and mists of copper salts may cause congestion of the nasal mucous membranes. If the salts reach the gastrointestinal tract, they act as irritants producing salivation, nausea, vomiting, gastric pain, hemorrhagic gastritis, and diarrhea. It is unlikely that poisoning by ingestion in industry would progress to a serious point as small amounts induce vomiting, emptying the stomach of copper salts. Chronic human intoxication occurs rarely and then only in individuals with Wilson's disease (hepatolenticular degeneration). This is a genetic condition caused by the pairing of abnormal autosomal recessive genes in which there is abnormally high absorption, retention, and storage of copper by the body. The disease is progressive and fatal if untreated.

Long Term Exposure: Copper may decrease fertility in both males and females. Repeated or prolonged contact may cause skin sensitization and allergy, thickening of the skin, and greenish color to the skin, teeth, and hair. Repeated exposure can cause chronic irritation of the nose and cause ulcers and hole in the septum dividing the inner nose. Repeated high exposure to copper can cause liver damage. There is evidence that workers in copper smelting plants have an increased risk of lung cancer, but this is thought to be due to arsenic trioxide and not copper.

Points of Attack: For copper dusts and mists: respiratory system, lungs, skin, liver, including risk with Wilson's disease, kidneys. For copper fume: respiratory system, skin, eyes, and risk with Wilson's disease.

Medical Surveillance: Serum and urine copper levels. Evaluation by a qualified allergist. Liver function tests. Copper often contains arsenic as an impurity. Wilson's disease is a rare hereditary condition which interferes with the body's ability to get rid of copper. If you have this condition, consult your doctor about copper exposure.

First Aid: If copper dust or powder gets into the eyes, remove any contact lenses at once and irrigate immediately for at least 15 minutes, occasionally lifting upper and lower lids. Seek medical attention immediately. If copper dusts or powder contacts the skin, remove contaminated clothing and wash immediately with soap and water. Seek medical attention immediately. If this chemical has been inhaled, remove from exposure, begin rescue breathing (using universal precautions) if breathing has stopped and CPR if heart action has stopped. Transfer promptly to a medical facility. When this chemical has been swallowed, get medical attention. Give large quantities of water and induce vomiting. Do not make an unconscious person vomit.

Note to Physician: In case of fume inhalation, treat pulmonary edema. Give prednisone or other corticosteroid orally to reduce tissue response to fume. Positive pressure ventilation may be necessary. Treat metal fume fever with bed rest, analgesics and antipyretics. The symptoms of metal fume fever may be delayed for 4 – 12 hours following exposure: it may last less than 36 hours.

Personal Protective Methods: For copper dusts, powder, or mists: Wear protective gloves and clothing to prevent any reasonable probability of skin contact. Safety equipment suppliers/manufacturers can provide recommendations on the most protective glove/clothing material for your operation. All protective clothing (suits, gloves, footwear, headgear) should be clean, available each day, and put on before work. Contact lenses should not be worn when working with this chemical. Wear dust-proof chemical goggles and face shield unless full facepiece respiratory protection is worn. Employees should wash immediately with soap when skin is wet or contaminated. Provide emergency showers and eyewash.

Respirator Selection: *Copper dusts and mists:* NIOSH/OSHA: *5 mg/m³:* DM (any dust and mist respirator). *10 mg/m³:* DMXSQ, if not present as a fume (any dust and mist respirator except single-use and quarter mask respirators); or SA (any supplied-air respirator). *25 mg/m³:* SA:CF (any supplied-air respirator operated in a continuous-flow mode); PAPRDM (any powered, air-purifying respirator with a dust and mist filter). *50 mg/m³:* HiEF (any air-purifying, full-facepiece respirator with a high-efficiency particulate filter); or PAPRTHiE (any powered, air-purifying respirator with a tight-fitting facepiece and a high-efficiency particulate filter); or SCBAF (any self-contained breathing apparatus with a full facepiece); or SAF (any supplied-air respirator with a full facepiece). *100 mg/m³:* SAF:PD,PP (any supplied-air respirator that has a full facepiece and is operated in a pressure-demand or other positive-pressure mode). *Emergency or* *planned entry into unknown concentrations or IDLH conditions:* SCBAF:PD,PP (any self-contained breathing apparatus that has a full facepiece and is operated in a pressure-demand or other positive-pressure mode); or SAF:PD,PP: ASCBA (any supplied-air respirator that has a full facepiece and is operated in a pressure-demand or other positive-pressure mode in combination with an auxiliary self-contained breathing apparatus operated in a pressure-demand or other positive- pressure mode). *Escape:* HiEF (any air-purifying, full-facepiece respirator with a high-efficiency particulate filter); or SCBAE (any appropriate escape-type, self-contained breathing apparatus).

Note: Substance reported to cause eye irritation or damage; may require eye protection.

Copper fume: NIOSH/OSHA: *Up to 1 mg/m³:* DMFu (any dust, mist, and fume respirator; or SA (any supplied-air respirator). *Up to 2.5 mg/m³:* SA:CF (any supplied-air respirator operated in a continuous-flow mode); or PAPRDMFu (any powered, air-purifying respirator with a dust, mist, and fume filter). *Up to 5 mg/m³:* HiEF (any air-purifying, full-facepiece respirator with a high-efficiency particulate filter); or SAT:CF (any supplied-air respirator that has a tight-fitting facepiece and is operated in a continuous-flow mode); or PAPRTHiE (any powered, air-purifying respirator with a tight-fitting facepiece and a high-efficiency particulate filter); or SCBAF (any self-contained breathing apparatus with a full facepiece); or SAF (any supplied-air respirator with a full facepiece). *Up to 100 mg/m³:* SAF:PD,PP (any supplied-air respirator that has a full facepiece and is operated in a pressure-demand or other positive-pressure mode). *Emergency or planned entry into unknown concentrations or IDLH conditions:* SCBAF:PD, PP (any self-contained breathing apparatus that has a full facepiece and is operated in a pressure-demand or other positive-pressure mode); or SAF:PD,PP:ASCBA (any supplied-air respirator that has a full facepiece and is operated in a pressure-demand or other positive-pressure mode in combination with an auxiliary self-contained breathing apparatus operated in a pressure-demand or other positive pressure mode). *Escape:* HiEF (any air-purifying, full-facepiece respirator with a high-efficiency particulate filter); or SCBAE (any appropriate escape-type, self-contained breathing apparatus).

Storage: Prior to working with copper you should be trained on its proper handling and storage. Store in tightly closed containers in a cool, well ventilated area away from acetylene gas, oxidizers, and other incompatible materials listed above. Use only non-sparking tools and equipment, especially when opening and closing containers of this chemical. Copper powder: sources of ignition such as smoking and open flames, are prohibited where copper powder is used, handled, or stored in a manner that could create a potential fire or explosion hazard. Store to avoid conditions which create fumes or fine dusts.

Shipping: Copper is not specifically cited in DOT's Performance-Oriented Packaging Standards.[19]

Spill Handling: Warn other workers of spill. Put on proper protective equipment and clothing. Sweep or vacuum up solids being careful not to raise dust levels. Evacuate persons not wearing protective equipment from area of spill or leak until clean-up is complete. Remove all ignition sources. Collect powdered material in the most convenient and safe manner and deposit in sealed containers. Ventilate area after clean-up is complete. It may be necessary to contain and dispose of copper dust and powder as a hazardous waste. If material or contaminated runoff enters waterways, notify downstream users of potentially contaminated waters. Contact your Department of Environmental Protection or your regional office of the federal EPA for specific recommendations. If employees are required to clean-up spills, they must be properly trained and equipped. OSHA 1910.120(q) may be applicable.

Fire Extinguishing: Copper powder can be a combustible solid. Copper metal does not burn. Use powdered dolomite, sodium chloride (common salt) or graphite. Do not use water. Copper may contain arsenic, poisonous gases may be produced in fire. If material or contaminated runoff enters waterways, notify downstream users of potentially contaminated waters. Notify local health and fire officials and pollution control agencies. If employees are expected to fight fires, they must be trained and equipped in OSHA 1910.156.

Disposal Method Suggested: Copper-containing wastes can be concentrated through the use of ion exchange, reverse osmosis, or evaporators to the point where copper can be electrolytically removed and sent to a reclaiming firm. If recovery is not feasible, the copper can be precipitated through the use of caustics and the sludge deposited in a chemical waste landfill.[22]

References

U.S. Environmental Protection Agency, Toxicology of Metals, Vol. II: Copper, Report EPA-600/1-77-022, pp 206–221, Research Triangle Park, NC (May 1977)

U.S. Environmental Protection Agency, Copper: ambient Water Quality Criteria, Washington, DC (1980)

National Academy of Sciences, Medical and Biologic Effects of Environmental Pollutants: Copper, Washington, DC (1977)

Sax, N. I., Ed., "Dangerous Properties of Industrial Materials Report," 1, No. 5, 48–49, New York, Van Nostrand Reinhold Co. (1981). (Copper)

New Jersey Department of Health and Senior Services and Senior Services, "Hazardous Substance Fact Sheet: Copper," Trenton, NJ (January 1999)

New York State Department of Health, "Chemical Fact Sheet: Copper," Albany, NY, Bureau of Toxic Substance Assessment (January 1986 and Version 3)

Copper Acetoarsenite

Molecular Formula: $C_4H_6As_6Cu_4O_{16}$

Synonyms: Acetato(trimetaarsenito)dicopper; Acetoarsenite de cuivre (French); Basle green; Cupric acetoarsenite; Emerald green; Imperial green; King's green; Meadow green; Mitis green; Moss green; Paris green; Parrot green; Patent green; Schweinfurth green; Vienna green

CAS Registry Number: 12002-03-8

RTECS Number: GL6475000

DOT ID: UN 1585

Regulatory Authority

- OSHA, 29CFR1910 Specifically Regulated Chemicals (See CFR 1910.1018)

Arsenic compounds:

- Carcinogen (arsenic compounds, n.o.s.) (IARC, Group I, carcinogenic to humans) (OSHA, select carcinogens)
- Clean Air Act, 42USC7412; Title I, Part A,§112 hazardous pollutants (arsenic compounds)
- Clean Water Act 40CFR401.15 Section 307, Toxic Pollutants, as arsenic and compounds
- RCRA, 40CFR261, Appendix 8 Hazardous Constituents, waste number not listed (arsenic compounds)
- SUPERFUND/EPCRA 40CFR302.4 Reportable Quantity (RQ): CERCLA, 1 lb (0.454 kg) (arsenic compounds)
- EPCRA Section 313: Form R *de minimis* concentration reporting level: 0.1% (inorganic arsenic)
- U.S. DOT 49CFR172.101, Appendix B, Regulated marine pollutant (arsenic compounds)
- Canada: Priority Substance List & Restricted Substances/Ocean Dumping Forbidden (CEPA), National Pollutant Release Inventory (NPRI) (arsenic compounds)

Copper compounds:

- CLEAN WATER ACT: Section 311 Hazardous Substances/RQ 40CFR117.3 (same as CERCLA, see below); Section 313 Water Priority Chemicals (57FR41331, 9/9/92); 40CFR401.15 Section 307 Toxic Pollutants, as copper and compounds
- RCRA 40CFR264, Appendix 9; TSD Facilities Ground Water Monitoring List. Suggested test method(s) (PQL μg/l): 6010 (60); 7210 (200) Note: All species in the ground water that contain copper are included
- SUPERFUND/EPCRA 40CFR302.4 Reportable Quantity (RQ): CERCLA, 100 lb (45.4 kg)
- EPCRA Section 313 Form R *de minimis* concentration reporting level: 1.0%
- Canada, WHMIS, Ingredients Disclosure List; National Pollutant Release Inventory (NPRI); CEPA Priority Substance List, Ocean dumping prohibited

Cited in U.S. State Regulations: California (A, G), Florida (G), Massachusetts (G), New Jersey (G), Pennsylvania (G)

Description: Copper acetoarsenite is an emerald-green, crystalline powder. Hazard Identification (based on NFPA-704 M Rating System): Health 3, Flammability 0, Reactivity 0. Insoluble in water.

Potential Exposure: It is used as an insecticide, wood preservative, and in paints for marine vessels.

Incompatibilities: Deomposes in water with prolonged heating. Incompatible with strong bases, strong acids.

Copper: The Federal standard (OSHA PEL 8-hour TWA)[58] for copper fume is 0.1 mg/m^3, and 1 mg/m^3 for copper dusts and mists. NIOSH recommends the same level for a 10-hour workshift. ACGIH recommends a TWA of 0.2 mg/m^3 for copper fume and 1 mg/m^3 for dusts and mists. The NIOSH IDLH is 100 mg/m^3.

Arsenic: The following exposure limits are for air levels only. When skin contact also occurs, overexposure is possible, even though air levels are less than the limits listed below. OSHA:[2] The legal airborne permissible exposure limit (PEL) is 0.010 mg/m^3 averaged over an 8-hour workshift. NIOSH:[2] The recommended airborne exposure limit is 0.002 mg/m^3 (ceiling), not to be exceeded during any 15 min. work period. ACGIH:[1] The recommended airborne exposure limit is 0.01 mg/m^3 averaged over an 8-hour workshift. The HSE (U.K.) Maximum Exposure Limit as As is 0.1 mg/m^3 TWA. California's workplace PEL is the same as ACGIH and an Action Level of 0.005 mg/m^3. The Australia limit is 0.05 mg/m^3 TWA (confirmed carcinogen); Israel 0.01 mg/m^3 TWA and Action Level 0.005 mg/m^3. Mexico level 0.2 mg/m^3 TWA. Canada: Alberta level 0.2 mg/m^3 TWA and STEL of 0.6 mg/m^3 (15 min); British Columbia level 0.5 mg/m^3 TWA; Ontario level 0.01 mg/m^3 TWAEV and STEV of 0.05; Quebec level 0.2 mg/m^3 TWAEV. The former USSR-UNEP/IRPTC project[43] has set a MAC of 0.003 mg/m^3 on an average daily basis for residential areas. In addition, several states have set guidelines or standards for arsenic in ambient air:[60] 0.06 mg/m^3 (California Prop. 65), 0.0002 μg/m^3 (Rhode Island), 0.00023 μg/m^3 (North Carolina), 0.024 μg/m^3 (Pennsylvania), 0.05 μg/m^3 (Connecticut), 0.07 – 0.39 μg/m^3 (Montana), 0.67 μg/m^3 (New York), 1.0 μg/m^3 (South Carolina), 2.0 μg/m^3 (North Dakota), 3.3 μg/m^3 (Virginia), 5 μg/m^3 (Nevada).

Determination in Air: Copper dusts and mists are collected on a filter, worked up with acid, measured by atomic absorption. See NIOSH Method #7029 for copper. For copper fume: filter collection, acid digestion, measurement by atomic absorption.

Routes of Entry: Inhalation, absorbed through the skin.

Harmful Effects and Symptoms

Short Term Exposure: Eye contact can cause severe irritation and burns. Skin contact can cause irritation, burning sensation, itching, thickening and color changes. This chemical can be absorbed through the skin, thereby increasing exposure.

Long Term Exposure: Repeated exposure can cause copper to deposit in the liver, kidneys, and other body organs, causing damage, atrophy of the inner lining of the nose and possible hole in the nasal septum, with a watery or bloody discharge. Metallic or garlic taste may also occur. Repeated skin exposure can cause skin allergy and possibly a green discoloration of the skin and hair. Some copper and arsenic compounds, but not this one, have been identified as carcinogens and certain arsenic compounds have been determined to be reproductive hazards. Therefore this chemical should be handled with extreme caution. May damage the nervous system.

Points of Attack: Skin, nervous system, liver, kidneys.

Medical Surveillance: Urine arsenic test. Examine skin for abnormal growths. Examination of the nose, skin, nails, and nervous system. Liver function tests, kidney function tests.

First Aid: If this chemical gets into the eyes, remove any contact lenses at once and irrigate immediately for at least 15 minutes, occasionally lifting upper and lower lids. Seek medical attention immediately. If this chemical contacts the skin, remove contaminated clothing and wash immediately with soap and water. Seek medical attention immediately. If this chemical has been inhaled, remove from exposure, begin rescue breathing (using universal precautions) if breathing has stopped and CPR if heart action has stopped. Transfer promptly to a medical facility. When this chemical has been swallowed, get medical attention. Give large quantities of water and induce vomiting. Do not make an unconscious person vomit.

Personal Protective Methods: Wear protective gloves and clothing to prevent any reasonable probability of skin contact. Safety equipment suppliers/manufacturers can provide recommendations on the most protective glove/clothing material for your operation. All protective clothing (suits, gloves, footwear, headgear) should be clean, available each day, and put on before work. Contact lenses should not be worn when working with this chemical. Wear dust-proof chemical goggles and face shield unless full facepiece respiratory protection is worn. Employees should wash immediately with soap when skin is wet or contaminated. Provide emergency showers and eyewash.

Respirator Selection: *Copper fume:* NIOSH/OSHA: *Up to 1 mg/m^3:* DMFu (any dust, mist, and fume respirator; or SA (any supplied-air respirator). *Up to 2.5 mg/m^3:* SA:CF (any supplied-air respirator operated in a continuous-flow mode); or PAPRDMFu (any powered, air-purifying respirator with a dust, mist, and fume filter). *Up to 5 mg/m^3:* HiEF (any air-purifying, full-facepiece respirator with a high-efficiency particulate filter); or SAT:CF (any supplied-air respirator that has a tight-fitting facepiece and is operated in a continuous-flow mode); or PAPRTHiE (any powered, air-purifying respirator with a tight-fitting facepiece and a high-efficiency particulate filter); or SCBAF (any self-contained breathing apparatus with a full facepiece); or SAF (any supplied-air respirator with a full facepiece). *Up to 100 mg/m^3:* SAF:PD,PP (any supplied-air respirator that has a full facepiece and is operated in a pressure-demand or other positive-pressure mode). *Emergency or planned entry into unknown concentrations or IDLH conditions:* SCBAF:PD,PP (any self-contained breathing apparatus that has a full facepiece and is operated in a pressure-demand or other positive-pressure mode); or SAF:PD,PP: ASCBA (any supplied-air respirator that has a full facepiece and is operated in a pressure-demand or other positive-pressure mode in combination with an auxiliary self-contained breathing apparatus operated in a pressure-demand or other positive pressure mode). *Escape:* HiEF (any air-purifying, full-facepiece respirator with a high-efficiency particulate filter); or SCBAE (any appropriate escape-type, self-contained breathing apparatus).

Arsenic: at any concentrations above the NIOSH REL, or where there is no REL, at any detectable concentration: SCBAF:PD,PP (any self-contained breathing apparatus that has a full faceplate and is operated in a pressure-demand or other positive-pressure mode); or SAF:PD,PP:ASCBA (any supplied-air respirator that has a full facepiece and is operated in a pressure-demand or other positive-pressure mode in combination with an auxiliary self-contained breathing apparatus operated in a pressure-demand or other positive-pressure mode). *Escape:* GMFAGHiE [any air-purifying, full-facepiece respirator (gas mask) with a chin-style, front-or back-mounted acid gas canister having a high-efficiency particulate filter]; or SCBAE (any appropriate escape-type, self-contained breathing apparatus).

Storage: Prior to working with copper acetoarsenite you should be trained on its proper handling and storage. Store in tightly closed containers in a cool, well ventilated area, away from strong bases, strong acids and moisture. Where possible, automatically pump liquid from drums or other storage containers to process containers

Shipping: DOT label requirement of "Poison." Copper acetoarsenite falls in UN/DOT Hazard Class 6.1 and Packing Group II.

Spill Handling: Evacuate persons not wearing protective equipment from area of spill or leak until clean-up is complete. Remove all ignition sources. Collect powdered material in the most convenient and safe manner and deposit in sealed containers. Ventilate area after clean-up is complete. It may be necessary to contain and dispose of this chemical as a hazardous waste. If material or contaminated runoff enters waterways, notify downstream users of potentially contaminated waters. Contact your Department of Environmental Protection or your regional office of the federal EPA for specific recommendations. If employees are required to clean-up spills, they must be properly trained and equipped. OSHA 1910.120(q) may be applicable.

Fire Extinguishing: This chemical may burn but does not easily ignite. Use dry chemical, carbon dioxide, water spray, or alcohol foam extinguishers. Poisonous fumes are produced in fire including arsenic oxide. If material or contaminated runoff enters waterways, notify downstream users of potentially contaminated waters. Notify local health and fire officials and pollution control agencies. Containers may explode in fire. From a secure, explosion-proof location, use water spray to cool exposed containers. If cooling streams are ineffective (venting sound increases in volume and pitch, tank discolors, or shows any signs of deforming), withdraw immediately to a secure position. If employees are expected to fight fires, they must be trained and equipped in OSHA 1910.156.

References

New Jersey Department of Health and Senior Services, "Hazardous Substance Fact Sheet, "Copper Acetoarsenite,"Trenton, NJ (January 1999)

Copper Chloride

Molecular Formula: Cl_2Cu

Synonyms: Copper bichloride; Copper(2+) chloride; Copper(II) chloride; Cupric chloride; Cupric chloride dihydrate; Cupric dichloride

CAS Registry Number: 7447-39-4

RTECS Number: GL7237000

DOT ID: UN 2802

Regulatory Authority

- CLEAN WATER ACT: Section 311 Hazardous Substances/RQ 40CFR117.3 (same as CERCLA, see below); Section 313 Water Priority Chemicals (57FR41331, 9/9/92); 40CFR401.15 Section 307 Toxic Pollutants, as copper and compounds
- RCRA 40CFR264, Appendix 9; TSD Facilities Ground Water Monitoring List. Suggested test method(s) (PQL μg/l): 6010 (60); 7210 (200) Note: All species in the ground water that contain copper are included
- SUPERFUND/EPCRA 40CFR302.4 Reportable Quantity (RQ): CERCLA, 10 lb (4.54 kg)
- EPCRA Section 313 Form R *de minimis* concentration reporting level: 1.0%
- U.S. DOT Regulated Marine Pollutant (49CFR172.101, Appendix B)
- Canada, WHMIS, Ingredients Disclosure List; National Pollutant Release Inventory (NPRI); CEPA Priority Substance List, Ocean dumping prohibited

Cited in U.S. State Regulations: California (A, G), Massachusetts (G), New Jersey (G), Pennsylvania (G)

Description: Copper chloride is a brownish-yellow powder. Boiling point = 993°C (decomposes below this point). Freezing/Melting point = 498°C. Hazard Identification (based on NFPA-704 M Rating System): Health 2, Flammability 0, Reactivity 0. Soluble in water.

Potential Exposure: It is used in petroleum, textiles, metallurgy, photography, agricultural products, feed additives, and wood preservation. It is also used in light sensitive paper manufacturing, glass pigments, and ceramics, and in making cyclonitrile.

Incompatibilities: Contact with strong acids forms monovalent copper salts and toxic hydrogen chloride gas. Forms shock-sensitive and explosive compounds with potassium, sodium, sodium hypobromite, nitromethane, acetylene. Keep away from moisture and alkali metals. Attacks metals in the presence of moisture. Reacts with moist air to form cupric chloride dihydrate. May be able to ignite combustible materials.

Permissible Exposure Limits in Air: The Federal standard (OSHA PEL 8-hour TWA)[58] for copper fume is 0.1 mg/m^3, and 1 mg/m^3 for copper dusts and mists. NIOSH recommends the same level for a 10-hour workshift. ACGIH recommends a TWA of 0.2 mg/m^3 for copper fume and 1 mg/m^3 for dusts and mists. The NIOSH IDLH is 100 mg/m^3.

Determination in Air: Copper dusts and mists are collected on a filter, worked up with acid, measured by atomic absorption. See NIOSH Method #7029 for copper. For copper fume: filter collection, acid digestion, measurement by atomic absorption.

Routes of Entry: Inhalation, ingestion.

Harmful Effects and Symptoms

Short Term Exposure: Irritates the eyes, skin, and respiratory tract. Skin or eye contact can cause corrosive burns and permanent damage. Irritates the stomach causing salivation, nausea, vomiting, stomach pain, and diarrhea.

Long Term Exposure: Repeated exposured may cause skin thickening and allergy. May cause a greenish color to the skin and hair. May damage the liver and kidneys. Repeated exposure can cause ulcers or a hole in the nasal septum with possible bleeding. May affect the blood.

Points of Attack: Kidneys, liver, skin, blood.

Medical Surveillance: Serum and urine copper levels. Liver and kidney function tests. Evaluation by a qualified allergist. Complete blood count (CBC). Wilson's disease is a rare hereditary condition which interferes with the body's ability to get rid of copper. If you have this condition, consult your doctor about copper exposure.

First Aid: If this chemical gets into the eyes, remove any contact lenses at once and irrigate immediately for at least 15 minutes, occasionally lifting upper and lower lids. Seek medical attention immediately. If this chemical contacts the skin, remove contaminated clothing and wash immediately with soap and water. Seek medical attention immediately. If this chemical has been inhaled, remove from exposure, begin rescue breathing (using universal precautions) if breathing has stopped and CPR if heart action has stopped. Transfer promptly to a medical facility. When this chemical has been swallowed, get medical attention. If victim is conscious, administer water or milk. Do not induce vomiting.

Personal Protective Methods: Wear protective gloves and clothing to prevent any reasonable probability of skin contact. Safety equipment suppliers/manufacturers can provide recommendations on the most protective glove/clothing material for your operation. All protective clothing (suits, gloves, footwear, headgear) should be clean, available each day, and put on before work. Contact lenses should not be worn when working with this chemical. Wear dust-proof chemical goggles and face shield unless full facepiece respiratory protection is worn. Employees should wash immediately with soap when skin is wet or contaminated. Provide emergency showers and eyewash.

Respirator Selection: NIOSH/OSHA: (copper dust and mist) *5 mg/m³:* DM (any dust and mist respirator). *10 mg/m³:* DMXSQ, if not present as a fume (any dust and mist respirator except single-use and quarter mask respirators); or SA (any supplied-air respirator). *25 mg/m³:* SA:CF (any supplied-air respirator operated in a continuous-flow mode); PAPRDM (any powered, air-purifying respirator with a dust and mist filter). *50 mg/m³:* HiEF (any air-purifying, full-facepiece respirator with a high-efficiency particulate filter); or PAPRTHiE (any powered, air-purifying respirator with a tight-fitting facepiece and a high-efficiency particulate filter); or SCBAF (any self-contained breathing apparatus with a full facepiece); or SAF (any supplied-air respirator with a full facepiece). *100 mg/m³:* SAF:PD,PP (any supplied-air respirator that has a full facepiece and is operated in a pressure-demand or other positive-pressure mode). *Emergency or planned entry into unknown concentrations or IDLH conditions:* SCBAF:PD,PP (any self-contained breathing apparatus that has a full facepiece and is operated in a pressure-demand or other positive-pressure mode); or SAF:PD,PP:ASCBA (any supplied-air respirator that has a full facepiece and is operated in a pressure-demand or other positive-pressure mode in combination with an auxiliary self-contained breathing apparatus operated in a pressure-demand or other positive- pressure mode). *Escape:* HiEF (any air-purifying, full-facepiece respirator with a high-efficiency particulate filter); or SCBAE (any appropriate escape-type, self-contained breathing apparatus).

Note: Substance reported to cause eye irritation or damage; may require eye protection.

NIOSH/OSHA: (copper fume) *Up to 1 mg/m³:* DMFu (any dust, mist, and fume respirator; or SA (any supplied-air respirator). *Up to 2.5 mg/m³:* SA:CF (any supplied-air respirator operated in a continuous-flow mode); or PAPRDMFu (any powered, air-purifying respirator with a dust, mist, and fume filter). *Up to 5 mg/m³:* HiEF (any air-purifying, full-facepiece respirator with a high-efficiency particulate filter); or SAT:CF (any supplied-air respirator that has a tight-fitting facepiece and is operated in a continuous-flow mode); or PAPRTHiE (any powered, air-purifying respirator with a tight-fitting facepiece and a high-efficiency particulate filter); or SCBAF (any self-contained breathing apparatus with a full facepiece); or SAF (any supplied-air respirator with a full facepiece). *Up to 100 mg/m³:* SAF:PD,PP (any supplied-air respirator that has a full facepiece and is operated in a pressure-demand or other positive-pressure mode). *Emergency or planned entry into unknown concentrations or IDLH conditions:* SCBAF:PD,PP (any self-contained breathing apparatus that has a full facepiece and is operated in a pressure-demand or other positive-pressure mode); or SAF:PD,PP:ASCBA (any supplied-air respirator that has a full facepiece and is operated in a pressure-demand or other positive-pressure mode in combination with an auxiliary self-contained breathing apparatus operated in a pressure-demand or other positive pressure mode). *Escape:* HiEF (any air-purifying, full-facepiece respirator with a high-efficiency particulate filter); or SCBAE (any appropriate escape-type, self-contained breathing apparatus).

Storage: Prior to working with this chemical you should be trained on its proper handling and storage. Store in tightly closed containers in a cool, well-ventilated area away from incompatible materials listed above, moisture, and heat.

Shipping: Label required is "Corrosive." Copper chloride is in DOT/UN Hazard Class 8 and Packing Group III.

Spill Handling: Evacuate persons not wearing protective equipment from area of spill or leak until clean-up is complete. Remove all ignition sources. Collect powdered material in the most convenient and safe manner and deposit in sealed containers. Ventilate area after clean-up is complete. It may be necessary to contain and dispose of this chemical as a hazardous waste. If material or contaminated runoff enters waterways, notify downstream users of potentially contaminated waters. Contact your Department of Environmental Protection or your regional office of the federal EPA for specific recommendations. If employees are required to clean-up spills, they must be properly trained and equipped. OSHA 1910.120(q) may be applicable.

Fire Extinguishing: This chemical does not burn, but may ignite combustible materials. Use any extinguisher suitable for surrounding fires. Poisonous gases are produced in fire including hydrogen chloride gas. If material or contaminated runoff enters waterways, notify downstream users of potentially contaminated waters. Notify local health and fire officials and pollution control agencies. Containers may explode in fire. From a secure, explosion-proof location, use water spray to cool exposed containers. If cooling streams are ineffective (venting sound increases in volume and pitch, tank discolors, or shows any signs of deforming), withdraw immediately to a secure position. If employees are expected to fight fires, they must be trained and equipped in OSHA 1910.156.

References

New Jersey Department of Health and Senior Services, "Hazardous Substance Fact Sheet, "Copper Chloride," Trenton, NJ (February 1999)

Copper Cyanide

Molecular Formula: CCuN, C_2CuN_2

Common Formula: CuCN, $Cu(CN)_2$

Synonyms: *cupric cyanide:* Copper(II) cyanide; Copper cynanamide; Cyanure de cuivre (French) *cuprous cyanide:* Cianuro de cobre (Spanish); Copper(1+) cyanide; Copper(I) cyanide; Cupricin

CAS Registry Number: 544-92-3 (cuprous cyanide); 14763-77-0 (cupric cyanide)

RTECS Number: GL7150000 (cuprous cyanide); GL7175000 (cupric cyanide)

DOT ID: UN 1587 (cupric cyanide)

Regulatory Authority

- Air Pollutant Standard Set (ACGIH)[1] (DFG)[3] (HSE)[33] (former USSR)[43] (OSHA)[58]

As cuprous cyanide:

- CLEAN AIR ACT: Hazardous Air Pollutants (Title I, Part A, Section 112) as cyanide compounds
- CLEAN WATER ACT: Toxic Pollutant (Section 401.15)
- EPA HAZARDOUS WASTE NUMBER (RCRA No.): P029
- RCRA, 40CFR261, Appendix 8 Hazardous Constituents
- SUPERFUND/EPCRA 40CFR302.4 Reportable Quantity (RQ): CERCLA, 10 lb (4.54 kg)
- RCRA 40CFR264, Appendix 9; TSD Facilities Ground Water Monitoring List. Suggested test method(s) (PQL μg/l): 6010 (60); 7210 (200) Note: All species in the ground water that contain copper are included
- EPCRA Section 313 Form R *de minimis* concentration reporting level: 1.0%. (copper)
- EPCRA Section 313 Form R *de minimis* concentration reporting level: 1.0%. (cyanide)
- U.S. DOT Regulated Marine Pollutant (49CFR172.101, Appendix B)
- Canada, WHMIS, Ingredients Disclosure List; National Pollutant Release Inventory (NPRI); CEPA Priority Substance List, Ocean dumping prohibited, as copper compounds, n.o.s.; Drinking Water Quality = 0.2 mg (CN)/l MAC as cyanide compounds

Cited in U.S. State Regulations: California (A, G), Kansas (G), Louisiana (G), Maine (G), Massachusetts (G), New Hampshire (G), New Jersey (G), Oklahoma (G), Pennsylvania (G), Vermont (G), Virginia (G), Washington (G), Wisconsin (G).

Description: Cuprous cyanide, CuCN, is a white crystalline substance. Cupric cyanide, $Cu(CN)_2$ is a yellowish-green powder which decomposes on heating. Freezing/Melting point = 473°C (in nitrogen). Hazard Identification (based on NFPA-704 M Rating System): Health 4, Flammability 0, Reactivity 0. Insoluble in water.

Potential Exposure: Copper Cyanide is used in electroplating copper on iron, and as an insecticide and a catalyst.

Incompatibilities: Contact with heat forms deadly hydrogen cyanide gas. May form hydrogen cyanide with water. Keep away from acetylene gas and chemically active metals such as potassium, sodium, magnesium, and zinc, strong oxidizers (chlorine, fluorine, peroxides, etc.).

Permissible Exposure Limits in Air: The legal airborne permissible exposure limit (OSHA) PEL[58] and the ACGIH recommended TLV is 1 mg/m^3 as copper dust and mist. The NIOSH IDLH is 100 mg (Cu)/m^3.

The limit set by the former USSR-UNEP/IRPTC project[43] is 0.3 mg/m^3 as a MAC in workplace air and 0.009 mg/m^3 as a momentary value in ambient air of residential areas; the daily average MAC allowable in residential areas is 0.004 mg/m^3.

Determination in Air: Collection by a filter and bubbler followed by measurement with an ion-specific electrode. See NIOSH Method #7904.[18]

Permissible Concentration in Water: The permissible concentration for copper set by USEPA to protect human health is 1 mg/liter and for cyanide is 0.2 mg (CN) per liter.[6] The Canadian MAC is the same. The former USSR-UNEP/IRPTC project[43] has set a MAC of 0.1 mg/l in water bodies used for domestic purposes and 0.05 mg/l in water bodies used for fishery purposes.

Determination in Water: Cyanide may be determined titrimetrically by EPA Methods 335.2 and 9010 which give total cyanide.

Routes of Entry: Inhalation, ingestion.

Harmful Effects and Symptoms

Short Term Exposure: Copper Cyanide can affect you when breathed in. Eye contact can cause severe burns with loss of vision. Skin contact can cause irritation or burns. Breathing Copper Cyanide causes irritation of respiratory tract, and may cause nose bleeds or sores in the nose.

Long Term Exposure: Repeated exposure can cause copper to deposit in the liver and other body organs, causing damage, atrophy of the inner lining of the nose, with a watery discharge. Metallic taste may also occur. Repeated skin exposure can cause skin allergy and possibly a green discoloration of the skin and hair. May be able to affect the lungs.

Points of Attack: Skin, lungs, possibly other body organs.

Medical Surveillance: For those with frequent or potentially high exposure (half the TLV or greater), the following are recommended before beginning work and at regular times after that: Lung function tests. If symptoms develop or overexposure is suspected, the following may be useful. Urine copper test.

First Aid: If this chemical gets into the eyes, remove any contact lenses at once and irrigate immediately for at least 15 minutes, occasionally lifting upper and lower lids. Seek medical attention immediately. If this chemical contacts the skin, remove contaminated clothing and wash immediately with soap and water. Seek medical attention immediately. If this chemical has been inhaled, remove from exposure, begin rescue breathing (using universal precautions) if breathing has stopped and CPR if heart action has stopped. Transfer promptly to a medical facility. When this chemical has been swallowed, get medical attention. Give large quantities of water and induce vomiting. Do not make an unconscious person vomit.

Personal Protective Methods: Wear protective gloves and clothing to prevent any reasonable probability of skin contact. Safety equipment suppliers/manufacturers can provide recommendations on the most protective glove/clothing material for your operation. All protective clothing (suits, gloves, footwear, headgear) should be clean, available each day, and put on before work. Contact lenses should not be worn when working with this chemical. Wear dust-proof chemical goggles and face shield unless full facepiece respiratory protection is worn. Employees should wash immediately with soap when skin is wet or contaminated. Provide emergency showers and eyewash.

Respirator Selection: NIOSH/OSHA: (copper dust and mist) *5 mg/m³:* DM (any dust and mist respirator). *10 mg/m³:* DMXSQ, if not present as a fume (any dust and mist respirator except single-use and quarter mask respirators); or SA (any supplied-air respirator). *25 mg/m³:* SA:CF (any supplied-air respirator operated in a continuous-flow mode); PAPRDM (any powered, air-purifying respirator with a dust and mist filter). *50 mg/m³:* HiEF (any air-purifying, full-facepiece respirator with a high-efficiency particulate filter); or PAPRTHiE (any powered, air-purifying respirator with a tight-fitting facepiece and a high-efficiency particulate filter); or SCBAF (any self-contained breathing apparatus with a full facepiece); or SAF (any supplied-air respirator with a full facepiece). *100 mg/m³:* SAF:PD,PP (any supplied-air respirator that has a full facepiece and is operated in a pressure-demand or other positive-pressure mode). *Emergency or planned entry into unknown concentrations or IDLH conditions:* SCBAF:PD,PP (any self-contained breathing apparatus that has a full facepiece and is operated in a pressure-demand or other positive-pressure mode); or SAF: PD,PP:ASCBA (any supplied-air respirator that has a full facepiece and is operated in a pressure-demand or other positive-pressure mode in combination with an auxiliary self-contained breathing apparatus operated in a pressure-demand or other positive- pressure mode). *Escape:* HiEF (any air-purifying, full-facepiece respirator with a high-efficiency particulate filter); or SCBAE (any appropriate escape-type, self-contained breathing apparatus).

Note: Substance reported to cause eye irritation or damage; may require eye protection.

Storage: Prior to working with Copper Cyanide you should be trained on its proper handling and storage. Copper Cyanide must be stored to avoid contact with chemically active metals (such as potassium, sodium, magnesium and zinc) since violent reactions occur. Store in tightly closed containers in a cool, well-ventilated area away from acetylene gas.

Shipping: Copper cyanide is classified by DOT[19] in Hazard Class 6.1 and Packing Group II. It must carry a "Poison" label.

Spill Handling: Evacuate persons not wearing protective equipment from area of spill or leak until clean-up is complete. Remove all ignition sources. Collect powdered material in the most convenient and safe manner and deposit in sealed containers. Ventilate area after clean-up is complete. It may be necessary to contain and dispose of this chemical as a hazardous waste. If material or contaminated runoff enters waterways, notify downstream users of potentially contaminated waters. Contact your Department of Environmental Protection or your regional office of the federal EPA for specific recommendations. If employees are required to clean-up spills, they must be properly trained and equipped. OSHA 1910.120(q) may be applicable.

Fire Extinguishing: This chemical is a non combustible solid. Use dry chemical, carbon dioxide, water spray, or foam extinguishers. Poisonous gases are produced in fire including cyanide gas and nitrous oxides. If material or contaminated runoff enters waterways, notify downstream users of potentially contaminated waters. Notify local health and fire officials and pollution control agencies. From a secure, explosion-proof location, use water spray to cool exposed containers. If cooling streams are ineffective (venting sound increases in volume and pitch, tank discolors, or shows any signs of deforming), withdraw immediately to a secure position. If employees are expected to fight fires, they must be trained and equipped in OSHA 1910.156.

Disposal Method Suggested: Copper containing wastes can be concentrated to the point where copper can be electrolytically removed and reclaimed.

If recovery is not feasible, the copper can be precipitated by alkali, the cyanide destroyed by alkaline oxidation yielding a sludge which can be sent to a chemical waste landfill.[22] In accordance with 40CFR165 recommendations for the disposal of pesticides and pesticide containers. Must be disposed properly by following package label directions or by contacting your state pesticide or environmental control agency or by contacting your regional EPA office.

References

New Jersey Department of Health and Senior Services and Senior Services, "Hazardous Substance Fact Sheet: Copper Cyanide," Trenton, NJ (February 1987)

Copper Sulfate

Molecular Formula: CuO_4S

Common Formula: $CuSO_4$

Synonyms: Basicop; BCS copper fungicide; Blue copper; Blue stone; Blue vitriol; Copper monosulfate; Copper sulfate (1:1); Copper(2+) sulfate; Copper(2+) sulfate (1:1); Copper(II) sulfate; Copper sulfate pentahydrate; CP basic sulfate; Cupric sulfate anhydrous; Cupric sulphate; EINECS No. 231-847-6; Griffin super Cu; Kupfersulfat (German); Phyto-Bordeaux; Roman vitriol; Sulfate of copper; Sulfate de cuivre (French); Sulfato de cobre (Spanish); Sulfuric acid, copper(2+) salt (1:1); TNCS 53; Triangle

CAS Registry Number: 7758-98-7

RTECS Number: GL8800000

DOT ID: No citation.

Regulatory Authority

- Air Pollutant Standard Set (Czechoslovakia)[35] (former USSR)[43]

As copper compounds:

- CLEAN WATER ACT: Toxic Pollutant (Section 401.15)
- RCRA 40CFR264, Appendix 9; TSD Facilities Ground Water Monitoring List. Suggested test method(s) (PQL μg/l): 6010 (60); 7210 (200) Note: All species in the ground water that contain copper are included
- EPCRA (Section 313): Includes any unique chemical substance that contains copper as part of that chemical's infrastructure. This category does not include copper phthalocyanide compounds that are substituted with only hydrogen, and/or chlorine, and/or bromine. Form R *de minimis* concentration reporting level: 1.0%
- Canada, WHMIS, Ingredients Disclosure List; National Pollutant Release Inventory (NPRI); CEPA Priority Substance List, Ocean dumping prohibited

Cited in U.S. State Regulations: California (A, G), Maine (G), New Jersey (G), New York (G), Oklahoma (G), Pennsylvania (G).

Description: Copper sulfate, $CuSO_4$, is a greenish-white crystalline solid which decomposes slightly above 200°C and decomposes to CuO at 650°C. Hazard Identification (based on NFPA-704 M Rating System): Health 2, Flammability 0, Reactivity 0. Highly soluble in water; forms bright blue solution.

Potential Exposure: Copper sulfate is used to detect and to remove trace amounts of water from alcohols and organic compounds; as a fungicide and algicide, in veterinary medicine and others.

Incompatibilities: Aqueous solution is an acid. May form explosive materials on contact with acetylene and nitromethane. Incompatible with strong bases, hydroxylamine, magnesium; zirconium, sodium hpobromite, hydrazine.

Permissible Exposure Limits in Air: The Federal standard (OSHA PEL 8-hour TWA)[58] for copper fume is 0.1 mg/m^3, and 1 mg/m^3 for copper dusts and mists. NIOSH recommends the same level for a 10-hour workshift. ACGIH recommends a TWA of 0.2 mg/m^3 for copper fume and 1 mg/m^3 for dusts and mists.

The former USSR-UNEP/IRPTC project[43] has set limits in the ambient air of residential areas of 0.009 mg/m^3 on a momentary basis and 0.004 mg/m^3 on a daily average basis. The NIOSH IDLH is 100 mg (Cu)/m^3

Czechoslovakia[35] has set a MAC of 0.1 mg/m^3 on a daily average basis; a MAC of 0.3 mg/m^3 on a 30-minute basis.

Determination in Air: Copper dusts and mists are collected on a filter, worked up with acid, measured by atomic absorption. See NIOSH Method #7029 for copper. For copper fume: filter collection, acid digestion, measurement by atomic absorption.

Permissible Concentration in Water: The former USSR-UNEP/IRPTC joint project[43] has set a MAC in water used for fishery purposes of 0.004 mg/l (0.001 as Cu). The EPA[6] has set a maximum of 1.0 mg/l in water to protect human health.

Routes of Entry: Inhalation, ingestion.

Harmful Effects and Symptoms

Short Term Exposure: *Inhalation:* May cause irritation to nose, throat and lungs causing coughing and wheezing.

Skin: May cause irritation of skin, localized coloration, itching and burns.

Eyes: May cause severe irritation, inflammation, burns, excessive tissue fluid and a cloudy cornea; possible permanent damage.

Ingestion: Poisonous if swallowed. May cause burning and metallic taste in mouth, blue skin coloration, intense inflammation of the stomach and intestines, abdominal pain, vomiting, diarrhea, blood in feces, headache, cold sweat, weak pulse, salivation, nausea, dehydration, low blood pressure, jaundice, and kidney failure. Death may result from a dose of a little as 1 teaspoon for a 150 pound person.

Long Term Exposure: May cause mutations in humans. May damage the testes and decrease fertility in both males and females. May cause skin allergy and thickening of the skin; copper deposits can cause discoloration in the skin and hair, leaving a green color. Repeated exposure can cause shrinking of the lining of the inner nose with watery discharge, liver damage. Individuals with Wilson's disease absorb, retain, and store copper excessively.

Points of Attack: Skin, reproductive system, liver.

Medical Surveillance: Serum and urine copper level. Liver and kidney tests. Examination by a qualified allergist. More than light alcohol consumption may exacerbate the liver damage caused by Copper sulfate.

First Aid: If this chemical gets into the eyes, remove any contact lenses at once and irrigate immediately for at least 15 minutes, occasionally lifting upper and lower lids. Seek medical attention immediately. If this chemical contacts the skin, remove contaminated clothing and wash immediately with soap and water. Seek medical attention immediately. If this chemical has been inhaled, remove from exposure, begin rescue breathing (using universal precautions) if breathing has stopped and CPR if heart action has stopped. Transfer promptly to a medical facility. When this chemical has been swallowed, get medical attention. Give large quantities of water and induce vomiting. Do not make an unconscious person vomit.

Note to Physician: Empty stomach by lavage with 0.1% solution of potassium ferrocyanide or milk. Liver or kidney function tests may be indicated. May result in methaemoglobinemia.

Personal Protective Methods: Wear protective gloves and clothing to prevent any reasonable probability of skin contact.

Safety equipment suppliers/manufacturers can provide recommendations on the most protective glove/clothing material for your operation. Neoprene and Polyvinyl Chloride are among the recommended protective materials. All protective clothing (suits, gloves, footwear, headgear) should be clean, available each day, and put on before work. Contact lenses should not be worn when working with this chemical. When working with liquids, wear splash-proof chemical goggles and face shield unless full facepiece respiratory protection is worn. When working with powders or dusts, wear dust-proof chemical goggles and face shield unless full facepiece respiratory protection is worn. Employees should wash immediately with soap when skin is wet or contaminated. Provide emergency showers and eyewash.

Respirator Selection: NIOSH/OSHA: (copper dust and mist) *5 mg/m³:* DM (any dust and mist respirator). *10 mg/m³:* DMXSQ, if not present as a fume (any dust and mist respirator except single-use and quarter mask respirators); or SA (any supplied-air respirator). *25 mg/m³:* SA:CF (any supplied-air respirator operated in a continuous-flow mode); PAPRDM (any powered, air-purifying respirator with a dust and mist filter). *50 mg/m³:* HiEF (any air-purifying, full-facepiece respirator with a high-efficiency particulate filter); or PAPRTHiE (any powered, air-purifying respirator with a tight-fitting facepiece and a high-efficiency particulate filter); or SCBAF (any self-contained breathing apparatus with a full facepiece); or SAF (any supplied-air respirator with a full facepiece). *100 mg/m³:* SAF:PD,PP (any supplied-air respirator that has a full facepiece and is operated in a pressure-demand or other positive-pressure mode). *Emergency or planned entry into unknown concentrations or IDLH conditions:* PD,PP (any self-contained breathing apparatus that has a full facepiece and is operated in a pressure-demand or other positive-pressure mode); or SAF:PD,PP:ASCBA (any supplied-air respirator that has a full facepiece and is operated in a pressure-demand or other positive-pressure mode in combination with an auxiliary self-contained breathing apparatus operated in a pressure-demand or other positive- pressure mode). *Escape:* HiEF (any air-purifying, full-facepiece respirator with a high-efficiency particulate filter); or SCBAE (any appropriate escape-type, self-contained breathing apparatus).

Note: Substance reported to cause eye irritation or damage; may require eye protection.

Storage: Prior to working with this chemical you should be trained on its proper handling and storage. Store in tightly closed containers in a cool, well-ventilated area away from incompatible materials.

Shipping: Copper sulfite is not specifically cited in DOT's Performance-Oriented Packaging Standards.[19]

Spill Handling: Evacuate persons not wearing protective equipment from area of spill or leak until clean-up is complete. Remove all ignition sources. Collect powdered material in the most convenient and safe manner and deposit in sealed containers. Ventilate area after clean-up is complete. It may be necessary to contain and dispose of this chemical as a hazardous waste. If material or contaminated runoff enters waterways, notify downstream users of potentially contaminated waters. Contact your Department of Environmental Protection or your regional office of the federal EPA for specific recommendations. If employees are required to clean-up spills, they must be properly trained and equipped. OSHA 1910.120(q) may be applicable.

Fire Extinguishing: This chemical is a non combustible solid. Use dry chemical, carbon dioxide, water spray, or foam extinguishers. Poisonous gases are produced in fire including oxides of sulfur. If material or contaminated runoff enters waterways, notify downstream users of potentially contaminated waters. Notify local health and fire officials and pollution control agencies. Containers may explode in fire. From a secure, explosion-proof location, use water spray to cool exposed containers. If cooling streams are ineffective (venting sound increases in volume and pitch, tank discolors, or shows any signs of deforming), withdraw immediately to a secure position. If employees are expected to fight fires, they must be trained and equipped in OSHA 1910.156.

Disposal Method Suggested: Add soda ash to waste $CuSO_4$ solution; let stand 24 hours. Decant and neutralize solution before flushing to sewer. Landfill sludge.[22]

References

New York State Department of Health, "Chemical Fact Sheet: Copper Sulfate," Albany, NY, Bureau of Toxic Substance Assessment (January 1986 and Version 3)

New Jersey Department of Health and Senior Services and Senior Services, "Hazardous Substance Fact Sheet: Cupric Sulfate," Trenton, NJ (January 1999)

Cotton Dust

Molecular Formula: $C_{6n}H_{10n}O_{5n}$

Common Formula: $(C_6H_{10}O_5)_n$

Synonyms: Cotton fiber (raw)

CAS Registry Number: None

RTECS Number: GN2275000

DOT ID: No citation.

Regulatory Authority

- Air Pollutant Standard Set (ACGIH)[1] (Australia) (DFG)[3] (HSE)[33] (Israel) (Mexico) (OSHA)[58] (Other Countries)[35] (Several Canadian Provinces)
- OSHA, 29CFR1910 Specifically Regulated Chemicals (See CFR 1910.1043)

Cited in U.S. State Regulations: Alaska (G), California (A), Florida (G), Illinois (G), Maine (G), Massachusetts (G), Minnesota (G), New Hampshire (G), New Jersey (G), Pennsylvania (G), Rhode Island (G), West Virginia (G).

Description: Cotton dust, $(C_6H_{10}O_5)_n$, is defined as dust generated into the atmosphere as a result of the processing of cotton fibers combined with any naturally occurring materials such as stems, leaves, bracts, and inorganic matter which may have accumulated on the cotton fibers during the growing or harvesting period. Any dust generated from processing of cotton through the weaving of fabric in textile mills and dust generated in other operations or manufacturing processes using new or waste cotton fibers or cotton fiber by-products from textile mills is considered cotton dust.

Potential Exposure: The Occupational Safety and Health Administration has estimated that 800,000 workers are involved in work with cotton fibers and thus are potentially exposed to cotton dust in the work place.

Incompatibilities: Strong oxidizers. Dust forms an explosive mixture with air.

Permissible Exposure Limits in Air: The legal OSHA PEL 8-hour TWA for cotton dust [as found in OSHA Table Z-1 (29CFR1910.1000)]: for cotton dust (raw) is 1 mg/m^3; for the cotton waste processing operations of waste recycling (sorting, blending, cleaning, and willowing) and garnetting. PELs for other sectors (as found in 29CFR1910.1043) are 0.200 mg/m^3 for yarn manufacturing and cotton washing operations, 0.500 mg/m^3 for textile mill waste house operations or for dust from "lower grade washed cotton" used during yarn manufacturing, and 0.750 mg/m^3 for textile slashing and weaving operations. The OSHA standard in 29CFR1910.1043 does not apply to cotton harvesting, ginning, or the handling and processing of woven or knitted materials and washed cotton. All PELs for cotton dust are mean concentrations of lint-free, respirable cotton dust collected by the vertical elutriator or an equivalent method and averaged over an 8-hour period. NIOSH recommends reducing exposures to cotton dust to the lowest feasible concentration to reduce the prevalence and severity of byssinosis; the REL is <0.200 mg/m^3 (as lint-free cotton dust). The NIOSH IDLH level is 100 mg/m^3. California set a PEL of 1 mg/m^3, prevent eye and skin contact [for cotton waste processing operations of waste recycling (See also §5190 for respiratory specifications)].

The TWA set by HSE is 0.5 mg/m^3 (total dust less fly) and 10 mg/m^3 (total inhalable dust) and 5 mg/m^3 (respirable dust). The MAK set by DFG[3] is 1.5 mg/m^3. Australia, Israel, Mexico, and the Canadian provinces of Alberta, BC, Ontario, and Quebec set TWA limits of 0.2 mg/m^3 (dust collected by the vertical elutriator cotton-dust sampler) and Mexico, Alberta, and BC set a STEL of 0.6 mg/m^3. The values set[35] in Argentina and Czechoslovakia for workplace air are 0.2 mg/m^3 (with a STEL of 0.6 mg/m^3 in Argentina). The TWA set in Japan[35] is 1.0 mg/m^3 for inhalable dust and 4 mg/m^3 for total dust. The former Soviet Union has set MAC values in ambient air in residential areas of 0.5 mg/m^3 on a momentary basis and 0.05 mg/m^3 on a daily average basis.

Determination in Air: Vertical elutriator; none; Gravimetric; OSHA (1910.1043).

Permissible Concentration in Water: No criteria set.

Routes of Entry: Inhalation of dust, ingestion, eye and skin contact.

Harmful Effects and Symptoms

Short Term Exposure: Human pulmonary effects. Breathing raw Cotton Fiber can cause coughing, fever (resembling metal fume fever), chills, and may cause nausea and vomiting, when first exposed. Allergens or fungi in the cotton or dust can cause illness.

Long Term Exposure: Repeated exposures can cause serious permanent lung damage ("brown lung" or byssinosis). The first symptoms are chest tightness and trouble breathing that occur a few hours after first starting work for the week. At first the symptoms go away after a day or two, but with continued exposure they can become constant.

Points of Attack: Cardiovascular system, respiratory system.

Medical Surveillance: *a:* Preplacement: A comprehensive physical examination shall be made available to include as a minimum: medical history, baseline forced vital capacity (FVC), and forced vital expiratory volume at 1 second (FEV_1). The history shall include administration of a questionnaire designed to elicit information regarding symptoms of chronic bronchitis, byssinosis, and dyspnea. If a positive personal history of respiratory allergy, chronic obstructive lung disease, or other diseases of the cardiopulmonary system are elicited, or where there is a positive history of smoking, the applicant shall be counseled on his increased risk from occupational exposure to cotton dust. At the time off this examination, the advisability of the workers using negative or positive pressure respirators shall be evaluated.

b: Each newly employed person shall be retested for ventilatory capacity (FVC and FEV_1) within 6 weeks of employment. This retest shall be performed on the first day at work after at least 40 hours' absence from exposure to cotton dust and shall be performed before and after at least 6 hours of exposure on the same day.

c: Each current employee exposed to cotton dust shall be offered a medical examination at least yearly that shall include administration of a questionnaire designed to elicit information regarding symptoms of chronic bronchitis, byssinosis, and dyspnea.

d: Each current employee exposed to cotton dust shall have measurement of forced vital capacity (FVC) and of forced expiratory volume at 1 sec (FEV_2). These tests of ventilatory function should be performed on the first day of work following at least 40 hours of absence from exposure to cotton dust, and shall be performed before and after at least 6 hours of exposure on the same day.

e: Ideally, the judgment of the employee's pulmonary function should be based on preplacement values (values taken before exposure to cotton dust). When preplacement values are not available then reference to standard pulmonary function value tables may be necessary. Note that these tables may not reflect normal values for different ethnic groups. For example, the average healthily black male may have an approximately 15% lower FVC than a Caucasian male of the same body build. A physician shall consider, in cases of significantly decreased pulmonary function, the impact of further exposure to cotton dust and evaluate the relative merits of a transfer to areas of less exposure or protective measures. A suggested plan for the management of cotton workers was proposed as a result of a conference on cotton workers' health.

f: Medical records, including information on all required medical examinations, shall be maintained for persons employed in work involving exposure to cotton dust. Medical records with pertinent supporting documents shall be maintained at least 20 years after the individual's termination of employment. These records shall be available to the medical representatives of the Secretary of Health, Education, and Welfare; of the Secretary of Labor; of the employee or former employee; and of the employer.

First Aid: If this chemical gets into the eyes, remove any contact lenses at once and irrigate immediately for at least 15 minutes, occasionally lifting upper and lower lids. Seek medical attention immediately. If this chemical contacts the skin, remove contaminated clothing and wash immediately with soap and water. Seek medical attention immediately. If this chemical has been inhaled, remove from exposure, begin rescue breathing (using universal precautions) if breathing has stopped and CPR if heart action has stopped. Transfer promptly to a medical facility. When this chemical has been swallowed, get medical attention. Give large quantities of water and induce vomiting. Do not make an unconscious person vomit.

Personal Protective Methods: Engineering control shall be used wherever feasible to maintain cotton dust concentrations below the prescribed limit. Administrative controls can also be used to reduce exposure. Wear protective gloves and clothing to prevent any reasonable probability of skin contact. Safety equipment suppliers/manufacturers can provide recommendations on the most protective glove/clothing material for your operation. All protective clothing (suits, gloves, footwear, headgear) should be clean, available each day, and put on before work. Contact lenses should not be worn when working with this chemical. Wear dust-proof goggles and face shield unless full facepiece respiratory protection is worn. Respirators shall also be provided and used for nonroutine operations (occasional brief exposures above the environmental limit and for emergencies) and shall be considered for use by employees who have symptoms even when exposed to concentrations below the established environmental limit. Employees should wash immediately with soap when skin is wet or contaminated. Provide emergency showers and eyewash.

Respirator Selection: NIOSH: *Up to 1 mg/m³:* D (any dust respirator). *Up to 2 mg/m³:* DXSQ (any dust respirator except single-use and quarter-mask respirators); or HiE (any air-purifying respirator with a high-efficiency particulate filter); or SA (any supplied-air respirator). *Up to 5 mg/m³:* SA:CF (any supplied-air respirator operated in a continuous-flow mode); or PAPRD (any powered, air-purifying respirator with a dust filter). *Up to 10 mg/m³:* HiEF (any air-purifying, full-facepiece respirator with a high-efficiency particulate filter); or SAT:CF (any supplied-air respirator that has a tight-fitting facepiece and is operated in a continuous-flow mode); or PAPRTHiE (any powered, air-purifying respirator with a tight-fitting facepiece and a high-efficiency particulate filter); or SCBAF (any self-contained breathing apparatus with a full facepiece); or SAF (any supplied-air respirator with a full facepiece).

Up to 100 mg/m³: SA:PD,PP (any supplied-air respirator operated in a pressure-demand or other positive-pressure mode). *Emergency or planned entry into unknown concentrations or IDLH conditions:* SCBAF:PD,PP (any self-contained breathing apparatus that has a full facepiece and is operated in a pressure-demand or other positive-pressure mode); or SAF:PD,PP:ASCBA (any supplied-air respirator that has a full facepiece and is operated in a pressure-demand or other positive-pressure mode in combination with an auxiliary self-contained breathing apparatus operated in a pressure-demand or other positive pressure mode). *Escape:* HiEF (any air-purifying, full-facepiece respirator with a high-efficiency particulate filter); or SCBAE (any appropriate escape-type, self-contained breathing apparatus).

Storage: Prior to working with this chemical you should be trained on its proper handling and storage. Cotton Fiber must be stored to avoid contact with strong oxidizers, (such as chlorine, bromine, and fluorine) since violent reactions occur. Store in tightly closed containers in a cool, well-ventilated area away from heat or flame.

Shipping: There is no specific DOT listing for cotton dust or fiber. Wet cotton and waste oily cotton fall in Hazard Class 4.2 and Packing Group III. They require a "Spontaneously Combustible" label and shipment by passenger aircraft or railcar or even by cargo aircraft is forbidden.

Spill Handling: Evacuate persons not wearing protective equipment from area of spill or leak until clean-up is complete. Remove all ignition sources. Collect cotton dust in the most convenient and safe manner and deposit in sealed containers. Ventilate area after clean-up is complete. It may be necessary to contain and dispose of this chemical as a hazardous waste. If material or contaminated runoff enters waterways, notify downstream users of potentially contaminated waters. Contact your Department of Environmental Protection or your regional office of the federal EPA for specific recommendations. If employees are required to clean-up spills, they must be properly trained and equipped. OSHA 1910.120(q) may be applicable.

Fire Extinguishing: Cotton fiber and dust is flammable. Use water only. Do not use chemical or carbon dioxide extinguishers. If employees are expected to fight fires, they must be trained and equipped in OSHA 1910.156.

Disposal Method Suggested: Cotton dust may be used as a cheap filling in quilt blankets. It contains a large organic fraction which can be digested anaerobically to produce manure plus biogas fuel.[22]

References

National Institute for Occupational Safety and Health, Criteria for a Recommended Standard: Occupational Exposure to Cotton Dust, NIOSH Document No. 75–118 (1975)

New Jersey Department of Health and Senior Services and Senior Services, "Hazardous Substance Fact Sheet: Cotton Fiber (Raw)," Trenton, NJ (January 1986)

Coumafuryl

Molecular Formula: $C_{17}H_{14}O_5$

Synonyms: 3-(α-Acetonylfurfuryl)-4-hydroxycoumarin; Cumafuryl (German); Foumarin®; Fumarin®; 3-(α-Furyl-b-acetylaethyl)-4-hydroxycumarin (German); 3-(1-Furyl-3-acetylethyl)-4-hydroxycoumarin; Krumkil; Ratafin®; Rat-A-Way®; Tomarin®

CAS Registry Number: 117-52-2

RTECS Number: GN4850000

DOT ID: UN 3027 (coumarin derivitive pesticide, solid poisonous)

Regulatory Authority

- Extremely Hazardous Substance (EPA-SARA) (TPQ = 10,000)[7] (Dropped from listing in 1988)

Cited in U.S. State Regulations: Massachusetts (G), New Jersey (G), Oklahoma (G).

Description: Coumafuryl, $C_{17}H_{14}O_2$, is a colorless, white, crystalline solid. Freezing/Melting point = 124°C. Insoluble in water.

Potential Exposure: This material is an anticoagulant rodenticide. Therefore, those involved in its manufacture, formulation and application are at risk.

Incompatibilities: Strong oxidizers may cause a fire and explosive hazard.

Permissible Exposure Limits in Air: No standards set.

Permissible Concentration in Water: No criteria set.

Routes of Entry: Ingestion, skin contact.

Harmful Effects and Symptoms

Short Term Exposure: Coumafuryl is very similar to warfarin, a hemorrhagic agent. Inhalation may cause symptoms described in long term exposure. With a single large ingested dose may cause hemorrhagic shock. The LD_{50} oral (mouse) is 14.7 mg/kg (highly toxic). This chemical can be absorbed through the skin, thereby increasing exposure or hemorrhagic effect. High exposure can cause death.

Long Term Exposure: Chronic exposure may cause death by hemorrhagic shock. Absorption by the lungs or after a few days or few weeks of repeated ingestion, may cause inhibition of prothrombin synthesis, nose bleeds and bleeding gums, hematoma, small reddish spots like a rash, bruises of the elbows, knees and buttocks, blood in urine and stools, anemia, occasional paralysis due to a stroke.

First Aid: If this chemical gets into the eyes, remove any contact lenses at once and irrigate immediately for at least 30 minutes, occasionally lifting upper and lower lids. Seek medical attention immediately. If this chemical contacts the skin, remove contaminated clothing and wash immediately with soap and water. Seek medical attention immediately. If this chemical has been inhaled, remove from exposure, begin rescue breathing (using universal precautions) if breathing has stopped and CPR if heart action has stopped. Transfer promptly to a medical facility. When this chemical has been swallowed, get medical attention. Give large quantities of water and induce vomiting. Do not make an unconscious person vomit. Speed in removing material from skin is of extreme importance. Remove and isolate contaminated clothing and shoes at the site. Keep victim quiet and maintain normal body temperature. Effects may be delayed. Keep victim under observation.

Personal Protective Methods: Wear protective gloves and clothing to prevent any reasonable probability of skin contact. Safety equipment suppliers/manufacturers can provide recommendations on the most protective glove/clothing material for your operation. All protective clothing (suits, gloves, footwear, headgear) should be clean, available each day, and put on before work. Contact lenses should not be worn when working with this chemical. Wear dust-proof chemical goggles and face shield unless full facepiece respiratory protection is worn. Employees should wash immediately with soap when skin is wet or contaminated. Provide emergency showers and eyewash.

Respirator Selection: *At any detectable concentration:* SCBAF:PD,PP (any MSHA/NIOSH approved self-contained breathing apparatus that has a full facepiece and is operated in a pressure-demand or other positive-pressure mode); or SAF:PD,PP:ASCBA (any supplied-air respirator that has a full facepiece and is operated in a pressure-demand or other positive-pressure mode in combination with an auxiliary, self-contained breathing apparatus operated in a pressure-demand or other positive pressure mode). *Escape:* HiEF [any air-purifying, full-facepiece respirator (gas mask) with a chin-style, front- or back-mounted organic vapor canister having a high-efficiency particulate filter]; or SCBAE (any appropriate escape-type, self-contained breathing apparatus).

Storage: Prior to working with this chemical you should be trained on its proper handling and storage. Store in tightly closed containers in a cool, well-ventilated area away from strong oxidizers.

Shipping: Coumarin derivative pesticide, solid, toxic, n.o.s., in Packing Group I require a "Poison" label. This material falls in the UN/DOT Hazard Class 6.1.[19][20]

Spill Handling: Evacuate persons not wearing protective equipment from area of spill or leak until clean-up is complete. Remove all ignition sources. Do not touch spilled material; stop leak if you can do so without risk. Use water spray to reduce vapors. For small spills, absorb with sand or other noncombustible absorbent material and place into containers for later disposal. For small dry spills, with clean shovel place material into clean, dry container and cover; move containers from spill area. For large spills, dike far ahead of spill for later disposal. Ventilate area after clean-up is complete. It may be necessary to contain and dispose of this chemical as a hazardous waste. If material or contaminated runoff enters waterways, notify downstream users of potentially contaminated waters. Contact your Department of Environmental Protection or your regional office of the federal EPA for specific recommendations. If employees are required to clean-up spills, they must be properly trained and equipped. OSHA 1910.120(q) may be applicable.

Fire Extinguishing: This chemical is a combustible solid. Stay upwind; keep out of low areas. Ventilate closed spaces before entering them. Small fires: dry chemical, carbon dioxide, water spray or foam. Large fires: water spray, fog or foam. Move container from fire area if you can do so without risk. Wear positive pressure breathing apparatus and special protective clothing. Remove an isolate contaminated clothing at the site. Fight fire from maximum distance. Dike fire control water for later disposal; do not scatter the material. Poisonous gases are produced in fire. If material or contaminated runoff enters waterways, notify downstream users of potentially contaminated waters. Notify local health and fire officials and pollution control agencies. Containers may explode in fire. From a secure, explosion-proof location, use water spray to cool exposed containers. If cooling streams are ineffective (venting sound increases in volume and pitch, tank discolors, or shows any signs of deforming), withdraw immediately to a secure position. If employees are expected to fight fires, they must be trained and equipped in OSHA 1910.156.

References

U.S. Environmental Protection Agency, "Chemical Profile: Coumafuryl," Washington, DC, Chemical Emergency Preparedness Program (October 31, 1985)

Coumaphos

Molecular Formula: $C_{14}H_{16}ClO_5PS$

Synonyms: Asuntol®; Azunthol; Bay 21/199; Bayer 21/199; Baymix; Baymix 50; 3-Chloro-7-hydroxy-4-methyl-coumarin *O,O*-diethyl phosphorothioate; 3-Chloro-7-hydroxy-4-methyl-coumarin *O,O*-diethyl phosphorothionate; 3-Chloro-7-hydroxy-4-methyl-coumarin *O*-ester with *O,O*-diethyl phosphorothioate; *O*-3-Chloro-4-methyl-7-coumarinyl *O,O*-diethyl phosphorothioate; 3-Chloro-4-methyl-7-coumarinyldiethyl phosphorothioate; 3-Chloro-4-methyl-7-hydroxycoumarindiethyl thiophosphoric acid ester; 3-Chloro-4-methylumbelliferone *O*-ester with *O,O*-diethyl phosphorothioate; Co-Ral®; Coumafos; Cumafos (Dutch, Spanish); *O,O*-Diaethyl-*O*-(3-chlor-4-methyl-cumarin-7-yl)-monothiophosphat (German); *o,o*-Diethyl-*o*-(3-chloor-4-methyl-cumarin-7-yl)monothiofosfaat (Dutch); *O,O*-Diethyl 3-chloro-4-methyl-7-umbelliferone thiophosphate; Diethyl3-chloro-4-methylumbelliferyl thionophosphate; *O,O*-Diethyl *O*-(3-chloro-4-methyl-7-coumarinyl) phosphorothioate; *O,O*-Diethylo (3-chloro-4-methylcoumarinyl-7) thiophosphate; *O,O*-Diethyl *O*-(3-chloro-4-methyl-2-oxo-2H-benzopyran-7-yl) phosphorothioate; *O,O*-Diethyl *O*-(3-chloro-4-methylumbelliferyl) phosphorothioate; Diethylthiophosphoric acid ester of 3-chloro-4-methyl-7-hydroxycoumarin; *o,o*-Dietil-*o*-(3-cloro-4-metil-cumarin-7-il-monotiofosfato) (Italian); Diolice; ENT 17,957; Meldane®; Meldone; Muscatox®; NCI-C08662; Negashunt®; Phosphorothioic acid, *O*-(3-chloro-4-methyl-2-oxo-2H-1-benzopyran-7-yl) *O,O*-diethyl ester; Phosphorothioic acid, *O,O*-diethyl ester, *O*-ester with 3-chloro-7-hydroxy-4-methylcoumarin; Resitox®; Suntol®; Thiophosphate de *O,O*-diethyle et de *O*-(3-chloro-4-methyl-7-coumarinyle) (French); Umbethion

CAS Registry Number: 56-72-4

RTECS Number: GN6300000

DOT ID: UN 2783

EEC Number: 015-038-00-3

Regulatory Authority

- CLEAN WATER ACT: Section 311 Hazardous Substances/RQ 40CFR117.3 (same as CERCLA, see below)
- SUPERFUND/EPCRA 40CFR355, Appendix B Extremely Hazardous Substances: TPQ = 100/10,000 lb (455/4,540 kg)
- SUPERFUND/EPCRA 40CFR302.4 Reportable Quantity (RQ): CERCLA, 10 lb (4.54 kg)
- U.S. DOT Regulated Marine Pollutant (49CFR172.101, Appendix B), severe pollutant
- Water Pollution Standard Proposed (Mexico)[35]

Cited in U.S. State Regulations: California (G), Florida (G), Illinois (G), Massachusetts (G), Michigan (G), New Hampshire (G), New Jersey (G), Pennsylvania (G).

Description: Coumaphos, $C_{14}H_{16}ClO_5PS$, is a white to brownish crystalline solid with a slight sulfurous odor. Freezing/Melting point = 91°C. Odor threshold = 0.02 ppm. Hazard Identification (based on NFPA-704 M Rating System): Health 3, Flammability 1, Reactivity 0. Insoluble in water.

Potential Exposure: Those involved in the manufacture, formulation and application of this material which is used for control of a wide variety of liver stock insects including cattle grubs, lice scabies, flies, and ticks; the common ectoparasites of sheep, goats, horse, swine, and poultry as well as for screw worms in all these animals.

Incompatibilities: Piperonyl butoxide, oxidizers, strong bases. Slowly reacts with caustics to be hydrolyzed. Keep away from water and heat.

Permissible Exposure Limits in Air: No standards set.

Determination in Air: OSHA versatile sampler-2; Toluene/Acetone; Gas chromatography/Flame photometric detection for sulfur, nitrogen, or phosphorus; NIOSH Method IV Method #5600, Organophosphorus Pesticides.

Permissible Concentration in Water: Mexico has set a maximum permissible concentration in estuaries of 0.02 mg/l.

Routes of Entry: Inhalation, ingestion, skin contact. This chemical can be absorbed through the skin, thereby increasing exposure.

Harmful Effects and Symptoms

Short Term Exposure: Contact may cause burns to skin and eyes. Fatal skin absorption can occur even if there is no feeling of irritation after contact. Cholinesterase inhibitor. Exposure can cause rapid, fatal organophosphate poisoning: with headache, sweating, nausea and vomiting, diarrhea, salivation, abdominal cramps, difficult breathing, stiffness of legs, blurring of vision, followed by loss of muscle coordination, muscle twitching, convulsions, coma, and death. The LD_{50} oral-rat is 13 mg/kg (highly toxic). The probable oral lethal dose is 50 – 500 mg/kg; or between 1 teaspoonful and 1 oz. For a 70 kg (150 lb) person. May be fatal if inhaled, swallowed, or absorbed through skin. The effects may be delayed.

Long Term Exposure: High or repeated exposure may cause nerve damage causing weakness, poor coordination in the arms and legs. May cause personality changes, depression, memory loss, or irritability. Cholinesterase inhibitor; cumulative effect is possible. This chemical may damage the nervous system with repeated exposure, resulting in convulsions, respiratory failure. May cause liver damage.

Points of Attack: Respiratory system, central nervous system, peripheral nervous system, plasma cholinesterase.

Medical Surveillance: Before employment and at regular times after that, the following are recommended: Plasma and red blood cell cholinesterase levels (tests for the enzyme poisoned by this chemical). If exposure stops, plasma levels return to normal in 1 – 2 weeks while red blood cell levels may be reduced for 1 – 3 months.

When cholinesterase enzyme levels are reduced by 25% or more below preemployment levels, risk of poisoning is increased, even if results are in lower ranges of "normal." Reassignment to work not involving organophosphate or carbamate pesticides is recommended until enzyme levels recover. If symptoms develop or overexposure occurs, repeat the above tests as soon as possible and get an exam of the nervous system. Also consider complete blood count. Consider chest x-ray following acute overexposure. Do not drink any alcoholic beverages before or during use. Alcohol promotes absorption of organic phosphates.

First Aid: If this chemical gets into the eyes, remove any contact lenses at once and irrigate immediately for at least 15 minutes, occasionally lifting upper and lower lids. Seek medical attention immediately. If this chemical contacts the skin, remove contaminated clothing and wash immediately with soap and water, followed by alcohol. Speed in removing material from skin is of extreme importance. Shampoo hair promptly if contaminated. Seek medical attention immediately. If this chemical has been inhaled, remove from exposure, begin rescue breathing (using universal precautions) if breathing has stopped and CPR if heart action has stopped. Transfer promptly to a medical facility. When this chemical has been swallowed, get medical attention. Give large quantities of water and induce vomiting. Do not make an unconscious person vomit. Effects may be delayed. Keep victim under observation.

Personal Protective Methods: Wear protective gloves and clothing to prevent any reasonable probability of skin contact. Safety equipment suppliers/manufacturers can provide recommendations on the most protective glove/clothing material for your operation. All protective clothing (suits, gloves, footwear, headgear) should be clean, available each day, and put on before work. Contact lenses should not be worn when working with this chemical. Wear dust-proof chemical goggles and face shield unless full facepiece respiratory protection is worn. Employees should wash immediately with soap when skin is wet or contaminated. Provide emergency showers and eyewash. In case of poisoning with this substance specific treatment is required; the appropriate means, including instructions, should be available.

Respirator Selection: Where the potential exists for exposure to Coumaphos, use a MSHA/NIOSH approved supplied-air respirator with a full facepiece operated in the positive pressure mode or with a full facepiece, hood, or helmet in the continuous flow mode, or use a MSHA/NIOSH approved self-contained breathing apparatus with a full facepiece operated in pressure-demand or other positive pressure mode.

Storage: Store in tightly closed containers in a cool, well-ventilated area.

Shipping: The DOT label required for Coumarin derivative pesticides, solid, n.o.s., is "Poison." The UN/DOT Hazard Class is 6.1 and the Packing Group is I. Regulated by US DOT as a severe marine pollutant.

Spill Handling: Do not touch spilled material; stop leak if you can do it without risk. Evacuate persons not wearing protective equipment from area of spill or leak until clean-up is complete. Remove all ignition sources. Use water spray to reduce vapors. *Small spills:* take up with sand or other noncombustible absorbent material and place into containers for later disposal. *Small dry spills:* with clean shove, place material into clean, dry container and cover; move containers from spill area. *Large spills:* dike far ahead of spill for later disposal. Ventilate area after clean-up is complete. It may be necessary to contain and dispose of this chemical as a hazardous waste. If material or contaminated runoff enters waterways, notify downstream users of potentially contaminated waters. Contact your Department of Environmental Protection or your regional office of the federal EPA for specific recommendations. If employees are required to clean-up spills, they must be properly trained and equipped. OSHA 1910.120(q) may be applicable.

Fire Extinguishing: This material may burn but does not ignite easily. Extinguish with water, foam, carbon dioxide, or dry chemicals. This chemical is a combustible solid. Use dry chemical, carbon dioxide, water spray, or alcohol foam extinguishers. Poisonous gases are produced in fire including chlorides, phosphorous oxides, sulfur oxides. If material or contaminated runoff enters waterways, notify downstream users of potentially contaminated waters. Notify local health and fire officials and pollution control agencies. From a secure, explosion-proof location, use water spray to cool exposed containers. If cooling streams are ineffective (venting sound increases in volume and pitch, tank discolors, or shows any signs of deforming), withdraw immediately to a secure position. If employees are expected to fight fires, they must be trained and equipped in OSHA 1910.156.

Disposal Method Suggested: Coumaphos can be decomposed by heating with concentrated alkali. Large amounts should be incinerated in a unit equipped with effluent gas scrubbing.[22]

References

U.S. Environmental Protection Agency, "Chemical Profile: Coumaphos," Washington, DC, Chemical Emergency Preparedness Program (October 31, 1985)

New Jersey Department of Health and Senior Services and Senior Services, "Hazardous Substance Fact Sheet: Coumaphos," Trenton, NJ (December 1998)

Sax, N. I., Ed., "Dangerous Properties of Industrial Materials Report," 4, No. 1, 53–56 (1984) and 9, No. 1, 19–29 (1989)

Coumatetralyl

Molecular Formula: $C_{19}H_{16}O_3$

Synonyms: Bay 25634; Bay ENE 11183B; Bayer 25,634; 2H-1-Benzopyran-2-one, 4-Hydroxy-3-(1,2,3,4-tetrahydro-1-naphthalenyl)-; Coumarin, 4-hydroxy-3-(1,2,3,4-tetrahydro-1-naphthyl)-; Cumatetralyl (German, Dutch); Endox; Endrocid; Endrocide; ENE 11183; 4-Hydroxy-3-(1,2,3,4-tetrahydro-1-naftyl)-4-cumarine (Dutch); 4-Hydroxy-3-(1,2,3,4-tetrahydro-1-napthalenyl)-2H-1-benzopyran-2-one; 4-Hydroxy-3-(1,2,3,4-tetrahydro-1-napthyl)cumarin; Racumin; Raucumin 57; Rodentin; 3-(α-Tetral)-4-oxycoumarin; 3-(α-Tetrayl)-4-hydroxycoumarin; 3-(D-Tetrayl)-4-hydroxycoumarin

CAS Registry Number: 5836-29-3

RTECS Number: GN7630000

DOT ID: UN 3027 (coumarin derivative pesticide, solid, poisonous)

Regulatory Authority

- SUPERFUND/EPCRA 40CFR355, Appendix B Extremely Hazardous Substances: TPQ = 500/10,000 lb (227/4,540 kg)
- SUPERFUND/EPCRA 40CFR302.4 Reportable Quantity (RQ): EHS, 1 lb (0.454 kg)

Cited in U.S. State Regulations: Massachusetts (G), New Jersey (G), Oklahoma (G), Pennsylvania (G).

Description: Coumatetralyl, $C_{19}H_{16}O_3$, is a yellowish-white. crystalline solid. Boiling point = 290°C. Freezing/Melting point = 69 – 70°C. Hazard Identification (based on NFPA-704 M Rating System): Health 4, Flammability 1, Reactivity 0. Soluble in hot water.

Potential Exposure: Coumatetralyl is used as a rodenticide, functioning as an anticoagulant that does not induce bait-shyness.

Incompatibilities: Oxidizers may cause fire and explosion hazard. Keep away from metals.

Permissible Exposure Limits in Air: No standards set.

Permissible Concentration in Water: No criteria set.

Routes of Entry: Inhalation, ingestion, skin contact.

Harmful Effects and Symptoms

Short Term Exposure: Inhalation may cause symptoms described in long term exposure. With a single large ingested dose may cause hemorrhagic shock. This chemical can be absorbed through the skin, thereby increasing exposure or hemorrhagic effect. High exposure can cause death. The LD_{50} oral (rat) is 16.5 mg/kg (highly toxic).

Long Term Exposure: Chronic exposure may cause death by hemorrhagic shock. Absorption by the lungs or after a few days or few weeks of repeated ingestion, may cause inhibition of prothrombin synthesis, nose bleeds and bleeding gums, hematoma, small reddish spots like a rash, bruises of the elbows, knees and buttocks, blood in urine and stools, anemia, occasional paralysis due to a stroke. Pre-existing blood clotting disease or liver disease are aggravated by Coumatetralyl exposure.

First Aid: If this chemical gets into the eyes, remove any contact lenses at once and irrigate immediately for at least 15 minutes, occasionally lifting upper and lower lids. Seek medical attention immediately. If this chemical contacts the skin, remove contaminated clothing and wash immediately with soap and water. Speed in removing material from skin is of extreme importance. Shampoo hair promptly if contaminated. Seek medical attention immediately. If this chemical has been inhaled, remove from exposure, begin rescue breathing (using universal precautions) if breathing has stopped and CPR if heart action has stopped. Transfer promptly to a medical facility. When this chemical has been swallowed, get medical attention. Give large quantities of water and induce vomiting. Do not make an unconscious person vomit. Medical observation is recommended for 24 – 36 hours following overexposure, as effects may be delayed.

Personal Protective Methods: Wear protective gloves and clothing to prevent any reasonable probability of skin contact. Safety equipment suppliers/manufacturers can provide recommendations on the most protective glove/clothing material for your operation. All protective clothing (suits, gloves, footwear, headgear) should be clean, available each day, and put on before work. Contact lenses should not be worn when working with this chemical. Wear splash or dust-proof chemical goggles and face shield unless full facepiece respiratory protection is worn. Employees should wash immediately with soap when skin is wet or contaminated. Provide emergency showers and eyewash.

Respirator Selection: *At any detectable concentration:* SCBAF:PD,PP (any MSHA/NIOSH approved self-contained breathing apparatus that has a full facepiece and is operated in a pressure-demand or other positive-pressure mode); or SAF:PD,PP:ASCBA (any supplied-air respirator that has a full facepiece and is operated in a pressure-demand or other positive-pressure mode in combination with an auxiliary, self-contained breathing apparatus operated in a pressure-demand or other positive pressure mode). *Escape:* HiEF [any air-purifying, full-facepiece respirator (gas mask) with a chin-style, front- or back-mounted organic vapor canister having a high-efficiency particulate filter]; or SCBAE (any appropriate escape-type, self-contained breathing apparatus).

Storage: Prior to working with this chemical you should be trained on its proper handling and storage. Store in tightly closed containers in a cool, well-ventilated area away from oxidizers.

Shipping: Coumarin derivative pesticides, solid, toxic, n.o.s., in Packing Group II require a "Poison" label. The UN/DOT Hazard Class is 6.1.[19][20]

Spill Handling: Evacuate persons not wearing protective equipment from area of spill or leak until clean-up is complete. Remove all ignition sources. Stay upwind; keep out of low areas. Ventilate closed spaces before entering them. Wear positive pressure breathing apparatus and special protective clothing. Remove and isolate contaminated clothing at the site. Do not touch spilled material. Use water spray to reduce vapors. With clean shovel place material into clean, dry container and cover. Dike far ahead of large spills for later disposal. Collect powdered material in the most convenient and safe manner and deposit in sealed containers. Ventilate area after clean-up is complete. It may be necessary to contain and dispose of this chemical as a hazardous waste. If material or contaminated runoff enters waterways, notify downstream users of potentially contaminated waters. Contact your Department of Environmental Protection or your regional office of the federal EPA for specific recommendations. If employees are required to clean-up spills, they must be properly trained and equipped. OSHA 1910.120(q) may be applicable.

Fire Extinguishing: This chemical is a noncombustible solid. Small fires: dry chemical, carbon dioxide, water spray or foam. Large fires: water spray, fog or foam. Stay upwind, keep out of low areas. Wear positive pressure breathing apparatus and special protective clothing. Fight fire from maximum distance. Dike fire control water for later disposal; do not scatter the material. Poisonous gases are produced in fire including oxides of carbon. If material or contaminated runoff enters waterways, notify downstream users of potentially contaminated waters. Notify local health and fire officials and pollution control agencies. From a secure, explosion-proof location, use water spray to cool exposed containers. If cooling streams are ineffective (venting sound increases in volume and pitch, tank discolors, or shows any signs of deforming), withdraw immediately to a secure position. If employees are expected to fight fires, they must be trained and equipped in OSHA 1910.156.

Disposal Method Suggested: Dissolve in a solvent and burn in a furnace by spraying in the solution.[22]

References

U.S. Environmental Protection Agency, "Chemical Profile: Coumatetralyl," Washington, DC, Chemical Emergency Preparedness Program (November 30, 1987)

p-Cresidine

Molecular Formula: $C_8H_{11}NO$

Common Formula: $C_6H_3(NH_2)(OCH_3)(CH_3)$

Synonyms: *m*-Amino-*p*-cresol, methyl ester; 3-Amino-*p*-cresol methyl ester; 1-Amino-2-methoxy-5-methylbenzene; 3-Amino-4-methoxytoluene; 2-Amino-4-methylanisole; *o*-Anisidine,5-methyl-; Azoic red 36; Benzeneamine, 2-methoxy-5-methyl-; C.I. azoic red 83; *p*-Cresidina (Spanish); Cresidine; Kresidine; Krezidin (German); Krezidine; 2-Methoxy-5-methylaniline; 2-Methoxy-5-methylbenzenamine (9CI); 4-Methoxy-*m*-toluidine; 4-Methyl-2-aminoanisole; 5-Methyl-*o*-anisidine; NCI-C02982; Paracresol

CAS Registry Number: 120-71-8

RTECS Number: BZ6720000

Regulatory Authority

- Carcinogen (Animal Positive) (DFG) (NCI) (NTP)[10] (Human Suspected) (IARC)[9]
- Air Pollutant Standard Set (North Dakota)[60]
- EPCRA Section 313 Form R *de minimis* concentration reporting level: 0.1%
- Canada, WHMIS, Ingredients Disclosure List

Cited in U.S. State Regulations: California (G), Florida (G), Illinois (G), Maine (G), Maryland (G), Massachusetts (G),

Michigan (G), Minnesota (G), New Jersey (G), North Dakota (A), Pennsylvania (G), West Virginia (G).

Description: p-Cresidine, $C_6H_3(NH_2)(OCH_3)(CH_3)$, is a white crystalline solid. Boiling point = 235°C. Freezing/Melting point = 52°C.

Slight solubility in water. Flash point ≥ 112°C. Hazard Identification (based on NFPA-704 M Rating System): Health 1, Flammability 2, Reactivity 0. Insoluble in water.

Potential Exposure: p-Cresidine appears to be used solely as an intermediate in the production of various azo dyes and pigments, including 11 dyes that are produced commercially in the United States.

Human exposure to p-Cresidine occurs primarily through inhalation of vapors or skin absorption of the liquid. Exposure to p-Cresidine is believed to be limited to workers in dye-production facilities. The Consumer Product Safety Commission staff believes it is possible that residual levels or trace impurities of p-Cresidine may be present in some dyes based on this chemical, and it may be present in the final consumer product. The presence of p-cresidine, even as a trace contaminant, may be cause for concern. However, data describing the actual levels of impurities in the final product and the potential for consumer exposure and uptake are currently lacking.

Incompatibilities: Contact with oxidizers may cause fire and explosions.

Permissible Exposure Limits in Air: Zero in ambient air set by North Dakota.[60]

Permissible Concentration in Water: No criteria set.

Harmful Effects and Symptoms

Short Term Exposure: Symptoms of exposure include skin and eye irritation, nausea, vomiting, liver damage, cyanosis, methemoglobinemia, weakness, drowsiness, somnolence and loss of consciousness. The oral LD_{50} rat is 1,450 mg/kg (slightly toxic).

Long Term Exposure: May cause cancer in humans. When administered in the diet, p-Cresidine was carcinogenic to rats, causing increased incidences of carcinomas and papillomas off the urinary bladder in both sexes, increased incidences of olfactory neuroblastomas in both sexes, increased incidences of olfactory neuroblastomas in both sexes, and of liver tumors in males. p-Cresidine was also carcinogenic in mice, causing carcinomas of the urinary bladders in both sexes and hepatocellular carcinomas in females. May cause anemia.

Points of Attack: Blood.

Medical Surveillance: Blood hemoglobin level. Complete blood count (CBC).

First Aid: If this chemical gets into the eyes, remove any contact lenses at once and irrigate immediately for at least 15 minutes, occasionally lifting upper and lower lids. Seek medical attention immediately. If this chemical contacts the skin, remove contaminated clothing and wash immediately with soap and water. Seek medical attention immediately. If this chemical has been inhaled, remove from exposure, begin rescue breathing (using universal precautions) if breathing has stopped and CPR if heart action has stopped. Transfer promptly to a medical facility. When this chemical has been swallowed, get medical attention. Give large quantities of water and induce vomiting. Do not make an unconscious person vomit.

Note to Physician: Treat for methemoglobinemia. Spectrophotometry may be required for precise determination of levels of methemoglobinemia in urine.

Personal Protective Methods: Wear protective gloves and clothing to prevent any reasonable probability of skin contact. Safety equipment suppliers/manufacturers can provide recommendations on the most protective glove/clothing material for your operation. All protective clothing (suits, gloves, footwear, headgear) should be clean, available each day, and put on before work. Contact lenses should not be worn when working with this chemical. Wear dust-proof chemical goggles and face shield unless full facepiece respiratory protection is worn. Employees should wash immediately with soap when skin is wet or contaminated. Provide emergency showers and eyewash.

Respirator Selection: *At any detectable concentration:* SCBAF:PD,PP (any MSHA/NIOSH approved self-contained breathing apparatus that has a full facepiece and is operated in a pressure-demand or other positive-pressure mode); or SAF: PD,PP:ASCBA (any supplied-air respirator that has a full facepiece and is operated in a pressure-demand or other positive-pressure mode in combination with an auxiliary, self-contained breathing apparatus operated in a pressure-demand or other positive pressure mode). *Escape:* HiEF [any air-purifying, full-facepiece respirator (gas mask) with a chin-style, front- or back-mounted organic vapor canister having a high-efficiency particulate filter]; or SCBAE (any appropriate escape-type, self-contained breathing apparatus).

Storage: Prior to working with this chemical you should be trained on its proper handling and storage. Before entering confined space where this chemical may be present, check to make sure that an explosive concentration does not exist. Store in tightly closed containers in a cool, well-ventilated area away from oxidizers. Store in a refrigerator under an inert atmosphere for prolonged storage. A regulated, marked area should be established where this chemical is handled, used, or stored in compliance with OSHA standard 1910.1045.

Shipping: p-Cresidine is not specifically cited in DOT's Performance-Oriented Packaging Standards.

Spill Handling: Evacuate persons not wearing protective equipment from area of spill or leak until clean-up is complete. Remove all ignition sources. Collect powdered material in the most convenient and safe manner and deposit in sealed containers. Ventilate area after clean-up is complete. It may be necessary to contain and dispose of this chemical as a hazardous waste. If material or contaminated runoff enters waterways, notify downstream users of potentially contaminated waters. Contact your Department of Environmental Protection or your regional office of the federal EPA for specific recommendations. If employees are required to clean-up spills, they must be properly trained and equipped. OSHA 1910.120(q) may be applicable.

Fire Extinguishing: This chemical is a combustible solid. Use dry chemical, carbon dioxide, water spray, or alcohol foam extinguishers. The substance decomposes on heating or on burning producing toxic and irritating fumes including oxides of carbon. If material or contaminated runoff enters waterways, notify downstream users of potentially contaminated waters. Notify local health and fire officials and pollution control agencies. From a secure, explosion-proof location, use water spray to cool exposed containers. If cooling streams are ineffective (venting sound increases in volume and pitch, tank discolors, or shows any signs of deforming), withdraw immediately to a secure position. If employees are expected to fight fires, they must be trained and equipped in OSHA 1910.156.

References

National Cancer Institute, Bioassay of p-Cresidine for Possible Carcinogenicity, DHHS Publication No. (NIH) 78-1394, National Technical Information Service, Springfield, Virginia (1979).

Cresols

Molecular Formula: C_7H_8O

Common Formula: $CH_3C_6H_4OH$

Synonyms: *m-cresol:* Benzene, 3-methyl-; *m*-Cresol; 3-Cresol; Cresol-*m*; *m*-Cresylic acid; EINECS No. 203-577-9; 1-Hydroxy-3-methylbenzene; *m*-Hydroxytoluene; 3-Hydroxytoluene; *m*-Kresol (German); Metacresol; *m*-Methylphenol; 3-Methylphenol; Phenol, 3-methyl-; *m*-Toluol

mixed isomers: Acede cresylique (French); AR-Toluenol; Bacillol; Cresol isomers; Cresoli (Italian); Cresols (*o*-; *m*-; *p*-); Cresols (all isomers); Cresols and cresylic acids, mixed; Cresylic acid; Hydroxytoluole (German); Kresole (German); Kresolen (Dutch); Krezol (Polish); Methylphenol; Phenol, methyl-; Tekresol; Tricresol

o-cresol: Benzene, 2-methyl-; *o*-Cresol; 2-Cresol; Cresol-*o*; cresol-*o*-; *o*-Cresylic acid; EINECS No. 202-432-8; 1-Hydroxy-2-methylbenzene; *o*-Hydroxytoluene; 2-Hydroxytoluene; *o*-Kresol (German); *o*-Methylphenol; 2-Methylphenol; Orthocresol; Phenol, 2-methyl; *o*-Toluol

p-cresol: Benzene, 4-methyl; *p*-Cresol; 4-Cresol; Cresol-*p*; *p*-Cresylic acid; EINECS No. 203-398-6; 1-Hydroxy-4-methylbenzene; *p*-Hydroxytoluene; 4-Hydroxytoluene; *p*-Kresol (German); *p*-Methylphenol; 4-Methylphenol; Phenol, 4-methyl; *p*-Toluol

CAS Registry Number: 1319-77-3 (mixed isomers, cresylic acid); 95-48-7 (o-); 108-39-4 (m-); 106-44-5 (p-)

RTECS Number: GO5950000 (mixed isomers); GO6300000 (o-); GO6125000 (m-); GO6475000 (p-)

DOT ID: UN 2076 (cresols)

EEC Number: 604-004-00-9 (all isomers)

Regulatory Authority

- Air Pollutant Standard Set (ACGIH)[1] (Australia) (DFG)[3] (HSE)[33] (Israel) (Mexico) (OSHA)[58] (Several States)[60] (Several Canadian Provinces)

Mixed isomers:

- CLEAN AIR ACT: Hazardous Air Pollutants (Title I, Part A, Section 112)
- CLEAN WATER ACT: Section 311 Hazardous Substances/RQ 100 lb (4.54 kg); Section 313 Water Priority Chemicals (57FR41331, 9/9/92)
- EPA HAZARDOUS WASTE NUMBER (RCRA No.): U052
- RCRA Toxicity Characteristic (Section 261.24), Maximum Concentration of Contaminants, regulatory level, 200 mg/l. *Note*: if o-, m- and p-Cresol concentrations cannot be differentiated, the total cresol (D026) concentration is used. The regulatory level of total cresol is 200 mg/l
- RCRA, 40CFR261, Appendix 8 Hazardous Constituents
- SUPERFUND/EPCRA 40CFR302.4 Reportable Quantity (RQ): CERCLA, 1,000 lb (454 kg)
- EPCRA Section 313 Form R *de minimis* concentration reporting level: 1.0%
- U.S. DOT Regulated Marine Pollutant (49CFR172.101, Appendix B) as cresols (o-; m; p-)
- Canada, WHMIS, Ingredients Disclosure List; National Pollutant Release Inventory (NPRI)

Cited in U.S. State Regulations: Alaska (G), California (A, G), Connecticut (A), Florida (G, A), Illinois (G), Indiana (A), Kansas (G), Louisiana (G), Maine (G), Maryland (G), Massachusetts (G, A), Nevada (A), New Hampshire (G), New Jersey (G), New York (G, A), North Carolina (A), North Dakota (A), Pennsylvania (G), Rhode Island (G), South Carolina (A), Vermont (G), Virginia (G, A), Washington (G), West Virginia (G), Wisconsin (G).

Description: Cresol, $CH_3C_6H_4OH$, is a mixture off the three isomeric cresols, o-, m-, and p-creosol. *o-:* Colorless or yellow liquid or crystals that turn dark on exposure to air and light, with characteristic phenolic odor. *m-:* Colorless or yellow liquid with characteristic odor. *p-:* Colorless to yellow crystals that turn dark on exposure to air and light, with characteristic odor. Cresols are slightly soluble in water. Some physical properties of the various isomers are as follows:

Explosive limits:

Isomer	Melting Point, °C	Boiling Point, °C	Flash Point, °C	Auto-ignition, °C	L.E.L. / U.E.L., %
mixed	11 – 36	185 – 205	43 – 82	ca. 560	1.1 / —
o-	30 – 31	191 – 192	81 – 83 (cc)	599	1.4 / —
m-	11 – 12	202	86 (cc)	558	1.1 / —
p-	36	201.8	86 (cc)	558 – 560	1.1 / —

Hazard Identification (based on NFPA-704 M Rating System): (cresols) Health 3, Flammability 2, Reactivity 0.

Potential Exposure: Cresol is used as a disinfectant, as an ore flotation agent, and as an intermediate in the manufacture of chemicals, dyes, plastics, and antioxidants. A mixture of isomers is generally used; the concentrations of the components are determined by the source of the cresol.

Incompatibilities: Forms explosive mixture with air. Incompatible with strong acids, oxidizers, alkalies, aliphatic amines, amides, chlorosulfonic acid, oleum. Decompose on heating producing strong acids and bases, causing fire and explosion hazard. Liquid attacks some plastics and rubber. Attacks many metals.

Permissible Exposure Limits in Air: The Federal standard OSHA PEL 8-hour TWA[58] and the ACGIH TLV 8-hour TWA limit,[1] HSE limit,[33] DFG MAK[3] Australia, Israel, Mexico are all 5 ppm (22 mg/m^3). There is no STEL value. The NIOSH IDLH value is 250 ppm. The notation "skin" is added indicating the possibility of cutaneous absorption. In Canada, the TWA is the same as above and Alberta has an STEL of 10 ppm (44 mg/m^3).

The former USSR has set a MAC in workplace air of 0.5 mg/m^3.[35] They have also set a MAC in ambient air in residential areas of 0.005 mg/m^3 both on a momentary and on a daily average basis.

A number of states have set guidelines or standards for cresols in ambient air[60] ranging from 12.0 μg/m^3 (Massachusetts) to 73.0 μg/m^3 (New York) to 110 μg/m^3 (Indiana) to 200 μg/m^3 (Connecticut) to 220 μg/m^3 (Florida, North Dakota, South Carolina) to 370 μg/m^3 (Virginia) to 524 μg/m^3 (Nevada) to 2,200 μg/m^3 (North Carolina).

Determination in Air: XAD-7® (tube); Methanol; Gas chromatography/Flame ionization detection; NIOSH IV Method #2546, Cresols and Phenol.

Permissible Concentration in Water: No criteria set, but EPA has suggested a permissible ambient concentration, based on health effects, of 304 μg/l. The former USSR-UNEP/IRPTC project[43] has set an MAC in water used for domestic purposes of 0.004 μg/l of m and p isomers. A limit of 0.003 mg/l of o-isomer is set for water bodies used for fishery purposes. Mexico[35] has set a MAC of 1.5 mg/l of cresols in the waters in estuaries.

Routes of Entry: Inhalation of percutaneous absorption of liquid or vapor, ingestion, eye and skin contact.

Harmful Effects and Symptoms

Short Term Exposure: Corrosive to the eyes, skin and respiratory tract. Inhalation can cause pulmonary edema, a medical emergency that can be delayed for several hours. This can cause death. Corrosive on ingestion. May affect the central nervous system. Exposure may result in death. The effects may be delayed. Inhalation: A level of 4 ppm was reported not to cause symptoms. Exposure to 48 ppm for 5 – 10 minutes may cause irritation to the nose, throat and lungs, nausea, vomiting and general muscle weakness. High blood pressure, tremors and convulsions have been reported in people exposed to Cresol at unspecified levels. Due to irritating effects, the inhalation of appreciable amounts of cresol vapor in the workplace under normal conditions is unlikely.

Skin: This is a major route of exposure and the one which causes most work-related cresol injuries. Contact with 20 ml of 90% cresol resulted in burns, swelling, internal bleeding and kidney damage, and has caused death in children; 10% cresol left on the skin for 2 hours has caused inflammation, blistering and scarring; 6% for 5 – 6 hours has caused drying and peeling, inflammation, facial tremors and tearing. A 4% solution tested on human skin produced no sensitization reaction.

Ingestion: 250 ml of 50% Cresol has caused pneumonia, irritation of the pancreas, kidney failure and unconsciousness. Symptoms reported from swallowing an unspecified amount of cresol include an immediate burning of the tissues, sweating, headache, dizziness, muscle weakness, ringing in the ears and pale skin appearance. Death has occurred from as little as 1 gram (about ¼ teaspoon).

Exposure may result in a burning pain in the mouth and throat; white necrotic lesions in the mouth, esophagus and stomach; abdominal pain, vomiting, diarrhea; paleness; sweating; weakness; headache; dizziness; ringing in ears; shallow respiration with "phenol" odor on the breath; scanty, dark-colored or "smoky" urine; and possible delirium followed by unconsciousness. Convulsions are rarely seen, except in children. Hypersensitivity develops in certain individuals.

The chemical is rated as a very toxic compound with a probable oral lethal dose in humans of 50 – 500 mg/kg, or between 1 teaspoon and 1 ounce for a 70 kg (150 lb person).

Long Term Exposure: Repeated or prolonged contact with skin may cause dermatitis. May cause lung damage Repeated or prolonged exposure may affect the central nervous system, kidneys, and liver. *p-:* Repeated or prolonged skin contact with the p- isomer may cause dermatitis.

Points of Attack: Central nervous system, respiratory system, liver, kidneys, skin, eyes.

Medical Surveillance: Consider the skin, eyes, respiratory system, and liver and kidney functions in placement or periodic examinations. Liver and kidney function tests. Evaluation by a qualified allergist. Consumption of alcohol may increase liver damage.

First Aid: If this chemical gets into the eyes, remove any contact lenses at once and irrigate immediately for at least 15 minutes, occasionally lifting upper and lower lids. Seek medical attention immediately. If this chemical contacts the skin, remove contaminated clothing and wash immediately with soap and water. Seek medical attention immediately. If this chemical has been inhaled, remove from exposure, begin rescue breathing (using universal precautions) if breathing has stopped and CPR if heart action has stopped. Transfer promptly to a medical facility. When this chemical has been swallowed, get medical attention. Give large quantities of water and induce vomiting. Do not make an unconscious person vomit. Medical observation is recommended for 24 – 48 hours after breathing overexposure, as pulmonary edema may be delayed. As first aid for pulmonary edema, a doctor or authorized paramedic may consider administering a corticosteroid spray.

Personal Protective Methods: Wear protective gloves and clothing to prevent any reasonable probability of skin contact. Safety equipment suppliers/manufacturers can provide recommendations on the most protective glove/clothing material for your operation. Saranex® has been recommended as a protective material for cresols. All protective clothing (suits, gloves, footwear, headgear) should be clean, available each day, and put on before work. Contact lenses should not be worn when working with this chemical. When working with liquids, wear splash-proof chemical goggles and face shield unless full facepiece respiratory protection is worn. When working with powders or dusts, wear dust-proof chemical goggles and face shield unless full facepiece respiratory protection is worn. Employees should wash immediately with soap when skin is wet or contaminated. Provide emergency showers and eyewash.

Respirator Selection: NIOSH as cresol: *23 ppm:* CCROVDM [any chemical cartridge respirator with organic vapor cartridge(s) in combination with a dust and mist filter]; SA (any supplied-air respirator). *57.5 ppm:* SA:CF (any supplied-air respirator operated in a continuous-flow mode); or PAPROVDM [any powered, air-purifying respirator with organic vapor cartridge(s) in combination with a dust and mist filter]. *115 ppm:* CCRFOVHiE [any chemical cartridge respirator with a full facepiece and organic vapor cartridge(s) in combination with a high efficiency particulate filter]; or GMFOVHiE [any air-purifying, full-facepiece respirator (gas mask) with a chin-style, front- or back-mounted organic vapor canister having a high-efficiency particulate filter]; or PAPRTHiE (any powered, air-purifying respirator with a tight-fitting facepiece and a high-efficiency particulate filter); or SAT:CF (any supplied-air respirator that has a tight-fitting facepiece and is operated in a continuous-flow mode); or SCBAF (any self-contained breathing apparatus with a full facepiece); or SAF (any supplied-air respirator with a full facepiece). *250 ppm:* SAF:PD,PP (any supplied-air respirator that has a full facepiece and is operated in a pressure-demand or other positive-pressure mode). *Emergency or planned entry into unknown concentrations or IDLH conditions:* SCBAF: PD,PP (any self-contained breathing apparatus that has a full facepiece and is operated in a pressure-demand or other positive-pressure mode); or SAF:PD,PP:ASCBA (any supplied-air respirator that has a full facepiece and is operated in a pressure-demand or other positive-pressure mode in combination with an auxiliary self-contained breathing apparatus operated in a pressure-demand or other positive pressure mode). *Escape:* GMFOVHiE [any air-purifying, full-facepiece respirator (gas mask) with a chin-style, front- or back-mounted organic vapor canister having a high-efficiency particulate filter]; or SCBAE (any appropriate escape-type, self-contained breathing apparatus).

Note: Substance reported to cause eye irritation or damage; may require eye protection.

Storage: Prior to working with cresols you should be trained on its proper handling and storage. Before entering confined space where Cresols may be present, check to make sure that an explosive concentration does not exist. Store in tightly closed containers in a cool, well ventilated area. Away from strong acids or oxidizers. Metal containers involving the transfer of this chemical should be grounded and bonded. Where possible, automatically pump liquid from drums or other storage containers to process containers. Drums must be equipped with self-closing valves, pressure vacuum bungs, and flame arresters. Use only non-sparking tools and equipment, especially when opening and closing containers of this chemical. Sources of ignition such as smoking and open flames, are prohibited where this chemical is used, handled, or stored in a manner that could create a potential fire or explosion hazard.

Shipping: The DOT label requirement is "Poison." The UN/DOT Hazard Class is 6.1 and the Packing Group is II.[19][20]

Spill Handling: Avoid inhalation. Wear proper respiratory protection and eye protection. Wear protective clothing. Do not touch spilled material. Stop leak if you can do so without risk. Use water spray to reduce vapors. For small spills: absorb the material with sand or other noncombustible absorbent material and place into containers for later disposal. For small dry spills, shovel up and place into clean, dry containers and cover. For large spills, dike far ahead of spill for later disposal. If clothing becomes contaminated remove immediately and isolate at the site. If a leak or spill has not ignited, use water spray to disperse the vapors and to provide protection for personnel attempting to stop a leak. Water spray may be used to flush spills away from exposures. It may be necessary to contain and dispose of this chemical as a hazardous waste. If material or contaminated runoff enters waterways, notify downstream users of potentially contaminated waters. Contact your Department of Environmental Protection or your regional office of the federal EPA for specific recommendations. If employees are required to clean-up spills, they must be properly trained and equipped. OSHA 1910.120(q) may be applicable.

Fire Extinguishing: Cresols are combustible. Stay upwind, keep out of low areas. Ventilate closed spaces before entering them. Use water to blanket the fire and dry chemical, foam, or carbon dioxide to extinguish the flames. Extinguish small fires with dry chemical, carbon dioxide, water spray, or foam. For large fires use water spray, fog, or foam. Move container from fire area if you can do so without risk. Poisonous gases are produced in fire. Vapors are heavier than air and will collect in low areas. Vapors may travel long distances to ignition sources and flashback. Vapors in confined areas may explode when exposed to fire. Containers may explode in fire. Storage containers and parts of containers may rocket great distances, in many directions. If material or contaminated runoff enters waterways, notify downstream users of potentially contaminated waters. Notify local health and fire officials and pollution control agencies. From a secure, explosion-proof location, use water spray to cool exposed containers. If cooling streams are ineffective (venting sound increases in volume and pitch, tank discolors, or shows any signs of deforming), withdraw immediately to a secure position. Dike fire control water for later disposal; don't scatter the material. If employees are expected to fight fires, they must be trained and equipped in OSHA 1910.156.

Disposal Method Suggested: Wastewaters may be subjected to biological treatment. Concentrations may be further reduced by ozone treatment. High concentration wastes may be destroyed in special waste incinerators.[22]

References

National Institute for Occupational Safety and Health, Criteria for a Recommended Standard: Occupational Exposure to Cresol, NIOSH Document No. 78–133, Washington, DC (1978)

U.S. Environmental Protection Agency, A Study of Industrial Data on Candidate Chemicals for Testing (Alkyl Phthalates and Cresols), Report EPA-560/5-78-002, Washington, DC (June 1978)

U.S. Environmental Protection Agency, Cresols and Cresylic Acid, Health and Environmental Effects Profile No. 54, Office of Solid Waste, Washington, DC (April 30, 1980)

U.S. Environmental Protection Agency, "Chemical Profile: Cresylic Acid," Washington, DC, Chemical Emergency Preparedness Program (October 31, 1985)

New Jersey Department of Health and Senior Services and Senior Services, "Hazardous Substance Fact Sheet: Cresols" Trenton, NJ (May 1998)

New Jersey Department of Health and Senior Services and Senior Services, "Hazardous Substance Fact Sheet: Cresylic Acid," Trenton, NJ (April 1985)

New York State Department of Health, "Chemical Fact Sheet: Cresol (Mixture)," Albany, NY, Bureau of Toxic Substance Assessment (March 1986). Also separate fact sheets on "ortho-Cresol," "meta-Cresol" and "para-Cresol."

Sax, N. I., Ed., "Dangerous Properties of Industrial Materials Report, for o-cresol: 5," No. 3, 30–34 (1985), for m-cresol: 1, No. 6, 44–46 (1981), and 6, No. 1, 41–46 (1986)

Crimidine

Molecular Formula: $C_7H_{10}ClN_3$

Common Formula: $C_7H_{10}N_3Cl$

Synonyms: Castrix®; 2-Chloor-4-dimethylamino-6-methyl-pyrimidine (Dutch); 2-Chloro-4-methyl-6-dimethylaminopyrimidine; 2-Cloro-4-dimetilamino-6-metil-pirimidina (Italian); Crimidin (German); Crimidina (Italian); Pyrimidine, 2-chloro-4-(dimethylamino)-6-methyl-; W 491

CAS Registry Number: 535-89-7

RTECS Number: UV8050000

DOT ID: UN 2588

Regulatory Authority

- Banned or Severely Restricted (In Agriculture) (Germany) (UN)[13]
- Very Toxic Substance (World Bank)[15]
- SUPERFUND/EPCRA 40CFR355, Appendix B Extremely Hazardous Substances: TPQ = 100/10,000 lb (4.54/4,540 kg)
- SUPERFUND/EPCRA 40CFR302.4 Reportable Quantity (RQ): EHS, 1 lb (0.454 kg)

Cited in U.S. State Regulations: California (G), Florida (G), Massachusetts (G), New Jersey (G), Oklahoma (G), Pennsylvania (G).

Description: Crimidine, $C_7H_{10}N_3Cl$, is a brown, waxy solid. Freezing/Melting point = 87°C. Hazard Identification (based on NFPA-704 M Rating System): Health 4, Flammability 1, Reactivity 0. Slightly soluble in water.

Potential Exposure: Crimidine is used as a rodenticide but is not registered in the U.S. as a pesticide.

Incompatibilities: Acids and acid fumes, strong bases.

Permissible Exposure Limits in Air: No standards set. This chemical can be absorbed through the skin, thereby increasing exposure.

Permissible Concentration in Water: No criteria set.

Routes of Entry: Ingestion, absorbed through the skin.

Harmful Effects and Symptoms

Short Term Exposure: Contact can cause eye and skin irritation and burns. Inhalation can irritate the nose and throat. Exposure may result in serious central nervous system damage with anxiety, restlessness, muscle stiffness, light sensitivity, noise sensitivity, touch sensitive, cold sweat, and leading to convulsions that may be fatal. If patient survives 5 – 6 hours there may not be serious problems. Extremely toxic; the LD_{50} oral (rat) is 1.25 mg/kg; probable oral lethal dose in humans is less than 5 mg/kg or less than 7 drops for a 70 kg (150 lb) person.

Long Term Exposure: Chronic health effects are unknown at this time.

Points of Attack: Central nervous system.

Medical Surveillance: There is no special test for this chemical. However, if illness occurs or overexposure is suspected, medical attention is recommended.

First Aid: *Eye Contact:* Immediately remove any contact lenses and flush with large amounts of water for at least 15 minutes, occasionally lifting upper and lower lids. Seek medical attention immediately.

Skin Contact: Quickly remove contaminated clothing. Immediately wash area with large amounts of water. Seek medical attention immediately.

Breathing: Remove the person from exposure, trying to avoid rapid, jerky motions or noise. Begin rescue breathing if breathing has stopped and CPR if heart action has stopped. If seizures occur, begin seizure first aid measures. Call for immediate medical attention to visit the patient prior to transfer if possible. Any facility using this chemical should have 24 hour rapid access to medical personnel with training and equipment for emergency treatment. All area employees should be trained in first aid measures for Castrix, including seizure management and CPR.

Personal Protective Methods: Wear protective gloves and clothing to prevent any reasonable probability of skin contact. Safety equipment suppliers/manufacturers can provide recommendations on the most protective glove/clothing material for your operation. All protective clothing (suits, gloves, footwear, headgear) should be clean, available each day, and put on before work. Contact lenses should not be worn when working with this chemical. Wear dust-proof chemical goggles and face shield unless full facepiece respiratory protection is worn. Employees should wash immediately with soap when skin is wet or contaminated. Provide emergency showers and eyewash.

Respirator Selection: Where the potential exists for exposure to Castrix, use a MSHA/NIOSH approved full facepiece respirator with a pesticide cartridge. Increased protection is obtained from full facepiece air purifying respirators. If while wearing a filter, cartridge or canister respirator, you can smell, taste, or otherwise detect Castrix, or in the case of a full facepiece respirator you experience eye irritation, leave the area immediately. Check to make sure the respirator-to-face seal is still good. If it is, replace the filter, cartridge, or canister. If the seal is no longer good, you may need a new respirator.

Where the potential for high exposures exists, use a MSHA/NIOSH approved supplied-air respirator with a full facepiece operated in the positive pressure mode or with a full facepiece, hood, or helmet in the continuous flow mode, or use a MSHA/NIOSH approved self-contained breathing apparatus with a full facepiece operated in pressure-demand or other positive pressure mode.

Storage: Prior to working with Castrix you should be trained on its proper handling and storage. Castrix must be stored to avoid contact with strong acids (such as hydrochloric, sulfuric and nitric) and acid fumes since violent reactions occur. Store in tightly closed containers in a cool, well-ventilated area.

Shipping: Pesticides, solid, toxic, n.o.s. require a "Poison" label. Crimidine falls in UN/DOT Hazard Class 6.1.[19][20]

Spill Handling: Evacuate and restrict persons not wearing protective equipment from area of spill or leak until cleanup is complete. Remove all ignition sources. Use organic vapor respiratory protection. Stay upwind; keep out of low areas. Wear self-contained (positive pressure if available) breathing apparatus and full protective clothing. Do not touch spilled material; stop leak if you can do so without risk. Small spills: absorb with sand or other noncombustible absorbent material and place into containers for later disposal. Small dry spills: with clean shovel place material into clean, dry container and cover; move containers from spill area. Large spills: dike far ahead of spill for later disposal. If water pollution occurs, notify appropriate authorities. Ventilate area of spill or leak. It may be necessary to contain and dispose of this chemical as a hazardous waste. If material or contaminated runoff enters waterways, notify downstream users of potentially contaminated waters. Contact your Department of Environmental Protection or your regional office of the federal EPA for specific recommendations. If employees are required to clean-up spills, they must be properly trained and equipped. OSHA 1910.120(q) may be applicable.

Fire Extinguishing: This chemical is a noncombustible solid. Use dry chemical, carbon dioxide, water spray, or alcohol foam extinguishers. Poisonous gases are produced in fire including chloride fumes and nitrogen oxides. If material or contaminated runoff enters waterways, notify downstream users of potentially contaminated waters. Notify local health and fire officials and pollution control agencies. From a secure, explosion-proof location, use water spray to cool exposed containers. If cooling streams are ineffective (venting sound increases in volume and pitch, tank discolors, or shows any signs of deforming), withdraw immediately to a secure position. If employees are expected to fight fires, they must be trained and equipped in OSHA 1910.156.

Disposal Method Suggested: In accordance with 40CFR 165 recommendations for the disposal of pesticides and pesticide containers. Must be disposed properly by following package label directions or by contacting your state pesticide or environmental control agency or by contacting your regional EPA office.

References

U.S. Environmental Protection Agency, "Chemical Profile: Crimidine," Washington, DC, Chemical Emergency Preparedness Program (November 30, 1987)

New Jersey Department of Health and Senior Services and Senior Services, "Hazardous Substance Fact Sheet: Castrix," Trenton, NJ (September 1999)

Crotonaldehyde

Molecular Formula: C_4H_6O

Common Formula: $CH_3CH=CHCHO$

Synonyms: *cis-:* Aldehido crotonico (Spanish); Aldehyde crotonique (French); 2-Butenal; Crotonic aldehyde; Krotonaldehyd (Czech); β-methylacrolein; Propylene aldehyde; Topenel

trans-: Aldehido crotonico, (E)- (Spanish); Aldehyde crotonique (E)- (French); (E)-2-Butenal; *trans*-2-Butenal; 2-Butenal, (E)-; Crotonal; Crotonaldehyde; Crotonaldehyde, (E)-; Crotonaldehyde, *trans-*; Crotonic aldehyde; Ethylene dipropionate (8CI); 3-Methylacroleine; NCI-C56279; Propylene aldehyde, (E)-; Propylene aldehyde, *trans-*; Propylene aldehyde-*trans*

CAS Registry Number: 123-73-9 (trans-); 4170-30-3 (cis-)

RTECS Number: GP9625000 (trans-); GP 9499000 (cis-)

DOT ID: UN 1143

EEC Number: 605-009-00-9

Regulatory Authority

- Carcinogen (Suspected) (DFG)[3] (ACGIH)[1]

trans-:

- Air Pollutant Standard Set (ACGIH)[1] (DFG)[3] (Mexico) (OSHA)[58] (Several Canadian Provinces)
- CLEAN AIR ACT: Accidental Release Prevention/Flammable substances, (Section 112[r], Table 3), TQ = 20,000 lb (9,080 kg)
- SUPERFUND/EPCRA 40CFR302.4 Reportable Quantity (RQ): CERCLA, 100 lb (45.4 kg)
- EPCRA Section 313 Form R *de minimis* concentration reporting level: 1.0%
- SUPERFUND/EPCRA 40CFR355, Appendix B Extremely Hazardous Substances: TPQ = 1,000 lb (454 kg)
- U.S. DOT Regulated Marine Pollutant (49CFR172.101, Appendix B)

cis-:

- Air Pollutant Standard Set (ACGIH)[1] (Australia) (DFG)[3] (HSE)[33] (Israel) (OSHA)[58] (Several States)[60] (Several Canadian Provinces0
- CLEAN AIR ACT: Accidental Release Prevention/Flammable substances, (Section 112[r], Table 3), TQ = 20,000 lb (9,080 kg)
- CLEAN WATER ACT: Section 311 Hazardous Substances/RQ 40CFR117.3 (same as CERCLA, see below)
- EPA HAZARDOUS WASTE NUMBER (RCRA No.): U053
- RCRA, 40CFR261, Appendix 8 Hazardous Constituents
- SUPERFUND/EPCRA 40CFR355, Appendix B Extremely Hazardous Substances: TPQ = 1,000 lb (454 kg)
- SUPERFUND/EPCRA 40CFR302.4 Reportable Quantity (RQ): CERCLA, 100 lb (45.4 kg)
- EPCRA Section 313 Form R *de minimis* concentration reporting level: 1.0%
- U.S. DOT Regulated Marine Pollutant (49CFR172.101, Appendix B)
- Canada, WHMIS, Ingredients Disclosure List

Cited in U.S. State Regulations: Alaska (G), California (G), Connecticut (A), Florida (G), Illinois (G), Louisiana (G), Maine (G), Massachusetts (G), Minnesota (G), Nevada (A), New Hampshire (G), New Jersey (G), North Dakota (A), Oklahoma (G), Pennsylvania (G), Rhode Island (G), Vermont (G), Virginia (G, A), Washington (G), West Virginia (G), Wisconsin (G).

Description: Crotonaldehyde, $CH_3CH=CHCHO$, is a water-white (turns pale-yellow on contact with air) with an irritating, pungent, suffocating odor. Odor threshold = 0.11 ppm. Boiling point = 104°C. Freezing/Melting point = (trans-) -77; (cis-) -69°C. Flash point = 13°C (oc). Autoignition temperature = 232°C. Explosive limits (trans-): LEL = 2.1%; UEL = 15.5%. Soluble in water. NFPA 704 M Hazard Identification (trans-): Health 4, Flammability 3, Reactivity 2.

Potential Exposure: Crotonaldehyde is used as an intermediate in the manufacture of n-butanol and crotonic and sorbic acids and in resin and rubber antioxidant manufacture; it is also used as a solvent in mineral oil purification, as a warning agent in fuel gas and as an alcohol denaturant.

Incompatibilities: A strong reducing agent. Forms explosive mixture with air. Readily converted by oxygen to peroxides and acids; heat or contact with alkalis and many other substances may cause polymerization. Incompatible with strong oxidizers, Strong acids including non-oxidizing mineral acids, ammonia, organic amines, aliphatic amines, aromatic amines, 1,3-butadiene, strong bases. Liquid attacks some plastics, rubber and coatings.

Permissible Exposure Limits in Air: The Federal limit (OSHA PEL 8-hour TWA)[58] and ACGIH recommended 8-hour TWA value is 2.0 ppm (6 mg/m^3). The HSE,[33] Israel, and Mexico TWA is the same and HSE, Mexico STEL value is 6.0 ppm (18 mg/m^3). The NIOSH IDLH level is 50 mg/m^3. The DFG[3] has set no numerical limits but has simply noted that this is a suspect carcinogen. The former USSR-UNEP/IRPTC project[43] has set an MAC in workplace air of 0.5 mg/m^3. Several states have set guidelines or standards for Crotonaldehyde in ambient air[60] ranging from 60 – 180 μg/m^3 (North Dakota) to 100 μg/m^3 (Virginia) to 120 μg/m^3 (Connecticut) to 143 μg/m^3 (Nevada).

Determination in Air: Bubbler; Reagent; Polarography; NIOSH IV Method #3516. Collection by charcoal tube, analysis by gas liquid chromatography. NIOSH: Crotonaldehyde, P&CAM 285.

Permissible Concentration in Water: No criteria set. Regulated by Clean Water Act.

Routes of Entry: Inhalation, ingestion, eye and skin contact.

Harmful Effects and Symptoms

Short Term Exposure: A lacramator (causing tearing). Contact or vapor can severe and painful irritation and burn eyes (can cause corneal damage) and skin. The vapor can irritate the respiratory tract causing cough and shortness of breath. The substance may affect the lungs, resulting in impaired function, coughing, and shortness of breath. Higher exposures can cause pulmonary edema, a medical emergency that can be delayed for several hours. This can cause death. Medical observation is indicated. Although slightly less toxic, Crotonaldehyde is similar chemically and toxicologically to acrolein which is rated as extremely toxic. The oral LD_{50} rat is 206 mg/kg (moderately toxic). A 15 minute exposure at 4.1 ppm is highly irritating to the nose and upper respiratory tract and causes tearing. Brief exposure at 45 ppm proved very disagreeable with prominent eye irritation. Toxic concentrations for human inhalation have been reported at 12 mg/m^3/10 min. As with acrolein, vapor exposures cause gastrointestinal distress when ingested.

Long Term Exposure: Crotonaldehyde causes mutations; such chemicals may have a cancer risk. There is limited evidence that this chemical causes cancer in animals, and may cause liver cancer. May cause skin allergy. Limited studies to date indicate that these substances have chemical reactivity and mutagenicity similar to acetaldehyde and malonaldehyde. Therefore, NIOSH recommends that careful consideration should be given to reducing exposures to these related aldehydes.

Points of Attack: Respiratory system, eyes, skin.

Medical Surveillance: Consider the skin, eyes and respiratory system in placement or periodic examinations. Evaluation by a qualified allergist.

First Aid: If this chemical gets into the eyes, remove any contact lenses at once and irrigate immediately for at least 15 minutes, occasionally lifting upper and lower lids. Seek medical attention immediately. If this chemical contacts the skin, remove contaminated clothing and wash immediately with soap and water. Seek medical attention immediately. If this chemical has been inhaled, remove from exposure, begin rescue breathing (using universal precautions) if breathing has stopped and CPR if heart action has stopped. Transfer promptly to a medical facility. When this chemical has been swallowed, get medical attention. Give large quantities of water and induce vomiting. Do not make an unconscious person vomit. Medical observation is recommended for 24 – 48 hours after breathing overexposure, as pulmonary edema may be delayed. As first aid for pulmonary edema, a doctor or authorized paramedic may consider administering a corticosteroid spray.

Personal Protective Methods: Wear protective gloves and clothing to prevent any reasonable probability of skin contact. Safety equipment suppliers/manufacturers can provide recommendations on the most protective glove/clothing material for your operation. Teflon and Butyl Rubber is among the recommended protective materials. All protective clothing (suits, gloves, footwear, headgear) should be clean, available each day, and put on before work. Contact lenses should not be worn when working with this chemical. Wear splash-proof chemical goggles and face shield unless full facepiece respiratory protection is worn. Employees should wash immediately with soap when skin is wet or contaminated. Provide emergency showers and eyewash.

Respirator Selection: NIOSH/OSHA: *20 ppm:* CCRFOV (any air-purifying, full-facepiece respirator (gas mask) with a chin-style, front- or back-mounted acid gas canister); or SA (any supplied-air respirator). *50 ppm:* SA:CF (any supplied-air respirator operated in a continuous-flow mode); or PAPROV [any powered, air-purifying respirator with organic vapor cartridge(s)]; or CCRFOV (any air-purifying, full-facepiece respirator (gas mask) with a chin-style, front- or back-mounted acid gas canister); or GMFOV [any air-purifying, full-facepiece respirator (gas mask) with a chin-style, front- or back-, mounted organic vapor canister]; or SCBAF (any self-contained breathing apparatus with a full facepiece); or SAF (any supplied-air respirator with a full facepiece) *Emergency or planned entry into unknown concentrations or IDLH conditions:* SCBAF:PD,PP (any self-contained breathing apparatus that has a full facepiece and is operated in a pressure-demand or other positive-pressure mode); or SAF:PD,PP:ASCBA (any supplied-air respirator that has a full facepiece and is operated in a pressure-demand or other positive-pressure mode in combination with an auxiliary self-contained breathing apparatus operated in a pressure-demand or other positive pressure mode). *Escape:* GMFOV [any air-purifying, full-facepiece respirator (gas mask) with a chin-style, front-or back-, mounted organic vapor canister]; or SCBAE (any appropriate escape-type, self-contained breathing apparatus).

Note: Substance reported to cause eye irritation or damage; may require eye protection.

Storage: Prior to working with Crotonaldehyde you should be trained on its proper handling and storage. Before entering confined space where this chemical may be present, check to make sure that an explosive concentration does not exist. Store in tightly closed containers in a cool, well ventilated area away from heat, caustics, ammonia, amines, oxidizing materials, mineral acids and 1,3-butadiene since violent reactions occur. Metal containers involving the transfer of this chemical should be grounded and bonded. Where possible, automatically pump liquid from drums or other storage containers to process containers. Drums must be equipped with self-closing valves, pressure vacuum bungs, and flame arresters. Use only non-sparking tools and equipment, especially when opening and closing containers of this chemical. Sources of ignition such as smoking and open flames, are prohibited where this chemical is used, handled, or stored in a manner that could create a potential fire or explosion hazard. A regulated, marked area should be established where this chemical is handled, used, or stored in compliance with OSHA standard 1910.1045.

Shipping: Stabilized Crotonaldehyde should be labeled: "Flammable Liquid, Poison." Shipment by passenger aircraft or railcar or even by cargo aircraft is forbidden. It falls in UN/DOT Hazard Class 3 and Packing Group I.

Spill Handling: Evacuate and restrict persons not wearing protective equipment from area of spill or leak until cleanup is complete. Remove all ignition sources. Keep unnecessary people away; isolate hazard area and deny entry. Stay upwind; keep out of low areas. Establish forced ventilation to keep levels below explosive limit. Wear positive pressure breathing apparatus and special protective clothing. Do not touch spilled material; stop leak if you can do so without risk. Use water spray to reduce vapors. Small spills: absorb with sand or other non-combustible absorbent material and place into containers for later disposal. Large spills: dike far ahead of spill for later disposal. It may be necessary to contain and dispose of this chemical as a hazardous waste. If material or contaminated runoff enters waterways, notify downstream users of potentially contaminated waters. Contact your Department of Environmental Protection or your regional office of the federal EPA for specific recommendations. If employees are required to clean-up spills, they must be properly trained and equipped. OSHA 1910.120(q) may be applicable.

Initial isolation and protective action distances

Distances shown are likely to be affected during the first 30 minutes after materials are spilled and could increase with time. If more than one tank car, cargo tank, portable tank, or large cylinder is involved in the incident is leaking, the protective action distance may need to be increased. You may need to seek emergency information from CHEMTREC at (800) 424-9300 or seek professional environmental engineering assistance from the U.S. EPA Environmental Response Team at (908) 548-8730 (24-hour response line).

Small spills (From a small package or a small leak from a large package)

First: Isolate in all directions (feet)..... 200

Then: Protect persons downwind (miles)

Day 0.1

Night 0.3

Large spills (From a large package or from many small packages)

First: Isolate in all directions (feet)..... 500

Then: Protect persons downwind (miles)

Day 0.3

Night 1.1

Fire Extinguishing: This chemical is a flammable liquid. Poisonous gases are produced in fire. Use dry chemical, carbon dioxide, or alcohol foam extinguishers. Vapors are heavier than air and will collect in low areas. Vapors may travel long distances to ignition sources and flashback. Vapors in confined areas may explode when exposed to fire. Containers may explode in fire. Storage containers and parts of containers may rocket great distances, in many directions. If material or contaminated runoff enters waterways, notify downstream users of potentially contaminated waters. Notify local health and fire officials and pollution control agencies. From a secure, explosion-proof location, use water spray to cool exposed containers. If cooling streams are ineffective (venting sound increases in volume and pitch, tank discolors, or shows any signs of deforming), withdraw immediately to a secure position. If employees are expected to fight fires, they must be trained and equipped in OSHA 1910.156.

Disposal Method Suggested: Incineration. May be absorbed on vermiculite and burned in open incinerator or dissolved in solvent and sprayed into incinerator.

References

U.S. Environmental Protection Agency, Crotonaldehyde, Health and Environmental Effects Profile No. 55, Office of Solid Waste, Washington, DC (April 30, 1980)

Sax, N. I., Ed., "Dangerous Properties of Industrial Materials Report," 4, No. 1, 56–59, New York, Van Nostrand Reinhold Co. (Jan./Feb. 1984)

U.S. Environmental Protection Agency, "Chemical Profile: Crotonaldehyde," Washington, DC, Chemical Emergency Preparedness Program (Nov. 30, 1987)

New Jersey Department of Health and Senior Services and Senior Services, "Hazardous Substance Fact Sheet: Crotonaldehyde," Trenton, NJ (September 1996)

Crotonic Acid

Molecular Formula: $C_4H_6O_2$

Synonyms: α-Butenoic acid; 2-Butenoic acid; α-Crotonic acid; β-Methylacrylic acid; 3-Methylacrylic acid

CAS Registry Number: 3724-65-0

DOT ID: UN 2823

Cited in U.S. State Regulations: Florida (G), Massachusetts (G), New Jersey (G), Pennsylvania (G)

Description: Crotonic acid, is a white, or colorless, crystalline solid with a pungent odor. May be transported as a molten liquid. Boiling point = 187 – 189°C. Freezing/Melting point = 72°C. Flash point = 88°C (oc). Soluble in water. This chemical is the commercially used trans- isomer. The cis- isomer is less stable and melts at 15°C.

Potential Exposure: Used to make plastics, resins, plasticizers, lacquers, and medicines.

Incompatibilities: Forms explosive mixture with air. A strong reducing agent. The aqueous solution is a weak acid. Violent reaction with oxidizers, combustibles, strong bases, peroxides. Moisture or strong sunlight (UV) may cause explosive polymerization. May accumulate static electrical charges, and may cause ignition of its vapors. Combustible when exposed to heat or flame.

Permissible Exposure Limits in Air: No OELs have been established.

Routes of Entry: Inhalation.

Harmful Effects and Symptoms

Short Term Exposure: Irritates the eyes, skin and respiratory tract. Corrosive: contact with the skin or eyes can cause burns and permanent damage. Inhalation can cause coughing and shortness of breath. Higher exposures can cause pulmonary edema, a medical emergency that can be delayed for several hours. This can cause death. Moderately toxic by ingestion, skin contact, and subcutaneous routes.

Long Term Exposure: Irritates the lungs; may cause bronchitis with coughing, phlegm, and/or shortness of breath.

Points of Attack: Lungs.

Medical Surveillance: Lung function tests. Consider chest x-ray following acute overexposure.

First Aid: If this chemical gets into the eyes, remove any contact lenses at once and irrigate immediately for at least 15 minutes, occasionally lifting upper and lower lids. Seek medical attention immediately. If this chemical contacts the skin, remove contaminated clothing and wash immediately with soap and water. Seek medical attention immediately. If this chemical has been inhaled, remove from exposure, begin rescue breathing (using universal precautions) if breathing has stopped and CPR if heart action has stopped. Transfer promptly to a medical facility. When this chemical has been swallowed, get medical attention. If victim is conscious, administer water or milk. Do not induce vomiting. Medical observation is recommended for 24 – 48 hours after breathing overexposure, as pulmonary edema may be delayed. As first aid for pulmonary edema, a doctor or authorized paramedic may consider administering a corticosteroid spray.

Personal Protective Methods: Wear protective gloves and clothing to prevent any reasonable probability of skin contact. Safety equipment suppliers/manufacturers can provide recommendations on the most protective glove/clothing material for your operation. All protective clothing (suits, gloves, footwear, headgear) should be clean, available each day, and put on before work. Contact lenses should not be worn when working with this chemical. Wear dust-proof chemical goggles and face shield unless full facepiece respiratory protection is worn. Employees should wash immediately with soap when skin is wet or contaminated. Provide emergency showers and eyewash.

Respirator Selection: *Where there is a potential for overexposure*: SCBAF:PD,PP (any MSHA/NIOSH approved self-contained breathing apparatus that has a full facepiece and is operated in a pressure-demand or other positive-pressure mode); or SAF:PD,PP:ASCBA (any supplied-air respirator that has a full facepiece and is operated in a pressure-demand or other positive-pressure mode in combination with an auxiliary, self-contained breathing apparatus operated in a pressure-demand or other positive pressure mode).

Storage: Prior to working with crotonic acid, you should be trained on its proper handling and storage. Store in tightly closed containers in a cool, well ventilated area away from oxidizers, strong bases, and reducing agents. Where possible, automatically pump liquid from drums or other storage containers to process containers

Shipping: Label required is "Corrosive." Crotonic acid is in DOT/UN Hazard Class 8 and Packing Group III.

Spill Handling: *Solid:* Evacuate persons not wearing protective equipment from area of spill or leak until clean-up is complete. Remove all ignition sources. Collect powdered material in the most convenient and safe manner and deposit in sealed containers. Ventilate area after clean-up is complete. It may be necessary to contain and dispose of this chemical as a hazardous waste. If material or contaminated runoff enters waterways, notify downstream users of potentially contaminated waters. Contact your Department of Environmental Protection or your regional office of the federal EPA for specific recommendations. If employees are required to clean-up spills, they must be properly trained and equipped. OSHA 1910.120(q) may be applicable.

Liquid: Evacuate and restrict persons not wearing protective equipment from area of spill or leak until cleanup is complete. Remove all ignition sources. Ventilate area of spill or leak. Absorb liquids in vermiculite, dry sand, earth, peat, carbon, or a similar material and deposit in sealed containers. Keep this chemical out of a confined space, such as a sewer, because of the possibility of an explosion, unless the sewer is designed to prevent the build-up of explosive concentrations. It may be necessary to contain and dispose of this chemical as a hazardous waste. If material or contaminated runoff enters waterways, notify downstream users of potentially contaminated waters. Contact your Department of Environmental Protection or your regional office of the federal EPA for specific recommendations. If employees are required to clean-up spills, they must be properly trained and equipped. OSHA 1910.120(q) may be applicable.

Fire Extinguishing: This chemical is a combustible solid. Use dry chemical, carbon dioxide, or alcohol or polymer foam extinguishers. Poisonous gases are produced in fire including acrid fumes and organic acids. If material or contaminated runoff enters waterways, notify downstream users of potentially contaminated waters. Notify local health and fire officials and pollution control agencies. Containers may explode in fire. From a secure, explosion-proof location, use water spray to cool exposed containers. If cooling streams are ineffective (venting sound increases in volume and pitch, tank discolors, or shows any signs of deforming), withdraw immediately to a secure position. If employees are expected to fight fires, they must be trained and equipped in OSHA 1910.156.

References

New Jersey Department of Health and Senior Services, Hazardous Substance Fact Sheet, "Crotonic Acid," Trenton NJ (January 1999)

Crufomate

Molecular Formula: $C_{12}H_{19}ClNO_3P$

Synonyms: *o*-(4-*terz*-Butil-2-cloro-fenil)-*o*-metil-fosforammide (Italian); 4-T-Butyl-2-chlorophenyl methyl methylphosphoramidate; 4-*tert*. Butyl 2-chlorophenyl methylphosphoramidate de methyle (French); *O*-(4-*tert*-Butyl-2-chlor-phenyl)-*O*-methyl-phosphorsaeure-*N*-methyl-amid (German); Dowco 132; ENT 25,602-X; *O*-Methyl *O*-2-chloro-4-*tert*-butylphenyl *N*-methylamidophosphate; Methylphosphoramidic acid, 4-*t*-butyl-2-chlorophenyl methyl ester; Montrel®; Phenol,4-*t*-butyl-2-chloro-, ester with methyl methylphosphoramidate; Phosphoramidic acid, 4-*tert*-butyl-2-chlorophenylphosphor amidate; Phosphoramidic acid, methyl-,4-*tert*-butyl-2-chlorophenyl; Phosphoramidic acid, methyl-,2-chloro-4-(1,1-dimethylethyl)phenyl methyl ester; Ruelene®; Ruelene® Drench; Rulene®; *o*-(4-Tertbutyl-2-chloor-fenyl)-*o*-methyl-fosforzuur-*N*-methyl-amide (Dutch)

CAS Registry Number: 299-86-5

RTECS Number: TB3850000

DOT ID: UN 2783 (solid); UN 3018 (liquid)

EEC Number: 015-074-00-X

Regulatory Authority

- Air Pollutant Standard Set (ACGIH)[1] (Australia) (Israel) (Mexico) (OSHA)[58] (Several States)[60] (Several Canadian Provinces)

Cited in U.S. State Regulations: Alaska (G), California (G), Connecticut (A), Florida (A), Illinois (G), Maine (G), Massachusetts (G), Nevada (A), New Hampshire (G), New Jersey (G), North Dakota (A), Pennsylvania (G), Rhode Island (G), Virginia (A), West Virginia (G).

Description: Crufomate, $C_{12}H_{19}ClNO_3P$, is a colorless crystalline compound. Commercial product is a yellow oil. Freezing/Melting point = 60°C (decomposes). Slightly soluble in water.

Potential Exposure: Those involved in the manufacture, formulation and application of this insecticide and anthelmintic for cattle.

Incompatibilities: The substance decomposes on heating forming corrosive and toxic fumes of hydrogen chloride, nitrogen oxides and phosphorous oxides. Incompatible with strongly alkaline and strongly acidic media. Unstable over long periods in aqueous preparations or above 140°F.

Permissible Exposure Limits in Air: There is no OSHA PEL. The NIOSH REL 10-hour TWA and ACGIH recommended 8-hour TWA is 5 mg/m^3 and the NIOSH STEL is 20 mg/m^3, not to be exceeded during any 15 minute work period. The Australian, Mexican, and Israeli TWA is the same as NIOSH and Mexico's STEL is 20 mg/m^3. The Canadian provinces of Alberta, BC, Ontario, and Quebec have adopted a TWA of 5 mg/m^3 and except for Quebec use the STEL of 20 mg/m^3. Several states have set guidelines or standards for crufomate in ambient air[60] ranging from 50 – 200 μg/m^3 (North Dakota) to 80 μg/m^3 (Virginia) to 100 μg/m^3 (Connecticut) to 119 μg/m^3 (Nevada).

Determination in Air: Filter; none; Gravimetric; NIOSH IV Method #0500, Particulates NOR (total). OSHA versatile sampler-2; Toluene/Acetone; Gas chromatography/Flame photometric detection for sulfur, nitrogen, or phosphorus; NIOSH Method IV Method #5600, Organophosphorus Pesticides.

Permissible Concentration in Water: No criteria set.

Routes of Entry: Skin absorption, inhalation of dust, ingestion.

Harmful Effects and Symptoms

Short Term Exposure: A cholinesterase inhibitor. Crufomate irritates the eyes, skin, and respiratory tract. Crufomate can affect you when breathed in and quickly enters the body by passing through the skin. Severe poisoning can occur from skin contact. It is a moderately toxic organophosphate chemical. Exposure can cause effects on the nervous system, rapid severe poisoning with headache, sweating, nausea and vomiting, diarrhea, loss of coordination, convulsions, respiratory failure, and death. The LD_{50} oral-rat is 460 mg/kg (slightly toxic). The health effects may be delayed

Long Term Exposure: Exposure may affect the developing fetus. Crufomate may damage the testes. High or repeated exposure may cause nerve damage and poor coordination in arms and legs. Repeated exposure may cause personality changes of depression, anxiety, or irritability.

Points of Attack: Respiratory system, central nervous system, peripheral nervous system, plasma cholinesterase.

Medical Surveillance: Before employment and at regular times after that, the following are recommended: Plasma and red blood cell cholinesterase levels (tests for the enzyme poisoned by this chemical). If exposure stops, plasma levels return to normal in 1 – 2 weeks while red blood cell levels may be reduced for 1 – 3 months.

When cholinesterase enzyme levels are reduced by 25% or more below preemployment levels, risk of poisoning is increased, even if results are in lower ranges of "normal." Reassignment to work not involving organophosphate or carbamate pesticides is recommended until enzyme levels recover. If symptoms develop or overexposure occurs, repeat the above tests as soon as possible and get an exam of the nervous system. Also consider complete blood count. Consider chest x-ray following acute overexposure. Do not drink any alcoholic beverages before or during use. Alcohol promotes absorption of organic phosphates.

First Aid: If this chemical gets into the eyes, remove any contact lenses at once and irrigate immediately for at least 15 minutes, occasionally lifting upper and lower lids. Seek medical attention immediately. If this chemical contacts the skin, remove contaminated clothing and wash immediately with soap and water. Speed in removing material from skin is of extreme importance. Shampoo hair promptly if contaminated. Seek medical attention immediately. If this chemical has been inhaled, remove from exposure, begin rescue breathing (using universal precautions) if breathing has stopped and CPR if heart action has stopped. Transfer promptly to a medical facility. When this chemical has been swallowed, get medical attention. Give large quantities of water and induce vomiting. Do not make an unconscious person vomit. Medical observation is recommended for 24 – 36 hours.

Personal Protective Methods: Wear protective gloves and clothing to prevent any reasonable probability of skin contact. Safety equipment suppliers/manufacturers can provide recommendations on the most protective glove/clothing material for your operation. All protective clothing (suits, gloves, footwear, headgear) should be clean, available each day, and put on before work. Contact lenses should not be worn when working with this chemical. Wear splash-proof chemical goggles when working with liquid or wear dust-proof goggles when working with powders or dusts unless full facepiece respiratory protection is worn. Employees should wash immediately with soap when skin is wet or contaminated. Provide emergency showers and eyewash.

Respirator Selection: Where the potential exists for exposures over 5 mg/m^3, use an MSHA/NIOSH approved full facepiece respirator with a pesticide cartridge. Greater protection is provided by a powered-air purifying respirator. Where the potential for high exposures exists, use an MSHA/NIOSH approved supplied-air respirator with a full facepiece operated in the positive pressure mode or with a full facepiece, hood, or helmet in the continuous flow mode, or use an MSHA/NIOSH approved self-contained breathing apparatus with a full facepiece operated in pressure-demand or other positive pressure mode.

Storage: Prior to working with this chemical you should be trained on its proper handling and storage. Store in tightly closed containers in a cool well-ventilated area away from heat. Also, avoid contact with oxidizers (such as peroxides, permanganates, chlorates, perchlorates, and nitrates).

Shipping: Organophosphate pesticides fall in Hazard Class 6.1 and Packing Group II. A "Keep Away From Food" label is required.

Spill Handling: Evacuate and restrict persons not wearing protective equipment from area of spill or leak until cleanup is complete. Remove all ignition sources. Ventilate area of spill or leak. Absorb liquids in vermiculite, dry sand, earth, peat, carbon, or a similar material and deposit in sealed containers. Collect powdered material in the most convenient and safe manner and deposit in sealed containers. It may be necessary to contain and dispose of this chemical as a hazardous waste. If material or contaminated runoff enters waterways, notify downstream users of potentially contaminated waters. Contact your Department of Environmental Protection or your regional office of the federal EPA for specific recommendations. If employees are required to clean up spills, they must be properly trained and equipped. OSHA 1910.120(q) may be applicable.

Fire Extinguishing: If Crufomate is in dry form, extinguish fire using an agent suitable for the type of surrounding fire; Crufomate itself does not burn. If an oily or liquid form is used, consider the flammability of the solvent in determining appropriate procedures. Use dry chemical, carbon dioxide, water spray, or alcohol foam extinguishers. Corrosive and toxic fumes of hydrogen chloride, nitrogen oxides and phosphorous oxides are produced in fire. If material or contaminated runoff enters waterways, notify downstream users of potentially contaminated waters. Notify local health and fire officials and pollution control agencies. Containers may explode in fire. From a secure, explosion-proof location, use water spray to cool exposed containers. If cooling streams are ineffective (venting sound increases in volume and pitch, tank discolors, or shows any signs of deforming), withdraw immediately to a secure position. If employees are expected to fight fires, they must be trained and equipped in OSHA 1910.156.

Disposal Method Suggested: Crufomate decomposes above pH 7.0 in alkaline media. In accordance with 40CFR165 recommendations for the disposal of pesticides and pesticide containers. Must be disposed properly by following package label directions or by contacting your state pesticide or environmental control agency or by contacting your regional EPA office.

References

New Jersey Department of Health and Senior Services and Senior Services, "Hazardous Substance Fact Sheet: Crufomate," Trenton, NJ (October, 1985)

Cumene

Molecular Formula: C_9H_{12}

Common Formula: $C_6H_5CH(CH_3)_2$

Synonyms: Benzene isopropyl; Benzene, (1-methylethyl-)-; Cum; Cumeen (Dutch); Cumeno (Spanish); Cumol; 2-Fenilpropano (Italian); 2-Fenyl-propan (Dutch); Isopropilbenzene (Italian); Isopropylbenzeen (Dutch); Isopropylbenzene; Isopropylbenzol; Isopropyl-benzol (German); 1-Methylethyl benzene; 2-Phenylpropane

CAS Registry Number: 98-82-8

RTECS Number: GR8575000

DOT ID: UN 1918

Regulatory Authority

- Air Pollutant Standard Set (ACGIH)[1] (Australia) (DFG)[3] (HSE)[33] (Israel) (Mexico) (OSHA)[58] (Several States)[60] (Several Canadian Provinces)
- CLEAN AIR ACT: Hazardous Air Pollutants (Title I, Part A, Section 112)
- EPA HAZARDOUS WASTE NUMBER (RCRA No.): U055
- RCRA, 40CFR261, Appendix 8 Hazardous Constituents
- SUPERFUND/EPCRA 40CFR302.4 Reportable Quantity (RQ): CERCLA, 5,000 lb (2,270 kg)
- EPCRA Section 313 Form R *de minimis* concentration reporting level: 1.0%
- U.S. DOT Regulated Marine Pollutant (49CFR172.101, Appendix B)
- Canada, WHMIS, Ingredients Disclosure List; National Pollutant Release Inventory (NPRI)

Cited in U.S. State Regulations: Alaska (G), California (A, G), Connecticut (A), Florida (G), Illinois (G), Kansas (G), Louisiana (G), Maine (G), Maryland (G), Massachusetts (G), Minnesota (G), Nevada (A), New Hampshire (G), New Jersey (G), New York (G), North Dakota (A), Oklahoma (G), Pennsylvania (G), Rhode Island (G), Vermont (G), Virginia (G, A), Washington (G), West Virginia (G), Wisconsin (G).

Description: Cumene, $C_6H_5CH(CH_3)_2$, is a colorless liquid with a sharp, penetrating, aromatic odor. The odor threshold is 0.008-0.132 ppm. Boiling point = 152°C. Flash point = 36°C. Hazard Identification (based on NFPA-704 M Rating System): Health 2, Flammability 3, Reactivity 1. Explosive limits are: LEL = 0.9%; UEL = 6.5%.

Potential Exposure: Cumene is used as a high octane gasoline component; it is used as a thinner for paints and lacquers; it is an important intermediate in phenol manufacture.

Incompatibilities: Vapor forms explosive mixture with air. Incompatible with strong acids (nitric, sulfuric acids), strong oxidizers. Air contact forms cumene hydroperoxide. Attacks rubber. May accumulate static electrical charges, and may cause ignition of its vapors.

Permissible Exposure Limits in Air: The OSHA PEL 8-hour TWA and ACGIH recommended TLV 8-hour TWA is 50 ppm (245 mg/m^3). DFG,[3] Australia, Israel, and the Canadian provinces of Ontario and Quebec have set the same limit. HSE,[33] Mexico and the Canadian provinces of Alberta and BC have set the same TWA and has set an STEL of 75 ppm (365 mg/m^3). They add the notation "skin" indicating the possibility of cutaneous absorption. The NIOSH IDLH level is 8,000 ppm. Some states have set guidelines or standards for Cumene in ambient air[60] ranging from 2.45 – 3.65 mg/m^3 (North Dakota) to 4.0 mg/m^3 (Virginia) to 4.9 mg/m^3 (Connecticut) to 5.83 mg/m^3 (Nevada).

Determination in Air: Charcoal adsorption, workup with CS_2, analysis by gas chromatography. See NIOSH Method #1501 for hydrocarbons, aromatic.

Permissible Concentration in Water: No criteria set.

Routes of Entry: Inhalation, ingestion, skin and eye contact. Passes through the unbroken skin.

Harmful Effects and Symptoms

Short Term Exposure: Irritates the eyes, skin and respiratory tract. Skin contact may cause a burning sensation and/or rash. Higher levels can cause dizziness, lightheadedness, headaches, unconsciousness, narcosis, coma. Levels of 4,000 ppm may cause unconsciousness. The LD_{50} oral-rat is 1,400 mg/kg (slightly toxic).

Long Term Exposure: Drying and cracking of the skin. May cause lung, liver, and kidney damage. Although cumene has not been adequately tested to determine whether brain or nerve damage could occur with repeated exposure, many solvents and other petroleum-based chemicals have been shown to cause such damage.

Points of Attack: Eyes, skin, respiratory system, central nervous system.

Medical Surveillance: Consider the points of attack in preplacement and periodic physical examinations. Interview for brain effects, including recent memory, mood, concentration, altered sleep patterns, and nervous system evaluation.

First Aid: If this chemical gets into the eyes, remove any contact lenses at once and irrigate immediately for at least 15 minutes, occasionally lifting upper and lower lids. Seek medical attention immediately. If this chemical contacts the skin, remove contaminated clothing and wash immediately with soap and water. Seek medical attention immediately. If this chemical has been inhaled, remove from exposure, begin rescue breathing (using universal precautions) if breathing has stopped and CPR if heart action has stopped. Transfer promptly to a medical facility. When this chemical has been swallowed, get medical attention. *Do not* induce vomiting.

Personal Protective Methods: Wear appropriate clothing to prevent repeated or prolonged skin contact. Chlorinated Polyethylene (CPE) is among the recommended protective materials. Wear eye protection to prevent any reasonable probability of eye contact. Employees should wash promptly when skin is wet or contaminated. Remove nonimpervious clothing promptly if contaminated or wet.

Respirator Selection: NIOSH/OSHA: *500 ppm:* CCROV [any chemical cartridge respirator with organic vapor cartridge(s)]; or SA (any supplied-air respirator). *1,000 ppm:* SA:CF (any supplied-air respirator operated in a continuous-flow mode); or PAPROV [any powered, air-purifying respirator with organic vapor cartridge(s)]; or CCRFOV [any chemical cartridge respirator with a full facepiece and organic vapor cartridge(s)]; or GMFOV [any air-purifying, full-facepiece respirator (gas mask) with a chin-style, front- or back-mounted acid gas canister]; or SCBAF (any self-contained breathing apparatus with a full facepiece) or SAF (any supplied-air respirator with a full facepiece). *Emergency or planned entry into unknown concentrations or IDLH conditions:* SCBAF:PD,PP (any self-contained breathing apparatus that has a full facepiece and is operated in a pressure-demand or other positive-pressure mode); or SAF:PD,PP: ASCBA (any supplied-air respirator that has a full facepiece and is operated in a pressure-demand or other positive-pressure mode in combination with an auxiliary self-contained breathing apparatus operated in a pressure-demand or other positive pressure mode). *Escape:* GMFOV [any air-purifying, full-facepiece respirator (gas mask) with a chin-style, front-or back-, mounted organic vapor canister]; or SCBAE (any appropriate escape-type, self-contained breathing apparatus).

Note: Substance reported to cause eye irritation or damage; may require eye protection.

Storage: Prior to working with this chemical you should be trained on its proper handling and storage. Before entering confined space where this chemical may be present, check to make sure that an explosive concentration does not exist. Cumene must be stored to avoid contact with oxidizers, such as permanganates, nitrites, peroxides, chlorates, and perchlorates, since violent reactions occur. Store in tightly closed containers in a cool well-ventilated area away from heat. Sources of ignition such as smoking and open flames are prohibited where Cumene is used, handled, or stored in a manner that could create a potential fire or explosion hazard. Sources of ignition such as smoking and open flames are prohibited where this chemical is used, handled, or stored in a manner that could create a potential fire or explosion hazard. Metal containers involving the transfer of 5 gallons or more of this chemical should be grounded and bonded. Drums must be equipped with self-closing valves, pressure vacuum bungs, and flame arresters. Use only nonsparking tools and equipment, especially when opening and closing containers of this chemical.

Shipping: Cumene requires a "Flammable Liquid" label. It falls in DOT Hazard Class 3 and Packing Group III.

Spill Handling: Evacuate and restrict persons not wearing protective equipment from area of spill or leak until cleanup is complete. Remove all ignition sources. Establish forced ventilation to keep levels below explosive limit. Absorb liquids in vermiculite, dry sand, earth, peat, carbon, or a similar material and deposit in sealed containers. It may be necessary to contain and dispose of this chemical as a hazardous waste. If material or contaminated runoff enters waterways, notify downstream users of potentially contaminated waters. Contact your Department of Environmental Protection or your regional office of the federal EPA for specific recommendations. If employees are required to clean-up spills, they must be properly trained and equipped. OSHA 1910.120(q) may be applicable.

Fire Extinguishing: This chemical is a combustible liquid. Poisonous gases are produced in fire. Use dry chemical, carbon dioxide, or alcohol foam extinguishers. Vapors are heavier than air and will collect in low areas. Vapors may travel long distances to ignition sources and flashback. Vapors in confined areas may explode when exposed to fire. Containers may explode in fire. Storage containers and parts of containers may rocket great distances, in many directions. If material or contaminated runoff enters waterways, notify downstream users of potentially contaminated waters. Notify local health and fire officials and pollution control agencies. From a secure, explosion-proof location, use water spray to cool exposed containers. If cooling streams are ineffective (venting sound increases in volume and pitch, tank discolors, or shows any signs of deforming), withdraw immediately to a secure position. If employees are expected to fight fires, they must be trained and equipped in OSHA 1910.156.

Disposal Method Suggested: Incineration.

References

Sax, N. I., Ed., "Dangerous Properties of Industrial Materials Report," 4, No. 1, 59–62, New York, Van Nostrand Reinhold Co. (Jan./Feb. 1984)

New Jersey Department of Health and Senior Services and Senior Services, "Hazardous Substance Fact Sheet: Cumene," Trenton, NJ (January 1986)

New York State Department of Health, "Chemical Fact Sheet: Cumene," Albany, NY, Bureau of Toxic Substance Assessment (March 1986)

Cumene Hydroperoxide

Molecular Formula: $C_9H_{12}O_2$

Common Formula: $C_6H_5C(CH_3)_2OOH$

Synonyms: Cumeenhydroperoxyde (Dutch); Cument hydroperoxide; Cumenyl hydroperoxide; Cumolhydroperoxid (German); α-Cumylhydroperoxide; Cumyl hydroperoxide; α,α-Dimethylbenzyl hydroperoxide; Hidroperoxido de cumeno (Spanish); Hydroperoxide, 1-methyl-1-phenylethyl-; Hydroperoxyde de cumene (French); Hydroperoxyde de cumyle (French); Hyperiz; Idroperossido di cumene (Italian); Idroperossido di cumolo (Italian); Isopropylbenzene hydroperoxide; Trigorox K 80

CAS Registry Number: 80-15-9

RTECS Number: MX2450000

DOT ID: UN 2116

EEC Number: 617-002-00-8

Regulatory Authority

- EPA HAZARDOUS WASTE NUMBER (RCRA No.): U096
- RCRA, 40CFR261, Appendix 8 Hazardous Constituents
- SUPERFUND/EPCRA 40CFR302.4 Reportable Quantity (RQ): CERCLA, 10 lb (4.54 kg)

- EPCRA Section 313 Form R *de minimis* concentration reporting level: 1.0%
- Canada, WHMIS, Ingredients Disclosure List; National Pollutant Release Inventory (NPRI)

Cited in U.S. State Regulations: California (G), Florida (G), Kansas (G), Louisiana (G), Maine (G), Massachusetts (G), New Hampshire (G), New Jersey (G), Oklahoma (G), Pennsylvania (G), Rhode Island (G), Vermont (G), Virginia (G), Washington (G), Wisconsin (G).

Description: Cumene hydroperoxide, $C_6H_5C(CH_3)_2OOH$, is a colorless to pale yellow liquid. Boiling point = 153°C. Freezing/Melting point = -10°C. It explodes on heating and its Flash point = 79°C. Its explosive limits are: LEL = 0.9%; UEL = 6.5%. Slightly soluble in water.

Potential Exposure: Cumene hydroperoxide is an intermediate in the process for making phenol plus acetone from Cumene. It also acts as a curing agent for unsaturated polyester resins.

Incompatibilities: May explode on heating at about 50°C. The substance is a strong oxidant and reacts violently with combustible and reducing materials, causing fire and explosion hazard. Contact with metallic salts of cobalt, copper or lead alloys and mineral acids, bases, amines may lead to violent decomposition. Vapor forms an explosive mixture with air. May accumulate static electrical charges, and may cause ignition of its vapors.

Permissible Exposure Limits in Air: No standards set.

Permissible Concentration in Water: No criteria set.

Routes of Entry: Inhalation, passing through the unbroken skin.

Harmful Effects and Symptoms

Short Term Exposure: Cumene hydroperoxide is corrosive to the eyes, skin, and respiratory tract. Eye contact can cause burns and permanent damage. Corrosive on ingestion. Inhalation may include nosebleeds, sore throat, hoarseness, cough with phlegm, increased saliva, and shortness of breath. Higher exposures can cause a build-up of fluid in the lungs (pulmonary edema), a medical emergency, which can cause death. Other exposure symptoms may include headache, dizziness, poor coordination and even unconsciousness. The oral LD_{50} rat is 382 mg/kg (moderately toxic).

Long Term Exposure: Cumene Hydroperoxide may cause mutations. Handle with extreme caution. May cause skin allergy. High or repeated overexposure may damage lungs, kidneys and liver.

Points of Attack: Lungs, skin, kidneys, liver.

Medical Surveillance: Before beginning employment and at regular times after that, for those with frequent or potentially high exposures, the following are recommended: Lung function tests. If symptoms develop or overexposure is suspected, the following also may be useful: Consider chest x-ray after acute overexposure. Evaluation by a qualified allergist, including careful exposure history and special testing, may help diagnose skin allergy. Tests for kidney and liver function. Complete blood count.

First Aid: If this chemical gets into the eyes, remove any contact lenses at once and irrigate immediately for at least 15 minutes, occasionally lifting upper and lower lids. Seek medical attention immediately. If this chemical contacts the skin, remove contaminated clothing and wash immediately with soap and water. Seek medical attention immediately. If this chemical has been inhaled, remove from exposure, begin rescue breathing (using universal precautions) if breathing has stopped and CPR if heart action has stopped. Transfer promptly to a medical facility. When this chemical has been swallowed, get medical attention. Give large quantities of water and induce vomiting. Do not make an unconscious person vomit. Medical observation is recommended for 24 – 48 hours after breathing overexposure, as pulmonary edema may be delayed. As first aid for pulmonary edema, a doctor or authorized paramedic may consider administering a corticosteroid spray.

Personal Protective Methods: Wear protective gloves and clothing to prevent any reasonable probability of skin contact. Safety equipment suppliers/manufacturers can provide recommendations on the most protective glove/clothing material for your operation. All protective clothing (suits, gloves, footwear, headgear) should be clean, available each day, and put on before work. Contact lenses should not be worn when working with this chemical. Wear splash-proof chemical goggles and face shield unless full facepiece respiratory protection is worn. Employees should wash immediately with soap when skin is wet or contaminated. Provide emergency showers and eyewash.

Respirator Selection: Where the potential for exposures to Cumene Hydroperoxide exists, use a MSHA/NIOSH approved supplied-air respirator with a full facepiece operated in the positive pressure mode or with a full facepiece, hood, or helmet in the continuous flow mode, or use a MSHA/NIOSH approved self-contained breathing apparatus with a full facepiece operated in pressure-demand or other positive pressure mode.

Storage: Prior to working with Cumend Hydroperoxide you should be trained on its proper handling and storage. Before entering confined space where this chemical may be present, check to make sure that an explosive concentration does not exist. Cumene Hydroperoxide must be stored to avoid contact with strong oxidizers (such as chlorine, bromine and fluorine), strong acids (such as hydrochloric, sulfuric and nitric) and organic materials, reducing agents, copper, copper or lead alloys, cobalt and mineral acids since violent reactions occur. Metal containers involving the transfer of this chemical should be grounded and bonded. Drums must be equipped with self-closing valves, pressure vacuum bungs, and flame arresters. Use only non-sparking tools and equipment, especially when opening and closing containers of this chemical. Sources of ignition such as smoking and open flames, are prohibited where this chemical is used, handled, or stored in a manner that could create a potential fire or explosion hazard.

Shipping: Cumene Hydroperoxide, technically pure, requires an "Organic Peroxide" label. It falls in DOT Hazard Class 5.2 and Packing Group I.

Spill Handling: Evacuate and restrict persons not wearing protective equipment from area of spill or leak until cleanup is complete. Remove all ignition sources. Establish forced ventilation to keep levels below explosive limit. Absorb liquids in vermiculite, dry sand, earth, peat, carbon, or a similar material and deposit in sealed containers. Keep Cumene Hydroperoxide out of a confined space, such as a sewer, because of the possibility of an explosion, unless the sewer is designed to prevent the build-up of explosive concentrations. It may be necessary to contain and dispose of this chemical as a hazardous waste. If material or contaminated runoff enters waterways, notify downstream users of potentially contaminated waters. Contact your Department of Environmental Protection or your regional office of the federal EPA for specific recommendations. If employees are required to clean-up spills, they must be properly trained and equipped. OSHA 1910.120(q) may be applicable.

Fire Extinguishing: This chemical is a flammable liquid. Poisonous gases including phenol are produced in fire. Use dry chemical, carbon dioxide, or alcohol foam extinguishers. Vapors are heavier than air and will collect in low areas. Vapors may travel long distances to ignition sources and flashback. Vapors in confined areas may explode when exposed to fire. Containers may explode in fire. Storage containers and parts of containers may rocket great distances, in many directions. If material or contaminated runoff enters waterways, notify downstream users of potentially contaminated waters. Notify local health and fire officials and pollution control agencies. From a secure, explosion-proof location, use water spray to cool exposed containers. If cooling streams are ineffective (venting sound increases in volume and pitch, tank discolors, or shows any signs of deforming), withdraw immediately to a secure position. If employees are expected to fight fires, they must be trained and equipped in OSHA 1910.156.

References

New Jersey Department of Health and Senior Services and Senior Services, "Hazardous Substance Fact Sheet: Cumene Hydroperoxide," Trenton, NJ (September 1996)

Cupferron

Molecular Formula: $C_6H_9N_3O_2$

Common Formula: $C_6H_5N(NO)ONH_4$

Synonyms: Ammonium-*N*-nitrosophenylhydroxylamine; Benzeneamine, *N*-Hydroxy-*N*-nitroso, ammonium salt; Hydroxylamine, *N*-nitroso-*N*-phenyl-, ammonium salt; Kupferron; NCI-C03258; *N*-Nitrosofenylhydroxylamin amonny (Czech); *N*-Nitrosophenylhydroxylamin ammonium salz (German); *N*-Nitroso-*N*-phenylhydroxylamine ammonium salt; *N*-Nitrosophenylhydroxylamine ammonium salt

CAS Registry Number: 135-20-6

RTECS Number: NC4725000

DOT ID: No citation.

Regulatory Authority

- Carcinogen (Animal Positive) (EPA) (NCI) (NTP)[9]
- EPCRA Section 313 Form R *de minimis* concentration reporting level: 0.1%
- Canada, WHMIS, Ingredients Disclosure List

Cited in U.S. State Regulations: California (A, G), Florida (G), Illinois (G), Maine (G), Maryland (G), Massachusetts (G), Michigan (G), Minnesota (G), New Jersey (G), Pennsylvania (G), West Virginia (G).

Description: Cupferron, $C_6H_9N_3O_2$, is a creamy-white crystalline compound. Freezing/Melting point = 163°C. Hazard Identification (based on NFPA-704 M Rating System): Health 2, Flammability 1, Reactivity 0. Soluble in water.

Potential Exposure: Cupferron is used to separate tin from zinc, and copper and iron from other metals in the laboratory. Cupferron also finds application as a quantitative reagent for vanadates and titanium and for the colorimetric determination of aluminum.

The potential for exposure appears to be greatest for those engaged in analytical or research studies involving use of the chemical. Workers may also be exposed to the compound during manufacturing processes.

Incompatibilities: Forms unstable and possibly explosive compounds with thorium salts, titanium, zirconium.

Permissible Exposure Limits in Air: No standards set.

Permissible Concentration in Water: No criteria set.

Routes of Entry: Human exposure to Cupferron occurs mainly through ingestion or inhalation of the dust from the dry salt. Skin absorption is a secondary route of exposure.

Harmful Effects and Symptoms

Long Term Exposure: Cupferron, given in the diet, was carcinogenic to Fisher 344 rats, causing hemangiosarcomas, hepatocellular carcinomas, and squamous-cell carcinomas of the forestomach in males and females, as well as carcinomas of the auditory sebaceous gland in females. The chemical was also carcinogenic to B6C3F1 mice, causing hemangiosarcomas in males; and hepatocellular carcinomas, carcinomas of the auditory sebaceous gland, a combination of hemangiosarcomas and hemangiomas, and adenomas of the Harderian gland in females. The oral LD_{50} rat is 257 mg/kg (moderately toxic).

First Aid: If this chemical gets into the eyes, remove any contact lenses at once and irrigate immediately for at least 15 minutes, occasionally lifting upper and lower lids. Seek medical attention immediately. If this chemical contacts the skin, remove contaminated clothing and wash immediately with soap and water. Seek medical attention immediately. If this chemical has been inhaled, remove from exposure, begin rescue breathing (using universal precautions) if breathing has stopped and CPR if heart action has stopped. Transfer promptly to a medical facility. When this chemical has been swallowed, get medical attention. Give large quantities of water and induce vomiting. Do not make an unconscious person vomit.

Personal Protective Methods: Wear protective gloves and clothing to prevent any reasonable probability of skin contact. Safety equipment suppliers/manufacturers can provide recommendations on the most protective glove/clothing material for your operation. All protective clothing (suits, gloves, footwear, headgear) should be clean, available each day, and put on before work. Contact lenses should not be worn when working with this chemical. Wear dust-proof chemical goggles and face shield unless full facepiece respiratory protection is worn. Employees should wash immediately with soap when skin is wet or contaminated. Provide emergency showers and eyewash.

Respirator Selection: *At any detectable concentration:* SCBAF:PD,PP (any MSHA/NIOSH approved self-contained breathing apparatus that has a full facepiece and is operated in a pressure-demand or other positive-pressure mode); or SAF:PD,PP:ASCBA (any supplied-air respirator that has a full facepiece and is operated in a pressure-demand or other positive-pressure mode in combination with an auxiliary, self-contained breathing apparatus operated in a pressure-demand or other positive pressure mode). *Escape:* HiEF [any air-purifying, full-facepiece respirator (gas mask) with a chin-style, front- or back-mounted organic vapor canister having a high-efficiency particulate filter]; or SCBAE (any appropriate escape-type, self-contained breathing apparatus).

Storage: Prior to working with cupferron you should be trained on its proper handling and storage. A regulated, marked area should be established where this chemical is handled, used, or stored in compliance with OSHA standard 1910.1045. Cupferron should be stored in a refrigerator or in a cool dry place and protected from exposure to moisture.

Shipping: Cupferron is not specifically cited in DOT's Performance-Oriented Packaging Standards.[19] It can be classified as a Hazardous Substance, solid, n.o.s. (UN 9188) which imposes no label requirements of maximum shipping quantity requirements. Such materials fall in Hazard Class ORM-E and Packing Group III.

Spill Handling: Evacuate persons not wearing protective equipment from area of spill or leak until clean-up is complete. Remove all ignition sources. Collect powdered material in the most convenient and safe manner and deposit in sealed containers. Ventilate area after clean-up is complete. It may be necessary to contain and dispose of this chemical as a hazardous waste. If material or contaminated runoff enters waterways, notify downstream users of potentially contaminated waters. Contact your Department of Environmental Protection or your regional office of the federal EPA for specific recommendations. If employees are required to clean-up spills, they must be properly trained and equipped. OSHA 1910.120(q) may be applicable.

Fire Extinguishing: Use dry chemical, carbon dioxide, water spray, or foam extinguishers. Poisonous gases including ammonia and nitrogen oxides are produced in fire. If material or contaminated runoff enters waterways, notify downstream users of potentially contaminated waters. Notify local health and fire officials and pollution control agencies. From a secure, explosion-proof location, use water spray to cool exposed containers. If cooling streams are ineffective (venting sound increases in volume and pitch, tank discolors, or shows any signs of deforming), withdraw immediately to a secure position. If employees are expected to fight fires, they must be trained and equipped in OSHA 1910.156.

References

National Cancer Institute, Bioassay of Cupferron for Possible Carcinogenicity, Technical Report Series No. 100, DHEW Publication No. (NIH) 78–1350, Bethesda, MD (1978)

Cupric Acetate

Molecular Formula: $C_4H_6CuO_4$

Common Formula: $Cu(OOCCH_3)_2$

Synonyms: Acetate de cuivre (French); Acetato de cobre (Spanish); Acetic acid, copper(2+) salt; Acetic acid, copper(II) salt; Acetic acid, cupric salt; Copper acetate; Copper(2+) acetate; Copper(II) acetate; Copper diacetate; Copper(2+) diacetate; Copper(II) diacetate; Crystallized verdigris; Crystals of Venus; Cupric diacetate; Neutral verdigris; Octan mednaty (Czech)

CAS Registry Number: 142-71-2

RTECS Number: AG3480000

DOT ID: UN 9106

EINECS Number: 205-553-3

Regulatory Authority

- Water Pollution Standard Proposed (EPA)[6] (former USSR)[35]
- CLEAN WATER ACT: Section 311 Hazardous Substances/RQ 40CFR117.3 (same as CERCLA, see below); Section 313 Water Priority Chemicals (57FR41331, 9/9/92); Toxic Pollutant (Section 401.15). as copper and compounds
- RCRA 40CFR264, Appendix 9; TSD Facilities Ground Water Monitoring List. Suggested test method(s) (PQL

µg/l): 6010 (60); 7210 (200) Note: All species in the ground water that contain copper are included

- SUPERFUND/EPCRA 40CFR302.4 Reportable Quantity (RQ): CERCLA, 100 lb (45.4 kg)
- EPCRA Section 313 Form R *de minimis* concentration reporting level: 1.0%
- Canada, WHMIS, Ingredients Disclosure List; National Pollutant Release Inventory (NPRI); CEPA Priority Substance List, Ocean dumping prohibited

Cited in U.S. State Regulations: California (A, G), Illinois (G), Massachusetts (G), New Hampshire (G), New Jersey (G), Pennsylvania (G).

Description: Cupric acetate, $Cu(OOCCH_3)_2$, is a greenish blue powder or small crystals. Boiling point = 240°C (decomposes). Freezing/Melting point = 115°C. Soluble in water.

Potential Exposure: Cupric acetate is used as a fungicide, as a catalyst for organic reactions, in textile dyeing and as a pigment for ceramics.

Incompatibilities: Forms explosive materials with acetylene gas, ammonia, caustic solutions, sodium hypobromite, notromethane. Keep away from chemically active metals, strong acids, nitrates. Decomposes above 240°C forming acetic acid fumes.

Permissible Exposure Limits in Air: The Federal standard (OSHA PEL 8-hour TWA)[58] for copper fume is 0.1 mg/m^3, and 1 mg/m^3 for copper dusts and mists. NIOSH recommends the same level for a 10-hour workshift. ACGIH recommends a TWA of 0.2 mg/m^3 for copper fume and 1 mg/m^3 for dusts and mists. The NIOSH IDLH is 100 mg/m^3.

Determination in Air: Copper dusts and mists are collected on a filter, worked up with acid, measured by atomic absorption. See NIOSH Method #7029 for copper. For copper fume: filter collection, acid digestion, measurement by atomic absorption.

Permissible Concentration in Water: See the entry on Copper.

Routes of Entry: Inhalation, ingestion.

Harmful Effects and Symptoms

Short Term Exposure: Inhaling Cupric acetate dust and vapors can irritate the respiratory tract causing coughing and wheezing. High levels may cause fluid to build up in the lungs (pulmonary edema), This can cause death. Corrosive: contact can irritate and may burn the skin and eyes. The LD_{50} oral-rat is 595 mg/kg (slightly toxic).

Long Term Exposure: Repeated exposure can cause skin allergy, thickening of the skin, and/or a green discoloration of the skin and hair. Repeated exposure can cause shrinking (atrophy) of the inner lining of the nose an may cause sores in the nose. Can cause liver and kidney damage.

Points of Attack: Skin, lung, liver, and kidney damage.

Medical Surveillance: For those with frequent or potentially high exposure (half the TLV or greater), the following are recommended before beginning work and at regular times after that: Lung function tests. If symptoms develop or overexposure is suspected, the following may be useful: Consider chest x-ray after acute overexposure. Serum and urine tests for copper can measure recent exposure. Liver and kidney function tests. Evaluation by a qualified allergist.

First Aid: If this chemical gets into the eyes, remove any contact lenses at once and irrigate immediately for at least 15 minutes, occasionally lifting upper and lower lids. Seek medical attention immediately. If this chemical contacts the skin, remove contaminated clothing and wash immediately with soap and water. Seek medical attention immediately. If this chemical has been inhaled, remove from exposure, begin rescue breathing (using universal precautions) if breathing has stopped and CPR if heart action has stopped. Transfer promptly to a medical facility. When this chemical has been swallowed, get medical attention. Give large quantities of water and induce vomiting. Do not make an unconscious person vomit. Medical observation is recommended for 24 – 48 hours after breathing overexposure, as pulmonary edema may be delayed. As first aid for pulmonary edema, a doctor or authorized paramedic may consider administering a corticosteroid spray.

Personal Protective Methods: Wear protective gloves and clothing to prevent any reasonable probability of skin contact. Safety equipment suppliers/manufacturers can provide recommendations on the most protective glove/clothing material for your operation. All protective clothing (suits, gloves, footwear, headgear) should be clean, available each day, and put on before work. Contact lenses should not be worn when working with this chemical. When working with liquids, wear splash-proof chemical goggles and face shield unless full facepiece respiratory protection is worn. When working with powders or dusts, wear dust-proof chemical goggles and face shield unless full facepiece respiratory protection is worn. Employees should wash immediately with soap when skin is wet or contaminated. Provide emergency showers and eyewash.

Respirator Selection: NIOSH/OSHA: (copper dust and mist) *5 mg/m^3:* DM (any dust and mist respirator). *10 mg/m^3:* DMXSQ, if not present as a fume (any dust and mist respirator except single-use and quarter mask respirators); or SA (any supplied-air respirator). *25 mg/m^3:* SA:CF (any supplied-air respirator operated in a continuous-flow mode); PAPRDM (any powered, air-purifying respirator with a dust and mist filter). *50 mg/m^3:* HiEF (any air-purifying, full-facepiece respirator with a high-efficiency particulate filter); or PAPRTHiE (any powered, air-purifying respirator with a tight-fitting facepiece and a high-efficiency particulate filter); or SCBAF (any self-contained breathing apparatus with a full facepiece); or SAF (any supplied-air respirator with a full facepiece). *100 mg/m^3:* SAF:PD,PP (any supplied-air respirator that has a full facepiece and is operated in a pressure-demand or other positive-pressure mode). *Emergency or planned entry into unknown concentrations or IDLH* *conditions:* SCBAF:PD,PP (any self-contained breathing apparatus that has a full facepiece and is operated in a pressure-demand or other positive-pressure mode); or SAF:PD, PP:ASCBA (any supplied-air respirator that has a full facepiece and is operated in a pressure-demand or other positive-pressure mode in combination with an auxiliary self-contained breathing apparatus operated in a pressure-demand or other positive- pressure mode). *Escape:* HiEF (any air-purifying, full-facepiece respirator with a high-efficiency particulate filter); or SCBAE (any appropriate escape-type, self-contained breathing apparatus).

Note: Substance reported to cause eye irritation or damage; may require eye protection.

NIOSH/OSHA: (copper fume) *Up to 1 mg/m³:* DMFu (any dust, mist, and fume respirator; or SA (any supplied-air respirator). *Up to 2.5 mg/m³:* SA:CF (any supplied-air respirator operated in a continuous-flow mode); or PAPRDMFu (any powered, air-purifying respirator with a dust, mist, and fume filter). *Up to 5 mg/m³:* HiEF (any air-purifying, full-facepiece respirator with a high-efficiency particulate filter); or SAT:CF (any supplied-air respirator that has a tight-fitting facepiece and is operated in a continuous-flow mode); or PAPRTHiE (any powered, air-purifying respirator with a tight-fitting facepiece and a high-efficiency particulate filter); or SCBAF (any self-contained breathing apparatus with a full facepiece); or SAF (any supplied-air respirator with a full facepiece). *Up to 100 mg/m³:* SAF:PD,PP (any supplied-air respirator that has a full facepiece and is operated in a pressure-demand or other positive-pressure mode). *Emergency or planned entry into unknown concentrations or IDLH conditions:* SCBAF:PD,PP (any self-contained breathing apparatus that has a full facepiece and is operated in a pressure-demand or other positive-pressure mode); or SAF: PD,PP:ASCBA (any supplied-air respirator that has a full facepiece and is operated in a pressure-demand or other positive-pressure mode in combination with an auxiliary self-contained breathing apparatus operated in a pressure-demand or other positive pressure mode). *Escape:* HiEF (any air-purifying, full-facepiece respirator with a high-efficiency particulate filter); or SCBAE (any appropriate escape-type, self-contained breathing apparatus).

Storage: Prior to working with this chemical you should be trained on its proper handling and storage. Store in tightly closed containers in a cool, well-ventilated area. Cupric acetate must be stored to avoid contact with acetylene gas, chemically active metals (such as potassium, sodium, magnesium, and zinc) since violent reactions occur.

Shipping: Cupric acetate is not specifically cited in DOT's Performance-Oriented Packaging Standards.

Spill Handling: Evacuate persons not wearing protective equipment from area of spill or leak until clean-up is complete. Remove all ignition sources. Collect powdered material in the most convenient and safe manner and deposit in sealed containers. Ventilate area after clean-up is complete. It may be necessary to contain and dispose of this chemical as a hazardous waste. If material or contaminated runoff enters waterways, notify downstream users of potentially contaminated waters. Contact your Department of Environmental Protection or your regional office of the federal EPA for specific recommendations. If employees are required to clean-up spills, they must be properly trained and equipped. OSHA 1910.120(q) may be applicable.

Fire Extinguishing: This chemical is a noncombustible solid. Use dry chemical, carbon dioxide, water spray, or foam extinguishers. Poisonous gases are produced in fire including acetic acid. If material or contaminated runoff enters waterways, notify downstream users of potentially contaminated waters. Notify local health and fire officials and pollution control agencies. Containers may explode in fire. From a secure, explosion-proof location, use water spray to cool exposed containers. If cooling streams are ineffective (venting sound increases in volume and pitch, tank discolors, or shows any signs of deforming), withdraw immediately to a secure position. If employees are expected to fight fires, they must be trained and equipped in OSHA 1910.156.

References

New Jersey Department of Health and Senior Services and Senior Services, "Hazardous Substance Fact Sheet: Cupric Acetate," Trenton, NJ (February 1999)

Cupric Nitrate

Molecular Formula: CuN_2O_6

Common Formula: $Cu(NO_3)_2$

Synonyms: Copper dinitrate; Copper(2+) nitrate; Copper(II) nitrate; Cupric dinitrate; Nitrato de cobre (Spanish); Nitric acid, copper(2+) salt; Nitric acid, copper(II) salt

CAS Registry Number: 3251-23-8

RTECS Number: WU7400000

DOT ID: UN 1479

Regulatory Authority

- Air Pollutant Standard Set (ACGIH)[1] (DFG)[3] (HSE)[33] (OSHA)[58]
- CLEAN WATER ACT: Section 311 Hazardous Substances/RQ 40CFR117.3 (same as CERCLA, see below); Section 313 Water Priority Chemicals (57FR41331, 9/9/92); Toxic Pollutant (Section 401.15). as copper and compounds
- RCRA 40CFR264, Appendix 9; TSD Facilities Ground Water Monitoring List. Suggested test method(s) (PQL μg/l): 6010 (60); 7210 (200) Note: All species in the ground water that contain copper are included
- SUPERFUND/EPCRA 40CFR302.4 Reportable Quantity (RQ): CERCLA, 100 lb (45.4 kg)
- EPCRA Section 313 Form R *de minimis* concentration reporting level: 1.0%

- Canada, WHMIS, Ingredients Disclosure List; National Pollutant Release Inventory (NPRI); CEPA Priority Substance List, Ocean dumping prohibited

Cited in U.S. State Regulations: California (A, G), Florida (G), Illinois (G), Massachusetts (G), New Hampshire (G), New Jersey (G), Oklahoma (G), Pennsylvania (G), Rhode Island (G).

Description: Cupric nitrate, $Cu(NO_3)_2$, is a blue crystalline solid. Boiling point = 172°C (decomposes below this point). Freezing/Melting point = 115°C. Soluble in water.

Potential Exposure: Cupric nitrate is used as an insecticide, in paint, varnish, enamel and in wood preservatives. Metal compounds are often used in "hot" operations in the workplace. These may include, but are not limited to, welding, brazing, soldering, plating, cutting, and metallizing. At the high temperatures reached in these operations, metals often form metal fumes which have different health effects and exposure standards than the original metal compound and require specialized controls.

Incompatibilities: A strong oxidizer. Aqueous solution is acidic; incompatible with bases. Violent reaction with potassium hexacyanoferrate; ammonia and potassium amide mixtures; acetic anhydrides, cyanides, ethers. Forms explosive materials with nitromethanes, sodium hypobromite, acetylene, chemically active metals such as potassium, sodium, etc. May ignite on contact with aluminum foil or tin. Risk of spontaneous combustion with combustibles (wood, cloth, etc.) organics, or reducing agents and readily oxidizable materials. Attacks metals in the presence of moisture.

Permissible Exposure Limits in Air: The Federal standard (OSHA PEL 8-hour TWA)[58] for copper fume is 0.1 mg/m^3, and 1 mg/m^3 for copper dusts and mists. NIOSH recommends the same level for a 10-hour workshift. ACGIH recommends a TWA of 0.2 mg/m^3 for copper fume and 1 mg/m^3 for dusts and mists. The NIOSH IDLH is 100 mg/m^3.

Determination in Air: Copper dusts and mists are collected on a filter, worked up with acid, measured by atomic absorption. See NIOSH Method #7029 for copper. For copper fume: filter collection, acid digestion, measurement by atomic absorption.

Permissible Concentration in Water: A limit of 1.0 μg/l in drinking water has been set for copper by EPA[6] and by the former USSR[35] as noted in the entry on copper.

Routes of Entry: Inhaltaion, ingestion.

Harmful Effects and Symptoms

Short Term Exposure: Skin and eye contact can cause irritation and burns. Inhalation can irritate the nose and throat, causing coughing and wheezing. Cupric nitrate may produce fumes that can cause "metal fume fever." Ingestion cause salivation, nausea, vomiting, stomach pain. May cause blood effects if swallowed. High exposure can cause unconsciousness. The LD_{50} oral-rat is 940 mg/kg (slightly toxic).

Long Term Exposure: Repeated exposure can cause copper to deposit in various parts of the body. Large deposits can make the skin and hair a green color. Repeated exposure can cause shrinking of the inner lining of the nose and may cause runny nose and sores. Excess deposits in the liver can cause liver damage. Metallic taste may also occur. Skin allergy with rash sometimes occurs. If allergy develops, even small future exposures can trigger rash. Repeated exposures can also cause thickening of the skin not caused by allergy.

Points of Attack: Skin, liver.

Medical Surveillance: For those with frequent or potentially high exposure (half the TLV or greater), the following are recommended before beginning work and at regular times after that: Lung function tests. If symptoms develop or overexposure is suspected, the following may be useful: Urine test for copper can measure recent exposure. Evaluation by a qualified allergist, including careful exposure history and special testing, may help diagnose skin allergy. Liver function tests.

First Aid: If this chemical gets into the eyes, remove any contact lenses at once and irrigate immediately for at least 15 minutes, occasionally lifting upper and lower lids. Seek medical attention immediately. If this chemical contacts the skin, remove contaminated clothing and wash immediately with soap and water. Seek medical attention immediately. If this chemical has been inhaled, remove from exposure, begin rescue breathing (using universal precautions) if breathing has stopped and CPR if heart action has stopped. Transfer promptly to a medical facility. When this chemical has been swallowed, get medical attention. Give large quantities of water and induce vomiting. Do not make an unconscious person vomit.

Note to Physician: In case of fume inhalation, treat pulmonary edema. Give prednisone or other corticosteroid orally to reduce tissue response to fume. Positive pressure ventilation may be necessary. Treat metal fume fever with bed rest, analgesics and antipyretics. The symptoms of metal fume fever may be delayed for 4 – 12 hours following exposure: it may last less than 36 hours.

Personal Protective Methods: Wear protective gloves and clothing to prevent any reasonable probability of skin contact. Safety equipment suppliers/manufacturers can provide recommendations on the most protective glove/clothing material for your operation. All protective clothing (suits, gloves, footwear, headgear) should be clean, available each day, and put on before work. Contact lenses should not be worn when working with this chemical. Wear dust-proof chemical goggles and face shield unless full facepiece respiratory protection is worn. Employees should wash immediately with soap when skin is wet or contaminated. Provide emergency showers and eyewash.

Respirator Selection: NIOSH/OSHA: (copper dust and mist) *5 mg/m^3:* DM (any dust and mist respirator). *10 mg/m^3:* DMXSQ, if not present as a fume (any dust and mist respirator except single-use and quarter mask respirators); or SA (any supplied-air respirator). *25 mg/m³:* SA:CF (any supplied-air respirator operated in a continuous-flow mode); PAPRDM (any powered, air-purifying respirator with a dust and mist filter). *50 mg/m³:* HiEF (any air-purifying, full-facepiece respirator with a high-efficiency particulate filter); or PAPRTHiE (any powered, air-purifying respirator with a tight-fitting facepiece and a high-efficiency particulate filter); or SCBAF (any self-contained breathing apparatus with a full facepiece); or SAF (any supplied-air respirator with a full facepiece). *100 mg/m³:* SAF:PD,PP (any supplied-air respirator that has a full facepiece and is operated in a pressure-demand or other positive-pressure mode). *Emergency or planned entry into unknown concentrations or IDLH conditions:* PD,PP (any self-contained breathing apparatus that has a full facepiece and is operated in a pressure-demand or other positive-pressure mode); or SAF:PD,PP:ASCBA (any supplied-air respirator that has a full facepiece and is operated in a pressure-demand or other positive-pressure mode in combination with an auxiliary self-contained breathing apparatus operated in a pressure-demand or other positive-pressure mode). *Escape:* HiEF (any air-purifying, full-facepiece respirator with a high-efficiency particulate filter); or SCBAE (any appropriate escape-type, self-contained breathing apparatus).

Note: Substance reported to cause eye irritation or damage; may require eye protection.

Storage: Prior to working with this chemical you should be trained on its proper handling and storage. Cupric nitrate must be stored to avoid contact with combustible, organic or other readily oxidizable materials, and chemically active metals (such as potassium, sodium, magnesium, and zinc) since violent reactions occur. Store in tightly closed containers in a cool well-ventilated area away from acetylene gas. Protect storage containers from physical damage. See OSHA standard 1910.104 and NFPA 43A *Code for the Storage of Liquid and Solid Oxidizers* for detailed handling and storage regulations.

Shipping: Oxidizing substances n.o.s., solid require an "Oxidizer" label. They fall in DOT Hazard Class 5.1 and Packing Group II.

Spill Handling: Evacuate persons not wearing protective equipment from area of spill or leak until clean-up is complete. Remove all ignition sources. Collect powdered material in the most convenient and safe manner and deposit in sealed containers. Ventilate area after clean-up is complete. Keep Cupric nitrate out of a confined space, such as a sewer, because of the possibility of an explosion, unless the sewer is designed to prevent the build-up of explosive concentrations. It may be necessary to contain and dispose of this chemical as a hazardous waste. If material or contaminated runoff enters waterways, notify downstream users of potentially contaminated waters. Contact your Department of Environmental Protection or your regional office of the federal EPA for specific recommendations. If employees are required to clean-up spills, they must be properly trained and equipped. OSHA 1910.120(q) may be applicable.

Fire Extinguishing: Copper nitrate itself is noncombustible, but it will increase the intensity of fire and may ignite combustible materials. Flooding amounts of water may be used to minimize its oxidizing effect on other materials. Caution: when large quantities are involved, application of water may cause scattering of molten material. Use dry chemical, carbon dioxide, water spray, or alcohol foam extinguishers. Decomposes below 170°C forming nitrogen oxides. If material or contaminated runoff enters waterways, notify downstream users of potentially contaminated waters. Notify local health and fire officials and pollution control agencies. From a secure, explosion-proof location, use water spray to cool exposed containers. If cooling streams are ineffective (venting sound increases in volume and pitch, tank discolors, or shows any signs of deforming), withdraw immediately to a secure position. If employees are expected to fight fires, they must be trained and equipped in OSHA 1910.156.

Disposal Method Suggested: Add slowly to water; stir in excess soda ash. Let stand then neutralize. Decant solution and flush to sewer; landfill sludge.

References

New Jersey Department of Health and Senior Services and Senior Services, "Hazardous Substance Fact Sheet: Cupric Nitrate," Trenton, NJ (February 1999)

Sax, N. I., Ed., "Dangerous Properties of Industrial Materials Report," 2, No. 5, 35–38 (1982) and 5, No. 6, 45–49 (1985)

Cupric Oxalate

Molecular Formula: $C_2H_2O_4 \cdot Cu$

Synonyms: Copper oxalate; Copper(II) oxalate; Ethanedioic acid, copper(2+) salt; Oxalic acid, copper(2+) salt

CAS Registry Number: 814-91-5

RTECS Number: RO2670000

DOT ID: UN 2775

Regulatory Authority

- Air Pollutant Standard Set (ACGIH) (OSHA) (NIOSH)
- CLEAN WATER ACT: Section 311 Hazardous Substances/RQ 40CFR117.3 (same as CERCLA, see below); Section 313 Water Priority Chemicals (57FR41331, 9/9/92); 40CFR401.15 Section 307 Toxic Pollutants, as copper and compounds
- SUPERFUND/EPCRA 40CFR302.4 Reportable Quantity (RQ): CERCLA, 100 lb (45.4 kg)
- RCRA 40CFR264, Appendix 9; TSD Facilities Ground Water Monitoring List. Suggested test method(s) (PQL μg/l): 6010 (60); 7210 (200) *Note:* All species in the ground water that contain copper are included
- EPCRA Section 313 Form R *de minimis* concentration reporting level: 1.0%

- Canada, WHMIS, Ingredients Disclosure List; National Pollutant Release Inventory (NPRI); CEPA Priority Substance List, Ocean dumping prohibited

Cited in U.S. State Regulations: California (A, G), New Jersey (G).

Description: Cupric oxalate is a bluish-white, odorless powder. Insoluble in water.

Potential Exposure: Used as a catalyst for organic reactions and in seed treatment as a repellent for birds and rodents.

Incompatibilities: Explosive materials are formed on contact with acetylene gas, ammonia, caustic solutions, sodium hypobromite, nitromethane. Slight heating can cause a weak explosion.

Permissible Exposure Limits in Air: The Federal standard (OSHA PEL 8-hour TWA)[58] for *copper fume* is 0.1 mg/m^3, and 1 mg/m^3 for *copper dusts and mists.* NIOSH recommends the same level for a 10-hour workshift. ACGIH recommends a TWA of 0.2 mg/m^3 for copper fume and 1 mg/m^3 for dusts and mists.

Determination in Air: Copper dusts and mists are collected on a filter, worked up with acid, measured by atomic absorption. See NIOSH Method #7029 for copper. For copper fume: filter collection, acid digestion, measurement by atomic absorption. See NIOSH Method #7200 for welding and brazing fume.

Permissible Concentration in Water: To protect freshwater aquatic life: 5.6 μg/l as a 24-hour average, never to exceed $e^{[0.94 \ln (\text{hardness}) - 1.23]}$μg/l. To protect human health: 1,000 μg/l.[6] Canada: Drinking Water Quality AO ≤ 1.0 mg/l. Mexico, Drinking Water = 1.0 mg/l

Determination in Water: Total copper may be determined by digestion followed by atomic absorption or by colorimetry (using neocuproine) or by Inductively Coupled Plasma (ICP) Optical Emission Spectrometry. Dissolved Copper may be determined by 0.45 μ filtration followed by the preceding methods.

Routes of Entry: Inhalation, ingestion.

Harmful Effects and Symptoms

Short Term Exposure: Powerful irritant. Contact with skin and/or eyes causes severe irritation, burns, and can cause permanent damage. Inhalation irritates the respiratory tract causing coughing and wheezing. Higher exposures can cause pulmonary edema, a medical emergency that can be delayed for several hours. This can cause death. If swallowed, Cupric oxalate has a caustic effect on the mouth, esophagus, and stomach, causing salivation, nausea, vomiting, diarrhea, and may cause damage to the kidneys.

Long Term Exposure: Repeated exposure can cause thickening of the skin, greenish color to the skin and hair, shrinking and perforation of the nasal septum with possible bleeding. May cause skin allergy and liver damage.

Points of Attack: Skin, liver.

Medical Surveillance: Liver function tests. Evaluation by a qualified allergist. Wilson's disease is a rare hereditary condition which interferes with the body's ability to get rid of copper. If you have this condition, consult your doctor about copper exposure.

First Aid: If this chemical gets into the eyes, remove any contact lenses at once and irrigate immediately for at least 15 minutes, occasionally lifting upper and lower lids. Seek medical attention immediately. If this chemical contacts the skin, remove contaminated clothing and wash immediately with soap and water. Seek medical attention immediately. If this chemical has been inhaled, remove from exposure, begin rescue breathing (using universal precautions) if breathing has stopped and CPR if heart action has stopped. Transfer promptly to a medical facility. When this chemical has been swallowed, get medical attention. Give large quantities of water and induce vomiting. Do not make an unconscious person vomit. Medical observation is recommended for 24 – 48 hours after breathing overexposure, as pulmonary edema may be delayed. As first aid for pulmonary edema, a doctor or authorized paramedic may consider administering a corticosteroid spray.

Personal Protective Methods: Wear protective gloves and clothing to prevent any reasonable probability of skin contact. Safety equipment suppliers/manufacturers can provide recommendations on the most protective glove/clothing material for your operation. All protective clothing (suits, gloves, footwear, headgear) should be clean, available each day, and put on before work. Contact lenses should not be worn when working with this chemical. Wear dust-proof chemical goggles and face shield unless full facepiece respiratory protection is worn. Employees should wash immediately with soap when skin is wet or contaminated. Provide emergency showers and eyewash.

Respirator Selection: *Copper dusts and mists:* NIOSH/OSHA: *5 mg/m^3:* DM (any dust and mist respirator). *10 mg/m^3:* DMXSQ, if not present as a fume (any dust and mist respirator except single-use and quarter mask respirators); or SA (any supplied-air respirator). *25 mg/m^3:* SA:CF (any supplied-air respirator operated in a continuous-flow mode); PAPRDM (any powered, air-purifying respirator with a dust and mist filter). *50 mg/m^3:* HiEF (any air-purifying, full-facepiece respirator with a high-efficiency particulate filter); or PAPRTHiE (any powered, air-purifying respirator with a tight-fitting facepiece and a high-efficiency particulate filter); or SCBAF (any self-contained breathing apparatus with a full facepiece); or SAF (any supplied-air respirator with a full facepiece). *100 mg/m^3:* SAF:PD,PP (any supplied-air respirator that has a full facepiece and is operated in a pressure-demand or other positive-pressure mode). *Emergency or planned entry into unknown concentrations or IDLH conditions:* SCBAF:PD,PP (any self-contained breathing apparatus that has a full facepiece and is operated in a pressure-demand or other positive-pressure mode); or SAF:PD,PP: ASCBA (any supplied-air respirator that has a full facepiece and is operated in a pressure-demand or other positive-pressure mode in combination with an auxiliary self contained breathing apparatus operated in a pressure-demand or other positive pressure mode).

Note: Substance reported to cause eye irritation or damage; may require eye protection.

Copper fume: NIOSH/OSHA: *Up to 1 mg/m³:* DMFu (any dust, mist, and fume respirator; or SA (any supplied-air respirator). *Up to 2.5 mg/m³:* SA:CF (any supplied-air respirator operated in a continuous-flow mode); or PAPRDMFu (any powered, air-purifying respirator with a dust, mist, and fume filter). *Up to 5 mg/m³:* HiEF (any air-purifying, full-facepiece respirator with a high-efficiency particulate filter); or SAT:CF (any supplied-air respirator that has a tight-fitting facepiece and is operated in a continuous-flow mode); or PAPRTHiE (any powered, air-purifying respirator with a tight-fitting facepiece and a high-efficiency particulate filter); or SCBAF (any self-contained breathing apparatus with a full facepiece); or SAF (any supplied-air respirator with a full facepiece). *Up to 100 mg/m³:* SAF:PD,PP (any supplied-air respirator that has a full facepiece and is operated in a pressure-demand or other positive-pressure mode *Emergency or planned entry into unknown concentrations or IDLH conditions:* SCBAF:PD,PP (any self-contained breathing apparatus that has a full facepiece and is operated in a pressure-demand or other positive-pressure mode); or SAF:PD,PP:ASCBA (any supplied-air respirator that has a full facepiece and is operated in a pressure-demand or other positive-pressure mode in combination with an auxiliary self-contained breathing apparatus operated in a pressure-demand or other positive pressure mode). *Escape:* HiEF (any air-purifying, full-facepiece respirator with a high-efficiency particulate filter); or SCBAE (any appropriate escape-type, self-contained breathing apparatus).

Storage: Prior to working with cupric oxalate you should be trained on its proper handling and storage. Store in tightly closed containers in a cool, well ventilated area away from acetylene gas, ammonia, caustic solutions, and nitromethane. Where possible, automatically pump liquid from drums or other storage containers to process containers.

Shipping: DOT label requirement of "Poison." Copper oxalate falls in UN/DOT Hazard Class 6.1 and Packing Group I or II.[19][20]

Spill Handling: Evacuate persons not wearing protective equipment from area of spill or leak until clean-up is complete. Remove all ignition sources. Collect powdered material in the most convenient and safe manner and deposit in sealed containers. Ventilate area after clean-up is complete. It may be necessary to contain and dispose of this chemical as a hazardous waste. If material or contaminated runoff enters waterways, notify downstream users of potentially contaminated waters. Contact your Department of Environmental Protection or your regional office of the federal EPA for specific recommendations. If employees are required to clean-up spills, they must be properly trained and equipped. OSHA 1910.120(q) may be applicable.

Fire Extinguishing: This chemical is a noncombustible solid. Use dry chemical, carbon dioxide, water spray, or foam extinguishers. Poisonous gases are produced in fire. If material or contaminated runoff enters waterways, notify downstream users of potentially contaminated waters. Notify local health and fire officials and pollution control agencies. Containers may explode in fire. From a secure, explosion-proof location, use water spray to cool exposed containers. If cooling streams are ineffective (venting sound increases in volume and pitch, tank discolors, or shows any signs of deforming), withdraw immediately to a secure position. If employees are expected to fight fires, they must be trained and equipped in OSHA 1910.156.

References

New Jersey Department of Health and Senior Services, "Hazardous Substance Fact Sheet, "Cupric Oxalate," Trenton, NJ (February 1999)

Cupriethylene Diamine

Molecular Formula: $C_2H_{10}CuN_2$

Common Formula: $Cu(C_2N_2H_{10})$

Synonyms: Complex; Copper-ethylenediamine complex; Cupriethylene diamine; Ethane, 1,2-diamino-, copper; Komeen®; Koplex® aquatic herbicide

CAS Registry Number: 13426-91-0

RTECS Number: KH8660000

DOT ID: UN 1761

Regulatory Authority

- CLEAN WATER ACT: Toxic Pollutant (Section 401.15) as copper and compounds
- RCRA 40CFR264, Appendix 9; TSD Facilities Ground Water Monitoring List. Suggested test method(s) (PQL µg/l): 6010 (60); 7210 (200) Note: All species in the ground water that contain copper are included
- EPCRA (Section 313): Includes any unique chemical substance that contains copper as part of that chemical's infrastructure. Form R *de minimis* concentration reporting level: 1.0%
- U.S. DOT 49CFR172.101, Appendix B, Regulated marine pollutant
- Canada, WHMIS, Ingredients Disclosure List; National Pollutant Release Inventory (NPRI); CEPA Priority Substance List, Ocean dumping prohibited

Cited in U.S. State Regulations: California (A, G), Maine (G), New Hampshire (G), New Jersey (G), Oklahoma (G), Pennsylvania (G).

Description: Cupriethylenediamine, $CCu(C_2N_2H_{10})$, is a purple liquid (may contain red or blue sediment), with an ammoniacal odor. Also described as a fishy odor. Boiling point = 100°C. It is non-flammable.

Potential Exposure: Cupriethylene diamine is used to dissolve cellulose products to give a cuprammonium-type solution.

Incompatibilities: Violent reaction with water. A powerful reducing agent. Violent reaction with oxidizers, organic materials, and many other substances. Forms unstable peroxides under normal conditions of temperature and storage. Dissolves wood, cotton and other cellulosic material.

Permissible Exposure Limits in Air: No standards set.

Permissible Concentration in Water: No criteria set but reference should be made to the entry on copper for possible application to copper ions introduced to water by Cupriethylene diamine.

Routes of Entry: Inhalation.

Harmful Effects and Symptoms

Short Term Exposure: Cupriethylene diamine is a corrosive chemical and can cause severe irritation and burns of the skin and eyes on contact. Exposure to Cupriethylene diamine can irritate the throat and air passages.

Long Term Exposure: Repeated exposure can cause thickening of the skin or a green color to form on the skin and hair. Repeated exposure can cause shrinking (atrophy) of the inner lining of the nose and may cause sores in the nose and watery discharge. Excessive buildup of cooper can cause liver damage.

Points of Attack: Skin, lungs, liver.

Medical Surveillance: Before beginning employment and at regular times after that, for those with frequent or potentially high exposures, the following are recommended: Lung function tests. If symptoms develop or overexposure is suspected, the following may be useful: Evaluation by a qualified allergist, including careful exposure history and special testing, may help diagnose skin allergy. Urine Copper test. Liver function tests.

First Aid: If this chemical gets into the eyes, remove any contact lenses at once and irrigate immediately for at least 15 minutes, occasionally lifting upper and lower lids. Seek medical attention immediately. If this chemical contacts the skin, remove contaminated clothing and wash immediately with soap and water. Seek medical attention immediately. If this chemical has been inhaled, remove from exposure, begin rescue breathing (using universal precautions) if breathing has stopped and CPR if heart action has stopped. Transfer promptly to a medical facility. When this chemical has been swallowed, get medical attention. If victim is conscious, administer water or milk. Do not induce vomiting.

Personal Protective Methods: Wear solvent-resistant gloves and clothing to prevent any reasonable probability of skin contact. Safety equipment suppliers/manufacturers can provide recommendations on the most protective glove/clothing material for your operation. All protective clothing (suits, gloves, footwear, headgear) should be clean, available each day, and put on before work. Contact lenses should not be worn when working with this chemical. Wear splash-proof chemical goggles and face shield unless full facepiece respiratory protection is worn. Employees should wash immediately with soap when skin is wet or contaminated. Remove nonimpervious clothing immediately if wet or contaminated. Provide emergency showers and eyewash.

Respirator Selection: NIOSH/OSHA: (copper dust and mist) *5 mg/m³:* DM (any dust and mist respirator). *10 mg/m³:* DMXSQ, if not present as a fume (any dust and mist respirator except single-use and quarter mask respirators); or SA (any supplied-air respirator). *25 mg/m³:* SA:CF (any supplied-air respirator operated in a continuous-flow mode); PAPRDM (any powered, air-purifying respirator with a dust and mist filter). *50 mg/m³:* HiEF (any air-purifying, full-facepiece respirator with a high-efficiency particulate filter); or PAPRTHiE (any powered, air-purifying respirator with a tight-fitting facepiece and a high-efficiency particulate filter); or SCBAF (any self-contained breathing apparatus with a full facepiece); or SAF (any supplied-air respirator with a full facepiece). *100 mg/m³:* SAF:PD,PP (any supplied-air respirator that has a full facepiece and is operated in a pressure-demand or other positive-pressure mode). *Emergency or planned entry into unknown concentrations or IDLH conditions:* PD,PP (any self-contained breathing apparatus that has a full facepiece and is operated in a pressure-demand or other positive-pressure mode); or SAF:PD,PP:ASCBA (any supplied-air respirator that has a full facepiece and is operated in a pressure-demand or other positive-pressure mode in combination with an auxiliary self-contained breathing apparatus operated in a pressure-demand or other positive- pressure mode). *Escape:* HiEF (any air-purifying, full-facepiece respirator with a high-efficiency particulate filter); or SCBAE (any appropriate escape-type, self-contained breathing apparatus).

Note: Substance reported to cause eye irritation or damage; may require eye protection.

Storage: Prior to working with this chemical you should be trained on its proper handling and storage. Store in tightly closed containers in a cool, well ventilated area away from incompatible materials listed above. Metal containers involving the transfer of this chemical should be grounded and bonded. Drums must be equipped with self-closing valves, pressure vacuum bungs, and flame arresters. Use only non-sparking tools and equipment, especially when opening and closing containers of this chemical. Sources of ignition such as smoking and open flames, are prohibited where this chemical is used, handled, or stored in a manner that could create a potential fire or explosion hazard.

Shipping: Cupriethylene diamine solution requires a "Corrosive, Poison" label according to DOT.[19] It falls in Hazard Class 8 and Packing Group II.

Spill Handling: Evacuate and restrict persons not wearing protective equipment from area of spill or leak until cleanup is complete. Remove all ignition sources. Ventilate area of spill or leak. Absorb liquids in vermiculite, dry sand, earth, peat, carbon, or a similar material and deposit in sealed containers. It may be necessary to contain and dispose of this chemical as a hazardous waste. If material or contaminated runoff enters waterways, notify downstream users of potentially contaminated waters. Contact your Department of Environmental Protection or your regional office of the federal EPA for specific recommendations. If employees are required to clean-up spills, they must be properly trained and equipped. OSHA 1910.120(q) may be applicable.

Fire Extinguishing: Cupriethylene diamine may burn, but does not readily ignite. Use dry chemical, carbon dioxide, or foam extinguishers. Poisonous gases are produced in fire including nitrous oxides. If material or contaminated runoff enters waterways, notify downstream users of potentially contaminated waters. Notify local health and fire officials and pollution control agencies. Containers may explode in fire. From a secure, explosion-proof location, use water spray to cool exposed containers. If cooling streams are ineffective (venting sound increases in volume and pitch, tank discolors, or shows any signs of deforming), withdraw immediately to a secure position. If employees are expected to fight fires, they must be trained and equipped in OSHA 1910.156.

References

New Jersey Department of Health and Senior Services and Senior Services, "Hazardous Substance Fact Sheet: Cupriethylene Diamine," Trenton, NJ (February 5, 1988)

Cyanamide

Molecular Formula: CH_2N_2

Common Formula: H_2NCN

Synonyms: Amidocyanogen; Carbamonitrile; Carbimide; Carbodiimide; Cyanogenamide; Cyanogen nitride

CAS Registry Number: 420-04-2

RTECS Number: GS5950000

DOT ID: UN 3276 (nitriles, toxic, n.o.s.)

EEC Number: 615-013-00-2

Regulatory Authority

- Air Pollutant Standard Set (ACGIH)[1] (Australia) (HSE)[33] (Israel) (Mexico) (former USSR)[43] (Several States)[60] (Several Canadian Provinces)

As cyanide compounds:

- CLEAN AIR ACT: Hazardous Air Pollutants (Title I, Part A, Section 112)
- CLEAN WATER ACT: 40CFR423, Appendix A, Priority Pollutants as cyanide, total
- EPCRA (Section 313): X+CN- where X = H+ or any other group where a formal dissociation may occur. For example, KCN or $Ca(CN)_2$. Form R *de minimis* concentration reporting level: 1.0%
- U.S. DOT Regulated Marine Pollutant (49CFR172.101, Appendix B) as cyanide mixtures, cyanide solutions
- Canada, WHMIS, Ingredients Disclosure List

Cited in U.S. State Regulations: Alaska (G), California (A, G), Connecticut (A), Florida (G, A), Illinois (G), Maine (G), Massachusetts (G), Minnesota (G), Nevada (A), New Hampshire (G), New Jersey (G), New York (A), North Dakota (A), Pennsylvania (G), Rhode Island (G), South Carolina (A), Virginia (A), West Virginia (G).

Description: Cyanamide, H_2NCN, is a combustible crystalline solid, but it is usually found as a 25% liquid solution. Boiling point = (decomposes) 260°C. Freezing/Melting point = 45°C. Flash point = 141°C. Hazard Identification (based on NFPA-704 M Rating System): Health 4, Flammability 1, Reactivity 3. Soluble in water.

Potential Exposure: Cyanamide may be melted to give a dimer, dicyandiamide or cyanoguanidine. At higher temperatures it gives the trimer, melamine, a raw material for melamine-formaldehyde resins.

Incompatibilities: Cyanamide may polymerize at temperatures above 122°C, or on evaporation of aqueous solutions. Reacts with acids, strong oxidants, strong reducing agents and water causing explosion and toxic hazard. Attacks various metals. Decomposes when heated above 49°C, on contact with acids, bases, 1,2-phenylene diamine salts, and moisture producing toxic fumes including nitrogen oxides and cyanides.

Permissible Exposure Limits in Air: There is no OSHA PEL. NIOSH and ACGIH[1] recommend a TWA of 2 mg/m^3 as does HSE,[33] Australia, Israel, and Mexico. The Canadian provinces of Alberta, BC, Ontario, and Quebec use the same TWA as well, and Alberta's STEL is 4 mg/m^3. The former USSR-UNEP/IRPTC project[43] has set an MAC of 0.5 mg/m^3 in workplace air and limits for ambient air in residential areas of 0.01 mg/m^3 on an average daily basis.

Several states have set guidelines or standards for Cyanamide in ambient air[60] ranging from 6.7 $\mu g/m^3$ (New York) to 20.0 $\mu g/m^3$ (Florida and North Dakota) to 35 $\mu g/m^3$ (Virginia) to 40 $\mu g/m^3$ (Connecticut) to 50 $\mu g/m^3$ (South Carolina).

Determination in Air: Filter; none; Gravimetric; NIOSH IV Method #0500, Particulates NOR (total).

Permissible Concentration in Water: No criteria set.

Routes of Entry: Inhalation, skin absorption, ingestion, skin and/or eye contact.

Harmful Effects and Symptoms

Short Term Exposure: Cyanamide is caustic and severely irritates the eyes, skin, and respiratory tract, and may affect the liver. Ingestion or inhalation may cause transitory intense redness of the face, headache, vertigo, increased respiration, tachycardia and hypotensions. The adverse effects off Cyanamide are potentiated by the ingestion of alcohol (beer, wine or liquor) within 1 – 2 days before or after exposure. Cyanamide is a highly reactive chemical and is a dangerous explosion hazard.

Long Term Exposure: Repeated or prolonged contact may cause skin sensitization and allergy. Exposure may cause liver and nervous system damage.

Points of Attack: Liver, skin.

Medical Surveillance: If overexposure occurs or if illness is suspected, the following are recommended: Liver function tests. Exam of the nervous system. Examination by a qualified allergist.

First Aid: If this chemical gets into the eyes, remove any contact lenses at once and irrigate immediately for at least 15 minutes, occasionally lifting upper and lower lids. Seek medical attention immediately. If this chemical contacts the skin, remove contaminated clothing and wash immediately with soap and water. Seek medical attention immediately. If this chemical has been inhaled, remove from exposure, begin rescue breathing (using universal precautions) if breathing has stopped and CPR if heart action has stopped. Transfer promptly to a medical facility. When this chemical has been swallowed, get medical attention. Give large quantities of water and induce vomiting. Do not make an unconscious person vomit.

Personal Protective Methods: Wear protective gloves and clothing to prevent any reasonable probability of skin contact. Safety equipment suppliers/manufacturers can provide recommendations on the most protective glove/clothing material for your operation. All protective clothing (suits, gloves, footwear, headgear) should be clean, available each day, and put on before work. Contact lenses should not be worn when working with this chemical. When working with liquids, wear splash-proof chemical goggles and face shield unless full facepiece respiratory protection is worn. When working with powders or dusts, wear dust-proof chemical goggles and face shield unless full facepiece respiratory protection is worn. Employees should wash immediately with soap when skin is wet or contaminated. Provide emergency showers and eyewash. See NIOSH Criteria Document 78–212 NITRILES.

Respirator Selection: Where the potential exists for exposures over 2 mg/m^3, use a MSHA/NIOSH approved supplied-air respirator with a full facepiece operated in the positive pressure mode or with a full facepiece, hood, or helmet in the continuous flow mode, or use a MSHA/NIOSH approved self-contained breathing apparatus with a full facepiece operated in pressure-demand or other positive pressure mode.

Storage: Prior to working with this chemical you should be trained on its proper handling and storage. Store in tightly closed containers in a cool well-ventilated area away from acids or acid fumes. Cyanamide can be stored in glass containers if it is stabilized with phosphoric, acetic, sulfuric, or boric acid; it attacks iron and steel, copper and brass.

Sources of ignition such as smoking and open flames are prohibited where Cyanamide is used, handled, or stored in a manner that could create a potential fire or explosion hazard. Wherever Cyanamide is used, handled, manufactured, or stored, use explosion-proof electrical equipment and fittings.

Shipping: Hazard Class: 6.1. Label: "Poison."

Spill Handling: Evacuate persons not wearing protective equipment from area of spill or leak until clean-up is complete. Remove all ignition sources. Ventilate area of spill or leak. Collect powdered material in the most convenient and safe manner and deposit in sealed containers. Absorb liquids in vermiculite, dry sand, earth, peat, carbon, or a similar material and deposit in sealed containers. It may be necessary to contain and dispose of this chemical as a hazardous waste. Keep Cyanamide out of a confined space, such as a sewer, because of the possibility of an explosion, unless the sewer is designed to prevent the build-up of explosive concentrations. If material or contaminated runoff enters waterways, notify downstream users of potentially contaminated waters. Contact your Department of Environmental Protection or your regional office of the federal EPA for specific recommendations. If employees are required to clean-up spills, they must be properly trained and equipped. OSHA 1910.120(q) may be applicable.

Fire Extinguishing: Cyanamide may burn, but does not readily ignite. Poisonous gas is produced in fire including nitrogen oxides and cyanides. Use dry chemical or CO_2 extinguishers. Containers may explode in fire. If material or contaminated runoff enters waterways, notify downstream users of potentially contaminated waters. Notify local health and fire officials and pollution control agencies. From a secure, explosion-proof location, use water spray to cool exposed containers. If cooling streams are ineffective (venting sound increases in volume and pitch, tank discolors, or shows any signs of deforming), withdraw immediately to a secure position. If employees are expected to fight fires, they must be trained and equipped in OSHA 1910.156.

Disposal Method Suggested: Add excess alkaline calcium hypochlorite with agitation. Flush to sewer after 24 hours. Cyanamide can also be destroyed in an incinerator equipped with afterburner and scrubber.

References

New Jersey Department of Health and Senior Services and Senior Services, "Hazardous Substance Fact Sheet: Cyanamide," Trenton, NJ (April 1986)

Sax, N. I., Ed., "Dangerous Properties of Industrial Materials Report," 8, No. 5, 65–68 (1988)

Cyanazine

Molecular Formula: $C_9H_{13}ClN_6$

Synonyms: Bladex®; Bladex® 80WP; 2-Chloro-4-(1-cyano-1-methylethylamino)-6-ethylamino-1,3,5-triazine; 2-Chloro-4-ethylamino-6-(1-cyano-1-methyl)ethylamino-*s*-triazine; 2-([4-Chloro-6-(ethylamino)-*s*-triazin-2-yl)amino)-2-methylpropanenitrile; 2-([4-Chloro-6-(ethylamino)-1,3,5-triazin-2-yl]amino)-2-methylpropanenitrile; 2-([4-Chloro-6-(ethylamino)-*s*-triazin-2-yl]amino)-2-methylpropionitrile; 2-(4-Chloro-6-ethylamino-1,3,5-triazin-2-ylamino)-2-methylpropionitrile; Cyanazine triazine pesticide; 1-Cyano-1-methylethyl)amino]-6-(ethylamino)-*s*-triazine; DW 3418®; EPA pesticide chemical code 100101; Fortrol®; Payze®; Propanenitrile, 2-([4-chloro-6-(ethylamino)-*s*-triazin-2-yl]amino)-2-methyl-; Propanenitrile, 2-([4-Chloro-6-(ethylamino)-1,3,5-triazin-2-yl]amino)-2-methyl-; SD 15418®; *s*-Triazine, 2-chloro-4-ethylamino-6-(1-cyano-1-methyl)ethylamino-; WL 19805®

CAS Registry Number: 21725-46-2

RTECS Number: UG1490000

DOT ID: UN 2588

EEC Number: 613-013-00-7

Regulatory Authority

- SAFE DRINKING WATER ACT: Priority List (55 FR 1470)
- EPCRA Section 313 Form R *de minimis* concentration reporting level: 1.0%
- CLEAN AIR ACT: Hazardous Air Pollutants (Title I, Part A, Section 112)
- CLEAN WATER ACT: 40CFR423, Appendix A, Priority Pollutants as cyanide, total
- EPA HAZARDOUS WASTE NUMBER (RCRA No.): P030 as cyanides soluble salts and complexes, n.o.s.
- RCRA, 40CFR261, Appendix 8 Hazardous Constituents. as cyanides, soluble salts and complexes, n.o.s.
- EPCRA (Section 313): X+CN- where X = H+ or any other group where a formal dissociation may occur. For example, KCN or $Ca(CN)_2$. Form R de minimis concentration reporting level: 1.0%
- U.S. DOT Regulated Marine Pollutant (49CFR172.101, Appendix B) as cyanide mixtures, cyanide solutions or cyanides, inorganic, n.o.s.
- Canada, Drinking Water Quality, IMAC = 0.01 mg/l

Cited in U.S. State Regulations: California (A, G), Kansas (G), Massachusetts (G), New Jersey (G), Pennsylvania (G)

Description: Cyanazine, $C_9H_{13}ClN_6$, is an off-white to tan crystalline solid. Freezing/Melting point = 167 – 169°C. Hazard Identification (based on NFPA-704 M Rating System): Health 2, Flammability 1, Reactivity 0. Soluble in water. Physical properties may be altered by carrier solvents used in commercial formulations.

Potential Exposure: Those involved in the manufacture, formulation and application of this herbicide.

Incompatibilities: Cyanazine decomposes in heat producing very toxic fumes and gases of hydrogen cyanide, hydrogen chloride, ethyl chloride, ammonia, acetone, and ethylene. Attacks metals in the presence of heat and moisture.

Permissible Exposure Limits in Air: No standards set. This chemical can be absorbed through the skin, thereby increasing exposure.

Determination in Air: See NIOSH Criteria Document 78-212 NITRILES.

Permissible Concentration in Water: A no-observed adverse effects level of 1 mg/kg/day has been determined by EPA. This has resulted in a drinking water equivalent level of 0.13 mg/l for a 10 kg child and a level of 0.046 mg/l on a long term basis. This also results in a lifetime health advisory of 0.009 mg/l for an adult. Kansas[61] has set a guideline for cyanazine in drinking water of 42 μg/l. Canada's IMAC for drinking water is 0.01 mg/l.

Determination in Water: High-performance liquid chromatography is applicable to the determination of Cyanazine in water according to EPA.

Routes of Entry: Inhalation, passing through the unbroken skin.

Harmful Effects and Symptoms

Short Term Exposure: This chemical can be absorbed through the skin, thereby increasing exposure. Exposure can irritate the nose, throat and bronchial tubes. Contact can irritate the skin or eyes. Overexposure can cause weakness, nausea and difficulty breathing. The oral LD_{50} rat is 149 mg/kg (moderately toxic). Toxicological properties may be altered by carrier solvents used in commercial formulations.

Long Term Exposure: Long-term effects are unknown. Related chemicals in the triazin chemical groups can cause liver damage, reduce thyroid function and/or cause skin allergy. May cause reproductive toxicity in humans.

Medical Surveillance: Liver function tests. Thyroid function tests. Evaluation by a qualified allergist.

First Aid: If this chemical gets into the eyes, remove any contact lenses at once and irrigate immediately for at least 15 minutes, occasionally lifting upper and lower lids. Seek medical attention immediately. If this chemical contacts the skin, remove contaminated clothing and wash immediately with soap and water. Seek medical attention immediately. If this chemical has been inhaled, remove from exposure, begin rescue breathing (using universal precautions) if breathing has stopped and CPR if heart action has stopped. Transfer promptly to a medical facility. When this chemical has been swallowed, get medical attention. Give large quantities of water and induce vomiting. Do not make an unconscious person vomit.

Personal Protective Methods: Wear protective gloves and clothing to prevent any reasonable probability of skin contact. Safety equipment suppliers/manufacturers can provide recommendations on the most protective glove/clothing material for your operation. All protective clothing (suits, gloves, footwear, headgear) should be clean, available each day, and put on before work. Contact lenses should not be worn when working with this chemical. Wear dust-proof chemical goggles and face shield unless full facepiece respiratory protection is worn. Employees should wash immediately with soap when skin is wet or contaminated. Provide emergency showers and eyewash. See NIOSH Criteria Document 78–212 NITRILES.

Respirator Selection: Engineering control must be effective to ensure that exposure to Cyanazine does not occur. Where the potential exists for exposures to Cyanazine use a MSHA/NIOSH approved full facepiece respirator with a pesticide cartridge. Increased protection is obtained from full facepiece air purifying respirators.

Where the potential for high exposures exists, use a MSHA/NIOSH approved supplied-air-respirator with a full facepiece operated in the positive pressure mode or with a full facepiece, hood, or helmet in the continuous flow mode, or use a MSHA/NIOSH approved self-contained breathing apparatus with a full facepiece operated in pressure-demand or other positive pressure mode.

Storage: Prior to working with Cyanazine you should be trained on its proper handling and storage. Store in tightly closed containers in a cool, well-ventilated area away from heat. Where possible, automatically pump liquid from drums or other storage containers to process containers.

Shipping: Pesticides, solid, toxic, n.o.s. are in Hazard Class 6.1 and cyanazine falls in Packing Group III. It requires a "Keep Away From Food" label.

Spill Handling: Evacuate persons not wearing protective equipment from area of spill or leak until clean-up is complete. Remove all ignition sources. If appropriate, moisten to prevent dust. Collect powdered material in the most convenient and safe manner and deposit in sealed containers. Ventilate area after clean-up is complete. It may be necessary to contain and dispose of this chemical as a hazardous waste. If material or contaminated runoff enters waterways, notify downstream users of potentially contaminated waters. Contact your Department of Environmental Protection or your regional office of the federal EPA for specific recommendations. If employees are required to clean-up spills, they must be properly trained and equipped. OSHA 1910.120(q) may be applicable.

Fire Extinguishing: Cyanazine may burn, but does not readily ignite. Use dry chemical, CO_2, water spray, or foam extinguishers. Poisonous gases are produced in fire including hydrogen cyanide, hydrogen chloride, ethyl chloride, ammonia, acetone and ethylene. If material or contaminated runoff enters waterways, notify downstream users of potentially contaminated waters. Notify local health and fire officials and pollution control agencies. From a secure, explosion-proof location, use water spray to cool exposed containers. If cooling streams are ineffective (venting sound increases in volume and pitch, tank discolors, or shows any signs of deforming), withdraw immediately to a secure position. If employees are expected to fight fires, they must be trained and equipped in OSHA 1910.156

Disposal Method Suggested: Incineration. In accordance with 40CFR165 recommendations for the disposal of pesticides and pesticide containers. Must be disposed properly by following package label directions or by contacting your state pesticide or environmental control agency or by contacting your regional EPA office.

References

Sax, N. I., Ed., "Dangerous Properties of Industrial Materials Report," 3, No. 1, 47–50, New York, Van Nostrand Reinhold Co. (1983)

U.S. Environmental Protection Agency, "Chemical Profile: Cyanazine," Washington, DC, Office of Drinking Water (August 1987)

New Jersey Department of Health and Senior Services and Senior Services, "Hazardous Substance Fact Sheet: Bladex," Trenton, NJ (November, 1986)

Cyanides

Molecular Formula: CKN, CNNa, KCN, NaCN

Synonyms: Carbon nitride ion (CN); Cianuro (Spanish); Cyanide anion; Cyanure (French); Isocyanide; KCN = potassium cyanide; NaCN = sodium cyanide (see the potassium and sodium cyanide entries for specific synonyms)

CAS Registry Number: 57-12-5 (cyanide ion); 151-50-8 (potassium cyanide); 143-33-9 (sodium cyanide)

RTECS Number: GS7175000 (cyanide ion); TS8750000 (potassium cyanide); VZ7520000 (sodium cyanide)

DOT ID: UN 1588 (cyanide ion); UN 1680 (potassium cyanide); UN 1689 (sodium cyanide)

EEC Number: 006-007-00-5

Regulatory Authority

- Banned or Severely Restricted (In Agriculture) (Germany) (In Consumer Products) (US)[13]
- Air Pollutant Standard Set (ACGIH)[1] (Australia) (DFG)[3] (HSE)[33] (Mexico) (OSHA)[58] (Several States)[60] (Several Canadian Provinces)

Cyanide:

- CLEAN WATER ACT: Section 313 Water Priority Chemicals (57FR41331, 9/9/92)
- RCRA 40CFR268.48; 61FR15654, Universal Treatment Standards: Wastewater (mg/l), 1.2; Nonwastewater (mg/kg), 590; RCRA 40CFR268.48; 61FR15654, Universal Treatment Standards as cyanides (total): Wastewater

(mg/l), 0.86; Nonwastewater (mg/kg), 30 as cyanides (amenable) *Note*: Both Cyanides (Total) and Cyanides (Amenable) for nonwastewaters are to be analyzed using Method 9010 or 9012, found in "Test Methods for Evaluating Solid Waste, Physical/Chemical Methods," EPA Publication SW-846, as incorporated by reference in 40 CFR 260.11, with a sample size of 10 grams and a distillation time of one hour and 15 minutes

- SAFE DRINKING WATER ACT: MCL, 0.2 mg/l; MCLG, 0.2 mg/l as free cyanide; Regulated chemical (47 FR 9352)
- U.S. DOT Regulated Marine Pollutant (49CFR172.101, Appendix B) as cyanides, inorganic, n.o.s.
- Canada, WHMIS, Ingredients Disclosure List; National Pollutant Release Inventory (NPRI); CEPA Priority Substance List, Ocean dumping prohibited

Potassium cyanide:

- CLEAN WATER ACT: Section 311 Hazardous Substances/RQ 40CFR117.3 (same as CERCLA, see below); Section 313 Water Priority Chemicals (57FR41331, 9/9/92)
- EPA HAZARDOUS WASTE NUMBER (RCRA No.): P098
- RCRA, 40CFR261, Appendix 8 Hazardous Constituents
- SUPERFUND/EPCRA 40CFR355, Appendix B Extremely Hazardous Substances: TPQ = 100 lb (45.4 kg)
- SUPERFUND/EPCRA 40CFR302.4 Reportable Quantity (RQ): CERCLA, 10 lb (4.54 kg)
- U.S. DOT Regulated Marine Pollutant (49CFR172.101, Appendix B)
- Canada, WHMIS, Ingredients Disclosure List; National Pollutant Release Inventory (NPRI); CEPA Priority Substance List, Ocean dumping prohibited

Sodium cyanide:

- CLEAN WATER ACT: Section 311 Hazardous Substances/RQ 40CFR117.3 (same as CERCLA, see below); Section 313 Water Priority Chemicals (57FR41331, 9/9/92)
- EPA HAZARDOUS WASTE NUMBER (RCRA No.): P106
- RCRA, 40CFR261, Appendix 8 Hazardous Constituents
- SUPERFUND/EPCRA 40CFR355, Appendix B Extremely Hazardous Substances: TPQ = 100 lb (45.4 kg)
- SUPERFUND/EPCRA 40CFR302.4 Reportable Quantity (RQ): CERCLA, 10 lb (4.54 kg)
- EPCRA Section 313: See Cyanide Compounds
- U.S. DOT Regulated Marine Pollutant (49CFR172.101, Appendix B)
- Canada, WHMIS, Ingredients Disclosure List; National Pollutant Release Inventory (NPRI); CEPA Priority Substance List, Ocean dumping prohibited

Cyanide compounds:

- CLEAN AIR ACT: Hazardous Air Pollutants (Title I, Part A, Section 112)
- CLEAN WATER ACT: 40CFR423, Appendix A, Priority Pollutants as cyanide, total
- EPA HAZARDOUS WASTE NUMBER (RCRA No.): P030 as cyanides soluble salts and complexes, n.o.s.
- RCRA, 40CFR261, Appendix 8 Hazardous Constituents. as cyanides, soluble salts and complexes, n.o.s.
- EPCRA (Section 313): X+CN- where X = H+ or any other group where a formal dissociation may occur. For example, KCN or $Ca(CN)_2$. Form R *de minimis* concentration reporting level: 1.0%
- U.S. DOT Regulated Marine Pollutant (49CFR172.101, Appendix B) as cyanide mixtures, cyanide solutions or cyanides, inorganic, n.o.s.
- Canada, WHMIS, Ingredients Disclosure List; National Pollutant Release Inventory (NPRI); CEPA Priority Substance List, Ocean dumping prohibited

Cyanides, soluble salts and compounds:

- CLEAN AIR ACT: Hazardous Air Pollutants (Title I, Part A, Section 112)
- CLEAN WATER ACT: 40CFR423, Appendix A, Priority Pollutants as cyanide, total; Toxic Pollutant (Section 401.15)
- EPA HAZARDOUS WASTE NUMBER (RCRA No.): P030
- RCRA, 40CFR261, Appendix 8 Hazardous Constituents
- RCRA 40CFR268.48; 61FR15654, Universal Treatment Standards: Wastewater (mg/l), 1.2 (total); 0.86 (amenable); Nonwastewater (mg/kg), 590 (total); 30 (amenable)
- RCRA 40CFR264, Appendix 9; TSD Facilities Ground Water Monitoring List. Suggested test method(s) (PQL μg/l): 9010 (40)
- SAFE DRINKING WATER ACT: MCL, 0.2 mg/l; MCLG, 0.2 mg/l; Regulated chemical (47 FR 9352)
- SUPERFUND/EPCRA 40CFR302.4 Reportable Quantity (RQ): CERCLA, 10 lb (4.54 kg)
- EPCRA Section 313 Form R *de minimis* concentration reporting level: 1.0%
- U.S. DOT Regulated Marine Pollutant (49CFR172.101, Appendix B) as cyanides, inorganic, n.o.s.
- Canada, WHMIS, Ingredients Disclosure List; National Pollutant Release Inventory (NPRI); CEPA Priority Substance List, Ocean dumping prohibited

Cited in U.S. State Regulations: Alaska (G), Arizona (W), California (A, G), Connecticut (A), Florida (G, A), Illinois (G), Kansas (G, W), Louisiana (G), Maine (G), Maryland (G), Massachusetts (G), Michigan (G), Minnesota (G), Nevada (A), New Hampshire (G), New Jersey (G), New York (A), North Dakota (A), Oklahoma (G), Pennsylvania (G), Rhode Island (G), South Carolina (A), South Dakota (A), Vermont (G), Virginia (G, A), Washington (G), West Virginia (G), Wisconsin (G).

Description: KCN and NaCN are white crystalline solids with a faint almond odor. Sodium cyanide also has a slight odor of hydrocyanic acid when damp. *KCN:* Boiling point = 1,625°C. Freezing/Melting point = 634°C. *NaCN:* Boiling point = 1,496°C. Freezing/Melting point = 564°C. NFPA 704 M

Hazard Identification (KCN and NaCN): Health 3, Flammability 0, Reactivity 0. Soluble in water.

Potential Exposure: Sodium and potassium cyanides are used primarily in the extraction of ores, electroplating, metal treatment, and various manufacturing processes.

Incompatibilities: The aqueous solution of potassium and sodium cyanide are highly corrosive, and strong bases. KCN and NaCN react violently with acids releases highly flammable hydrogen cyanide. Potassium and sodium cyanide are incompatible with strong oxidizers (such as acids, acid salts, chlorates, nitrates), organic anhydrides, isocyanates, alkylene oxides, epichlorohydrin, aldehydes, alcohols, glycols, phenols, cresols, caprolactum. Reacts with water forming hydrogen cyanide. Attacks aluminum, copper, zinc in the presence of moisture. KCN and NaCN absorb moisture from the air forming a corrosive syrup.

Permissible Exposure Limits in Air: The OSHA PEL 8-hour TWA is 5 mg/m^3 (4.7 ppm). The PEL does not include hydrogen cyanide. NIOSH recommended ceiling (10 minute) is also 5 mg/m^3 (4.7 ppm). ACGIH,[1] DFG,[3] HSE[33] have recommended 5 mg/m^3 as a TWA. The NIOSH IDLH level is 25 mg/m^3. ACGIH adds the notation "skin" indicating the possibility off cutaneous absorption. The former USSR-UNEP/IRPTC project[43] has set an MAC in workplace air of 0.3 mg/m^3. Further, they have set MAC values for ambient air in residential areas of 0.009 mg/m^3 on a momentary basis and 0.004 mg/m^3 on an average daily basis.

Several states have set guidelines or standards for cyanides in ambient air[60] ranging from 16.7 μg/m^3 (New York) to 50.0 μg/m^3 (Florida and North Dakota) to 80.0 μg/m^3 (Virginia) to 100 μg/m^3 (Connecticut and South Dakota) to 125 μg/m^3 (South Carolina) to 119.0 μg/m^3 (Nevada).

Determination in Air: Filter/Bubbler; Potassium hydroxide; Ion-specific electrode; NIOSH IV Method #7904, Cyanides. See also Method #6010, Hydrogen Cyanide.[18]

Permissible Concentration in Water: In 1976 the EPA criterion was 5.0 μg/l for freshwater and marine aquatic life and wildlife. As of 1980, the criteria are: To protect freshwater aquatic life: 3.5 μg/l as a 24-hour average, never to exceed 52.0 μg/l. To protect saltwater aquatic life: 30.0 μg/l on an acute toxicity basis; 2.0 μg/l on a chronic toxicity basis. To protect human health: 200 μg/l. The allowable daily intake for man is 8.4 mg/day.[6]

On the international scene, the South African Bureau of Standards has set 10 μg/l, the World Health Organization (WHO) 10 μg/l and the Federal Republic of Germany 50 μg/l as drinking water standards.

Other international limits[35] include an EEC limit of 50 μg/l; Mexican limits of 200 μg/l in drinking water and 1.0 μg/l in coastal waters and a Swedish limit of 100 μg/l. The former USSR-UNEP/IRPTC project[43] has set an MAC of 100 μg/l in water bodies used for domestic purposes and 50 μg/l in water for fishery purposes.

The U.S. E.P.A.[49] has determined a no-observed-adverse-effect-level (NOAEL) of 10.8 mg/kg/day which yields a lifetime health advisory of 154 μg/l. States which have set guidelines for cyanides in drinking water[61] include Arizona at 160 μg/l and Kansas at 220 μg/l.

Determination in Water: Distillation followed by silver nitrate titration or colorimetric analysis using pyridine pyrazolone (or barbituric acid).

Routes of Entry: Potassium cyanide can be absorbed through the skin, inhalation, ingestion.

Harmful Effects and Symptoms

Short Term Exposure: Potassium and sodium cyanide are corrosive to the eyes, the skin and the respiratory tract. Contact can cause eye and skin burns; may cause permanent damage to the eyes. Corrosive if swallowed. These substances may affect the central nervous system. Symptoms include headaches; confusion; nausea, pounding heart, weakness, unconsciousness and death.

Long Term Exposure: Repeated or prolonged contact with potassium or sodium cyanide may cause thyroid gland enlargement. May cause nosebleed and sores in the nose; changes in blood cell count. May cause central nervous system damage with headache, dizziness, confusion; nausea, vomiting, pounding heart, weakness in the arms and legs, unconsciousness and death. may affect liver and kidney function.

Points of Attack: Liver, kidneys, skin, cardiovascular system, central nervous system, thyroid.

Medical Surveillance: Consider the points of attack in preplacement and periodic physical examinations. Urine thiocyanate levels. Blood cyanide levels. Complete blood count (CBC). Evaluation of thyroid function. Liver function tests. Kidney function tests. Central nervous system tests. EKG.

First Aid: If this chemical gets into the eyes, remove any contact lenses at once and irrigate immediately for at least 15 minutes, occasionally lifting upper and lower lids. Seek medical attention immediately. If this chemical contacts the skin, remove contaminated clothing and wash immediately with soap and water. Speed in removing material from skin is of extreme importance. Shampoo hair promptly if contaminated. Seek medical attention immediately. If this chemical has been inhaled, remove from exposure, begin rescue breathing (using universal precautions) if breathing has stopped and CPR if heart action has stopped. Transfer promptly to a medical facility. When this chemical has been swallowed, get medical attention. Give large quantities of water and induce vomiting. Do not make an unconscious person vomit. Keep under observation for 24 – 48 hours as symptoms may return.

Use amyl nitrate capsules if symptoms develop. All area employees should be trained regularly in emergency measures for cyanide poisoning and in CPR. A cyanide antidote kit should be kept in the immediate work area and must be rapidly available. Kit ingredients should be replaced every 1 – 2 years to ensure freshness. Persons trained in the use of this kit, oxygen use, and CPR must be quickly available.

Note to Physician: Consider the administration of Ketocyanor (cobalt edetate) in dose of 300 – 600 mg i.v. initially. If recovery does not occur quickly (in 1 – 2 minutes) give a second dose of 300 mg, followed by i.v. glucose 5%. Alternatively, administer sodium nitrite (3%) in an i.v. dose of 10 ml over 3 minutes.

Personal Protective Methods: Wear protective gloves and clothing to prevent any reasonable probability of skin contact. Safety equipment suppliers/manufacturers can provide recommendations on the most protective glove/clothing material for your operation. All protective clothing (suits, gloves, footwear, headgear) should be clean, available each day, and put on before work. Contact lenses should not be worn when working with this chemical. Wear splash-proof chemical goggles and face shield when working with liquid, unless full facepiece respiratory protection is worn. Wear dust-proof goggles and face shield when working with powders or dust, unless full facepiece respiratory protection is worn. Employees should wash immediately with soap when skin is wet or contaminated. Provide emergency showers and eyewash.

Respirator Selection: NIOSH/OSHA: *up to 25 mg/m³:* SA (any supplied-air respirator); or SCBAF (any self-contained breathing apparatus with full facepiece). *Emergency or planned entry into unknown concentrations or IDLH conditions:* SCBAF:PD,PP (any self-contained breathing apparatus that has a full facepiece and is operated in a pressure-demand or other positive-pressure mode); or SAF:PD,PP:ASCBA (any supplied-air respirator that has a full facepiece and is operated in a pressure-demand or other positive-pressure mode in combination with an auxiliary self-contained breathing apparatus operated in a pressure-demand or other positive pressure mode). *Escape:* GMFSHiE (any air-purifying, full-facepiece respirator (gas mask) with a chin-style, front-or back-, mounted canister providing protection against the compound of concern and having a high efficiency particulate filter); or SCBAE (any appropriate escape-type, self-contained breathing apparatus).

Storage: Prior to working with cyanides you should be trained on its proper handling and storage. A regulated, marked area should be established where this chemical is handled, used, or stored in compliance with OSHA standard 1910.1045. Protect against physical damage. Store in cool dry place. Separate from other storage and protect from acids and oxidizing materials.

Shipping: Cyanides, inorganic, n.o.s. require a "Poison" label. They fall in Hazard Class 6.1 and in Packing Group II.

Spill Handling: Avoid contact with solids, dusts or solutions. Wear chemical protective suit with self-contained breathing apparatus. Evacuate persons not wearing protective equipment from area of spill or leak until clean-up is complete. Remove all ignition sources. Collect powdered material in the most convenient and safe manner and deposit in sealed containers. Ventilate area after clean-up is complete. Do not allow this chemical to enter the environment. It may be necessary to contain and dispose of this chemical as a hazardous waste. If material or contaminated runoff enters waterways, notify downstream users of potentially contaminated waters. Contact your Department of Environmental Protection or your regional office of the federal EPA for specific recommendations. If employees are required to clean-up spills, they must be properly trained and equipped. OSHA 1910.120(q) may be applicable.

Initial isolation and protective action distances

Distances shown are likely to be affected during the first 30 minutes after materials are spilled and could increase with time. If more than one tank car, cargo tank, portable tank, or large cylinder is involved in the incident is leaking, the protective action distance may need to be increased. You may need to seek emergency information from CHEMTREC at (800) 424-9300 or seek professional environmental engineering assistance from the U.S. EPA Environmental Response Team at (908)548-8730 (24-hour response line).

UN 1680 (Potassium cyanide), UN 1989 (Sodium cyanide) is on the DOT's list of dangerous water-reactive materials which create large amounts of toxic vapor when *spilled in water:* Dangerous from 0.5 – 10 km (0.3 – 6.0 miles) downwind.

Fire Extinguishing: Cyanides such as KCN and NaCN are not combustible themselves but contact with acid releases highly flammable hydrogen cyanide and oxides of nitrogen. Reacts with water releasing hydrogen cyanide. NO hydrous agents. NO water. NO carbon dioxide. Use dry chemical and foam on surrounding fires. Vapors are heavier than air and may collect in low areas. Containers may explode in fire. Storage containers and parts of containers may rocket great distances, in many directions. If material or contaminated runoff enters waterways, notify downstream users of potentially contaminated waters. Notify local health and fire officials and pollution control agencies. From a secure, explosion-proof location, use water spray to cool exposed containers. If cooling streams are ineffective (venting sound increases in volume and pitch, tank discolors, or shows any signs of deforming), withdraw immediately to a secure position. If employees are expected to fight fires, they must be trained and equipped in OSHA 1910.156.

Disposal Method Suggested: Consult with environmental regulatory agencies for guidance on acceptable disposal practices. Generators of waste containing this contaminant (≥ 100 kg/mo) must conform with EPA regulations governing storage, transportation, treatment, and waste disposal. In accordance with 40CFR165 recommendations for the disposal of pesticides and pesticide containers. Must be disposed properly by following package label directions or by contacting your state pesticide or environmental control agency or by contacting your regional EPA office. Add strong alkaline hypochlorite and react for 24 hours. Then flush to sewer with large volumes of water.[22]

References

U.S. Environmental Protection Agency, Cyanides: Ambient Water Quality Criteria, Washington, DC (1980)

National Institute for Occupational Safety and Health, Criteria for a Recommended Standard: Occupational Exposure to Hydrogen Cyanide and Cyanide Salts, NIOSH Document No. 77–108, Washington, DC (1977)

U.S. Environmental Protection Agency, Reviews of the Environmental Effects of Pollutants; V: Cyanide, Report No. EPA-600/1-78-027, Washington, DC (1978)

U.S. Environmental Protection Agency, Cyanides, Health and Environmental Effects Profile No. 56, Office of Solid Waste, Washington, DC (April 30, 1980)

Sax, N. I., Ed., "Dangerous Properties of Industrial Materials Report," 3, No. 6, 56–60 (1983) (Potassium Cyanide)

Sax, N. I., Ed., "Dangerous Properties of Industrial Materials Report," 3, No. 6, 60–63 (1983) (Sodium Cyanide)

U.S. Public Health Service, "Toxicological Profile for Cyanide," Atlanta, Georgia, Agency for Toxic Substances and Disease Registry (January, 1988)

U.S. Environmental Protection Agency, "Chemical Profile: Potassium Cyanide," Washington, DC, Chemical Emergency Preparedness Program (November 30, 1987)

U.S. Environmental Protection Agency, "Chemical Profile: Sodium Cyanide," Washington, DC, Chemical Emergency Preparedness Program (November 30, 1987)

New Jersey Department of Health and Senior Services and Senior Services, "Hazardous Substance Fact Sheet: Potassium Cyanide," Trenton, NJ (June 1998)

New Jersey Department of Health and Senior Services and Senior Services, "Hazardous Substance Fact Sheet: Sodium Cyanide," Trenton, NJ (June 1998)

Cyanogen

Molecular Formula: C_2N_2

Common Formula: $(CN)_2$

Synonyms: Carbon nitride; Cyanogene (French); Cyanogen gas; Dicyan; Dicyanogen; Ethanedinitrile; Monocyanogen; Nitriloacetonitrile; Oxalic acid dinitrile; Oxalic nitrile; Oxalonitrile; Oxalyl cyanide; Prussite

CAS Registry Number: 460-19-5

RTECS Number: GT1925000

DOT ID: UN 1026

Regulatory Authority

- Air Pollutant Standard Set (ACGIH)[1] (Australia) (DFG)[3] (HSE)[33] (Israel) (Mexico) (OSHA)[58] (Several States)[60] (Several Canadian Provinces)
- CLEAN AIR ACT: Accidental Release Prevention/Flammable substances, (Section 112[r], Table 3), TQ = 10,000 lb (4,540 kg)
- EPA HAZARDOUS WASTE NUMBER (RCRA No.): P031
- RCRA, 40CFR261, Appendix 8 Hazardous Constituents
- SUPERFUND/EPCRA 40CFR302.4 Reportable Quantity (RQ): CERCLA, 100 lb (45.4 kg)
- U.S. DOT Regulated Marine Pollutant (49CFR172.101, Appendix B) as cyanides, inorganic, n.o.s.
- U.S. DOT 49CFR172.101, Inhalation Hazardous Chemical
- Canada, WHMIS, Ingredients Disclosure List; National Pollutant Release Inventory (NPRI); CEPA Priority Substance List, Ocean dumping prohibited

Cited in U.S. State Regulations: Alaska (G), California (A, G), Connecticut (A), Florida (G, A), Illinois (G), Kansas (G), Louisiana (G), Maine (G), Massachusetts (G), Minnesota (G), Nevada (A), New Hampshire (G), New Jersey (G), New York (A), North Dakota (A), Oklahoma (G), Pennsylvania (G), Rhode Island (G, A, W), South Carolina (A), South Dakota (A, W), Tennessee (W), Utah (W), Vermont (G), Virginia (G, A), Washington (G), West Virginia (G), Wisconsin (G).

Description: Cyanogen, $(CN)_2$, is a colorless, flammable, compressed liquefied gas at room temperature. At deadly levels only, it has a pungent, almond-like odor.* Boiling point = –21°C. Freezing/Melting point = -28°C. Explosive limits: LEL = 6.6%; UEL = 32.0%. Hazard Identification (based on NFPA-704 M Rating System): Health 4, Flammability 4, Reactivity 2. Soluble in water.

* The irritant properties of cyanogen have been tested using both human male and female subjects, 21 – 65 years of age. The distinctive bitter almond smell of cyanogen could not be detected at concentrations of 50, 100 and 250 ppm. When exposed to 8 ppm for 6 minutes or 16 ppm for 6 – 8 minutes, immediate eye and nose irritation occurred.

Potential Exposure: Cyanogen is currently used as an intermediate in organic syntheses; at one time, it was used in poison gas warfare.

Incompatibilities: Explosive reaction with acids, strong oxidizers (e.g., dichlorine oxide, fluorine) Slowly hydrolyzed in water to form hydrogen cyanide, oxalic acid, ammonia.

Permissible Exposure Limits in Air: There is no OSHA PEL. NIOSH recommended TWA is 10 ppm (20 mg/m^3). The ACGIH. DFG,[3] HSE,[33] Australian, Israeli, and Mexican TWA is 10 ppm (20 mg/m^3). The Canadian TWA for Alberta, BC, Ontario, and Quebec are the same and Alberta's STEL is 20 ppm (43 mg/m^3). Some other foreign standards are as follows: Rumania (average), 3 mg/m^3, (ceiling) 5 mg/m^3; Australia, Belgium, Finland, the Netherlands, Switzerland, 2 mg/m^3.

A number of states have set guidelines or standards for Cyanogen in ambient air[60] ranging from 66.7 μg/m^3 (New York) to 200 μg/m^3 (Florida and North Dakota) to 350 μg/m^3 (Virginia) to 400 μg/m^3 (Connecticut) to 476 μg/m^3 (Nevada) to 500 μg/m^3 (South Carolina).

Determination in Air: No Method available.

Permissible Concentration in Water: No criteria set.

Routes of Entry: Inhalation, passing through the skin, and eye contact.

Harmful Effects and Symptoms

Short Term Exposure: A lacrimator (causing tearing). This chemical can be absorbed through the skin, thereby increasing exposure. Irritates eyes, nose, upper respiratory system. Skin contact with the liquid caused frostbite. Vision loss can occur following a high exposure. Cyanogen hydrolyzes to yield one molecule of hydrogen cyanide and one of cyanate; based on this, the toxic effects of $(CN)_2$ are thought to be comparable to HCN. The cyanide ion when released in the body causes a form of asphyxia by inhibiting many enzymes-especially those concerned with cellular respiration. Although the blood is saturated with oxygen, the tissues are not able to use it. Symptoms appear within a few seconds or minutes of ingesting or breathing vapors. Symptoms include cherry red lips, tachypnea, hypernea, bradycardia; headache, vertigo (an illusion of movement), convulsions; dizziness. The victims experience constriction of the chest, giddiness, confusion, headache, hypernea, palpitation, unconsciousness, convulsion, feeble and rapid respiration, and an extremely weak pulse. Death occurs within a few minutes after a large dose.

Long Term Exposure: Enlargement of the thyroid gland. There is some evidence that long term exposure can cause damage to the nervous system. Victims experience loss of appetite, weight loss.

Points of Attack: Eyes, respiratory system, central nervous system, cardiovascular system.

Medical Surveillance: Before beginning employment and at regular times after that, the following is recommended: Serum and urine thiocyanate levels. If symptoms develop or overexposure is suspected, the following may be useful: Exam of the thyroid.

First Aid: If this chemical gets into the eyes, remove any contact lenses at once and irrigate immediately for at least 15 minutes, occasionally lifting upper and lower lids. Seek medical attention immediately. If this chemical contacts the skin, remove contaminated clothing and wash immediately with soap and water. Seek medical attention immediately. If this chemical has been inhaled, remove from exposure, begin rescue breathing (using universal precautions) if breathing has stopped and CPR if heart action has stopped. Transfer promptly to a medical facility. When this chemical has been swallowed, get medical attention. Give large quantities of water and induce vomiting. Do not make an unconscious person vomit. Medical observation is advised for 24 – 48 hours.

If frostbite has occurred, seek medical attention immediately; do NOT rub the affected areas or flush them with water. In order to prevent further tissue damage, do NOT attempt to remove frozen clothing from frostbitten areas. If frostbite has NOT occurred, immediately and thoroughly wash contaminated skin with soap and water.

If cyanide poisoning is suspected: Use amyl nitrate capsules if symptoms develop. All area employees should be trained regularly in emergency measures for cyanide poisoning and in CPR. A cyanide antidote kit should be kept in the immediate work area and must be rapidly available. Kit ingredients should be replaced every 1 – 2 years to ensure freshness. Persons trained in the use of this kit, oxygen use, and CPR must be quickly available.

Personal Protective Methods: Lower exposure can cause irritation of the eyes, nose and throat. If these symptoms are noticed, immediately leave the work area.

Clothing: Avoid skin contact with Cyanogen. Wear protective gloves and clothing. Safety equipment suppliers/manufacturers can provide recommendations on the most protective glove/clothing material for your operation. All protective clothing (suits, gloves, footwear, headgear) should be clean, available each day, and put on before work.

Eye Protection: Wear air-tight gas-proof goggles, unless full facepiece respiratory protection is worn.

Where exposure to the liquefied compressed gas may occur, employees should be provided with special clothing designed to prevent frostbite. See NIOSH Criteria Document 78-212 NITRILES.

Respirator Selection: Where the potential exists for exposures over 10 ppm, use an MSHA/NIOSH approved supplied-air respirator with a full facepiece operated in the positive pressure mode or with a full facepiece, hood, or helmet in the continuous flow mode, or use an MSHA/NIOSH approved self-contained breathing apparatus with a full facepiece operated in pressure-demand or other positive pressure mode.

Storage: Prior to working with Cyanogen you should be trained on its proper handling and storage. Before entering confined space where this chemical may be present, check to make sure that an explosive concentration does not exist. Store in tightly closed containers in a cool well-ventilated area away from heat and light. Cyanogen must be stored to avoid contact with fluorine and oxygen, water or steam, acid or acid fumes, since violent reactions occur. Sources of ignition such as smoking and open flames are prohibited where Cyanogen is handled, used, or stored. Outside storage or storage in an area of non-combustible construction, is preferable. Use only non-sparking tools and equipment, especially when opening and closing containers of this chemical. Where Cyanogen is used, handled, manufactured, or stored, use explosion-proof electrical equipment and fittings. Sources of ignition such as smoking and open flames, are prohibited where this chemical is used, handled, or stored in a manner that could create a potential fire or explosion hazard. Procedures for the handling, use and storage of cylinders should be in compliance with OSHA 1910.101 and 1910.169 as with the recommendations of the Compressed Gas Association.

Shipping: Cyanogen, liquefied, falls in DOT Hazard Class 2.3 and Packing Group I. It must be labeled: "Poison Gas, Flammable Gas." Shipment by passenger aircraft or railcar or even by cargo aircraft is forbidden.[19]

Spill Handling: Evacuate and restrict persons not wearing protective equipment from area of spill or leak until cleanup is complete. Remove all ignition sources. Establish forced ventilation to keep levels below explosive limit. Stop the flow of gas if it can be done safely. If source of leak is a cylinder and the leak cannot be stopped in place, remove leaking cylinder to a safe place in the open air, and repair leak or allow cylinder to empty. Keep cyanogen out of confined space, such as sewer because of the possibility of explosion, unless the sewer is designed to prevent the buildup of explosive concentrations. It may be necessary to contain and dispose of cyanogen as a hazardous waste. Contact your Department of Environmental Protection or your regional office of the federal EPA for specific recommendations. If employees are required to clean-up spills, they must be properly trained and equipped. OSHA 1910.120(q) may be applicable.

Initial isolation and protective action distances

Distances shown are likely to be affected during the first 30 minutes after materials are spilled and could increase with time. If more than one tank car, cargo tank, portable tank, or large cylinder is involved in the incident is leaking, the protective action distance may need to be increased. You may need to seek emergency information from CHEMTREC at (800) 424-9300 or seek professional environmental engineering assistance from the U.S. EPA Environmental Response Team at (908)548-8730 (24-hour response line).

Small spills (From a small package or a small leak from a large package)

First: Isolate in all directions (feet)..... 200

Then: Protect persons downwind (miles)

Day 0.2

Night 0.6

Large spills (From a large package or from many small packages)

First: Isolate in all directions (feet)..... 700

Then: Protect persons downwind (miles)

Day 0.5

Night 2.2

Restrict persons not wearing protective equipment from areas of leaks until clean-up is complete. Remove all ignition sources. Ventilate area of leak to disperse the gas. Stop flow of gas. If source of leak is a cylinder and the leak cannot be stopped in place, remove the leaking cylinder to a safe place in the open air, and repair leak or allow cylinder to empty.

Fire Extinguishing: This chemical is a highly flammable gas; it burns with a purple-tinged flame. Poisonous gases including hydrogen cyanide are produced in fire. Stop the flow of gas if it can be done safely. Do not use water. Vapors are heavier than air and will collect in low areas. Vapors may travel long distances to ignition sources and flashback. Vapors in confined areas may explode when exposed to fire. Containers may explode in fire. Storage containers and parts of containers may rocket great distances, in many directions. If material or contaminated runoff enters waterways, notify downstream users of potentially contaminated waters. Notify local health and fire officials and pollution control agencies. From a secure, explosion-proof location, use water spray to cool exposed containers. If cooling streams are ineffective (venting sound increases in volume and pitch, tank discolors, or shows any signs of deforming), withdraw immediately to a secure position. If employees are expected to fight fires, they must be trained and equipped in OSHA 1910.156.

Disposal Method Suggested: Incineration; oxides or nitrogen are removed from the effluent gas by scrubbers and/or thermal devices.

References

National Institute for Occupational Safety and Health, Information Profiles on Potential Occupational Hazards-Single Chemicals: Cyanogen, Report TR 79-607, pp 39–44, Rockville, MD (December 1979)

Sax, N. I., Ed., "Dangerous Properties of Industrial Materials Report," 2, No. 1, 103–105, New York, Van Nostrand Reinhold Co. (1982)

New Jersey Department of Health and Senior Services and Senior Services, "Hazardous Substance Fact Sheet: Cyanogen," Trenton, NJ (January 1986)

Cyanogen Bromide

Molecular Formula: BrCN

Synonyms: Bromine cyanide; Bromocyan; Bromocyanogen; Bromure de cyanogen (French); Bromuro de cianogeno (Spanish); Campilit; Cyanobromide; Cyanogen monobromide

CAS Registry Number: 506-68-3

RTECS Number: GT2100000

DOT ID: UN 1889

EINECS Number: 208-051-2

Regulatory Authority

- EPA HAZARDOUS WASTE NUMBER (RCRA No.): U246
- RCRA, 40CFR261, Appendix 8 Hazardous Constituents
- SUPERFUND/EPCRA 40CFR355, Appendix B Extremely Hazardous Substances: TPQ = 500/10,000 lb (227/4,540 kg)
- SUPERFUND/EPCRA 40CFR302.4 Reportable Quantity (RQ): CERCLA, 1,000 lb (454 kg)
- MARINE POLLUTANT (49CFR, Subchapter 172.10)
- Canada, WHMIS, Ingredients Disclosure List; National Pollutant Release Inventory (NPRI); CEPA Priority Substance List, Canada, WHMIS, Ingredients Disclosure List;

National Pollutant Release Inventory (NPRI); CEPA Priority Substance List, Ocean dumping prohibited

Cited in U.S. State Regulations: California (A, G), Florida (G), Kansas (G), Louisiana (G), Maine (G), Massachusetts (G), New Hampshire (G), Oklahoma (G), Pennsylvania (G), Rhode Island (G), Vermont (G), Virginia (G), Washington (G), Wisconsin (G).

Description: Cyanogen Bromide, BrCN, is a colorless or white, volatile, crystalline solid with a penetrating odor. Boiling point = 61 – 62°C. Freezing/Melting point = 52°C. Soluble in water; decomposes slowly forming hydrogen cyanide and hydrogen bromide.

Potential Exposure: Those manufacturing this compound or using it in organic synthesis or as a fumigant, in textile treatment, in gold cyaniding or as a military poison gas.

Incompatibilities: May be unstable unless dry and pure. Violent reaction with acids, ammonia, amines. The substance decomposes on heating or on contact with water, acids, or acid vapors producing highly toxic and flammable hydrogen cyanide and corrosive hydrogen bromide.

Permissible Exposure Limits in Air: No standards set.

Determination in Air: No criteria set. However, inasmuch as it is a cyanide compound, the exposure limits are listed here: OSHA and ACGIH: 5 mg/m^3 TWA; NIOSH: Ceiling limit, 4.7 ppm; 5 mg/m^3 per 10 minutes as cyanides. All have notations that skin contact contributes significantly in overall exposure. IDLH = 25 mg/m^3 as CN.

Permissible Concentration in Water: No criteria set.

Routes of Entry: Inhalation, ingestion, skin absorption.

Harmful Effects and Symptoms

Short Term Exposure: Cyanogen bromide's toxic action resembles that of hydrocyanic acid. Cyanogen bromide is corrosive to the eyes, skin, and respiratory tract. Higher exposures can cause pulmonary edema, a medical emergency that can be delayed for several hours. This can cause death. Exposure may result in death. Super toxic; probable oral lethal dose in humans in less than 5 mg/kg or a taste (less than 7 drops) for a 70 kg (150 lb) person. Vapors are highly irritant and very poisonous.

High concentrations produce excessive respiration (causing increased uptake of cyanide), then labored breathing, paralysis, unconsciousness, convulsions and respiratory arrest. Headache, dizziness, nausea, and vomiting may occur with lesser concentrations.

Patients may experience confusion, anxiety, an initial rise in blood pressure with a decreased heart heat followed by an increased heart beat; cyanosis is not a consistent finding, in fact, the patient may be reddish. An odor of bitter almonds on the patient's breath may be present. Individuals with chronic diseases of the kidneys, respiratory tract, skin, or thyroid are at greater risk of developing toxic cyanide effects.

Long Term Exposure: Repeated or prolonged exposure to cyanogen bromide may cause thyroid gland enlargement. Chronic exposure may cause fatigue and weakness.

Points of Attack: Eyes, respiratory system, thyroid gland.

Medical Surveillance: Thyroid gland examination.

First Aid: If this chemical gets into the eyes, remove any contact lenses at once and irrigate immediately for at least 15 minutes, occasionally lifting upper and lower lids. Seek medical attention immediately. If this chemical contacts the skin, remove contaminated clothing and wash immediately with soap and water. Seek medical attention immediately. Do not perform direct mouth to mouth resuscitation; use bag/mask apparatus. If this chemical has been inhaled, remove from exposure, begin rescue breathing (using universal precautions) if breathing has stopped and CPR if heart action has stopped. Transfer promptly to a medical facility. When this chemical has been swallowed, get medical attention. If victim is conscious, administer water or milk. Do not induce vomiting. Medical observation is recommended for 24 – 48 hours after breathing overexposure, as pulmonary edema may be delayed. As first aid for pulmonary edema, a doctor or authorized paramedic may consider administering a corticosteroid spray.

Use amyl nitrate capsules if symptoms of cyanide poisoning develop. All area employees should be trained regularly in emergency measures for cyanide poisoning and in CPR. A cyanide antidote kit should be kept in the immediate work area and must be rapidly available. Kit ingredients should be replaced every 1 – 2 years to ensure freshness. Persons trained in the use of this kit, oxygen use, and CPR must be quickly available.

Personal Protective Methods: Wear protective gloves and clothing to prevent any reasonable probability of skin contact. Safety equipment suppliers/manufacturers can provide recommendations on the most protective glove/clothing material for your operation. All protective clothing (suits, gloves, footwear, headgear) should be clean, available each day, and put on before work. Contact lenses should not be worn when working with this chemical. Wear dust-proof chemical goggles and face shield unless full facepiece respiratory protection is worn. Employees should wash immediately with soap when skin is wet or contaminated. Provide emergency showers and eyewash.

Respirator Selection: Where there is a potential for overexposure: SCBAF:PD,PP (any MSHA/NIOSH approved self-contained breathing apparatus that has a full facepiece and is operated in a pressure-demand or other positive-pressure mode); or SAF:PD,PP:ASCBA (any supplied-air respirator that has a full facepiece and is operated in a pressure-demand or other positive-pressure mode in combination with an auxiliary, self-contained breathing apparatus operated in a pressure-demand or other positive pressure mode).

Storage: If dried over sodium, pure material may be stored in a desiccator for several months. Impure material decomposes and may explode. Prior to working with this chemical you should be trained on its proper handling and storage. Store in tightly closed containers in a cool, well-ventilated area away from moisture, acids, ammonia, amines, and incompatible materials listed above.

Shipping: DOT label required is "Poison, Corrosive." Shipment by passenger aircraft or railcar or even by cargo aircraft is forbidden. Falls in UN/DOT Hazard Class 6.1 and Packing Group I (19, 20). A DOT regulated marine pollutant.

Spill Handling: Evacuate persons not wearing protective equipment from area of spill or leak until clean up is complete. Remove all ignition sources. Protective clothing including impervious hand protection should be worn. Wear positive pressure breathing apparatus. Do not touch spilled material; stop leak if you can do it without risk. Use water spray to reduce vapors. Small spills: take up with sand or other noncombustible absorbent material and place into containers for later disposal. Large spills: dike far ahead of spill for later disposal. Collect powdered material in the most convenient and safe manner and deposit in sealed containers. Ventilate area after clean-up is complete. It may be necessary to contain and dispose of this chemical as a hazardous waste. If material or contaminated runoff enters waterways, notify downstream users of potentially contaminated waters. Contact your Department of Environmental Protection or your regional office of the federal EPA for specific recommendations. If employees are required to clean up spills, they must be properly trained and equipped. OSHA 1910.120(q) may be applicable.

Fire Extinguishing: Firefighting gear (including SCBA) does not provide adequate protection. If exposure occurs, remove and isolate gear immediately and thoroughly decontaminate personnel. If conditions permit, do not extinguish. Cool exposures using unattended monitors. (FEMA). If fire must be extinguished, use agent suitable for type of surrounding fire. Material itself does not burn or burns with difficulty. Do not use water on material itself. If large quantities of combustibles are involved, use water in flooding quantities as spray and fog. Use water spray to absorb vapors. Keep material out of water sources and sewers. Use water spray to knock down vapors. Vapors are heavier than air and will collect in low areas. Wear full protective clothing. Avoid direct water contact as it will cause cyanogen bromide to decompose, releasing toxic gases. Avoid breathing vapors; keep upwind; wear self-contained breathing apparatus. Poisonous gases are produced in fire. If material or contaminated runoff enters waterways, notify downstream users of potentially contaminated waters. Notify local health and fire officials and pollution control agencies. From a secure, explosion-proof location, use water spray to cool exposed containers. If cooling streams are ineffective (venting sound increases in volume and pitch, tank discolors, or shows any signs of deforming), withdraw immediately to a secure position. If employees are expected to fight fires, they must be trained and equipped in OSHA 1910.156.

Disposal Method Suggested: May be added to strong alkaline solution of calcium hypochlorite, let stand for 24 hours and flushed to sewer. May also be dissolved in flammable solvent and sprayed into an incinerator equipped with afterburner and scrubber.

References

Sax, N. I., Ed., "Dangerous Properties of Industrial Materials Report," 1, No. 8, 60–62, New York, Van Nostrand Reinhold Co. (1981)

U.S. Environmental Protection Agency, "Chemical Profile: Cyanogen Bromide," Washington, DC, Chemical Emergency Preparedness Program (November 30, 1987)

Cyanogen Chloride

Molecular Formula: CClN

Common Formula: CNCl

Synonyms: Chlorcyan; Chlorine cyanide; Chlorocyan; Chlorocyanide; Chlorocyanogen; Chlorure de cyanogene (French); Cloruro de cianogeno (Spanish); Cyanogen chloride [(CN)Cl]; Cyanogen chloride, containing less than 0.9% water

CAS Registry Number: 506-77-4

RTECS Number: GT2275000

DOT ID: UN 1589 (inhibited)

Regulatory Authority

- Air Pollutant Standard Set (ACGIH)[1] (Australia) (HSE)[33] (Israel) (OSHA)[58] (Several States)[60] (Several Canadian Provinces)
- OSHA Process Safety Management of Highly Hazardous Chemicals (29CFR, Part 1910.119, Appendix A): Threshold Quantity: 500 pounds
- CLEAN AIR ACT: Accidental Release Prevention/Flammable substances, (Section 112[r], Table 3), TQ = 10,000 lb (4,540 kg)
- CLEAN WATER ACT: Section 311 Hazardous Substances/RQ 40CFR117.3 (same as CERCLA, see below); Section 313 Water Priority Chemicals (57FR41331, 9/9/92)
- EPA HAZARDOUS WASTE NUMBER (RCRA No.): P033
- RCRA, 40CFR261, Appendix 8 Hazardous Constituents
- SAFE DRINKING WATER ACT: Priority List (55 FR 1470)
- SUPERFUND/EPCRA 40CFR302.4 Reportable Quantity (RQ): CERCLA, 10 lb (4.54 kg)
- U.S. DOT Regulated Marine Pollutant (49CFR172.101, Appendix B)
- U.S. DOT 49CFR172.101, Inhalation Hazardous Chemical
- Canada, WHMIS, Ingredients Disclosure List; National Pollutant Release Inventory (NPRI); CEPA Priority Substance List, Ocean dumping prohibited

Cited in U.S. State Regulations: Alaska (G), California (A, G), Florida (G), Illinois (G), Kansas (G), Louisiana (G), Maine (G), Massachusetts (G), Minnesota (G), Nevada (A), New Hampshire (G), New Jersey (G), North Dakota (A), Pennsylvania (G), Rhode Island (G), Vermont (G), Virginia (G, A), Washington (G), West Virginia (G), Wisconsin (G).

Description: Cyanogen chloride, CNCl, is a colorless gas or liquid (below 55°F/13°C) with a pungent, irritating odor. Shipped as a liquefied gas. A solid below 20°F/-7°C. Boiling point = 13°C. Freezing/Melting point = -6°C. Flash point = 51°C. Soluble in water (slowly decomposes).

Potential Exposure: Cyanaogen chloride is used as a fumigant, metal cleaner, in ore refining, production of synthetic rubber and in chemical synthesis. Cyanogen chloride can be used in the military as a poison gas.

Incompatibilities: May be stabilized to prevent polymerization. Cyanogen chloride may polymerize violently if contaminated with chlorine. In crude form chemical trimerizes violently if catalyzed by traces of hydrogen chloride or ammonium chloride. Contact with alcohols, acids, acid salts, amines, strong alkalis, olefins, strong oxidizers may cause fire and explosion. Heat causes decomposition producing toxic and corrosive fumes of hydrogen cyanide, hydrochloric acid, nitrogen oxides. Reacts slowly with water or water vapor to form hydrogen chloride. Attacks copper, brass, and bronze in the presence of moisture.

Permissible Exposure Limits in Air: No PEL has been set. NIOSH and ACGIH recommend a ceiling value of 0.3 ppm (0.6 mg/m^3). HSE and Israel set a STEL 0.3 ppm (0.6 mg/m^3). The Canadian provinces of Alberta, British Columbia, Ontario and Quebec set the same STEL value. Some states have set guidelines and standards for cyanogen chloride in ambient air[60] ranging from 5.0 $\mu g/m^3$ (Virginia) to 6.0 $\mu g/m^3$ (North Dakota) to 14.0 $\mu g/m^3$ (Nevada).

Determination in Air: No method available.

Permissible Concentration in Water: No criteria set.

Routes of Entry: Inhalation, skin absorption (liquid), ingestion (liquid), skin and/or eye contact (liquid).

Harmful Effects and Symptoms

Short Term Exposure: Cyanogen chloride is converted to cyanide in the body. A lacrimator. Cyanogen chloride severely irritates the eyes, skin, and respiratory tract. Inhalation can cause weakness, headache, giddiness, dizziness, confusion, nausea, vomiting; irregular/irregularities heartbeat, and pulmonary edema, a medical emergency that can be delayed for several hours. This can cause death. Skin contact with the liquid may cause frostbite and irritation. The toxicity of cyanogen chloride resides very largely on its pharmacokinetic property of yielding readily to hydrocyanic acid in vivo. Inhaling small amounts of Cyanogen chloride causes dizziness, weakness, congestion of the lungs, hoarseness, conjunctivitis, loss of appetite, weight loss, and mental deterioration. These effects are similar to those found from inhalation of cyanide.

Ingestion or inhalation of a lethal dose of cyanogen chloride (LD_{50} = 13 mg/kg), as for cyanide or other cyanogenic compounds, causes dizziness, rapid respiration, vomiting, flushing, headache, drowsiness, drop in blood pressure, rapid pulse, unconsciousness, convulsions, with death occurring within 4 hours.

Points of Attack: Eyes, skin, respiratory system, central nervous system, cardiovascular system.

Medical Surveillance: Lung function tests. EKG.

First Aid: If this chemical gets into the eyes, remove any contact lenses at once and irrigate immediately for at least 15 minutes, occasionally lifting upper and lower lids. Seek medical attention immediately. If this chemical contacts the skin, remove contaminated clothing and wash immediately with soap and water. Seek medical attention immediately. If this chemical has been inhaled, remove from exposure, begin rescue breathing (using universal precautions) if breathing has stopped and CPR if heart action has stopped. Transfer promptly to a medical facility. When this chemical has been swallowed, get medical attention. Give large quantities of water and induce vomiting. Do not make an unconscious person vomit. Medical observation is recommended for 24 – 48 hours after breathing overexposure, as pulmonary edema may be delayed. As first aid for pulmonary edema, a doctor or authorized paramedic may consider administering a corticosteroid spray.

If frostbite has occurred, seek medical attention immediately; do NOT rub the affected areas or flush them with water. In order to prevent further tissue damage, do NOT attempt to remove frozen clothing from frostbitten areas. If frostbite has NOT occurred, immediately and thoroughly wash contaminated skin with soap and water.

Use amyl nitrate capsules if symptoms develop. All area employees should be trained regularly in emergency measures for cyanide poisoning and in CPR. A cyanide antidote kit should be kept in the immediate work area and must be rapidly available. Kit ingredients should be replaced every 1 – 2 years to ensure freshness. Persons trained in the use of this kit, oxygen use, and CPR must be quickly available.

Personal Protective Methods: Wear protective gloves and clothing to prevent any reasonable probability of skin contact. Safety equipment suppliers/manufacturers can provide recommendations on the most protective glove/clothing material for your operation. All protective clothing (suits, gloves, footwear, headgear) should be clean, available each day, and put on before work. Contact lenses should not be worn when working with this chemical. Wear gas- and splash-proof chemical goggles and face shield unless full facepiece respiratory protection is worn. Employees should wash immediately with soap when skin is wet or contaminated. Provide emergency showers and eyewash.

Where exposure to the liquefied compressed gas may occur, employees should be provided with special clothing designed to prevent frostbite.

Respirator Selection: Where there is a potential for over-exposure: SCBAF:PD,PP (any MSHA/NIOSH approved self-contained breathing apparatus that has a full face-piece and is operated in a pressure-demand or other positive-pressure mode); or SAF:PD,PP:ASCBA (any supplied-air respirator that has a full facepiece and is operated in a pressure-demand or other positive-pressure mode in combination with an auxiliary, self-contained breathing apparatus operated in a pressure-demand or other positive pressure mode).

Storage: Prior to working with this chemical you should be trained on its proper handling and storage. Store in tightly closed containers in a cool, well-ventilated area Provide ventilation along the floor as the vapors are heavier than air. Procedures for the handling, use and storage of cylinders should be in compliance with OSHA 1910.101 and 1910.169 as with the recommendations of the Compressed Gas Association.

Shipping: Cyanogen chloride must be labeled: "Poison Gas, Flammable Gas." It falls in DOT Hazard Class 2.3 and Packing Group I. Shipment by passenger aircraft or railcar or even by cargo aircraft is forbidden.

Spill Handling: If in a building, evacuate building and confine vapors by closing doors and shutting down HVAC systems. Restrict persons not wearing protective equipment from area of spill or leak until cleanup is complete. Remove all ignition sources. Establish forced ventilation to keep levels below explosive limit and to disperse the gas. Wear chemical protective suit with self-contained breathing apparatus to combat spills. Stay upwind and use water spray to "knock down" vapor; contain runoff. Stop the flow of gas, if it can be done safely from a distance. If source is a cylinder and the leak cannot be stopped in place, remove the leaking cylinder to a safe place, and repair leak or allow cylinder to empty. Keep this chemical out of confined spaces, such as a sewer, because of the possibility of explosion, unless the sewer is designed to prevent the buildup of explosive concentrations. If employees are required to clean-up spills, they must be properly trained and equipped. OSHA 1910.120(q) may be applicable.

Initial isolation and protective action distances

Distances shown are likely to be affected during the first 30 minutes after materials are spilled and could increase with time. If more than one tank car, cargo tank, portable tank, or large cylinder is involved in the incident is leaking, the protective action distance may need to be increased. You may need to seek emergency information from CHEMTREC at (800) 424-9300 or seek professional environmental engineering assistance from the U.S. EPA Environmental Response Team at (908)548-8730 (24-hour response line).

Small spills (From a small package or a small leak from a large package)

First: Isolate in all directions (feet)..... 300

Then: Protect persons downwind (miles)

Day 0.3

Night 1.3

Large spills (From a large package or from many small packages)

First: Isolate in all directions (feet)..... 1,000

Then: Protect persons downwind (miles)

Day 1.1

Night 4.9

Fire Extinguishing: Firefighting gear (including SCBA) does not provide adequate protection. If exposure occurs, remove and isolate gear immediately and thoroughly decontaminate personnel. This material is not combustible. Use extinguishing agents suitable for surrounding fire. Heat causes decomposition producing toxic and corrosive fumes of hydrogen cyanide, hydrochloric acid, nitrogen oxides. If material or contaminated runoff enters waterways, notify downstream users of potentially contaminated waters. Notify local health and fire officials and pollution control agencies. From a secure, explosion-proof location, use water spray to cool exposed containers. If cooling streams are ineffective (venting sound increases in volume and pitch, tank discolors, or shows any signs of deforming), withdraw immediately to a secure position. If employees are expected to fight fires, they must be trained and equipped in OSHA 1910.156.

Disposal Method Suggested: React with strong calcium hypochlorite solution for 24 hours, then flush to sewer with large volumes of water.

References

U.S. Environmental Protection Agency, Cyanogen Chloride, Health and Environmental Effects Profile No. 57, Office of Solid Waste, Washington, DC (April 30, 1980)

Sax, N. I., Ed., "Dangerous Properties of Industrial Materials Report," 1, No. 8, 62–63 (1981) and 6, No. 1, 46–49 (1986)

Cyanogen Iodide

Molecular Formula: CIN

Common Formula: CNI

Synonyms: Cyanogen moniodide; Iodine cyanide; Jodcyan; NCI; Yoduro de cianogeno (Spanish)

CAS Registry Number: 506-78-5

RTECS Number: NN1750000

DOT ID: No citation.

Regulatory Authority

- SUPERFUND/EPCRA 40CFR355, Appendix B Extremely Hazardous Substances: TPQ = 1,000/10,000 lb (454/4,540 kg)
- SUPERFUND/EPCRA 40CFR302.4 Reportable Quantity (RQ): EHS, 1 lb (0.454 kg)
- U.S. DOT Regulated Marine Pollutant (49CFR172.101, Appendix B) as cyanides, inorganic, n.o.s.
- Canada, WHMIS, Ingredients Disclosure List; National Pollutant Release Inventory (NPRI); CEPA Priority Substance List, Ocean dumping prohibited

Cited in U.S. State Regulations: California (A, G), Florida (G), Massachusetts (G), New Jersey (G), Pennsylvania (G).

Description: Cyanogen Iodide, CNI, is a combustible, white crystalline solid with a very pungent odor. Freezing/Melting point = 146.5°C. Soluble in water.

Potential Exposure: Used by taxidermists as a preservative. Generally used for destroying all lower forms of life.

Incompatibilities: Incompatible with phosphorus.

Permissible Exposure Limits in Air: No standards set. However, inasmuch as it is a cyanide compound, the exposure limits are listed here: OSHA and ACGIH: 5 mg/m^3 TWA; NIOSH: Ceiling limit, 4.7 ppm; 5 mg/m^3 per 10 minutes as cyanides. All have notations that skin contact contributes significantly in overall exposure. IDLH = 25 mg/m^3 as CN

Determination in Air: No methods listed.

Permissible Concentration in Water: No criteria set.

Routes of Entry: Ingestion, absorbed through the skin.

Harmful Effects and Symptoms

Short Term Exposure: Highly irritation to eyes and skin. Converted to cyanide in the body. Causes convulsions, paralysis and death from respiratory failure. Poisonous, may be fatal if swallowed or absorbed through skin. Health effects are similar to cyanides and iodides. Upon ingestion, a bitter, acrid, burning taste is sometimes noted. Other symptoms are anxiety, confusion, dizziness, giddiness, rapid and difficult breathing, palpitations, tightness in chest, unconsciousness, violent convulsions and death.

Long Term Exposure: Long term contact with iodides can cause weakness, anemia, loss of appetite, loss of weight, and general depression.

Points of Attack: Blood.

Medical Surveillance: EKG, blood cyanide level.

First Aid: If this chemical gets into the eyes, remove any contact lenses at once and irrigate immediately for at least 15 minutes, occasionally lifting upper and lower lids. Seek medical attention immediately. If this chemical contacts the skin, remove contaminated clothing and wash immediately with soap and water. Seek medical attention immediately. If this chemical has been inhaled, remove from exposure, begin rescue breathing (using universal precautions) if breathing has stopped and CPR if heart action has stopped. Transfer promptly to a medical facility. When this chemical has been swallowed, get medical attention. Give large quantities of water and induce vomiting. Do not make an unconscious person vomit. Effects may be delayed; keep victim under observation.

Use amyl nitrate capsules if symptoms develop. All area employees should be trained regularly in emergency measures for cyanide poisoning and in CPR. A cyanide antidote kit should be kept in the immediate work area and must be rapidly available. Kit ingredients should be replaced every 1 – 2 years to ensure freshness. Persons trained in the use of this kit, oxygen use, and CPR must be quickly available.

Personal Protective Methods: Wear protective gloves and clothing to prevent any reasonable probability of skin contact. Safety equipment suppliers/manufacturers can provide recommendations on the most protective glove/clothing material for your operation. All protective clothing (suits, gloves, footwear, headgear) should be clean, available each day, and put on before work. Contact lenses should not be worn when working with this chemical. Wear dust-proof chemical goggles and face shield unless full facepiece respiratory protection is worn. Employees should wash immediately with soap when skin is wet or contaminated. Provide emergency showers and eyewash.

Respirator Selection: *Where there is a potential for overexposure:* SCBAF:PD,PP (any MSHA/NIOSH approved self-contained breathing apparatus that has a full facepiece and is operated in a pressure-demand or other positive-pressure mode); or SAF:PD,PP:ASCBA (any supplied-air respirator that has a full facepiece and is operated in a pressure-demand or other positive-pressure mode in combination with an auxiliary, self-contained breathing apparatus operated in a pressure-demand or other positive pressure mode).

Storage: Prior to working with cyanogen iodide you should be trained on its proper handling and storage. Store in tightly closed containers in a cool, well ventilated area. Metal containers involving the transfer of this chemical should be grounded and bonded. Drums must be equipped with self-closing valves, pressure vacuum bungs, and flame arresters. Use only non-sparking tools and equipment, especially when opening and closing containers of this chemical. Sources of ignition such as smoking and open flames, are prohibited where this chemical is used, handled, or stored in a manner that could create a potential fire or explosion hazard.

Shipping: Cyanides, inorganic, n.o.s. require a "Poison" label. Packing Group I. This material falls in UN/DOT Hazard Class 6.1.

Spill Handling: If outside, cover material to protect from wind, rain, or spray. Evacuate persons not wearing protective equipment from area of spill or leak until clean-up is complete. Remove all ignition sources. Collect powdered material in the most convenient and safe manner and deposit in sealed containers. Ventilate area after clean-up is complete. It may be necessary to contain and dispose of this chemical as a hazardous waste. If material or contaminated runoff enters waterways, notify downstream users of potentially contaminated waters. Contact your Department of Environmental Protection or your regional office of the federal EPA for specific recommendations. If employees are required to clean-up spills, they must be properly trained and equipped. OSHA 1910.120(q) may be applicable.

Fire Extinguishing: Firefighting gear (including SCBA) does not provide adequate protection. If exposure occurs, remove and isolate gear immediately and thoroughly decontaminate personnel. This chemical is a combustible solid. Small fires: dry chemical, carbon dioxide, water spray, or foam. Large fires: water spray, fog, or foam. Move container from fire area if you can do so without risk. Fight fire from maximum distance. Dike fire control water for later disposal; do not scatter the material. Keep unnecessary people away; isolate hazard area and deny entry. Stay upwind; keep out of low areas. Ventilate closed spaces before entering them. Wear positive pressure breathing apparatus and special protective clothing. Remove and isolate contaminated clothing at the site. Poisonous gases are produced in fire including cyanide gas, iodide gas and nitrogen oxides. If material or contaminated runoff enters waterways, notify downstream users of potentially contaminated waters. Notify local health and fire officials and pollution control agencies. From a secure, explosion-proof location, use water spray to cool exposed containers. If cooling streams are ineffective (venting sound increases in volume and pitch, tank discolors, or shows any signs of deforming), withdraw immediately to a secure position. If employees are expected to fight fires, they must be trained and equipped in OSHA 1910.156.

References

U.S. Environmental Protection Agency, "Chemical Profile: Cyanogen Iodine," Washington, DC, Chemical Emergency Preparedness Program (November 30, 1987)

Cyanophos

Molecular Formula: $C_8H_{16}N_5O_6P_2S_2$

Synonyms: BAY 34727; Bayer 34727; Ciafos; Cianofos (Spanish); *O*-(4-Cyanophenyl) *O,O*-dimethyl phosphorothioate; *O,p*-Cyanophenyl *O,O*-dimethyl phosphorothioate; Cyanophos organophosphate compound; Cyanox; Cyap; *O,O*-Dimethyl-*O*-(4-cyano-phenyl)-monothiophosphat (German); *O,O*-Dimethyl *O,p*-cyanophenyl phosphorothioate; *O,O*-Dimethyl *O*-4-cyanophenyl phosphorothioate; ENT 25,675; May & Baker S-4084; Phosphorothioic acid, *O*-(4-cyanophenyl)-*O,O*-dimethyl ester; Phosphorothioic acid, *O*-(4-cyanophenyl)-9,9-dimethyl ester; Phosphorothioic acid, *O,O*-dimethyl ester, *O*-ester with *p*-hydroxybenzonitrile; S 4084; Sumitomo S 4084; Sunitomo S 4084

CAS Registry Number: 2636-26-2

RTECS Number: TF7600000

Regulatory Authority

- SUPERFUND/EPCRA 40CFR355, Appendix B Extremely Hazardous Substances: TPQ = 1,000 lb (454 kg)
- SUPERFUND/EPCRA 40CFR302.4 Reportable Quantity (RQ): EHS, 1 lb (0.454 kg)
- U.S. DOT Regulated Marine Pollutant (49CFR172.101, Appendix B)
- Canada, WHMIS, Ingredients Disclosure List; National Pollutant Release Inventory (NPRI); CEPA Priority Substance List, Ocean dumping prohibited

Cited in U.S. State Regulations: California (A, G), Florida (G), Massachusetts (G), New Jersey (G), Pennsylvania (G).

Description:

Cyanophos is a yellow to reddish-yellow or amber liquid. Boiling point = 119°C (decomposes). Freezing/Melting point = 14 – 15°C. Slightly soluble in water.

Potential Exposure: Those involved in the manufacture, formulation and application of this insecticide which is used against rice stem borers and house flies. It is not registered as a pesticide in the U.S.

Incompatibilities: Alkaline materials and exposure to light can cause rapid decomposition.

Permissible Exposure Limits in Air: No standards set. However, inasmuch as it is a cyanide compound, the exposure limits are listed here: OSHA and ACGIH: 5 mg/m^3 TWA; NIOSH: Ceiling limit, 4.7 ppm; 5 mg/m^3 per 10 minutes as cyanides. All have notations that skin contact contributes significantly in overall exposure. IDLH = 25 mg/m^3 as CN.

Determination in Air: Filter/Bubbler; Potassium hydroxide; Ion-specific electrode; NIOSH IV Method #7904, Cyanides. OSHA versatile sampler-2; Toluene/Acetone; Gas chromatography/Flame photometric detection for sulfur, nitrogen, or phosphorus; NIOSH Method IV Method #5600, Organophosphorus Pesticides.

Permissible Concentration in Water: No criteria set.

Routes of Entry: Inhalation, ingestion, skin contact. Absorbed through the skin.

Harmful Effects and Symptoms

Short Term Exposure: Cyanophos is an organophosphorus insecticide. It is a cholinesterase inhibitor. Death may occur after a large oral dose; with smaller accidental doses, onset of illness may be delayed. The LD_{50} oral (rat) is 25 mg/kg (highly toxic).

Symptoms of organophosphorus pesticide poisoning include: headache, giddiness, nervousness, blurred vision, weakness, nausea, cramps, diarrhea, and discomfort in the chest. Signs include: sweating, pinpoint pupils, tearing, salivation and other excessive respiratory tract secretion, vomiting, cyanosis, papilledema, uncontrollable muscle twitches followed by muscular weakness, convulsions, coma, loss of sphincter control.

Long Term Exposure: Cholinesterase inhibitor; possible cumulative effect. Cyanophos may damage the nervous system, resulting in convulsions, respiratory failure. May cause liver damage.

Points of Attack: Respiratory system, central nervous system, peripheral nervous system, plasma cholinesterase.

Medical Surveillance: Before employment and at regular times after that, the following are recommended: Plasma and red blood cell cholinesterase levels (tests for the enzyme poisoned by this chemical). If exposure stops, plasma levels return to normal in 1 – 2 weeks while red blood cell levels may be reduced for 1 – 3 months.

When cholinesterase enzyme levels are reduced by 25% or more below preemployment levels, risk of poisoning is increased, even if results are in lower ranges of "normal." Reassignment to work not involving organophosphate or carbamate pesticides is recommended until enzyme levels recover. If symptoms develop or overexposure occurs, repeat the above tests as soon as possible and get an exam of the nervous system. Also consider complete blood count. Consider chest x-ray following acute overexposure. Do not drink any alcoholic beverages before or during use. Alcohol promotes absorption of organic phosphates.

First Aid: If this chemical gets into the eyes, remove any contact lenses at once and irrigate immediately for at least 15 minutes, occasionally lifting upper and lower lids. Seek medical attention immediately. If this chemical contacts the skin, remove contaminated clothing and wash immediately with soap and water. Speed in removing material from skin is of extreme importance. Shampoo hair promptly if contaminated. Seek medical attention immediately. If this chemical has been inhaled, remove from exposure, begin rescue breathing (using universal precautions) if breathing has stopped and CPR if heart action has stopped. Transfer promptly to a medical facility. When this chemical has been swallowed, get medical attention. Give large quantities of water and induce vomiting. Do not make an unconscious person vomit. Effects may be delayed; keep victim under observation.

Personal Protective Methods: Wear protective gloves and clothing to prevent any reasonable probability of skin contact. Safety equipment suppliers/manufacturers can provide recommendations on the most protective glove/clothing material for your operation. All protective clothing (suits, gloves, footwear, headgear) should be clean, available each day, and put on before work. Contact lenses should not be worn when working with this chemical. Wear splash-proof chemical goggles and face shield unless full facepiece respiratory protection is worn. Employees should wash immediately with soap when skin is wet or contaminated. Provide emergency showers and eyewash. See NIOSH also Criteria Document 78-212 NITRILES.

Respirator Selection: *Where there is a potential for overexposure:* SCBAF:PD,PP (any MSHA/NIOSH approved self-contained breathing apparatus that has a full facepiece and is operated in a pressure-demand or other positive-pressure mode); or SAF:PD,PP:ASCBA (any supplied-air respirator that has a full facepiece and is operated in a pressure-demand or other positive-pressure mode in combination with an auxiliary, self-contained breathing apparatus operated in a pressure-demand or other positive pressure mode).

Storage: Cyanophos is stable to storage for 2 years or more under normal conditions. Prior to working with this chemical you should be trained on its proper handling and storage. Store in tightly closed containers in a cool, well-ventilated area.

Shipping: Organophosphorus pesticides, liquid, toxic, n.o.s. require a "Poison" label. Cyanophos falls in UN/DOT Hazard Class 6.1.[19][20]

Spill Handling: Stay upwind; keep out of low areas. Ventilate closed spaces before entering them. Wear positive pressure breathing apparatus and special protective clothing. Remove and isolate contaminated clothing at the site. Do not touch spilled material. Use water spray to reduce vapors. Evacuate and restrict persons not wearing protective equipment from area of spill or leak until cleanup is complete. Remove all ignition sources. Ventilate area of spill or leak. Absorb liquids in vermiculite, dry sand, earth, peat, carbon, or a similar material and deposit in sealed containers. Dike far ahead of large spills for later disposal. It may be necessary to contain and dispose of this chemical as a hazardous waste. If material or contaminated runoff enters waterways, notify downstream users of potentially contaminated waters. Contact your Department of Environmental Protection or your regional office of the federal EPA for specific recommendations. If employees are required to clean-up spills, they must be properly trained and equipped. OSHA 1910.120(q) may be applicable.

Fire Extinguishing: This material may burn, but does not ignite readily. For small fires, use dry chemical, carbon dioxide, water spray, or foam. For large fires, use water spray, fog, or foam. Stay upwind; keep out of low areas. Move container from fire area if you can do it without risk. Fight fire from maximum distance. Dike fire control water for later disposal; do not scatter the material. Wear positive pressure breathing apparatus and special protective clothing. Poisonous gases are produced in fire including nitrogen oxides, phosphorous oxides, cyanide and sulfur oxides. If material or contaminated runoff enters waterways, notify downstream users of potentially contaminated waters. Notify local health and fire officials and pollution control agencies. From a secure, explosion-proof location, use water spray to cool exposed containers. If cooling streams are ineffective (venting sound increases in volume and pitch, tank discolors, or shows any signs of deforming), withdraw immediately to a secure position. If employees are expected to fight fires, they must be trained and equipped in OSHA 1910.156.

Disposal Method Suggested: In accordance with 40CFR 165 recommendations for the disposal of pesticides and pesticide containers. Must be disposed properly by following package label directions or by contacting your state pesticide or environmental control agency or by contacting your regional EPA office.

References

U.S. Environmental Protection Agency, "Chemical Profile: Cyanophos," Washington, DC, Chemical Emergency Preparedness Program (November 30, 1987)

Cyanopyridines

Molecular Formula: $C_6H_4N_2$

Common Formula: C_5H_4N-CN

Synonyms: *2-cyano-:* Picolinic acid nitrile; 2-Pyridinecarbonitrile

3-cyano-: 3-Azabenzonitrile; 3-Cyanopyridine; 3-Cyjanopirydyna; Nicotinic acid nitrile; Nicotinonitrile; Nitryl-kwasu-nikotynowego (Polish); 3-Pyridinecarbonitrile; 3-Pyridinenitrile; 3-Pyridylcarbonitrile

4-cyano-: 4-Azabenzonitrile; Isonicotinonitrile; 4-Pyridine carbonitrile

CAS Registry Number: 100-70-9 (2-cyano-); 100-54-9 (3-cyano-); 100-48-1 (4-cyano-)

RTECS Number: QT3030000 (3-cyano-)

DOT ID: UN 3276 (nitriles, toxic, n.o.s.)

Regulatory Authority

As cyanide compounds:

- CLEAN AIR ACT: Hazardous Air Pollutants (Title I, Part A, Section 112)
- CLEAN WATER ACT: 40CFR423, Appendix A, Priority Pollutants as cyanide, total
- EPA HAZARDOUS WASTE NUMBER (RCRA No.): P030 as cyanides soluble salts and complexes, n.o.s.
- RCRA, 40CFR261, Appendix 8 Hazardous Constituents. as cyanides, soluble salts and complexes, n.o.s.
- EPCRA (Section 313): X+CN- where X = H+ or any other group where a formal dissociation may occur. For example, KCN or $Ca(CN)_2$. Form R de minimis concentration reporting level: 1.0%
- U.S. DOT Regulated Marine Pollutant (49CFR172.101, Appendix B) as cyanide mixtures, cyanide solutions or cyanides, inorganic, n.o.s.

Cited in U.S. State Regulations: New York (G).

Description: The cyanopyridines, C_5H_4N-CN, are as follows:

Isomer	Freez./Melt. Point, °C	Boiling Point, °C	Flash Point, °C	Appearance
2-Cyano-	—	—	89	—
3-Cyano-	47 – 49	83 – 84	—	Colorless liquidor gray crystals
4-Cyano-	—	—	—	—

3- Cyano is soluble in water.

Potential Exposure: The cyanopyridines are used in the synthesis of organic compounds and as corrosion inhibitors for aluminum.

Incompatibilities: Oxidizing agents such as perchlorates, peroxides and permanganates.

Permissible Exposure Limits in Air: The NIOSH REL for nitriles is a ceiling limit of 6 mg/m^3, not to be exceeded in any 15-minute work period.

Determination in Air: See NIOSH Criteria Document 78-212 NITRILES.

Permissible Concentration in Water: No criteria set.

Harmful Effects and Symptoms

Short Term Exposure: *Inhalation:* May cause irritation to the nose and throat. *Skin:* May cause irritation. *Eyes:* May cause irritation. Animal data suggests eye damage can result from contact. *Ingestion:* Possible central nervous system damage due to cyanide content.

Long Term Exposure: No information found. The LD_{50} oral-rat for 3-cyanopiridines is 1,105 mg/kg (slightly toxic).

First Aid: If this chemical gets into the eyes, remove any contact lenses at once and irrigate immediately for at least 15 minutes, occasionally lifting upper and lower lids. Seek medical attention immediately. If this chemical contacts the skin, remove contaminated clothing and wash immediately with soap and water. Seek medical attention immediately. If this chemical has been inhaled, remove from exposure, begin rescue breathing (using universal precautions) if breathing has stopped and CPR if heart action has stopped. Transfer promptly to a medical facility. When this chemical has been swallowed, get medical attention. Give large quantities of water and induce vomiting. Do not make an unconscious person vomit.

Personal Protective Methods: Wear protective gloves and clothing to prevent any reasonable probability of skin contact. Safety equipment suppliers/manufacturers can provide recommendations on the most protective glove/clothing material for your operation. All protective clothing (suits, gloves, footwear, headgear) should be clean, available each day, and put on before work. Contact lenses should not be worn when working with this chemical. When working with liquids, wear splash-proof chemical goggles and face shield unless full facepiece respiratory protection is worn. When working with powders or dusts, wear dust-proof chemical goggles and face shield unless full facepiece respiratory protection is worn. Employees should wash immediately with soap when skin is wet or contaminated. Provide emergency showers and eyewash. See NIOSH Criteria Document 78-212 NITRILES.

Respirator Selection: Wear a chemical cartridge respirator with organic vapor or organic vapor/acid gas cartridges, if necessary.

Storage: Prior to working with this chemical you should be trained on its proper handling and storage. Store in tightly closed containers in a cool, well-ventilated area away from heat or flame and separate from oxidizing materials.

Shipping: The cyanopyridines are not specifically cited in DOT's Performance-Oriented Packaging Standards.[19]

Spill Handling: Evacuate and restrict persons not wearing protective equipment from area of spill or leak until cleanup is complete. Remove all ignition sources. Ventilate area of spill or leak. Absorb liquids in vermiculite, dry sand, earth, peat, carbon, or a similar material and deposit in sealed containers. Wash area of spill with soap and water. It may be necessary to contain and dispose of this chemical as a hazardous waste. If material or contaminated runoff enters waterways, notify downstream users of potentially contaminated waters. Contact your Department of Environmental Protection or your regional office of the federal EPA for specific recommendations. If employees are required to clean-up spills, they must be properly trained and equipped. OSHA 1910.120(q) may be applicable.

Fire Extinguishing: This chemical is a combustible liquid. Poisonous gases are produced in fire. Use dry chemical, carbon dioxide, or alcohol foam extinguishers. Vapors are heavier than air and will collect in low areas. Vapors may travel long distances to ignition sources and flashback. Vapors in confined areas may explode when exposed to fire. Containers may explode in fire. Storage containers and parts of containers may rocket great distances, in many directions. If material or contaminated runoff enters waterways, notify downstream users of potentially contaminated waters. Notify local health and fire officials and pollution control agencies. From a secure, explosion-proof location, use water spray to cool exposed containers. If cooling streams are ineffective (venting sound increases in volume and pitch, tank discolors, or shows any signs of deforming), withdraw immediately to a secure position. If employees are expected to fight fires, they must be trained and equipped in OSHA 1910.156.

References

New York State Department of Health, "Chemical Fact Sheet: Cyanopyridine(s)," Albany, NY, Bureau of Toxic Substance Assessment (June 1986)

Cycasin

Molecular Formula: $C_8H_{16}N_2O_8$

Synonyms: β-D-Glucopyranoside, (methyl-ONN-azoxy)methyl-; Side methylazoxymethanol β-D-glucoside

CAS Registry Number: 14901-08-7

RTECS Number: LZ5982000

DOT ID: No citation.

Regulatory Authority

- Carcinogen (Animal Positive) (IARC)[9] (NTP)[1]
- Air Pollutant Standard Set (North Dakota)[60]
- Hazardous Waste (EPA-RCRA) Hazardous Constituent Waste (EPA)

Cited in U.S. State Regulations: California (G), Florida (G), Illinois (G, W), Kansas (G), Louisiana (G), Maine (G), Massachusetts (G), Michigan (G), Minnesota (G), New Hampshire (G), New Jersey (G), North Dakota (A), Pennsylvania (G), Vermont (G), Virginia (G), Washington (G), West Virginia (G), Wisconsin (G).

Description: Cycasin, $C_8H_{16}N_2O_8$, is a crystalline solid. Freezing/Melting point = (decomposes) 154°C. Hazard Identification (based on NFPA-704 M Rating System): Health 2, Flammability 2, Reactivity 0.

Potential Exposure: Cycasin occurs naturally, in the seeds, roots, and leaves of cycad plants which are found in the tropical and subtropical regions of the world. Nuts from the cycads are used to make chips, flour, and starch.

Cycasin is not produced or used commercially. The major potential exposure is the ingestion of the foods containing Cycasin. It is estimated that about 50 – 55 percent of the inhabitants of Guam are potentially exposed (50,000 – 60,000 persons) to Cycasin. Wastewater from the preparation of the cycad nuts contain large amounts of Cycasin and represents a potential secondary exposure source.

Permissible Exposure Limits in Air: North Dakota[60] has set a guideline for Cycasin in ambient air of zero.

Permissible Concentration in Water: No criteria set.

Harmful Effects and Symptoms

Cycasin is carcinogenic in 5 animal species, inducing tumors in various organs. Following oral exposure, it is carcinogenic in the rat, hamster, guinea pig and fish. By this route, the data in the mouse is of borderline significance and the negative experiment in chickens only lasted 68 weeks. It is active in single-dose experiments and following prenatal exposure. The carcinogenicity of its metabolite, methylazoxymethanol, has been demonstrated in the rat and the hamster and that of a closely related synthetic substance, methylazoxymethanol acetate, in the rat. The LD_{50} oral-rat is 270 mg/kg (moderately toxic).

Short Term Exposure: Poisonous.

Long Term Exposure: A possible human carcinogen. Laboratory tests on animals suggest this chemical is a teratogen; mutation data has been reported.

First Aid: If this chemical gets into the eyes, remove any contact lenses at once and irrigate immediately for at least 15 minutes, occasionally lifting upper and lower lids. Seek medical attention immediately. If this chemical contacts the skin, remove contaminated clothing and wash immediately with soap and water. Seek medical attention immediately. If this chemical has been inhaled, remove from exposure, begin rescue breathing (using universal precautions) if breathing has stopped and CPR if heart action has stopped. Transfer promptly to a medical facility. When this chemical has been swallowed, get medical attention. Give large quantities of water and induce vomiting. Do not make an unconscious person vomit.

Personal Protective Methods: Wear protective gloves and clothing to prevent any reasonable probability of skin contact. Safety equipment suppliers/manufacturers can provide recommendations on the most protective glove/clothing material for your operation. All protective clothing (suits, gloves, footwear, headgear) should be clean, available each day, and put on before work. Contact lenses should not be worn when working with this chemical. When working with liquids, wear splash-proof chemical goggles and face shield unless full facepiece respiratory protection is worn. When working with powders or dusts, wear dust-proof chemical goggles and face shield unless full facepiece respiratory protection is worn. Employees should wash immediately with soap when skin is wet or contaminated. Provide emergency showers and eyewash.

Respirator Selection: *At any detectable concentration:* SCBAF:PD,PP (any MSHA/NIOSH approved self-contained breathing apparatus that has a full facepiece and is operated in a pressure-demand or other positive-pressure mode); or SAF:PD,PP:ASCBA (any supplied-air respirator that has a full facepiece and is operated in a pressure-demand or other positive-pressure mode in combination with an auxiliary, self-contained breathing apparatus operated in a pressure-demand or other positive pressure mode). *Escape:* HiEF [any air-purifying, full-facepiece respirator (gas mask) with a chin-style, front- or back-mounted organic vapor canister having a high-efficiency particulate filter]; or SCBAE (any appropriate escape-type, self-contained breathing apparatus).

Storage: Prior to working with Cycasin you should be trained on its proper handling and storage. A regulated, marked area should be established where this chemical is handled, used, or stored in compliance with OSHA standard 1910.1045. Store in tightly closed containers in a cool, well-ventilated area.

Shipping: Cycasin is not cited by DOT[19] in its Performance-Oriented Packaging Standards.

Spill Handling: Evacuate persons not wearing protective equipment from area of spill or leak until clean-up is complete. Remove all ignition sources. Collect powdered material in the most convenient and safe manner and deposit in sealed containers. Ventilate area after clean-up is complete. It may be necessary to contain and dispose of this chemical as a hazardous waste. If material or contaminated runoff enters waterways, notify downstream users of potentially contaminated waters. Contact your Department of Environmental Protection or your regional office of the federal EPA for specific recommendations. If employees are required to clean-up spills, they must be properly trained and equipped. OSHA 1910.120(q) may be applicable.

Fire Extinguishing: Use dry chemical, carbon dioxide, water spray, or alcohol foam extinguishers. Poisonous gases are produced in fire, including nitrogen oxides. If material or contaminated runoff enters waterways, notify downstream users of potentially contaminated waters. Notify local health and fire officials and pollution control agencies. From a secure, explosion-proof location, use water spray to cool exposed containers. If cooling streams are ineffective (venting sound increases in volume and pitch, tank discolors, or shows any signs of deforming), withdraw immediately to a secure position. If employees are expected to fight fires, they must be trained and equipped in OSHA 1910.156.

References

Sax, N. I., Ed., "Dangerous Properties of Industrial Materials Report," 1, No. 3, 48–49 (1981)

Cycloheptene

Molecular Formula: C_7H_{12}

Synonyms: Suberane; Suberylene

CAS Registry Number: 628-92-2

RTECS Number: GU4615000

DOT ID: UN 2242

Cited in U.S. State Regulations: New Hampshire (G), New Jersey (G).

Description:

Cycloheptene is a flammable, colorless, oily liquid. Boiling point = 115°C. Flash point = –7°C. Hazard Identification (based on NFPA-704 M Rating System): Health 1, Flammability 3, Reactivity 0. Insoluble in water.

Potential Exposure: Cycloheptene may be used in organic synthesis.

Incompatibilities: Contact with strong oxidizers may cause a fire or explosion hazard.

Permissible Exposure Limits in Air: No standards set.

Permissible Concentration in Water: No criteria set.

Routes of Entry: Inhalation, passing through the skin.

Harmful Effects and Symptoms

Short Term Exposure: Cycloheptene can affect you when breathed in and by passing through your skin. Exposure can cause you to feel dizzy, lightheaded and to pass out. Contact can irritate the skin.

Long Term Exposure: May cause drying and cracking of the skin.

Points of Attack: Skin.

First Aid: If this chemical gets into the eyes, remove any contact lenses at once and irrigate immediately for at least 15 minutes, occasionally lifting upper and lower lids. Seek medical attention immediately. If this chemical contacts the skin, remove contaminated clothing and wash immediately with soap and water. Seek medical attention immediately. If this chemical has been inhaled, remove from exposure, begin rescue breathing (using universal precautions) if breathing has stopped and CPR if heart action has stopped. Transfer promptly to a medical facility. When this chemical has been swallowed, get medical attention. Give large quantities of water and induce vomiting. Do not make an unconscious person vomit.

Personal Protective Methods: *Clothing:* Avoid skin contact with Cycloheptene. Wear solvent-resistant gloves and clothing. Safety equipment suppliers/manufacturers can provide recommendations on the most protective glove/clothing material for your operation. All protective clothing (suits, gloves, footwear, headgear) should be clean, available each day and put on before work.

Eye Protection: Wear splash-proof chemical goggles and face shield when working with liquid, unless full facepiece respiratory protection is worn.

Respirator Selection: Where the potential exists for exposures to Cycloheptene, use a MSHA/NIOSH approved supplied-air respirator with a full facepiece operated in the positive pressure mode or with a full facepiece, hood, or helmet in the continuous flow mode, or use a MSHA/NIOSH approved self-contained breathing apparatus with a full facepiece operated in pressure-demand or other positive pressure mode.

Storage: Cycloheptene must be stored to avoid contact with strong oxidizers (such as chlorine, bromine and fluorine) since violent reactions occur. Store in tightly closed containers in a cool, well-ventilated area. Sources of ignition, such as smoking and open flames, are prohibited where Cycloheptene is used, handled, or stored in a manner that could create a potential fire or explosion hazard.

Metal containers involving the transfer of 5 gallons or more of Cycloheptene should be grounded and bonded. Drums must be equipped with self-closing valves, pressure vacuum bungs and flame arresters. Use only non-sparking tools and equipment, especially when opening and closing containers of Cycloheptene. Wherever Cycloheptene is used, handled, manufactured, or stored, use explosion-proof electrical equipment and fittings.

Shipping: Cycloheptene falls in DOT Hazard Class 3 and Packing Group II. It must be labeled "Flammable Liquid."

Spill Handling: Evacuate and restrict persons not wearing protective equipment from area of spill or leak until cleanup is complete. Remove all ignition sources. Ventilate area of spill or leak. Absorb liquids in vermiculite, dry sand, earth, peat, carbon, or a similar material and deposit in sealed containers. Keep Cycloheptene out of a confined space, such as a sewer, because of the possibility of an explosion, unless the sewer is designed to prevent the build-up of explosive concentrations. It may be necessary to contain and dispose of this chemical as a hazardous waste. If material or contaminated runoff enters waterways, notify downstream users of potentially contaminated waters. Contact your Department of Environmental Protection or your regional office of the federal EPA for specific recommendations. If employees are required to clean-up spills, they must be properly trained and equipped. OSHA 1910.120(q) may be applicable.

Fire Extinguishing: This chemical is a flammable liquid. Poisonous gases are produced in fire. Use dry chemical, carbon dioxide, or foam extinguishers. Vapors are heavier than air and will collect in low areas. Vapors may travel long distances to ignition sources and flashback. Vapors in confined areas may explode when exposed to fire. Containers may explode in fire. Storage containers and parts of containers may rocket great distances, in many directions. If material or contaminated runoff enters waterways, notify downstream users of potentially contaminated waters. Notify local health and fire officials and pollution control agencies. From a secure, explosion-proof location, use water spray to cool exposed containers. If cooling streams are ineffective (venting sound increases in volume and pitch, tank discolors, or shows any signs of deforming), withdraw immediately to a secure position. If employees are expected to fight fires, they must be trained and equipped in OSHA 1910.156.

Disposal Method Suggested: Incineration.

References

New Jersey Department of Health and Senior Services and Senior Services, "Hazardous Substance Fact Sheet: Cycloheptene," Trenton, NJ (January 11, 1988)

Cyclohexane

Molecular Formula: C_6H_{12}

Synonyms: Benzene hexahydride; Benzene, hexahydro; Cicloesano (Italian); Ciclohexano (Spanish); Cyclohexaan (Dutch); Cyclohexan (German); Cykloheksan (Polish); Hexahydrobenzene; Hexamethylene; Hexanaphthene

CAS Registry Number: 110-82-7

RTECS Number: GU6300000

DOT ID: UN 1145

EEC Number: 601-017-00-1

EINECS Number: 203-806-2

Regulatory Authority

- Air Pollutant Standard Set (ACGIH)[1] (Australia) (DFG)[3] (HSE)[33] (Israel) (Mexico) (former USSR)[43] (OSHA)[58] (Several States)[60] (Several Canadian Provinces)
- CLEAN WATER ACT: Section 311 Hazardous Substances/RQ 40CFR117.3 (same as CERCLA, see below); Section 313 Water Priority Chemicals (57FR41331, 9/9/92)
- EPA HAZARDOUS WASTE NUMBER (RCRA No.): U056
- RCRA, 40CFR261, Appendix 8 Hazardous Constituents
- SUPERFUND/EPCRA 40CFR302.4 Reportable Quantity (RQ): CERCLA, 1,000 lb (454 kg)
- EPCRA Section 313 Form R *de minimis* concentration reporting level: 1.0%
- Canada, WHMIS, Ingredients Disclosure List; National Pollutant Release Inventory (NPRI)

Cited in U.S. State Regulations: Alaska (G), California (A, G), Connecticut (A), Florida (G, A), Illinois (G), Kansas (G), Louisiana (G), Maine (G), Maryland (G), Massachusetts (G, A), Minnesota (G), Nevada (A), New Hampshire (G), New Jersey (G), New York (G, A), North Dakota (A), Oklahoma (G), Pennsylvania (G), Rhode Island (G), Vermont (G), Virginia (G, A), Washington (G), West Virginia (G), Wisconsin (G).

Description: Cyclohexane, C_6H_{12}, is a colorless liquid with a mild, sweet odor. Odor threshold = 0.16 ppm. Boiling point = 81°C. Freezing/Melting point = 7°C. Flash point = -20°C. Autoignition temperature = 260°C. Hazard Identification (based on NFPA-704 M Rating System): Health 1, Flammability 3, Reactivity 0. The explosive limits are: LEL = 1.3%; UEL = 8%. Insoluble in water.

Potential Exposure: Cyclohexane is used as a chemical intermediate, as a solvent for fats, oils, waxes, resins, and certain synthetic rubbers, and as an extractant of essential oils in the perfume industry.

Incompatibilities: Forms explosive mixture with air. Contact with oxidizers, nitrogen dioxide, oxygen can cause fire and explosion hazard. Can explode in heat when mixed with dinitrogen tetraoxide liquid.

Permissible Exposure Limits in Air: The Federal limit (OSHA PEL 8-hour TWA),[58] and the DFG, HSE, Israeli, Australian, Mexican, and ACGIH value is 300 ppm (1,050 mg/m^3). The HSE[33] and Mexican STEL value is 375 ppm (1,300 mg/m^3). The NIOSH IDLH level is 1,300 ppm. [10% LEL. Canadian provincial TWA for Alberta, BC, Ontario, and Quebec are the same as OSHA and Alberta's STEL is 375 ppm. The former USSR-UNEP/IRPTC project[43] has set an MAC in workplace air of 80 mg/m^3. They have also set MAC values for ambient air in residential areas of 1.4 mg/m^3 on either a momentary or a daily average basis. Several states have set guidelines or standards for cyclohexane in ambient air[60] ranging from 1.4 mg/m^3 (Massachusetts) to 10.5 – 13.0 mg/m^3 (North Dakota) to 17.0 mg/m^3 (Virginia) to 21.0 mg/m^3 (Connecticut, Florida and New York) to 25.0 mg/m^3 (Nevada).

Determination in Air: Adsorption on charcoal. Workup with CS_2, analysis by gas chromatography. See NIOSH Method #1500 for hydrocarbons (B.P. 36 – 126°C).

Permissible Concentration in Water: The former USSR-UNEP/IRPTC joint project[43] has set an MAC of 0.1 mg/l in water bodies used for domestic purposes and 0.01 mg/l in water bodies used for fishery purposes.

Routes of Entry: Inhalation, ingestion, skin and/or eye contact.

Harmful Effects and Symptoms

Short Term Exposure: High concentrations (300 ppm): irritates eyes, nose and respiratory tract. Inhalation of high concentration (300 ppm) may cause irritation of the eyes, nose and throat. Higher concentrations may act as a narcotic resulting in dizziness, nausea, vomiting or loss of consciousness. Levels of 1,800 ppm can cause death. Vapor or liquid may cause skin irritation. This chemical destroys the skin's natural oils. If allowed to remain in contact with skin, may cause cracking, drying, chapping, smarting, and reddening. Ingestion of the liquid may cause aspiration into the lungs and chemical pneumonia. Animal studies suggest a lethal dose between one ounce and one pint for an adult. Exposure to high levels can cause nausea, dizziness, lightheadedness, and drowsiness. Unconsciousness and death may occur at levels far above the occupational exposure limit. Alcohol synergistically increases the toxic effects of cyclohexane.

Long Term Exposure: Prolonged or repeated exposure may cause skin drying, rash, and dermatitis. May cause damage to the liver, kidneys, brain, heart and circulatory system.

Points of Attack: Eyes, respiratory system, central nervous system, skin.

Medical Surveillance: Consider possible irritant effects to the skin and respiratory tract in any preplacement or periodic examination, as well as any renal or liver complications.

First Aid: If this chemical gets into the eyes, remove any contact lenses at once and irrigate immediately for at least 15 minutes, occasionally lifting upper and lower lids. Seek medical attention immediately. If this chemical contacts the skin, remove contaminated clothing and wash immediately with soap and water. Seek medical attention immediately. If this chemical has been inhaled, remove from exposure, begin rescue breathing (using universal precautions) if breathing has stopped and CPR if heart action has stopped. Transfer promptly to a medical facility. When this chemical has been swallowed, get medical attention. Do not induce vomiting.

Personal Protective Methods: Wear appropriate clothing to prevent repeated or prolonged skin contact. Nitrile,

Viton, and Teflon are among the recommended protective materials. Wear eye protection to prevent any reasonable probability of eye contact. Employees should wash promptly when skin is wet or contaminated. Remove clothing immediately if wet or contaminated to avoid flammability hazard.

Respirator Selection: NIOSH/OSHA: *1,000 ppm:* SA:CF (any supplied-air respirator operated in a continuous-flow mode); or PAPROV [any powered, air-purifying respirator with organic vapor cartridge(s)]; or CCRFOV [any chemical cartridge respirator with a full facepiece and organic vapor cartridge(s)]; or GMFOV [any air-purifying, full-facepiece respirator (gas mask) with a chin-style, front- or back-mounted acid gas canister]; or SCBAF (any self-contained breathing apparatus with a full facepiece; or SAF (any supplied-air respirator with a full facepiece). *Emergency or planned entry into unknown concentrations or IDLH conditions:* SCBAF:PD,PP (any self-contained breathing apparatus that has a full facepiece and is operated in a pressure-demand or other positive-pressure mode); or SAF:PD,PP:ASCBA (any supplied-air respirator that has a full facepiece and is operated in a pressure-demand or other positive-pressure mode in combination with an auxiliary self-contained breathing apparatus operated in a pressure-demand or other positive pressure mode). *Escape:* GMFOV [any air-purifying, full-facepiece respirator (gas mask) with a chin-style, front-or back-, mounted organic vapor canister]; or SCBAE (any appropriate escape-type, self-contained breathing apparatus).

Note: Substance causes eye irritation or damage; eye protection needed; shield, hydrocarbon-insoluble rubber or plastic apron.

Storage: Prior to working with cyclohexane you should be trained on its proper handling and storage. Before entering confined space where cyclohexane may be present, check to make sure that an explosive concentration does not exist. Cyclohexane must be stored to avoid contact with oxidizers (such as perchlorates, peroxides, permanganates, chlorates, and nitrates) since violent reactions occur. Store in tightly closed containers in a cool well-ventilated area away from heat. Metal containers involving the transfer of this chemical should be grounded and bonded. Where possible, automatically pump liquid from drums or other storage containers to process containers. Drums must be equipped with self-closing valves, pressure vacuum bungs, and flame arresters. Use only non-sparking tools and equipment, especially when opening and closing containers of this chemical. Sources of ignition such as smoking and open flames, are prohibited where this chemical is used, handled, or stored in a manner that could create a potential fire or explosion hazard. Wherever cyclohexane is used, handled, manufactured, or stored, use explosion-proof electrical equipment and fittings.

Shipping: Cyclohexane falls in DOT Hazard Class 3 and Packing Group II. The label required is "Flammable Liquid."

Spill Handling: Evacuate and restrict persons not wearing protective equipment from area of spill or leak until cleanup is complete. Remove all ignition sources. Establish forced ventilation to keep levels below explosive limit. Absorb liquids in vermiculite, dry sand, earth, peat, carbon, or a similar material and deposit in sealed containers. It may be necessary to contain and dispose of this chemical as a hazardous waste. If material or contaminated runoff enters waterways, notify downstream users of potentially contaminated waters. Contact your Department of Environmental Protection or your regional office of the federal EPA for specific recommendations. If employees are required to clean-up spills, they must be properly trained and equipped. OSHA 1910.120(q) may be applicable.

Fire Extinguishing: This chemical is a highly flammable liquid. Poisonous gases are produced in fire. Use dry chemical, carbon dioxide, or foam extinguishers. Vapors are heavier than air and will collect in low areas. Vapors may travel long distances to ignition sources and flashback. Vapors in confined areas may explode when exposed to fire. Containers may explode in fire. Storage containers and parts of containers may rocket great distances, in many directions. If material or contaminated runoff enters waterways, notify downstream users of potentially contaminated waters. Notify local health and fire officials and pollution control agencies. From a secure, explosion-proof location, use water spray to cool exposed containers. If cooling streams are ineffective (venting sound increases in volume and pitch, tank discolors, or shows any signs of deforming), withdraw immediately to a secure position. If employees are expected to fight fires, they must be trained and equipped in OSHA 1910.156.

Disposal Method Suggested: Incineration.

References

New Jersey Department of Health and Senior Services and Senior Services, "Hazardous Substance Fact Sheet: Cyclohexane," Trenton, NJ (January 1986)

New York State Department of Health, "Chemical Fact Sheet: Cyclohexane," Albany, NY, Bureau of Toxic Substance Assessment of Toxic Substance Assessment (Mar. 1986 and Version 2)

Cyclohexanol

Molecular Formula: $C_6H_{12}O$

Common Formula: $C_6H_{11}OH$

Synonyms: Adronal; Anol; Cicloesanolo (Italian); Ciclohexanol (Spanish); 1-Cyclohexanol; Cyclohexyl alcohol; Cykloheksanol (Polish); Hexahydrophenol; Hexalin; Hydralin; Hydrophenol; Hydroxycyclohexane; Naxol; Phenol, hexahydro-

CAS Registry Number: 108-93-0

RTECS Number: GB7875000

DOT ID: No citation.

EEC Number: 603-009-00-3

Regulatory Authority

- Air Pollutant Standard Set (ACGIH)[1] (Australia) (DFG)[3] (HSE)[33] (Israel) (Mexico) (former USSR)[43] (OSHA)[58] (Several States)[60] (Several Canadian Provinces)
- EPCRA Section 313 Form R *de minimis* concentration reporting level: 1.0%
- Canada, WHMIS, Ingredients Disclosure List

Cited in U.S. State Regulations: Alaska (G), California (A, G), Connecticut (A), Florida (G), Illinois (G), Kansas (A), Maine (G), Massachusetts (G), Minnesota (G), Nevada (A), New Hampshire (G), New Jersey (G), North Dakota (A), Pennsylvania (G), Rhode Island (G), Virginia (A), West Virginia (G).

Description: Cyclohexanol, $C_6H_{11}OH$, is a sticky solid (above 25°C/77°F) or colorless, viscous liquid with a faint camphor odor. Odor threshold = 3.5 ppm. Boiling point = 161°C. Freezing/Melting point = 24°C. Flash point = 68°C. Autoignition temperature = 300°C. Hazard Identification (based on NFPA-704 M Rating System): Health 1, Flammability 2, Reactivity 0. Slight solubility in water.

Potential Exposure: Cyclohexanol is used as a solvent for ethyl cellulose and other resins; it is used in soap manufacture; it is used as a raw material for adipic acid manufacture as a nylon intermediate.

Incompatibilities: Forms explosive mixture in air. Contact with strong oxidizers cause a fire and explosion hazard. Attacks some plastics.

Permissible Exposure Limits in Air: The Federal limit (OSHA PEL 8-hour TWA),[58] and the DFG MAK,[3] HSE,[33] Australian, Israeli, Mexican, and ACGIH[1] TWA value is 50 ppm (200 mg/m^3). The NIOSH IDLH level is 400 ppm. The Canadian provincial TWA level for Alberta, BC, Ontario, and Quebec is the same as OSHA and Alberta's STEL is 75 ppm. Nearly all of these values carry the notation "skin." Warning that this chemical can be absorbed through the skin, thereby increasing exposure. The former USSR-UNEP/IRPTC project[43] has also set an MAC for ambient air in residential areas of 0.06 mg/m^3 on either a momentary or a daily average basis. Several states have set guidelines or standards for Cyclohexanol in ambient air[60] ranging from 0.476 mg/m^3 (Kansas) to 2.0 mg/m^3 (North Dakota) to 3.3 mg/m^3 (Virginia) to 4.0 mg/m^3 (Connecticut) to 4.76 mg/m^3 (Nevada).

Determination in Air: Adsorption on charcoal, workup with 2-propanol in CS_2, analysis by gas chromatography/flame ionization. See NIOSH IV, Method # 1402 for Alcohols III.

Permissible Concentration in Water: The former USSR-UNEP/IRPTC joint project[43] has set an MAC in water bodies used for domestic purposes of 0.05 mg/l.

Routes of Entry: Iinhalation, skin absorption, ingestion, skin and/or eye contact.

Harmful Effects and Symptoms

Short Term Exposure: Cyclohexanol irritates the eyes, skin, and respiratory tract. May affect the central nervous system. In high concentrations it can cause headache, nausea, vomiting, dizziness, and unconsciousness.

Long Term Exposure: Removes the natural oils from the skin causing drying, cracking, and dermatitis. Prolonged or high exposures can cause liver, kidney and, lung damage.

Points of Attack: Eyes, skin respiratory system.

Medical Surveillance: Consider the points of attack in preplacement and periodic physical examinations. Lung function tests, liver and kidney function tests.

First Aid: If this chemical gets into the eyes, remove any contact lenses at once and irrigate immediately for at least 15 minutes, occasionally lifting upper and lower lids. Seek medical attention immediately. If this chemical contacts the skin, remove contaminated clothing and wash immediately with soap and water. Seek medical attention immediately. If this chemical has been inhaled, remove from exposure, begin rescue breathing (using universal precautions) if breathing has stopped and CPR if heart action has stopped. Transfer promptly to a medical facility. When this chemical has been swallowed, get medical attention. Give large quantities of water and induce vomiting. Do not make an unconscious person vomit.

Personal Protective Methods: Wear protective gloves and clothing to prevent any reasonable probability of skin contact. Safety equipment suppliers/manufacturers can provide recommendations on the most protective glove/clothing material for your operation. Nitrile, Polyvinyl Alcohol, Viton, and Silvershield are among the recommended protective materials. All protective clothing (suits, gloves, footwear, headgear) should be clean, available each day, and put on before work. Contact lenses should not be worn when working with this chemical. When working with liquids, wear splash-proof chemical goggles and face shield unless full facepiece respiratory protection is worn. When working with powders or dusts, wear dust-proof chemical goggles and face shield unless full facepiece respiratory protection is worn. Employees should wash immediately with soap when skin is wet or contaminated. Provide emergency showers and eyewash.

Respirator Selection: NIOSH/OSHA: *Up to 400 ppm:* CCROV [any chemical cartridge respirator with organic vapor cartridge(s)];* or PAPROV [any powered, air-purifying respirator with organic vapor cartridge(s)];* or GMFOV (any air-purifying, full-facepiece respirator (gas mask) with a chin-style, front- or back-mounted organic vapor canister); or SA (any supplied-air respirator);* or SCBAF (any self-contained breathing apparatus with a full facepiece *Emergency or planned entry into unknown concentrations or IDLH conditions:* SCBAF:PD,PP (any self-contained breathing apparatus that has a full facepiece and is operated in a pressure-demand or other positive-pressure mode); or SAF:PD,PP:ASCBA (any supplied-air respirator that has a full facepiece and is operated in a pressure-demand or other positive-pressure mode in combination with an auxiliary self-contained breathing apparatus operated in a pressure-demand or other positive-pressure mode). *Escape:* GMFOV [any air-purifying, full-facepiece respirator (gas mask) with a chin-style, front-or back-mounted organic vapor canister] or SCBAE (any appropriate escape-type, self-contained breathing apparatus).

* Substance reported to cause eye irritation or damage; eye protection required.

Storage: Prior to working with Cyclohexanol you should be trained on its proper handling and storage. Before entering confined space where this chemical may be present, check to make sure that an explosive concentration does not exist. Cyclohexanol must be stored to avoid contact with strong oxidizers (such as chlorine, bromine, and fluorine) since violent reactions occur. Metal containers involving the transfer of this chemical should be grounded and bonded. Where possible, automatically pump liquid from drums or other storage containers to process containers. Drums must be equipped with self-closing valves, pressure vacuum bungs, and flame arresters. Use only non-sparking tools and equipment, especially when opening and closing containers of this chemical. Sources of ignition such as smoking and open flames, are prohibited where this chemical is used, handled, or stored in a manner that could create a potential fire or explosion hazard. Wherever this chemical is used, handled, manufactured, or stored, use explosion-proof electrical equipment and fittings.

Shipping: Cyclohexanol is not specifically cited by DOT[19] but the combustible liquid n.o.s. category may be applied. This falls in Hazard Class 3 and Packing Group III. There are no label requirements.

Spill Handling: Evacuate and restrict persons not wearing protective equipment from area of spill or leak until cleanup is complete. Remove all ignition sources. Ventilate area of spill or leak. Absorb liquids in vermiculite, dry sand, earth, or a similar material and deposit in sealed containers. Collect powdered material in the most convenient and safe manner and deposit in sealed containers. It may be necessary to contain and dispose of this chemical as a hazardous waste. If material or contaminated runoff enters waterways, notify downstream users of potentially contaminated waters. Contact your Department of Environmental Protection or your regional office of the federal EPA for specific recommendations. If employees are required to clean-up spills, they must be properly trained and equipped. OSHA 1910.120(q) may be applicable.

Fire Extinguishing: This chemical is a combustible liquid or solid. Poisonous gases produced in fire. Use dry chemical, carbon dioxide, or alcohol foam extinguishers. Vapors are heavier than air and will collect in low areas. Vapors may travel long distances to ignition sources and flashback. Vapors in confined areas may explode when exposed to fire. Containers may explode in fire. Storage containers and parts of containers may rocket great distances, in many directions. If material or contaminated runoff enters waterways, notify downstream users of potentially contaminated waters. Notify local health and fire officials and pollution control agencies. From a secure, explosion-proof location, use water spray to cool exposed containers. If cooling streams are ineffective (venting sound increases in volume and pitch, tank discolors, or shows any signs of deforming), withdraw immediately to a secure position. If employees are expected to fight fires, they must be trained and equipped in OSHA 1910.156.

Disposal Method Suggested: Incineration.

References

New Jersey Department of Health and Senior Services and Senior Services, "Hazardous Substance Fact Sheet: Cyclohexanol," Trenton, NJ (January 1997)

Cyclohexanone

Molecular Formula: $C_6H_{10}O$

Synonyms: Anon; Anone; Cicloesanone (Italian); Ciclohexanona (Spanish); Cyclohexanon (Dutch); Cyclohexyl ketone; Cykloheksanon (Polish); Hexalin; Hexanon; Hydralin; Hytrol O; Ketohexamethylene; Nadone; NCI-C55005; Oxocyclohexane; Pimelic ketone; Pimelin ketone; Pomelic acetone; Sextone

CAS Registry Number: 108-94-1

RTECS Number: GW1050000

DOT ID: UN 1915

EEC Number: 606-010-00-7

EINECS Number: 203-631-1

Regulatory Authority

- Air Pollutant Standard Set (ACGIH)[1] (Australia) (DFG)[3] (HSE)[33] (Israel) (Mexico) (Russia)[43] (OSHA)[58] (Several States)[60]
- EPA HAZARDOUS WASTE NUMBER (RCRA No.): U057
- RCRA, 40CFR261, Appendix 8 Hazardous Constituents
- RCRA 40CFR268.48; 61FR15654, Universal Treatment Standards: Wastewater (mg/l), 0.36; Nonwastewater (mg/l), 0.75 TCLP
- SUPERFUND/EPCRA 40CFR302.4 Reportable Quantity (RQ): CERCLA, 5,000 lb (2,270 kg)
- Canada, WHMIS, Ingredients Disclosure List

Cited in U.S. State Regulations: Alaska (G), California (A, G), Connecticut (A), Florida (G), Illinois (G), Kansas (G), Louisiana (G), Maine (G), Massachusetts (G), Minnesota (G), Nevada (A), New Hampshire (G), New Jersey (G), North Dakota (A), Pennsylvania (G), Rhode Island (G), Vermont (G), Virginia (G, A), Washington (G), West Virginia (G), Wisconsin (G).

Description: Cyclohexanone, $C_6H_{10}O$, is a water-white to slightly-yellow liquid with a peppermint-like or acetone-like odor. The odor threshold is 0.12 – 0.24 ppm in air. Boiling point = 157°C. Freezing/Melting point = -32.1°C. Flash point = 44°C. Autoignition temperature = 420°C. The explosive limits are: LEL = 1.1%; UEL = 9.4%. Hazard Identification (based on NFPA-704 M Rating System): Health 1, Flammability 2, Reactivity 0 Soluble in water.

Potential Exposure: It is used in metal degreasing and as a solvent for lacquers, resins, and insecticides. It is an intermediate in adipic acid manufacture.

Incompatibilities: Forms explosive mixture with air. Contact with oxidizing agents, nitric acid may cause a violent reaction. Do not use brass, copper, bronze or lead fittings. Attacks many coatings and plastic materials.

Permissible Exposure Limits in Air: The Federal limit (OSHA PEL 8-hour TWA), DFG MAK, Mexican value[3] is 50 ppm (200 mg/m^3). The ACGIH and NIOSH recommend a TWA of 25 ppm (100 mg/m^3) as has Australia, Israel, and HSE,[33] and the HSE[33] STEL value is 100 ppm (400 mg/m^3). The NIOSH IDLH level is 700 ppm. The OSHA and NIOSH limits also bear the notation "skin" which indicates the possibility of cutaneous absorption. In Canada the provincial TWA is 25 ppm for Alberta, Ontario, and Quebec and Alberta's STEL is 100 ppm. The British Columbia TWA is 50 ppm. The former USSR-UNEP/IRPTC project[43] has set an MAC of 10 mg/m^3 in workplace air and 0.04 mg/m^3 I ambient air in residential areas on a momentary basis. Several states have set guidelines or standards for cyclohexanone in ambient air[60] ranging from 1.0 – 4.0 mg/m^3 (North Dakota) to 1.6 mg/m^3 (Virginia) to 2.0 mg/m^3 (Connecticut) to 2.38 mg/m^3 (Nevada).

Determination in Air: Charcoal adsorption, workup with CS_2, measurement by gas chromatography. See NIOSH Method #1300 for ketones.

Permissible Concentration in Water: The former USSR-UNEP/IRPTC project[43] has set an MAC in water bodies used for domestic purposes of 0.2 mg/l.

Routes of Entry: Inhalation, skin absorption, ingestion, skin and/or eye contact

Harmful Effects and Symptoms

Short Term Exposure: Cyclohexanone irritates the eyes, skin, and respiratory tract. Contact can burn the eyes. The LD_{50} oral-rat is 1,535 mg/kg (slightly toxic). Cyclohexanone may affect the central nervous system. Exposure of high concentrations can cause dizziness, lightheadedness and unconsciousness.

Long Term Exposure: Repeated or prolonged contact with skin may cause drying, cracking, and dermatitis. The following chronic (long term) health effects can occur at some time after exposure to cyclohexanone and can last for months or years: Cyclohexanone may damage the developing fetus. Long-term exposure may cause liver and kidney damage. Long-term exposure may cause clouding of the eye lenses (cataracts).

Points of Attack: Eyes, skin, respiratory system, central nervous system, liver, kidneys.

Medical Surveillance: For those with frequent or potentially high exposure (half the TLV or greater, or significant skin contact) the following are recommended before beginning work and at regular times after that: Liver function tests. If symptoms develop or overexposure is suspected, the following may also be useful: Kidney function tests. Exam of the eyes. Interview for brain effects including recent memory, mood, concentration, headaches, malaise and altered sleep patterns. Consider cerebellar, autonomic, and peripheral nervous system evaluation. Positive and borderline victims should be referred for neuropsychological testing.

First Aid: If this chemical gets into the eyes, remove any contact lenses at once and irrigate immediately for at least 15 minutes, occasionally lifting upper and lower lids. Seek medical attention immediately. If this chemical contacts the skin, remove contaminated clothing and wash immediately with soap and water. Seek medical attention immediately. If this chemical has been inhaled, remove from exposure, begin rescue breathing (using universal precautions) if breathing has stopped and CPR if heart action has stopped. Transfer promptly to a medical facility. When this chemical has been swallowed, get medical attention. Give large quantities of water and induce vomiting. Do not make an unconscious person vomit.

Personal Protective Methods: Wear protective gloves and clothing to prevent any reasonable probability of skin contact. Safety equipment suppliers/manufacturers can provide recommendations on the most protective glove/clothing material for your operation. Silvershield, Polyvinyl Alcohol, and Butyl Rubber are among the recommended protective materials. All protective clothing (suits, gloves, footwear, headgear) should be clean, available each day, and put on before work. Contact lenses should not be worn when working with this chemical. Wear splash-proof chemical goggles and face shield unless full facepiece respiratory protection is worn. Employees should wash immediately with soap when skin is wet or contaminated. Provide emergency showers and eyewash. Remove clothing immediately if set or contaminated to avoid flammability hazard.

Respirator Selection: NIOSH/OSHA: *625 ppm:* SA:CF (any supplied-air respirator operated in a continuous-flow mode); or PAPROV [any powered, air-purifying respirator with organic vapor cartridge(s)]. *700 ppm:* CCRFOV [any chemical cartridge respirator with a full facepiece and organic vapor cartridge(s)]; or GMFOV [any air-purifying, full-facepiece respirator (gas mask) with a chin-style, front-or back-, mounted organic vapor canister]; or PAPRTOV [any powered, air-purifying respirator with a tight-fitting facepiece and organic vapor cartridge(s)]; or SCBAF (any self-contained breathing apparatus with a full facepiece); or SAF (any supplied-air respirator with a full facepiece). *Emergency or planned entry into unknown concentra-*

tions or IDLH conditions: SCBAF:PD,PP (any self-contained breathing apparatus that has a full facepiece and is operated in a pressure-demand or other positive-pressure mode); or SAF:PD, PP:ASCBA (any supplied-air respirator that has a full facepiece and is operated in a pressure-demand or other positive-pressure mode in combination with an auxiliary self-contained breathing apparatus operated in a pressure-demand or other positive pressure mode). *Escape:* GMFOV [any air-purifying, full-facepiece respirator (gas mask) with a chin-style, front- or back-, mounted organic vapor canister]; or SCBAE (any appropriate escape-type, self-contained breathing apparatus).

Note: Substance causes eye irritation or damage; eye protection needed.

Storage: Prior to working with Cyclohexanone you should be trained on its proper handling and storage. Before entering confined space where this chemical may be present, check to make sure that an explosive concentration does not exist. Cyclohexanone must be stored to avoid contact with oxidizers (such as perchlorates, peroxides, chlorates, nitrates, and permangates) since violent reactions occur. Store in tightly closed containers in a cool well-ventilated area away from heat, sparks, and flames. Metal containers involving the transfer of this chemical should be grounded and bonded. Where possible, automatically pump liquid from drums or other storage containers to process containers. Drums must be equipped with self-closing valves, pressure vacuum bungs, and flame arresters. Use only non-sparking tools and equipment, especially when opening and closing containers of this chemical. Sources of ignition such as smoking and open flames, are prohibited where this chemical is used, handled, or stored in a manner that could create a potential fire or explosion hazard. Wherever this chemical is used, handled, manufactured, or stored, use explosion-proof electrical equipment and fittings.

Shipping: Cyclohexanone requires a "Flammable Liquid" label. It falls in DOT Hazard Class 3 and Packing Group III.

Spill Handling: Evacuate and restrict persons not wearing protective equipment from area of spill or leak until cleanup is complete. Remove all ignition sources. Establish forced ventilation to keep levels below explosive limit. Absorb liquids in vermiculite, dry sand, earth, peat, carbon, or a similar material and deposit in sealed containers. Keep this chemical out of a confined space, such as a sewer, because of the possibility of an explosion, unless the sewer is designed to prevent the build-up of explosive concentrations. It may be necessary to contain and dispose of this chemical as a hazardous waste. If material or contaminated runoff enters waterways, notify downstream users of potentially contaminated waters. Contact your Department of Environmental Protection or your regional office of the federal EPA for specific recommendations. If employees are required to clean-up spills, they must be properly trained and equipped. OSHA 1910.120(q) may be applicable.

Fire Extinguishing: This chemical is a combustible liquid. Poisonous gases produced in fire. Use dry chemical, carbon dioxide, or alcohol foam extinguishers. Vapors are heavier than air and will collect in low areas. Vapors may travel long distances to ignition sources and flashback. Vapors in confined areas may explode when exposed to fire. Containers may explode in fire. Storage containers and parts of containers may rocket great distances, in many directions. If material or contaminated runoff enters waterways, notify downstream users of potentially contaminated waters. Notify local health and fire officials and pollution control agencies. From a secure, explosion-proof location, use water spray to cool exposed containers. If cooling streams are ineffective (venting sound increases in volume and pitch, tank discolors, or shows any signs of deforming), withdraw immediately to a secure position. If employees are expected to fight fires, they must be trained and equipped in OSHA 1910.156.

Disposal Method Suggested: Incineration.

References

National Institute for Occupational Safety and Health, Criteria for a Recommended Standard: Occupational Exposure to Ketones, NIOSH Document No. 78–173 (1978)

New Jersey Department of Health and Senior Services and Senior Services, "Hazardous Substance Fact Sheet: Cyclohexanone," Trenton, NJ (February 1989)

Sax, N. I., Ed., "Dangerous Properties of Industrial Materials Report," 5, No. 6, 50–52 (1985)

Cyclohexene

Molecular Formula: C_6H_{10}

Synonyms: Benzene tetrahydride; Hexanaphthylene; Tetrahydrobenzene

CAS Registry Number: 110-83-8

RTECS Number: GW2500000

DOT ID: UN 2256

Regulatory Authority

- Air Pollutant Standard Set (ACGIH)[1] (Australia) (DFG)[3] (HSE)[33] (Israel) (Mexico) (OSHA)[58] (Several States)[60] (Several Canadian Provinces)
- Canada, WHMIS, Ingredients Disclosure List

Cited in U.S. State Regulations: Alaska (G), Connecticut (A), Florida (G), Illinois (G), Maine (G), Massachusetts (G), Nevada (A), New Hampshire (G), New Jersey (G), North Dakota (A), Pennsylvania (G), Rhode Island (G), Virginia (A), West Virginia (G).

Description: C_6H_{10} is a colorless liquid with a sweetish odor. Boiling point = 82 – 83°C. Freezing/Melting point = -104°C. Flash point = -6°C. Autoignition temperature = 310°C. Explosive limits in air: LEL = 1.2%; UEL = -4.8% (@ 100°C). Hazard Identification (based on NFPA-704 M Rating System): Health 1, Flammability 3, Reactivity 0. Insoluble in water.

Potential Exposure: May be used as an intermediate in making other chemicals (e.g., adipic acid, maleic acid, hexahydrobenzoic acid), oil extraction and as a catalyst solvent.

Incompatibilities: Forms explosive mixture with air. The substance can form explosive peroxides. The substance may polymerize under certain conditions. Reacts with strong oxidants causing fire and explosion hazard.

Permissible Exposure Limits in Air: Federal limit (OSHA PEL 8-hour TWA) and the ACGIH and NIOSH recommended level is 300 ppm (1,015 mg/m^3). The HSE,[33] Australian, Mexican, Israeli, DFG,[3] and Canadian provincial (Alberta, BC, Ontario, and Quebec) values are the same and Alberta's STEL is 375 ppm. The NIOSH IDLH level is 2,000 ppm. This chemical can be absorbed through the skin, thereby increasing exposure. In addition, several states have set guidelines or standards for Cyclohexene in ambient air[60] ranging from 10.15 mg/m^3 (North Dakota) to 17.0 mg/m^3 (Virginia) to 20.3 mg/m^3 (Connecticut) to 24.2 mg/m^3 (Nevada).

Determination in Air: Charcoal adsorption followed by workup with CS_2, and analysis by gas chromatography. See NIOSH Method #1500 for hydrocarbons BP36-126°C.

Permissible Concentration in Water: The former USSR-UNEP/IRPTC project[43] has set an MAC in water bodies used for domestic purposes of 0.02 mg/l.

Routes of Entry: Inhalation, ingestion, skin and/or eye contact

Harmful Effects and Symptoms

Short Term Exposure: Cyclohexene irritates the eyes, skin, and respiratory tract. This chemical can be absorbed through the skin, thereby increasing exposure. Swallowing the liquid may cause droplets to enter the lung and cause chemical pneumonia. Overexposure can cause dizziness, lightheadedness, loss of muscle coordination. Higher exposures can cause tremors, collapse and death. A closely related chemical, cyclopropane, can cause irregular heat beat, although it is not known if this chemical causes the same effect. High exposure can cause liver and brain damage.

Long Term Exposure: Repeated or high concentrations can cause dry skin and rash, liver and brain damage.

Points of Attack: Eyes, skin, respiratory system, central nervous system.

Medical Surveillance: Consider the points of attach in preplacement and periodic physical examinations. Liver function tests. Examination of the nervous system.

First Aid: If this chemical gets into the eyes, remove any contact lenses at once and irrigate immediately for at least 15 minutes, occasionally lifting upper and lower lids. Seek medical attention immediately. If this chemical contacts the skin, remove contaminated clothing and wash immediately with soap and water. Seek medical attention immediately. If this chemical has been inhaled, remove from exposure, begin rescue breathing (using universal precautions) if breathing has stopped and CPR if heart action has stopped. Transfer promptly to a medical facility. When this chemical has been swallowed, get medical attention. Do NOT induce vomiting.

Personal Protective Methods: Wear solvent-resistant gloves and clothing to prevent any reasonable probability of skin contact. Safety equipment suppliers/manufacturers can provide recommendations on the most protective glove/clothing material for your operation. All protective clothing (suits, gloves, footwear, headgear) should be clean, available each day, and put on before work. Contact lenses should not be worn when working with this chemical. Wear splash-proof chemical goggles and face shield unless full facepiece respiratory protection is worn. Employees should wash immediately with soap when skin is wet or contaminated. Remove nonimpervious clothing immediately if wet or contaminated. Provide emergency showers and eyewash. Remove clothing immediately if wet or contaminated to avoid flammability hazard.

Respirator Selection: NIOSH/OSHA: *Up to 2,000 ppm*: SA:CF (any supplied-air respirator operated in a continuous-flow mode); or PAPROV [any powered, air-purifying respirator with organic vapor cartridge(s)]; or CCRFOV [any chemical cartridge respirator with a full facepiece and organic vapor cartridge(s)]; or GMFOV [any air-purifying, full-facepiece respirator (gas mask) with a chin-style, front- or back-mounted organic vapor canister]; or SCBAF (any self-contained breathing apparatus with a full facepiece); or SAF (any supplied-air respirator with a full facepiece). *Emergency or planned entry into unknown concentrations or IDLH conditions:* SCBAF:PD,PP (any self-contained breathing apparatus that has a full facepiece and is operated in a pressure-demand or other positive-pressure mode); or SAF:PD,PP:ASCBA (any supplied-air respirator that has a full facepiece and is operated in a pressure-demand or other positive-pressure mode in combination with an auxiliary self-contained breathing apparatus operated in a pressure-demand or other positive-pressure mode). *Escape:* GMFOV [any air-purifying, full-facepiece respirator (gas mask) with a chin-style, front-or back-mounted organic vapor canister] or SCBAE (any appropriate escape-type, self-contained breathing apparatus).

Storage: Prior to working with this chemical you should be trained on its proper handling and storage. Before entering confined space where this chemical may be present, check to make sure that an explosive concentration does not exist. Store in tightly closed containers in a cool, well-ventilated area away from strong oxidizers (such as chlorine, bromine and fluorine). Sources of ignition, such as smoking and open flames, are prohibited where cyclohexene is handled, used, or stored. Metal containers involving the transfer of 5 gallons or more of cyclohexene should be grounded and bonded. Drums must be equipped with self-closing valves, pressure vacuum bungs, and flame arresters. Use only non-sparking tools and equipment, especially when opening and closing containers of cyclohexene. Wherever cyclohexene is used, handled, manufactured, or stored, use explosion-proof electrical equipment and fittings.

Shipping: Cyclohexene falls in Hazard Class 3 and Packing Group II. The label required is "Flammable Liquid."

Spill Handling: Evacuate and restrict persons not wearing protective equipment from area of spill or leak until cleanup is complete. Remove all ignition sources. Establish forced ventilation to keep levels below explosive limit. Absorb liquids in vermiculite, dry sand, earth, peat, carbon, or a similar material and deposit in sealed containers. Keep cyclohexene out of a confined space, such as a sewer, because of the possibility of an explosion, unless the sewer is designed to prevent the build-up of explosive concentrations. It may be necessary to contain and dispose of this chemical as a hazardous waste. If material or contaminated runoff enters waterways, notify downstream users of potentially contaminated waters. Contact your Department of Environmental Protection or your regional office of the federal EPA for specific recommendations. If employees are required to clean-up spills, they must be properly trained and equipped. OSHA 1910.120(q) may be applicable.

Fire Extinguishing: This chemical is a flammable liquid. Poisonous gases produced in fire. Use dry chemical, carbon dioxide, or foam extinguishers. Vapors are heavier than air and will collect in low areas. Vapors may travel long distances to ignition sources and flashback. Vapors in confined areas may explode when exposed to fire. Containers may explode in fire. Storage containers and parts of containers may rocket great distances, in many directions. If material or contaminated runoff enters waterways, notify downstream users of potentially contaminated waters. Notify local health and fire officials and pollution control agencies. From a secure, explosion-proof location, use water spray to cool exposed containers. If cooling streams are ineffective (venting sound increases in volume and pitch, tank discolors, or shows any signs of deforming), withdraw immediately to a secure position. If employees are expected to fight fires, they must be trained and equipped in OSHA 1910.156.

Disposal Method Suggested: Incineration.

References

New Jersey Department of Health and Senior Services and Senior Services, "Hazardous Substance Fact Sheet: Cyclohexene," Trenton, NJ (May 1986)

Cyclohexenyl Trichlorosilane

Molecular Formula: $C_9H_9Cl_3Si$

Common Formula: $C_6H_9SiCl_3$

Synonyms: Cyclohexene, 4-(trichlorosilyl)-; Trichloro-3-cyclohexenylsilane; 4-(Trichlorosilyl) cyclohexene

CAS Registry Number: 10137-69-6

RTECS Number: VV2800000

DOT ID: UN 1762

Regulatory Authority

- Canada, WHMIS, Ingredients Disclosure List

Cited in U.S. State Regulations: Maine (G), New Hampshire (G), New Jersey (G), Oklahoma (G); Pennsylvania (G).

Description: Cyclohexenyl Trichlorosilane, $C_6H_9SiCl_3$, is a colorless fuming liquid that smells like hydrogen chloride. Boiling point ≥ 149°C. Flash point ≥ 66°C. Hazard Identification (based on NFPA-704 M Rating System): Health 2, Flammability 2, Reactivity 0. Reacts with water.

Potential Exposure: This material is used to make silicone polymers.

Incompatibilities: Steam and moisture form toxic and corrosive chloride gases, including hydrogen chloride. Attacks metals in the presence of moisture. Some chlorosilanes self-ignite in air. Contact with ammonia can cause a self-igniting compound.

Permissible Exposure Limits in Air: No standards set.

Determination in Air: No OEL established.

Permissible Concentration in Water: No criteria set. Water reactive.

Routes of Entry: Inhalation.

Harmful Effects and Symptoms

Short Term Exposure: Cyclohexenyl Trichlorosilane can affect you when breathed in. Exposure can irritate the lungs causing coughing and/or shortness of breath. Higher exposures can cause a build-up of fluid (pulmonary edema), a medical emergency. This can cause death. Cyclohexenyl Trichlorosilane is a corrosive chemical and contact can cause severe skin and eye burns. Exposure can irritate the eyes, nose and throat. The oral LD_{50} rat is 2,830 mg/kg (slightly toxic).

Long Term Exposure: Repeated exposure may cause bronchitis with phlegm and shortness of breath.

Points of Attack: Lungs.

Medical Surveillance: Before beginning employment and at regular times after that, for those with frequent or potentially high exposures, the following are recommended: Lung function tests. If symptoms develop or overexposure is suspected, the following may be useful: Consider chest x-ray after acute overexposure.

First Aid: If this chemical gets into the eyes, remove any contact lenses at once and irrigate immediately for at least 15 minutes, occasionally lifting upper and lower lids. Seek medical attention immediately. If this chemical contacts the skin, remove contaminated clothing and wash immediately with soap and water. Seek medical attention immediately. If this chemical has been inhaled, remove from exposure, begin rescue breathing (using universal precautions) if breathing has stopped and CPR if heart action has stopped. Transfer promptly to a medical facility. When this chemical has

been swallowed, get medical attention. If victim is conscious, administer water or milk. Do not induce vomiting. Medical observation is recommended for 24 – 48 hours after breathing overexposure, as pulmonary edema may be delayed. As first aid for pulmonary edema, a doctor or authorized paramedic may consider administering a corticosteroid spray.

Personal Protective Methods: Wear protective gloves and clothing to prevent any reasonable probability of skin contact. Safety equipment suppliers/manufacturers can provide recommendations on the most protective glove/clothing material for your operation. All protective clothing (suits, gloves, footwear, headgear) should be clean, available each day, and put on before work. Contact lenses should not be worn when working with this chemical. Wear plash-proof chemical goggles and face shield unless full facepiece respiratory protection is worn. Employees should wash immediately with soap when skin is wet or contaminated. Provide emergency showers and eyewash.

Respirator Selection: Where the potential for exposure to Cyclohexenyl Trichlorosilane exists, use a MSHA/NIOSH approved supplied-air respirator with a full facepiece operated in the positive pressure mode or with a full facepiece, hood, or helmet in the continuous flow mode, or use a MSHA/NIOSH approved self-contained breathing apparatus with a full facepiece operated in pressure-demand or other positive pressure mode.

Storage: Prior to working with Cyclohexenyl Trichlorosilane you should be trained on its proper handling and storage. Store in tightly closed containers in a cool, well-ventilated area away from water, steam and moisture because toxic and corrosive chloride gases including hydrogen chloride can be produced. Sources of ignition, such as smoking and open flames, are prohibited where Cyclohexenyl Trichlorosilane is used, handled, or stored in a manner that could create a potential fire or explosion hazard.

Shipping: This material falls in DOT Hazard Class 8 and Packing Group II. It requires a "Corrosive" label. Shipment by passenger aircraft or railcar is forbidden.

Spill Handling: Evacuate and restrict persons not wearing protective equipment from area of spill or leak until cleanup is complete. Remove all ignition sources. Ventilate area of spill or leak. Absorb liquids in vermiculite, dry sand, earth, peat, carbon, or a similar material and deposit in sealed containers. Keep Cyclohexenyl Trichlorosilane out of a confined space, such as a sewer, because of the possibility of an explosion, unless the sewer is designed to prevent the build-up of explosive concentrations. It may be necessary to contain and dispose of this chemical as a hazardous waste. If material or contaminated runoff enters waterways, notify downstream users of potentially contaminated waters. Contact your Department of Environmental Protection or your regional office of the federal EPA for specific recommendations. If employees are required to clean-up spills, they must be properly trained and equipped. OSHA 1910.120(q) may be applicable.

Fire Extinguishing: This chemical is a combustible liquid. Poisonous gases including chlorine are produced in fire. Use dry chemical, carbon dioxide, or foam extinguishers. Fire may restart after it has been extinguished. Vapors are heavier than air and will collect in low areas. Vapors may travel long distances to ignition sources and flashback. Vapors in confined areas may explode when exposed to fire. Containers may explode in fire. Storage containers and parts of containers may rocket great distances, in many directions. If material or contaminated runoff enters waterways, notify downstream users of potentially contaminated waters. Notify local health and fire officials and pollution control agencies. From a secure, explosion-proof location, use water spray to cool exposed containers. If cooling streams are ineffective (venting sound increases in volume and pitch, tank discolors, or shows any signs of deforming), withdraw immediately to a secure position. If employees are expected to fight fires, they must be trained and equipped in OSHA 1910.156.

References

New Jersey Department of Health and Senior Services and Senior Services, "Hazardous Substance Fact Sheet: Cyclohexenyl Trichlorosilane," Trenton, NJ (December 1998)

Cycloheximide

Molecular Formula: $C_{15}H_{23}NO_4$

Synonyms: Acti-Aid®; Actidione; Actidione TGF; Actidone; Actispray; 3[2-(3,5-Dimethyl-2-oxocyclohexyl)-2-hydroxyethyl]glutarimide; Hizarocin®; Kaken®; Naramycin®; Naramycin A®; Neocycloheximide®; NSC-185; 2,6-Piperidinedione, 4-(2-3,5-dimethyl-2-oxocyclohexyl)-2-hydroxyethyl-, (IS)-[1α (S*),3α,5β]-

CAS Registry Number: 66-81-9

RTECS Number: MA4375000

DOT ID: UN 2588

Regulatory Authority

- Banned or Severely Restricted (In Agriculture) (Malaysia) (UN)[13]
- Very Toxic Substance (World Bank)[15]
- SUPERFUND/EPCRA 40CFR302.4 Reportable Quantity (RQ): EHS, 1 lb (0.454 kg)
- SUPERFUND/EPCRA 40CFR355, Appendix B Extremely Hazardous Substances: TPQ = 100/10,000 lb (45.4/4,540 kg)
- Canada, WHMIS, Ingredients Disclosure List

Cited in U.S. State Regulations: California (A, G), Florida (G), Illinois (G), Massachusetts (G), Michigan (G), New Jersey (G), New Hampshire (G), Oklahoma (G), Pennsylvania (G).

Description: Cycloheximide, $C_{15}H_{23}NO_4$, is a colorless crystalline substance. Freezing/Melting point = 119.5 – 121°C. Slightly soluble in water.

Potential Exposure: Those involved in the manufacture, formulation or application of this fungicide and pesticide. Used as an antibiotic, plant growth regulator, and protein synthesis inhibitor. Also used as a repellent for rodents and other animal pests, and in cancer therapy.

Incompatibilities: Incompatible with oxidizers, acid anhydrides, strong bases.

Permissible Exposure Limits in Air: No OEL has been established.

Permissible Concentration in Water: No criteria set.

Routes of Entry: Inhalation.

Harmful Effects and Symptoms

Short Term Exposure: Contact can cause eye and skin irritation. Exposure can cause excessive salivation, nausea, vomiting, diarrhea, and elevated blood urea nitrogen (BUN). High exposures can also cause imbalance, tremors, seizures and coma. Extremely toxic (LD_{50} value for rats is only 3.7 mg/kg). The probable oral lethal dose in humans is 5 – 50 mg/kg, or 7 drops to 1 teaspoonful for a 150 lb person. Signs of skin irritation may appear as much as 6 – 24 hours after exposure.

Long Term Exposure: May cause mutations and damage the developing fetus. May cause liver and kidney damage.

Points of Attack: Reproductive system. Liver and kidneys.

Medical Surveillance: Liver and kidney function tests.

First Aid: If this chemical gets into the eyes, remove any contact lenses at once and irrigate immediately for at least 15 minutes, occasionally lifting upper and lower lids. Seek medical attention immediately. If this chemical contacts the skin, remove contaminated clothing and wash immediately with soap and water. Seek medical attention immediately. If this chemical has been inhaled, remove from exposure, begin rescue breathing (using universal precautions) if breathing has stopped and CPR if heart action has stopped. Transfer promptly to a medical facility. When this chemical has been swallowed, get medical attention. Give large quantities of water and induce vomiting. Do not make an unconscious person vomit. Medical observation is recommended for 24 – 48 hours following skin contact.

Personal Protective Methods: Wear protective gloves and clothing to prevent any reasonable probability of skin contact. Safety equipment suppliers/manufacturers can provide recommendations on the most protective glove/clothing material for your operation. All protective clothing (suits, gloves, footwear, headgear) should be clean, available each day, and put on before work. Contact lenses should not be worn when working with this chemical. Wear dust-proof chemical goggles and face shield unless full facepiece respiratory protection is worn. Employees should wash immediately with soap when skin is wet or contaminated. Provide emergency showers and eyewash.

Respirator Selection: *Where there is a potential for overexposure:* SCBAF:PD,PP (any MSHA/NIOSH approved self-contained breathing apparatus that has a full facepiece and is operated in a pressure-demand or other positive-pressure mode); or SAF:PD,PP:ASCBA (any supplied-air respirator that has a full facepiece and is operated in a pressure-demand or other positive-pressure mode in combination with an auxiliary, self-contained breathing apparatus operated in a pressure-demand or other positive pressure mode).

Storage: Prior to working with this chemical you should be trained on its proper handling and storage. Store in tightly closed containers in a cool, well-ventilated area away from oxidizing agents, acid hydrides, and strong bases.

Shipping: This material may be classified as a poisonous solid n.o.s. (UN 2811) which puts it in Hazard Class 6.1 and Packing Group I. It then requires a "Poison" label.

Spill Handling: Avoid breathing dusts. Keep upwind. Wear self-contained breathing apparatus. Material is rapidly inactivated at room temperature by dilute alkali. Evacuate persons not wearing protective equipment from area of spill or leak until clean-up is complete. Remove all ignition sources. Collect powdered material in the most convenient and safe manner and deposit in sealed containers. Ventilate area after clean-up is complete. It may be necessary to contain and dispose of this chemical as a hazardous waste. If material or contaminated runoff enters waterways, notify downstream users of potentially contaminated waters. Contact your Department of Environmental Protection or your regional office of the federal EPA for specific recommendations. If employees are required to clean-up spills, they must be properly trained and equipped. OSHA 1910.120(q) may be applicable.

Fire Extinguishing: This chemical is a noncombustible solid. Wear SCBA and full protective clothing. Use dry chemical, carbon dioxide, water spray, or alcohol or polymer foam extinguishers. Poisonous gases are produced in fire including nitrogen oxides and carbon monoxide. If material or contaminated runoff enters waterways, notify downstream users of potentially contaminated waters. Notify local health and fire officials and pollution control agencies. From a secure, explosion-proof location, use water spray to cool exposed containers. If cooling streams are ineffective (venting sound increases in volume and pitch, tank discolors, or shows any signs of deforming), withdraw immediately to a secure position. If employees are expected to fight fires, they must be trained and equipped in OSHA 1910.156.

Disposal Method Suggested: High-temperature incinerator with flue gas scrubbing equipment.

References

Sax, N. I., Ed., "Dangerous Properties of Industrial Materials Report," 2, No. 5, 41–43 (1982) and 9, No. 1, 55–64 (1989)

U.S. Environmental Protection Agency, "Chemical Profile: Cycloheximide," Washington, DC, Chemical Emergency Preparedness Program (November 30, 1987)

New Jersey Department of Health and Senior Services, Hazardous Substance Fact Sheet, "Cycloheximide," Trenton NJ (January 1999)

Cyclohexylamine

Molecular Formula: $C_6H_{13}N$

Common Formula: $C_6H_{11}NH_2$

Synonyms: Aminocyclohexane; Aminohexahydrobenzene; Aniline, hexahydro-; CHA; Ciclohexilamina (Spanish); Cyclohexanamine; Cyclohexaneamine; Hexahydroaniline; Hexahydrobenzenamine

CAS Registry Number: 108-91-8

RTECS Number: GX0700000

DOT ID: UN 2357

EEC Number: 612-050-00-8

EINECS Number: 203-629-0

Regulatory Authority

- Air Pollutant Standard Set (ACGIH)[1] (Australia) (DFG)[3] (HSE)[33] (Israel) (Mexico) (former USSR)[43] (OSHA)[58] (Several States)[60] (Several Canadian Provinces)
- CLEAN AIR ACT: Accidental Release Prevention/Flammable substances, (Section 112[r], Table 3), TQ = 15,000 lb (6,810 kg)
- SUPERFUND/EPCRA 40CFR355, Appendix B Extremely Hazardous Substances: TPQ = 10,000 lb (4,540 kg)
- SUPERFUND/EPCRA 40CFR302.4 Reportable Quantity (RQ): EHS, 1 lb (0.454 kg)
- Canada, WHMIS, Ingredients Disclosure List

Cited in U.S. State Regulations: Alaska (G), California (A, G), Connecticut (A), Florida (G), Illinois (G), Kansas (A), Maine (G), Massachusetts (G), Minnesota (G), Nevada (A), New Hampshire (G), New Jersey (G), North Dakota (A), Pennsylvania (G), Rhode Island (G), Virginia (A), West Virginia (G).

Description: Cyclohexylamine, $C_6H_{11}NH_2$, is a colorless to yellow liquid with an unpleasant fishy odor. Boiling point = 135°C. Freezing/Melting point = -18°C. Flash point = 31°C. Autoignition temperature = 293°C. Explosive limits: LEL = 1.5%; UEL = 9.4%. Hazard Identification (based on NFPA-704 M Rating System): Health 3, Flammability 3, Reactivity 0. Soluble in water.

Potential Exposure: CHA is used in making dyes, chemicals, dry cleaning chemicals, insecticides, plasticizers, rubber chemicals, and as a chemical intermediate in the production on cyclamate sweeteners. It is also used as a boiler feedwater additive.

Incompatibilities: Forms explosive mixture with air. Cyclohexylamine is a strong base, it reacts violently with acid. Contact with strong oxidizers may cause fire and explosion hazard. Incompatible with organic anhydrides, isocyanates, vinyl acetate, acrylates, substituted allyls, alkylene oxides, epichlorohydrin, ketones, aldehydes, alcohols, glycols, phenols, cresols, caprolactum solution, lead. Corrosive to copper alloys, zinc or galvanized steel.

Permissible Exposure Limits in Air: There is no OSHA PEL. A TLV of 10 ppm (40 mg/m^3) (skin) has been recommended or adopted by NIOSH, ACGIH, DFG,[3] Israel, Mexico, Austalia, and HSE.[33] The notation "skin" is added by HSE indicating the possibility of cutaneous absorption. The Canadian provincial level for Alberta, BC, Ontario, and Quebec is the same, and Alberta's STEL is 20 ppm. Several states have set guidelines or standards for cyclohexylamine in ambient air[60] ranging from 95.238 μg/m^3 (Kansas) to 400 μg/m^3 (North Dakota) to 650 μg/m^3 (Virginia) to 800 μg/m^3 (Connecticut) to 952 μg/m^3 (Nevada). A value of 1.0 mg/m^3 has been set as an allowable MAC in workplace air by the former USSR-UNEP/IRPTC project.[43]

Determination in Air: Si gel; H2SO4; Gas chromatography/Flame ionization detection; NIOSH IV Method #2010, Amines, Aliphatic.

Permissible Concentration in Water: No criteria set but EPA[32] has suggested a permissible concentration based on health effects of 550 μg/l.

Routes of Entry: Inhalation, ingestion, skin absorption, eye and skin contact.

Harmful Effects and Symptoms

Short Term Exposure: This chemical can be absorbed through the skin, thereby increasing exposure. Cyclohexylamine is caustic to the skin, eyes, and respiaratory tract, and its systemic effects in humans include nausea, vomiting, anxiety, restlessness, drowsiness, lightheadedness, anxiety, apprehension, slurred speech, papillary dilation, severe skin irritation. Cyclohexylamine may also be a skin sensitizer. Inhalation can cause pulmonary edema, a medical emergency that can be delayed for several hours. This can cause death.

This material is classified as moderately toxic, the LD_{50} oral-rat is 156 mg/kg-probable oral lethal dose is 50 –500 mg/kg or between 1 teaspoon and 1 ounce for a 70 kg (150 lb) persons. It is considered a nerve poison, and is a weak methemoglobin-forming substance.

Long Term Exposure: Cyclohexylamine causes mutations (genetic changes). Such chemicals may have a cancer risk. Many scientists believe there is no safe level of exposure to a cancer-causing agent. It may damage the developing fetus. It may also damage the testes (male reproductive gland) and reduce the fertility of females. *Other Long-Term Effects:* Exposure may increase blood pressure. Repeated or severe exposures may damage vision and possible kidneys and liver. Cyclohexylamine may cause a skin allergy. Very low future exposures can cause itching and a skin rash. The Food and Drug Administration banned the use of cyclamates as artificial sweeteners in 1969 because of their metabolic conversion to cyclohexylamine, which was thought to be carcinogenic in rats. There is now no evidence that CHA is a carcinogen, however.

Points of Attack: Eyes, skin, respiratory system, central nervous system.

Medical Surveillance: Before beginning employment and at regular times after that, the following is recommended: Blood pressure check. If symptoms develop or overexposure is suspected, the following may be useful: Eye and vision exam. Kidney and liver function tests. Skin testing with dilute cyclohexylamine may help diagnose allergy, if done by a qualified allergist.

First Aid: If this chemical gets into the eyes, remove any contact lenses at once and irrigate immediately for at least 15 minutes, occasionally lifting upper and lower lids. Seek medical attention immediately. If this chemical contacts the skin, remove contaminated clothing and wash immediately with soap and water. Seek medical attention immediately. If this chemical has been inhaled, remove from exposure, begin rescue breathing (using universal precautions) if breathing has stopped and CPR if heart action has stopped. Transfer promptly to a medical facility. When this chemical has been swallowed, get medical attention. If victim is conscious, administer water or milk. Do not induce vomiting. Medical observation is recommended for 24 – 48 hours after breathing overexposure, as pulmonary edema may be delayed. As first aid for pulmonary edema, a doctor or authorized paramedic may consider administering a corticosteroid spray.

Note to Physician: Treat for methemoglobinemia. Spectrophotometry may be required for precise determination of levels of methemoglobinemia in urine.

Personal Protective Methods: Wear protective gloves and clothing to prevent any reasonable probability of skin contact. Safety equipment suppliers/manufacturers can provide recommendations on the most protective glove/clothing material for your operation. All protective clothing (suits, gloves, footwear, headgear) should be clean, available each day, and put on before work. Contact lenses should not be worn when working with this chemical. Wear splash-proof chemical goggles and face shield unless full facepiece respiratory protection is worn. Employees should wash immediately with soap when skin is wet or contaminated. Provide emergency showers and eyewash.

Respirator Selection: Where the potential exists for exposures over 10 ppm, use an MSHA/NIOSH approved full facepiece respirator with an organic vapor cartridge/canister. Increased protection is obtained from full facepiece powered air purifying respirators. Where the potential for high exposures exists, use an MSHA/NIOSH approved supplied-air respirator with a full facepiece operated in the positive pressure mode or with a full facepiece, hood, or helmet in the continuous flow mode, or use an MSHA/NIOSH approved self-contained breathing apparatus with a full facepiece operated in pressure-demand or other positive pressure mode.

Storage: Prior to working with cyclohexylamine you should be trained on its proper handling and storage. Before entering confined space where this chemical may be present, check to make sure that an explosive concentration does not exist. Protect containers from physical damage. Outdoor or detached storage is recommended. Cyclohexylamine must be stored to avoid contact with strong oxidizers (such as chlorine, bromine, and fluorine), since violent reactions occur. Store in tightly closed containers in a cool, well-ventilated area away from sources of heat. Sources of ignition such as smoking and open flames are prohibited where cyclohexylamine is handled, used, or stored. The vapor can form explosive mixtures in the air. Store in tightly closed containers in a cool, well ventilated area. Metal containers involving the transfer of this chemical should be grounded and bonded. Where possible, automatically pump liquid from drums or other storage containers to process containers. Drums must be equipped with self-closing valves, pressure vacuum bungs, and flame arresters. Use only non-sparking tools and equipment, especially when opening and closing containers of this chemical. Sources of ignition such as smoking and open flames, are prohibited where this chemical is used, handled, or stored in a manner that could create a potential fire or explosion hazard. Wherever this chemical is used, handled, manufactured, or stored, use explosion-proof electrical equipment and fittings.

Shipping: The label required is "Corrosive, Flammable Liquid." The UN/DOT Hazard Class is 8 and the Packing Group is II.

Spill Handling: Stay upwind; keep out of low areas. Wear self-contained (positive pressure if available) breathing apparatus and full protective clothing. Establish forced ventilation to keep levels below explosive limit. Shut off ignition sources; no flares, smoking or flames in hazard area. Do not touch spilled material; stop leak if you can do it without risk. Use water spray to reduce vapors. Small spills: take up with vermiculite, dry sand or other non-combustible absorbent material and place into containers for later disposal. Large spills: dike far ahead of spill for later disposal. Evacuate and restrict persons not wearing protective equipment from area of spill or leak until cleanup is complete. It may be necessary to contain and dispose of this chemical as a hazardous waste. If material or contaminated runoff enters waterways, notify downstream users of potentially contaminated waters. Contact your Department of Environmental Protection or your regional office of the federal EPA for specific recommendations. If employees are required to clean-up spills, they must be properly trained and equipped. OSHA 1910.120(q) may be applicable.

Fire Extinguishing: This chemical is a combustible liquid. Poisonous gases including nitrogen oxides are produced in fire. Use dry chemical, carbon dioxide, or foam extinguishers. Vapors are heavier than air and will collect in low areas. Vapors may travel long distances to ignition sources and flashback. Vapors in confined areas may explode when exposed to fire. Containers may explode in fire. Storage containers and parts of containers may rocket great distances, in many directions. If material or contaminated runoff enters waterways, notify downstream users of potentially contaminated waters.

Notify local health and fire officials and pollution control agencies. From a secure, explosion-proof location, use water spray to cool exposed containers. If cooling streams are ineffective (venting sound increases in volume and pitch, tank discolors, or shows any signs of deforming), withdraw immediately to a secure position. If employees are expected to fight fires, they must be trained and equipped in OSHA 1910.156.

Disposal Method Suggested: Incineration; incinerator is equipped with a scrubber or thermal unit to reduce NO_x emissions.

References

U.S. Environmental Protection Agency, "Chemical Profile: Cyclohexylamine," Washington, DC (October 21, 1977)

National Institute for Occupational Safety and Health, Information Profiles on Potential Occupational Hazards — Single Chemicals: Cyclohexylamine, Report TR 79-607, pp 45–55, Rockville, MD (December 1979)

U.S. Environmental Protection Agency, "Chemical Profile: Cyclohexylamine," Washington, DC, Chemical Emergency Preparedness Program (November 30, 1987)

New Jersey Department of Health and Senior Services and Senior Services, "Hazardous Substance Fact Sheet: Cyclohexylamine," Trenton, NJ (January 1986)

Cyclohexyl Isocyanate

Molecular Formula: $C_7H_{11}NO$

Common Formula: $C_6H_{11}NCO$

Synonyms: CHI; Cyclohexane, isocyanato-; Isocyanatocyclohexane; Isocyanic acid, cyclohexyl ester

CAS Registry Number: 3173-53-3

RTECS Number: NQ8650000

DOT ID: UN 2488

Regulatory Authority

- Canada, WHMIS, Ingredients Disclosure List

Cited in U.S. State Regulations: California (A, G), New Hampshire (G), New Jersey (G), Pennsylvania (G).

Description: Cyclohexyl Isocyanate, CHI, $C_6H_{11}NCO$, is a clear liquid with a sharp, pungent odor. Boiling point = 168°C. Flash point = 48°C (cc). Hazard Identification (based on NFPA-704 M Rating System): Health 2, Flammability 2, Reactivity 0. Soluble in water (reaction).

Potential Exposure: The material is used in the synthesis of agricultural chemicals.

Incompatibilities: Cyclohexyl Isocyanate polymerize due to heating above 93°C/200°F and under the influence of various chemicals including organometallic compounds. The substance decomposes in fire producing toxic fumes of hydrogen cyanide and nitrogen oxides. Reacts with oxidants and strong bases, water, alcohol, acids, amines, metal compounds, surface actives materials.

Permissible Exposure Limits in Air: No standards set.

Permissible Concentration in Water: No criteria set.

Routes of Entry: Inhalation, ingestion.

Harmful Effects and Symptoms

Short Term Exposure: A lacrimator (causes tearing). Exposure can severely irritate and may burn the skin, eyes, nose, throat and lungs. Very high levels may lead to a build-up of fluid in the lungs (pulmonary edema), a medical emergency that can be delayed for several hours. This can cause death.

Long Term Exposure: Repeated or prolonged contact may cause skin sensitization. Repeated or prolonged inhalation exposure may cause lung damage and /or asthma.

Points of Attack: Lungs, skin.

Medical Surveillance: For those with frequent or potentially high exposure (half the TLV or greater), the following are recommended before beginning work and at regular times after that: Lung function tests. If symptoms develop or overexposure is suspected, the following may be useful: Consider chest x-ray after acute overexposure.

First Aid: If this chemical gets into the eyes, remove any contact lenses at once and irrigate immediately for at least 30 minutes, occasionally lifting upper and lower lids. Seek medical attention immediately. If this chemical contacts the skin, remove contaminated clothing and wash immediately with soap and water. Seek medical attention immediately. If this chemical has been inhaled, remove from exposure, begin rescue breathing (using universal precautions) if breathing has stopped and CPR if heart action has stopped. Transfer promptly to a medical facility. When this chemical has been swallowed, get medical attention. Give large quantities of water and induce vomiting. Do not make an unconscious person vomit. Medical observation is recommended for 24 – 48 hours after breathing overexposure, as pulmonary edema may be delayed. As first aid for pulmonary edema, a doctor or authorized paramedic may consider administering a corticosteroid spray.

Personal Protective Methods: Wear protective gloves and clothing to prevent any reasonable probability of skin contact. Safety equipment suppliers/manufacturers can provide recommendations on the most protective glove/clothing material for your operation. Butyl rubber gloves have been recommended by the manufacturer as protection against this substance. All protective clothing (suits, gloves, footwear, headgear) should be clean, available each day, and put on before work. Contact lenses should not be worn when working with this chemical. Wear splash-proof chemical goggles and face shield unless full facepiece respiratory protection is worn. Employees should wash immediately with soap when skin is wet or contaminated. Provide emergency showers and eyewash.

Respirator Selection: Where the potential for exposure to Cyclohexyl Isocyanate exists, use a MSHA/NIOSH approved supplied-air respirator with a full facepiece operated in the positive pressure mode or with a full facepiece, hood, or helmet in the continuous flow mode, or use a MSHA/NIOSH approved self-contained breathing apparatus with a full facepiece operated in pressure-demand or other positive pressure mode.

Storage: Prior to working with Cyclohexyl Isocyanate you should be trained on its proper handling and storage. Cyclohexyl Isocyanate must be stored to avoid contact with moisture and temperatures above 200°F since violent reactions occur. Store in tightly closed containers in a cool, well-ventilated area away from water, strong bases, alcohol, metal compounds or surface active materials. Use only non-sparking tools and equipment, especially when opening and closing containers of this chemical. Sources of ignition such as smoking and open flames, are prohibited where this chemical is used, handled, or stored in a manner that could create a potential fire or explosion hazard.

Shipping: Cyclohexyl Isocyanate requires a "Poison" label. It falls in Hazard Class 6.1 and packing Group I. Shipment by passenger aircraft or railcar or even by cargo aircraft is forbidden.[19]

Spill Handling: Evacuate and restrict persons not wearing protective equipment from area of spill or leak until cleanup is complete. Remove all ignition sources. Ventilate area of spill or leak. Absorb liquids in vermiculite, dry sand, earth, peat, carbon, or a similar material and deposit in sealed containers. It may be necessary to contain and dispose of this chemical as a hazardous waste. If material or contaminated runoff enters waterways, notify downstream users of potentially contaminated waters. Contact your Department of Environmental Protection or your regional office of the federal EPA for specific recommendations. If employees are required to clean-up spills, they must be properly trained and equipped. OSHA 1910.120(q) may be applicable.

Initial isolation and protective action distances

Distances shown are likely to be affected during the first 30 minutes after materials are spilled and could increase with time. If more than one tank car, cargo tank, portable tank, or large cylinder is involved in the incident is leaking, the protective action distance may need to be increased. You may need to seek emergency information from CHEMTREC at (800) 424-9300 or seek professional environmental engineering assistance from the U.S. EPA Environmental Response Team at (908)548-8730 (24-hour response line).

Small spills (From a small package or a small leak from a large package)

First: Isolate in all directions (feet)..... 500

Then: Protect persons downwind (miles)

Day 0.6

Night 2.9

Large spills (From a large package or from many small packages)

First: Isolate in all directions (feet)..... 1500

Then: Protect persons downwind (miles)

Day 2.4

Night 7.0

Fire Extinguishing: Cyclohexyl Isocyanate may burn, but does not readily ignite. Containers may explode in fire. Poisonous gases are produced in fire, including hydrogen cyanide and nitrogen oxides. Use dry chemical, CO_2, water spray, or foam extinguishers. Vapors are heavier than air and will collect in low areas. Vapors in confined areas may explode when exposed to fire. Containers may explode in fire. Storage containers and parts of containers may rocket great distances, in many directions. If material or contaminated runoff enters waterways, notify downstream users of potentially contaminated waters. Notify local health and fire officials and pollution control agencies. From a secure, explosion-proof location, use water spray to cool exposed containers. If cooling streams are ineffective (venting sound increases in volume and pitch, tank discolors, or shows any signs of deforming), withdraw immediately to a secure position. If employees are expected to fight fires, they must be trained and equipped in OSHA 1910.156.

References

New Jersey Department of Health and Senior Services and Senior Services, "Hazardous Substance Fact Sheet: Cyclohexyl Isocyanate," Trenton, NJ (May 1986)

Cyclohexyl Trichlorosilane

Molecular Formula: $C_6H_{11}Cl_3Si$

Common Formula: $C_6H_{11}SiCl_3$

Synonyms: Silane, trichlorohexyl-; Trichlorocyclohexylsilane; 1-(Trichlorosilyl)cyclohexane

CAS Registry Number: 98-12-4

RTECS Number: VV2890000

DOT ID: UN 1763

Regulatory Authority

- Canada, WHMIS, Ingredients Disclosure List

Cited in U.S. State Regulations: Florida (G), Maine (G), Massachusetts (G), New Hampshire (G), New Jersey (G), Oklahoma (G), Pennsylvania (G).

Description: Cyclohexyl Trichlorosilane, $C_6H_{11}SiCl_3$, is a colorless to pale yellow liquid. Boiling point = 208°C. Flash point = 91°C. Hazard Identification (based on NFPA-704 M Rating System): Health 2, Flammability 2, Reactivity 1. Insoluble in water (reactive).

Potential Exposure: The material is used to make silicone polymers.

Incompatibilities: Contact with water or moisture forms hydrochloric acid. Incompatible with heat, strong oxidizers, strong bases, alcohols, metal compounds. Attacks metals in the presence of moisture.

Permissible Exposure Limits in Air: No standards set.

Permissible Concentration in Water: No criteria set.

Routes of Entry: Inhalation

Harmful Effects and Symptoms

Short Term Exposure: Cyclohexyl Trichlorosilane can affect you when breathed in. Exposure can irritate the eyes, nose, throat and lungs causing coughing, wheezing, and/or shortness of breath. Cyclohexyl Trichlorosilane is a corrosive chemical and can cause severe skin and eye burns leading to permanent eye damage. Higher exposures can cause pulmonary edema, a medical emergency that can be delayed for several hours. This can cause death. The oral LD_{50} rat is 2,830 mg/kg (slightly toxic).

Long Term Exposure: Repeated exposure may cause bronchitis to develop.

Points of Attack: Lungs.

Medical Surveillance: For those with frequent or potentially high exposure (half the TLV or greater), the following are recommended before beginning work and at regular times after that: Lung function tests. If symptoms develop or overexposure is suspected, the following may be useful: Consider chest x-ray after acute overexposure.

First Aid: If this chemical gets into the eyes, remove any contact lenses at once and irrigate immediately for at least 15 minutes, occasionally lifting upper and lower lids. Seek medical attention immediately. If this chemical contacts the skin, remove contaminated clothing and wash immediately with soap and water. Seek medical attention immediately. If this chemical has been inhaled, remove from exposure, begin rescue breathing (using universal precautions) if breathing has stopped and CPR if heart action has stopped. Transfer promptly to a medical facility. When this chemical has been swallowed, get medical attention. If victim is conscious, administer water or milk. Do not induce vomiting. Medical observation is recommended for 24 – 48 hours after breathing overexposure, as pulmonary edema may be delayed. As first aid for pulmonary edema, a doctor or authorized paramedic may consider administering a corticosteroid spray.

Personal Protective Methods: Wear protective gloves and clothing to prevent any reasonable probability of skin contact. Safety equipment suppliers/manufacturers can provide recommendations on the most protective glove/clothing material for your operation. All protective clothing (suits, gloves, footwear, headgear) should be clean, available each day, and put on before work. Contact lenses should not be worn when working with this chemical. Wear splash-proof chemical goggles and face shield unless full facepiece respiratory protection is worn. Employees should wash immediately with soap when skin is wet or contaminated. Provide emergency showers and eyewash.

Respirator Selection: Where the potential for exposure to Cyclohexyl Trichlorosilane exists, use a MSHA/NIOSH approved supplied-air respirator with a full facepiece operated in the positive pressure mode or with a full facepiece, hood, or helmet in the continuous flow mode, or use a MSHA/NIOSH approved self-contained breathing apparatus with a full facepiece operated in pressure-demand or other positive pressure mode.

Storage: Prior to working with this chemical you should be trained on its proper handling and storage. Store in tightly closed containers in a cool, well-ventilated area away from water, steam and moisture and away from combustible materials such as wood, oil and paper, and incompatible materials listed above. Metal containers involving the transfer of this chemical should be grounded and bonded. Where possible, automatically pump liquid from drums or other storage containers to process containers. Drums must be equipped with self-closing valves, pressure vacuum bungs, and flame arresters. Use only non-sparking tools and equipment, especially when opening and closing containers of this chemical. Sources of ignition such as smoking and open flames, are prohibited where this chemical is used, handled, or stored in a manner that could create a potential fire or explosion hazard. Wherever this chemical is used, handled, manufactured, or stored, use explosion-proof electrical equipment and fittings.

Shipping: Cyclohexyl Trichlorosilane requires a "Corrosive" label. It falls in DOT Hazard Class 8 and Packing Group II. Shipment by passenger aircraft or railcar is forbidden.

Spill Handling: Evacuate and restrict persons not wearing protective equipment from area of spill or leak until cleanup is complete. Remove all ignition sources. Ventilate area of spill or leak. Absorb liquids in vermiculite, dry sand, earth, peat, carbon, or a similar material and deposit in sealed containers. It may be necessary to contain and dispose of this chemical as a hazardous waste. If material or contaminated runoff enters waterways, notify downstream users of potentially contaminated waters. Contact your Department of Environmental Protection or your regional office of the federal EPA for specific recommendations. If employees are required to clean-up spills, they must be properly trained and equipped. OSHA 1910.120(q) may be applicable.

Fire Extinguishing: This chemical is a combustible liquid. Poisonous gases including chloride and hydrogen chloride are produced in fire. Do not use water or hydrous agents. Use dry chemical or carbon dioxide. Vapors are heavier than air and will collect in low areas. Vapors in confined areas may explode when exposed to fire. Containers may explode in fire. Storage containers and parts of containers may rocket great distances, in many directions. If material or contaminated runoff enters waterways, notify downstream users of potentially contaminated waters. Notify local health and fire officials and pollution control agencies. From a secure, explosion-proof location, use water spray to cool exposed containers. If cooling streams are ineffective (venting sound increases in volume and pitch, tank discolors, or shows any signs of deforming), withdraw immediately to a secure position. If employees are expected to fight fires, they must be trained and equipped in OSHA 1910.156.

References

New Jersey Department of Health and Senior Services and Senior Services, "Hazardous Substance Fact Sheet: Cyclohexyl Trichlorosilane," Trenton, NJ (May 1986)

Cyclonite

Molecular Formula: $C_3H_6N_6O_6$

Synonyms: Cyclotrimethylenetrinitramine; Hexahydro-1,3,5-trinitro-*s*-triazine; Hexogen; HMX; RDX; Trimethylenetrinitramine; 1,3,5-Trinitro-1,3,5-triazacyclohexane; Trinitrotrimethylenetriamine

CAS Registry Number: 121-82-4

RTECS Number: XY9450000

DOT ID: UN 0072

Regulatory Authority

- Explosive Substance (World Bank)[15]
- Air Pollutant Standard Set (ACGIH)[1] (Australia) (HSE)[33] (Israel) (Mexico) (OSHA)[58] (Several States)[60] (Several Canadian Provinces)

Cited in U.S. State Regulations: Alaska (G), California (A, G), Connecticut (A), Florida (G), Illinois (G), Maine (G), Massachusetts (G), Minnesota (G), Nevada (A), New Hampshire (G), New Jersey (G), North Dakota (A), Pennsylvania (G), Rhode Island (G), Virginia (A), West Virginia (G).

Description: Hexahydro-1,3,5-trinitro-s-triazine is a white crystalline compound. Freezing/Melting point = 203 – 204°C. Flash point = explodes; Insoluble in water.

Potential Exposure: Those involved in the manufacture of this material and its handling in munitions and solid-propellant manufacture. It is also used as a rat poison, a powerful military explosive, a base charge for detonators, and in plastic explosives.

Incompatibilities: Heat, physical damage, shock, and detonators. Detonates on contact with mercury fulminate. Keep away from other explosives, combustibles, strong oxidizers, and aqueous alkaline solutions.

Permissible Exposure Limits in Air: There is no OSHA PEL. The NIOSH recommended TWA is 1.5 mg/m^3 and STEL of 3.0 mg/m^3. HSE and the Canadian provinces of Alberta and British Columbia adopted the same values as NIOSH. ACGIH[1] recommends a TWA of 0.5 mg/m^3 Australia, Israel, Mexico, and the Canadian provinces of Ontario and Quebec set the TWA of 1.5 mg/m^3 with no STEL. Several states have set guidelines or standards for cyclonite in ambient air[60] ranging from 15 μg/m^3 (North Dakota) to 25 μg/m^3 (Virginia) to 30 μg/m^3 (Connecticut) to 36 μg/m^3 (Nevada). This chemical can be absorbed through the skin, thereby increasing exposure.

Determination in Air: Filter; none; Gravimetric; NIOSH IV Method #0500, Particulates NOR (total).

Permissible Concentration in Water: No criteria set.

Routes of Entry: Inhalation, skin absorption, ingestion, skin and/or eye contact.

Harmful Effects and Symptoms

Short Term Exposure: This chemical can be absorbed through the skin, thereby increasing exposure. Irritates the eyes, skin and respiratory tract. Skin contact causes a burning sensation and rash. Other symptoms include headache, nausea, vomiting, loss of appetite, weakness, confusion, dizziness. The oral LD_{50} rat is 100 mg/kg (moderately toxic).

Long Term Exposure: Repeated exposure can cause irritability, sleeplessness, seizures, anorexia, kidney damage. Exposure to high levels can damage the nervous system.

Points of Attack: Eyes, skin, central nervous system.

Medical Surveillance: Examination of the nervous system. Kidney function tests.

First Aid: If this chemical gets into the eyes, remove any contact lenses at once and irrigate immediately for at least 15 minutes, occasionally lifting upper and lower lids. Seek medical attention immediately. If this chemical contacts the skin, remove contaminated clothing and wash immediately with soap and water. Seek medical attention immediately. If this chemical has been inhaled, remove from exposure, begin rescue breathing (using universal precautions) if breathing has stopped and CPR if heart action has stopped. Transfer promptly to a medical facility. When this chemical has been swallowed, get medical attention. Give large quantities of water and induce vomiting. Do not make an unconscious person vomit.

Personal Protective Methods: Wear protective gloves and clothing to prevent any reasonable probability of skin contact. Safety equipment suppliers/manufacturers can provide recommendations on the most protective glove/clothing material for your operation. Neoprene gloves and plastic clothing are recommended. All protective clothing (suits, gloves, footwear, headgear) should be clean, available each day, and put on before work. Contact lenses should not be worn when working with this chemical. Wear dust-proof chemical goggles and face shield unless full facepiece respiratory protection is worn. Employees should wash immediately with soap when skin is wet or contaminated. Provide emergency showers and eyewash.

Respirator Selection: Use chemical cartridge respirator. *Where there is a potential for overexposure:* SCBAF:PD,PP (any MSHA/NIOSH approved self-contained breathing apparatus that has a full facepiece and is operated in a pressure-demand or other positive-pressure mode); or SAF:PD,PP: ASCBA (any supplied-air respirator that has a full facepiece and is operated in a pressure-demand or other positive-pressure mode in combination with an auxiliary, self-contained breathing apparatus operated in a pressure-demand or other positive pressure mode).

Storage: Prior to working with this chemical you should be trained on its proper handling and storage. Detached storage in a secure area is recommended. Store in tightly closed containers in a cool well-ventilated area away from other explosives, combustibles or strong oxidizers (such as chlorine, bromine, and fluorine). Also keep cyclonite away from shock and heat sources. Use only non-sparking tools and equipment, especially when opening and closing containers of this chemical. Sources of ignition such as smoking and open flames, are prohibited where this chemical is used, handled, or stored in a manner that could create a potential fire or explosion hazard. Wherever this chemical is used, handled, manufactured, or stored, use explosion-proof electrical equipment and fittings.

Shipping: Cyclonite falls in DOT Hazard Class 1.1 D and must bear an "Explosives" label.

Spill Handling: Evacuate and restrict persons not wearing protective equipment from area of spill or leak until cleanup is complete. Remove all ignition sources. Cover spill with soda ash, mixed and sprayed with water. Place into bucket of water and allow to stand for two hours. Do not operate radio transmitters or electronic detonator in spill area. Ventilate area of spill or leak after clean-up is complete. Keep Cyclonite out of a confined space, such as a sewer, because of the possibility of an explosion, unless the sewer is designed to prevent the build-up of explosive concentrations. It may be necessary to contain and dispose of this chemical as a hazardous waste. If material or contaminated runoff enters waterways, notify downstream users of potentially contaminated waters. Contact your Department of Environmental Protection or your regional office of the federal EPA for specific recommendations. If employees are required to clean-up spills, they must be properly trained and equipped. OSHA 1910.120(q) may be applicable.

Fire Extinguishing: Cyclonite is a highly dangerous explosion hazard, especially at high temperatures. In case of a fire, immediately evacuate area. An evacuation distance of one mile is recommended Poisonous gases are produced in fire including nitrogen oxides. If material or contaminated runoff enters waterways, notify downstream users of potentially contaminated waters. Notify local health and fire officials and pollution control agencies. From a secure, explosion-proof location, use water spray to cool exposed containers. If cooling streams are ineffective (venting sound increases in volume and pitch, tank discolors, or shows any signs of deforming), withdraw immediately to a secure position. If employees are expected to fight fires, they must be trained and equipped in OSHA 1910.156.

Disposal Method Suggested: Pour over soda ash, neutralize and flush to sewer with water. Also HMX may be recovered from solid propellant waste.

References

New Jersey Department of Health and Senior Services and Senior Services, "Hazardous Substance Fact Sheet: Cyclonite," Trenton, NJ (Aug. 1985)

Cyclopentadiene

Molecular Formula: C_5H_6

Synonyms: 1,3-Cyclopentadiene; Pentole; Pyropentylene; R-Pentene

CAS Registry Number: 542-92-7

RTECS Number: GY1000000

DOT ID: UN 1993

Regulatory Authority

- Air Pollutant Standard Set (ACGIH)[1] (Australia) (DFG)[3] (former USSR)[43] (Israel) (Mexico) (OSHA)[58] (Several States)[60] (Several Canadian Provinces)
- Canada, WHMIS, Ingredients Disclosure List

Cited in U.S. State Regulations: Alaska (G), California (A, G), Connecticut (A), Florida (A), Illinois (G), Maine (G), Massachusetts (G), Minnesota (G), Nevada (A), New Hampshire (G), New Jersey (G), North Dakota (A), Pennsylvania (G), Rhode Island (G), Virginia (A), West Virginia (G).

Description: Cyclopentadiene, C_5H_6, is a flammable, colorless liquid with a sweet odor, like turpentine. Boiling point = 42.0°C. Freezing/Melting point = -85°C. Flash point = 25°C (oc). Autoignition temperature = 640°C. Insoluble in water.

Potential Exposure: Cyclopentadiene is used as an intermediate in the manufacture of resins, insecticides, fungicides, and other chemicals.

Incompatibilities: Forms explosive mixture with air. Should be stored at -4 – 32°F/-20 – 0°C. Converted (dimerized) to higher-boiling dicyclopentadiene upon standing in air and at 32°F/0°C; this conversion may be violent and exothermic; this reaction is accelerated by peroxides or trichloroacetic acid. Reacts violently with potassium hydroxide. Violent reaction with strong oxidizers, strong acids, dinitrogen tetroxide, magnesium. Incompatible with alkaline earth metals, nitrogen oxides. May accumulate static electrical charges, and may cause ignition of its vapors.

Permissible Exposure Limits in Air: The Federal standard (OSHA PEL 8-hour TWA),[58] DFG MAK,[3] Australian, Israeli, Mexican, and ACGIH TLV value is 75 ppm (200 mg/m^3). The Canadian TWA for Alberta, BC, Ontario, and Quebec is the same as OSHA and the STEL fir Alberta and BC is 150 ppm. The NIOSH IDLH level is 750 ppm. The former USSR-UNEP/IRPTC project[43] has set an MAC in workplace air of 5 mg/m^3. Several states have set guidelines or standards for cyclopentadiene in ambient air[60] ranging from 2.0 – 4.0 mg/m^3

(North Dakota) to 3.3 mg/m^3 (Virginia) to 4.0 mg/m^3 (Connecticut) to 4.76 mg/m^3 (Nevada).

Determination in Air: Chromosorb tube-104*; Ethyl acetate; Gas chromatography/Flame ionization detection; NIOSH IV Method #2523.

Permissible Concentration in Water: No criteria set.

Routes of Entry: Inhalation, ingestion, skin and/or eye contact.

Harmful Effects and Symptoms

Short Term Exposure: Exposure can irritate the eyes, skin, and respiratory tract. Skin contact causes a burning sensation and rash.

Long Term Exposure: Exposure may damage the liver and kidneys. Exposure can cause a skin allergy to develop. If allergy develops, even low exposures may cause symptoms.

Points of Attack: Eyes, skin, respiratory system.

Medical Surveillance: If symptoms develop or overexposure is suspected, the following may be useful: Liver and kidney function tests. Evaluation by a qualified allergist, including careful exposure history and special testing, may help diagnose skin allergy.

First Aid: If this chemical gets into the eyes, remove any contact lenses at once and irrigate immediately for at least 15 minutes, occasionally lifting upper and lower lids. Seek medical attention immediately. If this chemical contacts the skin, remove contaminated clothing and wash immediately with soap and water. Seek medical attention immediately. If this chemical has been inhaled, remove from exposure, begin rescue breathing (using universal precautions) if breathing has stopped and CPR if heart action has stopped. Transfer promptly to a medical facility. When this chemical has been swallowed, get medical attention. Do NOT induce vomiting.

Personal Protective Methods: Wear appropriate clothing to prevent repeated or prolonged skin contact. Wear eye protection to prevent any reasonable probability of eye contact. Employees should wash promptly when skin is wet or contaminated. Remove clothing immediately if wet or contaminated to avoid flammability hazard.

Respirator Selection: *NIOSH/OSHA: Up to 750 ppm:* CCROV [any chemical cartridge respirator with organic vapor cartridge(s)]; or GMFOV [any air-purifying, full-facepiece respirator (gas mask) with a chin-style, front- or back-mounted organic vapor canister]; or PAPROV [any powered, air-purifying respirator with organic vapor cartridge(s)]; or SA (any supplied-air respirator); or SCBAF (any self-contained breathing apparatus with a full facepiece). *Emergency or planned entry into unknown concentrations or IDLH conditions:* SCBAF:PD,PP (any self-contained breathing apparatus that has a full facepiece and is operated in a pressure-demand or other positive-pressure mode); or SAF:PD,PP:ASCBA (any supplied-air respirator that has a full facepiece and is operated in a pressure-demand or other positive-pressure mode in combination with an auxiliary self-contained breathing apparatus operated in a pressure-demand or other positive-pressure mode). *Escape:* GMFOV [any air-purifying, full-facepiece respirator (gas mask) with a chin-style, front-or back-mounted organic vapor canister] or SCBAE (any appropriate escape-type, self-contained breathing apparatus).

Storage: Prior to working with Cyclopentadiene you should be trained on its proper handling and storage. Cyclopentadiene must be stored to avoid contact with strong oxidizing agents because violent reactions occur. Cyclopentadiene must be stored at temperatures below 32°F or it may undergo an explosive chemical reaction. Sources of ignition such as smoking and open flames are prohibited where Cyclopentadiene is used, handled, or stored in a manner that could create a potential fire or explosion hazard. Metal containers involving the transfer of this chemical should be grounded and bonded. Where possible, automatically pump liquid from drums or other storage containers to process containers. Drums must be equipped with self-closing valves, pressure vacuum bungs, and flame arresters. Use only non-sparking tools and equipment, especially when opening and closing containers of this chemical. Sources of ignition such as smoking and open flames, are prohibited where this chemical is used, handled, or stored in a manner that could create a potential fire or explosion hazard. Wherever this chemical is used, handled, manufactured, or stored, use explosion-proof electrical equipment and fittings.

Shipping: The DOT[19] makes no specific citation of cyclopentadiene in its Performance Oriented Packaging Standards.

Spill Handling: Evacuate and restrict persons not wearing protective equipment from area of spill or leak until cleanup is complete. Remove all ignition sources. Ventilate area of spill or leak. Absorb liquids in vermiculite, dry sand, earth, peat, carbon, or a similar material and deposit in sealed containers. Keep this chemical out of a confined space, such as a sewer, because of the possibility of an explosion, unless the sewer is designed to prevent the build-up of explosive concentrations. It may be necessary to contain and dispose of this chemical as a hazardous waste. If material or contaminated runoff enters waterways, notify downstream users of potentially contaminated waters. Contact your Department of Environmental Protection or your regional office of the federal EPA for specific recommendations. If employees are required to clean-up spills, they must be properly trained and equipped. OSHA 1910.120(q) may be applicable.

Fire Extinguishing: This chemical is a flammable liquid. Poisonous gases including carbon monoxide are produced in fire. Use dry chemical, carbon dioxide, or foam extinguishers. Vapors are heavier than air and will collect in low areas. Vapors may travel long distances to ignition sources and flashback. Vapors in confined areas may explode when exposed to fire. Containers may explode in fire. Storage containers and parts of containers may rocket great distances, in many directions. If material or contaminated runoff enters waterways, notify downstream users of potentially contaminated waters. Notify local health and fire officials and pollution control agencies. From a secure, explosion-proof location, use water spray to cool exposed containers. If cooling streams are ineffective (venting sound increases in volume and pitch, tank discolors, or shows any signs of deforming), withdraw immediately to a secure position. If employees are expected to fight fires, they must be trained and equipped in OSHA 1910.156.

Disposal Method Suggested: Incineration.

References

New Jersey Department of Health and Senior Services and Senior Services, "Hazardous Substance Fact Sheet: Cyclopentadiene," Trenton, NJ (January 1999)

Cyclopentane

Molecular Formula: C_5H_{10},

Synonyms: Pentamethylene

CAS Registry Number: 287-92-3

RTECS Number: GY2390000

DOT ID: UN 1146

EEC Number: 601-030-00-2

Regulatory Authority

- Air Pollutant Standard Set (ACGIH)[1] (Australia) (Israel) (Mexico) (OSHA)[58] (Several States)[60] (Several Canadian Provinces)
- Canada, WHMIS, Ingredients Disclosure List

Cited in U.S. State Regulations: Alaska (G), California (A, G), Connecticut (A), Florida (A), Illinois (G), Maine (G), Massachusetts (G), Minnesota (G), Nevada (A), New Hampshire (G), New Jersey (G), North Dakota (A), Oklahoma (G), Pennsylvania (G), Rhode Island (G), Vermont (W), Virginia (A), West Virginia (G).

Description:

Cyclopentane, C_5H_{10}, is a colorless liquid. Boiling point = 49°C. Freezing/Melting point = -94°C. Flash point ≤ -7°C. Autoignition temperature = 361°C Explosive limits in air: LEL = 1.5%; UEL = 8.7%. Hazard Identification (based on NFPA-704 M Rating System): Health 1, Flammability 3, Reactivity 0. Insoluble in water

Potential Exposure: Cyclopentane is used as a solvent.

Incompatibilities: Forms explosive mixture with air. May accumulate static electrical charges, and may cause ignition of its vapors. Contact with strong oxidizers may cause fire and explosion.

Permissible Exposure Limits in Air: There is no OSHA PEL. NIOSH and ACGIH[1] recommend a TWA of 600 ppm (1,720 mg/m^3). Australia, Israel, and the Canadian Provinces of Alberta, BC, Ontario, and Quebec set the same TWA as NIOSH, and Alberta's STEL is 900 ppm. Several states have set guidelines or standards for cyclopentane in ambient air[60] ranging from 17.0 mg/m^3 (Connecticut) to 17.2 – 25.8 mg/m^3 (North Dakota) to 29.0 mg/m^3 (Virginia) to 41.0 (Nevada).

Determination in Air: No method available.

Permissible Concentration in Water: Vermont[61] has set a guideline for cyclopentane in drinking water of 30.7 mg/l.

Routes of Entry: Inhalation, ingestion, skin and/or eye contact This chemical can be absorbed through the skin, thereby increasing exposure.

Harmful Effects and Symptoms

Short Term Exposure: Vapors are irritating to eyes, nose, and throat. If inhaled, will cause dizziness, nausea, vomiting, difficult breathing or loss of consciousness. Ingestion irritates the stomach. Contact with the liquid is irritating to eyes and skin. This chemical can be absorbed through the skin, thereby increasing exposure. This compound is moderately toxic by ingestion and inhalation. Ingestion causes irritation of the stomach, and aspiration produces severe lung irritation and rapidly developing pulmonary edema. Contact with liquid irritates eyes and skin. Cyclopentane is a nervous system depressant. Alcohol consumption synergistically increases the toxic effects of this compound.

Long Term Exposure: Cyclopentane can cause drying and cracking of the skin.

Points of Attack: Eyes, nose, throat, respiratory system, stomach and nervous system.

Medical Surveillance: Consider possible irritant effects to the skin and respiratory tract in preplacement and periodic examinations as well as any renal or liver complications.

First Aid: If this chemical gets into the eyes, remove any contact lenses at once and irrigate immediately for at least 15 minutes, occasionally lifting upper and lower lids. Seek medical attention immediately. If this chemical contacts the skin, remove contaminated clothing and wash immediately with soap and water. Seek medical attention immediately. If this chemical has been inhaled, remove from exposure, begin rescue breathing (using universal precautions) if breathing has stopped and CPR if heart action has stopped. Transfer promptly to a medical facility. When this chemical has been swallowed, get medical attention. Do not induce vomiting, guard against aspiration. Medical observation is recommended for 24 – 48 hours after breathing overexposure, as pulmonary edema may be delayed. As first aid for pulmonary edema, a doctor or authorized paramedic may consider administering a corticosteroid spray.

Personal Protective Methods: Wear solvent-resistant gloves and clothing to prevent any reasonable probability of skin contact. Safety equipment suppliers/manufacturers can provide recommendations on the most protective glove/

clothing material for your operation. All protective clothing (suits, gloves, footwear, headgear) should be clean, available each day, and put on before work. Contact lenses should not be worn when working with this chemical. Wear splash-proof chemical goggles and face shield unless full facepiece respiratory protection is worn. Employees should wash immediately with soap when skin is wet or contaminated. Remove nonimpervious clothing immediately if wet or contaminated. Provide emergency showers and eyewash.

Respirator Selection: *Where there is a potential for overexposure:* SCBAF:PD,PP (any MSHA/NIOSH approved self-contained breathing apparatus that has a full facepiece and is operated in a pressure-demand or other positive-pressure mode); or SAF:PD,PP:ASCBA (any supplied-air respirator that has a full facepiece and is operated in a pressure-demand or other positive-pressure mode in combination with an auxiliary, self-contained breathing apparatus operated in a pressure-demand or other positive pressure mode).

Storage: Prior to working with Cyclopentane you should be trained on its proper handling and storage. Before entering confined space where Cyclopentane may be present, check to make sure that an explosive concentration does not exist. Store in tightly closed containers in a cool well-ventilated area away from strong oxidizers (such as chlorine, bromine, and fluorine). Use only non-sparking tools and equipment, especially when opening and closing containers of Cyclopentane. Metal containers involving the transfer of this chemical should be grounded and bonded. Where possible, automatically pump liquid from drums or other storage containers to process containers. Drums must be equipped with self-closing valves, pressure vacuum bungs, and flame arresters. Use only non-sparking tools and equipment, especially when opening and closing containers of this chemical. Sources of ignition such as smoking and open flames, are prohibited where this chemical is used, handled, or stored in a manner that could create a potential fire or explosion hazard. Wherever this chemical is used, handled, manufactured, or stored, use explosion-proof electrical equipment and fittings.

Shipping: The label required by DOT is "Flammable Liquid." The UN/DOT Hazard Class is 3 and the Packing Group is II.[19][20]

Spill Handling: Evacuate and restrict persons not wearing protective equipment from area of spill or leak until cleanup is complete. Remove all ignition sources. Establish forced ventilation to keep levels below explosive limit. Absorb liquids in vermiculite, dry sand, earth, peat, carbon, or a similar material and deposit in sealed containers. It may be necessary to contain and dispose of this chemical as a hazardous waste. If material or contaminated runoff enters waterways, notify downstream users of potentially contaminated waters. Contact your Department of Environmental Protection or your regional office of the federal EPA for specific recommendations. If employees are required to clean-up spills, they must be properly trained and equipped. OSHA 1910.120(q) may be applicable.

Fire Extinguishing: This chemical is a flammable liquid. Poisonous gases are produced in fire. Use dry chemical, carbon dioxide, or foam extinguishers. Water may be ineffective because of low flash point. Do not extinguish fire unless flow of chemical can be stopped. Vapors are heavier than air and will collect in low areas. Vapors may travel long distances to ignition sources and flashback. Vapors in confined areas may explode when exposed to fire. Containers may explode in fire. Storage containers and parts of containers may rocket great distances, in many directions. If material or contaminated runoff enters waterways, notify downstream users of potentially contaminated waters. Notify local health and fire officials and pollution control agencies. From a secure, explosion-proof location, use water spray to cool exposed containers. If cooling streams are ineffective (venting sound increases in volume and pitch, tank discolors, or shows any signs of deforming), withdraw immediately to a secure position. If employees are expected to fight fires, they must be trained and equipped in OSHA 1910.156.

Disposal Method Suggested: Incineration.

References

New Jersey Department of Health and Senior Services and Senior Services, "Hazardous Substance Fact Sheet: Cyclopentane," Trenton, NJ (February 1989).

U.S. Environmental Protection Agency, "Chemical Profile: Cyclopentane," Washington, DC, Chemical Emergency Preparedness Program (October 31, 1985).

Cyclophosphamide

Molecular Formula: $C_7H_{15}Cl_2N_2O_2P$

Synonyms: ASTA; ASTA B 518; B 518; *N,N*-Bis-(β-chloraethyl)-*N'*,*O*-propylen-phosphorsaeure-ester-diamid (German); 2-[Bis(2-chloroethyl)amino]-1-oxa-3-aza-2-phosphocyclohexane 2-oxide monohydrate; 2-[Bis(2-chloroethyl)amino]-2H-1,3, 2-oxazaphosphorine 2-oxide; 1-Bis(2-chloroethyl)amino-1-oxo-2-aza-5-oxaphosphoridine monohydrate; [Bis(chloro-2-ethyl)amino]-2-tetrahydro-3,4,5,6-oxazaphosphorine-1,3, 2-oxide-2 hydrate; 2-[Bis(2-chloroethyl)amino]tetrahydro(2H)-1,3,2-oxazaphosphorine 2-oxide monohydrate; *N,N*-Bis(2-chloroethyl)-*N'*-(3-hydroxypropyl)phosphorodiamidic acid intramol ester hydrate; Bis(2-chloroethyl)phosphoramide-cyclic propanolamide ester; Bis(2-chloroethyl)phosphoramide cyclic propanolamide ester monohydrate; *N,N*-Bis(β-chloroethyl)-*N'*,*O*-propylene-phosphoric acid ester amidemonohydrate; *N,N*-Bis(2-chloroethyl)-*N'*,*O*-propylenephosphoric acid ester diamide; *N,N*-Bis(β-chloroethyl)-*N'*,*O*-propylenephosphoric acid ester diamidemonohydrate; *N,N*-Bis(2-chloroethyl)tetrahydro-2H-1,3,2-oxaphosphorin-2-amine,2-oxide monohydrate; *N,N*-Bis(β-chloroethyl)-*N'*,*O*-trimethylenephosphoric acid ester diamide; CB-4564; Clafen; Claphene (French); CP; CPA;

CTX; CY; Cyclic *N'*,*O*-propylene ester of *N*,*N*-bis(2-chloroethyl)phosphorodiamidic acid monohydrate; Cyclophosphamid; Cyclophosphamide; Cyclophosphamidum; Cyclophosphan; Cyclophosphane; Cyclophosphoramide; Cytophosphan; Cytoxan; 2-[Di(2-chloroethyl)amino]-1-oxa-3-aza-2-phosphacyclohexane-2-oxide monohydrate; 2-[Di(2-chloroethyl)amino)-2-oxide, *N*,*N*-di(2-chloroethyl)amino-*N*,*O*-propylenephosphoric acid ester diamide monohydrate; *N*,*N*-Di(2-chloroethyl)-*N*,*O*-propylenephosphoric acid ester diamide; Endoxan; Endoxana; Endoxanal; Endoxan-ASTA; Endoxane; Endoxan R; Enduxan; Genoxal; Mitoxan; NCI-C04900; NSC 26271; 2H-1,3,2-Oxazaphosphorin-2-amine, *N*,*N*-Bis(2-chloroethyl)tetrahydro-, 2-oxide; 2-H-1,3,2-Oxazaphosphorinane; 2H-1,3,2-Oxazaphosphorine,2-bis(2-chloroethyl)aminotetrahydro-2-oxide; Phosphorodiamidic acid, *N*,*N*-bis(2-chloroethyl)-*N'*-(3-hydroxypropyl)-, intramol. ester; Procytox; Semdoxan; Sendoxan; Senduxan; Zyklophosphamid (German)

CAS Registry Number: 50-18-0

RTECS Number: RP5950000

DOT ID: UN 3099

Regulatory Authority

- Carcinogen (Animal and Human Positive) (IARC)[9] (NTP)[10] (NCI)[9]
- EPA HAZARDOUS WASTE NUMBER (RCRA No.): U058
- RCRA, 40CFR261, Appendix 8 Hazardous Constituents
- SUPERFUND/EPCRA 40CFR302.4 Reportable Quantity (RQ): CERCLA, 10 lb (4.54 kg)

Cited in U.S. State Regulations: California (A, G), Florida (G), Illinois (G), Kansas (G), Louisiana (G), Maine (G), Massachusetts (G), Michigan (G), Minnesota (G), New Hampshire (G), New Jersey (G), Pennsylvania (G), Rhode Island (G), Vermont (G), Virginia (G), Washington (G), West Virginia (G), Wisconsin (G)

Description: Cyclophosphamide, $C_7H_{15}Cl_2N_2O_2P$, is a white crystalline powder (monohydrate). Darkens on exposure to light. Odorless. Freezing/Melting point = 41 – 45°C. Flash point 112°C. Soluble in water.

Potential Exposure: Cyclophosphamide is used in the treatment of malignant lymphoma, multiple meyloma, leukemias, and other malignant diseases. Cyclophosphamide has been tested as an insect chemosterilant and for use in the chemical shearing of sheep.

Cyclophosphamide is not produced in the United States. It is produced in Germany and exported to the United States where one company has formulated and marketed the drug since 1959. United States sales are approximately 1,300 lb annually.

The FDA estimates that 200,000 – 300,000 patients per year are treated with Cyclophosphamide. It is administered orally and through injection. The adult dosage is usually 1 – 5 mg/kg of body weight daily or 10 – 15 mg/kg administered intravenous every 7 – 10 days.

Incompatibilities: Heat. Protect from temperatures above 86°F/30°C. See storage recommendations.

Permissible Exposure Limits in Air: No standards set.

Permissible Concentration in Water: No criteria set.

Routes of Entry: Inhalation, skin and/or eye contact Absorbed through the skin.

Harmful Effects and Symptoms

Short Term Exposure: Cyclophosphamide irritates the eyes, skin, and respiratory tract. Eye contact can cause severe damage with possible loss of vision. This chemical can be absorbed through the skin, thereby increasing exposure. Cyclophosphamide may affect the blood, bladder, central nervous system, and heart. Symptoms of Cyclophosphamide exposure include G.I. disturbance, leukopenia, nausea, dizziness, liver dysfunction and hair loss.

Long Term Exposure: Cyclophosphamide is a carcinogen and probable teratogen in humans. It causes bladder and skin cancer. There is limited evidence that this chemical causes sterility in males and females. Repeated exposure may interfere with the body's ability to produce blood cells (anemia). May cause liver damage. The substance may have effects on the blood, bladder, lungs and bone marrow, resulting in leucopenia, cystitis, pulmonary fibrosis. There is sufficient evidence for the carcinogenicity of cyclophosphamide both in humans and in experimental animals. Cyclophosphamide was carcinogenic in rats following administration in drinking water and intravenous injection, and in mice following subcutaneous injection. Dosages were comparable to those used in clinical practice. The chemical produced benign and malignant tumors at various sites including bladder tumors in the rats.

Epidemiological studies are available in which persons treated with cyclophosphamide for a variety of medical conditions were compared with similarly affected controls. These studies consistently demonstrate an excess of various neoplasms and leukemias in the treated groups, although the number in all five studies was small. The oral LD_{50} mouse is 137 mg/kg (moderately toxic).

Points of Attack: Bladder, skin, liver.

Medical Surveillance: Complete blood count (CBC). Liver function tests.

First Aid: If this chemical gets into the eyes, remove any contact lenses at once and irrigate immediately for at least 30 minutes, occasionally lifting upper and lower lids. Seek medical attention immediately. If this chemical contacts the skin, remove contaminated clothing and wash immediately with soap and water. Seek medical attention immediately. If this chemical has been inhaled, remove from exposure, begin rescue breathing (using universal precautions) if breathing has stopped and CPR if heart action has stopped. Transfer promptly to a medical facility. When this chemical has been swallowed, get medical attention. Give large quantities of water and induce vomiting. Do not make an unconscious person vomit.

Personal Protective Methods: Wear protective gloves and clothing to prevent any reasonable probability of skin contact. Safety equipment suppliers/manufacturers can provide recommendations on the most protective glove/clothing material for your operation. All protective clothing (suits, gloves, footwear, headgear) should be clean, available each day, and put on before work. Contact lenses should not be worn when working with this chemical. Eye protection is included in the recommended respiratory protection. Employees should wash immediately with soap when skin is wet or contaminated. Provide emergency showers and eyewash.

Respirator Selection: *At any detectable concentration:* SCBAF:PD,PP (any MSHA/NIOSH approved self-contained breathing apparatus that has a full facepiece and is operated in a pressure-demand or other positive-pressure mode); or SAF:PD,PP:ASCBA (any supplied-air respirator that has a full facepiece and is operated in a pressure-demand or other positive-pressure mode in combination with an auxiliary, self-contained breathing apparatus operated in a pressure-demand or other positive pressure mode). *Escape:* HiEF [any air-purifying, full-facepiece respirator (gas mask) with a chin-style, front- or back-mounted organic vapor canister having a high-efficiency particulate filter]; or SCBAE (any appropriate escape-type, self-contained breathing apparatus).

Storage: Prior to working with Cyclophosphamide you should be trained on its proper handling and storage. Store in a refrigerator and protect from light and moisture. Storage at or below 77°F/25°C is recommended. Metal containers involving the transfer of this chemical should be grounded and bonded. Where possible, automatically pump liquid from drums or other storage containers to process containers. Drums must be equipped with self-closing valves, pressure vacuum bungs, and flame arresters. Use only non-sparking tools and equipment, especially when opening and closing containers of this chemical. Sources of ignition such as smoking and open flames, are prohibited where this chemical is used, handled, or stored in a manner that could create a potential fire or explosion hazard. Wherever this chemical is used, handled, manufactured, or stored, use explosion-proof electrical equipment and fittings. A regulated, marked area should be established where this chemical is handled, used, or stored in compliance with OSHA standard 1910.1045.

Shipping: Medicines, poisonous, solid, n.o.s. fall in Hazard Class 6.1. Cyclophosphamide falls in Packing Group III and requires a "Keep Away From Food" label.

Spill Handling: Evacuate persons not wearing protective equipment from area of spill or leak until clean-up is complete. Remove all ignition sources. Dampen spilled material with water to avoid dust. Collect dampened or powdered material in the most convenient and safe manner and deposit in sealed containers. Ventilate area after clean-up is complete. It may be necessary to contain and dispose of this chemical as a hazardous waste. If material or contaminated runoff enters waterways, notify downstream users of potentially contaminated waters. Contact your Department of Environmental Protection or your regional office of the federal EPA for specific recommendations. If employees are required to clean-up spills, they must be properly trained and equipped. OSHA 1910.120(q) may be applicable.

Fire Extinguishing: This chemical is a combustible solid. Use dry chemical, carbon dioxide, or foam extinguishers. Poisonous gases are produced in fire. If material or contaminated runoff enters waterways, notify downstream users of potentially contaminated waters. Notify local health and fire officials and pollution control agencies. Containers may explode in fire. From a secure, explosion-proof location, use water spray to cool exposed containers. If cooling streams are ineffective (venting sound increases in volume and pitch, tank discolors, or shows any signs of deforming), withdraw immediately to a secure position. If employees are expected to fight fires, they must be trained and equipped in OSHA 1910.156.

References

Sax, N. I., Ed., "Dangerous Properties of Industrial Materials Report," 1, No. 3, 62–64 (1981) (As Endoxan)

Cyclopropane

Molecular Formula: C_3H_6

Synonyms: Ciclopropano (Spanish); Cyclopropane, liquified; Trimethylene

CAS Registry Number: 75-19-4

RTECS Number: GZ0690000

DOT ID: UN 1027

Regulatory Authority

- CLEAN AIR ACT: Accidental Release Prevention/Flammable substances, (Section 112[r], Table 3), TQ = 10,000 lb (4,540 kg)

Cited in U.S. State Regulations: California (A, G), Florida (G), Maine (G), Massachusetts (G), New Jersey (G), Pennsylvania (G), Rhode Island (G).

Description:

Cyclopropane is a gas. Boiling point = -34°C. Freezing/Melting point = -127°C. Flash point = flammable gas. Flammable limits: LEL = 2.4%; UEL = 10.4%. Autoignition temperature = 498°C. Hazard Identification (based on NFPA-704 M Rating System): Health 1, Flammability 4, Reactivity 0.

Potential Exposure: Cyclopropane is used as an anesthetic and to make other chemicals.

Incompatibilities: Forms explosive mixture with air. Heat, flame or contact with oxidizers can cause fire and explosion hazard. May accumulate static electrical charges, and may cause ignition of its vapors.

Permissible Exposure Limits in Air: No standards set.

Determination in Air: Charcoal adsorption followed by workup with CS_2, and analysis by gas chromatography. See NIOSH Method #1500 for hydrocarbons

Permissible Concentration in Water: No criteria set.

Harmful Effects and Symptoms

Short Term Exposure: Cyclopropane can affect you when breathed in. Cyclopropane is used as a surgical anesthetic. High levels can cause you to feel dizzy, lightheaded, and to pass out. Very high levels can cause coma and death. Liquid can cause frostbite. May affect the nervous system, and cause heart-rate disorders.

Medical Surveillance: EKG, examination of the nervous system.

First Aid: If this chemical gets into the eyes, remove any contact lenses at once and irrigate immediately for at least 15 minutes, occasionally lifting upper and lower lids. Seek medical attention immediately. If this chemical contacts the skin, remove contaminated clothing and wash immediately with soap and water. Seek medical attention immediately. If this chemical has been inhaled, remove from exposure, begin rescue breathing (using universal precautions) if breathing has stopped and CPR if heart action has stopped. Transfer promptly to a medical facility. If frostbite has occurred, seek medical attention immediately; do *NOT* rub the affected areas or flush them with water. In order to prevent further tissue damage, do *NOT* attempt to remove frozen clothing from frostbitten areas. If frostbite has *NOT* occurred, immediately and thoroughly wash contaminated skin with soap and water.

Personal Protective Methods: *Clothing:* Avoid skin contact with Cyclopropane. Wear protective gloves and clothing. Safety equipment suppliers/manufacturers can provide recommendations on the most protective glove/clothing material for your operation. All protective clothing (suits, gloves, footwear, headgear) should be clean, available each day, and put on before work. Where exposure to cold equipment, vapors, or liquid may occur, employees should be provided with special clothing designed to prevent the freezing of body tissue.

Eye Protection: Wear gas-proof goggles, unless full facepiece respiratory protection is worn. Wear splash-proof chemical goggles when working with liquid, unless full facepiece respiratory protection is worn.

Respirator Selection: Where the potential exists for exposure to Cyclopropane, use a MSHA/NIOSH approved supplied-air respirator with a full facepiece operated in the positive pressure mode or with a full facepiece, hood, or helmet in the continuous flow mode, or use a MSHA/NIOSH approved self-contained breathing apparatus with a full facepiece operated in pressure-demand or other positive pressure mode.

Storage: Prior to working with Cyclopropane you should be trained on its proper handling and storage. Before entering confined space where this gas may be present, check to make sure that an explosive concentration does not exist. Cyclopropane must be stored to avoid contact with oxidizers (such as perchlorates, peroxides, permanganates, chlorates and nitrates) and oxygen since violent reactions occur. Store in tightly closed containers in a cool, well-ventilated area away from heat or flame. Outside or detached storage is recommended. Use only non-sparking tools and equipment, especially when opening and closing containers of this chemical. Sources of ignition such as smoking and open flames, are prohibited where this chemical is used, handled, or stored in a manner that could create a potential fire or explosion hazard. Wherever this chemical is used, handled, manufactured, or stored, use explosion-proof electrical equipment and fittings. Procedures for the handling, use and storage of cylinders should be in compliance with OSHA 1910.101 and 1910.169 as with the recommendations of the Compressed Gas Association.

Shipping: Cyclopropane, liquefied, falls in Hazard Class 2.1. It requires a "Flammable Gas" label. Shipment by passenger aircraft or railcar is forbidden.

Spill Handling: Evacuate and restrict persons not wearing protective equipment from area of spill or leak until cleanup is complete. Remove all ignition sources. Establish forced ventilation to keep levels below explosive limit and to disperse the gas. Stop the flow of gas if it can be done safely. If source of leak is a cylinder and the leak cannot be stopped in place, remove leaking cylinder to a safe place in the open air, and repair leak or allow cylinder to empty. Keep this chemical out of confined space, such as sewer because of the possibility of explosion, unless the sewer is designed to prevent the buildup of explosive concentrations. It may be necessary to contain and dispose of this chemical as a hazardous waste. Contact your Department of Environmental Protection or your regional office of the federal EPA for specific recommendations. If employees are required to clean-up spills, they must be properly trained and equipped. OSHA 1910.120(q) may be applicable.

Fire Extinguishing: Cyclopropane is a flammable gas. Poisonous gases are produced in fire. Do not extinguish the fire unless the flow of gas can be stopped and any remaining gas is out of the line. Specially trained personnel may use fog lines to cool exposures and let the fire burn itself out. Vapors are heavier than air and will collect in low areas. Vapors may travel long distances to ignition sources and flashback. Vapors in confined areas may explode when exposed to fire. Containers may explode in fire. Storage containers and parts of containers may rocket great distances, in many directions. If material or contaminated runoff enters waterways, notify downstream users of potentially contaminated waters. Notify local health and fire officials and pollution control agencies. From a secure, explosion-proof location, use water spray to cool exposed containers. If cooling streams are ineffective (venting sound increases in volume and pitch, tank discolors, or shows any signs of deforming), withdraw immediately to a secure position. If cylinders are exposed to excessive heat from fire or flame contact, withdraw immediately to a secure location. If employees are expected to fight fires, they must be trained and equipped in OSHA 1910.156.

References

New Jersey Department of Health and Senior Services and Senior Services, "Hazardous Substance Fact Sheet: Cyclopropane," Trenton, NJ (February 1987)

Cyhexatin

Molecular Formula: $C_{18}H_{34}O_4Sn$

Common Formula: $(C_6H_{11}O)_3SnOH$

Synonyms: Plictran®; Tricyclohexylhydroxystannane and ENT 27395-X; Tricyclohexyltin hydroxide

CAS Registry Number: 13121-70-5

RTECS Number: WH8750000

Regulatory Authority

- Air Pollutant Standard Set (ACGIH)[1] (Australia) (HSE)[33] (Israel) (Mexico) (OSHA)[58] (Several States)[60]
- U.S. DOT 49CFR172.101, Appendix B, Regulated marine pollutant, listed as "severe pollutant."
- Canada, CEPA Prohibited Export Substance List

Cited in U.S. State Regulations: Alaska (G), California (A, G), Connecticut (A), Florida (G), Illinois (G), Maine (G), Massachusetts (G), Minnesota (G), Nevada (A), New Hampshire (G), New Jersey (G), North Dakota (A), Pennsylvania (G), Rhode Island (G), Virginia (A), West Virginia (G).

Description: Cyhexatin, $(C_6H_{11}O)_3SnOH$, is a colorless to white, nearly odorless, crystalline powder. Boiling point = 227°C (decomposes). Freezing/Melting point = 195 – 198°C. Practically insoluble in water. An organotin compound.

Potential Exposure: Those involved in the manufacture, formulation and application of this acaricide (miticide).

Incompatibilities: Strong oxidizers, ultraviolet light

Permissible Exposure Limits in Air: The OSHA PEL for organotin compounds is 0.32 mg/m^3 [0.1 mg/m^3 (as Sn)]. The ACGIH recommended TWA value is 5 mg/m^3. Australia, Israel, Mexico, and the HSE[33] have set that same TWA as ACGIH and HSE's STEL value is 10 mg/m^3. In Canada the provinces of Alberta, BC, Ontario, and Quebec have set a TWA of 5 mg/m^3 and the STEL for Alberta and BC is 10 mg/m^3. Several states have set guidelines or standards for cyhexatin in ambient air[60] ranging from 50 μg/m^3 (North Dakota) to 80 μg/m^3 (Virginia) to 100 μg/m^3 (Connecticut) to 119 μg/m^3 (Nevada). The NIOSH IDLH is 80 mg/m^3 [25 mg/m^3 (as Sn)].

Determination in Air: Filter/XAD-2® (tube); CH_3COOH/CH_3CN; High-pressure liquid chromatography/Graphite furnace atomic absorption spectrometry; NIOSH IV Method #5504, Organotin.

Permissible Concentration in Water: No criteria set.

Routes of Entry: Inhalation, skin absorption, ingestion, skin and/or eye contact.

Harmful Effects and Symptoms

Short Term Exposure: Irritates eyes, skin, and respiratory system. Symptoms of exposure include headache, vertigo (an illusion of movement); sore throat, cough; abdominal pain, vomiting; skin burns, pruritus. Cyhexatin is moderate in acute oral toxicity to animals. This is in contrast to alkyl tin compounds with smaller (methyl and ethyl) radicals which are highly toxic. A diet including 6 mg/kg of body weight of cyhexatin for two years showed no effect in rats. The oral LD_{50} rat is 180 mg/kg (moderately toxic).

Long Term Exposure: This chemical has been shown to cause liver and kidney damage in animals.

Points of Attack: Eyes, skin, respiratory system, liver, kidneys.

Medical Surveillance: Liver function tests. Kidney function tests.

First Aid: If this chemical gets into the eyes, remove any contact lenses at once and irrigate immediately for at least 15 minutes, occasionally lifting upper and lower lids. Seek medical attention immediately. If this chemical contacts the skin, remove contaminated clothing and wash immediately with soap and water. Speed in removing material from skin is of extreme importance. Shampoo hair promptly if contaminated. Seek medical attention immediately. If this chemical has been inhaled, remove from exposure, begin rescue breathing (using universal precautions) if breathing has stopped and CPR if heart action has stopped. Transfer promptly to a medical facility. When this chemical has been swallowed, get medical attention. Give large quantities of water and induce vomiting. Do not make an unconscious person vomit.

Personal Protective Methods: Wear protective gloves and clothing to prevent any reasonable probability of skin contact. Safety equipment suppliers/manufacturers can provide recommendations on the most protective glove/clothing material for your operation. All protective clothing (suits, gloves, footwear, headgear) should be clean, available each day, and put on before work. Contact lenses should not be worn when working with this chemical. Wear dust-proof chemical goggles and face shield unless full facepiece respiratory protection is worn. Employees should wash immediately with soap when skin is wet or contaminated. Provide emergency showers and eyewash.

Respirator Selection: NIOSH/OSHA: *Up to 3.2 mg/m^3:* CCROVDM [any chemical cartridge respirator with organic vapor cartridge(s) in combination with a dust and mist filter]; or SA (any supplied-air respirator). *Up to 8 mg/m^3:* SA:CF (any supplied-air respirator operated in a continuous-flow mode); or PAPROVDM [any powered, air-purifying respirator with organic vapor cartridge(s) in combination with a dust and mist filter]. *Up to 16 mg/m^3:* CCRFOVHiE [any chemical cartridge respirator with a full facepiece and organic vapor cartridge(s) in combination with a high-efficiency particulate filter]; or GMFOVHiE [any air-purifying, full-facepiece respirator (gas mask) with a chin-style, front- or back-mounted organic vapor canister having a high-efficiency particulate filter]; or PAPRTOVHiE [any powered, air-purifying respirator with a tight-fitting face-piece and organic vapor cartridge(s) in combination with a high-efficiency particulate filter]; or SAT:CF (any supplied-air respirator that has a tight-fitting facepiece and is operated in a continuous-flow mode); or SCBAF (any self-contained breathing apparatus with a full facepiece); or SAF (any supplied-air respirator with a full facepiece). *Up to 80 mg/m³:* SAF:PD,PP (any supplied-air respirator that has a full facepiece and is operated in a pressure-demand or other positive-pressure mode). *Emergency or planned entry into unknown concentrations or IDLH conditions:* SCBAF: PD,PP (any self-contained breathing apparatus that has a full faceplate and is operated in a pressure-demand or other positive-pressure mode); or SAF:PD,PP: ASCBA (any supplied-air respirator that has a full facepiece and is operated in a pressure-demand or other positive-pressure mode in combination with an auxiliary, self-contained breathing apparatus operated in a pressure-demand or other positive-pressure mode). *Escape:* GMFOVHiE [any air-purifying, full-facepiece respirator (gas mask) with a chin-style, front- or back-mounted organic vapor canister having a high-efficiency particulate filter]; or SCBAE (any appropriate escape-type, self-contained breathing apparatus).

Storage: Prior to working with this chemical you should be trained on its proper handling and storage. Store in tightly closed containers in a cool, dark, well-ventilated area away from oxidizers.

Shipping: Cyhexatin is not specifically cited in DOT's Performance-Oriented Packaging Standards.[19]

Spill Handling: Evacuate persons not wearing protective equipment from area of spill or leak until clean-up is complete. Remove all ignition sources. Collect powdered material in the most convenient and safe manner and deposit in sealed containers. Ventilate area after clean-up is complete. It may be necessary to contain and dispose of this chemical as a hazardous waste. If material or contaminated runoff enters waterways, notify downstream users of potentially contaminated waters. Contact your Department of Environmental Protection or your regional office of the federal EPA for specific recommendations. If employees are required to clean-up spills, they must be properly trained and equipped. OSHA 1910.120(q) may be applicable.

Fire Extinguishing: This chemical is a combustible solid. Use dry chemical, carbon dioxide, water spray, or alcohol foam extinguishers. Poisonous gases are produced in fire. If material or contaminated runoff enters waterways, notify downstream users of potentially contaminated waters. Notify local health and fire officials and pollution control agencies. From a secure, explosion-proof location, use water spray to cool exposed containers. If cooling streams are ineffective (venting sound increases in volume and pitch, tank discolors, or shows any signs of deforming), withdraw immediately to a secure position. If employees are expected to fight fires, they must be trained and equipped in OSHA 1910.156.

References

National Institute for Occupational Safety and Health, Criteria for a Recommended Standard: Occupational Exposure to Organotin Compounds, NOSH Document No. 77–115 (1977)

D

2,4-D

Molecular Formula: $C_8H_6Cl_2O_3$

Common Formula: $Cl_2C_6H_3OCH_2COOH$

Synonyms: Acetic acid (2,4-dichlorophenoxy)-; Acide 2,4-dichloro phenoxyacetique (French); Acido (2,4-dicloro-fenossi)-acetico (Italian); Acido 2,4-diclorofenoxiacetico (Spanish); Agrotect; Amidox; Amoxone; Aqua-Kleen; Barrage; BH 2,4-D; Brush-Rhap; B-Selektonon; Bush killer; Chipco turf herbicide "D"; Chloroxone; Citrus fix; Crop rider; Crotilin; D 50; Dacamine; 2,4-d Acid; Debroussaillant 600; Decamine; Ded-Weed; Ded-Weed LV-69; Deherban; Desormone; (2,4-Dichloor-fenoxy)-azijnzuur (Dutch); Dichlorophenoxyacetic acid; 2,4-Dichlorophenoxyacetic acid, salts and esters; 2,4-Dichlorphenoxyacetic acid; (2,4-Dichlor-phenoxy)-essigsaeure (German); Dicopur; Dicotox; Dinoxol; DMA-4; Dormone; 2,4-D Phenoxy pesticide; 2,4-D, Salts and esters; 2,4-Dwuchlorofenoksyoctowy kwas (Polish); Emulsamine BK; Emulsamine E-3; ENT 8,538; Envert 171; Envert DT; Esteron; Esteron 44 weed killer; Esteron 76 BE; Esteron 99; Esteron 99 concentrate; Esteron brush killer; Esterone four; Estone; Farmco; Fernesta; Fernimine; Fernoxone; Ferxone; Foredex 75; Formula 40; Hedonal; Hedonal (herbicide); Herbidal; Ipaner; Krotiline; Kwas 2,4-dwuchlorofenoksyoctowy; Kwasu 2,4-dwuchlorofenoksoctowego; Kyselina 2,4-dichlorfenoxyoctova; Lawn-Keep; Macrondray; Miracle; Monosan; Mota maskros; Moxone; Netagrone; Netagrone 600; NSC 423; 2,4-PA (in Japan); Pennamine; Pennamine D; Phenox; Pielik; Planotox; Plantgard; Rhodia; Rtecs No. AG6825000; Salvo; Spritz-Hormin/2,4-D; Spritz-Hormit/2,4-D; Super D weedone; Superormone concentre; Transamine; Tributon; Trinoxol; U 46; U 46DP; U-5043; Vergemaster; Verton; Verton 2D; Verton D; Vertron 2D; Vidon 638; Visko; Visko-Rhap; Visko-Rhap low drift herbicides; Visko-Rhap low volatile 4L; Weed-AG-Bar; Weedar; Weedar-64; Weed-B-Gon; Weedez wonder bar; Weedone; Weedone LV4; Weed-Rhap; Weed tox; Weedtrol

CAS Registry Number: 94-75-7

RTECS Number: AG6825000

DOT ID: UN 2765

EEC Number: 607-039-00-8

Regulatory Authority

- Carcinogen (Human Suspected) (IARC)[9]
- Air Pollutant Standard Set (ACGIH)[1] (Australia) (DFG)[3] (HSE)[33] (Israel) (Mexico) (former USSR)[35] (OSHA)[58] (Several States)[60] (Several Canadian Provinces)
- Water Pollution Standard Proposed (EPA, Mexico)[35] (Maine, Minnesota)[61]
- Clean Air Act: Hazardous Air Pollutants (Title I, Part A, Section 112)
- Clean Water Act: Section 311 Hazardous Substances/RQ 40CFR117.3 (same as CERCLA, see below); Section 313 Water Priority Chemicals (57FR41331, 9/9/92)
- Reportable Quantity (RQ): CERCLA, 100 lb (45.5 kg)
- EPA Hazardous Waste Number (RCRA No.): U240, D016
- RCRA, 40CFR261, Appendix 8 Hazardous Constituents
- RCRA Toxicity Characteristic (Section 261.24), Maximum
- Concentration of Contaminants, regulatory level, 10.0 mg/l
- RCRA 40CFR268.48; 61FR15654, Universal Treatment Standards: Wastewater (mg/l), 0.72; Nonwastewater (mg/kg), 10
- RCRA 40CFR264, Appendix 9; TSD Facilities Ground Water Monitoring List. Suggested test method(s) (PQL μg/l): 8150 (10)
- Safe Drinking Water Act: MCL, 0.1 mg/l; MCGL, 0.07 mg/l; Regulated chemical (47 FR 9352) as 2,4-D
- CERCLA/SARA 313: Form R *de minimis* concentration reporting level: 1.0%
- U.S. DOT Regulated Marine Pollutant (49CFR172.101, Appendix B)
- Canada, Drinking Water Quality, 0.1 mg/l IMAC
- Mexico, Drinking Water Criteria, 0.1 mg/l

Cited in U.S. State Regulations: Alaska (G), California (A, G), Connecticut (A), Florida (A), Illinois (G), Kansas (G), Louisiana (G), Maine (G, W), Massachusetts (G), Michigan (G), Minnesota (W), Nevada (A), New Hampshire (G), New Jersey (G), New York (G), North Dakota (A), Pennsylvania (G, A), Rhode Island (G), Vermont (G),

Virginia (G, A), Washington (G), West Virginia (G), Wisconsin (G).

Description: 2,4-Dichlorophenoxyacetic acid, $Cl_2C_6H_3O$-CH_2COOH, is a white to yellow crystalline powder with a slight phenolic odor. Boiling point = (decomposes) Freezing/Melting point = 138°C. Slightly soluble in water. The taste and odor threshold in water is 3.13 mg/l.

Potential Exposure: 2,4-Dichlorophenoxyacetic acid, was introduced as a plant growth-regulator in 1942. It is registered in the United States as a herbicide for control of broadleaf plants and as a plant growth-regulator. Thus, workers engaged in manufacture, formulation or application are affected as may be citizens in areas of application.

Incompatibilities: A weak acid, incompatible with bases. Decomposes in sunlight or heat, forming hydrogen chloride and phosgene. Contact with strong oxidizers may cause fire and explosions.

Permissible Exposure Limits in Air: The legal limit (OSHA PEL), and ACGIH recommended TWA is 10 mg/m^3 as is the value for DFG,[3] HSE value,[33] Israel, Mexico, and Australia. The HSE[33] and Mexico adds a STEL of 20 mg/m^3. The NIOSH IDLH value is 100 mg/m^3. The former USSR[35] has set a MAC in workplace air of 1.0 mg/m^3. The Canadian provincial limit for Alberta, BC, Ontario, and Quebec is the dame as OSHA and Alberta and BC set a STEL of 20 mg/m^3. The former USSR-UNEP/IRPTC project[43] has set a MAC in ambient air in residential areas of 0.02 mg/m^3 on a momentary basis and 0.01 mg/m^3 on a daily average basis for the sodium salt of 2,4-D.

In addition, several states have set guidelines or standards for 2,4-D in ambient air[60] ranging from 100 μg/m^3 (North Dakota) to 105 μg/m^3 (Pennsylvania) to 160 μg/m^3 (Virginia) to 200 μg/m^3 (Connecticut) to 238 μg/m^3 (Nevada).

Determination in Air: Collection on a glass fiber filter and analysis by HPLC with UV detection. See NIOSH Method #5001.[18]

Permissible Concentration in Water: The U.S.[35] has set an MPC in bottled water intended for human consumption of 0.1 mg/l. Mexico[35] has set levels in ambient water of 0.1 mg/l in estuaries and 0.01 mg/l in coastal waters.

The former USSR-UNEP/IRPTC project[43] has set an MAV of 1.0 mg/l in water bodies used for drinking purposes for the sodium salt and 0.62 mg/l in water bodies used for fishery purposes.

A no-observed-adverse-effect-level (NOAEL) of 1 mg/kg/day has been determined[47] which results in the calculation of a lifetime health advisory of 0.070 mg/l. This level has been proposed by EPA[62] as a maximum level in drinking water.

Drinking water levels for Canada and Mexico are 0.1 mg/l. States which have set guidelines for 2,4-D in drinking water[61] include Minnesota at 70 μg/l and Maine at 100 μg/l.

Determination in Water: Filter; Methanol; High-pressure liquid chromatography/Ultraviolet detection; NIOSH IV Method #5001

Routes of Entry: Inhalation, skin absorption, ingestion, skin and eye contact.

Harmful Effects and Symptoms

Short Term Exposure: *Inhalation:* May cause irritation of the mouth, nose and throat, headache, nausea, vomiting, and diarrhea at levels above 10 mg/m^3. Nerve damage, which may be delayed, may include swelling of legs and feet, muscle twitch and stupor. Severe exposures may result in death.

Skin: Dust or liquid left in contact with the skin for several hours may be absorbed. This may result in severe delayed symptoms as listed above. These symptoms may last for months or years.

Eyes: Irritation may occur.

Ingestion: The oral dose required to produce symptoms is about 1/12 ounce (1/2 teaspoon). Increasing amounts may result in increasingly severe symptoms as listed above. Death has resulted from as little as 1/5 ounce. Survival for more than 48 hours is usually followed by complete recovery although symptoms may last for several months.

Long Term Exposure: Workers exposed to 2,4-D in the manufacturing process over a five to 10 year period at levels above 10 mg/m^3 complained of weakness, rapid fatigue, headache and vertigo. Liver damage, low blood pressure and slowed heartbeat were also found. Based on animal tests, 2,4-D may affects human reproduction

Points of Attack: Skin, central nervous system, liver, kidneys.

Medical Surveillance: If symptoms develop or overexposure is suspected, the following may be useful: Liver and kidney function tests. Exam of the nervous system.

First Aid: If this chemical gets into the eyes, remove any contact lenses at once and irrigate immediately for at least 15 minutes, occasionally lifting upper and lower lids. Seek medical attention immediately. If this chemical contacts the skin, remove contaminated clothing and wash immediately with soap and water. Seek medical attention immediately. If this chemical has been inhaled, remove from exposure, begin rescue breathing (using universal precautions) if breathing has stopped and CPR if heart action has stopped. Transfer promptly to a medical facility. When this chemical has been swallowed, get medical attention. Give large quantities of water and induce vomiting. Do not make an unconscious person vomit.

Personal Protective Methods: Wear protective gloves and clothing to prevent any reasonable probability of skin contact. Safety equipment suppliers/manufacturers can provide recommendations on the most protective glove/clothing material for your operation.. All protective clothing (suits, gloves, footwear, headgear) should be clean, available each day, and put on before work. Contact lenses should not be worn when working with this chemical. Wear dust-proof chemical

goggles and face shield unless full facepiece respiratory protection is worn. Employees should wash immediately with soap when skin is wet or contaminated. Provide emergency showers and eyewash.

Respirator Selection: NIOSH: *100 mg/m³:* CCROVDMFu ([any chemical cartridge respirator with organic vapor cartridge(s) in combination with a dust, mist, and fume filter]; or GMFOVHiE [any air-purifying, full-facepiece respirator (gas mask) with a chin-style, front- or back-mounted organic vapor canister having a high-efficiency particulate filter]; or PAPROVDMFu [any powered, air-purifying respirator with organic vapor cartridge(s) in combination with a dust, mist, and fume filter]; or SA (any supplied-air respirator); or SCBAF (any self-contained breathing apparatus with a full facepiece); or SAF (any supplied-air respirator with a full facepiece); or SAT:CF (any supplied-air respirator that has a tight-fitting facepiece and is operated in a continuous-flow mode). *Emergency or planned entry into unknown concentrations or IDLH conditions:* SCBAF:PD,PP (any self-contained breathing apparatus that has a full facepiece and is operated in a pressure-demand or other positive-pressure mode); or SAF:PD,PP:ASCBA (any supplied-air respirator that has a full facepiece and is operated in a pressure-demand or other positive-pressure mode in combination with an auxiliary self-contained breathing apparatus operated in a pressure-demand or other positive pressure mode). *Escape*: GMFOVHiE [any air-purifying, full-facepiece respirator (gas mask) with a chin-style, front- or back-mounted organic vapor canister having a high-efficiency particulate filter]; or SCBAE (any appropriate escape-type, self-contained breathing apparatus).

Storage: Prior to working with 2,4-D, you should be trained on its proper handling and storage. Store in tightly closed containers in a dark, cool, well-ventilated area. Keep away from oxidizers, heat, and sunlight. A regulated, marked area should be established where this chemical is stored in compliance with OSHA standard 1910.1045.

Shipping: Phenoxy pesticides, solid, toxic, n.o.s. fall in DOT Hazard Class 6.1 and 2,4-D in Packing Group III. The label required is "Keep Away From Food." A DOT regulated marine pollutant.

Spill Handling: Evacuate persons not wearing protective equipment from area of spill or leak until clean-up is complete. Remove all ignition sources. Collect powdered material in the most convenient and safe manner and deposit in sealed containers. Ventilate area after clean-up is complete. It may be necessary to contain and dispose of this chemical as a hazardous waste. If material or contaminated runoff enters waterways, notify downstream users of potentially contaminated waters. Contact your Department of Environmental Protection or your regional office of the federal EPA for specific recommendations. If employees are required to clean-up spills, they must be properly trained and equipped. OSHA 1910.120(q) may be applicable.

Fire Extinguishing: 2,4-Dichlorophenoxyacetic acid ester may burn, but does not readily ignite. Use dry chemical, CO_2, water spray, or foam extinguishers. Poisonous gases are produced in fire including hydrogen chloride, carbon monoxide, and phosgene. If material or contaminated runoff enters waterways, notify downstream users of potentially contaminated waters. Notify local health and fire officials and pollution control agencies. From a secure, explosion-proof location, use water spray to cool exposed containers. If cooling streams are ineffective (venting sound increases in volume and pitch, tank discolors, or shows any signs of deforming), withdraw immediately to a secure position. If employees are expected to fight fires, they must be trained and equipped in OSHA 1910.156.

Disposal Method Suggested: Incineration of phenoxys is effective in 1 second at 1,800°F using a straight combustion process or at 900°F using catalytic combustion. Over 99% decomposition was reported when small amounts of 2,4-D were burned in a polyethylene bag. See reference[22] for additional detail. In accordance with 40CFR165 recommendations for the disposal of pesticides and pesticide containers. Must be disposed properly by following package label directions or by contacting your state pesticide or environmental control agency or by contacting your regional EPA office. Consult with environmental regulatory agencies for guidance on acceptable disposal practices. Generators of waste containing this contaminant (≥100 kg/mo) must conform with EPA regulations governing storage, transportation, treatment, and waste disposal.

References

U.S. Environmental Protection Agency, 2,4-Dichlorophenoxy Acetic Acid, Health and Environmental Effects Profile No. 77, Washington, DC, Office of Solid Waste (April 30, 1980)

New Jersey Department of Health, "Hazardous Substance Fact Sheet: 2,4-Dichlorophenoxyacetic Acid Ester," Trenton, NJ (March 1987)

New York State Department of Health, "Chemical Fact Sheet: 2,4-D," Albany, NY, Bureau of Toxic Substance Assessment (March 1986 and Version 2)

Sax N. I., Ed., "Dangerous Properties of Industrial Materials Report," 1, No. 6, 49-52 (1981) and 7, No. 6, 11-46 (1987)

Dacarbazine

Molecular Formula: $C_6H_{10}N_6O$

Synonyms: Deticene; DIC; (Dimethyltriazeno)imidazole-carboxamide; 4-(3,3-Dimethyl-1-triazeno)imidazole-5-carboxamide; 4-(5)-(3,3-Dimethyl-1-triazeno)imidazole-5(4)-carboxamide; 4-(Dimethyltriazeno)imidazole-5-carboxamide; 5-(3,3-Dimethyl-1-triazeno)imidazole-4-carboxamide; 5-(3,3-Dimethyltriazeno)imidazole-4-carboxamide; 5-(Dimethyltriazeno)imidazole-4-carboxamide; 5-(3,3-Dimethyl-1-triazenyl)-

1H-imidazole-4-carboxamide; DTIC; DTIC-Dome; NCI-C04717; NSC-45388

CAS Registry Number: 4342-03-4

RTECS Number: NI3950000

DOT ID: UN 3249; UN 2811 (medicines, toxic, solid, n.o.s.)

Regulatory Authority

- Carcinogen (Animal Positive, Human Suspected) (IARC) (NCI) (NTP)[9]

Cited in U.S. State Regulations: California (A, G), Florida (G), Illinois (G), Maine (G), Massachusetts (G), Minnesota (G), New Jersey (G), Pennsylvania (G), Rhode Island (G).

Description: Dacarbazine, $C_6H_{10}N_6O$, is an ivory-colored crystalline solid. Freezing/Melting point = reported at 205°C.

Potential Exposure: Dacarbazine is used in cancer chemotherapy. Dacarbazine is used as an antineoplastic agent in the treatment of certain skin cancers and is occasionally used in the therapy of other neoplastic diseases which have become resistant to alternative treatment.

Health professionals who handle this drug (for example, pharmacists, nurses, and physicians) may possibly be exposed during drug preparation, administration, or cleanup; however, the risks can be avoided through use of appropriate containment equipment and work practices.[10] People receiving dacarbazine in treatment are also exposed.

Incompatibilities: Strong oxidizers. Explosive decomposition reported at 250 – 255°C.

Permissible Exposure Limits in Air: No standards set.

Permissible Concentration in Water: No criteria set.

Harmful Effects and Symptoms

Long Term Exposure: There is sufficient evidence that dacarbazine is carcinogenic in experimental animals. Rats given dacarbazine orally or intraperitoneally developed tumors of the breast, thymus, spleen, and brain in a minimum of 18 weeks after initial exposure. Intraperitoneal administration of dacarbazine to mice produced tumors in the lung, blood-producing tissue, and uterus.

First Aid: *Skin Contact:*[52] Flood all areas of body that have contacted the substance with water. Don't wait to remove contaminated clothing; do it under the water stream. Use soap to help assure removal. Isolate contaminated clothing when removed to prevent contact by others.

Eye Contact: Remove any contact lenses at once. Flush eyes well with copious quantities of water or normal saline for at least 20 – 30 minutes. Seek medical attention.

Inhalation: Leave contaminated area immediately; breathe fresh air. Proper respiratory protection must be supplied to any rescuers. If coughing, difficult breathing or any other symptoms develop, seek medical attention at once, even if symptoms develop many hours after exposure.

Ingestion: If convulsions are not present, give a glass or two of water or milk to dilute the substance. Assure that the person's airway is unobstructed and contact a hospital or poison center immediately for advice on whether or not to induce vomiting.

Personal Protective Methods: Wear protective gloves and clothing to prevent any reasonable probability of skin contact. Safety equipment suppliers/manufacturers can provide recommendations on the most protective glove/clothing material for your operation.. All protective clothing (suits, gloves, footwear, headgear) should be clean, available each day, and put on before work. Contact lenses should not be worn when working with this chemical. Wear dust-proof chemical goggles and face shield unless full facepiece respiratory protection is worn. Employees should wash immediately with soap when skin is wet or contaminated. Provide emergency showers and eyewash.

Respirator Selection: *At any detectable concentration:* SCBAF:PD,PP (any MSHA/NIOSH approved self-contained breathing apparatus that has a full facepiece and is operated in a pressure-demand or other positive-pressure mode); or SAF:PD,PP:ASCBA (any supplied-air respirator that has a full facepiece and is operated in a pressure-demand or other positive-pressure mode in combination with an auxiliary, self-contained breathing apparatus operated in a pressure-demand or other positive pressure mode). *Escape:* HiEF (any air-purifying, full-facepiece respirator (gas mask) with a chin-style, front- or back-mounted organic vapor canister having a high-efficiency particulate filter); or SCBAE (any appropriate escape-type, self-contained breathing apparatus).

Storage: Prior to working with dacarbazine you should be trained on its proper handling and storage. Store in a refrigerator or a cool, dry place and protect from light. A regulated, marked area should be established where this chemical is stored in compliance with OSHA standard 1910.1045.

Shipping: Medicines, solid, toxic, n.o.s. DOT Hazard Class 6.1and Packing Group II. Label "Poison."

Spill Handling: Evacuate persons not wearing protective equipment from area of spill or leak until clean-up is complete. Remove all ignition sources. Dampen spilled material with alcohol to avoid dust. Collect powdered material in the most convenient and safe manner and deposit in sealed containers. Ventilate area after clean-up is complete. It may be necessary to contain and dispose of this chemical as a hazardous waste. If material or contaminated runoff enters waterways, notify downstream users of potentially contaminated waters. Contact your Department of Environmental Protection or your regional office of the federal EPA for specific recommendations. If employees are required to clean-up spills, they must be properly trained and equipped. OSHA 1910.120(q) may be applicable.

Fire Extinguishing: Use dry chemical, carbon dioxide, water spray, or alcohol foam extinguishers. Poisonous gases are produced in fire including nitrogen oxides. If material or contaminated runoff enters waterways, notify downstream users of potentially contaminated waters. Notify local health and fire officials and pollution control agencies. From a secure, explosion-proof location, use water spray to cool exposed containers. If cooling streams are ineffective (venting sound increases in volume and pitch, tank discolors, or shows any signs of deforming), withdraw immediately to a secure position. If employees are expected to fight fires, they must be trained and equipped in OSHA 1910.156.

Daminozide

Molecular Formula: $C_6H_{12}N_2O_3$

Common Formula: $(CH_3)_2NNHCOCH_2CH_2COOH$

Synonyms: Alar®; Alar®-85; Aminozide®; B-9; B995; Bernsteinsaeure-2,2-dimethylhydrazid (German); B-Nine; Butanedioic acid mono(2,2-dimethylhydrazide); Daminozide (USDA); Dimas; *N*-Dimethyl amino-β-carbamyl propionic acid; *N*-(Dimethylamino)succinamic acid; *N*-Dimethylamino-succinamidsaeure (German); DMASA; DMSA; Kylar; NCI-C03827; SADH; Succinic acid 2,2-dimethylhydrazide; Succinic-1,1-dimethyl hydrazide

CAS Registry Number: 1596-84-5

RTECS Number: WM9625000

DOT ID: NA 9188

Regulatory Authority

- Carcinogen (Animal Positive) (NCI) (NTP)[9]

Cited in U.S. State Regulations: California (A, G), Massachusetts (G), New Jersey (G), Pennsylvania (G).

Description: Daminozide, $(CH_3)_2NNHCOCH_2CH_2COOH$, is a colorless crystalline solid. Freezing/Melting point = 157 – 164°C.

Potential Exposure: Daminozide is a plant growth regulator used on certain fruit (especially apples) to improve the balance between growth and fruit production and to improve fruit quality and synchronize maturity. U.S. sales were halted in 1989 because of health considerations.

Permissible Exposure Limits in Air: No standards set.

Permissible Concentration in Water: No criteria set.

Harmful Effects and Symptoms

The acute oral LD_{50} for rats is 8,400 mg/kg (insignificantly toxic). However, Daminozide metabolizes to diamethylhydrazine which is a proven carcinogen in animal tests. In has a low dermal irritation potential and it is neither teratogenic nor mutagenic.[55] It is not an acute toxicant to fish or wildlife.

First Aid: *Skin Contact:*[52] Flood all areas of body that have contacted the substance with water. Don't wait to remove contaminated clothing; do it under the water stream. Use soap to help assure removal. Isolate contaminated clothing when removed to prevent contact by others.

Eye Contact: Remove any contact lenses at once. Flush eyes well with copious quantities of water or normal saline for at least 20 – 30 minutes. Seek medical attention.

Inhalation: Leave contaminated area immediately; breathe fresh air. Proper respiratory protection must be supplied to any rescuers. If coughing, difficult breathing or any other symptoms develop, seek medical attention at once, even if symptoms develop many hours after exposure.

Ingestion: If convulsions are not present, give a glass or two of water or milk to dilute the substance. Assure that the person's airway is unobstructed and contact a hospital or poison center immediately for advice on whether or not to induce vomiting.

Personal Protective Methods: Wear protective gloves and clothing to prevent any reasonable probability of skin contact. Safety equipment suppliers/manufacturers can provide recommendations on the most protective glove/clothing material for your operation.. All protective clothing (suits, gloves, footwear, headgear) should be clean, available each day, and put on before work. Contact lenses should not be worn when working with this chemical. Wear dust-proof chemical goggles and face shield unless full facepiece respiratory protection is worn. Employees should wash immediately with soap when skin is wet or contaminated. Provide emergency showers and eyewash.

Respirator Selection: *At any detectable concentration:* SCBAF:PD,PP (any MSHA/NIOSH approved self-contained breathing apparatus that has a full facepiece and is operated in a pressure-demand or other positive-pressure mode); or SAF:PD,PP:ASCBA (any supplied-air respirator that has a full facepiece and is operated in a pressure-demand or other positive-pressure mode in combination with an auxiliary, self-contained breathing apparatus operated in a pressure-demand or other positive pressure mode). *Escape:* HiEF (any air-purifying, full-facepiece respirator (gas mask) with a chin-style, front- or back-mounted organic vapor canister having a high-efficiency particulate filter); or SCBAE (any appropriate escape-type, self-contained breathing apparatus).

Storage: Prior to working with Daminozide you should be trained on its proper handling and storage. Store in a cool, dry place or in a refrigerator. A regulated, marked area should be established where this chemical is handled, used, or stored in compliance with OSHA standard 1910.1045.

Shipping: Daminozide may be classified as a hazardous substance, solid, n.o.s. which falls in Hazard Class ORM-E, and Packing Group III which impose no label requirements or limits on quantities shipped by air or passenger railcar.

Spill Handling: Evacuate persons not wearing protective equipment from area of spill or leak until clean-up is complete. Remove all ignition sources. Collect powdered material in the most convenient and safe manner and deposit in sealed containers. Ventilate area after clean-up is complete. It may be necessary to contain and dispose of this chemical as a hazardous waste. If material or contaminated runoff enters waterways, notify downstream users of potentially contaminated waters. Contact your Department of Environmental Protection or your regional office of the federal EPA for specific recommendations. If employees are required to clean-up spills, they must be properly trained and equipped. OSHA 1910.120(q) may be applicable.

Fire Extinguishing: Use dry chemical, carbon dioxide, water spray, or alcohol foam extinguishers. Poisonous gases are produced in fire including nitrogen oxides. If material or contaminated runoff enters waterways, notify downstream users of potentially contaminated waters. Notify local health and fire officials and pollution control agencies. From a secure, explosion-proof location, use water spray to cool exposed containers. If cooling streams are ineffective (venting sound increases in volume and pitch, tank discolors, or shows any signs of deforming), withdraw immediately to a secure position. If employees are expected to fight fires, they must be trained and equipped in OSHA 1910.156.

Daunomycin

Molecular Formula: $C_{27}H_{29}NO_{10}$

Synonyms: 13,057 R.P.; Acetyladriamycin; Cerubidin; Daunamycin; Daunomicina (Spanish); Daunorubicin; Daunorubicine; DM; FI6339; Leukaemmycin C; 5,12-Naphthacenedione, 8-Acetyl-10-(3-amino-2,3,6-trideoxy-α-L-lyxohexopyranosyl)oxy-7,8,9,10-tetrahydro-6,8,11-trihydroxy-1-methoxy-, (8-cis)-; NCI-C04693; NSC-82151; RP 13057; Rubidomycin; Rubidomycine; Rubomycin C; Rubomycin C-1; Streptomyces

CAS Registry Number: 20830-81-3

RTECS Number: HB7875000

DOT ID: UN 3077

Regulatory Authority

- Carcinogen (animal positive) (IARC) (OSHA)
- EPA Hazardous Waste Number (RCRA No.): U059
- RCRA, 40CFR261, Appendix 8 Hazardous Constituents
- RCRA Land Ban Waste Restrictions
- Superfund/EPCRA 40CFR302.4 Reportable Quantity (RQ): CERCLA, 10 lb (4.54 kg)

Cited in U.S. State Regulations: California (A, G), Florida (G), Massachusetts (G), Minnesota (G), New Jersey (G), Pennsylvania (G)

Description: Daunomycin is a thin red, needle-shaped material. Freezing/Melting point = 190°C (decomposes). Soluble in water.

Potential Exposure: An antibiotic. It is used as a medicine for treating cancer.

Permissible Exposure Limits in Air: No limits set.

Harmful Effects and Symptoms

Short Term Exposure: Poisonous if swallowed.

Long Term Exposure: It may cause cancer in humans and may be teratogenic.

First Aid: If this chemical gets into the eyes, remove any contact lenses at once and irrigate immediately for at least 15 minutes, occasionally lifting upper and lower lids. Seek medical attention immediately. If this chemical contacts the skin, remove contaminated clothing and wash immediately with soap and water. Seek medical attention immediately. If this chemical has been inhaled, remove from exposure, begin rescue breathing (using universal precautions) if breathing has stopped and CPR if heart action has stopped. Transfer promptly to a medical facility. When this chemical has been swallowed, get medical attention. Give large quantities of water and induce vomiting. Do not make an unconscious person vomit.

Personal Protective Methods: Wear protective gloves and clothing to prevent any reasonable probability of skin contact. Safety equipment suppliers/manufacturers can provide recommendations on the most protective glove/clothing material for your operation.. All protective clothing (suits, gloves, footwear, headgear) should be clean, available each day, and put on before work. Contact lenses should not be worn when working with this chemical. Wear dust-proof chemical goggles and face shield unless full facepiece respiratory protection is worn. Employees should wash immediately with soap when skin is wet or contaminated. Provide emergency showers and eyewash.

Respirator Selection: Where the potential for exposure to this chemical, use a MSHA/NIOSH approved supplied-air respirator with a full facepiece operated in the positive pressure mode or with a full facepiece, hood, or helmet in the continuous flow mode, or use a MSHA/NIOSH approved self-contained breathing apparatus with a full facepiece operated in pressure-demand or other positive pressure mode.

Storage: Prior to working with Daunomycin you should be trained on its proper handling and storage. A regulated, marked area should be established where this chemical is handled, used, or stored in compliance with OSHA standard 1910.1045. Store in tightly closed containers in a cool, well ventilated area.

Shipping: Environmentally hazardous solid, n.o.s. Hazard Class: 9. Label: "Class 9." Packing Group: III.

Spill Handling: Evacuate persons not wearing protective equipment from area of spill or leak until clean-up is complete. Remove all ignition sources. Collect powdered material in the most convenient and safe manner and deposit in sealed containers. Ventilate area after clean-up is complete. It may be necessary to contain and dispose of this chemical as a hazardous waste. If material or contaminated runoff enters waterways, notify downstream users of potentially contaminated waters. Contact your Department of Environmental Protection or your regional office of the federal EPA for specific recommendations. If employees are required to clean-up spills, they must be properly trained and equipped. OSHA 1910.120(q) may be applicable.

Fire Extinguishing: This chemical may burn but does not easily ignite. Use dry chemical, carbon dioxide, water spray, or alcohol foam extinguishers. Poisonous gases are produced in fire including nitrogen oxides. If material or contaminated runoff enters waterways, notify downstream users of potentially contaminated waters. Notify local health and fire officials and pollution control agencies. From a secure, explosion-proof location, use water spray to cool exposed containers. If cooling streams are ineffective (venting sound increases in volume and pitch, tank discolors, or shows any signs of deforming), withdraw immediately to a secure position. If employees are expected to fight fires, they must be trained and equipped in OSHA 1910.156.

Disposal Method Suggested: Consult with environmental regulatory agencies for guidance on acceptable disposal practices. Generators of waste containing this contaminant (≥100 kg/mo) must conform with EPA regulations governing storage, transportation, treatment, and waste disposal.

References

New Jersey Department of Health and Senior Services, "Hazardous Substance Fact Sheet, Daunomycin," Trenton, NJ (January 1999)

DDT

Molecular Formula: $C_{14}H_9Cl_5$

Synonyms: Agritan; Anofex; Arkotine; Azotox; Benzene, 1,1'-(2,2,2-Trichloroethylidene)bis(4-chloro); α,α-Bis(*p*-chlorophenyl)-beta,beta,beta-trichlorethane; 1,1-Bis-(*p*-chlorophenyl)-2,2,2-trichloroethane; 2,2-Bis(*p*-chlorophenyl)-1,1-trichloroethane; Bosan supra; Bovidermol; Chlorophenothan; α-Chlorophenothane; Chlorophenothane; Chlorophenotoxum; Citox; Clofenotane; *p,p'*-DDT; 4,4' DDT; Dedelo; Deoval; Detox; Detoxan; Dibovan; *p,p'*-Dichlorodiphenyltrichloroethane; 4,4'-Dichlorodiphenyltrichloroethane; Dichlorodiphenyltrichloroethane; Dichlorodiphenyl trichloroethane 2,2-bis(*p*-chlorophenyl)-1,1,1-trichloroethane; Diclorodifeniltricloroetano (Spanish); Dicophane; Didigam; Didimac; Diphenyltrichloroethane; Dodat®; Dykol®; ENT1,506; Estonate; Ethane, 1,1,1-Trichloro-2,2-bis(*p*-chlorophenyl)-; Genitox®; Gesafid®; Gesapon®; Gesarex®; Gesarol®; Guesarol®; Gyron®; Havero-Extra®; Ivoran®; Ixodex®; Kopsol®; Mutoxin; NA 2761; NCI-C00464; Neocid®; OMS 16; Parachlorocidum; PEB1; Pentachlorin; Pentech; Pzeidan; Rukseam; Santobane; 1,1,1-Trichloor-2,2-bis(4-chloorfenyl)-ethaan (Dutch); 1,1,1-Trichlor-2,2-bis(4-chlor-phenyl)-aethan (German); 1,1,1-Trichloro-2,2-bis(*p*-chlorophenyl)ethane; Trichlorobis(4-chlorophenyl)ethane; 1,1,1-Trichloro-2,2-di(4-chlorophenyl)-ethane; 1,1,1-Tricloro-2,2-bis(4-cloro-fenil)-etano (Italian); Zeidane; Zerdane

CAS Registry Number: 50-29-3

RTECS Number: KJ3325000

DOT ID: UN/NA 2761

EEC Number: 602-045-00-7

Regulatory Authority

- Carcinogen (Animal Positive) (IARC)[9]
- Banned or Severely Restricted (Many Countries) (UN)[13]
- Air Pollutant Standard Set (ACGIH)[1] (Australia) (DFG)[3] (HSE)[33] (Israel) (Mexico) (former USSR)[43] (OSHA)[58] (Several States)[60] (Several Canadian Provinces)
- Clean Water Act: Section 311 Hazardous Substances/RQ 40CFR117.3 (same as CERCLA, see below); 40CFR423, Appendix A, Priority Pollutants; Section 313 Water Priority Chemicals (57FR41331, 9/9/92); Toxic Pollutant (Section 401.15)
- EPA Hazardous Waste Number (RCRA No.): U061
- RCRA, 40CFR261, Appendix 8 Hazardous Constituents
- RCRA 40CFR268.48; 61FR15654, Universal Treatment Standards: Wastewater (mg/l), 0.0039; Nonwastewater (mg/kg), 0.087
- RCRA 40CFR264, Appendix 9; TSD Facilities Ground Water Monitoring List. Suggested test method(s) (PQL μg/l): 8080 (0.1); 8270 (10)
- Superfund/EPCRA 40CFR302.4 Reportable Quantity (RQ): CERCLA, 1 lb (0.454 kg)
- U.S. DOT Regulated Marine Pollutant (49CFR172.101, Appendix B), severe pollutant
- Canada, Drinking Water Quality, 0.03 mg/l MAC
- Mexico, Drinking Water Criteria, 0.001 mg/l

Cited in U.S. State Regulations: Alaska (G), California (A, G), Connecticut (A), Florida (G), Illinois (G, W), Kansas (G, A, W), Louisiana (G), Maine (G, W), Massachusetts (G), Michigan (G), Minnesota (W), Nevada (A), New Hampshire (G), New Jersey (G), New York (G), North Dakota (A), Oklahoma (G), Pennsylvania (G, A), Rhode Island (G), Vermont (G), Virginia (G, A), Washington (G), West Virginia (G), Wisconsin (G).

Description: DDT is a waxy solid or slightly off-white powder of indefinite melting point with a weak, chemical odor. Freezing/Melting point = 107 – 109°C. Flash point = 72 – 75°C. Hazard Identification (based on NFPA-704 M Rating System): Health 2, Flammability 2, Reactivity 0. Insoluble in water.

Potential Exposure: DDT is a low-cost broad-spectrum insecticide. However, following an extensive review of health and environmental hazards of the use of DDT, U.S. EPA decided to ban further use of DDT in December 1972. This decision was based on several properties of DDT that had been well evidenced:[1] DDT and its metabolites are toxicants with long-term persistence in soil and water;[2] it is widely dispersed by erosion, runoff and volatization; and[3] the low-water solubility and high lipophilicity of DDT result in concentrated accumulation of DDT in the fat of wildlife and humans which may be hazardous.

Incompatibilities: Contact with strong oxidizers may cause fire and explosion hazard. Incompatible with salts of iron or aluminum, and bases. Do not store in iron containers.

Permissible Exposure Limits in Air: The Federal limit (OSHA PEL)[58] and the ACGIH recommended TWA value is 1 mg/m^3 as is the Australian, Israeli, Mexican, DFG,[3] HSE,[33] and Canadian provincial (Alberta, BC, Ontario, Quebec) value. The HSE, Mexican and Alberta, and BC STEL value is 3 mg/m^3. The NIOSH recommended REL is 0.5 mg/m^3. The former USSR-UNEP/IRPTC project[43] has set a MAC in workplace air of 0.1 mg/m^3 and values for ambient air in residential areas of 0.005 mg/m^3 on a momentary basis and 0.001 mg/m^3 on a daily average basis. Several states have set guidelines or standards for DDT in ambient air[60] ranging from 1.8 μg/m^3 (Pennsylvania) to 2.38 μg/m^3 (Kansas) to 5.0 μg/m^3 (Connecticut) to 10.0 μg/m^3 (North Dakota) to 16.0 μg/m^3 (Virginia) to 24 μg/m^3 (Nevada).

Determination in Air: Collection on a filter, workup with isooctane, analysis by gas chromatography. See NIOSH Method #S-274.

Permissible Concentration in Water: To protect freshwater aquatic life – 0.0010 μg/l as a 24 hr average; never to exceed 1.1 μg/l. To protect saltwater aquatic life – 0.0010 μg/l as a 24 hr average; never to exceed 0.13 μg/l. To protect human health – preferably zero. An additional lifetime cancer risk of 1 in 100,000 is imposed by a level of 0.24 ng/l (0.00024 μg/l).

Various states have set guidelines and standards for DDT in drinking water[61] ranging from guidelines of 0.42 μg/l (Kansas) to 0.83 μg/l (Maine) to 1.0 μg/l (Minnesota) and a standard of 50 μg/l (Illinois). The former USSR has set a MAC of 0.1 mg/l in water used for domestic purposes and zero in surface water for fishing.[35]

Canada has set a water quality MAC of 0.03 mg/l. Mexico[35] has set an MPC of 0.001 mg/l in drinking water supply; of 0.006 mg/l in estuaries and 0.6 μg/l in coastal waters.

Determination in Water: Gas chromatography (EPA Method 608) or gas chromatography plus mass spectrometry (EPA Method 625).

Routes of Entry: Inhalation, skin absorption, ingestion, eye and skin contact.

Harmful Effects and Symptoms

Short Term Exposure: *Inhalation:* Can cause irritation. 500 – 4,200 mg/m^3 has produced dizziness.

Skin: Can cause irritation in very high concentrations. DDT can be absorbed through the skin if dissolved in vegetable oils or other solvents.

Eyes: Can cause irritation.

Ingestion: 1/30 – 1/4 ounce has caused nausea, vomiting, headache and convulsions. Other symptoms include weakness, restlessness, dizziness, incoordination, numbness of face and extremities, abdominal pain, diarrhea, tremors, and death. Symptoms may be delayed from 1/2 – 3 hours. Estimated lethal dose is between 1 teaspoon and 1 ounce.

Can cause a prickling or tingling sensation in the mouth, tongue, lower face, nausea, vomiting, confusion, a sense of apprehension, weakness, loss of muscle control, and tremors, paresthesia tongue, lips, face; tremor; dizziness, confusion, malaise (vague feeling of discomfort), headache, fatigue; convulsions; paresis hands. High exposures can cause convulsions and death.

Long Term Exposure: DDT may cause liver and kidney damage. Prolonged or repeated exposure can cause irritation of the eyes, skin, and throat. Occupational exposure to DDT has been associated with changes in genetic material. DDT levels build up and stay in the body for long periods of time. Exposure to DDT and Aldrin may increase retention of DDT in the body. DDT causes cancer in laboratory animals. Whether it causes cancer in humans is unknown.

Points of Attack: Eyes, skin, central nervous system, kidneys, liver, peripheral nervous system. Cancer Site in animals: liver, lung, and lymphatic tumors.

Medical Surveillance: Serum DDT level. Urine *dichlorodiphenyl acetic acid* level. Liver and kidney function tests.

First Aid: If this chemical gets into the eyes, remove any contact lenses at once and irrigate immediately for at least 15 minutes, occasionally lifting upper and lower lids. Seek medical attention immediately. If this chemical contacts the skin, remove contaminated clothing and wash immediately with soap and water. Speed in removing material from skin is of extreme importance. Shampoo hair promptly if contaminated. Seek medical attention immediately. If this chemical has been inhaled, remove from exposure, begin rescue breathing (using universal precautions) if breathing has stopped and CPR if heart action has stopped. Transfer promptly to a medical facility. When this chemical has been swallowed, get medical attention. Give large quantities of water and induce vomiting. Do not make an unconscious person vomit.

Personal Protective Methods: Wear protective gloves and clothing to prevent any reasonable probability of skin contact. Safety equipment suppliers/manufacturers can provide recommendations on the most protective glove/clothing material for your operation.. All protective clothing (suits, gloves, footwear, headgear) should be clean, available each day, and put on before work. Contact lenses should not be worn when working with this chemical. Wear dust-proof chemical goggles and face shield unless full facepiece respiratory protection is worn. Employees should wash immediately with soap when skin is wet or contaminated. Provide emergency showers and eyewash.

Respirator Selection: NIOSH: *At any concentrations above the NIOSH REL, or where there is no REL, at any detectable concentration:* SCBAF:PD,PP (any self-contained breathing apparatus that has a full facepiece and is operated in a pressure-demand or other positive-pressure mode); or SAF:PD,PP:ASCBA (any supplied-air respirator that has a full facepiece and is operated in a pressure-demand or other positive-pressure mode in combination with an auxiliary self-contained breathing apparatus operated in a pressure-demand or other positive pressure mode). *Escape:* GMFOVHiE [any air-purifying, full-facepiece respirator (gas mask) with a chin-style, front- or back-mounted organic vapor canister having a high-efficiency particulate filter]; or SCBAE (any appropriate escape-type, self-contained breathing apparatus).

Storage: Prior to working with DDT you should be trained on its proper handling and storage. Store in tightly closed containers in a cool, well-ventilated area away from strong oxidizers, strong bases, and heat. Should not be stored in iron containers. A regulated, marked area should be established where this chemical is handled, used, or stored in compliance with OSHA standard 1910.1045.

Shipping: Organochlorine pesticides, solid, toxic, n.o.s. fall in DOT Hazard Class 6.1 and DDT in Packing Group III. The label required is "Keep Away From Food." The limit on passenger aircraft or railcar shipment is 100 kg; on cargo aircraft shipment is 200 kg.

Spill Handling: Evacuate persons not wearing protective equipment from area of spill or leak until clean-up is complete. Remove all ignition sources. Do not dry sweep. Use vacuum (use special HEPA vac, NOT a standard shop-vac) or a wet method to reduce dust during clean-up. Collect powdered material in the most convenient and safe manner and deposit in sealed containers. Ventilate area after clean-up is complete. It may be necessary to contain and dispose of this chemical as a hazardous waste. If material or contaminated runoff enters waterways, notify downstream users of potentially contaminated waters. Contact your Department of Environmental Protection or your regional office of the federal EPA for specific recommendations. If employees are required to clean-up spills, they must be properly trained and equipped. OSHA 1910.120(q) may be applicable.

Fire Extinguishing: This chemical is a combustible solid. Use dry chemical, carbon dioxide, water spray, or alcohol foam extinguishers. Poisonous gases are produced in fire including toxic chlorides. If material or contaminated runoff enters waterways, notify downstream users of potentially contaminated waters. Notify local health and fire officials and pollution control agencies. From a secure, explosion-proof location, use water spray to cool exposed containers. If cooling streams are ineffective (venting sound increases in volume and pitch, tank discolors, or shows any signs of deforming), withdraw immediately to a secure position. If employees are expected to fight fires, they must be trained and equipped in OSHA 1910.156.

Disposal Method Suggested: Incineration has been successfully used on a large scale for several years and huge incinerator equipment with scrubbers to catch HCl, a combustion product, are in use at several facilities such as Hooker Chemical, Dow Chemical and other producers of chlorinated hydrocarbon products. One incinerator operates at 900 – 1,400°C with air and steam added which precludes formation of Cl_2. A few companies also construct incinerator-scrubber combinations of smaller size, e.g., a system built by Garver-Davis, Inc., of Cleveland, Ohio, for the Canadian government can handle 200 – 500 lb DDT/day as a kerosene solution. In accordance with 40CFR165 recommendations for the disposal of pesticides and pesticide containers. Must be disposed properly by following package label directions or by contacting your state pesticide or environmental control agency or by contacting your regional EPA office. Consult with environmental regulatory agencies for guidance on acceptable disposal practices. Generators of waste containing this contaminant (≥100 kg/mo) must conform with EPA regulations governing storage, transportation, treatment, and waste disposal.

References

U.S. Environmental Protection Agency, DDT: Ambient Water Quality Criteria, Washington, DC (1980)

U.S. Environmental Protection Agency, DDT Health and Environmental Effects Profile No. 60, Washington, DC, Office of Solid Waste (April 30, 1980)

Sax N. I., Ed., "Dangerous Properties of Industrial Materials Report," 1, No. 3, 51-54 (1981) and 5, No. 1, 12-20 (1985)

New Jersey Department of Health and Senior Services, Hazardous Substance Fact Sheet, "DDT" Trenton, NJ (September 1996)

New York State Department of Health, "Chemical Fact Sheet: DDT." Albany, NY, Bureau of Toxic Substance Assessment (Mar. 1986 and Version 2)

Decaborane

Molecular Formula: $B_{10}H_{14}$

Synonyms: Boron hydride; Decaborane; Decaborano (Spanish); Decarboron tetradecahydride

CAS Registry Number: 17702-41-9

RTECS Number: HD1400000

DOT ID: UN 1868

Regulatory Authority

- Air Pollutant Standard Set (ACGIH)[1] (Australia) (DFG) (Israel) (Mexico) (OSHA)[58] (Several States)[60] (Several Canadian Provinces)
- Superfund/EPCRA 40CFR355, Appendix B Extremely Hazardous Substances: TPQ = 500/10,000 lb (227/4,540 kg)
- Superfund/EPCRA 40CFR302.4 Reportable Quantity (RQ): EHS, 1 lb (0.454 kg)
- Canada, WHMIS, Ingredients Disclosure List

Cited in U.S. State Regulations: Alaska (G), California (A, G), Connecticut (A), Florida (G), Illinois (G), Maine (G), Massachusetts (G), Minnesota (G), Nevada (A), New Hampshire

(G), New Jersey (G), North Dakota (A), Oklahoma (G), Pennsylvania (G), Rhode Island (G), Virginia (A), West Virginia (G).

Description: Decaborane, $B_{10}H_{14}$, is a colorless solid with a bitter odor. The odor threshold is 0.06 ppm. Boiling point = 213°C. Freezing/Melting point = 99.6°C. Flash point = 80°C (cc). Autoignition temperature = 149°C. Hazard Identification (based on NFPA-704 M Rating System): Health 3, Flammability 2, Reactivity 1. Very slightly soluble in cold water. Reacts with hot water.

Potential Exposure: Decaborane is used as a catalyst in olefin polymerization, in rocket propellants, in gasoline additives and as a vulcanizing agent for rubber.

Incompatibilities: May ignite SPONTANEOUSLY on exposure to air. Decomposes slowly in hot water. Incompatible with oxidizers, oxygenated solvents, dimethyl sulfoxide (reaction may be violent). Carbon tetrachloride, ethers, halocarbons, halogenated compounds form shock-sensitive mixtures. Attacks some plastics, rubber and coatings.

Permissible Exposure Limits in Air: The Federal limit (OSHA PEL)[58] and the NIOSH and ACGIH recommended TWA[1] is 0.05 ppm (0.3 mg/m^3). The STEL value from NIOSH and ACGIH is 0.15 ppm (0.9 mg/m^3). OSHA, NIOSH, and ACGIH add the notation "skin" indicating the possibility of cutaneous absorption. The NIOSH IDLH level is 15 mg/m^3. Australia, DFG, Israel, Mexico and Canadian provincial (Alberta, BC, Ontario, and Quebec) TWA level is the same as OSHA and Australia, Israel, Mexico, and Alberta, BC, Ontario, and Quebec STEL is 0.15 ppm. Several states have set guidelines or standards for decaborane in ambient air[60] ranging from 3-9 μg/m^3 (North Dakota) to 5 μg/m^3 (Virginia) to 6 μg/m^3 (Connecticut) and 7 μg/m^3 (Nevada).

Determination in Air: No method available.

Permissible Concentration in Water: No criteria set. (Decaborane hydrolyzes slowly in water.)

Routes of Entry: Inhalation, skin absorption, ingestion, eye and skin contact.

Harmful Effects and Symptoms

Short Term Exposure: Vapor exposure may cause clouding of the eyes with loss of vision. Contact can cause severe eye burns and may also irritate the skin. Sign and symptoms of acute exposure to decaborane may include tightness in the chest, dyspnea (shortness of breath), cough, and wheezing. Nausea and pulmonary edema may also occur. Neurological effects of acute exposure include dizziness, headache, weakness, incoordination, muscle spasms, tremor, and seizures. Exposure to decaborane may irritate or burn the skin, eyes, and mucous membranes. Exposure to decaborane may irritate or burn the skin, eyes, and mucous membranes.

Exposure can cause restlessness, headaches, dizziness, and nausea. High concentrations can cause muscle twitching, convulsions, unconsciousness, and death. High or repeated exposures may damage the liver and kidneys. Decaborane can be absorbed through the skin, thereby increasing exposure.

Long Term Exposure: May cause damage to the central nervous system, liver and kidneys.

Points of Attack: Central nervous system, liver, kidneys.

Medical Surveillance: Before beginning employment and at regular times after that, the following is recommended: Examination of the nervous system. If symptoms develop or overexposure is suspected, the following may be useful: Liver and kidney function tests. Exam of the eyes and vision.

First Aid: If this chemical gets into the eyes, remove any contact lenses at once and irrigate immediately for at least 15 minutes, occasionally lifting upper and lower lids. Seek medical attention immediately. If this chemical contacts the skin, remove contaminated clothing and wash immediately with soap and water. Seek medical attention immediately. If this chemical has been inhaled, remove from exposure, begin rescue breathing (using universal precautions) if breathing has stopped and CPR if heart action has stopped. Transfer promptly to a medical facility. When this chemical has been swallowed, get medical attention. Give large quantities of water and induce vomiting. Do not make an unconscious person vomit.

Personal Protective Methods: Wear protective gloves and clothing to prevent any reasonable probability of skin contact. Safety equipment suppliers/manufacturers can provide recommendations on the most protective glove/clothing material for your operation.. All protective clothing (suits, gloves, footwear, headgear) should be clean, available each day, and put on before work. Contact lenses should not be worn when working with this chemical. Wear dust-proof chemical goggles and face shield unless full facepiece respiratory protection is worn. Employees should wash immediately with soap when skin is wet or contaminated. Provide emergency showers and eyewash.

Respirator Selection: NIOSH/OSHA: *Up to 3 mg/m^3:* SA (any supplied-air respirator). *Up to 7.5 mg/m^3*: SA:CF (any supplied-air respirator operated in a continuous-flow mode). *Up to 15 mg/m^3*: SAT:CF (any supplied-air respirator that has a tight-fitting facepiece and is operated in a continuous-flow mode); or SCBAF (any self-contained breathing apparatus with a full facepiece); or SAF (any supplied-air respirator with a full facepiece). *Emergency or planned entry into unknown concentrations or IDLH conditions:* PD:PP (any self-contained breathing apparatus that has a full facepiece and is operated in a pressure-demand or other positive-pressure mode); or SAF:PD,PP:ASCUBA (any supplied-air respirator that has a full facepiece and is operated in a pressure-demand or other positive-pressure mode in combination with an auxiliary self-contained positive-pressure breathing apparatus). *Escape:* GMFOV [any air-purifying, full-facepiece respirator (gas mask) with a chin-style, front- or back-mounted organic vapor canister having a high-efficiency particulate filter]; or SCBAE (any appropriate escape-type, self-contained breathing apparatus).

Storage: Prior to working with decaborane you should be trained on its proper handling and storage. Decaborane must be stored to avoid contact with oxidizers, such as permanganates, nitrates, peroxides, chlorates, and perchlorates; or halogenated compounds, since violent reactions occur. Store in tightly closed containers in a cool well-ventilated area away from heat and water. Heat can cause an explosion. Contact with water can slowly produce flammable hydrogen gas. Detached storage is preferable. Sources of ignition such as smoking and open flames are prohibited where decaborane is handled, used or stored.

Shipping: Decaborane must be labeled: "Flammable, Solid, Poison." It falls in UN/DOT Hazard Class 4.1 and Packing Group II. Shipment by passenger aircraft or railcar is forbidden.

Spill Handling: Evacuate persons not wearing protective equipment from area of spill or leak until clean-up is complete. Remove all ignition sources. For small quantities, sweep into paper or other suitable material, place in appropriate container and burn in safe place (such as fume hood). Large quantities may be reclaimed. If reclamation is not practical, dissolve in flammable solvent (such as alcohol) and atomize in suitable combustion chamber.

Do not touch spilled material; stop leak if you can do so without risk. Use water spray to reduce vapors. Small spills: cover with water, sand or earth; shovel into metal container and keep material under water. Large spills: dike for later disposal and cover with wet sand or water. Clean up only under supervision of an expert. Ventilate area after clean-up is complete. It may be necessary to contain and dispose of this chemical as a hazardous waste. If material or contaminated runoff enters waterways, notify downstream users of potentially contaminated waters. Contact your Department of Environmental Protection or your regional office of the federal EPA for specific recommendations. If employees are required to clean-up spills, they must be properly trained and equipped. OSHA 1910.120(q) may be applicable.

Fire Extinguishing: This chemical is a combustible solid. The solid can self-ignite in oxygen, and mixtures with oxidizing materials can be explosive. Use dry chemical or CO_2 extinguishers. Avoid halogenated extinguishing agents, as they can react violently. Poisonous gases are produced in fire including hydrogen and boron oxide. If material or contaminated runoff enters waterways, notify downstream users of potentially contaminated waters. Notify local health and fire officials and pollution control agencies. Containers may explode in fire. From a secure, explosion-proof location, use water spray to cool exposed containers. If cooling streams are ineffective (venting sound increases in volume and pitch, tank discolors, or shows any signs of deforming), withdraw immediately to a secure position. If employees are expected to fight fires, they must be trained and equipped in OSHA 1910.156.

Disposal Method Suggested: Incineration with aqueous scrubbing of exhaust gases to remove B_2O_3 particulates.

References

Sax N. I., Ed., "Dangerous Properties of Industrial Materials Report," 1, No. 8, 64–65 (1981)

U.S. Environmental Protection Agency, "Chemical Profile: Decaborane,"[14] Washington, DC, Chemical Emergency Preparedness Program (November 30, 1987)

New Jersey Department of Health, "Hazardous Substance Fact Sheet: Decaborane," Trenton, NJ (February 1999)

Decabromodiphenyl Ether

Molecular Formula: $C_{12}Br_{10}O$

Synonyms: AFR 1021; Benzene, 1,1'-oxybis(2,3,4,5,6-pentabromo-); Berkflam B 10E; Bis(pentabromophenyl) ether; BR 55N; Bromkal 81; Bromkal 82-ODE; Bromkal 83-1ODE; DE83; DE 83R; Decabrom; Decabromobiphenyl ether; Decabromobiphenyl oxide; Decabromodiphenyl oxide (EPA); Decabromophenyl ether; EB 10FP; EBR 700; Ether, bis(pentabromophenyl); FR 300; FR 300BA; FRP 53; FR-PE; 1,1'-Oxybis(2,3,4,5,6-pentabromobenzene) (9CI); Pentabromophenyl ether; Planelon DB 100; Saytex 102; Saytex 102E; Tardex 100

CAS Registry Number: 1163-19-5

RTECS Number: KN3525000

DOT ID: UN 3077 (environmentally hazardous substances, n.o.s.)

Regulatory Authority

- EPCRA Section 313 Form R *de minimis* concentration reporting level: 1.0%
- Canada National Pollutant Release Inventory (NPRI)

Cited in U.S. State Regulations: California (A, G), Florida (G), Massachusetts (G), Minnesota (G), New Jersey (G), Pennsylvania (G).

Description: DBDPO is a white to off white, powder. Hazard Identification (based on NFPA-704 M Rating System): Health 2, Flammability 0, Reactivity 0. Slightly soluble in water.

Potential Exposure: It is used as a fire retardant for thermoplastics and man-made fibers.

Incompatibilities: Contact with strong oxidizers may cause a fire and explosion hazard. Ethers have a tendency to form unstable and explosive peroxides.

Permissible Exposure Limits in Air: The American Industrial Hygiene Association recommends a WEEL of 5 mg/m^3. No other exposure limits have been established.

Routes of Entry: Inhalation and through the skin.

Harmful Effects and Symptoms

Short Term Exposure: This chemical can be absorbed through the skin, thereby increasing exposure. DBDPO irritates the eyes, skin and respiratory tract.

Long Term Exposure: This chemical can accumulate in the body and may cause liver damage. Thyroid enlargement (goiter) may occur. There is limited evidence that this chemical causes cancer in animals; it may cause liver cancer (NTP). Some closely related polybrominated biphenyl compounds have been shown to damage the developing fetus.

Points of Attack: Liver, thyroid.

Medical Surveillance: Liver and thyroid function tests.

First Aid: If this chemical gets into the eyes, remove any contact lenses at once and irrigate immediately for at least 15 minutes, occasionally lifting upper and lower lids. Seek medical attention immediately. If this chemical contacts the skin, remove contaminated clothing and wash immediately with soap and water. Seek medical attention immediately. If this chemical has been inhaled, remove from exposure, begin rescue breathing (using universal precautions) if breathing has stopped and CPR if heart action has stopped. Transfer promptly to a medical facility. When this chemical has been swallowed, get medical attention. Give large quantities of water and induce vomiting. Do not make an unconscious person vomit.

Personal Protective Methods: Wear protective gloves and clothing to prevent any reasonable probability of skin contact. Safety equipment suppliers/manufacturers can provide recommendations on the most protective glove/clothing material for your operation.. All protective clothing (suits, gloves, footwear, headgear) should be clean, available each day, and put on before work. Contact lenses should not be worn when working with this chemical. Wear dust-proof chemical goggles and face shield unless full facepiece respiratory protection is worn. Employees should wash immediately with soap when skin is wet or contaminated. Provide emergency showers and eyewash.

Respirator Selection: Where the potential for exposure to this chemical, use a MSHA/NIOSH approved supplied-air respirator with a full facepiece operated in the positive pressure mode or with a full facepiece, hood, or helmet in the continuous flow mode, or use a MSHA/NIOSH approved self-contained breathing apparatus with a full facepiece operated in pressure-demand or other positive pressure mode.

Storage: Prior to working with DPDPO you should be trained on its proper handling and storage. Store in tightly closed containers in a cool, well ventilated area away from strong oxidizers. Where possible, automatically transfer this chemical from drums or other storage containers to process containers. Use only non-sparking tools and equipment, especially when opening and closing containers of this chemical. Sources of ignition such as smoking and open flames, are prohibited where this chemical is used, handled, or stored in a manner that could create a potential fire or explosion hazard. Wherever this chemical is used, handled, manufactured, or stored, use explosion-proof electrical equipment and fittings.

Shipping: This chemical is on the Community Right-to-Know list and is an environmentally hazardous substance, solid, n.o.s. It is in Hazard Class 9 and Packing Group III.

Spill Handling: Evacuate persons not wearing protective equipment from area of spill or leak until clean-up is complete. Remove all ignition sources. Collect powdered material in the most convenient and safe manner and deposit in sealed containers. Do not dry sweep. Ventilate area after clean-up is complete. It may be necessary to contain and dispose of this chemical as a hazardous waste. If material or contaminated runoff enters waterways, notify downstream users of potentially contaminated waters. Contact your Department of Environmental Protection or your regional office of the federal EPA for specific recommendations. If employees are required to clean-up spills, they must be properly trained and equipped. OSHA 1910.120(q) may be applicable.

Fire Extinguishing: This chemical may burn but does not easily ignite. Use dry chemical, carbon dioxide, water spray, or foam extinguishers. Poisonous gases are produced in fire including hydrogen bromide. If material or contaminated runoff enters waterways, notify downstream users of potentially contaminated waters. Notify local health and fire officials and pollution control agencies. From a secure, explosion-proof location, use water spray to cool exposed containers. If cooling streams are ineffective (venting sound increases in volume and pitch, tank discolors, or shows any signs of deforming), withdraw immediately to a secure position. If employees are expected to fight fires, they must be trained and equipped in OSHA 1910.156.

References

New Jersey Department of Health and Senior Services, "Hazardous Substance Fact Sheet, Decabromodiphenyl Ether," Trenton NJ (May 1998)

Demeton

Molecular Formula: $C_8H_{19}O_3PS_2$

Synonyms: *demeton:* Bay 10756; Bayer 10756; Bayer 8169; Demetona (Spanish); Demeton-O + Demeton-S; Demox; Denox; Diethoxy thiophosphoric acid ester of 2-ethylmercaptoethanol; *O,O*-Diethyl-2-ethylmercaptoethyl thiophosphate, diethoxythiophosphoric acid; *O,O*-Diethyl *S*-2-(ethylthio)ethyl phosphorothioate mixed with phosphorothioic acid, *O,O*-diethyl *O*-2-(ethylthio)ethyl ester; E-1059; ENT 17295; Mercaptophos (in former USSR); Phosphorothioic acid,*O,O*-diethyl *O*-2-(ethylthio)ethyl ester, mixed with *O,O*-diethyl *S*-2-(ethylthio)ethyl phosphorothioate; Systemox; Systox; UL

demeton-S: Bay 18436; Bayer 18436; Bayer 25/154; Demeton-*S*-metile (Italian); *O,O*-Dimethyl-*S*-(2-aethtyl-thio-aethyl)-

monothiophosphat (German); *O,O*-Dimethyl *S*-(2-eththio-ethyl) phosphorothioate; *O,O*-Dimethyl *S*-[2-(eththio)ethyl] phosphorthioate; Dimethyl *S*-(2-eththioethyl) thiophosphate; *O,O*-Dimethyl *S*-ethylmercaptoethyl thiophosphate; *O,O*-Dimethyl *S*-ethylmercaptoethyl thiophosphate, thiolo isomer; *O,O*-Dimethyl-*S*-(*S*-ethylthio-ethyl)-monothiofosfaat (Dutch); *O,O*-Dimethyl-*S*-(3-thia-pentyl)-monothiophosphat (German); *O,O*-Dimetil-*S*-(2-etilito-etil)-monotiofosfato (Italian); Duratox; *S*-[2-(Ethyl-thio)ethyl] *O,O*-dimethyl phosphorothioate; *S*-[2-(Ethyl-thio)ethyl]dimethyl phosphorothiolate; *S*-[2-(Ethyl-thio)ethyl] *O,O*-dimethyl thiophosphosphate; Isometa-systox; Isomethylsystox; Metaisoseptox; Metaisosystox; Metasystox Forte; Methyl demeton thioester; Methyl isosystox; Methyl-mercaptofos teolovy; Thiophosphate de *O,O*-dimethyle et de *S*-2-ethylthioethyle (French)

CAS Registry Number: 8065-48-3 (mixture); 298-03-3 (demeton-O); 126-75-0 (demeton-S)

RTECS Number: TF3150000 (mixture); TF3125000 (demeton-O); FT3130000 (demeton-S)

DOT ID: UN 3017 (mixture); UN 2783 (demeton-O); UN 2783 (demeton-S)

EEC Number: 015-031-00-5 (demeton-S); 015-030-00-X (demeton-O)

Regulatory Authority

- Banned or Severely Restricted (In Agriculture) (Germany and Russia)[13]
- Very Toxic Substance (World Bank)[15]
- Air Pollutant Standard Set (ACGIH)[1] (Australia) (DFG)[3] (Israel) (Mexico) (OSHA)[58] (Several States)[60] (Several Canadian Provinces)

For demeton and demeton-S:

- Superfund/EPCRA 40CFR355, Appendix B Extremely Hazardous Substances: TPQ = 500 lb (227 kg)
- Superfund/EPCRA 40CFR302.4 Reportable Quantity (RQ): EHS, 1 lb (0.454 kg)

Cited in U.S. State Regulations: Alaska (G), California (A, G), Connecticut (A), Florida (G), Illinois (G), Maine (G), Massachusetts (G), Michigan (G), Minnesota (G), Nevada (A), New Hampshire (G), New Jersey (G), New York (G), North Dakota (A), Oklahoma (G), Pennsylvania (G), Rhode Island (G), West Virginia (G).

Description: Is a light brown liquid with an odor of sulfur compounds. Freezing/Melting point ≤-13°C. Boiling point = 134°C at 2 mm Hg. Flash point = 45°C. Explosive limits: LEL = 1%; UEL = 5.3%. Hazard Identification (based on NFPA-704 M Rating System): Health 3, Flammability 2, Reactivity 0. Insoluble in water.

Potential Exposure: Those involved in the manufacture, formulation, and application of this systemic insecticide and acaricide.

Incompatibilities: Forms explosive mixture with air. Strong oxidizers, strong bases, soluble mercury, other pesticides, and water.

Permissible Exposure Limits in Air: The Federal limit (OSHA PEL)[58] is 0.1 mg/m^3. The ACGIH recommended TWA value is 0.01 ppm (0.11 mg/m^3). The DFG, Australian, Israeli, Mexican, and Canadian provincial (Alberta, BC, Ontario, Quebec) TWA value[3] is the same as OSHA and Mexico, Alberta, and BC set a STEL of 0.03 ppm (0.3 mg/m^3). OSHA, NIOSH and ACGIH add the notation "skin" indicating the possibility of cutaneous absorption. The NIOSH IDLH level is 20 mg/m^3. The MAC in workplace air set in the former USSR is 0.02 mg/m^3.[35][43] Brazil has set a MAC in workplace air of 0.08 mg/m^3.[35] States which have set guidelines or standards for Demeton in ambient air[60] include North Dakota at 1.0 μg/m^3 and Connecticut and Nevada at 2.0 μg/m^3.

Determination in Air: Filter/XAD-2® (tube); Toluene; Gas chromatography/Flame photometric detection for sulfur, nitrogen, or phosphorus; NIOSH IV Method #5514.

Permissible Concentration in Water: A MAC of 0.01 mg/l in water bodies used for domestic purposes has been set in the former USSR.[35][43]

Routes of Entry: Inhalation, skin absorption, ingestion, eye and skin contact.

Harmful Effects and Symptoms

Short Term Exposure: Demeton can be absorbed through the skin, thereby increasing exposure. Demeton may cause effects on the nervous system by cholinesterase inhibiting effect, causing convulsions, respiratory failure and possible death. High exposure (above OEL) may result in unconsciousness and death. Acute exposure to Demeton may produce the following symptoms of exposure: pinpoint pupils, blurred vision, headache, dizziness, muscle spasms, and profound weakness. Vomiting, diarrhea, abdominal pain, seizures, and coma may also occur. The heart rate may decrease following oral exposure or increase following dermal exposure. Chest pain may be noted. Hypotension (low blood pressure) may occur, although hypertension (high blood pressure) is not uncommon. Respiratory symptoms include dyspnea (shortness of breath), respiratory depression, and respiratory paralysis. Psychosis may occur.

This material is a cholinesterase inhibitor. It is readily absorbed through the skin and is extremely toxic. Probable human lethal oral dose is 5 – 50 mg/kg or 7 drops to 1 teaspoonful for 150 lb. person. Acute dose is believed to be 12 – 20 mg by oral route. The effects may be delayed. Medical observation is indicated.

Long Term Exposure: May cause mutations. May damage the developing fetus. May damage the nervous system, causing sensation of "pins and needles" in the hands and feet. May cause depression, irritability and personality changes. Cumulative effect is possible. Demeton may affect cholinesterase, causing significant depression of blood cholinesterase.

Points of Attack: Respiratory system, lungs, central nervous system, cardiovascular system, skin, eyes, plasma and red blood cell cholinesterase.

Medical Surveillance: Before employment and at regular times after that, the following are recommended: Plasma and red blood cell cholinesterase levels (tests for the enzyme poisoned by this chemical). If exposure stops, plasma levels return to normal in 1 – 2 weeks while red blood cell levels may be reduced for 1 – 3 months.

When cholinesterase enzyme levels are reduced by 25% or more below preemployment levels, risk of poisoning is increased, even if results are in lower ranges of "normal." Reassignment to work not involving organophosphate or carbamate pesticides is recommended until enzyme levels recover. If symptoms develop or overexposure occurs, repeat the above tests as soon as possible and get an exam of the nervous system. Also consider complete blood count. Consider chest x-ray following acute overexposure. Do not drink any alcoholic beverages before or during use. Alcohol promotes absorption of organic phosphates.

First Aid: If this chemical gets into the eyes, remove any contact lenses at once and irrigate immediately for at least 15 minutes, occasionally lifting upper and lower lids. Seek medical attention immediately. If this chemical contacts the skin, remove contaminated clothing and wash immediately with soap and water. Speed in removing material from skin is of extreme importance. Shampoo hair promptly if contaminated. Seek medical attention immediately. If this chemical has been inhaled, remove from exposure, begin rescue breathing (using universal precautions) if breathing has stopped and CPR if heart action has stopped. Transfer promptly to a medical facility. When this chemical has been swallowed, get medical attention. Give large quantities of water and induce vomiting. Do not make an unconscious person vomit. Effects may be delayed; medical observation is recommended.

Personal Protective Methods: Wear protective gloves and clothing to prevent any reasonable probability of skin contact. Safety equipment suppliers/manufacturers can provide recommendations on the most protective glove/clothing material for your operation.. All protective clothing (suits, gloves, footwear, headgear) should be clean, available each day, and put on before work. Contact lenses should not be worn when working with this chemical. Wear splash-proof chemical goggles and face shield unless full facepiece respiratory protection is worn. Employees should wash immediately with soap when skin is wet or contaminated. Provide emergency showers and eyewash.

Respirator Selection: O NIOSH/OSHA: *1 mg/m³*: SA (any supplied-air respirator). *2.5 mg/m³*: SA:CF (any supplied-air respirator operated in a continuous-flow mode). *5 mg/m³*: SAT:CF (any supplied-air respirator that has a tight-fitting facepiece and is operated in a continuous-flow mode); or SCBAF (any self-contained breathing apparatus with a full facepiece); or SAF (any supplied-air respirator with a full facepiece). 10 mg/m³: SA:PD,PP (any supplied-air respirator operated in a pressure-demand or other positive-pressure mode). *Emergency or planned entry into unknown concentrations or IDLH conditions:* SCBAF:PD,PP (any self-contained breathing apparatus that has a full faceplate and is operated in a pressure-demand or other positive-pressure mode); or SAF:PD,PP:ASCBA (any supplied-air respirator that has a full facepiece and is operated in a pressure-demand or other positive-pressure mode in combination with an auxiliary self-contained breathing apparatus operated in a pressure-demand or other positive pressure mode). *Escape:* GMFOVHiE [any air-purifying, full-facepiece respirator (gas mask) with a chin-style, front- or back-mounted organic vapor canister having a high-efficiency particulate filter]; or SCBAE (any appropriate escape-type, self-contained breathing apparatus).

Storage: Prior to working with demeton you should be trained on its proper handling and storage. A regulated, marked area should be established where this chemical is handled, used, or stored in compliance with OSHA standard 1910.1045. Before entering confined space where this chemical may be present, check to make sure that an explosive concentration does not exist. Store in tightly closed containers in a cool, well ventilated area away from oxidizers, strong bases, water, soluble mercury, and other pesticides. Where possible, automatically pump liquid from drums or other storage containers to process containers.

Shipping: Organophosphorus pesticides, liquid, toxic, flammable, n.o.s. with flash points above 23°C require a "Poison, Flammable Liquid" label in Packing Group II and Hazard Class 6.1.

Spill Handling: Evacuate and restrict persons not wearing protective equipment from area of spill or leak until cleanup is complete. Remove all ignition sources. Establish forced ventilation to keep levels below explosive limit. Absorb liquids in vermiculite, dry sand, earth, peat, carbon, or a similar material and deposit in sealed containers. Keep this chemical out of a confined space, such as a sewer, because of the possibility of an explosion, unless the sewer is designed to prevent the build-up of explosive concentrations. It may be necessary to contain and dispose of this chemical as a hazardous waste. If material or contaminated runoff enters waterways, notify downstream users of potentially contaminated waters. Contact your Department of Environmental Protection or your regional office of the federal EPA for specific recommendations. If employees are required to clean-up spills, they must be properly trained and equipped. OSHA 1910.120(q) may be applicable.

Fire Extinguishing: This chemical is a combustible liquid. Poisonous gases including sulfur oxides and carbon monoxide are produced in fire. Use dry chemical, carbon dioxide, or foam extinguishers. Vapors are heavier than air and will collect in low areas. Vapors may travel long distances to ignition sources and flashback. Vapors in confined areas may explode when exposed to fire. Containers may explode in fire. Storage containers and parts of containers may rocket great distances, in many directions. If material or contaminated runoff enters waterways, notify downstream users of potentially contaminated waters. Notify local health and fire officials and pollution control agencies. From a secure, explosion-proof location, use water spray to cool exposed containers. If cooling streams are ineffective (venting sound increases in volume and pitch, tank discolors, or shows any signs of deforming), withdraw immediately to a secure position. If employees are expected to fight fires, they must be trained and equipped in OSHA 1910.156.

Disposal Method Suggested: The thiono and thiolo isomers of this mixture are 50% hydrolyzed in 75 minutes and 0.85 minute, respectively at 20°C and pH 13. At pH 9 and 70°C, the half life of Demeton is 1.25 hr, but an pH 1 – 5 it is over 11 hr. Sand and rushed limestone may be added together with a flammable solvent; the resultant mixture may be burned in a furnace with afterburner and alkaline scrubber.[22] In accordance with 40CFR165 recommendations for the disposal of pesticides and pesticide containers. Must be disposed properly by following package label directions or by contacting your state pesticide or environmental control agency or by contacting your regional EPA office.

References

U.S. Environmental Protection Agency, "Chemical Profile: Demeton," Washington, DC, Chemical Emergency Preparedness Program (November 30, 1987)

New Jersey Department of Health, "Hazardous Substance Fact Sheet: Demeton," Trenton, NJ (April 1999)

New York State Department of Health, "Chemical Fact Sheet: Demeton," Albany, NY, Bureau of Toxic Substance Assessment (April 1986)

Demeton-Methyl

Molecular Formula: $C_6H_{15}O_3PS_2$

Synonyms: Bay 15203; Bayer 21/116; Demethon-methyl; Duratox®; ENT18,862; *S*(and *O*)-2-(Ethylthio)ethyl *O,O*-dimethyl phosphorothioate; Metasystox®; Methyl demeton; Methyl mercaptophos; Methyl systox; Phosphorothioic acid, *O*-2-(ethylthio)ethyl *O,O*-dimethyl ester mixed with *S*-2-(ethylthio)ethyl *O,O*-dimethyl phosphorothioate

CAS Registry Number: 8022-00-2

RTECS Number: TG1760000

DOT ID: UN2783

Regulatory Authority

- Banned or Severely Restricted (Restricted In Many Countries) (UN)[35]
- Air Pollutant Standard Set (DFG)[3] (ACGIH)

Cited in U.S. State Regulations: Florida (G), Massachusetts (G), Pennsylvania (G).

Description: Demeton-Methyl is a colorless to pale yellow oily liquid with an unpleasant odor. Hazard Identification (based on NFPA-704 M Rating System): Health 2, Flammability 2, Reactivity 0. Slightly soluble in water.

Potential Exposure: An organophosphate insecticide. Those engaged in the manufacture, formulation and application of the insecticide and acaricide on agricultural and horticultural crops.

Incompatibilities: Strong oxidizers such as chlorine, bromine, fluorine.

Permissible Exposure Limits in Air: ACGIH recommended TLV is 0.5 mg/m^3 TWA, with skin notation. The DFG MAK is 0.5 ppm (4.8 mg/m^3). This chemical can be absorbed through the skin, thereby increasing exposure.

Determination in Air: OSHA versatile sampler-2; Toluene/Acetone; Gas chromatography/Flame photometric detection for sulfur, nitrogen, or phosphorus; NIOSH Method IV Method #5600, Organophosphorus Pesticides.

Permissible Concentration in Water: No criteria set.

Routes of Entry: Inhalation and through the skin.

Harmful Effects and Symptoms

Short Term Exposure: Methyl demeton can be fatal by skin contact even if there is no feeling of irritation. Exposure can cause rapid, fatal organophosphate poisoning. Acute exposure to this chemical may produce the following signs and symptoms: pinpoint pupils, blurred vision, headache, dizziness, muscle spasms, and profound weakness, vomiting, diarrhea, abdominal pain, loss of coordination, seizures, coma and death. The heart rate may decrease following oral exposure or increase following dermal exposure. Hypotension (low blood pressure) is not uncommon. Respiratory symptoms include dyspnea (shortness of breath), respiratory depression, and respiratory paralysis. Psychosis may occur. Eye contact may cause irritation.

Long Term Exposure: May cause mutations. In animal studies this chemical causes a decrease in fertility and is toxic to the animal fetus. See also Demeton entry.

Points of Attack: Respiratory system, lungs, central nervous system, cardiovascular system, skin, eyes, plasma and red blood cell cholinesterase.

Medical Surveillance: Before employment and at regular times after that, the following are recommended: Plasma and red blood cell cholinesterase levels (tests for the enzyme poisoned by this chemical). If exposure stops, plasma levels return to normal in 1 – 2 weeks while red blood cell levels may be reduced for 1 – 3 months.

When cholinesterase enzyme levels are reduced by 25% or more below preemployment levels, risk of poisoning is increased, even if results are in lower ranges of "normal." Reassignment to work not involving organophosphate or carbamate pesticides is recommended until enzyme levels recover. If symptoms develop or overexposure occurs, repeat the above tests as soon as possible and get an exam of the nervous system. Also consider complete blood count. Consider chest x-ray following acute overexposure. Do not drink any alcoholic beverages before or during use. Alcohol promotes absorption of organic phosphates.

First Aid: If this chemical gets into the eyes, remove any contact lenses at once and irrigate immediately for at least 30 minutes, occasionally lifting upper and lower lids. Seek medical attention immediately. If this chemical contacts the skin, remove contaminated clothing and wash immediately with soap and water. Speed in removing material from skin is of extreme importance. Shampoo hair promptly if contaminated. Seek medical attention immediately. If this chemical has been inhaled, remove from exposure, begin rescue breathing (using universal precautions) if breathing has stopped and CPR if heart action has stopped. Transfer promptly to a medical facility. When this chemical has been swallowed, get medical attention. Give large quantities of water and induce vomiting. Do not make an unconscious person vomit. Effects may be delayed; medical observation is recommended.

Note to Physician: 1,1'-trimethylenebis(4-formylpyridinium bromide)dioxime (a.k.a TMB-4 DIBROMIDE and TMV-4) have been used as an antidote for organophosphate poisoning. Contact local poison control center for additional guidance.

Personal Protective Methods: Wear protective gloves and clothing to prevent any reasonable probability of skin contact. Safety equipment suppliers/manufacturers can provide recommendations on the most protective glove/clothing material for your operation.. All protective clothing (suits, gloves, footwear, headgear) should be clean, available each day, and put on before work. Contact lenses should not be worn when working with this chemical. Wear splash-proof chemical goggles and face shield unless full facepiece respiratory protection is worn. Employees should wash immediately with soap when skin is wet or contaminated. Provide emergency showers and eyewash.

Respirator Selection: See Demeton for guidance.

Storage: See Demeton for guidance.

Shipping: See Demeton for guidance.

Spill Handling: Evacuate and restrict persons not wearing protective equipment from area of spill or leak until cleanup is complete. Remove all ignition sources. Ventilate area of spill or leak. Absorb liquids in vermiculite, dry sand, earth, peat, carbon, or a similar material and deposit in sealed containers. Keep this chemical out of a confined space, such as a sewer, because of the possibility of an explosion, unless the sewer is designed to prevent the build-up of explosive concentrations. It may be necessary to contain and dispose of this chemical as a hazardous waste. If material or contaminated runoff enters waterways, notify downstream users of potentially contaminated waters. Contact your Department of Environmental Protection or your regional office of the federal EPA for specific recommendations. If employees are required to clean-up spills, they must be properly trained and equipped. OSHA 1910.120(q) may be applicable.

Fire Extinguishing: This chemical is a combustible liquid. Poisonous gases including oxides of phosphorus and sulfur are produced in fire. Use dry chemical, carbon dioxide, or alcohol foam extinguishers. Vapors are heavier than air and will collect in low areas. Containers may explode in fire. Storage containers and parts of containers may rocket great distances, in many directions. If material or contaminated runoff enters waterways, notify downstream users of potentially contaminated waters. Notify local health and fire officials and pollution control agencies. From a secure, explosion-proof location, use water spray to cool exposed containers. If cooling streams are ineffective (venting sound increases in volume and pitch, tank discolors, or shows any signs of deforming), withdraw immediately to a secure position. If employees are expected to fight fires, they must be trained and equipped in OSHA 1910.156.

Disposal Method Suggested: Alkaline hydrolysis or incineration.[22] In accordance with 40CFR165 recommendations for the disposal of pesticides and pesticide containers. Must be disposed properly by following package label directions or by contacting your state pesticide or environmental control agency or by contacting your regional EPA office.

References

Sax N. I., Ed., "Dangerous Properties of Industrial Materials Report," 1, No. 68–69 (1981). (As Meta-Systox)

U.S. Environmental Protection Agency, "Chemical Profile: Demeton-S-Methyl," Washington, DC, Chemical Emergency Preparedness Program (November 30, 1987)

New Jersey Department of Health and Senior Services, Hazardous Substance Fact Sheet, "Methyl Demeton," Trenton NJ (March 1989)

2,4-DES-Sodium

Molecular Formula: $C_8H_7Cl_2NaO_5$

Common Formula: $Cl_2C_6H_3O(CH_2)_2OSO_3Na$

Synonyms: Crag herbicide 1; Crag Sesone; 2,4-Des-Na; 2,4-Des-natrium (German); 2-(2,4-Dichlorophenoxy)ethanol hydrogen sulfate sodium salt; 2,4-Dichlorophenoxyethyl sulfate, sodium salt; Disul; Disul-NA; Disul-sodium; Natrium-2,4-dichlorphenoxyathylsulfat (German); SES; Sesone; Sodium-2-(2,4-dichlorophenoxy)ethyl sulfate; Sodium-2,4-dichlorophenoxyethyl sulphate; Sodium-2,4-dichlorophenyl cellosolve sulfate

CAS Registry Number: 136-78-7

RTECS Number: KK4900000

DOT ID: UN 2765

Regulatory Authority

- Air Pollutant Standard Set (ACGIH)[1] (OSHA)[58] (Several States)[60]

Cited in U.S. State Regulations: Alaska (G), California (G), Connecticut (A), Florida (G), Illinois (G), Maine (G), Massachusetts (G), Nevada (A), New Hampshire (G), New Jersey (G), North Dakota (A), Pennsylvania (G), Rhode Island (G), Virginia (A), West Virginia (G).

Description: Sesone, $Cl_2C_6H_3O(CH_2)_2OSO_3Na$, is a colorless, odorless, crystalline solid. Freezing/Melting point = 245°C (decomposes). Soluble in water. May be used in a carrier solvent which may change its physical properties.

Potential Exposure: Those involved in manufacture, formulation and application of this herbicide as well as citizens in the area of application.

Incompatibilities: Strong oxidizers, acids.

Permissible Exposure Limits in Air: The Federal Limit (OSHA PEL) is 10 mg/m^3, respirable fraction is 5 mg/m^3.[58] ACGIH as of has set a TWA of 10 mg/m^3. The NIOSH IDLH level is 500 mg/m^3. Several states have set guidelines or standards for sesone in ambient air[60] ranging from 100 μg/m^3 (North Dakota) to 160 μg/m^3 (Virginia) to 200 μg/m^3 (Connecticut) to 238 μg/m^3 (Nevada).

Determination in Air: See NIOSH Method # S-356.

Permissible Concentration in Water: No criteria set.

Routes of Entry: Inhalation, ingestion skin and eye contact.

Harmful Effects and Symptoms

Short Term Exposure: Irritates eyes, skin, and respiratory tract. High levels of exposure may cause central nervous system effects, convulsions. May affect the kidneys and liver.

Long Term Exposure: May cause liver and kidney damage.

Points of Attack: Eyes, skin, central nervous system, liver, kidneys.

Medical Surveillance: Liver function. Kidney function. Tests of nervous system.

First Aid: If this chemical gets into the eyes, remove any contact lenses at once and irrigate immediately for at least 15 minutes, occasionally lifting upper and lower lids. Seek medical attention immediately. If this chemical contacts the skin, remove contaminated clothing and wash immediately with soap and water. Seek medical attention immediately. If this chemical has been inhaled, remove from exposure, begin rescue breathing (using universal precautions) if breathing has stopped and CPR if heart action has stopped. Transfer promptly to a medical facility. When this chemical has been swallowed, get medical attention. Give large quantities of water and induce vomiting. Do not make an unconscious person vomit.

Personal Protective Methods: Wear protective gloves and clothing to prevent any reasonable probability of skin contact. Safety equipment suppliers/manufacturers can provide recommendations on the most protective glove/clothing material for your operation.. All protective clothing (suits, gloves, footwear, headgear) should be clean, available each day, and put on before work. Contact lenses should not be worn when working with this chemical. Wear dust-proof chemical goggles and face shield unless full facepiece respiratory protection is worn. Employees should wash immediately with soap when skin is wet or contaminated. Provide emergency showers and eyewash.

Respirator Selection: NIOSH: *50 mg/m^3:* DM (any dust and mist respirator). *100 mg/m^3*: DMXSQ (any dust and mist respirator except single-use and quarter-mask respirators); or SA (any supplied-air respirator). *250 mg/m^3*: PAPRDM (any supplied-air respirator operated in a continuous-flow mode); or SA:CF (any powered, air-purifying respirator with a dust and mist filter). *500 mg/m^3*: HiEF (any air-purifying, full-facepiece respirator with a high-efficiency particulate filter); or SCBAF (any powered, air-purifying respirator with a tight-fitting facepiece and a high-efficiency particulate filter*; or SAF (any supplied-air respirator that has a tight-fitting facepiece and is operated in a continuous-flow mode*; PAPRTHiE (any self-contained breathing apparatus with a full facepiece); or SAT:CF (any supplied-air respirator with a full facepiece). *Emergency or planned entry into unknown concentrations or IDLH conditions:* SCBAF:PD,PP (any self-contained breathing apparatus that has a full facepiece and is operated in a pressure-demand or other positive-pressure mode); or SAF:PD,PP:ASCBA (any supplied-air respirator that has a full facepiece and is operated in a pressure-demand or other positive-pressure mode in combination with an auxiliary self-contained breathing apparatus operated in a pressure-demand or other positive pressure mode.) *Escape:* HiEF (any air-purifying, full-facepiece respirator with a high-efficiency particulate filter); or SCBAE (any appropriate escape-type, self-contained breathing apparatus).

* Substance reported to cause eye irritation or damage; may require eye protection.

Storage: Prior to working with this chemical you should be trained on its proper handling and storage. Store in tightly closed containers in a cool, well ventilated area away from oxidizers and acids. Where possible, automatically pump liquid from drums or other storage containers to process containers.

Shipping: Phenoxy pesticides, solid, toxic, n.o.s. fall in Hazard Class 6.1. A "Keep Away From Food" label is required.

Spill Handling: Evacuate persons not wearing protective equipment from area of spill or leak until clean-up is complete. Remove all ignition sources. Collect powdered material in the most convenient and safe manner and deposit in sealed containers. Ventilate area after clean-up is complete. It may be necessary to contain and dispose of this chemical as a hazardous waste. If material or contaminated runoff enters waterways, notify downstream users of potentially contaminated waters. Contact your Department of Environmental Protection or your regional office of the federal EPA for specific recommendations. If employees are required to clean-up spills, they must be properly trained and equipped. OSHA 1910.120(q) may be applicable.

Fire Extinguishing: This chemical is a noncombustible solid. Use dry chemical, carbon dioxide, water spray, or alcohol foam extinguishers. Poisonous gases are produced in fire. If material or contaminated runoff enters waterways, notify downstream users of potentially contaminated waters. Notify local health and fire officials and pollution control agencies. From a secure, explosion-proof location, use water spray to cool exposed containers. If cooling streams are ineffective (venting sound increases in volume and pitch, tank discolors, or shows any signs of deforming), withdraw immediately to a secure position. If employees are expected to fight fires, they must be trained and equipped in OSHA 1910.156.

Disposal Method Suggested: Sesone is hydrolyzed by alkali to $NaHSO_4$ and apparently the dichlorophenoxyethanol. In accordance with 40CFR165 recommendations for the disposal of pesticides and pesticide containers. Must be disposed properly by following package label directions or by contacting your state pesticide or environmental control agency or by contacting your regional EPA office.

Diacetone Alcohol

Molecular Formula: $C_6H_{12}O_2$

Common Formula: $(CH_3)_2C(OH)CH_2COCH_3$

Synonyms: Acetonyldimethylcarbinol; Diacetonalcohol (Dutch); Diacetonalcool (Italian); Diacetonalkohol (German); Diacetone-alcool (French); Diacetone, 4-hydroxy-4-methyl-2-pentatone, 2-methyl-2-pentanol-4-one; Diketone alcohol; Dimethylacetonylcarbinol; EEC No. 603-016-00-1; 4-Hydroxy-2-keto-4-methylpentane; 4-Hydroxyl-2-keto-4-methylpentane; 4-Hydroxy-4-methyl-pentan-2-on (German, Dutch); 4-Hydroxy-4-methyl-2-pentanone; 4-Hydroxy-4-methylpentan-2-one; 4-Idrossi-4-metil-pentan-2-one (Italian); 4-Methyl-4-hydroxy-2-pentanone; 2-Methyl-2-pentanol-4-one; 2-Pentanone, 4-hydroxy-4-methyl-; tyranton

CAS Registry Number: 123-42-2

RTECS Number: SA9100000

DOT ID: UN 1148

Regulatory Authority

- Air Pollutant Standard Set (ACGIH)[1] (Australia) (DFG)[3] (HSE)[33] (Israel) (Mexico) (former USSR)[43] (OSHA)[58] (Several States)[60]
- Canada, WHMIS, Ingredients Disclosure List

Cited in U.S. State Regulations: Alaska (G), California (A, G), Connecticut (A), Florida (G), Illinois (G), Maine (G), Massachusetts (G), Minnesota (G), Nevada (A), New Hampshire (G), New Jersey (G), North Dakota (A), Pennsylvania (G), Rhode Island (G), Virginia (A), West Virginia (G).

Description: Diacetone alcohol, $(CH_3)_2C(OH)CH_2COCH_3$, is a colorless liquid with a mild, minty odor. Odor threshold = 0.28 ppm. Boiling point = 164°C. Freezing/Melting point = -44°C. Flash point = 64°C; 58°C (acetone free); 64°C (commercial grade). Explosive limits: LEL = 1.8%; UEL = 6.9%. Hazard Identification (based on NFPA-704 M Rating System): Health 1, Flammability 2, Reactivity 0. Soluble in water.

Potential Exposure: It is used as a solvent for pigments, cellulose esters, oils and fats. It is used in hydraulic brake fluids and in antifreeze formulations.

Incompatibilities: Strong oxidizers, strong alkalis.

Permissible Exposure Limits in Air: The Federal Limit (OSHA PEL),[58] DFG MAK,[3] HSE value,[33] Australian, Israeli, Mexican, Canadian provincial (Alberta, BC, Ontario, & Quebec), and the ACGIH TWA value is 50 ppm (240 mg/m^3). The Mexican, Alberta, BC, Ontario, and HSE STEL value is 75 ppm (360 mg/m^3). The NIOSH IDLH level is 1,800 ppm. The former USSR-UNEP/IRPTC project[43] has set a MAC in workplace air of 100 mg/m^3.

Several states have set guidelines or standards for diacetone alcohol in ambient air[60] ranging from 0.03 mg/m^3 (Nevada) to 2.4 – 3.6 mg/m^3 (North Dakota) to 4.0 mg/m^3 (Virginia) to 4.8 mg/m^3 (Connecticut).

Determination in Air: Adsorption on charcoal, workup with 2-propanol in CS_2, analysis by gas chromatography. See NIOSH Method # 1402 for Alcohols.

Permissible Concentration in Water: No criteria set.

Routes of Entry: Inhalation, ingestion, skin and eye contact.

Harmful Effects and Symptoms

Short Term Exposure: Irritates the eyes, skin, and respiratory tract. Eye contact can cause corneal tissue damage. Contact can irritate the skin, causing a burning sensation. It can cause you to become dizzy and lightheaded and to pass out.

Long Term Exposure: Repeated contact may lead to skin rash. Exposure may damage the liver, kidneys and the blood cells. Although there is no evidence involving this chemical, many similar solvents can cause nerve and brain damage.

Points of Attack: Eyes, skin, respiratory system.

Medical Surveillance: If symptoms develop or overexposure is suspected, the following may be useful: Liver and kidney function tests. Complete blood count.

First Aid: If this chemical gets into the eyes, remove any contact lenses at once and irrigate immediately for at least 15 minutes, occasionally lifting upper and lower lids. Seek

medical attention immediately. If this chemical contacts the skin, remove contaminated clothing and wash immediately with soap and water. Seek medical attention immediately. If this chemical has been inhaled, remove from exposure, begin rescue breathing (using universal precautions) if breathing has stopped and CPR if heart action has stopped. Transfer promptly to a medical facility. When this chemical has been swallowed, get medical attention. Give large quantities of water and induce vomiting. Do not make an unconscious person vomit.

Personal Protective Methods: Wear protective gloves and clothing to prevent any reasonable probability of skin contact. Safety equipment suppliers/manufacturers can provide recommendations on the most protective glove/clothing material for your operation. Neoprene and Polyvinyl Chloride are among the recommended protective materials. All protective clothing (suits, gloves, footwear, headgear) should be clean, available each day, and put on before work. Contact lenses should not be worn when working with this chemical. Wear splash-proof chemical goggles and face shield unless full facepiece respiratory protection is worn. Employees should wash immediately with soap when skin is wet or contaminated. Provide emergency showers and eyewash.

Respirator Selection: NIOSH/OSHA: *1,250 ppm:* SA:CF (any supplied-air respirator operated in a continuous-flow mode); or PAPROV [any powered, air-purifying respirator with organic vapor cartridge(s)]. *1,800 ppm:* CCRFOV [any chemical cartridge respirator with a full facepiece and organic vapor cartridge(s)]; or GMFOV [any air-purifying, full-facepiece respirator (gas mask) with a chin-style, front- or back-mounted acid gas canister]; or PAPRTOV [any powered, air-purifying respirator with a tight-fitting facepiece and organic vapor cartridge(s)]; or SCBAF (any self-contained breathing apparatus with a full facepiece); or SAF (any supplied-air respirator with a full facepiece). *Emergency or planned entry into unknown concentrations or IDLH conditions:* SCBAF:PD,PP (any self-contained breathing apparatus that has a full facepiece and is operated in a pressure-demand or other positive-pressure mode); or SAF:PD,PP:ASCBA (any supplied-air respirator that has a full facepiece and is operated in a pressure-demand or other positive-pressure mode in combination with an auxiliary self-contained breathing apparatus operated in a pressure-demand or other positive pressure mode). *Escape:* GMFOV [any air-purifying, full-facepiece respirator (gas mask) with a chin-style, front-or back-, mounted organic vapor canister]; or SCBAE (any appropriate escape-type, self-contained breathing apparatus).

Note: Substance causes eye irritation or damage; eye protection needed.

Storage: Prior to working with this chemical you should be trained on its proper handling and storage. Before entering confined space where this chemical may be present, check to make sure that an explosive concentration does not exist. Diacetone alcohol must be stored to avoid contact with strong oxidizers (such as chlorine, bromine, and fluorine) or strong alkalis (such as sodium hydroxide or potassium hydroxide) since violent reactions occur. Store in tightly closed containers in a cool well-ventilated area away from heat. Sources of ignition such as smoking and open flames are prohibited where diacetone alcohol is used, handled, or stored in a manner that could create a potential fire or explosion hazard. Metal containers involving the transfer of 5 gallons or more of diacetone alcohol should be grounded and bonded. Drums must be equipped with self-closing valves, pressure vacuum bungs, and flame arresters.

Shipping: Diacetone alcohol falls in DOT Hazard Class 3 and Packing Group III. It must be labeled: "Flammable Liquid."

Spill Handling: Evacuate and restrict persons not wearing protective equipment from area of spill or leak until cleanup is complete. Remove all ignition sources. Establish forced ventilation to keep levels below explosive limit. Absorb liquids in vermiculite, dry sand, earth, peat, carbon, or a similar material and deposit in sealed containers. Keep this chemical out of a confined space, such as a sewer, because of the possibility of an explosion, unless the sewer is designed to prevent the build-up of explosive concentrations. It may be necessary to contain and dispose of this chemical as a hazardous waste. If material or contaminated runoff enters waterways, notify downstream users of potentially contaminated waters. Contact your Department of Environmental Protection or your regional office of the federal EPA for specific recommendations. If employees are required to clean-up spills, they must be properly trained and equipped. OSHA 1910.120(q) may be applicable.

Fire Extinguishing: This chemical is a combustible liquid. Poisonous gases are produced in fire. Use dry chemical, carbon dioxide, or alcohol foam extinguishers. Vapors are heavier than air and will collect in low areas. Vapors may travel long distances to ignition sources and flashback. Vapors in confined areas may explode when exposed to fire. Containers may explode in fire. Storage containers and parts of containers may rocket great distances, in many directions. If material or contaminated runoff enters waterways, notify downstream users of potentially contaminated waters. Notify local health and fire officials and pollution control agencies. From a secure, explosion-proof location, use water spray to cool exposed containers. If cooling streams are ineffective (venting sound increases in volume and pitch, tank discolors, or shows any signs of deforming), withdraw immediately to a secure position. If employees are expected to fight fires, they must be trained and equipped in OSHA 1910.156.

Disposal Method Suggested: Incineration.

References

National Institute for Occupational Safety and Health, "Criteria for a Recommended Standard: Occupational Exposure to Ketones," NIOSH Doc. No. 78–173 (1978)

New Jersey Department of Health, "Hazardous Substance Fact Sheet: Diacetone Alcohol," Trenton, NJ (October 1996)

Dialifor

Molecular Formula: $C_{14}H_{17}ClNO_4PS_2$

Synonyms: *N*-[2-Chloro-1-(diethoxyphosphinpthioyl-thio)ethyl]phthalimide; *S*-[2-Chloro-1-(1,3-dihydro-1,3-dioxo-2H-isoindol-2-yl)ethyl] *O,O*-diethyl phosphorodithioate; *S*-(2-Chloro-1-phthalimidoethyl) *O,O*-diethyl phosphorodithioate; Dialifos; *O,O*-Diethyl *S*-(2-chloro-1-phthalimidoethyl) phosphorodithioate; *O,O*-Diethyl phosphorodithioate *S*-ester with *N*-(2-chloro-1-mercaptoethyl) phthalimide; ENT 27,320; Hercules 14503; Phosphorodithioic acid, *S*-[2-chloro-1-(1,3-dihydro-1,3-dioxo-2H-isoindol-2-yl)ethyl] *O,O*-diethyl ester; Phosphorodithioic acid, 5-[2-chloro-1-(1,3-dihydro-1,3-dioxo-2H-isoindol-2-yl)ethyl] *O,O*-diethyl ester; Phosphorodithioic acid, *S*-(2-chloro-1-phthalimidoethyl) *O,O*-diethyl ester; Torak

CAS Registry Number: 10311-84-9

RTECS Number: TD5165000

DOT ID: UN 2783

Regulatory Authority

- Banned or Severely Restricted (Malaysia, DDR)[13]
- Very Toxic Substance (World Bank)[15]
- Superfund/EPCRA 40CFR355, Appendix B Extremely Hazardous Substances: TPQ = 100/10,000 lb (45.4/4,540 kg)
- U.S. DOT Regulated Marine Pollutant (49CFR172.101, Appendix B)

Cited in U.S. State Regulations: California (G), Illinois (G), Massachusetts (G), New Hampshire (G), New Jersey (G), Pennsylvania (G).

Description: Dialifor, $C_{14}H_{17}ClNO_4PS_2$, is a crystalline solid. Also reported as an oil. Freezing/Melting point (solid) = 67 – 69°C. Insoluble in water.

Potential Exposure: Those involved in the manufacture, formulation and application of this insecticide.

Incompatibilities: Strong bases.

Permissible Exposure Limits in Air: No standards set.

Determination in Air: OSHA versatile sampler-2; Toluene/Acetone; Gas chromatography/Flame photometric detection for sulfur, nitrogen, or phosphorus; NIOSH Method IV Method #5600, Organophosphorus Pesticides.

Permissible Concentration in Water: No criteria set.

Routes of Entry: Inhalation, passing through the skin, ingestion.

Harmful Effects and Symptoms

Short Term Exposure: This material is highly toxic (the LD_{50} for rats is 5 mg/kg). This material can cause serious symptoms and in extreme cases death by respiratory arrest. Organic phosphorus insecticides are absorbed by the skin, as well as by the respiratory and gastrointestinal tracts. They are cholinesterase inhibitors. Symptoms of exposure include headache, giddiness, blurred vision, nervousness, weakness, nausea, cramps, diarrhea, and discomfort in the chest. Signs include sweating, tearing, salivation, vomiting, cyanosis, convulsions, coma, loss of reflexes and loss of sphincter control.

Long Term Exposure: Cholinesterase inhibitor; cumulative effect is possible. This chemical may damage the nervous system with repeated exposure, resulting in convulsions, respiratory failure. May cause liver damage.

Points of Attack: Respiratory system, lungs, central nervous system, cardiovascular system, skin, eyes, plasma and red blood cell cholinesterase.

Medical Surveillance: Before employment and at regular times after that, the following are recommended: Plasma and red blood cell cholinesterase levels (tests for the enzyme poisoned by this chemical). If exposure stops, plasma levels return to normal in 1 – 2 weeks while red blood cell levels may be reduced for 1 – 3 months.

When cholinesterase enzyme levels are reduced by 25% or more below preemployment levels, risk of poisoning is increased, even if results are in lower ranges of "normal." Reassignment to work not involving organophosphate or carbamate pesticides is recommended until enzyme levels recover. If symptoms develop or overexposure occurs, repeat the above tests as soon as possible and get an exam of the nervous system. Also consider complete blood count. Consider chest x-ray following acute overexposure. Do not drink any alcoholic beverages before or during use. Alcohol promotes absorption of organic phosphates.

First Aid: If this chemical gets into the eyes, remove any contact lenses at once and irrigate immediately for at least 15 minutes, occasionally lifting upper and lower lids. Seek medical attention immediately. If this chemical contacts the skin, remove contaminated clothing and wash immediately with soap and water. Speed in removing material from skin is of extreme importance. Shampoo hair promptly if contaminated. Seek medical attention immediately. If this chemical has been inhaled, remove from exposure, begin rescue breathing (using universal precautions) if breathing has stopped and CPR if heart action has stopped. Transfer promptly to a medical facility. When this chemical has been swallowed, get medical attention. Give large quantities of water and induce vomiting. Do not make an unconscious person vomit. Effects may be delayed; medical observation is recommended.

Personal Protective Methods: Wear protective gloves and clothing to prevent any reasonable probability of skin contact. Safety equipment suppliers/manufacturers can provide recommendations on the most protective glove/clothing material for your operation.. All protective clothing (suits, gloves, footwear, headgear) should be clean, available each day, and put on before work. Contact lenses should not be worn when working with this chemical. Wear dust-proof chemical goggles and face shield unless full facepiece respiratory protection is worn. Employees should wash immediately with soap when skin is wet or contaminated. Provide emergency showers and eyewash.

Respirator Selection: *At any detectable concentration:* SCBAF:PD,PP (any MSHA/NIOSH approved self-contained breathing apparatus that has a full facepiece and is operated in a pressure-demand or other positive-pressure mode); or SAF:PD,PP:ASCBA (any supplied-air respirator that has a full facepiece and is operated in a pressure-demand or other positive-pressure mode in combination with an auxiliary, self-contained breathing apparatus operated in a pressure-demand or other positive pressure mode). *Escape:* HiEF (any air-purifying, full-facepiece respirator (gas mask) with a chin-style, front- or back-mounted organic vapor canister having a high-efficiency particulate filter); or SCBAE (any appropriate escape-type, self-contained breathing apparatus).

Storage: Prior to working with dialifor you should be trained on its proper handling and storage. Store in tightly closed containers in a cool, well ventilated area away from strong bases.

Shipping: Organophosphorus pesticides, solid, toxic, n.o.s. fall in DOT Hazard Class 6.1 and dialifor falls in Packing Group I. The label required is "Poison."

Spill Handling: Evacuate persons not wearing protective equipment from area of spill or leak until clean-up is complete. Remove all ignition sources. Collect powdered material in the most convenient and safe manner and deposit in sealed containers. Ventilate area after clean-up is complete. It may be necessary to contain and dispose of this chemical as a hazardous waste. If material or contaminated runoff enters waterways, notify downstream users of potentially contaminated waters. Contact your Department of Environmental Protection or your regional office of the federal EPA for specific recommendations. If employees are required to clean-up spills, they must be properly trained and equipped. OSHA 1910.120(q) may be applicable.

Fire Extinguishing: Use dry chemical, carbon dioxide, water spray, or foam extinguishers. Poisonous gases are produced in fire including phosphorus, sulfur, and nitrogen oxides. If material or contaminated runoff enters waterways, notify downstream users of potentially contaminated waters. Notify local health and fire officials and pollution control agencies. From a secure, explosion-proof location, use water spray to cool exposed containers. If cooling streams are ineffective (venting sound increases in volume and pitch, tank discolors, or shows any signs of deforming), withdraw immediately to a secure position. If employees are expected to fight fires, they must be trained and equipped in OSHA 1910.156.

Disposal Method Suggested: Alkaline hydrolysis or incineration. In accordance with 40CFR165 recommendations for the disposal of pesticides and pesticide containers. Must be disposed properly by following package label directions or by contacting your state pesticide or environmental control agency or by contacting your regional EPA office.

References

Sax N. I., Ed., "Dangerous Properties of Industrial Materials Report," 2, No. 5, 43-45 (1982)

U.S. Environmental Protection Agency, "Chemical Profile: Dialifor," Washington, DC, Chemical Emergency Preparedness Program (November 30, 1987)

Diallate

Molecular Formula: $C_{10}H_{17}Cl_2NOS$

Common Formula: $[(CH_3)_2CH]_2NCOSCH_2CCl{=}CHCl$

Synonyms: Avadex; Bis(1-methylethyl)carbamothioic acid, *S*-(2,3-dichloro-2-propenyl) ester; Carbamothioic acid, bis(1-methylethyl) *S*-(2,3-dichloro-2-propenyl) ester; CP 15,336; DATC; 2,3-DCDT; Diallaat (Dutch); Diallat (German); Di-allate; Diallate carbamate herbicide; *S*-(2,3-Dichlor-allyl)-*N,N*-diisopropyl-monothiocarbamaat (Dutch); *S*-(2,3-Dichloro-allil)-*N,N*-diisopropil-monotiocarbammato (Italian); *S*-(2,3-Dichloroallyl) diisopropylthiocarbamate]; *S*-2,3-Dichloroallyl di-isopropyl(thiocarbamate); *S*-2,3-Dichloroallyl diisopropyl-thiocarbamate; *S*-2,3-Dichloroallyldiisopropyl thiocarbamate; Dichloroallyldiisopropyl thiocarbamate; 2,3-Dichloroallyl *N,N*-Diisopropylthiolcarbamate; 2,3-Dichloro-2-propene-1-thiol, iisopropylcarbamate; *S*-(2,3-Dichloro-2-propenyl)bis(1-methylethyl) carbamothioate; Diisopropylthiocarbamic acid, -(2,3-dichloroallyl) ester; Di-isopropylthiolocarbamate des-(2,3-dichloro allyle) (French); 2-Propene-1-thiol, 2,3-dichloro-, diisopropylcarbamate

CAS Registry Number: 2303-26-4

RTECS Number: EX8225000

DOT ID: UN 2757 (carbamate pesticides, solid, toxic); UN 2992 (carbamate pesticides, liquid, toxic)

Regulatory Authority

- Carcinogen (Animal Positive) (IARC)[9]
- EPA Hazardous Waste Number (RCRA No.): U062
- RCRA, 40CFR261, Appendix 8 Hazardous Constituents
- RCRA 40CFR264, Appendix 9; TSD Facilities Ground Water Monitoring List. Suggested test method(s) (PQL μg/l): 8270 (10)
- Superfund/EPCRA 40CFR302.4 Reportable Quantity (RQ): CERCLA, 100 lb (45.4 kg)
- EPCRA Section 313 Form R *de minimis* concentration reporting level: 1.0%
- U.S. DOT Regulated Marine Pollutant (49CFR172.101, Appendix B)

Cited in U.S. State Regulations: California (A, G), Florida (G), Kansas (G), Louisiana (G), Massachusetts (G), Michigan (G), New Jersey (G), Pennsylvania (G), Rhode Island (G), Vermont (G), Virginia (G), Washington (G), Wisconsin (G).

Description: Diallate, $C_{10}H_{17}Cl_2NOS$, is a brown liquid. Boiling point = 150°C @ 9 mm Hg pressure. Freezing/Melting point = 25 – 30°C. Hazard Identification (based on NFPA-704 M Rating System): Health 2, Flammability 0, Reactivity 0. Slightly soluble in water.

Potential Exposure: Those involved in the manufacture, formulation and application of this re-emergence herbicide.

Incompatibilities: Alkalis.

Permissible Exposure Limits in Air: No standards set.

Permissible Concentration in Water: The former USSR has set a MAC of 0.03 mg/l in water used for domestic purposes.[35][43]

Routes of Entry: Inhalation, passing through the skin, ingestion.

Harmful Effects and Symptoms

Short Term Exposure: Eye contact can irritate and possibly cause burns. Inhalation caused irritation of the respiratory tract with chest tightness and/or difficulty breathing. Higher levels can affect the nervous system. With nausea, vomiting, diarrhea, abdominal pain, reduced muscle coordination, blurred vision, muscle twitching, convulsions, coma and possible death.

Long Term Exposure: High or repeated exposures can cause liver and kidney damage. There is limited evidence that diallate causes liver cancer in animals.

Points of Attack: Skin, eyes, nervous system.

Medical Surveillance: Lung function tests. Kidney and liver function tests. Examination of the nervous system. Interview exposed person for brain effects, including memory, mood, concentration, headaches, malaise, and altered sleep patterns.

First Aid: If this chemical gets into the eyes, remove any contact lenses at once and irrigate immediately for at least 15 minutes, occasionally lifting upper and lower lids. Seek medical attention immediately. If this chemical contacts the skin, remove contaminated clothing and wash immediately with soap and water. Seek medical attention immediately. If this chemical has been inhaled, remove from exposure, begin rescue breathing (using universal precautions) if breathing has stopped and CPR if heart action has stopped. Transfer promptly to a medical facility. When this chemical has been swallowed, get medical attention. Give large quantities of water and induce vomiting. Do not make an unconscious person vomit.

Personal Protective Methods: Wear protective gloves and clothing to prevent any reasonable probability of skin contact. Safety equipment suppliers/manufacturers can provide recommendations on the most protective glove/clothing material for your operation.. All protective clothing (suits, gloves, footwear, headgear) should be clean, available each day, and put on before work. Contact lenses should not be worn when working with this chemical. Wear splash-proof chemical goggles and face shield unless full facepiece respiratory protection is worn. Employees should wash immediately with soap when skin is wet or contaminated. Provide emergency showers and eyewash.

Respirator Selection: *At any detectable concentration:* SCBAF:PD,PP (any MSHA/NIOSH approved self-contained breathing apparatus that has a full facepiece and is operated in a pressure-demand or other positive-pressure mode); or SAF:PD,PP:ASCBA (any supplied-air respirator that has a full facepiece and is operated in a pressure-demand or other positive-pressure mode in combination with an auxiliary, self-contained breathing apparatus operated in a pressure-demand or other positive pressure mode). *Escape:* HiEF (any air-purifying, full-facepiece respirator (gas mask) with a chin-style, front- or back-mounted organic vapor canister having a high-efficiency particulate filter); or SCBAE (any appropriate escape-type, self-contained breathing apparatus).

Storage: Prior to working with diallate you should be trained on its proper handling and storage. Store in tightly closed containers in a cool, well ventilated area away from alkalies. Where possible, automatically pump liquid from drums or other storage containers to process containers. A regulated, marked area should be established where this chemical is handled, used, or stored in compliance with OSHA standard 1910.1045.

Shipping: Pesticides, solid, toxic, n.o.s. fall in DOT Hazard Class 6.1 and diallate falls in Packing Group III. It must bear a "Keep Away From Food" label.

Spill Handling: Evacuate and restrict persons not wearing protective equipment from area of spill or leak until cleanup is complete. Remove all ignition sources. Ventilate area of spill or leak. Absorb liquids in vermiculite, dry sand, earth, peat, carbon, or a similar material and deposit in sealed containers. Keep this chemical out of a confined space, such as a sewer, because of the possibility of an explosion, unless the sewer is designed to prevent the build-up of explosive concentrations. It may be necessary to contain and dispose of this chemical as a hazardous waste. If material or contaminated runoff enters waterways, notify downstream users of potentially contaminated waters. Contact your Department of Environmental Protection or your regional office of the federal EPA for specific recommendations. If employees are required to clean-up spills, they must be properly trained and equipped. OSHA 1910.120(q) may be applicable.

Fire Extinguishing: Poisonous gases including sulfur oxides, nitrogen oxides, and chlorides are produced in fire. Use dry chemical, carbon dioxide, or alcohol foam extinguishers. Vapors are heavier than air and will collect in low areas.

Containers may explode in fire. Storage containers and parts of containers may rocket great distances, in many directions. If material or contaminated runoff enters waterways, notify downstream users of potentially contaminated waters. Notify local health and fire officials and pollution control agencies. From a secure, explosion-proof location, use water spray to cool exposed containers. If cooling streams are ineffective (venting sound increases in volume and pitch, tank discolors, or shows any signs of deforming), withdraw immediately to a secure position. If employees are expected to fight fires, they must be trained and equipped in OSHA 1910.156.

Disposal Method Suggested: Land burial is acceptable for small quantities. Larger quantities can be incinerated.[22] In accordance with 40CFR165 recommendations for the disposal of pesticides and pesticide containers. Must be disposed properly by following package label directions or by contacting your state pesticide or environmental control agency or by contacting your regional EPA office. Consult with environmental regulatory agencies for guidance on acceptable disposal practices. Generators of waste containing this contaminant (≥100 kg/mo) must conform with EPA regulations governing storage, transportation, treatment, and waste disposal.

References

New Jersey Department of Health and Senior Services, Hazardous Substance Fact Sheet, "Diallate," Trenton NJ (April 1997)

Sax N. I., Ed., "Dangerous Properties of Industrial Materials Report," 3, No. 1, 50–53 (1983)

2,4-Diaminoanisole

Molecular Formula: $C_7H_{10}N_2O$

Common Formula: $H_3COC_6H_3(NH_2)_2$

Synonyms: 3-Amino-4-methoxyaniline; 1,3-Benzenediamine, 4-methoxy-; C.I. 76050; C.I. oxidation base 12; 2,4-DAA; 2,4-Diamineanisole; *m*-Diaminoanisole; 2,4-Diaminoanisole; 1,3-Diamino-4-methoxybenzene; 2,4-Diamino-1-methoxybenzene; 2,4-Diaminophenyl methyl ether; Furro L; 4-Methoxy-1,3-benzenediamine; *p*-Methoxy-*m*-phenylenediamine; 4-Methoxy-*m*-phenylenediamine; 4-Methoxy-1,3-phenylenediamine; 4-MMPD; Pelagol DA; Pelagol grey L; Pelagol L; *m*-Phenylenediamine, 4-methoxy-

sulfate: Anisole, 2,4-diamino-, hydrogen sulfate; Anisole, 2,4-diamino-, sulfate; 1,3-Benzenediamine, 4-methoxy, sulfate (1:1); C.I. 76051; C.I. Oxidation base 12A; 2,4-DAA Sulfate; 2,4-Diaminoanisole sulphate; 2,4-Diamino-anisol sulphate; 2,4-Diamino-1-methoxybenzene; 1,3-Diamino-4-methoxybenzene sulphate; 2,4-Diamino-1-methoxybenzene sulphate; 2,4-Diaminosole sulphate; Durafur brown MN; Fouramine BA; Fourrine 76; Fourrine SLA; Furro SLA; 4-Methoxy-1,3-benzenediamine sulfate; 4-Methoxy-1,3-benzenediamine sulfate (1:1); 4-Methoxy-1,3-benzenediamine sulphate; 4-Methoxy-*m*-phenylenediamine sulfate; *p*-Methoxy-*m*-phenylenediamine sulphate; 4-Methoxy-*m*-phenylenediamine sulphate; 4-MMPD Sulphate; NAKO TSA; NCI-C01989; Oxidation base 12A; Pelagol BA; Pelagol grey; Pelagol grey SLA; Pelagol SLA; Renal SLA; Ursol SLA; Zoba SLE

CAS Registry Number: 615-05-4; 39156-41-7 (sulfate)

RTECS Number: ST2690000

DOT ID: UN 3077

Regulatory Authority

- Carcinogen (Human Suspected) (IARC)[9] (Animal Positive) (NTP) (DFG)[3]
- Banned or Severely Restricted (Sweden)[35]
- EPCRA Section 313 Form R *de minimis* concentration reporting level: 0.1%
- Canada, WHMIS, Ingredients Disclosure List

Cited in U.S. State Regulations: California (A, G), Florida (G), Illinois (G), Maine (G), Massachusetts (G), Maryland (G), Michigan (G), Minnesota (G), New Jersey (G), Pennsylvania (G), West Virginia (G).

Description: 2,4-Diaminoanisole, $H_3COC_6H_3(NH_2)_2$, is a needle-like solid. Freezing/Melting point 67 – 68°C. The sulfate is an off-white to violet powder. Soluble in water.

Potential Exposure: The principal use of 2,4-diaminoanisole is as a component of oxidation (permanent) hair and fur dye formulations. Human exposure to 2,4-diaminoanisole sulfate may possibly occur through skin absorption at chemical and dye production facilities, as well as through dermal contact in persons using hair dyes containing the chemical.

Incompatibilities: Sulfates react violently with aluminum and magnesium.

Permissible Exposure Limits in Air: No standards set. Can be absorbed through the skin thereby increasing exposure.

Permissible Concentration in Water: No criteria set.

Routes of Entry: Inhalation, passing through the skin.

Harmful Effects and Symptoms

Short Term Exposure: This chemical can be absorbed through the skin, thereby increasing exposure. Contact may cause irritation and possible eye damage. High exposures to the sulfate can cause poisoning with trembling, diarrhea, trouble breathing and even death.

Long Term Exposure: The sulfate can cause both skin and lung allergies to develop. There is sufficient evidence for the carcinogenicity of 2,4-diaminoanisole sulfate in experimental animals. In rats, dietary administration of the technical grade 2,4-diaminoanisole sulfate increased the incidence of cancers of the skin and the associated glands, and of thyroid cancers in each sex. In mice, dietary administration of 2,4-diaminoanisole sulfate induced thyroid tumors in each sex. Female rats exposed to technical-grade 2,4-diamonoanisole sulfate in the feed developed tumors of the thyroid, mammary, clitoral, and pituitary glands.

Points of Attack: Skin, eyes, lungs.

Medical Surveillance: Evaluation by a qualified allergist. Lung function tests.

First Aid: If this chemical gets into the eyes, remove any contact lenses at once and irrigate immediately for at least 15 minutes, occasionally lifting upper and lower lids. Seek medical attention immediately. If this chemical contacts the skin, remove contaminated clothing and wash immediately with soap and water. Seek medical attention immediately. If this chemical has been inhaled, remove from exposure, begin rescue breathing (using universal precautions) if breathing has stopped and CPR if heart action has stopped. Transfer promptly to a medical facility. When this chemical has been swallowed, get medical attention. Give large quantities of water and induce vomiting. Do not make an unconscious person vomit.

Personal Protective Methods: Gloves are usually worn by hairdressers when applying hair dyes. Beyond that, NIOSH recommends minimization of exposure. Wear protective gloves and clothing to prevent any reasonable probability of skin contact. Safety equipment suppliers/manufacturers can provide recommendations on the most protective glove/clothing material for your operation.. All protective clothing (suits, gloves, footwear, headgear) should be clean, available each day, and put on before work. Contact lenses should not be worn when working with this chemical. Wear dust-proof chemical goggles and face shield unless full facepiece respiratory protection is worn. Employees should wash immediately with soap when skin is wet or contaminated. Provide emergency showers and eyewash.

Respirator Selection: *At any detectable concentration:* SCBAF:PD,PP (any MSHA/NIOSH approved self-contained breathing apparatus that has a full facepiece and is operated in a pressure-demand or other positive-pressure mode); or SAF:PD,PP:ASCBA (any supplied-air respirator that has a full facepiece and is operated in a pressure-demand or other positive-pressure mode in combination with an auxiliary, self-contained breathing apparatus operated in a pressure-demand or other positive pressure mode). *Escape:* HiEF (any air-purifying, full-facepiece respirator (gas mask) with a chin-style, front- or back-mounted organic vapor canister having a high-efficiency particulate filter); or SCBAE (any appropriate escape-type, self-contained breathing apparatus).

Storage: Prior to working with 2,4-DAA sulfate you should be trained on its proper handling and storage. A regulated, marked area should be established where this chemical is handled, used or stored in compliance with OSHA standard 1910.1045. Store in tightly closed containers in a cool, well ventilated area away from aluminum and magnesium.

Shipping: Environmentally hazardous solid, n.o.s. Hazard Class: 9. Label: "Class 9."

Spill Handling: Evacuate persons not wearing protective equipment from area of spill or leak until clean-up is complete. Remove all ignition sources. Collect powdered material in the most convenient and safe manner and deposit in sealed containers. Ventilate area after clean-up is complete. It may be necessary to contain and dispose of this chemical as a hazardous waste. If material or contaminated runoff enters waterways, notify downstream users of potentially contaminated waters. Contact your Department of Environmental Protection or your regional office of the federal EPA for specific recommendations. If employees are required to clean-up spills, they must be properly trained and equipped. OSHA 1910.120(q) may be applicable.

Fire Extinguishing: This chemical is a noncombustible solid. Use dry chemical, carbon dioxide, water spray, or alcohol foam extinguishers. Poisonous gases are produced in fire including nitrogen and sulfur oxides. If material or contaminated runoff enters waterways, notify downstream users of potentially contaminated waters. Notify local health and fire officials and pollution control agencies. From a secure, explosion-proof location, use water spray to cool exposed containers. If cooling streams are ineffective (venting sound increases in volume and pitch, tank discolors, or shows any signs of deforming), withdraw immediately to a secure position. If employees are expected to fight fires, they must be trained and equipped in OSHA 1910.156.

Disposal Method Suggested: Incineration.[22]

References

National Institute for Occupational Safety and Health, 2,4-Diaminoanisole in Hair and Fur Dyes, Current Intelligence Bulletin No. 19, Washington, DC (January 13, 1978)

New Jersey Department of Health and Senior Services, Hazardous Substance Fact Sheet, "2,4-Diaminoanisole Sulfate," Trenton NJ (January 1989)

4,4'-Diaminodiphenyl Ether

Molecular Formula: $C_{12}H_{12}N_2O$

Synonyms: *p*-Aminophenyl ether; 4-Aminophenyl ether; Aniline, 4,4'-oxydi-; Benzenamine, 4,4'-oxybis-; Bis(*p*-aminophenyl) ether; Bis(4-aminophenyl) ether; 4,4-Diaminodiphenyl ether; Diaminodiphenyl ether; 4,4'-Diaminofenol eter (Spanish); 4,4'-Diaminophenyl ether; NCI-C50146; Oxybis(4-aminobenzene); *p,p*'-Oxybis(aniline); 4,4'-Oxybis(aniline); *p,p*'-Oxydianiline; 4,4'-Oxydianiline; 4,4'-Oxydiphenylamine; Oxydi-*p*-phenylenediamine

CAS Registry Number: 101-80-4

RTECS Number: BY7900000

DOT ID: UN 3077 (environmentally hazardous substances, n.o.s.)

Regulatory Authority

- Carcinogen (animal positive) (IARC) (NTP) (DFG)
- EPCRA Section 313 Form R *de minimis* concentration reporting level: 0.1%

Cited in U.S. State Regulations: California (A, G), Florida (G), Massachusetts (G), Minnesota (G), New Jersey (G), Pennsylvania (G).

Description: 4,4'-Diaminophenyl ether, is a colorless or light pink crystalline solid. Freezing/Melting point = 187°C. Boiling point ≥300°C. Hazard Identification (based on NFPA-704 M Rating System): Health 1, Flammability 1, Reactivity 0. Insoluble in water.

Potential Exposure: Used to make special high temperature-resistant resins.

Incompatibilities: Contact with oxidizers may cause fire and explosion hazard. Ethers have a tendency to form explosive peroxides.

Permissible Exposure Limits in Air: No exposure limits have been established.

Routes of Entry: Inhalation, passing through the skin.

Harmful Effects and Symptoms

Short Term Exposure: This chemical can be absorbed through the skin, thereby increasing exposure. Contact can irritate the eyes, skin and respiratory tract.

Long Term Exposure: May cause skin allergy. Repeated exposure may cause liver damage and enlarged thyroid (goiter). This chemical has been shown to cause liver, thyroid cancer and lymphoma in animals, and may be a human carcinogen.

Points of Attack: Liver, thyroid, skin.

Medical Surveillance: Liver and thyroid function tests. Evaluation by a qualified allergist. Regular consumption of alcohol may increase the liver damage caused by 4,4'-diaminodiphenyl ether.

First Aid: If this chemical gets into the eyes, remove any contact lenses at once and irrigate immediately for at least 15 minutes, occasionally lifting upper and lower lids. Seek medical attention immediately. If this chemical contacts the skin, remove contaminated clothing and wash immediately with soap and water. Seek medical attention immediately. If this chemical has been inhaled, remove from exposure, begin rescue breathing (using universal precautions) if breathing has stopped and CPR if heart action has stopped. Transfer promptly to a medical facility. When this chemical has been swallowed, get medical attention. Give large quantities of water and induce vomiting. Do not make an unconscious person vomit.

Personal Protective Methods: Wear protective gloves and clothing to prevent any reasonable probability of skin contact. Safety equipment suppliers/manufacturers can provide recommendations on the most protective glove/clothing material for your operation.. All protective clothing (suits, gloves, footwear, headgear) should be clean, available each day, and put on before work. Contact lenses should not be worn when working with this chemical. Wear dust-proof chemical goggles and face shield unless full facepiece respiratory protection is worn. Employees should wash immediately with soap when skin is wet or contaminated. Provide emergency showers and eyewash.

Respirator Selection: Where the potential for exposure to this chemical, use a MSHA/NIOSH approved supplied-air respirator with a full facepiece operated in the positive pressure mode or with a full facepiece, hood, or helmet in the continuous flow mode, or use a MSHA/NIOSH approved self-contained breathing apparatus with a full facepiece operated in pressure-demand or other positive pressure mode.

Storage: Prior to working with this chemical you should be trained on its proper handling and storage. Store in tightly closed containers in a cool, well ventilated area away from strong oxidizers. Where possible, automatically transfer this chemical from drums or other storage containers to process containers. Use only non-sparking tools and equipment, especially when opening and closing containers of this chemical. Sources of ignition such as smoking and open flames, are prohibited where this chemical is used, handled, or stored in a manner that could create a potential fire or explosion hazard. Wherever this chemical is used, handled, manufactured, or stored, use explosion-proof electrical equipment and fittings. A regulated, marked area should be established where this chemical is handled, used, or stored in compliance with OSHA standard 1910.1045.

Shipping: This chemical is on the Community Right-to-Know list and is an environmentally hazardous substance, solid, n.o.s. It is in Hazard Class 9 and Packing Group III.

Spill Handling: Evacuate persons not wearing protective equipment from area of spill or leak until clean-up is complete. Remove all ignition sources. Collect powdered material in the most convenient and safe manner and deposit in sealed containers. Do not dry sweep. Ventilate area after clean-up is complete. It may be necessary to contain and dispose of this chemical as a hazardous waste. If material or contaminated runoff enters waterways, notify downstream users of potentially contaminated waters. Contact your Department of Environmental Protection or your regional office of the federal EPA for specific recommendations. If employees are required to clean-up spills, they must be properly trained and equipped. OSHA 1910.120(q) may be applicable.

Fire Extinguishing: This chemical may burn but does not easily ignite. Use dry chemical, carbon dioxide, water spray, or alcohol foam extinguishers. Poisonous gases are produced in fire including nitrogen oxides and carbon monoxide. If material or contaminated runoff enters waterways, notify downstream users of potentially contaminated waters. Notify local health and fire officials and pollution control agencies. From a secure, explosion-proof location, use water spray to cool exposed containers. If cooling streams are ineffective (venting sound increases in volume and pitch, tank discolors, or shows any signs of deforming), withdraw immediately to a secure position. If employees are expected to fight fires, they must be trained and equipped in OSHA 1910.156.

References

New Jersey Department of Health and Senior Services, "Hazardous Substance Fact Sheet, 4,4'-Diaminodiphenyl Ether," Trenton NJ (January 1999)

4,4'-Diaminodiphenylmethane

Molecular Formula: $C_{13}H_{14}N_2$

Common Formula: $H_2NC_6H_4CH_2C_6H_4NH_2$

Synonyms: 4-(4-Aminobenzyl)aniline; Ancamine TL; Aniline, 4,4'-methylenedi-; Araldite hardener 972; Benzenamine, 4,4'-methylenebis-; Benzenamine, 4,4'-methylenebis-(aniline); Bis-*p*-aminofenylmethan; Bis(*p*-aminophenyl)methane; Bis(4-aminophenyl)methane; Bis(aminophenyl)methane; Curithane; DADPM; DAPM; DDM; DDV; *p,p'*-Diaminodifenylmethan; *p,p'*-Diaminodiphenylmethane; 4,4'-Diaminodiphenylmethane; Diaminodiphenylmethane; Di-(4-aminophenyl)methane; Dianilinemethane; 4,4'-Diphenylmethanediamine; Epicure DDM; Epikure DDM; HT 972; Jeffamine AP-20; MDA; 4,4'-Methylenebis(aniline); Methylenebis(aniline); 4,4'-Methylenebis(Benzeneamine); *p,p'*-Methylenedianiline; Methylenedianiline; 4,4'-Methylenedibenzenamine; 4,4'-Metilendianilina (Spanish); Sumicure M; Tonox

CAS Registry Number: 101-77-9

RTECS Number: BY5425000

DOT ID: UN 2651

EEC Number: 612-051-00-1

Regulatory Authority

- Carcinogen (Animal Positive) (IARC)[9] (ACGIH)[1] (DFG)[3]
- OSHA, 29CFR1910 Specifically Regulated Chemicals (See CFR 1910.1050)
- Air Pollutant Standard Set (ACGIH)[1] (HSE)[33] (Several States)[60]
- Clean Air Act: Hazardous Air Pollutants (Title I, Part A, Section 112)
- Superfund/EPCRA 40CFR302.4 Reportable Quantity (RQ): CERCLA, 1 lb (0.454 kg)
- EPCRA Section 313 Form R *de minimis* concentration reporting level: 0.1%

Cited in U.S. State Regulations: California (G), Connecticut (A), Florida (G), Kansas (A), Maryland (G), Massachusetts (G), Nevada (A), New Hampshire (G), New Jersey (G), New York (A), North Dakota (A), Pennsylvania (G), Rhode Island (G), South Carolina (A), Virginia (A), West Virginia (G).

Description: 4,4'-Diaminodiphenylmethane, $H_2NC_6H_4CH_2C_6H_4NH_2$, is a pale yellow crystalline solid (turns light brown on contact with air) with a faint amine-like odor. Boiling point = 398°C. Freezing/Melting point = 92°C. Flash point = 221°C (cc). Slightly soluble in water.

Potential Exposure: Approximately 99% of the DDM produced is consumed in its crude form (occasionally containing not more than 50% DDM and ply-DDM) at its production site by reaction with phosgene in the preparation of isocyanates and polyisocyanates. These isocyanates and polyisocyanates are employed in the manufacture of rigid polyurethane foams which find application as thermal insulation. Polyisocyanates are also used in the preparation of the semiflexible polyurethane foams used for automotive safety cushioning.

DDM is also used as: an epoxy hardening agent; a raw material in the production of polyurethane elastomers; in the rubber industry as a curative for neoprene and as an antifrosting agent (antioxidant) in footwear; a raw material in the production of Quana® nylon; and a raw material in the preparation of poly(amide-imide) resins (used in magnet wire enamels).

Incompatibilities: A weak base. Strong oxidizers may cause a fire and explosion hazard.

Permissible Exposure Limits in Air: The OSHA PEL (1910.1050) TWA is 0.010 ppm and STEL of 0.100 ppm. ACGIH recommends a TWA value of 0.1 ppm (0.8 mg/m^3). HSE[33] has adopted this same TWA and added a STEL of 0.5 ppm (4.0 mg/m^3). Several states have set guidelines or standards for MDA in ambient air[60] ranging from zero (North Dakota) to 0.4 μg/m^3 (Kansas) to 0.8 μg/m^3 (Virginia) to 2.67 μg/m^3 (New York) to 4.0 μg/m^3 (South Carolina) to 8.0 μg/m^3 (Connecticut) to 19.0 μg/m^3 (Nevada).

Determination in Air: Use OSHA Method #ID-57; or, Special Filter; Potassium hydroxide/Methanol; High-pressure liquid chromatography/Ultraviolet/Electrochemical detection; NIOSH IV Method #5029.

Permissible Concentration in Water: No criteria set.

Routes of Entry: Inhalation, passing through the skin.

Harmful Effects and Symptoms

Short Term Exposure: Irritates the eyes. A single large exposure or repeated smaller exposures can cause serious liver disease (toxic hepatitis) with symptoms of fever, upper abdominal pain, jaundice, dark urine, fatigue, and loss of appetite.

Long Term Exposure: 4,4'-Methylene Dianiline can cause liver damage and may damage the kidneys. Repeated or prolonged contact with skin may cause skin sensitization and dermatitis. Causes thyroid and other cancers in animals; a possible carcinogen in humans.

Points of Attack: Liver, kidneys, skin.

Medical Surveillance: Before beginning employment and at regular times after that, the following are recommended: Liver function tests. If symptoms develop or overexposure is suspected, the following may be useful: Kidney function tests. Evaluation by a qualified allergist.

First Aid: If this chemical gets into the eyes, remove any contact lenses at once and irrigate immediately for at least 15 minutes, occasionally lifting upper and lower lids. Seek medical attention immediately. If this chemical contacts the skin, remove contaminated clothing and wash immediately with soap and water. Seek medical attention immediately. If this chemical has been inhaled, remove from exposure, begin rescue breathing (using universal precautions) if breathing has stopped and CPR if heart action has stopped. Transfer promptly to a medical facility. When this chemical has been swallowed, get medical attention. Give large quantities of water and induce vomiting. Do not make an unconscious person vomit.

Personal Protective Methods: Wear protective gloves and clothing to prevent any reasonable probability of skin contact. Safety equipment suppliers/manufacturers can provide recommendations on the most protective glove/clothing material for your operation.. All protective clothing (suits, gloves, footwear, headgear) should be clean, available each day, and put on before work. Contact lenses should not be worn when working with this chemical. Wear dust-proof chemical goggles and face shield unless full facepiece respiratory protection is worn. Employees should wash immediately with soap when skin is wet or contaminated. Provide emergency showers and eyewash.

Respirator Selection: Where the potential exists for exposures over 0.1 ppm, use a MSHA/NIOSH approved supplied-air respirator with a full facepiece operated in the positive pressure mode or with a full facepiece, hood, or helmet in the continuous flow mode, or use a MSHA/NIOSH approved self-contained breathing apparatus with a full facepiece operated in pressure-demand or other positive pressure mode.

Storage: Prior to working with this chemical you should be trained on its proper handling and storage. Store in tightly closed containers in a cool, well-ventilated area away from strong oxidizers (such as chlorine, bromine, and fluorine). A regulated, marked area should be established where this chemical is handled, used, or stored in compliance with OSHA standard 1910.1045.

Shipping: This material falls in Hazard Class 6.1 and Packing Group III. The label required is "Keep Away From Food."

Spill Handling: Evacuate persons not wearing protective equipment from area of spill or leak until clean-up is complete. Remove all ignition sources. Collect powdered material in the most convenient and safe manner and deposit in sealed containers. Ventilate area after clean-up is complete. It may be necessary to contain and dispose of this chemical as a hazardous waste. If material or contaminated runoff enters waterways, notify downstream users of potentially contaminated waters. Contact your Department of Environmental Protection or your regional office of the federal EPA for specific recommendations. If employees are required to clean-up spills, they must be properly trained and equipped. OSHA 1910.120(q) may be applicable.

Fire Extinguishing: This chemical is a combustible solid. Use dry chemical, carbon dioxide, water spray, or alcohol foam extinguishers. Poisonous gases are produced in fire including aniline and nitrogen oxides. If material or contaminated runoff enters waterways, notify downstream users of potentially contaminated waters. Notify local health and fire officials and pollution control agencies. From a secure, explosion-proof location, use water spray to cool exposed containers. If cooling streams are ineffective (venting sound increases in volume and pitch, tank discolors, or shows any signs of deforming), withdraw immediately to a secure position. If employees are expected to fight fires, they must be trained and equipped in OSHA 1910.156.

Disposal Method Suggested: Controlled incineration (oxides of nitrogen are removed from the effluent gas by scrubbers and/or thermal devices).

References

New Jersey Department of Health, "Hazardous Substance Fact Sheet: 4,4'-Methylene Dianiline," Trenton, NJ (March 1986)

National Institute for Occupational Safety and Health, "Current Intelligence Bulletin 47:4,4'-Methylenedianiline (MDA)," Cincinnati, Ohio (July 25, 1986)

Occupational Health and Safety Administration, "Occupational Exposure to 4,4'-Methylenedianiline (MDA)," Federal Register 54, No. 91, 20672–20741 (May 12, 1989)

Diatomaceous Earth

Molecular Formula: SiO_2

Synonyms: Amorphous silica; Diatomaceous silica; Diatomite, uncalcined; Precipitated amorphous silica; Silica, amorphous diatomaceous earth; Silicon dioxide (amorphous)

CAS Registry Number: 61790-53-2; 7631-86-9 (Silica, amorphous)

RTECS Number: HL8600000

DOT ID: No citation.

Regulatory Authority

- Air Pollutant Standard Set (ACGIH)[1] (former USSR)[43] (OSHA)[58]

Cited in U.S. State Regulations: Illinois (G), New Hampshire (G), New Jersey (G), Rhode Island (G), West Virginia (G).

Description: Diatomaceous earth is a transparent to gray, odorless amorphous powder. Boiling point = 2,230°C. Freezing/Melting point = 1,710°C. Insoluble in water.

Potential Exposure: Diatomaceous earth is used as a filtering agent and as a filler in construction materials, pesticides, paints, and varnishes. The calcined version (which has been heat treated) is the most dangerous and contains crystallized silica, and should be handled as silica. See also other entries on silica.

Incompatibilities: High temperatures causes the formation of crystalline silica. Incompatible with oxygen difluoride, chlorine difluoride.

Permissible Exposure Limits in Air: The OSHA PEL TWA 20 MPPCF [80 mg/m^3/%SiO_2]. NIOSH recommends 6 mg/m^3 as the limit. The ACGIH recommended airborne exposure limit for respirable (small particles) diatomaceous earth is 1.0 mg/m^3 of total dust (large and small particles containing no asbestos and less than 1% free silica) averaged over an 8-hour workshift. The NIOSH IDLH is 3,000 mg/m^3 The former USSR-UNEP/IRPTC project has set a MAC of 2 mg/m^3 in workplace air.[43]

Determination in Air: Filter; Low-temperature ashing; X-ray diffraction spectrometry; NIOSH IV, Method #7501; or, Gravimetric plus OSHA Method ID/42.[58]

Permissible Concentration in Water: No criteria set.

Routes of Entry: Inhalation.

Harmful Effects and Symptoms

Short Term Exposure: Unknown at this time.

Long Term Exposure: Exposure can cause permanent scarring of the lungs, especially if diatomaceous earth has been calcined (heat treated). Symptoms include shortness of breath and cough. This can begin anywhere from months to years after exposure. The name of this disease is silicosis. With heavy exposure, individuals may become respiratory cripples. This can be fatal.

Points of Attack: Lungs.

Medical Surveillance: Before first exposure to calcined diatomaceous earth and at regular times after, the following are recommended: Medical exam of the lungs. Lung function tests. Chest x-ray (every 2 – 5 years).

First Aid: If this chemical gets into the eyes, remove any contact lenses at once and irrigate immediately for at least 15 minutes, occasionally lifting upper and lower lids. Seek medical attention immediately. If this chemical has been inhaled, remove from exposure, begin rescue breathing (using universal precautions) if breathing has stopped and CPR if heart action has stopped. Transfer promptly to a medical facility.

Personal Protective Methods: Wear protective gloves and clothing. Safety equipment suppliers/manufacturers can provide recommendations on the most protective glove/clothing material for your operation.. All protective clothing (suits, gloves, footwear, headgear) should be clean, available each day, and put on before work. Contact lenses should not be worn when working with this chemical. Wear dust-proof chemical goggles and face shield unless full facepiece respiratory protection is worn. Employees should wash immediately with soap when skin is wet or contaminated. Provide emergency showers and eyewash.

Respirator Selection: NIOSH: *30 mg/m^3:* DM (any dust and mist respirator). *60 mg/m^3:* DMXSQ (any dust and mist respirator except single-use and quarter mask respirators); or SA (any supplied-air respirator). *150 mg/m^3:* SA:CF (any supplied-air respirator operated in a continuous-flow mode); or PAPRDM (any powered, air-purifying respirator with a dust and mist filter). *300 mg/m^3:* HiEF (any air-purifying, full-facepiece respirator with a high-efficiency particulate filter); or SAT:CF (any supplied-air respirator that has a tight-fitting facepiece and is operated in a continuous-flow mode); or PAPRTHiE (any powered, air-purifying respirator with a tight-fitting facepiece and a high-efficiency particulate filter); or SCBAF (any self-contained breathing apparatus with a full facepiece); or SAF (any supplied-air respirator with a full facepiece). *3,000 mg/m^3:* SAF:PD,PP (any supplied-air respirator that has a full facepiece and is operated in a pressure-demand or other positive-pressure mode). *Emergency or planned entry into unknown concentrations or IDLH conditions:* SCBAF:PD,PP (any self-contained breathing apparatus that has a full facepiece and is operated in a pressure-demand or other positive-pressure mode); or SAF:PD,PP:ASCBA (any supplied-air respirator that has a full facepiece and is operated in a pressure-demand or other positive-pressure mode in combination with an auxiliary, self-contained breathing apparatus operated in a pressure-demand or other positive-pressure mode). *Escape:* HiEF (any air-purifying, full-facepiece respirator with a high-efficiency particulate filter); or SCBAE (any appropriate escape-type, self-contained breathing apparatus).

Storage: Store in tightly closed containers in a cool well-ventilated area.

Shipping: This material is not singled out by DOT[19] in its Performance-Oriented Packaging Standards.

Spill Handling: Collect powdered material in the most convenient and safe manner and deposit in sealed containers. Ventilate area after clean-up is complete. It may be necessary to contain and dispose of this chemical as a hazardous waste. If material or contaminated runoff enters waterways, notify downstream users of potentially contaminated waters. Contact your Department of Environmental Protection or your regional office of the federal EPA for specific recommendations. If employees are required to clean-up spills, they must be properly trained and equipped. OSHA 1910.120(q) may be applicable.

Fire Extinguishing: Use any extinguishing agent suitable for surrounding fire. Poisonous gases are produced in fire. If material or contaminated runoff enters waterways, notify downstream users of potentially contaminated waters. Notify local health and fire officials and pollution control agencies. From a secure, explosion-proof location, use water spray to cool exposed containers. If cooling streams are ineffective (venting sound increases in volume and pitch, tank discolors, or shows any signs of deforming), withdraw immediately to a secure position. If employees are expected to fight fires, they must be trained and equipped in OSHA 1910.156.

References

New Jersey Department of Health, "Hazardous Substance Fact Sheet: Diatomaceous Earth," Trenton, NJ (August 1985)

Diazepam

Molecular Formula: $C_{16}H_{13}ClN_2O$

Synonyms: Alboral; Aliseum; Amiprol; Ansiolin; Ansiolisina; Apaurin; Apozepam; Assival; Atensine; Atilen; 2H-1,4-Benzodiazepin-2-one,7-chloro-1,3-dihydro-1-methyl-5-phenyl-bialzepam; Calmocitene; Calmpose; Cercine; Ceregulart; 7-Chloro-1,3-dihydro-1-methyl-5-phenyl-2H-1,4-benzodiazepin-2-one; 7-Chloro-1-methyl-5-3H-1,4-benzodiazepin-2(1H)-one; 7-Chloro-1-methyl-2-oxo-5-phenyl-3H-1,4-benzodiazepine; 7-Chloro-1-methyl-5-phenyl-2H-1,4-benzodiazepin-2-one; 7-Chloro-1-methyl-5-phenyl-3H-1,4-benzodiazepin-2(1H)-one; 7-Chloro-1-methyl-5-phenyl-1,3-dihydro-2H-1,4-benzodiazepin-2-one; Condition; DAP; Diacepan; Diapam; Diazemuls; Diazepam; Diazepamu (Polish); Diazetard; Dienpax; Dipam; Dipezona; Domalium; Duksen; Duxen; E-Pam; Eridan; Faustan; Freudal; Frustan; Gihitan; Horizon; Kabivitrum; Kiatrium; LA-III; Lembrol; Levium; Liberetas; Methyl diazepinone; 1-Methyl-5-phenyl-7-chloro-1,3-dihydro-2H-1,4-benzodiazepin-2-one; Morosan; Noan; NSC-77518; Pacitran; Paranten; Paxate; Paxel; Plidan; Quetinil; Quiatril; Quievita; Relaminal; Relanium; Relax; Renborin; RO 5-2807; S.A.R.L.; Saromet; Sedipam; Seduksen; Seduxen; Serenack; Serenamin; Serenzin; Setonil; Sibazon; Sonacon; Stesolid; Stesolin; Tensopam; Tranimul; Tranqdyn; Tranquirit; Umbrium; Unisedil; Usempax AP; Valeo; Valitran; Valium; Valium R; Vatran; Velium; Vival; Vivol; WY-3467; Zipan

CAS Registry Number: 439-14-5

RTECS Number: DF1575000

DOT ID: UN 3249

Cited in U.S. State Regulations: California (G), Massachusetts (G), New Jersey (G). Pennsylvania (G).

Description: Diazepam, $C_{16}H_{13}ClN_2O$, is a yellow crystalline powder. Freezing/Melting point = 125 – 126°C. Insoluble in water.

Potential Exposure: Those involved in the manufacture, packaging or consumption of this widely used tranquilizing drug.

Permissible Exposure Limits in Air: No standards set.

Permissible Concentration in Water: No criteria set.

Routes of Entry: Inhalation.

Harmful Effects and Symptoms

Short Term Exposure: When used as a medial drug, diazepam can cause drowsiness and difficulty with coordination, concentration and balance. It may also cause irritability, anxiety, weakness, headaches, upset stomach and joint pains. Less common side effects include jaundice, skin rashes, or a drop in the white blood cell count. These effects might also occur from workplace exposure due to breathing in dust during packaging or manufacture. The oral LD_{50} rat is 352 mg/kg (moderately toxic).

Long Term Exposure: Diazepam is a probable teratogen and may be a mutagen. Handle with extreme caution. Sudden discontinuing the exposure following high exposure for at least 3 months may cause shakiness, irritability, and possible convulsions.

Medical Surveillance: If symptoms develop or overexposure is suspected, the following may be useful: Blood test for diazepam level.

First Aid: If this chemical gets into the eyes, remove any contact lenses at once and irrigate immediately for at least 15 minutes, occasionally lifting upper and lower lids. If this chemical contacts the skin, remove contaminated clothing and wash immediately with soap and water.

Personal Protective Methods: Wear protective gloves and clothing to prevent any reasonable probability of skin contact. Safety equipment suppliers/manufacturers can provide recommendations on the most protective glove/clothing material for your operation.. All protective clothing (suits, gloves, footwear, headgear) should be clean, available each day, and put on before work. Contact lenses should not be worn when working with this chemical. Wear dust-proof chemical goggles and face shield unless full facepiece respiratory protection is worn. Employees should wash immediately with soap when skin is wet or contaminated. Provide emergency showers and eyewash.

Respirator Selection: Where the potential exists for exposure to diazepam, use a MSHA/NIOSH approved full facepiece respirator with a high efficiency particulate filter. Greater protection is provided by a powered-air purifying respirator. Where the potential for high exposures exists, use a MSHA/NIOSH approved supplied-air respirator with a full facepiece operated in the positive pressure mode or with a full facepiece, hood, or helmet in the continuous flow mode, or use a MSHA/NIOSH approved self-contained breathing apparatus with a full facepiece operated in the pressure-demand or other positive pressure mode.

Storage: Prior to working with diazepam you should be trained on its proper handling and storage. Store in tightly closed containers in a cool, well-ventilated area. If you are required to work in a sterile environment you require specific training.

Shipping: Diazepam is not specifically cited but medicines, poisonous, solid, n.o.s. fall in DOT Hazard Class 6.1 and diazepam would fall in Packing Group III. It requires a "Keep Away From Food" label.

Spill Handling: Evacuate persons not wearing protective equipment from area of spill or leak until clean-up is complete. Remove all ignition sources. Collect powdered material in the most convenient and safe manner and deposit in sealed containers. Ventilate area after clean-up is complete. It may be necessary to contain and dispose of this chemical as a hazardous waste. If material or contaminated runoff enters waterways, notify downstream users of potentially contaminated waters. Contact your Department of Environmental Protection or your regional office of the federal EPA for specific recommendations. If employees are required to clean-up spills, they must be properly trained and equipped. OSHA 1910.120(q) may be applicable.

Fire Extinguishing: Use extinguishing agents suitable for surrounding fire. Poisonous gases are produced in fire including chlorine, and oxides of nitrogen. If material or contaminated runoff enters waterways, notify downstream users of potentially contaminated waters. Notify local health and fire officials and pollution control agencies. From a secure, explosion-proof location, use water spray to cool exposed containers. If cooling streams are ineffective (venting sound increases in volume and pitch, tank discolors, or shows any signs of deforming), withdraw immediately to a secure position. If employees are expected to fight fires, they must be trained and equipped in OSHA 1910.156.

References

New Jersey Department of Health, "Hazardous Substance Fact Sheet: Diazepam," Trenton, NJ (May 1986)

Diazinon

Molecular Formula: $C_{12}H_{21}O_3N_2SP$

Synonyms: AG-500; AI3-19507; Alfa-Tox; Antigal; Antlak; Basudin; Basudin 10 G; Basudin E; Bazuden; Caswell No. 342; Dazzel; *O,O*-Diaethyl-*O*-(2-isopropyl-4-methyl-pyrimidin-6-yl)-monothiophosphat (German); *O,O*-Diaethyl-*O*-(2-isopropyl-4-methyl-6-pyrimidyl)-thionophosphat (German); Dianon; Diaterr-Fos; Diazajet; Diazatol; Diazide; Diazinon AG 500; Diazinone; Diazitol; Diazol; Dicid; Diethyl 2-isopropyl-4-methyl-6-pyrimidinl phosphorothionate; Diethyl 4-(2-isopropyl-6-methyl-pyrimidinl)phosphorothionate; *o,o*-Diethyl-*o*-(2-isopropyl-4-methyl-pyrimidin-6-yl)-monothiofospaat (Dutch); *O,O*-Diethyl *O*-2-isopropyl-6-methylpyrimidin-4-yl phosphorothionate; *O,O*-Diethyl *O*-(2-isopropyl-4-methyl-6-pyrimidyl) phosphorothionate; *O,O*-Diethyl *O*-(2-isopropyl-4-methyl-6-pyrimidyl) thionophosphate; Diethyl 2-isopropyl-4-methyl-6-pyrimidylthionophosphate; *O,O*-Diethyl 2-isopropyl-4-methylpyrimidyl-6-thiophosphate; *O,O*-Diethyl *O*-6-methyl-2-isopropyl-4-pyrimidinyl phosphorthioate; *O,O*-Diethyl *O*-[6-methyl-2-(1-methylethyl)-4-pyrimidinyl] phosphorthioate; Dimpylate; Dipofene; Diziktol; Dizinon; Dyzol; ENT 19,507; EPA pesticide chemical code 057801; Exodin; G-24480; G 301; Gardentox; Geigy 24480; Isopropylmethylpyrimidyl diethyl thiophosphate; *O*-2-Isopropyl-4-methylpyrimyl *O,O*-diethyl phosphorothioate; Kayazinon; Kayazol; NA 2783 (DOT); NCI-C08673; Neocidol; Neocidol (oil); Nipsan; Nucidol; Oleodiazinon; Phosphoric acid, *O,O*-diethyl *O*-6-methyl-2-(1-methylethyl)-4-pyrimidinyl ester; Phosphorothioate, *O,O*-diethyl *O*-6-(2-isopropyl-4-methylpyrimidyl); Phosphorothioic acid, *O,O*-diethyl *O*-(2-isopropyl-6-methyl-4-pyrimidinyl) ester; Phosphorothioic acid, *O,O*-diethyl *O*-(isopropylmethylpyrimidyl) ester; Phosphorothioic acid, *O,O*-diethyl *O*-[6-methyl-2-(1-methylethyl)-4-pyrimidinyl] ester; 4-Pyrimidinol, 2-isopropyl-6-methyl-, *O*-ester with *O,O*-diethyl phosphorothioate; Root Guard; RTECS No. TF33450000; Sarolex; Spectracide; Spectracide 25EC; Srolex; Thiophosphate de *O,O*-diethyle et de *O*-2-isopropyl-4-methyl 6-pyrimidyle (French); Thiophosphoric acid 2-isopropyl-4-methyl-6-pyrimidyl diethyl ester

CAS Registry Number: 333-41-5

RTECS Number: TF3325000

DOT ID: UN/NA 2783

EEC Number: 015-040-00-4

Regulatory Authority

- Air Pollutant Standard Set (ACGIH)[1] (Argentina) (Australia) (DFG)[3] (HSE)[33] (Israel) (Mexico) (former USSR)[35] (Several States)[60] (Several Canadian Provinces)
- Clean Water Act: Section 311 Hazardous Substances/RQ 40CFR117.3 (same as CERCLA, see below)
- Superfund/EPCRA 40CFR302.4 Reportable Quantity (RQ): CERCLA, 1 lb (0.454 kg)
- EPCRA Section 313 Form R *de minimis* concentration reporting level: 1.0%
- U.S. DOT Regulated Marine Pollutant (49CFR172.101, Appendix B), severe pollutant
- Canada, Drinking Water Quality, 0.02 mg/l MAC

Cited in U.S. State Regulations: Alaska (G), California (A, W), Connecticut (A), Florida (G), Illinois (G), Kansas (W), Maine (G, W), Massachusetts (G), Michigan (G), Minnesota (G), Nevada (A), New Hampshire (G, W), New Jersey (G), North Dakota (A), Pennsylvania (G), Rhode Island (G), Virginia (A), West Virginia (G).

Description: Diazinon, $C_{12}H_{21}O_3N_2SP$, is a combustible, colorless, oily liquid with a faint amine odor. Technical grade is pale to dark brown. Boiling point = (decomposes)120°C; 83 – 84°C (at 0.002 mm). Flash point = 82°C. Hazard Identification (based on NFPA-704 M Rating System): Health 3, Flammability 1, Reactivity 0. Slightly soluble in water.

Potential Exposure: To producers, formulators and applicators of this nonsystemic pesticide and acaricide. Diazinon is used in the United States on a wide variety of agricultural crops, ornamentals, domestic animals, lawns and gardens, and household pests.

Incompatibilities: Hydrolyzes slowly in water and dilute acid. Reacts with strong acids and alkalis with possible formation of highly toxic tetraethyl thiopyrophosphates. Incompatible with copper-containing compounds.

Permissible Exposure Limits in Air: There is no OSHA PEL. NIOSH and ACGIH recommend an airborne exposure limit of 0.1 mg/m^3 TWA with the notation "skin" indicating the possibility of cutaneous absorption. Argentina, Australia, Israel, Mexico, HSE,[33] and the Canadian provinces of Alberta, BC, Ontario, and Quebec use this same value but HSE, Mexico, Alberta, and BC add the STEL of 0.3 mg/m^3 as does Argentina.[35] The DFG, however, has set an MAK of 1.0 mg/m^3 and a STEL of 10 mg/m^3.[3] The former USSR[35] has set a MAC of 0.2 mg/m^3 in workplace air and a MAC for ambient air in residential areas of 0.01 mg/m^3 on either a momentary or a daily average basis. In addition, several states have set guidelines or standards for Diazinon in ambient air[60] ranging from 1.0 μg/m^3 (North Dakota) to 1.6 μg/m^3 (Virginia) to 2.0 μg/m^3 (Connecticut and Nevada).

Determination in Air: OSHA versatile sampler-2; Toluene/Acetone; Gas chromatography/Flame photometric detection for sulfur, nitrogen, or phosphorus; NIOSH IV Method #5600, Organophosphorus Pesticides. See also OSHA Method #62.[58]

Permissible Concentration in Water: Canada set a MAC of 0.02 mg/l. The former USSR[35] has set a MAC in water bodies for domestic purposes of 0.3 mg/l. The USEPA has determined a NOAEL of 0.05 mg/kg/day which gives a long-term health advisory of 0.0175 mg/l and a lifetime health advisory of 0.00063 mg/l. Several states have set guidelines for Diazinon in drinking water[61] ranging from 4 μg/l (Maine) to 14 μg/l (California and Kansas).

Determination in Water: By Methylene Chloride extraction followed by gas chromatography.

Routes of Entry: Inhalation, skin absorption, ingestion, skin and/or eye contact.

Harmful Effects and Symptoms

Short Term Exposure: Diazinon can affect you when breathed in and quickly enters the body by passing through the skin. May cause skin and eye irritation. Exposure can cause organophosphate poisoning with headache, sweating, nausea, and vomiting, diarrhea, muscle twitching and possible death. It is a moderately toxic organophosphate chemical. The LD_{50} oral-rat is 66 mg/kg (moderately toxic).

Long Term Exposure: Diazinon may damage the developing fetus. Exposure can cause severe organophosphate poisoning with headache, sweating, nausea and vomiting, diarrhea, loss of coordination, and death. Diazinon may affect the liver.

Points of Attack: Eyes, respiratory system, central nervous system, cardiovascular system, blood cholinesterase.

Medical Surveillance: Before employment and at regular times after that, the following are recommended: Plasma and red blood cell cholinesterase levels (tests for the enzyme poisoned by this chemical). If exposure stops, plasma levels return to normal in 1 – 2 weeks while red blood cell levels may be reduced for 1 – 3 months.

When cholinesterase enzyme levels are reduced by 25% or more below preemployment levels, risk of poisoning is increased, even if results are in lower ranges of "normal." Reassignment to work not involving organophosphate or carbamate pesticides is recommended until enzyme levels recover. If symptoms develop or overexposure occurs, repeat the above tests as soon as possible and get an exam of the nervous system. Also consider complete blood count. Consider chest x-ray following acute overexposure. Do not drink any alcoholic beverages before or during use. Alcohol promotes absorption of organic phosphates. Liver function tests. Exam of the nervous system. Complete blood count.

First Aid: If this chemical gets into the eyes, remove any contact lenses at once and irrigate immediately for at least 15 minutes, occasionally lifting upper and lower lids. Seek medical attention immediately. If this chemical contacts the skin, remove contaminated clothing and wash immediately with soap and water. Speed in removing material from skin is of extreme importance. Shampoo hair promptly if contaminated. Seek medical attention immediately. If this chemical has been inhaled, remove from exposure, begin rescue breathing (using universal precautions) if breathing has stopped and CPR if heart action has stopped. Transfer promptly to a medical facility. When this chemical has been swallowed, get medical attention. Give large quantities of water and induce vomiting. Do not make an unconscious person vomit. Effects may be delayed; medical observation is recommended.

Personal Protective Methods: Wear protective gloves and clothing to prevent any reasonable probability of skin contact. Safety equipment suppliers/manufacturers can provide recommendations on the most protective glove/clothing material for your operation.. All protective clothing (suits, gloves, footwear, headgear) should be clean, available each day, and put on before work. Contact lenses should not be worn when working with this chemical. Wear splash-proof chemical goggles and face shield unless full facepiece respiratory protection is worn. Employees should wash immediately with soap when skin is wet or contaminated. Provide emergency showers and eyewash.

Respirator Selection: Where the potential exists for exposures over 0.1 mg/m^3 use a MSHA/NIOSH approved respirator with a pesticide cartridge. The prefilter should be a high efficiency particulate filter. More protection is provided by a full facepiece respirator than by a half-mask respirator, and even greater protection is provided by a powered-air purifying respirator.

Where the potential for high exposures exists, use an MSHA/NIOSH approved supplied-air respirator with a full facepiece operated in the positive pressure mode or with a full facepiece, hood, or helmet in the continuous flow mode, or use an MSHA/NIOSH approved self-contained breathing apparatus with a full facepiece operated in the pressure-demand or other positive pressure mode.

Storage: Store in tightly closed containers in a cool well-ventilated area away from water, and oxidizers such as (peroxides, nitrated, permanganates, chlorates, and perchlorates).

Shipping: Organophosphorus pesticides, liquid, toxic n.o.s. fall in DOT Hazard Class 6.1 and Diazinon in Packing Group III. The label required is "Keep Away From Food."

Spill Handling: Evacuate and restrict persons not wearing protective equipment from area of spill or leak until cleanup is complete. Remove all ignition sources. Ventilate area of spill or leak. Collect for reclamation or absorb in vermiculate, dry sand, earth or a similar material. Dispose by absorbing in vermiculite, dry sand, earth or a similar material, and depositing in an approved facility. Do not use water, as toxic gases may be produced. It may be necessary to contain and dispose of this chemical as a hazardous waste. If material or contaminated runoff enters waterways, notify downstream users of potentially contaminated waters. Contact your Department of Environmental Protection or your regional office of the federal EPA for specific recommendations. If employees are required to clean-up spills, they must be properly trained and equipped. OSHA 1910.120(q) may be applicable.

Fire Extinguishing: This chemical is a combustible liquid. Diazinon decomposes on heating above 120°C producing toxic fumes including phosphorous oxides and sulfur oxides. Use dry chemical, carbon dioxide, or foam extinguishers. Vapors are heavier than air and will collect in low areas. Containers may explode in fire. Storage containers and parts of containers may rocket great distances, in many directions. If material or contaminated runoff enters waterways, notify downstream users of potentially contaminated waters. Notify local health and fire officials and pollution control agencies. From a secure, explosion-proof location, use water spray to cool exposed containers. If cooling streams are ineffective (venting sound increases in volume and pitch, tank discolors, or shows any signs of deforming), withdraw immediately to a secure position. If employees are expected to fight fires, they must be trained and equipped in OSHA 1910.156.

Disposal Method Suggested: Diazinon is hydrolyzed in acid media about 12 times as rapidly as parathion, and at about the same rate as parathion in alkaline media. In excess water this compound yields diethylthiophosphoric acid and 2-isopropyl-4-methyl-6-hydroxypyrimidine. With insufficient water, highly toxic tetraethyl monothiopyrophosphate is formed. Therefore, incineration would be a preferable ultimate disposal method with caustic scrubbing of the incinerator effluent.[22] In accordance with 40CFR165 recommendations for the disposal of pesticides and pesticide containers. Must be disposed properly by following package label directions or by contacting your state pesticide or environmental control agency or by contacting your regional EPA office.

References

New Jersey Department of Health, "Hazardous Substance Fact Sheet: Diazinon," Trenton, NJ (March 1998)

U.S. Environmental Protection Agency, "Health Advisory: Diazinon," Washington, DC, Office of Drinking Water (August 1987)

Sax N. I., Ed., "Dangerous Properties of Industrial Materials Report," 7, No. 5, 36–43 (1987)

Diazomethane

Molecular Formula: CH_2N_2

Synonyms: Azimethylene; Azomethylene; Diazirine; Diazometano (Spanish); Diazonium methylide; Methane, diazo-

CAS Registry Number: 334-88-3

RTECS Number: PA7000000

EEC Number: 006-068-00-8

Regulatory Authority

- Carcinogen (Animal Positive) (IARC)[9] (DFG)[3]
- Banned or Severely Restricted (Belgium, Sweden) (UN)[13]
- OSHA 29CFR1910.119, Appendix A. Process Safety List of Highly Hazardous Chemicals, TQ = 500 lb (227 kg)
- Air Pollutant Standard Set (ACGIH)[1] (HSE)[33] (OSHA)[58] (Several States)[60]
- Clean Air Act: Hazardous Air Pollutants (Title I, Part A, Section 112)
- Superfund/EPCRA 40CFR302.4 Reportable Quantity (RQ): CERCLA, 1 lb (0.454 kg)
- EPCRA Section 313 Form R *de minimis* concentration reporting level: 1.0%
- Canada, WHMIS, Ingredients Disclosure List

Cited in U.S. State Regulations: Alaska (G), California (A, G), Connecticut (A), Florida (G, A), Maine (G), Maryland (G), Massachusetts (G), Minnesota (G), Nevada (A), New Hampshire (G, W), New Jersey (G), New York (A), North Dakota (A), Pennsylvania (G), Rhode Island (G), South Carolina (A), Virginia (A), West Virginia (G).

Description: Diazomethane, CH_2N_2, is a flammable, yellow gas or a liquid under pressure. Boiling point = -23°C. Freezing/Melting point = -145°C. Autoignition temperature = 100°C (explodes). Hazard Identification (based on NFPA-704 M Rating System): Health 4, Flammability 3, Reactivity 3. Decomposes in water.

Potential Exposure: Diazomethane is a powerful methylating agent for acidic compounds such as carboxylic acids, phenols and enols. It is used in pesticide manufacture and pharmaceutical manufacture.

Incompatibilities: Heat (at about or above 100°C), shock, friction, concussion, sunlight or other intense illuminations may cause explosions. Contact with alkali metals, drying agents such as calcium sulfate, or rough edges (such as ground glass) may cause explosions.

Permissible Exposure Limits in Air: The Federal Limit (OSHA PEL)[58] and the recommended ACGIH TWA value is 0.2 ppm (0.4 mg/m^3). The Australian, HSE, Israeli, and Canadian provincial (Alberta, BC, Ontario, Quebec) value is the same and Alberta's STEL is 0.6 ppm. The NIOSH IDLH level is 2 ppm. Several states have set guidelines or standards for diazomethane in ambient air[60] ranging from 1.3 μg/m^3 (New York) to 2.0 μg/m^3 (South Carolina) to 4.0 μg/m^3 (Florida and North Dakota) to 7.0 μg/m^3 (Virginia) to 8.0 μg/m^3 (Connecticut) to 10.0 μg/m^3 (Nevada).

Determination in Air: XAD-2® (tube); CS2; Gas chromatography/Flame ionization detection; NIOSH IV Method #2515.[18]

Permissible Concentration in Water: No criteria set, but EPA[32] has suggested a permissible ambient goal of 5.5 μg/l, based on health effects.

Routes of Entry: Inhalation, ingestion, skin and eye contact (liquid).

Harmful Effects and Symptoms

Short Term Exposure: Cough, shortness of breath; headaches; flushed skin, fever; chest pain, pulmonary edema, pneumonitis; asthma; eye irritation. It is extremely toxic. Inhalation can cause pulmonary edema, a medical emergency that can be delayed for several hours. This can cause death. Exposure can cause severe lung damage with symptoms of coughing, chest pain, shortness of breath, fever and fatigue. Exposure to the gas or liquid can cause severe skin burns and eye damage. Contact with the liquid can cause frostbite.

Long Term Exposure: Repeated exposures, even at low levels, may cause an asthma-like lung allergy. This chemical is a possible human carcinogen.

Points of Attack: Respiratory system, lungs, eyes, skin.

Medical Surveillance: Before beginning employment and at regular times after that, the following are recommended: Lung function tests. These may be normal at first if the person is not having an attack at the time. If symptoms develop or overexposure is suspected, the following may be useful: Evaluation by a qualified allergist, including careful exposure history and special testing, may help diagnose allergy. Consider chest x-ray after acute overexposure.

First Aid: If this chemical gets into the eyes, remove any contact lenses at once and irrigate immediately for at least 15 minutes, occasionally lifting upper and lower lids. Seek medical attention immediately. If this chemical contacts the skin, remove contaminated clothing and wash immediately with soap and water. Seek medical attention immediately. If this chemical has been inhaled, remove from exposure, begin rescue breathing (using universal precautions) if breathing has stopped and CPR if heart action has stopped. Transfer promptly to a medical facility. When this chemical has been swallowed, get medical attention. Give large quantities of water and induce vomiting. Do not make an unconscious person vomit. Medical observation is recommended for 24 – 48 hours after breathing overexposure, as pulmonary edema may be delayed. As first aid for pulmonary edema, a doctor or authorized paramedic may consider administering a corticosteroid spray. If frostbite has occurred, seek medical attention immediately; do *NOT* rub the affected areas or flush them with water. In order to prevent further tissue damage, do *NOT* attempt to remove frozen clothing from frostbitten areas. If frostbite has *NOT* occurred, immediately and thoroughly wash contaminated skin with soap and water.

Personal Protective Methods: Wear protective gloves and clothing to prevent any reasonable probability of skin contact. Safety equipment suppliers/manufacturers can provide recommendations on the most protective glove/clothing material for your operation.. All protective clothing (suits, gloves, footwear, headgear) should be clean, available each day, and put on before work. Contact lenses should not be worn when working with this chemical. Wear eye protection to prevent any possibility of contact. Employees should wash immediately with soap when skin is wet or contaminated. Provide emergency showers and eyewash.

Where exposure to the liquefied compressed gas may occur, employees should be provided with special clothing designed to prevent frostbite.

Respirator Selection: NIOSH/OSHA: *up to 2 ppm:* SA (any supplied-air respirator);* or SCBA (any self-contained breathing apparatus with a full facepiece). *Emergency or planned entry into unknown concentrations or IDLH conditions*: SCBAF:PD,PP (any self-contained breathing apparatus that has a full facepiece and is operated in a pressure-demand or other positive-pressure mode); or SAF:PD,PP:ASCBA (any supplied-air respirator that has a full facepiece and is operated in a pressure-demand or other positive-pressure mode in combination with an auxiliary self-contained breathing apparatus operated in a pressure-demand or other positive-pressure mode). *Escape:* GMFOV [any air-purifying, full-facepiece respirator (gas mask) with a chin-style, front-or back-mounted organic vapor canister] or SCBAE (any appropriate escape-type, self-contained breathing apparatus).

* Substance causes eye irritation or damage; eye protection needed

Storage: Prior to working with this chemical you should be trained on its proper handling and storage. Diazomethane must be stored to avoid contact with alkali metals, such as lithium, sodium, or potassium; or drying agents, such as calcium sulfate, since violent reactions occur. Safety barriers or shields should be used to protect workers from accidental explosions. Sources of ignition such as smoking and open flames are prohibited where diazomethane is handled, used, or stored. Metal containers used in the transfer of 5 gallons or more diazomethane should be grounded and bonded. Drums must be equipped with self-closing valves, pressure vacuum bungs, and flame arresters. Use only non-sparking tools and equipment, especially when opening and closing containers of diazomethane. Wherever diazomethane is used, handled, manufactured, or stored, use explosion-proof electrical equipment and fittings. A regulated, marked area should be established where this chemical is handled, used, or stored in compliance with OSHA standard 1910.1045.

Shipping: Diazomethane is not specifically covered in DOT's Performance-Oriented Packaging Standards.[19]

Spill Handling: If in a building, evacuate building and confine vapors by closing doors and shutting down HVAC systems. Restrict persons not wearing protective equipment from area of spill or leak until cleanup is complete. Remove all ignition sources. Ventilate area of spill or leak to disperse the gas. Wear chemical protective suit with self-contained breathing apparatus to combat spills. Stay upwind and use water spray to "knock down" vapor; contain runoff. Stop the flow of gas, if it can be done safely from a distance. If source is a cylinder and the leak cannot be stopped in place, remove the leaking cylinder to a safe place, and repair leak or allow cylinder to empty. If in liquid form, allow to vaporize or absorb the spilled chemical by using a sponge and water. Decompose chemically with a 10% ceric ammonium nitrate solution. Keep this chemical out of confined spaces, such as a sewer, because of the possibility of explosion, unless the sewer is designed to prevent the buildup of explosive concentrations. If employees are required to clean-up spills, they must be properly trained and equipped. OSHA 1910.120(q) may be applicable.

Fire Extinguishing: This chemical is a flammable and explosive gas. Poisonous gases including nitrogen oxides are produced in fire. Use dry chemical or sand to extinguish fire. Vapors are heavier than air and will collect in low areas. Vapors may travel long distances to ignition sources and flashback. Vapors in confined areas may explode when exposed to fire. Containers may explode in fire. Storage containers and parts of containers may rocket great distances, in many directions. If material or contaminated runoff enters waterways, notify downstream users of potentially contaminated waters. Notify local health and fire officials and pollution control agencies. From a secure, explosion-proof location, use water spray to cool exposed containers. If cooling streams are ineffective (venting sound increases in volume and pitch, tank discolors, or shows any signs of deforming), withdraw immediately to a secure position. If employees are expected to fight fires, they must be trained and equipped in OSHA 1910.156.

Disposal Method Suggested: Decompose chemically with ceric ammonium nitrate under constant agitation and cooling.[24]

References

Sax N. I., Ed., "Dangerous Properties of Industrial Materials Report," 1, No. 3, 55 (1981)

New Jersey Department of Health, "Hazardous Substance Fact Sheet: Diazomethane," Trenton, NJ (October 1998)

Dibenz(a,h)Anthracene

Molecular Formula: $C_{22}H_{14}$

Synonyms: AI3-18996; 1,2:5,6-Benzanthracene; 1,2,5,6-DBA; DBA; 1,2,5,6-Dibenzanthracene; 1,2:5,6-Dibenz(a) anthracene; 1,2:5,6-Dibenzanthracene; Dibenzanthracene; Dibenz(a,h)antraceno (Spanish); 1,2:5,6-Dibenzoanthracene

CAS Registry Number: 53-70-3

RTECS Number: HN2625000

DOT ID: UN 2811

EEC Number: 601-041-00-2

Regulatory Authority

- Carcinogen (Limited human data) (IARC)[9] (Animal Positive) (DFG)
- OSHA, 29CFR1910 Specifically Regulated Chemicals (See CFR 1910.1002) as coal tar pitch volatiles
- Air Pollutant Standard Set (ACGIH)[1] (HSE)[33] (OSHA)[58]
- Clean Water Act: 40CFR423, Appendix A, Priority Pollutants; Section 313 Water Priority Chemicals (57FR41331, 9/9/92); 40CFR401.15 Section 307 Toxic Pollutants as polynuclear aromatic hydrocarbons
- EPA Hazardous Waste Number (RCRA No.): U063
- RCRA, 40CFR261, Appendix 8 Hazardous Constituents
- RCRA 40CFR268.48; 61FR15654, Universal Treatment Standards: Wastewater (mg/l), 0.055; Nonwastewater (mg/kg), 8.2
- RCRA 40CFR264, Appendix 9; TSD Facilities Ground Water Monitoring List. Suggested test method(s) (PQL µg/l): 8100 (200); 8270 (10)
- Superfund/EPCRA 40CFR302.4 Reportable Quantity (RQ): CERCLA, 1 lb (0.454 kg)
- EPCRA Section 313 Form R *de minimis* concentration reporting level: 0.1
- Canada, WHMIS, Ingredients Disclosure List

Cited in U.S. State Regulations: California (A, G), Florida (G), Illinois (G), Kansas (G, W), Louisiana (G), Massachusetts (G), Michigan (G), Minnesota (G), New Hampshire (G), New Jersey (G), Pennsylvania (G), Vermont (G), Virginia (G), Washington (G), West Virginia (G), Wisconsin (G).

Description: Dibenz(a,h)anthracene is a polynuclear aromatic hydrocarbon consisting of 5 benzene rings fused together. It has the formula $C_{22}H_{14}$. It is a colorless, crystalline solid. Boiling point = 524°C. Freezing/Melting point = 267 – 270°C. Hazard Identification (based on NFPA-704 M Rating System): Health 1, Flammability 1, Reactivity 0. Insoluble in water.

Potential Exposure: Dibenz(a,h)anthracene [DB(a,h)A] is a chemical substance formed during the incomplete burning of fossil duel, garbage, or any organic matter and is found in smoke in general; it condenses on dust particles and is distributed into water and soil and on crops. DB(a,h)A is a polycyclic aromatic hydrocarbon (PAH) and is also a component of coal tar pitch, which is used in industry as a binder for electrodes, and of creosote used to preserve wood. PAHs are also found in limited amounts in bitumens and asphalt used in industry and for paving.

Incompatibilities: Strong oxidizers.

Permissible Exposure Limits in Air: Dibenz(a,h)anthracene falls in the category of coal tar pitch volatiles which has a TWA set by OSHA: PEL 0.2 mg/m^3 (benzene-soluble fraction) (1910.1002).

Determination in Air: Sample preparation by solvent extraction may be followed by measurement by HPLC:NIOSH Method #5506 or gas chromatography: NIOSH Method 5515.[18] For coal tar volatiles NIOSH recommends: Filter; Benzene; Gravimetric; OSHA Method #58.

Permissible Concentration in Water: Under the priority toxic pollutant criteria, the recommended level for the protection of human health is zero. Various levels have been set forth for various lifetime cancer risks.[6] Kansas has set a guideline for BDA in drinking water of 0.029 μg/l.[61]

Determination in Water: Extraction with Methylene chloride may be followed by measurement by gas chromatography coupled with mass spectrometry.

Routes of Entry: Inhalation, skin and/or eye contact.

Harmful Effects and Symptoms

Long Term Exposure: DB(a,h)A is a toxic chemical and is a probable carcinogen in humans. It has caused cancer in laboratory animals when it is ingested or applied to their skin. Because DB(a,h)A causes cancer in animals, it is likely that humans exposed in the same manner would develop cancer as well. DB(a,h)A may affect the skin, resulting in photosensitization.

Points of Attack: Respiratory system, skin, bladder, kidneys.

Medical Surveillance: See the entry on Coal Tar Pitch Volatiles.

First Aid: If this chemical gets into the eyes, remove any contact lenses at once and irrigate immediately for at least 15 minutes, occasionally lifting upper and lower lids. Seek medical attention immediately. If this chemical contacts the skin, remove contaminated clothing and wash immediately with soap and water. Seek medical attention immediately. If this chemical has been inhaled, remove from exposure, begin rescue breathing (using universal precautions) if breathing has stopped and CPR if heart action has stopped. Transfer promptly to a medical facility. When this chemical has been swallowed, get medical attention. Give large quantities of water and induce vomiting. Do not make an unconscious person vomit.

Personal Protective Methods: Wear protective gloves and clothing to prevent any reasonable probability of skin contact. Safety equipment suppliers/manufacturers can provide recommendations on the most protective glove/clothing material for your operation.. All protective clothing (suits, gloves, footwear, headgear) should be clean, available each day, and put on before work. Contact lenses should not be worn when working with this chemical. Wear dust-proof chemical goggles and face shield unless full facepiece respiratory protection is worn.

Respirator Selection: See the entry on Coal Tar Pitch Volatiles.

Storage: See the entry on Coal Tar Pitch Volatiles. A regulated, marked area should be established where this chemical is handled, used, or stored in compliance with OSHA standard 1910.1045.

Shipping: See the entry on Coal Tar Pitch Volatiles.

Spill Handling: Evacuate persons not wearing protective equipment from area of spill or leak until clean-up is complete. Remove all ignition sources. Collect powdered material in the most convenient and safe manner and deposit in sealed containers. Ventilate area after clean-up is complete. It may be necessary to contain and dispose of this chemical as a hazardous waste. If material or contaminated runoff enters waterways, notify downstream users of potentially contaminated waters. Contact your Department of Environmental Protection or your regional office of the federal EPA for specific recommendations. If employees are required to clean-up spills, they must be properly trained and equipped. OSHA 1910.120(q) may be applicable.

Fire Extinguishing: This chemical is a combustible solid. Use dry chemical, carbon dioxide, water spray, or alcohol extinguishers. Poisonous gases are produced in fire. If material or contaminated runoff enters waterways, notify downstream users of potentially contaminated waters. Notify local health and fire officials and pollution control agencies. From a secure, explosion-proof location, use water spray to cool exposed containers. If cooling streams are ineffective (venting sound increases in volume and pitch, tank discolors, or shows any signs of deforming), withdraw immediately to a secure position. If employees are expected to fight fires, they must be trained and equipped in OSHA 1910.156.

Disposal Method Suggested: See the entry on Coal Tar Pitch Volatiles. Consult with environmental regulatory agencies for guidance on acceptable disposal practices. Generators of waste containing this contaminant (≥100 kg/mo) must conform with EPA regulations governing storage, transportation, treatment, and waste disposal.

References

U.S. Public Health Service, "Toxicological Profile for Dibenz(a,h)-Anthracene," Atlanta, Georgia, Agency for Toxic Substances & Disease Registry (October 1987)

Sax N. I., Ed., "Dangerous Properties of Industrial Materials Report," 4, No. 6, 94–104 (1984)

Dibenzofuran

Molecular Formula: $C_{12}H_8O$

Synonyms: (1,1'-Biphenyl)-2,2'-diyl oxide; 2,2'-Biphenylene oxide; 2,2'-Biphenylyleme oxide; Dibenzo(b,d)furan; Dibenzofurano (Spanish); Diphenylene oxide

CAS Registry Number: 132-64-9

RTECS Number: HP4450000

DOT ID: UN 3077

Regulatory Authority

- Clean Air Act: Hazardous Air Pollutants (Title I, Part A, Section 112) as dibenzofurans
- RCRA 40CFR264, Appendix 9; TSD Facilities Ground Water Monitoring List. Suggested test method(s) (PQL μg/l): 8270 (10)
- Superfund/EPCRA 40CFR302.4 Reportable Quantity (RQ): CERCLA, 1 lb (0.454 kg)
- EPCRA Section 313 Form R *de minimis* concentration reporting level: 1.0%
- Canada, CEPA Toxic Substances

Cited in U.S. State Regulations: California (A, G), Maryland (G), Massachusetts (G), New Jersey (G), Pennsylvania (G).

Description: Dibenzofuran, $C_{12}H_8O$ is a white crystalline powder. Boiling point = 285 – 288°C. Freezing/Melting point = 85 – 87°C. Hazard Identification (based on NFPA-704 M Rating System): Health 1, Flammability 1, Reactivity 0. Very slightly soluble in water.

Potential Exposure: This material is used as an insecticide and in organic synthesis to make other chemicals. It is derived from coal tar creosote.

Incompatibilities: Strong oxidizers.

Permissible Exposure Limits in Air: No standards set.

Permissible Concentration in Water: No criteria set.

Routes of Entry: Inhalation, passing through the skin.

Harmful Effects and Symptoms

Short Term Exposure: Dibenzofuran can be absorbed through the skin, thereby increasing exposure. Exposure irritates the eyes, skin and respiratory tract. Poisonous if ingested. See also entry for coal tar.

Long Term Exposure: Repeated contact may cause skin growths, rashes, and changes in skin color. Exposure to sunlight may make rash worse.

Points of Attack: Skin.

Medical Surveillance: Evaluation by a qualified allergist.

First Aid: *Skin Contact:*[52] Flood all areas of body that have contacted the substance with water. Don't wait to remove contaminated clothing; do it under the water stream. Use soap to help assure removal. Isolate contaminated clothing when removed to prevent contact by others.

Eye Contact: Remove any contact lenses at once. Flush eyes well with copious quantities of water or normal saline for at least 20 – 30 minutes. Seek medical attention.

Inhalation: Leave contaminated area immediately; breathe fresh air. Proper respiratory protection must be supplied to any rescuers. If coughing, difficult breathing or any other symptoms develop, seek medical attention at once, even if symptoms develop many hours after exposure.

Ingestion: If convulsions are not present, give a glass or two of water or milk to dilute the substance. Assure that the person's airway is unobstructed and contact a hospital or poison center immediately for advice on whether or not to induce vomiting.

Personal Protective Methods: Wear protective gloves and clothing to prevent any reasonable probability of skin contact. Safety equipment suppliers/manufacturers can provide recommendations on the most protective glove/clothing material for your operation.. All protective clothing (suits, gloves, footwear, headgear) should be clean, available each day, and put on before work. Contact lenses should not be worn when working with this chemical. Wear dust-proof chemical goggles and face shield unless full facepiece respiratory protection is worn. Employees should wash immediately with soap when skin is wet or contaminated. Provide emergency showers and eyewash.

Respirator Selection: *At any detectable concentration:* SCBAF:PD,PP (any MSHA/NIOSH approved self-contained breathing apparatus that has a full facepiece and is operated in a pressure-demand or other positive-pressure mode); or SAF:PD,PP:ASCBA (any supplied-air respirator that has a full facepiece and is operated in a pressure-demand or other positive-pressure mode in combination with an auxiliary, self-contained breathing apparatus operated in a pressure-demand or other positive pressure mode). *Escape:* HiEF (any air-purifying, full-facepiece respirator (gas mask) with a chin-style, front- or back-mounted organic vapor canister having a high-efficiency particulate filter); or SCBAE (any appropriate escape-type, self-contained breathing apparatus).

Storage: Prior to working with Dibenzofuran you should be trained on its proper handling and storage. Store in tightly closed containers in a refrigerator or cool, well ventilated area away from oxidizers.

Shipping: Environmentally hazardous solid, n.o.s. Hazard Class: 9. Label: "Class 9." Packing Group: III.

Spill Handling: Evacuate persons not wearing protective equipment from area of spill or leak until clean-up is complete. Remove all ignition sources. Collect powdered material in the most convenient and safe manner and deposit in sealed containers. Ventilate area after clean-up is complete. It may be necessary to contain and dispose of this chemical as a hazardous waste. If material or contaminated runoff enters waterways, notify downstream users of potentially contaminated waters. Contact your Department of Environmental

Protection or your regional office of the federal EPA for specific recommendations. If employees are required to clean-up spills, they must be properly trained and equipped. OSHA 1910.120(q) may be applicable.

Fire Extinguishing: Use dry chemical, carbon dioxide, water spray, or foam extinguishers. Poisonous gases are produced in fire. If material or contaminated runoff enters waterways, notify downstream users of potentially contaminated waters. Notify local health and fire officials and pollution control agencies. From a secure, explosion-proof location, use water spray to cool exposed containers. If cooling streams are ineffective (venting sound increases in volume and pitch, tank discolors, or shows any signs of deforming), withdraw immediately to a secure position. If employees are expected to fight fires, they must be trained and equipped in OSHA 1910.156.

Disposal Method Suggested: In accordance with 40CFR165 recommendations for the disposal of pesticides and pesticide containers. Must be disposed properly by following package label directions or by contacting your state pesticide or environmental control agency or by contacting your regional EPA office.

References

New Jersey Department of Health and Senior Services, Hazardous Substance Fact Sheet, "Dibenzofuran," Trenton NJ (May, 1998)

Dibenzyl Dichlorosilane

Molecular Formula: $C_{14}H_{14}Cl_2Si$

Common Formula: $(C_6H_5CH_2)_2SiCl_2$

Synonyms: Dichlorobis(phenylmethye) silane

CAS Registry Number: 18414-36-3

RTECS Number: VV2977000

DOT ID: UN 2434

Cited in U.S. State Regulations: New Jersey (G).

Description: $(C_6H_5CH_2)_2siCl_2$ is a colorless liquid.

Potential Exposure: This material is used as an intermediate in the production of silicone polymers.

Incompatibilities: Combustible materials. Contact with water, steam or moisture produces corrosive hydrogen chloride gas.

Permissible Exposure Limits in Air: No standards set.

Permissible Concentration in Water: No criteria set.

Routes of Entry: Inhalation.

Harmful Effects and Symptoms

Short Term Exposure: The health effects are not well known at this time. However, closely related chemicals cause irritation of the eyes, nose, throat and lungs. Dibenzyl dichlorosilane is a corrosive chemical and can cause severe eye and skin burns. This substance can give off corrosive hydrogen chloride gas on contact with water, steam, or moisture. It is possible that higher exposures may cause pulmonary edema, a medical emergency that can be delayed for several hours. This can cause death.

Long Term Exposure: Although it is not known if this chemical causes lung problems, similar corrosive or highly irritating substances may affect the lungs.

Points of Attack: Skin, lungs.

Medical Surveillance: Before beginning employment and at regular times after that, for those with frequent or potentially high exposures, the following is recommended: Lung function tests. If symptoms develop or overexposure is suspected, the following may be useful: Consider chest x-ray after acute overexposure.

First Aid: If this chemical gets into the eyes, remove any contact lenses at once and irrigate immediately for at least 30 minutes, occasionally lifting upper and lower lids. Seek medical attention immediately. If this chemical contacts the skin, remove contaminated clothing and wash immediately with soap and water. Seek medical attention immediately. If this chemical has been inhaled, remove from exposure, begin rescue breathing (using universal precautions) if breathing has stopped and CPR if heart action has stopped. Transfer promptly to a medical facility. When this chemical has been swallowed, get medical attention. If victim is conscious, administer water or milk. Do not induce vomiting. Medical observation is recommended for 24 – 48 hours after breathing overexposure, as pulmonary edema may be delayed. As first aid for pulmonary edema, a doctor or authorized paramedic may consider administering a corticosteroid spray.

Personal Protective Methods: Wear protective gloves and clothing to prevent any reasonable probability of skin contact. Safety equipment suppliers/manufacturers can provide recommendations on the most protective glove/clothing material for your operation.. All protective clothing (suits, gloves, footwear, headgear) should be clean, available each day, and put on before work. Contact lenses should not be worn when working with this chemical. Wear splash-proof chemical goggles and face shield unless full facepiece respiratory protection is worn. Employees should wash immediately with soap when skin is wet or contaminated. Provide emergency showers and eyewash.

Respirator Selection: Where the potential exists for exposure to Dibenzyl dichlorosilane, use a MSHA/NIOSH approved supplied-air respirator with a full facepiece operated in the positive pressure mode or with a full facepiece, hood, or helmet in the continuous flow mode, or use a MSHA/NIOSH approved self-contained breathing apparatus with a full facepiece operated in pressure-demand or other positive pressure mode.

Storage: Prior to working with this chemical you should be trained on its proper handling and storage. Store in tightly closed containers in a cool, well ventilated area away from combustible materials and any form of moisture. Where possible, automatically pump liquid from drums or other storage containers to process containers.

Shipping: Dibenzyl dichlorosilane requires a "Corrosive" label according to DOT.[19] It falls in Hazard Class 8 and Packing Group II.

Spill Handling: Evacuate and restrict persons not wearing protective equipment from area of spill or leak until cleanup is complete. Remove all ignition sources. Ventilate area of spill or leak. Absorb liquids in vermiculite, dry sand, earth, peat, carbon, or a similar material and deposit in sealed containers. It may be necessary to contain and dispose of this chemical as a hazardous waste. If material or contaminated runoff enters waterways, notify downstream users of potentially contaminated waters. Contact your Department of Environmental Protection or your regional office of the federal EPA for specific recommendations. If employees are required to clean-up spills, they must be properly trained and equipped. OSHA 1910.120(q) may be applicable.

Fire Extinguishing: Dibenzyl dichlorosilane may burn, but does not readily ignite. Use dry chemical, CO_2, or foam extinguishers. Poisonous gases are produced in fire. Vapors are heavier than air and will collect in low areas. If material or contaminated runoff enters waterways, notify downstream users of potentially contaminated waters. Notify local health and fire officials and pollution control agencies. From a secure, explosion-proof location, use water spray to cool exposed containers. If cooling streams are ineffective (venting sound increases in volume and pitch, tank discolors, or shows any signs of deforming), withdraw immediately to a secure position. If employees are expected to fight fires, they must be trained and equipped in OSHA 1910.156.

References

New Jersey Department of Health, "Hazardous Substance Fact Sheet: "Dibenzyl Dichlorosilane," Trenton, NJ (May 1986)

Diborane

Molecular Formula: B_2H_6

Synonyms: Boroethane; Boron hydride; Diborane (6); Diborane hexanhydride; Diborano (Spanish); Diboron hexahydride

CAS Registry Number: 19287-45-7

RTECS Number: HQ9275000

DOT ID: UN 1911

Regulatory Authority

- Air Pollutant Standard Set (ACGIH)[1] (Australia) (DFG)[3] (HSE)[33] (Israel) (Mexico) (OSHA)[58] (Several States)[60] (Several Canadian Provinces)
- OSHA 29CFR1910.119, Appendix A, Process Safety List of Highly Hazardous Chemicals, TQ = 100 lb (45 kg)
- Clean Air Act: Accidental Release Prevention/Flammable substances, (Section 112[r], Table 3), TQ = 2,500 lb (1,135 kg)
- Superfund/EPCRA 40CFR355, Appendix B Extremely Hazardous Substances: TPQ = 100 lb (45.4 kg)
- Superfund/EPCRA 40CFR302.4 Reportable Quantity (RQ): EHS, 1 lb (0.454 kg)
- U.S. DOT 49CFR172.101, Inhalation Hazardous Chemical
- Canada, WHMIS, Ingredients Disclosure List

Cited in U.S. State Regulations: Alaska (G), California (A, G), Connecticut (A), Florida (G), Illinois (G), Maine (G), Massachusetts (G), Minnesota (G), Nevada (A), New Hampshire (G), New Jersey (G), North Dakota (A), Oklahoma (G), Pennsylvania (G), Rhode Island (G), Virginia (A), West Virginia (G).

Description: Diborane, B_2H_6, is a compressed, colorless, flammable gas with a nauseating, sickly-sweet odor. The odor threshold = 2.5 ppm. Boiling point = -92.5°C. Freezing/Melting point = -165.5°C. Flash point = -90°C (flammable gas). Autoignition temperature = 40 – 50°C. Explosive limits: LEL = 0.8%; UEL = 88%. Hazard Identification (based on NFPA-704 M Rating System): Health 4, Flammability 4, Reactivity 3. Reacts with water.

Potential Exposure: Diborane is used as a catalyst for olefin polymerization, a rubber vulcanizer, a reducing agent, a flame-speed accelerator, a chemical intermediate for other boron hydrides, and as a doping agent; and in rocket propellants and in the conversion of olefins to trialkyl boranes and primary alcohols.

Incompatibilities: A strong reducing agent. Unstable above 8°C. Diborane can polymerize forming liquid pentaborane. It ignites spontaneously in moist air, and on contact with water, hydrolyzes exothermically forming hydrogen and boric acid. Contact halogenated compounds may cause fire and explosion. Contact with aluminum, lithium and other active metals form hydrides which may ignite spontaneously. Incompatible with aluminum, carbon tetrachloride, nitric acid, nitrogen trifluoride and many other chemicals. Reacts with oxidized surfaces. Attacks some plastics, rubber or coatings.

Permissible Exposure Limits in Air: The Federal Limit (OSHA PEL),[58] the DFG MAK[3] and the HSE, Australian, Israeli, Mexican, and Canadian provinces of Alberta, BC, Ontario, and Quebec TWA[33] is 0.1 ppm (0.1 mg/m^3) and Alberta set the STEL at 0.3 ppm. The NIOSH IDLH level is 15 ppm. In addition, some states have set guidelines or standards for Diborane in ambient air[60] ranging from 30 μg/m^3 (North Dakota) to 50 μg/m^3 (Virginia) to 60 μg/m^3 (Connecticut) to 71 μg/m^3 (Nevada).

Determination in Air: Filter/Special charcoal tube; H_2O_2; Plasma emission spectroscopy; NIOSH IV Method #6006.

Permissible Concentration in Water: No criteria set. (Diborane reacts on contact with water as noted above.)

Routes of Entry: Inhalation, skin and eye contact.

Harmful Effects and Symptoms

Short Term Exposure: Diborane is the least toxic of the boron hydrides. In acute poisoning, the symptoms are similar to "metal fume fever:" tightness, heaviness and burning in chest, coughing, shortness of breath, chills, fever, pericardial pain, nausea, shivering, and drowsiness. Signs appear soon after exposure or after a latent period of up to 24 hours and persist for 1 – 3 days or more. Pneumonia may develop later. Reversible liver and kidney changes were seen in rats exposed to very high gas levels. This has not been noted in man.

Subacute poisoning is characterized by pulmonary irritation symptoms, and if this is prolonged, central nervous system symptoms such as headaches, dizziness, vertigo, chills, fatigue, muscular weakness, and only infrequent transient tremors, appear. Convulsions do not occur. NIOSH lists symptoms as; chest tightness, precordial pain, shortness breathing, nonproductive cough, nausea; headache, lightheadedness, vertigo (an illusion of movement), chills, fever, fatigue, weakness, tremor, muscle fasiculation. In animals: liver, kidney damage; pulmonary edema; hemorrhage.

Long Term Exposure: Prolonged exposure may cause lungs damage. Chronic exposure leads to wheezing, dyspnea, tightness, dry cough, rales, and hyperventilation which persist for several years.

Points of Attack: Respiratory system, central nervous system, liver, kidneys.

Medical Surveillance: Before beginning employment and at regular times after that, the following are recommended: Lung function tests. If symptoms develop or overexposure has occurred, the following may be useful: Liver and kidney function tests. Examination of the nervous system. Consider chest x-ray after acute overexposure.

First Aid: If this chemical gets into the eyes, remove any contact lenses at once and irrigate immediately for at least 15 minutes, occasionally lifting upper and lower lids. Seek medical attention immediately. If this chemical contacts the skin, remove contaminated clothing and wash immediately with soap and water. Seek medical attention immediately. If this chemical has been inhaled, remove from exposure, begin rescue breathing (using universal precautions) if breathing has stopped and CPR if heart action has stopped. Transfer promptly to a medical facility. When this chemical has been swallowed, get medical attention. Give large quantities of water and induce vomiting. Do not make an unconscious person vomit. Medical observation is recommended for 24 – 48 hours after breathing overexposure, as pulmonary edema may be delayed. As first aid for pulmonary edema, a doctor or authorized paramedic may consider administering a corticosteroid spray.

Personal Protective Methods: Wear protective gloves and clothing to prevent any reasonable probability of skin contact. For emergency situations, wear a positive pressure, pressure-demand, full facepiece self-contained breathing apparatus (SCBA) or pressure-demand supplied air respirator with escape SCBA and a fully-encapsulating, chemical resistant suit. Safety equipment suppliers/manufacturers can provide recommendations on the most protective glove/clothing material for your operation.. All protective clothing (suits, gloves, footwear, headgear) should be clean, available each day, and put on before work. Contact lenses should not be worn when working with this chemical. Wear gas-proof chemical goggles and face shield unless full facepiece respiratory protection is worn. Employees should wash immediately with soap when skin is wet or contaminated. Provide emergency showers and eyewash.

Respirator Selection: NIOSH/OSHA: *Up to 1 ppm*: SA (any supplied-air respirator). *Up to 2.5 ppm:* SA:CF (any supplied-air respirator operated in a continuous-flow mode). *Up to 5 ppm:* SCBAF (any supplied-air respirator that has a tight-fitting facepiece and is operated in a continuous-flow mode); or SAF (any self-contained breathing apparatus with a full facepiece); or SAT:CF (any supplied-air respirator with a full facepiece). *Up to 50 ppm:* (any supplied-air respirator operated in a pressure-demand or other positive-pressure mode). *Emergency or planned entry into unknown concentrations or IDLH conditions*: SCBAF:PD,PP (any self-contained breathing apparatus that has a full facepiece and is operated in a pressure-demand or other positive-pressure mode); or SAF:PD,PP:ASCBA (any supplied-air respirator that has a full facepiece and is operated in a pressure-demand or other positive-pressure mode in combination with an auxiliary self-contained breathing apparatus operated in a pressure-demand or other positive-pressure mode). *Escape:* GMFS [any air-purifying, full-facepiece respirator (gas mask) with a chin-style, front- or back-mounted canister providing protection against the compound of concern]; or SCBAE (any appropriate escape-type, self-contained breathing apparatus).

Storage: Prior to working with this chemical you should be trained on its proper handling and storage. Before entering confined space where this chemical may be present, check to make sure that an explosive concentration does not exist. Diborane must be stored to avoid contact with: air, active metals such as aluminum and lithium; halogenated compounds such as chlorine; and oxidizing agents such as permanganates, nitrates, peroxides, chlorates, and perchlorates, since violent reactions occur. Containers should be dry, clean and free of oxygen. Store in tightly closed containers in a cool well-ventilated area away from heat and moisture. Containers are usually stored in "dry ice" or are refrigerated in some other way. Diborane can ignite spontaneously in moist air at room temperature. Use dry nitrogen purge in any transfer. Sources of ignition such as smoking and open flames are prohibited where Diborane is handled, used, or stored. Use only non-sparking tools and equipment, especially when opening and closing containers of Diborane. Wherever Diborane is used, handled, manufactured, or stored, use explosion-proof electrical equipment and fittings. Procedures for the handling, use and storage of cylinders should be in compliance with OSHA 1910.101 and 1910.169 as with the recommendations of the Compressed Gas Association.

Shipping: Diborane must be labeled: "Poison Gas, Flammable Gas." It falls in DOT Hazard Class 2.3 and Packing Group I. Shipment by passenger aircraft or railcar or even by cargo aircraft is forbidden.

Spill Handling: If in a building, evacuate building and confine vapors by closing doors and shutting down HVAC systems. Restrict persons not wearing protective equipment from area of spill or leak until cleanup is complete. Remove all ignition sources. Establish forced ventilation to keep levels below explosive limit. Wear chemical protective suit with self-contained breathing apparatus to combat spills. Stay upwind and use water spray to "knock down" vapor; contain runoff. Stop the flow of gas, if it can be done safely from a distance. If possible dilute the leak with an inert gas and exhaust through a fume hood. If source is a cylinder and the leak cannot be stopped in place, specially trained personnel may be able to remove the leaking cylinder to a safe place in the open air and repair leak or allow cylinder to empty. Keep this chemical out of confined spaces, such as a sewer, because of the possibility of explosion, unless the sewer is designed to prevent the buildup of explosive concentrations. If employees are required to clean-up spills, they must be properly trained and equipped. OSHA 1910.120(q) may be applicable.

Initial isolation and protective action distances

Distances shown are likely to be affected during the first 30 minutes after materials are spilled and could increase with time. If more than one tank car, cargo tank, portable tank, or large cylinder is involved in the incident is leaking, the protective action distance may need to be increased. You may need to seek emergency information from CHEMTREC at (800) 424-9300 or seek professional environmental engineering assistance from the U.S. EPA Environmental Response Team at (908) 548-8730 (24-hour response line).

Small spills (From a small package or a small leak from a large package)

First: Isolate in all directions (feet)..... 400

Then: Protect persons downwind (miles)

Day 0.3

Night 1.4

Large spills (From a large package or from many small packages)

First: Isolate in all directions (feet) .. 1,000

Then: Protect persons downwind (miles)

Day 1.2

Night 5.2

Fire Extinguishing: Firefighting gear (including SCBA) does not provide adequate protection. If exposure occurs, remove and isolate gear immediately and thoroughly decontaminate personnel. This chemical is a highly flammable and reactive gas; it will ignite without warning in moist air at room temperature. Poisonous gases produced in fire including boron and hydrogen at high heat, and hydrogen and boron hydrides at lower temperatures. Approach fire with extreme caution; consider letting it burn. Do not extinguish fire unless the flow of gas can be stopped and any remaining gas is out of the line. Specially trained personnel may use fog lines to cool exposures and let the fire burn itself out. Stop the flow of gas and use water spray to protect personnel during the shut-off. Use water from an unmanned source to keep fire-exposed containers cool. Diborane may react violently with halogenated extinguishing agents. Vapors are heavier than air and will collect in low areas. Vapors may travel long distances to ignition sources and flashback. Vapors in confined areas may explode when exposed to fire. Containers may explode in fire. Storage containers and parts of containers may rocket great distances, in many directions. If material or contaminated runoff enters waterways, notify downstream users of potentially contaminated waters. Notify local health and fire officials and pollution control agencies. From a secure, explosion-proof location, use water spray to cool exposed containers. If cooling streams are ineffective (venting sound increases in volume and pitch, tank discolors, or shows any signs of deforming), withdraw immediately to a secure position. If employees are expected to fight fires, they must be trained and equipped in OSHA 1910.156.

Disposal Method Suggested: Incineration with aqueous scrubbing of exhaust gases to remove B_2O_3 particulates.

References

Sax N. I., Ed., "Dangerous Properties of Industrial Materials Report," 2, No. 1, 105–107 (1982)

U.S. Environmental Protection Agency, "Chemical Profile: Diborane," Washington, DC, Chemical Emergency Preparedness Program (November 30, 1987)

New Jersey Department of Health, "Hazardous Substance Fact Sheet: Diborane," Trenton, NJ (April 1986)

Dibromobenzene

Molecular Formula: $C_6H_4Br_2$

Synonyms: Benzene dibromide; Benzene, dibromo-; *o*-Dibromobenzene

CAS Registry Number: 26249-12-7

RTECS Number: CZ1780000

DOT ID: UN 2711

Regulatory Authority

- Air Pollutant Standard Set[43]

Cited in U.S. State Regulations: New Hampshire (G), New Jersey (G).

Description: Dibromobenzene, $C_6H_4Br_2$, is a heavy colorless liquid with a pleasant aromatic odor. Boiling point = 225.5°C; Freezing/Melting point = 7°C. Flash point = 47°C. Insoluble in water.

Potential Exposure: Dibromobenzene is used as a solvent for oils and in organic synthesis.

Incompatibilities: Strong oxidizers.

Permissible Exposure Limits in Air: No workplace limits were found but the former USSR-UNEP/IRPTC project[43] has set a MAC value for ambient air in residential areas of 0.2 mg/m^3 on a momentary basis.

Permissible Concentration in Water: No criteria set.

Routes of Entry: Inhalation, passing through the skin.

Harmful Effects and Symptoms

Short Term Exposure: Exposure can irritate the eyes, nose and throat. High levels can cause you to feel dizzy, lightheaded and to pass out. Contact can irritate the eyes and skin.

Long Term Exposure: Similar chemicals can cause liver damage.

Medical Surveillance: If symptoms develop or overexposure is suspected, the following may be useful: Liver function tests.

First Aid: If this chemical gets into the eyes, remove any contact lenses at once and irrigate immediately for at least 15 minutes, occasionally lifting upper and lower lids. Seek medical attention immediately. If this chemical contacts the skin, remove contaminated clothing and wash immediately with soap and water. Seek medical attention immediately. If this chemical has been inhaled, remove from exposure, begin rescue breathing (using universal precautions) if breathing has stopped and CPR if heart action has stopped. Transfer promptly to a medical facility. When this chemical has been swallowed, get medical attention. Give large quantities of water and induce vomiting. Do not make an unconscious person vomit.

Personal Protective Methods: Wear protective gloves and clothing to prevent any reasonable probability of skin contact. Safety equipment suppliers/manufacturers can provide recommendations on the most protective glove/clothing material for your operation.. All protective clothing (suits, gloves, footwear, headgear) should be clean, available each day, and put on before work. Contact lenses should not be worn when working with this chemical. Wear splash-proof chemical goggles and face shield unless full facepiece respiratory protection is worn. Employees should wash immediately with soap when skin is wet or contaminated. Provide emergency showers and eyewash.

Respirator Selection: Where the potential exists for exposures to Dibromobenzene, use a MSHA/NIOSH approved supplied-air respirator with a full facepiece operated in the positive pressure mode or with a full facepiece, hood, or helmet in the continuous flow mode, or use a MSHA/NIOSH approved self-contained breathing apparatus with a full facepiece operated in pressure-demand or other positive pressure mode.

Storage: Prior to working with dibromobenzene you should be trained on its proper handling and storage. Store in tightly closed containers in a cool, well-ventilated area away from strong oxidizers. Sources of ignition, such as smoking and open flames, are prohibited where Dibromobenzene is used, handled, or stored, in a manner that could create a potential fire or explosion hazard.

Shipping: Dibromobenzene requires a "Flammable Liquid" label. It falls in DOT Hazard Class 3 and Packing Group III.

Spill Handling: Evacuate and restrict persons not wearing protective equipment from area of spill or leak until cleanup is complete. Remove all ignition sources. Ventilate area of spill or leak. Absorb liquids in vermiculite, dry sand, earth, peat, carbon, or a similar material and deposit in sealed containers. Keep this chemical out of a confined space, such as a sewer, because of the possibility of an explosion, unless the sewer is designed to prevent the build-up of explosive concentrations. It may be necessary to contain and dispose of this chemical as a hazardous waste. If material or contaminated runoff enters waterways, notify downstream users of potentially contaminated waters. Contact your Department of Environmental Protection or your regional office of the federal EPA for specific recommendations. If employees are required to clean-up spills, they must be properly trained and equipped. OSHA 1910.120(q) may be applicable.

Fire Extinguishing: This chemical is a combustible liquid. Poisonous gases are produced in fire. Use dry chemical, carbon dioxide, or alcohol foam extinguishers. Vapors are heavier than air and will collect in low areas. Vapors may travel long distances to ignition sources and flashback. Vapors in confined areas may explode when exposed to fire. Containers may explode in fire. Storage containers and parts of containers may rocket great distances, in many directions. If material or contaminated runoff enters waterways, notify downstream users of potentially contaminated waters. Notify local health and fire officials and pollution control agencies. From a secure, explosion-proof location, use water spray to cool exposed containers. If cooling streams are ineffective (venting sound increases in volume and pitch, tank discolors, or shows any signs of deforming), withdraw immediately to a secure position. If employees are expected to fight fires, they must be trained and equipped in OSHA 1910.156.

Disposal Method Suggested: Incineration with flue gas scrubbing.

References

New Jersey Department of Health, "Hazardous Substance Fact Sheet: Dibromobenzene," Trenton, NJ (August 5, 1987)

Dibromochloromethane

Molecular Formula: $CHBr_2Cl$

Synonyms: CDBM; Chlorodibromomethane; Clorodibromometano (Spanish); NCI-C55254

CAS Registry Number: 124-48-1

RTECS Number: PA6360000

DOT ID: UN 3082

Regulatory Authority

- Carcinogen (Animal Suspected) (NTP)[9]
- Water Pollution Standard Proposed (EPA) (Illinois)[61]
- Clean Water Act: 40CFR423, Appendix A, Priority Pollutants
- Safe Drinking Water Act: Priority List (55 FR 1470)
- Safe Drinking Water Act: MCL 0.10 mg/l, as trihalomethanes
- RCRA 40CFR268.48; 61FR15654, Universal Treatment Standards: Wastewater (mg/l), 0.057; Nonwastewater (mg/kg), 15
- RCRA 40CFR264, Appendix 9; TSD Facilities Ground Water Monitoring List. Suggested test method(s) (PQL μg/l): 8010 (1); 8240 (5)
- Superfund/EPCRA 40CFR302.4 Reportable Quantity (RQ): CERCLA, 100 lb (45.4 kg)

Cited in U.S. State Regulations: California (A, G), Illinois (W), Massachusetts (G), New Jersey (G), Pennsylvania (G).

Description: Dibromochloromethane, $CHBr_2Cl$, is a clear colorless liquid. Boiling point = 119 – 120°C. Freezing/Melting point = -22°C. Also reported as ≤20°C.

Potential Exposure: Dibromochloromethane is used as a chemical intermediate in the manufacture o fire extinguishing agents, aerosol propellants, refrigerants, and pesticides. Dibromochloromethane has been detected in drinking water in the United States. It is believed to be formed by the haloform reaction that may occur during water chlorination. Dibromochloromethane can be removed from drinking water via treatment with activated carbon. There is a potential for dibromochloromethane to accumulate in the aquatic environment because of its resistance to degradation. Volatilization is likely to be an important means of environmental transport.

Incompatibilities: Strong oxidizers.

Permissible Exposure Limits in Air: No standards set.

Permissible Concentration in Water: The Maximum Contaminant Level (MCL) for total trihalomethanes (including dibromochloromethane) in drinking water has been set by the U.S. EPA at 0.10 mg/l (44 RF 68624). Illinois has set a guideline for CDBN in drinking water[61] of 1.0 μg/l.

Harmful Effects and Symptoms

Short Term Exposure: Very little toxicity information is available. It is however an irritant and narcotic. Symptoms include dizziness, headache and liver and kidney damage.

Long Term Exposure: May cause liver and kidney damage. Dibromochloromethane gave positive results in mutagenicity tests with salmonella typhimurium TA 100. It is currently under test by the National Cancer Institute. The LD_{50} oral-rat is 848 mg/kg (slightly toxic).

Points of Attack: Liver, kidneys, skin.

Medical Surveillance: Liver and kidney function tests.

First Aid: *Skin Contact:*[52] Flood all areas of body that have contacted the substance with water. Don't wait to remove contaminated clothing; do it under the water stream. Use soap to help assure removal. Isolate contaminated clothing when removed to prevent contact by others.

Eye Contact: Remove any contact lenses at once. Flush eyes well with copious quantities of water or normal saline for at least 20 – 30 minutes. Seek medical attention.

Inhalation: Leave contaminated area immediately; breathe fresh air. Proper respiratory protection must be supplied to any rescuers. If coughing, difficult breathing or any other symptoms develop, seek medical attention at once, even if symptoms develop many hours after exposure.

Ingestion: If convulsions are not present, give a glass or two of water or milk to dilute the substance. Assure that the person's airway is unobstructed and contact a hospital or poison center immediately for advice on whether or not to induce vomiting.

Personal Protective Methods: Wear protective gloves (nitrile gloves may provide protection[52]) and clothing to prevent any reasonable probability of skin contact. Safety equipment suppliers/manufacturers can provide recommendations on the most protective glove/clothing material for your operation.. All protective clothing (suits, gloves, footwear, headgear) should be clean, available each day, and put on before work. Contact lenses should not be worn when working with this chemical. Wear splash-proof chemical goggles and face shield unless full facepiece respiratory protection is worn. Employees should wash immediately with soap when skin is wet or contaminated. Provide emergency showers and eyewash.

Respirator Selection: *At any detectable concentration:* SCBAF:PD,PP (any MSHA/NIOSH approved self-contained breathing apparatus that has a full facepiece and is operated in a pressure-demand or other positive-pressure mode); or SAF:PD,PP:ASCBA (any supplied-air respirator that has a full facepiece and is operated in a pressure-demand or other positive-pressure mode in combination with an auxiliary, self-contained breathing apparatus operated in a pressure-demand or other positive pressure mode). *Escape:* HiEF (any air-purifying, full-facepiece respirator (gas mask) with a chin-style, front- or back-mounted organic vapor canister having a high-efficiency particulate filter); or SCBAE (any appropriate escape-type, self-contained breathing apparatus).

Storage: Prior to working with this chemical you should be trained on its proper handling and storage. Store in tightly closed containers under an inert atmosphere, away from light, in a refrigerator. Where possible, automatically pump liquid from drums or other storage containers to process containers. A regulated, marked area should be established where this chemical is handled, used, or stored in compliance with OSHA standard 1910.1045.

Shipping: Environmentally hazardous liquid, n.o.s. Hazard Class: 9. Label: "Class 9." Packing Group: III.

Spill Handling: Evacuate and restrict persons not wearing protective equipment from area of spill or leak until cleanup is complete. Remove all ignition sources. Ventilate area of spill or leak. Absorb liquids in vermiculite, dry sand, earth, peat, carbon, or a similar material and deposit in sealed containers. Keep this chemical out of a confined space, such as a sewer, because of the possibility of an explosion, unless the sewer is designed to prevent the build-up of explosive concentrations. It may be necessary to contain and dispose of this chemical as a hazardous waste. If material or contaminated runoff enters waterways, notify downstream users of potentially contaminated waters. Contact your Department of Environmental Protection or your regional office of the federal EPA for specific recommendations. If employees are required to clean-up spills, they must be properly trained and equipped. OSHA 1910.120(q) may be applicable.

Fire Extinguishing: Poisonous gases including chlorine and bromine are produced in fire. Use dry chemical, carbon dioxide, or foam extinguishers. Vapors are heavier than air and will collect in low areas. Storage containers and parts of containers may rocket great distances, in many directions. If material or contaminated runoff enters waterways, notify downstream users of potentially contaminated waters. Notify local health and fire officials and pollution control agencies. From a secure, explosion-proof location, use water spray to cool exposed containers. If cooling streams are ineffective (venting sound increases in volume and pitch, tank discolors, or shows any signs of deforming), withdraw immediately to a secure position. If employees are expected to fight fires, they must be trained and equipped in OSHA 1910.156.

Disposal Method Suggested: May be destroyed by high-temperature incinerator equipped with an HCl scrubber.

References

U.S. Environmental Protection Agency, Dibromochloromethane, Health and Environmental Effects Profile No. 61, Washington, DC, Office of Solid Waste (April 30, 1980)

Sax N. I., Ed., "Dangerous Properties of Industrial Materials Report," 5, No. 2, 61–63 (1985)

Dibromochloropropane

Molecular Formula: $C_3H_5Br_2Cl$

Common Formula: $CH_2BrCHBrCH_2Cl$

Synonyms: BBC 12; 1-Chloro-2,3-dibromopropane; 3-Chloro-1,2-dibromopropane; DBCP; 1,2-Dibrom-3-chlor-propan (German); Dibromchlorpropan (German); Dibromo-chloropropane; 1,2-Dibromo-3-cloro-propano (Italian); 1,2-Dibromo-3-cloropropano (Spanish); 1,2-Dibroom-3-chloor-propaan (Dutch); Fumagone; Fumazone; NCI-C00500; Nemabrom; Nemafume; Nemagon; Nemagon 20; Nemagon 90; Nemagone; Nemagone 20G; Nemagon soil fumigant; Nemanax; Nemapaz; Nemaset; Nematocide; Nematox; Nemazon; OS 1897; Oxy DBCP; Propane, 1,2-dibromo-3-chloro-; SD 1897

CAS Registry Number: 96-12-8

RTECS Number: TX8750000

DOT ID: UN 2872

EEC Number: 602-021-00-6

Regulatory Authority

- Carcinogen (Animal Positive) (IARC) (NCI) (NTP)[9] (DFG)[3]
- Banned or Severely Restricted (Several Countries) (UN)[35]
- OSHA, 29CFR1910 Specifically Regulated Chemicals (See CFR 1910.1044)
- Air Pollutant Standard Set (US, Argentina)[35] (former USSR)[43] (Several States)[60]
- Clean Air Act: Hazardous Air Pollutants (Title I, Part A, Section 112)
- EPA Hazardous Waste Number (RCRA No.): U066
- RCRA, 40CFR261, Appendix 8 Hazardous Constituents
- RCRA 40CFR268.48; 61FR15654, Universal Treatment Standards: Wastewater (mg/l), 0.11; Nonwastewater (mg/kg), 15
- RCRA 40CFR264, Appendix 9; TSD Facilities Ground Water Monitoring List. Suggested test method(s) (PQL μg/l): 8010 (100); 8240 (5); 8270 (10)
- Safe Drinking Water Act: MCL, 0.0002 mg/l; MCLG, zero; Regulated chemical (47FR9352)
- Superfund/EPCRA 40CFR302.4 Reportable Quantity (RQ): CERCLA, 1 lb (0.454 kg)
- EPCRA Section 313 Form R *de minimis* concentration reporting level: 0.1%
- Canada, WHMIS, Ingredients Disclosure List

Cited in U.S. State Regulations: Arizona (W), California (A, G, W), Connecticut (A), Florida (G), Kansas (G), Louisiana (G), Maine (G), Maryland (G), Massachusetts (G), Michigan (G), Minnesota (W), New Hampshire (G), New Jersey (G), North Dakota (A), Pennsylvania (G, A), Vermont (G),

Virginia (G), Washington (G), West Virginia (G), Wisconsin (G, W).

Description: DBCP, $CH_2BrCHBrCH_2Cl$, is an amber to brown liquid (a solid below 6°C/43°F) with a strong, pungent odor. It has an odor and taste threshold at 0.01 mg/l in water. Boiling point = 196°C (decomposes). Freezing/Melting point = 6.1°C. Flash point = 77°C. Hazard Identification (based on NFPA-704 M Rating System): Health 2, Flammability 2, Reactivity 0. Slightly soluble in water.

Potential Exposure: DBCP has been used in agriculture as a menatocide since 1955, being supplied for such use in the forms of liquid concentrate, emulsifiable concentrate, powder, granules, and solid material. A rebuttable presumption against registration for pesticide uses was issued by U.S. EPA on September 22, 1977, on the basis of oncogenicity and reproductive effects. Then, as of November 3, 1977, EPA in a further action suspended all registrations of end use products, subject to various specific restrictions.

Incompatibilities: Reacts with oxidizers and chemically active metals (i.e., aluminum, magnesium and tin alloys). Attacks some rubber materials and coatings. Corrosive to metals.

Permissible Exposure Limits in Air: The OSHA PEL is 0.001 ppm TWA and that no exposure to eyes or skin should occur.[35] Argentina has set a TWA of 0.25 mg/m^3 with a STEL of 0.75 mg/m^3. Sweden and Germany have set no limits; simply stated that DBCP is a carcinogenic substance and should be avoided. The former USSR-UNEP/IRPTC project[43] has set a MAC in ambient air in residential areas of 0.0004 mg/m^3 on a momentary basis and 0.00003 mg/m^3 on a daily average basis.

In addition, several states have set guidelines or standards for DBCP in ambient air[60] ranging from zero (North Dakota) to 0.05 μg/m^3 (Connecticut) to 1.0 μg/m^3 (Pennsylvania).

Permissible Concentration in Water: The former USSR-UNEP/IRPTC project[43] has set a MAC of 0.01 mg/l in water bodies used for domestic purposes. The USEPA[47] has set a one-day health advisory of 0.2 mg/l and a ten-day health advisory of 0.02 mg/l both for a 10 kg child. Longer term health advisories could not be calculated because of the carcinogenicity of DBCP. EPA has recently proposed[62] a maximum drinking water level of 0.0002 mg/l (0.2 μg/l).

Several states have set guidelines for DBCP in drinking water[61] ranging from 0.025 μg/l (Arizona) to 0.25 μg/l (Minnesota) to 0.5 μg/l (Wisconsin) to 1.0 μg/l (California).

Determination in Water: By purge-and-trap gas chromatography.[47]

Routes of Entry: Inhalation, skin absorption, ingestion, skin and/or eye contact.

Harmful Effects and Symptoms

Short Term Exposure: Symptoms include severe local irritation to eyes, skin and mucous membranes. Nausea and vomiting may occur after ingestion. Exposure to DBCP can cause headache, nausea, vomiting, weakness, lightheadedness, unconsciousness, and possible death. Higher exposures can cause pulmonary edema, a medical emergency that can be delayed for several hours. This can cause death.

Long Term Exposure: The possible effects on the health of employees chronically exposed to repeated or lower exposures of DBCP may include sterility, diminished renal function, and degeneration and cirrhosis of the liver. DBCP is a probable carcinogen in humans. It has been shown to cause stomach, breast, tongue, and nasal cavity cancer in animals. May damage the testes and decrease fertility in males and females. Repeated exposure can damage the eyes causing clouding of lens or cornea, opens sores on the skin, liver and kidney damage.

Points of Attack: Eyes, skin, respiratory system, central nervous system, liver, kidneys, spleen, reproductive system, digestive system.

Medical Surveillance: Medical surveillance shall be made available to employees as outlined below:

- Comprehensive preplacement or initial medical and work histories with emphasis on reproductive experience and menstrual history.
- Comprehensive physical examination with emphasis on the genito-urinary tract including testicle size and consistency in males.
- Semen analysis to include sperm count, motility and morphology.
- Other tests, such as serum testosterone, serum follicle stimulating hormone (FSH), and serum lutenizing hormone (LH) may be carried out if, in the opinion of the responsible physician, they are indicated. In addition, screening tests of the renal and hepatic systems may be considered.
- A judgment of the worker's ability to use positive pressure respirators.
- Employees shall be counseled by the physician to ensure that each employees is aware that DBCP has been implicated in the production of effects on the reproductive system including sterility in male workers. In addition, they should be made aware that cancer was produced in some animals. While the relevancy of these findings is not yet clearly defined, they do indicate that both employees and employers should do everything possible to minimize exposure to DBCP.
- Periodic examinations containing the elements of the preplacement or initial examination shall be made available on at least an annual basis.
- Examinations of current employees shall be made available as soon as practicable after the promulgation of a standards for DBCP.
- Medical surveillance shall be made available to any worker suspected of having been exposed to DBCP.

- Pertinent medical records shall be maintained for all employees subject to exposure to DBCP in the workplace. Such records shall be maintained for 30 years and shall be available to medical representatives of the U.S. Government, the employer and the employee.

First Aid: If this chemical gets into the eyes, remove any contact lenses at once and irrigate immediately for at least 20 – 30 minutes, occasionally lifting upper and lower lids. Seek medical attention immediately. If this chemical contacts the skin, remove contaminated clothing and wash immediately with soap and water. Seek medical attention immediately. If this chemical has been inhaled, remove from exposure, begin rescue breathing (using universal precautions) if breathing has stopped and CPR if heart action has stopped. Transfer promptly to a medical facility. When this chemical has been swallowed, get medical attention. Contact local poison control center for advice about inducing vomiting. Medical observation is recommended for 24 – 48 hours after breathing overexposure, as pulmonary edema may be delayed. As first aid for pulmonary edema, a doctor or authorized paramedic may consider administering a corticosteroid spray.

Personal Protective Methods: Wear protective gloves and clothing to prevent any reasonable probability of skin contact. Safety equipment suppliers/manufacturers can provide recommendations on the most protective glove/clothing material for your operation.. All protective clothing (suits, gloves, footwear, headgear) should be clean, available each day, and put on before work. Contact lenses should not be worn when working with this chemical. Wear splash-proof chemical goggles and face shield unless full facepiece respiratory protection is worn. Employees should wash immediately with soap when skin is wet or contaminated. Provide emergency showers and eyewash.

Note: Protective clothing shall be resistant to the penetration and to the chemical action of dibromochloropropane. Additional protection, including gloves, bib-type aprons, boots, and overshoes, shall be provided for, and worn by, each employee during any operation that may cause direct contact with liquid dibromochloropropane. Supplied-air hoods or suits resistant to penetration by dibromochloropropane shall be worn when entering confined spaces, such as pits, or storage tanks. In situations where heat stress is likely to occur, supplied-air suits, preferably cooled, are recommended. The employer shall ensure that all personal protective clothing is inspected regularly for defects and is maintained in a clean and satisfactory condition by the employee.

Respirator Selection: Engineering controls shall be used wherever needed to keep airborne dibromochloropropane concentrations below the recommended occupational exposure limit. Compliance with this limit may be achieved by the use of respirators under the following conditions only:

- During the time necessary to install or test the required engineering controls.
- For nonroutine operations, such as emergency maintenance or repair activities.
- During emergencies when air concentrations of dibromochloropropane may exceed the recommended occupational exposure limit.

When a respirator is permitted, NIOSH recommends the following:

At any detectable concentration: SCBAF:PD,PP (any MSHA/NIOSH approved self-contained breathing apparatus that has a full facepiece and is operated in a pressure-demand or other positive-pressure mode); or SAF:PD,PP:ASCBA (any supplied-air respirator that has a full facepiece and is operated in a pressure-demand or other positive-pressure mode in combination with an auxiliary, self-contained breathing apparatus operated in a pressure-demand or other positive pressure mode). *Escape:* GMFOVHiE [any air-purifying, full-facepiece respirator (gas mask) with a chin-style, front- or back-mounted organic vapor canister having a high-efficiency particulate filter]; or SCBAE (any appropriate escape-type, self-contained breathing apparatus).

Storage: Prior to working with DBCP you should be trained on its proper handling and storage. Store in a refrigerator. Protect from alkalis and reactive metals. Protection from light is recommended for long term storage. Where possible, automatically pump liquid from drums or other storage containers to process containers. A regulated, marked area should be established where this chemical is handled, used, or stored in compliance with OSHA standard 1910.1045.

Shipping: DBCP requires a "Keep Away From Food" label. It falls in DOT Hazard Class 6.1 and Packing Group III.

Spill Handling: Evacuate and restrict persons not wearing protective equipment from area of spill or leak until cleanup is complete. Remove all ignition sources. Ventilate area of spill or leak. Absorb liquids in vermiculite, dry sand, earth, peat, carbon, or a similar material and deposit in sealed containers. Keep this chemical out of a confined space, such as a sewer, because of the possibility of an explosion, unless the sewer is designed to prevent the build-up of explosive concentrations. It may be necessary to contain and dispose of this chemical as a hazardous waste. If material or contaminated runoff enters waterways, notify downstream users of potentially contaminated waters. Contact your Department of Environmental Protection or your regional office of the federal EPA for specific recommendations. If employees are required to clean-up spills, they must be properly trained and equipped. OSHA 1910.120(q) may be applicable.

Fire Extinguishing: This chemical is a combustible liquid. Poisonous gases are produced in fire. Use dry chemical, carbon dioxide, or foam extinguishers. Vapors are heavier than air and will collect in low areas. Vapors may travel long distances to ignition sources and flashback. Vapors in confined areas may explode when exposed to fire. Containers may explode in fire. Storage containers and parts of containers may rocket great distances, in many directions. If material or contaminated runoff enters waterways, notify downstream users of potentially contaminated waters. Notify local health and fire officials and pollution control agencies. From a secure, explosion-proof location, use water spray to cool exposed containers. If cooling streams are ineffective (venting sound increases in volume and pitch, tank discolors, or shows any signs of deforming), withdraw immediately to a secure position. If employees are expected to fight fires, they must be trained and equipped in OSHA 1910.156.

Disposal Method Suggested: Use incinerator equipped with afterburner and scrubber.[22] Consult with environmental regulatory agencies for guidance on acceptable disposal practices. Generators of waste containing this contaminant (≥100 kg/mo) must conform with EPA regulations governing storage, transportation, treatment, and waste disposal.

References

National Institute for Occupational Safety and Health, Criteria for a Recommended Standard: Occupational Exposure to Dibromochloropropane, NIOSH Doc. No. 78–115 (1978)

Sax N. I., Ed., "Dangerous Properties of Industrial Materials Report," 1, No. 3, 55–57 (1981)

New Jersey Department of Health and Senior Services, Hazardous Substance Fact Sheet, "DBCP." Trenton NJ (June 1998)

Dibutylamine

Molecular Formula: $C_8H_{19}N$

Synonyms: 1-Butanamine, *N*-butyl; *N*-butyl-1-butanamine; DBA; *N*-Dibutylamine; Di(*N*-butyl)amine; Dibutylamine; DNBA

CAS Registry Number: 111-92-2

RTECS Number: HR7780000

DOT ID: UN 2248

EEC Number: 612-049-00-0

Regulatory Authority

- Canada, WHMIS, Ingredients Disclosure List

Cited in U.S. State Regulations: California (A, G), Florida (G), Massachusetts (G), Minnesota (G), New Jersey (G), Pennsylvania (G).

Description: Dibutylamine is a colorless liquid with an odor of ammonia. Boiling point = 159 – 161°C. Freezing/Melting point = -59 – -61.9°C. Flash point = 42 – 47°C. Autoignition temperature = 260°C. Explosive Limits: LEL = 1.1%; UEL – unknown. Hazard Identification (based on NFPA-704 M Rating System): Health 3, Flammability 2, Reactivity 0. Slightly soluble in water.

Potential Exposure: Used as a corrosion inhibitor, and intermediate for emulsifiers, rubber products, dyes, and insecticides.

Incompatibilities: Forms explosive mixture with air. Aqueous solution is a strong base. Incompatible with acids, acid chlorides, acid anhydrides, halogens, isocyanates, vinyl acetate, acrylates, substituted allyls, alkylene oxides, epichlorohydrin, ketones, aldehydes, alcohols, glycols, phenols, cresols, caprolactum solution, strong oxidizers, reactive organic compounds. Attacks copper alloys, zinc, tin, tin alloys, galvanized steel. Also, carbon dioxide is listed as incompatible by the state of New Jersey.

Permissible Exposure Limits in Air: There is no OSHA PEL. The American Industrial Hygiene Association (AIHA) recommends a WEEL ceiling of 5 ppm (26.5 mg/m^3).

Routes of Entry: Inhalation, through the skin, ingestion.

Harmful Effects and Symptoms

Short Term Exposure: This chemical is corrosive. This chemical can be absorbed through the skin, thereby increasing exposure. Skin or eye contact can cause severe irritation and burns. Inhalation can cause irritation of the respiratory tract and/or shortness of breath. Higher exposures can cause pulmonary edema, a medical emergency that can be delayed for several hours. This can cause death.

Long Term Exposure: May cause lung irritation; bronchitis may develop with coughing, phlegm and/or shortness of breath.

Points of Attack: Lungs.

Medical Surveillance: Lung function tests. Consider chest x-ray following acute over exposure.

First Aid: If this chemical gets into the eyes, remove any contact lenses at once and irrigate immediately for at least 15 minutes, occasionally lifting upper and lower lids. Seek medical attention immediately. If this chemical contacts the skin, remove contaminated clothing and wash immediately with soap and water. Seek medical attention immediately. If this chemical has been inhaled, remove from exposure, begin rescue breathing (using universal precautions) if breathing has stopped and CPR if heart action has stopped. Transfer promptly to a medical facility. When this chemical has been swallowed, get medical attention. If victim is *conscious*, administer water or milk. Do not induce vomiting. Medical observation is recommended for 24 – 48 hours after breathing overexposure, as pulmonary edema may be delayed. As first aid for pulmonary edema, a doctor or authorized paramedic may consider administering a corticosteroid spray.

Personal Protective Methods: Wear protective gloves and clothing to prevent any reasonable probability of skin contact. Safety equipment suppliers/manufacturers can provide recommendations on the most protective glove/clothing material for your operation. Viton and Polyvinyl Alcohol are among the recommended protective materials. All protective clothing (suits, gloves, footwear, headgear) should be clean, available each day, and put on before work. Contact lenses should not be worn when working with this chemical. Wear splash-proof chemical goggles and face shield unless full facepiece respiratory protection is worn. Employees should wash immediately with soap when skin is wet or contaminated. Provide emergency showers and eyewash.

Respirator Selection: Where the potential for exposure to this chemical, use a MSHA/NIOSH approved supplied-air respirator with a full facepiece operated in the positive pressure mode or with a full facepiece, hood, or helmet in the continuous flow mode, or use a MSHA/NIOSH approved self-contained breathing apparatus with a full facepiece operated in pressure-demand or other positive pressure mode.

Storage: Prior to working with dibutylamine you should be trained on its proper handling and storage. Before entering confined space where this chemical may be present, check to make sure that an explosive concentration does not exist. Store in tightly closed containers in a cool, well ventilated area away from incompatible materials listed above. Metal containers involving the transfer of this chemical should be grounded and bonded. Where possible, automatically pump liquid from drums or other storage containers to process containers. Drums must be equipped with self-closing valves, pressure vacuum bungs, and flame arresters. Use only non-sparking tools and equipment, especially when opening and closing containers of this chemical. Sources of ignition such as smoking and open flames, are prohibited where this chemical is used, handled, or stored in a manner that could create a potential fire or explosion hazard. Wherever this chemical is used, handled, manufactured, or stored, use explosion-proof electrical equipment and fittings.

Shipping: Label as a "Corrosive, Flammable Liquid." Hazard Class 8 and Packing Group II.

Spill Handling: Evacuate and restrict persons not wearing protective equipment from area of spill or leak until cleanup is complete. Remove all ignition sources. Establish forced ventilation to keep levels below explosive limit. Absorb liquids in vermiculite, dry sand, earth, peat, carbon, or a similar material and deposit in sealed containers. Keep this chemical out of a confined space, such as a sewer, because of the possibility of an explosion, unless the sewer is designed to prevent the build-up of explosive concentrations. It may be necessary to contain and dispose of this chemical as a hazardous waste. If material or contaminated runoff enters waterways, notify downstream users of potentially contaminated waters. Contact your Department of Environmental Protection or your regional office of the federal EPA for specific recommendations. If employees are required to clean-up spills, they must be properly trained and equipped. OSHA 1910.120(q) may be applicable.

Fire Extinguishing: This chemical is a flammable liquid. Poisonous gases including nitrogen oxides, hydrocarbons, amines, and carbon monoxide are produced in fire. Use dry chemical, water spray, or alcohol foam extinguishers. Vapors are heavier than air and will collect in low areas. Vapors may travel long distances to ignition sources and flashback. Vapors in confined areas may explode when exposed to fire. Containers may explode in fire. Storage containers and parts of containers may rocket great distances, in many directions. If material or contaminated runoff enters waterways, notify downstream users of potentially contaminated waters. Notify local health and fire officials and pollution control agencies. From a secure, explosion-proof location, use water spray to cool exposed containers. If cooling streams are ineffective (venting sound increases in volume and pitch, tank discolors, or shows any signs of deforming), withdraw immediately to a secure position. If employees are expected to fight fires, they must be trained and equipped in OSHA 1910.156.

Disposal Method Suggested: Incineration.

References

New Jersey Department of Health and Senior Services, "Hazardous Substance Fact Sheet, DIBUTYLAMINE," Trenton NJ (October 1998)

Dibutylaminoethanol

Molecular Formula: $C_{10}H_{23}NO$

Common Formula: $(C_4H_9)_2NCH_2CH_2OH$

Synonyms: DBAE; β-Di-*N*-butylaminoethanol; 2-Di-*N*-butylaminoethanol; 2-Dibutylaminoethanol; Dibutylaminoethanol; 2-Di-*N*-butylaminoethyl alcohol; *N,N*-Dibutylethanolamine; *N,N*-Dibutyl-*N*-(2-hydroxyethyl)amine

CAS Registry Number: 102-8-18

RTECS Number: KK3850000

DOT ID: UN 2873

Regulatory Authority

- Air Pollutant Standard Set (ACGIH)[1] (Australia) (Israel) (Mexico) (OSHA)[58] (Several States)[60] (Several Canadian Provinces)

Cited in U.S. State Regulations: Alaska (G), California (A, G), Connecticut (A), Florida (G), Maine (G), Massachusetts (G), Minnesota (G), Nevada (A), New Hampshire (G), New Jersey (G), North Dakota (A), Pennsylvania (G), Rhode Island (G), Virginia (A), West Virginia (G).

Description: 2-Di-n-butylaminoethanol, $(C_4H_9)_2$-NCH_2CH_2OH, is a colorless liquid with a faint amine-like odor. Boiling point = 224 – 232°C. Flash point = 90°C. Insoluble in water.

Potential Exposure: This material is used in organic synthesis.

Incompatibilities: Oxidizers.

Permissible Exposure Limits in Air: The Federal Limit (OSHA PEL)[58] and the ACGIH TWA value is 2 ppm (14 mg/m^3) with the notation "skin" indicating the possibility of cutaneous absorption. Australia, Israel, Mexico, and the Canadian provinces of Alberta, BC, Ontario, and Quebec use the same TWA and Alberta, BD, and Mexico set a STEL of 4 ppm (28 mg/m^3). Several states have set guidelines or standards for dibutylaminoethanol in ambient air[60] ranging from 0.14 mg/m^3 (North Dakota) to 0.23 mg/m^3 (Virginia) to 0.28 mg/m^3 (Connecticut) to 0.333 mg/m^3 (Nevada).

Determination in Air: Si gel; Methanol/Water; Gas chromatography/Flame ionization detection; NIOSH IV Method #2007, Aminoethanol Compounds.

Permissible Concentration in Water: No criteria set.

Routes of Entry: Inhalation, skin absorption, ingestion, skin and/or eye contact.

Harmful Effects and Symptoms

Short Term Exposure: Contact may burn the eyes and irritate the skin. The vapor can irritate the nose, throat and bronchial tubes. The oral LD_{50} rat is 1,070 mg/kg (slightly toxic).

Long Term Exposure: High or repeated exposure may damage the liver and kidneys. Related chemicals can cause lung allergy (asthma) or skin allergy, with rash. They can also cause a fluid build-up in the lungs with high exposures, a medical emergency. It its not known whether 2-n-dibutylaminoethanol has these effects.

Points of Attack: Eyes, skin, respiratory system.

Medical Surveillance: If symptoms develop or overexposure is suspected, the following may be useful: Liver and kidney function tests. Lung function tests. These may be normal if the person is not having an attack at the time of the test. Evaluation by a qualified allergist, including careful exposure history and special testing, may help diagnose skin allergy. Consider chest x-ray after acute overexposure.

First Aid: If this chemical gets into the eyes, remove any contact lenses at once and irrigate immediately for at least 15 minutes, occasionally lifting upper and lower lids. Seek medical attention immediately. If this chemical contacts the skin, remove contaminated clothing and wash immediately with soap and water. Seek medical attention immediately. If this chemical has been inhaled, remove from exposure, begin rescue breathing (using universal precautions) if breathing has stopped and CPR if heart action has stopped. Transfer promptly to a medical facility. When this chemical has been swallowed, get medical attention. Give large quantities of water and induce vomiting. Do not make an unconscious person vomit.

Personal Protective Methods: Wear protective gloves and clothing to prevent any reasonable probability of skin contact. Safety equipment suppliers/manufacturers can provide recommendations on the most protective glove/clothing material for your operation.. All protective clothing (suits, gloves, footwear, headgear) should be clean, available each day, and put on before work. Contact lenses should not be worn when working with this chemical. Wear splash-proof chemical goggles and face shield unless full facepiece respiratory protection is worn. Employees should wash immediately with soap when skin is wet or contaminated. Provide emergency showers and eyewash.

Respirator Selection: Where the potential exists for exposures over 2 ppm, use a MSHA/NIOSH approved supplied-air respirator with a full facepiece operated in the positive pressure mode or with a full facepiece, hood, or helmet in the continuous flow mode, or use a MSHA/NIOSH approved self-contained breathing apparatus with a full facepiece operated in the pressure-demand or other positive pressure mode.

Storage: Prior to working with DBAE you should be trained on its proper handling and storage. 2-n-dibutylamineothanol must be stored to avoid contact with oxidizers (such as perchlorates, peroxides, permanganates, chlorates, and nitrates) since violent reactions may occur. Sources of ignition such as smoking and open flames are prohibited where 2-n-dibutylaminoethanol is used, handled, or stored in a manner that could create a potential fire or explosion hazard. Where possible, automatically pump liquid from drums or other storage containers to process containers.

Shipping: Dibutylaminoethanol must carry a "Keep Away From Food" label. It falls in DOT Hazard Class 6.1 and packing Group III.

Spill Handling: Evacuate and restrict persons not wearing protective equipment from area of spill or leak until cleanup is complete. Remove all ignition sources. Ventilate area of spill or leak. Absorb liquids in vermiculite, dry sand, earth, peat, carbon, or a similar material and deposit in sealed containers. Keep this chemical out of a confined space, such as a sewer, because of the possibility of an explosion, unless the sewer is designed to prevent the build-up of explosive concentrations. It may be necessary to contain and dispose of this chemical as a hazardous waste. If material or contaminated runoff enters waterways, notify downstream users of potentially contaminated waters. Contact your Department of Environmental Protection or your regional office of the federal EPA for specific recommendations. If employees are required to clean-up spills, they must be properly trained and equipped. OSHA 1910.120(q) may be applicable.

Fire Extinguishing: This chemical is a combustible liquid. Poisonous gases including nitrogen oxides are produced in fire. Use dry chemical, carbon dioxide, or foam extinguishers. Vapors are heavier than air and will collect in low areas. Vapors may travel long distances to ignition sources and flashback. Vapors in confined areas may explode when exposed to fire. Containers may explode in fire. Storage containers and parts of containers may rocket great distances, in many directions. If material or contaminated runoff enters waterways, notify downstream users of potentially contaminated waters. Notify local health and fire officials and pollution control agencies. From a secure, explosion-proof location, use water spray to cool exposed containers. If cooling

streams are ineffective (venting sound increases in volume and pitch, tank discolors, or shows any signs of deforming), withdraw immediately to a secure position. If employees are expected to fight fires, they must be trained and equipped in OSHA 1910.156.

References

New Jersey Department of Health, "Hazardous Substance Fact Sheet: "2-n-Dibutylaminoethanol," Trenton, NJ (October 1998)

Di-tert-Butyl-p-Cresol

Molecular Formula: $C_{15}H_{24}O$

Common Formula: $[(CH_3)_3C]_2C_6H_2(CH_3)OH$

Synonyms: BHT; Butylated hydroxytoluene; DBPC; Dibutylated hydroxytoluene; 4-Methyl-2,6-di-*tert*-butylphenol

CAS Registry Number: 128-37-0

RTECS Number: GO7875000

DOT ID: NA 9188

Regulatory Authority

- Carcinogen (Animal Suspected) (IARC)[9]
- Air Pollutant Standard Set (ACGIH)[1] (HSE)

Cited in U.S. State Regulations: Alaska (G), California (G), Maine (G), Massachusetts (G), New Hampshire (G), New Jersey (G), Pennsylvania (G), West Virginia (G).

Description: DBPC, $[(CH_3)_3C]_2C_6H_2(CH_3)OH$, is a white to pale yellow crystalline solid or powder. Boiling point = 265°C. Freezing/Melting point = 70°C. Flash point = 127°C (cc). Insoluble in water.

Potential Exposure: DBPC is used as an antioxidant to stabilize petroleum fuels, rubber and vinyl plastics. It is also used as an antioxidant in human foods and animal feeds.

Incompatibilities: Oxidizers may cause fire and explosion hazard.

Permissible Exposure Limits in Air: There is no OSHA PEL, but ACGIH and NIOSH recommend a TWA value of 10 mg/m^3. HSE set the same TWA.

Determination in Air: Si gel; Methanol/CS2; Gas chromatography/Flame ionization detection; NIOSH II,[1] P&CAM Method #226.

Permissible Concentration in Water: No criteria set.

Harmful Effects and Symptoms

Short Term Exposure: Irritates the eyes and skin. This compound has an acute oral LD_{50} for rats of 890 mg/kg which is classified as slightly toxic.

Long Term Exposure: 2,6-di-tert-butyl-p-cresol may damage the developing fetus. Repeated exposure to high levels may affect the liver. DBPC may cause changes in behavior and learning ability, and reduce the blood's ability to clot, but this is not known for sure at this time.

Points of Attack: Eyes, skin, liver.

Medical Surveillance: Liver function tests.

First Aid: If this chemical gets into the eyes, remove any contact lenses at once and irrigate immediately for at least 15 minutes, occasionally lifting upper and lower lids. Seek medical attention immediately. If this chemical contacts the skin, remove contaminated clothing and wash immediately with soap and water. Seek medical attention immediately. If this chemical has been inhaled, remove from exposure, begin rescue breathing (using universal precautions) if breathing has stopped and CPR if heart action has stopped. Transfer promptly to a medical facility. When this chemical has been swallowed, get medical attention. Give large quantities of water and induce vomiting. Do not make an unconscious person vomit.

Personal Protective Methods: Wear protective gloves and clothing to prevent any reasonable probability of skin contact. Safety equipment suppliers/manufacturers can provide recommendations on the most protective glove/clothing material for your operation.. All protective clothing (suits, gloves, footwear, headgear) should be clean, available each day, and put on before work. Contact lenses should not be worn when working with this chemical. Wear dust-proof chemical goggles and face shield unless full facepiece respiratory protection is worn. Employees should wash immediately with soap when skin is wet or contaminated. Provide emergency showers and eyewash.

Respirator Selection: NIOSH: *At any detectable concentration:* SCBAF:PD,PP (any MSHA/NIOSH approved self-contained breathing apparatus that has a full facepiece and is operated in a pressure-demand or other positive-pressure mode); or SAF:PD,PP:ASCBA (any supplied-air respirator that has a full facepiece and is operated in a pressure-demand or other positive-pressure mode in combination with an auxiliary, self-contained breathing apparatus operated in a pressure-demand or other positive pressure mode). *Escape:* GMFOV [any air-purifying, full-facepiece respirator (gas mask) with a chin-style, front-or back-mounted organic vapor canister] or SCBAE (any appropriate escape-type, self-contained breathing apparatus).

Storage: Prior to working with DBPC you should be trained on its proper handling and storage. A regulated, marked area should be established where this chemical is handled, used, or stored in compliance with OSHA standard 1910.1045. Store in tightly closed containers in a cool well-ventilated area away from oxidizing agents (such as peroxides, permanganates, chlorates, perchlorates, and nitrates).

Shipping: DBPC may be treated as a hazardous substance, solid, n.o.s. which falls in DOT Hazards Class ORM-E and Packing Group III. There are no label requirements.

Spill Handling: Evacuate persons not wearing protective equipment from area of spill or leak until clean-up is complete. Remove all ignition sources. Collect powdered material in the most convenient and safe manner and deposit in sealed containers. Ventilate area after clean-up is complete. It may be necessary to contain and dispose of this chemical as a hazardous waste. If material or contaminated runoff enters waterways, notify downstream users of potentially contaminated waters. Contact your Department of Environmental Protection or your regional office of the federal EPA for specific recommendations. If employees are required to clean-up spills, they must be properly trained and equipped. OSHA 1910.120(q) may be applicable.

Fire Extinguishing: DBPC is a combustible solid, but does not readily ignite. Use dry chemical, carbon dioxide, water spray, or alcohol foam extinguishers. Water or foam may cause frothing. Poisonous gases are produced in fire. If material or contaminated runoff enters waterways, notify downstream users of potentially contaminated waters. Notify local health and fire officials and pollution control agencies. From a secure, explosion-proof location, use water spray to cool exposed containers. If cooling streams are ineffective (venting sound increases in volume and pitch, tank discolors, or shows any signs of deforming), withdraw immediately to a secure position. If employees are expected to fight fires, they must be trained and equipped in OSHA 1910.156.

References

New Jersey Department of Health, "Hazardous Substance Fact Sheet: 2,6-Di-tert-Butyl-p-Cresol," Trenton, NJ (April 1997)

Dibutyl Phosphate

Molecular Formula: $C_8H_{18}NO_4P$

Common Formula: $(n\text{-}C_4H_9O)_2PO(ON)$

Synonyms: Dibutyl acid *o*-phosphate; Dibutyl acid phosphate; Di-*N*-butyl hydrogen phosphate; Di-*N*-butyl phosphate; Dibutyl phosphoric acid; Phosphoric acid, dibutyl ester

CAS Registry Number: 107-66-4

RTECS Number: TB9605000

DOT ID: UN 1760

Regulatory Authority

- Air Pollutant Standard Set (ACGIH)[1] (Australia) (Israel) (Mexico) (HSE)[33] (OSHA)[58] (Several States)[60] (Several Canadian Provinces)
- Canada, WHMIS, Ingredients Disclosure List

Cited in U.S. State Regulations: Alaska (G), California (A, G), Connecticut (A), Florida (G), Illinois (G), Maine (G), Massachusetts (G), Minnesota (G), Nevada (A), New Hampshire (G), New Jersey (G), North Dakota (A), Pennsylvania (G), Rhode Island (G), Virginia (A), West Virginia (G).

Description: Dibutyl Phosphate, $(n\text{-}C_4H_9O)_2PO(ON)$, is a pale amber to brown, odorless liquid. Boiling point = 135 – 138°C. Freezing/Melting point = -13°C. Flash point = 188°C (oc). Auto-ignition temperature = 420°C. Insoluble in water.

Potential Exposure: This material is used as a catalyst in organic synthesis.

Incompatibilities: Dibutyl phosphate is a medium strong acid. Reacts with strong oxidizers. Attacks many metals forming flammable and explosive hydrogen gas. Attacks some plastics, rubber and coatings.

Permissible Exposure Limits in Air: The Federal Limit (OSHA PEL),[58] the HSE value[33] and the ACGIH TWA value is 1.0 ppm (5 mg/m^3) and the STEL value is 2.0 ppm (10 mg/m^3). The NIOSH IDLH level is 30 ppm. Australia, Israel, Mexico, and the Canadian provinces of Alberta, BC, Ontario, and Quebec set the same levels as OSHA. Some states have set guidelines or standards for dibutyl phosphate in ambient air[60] ranging from 50 – 100 μg/m^3 (North Dakota) to 80 μg/m^3 (Virginia) to 100 μg/m^3 (Connecticut) to 119 μg/m^3 (Nevada).

Determination in Air: Filter; CH3CN; Gas chromatography/Flame photometric detection for sulfur, nitrogen, or phosphorus; NIOSH IV Method #5017.

Permissible Concentration in Water: No criteria set.

Routes of Entry: Inhalation, ingestion, eye and/or skin contact. Passes through the skin.

Harmful Effects and Symptoms

Short Term Exposure: Exposure can irritate and burn the eyes. Inhalation can irritate the respiratory tract causing coughing, wheezing, and shortness of breath. Skin contact irritates the skin causing rash or burning sensation.

Long Term Exposure: Can cause drying and cracking of skin. Can cause lung irritation.

Points of Attack: Respiratory system, skin, eyes.

Medical Surveillance: For those with frequent or potentially high exposure (half the TLV or greater) the following are recommended before beginning work and at regular times after that: Lung function tests.

First Aid: If this chemical gets into the eyes, remove any contact lenses at once and irrigate immediately for at least 15 minutes, occasionally lifting upper and lower lids. Seek medical attention immediately. If this chemical contacts the skin, remove contaminated clothing and wash immediately with soap and water. Seek medical attention immediately. If this chemical has been inhaled, remove from exposure, begin rescue breathing (using universal precautions) if breathing has stopped and CPR if heart action has stopped. Transfer promptly to a medical facility. When this chemical has been swallowed, get medical attention. If victim is conscious, administer water or milk. Do not induce vomiting.

Personal Protective Methods: Wear protective gloves and clothing to prevent any reasonable probability of skin contact. Safety equipment suppliers/manufacturers can provide recommendations on the most protective glove/clothing material for your operation.. All protective clothing (suits, gloves, footwear, headgear) should be clean, available each day, and put on before work. Contact lenses should not be worn when working with this chemical. Wear splash-proof chemical goggles and face shield unless full facepiece respiratory protection is worn. Employees should wash immediately with soap when skin is wet or contaminated. Provide emergency showers and eyewash.

Respirator Selection: NIOSH/OSHA: *Up to 10 ppm:* SA (any supplied-air respirator). *Up to 25 ppm:* SA:CF (any supplied-air respirator operated in a continuous-flow mode. *Up to 30 ppm:* SAT:CF (any supplied-air respirator that has a tight-fitting facepiece and is operated in a continuous-flow mode); or SCBAF (any self-contained breathing apparatus with a full facepiece); or SAF (any supplied-air respirator with a full facepiece). *Emergency or planned entry into unknown concentrations or IDLH conditions:* SCBAF:PD,PP (any self-contained breathing apparatus that has a full facepiece and is operated in a pressure-demand or other positive-pressure mode); or SAF:PD,PP:ASCBA (any supplied-air respirator that has a full facepiece and is operated in a pressure-demand or other positive-pressure mode in combination with an auxiliary, self-contained breathing apparatus operated in a pressure-demand or other positive-pressure mode). *Escape:* GMFOVHiE [any air-purifying, full-facepiece respirator (gas mask) with a chin-style, front- or back-mounted organic vapor canister having a high-efficiency particulate filter]; or SCBAE (any appropriate escape-type, self-contained breathing apparatus).

Storage: Prior to working with this chemical you should be trained on its proper handling and storage. Dibutyl phosphate must be stored to avoid contact with strong oxidizers such as chlorine, chlorine dioxide, and bromine, since violent reactions occur.

Shipping: There is no specific citation for dibutyl phosphate, but as a corrosive liquid, n.o.s. subject to Hazard Class 8. Label: "Corrosive," Packing Group I or II.

Spill Handling: Evacuate and restrict persons not wearing protective equipment from area of spill or leak until cleanup is complete. Remove all ignition sources. Ventilate area of spill or leak. Absorb liquids in vermiculite, dry sand, earth, peat, carbon, or a similar material and deposit in sealed containers. Keep this chemical out of a confined space, such as a sewer, because of the possibility of an explosion, unless the sewer is designed to prevent the build-up of explosive concentrations. It may be necessary to contain and dispose of this chemical as a hazardous waste. If material or contaminated runoff enters waterways, notify downstream users of potentially contaminated waters. Contact your Department of Environmental Protection or your regional office of the federal EPA for specific recommendations. If employees are required to clean-up spills, they must be properly trained and equipped. OSHA 1910.120(q) may be applicable.

Fire Extinguishing: Dibutyl phosphate is a combustible liquid. Poisonous gases including oxides of phosphorus, carbon monoxide, phosphine, and phosphoric acid are produced in fire. Vapors are heavier than air and will collect in low areas. Vapors may travel long distances to ignition sources and flashback. Vapors in confined areas may explode when exposed to fire. Containers may explode in fire. Storage containers and parts of containers may rocket great distances, in many directions. If material or contaminated runoff enters waterways, notify downstream users of potentially contaminated waters. Notify local health and fire officials and pollution control agencies. From a secure, explosion-proof location, use water spray to cool exposed containers. If cooling streams are ineffective (venting sound increases in volume and pitch, tank discolors, or shows any signs of deforming), withdraw immediately to a secure position. If employees are expected to fight fires, they must be trained and equipped in OSHA 1910.156.

References

New Jersey Department of Health, "Hazardous Substance Fact Sheet: Dibutyl Phosphate," Trenton, NJ (February 1998)

Dibutyl Phthalate

Molecular Formula: $C_{16}H_{22}O_4$

Common Formula: $C_6H_4(COOC_4H_9)_2$

Synonyms: *o*-Benzenedicarboxylic acid, dibutyl ester; 1,2-Benzenedicarboxylic acid, dibutyl ester; Benzene-*o*-dicarboxylic acid di-*n*-butyl ester; Bis-*n*-butyl phthalate; BUFA; Butyl phthalate; *n*-Butyl phthalate (DOT); Celluflex DPB; DBP; DBP (Ester); Di(*n*-butyl) 1,2-benzenedicarboxylate; Dibutyl 1,2-benzene dicarboxylate; Dibutyl 1,2-benzenedicarboxylate; Di-*n*-butyl phthalate; Dibutyl *o*-phthalate; EINECS No. 201-557-4; Elaol; Ergoplast FDB; Ftalato de *n*-butilo (Spanish); Genoplast B; Hexaplas M/B; Kodaflex dibutyl phthalate (DBP); Morflex-240; NLA-10; Palatinol C; Palatinol DBP; Phthalic acid, dibutyl ester; Polycizer DBP; PX 104; RC Plasticizer DBP Staflex DBP; Uniplex 150; Witcizer 300

CAS Registry Number: 84-74-2

RTECS Number: TI0875000

DOT ID: UN/NA 9095 (n-butyl phthalate); UN 3082

Regulatory Authority

- Air Pollutant Standard Set (ACGIH)[1] (HSE) (Australia) (Israel) (Mexico)[33] (OSHA)[68] (Other Countries)[35] (Several Canadian Provinces)
- Clean Air Act: Hazardous Air Pollutants (Title I, Part A, Section 112)
- Clean Water Act: 40CFR423, Appendix A, Priority Pollutants; Section 313 Water Priority Chemicals (57FR41331, 9/9/92)
- EPA Hazardous Waste Number (RCRA No.): U069
- RCRA, 40CFR261, Appendix 8 Hazardous Constituents

- RCRA 40CFR268.48; 61FR15654, Universal Treatment Standards: Wastewater (mg/l), 0.057; Nonwastewater (mg/kg), 28
- RCRA 40CFR264, Appendix 9; TSD Facilities Ground Water Monitoring List. Suggested test method(s) (PQL μg/l): 8060 (5); 8270 (10)
- Superfund/EPCRA 40CFR302.4 Reportable Quantity (RQ): CERCLA, 10 lb(4.55 kg)
- EPCRA Section 313 Form R *de minimis* concentration reporting level: 1.0%
- U.S. DOT Regulated Marine Pollutant (49CFR172.101, Appendix B)
- Canada, WHMIS, Ingredients Disclosure List, CEPA Priority Substance List, National Priority Release Inventory

Cited in U.S. State Regulations: Alaska (G), Florida (G), Illinois (G), Kansas (G, W), Louisiana (G), Maine (G, W), Maryland (G), Massachusetts (G), Michigan (G), New Hampshire (G), New Jersey (G), Pennsylvania (G), Rhode Island (G), Vermont (G), Virginia (G), Washington (G), West Virginia (G), Wisconsin (G).

Description: Dibutyl phthalate, $C_6H_4(COOC_4H_9)_2$, is a colorless oily liquid with a very weak aromatic odor. Boiling point = 340°C. Freezing/Melting point = -35°C. Flash point = 157°C. Auto-ignition temperature = 402°C. Hazard Identification (based on NFPA-704 M Rating System): Health 0, Flammability 1, Reactivity 0. Explosive limits: LEL = 0.5% @ 235°C; UEL = unknown. Practically insoluble in water.

Potential Exposure: Use in plasticizing vinyl acetate emulsion systems and in plasticizing cellulose esters. Also used as a lacquer solvent and insect repellent.

Incompatibilities: DBP is a medium strong acid. Reacts with strong oxididizers, strong alkalis, strong acids, nitrates. Attacks many metals forming flammable and explosive hydrogen gas.

Permissible Exposure Limits in Air: The Federal legal limit (OSHA PEL) and ACGIH recommended TWA is 5 mg/m^3. Australia, Israel, Mexico, and the Canadian provinces of Alberta, BC, Ontario, and Quebec, and the HSE[33] use this value and HSE, Mexico, Alberta, and BC add the STEL of 10 mg/m^3. The former USSR has set a MAC in workplace air of 0.5 mg/m^3.[35][43] The former USSR[35] has set a limit for ambient air of 0.1 mg/m^3 on a once daily basis; Czechoslovakia[35] has set ambient air limits at 0.8 mg/m^3 on a daily average basis and 2.4 mg/m^3 on a half-hour exposure basis. The NIOSH IDLH level is 9,300 mg/m^3.

Determination in Air: Collection on a filter; Work-up with CS_2, Gas chromatography/Flame ionization detection; NIOSH IV Method #5020.

Permissible Concentration in Water: To protect freshwater aquatic life – 940 μg/l on an acute basis and 3 μg/l on a chronic basis for all phthalate esters. On a chronic basis, as low as 2,944 μg/l. To protect human health – 34,000 μg/l.[6] Some states have set guidelines for DBP in drinking water[61] ranging from 770 μg/l in Kansas to 2,200 μg/l in Maine. The former USSR-UNEP/IRPTC project[43] has set a MAC in water bodies used for domestic purposes of 0.2 mg/l.

Determination in Water: Gas chromatography (EPA Method 606) or gas chromatography plus mass spectrometry (EPA Method 625).

Routes of Entry: Inhalation, ingestion, eye and skin contact.

Harmful Effects and Symptoms

Short Term Exposure: The substance irritates the eyes, skin, and nasal passages and upper respiratory system. May cause stomach irritation; light sensitivity.

Long Term Exposure: Unknown at this time. However this chemical may cause lung problems. Di-n-butyl phthalate may also damage the developing fetus and may also damage the testes (male reproductive glands).

Points of Attack: Eyes, respiratory system, gastrointestinal system.

Medical Surveillance: Consider the points of attack in preplacement and periodic physical examinations. Lung function tests.

First Aid: If this chemical gets into the eyes, remove any contact lenses at once and irrigate immediately for at least 15 minutes, occasionally lifting upper and lower lids. Seek medical attention immediately. If this chemical contacts the skin, remove contaminated clothing and wash immediately with soap and water. Seek medical attention immediately. If this chemical has been inhaled, remove from exposure, begin rescue breathing (using universal precautions) if breathing has stopped and CPR if heart action has stopped. Transfer promptly to a medical facility. When this chemical has been swallowed, get medical attention. Give large quantities of water and induce vomiting. Do not make an unconscious person vomit.

Personal Protective Methods: Wear protective gloves and clothing to prevent any reasonable probability of skin contact. Safety equipment suppliers/manufacturers can provide recommendations on the most protective glove/clothing material for your operation. ACGIH recommends butyl rubber, neoprene, nitrile rubber, and viton as good to excellent protective materials. All protective clothing (suits, gloves, footwear, headgear) should be clean, available each day, and put on before work. Contact lenses should not be worn when working with this chemical. Wear splash-proof chemical goggles and face shield unless full facepiece respiratory protection is worn. Employees should wash immediately with soap when skin is wet or contaminated. Provide emergency showers and eyewash.

Respirator Selection: NIOSH/OSHA: *50 mg/m^3:* DMF (any dust and mist respirator with a full facepiece). *125 mg/m^3:* SA:CF (any supplied-air respirator operated in a continuous-flow mode); or PAPRDM (any powered, air-purifying respirator with a dust and mist filter). *250 mg/m³:* HiEF (any air-purifying, full-facepiece respirator with a high-efficiency particulate filter); or SCBAF (any self-contained breathing apparatus with a full facepiece); or SAF (any supplied-air respirator with a full facepiece). *4,000 mg/m³:* SAF:PD,PP (any supplied-air respirator that has a full facepiece and is operated in a pressure-demand or other positive-pressure mode). Emergency or planned entry into unknown concentrations or IDLH conditions: SCBAF:PD,PP (any self-contained breathing apparatus that has a full facepiece and is operated in a pressure-demand or other positive-pressure mode); or SAF:PD,PP:ASCBA (any supplied-air respirator that has a full facepiece and is operated in a pressure-demand or other positive-pressure mode in combination with an auxiliary self-contained breathing apparatus operated in a pressure-demand or other positive pressure mode). *Escape:* HiEF (any air-purifying, full-facepiece respirator with a high-efficiency particulate filter); or SCBAE (any appropriate escape-type, self-contained breathing apparatus).

Note: Substance causes eye irritation or damage; eye protection needed.

Storage: Prior to working with this chemical you should be trained on its proper handling and storage. Before entering confined space where this chemical may be present, check to make sure that an explosive concentration does not exist. Di-n-butyl phthalate must be stored to avoid contact with strong oxidizers (such as chlorine, bromine, or chlorine dioxide); strong alkalis (such as sodium hydroxide, potassium hydroxide and lithium hydroxide); and strong acids (such as sulfuric acid, hydrochloric acid, and nitric acid) since violent reactions occur. Store in tightly closed containers in a cool well-ventilated area away from heat. Sources of ignition such as smoking and open flames are prohibited where di-n-butyl phthalate is used, handled, or stored in a manner that could create a potential fire or explosion hazard.

Shipping: DBP is an environmentally hazardous liquid, n.o.s. It is in Hazard Class 9. Label: "Class 9." Packing Group III.

Spill Handling: Evacuate and restrict persons not wearing protective equipment from area of spill or leak until cleanup is complete. Remove all ignition sources. Establish forced ventilation to keep levels below explosive limit. Absorb liquids in vermiculite, dry sand, earth, peat, carbon, or a similar material and deposit in sealed containers. Keep this chemical out of a confined space, such as a sewer, because of the possibility of an explosion, unless the sewer is designed to prevent the build-up of explosive concentrations. It may be necessary to contain and dispose of this chemical as a hazardous waste. If material or contaminated runoff enters waterways, notify downstream users of potentially contaminated waters. Contact your Department of Environmental Protection or your regional office of the federal EPA for specific recommendations. If employees are required to clean-up spills, they must be properly trained and equipped. OSHA 1910.120(q) may be applicable.

Fire Extinguishing: This chemical is a combustible liquid. Poisonous gases including phthalic anhydride are produced in fire. Use dry chemicals or carbon dioxide. Vapors are heavier than air and will collect in low areas. Vapors may travel long distances to ignition sources and flashback. Vapors in confined areas may explode when exposed to fire. Containers may explode in fire. Storage containers and parts of containers may rocket great distances, in many directions. If material or contaminated runoff enters waterways, notify downstream users of potentially contaminated waters. Notify local health and fire officials and pollution control agencies. From a secure, explosion-proof location, use water spray to cool exposed containers. If cooling streams are ineffective (venting sound increases in volume and pitch, tank discolors, or shows any signs of deforming), withdraw immediately to a secure position. If employees are expected to fight fires, they must be trained and equipped in OSHA 1910.156.

Disposal Method Suggested: May be absorbed on vermiculite, sand or earth and disposed of in a sanitary landfill. Alternatively, it may be incinerated.[22] Consult with environmental regulatory agencies for guidance on acceptable disposal practices. Generators of waste containing this contaminant (≥100 kg/mo) must conform with EPA regulations governing storage, transportation, treatment, and waste disposal.

References

U.S. Environmental Protection Agency, Phthalate Esters: Ambient Water Quality Criteria, Washington, DC (1980)

National Institute for Occupational Safety and Health, Profiles on Occupational Hazards for Criteria Document Priorities: Phthalates, 97–103, Report PB-274-273, Cincinnati, OH (1977)

U.S. Environmental Protection Agency, Di-n-butyl Phthalate, Health and Environmental Effects Profile No. 62, Washington, DC, Office of Solid Waste (April 30, 1980)

New Jersey Department of Health, "Hazardous Substance Fact Sheet: Di-n-butyl Phthalate," Trenton, NJ (February 1989)

Sax N. I., Ed., "Dangerous Properties of Industrial Materials Report," 5, No. 4, 40–44 (1985)

Dicamba

Molecular Formula: $C_8H_6Cl_2O_3$

Common Formula: $Cl_2C_6H_2(OCH_3)COOH$

Synonyms: Acido (3,6-dichloro-2-metossi)-benzoico (Italian); AI3-27556; *o*-Anisic acid, 3,6-dichloro-; Banex; Banlen; Banvel; Banvel 4S; Banvel 4WS; Banvel CST; Banvel herbicide; Banvel II herbicide; Benzoic acid, 3,6-dichloro-2-methoxy-; Brush Buster; Caswell No. 295; Compound B dicamba; Dianat (Russian); Dianate; Dicamba benzoic acid herbicide; Dicambra; 3,6-Dichloor-2-methoxy-benzoeizuur (Dutch); 3,6-Dichlor-3-methoxy-benzoesaeure (German); 3,6-Dichloro-*o*-anisic acid; 2,5-Dichloro-

6-methoxybenzoic acid; 3,6-Dichloro-2-methoxybenzoic acid; EPA pesticide chemical code 029801; MDBA; Mediben; 2-Methoxy-3,6-dichlorobenzoic acid; Velsicol 58-CS-11; Velsicol compound R

CAS Registry Number: 1918-00-9

RTECS Number: DG7525000

DOT ID: UN 2769

EEC Number: 607-043-00-X

Regulatory Authority

- Air Pollutant Standard Set (former USSR)[35]
- Clean Water Act: Section 311 Hazardous Substances/RQ 40CFR117.3 (same as CERCLA, see below)
- Safe Drinking Water Act: Priority List (55 FR 1470)
- Superfund/EPCRA 40CFR302.4 Reportable Quantity (RQ): CERCLA, 1,000 lb (454 kg)
- EPCRA Section 313 Form R *de minimis* concentration reporting level: 1.0%
- Canada, Drinking Water Quality, 0.12 mg/l MAC

Cited in U.S. State Regulations: California (G, W), Illinois (G), Maine (W), Massachusetts (G), New Hampshire (G), New Jersey (G), Pennsylvania (G), Wisconsin (W).

Description: Dicamba, $Cl_2C_6H_2(OCH_3)COOH$, is a white or brown nonflammable, colorless, odorless solid. Boiling point = 200°C (decomposes below B.P.). Freezing/Melting point = 114 – 116°C. Hazard Identification (based on NFPA-704 M Rating System): Health 1, Flammability 0, Reactivity 0. Slightly soluble in water.

Potential Exposure: Those involved in manufacture, formulation and application of this postemergence herbicide. Used to control allual and perennial broad leaf weeds in corn, small grain pastures, and non-croplands.

Incompatibilities: Incompatible with sulfuric acid, bases, ammonia, aliphatic amines, alkanolamines, isocyanates, alkylene oxides, epichlorohydrin. Dicamba decomposes in heat producing toxic and corrosive fumes including hydrogen chloride.

Permissible Exposure Limits in Air: The former USSR[35] has set a MAC in workplace air of 1.0 mg/m^3. Although no U.S. exposure limits have been established, this chemical can be absorbed through the skin, thereby increasing exposure.

Permissible Concentration in Water: A no-adverse effect level in drinking water has been calculated by NSA/NRC[46] at 0.009 mg/l. States which have set guidelines for dicamba in drinking water[61] include Maine at 9.0 µg/l and Wisconsin at 12.5 µg/l. Canada's MAC in drinking water is 0.12 mg/l.

Determination in Water: A detection limit of 1 ppb for dicamba by electron-capture gas chromatography has been reported by NAS/NRC.[46]

Routes of Entry: Ingestion, inhalation, and through the skin.

Harmful Effects and Symptoms

Short Term Exposure: Dicamba irritates the eyes, skin, and respiratory tract. Exposure can cause nausea, vomiting, loss of appetite and weight, muscle weakness, and exhaustion. The acute toxicity of dicamba is relatively low. Dicamba produced no adverse effect when fed to rats at up to 19.3 mg/kg/day and 25 mg/kg/day in subchronic and chronic studies. The no-adverse-effect dose in dogs was 1.25 mg/kg/day in a 2-year feeding study. Based on these data, an ADI was calculated at 0.0012 mg/kg/day. The LD_{50} oral-rat is 1,037 mg/kg (slightly toxic).

Long Term Exposure: May affect the liver.

Points of Attack: Liver.

Medical Surveillance: Liver function tests.

First Aid: If this chemical gets into the eyes, remove any contact lenses at once and irrigate immediately for at least 15 minutes, occasionally lifting upper and lower lids. Seek medical attention immediately. If this chemical contacts the skin, remove contaminated clothing and wash immediately with soap and water. Seek medical attention immediately. If this chemical has been inhaled, remove from exposure, begin rescue breathing (using universal precautions) if breathing has stopped and CPR if heart action has stopped. Transfer promptly to a medical facility. When this chemical has been swallowed, get medical attention. Give large quantities of water and induce vomiting. Do not make an unconscious person vomit.

Personal Protective Methods: Wear protective gloves and clothing to prevent any reasonable probability of skin contact. Safety equipment suppliers/manufacturers can provide recommendations on the most protective glove/clothing material for your operation.. All protective clothing (suits, gloves, footwear, headgear) should be clean, available each day, and put on before work. Contact lenses should not be worn when working with this chemical. Wear dust-proof chemical goggles and face shield unless full facepiece respiratory protection is worn. Employees should wash immediately with soap when skin is wet or contaminated. Provide emergency showers and eyewash.

Respirator Selection: *At any detectable concentration:* SCBAF:PD,PP (any MSHA/NIOSH approved self-contained breathing apparatus that has a full facepiece and is operated in a pressure-demand or other positive-pressure mode); or SAF:PD,PP:ASCBA (any supplied-air respirator that has a full facepiece and is operated in a pressure-demand or other positive-pressure mode in combination with an auxiliary, self-contained breathing apparatus operated in a pressure-demand or other positive pressure mode). *Escape:* HiEF (any air-purifying, full-facepiece respirator (gas mask) with a chin-style, front- or back-mounted organic vapor canister having a high-efficiency particulate filter); or SCBAE (any appropriate escape-type, self-contained breathing apparatus).

Storage: Prior to working with dicamba you should be trained on its proper handling and storage. Store in tightly closed containers in a cool, well ventilated area away from incompatible materials listed above, heat and water.

Shipping: Benzoic derivative pesticides, solid, toxic, n.o.s. could be applied to dicamba in which case the material falls in Hazard Class 6.1 and Packing Group III. The label required is "Keep Away From Food."

Spill Handling: Evacuate persons not wearing protective equipment from area of spill or leak until clean-up is complete. Remove all ignition sources. Collect powdered material in the most convenient and safe manner and deposit in sealed containers. Ventilate area after clean-up is complete. It may be necessary to contain and dispose of this chemical as a hazardous waste. If material or contaminated runoff enters waterways, notify downstream users of potentially contaminated waters. Contact your Department of Environmental Protection or your regional office of the federal EPA for specific recommendations. If employees are required to clean-up spills, they must be properly trained and equipped. OSHA 1910.120(q) may be applicable.

Fire Extinguishing: This chemical is a noncombustible solid. Use extinguishers suitable for surrounding fire. Poisonous gases are produced in fire, including chlorine and hydrogen chloride. If material or contaminated runoff enters waterways, notify downstream users of potentially contaminated waters. Notify local health and fire officials and pollution control agencies. From a secure, explosion-proof location, use water spray to cool exposed containers. If cooling streams are ineffective (venting sound increases in volume and pitch, tank discolors, or shows any signs of deforming), withdraw immediately to a secure position. If employees are expected to fight fires, they must be trained and equipped in OSHA 1910.156.

Disposal Method Suggested: Land disposal or incineration are disposal options.[22] In accordance with 40CFR165 recommendations for the disposal of pesticides and pesticide containers. Must be disposed properly by following package label directions or by contacting your state pesticide or environmental control agency or by contacting your regional EPA office.

References

U.S. Environmental Protection Agency, "Health Advisory: Dicamba," Washington, DC, Office of Drinking Water (August 1987)

New Jersey Department of Health and Senior Services, Hazardous Substance Fact Sheet, "Dicamba," Trenton, NJ (January 1999)

Dichloroacetic Acid

Molecular Formula: $C_2H_2C_{12}O_2$

Synonyms: Acetic acid, bichloro-; Acetic acid, dichloro-; Bichloroacetic acid; DCA; Dichlorethanoic acid

CAS Registry Number: 79-43-6

RTECS Number: AG6125000

DOT ID: UN 1764

EEC Number: 607-066-00-5

Regulatory Authority

- Air Pollutant Standard Set (former USSR)

Cited in U.S. State Regulations: New Jersey (G).

Description: DCA is a corrosive, combustible, colorless liquid with a pungent odor. Boiling point = 193 – 194°C. Freezing/Melting point = 9.7°C. Flash point = 110°C. Hazard Identification (based on NFPA-704 M Rating System): Health 3, Flammability 1, Reactivity 0. Soluble in water.

Potential Exposure: Used as a fungicide, a medication, and a chemical intermediate in pharmaceuticals.

Incompatibilities: DCA is a medium strong acid; incompatible with non-oxidizing mineral acids, organic acids, bases, acrylates, aldehydes, alcohols, alkylene oxides, ammonia, aliphatic amines, alkanolamines, aromatic amines, amides, glycols, isocyanates, ketones. Attacks metals generating flammable hydrogen gas. Attacks some plastics, rubber and coatings.

Permissible Exposure Limits in Air: The former USSR set an airborne exposure limit of 4 mg/m^3.

Routes of Entry: Inhalation, ingestion.

Harmful Effects and Symptoms

Short Term Exposure: Corrosive to the eyes, skin, and respiratory tract; causes severe irritation and burns. Eye contact may cause permanent damage. Higher exposures can cause pulmonary edema, a medical emergency that can be delayed for several hours. This can cause death. Corrosive if swallowed.

Long Term Exposure: May cause damage to the developing fetus. May affect the liver and kidneys. May damage the nervous system causing numbness, "pins and needles," and/or weakness in the hands and feet. Repeated exposure may cause lung irritation, bronchitis. There is limited evidence that DCA causes liver cancer in animals.

Points of Attack: Lungs, liver, kidneys, nervous system.

Medical Surveillance: Liver and kidney function tests. Lung function tests. Examination of the nervous system. Consider chest x-ray following acute overexposure.

First Aid: If this chemical gets into the eyes, remove any contact lenses at once and irrigate immediately for at least 15 minutes, occasionally lifting upper and lower lids. Seek medical attention immediately. If this chemical contacts the skin, remove contaminated clothing and wash immediately with soap and water. Seek medical attention immediately. If this chemical has been inhaled, remove from exposure, begin rescue breathing (using universal precautions) if breathing has stopped and CPR if heart action has stopped. Transfer promptly to a medical facility. When this chemical has been swallowed, get medical attention. If victim is conscious, administer water or milk. Do not induce vomiting. Medical observation is recommended for 24 – 48 hours after breathing overexposure, as pulmonary edema may be delayed. As first aid for pulmonary edema, a doctor or authorized paramedic may consider administering a corticosteroid spray.

Personal Protective Methods: Wear protective gloves and clothing to prevent any reasonable probability of skin contact. Safety equipment suppliers/manufacturers can provide recommendations on the most protective glove/clothing material for your operation.. All protective clothing (suits, gloves, footwear, headgear) should be clean, available each day, and put on before work. Contact lenses should not be worn when working with this chemical. Wear splash-proof chemical goggles and face shield unless full facepiece respiratory protection is worn. Employees should wash immediately with soap when skin is wet or contaminated. Provide emergency showers and eyewash.

Respirator Selection: Where the potential for exposure to this chemical, use a MSHA/NIOSH approved supplied-air respirator with a full facepiece operated in the positive pressure mode or with a full facepiece, hood, or helmet in the continuous flow mode, or use a MSHA/NIOSH approved self-contained breathing apparatus with a full facepiece operated in pressure-demand or other positive pressure mode.

Storage: Prior to working with DCA you should be trained on its proper handling and storage. Store in tightly closed containers in a cool, well ventilated area away from incompatible materials listed above. Metal containers involving the transfer of this chemical should be grounded and bonded. Where possible, automatically pump liquid from drums or other storage containers to process containers. Drums must be equipped with self-closing valves, pressure vacuum bungs, and flame arresters. Use only non-sparking tools and equipment, especially when opening and closing containers of this chemical. Sources of ignition such as smoking and open flames, are prohibited where this chemical is used, handled, or stored in a manner that could create a potential fire or explosion hazard. Wherever this chemical is used, handled, manufactured, or stored, use explosion-proof electrical equipment and fittings.

Shipping: Label "Corrosive." Hazard Class 8. Packing Group II.

Spill Handling: Evacuate and restrict persons not wearing protective equipment from area of spill or leak until cleanup is complete. Remove all ignition sources. Ventilate area of spill or leak. Absorb liquids in vermiculite, dry sand, earth, peat, carbon, or a similar material and deposit in sealed containers. Keep this chemical out of a confined space, such as a sewer, because of the possibility of an explosion, unless the sewer is designed to prevent the build-up of explosive concentrations. It may be necessary to contain and dispose of this chemical as a hazardous waste. If material or contaminated runoff enters waterways, notify downstream users of potentially contaminated waters. Contact your Department of Environmental Protection or your regional office of the federal EPA for specific recommendations. If employees are required to clean-up spills, they must be properly trained and equipped. OSHA 1910.120(q) may be applicable.

Fire Extinguishing: This chemical is a combustible liquid, but does not readily ignite. Poisonous gases including phosgene, hydrogen chloride, carbon monoxide are produced in fire. Use dry chemical, carbon dioxide, or alcohol foam extinguishers. Vapors are heavier than air and will collect in low areas. Vapors in confined areas may explode when exposed to fire. Containers may explode in fire. Storage containers and parts of containers may rocket great distances, in many directions. If material or contaminated runoff enters waterways, notify downstream users of potentially contaminated waters. Notify local health and fire officials and pollution control agencies. From a secure, explosion-proof location, use water spray to cool exposed containers. If cooling streams are ineffective (venting sound increases in volume and pitch, tank discolors, or shows any signs of deforming), withdraw immediately to a secure position. If employees are expected to fight fires, they must be trained and equipped in OSHA 1910.156.

References

New Jersey Department of Health and Senior Services, "Hazardous Substance Fact Sheet, Dichloroacetic Acid," Trenton NJ (February 1999)

Dichloroacetylene

Molecular Formula: C_2Cl_2

Common Formula: ClC≡CCl

Synonyms: DCA; Dichlorlethyne

CAS Registry Number: 7572-29-4

RTECS Number: AP1080000

DOT ID: None; "Forbidden"

Regulatory Authority

- Carcinogen (Animal Suspected) (IARC)[9] (Animal Positive) (DFG)[3]
- Air Pollutant Standard Set (ACGIH)[1] (HSE)[33] (Several States)[60]
- OSHA 29CFR1910.119, Appendix A. Process Safety List of Highly Hazardous Chemicals, TQ = 250 lb (114 kg)
- Canada, WHMIS, Ingredients Disclosure List

Cited in U.S. State Regulations: Alaska (G), California (A, G), Connecticut (A), Florida (A), Massachusetts (G), Minnesota (G), New Jersey (G), North Dakota (A), Pennsylvania (G), Rhode Island (G), Virginia (A), West Virginia (G).

Description: DCA, ClC≡CCl, is a volatile oil with a disagreeable, sweetish odor. A gas above 32°C/90°F. Boiling point = 32°C (explodes). Freezing/Melting point = -50 – -66°C.

Hazard Identification (based on NFPA-704 M Rating System): Health 3, Flammability 0, Reactivity 0. Insoluble in water.

Potential Exposure: DCA is not produced commercially. Dichloroacetylene may be formed by incineration of trichloroethylene below optimal furnace temperatures. Also in closed circuit anesthesia with trichloroethylene, heat and moisture produced by soda-lime absorption of CO_2 may produce dichloroacetylene along with phosgene and CO.

Incompatibilities: Heat or shock may cause explosion. Violent reaction with oxidizers and strong acids (forms poisonous gases of phosgene and hydrogen chloride).

Permissible Exposure Limits in Air: There is no Federal Limit (OSHA PEL). NIOSH recommends an exposure limit of 0.1 ppm TWA, and the ACGIH recommends a ceiling value of 0.1 ppm (0.4 mg/m^3) not to be exceeded at any time. Mexico, Israel, Australia, and the Canadian provinces of Alberta, BC, Ontario, and Quebec set a peak limitation or ceiling of 0.1 ppm. The HSE[33] has set a STEL of 0.1 ppm. DFG set no numerical value and has listed DCA as a carcinogen. Some states have set guidelines or standards for dichloroacetylene in ambient air[60] ranging from 3.2 $\mu g/m^3$ (Virginia) to 4.0 $\mu g/m^3$ (North Dakota) to 8.0 $\mu g/m^3$ (Connecticut) to 10.0 $\mu g/m^3$ (Nevada).

Determination in Air: The NIOSH pocket guide lists no methods. However, see NIOSH Method 1003 for hydrocarbons, halogenated.

Permissible Concentration in Water: No criteria set.

Routes of Entry: Inhalation, skin absorption, ingestion, skin and/or eye contact.

Harmful Effects and Symptoms

Short Term Exposure: Eye contact can cause irritation. Inhalation can irritate the respiratory tract with coughing, wheezing, and/or shortness of breath. Exposure can cause headache, loss of appetite, extreme nausea, vomiting, involvement of the trigeminal nerve and facial muscles causing paralysis of the face, and the development of facial herpes. Higher exposure may cause a build-up of fluid in the lungs (pulmonary edema). This can cause death.

Long Term Exposure: There is limited evidence that DCA causes kidney cancer in animals. Dichloroacetylene can cause nervous system damage leading to weakness and behavioral changes, and may affect the kidneys. May cause lung irritation and the development of bronchitis with coughing, phlegm, and/or shortness of breath.

Points of Attack: Central nervous system, lungs, kidneys.

Medical Surveillance: For those with frequent or potentially high exposure (half the TLV or greater), the following are recommended before beginning work and at regular times after that: Lung function tests. If symptoms develop or overexposure is suspected, the following may be useful: Liver and kidney function tests. Exam of the nervous system. Consider chest x-ray after acute overexposure.

First Aid: If this chemical gets into the eyes, remove any contact lenses at once and irrigate immediately for at least 15 minutes, occasionally lifting upper and lower lids. Seek medical attention immediately. If this chemical contacts the skin, remove contaminated clothing and wash immediately with soap and water. Seek medical attention immediately. If this chemical has been inhaled, remove from exposure, begin rescue breathing (using universal precautions) if breathing has stopped and CPR if heart action has stopped. Transfer promptly to a medical facility. When this chemical has been swallowed, get medical attention. Give large quantities of water and induce vomiting. Do not make an unconscious person vomit. Medical observation is recommended for 24 – 48 hours after breathing overexposure, as pulmonary edema may be delayed. As first aid for pulmonary edema, a doctor or authorized paramedic may consider administering a corticosteroid spray.

Personal Protective Methods: Avoid skin contact. Wear protective gloves and clothing to prevent any reasonable probability of skin contact. Safety equipment suppliers/ manufacturers can provide recommendations on the most protective glove/clothing material for your operation. All protective clothing (suits, gloves, footwear, headgear) should be clean, available each day, and put on before work. Contact lenses should not be worn when working with this chemical. Wear splash or dust-proof chemical goggles and face shield unless full facepiece respiratory protection is worn. Employees should wash immediately with soap when skin is wet or contaminated. Provide emergency showers and eyewash. Wear protective gloves and clothing. Safety equipment suppliers/manufacturers can provide recommendations on the most protective glove/clothing material for your operation. All protective clothing (suits, gloves, footwear, headgear) should be clean, available each day, and put on before work.

Eye Protection: Wear splash-proof chemical goggles when working with liquid, unless full facepiece respiratory protection is worn. Wear gas-proof goggles, unless full facepiece respiratory protection is worn.

Respirator Selection: NIOSH: *At any detectable concentration:* SCBAF:PD,PP (any MSHA/NIOSH approved self-contained breathing apparatus that has a full facepiece and is operated in a pressure-demand or other positive-pressure mode); or SAF:PD,PP:ASCBA (any supplied-air respirator that has a full facepiece and is operated in a pressure-demand or other positive-pressure mode in combination with an auxiliary, self-contained breathing apparatus operated in a pressure-demand or other positive pressure mode). *Escape:* GMFOV [any air-purifying, full-facepiece respirator (gas mask) with a chin-style, front-or back-mounted organic vapor canister] or SCBAE (any appropriate escape-type, self-contained breathing apparatus).

Storage: Prior to working with DCA you should be trained on its proper handling and storage. Dichloroacetylene must be stored to avoid contact with oxidizers (such as perchlorates, peroxides, permanganates, chlorates, and nitrates) since violent reactions occur. Keep dichloroacetylene away from strong acids (such as hydrochloric, sulfuric, and nitric), because poisonous gases may be given off including phosgene and hydrogen chloride. Store in tightly closed containers in a cool, well-ventilated area away from heat, potassium, sodium and aluminum powders. Sources of ignition such as smoking and open flames are prohibited where dichloroacetylene is used, handled, or stored in a manner that could create a potential fire or explosion hazard. Use only non-sparking tools and equipment, especially when opening and closing containers of dichloroacetylene. Wherever dichloroacetylene is used, handled, manufactured, or stored, use explosion-proof electrical equipment and fittings. A regulated, marked area should be established where this chemical is handled, used, or stored in compliance with OSHA standard 1910.1045.

Shipping: Dichloroacetylene is cited by DOT[19] as "Forbidden" to transport.

Spill Handling: Evacuate and restrict persons not wearing protective equipment from area of spill or leak until cleanup is complete. Remove all ignition sources. Ventilate area of spill or leak. Absorb liquids in vermiculite, dry sand, earth, peat, carbon, or a similar material and deposit in sealed containers. Keep this chemical out of a confined space, such as a sewer, because of the possibility of an explosion, unless the sewer is designed to prevent the build-up of explosive concentrations. It may be necessary to contain and dispose of this chemical as a hazardous waste. If material or contaminated runoff enters waterways, notify downstream users of potentially contaminated waters. Contact your Department of Environmental Protection or your regional office of the federal EPA for specific recommendations. If employees are required to clean-up spills, they must be properly trained and equipped. OSHA 1910.120(q) may be applicable.

Fire Extinguishing: Dichloroacetylene is a combustible liquid; becomes a gas above 32°C/90°F. It will explode before it reaches a temperature that is hot enough to burn. Poisonous gas is produced in fire. Containers may explode in fire. Storage containers and parts of containers may rocket great distances, in many directions. If material or contaminated runoff enters waterways, notify downstream users of potentially contaminated waters. Notify local health and fire officials and pollution control agencies. From a secure, explosion-proof location, use water spray to cool exposed containers. If cooling streams are ineffective (venting sound increases in volume and pitch, tank discolors, or shows any signs of deforming), withdraw immediately to a secure position. If employees are expected to fight fires, they must be trained and equipped in OSHA 1910.156.

References

New Jersey Department of Health, "Hazardous Substance Fact Sheet: Dichloroacetylene," Trenton, NJ (April 1997)

Dichlorobenzalkonium Chloride

Molecular Formula: C_8H_{17} to $C_{18}H_{37}$

Common Formula: $C_9H_{11}Cl_2N{\cdot}Cl$

Synonyms: Alkyl(C_6H_{18})dimethyl-3,4-dichlorobenzyl-ammonium chloride; Alkyl(C_8H_{17} to $C_{18}H_{37}$) dimethyl-3,4-dichlorobenzyl ammonium chloride; Dichlorobenzalkonium chloride; Tetrosan®

CAS Registry Number: 8023-53-8

RTECS Number: BO3200000

Regulatory Authority

- Extremely Hazardous Substance (EPA-SARA) (TPQ = 10,000) (Dropped form listing in 1988)

Cited in U.S. State Regulations: Massachusetts (G).

Description: Dichlorobenzalkonium chloride, $C_9H_{11}Cl_2N{\cdot}Cl$, is a colorless, crystalline solid. Soluble in water.

Potential Exposure: This material is used as an antiseptic, germicide, algicide, sterilizer, and deodorant.

Incompatibilities: Chlorides may be incompatible with acids, acid fumes. Esters may be incompatible with moisture and nitrates.

Permissible Exposure Limits in Air: No standards set.

Permissible Concentration in Water: No criteria set.

Harmful Effects and Symptoms

Short Term Exposure: Ingestion causes burning pain in the mouth, throat and abdomen with spitting of blood; drooling; vomiting; ulcers in the mouth and throat; shock; restlessness; confusion; weakness; apprehension; muscle weakness; difficulty in breathing; depression; bluing of the skin; and death from shock or asphyxiation.

Ten percent concentrated aqueous solutions are irritating to the skin and concentrations as low as 0.1 – 0.5% are irritating to the eyes and mucous membranes. Ingestion can cause corrosion of upper intestinal tract. Swelling of throat and lungs with fluid (edema) also can occur. Death can occur due to paralysis of respiratory muscles or circulatory collapse. The oral LD_{50} rat is 730 mg/kg (slightly toxic).

Long Term Exposure: May cause liver and kidney damage. Repeated contact may cause allergic reaction to skin and lungs.

Points of Attack: Liver, kidneys, lungs, skin.

Medical Surveillance: Liver and kidney function tests. Lung function tests. Examination by a qualified allergist.

First Aid: If this chemical gets into the eyes, remove any contact lenses at once and irrigate immediately for at least 15 minutes, occasionally lifting upper and lower lids. Seek medical attention immediately. If this chemical contacts the skin, remove contaminated clothing and wash immediately with soap and water. Seek medical attention immediately. If this chemical has been inhaled, remove from exposure, begin

rescue breathing (using universal precautions) if breathing has stopped and CPR if heart action has stopped. Transfer promptly to a medical facility. When this chemical has been swallowed, get medical attention. If concentrated (10% or greater) solution is ingested, patient should swallow a large quantity of milk, egg whites or gelatin solution. If breathing is difficult, give oxygen.

Personal Protective Methods: Wear protective gloves and clothing to prevent any reasonable probability of skin contact. Safety equipment suppliers/manufacturers can provide recommendations on the most protective glove/clothing material for your operation.. All protective clothing (suits, gloves, footwear, headgear) should be clean, available each day, and put on before work. Contact lenses should not be worn when working with this chemical. Wear dust-proof chemical goggles and face shield unless full facepiece respiratory protection is worn. Employees should wash immediately with soap when skin is wet or contaminated. Provide emergency showers and eyewash.

Respirator Selection: *At any detectable concentration:* SCBAF:PD,PP (any MSHA/NIOSH approved self-contained breathing apparatus that has a full facepiece and is operated in a pressure-demand or other positive-pressure mode); or SAF:PD,PP:ASCBA (any supplied-air respirator that has a full facepiece and is operated in a pressure-demand or other positive-pressure mode in combination with an auxiliary, self-contained breathing apparatus operated in a pressure-demand or other positive pressure mode). *Escape:* HiEF (any air-purifying, full-facepiece respirator (gas mask) with a chin-style, front- or back-mounted organic vapor canister having a high-efficiency particulate filter); or SCBAE (any appropriate escape-type, self-contained breathing apparatus).

Storage: Prior to working with this chemical you should be trained on its proper handling and storage. Store in tightly closed containers in a cool, well ventilated area away from acids, acid fumes, moisture and nitrates.

Spill Handling: Evacuate persons not wearing protective equipment from area of spill or leak until clean-up is complete. Remove all ignition sources. Collect powdered material in the most convenient and safe manner and deposit in sealed containers. Ventilate area after clean-up is complete. It may be necessary to contain and dispose of this chemical as a hazardous waste. If material or contaminated runoff enters waterways, notify downstream users of potentially contaminated waters. Contact your Department of Environmental Protection or your regional office of the federal EPA for specific recommendations. If employees are required to clean-up spills, they must be properly trained and equipped. OSHA 1910.120(q) may be applicable.

Fire Extinguishing: Use dry chemical, carbon dioxide, or foam extinguishers. Poisonous gases are produced in fire including nitrogen oxides, ammonia, and chlorides. If material or contaminated runoff enters waterways, notify downstream users of potentially contaminated waters. Notify local health and fire officials and pollution control agencies. From a secure, explosion-proof location, use water spray to cool exposed containers. If cooling streams are ineffective (venting sound increases in volume and pitch, tank discolors, or shows any signs of deforming), withdraw immediately to a secure position. If employees are expected to fight fires, they must be trained and equipped in OSHA 1910.156.

References

U.S. Environmental Protection Agency, "Chemical Profile: Dichlorobenzalkonium Chloride," Washington, DC, Chemical Emergency Preparedness Program (October 31, 1985)

Dichlorobenzenes

Molecular Formula: $C_6H_4Cl_2$

Synonyms: *1,2-DCB:* Benzene, 1,2-dichloro-; Chloroben; Chloroden; Cloroben; DCB; *o*-Dichlorbenzol; Dichloricide; *o*-Dichlorobenzene; 1,2-Dichlorobenzene; Dichlorobenzene, *o*-; *o*-Dichlorobenzol; *o*-Diclorobenceno (Spanish); 1,2-Diclorobenceno (Spanish); Dilantin DB; Dilatin DB; Dizene; Dowtherm E; J100; NCI-C54944; ODB; ODCB; Orthodichlorobenzene; Orthodichlorobenzol; Special termite fluid; Termitkil; Ultramac S40

1,3-DCB: Benzene, *m*-dichloro-; Benzene, 1,3-dichloro-; *m*-Dichlorobenzene; *m*-Dichlorobenzol; *m*-Diclorobenceno (Spanish); 1,3-Diclorobenceno (Spanish); Metadichlorobenzene; *m*-Phenylene dichloro

1,4-DCB: Benzene, *p*-dichloro-; Benzene, 1,4-dichloro-; *p*-Chlorophenyl chloride; Di-chloricide; *p*-Dichlorobenzene; *p*-Diclorobenceno (Spanish); 1,4-Diclorobenceno (Spanish); EINECS No. 203-400-5; Evola; Paracide; Para Crystals; Paradi; Paradichlorobenzene; Paradow; Paramoth; Paranuggets; Parazene; PDB; Persia-Perazol; Santochlor; Santoclor

mixed isomers: Amisia-Mottenschutz; Benzene, dichloro-; DCB; Dichlorobenzene (mixed isomers); Diclorobenceno (Spanish); Dilatin DBI; Mottenschutzmittel evau P; Mott-EX; Totamott

CAS Registry Number: 95-50-1 (o-DCB); 541-73-1 (m-DCB); 106-46-7 (p-DCB); 25321-22-6 (mixed isomers)

RTECS Number: CZ4500000 (o-DCB); CZ4499000 (m-DCB); CZ4550000 (p-DCB); CZ4430000 (mixed isomers)

DOT ID: UN 1591 (o-DCB); NA 1993 (m-DCB); UN 1592 (p-DCB)

EEC Number: 602-034-00-7 (1,2-DCB); 602-035-00-2 (1,4-DCB)

EINECS Number: 202-425-9

Regulatory Authority

- Air Pollutant Standard Set (ACGIH)[1] (Australia) (DFG)[3] (HSE)[33] (Israel) (Mexico) (OSHA)[58] (former USSR)[35][43] (Several States)[60] (Several Canadian Provinces)

1,2-DCB:

- Clean Water Act: Section 311 Hazardous Substances/RQ 40CFR117.3 (same as CERCLA, see below); 40CFR423, Appendix A, Priority Pollutants; Section 313 Water Priority Chemicals (57FR41331, 9/9/92); Toxic Pollutant (Section 401.15)
- EPA Hazardous Waste Number (RCRA No.): U070
- RCRA, 40CFR261, Appendix 8 Hazardous Constituents
- RCRA 40CFR268.48; 61FR15654, Universal Treatment Standards: Wastewater (mg/l), 0.088; Nonwastewater (mg/kg), 6.0
- RCRA 40CFR264, Appendix 9; TSD Facilities Ground Water Monitoring List. Suggested test method(s) (PQL μg/l): 8010 (2); 8020 (5); 8120 (10); 8270 (10)
- Safe Drinking Water Act: MCL, 0.6 mg/l; MCLG, 0.6 mg/l; Regulated chemical (47 FR 9352)
- Superfund/EPCRA 40CFR302.4 Reportable Quantity (RQ): CERCLA, 100 lb (45.4 kg)
- EPCRA Section 313 Form R *de minimis* concentration reporting level: 1.0%
- U.S. DOT Regulated Marine Pollutant (49CFR172.101, Appendix B)
- Canada, WHMIS, Ingredients Disclosure List, CEPA Priority Substance List, National Pollutant Release Inventory (NPRI); Drinking Water Quality 0.2 mg/l MAC and ≤0.003 mg/l AO

1,3-DCB:

- Clean Water Act: Section 311 Hazardous Substances/RQ 40CFR117.3 (same as CERCLA, see below); 40CFR423, Appendix A, Priority Pollutants; Section 313 Water Priority Chemicals (57FR41331, 9/9/92); Toxic Pollutant (Section 401.15)
- EPA Hazardous Waste Number (RCRA No.): U071
- RCRA, 40CFR261, Appendix 8 Hazardous Constituents
- RCRA 40CFR268.48; 61FR15654, Universal Treatment Standards: Wastewater (mg/l), 0.036; Nonwastewater (mg/kg), 6.0
- RCRA 40CFR264, Appendix 9; TSD Facilities Ground Water Monitoring List. Suggested test method(s) (PQL μg/l): 8010 (5); 8020 (5); 8120 (10); 8270 (10)
- Safe Drinking Water Act: Regulated chemical (47FR9352) as dichlorobenzene; Priority List (55FR1470)
- Superfund/EPCRA 40CFR302.4 Reportable Quantity (RQ): CERCLA, 100 lb (45.4 kg)
- EPCRA Section 313 Form R *de minimis* concentration reporting level: 1.0%
- U.S. DOT Regulated Marine Pollutant (49CFR172.101, Appendix B)
- Canada, WHMIS, Ingredients Disclosure List

1,4-DCB:

- Carcinogen, (Animal Positive) (IARC), (NTP), (ACGIH)
- Clean Air Act: Hazardous Air Pollutants (Title I, Part A, Section 112)
- Clean Water Act: Section 311 Hazardous Substances/RQ 40CFR117.3 (same as CERCLA, see below); 40CFR423, Appendix A, Priority Pollutants; Section 313 Water Priority Chemicals (57FR41331, 9/9/92); Toxic Pollutant (Section 401.15)
- EPA Hazardous Waste Number (RCRA No.): U072; D027
- RCRA, 40CFR261, Appendix 8 Hazardous Constituents
- RCRA Toxicity Characteristic (Section 261.24), Maximum Concentration of Contaminants, regulatory level, 7.5 mg/l
- RCRA 40CFR268.48; 61FR15654, Universal Treatment Standards: Wastewater (mg/l), 0.090; Nonwastewater (mg/kg), 6.0
- RCRA 40CFR264, Appendix 9; TSD Facilities Ground Water Monitoring List. Suggested test method(s) (PQL μg/l): 8010 (2); 8020 (5); 8120 (15); 8270 (10)
- Safe Drinking Water Act: MCL, 0.075 mg/l; MCLG, 0.075 mg/l; Regulated chemical (47FR9352)
- Superfund/EPCRA 40CFR302.4 Reportable Quantity (RQ): CERCLA, 100 lb (45.4 kg)
- EPCRA Section 313 Form R *de minimis* concentration reporting level: 0.1%
- U.S. DOT Regulated Marine Pollutant (49CFR172.101, Appendix B)
- Canada, WHMIS, Ingredients Disclosure List, CEPA Priority Substance List, National Pollutant Release Inventory (NPRI); Drinking Water Quality 0.005 mg/l MAC and ≤0.001 mg/l AO

Mixed isomers:

- Clean Water Act: Section 311 Hazardous Substances/RQ 40CFR117.3 (same as CERCLA, see below); 40CFR423, Appendix A, Priority Pollutants; Section 313 Water Priority Chemicals (57FR41331, 9/9/92); Toxic Pollutant (Section 401.15)
- RCRA, 40CFR261, Appendix 8 Hazardous Constituents, waste number not listed
- Safe Drinking Water Act: Regulated chemical (47 FR 9352)
- Superfund/EPCRA 40CFR302.4 Reportable Quantity (RQ): CERCLA, 100 lb (45.4 kg)
- EPCRA Section 313 Form R *de minimis* concentration reporting level: 0.1%
- U.S. DOT Regulated Marine Pollutant (49CFR172.101, Appendix B)
- Canada, WHMIS, Ingredients Disclosure List, CEPA Priority Substance List, National Pollutant Release Inventory (NPRI)
- Mexico, Drinking Water, 0.4 mg/l

Cited in U.S. State Regulations: Alaska (G), Arizona (W), California (A, W), Connecticut (A), Florida (G), Illinois (G), Indiana (A), Kansas (G, W), Louisiana (G), Maine (G, W),

Maryland (G), Massachusetts (G, A), Michigan (G), Minnesota (W), Nevada (A), New Hampshire (G, W), New Jersey (G, W), New York (G, A), North Carolina (A), North Dakota (A), Oklahoma (G), Pennsylvania (G), Rhode Island (G), South Carolina (A), Vermont (G, W), Virginia (G, A), Washington (G), West Virginia (G), Wisconsin (G, W).

Description: There are three isomeric forms of dichlorobenzene, $C_6H_4Cl_2$: o-DCB is a colorless to pale yellow liquid with a pleasant, aromatic odor. Odor threshold = 0.30 ppm. Boiling point = 180°. Freezing/Melting point = -17°C. Flash point = 60°C. Flash point = 66°C c.c. Autoignition temperature: 648°C. Explosive limits: LEL = 2.2%; UEL = 9.2%. Hazard Identification (based on NFPA-704 M Rating System): Health 2, Flammability 2, Reactivity 0. Insoluble in water.

m-DCB is a liquid. Boiling point =172°C. p-DCB is a colorless or white solid with a mothball-like odor. Odor threshold = 0.18 ppm. Boiling point = 174°C. Freezing/Melting point = 53°C. Flash point = 66°C. Insoluble in water. Explosive limits: LEL = 2.5%; UEL = unknown. Hazard Identification (based on NFPA-704 M Rating System): Health 2, Flammability 2, Reactivity 0. Insoluble in water.

Potential Exposure: The major uses of o-DCB are as a process solvent in the manufacturing of toluene diisocyanate and as an intermediate in the synthesis of dyestuffs, herbicides, and degreasers. p-Dichlorbenzene is used primarily as an air deodorant and in insecticide, which accounts for 90% of the total production of this isomer. Information is not available concerning the production and use of m-DCB. However, it may occur as a contaminant of o- or p-DCB formulations. Both o- and p-isomers are produced almost entirely as by-products during the production of monochlorobenzene.

Incompatibilities: For o-DCB and m-DCB: acid fumes, chlorides, strong oxidizers, hot aluminum or aluminum alloys. For p-DCB: Strong oxidizers; although, incompatibilities for this chemical may also include other materials listed for o-DCB.

Permissible Exposure Limits in Air:

	OSHA PEL	ACGIH TLV	NIOSH IDLH
o-DCB	50 ppm (300 mg/m³)*	25 ppm TWA and (STEL: 50 ppm*)	200 ppm
p-DCB	75 ppm TWA (450 mg/m³)	10 ppm TWA (60 mg/m³)	150 ppm (Carcinogen)

* As a ceiling value, not to be exceeded at any time.

Foreign Regulations (ppm) TWA (except as noted)

	Australia	DFG	HSE	Israel	Mexico	Canada**
o-DCB	50*	50	50*	50*	50	50*
p-DCB	75	50	75	75	75	75
	110*	100*	110*	110*	110*	110*

* As a ceiling or peak limitation value, not to be exceeded at any time, except for DFG Peak Limitation (30 minute) which is not to be exceeded 4 times during a normal workshift.

** Alb, BC, Ont, Que

No occupational exposure limits have been established for m-DCB. The former USSR-UNEP/IRPTC project gives a MAC in workplace air of 20 mg/m³.[43] The former USSR gives a MAC for ambient air in residential areas of 0.035 mg/m³ on a once daily basis for p-DCB.[35]

A number of states have set guidelines or standards for chlorobenzenes in ambient air[60] as follows:

State	o-DCB- (mg/m³)	p-DCB (mg/m³)
Connecticut	0	9.0
Indiana	1.5	—
Massachusetts	0.082	0.0042
Nevada	7.14	10.7
New York	1.0	—
North Carolina	—	67.5
North Dakota	3.0	6.75
South Carolina	—	4.5
Virginia	2.5	7.5

Determination in Air: Charcoal adsorption followed by CS_2 workup and gas chromatographic analysis. See NIOSH Method 1003 for halogenated hydrocarbons.

Permissible Concentration in Water: To protect freshwater aquatic life – 1,120 μg/l on an acute toxicity basis and 763 μg/l on a chronic basis. To protect saltwater aquatic life: 1,970 μg/l on an acute toxicity basis. To protect human health: 400 μg/l for all isomers.[6] The former USSR established a MAC of 0.002 mg/l for water bodies used for domestic purposes.[35][43] The USEPA[48] has derived lifetime health advisories for o- and m-DCB as 0.62 mg/l (620 μg/l) and for p-DCB of 0.075 mg/l (75 μg/l). EPA[62] has recently proposed a maximum level for o-DCB o 0.6 mg/l in drinking water. See also Regulatory section for drinking water criteria for Mexico and Canada.

A number of states have set guidelines and standards for chlorobenzenes in drinking water[61] as follows:

State	o-DCB- (μg/l)	m-DCB- (μg/l)	p-DCB- (μg/l)
Arizona	620	—	750
California	130	130	0.3 – 0.5
Kansas	620	620	—
Maine	85	—	27
Minnesota	620	—	750
New Jersey	600	600	6
Vermont	620	—	—
Wisconsin	1,250	1,250	5

Determination in Water: Gas chromatography (EPA Methods 601, 602, 612) or gas chromatography plus mass spectrometry (EPA Method 625). Gas-chromatographic methods have been developed for p-PDB with a sensitivity of 380 pg/cm peak high, and p-PDB concentrations as low as 1.0 ppb in water have been analyzed according to NAS/NRC.

Routes of Entry: o-DCB: inhalation, skin absorption, ingestion, skin and/or eye contact. p-DCB: Inhalation, ingestion, eye and skin contact.

Harmful Effects and Symptoms

Human exposure to dichlorobenzene is reported to cause hemolytic anemia and liver necrosis, and 1,4-dichlorobenzene has been found in human adipose tissue. In addition, the dichlorobenzenes are toxic to nonhuman mammals, birds, and aquatic organisms and impart an offensive taste and odor to water. The dichlorobenzenes are metabolized by mammals, including humans, to various dichlorophenols, some of which are as toxic as the dichlorobenzenes.

Exposure can damage blood cells. Contact can cause irritation of the skin and eyes. Prolonged contact can cause severe burns. It may damage the liver, kidneys and lungs. Exposure can cause headache, dizziness, swelling of the eyes, hands and feet, and nausea. Higher levels can cause severe liver damage and death. Persons with preexisting pathology (hepatic, renal, central nervous system, blood) or metabolic disorders, who are taking certain drugs (hormones or otherwise metabolically active), or who are otherwise exposed to DCBs or related (chemically or biologically) chemicals by such means as occupation, or domestic use or abuse (e.g., pica or "sniffing") of DCB products, might well be considered at increased risk from exposure to DCBs.

Short Term Exposure: *o-DCB:* Can be absorbed through the skin, thereby increasing exposure. Irritates the eyes, skin, and respiratory tract. Prolonged skin contact may cause blisters. May affect the central nervous system. Exposure can cause headache and nausea. Higher exposure can cause dizziness, lightheadedness, and unconsciousness.

m-DCB: Can be absorbed through the skin, thereby increasing exposure. Symptoms are similar to o-DCB and m-DCB may damage the red blood cells leading to low blood count.

p-DCB: Can be absorbed through the skin, thereby increasing exposure. Exposure can cause headache, dizziness, nausea, swelling of the hands and feet. Contact with the dust can irritate and burn the eyes and skin. Skin allergy may develop.

Long Term Exposure: *o-DCB:* Repeated or prolonged contact may cause skin sensitization and allergy. Long term exposure may cause damage to the blood cells, liver, kidneys, and lungs.

m-DCB: Repeated or prolonged contact may cause skin sensitization and allergy. May affect the liver and kidneys.

p-DCB: May be carcinogenic to humans; it causes kidney and liver cancer in animals. There is a suggested association between this chemical and leukemia. There is evidence that p-DCB can damage the developing animal fetus. Repeated exposure can damage the nervous system, skin allergy and damage the lungs, liver, and kidneys. p-DCG may affect the blood and cause hemolytic anemia.

Points of Attack: o-DCB and m-DCB: liver, kidneys, skin, eyes. p-DCB: liver, respiratory system, eyes, kidneys, skin.

Medical Surveillance: For those with frequent or potentially high exposure (half the TLV or greater, or significant skin contact) the following are recommended before beginning work and at regular times after that: Liver, kidney and lung function tests. Complete blood count. If symptoms develop or overexposure is suspected, the following may be useful: Evaluation by a qualified allergist, including careful exposure history and special testing, may help diagnose skin allergy.

First Aid: If this chemical gets into the eyes, remove any contact lenses at once and irrigate immediately for at least 15 minutes, occasionally lifting upper and lower lids. Seek medical attention immediately. If this chemical contacts the skin, remove contaminated clothing and wash immediately with soap and water. Seek medical attention immediately. If this chemical has been inhaled, remove from exposure, begin rescue breathing (using universal precautions) if breathing has stopped and CPR if heart action has stopped. Transfer promptly to a medical facility. When this chemical has been swallowed, get medical attention. Give large quantities of water and induce vomiting. Do not make an unconscious person vomit.

Personal Protective Methods: Wear protective gloves and clothing to prevent any reasonable probability of skin contact. Safety equipment suppliers/manufacturers can provide recommendations on the most protective glove/clothing material for your operation.. All protective clothing (suits, gloves, footwear, headgear) should be clean, available each day, and put on before work. Contact lenses should not be worn when working with this chemical. Wear splash-proof (o-DCB or o-DCB) or dust-proof (p-DCB) chemical goggles and face shield unless full facepiece respiratory protection is worn. Employees should wash immediately with soap when skin is wet or contaminated. Provide emergency showers and eyewash.

Respirator Selection: *For o-DCB:* NIOSH/OSHA: *2,000 ppm:* CCRFOV (any air-purifying, full-facepiece respirator (gas mask) with a chin-style, front- or back-mounted acid gas canister); or PAPROV [any powered, air-purifying respirator with organic vapor cartridge(s)]; or SCBAF (any self-contained breathing apparatus with a full facepiece); or SAF (any supplied-air respirator with a full facepiece). Emergency or planned entry into unknown concentrations or IDLH conditions: SCBAF:PD,PP (any self-contained breathing apparatus that has a full facepiece and is operated in a pressure-demand or other positive-pressure mode); or SAF:PD,PP:ASCBA (any supplied-air respirator that has a full facepiece and is operated in a pressure-demand or other positive-pressure mode in combination with an auxiliary self-contained breathing apparatus operated in a pressure-demand or other positive pressure mode). *Escape:* GMFOV [any air-purifying, full-facepiece respirator (gas mask) with a chin-style, front-or back-, mounted organic vapor canister]; or SCBAE (any appropriate escape-type, self-contained breathing apparatus).

Note: Substance causes eye irritation or damage; eye protection needed.

For p-DCB: NIOSH: *At any detectable concentration:* SCBAF:PD,PP (any MSHA/NIOSH approved self-contained breathing apparatus that has a full facepiece and is operated in a pressure-demand or other positive-pressure mode); or SAF:PD,PP:ASCBA (any supplied-air respirator that has a full facepiece and is operated in a pressure-demand or other positive-pressure mode in combination with an auxiliary, self-contained breathing apparatus operated in a pressure-demand or other positive pressure mode). *Escape:* GMFOV [any air-purifying, full-facepiece respirator (gas mask) with a chin-style, front-or back-mounted organic vapor canister] or SCBAE (any appropriate escape-type, self-contained breathing apparatus).

Storage: Prior to working with any DCB you should be trained on its proper handling and storage. Before entering confined space where this chemical may be present, check to make sure that an explosive concentration does not exist. Dichlorobenzene must be stored to avoid contact with strong oxidizers, such as permanganates, nitrates, peroxides, chlorates, and perchlorates, hot aluminum or aluminum alloys, since violent reactions occur. Store in tightly closed containers in a cool well-ventilated area away from heat and direct light. Sources of ignition such as smoking and open flames are prohibited where dichlorobenzene is used, handled, or stored in a manner that could create a potential fire or explosion hazard. A regulated, marked area should be established where p-DCB is handled, used, or stored in compliance with OSHA standard 1910.1045.

Shipping: Both o-DCB and p-DCB isomers require a "Keep Away From Food" label. They fall in DOT Hazard Class 6.1 and Packing Group III.

Spill Handling: *p-DCB:* Evacuate persons not wearing protective equipment from area of spill or leak until clean-up is complete. Remove all ignition sources. Collect powdered material in the most convenient and safe manner and deposit in sealed containers. Ventilate area after clean-up is complete. It may be necessary to contain and dispose of this chemical as a hazardous waste. If material or contaminated runoff enters waterways, notify downstream users of potentially contaminated waters. Contact your Department of Environmental Protection or your regional office of the federal EPA for specific recommendations. If employees are required to clean-up spills, they must be properly trained and equipped. OSHA 1910.120(q) may be applicable.

o-DCB and m-DCB: Evacuate and restrict persons not wearing protective equipment from area of spill or leak until cleanup is complete. Remove all ignition sources. Establish forced ventilation to keep levels below explosive limit. Absorb liquids in vermiculite, dry sand, earth, peat, carbon, or a similar material and deposit in sealed containers. Keep this chemical out of a confined space, such as a sewer, because of the possibility of an explosion, unless the sewer is designed to prevent the build-up of explosive concentrations. It may be necessary to contain and dispose of this chemical as a hazardous waste. If material or contaminated runoff enters waterways, notify downstream users of potentially contaminated waters. Contact your Department of Environmental Protection or your regional office of the federal EPA for specific recommendations. If employees are required to clean-up spills, they must be properly trained and equipped. OSHA 1910.120(q) may be applicable.

Fire Extinguishing: *p-DCB:* This chemical is a combustible solid. Use dry chemical, carbon dioxide, or foam extinguishers. Poisonous gases are produced in fire including hydrogen chloride. If material or contaminated runoff enters waterways, notify downstream users of potentially contaminated waters. Notify local health and fire officials and pollution control agencies. From a secure, explosion-proof location, use water spray to cool exposed containers. If cooling streams are ineffective (venting sound increases in volume and pitch, tank discolors, or shows any signs of deforming), withdraw immediately to a secure position. If employees are expected to fight fires, they must be trained and equipped in OSHA 1910.156.

o-DCB and p-DCB: These chemicals are combustible liquid. Poisonous gases produced in fire including hydrogen chloride and chlorine. Use dry chemical, carbon dioxide, or foam extinguishers. Vapors are heavier than air and will collect in low areas. Vapors may travel long distances to ignition sources and flashback. Vapors in confined areas may explode when exposed to fire. Containers may explode in fire. Storage containers and parts of containers may rocket great distances, in many directions. If material or contaminated runoff enters waterways, notify downstream users of potentially contaminated waters. Notify local health and fire officials and pollution control agencies. From a secure, explosion-proof location, use water spray to cool exposed containers. If cooling streams are ineffective (venting sound increases in volume and pitch, tank discolors, or shows any signs of deforming), withdraw immediately to a secure position. If employees are expected to fight fires, they must be trained and equipped in OSHA 1910.156.

Disposal Method Suggested: Incineration, preferably after mixing with another combustible fuel. Care must be exercised to assure complete combustion to prevent the formation of phosgene. An acid scrubber is necessary to remove the halo acids produced.[22] Consult with environmental regulatory agencies for guidance on acceptable disposal practices. Generators of waste containing this contaminant (≥100 kg/mo) must conform with EPA regulations governing storage, transportation, treatment, and waste disposal.

References

U.S. Environmental Protection Agency, Dichlorobenzene: Ambient Water Quality Criteria, Washington, DC (1980)

U.S. Environmental Protection Agency, 1,2-Dichlorobenzene, Health and Environmental Effects Profile No. 64, Washington, DC, Office of Solid Waste (April 30, 1980)

U.S. Environmental Protection Agency, 1,3-Dichlorobenzene, Health and Environmental Effects Profile No. 65, Washington, DC, Office of Solid Waste (April 30, 1980)

U.S. Environmental Protection Agency, 1,4-Dichlorobenzene, Health and Environmental Effects Profile No. 66, Washington, DC, Office of Solid Waste (April 30, 1980)

U.S. Environmental Protection Agency, Dichlorobenzenes, Health and Environmental Effects Profile No. 67, Washington, DC, Office of Solid Waste (April 30, 1980)

Sax N. I., Ed., "Dangerous Properties of Industrial Materials Report," 4, No. 2, 45–48 (1984) (1,3-Dichlorobenzene); 4, No. 2, 49–52, and 6, No. 2, 50-57 (1986) (Mixed isomers)

U.S. Public Health Service, "Toxicological Profile for 1,4-Dichlorobenzene," Atlanta, Georgia, Agency for Toxic Substances & Disease Registry (December 1987)

New Jersey Department of Health, "Hazardous Substance Fact Sheet: 1,2-Dichlorobenzene," Trenton, NJ (June 1998)

New Jersey Department of Health, "Hazardous Substance Fact Sheet: 1,3-Dichlorobenzene," Trenton, NJ (February 1999)

New Jersey Department of Health, "Hazardous Substance Fact Sheet: 1,4-Dichlorobenzene," Trenton, NJ (June 1998)

New York State Department of Health, "Chemical Fact Sheet: ortho-Dichlorobenzene," Albany, NY, Bureau of Toxic Substance Assessment (March 1986)

New York State Department of Health, "Chemical Fact Sheet: para-Dichlorobenzene," Albany, NY, Bureau of Toxic Substance Assessment (April 1986)

3,3'-Dichlorobenzidine (and its Salts)

Molecular Formula: $C_{12}H_{10}Cl_2N_2$

Common Formula: $C_6H_3ClNH_2C_6H_3ClNH_2$

Synonyms: Benzidine, 3,3'-dichloro-; (1,1'-biphenyl)-4,4'-diamine, 3,3'-dichloro-; C.I. 23060; Curithane C 126; DCB; 4,4'-Diamino-3,3'-dichlorobiphenyl; 4,4'-Diamino-3,3'-dichlorodiphenyl; 3,3'-Dichlorobenzidin (Czech); *o,o'*-Dichlorobenzidine; Dichlorobenzidine; 3,3'-Dichloro-4,4'-biphenyldiamine; 3,3'-Dichlorobiphenyl-4,4'-diamine; 3,3'-Dichloro-4,4'-diamino(1,1-biphenyl); 3,3'-Dichloro-4,4'-diaminobiphenyl; 3,3-Diclorobencidina (Spanish)

dihydrochloride: A13-22046; Benzidine, 3,3'-dichloro-, dihydrochloride; (1,1'-Biphenyl)-4,4'-diamine, 3,3'-dichloro-, dihydrochloride; 3,3'-Dichlorobenzidine hydrochloride; 3,3'-Dichloro-(1,1'-biphenyl)-4,4'-diamine dihydrochloride

sulfate: (1,1'-Biphenyl)-4,4'-diamine, 3,3'-dichloro-, sulfate (1:2); 3,3'-Dichlorobenzidine dihydrogen bis(sulfate); 3,3'-Dichlorobenzidine sulphate; Sulfato de 3,3-diclorobenzidina (Spanish)

CAS Registry Number: 91-94-1; 612-83-9 (dihydrochloride); 64969-34-2 (sulfate)

RTECS Number: DD0525000; DD0550000 (dihydrochloride)

DOT ID: UN 3077

EEC Number: 612-068-00-4

Regulatory Authority

3,3-isomer:

- Carcinogen (Animal Positive) (IARC)[9] (suspect) NTP
- Banned or Severely Restricted (Several Countries) (UN)[13][35]
- OSHA, 29CFR1910 Specifically Regulated Chemicals (See CFR 1910.1007)
- Air Pollutant Standard Set (Several States)[60]
- Clean Air Act: Hazardous Air Pollutants (Title I, Part A, Section 112)
- Clean Water Act: 40CFR423, Appendix A, Priority Pollutants; Section 313 Water Priority Chemicals (57FR41331, 9/9/92); Toxic Pollutant (Section 401.15)
- EPA Hazardous Waste Number (RCRA No.): U073
- RCRA, 40CFR261, Appendix 8 Hazardous Constituents.
- RCRA 40CFR264, Appendix 9; TSD Facilities Ground Water Monitoring List. Suggested test method(s) (PQL μg/l): 8270 (20)
- Superfund/EPCRA 40CFR302.4 Reportable Quantity (RQ): CERCLA, 1 lb (0.454 kg)
- EPCRA Section 313 Form R *de minimis* concentration reporting level: 0.1%
- Canada, WHMIS, Ingredients Disclosure List

dyhydrochloride and sulfate:

- OSHA, 29CFR1910 Specifically Regulated Chemicals (See CFR 1910.1007)
- EPCRA Section 313 Form R *de minimis* concentration reporting level: 0.1%

Cited in U.S. State Regulations: Alaska (G), California (A, G), Florida (G), Illinois (G), Kansas (G, W), Louisiana (G), Maine (G), Maryland (G), Massachusetts (G), Michigan (G), Minnesota (W), New Hampshire (G), New Jersey (G), New York (A), North Dakota (A), Pennsylvania (G, A), Rhode Island (G, A), South Carolina (A), Vermont (G), Virginia (G, A), Washington (G), West Virginia (G), Wisconsin (G).

Description: 3,3'-Dichlorobenzidine, $C_6H_3ClNH_2C_6H_3ClNH_2$, is a gray or purple crystalline solid. Freezing/Melting point = 132 – 133°C. Boiling point = 368°C. Autoignition temperature = 350°C. Hazard Identification (based on NFPA-704 M Rating System): Health 0, Flammability 1, Reactivity 0. Insoluble in water.

Potential Exposure: The major uses of dichlorobenzidine are in the manufacture of pigments for printing ink, textiles, plastics, and crayons and as a curing agent for solid urethane plastics. There are no substitutes for many of its uses. Additional groups that may be at risk include workers in the printing or graphic arts professions handling the 3,3'-DCB-based azo pigments. 3,3'-DCB may be present as an impurity in the pigments, and there is some evidence that 3,3'-DCB may be metabolically liberated from the azo pigment.

Incompatibilities: None reported.

Permissible Exposure Limits in Air: 3,3'-Dichlorobenzidine and its salts are included in a Federal standard for carcinogens; all contact with it should be avoided. Skin absorption is possible. ACGIH has categorized DCB as an "Industrial Substance Suspect of Carcinogenic Potential for Man" as has DFG.[3] Several states have set guidelines or standards for dichlorobenzidine in ambient air[60] ranging from zero (North Dakota, Pennsylvania, Virginia) to 0.002 $\mu g/m^3$ (Rhode Island) to 0.10 $\mu g/m^3$ (New York) to 0.15 $\mu g/m^3$ (South Carolina).

Determination in Air: Collection on a filter, elution with triethylamine in methanol, analysis by high performance liquid chromatography. See NIOSH Method 5509.[18]

Permissible Concentration in Water: To protect freshwater and saltwater aquatic life – no criteria developed due to insufficient data. To protect human health – preferably zero. An additional life-time cancer risk of 1 in 100,000 results at a level of 0.103 μg/l.[6] States which have set guidelines for dichlorobenzidine in drinking water[61] include Kansas and Minnesota – both at 0.21 μg/l.

Determination in Water: Chloroform extraction followed by concentration and high performance liquid chromatography (EPA Method 605) or gas chromatography plus mass spectrometry (EPA Method 625).

Routes of Entry: Inhalation, skin absorption, ingestion, skin and/or eye contact.

Harmful Effects and Symptoms

Short Term Exposure: Skin allergic sensitization, dermatitis; headache, dizziness; caustic burns; frequent urination, dysuria; hematuria (blood in the urine); gastrointestinal upset; upper respiratory infection.

Long Term Exposure: 3,3'-Dichlorobenzidine was shown to be a potent carcinogen in rats and mice in feeding and injection experiments, but no bladder tumors were produced. The LD_{50} oral-rat is 5,250 mg/kg (insignificantly toxic).

Points of Attack: Bladder, liver, lung, skin, gastrointestinal tract.

Medical Surveillance: Preplacement and periodic examinations should include history of exposure to other carcinogens, smoking, alcohol, medication, and family history. The skin, lung, kidney, bladder, and liver should be evaluated; sputum or urinary cytology may be helpful. Examination by a qualified allergist.

First Aid: If this chemical gets into the eyes, remove any contact lenses at once and irrigate immediately for at least 15 minutes, occasionally lifting upper and lower lids. Seek medical attention immediately. If this chemical contacts the skin, remove contaminated clothing and wash immediately with soap and water. Seek medical attention immediately. If this chemical has been inhaled, remove from exposure, begin rescue breathing (using universal precautions) if breathing has stopped and CPR if heart action has stopped. Transfer promptly to a medical facility. When this chemical has been swallowed, get medical attention. Give large quantities of water and induce vomiting. Do not make an unconscious person vomit.

Personal Protective Methods: Wear protective gloves and clothing to prevent any reasonable probability of skin contact. Full body protective clothing and gloves should be used by those employed in handling operations. Fullface supplied air respirators of continuous flow or pressure demand type should also be used. On exit from a regulated area, employees should shower and change into street clothes, leaving their protective clothing and equipment at the point of exit to be placed in impervious containers at the end of the work shift for decontamination or disposal. Effective methods should be used to clean and decontaminate gloves and clothing. Safety equipment suppliers/manufacturers can provide recommendations on the most protective glove/clothing material for your operation.. All protective clothing (suits, gloves, footwear, headgear) should be clean, available each day, and put on before work. Contact lenses should not be worn when working with this chemical. Wear dust-proof chemical goggles and face shield unless full facepiece respiratory protection is worn. Employees should wash immediately with soap when skin is wet or contaminated. Provide emergency showers and eyewash.

Respirator Selection: NIOSH *At any detectable concentration:* SCBAF:PD,PP (any MSHA/NIOSH approved self-contained breathing apparatus that has a full facepiece and is operated in a pressure-demand or other positive-pressure mode); or SAF:PD,PP:ASCBA (any supplied-air respirator that has a full facepiece and is operated in a pressure-demand or other positive-pressure mode in combination with an auxiliary, self-contained breathing apparatus operated in a pressure-demand or other positive pressure mode). *Escape:* HiEF (any air-purifying, full-facepiece respirator (gas mask) with a chin-style, front- or back-mounted organic vapor canister having a high-efficiency particulate filter); or SCBAE (any appropriate escape-type, self-contained breathing apparatus).

Storage: Prior to working with this chemical you should be trained on its proper handling and storage. Store in tightly closed containers in a cool, well-ventilated area. A regulated, marked area should be established where this chemical is handled, used, or stored in compliance with OSHA standard 1910.1045.

Shipping: Environmentally hazardous solid, n.o.s. Hazard Class: 9. Label: "Class 9."

Spill Handling: Evacuate persons not wearing protective equipment from area of spill or leak until clean-up is complete. Remove all ignition sources. Collect powdered material in the most convenient and safe manner and deposit in sealed containers. Ventilate area after clean-up is complete. It may be necessary to contain and dispose of this chemical as a hazardous waste. If material or contaminated runoff enters waterways, notify downstream users of potentially contaminated waters. Contact your Department of Environmental Protection or your regional office of the federal EPA for specific recommendations. If employees are required to clean-up spills, they must be properly trained and equipped. OSHA 1910.120(q) may be applicable.

Fire Extinguishing: This chemical is a combustible solid. Use dry chemical, carbon dioxide, water spray, or alcohol foam extinguishers. Poisonous gases are produced in fire. If material or contaminated runoff enters waterways, notify downstream users of potentially contaminated waters. Notify local health and fire officials and pollution control agencies. From a secure, explosion-proof location, use water spray to cool exposed containers. If cooling streams are ineffective (venting sound increases in volume and pitch, tank discolors, or shows any signs of deforming), withdraw immediately to a secure position. If employees are expected to fight fires, they must be trained and equipped in OSHA 1910.156.

Disposal Method Suggested: Incineration (1,500°F, 0.5 second for primary combustion; 2,200°F, 1.0 second for secondary combustion). The formation of elemental chlorine can be prevented through injection of steam or methane into the combustion process. No_x may be abated through the use of thermal or catalytic devices.[22] Consult with environmental regulatory agencies for guidance on acceptable disposal practices. Generators of waste containing this contaminant (≥100 kg/mo) must conform with EPA regulations governing storage, transportation, treatment, and waste disposal.

References

U.S. Environmental Protection Agency, 3,3'-Dichlorobenzidine: Ambient Water Quality Criteria, Washington, DC (1980)

U.S. Environmental Protection Agency, 3,3'-Dichlorobenzidine, Health and Environmental Effects Profile No. 68, Washington, DC, Office o Solid Waste (April 30, 1980)

Sax N. I., Ed., "Dangerous Properties of Industrial Materials Report," 2, No. 5, 45–49 (1982) and 3, No. 2, 79–82 (1983)

1,4-Dichloro-2-Butene

Molecular Formula: $C_4H_6Cl_2$

Synonyms: *cis-:* 2-Butene, 1,4-dichloro-; 1,4-DCB; DCB; 1,4-Dichloro-2-butene; 1,4-Dichlorobutene-2; 1,4-Dicloro-2-butano (Spanish)

trans-: AI3-52332; 2-Butene, 1,4-dichloro-, (E)-; 2-Butene, 1,4-dichloro-, *trans*-; 2-Butylene dichloride; (E)-1,4-Dichloro-2-butene; (E)-1,4-Dichlorobutene; *trans*-1,4-Dichloro-2-butene; *trans*-1,4-dichlorobutene; *trans*-2,3-Dichlorobut-2-ene; 1,4-Dichloro-*trans*-2-butene; 1,4-Dichlorobutene-2, (E)-; 1,4-Dichlorobutene-2, *trans*-

CAS Registry Number: 764-41-0 (cis-); 110-57-6 (trans-); 11069-19-5 (mixed isomers)

RTECS Number: EM4900000 (cis-); EM4903000 (trans-); EM4730000 (mixed isomers)

DOT ID: UN 2924 (mixed)

Regulatory Authority

- Carcinogen (DFG)[3]

cis-:

- Air Pollutant Standard Set[43] (ACGIH)
- EPA Hazardous Waste Number (RCRA No.): U074
- RCRA, 40CFR261, Appendix 8 Hazardous Constituents
- RCRA Land Ban Waste Restrictions
- Superfund/EPCRA 40CFR302.4 Reportable Quantity (RQ): CERCLA, 1 lb (0.454 kg)
- EPCRA Section 313 Form R *de minimis* concentration reporting level: 1.0%
- Canada, WHMIS, Ingredients Disclosure List

trans-:

- RCRA 40CFR264, Appendix 9; TSD Facilities Ground Water Monitoring List. Suggested test method(s) (PQL µg/l): 8240 (5)
- Superfund/EPCRA 40CFR355, Appendix B Extremely Hazardous Substances: TPQ = 500 lb (227 kg)
- EPCRA Section 313 Form R *de minimis* concentration reporting level: 1.0%

Cited in U.S. State Regulations: California (A, G), Florida (G), Kansas (G), Louisiana (G), Maine (G), Massachusetts (G), New Hampshire (G), New Jersey (G), Pennsylvania (G), Vermont (G), Virginia (A), Washington (G), Wisconsin (G).

Description: The 1,4-dichloro-2-butenes are colorless liquids. The *cis*-isomer: Boiling point = 153°C. Freezing/Melting point = -48°C. The *trans*-isomer: Boiling point = 155.5°C. Freezing/Melting point = 1 – 3°C. This material has a sweet, pungent odor. The explosive limits for the *trans*-isomer are LEL = 1.5%; UEL = 4%. Soluble in water.

Potential Exposure: DC occurs as a by-product in chloroprene manufacture and may be used as a chemical intermediate.

Incompatibilities: Strong oxidizers.

Permissible Exposure Limits in Air: ACGIH set a TWA of 0.005 ppm (0.025 mg/m^3) for the cis- isomer (764-41-0). The former USSR-UNEP/IRPTC project[43] has set a MAC in workplace air of 1.0 mg/m^3 for 1,3-dichloro-2-butene but has no value for 1,4-dichloro-2-butene.

Permissible Concentration in Water: The former USSR-UNEP/IRPTC project[43] has set a MAC in water bodies used for domestic purposes of 0.05 mg/l for 1,3-dichloro-2-butene but has no value for 1,4-dichloro-2-butene.

Routes of Entry: Inhalation, absorbed through the skin.

Harmful Effects and Symptoms

Short Term Exposure: Symptoms of exposure include respiratory distress and burns to skin and eyes. Inhalation of vapor irritates nose and throat. Contact with eyes causes intense irritation and tears. Contact of liquid with skin causes severe blistering. Ingestion causes severe irritation of mouth and stomach. Liquid and vapors from the material are highly corrosive and may damage skin, eyes, lungs, and internal organs.

Long Term Exposure: The material is a carcinogen according to DFG[3] and ACGIH. Corrosive substances may affect the lungs.

Points of Attack: Lungs, skin.

Medical Surveillance: Lung function tests. Consider chest x-ray following acute overexposure.

First Aid: If this chemical gets into the eyes, remove any contact lenses at once and irrigate immediately for at least 15 minutes, occasionally lifting upper and lower lids. Seek medical attention immediately. If this chemical contacts the skin, remove contaminated clothing and wash immediately with soap and water. Seek medical attention immediately. If this chemical has been inhaled, remove from exposure, begin rescue breathing (using universal precautions) if breathing has stopped and CPR if heart action has stopped. Transfer promptly to a medical facility. When this chemical has been swallowed, get medical attention. If victim is conscious, administer water or milk. Do not induce vomiting.

Personal Protective Methods: Wear protective gloves and clothing to prevent any reasonable probability of skin contact. Safety equipment suppliers/manufacturers can provide recommendations on the most protective glove/clothing material for your operation. Viton and Saranex are among the recommended protective materials. All protective clothing (suits, gloves, footwear, headgear) should be clean, available each day, and put on before work. Contact lenses should not be worn when working with this chemical. Wear splash-proof chemical goggles and face shield unless full facepiece respiratory protection is worn. Employees should wash immediately with soap when skin is wet or contaminated. Provide emergency showers and eyewash.

Respirator Selection: *At any detectable concentration:* SCBAF:PD,PP (any MSHA/NIOSH approved self-contained breathing apparatus that has a full facepiece and is operated in a pressure-demand or other positive-pressure mode); or SAF:PD,PP:ASCBA (any supplied-air respirator that has a full facepiece and is operated in a pressure-demand or other positive-pressure mode in combination with an auxiliary, self-contained breathing apparatus operated in a pressure-demand or other positive pressure mode). *Escape:* HiEF (any air-purifying, full-facepiece respirator (gas mask) with a chin-style, front- or back-mounted organic vapor canister having a high-efficiency particulate filter); or SCBAE (any appropriate escape-type, self-contained breathing apparatus).

Storage: Prior to working with this chemical you should be trained on its proper handling and storage. Before entering confined space where this chemical may be present, check to make sure that an explosive concentration does not exist. Store in tightly closed containers in a cool, well ventilated area away from oxidizers. Where possible, automatically pump liquid from drums or other storage containers to process containers. A regulated, marked area should be established where this chemical is handled, used, or stored in compliance with OSHA standard 1910.1045.

Shipping: Dichlorobutene falls in DOT Hazard Class 8 and Packing Group I. It must be labeled: "Corrosive, Flammable Liquid."

Spill Handling: Evacuate and restrict persons not wearing protective equipment from area of spill or leak until cleanup is complete. Remove all ignition sources. Establish forced ventilation to keep levels below explosive limit. Stay upwind; keep out of low areas. Shut off ignition sources. Do not touch spilled material. Use water spray to reduce vapors, but do not get water inside containers. For small spills, absorb with sand or other non-combustible absorbent material. For large spills, dike far ahead o spill for later disposal. Keep this chemical out of a confined space, such as a sewer, because of the possibility of an explosion, unless the sewer is designed to prevent the build-up of explosive concentrations. It may be necessary to contain and dispose of this chemical as a hazardous waste. If material or contaminated runoff enters waterways, notify downstream users of potentially contaminated waters. Contact your Department of Environmental Protection or your regional office of the federal EPA for specific recommendations. If employees are required to clean-up spills, they must be properly trained and equipped. OSHA 1910.120(q) may be applicable.

Fire Extinguishing: The material will burn, though it may require some effort to ignite. Fire produces irritating and poisonous gases. When heated to decomposition, it emits toxic fumes of chlorine-containing compounds. For small fires, use dry chemical, carbon dioxide, spray or foam. For large fires, use water spray, fog, or foam. Wear positive pressure breathing apparatus and full protective clothing. Move containers from fire area if you can do so without risk. Spray containers with cooling water until well after fire is out. Isolate for one-half mile in all directions if tank car or truck is involved in a fire. Vapors are heavier than air and will collect in low areas. Vapors in confined areas may explode when exposed to fire. Containers may explode in fire. Storage containers and parts of containers may rocket great distances, in many directions. If material or contaminated runoff enters waterways, notify downstream users of potentially contaminated waters. Notify local health and fire officials and pollution control agencies. From a secure, explosion-proof location, use water spray to cool exposed containers. If cooling streams are ineffective (venting sound increases in volume and pitch, tank discolors, or shows any signs of deforming), withdraw immediately to a secure position. If employees are expected to fight fires, they must be trained and equipped in OSHA 1910.156.

Disposal Method Suggested: High-temperature incineration with hydrochloric acid scrubbing. Consult with environmental regulatory agencies for guidance on acceptable disposal practices. Generators of waste containing this contaminant (≥100 kg/mo) must conform with EPA regulations governing storage, transportation, treatment, and waste disposal.

References

U.S. Environmental Protection Agency, "Chemical Profile: 1,4-Dichlorobutene, Trans-)," Washington, DC, Chemical Emergency Preparedness Program (November 30, 1987)

Sax N. I., Ed., "Dangerous Properties of Industrial Materials Report," 4, No. 3, 41–44 (1984)

Dichlorodifluoro-Ethylene

Molecular Formula: $C_2Cl_2F_2$

Common Formula: CFCl=CFCl

Synonyms: Dichlorodifluoroethene; Ethene, Dichlorofluoro-

CAS Registry Number: 27156-03-2

RTECS Number: KV9460000

DOT ID: NA 9018

Cited in U.S. State Regulations: Maine (G), New Jersey (G).

Description: Dichlorodifluoro-ethylene, CFCl=CFCl, is a colorless gas or liquid. Boiling point = 21.1°C. Insoluble in water.

Potential Exposure: Those involved in the manufacture of this compound or its use in the synthesis of fluorochemicals. Used for chemical research and development purposes.

Permissible Exposure Limits in Air: No standards set.

Permissible Concentration in Water: No criteria set.

Routes of Entry: Inhalation.

Harmful Effects and Symptoms

Short Term Exposure: Dichlorodifluoro-ethylene can affect you when breathed in. Exposure can irritate the eyes, nose and throat. High levels can cause you to become dizzy, lightheaded and pass out. Similar compounds can cause the heart to beat irregularly, or stop, which can cause death.

Long Term Exposure: May cause liver and kidney damage.

Points of Attack: Liver, kidneys.

Medical Surveillance: If symptoms develop or overexposure has occurred, the following tests may be useful: Liver function tests. Kidney function tests. Holter monitor (a special 24 hour EKG to look for irregular heart beat).

First Aid: If this chemical gets into the eyes, remove any contact lenses at once and irrigate immediately for at least 15 minutes, occasionally lifting upper and lower lids. Seek medical attention immediately. If this chemical contacts the skin, remove contaminated clothing and wash immediately with soap and water. Seek medical attention immediately. If this chemical has been inhaled, remove from exposure, begin rescue breathing (using universal precautions) if breathing has stopped and CPR if heart action has stopped. Transfer promptly to a medical facility. When this chemical has been swallowed, get medical attention. Give large quantities of water and induce vomiting. Do not make an unconscious person vomit.

Personal Protective Methods: Wear protective gloves and clothing to prevent any reasonable probability of skin contact. Safety equipment suppliers/manufacturers can provide recommendations on the most protective glove/clothing material for your operation.. All protective clothing (suits, gloves, footwear, headgear) should be clean, available each day, and put on before work. Contact lenses should not be worn when working with this chemical. Wear splash-proof chemical goggles and face shield unless full facepiece respiratory protection is worn. Employees should wash immediately with soap when skin is wet or contaminated. Provide emergency showers and eyewash.

Respirator Selection: Where the potential for exposure to Dichlorodifluoro-ethylene exists, use a MSHA/NIOSH approved supplied-air respirator with a full facepiece operated in the positive pressure mode or with a full facepiece, hood, or helmet in the continuous flow mode or use a MSHA/NIOSH approved self-contained breathing apparatus with a full facepiece operated in pressure-demand or other positive pressure mode.

Storage: Prior to working with this chemical you should be trained on its proper handling and storage. In contact with water or steam, dichlorodifluoro-ethylene produces highly toxic and corrosive fumes. Store in tightly closed containers in a cool, dry area.

Shipping: In its performance-oriented packaging standards,[19] this material is not specifically cited.

Spill Handling: Evacuate and restrict persons not wearing protective equipment from area of spill or leak until cleanup is complete. Remove all ignition sources. Ventilate area of spill or leak. If dichlorodifluoro-ethylene gas is leaked, take the following steps: Restrict persons not wearing protective equipment from area of leak until clean-up is complete. Ventilate area of leak to disperse the gas. Stop flow of gas. If source of leak is a cylinder and the leak cannot be stopped in place, remove the leaking cylinder to a safe place in the open air and repair leak or allow cylinder to empty. Absorb liquids in vermiculite, dry sand, earth, peat, carbon, or a similar material and deposit in sealed containers. Keep this chemical out of a confined space, such as a sewer, because of the possibility of an explosion, unless the sewer is designed to prevent the build-up of explosive concentrations. It may be necessary to contain and dispose of this chemical as a hazardous waste. If material or contaminated runoff enters waterways, notify downstream users of potentially contaminated waters. Contact your Department of Environmental Protection or your regional office of the federal EPA for specific recommendations. If employees are required to clean-up spills, they must be properly trained and equipped. OSHA 1910.120(q) may be applicable.

Fire Extinguishing: Dichlorodifluoro-ethylene may burn, but does not readily ignite. Poisonous gases are produced in the fire, including highly toxic fumes of fluorides and chlorides. Containers may explode in fire. Use dry chemical or CO_2 extinguishers. Use water spray to keep fire-exposed containers cool. Vapors are heavier than air and will collect in low areas. Vapors in confined areas may explode when exposed to fire. Containers may explode in fire. Storage containers and parts of containers may rocket great distances, in many directions. If material or contaminated runoff enters waterways, notify downstream users of potentially contaminated waters. Notify local health and fire officials and pollution control agencies. From a secure, explosion-proof location, use water spray to cool exposed containers. If cooling streams are ineffective (venting sound increases in volume and pitch, tank discolors, or shows any signs of deforming), withdraw immediately to a secure position. If employees are expected to fight fires, they must be trained and equipped in OSHA 1910.156.

References

New Jersey Department of Health, "Hazardous Substance Fact Sheet: Dichlorodifluoro-Ethylene," Trenton, NJ (February 1987)

Dichlorodifluoromethane

Molecular Formula: CCl_2F_2

Synonyms: Algofrene Type 2; Arcton 6; Arcton 12; CFC-12; Diclorodifluometano (Spanish); Difluorodichloromethane; Dwuchlorodwufluorometan (Polish); Electro-CF 12; Eskimon 12; F 12; FC 12; Fluorocarbon 12; Freon 12; Freon F-12; Frigen 12; Genetron 12; Halocarbon 12/Ucon 12; Halon; Halon 122; Isceon 122; Isotron 2; Isotron 12; Ledon 12; Methane, Dichlorodifluoro-; Propellant 12; R 12; Refrigerant 12; Ucon 12; Ucon 12/Halocarbon 12

CAS Registry Number: 75-71-8

RTECS Number: PA8200000

DOT ID: UN 1028

Regulatory Authority

- Air Pollutant Standard Set (Australia) (OSHA)[58] (HSE)[33] (Israel) (Sweden)[35] (former USSR)[35][43] (Several States)[60] (Several Canadian Provinces)
- Clean Air Act: Stratospheric ozone protection (Title VI, Subpart A, Appendix A), Class I, Ozone Depletion Potential = 1.0
- EPA Hazardous Waste Number (RCRA No.): U075
- RCRA, 40CFR261, Appendix 8 Hazardous Constituents
- RCRA 40CFR268.48; 61FR15654, Universal Treatment Standards: Wastewater (mg/l), 0.23; Nonwastewater (mg/kg), 7.2
- RCRA 40CFR264, Appendix 9; TSD Facilities Ground Water Monitoring List. Suggested test method(s) (PQL μg/l): 8010 (10); 8240 (5)
- Safe Drinking Water Act: Priority List (55 FR 1470) (Removed January 1981)
- Superfund/EPCRA 40CFR302.4 Reportable Quantity (RQ): CERCLA, 5,000 lb (2,270 kg)
- EPCRA Section 313 Form R *de minimis* concentration reporting level: 1.0%
- Canada, WHMIS, Ingredients Disclosure List

Cited in U.S. State Regulations: Alaska (G), Arizona (W), California (A, G), Connecticut (A), Florida (G), Illinois (G), Kansas (G, W), Louisiana (G), Maine (G, W), Massachusetts (G), Minnesota (G), Nevada (A), New Jersey (G), New York (G), North Carolina (A), North Dakota (A), Pennsylvania (G), Rhode Island (G), Vermont (G), Virginia (G, A), Washington (G), West Virginia (G), Wisconsin (G).

Description: Dichlorodifluoromethane, CCl_2F_2, is a colorless, nonflammable gas with a characteristic ether-like odor at >20% by volume. Shipped as a compressed gas. Boiling point = -30°C. Freezing/Melting point = -158°C. Hazard Identification (based on NFPA-704 M Rating System): Health 1, Flammability 0, Reactivity 0. Soluble in water.

Potential Exposure: Dichlorodifluoromethane is used as an aerosol propellant, refrigerant and foaming agent.

Incompatibilities: Chemically active metals-sodium, potassium, calcium, powdered aluminum, zinc, magnesium. Attacks magnesium and its alloys.

Permissible Exposure Limits in Air: The Federal Limit (OSHA PEL),[58] the Australian, Israeli, DFG MAK, HSE values, and the ACGIH TWA value is 1,000 ppm (4,950 mg/m^3). The Canadian provinces of Alberta, BC, Ontario, and Quebec use the same TWA values as OSHA and HSE, Alberta, and British Columbia set the STEL value at 1,250 ppm (6,200 mg/m^3).[33] Sweden[35] has set a TWA of 500 ppm (2,500 mg/m^3) and a STEL of 750 ppm (4,000 mg/m^3). The former USSR has set a MAC in workplace air of 3,000 mg/m^3 as well as values for ambient air in residential areas of 100 mg/m^3 on a momentary basis and 10 mg/m^3 on a daily average basis.[35][43] The NIOSH IDLH level is 15,000 ppm.

Several states have set guidelines or standards for R-12 in ambient air[60] ranging from 49.5 mg/m^3 (North Dakota) to 82.5 mg/m^3 (Virginia) to 99.0 mg/m^3 (Connecticut) to 118.0 mg/m^3 (Nevada) to 247.0 mg/m^3 (North Carolina). Those values may well be modified in the future because of concern over the effect of chlorofluorocarbons on the depletion of the ozone layer in the atmosphere.

Determination in Air: Charcoal tube;[2] CH_2C_{12}; Gas chromatography/Flame ionization detection; NIOSH IV, Method #1018.

Permissible Concentration in Water: Human health protection-preferably zero. Additional lifetime cancer risk of 1 in 100,000 results at a level of 1.9 µg/l. In Jan. 1981 EPA (46FR2266) removed F-12 from the priority toxic pollutant list.[6] The former USSR[35][43] has set a MAC in water used for domestic purposes of 10 mg/l. Several states have set guidelines for R-12 in drinking water[61] ranging from 1.0 µg/l in Arizona to 160 µg/l (Maine) to 5,600 µg/l (Kansas).

Determination in Water: Inert gas purge followed by gas chromatography with halide specific detector (EPA Method 601).

Routes of Entry: Inhalation, eye and/or skin contact (liquid).

Harmful Effects and Symptoms

Short Term Exposure: Exposure can cause you to become dizzy and lightheaded and to have trouble concentrating. Exposure can cause the heart to beat irregularly (cardiac arrythmia) or cause heart arrest. This can cause death. Contact with the liquid can cause severe eye and skin burns from frostbite. Breathing the gas can irritate the mouth, nose and throat. High levels can cause asphyxiation.

Points of Attack: Cardiovascular system, peripheral nervous system.

Medical Surveillance: If symptoms develop or overexposure is suspected, the following may be useful: Special 24-hour EKG (Holter monitor) to look for irregular heartbeat.

First Aid: If this chemical gets into the eyes, remove any contact lenses at once and irrigate immediately for at least 15 minutes, occasionally lifting upper and lower lids. Seek medical attention immediately. If this chemical contacts the skin, remove contaminated clothing and wash immediately with soap and water. Seek medical attention immediately. If this chemical has been inhaled, remove from exposure, begin rescue breathing (using universal precautions) if breathing has stopped and CPR if heart action has stopped. Transfer promptly to a medical facility.

Personal Protective Methods: Wear protective gloves and clothing to prevent any reasonable probability of skin contact. Safety equipment suppliers/manufacturers can provide recommendations on the most protective glove/clothing material for your operation. ACGIH recommends neoprene rubber as a good to excellent protective material. All protective clothing (suits, gloves, footwear, headgear) should be clean, available each day, and put on before work. Contact lenses should not be worn when working with this chemical. Wear splash-proof chemical goggles and face shield unless full facepiece respiratory protection is worn. Employees should wash immediately with soap when skin is wet or contaminated. Provide emergency showers and eyewash. Where cold equipment, vapors, or liquid may occur, employees should be provided with special clothing designed to prevent the freezing of body tissues.

Respirator Selection: NIOSH/OSHA: *10,000 ppm:* SA (any supplied-air respirator). *15,000 ppm*: SA:CF (any supplied-air respirator operated in a continuous-flow mode); or SCBAF (any self-contained breathing apparatus with a full facepiece); or SAF (any supplied-air respirator with a full facepiece). *Emergency or planned entry into unknown concentrations or IDLH conditions*: SCBAF:PD,PP (any self-contained breathing apparatus that has a full facepiece and is operated in a pressure-demand or other positive-pressure mode); or SAF:PD,PP:ASCBA (any supplied-air respirator that has a full facepiece and is operated in a pressure-demand or other positive-pressure mode in combination with an auxiliary self-contained breathing apparatus operated in a pressure-demand or other positive pressure mode). *Escape:* GMFOV [any air-purifying, full-facepiece respirator (gas mask) with a chin-style, front-or back-, mounted organic vapor canister]; or SCBAE (any appropriate escape-type, self-contained breathing apparatus).

Storage: Prior to working with this chemical you should be trained on its proper handling and storage. Store in tightly closed containers in a cool well-ventilated area away from heat. Dichlorodifluoromethane must be stored to avoid contact with chemically active metals (such as sodium, potassium, calcium, powdered aluminum, zinc, and magnesium) since violent reactions occur. Procedures for the handling, use and storage of cylinders should be in compliance with OSHA 1910.101 and 1910.169 as with the recommendations of the Compressed Gas Association.

Shipping: R-12 must be labeled "Nonflammable Gas." It falls in DOT Hazard Class 2.2.

Spill Handling: Evacuate and restrict persons not wearing protective equipment from area of spill or leak until cleanup is complete. Remove all ignition sources. Ventilate area of spill or leak to disperse the gas. Stop flow of gas. If source of leak is a cylinder and the leak cannot be stopped in place, remove the leaking cylinder to a safe place in the open air, and repair leak or allow cylinder to empty. It may be necessary to contain and dispose of this chemical as a hazardous waste. Contact your Department of Environmental Protection or your regional office of the federal EPA for specific recommendations. If employees are required to clean-up spills, they must be properly trained and equipped. OSHA 1910.120(q) may be applicable.

Fire Extinguishing: Extinguish fire using an agent suitable for type of surrounding fire. The material itself does not burn. Poisonous gases including hydrogen chloride are produced in fire. Vapors are heavier than air and will collect in low areas. Containers may explode in fire. Storage containers and parts of containers may rocket great distances, in many directions. If material or contaminated runoff enters waterways, notify downstream users of potentially contaminated waters. Notify local health and fire officials and pollution control agencies. From a secure, explosion-proof location, use water spray to cool exposed containers. If cooling streams are ineffective (venting sound increases in volume and pitch, tank discolors, or shows any signs of deforming), withdraw immediately to a secure position. If employees are expected to fight fires, they must be trained and equipped in OSHA 1910.156.

Disposal Method Suggested: Incineration, preferably after mixing with another combustible fuel. Care must be exercised to assure complete combustion to prevent the formation of phosgene. An acid scrubber is necessary to remove the halo acids produced. Because of potential ozone decomposition in the stratosphere, R-12 should be released to the atmosphere only as a last resort.[22] Consult with environmental regulatory agencies for guidance on acceptable disposal practices. Generators of waste containing this contaminant (≥100 kg/mo) must conform with EPA regulations governing storage, transportation, treatment, and waste disposal.

References

U.S. Environmental Protection Agency, Halomethanes: Ambient Water Quality Criteria, Washington, DC (1980)

U.S. Environmental Protection Agency, Trichlorofluoromethane and Dichlorodifluoromethane, Health and Environmental Effects Profile No. 167, Washington, DC, Office of Solid Waste (April 30, 1980)

New Jersey Department of Health, "Hazardous Substance Fact Sheet: Dichlorodifluoromethane," Trenton, NJ (May 1998)

New York State Department of Health, "Chemical Fact Sheet: Dichlorodifluoromethane," Albany, NY, Bureau of Toxic Substance Assessment (June 1986)

1,3-Dichloro-5,5-Dimethylhydantoin

Molecular Formula: $C_5H_6Cl_2N_2O_2$

Synonyms: Dactin®; Dantion; DCDMH; DDH; 1,3-Dicloro-5,5-dimehyl-2,4-imidazolinedione; Halane®; 2,4-Imidazolidinedione, 1,3-dichloro-5,5-dimethyl

CAS Registry Number: 118-52-5

RTECS Number: MU0700000

DOT ID: No citation.

Regulatory Authority

- Air Pollutant Standard Set (ACGIH)[1] (Australia) (HSE)[33] (Israel) (Mexico) (OSHA)[58] (Several States)[60] (Several Canadian Provinces)

Cited in U.S. State Regulations: Alaska (G), California (A, G), Connecticut (A), Florida (G), Illinois (G), Maine (G), Massachusetts (G), Michigan (G), Minnesota (G), Nevada (A), New Hampshire (G), New Jersey (G), North Dakota (A), Pennsylvania (G), Rhode Island (G), Virginia (A), West Virginia (G).

Description: DCDMH, $C_5H_6Cl_2N_2O_2$, is a combustible, white powder with a chlorine-like odor and a structural formula.

Freezing/Melting point = 130°C. Flash point = 175°C. Slightly soluble in water.

Potential Exposure: It is used as a chlorinating agent, disinfectant and laundry bleach. It is used as a polymerization catalyst in making vinyl chloride and in drug and pesticide synthesis.

Incompatibilities: A strong oxidizer. Contact with water forms poisonous and corrosive gases. Mixtures with xylene may explode. Not compatible with moisture (especially hot water, steam), strong acids, easily oxidized materials (such as ammonia salts, sulfides, etc), reducing agents, strong bases, ammonium salts, sulfides.

Permissible Exposure Limits in Air: The Federal Limit (OSHA PEL)[58] and ACGIH TLV is 0.2 mg/m^3 TWA as is the HSE, Australian, Israeli, Mexican and Canadian provincial (Alberta, BC, Ontario, and Quebec) value. The ACGIH, HSE, Israeli, Mexican, Australian and Canadian provincial STEL is 0.4 mg/m^3 The NIOSH IDLH level is 5 mg/m^3. Several states have set guidelines or standards for this material in ambient air[60] ranging from 2.0 – 4.0 μg/m^3 (North Dakota) to 3.2 μg/m^3 (Virginia) to 4.0 μg/m^3 (Connecticut) to 5.0 μg/m^3 (Nevada).

Determination in Air: No method available.

Permissible Concentration in Water: No criteria set.

Routes of Entry: Inhalation, ingestion, eye and/or skin contact.

Harmful Effects and Symptoms

Short Term Exposure: Irritation of skin, eyes, mucous membrane and respiratory system. The oral LD_{50} rat is 542 mg/kg (slightly toxic). Higher levels can irritate the lungs, causing a build-up of fluid (pulmonary edema). This can cause death.

Long Term Exposure: Can irritate the lungs causing bronchitis with coughing, phlegm, and/or shortness of breath.

Points of Attack: Eyes, respiratory system.

Medical Surveillance: For those with frequent or potentially high exposure (half the TLV or greater), the following are recommended before beginning work and at regular times after that: Lung function tests. If symptoms develop or overexposure is suspected, the following may be useful: Consider chest x-ray after acute overexposure.

First Aid: If this chemical gets into the eyes, remove any contact lenses at once and irrigate immediately for at least 15 minutes, occasionally lifting upper and lower lids. Seek medical attention immediately. If this chemical contacts the skin, remove contaminated clothing and wash immediately with soap and water. Seek medical attention immediately. If this chemical has been inhaled, remove from exposure, begin rescue breathing (using universal precautions) if breathing has stopped and CPR if heart action has stopped. Transfer promptly to a medical facility. When this chemical has been swallowed, get medical attention. Give large quantities of water and induce vomiting. Do not make an unconscious person vomit. Medical observation is recommended for 24 – 48 hours after breathing overexposure, as pulmonary edema may be delayed. As first aid for pulmonary edema, a doctor or authorized paramedic may consider administering a corticosteroid spray.

Personal Protective Methods: Wear protective gloves and clothing to prevent any reasonable probability of skin contact. Safety equipment suppliers/manufacturers can provide recommendations on the most protective glove/clothing material for your operation.. All protective clothing (suits, gloves, footwear, headgear) should be clean, available each day, and put on before work. Contact lenses should not be worn when working with this chemical. Wear dust-proof chemical goggles and face shield unless full facepiece respiratory protection is worn. Employees should wash immediately with soap when skin is wet or contaminated. Provide emergency showers and eyewash.

Respirator Selection: NIOSH/OSHA: Up to 2 mg/m^3: SA (any supplied-air respirator). Up to 5 mg/m^3: SA:CF (any supplied-air respirator operated in a continuous-flow mode); or SCBAF (any self-contained breathing apparatus with a full facepiece); or SAF (any supplied-air respirator with a full facepiece). *Emergency or planned entry into unknown concentrations or IDLH conditions:* SCBAF:PD,PP (any self-contained breathing apparatus that has a full facepiece and is operated in a pressure-demand or other positive-pressure mode); or SAF:PD,PP:ASCBA (any supplied-air respirator that has a full facepiece and is operated in a pressure-demand or other positive-pressure mode in combination with an auxiliary self-contained breathing apparatus operated in a pressure-demand or other positive pressure mode). *Escape:* GMFSHiE [any air-purifying, full-facepiece respirator (gas mask) with a chin-style, front-or back-, mounted canister providing protection against the compound of concern and having a high efficiency particulate filter]; or SCBAE (any appropriate escape-type, self-contained breathing apparatus).

Storage: Prior to working with this chemical you should be trained on its proper handling and storage. 1,3-Dichloro-5,5-Dimethyl Hydantoin must be stored to avoid contact with strong acids (such as sulfuric acid, nitric acid, or hydrochloric acid) and easily oxidized materials (such as ammonium salts and sulfides) since violent reactions occur and poisonous gases can be produced. Store in tightly closed containers in a cool well-ventilated area away from water or steam. 1,3-Dichloro-5,5-dimethylhydantoin decomposes with formation of poisonous gases at 395 – 410°F. If 1,3-dichloro-5,5-dimethylhydantoin contacts water or steam, it decomposes at lower temperatures and produces poisonous gases including chlorine. Sources of ignition such as smoking and open flames are prohibited where 1,3-dichloro-5,5-dimethylhydantoin is used, handled, or stored in a manner that could create a potential fire or explosion hazard.

Shipping: This material is not specifically cited in DOT's Performance-Oriented Packaging Standards.[19]

Spill Handling: Evacuate persons not wearing protective equipment from area of spill or leak until clean-up is complete. Remove all ignition sources. Collect powdered material in the most convenient and safe manner and deposit in sealed containers. Ventilate area after clean-up is complete. It may be necessary to contain and dispose of this chemical as a hazardous waste. If material or contaminated runoff enters waterways, notify downstream users of potentially contaminated waters. Contact your Department of Environmental Protection or your regional office of the federal EPA for specific recommendations. If employees are required to clean-up spills, they must be properly trained and equipped. OSHA 1910.120(q) may be applicable.

Fire Extinguishing: This chemical is a combustible solid. Use dry chemical or carbon dioxide. DO NOT use water. Poisonous gases are produced in fire including chlorine, phosgene, and nitrogen oxides. If material or contaminated runoff enters waterways, notify downstream users of potentially contaminated waters. Notify local health and fire officials and pollution control agencies. From a secure, explosion-proof location, use water spray to cool exposed containers. If cooling streams are ineffective (venting sound increases in volume and pitch, tank discolors, or shows any signs of deforming), withdraw immediately to a secure position. If employees are expected to fight fires, they must be trained and equipped in OSHA 1910.156.

Disposal Method Suggested: Incineration (1,500°F, 0.5 sec for primary combustion; 220°F, 1.0 sec for secondary combustion). The formation of elemental chlorine can be prevented by injection of steam or methane into the combustion process. No_x may be abated by the use of thermal or catalytic devices.

References

New Jersey Department of Health, "Hazardous Substance Fact Sheet: 1,3-Dichloro-5,5-Dimethyl Hydantoin," Trenton, NJ (October 1998)

1,1-Dichloroethane

Molecular Formula: $C_2H_4Cl_2$

Common Formula: CH_3CHCl_2

Synonyms: Aethylidenchlorid (German); *asym*-Dichloroethane; Chlorinated hydrochloric ether; Chlorure d'ethylidene (French); Cloruro di etilidene (Italian); 1,1-Dichloorethaan (Dutch); 1,1-Dichloraethan (German); Dichloromethylethane; 1,1-Dicloroetano (Italian); 1,1-Dicloroetano (Spanish); Ethane, 1,1-dichloro-; 1,1-Ethylidene chloride; Ethylidene chloride; Ethylidene dichloride; NCI-C04535; Vinylidene chloride

CAS Registry Number: 75-34-3

RTECS Number: KI0175000

DOT ID: UN 2362

EEC Number: 602-011-00-1

Regulatory Authority

- Air Pollutant Standard Set (ACGIH)[1] (DFG)[3] (HSE)[33] (OSHA)[58] (Several States)[60]
- Clean Air Act: Hazardous Air Pollutants (Title I, Part A, Section 112)
- Clean Water Act: 40CFR423, Appendix A, Priority Pollutants
- EPA Hazardous Waste Number (RCRA No.): U076
- RCRA, 40CFR261, Appendix 8 Hazardous Constituents
- RCRA 40CFR268.48; 61FR15654, Universal Treatment Standards: Wastewater (mg/l), 0.059; Nonwastewater (mg/kg), 6.0
- RCRA 40CFR264, Appendix 9; TSD Facilities Ground Water Monitoring List. Suggested test method(s) (PQL μg/l): 8010 (1); 8240 (5)
- Safe Drinking Water Act: Priority List (55 FR 1470)
- Superfund/EPCRA 40CFR302.4 Reportable Quantity (RQ): CERCLA, 1,000 lb (454 kg)
- EPCRA Section 313 Form R *de minimis* concentration reporting level: 1.0%
- U.S. DOT Regulated Marine Pollutant (49CFR172.101, Appendix B)
- Canada, WHMIS, Ingredients Disclosure List

Cited in U.S. State Regulations: Alaska (G), California (A, W), Connecticut (A), Florida (G), Illinois (G, W), Kansas (G), Louisiana (G), Maine (G), Massachusetts (G), Minnesota (G), Nevada (A), New Hampshire (G), New Jersey (G), New Mexico (W), New York (G), North Dakota (A), Pennsylvania (G), Rhode Island (G), Vermont (G, W), Virginia (G, A), Washington (G), West Virginia (G), Wisconsin (G, W).

Description: 1,1-Didichloroethane, CH_3CHCl_2, is a colorless, oily liquid with a chloroform-like odor. Odor threshold = 50 – 1,350 ppm. Boiling point = 57°C. Freezing/Melting point = -98°C. Flash point = -17°C (cc). Autoignition temperature = 458°C. Hazard Identification (based on NFPA-704 M Rating System): Health 2, Flammability 3, Reactivity 0. Explosive limits: LEL = 5.4%; UEL = 11.4%. Slightly soluble in water.

Potential Exposure: It is used as a solvent and cleaning and degreasing agent as well as in organic synthesis as an intermediate.

Incompatibilities: Forms explosive mixture with air. Reacts violently with strong oxidizers, alkali metals, earth-alkali metals, powdered metals, causing fire and explosion hazard. Contact with strong caustic will produce flammable and toxic acetaldehyde gas. Attacks aluminum, iron. Attacks some plastics (including polyethylene) and coatings.

Permissible Exposure Limits in Air: The Federal Limit (OSHA PEL)[58] is 100 ppm (400 mg/m^3) as is the Quebec and DFG MAK value.[3] The ACGIH TWA value[1] is 200 ppm (810 mg/m^3) and the STEL value is 250 ppm (1,010 mg/m^3). Australia, Israel, Mexico, and the Canadian provinces of Alberta, BC, and Ontario have the same limits as ACGIH. The HSE TWA value is 200 ppm and STEL is 400 ppm. The NIOSH IDLH level is 3,000 ppm. Several states have set guidelines or standards for ethylidene chloride in ambient air[60] ranging from 8.0 mg/m^3 (Connecticut) to 8.1 – 10.1 mg/m^3 (North Dakota) to 13.5 mg/m^3 (Virginia) to 19.3 mg/m^3 (Nevada).

Determination in Air: Charcoal tube; CS_2; Gas chromatography/Flame ionization detection; NIOSH IV, Method#1003, Halogenated Hydrocarbons.

Permissible Concentration in Water: No criteria set for aquatic life or human health due to insufficient data.[6] Several states have set guidelines or standards for ethylidene chloride in drinking water[61] ranging from 1.0 μg/l (Illinois) to 20.0 μg/l (California) to 25 μg/l (New Mexico) to 70.0 μg/l (Vermont) to 850 μg/l (Wisconsin).

Determination in Water: Inert gas purge followed be gas chromatography with halide specific detection (EPA Method 601) or gas chromatography plus mass spectrometry (EPA Method 624).

Routes of Entry: Inhalation, ingestion, eye and skin contact.

Harmful Effects and Symptoms

Short Term Exposure: This chemical may affect the central nervous system. Exposure can cause drowsiness, unconsciousness, and death. High exposures may damage the liver or kidneys. Contact can cause eye and skin irritation with eye burns. Long-term exposure can cause thickening and cracking of skin. 1,1-Dichloroethane is a highly flammable liquid and a dangerous fire hazard. Never use near combustion sources. Do not use 1,1-dichlorethane where welding is being done because deadly phosgene gas can be formed.

Long Term Exposure: It may damage the developing fetus. The liquid destroys the skin's natural oils. May affects on the kidneys and liver. A chloroethane, this chemical may be a potential occupational carcinogen. Prolonged skin contact can cause thickening and cracking of the skin and mild burns. Although not adequately evaluated, similar petroleum-based chemicals can cause brain or other nerve damage.

Points of Attack: Skin, liver, kidneys, lungs, central nervous system.

Medical Surveillance: If overexposure or illness is suspected, consider: Liver and kidney function tests. Interview for brain effects, including recent, memory, mood, concentration, headaches, malaise, altered sleep patterns. Consider autonomic and peripheral nervous system evaluation. Positive and borderline cases should be referred for neuropsychological testing.

First Aid: If this chemical gets into the eyes, remove any contact lenses at once and irrigate immediately for at least 15 minutes, occasionally lifting upper and lower lids. Seek medical attention immediately. If this chemical contacts the skin, remove contaminated clothing and wash immediately with soap and water. Seek medical attention immediately. If this chemical has been inhaled, remove from exposure, begin rescue breathing (using universal precautions) if breathing has stopped and CPR if heart action has stopped. Transfer promptly to a medical facility. When this chemical has been swallowed, get medical attention. If victim is conscious, administer water or milk. Do not induce vomiting.

Personal Protective Methods: Wear solvent-resistant protective gloves and clothing to prevent any reasonable probability of skin contact. Safety equipment suppliers/manufacturers can provide recommendations on the most protective glove/clothing material for your operation. Teflon is among the recommended protective materials. All protective clothing (suits, gloves, footwear, headgear) should be clean, available each day, and put on before work. Contact lenses should not be worn when working with this chemical. Wear splash-proof chemical goggles and face shield unless full facepiece respiratory protection is worn. Employees should wash immediately with soap when skin is wet or contaminated. Provide emergency showers and eyewash.

Respirator Selection: NIOSH/OSHA: *1,000 ppm:* SA (any supplied-air respirator). *2,500 ppm:* SA:CF (any supplied-air respirator operated in a continuous-flow mode). *3,000 ppm:* SCBAF (any self-contained breathing apparatus with a full facepiece); or SAF (any supplied-air respirator with a full facepiece). Emergency or planned entry into unknown concentrations or IDLH conditions: SCBAF:PD,PP (any self-contained breathing apparatus that has a full facepiece and is operated in a pressure-demand or other positive-pressure mode); or SAF:PD,PP:ASCBA (any supplied-air respirator that has a full facepiece and is operated in a pressure-demand or other positive-pressure mode in combination with an auxiliary self-contained breathing apparatus operated in a pressure-demand or other positive pressure mode). *Escape:* GMFOV [any air-purifying, full-facepiece respirator (gas mask) with a chin-style, front- or back- mounted organic vapor canister]; or SCBAE (any appropriate escape-type, self-contained breathing apparatus).

Storage: Prior to working with this chemical you should be trained on its proper handling and storage. Before entering confined space where this chemical may be present, check to make sure that an explosive concentration does not exist. 1,1-Dichloroethane must be stored to avoid contact with strong oxidizers such as chlorine, bromine and fluorine since violent reactions occur. Store in tightly closed containers in a cool well-ventilated area away from heat. Sources of ignition such as smoking and open flames are prohibited where 1,1-dichlorethane is used, handled, or stored in a manner that could create a potential fire or explosion hazard. Metal containers used in the transfer of 5 gallons or more of 1,1-dichloroethane should be grounded and bonded. Drums must be quipped with self-closing valves, pressure vacuum bungs, and flame arresters. Use only non-sparking tools and equipment, especially when opening and closing containers of 1,1-dichloroethane.

Shipping: This material requires a "Flammable Liquid" label. It falls in DOT Hazard Class 3 and Packing Group II.

Spill Handling: Evacuate and restrict persons not wearing protective equipment from area of spill or leak until cleanup is complete. Remove all ignition sources. Establish forced ventilation to keep levels below explosive limit. Absorb liquids in vermiculite, dry sand, earth, peat, carbon, or a similar material and deposit in sealed containers. Keep this chemical out of a confined space, such as a sewer, because of the possibility of an explosion, unless the sewer is designed to prevent the build-up of explosive concentrations. It may be necessary to contain and dispose of this chemical as a hazardous waste. If material or contaminated runoff enters waterways, notify downstream users of potentially contaminated waters. Contact your Department of Environmental Protection or your regional office of the federal EPA for specific recommendations. If employees are required to clean-up spills, they must be properly trained and equipped. OSHA 1910.120(q) may be applicable.

Fire Extinguishing: This chemical is a flammable liquid. Poisonous and corrosive fumes including phosgene and hydrogen chloride are produced in fire. Use dry chemical, carbon dioxide, or foam extinguishers. Vapors are heavier than air and will collect in low areas. Vapors may travel long distances to ignition sources and flashback. Vapors in confined areas may explode when exposed to fire. Containers may explode in fire. Storage containers and parts of containers may rocket great distances, in many directions. If material or contaminated runoff enters waterways, notify downstream users of potentially contaminated waters. Notify local health and fire officials and pollution control agencies. From a secure, explosion-proof location, use water spray to cool exposed containers. If cooling streams are ineffective (venting sound increases in volume and pitch, tank discolors, or shows any signs of deforming), withdraw immediately to a secure position. If employees are expected to fight fires, they must be trained and equipped in OSHA 1910.156.

Disposal Method Suggested: Incineration; preferably after mixing with another combustible fuel. Care must be exercised to assure complete combustion to prevent the formation of phosgene. An acid scrubber is necessary to remove the halo acids produced. Consult with environmental regulatory agencies for guidance on acceptable disposal practices. Generators of waste containing this contaminant (≥100 kg/mo) must conform with EPA regulations governing storage, transportation, treatment, and waste disposal.

References

U.S. Environmental Protection Agency, Chloroethanes: Ambient Water Quality Criteria, Washington, DC (1980)

U.S. Environmental Protection Agency, 1,1-Dichloroethane, Health and Environmental Effects Profile No. 69, Washington, DC, Office of Solid Waste (April 30, 1980)

Sax N. I., Ed., "Dangerous Properties of Industrial Materials Report," 4, No. 3, 44–47 (1984)

New York State Department of Health, "Chemical Fact Sheet: 1,1-Dichloroethane," Albany, NY, Bureau of Toxic Substance Assessment (May 1986)

New Jersey Department of Health, "Hazardous Substance Fact Sheet: 1,1-Dichloroethane," Trenton, NJ (February 1989)

Dichloroethyl Acetate

Molecular Formula: $C_4H_6Cl_2O_2$

Common Formula: $CH_3COOCHClCH_2Cl$

Synonyms: Aceto de 1,2-dicloroetilo (Spanish); 1,2-Dichloroethanol acetate; 1,2-Dichloroethyl acetate; Ethanol, 1,2-dichloro-, acetate

CAS Registry Number: 10140-87-1

RTECS Number: KK4200000

Regulatory Authority

- Superfund/EPCRA 40CFR355, Appendix B Extremely Hazardous Substances: TPQ = 1,000 lb (454 kg)
- Superfund/EPCRA 40CFR302.4 Reportable Quantity (RQ): EHS, 1 lb (0.454 kg)

Cited in U.S. State Regulations: California (G), Massachusetts (G), New Jersey (G).

Description: Dichloroethyl acetate, $C_4H_6Cl_2O_2$, is a water-white liquid. Boiling point = 58 – 65°C @ 13 mm Hg pressure. Flash point = 152°C. Insoluble in water.

Potential Exposure: This material is used in organic synthesis.

Incompatibilities: Strong oxidizers. May explode on contact with nitrates and heat.

Permissible Exposure Limits in Air: No standards set.

Permissible Concentration in Water: No criteria set.

Routes of Entry: Inhalation.

Harmful Effects and Symptoms

Short Term Exposure: EPA states that the material is toxic by inhalation.

First Aid: If this chemical gets into the eyes, remove any contact lenses at once and irrigate immediately for at least 15 minutes, occasionally lifting upper and lower lids. Seek medical attention immediately. If this chemical contacts the skin, remove contaminated clothing and wash immediately with soap and water. Seek medical attention immediately. If this chemical has been inhaled, remove from exposure, begin rescue breathing (using universal precautions) if breathing has stopped and CPR if heart action has stopped. Transfer promptly to a medical facility. When this chemical has been swallowed, get medical attention. Give large quantities of water and induce vomiting. Do not make an unconscious person vomit.

Personal Protective Methods: Wear protective gloves and clothing to prevent any reasonable probability of skin contact. Safety equipment suppliers/manufacturers can provide recommendations on the most protective glove/clothing material for your operation.. All protective clothing (suits, gloves, footwear, headgear) should be clean, available each day, and put on before work. Contact lenses should not be worn when working with this chemical. Wear splash-proof chemical goggles and face shield unless full facepiece respiratory protection is worn. Employees should wash immediately with soap when skin is wet or contaminated.

Respirator Selection: *At any detectable concentration:* SCBAF:PD,PP (any MSHA/NIOSH approved self-contained breathing apparatus that has a full facepiece and is operated in a pressure-demand or other positive-pressure mode); or SAF:PD,PP:ASCBA (any supplied-air respirator that has a full facepiece and is operated in a pressure-demand or other positive-pressure mode in combination with an auxiliary, self-contained breathing apparatus operated in a pressure-demand or other positive pressure mode). *Escape:* HiEF (any air-purifying, full-facepiece respirator (gas mask) with a chin-style, front- or back-mounted organic vapor canister having a high-efficiency particulate filter); or SCBAE (any appropriate escape-type, self-contained breathing apparatus).

Storage: Prior to working with this chemical you should be trained on its proper handling and storage. Store in tightly closed containers in a cool, well ventilated area away from oxidizers and nitrates. Where possible, automatically pump liquid from drums or other storage containers to process containers.

Shipping: Combustible liquids, n.o.s. fall in DOT Hazard Class 3 and Packing Group III. There is no label requirement.

Spill Handling: Evacuate and restrict persons not wearing protective equipment from area of spill or leak until cleanup is complete. Remove all ignition sources. Stay upwind; keep out of low areas. Shut off ignition sources; no flares, smoking or flames in hazard area. Stop leak if you can do it without risk. Use water spray to reduce vapors. Small spills: take up with sand or other non-combustible absorbent material and place into containers for later disposal. Large spills: dike far ahead of spill for later disposal. Ventilate area of spill or leak. Keep this chemical out of a confined space, such as a sewer, because of the possibility of an explosion, unless the sewer is designed to prevent the build-up of explosive concentrations. It may be necessary to contain and dispose of this chemical as a hazardous waste. If material or contaminated runoff enters waterways, notify downstream users of potentially contaminated waters. Contact your Department of Environmental Protection or your regional office of the federal EPA for specific recommendations. If employees are required to cleanup spills, they must be properly trained and equipped. OSHA 1910.120(q) may be applicable.

Fire Extinguishing: This chemical is a combustible liquid. Poisonous gases including phosgene and chlorine are produced in fire. Use dry chemical, carbon dioxide, or foam extinguishers. Vapors are heavier than air and will collect in low areas. Vapors may travel long distances to ignition sources and flashback. Vapors in confined areas may explode when exposed to fire. Containers may explode in fire. Storage containers and parts of containers may rocket great distances, in many directions. If material or contaminated runoff enters waterways, notify downstream users of potentially contaminated waters. Notify local health and fire officials and pollution control agencies. From a secure, explosion-proof location, use water spray to cool exposed containers. If cooling streams are ineffective (venting sound increases in volume and pitch, tank discolors, or shows any signs of deforming), withdraw immediately to a secure position. If employees are expected to fight fires, they must be trained and equipped in OSHA 1910.156.

References

U.S. Environmental Protection Agency, "Chemical Profile: Ethanol, 1,2-Dichloro, Acetate," Washington, DC, Chemical Emergency Preparedness Program (November 30, 1987)

1,2-Dichloroethylene

Molecular Formula: $C_2H_2Cl_2$

Common Formula: ClCH=CHCl

Synonyms: Acetylene dichloride; 1,2-Dichloraethen (German); 1,2-Dichloroethene; *sym*-Dichloroethylene; Dichloro-1,2-ethylene; 1,2-Dicloroeteno (Spanish); Dioform; Ethene, 1,2-dichloro-; Ethylene, 1,2-dichloro-; NCI-C56031

cis-isomer: Acetyalyne-dichloride; *cis*-1,2-Dichlorethylene; *cis*-1,2-Dichloroethene; (Z)-1,2-Dichloroethylene; *cis*-Dichloroethylene; 1,2-*cis*-Dichloroethylene; Ethene, 1,2-dichloro-, (Z)-ethene; Ethylene, 1,2-dichloro-, (Z)

trans-isomer: trans-Acetylene dichloride; *trans*-1,2-Dichloroethylene; *trans*-Dichloroethylene; *trans*-1,2-Dicloroeteno (Spanish); Dioform; Ethene, *trans*-1,2-dichloro-; Ethene, 1,2-dichloro-, (E)-

CAS Registry Number: 540-59-0; 156-59-2 (cis-); 156-60-5 (trans-)

RTECS Number: KV9360000; KV9420000 (cis-); RV9400000 (trans-)

DOT ID: UN 1150

EEC Number: 602-026-00-3

Regulatory Authority ;

- Clean Water Act: 40CFR423, Appendix A, Priority Pollutants; Section 313 Water Priority Chemicals (57FR41331, 9/9/92); Toxic Pollutant (Section 401.15)
- EPCRA Section 313 Form R *de minimis* concentration reporting level: 1.0%
- Canada, WHMIS, Ingredients Disclosure List

trans-isomer:

- Clean Water Act: 40CFR423, Appendix A, Priority Pollutants; Section 313 Water Priority Chemicals (57FR41331, 9/9/92)
- EPA Hazardous Waste Number (RCRA No.): U079
- RCRA, 40CFR261, Appendix 8 Hazardous Constituents
- RCRA 40CFR268.48; 61FR15654, Universal Treatment Standards: Wastewater (mg/l), 0.054; Nonwastewater (mg/kg), 30
- RCRA 40CFR264, Appendix 9; TSD Facilities Ground Water Monitoring List. Suggested test method(s) (PQL μg/l): 8010 (1); 8240 (5)
- Safe Drinking Water Act: MCL, 0.1 mg/l; MCLG, 0.1 mg/l; Regulated chemical (47 FR 9352)
- Superfund/EPCRA 40CFR302.4 Reportable Quantity (RQ): CERCLA, 1,000 lb (454 kg)

cis-isomer:

- Safe Drinking Water Act: MCL, 0.07 mg/l; MCLG, 0.07 mg/l; Regulated chemical (47 FR 9352)

Cited in U.S. State Regulations: Alaska (G), Arizona (W), California (W), Connecticut (A), Florida (G), Illinois (G), Kansas (G, W), Louisiana (G), Maine (G, W), Maryland (G), Massachusetts (G, A), Minnesota (W), Nevada (A), New Hampshire (G), New Jersey (G, W), New York (G), Oklahoma (G), Pennsylvania (G), Rhode Island (G), Vermont (G, W), Virginia (G, A), Washington (G), West Virginia (G), Wisconsin (G, W).

Description: 1,2-Dichloroethylene, ClCH=CHCl, exists in two isomers, cis- 60% and trans- 40%. There are variations in toxicity between these two forms. At room temperature, these chemicals are colorless liquids with a slightly acrid, ethereal odor. The odor threshold in air is 17 ppm. Mixed isomers: Boiling point = 48°C. Explosive Limits: LEL = 9.7%; UEL = 12.8%. Autoignition temperature = 460°C. *cis-isomer:* Boiling point = 60.3°C. Freezing/Melting point = -82°C. Flash point = 6°C. *trans-isomer:* Boiling point = 47.5°C. Freezing/Melting point = -49°C. Flash point = 2 – 4°C. The explosive limits are: LEL = 5.6% or 9.7%;[52] UEL = 12.8%. Slightly soluble in water.

Potential Exposure: 1,2-Dichloroethylene is used as a solvent for waxes, resins, and acetylcellulose. It is also used in the extraction of rubber, as a refrigerant, in the manufacture of pharmaceuticals and artificial pearls, and in the extraction of oils and fats from fish and meat.

Incompatibilities: Forms explosive mixture with air. Attacks some plastics, rubber and coatings. Strong oxidizers. Gradual decomposition results in hydrochloric acid formation in the presence of ultraviolet light or upon contact with hot metal or other hot surfaces. Reacts with strong oxidizers, strong bases, potassium hydroxide, difluoromethylene, dihypofluoride, nitrogen tetroxide (explosive), or copper (and its alloys) producing toxic chloroacetylene which is spontaneously flammable on contact with air. Attacks some plastics and coatings.

Permissible Exposure Limits in Air: For 1,2-DCE (540-59-0), the Federal Limit (OSHA PEL)[58] and ACGIH TLV is 200 ppm (790 mg/m^3) TWA. The DFG MAK, HSE, Australian, Israeli, and Canadian provincial (Alberta, BC, Ontario, Quebec) TWA values are the same as OSHA. The HSE, Alberta, BC, and Ontario STEL value is 250 ppm (1,000 mg/m^3). The NIOSH ILDH level is 1,000 ppm. Japan[35] has set a TWA value in workplace air of 150 ppm (596 mg/m^3). Several States have set guidelines or standards for *sym*-dichloroethylene in ambient air[60] ranging from 0.110 mg/m^3 (Massachusetts) to 13.0 mg/m^3 (Virginia) to 15.8 mg/m^3 (Connecticut) to 18.8 mg/m^3 (Nevada).

Determination in Air: Charcoal absorption workup with CS_2, analysis by gas chromatography. See NIOSH Method #1003 for hydrocarbons, halogenated.

Permissible Concentration in Water: To protect freshwater aquatic life – 11,600 μg/l on an acute toxicity basis for dichloroethylenes in general. To protect saltwater aquatic life: 224,000 μg/l on an acute toxicity basis for dichloroethylenes as a class. To protect human health-no criteria developed due to insufficient data.[6] A long term health advisory for cis-1,2-DCE has been determined by EPA[48] as 3.5 mg/l and a lifetime health advisory as 0.07 mg/l. Maximum levels in drinking water have been set by EPA[62] at 0.07 mg/l for the cis-isomer and 0.10 mg/l for the trans-isomer.

Several states have set guidelines or standards for the dichloroethylenes in drinking water[61] as follows:

State	Cis- (μg/l)	Trans- (μg/l)	Sym- (μg/l)
Arizona	—	—	70
California	16	16	—
Kansas	70	70	—
Maine	400	270	—
Minnesota	70	70	—
New Jersey	—	—	10
Vermont	—	—	70
Wisconsin	100	100	—

Determination in Water: trans-1,2-dichloroethylene may be determined by inert gas purge followed by gas chromatography with halide specific detection (EPA Method 601) or gas chromatography plus mass spectrometry (EPA Method 624).

Routes of Entry: Inhalation of the vapor, ingestion, skin and eye contact.

Short Term Exposure: This liquid can act as a primary irritant producing dermatitis and irritation of mucous membranes. 1,2-DCE irritates the eyes, skin, and respiratory tract. Skin contact can cause a burning sensation and rash. 1,2-Dichloroethylene acts principally as a narcotic, causing central nervous system depression. Symptoms of acute exposure include dizziness, nausea and frequent vomiting, and central nervous system intoxication similar to that caused by alcohol. High levels can cause unconsciousness. The oral LD_{50} rat is 770 mg/kg (slightly toxic).

Long Term Exposure: Destroys skin's natural oils. Repeated exposure may damage the liver and kidneys. Renal effects, when they do occur, are transient.

Points of Attack: Respiratory system, eyes, central nervous system.

Medical Surveillance: Consider possible irritant effects on skin or respiratory tract as well as liver and renal function is preplacement or periodic examinations. Expired air analysis may be useful in detecting exposure. For those with frequent or potentially high exposure (half the TLV or greater), the following are recommended before beginning work and at regular times after that: Lung function tests. If symptoms develop or overexposure is suspected, the following may be useful: Liver and kidney function tests. Complete blood count.

First Aid: If this chemical gets into the eyes, remove any contact lenses at once and irrigate immediately for at least 15 minutes, occasionally lifting upper and lower lids. Seek medical attention immediately. If this chemical contacts the skin, remove contaminated clothing and wash immediately with soap and water. Seek medical attention immediately. If this chemical has been inhaled, remove from exposure, begin rescue breathing (using universal precautions) if breathing has stopped and CPR if heart action has stopped. Transfer promptly to a medical facility. When this chemical has

been swallowed, get medical attention. Give large quantities of water and induce vomiting. Do not make an unconscious person vomit.

Personal Protective Methods: Wear protective gloves and clothing to prevent any reasonable probability of skin contact. Safety equipment suppliers/manufacturers can provide recommendations on the most protective glove/clothing material for your operation. Viton is among the recommended protective materials for the *cis-* and *trans-* isomers. All protective clothing (suits, gloves, footwear, headgear) should be clean, available each day, and put on before work. Contact lenses should not be worn when working with this chemical. Wear splash-proof chemical goggles and face shield unless full facepiece respiratory protection is worn. Employees should wash immediately with soap when skin is wet or contaminated. Provide emergency showers and eyewash.

Respirator Selection: NIOSH/OSHA: *1,000 ppm:* SA:CF (any supplied-air respirator operated in a continuous-flow mode); or PAPROV [any powered, air-purifying respirator with organic vapor cartridge(s)]; or CCRFOV [any chemical cartridge respirator with a full facepiece and organic vapor cartridge(s)]; or GMFOV [any air-purifying, full-facepiece respirator (gas mask) with a chin-style, front- or back-mounted acid gas canister]; or SCBAF (any self-contained breathing apparatus with a full facepiece); or SAF (any supplied-air respirator with a full facepiece). *Emergency or planned entry into unknown concentrations or IDLH conditions*: SCBAF:PD,PP (any self-contained breathing apparatus that has a full facepiece and is operated in a pressure-demand or other positive-pressure mode); or SAF:PD,PP:ASCBA (any supplied-air respirator that has a full facepiece and is operated in a pressure-demand or other positive-pressure mode in combination with an auxiliary self-contained breathing apparatus operated in a pressure-demand or other positive pressure mode). *Escape:* GMFOV [any air-purifying, full-facepiece respirator (gas mask) with a chin-style, front-or back-, mounted organic vapor canister]; or SCBAE (any appropriate escape-type, self-contained breathing apparatus).

Note: Substance causes eye irritation or damage; eye protection needed.

Storage: Prior to working with 1,2-DCE you should be trained on its proper handling and storage. Before entering confined space where this chemical may be present, check to make sure that an explosive concentration does not exist. 1,2-Dichloroethylene must be stored to avoid contact with strong oxidizers (such as chlorine, bromine and fluorine) since violent reactions occur. Store in tightly closed containers in a cool, well-ventilated area away from heat. Sources of ignition, such as smoking and open flames, are prohibited where 1,2-dichloroethylene is used, handled, or stored. Metal containers involving the transfer of 5 gallons or more of 1,2-dichloroethylene should be grounded and bonded. Drums must be equipped with self-closing valves, pressure vacuum bungs and flame arresters. Use only non-sparking tools and equipment, especially when opening and closing containers of 1,2-dichloroethylene. Wherever 1,2-dichloroethylene is used, handled, manufactured, or stored, use explosion-proof electrical equipment and fittings.

Shipping: Dichloroethylene must carry a "Flammable Liquid" label. It falls in DOT Hazard Class 3 and Packing Group II.

Spill Handling: Evacuate and restrict persons not wearing protective equipment from area of spill or leak until cleanup is complete. Remove all ignition sources. Establish forced ventilation to keep levels below explosive limit. Absorb liquids in vermiculite, dry sand, earth, peat, carbon, or a similar material and deposit in sealed containers. Keep this chemical out of a confined space, such as a sewer, because of the possibility of an explosion, unless the sewer is designed to prevent the build-up of explosive concentrations. It may be necessary to contain and dispose of this chemical as a hazardous waste. If material or contaminated runoff enters waterways, notify downstream users of potentially contaminated waters. Contact your Department of Environmental Protection or your regional office of the federal EPA for specific recommendations. If employees are required to clean-up spills, they must be properly trained and equipped. OSHA 1910.120(q) may be applicable.

Fire Extinguishing: 1,2-DCE is a flammable liquid. Poisonous gases including phosgene and hydrogen chloride are produced in fire. Use dry chemical, carbon dioxide, or foam extinguishers. Vapors are heavier than air and will collect in low areas. Vapors may travel long distances to ignition sources and flashback. Vapors in confined areas may explode when exposed to fire. Containers may explode in fire. Storage containers and parts of containers may rocket great distances, in many directions. If material or contaminated runoff enters waterways, notify downstream users of potentially contaminated waters. Notify local health and fire officials and pollution control agencies. From a secure, explosion-proof location, use water spray to cool exposed containers. If cooling streams are ineffective (venting sound increases in volume and pitch, tank discolors, or shows any signs of deforming), withdraw immediately to a secure position. If employees are expected to fight fires, they must be trained and equipped in OSHA 1910.156.

Disposal Method Suggested: Incineration, preferably after mixing with another combustible fuel. Care must be exercised to assure complete combustion to prevent the formation of phosgene. An Acid scrubber is necessary to remove the halo acids produced.[22] Consult with environmental regulatory agencies for guidance on acceptable disposal practices. Generators of waste containing this contaminant (≥100 kg/mo) must conform with EPA regulations governing storage, transportation, treatment, and waste disposal.

References

U.S. Environmental Protection Agency, Dichloroethylenes: Ambient Water Quality Criteria, Washington, DC (1980) sufficient data. For the protection of human rably zero. An additional lifetime cancer risk of is posed by a concentration of 0.3 μg/l.[6] Some set guidelines for BCEE in drinking water[61] n 0.31 μg/l (Minnesota) to 4.2 μg/l (Kansas) to aine) to 10.0 μg/l (Arizona).

ion in Water: CH_2Cl_2 extraction followed by gas raphy with halogen; specific detector (EPA Method s chromatography plus mass spectrometry (EPA th25).

ute Entry: Inhalation, skin absorption, ingestion, skin d/ore contact.

arm/Effects and Symptoms

hort Tm Exposure: BCEE can be absorbed through the kin, thereby increasing exposure. Exposure irritates the eyes, skin, and respiratory tract. Skin and eye contact may cause burns. Higher exposures can cause pulmonary edema, a medical emergency that can be delayed for several hours. This can cause death. At concentrations above 500 ppm, coughing, retching, and vomiting may occur, as well as profuse tearing. There can be irritation at lower concentrations. This material is very toxic; the probable oral lethal dose is 50 – 500 mg/kg, or between 1 teaspoon and 1 ounce for a 150 pound person. It can be a central nervous system depressant in high concentrations. It is extremely irritating to the eyes, nose, and respiratory passages. It can penetrate the skin to cause serious and even fatal poisoning. Poisonous; may be fatal if inhaled, swallowed or absorbed through skin.

Long Term Exposure: BCEE may damage the liver and kidneys. Can irritate the lungs; repeated exposures may cause bronchitis.

Points of Attack: Respiratory system, skin, eyes, liver and kidneys. This chemical causes liver cancer in animals and may be a potential human carcinogen.

Medical Surveillance: Before beginning employment and at regular times after that, the following are recommended: Lung function tests. If symptoms develop or overexposure is suspected, the following may also be useful: Liver, kidney, and lung function tests. Consider chest x-ray following acute overexposure.

First Aid: If this chemical gets into the eyes, remove any contact lenses at once and irrigate immediately for at least 15 minutes, occasionally lifting upper and lower lids. Seek medical attention immediately. If this chemical contacts the skin, remove contaminated clothing and wash immediately with soap and water. Seek medical attention immediately. If this chemical has been inhaled, remove from exposure, begin rescue breathing (using universal precautions) if breathing has stopped and CPR if heart action has stopped. Transfer promptly to a medical facility. When this chemical has been swallowed, get medical attention. Give large quantities of water and induce vomiting. Do not make an unconscious person vomit. Medical observation is recommended for 24 – 48 hours after breathing overexposure, as pulmonary edema may be delayed. As first aid for pulmonary edema, a doctor or authorized paramedic may consider administering a corticosteroid spray.

Personal Protective Methods: Wear protective gloves and clothing to prevent any reasonable probability of skin contact. Safety equipment suppliers/manufacturers can provide recommendations on the most protective glove/clothing material for your operation. Teflon and Chlorinated Polyethylene are among the recommended protective materials. All protective clothing (suits, gloves, footwear, headgear) should be clean, available each day, and put on before work. Contact lenses should not be worn when working with this chemical. Wear splash-proof chemical goggles and face shield unless full facepiece respiratory protection is worn. Employees should wash immediately with soap when skin is wet or contaminated. Provide emergency showers and eyewash.

Respirator Selection: NIOSH: *At any detectable concentration:* SCBAF:PD,PP (any MSHA/NIOSH approved self-contained breathing apparatus that has a full facepiece and is operated in a pressure-demand or other positive-pressure mode); or SAF:PD,PP:ASCBA (any supplied-air respirator that has a full facepiece and is operated in a pressure-demand or other positive-pressure mode in combination with an auxiliary, self-contained breathing apparatus operated in a pressure-demand or other positive pressure mode). *Escape:* GMFOV [any air-purifying, full-facepiece respirator (gas mask) with a chin-style, front-or back-mounted organic vapor canister] or SCBAE (any appropriate escape-type, self-contained breathing apparatus).

Storage: Prior to working with BCEE you should be trained on its proper handling and storage. Before entering confined space where this chemical may be present, check to make sure that an explosive concentration does not exist. Bis (2-chloroethyl) ether must be stored to avoid contact with strong oxidizers such as chlorine, bromine, and chlorine dioxide since violent reactions occur, and moisture. Store in tightly closed containers in a cool, dry, well-ventilated area away from heat. Sources of ignition such as smoking and open flames are prohibited where bis (2-chloroethyl) ether is used, handled, or stored in a manner that could create a potential fire or explosion hazard. A regulated, marked area should be established where this chemical is handled, used, or stored in compliance with OSHA standard 1910.1045.

Shipping: Dichloroethyl ether must carry a "Poison" label. It falls in DOT Hazard Class 6.1 and Packing Group II.

Spill Handling: Evacuate and restrict persons not wearing protective equipment from area of spill or leak until cleanup is complete. Remove all ignition sources. Establish forced ventilation to keep levels below explosive limit. Absorb spills in vermiculite, dry sand, earth, or similar material. Keep material out of water sources and sewers. Build dikes to contain flow as necessary. Use water spray to

U.S. Environmental Protection Agency, Trans-1,2-Dichloroethylenes, Health and Environmental Effects Profile No. 72, Washington, DC, Office of Solid Waste (April 30, 1980)

U.S. Environmental Protection Agency, Dichloroethylenes, Health and Environmental Effects Profile N0. 73, Washington, DC, Office of Solid Waste (April 30, 1980)

Sax N. I., Ed., "Dangerous Properties of Industrial Materials Report," 4, No. 3, 48–53 (1984)

New Jersey Department of Health, "Hazardous Substance Fact Sheet: 1,2-Dichloroethylene," Trenton, NJ (September 1996)

Dichloroethyl Ether

Molecular Formula: $C_4H_8Cl_2O$

Common Formula: $ClCH_2CH_2OCH_2CH_2Cl$

Synonyms: BCEE; Bis(β-chloroethyl) ether; Bis(2-chloroethyl) ether; Bis(2-cloroetil)eter (Spanish); Chlorex; 1-Chloro-2-(β-chloroethoxy)ethane; Chloroethyl ether (DOT); Clorex; DCEE; 2,2'-Dichloorethylether (Dutch); 2,2'-Dichlor-diaethylaether (German); 2,2'-Dichlorethyl ether (DOT); β,β'-Dichlorodiethyl ether; 2,2'-Dichlorodiethyl ether; Dichloroether; β,β'-Dichloroethyl ether; *sym*-Dichloroethyl ether; 2,2'-Dichloroethyl ether; Di(β-chloroethyl) ether; Di(2-chloroethyl) ether; Dichloroethyl ether; Dichloroethyl oxide; 2,2'-Dicloroetiletere (Italian); Dwuchlorodwuetylowy eter (Polish); ENT 4,504; Ethane, 1,1'-oxybis 2-chloro-; Ether dichlore (French); 1,1'-Oxybis(2-chloro)ethane; Oxyde de chlorethyle (French)

CAS Registry Number: 111-44-4

RTECS Number: KN0875000

DOT ID: UN 1916

Regulatory Authority

- Carcinogen (Animal Positive) (IARC)[9]
- Banned or Severely Restricted (Finland, Sweden) (UN)[13]
- Air Pollutant Standard Set (ACGIH)[1](Australia) (DFG) (Israel) (former USSR)[43] (OSHA)[58] (Several States)[60]
- Clean Air Act: Hazardous Air Pollutants (Title I, Part A, Section 112)
- Clean Water Act: 40CFR423, Appendix A, Priority Pollutants; Section 313 Water Priority Chemicals (57FR41331, 9/9/92)
- EPA Hazardous Waste Number (RCRA No.): U025
- RCRA, 40CFR261, Appendix 8 Hazardous Constituents
- RCRA 40CFR268.48; 61FR15654, Universal Treatment Standards: Wastewater (mg/l), 0.033; Nonwastewater (mg/kg), 6.0
- RCRA 40CFR264, Appendix 9; TSD Facilities Ground Water Monitoring List. Suggested test method(s) (PQL µg/l): 8270 (10)
- Superfund/EPCRA 40CFR302.4 Reportable Quantity (RQ): CERCLA, 10 lb (4.54 kg)
- Superfund/EPCRA ardous Substances: 1
- EPCRA Section 313 F porting level: 0.1%
- Canada, WHMIS, Ingredie
- Mexico, drinking water criteri

Cited in U.S. State Regulations: Connecticut (A), Illinois (G), Kansas Maine (G, W), Massachusetts (G, vada (A), New Hampshire (G), New kota (A), Oklahoma (G), Pennsylvani land (G), Vermont (G), Virginia (G, A), West Virginia (G), Wisconsin (G).

Description: Dichloroethyl ether, $ClCH_2CH_2$ a clear, colorless liquid with a pungent, fruity described as having a chlorinated solvent-like point = 176 – 178°C. Flash point = 55°C.[17] Autoig perature = 369°C. Explosive limits: LEL = 2.7%; known. Hazard Identification (based on NFPA-704 System): Health 3, Flammability 2, Reactivity 1. Inso water.

Potential Exposure: Dichloroethyl ether is used in manufacture of paint, varnish, lacquer, soap, and finish mover. It is also used as a solvent for cellulose esters, naph thalenes, oils, fats, greases, pectin, tar, and gum; in dry cleaning, in textile scouring; and in soil fumigation.

Incompatibilities: Contact with moisture caused decomposition producing hydrochloric acid. Can form peroxides. Forms explosive mixture with air. Contact with strong oxidizers may cause fire and explosion hazard. Attacks some plastics, rubber and coatings. Attacks metals in the presence of moisture.

Permissible Exposure Limits in Air: The Federal Limit (OSHA PEL) and the ACGIH recommended TLV is 5 ppm (30 mg/m^3) and the STEL value in both cases is 10 ppm (60 mg/m^3). They both add the notation "skin" indicating the possibility of cutaneous absorption. Australia, Israel and the Canadian provinces of Alberta, BC, Ontario, and Quebec set the same TWA and STEL as OSHA. The DFG MAK value is 10 ppm. NIOSH IDLH = [Ca]100 ppm. The former USSR-UNEP/IRPTC project[43] has set a MAC in workplace air of 2 mg/m^3. Several states have set guidelines have set guidelines or standards for BCEE in ambient air[60] ranging from 0.0714 mg/m^3 (Kansas) to 0.3 – 0.6 mg/m^3 (North Dakota) to 0.5 mg/m^3 (Virginia) to 0.6 mg/m^3 (Connecticut) to 0.714 mg/m^3 (Nevada) to 0.72 mg/m^3 (Pennsylvania).

Determination in Air: Charcoal tube; CS_2; Gas chromatography/Flame ionization detection; NIOSH IV Method #1004.

Permissible Concentration in Water: To protect freshwater aquatic life – 238,000 µg/l, for chloroalkyl ethers in general. No criteria developed for protection of saltwater aquati knock down vapors. Spill or leak: shut off ignition sources; no flares, smoking or flames in hazard area. Do not touch spilled material; stop leak if you can do so without risk. Use water spray to reduce vapors. Small spills: absorb with sand or other noncombustible absorbent material and place into containers for later disposal. Large spills: dike far ahead of spill for later disposal. Keep unnecessary people away; isolate hazard area and deny entry. Keep this chemical out of a confined space, such as a sewer, because of the possibility of an explosion, unless the sewer is designed to prevent the build-up of explosive concentrations. It may be necessary to contain and dispose of this chemical as a hazardous waste. If material or contaminated runoff enters waterways, notify downstream users of potentially contaminated waters. Contact your Department of Environmental Protection or your regional office of the federal EPA for specific recommendations. If employees are required to clean-up spills, they must be properly trained and equipped. OSHA 1910.120(q) may be applicable.

Fire Extinguishing: This chemical is a combustible liquid. Poisonous gases are produced in fire. Use dry chemical, carbon dioxide, or foam extinguishers. Dike fire control water for later disposal; do not scatter the material. Vapors are heavier than air and will collect in low areas. Vapors may travel long distances to ignition sources and flashback. Vapors in confined areas may explode when exposed to fire. Containers may explode in fire. Storage containers and parts of containers may rocket great distances, in many directions. If material or contaminated runoff enters waterways, notify downstream users of potentially contaminated waters. Notify local health and fire officials and pollution control agencies. From a secure, explosion-proof location, use water spray to cool exposed containers. If cooling streams are ineffective (venting sound increases in volume and pitch, tank discolors, or shows any signs of deforming), withdraw immediately to a secure position. If employees are expected to fight fires, they must be trained and equipped in OSHA 1910.156.

Disposal Method Suggested: Incineration, preferably after mixing with another combustible fuel. Care must be exercised to assure complete combustion to prevent the formation of phosgene. An acid scrubber is necessary to remove the halo acids produced. Consult with environmental regulatory agencies for guidance on acceptable disposal practices. Generators of waste containing this contaminant (≥100 kg/mo) must conform with EPA regulations governing storage, transportation, treatment, and waste disposal.

References

U.S. Environmental Protection Agency, Chloroalkyl Ethers: Ambient Water Quality Criteria, Washington, DC (1980)

U.S. Environmental Protection Agency, Bis (2-Chloroethyl) Ether, Health and Environmental Effects Profile No. 24, Washington, DC, Office of Solid Waste (April 30, 1980)

International Agency for Research on Cancer, IARC Monographs on the Carcinogenic Risks of Chemicals to Humans, Lyon, France 9, 117 (1975)

U.S. Environmental Protection Agency, "Chemical Profile: Dichloroethyl Ether," Washington, DC, Chemical Emergency Preparedness Program (November 30, 1987)

New Jersey Department of Health, "Hazardous Substance Fact Sheet: Bis (2-Chloroethyl) Ether," Trenton, NJ (January 1986)

Sax N. I., Ed., "Dangerous Properties of Industrial Materials Report," 1, No. 4, 62–76 (1987)

Dichloromethylphenylsilane

Molecular Formula: $C_7H_8Cl_2Si$

Common Formula: $C_6H_5SiCl_2CH_3$

Synonyms: Dichloromethylphenylsilane; Methylphenyldichlorosilane; Phenylmethyldichlorosilane; Silane, dichloromethylphenyl-

CAS Registry Number: 149-74-6

RTECS Number: VV3530000

DOT ID: UN 2437

Regulatory Authority

- Air Pollutant Standard Set (former USSR)[43]
- Superfund/EPCRA 40CFR355, Appendix B Extremely Hazardous Substances: TPQ = 1,000 lb (454 kg)
- Superfund/EPCRA 40CFR302.4 Reportable Quantity (RQ): EHS, 1 lb (0.454 kg)

Cited in U.S. State Regulations: California (G), Florida (G), Massachusetts (G), New Jersey (G), Pennsylvania (G).

Description: Dichloromethylphenylsilane, $C_6H_5SiCl_2$-CH_3, is a flammable, colorless liquid. Boiling point = 205°C. Flash point = 28°C. Hazard Identification (based on NFPA-704 M Rating System): Health 2, Flammability 3, Reactivity 1. Reacts with water.

Potential Exposure: Used in the manufacture of silicones; and as a chemical intermediate for silicone fluids, resins, and elastomers.

Incompatibilities: Forms explosive mixture with air. Contact with moisture causes decomposition, forming HCl and explosive hydrogen. Contact with ammonia may form a self-igniting material. Reacts strongly with oxidizing materials. Attacks some metals in the presence of moisture.

Permissible Exposure Limits in Air: The former USSR-UNEP/IRPTC joint project[43] has set a MAC in workplace air of 1 mg/m^3.

Permissible Concentration in Water: No criteria set.

Routes of Entry: Inhalation, ingestion, skin contact.

Harmful Effects and Symptoms

Short Term Exposure: The chemical is toxic and a corrosive irritant. Contact with the skin or eyes causes burns and permanent damage. Inhalation may cause irritation. Higher exposures may cause pulmonary edema, a medical emergency that can be delayed for several hours. This can cause death.

Long Term Exposure: Repeated exposure may cause lung irritation and bronchitis.

Points of Attack: Lungs.

Medical Surveillance: Lung function tests. Consider chest x-ray following acute overexposure.

First Aid: If this chemical gets into the eyes, remove any contact lenses at once and irrigate immediately for at least 15 minutes, occasionally lifting upper and lower lids. Seek medical attention immediately. If this chemical contacts the skin, remove contaminated clothing and wash immediately with soap and water. Seek medical attention immediately. If this chemical has been inhaled, remove from exposure, begin rescue breathing (using universal precautions) if breathing has stopped and CPR if heart action has stopped. Transfer promptly to a medical facility. When this chemical has been swallowed, get medical attention. If victim is conscious, administer water or milk. Do not induce vomiting.

Personal Protective Methods: Wear protective gloves and clothing to prevent any reasonable probability of skin contact. Safety equipment suppliers/manufacturers can provide recommendations on the most protective glove/clothing material for your operation.. All protective clothing (suits, gloves, footwear, headgear) should be clean, available each day, and put on before work. Contact lenses should not be worn when working with this chemical. Wear splash-proof chemical goggles and face shield unless full facepiece respiratory protection is worn. Employees should wash immediately with soap when skin is wet or contaminated. Provide emergency showers and eyewash.

Respirator Selection: *At any detectable concentration:* SCBAF:PD,PP (any MSHA/NIOSH approved self-contained breathing apparatus that has a full facepiece and is operated in a pressure-demand or other positive-pressure mode); or SAF:PD,PP:ASCBA (any supplied-air respirator that has a full facepiece and is operated in a pressure-demand or other positive-pressure mode in combination with an auxiliary, self-contained breathing apparatus operated in a pressure-demand or other positive pressure mode). *Escape:* GMFOV [any air-purifying, full-facepiece respirator (gas mask) with a chin-style, front-or back-mounted organic vapor canister] or SCBAE (any appropriate escape-type, self-contained breathing apparatus).

Storage: Prior to working with methylphenyldichlorosilane you should be trained on its proper handling and storage. Store in tightly closed containers in a cool, well ventilated area away from moisture, oxidizers, and ammonia. Where possible, automatically pump liquid from drums or other storage containers to process containers.

Shipping: Methylphenyldichlorosilane requires a "Corrosive" label. It falls in DOT Hazard Class 8 and Packing Group II.

Spill Handling: Evacuate and restrict persons not wearing protective equipment from area of spill or leak until cleanup is complete. Remove all ignition sources. Ventilate area of spill or leak. Absorb liquids in vermiculite, dry sand, earth, peat, carbon, or a similar material and deposit in sealed containers. Keep this chemical out of a confined space, such as a sewer, because of the possibility of an explosion, unless the sewer is designed to prevent the build-up of explosive concentrations. It may be necessary to contain and dispose of this chemical as a hazardous waste. If material or contaminated runoff enters waterways, notify downstream users of potentially contaminated waters. Contact your Department of Environmental Protection or your regional office of the federal EPA for specific recommendations. If employees are required to clean-up spills, they must be properly trained and equipped. OSHA 1910.120(q) may be applicable.

Fire Extinguishing: This chemical is a flammable liquid. Poisonous gases including chlorine are produced in fire. For small fires, use dry chemical, carbon dioxide. Do not use water. Wear self-contained (positive pressure) breathing apparatus with full protective clothing. Vapors are heavier than air and will collect in low areas. Vapors may travel long distances to ignition sources and flashback. Vapors in confined areas may explode when exposed to fire. Containers may explode in fire. Storage containers and parts of containers may rocket great distances, in many directions. If material or contaminated runoff enters waterways, notify downstream users of potentially contaminated waters. Notify local health and fire officials and pollution control agencies. From a secure, explosion-proof location, use water spray to cool exposed containers. Do not get water inside containers. If cooling streams are ineffective (venting sound increases in volume and pitch, tank discolors, or shows any signs of deforming), withdraw immediately to a secure position. If employees are expected to fight fires, they must be trained and equipped in OSHA 1910.156.

References

U.S. Environmental Protection Agency, "Chemical Profile: Dichloromethylphenylsilane," Washington, DC, Chemical Emergency Preparedness Program (November 30, 1987)

Dichloromonofluoromethane

Molecular Formula: $CHCl_2F$

Synonyms: Algofrene type 5; Dichloromonofluorometha NE; F 21; FC 21; Fluorodichloromethane; Freon F 21; Genetron 21; Halon 112; HCFC-21; Methane, dichlorofluoro-; Monofluorodichlorometha NE; R 21 (refrigerant); Refrigerant 21

CAS Registry Number: 75-43-4

RTECS Number: PA8400000

DOT ID: UN 1029

Regulatory Authority

- Air Pollutant Standard Set (ACGIH)[1] (Australia) (DFG)[3] (HSE)[33] (Israel) (former USSR)[43] (OSHA)[58] (Several States)[60] (Several Canadian Provinces)
- Canada, WHMIS, Ingredients Disclosure List

Cited in U.S. State Regulations: Alaska (G), California (A, G), Connecticut (A), Florida (G), Illinois (G), Maine (G), Massachusetts (G), Minnesota (G), Nevada (A), New Hampshire (G), New Jersey (G), North Carolina (A), North Dakota (A), Pennsylvania (G), Rhode Island (G), Virginia (A), West Virginia (G).

Description: Dichlorofluoromethane, $CHCl_2F$, is a heavy, colorless gas or liquid (below 9°C) with a slight ether-like odor. Boiling point = 9°C. Freezing/Melting point = -135°C. Autoignition temperature = 522°C. Hazard Identification (based on NFPA-704 M Rating System): Health 1, Flammability 0, Reactivity 0.

Potential Exposure: This material is used as a refrigerant and a propellant gas.

Incompatibilities: Reacts violently with chemically active metals: sodium, potassium, calcium, powdered aluminum, zinc, magnesium; alkali, alkaline earth. Reacts with acids or acid fumes producing highly toxic chlorine and fluorine fumes. Attacks some forms of plastics, rubber, and coatings.

Permissible Exposure Limits in Air: The Federal limit (OSHA PEL) is 1,000 ppm (4,200 mg/m^3) TWA. The Mexico limit is 500 ppm. The ACGIH recommended TLV, DFG MAK,[3] HSE,[33] Australian, Israeli, and Canadian provincial (Alberta, Ontario, and Quebec) TWA value is 10 ppm (40 mg/m^3) and Alberta's STEL is 20 ppm. British Columbia's TWA is the same as OSHA's. The NIOSH IDLH level is 5,000 ppm. The former USSR-UNEP/IRPTC joint project[43] has set a MAC in ambient air of residential areas of 100 mg/m^3 on a momentary basis and 10 mg/m^3 on an average daily basis. Several states have set guidelines or standards for R-21 in ambient air[60] ranging from 0.4 mg/m^3 (North Dakota) to 0.5 mg/m^3 (North Carolina) to 0.65 mg/m^3 (Virginia) to 0.8 mg/m^3 (Connecticut) to 0.952 mg/m^3 (Nevada).

Determination in Air: Charcoal tube;[2] Workup with CS_2; analysis by gas chromatography/Flame ionization detection; NIOSH IV Method #2516.

Permissible Concentration in Water: No criteria set.

Routes of Entry: Inhalation, ingestion, eye and skin contact.

Harmful Effects and Symptoms

Short Term Exposure: Contact with the liquid may cause frostbite. This chemical affects the central nervous system at high concentrations. High exposure could cause asphyxia and/or cardiac arrhythmia (irregular heartbeat). This might lead to cardiac arrest.

Long Term Exposure: May damage the developing fetus. May damage the liver. May cause irregular heartbeat.

Points of Attack: Respiratory system, lungs, cardiovascular system.

Medical Surveillance: Consider the points of attack in preplacement and periodic physical examinations. EKG, Liver function tests.

First Aid: If this chemical gets into the eyes, remove any contact lenses at once and irrigate immediately for at least 15 minutes, occasionally lifting upper and lower lids. Seek medical attention immediately. If this chemical contacts the skin, remove contaminated clothing and wash immediately with soap and water. Seek medical attention immediately. If this chemical has been inhaled, remove from exposure, begin rescue breathing (using universal precautions) if breathing has stopped and CPR if heart action has stopped. Transfer promptly to a medical facility. When this chemical has been swallowed, get medical attention. Give large quantities of water and induce vomiting. Do not make an unconscious person vomit. If frostbite has occurred, seek medical attention immediately; do NOT rub the affected areas or flush them with water. In order to prevent further tissue damage, do NOT attempt to remove frozen clothing from frostbitten areas. If frostbite has NOT occurred, immediately and thoroughly wash contaminated skin with soap and water.

Personal Protective Methods: Wear protective gloves and clothing to prevent any reasonable probability of skin contact. Safety equipment suppliers/manufacturers can provide recommendations on the most protective glove/clothing material for your operation.. All protective clothing (suits, gloves, footwear, headgear) should be clean, available each day, and put on before work. Contact lenses should not be worn when working with this chemical. Wear splash-proof chemical goggles and face shield unless full facepiece respiratory protection is worn. Employees should wash immediately with soap when skin is wet or contaminated. Provide emergency showers and eyewash. Where exposure to the liquefied compressed gas may occur, employees should be provided with special clothing designed to prevent frostbite.

Respirator Selection: NIOSH/OSHA: *100 ppm*: SA (any supplied-air respirator). *250 ppm:* SA:CF (any supplied-air respirator operated in a continuous-flow mode). *500 ppm*: SCBAF (any self-contained breathing apparatus with a full facepiece); or SAF (any supplied-air respirator with a full facepiece). *5,000 ppm:* SA:PD,PP (any supplied-air respirator operated in a pressure-demand or other positive-pressure mode). *Emergency or planned entry into unknown concentrations or IDLH conditions*: SCBAF:PD,PP (any self-contained breathing apparatus that has a full facepiece and is operated in a pressure-demand or other positive-pressure mode); or SAF:PD,PP:ASCBA (any supplied-air respirator that has a full facepiece and is operated in a pressure-demand or other positive-pressure mode in combination with an auxiliary self-contained breathing apparatus operated in a pressure-demand or other positive pressure mode). *Escape:* GMFOV [any air-purifying, full-facepiece respirator (gas mask) with a chin-style, front-or back-, mounted organic vapor canister]; or SCBAE (any appropriate escape-type, self-contained breathing apparatus).

Storage: Prior to working with dichlorofluoromethane you should be trained on its proper handling and storage. Store in tightly closed containers in a cool, well ventilated area away from chemically active metals, acids, acid fumes, alkali, alkaline earth since violent reaction occur. Procedures for the handling, use and storage of cylinders should be in compliance with OSHA 1910.101 and 1910.169 as with the recommendations of the Compressed Gas Association.

Shipping: Dichloromonofluoromethane must be labeled "Nonflammable Gas." It falls in DOT Hazard Class 2.2.

Spill Handling: Evacuate and restrict persons not wearing protective equipment from area of spill or leak until cleanup is complete. Remove all ignition sources. Ventilate area of spill or leak. Absorb liquids in vermiculite, dry sand, earth, peat, carbon, or a similar material and deposit in sealed containers. Keep this chemical out of a confined space, such as a sewer, because of the possibility of an explosion, unless the sewer is designed to prevent the build-up of explosive concentrations. It may be necessary to contain and dispose of this chemical as a hazardous waste. If material or contaminated runoff enters waterways, notify downstream users of potentially contaminated waters. Contact your Department of Environmental Protection or your regional office of the federal EPA for specific recommendations. If employees are required to clean-up spills, they must be properly trained and equipped. OSHA 1910.120(q) may be applicable.

Fire Extinguishing: This chemical is a combustible liquid. Poisonous gases including chlorine, fluorine are produced in fire. Use dry chemical, carbon dioxide, or alcohol foam extinguishers. Vapors are heavier than air and will collect in low areas. Vapors may travel long distances to ignition sources and flashback. Vapors in confined areas may explode when exposed to fire. Containers may explode in fire. Storage containers and parts of containers may rocket great distances, in many directions. If material or contaminated runoff enters waterways, notify downstream users of potentially contaminated waters. Notify local health and fire officials and pollution control agencies. From a secure, explosion-proof location, use water spray to cool exposed containers. If cooling streams are ineffective (venting sound increases in volume and pitch, tank discolors, or shows any signs of deforming), withdraw immediately to a secure position. If employees are expected to fight fires, they must be trained and equipped in OSHA 1910.156.

Disposal Method Suggested: Incineration, preferably after mixing with another combustible fuel. Care must be exercised to assure complete combustion to prevent the formation of phosgene. An acid scrubber is necessary to remove the halo acids produced. Because of recent discovery of potential ozone decomposition in the stratosphere, this material should be released to the atmosphere only as a last resort.

References

New Jersey Department of Health and Senior Services, Hazardous Substance Fact Sheet, "Dichlorofluoromethane," Trenton, NJ (January 1999)

1,1-Dichloro-1-Nitroethane

Molecular Formula: $C_2H_3Cl_2NO_2$

Common Formula: $CH_3CCl_2NO_2$

Synonyms: Dichloronitroethane; Ethide®

CAS Registry Number: 594-72-9

RTECS Number: KI1050000

DOT ID: UN 2650

Regulatory Authority

- Air Pollutant Standard Set (ACGIH)[1] (Australia) (DFG)[3] (Israel) (Mexico) (OSHA)[58] (Several States)[60] (Several Canadian Provinces)

Cited in U.S. State Regulations: Alaska (G), California (A, G), Connecticut (A), Florida (G), Illinois (G), Massachusetts (G), Minnesota (G), Nevada (A), New Hampshire (G), New Jersey (G), North Dakota (A), Pennsylvania (G), Rhode Island (G), Virginia (A), West Virginia (G).

Description: 1,1-Dichloro-1-nitroethane, $CH_3CCl_2NO_2$, is a colorless liquid with an unpleasant odor; causes tears. Boiling point = 124°C. Flash point = 76°C. Hazard Identification (based on NFPA-704 M Rating System): Health 2, Flammability 2, Reactivity 3. Insoluble in water.

Potential Exposure: This material is used as a fumigant insecticide. Therefore, those engaged in the manufacture, formulation and application of this material may be exposed.

Incompatibilities: Strong oxidizers. Corrosive to iron in the presence of moisture.

Permissible Exposure Limits in Air: The Federal Limit (OSHA PEL)[58] and the DFG MAC value[3] is a ceiling value of 10 ppm (60 mg/m^3), not to exceeded at any time. British Columbia set the same ceiling limit as OSHA. NIOSH and ACGIH recommend a TWA of 2 ppm (10 mg/m^3). Ontario and Quebec use the same TWA as NIOSH. Australia, Israel, Mexico and Alberta set a TWA of 2 ppm, and Mexico's STEL is 10 ppm. The NIOSH IDLH level is 25 ppm. Several states have set guidelines or standards for dichloronitroehtane in ambient air[60] ranging from 100 μg/m^3 (North Dakota) to 160 μg/m^3 (Virginia) to 200 μg/m^3 (Connecticut) to 238 μg/m^3 (Nevada).

Determination in Air: Charcoal tube (petroleum-based); CS_2; Gas chromatography/Flame ionization detection; NIOSH IV Method #1601.

Permissible Concentration in Water: No criteria set.

Routes of Entry: Inhalation, ingestion, eye and/or skin contact.

Harmful Effects and Symptoms

Short Term Exposure: In animals: lung, skin, eye irritation. Higher exposures can cause pulmonary edema, a medical emergency that can be delayed for several hours. This can cause death. The oral LD_{50} rat is 410 mg/kg (moderately toxic).

Long Term Exposure: This chemical causes liver, heart, kidney, and blood vessel damage in animals.

Points of Attack: Lungs.

Medical Surveillance: Consider the points of attack in preplacement and periodic physical examinations.

First Aid: If this chemical gets into the eyes, remove any contact lenses at once and irrigate immediately for at least 15 minutes, occasionally lifting upper and lower lids. Seek medical attention immediately. If this chemical contacts the skin, remove contaminated clothing and wash immediately with soap and water. Seek medical attention immediately. If this chemical has been inhaled, remove from exposure, begin rescue breathing (using universal precautions) if breathing has stopped and CPR if heart action has stopped. Transfer promptly to a medical facility. When this chemical has been swallowed, get medical attention. Give large quantities of water and induce vomiting. Do not make an unconscious person vomit. Medical observation is recommended for 24 – 48 hours after breathing overexposure, as pulmonary edema may be delayed. As first aid for pulmonary edema, a doctor or authorized paramedic may consider administering a corticosteroid spray.

Personal Protective Methods: Wear protective gloves and clothing to prevent any reasonable probability of skin contact. Safety equipment suppliers/manufacturers can provide recommendations on the most protective glove/clothing material for your operation.. All protective clothing (suits, gloves, footwear, headgear) should be clean, available each day, and put on before work. Contact lenses should not be worn when working with this chemical. Wear splash-proof chemical goggles and face shield unless full facepiece respiratory protection is worn. Employees should wash immediately with soap when skin is wet or contaminated. Provide emergency showers and eyewash.

Respirator Selection: NIOSH: *Up to 20 ppm:* SA (any supplied-air respirator). *Up to 25 ppm:* SA:CF (any supplied-air respirator operated in a continuous-flow mode); or SCBAF (any self-contained breathing apparatus with a full facepiece). SAF (any supplied-air respirator with a full facepiece). *Emergency or planned entry into unknown concentrations or IDLH conditions*: SCBAF:PD,PP (any self-contained breathing apparatus that has a full facepiece and is operated in a pressure-demand or other positive-pressure mode); or SAF:PD,PP:ASCBA (any supplied-air respirator that has a full facepiece and is operated in a pressure-demand or other positive-pressure mode in combination with an auxiliary self-contained breathing apparatus operated in a pressure-demand or other positive-pressure mode). *Escape:* GMFOV [any air-purifying, full-facepiece respirator (gas mask) with a chin-style, front-or back-mounted organic vapor canister] or SCBAE (any appropriate escape-type, self-contained breathing apparatus).

Storage: Prior to working with this chemical you should be trained on its proper handling and storage. Store in tightly closed containers in a refrigerator away from oxidizers and sources of ignition. Where possible, automatically pump liquid from drums or other storage containers to process containers.

Shipping: This material must carry a "Poison" label. It falls in DOT Hazard Class 6.1 and Packing Group II.

Spill Handling: Evacuate and restrict persons not wearing protective equipment from area of spill or leak until cleanup is complete. Remove all ignition sources. Ventilate area of spill or leak. Absorb liquids in vermiculite, dry sand, earth, peat, carbon, or a similar material and deposit in sealed containers. Keep this chemical out of a confined space, such as a sewer, because of the possibility of an explosion, unless the sewer is designed to prevent the build-up of explosive concentrations. It may be necessary to contain and dispose of this chemical as a hazardous waste. If material or contaminated runoff enters waterways, notify downstream users of potentially contaminated waters. Contact your Department of Environmental Protection or your regional office of the federal EPA for specific recommendations. If employees are required to clean-up spills, they must be properly trained and equipped. OSHA 1910.120(q) may be applicable.

Fire Extinguishing: This chemical is a flammable liquid. Poisonous gases are produced in fire. Water may be used to blanket fire since liquid is heavier than water (sp gr = 1.4). Apply water gently to the surface of the liquid.[17] Vapors are heavier than air and will collect in low areas. Vapors may travel long distances to ignition sources and flashback. Vapors in confined areas may explode when exposed to fire. Containers may explode in fire. Storage containers and parts of containers may rocket great distances, in many directions. If material or contaminated runoff enters waterways, notify downstream users of potentially contaminated waters. Notify local health and fire officials and pollution control agencies. From a secure, explosion-proof location, use water spray to cool exposed containers. If cooling streams are ineffective (venting sound increases in volume and pitch, tank discolors, or shows any signs of deforming), withdraw immediately to a secure position. If employees are expected to fight fires, they must be trained and equipped in OSHA 1910.156.

Disposal Method Suggested: Incineration (1,500°F, 0.5 second for primary combustion; 2,200°F, 1.0 second for secondary combustion). The formation of elemental chlorine can be prevented through injection of steam or methane into the combustion process. No_x may be abated through the use of thermal or catalytic devices.

2,4-Dichlorophenol

Molecular Formula: $C_6H_4Cl_2O$

Common Formula: $Cl_2C_6H_3OH$

Synonyms: 2,4-DCP; 1,3-Dichloro-4-hydroxybenzene; 4,6-Dichlorophenol; 2,4-Diclorofenol (Spanish); Phenol, 2,4-dichloro-

CAS Registry Number: 120-83-2

RTECS Number: SK8575000

DOT ID: UN 2020

EEC Number: 604-011-00-7

Regulatory Authority

- Carcinogen (Human Suspected) (IARC)[9]
- Clean Water Act: 40CFR401.15 Section 307 Toxic Pollutants; Section 313 Water Priority Chemicals (57FR41331, 9/9/92)
- EPA Hazardous Waste Number (RCRA No.): U081
- RCRA, 40CFR261, Appendix 8 Hazardous Constituents
- RCRA 40CFR268.48; 61FR15654, Universal Treatment Standards: Wastewater (mg/l), 0.044; Nonwastewater (mg/kg), 14
- RCRA 40CFR264, Appendix 9; TSD Facilities Ground Water Monitoring List. Suggested test method(s) (PQL μg/l): 8040 (5); 8270 (10)
- Superfund/EPCRA 40CFR302.4 Reportable Quantity (RQ): CERCLA, 100 lb (45.4 kg)
- EPCRA Section 313 Form R *de minimis* concentration reporting level: 1.0%
- U.S. DOT Regulated Marine Pollutant (49CFR172.101, Appendix B)
- Canada, Drinking Water, ≤0.0003 mg/l MAC, National Pollutant Release Inventory (NPRI)
- Mexico Drinking water, 0.03 mg/l

Cited in U.S. State Regulations: California (A, G), Illinois (G), Kansas (G, W), Louisiana (G), Maine (W), Maryland (G), Massachusetts (G), Michigan (A, G), New Hampshire (G), New Jersey (G), Pennsylvania (G), Vermont (G), Virginia (G), Washington (G), Wisconsin (G).

Description: 2,4-DCP, $Cl_2C_6H_3OH$, is a colorless crystalline solid with a characteristic odor. Boiling point = 210°C. Freezing/Melting point = 45.0°C. Flash point = 113°C. Hazard Identification (based on NFPA-704 M Rating System): Health 1, Flammability 1, Reactivity 0. Slightly soluble in water.

Potential Exposure: 2,4-Dichlorophenol is a commercially produced substituted phenol used entirely in the manufacture of industrial and agricultural products. As an intermediate in the chemical industry, 2,4-DCP is utilized as the feedstock for the manufacture of 2,4-dichlorophenoxyacetic acid (2,4-D), and 2,4-D derivatives (germicides, soil sterilants, etc.) and certain methyl compounds used in mothproofing, antiseptics and seed disinfectants. 2,4-DCP is also reacted with benzene sulfonyl chloride to produce miticides or further chlorinated to pentachlorophenol, a wood preservative. It is thus a widely used pesticide intermediate. The only group expected to be at risk for high exposure to 2,4-DCP is industrial workers involved in the manufacturing or handling of 2,4-DCP and 2,4-D.

Incompatibilities: Violent reaction with strong oxidizers. Contact with acids or acid fumes causes decomposition releasing poisonous chlorine gas. Incompatible with caustics, acid anhydrides, acid chlorides. Quickly corrodes aluminum; slowly corrodes zinc, tin, brass, bronze, copper and its alloys. May accumulate static electrical charges, and may cause ignition of its vapors.

Permissible Exposure Limits in Air: No occupational exposure limits have been established for 2,4-DCP. Michigan[60] has set a guideline for 2,4-dichlorophenol in ambient air of 77.0 μg/m³.

Permissible Concentration in Water: To protect freshwater aquatic life – 2,020 μg/l on an acute toxicity basis and 365 μg/l on a chronic toxicity basis. To protect saltwater aquatic life-no criteria because of insufficient data. To protect human health – 0.3 μg/l based on organoleptic effects and 3,090 μg/l based on toxicity data.[6]

Canada, Drinking Water, ≤0.0003 mg/l MAC. Mexico Drinking water, 0.03 mg/l. The former USSR[35][43] has set a MAC in water bodies used for domestic purposes of 2 μg/l. This applies to all dichlorophenols isomers. Two states have set guidelines for 2,4-dichlophenol in drinking water: Maine at 200 μg/l and Kansas at 700 μg/l.

Determination in Water: Methylene chloride extraction followed by gas chromatography with flame ionization or electron capture detection (EPA Method 604) or gas chromatography plus mass spectrometry (EPA Method 625).

Routes of Entry: Inhalation, ingestion, through the skin.

Harmful Effects and Symptoms

Although a paucity of aquatic toxicity data exists, 2,4-DCP appears to be less toxic than the higher chlorinated phenols. 2,4-DCP's toxicity to certain microorganisms and plant life has been demonstrated and its tumor promoting potential in mice has been reported. In addition, it has been demonstrated that 2,4-DCP can produce objectionable odors when present in water at extremely low levels. These findings, in conjunction with potential 2,4-DCP pollution by waste sources from commercial processes or the inadvertent production of 2,4-DCP due to chlorination of waters containing phenol, lead to the conclusion that 2,4-DCP represents a potential threat to aquatic and terrestrial life, including man, 2,4-DCP can irritate tissue and mucous membranes. The LD_{50} oral-rat is 580 mg/kg (slightly toxic).

Short Term Exposure: 2,4-DCP can be absorbed through the skin, thereby increasing exposure. Skin or eye contact can cause irritation and burns. Inhalation can cause respiratory irritation, coughing and wheezing.

Long Term Exposure: May cause liver and kidney damage. May affect the nervous system causing headache, dizziness, nausea, vomiting, weakness, and possible coma. Several other chlorophenols are carcinogenic, although 2,4-DCP has not been identified specifically as a carcinogen.

Points of Attack: Nervous system, liver and kidneys.

Medical Surveillance: Examination of the nervous system. Liver and kidney function tests.

First Aid: If this chemical gets into the eyes, remove any contact lenses at once and irrigate immediately for at least 15 minutes, occasionally lifting upper and lower lids. Seek medical attention immediately. If this chemical contacts the skin, remove contaminated clothing and wash immediately with soap and water. Speed in removing material from skin is of extreme importance. Shampoo hair promptly if contaminated. Seek medical attention immediately. If this chemical has been inhaled, remove from exposure, begin rescue breathing (using universal precautions) if breathing has stopped and CPR if heart action has stopped. Transfer promptly to a medical facility. When this chemical has been swallowed, get medical attention. Give large quantities of water and induce vomiting. Do not make an unconscious person vomit.

Personal Protective Methods: Wear protective gloves and clothing to prevent any reasonable probability of skin contact. Safety equipment suppliers/manufacturers can provide recommendations on the most protective glove/clothing material for your operation.. All protective clothing (suits, gloves, footwear, headgear) should be clean, available each day, and put on before work. Contact lenses should not be worn when working with this chemical. Wear dust-proof chemical goggles and face shield unless full facepiece respiratory protection is worn. Employees should wash immediately with soap when skin is wet or contaminated. Provide emergency showers and eyewash.

Respirator Selection: *At any detectable concentration:* SCBAF:PD,PP (any MSHA/NIOSH approved self-contained breathing apparatus that has a full facepiece and is operated in a pressure-demand or other positive-pressure mode); or SAF:PD,PP:ASCBA (any supplied-air respirator that has a full facepiece and is operated in a pressure-demand or other positive-pressure mode in combination with an auxiliary, self-contained breathing apparatus operated in a pressure-demand or other positive pressure mode). *Escape:* HiEF (any air-purifying, full-facepiece respirator (gas mask) with a chin-style, front- or back-mounted organic vapor canister having a high-efficiency particulate filter); or SCBAE (any appropriate escape-type, self-contained breathing apparatus).

Storage: Prior to working with 2,4-DCP you should be trained on its proper handling and storage. Store in tightly closed containers in a refrigerator away from oxidizers, acid, acid fumes, acid chlorides, acid anhydrides, caustics. A regulated, marked area should be established where this chemical is handled, used, or stored in compliance with OSHA standard 1910.1045.

Shipping: Chlorophenols, solid require a "Keep Away From Food" label. They fall in DOT Hazard Class 6.1 and Packing Group III.

Spill Handling: Evacuate persons not wearing protective equipment from area of spill or leak until clean-up is complete. Remove all ignition sources. Collect powdered material in the most convenient and safe manner and deposit in sealed containers. Ventilate area after clean-up is complete. It may be necessary to contain and dispose of this chemical as a hazardous waste. If material or contaminated runoff enters waterways, notify downstream users of potentially contaminated waters. Contact your Department of Environmental Protection or your regional office of the federal EPA for specific recommendations. If employees are required to clean-up spills, they must be properly trained and equipped. OSHA 1910.120(q) may be applicable.

Fire Extinguishing: This chemical is a combustible solid. Use dry chemical, carbon dioxide, water spray, or alcohol foam extinguishers. Dry chemical or CO_2 are preferred extinguishers. Water or foam may cause frothing. Poisonous gases are produced in fire including corrosive hydrogen chloride. If material or contaminated runoff enters waterways, notify downstream users of potentially contaminated waters. Notify local health and fire officials and pollution control agencies. Containers may explode in fire. From a secure, explosion-proof location, use water spray to cool exposed containers. If cooling streams are ineffective (venting sound increases in volume and pitch, tank discolors, or shows any signs of deforming), withdraw immediately to a secure position. If employees are expected to fight fires, they must be trained and equipped in OSHA 1910.156.

Disposal Method Suggested: Dissolve in a combustible solvent and incinerate in a furnace equipped with afterburner and scrubber.[22] In accordance with 40CFR165 recommendations for the disposal of pesticides and pesticide containers. Must be disposed properly by following package label directions or by contacting your state pesticide or environmental control agency or by contacting your regional EPA office. Consult with environmental regulatory agencies for guidance on acceptable disposal practices. Generators of waste containing this contaminant ($\geq$100 kg/mo) must conform with EPA regulations governing storage, transportation, treatment, and waste disposal.

References

U.S. Environmental Protection Agency, 2,4-Dichlorophenol: Ambient Water Quality Criteria, Washington, DC (1980)

U.S. Environmental Protection Agency, "Chemical Hazard Information Profile: Mono/Dichlorophenols, Washington, DC (1979)

U.S. Environmental Protection Agency, 2,4-Dichlorophenol, Health and Environmental Effects Profile No. 75, Washington, DC, Office of Solid Waste (April 30, 1980)

Sax N. I., Ed., "Dangerous Properties of Industrial Materials Report," 1, No. 7, 50–52 (1981), & 7, No. 3, 70–86 (1987)

New Jersey Department of Health and Senior Services, "Hazardous Substance Fact Sheet, 2,4-Dichlorophenol."Trenton, NJ (June 1998)

2,6-Dichlorophenol

Molecular Formula: $C_6H_4Cl_2O$

Common Formula: $Cl_2C_6H_3OH$

Synonyms: 2,6-DCP; 2,6-Dichlorofenol (Czech); 2,6-Diclorofenol (Spanish)

CAS Registry Number: 87-65-0

RTECS Number: SK8750000

DOT ID: UN 2021

Regulatory Authority

- Clean Water Act: 40CFR423, Appendix A, Priority Pollutants; Section 313 Water Priority Chemicals (57FR41331, 9/9/92)
- EPA Hazardous Waste Number (RCRA No.): U082
- RCRA, 40CFR261, Appendix 8 Hazardous Constituents
- RCRA 40CFR268.48; 61FR15654, Universal Treatment Standards: Wastewater (mg/l), 0.044; Nonwastewater (mg/kg), 14
- RCRA 40CFR264, Appendix 9; TSD Facilities Ground Water Monitoring List. Suggested test method(s) (PQL μg/l): 8270 (10)
- Superfund/EPCRA 40CFR302.4 Reportable Quantity (RQ): CERCLA, 100 lb (45.4 kg)
- EPCRA Section 313 Form R *de minimis* concentration reporting level: 1.0%
- U.S. DOT Regulated Marine Pollutant (49CFR172.101, Appendix B)
- Canada, WHMIS, Ingredients Disclosure List

Cited in U.S. State Regulations: Kansas (G, W), Louisiana (G), Massachusetts (G), New Jersey (G), Pennsylvania (G), Vermont (G), Virginia (G), Washington (G), Wisconsin (G).

Description: 2,6-Dichlorophenol, $Cl_2C_6H_3OH$, is a white crystalline solid having a strong odor similar to o-chlorophenol. Freezing/Melting point = 68 – 69°C.

Potential Exposure: 2,6-Dichlorophenol is produced as a by-product from the direct chlorination of phenol. It is used primarily as a starting material for the manufacture of trichlorophenols, tetrachlorophenols, and pentachlorophenols. It also acts as a sex pheromone for the Lone Star tick.

Incompatibilities: Incompatible with strong oxidizers, acids, acid fumes, acid anhydrides, acid chlorides.

Permissible Exposure Limits in Air: No occupational exposure limits have been established for 2,6-DCP

Permissible Concentration in Water: The former USSR[35][43] has set a MAC in water bodies used for domestic purposes o 2.0 μg/l (applies to all chlorophenol isomers). Kansas[61] has set a guideline of 0.2 μg/l of 2,6-dichlorophenol in drinking water.

Routes of Entry: Inhalation, through the skin.

Harmful Effects and Symptoms

Short Term Exposure: 2,6 DCP can be absorbed through the skin, thereby increasing exposure. Severe local irritation of eyes, skin, and mucous membranes; causes burns. Other symptoms are tremors, convulsions and respiratory inhibition. The oral LD_{50} rat is 2,940 mg/kg (slightly toxic).

Long Term Exposure: May cause liver and kidney damage. May affect the nervous system causing headache, dizziness, nausea, vomiting, weakness, and possible coma. Several other chlorophenols are carcinogenic, although 2,6-DCP has not been identified specifically as a carcinogen.

Points of Attack: Nervous system, liver and kidneys.

Medical Surveillance: Examination of the nervous system. Liver and kidney function tests.

First Aid: If this chemical gets into the eyes, remove any contact lenses at once and irrigate immediately for at least 15 minutes, occasionally lifting upper and lower lids. Seek medical attention immediately. If this chemical contacts the skin, remove contaminated clothing and wash immediately with soap and water. Seek medical attention immediately. If this chemical has been inhaled, remove from exposure, begin rescue breathing (using universal precautions) if breathing has stopped and CPR if heart action has stopped. Transfer promptly to a medical facility. When this chemical has been swallowed, get medical attention. Give large quantities of water and induce vomiting. Do not make an unconscious person vomit.

Personal Protective Methods: Wear protective gloves and clothing to prevent any reasonable probability of skin contact. Safety equipment suppliers/manufacturers can provide recommendations on the most protective glove/clothing material for your operation.. All protective clothing (suits, gloves, footwear, headgear) should be clean, available each day, and put on before work. Contact lenses should not be worn when working with this chemical. Wear dust-proof chemical goggles and face shield unless full facepiece respiratory protection is worn. Employees should wash immediately with soap when skin is wet or contaminated. Provide emergency showers and eyewash.

Respirator Selection: *At any detectable concentration:* SCBAF:PD,PP (any MSHA/NIOSH approved self-contained breathing apparatus that has a full facepiece and is operated in a pressure-demand or other positive-pressure mode); or

SAF:PD,PP:ASCBA (any supplied-air respirator that has a full facepiece and is operated in a pressure-demand or other positive-pressure mode in combination with an auxiliary, self-contained breathing apparatus operated in a pressure-demand or other positive pressure mode). *Escape:* HiEF (any air-purifying, full-facepiece respirator (gas mask) with a chin-style, front- or back-mounted organic vapor canister having a high-efficiency particulate filter); or SCBAE (any appropriate escape-type, self-contained breathing apparatus).

Storage: Prior to working with 2,6-DCP you should be trained on its proper handling and storage. Store in tightly closed containers in a refrigerator away from oxidizers and other incompatible materials listed above.

Shipping: Chlorophenols, solid require a "Keep Away From Food" label. They fall in DOT Hazard Class 6.1 and Packing Group III.

Spill Handling: Evacuate persons not wearing protective equipment from area of spill or leak until clean-up is complete. Remove all ignition sources. Collect powdered material in the most convenient and safe manner and deposit in sealed containers. Ventilate area after clean-up is complete. It may be necessary to contain and dispose of this chemical as a hazardous waste. If material or contaminated runoff enters waterways, notify downstream users of potentially contaminated waters. Contact your Department of Environmental Protection or your regional office of the federal EPA for specific recommendations. If employees are required to clean-up spills, they must be properly trained and equipped. OSHA 1910.120(q) may be applicable.

Fire Extinguishing: This chemical is a combustible solid. Use dry chemical, carbon dioxide, water spray, or alcohol foam extinguishers. Poisonous gases are produced in fire. If material or contaminated runoff enters waterways, notify downstream users of potentially contaminated waters. Notify local health and fire officials and pollution control agencies. From a secure, explosion-proof location, use water spray to cool exposed containers. If cooling streams are ineffective (venting sound increases in volume and pitch, tank discolors, or shows any signs of deforming), withdraw immediately to a secure position. If employees are expected to fight fires, they must be trained and equipped in OSHA 1910.156.

Disposal Method Suggested: Dissolve in a combustible solvent and incinerate in a furnace equipped with afterburner and scrubber.[22] In accordance with 40CFR165 recommendations for the disposal of pesticides and pesticide containers. Must be disposed properly by following package label directions or by contacting your state pesticide or environmental control agency or by contacting your regional EPA office. Consult with environmental regulatory agencies for guidance on acceptable disposal practices. Generators of waste containing this contaminant (≥100 kg/mo) must conform with EPA regulations governing storage, transportation, treatment, and waste disposal.

References

U.S. Environmental Protection Agency, 2,6-Dichlorophenol, Health and Environmental Effects Profile No. 76, Washington, DC, Office of Solid Waste (April 30, 1980)

Sax N. I., Ed., "Dangerous Properties of Industrial Materials Report," 4, No. 5, 35–38 (1984)

2-(2,4-Dichlorophenoxy) Propionic Acid

Molecular Formula: $C_9H_8C_{12}O_3$

Synonyms: Acide-2-(2,4-dichloro-phenoxy)propionique (French); Acido-2-(2,4-dicloro-fenossi)propionico (Italian); Acido 2-(2,4-diclorofenoxi)propionico (Spanish); BH 2,4-DP; Celatox-DP; Cornox RD; Cornox RK; Desormone; 2(2,4-Dichloor-fenoxy)propionzuur (Dutch); α-(2,4-Dichlorophenoxy)propionic acid; (+-)-2-(2,4-Dichlorophenoxy)propionic acid; 2-(2,4-Dichlorophenoxy)propionic acid; 2,4-Dichlorophenoxy-α-propionic acid; 2,4-Dichlorophenoxypropionic acid; Dichloroprop; 2-(2,4-Dichlor-phenoxy)-propionsaeure (German); Dichlorprop; 2-(2,4-DP); 2,4-DP (EPA); Embutox; Graminon-plus; Hedonal; Hedonal DP; Herbizid DP; Hormatox; Kildip; NSC 39624; Polyclene; Polymone; Polytox; Propanoic acid, 2-(2,4-dichlorophenoxy)-; Propionic acid, 2-(2,4-dichlorophenoxy)-; RD 406; Seritox 50; U 46; U46 DP-Fluid; Visko-Rhap; Weedone 170; Weedone DP

CAS Registry Number: 120-36-5

RTECS Number: UF1050000

DOT ID: UN 2765 (phenoxy pesticides, solid, toxic)

EEC Number: 607-045-00-0

Regulatory Authority

- EPCRA Section 313 Form R *de minimis* concentration reporting level: 0.1%
- U.S. DOT 49CFR172.101, Appendix B, Regulated marine pollutant

Cited in U.S. State Regulations: California (G), Florida (G), New Jersey (G), Pennsylvania (G).

Description: 2,4-DP is a combustible, colorless to yellowish to tan crystalline solid with a faint phenolic odor. Freezing/Melting point = 117 – 118°C. Insoluble in water. May be applied as a liquid containing a flammable carrier.

Potential Exposure: A phenoxy herbicide.

Incompatibilities: Contact with oxidizers may cause a fire and explosion hazard. The aqueous solution is a weak acid. Attacks many metals in presence of moisture.

Permissible Exposure Limits in Air: No occupational exposure limits have been established.

Routes of Entry: Inhalation and passing through the skin.

Harmful Effects and Symptoms

Short Term Exposure: Irritates the eyes, skin, and respiratory tract. Exposure can cause headache, fatigue, muscle twitching, fever, nausea, vomiting, diarrhea, stomach pain, and poor appetite, convulsions.

Long Term Exposure: There is limited evidence that related phenoxy herbicide compounds cause cancer. May cause liver damage.

Points of Attack: Liver.

Medical Surveillance: Liver function tests.

First Aid: If this chemical gets into the eyes, remove any contact lenses at once and irrigate immediately for at least 15 minutes, occasionally lifting upper and lower lids. Seek medical attention immediately. If this chemical contacts the skin, remove contaminated clothing and wash immediately with soap and water. Seek medical attention immediately. If this chemical has been inhaled, remove from exposure, begin rescue breathing (using universal precautions) if breathing has stopped and CPR if heart action has stopped. Transfer promptly to a medical facility. When this chemical has been swallowed, get medical attention. Give large quantities of water and induce vomiting. Do not make an unconscious person vomit.

Personal Protective Methods: Wear protective gloves and clothing to prevent any reasonable probability of skin contact. Safety equipment suppliers/manufacturers can provide recommendations on the most protective glove/clothing material for your operation.. All protective clothing (suits, gloves, footwear, headgear) should be clean, available each day, and put on before work. Contact lenses should not be worn when working with this chemical. Wear dust-proof chemical goggles and face shield unless full facepiece respiratory protection is worn. Employees should wash immediately with soap when skin is wet or contaminated. Provide emergency showers and eyewash.

Respirator Selection: Where the potential for exposure to this chemical, use a MSHA/NIOSH approved supplied-air respirator with a full facepiece operated in the positive pressure mode or with a full facepiece, hood, or helmet in the continuous flow mode, or use a MSHA/NIOSH approved self-contained breathing apparatus with a full facepiece operated in pressure-demand or other positive pressure mode.

Storage: Prior to working with 2,4-DP you should be trained on its proper handling and storage. Store in tightly closed containers in a cool, well ventilated area away from strong oxidizers and bases. Metal containers involving the transfer of this chemical should be grounded and bonded. Where possible, automatically transfer material from drums or other storage containers to process containers. Drums must be equipped with self-closing valves, pressure vacuum bungs, and flame arresters. Use only non-sparking tools and equipment, especially when opening and closing containers of this chemical. Sources of ignition such as smoking and open flames, are prohibited where this chemical is used, handled, or stored in a manner that could create a potential fire or explosion hazard. Wherever this chemical is used, handled, manufactured, or stored, use explosion-proof electrical equipment and fittings.

Shipping: Label "Poison," UN Hazard Class: 6.1. This chemical is a DOT regulated marine pollutant.

Spill Handling: *Solid material:* Evacuate persons not wearing protective equipment from area of spill or leak until clean-up is complete. Remove all ignition sources. Collect powdered material in the most convenient and safe manner and deposit in sealed containers. Ventilate area after clean-up is complete. It may be necessary to contain and dispose of this chemical as a hazardous waste. If material or contaminated runoff enters waterways, notify downstream users of potentially contaminated waters. Contact your Department of Environmental Protection or your regional office of the federal EPA for specific recommendations. If employees are required to clean-up spills, they must be properly trained and equipped. OSHA 1910.120(q) may be applicable.

Liquid formulations containing organic solvents: Evacuate and restrict persons not wearing protective equipment from area of spill or leak until cleanup is complete. Remove all ignition sources. Ventilate area of spill or leak. Absorb liquids in vermiculite, dry sand, earth, peat, carbon, or a similar material and deposit in sealed containers. Keep this chemical out of a confined space, such as a sewer, because of the possibility of an explosion, unless the sewer is designed to prevent the build-up of explosive concentrations. It may be necessary to contain and dispose of this chemical as a hazardous waste. If material or contaminated runoff enters waterways, notify downstream users of potentially contaminated waters. Contact your Department of Environmental Protection or your regional office of the federal EPA for specific recommendations. If employees are required to clean-up spills, they must be properly trained and equipped. OSHA 1910.120(q) may be applicable.

Fire Extinguishing: *Solid material:* This chemical is a combustible solid. Use dry chemical, carbon dioxide, water spray, or alcohol foam extinguishers. Poisonous gases are produced in fire including phosgene, hydrogen chloride and carbon monoxide. If material or contaminated runoff enters waterways, notify downstream users of potentially contaminated waters. Notify local health and fire officials and pollution control agencies. From a secure, explosion-proof location, use water spray to cool exposed containers. If cooling streams are ineffective (venting sound increases in volume and pitch, tank discolors, or shows any signs of deforming), withdraw immediately to a secure position. If employees are expected to fight fires, they must be trained and equipped in OSHA 1910.156.

Liquid formulations containing organic solvents: This chemical is a combustible liquid. Poisonous gases including phosgene, hydrogen chloride and carbon monoxide are produced in fire. Use dry chemical, carbon dioxide, or alcohol foam extinguishers. Vapors are heavier than air and will collect in low areas. Vapors may travel long distances to ignition sources and flashback. Vapors in confined areas may explode when exposed to fire. Containers may explode in fire. Storage containers and parts of containers may rocket great distances, in many directions. If material or contaminated runoff enters waterways, notify downstream users of potentially contaminated waters. Notify local health and fire officials and pollution control agencies. From a secure, explosion-proof location, use water spray to cool exposed containers. If cooling streams are ineffective (venting sound increases in volume and pitch, tank discolors, or shows any signs of deforming), withdraw immediately to a secure position. If employees are expected to fight fires, they must be trained and equipped in OSHA 1910.156.

Disposal Method Suggested: In accordance with 40CFR165 recommendations for the disposal of pesticides and pesticide containers. Must be disposed properly by following package label directions or by contacting your state pesticide or environmental control agency or by contacting your regional EPA office.

References

New Jersey Department of Health and Senior Services, "Hazardous Substance Fact Sheet, 2-(2,4-Dichlorophenoxy) Propionic Acid," Trenton NJ (February 1999).

Dichlorophenyl Isocyanates

Molecular Formula: $C_7H_3C_{12}NO$ (102-36-3)

Synonyms: *102-36-3:* Benzene, 1,2-dichloro-4-isocyanato-; 3,4-Dichlorfenylisokyanat; 1,2-Dichloro-4-phenyl isocyanate; 3,4-Dichlorophenyl isocyanate; Dichlorophenyl isocyanate

other dichlorophenyl isocyanates: Benzene, 2,4-dichloro-1-isocyanato-; 1,2-Dichloro-3-phenyl isocyanate; 1,3-Dichloro-2-phenyl isocyanate; 1,4-Dichloro-2-phenyl isocyanate; 2,4-Dichloro-1-phenyl isocyanate; Dichlorophenyl isocyanate

CAS Registry Number: *General group:* 102-36-3; 41195-90-8; 39920-37-1; 5392-82-5; 2612-57-9; 34893-92-0

RTECS Number: NQ8755000 (General group); NQ8760000 (102-36-3)

DOT ID: UN 2250

Regulatory Authority

- Air Pollutant Standard Set (former USSR)[43]
- Superfund/EPCRA 40CFR355, Appendix B Extremely Hazardous Substances: TPQ = 500/10,000 lb (227/4,540 kg) (102-36-3)
- Superfund/EPCRA 40CFR302.4 Reportable Quantity (RQ): EHS, 1 lb (0.454 kg) (102-36-3)
- Canada, WHMIS, Ingredients Disclosure List (all isomers)

Cited in U.S. State Regulations: California (G), Massachusetts (G), New Hampshire (G), New Jersey (G).

Description: The dichlorophenyl isocyanates are combustible, crystalline (sugar or sand-like) solids. In general, they are white to yellow in color, but the 1,4-dichloro-2-phenyl isomer is white to light green. Their flash points are generally over 113°C but that of the 1,3-dichloro-2-phenyl isomer is reported as 77°C. These chemicals are insoluble in water, and some may be reactive. *1,2-dichloro-4-isomer 102-36-3:* is the isomer of regulatory focus: Freezing/Melting point = 42 – 43°C. Boiling point = 112° @ 12 mm. Reactive in water.

Potential Exposure: Those materials used as chemical intermediates.

Incompatibilities: Water (forms carbon dioxide), alcohols and glycols, acids, ammonia, strong bases, carboxylic acids, amines, caprolactum solutions, and metals.

Permissible Exposure Limits in Air: The former USSR-UNEP/IRPTC project[43] has set a MAC in workplace air of 0.3 mg/m^3 for the 3,4-dichlorophenyl isomer.

Determination in Air: Impinger; Reagent; High-pressure liquid chromatography/Fluorescence/Electrochemical detection; NIOSH IV Method #5522, Isocyanates. See also Method #5521.

Permissible Concentration in Water: No criteria set. Reacts with water forming carbon dioxide.

Harmful Effects and Symptoms

Short Term Exposure: Exposure can irritate the eyes, nose, throat, air passages and lungs causing coughing, shortness of breath and tightness in the chest.

Long Term Exposure: Dichlorophenyl isocyanates may cause an asthma-like allergy. Future exposures can cause asthma attacks with shortness of breath, wheezing, cough, and/or chest tightness.

Points of Attack: Lungs.

Medical Surveillance: Before beginning employment and at regular times after that, for those with frequent or potentially high exposures, the following are recommended: Lung function tests. These may be normal if the person is not having an attack at the time of the test. If symptoms develop or overexposure is suspected, the following may be useful: Evaluation by a qualified allergist, including careful exposure history and special testing, may help diagnose allergy.

First Aid: If this chemical gets into the eyes, remove any contact lenses at once and irrigate immediately for at least 15 minutes, occasionally lifting upper and lower lids. Seek medical attention immediately. If this chemical contacts the skin, remove contaminated clothing and wash immediately with soap and water. Seek medical attention immediately. If this chemical has been inhaled, remove from exposure, begin

rescue breathing (using universal precautions) if breathing has stopped and CPR if heart action has stopped. Transfer promptly to a medical facility. When this chemical has been swallowed, get medical attention. Give large quantities of water and induce vomiting. Do not make an unconscious person vomit.

Personal Protective Methods: Wear protective gloves and clothing to prevent any reasonable probability of skin contact. Safety equipment suppliers/manufacturers can provide recommendations on the most protective glove/clothing material for your operation.. All protective clothing (suits, gloves, footwear, headgear) should be clean, available each day, and put on before work. Contact lenses should not be worn when working with this chemical. Wear dust-proof chemical goggles and face shield unless full facepiece respiratory protection is worn. Employees should wash immediately with soap when skin is wet or contaminated. Provide emergency showers and eyewash.

Respirator Selection: Where the potential for exposure to dichlorophenyl isocyanates exists, use a MSHA/NIOSH approved supplied-air respirator with a full facepiece operated in the positive pressure mode or with a full facepiece, hood, or helmet in the continuous flow mode, or use a MSHA/NIOSH approved self-contained breathing apparatus with a full facepiece operated in pressure-demand or other positive pressure mode.

Storage: Prior to working with these chemicals you should be trained on its proper handling and storage. Store to avoid contact with alcohols, strong bases (such as potassium hydroxide and sodium hydroxide), carboxylic acids, amines, and metals since violent reactions occur. Store in tightly closed containers in a dry, cool, well-ventilated area away from moisture and temperatures above 40°C.

Shipping: Dichlorophenyl isocyanates must carry a "Poison" label. They fall in DOT Hazard Class 6.1 and Packing Group II.

Spill Handling: Evacuate persons not wearing protective equipment from area of spill or leak until clean-up is complete. Remove all ignition sources. Collect powdered material in the most convenient and safe manner and deposit in sealed containers. Ventilate area after clean-up is complete. It may be necessary to contain and dispose of this chemical as a hazardous waste. If material or contaminated runoff enters waterways, notify downstream users of potentially contaminated waters. Contact your Department of Environmental Protection or your regional office of the federal EPA for specific recommendations. If employees are required to clean-up spills, they must be properly trained and equipped. OSHA 1910.120(q) may be applicable.

Fire Extinguishing: These chemicals are combustible solids. Use dry chemical, carbon dioxide, or water spray extinguishers. Poisonous gases are produced in fire, including hydrogen cyanide, oxides of nitrogen and hydrogen chloride gas. If material or contaminated runoff enters waterways, notify downstream users of potentially contaminated waters. Notify local health and fire officials and pollution control agencies. Containers may explode in fire. From a secure, explosion-proof location, use water spray to cool exposed containers. If cooling streams are ineffective (venting sound increases in volume and pitch, tank discolors, or shows any signs of deforming), withdraw immediately to a secure position. If employees are expected to fight fires, they must be trained and equipped in OSHA 1910.156.

Disposal Method Suggested: Combustion in an incinerator equipped with afterburner and fume scrubber.

References

New Jersey Department of Health, "Hazardous Substance Fact Sheets: 1,2-Dichloro-4-Phenyl Isocyanate," Trenton, NJ (April 1986)

New Jersey Department of Health, "Hazardous Substance Fact Sheets: 1,3-Dichloro-2-Phenyl Isocyanate," Trenton, NJ (April 1986)

New Jersey Department of Health, "Hazardous Substance Fact Sheets: 1,4-Dichloro-2-Phenyl Isocyanate," Trenton, NJ (April 1986)

New Jersey Department of Health, "Hazardous Substance Fact Sheets: 2,4-Dichloro-1-Phenyl Isocyanate," Trenton, NJ (April 1986)

New Jersey Department of Health, "Hazardous Substance Fact Sheets: 1,2-Dichloro-3-Phenyl Isocyanate," Trenton, NJ (April 1986)

Dichlorophenyl Trichlorosilane

Molecular Formula: $C_6H_3Cl_5Si$

Common Formula: $Cl_2C_6H_3SiCl_3$

Synonyms: Silane, trichloro(dichlorophenyl)-; Trichloro(dichlorophenyl)silane

CAS Registry Number: 27137-85-5

RTECS Number: VV3540000

DOT ID: UN 1766

Regulatory Authority

- Air Pollutant Standard Set (former USSR)[43]

Cited in U.S. State Regulations: Maine (G), Massachusetts (G), New Hampshire (G), New Jersey (G).

Description: Dichlorophenyl trichlorosilane, $Cl_2C_6H_3SiCl_3$, is a straw-colored liquid. Flash point = 141°C. Decomposes in water.

Potential Exposure: This material is used in silicone polymer manufacture.

Incompatibilities: Water, combustible materials.

Permissible Exposure Limits in Air: The former USSR-UNEP/IRPTC project[43] has set a MAC in workplace air of 1.0 mg/m^3.

Permissible Concentration in Water: No criteria set.

Routes of Entry: Inhalation.

Harmful Effects and Symptoms

Short Term Exposure: Exposure can irritate the lungs causing coughing and/or shortness of breath. Higher exposures can cause a build-up of fluid in the lungs (pulmonary edema). This can cause death. This substance is a corrosive chemical and contact can cause severe skin and eye burns. Exposure can irritate the eyes, nose and throat.

Long Term Exposure: Repeated exposure may affect the lungs.

Points of Attack: Lungs.

Medical Surveillance: Before beginning employment and at regular times after that, for those with frequent or potentially high exposures, the following are recommended: Lung function tests. If symptoms develop or overexposure is suspected, the following may be useful: Consider chest x-ray following acute overexposure.

First Aid: If this chemical gets into the eyes, remove any contact lenses at once and irrigate immediately for at least 15 minutes, occasionally lifting upper and lower lids. Seek medical attention immediately. If this chemical contacts the skin, remove contaminated clothing and wash immediately with soap and water. Seek medical attention immediately. If this chemical has been inhaled, remove from exposure, begin rescue breathing (using universal precautions) if breathing has stopped and CPR if heart action has stopped. Transfer promptly to a medical facility. When this chemical has been swallowed, get medical attention. If victim is conscious, administer water or milk. Do not induce vomiting. Medical observation is recommended for 24 – 48 hours after breathing overexposure, as pulmonary edema may be delayed. As first aid for pulmonary edema, a doctor or authorized paramedic may consider administering a corticosteroid spray.

Personal Protective Methods: Wear protective gloves and clothing to prevent any reasonable probability of skin contact. Safety equipment suppliers/manufacturers can provide recommendations on the most protective glove/clothing material for your operation.. All protective clothing (suits, gloves, footwear, headgear) should be clean, available each day, and put on before work. Contact lenses should not be worn when working with this chemical. Wear splash-proof chemical goggles and face shield unless full facepiece respiratory protection is worn. Employees should wash immediately with soap when skin is wet or contaminated. Provide emergency showers and eyewash.

Respirator Selection: Where the potential for exposures to dichlorophenyl Trichlorosilane exists, use a MSHA/NIOSH approved supplied-air respirator with a full facepiece operated in the positive pressure mode or with a full facepiece, hood, or helmet in the continuous flow mode, or use a MSHA/NIOSH approved self-contained breathing apparatus with a full facepiece operated in pressure demand or other positive pressure mode.

Storage: Prior to working with this chemical you should be trained on its proper handling and storage. Dichlorophenyl trichlorosilane should be stored to avoid contact with moisture or with combustible materials such as wood, paper and oil.

Shipping: This material must carry a "Corrosive" label. It falls in DOT Hazard Class 8 and Packing Group II.

Spill Handling: Evacuate and restrict persons not wearing protective equipment from area of spill or leak until cleanup is complete. Remove all ignition sources. Ventilate area of spill or leak. Absorb liquids in vermiculite, dry sand, earth, peat, carbon, or a similar material and deposit in sealed containers. Keep this chemical out of a confined space, such as a sewer, because of the possibility of an explosion, unless the sewer is designed to prevent the build-up of explosive concentrations. It may be necessary to contain and dispose of this chemical as a hazardous waste. If material or contaminated runoff enters waterways, notify downstream users of potentially contaminated waters. Contact your Department of Environmental Protection or your regional office of the federal EPA for specific recommendations. If employees are required to clean-up spills, they must be properly trained and equipped. OSHA 1910.120(q) may be applicable.

Fire Extinguishing: Dichlorophenyl Trichlorosilane may burn, but does not readily ignite. Poisonous gas is produced in fire. Use dry chemical, CO_2, or foam extinguishers. Vapors are heavier than air and will collect in low areas. Vapors may travel long distances to ignition sources and flashback. Storage containers and parts of containers may rocket great distances, in many directions. If material or contaminated runoff enters waterways, notify downstream users of potentially contaminated waters. Notify local health and fire officials and pollution control agencies. From a secure, explosion-proof location, use water spray to cool exposed containers. If cooling streams are ineffective (venting sound increases in volume and pitch, tank discolors, or shows any signs of deforming), withdraw immediately to a secure position. If employees are expected to fight fires, they must be trained and equipped in OSHA 1910.156.

References

New Jersey Department of Health, "Hazardous Substance Fact Sheet: Dichloro Phenyl Trichlorosilane," Trenton, NJ (September 1987)

1,2-Dichloropropane

Molecular Formula: $C_3H_6Cl_2$

Common Formula: $ClCH_2CHClCH_3$

Synonyms: Bichlorure de propylene (French); α,β-Dichloropropane; 1,2-Dicloroprpano (Spanish); Dwuchloropropan (Polish); ENT 15,406; NCI-C55141; Propane, 1,2-dichloro-; Propylene chloride; α,β-Propylene dichloride; Propylene dichloride

CAS Registry Number: 78-87-5

RTECS Number: TX9625000

DOT ID: UN 1279

EEC Number: 602-020-00-0

EINECS Number: 201-152-2

Regulatory Authority

- Carcinogen (Animal Suspected) (IARC)[9]
- Air Pollutant Standard Set (ACGIH)[1] (Australia) (DFG)[3] (Israel) (Mexico) (former USSR)[35][43] (OSHA)[58] (Several States)[60] (Several Canadian Provinces)
- Clean Air Act: Hazardous Air Pollutants (Title I, Part A, Section 112)
- Clean Water Act: Section 311 Hazardous Substances/ RQ 40CFR117.3 (same as CERCLA, see below); 40CFR423, Appendix A, Priority Pollutants; Section 313 Water Priority Chemicals (57FR41331, 9/9/92)
- EPA Hazardous Waste Number (RCRA No.): U083
- RCRA, 40CFR261, Appendix 8 Hazardous Constituents
- RCRA 40CFR268.48; 61FR15654, Universal Treatment Standards: Wastewater (mg/l), 0.85; Nonwastewater (mg/kg), 18
- RCRA 40CFR264, Appendix 9; TSD Facilities Ground Water Monitoring List. Suggested test method(s) (PQL μg/l): 8010 (0.5); 8240 (5)
- Safe Drinking Water Act: MCL, 0.005 mg/l; MCLG, zero; Regulated chemical (47 FR 9352); Priority List (55 FR 1470)
- Superfund/EPCRA 40CFR302.4 Reportable Quantity (RQ): CERCLA, 1,000 lb (454 kg)
- EPCRA Section 313 Form R *de minimis* concentration reporting level: 1.0%
- U.S. DOT Regulated Marine Pollutant (49CFR172.101, Appendix B)
- Canada, WHMIS, Ingredients Disclosure List, National Pollution Release Inventory (NPRI)

Cited in U.S. State Regulations: Alaska (G), Arizona (W), California (A, W), Connecticut (A, W), Florida (G), Illinois (G), Kansas (G, A, W), Louisiana (G), Maine (G), Maryland (G), Massachusetts (G, A, W), Minnesota (W), Nevada (A), New Jersey (G), North Dakota (A), Oklahoma (G), Pennsylvania (G), Rhode Island (G), Vermont (G), Virginia (G, A), Washington (G), West Virginia (G), Wisconsin (G).

Description: Dichloropropane, $CH_2CHClCH_3$, is a colorless stable liquid with an odor similar to chloroform. The odor threshold in air is 0.25 ppm. Boiling point = 96°C. Freezing/ Melting point = -100°C. Flash point = 16°C (cc). Autoignition temperature = 557°C. Explosive limits: LEL= 3.4%; UEL = 14.5%. Hazard Identification (based on NFPA-704 M Rating System): Health 2, Flammability 3, Reactivity 0. Very slightly soluble in water.

Potential Exposure: Dichloropropane is used as a chemical intermediate in perchloroethylene and carbon tetrachloride synthesis and as a lead scavenger for antiknock fluids. It is also used as a solvent for fats, oils, waxes, gums and resins, and in solvent mixtures for cellulose esters and ethers. Other applications include the use of dichloropropane as a fumigant, alone and in combination with dichloropropane, as a scouring compound, and a metal degreasing agent. It is also used as an insecticidal fumigant.

Incompatibilities: Forms explosive mixture with air. May accumulate static electrical charges, and may cause ignition of its vapors. Contact with strong oxidizers, powdered aluminum may cause fire and explosion hazard. Strong acids can cause decomposition and the formation of hydrogen chloride vapors. Reacts with strong bases, o-dichlorobenzene, 1,2-dichloroethane. Corrosive to aluminum and its alloys. Attacks some plastics, rubber and coatings.

Permissible Exposure Limits in Air: The Federal Limit (OSHA PEL),[58] DFG MAK and the ACGIH TWA recommended value is 75 ppm (350 mg/m^3) TWA and the ACGIH STEL is 110 ppm (510 mg/m^3). Australia, Israel, Mexico, and the Canadian provinces of Alberta, BC, Ontario, and Quebec use the same values as ACGIH. The NIOSH IDLH level = (Ca) 400 ppm. Brazil[35] has set a workplace limit of 59 ppm (275 mg/m^3). Russia[35][43] has set a MAC in workplace air of 10 mg/m^3 and a limit in ambient air in residential areas of 0.18 mg/m^3 on a daily average basis.

Several states have set guidelines or standards for propylene dichloride in ambient air[60] ranging from 5.1 μg/m^3 (Massachusetts) to 13.89 μg/m^3 (Kansas) to 3,500 – 5,100 μg/m^3 (North Dakota) to 5,800 μg/m^3 (Virginia) to 7,000 μg/m^3 (Connecticut) to 8,330 μg/m^3 (Nevada).

Determination in Air: Charcoal tube (petroleum-based); Acetone/Cyclohexane; Gas chromatography/Electrochemical detection; NIOSH IV Method #1013, 1,2-Dichloropropane.

Permissible Concentration in Water: To protect freshwater aquatic life – 23,000 μg/l on an acute toxicity basis and 5,700 μg/l on a chronic basis. To protect saltwater aquatic life – 10,300 μg/l on an acute toxicity basis and 3,040 μg/l on a chronic basis. To protect human health – no value set because of insufficient data.[6] EPA[62] has recently proposed a limit in drinking water of 0.005 mg/l (5 μg/l).

Several states have set guidelines for propylene dichloride in drinking water[61] ranging from 1 μg/l (Arizona, Massachusetts) to 6 μg/l (Kansas and Minnesota) to 10 μg/l (California and Connecticut). The former USSR[35][43] has set a MAC in water bodies used for domestic purposes of 0.4 mg/l. The USEPA[47] has derived a no-observed adverse effects level (NOAEL) of 8.8 mg/kg/day which gives a 10-day health advisory for a 10-kg child of 0.09 mg/l.

Determination in Water: Inert gas purge followed by gas chromatography with halide specific detection (EPA Method 601) or gas chromatography plus mass spectrometry (EPA Method 624).

Routes of Entry: Inhalation of vapor, ingestion, eye and skin contact.

Harmful Effects and Symptoms

Short Term Exposure: 1,2-Dichloropropane irritates the eyes, skin, and respiratory tract. It may affect the nervous system. High concentration exposure can result in lightheadedness, dizziness, and unconsciousness. Propylene dichloride may cause dermatitis and defatting of the skin. More severe irritation may occur if it is confined against the skin by clothing. Undiluted, it is moderately irritating to the eyes, but does not cause permanent injury. The vapor can irritate the nose, throat, eyes and air passages. Repeated or prolonged skin contact can cause rash.

Long Term Exposure: Repeated exposure can cause skin drying and dermatitis. There is limited evidence that this chemical causes cancer in animals. It may cause liver cancer. In animal experiments, acute exposure to propylene dichloride produced central nervous system narcosis, fatty degeneration of the liver and kidneys. High or repeated exposure can damage the liver, kidneys and brain, Early symptoms include headaches, nausea, personality changes. Based on animal tests this chemical may affect reproduction and may cause malformations in the human fetus.

Points of Attack: Eyes, skin, respiratory system, liver, kidneys, central nervous system. *Cancer Site:* (in animals) liver & mammary gland tumors.

Medical Surveillance: Liver and kidney function tests. Examination of the nervous system. Evaluate the skin condition.

First Aid: If this chemical gets into the eyes, remove any contact lenses at once and irrigate immediately for at least 15 minutes, occasionally lifting upper and lower lids. Seek medical attention immediately. If this chemical contacts the skin, remove contaminated clothing and wash immediately with soap and water. Seek medical attention immediately. If this chemical has been inhaled, remove from exposure, begin rescue breathing (using universal precautions) if breathing has stopped and CPR if heart action has stopped. Transfer promptly to a medical facility. When this chemical has been swallowed, get medical attention. Give large quantities of water and induce vomiting. Do not make an unconscious person vomit.

Personal Protective Methods: Wear protective gloves and clothing to prevent any reasonable probability of skin contact. Safety equipment suppliers/manufacturers can provide recommendations on the most protective glove/clothing material for your operation. Teflon is among the recommended protective materials for all dichloropropane isomers. All protective clothing (suits, gloves, footwear, headgear) should be clean, available each day, and put on before work. Contact lenses should not be worn when working with this chemical. Wear splash-proof chemical goggles and face shield unless full facepiece respiratory protection is worn. Employees should wash immediately with soap when skin is wet or contaminated. Provide emergency showers and eyewash.

Respirator Selection: NIOSH: At *any concentrations above the NIOSH REL, or where there is no REL, at any detectable concentration:* SCBAF:PD,PP (any self-contained breathing apparatus that has a full facepiece and is operated in a pressure-demand or other positive-pressure mode); or SAF:PD,PP:ASCBA (any supplied-air respirator that has a full facepiece and is operated in a pressure-demand or other positive-pressure mode in combination with an auxiliary self-contained breathing apparatus operated in a pressure-demand or other positive pressure mode). *Escape:* GMFOV [any air-purifying, full-facepiece respirator (gas mask) with a chin-style, front- or back-, mounted organic vapor canister]; or SCBAE (any appropriate escape-type, self-contained breathing apparatus).

Storage: Prior to working with 1,2-dichlropropane you should be trained on its proper handling and storage. Before entering confined space where this chemical may be present, check to make sure that an explosive concentration does not exist. 1,2-dichlropropane must be stored to avoid contact with aluminum since violent reactions occur. Store in tightly closed containers in a cool, well-ventilated area away from strong oxidizers (such as chlorine, bromine and fluorine), strong acids (such as hydrochloric, sulfuric and nitric), o-dichlorobenzene and 1,2-dichlroethane. Sources of ignition, such as smoking and open flames, are prohibited where 1,2-dichlropropane, is handled, used, or stored. Metal containers involving the transfer of 5 gallons or more of 1,2-dichlropropane should be grounded and bonded. Drums must be equipped with self-closing valves, pressure vacuum bungs, and flame arresters. Use only non-sparking tools and equipment, especially when opening and closing containers of 1,2-dichloropropane. Wherever 1,2-dichloropropane is used, handled, manufactured, or stored, use explosion-proof electrical equipment and fittings. A regulated, marked area should be established where this chemical is handled, used, or stored in compliance with OSHA standard 1910.1045.

Shipping: Propylene dichloride must be labeled: "Flammable Liquid." It falls in DOT Hazard Class 3 and Packing Group II.

Spill Handling: Evacuate and restrict persons not wearing protective equipment from area of spill or leak until cleanup is complete. Remove all ignition sources. Stay upwind; keep out of low areas. Establish forced ventilation to keep levels below explosive limit. Absorb liquids in vermiculite, dry sand, earth, peat, carbon, or a similar material and deposit in sealed containers. Keep this chemical out of a confined space, such as a sewer, because of the possibility of an explosion, unless the sewer is designed to prevent the build-up of explosive concentrations. It may be necessary to contain and dispose of this chemical as a hazardous waste. If material or contaminated runoff enters waterways, notify downstream users of potentially contaminated waters. Contact your Department of Environmental Protection or your regional office of the federal EPA for specific recommendations. If employees are required to clean-up spills, they must be properly trained and equipped. OSHA 1910.120(q) may be applicable.

Fire Extinguishing: This chemical is a flammable liquid. Poisonous gases including chlorine are produced in fire. Use dry chemical, CO_2, or foam extinguishers. Water may be ineffective, except to blanket fire. Vapors are heavier than air and will collect in low areas. Vapors may travel long distances to ignition sources and flashback. Vapors in confined areas may explode when exposed to fire. Containers may explode in fire. Storage containers and parts of containers may rocket great distances, in many directions. If material or contaminated runoff enters waterways, notify downstream users of potentially contaminated waters. Notify local health and fire officials and pollution control agencies. From a secure, explosion-proof location, use water spray to cool exposed containers. If cooling streams are ineffective (venting sound increases in volume and pitch, tank discolors, or shows any signs of deforming), withdraw immediately to a secure position. If employees are expected to fight fires, they must be trained and equipped in OSHA 1910.156.

Disposal Method Suggested: Incineration, preferable after mixing with another combustible fuel. Care must be exercised to assure complete combustion to prevent the formation of phosgene. An acid scrubber is necessary to remove the halo acids produced.[22] Consult with environmental regulatory agencies for guidance on acceptable disposal practices. Generators of waste containing this contaminant (≥100 kg/mo) must conform with EPA regulations governing storage, transportation, treatment, and waste disposal.

References

U.S. Environmental Protection Agency, Dichloropropanes/ Dichloropopenes: Ambient Water Quality Criteria, Washington, DC (1980)

National Institute for Occupational Safety and Health, Profiles on Occupational Hazards for Criteria Document Priorities: Dichloropropane, pp 292–294, Report PB-274, -73, Cincinnati, OH (1977)

U.S. Environmental Protection Agency, 1,2-Dichloropopane, Health and Environmental Effects Profile No. 78, Washington, DC, Office of Solid Waste (April 30, 1980)

U.S. Environmental Protection Agency, Dichloropropanes/ Dichloropropenes, Health and Environmental Effects Profile No. 79, Washington, DC, Office of Solid Waste (April, 30, 1980)

New Jersey Department of Health, "Hazardous Substance Fact Sheet: 1,2-Dichloropropane," Trenton, NJ (May 1986)

Dichloropropanols

Molecular Formula: $C_3H_6Cl_2O$

Common Formula: $C_3H_6OCl_2$

Synonyms: *96-23-1:* α-Dichlorohydrin; Dichlorohydrin; *sym*-Dichloroisopropyl alcohol; 1,3-Dichloropropanol-2; *sym*-Glycerol dichlorohydrin; Glycerol α,β-dichlorohydrin; U 25,354

616-23-9: 1,2-Dichloro-3-propanol; 1,2-Dichloropropanol-3; 1,3-Dichloro-2-propanol; 2,3-Dichloro-1-propanol; 2,3-Dichloropropanol; Glycerol-α,β-dichlorohydrin

CAS Registry Number: 96-23-1; 616-23-9; 26545-73-3 (dichloropropanols)

RTECS Number: UB1400000 (96-23-1); UB1225000 (616-23-9)

DOT ID: UN 2750 (1,3-Dichloro-2-propanol)

Regulatory Authority

- Carcinogen (Animal Positive) (DFG) (96-23-1)
- Hazardous Constituent Waste (EPA-RCRA), as dichloropropanols
- Canada, WHMIS, Ingredients Disclosure List

Cited in U.S. State Regulations: Florida (G), Kansas (G), Louisiana (G), Massachusetts (G), New Hampshire (G), New Jersey (G), Pennsylvania (G), Vermont (G), Virginia (G), Washington (G), Wisconsin (G).

Description: There are 4 isomers of dichloropropanols 1,3-dichloro-2-propanol (96-23-1) and "dichloropropanols" (26545-73-3) are citations in environmental regulations: $C_3H_6OCl_2$ is a colorless viscous liquid with a chloroform-like odor. Slightly soluble in water. 1,3-Dichloro-2-propanol, BP 174°C. Flash point = 74°C. Freezing/Melting point = -4°C. Hazard Identification (based on NFPA-704 M Rating System): Health 2, Flammability 2, Reactivity 0. Soluble in water. 2,3-Dichloro-1-propanol, Boiling point =182°C. 3,3-Dichloro-1-propanol, Boiling point = 82 – 83°C. 1,1-Dichloro-2-propanol, Boiling point = 146 – 148°C.

Potential Exposure: It is used as a solvent for hard resins and nitrocellulose, in the manufacture of photographic chemicals and lacquer, as a cement for Celluloid, and as a binder of water colors. It occurs in effluents from glycerol and halohydrin production plants.

Incompatibilities: Oxidizers.

Permissible Exposure Limits in Air: No standards set.

Permissible Concentration in Water: No criteria set.

Harmful Effects and Symptoms

Short Term Exposure: *2,3-dichloro-1-propanol:*[52] Irritation of the eyes, skin, and mucous membranes; dyspnea, coughing, nausea, vomiting, diarrhea, abdominal pain, gastrointestinal hemorrhage, toxic hepatitis, jaundice, hemolytic anemia; decreased urinary output from nephritis and renal failure; somnolence, cerebral hemorrhage, central nervous system depression and coma. High exposures can cause pulmonary edema, a medical emergency that can be delayed for several hours. This can cause death.

1,3-dichloro-2-propanol:[52] Irritation of skin, eyes, and respiratory system; inhalation may cause headache, vertigo, nausea, vomiting; may also cause coma and liver damage. High exposures can cause pulmonary edema, a medical emergency that can be delayed for several hours. This can cause death.

Long Term Exposure: Causes cancer in animals; may be a potential human carcinogen. May cause liver, kidney, and /or lung damage.

Points of Attack: Central nervous system, liver, kidneys, lungs.

Medical Surveillance: Kidney function tests. Liver function tests. Lung function tests. Consider chest x-ray following acute overexposure.

First Aid: If this chemical gets into the eyes, remove any contact lenses at once and irrigate immediately for at least 15 minutes, occasionally lifting upper and lower lids. Seek medical attention immediately. If this chemical contacts the skin, remove contaminated clothing and wash immediately with soap and water. Seek medical attention immediately. If this chemical has been inhaled, remove from exposure, begin rescue breathing (using universal precautions) if breathing has stopped and CPR if heart action has stopped. Transfer promptly to a medical facility. When this chemical has been swallowed, get medical attention. Give large quantities of water and induce vomiting. Do not make an unconscious person vomit. Medical observation is recommended for 24 – 48 hours after breathing overexposure, as pulmonary edema may be delayed. As first aid for pulmonary edema, a doctor or authorized paramedic may consider administering a corticosteroid spray.

Personal Protective Methods: Wear protective gloves and clothing to prevent any reasonable probability of skin contact. Safety equipment suppliers/manufacturers can provide recommendations on the most protective glove/clothing material for your operation.. All protective clothing (suits, gloves, footwear, headgear) should be clean, available each day, and put on before work. Contact lenses should not be worn when working with this chemical. Wear splash-proof chemical goggles and face shield unless full facepiece respiratory protection is worn. Employees should wash immediately with soap when skin is wet or contaminated. Provide emergency showers and eyewash.

Respirator Selection: *At any detectable concentration:* SCBAF:PD,PP (any MSHA/NIOSH approved self-contained breathing apparatus that has a full facepiece and is operated in a pressure-demand or other positive-pressure mode); or SAF:PD,PP:ASCBA (any supplied-air respirator that has a full facepiece and is operated in a pressure-demand or other positive-pressure mode in combination with an auxiliary, self-contained breathing apparatus operated in a pressure-demand or other positive pressure mode). *Escape:* HiEF (any air-purifying, full-facepiece respirator (gas mask) with a chin-style, front- or back-mounted organic vapor canister having a high-efficiency particulate filter); or SCBAE (any appropriate escape-type, self-contained breathing apparatus).

Storage: Prior to working with 1,3-dichloro-2-propanol you should be trained on its proper handling and storage. Before entering confined space where this chemical may be present, check to make sure that an explosive concentration does not exist. Store in tightly closed containers in a refrigerator under inert atmosphere away from oxidizers. Metal containers involving the transfer of this chemical should be grounded and bonded. Where possible, automatically pump liquid from drums or other storage containers to process containers. Drums must be equipped with self-closing valves, pressure vacuum bungs, and flame arresters. Use only non-sparking tools and equipment, especially when opening and closing containers of this chemical. Sources of ignition such as smoking and open flames, are prohibited where this chemical is used, handled, or stored in a manner that could create a potential fire or explosion hazard. Wherever this chemical is used, handled, manufactured, or stored, use explosion-proof electrical equipment and fittings. A regulated, marked area should be established where this chemical is handled, used, or stored in compliance with OSHA standard 1910.1045.

Shipping: 1,3-Dichloro-2-propanol requires a "Poison" label. It falls in DOT Hazard Class 6.1 and Packing Group II.

Spill Handling: Evacuate and restrict persons not wearing protective equipment from area of spill or leak until cleanup is complete. Remove all ignition sources. Ventilate area of spill or leak. Absorb liquids in vermiculite, dry sand, earth, peat, carbon, or a similar material and deposit in sealed containers. Keep this chemical out of a confined space, such as a sewer, because of the possibility of an explosion, unless the sewer is designed to prevent the build-up of explosive concentrations. It may be necessary to contain and dispose of this chemical as a hazardous waste. If material or contaminated runoff enters waterways, notify downstream users of potentially contaminated waters. Contact your Department of Environmental Protection or your regional office of the federal EPA for specific recommendations. If employees are required to clean-up spills, they must be properly trained and equipped. OSHA 1910.120(q) may be applicable.

Fire Extinguishing: This chemical is a combustible liquid. Poisonous gases including chlorine and phosgene are produced in fire. Use alcohol foam extinguishers. Vapors are heavier than air and will collect in low areas. Vapors may travel long distances to ignition sources and flashback. Vapors in confined areas may explode when exposed to fire. Containers may explode in fire. Storage containers and parts of containers may rocket great distances, in many directions. If material or contaminated runoff enters waterways, notify downstream users of potentially contaminated waters. Notify local health and fire officials and pollution control agencies. From a secure, explosion-proof location, use water spray to cool exposed containers. If cooling streams are ineffective (venting sound increases in volume and pitch, tank discolors, or shows any signs of deforming), withdraw immediately to a secure position. If employees are expected to fight fires, they must be trained and equipped in OSHA 1910.156.

References

U.S. Environmental Protection Agency, Dichloropropanol, Health and Environmental Effects Profile No. 80, Washington, DC, Office of Solid Waste (April 30, 1980)

1,3-Dichloropropene

Molecular Formula: $C_3H_4Cl_2$

Common Formula: $CHCl{=}CHCH_2Cl$

Synonyms: β-Chloroallyl chloride; 3-Chloroallyl chloride; 3-Chloropropenyl chloride; 1,3-D; 1,3-Dichloro-1-propene; 1,3-Dichloro-2-propene; α,β-Dichloropropylene; 1,3-Dichloropropylene; 1,3-Dicloropropeno (Spanish); 1-Propene, 1,3-dichloro-; Propene, 1,3-dichloro-; Telone; Telone II

CAS Registry Number: 542-75-6; 10061-01-5 (cis-); 10061-02-6 (trans-)

RTECS Number: UC8310000

DOT ID: UN 2047

Regulatory Authority

- Carcinogen (Animal Positive) (IARC)[9] (Australia) (NTP)[10] (Israel) (DFG, cis- and trans-isomers)[3] (Several Canadian Provinces)
- Air Pollutant Standard Set (ACGIH)[1] (HSE)[33] (former USSR)[35][43] (Several States)[60]
- Clean Air Act: Hazardous Air Pollutants (Title I, Part A, Section 112)
- Clean Water Act: Section 311 Hazardous Substances/RQ 40CFR117.3 (same as CERCLA, see below); 40CFR423, Appendix A, Priority Pollutants; Section 313 Water Priority Chemicals (57FR41331, 9/9/92)
- EPA Hazardous Waste Number (RCRA No.): U084
- RCRA, 40CFR261, Appendix 8 Hazardous Constituents
- Superfund/EPCRA 40CFR302.4 Reportable Quantity (RQ): CERCLA, 100 lb (45.4 kg)
- EPCRA Section 313 Form R *de minimis* concentration reporting level: 0.1%
- U.S. DOT Regulated Marine Pollutant (49CFR172.101, Appendix B)
- Canada, WHMIS, Ingredients Disclosure List
- Mexico, drinking water, 0.09 mg/l

trans-isomer:

- RCRA 40CFR268.48; 61FR15654, Universal Treatment Standards: Wastewater (mg/l), 0.036; Nonwastewater (mg/kg), 18
- RCRA 40CFR264, Appendix 9; TSD Facilities Ground Water Monitoring List. Suggested test method(s) (PQL μg/l): 8010 (5); 8240 (5)
- EPCRA Section 313 Form R *de minimis* concentration reporting level: 0.1%

Cited in U.S. State Regulations: Alaska (G), Arizona (W), California (A, G), Connecticut (A, W), Florida (G), Kansas (G, W), Louisiana (G), Massachusetts (G), Minnesota (G), Nevada (A), New Hampshire (G), New Jersey (G), North Dakota (A), Oklahoma (G), Pennsylvania (G), Rhode Island (G), Vermont (G, W), Virginia (G, A), Washington (G), West Virginia (G), Wisconsin (G).

Description: 1,3-Dichloropropene, $CHCl{=}CHCH_2Cl$, is a colorless to straw-colored liquid with a sharp, sweet, irritating, chloroform-like odor. Boiling point = 106°C; 103 – 110°C (mixed cis- and trans- isomers). Freezing/Melting point = -84°C. Flash point = 35°C. The explosive limits are: LEL = 5.3%; UEL = 14.5%. Hazard Identification (based on NFPA-704 M Rating System): Health 2, Flammability 3, Reactivity 0. Practically insoluble in water.

Potential Exposure: Workers engaged in manufacture, formulation and application of this soil fumigant and nematocide. It is used in combinations with dichloropropanes as a soil fumigant also.

Incompatibilities: Forms explosive mixture with air. Violent reaction with strong oxidizers. May accumulate static electrical charges, and may cause ignition of its vapors. Incompatible with strong acids, oxidizers, aluminum or magnesium compounds, aliphatic amines, alkanolamines, alkaline materials, halogens, or corrosives.

Note: Epichlorohydrin may be added as a stabilizer.

Permissible Exposure Limits in Air: There is no OSHA PEL. NIOSH and ACGIH recommend a TWA of 1.0 ppm (5 mg/m^3). HSE, Australia, Israel, and the Canadian provinces of Alberta, Ontario, and Quebec set the same TWA as ACGIH and HSE,[33] and Alberta add a STEL of 10 ppm (50 mg/m^3). DFG has no numerical limit, but notes carcinogenic effect in animals. The additional notation "skin" indicates the possibility of cutaneous absorption. The former USSR[35][43] has set a MAC in workplace air of 5.0 mg/m^3 and MAC values for ambient air in residential areas of 0.1 mg/m^3 on a momentary basis and 0.01 mg/m^3 on a daily average basis.

Several states have set guidelines or standards for 1,3-dichloropropene in ambient air[60] ranging from 50 μg/m³ (North Dakota) to 80 μg/m³ (Virginia) to 100 μg/m³ (Connecticut) to 119 μg/m³ (Nevada).

Determination in Air: No method listed.

Permissible Concentration in Water: To protect freshwater aquatic life – 6,060 μg/l on an acute toxicity basis and 244 μg/l on a chronic basis. To protect saltwater aquatic life – 790 μg/l on an acute toxicity basis. To protect human health – 87.0 μg/l.[6] The former USSR[35][43] has set a MAC in water bodies used for domestic purposes of 0.4 mg/l. Mexico's drinking water criteria is 0.09 mg/l.

A no-observed-adverse-effects level (NOAEL) of 3.0 mg/kg/day has been determined by the EPA (see Health Advisory reference below). This results in a drinking water level on a lifetime basis of 0.011 mg/l (11.0 μg/l). Several states have set guidelines for 1,3-dichloropopene in drinking water[61] ranging from 10 μg/l (Connecticut) to 87 μg/l (Arizona and Kansas) to 89 μg/l (Vermont).

Determination in Water: Inert gas purge followed by gas chromatography with halide specific detection (EPA Method 601) or gas chromatography plus mass spectrometry (EPA Method 624).

Routes of Entry: Inhalation, skin absorption, ingestion, skin and/or eye contact.

Harmful Effects and Symptoms

Short Term Exposure: This chemical can be absorbed through the skin, thereby increasing exposure. Exposure can cause headaches, chest pain, and dizziness. High levels can cause you to pass out. Contact can severely burn the eyes and skin, with permanent damage. High exposures can damage the kidneys, liver and lungs.

Long Term Exposure: There is evidence that 1,3-dichloropropene causes cancer in animals and humans. May damage the kidneys, liver and lungs. May cause chronic headache, and personality changes.

Points of Attack: Eyes, skin, respiratory system, central nervous system, liver, kidneys. *Cancer Site*: (in animals) cancer of the bladder, liver, lung & stomach.

Medical Surveillance: Before beginning employment and at regular times after that, the following are recommended: Liver function tests. Lung function tests. Kidney function tests.

First Aid: If this chemical gets into the eyes, remove any contact lenses at once and irrigate immediately for at least 15 minutes, occasionally lifting upper and lower lids. Seek medical attention immediately. If this chemical contacts the skin, remove contaminated clothing and wash immediately with soap and water. Seek medical attention immediately. If this chemical has been inhaled, remove from exposure, begin rescue breathing (using universal precautions) if breathing has stopped and CPR if heart action has stopped. Transfer promptly to a medical facility. When this chemical has been swallowed, get medical attention. Give large quantities of water and induce vomiting. Do not make an unconscious person vomit.

Personal Protective Methods: Wear protective gloves and clothing to prevent any reasonable probability of skin contact. Safety equipment suppliers/manufacturers can provide recommendations on the most protective glove/clothing material for your operation. Viton and Polyvinyl Alcohol are among the recommended protective materials. All protective clothing (suits, gloves, footwear, headgear) should be clean, available each day, and put on before work. Contact lenses should not be worn when working with this chemical. Wear splash-proof chemical goggles and face shield unless full facepiece respiratory protection is worn. Employees should wash immediately with soap when skin is wet or contaminated. Provide emergency showers and eyewash.

Respirator Selection: *At any concentrations above the NIOSH REL, or where there is no REL, at any detectable concentration:* SCBAF:PD,PP (any MSHA/NIOSH approved self-contained breathing apparatus that has a full facepiece and is operated in a pressure-demand or other positive-pressure mode) or SAF:PD,PP:ASCBA (any supplied-air respirator that has a full facepiece and is operated in a pressure-demand or other positive-pressure mode in combination with an auxiliary self-contained breathing apparatus operated in a pressure-demand or other positive pressure mode). *Escape:* GMFOV [Any air-purifying, full-facepiece respirator (gas mask) with a chin-style, front- or back-mounted organic vapor canister]; or SCBAE (any appropriate escape-type, self-contained breathing apparatus).

Storage: Prior to working with 1,3-dichloropropene you should be trained on its proper handling and storage. Before entering confined space where this chemical may be present, check to make sure that an explosive concentration does not exist. 1,3-dichloropropene must be stored to avoid contact with aluminum or magnesium compounds; substances containing fluorine, chlorine, bromine or iodine; and alkaline or corrosive materials since violent reactions occur. Store in tightly closed containers in a cool, well-ventilated area away from heat. Separate outside storage is preferred. Sources of ignition such as smoking and open flames are prohibited where 1,3-dichloropropene is handled, used, or stored. Metal containers involving the transfer of 5 gallons or more of 1,3-dichloropropene should be grounded and bonded. Drums must be equipped with self-closing valves, pressure vacuum bungs, and flame arresters. Use only non-sparking tools and equipment, especially when opening and closing containers or 1,3-dichloropropene. A regulated, marked area should be established where this chemical is handled, used, or stored in compliance with OSHA standard 1910.1045.

Shipping: Dichloropropene must be labeled "Flammable Liquid." It falls in DOT Hazard Class 3 and Packing Group II.

Spill Handling: Evacuate and restrict persons not wearing protective equipment from area of spill or leak until cleanup is complete. Remove all ignition sources. Stay upwind; keep out of low areas. Establish forced ventilation to keep levels below explosive limit. Absorb liquids in vermiculite, dry sand, earth, peat, carbon, or a similar material and deposit in sealed containers. Keep this chemical out of a confined space, such as a sewer, because of the possibility of an explosion, unless the sewer is designed to prevent the build-up of explosive concentrations. It may be necessary to contain and dispose of this chemical as a hazardous waste. If material or contaminated runoff enters waterways, notify downstream users of potentially contaminated waters. Contact your Department of Environmental Protection or your regional office of the federal EPA for specific recommendations. If employees are required to clean-up spills, they must be properly trained and equipped. OSHA 1910.120(q) may be applicable.

Fire Extinguishing: This chemical is a flammable liquid. Poisonous gases including hydrogen chloride are produced in fire. Use dry chemical, carbon dioxide, or foam extinguishers. Vapors are heavier than air and will collect in low areas. Vapors may travel long distances to ignition sources and flashback. Vapors in confined areas may explode when exposed to fire. Containers may explode in fire. Storage containers and parts of containers may rocket great distances, in many directions. If material or contaminated runoff enters waterways, notify downstream users of potentially contaminated waters. Notify local health and fire officials and pollution control agencies. From a secure, explosion-proof location, use water spray to cool exposed containers. If cooling streams are ineffective (venting sound increases in volume and pitch, tank discolors, or shows any signs of deforming), withdraw immediately to a secure position. If employees are expected to fight fires, they must be trained and equipped in OSHA 1910.156.

Disposal Method Suggested: Incineration, preferably after mixing with another combustible fuel. Care must be exercised to assure complete combustion to prevent the formation of phosgene. An acid scrubber is necessary to remove the halo acids produced.[22] In accordance with 40CFR165 recommendations for the disposal of pesticides and pesticide containers. Must be disposed properly by following package label directions or by contacting your state pesticide or environmental control agency or by contacting your regional EPA office. Consult with environmental regulatory agencies for guidance on acceptable disposal practices. Generators of waste containing this contaminant (≥100 kg/mo) must conform with EPA regulations governing storage, transportation, treatment, and waste disposal.

References

U.S. Environmental Protection Agency, Dichloropropanes/Dichloropropenes: Ambient Water Quality Criteria, Washington, DC (1980)

U.S. Environmental Protection Agency, 1,3-Dichloropropene, Health and Environmental Effects Profile No. 81, Washington, DC, Office of Solid Waste (April 30, 1980)

U.S. Environmental Protection Agency, Dichloropropanes/Dichloropropenes: Health and Environmental Effects Profile No. 79, Washington, DC, Office of Solid Waste (April 30, 1980)

U.S. Environmental Protection Agency, "Health Advisory: 1,3-Dichloropropene," Washington, DC, Office of Drinking Water (August 1987)

New Jersey Department of Health, "Hazardous Substance Fact Sheet: 1,3-Dichloropropene," Trenton, NJ (April 1986)

Sax N. I., Ed., "Dangerous Properties of Industrial Materials Report," 6, No. 5, 88–93 (1986)

2,2-Dichloropropionic Acid

Molecular Formula: $C_3H_4Cl_2O_2$

Common Formula: CH_3CCl_2COOH

Synonyms: Acido 2,2-dicloropropionico (Spanish); Atlas Lignum (formulation); Basapon; Basapon B; Basapon/Basapon N; Basinex; BH Dalapon; BH Rasinox R (formulation); BH Total (formulation); Crisapon; Dalapon; Dalapon 85; Dalapon aliphatic acid herbicide; Ded-Weed; Destral; Devipon; α,α-Dichloropropionic acid; α-Dichloropropionic acid; Dowpon, Dowpon M; Fydulan (Formulation); Gramevin; Kenapon; Liropon; Proprop (South Africa); Radapon; Revenge; Synchemicals couch and grass killer; Unipon; Volunteered

CAS Registry Number: 75-99-0

RTECS Number: UF0690000

DOT ID: UN/NA 1760

Regulatory Authority

- Air Pollutant Standard Set (ACGIH)
- Clean Water Act: Section 311 Hazardous Substances/RQ 40CFR117.3 (same as CERCLA, see below)
- Safe Drinking Water Act: MCL, 0.2 mg/l; MCLG, 0.2 mg/l
- Superfund/EPCRA 40CFR302.4 Reportable Quantity (RQ): CERCLA, 5,000 lb (2,270 kg)

Cited in U.S. State Regulations: Alaska (G), California (A, G), Florida (G), Maine (G), Massachusetts (G), Minnesota (G), New Hampshire (G), New Jersey (G), Pennsylvania (G), Rhode Island (G), West Virginia (G).

Description: 2,2-Dichloropropionic acid, CH_3CCl_2COOH, is a colorless liquid with an acrid odor or a white to tan powder below 8°C. The sodium salt, a white powder, is often used. Boiling point = 185 – 190°C. Freezing/Melting point = 8°C. Hazard Identification (based on NFPA-704 M Rating System): Health 1, Flammability 0, Reactivity 0. Highly soluble in water (slowly reactive).

Potential Exposure: Those involved in the manufacture, formulation and application of the herbicide.

Incompatibilities: *Metals:* highly corrosive to aluminum and copper alloys. Reacts slowly in water to form hydrochloric and pyruvic acids.

Permissible Exposure Limits in Air: There is no OSHA PEL.[58] NIOSH and ACGIH recommend a TWA value of 1 ppm (6 mg/m^3). Israel, Mexico, DFG,[3] and the Canadian provinces of Alberta, Ontario, and Quebec set the same value as ACGIH and Alberta's STEL is 2 ppm. The former USSR[35][43] has set a MAC in workplace air of 10 mg/m^3 and a value for ambient air in residential areas of 0.03 mg/m^3.[35]

Determination in Air: No method available.

Permissible Concentration in Water: The no-observed-adverse effects level (NOAEL) for the acid is 8 mg/kg/day according to the EPA Health Advisory. This gives a lifetime drinking water equivalent level of 2.8 mg/l.

Determination in Water: By gas chromatography. See Health Advisory referenced below for details.

Routes of Entry: Inhalation, ingestion, skin and/or eye contact.

Harmful Effects and Symptoms

Short Term Exposure: 2,2-dichloropropionic acid is a corrosive chemical and can cause irritation and burn the skin and eyes, causing permanent damage. Exposure may cause symptoms of throat pain, loss of appetite, nausea, and sweating, lassitude (weakness, exhaustion), diarrhea, vomiting, slowing of pulse; central nervous system depressant/depression. 2,2-dichloropropionic acid can affect you when breathed in. Exposure can irritate the upper respiratory tract. 2,2-dichloropropionic acid may cause a skin allergy. If allergy develops, very low future exposures can cause itching and a skin rash.

Long Term Exposure: May cause liver and kidney damage. May cause skin allergy.

Points of Attack: Eyes, skin, respiratory system, gastrointestinal tract, central nervous system.

Medical Surveillance: If overexposure or illness is suspected, consider: Lung function tests. Kidney and liver function tests.

First Aid: If this chemical gets into the eyes, remove any contact lenses at once and irrigate immediately for at least 15 minutes, occasionally lifting upper and lower lids. Seek medical attention immediately. If this chemical contacts the skin, remove contaminated clothing and wash immediately with soap and water. Seek medical attention immediately. If this chemical has been inhaled, remove from exposure, begin rescue breathing (using universal precautions) if breathing has stopped and CPR if heart action has stopped. Transfer promptly to a medical facility. When this chemical has been swallowed, get medical attention. If victim is conscious, administer water or milk. Do not induce vomiting.

Personal Protective Methods: Wear acid resistant protective gloves and clothing to prevent any reasonable probability of skin contact. Safety equipment suppliers/manufacturers can provide recommendations on the most protective glove/clothing material for your operation.. All protective clothing (suits, gloves, footwear, headgear) should be clean, available each day, and put on before work. Contact lenses should not be worn when working with this chemical. Wear splash-proof chemical goggles and face shield when working with liquids or dust-proof goggles and faceshield when working with powders or dusts, unless full facepiece respiratory protection is worn. Employees should wash immediately with soap when skin is wet or contaminated. Provide emergency showers and eyewash.

Respirator Selection: Where the potential exists for exposures over 1 ppm, use an MSHA/NIOSH approved supplied-air respirator with a full facepiece operated in the positive pressure mode or with a full facepiece, hood, or helmet in the continuous flow mode, or use an MSHA/NIOSH approved self-contained breathing apparatus with a full facepiece operated in pressure-demand or other positive pressure mode.

Storage: Prior to working with this chemical you should be trained on its proper handling and storage. Store in tightly closed containers in a cool, well ventilated area away from metals and moisture.

Shipping: Acids, liquid, n.o.s. must carry a "Corrosive" label. They fall in DOT Hazard Class 8 and Packing Group II.

Spill Handling: *Liquid:* Evacuate and restrict persons not wearing protective equipment from area of spill or leak until cleanup is complete. Remove all ignition sources. Ventilate area of spill or leak. Absorb liquids in vermiculite, dry sand, earth, peat, carbon, or a similar material and deposit in sealed containers. Keep this chemical out of a confined space, such as a sewer, because of the possibility of an explosion, unless the sewer is designed to prevent the build-up of explosive concentrations. It may be necessary to contain and dispose of this chemical as a hazardous waste. If material or contaminated runoff enters waterways, notify downstream users of potentially contaminated waters. Contact your Department of Environmental Protection or your regional office of the federal EPA for specific recommendations. If employees are required to clean-up spills, they must be properly trained and equipped. OSHA 1910.120(q) may be applicable.

Powder: Evacuate persons not wearing protective equipment from area of spill or leak until clean-up is complete. Remove all ignition sources. Collect powdered material in the most convenient and safe manner and deposit in sealed containers. Ventilate area after clean-up is complete. It may be necessary to contain and dispose of this chemical as a hazardous waste. If material or contaminated runoff enters waterways, notify downstream users of potentially contaminated waters. Contact your Department of Environmental Protection or your regional office of the federal EPA for specific recommendations. If employees are required to clean-up spills, they must be properly trained and equipped. OSHA 1910.120(q) may be applicable.

Fire Extinguishing: This chemical may burn, but does not easily ignite. Poisonous gases including chlorine are produced in fire. Use dry chemical, carbon dioxide, or foam extinguishers. Vapors are heavier than air and will collect in low areas. Containers may explode in fire. Storage containers and parts of containers may rocket great distances, in many directions. If material or contaminated runoff enters waterways, notify downstream users of potentially contaminated waters. Notify local health and fire officials and pollution control agencies. From a secure, explosion-proof location, use water spray to cool exposed containers. If cooling streams are ineffective (venting sound increases in volume and pitch, tank discolors, or shows any signs of deforming), withdraw immediately to a secure position. If employees are expected to fight fires, they must be trained and equipped in OSHA 1910.156.

Disposal Method Suggested: Incineration in a unit with efficient gas scrubbing.[22] In accordance with 40CFR165 recommendations for the disposal of pesticides and pesticide containers. Must be disposed properly by following package label directions or by contacting your state pesticide or environmental control agency or by contacting your regional EPA office.

References

Sax N. I., Ed., "Dangerous Properties of Industrial Materials Report," 3, No. 2, 74–77 (1983)

U.S. Environmental Protection Agency, "Health Advisory: Dalapon," Washington, DC, Office of Drinking Water (August 1987)

New Jersey Department of Health, "Hazardous Substance Fact Sheet: 2,2-Dichloropropionic Acid," Trenton, NJ (February 1989)

Dichlorotetrafluoroethane

Molecular Formula: $C_2Cl_2F_4$

Common Formula: $F_2ClCCClF_2$

Synonyms: Arcton 114; Arcton 33; CFC-114; Cryofluoran; Cryofluorane; *sym*-Dichlorotetrafluoroethane; 1,2-Dichloro-1,1,2,2-tetrafluoroethane; *sim*-Diclorotetrafluoetano (Spanish); Ethane, 1,2-dichloro-1,1,2,2-tetrafluoro-; Ethane, 1,2-dichlorotetrafluoro-; F 114; FC 114; Fluorane 114; Fluorocarbon 114; Freon 114; Frigen 114; Frigiderm; Genetron 114; Genetron 316; Halocarbon 114; Halon 242; Ledon 114; Propellant 114; R 114; 1,1,2,2-Tetrafluoro-1,2-dichloroethane; Ucon 114

CAS Registry Number: 76-14-2

RTECS Number: KI1101000

DOT ID: UN 1958

Regulatory Authority

- Air Pollutant Standard Set (ACGIH)[1] (Australia) (DFG)[3] (HSE)[33] (Israel) (Mexico) (Russia)[43] (OSHA)[58] (Several States)[60] (Several Canadian Provinces)
- Clean Air Act: Stratospheric ozone protection (Title VI, Subpart A, Appendix A), Class I, Ozone Depletion Potential = 1.0
- EPCRA Section 313 Form R *de minimis* concentration reporting level: 1.0%
- Canada, WHMIS, Ingredients Disclosure List

Cited in U.S. State Regulations: Alaska (G), California (A), Connecticut (A), Florida (G), Illinois (G), Maine (G), Massachusetts (G), Minnesota (G), Nevada (A), New Hampshire (G), New Jersey (G), North Dakota (A), Rhode Island (G), Virginia (A), West Virginia (G).

Description: CFC 114, $F_2ClCCClF_2$, colorless gas with a faint, ether-like odor at high concentrations. A liquid below 3.3°C/38°F. Shipped as a liquefied compressed gas. Boiling point = 3°C. Freezing/Melting point = -94°C. Slightly soluble in water.

Potential Exposure: This material is used as a refrigerant and also as a propellant gas.

Incompatibilities: Reacts with acids and acid fumes forming highly toxic chloride gases. Chemically active metals: sodium, potassium, calcium, powdered aluminum, zinc, and magnesium. Attacks some plastics and coatings.

Permissible Exposure Limits in Air: The Federal Limit (OSHA PEL),[58] DFG MAK,[3] Australian, Israeli, Mexican, Canadian provincial (Alberta, BC, Ontario, Quebec), and the recommended ACGIH TWA value is 1,000 ppm (7,000 mg/m^3) and HSE,[33] Mexico, Alberta, BC STEL value is 1,250 ppm (8,750 mg/m^3). The NIOSH IDLH level is 15,000 ppm. The former USSR-UNEP/IRPTC project[43] has set a MAC in workplace air of 3,000 mg/m^3. Several states have set guidelines or standards for R-114 in ambient air[60] ranging from 70 mg/m^3 (North Dakota) to 115 mg/m^3 (Virginia) to 140 mg/m^3 (Connecticut) to 167 mg/m^3 (Nevada).

Determination in Air: Charcoal tube;[2] CH_2Cl_2; Gas chromatography/Flame ionization detection; NIOSH IV, Method #1018.

Permissible Concentration in Water: No criteria set.

Routes of Entry: Inhalation, skin and/or eye contact (liquid).

Harmful Effects and Symptoms

Short Term Exposure: Dichlorotetrafluoroethane can affect you when breathed in. Irritates the eyes and upper respiratory tract. Inhalation of vapors can cause you to become dizzy and lightheaded, drowsy, and pass out. It can cause the heart to beat irregularly or stop, which can cause death. Contact with the liquid can cause frostbite, burning the eyes and skin.

Long Term Exposure: Can irritate the lungs and cause bronchitis with coughing, phlegm, and/or shortness of breath.

Points of Attack: Respiratory system, cardiovascular system.

Medical Surveillance: For those with frequent or potentially high exposure (half the TLV or greater), the following are recommended before beginning work and at regular times after that: Lung function tests. If symptoms develop or overexposure is suspected, the following may be useful: Holter monitor (a special 24 hour EKG to look for irregular heart rhythms).

First Aid: If this chemical gets into the eyes, remove any contact lenses at once and irrigate immediately for at least 15 minutes, occasionally lifting upper and lower lids. Seek medical attention immediately. If this chemical contacts the skin, remove contaminated clothing and wash immediately with soap and water. Seek medical attention immediately. If this chemical has been inhaled, remove from exposure, begin rescue breathing (using universal precautions) if breathing has stopped and CPR if heart action has stopped. Transfer promptly to a medical facility. When this chemical has been swallowed, get medical attention. Give large quantities of water and induce vomiting. Do not make an unconscious person vomit. If frostbite has occurred, seek medical attention immediately; do *NOT* rub the affected areas or flush them with water. In order to prevent further tissue damage, do *NOT* attempt to remove frozen clothing from frostbitten areas. If frostbite has *NOT* occurred, immediately and thoroughly wash contaminated skin with soap and water.

Personal Protective Methods: Wear protective gloves and clothing to prevent any reasonable probability of skin contact. Safety equipment suppliers/manufacturers can provide recommendations on the most protective glove/clothing material for your operation.. All protective clothing (suits, gloves, footwear, headgear) should be clean, available each day, and put on before work. Contact lenses should not be worn when working with this chemical. Wear splash-proof chemical goggles and face shield when working with the liquid, wear gas-proof goggles and face shield when working with the gas unless full facepiece respiratory protection is worn. Employees should wash immediately with soap when skin is wet or contaminated. Provide emergency showers and eyewash. Cold equipment, vapors, or liquid may occur, special gloves and clothing designed to prevent freezing of body tissues should be used. Where exposure to the liquefied compressed gas may occur, employees should be provided with special clothing designed to prevent frostbite.

Respirator Selection: NIOSH/OSHA: *10,000 ppm:* SA (any supplied-air respirator). *15,000 ppm:* SA:CF (any supplied-air respirator operated in a continuous-flow mode); or SCBAF (any self-contained breathing apparatus with a full facepiece); or SAF (any supplied-air respirator with a full facepiece). Emergency or planned entry into unknown concentrations or IDLH conditions: SCBAF:PD,PP (any self-contained breathing apparatus that has a full facepiece and is operated in a pressure-demand or other positive-pressure mode); or SAF:PD,PP:ASCBA (any supplied-air respirator that has a full facepiece and is operated in a pressure-demand or other positive-pressure mode in combination with an auxiliary self-contained breathing apparatus operated in a pressure-demand or other positive pressure mode). *Escape:* GMFOV [any air-purifying, full-facepiece respirator (gas mask) with a chin-style, front-or back-, mounted organic vapor canister]; or SCBAE (any appropriate escape-type, self-contained breathing apparatus).

Storage: Prior to working with CFC 114 you should be trained on its proper handling and storage. Dichlorotetrafluoroethane must be stored to avoid contact with acids, acid fumes, chemically active metals, such as sodium, potassium, calcium, powdered aluminum, zinc and magnesium, since violent reactions occur. Store in tightly closed containers in a cool, well-ventilated area away from heat. Procedures for the handling, use and storage of cylinders should be in compliance with OSHA 1910.101 and 1910.169 as with the recommendations of the Compressed Gas Association.

Shipping: This material must carry a "Nonflammable Gas" label. It falls in DOT Hazard Class 2.2.

Spill Handling: Evacuate and restrict persons not wearing protective equipment from area of spill or leak until cleanup is complete. Remove all ignition sources. Ventilate area of spill or leak. Ventilate area of leak to disperse the gas. Stop flow of gas. If source of leak is a cylinder and the leak cannot be stopped in place, remove the leaking cylinder to a safe place in the open air, and repair leak or allow cylinder to empty. If the liquid is spilled or leaked, allow it to vaporize. It may be necessary to contain and dispose of this chemical as a hazardous waste. If material or contaminated runoff enters waterways, notify downstream users of potentially contaminated waters. Contact your Department of Environmental Protection or your regional office of the federal EPA for specific recommendations. If employees are required to clean-up spills, they must be properly trained and equipped. OSHA 1910.120(q) may be applicable.

Fire Extinguishing: Dichlorotetrafluoroethane is a nonflammable liquid or gas. Extinguish fire using an agent suitable for type of surrounding fire. Poisonous gases are produced in fire including hydrogen chloride, phosgene, and hydrogen fluoride. Vapors are heavier than air and will collect in low areas. Containers may explode in fire. Storage containers and parts of containers may rocket great distances, in many directions. Notify local health and fire officials and pollution control agencies. From a secure, explosion-proof location, use water spray to cool exposed containers. If cooling streams are ineffective (venting sound increases in volume and pitch, tank discolors, or shows any signs of deforming), withdraw immediately to a secure position. If employees are expected to fight fires, they must be trained and equipped in OSHA 1910.156.

Disposal Method Suggested: Incineration after mixing with combustible fuel. Use flue gas scrubber.[22]

References

New Jersey Department of Health, "Hazardous Substance Fact Sheet: Dichlorotetrafluoroethane," Trenton, NJ (June 1998)

Dichlorvos

Molecular Formula: $C_4H_7Cl_2O_4P$

Synonyms: Apavap; Astrobot; Atgard; Atgard V; Bay 19149; Bayer 19149; Benfos; Bibesol; Brevinyl; Brevinyl E 50; Canogard; Cekusan; Chlorvinphos; Cyanophos; Cypona; DDVF; DDVP (Insecticide); Dedevap; Deriban; Derribante; DES; Devikol; Dichlofos; (2,2-Dichloor-vinyl)-dimethyl-fosfaat (Dutch); Dichloorvo (Dutch); Dichlorfos (Polish); Dichlorman; 2,2-Dichloroethenol dimethyl phosphate; 2,2-Dichloroethenyl dimethyl phosphate; 2,2-Dichlorovinyl dimethyl phosphate; Dichlorovos; (2,2-Dichlorvinyl)-dimethyl-phosphat (German); *O*-(2,2-Dichlorvinyl)*O,O*-dimethylphosphat (German); (2,2-Dicloro-vinil)dimetilfosfato (Italian); Dimethyl 2,2-dichloroethenyl phosphate; *O,O*-Dimethyl 2,2-dichlorovinyl phosphate; Dimethyl 2,2-dichlorovinyl phosphate; Dimethyl dichlorovinyl phosphate; Divipan; Dquigard; Duo-Kill; Duravos; ENT 20,738; Equigard; Equigel; Estrosel; Estrosol; Ethenol, 2,2-dichloro-, dimethyl phosphate; Fecama; Fekama; Fly-Die; Fly fighter; Herkal; Insectigas D; Krecalvin; Lindan; Mafu; Marvex; Mopari; NCI-C00113; Nefrafos; Nerkol; Nogos; Nogos 50; Nogos G; No-Pest; No-Pest Strip; Novotox; NSC-6738; Nuva; Nuvan; Nuvan 100EC; Nuvan 7; OKO; OMS 14; Panaplate; Phosphate de dimethyle et de 2,2-dichlorovinyle (French); Phosphoric acid, 2,2-dichloroethenyl dimethyl ester; Phosphoric acid, 2-dichloroethenyl dimethyl ester; Phosphoric acid, 2,2-dichlorovinyl dimethyl ester; Phosvit; SD 1750; Szklarniak; Tap 9VP; Task; Task Tabs; Tenac; Tetravos; UN 2783; Unifos (Pesticide); Unitox; Vapona; Vapona insecticide; Vaponite; Verdican; Verdipor; Vinylofos; Vinylophos; Winylophos

CAS Registry Number: 62-73-7

RTECS Number: TC0350000

DOT ID: UN 3017

EEC Number: 015-019-00-X

Regulatory Authority

- Air Pollutant Standard Set (ACGIH)[1] (Australia) (DFG)[3] (HSE)[33] (Israel) (Mexico) (Argentina) (OSHA) (former USSR)[35] (Several Canadian Provinces)
- Clean Air Act: Hazardous Air Pollutants (Title I, Part A, Section 112)
- Clean Water Act: Section 311 Hazardous Substances/RQ 40CFR117.3 (same as CERCLA, see below); Section 313 Water Priority Chemicals (57FR41331, 9/9/92)
- Superfund/EPCRA 40CFR355, Appendix B Extremely Hazardous Substances: TPQ = 1,000 lb (454 kg)
- Superfund/EPCRA 40CFR302.4 Reportable Quantity (RQ): CERCLA, 10 lb (4.54 kg)
- EPCRA Section 313 Form R *de minimis* concentration reporting level: 1.0%
- U.S. DOT Regulated Marine Pollutant (49CFR172.101, Appendix B)

Cited in U.S. State Regulations: Alaska (G), California (A, G), Florida (G), Illinois (G), Maine (G), Massachusetts (G), Michigan (G), Minnesota (G), New Hampshire (G), New Jersey (G), Pennsylvania (G), Rhode Island (G), West Virginia (G).

Description: Dichlorvos is a colorless to amber liquid with a mild aromatic odor. Boiling point = 140°C under 20 mm Hg pressure. Flash point ≥79°C. Hazard Identification (based on NFPA-704 M Rating System): Health 3, Flammability 1, Reactivity 0. Moderate solubility in water.

Potential Exposure: Those involved in manufacture, formulation and application of this fumigant insecticide in household, public health and agricultural uses.

Incompatibilities: Strong acids, strong alkalis. Corrosive to iron, mild steel, and some forms of plastics, rubber, and coatings.

Permissible Exposure Limits in Air: The Federal limit (OSHA PEL), Australian, Israeli, DFG MAK, HSE, Canadian provincial (Alberta, BC, Ontario, Quebec), and ACGIH value is 0.1 ppm (1 mg/m^3) TWA. And HSE,[33] Alberta, BC, and STEL value of 0.3 mg/m^3. The Mexican TWA is 0.16 ppm. The NIOSH IDLH level is 100 mg/m^3. The notation "skin" is added indicating the possibility of cutaneous absorption. Argentina[35] uses the same TWA limits set forth above. The former USSR[35] has set a MAC in workplace air of 0.2 mg/m^3 and a MAC for ambient air in residential areas of 0.07 mg/m^3 on a once-a-day basis and 0.002 mg/m^3 on a daily average basis.

Determination in Air: XAD-2® (tube); Toluene; Gas chromatography/Flame photometric detection for sulfur, nitrogen, or phosphorus; NIOSH II(5), P&CAM Method #295.

Permissible Concentration in Water: The former USSR[35] has set a MAC in water bodies used for domestic purposes of 1.0 mg/l and in water for fishing of zero.

Routes of Entry: Inhalation, skin absorption, ingestion, eye and/or skin contact.

Harmful Effects and Symptoms

Symptoms of exposure include sweating, twitching, contracted pupils, respiratory distress (tightness in the chest and wheezing), salivation (drooling), lacrimation (tearing), nausea, vomiting, abdominal cramps, diarrhea, involuntary defecation and urination, slurred speech, coma, apnea (cessation of breathing), and death.

Dichlorvos is a very toxic compound with a probable lethal oral dose in humans between 50 and 500 mg/kg, or between 1 teaspoon and 1 oz. for a 70 kg (150 lb.) person. However, brief exposure (30 – 60 minutes) to vapor concentrations as high as 6.9 mg/liter did not result in clinical signs or depressed serum cholinesterase levels. Toxic changes are typical of organophosphate insecticide poisoning with progression to respiratory distress, respiratory paralysis, and death if there is no clinical intervention.

Short Term Exposure: Dichlorvos irritates the eyes and skin. Symptoms include miosis, aching eyes; rhinorrhea (discharge of thin nasal mucous); headache; chest tightness, wheezing, laryngeal spasm, salivation; cyanosis; anorexia, nausea, vomiting, diarrhea; sweating; muscle fasiculation, paralysis, giddiness, ataxia; convulsions; low blood pressure, cardiac irregular/irregularities. The substance may cause effects on the central nervous system. Cholinesterase inhibitor. High levels of exposure may result in death.

Long Term Exposure: Repeated or prolonged contact with skin may cause skin sensitization and dermatitis. Cholinesterase inhibitor; cumulative effect is possible: see short term exposure. This substance may be carcinogenic to humans; it has been shown to cause cancer of the pancreas in animals. There is limited evidence that Dichlorvos is a teratogen in animals, and may cause birth defects or damage the fetus in humans.

Points of Attack: Eyes, skin, respiratory system, cardiovascular system, central nervous system, blood cholinesterase.

Medical Surveillance: Before employment and at regular times after that, the following are recommended: Plasma and red blood cell cholinesterase levels (tests for the enzyme poisoned by this chemical). If exposure stops, plasma levels return to normal in 1 – 2 weeks while red blood cell levels may be reduced for 1 – 3 months.

When cholinesterase enzyme levels are reduced by 25% or more below preemployment levels, risk of poisoning is increased, even if results are in lower ranges of "normal." Reassignment to work not involving organophosphate or carbamate pesticides is recommended until enzyme levels recover. If symptoms develop or overexposure occurs, repeat the above tests as soon as possible and get an exam of the nervous system. Also consider complete blood count. Consider chest x-ray following acute overexposure. Do not drink any alcoholic beverages before or during use. Alcohol promotes absorption of organic phosphates.

First Aid: If this chemical gets into the eyes, remove any contact lenses at once and irrigate immediately for at least 15 minutes, occasionally lifting upper and lower lids. Seek medical attention immediately. If this chemical contacts the skin, remove contaminated clothing and wash immediately with soap and water. Speed in removing material from skin is of extreme importance. Shampoo hair promptly if contaminated. Seek medical attention immediately. If this chemical has been inhaled, remove from exposure, begin rescue breathing (using universal precautions) if breathing has stopped and CPR if heart action has stopped. Transfer promptly to a medical facility. When this chemical has been swallowed, get medical attention. Give large quantities of water and induce vomiting. Do not make an unconscious person vomit. Medical observation is recommended for 24 – 48 hours following overexposure.

Personal Protective Methods: Wear protective gloves and clothing to prevent any reasonable probability of skin contact. Safety equipment suppliers/manufacturers can provide recommendations on the most protective glove/clothing material for your operation.. All protective clothing (suits, gloves, footwear, headgear) should be clean, available each day, and put on before work. Contact lenses should not be worn when working with this chemical. Wear splash-proof chemical goggles and face shield unless full facepiece respiratory protection is worn. Employees should wash immediately with soap when skin is wet or contaminated. Provide emergency showers and eyewash.

Respirator Selection: NIOSH/OSHA: 10 mg/m^3: SA (any supplied-air respirator). *25 mg/m^3*: SA:CF (any supplied-air respirator operated in a continuous-flow mode). *50 mg/m^3*: SAT:CF (any supplied-air respirator that has a tight-fitting facepiece and is operated in a continuous-flow mode); or SCBAF (any self-contained breathing apparatus with a full facepiece); or SAF (any supplied-air respirator with a full facepiece). *100 mg/m^3*: SA:PD,PP (any supplied-air respirator operated in a pressure-demand or other positive-pressure mode). *Emergency or planned entry into unknown concentrations or IDLH conditions*: SCBAF:PD,PP (any self-contained breathing apparatus that has a full facepiece and is operated in a pressure-demand or other positive-pressure mode); or SAF:PD,PP:ASCBA (any supplied-air respirator that has a full facepiece and is operated in a pressure-demand or other positive-pressure mode in combination with an auxiliary self-contained breathing apparatus operated in a pressure-demand or other positive pressure mode). *Escape:* GMFOVHiE [any air-purifying, full-facepiece respirator (gas mask) with a chin-style, front- or back-mounted organic vapor canister having a high-efficiency particulate filter]; or SCBAE (any appropriate escape-type, self-contained breathing apparatus).

Storage: Prior to working with Dichlorvos you should be trained on its proper handling and storage. Store in tightly closed containers in a cool, well-ventilated area away from strong acids, strong alkalis. Dichlorvos will attack some forms of mild iron, plastics, rubber and coatings.

Shipping: Organophosphorus pesticides, liquid, toxic, flammable, n.o.s. fall in Hazard Class 6.1 and Dichlorvos in Packing Group II. It requires a "Poison, Flammable Liquid" label.

Spill Handling: Evacuate and restrict persons not wearing protective equipment from area of spill or leak until cleanup is complete. Remove all ignition sources. Ventilate area of spill or leak. Remove and isolate contaminated clothing at the site. Do not touch spilled material; stop leaks if you can do it without risk. Reduce vapors with water spray. Take up small spills with sand or other noncombustible absorbent material for later disposal in canisters. Dike large spills far ahead of spill for later disposal. Keep this chemical out of a confined space, such as a sewer, because of the possibility of an explosion, unless the sewer is designed to prevent the build-up of explosive concentrations. It may be necessary to contain and dispose of this chemical as a hazardous waste. If material or contaminated runoff enters waterways, notify downstream users of potentially contaminated waters. Contact your Department of Environmental Protection or your regional office of the federal EPA for specific recommendations. If employees are required to clean-up spills, they must be properly trained and equipped. OSHA 1910.120(q) may be applicable.

Fire Extinguishing: This chemical is a combustible liquid. Poisonous gases including hydrogen chloride and phosphoric acid are produced in fire. Use dry chemical, carbon dioxide, or foam extinguishers. Use self-contained breathing apparatus with a full face piece operated on pressure-demand or other positive pressure mode. Prevent skin contact with protective clothing. Isolate area and deny entry. Fight fire from maximum distance. Dike fire control water for future disposal. Vapors are heavier than air and will collect in low areas. Vapors in confined areas may explode when exposed to fire. Containers may explode in fire. Storage containers and parts of containers may rocket great distances, in many directions. If material or contaminated runoff enters waterways, notify downstream users of potentially contaminated waters. Notify local health and fire officials and pollution control agencies. From a secure, explosion-proof location, use water spray to cool exposed containers. If cooling streams are ineffective (venting sound increases in volume and pitch, tank discolors, or shows any signs of deforming), withdraw immediately to a secure position. If employees are expected to fight fires, they must be trained and equipped in OSHA 1910.156.

Disposal Method Suggested: 50% hydrolysis is obtained in pure water in 25 minutes at 70°C and in 61.5 days at 20°C. A buffered solution yields 50% hydrolysis (37.5°C) in 301 minutes at pH 8, 462 minutes at pH 7, 620 minutes at pH 5.4. Hydrolysis yields no toxic residues. Incineration in a furnace equipped with an afterburner and alkaline scrubber is recommended as is alkaline hydrolysis followed by soil burial.[22] In accordance with 40CFR165 recommendations for the disposal of pesticides and pesticide containers. Must be disposed properly by following package label directions or by contacting your state pesticide or environmental control agency or by contacting your regional EPA office.

References

U.S. Environmental Protection Agency, Investigation of Selected Potential Environmental Contaminants: Haloalkyl Phosphates, Report EPA-560/2076-007, Washington, DC (August 1976)

Sax N. I., Ed., "Dangerous Properties of Industrial Materials Report," 1, No. 3, 57–59 (1981)

U.S. Environmental Protection Agency, "Chemical Profile: Dichlorvos," Washington, DC, Chemical Emergency Preparedness Program (November 30, 1987)

New Jersey Department of Health, "Hazardous Substance Fact Sheet: Dichlorvos," Trenton, NJ (October 1996)

Dicofol

Molecular Formula: $C_{14}H_9Cl_5$

Synonyms: Acarin; Benzenemethanol, 4-chloro-α(-4-chlorophenyl)-α-(trichloromethyl)-; Benzhydrol, 4,4'-dichloro-α-(trichloromethyl)-; 1,1-Bis(*p*-chlorophenyl)-2,2,2-trichloroethanol; 1,1-Bis(4-chlorophenyl)-2,2,2-trichloroethanol; 4-Chloro-α-(4-chlorophenyl)-α-(trichloromethyl) benzene methanol; CPCA; Decofol; Dichlorokelthane; Di(*p*-chlorophenyl) trichloromethyl carbinol; 4,4'-Dichloro-α-(trichloromethyl)benzhydrol; DTMC; ENT 23,648; Ethanol, 2,2,2-trichloro-1,1-bis(4-chlorophenyl)-; Fumite dicofol; FW 293; Keltane; *p,p'*-Kelthane; Kelthane; Kelthane A; Kelthanethanol; Milbol; Mitigan; NCI-C00486; 2,2,2-Trichloro-1,1-bis(*p*-chlorophenyl)ethanol; 2,2,2-Trichloro-1,1-bis(4-chlorophenyl)ethanol; 2,2,2-Trichloro-1,1-di(4-chlorophenyl)ethanol

CAS Registry Number: 115-32-2

RTECS Number: DC8400000

DOT ID: UN2761

EEC Number: 603-044-00-4

Regulatory Authority

- Carcinogen (animal positive) NTP
- Clean Water Act: Section 311 Hazardous Substances/RQ 40CFR117.3 (same as CERCLA, see below); Section 313 Water Priority Chemicals (57FR41331, 9/9/92)
- Superfund/EPCRA 40CFR302.4 Reportable Quantity (RQ): CERCLA, 10 lb (4.54 kg)
- EPCRA Section 313 Form R *de minimis* concentration reporting level: 1.0%

Cited in U.S. State Regulations: California (A, G), Massachusetts (G), New Jersey (G), Pennsylvania (G).

Description: Dicofol is a white or brown waxy solid. Flash point = 120°C. Hazard Identification (based on NFPA-704 M Rating System): Health 2, Flammability 1, Reactivity 0. Slightly soluble in water.

Potential Exposure: An organochlorine pesticide.

Incompatibilities: Incompatible with alkaline pesticides, strong acids, acid fumes, aliphatic amines, isocyanates, and steel.

Permissible Exposure Limits in Air: No occupational exposure limits have been established.

Routes of Entry: Inhalation, passing through the skin, ingestion.

Harmful Effects and Symptoms

Short Term Exposure: Dicofol can be absorbed through the skin, thereby increasing exposure. It irritates the skin and the respiratory tract. Exposure can cause headache, nausea, vomiting and poor appetite. Dicofol may affect the central nervous system causing numbness and weakness in the hands and feet, muscle twitching, seizures, unconsciousness and death.

Long Term Exposure: May affect the liver and kidneys. May cause personality changes with depression, anxiety, and irritability. May decrease fertility in females. Prolonged or repeated skin contact may cause dermatitis. There is limited evidence that Dicofol causes liver cancer in animals.

Points of Attack: Skin, nervous system, liver, kidneys.

Medical Surveillance: Liver and kidney function tests. Examination of the nervous system. Dermatological examination.

First Aid: If this chemical gets into the eyes, remove any contact lenses at once and irrigate immediately for at least 15 minutes, occasionally lifting upper and lower lids. Seek medical attention immediately. If this chemical contacts the skin, remove contaminated clothing and wash immediately with soap and water. Seek medical attention immediately. If this chemical has been inhaled, remove from exposure, begin rescue breathing (using universal precautions) if breathing has stopped and CPR if heart action has stopped. Transfer promptly to a medical facility. When this chemical has been swallowed, get medical attention. Give large quantities of water and induce vomiting. Do not make an unconscious person vomit.

Personal Protective Methods: Wear protective gloves and clothing to prevent any reasonable probability of skin contact. Safety equipment suppliers/manufacturers can provide recommendations on the most protective glove/clothing material for your operation.. All protective clothing (suits, gloves, footwear, headgear) should be clean, available each day, and put on before work. Contact lenses should not be worn when working with this chemical. Wear dust-proof chemical goggles and face shield unless full facepiece respiratory protection is worn. For liquid solutions containing Dicofol wear indirect-vent, impact and splash-resistant goggles. Employees should wash immediately with soap when skin is wet or contaminated. Provide emergency showers and eyewash.

Respirator Selection: Where the potential for exposure to this chemical, use a MSHA/NIOSH approved supplied-air respirator with a full facepiece operated in the positive pressure mode or with a full facepiece, hood, or helmet in the continuous flow mode, or use a MSHA/NIOSH approved self-contained breathing apparatus with a full facepiece operated in pressure-demand or other positive pressure mode.

Storage: Prior to working with Dicofol you should be trained on its proper handling and storage. Store in tightly closed containers in a cool, well ventilated area away from alkaline pesticides, strong acids, acid fumes and steel. Where possible, automatically transfer material from drums or other storage containers to process containers. A regulated, marked area should be established where this chemical is handled, used, or stored in compliance with OSHA standard 1910.1045.

Shipping: Organochlorine pesticides, solid, toxic, n.o.s. fall in DOT Hazard Class 6.1 and DDT in Packing Group III. The label required is "Keep Away From Food."

Spill Handling: Evacuate persons not wearing protective equipment from area of spill or leak until clean-up is complete. Remove all ignition sources. Collect powdered material in the most convenient and safe manner and deposit in sealed containers. Ventilate area after clean-up is complete. It may be necessary to contain and dispose of this chemical as a hazardous waste. If material or contaminated runoff enters waterways, notify downstream users of potentially contaminated waters. Contact your Department of Environmental Protection or your regional office of the federal EPA for specific recommendations. If employees are required to clean-up spills, they must be properly trained and equipped. OSHA 1910.120(q) may be applicable.

Fire Extinguishing: This chemical is a combustible solid. Use dry chemical, carbon dioxide, water spray, or alcohol foam extinguishers. Poisonous gases are produced in fire including hydrogen chloride. If material or contaminated runoff enters waterways, notify downstream users of potentially contaminated waters. Notify local health and fire officials and pollution control agencies. From a secure, explosion-proof location, use water spray to cool exposed containers. If cooling streams are ineffective (venting sound increases in volume and pitch, tank discolors, or shows any signs of deforming), withdraw immediately to a secure position. If employees are expected to fight fires, they must be trained and equipped in OSHA 1910.156.

Disposal Method Suggested: In accordance with 40CFR165 recommendations for the disposal of pesticides and pesticide containers. Must be disposed properly by following package label directions or by contacting your state pesticide or environmental control agency or by contacting your regional EPA office.

References

New Jersey Department of Health and Senior Services, "Hazardous Substance Fact Sheet, DICOFOL," Trenton NJ (October 1998).

Dicrotophos

Molecular Formula: $C_8H_{16}NO_5P$

Synonyms: Bidirl; Bidrin; Bidrin (Shell); Bidrin-R; C-709; C-709 (Ciba-Geigy); Carbicrin; Carbicron; Carbomicron; Ciba 709; Crotonamide, 3-hydroxy-*N*,*N*-dimethyl-, *cis*-, dimethyl phosphate; Crotonamide, 3-hydroxy-*N*,*N*-dimethyl-, dimethylphosphate, (E)-; Crotonamide, 3-hydroxy-*N*,*N*-dimethyl-, dimethylphosphate, *cis*-; Diapadrin; Dicroptophos; Dicrotofos (Dutch); Didrin; 3-(Dimethoxyphosphinyloxy)-*N*,*N*-dimethyl[e]crotonamide; 3-(Dimethoxyphosphinyloxy)-*N*,*N*-dimethyl-*cis*-crotonamide; 3-(Dimethoxyphosphinyloxy)-*N*,*N*-dimethylisocrotonamide; 3-(Dimethylamino)-1-methyl-3-oxo-1-propenyl dimethyl phosphate; (E)-2-Dimethylcarbamoyl-1-methylvinyl dimethylphosphate; *cis*-2-Dimethylcarbamoyl-1-methylvinyl dimethylphosphate; *O,O*-Dimethyl-*O*-(2-dimethylcarbamoyl-1-methyl-vinyl)phosphat (German); *o,o*-Dimethyl-*o*-(1,4-dimethyl-3-oxo-4-aza-pent-1-enyl)fosfaat (Dutch); *O,O*-Dimethyl *O*-(1,4-dimethyl-3-oxo-4-azapent-1-enyl) phosphate; *O,O*-Dimethyl *O*-(*N*,*N*-dimethylcarbamoyl-1-methylvinyl) phosphate; Dimethyl phosphate ester with 3-hydroxy-*N*,*N*-dimethyl-*cis*-crotonamide; Dimethyl phosphate of 3-hydroxy-*N*,*N*-dimethyl-*cis*-crotonamide; *o,o*-Dimetil-*o*-(1,4-dimetil-3-oxo-4-aza-pent-1-enil)-fosfato (Italian); Ektafos; Ektofos; ENT 24,482; 3-Hydroxy-*N*,*N*-dimethyl-(E)-crotonamide dimethyl phosphate; 3-Hydroxy-*N*,*N*-dimethyl-*cis*-crotonamide dimethyl phosphate; 3-Hydroxydimethyl crotonamide dimethyl phosphate; Karbicron; Phosphatede dimethyle et de 2-dimethylcarbamoyl 1-methyl vinyle (French); Phosphoric acid, 3-(dimethylamino)-1-methyl-3-oxo-1-propenyl dimethyl ester, (E)-; Phosphoric acid, dimethyl ester, ester with (E)-3-hydroxy-*N*,*N*-dimethylcrotonamide; Phosphoric acid, dimethyl ester, ester with *cis*-3-hydroxy-*N*,*N*-dimethylcrotonamide; SD 3562; Shell SD-3562

CAS Registry Number: 141-66-2

RTECS Number: TC3850000

DOT ID: UN3018

EEC Number: 015-073-00-4

Regulatory Authority

- Banned or Severely Restricted (DDR and Malaysia) (UN)[13]
- Air Pollutant Standard Set (ACGIH)[1] (Australia) (Israel) (Mexico) (Several States)[60] (Several Canadian Provinces)
- Superfund/EPCRA 40CFR355, Appendix B Extremely Hazardous Substances: TPQ = 100 lb (45.4 kg)
- Superfund/EPCRA 40CFR302.4 Reportable Quantity (RQ): EHS, 1 lb (0.454 kg)
- U.S. DOT Regulated Marine Pollutant (49CFR172.101, Appendix B)

Cited in U.S. State Regulations: Alaska (G), California (A, G), Connecticut (A), Florida (G), Illinois (G), Maine (G), Massachusetts (G), Michigan (G), Minnesota (G), Nevada (A), New Hampshire (G), New Jersey (G), North Dakota (A), Pennsylvania (G), Rhode Island (G), Virginia (A), West Virginia (G).

Description: 3-(dimethylamino)-1-methyl-3-oxo-1-propenyl dimethyl phosphate, is an amber liquid with a mild ester odor. Boiling point = 400°C. Decomposes below boiling point @ 75°C after storage for 31 days. Flash point ≥94°C. Hazard Identification (based on NFPA-704 M Rating System): Health 3, Flammability 1, Reactivity 0. Soluble in water.

Potential Exposure: Those involved in the manufacture, formulation and application of this insecticide and acaricide.

Incompatibilities: Attacks some metals: Corrosive to cast iron, mild steel, brass, and stainless steel304. Decomposes after prolonged storage, but is stable when stored in glass or polyethylene containers with temperatures to 40°C.

Permissible Exposure Limits in Air: There is no OSHA PEL. The ACGIH recommended airborne exposure limit 0.25 mg/m^3 TWA. NIOSH recommends the same exposure limit. Australia, Israel, Mexico, and the Canadian provinces of Alberta, BC, Ontario, and Quebec set the same limit as ACGIH and Alberta's STEL is 0.75 mg/m^3. The TWA bears the notation "skin" indicating the possibility of cutaneous absorption. Several states have set guidelines or standards for dicrotophos in ambient air[60] ranging from 2.5 μ/m^3 (North Dakota) to 4.0 μg/m^3 (Virginia) to 5.0 μg/m^3 (Connecticut) to 6.0 μg/m^3 (Nevada).

Determination in Air: OSHA versatile sampler-2; Toluene/Acetone; Gas chromatography/Flame photometric detection for sulfur, nitrogen, or phosphorus; NIOSH IV, Method #5600, Organophosphorus Pesticides.

Permissible Concentration in Water: No criteria set.

Routes of Entry: Inhalation, skin absorption, ingestion, skin and/or eye contact.

Harmful Effects and Symptoms

Short Term Exposure: Dicrotophos may affects the nervous system, causing convulsions, respiratory failure. Dicrotophos is a cholinesterase inhibitor which can penetrate the skin. Effects may be cumulative. It is extremely toxic. Probable human oral lethal dose is 5 – 50 mg/kg, 7 drops to 1 teaspoonful for a 70 kg (150 lb.) person. Closely related in toxicity to azodrin.

Acute exposure to dicrotophos may produce the following signs and symptoms; pinpoint pupils, blurred vision, headache, dizziness, muscle spasms, and profound weakness. Vomiting, diarrhea, abdominal pain, seizures, and coma may also occur. The heart rate may decrease following oral exposure or increase following dermal exposure. Hypotension (low blood pressure) is not uncommon. Respiratory symptoms include dyspnea (shortness of breath), respiratory depression, and respiratory paralysis. Psychosis may occur. The effects may be delayed.

Long Term Exposure: Dicrotophos is a cholinesterase inhibitor; cumulative effect is possible. May damage the nervous system causing numbness, "pins and needles," sensation and/or weakness of the hands and feet. Repeated exposure may cause personality changes of depression, anxiety or irritability.

Points of Attack: Respiratory system, lungs, central nervous system, cardiovascular system, skin, eyes, plasma and red blood cell cholinesterase.

Medical Surveillance: Before employment and at regular times after that, the following are recommended: Plasma and red blood cell cholinesterase levels (tests for the enzyme poisoned by this chemical). If exposure stops, plasma levels return to normal in 1 – 2 weeks while red blood cell levels may be reduced for 1 – 3 months.

When cholinesterase enzyme levels are reduced by 25% or more below preemployment levels, risk of poisoning is increased, even if results are in lower ranges of "normal." Reassignment to work not involving organophosphate or carbamate pesticides is recommended until enzyme levels recover. If symptoms develop or overexposure occurs, repeat the above tests as soon as possible and get an exam of the nervous system. Also consider complete blood count. Consider chest x-ray following acute overexposure. Do not drink any alcoholic beverages before or during use. Alcohol promotes absorption of organic phosphates.

First Aid: If this chemical gets into the eyes, remove any contact lenses at once and irrigate immediately for at least 15 minutes, occasionally lifting upper and lower lids. Seek medical attention immediately. If this chemical contacts the skin, remove contaminated clothing and wash immediately with soap and water. Speed in removing material from skin is of extreme importance. Shampoo hair promptly if contaminated. Seek medical attention immediately. If this chemical has been inhaled, remove from exposure, begin rescue breathing (using universal precautions) if breathing has stopped and CPR if heart action has stopped. Transfer promptly to a medical facility. Medical observation is recommended; effect may be delayed.

Note to Physician: 1,1'-trimethylenebis(4-formylpyridinium bromide) dioxime (a.k.a TMB-4 DIBROMIDE and TMV-4) have been used as an antidote for organophosphate poisoning.

Personal Protective Methods: Wear protective gloves and clothing to prevent any reasonable probability of skin contact. Safety equipment suppliers/manufacturers can provide recommendations on the most protective glove/clothing material for your operation.. All protective clothing (suits, gloves, footwear, headgear) should be clean, available each day, and put on before work. Contact lenses should not be worn when working with this chemical. Wear splash-proof chemical goggles and face shield unless full facepiece respiratory protection is worn. Employees should wash immediately with soap when skin is wet or contaminated. Provide emergency showers and eyewash.

Respirator Selection: *At any detectable concentration:* SCBAF:PD,PP (any MSHA/NIOSH approved self-contained breathing apparatus that has a full facepiece and is operated in a pressure-demand or other positive-pressure mode); or SAF:PD,PP:ASCBA (any supplied-air respirator that has a full facepiece and is operated in a pressure-demand or other positive-pressure mode in combination with an auxiliary, self-contained breathing apparatus operated in a pressure-demand or other positive pressure mode).

Storage: Prior to working with this chemical you should be trained on its proper handling and storage. Store in tightly closed containers in a cool, well ventilated area away from strong oxidizers, metals, strong bases and heat. Where possible, automatically pump liquid from drums or other storage containers to process containers.

Shipping: Organophosphorus pesticides, liquid, toxic, n.o.s. fall in Hazard Class 6.1 and dicrotophos in Packing Group I. It requires a "Poison" label.

Spill Handling: Evacuate and restrict persons not wearing protective equipment from area of spill or leak until cleanup is complete. Do not touch spilled material; stop leak if you can do so without risk. Use water spray to reduce vapors. Small spills: absorb with sand or other noncombustible absorbent material and place into containers for later disposal. Small dry spills: with clean shovel place material into clean, dry container and cover; move containers from spill area. Large spills: dike far ahead of spill for later disposal. Stay upwind; keep out of low areas. Ventilate closed spaces before entering them. Remove all ignition sources. It may be necessary to contain and dispose of this chemical as a hazardous waste. If material or contaminated runoff enters waterways, notify downstream users of potentially contaminated waters. Contact your Department of Environmental Protection or your regional office of the federal EPA for specific recommendations. If employees are required to clean-up spills, they must be properly trained and equipped. OSHA 1910.120(q) may be applicable.

Fire Extinguishing: This material may burn but does not ignite readily. Poisonous gases including nitrogen oxides, phosphorus oxides, carbon monoxide are produced in fire. Fire and runoff from fire control water may produce irritating or poisonous gases. Extinguish with dry chemical, carbon dioxide, water spray, fog, or foam. Move container from fire area if you can do so without risk. Fight fire from maximum distance. Dike fire control water for later disposal; do not scatter the material. Wear positive pressure breathing apparatus and special protective clothing. Containers may explode in fire. Storage containers and parts of containers may rocket great distances, in many directions. If material or contaminated runoff enters waterways, notify downstream users of potentially contaminated waters. Notify local health and fire officials and pollution control agencies. From a secure, explosion-proof location, use water spray to cool exposed containers. If cooling streams are ineffective (venting sound increases in volume and pitch, tank discolors, or shows any signs of deforming), withdraw immediately to a secure position. If employees are expected to fight fires, they must be trained and equipped in OSHA 1910.156.

Disposal Method Suggested: Dicrotophos decomposes after 7 days at 90°C and 31 days at 75°C. Hydrolysis is 50% complete in aqueous solutions at 38°C after 50 days at pH 9.1 (100 days are required at pH 1.1). Alkaline hydrolysis (NaOH) yields $(CH_3)_2NH$. Incineration is also recommended as a disposal method.[22] In accordance with 40CFR165 recommendations for the disposal of pesticides and pesticide containers. Must be disposed properly by following package label directions or by contacting your state pesticide or environmental control agency or by contacting your regional EPA office.

References

Sax N. I., Ed., "Dangerous Properties of Industrial Materials Report," 2, No. 5, 49–54 (1982)

U.S. Environmental Protection Agency, "Chemical Profile: Dicrotophos," Washington, DC, Chemical Emergency Preparedness Program (October 1998)

Dicyclohexylamine

Molecular Formula: $C_{12}H_{23}N$

Common Formula: C_6H_{11}-NH-C_6H_{11}

Synonyms: *N*-Cyclohexylcyclohexanamine; DCHA; Dicha; Di-Cha; Dicyklohexylamin (Czech); Dodecahydrodiphenylamine

CAS Registry Number: 101-83-7

RTECS Number: HY4025000

DOT ID: UN2565

Regulatory Authority

- Air Pollutant Standard Set (former USSR)[43]
- Canada, WHMIS, Ingredients Disclosure List

Cited in U.S. State Regulations: Florida (G), Massachusetts (G), New Hampshire (G), New Jersey (G), Pennsylvania (G).

Description: Dicyclohexylamine, C_6H_{11}-NH-C_6H_{11}, is a combustible, colorless liquid with a faint amine odor. Boiling point = 256°C. Flash point ≥99°C. Hazard Identification (based on NFPA-704 M Rating System): Health 3, Flammability 1, Reactivity 0. Slightly soluble in water.

Potential Exposure: Dicyclohexylamine salts of fatty acids and sulfuric acid have soap and detergent properties useful to the printing and textile industries. Metal complexes of DI-CHA are used as catalysts in the paint, varnish, and the ink industries. Several vapor-phase corrosion inhibitors are solid DI-CHA derivatives. These compounds are slightly colatile at normal temperatures and are used to protect packaged or stored ferrous metals from atmospheric corrosion. Dicyclohexylamine is also used for a number of other purposes: plasticizers; insecticidal formulations; antioxidant in lubricating oils, fuels, and rubber; and extractant.

Incompatibilities: Contact with strong oxidizers can cause fire and explosion hazard.

Permissible Exposure Limits in Air: The former USSR-UNEP/IRPTC project[43] has set a MAC of 1.0 mg/m³ in workplace air and a MAC for ambient air in residential areas of 0.008 mg/m³ on a momentary basis.

Permissible Concentration in Water: The former USSR-UNEP/IRPTC project[43] has set a MAC of 0.01 mg/l in water bodies used for domestic purposes.

Routes of Entry: Inhalation, ingestion.

Harmful Effects and Symptoms

Short Term Exposure: Corrosive to the skin and eyes; may cause burns and permanent damage. Dicyclohexylamine is somewhat more toxic that cyclohexylamine. Poisoning symptoms and death appear earlier in rabbits injected with 0.5 g/kg DI-CHA (as opposed to CHA). Doses of 0.25 g/kg are just sublethal, causing convulsions and reversible paralysis. Dicyclohexylamine is a skin irritant. The LD_{50} oral-rat is 373 mg/kg (moderately toxic). Symptoms of exposure include severe irritation of the eyes, skin and mucous membranes. Also nausea, vomiting, weakness and irritation of the gastrointestinal tract.

First Aid: *Skin Contact:*[52] Flood all areas of body that have contacted the substance with water. Don't wait to remove contaminated clothing; do it under the water stream. Use soap to help assure removal. Isolate contaminated clothing when removed to prevent contact by others.

Eye Contact: Remove any contact lenses at once. Flush eyes well with copious quantities of water or normal saline for at lest 20 – 30 minutes. Seek medical attention.

Inhalation: Leave contaminated area immediately; breathe fresh air. Proper respiratory protection must be supplied to any rescuers. If coughing, difficult breathing or any other symptoms develop, seek medical attention at once, even if symptoms develop many hours after exposure.

Ingestion: If convulsions are not present, give a glass or two of water or milk to dilute the substance. Assure that the person's airway is unobstructed and contact a hospital or poison center immediately for advice on whether or not to induce vomiting.

Personal Protective Methods: Wear protective gloves and clothing to prevent any reasonable probability of skin contact. Safety equipment suppliers/manufacturers can provide recommendations on the most protective glove/clothing material for your operation.. All protective clothing (suits, gloves, footwear, headgear) should be clean, available each day, and put on before work. Contact lenses should not be worn when working with this chemical. Wear splash-proof chemical goggles and face shield unless full facepiece respiratory protection is worn. Employees should wash immediately with soap when skin is wet or contaminated. Provide emergency showers and eyewash.

Respirator Selection: *At any detectable concentration:* SCBAF:PD,PP (any MSHA/NIOSH approved self-contained *st Aid:* *Skin Contact:*[52] Flood all areas of body that e contacted the substance with water. Don't wait to re-ve contaminated clothing; do it under the water stream. soap to help assure removal. Isolate contaminated cloth- when removed to prevent contact by others.

Contact: Remove any contact lenses at once. Immedi-ly flush eyes well with copious quantities of water or mal saline for at least 20 – 30 minutes. Seek medical ention.

alation: Leave contaminated area immediately; breathe sh air. Proper respiratory protection must be supplied to rescuers. If coughing, difficult breathing or any other nptoms develop, seek medical attention at once, even if nptoms develop many hours after exposure.

estion: Contact a physician, hospital or poison center at ce. If the victim is unconscious or convulsing, do not in-ce vomiting or give anything by mouth. Assure that his way is open and lay him on his side with his head lower n his body and transport immediately to a medical facility. conscious and not convulsing, give a glass of water to ute the substance. Vomiting should not be induced without hysician's advice.

te to Physician: Treat for methemoglobinemia. Spectro-otometry may be required for precise determination of lev-of methemoglobinemia in urine.

rsonal Protective Methods: Wear protective gloves and thing to prevent any reasonable probability of skin con-t. Safety equipment suppliers/manufacturers can provide commendations on the most protective glove/clothing ma-ial for your operation.. All protective clothing (suits, gloves, otwear, headgear) should be clean, available each day, and t on before work. Contact lenses should not be worn when rking with this chemical. Wear dust-proof chemical goggles d face shield unless full facepiece respiratory protection is rn. Employees should wash immediately with soap when in is wet or contaminated. Provide emergency showers and ewash.

spirator Selection: *At any detectable concentration:* CBAF:PD,PP (any MSHA/NIOSH approved self-con-ined breathing apparatus that has a full facepiece and operated in a pressure-demand or other positive-pres-re mode); or SAF:PD,PP:ASCBA (any supplied-air res-rator that has a full facepiece and is operated in a pres-re-demand or other positive-pressure mode in combi-tion with an auxiliary, self-contained breathing appara-s operated in a pressure-demand or other positive pres-re mode.

orage: Prior to working with this chemical you should be ained on its proper handling and storage Store in a refrigera-r under an inert atmosphere for long-term storage.

ipping: Dicyclohexylammonium nitrite requires a "Flam-able Solid" label. It falls in DOT Hazard Class 4.1 and Pack-g Group III.

Spill Handling: Evacuate persons not wearing protective equipment from area of spill or leak until clean-up is complete. Remove all ignition sources. Dampen spilled material with water to avoid airborne dust. Collect powdered material in the most convenient and safe manner and deposit in sealed containers. Ventilate area after clean-up is complete. It may be necessary to contain and dispose of this chemical as a hazardous waste. If material or contaminated runoff enters waterways, notify downstream users of potentially contaminated waters. Contact your Department of Environmental Protection or your regional office of the federal EPA for specific recommendations. If employees are required to clean-up spills, they must be properly trained and equipped. OSHA 1910.120(q) may be applicable.

Fire Extinguishing: This chemical is a flammable solid. Use dry chemical, carbon dioxide, water spray, or foam extinguishers. Poisonous gases are produced in fire including nitrogen oxides and nitrous acid. If material or contaminated runoff enters waterways, notify downstream users of potentially contaminated waters. Notify local health and fire officials and pollution control agencies. From a secure, explosion-proof location, use water spray to cool exposed containers. If cooling streams are ineffective (venting sound increases in volume and pitch, tank discolors, or shows any signs of deforming), withdraw immediately to a secure position. If employees are expected to fight fires, they must be trained and equipped in OSHA 1910.156.

Disposal Method Suggested: Incineration; incinerator is equipped with a scrubber or thermal unit to reduce No_x emissions.

References

U.S. Environmental Protection Agency, "Chemical Hazard Information Profile: Cyclohexylamine," Washington, DC (October 21, 1977)

Dicyclopentadiene

Molecular Formula: $C_{10}H_{12}$

Synonyms: Bicyclopentadiene; Biscyclopentadiene; 1,3-Cyclopentadiene, dimer; DCPD; Diciclopentadieno (Spanish); Dicyklopentadien (Czech); Dimer cyklopentadienu (Czech); 4,7-Methano-1H-indene; 4,7-Methano-1H-indene, 3a,4,7,7a-tetrahydro-; 3a,4,7,7a-Tetrahydro-4,7-methanoindene

CAS Registry Number: 77-73-6

RTECS Number: PC1050000

DOT ID: UN 2048

Regulatory Authority

- Air Pollutant Standard Set (ACGIH)[1] (Australia) (DFG) (Israel) (Mexico) (former USSR)[43] (Several States)[60] (Several Canadian Provinces)
- EPCRA Section 313 Form R *de minimis* concentration reporting level: 1.0%
- Canada, WHMIS, Ingredients Disclosure List

breathing apparatus that has a full facepiece and is operated in a pressure-demand or other positive-pressure mode); or SAF:PD,PP:ASCBA (any supplied-air respirator that has a full facepiece and is operated in a pressure-demand or other positive-pressure mode in combination with an auxiliary, self-contained breathing apparatus operated in a pressure-demand or other positive pressure mode).

Storage: Prior to working with DI-CHA you should be trained on its proper handling and storage. Store in tightly closed containers in a refrigerator away from oxidizers. Where possible, automatically pump liquid from drums or other storage containers to process containers. Protection from air is recommended for long term storage.

Shipping: Dicyclohexylamine requires a "Corrosive" label. If falls in DOT Hazard Class 8 and Packing Group III.

Spill Handling: Evacuate and restrict persons not wearing protective equipment from area of spill or leak until cleanup is complete. Remove all ignition sources. Ventilate area of spill or leak. Absorb liquids in vermiculite, dry sand, earth, peat, carbon, or a similar material and deposit in sealed containers. Keep this chemical out of a confined space, such as a sewer, because of the possibility of an explosion, unless the sewer is designed to prevent the build-up of explosive concentrations. It may be necessary to contain and dispose of this chemical as a hazardous waste. If material or contaminated runoff enters waterways, notify downstream users of potentially contaminated waters. Contact your Department of Environmental Protection or your regional office of the federal EPA for specific recommendations. If employees are required to clean-up spills, they must be properly trained and equipped. OSHA 1910.120(q) may be applicable.

Fire Extinguishing: This chemical is a combustible liquid. Poisonous gases are produced in fire. Use dry chemical, carbon dioxide, or alcohol foam extinguishers. Vapors are heavier than air and will collect in low areas. Vapors may travel long distances to ignition sources and flashback. Vapors in confined areas may explode when exposed to fire. Containers may explode in fire. Storage containers and parts of containers may rocket great distances, in many directions. If material or contaminated runoff enters waterways, notify downstream users of potentially contaminated waters. Notify local health and fire officials and pollution control agencies. From a secure, explosion-proof location, use water spray to cool exposed containers. If cooling streams are ineffective (venting sound increases in volume and pitch, tank discolors, or shows any signs of deforming), withdraw immediately to a secure position. If employees are expected to fight fires, they must be trained and equipped in OSHA 1910.156.

Disposal Method Suggested: Incineration; incinerator is equipped with a scrubber or thermal unit to reduce No_x emissions.

References

U.S. Environmental Protection Agency, "Chemic[al In]formation Profile: Cyclohexylamine," Washing[ton, DC (Oc]tober 21, 1977)

Dicyclohexylamine Nitrite

Molecular Formula: $C_{12}H_{24}N_2O_2$

Common Formula: C_6H_{11}-NH-$C_6H_{11}HNO_2$

Synonyms: Dechan; Di-Chan; Dichan (Czec[h]; ... hexylaminonitrite; Dicyclohexylammonium ... cyklohexylamin nitrit (Czech); Dicynit (Czec[h]); ... hydrophenylamine nitrite; Dusitan dicyklol... (Czech)

CAS Registry Number: 3129-91-7

RTECS Number: HY4200000

DOT ID: UN 2687

Regulatory Authority

- Air Pollutant Standard Set (former USSR)[43]

Description: Dicyclohexylamine nitrite, ... $C_6H_{11}HNO_2$, is a flammable white powder whi[ch] ... volatility at room temperature and higher.

Potential Exposure: It is used s a vapor-phas[e] ... inhibitor whereby it vaporizes either from the s... from solution and offers protection against atmos... ing. Wrapping paper, plastic wraps, and other m... be impregnated with di-Chan to protect metal ... packaging and storage.

Incompatibilities: Nitrates are strong oxidizers. C... reducing materials and easily oxidized materials ... fire and explosion hazard.

Permissible Exposure Limits in Air: The for[mer] ... UNEP/IRPTC joint project[43] has set a MAC of ... in workplace air and a MAC in ambient air in res... eas of 0.02 mg/m^3.

Permissible Concentration in Water: The for[mer] ... UNEP/IRPTC project[43] has set a MAC in water ... for domestic purposes of 0.01 mg/l.

Routes of Entry: Inhalation, ingestion.

Harmful Effects and Symptoms

Short Term Exposure: The oral LD_{50} rat is ... (moderately toxic). It decreases blood pressure.

Long Term Exposure: Prolonged exposure t[o] ... hexylamine nitrite vapor is reported to lead to ... the CNS, erythrocytes, and methemoglobinemia, ... turb the functional state of the liver and kidneys ... workers.

Points of Attack: Blood, liver, kidneys.

Medical Surveillance: Blood pressure. Compl[ete blood] count. Liver and kidney function tests.

Cited in U.S. State Regulations: Alaska (G), California (A, G), Connecticut (A), Florida (G), Illinois (G), Maine (G), Massachusetts (G), Minnesota (G), Nevada (A), New Hampshire (G), New Jersey (G), North Dakota (A), Pennsylvania (G), Rhode Island (G), Virginia (A), West Virginia (G).

Description: Cyclopentadiene, $C_{10}H_{12}$, is a crystalline solid or a liquid (above 32°C) with a disagreeable, camphor-like odor. The odor threshold is 0.011 (detectable); 0.020 ppm (recognizable) (AIHA). Boiling point = (decomposes) 172°C. Freezing/Melting point = 32°C. Flash point = 32°C. Autoignition temperature = 503°C. Hazard Identification (based on NFPA-704 M Rating System): Health 1, Flammability 3, Reactivity 1. Explosive limits: LEL = 0.8; UEL = 6.3. Insoluble in water.

Potential Exposure: This compound is used in the manufacture of cyclopentadiene as a pesticide intermediate. It is used in the production of ferrocene compounds. It is used in paints and varnishes and resin manufacture.

Incompatibilities: Forms explosive mixture with air above flash point. Depolymerizes at boiling point and forms two molecules of cyclopentadiene; unless inhibited and maintained under inert atmosphere to prevent polymerization. Violent reaction with strong oxidizers, strong acids, strong bases. Can accumulate static electrical charges, and may cause ignition of its vapors.

Permissible Exposure Limits in Air: There is no Federal Limit (OSHA PEL).[58] The NIOSH and ACGIH recommended TWA value is 5 ppm (30 mg/m^3). Australia, Mexico, Israel, and the Canadian provinces of Alberta, BC, Ontario, and Quebec set the same TWA level as ACGIH and the Alberta STEL is 10 ppm. The DFG MAK is 0.5 ppm (2.7 mg/m^3). The former USSR-UNEP/IRPTC project[43] has set a MAC of 1.0 mg/m^3 in workplace air. Several states have set guidelines or standards for dicyclopentadiene in ambient air[60] ranging from 0.3 mg/m^3 (North Dakota) to 0.5 mg/m^3 (Virginia) to 0.6 mg/m^3 (Connecticut) to 0.714 mg/m^3 (Nevada).

Determination in Air: No method available.

Permissible Concentration in Water: No criteria set.

Routes of Entry: Inhalation, ingestion, skin and/or eye contact.

Harmful Effects and Symptoms

Short Term Exposure: Dicyclopentadiene irritates the skin and eyes. Human subjects reported an odor threshold at 0.003 ppm, slight eye or throat irritation at 1 – 5 ppm. May cause headache, loss of balance and coordination, and even convulsions. Increased urinary frequency have been reported in exposed workers.

Long Term Exposure: May affect the liver, kidneys, nervous system, and gastro-intestinal tract. May cause lung irritation with possible cough and shortness of breath. Animal experiments produced leukocytosis and kidney lesions.

Points of Attack: Eyes, skin, respiratory system, central nervous system, kidneys.

Medical Surveillance: Before beginning employment and at regular times after that, for those with frequent or potentially high exposures, the following is recommended: Lung and kidney function tests. If symptoms develop or overexposure is suspected, the following may be useful: Exam of the nervous system and kidney function tests.

First Aid: If this chemical gets into the eyes, remove any contact lenses at once and irrigate immediately for at least 15 minutes, occasionally lifting upper and lower lids. Seek medical attention immediately. If this chemical contacts the skin, remove contaminated clothing and wash immediately with soap and water. Seek medical attention immediately. If this chemical has been inhaled, remove from exposure, begin rescue breathing (using universal precautions) if breathing has stopped and CPR if heart action has stopped. Transfer promptly to a medical facility. When this chemical has been swallowed, get medical attention. Give large quantities of water and induce vomiting. Do not make an unconscious person vomit.

Personal Protective Methods: Wear protective gloves and clothing to prevent any reasonable probability of skin contact. Safety equipment suppliers/manufacturers can provide recommendations on the most protective glove/clothing material for your operation.. All protective clothing (suits, gloves, footwear, headgear) should be clean, available each day, and put on before work. Contact lenses should not be worn when working with this chemical. Wear dust-proof chemical goggles when working with powders or dusts and splash-proof goggles when working with liquids unless full facepiece respiratory protection is worn. Employees should wash immediately with soap when skin is wet or contaminated. Provide emergency showers and eyewash.

Respirator Selection: Where the potential exists for exposures over 5 ppm, use an MSHA/NIOSH approved respirator with a combination cartridge offering protection against organic vapors as well as against dust, fume, and mist. More protection is provided by a full facepiece respirator than by a half mask respirator, and even greater protection is provided by a powered-air purifying respirator. If while wearing a filter, cartridge or canister respirator, you can smell, taste, or otherwise detect dicyclopentadiene, or in the case of a full facepiece respirator you experience eye irritation, leave the area immediately. Check to make sure the respirator-to-face seal is still good. If it is, replace the filter, cartridge, or canister. If the seal is no longer good, you may need a new respirator.

Where the potential for high exposures exists, use an MSHA/NIOSH approved supplied-air respirator with a full facepiece operated in the positive pressure mode or with a full facepiece, hood, or helmet in the continuous flow mode, or use an MSHA/NIOSH approved self-contained breathing apparatus with a full facepiece operated in pressure-demand or other positive pressure mode.

Storage: Prior to working with this chemical you should be trained on its proper handling and storage. Before entering confined space where this chemical may be present, check to make sure that an explosive concentration does not exist. Store in tightly closed containers in a cool, well-ventilated area away from oxidizing materials, strong acids, and strong bases. Sources of ignition such as smoking and open flames are prohibited where this chemical is used, handled, or stored in a manner that could create a potential fire or explosion hazard. Metal containers involving the transfer of 5 gallons or more of this chemical should be grounded and bonded. Drums must be equipped with self-closing valves, pressure vacuum bungs, and flame arresters. Use only nonsparking tools and equipment, especially when opening and closing containers of this chemical.

Shipping: Dicyclopentadiene must carry a "Flammable Liquid" label. It falls in DOT Hazard Class 3 and Packing Group III.

Spill Handling: *Powders and dusts:* Evacuate persons not wearing protective equipment from area of spill or leak until clean-up is complete. Remove all ignition sources. Collect powdered material in the most convenient and safe manner and deposit in sealed containers. Ventilate area after clean-up is complete. It may be necessary to contain and dispose of this chemical as a hazardous waste. If material or contaminated runoff enters waterways, notify downstream users of potentially contaminated waters. Contact your Department of Environmental Protection or your regional office of the federal EPA for specific recommendations. If employees are required to clean-up spills, they must be properly trained and equipped. OSHA 1910.120(q) may be applicable.

Liquid: Evacuate and restrict persons not wearing protective equipment from area of spill or leak until cleanup is complete. Remove all ignition sources. Establish forced ventilation to keep levels below explosive limit. Absorb liquids in vermiculite, dry sand, earth, peat, carbon, or a similar material and deposit in sealed containers. Keep this chemical out of a confined space, such as a sewer, because of the possibility of an explosion, unless the sewer is designed to prevent the build-up of explosive concentrations. It may be necessary to contain and dispose of this chemical as a hazardous waste. If material or contaminated runoff enters waterways, notify downstream users of potentially contaminated waters. Contact your Department of Environmental Protection or your regional office of the federal EPA for specific recommendations. If employees are required to clean-up spills, they must be properly trained and equipped. OSHA 1910.120(q) may be applicable.

Fire Extinguishing: This chemical is a flammable liquid. Poisonous gases are produced in fire. Use dry chemical, carbon dioxide, or foam extinguishers. Vapors are heavier than air and will collect in low areas. Vapors may travel long distances to ignition sources and flashback. Vapors in confined areas may explode when exposed to fire. Containers may explode in fire. Storage containers and parts of containers may rocket great distances, in many directions. If material or contaminated runoff enters waterways, notify downstream users of potentially contaminated waters. Notify local health and fire officials and pollution control agencies. From a secure, explosion-proof location, use water spray to cool exposed containers. If cooling streams are ineffective (venting sound increases in volume and pitch, tank discolors, or shows any signs of deforming), withdraw immediately to a secure position. If employees are expected to fight fires, they must be trained and equipped in OSHA 1910.156.

Disposal Method Suggested: Incineration.

References

New Jersey Department of Health, "Hazardous Substance Fact Sheet: Dicyclopentadiene," Trenton, NJ (February 1989)

Dicyclopentadienyl Iron

Molecular Formula: $C_{10}H_{10}Fe$

Common Formula: $(C_5H_5)_2Fe$

Synonyms: Bis (cyclopentadienyl) iron; Bis-cyclopentadienyl iron; Ferrocene; Iron dicyclopentadienyl

CAS Registry Number: 102-54-5

RTECS Number: LK0700000

DOT ID: NA 9188

Regulatory Authority

- Air Pollutant Standard Set (ACGIH)[1] (Australia) (Israel) (Mexico) (HSE)[33] (Several Canadian Provinces)
- Canada, WHMIS, Ingredients Disclosure List

Cited in U.S. State Regulations: Alaska (G), California (A, G), Florida (G), Illinois (G), Maine (G), Massachusetts (G), Minnesota (G), New Hampshire (G), New Jersey (G), Pennsylvania (G), Rhode Island (G), West Virginia (G).

Description: Dicyclopentadienyl iron, $(C_5H_5)_2Fe$, is a bright orange crystalline solid with a camphor-like odor. Boiling point = 249°C. Freezing/Melting point = 173 – 174°C. Insoluble in water.

Potential Exposure: It is used as an additive for furnace oils, gasoline, and jet fuels to reduce combustion smoke. It is used in making rubber, silicone resins and high temperature polymers, and in coatings for missiles and satellites.

Incompatibilities: Violent reaction with ammonium perchlorate, tetranitromethane, mercury(II) nitrate. Incomparable with oxidizers.

Permissible Exposure Limits in Air: OSHA[58] has set a TWA for the respirable fraction of 5 mg/m^3 and a TWA for total dust of 15 mg/m^3. NIOSH recommends a TWA for the respirable fraction of 5 mg/m^3 and a TWA for total dust of 10 mg/m^3. ACGIH recommends a TWA value of 10 mg/m^3. HSE,[33] Israel, Mexico, and the Canadian provinces of

Alberta, BC, Ontario, and Quebec have adopted the same TWA value as ACGIH, and HSE, Alberta, British Columbia, and Mexico set a STEL of 20 mg/m^3.

Determination in Air: Filter; H_2SO_4/H_2O_2/Hydrochloric acid; Inductively coupled plasma; OSHA Method #ID125G, Iron Oxide Fume.

Permissible Concentration in Water: No criteria set, but EPA[32] has suggested a permissible ambient goal of 530 μg/l based on health effects.

Routes of Entry: Inhalation, ingestion, skin and/or eye contact.

Harmful Effects and Symptoms

Short Term Exposure: Irritates the eyes, skin, and respiratory system. Dicyclopentadienyl iron is classified as a slightly toxic material, but the toxicological properties have not been extensively investigated. The LD_{50} oral-rat is 1,320 mg/kg (slightly toxic). Dicyclopentadienyl iron can affect you when breathed in.

Long Term Exposure: Repeated high exposures may cause mood changes such as irritability. Dicyclopentadienyl iron may cause mutations. Handle with extreme caution. Exposure may affect liver and lung functions. In animals: liver, red blood cell, testicular changes. (NIOSH)

Points of Attack: Eyes, skin, respiratory system, liver, blood, reproductive system.

Medical Surveillance: For those with frequent or potentially high exposure (half the TLV or greater), the following are recommended before beginning work and at regular times after that: Liver function tests. Lung function tests. Complete blood count (CBC). Serum iron level.

First Aid: If this chemical gets into the eyes, remove any contact lenses at once and irrigate immediately for at least 15 minutes, occasionally lifting upper and lower lids. Seek medical attention immediately. If this chemical contacts the skin, remove contaminated clothing and wash immediately with soap and water. Seek medical attention immediately. If this chemical has been inhaled, remove from exposure, begin rescue breathing (using universal precautions) if breathing has stopped and CPR if heart action has stopped. Transfer promptly to a medical facility. When this chemical has been swallowed, get medical attention. Give large quantities of water and induce vomiting. Do not make an unconscious person vomit.

Personal Protective Methods: Wear protective gloves and clothing to prevent any reasonable probability of skin contact. Safety equipment suppliers/manufacturers can provide recommendations on the most protective glove/clothing material for your operation.. All protective clothing (suits, gloves, footwear, headgear) should be clean, available each day, and put on before work. Contact lenses should not be worn when working with this chemical. Wear dust-proof chemical goggles and face shield unless full facepiece respiratory protection is worn. Employees should wash immediately with soap when skin is wet or contaminated. Provide emergency showers and eyewash.

Respirator Selection: Where the potential exists for exposures over 10 mg/m^3, use a MSHA/NIOSH approved respirator equipped with particulate (dust/fume/mist) filters. Particulate filters must be checked every day before work for physical damage, such as rips or tears, and replaced as needed. If while wearing a filter, cartridge or canister respirator, you can smell, taste, or otherwise detect Dicyclopentadienyl iron, or in the case of a full facepiece respirator you experience eye irritation, leave the area immediately. Check to make sure the respirator-to-face seal is still good. If it is, replace the filter, cartridge, or canister. If the seal is no longer good, you may need a new respirator.

Be sure to consider all potential exposures in your workplace. You may need a combination of filters, profilers, cartridges, or canisters to protect against different forms of a chemical (such as vapor and mist) or against a mixture of chemicals. Where the potential for high exposures exists, use a MSHA/NIOSH approved supplied-air respirator with a full facepiece operated in the positive pressure mode or with a full facepiece, hood, or helmet in the continuous flow mode, or use a MSHA/ NIOSH approved self-contained breathing apparatus with a full facepiece operated in pressure-demand or other positive pressure mode.

Storage: Prior to working with this chemical you should be trained on its proper handling and storage. Store in tightly closed containers in a cool, well-ventilated area away from oxidizers, ammonium perchlorate, tetranitromethane, mercury(II) nitrate, and heat. Sources of ignition such as smoking and open flames, are prohibited where this chemical is used, handled, or stored in a manner that could create a potential fire or explosion hazard.

Shipping: Ferrocene is not specifically cited by DOT in its Performance-Oriented Packaging Standards but it can be considered under Hazardous Substances, Solid, n.o.s. which imposes no label requirements or maximum shipping quantity limits.

Spill Handling: Evacuate persons not wearing protective equipment from area of spill or leak until clean-up is complete. Remove all ignition sources. Collect powdered material in the most convenient and safe manner and deposit in sealed containers. Ventilate area after clean-up is complete. It may be necessary to contain and dispose of this chemical as a hazardous waste. If material or contaminated runoff enters waterways, notify downstream users of potentially contaminated waters. Contact your Department of Environmental Protection or your regional office of the federal EPA for specific recommendations. If employees are required to clean-up spills, they must be properly trained and equipped. OSHA 1910.120(q) may be applicable.

Fire Extinguishing: Dicyclopentadienyl iron is a combustible solid. Extinguish fire using an agent suitable for type of surrounding fire. Dicyclopentadienyl iron itself does not burn. Poisonous gases are produced in fire. If material or contaminated runoff enters waterways, notify downstream users of potentially contaminated waters. Notify local health and fire officials and pollution control agencies. From a secure, explosion-proof location, use water spray to cool exposed containers. If cooling streams are ineffective (venting sound increases in volume and pitch, tank discolors, or shows any signs of deforming), withdraw immediately to a secure position. If employees are expected to fight fires, they must be trained and equipped in OSHA 1910.156.d in OSHA 1910.156.

References

Sax N. I., Ed., "Dangerous Properties of Industrial Materials Report," 1, No. 4, 67–68 (1981)

New Jersey Department of Health, "Hazardous Substance Fact Sheet: Dicyclopentadienyl Iron," Trenton, NJ (October 1998)

Sax N. I., Ed., "Dangerous Properties of Industrial Materials Report," 1, No. 4, 67–68 (1981) (as Ferrocene)

Dieldrin

Molecular Formula: $C_{12}H_8Cl_6O$

Synonyms: Alvit; Compound 497; Dieldrex; Dieldrina (Spanish); Dieldrine (French); Dieldrite; 2,7:3,6-Dimethanonaphtha(2,3b)oxirene,3,4,5,6,9,9-hexachloro-1a,2,2a,3,6,6a,7,7a-octahydro-(1a α,2.β,2a.α,3β,6.β,6aα,7β,7aα); ENT 16,225; HEOD; 1,2,3,4,10,10-Hexachloro-6,7-epoxy-1,4,4a,5,6,7,8,8a-octahydro-1,4-endo,exo-5,8-di-methanonaphthalene; Hexachloroepoxyoctahydro-endoexo-dimethanonaphthalene; 3,4,5,6,9,9-Hexachloro-1a,2,2a,3,6,6a,7,7a-octahydro-2,7:3,6-dimethano; 3,4,5,6,9,9-Hexachloro-1a, 2, 2a, 3, 6, 6a, 7, 7a-octahydro-2,7:3,6-dimethanonaphth(2,3-b)oxirene; (1r,4s,4as,5r,6r,7s,8s,8ar)1,2,3,4,10,10-Hexachloro-1,4,4a,5,6,7,8,8a-octahydro-6,7-epoxy-1,4:5,8-dimethanonaphthalene; Illoxol; Killgerm dethlac insecticidal laquer; NCI-C00124; Octalox; Oxralox; Panoram; Panoram D-31; Quintox

CAS Registry Number: 60-57-1

RTECS Number: IO1750000

DOT ID: UN/NA 2761

EEC Number: 602-049-00-9

Regulatory Authority

- Carcinogen (Animal Positive) (IARC) (NCI)[9]
- Banned or Severely Restricted (Many Countries) (UN)[13]
- Air Pollutant Standard Set (ACGIH)[1] (Australia) (DFG)[3] (HSE)[33] (Israel) (Mexico) (former USSR)[35][43] (OSHA)[58] (Several States)[60]
- Clean Water Act: Section 311 Hazardous Substances/RQ 40CFR117.3 (same as CERCLA, see below); 40CFR423, Appendix A, Priority Pollutants; Section 313 Water Priority Chemicals (57FR41331, 9/9/92); Toxic Pollutant (Section 401.15)
- EPA Hazardous Waste Number (RCRA No.): P037
- RCRA, 40CFR261, Appendix 8 Hazardous Constituents
- RCRA 40CFR268.48; 61FR15654, Universal Treatment Standards: Wastewater (mg/l), 0.017; Nonwastewater (mg/kg), 0.13
- RCRA 40CFR264, Appendix 9; TSD Facilities Ground Water Monitoring List. Suggested test method(s) (PQL μg/l): 8080 (0.05); 8270 (10)
- Superfund/EPCRA 40CFR302.4 Reportable Quantity (RQ): CERCLA, 1 lb (0.454 kg)
- U.S. DOT Regulated Marine Pollutant (49CFR172.101, Appendix B)
- Mexico, Drinking Water Criteria, 0.0000007 mg/l

Cited in U.S. State Regulations: Alaska (G), California (A, G, W), Connecticut (A), Florida (G), Illinois (G, W), Kansas (G, A, W), Louisiana (G), Maine (G), Massachusetts (G), Michigan (G), Minnesota (W), Nevada (A), New Hampshire (G, W), New Jersey (G), New York (G), North Dakota (A), Oklahoma (G), Pennsylvania (G, A), Rhode Island (G), Vermont (G), Virginia (G, A), Washington (G), West Virginia (G), Wisconsin (G).

Description: Dieldrin, $C_{12}H_8Cl_6O$, is a colorless to light tan solid with a mild chemical odor. The odor threshold in water is 0.04 mg/l. Boiling point = (decomposes). Freezing/Melting point = 175 – 176°C. Hazard Identification (based on NFPA-704 M Rating System): Health 3, Flammability 0, Reactivity 0. Insoluble in water.

Potential Exposure: Aldrin and dieldrin are manmade compounds belonging to the group of cyclodiene insecticides. They are a subgroup of the chlorinated cyclic hydrocarbon insecticides which include DDT, BHC, etc. They were manufactured in the United States by Shell Chemical Co. until the U.S. EPA prohibited their manufacture in 1974 under the Federal Insecticide, Fungicide, and Rodenticide Act. The primary use off the chemicals in the past was for control of corn pests, although they were also used by the citrus industry.

Dieldrin's persistence in the environment is due to its extremely low volatility (i.e., a vapor pressure of 1.78×10^{-7} mm mercury at 20°C), and low solubility in water (186 μg/l at 25 – 29°C). In addition, dieldrin is extremely apolar, resulting in a high affinity for fat which accounts for its retention in animal fats, plant waxes, and other such organic matter in the environment. The fat solubility of dieldrin results in the progressive accumulation in the food chain which may result in a concentration in an organism which would exceed the lethal limit for a consumer.

Incompatibilities: Incompatible with strong acids: concentrated mineral acids, acid catalysts, phenols, strong oxidizers, phenols, active metals, like sodium, potassium, magnesium, and zinc. Keep away from copper, iron, and their salts.

Permissible Exposure Limits in Air: The Federal Limit (OSHA PEL),[58] as well as the recommended ACGIH TWA value is 0.25 mg/m^3. The DFG MAK, HSE, Australian, Israeli,

Mexican and Canadian provincial (Alberta, BC, Ontario, and Quebec) TWA is the same as OSHA and the STEL value set by Mexico, Alberta, British Columbia, and HSE[33] is 0.75 mg/m^3. The ceiling value in Germany[35] is 2.5 mg/m^3. The notation "skin" indicates the possibility of cutaneous absorption. The NIOSH IDLH level is [Ca] 50 mg/m^3. The former USSR[35][43] has set a MAC in workplace air of 0.01 mg/m^3.

Several states have set guidelines or standards for dieldrin in ambient air[60] ranging from 0.035 μg/m^3 (Pennsylvania) to 0.595 μg/m^3 (Kansas) to 2.5 μg/m^3 (North Dakota) to 4.0 μg/m^3 (Virginia) to 5.0 μg/m^3 (Connecticut) to 6.0 μg/m^3 (Nevada).

Determination in Air: Filter; Isooctane; Gas chromatography/Electrochemical detection; NIOSH II(3), Method #S283.

Permissible Concentration in Water: To protect freshwater aquatic life – 0.0019 μg/l as a 24 hr average, never to exceed 2.5 μg/l. To protect saltwater aquatic life – 0.0019 μg/l as a 24 hr average never to exceed 0.71 μg/l. To protect human health – preferably zero. An additional lifetime cancer risk of 1 in 100,000 results at a level of 0.71 ng/l (0.00071 μg/l).[6]

Mexico[35] has set MAC values for dieldrin of 0.0000007 mg/l in water used for drinking water supply; of 0.003 mg/l in estuaries and 0.03 μg/l in estuaries. WHO[35] has set a limit of 0.03 μg/l in drinking water. A NOAEL (no observed adverse effects level) of 0.005 mg/kg/day has been calculated by EPA which results in the calculation of a drinking water equivalent of 1.75 μg/l. No lifetime health advisory could be calculated in view of the cancer risk. Several states have set standards or guidelines for dieldrin in drinking water[61] ranging from 0.01 μg/l (Minnesota) to 0.019 μg/l (Kansas) to 0.05 μg/l (California) to 1.0 μg/l (Illinois).

Determination in Water: Methylene chloride extraction followed by gas chromatography with electron capture or halogen specific detection (EPA Method 608) or gas chromatography plus mass spectrometry (EPA Method 625).

Routes of Entry: Inhalation, skin absorption, ingestion, eye and/or skin contact.

Harmful Effects and Symptoms

During the past decade, considerable information has been generated concerning the toxicity and potential carcinogenicity of the two organochlorine pesticides, aldrin and dieldrin. These two pesticides are usually considered together since Aldrin is readily epoxidized to dieldrin in the environment. Both are acutely toxic to most forms of life including arthropods, mollusks, invertebrates, amphibians, reptiles, fish, birds and mammals. Dieldrin is extremely persistent in the environment. By means of bioaccumulation it is concentrated many times as it moves up the food chain.

Short Term Exposure: *Inhalation:* May cause nausea, drowsiness, loss of appetite, visual disturbances and insomnia. Sprays of 1 – 2½% have caused giddiness, headache, muscle twitching, convulsions and loss of consciousness.

Skin: Can be absorbed to cause or increase the severity of symptoms as listed under ingestion. Contact may cause skin rash.

Eyes: May cause irritation, redness, and affect vision.

Ingestion: Can cause headache, nausea, irritability, insomnia, high blood pressure, vision problems, loss of coordination, profuse sweating, dizziness, frothing at the mouth, convulsions and loss of consciousness. Death may occur from as little as 1/20 ounce (1.4 gram). Some symptoms may be delayed up to 12 hours.

Exposure to dieldrin may affects the central nervous system, resulting in convulsions.

Long Term Exposure: May cause liver damage. Dieldrin accumulates in the human body. Dieldrin has caused cancer in laboratory animals. It is considered a suspect occupational carcinogen. May damage the developing fetus. May reduce fertility in males and females. Dieldrin concentrates in breast milk, and therefore, may be transferred to breast feeding infants. Repeated higher exposure can cause tremors, muscle twitching and seizures (convulsions) and may lead to coma and death. Convulsions are somewhat delayed and may occur weeks or months following exposure. Repeated exposure may cause personality changes of depression, anxiety or irritability.

Points of Attack: Central nervous system, liver, kidneys, skin. *Cancer Site* (in animals): lung, liver, thyroid and adrenal gland tumors.

Medical Surveillance: Before employment and at regular times after that, the following are recommended: Plasma and red blood cell cholinesterase levels (tests for the enzyme poisoned by this chemical). If exposure stops, plasma levels return to normal in 1 – 2 weeks while red blood cell levels may be reduced for 1 – 3 months.

When cholinesterase enzyme levels are reduced by 25% or more below preemployment levels, risk of poisoning is increased, even if results are in lower ranges of "normal." Reassignment to work not involving organophosphate or carbamate pesticides is recommended until enzyme levels recover. If symptoms develop or overexposure occurs, repeat the above tests as soon as possible and get an exam of the nervous system. Also consider complete blood count. Consider chest x-ray following acute overexposure. Do not drink any alcoholic beverages before or during use. Alcohol promotes absorption of organic phosphates. Blood dieldrin level. Examination of the nervous system. If symptoms develop or overexposure is suspected, the following may be useful: Liver function tests. EEG. Blood dieldrin levels (Normal = less than 1 mg/100 ml; level should not exceed 15 mg/100 ml). Examination of the nervous system.

First Aid: If this chemical gets into the eyes, remove any contact lenses at once and irrigate immediately for at least 15 minutes, occasionally lifting upper and lower lids. Seek medical attention immediately. If this chemical contacts the

skin, remove contaminated clothing and wash immediately with soap and water. Seek medical attention immediately. If this chemical has been inhaled, remove from exposure, begin rescue breathing (using universal precautions) if breathing has stopped and CPR if heart action has stopped. Transfer promptly to a medical facility. When this chemical has been swallowed, get medical attention. Give large quantities of water and induce vomiting. Do not make an unconscious person vomit. Medical observation is recommended for 12 hours after overexposure.

Personal Protective Methods: Wear protective gloves and clothing to prevent any reasonable probability of skin contact. Safety equipment suppliers/manufacturers can provide recommendations on the most protective glove/clothing material for your operation.. All protective clothing (suits, gloves, footwear, headgear) should be clean, available each day, and put on before work. Contact lenses should not be worn when working with this chemical. Wear dust-proof chemical goggles and face shield unless full facepiece respiratory protection is worn. Employees should wash immediately with soap when skin is wet or contaminated. Provide emergency showers and eyewash.

Respirator Selection: NIOSH: At *any concentrations above the NIOSH REL, or where there is no REL, at any detectable concentration:* SCBAF:PD,PP (any self-contained breathing apparatus that has a full facepiece and is operated in a pressure-demand or other positive-pressure mode); or SAF:PD,PP:ASCBA (any supplied-air respirator that has a full facepiece and is operated in a pressure-demand or other positive-pressure mode in combination with an auxiliary self-contained breathing apparatus operated in a pressure-demand or other positive pressure mode). *Escape:* GMFOVHiE [any air-purifying, full-facepiece respirator (gas mask) with a chin-style, front- or back-mounted organic vapor canister having a high-efficiency particulate filter]; or SCBAE (any appropriate escape-type, self-contained breathing apparatus).

Storage: Prior to working with dieldrin you should be trained on its proper handling and storage. Dieldrin must be stored to avoid contact with oxidizers (such as perchlorates, peroxides, permanganates, chlorates and nitrates); strong acids (such as hydrochloric, sulfuric and nitric); chemically active metals (such as potassium, sodium, magnesium and zinc) since violent reactions occur. A regulated, marked area should be established where this chemical is handled, used, or stored in compliance with OSHA standard 1910.1045.

Shipping: Dieldrin must carry a "Poison" label. It falls in Hazard Class 6.1 and Packing Group II.

Spill Handling: Evacuate persons not wearing protective equipment from area of spill or leak until clean-up is complete. Remove all ignition sources. Collect powdered material in the most convenient and safe manner and deposit in sealed containers. Ventilate area after clean-up is complete. It may be necessary to contain and dispose of this chemical as a hazardous waste. If material or contaminated runoff enters waterways, notify downstream users of potentially contaminated waters. Contact your Department of Environmental Protection or your regional office of the federal EPA for specific recommendations. If employees are required to clean-up spills, they must be properly trained and equipped. OSHA 1910.120(q) may be applicable.

Fire Extinguishing: Dieldrin is a non-combustible solid. Commercial solutions may contain flammable or combustible liquids. Use dry chemical, carbon dioxide, water spray, alcohol or polymer foam extinguishers. Poisonous gases are produced in fire including hydrogen chloride and carbon monoxide. If material or contaminated runoff enters waterways, notify downstream users of potentially contaminated waters. Notify local health and fire officials and pollution control agencies. Containers may explode in fire. From a secure, explosion-proof location, use water spray to cool exposed containers. If cooling streams are ineffective (venting sound increases in volume and pitch, tank discolors, or shows any signs of deforming), withdraw immediately to a secure position. If employees are expected to fight fires, they must be trained and equipped in OSHA 1910.156.

Disposal Method Suggested: Incineration (1,500°F, 0.5 second minimum for primary combustion; 3,200°F, 1.0 second for secondary combustion) with adequate scrubbing and ash disposal facilities.[22] In accordance with 40CFR165 recommendations for the disposal of pesticides and pesticide containers. Must be disposed properly by following package label directions or by contacting your state pesticide or environmental control agency or by contacting your regional EPA office. Consult with environmental regulatory agencies for guidance on acceptable disposal practices. Generators of waste containing this contaminant (≥100 kg/mo) must conform with EPA regulations governing storage, transportation, treatment, and waste disposal.

References

U.S. Environmental Protection Agency, Aldrin/Dieldrin: Ambient Water Quality Criteria, Washington, DC (1980)

U.S. Environmental Protection Agency, Dieldrin, Health and Environmental Effects Profile No. 82, Washington, DC, Office of Solid Waste (April 30, 1980)

Sax N. I., Ed., "Dangerous Properties of Industrial Materials Report," 1, No. 4, 52–55 (1981) an 6, No. 1, 9–16 (1986)

U.S. Environmental Protection Agency, "Health Advisory: Dieldrin," Washington, DC, Office of Drinking Water (August 1987)

New Jersey Department of Health, "Hazardous Substance Fact Sheet: Dieldrin," Trenton, NJ (November 1998)

New York State Department of Health, "Chemical Fact Sheet: Dieldrin," Albany, NY, Bureau of Toxic Substance Assessment (January 1986 & Version 2)

U.S. Public Health Service, "Toxicological Profile for Aldrin/Dieldrin," Atlanta, Georgia, Agency for Toxic Substances & Disease Registry (November 1987)

Diepoxybutane

Molecular Formula: $C_4H_6O_2$

Synonyms: 1,1'-Bi(Ethylene Oxide); Bioxirane; 2,2'-Bioxirane; Butadiendioxyd (German); Butadiene Diepoxide; 1,3-Butadiene Diepoxide; Butadiene Dioxide; Butane Diepoxide; Butane, 1,2:3,4-Diepoxy-; DEB; Dioxybutadiene; ENT 26,592; Erythritol Anhydride

CAS Registry Number: 1464-53-5

RTECS Number: EJ8225000

Regulatory Authority

- Carcinogen (Animal Positive) (IARC)[9] (suspect carcinogen) NTP
- Air Pollutant Standard Set (North Dakota)[60]
- EPA Hazardous Waste Number (RCRA No.): U085
- RCRA, 40CFR261, Appendix 8 Hazardous Constituents
- EPCRA Section 313 Form R *de minimis* concentration reporting level: 0.1%
- CALIFORNIA'S PROPOSITION 65: Carcinogen
- Superfund/EPCRA 40CFR355, Appendix B Extremely Hazardous Substances: TPQ = 500 lb (227 kg)
- Superfund/EPCRA 40CFR302.4 Reportable Quantity (RQ): CERCLA, 10 lb (4.54 kg)
- Canada, WHMIS, Ingredients Disclosure List

Cited in U.S. State Regulations: California (A, G), Florida (G), Illinois (G), Kansas (G), Louisiana (G), Maine (G), Maryland (G), Massachusetts (G), Michigan (G), Minnesota (G), New Jersey (G), North Dakota (A), Pennsylvania (G), Vermont (G), Virginia (G), Washington (G), West Virginia (G), Wisconsin (G).

Description: Diepoxybutane is a colorless liquid. Boiling point = 138°C. Freezing/Melting point = 19°C. Mixes with water.

Potential Exposure: DEB is primarily used in research and experimental work, as a curing agent for polymers, as a crosskicking agent for textile fabrics, and in preventing microbial spoilage in substances. DEB is also used commercially as mixed stereoisomers and as individual isomers in the preparation of erythritol and other pharmaceuticals.

Permissible Exposure Limits in Air: North Dakota[60] has set a guideline of zero in ambient air.

Permissible Concentration in Water: No criteria set.

Routes of Entry: Human exposure to DEB is principally through inhalation and skin absorption.

Harmful Effects and Symptoms

Short Term Exposure: Eye and skin contact can cause severe irritation. Accidental minor exposure caused swelling of the eyelids, upper respiratory tract irritation and painful eye irritation 6 hours after exposure.

Long Term Exposure: There is sufficient evidence for the carcinogenicity of diepoxybutane in experimental animals. Two forms of 1,2:3,4-diepoxybutane (DL and meso) were carcinogenic in mice by skin application. Both compounds produced squamous-cell skin carcinomas. The D,L-racemate also produced local sarcomas in mice and rats by subcutaneous injection.

First Aid: If this chemical gets into the eyes, remove any contact lenses at once and irrigate immediately for at least 15 minutes, occasionally lifting upper and lower lids. Seek medical attention immediately. If this chemical contacts the skin, remove contaminated clothing and wash immediately with soap and water. Seek medical attention immediately. If this chemical has been inhaled, remove from exposure, begin rescue breathing (using universal precautions) if breathing has stopped and CPR if heart action has stopped. Transfer promptly to a medical facility. When this chemical has been swallowed, get medical attention. Give large quantities of water and induce vomiting. Do not make an unconscious person vomit.

Personal Protective Methods: Wear protective gloves and clothing to prevent any reasonable probability of skin contact. Safety equipment suppliers/manufacturers can provide recommendations on the most protective glove/clothing material for your operation.. All protective clothing (suits, gloves, footwear, headgear) should be clean, available each day, and put on before work. Contact lenses should not be worn when working with this chemical. Wear splash-proof chemical goggles and face shield unless full facepiece respiratory protection is worn. Employees should wash immediately with soap when skin is wet or contaminated. Provide emergency showers and eyewash.

Respirator Selection: *At any detectable concentration:* SCBAF:PD,PP (any MSHA/NIOSH approved self-contained breathing apparatus that has a full facepiece and is operated in a pressure-demand or other positive-pressure mode); or SAF:PD,PP:ASCBA (any supplied-air respirator that has a full facepiece and is operated in a pressure-demand or other positive-pressure mode in combination with an auxiliary, self-contained breathing apparatus operated in a pressure-demand or other positive pressure mode). *Escape:* GMFOV [any air-purifying, full-facepiece respirator (gas mask) with a chin-style, front-or back-mounted organic vapor canister] or SCBAE (any appropriate escape-type, self-contained breathing apparatus).

Storage: Prior to working with diepoxybutane you should be trained on its proper handling and storage. Store in tightly closed containers in a cool, well-ventilated area. A regulated, marked area should be established where this chemical is handled, used, or stored in compliance with OSHA standard 1910.1045.

Spill Handling: Evacuate and restrict persons not wearing protective equipment from area of spill or leak until cleanup is complete. Remove all ignition sources. Ventilate area of spill or leak. Absorb the material on carbon, clay, bentonite or sawdust. For spills on water bodies, isolate the contaminated water and add dilute HCl or acetic acid to detoxify; then treat the water by biological treatment. It may be necessary to contain and dispose of this chemical as a hazardous waste. If material or contaminated runoff enters waterways, notify downstream users of potentially contaminated waters. Contact your Department of Environmental Protection or your regional office of the federal EPA for specific recommendations. If employees are required to clean-up spills, they must be properly trained and equipped. OSHA 1910.120(q) may be applicable.

Fire Extinguishing: Poisonous gases are produced in fire. Use dry chemical, carbon dioxide, or foam extinguishers. Vapors are heavier than air and will collect in low areas. Containers may explode in fire. Storage containers and parts of containers may rocket great distances, in many directions. If material or contaminated runoff enters waterways, notify downstream users of potentially contaminated waters. Notify local health and fire officials and pollution control agencies. From a secure, explosion-proof location, use water spray to cool exposed containers. If cooling streams are ineffective (venting sound increases in volume and pitch, tank discolors, or shows any signs of deforming), withdraw immediately to a secure position. If employees are expected to fight fires, they must be trained and equipped in OSHA 1910.156.

Disposal Method Suggested: Detoxify by HCl or acetic acid treatment followed by secure landfill disposal, landfarming or biological treatment. Incineration is another alternative. Consult with environmental regulatory agencies for guidance on acceptable disposal practices. Generators of waste containing this contaminant (≥100 kg/mo) must conform with EPA regulations governing storage, transportation, treatment, and waste disposal.

References

Sax N. I., Ed., "Dangerous Properties of Industrial Materials Report," 4, No. 3, 56–60 (1984)

U.S. Environmental Protection Agency, "Chemical Profile: Diepoxybutane," Washington, DC, Chemical Emergency Preparedness Program (November 30, 1987)

Diethanolamine

Molecular Formula: $C_4H_{11}NO_2$

Synonyms: *N,N*-Bis(2-hydroxyethyl)amine; Bis(2-hydroxyethyl)amine; Bis(hydroxyethyl)amine; DEA; Diaethanolamin (German); Dietanolamina (Spanish); Diethanolamin (Czech); *N,N*-Diethanolamine; Diethylolamine; 2,2'-Dihydroxydiethylamine; Di(2-hydroxyethyl)amine; Diolamine; EINECS No. 203-868-0; Ethanol, 2,2'-iminobis-; Ethanol, 2,2'-iminodi-; 2-[(2-Hydroxyethyl)amino]ethanol; 2,2'-Iminobis(ethanol); 2,2'-Iminodi-1-ethanol; 2,2'-Iminodiethanol; Iminodiethanol; NCI C55174

CAS Registry Number: 111-42-2

RTECS Number: KL2975000

DOT ID: UN 3077 (solid); UN 3082 (liquid)

EEC Number: 603-071-00-1

Regulatory Authority

- Air Pollutant Standard Set (ACGIH)[1] (Australia) (DFG)[3] (HSE)[33] (Israel) (former USSR)[35] (Several Canadian Provinces)
- Clean Air Act: Hazardous Air Pollutants (Title I, Part A, Section 112)
- Superfund/EPCRA 40CFR302.4 Reportable Quantity (RQ): CERCLA, 1 lb. (0.454 kg.).
- EPCRA Section 313 Form R *de minimis* concentration reporting level: 1.0%
- Canada, WHMIS, Ingredients Disclosure List; National Pollutant Release Inventory (NPRI)

Cited in U.S. State Regulations: California (A, G), Florida (G), Massachusetts (G), Minnesota (G), New Jersey (G), Pennsylvania (G).

Description: Diethanolamine, is a colorless crystalline solid or a syrupy, white liquid (above 82°F) with a mild, ammonia-like odor. Odor threshold = 0.27 ppm. Boiling point = 269°C (decomposes). Freezing/Melting point = 28°C. Flash point = 172°C (oc). Autoignition temperature = 662°C. Explosive limits: LEL = 1.6%; UEL = 9.8%. Hazard Identification (based on NFPA-704 M Rating System): Health 1, Flammability 1, Reactivity 0. Highly soluble in water.

Potential Exposure: Used as a detergent, as and intermediate for making other chemicals, and as an absorbent for acid gases.

Incompatibilities: The aqueous solution is a medium strong base. Violent reaction with strong acids, oxidizers, acid anhydrides, halides. Incompatible with isocyanates, vinyl acetate, acrylates, substituted allyls, alkylene oxides, epichlorohydrin, aldehydes. Reacts with CO_2 in the air. Hygroscopic (i.e., absorbs moisture from the air). Corrosive to copper, zinc, aluminum and their alloys, and galvanized iron.

Permissible Exposure Limits in Air: There is no OSHA PEL. NIOSH recommends a limit of 3 ppm (15 mg/m^3). ACGIH recommends a limit of 0.46 ppm (2 mg/m^3). Australia, Israel, and the Canadian provinces of Alberta, BC, Ontario, and Quebec set a limit of 3 ppm and Alberta set an STEL of 6 ppm.

Determination in Air: Impinger; Reagent; Ion chromatography; NIOSH IV Method #3509, Aminoethanol Compounds II.

Routes of Entry: Inhalation, ingestion, skin and/or eye contact.

Harmful Effects and Symptoms

Short Term Exposure: A lacrimator. Irritates the eyes, skin, nose, throat. Contact can cause severe irritation and burns of the skin and eyes. Inhalation can cause headache, nausea, and vomiting; coughing, wheezing and/or shortness of breath.

Long Term Exposure: Diethanolamine may affect the kidneys and liver. Prolonged or repeated skin exposure may cause a skin allergy and rash.

Points of Attack: Eyes, skin, respiratory system.

Medical Surveillance: Liver and kidney function tests. Lung function tests. Evaluation by a qualified allergist.

First Aid: If this chemical gets into the eyes, remove any contact lenses at once and irrigate immediately for at least 15 minutes, occasionally lifting upper and lower lids. Seek medical attention immediately. If this chemical contacts the skin, remove contaminated clothing and wash immediately with soap and water. Seek medical attention immediately. If this chemical has been inhaled, remove from exposure, begin rescue breathing (using universal precautions) if breathing has stopped and CPR if heart action has stopped. Transfer promptly to a medical facility. When this chemical has been swallowed, get medical attention. Give large quantities of water and induce vomiting. Do not make an unconscious person vomit.

Personal Protective Methods: Wear protective gloves and clothing to prevent any reasonable probability of skin contact. Safety equipment suppliers/manufacturers can provide recommendations on the most protective glove/clothing material for your operation. Butyl rubber, Neoprene, Viton, and Teflon are among the recommended protective materials. All protective clothing (suits, gloves, footwear, headgear) should be clean, available each day, and put on before work. Contact lenses should not be worn when working with this chemical. Wear splash or dust-proof chemical goggles and face shield when working with the liquid or solid material, unless full facepiece respiratory protection is worn. Employees should wash immediately with soap when skin is wet or contaminated. Provide emergency showers and eyewash.

Respirator Selection: Where the potential for exposure to this chemical, use a MSHA/NIOSH approved supplied-air respirator with a full facepiece operated in the positive pressure mode or with a full facepiece, hood, or helmet in the continuous flow mode, or use a MSHA/NIOSH approved self-contained breathing apparatus with a full facepiece operated in pressure-demand or other positive pressure mode.

Storage: Prior to working with diethanolamine you should be trained on its proper handling and storage. Before entering confined space where this chemical may be present, check to make sure that an explosive concentration does not exist. Store in tightly closed containers in a cool, well ventilated area. Metal containers involving the transfer of this chemical should be grounded and bonded. Where possible, automatically pump liquid from drums or other storage containers to process containers. Drums must be equipped with self-closing valves, pressure vacuum bungs, and flame arresters. Use only non-sparking tools and equipment, especially when opening and closing containers of this chemical. Sources of ignition such as smoking and open flames, are prohibited where this chemical is used, handled, or stored in a manner that could create a potential fire or explosion hazard. Wherever this chemical is used, handled, manufactured, or stored, use explosion-proof electrical equipment and fittings.

Shipping: Environmentally hazardous solid, (or liquid), n.o.s. Hazard Class: 9. Label: "Class 9."

Spill Handling: *Dry material:* Evacuate persons not wearing protective equipment from area of spill or leak until clean-up is complete. Remove all ignition sources. Collect powdered material in the most convenient and safe manner and deposit in sealed containers. Ventilate area after clean-up is complete. It may be necessary to contain and dispose of this chemical as a hazardous waste. If material or contaminated runoff enters waterways, notify downstream users of potentially contaminated waters. Contact your Department of Environmental Protection or your regional office of the federal EPA for specific recommendations. If employees are required to clean-up spills, they must be properly trained and equipped. OSHA 1910.120(q) may be applicable.

Liquid: Evacuate and restrict persons not wearing protective equipment from area of spill or leak until cleanup is complete. Remove all ignition sources. Establish forced ventilation to keep levels below explosive limit. Absorb liquids in vermiculite, dry sand, earth, peat, carbon, or a similar material and deposit in sealed containers. Keep this chemical out of a confined space, such as a sewer, because of the possibility of an explosion, unless the sewer is designed to prevent the build-up of explosive concentrations. It may be necessary to contain and dispose of this chemical as a hazardous waste. If material or contaminated runoff enters waterways, notify downstream users of potentially contaminated waters. Contact your Department of Environmental Protection or your regional office of the federal EPA for specific recommendations. If employees are required to clean-up spills, they must be properly trained and equipped. OSHA 1910.120(q) may be applicable.

Fire Extinguishing: *Dry material:* This chemical is a combustible solid. Use dry chemical, carbon dioxide, water spray, or alcohol foam extinguishers. Poisonous gases are produced in fire. If material or contaminated runoff enters waterways, notify downstream users of potentially contaminated waters. Notify local health and fire officials and pollution control agencies. From a secure, explosion-proof location, use water spray to cool exposed containers. If cooling streams are ineffective (venting sound increases in volume and pitch, tank discolors, or shows any signs of deforming), withdraw immediately to a secure position. If employees are expected to fight fires, they must be trained and equipped in OSHA 1910.156.

Liquid: This chemical is a combustible liquid. Poisonous gases are produced in fire. Use dry chemical, carbon dioxide, or alcohol foam extinguishers. Vapors are heavier than air and will collect in low areas. Vapors in confined areas may explode when exposed to fire. Containers may explode in fire. Storage containers and parts of containers may rocket great distances, in many directions. If material or contaminated runoff enters waterways, notify downstream users of potentially contaminated waters. Notify local health and fire officials and pollution control agencies. From a secure, explosion-proof location, use water spray to cool exposed containers. If cooling streams are ineffective (venting sound increases in volume and pitch, tank discolors, or shows any signs of deforming), withdraw immediately to a secure position. If employees are expected to fight fires, they must be trained and equipped in OSHA 1910.156.

Disposal Method Suggested: Incineration.

References

New Jersey Department of Health and Senior Services, "Hazardous Substance Fact Sheet, Diethanolamine." Trenton NJ (October 1996)

Diethoxypropene

Molecular Formula: $C_7H_{14}O_2$

Common Formula: $(C_2H_5O)_2CHCH{=}CH_2$

Synonyms: Acrolein acetal; Acrylaldehyde diethyl; 3,3-Diethoxy-1-propene; 3,3-Diethoxypropene; Propenal diethyl acetal; 1-Propene, 3,3-diethoxy-

CAS Registry Number: 3054-95-3

RTECS Number: AS1370000

DOT ID: UN 2374

Cited in U.S. State Regulations: New Hampshire (G), New Jersey (G).

Description: 3,3-Diethoxypropene, $(C_2H_5O)_2CHCH{=}CH_2$ is a colorless liquid. Flash point = 40°C. Boiling point = 123.5°C.

Potential Exposure: Used in organic synthesis.

Incompatibilities: Forms explosive mixture with air. Contact with acids and strong oxidizers may cause fire and explosion hazard.

Permissible Exposure Limits in Air: No standards set.

Permissible Concentration in Water: No criteria set.

Harmful Effects and Symptoms

There is no information about the health effects of diethoxy propene known at this time. Diethoxypropene is a flammable liquid and a fire hazard.

First Aid: If this chemical gets into the eyes, remove any contact lenses at once and irrigate immediately for at least 15 minutes, occasionally lifting upper and lower lids. Seek medical attention immediately. If this chemical contacts the skin, remove contaminated clothing and wash immediately with soap and water. Seek medical attention immediately. If this chemical has been inhaled, remove from exposure, begin rescue breathing (using universal precautions) if breathing has stopped and CPR if heart action has stopped. Transfer promptly to a medical facility. When this chemical has been swallowed, get medical attention. Give large quantities of water and induce vomiting. Do not make an unconscious person vomit.

Personal Protective Methods: Wear protective gloves and clothing to prevent any reasonable probability of skin contact. Safety equipment suppliers/manufacturers can provide recommendations on the most protective glove/clothing material for your operation.. All protective clothing (suits, gloves, footwear, headgear) should be clean, available each day, and put on before work. Contact lenses should not be worn when working with this chemical. Wear splash-proof chemical goggles and face shield unless full facepiece respiratory protection is worn. Employees should wash immediately with soap when skin is wet or contaminated. Provide emergency showers and eyewash.

Respirator Selection: Where the potential for high exposures exists, use a MSHA/NIOSH approved supplied-air respirator with a full facepiece operated in the positive pressure mode or with a full facepiece, hood, or helmet in the continuous flow mode, or use a MSHA/NIOSH approved self-contained breathing apparatus with a full facepiece operated in pressure-demand or other positive pressure mode.

Storage: Prior to working with this chemical you should be trained on its proper handling and storage. Store in tightly closed containers in a cool, well-ventilated area away from acids and strong oxidizers. Sources of ignition, such as smoking and open flames, are prohibited where Diethoxypropene is handled, used, or stored. Metal containers involving the transfer of 5 gallons or more of Diethoxypropene should be grounded and bonded. Drums must be equipped with self-closing valves, pressure vacuum bungs and flame arresters. Use only non-sparking tools and equipment, especially when opening and closing containers of Diethoxypropene. Wherever Diethoxypropene is used, handled, manufactured, or stored, use explosion-proof electrical equipment and fittings.

Shipping: 3,3-Diethoxypropene must bear a "Flammable Liquid" label. It falls in DOT Hazard Class 3 and Packing Group II.

Spill Handling: Evacuate and restrict persons not wearing protective equipment from area of spill or leak until cleanup is complete. Remove all ignition sources. Ventilate area of spill or leak. Absorb liquids in vermiculite, dry sand, earth, peat, carbon, or a similar material and deposit in sealed containers. Keep this chemical out of a confined space, such as a sewer, because of the possibility of an explosion, unless the sewer is designed to prevent the build-up of explosive concentrations. It may be necessary to contain and dispose of this chemical as a hazardous waste. If material or contaminated runoff enters waterways, notify downstream users of potentially contaminated waters. Contact your Department of Environmental Protection or your regional office of the federal EPA for specific recommendations. If employees are required to clean-up spills, they must be properly trained and equipped. OSHA 1910.120(q) may be applicable.

Fire Extinguishing: This chemical is a flammable liquid. Poisonous gases are produced in fire. Use dry chemical, carbon dioxide, or alcohol foam extinguishers. Vapors are heavier than air and will collect in low areas. Vapors may travel long distances to ignition sources and flashback. Vapors in confined areas may explode when exposed to fire. Containers may explode in fire. Storage containers and parts of containers may rocket great distances, in many directions. If material or contaminated runoff enters waterways, notify downstream users of potentially contaminated waters. Notify local health and fire officials and pollution control agencies. From a secure, explosion-proof location, use water spray to cool exposed containers. If cooling streams are ineffective (venting sound increases in volume and pitch, tank discolors, or shows any signs of deforming), withdraw immediately to a secure position. If employees are expected to fight fires, they must be trained and equipped in OSHA 1910.156.

References

New Jersey Department of Health, "Hazardous Substance Fact Sheet: Diethoxypropene," Trenton, NJ (September 23, 1987)

Diethylamine

Molecular Formula: $C_4H_{11}N$

Common Formula: $(C_2H_5)_2NH$

Synonyms: 2-Aminopentane; DEN; Diaethyamin (German); Diethamine; *N,N*-Diethylamine; Dietilamina (Italian, Spanish); Dwuetyloamina (Polish); *N*-Ethyl ethanamine

CAS Registry Number: 109-89-7

RTECS Number: HZ8750000

DOT ID: UN 1154

EEC Number: 612-003-00-X

EINECS Number: 203-716-3

Regulatory Authority

- Air Pollutant Standard Set (ACGIH)[1] (Australia) (DFG)[3] (HSE)[33] (Israel) (Mexico) (OSHA)[58] (former USSR)[43] (Several States)[60] (Several Canadian Provinces)
- Clean Water Act: Section 311 Hazardous Substances/RQ 100 lb (45.4 kg)
- Superfund/EPCRA 40CFR302.4 Reportable Quantity (RQ): CERCLA, 1,000 lb (454 kg)
- Canada, WHMIS, Ingredients Disclosure List

Cited in U.S. State Regulations: Alaska (G), California (A, G), Connecticut (A), Florida (G), Illinois (G), Maine (G), Massachusetts (G, A), Minnesota (G), Nevada (A), New Jersey (G), North Dakota (A), Oklahoma (G), Pennsylvania (G), Rhode Island (G), Virginia (A), West Virginia (G).

Description: DEA, $(C_2H_5)_2NH$, is a colorless liquid with a fishy ammonia-like odor. The odor threshold is 0.14 ppm. Boiling point = 55.6°C. Freezing/Melting point = -50°C. Flash point = -23°C. Autoignition temperature = 312°C. Explosive limits: LEL = 1.8%; UEL = 10.1%. Hazard Identification (based on NFPA-704 M Rating System): Health 3, Flammability 3, Reactivity 0. Soluble in water.

Potential Exposure: Diethylamine (DEA) is used in the manufacture of the following chemicals: diethyldithiocarbamate and thiurams (rubber processing accelerators), diethylaminoethanol (medicinal intermediate), diethylaminopropylamine (epoxy curing agent), N,N-diethyl-m-toluamide and other pesticides, and 2-diethylaminoethylmethacrylate. It is used in the manufacture of several drugs.

Incompatibilities: Forms explosive mixture with air. May accumulate static electrical charges, and may cause ignition of its vapors. Violent reaction with strong oxidizers, strong acids, cellulose nitrate. Incompatible with organic anhydrides, isocyanates, vinyl acetate, acrylates, substituted allyls, alkylene oxides, epichlorohydrin, ketones, aldehydes, alcohols, glycols, mercury, phenols, cresols, caprolactum solution. Attacks aluminum, copper, lead, tin, zinc and alloys.

Permissible Exposure Limits in Air: The Federal Limit (OSHA PEL)[58] is 25 ppm (75 mg/m^3). TWA. Australia, Israel, Mexico, HSE TWA, recommended ACGIH TLV, and Canadian provinces of Alberta, BC, Ontario, and Quebec TWA is 10 ppm (30 mg/m^3) and all have set a STEL value of 25 ppm (75 mg/m^3). The DFG MAK is 10 ppm and peak limitation is 20 ppm (10min) not to be exceeded 4 times per workshift. The NIOSH IDLH level is 200 ppm. The former USSR-UNEP/IRPTC joint project[43] has set a MAC in workplace air of 30 mg/m^3 as well as MAC values for the ambient air in residential areas of 0.05 mg/m^3 both on a momentary and a daily average basis. Several states have set guidelines or standards for Diethylamine in ambient air[60] ranging from 4.1 μg/m^3 (Massachusetts) to 300 – 750 μg/m^3 (North Dakota) to 500 μg/m^3 (Virginia) to 600 μg/m^3 (Connecticut) to 715 μg/m^3 (Nevada).

Determination in Air: Adsorption on silica, workup with H_2SO_4 in CH_3OH (Methanol); Gas chromatography/Flame ionization detection; NIOSH IV, Method #2010.

Permissible Concentration in Water: The former USSR-UNEP/IRPTC joint project[43] has set a MAC of 2 mg/l in water bodies used for domestic purposes.

Routes of Entry: Inhalation, ingestion, skin absorption, eye and skin contact.

Harmful Effects and Symptoms

Short Term Exposure: It is irritant and corrosive to eyes, skin, mucous membranes and the respiratory tract. Dyspnea (difficult breathing) results on inhalation and it may be fatal.[24] Higher exposures can cause pulmonary edema, a medical emergency that can be delayed for several hours. This can cause death. DEA can be absorbed through the skin, thereby increasing exposure. The oral LD_{50} rat is 540 mg/kg (slightly toxic).

Long Term Exposure: Repeated exposure may affect the liver and kidneys and result in swelling of the eyes, impaired vision, and the appearance of "circles" or halos around lights. Repeated exposures may cause bronchitis to develop with coughing, phlegm, and/or shortness of breath. In animals; myocardial degeneration.

Points of Attack: Eyes, skin, respiratory system, cardiovascular system.

Medical Surveillance: Consider the points of attack in preplacement and periodic physical examinations. Lung function tests. Consider chest x-ray following acute overexposure. Consider vision tests, especially if symptoms have been noted. Liver and kidney function tests.

First Aid: If this chemical gets into the eyes, remove any contact lenses at once and irrigate immediately for at least 15 minutes, occasionally lifting upper and lower lids. Seek medical attention immediately. If this chemical contacts the skin, remove contaminated clothing and wash immediately with soap and water. Seek medical attention immediately. If this chemical has been inhaled, remove from exposure, begin rescue breathing (using universal precautions) if breathing has stopped and CPR if heart action has stopped. Transfer promptly to a medical facility. When this chemical has been swallowed, get medical attention. Give large quantities of water and induce vomiting. Do not make an unconscious person vomit. Medical observation is recommended for 24 – 48 hours after breathing overexposure, as pulmonary edema may be delayed. As first aid for pulmonary edema, a doctor or authorized paramedic may consider administering a corticosteroid spray.

Personal Protective Methods: Wear protective gloves and clothing to prevent any reasonable probability of skin contact. Safety equipment suppliers/manufacturers can provide recommendations on the most protective glove/clothing material for your operation. Use butyl rubber gloves. All protective clothing (suits, gloves, footwear, headgear) should be clean, available each day, and put on before work. Contact lenses should not be worn when working with this chemical. Wear splash-proof chemical goggles and face shield unless full facepiece respiratory protection is worn. Employees should wash immediately with soap when skin is wet or contaminated. Provide emergency showers and eyewash.

Respirator Selection: NIOSH/OSHA: *200 ppm:* SA:CF (any supplied-air respirator operated in a continuous-flow mode); or PAPRS [any powered, air-purifying respirator with cartridge(s) providing protection against the compound of concern]; or CCRFS [any chemical cartridge respirator with a full facepiece and cartridge(s) providing protection against the compound of concern organic vapor and acid gas cartridge(s)]; or GMFS [any air-purifying, full-facepiece respirator (gas mask) with a chin-style, front- or back-mounted canister providing protection against the compound of concern]; or SCBAF (any self-contained breathing apparatus with a full facepiece); or SAF (any supplied-air respirator with a full facepiece). *Emergency or planned entry into unknown concentrations or IDLH conditions:* SCBAF:PD,PP (any self-contained breathing apparatus that has a full facepiece and is operated in a pressure-demand or other positive-pressure mode); or SAF:PD,PP:ASCBA (any supplied-air respirator that has a full facepiece and is operated in a pressure-demand or other positive-pressure mode in combination with an auxiliary self-contained breathing apparatus operated in a pressure-demand or other positive pressure mode). *Escape:* GMFS [any air-purifying, full-facepiece respirator (gas mask) with a chin-style, front- or back-mounted canister providing protection against the compound of concern]; or SCBAE (any appropriate escape-type, self-contained breathing apparatus).

Note: Substance causes eye irritation or damage; eye protection needed.

Storage: Prior to working with DEA you should be trained on its proper handling and storage. Before entering confined space where this chemical may be present, check to make sure that an explosive concentration does not exist. Keep containers well closed. Protect against direct sunlight. Store in a cool place. Protect containers against physical damage. Outdoors or detached storage is preferred. In case of indoor storage, store in a standard combustible liquid storage room or cabinet. Good ventilation. Separate from oxidizing materials, strong acids, cellulose nitrate, and other incompatible materials listed above.

Shipping: This compound requires a shipping label of: "Flammable Liquid." It falls in DOT Hazard Class 3 and Packing Group II.

Spill Handling: Evacuate and restrict persons not wearing protective equipment from area of spill or leak until cleanup is complete. Remove all ignition sources. Establish forced ventilation to keep levels below explosive limit. Spread sodium bisulfate and sprinkle water upon a spill,[24] then flush to sewer. Alternatively, absorb liquids in vermiculite, dry sand, earth, peat, carbon, or a similar material and deposit in sealed containers. Keep this chemical out of a confined space, such as a sewer, because of the possibility of an explosion, unless the sewer is designed to prevent the build-up of explosive concentrations. It may be necessary to contain and dispose of this chemical as a hazardous waste. If material or contaminated runoff enters waterways, notify downstream users of potentially contaminated waters. Contact your Department of Environmental Protection or your regional office of the federal EPA for specific recommendations. If employees are required to clean-up spills, they must be properly trained and equipped. OSHA 1910.120(q) may be applicable.

Fire Extinguishing: This chemical is a flammable liquid. Poisonous gases are produced in fire. Use dry chemical, foam, or carbon dioxide fire extinguishing agent. Water may be ineffective, but use water spray to keep fire-exposed containers cool, to disperse the vapor, to flush spills away from exposures, to dilute spills to nonflammable mixtures and thus to prevent the spread of fires. Vapors are heavier than air and will collect in low areas. Vapors may travel long distances to ignition sources and flashback. Vapors in confined areas may explode when exposed to fire. Containers may explode in fire. Storage containers and parts of containers may rocket great distances, in many directions. If material or contaminated runoff enters waterways, notify downstream users of potentially contaminated waters. Notify local health and fire officials and pollution control agencies. From a secure, explosion-proof location, use water spray to cool exposed containers. If cooling streams are ineffective (venting sound increases in volume and pitch, tank discolors, or shows any signs of deforming), withdraw immediately to a secure position. If employees are expected to fight fires, they must be trained and equipped in OSHA 1910.156.

Disposal Method Suggested: Incineration; incinerator is equipped with a scrubber or thermal unit to reduce No_x emissions.

References

National Institute for Occupational Safety and Health, Profiles on Occupational Hazards for Criteria Document Priorities: Primary Aliphatic Amines, 154-166, Report PB-274,073, Cincinnati, OH (1977)

U.S. Environmental Protection Agency, "Chemical Hazard Information Profile: Ethylamines," Washington, DC (April 1, 1978)

New Jersey Department of Health and Senior Services, "Hazardous Substance Fact Sheet, Diethylamine," Trenton, NJ (May 1998)

Diethylaminoethanol

Molecular Formula: $C_6H_{15}NO$

Common Formula: $(C_2H_5)_2NCH_2CH_2OH$

Synonyms: DEAE; Diaethylaminoaethanol (German); 2-Diethylamino-; β-Diethylaminoethanol; *N*-Diethylaminoethanol; 2-(Diethylamino)ethanol; 2-*N*-Diethylaminoethanol; β-Diethylaminoethyl alcohol; 2-(Diethylamino) ethyl alcohol; *N,N*-Diethylethanolamine; *N,N*-Diethyl-*N*-(β-hydroxyethyl) amine; *N,N*-Diethyl-2-hydroxyethylamine; *N*-1,1-Diethyl-*N*-(2-hydroxyethyl)amine; Ethanol, 2-(diethylamino)-; 2-Hydroxytriethylamine

CAS Registry Number: 100-37-8

RTECS Number: KK5075000

DOT ID: UN 2686

EEC Number: 603-048-00-6

Regulatory Authority

- Air Pollutant Standard Set (ACGIH)[1] (Australia) (DFG)[3] (HSE)[33] (Israel) (Mexico) (OSHA)[58] (Several States)[60] (Several Canadian Provinces)
- Canada, WHMIS, Ingredients Disclosure List

Cited in U.S. State Regulations: Alaska (G), California (A, G), Connecticut (A), Florida (G), Illinois (G), Maine (G), Massachusetts (G), Minnesota (G), Nevada (A), New Hampshire (G), New Jersey (G), North Dakota (A), Pennsylvania (G), Rhode Island (G), Virginia (A), West Virginia (G).

Description: DEAE, $(C_2H_5)_2NCH_2CH_2OH$, is a colorless liquid with a weak ammoniacal odor. The odor perception limit in air is 0.011 ppm; the odor recognition level is 0.04 ppm.[41] Boiling point = 161°C. Freezing/Melting point = -70°C. Flash point = 60°C.[17] Autoignition temperature = 320°C. Explosive limits: LEL= 6.7; UEL = 11.7. Hazard Identification (based on NFPA-704 M Rating System): Health 3, Flammability 2, Reactivity 0. Soluble in water.

Potential Exposure: This compound is used as a chemical intermediate for the production of emulsifiers, detergents, solubilizers, cosmetics, and textile finishing agents. It is also used in drug manufacture.

Incompatibilities: Forms explosive mixture with air. Violent reaction with oxidizers, strong acids, acid chlorides, and isocyanates. Attacks light metals and copper. Attacks some plastics and rubber.

Permissible Exposure Limits in Air: The Federal Limit (OSHA PEL)[58] is 10 ppm (50 mg/m^3). The Australian, Israeli, Mexican, Canadian provincial (Alberta, BC, Ontario, Quebec), HSE[33] have the same TWA and Alberta's STEL is 20 ppm. The DFG MAK is 5 ppm (24 mg/m^3). The ACGIH recommended limit is 2 ppm TWA. The notation "skin" indicates the possibility of cutaneous absorption. The NIOSH IDLH level is 100 ppm. Several states have set guidelines or standards for diethylaminoethanol in ambient air[60] ranging from 0.5 mg/m^3 (North Dakota) to 0.8 mg/m^3 (Virginia) to 1.0 mg/m^3 (Connecticut) to 1.19 mg/m^3 (Nevada).

Determination in Air: Adsorption on silica gel; Methanol/Water; Gas chromatography/Flame ionization detection; NIOSH IV, Method #2007, Aminoethanol Compounds I.

Permissible Concentration in Water: No criteria set.

Routes of Entry: Inhalation, skin absorption, ingestion, eye and/or skin contact.

Harmful Effects and Symptoms

Short Term Exposure: Diethylaminoethanol can affect you when breathed in by passing through your skin. Contact can cause very severe burns of the eyes, leading to permanent damage. It can also irritate the skin, causing a rash and burning sensation upon contact. Breathing the vapor may irritate the lungs, causing coughing and/or shortness of breath. Inhalation of vapors can cause pulmonary edema, a medical emergency that can be delayed for several hours. This can cause death. Symptoms include nausea, vomiting; eye, skin and respiratory irritation. The oral LD_{50} rat is 1,300 mg/kg (slightly toxic).

Long Term Exposure: Diethylaminoethanol can cause a skin allergy to develop and may affect the nervous system. May cause lung irritation.

Points of Attack: Eyes, skin, respiratory system.

Medical Surveillance: For those with frequent or potentially high exposure (half the TLV or greater) the following are recommended before beginning work and at regular times after that: Lung function tests. If symptoms develop or overexposure is suspected, the following may be useful: Evaluation by a qualified allergist, including careful exposure history and special testing, may help diagnose skin allergy.

First Aid: If this chemical gets into the eyes, remove any contact lenses at once and irrigate immediately for at least 15 minutes, occasionally lifting upper and lower lids. Seek medical attention immediately. If this chemical contacts the skin, remove contaminated clothing and wash immediately with soap and water. Seek medical attention immediately. If this chemical has been inhaled, remove from exposure, begin rescue breathing (using universal precautions) if breathing has stopped and CPR if heart action has stopped. Transfer promptly to a medical facility. When this chemical has been swallowed, get medical attention. Give large quantities of water and induce vomiting. Do not make an unconscious person vomit. Medical observation is recommended for 24 – 48 hours after breathing overexposure, as pulmonary edema may be delayed. As first aid for pulmonary edema, a doctor or authorized paramedic may consider administering a corticosteroid spray.

Personal Protective Methods: Wear protective gloves and clothing to prevent any reasonable probability of skin contact. Safety equipment suppliers/manufacturers can provide recommendations on the most protective glove/clothing material for your operation. Viton, Polyvinyl Alcohol, and Butyl Rubber are among the recommended protective materials. All protective clothing (suits, gloves, footwear, headgear) should be clean, available each day, and put on before work. Contact lenses should not be worn when working with this chemical. Wear splash-proof chemical goggles and face shield unless full facepiece respiratory protection is worn. Employees should wash immediately with soap when skin is wet or contaminated. Provide emergency showers and eyewash.

Respirator Selection: NIOSH/OSHA: *100 ppm:* CCROV [any chemical cartridge respirator with organic vapor cartridge(s)]; or GMFOV [any air-purifying, full-facepiece respirator (gas mask) with a chin-style, front- or back-mounted acid gas canister]; or PAPROV [any powered, air-purifying respirator with organic vapor cartridge(s)]; or SA (any supplied-air respirator); or SCBAF (any self-contained breathing apparatus with a full facepiece). *Emergency or planned entry into unknown concentrations or IDLH conditions:* SCBAF:PD,PP (any self-contained breathing apparatus that has a full facepiece and is operated in a pressure-demand or other positive-pressure mode); or SAF:PD,PP: ASCBA (any supplied-air respirator that has a full facepiece and is operated in a pressure-demand or other positive-pressure mode in combination with an auxiliary self-contained breathing apparatus operated in a pressure-demand or other positive pressure mode). *Escape:* GMFOV [any air-purifying, full-facepiece respirator (gas mask) with a chin-style, front- or back-mounted organic vapor canister]; or SCBAE (any appropriate escape-type, self-contained breathing apparatus).

Note: Substance reported to cause eye irritation or damage; may require eye protection.

Storage: Prior to working with DEAE you should be trained on its proper handling and storage. Before entering confined space where this chemical may be present, check to make sure that an explosive concentration does not exist. Diethylaminoethanol must be stored to avoid contact with strong acids (such as hydrochloric, sulfuric, and nitric), and strong oxidizers (such as chlorine, bromine, and fluorine), because violent reactions occur. Store in tightly closed containers in a cool, well-ventilated area away from heat. Sources of ignition such as smoking and open flames are prohibited where diethylaminoethanol is used, handled, or stored in a manner that could create a potential fire or explosion hazard.

Shipping: Diethylaminoethanol must carry a "Flammable Liquid" label. It falls in DOT Hazard Class 3 and Packing Group III.

Spill Handling: Evacuate and restrict persons not wearing protective equipment from area of spill or leak until cleanup is complete. Remove all ignition sources. Establish forced ventilation to keep levels below explosive limit. Absorb liquids in vermiculite, dry sand, earth, peat, carbon, or a similar material and deposit in sealed containers. Keep this chemical out of a confined space, such as a sewer, because of the possibility of an explosion, unless the sewer is designed to prevent the build-up of explosive concentrations. It may be necessary to contain and dispose of this chemical as a hazardous waste. If material or contaminated runoff enters waterways, notify downstream users of potentially contaminated waters. Contact your Department of Environmental Protection or your regional office of the federal EPA for specific recommendations. If employees are required to clean-up spills, they must be properly trained and equipped. OSHA 1910.120(q) may be applicable.

Fire Extinguishing: This chemical is a combustible liquid. Poisonous gases are produced in fire. Use dry chemical, carbon dioxide, or foam extinguishers. Vapors are heavier than air and will collect in low areas. Vapors may travel long distances to ignition sources and flashback. Vapors in confined areas may explode when exposed to fire. Containers may explode in fire. Storage containers and parts of containers may rocket great distances, in many directions. If material or contaminated runoff enters waterways, notify downstream users of potentially contaminated waters. Notify local health and fire officials and pollution control agencies. From a secure, explosion-proof location, use water spray to cool exposed containers. If cooling streams are ineffective (venting sound increases in volume and pitch, tank discolors, or shows any signs of deforming), withdraw immediately to a secure position. If employees are expected to fight fires, they must be trained and equipped in OSHA 1910.156.

Disposal Method Suggested: Controlled incineration; incinerator is equipped with a scrubber or thermal unit to reduce No_x emissions.

References

New Jersey Department of Health, "Hazardous Substance Fact Sheet: Diethylaminoethanol," Trenton, NJ (September 1996)

N,N-Diethyl Aniline

Molecular Formula: $C_{10}H_{15}N$

Common Formula: $C_6H_5N(C_2H_5)_2$

Synonyms: Benzenamine, *N,N*-diethyl-; DEA; Diaethyanilin (German); *N,N*-Diethylaminobenzene; *N,N*-Diethylanilin (Czech); Diethylaniline; *N,N*-Diethylbenzenamine; Diethylphenylamine; *N,N*-Dietilanilina (Spanish)

CAS Registry Number: 91-66-7

RTECS Number: BX3400000

DOT ID: UN 2432

Regulatory Authority

- Superfund/EPCRA 40CFR302.4 Reportable Quantity (RQ): CERCLA, 1 lb (0.454 kg)
- Canada, WHMIS, Ingredients Disclosure List

Cited in U.S. State Regulations: Florida (G), Massachusetts (G), New Hampshire (G), New Jersey (G), Pennsylvania (G).

Description: Diethyl Aniline, $C_6H_5N(C_2H_5)_2$, is a colorless to yellow liquid. Boiling point = 216°C. Flash point = 85°C. Autoignition temperature = 630°C. Hazard Identification (based on NFPA-704 M Rating System): Health 3, Flammability 2, Reactivity 0. Slightly soluble in water.

Potential Exposure: Diethyl Aniline is used in organic synthesis and as diestuff intermediate.

Incompatibilities: Forms explosive mixture with air. Violent reaction with strong oxidizers, halogens. Incompatible with acids, organic anhydrides, isocyanates, aldehydes.

Permissible Exposure Limits in Air: No standards set. DEA can be absorbed through the skin, thereby increasing exposure.

Permissible Concentration in Water: No criteria set.

Routes of Entry: Inhalation, passing through the skin.

Harmful Effects and Symptoms

Short Term Exposure: Exposure by skin contact or breathing can interfere with the ability of the blood to carry oxygen (a condition called methemoglobinemia). This can cause headaches, dizziness, a bluish color to the skin and lips, trouble breathing, and even collapse and death. Repeated exposures can cause a low red blood count (anemia). Very high single or repeated high exposures to diethyl aniline can damage the liver.

Long Term Exposure: May cause anemia and liver damage.

Points of Attack: Blood, liver.

Medical Surveillance: Before beginning employment and at regular times after that, for those with frequent or potentially high exposures, the following are recommended: Liver function tests. Complete blood count (CBC). If symptoms develop or overexposure is suspected, the following may be useful: Blood methemoglobin level.

First Aid: If this chemical gets into the eyes, remove any contact lenses at once and irrigate immediately for at least 15 minutes, occasionally lifting upper and lower lids. Seek medical attention immediately. If this chemical contacts the skin, remove contaminated clothing and wash immediately with soap and water. Seek medical attention immediately. If this chemical has been inhaled, remove from exposure, begin rescue breathing (using universal precautions) if breathing has stopped and CPR if heart action has stopped. Transfer promptly to a medical facility. When this chemical has been swallowed, get medical attention. Give large quantities of water and induce vomiting. Do not make an unconscious person vomit.

Note to Physician: Treat for methemoglobinemia. Spectrophotometry may be required for precise determination of levels of methemoglobinemia in urine.

Personal Protective Methods: Wear protective gloves and clothing to prevent any reasonable probability of skin contact. Safety equipment suppliers/manufacturers can provide recommendations on the most protective glove/clothing material for your operation.. All protective clothing (suits, gloves, footwear, headgear) should be clean, available each day, and put on before work. Contact lenses should not be worn when working with this chemical. Wear splash-proof chemical goggles and face shield unless full facepiece respiratory protection is worn. Employees should wash immediately with soap when skin is wet or contaminated. Provide emergency showers and eyewash.

Respirator Selection: Where the potential exists for exposures to diethyl aniline, use a MSHA/NIOSH approved supplied-air respirator with a full facepiece operated in the positive pressure mode or with a full facepiece, hood, or helmet in the continuous flow mode, or use a MSHA/NIOSH approved self-contained breathing apparatus with a full facepiece operated in pressure-demand or other positive pressure mode.

Storage: Prior to working with DEA you should be trained on its proper handling and storage. Diethyl aniline must be stored to avoid contact with strong oxidizers (such as chlorine, bromine and fluorine) and strong acids since violent reactions occur. Store in tightly closed containers in a cool, well-ventilated area away from strong acids (such as hydrochloric, sulfuric and nitric) and direct sunlight.

Shipping: N,N-Diethyl aniline must be labeled: "Keep Away From Food." It falls in DOT Hazard Class 6.1 and Packing Group III.

Spill Handling: Evacuate and restrict persons not wearing protective equipment from area of spill or leak until cleanup is complete. Remove all ignition sources. Ventilate area of spill or leak. Absorb liquids in vermiculite, dry sand, earth, peat, carbon, or a similar material and deposit in sealed containers. Keep this chemical out of a confined space, such as a sewer, because of the possibility of an explosion, unless the sewer is designed to prevent the build-up of explosive concentrations. It may be necessary to contain and dispose of this chemical as a hazardous waste. If material or contaminated runoff enters waterways, notify downstream users of potentially contaminated waters. Contact your Department of Environmental Protection or your regional office of the federal EPA for specific recommendations. If employees are required to clean-up spills, they must be properly trained and equipped. OSHA 1910.120(q) may be applicable.

Fire Extinguishing: This chemical is a combustible liquid. Poisonous gases including nitrogen oxides are produced in fire. Use dry chemical, carbon dioxide, or foam extinguishers. Vapors are heavier than air and will collect in low areas. Vapors may travel long distances to ignition sources and flashback. Vapors in confined areas may explode when exposed to fire. Containers may explode in fire. Storage containers and parts of containers may rocket great distances, in many directions. If material or contaminated runoff enters waterways, notify downstream users of potentially contaminated waters. Notify local health and fire officials and pollution control agencies. From a secure, explosion-proof location, use water spray to cool exposed containers. If cooling streams are ineffective (venting sound increases in volume and pitch, tank discolors, or shows any signs of deforming), withdraw immediately to a secure position. If employees are expected to fight fires, they must be trained and equipped in OSHA 1910.156.

References

New Jersey Department of Health, "Hazardous Substance Fact Sheet: Diethyl Aniline," Trenton, NJ (September 1996)

Diethylcarbamazine Citrate

Molecular Formula: $C_{10}H_{21}N_3O{\cdot}C_6H_8O_7$

Synonyms: Banocide; Caricide; Caritrol; Dicarocide; Diethylcarbamazane citrate; Diethylcarbamazine citrate; Diethylcarbamazine hydrogen citrate; 1-Diethylcarbamoyl-4-methylpiperazine dihydrogen citrate; *N,N*-Diethyl-4-methyl-1-piperazine carboxamide citrate; *N,N*-diethyl-4-methyl-1-piperazinecarboxamide dihydrogen citrate; *N,N*-Diethyl-4-methyl-1-piperazinecarboxamide 2-hydroxy-1,2,3-propanetricarboxylate; Ditrazin; Ditrazin citrate; Ditrazine; Ditrazine citrate; Ethodryl citrate; Franocide; Franozan; Hetrazan; Loxuran; 1-Methyl-4-diethylcarbamoylpiperazine citrate

CAS Registry Number: 1642-54-2

RTECS Number: TL1225000

Cited in U.S. State Regulations: Florida (G), Massachusetts (G), New Jersey (G), Pennsylvania (G).

Description: Diethylcarbamazine Citrate is an odorless, white crystalline powder. Freezing/Melting point = 135 – 143°C. Hazard Identification (based on NFPA-704 M Rating System): Health 2, Flammability 1, Reactivity 0. Soluble in water.

Potential Exposure: Used against filariasis in man and animals. Especially popular in veterinary medicine (anthelmintic).

Routes of Entry: Inhalation, ingestion, absorbed through the skin.

Harmful Effects and Symptoms

Short Term Exposure: Poisonous if inhaled. Symptoms of exposure include nausea, vomiting, headache, weakness, and (as seen in dogs) muscle tremors and convulsions. May be fatal if inhaled, swallowed or absorbed through skin. Contact may cause burns to skin and eyes. The average adult man may tolerate a single dose of 1.5 g without ill effects.

Long Term Exposure: May have reproductive affects.

First Aid: If this chemical gets into the eyes, remove any contact lenses at once and irrigate immediately for at least 15 minutes, occasionally lifting upper and lower lids. Seek medical attention immediately. If this chemical contacts the skin, remove contaminated clothing and wash immediately with soap and water. Seek medical attention immediately. If this chemical has been inhaled, remove from exposure, begin rescue breathing (using universal precautions) if breathing has stopped and CPR if heart action has stopped. Transfer promptly to a medical facility. When this chemical has been swallowed, get medical attention. Give large quantities of water and induce vomiting. Do not make an unconscious person vomit.

Personal Protective Methods: For emergency situations, wear protective gloves and clothing to prevent any reasonable probability of skin contact. Safety equipment suppliers/manufacturers can provide recommendations on the most protective glove/clothing material for your operation.. All protective clothing (suits, gloves, footwear, headgear) should be clean, available each day, and put on before work. Contact lenses should not be worn when working with this chemical. Wear dust-proof chemical goggles and face shield unless full facepiece respiratory protection is worn. Employees should wash immediately with soap when skin is wet or contaminated. Provide emergency showers and eyewash, or a positive pressure, pressure-demand, full facepiece self-contained breathing apparatus (SCBA) or pressure-demand supplied air respirator with escape SCBA and a fully-encapsulating, chemical resistant suit.

Respirator Selection: *At any detectable concentration:* SCBAF:PD,PP (any MSHA/NIOSH approved self-contained breathing apparatus that has a full facepiece and is operated in a pressure-demand or other positive-pressure mode); or SAF:PD,PP:ASCBA (any supplied-air respirator that has a full facepiece and is operated in a pressure-demand or other positive-pressure mode in combination with an auxiliary, self-contained breathing apparatus operated in a pressure-demand or other positive pressure mode).

Storage: Prior to working with this chemical you should be trained on its proper handling and storage. Store in tightly closed containers in a cool, well ventilated area.

Spill Handling: Evacuate persons not wearing protective equipment from area of spill or leak until clean-up is complete. Remove all ignition sources. Collect powdered material in the most convenient and safe manner and deposit in sealed containers. Ventilate area after clean-up is complete. It may be necessary to contain and dispose of this chemical as a hazardous waste. If material or contaminated runoff enters waterways, notify downstream users of potentially contaminated waters. Contact your Department of Environmental Protection or your regional office of the federal EPA for specific recommendations. If employees are required to clean-up spills, they must be properly trained and equipped. OSHA 1910.120(q) may be applicable.

Fire Extinguishing: Use dry chemical, carbon dioxide, water spray, or alcohol foam extinguishers. Poisonous gases are produced in fire including nitrogen oxides. If material or contaminated runoff enters waterways, notify downstream users of potentially contaminated waters. Notify local health and fire officials and pollution control agencies. From a secure, explosion-proof location, use water spray to cool exposed containers. If cooling streams are ineffective (venting sound increases in volume and pitch, tank discolors, or shows any signs of deforming), withdraw immediately to a secure position. If employees are expected to fight fires, they must be trained and equipped in OSHA 1910.156.

References

U.S. Environmental Protection Agency, "Chemical Profile: Diethylcarbamazine Citrate," Washington, DC, Chemical Emergency Preparedness Program (November 30, 1987)

Diethyl Carbamyl Chloride

Molecular Formula: $C_5H_{10}ClNO$

Common Formula: $(C_2H_5)_2NCOCl$

Synonyms: Diethylcarbamic chloride; Diethylcarbamidoyl chloride; *N,N*-Diethylcarbamoyl chloride; Diethylcarbamyl chloride

CAS Registry Number: 88-10-8

RTECS Number: FD4025000

Regulatory Authority

- Carcinogen (Animal Positive) (DFG)[3]

Cited in U.S. State Regulations: Florida (G), Massachusetts (G), New Hampshire (G), New Jersey (G), Pennsylvania (G).

Description: Diethyl carbamoyl chloride, $(C_2H_5)_2NCOCl$ is a colorless liquid. Boiling point = 187 – 190°C. Flash point = 163 – 172°C. Hazard Identification (based on NFPA-704 M Rating System): Health 2, Flammability 1, Reactivity 2. Soluble in water (reactive).

Potential Exposure: In the synthesis of the pharmaceutical Diethylcarbamazine citrate, an anthelmintic (working agent) produced and marketed under the trade names Hetrazan and Caricide.

Incompatibilities: Will react with water or steam to produce toxic and corrosive fumes.

Permissible Exposure Limits in Air: No standards set.

Permissible Concentration in Water: No criteria set.

Harmful Effects and Symptoms

Long Term Exposure: A study has shown DECC to be mutagenic in two E.coli strains (WP2 and WP2S from Witkin). However, DECC was not as mutagenic as its close analog, dimethylcarbamoyl chloride (DMCC). A suspected carcinogen.

First Aid: If this chemical gets into the eyes, remove any contact lenses at once and irrigate immediately for at least 15 minutes, occasionally lifting upper and lower lids. Seek medical attention immediately. If this chemical contacts the skin, remove contaminated clothing and wash immediately with soap and water. Seek medical attention immediately. If this chemical has been inhaled, remove from exposure, begin rescue breathing (using universal precautions) if breathing has stopped and CPR if heart action has stopped. Transfer promptly to a medical facility. When this chemical has been swallowed, get medical attention. If victim is conscious, administer water or milk. Do not induce vomiting.

Personal Protective Methods: Wear protective gloves and clothing to prevent any reasonable probability of skin contact. Safety equipment suppliers/manufacturers can provide recommendations on the most protective glove/clothing material for your operation.. All protective clothing (suits, gloves, footwear, headgear) should be clean, available each day, and put on before work. Contact lenses should not be worn when working with this chemical. Wear splash-proof chemical goggles and face shield unless full facepiece respiratory protection is worn. Employees should wash immediately with soap when skin is wet or contaminated. Provide emergency showers and eyewash.

Respirator Selection: *At any detectable concentration:* SCBAF:PD,PP (any MSHA/NIOSH approved self-contained breathing apparatus that has a full facepiece and is operated in a pressure-demand or other positive-pressure mode); or SAF:PD,PP:ASCBA (any supplied-air respirator that has a full facepiece and is operated in a pressure-demand or other positive-pressure mode in combination with an auxiliary, self-contained breathing apparatus operated in a pressure-demand or other positive pressure mode).

Storage: Prior to working with this chemical you should be trained on its proper handling and storage. Store in tightly closed containers in a cool, well ventilated area away from oxidizers and reducing agents. Where possible, automatically pump liquid from drums or other storage containers to process containers. A regulated, marked area should be established where this chemical is handled, used, or stored in compliance with OSHA standard 1910.1045.

Spill Handling: Evacuate and restrict persons not wearing protective equipment from area of spill or leak until cleanup is complete. Remove all ignition sources. Ventilate area of spill or leak. Absorb liquids in vermiculite, dry sand, earth, peat, carbon, or a similar material and deposit in sealed containers. Keep this chemical out of a confined space, such as a sewer, because of the possibility of an explosion, unless the sewer is designed to prevent the build-up of explosive concentrations. It may be necessary to contain and dispose of this chemical as a hazardous waste. If material or contaminated runoff enters waterways, notify downstream users of potentially contaminated waters. Contact your Department of Environmental Protection or your regional office of the federal EPA for specific recommendations. If employees are required to clean-up spills, they must be properly trained and equipped. OSHA 1910.120(q) may be applicable.

Fire Extinguishing: This chemical is a combustible liquid. Poisonous gases including nitrogen oxides and chlorides are produced in fire. Use dry chemical, carbon dioxide extinguishers. Reacts with water forming corrosive and toxic fumes. Vapors are heavier than air and will collect in low areas. Vapors in confined areas may explode when exposed to fire. Containers may explode in fire. Storage containers and parts of containers may rocket great distances, in many directions. If material or contaminated runoff enters waterways, notify downstream users of potentially contaminated waters. Notify local health and fire officials and pollution control agencies. From a secure, explosion-proof location, use water spray to cool exposed containers. If cooling streams are ineffective (venting sound increases in volume and pitch, tank discolors, or shows any signs of deforming), withdraw immediately to a secure position. If employees are expected to fight fires, they must be trained and equipped in OSHA 1910.156.

Diethyl Chlorophosphate

Molecular Formula: $C_4H_{10}ClO_3P$

Common Formula: $P(OC_2H_5)_2OCl$

Synonyms: Chlorophosphoric acid diethyl ester; Clorofosfato de dietilo (Spanish); Diethoxyphosphorous oxychloride; Phosphorochloridic acid, diethyl ester

CAS Registry Number: 814-49-3

RTECS Number: TD1400000

DOT ID: UN 3082

Regulatory Authority

- Superfund/EPCRA 40CFR355, Appendix B Extremely Hazardous Substances: TPQ = 500 lb (227 kg)
- Superfund/EPCRA 40CFR302.4 Reportable Quantity (RQ): EHS, 1 lb (0.454 kg)

Cited in U.S. State Regulations: Florida (G), Massachusetts (G), New Jersey (G), Pennsylvania.

Description: Diethyl chlorophosphate, $P(OC_2H_5)_2OCl$, is a combustible, clear liquid with an unpleasant odor. Boiling point = 60°C under 2.0 mm pressure. Hazard Identification (based on NFPA-704 M Rating System): Health 4, Flammability 1, Reactivity 0.

Potential Exposure: This material may be used as an intermediate in the manufacture of pesticides and chemical warfare agents.

Incompatibilities: Strong oxidizers.

Permissible Exposure Limits in Air: No standards set. A deadly poison by dermal contact.

Determination in Air: OSHA versatile sampler-2; Toluene/Acetone; Gas chromatography/Flame photometric detection for sulfur, nitrogen, or phosphorus; NIOSH Method IV Method #5600, Organophosphorus Pesticides.

Permissible Concentration in Water: No criteria set.

Routes of Entry: Inhalation, skin contact.

Harmful Effects and Symptoms

Short Term Exposure: Diethyl chlorophosphate may severely irritate or burn the skin, eyes, or mucous membranes. This material is a cholinesterase inhibitor. It has high oral and very high dermal toxicity. It is a skin irritant. It is also toxic by inhalation. Acute exposure to diethyl chlorophosphate may produce the following signs and symptoms: pinpoint pupils, blurred vision, headache, dizziness, muscle spasms, and profound weakness. Vomiting, diarrhea, abdominal pain, seizures, and coma may also occur. The heart rate may either decrease following oral exposure or increase following dermal exposure. Chest pain may be noted. Hypotension (low blood pressure) may be observed, although hypertension (high blood pressure) is not uncommon. Respiratory symptoms include dyspnea (shortness of breath), respiratory depression, and respiratory paralysis.

Long Term Exposure: Cholinesterase inhibitor; cumulative effect is possible. Diethyl chlorophosphate may damage the nervous system with repeated exposure, resulting in convulsions, respiratory failure. May cause liver damage.

Points of Attack: Respiratory system, lungs, central nervous system, cardiovascular system, skin, eyes, plasma and red blood cell cholinesterase.

Medical Surveillance: Before employment and at regular times after that, the following are recommended: Plasma and red blood cell cholinesterase levels (tests for the enzyme poisoned by this chemical). If exposure stops, plasma levels return to normal in 1 – 2 weeks while red blood cell levels may be reduced for 1 – 3 months.

When cholinesterase enzyme levels are reduced by 25% or more below preemployment levels, risk of poisoning is increased, even if results are in lower ranges of "normal." Reassignment to work not involving organophosphate or carbamate pesticides is recommended until enzyme levels recover. If symptoms develop or overexposure occurs, repeat the above tests as soon as possible and get an exam of the nervous system. Also consider complete blood count. Consider chest x-ray following acute overexposure. Do not drink any alcoholic beverages before or during use. Alcohol promotes absorption of organic phosphates.

First Aid: If this chemical gets into the eyes, remove any contact lenses at once and irrigate immediately for at least 15 minutes, occasionally lifting upper and lower lids. Seek medical attention immediately. If this chemical contacts the skin, remove contaminated clothing and wash immediately with soap and water. Speed in removing material from skin is of extreme importance. Shampoo hair promptly if contaminated. Seek medical attention immediately. If this chemical has been inhaled, remove from exposure, begin rescue breathing (using universal precautions) if breathing has stopped and CPR if heart action has stopped. Transfer promptly to a medical facility. When this chemical has been swallowed, get medical attention. Give conscious victims water or milk. Promote excretion by administering saline cathartic or sorbitol. Do not make an unconscious person vomit.

Personal Protective Methods: For emergency situations, wear protective gloves and clothing to prevent any reasonable probability of skin contact. Safety equipment suppliers/manufacturers can provide recommendations on the most protective glove/clothing material for your operation. All protective clothing (suits, gloves, footwear, headgear) should be clean, available each day, and put on before work. Contact lenses should not be worn when working with this chemical. Wear splash-proof chemical goggles and face shield unless full facepiece respiratory protection is worn. Employees should wash immediately with soap when skin is wet or contaminated. Provide emergency showers and eyewash. For emergency situations wear a positive pressure, pressure-demand, full facepiece self-contained breathing apparatus (SCBA) or pressure-demand supplied air respirator with escape (SCBA) and a fully-encapsulating, chemical resistant suit.

Respirator Selection: *At any detectable concentration:* SCBAF:PD,PP (any MSHA/NIOSH approved self-contained breathing apparatus that has a full facepiece and is operated in a pressure-demand or other positive-pressure mode); or SAF:PD,PP:ASCBA (any supplied-air respirator that has a full facepiece and is operated in a pressure-demand or other positive-pressure mode in combination with an auxiliary, self-contained breathing apparatus operated in a pressure-demand or other positive pressure mode). *Escape:* HiEF (any air-purifying, full-facepiece respirator (gas mask) with a chin-style, front- or back-mounted organic vapor canister having a high-efficiency particulate filter); or SCBAE (any appropriate escape-type, self-contained breathing apparatus).

Storage: Prior to working with this chemical you should be trained on its proper handling and storage. Store in tightly closed containers in a cool, well ventilated area away from oxidizers and reducing agents. Where possible, automatically pump liquid from drums or other storage containers to process containers.

Shipping: Environmentally hazardous liquid, n.o.s. Hazard Class: 9. Label: "Class 9."

Spill Handling: Evacuate and restrict persons not wearing protective equipment from area of spill or leak until cleanup is complete. Do not touch spilled material. Do not breathe vapors. Stay upwind; keep out of low areas. Remove all ignition sources. Ventilate area of spill or leak. This material is a combustible liquid. For a spill or leak of a combustible liquid, shut off ignition sources; no flares, smoking or flames in hazard area. Stop leak if you can do so without risk. Use water spray to reduce vapors. Small spills: absorb with sand or other noncombustible absorbent material and place into containers for later disposal. Large spills: dike far ahead of spill for later disposal. Keep this chemical out of a confined space, such as a sewer, because of the possibility of an explosion, unless the sewer is designed to prevent the build-up of explosive concentrations. It may be necessary to contain and dispose of this chemical as a hazardous waste. If material or contaminated runoff enters waterways, notify downstream users of potentially contaminated waters. Contact your Department of Environmental Protection or your regional office of the federal EPA for specific recommendations. If employees are required to clean-up spills, they must be properly trained and equipped. OSHA 1910.120(q) may be applicable.

Fire Extinguishing: This chemical is a combustible liquid. Poisonous gases including phosphorus oxides and chlorides are produced in fire. Small fires: dry chemical, carbon dioxide, water spray, or foam. Large fires: water spray, fog, or foam. Move container from fire area if you can do it without risk. Cool containers that are exposed to flames with water from the side until well after fire is out. For massive fire in cargo area, use unmanned hose holder or monitor nozzles; if this is impossible, withdraw form area and let fire burn. Stay upwind; keep out of low areas. Wear self-contained (positive pressure if available) breathing apparatus and full protective clothing. Isolate for 1/2 mile in all directions if tank car or truck is involved in fire. Vapors are heavier than air and will collect in low areas. Vapors in confined areas may explode when exposed to fire. Containers may explode in fire. Storage containers and parts of containers may rocket great distances, in many directions. If material or contaminated runoff enters waterways, notify downstream users of potentially contaminated waters. Notify local health and fire officials and pollution control agencies. From a secure, explosion-proof location, use water spray to cool exposed containers. If cooling streams are ineffective (venting sound increases in volume and pitch, tank discolors, or shows any signs of deforming), withdraw immediately to a secure position. If employees are expected to fight fires, they must be trained and equipped in OSHA 1910.156.

References

U.S. Environmental Protection Agency, "Chemical Profile: Diethyl Chlorophosphate," Washington, DC, Chemical Emergency Preparedness Program (November 30, 1987)

Diethylene Triamine

Molecular Formula: $C_4H_{13}N_3$

Common Formula: $(NH_2CH_2CH_2)_2NH$

Synonyms: Aminoethandiamine; *N*-(2-Aminoethyl); Aminoethylethandiamine; 3-Azapentane-1,5-diamine; Bis (2-aminoethyl) amine; Bis(β-aminoethyl)amine; Bis(2-aminoethyl)amine; DETA; 2,2'-Diaminodiethylamine; Diethylenetriamine; 1,2-Ethanediamine, *N*-(2-aminoethyl)-; Ethylenediamine; 2,2'-Iminobisethylamine; 3-a2a pentane-1, 5-diamine; 1,4,7-Triazaheptane

CAS Registry Number: 111-40-0

RTECS Number: IE1225000

DOT ID: UN 2079

EEC Number: 612-058-00-X

Regulatory Authority

- Air Pollutant Standard Set (ACGIH)[1] (Australia) (HSE)[33] Israel) (Mexico) (Sweden)[35] former (former USSR)[43] (OSHA)[58] (Several States)[60] (Several Canadian Provinces)
- Canada, WHMIS, Ingredients Disclosure List

Cited in U.S. State Regulations: Alaska (G), California (A, G), Connecticut (A), Florida (G), Illinois (G), Maine (G), Massachusetts (G), Minnesota (G), Nevada (A), New Hampshire (G), New Jersey (G), North Dakota (A), Pennsylvania (G), Rhode Island (G), Virginia (A), West Virginia (G).

Description: Diethylene triamine, $(NH_2CH_2CH_2)_2NH$, is a flammable, thick yellow liquid with an ammoniacal odor. The odor threshold is 100 ppm.[41] Boiling point = 207°C. Freezing/Melting point = -39°C. Flash point = 98°C. Explosive limits: LEL = 2%; UEL = 6.7%.[17] Hazard Identification (based on NFPA-704 M Rating System): Health 3, Flammability 1, Reactivity 0. Soluble in water.

Potential Exposure: This material is used as a solvent for sulfur, acid gases, resins and dyes.

Incompatibilities: Forms explosive mixture with air. Ignites spontaneously on contact with cellulose nitrate. Contact with silver, cobalt, or chromium compounds may cause explosions. Incompatible with acids, halogenated organics, organic anhydrides, isocyanates, vinyl acetate, acrylates, substituted allyls, alkylene oxides, epichlorohydrin, ketones, aldehydes, alcohols, glycols, mercury, phenols, cresols, caprolactum solution, strong oxidizers. Attacks aluminum, copper, copper alloys, lead, tin, zinc and alloys.

Permissible Exposure Limits in Air: There is no OSHA PEL.[58] HSE,[33] Israel, Australia, Mexico, ACGIH TLV, and the Canadian provinces of Alberta, BC, Ontario, and Quebec TWA is 1 ppm (4.2 mg/m^3) and the Alberta STEL is 3 ppm. All warn about the potential for skin absorbtion. The former USSR[35] has set a MAC in ambient air in residential areas of 0.01 mg/m^3 on a once-daily basis. Several states have set guidelines or standards for diethylene triamine in ambient air[60] ranging from 40.0 μg/m^3 (North Dakota) to 65 μg/m^3 (Virginia) to 80.0 μg/m^3 (Connecticut) to 95 μg/m^3 (Nevada).

Determination in Air: XAD-2® (tube); Wear DMF (any dust and mist respirator with a full facepiece); High-pressure liquid chromatography/Ultraviolet detection; NIOSH IV Method #2540.

Permissible Concentration in Water: The former USSR[35][43] has set a MAC in water bodies used for domestic purposes of 0.2 mg/l.

Routes of Entry: Inhalation, skin absorption, ingestion, skin and/or eye contact.

Harmful Effects and Symptoms

Short Term Exposure: Diethylene triamine can affect you when breathed in and by passing through you skin. Exposure can severely irritate the eyes, skin, and respiratory system. Diethylene triamine is a corrosive liquid and contact can irritate the skin and may irritate and burn the eyes. Skin sensitization, dermatitis and pulmonary sensitization and irritation, possibly leading to bronchial asthma. Also conjunctivitis, keratitis; nausea, vomiting; eye, skin necrosis; cough, dyspnea (breathing difficulty). The oral LD_{50} rat is 1,080 mg/kg (slightly toxic).

Long Term Exposure: Diethylene triamine can cause both skin and lung allergies to develop. Once allergy develops, further low exposures can trigger allergic effects.

Points of Attack: Eyes, skin, respiratory system.

Medical Surveillance: Before beginning employment and at regular times after that, for those with frequent or potentially high exposures, the following is recommended: Lung function tests. If symptoms develop or overexposure is suspected, the following may be useful: Skin testing with diluted diethylene triamine may help diagnose allergy if done by a qualified allergist.

First Aid: If this chemical gets into the eyes, remove any contact lenses at once and irrigate immediately for at least 15 minutes, occasionally lifting upper and lower lids. Seek medical attention immediately. If this chemical contacts the skin, remove contaminated clothing and wash immediately with soap and water. Seek medical attention immediately. If this chemical has been inhaled, remove from exposure, begin rescue breathing (using universal precautions) if breathing has stopped and CPR if heart action has stopped. Transfer promptly to a medical facility. When this chemical has been swallowed, get medical attention. If victim is conscious, administer water or milk. Do not induce vomiting.

Personal Protective Methods: Wear protective gloves and clothing to prevent any reasonable probability of skin contact. Safety equipment suppliers/manufacturers can provide recommendations on the most protective glove/clothing material for your operation.. All protective clothing (suits, gloves, footwear, headgear) should be clean, available each day, and put on before work. Contact lenses should not be worn when working with this chemical. Wear splash-proof chemical goggles and face shield unless full facepiece respiratory protection is worn. Employees should wash immediately with soap when skin is wet or contaminated. Provide emergency showers and eyewash.

Respirator Selection: Where the potential exists for exposures over 1 ppm, use an MSHA/NIOSH approved supplied-air respirator with a full facepiece operated in the positive pressure mode or with a full facepiece, hood, or helmet in the continuous flow mode, or use an MSHA/NIOSH approved self-contained breathing apparatus with a full facepiece operated in pressure-demand or other positive pressure mode.

Storage: Prior to working with this chemical you should be trained on its proper handling and storage. Before entering confined space where this chemical may be present, check to make sure that an explosive concentration does not exist. Store in tightly closed containers in a cool, well-ventilated area away from acids, halogenated organics, and oxidizers (such as perchlorates, peroxides, chlorates, nitrates and permanganates). Protect containers from physical damage. Sources of ignition such as smoking and open flames are prohibited where this chemical is used, handled, or stored in a manner that could create a potential fire or explosion hazard. Metal containers involving the transfer of 5 gallons or more of this chemical should be grounded and bonded. Drums must be equipped with self-closing valves, pressure vacuum bungs, and flame arresters. Use only nonsparking tools and equipment, especially when opening and closing containers of this chemical.

Shipping: This material requires a "Corrosive" label. It falls in DOT Hazard Class 8 and Packing Group II.

Spill Handling: Evacuate and restrict persons not wearing protective equipment from area of spill or leak until cleanup is complete. Establish forced ventilation to keep levels below explosive limit. Absorb liquids in vermiculite, dry sand, earth, peat, carbon, or a similar material and deposit in sealed containers. Keep this chemical out of a confined space, such as a sewer, because of the possibility of an explosion, unless the sewer is designed to prevent the build-up of explosive concentrations. It may be necessary to contain and dispose of this chemical as a hazardous waste. If material or contaminated runoff enters waterways, notify downstream users of potentially contaminated waters. Contact your Department of Environmental Protection or your regional office of the federal EPA for specific recommendations. If employees are required to clean-up spills, they must be properly trained and equipped. OSHA 1910.120(q) may be applicable.

Fire Extinguishing: Diethylene triamine may burn, but does not readily ignite. Use dry chemical, CO_2, water spray, or foam extinguishers. Poisonous gases are produced in fire. Vapors are heavier than air and will collect in low areas. Containers may explode in fire. Storage containers and parts of containers may rocket great distances, in many directions. If material or contaminated runoff enters waterways, notify downstream users of potentially contaminated waters. Notify local health and fire officials and pollution control agencies. From a secure, explosion-proof location, use water spray to cool exposed containers. If cooling streams are ineffective (venting sound increases in volume and pitch, tank discolors, or shows any signs of deforming), withdraw immediately to a secure position. If employees are expected to fight fires, they must be trained and equipped in OSHA 1910.156.

Disposal Method Suggested: Incinerate in admixture with flammable solvent in furnace equipped with afterburner and scrubber.[22]

References

New Jersey Department of Health, "Hazardous Substance Fact Sheet: Diethylene Triamine," Trenton, NJ (February 1989)

Di(2-Ethylhexyl) Phthalate

Molecular Formula: $C_{24}H_{38}O_4$

Synonyms: BEHP; 1,2-Benzenedicarboxylic acid, bis(2-ethylhexyl) ester; 1,2-Benzenedicarboxylic acid, dioctyl ester; Bis(2-ethylhexyl) 1,2-benzenedicarboxylate; Bis(2-ethylhexyl) phthalate; Bis(2-etilhexil)ftalato (Spanish); Bisoflex 81; Bisoflex 82; Bisoflex DOP; Compound 889; DAF 68; DEHP; Diester of 2-ethylhexyl alcohol and phthalic acid; Di(2-ethylhexyl) *o*-phthalate; Di(2-ethylhexyl)phthalate; Di-*s*-octyl phthalate; Di-*sec*-octyl phthalate; Dioctyl phthalate; DOF; DOP; Ergoplast FDO; Ergoplast FDO-S; 2-Ethylhexyl phthalate; Ethylhexyl phthalate; Eviplast 80; Eviplast 81; Fleximel; Flexol DOP; Flexol plasticizer DOP; Ftalato de(2-etilhexilo) (Spanish); Good-Rite GP 264; Hatcol DOP; Hercoflex 260; Kodaflex DOP; Mitsubishi DOP; Mollano; NCI-C52733; Nuoplaz DOP; Octoil; Octyl phthalate; Octyl phthalate, di-*sec*; Palatinol AH; Phthalic acid dioctyl ester; Pittsburgh PX-138; Plasthall DOP; Plasticizer 28P; Platinol AH; Platinol DOP; Polycizer 162; PX-138; RC Plasticizer DOP; Reomol D 79P; Reomol DOP; Sicol 150; Staflex DOP; Truflex DOP; Vestinol AH; Vinicizer 80; Witcizer 312

CAS Registry Number: 117-81-7

RTECS Number: TI0350000

DOT ID: UN 3082

EEC Number: 204-211-0

Regulatory Authority

- Carcinogen (Animal Positive) (IARC) (NTP)[9] (ACGIH)[1]
- Air Pollutant Standard Set (ACGIH)[1] (DFG)[3] (Australia) (HSE)[33] (Israel) (Mexico) (former USSR)[35] (Several States)[60] (Several Canadian Provinces)

- Clean Air Act: Hazardous Air Pollutants (Title I, Part A, Section 112)
- Clean Water Act: 40CFR423, Appendix A, Priority Pollutants
- EPA Hazardous Waste Number (RCRA No.): U028
- RCRA, 40CFR261, Appendix 8 Hazardous Constituents
- RCRA 40CFR268.48; 61FR15654, Universal Treatment Standards: Wastewater (mg/l), 0.28; Nonwastewater (mg/kg), 28
- RCRA 40CFR264, Appendix 9; TSD Facilities Ground Water Monitoring List. Suggested test method(s) (PQL μg/l): 8060 (20); 8270 (10)
- Safe Drinking Water Act: MCL, 0.006 mg/l; MCLG, zero
- Superfund/EPCRA 40CFR302.4 Reportable Quantity (RQ): CERCLA, 100 lb (45.4 kg)
- EPCRA Section 313 Form R *de minimis* concentration reporting level: 0.1%
- Canada, WHMIS, Ingredients Disclosure List, National Pollutant Release Inventory (NPRI)

Cited in U.S. State Regulations: California (G), Connecticut (A, G), Illinois (G), Kansas (G), Louisiana (G), Maine (G), Maryland (G), Massachusetts (G, A), Minnesota (G), Nevada (A), New Hampshire (G), New Jersey (G), New York (A), North Carolina (A), North Dakota (A), Pennsylvania (G, A), Rhode Island (G, A), Vermont (G), Virginia (G, A), Washington (G), Wisconsin (G).

Description: Di(2-Ethylhexyl) Phthalate is a colorless oily liquid with almost no odor. Boiling point = 387°C. Freezing/Melting point = -50°C. Flash point = 215°C. Explosive limits: LEL = 0.3%; UEL – unknown. Slightly soluble in water.

Potential Exposure: Di (2-ethylhexyl) phthalate (DEHP) is commercially produced by the reaction of 2-ethylhexyl alcohol and phthalic anhydride. It is used as a plasticizers for resins and in the manufacture of organic pump fluids. Two groups are at risk in regard to phthalic acid esters. These are workers in the industrial environment in which the phthalates are manufactured or used and patients receiving chronic transfusion of blood and blood products stored in PVC blood bags.

Incompatibilities: Nitrates, strong oxidizers, strong acids, strong alkalis.

Permissible Exposure Limits in Air: There is no OSHA PEL. The NIOSH and ACGIH recommended TWA value is 5 mg/m^3 and STEL is 10 mg/m^3. Australia, Israel, Mexico, and the Canadian provinces of Alberta, BC, and Quebec set the same airborne levels. Ontario set a TWA of 3 mg/m^3 and STEL is 10 mg/m^3. The DFG, set an MAK of 10 mg/m^3. The NIOSH IDLH level is [Ca] 5,000 mg/m^3. The former USSR[35] has set a MAC in workplace air of 0.1 mg/m^3. A number of states have set guidelines or standards for bis (2-ethylhexyl) phthalate in ambient air[60] – 0.5 μg/m^3 (Rhode Island) to 0.68 μg/m^3 (Massachusetts) to 16.0 μg/m^3 (New York) to 25.0 μg/m^3 (North Carolina) to 80.0 μg/m^3 (Virginia) to 100 μg/m^3 (Connecticut) to 119 μg/m^3 (Nevada) to 120 μg/m^3 (Pennsylvania).

Determination in Air: Collection on a filter, workup with CS_2; Gas chromatography/Flame ionization detection; NIOSH IV, Method #5020, di(2-Ethylhexyl) Phthalate.

Permissible Concentration in Water: For freshwater and saltwater aquatic life, no criteria have been set due to lack of data. For protection of human health, the ambient water criterion is 15 mg/l (15,000 μg/l).[6] A no-adverse effect level in drinking water has been calculated by NAS/NRC to be 4.2 mg/l. An acceptable daily intake (ADI) value of 0.6 μg/kg/day has been calculated by NAS/NRC.

Determination in Water: Gas chromatography (EPA Method 606) or gas chromatography plus mass spectrometry (EPA Method 625).

Routes of Entry: Inhalation, ingestion, skin and/or eye contact.

Harmful Effects and Symptoms

Short Term Exposure: DEHP irritates the eyes, skin, and respiratory tract, and may affect the gastrointestinal tract. Symptoms also include irritation of the mucous membranes; nausea; diarrhea.

Long Term Exposure: Repeated or prolonged contact with skin may cause dermatitis. The very low levels of DEHP to which humans are normally exposed have not been shown to cause adverse health effects. But DEHP causes cancer in rats and mice. It is also known to produce liver damage and male reproductive system damage, affect reproduction, and produce birth defects in laboratory animals. However, none of these effects have been documented in humans. This complicates estimating which kinds of health effects and exposure levels will actually affect humans. However, it is prudent to regard the animal data as indicating some degree of concern for harmful human effects (a potential occupational carcinogen) until research can more reasonably conclude that no harm can occur.

Points of Attack: Eyes, respiratory system, central nervous system, liver, reproductive system, gastrointestinal tract. Cancer Site [in animals: liver tumors].

Medical Surveillance: If symptoms develop or overexposure is suspected, the following may be useful: Liver and kidney function tests. Examination of the nervous system. Any evaluation should include a careful history of past and present symptoms.

First Aid: If this chemical gets into the eyes, remove any contact lenses at once and irrigate immediately for at least 15 minutes, occasionally lifting upper and lower lids. Seek medical attention immediately. If this chemical contacts the skin, remove contaminated clothing and wash immediately with soap and water. Seek medical attention immediately. If this chemical has been inhaled, remove from exposure, begin rescue breathing (using universal precautions) if breathing has stopped and CPR if heart action has stopped. Transfer promptly to a medical facility. When this chemical has been swallowed, get medical attention. Give large quantities of water and induce vomiting. Do not make an unconscious person vomit.

Personal Protective Methods: Wear protective gloves and clothing to prevent any reasonable probability of skin contact. Safety equipment suppliers/manufacturers can provide recommendations on the most protective glove/clothing material for your operation.. All protective clothing (suits, gloves, footwear, headgear) should be clean, available each day, and put on before work. Contact lenses should not be worn when working with this chemical. Wear splash-proof chemical goggles and face shield unless full facepiece respiratory protection is worn. Employees should wash immediately with soap when skin is wet or contaminated. Provide emergency showers and eyewash.

Respirator Selection: *At any detectable concentration:* SCBAF:PD,PP (any MSHA/NIOSH approved self-contained breathing apparatus that has a full facepiece and is operated in a pressure-demand or other positive-pressure mode); or SAF:PD,PP:ASCBA (any supplied-air respirator that has a full facepiece and is operated in a pressure-demand or other positive-pressure mode in combination with an auxiliary, self-contained breathing apparatus operated in a pressure-demand or other positive pressure mode). *Escape:* HiEF (any air-purifying, full-facepiece respirator (gas mask) with a chin-style, front- or back-mounted organic vapor canister having a high-efficiency particulate filter); or SCBAE (any appropriate escape-type, self-contained breathing apparatus).

Storage: Prior to working with DEHP you should be trained on its proper handling and storage. Before entering confined space where this chemical may be present, check to make sure that an explosive concentration does not exist. Bis (2-ethylhexyl) phthalate must be stored to avoid contact with oxidizing materials, such as permanganates, nitrates, peroxides, chlorates, and perchlorates, since violent reactions occur. Store in tightly closed containers in a cool, well-ventilated area away from heat. Sources of ignition such as smoking and open flames are prohibited where bis (2-ethylhexyl) phthalate is used, handled, or stored in a manner that could create a potential fire or explosion hazard. Sources of ignition such as smoking and open flames are prohibited where this chemical is used, handled, or stored in a manner that could create a potential fire or explosion hazard. Metal containers involving the transfer of 5 gallons or more of this chemical should be grounded and bonded. Drums must be equipped with self-closing valves, pressure vacuum bungs, and flame arresters. Use only nonsparking tools and equipment, especially when opening and closing containers of this chemical. A regulated, marked area should be established where this chemical is handled, used, or stored in compliance with OSHA standard 1910.1045.

Shipping: Environmentally hazardous liquid, n.o.s. Hazard Class: 9. Label: "Class 9."

Spill Handling: Evacuate and restrict persons not wearing protective equipment from area of spill or leak until cleanup is complete. Remove all ignition sources. Establish forced ventilation to keep levels below explosive limit. Absorb liquids in vermiculite, dry sand, earth, peat, carbon, or a similar material and deposit in sealed containers. Keep this chemical out of a confined space, such as a sewer, because of the possibility of an explosion, unless the sewer is designed to prevent the build-up of explosive concentrations. It may be necessary to contain and dispose of this chemical as a hazardous waste. If material or contaminated runoff enters waterways, notify downstream users of potentially contaminated waters. Contact your Department of Environmental Protection or your regional office of the federal EPA for specific recommendations. If employees are required to clean-up spills, they must be properly trained and equipped. OSHA 1910.120(q) may be applicable.

Fire Extinguishing: This chemical is a combustible liquid. It may burn but does not easily ignite. Poisonous gases are produced in fire. Use dry chemical, carbon dioxide, or foam extinguishers. Vapors in confined areas may explode when exposed to fire. Containers may explode in fire. Storage containers and parts of containers may rocket great distances, in many directions. If material or contaminated runoff enters waterways, notify downstream users of potentially contaminated waters. Notify local health and fire officials and pollution control agencies. From a secure, explosion-proof location, use water spray to cool exposed containers. If cooling streams are ineffective (venting sound increases in volume and pitch, tank discolors, or shows any signs of deforming), withdraw immediately to a secure position. If employees are expected to fight fires, they must be trained and equipped in OSHA 1910.156.

Disposal Method Suggested: Incineration.[22] Consult with environmental regulatory agencies for guidance on acceptable disposal practices. Generators of waste containing this contaminant ($\geq$100 kg/mo) must conform with EPA regulations governing storage, transportation, treatment, and waste disposal.

References

U.S. Environmental Protection Agency, Phthalate Esters: Ambient Water Quality Criteria, Washington, DC (1980)

National Institute for Occupational Safety and Health, Profiles on Occupational Hazards for Criteria Document Priorities: Phthalates, 97–103, Report PB-274,073, Rockville, MD (1977)

U.S. Environmental Protection Agency, Bis (2-Ethylhexyl) Phthalate, Health and Environmental Effects Profile No. 27, Washington, DC, Office of Solid Waste (April 30, 1980)

Sax N. I., Ed., "Dangerous Properties of Industrial Materials Report," 1, No. 7, 52–54 (1981) and 2, No. 2, 22–25 (1982)

U.S. Public Health Service, "Toxicological Profile for Di-(2-Ethylhexyl) Phthalate," Atlanta, Georgia, Agency for Toxic Substances and Disease Registry (December 1987)

New Jersey Department of Health, "Hazardous Substance Fact Sheet: Bis (2-Ethylhexyl) Phthalate," Trenton, NJ (January 1989)

Diethyl Ketone

Molecular Formula: $C_5H_{10}O$

Common Formula: $C_2H_5COC_2H_5$

Synonyms: DEK; Diethylcetone (French); Dimethylacetone; Ethyl ketone; Ethyl propionyl; Metacetone; Methacetone; 3-Pentanone; Pentanone-3; 3-Pentanone dimethyl acetone; Propione

CAS Registry Number: 96-22-0

RTECS Number: SA8050000

DOT ID: UN 1156

EEC Number: 606-006-00-5

Regulatory Authority

- Air Pollutant Standard Set (ACGIH)[1] (Australia) (DFG)[3] (HSE)[33] (Israel) (Several States)[60] (Several Canadian Provinces)
- Canada, WHMIS, Ingredients Disclosure List

Cited in U.S. State Regulations: Alaska (G), California (A, G), Connecticut (A), Florida (G), Illinois (G), Maine (G), Massachusetts (G), Minnesota (G), Nevada (A), New Hampshire (G), New Jersey (G), North Dakota (A), Oklahoma (G), Pennsylvania (G), Rhode Island (G), Virginia (A), West Virginia (G).

Description: Diethyl Ketone, $C_2H_5COC_2H_5$, is a colorless liquid with an acetone-like odor (smells like nail-polish remover). Odor threshold 2.8 ppm. Boiling point = 101°C. Freezing/Melting point = -42°C. Flash point = 12.8°C (oc). Autoignition temperature = 452°C. Explosive limits: LEL = 1.6%; UEL = 6.4%. Hazard Identification (based on NFPA-704 M Rating System): Health 1, Flammability 3, Reactivity 0. Moderately soluble in water.

Potential Exposure

This compound is used as a solvent and in organic synthesis and making medicines.

Incompatibilities: Violent reaction with oxidizers, causing fire and explosion hazard. Forms explosive mixture with air. Incompatible with strong acids, aliphatic amines. Attacks many plastics, rubber and coatings. May accumulate static electrical charges, and may cause ignition of its vapors.

Permissible Exposure Limits in Air: There is no OSHA PEL.[58] NIOSH and ACGIH recommend a TWA of 200 ppm (705 mg/m^3). Australia, Israel, HSE,[33] and the Canadian provinces of Alberta, BC, Ontario, and Quebec set the same TWA as ACGIH and HSE's and Alberta's STEL is 250 ppm (875 mg/m^3). Several states have set guidelines or standards for diethyl ketone in ambient air[60] ranging from 7.05 mg/m^3 (North Dakota) to 12.0 mg/m^3 (Virginia) to 14.1 mg/m^3 (Connecticut) to 16.8 mg/m^3 (Nevada).

Determination in Air: No method available.

Permissible Concentration in Water: No criteria set.

Routes of Entry: Inhalation, ingestion, skin and/or eye contact.

Harmful Effects and Symptoms

Short Term Exposure: DEK can be absorbed through the skin, thereby increasing exposure DEK irritates the eyes, skin, and respiratory tract. Higher exposures can cause dizziness, lightheadedness, and unconsciousness. This material is slightly toxic. The oral LD_{50} rat is 2,140 mg/kg.

Long Term Exposure: Removes the skins natural oils. Some, but not all, ketones can cause nerve damage.

Points of Attack: Eyes, skin, respiratory system.

Medical Surveillance: Interview for brain effects, including memory, mood, concentration, headaches, altered sleep patterns.

First Aid: If this chemical gets into the eyes, remove any contact lenses at once and irrigate immediately for at least 15 minutes, occasionally lifting upper and lower lids. Seek medical attention immediately. If this chemical contacts the skin, remove contaminated clothing and wash immediately with soap and water. Seek medical attention immediately. If this chemical has been inhaled, remove from exposure, begin rescue breathing (using universal precautions) if breathing has stopped and CPR if heart action has stopped. Transfer promptly to a medical facility. When this chemical has been swallowed, get medical attention. Give large quantities of water and induce vomiting. Do not make an unconscious person vomit.

Personal Protective Methods: Wear protective gloves and clothing to prevent any reasonable probability of skin contact. Safety equipment suppliers/manufacturers can provide recommendations on the most protective glove/clothing material for your operation.. All protective clothing (suits, gloves, footwear, headgear) should be clean, available each day, and put on before work. Contact lenses should not be worn when working with this chemical. Wear splash-proof chemical goggles and face shield unless full facepiece respiratory protection is worn. Employees should wash immediately with soap when skin is wet or contaminated. Provide emergency showers and eyewash.

Respirator Selection: Where the potential exists for exposures over 200 ppm, use an MSHA/NIOSH approved respirator with an organic vapor cartridge/canister. More protection is provided by a full facepiece respirator than by a half-mask respirator, and even greater protection is provided by a powered-air purifying respirator. If while wearing a filter, cartridge or canister respirator, you can smell, taste, or otherwise detect diethyl ketone, or in the case of a full facepiece respirator you experience eye irritation, leave the area immediately. Check to make sure the respirator-to-face seal is still good. If it is, replace the filter, cartridge, or canister. If the seal is no longer good, you may need a new respirator.

Where the potential for high exposures exists, use an MSHA/NIOSH approved supplied-air respirator with a full facepiece operated in the positive pressure mode or with a full facepiece, hood, or helmet in the continuous flow mode, or use an MSHA/NIOSH approved self-contained breathing apparatus with a full facepiece operated in pressure-demand or other positive pressure mode.

Storage: Prior to working with DEK you should be trained on its proper handling and storage. Before entering confined space where this chemical may be present, check to make sure that an explosive concentration does not exist. Diethyl ketone must be stored to avoid contact with oxidizing materials (such as peroxides, perchlorates, chlorates, permanganates, and nitrates) since violent reactions occur. Store in tightly closed containers in a cool, well-ventilated area away from sources of heat. Sources of ignition such as smoking and open flames are prohibited where diethyl ketone is used, handled, or stored in a manner that could create a potential fire or explosion hazard, Metal containers involving the transfer of 5 gallons or more of diethyl ketone should be grounded and bonded. Drums must be equipped with self-closing valves, pressure vacuum bungs, and flame arresters. Use only non-sparking tools and equipment, especially when opening and closing containers of diethyl ketone. Wherever diethyl ketone is used, handled, manufactured, or stored, use explosion-proof electrical equipment and fittings.

Shipping: This compound requires a shipping label of: "Flammable Liquid." It falls in DOT Hazard Class 3 and Packing Group II.

Spill Handling: Evacuate and restrict persons not wearing protective equipment from area of spill or leak until cleanup is complete. Remove all ignition sources. Establish forced ventilation to keep levels below explosive limit. Absorb liquids in vermiculite, dry sand, earth, peat, carbon, or a similar material and deposit in sealed containers. Keep this chemical out of a confined space, such as a sewer, because of the possibility of an explosion, unless the sewer is designed to prevent the build-up of explosive concentrations. It may be necessary to contain and dispose of this chemical as a hazardous waste. If material or contaminated runoff enters waterways, notify downstream users of potentially contaminated waters. Contact your Department of Environmental Protection or your regional office of the federal EPA for specific recommendations. If employees are required to clean-up spills, they must be properly trained and equipped. OSHA 1910.120(q) may be applicable.

Fire Extinguishing: This chemical is a flammable liquid. Poisonous gases are produced in fire. Use alcohol foam extinguishers. Vapors are heavier than air and will collect in low areas. Vapors may travel long distances to ignition sources and flashback. Vapors in confined areas may explode when exposed to fire. Containers may explode in fire. Storage containers and parts of containers may rocket great distances, in many directions. If material or contaminated runoff enters waterways, notify downstream users of potentially contaminated waters. Notify local health and fire officials and pollution control agencies. From a secure, explosion-proof location, use water spray to cool exposed containers. If cooling streams are ineffective (venting sound increases in volume and pitch, tank discolors, or shows any signs of deforming), withdraw immediately to a secure position. If employees are expected to fight fires, they must be trained and equipped in OSHA 1910.156.

Disposal Method Suggested: Incineration.

References

National Institute for Occupational Safety and Health, Criteria for a Recommended Standard: Occupational Exposure to Ketones, NIOSH Doc. No. 78–173 (1978)

New Jersey Department of Health, "Hazardous Substance Fact Sheet: Diethyl Ketone," Trenton, NJ (February 1989)

O,O-Diethyl-S-Methyl Phosphorodithioate

Molecular Formula: $C_5H_{13}O_2PS_2$

Synonyms: Phosphorodithioc acid, O,O-diethyl *S*-methyl ester

CAS Registry Number: 3288-58-2

RTECS Number: TD9670000

DOT ID: UN 2783

Regulatory Authority

- EPA Hazardous Waste Number (RCRA No.): U087
- RCRA, 40CFR261, Appendix 8 Hazardous Constituents
- RCRA Land Ban Waste Restrictions
- Superfund/EPCRA 40CFR302.4 Reportable Quantity (RQ): CERCLA, 5,000 lb (2,270 kg)

Cited in U.S. State Regulations: New Jersey (G), Massachusetts (G), Pennsylvania (G).

Description: O,O-Diethyl-S-Methyl Phosphorodithioate is an organophosphate and the methyl derivative of O,O-diethyl dithiophosphoric acid (which see).

The compound has partly insecticidal, acaricidal and fungicidal activity and is useful as an intermediate for organic synthesis.

Permissible Exposure Limits in Air: No standards set.

Determination in Air: OSHA versatile sampler-2; Toluene/Acetone; Gas chromatography/Flame photometric detection for sulfur, nitrogen, or phosphorus; NIOSH Method IV Method #5600, Organophosphorus Pesticides.

Permissible Concentration in Water: No criteria set.

Routes of Entry: Inhalation, skin contact.

Harmful Effects and Symptoms

Short Term Exposure: Many organic phosphorus insecticides are absorbed by the skin, as well as by the respiratory and gastrointestinal tracts. They are cholinesterase inhibitors. Symptoms of exposure include headache, giddiness, blurred vision, nervousness, weakness, nausea, cramps, diarrhea, and discomfort in the chest. Signs include sweating, tearing, salivation, vomiting, cyanosis, convulsions, coma, loss of reflexes and loss of sphincter control.

Long Term Exposure: There is no available information on the possible carcinogenic, mutagenic, teratogenic or adverse reproductive effects of O,O-diethyl-S-methyl phosphorodithioate. It, like other organophosphate compounds, is expected to produce cholinesterase inhibition in humans. There is no available data on the aquatic toxicity of this compound. The oral LD_{50} mouse is 156 mg/kg (Moderately toxic).

Points of Attack: Respiratory system, lungs, central nervous system, cardiovascular system, skin, eyes, plasma and red blood cell cholinesterase.

Medical Surveillance: Before employment and at regular times after that, the following are recommended: Plasma and red blood cell cholinesterase levels (tests for the enzyme poisoned by this chemical). If exposure stops, plasma levels return to normal in 1 – 2 weeks while red blood cell levels may be reduced for 1 – 3 months.

When cholinesterase enzyme levels are reduced by 25% or more below preemployment levels, risk of poisoning is increased, even if results are in lower ranges of "normal." Reassignment to work not involving organophosphate or carbamate pesticides is recommended until enzyme levels recover. If symptoms develop or overexposure occurs, repeat the above tests as soon as possible and get an exam of the nervous system. Also consider complete blood count. Consider chest x-ray following acute overexposure. Do not drink any alcoholic beverages before or during use. Alcohol promotes absorption of organic phosphates.

First Aid: If this chemical gets into the eyes, remove any contact lenses at once and irrigate immediately for at least 15 minutes, occasionally lifting upper and lower lids. Seek medical attention immediately. If this chemical contacts the skin, remove contaminated clothing and wash immediately with soap and water. Speed in removing material from skin is of extreme importance. Shampoo hair promptly if contaminated. Seek medical attention immediately. If this chemical has been inhaled, remove from exposure, begin rescue breathing (using universal precautions) if breathing has stopped and CPR if heart action has stopped. Transfer promptly to a medical facility. When this chemical has been swallowed, get medical attention. Give large quantities of water and induce vomiting. Do not make an unconscious person vomit. Medical observation is recommended.

Personal Protective Methods: Wear protective gloves and clothing to prevent any reasonable probability of skin contact. Safety equipment suppliers/manufacturers can provide recommendations on the most protective glove/clothing material for your operation.. All protective clothing (suits, gloves, footwear, headgear) should be clean, available each day, and put on before work. Contact lenses should not be worn when working with this chemical. Wear splash-proof goggles and face shield if working with the liquid or dust-proof chemical goggles and face shield if working with dry material unless full facepiece respiratory protection is worn. Employees should wash immediately with soap when skin is wet or contaminated. Provide emergency showers and eyewash.

Respirator Selection: *At any detectable concentration:* SCBAF:PD,PP (any MSHA/NIOSH approved self-contained breathing apparatus that has a full facepiece and is operated in a pressure-demand or other positive-pressure mode); or SAF:PD,PP:ASCBA (any supplied-air respirator that has a full facepiece and is operated in a pressure-demand or other positive-pressure mode in combination with an auxiliary, self-contained breathing apparatus operated in a pressure-demand or other positive pressure mode).

Storage: Prior to working with this chemical you should be trained on its proper handling and storage. Store in tightly closed containers in a cool, well ventilated area away from oxidizers and reducing agents.

Spill Handling: *Dry material:* Evacuate persons not wearing protective equipment from area of spill or leak until clean-up is complete. Remove all ignition sources. Collect powdered material in the most convenient and safe manner and deposit in sealed containers. Ventilate area after clean-up is complete. It may be necessary to contain and dispose of this chemical as a hazardous waste. If material or contaminated runoff enters waterways, notify downstream users of potentially contaminated waters. Contact your Department of Environmental Protection or your regional office of the federal EPA for specific recommendations. If employees are required to clean-up spills, they must be properly trained and equipped. OSHA 1910.120(q) may be applicable.

Liquid: Evacuate and restrict persons not wearing protective equipment from area of spill or leak until cleanup is complete. Remove all ignition sources. Ventilate area of spill or leak. Absorb liquids in vermiculite, dry sand, earth, peat, carbon, or a similar material and deposit in sealed containers. Keep this chemical out of a confined space, such as a sewer, because of the possibility of an explosion, unless the sewer is designed to prevent the build-up of explosive concentrations. It may be necessary to contain and dispose of this chemical as a hazardous waste. If material or contaminated runoff enters waterways, notify downstream users of potentially contaminated waters. Contact your Department of Environmental Protection or your regional office of the federal EPA for specific recommendations. If employees are required to clean-up spills, they must be properly trained and equipped. OSHA 1910.120(q) may be applicable.

Fire Extinguishing: Use dry chemical, carbon dioxide, water spray, or alcohol foam extinguishers. Poisonous gases are produced in fire. If material or contaminated runoff enters waterways, notify downstream users of potentially contaminated waters. Notify local health and fire officials and pollution control agencies. From a secure, explosion-proof location, use water spray to cool exposed containers. If cooling streams are ineffective (venting sound increases in volume and pitch, tank discolors, or shows any signs of deforming), withdraw immediately to a secure position. If employees are expected to fight fires, they must be trained and equipped in OSHA 1910.156.

Disposal Method Suggested: In accordance with 40CFR165 recommendations for the disposal of pesticides and pesticide containers. Must be disposed properly by following package label directions or by contacting your state pesticide or environmental control agency or by contacting your regional EPA office. Consult with environmental regulatory agencies for guidance on acceptable disposal practices. Generators of waste containing this contaminant (≥100 kg/mo) must conform with EPA regulations governing storage, transportation, treatment, and waste disposal.

References

U.S. Environmental Protection Agency, O,O-Diethyl-S-Methyl Phosphorodithioate, Health and Environmental Effects Profile No. 84, Washington, DC, Office of Solid Waste (April 30, 1980)

Diethyl-p-Phenylenediamine

Molecular Formula: $C_{10}H_{16}N_2$

Common Formula: $C_6H_4NH_2N(C_2H_5)_2$

Synonyms: *p*-Aminodiethylaniline; 4-Amino-*N,N*-diethylaniline; *p*-(Diethylamino)aniline; 4-(Diethylamino)aniline; Diethylaminoaniline; *N,N'*-Diethyl-*p*-fenylendiamin; *N,N'*-Diethyl-*p*-phenylenediamine; Diethyl-*p*-phenylenediamine; *N,N*-Diethyl-*p*-phosphoric acid; DPD

CAS Registry Number: 93-05-0

RTECS Number: SS9275000

Regulatory Authority

- Extremely Hazardous Substance (EPA-SARA) (Dropped from listing in 1988)

Cited in U.S. State Regulations: Massachusetts (G), New Jersey (G).

Description: DPD, $C_6H_4NH_2N(C_2H_5)_2$, is a clear liquid. Boiling point = 260 – 262°C. Flash point ≥110°C.[52]

Potential Exposure: Used as a dye intermediate and in color photography.

Incompatibilities: Contact with strong oxidizers may cause fire and explosion hazard.

Permissible Exposure Limits in Air: No standards set.

Permissible Concentration in Water: No criteria set.

Harmful Effects and Symptoms

Short Term Exposure: DPD is poisonous and a skin irritant. The lowest toxic dermal dose reported in humans is 73 μg/kg. The LD_{50} low-oral-cat is 300 mg/kg and the LD_{50} low dermal rabbit is 5 mg/kg.

Long Term Exposure: May cause allergic dermatitis.

Points of Attack: Skin.

Medical Surveillance: Examination by a qualified allergist.

First Aid: If this chemical gets into the eyes, remove any contact lenses at once and irrigate immediately for at least 15 minutes, occasionally lifting upper and lower lids. Seek medical attention immediately. If this chemical contacts the skin, remove contaminated clothing and wash immediately with soap and water. Seek medical attention immediately. If this chemical has been inhaled, remove from exposure, begin rescue breathing (using universal precautions) if breathing has stopped and CPR if heart action has stopped. Transfer promptly to a medical facility. When this chemical has been swallowed, get medical attention. Give large quantities of water and induce vomiting. Do not make an unconscious person vomit.

Personal Protective Methods: Wear protective gloves and clothing to prevent any reasonable probability of skin contact. Safety equipment suppliers/manufacturers can provide recommendations on the most protective glove/clothing material for your operation.. All protective clothing (suits, gloves, footwear, headgear) should be clean, available each day, and put on before work. Contact lenses should not be worn when working with this chemical. Wear splash-proof chemical goggles and face shield unless full facepiece respiratory protection is worn. Employees should wash immediately with soap when skin is wet or contaminated. Provide emergency showers and eyewash.

Storage: Prior to working with DPD you should be trained on its proper handling and storage. Store in tightly closed containers in a cool, well ventilated area. Metal containers involving the transfer of this chemical should be grounded and bonded. Where possible, automatically pump liquid from drums or other storage containers to process containers. Drums must be equipped with self-closing valves, pressure vacuum bungs, and flame arresters. Use only non-sparking tools and equipment, especially when opening and closing containers of this chemical. Sources of ignition such as smoking and open flames, are prohibited where this chemical is used, handled, or stored in a manner that could create a potential fire or explosion hazard. Wherever this chemical is used, handled, manufactured, or stored, use explosion-proof electrical equipment and fittings.

Spill Handling: Evacuate and restrict persons not wearing protective equipment from area of spill or leak until cleanup is complete. Remove all ignition sources. Ventilate area of spill or leak. Absorb liquids in vermiculite, dry sand, earth, peat, carbon, or a similar material and deposit in sealed containers. Large spills: dike far ahead of spill for later disposal. Keep this chemical out of a confined space, such as a sewer, because of the possibility of an explosion, unless the sewer is designed to prevent the build-up of explosive concentrations. It may be necessary to contain and dispose of this chemical as a hazardous waste. If material or contaminated runoff enters waterways, notify downstream users of potentially contaminated waters. Contact your Department of Environmental Protection or your regional office of the federal EPA for specific recommendations. If employees are required to clean-up spills, they must be properly trained and equipped. OSHA 1910.120(q) may be applicable.

Fire Extinguishing: This chemical is a combustible liquid. Poisonous gases including nitrogen oxides are produced in fire. Extinguish with dry chemical, carbon dioxide, water spray, fog, or foam. Stay upwind; keep out of low areas. Wear self-contained (positive pressure if available) breathing apparatus and full protective clothing. Move container from fire area if you can do it without risk. Vapors are heavier than air and will collect in low areas. Vapors may travel long distances to ignition sources and flashback. Vapors in confined areas may explode when exposed to fire. Containers may explode in fire. Storage containers and parts of containers may rocket great distances, in many directions. If material or contaminated runoff enters waterways, notify downstream users of potentially contaminated waters. Notify local health and fire officials and pollution control agencies. From a secure, explosion-proof location, use water spray to cool exposed containers. If cooling streams are ineffective (venting sound increases in volume and pitch, tank discolors, or shows any signs of deforming), withdraw immediately to a secure position. If employees are expected to fight fires, they must be trained and equipped in OSHA 1910.156.

References

U.S. Environmental Protection Agency, "Chemical Profile: Diethyl-p-Phenylenediamine," Washington, DC, Chemical Emergency Preparedness Program (October 31, 1985)

Diethyl Phthalate

Molecular Formula: $C_{12}H_{14}O_4$

Common Formula: $C_6H_4(OCOC_2H_5)_2$

Synonyms: Anozol; *o*-Benzenedicarboxylic acid diethyl ester; 1,2-Benzenedicarboxylic acid, diethyl ester; DEP; Diethyl 1,2-benzenedicarboxylate; Diethyl *p*-phthalate; Estol 1550; Ethyl phthalate; Ftalato de dietilo (Spanish); NCI-C60048; Neantine; Palatinol A; Phthalic acid, diethyl ester; Phthalol; Phthalsaeurediaethylester (German); Placidol E; Solvanol; Unimoll DA

CAS Registry Number: 84-66-2

RTECS Number: TI1050000

DOT ID: UN 3082

EINECS Number: 201-550-6

Regulatory Authority

- Air Pollutant Standard Set (ACGIH)[1] (Australia) (HSE)[33] (Israel) (Mexico) (OSHA)[58] (former USSR)[35] (Several States)[60] (Several Canadian Provinces)
- Clean Water Act: 40CFR423, Appendix A, Priority Pollutants; Section 313 Water Priority Chemicals (57FR41331, 9/9/92)
- EPA Hazardous Waste Number (RCRA No.): U088
- RCRA, 40CFR261, Appendix 8 Hazardous Constituents
- RCRA 40CFR268.48; 61FR15654, Universal Treatment Standards: Wastewater (mg/l), 0.20; Nonwastewater (mg/kg), 28
- RCRA 40CFR264, Appendix 9; TSD Facilities Ground Water Monitoring List. Suggested test method(s) (PQL μg/l): 8060 (5); 8270 (10)
- Superfund/EPCRA 40CFR302.4 Reportable Quantity (RQ): CERCLA, 1,000 lb (454 kg)
- EPCRA Section 313: Deleted from EPCRA/SARA 313 July 29, 1996 (FR Vol. 61, No. 146, p. 39356–39357)
- Canada, WHMIS, Ingredients Disclosure List; National Pollutant Release Inventory (NPRI)
- Mexico, Drinking Water Criteria, Wastewater

Cited in U.S. State Regulations: Alaska (G), California (A, G), Connecticut (A), Florida (G, A), Illinois (G), Kansas (G, W), Louisiana (G), Maine (G), Maryland (G), Massachusetts (G), Minnesota (G), Nevada (A), New Hampshire (G), New Jersey (G), New York (G, A), North Dakota (A), Pennsylvania (G), Rhode Island (G), South Carolina (A), Vermont (G), Virginia (G, A), Washington (G), West Virginia (G), Wisconsin (G).

Description: Diethyl phthalate, $C_6H_4(OCOC_2H_5)_2$, is a water-white odorless liquid. Boiling point = 296°C. Flash point = 161°C (cc). Autoignition temperature = 457°C. Explosive limits: LEL = 0.7% @ 186°C; UEL – unknown. Hazard Identification (based on NFPA-704 M Rating System): Health 0, Flammability 1, Reactivity 0. Insoluble in water.

Potential Exposure: This material is used as a solvent for cellulose esters, as a vehicle in pesticidal sprays, as a fixative and solvent in perfumery, as an alcohol denaturant and as plasticizers in solid rocket propellants.

Incompatibilities: Violent reaction with strong acids, strong oxidizers including permanganates, water. Attacks some forms of plastic.

Permissible Exposure Limits in Air: There is no OSHA PEL.[58] Australia, Mexico, the Canadian provinces of Alberta, BC, Ontario, and Quebec, the HSE[58] and ACGIH TWA value is

5 mg/m^3 and the HSE, Mexico, Alberta, and British Columbia STEL is 10 mg/m^3. Ontario has proposed a TWA of 3 mg/m^3 and STEL of 5 mg/m^3. The former USSR[35] has set a MAC value in workplace air of 0.5 mg/m^3. Several states have set guidelines or standards for diethyl phthalate in ambient air[60] ranging from 16.7 μg/m^3 (New York) to 25.0 μg/m^3 (South Carolina) to 50.0 μg/m^3 (Connecticut and Florida) to 50 – 100 μg/m^3 (North Dakota) to 80.0 μg/m^3 (Virginia) to 119 μg/m^3 (Nevada).

Determination in Air: Collection by OSHA versatile sampler-Tenax; Toluene; Gas chromatography/Flame ionization detection; OSHA Method #104.

Permissible Concentration in Water: Data are insufficient to draft criterion for the protection o either freshwater or marine organisms. The recommended water quality criterion level for protection of human health is 350 mg/l (350,000 μg/l).[6] Kansas has set guideline for drinking water of 350 mg/l.[61] The Mexican drinking water criteria is 350.0 mg/l.

Determination in Water: Methylene chloride extraction followed by gas chromatography with flame ionization or electron capture detection (EPA Method #606) or gas chromatography plus mass spectrometry (EPA Method #625).

Routes of Entry: Inhalation, ingestion, skin and/or eye contact.

Harmful Effects and Symptoms

Short Term Exposure: Inhalation: Levels above 500 mg/m^3 may cause irritation. Exposure to heated vapors may cause irritation of the nose and throat, dizziness and nausea. Sensitive individuals may develop an allergic reaction similar to asthma.

Skin: Can be absorbed. May cause irritation and allergy.

Eyes: Contact can cause irritation and damage to the cornea.

Ingestion: Animal studies suggest that about 2.5 liquid ounces may cause death in a 150 pound adult.

Symptoms include headache, dizziness, nausea; lacrimation (discharge of tears); possible polyneuropathy, vestibular dysfunction; pain, numbness, weakness, spasms in arms and legs.

Long Term Exposure: Prolonged inhalation may cause irritation of the nose, throat and respiratory system. Repeated exposure may cause nerve damage, with pain, numbness, or weakness in the arms and legs. In animals: reproductive effects.

Points of Attack: Eyes, skin, respiratory system, central nervous system, peripheral nervous system, reproductive system.

Medical Surveillance: If symptoms develop or overexposure is suspected, the following are recommended: Medical exam of the nervous system. Complete blood count and differential.

First Aid: If this chemical gets into the eyes, remove any contact lenses at once and irrigate immediately for at least 15 minutes, occasionally lifting upper and lower lids. Seek medical attention immediately. If this chemical contacts the skin, remove contaminated clothing and wash immediately with soap and water. Seek medical attention immediately. If this chemical has been inhaled, remove from exposure, begin rescue breathing (using universal precautions) if breathing has stopped and CPR if heart action has stopped. Transfer promptly to a medical facility. When this chemical has been swallowed, get medical attention. Give large quantities of water and induce vomiting. Do not make an unconscious person vomit.

Personal Protective Methods: Wear solvent-resistant protective gloves and clothing to prevent any reasonable probability of skin contact. Safety equipment suppliers/manufacturers can provide recommendations on the most protective glove/clothing material for your operation.. All protective clothing (suits, gloves, footwear, headgear) should be clean, available each day, and put on before work. Contact lenses should not be worn when working with this chemical. Wear splash-proof chemical goggles and face shield unless full facepiece respiratory protection is worn. Employees should wash immediately with soap when skin is wet or contaminated. Provide emergency showers and eyewash.

Respirator Selection: Where the potential exists for exposure over 5 mg/m^3, use an MSHA/NIOSH approved full facepiece respirator with an organic vapor cartridge/canister. Increased protection is obtained from full facepiece powered air purifying respirators. If while wearing a filter, cartridge or canister respirator, you can smell, taste, or otherwise detect diethyl phthalate, or in the case of a full facepiece respirator you experience eye irritation, leave the area immediately.

Storage: Prior to working with DEP you should be trained on its proper handling and storage. Before entering confined space where this chemical may be present, check to make sure that an explosive concentration does not exist. Store in tightly closed containers in a cool, well-ventilated area away from heat and oxidizing agents (such as permanganates, nitrates, chlorates, perchlorates, and peroxides). Metal containers involving the transfer of this chemical should be grounded and bonded. Where possible, automatically pump liquid from drums or other storage containers to process containers. Drums must be equipped with self-closing valves, pressure vacuum bungs, and flame arresters. Use only non-sparking tools and equipment, especially when opening and closing containers of this chemical. Sources of ignition such as smoking and open flames, are prohibited where this chemical is used, handled, or stored in a manner that could create a potential fire or explosion hazard. Wherever this chemical is used, handled, manufactured, or stored, use explosion-proof electrical equipment and fittings.

Shipping: Environmentally hazardous liquid, n.o.s. Hazard Class: 9. Label: "Class 9." Packing Group: III.

Spill Handling: Evacuate and restrict persons not wearing protective equipment from area of spill or leak until cleanup is complete. Remove all ignition sources. Establish forced ventilation to keep levels below explosive limit. Absorb liquids in vermiculite, dry sand, earth, peat, carbon, or a similar material and deposit in sealed containers. Keep this chemical out of a confined space, such as a sewer, because of the possibility of an explosion, unless the sewer is designed to prevent the build-up of explosive concentrations. It may be necessary to contain and dispose of this chemical as a hazardous waste. If material or contaminated runoff enters waterways, notify downstream users of potentially contaminated waters. Contact your Department of Environmental Protection or your regional office of the federal EPA for specific recommendations. If employees are required to clean-up spills, they must be properly trained and equipped. OSHA 1910.120(q) may be applicable.

Fire Extinguishing: This chemical is a combustible liquid. However, ignition is difficult. Poisonous gases are produced in fire. Use dry chemical, carbon dioxide, or foam extinguishers. Water or foam may cause frothing. Vapors are heavier than air and will collect in low areas. Vapors may travel long distances to ignition sources and flashback. Vapors in confined areas may explode when exposed to fire. Containers may explode in fire. Storage containers and parts of containers may rocket great distances, in many directions. If material or contaminated runoff enters waterways, notify downstream users of potentially contaminated waters. Notify local health and fire officials and pollution control agencies. From a secure, explosion-proof location, use water spray to cool exposed containers. If cooling streams are ineffective (venting sound increases in volume and pitch, tank discolors, or shows any signs of deforming), withdraw immediately to a secure position. If employees are expected to fight fires, they must be trained and equipped in OSHA 1910.156.

Disposal Method Suggested: Incineration. In accordance with 40CFR165 recommendations for the disposal of pesticides and pesticide containers. Must be disposed properly by following package label directions or by contacting your state pesticide or environmental control agency or by contacting your regional EPA office. Consult with environmental regulatory agencies for guidance on acceptable disposal practices. Generators of waste containing this contaminant (≥100 kg/mo) must conform with EPA regulations governing storage, transportation, treatment, and waste disposal.

References

U.S. Environmental Protection Agency, Phthalate Esters: Ambient Water Quality Criteria, Washington, DC (1980)

U.S. Environmental Protection Agency, Diethyl Phthalate, Health and Environmental Effects Profile No. 85, Washington, DC, Office of Solid Waste (April 30, 1980)

New Jersey Department of Health, "Hazardous Substance Fact Sheet: Diethyl Phthalate," Trenton, NJ (July 1996)

Diethylstilbestrol

Molecular Formula: $C_{18}H_{20}O_2$

Synonyms: Acnestrol; Agostilben; Antigestil; Bio-DES; 3,4-Bis(*p*-hydrophenyl)-3-hexene; Bufton; Climaterine; Comestrol; Comestrol estrobene; Cyren; Dawe's destrol; DEB; DES; Desma; DES (synthetic estrogen); Destrol; Diastyl; Dibestrol; Dicorvin; Di-estryl; (E)-4,4'-(1,2-Diethyl-1,2-ethenediyl)bisphenol; *trans*-4,4'-(1,2-Diethyl-1,2-ethenediyl)bisphenol; 4,4'-(1,2-Diethyl-1,2-ethenediyl)bis-phenol; α,α'-Diethyl-(E)-4,4'd-stilbenediol; α,α'-Diethylstilbenediol; *trans*-α,α'-Diethyl-stilbenediol; 2,2'-Diethyl-4,4'-stilbenediol; *trans*-Diethylstilbestrol; Dietilestilbestrol (Spanish); 4,4'-Dihydroxy-α,β-diethylstilbene; 4,4'-Dihydroxydiethylstilbene; 3,4'(4,4'-Dihydroxyphenyl)hex-3-ene; Distilbene; Domestrol; Dyestrol; Estilben; Estril; Estrobene; Estrogen; Estromenin; Estrosyn; Follidiene; Fonatol; Grafestrol; Gynopharm; Hibestrol; Idroestril; Iscovesco; Makarol; Menosyilbeen; Micrest; Microest; Milestrol; Neo-Oestranol 1; NSC-3070; Oekolp; Oestrogenine; Oestrol; Oestromenin; Oestromensil; Oestromensyl; Oestromienin; Oestromon; Pabestrol; Palestrol; Percutatrine oestrogenique iscovesco; Phenol, 4,4'-(1,2-diethyl-1,2-ethenediyl)bis-, (E)-; Protectona; Rumestrol 2; Sedestran; Serral; Sexocretin; Sibol; Sintestrol; Stibilium; Stil; Stilbestrol; Stilbestrone; Stilbetin; Stilboefral; Stilboestroform; Stilboestrol; Stilbofollin; Stilbol; Stilkap; Stil-Rol; Synestrin; Synthoestrin; Synthofolin; Syntofolin; Tampovagan stilboestrol; Tylosterone; Vagestrol; Vetag

CAS Registry Number: 56-53-1

RTECS Number: WJ5600000

DOT ID: UN 3077

Regulatory Authority

- Carcinogen (Carcinogenic to humans) (IARC)[9] (NTP)
- Banned or Severely Restricted (Many Countries) (UN)[13]
- EPA Hazardous Waste Number (RCRA No.): U089
- RCRA, 40CFR261, Appendix 8 Hazardous Constituents
- Superfund/EPCRA 40CFR302.4 Reportable Quantity (RQ): CERCLA, 1 lb (0.454 kg)

Cited in U.S. State Regulations: California (A, G), Florida (G), Illinois (G), Kansas (G), Louisiana (G), Maine (G), Massachusetts (G), Michigan (G), Minnesota (G), New Hampshire (G), New Jersey (G), Pennsylvania (G), Vermont (G), Virginia (G), Washington (G), West Virginia (G), Wisconsin (G).

Description: DES is a white, odorless, crystalline powder or plates. Freezing/Melting point = 169 – 172°C. Hazard Identification (based on NFPA-704 M Rating System): Health 3, Flammability 1, Reactivity 0. Soluble in water.

Potential Exposure: DES had been used extensively as a growth promoter for cattle and sheep and is used in veterinary medicine to treat estrogen-deficiency disorders. It has been used in humans to prevent spontaneous abortions, to treat symptoms associated with menopause, menstrual disorders, postpartum breast engorgement, primary ovarian failure, breast cancer, and prostate cancers in males.

In 1979, the DFA revoked all use of DES in food-producing animals. The Court of Appeals upheld the Commissioner's decision to revoke the use of DES in all food-producing animals on November 24, 1980, the motion to reconsider was denied on December 24, 1980.

Permissible Exposure Limits in Air: No standards set.

Permissible Concentration in Water: No criteria set.

Routes of Entry: Inhalation.

Harmful Effects and Symptoms

Short Term Exposure: Nausea, headache, vomiting; local irritation of eyes, skin and mucous membranes.

Long Term Exposure: A confirmed human carcinogen. May affect the glands. A possible teratogen. May cause vaginal bleeding. In men, it may cause impotence, breast enlargement, and other feminizing effects.

Points of Attack: Cancer sites: Skin, liver, and lung tumors; uterine and other reproductive system tumors in the female newborn of exposed women. Glandular system.

First Aid: *Skin Contact:*[52] Flood areas of body that have contacted the substance with water. Don't wait to remove contaminated clothing; do it under the water stream. Use soap to help assure removal. Isolate contaminated clothing when removed to prevent contact by others.

Eye Contact: Remove any contact lenses at once. Flush eyes well with copious quantities of water or normal saline for at least 20 – 30 minutes. Seek medical attention.

Inhalation: Leave contaminated area immediately, breathe fresh air. Proper respiratory protection must be supplied to any rescuers. If coughing, difficult breathing or any other symptoms develop, seek medical attention at once even if symptoms develop many hours after exposure.

Ingestion: If convulsions are not present, give a glass or two of water or milk to dilute the substance. Assure that the person's airway is unobstructed and contact a hospital or poison center immediately for advice on whether or not to induce vomiting.

Personal Protective Methods: Wear protective gloves and clothing to prevent any reasonable probability of skin contact. Safety equipment suppliers/manufacturers can provide recommendations on the most protective glove/clothing material for your operation.. All protective clothing (suits, gloves, footwear, headgear) should be clean, available each day, and put on before work. Contact lenses should not be worn when working with this chemical. Wear dust-proof chemical goggles and face shield unless full facepiece respiratory protection is worn. Employees should wash immediately with soap when skin is wet or contaminated. Provide emergency showers and eyewash.

Respirator Selection: *At any detectable concentration:* SCBAF:PD,PP (any MSHA/NIOSH approved self-contained breathing apparatus that has a full facepiece and is operated in a pressure-demand or other positive-pressure mode); or SAF:PD,PP:ASCBA (any supplied-air respirator that has a full facepiece and is operated in a pressure-demand or other positive-pressure mode in combination with an auxiliary, self-contained breathing apparatus operated in a pressure-demand or other positive pressure mode). *Escape:* HiEF (any air-purifying, full-facepiece respirator (gas mask) with a chin-style, front- or back-mounted organic vapor canister having a high-efficiency particulate filter/Any appropriate escape-type, self-contained breathing apparatus); or SCBAE (any appropriate escape-type, self-contained breathing apparatus).

Storage: Prior to working with DES you should be trained on its proper handling and storage. Store in a refrigerator under an inert atmosphere and protect from light. A regulated, marked area should be established where this chemical is handled, used, or stored in compliance with OSHA standard 1910.1045.

Shipping: Environmentally hazardous solid, n.o.s. Hazard Class: 9. Label: "Class 9."

Spill Handling: Evacuate persons not wearing protective equipment from area of spill or leak until clean-up is complete. Remove all ignition sources. Collect powdered material in the most convenient and safe manner and deposit in sealed containers. Ventilate area after clean-up is complete. It may be necessary to contain and dispose of this chemical as a hazardous waste. If material or contaminated runoff enters waterways, notify downstream users of potentially contaminated waters. Contact your Department of Environmental Protection or your regional office of the federal EPA for specific recommendations. If employees are required to clean-up spills, they must be properly trained and equipped. OSHA 1910.120(q) may be applicable.

Fire Extinguishing: Use dry chemical, carbon dioxide, water spray, or foam extinguishers. Poisonous gases are produced in fire. If material or contaminated runoff enters waterways, notify downstream users of potentially contaminated waters. Notify local health and fire officials and pollution control agencies. From a secure, explosion-proof location, use water spray to cool exposed containers. If cooling streams are ineffective (venting sound increases in volume and pitch, tank discolors, or shows any signs of deforming), withdraw immediately to a secure position. If employees are expected to fight fires, they must be trained and equipped in OSHA 1910.156.

Disposal Method Suggested: Consult with environmental regulatory agencies for guidance on acceptable disposal practices. Generators of waste containing this contaminant (≥100 kg/mo) must conform with EPA regulations governing storage, transportation, treatment, and waste disposal.

References

Sax N. I., Ed., "Dangerous Properties of Industrial Materials Report," 1, No. 3, 59–61 (1981) and 6, No. 2, 57–62 (1986)

Diethyl Sulfate

Molecular Formula: $C_4H_{10}O_4S$

Common Formula: $(C_2H_5)_2SO_4$

Synonyms: DES; Diaethylsulfat (German); Diethylester kyseliny sirove (Czech); Diethyl ester sulfuric acid; Diethyl sulphate; Diethyl tetraoxosulfate; Diethyl tetraoxosulphate; DS; Ethyl sulfate; Ethyl sulphate; Sulfato de dietilo (Spanish); Sulfuric acid, diethyl ester

CAS Registry Number: 64-67-5

RTECS Number: WS7875000

DOT ID: UN 1594

EEC Number: 016-027-00-6

Regulatory Authority

- Carcinogen (Animal Positive) (IARC)[9] (NTP)[10] (DFG)[3]
- Banned or Severely Restricted (Sweden) (UN)[13]
- Air Pollutant Standard Set (North Dakota)[60]
- Clean Air Act: Hazardous Air Pollutants (Title I, Part A, Section 112)
- Superfund/EPCRA 40CFR302.4 Reportable Quantity (RQ): CERCLA, 1 lb (0.454 kg)
- EPCRA Section 313 Form R *de minimis* concentration reporting level: 0.1%
- Canada, WHMIS, Ingredients Disclosure List; National Pollutant Release Inventory (NPRI)

Cited in U.S. State Regulations: California (G), Florida (G), Illinois (G), Maine (G), Maryland (G), Massachusetts (G), New Hampshire (G), Pennsylvania (G), Rhode Island (G).

Description: Diethyl sulfate, $(C_2H_5)_2SO_4$, is a colorless, oily liquid with a faint peppermint-like or ether-like odor. Turns brown on contact with air. Boiling point = 209°C (decomposes). Freezing/Melting point = -25°C. Flash point = 104°C. Autoignition temperature = 435°C. Hazard Identification (based on NFPA-704 M Rating System): Health 3, Flammability 1, Reactivity 1. Insoluble in water with slight decomposition.

Potential Exposure: Ehylation agent (synthesis of dye and pharmaceuticals), synthesizing agent of ammonium chloride compounds. Making isopropyl alcohol, ethyl alcohol and other chemicals. Dehydrating agent. Extractant for gasoline.

Incompatibilities: Vigorous reaction with strong oxidizers or water (forms sulfuric acid). The aqueous solution is a strong acid; incompatible with non-oxidizing mineral acids, organic acids, bases, acrylates, aldehydes, alcohols, alkylene oxides, ammonia, aliphatic amines, alkanolamines, aromatic amines, amides, chlorates, epichlorohydrin, fulminates, glycols, isocyanates, ketones, metals (powdered), organic anhydrides, perchlorates, picrates, substituted allyls, phenols and cresols. DS decomposes when heated producing diethyl ether and fumes of sulfur oxides. DS is a strong oxidizer; reacts with combustible and reducing materials. DS is a strong reducing agent; reacts with oxidizing materials. Forms explosive hydrogen gas on contact with iron in the presence of water.

Permissible Exposure Limits in Air: There are no occupational exposure limits. North Dakota[60] has set a guideline of zero concentration in ambient air.

Permissible Concentration in Water: No criteria set.

Harmful Effects and Symptoms

Short Term Exposure: DS is corrosive to the eyes, skin, and respiratory tract. Inhalation of the aerosol can cause pulmonary edema, a medical emergency that can be delayed for several hours. This can cause death. A poison by subcutaneous and inhalation routes. Moderate acute toxicity by ingestion and skin routes. An experimental tumorigen, transplacental brain carcinogen.

Long Term Exposure: Repeated or prolonged contact with skin may cause dermatitis. DS is a probable human carcinogen. May cause heritable genetic damage in humans. May cause lung damage.

Points of Attack: Lungs, central nervous system, skin, reproductive system. Cancer sites: lungs and stomach in animals; throat cancer in humans

Medical Surveillance: Preclude from exposure those individuals with diseases of central nervous system, kidneys and liver. Lung function tests. Consider chest x-ray following acute overexposure.

First Aid: If this chemical gets into the eyes, remove any contact lenses at once and irrigate immediately for at least 15 minutes, occasionally lifting upper and lower lids. Seek medical attention immediately. If this chemical contacts the skin, remove contaminated clothing and wash immediately with soap and water. Seek medical attention immediately. If this chemical has been inhaled, remove from exposure, begin rescue breathing (using universal precautions) if breathing has stopped and CPR if heart action has stopped. Transfer promptly to a medical facility. When this chemical has been swallowed, get medical attention. Gastric lavage (stomach wash), if swallowed, followed by saline catharsis. Do not make an unconscious person vomit. Medical observation is recommended for 24 – 48 hours after breathing overexposure, as pulmonary edema may be delayed. As first aid for pulmonary edema, a doctor or authorized paramedic may consider administering a corticosteroid spray.

Personal Protective Methods: Wear protective gloves and clothing to prevent any reasonable probability of skin contact. Safety equipment suppliers/manufacturers can provide recommendations on the most protective glove/clothing material for your operation.. All protective clothing (suits, gloves, footwear, headgear) should be clean, available each day, and put on before work. Contact lenses should not be worn when working with this chemical. Wear splash-proof chemical goggles and face shield unless full facepiece respiratory protection is worn. Employees should wash immediately with

soap when skin is wet or contaminated. Provide emergency showers and eyewash.

Respirator Selection: *At any detectable concentration:* SCBAF:PD,PP (any MSHA/NIOSH approved self-contained breathing apparatus that has a full facepiece and is operated in a pressure-demand or other positive-pressure mode); or SAF:PD,PP:ASCBA (any supplied-air respirator that has a full facepiece and is operated in a pressure-demand or other positive-pressure mode in combination with an auxiliary, self-contained breathing apparatus operated in a pressure-demand or other positive pressure mode).

Storage: Protect against physical damage. Store in a cool, dry well-ventilated location away from any area where the fire hazard may be acute. Outside or detached storage is preferred. Separate from other storage. Before entering confined space where DS may be present, check to make sure that an explosive concentration does not exist. Store in tightly closed containers in a cool, well ventilated area. Metal containers involving the transfer of this chemical should be grounded and bonded. Where possible, automatically pump liquid from drums or other storage containers to process containers. Drums must be equipped with self-closing valves, pressure vacuum bungs, and flame arresters. Use only non-sparking tools and equipment, especially when opening and closing containers of this chemical. Sources of ignition such as smoking and open flames, are prohibited where this chemical is used, handled, or stored in a manner that could create a potential fire or explosion hazard. Wherever this chemical is used, handled, manufactured, or stored, use explosion-proof electrical equipment and fittings. A regulated, marked area should be established where this chemical is handled, used, or stored in compliance with OSHA standard 1910.1045.

Shipping: This compound requires a shipping label of: "Poison." It falls in DOT Hazard Class 6.1 and Packing Group II.

Spill Handling: Evacuate and restrict persons not wearing protective equipment from area of spill or leak until cleanup is complete. Remove all ignition sources. Ventilate area of spill or leak. Absorb liquids in vermiculite, dry sand, earth, peat, carbon, or a similar material and deposit in sealed containers. Keep this chemical out of a confined space, such as a sewer, because of the possibility of an explosion, unless the sewer is designed to prevent the build-up of explosive concentrations. It may be necessary to contain and dispose of this chemical as a hazardous waste. If material or contaminated runoff enters waterways, notify downstream users of potentially contaminated waters. Contact your Department of Environmental Protection or your regional office of the federal EPA for specific recommendations. If employees are required to clean-up spills, they must be properly trained and equipped. OSHA 1910.120(q) may be applicable.

Fire Extinguishing: This chemical is a combustible liquid. Poisonous gases including sulfur oxides are produced in fire. Use dry chemical, carbon dioxide, or alcohol foam extinguishers. Water or foam may cause frothing and sulfuric acid. Vapors are heavier than air and will collect in low areas. Vapors may travel long distances to ignition sources and flashback. Vapors in confined areas may explode when exposed to fire. Containers may explode in fire. Storage containers and parts of containers may rocket great distances, in many directions. If material or contaminated runoff enters waterways, notify downstream users of potentially contaminated waters. Notify local health and fire officials and pollution control agencies. From a secure, explosion-proof location, use water spray to cool exposed containers. If cooling streams are ineffective (venting sound increases in volume and pitch, tank discolors, or shows any signs of deforming), withdraw immediately to a secure position. If employees are expected to fight fires, they must be trained and equipped in OSHA 1910.156.

Disposal Method Suggested: Dissolve in a combustible solvent. Scatter the spray of the solvent into the furnace with afterburner and alkali scrubber.

References

New Jersey Department of Health and Senior Services, "Hazardous Substance Fact Sheet, Diethyl Sulfate," Trenton, NJ (September 1988)

Diethyl Zinc

Molecular Formula: $C_4H_{10}Zn$

Common Formula: $(C_2H_5)_2Zn$

Synonyms: Ethyl zinc; Zinc, diethyl-; Zinc ethide; Zinc ethyl

CAS Registry Number: 557-20-0

RTECS Number: ZH2070000

DOT ID: UN 1366

Regulatory Authority

- OSHA 29CFR1910.119, Appendix A, Process Safety List of Highly Hazardous Chemicals, TQ = 10,000 lb (4,540 kg)

Cited in U.S. State Regulations: Florida (G), Massachusetts (G), New Hampshire (G), New Jersey (G), Oklahoma (G), Pennsylvania (G), Rhode Island (G).

Description: Diethyl Zinc, $(C_2H_5)_2Zn$, is a colorless liquid. Boiling point = 124°C. Flash point = Ignites spontaneously in air at room temperature. Hazard Identification (based on NFPA-704 M Rating System): Health 3, Flammability 4, Reactivity 3. Water reactive. Do not use water, foam or halogenated extinguishing agents. Contact with water causes violent decomposition releasing a flammable gas. It should be used under a dry inert gas or kerosene blanket or in an evacuated system.

Potential Exposure: Diethyl zinc is used in organic syntheses, as a catalyst in the manufacture of olefin polymers and as a high-energy aircraft and missile fuel.

Incompatibilities: Ignites spontaneously on contact with air or strong oxidizers. Explosive decomposition at 245°F/120°C.

Violent reaction with hydrazine, sulfur dioxide, halogens, some alcohols, ozone; possible fire and explosions. Contact with water forms ethane gas.

Permissible Exposure Limits in Air: No standards set.

Permissible Concentration in Water: No criteria set.

Harmful Effects and Symptoms

Short Term Exposure: Contact can irritate or burn the eyes and skin.

Long Term Exposure: Unknown at this time.

First Aid: If this chemical gets into the eyes, remove any contact lenses at once and irrigate immediately for at least 15 minutes, occasionally lifting upper and lower lids. Seek medical attention immediately. If this chemical contacts the skin, remove contaminated clothing and wash immediately with soap and water. Seek medical attention immediately. If this chemical has been inhaled, remove from exposure, begin rescue breathing (using universal precautions) if breathing has stopped and CPR if heart action has stopped. Transfer promptly to a medical facility. When this chemical has been swallowed, get medical attention. Give large quantities of water and induce vomiting. Do not make an unconscious person vomit.

Personal Protective Methods: Wear protective gloves and clothing to prevent any reasonable probability of skin contact. Because diethyl zinc can ignite spontaneously, protective clothing should be fire-retardant or fire-proof. Safety equipment suppliers/manufacturers can provide recommendations on the most protective glove/clothing material for your operation.. All protective clothing (suits, gloves, footwear, headgear) should be clean, available each day, and put on before work. Contact lenses should not be worn when working with this chemical. Wear splash-proof chemical goggles and face shield unless full facepiece respiratory protection is worn. Employees should wash immediately with soap when skin is wet or contaminated. Provide emergency showers and eyewash.

Respirator Selection: Where the potential for exposures exists to diethyl zinc, use a MSHA/NIOSH approved supplied-air respirator with a full facepiece operated in the positive pressure mode or with a full facepiece, hood, or helmet in the continuous flow mode, or use a MSHA/NIOSH approved self-contained breathing apparatus with a full facepiece operated in pressure-demand or other positive pressure mode.

Storage: Prior to working with diethyl zinc you should be trained on its proper handling and storage. Before entering confined space where diethyl zinc may be present, check to make sure that an explosive concentration does not exist. Diethyl zinc must be stored to avoid contact with water, chlorine, hydrazine and oxidizers since violent reactions occur. Store in sealed tubes or cylinders under a dry, inert gas blanket or in an evacuated system. Shade from radiant heat and protect from rain. Metal containers involving the transfer of 5 gallons or more of diethyl zinc should due grounded and bonded. Wherever diethyl zinc is used, handled, manufactured, or stored, use explosion proof electrical equipment and fittings. Use only non-sparking tools and equipment, especially when opening and closing containers of this chemical. Sources of ignition such as smoking and open flames, are prohibited where this chemical is used, handled, or stored in a manner that could create a potential fire or explosion hazard.

Shipping: This compound requires a shipping label of: "Spontaneously Combustible, Dangerous When Wet." It falls in DOT Hazard 4.2 and Packing Group I. Passenger aircraft or railcar shipment is forbidden as is cargo aircraft shipment.

Spill Handling: Evacuate and restrict persons not wearing protective equipment from area of spill or leak until cleanup is complete. Remove all ignition sources. Ventilate area of spill or leak. Absorb liquids in vermiculite, dry sand, earth, peat, carbon, or a similar material and deposit in sealed containers. Keep this chemical out of a confined space, such as a sewer, because of the possibility of an explosion, unless the sewer is designed to prevent the build-up of explosive concentrations. It may be necessary to contain and dispose of this chemical as a hazardous waste. If material or contaminated runoff enters waterways, notify downstream users of potentially contaminated waters. Contact your Department of Environmental Protection or your regional office of the federal EPA for specific recommendations. If employees are required to clean-up spills, they must be properly trained and equipped. OSHA 1910.120(q) may be applicable.

Fire Extinguishing: This chemical is a flammable and highly reactive liquid. It can ignite spontaneously if it is exposed to air or moisture at room temperature. Poisonous gases including zinc oxide are produced in fire. Use dry chemical, soda ash, or lime extinguishers. Do not use water, foam, or halogenated extinguishers. Fire may restart after it has been extinguished. Vapors are heavier than air and will collect in low areas. Vapors may travel long distances to ignition sources and flashback. Vapors in confined areas may explode when exposed to fire. Containers may explode in fire. Storage containers and parts of containers may rocket great distances, in many directions. If material or contaminated runoff enters waterways, notify downstream users of potentially contaminated waters. Notify local health and fire officials and pollution control agencies. From a secure, explosion-proof location, use water spray to cool exposed containers. If cooling streams are ineffective (venting sound increases in volume and pitch, tank discolors, or shows any signs of deforming), withdraw immediately to a secure position. If employees are expected to fight fires, they must be trained and equipped in OSHA 1910.156.

References

New Jersey Department of Health, "Hazardous Substance Fact Sheet: Diethyl Zinc," Trenton, NJ (June 1986)

Difluorodibromomethane

Molecular Formula: CBr_2F_2

Synonyms: Dibromodifluoromethane; Freon® 12B2; Halon® 1202; Methane, dibromofluoro-

CAS Registry Number: 75-61-6

RTECS Number: PA7525000

DOT ID: UN 1941

Regulatory Authority

- Air Pollutant Standard Set (ACGIH)[1] (Australia) (DFG)[3] (HSE)[33] (Israel) (Mexico) (OSHA) (Several States)[60] (Several Canadian Provinces)

Cited in U.S. State Regulations: Alaska (G), California (G), Connecticut (A), Florida (G), Illinois (G), Maine (G), Massachusetts (G), Minnesota (G), Nevada (A), New Hampshire (G), New Jersey (G), North Dakota (A), Pennsylvania (G), Rhode Island (G), Virginia (A), West Virginia (G).

Description: Difluorodibromomethane, CBr_2F_2, is a colorless heavy liquid or gas (above 24°C/76°F) with a characteristic odor. Boiling point = 24°C. Freezing/Melting point = -146°C. Insoluble in water.

Potential Exposure: This material is used as a fire-extinguishing agent and in making dyes and pharmaceuticals.

Incompatibilities: Reacts with chemically active metals such as sodium, potassium, calcium, powdered aluminum, zinc, magnesium. Attacks some plastics, rubbers and coatings.

Permissible Exposure Limits in Air: The Federal Limit (OSHA PEL)[58] is 100 ppm (860 mg/m^3). The HSE, DFG, ACGIH, Australia, Israel, Mexico and the Canadian provinces of Alberta, BC, Ontario, and Quebec use the same TWA limit as OSHA and HSE, Mexico, Alberta, British Columbia set a STEL value is 150 ppm (1,290 mg/m^3). The NIOSH IDLH level is 2,000 ppm. Several states have set guidelines or standards for this compound in ambient air[60] ranging from 8.6 mg/m^3 (North Dakota) to 14.0 mg/m^3 (Virginia) to 17.2 mg/m^3 (Connecticut) to 29.5 mg/m^3 (Nevada).

Determination in Air: Charcoal adsorption, workup with isopropanol, analysis by gas chromatography. See NIOSH Method 1012.[18]

Permissible Concentration in Water: No criteria set.

Routes of Entry: Inhalation, ingestion, eye and/or skin contact.

Harmful Effects and Symptoms

Short Term Exposure: Contact may cause eye and skin irritation. Inhalation can irritate the nose and throat. High exposures can cause you to become dizzy, lightheaded, and to pass out. Also, high exposures can cause pulmonary edema, a medical emergency that can be delayed for several hours. This can cause death. Contact with the liquid can cause frostbite.

Long Term Exposure: May cause liver damage. In animals: central nervous system symptoms; liver damage.

Points of Attack: Respiratory system, central nervous system, liver.

Medical Surveillance: If symptoms develop or overexposure has occurred, the following may be useful: Consider chest x-ray following acute overexposure. Liver and lung function tests. Examination of the nervous system.

First Aid: If this chemical gets into the eyes, remove any contact lenses at once and irrigate immediately for at least 15 minutes, occasionally lifting upper and lower lids. Seek medical attention immediately. If this chemical contacts the skin, remove contaminated clothing and wash immediately with soap and water. Seek medical attention immediately. If this chemical has been inhaled, remove from exposure, begin rescue breathing (using universal precautions) if breathing has stopped and CPR if heart action has stopped. Transfer promptly to a medical facility. When this chemical has been swallowed, get medical attention. Give large quantities of water and induce vomiting. Do not make an unconscious person vomit. If frostbite has occurred, seek medical attention immediately; do *NOT* rub the affected areas or flush them with water. In order to prevent further tissue damage, do *NOT* attempt to remove frozen clothing from frostbitten areas. If frostbite has *NOT* occurred, immediately and thoroughly wash contaminated skin with soap and water.

Personal Protective Methods: Wear protective gloves and clothing to prevent any reasonable probability of skin contact. Safety equipment suppliers/manufacturers can provide recommendations on the most protective glove/clothing material for your operation.. All protective clothing (suits, gloves, footwear, headgear) should be clean, available each day, and put on before work. Contact lenses should not be worn when working with this chemical. Wear splash-proof chemical goggles and face shield when working with liquid, unless full facepiece respiratory protection is worn. When working with Difluorodibromomethane gas wear gas-proof goggles and face shield, unless full facepiece respiratory protection is worn. Employees should wash immediately with soap when skin is wet or contaminated. Provide emergency showers and eyewash. Where exposure to the liquefied compressed gas may occur, employees should be provided with special clothing designed to prevent frostbite.

Respirator Selection: OSHA: up to *1,000 ppm:* SA (any supplied respirator). *Up to 2,000 ppm:* SA:CF (any supplied-air respirator operated in a continuous-flow mode); or SCBAF (any self-contained breathing apparatus with a full facepiece); SAF (any supplied-air respirator with a full facepiece). *Emergency or planned entry into unknown concentrations or IDLH conditions*: SCBAF:PD,PP (any self-contained breathing apparatus that has a full facepiece and is operated in a pressure-demand or other positive-pressure mode); or SAF:PD,PP:ASCBA (any supplied-air respirator that has a full facepiece and is operated in a pressure-demand or other positive-pressure mode in combination with an auxiliary self-contained breathing apparatus operated in a pressure-demand or other positive-pressure mode). *Escape:* GMFOV [any air-purifying, full-facepiece respirator (gas mask) with a chin-style, front- or back-mounted organic vapor canister] or SCBAE (any appropriate escape-type, self-contained breathing apparatus).

Storage: Difluorodibromomethane must be stored to avoid contact with chemically active metals such as sodium, potassium, calcium, powdered aluminum, zinc and magnesium, since violent reactions occur. Store in tightly closed containers in a cool, well-ventilated area away from heat.

Shipping: This compound requires a shipping label of: Class 9. It falls in DOT Hazard Class 9 and Packing III.

Spill Handling: *Liquid:* Evacuate and restrict persons not wearing protective equipment from area of spill or leak until cleanup is complete. Remove all ignition sources. Ventilate area of spill or leak. Absorb liquids in vermiculite, dry sand, earth, peat, carbon, or a similar material and deposit in sealed containers. Keep this chemical out of a confined space, such as a sewer, because of the possibility of an explosion, unless the sewer is designed to prevent the build-up of explosive concentrations. It may be necessary to contain and dispose of this chemical as a hazardous waste. If material or contaminated runoff enters waterways, notify downstream users of potentially contaminated waters. Contact your Department of Environmental Protection or your regional office of the federal EPA for specific recommendations. If employees are required to clean-up spills, they must be properly trained and equipped. OSHA 1910.120(q) may be applicable.

Gas: Restrict persons not wearing protective equipment from area of spill or leak until cleanup is complete. Ventilate area of spill or leak. If the gas is leaking, stop flow of gas. If source of leak is a cylinder and the leak cannot be stopped in place, remove the leaking cylinder to a safe place in the open air, and repair leak or allow cylinder to empty. If the liquid is spilled or leaked, allow it to vaporize.

Fire Extinguishing: This chemical is a noncombustible liquid or gas. Poisonous gases including hydrogen bromide, hydrogen fluoride, carbon monoxide are produced in fire. Use dry chemical, carbon dioxide, or alcohol or polymer foam extinguishers. Vapors are heavier than air and will collect in low areas. Containers may explode in fire. Storage containers and parts of containers may rocket great distances, in many directions. If material or contaminated runoff enters waterways, notify downstream users of potentially contaminated waters. Notify local health and fire officials and pollution control agencies. From a secure, explosion-proof location, use water spray to cool exposed containers. If cooling streams are ineffective (venting sound increases in volume and pitch, tank discolors, or shows any signs of deforming), withdraw immediately to a secure position. If employees are expected to fight fires, they must be trained and equipped in OSHA 1910.156.

Disposal Method Suggested: Venting to the atmosphere was formerly the preferred method. Now, concern for destruction of the ozone layer has made this method unacceptable. Return to supplier (where possible) or high-temperature incineration must now be used.

References

New Jersey Department of Health, "Hazardous Substance Fact Sheet: Difluorodibromomethane," Trenton, NJ (December 1998)

Digitoxin

Molecular Formula: $C_{41}H_{64}O_{13}$

Synonyms: Acedoxin; Asthenthilo; Cardidigin; Cardigin; Carditoxin; Cristapurat; Crystalline digitalin; Crystodigin; (3β,5β)-3-[(*O*-2,6-Dideoxy-β-D-ribohexopyranosyl-(1→4)-*O*-2,6-dideoxy-β-D-ribohexopyranosyl-(1→4)-2,6-dideoxy-β-D-ribo-hexopyranosyl)oxy]-14-hydroxycard-20(22)-enolide; Digilong; Digimed; Digimerck; Digisidin; Digitalin; Digitaline (French); Digitaline cristalliseel digitaline nativelle; Digitalinum verum; Digitophyllin; Digitoxigenin-tridigitoxosid (German); Digitoxigenin tridigitoxoside; Digitoxina (Spanish); Ditaven; Glucodigin; Lanatoxin; Monoglycocoard; Myodigin; Purodigin; Purpurid; Tardigal; Tri-digitoxoside (German); Unidigin

CAS Registry Number: 71-63-6

RTECS Number: IH2275000

Regulatory Authority

- Superfund/EPCRA 40CFR355, Appendix B Extremely Hazardous Substances: TPQ = 100/10,000 lb (45.4/4,540 kg)
- Superfund/EPCRA 40CFR302.4 Reportable Quantity (RQ): EHS, 1 lb (0.454 kg)

Cited in U.S. State Regulations: California (G), Florida (G), Massachusetts (G), New Jersey (G), Pennsylvania (G).

Description: Digitoxin, $C_{41}H_{64}O_{13}$, is a white crystalline solid. Freezing/Melting point = 256 – 257°C. Slightly soluble in water.

Potential Exposure: This material is used as a cardiotonic drug. Digitoxin is the most toxic component of digitalis.

Permissible Exposure Limits in Air: No standards set.

Permissible Concentration in Water: No criteria set.

Harmful Effects and Symptoms

Short Term Exposure: Eye contact can cause irritation. Symptoms of exposure include nausea and vomiting. Headache, malaise, fatigue, weakness, drowsiness, abdominal discomfort are symptomatic of toxicity. Visual disturbances (reduction of visual acuity, illusions of flickering or shimmering lights, abnormal color vision) and emotional disorders (including confusion, disorientation, aphasia, delirium, hallucinations, and rarely convulsions) are also possible toxic effects.

Material is bioactive and capable of causing cardiac arrhythmias and electrolyte imbalances that may be fatal. Death is due to ventricular fibrillation or cardiac arrest. Material has a high toxicity hazard rating; it may cause death or permanent injury after a very short exposure. It is classified as super toxic; an estimated single lethal dose is 3 – 10 mg. The LD_{50} rat is 56 mg/kg.

Long Term Exposure: May cause reproductive effects.

Points of Attack: Heart.

Medical Surveillance: EKG

First Aid: If this chemical gets into the eyes, remove any contact lenses at once and irrigate immediately for at least 15 minutes, occasionally lifting upper and lower lids. Seek medical attention immediately. If this chemical contacts the skin, remove contaminated clothing and wash immediately with soap and water. Seek medical attention immediately. If this chemical has been inhaled, remove from exposure, begin rescue breathing (using universal precautions) if breathing has stopped and CPR if heart action has stopped. Transfer promptly to a medical facility. When this chemical has been swallowed, get medical attention. Give large quantities of water and induce vomiting. Do not make an unconscious person vomit.

Personal Protective Methods: Wear protective gloves and clothing to prevent any reasonable probability of skin contact. Safety equipment suppliers/manufacturers can provide recommendations on the most protective glove/clothing material for your operation.. All protective clothing (suits, gloves, footwear, headgear) should be clean, available each day, and put on before work. Contact lenses should not be worn when working with this chemical. Wear dust-proof chemical goggles and face shield unless full facepiece respiratory protection is worn. Employees should wash immediately with soap when skin is wet or contaminated. Provide emergency showers and eyewash.

Respirator Selection: *At any detectable concentration:* SCBAF:PD,PP (any MSHA/NIOSH approved self-contained breathing apparatus that has a full facepiece and is operated in a pressure-demand or other positive-pressure mode); or SAF:PD,PP:ASCBA (any supplied-air respirator that has a full facepiece and is operated in a pressure-demand or other positive-pressure mode in combination with an auxiliary, self-contained breathing apparatus operated in a pressure-demand or other positive pressure mode). *Escape:* HiEF (any air-purifying, full-facepiece respirator (gas mask) with a chin-style, front- or back-mounted organic vapor canister having a high-efficiency particulate filter); or SCBAE (any appropriate escape-type, self-contained breathing apparatus).

Storage: Prior to working with digitoxin you should be trained on its proper handling and storage. Store in tightly closed containers in a refrigerator.

Shipping: This compound can be classed under medicines, poisonous, solid, n.o.s. It requires a shipping label of: "Poison." It falls in DOT Hazard Class 6.1 and Packing Group II.

Spill Handling: Evacuate persons not wearing protective equipment from area of spill or leak until clean-up is complete. Remove all ignition sources. Collect powdered material in the most convenient and safe manner and deposit in sealed containers. Ventilate area after clean-up is complete. It may be necessary to contain and dispose of this chemical as a hazardous waste. If material or contaminated runoff enters waterways, notify downstream users of potentially contaminated waters. Contact your Department of Environmental Protection or your regional office of the federal EPA for specific recommendations. If employees are required to clean-up spills, they must be properly trained and equipped. OSHA 1910.120(q) may be applicable.

Fire Extinguishing: Use dry chemical, carbon dioxide, water spray, or alcohol foam extinguishers. Poisonous gases are produced in fire. If material or contaminated runoff enters waterways, notify downstream users of potentially contaminated waters. Notify local health and fire officials and pollution control agencies. From a secure, explosion-proof location, use water spray to cool exposed containers. If cooling streams are ineffective (venting sound increases in volume and pitch, tank discolors, or shows any signs of deforming), withdraw immediately to a secure position. If employees are expected to fight fires, they must be trained and equipped in OSHA 1910.156.

References

U.S. Environmental Protection Agency, "Chemical Profile: Digitoxin," Washington, DC, Chemical Emergency Preparedness Program (November 30, 1987)

Diglycidyl Ether

Molecular Formula: $C_6H_{10}O_3$

Synonyms: Bis(2,3-epoxypropyl) ether; Bis(2-3-epoxypropyl) ether; DGE; Diallyl ether dioxide; Di(2,3-epoxy)propyl ether; Di(epoxypropyl) ether; 2-Epoxypropyl ether; Ether, bis(2,3-epoxypropyl); Ether, diglycidyl; NSV 54739; Oxirane, 2,2'-oxybis (methylene) bis-; 2,2'-Oxybis(methylene)bisoxirane

CAS Registry Number: 2238-07-5

RTECS Number: KN2350000

Regulatory Authority

- Air Pollutant Standard Set (ACGIH)[1] (Australia) (DFG)[3] (HSE)[33] (Israel) (Mexico) (OSHA)[58] (Several States)[60] (Several Canadian Provinces)
- Superfund/EPCRA 40CFR355, Appendix B Extremely Hazardous Substances: TPQ = 1,000 lb (454 kg)

- Superfund/EPCRA 40CFR302.4 Reportable Quantity (RQ): EHS, 1 lb (0.454 kg)
- Canada, WHMIS, Ingredients Disclosure List

Cited in U.S. State Regulations: Alaska (G), California (A, G), Connecticut (A), Florida (G), Illinois (G), Maine (G), Massachusetts (G), Minnesota (G), Nevada (A), New Hampshire (G, W), New Jersey (G), North Dakota (A), Oklahoma (G), Pennsylvania (G), Rhode Island (G), Virginia (A), West Virginia (G).

Description: $C_6H_{10}O_3$ is a colorless liquid with a strong, irritant odor. Boiling point = 260°C. Flash point = 64°C. Hazard Identification (based on NFPA-704 M Rating System): Health 4, Flammability 2, Reactivity 0.

Potential Exposure: This material is used as a reactive diluent for epoxy resins, as a textile treating agent and as a stabilizer for chlorinated organic compounds.

Incompatibilities: Forms explosive mixture with air. May explode when heated. Contact with strong oxidizers may cause fire and explosions. Ethers, as a class, tend to form peroxides upon contact with air and exposure to light. Attacks some forms of plastics, coatings and rubber.

Permissible Exposure Limits in Air: The OSHA PEL is 0.5 ppm (Ceiling) not to be exceeded at any time. The Canadian province of British Columbia set the same ceiling limit as OSHA. The ACGIH recommends a TWA of 0.1 ppm (0.53 mg/m^3) but no STEL value. Australia, Israel, Mexico, the DFG,[3] HSE[33] and Canadian provinces of Alberta, Ontario, and Quebec set the same TWA as ACGIH and Alberta set a STEL of 0.3 ppm. The NIOSH IDLH level is [Ca] 10 ppm. Several states have set guidelines or standards for DFG in ambient air[60] ranging from 5.0 μg/m^3 (North Dakota) to 5 – 10 μg/m^3 (Connecticut) to 8.0 μg/m^3 (Virginia) to 2,380 μg/m^3 (Nevada).

Determination in Air: Adsorption by charcoal tube, analysis by gas liquid chromatography.

Permissible Concentration in Water: No criteria set.

Routes of Entry: Inhalation, ingestion, eye and/or skin contact.

Harmful Effects and Symptoms

Short Term Exposure: Diglycidyl ether can affect you when breathed in and by passing through your skin. Higher levels can cause you to feel dizzy, lightheaded and to pass out. Death can occur. Inhalation of the vapors can cause pulmonary edema, a medical emergency that can be delayed for several hours. This can cause death. DGE may affect the blood, kidneys, liver, testes. Contact can irritate and burn the eyes and skin and cause skin allergy. Because this is a mutagen, handle it as a possible carcinogen – with extreme caution.

Long Term Exposure: Repeated exposure can damage the liver, kidneys and lower blood cell count. Diglycidyl ether may damage the lungs and cause clouding of the eyes. Repeated or prolonged contact with skin may cause skin sensitization and dermatitis. Animal tests show that DGE may damage the testes. Designated a potential human carcinogen by NIOSH (as glycidyl ethers).

Points of Attack: Eyes, skin, respiratory system, reproductive system. Cancer site in animals: skin tumors.

Medical Surveillance: Before beginning employment and at regular times after that, for those with frequent or potentially high exposures, the following are recommended: Liver and kidney function tests Lung function tests. If symptoms develop or overexposure is suspected, the following may be useful: Evaluation by a qualified allergist, including careful exposure history and special testing, may help diagnose skin allergy. Exam of the eyes and vision.

First Aid: If this chemical gets into the eyes, remove any contact lenses at once and irrigate immediately for at least 15 minutes, occasionally lifting upper and lower lids. Seek medical attention immediately. If this chemical contacts the skin, remove contaminated clothing and wash immediately with soap and water. Seek medical attention immediately. If this chemical has been inhaled, remove from exposure, begin rescue breathing (using universal precautions) if breathing has stopped and CPR if heart action has stopped. Transfer promptly to a medical facility. When this chemical has been swallowed, get medical attention. Give large quantities of water and induce vomiting. Do not make an unconscious person vomit.

Personal Protective Methods: Wear protective gloves and clothing to prevent any reasonable probability of skin contact. Safety equipment suppliers/manufacturers can provide recommendations on the most protective glove/clothing material for your operation.. All protective clothing (suits, gloves, footwear, headgear) should be clean, available each day, and put on before work. Contact lenses should not be worn when working with this chemical. Wear splash-proof chemical goggles and face shield unless full facepiece respiratory protection is worn. Employees should wash immediately with soap when skin is wet or contaminated. Provide emergency showers and eyewash.

Respirator Selection: NIOSH: At *any concentrations above the NIOSH REL, or where there is no REL, at any detectable concentration:* SCBAF:PD,PP (any MSHA/NIOSH approved self-contained breathing apparatus that has a full facepiece and is operated in a pressure-demand or other positive-pressure mode) or SAF:PD,PP:ASCBA (any supplied-air respirator that has a full facepiece and is operated in a pressure-demand or other positive-pressure mode in combination with an auxiliary self-contained breathing apparatus operated in a pressure-demand or other positive pressure mode). *Escape:* GMFOV [Any air-purifying, full-facepiece respirator (gas mask) with a chin-style, front-or back-mounted organic vapor canister]; or SCBAE (any appropriate escape-type, self-contained breathing apparatus).

Storage: Prior to working with DGE you should be trained on its proper handling and storage. Diglycidyl ether must be stored to avoid contact with strong oxidizers, such as chlorine, chlorine dioxide, bromine, nitrates and permanganates since violent reactions occur. Store in tightly closed containers in a cool, well-ventilated area away from heat. Sources of ignition, such as smoking and open flames, are prohibited where Diglycidyl ether is used, handled, or stored in a manner that could create a potential fire o explosion hazard.

Shipping: This material is not covered in DOT's Performance-Oriented Packaging Standards.[19]

Spill Handling: Evacuate and restrict persons not wearing protective equipment from area of spill or leak until cleanup is complete. Remove all ignition sources. Ventilate area of spill or leak. Absorb liquids in vermiculite, dry sand, earth, peat, carbon, or a similar material and deposit in sealed containers. Keep this chemical out of a confined space, such as a sewer, because of the possibility of an explosion, unless the sewer is designed to prevent the build-up of explosive concentrations. It may be necessary to contain and dispose of this chemical as a hazardous waste. If material or contaminated runoff enters waterways, notify downstream users of potentially contaminated waters. Contact your Department of Environmental Protection or your regional office of the federal EPA for specific recommendations. If employees are required to clean-up spills, they must be properly trained and equipped. OSHA 1910.120(q) may be applicable.

Fire Extinguishing: This chemical is a flammable liquid that may explode on heating. Poisonous gases are produced in fire. Use dry chemical, carbon dioxide, or alcohol foam extinguishers. Vapors are heavier than air and will collect in low areas. Vapors may travel long distances to ignition sources and flashback. Vapors in confined areas may explode when exposed to fire. Containers may explode in fire. Storage containers and parts of containers may rocket great distances, in many directions. If material or contaminated runoff enters waterways, notify downstream users of potentially contaminated waters. Notify local health and fire officials and pollution control agencies. From a secure, explosion-proof location, use water spray to cool exposed containers. If cooling streams are ineffective (venting sound increases in volume and pitch, tank discolors, or shows any signs of deforming), withdraw immediately to a secure position. If employees are expected to fight fires, they must be trained and equipped in OSHA 1910.156.

Disposal Method Suggested: Incineration.

References

National Institute for Occupational Safety and Health, Information Profiles on Potential Occupational Hazards: Glycidyl Ethers, 116–123, Report PB-276–678, Rockville, MD (October 1977)

National Institute for Occupational Safety and Health, Criteria for a Recommended Standard: Occupational Exposure to Glycidyl Ethers, NIOSH Doc. No. 78–166, Washington, DC (1978)

U.S. Environmental Protection Agency, "Chemical Profile: Diglycidyl Ether," Washington, DC, Chemical Emergency Preparedness Program (November 30, 1987)

New Jersey Department of Health, "Hazardous Substance Fact Sheet: Diglycidyl Ether," Trenton, NJ (November 1986)

Diglycidyl Ether of Bisphenol A

Molecular Formula: $C_{21}H_{24}O_4$

Synonyms: 2,2'-Bis [(*p*-2,3-epoxy propoxy) phenyl] propane; 2,2-Bis[4-(2,3-epoxypropyloxy)phenyl]propane; Bis(4-glycidyloxyphenyl)dimethylamethane; 2,2-Bis(*p*-glycidyloxyphenyl)propane; Bis(4-hydroxyphenyl)dimethylmethane diglycidyl ether; 2,2-Bis(*p*-hydroxyphenyl)propane, diglycidyl ether; 2,2-Bis(4-hydroxyphenyl)propane, diglycidyl ether; D.E.R. 332; Diglycidyl bisphenol A; Diglycidyl ether of 2,2-bis(*p*-hydroxyphenyl)propane; Diglycidyl ether of 2,2-bis(4-hydroxyphenyl)propane; Diglycidyl ether of bisphenol A; Diglycidyl ether of 4,4'-isopropylidenediphenol; *p,p*'-Dihydroxydiphenyldimethylmethane diglycidyl ether; 4,4'-Dihydroxydiphenyldimethylmethane diglycidyl ether; EPI-REZ 508; EPI-REZ 510; Epon 828; Epoxide A; ERL-2774; 4,4'-Isopropylidenediphenol diglycidyl ether; 2,2'-[(1-Methylethylidene)bis(4,1-phenyleneoxymethylene)] bisoxirane

CAS Registry Number: 1675-54-3

RTECS Number: TX3800000

Regulatory Authority

- Canada, WHMIS, Ingredients Disclosure List

Cited in U.S. State Regulations: New Jersey (G).

Description: Diglycidyl Ether of Bisphenol A is an odorless amber liquid. Freezing/Melting point = 8 – 12°C and decomposes in lieu of boiling. Flash point = 79°C.

Potential Exposure: Diglycidyl ether of Bisphenol A is used as a basic active ingredient of epoxy resins.

Incompatibilities: Incompatible with strong acids, strong oxidizers. Presumed to form explosive peroxides.

Permissible Exposure Limits in Air: No standards set.

Permissible Concentration in Water: No criteria set.

Routes of Entry: Inhalation, passing through the skin.

Harmful Effects and Symptoms

Short Term Exposure: Diglycidyl ether of Bisphenol A can affect you when breathed in and by passing through your skin. Skin contact can cause irritation and rash. It can also lead to the development of skin allergy. Once allergy develops even low future exposures can trigger rash. Eye contact causes irritation. The vapors can irritate the eyes, nose, throat and bronchial tubes, causing nosebleeds, coughing and tightness in the chest.

Long Term Exposure: There is limited evidence that this chemical causes skin cancer in animals. May cause skin sensitization and allergy.

Points of Attack: Skin, lungs.

Medical Surveillance: If symptoms develop or overexposure is suspected, the following may be useful: Consider lung function tests. Evaluation by a qualified allergist, including careful exposure history and special testing, may help diagnose skin allergy.

First Aid: If this chemical gets into the eyes, remove any contact lenses at once and irrigate immediately for at least 15 minutes, occasionally lifting upper and lower lids. Seek medical attention immediately. If this chemical contacts the skin, remove contaminated clothing and wash immediately with soap and water. Seek medical attention immediately. If this chemical has been inhaled, remove from exposure, begin rescue breathing (using universal precautions) if breathing has stopped and CPR if heart action has stopped. Transfer promptly to a medical facility. When this chemical has been swallowed, get medical attention. Give large quantities of water and induce vomiting. Do not make an unconscious person vomit.

Personal Protective Methods: Wear protective gloves and clothing to prevent any reasonable probability of skin contact. Safety equipment suppliers/manufacturers can provide recommendations on the most protective glove/clothing material for your operation.. All protective clothing (suits, gloves, footwear, headgear) should be clean, available each day, and put on before work. Contact lenses should not be worn when working with this chemical. Wear splash-proof chemical goggles and face shield unless full facepiece respiratory protection is worn. Employees should wash immediately with soap when skin is wet or contaminated. Provide emergency showers and eyewash.

Respirator Selection: Where the potential exists for exposures to Diglycidyl ether of Bisphenol A, use a MSHA/NOSH approved supplied-air respirator with a full facepiece operated in the positive pressure mode or with a full facepiece, hood, or helmet in the continuous flow mode, or use a MSHA/NIOSH approved self-contained breathing apparatus with a full facepiece operated in pressure-demand or other positive pressure mode.

Storage: Prior to working with this chemical you should be trained on its proper handling and storage. Store in tightly closed containers in a dark, cool, well-ventilated area away from strong oxidizers, acids and heat.

Shipping: This material is not covered in DOT's Performance-Oriented Packaging Standards.[19] It may however be classified as a combustible liquid n.o.s. This class requires no shipping label. It falls in DOT Hazard Class 3 and Packing Group III.

Spill Handling: Evacuate and restrict persons not wearing protective equipment from area of spill or leak until cleanup is complete. Remove all ignition sources. Ventilate area of spill or leak. Absorb liquids in vermiculite, dry sand, earth, peat, carbon, or a similar material and deposit in sealed containers. Keep this chemical out of a confined space, such as a sewer, because of the possibility of an explosion, unless the sewer is designed to prevent the build-up of explosive concentrations. It may be necessary to contain and dispose of this chemical as a hazardous waste. If material or contaminated runoff enters waterways, notify downstream users of potentially contaminated waters. Contact your Department of Environmental Protection or your regional office of the federal EPA for specific recommendations. If employees are required to clean-up spills, they must be properly trained and equipped. OSHA 1910.120(q) may be applicable.

Fire Extinguishing: This chemical is a combustible liquid. Poisonous gases are produced in fire. Use dry chemical, carbon dioxide, or alcohol foam extinguishers. Vapors are heavier than air and will collect in low areas. Vapors may travel long distances to ignition sources and flashback. Vapors in confined areas may explode when exposed to fire. Containers may explode in fire. Storage containers and parts of containers may rocket great distances, in many directions. If material or contaminated runoff enters waterways, notify downstream users of potentially contaminated waters. Notify local health and fire officials and pollution control agencies. From a secure, explosion-proof location, use water spray to cool exposed containers. If cooling streams are ineffective (venting sound increases in volume and pitch, tank discolors, or shows any signs of deforming), withdraw immediately to a secure position. If employees are expected to fight fires, they must be trained and equipped in OSHA 1910.156.

Disposal Method Suggested: Incineration.

References

New Jersey Department of Health, "Hazardous Substance Fact Sheet: Diglycidyl Ether of Bisphenol A," Trenton, NJ (January 11, 1988)

National Institute for Occupational Safety and Health, "Glycidyl Ethers," NIOSH Current Intelligence Bulletin, No. 29, Cincinnati, Ohio (October 12, 1978)

Digoxin

Molecular Formula: $C_{41}H_{64}O_{14}$

Synonyms: Chloroformic digitalin; (3β,5β,12β)-3-[(*O*-2,6-Dideoxy-β-D-ribohexopyranosyl-(1→4)-*O*-2,6-dideoxy-β-D-ribohexopyranosyl-(1→4)-2,6-dideoxy-β-D-ribo-hexopyranosyl)oxy]-12,14-dihydroxycarD-20(22)-enolide; Digacin; Digitalis glycoside; Digoxigenintridigitoxosid (German); Digoxina (Spanish); Digoxine; Homolle's Digitalin; Lanicor; Lanoxin; Rougoxin; SK-Digoxin

CAS Registry Number: 20830-75-5

RTECS Number: IH6125000

Regulatory Authority

- Superfund/EPCRA 40CFR355, Appendix B Extremely Hazardous Substances: TPQ = 10/10,000 lb (0.454/4,540 kg)
- Superfund/EPCRA 40CFR302.4 Reportable Quantity (RQ): EHS, 1 lb (0.454 kg)

Cited in U.S. State Regulations: Massachusetts (G).

Description: Digoxin, $C_{41}H_{64}O_{14}$, is a white crystalline solid. Freezing/Melting point = 230 – 265°C (decomposes).

Potential Exposure: Digoxin is used as a cardiotonic drug.

Permissible Exposure Limits in Air: No standard set.

Permissible Concentration in Water: No criteria set.

Routes of Entry: Ingestion, inhalation, skin and/or eyes.

Harmful Effects and Symptoms

Short Term Exposure: Symptoms of exposure refer specifically to digitalis and include nausea and vomiting, headache, fatigue, weakness, drowsiness, and abdominal discomfort are symptomatic of toxicity. Visual disturbances (including blurring, halos, and aberrations of color), emotional disorders (including confusion, disorientation, aphasia, delirium, and hallucinations) and convulsions. Material is a digitalis glycoside. Ingestion can cause death. Material is considered super toxic; probable human oral lethal dose is less than 5 mg/kg, a taste (less than 7 drops) for a 70 kg (150 lb.) person.

Persons at risk include those taking drugs for thyroid and renal diseases. Quinidine and diuretics taken concurrently with digoxin can be hazardous. It should be used with extreme care during pregnancy and in nursing mothers.

First Aid: If this chemical gets into the eyes, remove any contact lenses at once and irrigate immediately for at least 15 minutes, occasionally lifting upper and lower lids. Seek medical attention immediately. If this chemical contacts the skin, remove contaminated clothing and wash immediately with soap and water. Seek medical attention immediately. If this chemical has been inhaled, remove from exposure, begin rescue breathing (using universal precautions) if breathing has stopped and CPR if heart action has stopped. Transfer promptly to a medical facility. When this chemical has been swallowed, get medical attention. Give large quantities of water and induce vomiting. Do not make an unconscious person vomit.

Personal Protective Methods: Wear protective gloves and clothing to prevent any reasonable probability of skin contact. Safety equipment suppliers/manufacturers can provide recommendations on the most protective glove/clothing material for your operation.. All protective clothing (suits, gloves, footwear, headgear) should be clean, available each day, and put on before work. Contact lenses should not be worn when working with this chemical. Wear dust-proof chemical goggles and face shield unless full facepiece respiratory protection is worn. Employees should wash immediately with soap when skin is wet or contaminated. Provide emergency showers and eyewash.

Respirator Selection: *At any detectable concentration:* SCBAF:PD,PP (any MSHA/NIOSH approved self-contained breathing apparatus that has a full facepiece and is operated in a pressure-demand or other positive-pressure mode); or SAF:PD,PP:ASCBA (any supplied-air respirator that has a full facepiece and is operated in a pressure-demand or other positive-pressure mode in combination with an auxiliary, self-contained breathing apparatus operated in a pressure-demand or other positive pressure mode).

Storage: Prior to working with this chemical you should be trained on its proper handling and storage. Avoid light. Store in tightly closed containers in a dark, cool, well ventilated area away from oxidizers and reducing agents.

Shipping: This compound falls in the medicines, poisonous, solid n.o.s. category and requires a shipping label of: "Poison." It falls in DOT Hazard Class 6.1 and Packing Group I.

Spill Handling: Evacuate persons not wearing protective equipment from area of spill or leak until clean-up is complete. Remove all ignition sources. Collect powdered material in the most convenient and safe manner and deposit in sealed containers. Ventilate area after clean-up is complete. It may be necessary to contain and dispose of this chemical as a hazardous waste. If material or contaminated runoff enters waterways, notify downstream users of potentially contaminated waters. Contact your Department of Environmental Protection or your regional office of the federal EPA for specific recommendations. If employees are required to clean-up spills, they must be properly trained and equipped. OSHA 1910.120(q) may be applicable.

Fire Extinguishing: This chemical is a combustible solid. Use dry chemical, carbon dioxide, water spray, or alcohol foam extinguishers. Poisonous gases are produced in fire. If material or contaminated runoff enters waterways, notify downstream users of potentially contaminated waters. Notify local health and fire officials and pollution control agencies. From a secure, explosion-proof location, use water spray to cool exposed containers. If cooling streams are ineffective (venting sound increases in volume and pitch, tank discolors, or shows any signs of deforming), withdraw immediately to a secure position. If employees are expected to fight fires, they must be trained and equipped in OSHA 1910.156.

References

U.S. Environmental Protection Agency, "Chemical Profile: Digoxin," Washington, DC, Chemical Emergency Preparedness Program (November 30, 1987)

Dihydrosafrole

Molecular Formula: $C_{10}H_{12}O_2$

Synonyms: AI3-03435; Benzene, 1,2-(methylenedioxy)-4-propyl-; 1,3-Benzodioxole, 5-propyl-; Dihidrosafrol (Spanish); Dihydroisosafrole; Dihydrosafrol; 2',3'-Dihydrosafrole; [1,2-(Methylenedioxy)-4-propyl]benzene; NSC 27867; 5-Propyl-1,3-benzodioxole; 4-Propyl-1,2-(methyle nedioxy)benzene; Safrole, dihydro-

CAS Registry Number: 94-58-6

RTECS Number: DA6125000

Regulatory Authority

- Carcinogen (Animal Positive) (IARC)[9]

- EPA Hazardous Waste Number (RCRA No.): U090
- RCRA, 40CFR261, Appendix 8 Hazardous Constituents
- Superfund/EPCRA 40CFR302.4 Reportable Quantity (RQ): CERCLA, 10 lb (4.54 kg)
- EPCRA Section 313 Form R *de minimis* concentration reporting level: 0.1%

Cited in U.S. State Regulations: California (A, G), Florida (G), Illinois (G), Kansas (G), Louisiana (G), Massachusetts (G), Michigan (G), Minnesota (G), New Hampshire (G), New Jersey (G), Pennsylvania (G), Vermont (G), Virginia (G), Washington (G), Wisconsin (G).

Description: Dihydrosafrole, $C_{10}H_{12}O_2$, is an oily liquid. Boiling point = 228°C.

Potential Exposure: Workers may be exposed to dihydrosafrole during its use as a chemical intermediate in the production of piperonyl butoxide and related insecticidal synergists, and in the production of fragrances for cosmetic products.

Permissible Exposure Limits in Air: No limits set.

Permissible Concentration in Water: No criteria set.

Determination in Water: By chromatography.

Harmful Effects and Symptoms

Short Term Exposure: Skin contact can cause irritation. An LD_{50} value of 2,260 mg/kg has been reported for rats (slightly toxic).

Long Term Exposure: Dihydrosafrole is recognized as a carcinogen in experimental animals by the International Agency for Research on Cancer (IARC) and the National Cancer Institute Bioassay Program.

Points of Attack: Liver, spleen.

Medical Surveillance: Liver function tests.

First Aid: If this chemical gets into the eyes, remove any contact lenses at once and irrigate immediately for at least 15 minutes, occasionally lifting upper and lower lids. Seek medical attention immediately. If this chemical contacts the skin, remove contaminated clothing and wash immediately with soap and water. Seek medical attention immediately. If this chemical has been inhaled, remove from exposure, begin rescue breathing (using universal precautions) if breathing has stopped and CPR if heart action has stopped. Transfer promptly to a medical facility. When this chemical has been swallowed, get medical attention. Give large quantities of water and induce vomiting. Do not make an unconscious person vomit.

Personal Protective Methods: Wear protective gloves and clothing to prevent any reasonable probability of skin contact. Safety equipment suppliers/manufacturers can provide recommendations on the most protective glove/clothing material for your operation.. All protective clothing (suits, gloves, footwear, headgear) should be clean, available each day, and put on before work. Contact lenses should not be worn when working with this chemical. Wear splash-proof chemical goggles and face shield unless full facepiece respiratory protection is worn. Employees should wash immediately with soap when skin is wet or contaminated. Provide emergency showers and eyewash.

Respirator Selection: *At any detectable concentration:* SCBAF:PD,PP (any MSHA/NIOSH approved self-contained breathing apparatus that has a full facepiece and is operated in a pressure-demand or other positive-pressure mode); or SAF:PD,PP:ASCBA (any supplied-air respirator that has a full facepiece and is operated in a pressure-demand or other positive-pressure mode in combination with an auxiliary, self-contained breathing apparatus operated in a pressure-demand or other positive pressure mode). *Escape:* HiEF (any air-purifying, full-facepiece respirator (gas mask) with a chin-style, front- or back-mounted organic vapor canister having a high-efficiency particulate filter); or SCBAE (any appropriate escape-type, self-contained breathing apparatus).

Storage: Prior to working with this chemical you should be trained on its proper handling and storage. Store in tightly closed containers in a cool, well ventilated area away from oxidizers and reducing agents. Where possible, automatically pump liquid from drums or other storage containers to process containers. A regulated, marked area should be established where this chemical is handled, used, or stored in compliance with OSHA standard 1910.1045.

Spill Handling: Evacuate and restrict persons not wearing protective equipment from area of spill or leak until cleanup is complete. Remove all ignition sources. Ventilate area of spill or leak. Absorb liquids in vermiculite, dry sand, earth, peat, carbon, or a similar material and deposit in sealed containers. It may be necessary to contain and dispose of this chemical as a hazardous waste. If material or contaminated runoff enters waterways, notify downstream users of potentially contaminated waters. Contact your Department of Environmental Protection or your regional office of the federal EPA for specific recommendations. If employees are required to clean-up spills, they must be properly trained and equipped. OSHA 1910.120(q) may be applicable.

Fire Extinguishing: Poisonous gases are produced in fire. Use dry chemical, carbon dioxide, or foam extinguishers. Vapors are heavier than air and will collect in low areas. Containers may explode in fire. Storage containers and parts of containers may rocket great distances, in many directions. If material or contaminated runoff enters waterways, notify downstream users of potentially contaminated waters. Notify local health and fire officials and pollution control agencies. From a secure, explosion-proof location, use water spray to cool exposed containers. If cooling streams are ineffective (venting sound increases in volume and pitch, tank discolors, or shows any signs of deforming), withdraw immediately to a secure position. If employees are expected to fight fires, they must be trained and equipped in OSHA 1910.156.

Disposal Method Suggested: The U.S.E.P.A. recommends packaging of product residues and sorbent media in epoxy-lined drums and disposal at an EPA-approved site. The compound may be destroyed by permanganate oxidation, high temperature incineration with scrubbing equipment, or microwave plasma treatment. Consult with environmental regulatory agencies for guidance on acceptable disposal practices. Generators of waste containing this contaminant (≥100 kg/mo) must conform with EPA regulations governing storage, transportation, treatment, and waste disposal.

References

U.S. Environmental Protection Agency, "Chemical Hazard Information Profile Draft Report: Dihydrosafrole," Washington, DC (May 17, 1984)

Sax N. I., Ed., "Dangerous Properties of Industrial Materials Report," 7, No. 2, 51–53 (1987)

Diisobutyl Ketone

Molecular Formula: $C_9H_{18}O$

Common Formula: $[(CH_3)_2CHCH_2]_2CO$

Synonyms: DIBK; Diisipropyl-acetone; Diisobutilchetone (Italian); Di-isobutylcetone (French); Diisobutylketon (Dutch, German); 5-Diisopropylacetone; Diisopropylacetone; 2,6-Dimethyl-4-heptane; 2,6-Dimethyl-heptan-4-on (Dutch); 2,6-Dimethyl-heptan-4-on (German); 2,6-Dimethyl-4-heptanone; 2,6-Dimethylheptan-4-one; 2,6-Dimethylheptanone; 2,6-Dimetileptan-4-one (Italian); 4-Heptanone,2,6-dimethyl-; Isobutyl ketone; Isovalerone; Valerone

CAS Registry Number: 108-83-8

RTECS Number: MJ5775000

DOT ID: UN1157

EEC Number: 606-005-00-X

Regulatory Authority

- Air Pollutant Standard Set (ACGIH)[1] (Australia) (DFG)[3] (HSE)[33] (Israel) (Mexico) (OSHA)[58] (Several States)[60] (Several Canadian Provinces)
- Canada, WHMIS, Ingredients Disclosure List

Cited in U.S. State Regulations: Alaska (G), California (A, G), Connecticut (A), Florida (G), Illinois (G), Maine (G), Massachusetts (G), Minnesota (G), Nevada (A), New Hampshire (G), New Jersey (G), North Dakota (A), Pennsylvania (G), Rhode Island (G), Virginia (A), West Virginia (G).

Description: Diisobutyl ketone, $[(CH_3)_2CHCH_2]_2CO$, is a colorless liquid with a mild, sweet odor. Boiling point = 168°C. Freezing/Melting point = -42°C. Flash point = 49°C. Autoignition temperature = 396°C. The explosive limits are LEL = 0.8%; UEL = 7.1% @ 93°C.[17] Hazard Identification (based on NFPA-704 M Rating System): Health 1, Flammability 2, Reactivity 0. Insoluble in water.

Potential Exposure: This material is used as a solvent, as a dispersant for resins, and as an intermediate in the synthesis of pharmaceuticals and pesticides.

Incompatibilities: Forms explosive mixture with air. Incompatible with strong acids, aliphatic amines, strong oxidizers. Attacks some forms of plastics, coatings and rubber.

Permissible Exposure Limits in Air: The OSHA PEL and DFG MAK[3] is 50 ppm (290 mg/m³). NIOSH and ACGIH recommend a TWA of 25 ppm (150 mg/m³). The HSE[33] Australia, Israel, Mexico and the Canadian provinces of Alberta, BC, Ontario, and Quebec have set the same TWA and Alberta set a STEL of 38 ppm. The NIOSH IDLH level is 500 ppm. Several states have set guidelines or standards for diisobutyl ketone in ambient air[60] ranging from 2.5 mg/m³ (North Dakota and Virginia) to 2.8 mg/m³ (Connecticut) to 3.57 mg/m³ (Nevada).

Determination in Air: Charcoal adsorption, workup with CS_2, analysis by gas chromatography. See NIOSH Method 1300.[18]

Permissible Concentration in Water: No criteria set.

Routes of Entry: Inhalation, ingestion, eye and/or skin contact.

Harmful Effects and Symptoms

Short Term Exposure: DIBK irritates the eyes, skin, respiratory tract. Exposure to high concentrations could cause headaches, dizziness, unconsciousness.

Long Term Exposure: Repeated or prolonged contact with skin may cause dermatitis, liver, and kidney damage.

Points of Attack: Eyes, skin, respiratory system, central nervous system, liver, kidneys.

Medical Surveillance: If symptoms develop or overexposure has occurred, the following may be useful: Liver and kidney function tests.

First Aid: If this chemical gets into the eyes, remove any contact lenses at once and irrigate immediately for at least 15 minutes, occasionally lifting upper and lower lids. Seek medical attention immediately. If this chemical contacts the skin, remove contaminated clothing and wash immediately with soap and water. Seek medical attention immediately. If this chemical has been inhaled, remove from exposure, begin rescue breathing (using universal precautions) if breathing has stopped and CPR if heart action has stopped. Transfer promptly to a medical facility. When this chemical has been swallowed, get medical attention. Give large quantities of water and induce vomiting. Do not make an unconscious person vomit.

Personal Protective Methods: Wear solvent-resistant protective gloves and clothing to prevent any reasonable probability of skin contact. Safety equipment suppliers/manufacturers can provide recommendations on the most protective glove/clothing material for your operation. Nitrile and Polyvinyl Alcohol are among the recommended protective materials. All protective clothing (suits, gloves, footwear, headgear) should be clean, available each day, and put on before work. Contact lenses should not be worn when working with this chemical. Wear splash-proof goggles and face shield, unless full facepiece respiratory protection is worn. Wear splash-proof chemical goggles and face shield when working with liquid, unless full facepiece respiratory protection is worn. Employees should wash immediately with soap when skin is wet or contaminated. Provide emergency showers and eyewash.

Respirator Selection: NIOSH/OSHA: *up to 500 ppm:* SA:CF (any supplied-air respirator operated in a continuous-flow mode); or PAPROV [any powered, air-purifying respirator with organic vapor cartridge(s)]; or CCRFOV [any air-purifying, full-facepiece respirator (gas mask) with a chin-style, front- or back-mounted acid gas canister]; or GMFOV [any air-purifying, full-facepiece respirator (gas mask) with a chin-style, front- or back-, mounted organic vapor canister]; or SCBAF (any self-contained breathing apparatus with a full facepiece); or SAF (any supplied-air respirator with a full facepiece). *Emergency or planned entry into unknown concentrations or IDLH conditions:* SCBAF:PD,PP (any self-contained breathing apparatus that has a full facepiece and is operated in a pressure-demand or other positive-pressure mode); or SAF:PD,PP:ASCBA (any supplied-air respirator that has a full facepiece and is operated in a pressure-demand or other positive-pressure mode in combination with an auxiliary self-contained breathing apparatus operated in a pressure-demand or other positive pressure mode). *Escape:* GMFOV [any air-purifying, full-facepiece respirator (gas mask) with a chin-style, front- or back-, mounted organic vapor canister]; or SCBAE (any appropriate escape-type, self-contained breathing apparatus).

Note: Substance causes eye irritation or damage; eye protection needed.

Storage: Prior to working with DIBK you should be trained on its proper handling and storage. Before entering confined space where DIBK may be present, check to make sure that an explosive concentration does not exist. Store in tightly closed containers in a cool, well ventilated area. Metal containers involving the transfer of this chemical should be grounded and bonded. Where possible, automatically pump liquid from drums or other storage containers to process containers. Drums must be equipped with self-closing valves, pressure vacuum bungs, and flame arresters. Use only non-sparking tools and equipment, especially when opening and closing containers of this chemical. Sources of ignition such as smoking and open flames, are prohibited where this chemical is used, handled, or stored in a manner that could create a potential fire or explosion hazard. Wherever this chemical is used, handled, manufactured, or stored, use explosion-proof electrical equipment and fittings.

Shipping: This compound requires a shipping label of: "Flammable Liquid." It falls in DOT Hazard Class 3 and Packing Group III.

Spill Handling: Evacuate and restrict persons not wearing protective equipment from area of spill or leak until cleanup is complete. Remove all ignition sources. Establish forced ventilation to keep levels below explosive limit. Absorb liquids in vermiculite, dry sand, earth, peat, carbon, or a similar material and deposit in sealed containers. Keep this chemical out of a confined space, such as a sewer, because of the possibility of an explosion, unless the sewer is designed to prevent the build-up of explosive concentrations. It may be necessary to contain and dispose of this chemical as a hazardous waste. If material or contaminated runoff enters waterways, notify downstream users of potentially contaminated waters. Contact your Department of Environmental Protection or your regional office of the federal EPA for specific recommendations. If employees are required to clean-up spills, they must be properly trained and equipped. OSHA 1910.120(q) may be applicable.

Fire Extinguishing: This chemical is a combustible liquid. Poisonous gases are produced in fire. Use dry chemical, carbon dioxide, or alcohol foam extinguishers. Vapors are heavier than air and will collect in low areas. Vapors may travel long distances to ignition sources and flashback. Vapors in confined areas may explode when exposed to fire. Containers may explode in fire. Storage containers and parts of containers may rocket great distances, in many directions. If material or contaminated runoff enters waterways, notify downstream users of potentially contaminated waters. Notify local health and fire officials and pollution control agencies. From a secure, explosion-proof location, use water spray to cool exposed containers. If cooling streams are ineffective (venting sound increases in volume and pitch, tank discolors, or shows any signs of deforming), withdraw immediately to a secure position. If employees are expected to fight fires, they must be trained and equipped in OSHA 1910.156.

Disposal Method Suggested: Incineration.[22]

References

National Institute for Occupational Safety and Health, Criteria for a Recommended Standard: Occupational Exposure to Ketones, NIOSH Do. No. 78–113, Washington, DC (1978)

New Jersey Department of Health, "Hazardous Substance Fact Sheet: 2,6-Dimethylheptanone," Trenton, NJ (September 1996)

Sax N. I., Ed., "Dangerous Properties of Industrial Materials Report," 1, No. 6, 51–52

Diisopropylamine

Molecular Formula: $C_6H_{15}N$

Common Formula: $(CH_3)_2CHNHCH(CH_3)_2$

Synonyms: Bis(isopropyl)amine; DIPA; *N*-(1-Methyethyl)-2-propanamine; 2-Propanamine, *N*-(1-methylethyl)-

CAS Registry Number: 108-18-9

RTECS Number: IM4025000

DOT ID: UN 1158

EEC Number: 612-048-00-5

Regulatory Authority

- Air Pollutant Standard Set (ACGIH)[1] (Australia) (HSE)[33] (Israel) (Mexico) (former USSR)[43] (OSHA)[58] (Several States)[60] (Several Canadian Provinces)
- Canada, WHMIS, Ingredients Disclosure List

Cited in U.S. State Regulations: Alaska (G), California (A, G), Connecticut (A), Florida (G), Illinois (G), Maine (G), Massachusetts (G), Minnesota (G), Nevada (A), New Jersey (G), North Dakota (A), Pennsylvania (G), Rhode Island (G), Virginia (A), West Virginia (G).

Description: Diisopropylamine, $(CH_3)_2CHNHCH(CH_3)_2$, is a colorless liquid with an ammoniacal odor. Boiling point = 83°C. There are other reported boiling points in the literature ranging from 90 – 96°C. Flash point = -1°C. Autoignition temperature = 316°C. Explosive limits: LEL = 1.1%; UEL = 7.1%. Insoluble in water.

Potential Exposure: This material is used as a chemical intermediate in the synthesis of pharmaceuticals and pesticides (Diallate, Fenamiphos and Triallate, for example).

Incompatibilities: Forms explosive mixture with air. This chemical is a strong base; reacts violently with strong oxidizers, strong acids. Attacks copper, zinc and their alloys, aluminum and galvanized steel. Attacks some forms of plastics and coatings.

Permissible Exposure Limits in Air: The Federal Limit (OSHA PEL),[58] the HSE TWA[33] and ACGIH TWA value is 5 ppm (20 mg/m^3). Australia, Israel, Mexico, and the Canadian provinces of Alberta, BC, Ontario, and Quebec set the same TWA level and Alberta set a STEL of 10 ppm. The notation "skin" indicates the possibility of cutaneous absorption. The NIOSH IDLH level is 200 ppm. The former USSR-UNEP/IRPTC project[43] has set a MAC in workplace air of 5 mg/m^3. Several states have set guidelines or standards for diisopropylamine in ambient air[60] ranging from 200 μg/m^3 (North Dakota) to 350 μg/m^3 (Virginia) to 400 μg/m^3 (Connecticut) to 476 μg/m^3 (Nevada).

Determination in Air: Impinger; Potassium hydroxide; Gas chromatography/Flame ionization detection; NIOSH II(4), Method #S141.

Permissible Concentration in Water: The former USSR-UNEP/IRPTC project[43] has set a MAC in water bodies used for domestic purposes of 0.5 mg/l.

Routes of Entry: Inhalation, ingestion, skin absorption, eye and/or skin contact.

Harmful Effects and Symptoms

Short Term Exposure: A lacrimator. The vapor is corrosive to the eyes and respiratory tract. Higher exposures can cause pulmonary edema, a medical emergency that can be delayed for several hours. This can cause death. Nausea vomiting; headaches; eye irritation, visual disturbance. Contact may cause eye damage, causing cloudy vision. Exposure may cause a skin allergy to develop, so that even very small future exposures will result in itching and skin rash.

Long Term Exposure: Repeated or prolonged contact with skin may cause skin allergy, dermatitis, lung damage and asthma-like allergy.

Points of Attack: Eyes, skin, respiratory system.

Medical Surveillance: Before beginning employment and at regular times after that, the following are recommended: Lung function tests. These may be normal if person is not having an attack at the time of test. If symptoms develop or overexposure is suspected, the following may be useful: Evaluation by a qualified allergist, including careful exposure history and special testing, may help diagnose skin allergy. Consider chest x-ray after acute overexposure.

First Aid: If this chemical gets into the eyes, remove any contact lenses at once and irrigate immediately for at least 15 minutes, occasionally lifting upper and lower lids. Seek medical attention immediately. If this chemical contacts the skin, remove contaminated clothing and wash immediately with soap and water. Seek medical attention immediately. If this chemical has been inhaled, remove from exposure, begin rescue breathing (using universal precautions) if breathing has stopped and CPR if heart action has stopped. Transfer promptly to a medical facility. When this chemical has been swallowed, get medical attention. Give large quantities of water and induce vomiting. Do not make an unconscious person vomit. Medical observation is recommended for 24 – 48 hours after breathing overexposure, as pulmonary edema may be delayed. As first aid for pulmonary edema, a doctor or authorized paramedic may consider administering a corticosteroid spray.

Personal Protective Methods: Wear protective gloves and clothing to prevent any reasonable probability of skin contact. Safety equipment suppliers/manufacturers can provide recommendations on the most protective glove/clothing material for your operation. Teflon, Viton, and Nitrile are among the recommended protective materials. All protective clothing (suits, gloves, footwear, headgear) should be clean, available each day, and put on before work. Contact lenses should not be worn when working with this chemical. Wear splash-proof chemical goggles and face shield unless full facepiece respiratory protection is worn. Employees should wash immediately with soap when skin is wet or contaminated. Provide emergency showers and eyewash.

Respirator Selection: NIOSH/OSHA: *125 ppm:* SA:CF (any supplied-air respirator operated in a continuous-flow mode); or PAPROV [any powered, air-purifying respirator with organic vapor cartridge(s)]. *200 ppm:* CCRFOV [any air-purifying, full-facepiece respirator (gas mask) with a chin-style, front- or back-mounted acid gas canister]; or GMFOV [any air-purifying, full-facepiece respirator (gas mask) with a chin-style, front- or back-, mounted organic vapor canister]; or PAPRTOV [any powered, air-purifying respirator with a tight-fitting facepiece and organic vapor cartridge(s)]; or

SCBAF (any self-contained breathing apparatus with a full facepiece); or SAF (any supplied-air respirator with a full facepiece). *Emergency or planned entry into unknown concentrations or IDLH conditions:* SCBAF:PD,PP (any self-contained breathing apparatus that has a full facepiece and is operated in a pressure-demand or other positive-pressure mode); or SAF:PD,PP:ASCBA (any supplied-air respirator that has a full facepiece and is operated in a pressure-demand or other positive-pressure mode in combination with an auxiliary self-contained breathing apparatus operated in a pressure-demand or other positive pressure mode). *Escape:* GMFOV [any air-purifying, full-facepiece respirator (gas mask) with a chin-style, front- or back-, mounted organic vapor canister]; or SCBAE (any appropriate escape-type, self-contained breathing apparatus).

Note: Substance causes eye irritation or damage; eye protection needed.

Storage: Prior to working with DIPA you should be trained on its proper handling and storage. Before entering confined space where this chemical may be present, check to make sure that an explosive concentration does not exist. Diisopropylamine must be stored to avoid contact with strong acids (such as hydrochloric, sulfuric, and nitric) or oxidizers (such as perchlorates, peroxides, permanganates, chlorates, and nitrates), because violent reactions occur. Store in tightly closed containers in a cool, well-ventilated area away from heat. Sources of ignition such as smoking and open flames are prohibited where diisopropylamine is used, handled, or stored in a manner that could create a potential fire or explosion hazard. Metal containers involving the transfer of 5 gallons or more of diisopropylamine should be grounded and bonded. Drums must be equipped with self-closing valves, pressure vacuum bungs, and flame arresters. Use only non-sparking tools and equipment, especially when opening and closing containers of diisopropylamine.

Shipping: This compound requires a shipping label of "Flammable Liquid." It falls in DOT Hazard Class 3 and Packing Group II.

Spill Handling: Evacuate and restrict persons not wearing protective equipment from area of spill or leak until cleanup is complete. Remove all ignition sources. Establish forced ventilation to keep levels below explosive limit. Absorb liquids in vermiculite, dry sand, earth, peat, carbon, or a similar material and deposit in sealed containers. Keep this chemical out of a confined space, such as a sewer, because of the possibility of an explosion, unless the sewer is designed to prevent the build-up of explosive concentrations. It may be necessary to contain and dispose of this chemical as a hazardous waste. If material or contaminated runoff enters waterways, notify downstream users of potentially contaminated waters. Contact your Department of Environmental Protection or your regional office of the federal EPA for specific recommendations. If employees are required to clean-up spills, they must be properly trained and equipped. OSHA 1910.120(q) may be applicable.

Fire Extinguishing: This chemical is a flammable liquid. Poisonous gases including nitrogen oxides are produced in fire. Use dry chemical, carbon dioxide, or foam extinguishers. Vapors are heavier than air and will collect in low areas. Vapors may travel long distances to ignition sources and flashback. Vapors in confined areas may explode when exposed to fire. Containers may explode in fire. Storage containers and parts of containers may rocket great distances, in many directions. If material or contaminated runoff enters waterways, notify downstream users of potentially contaminated waters. Notify local health and fire officials and pollution control agencies. From a secure, explosion-proof location, use water spray to cool exposed containers. If cooling streams are ineffective (venting sound increases in volume and pitch, tank discolors, or shows any signs of deforming), withdraw immediately to a secure position. If employees are expected to fight fires, they must be trained and equipped in OSHA 1910.156.

Disposal Method Suggested: Incineration; incinerator is equipped with a scrubber or thermal unit to reduce No_x emissions.

References

New Jersey Department of Health, "Hazardous Substance Fact Sheet: Diisopropylamine," Trenton, NJ (March 1986)

Diisopropyl Ether

Molecular Formula: $C_6H_{14}O$

Common Formula: $[(CH_3)_2CH]_2O$

Synonyms: Diisopropyl ether; Diisopropyl oxide; 2-Isopropoxy propane; 2-Isopropoxypropane; Isopropyl ether; 2,2'-Oxybispropane; Propane, 2,2'-oxybis-

CAS Registry Number: 108-20-3

RTECS Number: TZ5425000

DOT ID: UN 1159

EEC Number: 603-045-00-X

Regulatory Authority

- Air Pollutant Standard Set (ACGIH)[1] (Australia) (DFG)[3] (HSE)[33] (Israel) (Mexico) (OSHA)[58] (Several States)[60] (Several Canadian Provinces)
- Canada, WHMIS, Ingredients Disclosure List

Cited in U.S. State Regulations: Alaska (G), California (A, G), Connecticut (A), Florida (G), Maine (G), Massachusetts (G), Minnesota (G), Nevada (A), New Hampshire (G), New Jersey (G), North Dakota (A), Oklahoma (G), Pennsylvania (G), Rhode Island (G), Virginia (A), West Virginia (G).

Description: Diisopropyl ether, $[(CH_3)_2CH]_2O$, is a colorless liquid with a sharp sweet ether-like odor. Boiling point = 69°C. Freezing/Melting point = -60°C. Flash point = -28°C. Auto-ignition temperature = 443°C. Explosive limits: LEL = 1.4%; UEL = 7.9%.[17] Hazard Identification (based on NFPA-704 M Rating System): Health 1, Flammability 3, Reactivity 1. Slightly soluble in water.

Potential Exposure: This material is used as a solvent and chemical intermediate.

Incompatibilities: Forms explosive mixture with air. Air contact forms explosive peroxides that may explode with heat or shock. Contact with strong oxidizers or strong acids may cause a fire and explosion hazard. Attacks some plastics, rubber and coatings.

Permissible Exposure Limits in Air: The Federal Limit (OSHA PEL)[58] and the DFG MAK value[3] is 500 ppm (2,100 mg/m^3). The ACGIH recommends a TWA value of 250 ppm (1,050 mg/m^3) and a STEL of 310 ppm (1,320 mg/m^3) as has HSE.[33] The NIOSH ILDH level is 1,400 ppm [10%LEL]. Several states have set guidelines or standards for diisopropyl ether in ambient air[60] ranging from 10.5-13.2 mg/m^3 (North Dakota) to 17.5 mg/m^3 (Virginia) to 21.0 mg/m^3 (Connecticut) to 25.0 mg/m^3 (Nevada).

Determination in Air: Charcoal adsorption, workup with CS_2, analysis by gas chromatography/Flame ionization detection; NIOSH IV, Method #1618.

Permissible Concentration in Water: No criteria set.

Routes of Entry: Inhalation, ingestion, eye and/or skin contact.

Harmful Effects and Symptoms

Short Term Exposure: Diisopropyl ether irritates the eyes, skin, and respiratory tract. May affect the central nervous system. Inhaling Breathing diisopropyl ether may cause drowsiness, dizziness, and nausea. Higher levels may cause unconsciousness and even death.

Long Term Exposure: Prolonged or repeated exposures may cause drying and cracking of the skin.

Points of Attack: Respiratory system, skin.

Medical Surveillance: Consider the points of attack in preplacement and periodic physical examinations.

First Aid: If this chemical gets into the eyes, remove any contact lenses at once and irrigate immediately for at least 15 minutes, occasionally lifting upper and lower lids. Seek medical attention immediately. If this chemical contacts the skin, remove contaminated clothing and wash immediately with soap and water. Seek medical attention immediately. If this chemical has been inhaled, remove from exposure, begin rescue breathing (using universal precautions) if breathing has stopped and CPR if heart action has stopped. Transfer promptly to a medical facility. When this chemical has been swallowed, get medical attention. Give large quantities of water and induce vomiting. Do not make an unconscious person vomit.

Personal Protective Methods: Wear protective gloves and clothing to prevent any reasonable probability of skin contact. Safety equipment suppliers/manufacturers can provide recommendations on the most protective glove/clothing material for your operation. ACGIH recommends nitrile rubber, polyvinyl alcohol, and viton materials as providing good to excellent protection in gloves and clothing. All protective clothing (suits, gloves, footwear, headgear) should be clean, available each day, and put on before work. Contact lenses should not be worn when working with this chemical. Wear splash-proof chemical goggles and face shield unless full facepiece respiratory protection is worn. Employees should wash immediately with soap when skin is wet or contaminated. Provide emergency showers and eyewash.

Respirator Selection: NIOSH/OSHA: *up to 1,400 ppm:* CCRFOV [any air-purifying, full-facepiece respirator (gas mask) with a chin-style, front- or back-mounted acid gas canister]; or PAPROV [any powered, air-purifying respirator with organic vapor cartridge(s)]; or GMFOV [any air-purifying, full-facepiece respirator (gas mask) with a chin-style, front- or back-, mounted organic vapor canister]; or SA (any supplied-air respirator); or SCBAF (any self-contained breathing apparatus with a full facepiece). *Emergency or planned entry into unknown concentrations or IDLH conditions:* SCBAF:PD,PP (any self-contained breathing apparatus that has a full facepiece and is operated in a pressure-demand or other positive-pressure mode); or SAF:PD,PP: ASCBA (any supplied-air respirator that has a full facepiece and is operated in a pressure-demand or other positive-pressure mode in combination with an auxiliary self-contained breathing apparatus operated in a pressure-demand or other positive pressure mode). *Escape:* GMFOV [any air-purifying, full-facepiece respirator (gas mask) with a chin-style, front- or back-, mounted organic vapor canister]; or SCBAE (any appropriate escape-type, self-contained breathing apparatus).

Note: Substance reported to cause eye irritation or damage; may require eye protection.

Storage: Prior to working with diisopropyl ether you should be trained on its proper handling and storage. Before entering confined space where this chemical may be present, check to make sure that an explosive concentration does not exist. Diisopropyl ether must be stored to avoid contact with strong oxidizers, such as bromine, chlorine, chlorine dioxide, and nitrates, since violent reactions occur. Store in tightly closed containers in a cool, well-ventilated area away from heat, spark, or direct sunlight. Unstable peroxides may form when diisopropyl ether is in contact with the air for a long time. It may then explode by itself or when heated or subjected to shock. Sources of ignition such as smoking and open flames are prohibited where diisopropyl ether is used, handled, or stored in a manner that could create a potential fire or explosion hazard. Metal containers involving the transfer of 5 gallons or more of diisopropyl ether should be grounded and bonded. Drums must be equipped with self-closing valves, pressure vacuum bungs, and flame arresters.

Shipping: This compound requires a shipping label of: "Flammable Liquid." It falls in DOT Hazard Class 3 and Packing Group II.

Spill Handling: Evacuate and restrict persons not wearing protective equipment from area of spill or leak until cleanup is complete. Remove all ignition sources. Establish forced ventilation to keep levels below explosive limit. Absorb liquids in vermiculite, dry sand, earth, peat, carbon, or a similar material and deposit in sealed containers. Keep this chemical out of a confined space, such as a sewer, because of the possibility of an explosion, unless the sewer is designed to prevent the build-up of explosive concentrations. It may be necessary to contain and dispose of this chemical as a hazardous waste. If material or contaminated runoff enters waterways, notify downstream users of potentially contaminated waters. Contact your Department of Environmental Protection or your regional office of the federal EPA for specific recommendations. If employees are required to clean-up spills, they must be properly trained and equipped. OSHA 1910.120(q) may be applicable.

Fire Extinguishing: This chemical is a flammable liquid. Poisonous gases are produced in fire. Use dry chemical, carbon dioxide, or alcohol foam extinguishers. Vapors are heavier than air and will collect in low areas. Vapors may travel long distances to ignition sources and flashback. Vapors in confined areas may explode when exposed to fire. Containers may explode in fire. Storage containers and parts of containers may rocket great distances, in many directions. If material or contaminated runoff enters waterways, notify downstream users of potentially contaminated waters. Notify local health and fire officials and pollution control agencies. From a secure, explosion-proof location, use water spray to cool exposed containers. If cooling streams are ineffective (venting sound increases in volume and pitch, tank discolors, or shows any signs of deforming), withdraw immediately to a secure position. If employees are expected to fight fires, they must be trained and equipped in OSHA 1910.156.

Disposal Method Suggested: Concentrated waste containing no peroxides-discharge liquid at a controlled rate near a pilot flame. Concentrated waste containing peroxide-perforation of a container of the waste from a safe distance followed by open burning.

References

New Jersey Department of Health, "Hazardous Substance Fact Sheet: Diisopropyl Ether," Trenton, NJ (February 1986)

Dimefox

Molecular Formula: $C_4H_{12}FN_2OP$

Synonyms: BFPO; Bis(dimethylamido) fluorophosphate; Bis(dimethylamido)fluorophosphine oxide; Bis(dimethylamido)phosphoryl fluoride; Bis(dimethylamino) fluorophosphate; Bisdimethylaminofluorophosphine oxide; BPF; CR 409; DIFO; DMF; ENT 19,109; Fluophosphoric acid di(dimethylamide); Fluorure de *N,N,N',N'*-tetramethyle phosphoro-diamide (French); Hanane; Pestox 14; Pestox IV; Pestox XIV; T-2002; Terra-Systam; Terra-Sytam; Terrasytum; *N,N,N',N'*-Tetramethyl-diamido-fosforzuur-fluoride (Dutch); Tetramethyldiamidophosphoric fluoride; *N,N,N',N'*-Tetramethyl-diamido-phosphorsaeure-fluorid (German); *N,N,N,N*-Tetramethylphosphorodiamidic fluoride; Tetramethylphosphorodiamidic fluoride; *N,N,N',N'*-Tetrametil-fosforodiammido-fluoruro (Italian); Tetra systam; TL 792 Wacker 14/10

CAS Registry Number: 115-26-4

RTECS Number: TD4025000

DOT ID: UN 3018

Regulatory Authority

- Very Toxic Substance (World Bank)[15]
- Superfund/EPCRA 40CFR355, Appendix B Extremely Hazardous Substances: TPQ = 500 lb (227 kg)
- Superfund/EPCRA 40CFR302.4 Reportable Quantity (RQ): EHS, 1 lb (0.454 kg)

Cited in U.S. State Regulations: California (A, G), Florida (G), Massachusetts (G), New Jersey (G), Oklahoma (G), Pennsylvania (G).

Description: Dimefox, $C_4H_{12}FN_2OP$, is a clear liquid. Boiling point = 86°C under 15 mm Hg pressure. Hazard Identification (based on NFPA-704 M Rating System): Health 4, Flammability 1, Reactivity 1. Soluble in water.

Potential Exposure: Those involved in the manufacture, formulation and application of this pesticide.

Incompatibilities: Strong oxidants, strong acids, halogens.

Permissible Exposure Limits in Air: No standards set.

Determination in Air: OSHA versatile sampler-2; Toluene/Acetone; Gas chromatography/Flame photometric detection for sulfur, nitrogen, or phosphorus; NIOSH Method IV Method #5600, Organophosphorus Pesticides.

Permissible Concentration in Water: No criteria set.

Harmful Effects and Symptoms

Short Term Exposure: Organic phosphorus insecticides are absorbed by the skin, as well as by the respiratory and gastrointestinal tracts. They are cholinesterase inhibitors. Symptoms of exposure include headache, giddiness, blurred vision, nervousness, weakness, nausea, cramps, diarrhea, and discomfort in the chest. Signs include sweating, tearing, salivation, vomiting, cyanosis, convulsions, coma, loss of reflexes and loss of sphincter control. This material is extremely toxic; the probable oral lethal dose (human) is 5 – 50 mg/kg, or 7 drops to 1 teaspoonful for a 150-lb. person. Death may occur from respiratory arrest. Hazards of vapor toxicity are high.

Long Term Exposure: Cholinesterase inhibitor; cumulative effect is possible. Dimefox may damage the nervous system with repeated exposure, resulting in convulsions, respiratory failure. May cause liver damage.

Points of Attack: Respiratory system, lungs, central nervous system, cardiovascular system, skin, eyes, plasma and red blood cell cholinesterase.

Medical Surveillance: Before employment and at regular times after that, the following are recommended: Plasma and red blood cell cholinesterase levels (tests for the enzyme poisoned by this chemical). If exposure stops, plasma levels return to normal in 1 – 2 weeks while red blood cell levels may be reduced for 1 – 3 months.

When cholinesterase enzyme levels are reduced by 25% or more below preemployment levels, risk of poisoning is increased, even if results are in lower ranges of "normal." Reassignment to work not involving organophosphate or carbamate pesticides is recommended until enzyme levels recover. If symptoms develop or overexposure occurs, repeat the above tests as soon as possible and get an exam of the nervous system. Also consider complete blood count. Consider chest x-ray following acute overexposure. Do not drink any alcoholic beverages before or during use. Alcohol promotes absorption of organic phosphates.

First Aid: If this chemical gets into the eyes, remove any contact lenses at once and irrigate immediately for at least 15 minutes, occasionally lifting upper and lower lids. Seek medical attention immediately. If this chemical contacts the skin, remove contaminated clothing and wash immediately with soap and water. Speed in removing material from skin is of extreme importance. Shampoo hair promptly if contaminated. Seek medical attention immediately. If this chemical has been inhaled, remove from exposure, begin rescue breathing (using universal precautions) if breathing has stopped and CPR if heart action has stopped. Transfer promptly to a medical facility. When this chemical has been swallowed, get medical attention. Give large quantities of water and induce vomiting. Do not make an unconscious person vomit. Keep victim quiet and maintain normal body temperature. Effects may be delayed; keep victim under observation.

Personal Protective Methods: Wear protective gloves and clothing to prevent any reasonable probability of skin contact. Safety equipment suppliers/manufacturers can provide recommendations on the most protective glove/clothing material for your operation.. All protective clothing (suits, gloves, footwear, headgear) should be clean, available each day, and put on before work. Contact lenses should not be worn when working with this chemical. Wear splash-proof chemical goggles and face shield unless full facepiece respiratory protection is worn. Employees should wash immediately with soap when skin is wet or contaminated. Provide emergency showers and eyewash.

Respirator Selection: *At any detectable concentration:* SCBAF:PD,PP (any MSHA/NIOSH approved self-contained breathing apparatus that has a full facepiece and is operated in a pressure-demand or other positive-pressure mode); or SAF:PD,PP:ASCBA (any supplied-air respirator that has a full facepiece and is operated in a pressure-demand or other positive-pressure mode in combination with an auxiliary, self-contained breathing apparatus operated in a pressure-demand or other positive pressure mode).

Storage: Prior to working with dimefox you should be trained on its proper handling and storage. Store in tightly closed containers in a cool, well ventilated area away from strong oxidizers, strong acids, halogens. Metal containers involving the transfer of this chemical should be grounded and bonded. Where possible, automatically pump liquid from drums or other storage containers to process containers. Drums must be equipped with self-closing valves, pressure vacuum bungs, and flame arresters. Use only non-sparking tools and equipment, especially when opening and closing containers of this chemical. Sources of ignition such as smoking and open flames, are prohibited where this chemical is used, handled, or stored in a manner that could create a potential fire or explosion hazard. Wherever this chemical is used, handled, manufactured, or stored, use explosion-proof electrical equipment and fittings.

Shipping: Organophosphorus pesticides, liquid, toxic, n.o.s. falls in Hazard Class 6.1 and Dimefox in Packing Group I. It requires a "Poison" label.

Spill Handling: Evacuate and restrict persons not wearing protective equipment from area of spill or leak until cleanup is complete. Remove all ignition sources. Ventilate area of spill or leak. Absorb liquids in vermiculite, dry sand, earth, peat, carbon, or a similar material and deposit in sealed containers. Keep this chemical out of a confined space, such as a sewer, because of the possibility of an explosion, unless the sewer is designed to prevent the build-up of explosive concentrations. It may be necessary to contain and dispose of this chemical as a hazardous waste. If material or contaminated runoff enters waterways, notify downstream users of potentially contaminated waters. Contact your Department of Environmental Protection or your regional office of the federal EPA for specific recommendations. If employees are required to clean-up spills, they must be properly trained and equipped. OSHA 1910.120(q) may be applicable.

Fire Extinguishing: This chemical is a combustible liquid. Poisonous gases including phosphorus oxides, nitrogen oxides, and fluorine are produced in fire. Use dry chemical, carbon dioxide, or foam extinguishers. DO NOT use halogen extinguishers. Vapors are heavier than air and will collect in low areas. Vapors may travel long distances to ignition sources and flashback. Vapors in confined areas may explode when exposed to fire. Containers may explode in fire. Storage containers and parts of containers may rocket great distances, in many directions. If material or contaminated runoff enters waterways, notify downstream users of potentially contaminated waters. Notify local health and fire officials and pollution control agencies. From a secure, explosion-proof location, use water spray to cool exposed containers. If cooling

streams are ineffective (venting sound increases in volume and pitch, tank discolors, or shows any signs of deforming), withdraw immediately to a secure position. If employees are expected to fight fires, they must be trained and equipped in OSHA 1910.156.

Disposal Method Suggested: Small batches may be mixed with sand and acid added in a pit or trench in clay soil. Larger quantities may be incinerated in a unit with effluent flue gas scrubbing.[22] In accordance with 40CFR165 recommendations for the disposal of pesticides and pesticide containers. Must be disposed properly by following package label directions or by contacting your state pesticide or environmental control agency or by contacting your regional EPA office.

References

U.S. Environmental Protection Agency, "Chemical Profile: Dimefox," Washington, DC, Chemical Emergency Preparedness Program (November 30, 1987)

New Jersey Department of Health and Senior Services, "Hazardous Substance Fact Sheet, Dimefox." Trenton, NJ (February 1999)

Dimethoate

Molecular Formula: $C_5H_{12}NO_3PS_2$

Common Formula: $H_3COP(S)(OCH_3)SCH_2CONHCH_3$

Synonyms: Acetic acid, *O,O*-dimethyldithiophosphoryl-, *N*-monomethylamide salt; *O,O*-Dimethyl *S*-(*N*-methylcarbamoylmethyl) dithiophosphate; Phosphamide; Phosphorodithioic acid, *O,O*-dimethyl *S*-[2-(methylamino)-2-oxoethyl] ester; Rogor

CAS Registry Number: 60-51-5

RTECS Number: TE1750000

DOT ID: NA 2783

EEC Number: 015-051-00-4

Regulatory Authority

- Banned or Severely Restricted (USEPA) (UN)[13]
- Air Pollutant Standard Set (former USSR)[35][43]
- EPA Hazardous Waste Number (RCRA No.): P044
- RCRA, 40CFR261, Appendix 8 Hazardous Constituents
- RCRA 40CFR264, Appendix 9; TSD Facilities Ground Water Monitoring List. Suggested test method(s) (PQL μg/l): 8270 (10)
- Superfund/EPCRA 40CFR355, Appendix B Extremely Hazardous Substances: TPQ = 500/10,000 lb (227/4,540 kg)
- Superfund/EPCRA 40CFR302.4 Reportable Quantity (RQ): CERCLA, 10 lb (4.54 kg)
- EPCRA Section 313 Form R *de minimis* concentration reporting level: 1.0%
- U.S. DOT Regulated Marine Pollutant (49CFR172.101, Appendix B)
- Canada Drinking Water Quality, 0.02 mg/l MAC

Cited in U.S. State Regulations: California (W), Florida (G), Kansas (G), Louisiana (G), Massachusetts (G), Michigan (G), New Jersey (G), Pennsylvania (G), Vermont (G), Virginia (G), Washington (G), Wisconsin (G, W).

Description: Dimethoate, $C_5H_{12}NO_3PS_2$, is a white crystalline solid with a camphor-like odor. Boiling point = 117°C. Freezing/Melting point = 52°C. Hazard Identification (based on NFPA-704 M Rating System): Health 3, Flammability 2, Reactivity 0. Slightly soluble in water.

Potential Exposure: Dimethoate is a contact and systemic organophosphate insecticide effective against a broad range of insects and mites when applied on a wide range of crops. It has not been produced in the U.S. since 1982.

Incompatibilities: Strong bases (alkalis). Do not store solid above 77 – 86°F (25 – 30°C). Liquid solutions must be stored above 45°F (7°C).

Permissible Exposure Limits in Air: The former USSR[35][43] has set air MAC in workplace air of 0.5 mg/m^3 and limits in the ambient air of residential areas of 0.003 mg/m^3 on either a momentary or a daily average basis.

Determination in Air: OSHA versatile sampler-2; Toluene/Acetone; Gas chromatography/Flame photometric detection for sulfur, nitrogen, or phosphorus; NIOSH Method IV Method #5600, Organophosphorus Pesticides.

Permissible Concentration in Water: A MAC in water bodies used for domestic purposes of 0.03 mg/l has been set by the former USSR.[35][43] States which have set guidelines for Dimethoate in drinking water include California at 140 μg/l and Wisconsin at 10 μg/l.

Harmful Effects and Symptoms

Short Term Exposure: Acute exposure to Dimethoate may produce the following signs and symptoms: pinpoint pupils, blurred vision, headache, dizziness, muscle spasms, and profound weakness. Vomiting, diarrhea, abdominal pain, seizures, and coma may also occur. The heart rate may decrease following oral exposure or increase following dermal exposure. Hypotension (low blood pressure) and chest pain may be noted. Hypertension (high blood pressure) is not uncommon. Respiratory effects may include dyspnea (shortness of breath), respiratory depression, and respiratory paralysis. Psychosis may occur.

Dimethoate is very toxic; the probably oral lethal dose in humans is between 50 – 500 mg/kg, or between 1 teaspoon and 1 ounce for a 70 kg (150 lb.) person. Dimethoate is a cholinesterase inhibitor, meaning it affects the central nervous system. Death is due to respiratory arrest arising from failure of respiratory center, paralysis of respiratory muscles, intense bronchoconstriction or all three. Dimethoate is a mutagen. Mutagens may have a cancer risk. All contact with this chemical should be reduced to the lowest possible level.

Long Term Exposure: Repeated or prolonged contact with skin may cause dermatitis. Cholinesterase inhibitor;

cumulative effect is possible. This chemical may damage the nervous system with repeated exposure, resulting in convulsions, respiratory failure. May cause liver damage. Animal tests indicate that this chemical possibly causes toxic effects upon human reproduction.

Points of Attack: Respiratory system, lungs, central nervous system, cardiovascular system, skin, eyes, plasma and red blood cell cholinesterase.

Medical Surveillance: Before employment and at regular times after that, the following are recommended: Plasma and red blood cell cholinesterase levels (tests for the enzyme poisoned by this chemical). If exposure stops, plasma levels return to normal in 1 – 2 weeks while red blood cell levels may be reduced for 1 – 3 months.

When cholinesterase enzyme levels are reduced by 25% or more below preemployment levels, risk of poisoning is increased, even if results are in lower ranges or "normal." Reassignment to work not involving organophosphate or carbamate pesticides is recommended until enzyme levels recover. If symptoms develop or overexposure occurs, repeat the above tests as soon as possible and get an exam of the nervous system. Do not drink any alcoholic beverages before or during use. Alcohol promotes absorption of organic phosphates.

First Aid: If this chemical gets into the eyes, remove any contact lenses at once and irrigate immediately for at least 15 minutes, occasionally lifting upper and lower lids. Seek medical attention immediately. If this chemical contacts the skin, remove contaminated clothing and wash immediately with soap and water. Speed in removing material from skin is of extreme importance. Shampoo hair promptly if contaminated. Seek medical attention immediately. If this chemical has been inhaled, remove from exposure, begin rescue breathing (using universal precautions) if breathing has stopped and CPR if heart action has stopped. Transfer promptly to a medical facility. When this chemical has been swallowed, get medical attention. Give large quantities of water and induce vomiting. Do not make an unconscious person vomit. Effects may be delayed; keep victim under medical observation. Obtain authorization and/or further instructions from the local hospital for administration of an antidote or performance of other invasive procedures.

Note to Physician: 1,1'-trimethylenebis(4-formylpyridinium bromide)dioxime (a.k.a TMB-4 DIBROMIDE and TMV-4) have been used as an antidote for organophosphate poisoning.

Personal Protective Methods: Wear protective gloves and clothing to prevent any reasonable probability of skin contact. Safety equipment suppliers/manufacturers can provide recommendations on the most protective glove/clothing material for your operation.. All protective clothing (suits, gloves, footwear, headgear) should be clean, available each day, and put on before work. Contact lenses should not be worn when working with this chemical. Wear dust-proof chemical goggles and face shield unless full facepiece respiratory protection is worn. Employees should wash immediately with soap when skin is wet or contaminated. Provide emergency showers and eyewash.

Respirator Selection: Where the potential exists for exposure to Dimethoate, use a MSHA/NIOSH approved supplied-air respirator with a full facepiece operated in the positive pressure mode or with a full facepiece, hood, or helmet in the continuous flow mode, or use a MSHA/NIOSH approved self-contained breathing apparatus with a full facepiece operated in pressure-demand or other positive pressure mode.

Storage: Store in tightly closed containers in a cool, well-ventilated area away from heat and strong bases. Do not store *solid* Dimethoate above 25 – 30°C/77 – 86°F. However, *liquid* solutions of Dimethoate must be stored above 7°C/45°F. Keep away from sources of heat, flames, or spark-generating equipment. Unstable in alkaline solution. Hydrolyzed by aqueous alkali. Stable in aqueous solutions. The compound is stable for 2 years under environmental conditions if stored in undamaged, original containers.

Shipping: "Organophosphorus pesticides, solid, toxic, n.o.s." would cover Dimethoate and fall in DOT Hazard Class 6.1. Dimethoate would fall in Packing Group I. It requires a "Poison" label. The limits on passenger aircraft or railcar shipment is 5 kg; on cargo aircraft shipment is 50 kg.

Spill Handling: Evacuate persons not wearing protective equipment from area of spill or leak until clean-up is complete. Remove all ignition sources. Collect powdered material in the most convenient and safe manner and deposit in sealed containers. Ventilate area after clean-up is complete. It may be necessary to contain and dispose of this chemical as a hazardous waste. If material or contaminated runoff enters waterways, notify downstream users of potentially contaminated waters. Contact your Department of Environmental Protection or your regional office of the federal EPA for specific recommendations. If employees are required to clean-up spills, they must be properly trained and equipped. OSHA 1910.120(q) may be applicable.

Fire Extinguishing: This chemical is a combustible solid. Use dry chemical, carbon dioxide, water spray, or alcohol foam extinguishers. Poisonous gases are produced in fire including nitrogen oxides, sulfur oxides, and phosphorus oxides. If material or contaminated runoff enters waterways, notify downstream users of potentially contaminated waters. Notify local health and fire officials and pollution control agencies. Containers may explode in fire. From a secure, explosion-proof location, use water spray to cool exposed containers. If cooling streams are ineffective (venting sound increases in volume and pitch, tank discolors, or shows any signs of deforming), withdraw immediately to a secure position. If employees are expected to fight fires, they must be trained and equipped in OSHA 1910.156.

Disposal Method Suggested: May be mixed with alkalis and then buried.[22] In accordance with 40CFR165 recommendations for the disposal of pesticides and pesticide containers. Must be disposed properly by following package label directions or by contacting your state pesticide or environmental control agency or by contacting your regional EPA office. Consult with environmental regulatory agencies for guidance on acceptable disposal practices. Generators of waste containing this contaminant (≥100 kg/mo) must conform with EPA regulations governing storage, transportation, treatment, and waste disposal.

References

U.S. Environmental Protection Agency, "Chemical Profile: Dimethoate," Washington, DC, Chemical Emergency Preparedness Program (November 30, 1987)

New Jersey Department of Health, "Hazardous Substance Fact Sheet: Dimethoate," Trenton, NJ (February 1999)

3,3'-Dimethoxybenzidine

Molecular Formula: $C_{14}H_{16}N_2O_2$

Common Formula: $H_2N(OCH_3)C_6H_3{\cdot}C_6H_3(OCH_3)NH_2$

Synonyms: Acetamine diazo black RD; Amacel developed navy SD; Azofix blue B salt; Azogene fast blue B; Azogene fast blue base; Benzidine, 3,3'-dimethoxy-; 4,4'-Bi-*o*-anisidine; (1,1'-Biphenyl)-4,4'-diamine, 3,3'-dimethoxy-; Blue base IRGA B; Blue base NB; Blue BN base; Brentamine fast blue B base; Cellitazol B; C.I. 24110; C.I. Azoic diazo component 48; Cibacete diazo navy blue 2B; C.I. Disperse Black 6; Diacelliton Fast Grey G; Diacel navy DC; 4,4'-Diamino-3,3'-dimethoxy-1,1'-biphenyl; *o*-Dianisidin (Czech, German); *o*-Dianisidina (Italian); Dianisidina (Italian, Spanish); *o,o*'-Dianisidine; *o*-Dianisidine; 3,3'-Dianisidine; Dianisidine; Diato blue base B; 3,3'-Dimethoxybenzidin (Czech); 3,3'-Dimethoxybezidine; 3,3'-Dimethoxy-4,4'-diaminodiphenyl; 3,3'-Dimetossibenzodina (Italian); 3,3'-Dimetoxibenzidina (Spanish); Fast blue base B; Fast blue B base; Fast blue DSC base; Hiltonil fast blue B base; Hiltosal fast blue B salt; Hindasol blue B salt; Kako blue B salt; kayaku blue B base; Lake blue B base; Meisei teryl diazo blue HR; Mitsui blue B base; Naphthani! blue B base; Neutrosel navy BN; Sanyo fast blue salt B; Setacyl diazo navy R; Spectrolene blue B; DMOB.

CAS Registry Number: 119-90-4

RTECS Number: DD0875000

DOT ID: UN 3077

Regulatory Authority

- Carcinogen (Animal Positive) (IARC)[9] (DFG)[3]
- Banned or Severely Restricted (Germany, Sweden) (UN)[35]
- Air Pollutant Standard Set (Several States)[60]
- Clean Air Act: Hazardous Air Pollutants (Title I, Part A, Section 112)
- EPA Hazardous Waste Number (RCRA No.): U091
- RCRA, 40CFR261, Appendix 8 Hazardous Constituents
- Superfund/EPCRA 40CFR302.4 Reportable Quantity (RQ): CERCLA, 100 lb (45.4 kg)
- EPCRA Section 313 Form R *de minimis* concentration reporting level: 0.1%
- Canada, WHMIS, Ingredients Disclosure List

Cited in U.S. State Regulations: California (A, G), Florida (G), Illinois (G), Kansas (G), Louisiana (G), Maryland (G), Massachusetts (G), Minnesota (G), New Hampshire (G), New Jersey (G), New York (A), North Dakota (A), Pennsylvania (G), South Carolina (A), Vermont (G), Virginia (G), Washington (G), Wisconsin (G).

Description: DMOB, $C_{14}H_{16}N_2O_2$, is a colorless crystalline material which may turn purple on standing in air. Freezing/Melting point = 137 – 138°C. Flash point = 206°C. Hazard Identification (based on NFPA-704 M Rating System): Flammability 1, Reactivity 0. Slightly soluble in water.

Potential Exposure: DMOB or its dihydrochloride is used principally as a chemical intermediate for the production of azo dyes. About 30% of DMOB is used as a chemical intermediate in the production of o-dianisidine diisocyanate (3,3'-dimethoxy-4,4'-diisocyanate-biphenyl) that is used in adhesive systems and also as a component of polyurethane elastomers and resins. DMOB is used as a dye itself for leather, paper, plastics, rubber, and textiles. DMOB has also been used in the detection of metals, thiocyanates, and nitrites.

Incompatibilities: Oxidizers.

Permissible Exposure Limits in Air: Guidelines or standards for dimethoxybenzidine in ambient air have been set[60] by North Dakota at zero, by New York at 0.2 μg/m³ and by South Carolina at 0.3 μg/m³.

Permissible Concentration in Water: No criteria set.

Routes of Entry: Human exposure to DMOB is possible through inhalation of dye particles from equipment vent systems and through skin absorption from the finished dye product, textile processing, mixing operations, or packaging process.

Harmful Effects and Symptoms

Short Term Exposure: May cause eye and skin irritation.

Long Term Exposure: There is sufficient evidence for the carcinogenicity of 3,3'-dimethoxybenzidine (o-dianisidine) in experimental animals.[1] 3,3'-Dimethoxydenzidine administered by stomach tube was carcinogenic in rats, producing tumors at various sites, including intestinal, skin and Zymbal's gland carcinomas. The findings in hamsters exposed to o-dianisidine in the feed also suggest carcinogenicity. The evidence for the carcinogenicity of o-dianisidine in humans in inadequate. Most of the workers exposed to this substance were also exposed to related amines, such as benzidine, which are strongly associated with urinary bladder cancer in humans.

Points of Attack: Bladder, breast, skin.

Medical Surveillance: Before beginning employment and at regular times after that, for those with frequent or potentially high exposures, the following are recommended: Exam of the breast, bladder, intestine, skin, stomach and ovary.

First Aid: If this chemical gets into the eyes, remove any contact lenses at once and irrigate immediately for at least 15 minutes, occasionally lifting upper and lower lids. Seek medical attention immediately. If this chemical contacts the skin, remove contaminated clothing and wash immediately with soap and water. Seek medical attention immediately. If this chemical has been inhaled, remove from exposure, begin rescue breathing (using universal precautions) if breathing has stopped and CPR if heart action has stopped. Transfer promptly to a medical facility. When this chemical has been swallowed, get medical attention. Give large quantities of water and induce vomiting. Do not make an unconscious person vomit.

Personal Protective Methods: Wear protective gloves and clothing to prevent any reasonable probability of skin contact. Safety equipment suppliers/manufacturers can provide recommendations on the most protective glove/clothing material for your operation.. All protective clothing (suits, gloves, footwear, headgear) should be clean, available each day, and put on before work. Contact lenses should not be worn when working with this chemical. Wear dust-proof chemical goggles and face shield unless full facepiece respiratory protection is worn. Employees should wash immediately with soap when skin is wet or contaminated. Provide emergency showers and eyewash.

Respirator Selection: At any exposure level, use a MSHA/NIOSH approved supplied-air respirator with a full facepiece operated in the positive pressure mode or with a full facepiece, hood, or helmet in the continuous flow mode, or use a MSHA/NIOSH approved self-contained breathing apparatus with a full facepiece operated in pressure-demand or other positive pressure mode.

Storage: 3,3'-Dimethoxybenzidine may react with oxidizers (such as perchlorates, peroxides, permanganates, chlorates and nitrates). Store in tightly closed containers in a cool, well-ventilated area. Avoid exposure to light. A regulated, marked area should be established where this chemical is handled, used, or stored in compliance with OSHA standard 1910.1045.

Shipping: Environmentally hazardous solid, n.o.s. Hazard Class: 9. Label: "Class 9." Packing Group: III.

Spill Handling: Evacuate persons not wearing protective equipment from area of spill or leak until clean-up is complete. Remove all ignition sources. Collect powdered material in the most convenient and safe manner and deposit in sealed containers. Ventilate area after clean-up is complete. It may be necessary to contain and dispose of this chemical as a hazardous waste. If material or contaminated runoff enters waterways, notify downstream users of potentially contaminated waters. Contact your Department of Environmental Protection or your regional office of the federal EPA for specific recommendations. If employees are required to clean-up spills, they must be properly trained and equipped. OSHA 1910.120(q) may be applicable.

Fire Extinguishing: 3,3'-Dimethoxybenzidine may burn, but does not readily ignite. Use dry chemical, CO_2, water spray, foam extinguishers or water spray extinguishers. Poisonous gases are produced in fire including nitrogen oxides and hydrogen chloride. If material or contaminated runoff enters waterways, notify downstream users of potentially contaminated waters. Notify local health and fire officials and pollution control agencies. From a secure, explosion-proof location, use water spray to cool exposed containers. If cooling streams are ineffective (venting sound increases in volume and pitch, tank discolors, or shows any signs of deforming), withdraw immediately to a secure position. If employees are expected to fight fires, they must be trained and equipped in OSHA 1910.156.

Disposal Method Suggested: Incineration.[22] Consult with environmental regulatory agencies for guidance on acceptable disposal practices. Generators of waste containing this contaminant (≥100 kg/mo) must conform with EPA regulations governing storage, transportation, treatment, and waste disposal.

References

Sax N. I., Ed., "Dangerous Properties of Industrial Materials Report," 7, No. 2, 44–47 (1987)

New Jersey Department of Health, "Hazardous Substance Fact Sheet: 3,3'-Dimethoxybenzidine," Trenton, NJ (January 15, 1988)

N,N-Dimethylacetamide

Molecular Formula: C_4H_9NO

Common Formula: $CH_3CON(CH_3)_2$

Synonyms: Acetamide *N,N*-dimethyl; Acetdimethylamide; Acetic acid, dimethylamide; Acetyldimethylamine; *N,N*-Dimethyl acetamide; Dimethylacetamide; Dimethylacetone amide; Dimethylamide acetate; DMA; DMAC

CAS Registry Number: 127-19-5

RTECS Number: AB7700000

DOT ID: No citation.

EEC Number: 616-011-00-4

Regulatory Authority

- Air Pollutant Standard Set (ACGIH)[1] (Australia) (DFG)[3] (HSE)[33] (Israel) (Mexico) (OSHA)[58] (Several States)[60] (Several Canadian Provinces)
- Canada, WHMIS, Ingredients Disclosure List

Cited in U.S. State Regulations: Alaska (G), Connecticut (A), Florida (G), Illinois (G), Maine (G), Massachusetts (G), Nevada (A), New Hampshire (G, W), New Jersey (G), New York (G), North Dakota (A), Pennsylvania (G), Rhode Island (G), Virginia (A), West Virginia (G).

Description: Dimethylacetamide, $CH_3CON(CH_3)_2$, is a colorless, nonvolatile liquid with a faint ammonia-like odor. The odor threshold is 47 ppm. Boiling point = 165°C. Freezing/

Melting point = -20°C. Flash point = 70°C (oc). Autoignition temperature = 490°C. The explosive limits are: LEL = 1.8% @ 100°C; UEL = 11.5% @ 160°C. Hazard Identification (based on NFPA-704 M Rating System): Health 2, Flammability 2, Reactivity 0. Soluble in water.

Potential Exposure: Dimethylacetamide is used primarily as a solvent for synthetic and natural resins, especially acrylic fibers and spandex. About 15% of dimethylacetamide production is used to make alkyl (C_{12-14}) dimethylamine oxide (a surfactant) and rubber chemicals. Dimethylacetamide is also used as an extraction solvent for butadiene manufacture.

Incompatibilities: Forms explosive mixture with air. Incompatible with non-oxidizing mineral acids, strong acids, ammonia, isocyanates, phenols, cresols, halogenated compounds above 185°F/85°C. Incompatible with carbon tetrachloride, other halogenated compounds when in contact with iron and oxidizers. Attacks some plastics, rubber and coatings.

Permissible Exposure Limits in Air: The Federal Limit (OSHA PEL),[58] the DFG MAK,[3] the HSE TWA[33] and the recommended TWA value is 10 ppm (35 mg/m^3) and the STEL value set by HSE[33] is 15 ppm (50 mg/m^3). Australia, Mexico, Israel, and the Canadian provinces of Alberta, BC, Ontario, and Quebec set the same TWA as OSHA and Alberta, BC, Mexico set a STEL of 15 ppm. The NIOSH IDLH level is 300 ppm. The notation "skin" indicates the possibility of cutaneous absorption. Several states have set guidelines or standards for DMA in ambient air[60] ranging from 350 μg/m^3 (North Dakota) to 429 μg/m^3 (Nevada) to 600 μg/m^3 (Virginia) to 700 μg/m^3 (Connecticut).

Determination in Air: Adsorption on silica, workup with methanol, analysis by gas chromatography/Flame ionization; NIOSH IV Method #2004.

Permissible Concentration in Water: No criteria set.

Routes of Entry: Inhalation of vapor and absorption through intact skin, ingestion and eye and skin contact.

Harmful Effects and Symptoms

Short Term Exposure: Irritates the eyes, skin, nose and throat.

Long Term Exposure: Repeated skin contact with the liquid can cause irritation and destroys the skin's natural oils. DMAC may affect the central nervous system and liver. Liver damage with nausea and/or jaundice can occur from skin or breathing exposure @ 400 ppm levels. Repeated or high exposures can cause depression, lethargy, hallucinations and personality changes. There is limited evidence that DMAC damages the developing fetus.

Points of Attack: Skin, liver, central nervous system.

Medical Surveillance: If symptoms develop or overexposure is suspected, the following may be useful: Liver function tests. An exam of the nervous system emphasizing personality changes.

First Aid: If this chemical gets into the eyes, remove any contact lenses at once and irrigate immediately for at least 15 minutes, occasionally lifting upper and lower lids. Seek medical attention immediately. If this chemical contacts the skin, remove contaminated clothing and wash immediately with soap and water. Seek medical attention immediately. If this chemical has been inhaled, remove from exposure, begin rescue breathing (using universal precautions) if breathing has stopped and CPR if heart action has stopped. Transfer promptly to a medical facility. When this chemical has been swallowed, get medical attention. Give large quantities of water and induce vomiting. Do not make an unconscious person vomit.

Personal Protective Methods: Wear protective gloves and clothing to prevent any reasonable probability of skin contact. Safety equipment suppliers/manufacturers can provide recommendations on the most protective glove/clothing material for your operation.. Saranex is among the recommended protective materials. All protective clothing (suits, gloves, footwear, headgear) should be clean, available each day, and put on before work. Contact lenses should not be worn when working with this chemical. Wear splash-proof chemical goggles and face shield unless full facepiece respiratory protection is worn. Employees should wash immediately with soap when skin is wet or contaminated. Provide emergency showers and eyewash.

Respirator Selection: NIOSH/OSHA: *100 ppm:* SA (any supplied-air respirator). *250 ppm*: SA:CF (any supplied-air respirator operated in a continuous-flow mode). *400 ppm*: SCBAF (any self-contained breathing apparatus with a full facepiece); or SAF (any supplied-air respirator with a full facepiece). *Emergency or planned entry into unknown concentrations or IDLH conditions*: SCBAF:PD,PP (any self-contained breathing apparatus that has a full facepiece and is operated in a pressure-demand or other positive-pressure mode); or SAF:PD,PP:ASCBA (any supplied-air respirator that has a full facepiece and is operated in a pressure-demand or other positive-pressure mode in combination with an auxiliary self-contained breathing apparatus operated in a pressure-demand or other positive pressure mode). *Escape:* GMFOV [any air-purifying, full-facepiece respirator (gas mask) with a chin-style, front- or back-, mounted organic vapor canister]; or SCBAE (any appropriate escape-type, self-contained breathing apparatus).

Note: Substance causes eye irritation or damage; eye protection needed.

Storage: Prior to working with DMAC you should be trained on its proper handling and storage. Before entering confined space where this chemical may be present, check to make sure that an explosive concentration does not exist. Dimethyl acetamide must be stored to avoid contact with carbon tetrachloride and other halogenated compounds when in contact with iron since violent reactions occur. Store in tightly closed containers in a cool, well-ventilated area. Sources of ignition such as smoking and open flames are prohibited where this chemical is used, handled, or stored in a manner that could create a potential fire or explosion hazard. Metal containers involving the transfer of 5 gallons or more of this chemical should be grounded and bonded. Drums must be equipped with self-closing valves, pressure vacuum bungs, and flame arresters. Use only nonsparking tools and equipment, especially when opening and closing containers of this chemical.

Shipping: This material is not cited in DOT's Performance-Oriented Packaging Standards.[19]

Spill Handling: Evacuate and restrict persons not wearing protective equipment from area of spill or leak until cleanup is complete. Remove all ignition sources. Establish forced ventilation to keep levels below explosive limit. Absorb liquids in vermiculite, dry sand, earth, peat, carbon, or a similar material and deposit in sealed containers. Keep this chemical out of a confined space, such as a sewer, because of the possibility of an explosion, unless the sewer is designed to prevent the build-up of explosive concentrations. It may be necessary to contain and dispose of this chemical as a hazardous waste. If material or contaminated runoff enters waterways, notify downstream users of potentially contaminated waters. Contact your Department of Environmental Protection or your regional office of the federal EPA for specific recommendations. If employees are required to clean-up spills, they must be properly trained and equipped. OSHA 1910.120(q) may be applicable.

Fire Extinguishing: This chemical is a combustible liquid. Poisonous gases including nitrogen oxides and carbon monoxide are produced in fire. Use dry chemical, carbon dioxide, or polymer foam extinguishers. Vapors are heavier than air and will collect in low areas. Vapors may travel long distances to ignition sources and flashback. Vapors in confined areas may explode when exposed to fire. Containers may explode in fire. Storage containers and parts of containers may rocket great distances, in many directions. If material or contaminated runoff enters waterways, notify downstream users of potentially contaminated waters. Notify local health and fire officials and pollution control agencies. From a secure, explosion-proof location, use water spray to cool exposed containers. If cooling streams are ineffective (venting sound increases in volume and pitch, tank discolors, or shows any signs of deforming), withdraw immediately to a secure position. If employees are expected to fight fires, they must be trained and equipped in OSHA 1910.156.

Disposal Method Suggested: Controlled incineration (incinerator is equipped with a scrubber or thermal unit to reduce nitrogen oxide emissions).

References

National Institute for Occupational Safety and Health, Information Profiles on Potential Occupational Hazards – Single Chemicals: N,N-Dimethyl Acetamide, 56–64, Report RT 79-607, Rockville, MD (December 1979)

Sax N. I., Ed., "Dangerous Properties of Industrial Materials Report," 1, No. 5, 50–51 (1981). New York, Van Nostrand Reinhold Co. (1981)

New Jersey Department of Health, "Hazardous Substance Fact Sheet: Dimethylacetamide," Trenton, NJ (December 1998)

New York State Department of Health, "Chemical Fact Sheet: N,N-Dimethylacetamide," Albany, NY, Bureau of Toxic Substance Assessment (March 1986)

Dimethylamine

Molecular Formula: C_2H_7N

Common Formula: $(CH_3)_2NH$

Synonyms: AI3-15638-X; *N,N*-Dimethylamine; Dimethylamine, anhydrous; DMA; Methanamine, *N*-methyl-; *N*-Methylmethanamine

CAS Registry Number: 124-40-3

RTECS Number: IP8750000

DOT ID: UN 1032 (anhydrous); UN 1160 (solution)

EEC Number: 612-001-00-9

Regulatory Authority

- Air Pollutant Standard Set (ACGIH)[1] (Australia) (DFG)[3] (HSE)[33] (Israel) (Mexico) (OSHA)[58] (Several States)[60] (Several Canadian Provinces)
- OSHA 29CFR1910.119, Appendix A, Process Safety List of Highly Hazardous Chemicals, TQ = 2,500 lb (1,135 kg)
- Clean Air Act: Accidental Release Prevention/Flammable substances, (Section 112[r], Table 3), TQ = 10,000 lb (4,540 kg)
- Clean Water Act: Section 311 Hazardous Substances/RQ 40CFR117.3 (same as CERCLA, see below)
- EPA Hazardous Waste Number (RCRA No.): U092
- RCRA, 40CFR261, Appendix 8 Hazardous Constituents
- Superfund/EPCRA 40CFR302.4 Reportable Quantity (RQ): CERCLA, 1,000 lb (454 kg)
- EPCRA Section 313 Form R *de minimis* concentration reporting level: 1.0%
- Canada, WHMIS, Ingredients Disclosure List

Cited in U.S. State Regulations: Alaska (G), California (A, G), Connecticut (A), Florida (G), Illinois (G), Kansas (G, A), Louisiana (G), Maine (G), Massachusetts (G), Minnesota (G), Nevada (A), New Hampshire (G), New Jersey (G), North Dakota (A), Oklahoma (G), Pennsylvania (G), Rhode Island (G), Vermont (G), Virginia (G, A), Washington (G), West Virginia (G), Wisconsin (G).

Description: DMA, $(CH_3)_2NH$, is a colorless, compressed gas or liquid (below 44°F) with a pungent, fishy, or ammonia-like odor. The odor threshold is 0.34 ppm. Boiling point = 7°C. Freezing/Melting point = -92°C. Flash point = flammable gas. Autoignition temperature = 401°C. Explosive limits: LEL = 2.8%; UEL = 14.4%. Hazard Identification (based on NFPA-

704 M Rating System): Health 3, Flammability 4, Reactivity 0. Soluble in water.

Potential Exposure: This material is used in leather tanning, as an accelerator in rubber vulcanization, in the manufacture of detergents and in drug synthesis and pesticide manufacture.

Incompatibilities: Dimethylamine is a medium-strong base. Reacts violently with strong oxidizers; with mercury causing fire and explosion hazard. Incompatible with acids, organic anhydrides, isocyanates, vinyl acetate, acrylates, substituted allyls, alkylene oxides, epichlorohydrin, ketones, aldehydes, alcohols, glycols, phenols, cresols, caprolactum solution. Attacks aluminum, copper, lead, tin, zinc and alloys; some plastics, rubbers and coatings.

Permissible Exposure Limits in Air: The Federal Limit (OSHA PEL)[58] and the ACGIH recommended TWA value is 10 ppm (18 mg/m^3) as is the DFG[3] and HSE[33] limit. No STEL value has been set. The NIOSH IDLH level is 500 ppm. The limit set in Brazil for workplace air[35] is 8 ppm (14 mg/m^3). The former USSR[35][43] has set a MAC in workplace air of 1.0 mg/m^3 and MAC values in ambient air in residential areas of 0.005 mg/m^3 on either a momentary or a daily average basis. In addition, several states have set guidelines or standards for dimethylamine in ambient air[60] ranging from 42.857 μg/m^3 (Kansas) to 180 μg/m^3 (North Dakota) to 300 μg/m^3 (Virginia) to 360 μg/m^3 (Connecticut) to 429 μg/m^3 (Nevada).

Determination in Air: Adsorption on silica, workup with H_2SO_4 in CH_3OH, analysis by gas chromatography/Flame ionization; NIOSH IV, Method #2010.

Permissible Concentration in Water: No criterion set, but EPA[32] has suggested a permissible ambient goal of 248 μg/l, based on health effects. The former USSR[35][43] has set a MAC in water bodies used for domestic purposes of 0.1 mg/l.

Routes of Entry: Inhalation, ingestion, eye and/or skin contact (liquid).

Harmful Effects and Symptoms

Short Term Exposure: Corrosive to the eyes, skin, respiratory tract. Eye or skin contact with the liquid can cause burns and permanent damage. Inhalation of the vapors can cause coughing, and/or shortness of breath, and higher levels can cause pulmonary edema, a medical emergency that can be delayed for several hours. This can cause death. Symptom include sneezing, coughing and dyspnea; pulmonary edema; conjunctivitis; dermatitis; burns of skin and mucous membranes. The oral LD_{50} rat is 698 mg/kg (slightly toxic). Rapid evaporation of the liquid may cause frostbite.

Long Term Exposure: Repeated or prolonged contact with skin may cause irritation, redness, itching; dermatitis. Repeated exposures may cause bronchitis with cough, phlegm, and/or shortness of breath.

Points of Attack: Respiratory system, lungs, skin, eyes.

Medical Surveillance: Before beginning employment and at regular times after that, the following are recommended: Liver function tests. Lung function tests. If symptoms develop or overexposure is suspected, the following may be useful: Consider chest x-ray after acute overexposure. If symptoms develop or overexposure has occurred, repeat these tests.

First Aid: If this chemical gets into the eyes, remove any contact lenses at once and irrigate immediately for at least 15 minutes, occasionally lifting upper and lower lids. Seek medical attention immediately. If this chemical contacts the skin, remove contaminated clothing and wash immediately with soap and water. Seek medical attention immediately. If this chemical has been inhaled, remove from exposure, begin rescue breathing (using universal precautions) if breathing has stopped and CPR if heart action has stopped. Transfer promptly to a medical facility. When this chemical has been swallowed, get medical attention. Give large quantities of water and induce vomiting. Do not make an unconscious person vomit. Medical observation is recommended for 24 – 48 hours after breathing overexposure, as pulmonary edema may be delayed. As first aid for pulmonary edema, a doctor or authorized paramedic may consider administering a corticosteroid spray.

If frostbite has occurred, seek medical attention immediately; do *NOT* rub the affected areas or flush them with water. In order to prevent further tissue damage, do *NOT* attempt to remove frozen clothing from frostbitten areas. If frostbite has *NOT* occurred, immediately and thoroughly wash contaminated skin with soap and water.

Personal Protective Methods: Wear protective gloves and clothing to prevent any reasonable probability of skin contact. Safety equipment suppliers/manufacturers can provide recommendations on the most protective glove/clothing material for your operation. Butyl Rubber is among the recommended protective materials. All protective clothing (suits, gloves, footwear, headgear) should be clean, available each day, and put on before work. Contact lenses should not be worn when working with this chemical. Wear splash-proof chemical goggles and face shield unless full facepiece respiratory protection is worn. Employees should wash immediately with soap when skin is wet or contaminated. Provide emergency showers and eyewash. Where exposure to the liquefied compressed gas may occur, employees should be provided with special clothing designed to prevent frostbite.

Respirator Selection: NIOSH/OSHA: *250 ppm:* SA:CF (any supplied-air respirator operated in a continuous-flow mode). *500 ppm:* SCBAF (any self-contained breathing apparatus with a full facepiece); or SAF (any supplied-air respirator with a full facepiece). *Emergency or planned entry into unknown concentrations or IDLH conditions:* SCBAF:PD,PP (any self-contained breathing apparatus that has a full facepiece and is operated in a pressure-demand or other positive-pressure mode); or SAF:PD,PP:ASCBA (any supplied-air respirator that has a full facepiece and is operated in a pressure-demand or other positive-pressure mode in combination with an auxiliary self-contained breathing apparatus operated in a pressure-demand or other positive pressure mode). *Escape:* GMFS [any air-purifying, full-facepiece respirator (gas mask) with a chin-style, front- or back-mounted canister providing protection against the compound of concern]; or SCBAE (any appropriate escape-type, self-contained breathing apparatus).

Note: Substance causes eye irritation or damage; eye protection needed.

Storage: Prior to working with dimethylamine you should be trained on its proper handling and storage. Before entering confined space where this chemical may be present, check to make sure that an explosive concentration does not exist. Dimethylamine must be stored to avoid contact with strong oxidizers (such as chlorine, bromine, and fluorine) and mercury because violent reactions occur. Store in tightly closed containers in a cool, well-ventilated area away from heat. Sources of ignition such as smoking and open flames are prohibited where dimethylamine is handled, used, or stored. Metal containers used in the transfer of 5 gallons or more of dimethylamine should be grounded and bonded. Drums must be equipped with self-closing valves, pressure vacuum bungs, and flame arresters. Use only non-sparking tools and equipment, especially when opening and closing containers o dimethylamine. Wherever dimethylamine is used, handled, manufactured, or stored, use explosion-proof electrical equipment and fittings. Procedures for the handling, use and storage of cylinders should be in compliance with OSHA 1910.101 and 1910.169 as with the recommendations of the Compressed Gas Association.

Shipping: This compound requires a shipping label of "Poison Gas, Flammable Gas." It falls in DOT Hazard Class 2.1. Passenger aircraft or railcar shipment is forbidden and cargo aircraft shipment is forbidden as well.

Spill Handling: Restrict persons not wearing protective equipment from area of spill or leak until clean-up is complete. Remove all ignition sources. Establish forced ventilation to keep levels below explosive limit and to disperse the gas. Stop flow of gas. If source of leak is a cylinder and the leak cannot be stopped in place, remove the leaking cylinder to a safe place in the open air, and repair leak or allow cylinder to empty. If in liquid form, allow to vaporize. Keep dimethylamine out of a confined space, such as a sewer, because of the possibility of an explosion, unless the sewer is designed to prevent the build-up of explosive concentrations. It may be necessary to contain and dispose of this chemical as a hazardous waste. If material or contaminated runoff enters waterways, notify downstream users of potentially contaminated waters. Contact your Department of Environmental Protection or your regional office of the federal EPA for specific recommendations. If employees are required to clean-up spills, they must be properly trained and equipped. OSHA 1910.120(q) may be applicable.

Fire Extinguishing: Dimethylamine is a flammable gas or liquid. Poisonous gases including nitrogen oxides are produced in fire. Use dry chemical, carbon dioxide, or foam extinguishers. Vapors are heavier than air and will collect in low areas. Vapors may travel long distances to ignition sources and flashback. Vapors in confined areas may explode when exposed to fire. Containers may explode in fire. Storage containers and parts of containers may rocket great distances, in many directions. If material or contaminated runoff enters waterways, notify downstream users of potentially contaminated waters. Notify local health and fire officials and pollution control agencies. From a secure, explosion-proof location, use water spray to cool exposed containers. If cooling streams are ineffective (venting sound increases in volume and pitch, tank discolors, or shows any signs of deforming), withdraw immediately to a secure position. If employees are expected to fight fires, they must be trained and equipped in OSHA 1910.156.

Disposal Method Suggested: Incineration; incinerator is equipped with a scrubber or thermal unit to reduce No_x emissions.[22] Consult with environmental regulatory agencies for guidance on acceptable disposal practices. Generators of waste containing this contaminant (≥100 kg/mo) must conform with EPA regulations governing storage, transportation, treatment, and waste disposal.

References

New Jersey Department of Health, "Hazardous Substance Fact Sheet: Dimethylamine," Trenton, NJ (April 1986)

4-Dimethylaminoazobenzene

Molecular Formula: $C_{14}H_{15}N_3$

Common Formula: $C_6H_5NNC_6H_4N(CH_3)_2$

Synonyms: Benzenamine, *N,N*-dimethyl-4-(phenylazo)-; Brilliant fast oil yellow; Brilliant fast spirit yellow; Brilliant oil yellow; Butter yellow; Cerasine yellow GG; C.I. 11020; C.I. solvent yellow 2; DAB; *p*-(Dimethylamino)azobenzene; 4-(*N,N*-Dimethylamino)azob enzene; Dimethylaminoazobenzene; *N,N*-Dimethyl-*p*-(phenylazo)aniline; *N,N*-Dimethyl-4-phenylazo aniline; Dimethyl yellow; DMAB; Enial yellow 2G; Fast oil yellow B; Fat yellow; Fat yellow A; Fat yellow AD OO; Fat yellow ES; Fat yellow ES extra; Fat yellow extra conc.; Fat yellow R; Grasal brilliant yellow; Iketon yellow extra; Methyl yellow; Oil yellow 20; Oil yellow 2625; Oil yellow 2G; Oil yellow BB; Oil yellow D; Oil yellow FN; Oil yellow G; Oil yellow GG; Oil yellow II; Oil yellow N; Oil yellow Pel; Oil yellow S; Oleal yellow 2G; Organol yellow ADM; Orient oil yellow GG; Petrol yellow WT; 4-(Phenylazo)-*N,N*-dimethylaniline; Resinol yellow GR; Silotras yellow T 2G; Somalia yellow A; Stear yellow JB; Sudan yellow GG; Sudan yellow GGA; Toyo oil yellow G; Waxoline yellow ADS; Yellow G soluble in grease

CAS Registry Number: 60-11-7

RTECS Number: BX7350000

DOT ID: UN 3077

Regulatory Authority

- Carcinogen (Animal Positive) (IARC)[9] (NTP)[10]
- OSHA Specifically Regulated Chemical (29CFR 1910.1015)
- Air Pollutant Standard Set (Several States)[60]
- OSHA, 29CFR1910 Specifically Regulated Chemicals (See CFR 1910.1015)
- Clean Air Act: Hazardous Air Pollutants (Title I, Part A, Section 112)
- EPA Hazardous Waste Number (RCRA No.): U093
- RCRA, 40CFR261, Appendix 8 Hazardous Constituents
- RCRA 40CFR268.48; 61FR15654, Universal Treatment Standards: Wastewater (mg/l), 0.13; Nonwastewater (mg/kg), N/A
- RCRA 40CFR264, Appendix 9; TSD Facilities Ground Water Monitoring List. Suggested test method(s) (PQL μg/l): 8270 (10)
- Superfund/EPCRA 40CFR302.4 Reportable Quantity (RQ): CERCLA, 10 lb (4.54 kg)
- EPCRA Section 313 Form R *de minimis* concentration reporting level: 0.1%
- Canada, WHMIS, Ingredients Disclosure List

Cited in U.S. State Regulations: Alaska (G), California (A, G), Florida (G), Illinois (G), Kansas (G), Louisiana (G), Maine (G), Massachusetts (G), Michigan (G), Minnesota (G), New Hampshire (G), New Jersey (G), New York (A), North Dakota (A), Oklahoma (G), Pennsylvania (G), Rhode Island (G), South Carolina (A), Vermont (G), Virginia (G), Washington (G), West Virginia (G), Wisconsin (G).

Description: DAB, $C_6H_5NNC_6H_4N(CH_3)_2$, is a flaky yellow crystal. Boiling point = (sublimes). Freezing/Melting point = 114 – 117°C. Very slightly soluble.

Potential Exposure: DAB is used for coloring polishes and other wax products, polystyrene, soap, and as a chemical indicator. Human exposure to DAB can occur through either inhalation or skin absorption.

Incompatibilities: None reported.

Permissible Exposure Limits in Air: 4-Dimethylaminoazobenzene is included in the Federal standard for carcinogens; all contact with it should be avoided. Exposure to this chemical is to be controlled through the required use of engineering controls, work practices, and personal protective equipment, including respirators. See 29 CFR 1910.1003-1910.1016 for specific details of these requirements. States which have set guidelines or standards for this compound in ambient air[60] include North Dakota at zero, New York at 0.03 μg/m³ and South Carolina at 125.0 μg/m³.

Determination in Air: G-Chromosorb tube P; 2-Propanol; Gas chromatography/Flame ionization detection; NIOSH II[4] P&CAM Method #284.

Permissible Concentration in Water: No criteria set.

Routes of Entry: Inhalation, skin absorption, ingestion, skin and/or eye contact.

Harmful Effects and Symptoms

Short Term Exposure: This compound can cause irritation of the eyes, skin, and respiratory system. It can cause contact dermatitis.

Long Term Exposure: Enlarged liver; liver, kidney dysfunction; contact dermatitis; cough, wheezing, dyspnea (breathing difficulty); bloody sputum; bronchial secretions; frequent urination, hematuria (blood in the urine), dysuria; and may cause liver or bladder cancer.

Points of Attack: Skin, respiratory system, liver, kidneys, bladder. Cancer Site in animals: liver and bladder tumors.

Medical Surveillance: Preplacement and periodic examinations should include a history of exposure to other carcinogens; use of alcohol, smoking, and medications; and family history. Special attention should be given to liver size and liver function tests.

First Aid: If this chemical gets into the eyes, remove any contact lenses at once and irrigate immediately for at least 15 minutes, occasionally lifting upper and lower lids. Seek medical attention immediately. If this chemical contacts the skin, remove contaminated clothing and wash immediately with soap and water. Seek medical attention immediately. If this chemical has been inhaled, remove from exposure, begin rescue breathing (using universal precautions) if breathing has stopped and CPR if heart action has stopped. Transfer promptly to a medical facility. Contact a physician, hospital or poison center at once. If the victim is unconscious or convulsing, do not induce vomiting or give anything by mouth. Assure that his airway is open and if conscious and not convulsing, give a glass of water to dilute the substance. Vomiting should not be induced without a physician's advice.

Personal Protective Methods: Wear protective gloves and clothing to prevent any reasonable probability of skin contact. Safety equipment suppliers/manufacturers can provide recommendations on the most protective glove/clothing material for your operation.. All protective clothing (suits, gloves, footwear, headgear) should be clean, available each day, and put on before work. Contact lenses should not be worn when working with this chemical. Wear dust-proof chemical goggles and face shield unless full facepiece respiratory protection is worn. Employees should wash immediately with soap when skin is wet or contaminated. Provide emergency showers and eyewash.

Respirator Selection: NIOSH: *At any detectable concentration:* SCBAF:PD,PP (any MSHA/NIOSH approved self-contained breathing apparatus that has a full facepiece and is operated in a pressure-demand or other positive-pressure mode); or SAF:PD,PP:ASCBA (any supplied-air respirator that has a full facepiece and is operated in a pressure-demand or other positive-pressure mode in combination with an auxiliary, self-contained breathing apparatus operated in a pressure-demand or other positive pressure mode). *Escape:* HiEF (any air-purifying, full-facepiece respirator (gas mask) with a chin-style, front- or back-mounted organic vapor canister having a high-efficiency particulate filter); or SCBAE (any appropriate escape-type, self-contained breathing apparatus).

Storage: Prior to working with DAB you should be trained on its proper handling and storage. Store in a cool, dry place and protect from exposure to light and air. A regulated, marked area should be established where this chemical is handled, used, or stored in compliance with OSHA standard 1910.1045.

Shipping: Environmentally hazardous solid, n.o.s. Hazard Class: 9. Label: "Class 9." Packing Group: III.

Spill Handling: *Dry material:* Evacuate persons not wearing protective equipment from area of spill or leak until clean-up is complete. Remove all ignition sources. Collect powdered material in the most convenient and safe manner and deposit in sealed containers. Ventilate area after clean-up is complete. It may be necessary to contain and dispose of this chemical as a hazardous waste. If material or contaminated runoff enters waterways, notify downstream users of potentially contaminated waters. Contact your Department of Environmental Protection or your regional office of the federal EPA for specific recommendations. If employees are required to clean-up spills, they must be properly trained and equipped. OSHA 1910.120(q) may be applicable.

Liquid: Evacuate and restrict persons not wearing protective equipment from area of spill or leak until cleanup is complete. Remove all ignition sources. Ventilate area of spill or leak. Absorb liquids in vermiculite, dry sand, earth, peat, carbon, or a similar material and deposit in sealed containers. It may be necessary to contain and dispose of this chemical as a hazardous waste. If material or contaminated runoff enters waterways, notify downstream users of potentially contaminated waters. Contact your Department of Environmental Protection or your regional office of the federal EPA for specific recommendations. If employees are required to clean-up spills, they must be properly trained and equipped. OSHA 1910.120(q) may be applicable.

Fire Extinguishing: Extinguish fire using an agent suitable for type of surrounding fire. 4-Dimethylaminoazobenzene itself does not burn. Poisonous gases are produced in fire. If material or contaminated runoff enters waterways, notify downstream users of potentially contaminated waters. Notify local health and fire officials and pollution control agencies. From a secure, explosion-proof location, use water spray to cool exposed containers. If cooling streams are ineffective (venting sound increases in volume and pitch, tank discolors, or shows any signs of deforming), withdraw immediately to a secure position. If employees are expected to fight fires, they must be trained and equipped in OSHA 1910.156.

Disposal Method Suggested: Consult with environmental regulatory agencies for guidance on acceptable disposal practices. Generators of waste containing this contaminant (≥100 kg/mo) must conform with EPA regulations governing storage, transportation, treatment, and waste disposal.

References

New Jersey Department of Health, "Hazardous Substance Fact Sheet: 4-Dimethylaminoazobenzene," Trenton, NJ (February 1986)

Dimethylaminoethanol

Molecular Formula: $C_4H_{11}N$

Synonyms: Deanol; 2-Dimethylaminoethanol; β-Dimethylaminoethyl alcohol; Dimethylethanolamine; *N,N*-Dimethyl-*N*-(2-hydroxyethyl)amine; *N,N*-Dimethyl-2-hydroxyethylamine; Ethanol, 2-dimethylamino-

CAS Registry Number: 108-01-0

RTECS Number: KK6125000

DOT ID: UN 2051

EEC Number: 603-047-00-0

Cited in U.S. State Regulations: Florida (G), New Jersey (G), Pennsylvania (G).

Description: Dimethylaminoethanol is a colorless liquid with a pungent odor. Odor threshold 0.25 ppm. Boiling point = 133°C. Freezing/Melting point = -59°C. Flash point = 41°C. Autoignition temperature = 295°C. Explosive limits: LEL = 1.6%; UEL = 11.9%. Hazard Identification (based on NFPA-704 M Rating System): Health 2, Flammability 2, Reactivity 0. Soluble in water.

Potential Exposure: Used in making dyestuffs, textiles, pharmaceuticals, and emulsifiers in paints and coatings. Also used as a medication in the treatment of behavioral problems of children.

Incompatibilities: Forms explosive mixture with air. Violent reaction with oxidizers, strong acids, acid chlorides, and isocyanates. Attacks copper and its alloys, galvanized steel, zinc and zinc alloys.

Routes of Entry: Inhalation, through the skin.

Harmful Effects and Symptoms

Short Term Exposure: Contact can cause severe irritation and burns to the eyes and skin, with possible permanent damage. Breathing the aerosol can cause lung irritation, coughing and/or shortness of breath. Higher exposures can cause pulmonary edema, a medical emergency that can be delayed for several hours. This can cause death. Exposure can cause headache, muscle tenderness, restlessness, increased irritability, lack of sleep and weight loss.

Long Term Exposure: Repeated skin contact may cause dermatitis. May cause an asthma-like allergy. May affect the nervous system. Corrosive substances may cause lung irritation and bronchitis.

Points of Attack: Nervous system, skin, lungs.

Medical Surveillance: Examination of the nervous system. Lung function tests. Consider chest x-ray following acute overexposure.

First Aid: If this chemical gets into the eyes, remove any contact lenses at once and irrigate immediately for at least 15 minutes, occasionally lifting upper and lower lids. Seek medical attention immediately. If this chemical contacts the skin, remove contaminated clothing and wash immediately with soap and water. Seek medical attention immediately. If this chemical has been inhaled, remove from exposure, begin rescue breathing (using universal precautions) if breathing has stopped and CPR if heart action has stopped. Transfer promptly to a medical facility. When this chemical has been swallowed, get medical attention. If victim is conscious, administer water or milk. Do not induce vomiting. Medical observation is recommended for 24 – 48 hours after breathing overexposure, as pulmonary edema may be delayed. As first aid for pulmonary edema, a doctor or authorized paramedic may consider administering a corticosteroid spray.

Personal Protective Methods: Wear protective gloves and clothing to prevent any reasonable probability of skin contact. Safety equipment suppliers/manufacturers can provide recommendations on the most protective glove/clothing material for your operation.. All protective clothing (suits, gloves, footwear, headgear) should be clean, available each day, and put on before work. Contact lenses should not be worn when working with this chemical. Wear splash-proof chemical goggles and face shield unless full facepiece respiratory protection is worn. Employees should wash immediately with soap when skin is wet or contaminated. Provide emergency showers and eyewash.

Respirator Selection: Where the potential for exposure to this chemical, use a MSHA/NIOSH approved supplied-air respirator with a full facepiece operated in the positive pressure mode or with a full facepiece, hood, or helmet in the continuous flow mode, or use a MSHA/NIOSH approved self-contained breathing apparatus with a full facepiece operated in pressure-demand or other positive pressure mode.

Storage: Prior to working with dimethylaminoethanol you should be trained on its proper handling and storage. Before entering confined space where may be present, check to make sure that an explosive concentration does not exist. Store in tightly closed containers in a cool, well ventilated area away from oxidizers, strong acids and chemically active metals. Metal containers involving the transfer of this chemical should be grounded and bonded. Where possible, automatically pump liquid from drums or other storage containers to process containers. Drums must be equipped with self-closing valves, pressure vacuum bungs, and flame arresters. Use only non-sparking tools and equipment, especially when opening and closing containers of this chemical. Sources of ignition such as smoking and open flames, are prohibited where this chemical is used, handled, or stored in a manner that could create a potential fire or explosion hazard. Wherever this chemical is used, handled, manufactured, or stored, use explosion-proof electrical equipment and fittings.

Shipping: Label "Corrosive, Flammable Liquid, Poison." Hazard Class 8. Packing Group II.

Spill Handling: Evacuate and restrict persons not wearing protective equipment from area of spill or leak until cleanup is complete. Remove all ignition sources. Establish forced ventilation to keep levels below explosive limit. Absorb liquids in vermiculite, dry sand, earth, peat, carbon, or a similar material and deposit in sealed containers. Keep this chemical out of a confined space, such as a sewer, because of the possibility of an explosion, unless the sewer is designed to prevent the build-up of explosive concentrations. It may be necessary to contain and dispose of this chemical as a hazardous waste. If material or contaminated runoff enters waterways, notify downstream users of potentially contaminated waters. Contact your Department of Environmental Protection or your regional office of the federal EPA for specific recommendations. If employees are required to clean-up spills, they must be properly trained and equipped. OSHA 1910.120(q) may be applicable.

Fire Extinguishing: This chemical is a flammable liquid. Poisonous gases including nitrogen oxides and carbon monoxide are produced in fire. Use dry chemical, carbon dioxide, or alcohol or polymer resistant foam extinguishers. Water may be ineffective. Vapors are heavier than air and will collect in low areas. Vapors may travel long distances to ignition sources and flashback. Vapors in confined areas may explode when exposed to fire. Containers may explode in fire. Storage containers and parts of containers may rocket great distances, in many directions. If material or contaminated runoff enters waterways, notify downstream users of potentially contaminated waters. Notify local health and fire officials and pollution control agencies. From a secure, explosion-proof location, use water spray to cool exposed containers. If cooling streams are ineffective (venting sound increases in volume and pitch, tank discolors, or shows any signs of deforming), withdraw immediately to a secure position. If employees are expected to fight fires, they must be trained and equipped in OSHA 1910.156.

Disposal Method Suggested: Incineration.

References

New Jersey Department of Health and Senior Services, "Hazardous Substance Fact Sheet, Dimethylaminoethanol," Trenton NJ (March 1999)

N,N-Dimethylaniline

Molecular Formula: $C_8H_{11}N$

Common Formula: $C_6H_5N(CH_3)_2$

Synonyms: Aniline, *N,N*-dimethyl-; Benzenamine, *N,N*-dimethyl-; (Dimethylamino)benzene; *N,N*-dimethylaminobenzene; *N*-dimethyl-aniline; Dimethylaniline; *N,N*-Dimethylbenzenamine; *N,N*-Dimethylphenylamine; Dimethylphenylamine; *N,N*-Dimetilanilina (Spanish); DMA; Dwumetyloanilina (Polish); NCI-C56428; Versneller NL 63/10

CAS Registry Number: 121-69-7

RTECS Number: BX4725000

DOT ID: UN 2253

EEC Number: 612-016-00-0

EINECS Number: 204-493-5

Regulatory Authority

- Carcinogen (some animal evidence) NTP
- Air Pollutant Standard Set (ACGIH)[1] (Australia) (DFG)[3] (HSE) (Israel) (Mexico) (former USSR)[43] (OSHA)[58] (Several States)[60] (Several Canadian Provinces)
- Clean Air Act: Hazardous Air Pollutants (Title I, Part A, Section 112)
- Superfund/EPCRA 40CFR302.4 Reportable Quantity (RQ): CERCLA, 1 lb (0.454 kg)
- EPCRA Section 313 Form R *de minimis* concentration reporting level: 1.0%
- Canada, WHMIS, Ingredients Disclosure List

Cited in U.S. State Regulations: Alaska (G), California (A, G), Connecticut (A), Florida (G), Illinois (G), Maine (G), Maryland (G), Massachusetts (G), Minnesota (G), Nevada (A), New Hampshire (G), New Jersey (G), North Dakota (A), Pennsylvania (G), Rhode Island (G), West Virginia (G).

Description: DMA, $C_6H_5N(CH_3)_2$, is a straw-colored liquid with a characteristic amine-like odor. It turns brown on contact with air. Boiling point = 192°C. Freezing/Melting point = 2.5°C. Flash point = 62 – 74°C. Autoignition temperature = 371°C. Explosive limits: LEL – unknown; UEL = 7%.

Potential Exposure: This material is used as an intermediate in the manufacture of many dyes and rubber chemicals. It is also used as an analytical reagent.

Incompatibilities: Forms explosive mixture with air. Contact with strong oxidizers, strong acids, benzoyl peroxide may cause fire and explosion hazard. Contact with hypochlorite bleaches form explosive chloroamines. Incompatible with anhydrides, isocyanates, aldehydes.

Permissible Exposure Limits in Air: The Federal Limit (OSHA PEL) and recommended ACGIH TLV is 5 ppm (25 mg/m^3) TWA and ACGIH set a STEL of 10 ppm. The notation "skin" indicates the possibility of cutaneous absorption. The TWA set by Australia, DFG, HSE, Israel, Mexico and the Canadian provinces of Alberta, BC, Ontario, and Quebec is 5 ppm and all have set a STEL of 10 ppm (50 mg/m^3). The NIOSH IDLH level is 100 ppm. The former USSR-UNEP/IRPTC joint project[43] has set a MAC in workplace air of 0.2 mg/m^3 and a MAC in ambient air in residential area (on both a momentary and a daily average basis of 0.0055 mg/m^3).

Some states have set guidelines or standards for *N,N*-dimethylaniline in ambient air[60] ranging from 250 – 500 μg/m^3 (North Dakota) to 500 μg/m^3 (Connecticut) to 560 μg/m^3 (Nevada).

Determination in Air: Adsorption on charcoal, workup with CS_2, analysis by gas chromatography. See NIOSH Method #2002 on Amines, Aromatic.

Permissible Concentration in Water: No criteria set, but EPA[32] has suggested permissible ambient concentrations of 345 μg/l.

Routes of Entry: Inhalation, skin absorption, ingestion, eye and/or skin contact.

Harmful Effects and Symptoms

Exposure: Contact may irritate or burn the eyes, and may irritate the nose and throat. Dimethylaniline can be absorbed through the skin, thereby increasing exposure. Swallowing the liquid may cause aspiration into the lungs with the risk of chemical pneumonia. May affect the blood, resulting in formation of methemoglobin. Symptoms of anoxia, cyanosis, weakness, dizziness, ataxia. Symptoms may be delayed.

Long Term Exposure: Repeated or prolonged contact with skin may cause dermatitis. While this chemical has not been identified as a reproductive hazard, there is some evidence that women working with aniline, a related compound, have a higher incidence of reproductive cycle disorders and abortions.

Points of Attack: Blood, kidneys, liver, cardiovascular system.

Medical Surveillance: A blood test for methemoglobin level should be done periodically and after any overexposure or if any symptoms develop. Any evaluation should include a careful history of past and present symptoms.

First Aid: If this chemical gets into the eyes, remove any contact lenses at once and irrigate immediately for at least 15 minutes, occasionally lifting upper and lower lids. Seek medical attention immediately. If this chemical contacts the skin, remove contaminated clothing and wash immediately with soap and water. Seek medical attention immediately. If this chemical has been inhaled, remove from exposure, begin rescue breathing (using universal precautions) if breathing has stopped and CPR if heart action has stopped. Transfer promptly to a medical facility. When this chemical has been swallowed, get medical attention. Give large quantities of water and induce vomiting. Do not make an unconscious person vomit. Medical observation is recommended; effects may be delayed.

Note to Physician: Treat for methemoglobinemia. Spectrophotometry may be required for precise determination of levels of methemoglobinemia in urine.

Personal Protective Methods: Wear protective gloves and clothing to prevent any reasonable probability of skin contact. Safety equipment suppliers/manufacturers can provide recommendations on the most protective glove/clothing material for your operation.. All protective clothing (suits, gloves, footwear, headgear) should be clean, available each day, and put on before work. Contact lenses should not be worn when working with this chemical. Wear splash-proof chemical goggles and face shield unless full facepiece respiratory protection is worn. Employees should wash immediately with soap when skin is wet or contaminated. Provide emergency showers and eyewash.

Respirator Selection: NIOSH: *Up to 50 ppm*: SA (any supplied-air respirator). *Up to 100 ppm:* SA: CF (any supplied-air respirator operated in a continuous-flow mode); or SCBAF (any self-contained breathing apparatus with a full facepiece); SAF (any supplied-air respirator with a full facepiece). *Emergency or planned entry into unknown concentrations or IDLH conditions*: SCBAF:PD,PP (any self-contained breathing apparatus that has a full facepiece and is operated in a pressure-demand or other positive-pressure mode); or SAF:PD,PP:ASCBA (any supplied-air respirator that has a full facepiece and is operated in a pressure-demand or other positive-pressure mode in combination with an auxiliary self-contained breathing apparatus operated in a pressure-demand or other positive-pressure mode). *Escape:* GMFOV [any air-purifying, full-facepiece respirator (gas mask) with a chin-style, front- or back-mounted organic vapor canister] or SCBAE (any appropriate escape-type, self-contained breathing apparatus).

Storage: Prior to working with dimethylaniline you should be trained on its proper handling and storage. Before entering confined space where this chemical may be present, check to make sure that an explosive concentration does not exist. Dimethylaniline must be stored to avoid contact with oxidizers (such as perchlorates, peroxides, permanganates, chlorates, and nitrates), strong acids, and benzoyl peroxide since violent reactions occur. Metal containers involving the transfer of this chemical should be grounded and bonded. Where possible, automatically pump liquid from drums or other storage containers to process containers. Drums must be equipped with self-closing valves, pressure vacuum bungs, and flame arresters. Use only non-sparking tools and equipment, especially when opening and closing containers of this chemical. Sources of ignition such as smoking and open flames, are prohibited where this chemical is used, handled, or stored in a manner that could create a potential fire or explosion hazard. Wherever this chemical is used, handled, manufactured, or stored, use explosion-proof electrical equipment and fittings. A regulated, marked area should be established where this chemical is handled, used, or stored in compliance with OSHA standard 1910.1045.

Shipping: This compound requires a shipping label of: "Poison." It falls in DOT Hazard Class 6.1 and Packing Group II.

Spill Handling: Evacuate and restrict persons not wearing protective equipment from area of spill or leak until cleanup is complete. Remove all ignition sources. Establish forced ventilation to keep levels below explosive limit. Cover and mix with a 9:1 mixture of sand and soda ash and deposit in sealed containers. Keep this chemical out of a confined space, such as a sewer, because of the possibility of an explosion, unless the sewer is designed to prevent the build-up of explosive concentrations. It may be necessary to contain and dispose of this chemical as a hazardous waste. If material or contaminated runoff enters waterways, notify downstream users of potentially contaminated waters. Contact your Department of Environmental Protection or your regional office of the federal EPA for specific recommendations. If employees are required to clean-up spills, they must be properly trained and equipped. OSHA 1910.120(q) may be applicable.

Fire Extinguishing: This chemical is a combustible liquid. Poisonous gases including nitrogen oxides and carbon monoxide are produced in fire. Use dry chemical, carbon dioxide, or foam extinguishers. Vapors are heavier than air and will collect in low areas. Vapors may travel long distances to ignition sources and flashback. Vapors in confined areas may explode when exposed to fire. Containers may explode in fire. Storage containers and parts of containers may rocket great distances, in many directions. If material or contaminated runoff enters waterways, notify downstream users of potentially contaminated waters. Notify local health and fire officials and pollution control agencies. From a secure, explosion-proof location, use water spray to cool exposed containers. If cooling streams are ineffective (venting sound increases in volume and pitch, tank discolors, or shows any signs of deforming), withdraw immediately to a secure position. If employees are expected to fight fires, they must be trained and equipped in OSHA 1910.156.

Disposal Method Suggested: Incineration in a furnace equipped with afterburner and scrubber.

References

New Jersey Department of Health, "Hazardous Substance Fact Sheet: Dimethylaniline," Trenton, NJ (January 1986)

Sax N. I., Ed., "Dangerous Properties of Industrial Materials Report," 5, No. 3, 34–41 (1985)

7,12-Dimethylbenz[a]anthracene

Molecular Formula: $C_{20}H_{16}$

Synonyms: AI3-50460; Benz(a)anthracene, 7,12-dimethyl-; Benz(a)anthracene, 9,10-dimethyl-; DBA; 6,7-Dimethyl-1,2-benzanthracene; 7,12-Dimethyl-1,2-benzanthracene; 7,12-Dimethylbenzanthracene; 9,10-Dimethyl-1,2-benzanthracene; 9,10-Dimethylbenz(a)anthracene; 9,10-Dimethylbenzanthracene; Dimethylbenz(a)anthracene; Dimethylbenzanthracene; 9,10-Dimethylbenz-1,2-benzanthracene; 9,10-Dimethylbenz-1,2-benzanthrazen (German); 7,12-Dimethylbenzo(a)anthracene; 1,4-Dimethyl-2,3-benzphenanthrene; 7,12-Dimetilbenz(a)antraceno (Spanish); 7,12-DMBA; DMBA; NCI-C03918; NSC 40823

CAS Registry Number: 57-97-6

RTECS Number: CW3850000

DOT ID: UN 3077

Regulatory Authority

- Carcinogen (Animal Positive) (numerous studies)
- Clean Water Act: 40CFR401.15 Section 307 Toxic Pollutants as polynuclear aromatic hydrocarbons
- EPA Hazardous Waste Number (RCRA No.): U094
- RCRA, 40CFR261, Appendix 8 Hazardous Constituents
- RCRA 40CFR264, Appendix 9; TSD Facilities Ground Water Monitoring List. Suggested test method(s) (PQL μg/l): 8270 (10)
- Superfund/EPCRA 40CFR302.4 Reportable Quantity (RQ): CERCLA, 1 lb. (0.454 kg.)
- EPCRA Section 313 Form R *de minimis* concentration reporting level: 0.1%
- Canada, WHMIS, Ingredients Disclosure List

Cited in U.S. State Regulations: California (A, G), Florida (G), Massachusetts (G), Minnesota (G), New Jersey (G), Pennsylvania (G).

Description: DBA is a greenish-yellow solid. Freezing/Melting point 121 – 123°C. Insoluble in water.

Potential Exposure: DBA is a polycyclic aromatic hydrocarbon (PAH) that is present in the smoke of cigarettes, burned wood and coal, coal tar, gasoline, and diesel exhaust. It is a medical and pharmaceutical research chemical.

Incompatibilities: Oxidizing agents.

Permissible Exposure Limits in Air: There is no established exposure limits.

Permissible Concentration in Water: See regulatory authority above.

Routes of Entry: Inhalation, skin and/or eye contact. Absorbed through the skin.

Harmful Effects and Symptoms

Short Term Exposure: Contact can irritate and burn the skin and eyes. Irritates the respiratory tract. High exposure may damage the blood cells causing anemia.

Long Term Exposure: Repeated exposure can irritate the lungs causing coughing, wheezing and/or difficult breathing. This chemical has been shown to cause skin and lung cancer in animals and has caused cancer in the offsprings of animals exposed during pregnancy. Believed to be a transplacental carcinogen. May damage the male reproductive glands. May cause liver and kidney damage. May cause anemia.

Points of Attack: Lungs, testes, liver, kidneys, blood. Cancer site in animals: lungs, skin.

Medical Surveillance: Liver and kidney function tests. Complete blood count (CBC). Lung function tests.

First Aid: If this chemical gets into the eyes, remove any contact lenses at once and irrigate immediately for at least 15 minutes, occasionally lifting upper and lower lids. If this chemical contacts the skin, remove contaminated clothing and wash immediately with soap and water. If this chemical has been inhaled, remove from exposure, begin rescue breathing (using universal precautions) if breathing has stopped and CPR if heart action has stopped. Transfer promptly to a medical facility.

Personal Protective Methods: Wear protective gloves and clothing to prevent any reasonable probability of skin contact. Safety equipment suppliers/manufacturers can provide recommendations on the most protective glove/clothing material for your operation.. All protective clothing (suits, gloves, footwear, headgear) should be clean, available each day, and put on before work. Contact lenses should not be worn when working with this chemical. Wear dust-proof chemical goggles and face shield unless full facepiece respiratory protection is worn. Employees should wash immediately with soap when skin is wet or contaminated. Provide emergency showers and eyewash.

Respirator Selection: Where the potential for exposure to this chemical, use a MSHA/NIOSH approved supplied-air respirator with a full facepiece operated in the positive pressure mode or with a full facepiece, hood, or helmet in the continuous flow mode, or use a MSHA/NIOSH approved self-contained breathing apparatus with a full facepiece operated in pressure-demand or other positive pressure mode.

Storage: Prior to working with DBA you should be trained on its proper handling and storage. Store in tightly closed containers in a cool, well ventilated area away from oxidizers. A regulated, marked area should be established where this chemical is handled, used, or stored in compliance with OSHA standard 1910.1045.

Shipping: Environmentally hazardous solid, n.o.s. Hazard Class: 9. Label: "Class 9." Packing Group: III.

Spill Handling: Evacuate persons not wearing protective equipment from area of spill or leak until clean-up is complete. Remove all ignition sources. Collect powdered material in the most convenient and safe manner and deposit in sealed containers. Ventilate area after clean-up is complete. It may be necessary to contain and dispose of this chemical as a hazardous waste. If material or contaminated runoff enters waterways, notify downstream users of potentially contaminated waters. Contact your Department of Environmental Protection or your regional office of the federal EPA for specific recommendations. If employees are required to clean-up spills, they must be properly trained and equipped. OSHA 1910.120(q) may be applicable.

Fire Extinguishing: Use dry chemical, carbon dioxide, water spray, or alcohol or polymer foam extinguishers. Poisonous gases are produced in fire including carbon monoxide. If material or contaminated runoff enters waterways, notify downstream users of potentially contaminated waters. Notify local health and fire officials and pollution control agencies. From a secure, explosion-proof location, use water spray to cool exposed containers. If cooling streams are ineffective (venting sound increases in volume and pitch, tank discolors, or shows any signs of deforming), withdraw immediately to a secure position. If employees are expected to fight fires, they must be trained and equipped in OSHA 1910.156.

Disposal Method Suggested: Consult with environmental regulatory agencies for guidance on acceptable disposal practices. Generators of waste containing this contaminant (≥100 kg/mo) must conform with EPA regulations governing storage, transportation, treatment, and waste disposal.

References

New Jersey Department of Health and Senior Services, "Hazardous Substance Fact Sheet, Benz[a]anthracene, 7,12-dimethyl," Trenton NJ (March 1999)

2,3-Dimethylbutane

Molecular Formula: C_6H_{14}

Common Formula: $CH_3CH(CH_3)CH(CH_3)CH_3$

Synonyms: Biisopropyl; Butane, 2,3-dimethyl-; Diisopropyl; Isohexane diisopropyl; Neohexane; 1,1,2,2-Tetramethylethane

CAS Registry Number: 79-29-8

RTECS Number: EJ9350000

DOT ID: UN 2457

Regulatory Authority

- Air Pollutant Standard Set (ACGIH)
- Canada, WHMIS, Ingredients Disclosure List

Cited in U.S. State Regulations: Florida (G), Maine (G), Massachusetts (G), New Hampshire (G), New Jersey (G), Pennsylvania (G).

Description: 2,3-Dimethylbutane, $CH_3CH(CH_3)CH(CH_3)CH_3$, is a colorless liquid. Boiling point = 58°C. Flash point = -29°C. Autoignition temperature = 405°C. Explosive limits: LEL = 1.2%; UEL = 7.0%. Hazard Identification (based on NFPA-704 M Rating System): Health 1, Flammability 3, Reactivity 0. Insoluble in water.

Potential Exposure: 2,3-Dimethylbutane is used in high octane fuel and to make organic chemicals.

Incompatibilities: Forms explosive mixture with air. Violent reaction with oxygen, strong oxidizers.

Permissible Exposure Limits in Air: Measured as hexane isomer, NIOSH recommends and airborne exposure limit of 100 ppm TWA for a 10-hour workshift and STEL of 510 ppm not to be exceeded during any 15 minute work period. ACGIH recommends a TWA limit of 500 ppm (8-hour workshift) and STEL of 1,000 ppm.

Determination in Air: No method available.

Permissible Concentration in Water: No criteria set.

Routes of Entry: Inhalation, ingestion, skin and/or eye contact.

Harmful Effects and Symptoms

Short Term Exposure: Contact can irritate the skin or eyes. Exposure can irritate the eyes, nose and throat. Exposure can also cause you to feel dizzy, lightheaded, giddy, and to pass out. Swallowing the liquid may cause aspiration into the lungs with the risk of chemical pneumonia. Prolonged contact may cause dermatitis.

Long Term Exposure: Repeated exposure may damage the kidneys. May cause dermatitis.

Points of Attack: Eyes, skin, respiratory system, central nervous system.

Medical Surveillance: If symptoms develop or overexposure is suspected, the following may be useful: Kidney function tests.

First Aid: If this chemical gets into the eyes, remove any contact lenses at once and irrigate immediately for at least 15 minutes, occasionally lifting upper and lower lids. Seek medical attention immediately. If this chemical contacts the skin, remove contaminated clothing and wash immediately with soap and water. Seek medical attention immediately. If this chemical has been inhaled, remove from exposure, begin rescue breathing (using universal precautions) if breathing has stopped and CPR if heart action has stopped. Transfer promptly to a medical facility. When this chemical has been swallowed, get medical attention. Give large quantities of water and induce vomiting. Do not make an unconscious person vomit.

Personal Protective Methods: Wear protective gloves and clothing to prevent any reasonable probability of skin contact. Safety equipment suppliers/manufacturers can provide recommendations on the most protective glove/clothing material for your operation.. All protective clothing (suits, gloves, footwear, headgear) should be clean, available each day, and put on before work. Contact lenses should not be worn when working with this chemical. Wear splash-proof chemical goggles and face shield unless full facepiece respiratory protection is worn. Employees should wash immediately with soap when skin is wet or contaminated. Provide emergency showers and eyewash.

Respirator Selection: NIOSH: *Up to 1,000 ppm:* SA (any supplied-air respirator).* *Up to 2,500 ppm:* SA: CF (any supplied-air respirator operated in a continuous-flow mode).* *Up to 5,000 ppm*: SAT: CF (any supplied-air respirator that has a tight-fitting facepiece and is operated in a continuous-flow mode);* or SCBAF (any self-contained breathing apparatus with a full facepiece); or SAF (any supplied-air respirator with a full facepiece). *Emergency or planned entry into unknown concentrations or IDLH conditions:* SCBAF:PD,PP (any self-contained breathing apparatus that has a full facepiece and is operated in a pressure-demand or other positive-pressure mode); or SAF:PD,PP:ASCBA (any supplied-air respirator that has a full facepiece and is operated in a pressure-demand or other positive-pressure mode in combination with an auxiliary self-contained breathing apparatus operated in a pressure-demand or other positive pressure mode). *Escape:* GMFOV [any air-purifying, full-facepiece respirator (gas mask) with a chin-style, front- or back-, mounted organic vapor canister]; or SCBAE (any appropriate escape-type, self-contained breathing apparatus).

* Substance reported to cause eye irritation or damage; may require eye protection.

Storage: Prior to working with 2,3-dimethylbutane you should be trained on its proper handling and storage. Before entering confined space where this chemical may be present, check to make sure that an explosive concentration does not exist. 2,3-Dimethylbutane must be stored to avoid contact with oxidizers (such as perchlorates, peroxides, permanganates, chlorates and nitrates) since violent reactions occur. Sources of ignition, such as smoking and open flames, are prohibited where 2,3-dimethylbutane is handled, used, or stored. Metal containers involving the transfer of 5 gallons or more of 2,3-diemthylbutane should be grounded and bonded. Drums must be equipped with self-closing valves, pressure vacuum bungs, and flame arresters. Use only non-sparking tools and equipment, especially when opening and closing containers of 2,3-dimethylbutane. Wherever 2,3-diemthylbutane is used, handled, manufactured, or stored, use explosion-proof electrical equipment and fittings.

Shipping: This compound requires a shipping label of: "Flammable Liquid." It falls in DOT Hazard Class 3 and Packing Group II.

Spill Handling: Evacuate and restrict persons not wearing protective equipment from area of spill or leak until cleanup is complete. Remove all ignition sources. Establish forced ventilation to keep levels below explosive limit. Absorb liquids in vermiculite, dry sand, earth, peat, carbon, or a similar material and deposit in sealed containers. Keep this chemical out of a confined space, such as a sewer, because of the possibility of an explosion, unless the sewer is designed to prevent the build-up of explosive concentrations. It may be necessary to contain and dispose of this chemical as a hazardous waste. If material or contaminated runoff enters waterways, notify downstream users of potentially contaminated waters. Contact your Department of Environmental Protection or your regional office of the federal EPA for specific recommendations. If employees are required to clean-up spills, they must be properly trained and equipped. OSHA 1910.120(q) may be applicable.

Fire Extinguishing: This chemical is a flammable. Poisonous gases including hydrocarbons and carbon monoxide are produced in fire. Use dry chemical, carbon dioxide, or alcohol or polymer foam extinguishers. Vapors are heavier than air and will collect in low areas. Vapors may travel long distances to ignition sources and flashback. Vapors in confined areas may explode when exposed to fire. Containers may explode in fire. Storage containers and parts of containers may rocket great distances, in many directions. If material or contaminated runoff enters waterways, notify downstream users of potentially contaminated waters. Notify local health and fire officials and pollution control agencies. From a secure, explosion-proof location, use water spray to cool exposed containers. If cooling streams are ineffective (venting sound increases in volume and pitch, tank discolors, or shows any signs of deforming), withdraw immediately to a secure position. If employees are expected to fight fires, they must be trained and equipped in OSHA 1910.156.

Disposal Method Suggested: Incineration.

References

New Jersey Department of Health, "Hazardous Substance Fact Sheet: 2,3-Dimethylbutane," Trenton, NJ (September 7, 1987)

Dimethyl Carbamoyl Chloride

Molecular Formula: C_3H_6ClNO

Common Formula: $(CH_3)_2NCOCl$

Synonyms: Carbamic chloride, dimethyl-; Carbamoyl chloride, *N,N*-dimethylaminocarbonyl chloride; Carbamyl chloride, *N,N*-dimethyl-; Chlorid kyseliny dimethylkarbaminove; Cloroformic acid dimethylamide; Cloruro de dimetilcarbamolilo (Spanish); DDC; Dimethylamid kyseliny chlormravenci (Czech); (Dimethylamino)carbonyl chloride; *N,N*-Dimethylaminocarbonyl chloride; *N,N*-Dimethylcarbamic acid chloride; Dimethylcarbamic acid chloride; Dimethylcarbamic chloride; *N,N*-Dimethylcarbamidoyl chloride; Dimethylcarbamidoyl chloride; *N,N*-Dimethylcarbamoyl chloride; Dimethylcarbamoyl chloride; *N,N*-Dimethylcarbamyl chloride; Dimethylcarbamyl chloride; Dimethylchloroformamide; Dimethylkarbamoylchlorid (German); DMCC

CAS Registry Number: 79-44-7

RTECS Number: FD4200000

DOT ID: UN 2262

Regulatory Authority

- Carcinogen (Animal Positive) (IARC)[9] (DFG)[3] (Suspected Carcinogen) NTP
- Very Toxic Substance (World Bank)[15]
- Air Pollutant Standard Set (Several States)[60]
- Clean Air Act: Hazardous Air Pollutants (Title I, Part A, Section 112)
- EPA Hazardous Waste Number (RCRA No.): U097
- RCRA, 40CFR261, Appendix 8 Hazardous Constituents
- Superfund/EPCRA 40CFR302.4 Reportable Quantity (RQ): CERCLA, 1 lb (0.454 kg)
- EPCRA Section 313 Form R *de minimis* concentration reporting level: 0.1%
- Canada, WHMIS, Ingredients Disclosure List

Cited in U.S. State Regulations: Alaska (G), California (A, G), Florida (G), Illinois (G), Kansas (G), Louisiana (G), Maine (G), Maryland (G), Massachusetts (G), Minnesota (G), New Hampshire (G), New Jersey (G), New York (A), North Dakota (A), Pennsylvania (G, A), Rhode Island (G), Vermont (G), Virginia (G, A), Washington (G), West Virginia (G), Wisconsin (G).

Description: DMCC, $(CH_3)_2NCOCl$, is a liquid. Boiling point = 165 – 167°C. Freezing/Melting point = -33°C. Flash point = 68°C. Reactive with water.

Potential Exposure: DMCC is used as a chemical intermediate in the production of pharmaceuticals, pesticides, rocket fuel, and in dye synthesis. Human exposure is limited to but not restricted to chemical workers, pesticide formulators, dye makers, and pharmaceutical workers. DMCC has been found at levels up to 6 ppm during the production of phthaloyl chlorides. It is possible that levels of exposure might be higher in facilities in which the chemical is used for further synthesis. When DMCC is used as a dye intermediate, exposure can occur from the amount of residue in the product and its ability to migrate.

Incompatibilities: Strong oxidizers, strong acids, strong bases, water (rapidly hydrolyzes in water to dimethylamine, carbon dioxide, and hydrogen chloride). Attacks metals in the presence of moisture.

Permissible Exposure Limits in Air: There are no OSHA PEL. NIOSH recommends, "no detectable exposure levels for proven carcinogenic substances" (Annals of the New York Academy of Sciences, 271:200-207, 1976). ACGIH and DFG classify DMCC as an "industrial substance suspect of carcinogenic potential for man with no assigned TWA." Several states have set guidelines or standards for DMCC in ambient air[60] ranging from zero in North Dakota to 0.03 $\mu g/m^3$ (New York) to 0.24 $\mu g/m^3$ (Pennsylvania) to 3.0 $\mu g/m^3$ (Virginia).

Determination in Air: No method available.

Permissible Concentration in Water: No criteria set.

Determination in Water: No method available.

Routes of Entry: Inhalation, skin absorption, ingestion, skin and/or eye contact.

Harmful Effects and Symptoms

Short Term Exposure: Contact can irritate and burn the skin and eyes. Breathing dimethyl carbamoyl chloride can irritate the nose, throat and lungs causing cough, wheezing, shortness of breath, dyspnea (breathing difficulty); headache, nausea, vomiting.

Long Term Exposure: Repeated exposure may cause laryngitis, bronchitis, liver damage. A suspected human carcinogen.

Points of Attack: Eyes, skin, respiratory system, liver. Cancer site in animals: nasal cancer.

Medical Surveillance: Before beginning employment and at regular times after that, the following are recommended: Lung function tests. If symptoms develop or overexposure is suspected, the following may be useful: Liver function tests. Exposed workers should check their skin from time to time for new growths, changes in moles and sores that won't heal. In case skin exposure occurs, it is usually easily cured when treated promptly.

First Aid: If this chemical gets into the eyes, remove any contact lenses at once and irrigate immediately for at least 15 minutes, occasionally lifting upper and lower lids. Seek medical attention immediately. If this chemical contacts the skin, remove contaminated clothing and wash immediately with soap and water. Seek medical attention immediately. If this chemical has been inhaled, remove from exposure, begin rescue breathing (using universal precautions) if breathing has stopped and CPR if heart action has stopped. Transfer promptly to a medical facility. When this chemical has been swallowed, get medical attention. If victim is conscious, administer water or milk. Do not induce vomiting.

Personal Protective Methods: Wear protective gloves and clothing to prevent any reasonable probability of skin contact. Safety equipment suppliers/manufacturers can provide recommendations on the most protective glove/clothing material for your operation.. All protective clothing (suits, gloves, footwear, headgear) should be clean, available each day, and put on before work. Contact lenses should not be worn when working with this chemical. Wear splash-proof chemical goggles and face shield unless full facepiece respiratory protection is worn. Employees should wash immediately with soap when skin is wet or contaminated. Provide emergency showers and eyewash.

Respirator Selection: *At any detectable concentration:* SCBAF:PD,PP (any MSHA/NIOSH approved self-contained breathing apparatus that has a full facepiece and is operated in a pressure-demand or other positive-pressure mode); or SAF:PD,PP:ASCBA (any supplied-air respirator that has a full facepiece and is operated in a pressure-demand or other positive-pressure mode in combination with an auxiliary, self-contained breathing apparatus operated in a pressure-demand or other positive pressure mode). *Escape:* GMFOV [any air-purifying, full-facepiece respirator (gas mask) with a chin-style, front- or back-mounted organic vapor canister] or SCBAE (any appropriate escape-type, self-contained breathing apparatus).

Storage: Prior to working with DMCC you should be trained on its proper handling and storage. Store in tightly closed containers in a cool, well ventilated area away from strong oxidizers, strong acids, strong bases and moisture since violent reactions can occur. Metal containers involving the transfer of this chemical should be grounded and bonded. Where possible, automatically pump liquid from drums or other storage containers to process containers. Drums must be equipped with self-closing valves, pressure vacuum bungs, and flame arresters. Use only non-sparking tools and equipment, especially when opening and closing containers of this chemical. Sources of ignition such as smoking and open flames, are prohibited where this chemical is used, handled, or stored in a manner that could create a potential fire or explosion hazard.

Wherever this chemical is used, handled, manufactured, or stored, use explosion-proof electrical equipment and fittings. A regulated, marked area should be established where this chemical is handled, used, or stored in compliance with OSHA standard 1910.1045.

Shipping: This compound requires a shipping label of: "Corrosive." It falls in DOT Hazard Class 8 and Packing Group II.

Spill Handling: Evacuate and restrict persons not wearing protective equipment from area of spill or leak until cleanup is complete. Remove all ignition sources. Ventilate area of spill or leak. Absorb liquids in vermiculite, dry sand, earth, peat, carbon, or a similar material and deposit in sealed containers. *DO NOT USE WATER* or wet method. Keep this chemical out of a confined space, such as a sewer, because of the possibility of an explosion, unless the sewer is designed to prevent the build-up of explosive concentrations. It may be necessary to contain and dispose of this chemical as a hazardous waste. If material or contaminated runoff enters waterways, notify downstream users of potentially contaminated waters. Contact your Department of Environmental Protection or your regional office of the federal EPA for specific recommendations. If employees are required to clean-up spills, they must be properly trained and equipped. OSHA 1910.120(q) may be applicable.

Fire Extinguishing: Dimethyl carbamoyl chloride is a combustible liquid. Do not allow water to get inside containers. Use dry chemical, CO_2, or foam extinguishers. Poisonous gases including carbon monoxide, hydrochloric acid, and nitrogen oxides are produced in fire. Vapors are heavier than air and will collect in low areas. Vapors may travel long distances to ignition sources and flashback. Vapors in confined areas may explode when exposed to fire. Containers may explode in fire. Storage containers and parts of containers may rocket great distances, in many directions. If material or contaminated runoff enters waterways, notify downstream users of potentially contaminated waters. Notify local health and fire officials and pollution control agencies. From a secure, explosion-proof location, use water spray to cool exposed containers. If cooling streams are ineffective (venting sound increases in volume and pitch, tank discolors, or shows any signs of deforming), withdraw immediately to a secure position. If employees are expected to fight fires, they must be trained and equipped in OSHA 1910.156.

Disposal Method Suggested: High-temperature incineration with scrubbing of flue gas for HCl and No_x removal. Consult with environmental regulatory agencies for guidance on acceptable disposal practices. Generators of waste containing this contaminant (≥100 kg/mo) must conform with EPA regulations governing storage, transportation, treatment, and waste disposal.

References

New Jersey Department of Health, "Hazardous Substance Fact Sheet: Dimethylcarbamoyl Chloride," Trenton, NJ (December 1998)

Sax N. I., Ed., "Dangerous Properties of Industrial Materials Report," 7, No. 1, 51–54 (1987)

2,5-Dimethyl-2,5-Di (tert-Butyl Peroxy) Hexane

Molecular Formula: $C_{16}H_{34}O_2$

Common Formula: $[(H_3C)_3COOC(CH_3)_2\text{-}CH_2]_2$

Synonyms: Aztec® 2,5-di; 2,5-Dimethyl-2,5-di(*t*-butylperoxy)hexane; Esperal®120; Luperco 101-P20; Lupersol 101; Peroxide, (1,1,4,4-tetramethyl-1,4-butanediyl)bis(1,1-dimethylethyl); Peroxide, (1,1,4,4-tetramethyltetramethylene)bis(*tert*-butyl); Polyvel CR-5F; Trigonox® 101-101/45; Varox

CAS Registry Number: 78-63-7

RTECS Number: MO1835000

DOT ID: UN 2155; UN 3105 (see organic peroxide tables)

EINECS Number: 201-128-1

Cited in U.S. State Regulations: New Jersey (G).

Description: 2,5-Dimethyl-2,5-di (tert-butyl peroxy) hexane,$[(H_3C)_3COOC(CH_3)_2\text{-}CH_2]_2$, is a colorless to yellow substance. Freezing/Melting point = 8°C and. Boiling point = 50 – 52°C under 0.1 mm pressure. Flash point = 85°C. Insoluble in water.

Potential Exposure: It is used as a catalyst in making polyethylene, polystyrene and polyester resins.

Incompatibilities: Forms explosive mixture with air. Reducing agents, strong oxidizers, strong acids, combustible materials.

Permissible Exposure Limits in Air: No standards set.

Permissible Concentration in Water: No criteria set.

Harmful Effects and Symptoms

Short Term Exposure: Little is known about the health effects of this substance; however, similar chemicals can irritate, or may burn, the skin and eyes. Exposure to high levels may cause you to feel dizzy, lightheaded and to pass out.

Long Term Exposure: Unknown at this time.

First Aid: If this chemical gets into the eyes, remove any contact lenses at once and irrigate immediately for at least 15 minutes, occasionally lifting upper and lower lids. Seek medical attention immediately. If this chemical contacts the skin, remove contaminated clothing and wash immediately with soap and water. Seek medical attention immediately. If this chemical has been inhaled, remove from exposure, begin rescue breathing (using universal precautions) if breathing has stopped and CPR if heart action has stopped. Transfer promptly to a medical facility. When this chemical has been swallowed, get medical attention. Give large quantities of water and induce vomiting. Do not make an unconscious person vomit.

Personal Protective Methods: Wear protective gloves and clothing to prevent any reasonable probability of skin contact.

Safety equipment suppliers/manufacturers can provide recommendations on the most protective glove/clothing material for your operation.. All protective clothing (suits, gloves, footwear, headgear) should be clean, available each day, and put on before work. Contact lenses should not be worn when working with this chemical. Wear splash-proof chemical goggles and face shield unless full facepiece respiratory protection is worn. Employees should wash immediately with soap when skin is wet or contaminated. Provide emergency showers and eyewash.

Respirator Selection: Where the potential exists for exposure to 2,5-dimethyl-2,5-di (tert-butyl peroxy) hexane, use a MSHA/NIOSH approved supplied-air respirator with a full facepiece operated in the positive pressure mode or with a full facepiece, hood, or helmet in the continuous flow mode, or use a MSHA/NIOSH approved self-contained breathing apparatus with a full facepiece operated in pressure-demand or other positive pressure mode.

Storage: Prior to working with this chemical you should be trained on its proper handling and storage. Store in tightly closed containers in a cool, well-ventilated area away from strong oxidizers (such as chlorine, bromine and fluorine), strong acids (such as hydrochloric, sulfuric and nitric) and combustibles. Sources of ignition, such as smoking and open flames, are prohibited where 2,5-dimethyl-2,5-di (tert-butyl peroxy) hexane is used, handled, or stored in a manner that could create a potential fire or explosion hazard. Use only non-sparking tools and equipment, especially when opening and closing containers of 2,5-dimethyl-2,5-di (tert-butyl peroxy) hexane. See OSHA standard 1910.104 and NFPA 43A *Code for the Storage of Liquid and Solid Oxidizers* for detailed handling and storage regulations.

Shipping: This compound requires a shipping label of: "Organic Peroxide." It falls in DOT Hazard Class 5.2 and Packing Group II.

Spill Handling: Evacuate and restrict persons not wearing protective equipment from area of spill or leak until cleanup is complete. Remove all ignition sources. Ventilate area of spill or leak. Absorb liquids in vermiculite, dry sand, earth, peat, carbon, or a similar material and deposit in sealed containers. Keep this chemical out of a confined space, such as a sewer, because of the possibility of an explosion, unless the sewer is designed to prevent the build-up of explosive concentrations. It may be necessary to contain and dispose of this chemical as a hazardous waste. If material or contaminated runoff enters waterways, notify downstream users of potentially contaminated waters. Contact your Department of Environmental Protection or your regional office of the federal EPA for specific recommendations. If employees are required to clean-up spills, they must be properly trained and equipped. OSHA 1910.120(q) may be applicable.

Fire Extinguishing: This chemical is a flammable liquid. Poisonous gases are produced in fire. Use dry chemical, carbon dioxide, or foam extinguishers. Vapors are heavier than air and will collect in low areas. Vapors may travel long distances to ignition sources and flashback. Vapors in confined areas may explode when exposed to fire. Containers may explode in fire. Storage containers and parts of containers may rocket great distances, in many directions. If material or contaminated runoff enters waterways, notify downstream users of potentially contaminated waters. Notify local health and fire officials and pollution control agencies. From a secure, explosion-proof location, use water spray to cool exposed containers. If cooling streams are ineffective (venting sound increases in volume and pitch, tank discolors, or shows any signs of deforming), withdraw immediately to a secure position. If employees are expected to fight fires, they must be trained and equipped in OSHA 1910.156.

References

New Jersey Department of Health, "Hazardous Substance Fact Sheet: 2,5-Dimethyl-2,5-Di (tert-Butyl Peroxy) Hexane," Trenton, NJ (March 1987)

Dimethyldichlorosilane

Molecular Formula: $C_2H_6Cl_2Si$

Common Formula: $SiCl_2(CH_3)_2$

Synonyms: A13-51462; Dichlorodimethylsilane; Dichlorodimethylsilicone; Dimetildiclorosilano (Spanish); DMCS; Inerton AW-DMCS; Inerton DW-DMC; NSC 77070; Silane, dichlorodimethyl-

CAS Registry Number: 75-78-5

RTECS Number: VV3150000

DOT ID: UN 1162

Regulatory Authority

- OSHA 29CFR1910.119, Appendix A, Process Safety List of Highly Hazardous Chemicals, TQ = 1,000 lb (454 kg)
- Clean Air Act: Accidental Release Prevention/Flammable substances, (Section 112[r], Table 3), TQ = 5,000 lb (2,270 kg)
- Superfund/EPCRA 40CFR355, Appendix B Extremely Hazardous Substances: TPQ = 500 lb (227 kg)
- Superfund/EPCRA 40CFR302.4 Reportable Quantity (RQ): EHS, 1 lb (0.454 kg)
- EPCRA Section 313 Form R *de minimis* concentration reporting level: 1.0%
- Canada, WHMIS, Ingredients Disclosure List

Cited in U.S. State Regulations: Florida (G), Maine (G), Massachusetts (G), New Hampshire (G), New Jersey (G), Oklahoma (G), Pennsylvania (G).

Description: DMCS, $SiCl_2(CH_3)_2$, is a colorless liquid with sharp, irritating odor. Boiling point = 70°C. Flash point ≤21°C. Explosive limits: LEL = 3.4%; UEL ≥9.5°C. Hazard Identification (based on NFPA-704 M Rating System): Health 3, Flammability 3, Reactivity 1. Decomposes in water.

Potential Exposure: Used as an intermediate in the manufacture of silicone polymers.

Incompatibilities: Forms explosive gas mixture with air. Water, steam, moisture forms toxic and corrosive hydrogen chloride gas. Incompatible with acetone, amines, ammonia, alcohols, strong oxidizers, caustics. Attacks most metals. Do not store in temperatures above 122°F/50°C.

Permissible Exposure Limits in Air: No standards set.

Determination in Air: Not established. However, this chemical is highly corrosive and a dangerous fire hazard.

Permissible Concentration in Water: Reacts with water. No criteria set.

Routes of Entry: Inhalation.

Harmful Effects and Symptoms

Short Term Exposure: Acute inhalation exposure may result in sneezing, choking, laryngitis, dyspnea (shortness of breath), respiratory tract irritation, and chest pain. Higher exposures can cause pulmonary edema, a medical emergency that can be delayed for several hours. This can cause death. Bleeding of nose and gums, ulceration of the nasal and oral mucosa, pulmonary edema, chronic bronchitis, and pneumonia may also occur. If the eyes have come in contact with dimethyldichlorosilane, irritation, pain, swelling, corneal erosion, and blindness may result. Dermatitis (red, inflamed skin), severe burns, pain, and shock generally follow dermal exposure. Inhalation irritates mucous membranes. Severe gastrointestinal damage may occur. Vapors cause severe eye and lung injury. Upon short contact, second and third degree burns may occur. Signs and symptoms of acute ingestion of dimethyldichlorosilane may be severe and include increased salivation, intense thirst, difficulty swallowing, chills, pain, and shock. Oral, esophageal, and stomach burns are common. Vomitus generally has a coffee-ground appearance. The potential for circulatory collapse is high following ingestion of dimethyldichlorosilane.

Long Term Exposure: Very irritating substances may affect the lungs.

Points of Attack: Lungs.

Medical Surveillance: Before beginning employment and at regular times after that, for those with frequent or potentially high exposures, the following is recommended: Lung function tests. If symptoms develop or overexposure is suspected, the following may be useful: Consider chest x-ray after acute overexposure.

First Aid: If this chemical gets into the eyes, remove any contact lenses at once and irrigate immediately for at least 15 minutes, occasionally lifting upper and lower lids. Seek medical attention immediately. If this chemical contacts the skin, remove contaminated clothing and wash immediately with soap and water. Seek medical attention immediately. If this chemical has been inhaled, remove from exposure, begin rescue breathing (using universal precautions) if breathing has stopped and CPR if heart action has stopped. Transfer promptly to a medical facility. When this chemical has been swallowed, get medical attention. If victim *is conscious*, administer water or milk. Do not induce vomiting. Medical observation is recommended for 24 – 48 hours after breathing overexposure, as pulmonary edema may be delayed. As first aid for pulmonary edema, a doctor or authorized paramedic may consider administering a corticosteroid spray.

Personal Protective Methods: Wear protective gloves and clothing to prevent any reasonable probability of skin contact. Safety equipment suppliers/manufacturers can provide recommendations on the most protective glove/clothing material for your operation.. All protective clothing (suits, gloves, footwear, headgear) should be clean, available each day, and put on before work. Contact lenses should not be worn when working with this chemical. Wear splash-proof chemical goggles and face shield unless full facepiece respiratory protection is worn. Employees should wash immediately with soap when skin is wet or contaminated. Provide emergency showers and eyewash.

Respirator Selection: Where the potential exists for exposure to dimethyl dichlorosilane, use a MSHA/NIOSH approved supplied-air respirator with a full facepiece operated in the positive pressure mode or with a full facepiece, hood, or helmet in the continuous flow mode, or use a MSHA/NIOSH approved self-contained breathing apparatus with a full facepiece operated in pressure-demand or other positive pressure mode.

Storage: Dimethyl dichlorosilane must be stored to avoid contact with oxidizers (such as perchlorates, peroxides, permanganates, chlorates and nitrates) since violent reactions occur. Before entering confined space where this chemical may be present, check to make sure that an explosive concentration does not exist. Store in tightly closed containers in a cool, well-ventilated area away from water, steam or moisture because toxic and corrosive hydrogen chloride gas can be produced. Do not store at temperatures above 50°C/122°F. Sources of ignition, such as smoking and open flames, are prohibited where dimethyl dichlorosilane is handled, used, or stored. Metal containers involving the transfer of 5 gallons or more of dimethyl dichlorosilane should be grounded and bonded. Drums must be equipped with self-closing valves, pressure vacuum bungs, and flame arresters. Use only nonsparking tools and equipment, especially when opening and closing containers of dimethyl dichlorosilane.

Shipping: This compound requires a shipping label of: "Flammable Liquid, Poison, Corrosive." It falls in DOT Hazard Class 3 and Packing Group I.

Spill Handling: Evacuate and restrict persons not wearing protective equipment from area of spill or leak until cleanup is complete. Remove all ignition sources. Establish forced ventilation to keep levels below explosive limit. Use water spray to reduce vapors; however do not get water inside containers. Small spills: Absorb liquids in vermiculite, dry sand, earth, peat, carbon, or a similar material and deposit in sealed containers. Large spills: dike far ahead of spill for later disposal. Use effective fume removal device. Keep this chemical out of a confined space, such as a sewer, because of the possibility of an explosion, unless the sewer is designed to prevent the build-up of explosive concentrations. It may be necessary to contain and dispose of this chemical as a hazardous waste. If material or contaminated runoff enters waterways, notify downstream users of potentially contaminated waters. Contact your Department of Environmental Protection or your regional office of the federal EPA for specific recommendations. If employees are required to clean-up spills, they must be properly trained and equipped. OSHA 1910.120(q) may be applicable.

Fire Extinguishing: This chemical is a combustible liquid. Poisonous gases including phosgene and hydrogen chloride are produced in fire. Use dry chemical, carbon dioxide, water spray, or foam extinguishers. Move container from fire area if it can be done without risk. Do not get water inside container. Isolate for one-half mile in all directions if tank car or truck involved in fire. Vapors are heavier than air and will collect in low areas. Vapors may travel long distances to ignition sources and flashback. Vapors in confined areas may explode when exposed to fire. Containers may explode in fire. Storage containers and parts of containers may rocket great distances, in many directions. If material or contaminated runoff enters waterways, notify downstream users of potentially contaminated waters. Notify local health and fire officials and pollution control agencies. From a secure, explosion-proof location, use water spray to cool exposed containers. If cooling streams are ineffective (venting sound increases in volume and pitch, tank discolors, or shows any signs of deforming), withdraw immediately to a secure position. If employees are expected to fight fires, they must be trained and equipped in OSHA 1910.156.

References

New Jersey Department of Health, "Hazardous Substance Fact Sheet: Dimethyl Dichlorosilane," Trenton, NJ (January 1996)

U.S. Environmental Protection Agency, "Chemical Profile: Dimethyl Dichlorosilane," Washington, DC, Chemical Emergency Preparedness Program (November 30, 1987)

Dimethyldiethoxysilane

Molecular Formula: $C_6H_{16}O_2Si$

Common Formula: $(CH_3)_2Si(OC_2H_5)_2$

Synonyms: Diethoxydimethylsilane; Silane, diethoxydimethyl-

CAS Registry Number: 78-62-6

RTECS Number: VV3590000

DOT ID: UN 2380

Regulatory Authority

- Canada, WHMIS, Ingredients Disclosure List

Cited in U.S. State Regulations: Florida (G), New Hampshire (G), New Jersey (G), Pennsylvania (G).

Description: Dimethyldiethoxysilane, $(CH_3)_2Si(OC_2H_5)_2$, is a colorless liquid. Boiling point = 114°C. Flash point = 23°C.

Potential Exposure: This material is used in water repellent formulations.

Incompatibilities: Forms explosive mixture with air. Strong oxidizers.

Permissible Exposure Limits in Air: No standards set.

Permissible Concentration in Water: No criteria set.

Routes of Entry: Inhalation.

Harmful Effects and Symptoms

Short Term Exposure: Exposure can irritate the eyes, nose and throat. Contact may irritate the skin causing a rash or burning feeling on contact.

Long Term Exposure: Unknown at this time.

First Aid: If this chemical gets into the eyes, remove any contact lenses at once and irrigate immediately for at least 15 minutes, occasionally lifting upper and lower lids. Seek medical attention immediately. If this chemical contacts the skin, remove contaminated clothing and wash immediately with soap and water. Seek medical attention immediately. If this chemical has been inhaled, remove from exposure, begin rescue breathing (using universal precautions) if breathing has stopped and CPR if heart action has stopped. Transfer promptly to a medical facility. When this chemical has been swallowed, get medical attention. Give large quantities of water and induce vomiting. Do not make an unconscious person vomit.

Personal Protective Methods: Wear protective gloves and clothing to prevent any reasonable probability of skin contact. Safety equipment suppliers/manufacturers can provide recommendations on the most protective glove/clothing material for your operation.. All protective clothing (suits, gloves, footwear, headgear) should be clean, available each day, and put on before work. Contact lenses should not be worn when working with this chemical. Wear splash-proof chemical goggles and face shield unless full facepiece respiratory protection is worn. Employees should wash immediately with soap when skin is wet or contaminated. Provide emergency showers and eyewash.

Respirator Selection: Where the potential exists for exposure to Dimethyldiethoxysilane, use a MSHA/NIOSH approved supplied-air respirator with a full facepiece operated in the positive pressure mode or with a full facepiece, hood, or helmet in the continuous flow mode, or use a MSHA/NIOSH approved self-contained breathing apparatus with a full facepiece operated in pressure-demand or other positive pressure mode.

Storage: Prior to working with this chemical you should be trained on its proper handling and storage. Store in tightly closed containers in a cool, well-ventilated area away from heat. Sources of ignition, such as smoking and open flames, are prohibited where Dimethyldiethoxysilane is handled, used or stored. Metal containers involving the transfer of 5 gallons or more of Dimethyldiethoxysilane should be grounded and bonded. Drums must be equipped with well-closing valves, pressure vacuum bungs, and flame arresters. Use only non-sparking tools and equipment, especially when opening and closing containers of Dimethyldiethoxysilane.

Shipping: This compound requires a shipping label of: "Flammable Liquid." It falls in DOT Hazard Class 3 and Packing Group II.

Spill Handling: Evacuate and restrict persons not wearing protective equipment from area of spill or leak until cleanup is complete. Remove all ignition sources. Ventilate area of spill or leak. Absorb liquids in vermiculite, dry sand, earth, peat, carbon, or a similar material and deposit in sealed containers. Keep this chemical out of a confined space, such as a sewer, because of the possibility of an explosion, unless the sewer is designed to prevent the build-up of explosive concentrations. It may be necessary to contain and dispose of this chemical as a hazardous waste. If material or contaminated runoff enters waterways, notify downstream users of potentially contaminated waters. Contact your Department of Environmental Protection or your regional office of the federal EPA for specific recommendations. If employees are required to clean-up spills, they must be properly trained and equipped. OSHA 1910.120(q) may be applicable.

Fire Extinguishing: This chemical is a flammable liquid. Poisonous gases are produced in fire. Use dry chemical, carbon dioxide, water spray, or alcohol foam extinguishers. Vapors are heavier than air and will collect in low areas. Vapors may travel long distances to ignition sources and flashback. Vapors in confined areas may explode when exposed to fire. Containers may explode in fire. Storage containers and parts of containers may rocket great distances, in many directions. If material or contaminated runoff enters waterways, notify downstream users of potentially contaminated waters. Notify local health and fire officials and pollution control agencies. From a secure, explosion-proof location, use water spray to cool exposed containers. If cooling streams are ineffective (venting sound increases in volume and pitch, tank discolors, or shows any signs of deforming), withdraw immediately to a secure position. If employees are expected to fight fires, they must be trained and equipped in OSHA 1910.156.

References

New Jersey Department of Health, "Hazardous Substance Fact Sheet: Dimethyldiethoxysilane," Trenton, NJ (June 1986)

Dimethyl Disulfide

Molecular Formula: $C_2H_6S_2$

Common Formula: CH_3SSCH_3

Synonyms: Methyl disulfide

CAS Registry Number: 624-92-0

RTECS Number: JO1927500

DOT ID: UN 2381

Regulatory Authority

- Air Pollutant Standard Set (former USSR)[43]

Cited in U.S. State Regulations: Massachusetts (G), New Hampshire (G), New Jersey (G), Pennsylvania (G).

Description: Dimethyldisulfide, CH_3SSCH_3, is a flammable liquid. The odor threshold in water is 0.3 – 1.2 parts per billion. Boiling point = 110°C. Flash point = 7°C.

Potential Exposure: This material may be used as an organic intermediate.

Incompatibilities: Forms explosive mixture with air. Contact with water forms hydrogen sulfide. Oxidizing materials, acids can cause a violent reaction.

Permissible Exposure Limits in Air: The former USSR-UNEP/IRPTC project[43] has set a MAC in ambient air in residential areas of 0.7 mg/m^3 on a momentary basis.

Permissible Concentration in Water: The former USSR-UNEP/IRPTC project[43] has set a MAC in water bodies used for domestic purposes of 0.04 mg/l.

Harmful Effects and Symptoms

Short Term Exposure: Eye contact can cause severe irritation and burns. This material is highly irritating and toxic by inhalation. Health hazards resemble those of sulfides and alkyl disulfides. The related dimethyl sulfide causes softening and irritation of the skin.

Long Term Exposure: Highly irritating substances may affect the lungs.

Points of Attack: Lungs.

Medical Surveillance: Lung function tests. Consider chest x-ray following acute overexposure.

First Aid: If this chemical gets into the eyes, remove any contact lenses at once and irrigate immediately for at least 15 minutes, occasionally lifting upper and lower lids. Seek medical attention immediately. If this chemical contacts the skin, remove contaminated clothing and wash immediately with soap and water. Seek medical attention immediately. If this chemical has been inhaled, remove from exposure, begin rescue breathing (using universal precautions) if breathing has stopped and CPR if heart action has stopped. Transfer promptly to a medical facility. When this chemical has been swallowed, get medical attention. Give large quantities of water and induce vomiting. Do not make an unconscious person vomit.

Personal Protective Methods: Wear protective gloves and clothing to prevent any reasonable probability of skin contact. Safety equipment suppliers/manufacturers can provide recommendations on the most protective glove/clothing material for your operation.. All protective clothing (suits, gloves, footwear, headgear) should be clean, available each day, and put on before work. Contact lenses should not be worn when working with this chemical. Wear splash-proof chemical goggles and face shield unless full facepiece respiratory protection is worn. Employees should wash immediately with soap when skin is wet or contaminated. Provide emergency showers and eyewash. For emergency situations, wear a positive pressure, pressure-demand, full facepiece self-contained breathing apparatus (SCBA) or pressure-demand supplied air respirator with escape SCBA and a fully-encapsulating, chemical resistant suit.

Respirator Selection: Use any MSHA/NIOSH approved self-contained breathing apparatus that has a full facepiece and is operated in a pressure-demand or other positive-pressure mode); or SAF:PD,PP:ASCBA (any supplied-air respirator that has a full facepiece and is operated in a pressure-demand or other positive-pressure mode in combination with an auxiliary, self-contained breathing apparatus operated in a pressure-demand or other positive pressure mode).

Storage: Prior to working with Dimethyl disulfide you should be trained on its proper handling and storage. Before entering confined space where this chemical may be present, check to make sure that an explosive concentration does not exist. Store in tightly closed containers in a cool, well ventilated area away from any form of moisture, oxidizers, acids. Metal containers involving the transfer of this chemical should be grounded and bonded. Where possible, automatically pump liquid from drums or other storage containers to process containers. Drums must be equipped with self-closing valves, pressure vacuum bungs, and flame arresters. Use only non-sparking tools and equipment, especially when opening and closing containers of this chemical. Sources of ignition such as smoking and open flames, are prohibited where this chemical is used, handled, or stored in a manner that could create a potential fire or explosion hazard. Wherever this chemical is used, handled, manufactured, or stored, use explosion-proof electrical equipment and fittings.

Shipping: This compound requires a shipping label of: "Flammable Liquid." It falls in DOT Hazard Class 3 and Packing Group II.

Spill Handling: Evacuate and restrict persons not wearing protective equipment from area of spill or leak until cleanup is complete. Remove all ignition sources. Ventilate area of spill or leak. Use water spray to reduce vapors.

Small spills: Absorb liquids in vermiculite, dry sand, earth, peat, carbon, or a similar material and deposit in sealed containers.

Large spills: dike far ahead of spill for later disposal. Keep this chemical out of a confined space, such as a sewer, because of the possibility of an explosion, unless the sewer is designed to prevent the build-up of explosive concentrations. It may be necessary to contain and dispose of this chemical as a hazardous waste. If material or contaminated runoff enters waterways, notify downstream users of potentially contaminated waters. Contact your Department of Environmental Protection or your regional office of the federal EPA for specific recommendations. If employees are required to clean-up spills, they must be properly trained and equipped. OSHA 1910.120(q) may be applicable.

Fire Extinguishing: This material may be ignited by heat, sparks or flames. Poisonous gases including sulfur oxides are produced in fire. Small fires: dry chemical, carbon dioxide, water spray, or foam. Large fires: water spray, fog, or foam. Move container from fire area if you can do so without risk. Spray cooling water on containers that are exposed to flames until well after fire is out. Fight fire from maximum distance. Dike fire control water for later disposal; do not scatter the material. Stay upwind; keep out of low areas. Wear positive pressure breathing apparatus and special protective clothing. Vapor explosion and poison hazard indoors, outdoors or in sewers. Vapors may travel long distances to ignition sources and flashback. Vapors in confined areas may explode when exposed to fire. Storage containers and parts of containers may rocket great distances, in many directions. If material or contaminated runoff enters waterways, notify downstream users of potentially contaminated waters. Notify local health and fire officials and pollution control agencies. From a secure, explosion-proof location, use water spray to cool exposed containers. If cooling streams are ineffective (venting sound increases in volume and pitch, tank discolors, or shows any signs of deforming), withdraw immediately to a secure position. If employees are expected to fight fires, they must be trained and equipped in OSHA 1910.156.

References

U.S. Environmental Protection Agency, "Chemical Profile: Methyl Disulfide," Washington, DC, Chemical Emergency Preparedness Program (November 30, 1987)

Dimethyl Ether

Molecular Formula: C_2H_6O

Synonyms: Dimethyl ether; Dimethyl oxide; DME; Eter metilico (Spanish); Methane oxybis-; Methyl ether; Oxybismethane; Wood ether

CAS Registry Number: 115-10-6

RTECS Number: PM4780000

DOT ID: UN 1033

EINECS Number: 204-065-8

Regulatory Authority

- Air Pollutant Standard Set (DFG) (AIHA)

- Clean Air Act: Accidental Release Prevention/Flammable substances, (Section 112[r], Table 3), TQ = 10,000 lb (4,540 kg)

Cited in U.S. State Regulations: California (A, G), Florida (G), Massachusetts (G), Minnesota (G), New Jersey (G), Pennsylvania (G).

Description: Dimethyl ether is a colorless compressed liquefied gas or liquid. Boiling point = -24°C. Freezing/Melting point = -141°C. Flash point = flammable gas (-41°C). Autoignition temperature = 350°C. Hazard Identification (based on NFPA-704 M Rating System): Health 1, Flammability 4, Reactivity 1. Explosive limits: LEL = 3.4%; UEL = 27.0%. Soluble in water.

Potential Exposure: Used as a refrigerant, solvent, propellent for aerosol sprays, and in making plastics.

Incompatibilities: Flammable gas. Forms explosive mixture with air. Forms unstable peroxides in containers that have been opened or remain in storage for more than 6 months. Peroxides can be detonated by friction, impact or heating. Violent reaction with strong oxidizers, aluminum hydride, lithium aluminum hydride. Keep away from heat, air, sunlight.

Permissible Exposure Limits in Air: There is no OSHA PEL. The DFG set a MAK of 1,000 ppm (1,900 mg/m^3) with Peak limitation of 2 times the MAK (60 min.) not to be exceeded 3 times per workshift. The AIHA recommends a level of 500 ppm (942 mg/m^3) TWA.

Routes of Entry: Inhalation.

Harmful Effects and Symptoms

Short Term Exposure: Irritates the eyes and respiratory tract. High exposure can cause headache, dizziness, lightheadedness and unconsciousness. Rapid evaporation of the liquid can cause severe frostbite. May affect the nervous system.

Points of Attack: Nervous system.

First Aid: If this chemical gets into the eyes, remove any contact lenses at once and irrigate immediately for at least 15 minutes, occasionally lifting upper and lower lids. Seek medical attention immediately. If this chemical contacts the skin, remove contaminated clothing and wash immediately with soap and water. Seek medical attention immediately. If this chemical has been inhaled, remove from exposure, begin rescue breathing (using universal precautions) if breathing has stopped and CPR if heart action has stopped. Transfer promptly to a medical facility. When this chemical has been swallowed, get medical attention. Give large quantities of water and induce vomiting. Do not make an unconscious person vomit. If frostbite has occurred, seek medical attention immediately; do *NOT* rub the affected areas or flush them with water. In order to prevent further tissue damage, do *NOT* attempt to remove frozen clothing from frostbitten areas. If frostbite has *NOT* occurred, immediately and thoroughly wash contaminated skin with soap and water.

Personal Protective Methods: Wear protective gloves and clothing to prevent any reasonable probability of skin contact. Safety equipment suppliers/manufacturers recommend butyl and neoprene as protective materials. Ali protective clothing (suits, gloves, footwear, headgear) should be clean, available each day, and put on before work. Contact lenses should not be worn when working with this chemical. For gas wear gas-proof goggles and face shield, for liquid wear splash-proof chemical goggles and face shield unless full facepiece respiratory protection is worn. Employees should wash immediately with soap when skin is wet or contaminated. Provide emergency showers and eyewash. Where exposure to the liquefied compressed gas may occur, employees should be provided with special clothing designed to prevent frostbite.

Respirator Selection: Where the potential for exposure to this chemical, use a MSHA/NIOSH approved supplied-air respirator with a full facepiece operated in the positive pressure mode or with a full facepiece, hood, or helmet in the continuous flow mode, or use a MSHA/NIOSH approved self-contained breathing apparatus with a full facepiece operated in pressure-demand or other positive pressure mode.

Storage: Prior to working with DME you should be trained on its proper handling and storage. Before entering confined space where DME may be present, check to make sure that an explosive concentration does not exist. Store in tightly closed containers in a cool, dark, well ventilated area. Metal containers involving the transfer of this chemical should be grounded and bonded. Where possible, automatically pump liquid from drums or other storage containers to process containers. Drums must be equipped with self-closing valves, pressure vacuum bungs, and flame arresters. Use only non-sparking tools and equipment, especially when opening and closing containers of this chemical. Sources of ignition such as smoking and open flames, are prohibited where this chemical is used, handled, or stored in a manner that could create a potential fire or explosion hazard. Wherever this chemical is used, handled, manufactured, or stored, use explosion-proof electrical equipment and fittings.

Note: Forms unstable peroxides in either containers that have been opened and remain in storage for more than 6 months. Procedures for the handling, use and storage of cylinders should be in compliance with OSHA 1910.101 and 1910.169 as with the recommendations of the Compressed Gas Association.

Shipping: Label "Flammable Gas." Hazard class 2.1. Packing Group II.

Spill Handling: If in a building, evacuate building and confine vapors by closing doors and shutting down HVAC systems. Restrict persons not wearing protective equipment from area of spill or leak until cleanup is complete. Remove all ignition sources. Establish forced ventilation to keep levels below explosive limit and to disperse the gas. Wear chemical protective suit with self-contained breathing apparatus to combat spills. Stay upwind and use water spray to "knock down" vapor; contain runoff. Stop the flow of gas, if it can be done safely from a distance. If source is a cylinder and the leak cannot be stopped in place, remove the leaking cylinder to a safe place, and repair leak or allow cylinder to empty. Keep this chemical out of confined spaces, such as a sewer, because of the possibility of explosion, unless the sewer is designed to prevent the buildup of explosive concentrations. If employees are required to clean-up spills, they must be properly trained and equipped. OSHA 1910.120(q) may be applicable.

Fire Extinguishing: This chemical is a flammable gas or liquid under pressure. Poisonous gases are produced in fire. Do not extinguish the fire unless the flow of gas can be stopped and any remaining gas is out of the line. Specially trained personnel may use fog lines to cool exposures and let the fire burn itself out. Vapors are heavier than air and will collect in low areas. Vapors may travel long distances to ignition sources and flashback. Vapors in confined areas may explode when exposed to fire. Containers may explode in fire. Storage containers and parts of containers may rocket great distances, in many directions. If material or contaminated runoff enters waterways, notify downstream users of potentially contaminated waters. Notify local health and fire officials and pollution control agencies. From a secure, explosion-proof location, use water spray to cool exposed containers. If cooling streams are ineffective (venting sound increases in volume and pitch, tank discolors, or shows any signs of deforming), withdraw immediately to a secure position. If cylinders are exposed to excessive heat from fire or flame contact, withdraw immediately to a secure location. If employees are expected to fight fires, they must be trained and equipped in OSHA 1910.156.

References

New Jersey Department of Health and Senior Services, "Hazardous Substance Fact Sheet, Dimethyl Ether," Trenton, NJ (June 1996)

N,N-Dimethylformamide

Molecular Formula: C_3H_7NO

Common Formula: $HCON(CH_3)_2$

Synonyms: AI3-03311; Dimethylformamid (German); *N,N*-Dimethylformamide; *N*-Dimethylformamide; *N,N*-Dimethylmethanamide; Dimetilformamida (Spanish); Dimetilformamide (Italian); Dimetylformamidu (Czech); DMF; DMFA; Dwumethyloformamid (Polish); Dynasolve 100; Formamide, *N,N*-dimethyl-; Formic acid, amide, *N,N*-dimethyl-; *N*-Formyldimethylamine; NCI-C60913; NSC-5356; U-4224; Weld-On P-70 primer

CAS Registry Number: 68-12-2

RTECS Number: LO2100000

DOT ID: UN 2265

EEC Number: 616-001-00-X

EINECS Number: 200-679-5

Regulatory Authority

- Air Pollutant Standard Set (ACGIH)[1] (Australia) (DFG)[3] (HSE)[33] (Israel) (Mexico) (former USSR)[35][43] (OSHA)[58] (Several States)[60] (Several Canadian Provinces)
- Clean Air Act: Hazardous Air Pollutants (Title I, Part A, Section 112)
- Superfund/EPCRA 40CFR302.4 Reportable Quantity (RQ): CERCLA, 1 lb (0.454 kg)
- EPCRA Section 313 Form R *de minimis* concentration reporting level: 1.0%
- Canada, WHMIS, Ingredients Disclosure List

Cited in U.S. State Regulations: Alaska (G), California (A, G), Connecticut (A), Florida (G), Illinois (G), Maine (G), Massachusetts (G, A), Minnesota (G), Nevada (A), New Hampshire (G), New Jersey (G), New York (G), North Dakota (A), Pennsylvania (G), Rhode Island (G), Virginia (A), West Virginia (G).

Description: Dimethylformamide, $HCON(CH_3)_2$, is a flammable, colorless liquid with a fishy, unpleasant odor at relatively low concentrations. The odor threshold is 0.47 – 100 ppm. Boiling point = 153°C. Freezing/Melting point = -61°C. Flash point = 58°C. Explosive limits are: LEL = 2.2%; UEL = 15.2% @ 100°C. Hazard Identification (based on NFPA-704 M Rating System): Health 1, Flammability 2, Reactivity 0. Soluble in water.

Potential Exposure: Dimethylformamide has powerful solvent properties for a wide range of organic compounds. Because of Dimethylformamide's physical properties, it has been used when solvents with a slow rate of evaporation are required. It finds particular usage in the manufacturer of polyacrylic fibers, butadiene, purified acetylene, pharmaceuticals, dyes, petroleum products, and other organic chemicals.

Incompatibilities: Forms explosive mixture with air. Contact with carbon tetrachloride and other halogenated compounds, particularly in contact with iron or strong oxidizers may cause fire and explosions. Vigorous reaction with alkylaluminums. Incompatible with non-oxidizing mineral acids, strong acids, chlorinated hydrocarbons, isocyanates, inorganic nitrates, phenols, cresols, ammonia, bromine, chromic anhydride, magnesium nitrate, methylene diisocyanate, phosphorous trioxide, triethylaluminum. Attacks some plastics, rubber and coatings.

Permissible Exposure Limits in Air: The Federal standard (OSHA PEL), the ACGIH TWA,[1] DFG, Mexico, the Canadian provinces of Alberta, BC, Ontario, and Quebec, and HSE value[33] is 10 ppm (30 mg/m^3). The notation "skin" indicates the possibility of cutaneous absorption. The STEL set by HSE,[33] Alberta, BC, and Mexico is 20 ppm (60 mg/m^3). The ILDH level is 500 ppm. Australia, Israel, Japan, Sweden and Czechoslovakia[35] use the same TWA. The former USSR[35][43] has set a MAC in workplace air of 10 mg/m^3. The former USSR has also set MAC values for ambient air in residential areas of 0.03 mg/m^3 on either a momentary or a daily average basis. Several states have set guidelines or standards for DMF in ambient air[60] ranging from 8.1 μg/m^3 (Massachusetts) to 300 μg/m^3 (North Dakota) to 500 μg/m^3 (Virginia) to 600 μg/m^3 (Connecticut) to 714 μg/m^3 (Nevada).

Determination in Air: Si gel; Methanol; Gas chromatography/Flame ionization detection; NOIOSH IV, Method #2004.[18]

Permissible Concentration in Water: A MAC value in water bodies used for domestic purposes has been set by The former USSR[35][43] at 10.0 mg/l. Beyond that a MAC in water bodies used for fishery purposes is 0.28 mg/l.

Routes of Entry: Inhalation of vapor, and absorption through intact skin, ingestion and skin and eye contact.

Harmful Effects and Symptoms

Short Term Exposure: DMF irritates the eyes, skin, and respiratory tract.

Inhalation: Thirteen workers exposed to concentrations below 20 ppm and occasionally to higher levels for up to 32 weeks complained of nausea, vomiting, and colicky abdominal pain; some cases of liver enlargement were detected. A worker who was splashed over 20% of his body surface and simultaneously exposed to high concentrations, initially suffered only skin irritation; abdominal pain began 62 hours after the exposure and became progressively more severe, with vomiting, and high blood pressure; the effects were gone by the 7th day after exposure. Some workers have noted facial flushing (especially after alcohol ingestion). May also cause loss of appetite, stomach pain, constipation, diarrhea, nausea, and vomiting; liver injury.

Skin: Rapidly penetrates the skin. May cause or increase the severity of effects reported above. It is also highly irritating to skin.

Eyes: Highly irritating to eyes and mucous membranes.

Ingestion: Fatal dose for humans has been estimated at 10 grams (about 1/3-ounce).

Long Term Exposure: Prolonged or repeated skin contact with the liquid defats the skin and may cause irritation and rash. Prolonged inhalation at 100 ppm has caused liver damage in animals. Kidney and liver damage in animals has also been reported. May cause damage to the developing fetus and there is limited evidence that DMF is a teratogen in animals.

Points of Attack: Eyes, skin, respiratory system, liver, kidneys, cardiovascular system.

Medical Surveillance: Preplacement and periodic examinations should be concerned particularly with liver and kidney function and with possible effects on the skin. Liver function tests. Kidney function tests. Urine N-methylformamide level (should not be above 40 mg/gm creatinine).

First Aid: If this chemical gets into the eyes, remove any contact lenses at once and irrigate immediately for at least 15 minutes, occasionally lifting upper and lower lids. Seek medical attention immediately. If this chemical contacts the skin, remove contaminated clothing and wash immediately with soap and water. Seek medical attention immediately. If this chemical has been inhaled, remove from exposure, begin rescue breathing (using universal precautions) if breathing has stopped and CPR if heart action has stopped. Transfer promptly to a medical facility. When this chemical has been swallowed, get medical attention. Give large quantities of water and induce vomiting. Do not make an unconscious person vomit.

Personal Protective Methods: Wear protective gloves and clothing to prevent any reasonable probability of skin contact. Safety equipment suppliers/manufacturers can provide recommendations on the most protective glove/clothing material for your operation. Butyl Rubber, Teflon, Silvershield, and Viton/Chlorobutyl Rubber are among the recommended protective materials. All protective clothing (suits, gloves, footwear, headgear) should be clean, available each day, and put on before work. Contact lenses should not be worn when working with this chemical. Wear splash-proof chemical goggles and face shield unless full facepiece respiratory protection is worn. Employees should wash immediately with soap when skin is wet or contaminated. Provide emergency showers and eyewash.

Respirator Selection: NIOSH/OSHA: *100 ppm:* SA (any supplied-air respirator). *250 ppm:* SA:CF (any supplied-air respirator operated in a continuous-flow mode). *500 ppm:* SAT:CF (any supplied-air respirator that has a tight-fitting facepiece and is operated in a continuous-flow mode); or SCBAF (any self-contained breathing apparatus with a full facepiece); or SAF (any supplied-air respirator with a full facepiece). *Emergency or planned entry into unknown concentrations or IDLH conditions:* SCBAF:PD,PP (any self-contained breathing apparatus that has a full facepiece and is operated in a pressure-demand or other positive-pressure mode); or SAF:PD,PP:ASCBA (any supplied-air respirator that has a full facepiece and is operated in a pressure-demand or other positive-pressure mode in combination with an auxiliary self-contained breathing apparatus operated in a pressure-demand or other positive pressure mode). *Escape:* GMFOV [any air-purifying, full-facepiece respirator (gas mask) with a chin-style, front- or back-, mounted organic vapor canister]; or SCBAE (any appropriate escape-type, self-contained breathing apparatus).

Note: Substance reported to cause eye irritation or damage; may require eye protection.

Storage: Prior to working with DMF you should be trained on its proper handling and storage. Before entering confined space where this chemical may be present, check to make sure that an explosive concentration does not exist. Dimethylformamide must be stored to avoid contact with carbon tetrachloride and other halogenated compounds, particularly when in contact with iron; with strong oxidizers (such as chlorine, chlorine dioxide and bromine; and with alkyl aluminums, since violent reactions occur. Store in tightly closed containers in a cool, well-ventilated area away from heat. Sources of ignition such as smoking and open flames are prohibited where this chemical is used, handled, or stored in a manner that could create a potential fire or explosion hazard. Metal containers involving the transfer of 5 gallons or more of this chemical should be grounded and bonded. Drums must be equipped with self-closing valves, pressure vacuum bungs, and flame arresters. Use only nonsparking tools and equipment, especially when opening and closing containers of this chemical.

Shipping: This compound requires a shipping label of: "Flammable Liquid." It falls in DOT Hazard Class 3 and Packing Group III.

Spill Handling: Evacuate and restrict persons not wearing protective equipment from area of spill or leak until cleanup is complete. Remove all ignition sources. Establish forced ventilation to keep levels below explosive limit. Absorb liquids in vermiculite, dry sand, earth, peat, carbon, or a similar material and deposit in sealed containers. Keep this chemical out of a confined space, such as a sewer, because of the possibility of an explosion, unless the sewer is designed to prevent the build-up of explosive concentrations. It may be necessary to contain and dispose of this chemical as a hazardous waste. If material or contaminated runoff enters waterways, notify downstream users of potentially contaminated waters. Contact your Department of Environmental Protection or your regional office of the federal EPA for specific recommendations. If employees are required to clean-up spills, they must be properly trained and equipped. OSHA 1910.120(q) may be applicable.

Fire Extinguishing: This chemical is a combustible liquid. Poisonous gases including dimethylamine and nitrogen oxides are produced in fire. DO NOT use halogenated extinguishing media. Use dry chemical, carbon dioxide, water spray, or alcohol foam extinguishers. Vapors are heavier than air and will collect in low areas. Vapors may travel long distances to ignition sources and flashback. Vapors in confined areas may explode when exposed to fire. Containers may explode in fire. Storage containers and parts of containers may rocket great distances, in many directions. If material or contaminated runoff enters waterways, notify downstream users of potentially contaminated waters. Notify local health and fire officials and pollution control agencies. From a secure, explosion-proof location, use water spray to cool exposed containers. If cooling streams are ineffective (venting sound increases in volume and pitch, tank discolors, or shows any signs of deforming), withdraw immediately to a secure position. If employees are expected to fight fires, they must be trained and equipped in OSHA 1910.156.

Disposal Method Suggested: Burn in solution in flammable solvent in furnace equipped with alkali scrubber.[22] Recovery and recycle is an alternative to disposal for DMF from fiber spin baths and from PVC reactor cleaning solvents.

References

U.S. Environmental Protection Agency, Chemical Hazard Information Profile: N,N-Dimethylformamide, Washington, DC (April 13, 1978)

National Institute for Occupational Safety and Health, Information Profiles on Potential Occupational Hazards – Single Chemicals: N,N-Dimethyl Formamide, 65–73, Report TR 79-607, Rockville, MD (December 1979)

Sax N. I., Ed., "Dangerous Properties of Industrial Materials Report," 1, No. 3, 61–62 (1981)

New Jersey Department of Health, "Hazardous Substance Fact Sheet: N,N-Dimethylformamide," Trenton, NJ, (September 1996)

1,1-Dimethylhydrazine

Molecular Formula: $C_2H_8N_2$

Common Formula: $(CH_3)_2NNH_2$

Synonyms: Dimazin; Dimazine®; Dimethylhydrazine; 1,1-Dimethyl hydrazine; *asym*-Dimethylhydrazine *N,N*-Dimethylhydrazine; U-Dimethylhydrazine; Hydrazine, 1,1-dimethyl-; UDMH;

CAS Registry Number: 57-14-7

RTECS Number: MV2450000

DOT ID: UN 1163

EEC Number: 007-012-00-5

Regulatory Authority

- Carcinogen (Animal Positive) (IARC)[9] (ACGIH)[1] (DFG)[3] (suspect carcinogen) NTP
- Air Pollutant Standard Set (ACGIH)[1] (Australia) (DFG) (Israel) (Mexico) (OSHA)[58] (Several States)[60] (Several Canadian Provinces)
- OSHA 29CFR1910.119, Appendix A, Process Safety List of Highly Hazardous Chemicals, TQ = 1,000 lb (454 kg)
- Clean Air Act: Hazardous Air Pollutants (Title I, Part A, Section 112); Accidental Release Prevention/Flammable substances, (Section 112[r], Table 3), TQ = 15,000 lb (6,810 kg)
- EPA Hazardous Waste Number (RCRA No.): U098
- RCRA, 40CFR261, Appendix 8 Hazardous Constituents
- EPCRA Section 313 Form R *de minimis* concentration reporting level: 0.1%
- Superfund/EPCRA 40CFR355, Appendix B Extremely Hazardous Substances: TPQ = 1,000 lb (454 kg)
- Superfund/EPCRA 40CFR302.4 Reportable Quantity (RQ): CERCLA, 10 lb (4.54 kg)
- Canada, WHMIS, Ingredients Disclosure List

Cited in U.S. State Regulations: Alaska (G), California (A, G), Connecticut (A), Florida (G, A), Illinois (G), Kansas (G), Louisiana (G), Maine (G), Maryland (G), Massachusetts (G), Michigan (G), Minnesota (G), Nevada (A), New Hampshire (G), New Jersey (G), New York (A), North Dakota (A), Oklahoma (G), Pennsylvania (G, A), Rhode Island (G), South

Carolina (A), Vermont (G), Virginia (G, A), Washington (G), West Virginia (G), Wisconsin (G).

Description: UDMH, $(CH_3)_2NNH_2$, is a fuming colorless liquid that turns yellow on contact with air, with a fishy, amine-like odor. The odor threshold is 6.1 – 14 ppm.[41] Boiling point = 63°C. Freezing/Melting point = -58°C. Flash point = -15°C. Autoignition temperature = 249°C. Explosive limits: LEL = 2.0%; UEL = 95%.[17] Hazard Identification (based on NFPA-704 M Rating System): Health 1, Flammability 2, Reactivity 0. Highly soluble in water.

Potential Exposure: This material is used as a component in liquid rocket propellant combinations; it is also used in photography, as an absorbent, and to make other chemicals.

Incompatibilities: Forms explosive mixture with air; may spontaneously ignite on contact with air and porous materials such as asbestos, wood, earth, cloth, etc. Reacts with oxygen causing fire and explosion hazard. A strong reducing agent, this chemical is incompatible with strong acids, halogens, metallic mercury, copper alloys, brass, iron, iron salts. Contact with strong oxidizers such as nitric acid, nitrogen tetroxide, hydrogen peroxide may cause spontaneous ignition. A strong base, this chemical is corrosive and reacts violently with acids. Attacks some plastics, rubber and coatings. May accumulate static electrical charges, and may cause ignition of its vapors.

Permissible Exposure Limits in Air: The Federal Limit (OSHA PEL) is 0.5 ppm (1 mg/m^3) TWA. NIOSH recommends an exposure limit of (Ceiling) 0.06 ppm that should not be exceeded at any time. ACGIH recommends an exposure limit of 0.01 ppm TWA. Australia, Israel, Mexico and the Canadian provinces of Alberta, BC, Ontario, and Quebec use the same TWA as OSHA and Alberta, BC, and Mexico set a STEL of 1 ppm. DFG set no numerical limit and warns of carcinogenicity and danger of skin and respiratory sensitization. The notation "skin" indicates the possibility of cutaneous absorption The ACGIH has the notation that UMDH is an "industrial substance suspect of carcinogenic potential for man." The NIOSH IDLH level is [Ca] 15 ppm. A number of states have set guidelines or standards for UDMH in ambient air[60] ranging from zero (North Dakota) to 2.4 μg/m^3 (Pennsylvania) to 3.3 μg/m^3 (New York) to 5.0 μg/m^3 (South Carolina) to 10.0 μg/m^3 (Connecticut, Florida, Virginia) to 24.0 μg/m^3 (Nevada).

Determination in Air: Collection by a bubbler, colorimetric determination with phosphomolybdic acid; Visible spectrophotometry; NIOSH IV Method #3515.

Permissible Concentration in Water: No criteria set, but EPA[32] has suggested a permissible ambient goal of 13.8 μg/l based on health effects.

Routes of Entry: Inhalation, skin absorption, ingestion, eye and skin contact.

Harmful Effects and Symptoms

This compound exhibits high acute toxicity as a result of exposure be all routes. Death or permanent injury may result after very short exposure to small quantities. Chronic exposure may cause pneumonia, liver damage, and kidney damage. Signs and symptoms of acute exposure to dimethylhydrazine may include eye irritation, facial numbness, facial swelling, and increased salivation. Headache, twitching, seizures, convulsions, and coma may also occur. Gastrointestinal effects include anorexia, nausea, and vomiting. Pulmonary edema and hypotension (low blood pressure) are common. Dimethylhydrazine is toxic to the liver, ruptures red blood cells, and may cause kidney damage. Dermal contact may result in strong skin and mucous membrane irritation.

Short Term Exposure: UDMH can be absorbed through the skin, thereby increasing exposure. It is corrosive to the eyes, skin, respiratory tract. Contact with the liquid may cause permanent eye damage. Inhalation of vapors can cause headache, nausea, dizziness, coughing, and pulmonary edema, a medical emergency that can be delayed for several hours. This can cause death. UDMH may affect the central nervous system, kidneys, and liver.

Long Term Exposure: UDMH may be carcinogenic to humans. It may affect the nervous system, liver, kidneys, and blood.

Points of Attack: Central nervous system, liver, gastrointestinal tract, blood, respiratory system, eyes, skin. Cancer site in animals: tumors of the lungs, liver, blood vessels and intestines. May sensitize the skin and lungs causing dermatitis and asthma-like symptoms.

Medical Surveillance: Liver and kidney function tests. Examination of the nervous system. Examination by a qualified allergist. Consider chest x-ray following acute overexposure.

First Aid: If this chemical gets into the eyes, remove any contact lenses at once and irrigate immediately for at least 15 minutes, occasionally lifting upper and lower lids. Seek medical attention immediately. If this chemical contacts the skin, remove contaminated clothing and wash immediately with soap and water. Seek medical attention immediately. If this chemical has been inhaled, remove from exposure, begin rescue breathing (using universal precautions) if breathing has stopped and CPR if heart action has stopped. Transfer promptly to a medical facility. When this chemical has been swallowed, get medical attention. If victim is conscious, administer water or milk. Do not induce vomiting. Medical observation is recommended for 24 – 48 hours after breathing overexposure, as pulmonary edema may be delayed. As first aid for pulmonary edema, a doctor or authorized paramedic may consider administering a corticosteroid spray.

Personal Protective Methods: Wear protective gloves and clothing to prevent any reasonable probability of skin contact. Safety equipment suppliers/manufacturers can provide recommendations on the most protective glove/clothing material for your operation. Butyl Rubber and Chlorobutyl Rubber are among the recommended protective material. All protective clothing (suits, gloves, footwear, headgear) should be clean, available each day, and put on before work. Contact lenses should not be worn when working with this chemical. Wear splash-proof chemical goggles and face shield unless full facepiece respiratory protection is worn. Employees should wash immediately with soap when skin is wet or contaminated. Remove clothing immediately if wet or contaminated to avoid flammability hazard. Provide emergency showers and eyewash.

Respirator Selection: NIOSH: *At any detectable concentration:* SCBAF:PD,PP (any MSHA/NIOSH approved self-contained breathing apparatus that has a full facepiece and is operated in a pressure-demand or other positive-pressure mode); or SAF:PD,PP:ASCBA (any supplied-air respirator that has a full facepiece and is operated in a pressure-demand or other positive-pressure mode in combination with an auxiliary, self-contained breathing apparatus operated in a pressure-demand or other positive pressure mode). *Escape:* GMFS [any air-purifying, full-facepiece respirator (gas mask) with a chin-style, front- or back-mounted canister providing protection against the compound of concern]; or SCBAE (any appropriate escape-type, self-contained breathing apparatus).

Storage: Prior to working with UDMH you should be trained on its proper handling and storage. Outside or detached storage is preferred. Before entering confined space where this chemical may be present, check to make sure that an explosive concentration does not exist. Store in tightly closed containers in a cool, well ventilated area. Keep dry and separate from porous materials, oxidizing agents, and other incompatible materials, some of which are listed above. Store in an inert atmosphere below 50°C. Do not use copper containers. Metal containers involving the transfer of this chemical should be grounded and bonded. Where possible, automatically pump liquid from drums or other storage containers to process containers. Drums must be equipped with self-closing valves, pressure vacuum bungs, and flame arresters. Use only non-sparking tools and equipment, especially when opening and closing containers of this chemical. Sources of ignition such as smoking and open flames, are prohibited where this chemical is used, handled, or stored in a manner that could create a potential fire or explosion hazard. Wherever this chemical is used, handled, manufactured, or stored, use explosion-proof electrical equipment and fittings. A regulated, marked area should be established where this chemical is handled, used, or stored in compliance with OSHA standard 1910.1045.

Shipping: This compound requires a shipping label of: "Flammable Liquid, Poison, Corrosive." It falls in DOT Hazard Class 6.1 and Packing Group I. Passenger, cargo aircraft or railcar shipment is forbidden.

Spill Handling: Evacuate and restrict persons not wearing protective equipment from area of spill or leak until cleanup is complete. Remove all ignition sources. Establish forced ventilation to keep levels below explosive limit. Do not touch spilled material. Stop leak if this can be done without risk. Use water spray to reduce vapors. Take up small spills vermiculite, dry sand, earth, peat, carbon, or a similar material and deposit in sealed containers. Dike far ahead of large spills for later disposal. Spills also may be removed with an aspirator. Transfer to glass container and neutralize with dilute sulfuric acid. Drain with copious amounts of water. Keep this chemical out of a confined space, such as a sewer, because of the possibility of an explosion, unless the sewer is designed to prevent the build-up of explosive concentrations. It may be necessary to contain and dispose of this chemical as a hazardous waste. If material or contaminated runoff enters waterways, notify downstream users of potentially contaminated waters. Contact your Department of Environmental Protection or your regional office of the federal EPA for specific recommendations. If employees are required to clean-up spills, they must be properly trained and equipped. OSHA 1910.120(q) may be applicable.

Initial isolation and protective action distances

Distances shown are likely to be affected during the first 30 minutes after materials are spilled and could increase with time. If more than one tank car, cargo tank, portable tank, or large cylinder is involved in the incident is leaking, the protective action distance may need to be increased. You may need to seek emergency information from CHEMTREC at (800) 424-9300 or seek professional environmental engineering assistance from the U.S. EPA Environmental Response Team at (908) 548-8730 (24-hour response line).

Small spills (From a small package or a small leak from a large package)

First: Isolate in all directions (feet) 400

Then: Protect persons downwind (miles)

Day 0.4

Night 1.9

Large spills (From a large package or from many small packages)

First: Isolate in all directions (feet) 1,200

Then: Protect persons downwind (miles)

Day 1.6

Night .... 7.0+

Fire Extinguishing: This chemical is a flammable liquid. Poisonous gases including nitrogen oxides, hydrogen, ammonia, dimethylamine and hydrazoic acid are produced in fire. Use dry chemical, carbon dioxide, or foam for small fires. Water may be ineffective. For large fires water fog, carbon dioxide, and bicarbonate agents may allow flashback and explosive re-ignition. Move containers from fire area if it can be done without risk. Dike fire control water for later disposal, do not scatter the material. Wear positive pressure breathing apparatus and special protective clothing. Isolation distances are listed above. Vapors are heavier than air and will collect in low areas. Vapors may travel long distances to ignition sources and flashback. Vapors in confined areas may explode when exposed to fire. Containers may explode in fire. Storage containers and parts of containers may rocket great distances, in many directions. If material or contaminated runoff enters waterways, notify downstream users of potentially contaminated waters. Notify local health and fire officials and pollution control agencies. From a secure, explosion-proof location, use water spray to cool exposed containers. If cooling streams are ineffective (venting sound increases in volume and pitch, tank discolors, or shows any signs of deforming), withdraw immediately to a secure position. If employees are expected to fight fires, they must be trained and equipped in OSHA 1910.156.

Disposal Method Suggested: Controlled incineration (oxides of nitrogen are removed from the effluent gas by scrubbers and/or thermal devices). Consult with environmental regulatory agencies for guidance on acceptable disposal practices. Generators of waste containing this contaminant (≥100 kg/mo) must conform with EPA regulations governing storage, transportation, treatment, and waste disposal.

References

National Institute for Occupational Safety and Health, Criteria for a Recommended Standard: Occupational Exposure to Hydrazines, NIOSH Doc. No. 78–172, Washington, DC (1978)

See Reference (A-60)

Sax N. I., Ed., "Dangerous Properties of Industrial Materials Report," 4, No. 3, 60–67 (1984)

U.S. Environmental Protection Agency, "Chemical Profile: 1,1-Dimethylhydrazine," Washington, DC, Chemical Emergency Preparedness Program (November 30, 1987)

New Jersey Department of Health and Senior Services, "Hazardous Substance Fact Sheet, 1,1-Dimethylhydrazine," Trenton, NJ (June 1998)

1,2-Dimethylhydrazine

Molecular Formula: $C_2H_8N_2$

Common Formula: $H_3CNHNHCH_3$

Synonyms: *N,N'*-Dimethyhydrazine; 1,2-Dimethylhydrazin (German); *sym*-Dimethylhydrazine; 1,2-Dimethylhydrazine; *sim*-Dimetilhidrazina (Spanish); 1,2-Dimetilhidrazina (Spanish); DMH; *symetryczna*-Dwumetylohydrazyna (Polish); Hydrazomethane; Hydroazomethane; SDMH

CAS Registry Number: 540-73-8

RTECS Number: MV2625000

DOT ID: UN 2382

Regulatory Authority

- Carcinogen (Animal Positive) (IARC)[9] (DFG)[3]
- Air Pollutant Standard Set (South Carolina)[60]
- EPA Hazardous Waste Number (RCRA No.): U099
- RCRA, 40CFR261, Appendix 8 Hazardous Constituents
- Superfund/EPCRA 40CFR302.4 Reportable Quantity (RQ): CERCLA, 1 lb (0.454 kg)
- Canada, WHMIS, Ingredients Disclosure List

Cited in U.S. State Regulations: California (A, G), Florida (G), Kansas (G), Louisiana (G), Massachusetts (G), Michigan (G), Minnesota (G), New Hampshire (G), New Jersey (G), Pennsylvania (G), Vermont (G), Virginia (G), Washington (G), Wisconsin (G).

Description: SDMH, $H_3CNHNHCH_3$, is a flammable, colorless, fuming liquid with a fishy, amine-like odor. Boiling point = 80 – 81°C. Freezing/Melting point = –9°C. Flash point ≤23°C. Soluble in water.

Potential Exposure: SDMH is an experimental rocket fuel and is used in chemical synthesis; a laboratory chemical.

Incompatibilities: Forms explosive mixture with air. A strong reducing agent and strong base. Violent reaction with strong oxidizers, strong acids, metallic oxides. Attacks some plastics, rubber and coatings. May accumulate static electrical charges, and may cause ignition of its vapors.

Permissible Exposure Limits in Air: South Carolina has set a guideline for ambient air of 5.0 $\mu g/m^3$. DFG has set no numerical limit but warns of cutaneous absorption, skin and respiratory sensitization and carcinogenicity.

Determination in Air: Bubbler; Phosphomolybdic-acid; Visible spectrophotometry; NIOSH IV Method #3515 (for 1,1-dimethylhydrazine).

Permissible Concentration in Water: No criteria set.

Routes of Entry: Inhalation, skin absorption, ingestion, skin and/or eye contact.

Harmful Effects and Symptoms

Short Term Exposure: SDMH is corrosive to the skin, eyes and respiratory system. It is probably a hemolytic agent.

Long Term Exposure: SDMH is a mutagen, teratogen and carcinogen.

First Aid: If this chemical gets into the eyes, remove any contact lenses at once and irrigate immediately for at least 15 minutes, occasionally lifting upper and lower lids. Seek medical attention immediately. If this chemical contacts the skin, remove contaminated clothing and wash immediately with soap and water. Seek medical attention immediately. If this chemical has been inhaled, remove from exposure, begin rescue breathing (using universal precautions) if breathing has stopped and CPR if heart action has stopped. Transfer promptly to a medical facility. When this chemical has been swallowed, get medical attention. If victim is conscious, administer water or milk. Do not induce vomiting.

Personal Protective Methods: Wear protective gloves and clothing to prevent any reasonable probability of skin contact. Safety equipment suppliers/manufacturers can provide recommendations on the most protective glove/clothing material for your operation.. All protective clothing (suits, gloves, footwear, headgear) should be clean, available each day, and put on before work. Contact lenses should not be worn when working with this chemical. Wear splash-proof chemical

goggles and face shield unless full facepiece respiratory protection is worn. Employees should wash immediately with soap when skin is wet or contaminated. Provide emergency showers and eyewash.

Respirator Selection: *At any detectable concentration:* SCBAF:PD,PP (any MSHA/NIOSH approved self-contained breathing apparatus that has a full facepiece and is operated in a pressure-demand or other positive-pressure mode); or SAF:PD,PP:ASCBA (any supplied-air respirator that has a full facepiece and is operated in a pressure-demand or other positive-pressure mode in combination with an auxiliary, self-contained breathing apparatus operated in a pressure-demand or other positive pressure mode). *Escape:* GMFS [any air-purifying, full-facepiece respirator (gas mask) with a chin-style, front- or back-mounted canister providing protection against the compound of concern/Any appropriate escape-type, self-contained breathing apparatus]; or SCBAE (any appropriate escape-type, self-contained breathing apparatus).

Storage: Prior to working with SDMH you should be trained on its proper handling and storage. Store in tightly closed containers in a dark, cold, well ventilated area. Metal containers involving the transfer of this chemical should be grounded and bonded. Where possible, automatically pump liquid from drums or other storage containers to process containers. Drums must be equipped with self-closing valves, pressure vacuum bungs, and flame arresters. Use only non-sparking tools and equipment, especially when opening and closing containers of this chemical. Sources of ignition such as smoking and open flames, are prohibited where this chemical is used, handled, or stored in a manner that could create a potential fire or explosion hazard. Wherever this chemical is used, handled, manufactured, or stored, use explosion-proof electrical equipment and fittings. A regulated, marked area should be established where this chemical is handled, used, or stored in compliance with OSHA standard 1910.1045.

Shipping: This compound requires a shipping label of: "Flammable Liquid, Poison, Corrosive." It falls in DOT Hazard Class 3 and Packing Group I. Passenger, cargo aircraft or railcar shipment is forbidden.

Spill Handling: Evacuate and restrict persons not wearing protective equipment from area of spill or leak until cleanup is complete. Remove all ignition sources. Ventilate area of spill or leak. Absorb liquids in vermiculite, dry sand, earth, peat, carbon, or a similar material and deposit in sealed containers. Keep this chemical out of a confined space, such as a sewer, because of the possibility of an explosion, unless the sewer is designed to prevent the build-up of explosive concentrations. It may be necessary to contain and dispose of this chemical as a hazardous waste. If material or contaminated runoff enters waterways, notify downstream users of potentially contaminated waters. Contact your Department of Environmental Protection or your regional office of the federal EPA for specific recommendations. If employees are required to clean-up spills, they must be properly trained and equipped. OSHA 1910.120(q) may be applicable.

Fire Extinguishing: This chemical is a combustible liquid. Poisonous gases including nitrogen oxides are produced in fire. Use dry chemical, carbon dioxide, or alcohol foam extinguishers. Vapors are heavier than air and will collect in low areas. Vapors may travel long distances to ignition sources and flashback. Vapors in confined areas may explode when exposed to fire. Containers may explode in fire. Storage containers and parts of containers may rocket great distances, in many directions. If material or contaminated runoff enters waterways, notify downstream users of potentially contaminated waters. Notify local health and fire officials and pollution control agencies. From a secure, explosion-proof location, use water spray to cool exposed containers. If cooling streams are ineffective (venting sound increases in volume and pitch, tank discolors, or shows any signs of deforming), withdraw immediately to a secure position. If employees are expected to fight fires, they must be trained and equipped in OSHA 1910.156.

Disposal Method Suggested: See entry for 1,1-Dimethylhydrazine. Consult with environmental regulatory agencies for guidance on acceptable disposal practices. Generators of waste containing this contaminant (≥100 kg/mo) must conform with EPA regulations governing storage, transportation, treatment, and waste disposal.

References

Sax N. I., Ed., "Dangerous Properties of Industrial Materials Report," 4, No. 3, 67–70 (1984)

2,4-Dimethylphenol

Molecular Formula: $C_8H_{10}O$

Common Formula: $HOC_6H_3(CH_3)_2$

Synonyms: 4,6-Dimethylphenol; 2,4-Dimetilfenol (Spanish); 2,4-DMP; 1-Hydroxy-2,4-dimethylbenzene; 4-Hydroxy-1,3-dimethylbenzene; Phenol, 2,4-dimethyl-; *m*-Xylenol; 2,4-Xylenol

CAS Registry Number: 105-67-9

RTECS Number: ZE5600000

DOT ID: UN 2261

Regulatory Authority

- Water Pollution Standard Proposed (EPA)[6] (California, Kansas)[61]
- Clean Water Act: 40CFR401.15 Section 307 Toxic Pollutants; 40CFR423, Appendix A, Priority Pollutants
- EPA Hazardous Waste Number (RCRA No.): U101
- RCRA, 40CFR261, Appendix 8 Hazardous Constituents
- RCRA 40CFR268.48; 61FR15654, Universal Treatment Standards: Wastewater (mg/l), 0.036; Nonwastewater (mg/kg), 1.4

- RCRA 40CFR264, Appendix 9; TSD Facilities Ground Water Monitoring List. Suggested test method(s) (PQL μg/l): 8040 (5); 8270 (10)
- Superfund/EPCRA 40CFR302.4 Reportable Quantity (RQ): CERCLA, 100 lb (45.4 kg)
- EPCRA Section 313 Form R *de minimis* concentration reporting level: 1.0%

Cited in U.S. State Regulations: California (W), Kansas (G, W), Louisiana (G), Maryland (G), New Jersey (G), Pennsylvania (G), Vermont (G), Virginia (G), Washington (G), Wisconsin (G).

Description: 2,4-DMP, $C_8H_{10}O$, $HOC_6H_3(CH_3)_2$, is a combustible, colorless crystalline solid. Boiling point 212°C. Freezing/Melting point = 27 – 28°C. Flash point = 110°C. The 2,4-isomer is 1 of 5 isomers of this formula. Hazard Identification (based on NFPA-704 M Rating System): Health 1, Flammability 2, Reactivity 0. Soluble in water.

Potential Exposure: 2,4-DMP finds use commercially as an important chemical feedstock or constituent for the manufacture of a wide range of commercial products for industry and agriculture. 2,4-Dimethylphenol is used in the manufacture of phenolic antioxidants, disinfectants, solvents, pharmaceuticals, insecticides, fungicides, plasticizers, rubber chemicals, polyphenylene oxide wetting agents, and dyestuffs, and is a additive or constituent of lubricants, gasolines, and Cresylic acid. 2,4-Dimethylphenol (2,4-DMP) is a naturally occurring, substituted phenol derived from the cresol fraction of petroleum or coal tars by fractional distillation and extraction with aqueous alkaline solutions. It is the Cresylic acid or tar acid fraction of coal tar.

Workers involved in the fractionation and distillation of petroleum or coal and coal tar products comprise one group at risk. Workers who are intermittently exposed to certain commercial degreasing agents containing cresol may also be at risk. Cigarette and marijuana smoking groups and those exposed to cigarette smoke inhale μg quantities of 2,4-demthylphenol.

Incompatibilities: Strong oxidizers.

Permissible Exposure Limits in Air: No standards set.

Permissible Concentration in Water: To protect freshwater aquatic life – 2,120 μg/l on an acute toxicity basis. To protect saltwater aquatic life – no criterion established due to insufficient data. To protect human health – in view of the relative paucity of data on the mutagenicity, carcinogenicity, teratogenicity and long term oral toxicity of 2,4-dimethylphenol, estimates of the effects of chronic oral exposure at low levels cannot be made with any confidence. It is recommended that studies to produce such information be conducted before limits in drinking water are established. A criterion of 400 μg/l is suggested by EPA on an organoleptic basis.[6] Kansas and California have set guidelines[61] at 400 μg/l also.

Determination in Water: Methyl chloride extraction followed by gas chromatography with flame ionization or electron capture detection (EPA Method 604) or gas chromatography plus mass spectrometry (EPA Method 625).

Routes of Entry: Inhalation, skin absorption, ingestion, skin and/or eye contact.

Harmful Effects and Symptoms

Short Term Exposure: m-Xylenol can be absorbed through the skin, thereby increasing exposure. Causes severe skin and eye irritation. Inhalation can cause lung irritation, with coughing and shortness of breath. May cause headaches, dizziness, nausea, vomiting, stomach pain and exhaustion.

Long Term Exposure: High or repeated exposure may affect the liver and kidneys.

Points of Attack: Liver, kidneys.

Medical Surveillance: Liver and kidney function tests.

First Aid: If this chemical gets into the eyes, remove any contact lenses at once and irrigate immediately for at least 15 minutes, occasionally lifting upper and lower lids. Seek medical attention immediately. If this chemical contacts the skin, remove contaminated clothing and wash immediately with soap and water. Seek medical attention immediately. If this chemical has been inhaled, remove from exposure, begin rescue breathing (using universal precautions) if breathing has stopped and CPR if heart action has stopped. Transfer promptly to a medical facility. When this chemical has been swallowed, get medical attention. If convulsions are not present, give a glass or two of water or milk to dilute the substance. Assure that the person's airway is unobstructed and contact a hospital or poison center immediately for advice on whether or not to induce vomiting.

Personal Protective Methods: Wear protective gloves and clothing to prevent any reasonable probability of skin contact. Safety equipment suppliers/manufacturers can provide recommendations on the most protective glove/clothing material for your operation.. All protective clothing (suits, gloves, footwear, headgear) should be clean, available each day, and put on before work. Contact lenses should not be worn when working with this chemical. Wear dust-proof chemical goggles and face shield unless full facepiece respiratory protection is worn. Employees should wash immediately with soap when skin is wet or contaminated. Provide emergency showers and eyewash. and self-contained breathing apparatus.

Respirator Selection: Use any MSHA/NIOSH approved self-contained breathing apparatus that has a full facepiece and is operated in a pressure-demand or other positive-pressure mode; or any supplied-air respirator that has a full facepiece and is operated in a pressure-demand or other positive-pressure mode in combination with an auxiliary, self-contained breathing apparatus operated in a pressure-demand or other positive pressure mode.

Storage: Prior to working with this chemical you should be trained on its proper handling and storage. Store in tightly closed containers in a cool, well ventilated area away from oxidizers and sources of ignition.

Shipping: This compound requires a shipping label of: "Poison" for xylenols. It falls in DOT Hazard Class 6.1 and Packing Group II.

Spill Handling: Evacuate persons not wearing protective equipment from area of spill or leak until clean-up is complete. Remove all ignition sources. Dampen spilled material with alcohol to avoid dust. *DO NOT DRY SWEEP.* Collect powdered material in the most convenient and safe manner and deposit in sealed containers. Ventilate area after clean-up is complete. It may be necessary to contain and dispose of this chemical as a hazardous waste. If material or contaminated runoff enters waterways, notify downstream users of potentially contaminated waters. Contact your Department of Environmental Protection or your regional office of the federal EPA for specific recommendations. If employees are required to clean-up spills, they must be properly trained and equipped. OSHA 1910.120(q) may be applicable.

Fire Extinguishing: This chemical is a combustible solid. Use dry chemical, carbon dioxide, water spray, or alcohol or polymer foam extinguishers. Poisonous gases are produced in fire including carbon monoxide. If material or contaminated runoff enters waterways, notify downstream users of potentially contaminated waters. Notify local health and fire officials and pollution control agencies. From a secure, explosion-proof location, use water spray to cool exposed containers. If cooling streams are ineffective (venting sound increases in volume and pitch, tank discolors, or shows any signs of deforming), withdraw immediately to a secure position. If employees are expected to fight fires, they must be trained and equipped in OSHA 1910.156.

Disposal Method Suggested: Incineration. Consult with environmental regulatory agencies for guidance on acceptable disposal practices. Generators of waste containing this contaminant (≥100 kg/mo) must conform with EPA regulations governing storage, transportation, treatment, and waste disposal.

References

U.S. Environmental Protection Agency, 2,4-Dimethylphenol: Ambient Water Quality Criteria, Washington, DC (1980)

U.S. Environmental Protection Agency, 2,4-Dimethylphenol, Health and Environmental Effects Profile No. 87, Washington, DC, Office of Solid Waste (April 30, 1980)

Sax N. I., Ed., "Dangerous Properties of Industrial Materials Report," 7, No. 3, 87–90 (1987)

New Jersey Department of Health and Senior Services, Hazardous Substance Fact Sheet, "2,4-Dimethylphenol," Trenton NJ (November 1998)

Dimethyl-p-Phenylenediamine

Molecular Formula: $C_8H_{12}N_2$

Common Formula: $(H_3C)_2NC_6H_4NH_2$

Synonyms: 4-(Dimethylamino) aniline; *N,N*-Dimethyl-*p*-phenylenediamine; Dimetil-*p*-fenilendiamina (Spanish)

CAS Registry Number: 99-98-9

RTECS Number: ST0874000

DOT ID: UN 2811

Regulatory Authority

- Superfund/EPCRA 40CFR355, Appendix B Extremely Hazardous Substances: TPQ = 10/10,000 lb (4.5/4,540 kg)
- Superfund/EPCRA 40CFR302.4 Reportable Quantity (RQ): EHS, 1 lb (0.454 kg)

Cited in U.S. State Regulations: Massachusetts (G), New Jersey (G), Pennsylvania (G).

Description: Dimethyl-p-phenylenediamine, $(H_3C)_2NC_6H_4NH_2$, is a colorless to reddish violet crystalline solid. Freezing/Melting point = 41°C. Boiling point = 262°C. Flash point = 90°C.

Potential Exposure: Used in the production of Methylene blue and photodeveloper. It is a reagent for hydrogen sulfide, cellulose, organic synthesis. Chemical intermediate for dyes and diazonium chloride salts; analytical reagent for chloroamine detection in water.

Incompatibilities: Strong oxidizers.

Permissible Exposure Limits in Air: No standards set.

Permissible Concentration in Water: No criteria set.

Harmful Effects and Symptoms

Short Term Exposure: Irritant to skin and eyes. Lowest toxic dose with skin effect is 14 mg/kg. The LD_{50} oral-rat is 50 mg/kg (moderately toxic).

First Aid: If this chemical gets into the eyes, remove any contact lenses at once and irrigate immediately for at least 15 minutes, occasionally lifting upper and lower lids. Seek medical attention immediately. If this chemical contacts the skin, remove contaminated clothing and wash immediately with soap and water. Seek medical attention immediately. If this chemical has been inhaled, remove from exposure, begin rescue breathing (using universal precautions) if breathing has stopped and CPR if heart action has stopped. Transfer promptly to a medical facility. When this chemical has been swallowed, get medical attention. Give large quantities of water and induce vomiting. Do not make an unconscious person vomit.

Personal Protective Methods: Wear protective gloves and clothing to prevent any reasonable probability of skin contact. Safety equipment suppliers/manufacturers can provide recommendations on the most protective glove/clothing material for your operation.. All protective clothing (suits, gloves, footwear, headgear) should be clean, available each day, and put on before work. Contact lenses should not be worn when working with this chemical. Wear dust-proof chemical goggles and face shield unless full facepiece respiratory protection is worn. Employees should wash immediately with soap when skin is wet or contaminated. Provide emergency showers and eyewash.

Storage: Store in cool dry place.

Shipping: Can be considered a poisonous solid n.o.s. This compound requires a shipping label of "Poison." It falls in DOT Hazard Class 6.1 and Packing Group II.

Spill Handling: Evacuate persons not wearing protective equipment from area of spill or leak until clean-up is complete. Remove all ignition sources. Collect powdered material in the most convenient and safe manner and deposit in sealed containers. Ventilate area after clean-up is complete. It may be necessary to contain and dispose of this chemical as a hazardous waste. If material or contaminated runoff enters waterways, notify downstream users of potentially contaminated waters. Contact your Department of Environmental Protection or your regional office of the federal EPA for specific recommendations. If employees are required to clean-up spills, they must be properly trained and equipped. OSHA 1910.120(q) may be applicable.

Fire Extinguishing: This chemical is a combustible solid. Use dry chemical, carbon dioxide, water spray, or alcohol foam extinguishers. Poisonous gases are produced in fire. If material or contaminated runoff enters waterways, notify downstream users of potentially contaminated waters. Notify local health and fire officials and pollution control agencies. Containers may explode in fire. From a secure, explosion-proof location, use water spray to cool exposed containers. If cooling streams are ineffective (venting sound increases in volume and pitch, tank discolors, or shows any signs of deforming), withdraw immediately to a secure position. If employees are expected to fight fires, they must be trained and equipped in OSHA 1910.156.

References

U.S. Environmental Protection Agency, "Chemical Profile: Demethyl-p-Pehnylenediamine," Washington, DC, Chemical Emergency Preparedness Program (November 30, 1987)

Dimethyl Phosphorochloridothioate

Molecular Formula: $C_2H_6ClO_2PS$

Common Formula: $(CH_3O)_2PSCl$

Synonyms: Chlorodimethoxyphosphine sulfide; Dimethoxythiophosphonyl chloride; *O,O*-Dimethyl chlorothionophosphate; Dimethyl chlorothionophosphate; *O,O*-Dimethyl chlorothiophosphate; Dimethyl chlorothiophosphonate; Dimethylchlorthiofosfat (Czech); *O,O*-Dimethylester kyseliny chlorthiofosforecne (Czech); *O,O*-Dimethyl phosphorochloridothioate; Dimethyl phosphorochloridothioate; Dimethyl phosphorochloridothionate; *O,O*-Dimethyl phosphorochlorothioate; Dimethyl phosphorochlorothioate; *O,O*-Dimethyl phosphorothionochloridate; Dimethyl thionochlorophosphate; *O,O*-Dimethyl thionophosphorochloridate; Dimethyl thionophosphorochloridate; *O,O*-Dimethyl thionophosphoryl chloride; *O,O*-Dimethyl thiophosphoric acid chloride; Dimethyl thiophosphorochloridate; *O,O*-Dimethyl thiophosphoryl chloride; Dimethyl thiophosphoryl chloride; Methyl PCT; NSC 132984; Phosphonothioic acid, chloro-, *O,O*-dimethyl ester; Phosphorochlorid othioic acid, *O,O*-dimethyl ester

CAS Registry Number: 2524-03-0

RTECS Number: TD1830000

DOT ID: UN 2923

Regulatory Authority

- Air Pollutant Standard Set (former USSR)[43]
- Superfund/EPCRA 40CFR355, Appendix B Extremely Hazardous Substances: TPQ = 500 lb (227 kg)
- Superfund/EPCRA 40CFR302.4 Reportable Quantity (RQ): EHS, 1 lb (0.454 kg)
- EPCRA Section 313 Form R *de minimis* concentration reporting level: 1.0%

Cited in U.S. State Regulations: California (A, G), Massachusetts (G), New Jersey (G), Pennsylvania (G).

Description: Dimethyl chlorothiophosphate, $(CH_3O)_2PSCl$, is a colorless to light amber liquid. Boiling point = 66 – 67°C under 16 mm Hg pressure.

Potential Exposure: This material is used as a chemical intermediate for insecticides, pesticides, and fungicides; oil and gasoline additives; plasticizers; corrosion inhibitors; flame retardants; and flotation agents. Not registered as a pesticide in the U.S.

Incompatibilities: Water.

Permissible Exposure Limits in Air: The former USSR-UNEP/IRPTC project[43] has set a MAC in workplace air of 0.5 mg/m^3.

Permissible Concentration in Water: The former USSR-UNEP/IRPTC project[43] has set a MAC in water bodies used for domestic purposes of 0.07 mg/l.

Harmful Effects and Symptoms

Short Term Exposure: Contact with the skin, eyes, or mucous membranes may result in severe irritation, burns, and pain. Acute exposure to methyl PCT may produce the following signs and symptoms: pinpoint pupils, blurred vision, headache, dizziness, muscle spasms and profound weakness. Vomiting, diarrhea, abdominal pain, seizures, and coma may also occur. The heart rate may either decrease following oral exposure or increase following dermal exposure. Hypotension (low blood pressure) may occur although hypertension (high blood pressure) in not uncommon. Chest pain may be noted. Respiratory symptoms include dyspnea (shortness of breath), respiratory depression, and respiratory paralysis. Psychosis may occur.

Long Term Exposure: Corrosive materials may affect the lungs.

Points of Attack: Lungs.

Medical Surveillance: Lung function tests. EKG. Consider chest x-ray following acute overexposure.

First Aid: If this chemical gets into the eyes, remove any contact lenses at once and irrigate immediately for at least 15 minutes, occasionally lifting upper and lower lids. Seek medical attention immediately. If this chemical contacts the skin, remove contaminated clothing and wash immediately with soap and water. Seek medical attention immediately. If this chemical has been inhaled, remove from exposure, begin rescue breathing (using universal precautions) if breathing has stopped and CPR if heart action has stopped. Transfer promptly to a medical facility. When this chemical has been swallowed, get medical attention. If victim is conscious, administer water or milk. Do not induce vomiting.

Personal Protective Methods: Wear protective gloves and clothing to prevent any reasonable probability of skin contact. Safety equipment suppliers/manufacturers can provide recommendations on the most protective glove/clothing material for your operation.. All protective clothing (suits, gloves, footwear, headgear) should be clean, available each day, and put on before work. Contact lenses should not be worn when working with this chemical. Wear splash-proof chemical goggles and face shield unless full facepiece respiratory protection is worn. Employees should wash immediately with soap when skin is wet or contaminated. Provide emergency showers and eyewash.

Storage: Prior to working with this chemical you should be trained on its proper handling and storage. Store in tightly closed containers in a cool, well ventilated area. Metal containers involving the transfer of this chemical should be grounded and bonded. Where possible, automatically pump liquid from drums or other storage containers to process containers. Drums must be equipped with self-closing valves, pressure vacuum bungs, and flame arresters. Sources of ignition such as smoking and open flames, are prohibited where this chemical is used, handled, or stored in a manner that could create a potential fire or explosion hazard. Wherever this chemical is used, handled, manufactured, or stored, use explosion-proof electrical equipment and fittings.

Shipping: Is considered as corrosive liquid, poisonous, n.o.s. by DOT. This compound requires a shipping label of: "Corrosive, Poison." It falls in DOT Hazard Class 8 and Packing Group II.

Spill Handling: Evacuate and restrict persons not wearing protective equipment from area of spill or leak until cleanup is complete. Remove all ignition sources. Ventilate area of spill or leak. Avoid inhalation. Apply powdered limestone, slaked lime, soda ash, or sodium bicarbonate. Do not touch spilled material; stop leak if you can do so without risk. Use water spray to reduce vapors. Small spills: Absorb liquids in vermiculite, dry sand, earth, peat, carbon, or a similar material and deposit in sealed containers. Keep this chemical out of a confined space, such as a sewer, because of the possibility of an explosion, unless the sewer is designed to prevent the build-up of explosive concentrations. It may be necessary to contain and dispose of this chemical as a hazardous waste. If material or contaminated runoff enters waterways, notify downstream users of potentially contaminated waters. Contact your Department of Environmental Protection or your regional office of the federal EPA for specific recommendations. If employees are required to clean-up spills, they must be properly trained and equipped. OSHA 1910.120(q) may be applicable.

Fire Extinguishing: This material may burn but does not ignite readily. This material may ignite combustibles (wood, paper, oil, etc.). May react violently with water. Extinguish with dry chemical, carbon dioxide, water spray, fog, or foam. Wear positive pressure breathing apparatus and special protective clothing. Move container from fire area if you can do so without risk. Poisonous gases are produced in fire. Vapors are heavier than air and will collect in low areas. Containers may explode in fire. Storage containers and parts of containers may rocket great distances, in many directions. If material or contaminated runoff enters waterways, notify downstream users of potentially contaminated waters. Notify local health and fire officials and pollution control agencies. From a secure, explosion-proof location, use water spray to cool exposed containers. If cooling streams are ineffective (venting sound increases in volume and pitch, tank discolors, or shows any signs of deforming), withdraw immediately to a secure position. If employees are expected to fight fires, they must be trained and equipped in OSHA 1910.156.

References

U.S. Environmental Protection Agency, "Chemical Profile: Dimethyl Phosphorochloridothioate," Washington, DC, Chemical Emergency Preparedness Program (November 30, 1987)

Dimethyl Phthalate

Molecular Formula: $C_{10}H_{10}O_4$

Common Formula: $C_6H_4(COOCH_3)_2$

Synonyms: Avolin; 1,2-Benzenedicarboxylic acid, dimethyl ester; Dimethyl 1,2-benzenedicarboxylate; Dimethyl benzene-*o*-dicarboxylate; Dimethyl *o*-phthalate; DMF (insect repellent); DMP; ENT 262; Fermine; Ftalato de dimetilo (Spanish); Kemester DMP; Kodaflex DMP; Mipax; NTM; Palatinol M; Phthalic acid, dimet; Phthalic acid, methyl ester; Phthalsaeuredimethylester (German); Repeftal; Solvanom; Solvarone; Unimoll DM; Uniplex 110

CAS Registry Number: 131-11-3

RTECS Number: TI1575000

DOT ID: UN 3082

EINECS Number: 205-011-6

Regulatory Authority

- Air Pollutant Standard Set (ACGIH)[1] (Australia) (HSE)[33] (Israel) (Mexico) (former USSR)[35] (OSHA)[58] (Several States)[60] (Several Canadian Provinces)
- Clean Air Act: Hazardous Air Pollutants (Title I, Part A, Section 112)
- Clean Water Act: 40CFR423, Appendix A, Priority Pollutants; Section 313 Water Priority Chemicals (57FR41331, 9/9/92); Toxic Pollutant (Section 401.15)
- EPA Hazardous Waste Number (RCRA No.): U102
- RCRA, 40CFR261, Appendix 8 Hazardous Constituents
- RCRA 40CFR268.48; 61FR15654, Universal Treatment Standards: Wastewater (mg/l), 0.047; Nonwastewater (mg/kg), 28
- RCRA 40CFR264, Appendix 9; TSD Facilities Ground Water Monitoring List. Suggested test method(s) (PQL µg/l): 8060 (5); 8270 (10)
- Superfund/EPCRA 40CFR302.4 Reportable Quantity (RQ): CERCLA, 5,000 lb (2,270 kg)
- EPCRA Section 313 Form R *de minimis* concentration reporting level: 1.0%
- Canada, WHMIS, Ingredients Disclosure List

Cited in U.S. State Regulations: Alaska (G), California (A, G), Connecticut (A), Illinois (G), Kansas (G, W), Louisiana (G), Maine (G), Maryland (G), Massachusetts (G), Minnesota (G), Nevada (A), New Hampshire (G), New Jersey (G), New York (G), North Dakota (A), Pennsylvania (G), Rhode Island (G), Vermont (G), Virginia (G, A), Washington (G), West Virginia (G), Wisconsin (G).

Description: Dimethyl phthalate, $C_6H_4(COOCH_3)_2$, is a colorless, oily liquid with a slight ester odor. Boiling point = 285°C. Freezing/Melting point = 5.5°C. Flash point = 146°C. Autoignition temperature = 490°C. The LEL is 0.9% @ 180°C. Hazard Identification (based on NFPA-704 M Rating System): Health 0, Flammability 1, Reactivity 0. Insoluble in water.

Potential Exposure: Dimethyl phthalate is used as a solvent, plasticizers for cellulose ester plastics and as an insect repellent.

Incompatibilities: Nitrates, strong oxidizers, strong alkalis, strong acids.

Permissible Exposure Limits in Air: The Federal Limit (OSHA PEL)[58] and the recommended ACGIH TLV is 5 mg/m^3 TWA. The HSE, Australia, Israel, Mexico, and the Canadian provinces of Alberta, BC, Ontario, and Quebec use the same TWA and the STEL set by HSE, Alberta, BC and Mexico is 10 mg/m^3. The NIOSH IDLH level is 2,000 mg/m^3. The former USSR[35] has set a MAC in workplace air of 0.3 mg/m^3. Several states have set guidelines or standards for DMP in ambient air[60] ranging from 50 µg/m^3 (North Dakota) to 80 µg/m^3 (Virginia) to 100 µg/m^3 (Connecticut) to 119 µg/m^3 (Nevada).

Determination in Air: Collection by OSHA versatile sampler-Tenax; Toluene; analysis by gas chromatography/flame ionization; OSHA Method #104.

Permissible Concentration in Water: To protect freshwater and saltwater aquatic life – no criteria developed due to insufficient data. To protect human health – 313 mg/l.[6] Kansas[61] has set 313 mg/l as a guideline for drinking water also.

Determination in Water: Methylene chloride extraction followed by gas chromatography with flame ionization or electron capture detection (EPA Method 606) or gas chromatography plus mass spectrometry (EPA Method 626).

Routes of Entry: Inhalation, ingestion, skin and/or eye contact.

Harmful Effects and Symptoms

Short Term Exposure: *Inhalation:* May cause irritation of the nose and throat with coughing. Prolonged inhalation of high levels may cause dizziness, disorientation, loss of coordination and slowing of heart and respiratory rate. Animal studies suggest that irritation may occur at 2,000 mg/m^3 and may cause death at 9,000 mg/m^3 for 6.5 hours.

Skin: No effects reported.

Eyes: May cause irritation and chemical burns.

Ingestion: May cause irritation of the lips, tongue, mouth and stomach, vomiting, diarrhea, dizziness, unconsciousness and coma. Animal studies suggest that the lethal dose for an adult is 6 – 18 ounces.

Long Term Exposure: No information available. Dimethyl phthalate is a possible teratogen in humans; there is limited evidence that it is a teratogen in animals. It may reduce fertility in males and females.

Points of Attack: Eyes, respiratory system, gastrointestinal tract.

Medical Surveillance: There is no special tests for this chemical. However, if illness occurs or overexposure is suspected, medical attention is recommended.

First Aid: If this chemical gets into the eyes, remove any contact lenses at once and irrigate immediately for at least 15 minutes, occasionally lifting upper and lower lids. Seek medical attention immediately. If this chemical contacts the skin, remove contaminated clothing and wash immediately with soap and water. Seek medical attention immediately. If this chemical has been inhaled, remove from exposure, begin rescue breathing (using universal precautions) if breathing has stopped and CPR if heart action has stopped. Transfer promptly to a medical facility. When this chemical has been swallowed, get medical attention. Give large quantities of water and induce vomiting. Do not make an unconscious person vomit.

Personal Protective Methods: Wear protective gloves and clothing to prevent any reasonable probability of skin contact. Safety equipment suppliers/manufacturers can provide recommendations on the most protective glove/clothing material for your operation.. All protective clothing (suits, gloves, footwear, headgear) should be clean, available each day, and put on before work. Contact lenses should not be worn when working with this chemical. Wear splash-proof chemical goggles and face shield unless full facepiece respiratory protection is worn. Employees should wash immediately with soap when skin is wet or contaminated. Provide emergency showers and eyewash.

Respirator Selection: NIOSH/OSHA: *50 mg/m³:* DMF (any dust and mist respirator with a full facepiece). *125 mg/m³:* SA:CF (any supplied-air respirator operated in a continuous-flow mode); or PAPRDM (any powered, air-purifying respirator with a dust and mist filter). *250 mg/m³:* HiEF (any air-purifying, full-facepiece respirator with a high-efficiency particulate filter); or SCBAF (any self-contained breathing apparatus with a full facepiece); or SAF (any supplied-air respirator with a full facepiece). *2,000 mg/m³:* SAF:PD,PP (any supplied-air respirator that has a full facepiece and is operated in a pressure-demand or other positive-pressure mode). *Emergency or planned entry into unknown concentrations or IDLH conditions:* SCBAF:PD,PP (any self-contained breathing apparatus that has a full facepiece and is operated in a pressure-demand or other positive-pressure mode); or SAF:PD,PP:ASCBA (any supplied-air respirator that has a full facepiece and is operated in a pressure-demand or other positive-pressure mode in combination with an auxiliary self-contained breathing apparatus operated in a pressure-demand or other positive pressure mode). *Escape:* HiEF (any air-purifying, full-facepiece respirator with a high-efficiency particulate filter); or SCBAE (any appropriate escape-type, self-contained breathing apparatus).

Note: Substance causes eye irritation or damage; eye protection needed.

Storage: Prior to working with dimethyl phthalate you should be trained on its proper handling and storage. Before entering confined space where this chemical may be present, check to make sure that an explosive concentration does not exist. Dimethyl phthalate must be stored to avoid contact with nitrates; strong alkalis such as sodium hydroxide, potassium hydroxide, and lithium hydroxide; strong oxidizers such as chlorine, chlorine dioxide, and bromine; and strong acids such as sulfuric acid, hydrochloric acid, and nitric acid, since violent reactions occur. Store in tightly closed containers in a cool, well-ventilated area away from heat or flame.

Shipping: Environmentally hazardous liquid, n.o.s. Hazard Class: 9. Label: "Class 9." Packing Group: III.

Spill Handling: Evacuate and restrict persons not wearing protective equipment from area of spill or leak until cleanup is complete. Remove all ignition sources. Establish forced ventilation to keep levels below explosive limit. Absorb liquids in vermiculite, dry sand, earth, peat, carbon, or a similar material and deposit in sealed containers. Keep this chemical out of a confined space, such as a sewer, because of the possibility of an explosion, unless the sewer is designed to prevent the build-up of explosive concentrations. It may be necessary to contain and dispose of this chemical as a hazardous waste. If material or contaminated runoff enters waterways, notify downstream users of potentially contaminated waters. Contact your Department of Environmental Protection or your regional office of the federal EPA for specific recommendations. If employees are required to clean-up spills, they must be properly trained and equipped. OSHA 1910.120(q) may be applicable.

Fire Extinguishing: This chemical is a combustible liquid. Poisonous gases are produced in fire. Use dry chemical, carbon dioxide, or foam extinguishers. Vapors are heavier than air and will collect in low areas. Vapors may travel long distances to ignition sources and flashback. Vapors in confined areas may explode when exposed to fire. Containers may explode in fire. Storage containers and parts of containers may rocket great distances, in many directions. If material or contaminated runoff enters waterways, notify downstream users of potentially contaminated waters. Notify local health and fire officials and pollution control agencies. From a secure, explosion-proof location, use water spray to cool exposed containers. If cooling streams are ineffective (venting sound increases in volume and pitch, tank discolors, or shows any signs of deforming), withdraw immediately to a secure position. If employees are expected to fight fires, they must be trained and equipped in OSHA 1910.156.

Disposal Method Suggested: Incineration.[22] Consult with environmental regulatory agencies for guidance on acceptable disposal practices. Generators of waste containing this contaminant (≥100 kg/mo) must conform with EPA regulations governing storage, transportation, treatment, and waste disposal.

References

U.S. Environmental Protection Agency, Phthalate Esters: Ambient Water Quality Criteria, Washington, DC (1980)

National Institute for Occupational Safety and Health, Profiles on Occupational Hazards for Criteria Document Priorities – Phthalates, 97–103, Report PB-274, 073, Cincinnati, OH (1977)

U.S. Environmental Protection Agency, Dimethyl Phthalate, Health and Environmental Effects Profile No. 88, Washington, DC, Office of Solid Waste (April 30, 1980)

Sax N. I., Ed., "Dangerous Properties of Industrial Materials Report," 2, No. 4, 80–84 (1982)

U.S. Environmental Protection Agency, "Chemical Profile: Dimethyl Phthalate," Washington, DC, Chemical Emergency Preparedness Program (October 31, 1985)

New Jersey Department of Health, "Hazardous Substance Fact Sheet: Dimethyl Phthalate," Trenton, NJ (October 1996)

New York State Department of Health, "Chemical Fact Sheet: Dimethylphthalate (DMP)," Albany, NY, Bureau of Toxic Substance Assessment (January 1986 and Version 2)

Dimethyl Sulfate

Molecular Formula: $C_2H_6O_4S$

Common Formula: $(CH_3)_2SO_4$

Synonyms: Dimethylester kyseliny sirove (Czech); Dimethyl monosulfate; Dimethyl sulfaat (Dutch); Dimethylsulfat (Czech); Dimethyl sulphate; Dimetilsulfato (Italian); DMS; Dwumetylowy siarczan (Polish); Methyle (sulfate de) (French); Methyl sulfate; Sulfate dimethylique (French); Sulfate de methyle (French); Sulfato de dimetilo (Spanish); Sulfuric acid, dimethyl ester

CAS Registry Number: 77-78-1

RTECS Number: WS8225000

DOT ID: UN 1595

EEC Number: 016-023-00-4

Regulatory Authority

- Carcinogen (Animal Positive) (IARC)[9] (ACGIH)[1] (DFG)[3]
- Air Pollutant Standard Set (ACGIH)[1] (Australia) (DFG)[3] (HSE)[33] (Israel) (Mexico) (OSHA)[58] (Other Countries)[35] (Several States)[60] (Several Canadian Provinces)
- Clean Air Act: Hazardous Air Pollutants (Title I, Part A, Section 112)
- EPA Hazardous Waste Number (RCRA No.): U103
- RCRA, 40CFR261, Appendix 8 Hazardous Constituents
- EPCRA Section 313 Form R *de minimis* concentration reporting level: 0.1%
- Superfund/EPCRA 40CFR355, Appendix B Extremely Hazardous Substances: TPQ = 500 lb (227 kg)
- Superfund/EPCRA 40CFR302.4 Reportable Quantity (RQ): CERCLA, 100 lb (45.4 kg)
- Canada, WHMIS, Ingredients Disclosure List

Cited in U.S. State Regulations: Alaska (G), California (A, G), Connecticut (A), Florida (G, A), Illinois (G), Kansas (G), Louisiana (G), Maine (G), Maryland (G), Massachusetts (G), Michigan (G), Minnesota (G), Nevada (A), New Hampshire (G), New Jersey (G), New York (A), North Carolina (A), North Dakota (A), Oklahoma (G), Pennsylvania (G, A), Rhode Island (G), South Carolina (A), Vermont (G), Virginia (G, A), Washington (G), West Virginia (G), Wisconsin (G).

Description: Dimethyl sulfate, $(CH_3)_2SO_4$, is an oily, colorless liquid with a faint onion-like odor. Boiling point = 188°C (decomposes). Freezing/Melting point = -32°C. Flash point = 83°C (oc). Autoignition temperature = 188°C. Hazard Identification (based on NFPA-704 M Rating System): Health 4, Flammability 2, Reactivity 0. Slightly soluble in water (hydrolysis above 18°C).

Potential Exposure: Industrial use of dimethyl sulfate is based upon its methylating properties. It is used in the manufacture of methyl esters, ethers and amines, in dyes, drugs, perfume, phenol derivatives, and other organic chemicals. It is also used as a solvent in the separation of mineral oils. It is used as an intermediate in the manufacture of many pharmaceuticals and pesticides.

Incompatibilities: The aqueous solution is a medium strong acid. Forms explosive mixture with air. Violent reaction with strong oxidizers, strong acids, strong alkalies, concentrated ammonia solutions with risks of fire and explosions. Reacts with water evolving heat and forming sulfuric acid. Attacks some plastics, rubber and coatings. Attacks metals in the presence of moisture.

Permissible Exposure Limits in Air: The Federal Limit (OSHA PEL)[58] is 1 ppm (5 mg/m^3) TWA. The ACGIH recommended TLV value is 0.1 ppm (0.5 mg/m^3). There is no DFG numerical value. ACGIH[1] and DFG[3] have the note that dimethyl sulfate is an "industrial substance suspect of carcinogenic potential for man." Skin absorption is possible. HSE[33] uses 0.1 ppm as the STEL. The NIOSH IDLH is [Ca] 7 ppm. In other countries,[35] MAC values for workplace air have been set at 0.4 mg/m^3 in Brazil at 0.1 mg/m^3 in the former USSR and 0.05 mg/m^3 in Czechoslovakia. Several states have set guidelines or standards for dimethyl sulfate in ambient air[60] ranging from zero (North Carolina and North Dakota) to 1.67 μg/m^3 (New York) to 2.5 μg/m^3 (Connecticut and South Carolina) to 5.0 μg/m^3 (Florida and Virginia) to 12.0 μg/m^3 (Nevada and Pennsylvania).

Determination in Air: Sampling by solid sorbent Porapak® tube-P; Diethyl ether; Gas chromatography/Electrolytic conductivity detection; NIOSH IV Method #2524.

Permissible Concentration in Water: No criteria set.

Routes of Entry: Inhalation of vapor, percutaneous absorption of liquid, ingestion, eye and skin contact.

Harmful Effects and Symptoms

Short Term Exposure: Dimethyl sulfate is corrosive to the eyes, skin, and the respiratory tract. Corrosive if ingested. Ingestion of the substance may cause oedema of lips, tongue and pharynx. Inhalation can cause pulmonary edema, a medical emergency that can be delayed for several hours. This can cause death. Extremely toxic vapors and liquid: a few whiffs or contact on skin could be fatal. Also acutely toxic if ingested. Delayed effects which are ultimately fatal may also occur. Lethal concentrations as low as 97 ppm/10 min have been reported in humans. DNA inhibitions and damage to human somatic cells, and sister chromatid exchange in human fibroblast cells were observed. Delayed appearance of symptoms may permit unnoticed exposure to lethal quantities. Immediate effects of vapor exposure are eye irritation, cough, swelling of tongue, lips, and larynx, and lungs (later). Ingestion or direct contact with mucous membranes causes corrosion. Once absorbed, lung damage and liver and kidney injury will occur. Liquid dermal exposure causes blistering, followed by convulsions, delirium, coma, and death in severe cases.

Long Term Exposure: Lungs may be affected. May affect the liver, kidneys and central nervous system. Dimethyl sulfate is probably carcinogenic to humans. May damage the developing fetus.

Points of Attack: Eyes, skin, respiratory system, liver, kidneys, central nervous system. Cancer site in animals: nasal and lung cancer.

Medical Surveillance: Preplacement and periodic medical examinations should give special consideration to the skin, eyes, central nervous system, lungs. Chest x-rays should be taken and lung, liver, and kidney functions evaluated. Sputum and urinary cytology may be useful in detecting the presence or absence of carcinogenic effects.

First Aid: If this chemical gets into the eyes, remove any contact lenses at once and irrigate immediately for at least 30 minutes, occasionally lifting upper and lower lids. Seek medical attention immediately. If this chemical contacts the skin, remove contaminated clothing and wash immediately with soap and water. Seek medical attention immediately. If this chemical has been inhaled, remove from exposure, begin rescue breathing (using universal precautions) if breathing has stopped and CPR if heart action has stopped. Transfer promptly to a medical facility. When this chemical has been swallowed, get medical attention. If victim is conscious, administer water or milk. Do not induce vomiting. Medical observation is recommended for 24 – 48 hours after breathing overexposure, as pulmonary edema may be delayed. As first aid for pulmonary edema, a doctor or authorized paramedic may consider administering a corticosteroid spray.

Personal Protective Methods: Wear protective gloves and clothing to prevent any reasonable probability of skin contact. Safety equipment suppliers/manufacturers can provide recommendations on the most protective glove/clothing material for your operation.. All protective clothing (suits, gloves, footwear, headgear) should be clean, available each day, and put on before work. Contact lenses should not be worn when working with this chemical. Wear splash-proof chemical goggles and face shield unless full facepiece respiratory protection is worn. Employees should wash immediately with soap when skin is wet or contaminated. Provide emergency showers and eyewash.

Respirator Selection: NIOSH: At *any concentrations above the NIOSH REL, or where there is no REL, at any detectable concentration:* SCBAF:PD,PP (any self-contained breathing apparatus that has a full facepiece and is operated in a pressure-demand or other positive-pressure mode); or SAF: PD,PP:ASCBA (any supplied-air respirator that has a full facepiece and is operated in a pressure-demand or other positive-pressure mode in combination with an auxiliary self-contained breathing apparatus operated in a pressure-demand or other positive pressure mode). Escape: GMFS [any air-purifying, full-facepiece respirator (gas mask) with a chin-style, front- or back-mounted canister providing protection against the compound of concern]; or SCBAE (any appropriate escape-type, self-contained breathing apparatus).

Storage: Prior to working with this chemical you should be trained on its proper handling and storage. Dimethyl sulfate must be stored to avoid contact with water, strong oxidizers (such as chlorine, bromine and fluorine) and strong ammonia solutions, since violent reactions occur. Store in tightly closed containers in a cool, well-ventilated area away from heat or flame. Do not allow this chemical to contact water. Sources of ignition, such as smoking and open flames, are prohibited where dimethyl sulfate is used, handled, or stored in a manner that could create a potential fire or explosion hazard. A regulated, marked area should be established where this chemical is handled, used, or stored in compliance with OSHA standard 1910.1045.

Shipping: This compound requires a shipping label of: "Poison, Corrosive." It falls in DOT Hazard Class 6.1 and Packing Group I.

Spill Handling: Evacuate and restrict persons not wearing protective equipment from area of spill or leak until cleanup is complete. Remove all ignition sources. Ventilate area of spill or leak. Absorb liquids in vermiculite, dry sand, earth, peat, carbon, or a similar material and deposit in sealed containers. Keep this chemical out of a confined space, such as a sewer, because of the possibility of an explosion, unless the sewer is designed to prevent the build-up of explosive concentrations. It may be necessary to contain and dispose of this chemical as a hazardous waste. If material or contaminated runoff enters waterways, notify downstream users of potentially contaminated waters. Contact your Department of Environmental Protection or your regional office of the federal EPA for specific recommendations. If employees are required to clean-up spills, they must be properly trained and equipped. OSHA 1910.120(q) may be applicable.

Fire Extinguishing: This chemical is a combustible liquid. Poisonous gases including sulfur oxides are produced in fire. Firefighting gear (including SCBA) does not provide adequate protection. If exposure occurs, remove and isolate gear immediately and thoroughly decontaminate personnel. Use dry chemical, carbon dioxide, or foam extinguishers. Vapors are heavier than air and will collect in low areas. Vapors may travel long distances to ignition sources and flashback. Vapors in confined areas may explode when exposed to fire. Containers may explode in fire. Storage containers and parts of containers may rocket great distances, in many directions. If material or contaminated runoff enters waterways, notify downstream users of potentially contaminated waters. Notify local health and fire officials and pollution control agencies. From a secure, explosion-proof location, use water spray to cool exposed containers. If cooling streams are ineffective (venting sound increases in volume and pitch, tank discolors, or shows any signs of deforming), withdraw immediately to a secure position. If employees are expected to fight fires, they must be trained and equipped in OSHA 1910.156.

Disposal Method Suggested: Incineration (1,800°F, 1.5 seconds minimum) of dilute, neutralized dimethyl sulfate waste is recommended. The incinerator must be equipped with efficient scrubbing devices for oxides of sulfur. Alkaline hydrolysis may also be used as may landfill burial.[22] Consult with environmental regulatory agencies for guidance on acceptable disposal practices. Generators of waste containing this contaminant (≥100 kg/mo) must conform with EPA regulations governing storage, transportation, treatment, and waste disposal.

References

National Institute for Occupational Safety and Health, Information Profiles on Potential Occupational Hazards – Single Chemicals: Dimethyl Sulfate, 74–84, Report TR79-607, Rockville, MD (December 1979)

Sax N. I., Ed., "Dangerous Properties of Industrial Materials Report," 1, No. 5, 51–53 (1981)

U.S. Environmental Protection Agency, "Chemical Profile: Dimethyl Sulfate," Washington, DC, Chemical Emergency Preparedness Program (November 30, 1987)

New Jersey Department of Health, "Hazardous Substance Fact Sheet: Dimethyl Sulfate," Trenton, NJ (October 1996)

Dimethyl Sulfide

Molecular Formula: C_2H_6S

Common Formula: $(CH_3)_2S$

Synonyms: Dimethyl monosulfide; Dimethyl sulphide; Dimethyl thioether; DMS; Methanethiomethane; Methyl monosulfide; Methyl sulphide; Methylthiomethane; 2-Thiapropane; Thiobismethane; 2-Thiopropane

CAS Registry Number: 75-18-3

RTECS Number: PV5075000

DOT ID: UN 1164

Regulatory Authority

- Air Pollutant Standard Set (UNEP)[43]

Cited in U.S. State Regulations: Florida (G), Maine (G), Massachusetts (G), New Hampshire (G), New Jersey (G), Oklahoma (G), Pennsylvania (G), Rhode Island (G).

Description: Dimethyl sulfide, $(CH_3)_2S$, is a colorless to yellow liquid with an unpleasant wild radish or cabbage-like odor. Boiling point = 37°C. Freezing/Melting point = -98°C. Flash point ≤-18°C. Autoignition temperature = 206°C. Explosive limits: LEL = 2.2%; UEL = 19.7%.[17] Hazard Identification (based on NFPA-704 M Rating System): Health 1, Flammability 4, Reactivity 0. Slightly soluble in water.

Potential Exposure: It is a gas odorant, catalyst impregnator, solvent for anhydrous mineral salts, flavoring ingredient in goods and beverages, chemical intermediate for solvents and dimethyl sulfoxide.

Incompatibilities: Reacts violently with strong oxidizing materials causing fire and explosion hazard.

Permissible Exposure Limits in Air: The former USSR-UNEP/IRPTC project[43] has set a MAC in workplace air of 50 mg/m^3 and a MAC in ambient air of residential areas of 0.08 mg/m^3 on a momentary basis.

Permissible Concentration in Water: The former USSR-UNEP/IRPTC project[43] has set a MAC in water bodies used for domestic purposes of 0.01 mg/l.

Routes of Entry: Inhalation, ingestion, skin contact.

Harmful Effects and Symptoms

Short Term Exposure: Dimethyl sulfide causes softening and irritation of the skin. Orally it is an irritant. Inhalation can cause dizziness and unconsciousness.

Long Term Exposure: Unknown at this time.

First Aid: If this chemical gets into the eyes, remove any contact lenses at once and irrigate immediately for at least 15 minutes, occasionally lifting upper and lower lids. Seek medical attention immediately. If this chemical contacts the skin, remove contaminated clothing and wash immediately with soap and water. Seek medical attention immediately. If this chemical has been inhaled, remove from exposure, begin rescue breathing (using universal precautions) if breathing has stopped and CPR if heart action has stopped. Transfer promptly to a medical facility. When this chemical has been swallowed, get medical attention. Give large quantities of water and induce vomiting. Do not make an unconscious person vomit.

Personal Protective Methods: Wear protective gloves and clothing to prevent any reasonable probability of skin contact. Safety equipment suppliers/manufacturers can provide recommendations on the most protective glove/clothing material for your operation.. All protective clothing (suits, gloves, footwear, headgear) should be clean, available each day, and put on before work. Contact lenses should not be worn when working with this chemical. Wear splash-proof chemical goggles and face shield unless full facepiece respiratory protection is worn. Employees should wash immediately with soap when skin is wet or contaminated. Provide emergency showers and eyewash.

Respirator Selection: *Use* any MSHA/NIOSH approved self-contained breathing apparatus that has a full facepiece and is operated in a pressure-demand or other positive-pressure mode); or any supplied-air respirator that has a full facepiece and is operated in a pressure-demand or other positive-pressure mode in combination with an auxiliary, self-contained breathing apparatus operated in a pressure-demand or other positive pressure mode.

Storage: Prior to working with Dimethyl sulfide you should be trained on its proper handling and storage. Before entering confined space where this chemical may be present, check to make sure that an explosive concentration does not exist. Store in tightly closed containers in a cool, well ventilated area. Metal containers involving the transfer of this chemical should be grounded and bonded. Where possible, automatically pump liquid from drums or other storage containers to process containers. Drums must be equipped with self-closing valves, pressure vacuum bungs, and flame arresters. Use only non-sparking tools and equipment, especially when opening and closing containers of this chemical. Sources of ignition such as smoking and open flames, are prohibited where this chemical is used, handled, or stored in a manner that could create a potential fire or explosion hazard. Wherever this chemical is used, handled, manufactured, or stored, use explosion-proof electrical equipment and fittings.

Shipping: This compound requires a shipping label of: "Flammable Liquid." It falls in DOT Hazard Class 3 and Packing Group I.

Spill Handling: Evacuate and restrict persons not wearing protective equipment from area of spill or leak until cleanup is complete. Remove all ignition sources. Establish forced ventilation to keep levels below explosive limit. Absorb liquids in vermiculite, dry sand, earth, peat, carbon, or a similar material and deposit in sealed containers. Keep this chemical out of a confined space, such as a sewer, because of the possibility of an explosion, unless the sewer is designed to prevent the build-up of explosive concentrations. It may be necessary to contain and dispose of this chemical as a hazardous waste. If material or contaminated runoff enters waterways, notify downstream users of potentially contaminated waters. Contact your Department of Environmental Protection or your regional office of the federal EPA for specific recommendations. If employees are required to clean-up spills, they must be properly trained and equipped. OSHA 1910.120(q) may be applicable.

Fire Extinguishing: This chemical is a combustible liquid. Poisonous gases including hydrogen sulfide and sulphur oxides are produced in fire. Use dry chemical, carbon dioxide, water spray, fog, or foam extinguishers. Wear self-contained breathing apparatus and full protective clothing. If it can be done safely, move container from fire area. For massive fire in cargo area, used unmanned hose holder or monitor nozzles; if this is impossible withdraw from area and let fire burn. Isolate for 1/2 mile in all direction if tank car or truck is involved in fire. Vapors are heavier than air and will collect in low areas. Vapors may travel long distances to ignition sources and flashback. Vapors in confined areas may explode when exposed to fire. Containers may explode in fire. Storage containers and parts of containers may rocket great distances, in many directions. If material or contaminated runoff enters waterways, notify downstream users of potentially contaminated waters. Notify local health and fire officials and pollution control agencies. From a secure, explosion-proof location, use water spray to cool exposed containers. If cooling streams are ineffective (venting sound increases in volume and pitch, tank discolors, or shows any signs of deforming), withdraw immediately to a secure position. If employees are expected to fight fires, they must be trained and equipped in OSHA 1910.156.

References

U.S. Environmental Protection Agency, "Chemical Profile: Dimethyl Sulfide," Washington, DC, Chemical Emergency Preparedness Program (November 30, 1987)

Dimethyl Sulfoxide

Molecular Formula: C_2H_6O

Common Formula: $(CH_3)_2SO$

Synonyms: A-10846; Deltan; Demasorb; Demavet; Demeso; Demsodrox; Dermasorb; Dimethyl sulphoxide; Dimexide; Dipiratril-Tropico; DMS-70; DMS-90; DMSO; Dolicur; Doligur; Domoso; Durasorb; Gamasol-90; Hyadur; Infiltrina; M-176; Methyl sulfoxide; NSC-763; Rimso-50; Somi-Pront; SQ 9453; Sulficyl bis(methane); Syntexan; Topsym

CAS Registry Number: 67-68-5

RTECS Number: PV6210000

DOT ID: No citation.

Cited in U.S. State Regulations: New Hampshire (G), New York (G).

Description: DMSO, $(CH_3)_2SO$, is a hygroscopic colorless to yellow liquid. Boiling point = 189°C. Freezing/Melting point = 18 – 19°C. Also reported as 8°C in the literature. Flash point = 95°C. Autoignition temperature = 215°C. Explosive limits: LEL = 2.6%; UEL = 42%.[17] Hazard Identification (based on NFPA-704 M Rating System): Health 1, Flammability 1, Reactivity 0. Slight solubility in water.

Potential Exposure: Used as a solvent, as a pharmaceutical, in chemicals production.

Incompatibilities: Violent reaction with strong oxidizers. Reacts with ethanoyl chloride, boron compounds, halides, metal alkoxides.

Permissible Exposure Limits in Air: No standards set.

Permissible Concentration in Water: No criteria set.

Harmful Effects and Symptoms

Short Term Exposure: Systemically, it produces anesthesia, vomiting, chills, cramps and lethargy. The LD_{50} oral-rat is 14,500 mg/kg (insignificantly toxic). It reportedly irritates the eyes and skin and the respiratory tract, however.[57] It also affects the blood. All routes of exposures can produce an intense garlic-like taste and breath odor. *Eyes:* A 7.5% solution can cause irritation and burning. *Inhalation:* Exposure to high concentrations of DMSO can cause some sedation and lowering of consciousness. Animal studies indicate that exposures of 900 ppm for 24 hours produced no ill effects. Skin: A 10% solution applied for 14 consecutive days produced some sedation, headache, nausea and dizziness along with such skin effects as irritation, drying and scaling. Contact with a 70 – 90% solution has resulted in immediate stinging and burning. DMSO is very readily absorbed through the skin and may increase the absorption of other substances dissolved in it or on the surface of the skin. *Ingestion:* No human data available. Animal studies indicate that symptoms from swallowing DMSO include throat and stomach irritation, vomiting and sedation.

Long Term Exposure: Repeated or prolonged contact with skin may cause irritation and dermatitis. DMSO may affect the liver, resulting in impaired function.

Points of Attack: Skin, liver.

Medical Surveillance: Liver function tests.

First Aid: If this chemical gets into the eyes, remove any contact lenses at once and irrigate immediately for at least 15 minutes, occasionally lifting upper and lower lids. Seek medical attention immediately. If this chemical contacts the skin, remove contaminated clothing and wash immediately with soap and water. Seek medical attention immediately. If this chemical has been inhaled, remove from exposure, begin rescue breathing (using universal precautions) if breathing has stopped and CPR if heart action has stopped. Transfer promptly to a medical facility. When this chemical has been swallowed, get medical attention. Give large quantities of water and induce vomiting. Do not make an unconscious person vomit.

Personal Protective Methods: Wear protective gloves and clothing to prevent any reasonable probability of skin contact. Safety equipment suppliers/manufacturers can provide recommendations on the most protective glove/clothing material for your operation.. All protective clothing (suits, gloves, footwear, headgear) should be clean, available each day, and put on before work. Contact lenses should not be worn when working with this chemical. Wear splash-proof chemical goggles and face shield unless full facepiece respiratory protection is worn. Employees should wash immediately with soap when skin is wet or contaminated. Provide emergency showers and eyewash.

Storage: Prior to working with this chemical you should be trained on its proper handling and storage. Protect from physical damage. Store in cool, dry place away from oxidizers and ignition sources. Before entering confined space where this chemical may be present, check to make sure that an explosive concentration does not exist. Store in tightly closed containers in a cool, well-ventilated area away from heat. Sources of ignition such as smoking and open flames are prohibited where this chemical is used, handled, or stored in a manner that could create a potential fire or explosion hazard. Metal containers involving the transfer of 5 gallons or more of this chemical should be grounded and bonded. Drums must be equipped with self-closing valves, pressure vacuum bungs, and flame arresters. Use only nonsparking tools and equipment, especially when opening and closing containers of this chemical.

Shipping: DMSO is not specifically cited in DOT's Performance-Oriented Packaging Standards.

Spill Handling: Evacuate and restrict persons not wearing protective equipment from area of spill or leak until cleanup is complete. Remove all ignition sources. Establish forced ventilation to keep levels below explosive limit. Absorb liquids in vermiculite, dry sand, earth, peat, carbon, or a similar material and deposit in sealed containers. Keep this chemical out of a confined space, such as a sewer, because of the possibility of an explosion, unless the sewer is designed to prevent the build-up of explosive concentrations. It may be necessary to contain and dispose of this chemical as a hazardous waste. If material or contaminated runoff enters waterways, notify downstream users of potentially contaminated waters. Contact your Department of Environmental Protection or your regional office of the federal EPA for specific recommendations. If employees are required to cleanup spills, they must be properly trained and equipped. OSHA 1910.120(q) may be applicable.

Fire Extinguishing: This chemical is a combustible liquid. Poisonous gases including sulfur dioxide are produced in fire. Use dry chemical, carbon dioxide, or foam extinguishers. Vapors are heavier than air and will collect in low areas. Vapors may travel long distances to ignition sources and flashback. Vapors in confined areas may explode when exposed to fire. Containers may explode in fire. Storage containers and parts of containers may rocket great distances, in many directions. If material or contaminated runoff enters waterways, notify downstream users of potentially contaminated waters. Notify local health and fire officials and pollution control agencies. From a secure, explosion-proof location, use water spray to cool exposed containers. If cooling streams are ineffective (venting sound increases in volume and pitch, tank discolors, or shows any signs of deforming), withdraw immediately to a secure position. If employees are expected to fight fires, they must be trained and equipped in OSHA 1910.156.

References

Sax N. I., Ed., "Dangerous Properties of Industrial Materials Report," 1, No. 1, 42–43 (1980)

New York State Department of Health, "Chemical Fact Sheet: DMSO," Albany, NY, Bureau of Toxic Substance Assessment (March 1986)

Dimethyl Terephthalate

Molecular Formula: $C_{10}H_{10}O_4$

Common Formula: $CH_3OCOC_6H_4COOCH_3$

Synonyms: 1,4-Benzenedicarboxylic acid dimethyl ester; Dimethyl *p*-phthalate; DMT; Terephthalic acid dimethyl ester

CAS Registry Number: 120-61-6

RTECS Number: WZ1225000

Regulatory Authority

- Carcinogen (Animal Positive) (NCI)[9]
- Air Pollutant Standard Set (UNEP)[43]
- Water Pollution Standard Proposed (UNEP)[43]

Cited in U.S. State Regulations: New Hampshire (G), New York (G).

Description: DMT, $CH_3OCOC_6H_4COOCH_3$, is a combustible, white, flaky solid. Freezing/Melting point = 141°C. Boiling point = 284 – 288°C. Flash point = 153°C (oc). Autoignition temperature = 515 – 518°C. Hazard Identification (based on NFPA-704 M Rating System): Health 1, Flammability 1, Reactivity 0. Insoluble in water.

Potential Exposure: Essentially all DMT is consumed in the production of polyethylene terephthalate, the polymer for polyester fibers and polyester films. Less than 2% of production is used to make polybutylene terephthalate resins and other specialty products.

Incompatibilities: Incompatible with strong acids, nitrates, strong oxidizers.

Permissible Exposure Limits in Air: The former USSR-UNEP/IRPTC project[43] has set a MAC in workplace air of 0.1 mg/m^3.

Permissible Concentration in Water: The former USSR-UNEP/IRPTC project[43] has set a MAC in water bodies used for domestic purposes of 1.5 mg/l.

Routes of Entry: Inhalation of dust or vapor, ingestion, skin and eye contact. Absorbed through the skin.

Harmful Effects and Symptoms

Short Term Exposure: DMT appears to have a very low order of toxicity. Acute animal studies indicate oral, i.p., and dermal LD_{50} values in excess of 3,400 mg/kg, and subchronic oral (10,000 ppm DMT in the diet for 96 days) and inhalation exposures (2 – 10 ppm, 4 hr/day × 58 days) have not resulted in any hematologic, blood chemical, or pathologic alterations attributable to DMT. Results of other experiments with rats and rabbits demonstrated that DMT is rapidly absorbed and excreted (primarily in the urine), and that no significant quantities accumulate in tissues following single or repeated oral, intratracheal, dermal, or ocular administration, DMT does not appear to irritate or sensitize rodent skin.

First Aid: If this chemical gets into the eyes, remove any contact lenses at once and irrigate immediately for at least 15 minutes, occasionally lifting upper and lower lids. Seek medical attention immediately. If this chemical contacts the skin, remove contaminated clothing and wash immediately with soap and water. Seek medical attention immediately. If this chemical has been inhaled, remove from exposure, begin rescue breathing (using universal precautions) if breathing has stopped and CPR if heart action has stopped. Transfer promptly to a medical facility. When this chemical has been swallowed, get medical attention. Give large quantities of water and induce vomiting. Do not make an unconscious person vomit.

Personal Protective Methods: Wear protective gloves and clothing to prevent any reasonable probability of skin contact. Safety equipment suppliers/manufacturers can provide recommendations on the most protective glove/clothing material for your operation.. All protective clothing (suits, gloves, footwear, headgear) should be clean, available each day, and put on before work. Contact lenses should not be worn when working with this chemical. Wear dust-proof chemical goggles and face shield unless full facepiece respiratory protection is worn. Employees should wash immediately with soap when skin is wet or contaminated. Provide emergency showers and eyewash.

Storage: Prior to working with DMT you should be trained on its proper handling and storage. Store in tightly closed containers in a cool, well ventilated area away from oxidizers, nitrates, acids and sources of ignition. A regulated, marked area should be established where this chemical is handled, used, or stored in compliance with OSHA standard 1910.1045.

Shipping: DMT is not specifically cited in DOT'S Performance- Oriented Packaging Standards. It could be classified as a Hazardous Substance, solid, n.o.s. in NA 9188. However this imposes no label requirements or shipping weight maximums.

Spill Handling: Evacuate persons not wearing protective equipment from area of spill or leak until clean-up is complete. Remove all ignition sources. Collect powdered material in the most convenient and safe manner and deposit in sealed containers. Ventilate area after clean-up is complete. It may be necessary to contain and dispose of this chemical as a hazardous waste. If material or contaminated runoff enters waterways, notify downstream users of potentially contaminated waters. Contact your Department of Environmental Protection or your regional office of the federal EPA for specific recommendations. If employees are required to clean-up spills, they must be properly trained and equipped. OSHA 1910.120(q) may be applicable.

Fire Extinguishing: Molten DMT will burn if ignited. It ignites at 153°C. Vapor or dust can form explosive mixtures in air. Use dry chemical, carbon dioxide, water spray, or foam extinguishers. Poisonous gases are produced in fire. If material or contaminated runoff enters waterways, notify downstream users of potentially contaminated waters. Notify local health and fire officials and pollution control agencies. From a secure, explosion-proof location, use water spray to cool exposed containers. If cooling streams are ineffective (venting sound increases in volume and pitch, tank discolors, or shows any signs of deforming), withdraw immediately to a secure position. If employees are expected to fight fires, they must be trained and equipped in OSHA 1910.156.

Disposal Method Suggested: Incineration.

References

National Institute for Occupational Safety and Health, Information Profiles on Potential Occupational Hazards – Single Chemicals: Dimethyl Terephthalate, pp 84–97, Report TR 79-607, Rockville, MD (December 1979)

New York State Department of Health, "Chemical Fact Sheet: Dimethyl Terephthalate," Albany, NY, Bureau of Toxic Substance Assessment (January 1986)

Dimetilan

Molecular Formula: $C_{10}H_{16}N_4O_3$

Synonyms: Carbamic acid, dimethyl-, 1-[(dimethylamino)carbonyl]-5-methyl-1H-pyrazol-2-yl ester; Carbamic acid, dimethyl-, ester with 3-hydroxy-*N*,*N*-5-trimethylpyrazole-1-carboxamide; Dimethyl carbamate ester of 3-hydroxy-*N*,*N*-5-trimethylpyrazole-1-carboxamide; Dimethylcarbamic acid 1-[(dimethylamino)carbonyl]-5-methyl-1H-pyrazol-3-yl ester; Dimethylcarbamic acid ester with 3-hydroxy-*N*,*N*,5-trimethylpyrazole-1-carboxamide; Dimethylcarbamic acid 5-methyl-1H-carboxamine; Dimethylcarbamic acid 5-methyl-1H-pyrazol-3-yl ester; 2-Dimethylcarbamoyl-3-methylpyrazolyl-(5)-*N*,*N*-dimethylcarbamat; 1-Dimethylcarbamoyl-5-methylpyrazol-3-yl dimethylcarbamate; Dimethylcarbamoyl-3-methyl-5-pyrazolyl dimethylcarbamate; Dimetilane; ENT25,595-X; ENT 25,922; Geigy 22870; 3-Hydroxy-*N*,*N*,5-trimethylpyrazole-1-carboxamidedimethylcarbamate (ester); 5-Methyl-1H-pyrazol-3-yl dimethylcarbamate; Snip; Snip fly

CAS Registry Number: 644-64-4

RTECS Number: EZ9084000

DOT ID: UN 2757

Regulatory Authority

- Banned or Severely Restricted (Portugal) (UN)[13]
- RCRA, 40CFR261, Appendix 8 Hazardous Constituents
- RCRA 40CFR268.48; 61FR15654, Universal Treatment Standards: Wastewater (mg/l), 0.056; Nonwastewater (mg/kg), 1.4
- Superfund/EPCRA 40CFR355, Appendix B Extremely Hazardous Substances: TPQ = 500/10,000 lb (227/4,540 kg)
- Superfund/EPCRA 40CFR302.4 Reportable Quantity (RQ): EHS, 1 lb (0.454 kg)

Cited in U.S. State Regulations: Massachusetts (G).

Description: Dimetilan, $C_{10}H_{16}N_4O_3$, is a yellow to reddish-brown solid. Freezing/Melting point = 68 – 71°C (the technical grade at 55 – 65°C). Hazard Identification (based on NFPA-704 M Rating System): Health 3, Flammability 1, Reactivity 0. Highly soluble in water.

Potential Exposure: Formerly an insecticide for insect control on livestock, especially housefly control. It is no longer produced commercially in the U.S.

Incompatibilities: Hydrolyzed by acids and alkalis.

Permissible Exposure Limits in Air: No standards set.

Permissible Concentration in Water: No criteria set.

Routes of Entry: Skin contact, inhalation.

Harmful Effects and Symptoms

Short Term Exposure: Very toxic; probable oral lethal dose for humans is 50 – 500 mg/kg or between 1 teaspoon and 1 oz for a 70 kg (150 lb) person. Dimetilan is highly toxic by ingestion and moderately toxic by contact with the skin. Death is primarily due to respiratory arrest of central origin, paralysis off the respiratory muscles, intense bronchoconstriction, or all three. This compound is a cholinesterase inhibitor. Symptoms are similar to carbaryl poisoning: nausea, vomiting, abdominal cramps, diarrhea, pinpoint pupils, excessive salivation, and sweating are common symptoms. Running nose and tightness in chest are common in inhalation exposures. Difficulty in breathing, raspy breathing, and loss of muscle coordination may also be seen. Exposure may also result in random jerky movements, incontinence, convulsions, and coma and death.

Long Term Exposure: Many carbamates affect the central nervous system.

Medical Surveillance: Before employment and at regular times after that, the following are recommended: Plasma and red blood cell cholinesterase levels (tests for the enzyme poisoned by this chemical). If exposure stops, plasma levels return to normal in 1 – 2 weeks while red blood cell levels may be reduced for 1 – 3 months.

When cholinesterase enzyme levels are reduced by 25% or more below preemployment levels, risk of poisoning is increased, even if results are in lower ranges of "normal." Reassignment to work not involving organophosphate or carbamate pesticides is recommended until enzyme levels recover. If symptoms develop or overexposure occurs, repeat the above tests as soon as possible and get an exam of the nervous system. Also consider complete blood count. Consider chest x-ray following acute overexposure. Do not drink any alcoholic beverages before or during use. Alcohol promotes absorption of organic phosphates.

First Aid: If this chemical gets into the eyes, remove any contact lenses at once and irrigate immediately for at least 15 minutes, occasionally lifting upper and lower lids. Seek medical attention immediately. If this chemical contacts the skin, remove contaminated clothing and wash immediately with soap and water. Speed in removing material from skin is of extreme importance. Shampoo hair promptly if contaminated. Seek medical attention immediately. If this chemical has been inhaled, remove from exposure, begin rescue breathing (using universal precautions) if breathing has stopped and CPR if heart action has stopped. Transfer promptly to a medical facility. When this chemical has been swallowed, get medical attention. Give large quantities of water and induce vomiting. Do not make an unconscious person vomit. Speed in removing material from skin is of extreme importance. Remove and isolate contaminated clothing and shoes at the site. Keep victim quiet and maintain normal body temperature. Effects may be delayed; keep victim under observation.

Personal Protective Methods: Wear protective gloves and clothing to prevent any reasonable probability of skin contact. Safety equipment suppliers/manufacturers can provide recommendations on the most protective glove/clothing material for your operation.. All protective clothing (suits, gloves, footwear, headgear) should be clean, available each day, and put on before work. Contact lenses should not be worn when working with this chemical. Wear dust-proof chemical goggles and face shield unless full facepiece respiratory protection is worn. Employees should wash immediately with soap when skin is wet or contaminated. Provide emergency showers and eyewash.

Storage: Prior to working with this chemical you should be trained on its proper handling and storage. Store in tightly closed containers in a cool, well-ventilated area.

Shipping: Carbamate pesticides, solid, toxic, n.o.s. are in Hazard Class 6.1 and Dimetilan is in Packing Group II. It requires a "Poison" label.

Spill Handling: Evacuate persons not wearing protective equipment from area of spill or leak until clean-up is complete. Remove all ignition sources. Collect powdered material in the most convenient and safe manner and deposit in sealed containers. Ventilate area after clean-up is complete. It may be necessary to contain and dispose of this chemical as a hazardous waste. If material or contaminated runoff enters waterways, notify downstream users of potentially contaminated waters. Contact your Department of Environmental Protection or your regional office of the federal EPA for specific recommendations. If employees are required to clean-up spills, they must be properly trained and equipped. OSHA 1910.120(q) may be applicable.

Fire Extinguishing: Extinguish fire using agent suitable for types of surrounding fire, as the material itself burns with difficulty. Use water in flooding quantities as a fog. Wear positive pressure breathing apparatus and special protective clothing. Move container from fire area. Fight fire from maximum distance. Dike fire control water for later disposal; do not scatter the material. Poisonous gases are produced in fire. If material or contaminated runoff enters waterways, notify downstream users of potentially contaminated waters. Notify local health and fire officials and pollution control agencies. From a secure, explosion-proof location, use water spray to cool exposed containers. If cooling streams are ineffective (venting sound increases in volume and pitch, tank discolors, or shows any signs of deforming), withdraw immediately to a secure position. If employees are expected to fight fires, they must be trained and equipped in OSHA 1910.156.

Disposal Method Suggested: Alkali treatment followed by soil burial.[22] Large amounts should be incinerated in a unit equipped with efficient gas scrubbing. In accordance with 40CFR165 recommendations for the disposal of pesticides and pesticide containers. Must be disposed properly by following package label directions or by contacting your state pesticide or environmental control agency or by contacting your regional EPA office.

References

U.S. Environmental Protection Agency, "Chemical Profile: Dimetilan," Washington, DC, Chemical Emergency Preparedness Program (November 30, 1987)

Dinitolmide

Molecular Formula: $C_8H_7N_3O_5$

Common Formula: $H_3CC_6H_2(NO_2)_2CONH_2$

Synonyms: 2-Methyl-3,5-dinitrobenzamide, *o*-Dinitro-toluamide; Zoalene®; Zoamix®

CAS Registry Number: 148-01-6

RTECS Number: XS4200000

DOT ID: No citation.

Regulatory Authority

- Air Pollutant Standard Set (ACGIH)[1] (Several States)[60]

Cited in U.S. State Regulations: Alaska (G), California (A, G), Connecticut (A), Florida (G), Illinois (G), Maine (G), Massachusetts (G), Nevada (A), New Hampshire (G), New Jersey (G), North Dakota (A), Pennsylvania (G), Rhode Island (G), Virginia (A), West Virginia (G).

Description: Dinitolmide, $H_3CC_6H_2(NO_2)_2CONH_2$, is a yellowish crystalline substance. Freezing/Melting point = 177 – 181°C. Very slightly soluble in water.

Potential Exposure: Those involved in the manufacture, formulation and application of this veterinary coccidiostat.

Incompatibilities: A weak oxidizing agent, but high temperatures and pressures may cause violent reactions. Heat can cause a violent exothermic reaction above 248°C. Contact with alkalies may form explosive metal salts.

Permissible Exposure Limits in Air: There is no OSHA PEL. ACGIH recommends a TWA value of 5 mg/m^3 and STEL of 10 mg/m^3. Several states have set standards or guidelines for dinitolmide in ambient air[60] ranging from 50 $\mu g/m^3$ (North Dakota) to 80 $\mu g/m^3$ (Virginia) to 100 $\mu g/m^3$ (Connecticut) to 119 $\mu g/m^3$ (Nevada).

Determination in Air: Filter; none; Gravimetric; NIOSH IV Method #0500, Particulates NOR (total).

Permissible Concentration in Water: No criteria set.

Routes of Entry: Inhalation, ingestion, skin and/or eye contact.

Harmful Effects and Symptoms

Short Term Exposure: Contact eczema; in animals. The oral LD_{50} rat is 600 mg/kg (slightly toxic).

Long Term Exposure: Dinitolmide may cause mutations. All contact with this chemical should be reduced to the lowest possible level. Prolonged exposure may have some effect on the liver; methemoglobinemia.

Points of Attack: Liver, blood, skin.

Medical Surveillance: If symptoms develop or overexposure is suspected, the following may be useful: Liver function tests. Evaluation by a qualified allergist. Complete blood count. Methemoglobin level.

First Aid: If this chemical gets into the eyes, remove any contact lenses at once and irrigate immediately for at least 15 minutes, occasionally lifting upper and lower lids. Seek medical attention immediately. If this chemical contacts the skin, remove contaminated clothing and wash immediately with soap and water. Seek medical attention immediately. If this chemical has been inhaled, remove from exposure, begin rescue breathing (using universal precautions) if breathing has stopped and CPR if heart action has stopped. Transfer promptly to a medical facility. When this chemical has been swallowed, get medical attention. Give large quantities of water and induce vomiting. Do not make an unconscious person vomit.

Note to Physician: Treat for methemoglobinemia. Spectrophotometry may be required for precise determination of levels of methemoglobinemia in urine.

Personal Protective Methods: Wear protective gloves and clothing to prevent any reasonable probability of skin contact. Safety equipment suppliers/manufacturers can provide recommendations on the most protective glove/clothing material for your operation.. All protective clothing (suits, gloves, footwear, headgear) should be clean, available each day, and put on before work. Contact lenses should not be worn when working with this chemical. Wear dust-proof chemical goggles and face shield unless full facepiece respiratory protection is worn. Employees should wash immediately with soap when skin is wet or contaminated. Provide emergency showers and eyewash.

Respirator Selection: Where the potential exists for exposures over 5 mg/m^3, use an MSHA/NIOSH approved respirator equipped with particulate (dust/fume/mist) filters. Particulate filters must be checked every day before work for physical damage, such as rips or tears, and replaced as needed. If while wearing a filter, cartridge or canister respirator, you can smell, taste, or otherwise detect dinitolmide, or in the case of a full facepiece respirator you experience eye irritation, leave the area immediately. Check to make sure the respirator to face seal is still good. If it is, replace the filter, cartridge, or canister. If the seal is no longer good, you may need a new respirator.

Where the potential for high exposures exists, use an MSHA/NIOSH approved supplied-air respirator with a full facepiece operated in the positive pressure mode or with a full facepiece, hood, or helmet in the continuous flow mode, or use an MSHA/NIOSH approved self-contained breathing apparatus with a full facepiece operated in the pressure-demand or other positive pressure mode.

Storage: Prior to working with dinitolmide you should be trained on its proper handling and storage. Store in tightly closed containers in a cool, well ventilated area away from alkalies and heat.

Shipping: Pesticides solid, toxic, n.o.s. fall in Hazard Class 6.1 and dinitolmide in Packing Group III. It requires a "Keep Away From Food" label.

Spill Handling: Evacuate persons not wearing protective equipment from area of spill or leak until clean-up is complete. Remove all ignition sources. Collect powdered material in the most convenient and safe manner and deposit in sealed containers. Ventilate area after clean-up is complete. It may be necessary to contain and dispose of this chemical as a hazardous waste. If material or contaminated runoff enters waterways, notify downstream users of potentially contaminated waters. Contact your Department of Environmental Protection or your regional office of the federal EPA for specific recommendations. If employees are required to clean-up spills, they must be properly trained and equipped. OSHA 1910.120(q) may be applicable.

Fire Extinguishing: Extinguish fire using an agent suitable for type of surrounding fire. Dinitolmide itself does not burn. Poisonous gases are produced in fire. If material or contaminated runoff enters waterways, notify downstream users of potentially contaminated waters. Notify local health and fire officials and pollution control agencies. From a secure, explosion-proof location, use water spray to cool exposed containers. If cooling streams are ineffective (venting sound increases in volume and pitch, tank discolors, or shows any signs of deforming), withdraw immediately to a secure position. If employees are expected to fight fires, they must be trained and equipped in OSHA 1910.156.

Disposal Method Suggested: In accordance with 40CFR165 recommendations for the disposal of pesticides and pesticide containers. Must be disposed properly by following package label directions or by contacting your state pesticide or environmental control agency or by contacting your regional EPA office.

References

New Jersey Department of Health, "Hazardous Substance Fact Sheet: Dinitolmide," Trenton, NJ (February 1989)

Dinitroanilines

Molecular Formula: $C_6H_5N_3O_4$

Synonyms: 2,4-Dinitraniline; 2,4-Dinitroanilin (German); 2,4-Dinitroanilina (Italian); 2,4-Dinitrobenzenamime; DNA; NCI-C60753

CAS Registry Number: 97-02-9 (2,4-); 602-03-9 (2,3-); 606-22-4 (2,6-); 618-87-1 (3,5-); 26471-56-7 (mixed isomers)

RTECS Number: BX9100000 (2,4-)

DOT ID: UN 1596

EEC Number: 612-040-00-1

Regulatory Authority

2,4-dinitroaniline:

- OSHA 29CFR1910.119, Appendix A. Process Safety List of Highly Hazardous Chemicals, TQ = 5,000 lb (2,270 kg)

Cited in U.S. State Regulations: New Jersey (G), Pennsylvania (G).

Description: Dinitroanilines are yellow to greenish-yellow, needle-like crystals with a musty odor. Boiling Point = 57°C. Freezing/Melting point = 188°C. Flash point = 224°C. Hazard Identification (based on NFPA-704 M Rating System): Health 3, Flammability 1, Reactivity 3. Insoluble in water.

Potential Exposure: Used as a corrosion inhibitor and to make azo dyes and toner pigment in printing inks.

Incompatibilities: Decomposes in moderate heat forming toxic vapors that form an explosive mixture with air. Violent reaction with strong oxidizers, strong acids, strong bases, acid chlorides, acid anhydrides and chloroformates.

Permissible Exposure Limits in Air: No OELs have been established.

Routes of Entry: Inhalation.

Harmful Effects and Symptoms

Short Term Exposure: Irritates the eyes, skin and respiratory tract. Inhalation can cause coughing and wheezing. High levels can cause methemoglobinemia, causing headache, dizziness, and blue color to the skin and lips. Higher levels can cause difficult breathing, collapse and possible death.

Long Term Exposure: See entry above. May cause liver and kidney damage.

Points of Attack: Blood, liver, kidneys.

Medical Surveillance: Blood methemoglobin level. Liver and kidney function tests.

First Aid: If this chemical gets into the eyes, remove any contact lenses at once and irrigate immediately for at least 15 minutes, occasionally lifting upper and lower lids. Seek medical attention immediately. If this chemical contacts the skin, remove contaminated clothing and wash immediately with soap and water. Seek medical attention immediately. If this chemical has been inhaled, remove from exposure, begin rescue breathing (using universal precautions) if breathing has stopped and CPR if heart action has stopped. Transfer promptly to a medical facility. When this chemical has been swallowed, get medical attention. Give large quantities of water and induce vomiting. Do not make an unconscious person vomit. In the case of poisoning, special first aid is required; antidotes for the formation of met hemoglobin should be available, including instructions.

Note to Physician: Treat for methemoglobinemia. Spectrophotometry may be required for precise determination of levels of methemoglobinemia in urine.

Personal Protective Methods: Wear protective gloves and clothing to prevent any reasonable probability of skin contact. Safety equipment suppliers/manufacturers can provide recommendations on the most protective glove/clothing material for your operation.. All protective clothing (suits, gloves, footwear, headgear) should be clean, available each day, and put on before work. Contact lenses should not be worn when working with this chemical. Wear dust-proof chemical goggles and face shield unless full facepiece respiratory protection is worn. Employees should wash immediately with soap when skin is wet or contaminated. Provide emergency showers and eyewash.

Respirator Selection: Where the potential for exposure to this chemical, use a MSHA/NIOSH approved supplied-air respirator with a full facepiece operated in the positive pressure mode or with a full facepiece, hood, or helmet in the continuous flow mode, or use a MSHA/NIOSH approved self-contained breathing apparatus with a full facepiece operated in pressure-demand or other positive pressure mode.

Storage: Prior to working with DNA you should be trained on its proper handling and storage. Before entering confined space where DNA may be present, check to make sure that an explosive concentration does not exist. Store in tightly closed containers in a cool, well ventilated area away from oxidizers, strong acids, strong bases and other incompatible materials listed above. Metal containers involving the transfer of this chemical should be grounded and bonded. Where possible, automatically pump liquid from drums or other storage containers to process containers. Drums must be equipped with self-closing valves, pressure vacuum bungs, and flame arresters. Use only non-sparking tools and equipment, especially when opening and closing containers of this chemical. Sources of ignition such as smoking and open flames, are prohibited where this chemical is used, handled, or stored in a manner that could create a potential fire or explosion hazard. Wherever this chemical is used, handled, manufactured, or stored, use explosion-proof electrical equipment and fittings.

Shipping: Label "Poison." Hazard Group 6.1. Packing Group II.

Spill Handling: Evacuate persons not wearing protective equipment from area of spill or leak until clean-up is complete. Remove all ignition sources. Do not dry sweep. Collect powdered material in the most convenient and safe manner and deposit in sealed containers. Ventilate area after clean-up is complete. It may be necessary to contain and dispose of this chemical as a hazardous waste. If material or contaminated runoff enters waterways, notify downstream users of potentially contaminated waters. Contact your Department of Environmental Protection or your regional office of the federal EPA for specific recommendations. If employees are required to clean-up spills, they must be properly trained and equipped. OSHA 1910.120(q) may be applicable.

Fire Extinguishing: This chemical is a combustible solid but does not readily ignite. However, moderate heat causes decomposition that produces toxic vapors that form an explosive mixture with air. Use dry chemical, carbon dioxide, water spray, or alcohol or polymer foam extinguishers. Poisonous gases are produced in fire including carbon monoxide and nitrogen oxides. Heated vapors in confined spaces can explode. If material or contaminated runoff enters waterways, notify downstream users of potentially contaminated waters. Notify local health and fire officials and pollution control agencies. Containers can explode in fire. From a secure, explosion-proof location, use water spray to cool exposed containers. If cooling streams are ineffective (venting sound increases in volume and pitch, tank discolors, or shows any signs of deforming), withdraw immediately to a secure position. If employees are expected to fight fires, they must be trained and equipped in OSHA 1910.156.

References

New Jersey Department of Health and Senior Services, "Hazardous Substance Fact Sheet, Dinitroanilines," Trenton, NJ (May 1999)

Dinitrobenzenes

Molecular Formula: $C_6H_4N_2O_4$

Common Formula: $C_6H_4(NO_2)_2$

Synonyms: *o-isomer:* Benzene, *o*-dinitro-; Benzene, 1,2-dinitro-; *o*-Dinitrobenceno (Spanish); 1,2-Dinitrobenzene; 1,2-Dinitrobenzol; 1,2-DNB

m-isomer: Benzene, *m*-dinitro-; Benzene, 1,3-dinitro-; *m*-Dinitrobenceno (Spanish); 1,3-Dinitrobenzene; 1,3-Dinitrobenzol; 1,3-DNB

p-isomer: Benzene, *p*-dinitro-; Benzene, 1,4-dinitro-; *p*-Dinitrobenceno (Spanish); 1,4-Dinitrobenzene; 1,4-Dinitrobenzol; Dithane A-4; 1,4-DNB

CAS Registry Number: 528-29-0 (ortho-); 99-65-0 (meta-); 100-25-4 (para-); 25154-54-5 (mixed isomers)

RTECS Number: CZ7450000 (ortho-); CZ7350000 (meta-); CZ7525000 (para-); CZ7340000 (mixed isomers)

DOT ID: UN 1597

EEC Number: 609-004-00-2

Regulatory Authority

- Air Pollutant Standard Set (ACGIH)[1] (DFG)[3] (HSE)[33] (former USSR)[43] (OSHA)[58] (Several States)[60]
- Water Pollution Standard Proposed (UNEP)[43]

mixed isomers:

- Clean Water Act: Section 311 Hazardous Substances/RQ 40CFR117.3 (same as CERCLA, see below)
- RCRA, 40CFR261, Appendix 8 Hazardous Constituents, AS dinitrobenzene, n.o.s., waste number not listed
- Superfund/EPCRA 40CFR302.4 Reportable Quantity (RQ): CERCLA, 100 lb (45.4 kg)
- Canada, WHMIS, Ingredients Disclosure List

meta-:

- Clean Water Act: Section 311 Hazardous Substances/RQ 40CFR117.3 (same as CERCLA, see below)
- RCRA 40CFR264, Appendix 9; TSD Facilities Ground Water Monitoring List. Suggested test method(s) (PQL μg/l): 8270 (10)
- Superfund/EPCRA 40CFR302.4 Reportable Quantity (RQ): CERCLA, 100 lb (45.4 kg)
- EPCRA Section 313 Form R *de minimis* concentration reporting level: 1.0%

ortho-:

- Clean Water Act: Section 311 Hazardous Substances/RQ 40CFR117.3 (same as CERCLA, see below)
- Superfund/EPCRA 40CFR302.4 Reportable Quantity (RQ): CERCLA, 100 lb (45.4 kg)
- EPCRA Section 313 Form R *de minimis* concentration reporting level: 1.0%

para-:

- Clean Water Act: Section 311 Hazardous Substances/RQ 40CFR117.3 (same as CERCLA, see below)
- RCRA 40CFR268.48; 61FR15654, Universal Treatment Standards: Wastewater (mg/l), 0.32; Nonwastewater (mg/kg), 2.3
- Superfund/EPCRA 40CFR302.4 Reportable Quantity (RQ): CERCLA, 100 lb (45.4 kg)
- EPCRA Section 313 Form R *de minimis* concentration reporting level: 1.0%

Cited in U.S. State Regulations: Alaska (G), Connecticut (A), Florida (G, A), Illinois (G), Kansas (G), Louisiana (G), Maine (G), Massachusetts (G), Nevada (A), New Hampshire (G), New Jersey (G), New York (A), South Carolina (A), Vermont (G), Virginia (G, A), Washington (G), West Virginia (G), Wisconsin (G).

Description: Dinitrobenzene, $C_6H_4(NO_2)_2$, exists in three isomers (o-, m-, p-); the meta form is the most widely used. All are white to yellow crystalline solids having a characteristic odor. Boiling point = (ortho-) 318°C; (meta-) 319°C; (para-) 299°C; (mixed-) @ 305°C. Freezing/Melting point = (ortho-) 117 – 118°C; (meta-) 90°C; (para-) 173 – 174°C; (mixed) 75 – 85°C. Flash point = (ortho-, meta-, para-) 150°C. NFPA 704 M Hazard Identification (ortho-): Health 3, Flammability 1,

Reactivity 4. Solubility in water is poor for o-, m-, and mixed isomers; none for p-.

Potential Exposure: Dinitrobenzene is used in the synthesis of dyestuffs, dyestuff intermediates, and explosives and in celluloid production.

Incompatibilities: Dinitrobenzene is impact, friction, and heat sensitive, and may explode. Dust explosion possible if mixed with air. Reacts violently with strong oxidizers, strong bases, and chemically active metals, causing fire and explosion hazard. Mixtures with nitric acid are highly explosive. Attacks some plastics.

Permissible Exposure Limits in Air: The Federal Limit (OSHA PEL)[58] for all isomers of dinitrobenzene is 1 mg/m^3 (0.15 ppm). The notation "skin" indicates the possibility of cutaneous absorption. This is also the ACGIH,[1] DFG[3] and HSE[33] values as well as the MAC value set by the former USSR-UNEP/IRPTC project[43] for workplace air. The STEL value adopted by HSE[33] is 0.5 ppm (3 mg/m^3). The NIOS ILDH level is 50 mg/m^3.

A number of states have set guidelines or standards for dinitrobenzenes in ambient air[60] as follows (all values in $\mu g/m^3$):

State	ortho-	meta-	para-
Connecticut	20.0	20.0	100.0
Florida	—	10.0	—
Nevada	—	24.0	—
New York	—	3.3	—
North Dakota	10.0	10.0	10.0
South Carolina	—	10.0	—
Virginia	16.0	16.0	16.0

Determination in Air: Collection by filter in series with an ethylene glycol bubbler, analysis by high pressure liquid chromatography/Ultraviolet detection; NIOSH II(4) Method #S214.

Permissible Concentration in Water: The MAC in water bodies used for domestic purposes has been set by the former USSR-UNEP/IRPTC project.[43]

Routes of Entry: Inhalation, percutaneous absorption of liquid, ingestion, eye and skin contact.

Harmful Effects and Symptoms

Short Term Exposure: Dinitrobenzene irritates the eyes, skin, and respiratory tract. Dinitrobenzene may affect the lungs and the ability of the blood to carry oxygen, resulting in the formation of methaemoglobin. This can cause headache, weakness, fatigue, dizziness and blue color to the skin and lips. Higher levels may cause difficulty in breathing, collapse, unconsciousness, and may result in death. The effects may be delayed. Consuming alcohol, exposure to sunlight, or hot baths may make symptoms worse. Exposure to dinitrobenzene may produce yellowish coloration of the skin, eyes and hair. Exposure to any isomer of dinitrobenzene may produce methemoglobinemia, symptoms of which are headaches, irritability, dizziness, weakness, nausea, vomiting, dyspnea, drowsiness, and unconsciousness. If treatment is not given promptly, death may occur. Dinitrobenzene may also cause a bitter almond taste or burning sensation in the mouth, dry throat, and thirst. Reduced vision may occur. In addition liver damage, hearing loss, and ringing of the ears may be produced. Repeated or prolonged exposure may cause anemia.

Long Term Exposure: Repeated exposure may cause hearing loss and changes of vision. Prolonged exposure may lead to liver damage and may cause anemia. Can cause serious reproductive toxicity in humans. The substance may have effects on the respiratory tract formation of methaemoglobin.

Points of Attack: Blood, liver, cardiovascular system, eyes, central nervous system.

Medical Surveillance: Preemployment and periodic examinations should be concerned particularly with a history of blood dyscrasias, reactions to medications, alcohol intake, eye disease, and skin and cardiovascular status. Liver and renal functions should be evaluated periodically as well as blood and general health. Blood methemoglobin levels should be followed until normal in all cases of suspected cyanosis. Dinitrobenzene can be determined in the urine; levels greater than 25 mg/l may indicate significant absorption. Complete blood count (CBC). Examination of the eyes and color vision. People with a medical condition called "G-6-P-D Deficiency" may have worse problems if exposed to dinitrobenzene.

First Aid: If this chemical gets into the eyes, remove any contact lenses at once and irrigate immediately for at least 15 minutes, occasionally lifting upper and lower lids. Seek medical attention immediately. If this chemical contacts the skin, remove contaminated clothing and wash immediately with soap and water. Seek medical attention immediately. If this chemical has been inhaled, remove from exposure, begin rescue breathing (using universal precautions) if breathing has stopped and CPR if heart action has stopped. Transfer promptly to a medical facility. When this chemical has been swallowed, get medical attention. Give large quantities of water and induce vomiting. Do not make an unconscious person vomit. Effects may be delayed; medical observation is recommended.

Note to Physician: Treat for methemoglobinemia. Spectrophotometry may be required for precise determination of levels of methemoglobinemia in urine.

Personal Protective Methods: Wear protective gloves and clothing to prevent any reasonable probability of skin contact. Safety equipment suppliers/manufacturers can provide recommendations on the most protective glove/clothing material for your operation.. All protective clothing (suits, gloves, footwear, headgear) should be clean, available each day, and put on before work. Contact lenses should not be worn when working with this chemical. Wear dust-proof chemical goggles and face shield unless full facepiece respiratory protection is worn. Employees should wash immediately with soap when skin is wet or contaminated. Provide emergency showers and eyewash.

Respirator Selection: NIOSH/OSHA: *5 mg/m³:* DM (any dust and mist respirator). *10 mg/m³*: DMXSQ (any dust and mist respirator except single-use and quarter mask respirators); or SA (any supplied-air respirator); or HiE (any air-purifying, respirator with a high-efficiency particulate filter). *25 mg/m³:* SA:CF (any supplied-air respirator operated in a continuous-flow mode); or PAPRDM (any powered, air-purifying respirator with a dust and mist filter). *50 mg/m³:* HiEF (any air-purifying, full-facepiece respirator with a high-efficiency particulate filter); or SAT:CF (any supplied-air respirator that has a tight-fitting facepiece and is operated in a continuous-flow mode); or PAPRTHiE (any powered, air-purifying respirator with a tight-fitting facepiece and a high-efficiency particulate filter); or SCBAF (any self-contained breathing apparatus with a full facepiece); or SAF (any supplied-air respirator with a full facepiece). *Emergency or planned entry into unknown concentrations or IDLH conditions:* SCBAF:PD,PP (any self-contained breathing apparatus that has a full facepiece and is operated in a pressure-demand or other positive-pressure mode); or SAF:PD,PP:ASCBA (any supplied-air respirator that has a full facepiece and is operated in a pressure-demand or other positive-pressure mode in combination with an auxiliary self-contained breathing apparatus operated in a pressure-demand or other positive pressure mode). *Escape:* HiEF (any air-purifying, full-facepiece respirator with a high-efficiency particulate filter); or SCBAE (any appropriate escape-type, self-contained breathing apparatus).

Storage: Prior to working with dinitrobenzene you should be trained on its proper handling and storage. Dinitrobenzene must be stored to avoid contact with strong oxidizers (such as chloride, bromine, chlorine dioxide, nitrates, and permanganates) since violent reactions occur. Contact with caustics and chemically active metals (such as tin and zinc) may evolve heat, causing a build-up in pressure. Store in tightly closed containers in a cool, well-ventilated area away from shock or heat, which may cause this chemical to explode. Storage outdoors or in explosion-proof areas is preferred. Sources of ignition such as smoking and open flames are prohibited where dinitrobenzene is handled, used, or stored. Metal containers used in the transfer of 5 gallons or more of dinitrobenzene should be grounded and bonded. Drums must be equipped with self-closing valves, pressure vacuum bungs, and flame arresters. Use only non-sparking tools and equipment, especially when opening and closing containers of dinitrobenzene. Wherever dinitrobenzene is used, handled, manufactured, or stored, use explosion-proof electrical equipment and fittings.

Shipping: Dinitrobenzenes fall in Hazard Class 6.1 and Packing Group II. They all require a "Poison" label.

Spill Handling: Evacuate persons not wearing protective equipment from area of spill or leak until clean-up is complete. Remove all ignition sources. Collect powdered material in the most convenient and safe manner and deposit in sealed containers. Ventilate area after clean-up is complete. It may be necessary to contain and dispose of this chemical as a hazardous waste. If material or contaminated runoff enters waterways, notify downstream users of potentially contaminated waters. Contact your Department of Environmental Protection or your regional office of the federal EPA for specific recommendations. If employees are required to clean-up spills, they must be properly trained and equipped. OSHA 1910.120(q) may be applicable.

Fire Extinguishing: Dinitrobenzene is an extremely explosive solid. Prolonged exposure to fire and heat may result in an explosion due to SPONTANEOUS decomposition. May explode on heating under confinement. Fight fires from an explosions-resistant location. Cool fire-exposed containers of dinitrobenzene with water. In advanced or massive fires, evacuate the area. Use dry chemical, carbon dioxide, water spray, or alcohol foam extinguishers. Poisonous gases are produced in fire including nitrogen oxides. If material or contaminated runoff enters waterways, notify downstream users of potentially contaminated waters. Notify local health and fire officials and pollution control agencies. From a secure, explosion-proof location, use water spray to cool exposed containers. If cooling streams are ineffective (venting sound increases in volume and pitch, tank discolors, or shows any signs of deforming), withdraw immediately to a more secure position. If employees are expected to fight fires, they must be trained and equipped in OSHA 1910.156.

Disposal Method Suggested: Incineration (1,800°F, 2.0 seconds minimum) followed by removal of the oxides of nitrogen that are formed using scrubbers and/or catalytic or thermal devices. The dilute wastes should be concentrated before incineration.

References

U.S. Environmental Protection Agency, Dinitrobenzenes, Health and Environmental Effects Profile No. 89, Wash, DC, Office of Solid Waste (April 30, 1980)

Sax N. I., Ed., "Dangerous Properties of Industrial Materials Report," 3, No. 3, 80–82 (1983) (p-Dinitrobenzene) and 5, No. 3, 51–53 (1985) (o-Dinitrobenzene) and 6, No. 1, 49–52 (1986) (m-Dinitrobenzene)

New Jersey Department of Health, "Hazardous Substance Fact Sheet: Dinitrobenzene," Trenton, NJ (February 1996)

Dinitro-o-Cresol

Molecular Formula: $C_7H_6N_2O_5$

Common Formula: $CH_3C_6H_2(NO_2)_2OH$

Synonyms: Antinonin; Antinonnin; Arborol; *o*-Cresol, 4,6-dinitro-; Degrassan; Dekrysil; Detal; Dillex; Dinitro; 3,5-Dinitro-*o*-cresol; Dinitro-*o*-cresol; Dinitrocresol; 4,6-Dinitro-*o*-cresol and salts; Dinitrodendtroxal; 3,5-Dinitro-2-hydroxytoluene; Dinitrol; 2,4-Dinitro-6-methylphenol; 4,6-Dinitro-2-methylphenol; DINOC; Dinurania; Ditrosol; DNOC; Effusan; Effusan 3436; Elgetol; Elgetol 30; Elipol; Extrar; Flavin-Sandoz; Hedolit; Hedolite; K III; K IV; Kreozan; Krezotol 50; Lipan; 2-Methyl-4,6-dinitrophenol; 6-Methyl-2,4-dinitrophenol; Neudorff DN 50; Nitrofan; Phenol, 2-methyl-4,6-dinitro-; Prokarbol; Rafex; Rafex 35; Raphatox; Sandolin; Sandolin A; Selinon; Sinox; Winterwash

CAS Registry Number: 534-52-1

RTECS Number: GO9625000

DOT ID: UN 1598

Regulatory Authority

- Banned or Severely Restricted (Sweden) (UN)[13]
- Air Pollutant Standard Set (ACGIH)[1] (DFG)[3] (HSE)[33] ()SHA)[58] (former USSR)[43] (Several States)[60]
- Clean Air Act: Hazardous Air Pollutants (Title I, Part A, Section 112)
- Clean Water Act: Section 313 Water Priority Chemicals (57FR41331, 9/9/92)
- EPA Hazardous Waste Number (RCRA No.): P047
- RCRA, 40CFR261, Appendix 8 Hazardous Constituents
- RCRA 40CFR268.48; 61FR15654, Universal Treatment Standards: Wastewater (mg/l), 0.28; Nonwastewater (mg/kg), 160
- RCRA 40CFR264, Appendix 9; TSD Facilities Ground Water Monitoring List. Suggested test method(s) (PQL μg/l): 8040 (150); 8270 (50)
- Superfund/EPCRA 40CFR355, Appendix B Extremely Hazardous Substances: TPQ = 10/10,000 lb (4.54/4,540 kg)
- Superfund/EPCRA 40CFR302.4 Reportable Quantity (RQ): CERCLA, 10 lb (4.54 kg)
- EPCRA Section 313 Form R *de minimis* concentration reporting level: 1.0%
- U.S. DOT Regulated Marine Pollutant (49CFR172.101, Appendix B)
- Canada, WHMIS, Ingredients Disclosure List

Cited in U.S. State Regulations: Alaska (G), Connecticut (A), Florida (G), Illinois (G), Kansas (G, W), Louisiana (G), Maine (G), Maryland (G), Massachusetts (G), Michigan (G), Nevada (A), New Hampshire (G), New Jersey (G), North Dakota (A), Oklahoma (G), Pennsylvania (G), Rhode Island (G), Vermont (G), Virginia (G, A), Washington (G), West Virginia (G), Wisconsin (G).

Description: DNOC, $CH_3C_6H_2(NO_2)_2OH$, exists in 9 isomeric forms of which 4,6-dinitro-o-cresol is the most important commercially, and heavily regulated. It is a noncombustible, yellow crystalline solid. Boiling point = 312°C. Freezing/Melting point = 88°C. Slightly soluble in water.

Potential Exposure: DNOC is widely used in agriculture as a herbicide and pesticide; it is also used in the dyestuff industry. Although 4,6-dinitro-o-cresol (DNOC) is no longer manufactured in the United States, a limited quantity is imported and used as a blossom-thinning agent on fruit trees and as a fungicide, insecticide, and miticides on fruit trees during the dormant season. Hence, individuals formulating or spraying the compound incur the highest risk of exposure to the compound.

Incompatibilities: Dust can form an explosive mixture with air. Strong oxidizers, oxidizers, strong bases. Protect from heat and shock.

Permissible Exposure Limits in Air: The Federal Limit (OSHA PEL),[58] the DFG[3] and HSE[33] values and recommended ACGIH TWA value for all isomers of DNOC is 0.2 mg/m^3. The notation "skin" indicates the possibility of cutaneous absorption. The HSE STEL value is 0.6 mg/m^3. The NIOSH IDLH level is 5.0 mg/m^3. The MAC in workplace air set by the former USSR-UNEP/IRPTC project[43] is 0.05 mg/m^3. They have also set MAC values for the ambient air in residential areas of 0.003 mg/m^3 on a momentary basis and 0.0008 mg/m^3 on a daily average basis. Several states have set guidelines or standards for DNOC in ambient air[60] ranging from 2.0 μg/m^3 (North Dakota) to 3.0 μg/m^3 (Virginia) to 4.0 μg/m^3 (Connecticut) to 5.0 μg/m^3 (Nevada).

Determination in Air: Collection by charcoal tube, analysis by gas liquid chromatography.

Permissible Concentration in Water: To protect human health, 13.4 μg/l.[6] The former USSR[35] has set a MAC in water bodies used for domestic purposes of 0.05 mg/l and in water bodies used for fishery purposes of 0.002 mg/l.[43]

Determination in Water: Filter/Bubbler; 2-Propanol; High-pressure liquid chromatography/Ultraviolet detection; NIOSH II(5) Method #S166. Methylene chloride extraction followed by gas chromatography with flame ionization or electron capture detection. EPA Method 604, or gas chromatography plus mass spectrometry EPA Method 625.

Routes of Entry: Inhalation, percutaneous absorption, ingestion, eye and/or skin contact.

Harmful Effects and Symptoms

Short Term Exposure: Early manifestations of acute dinitrocresol exposure include fever, sweating, headache, and confusion. Blood pressure, pulse, and respiratory rate are often elevated. Severe exposure may result in restlessness, seizures, and coma. Other signs and symptoms include dyspnea (shortness of breath), cyanosis (blue tint to skin and mucous membranes), pulmonary edema, nausea, vomiting, and abdominal pain. Liver injury with associated jaundice, kidney failure, and cardiac arrhythmias are commonly noted. Dermal exposure results in yellow staining of the skin and may produce burns. Dinitrocresol may irritate and burn the eyes and mucous membranes. DNOC is an extremely toxic material; probable oral lethal dose is 5 – 50 mg/kg in humans or between 7 drops and 1 teaspoon for a 70 kg (150 lb) person.

Long Term Exposure: May damage the liver, kidneys and blood cells. May stain yellow the skin, eyes, and fingernails. Repeated exposure can cause anxiety, fatigue, insomnia, excessive perperation, unusual thirst, weight loss and cataracts in the eyes.

Points of Attack: Cardiovascular system, endocrine system.

Medical Surveillance: Before beginning employment, at regular times after that and if symptoms develop or overexposure has occurred, the following may be useful: Exam of eyes for cataracts. Exam of skin and nails for staining. Blood tests for 4,6-dinitro-o-cresol. Persons with blood levels over 10 ppm (10 mg/l) should be kept away from further exposure until levels return to normal. If symptoms develop or overexposure is suspected, the following may be useful: Liver and kidney function tests. Complete blood count.

First Aid: If this chemical gets into the eyes, remove any contact lenses at once and irrigate immediately for at least 15 minutes, occasionally lifting upper and lower lids. Seek medical attention immediately. If this chemical contacts the skin, remove contaminated clothing and wash immediately with soap and water. Seek medical attention immediately. If this chemical has been inhaled, remove from exposure, begin rescue breathing (using universal precautions) if breathing has stopped and CPR if heart action has stopped. Transfer promptly to a medical facility. When this chemical has been swallowed, get medical attention. Give large quantities of water and induce vomiting. Do not make an unconscious person vomit. If high fever is present, drench victim's clothes in cool water, or immerse person in cool bath before transfer.

Personal Protective Methods: Wear protective gloves and clothing to prevent any reasonable probability of skin contact. Safety equipment suppliers/manufacturers can provide recommendations on the most protective glove/clothing material for your operation. ACGIH recommends SARANEX, natural rubber, neoprene and chlorinated polyethylene as providing good to excellent protection. All protective clothing (suits, gloves, footwear, headgear) should be clean, available each day, and put on before work. Contact lenses should not be worn when working with this chemical. Wear dust-proof chemical goggles and face shield unless full facepiece respiratory protection is worn. Employees should wash immediately with soap when skin is wet or contaminated. Provide emergency showers and eyewash.

Respirator Selection: NIOSH/OSHA: *Up to 2 mg/m^3:* DMF (any dust and mist respirator with a full facepiece). *Up to 5 mg/m^3:* PAPRDM (APF = 50) (any air-purifying, full-facepiece respirator with a high-efficiency particulate filter); or SA:CF (APF = 25) Any supplied-air respirator operated in a continuous-flow mode; or HiEF (APF = 25) (any powered, air-purifying respirator with a dust and mist filter); or SF (APF = 50) (any self-contained breathing apparatus with a full facepiece); or SCBAF (APF = 50) (any supplied-air respirator with a full facepiece). *At any concentrations above the NIOSH REL, or where there is no REL, at any detectable concentration:* SCBAF:PD,PP (any self-contained breathing apparatus that has a full facepiece and is operated in a pressure-demand or other positive-pressure mode); or SAF:PD,PP:ASCBA (any supplied-air respirator that has a full facepiece and is operated in a pressure-demand or other positive-pressure mode in combination with an auxiliary, self-contained breathing apparatus operated in a pressure-demand or other positive-pressure mode). *Escape:* HiEF (any air-purifying, full-facepiece respirator with a high-efficiency particulate filter); or SCBAE (any appropriate escape-type, self-contained breathing apparatus).

Storage: Prior to working with this chemical you should be trained on its proper handling and storage. 4,6-Dinitro-o-cresol must be stored to avoid contact with strong oxidizers (such as bromine, chlorine, chlorine dioxide and nitrates) since violent reactions occur. Store in tightly closed containers in a cool, well-ventilated area away from heat. Sources of ignition, such as smoking and open flames, are prohibited where 4,6-dinitro-o-cresol is used, handled, or stored in a manner that could create a potential fire or explosion hazard. Use only non-sparking tools and equipment, especially when opening and closing containers of 4,6-dinitro-o-cresol. Wherever 4,6-dinitro-o-cresol is used, handled, manufactured, or stored, use explosion-proof electrical equipment and fittings.

Shipping: This compound requires a shipping label of: "Poison." It falls in DOT Hazard Class 6.1 and Packing Group II.

Spill Handling: Evacuate persons not wearing protective equipment from area of spill or leak until clean-up is complete. Remove all ignition sources. Do not touch spilled material; stop leak if it can be done without risk. Take up small spills with sand or other noncombustible absorbent material; place into containers for later disposal. Small dry spills: collect powdered material in the most convenient and safe manner and deposit in sealed containers. Large spills should be diked for later disposal. Ventilate area after clean-up is complete. It may be necessary to contain and dispose of this chemical as a hazardous waste. If material or contaminated runoff enters waterways, notify downstream users of potentially contaminated waters. Contact your Department of Environmental Protection or your regional office of the federal EPA for specific recommendations. If employees are required to clean-up spills, they must be properly trained and equipped. OSHA 1910.120(q) may be applicable.

Fire Extinguishing: 4,6-Dinitro-o-cresol may burn but does not easily ignite. The dust can form an explosive mixture with air. Use dry chemical, carbon dioxide, water spray, or alcohol foam extinguishers. Poisonous gases are produced in fire including nitrogen oxides. If material or contaminated runoff enters waterways, notify downstream users of potentially contaminated waters. Notify local health and fire officials and pollution control agencies. From a secure, explosion-proof location, use water spray to cool exposed containers. If cooling streams are ineffective (venting sound increases in volume and pitch, tank discolors, or shows any signs of deforming), withdraw immediately to a secure position. If employees are expected to fight fires, they must be trained and equipped in OSHA 1910.156.

Disposal Method Suggested: Incineration (1,100°F minimum) with adequate scrubbing and ash disposal facilities.[22] In accordance with 40CFR165 recommendations for the disposal of pesticides and pesticide containers. Must be disposed properly by following package label directions or by contacting your state pesticide or environmental control agency or by contacting your regional EPA office. Consult with environmental regulatory agencies for guidance on acceptable disposal practices. Generators of waste containing this contaminant (≥100 kg/mo) must conform with EPA regulations governing storage, transportation, treatment, and waste disposal.

References

National Institute for Occupational Safety and Health, Criteria for a Recommended Standards: Occupational Exposure to Dinitro-ortho-Cresol, NIOSH Publication No. 78–131, Washington, DC (1978)

U.S. Environmental Protection Agency, Nitrophenols: Ambient Water Quality Criteria, Washington, DC (1980)

U.S. Environmental Protection Agency, 4,6-Dinitro-o-Cresol, Health and Environmental Effects Profile No. 90, Washington, DC, Office of Solid Waste (April 30, 1980)

Sax N. I., Ed., "Dangerous Properties of Industrial Materials Report," 2, No. 5, 54–59 (1982) and 4, No. 1, 62–66 (1984)

U.S. Environmental Protection Agency, "Chemical Profile: Dinitrocresol," Washington, DC, Chemical Emergency Preparedness Program (November 30, 1987)

New Jersey Department of Health, "Hazardous Substance Fact Sheet: 4,6-Dinitro-o-Cresol," Trenton, NJ (June 1986)

Dinitronaphthalenes

Molecular Formula: $C_{10}H_6N_2O_4$

Common Formula: $C_{10}H_6(NO_2)_2$

Synonyms: 1,3-Dinitronaphthalene; 1,5-Dinitronaphthalene; Naphthalene, 1,3-dinitro-; Naphthalene, 1,8-dinitro-

CAS Registry Number: 606-37-1 (1,3-); 605-71-1 (1,5-); 602-38-0 (1,8-); 27478-34-8 (all)

RTECS Number: QJ4550800 (1,3-); QJ4551000 (1,5-); QJ4552000 (1,8-); QJ4550000 (all)

DOT ID: UN 2538 (nitronaphthalene)

Regulatory Authority

- Carcinogen (Suspected) (DFG)[3]
- Water Pollution Standard Proposed (former USSR)[43]

Description: Dinitronaphthalene, $C_{10}H_6(NO_2)_2$ is a yellowish crystalline solid. The various isomers have the following Freezing/Melting points: 1,3-dinitro: 146 – 148°C; 1,5-dinitro: 218°C, and 1,8-dinitro: 173°C. These substances are highly flammable, and shock and heat sensitive.

Potential Exposure: Used in dye intermediate manufacture and explosives manufacture.

Incompatibilities: May form explosive metal salts with alkalies. Aluminum in the presence of heat. Sulfur, sulfuric acid in the presence of heat.

Permissible Exposure Limits in Air: No standards set.

Permissible Concentration in Water: The former USSR-UNEP/IRPTC project[43] has set a MAC in water bodies used for domestic purposes of 1 mg/l.

Harmful Effects and Symptoms

Short Term Exposure: Irritation of the skin, eyes, and mucous membranes; headache, dizziness, nausea, vomiting.[52]

Long Term Exposure: A suspected carcinogen. Insomnia, fatigue, weight loss, central nervous system depression, skin pigmentation, liver damage, toxic hepatitis, kidney damage, anemia, and cyanosis.

First Aid: *Skin Contact:*[52] Flood all areas of body that have contacted the substance with water. Don't wait to remove contaminated clothing; do it under the water stream. Use soap to help assure removal. Isolate contaminated clothing when removed to prevent contact by others.

Eye Contact: Remove any contact lenses at once. Immediately flush eyes well with copious quantities of water or normal saline for at least 20 – 30 minutes. Seek medical attention.

Inhalation: Leave contaminated area immediately; breathe fresh air. Proper respiratory protection must be supplied to any rescuers. If coughing, difficult breathing or any other symptoms develop, seek medical attention at once, even if symptoms develop may hours after exposure.

Ingestion: If unconscious or convulsing, do not induce vomiting or give anything by mouth. Assure that victim's airway is open and lay him on his side with his head lower than his body and transport a once to a medical facility. If conscious and not convulsing, give a glass of water to dilute the substance. If medical advice is not readily available, do not induce vomiting, and rush the victim to the nearest medical facility.

Personal Protective Methods: Wear protective gloves and clothing to prevent any reasonable probability of skin contact. Safety equipment suppliers/manufacturers can provide recommendations on the most protective glove/clothing material for your operation.. All protective clothing (suits, gloves, footwear, headgear) should be clean, available each day, and put on before work. Contact lenses should not be worn when working with this chemical. Wear dust-proof chemical goggles and face shield unless full facepiece respiratory protection is worn. Employees should wash immediately with soap when skin is wet or contaminated. Provide emergency showers and eyewash.

Respirator Selection: *At any detectable concentration:* SCBAF:PD,PP (any MSHA/NIOSH approved self-contained breathing apparatus that has a full facepiece and is operated in a pressure-demand or other positive-pressure mode); or SAF:PD,PP:ASCBA (any supplied-air respirator that has a full facepiece and is operated in a pressure-demand or other positive-pressure mode in combination with an auxiliary, self-contained breathing apparatus operated in a pressure-demand or other positive pressure mode). Escape: HiEF (any air-purifying, full-facepiece respirator (gas mask) with a chin-style, front- or back-mounted organic vapor canister having a high-efficiency particulate filter); or SCBAE (any appropriate escape-type, self-contained breathing apparatus).

Storage: Prior to working with dinitronaphthalenes you should be trained on its proper handling and storage. Store in tightly closed containers in a refrigerator or cool, well ventilated area away from alkalies, aluminum, sulfur, sulfuric acid, and heat. A regulated, marked area should be established where this chemical is handled, used, or stored in compliance with OSHA standard 1910.1045.

Shipping: Nitronaphthalene falls in UN/DOT class 2538, Hazard Class 4.1 and Packing Group III. The label required is "Flammable Solid."

Spill Handling: Evacuate persons not wearing protective equipment from area of spill or leak until clean-up is complete. Remove all ignition sources. Collect powdered material in the most convenient and safe manner and deposit in sealed containers. Ventilate area after clean-up is complete. It may be necessary to contain and dispose of this chemical as a hazardous waste. If material or contaminated runoff enters waterways, notify downstream users of potentially contaminated waters. Contact your Department of Environmental Protection or your regional office of the federal EPA for specific recommendations. If employees are required to clean-up spills, they must be properly trained and equipped. OSHA 1910.120(q) may be applicable.

Fire Extinguishing: These are very flammable materials. They are sensitive to heat and shock. Hence, extreme care must be exercised in fire fighting. Use dry chemical, carbon dioxide, water spray, or alcohol foam extinguishers. Poisonous gases are produced in fire including nitrogen oxides. If material or contaminated runoff enters waterways, notify downstream users of potentially contaminated waters. Notify local health and fire officials and pollution control agencies. From a secure, explosion-proof location, use water spray to cool exposed containers. If cooling streams are ineffective (venting sound increases in volume and pitch, tank discolors, or shows any signs of deforming), withdraw immediately to a secure position. If employees are expected to fight fires, they must be trained and equipped in OSHA 1910.156.

Dinitrophenols

Molecular Formula: $C_6H_4N_2O_5$

Common Formula: $C_6H_3(NO_2)_2OH$

Synonyms: *2,4-isomer:* Aldifen; Chemox PE; 2,4-Dinitrofenol (Dutch, Spanish); 2,4-Dinitrofenolo (Italian); α-Dinitrophenol; Dinofan 51285; 2,4-DNP; Fenoxyl carbon N; 1-Hydroxy-2,4-dinitrobenzene; Maroxol-50; Nitro kleenup; NSC-1532; Phenol, α-dinitro-; Phenol, 2,4-dinitro-; Solfo black 2B supra; Solfo black B; Solfo black BB; Solfo black G; Solfo black SB; Tertrosulphur black PB; Tertrosulphur PBR

2,5-isomer: 2,5-Dinitrofenol (Dutch, Spanish); 2,5-Dinitrofenolo (Italian); 2,5-DNP; γ-Dinitrophenol; 1-Hydroxy-2,5-dinitrobenzene; Phenol, 2,5-dinitro-

2,6-isomer: 2,6-Dinitrofenol (Dutch, Spanish); 2,6-Dinitrofenolo (Italian); β-Dinitrophenol; *o,o*-Dinitrophenol; 2,6-DNP; 1-Hydroxy-2,6-dinitrobenzene; Phenol, 2,6-dinitro-

mixed isomers: Dinitrofenol (Dutch, Spanish); Dinitrofenolo (Italian); Dinitrophenol (mixed isomers); Hydroxydinitrobenzene; Phenol, dinitro-

CAS Registry Number: 66-56-8 (2,3-); 51-28-5 (2,4-); 329-71-5 (2,5-); 573-56-8 (2,6-); 577-71-9 (3,4-); 586-11-6 (3,5-); 25550-58-7 (mixed isomers)

RTECS Number: SL2700000 (2,3-); SL2800000 (2,4-); SL2900000 (2,5); SL2795000 (2,6); SL3000000 (3,4); SL3050000 (3,5)

DOT ID: UN 1320 (dinitrophenol, wetted with not less than 15% water); UN 1599 (solution)

EEC Number: 609-016-00-8

Regulatory Authority

- Explosive Substances (World Bank)[15]
- Air Pollutant Standard Set (former USSR)[43]

2,4-isomer:

- Clean Air Act: Hazardous Air Pollutants (Title I, Part A, Section 112)
- Clean Water Act: Section 311 Hazardous Substances/RQ 40CFR117.3 (same as CERCLA, see below); Section 313 Water Priority Chemicals (57FR41331, 9/9/92)
- EPA Hazardous Waste Number (RCRA No.): P048
- RCRA, 40CFR261, Appendix 8 Hazardous Constituents
- RCRA 40CFR268.48; 61FR15654, Universal Treatment Standards: Wastewater (mg/l), 0.12; Nonwastewater (mg/kg), 160
- RCRA 40CFR264, Appendix 9; TSD Facilities Ground Water Monitoring List. Suggested test method(s) (PQL μg/l): 8040 (150); 8270 (50)
- Safe Drinking Water Act: Priority List (55FR1470)
- Superfund/EPCRA 40CFR302.4 Reportable Quantity (RQ): CERCLA, 10 lb (4.54 kg)

- EPCRA Section 313 Form R *de minimis* concentration reporting level: 1.0%
- U.S. DOT Regulated Marine Pollutant (49CFR172.101, Appendix B)

2,5-isomer:

- Clean Water Act: Section 311 Hazardous Substances/RQ 40CFR117.3 (same as CERCLA, see below)
- Superfund/EPCRA 40CFR302.4 Reportable Quantity (RQ): CERCLA, 10 lb (4.54 kg)
- U.S. DOT Regulated Marine Pollutant (49CFR172.101, Appendix B)

2,6-isomer:

- Clean Water Act: Section 311 Hazardous Substances/RQ 40CFR117.3 (same as CERCLA, see below)
- Superfund/EPCRA 40CFR302.4 Reportable Quantity (RQ): CERCLA, 10 lb (4.54 kg)
- U.S. DOT Regulated Marine Pollutant (49CFR172.101, Appendix B)

Mixed isomers:

- Superfund/EPCRA 40CFR302.4 Reportable Quantity (RQ): CERCLA, 10 lb (4.54 kg)
- U.S. DOT Regulated Marine Pollutant (49CFR172.101, Appendix B)

Cited in U.S. State Regulations: California (G), Illinois (G), Kansas (G, W), Louisiana (G), Maine (G, W), Maryland (G), Michigan (G), New Hampshire (G), New Jersey (G), New York (G), Oklahoma (G), Pennsylvania (G), Vermont (G), Virginia (G), Washington (G), Wisconsin (G).

Description: The dinitrophenols, $C_6H_3(NO_2)_2OH$ are yellow crystalline solids with a sweet, musty odor. Very slightly soluble in water. Properties:

Isomer	Melting Point °C	Boiling Point °C
2,3-	144	
2,4-	112 – 115	sublimes
2,5-	104	
2,6-	63.5	
3,4-	134	
3,5-	122 – 123	

As noted in the Regulatory section above, 2,4-, 2,5-, 2,6-, and mixed isomers of dinitrophenol are the most heavily regulated of this group. Due to its explosive properties, dinitrophenol is used in the form of a wetted solution or water paste.

Potential Exposure: 2,4-DNP is used in the manufacturing of dyestuff intermediates, wood preservatives, pesticides, herbicides, explosives, chemical indicators, photographic developers, and also in chemical synthesis.

Incompatibilities: Dust forms an explosive mixture with air. Explosion can be caused by heat, friction or shock. Contact with reducing agents, combustibles may cause fire and explosions. Forms shock-sensitive explosive salts with ammonia, strong bases, and most metals. May accumulate static electrical charges, and may cause ignition of its vapors.

Permissible Exposure Limits in Air: The former USSR-UNEP/IRPTC joint project[43] has set a MAC in workplace air of 0.05 mg/m^3.

Permissible Concentration in Water: To protect freshwater aquatic life – 230 μg/l on an acute toxicity basis for nitrophenols as a class. To protect saltwater aquatic life 4,850 μg/l on an acute toxicity basis for nitrophenols as a class. To protect human health – 70.0 μg/l.[6] This compares to a limit of 30 μg/l set in the USSR.[43] Kansas and Maine have set guidelines for 2,4-dinitrophenol in drinking water of 31 μg/l and 110 μg/l, respectively.

Determination in Water: Methylene chloride extraction followed by gas chromatography with flame ionization or electron capture detection (EPA Method 604) or gas chromatography plus mass spectrometry (EPA Method 625).

Routes of Entry: Inhalation, through the skin, ingestion.

Harmful Effects and Symptoms

Short Term Exposure: Dinitrophenol can affect you when breathed and by passing through skin. Contact with the 2,4-isomer can cause severe irritation and burns to the eyes and skin. May affect the metabolism, causing very high body temperature. May affects the peripheral nervous system causing numbness, "pins and needles," and/or weakness of the hands and feet. Exposure can cause a bluish color to skin and lips, headaches, temperature rise, dizziness, collapse, convulsions (fits), coma and even death. Exposure may irritate the lungs causing coughing and shortness of breath. Higher levels can cause a build-up of fluid in the lungs, a medical emergency, which can cause death.

Long Term Exposure: Repeated or prolonged contact with skin may cause dermatitis with rash and drying and itching of the skin. Dinitrophenol may have affects the eyes, causing cataracts. Exposure can damage the liver and kidneys, and affect the thyroid gland. May cause lung irritation and the development of bronchitis with coughing and shortness of breath. May damage the nervous system. High exposure may damage the developing fetus. Repeated exposure to the 2,4- isomer can damage blood cells, causing anemia.

Points of Attack: Skin, liver, kidneys, lungs, peripheral nervous system, eyes, thyroid gland, blood.

Medical Surveillance: Before beginning employment and at regular times after that, the following are recommended: Eye exam for cataracts. Lung function tests. If symptoms develop or overexposure is suspected, the following may be useful: Blood test for methemoglobin level. Complete blood count (CBC). Lung function tests. Liver and kidney function tests. Thyroid function tests. Evaluation by a qualified allergist, including careful exposure history and special testing, may help diagnose skin allergy. Consider chest x-ray after acute exposure.

First Aid: If this chemical gets into the eyes, remove any contact lenses at once and irrigate immediately for at least 30 minutes, occasionally lifting upper and lower lids. Seek medical attention immediately. If this chemical contacts the skin, remove contaminated clothing and wash immediately with soap and water. Seek medical attention immediately. If this chemical has been inhaled, remove from exposure, begin rescue breathing (using universal precautions) if breathing has stopped and CPR if heart action has stopped. Transfer promptly to a medical facility. When this chemical has been swallowed, get medical attention. Give large quantities of water and induce vomiting. Do not make an unconscious person vomit.

Note to Physician: Treat for methemoglobinemia. Spectrophotometry may be required for precise determination of levels of methemoglobinemia in urine.

Personal Protective Methods: Wear protective gloves and clothing to prevent any reasonable probability of skin contact. Safety equipment suppliers/manufacturers can provide recommendations on the most protective glove/clothing material for your operation.. All protective clothing (suits, gloves, footwear, headgear) should be clean, available each day, and put on before work. Contact lenses should not be worn when working with this chemical. Wear dust-proof chemical goggles and face shield unless full facepiece respiratory protection is worn. Employees should wash immediately with soap when skin is wet or contaminated. Provide emergency showers and eyewash.

Respirator Selection: Where the potential for exposure to Dinitrophenol exists, use a MSHA/NIOSH approved supplied-air respirator with a full facepiece operated in the positive pressure mode or with a full facepiece, hood, or helmet in the continuous flow mode, or use a MSHA/NIOSH approved self-contained breathing apparatus with a full facepiece operated in pressure-demand or other positive pressure mode.

Storage: Prior to working with this chemical you should be trained on its proper handling and storage. Dinitrophenol should be kept wet and protected from thermal and mechanical shock. Dinitrophenol must be stored to avoid contact with strong oxidizers (such as chlorine, bromine and fluorine) and metals and metal compounds since violent reactions occur. Sources of ignition, such as smoking and open flames, are prohibited where Dinitrophenol is handled, used, or stored. Use only non-sparking tools and equipment, especially when opening and closing containers of Dinitrophenol. Wherever Dinitrophenol is used, handled, manufactured, or stored, use explosion-proof electrical equipment and fittings.

Shipping: Dinitrophenol wetted with at least 15% water requires a shipping label of: "Flammable Solid, Poison." It falls in DOT Hazard Class 4.1 and Packing Group I.

Spill Handling: Evacuate persons not wearing protective equipment from area of spill or leak until clean-up is complete. Remove all ignition sources. Wet spilled material with water. Collect powdered material in the most convenient and safe manner and deposit in sealed containers. Ventilate area after clean-up is complete. It may be necessary to contain and dispose of this chemical as a hazardous waste. Keep dinitrophenol out of a confined space, such as a sewer, because of the possibility of an explosion, unless the sewer is designed to prevent the build-up of explosive concentrations. If material or contaminated runoff enters waterways, notify downstream users of potentially contaminated waters. Contact your Department of Environmental Protection or your regional office of the federal EPA for specific recommendations. If employees are required to clean-up spills, they must be properly trained and equipped. OSHA 1910.120(q) may be applicable.

Fire Extinguishing: Dry dinitrophenol is a severe explosion hazard when exposed to heat or shock. Wetted dinitrophenol in solution may be combustible or flammable, depending on the solvent. Poisonous gases are produced in fire, including oxides of nitrogen and carbon monoxide. Use flooding quantities of water to extinguish fire. If water is not available use dry chemical or dirt. If material or contaminated runoff enters waterways, notify downstream users of potentially contaminated waters. Notify local health and fire officials and pollution control agencies. Containers may explode in fire. From a secure, explosion-proof location, use water spray to cool exposed containers. If cooling streams are ineffective (venting sound increases in volume and pitch, tank discolors, or shows any signs of deforming), withdraw immediately to a secure position. If employees are expected to fight fires, they must be trained and equipped in OSHA 1910.156.

Disposal Method Suggested: Incineration (1,800°F, 2.0 seconds minimum) with adequate scrubbing equipment for the removal of No_x. Consult with environmental regulatory agencies for guidance on acceptable disposal practices. Generators of waste containing this contaminant (≥100 kg/mo) must conform with EPA regulations governing storage, transportation, treatment, and waste disposal.

References

U.S. Environmental Protection Agency, Nitrophenols: Ambient Water Quality Criteria, Washington, DC (1980)

U.S. Environmental Protection Agency, 2,4-Dinitrophenol, Health and Environmental Effects Profile No. 91, Washington, DC, Office of Solid Waste (April 30, 1980)

Sax N. I., Ed., "Dangerous Properties of Industrial Materials Report," 2, No. 2, 25–27 (1982) and 3, No. 2, 38–44 (1983)

New York State Department of Health, "Chemical Fact Sheet: 2,4-Dinitrophenol," Albany, NY, Bureau of Toxic Substance Assessment (March 1986)

New Jersey Department of Health, "Hazardous Substance Fact Sheet: Dinitrophenol," Trenton, NJ (October 1996)

New Jersey Department of Health, "Hazardous Substance Fact Sheet: 2,4-Dinitrophenol," Trenton, NJ (March 1999)

2,4-Dinitrotoluene

Molecular Formula: $C_7H_6N_2O_4$

Common Formula: $C_6H_3(NO_2)_2CH_3$

Synonyms: Benzene, 2,4-DNT; Benzene, 1-methyl-2,4-dinitro-; 2,4-Dinitrotolueno (Spanish); 2,4-DNT; 1-Methyl-2,4-dinitobenzene; NCI-C01865; Toluene, 2,4-dinitro-

CAS Registry Number: 121-14-2 (2,4-); 25321-14-6 (dinitrotoluene)

RTECS Number: XT1575000

DOT ID: UN 1600 (molten); UN 2038 (solid)

EEC Number: 609-007-00-9

Regulatory Authority

- Carcinogen (Animal Suspected) (IARC)[9] (Animal Positive) (DFG)[3]
- Air Pollutant Standard Set (ACGIH)[1] (HSE)[33] (former USSR)[43] (OSHA)[58] (Several States)[60]
- Clean Air Act: Hazardous Air Pollutants (Title I, Part A, Section 112)
- Clean Water Act: Section 311 Hazardous Substances/RQ 40CFR117.3 (same as CERCLA, see below); 40CFR423, Appendix A, Priority Pollutants; Section 313 Water Priority Chemicals (57FR41331, 9/9/92); Toxic Pollutant (Section 401.15) as dinitrotoluene
- EPA Hazardous Waste Number (RCRA No.): U105; D030
- RCRA, 40CFR261, Appendix 8 Hazardous Constituents
- RCRA Toxicity Characteristic (Section 261.24), Maximum
- Concentration of Contaminants, regulatory level, 0.13 mg/l
- RCRA 40CFR268.48; 61FR15654, Universal Treatment Standards: Wastewater (mg/l), 0.32; Nonwastewater (mg/kg), 140
- RCRA 40CFR264, Appendix 9; TSD Facilities Ground Water Monitoring List. Suggested test method(s) (PQL μg/l): 8090 (0.2); 8270 (10)
- Safe Drinking Water Act: Priority List (55FR1470)
- Superfund/EPCRA 40CFR302.4 Reportable Quantity (RQ): CERCLA, 10 lb (4.54 kg)
- EPCRA Section 313 Form R *de minimis* concentration reporting level: 1.0%
- Canada, WHMIS, Ingredients Disclosure List

Cited in U.S. State Regulations: Alaska (G), California (A, G), Connecticut (A), Florida (G), Illinois (G), Kansas (G, W), Louisiana (G), Maine (G), Maryland (G), Massachusetts (G), Michigan (G), Minnesota (W), Nevada (A), New Hampshire (G), New Jersey (G), New York (G), North Dakota (A), Oklahoma (G), Pennsylvania (G), Rhode Island (G), Vermont (G), Virginia (G, A), Washington (G), West Virginia (G), Wisconsin (G).

Description: Six isomers of dinitrotoluene, $C_6H_3(NO_2)_2CH_3$, exist, the most important being 2,4-dinitrotoluene, an orange-yellow crystalline solid. Boiling point = 300°C. Freezing/Melting point = 70°C. Flash point = 207°C. Hazard Identification (based on NFPA-704 M Rating System): Health 3, Flammability 1, Reactivity 3. Insoluble in water.

Potential Exposure: DNT is used in the preparation of polyurethane foams and manufacture of toluene diisocyanate for the production of polyurethane plastics, in the production of military and commercial explosives, to plasticize cellulose nitrate in explosives, to moderate the burning rate of propellants and explosives, in the manufacture of gelatin explosives, as a water-proofing coating for smokeless powders, as an intermediate in TNT manufacture, and in the manufacture of azo dye intermediate.

Incompatibilities: Dust forms an explosive mixture with air. Heat forms corrosive nitrogen oxide fumes and may cause explosion. Commercial grades will decompose at 250°C/482°F, with self-sustaining decomposition at 280°C/536°F. Contact with strong oxidizers, caustics, and reducing agents may cause fire and explosions. Contact with nitric acid forms an explosive material. Contact with sodium oxide causes ignition. Not compatible with chemically active metals such as tin and zinc.

Permissible Exposure Limits in Air: The Federal Limit (OSHA PEL)[58] is 1.5 mg/m^3 as is the HSE TWA.[33] The notation "skin" indicates the possibility of cutaneous absorption. The HSE STEL is 5 mg/m^3. The ACGIH recommended airborne exposure limit is 0.2 mg/m^3. The NIOSH IDLH level is (Ca) 50 mg/m^3. The former USSR-UNEP/IRPTC joint project[43] has set a MAC in workplace air of 1.0 mg/m^3. Several states have set guidelines or standards for DNT in ambient air[60] ranging from 15 μg/m^3 (Connecticut and North Dakota) to 25 μg/m^3 (Virginia) to 36 μg/m^3 (Nevada).

Determination in Air: Filter/Tenax; Acetone; Gas chromatography/Thermal energy analyzer detection-Explosives package; OSHA Method #44.

Permissible Concentration in Water: To protect freshwater aquatic life – 300 μg/l on an acute toxicity basis and 230 μg/l on a chronic toxicity basis. To protect saltwater aquatic life – 590 μg/l on an acute toxicity basis. To protect human health – preferably zero. An additional lifetime cancer risk of 1 in 100,000 results at a concentration of 1.1 μg/l of 2,4-DNT.[6] The former USSR-UNEP/IRPTC joint project[43] has set a MAC in water bodies uses for domestic purposes of 0.5 mg/l. In addition, the states o Kansas and Minnesota have set guidelines for DNT in drinking water,[61] both at a level of 1.1 μg/l.

Determination in Water: Methylene chloride extraction followed by exchange to toluene and gas chromatography with flame ionization detection (EPA Method 609) or gas chromatography plus mass spectrometry (EPA Method 625).

Routes of Entry: Inhalation of vapor, percutaneous absorption of liquid, ingestion and eye and/or skin contact.

Harmful Effects and Symptoms

Short Term Exposure: *Inhalation:* May affect the central nervous system, cardiovascular system, and the blood, resulting in the formation of methemoglobin. Symptoms may be delayed for up to 4 hours. Both the nervous system and the blood are affected. Nervous system effects may include confusion, disorientation, dizziness, weakness, drowsiness and coma. Convulsions may occur. Blood effects are from a decreased ability to carry oxygen and may include moderate to severe headache, nausea, vomiting, blue coloring of skin, fall in blood pressure and an irregular heartbeat. Ingestion of alcohol is reported to aggravate toxic effects.

Skin: Readily absorbed through the skin. Small amounts absorbed from clothes or shoes may cause or increase the severity of the symptoms listed above. Irritates the skin.

Eyes: Hot fumes may cause severe burning of eyelids and cornea, resulting in permanent scarring.

Ingestion: Animal studies suggest that ingestion causes symptoms listed under Inhalation.

Long Term Exposure: Reported or prolonged exposure may cause anemia, decrease of oxygen-carrying capacity of the blood, blue skin coloration, liver damage and jaundice, anemia, weight loss, and skin rash. May affect the nervous system causing fatigue, nausea, vomiting, personality changes such as irritability, anxiety, confusion, and depression. May be a cancer causing agent in humans since it has been shown to cause liver cancer in animals. The 2,6- isomer caused mutations, liver cancer in animals, and may decrease fertility in males and females.

Points of Attack: Blood, liver, cardiovascular system, reproductive system. Cancer site in animals: liver, skin and kidney tumors.

Medical Surveillance: Before beginning employment and at regular times after that, the following are recommended: Complete blood count (CBC). Urinary dinitrotoluene level. Liver and kidney function tests. Blood methemoglobin level.

First Aid: If this chemical gets into the eyes, remove any contact lenses at once and irrigate immediately for at least 15 minutes, occasionally lifting upper and lower lids. Seek medical attention immediately. If this chemical contacts the skin, remove contaminated clothing and wash immediately with soap and water. Seek medical attention immediately. If this chemical has been inhaled, remove from exposure, begin rescue breathing (using universal precautions) if breathing has stopped and CPR if heart action has stopped. Transfer promptly to a medical facility. When this chemical has been swallowed, get medical attention. Give large quantities of water and induce vomiting. Do not make an unconscious person vomit. The formation of methemoglobin may be delayed; medical observation is recommended.

Note to Physician: Treat for methemoglobinemia. Spectrophotometry may be required for precise determination of levels of methemoglobinemia in urine.

Personal Protective Methods: Wear protective gloves and clothing to prevent any reasonable probability of skin contact. Safety equipment suppliers/manufacturers can provide recommendations on the most protective glove/clothing material for your operation.. All protective clothing (suits, gloves, footwear, headgear) should be clean, available each day, and put on before work. Contact lenses should not be worn when working with this chemical. Wear dust-proof chemical goggles and face shield unless full facepiece respiratory protection is worn. Employees should wash immediately with soap when skin is wet or contaminated. Provide emergency showers and eyewash.

Respirator Selection: NIOSH: *At any detectable concentration:* SCBAF:PD,PP (any self-contained breathing apparatus that has a full facepiece and is operated in a pressure-demand or other positive-pressure mode); or SAF:PD,PP:ASCBA (any supplied-air respirator that has a full facepiece and is operated in a pressure-demand or other positive-pressure mode in combination with an auxiliary, self-contained breathing apparatus operated in a pressure-demand or other positive-pressure mode). *Escape:* GMFOVHiE [any air-purifying, full-facepiece respirator (gas mask) with a chin-style, front- or back-mounted organic vapor canister having a high-efficiency particulate filter]; or SCBAE (any appropriate escape-type, self-contained breathing apparatus).

Storage: Prior to working with dinitrotoluenes you should be trained on its proper handling and storage. 2,4-Dinitrotoluene must be stored to avoid contact with strong oxidizers, such as chlorine, chlorine dioxide, bromine, nitrates, and permanganates; caustics, such as sodium hydroxide and potassium hydroxide; and chemically active metals, such as tin or zinc, since violent reactions occur. Contact with strong oxidizers can cause fire or explosions. Also, striking it or dropping it may cause detonation and explosion. Store in tightly closed containers in a cool, well-ventilated area away from heat (temperatures above 482°F). Sources of ignition such as smoking and open flames are prohibited where 2,4-dinitrotoluene is used, handled, or stored in a manner that could create a potential fire or explosion hazard. Use only non-sparking tools and equipment, especially when opening and closing containers of 2,4-dinitrotoluene. Wherever 2,4-dinitrotoluene is used, handled, manufactured, or stored, use explosion-proof electrical equipment and fittings. A regulated, marked area should be established where this chemical is handled, used, or stored in compliance with OSHA standard 1910.1045.

Shipping: Dinitrotoluenes, solid require a shipping label of: "Poison." It falls in DOT Hazard Class 6.1 and Packing Group II.

Spill Handling: *Solid:* Evacuate persons not wearing protective equipment from area of spill or leak until clean-up is complete. Remove all ignition sources. Collect powdered material in the most convenient and safe manner and deposit in sealed containers. Ventilate area after clean-up is complete. It may be necessary to contain and dispose of this chemical as a hazardous waste. If material or contaminated runoff enters waterways, notify downstream users of potentially contaminated waters. Contact your Department of Environmental Protection or your regional office of the federal EPA for specific recommendations. If employees are required to clean-up spills, they must be properly trained and equipped. OSHA 1910.120(q) may be applicable.

Liquid: Evacuate and restrict persons not wearing protective equipment from area of spill or leak until cleanup is complete. Remove all ignition sources. Ventilate area of spill or leak. Absorb liquids in vermiculite, dry sand, earth, peat, carbon, or a similar material and deposit in sealed containers. Keep this chemical out of a confined space, such as a sewer, because of the possibility of an explosion, unless the sewer is designed to prevent the build-up of explosive concentrations. It may be necessary to contain and dispose of this chemical as a hazardous waste. If material or contaminated runoff enters waterways, notify downstream users of potentially contaminated waters. Contact your Department of Environmental Protection or your regional office of the federal EPA for specific recommendations. If employees are required to clean-up spills, they must be properly trained and equipped. OSHA 1910.120(q) may be applicable.

Fire Extinguishing: Molten 2,4-dinitrotoluene is combustible. It may burn, but does not readily ignite. Use extreme caution when fighting a fire, since 2,4-dinitrotoluene could explode. Use dry chemical, CO_2, or water spray extinguishers. If the fire is advanced, evacuate the area. Poisonous gases are produced in fire. If material or contaminated runoff enters waterways, notify downstream users of potentially contaminated waters. Notify local health and fire officials and pollution control agencies. From a secure, explosion-proof location, use water spray to cool exposed containers. If cooling streams are ineffective (venting sound increases in volume and pitch, tank discolors, or shows any signs of deforming), withdraw immediately to a secure position. If employees are expected to fight fires, they must be trained and equipped in OSHA 1910.156.

Disposal Method Suggested: Pretreatment involves contact of the dinitrotoluene contaminated waste with $NaHCO_3$ and solid combustibles followed by incineration in an alkaline scrubber equipped incinerator unit. Consult with environmental regulatory agencies for guidance on acceptable disposal practices. Generators of waste containing this contaminant (≥100 kg/mo) must conform with EPA regulations governing storage, transportation, treatment, and waste disposal.

References

U.S. Environmental Protection Agency, Dinitrotoluenes: Ambient Water Quality Criteria, Washington, DC (1980)

U.S. Environmental Protection Agency, "Chemical Hazard Information Profile: 2,4-Dinitrotoluene," Washington, DC (March 9, 1978)

U.S. Environmental Protection Agency, Dinitrotoluenes, Health and Environmental Effects Profile No. 92, Washington, DC, Office of Solid Waste (April 30, 1980)

U.S. Environmental Protection Agency, 2,4-Dinitrotoluene, Health and Environmental Effects Profile No. 93, Washington, DC, Office of Solid Waste (April 30, 1980)

Sax N. I., Ed., "Dangerous Properties of Industrial Materials Report," 3, No. 2, 70–72 (1983)

New York State Department of Health, "Chemical Fact Sheet: 2,4-Dinitrotoluene," Albany, NY, Bureau of Toxic Substance Assessment (May 1986)

New Jersey Department of Health, "Hazardous Substance Fact Sheet: 2,4-Dinitrotoluene," Trenton, NJ (April 1986)

Dinoseb

Molecular Formula: $C_{10}H_{12}N_2O_5$

Common Formula: $C_6H_2(NO_2)_2(C_4H_9)OH$

Synonyms: Aatox; AI3-01122; Aretit; Basanite; BNP 20; BNP 30; Butaphene; 2-*sec*-Butyl-4,6-dinitrophenol; Caldon; Caswell No. 392DD; Chemox; Chemox general; Chemox P.E.; DBNF; Dinitrall; Dinitro; Dinitro-3; 4,6-Dinitro-2-*sec*-butylfenol (Czech); 2,4-Dinitro-6-*sec*-butylphenol; 4,6-Dinitro-*o,sec*-butylphenol; 4,6-Dinitro-2-*sec*-butylphenol; Dinitro-*o,sec*-butylphenol; Dinitro-butylphenol; 2,4-Dinitro-6-(1-methylpropyl)phenol; 4,6-Dinitro-2-(1-methyl-*N*-propyl)phenol; 4,6-Dinitro-2-(1-methyl-propyl)phenol; Dinoseb; DN 289; DNBP; DNOSBP; DNPB; DNSBP; Dow general; Dow general weed killer; Dow selective weed killer; Dynamyte; Dytop; Elgetol 318; ENT 1,122; EPA pesticide chemical code 037505; Gebutox; Hel-Fire; Ivosit; Kiloseb; Knox-Weed; Ladob; Laseb; Liro DNBP; 6-(1-Methyl-propyl)-2,4-dinitrofenol (Dutch); 2-(1-Methylpropyl)-4,6-dinitrophenol; 6-(1-Metil-propil)-2,4-dinitrnolo (Italian); Nitropone C; NSC 202753; Phenol, 2-*sec*-butyl-4,6-dinitro-; Phenol, 2-(1-methylpropyl)-4,6-dinitro-; Phenotan; Premerge; Premerge 3; Sinox general; Sparic; Spurge; Subitex; Unicrop DNBP; Vertac dinitro weed killer; Vertac general weed killer; Vertac selective weed killer

CAS Registry Number: 88-85-7

RTECS Number: SJ9800000

DOT ID: UN 2765

EEC Number: 609-025-00-7

Regulatory Authority

- Banned or Severely Restricted (Several Countries) (UN)[13]
- Air Pollutant Standard Set (former USSR)[43]
- EPA Hazardous Waste Number (RCRA No.): P020
- RCRA, 40CFR261, Appendix 8 Hazardous Constituents
- RCRA 40CFR268.48; 61FR15654, Universal Treatment Standards: Wastewater (mg/l), 0.066; Nonwastewater (mg/kg), 2.515
- RCRA 40CFR264, Appendix 9; TSD Facilities Ground Water Monitoring List. Suggested test method(s) (PQL μg/l): 8150 (1); 8270 (10)

- Safe Drinking Water Act: MCL, 0.007 mg/l; MCLG, 0.007 mg/l; Regulated chemical (47FR9352)
- Superfund/EPCRA 40CFR355, Appendix B Extremely Hazardous Substances: TPQ = 100/10,000 lb (45.4/4,540 kg)
- Superfund/EPCRA 40CFR302.4 Reportable Quantity (RQ): CERCLA, 1,000 lb (454 kg)
- EPCRA Section 313 Form R *de minimis* concentration reporting level: 1.0%
- U.S. DOT Regulated Marine Pollutant (49CFR172.101, Appendix B)

Cited in U.S. State Regulations: California (G), Connecticut (W), Illinois (G), Kansas (G, W), Louisiana (G), Maine (W), Massachusetts (G, W), Michigan (G), New Hampshire (G), New Jersey (G), Oklahoma (G), Pennsylvania (G), Vermont (G), Virginia (G), Washington (G), Wisconsin (G, W).

Description: Dinoseb, $C_6H_2(NO_2)_2(C_4H_9)OH$, is an orange-brown viscous liquid with a pungent odor or an orange crystalline solid. Freezing/Melting point = 38 – 42°C. Flash point = 16 – 29°C (for 3 commercial products). Insoluble in water.

Potential Exposure: This material is used as a plant growth regulator, insecticide and herbicide.

Incompatibilities: The solution in water is a weak acid. Attacks many metals in presence of water.

Permissible Exposure Limits in Air: The former USSR-UNEP/IRPTC project[43] has set a MAC in workplace air of 0.05 mg/m^3.

Permissible Concentration in Water: A health advisory of 3,5 μg/l has been developed by EPA based on possible teratogenic action of dinoseb as described in the EPA document referred to below. In addition, several states have set guidelines for dinoseb in drinking water[61] ranging from 2.0 μg/l (Maine) to 5.0 μg/l (Massachusetts) to 13.0 μg/l (Wisconsin) to 39.0 μg/l (Kansas). The former USSR[35] has set a MAC in surface water of 0.1 mg/l.

Determination in Water: Extraction with ether, conversion to methyl ester and determination by electron capture gas chromatography.

Harmful Effects and Symptoms

Short Term Exposure: Dinoseb causes eye irritation. May affect the gastrointestinal tract and central nervous system. Early manifestations of dinoseb exposure include fever, sweating, headache, and confusion. Elevations of blood pressure, pulse, and respiratory rate are common. Severe exposure may result in restlessness, seizures, and coma. Other signs and symptoms include dyspnea (shortness of breath), nausea, vomiting, and abdominal pain. Liver injury with associated jaundice, kidney failure, and cardiac arrhythmias may be noted. Inhalation of the aerosol may cause pulmonary edema, a medical emergency that can be delayed for several hours. This can cause death. Muscle weakness may be pronounced. Dermal exposure results in yellow staining of the skin and may produce burns.

Warning: Exposure to dinoseb fumes or aerosol in hot environment may cause death. Effects may be delayed from several hours to 2 days. Caution is advised. Toxicity of dinoseb is enhanced by high ambient temperature and physical activity. Dinoseb is extremely toxic: Probable oral lethal dose is 5 – 50 mg/kg; between 7 drops and 1 teaspoonful for 70 kg person (150 lb).

Long Term Exposure: Dinoseb may affect the kidneys, liver, blood, immune system, and eyes; may cause cataracts. May cause reproductive toxicity in humans.

Points of Attack: Liver, kidneys, blood, cardiovascular system, immune system, eyes.

Medical Surveillance: Liver and kidney function tests. Complete blood count (CBC). Eye examination. EKG.

First Aid: If this chemical gets into the eyes, remove any contact lenses at once and irrigate immediately for at least 15 minutes, occasionally lifting upper and lower lids. Seek medical attention immediately. If this chemical contacts the skin, remove contaminated clothing and wash immediately with soap and water. Seek medical attention immediately. If this chemical has been inhaled, remove from exposure, begin rescue breathing (using universal precautions) if breathing has stopped and CPR if heart action has stopped. Transfer promptly to a medical facility. When this chemical has been swallowed, get medical attention. Give large quantities of water and induce vomiting. Do not make an unconscious person vomit. Consult poison center on use of antidotes.

Personal Protective Methods: Wear protective gloves and clothing to prevent any reasonable probability of skin contact. Safety equipment suppliers/manufacturers can provide recommendations on the most protective glove/clothing material for your operation.. All protective clothing (suits, gloves, footwear, headgear) should be clean, available each day, and put on before work. Contact lenses should not be worn when working with this chemical. If working with liquid wear splash-proof chemical goggles, if working dry material wear dust-proof chemical goggles and face shield unless full facepiece respiratory protection is worn. Employees should wash immediately with soap when skin is wet or contaminated. Provide emergency showers and eyewash.

Respirator Selection: ***Storage:*** Prior to working with dinoseb you should be trained on its proper handling and storage. Store in tightly closed containers in a cool, well ventilated area.

Shipping: Phenoxy pesticides, solid, toxic, n.o.s. require a "Poison" label. They fall in Hazard Class 6.1 and dinoseb in Packing Group II.

Spill Handling: Do not handle broken packages without protective equipment. Wash away any material which may have contacted the body with copious amounts of water.

Dry material: Evacuate persons not wearing protective equipment from area of spill or leak until clean-up is complete. Remove all ignition sources. Keep spilled material wet. Do not attempt to sweep up dry material. Use HEPA vacuum or wet method to reduce dust during clean-up. DO NOT DRY SWEEP. Collect powdered material in the most convenient and safe manner and deposit in sealed containers. Ventilate area after clean-up is complete. Keep material out of water sources and sewers. It may be necessary to contain and dispose of this chemical as a hazardous waste. If material or contaminated runoff enters waterways, notify downstream users of potentially contaminated waters. Contact your Department of Environmental Protection or your regional office of the federal EPA for specific recommendations. If employees are required to clean-up spills, they must be properly trained and equipped. OSHA 1910.120(q) may be applicable.

Liquid: Evacuate and restrict persons not wearing protective equipment from area of spill or leak until cleanup is complete. Remove all ignition sources. Ventilate area of spill or leak. Absorb liquids in vermiculite, dry sand, earth, peat, carbon, or a similar material and deposit in sealed containers. Keep material out of water sources and sewers. It may be necessary to contain and dispose of this chemical as a hazardous waste. If material or contaminated runoff enters waterways, notify downstream users of potentially contaminated waters. Contact your Department of Environmental Protection or your regional office of the federal EPA for specific recommendations. If employees are required to clean-up spills, they must be properly trained and equipped. OSHA 1910.120(q) may be applicable.

Fire Extinguishing: Extinguish by flooding with water. Cool all affected containers with flooding quantities of water. Wear self-contained breathing apparatus and full protective clothing. If fire becomes uncontrollable, evacuate for a radius of 1 mile. It is dangerously explosive. When not water-wet it is a high explosive. Dry, the material is easily ignited and it will burn very vigorously. Poisonous gases are produced in fire. If material or contaminated runoff enters waterways, notify downstream users of potentially contaminated waters. Notify local health and fire officials and pollution control agencies. From a secure, explosion-proof location, use water spray to cool exposed containers. If cooling streams are ineffective (venting sound increases in volume and pitch, tank discolors, or shows any signs of deforming), withdraw immediately to a secure position. If employees are expected to fight fires, they must be trained and equipped in OSHA 1910.156.

Disposal Method Suggested: Incineration.[22] Conduct at 1,000°C for 2.0 seconds minimum with scrubber for No_x removal is recommended. In accordance with 40CFR165 recommendations for the disposal of pesticides and pesticide containers. Must be disposed properly by following package label directions or by contacting your state pesticide or environmental control agency or by contacting your regional EPA office. Consult with environmental regulatory agencies for guidance on acceptable disposal practices. Generators of waste containing this contaminant (≥100 kg/mo) must conform with EPA regulations governing storage, transportation, treatment, and waste disposal.

References

U.S. Environmental Protection Agency, "Chemical Profile: Dinoseb," Washington, DC, Chemical Emergency Preparedness Program (November 30, 1987)

U.S. Environmental Protection Agency, "Health Advisory: Dinoseb," Washington, DC, Office of Drinking Water (August 1987)

Dinoterb

Molecular Formula: $C_{10}H_{12}N_2O_5$

Common Formula: $C_6H_2(NO_2)_2(C_4H_9)(OH)$

Synonyms: *o,tert*-Butyl-4,6-dinitrophenol; 2-(1,1-Dimethylethyl)-4,6-dinitrophenol; 2,4-Dinitro-6-*tert*-butylphenol; Dinitroterb; DNTBP; Herbogil; Phenol-2-*tert*-butyl-4,6-dinitro-; Phenol, 2-(1,1-dimethylethyl)4,6-dinitro-

CAS Registry Number: 1420-07-1

RTECS Number: SK0160000

Regulatory Authority

- Superfund/EPCRA 40CFR355, Appendix B Extremely Hazardous Substances: TPQ = 500/10,000 lb (227/4,540 kg)
- Superfund/EPCRA 40CFR302.4 Reportable Quantity (RQ): EHS, 1 lb (0.454 kg) (TPQ = 500)[7]

Cited in U.S. State Regulations: California (G), Massachusetts (G), New Jersey (G), Pennsylvania (G).

Description: Dinoterb, $C_6H_2(NO_2)_2(C_4H_9)(OH)$, is a yellow solid. Freezing/Melting point = 126°C. Hazard Identification (based on NFPA-704 M Rating System): Health 3, Flammability 2, Reactivity 3.

Potential Exposure: This material is a herbicide and a rodenticide.

Incompatibilities: Strong caustics. Heat may cause material to explode.

Permissible Exposure Limits in Air: No standards set.

Permissible Concentration in Water: No criteria set.

Harmful Effects and Symptoms

Short Term Exposure: Symptoms of poisoning are similar to other dinitrophenols and may include nausea, gastric distress, restlessness, sensation of heat, flushed skin, sweating, thirst, deep and rapid breathing, rapid heart rate, fever, and lack of oxygen to tissues (blueness of skin). This compound is toxic by all routes of exposure. The dangerous single oral dose of dinitro-o-cresol, a structurally similar compound, is estimated to be about 29 mg/kg.

First Aid: If this chemical gets into the eyes, remove any contact lenses at once and irrigate immediately for at least 15 minutes, occasionally lifting upper and lower lids. Seek medical attention immediately. If this chemical contacts the skin, remove contaminated clothing and wash immediately with soap and water. Seek medical attention immediately. If this chemical has been inhaled, remove from exposure, begin rescue breathing (using universal precautions) if breathing has stopped and CPR if heart action has stopped. Transfer promptly to a medical facility. When this chemical has been swallowed, get medical attention. Give large quantities of water and induce vomiting. Do not make an unconscious person vomit.

Personal Protective Methods: Wear protective gloves and clothing to prevent any reasonable probability of skin contact. Safety equipment suppliers/manufacturers can provide recommendations on the most protective glove/clothing material for your operation.. All protective clothing (suits, gloves, footwear, headgear) should be clean, available each day, and put on before work. Contact lenses should not be worn when working with this chemical. Wear dust-proof chemical goggles and face shield unless full facepiece respiratory protection is worn. Employees should wash immediately with soap when skin is wet or contaminated. Provide emergency showers and eyewash.

Respirator Selection: Where the potential for exposure to dinoterb exists, use a MSHA/NIOSH approved supplied-air respirator with a full facepiece operated in the positive pressure mode or with a full facepiece, hood, or helmet in the continuous flow mode, or use a MSHA/NIOSH approved self-contained breathing apparatus with a full facepiece operated in pressure-demand or other positive pressure mode.

Storage: Prior to working with dinoterb you should be trained on its proper handling and storage. Store in tightly closed containers in a cool, well ventilated area.

Shipping: Phenoxy pesticides, solid, toxic, n.o.s. require a "Poison" label. They fall in Hazard Class 6.1 and dinoterb in Packing Group II.

Spill Handling: *Dry material:* Evacuate persons not wearing protective equipment from area of spill or leak until clean-up is complete. Remove all ignition sources. Collect powdered material in the most convenient and safe manner and deposit in sealed containers. Ventilate area after clean-up is complete. It may be necessary to contain and dispose of this chemical as a hazardous waste. If material or contaminated runoff enters waterways, notify downstream users of potentially contaminated waters. Contact your Department of Environmental Protection or your regional office of the federal EPA for specific recommendations. If employees are required to clean-up spills, they must be properly trained and equipped. OSHA 1910.120(q) may be applicable.

Wet spills: Evacuate and restrict persons not wearing protective equipment from area of spill or leak until cleanup is complete. Remove all ignition sources. Ventilate area of spill or leak. Absorb liquids in vermiculite, dry sand, earth, peat, carbon, or a similar material and deposit in sealed containers. Keep this chemical out of a confined space, such as a sewer, because of the possibility of an explosion, unless the sewer is designed to prevent the build-up of explosive concentrations. It may be necessary to contain and dispose of this chemical as a hazardous waste. If material or contaminated runoff enters waterways, notify downstream users of potentially contaminated waters. Contact your Department of Environmental Protection or your regional office of the federal EPA for specific recommendations. If employees are required to clean-up spills, they must be properly trained and equipped. OSHA 1910.120(q) may be applicable.

Fire Extinguishing: Heat may cause material to explode. Use dry chemical, carbon dioxide, water spray, or foam for small fires, and water spray, fog, or foam for large fires. Move container from fire area if it can be done safely. Isolate hazard area, stay upwind, and keep out of low areas. Wear self-contained breathing apparatus and full protective clothing. Poisonous gases are produced in fire including nitrogen oxides. If material or contaminated runoff enters waterways, notify downstream users of potentially contaminated waters. Notify local health and fire officials and pollution control agencies. Containers may explode in fire. From a secure, explosion-proof location, use water spray to cool exposed containers. If cooling streams are ineffective (venting sound increases in volume and pitch, tank discolors, or shows any signs of deforming), withdraw immediately to a secure position. If employees are expected to fight fires, they must be trained and equipped in OSHA 1910.156.

References

U.S. Environmental Protection Agency, "Chemical Profile: Dinoterb," Washington, DC, Chemical Emergency Preparedness Program (November 30, 1987)

Di-n-Octyl Phthalate

Molecular Formula: $C_{24}H_{38}O_4$

Common Formula: $C_6H_4(COOC_8H_{17})_2$

Synonyms: 1,2-Benzenedicarboxylic acid, di-*n*-octyl ester; Bis(2-ethylhexyl) phthalate; Cellulex DOP; Di-*sec*-(2-ethylhexyl) phthalate; Dinopol NOP; Di-*n*-octyl phthalate; Di-*sec*-octyl phthalate; DNOP; DOP; *n*-Octyl phthalate; Octyl phthalate; Phthalic acid, dioctyl ester; PX-138; Vinicizer-85

CAS Registry Number: 117-84-0

RTECS Number: TI1925000

DOT ID: UN 3082

Regulatory Authority

- Clean Water Act: 40CFR423, Appendix A, Priority Pollutants; Section 313 Water Priority Chemicals (57FR41331, 9/9/92)
- EPA Hazardous Waste Number (RCRA No.): U107
- RCRA, 40CFR261, Appendix 8 Hazardous Constituents
- RCRA 40CFR268.48; 61FR15654, Universal Treatment Standards: Wastewater (mg/l), 0.017; Nonwastewater (mg/kg), 28

- RCRA 40CFR264, Appendix 9; TSD Facilities Ground Water Monitoring List. Suggested test method(s) (PQL μg/l): 8060 (30); 8270 (10)
- Superfund/EPCRA 40CFR302.4 Reportable Quantity (RQ): CERCLA, 5,000 lb (2,270 kg)
- Canada, WHMIS, Ingredients Disclosure List

Cited in U.S. State Regulations: California (G), Kansas (G), Louisiana (G), Maryland (G), Massachusetts (G), New Jersey (G), Pennsylvania (G), Rhode Island (G), Vermont (G), Virginia (G), Washington (G), Wisconsin (G).

Description: DNOP, $C_6H_4(COOC_8H_{17})_2$, is a liquid. Boiling point = 230°C under 5 mm Hg pressure. Flash point = 215°C. Autoignition temperature = 390°C. Explosive Limits: LEL = 0.3% @ 245°C; UEL – unknown. Hazard Identification (based on NFPA-704 M Rating System): Health 0, Flammability 1, Reactivity 0. Insoluble in water.

Potential Exposure: DNOP is used as plasticizers in plastics product manufacture.

Incompatibilities: Water contact causes foaming. Incompatible with oxidizers, strong acids, alkyl-halides since violent reactions occur.

Permissible Exposure Limits in Air: No standards set.

Permissible Concentration in Water: Insufficient data are available to permit EPA to define criteria to prevent damage to aquatic life or harm to humans.[6] The former USSR-UNEP/IRPTC project[43] has set a MAC in water bodies used for domestic purposes of 1.0 mg/l.

Determination in Water: Methylene chloride extraction followed by gas chromatography with flame ionization or electron capture detection (EPA Method 606) or gas chromatography plus mass spectrometry (EPA Method 625).

Routes of Entry: Inhalation.

Harmful Effects and Symptoms

Short Term Exposure: Di-n-octyl phthalate can affect you when breathed in and may enter the body through the skin. Eye contact may cause irritation. Repeated skin contact may cause dryness, cracking and rash. Breathing the vapor may irritate the nose, throat and bronchial tubes. High exposure levels can irritate the lungs and, if prolonged, could cause death.

Long Term Exposure: High or repeated exposure may affect the liver or kidneys. Repeated skin contact can cause dryness, cracking, and rash. This chemical has been shown to be a teratogen in animals.

Points of Attack: Skin, liver, kidneys.

Medical Surveillance: Liver, kidney, and lung function tests. Evaluation by a qualified allergist. Consider chest x-ray following acute overexposure.

First Aid: If this chemical gets into the eyes, remove any contact lenses at once and irrigate immediately for at least 15 minutes, occasionally lifting upper and lower lids. Seek medical attention immediately. If this chemical contacts the skin, remove contaminated clothing and wash immediately with soap and water. Seek medical attention immediately. If this chemical has been inhaled, remove from exposure, begin rescue breathing (using universal precautions) if breathing has stopped and CPR if heart action has stopped. Transfer promptly to a medical facility. When this chemical has been swallowed, get medical attention. Give large quantities of water and induce vomiting. Do not make an unconscious person vomit.

Personal Protective Methods: Wear protective gloves and clothing to prevent any reasonable probability of skin contact. Safety equipment suppliers/manufacturers can provide recommendations on the most protective glove/clothing material for your operation. ACGIH recommends butyl rubber, nitrile rubber and Viton as a protective material. All protective clothing (suits, gloves, footwear, headgear) should be clean, available each day, and put on before work. Contact lenses should not be worn when working with this chemical. Wear splash-proof chemical goggles and face shield unless full facepiece respiratory protection is worn. Employees should wash immediately with soap when skin is wet or contaminated. Provide emergency showers and eyewash.

Respirator Selection: Where the potential exists for exposures to di-n-octyl phthalate, use a MSHA/NIOSH approved supplied-air respirator with a full facepiece operated in the positive pressure mode or with a full facepiece, hood, or helmet in the continuous flow mode, or use a MSHA/NIOSH approved self-contained breathing apparatus with a full facepiece operated in pressure-demand or other positive pressure mode.

Storage: Prior to working with this chemical you should be trained on its proper handling and storage. Before entering confined space where this chemical may be present, check to make sure that an explosive concentration does not exist. Di-n-octyl phthalate must be stored to avoid contact with strong oxidizers (such as chlorine, bromine and fluorine) and strong acids (such as hydrochloric, sulfuric and nitric) and alkyl-halides since violent reactions occur. Store in tightly closed containers in a cool, well-ventilated area away from heat. Sources of ignition such as smoking and open flames are prohibited where this chemical is used, handled, or stored in a manner that could create a potential fire or explosion hazard. Metal containers involving the transfer of 5 gallons or more of this chemical should be grounded and bonded. Drums must be equipped with self-closing valves, pressure vacuum bungs, and flame arresters. Use only nonsparking tools and equipment, especially when opening and closing containers of this chemical.

Shipping: Environmentally hazardous solid, n.o.s. Hazard Class: 9. Label: "Class 9."

Spill Handling: Evacuate and restrict persons not wearing protective equipment from area of spill or leak until cleanup is complete. Remove all ignition sources. Establish forced ventilation to keep levels below explosive limit.

Absorb liquids in vermiculite, dry sand, earth, peat, carbon, or a similar material and deposit in sealed containers. Keep this chemical out of a confined space, such as a sewer, because of the possibility of an explosion, unless the sewer is designed to prevent the build-up of explosive concentrations. It may be necessary to contain and dispose of this chemical as a hazardous waste. If material or contaminated runoff enters waterways, notify downstream users of potentially contaminated waters. Contact your Department of Environmental Protection or your regional office of the federal EPA for specific recommendations. If employees are required to clean-up spills, they must be properly trained and equipped. OSHA 1910.120(q) may be applicable.

Fire Extinguishing: This chemical is a combustible liquid. Poisonous gases are produced in fire. Use dry chemical, carbon dioxide, or alcohol foam extinguishers. Vapors are heavier than air and will collect in low areas. Vapors may travel long distances to ignition sources and flashback. Vapors in confined areas may explode when exposed to fire. Containers may explode in fire. Storage containers and parts of containers may rocket great distances, in many directions. If material or contaminated runoff enters waterways, notify downstream users of potentially contaminated waters. Notify local health and fire officials and pollution control agencies. From a secure, explosion-proof location, use water spray to cool exposed containers. If cooling streams are ineffective (venting sound increases in volume and pitch, tank discolors, or shows any signs of deforming), withdraw immediately to a secure position. If employees are expected to fight fires, they must be trained and equipped in OSHA 1910.156.

Disposal Method Suggested: Incineration. Consult with environmental regulatory agencies for guidance on acceptable disposal practices. Generators of waste containing this contaminant (≥100 kg/mo) must conform with EPA regulations governing storage, transportation, treatment, and waste disposal.

References

U.S. Environmental Protection Agency, Phthalate Esters: Ambient Water Quality Criteria, Washington, DC (1980)

U.S. Environmental Protection Agency, Di-n-Octyl Phthalate, Health and Environmental Effects Profile No. 95, Washington, DC, Office of Solid Waste (April 30, 1980)

Sax N. I., Ed., "Dangerous Properties of Industrial Materials Report," 6, No. 1, 52–56 (1986)

U.S. Environmental Protection Agency, "Chemical Profile: Dioctyl Phthalate," Washington, DC, Chemical Emergency Preparedness Program (October 31, 1985)

New Jersey Department of Health, "Hazardous Substance Fact Sheet: Di-n-Octyl Phthalate," Trenton, NJ (January 1996)

Dioxane

Molecular Formula: $C_4H_8O_2$

Synonyms: 6200 drum cleaning solvent; 6500 drum cleaning solvent; Chlorothene SM solvent; 1,4-Diethylene dioxide; Diethylene dioxide; Diethylene ether; Di(ethylene oxide); Diethylene oxide; Diokan; Dioksan (Polish); Diossaoxan (Czech); 1,4-Dioxacyclohexane; 1,4-Dioxan (German); Dioxan; Dioxan-1,4 (German); *p*-Dioxane; 1,4-Dioxane; Dioxane; Dioxanne (French); 1,4-Dioxin, tetrahydro-; Dioxyethylene ether; Glycol ethylene ether; NCI-C03689; NE 220; Solvent 111; STCC 4909155; Tetrahydro-*p*-dioxin; Tetrahydro-1,4-dioxin

CAS Registry Number: 123-91-1

RTECS Number: JG8225000

DOT ID: UN 1165

EEC Number: 603-024-00-5

Regulatory Authority

- Carcinogen (Animal Positive) (IARC) (NCI)[9] (DFG)[3]
- Air Pollutant Standard Set (ACGIH)[1] (DFG)[3] (HSE)[33] (OSHA)[58] (former USSR)[35] (Several States)[60]
- Clean Air Act: Hazardous Air Pollutants (Title I, Part A, Section 112)
- EPA Hazardous Waste Number (RCRA No.): U108
- RCRA, 40CFR261, Appendix 8 Hazardous Constituents
- RCRA 40CFR268.48; 61FR15654, Universal Treatment Standards: Wastewater (mg/l), 12.0; Nonwastewater (mg/kg), 170
- RCRA 40CFR264, Appendix 9; TSD Facilities Ground Water Monitoring List. Suggested test method(s) (PQL μg/l): 8015 (150)
- Superfund/EPCRA 40CFR302.4 Reportable Quantity (RQ): CERCLA, 100 lb (45.4 kg)
- EPCRA Section 313 Form R *de minimis* concentration reporting level: 0.1%
- Canada, WHMIS, Ingredients Disclosure List

Cited in U.S. State Regulations: Alaska (G), California (A, G), Connecticut (A, W), Florida (G, A), Illinois (G), Kansas (G, A), Louisiana (G), Maine (G), Maryland (G), Massachusetts (G, A), Michigan (G), Minnesota (G), Nevada (A), New Hampshire (G, W), New Jersey (G), New York (G, A), North Carolina (A), North Dakota (A), Oklahoma (G), Pennsylvania (G, A), Rhode Island (G), South Carolina (A), South Dakota (A), Vermont (G), Virginia (G, A), Washington (G), West Virginia (G), Wisconsin (G).

Description: Dioxane, is a volatile, colorless liquid that may form explosive peroxides during storage. Boiling point = 101°C. Freezing/Melting point = 12°C. Flash point = 12°C. Autoignition temperature = 180°C. Explosive limits: LEL = 2.0%; UEL = 22%. Hazard Identification (based on NFPA-704 M Rating System): Health 2, Flammability 3, Reactivity 1. Soluble in water.

Potential Exposure: Dioxane finds its primary use as a solvent for cellulose acetate, dyes, fats, greases, lacquers, mineral oil, paints, polyvinyl polymers, resins, varnishes, and waxes. It finds particular usage in paint and varnish strippers, as a wetting agent and dispersing agent in textile processing, dye baths, stain and printing compositions, and in the preparation of histological slides.

Incompatibilities: Forms explosive mixture with air. Moisture forms unstable and explosive peroxides. Incompatible with strong oxidizers, strong acids, decaborane, triethyl aluminum, oxygen, halogens, reducingagnts, moisture, and heat. Reacts explosively with catalysts such as Raney-nickel (above 210°C). Attacks many plastics. May accumulate static electrical charges, and may cause ignition of its vapors.

Permissible Exposure Limits in Air: The Federal Limit (OSHA PEL)[58] is 100 ppm TWA. The HSE TWA,[33] and Sweden,[35] the ACGIH TLV is 25 ppm (90 mg/m^3). The DFG MAK is 20 ppm (73 mg/m^3) The notation "skin" indicates the possibility off cutaneous absorption. NIOSH recommended airborne exposure limit is 1 ppm (3.6 mg/m^3) based on a 30-minute sampling period. The STEL value adopted by HSE[33] is 100 ppm (360 mg/m^3). The NIOSH IDLH level is 500 ppm. The former USSR[35][43] has set a MAC in workplace air of 10 mg/m^3 and a MAC in the ambient air of residential areas[35] of 0.7 mg/m^3 on a once-daily basis.

In addition, several states have set guidelines or standards for Dioxane in ambient air[60] ranging from zero (North Carolina) to 2.4 μg/m^3 (Massachusetts) to 214.286 μg/m^3 (Kansas) to 300 μg/m^3 (New York) to 311 μg/m^3 (Pennsylvania) to 450 μg/m^3 (Connecticut and South Carolina) to 900 μg/m^3 (Florida, South Dakota) to 1,500 μg/m^3 (Virginia) to 2,140 μg/m^3 (Nevada).

Determination in Air: Charcoal adsorption, CS_2 workup, gas chromatography analysis. See NIOSH Method 1602.[18]

Permissible Concentration in Water: EPA[32] has suggested a permissible ambient goal of 2,480 μg/l based on health effects. The EPA[48] has developed a one-day health advisory of 4 mg/l and a ten-day health advisory of 0.4 mg/l but notes that carcinogenicity makes formulation of a longer term advisory inadvisable. Connecticut[61] has set a guideline for Dioxane in drinking water of 20 μg/l.

Determination in Water: Purge and trap gas chromatography-mass spectrometry may be used.[48]

Routes of Entry: Inhalation of vapor, percutaneous absorption, ingestion, eye and skin contact.

Harmful Effects and Symptoms

Short Term Exposure: Contact with dioxane can cause eye and skin irritation and burns. Inhalation can cause irritation of the respiratory tract with coughing and shortness of breath. Higher exposure to dioxane vapor can cause headache, lightheadedness, dizziness, loss of appetite, headache, nausea, vomiting, stomach pain, drowsiness, and unconsciousness. Overexposure may cause liver and kidney damage. Prolonged skin exposure to the liquid may cause drying and cracking.

Long Term Exposure: Prolonged or repeated exposure may cause liver and kidney damage. The liquid defats the skin, causing drying and cracking. This chemical is a potential occupational carcinogen.

Points of Attack: Eyes, skin, respiratory system, liver, kidneys. Cancer site in animals: lung, liver, and nasal cavity tumors.

Medical Surveillance: Preplacement and periodic examinations should be directed to symptoms of headache and dizziness, as well as nausea and other gastrointestinal disturbances. The condition of the skin should be considered. For those with frequent or potentially high exposure (half the TLV or greater, or significant skin contact), the following are recommended before beginning work and at regular times after that: Liver function tests. Kidney function tests, including routine urine test.

First Aid: If this chemical gets into the eyes, remove any contact lenses at once and irrigate immediately for at least 15 minutes, occasionally lifting upper and lower lids. Seek medical attention immediately. If this chemical contacts the skin, remove contaminated clothing and wash immediately with soap and water. Seek medical attention immediately. If this chemical has been inhaled, remove from exposure, begin rescue breathing (using universal precautions) if breathing has stopped and CPR if heart action has stopped. Transfer promptly to a medical facility. When this chemical has been swallowed, get medical attention. Give large quantities of water and induce vomiting. Do not make an unconscious person vomit.

Personal Protective Methods: Wear protective gloves and clothing to prevent any reasonable probability of skin contact. Safety equipment suppliers/manufacturers can provide recommendations on the most protective glove/clothing material for your operation. Butyl rubber and Teflon are among the recommended protective materials. All protective clothing (suits, gloves, footwear, headgear) should be clean, available each day, and put on before work. Contact lenses should not be worn when working with this chemical. Wear splash-proof chemical goggles and face shield unless full facepiece respiratory protection is worn. Employees should wash immediately with soap when skin is wet or contaminated. Provide emergency showers and eyewash. Remove clothing immediately if wet or contaminated to avoid flammability hazard.

Respirator Selection: *NIOSH: At any concentrations above the NIOSH REL, or where there is no REL, at any detectable concentration:* SCBAF:PD,PP (any self-contained breathing apparatus that has a full facepiece and is operated in a pressure-demand or other positive-pressure mode); or SAF:PD,PP:ASCBA (any supplied-air respirator that has a full facepiece and is operated in a pressure-demand or other positive-pressure mode in combination with an auxiliary self-contained breathing apparatus operated in a pressure-demand or other positive pressure mode). *Escape:* GMFOV [any air-purifying, full-facepiece respirator (gas mask) with a chin-style, front- or back-, mounted organic vapor canister]; or SCBAE (any appropriate escape-type, self-contained breathing apparatus).

Storage: Prior to working with dioxane you should be trained on its proper handling and storage. Before entering confined space where this chemical may be present, check to make sure that an explosive concentration does not exist. 1,4-dioxane must be stored to avoid contact with strong oxidizers (such as chlorine, bromine and fluorine) since violent reactions occur. Sources of ignition, such as smoking and open flames, are prohibited where 1,4-dioxane is handled, used, or stored. Metal containers involving the transfer of 5 gallons or more of 1,4-dioxane should be grounded and bonded. Drums must be equipped with self-closing valves, pressure vacuum bungs and flame arresters. Use only non-sparking tools and equipment, especially when opening and closing containers of 1,4-dioxane. A regulated, marked area should be established where this chemical is handled, used, or stored in compliance with OSHA standard 1910.1045.

Shipping: This compound requires a shipping label of: "Flammable Liquid." It falls in DOT Hazard Class 3 and Packing Group II.

Spill Handling: Evacuate and restrict persons not wearing protective equipment from area of spill or leak until cleanup is complete. Remove all ignition sources. Establish forced ventilation to keep levels below explosive limit. Absorb liquids in vermiculite, dry sand, earth, peat, carbon, or a similar material and deposit in sealed containers. Keep this chemical out of a confined space, such as a sewer, because of the possibility of an explosion, unless the sewer is designed to prevent the build-up of explosive concentrations. It may be necessary to contain and dispose of this chemical as a hazardous waste. If material or contaminated runoff enters waterways, notify downstream users of potentially contaminated waters. Contact your Department of Environmental Protection or your regional office of the federal EPA for specific recommendations. If employees are required to clean-up spills, they must be properly trained and equipped. OSHA 1910.120(q) may be applicable.

Fire Extinguishing: This chemical is a flammable liquid. Poisonous gases are produced in fire. Use dry chemical, carbon dioxide, or foam extinguishers. Do not use halogens. Vapors are heavier than air and will collect in low areas. Vapors may travel long distances to ignition sources and flashback. Vapors in confined areas may explode when exposed to fire. Containers may explode in fire. Storage containers and parts of containers may rocket great distances, in many directions. If material or contaminated runoff enters waterways, notify downstream users of potentially contaminated waters. Notify local health and fire officials and pollution control agencies. From a secure, explosion-proof location, use water spray to cool exposed containers. If cooling streams are ineffective (venting sound increases in volume and pitch, tank discolors, or shows any signs of deforming), withdraw immediately to a secure position. If employees are expected to fight fires, they must be trained and equipped in OSHA 1910.156.

Disposal Method Suggested: Concentrated waste containing no peroxides-discharge liquid at a controlled rate near a pilot flame. Concentrated waste containing peroxides-perforation of a container of the waste from a safe distance followed by open burning.[22] Consult with environmental regulatory agencies for guidance on acceptable disposal practices. Generators of waste containing this contaminant (≥100 kg/mo) must conform with EPA regulations governing storage, transportation, treatment, and waste disposal.

References

National Institute for Occupational Safety and Health, Criteria for a Recommended Standard: Occupational Exposure to Dioxane, NIOSH Doc. No. 77–226 (1977)

U.S. Environmental Protection Agency, "Chemical Hazard Information Profile: Dioxane," Washington, DC (1979)

Sax N. I., Ed., "Dangerous Properties of Industrial Materials Report," 8, No.1, 32–42 (1988)

New Jersey Department of Health, "Hazardous Substance Fact Sheet: 1,4-Dioxane," Trenton, NJ (February 1996)

New York State Department of Health, "Chemical Fact Sheet: 1,4-Dioxane," Albany, NY, Bureau of Toxic Substance Assessment (January 1986)

Dioxathion

Molecular Formula: $C_{12}H_{26}O_6P_2S_4$

Synonyms: AC 528; Bis(dithiophospate de *o,o*-diethyle) de *S,S*'-(1,4-dioxanne-2,3-diyle) (French); Delnatex; Delnav; 1,4-Dioxan-2,3-diyl *S,S*-di(*O,O*-diethyl phosphorodithioate); 2,3-Dioxanedithiol *S,S*-bis(*O,O*-diethylphosphorodithioate); *S,S*'-*p*-Dioxane-2,3-diyl bis(*O,O*-diethylphosphorodithioate); *S,S*'-1,4-Dioxane-2,3-diyl bis(*O,O*-diethyl phosphorodithioate); *S,S*'-(1,4-Dioxane-2,3-diyl) *O,O,O*',*O*'-tetraethylbis(phosphorodithioate); *S,S*'-1,4-Dioxane-2,3-diyl *O,O,O*-tetraethyl ester; ENT 22879; Hercules AC528; Kavadel; Navadel; NCI-C00395; Phosphorodithioic acid, *O,O*-diethyl ester, *S,S*-diester with *p*-dioxane-2,3-dithiol; Phosphorodithioic acid, *S,S*'-*p*-dioxane-2,3-diyl *O,O,O*',*O*'-tetraethyl ester; Phosphorodithioic acid, *S,S*'-1,4-dioxane-2,3-diyl *O,O,O*',*O*'-tetraethyl ester; Phosphorodithioic acid, 5,5'-, 1,4-dioxane-2,3-diyl *O,O,O*',*O*'-tetraethyl ester

CAS Registry Number: 78-34-2

RTECS Number: TE3350000

DOT ID: UN 2783; UN 3018

EEC Number: 015-063-00-X

Regulatory Authority

- Air Pollutant Standard Set (ACGIH)[1] (HSE)[33] (OSHA)[58] (Several States)[60]
- Superfund/EPCRA 40CFR355, Appendix B Extremely Hazardous Substances: TPQ = 500 lb (227 kg)
- Superfund/EPCRA 40CFR302.4 Reportable Quantity (RQ): EHS, 1 lb (0.454 kg)

- U.S. DOT Regulated Marine Pollutant (49CFR172.101, Appendix B)

Cited in U.S. State Regulations: Alaska (G), California (G), Connecticut (A), Florida (G), Illinois (G), Maine (G), Massachusetts (G), Michigan (G), Nevada (A), New Hampshire (G), New Jersey (G), New York (A), North Dakota (A), Oklahoma (G), Pennsylvania (G), Rhode Island (G), Virginia (A), West Virginia (G).

Description: Dioxathion, $C_{12}H_{26}O_6P_2S_4$, is a viscous, reddish-brown liquid or powder with a garlic-like odor. Freezing/Melting point = -20°C. Hazard Identification (based on NFPA-704 M Rating System): Health 3, Flammability 0, Reactivity 0. Insoluble in water.

Potential Exposure: Those involved in the manufacture, formulation or application of this insecticide and acaricide.

Incompatibilities: Incompatible with strong acids. Dioxathion is hydrolyzed by strong bases, attacks iron and tin surfaces.

Permissible Exposure Limits in Air: There is no OSHA PEL. NIOSH, ACGIH TLV and HSE[33] TWA value is 0.2 mg/m^3 with the notation "skin" indicating the possibility of cutaneous absorption. There is no NIOSH ILDH value. Several states have set guidelines or standards for Dioxathion in ambient air[60] ranging from zero (New York) to 2.0 μg/m^3 (North Dakota) to 3.0 μg/m^3 (Virginia) to 4.0 μg/m^3 (Connecticut) to 5.0 μg/m^3 (Nevada).

Determination in Air: Although NIOSH lists "none available," the organophosphate method is listed: OSHA versatile sampler-2; Toluene/Acetone; Gas chromatography/Flame photometric detection for sulfur, nitrogen, or phosphorus; NIOSH Method IV Method #5600, Organophosphorus Pesticides.

Permissible Concentration in Water: No criteria set.

Routes of Entry: Inhalation, skin absorption, ingestion, skin and/or eye contact.

Harmful Effects and Symptoms

Short Term Exposure: Dioxathion is a cholinesterase inhibitor. It may affect the nervous system, resulting in convulsions, respiratory failure. Exposure to high level may cause death. Acute exposure to Dioxathion may produce the following signs and symptoms: pinpoint pupils, blurred vision, headache, dizziness, muscle spasms, and profound weakness. Vomiting, diarrhea, abdominal pain, seizures, and coma may also occur. The heart rate may decrease following oral exposure or increase following dermal exposure. Hypotension (low blood pressure) may occur although hypertension (high blood pressure) is not uncommon. Chest pain may be noted. Respiratory symptoms include dyspnea (shortness of breath), respiratory depression, and respiratory paralysis. Psychosis may occur.

Dioxathion is very toxic. Probable oral lethal dose for humans is 50 – 500 mg/kg or between 1 teaspoonful and 1 oz for a 70 kg (150 lb) person. It is a cholinesterase inhibitor. Death is primarily due to respiratory arrest arising from failure of the respiratory center, paralysis of respiratory muscles, intense bronchoconstriction, or all three.

Long Term Exposure: Cholinesterase inhibitor; cumulative effect is possible. This chemical may damage the nervous system with repeated exposure, resulting in convulsions, respiratory failure. May cause liver damage.

Points of Attack: Respiratory system, lungs, central nervous system, cardiovascular system, skin, eyes, plasma and red blood cell cholinesterase.

Medical Surveillance: Before employment and at regular times after that, the following are recommended: Plasma and red blood cell cholinesterase levels (tests for the enzyme poisoned by this chemical). If exposure stops, plasma levels return to normal in 1 – 2 weeks while red blood cell levels may be reduced for 1 – 3 months.

When cholinesterase enzyme levels are reduced by 25% or more below preemployment levels, risk of poisoning is increased, even if results are in lower ranges of "normal." Reassignment to work not involving organophosphate or carbamate pesticides is recommended until enzyme levels recover. If symptoms develop or overexposure occurs, repeat the above tests as soon as possible and get an exam of the nervous system. Also consider complete blood count. Consider chest x-ray following acute overexposure. Do not drink any alcoholic beverages before or during use. Alcohol promotes absorption of organic phosphates.

First Aid: If this chemical gets into the eyes, remove any contact lenses at once and irrigate immediately for at least 15 minutes, occasionally lifting upper and lower lids. Seek medical attention immediately. If this chemical contacts the skin, remove contaminated clothing and wash immediately with soap and water. Speed in removing material from skin is of extreme importance. Shampoo hair promptly if contaminated. Seek medical attention immediately. If this chemical has been inhaled, remove from exposure, begin rescue breathing (using universal precautions) if breathing has stopped and CPR if heart action has stopped. Transfer promptly to a medical facility. When this chemical has been swallowed, get medical attention. Give large quantities of water and induce vomiting. Do not make an unconscious person vomit. Effects may be delayed; medical observation is recommended. Obtain authorization and/or further instructions from the local hospital for administration of an antidote or performance of other invasive procedures.

Personal Protective Methods: Wear protective gloves and clothing to prevent any reasonable probability of skin contact. Safety equipment suppliers/manufacturers can provide recommendations on the most protective glove/clothing material for your operation.. All protective clothing (suits, gloves, footwear, headgear) should be clean, available each day, and put on before work. Contact lenses should not be worn when working with this chemical. Wear splash-proof chemical goggles and face shield unless full facepiece respiratory protection is worn. Employees should wash immediately with soap when skin is wet or contaminated. Provide emergency showers and eyewash.

Respirator Selection: Where the potential exists for exposures over 0.2 mg/m^3, use an MSHA/NIOSH approved supplied-air respirator with a full facepiece operated in the positive pressure mode or with a full facepiece, hood, or helmet in the continuous flow mode, or use an MSHA/NIOSH approved self-contained breathing apparatus with a full facepiece operated in pressure-demand or other positive pressure mode.

Storage: Prior to working with this chemical you should be trained on its proper handling and storage. Store in tightly closed containers in a cool, well-ventilated area.

Shipping: Dioxathion is not specifically cited by DOT but falls in the Organophosphorus Pesticides, Liquid, Toxic n.o.s. Category which is in Hazard Class 6.1. Dioxathion falls in Shipping Group II which requires a "Poison" label. This chemical is a regulated marine pollutant.

Spill Handling: Evacuate and restrict persons not wearing protective equipment from area of spill or leak until cleanup is complete. Remove all ignition sources. Ventilate area of spill or leak. Small spill: absorb with sand of other noncombustible absorbent material and place into containers for later disposal. Large spills: dike far ahead of spill for later disposal. It may be necessary to contain and dispose of this chemical as a hazardous waste. If material or contaminated runoff enters waterways, notify downstream users of potentially contaminated waters. Contact your Department of Environmental Protection or your regional office of the federal EPA for specific recommendations. If employees are required to clean-up spills, they must be properly trained and equipped. OSHA 1910.120(q) may be applicable.

Fire Extinguishing: Dioxathion does not burn. Use fire extinguishing agents suitable for surrounding fire. Decomposes above 135°C; poisonous gases including nitrogen, phosphorus, and sulfur oxides are produced. Move container from fire area if you can do so without risk. Fight fire from maximum distance. Dike fire control water for later disposal; do not scatter the material. Vapors are heavier than air and will collect in low areas. Containers may explode in fire. Storage containers and parts of containers may rocket great distances, in many directions. If material or contaminated runoff enters waterways, notify downstream users of potentially contaminated waters. Notify local health and fire officials and pollution control agencies. From a secure, explosion-proof location, use water spray to cool exposed containers. If cooling streams are ineffective (venting sound increases in volume and pitch, tank discolors, or shows any signs of deforming), withdraw immediately to a secure position. If employees are expected to fight fires, they must be trained and equipped in OSHA 1910.156.

Disposal Method Suggested: For small quantities: mix with lime and bury. For large amounts, use incineration with effluent gas scrubbing.[22] In accordance with 40CFR165 recommendations for the disposal of pesticides and pesticide containers. Must be disposed properly by following package label directions or by contacting your state pesticide or environmental control agency or by contacting your regional EPA office.

References

Sax N. I., Ed., "Dangerous Properties of Industrial Materials Report," 2, No. 5, 60–63 (1982) New York, Van Nostrand Reinhold Co. (1982)

U.S. Environmental Protection Agency, "Chemical Profile: Dioxathion," Washington, DC, Chemical Emergency Preparedness Program (November 30, 1987)

New Jersey Department of Health, "Hazardous Substance Fact Sheet: Dioxathion," Trenton, NJ (December 1998)

Dioxolane

Molecular Formula: $C_3H_6O_2$

Synonyms: 1,3-Dioxacyclopentane; 1,3-Dioxolan; Ethylene glycol formal; Formal glycol; Glycol formal

CAS Registry Number: 646-06-0

RTECS Number: JH6760000

DOT ID: UN 1166

Regulatory Authority

- Air Pollutant Standard Set (former USSR)[43]

Cited in U.S. State Regulations: Florida (G), Maine (G), Massachusetts (G), New Hampshire (G), New Jersey (G), Pennsylvania (G).

Description: Dioxolane is a water-white liquid. Boiling point = 74°C. Flash point = 2°C (oc). Hazard Identification (based on NFPA-704 M Rating System): Health 2, Flammability 3, Reactivity 2. Soluble in water.

Potential Exposure: Is a low-boiling solvent used for extraction of oils, fats, waxes, dyes and cellulose derivatives. May be used as a cross-linking agent in phenolic resins.

Incompatibilities: Forms explosive mixture with air. Contact with oxidizers may cause fire and explosion hazard.

Permissible Exposure Limits in Air: The former USSR-UNEP/IRPTC project[43] has set a MAC in workplace air of 50 mg/m^3.

Determination in Air: No method available.

Permissible Concentration in Water: No criteria set.

Routes of Entry: Inhalation, absorbed through the skin.

Harmful Effects and Symptoms

Short Term Exposure: Moderately toxic by inhalation and ingestion. May be poisonous if inhaled or absorbed through skin. Contact may irritate or burn skin and eyes. Vapors may cause dizziness or suffocation.

First Aid: If this chemical gets into the eyes, remove any contact lenses at once and irrigate immediately for at least 15 minutes, occasionally lifting upper and lower lids. Seek medical attention immediately. If this chemical contacts the skin, remove contaminated clothing and wash immediately with soap and water. Seek medical attention immediately. If this chemical has been inhaled, remove from exposure, begin rescue breathing (using universal precautions) if breathing has stopped and CPR if heart action has stopped. Transfer promptly to a medical facility. When this chemical has been swallowed, get medical attention. Give large quantities of water and induce vomiting. Do not make an unconscious person vomit.

Personal Protective Methods: Wear protective gloves and clothing to prevent any reasonable probability of skin contact. Safety equipment suppliers/manufacturers can provide recommendations on the most protective glove/clothing material for your operation.. All protective clothing (suits, gloves, footwear, headgear) should be clean, available each day, and put on before work. Contact lenses should not be worn when working with this chemical. Wear splash-proof chemical goggles and face shield unless full facepiece respiratory protection is worn. Employees should wash immediately with soap when skin is wet or contaminated. Provide emergency showers and eyewash.

Respirator Selection: Where the potential exists for exposure to this chemical, use a MSHA/NIOSH approved supplied-air respirator with a full facepiece operated in the positive pressure mode or with a full facepiece, hood, or helmet in the continuous flow mode, or use a MSHA/NIOSH approved self-contained breathing apparatus with a full facepiece operated in pressure-demand or other positive pressure mode.

Storage: Prior to working with this chemical you should be trained on its proper handling and storage. Store in tightly closed containers in a cool, well ventilated area away from oxidizers and reducing agents. Where possible, automatically pump liquid from drums or other storage containers to process containers.

Shipping: This compound requires a shipping label of: "Flammable Liquid." It falls in DOT Hazard Class 3 and Packing Group II.

Spill Handling: This chemical is a flammable liquid. Evacuate and restrict persons not wearing protective equipment from area of spill or leak until cleanup is complete. Remove all ignition sources. Ventilate area of spill or leak. If material is not on fire and not involved in fire, then keep sparks, flames, and other sources of ignition away. Build dikes to contain flow as necessary. Attempt to stop leak if this can be done without hazard. Use water spray to disperse vapors, and dilute standing pools of liquid. Avoid breathing vapors. Keep upwind. Wear boots, protective goggles, and gloves. Do not handle broken packages without protective equipment. Wash away any material which may have contacted the body with copious amounts of water or soap and water. Small spills: absorb with sand or other noncombustible absorbent material and place into containers for later disposal. Large spills: dike far ahead of spill for later disposal. Keep this chemical out of a confined space, such as a sewer, because of the possibility of an explosion, unless the sewer is designed to prevent the build-up of explosive concentrations. It may be necessary to contain and dispose of this chemical as a hazardous waste. If material or contaminated runoff enters waterways, notify downstream users of potentially contaminated waters. Contact your Department of Environmental Protection or your regional office of the federal EPA for specific recommendations. If employees are required to clean-up spills, they must be properly trained and equipped. OSHA 1910.120(q) may be applicable.

Fire Extinguishing: This chemical is a flammable liquid. Do not extinguish unless flow can be stopped. Use water in flooding quantities as fog. Solid streams of water may be ineffective. Use "alcohol" foam, carbon dioxide, or dry chemical. Small fires: dry chemical, carbon dioxide, water spray or alcohol foam. Large fires: water spray, fog or alcohol foam. Move container from fire area if you can do so without risk. Spray cooling water on containers that are exposed to flames until well after fire is out. For massive fire in cargo area, use unmanned hose holder or monitor nozzles; if this is impossible, withdraw from area and let fire burn. Poisonous gases are produced in fire. Vapors are heavier than air and will collect in low areas. Vapors may travel long distances to ignition sources and flashback. Vapors in confined areas may explode when exposed to fire. Containers may explode in fire. Storage containers and parts of containers may rocket great distances, in many directions. If material or contaminated runoff enters waterways, notify downstream users of potentially contaminated waters. Notify local health and fire officials and pollution control agencies. From a secure, explosion-proof location, use water spray to cool exposed containers. If cooling streams are ineffective (venting sound increases in volume and pitch, tank discolors, or shows any signs of deforming), withdraw immediately to a secure position. If employees are expected to fight fires, they must be trained and equipped in OSHA 1910.156.

Disposal Method Suggested: Incineration.

References

U.S. Environmental Protection Agency, "Chemical Profile: Dioxolane," Washington, DC, Chemical Emergency Preparedness Program (October 31, 1985)

Dipentene

Molecular Formula: $C_{10}H_{16}$

Synonyms: Cajeputene; Cinene; DL-*p*-Mentha-1,8-diene; Limonene; *p*-Mentha-1,8-diene; 1-Methyl-4-(1-methylethenyl) cyclohexane; Terpinene

CAS Registry Number: 138-86-3

RTECS Number: OS8100000

DOT ID: UN 2052

Cited in U.S. State Regulations: New Hampshire (G), New Jersey (G), Pennsylvania (G).

Description: Dipentene, $C_{10}H_{16}$, is a colorless liquid with a lemon-like odor. Boiling point = 170°C. Flash point = 45°C. Autoignition temperature = 237°C. Explosive limits: LEL = 0.7%; UEL = 6.1%, both at 150°C. Hazard Identification (based on NFPA-704 M Rating System): Health 0, Flammability 2, Reactivity 0. Insoluble in water.

Potential Exposure: Dipentene is used as a solvent, in rubber compounding and reclamation, and to make paints, enamels, lacquer and perfumes.

Incompatibilities: Forms explosive mixture with air. Contact with oxidizers may cause fire and explosion hazard.

Permissible Exposure Limits in Air: No standards set.

Permissible Concentration in Water: No criteria set.

Routes of Entry: Inhalation, passing through the skin.

Harmful Effects and Symptoms

Short Term Exposure: Dipentene can affect you when breathed in and by passing through your skin. Contact can irritate the eyes and skin. High exposures may damage the kidneys.

Long Term Exposure: Dipentene may cause a skin allergy. If allergy develops, very low future exposures can cause itching and a skin rash. There is limited evidence that dipentene causes kidney cancer in male rats (NJ).

Points of Attack: Skin.

Medical Surveillance: If symptoms develop or overexposure is suspected, the following may be useful: Evaluation by a qualified allergist, including careful exposure history and special testing, may help diagnose skin allergy. Kidney function tests.

First Aid: If this chemical gets into the eyes, remove any contact lenses at once and irrigate immediately for at least 15 minutes, occasionally lifting upper and lower lids. Seek medical attention immediately. If this chemical contacts the skin, remove contaminated clothing and wash immediately with soap and water. Seek medical attention immediately. If this chemical has been inhaled, remove from exposure, begin rescue breathing (using universal precautions) if breathing has stopped and CPR if heart action has stopped. Transfer promptly to a medical facility. When this chemical has been swallowed, get medical attention. Give large quantities of water and induce vomiting. Do not make an unconscious person vomit.

Personal Protective Methods: Wear protective gloves and clothing to prevent any reasonable probability of skin contact. Safety equipment suppliers/manufacturers can provide recommendations on the most protective glove/clothing material for your operation.. All protective clothing (suits, gloves, footwear, headgear) should be clean, available each day, and put on before work. Contact lenses should not be worn when working with this chemical. Wear splash-proof chemical goggles and face shield unless full facepiece respiratory protection is worn. Employees should wash immediately with soap when skin is wet or contaminated. Provide emergency showers and eyewash.

Respirator Selection: Where the potential exists for exposure to Dipentene, use a MSHA/NIOSH approved supplied-air respirator with a full facepiece operated in the positive pressure mode or with a full facepiece, hood, or helmet in the continuous flow mode, or use a MSHA/NIOSH approved self-contained breathing apparatus with a full facepiece operated in pressure-demand or other positive pressure mode.

Storage: Store in tightly closed containers in a cool, well-ventilated area. Before entering confined space where this chemical may be present, check to make sure that an explosive concentration does not exist. Store in tightly closed containers in a cool, well-ventilated area away from heat and incompatible materials. Sources of ignition such as smoking and open flames are prohibited where this chemical is used, handled, or stored in a manner that could create a potential fire or explosion hazard. Metal containers involving the transfer of 5 gallons or more of this chemical should be grounded and bonded. Drums must be equipped with self-closing valves, pressure vacuum bungs, and flame arresters. Use only nonsparking tools and equipment, especially when opening and closing containers of this chemical.

Shipping: This compound requires a shipping label of: "Flammable Liquid." It falls in DOT Hazard Class 3 and Packing Group III.

Spill Handling: Evacuate and restrict persons not wearing protective equipment from area of spill or leak until cleanup is complete. Remove all ignition sources. Establish forced ventilation to keep levels below explosive limit. Absorb liquids in vermiculite, dry sand, earth, peat, carbon, or a similar material and deposit in sealed containers. Keep this chemical out of a confined space, such as a sewer, because of the possibility of an explosion, unless the sewer is designed to prevent the build-up of explosive concentrations. It may be necessary to contain and dispose of this chemical as a hazardous waste. If material or contaminated runoff enters waterways, notify downstream users of potentially contaminated waters. Contact your Department of Environmental Protection or your regional office of the federal EPA for specific recommendations. If employees are required to clean-up spills, they must be properly trained and equipped. OSHA 1910.120(q) may be applicable.

Fire Extinguishing: This chemical is a combustible liquid. Poisonous gases are produced in fire. Use dry chemical, carbon dioxide, or alcohol foam extinguishers. Vapors are heavier than air and will collect in low areas. Vapors may travel long distances to ignition sources and flashback. Vapors in confined areas may explode when exposed to fire. Containers may explode in fire. Storage containers and parts of containers may rocket great distances, in many directions. If material or contaminated runoff enters waterways, notify downstream users of potentially contaminated waters. Notify local health and fire officials and pollution control agencies. From a secure, explosion-proof location, use water spray to cool exposed containers. If cooling streams are ineffective (venting sound increases in volume and pitch, tank discolors, or shows any signs of deforming), withdraw immediately to a secure position. If employees are expected to fight fires, they must be trained and equipped in OSHA 1910.156.

Disposal Method Suggested: Incineration.

References

Sax N. I., Ed., "Dangerous Properties of Industrial Materials Report," 2, No. 3, 78–79 (1982)

New Jersey Department of Health, "Hazardous Substance Fact Sheet: Dipentene," Trenton, NJ (December 1996)

Diphacinone

Molecular Formula: $C_{23}H_{16}O_3$

Synonyms: Didandin; Dipaxin; Diphacin; Diphacinon; Diphenacin; Diphenadion; Diphenadione; 2-Diphenylacetyl-1,3-diketohydrindene; 2-(Diphenylacetyl)indan-1,3-indandione; 2-(Diphenylacetyl)-1H-indene-1,3(2H)-dione; Oragulant; PID; Promar; Ramik; Ratindan 1; Solvan; U 1363

CAS Registry Number: 82-66-6

RTECS Number: NK5600000

DOT ID: UN 3027

Regulatory Authority

- Very Toxic Substance (World Bank)[15]
- Superfund/EPCRA 40CFR355, Appendix B Extremely Hazardous Substances: TPQ = 10/10,000 lb (4.54/4,540 kg)
- Superfund/EPCRA 40CFR302.4 Reportable Quantity (RQ): EHS, 1 lb (0.454 kg)

Cited in U.S. State Regulations: California (G), Massachusetts (G), New Jersey (G), Pennsylvania (G).

Description: Diphacione, $C_{23}H_{16}O_3$, is an odorless, pale yellow, crystalline solid. Freezing/Melting point = 146 – 147°C. Hazard Identification (based on NFPA-704 M Rating System): Health 3, Flammability 1, Reactivity 0. Very slightly soluble in water.

Potential Exposure: This material is used as a rodenticide and as an anticoagulant medication.

Permissible Exposure Limits in Air: No standards set.

Permissible Concentration in Water: No criteria set.

Harmful Effects and Symptoms

Short Term Exposure: This material is extremely toxic; probable oral lethal dose in humans is 5 – 50 mg/kg, or between 7 drops and 1 teaspoonful for a 150 lb persons. Diphacinone is an anticoagulant (inhibits blood clotting). Hemorrhage is the most common effect and may be manifested by nose bleeding, gum bleeding, bloody stools and urine, ecchymoses (extravasations of blood into skin), and hemoptysis (coughing up blood). Bruising is heightened. Abdominal and flank pains are also common. Other signs and symptoms include flushing, dizziness, hypotension (low blood pressure), dyspnea (shortness of breath), cyanosis (blue tint to the skin and mucous membranes), fever, and diarrhea.

Long Term Exposure: May affect the liver and kidneys. Repeated exposure may cause low white blood cell count and affect the brain.

Points of Attack: Blood, liver, kidneys.

Medical Surveillance: Blood test for clotting time (PT, INR, or PTT). Stool and urine tests for blood. Liver and kidney function tests. Complete blood count (CBC). EEG.

First Aid: If this chemical gets into the eyes, remove any contact lenses at once and irrigate immediately for at least 15 minutes, occasionally lifting upper and lower lids. Seek medical attention immediately. If this chemical contacts the skin, remove contaminated clothing and wash immediately with soap and water. Seek medical attention immediately. If this chemical has been inhaled, remove from exposure, begin rescue breathing (using universal precautions) if breathing has stopped and CPR if heart action has stopped. Transfer promptly to a medical facility. When this chemical has been swallowed, get medical attention. Give large quantities of water and induce vomiting. Do not make an unconscious person vomit. Obtain authorization and/or further instructions from the local hospital for administration of an antidote or performance of other invasive procedures. Rush to a health care facility. Acute exposure to Diphacinone may require decontamination and life support for the victims. Emergency personnel should wear protective clothing appropriate to the type and degree of contamination. Air-purifying or supplied-air respiratory equipment should also be worn, as necessary. Rescue vehicles should carry supplies such as plastic sheeting and disposable plastic bags to assist in preventing spread of contamination.

Personal Protective Methods: For emergency situations, wear a positive pressure, pressure-demand, full facepiece self-contained breathing apparatus (SCBA) or pressure-demand supplied air respirator with escape SCBA and a fully-encapsulating, chemical resistant suit. Safety equipment suppliers/manufacturers can provide recommendations on the most protective glove/clothing material for your operation.. All protective clothing (suits, gloves, footwear, headgear) should be clean, available each day, and put on before work. Contact lenses should not be worn when working with this chemical. Employees should wash immediately with soap when skin is wet or contaminated. Provide emergency showers and eyewash.

Respirator Selection: Where the potential exists for exposure to Diphacinone, use a MSHA/NIOSH approved supplied-air respirator with a full facepiece operated in the positive pressure mode or with a full facepiece, hood, or helmet in the continuous flow mode, or use a MSHA/NIOSH approved self-contained breathing apparatus with a full facepiece operated in pressure-demand or other positive pressure mode.

Storage: Prior to working with this chemical you should be trained on its proper handling and storage. Store in tightly closed containers in a cool, well ventilated area.

Shipping: Coumarin derivative pesticides, solid, toxic, n.o.s. fall in Hazard Class 6.1 and Diphacinone in Packing Group I. The label required is "Poison."

Spill Handling: Evacuate persons not wearing protective equipment from area of spill or leak until clean-up is complete. Remove all ignition sources. Do not touch spilled material; stop leak if you can do it without risk. Use water spray to reduce vapors. Small spills: Take up with sand or other noncombustible absorbent material and place into containers for later disposal. Small dry spills: with clean shovel place material into clean, dry container and cover; move containers from spill area. Large spills: dike far ahead of spill for later disposal. Ventilate area after clean-up is complete. It may be necessary to contain and dispose of this chemical as a hazardous waste. If material or contaminated runoff enters waterways, notify downstream users of potentially contaminated waters. Contact your Department of Environmental Protection or your regional office of the federal EPA for specific recommendations. If employees are required to clean-up spills, they must be properly trained and equipped. OSHA 1910.120(q) may be applicable.

Fire Extinguishing: Use dry chemical, carbon dioxide, water spray, or foam extinguishers. Poisonous gases are produced in fire. If material or contaminated runoff enters waterways, notify downstream users of potentially contaminated waters. Notify local health and fire officials and pollution control agencies. From a secure, explosion-proof location, use water spray to cool exposed containers. If cooling streams are ineffective (venting sound increases in volume and pitch, tank discolors, or shows any signs of deforming), withdraw immediately to a secure position. If employees are expected to fight fires, they must be trained and equipped in OSHA 1910.156.

Disposal Method Suggested: In accordance with 40CFR165 recommendations for the disposal of pesticides and pesticide containers. Must be disposed properly by following package label directions or by contacting your state pesticide or environmental control agency or by contacting your regional EPA office.

References

U.S. Environmental Protection Agency, "Chemical Profile: Diphacinone," Washington, DC, Chemical Emergency Preparedness Program (November 30, 1987)

Diphenamid

Molecular Formula: $C_{16}H_{17}NO$

Common Formula: $(CH_3)_2NCOCH(C_6H_5)_2$

Synonyms: 80W; Acetamide, *N,N*-dimethyl-2,2-diphenyl-; Benzeneacetamide, *N,N*-dimethyl-α-phenyl-; DIF 4; Difenamid (Spanish); *N,N*-Dimethyl-α,α-diphenylacetamide; *N,N*-Dimethyl-2,2-diphenylacetamide; *N,N*-Dimethyldiphenylacetamide; *N,N*-Dimethyl-α-phenylbenzeneacetamide; Dimid; Diphenamide; Diphenylamide; 2,2-Diphenyl-*N,N*-dimethylacetamide; Dymid; Enide; Enide 50W; FDN; Fenam; L 34314; Lilly 34314; Rideon; U 4513; Zarur

CAS Registry Number: 957-51-7

RTECS Number: AB8050000

Regulatory Authority

- Air Pollutant Standard Set (former USSR)[35]
- EPCRA Section 313 Form R *de minimis* concentration reporting level: 1.0%

Cited in U.S. State Regulations: California (G, W), New Jersey (G), Pennsylvania (G).

Description: Diphenamid, $(CH_3)_2NCOCH(C_6H_5)_2$, is a white crystalline solid in various forms. Freezing/Melting point = 135°C. Hazard Identification (based on NFPA-704 M Rating System): Health 1, Flammability 0, Reactivity 0. Very slightly soluble in water.

Potential Exposure: This material is used as a pre-emergent and selective herbicide for tomatoes, peanuts, alfalfa, soybeans, cotton and other crops.

Incompatibilities: Reacts with strong oxidants, strong acids and alkalies.

Permissible Exposure Limits in Air: The former USSR[35] has set a MAC in workplace air of 5.0 mg/m^3.

Permissible Concentration in Water: The former USSR[35] has set a MAC in water bodies used for domestic purposes of 1.2 mg/l. A lifetime health advisory of 0.2 mg/l has been calculated by EPA. California[61] has set a guideline for diphenamid in drinking water of 40 μg/l.

Harmful Effects and Symptoms

Short Term Exposure: The oral LD_{50} rat is 685 mg/kg (slightly toxic).

Long Term Exposure: A slight increase in liver weights was observed in long-term animal feeding studies. Mutation data reported.

First Aid: If this chemical gets into the eyes, remove any contact lenses at once and irrigate immediately for at least 15 minutes, occasionally lifting upper and lower lids. Seek medical attention immediately. If this chemical contacts the skin, remove contaminated clothing and wash immediately with soap and water. Seek medical attention immediately. If

this chemical has been inhaled, remove from exposure, begin rescue breathing (using universal precautions) if breathing has stopped and CPR if heart action has stopped. Transfer promptly to a medical facility. When this chemical has been swallowed, get medical attention. Give large quantities of water and induce vomiting. Do not make an unconscious person vomit.

Personal Protective Methods: Wear protective gloves and clothing to prevent any reasonable probability of skin contact. Safety equipment suppliers/manufacturers can provide recommendations on the most protective glove/clothing material for your operation.. All protective clothing (suits, gloves, footwear, headgear) should be clean, available each day, and put on before work. Contact lenses should not be worn when working with this chemical. Wear dust-proof chemical goggles and face shield unless full facepiece respiratory protection is worn. Employees should wash immediately with soap when skin is wet or contaminated. Provide emergency showers and eyewash.

Respirator Selection: Where the potential for exposure to diphenamid exists, use a MSHA/NIOSH approved supplied-air respirator with a full facepiece operated in the positive pressure mode or with a full facepiece, hood, or helmet in the continuous flow mode, or use a MSHA/NIOSH approved self-contained breathing apparatus with a full facepiece operated in pressure-demand or other positive pressure mode.

Storage: Prior to working with this chemical you should be trained on its proper handling and storage. Store in tightly closed containers in a cool, well ventilated area away from strong oxidizers, strong acids, strong alkalies.

Spill Handling: Evacuate persons not wearing protective equipment from area of spill or leak until clean-up is complete. Remove all ignition sources. Collect powdered material in the most convenient and safe manner and deposit in sealed containers. Ventilate area after clean-up is complete. It may be necessary to contain and dispose of this chemical as a hazardous waste. If material or contaminated runoff enters waterways, notify downstream users of potentially contaminated waters. Contact your Department of Environmental Protection or your regional office of the federal EPA for specific recommendations. If employees are required to clean-up spills, they must be properly trained and equipped. OSHA 1910.120(q) may be applicable.

Fire Extinguishing: This chemical is a combustible solid. Use dry chemical, carbon dioxide, water spray, or alcohol foam extinguishers. Diphenamid decomposes at 210°C forming toxic and corrosive gases including nitrogen oxides. If material or contaminated runoff enters waterways, notify downstream users of potentially contaminated waters. Notify local health and fire officials and pollution control agencies. From a secure, explosion-proof location, use water spray to cool exposed containers. If cooling streams are ineffective (venting sound increases in volume and pitch, tank discolors, or shows any signs of deforming), withdraw immediately to a secure position. If employees are expected to fight fires, they must be trained and equipped in OSHA 1910.156.

Disposal Method Suggested: Small amounts may be destroyed by alkaline hydrolysis. Admixture with alkali can be followed by soil burial.[22] Larger quantities can be disposed of by incineration in admixture with acetone or xylene and using effluent gas scrubbing.[22]

References

U.S. Environmental Protection Agency, "Chemical Profile: Diphenamid," Washington, DC, Office of Drinking Water (August 1987)

Diphenylamine

Molecular Formula: $C_{12}H_{11}N$

Common Formula: $(C_6H_5)_2NH$

Synonyms: Acetamide, 2-Biphenylyl-*N*-pyridyl-; AI3-00781; Aniline, *N*-phenyl-; Anilinobenzene; Benzenamine, *N*-phenyl-; Benzene, anilino-; Benzene, (phenylamino)-; Big Dipper; 2-Biphenylyl-*N*-pyridylacetamide; Caswell No. 398; C.I. 10355; Deccoscald 282; DFA; Difenilamina (Spanish); Diphenpyramide; *N,N*-Diphenylamine; DPA; EPA pesticide chemical code 038501; No scald DPA 283; NSC 215210; *N*-Phenylaniline; Phenylaniline; *N*-Phenylbenzenamine; *N*-Phenylbenzeneamine; Poly(diphenylamine); Pyridyl-biphenylyl-acetamide; Scaldip; Z-876

CAS Registry Number: 122-39-4

RTECS Number: JJ7800000

DOT ID: UN 2811

EEC Number: 612-026-00-5

Regulatory Authority

- Air Pollutant Standard Set (ACGIH)[1] (Australia) (HSE)[33] (Israel) (Mexico) (OSHA)[58] (Several States)[60] (Several Canadian Provinces)
- RCRA Hazardous Constituent Waste[5]

Cited in U.S. State Regulations: Alaska (G), California (A, G), Connecticut (A), Florida (G), Illinois (G), Kansas (G), Louisiana (G), Maine (G), Massachusetts (G, A), Minnesota (G), Nevada (A), New Hampshire (G), New Jersey (G), North Dakota (A), Oklahoma (G), Pennsylvania (G), Rhode Island (G, A), Vermont (G), Virginia (G, A), Washington (G), West Virginia (G), Wisconsin (G).

Description: DPA, $(C_6H_5)_2NH$, is a colorless to amber to brown crystalline solid or liquid with a pungent, floral odor. Contact with light causes a color change. Freezing/Melting point = 53°C. Boiling point = 302°C. Flash point = 153°C. Autoignition temperature = 634°C. Hazard Identification (based on NFPA-704 M Rating System): Health 3, Flammability 1, Reactivity 0. Insoluble in water.

Potential Exposure: DPA is used in antioxidants and in stabilizers for plastics including solid rocket propellants. It is used in the manufacture of pharmaceuticals, pesticides, explosives and dyes.

Incompatibilities: Dust may form an explosive mixture with air. Incompatible with strong acids, strong oxidizers, aldehydes, organic anhydrides, isocyanates, hexachloromelamine, trichloromelamine. Reacts with nitrogen oxides forming N-nitrosodiphenylamine and heat-, friction-, and shock-sensitive nitro products.

Permissible Exposure Limits in Air: There is no OSHA PEL.[58] The ACGIH and HSE[33] TWA value is 10 mg/m^3 and HSE[33] has set a STEL of 20 mg/m^3. Czechoslovakia[35] has set a TWA of 5 mg/m^3 with a ceiling value of 10 mg/m^3. Some states have set guidelines or standards for diphenylamine in ambient air[60] ranging from 0.68 $\mu g/m^3$ (Massachusetts) to 100.0 $\mu g/m^3$ (North Dakota) to 160.0 $\mu g/m^3$ (Virginia) to 200.0 $\mu g/m^3$ (Connecticut and Rhode Island) to 238.0 $\mu g/m^3$ (Nevada).

Determination in Air: Filter (special);[2] Methanol; High-pressure liquid chromatography/Ultraviolet; OSHA Method #78.

Permissible Concentration in Water: The former USSR[35][43] has set a MAC in water bodies used for domestic purposes of 0.05 mg/l.

Routes of Entry: Inhalation of dust, ingestion, skin absorption, eye and/or skin contact.

Harmful Effects and Symptoms

Short Term Exposure: DFA irritates the eyes, skin, and respiratory tract. It may affect the blood (resulting in formation of methemoglobin). Industrial poisoning has been encountered and was manifested clinically by bladder symptoms, tachycardia, hypertension and skin trouble. Overexposure can damage the liver, kidneys (polycystic kidneys) and may cause bladder symptoms and blood in the urine.

Long Term Exposure: DFA may affect the liver, kidneys and blood, forming methemoglobin. NIOSH warns that the carcinogen 4-Aminodiphenyl may be present as an impurity in the commercial product. In animals: teratogenic effects.

Points of Attack: Eyes, skin, respiratory system, cardiovascular system, blood, bladder, reproductive system.

Medical Surveillance: If symptoms develop or overexposure is suspected, the following may be useful: Evaluation of the kidneys and urinary system. Urine cytology (to look for abnormal cells). Liver function tests.

First Aid: If this chemical gets into the eyes, remove any contact lenses at once and irrigate immediately for at least 15 minutes, occasionally lifting upper and lower lids. Seek medical attention immediately. If this chemical contacts the skin, remove contaminated clothing and wash immediately with soap and water. Seek medical attention immediately. If this chemical has been inhaled, remove from exposure, begin rescue breathing (using universal precautions) if breathing has stopped and CPR if heart action has stopped. Transfer promptly to a medical facility. When this chemical has been swallowed, get medical attention. Give large quantities of water and induce vomiting. Do not make an unconscious person vomit. Medical observation is recommended.

Personal Protective Methods: Wear protective gloves and clothing to prevent any reasonable probability of skin contact. Safety equipment suppliers/manufacturers can provide recommendations on the most protective glove/clothing material for your operation.. All protective clothing (suits, gloves, footwear, headgear) should be clean, available each day, and put on before work. Contact lenses should not be worn when working with this chemical. Wear dust-proof chemical goggles and face shield unless full facepiece respiratory protection is worn. Employees should wash immediately with soap when skin is wet or contaminated. Provide emergency showers and eyewash.

Respirator Selection: Where the potential exists for exposures over 10 mg/m^3, use a MSHA/NIOSH approved full facepiece respirator with a high efficiency particulate filter. Greater protection is provided by a powered-air purifying respirator. Particulate filters must be checked every day before work for physical damage, such as rips or tears, and replaced as needed. If while wearing a filter, cartridge or canister respirator, you can smell, taste, or otherwise detect diphenylamine, or in the case of a full facepiece respirator you experience eye irritation, leave the area immediately. Check to make sure the respirator-to-face seal is still good. If it is, replace the filter, cartridge, or canister. If the seal is no longer good, you may need a new respirator.

Where the potential for high exposures exists, use a MSHA/NIOSH approved supplied-air respirator with a full facepiece operated in the positive pressure mode or with a full facepiece, hood, or helmet in the continuous flow mode, or use a MSHA/NIOSH approved self-contained breathing apparatus with a full facepiece operated in pressure-demand or other positive pressure mode.

Storage: Store in tightly closed containers in a cool, well-ventilated area away from light. Sources of ignition, such as smoking and open flames, are prohibited where diphenylamine is used, handled, or stored in a manner that could create a potential fire or explosion hazard.

Spill Handling: Evacuate persons not wearing protective equipment from area of spill or leak until clean-up is complete. Remove all ignition sources. Collect powdered material in the most convenient and safe manner and deposit in sealed containers. Ventilate area after clean-up is complete. It may be necessary to contain and dispose of this chemical as a hazardous waste. If material or contaminated runoff enters waterways, notify downstream users of potentially contaminated waters. Contact your Department of Environmental Protection or your regional office of the federal EPA for specific recommendations. If employees are required to clean-up spills, they must be properly trained and equipped. OSHA 1910.120(q) may be applicable.

Fire Extinguishing: This chemical is a combustible solid. Use dry chemical, carbon dioxide, water spray, or foam extinguishers. Poisonous gases are produced in fire. If material or contaminated runoff enters waterways, notify downstream users of potentially contaminated waters. Notify local health and fire officials and pollution control agencies. From a secure, explosion-proof location, use water spray to cool exposed containers. If cooling streams are ineffective (venting sound increases in volume and pitch, tank discolors, or shows any signs of deforming), withdraw immediately to a secure position. If employees are expected to fight fires, they must be trained and equipped in OSHA 1910.156.

Disposal Method Suggested: Burn in admixture with flammable solvent in furnace equipped with afterburner and scrubber.[22]

References

Sax N. I., Ed., "Dangerous Properties of Industrial Materials Report," 2, No. 5, 63–66 (1982)

New Jersey Department of Health, "Hazardous Substance Fact Sheet: Diphenylamine," Trenton, NJ (December 1996)

Diphenyl Dichlorosilane

Molecular Formula: $C_{12}H_{10}Cl_2Si$

Common Formula: $(C_6H_5)_2SiCl_2$

Synonyms: Dichlorodiphenylsilane; Diphenylsilicon dichloride

CAS Registry Number: 80-10-4

RTECS Number: VV3190000

DOT ID: UN 1769

Cited in U.S. State Regulations: California (G), Florida (G), Maine (G), Massachusetts (G), New Jersey (G), Oklahoma (G), Pennsylvania (G).

Description: Diphenyl dichlorosilane, $(C_6H_5)_2SiCl_2$, is a colorless liquid. It has a sharp, pungent HCl-like odor. Boiling point = 305°C. Flash point = 142°C.Hazard Identification (based on NFPA-704 M Rating System): Health 3, Flammability 1, Reactivity 0. Insoluble in water.

Potential Exposure: This material is used in the synthesis of silicone lubricants.

Incompatibilities: Water, steam, strong oxidizers.

Permissible Exposure Limits in Air: No standards set.

Permissible Concentration in Water: No criteria set.

Harmful Effects and Symptoms

Short Term Exposure: Vapors can irritate the eyes, nose, and throat. Higher levels could irritate the lungs and even lead to pulmonary edema, a medical emergency. This can cause death. This substance is a corrosive chemical and contact can cause severe skin burns and severe eye burns leading to permanent damage.

Long Term Exposure: Corrosive materials may cause lung effects, bronchitis with coughing and shortness of breath.

Points of Attack: Lungs, eyes, skin.

Medical Surveillance: For those with frequent or potentially high exposure the following are recommended before beginning work and at regular times after that: Lung function tests. If symptoms develop or overexposure is suspected, the following may be useful: Consider chest x-ray after acute overexposure.

First Aid: If this chemical gets into the eyes, remove any contact lenses at once and irrigate immediately for at least 15 minutes, occasionally lifting upper and lower lids. Seek medical attention immediately. If this chemical contacts the skin, remove contaminated clothing and wash immediately with soap and water. Seek medical attention immediately. If this chemical has been inhaled, remove from exposure, begin rescue breathing (using universal precautions) if breathing has stopped and CPR if heart action has stopped. Transfer promptly to a medical facility. When this chemical has been swallowed, get medical attention. If victim is conscious, administer water or milk. Do not induce vomiting. Medical observation is recommended for 24 – 48 hours after breathing overexposure, as pulmonary edema may be delayed. As first aid for pulmonary edema, a doctor or authorized paramedic may consider administering a corticosteroid spray.

Personal Protective Methods: Wear protective gloves and clothing to prevent any reasonable probability of skin contact. Safety equipment suppliers/manufacturers can provide recommendations on the most protective glove/clothing material for your operation.. All protective clothing (suits, gloves, footwear, headgear) should be clean, available each day, and put on before work. Contact lenses should not be worn when working with this chemical. Wear splash-proof chemical goggles and face shield unless full facepiece respiratory protection is worn. Employees should wash immediately with soap when skin is wet or contaminated. Provide emergency showers and eyewash.

Respirator Selection: Where the potential for exposure to diphenyl dichlorosilane exists, use a MSHA/NIOSH approved supplied-air respirator with a full facepiece operated in the positive pressure mode or with a full facepiece, hood, or helmet in the continuous flow mode, or use a MSHA/NIOSH approved self-contained breathing apparatus with a full facepiece operated in pressure-demand or other positive pressure mode.

Storage: Diphenyl dichlorosilane must be stored to avoid contact with oxidizers (such as perchlorates, peroxides, permanganates, chlorates, and nitrates), bases, alcohols and acids since violent reactions occur. Store in tightly closed containers in a cool, well-ventilated area away from water, steam and moisture because toxic and corrosive chloride gases including hydrogen chloride can be produced. Sources of ignition such as smoking and open flames are prohibited where diphenyl dichlorosilane is used, handled, or stored in a manner that could create a potential fire or explosion hazard.

Shipping: This compound requires a shipping label of: "Corrosive." It falls in DOT Hazard Class 8 and Packing Group II.

Spill Handling: Evacuate and restrict persons not wearing protective equipment from area of spill or leak until cleanup is complete. Remove all ignition sources. Ventilate area of spill or leak. Absorb liquids in vermiculite, dry sand, earth, peat, carbon, or a similar material and deposit in sealed containers. Keep this chemical out of a confined space, such as a sewer, because of the possibility of an explosion, unless the sewer is designed to prevent the build-up of explosive concentrations. It may be necessary to contain and dispose of this chemical as a hazardous waste. If material or contaminated runoff enters waterways, notify downstream users of potentially contaminated waters. Contact your Department of Environmental Protection or your regional office of the federal EPA for specific recommendations. If employees are required to clean-up spills, they must be properly trained and equipped. OSHA 1910.120(q) may be applicable.

Fire Extinguishing: Diphenyl dichlorosilane is a combustible liquid. Diphenyl dichlorosilane may burn, but does not readily ignite. Poisonous gases are produced in fire, including hydrogen chloride and phosgene. Fire may restart after it has been extinguished. Using water or foam extinguishers may cause frothing. Water spray may be useful. Use dry chemical or CO_2 as extinguishing agents. Vapors are heavier than air and will collect in low areas. Containers may explode in fire. Storage containers and parts of containers may rocket great distances, in many directions. If material or contaminated runoff enters waterways, notify downstream users of potentially contaminated waters. Notify local health and fire officials and pollution control agencies. From a secure, explosion-proof location, use water spray to cool exposed containers. If cooling streams are ineffective (venting sound increases in volume and pitch, tank discolors, or shows any signs of deforming), withdraw immediately to a secure position. If employees are expected to fight fires, they must be trained and equipped in OSHA 1910.156.

References

New Jersey Department of Health, "Hazardous Substance Fact Sheet: Diphenyl Dichlorosilane," Trenton, NJ (April 1986)

1,2-Diphenylhydrazine

Molecular Formula: $C_{12}H_{12}N_2$

Common Formula: $C_6H_5NHNHC_6H_5$

Synonyms: Benzene, 1,1'-hydrazobis-; *N,N'*-Bianiline; *N,N'*-Difenilhidracina (Spanish); 1,2-Difenilhidracina (Spanish); *N,N'*-Diphenylhydrazine; *sym*-Diphenylhydrazine; DPH; Hydrazobenzen (Czech); Hydrazobenzene; Hydrazodibenzene; NCI-CO1854

CAS Registry Number: 122-66-7

RTECS Number: MW2625000

DOT ID: UN 3077

EEC Number: 007-021-00-4

Regulatory Authority

- Carcinogen (Animal Positive) (NCI)[9]
- Clean Air Act: Hazardous Air Pollutants (Title I, Part A, Section 112)
- Clean Water Act: 40CFR423, Appendix A, Priority Pollutants; Section 313 Water Priority Chemicals (57FR41331, 9/9/92); 40CFR401.15 Section 307 Toxic Pollutants, as diphenylhydrazine
- EPA Hazardous Waste Number (RCRA No.): U109
- RCRA, 40CFR261, Appendix 8 Hazardous Constituents
- RCRA 40CFR268.48; 61FR15654, Universal Treatment Standards: Wastewater (mg/l), 0.087; Nonwastewater (mg/kg), N/A
- Safe Drinking Water Act: Priority List (55 FR 1470)
- Superfund/EPCRA 40CFR302.4 Reportable Quantity (RQ): CERCLA, 10 lb (4.54 kg)
- EPCRA Section 313 Form R *de minimis* concentration reporting level: 0.1%

Cited in U.S. State Regulations: California (G), Florida (G), Illinois (G), Kansas (G, W), Louisiana (G), Maine (G), Maryland (G), Massachusetts (G), Michigan (G), Minnesota (W), Pennsylvania (G), Vermont (G), Virginia (G), Washington (G), West Virginia (G), Wisconsin (G).

Description: DPH, $C_6H_5NHNHC_6H_5$, is a white to yellow crystalline compound. Freezing/Melting point = 131°C. Slightly soluble in water.

Potential Exposure: 1,2-diphenylhydrazine (DPH) is a precursor in the manufacture of benzidine, an intermediate in the production of dyes. 1,2-diphenylhydrazine is used in the synthesis of phenylbutazone, a potent anti-inflammatory (antiarthritic) drug. Manufacturers of dyes and pharmaceuticals are subject to occupational exposure. Groups working in the laboratory and forensic medicine may also be subject to 1,2-diphenylhydrazine exposure.

Incompatibilities: Not compatible with oxidizers, strong acids, acid chlorides, acid anhydrides, and mineral acids forming benzidine. Store under nitrogen.

Permissible Exposure Limits in Air: No limits set.

Permissible Concentration in Water: To protect freshwater aquatic life – 270 μg/l on an acute toxicity basis. To protect saltwater aquatic life – no criterion developed due to insufficient data. To protect human health – preferably zero. An additional lifetime cancer risk of 1 in 100,000 is presented by a concentration of 0.4 μg/l.[6] Two states have set guidelines for 1,2-diphenylhydrazine in drinking water both at the level of 0.45 μg/l.

Determination in Water: Gas chromatography plus mass spectrometry (EPA Method 625).

Routes of Entry: Inhalation, passing through the skin.

Harmful Effects and Symptoms

Short Term Exposure: Irritates the eyes, skin, and respiratory tract causing coughing and wheezing.

Long Term Exposure: This substance is possibly carcinogenic to humans. It has been shown to cause liver and breast cancer in animals. Diphenylhydrazine is a suspected carcinogen in humans because of its structural relationship to benzidine, which is an established human bladder carcinogen. Recent studies in rats and mice have shown that diphenylhydrazine produces both benign and malignant tumors when administered subcutaneously. Carcinogenicity in both rats of both sexes and female mice has been established. In view of the relative paucity of data on the mutagenicity, teratogenicity and long-term oral toxicity of diphenylhydrazine, estimates of the effects of chronic oral exposure at low levels cannot be made with any confidence according to NAS/NRC.[46] May cause skin allergy with itching and skin rash. Repeated exposure may cause anemia and may damage the liver.

Points of Attack: Liver, blood, skin.

Medical Surveillance: Liver function tests. Evaluation by a qualified allergist. Complete Blood count (CBC).

First Aid: *Skin Contact:*[52] Flood all areas of body that have contacted the substance with water. Don't wait to remove contaminated clothing; do it under the water stream. Use soap to help assure removal. Isolate contaminated clothing when removed to prevent contact by others.

Eye Contact: Remove any contact lenses at once. Flush eyes well with copious quantities of water or normal saline for at least 20 – 30 minutes. Seek medical attention.

Inhalation: Leave contaminated area immediately; breathe fresh air. Proper respiratory protection must be supplied to any rescuers. If coughing, difficult breathing or any other symptoms develop, seek medical attention at once, even if symptoms develop many hours after exposure.

Ingestion: If convulsions are not present, give a glass or two of water or milk to dilute the substance. Assure that the person's airway is unobstructed and contact a hospital or poison center immediately for advice on whether or not to induce vomiting.

Personal Protective Methods: Wear protective gloves and clothing to prevent any reasonable probability of skin contact. Safety equipment suppliers/manufacturers can provide recommendations on the most protective glove/clothing material for your operation.. All protective clothing (suits, gloves, footwear, headgear) should be clean, available each day, and put on before work. Contact lenses should not be worn when working with this chemical. Wear dust-proof chemical goggles and face shield unless full facepiece respiratory protection is worn. Employees should wash immediately with soap when skin is wet or contaminated. Provide emergency showers and eyewash.

Respirator Selection: Where the potential exists for exposure to 1,2-diphenylhydrazine, use a MSHA/NIOSH approved supplied-air respirator with a full facepiece operated in the positive pressure mode or with a full facepiece, hood, or helmet in the continuous flow mode, or use a MSHA/NIOSH approved self-contained breathing apparatus with a full facepiece operated in pressure-demand or other positive pressure mode.

Storage: Prior to working with this chemical you should be trained on its proper handling and storage. Store under nitrogen away from oxidizers, strong acids, acid chlorides, acid anhydrides. A regulated, marked area should be established where this chemical is handled, used, or stored in compliance with OSHA standard 1910.1045.

Shipping: Environmentally hazardous solid, n.o.s. Hazard Class: 9. Label: "Class 9." Packing Group: III.

Spill Handling: Evacuate persons not wearing protective equipment from area of spill or leak until clean-up is complete. Remove all ignition sources. Dampen spilled material with alcohol to avoid dust. Collect powdered material in the most convenient and safe manner and deposit in sealed containers. Ventilate area after clean-up is complete. It may be necessary to contain and dispose of this chemical as a hazardous waste. If material or contaminated runoff enters waterways, notify downstream users of potentially contaminated waters. Contact your Department of Environmental Protection or your regional office of the federal EPA for specific recommendations. If employees are required to clean-up spills, they must be properly trained and equipped. OSHA 1910.120(q) may be applicable.

Fire Extinguishing: Use dry chemical, carbon dioxide, water spray, or alcohol or polymer foam extinguishers. Poisonous gases are produced in fire including nitrogen oxides and carbon monoxide. If material or contaminated runoff enters waterways, notify downstream users of potentially contaminated waters. Notify local health and fire officials and pollution control agencies. From a secure, explosion-proof location, use water spray to cool exposed containers. If cooling streams are ineffective (venting sound increases in volume and pitch, tank discolors, or shows any signs of deforming), withdraw immediately to a secure position. If employees are expected to fight fires, they must be trained and equipped in OSHA 1910.156.

Disposal Method Suggested: Controlled incineration whereby oxides of nitrogen are removed from the effluent gas by scrubber, catalytic or thermal device.[22] Consult with environmental regulatory agencies for guidance on acceptable disposal practices. Generators of waste containing this contaminant (≥100 kg/mo) must conform with EPA regulations governing storage, transportation, treatment, and waste disposal.

References

U.S. Environmental Protection Agency, Diphenylhydrazine: Ambient Water Quality Criteria, Washington, DC (1980)

U.S. Environmental Protection Agency, 1,2-Diphenylhydrazine, Health and Environmental Effects Profile No. 96, Washington, DC, Office of Solid Waste (April 30, 1980)

Sax N. I., Ed., "Dangerous Properties of Industrial Materials Report," 2, No. 5, 68–70 (1982) and 3, No. 2, 45–46 (1983) and 6, No. 1, 61–68 (1986)

Diphenyl Oxide

Molecular Formula: $C_{12}H_{10}O$

Common Formula: $(C_6H_5)_2O$

Synonyms: Diphenyl ether; Dowtherm A (diphenyl/diphenyl oxide mixture); 1,1'-Oxybisbenzene; Oxydiphenyl; Phenoxy benzene; Phenoxybenzene; Phenyl ether; Phenyl oxide

CAS Registry Number: 101-84-8

RTECS Number: KN8970000

DOT ID

NA 9188

Regulatory Authority

- Air Pollutant Standard Set (ACGIH)[1] (Australia) (DFG)[3] (former USSR)[35][43] (Israel) (Mexico) (Several States)[60] (Several Canadian Provinces)
- Canada, WHMIS, Ingredients Disclosure List

Cited in U.S. State Regulations: Connecticut (A), Massachusetts (G), Nevada (A), New Hampshire (G), North Dakota (A), Pennsylvania (G), Virginia (A).

Description: Diphenyl Oxide, $(C_6H_5)_2O$, is a colorless solid or liquid with a highly disagreeable odor. The odor threshold is 0.1 ppm. Boiling point = 258°C. Freezing/Melting point = 27°C. Flash point = 115°C. Explosive limits: LEL = 0.7%; UEL = 6.0%. Insoluble in water. It is often used commercially in admixture with diphenyl also known as biphenyl. See also Biphenyl record.

Potential Exposure: This material is used as a heat transfer medium; it is used in perfuming soaps and in organic synthesis.

Incompatibilities: Strong oxidizers may cause fire and explosions. Attacks some plastics, rubber and coatings.

Permissible Exposure Limits in Air: The Federal Limit (OSHA PEL),[58] the DFG MAK value[3] and the ACGIH TWA value is 1.0 ppm (7 mg/m^3). The ACGIH STEL is 2 ppm (14 mg/m^3). The NIOSH IDLH is 100 ppm. The former USSR[35][43] has set a MAC in workplace air of 5.0 mg/m^3 and also[35] has set a MAC in ambient air of residential areas of 0.03 mg/m^3 on a once-daily basis. In addition, several states have set guidelines or standards for diphenyl oxide in ambient air ranging from 70 – 140 μg/m^3 (North Dakota) to 115 μg/m^3 (Virginia) to 140 μg/m^3 (Connecticut) to 168 μg/m^3 (Nevada).

Determination in Air: Charcoal adsorption, CS_2, workup, Gas chromatography/Flame ionization detection; NIOSH IV Method #1617.

Permissible Concentration in Water: No criteria set.

Routes of Entry: Inhalation, eye and skin contact, ingestion.

Harmful Effects and Symptoms

Short Term Exposure: Inhalation may cause irritation of eyes, skin, and respiratory system. The odor can cause nausea.

Long Term Exposure: May cause dermatitis.

Points of Attack: Eyes, skin, respiratory system.

Medical Surveillance: Consider the points of attack in preplacement and periodic physical examinations.

First Aid: If this chemical gets into the eyes, remove any contact lenses at once and irrigate immediately for at least 15 minutes, occasionally lifting upper and lower lids. Seek medical attention immediately. If this chemical contacts the skin, remove contaminated clothing and wash immediately with soap and water. Seek medical attention immediately. If this chemical has been inhaled, remove from exposure, begin rescue breathing (using universal precautions) if breathing has stopped and CPR if heart action has stopped. Transfer promptly to a medical facility. When this chemical has been swallowed, get medical attention. Give large quantities of water and induce vomiting. Do not make an unconscious person vomit.

Personal Protective Methods: Wear protective gloves and clothing to prevent any reasonable probability of skin contact. Safety equipment suppliers/manufacturers can provide recommendations on the most protective glove/clothing material for your operation.. All protective clothing (suits, gloves, footwear, headgear) should be clean, available each day, and put on before work. Contact lenses should not be worn when working with this chemical. Wear splash-proof goggles and face shield if working with the liquid or dust-proof chemical goggles and face shield if working with the crystalline material, unless full facepiece respiratory protection is worn. Employees should wash immediately with soap when skin is wet or contaminated. Provide emergency showers and eyewash.

Respirator Selection: NIOSH/OSHA: *25 ppm:* SA:CF (any supplied-air respirator operated in a continuous-flow mode); PAPROVDM [any powered, air-purifying respirator with organic vapor cartridge(s) in combination with a dust and mist filter]. *50 ppm:* CCRFOVHiE [any chemical cartridge respirator with a full facepiece and organic vapor cartridge(s) in combination with a high efficiency particulate filter]; or GMFOVHiE [any air-purifying, full-facepiece respirator (gas mask) with a chin-style, front- or back-mounted organic vapor canister having a high-efficiency particulate filter]; or SCBAF (any self-contained breathing apparatus with a full facepiece); or SAF (any supplied-air respirator with a full facepiece. *100 ppm:* SAF:PD,PP (any supplied-air respirator that has a full facepiece and is operated in a pressure-demand or other positive-pressure mode). *Emergency or planned entry into* *unknown concentrations or IDLH conditions*: SCBAF: PD,PP (any self-contained breathing apparatus that has a full facepiece and is operated in a pressure-demand or other positive-pressure mode); or SAF:PD,PP:ASCBA (any supplied-air respirator that has a full facepiece and is operated in a pressure-demand or other positive-pressure mode in combination with an auxiliary self-contained breathing apparatus operated in a pressure-demand or other positive pressure mode). *Escape:* GMFOVHiE [any air-purifying, full-facepiece respirator (gas mask) with a chin-style, front- or back-mounted organic vapor canister having a high-efficiency particulate filter]; or SCBAE (any appropriate escape-type, self-contained breathing apparatus).

Note: Substance causes eye irritation or damage; eye protection needed.

Storage: Prior to working with diphenyl oxide you should be trained on its proper handling and storage. Before entering confined space where this chemical may be present, check to make sure that an explosive concentration does not exist. Store in tightly closed containers in a cool, well ventilated area away from oxidizers. Store in tightly closed containers in a cool, well-ventilated area away from heat. Sources of ignition such as smoking and open flames are prohibited where this chemical is used, handled, or stored in a manner that could create a potential fire or explosion hazard. Metal containers involving the transfer of 5 gallons or more of this chemical should be grounded and bonded. Drums must be equipped with self-closing valves, pressure vacuum bungs, and flame arresters. Use only nonsparking tools and equipment, especially when opening and closing containers of this chemical.

Shipping: Diphenyl oxide is not specifically cited by DOT[19] but may be classed as a hazardous substance, solid, n.o.s. The Hazard Class is ORM-E, the Packing Group is III.

Spill Handling: *Dry material:* Evacuate persons not wearing protective equipment from area of spill or leak until clean-up is complete. Remove all ignition sources. Dampen dry material with alcohol to avoid any dust. Collect powdered material in the most convenient and safe manner and deposit in sealed containers. Ventilate area after clean-up is complete. It may be necessary to contain and dispose of this chemical as a hazardous waste. If material or contaminated runoff enters waterways, notify downstream users of potentially contaminated waters. Contact your Department of Environmental Protection or your regional office of the federal EPA for specific recommendations. If employees are required to clean-up spills, they must be properly trained and equipped. OSHA 1910.120(q) may be applicable.

Liquid: Evacuate and restrict persons not wearing protective equipment from area of spill or leak until cleanup is complete. Remove all ignition sources. Establish forced ventilation to keep levels below explosive limit. Absorb liquids in vermiculite, dry sand, earth, peat, carbon, or a similar material and deposit in sealed containers. Keep this chemical out of a confined space, such as a sewer, because of the possibility of an explosion, unless the sewer is designed to prevent the build-up of explosive concentrations. It may be necessary to contain and dispose of this chemical as a hazardous waste. If material or contaminated runoff enters waterways, notify downstream users of potentially contaminated waters. Contact your Department of Environmental Protection or your regional office of the federal EPA for specific recommendations. If employees are required to clean-up spills, they must be properly trained and equipped. OSHA 1910.120(q) may be applicable.

Fire Extinguishing: *Dry material:* This chemical is a combustible solid. Use dry chemical, carbon dioxide, water spray, or alcohol foam extinguishers. Poisonous gases are produced in fire. If material or contaminated runoff enters waterways, notify downstream users of potentially contaminated waters. Notify local health and fire officials and pollution control agencies. From a secure, explosion-proof location, use water spray to cool exposed containers. If cooling streams are ineffective (venting sound increases in volume and pitch, tank discolors, or shows any signs of deforming), withdraw immediately to a secure position. If employees are expected to fight fires, they must be trained and equipped in OSHA 1910.156.

Liquid: This chemical is a combustible liquid. Poisonous gases are produced in fire. Use dry chemical, carbon dioxide, or alcohol foam extinguishers. Vapors are heavier than air and will collect in low areas. Vapors may travel long distances to ignition sources and flashback. Vapors in confined areas may explode when exposed to fire. Containers may explode in fire. Storage containers and parts of containers may rocket great distances, in many directions. If material or contaminated runoff enters waterways, notify downstream users of potentially contaminated waters. Notify local health and fire officials and pollution control agencies. From a secure, explosion-proof location, use water spray to cool exposed containers. If cooling streams are ineffective (venting sound increases in volume and pitch, tank discolors, or shows any signs of deforming), withdraw immediately to a secure position. If employees are expected to fight fires, they must be trained and equipped in OSHA 1910.156.

Disposal Method Suggested: Incineration.[22]

Dipropylamine

Molecular Formula: $C_6H_{15}N$

Synonyms: Di-*N*-propilamina (Spanish); *N*-Dipropylamine; Di-*N*-propylamine; 1-Propylamine, *N*-propyl; *N*-Propyl-1-propanamine

CAS Registry Number: 142-84-7

RTECS Number: JL9200000

DOT ID: UN 2383

Regulatory Authority

- EPA Hazardous Waste Number (RCRA No.): U110
- RCRA, 40CFR261, Appendix 8 Hazardous Constituents
- Superfund/EPCRA 40CFR302.4 Reportable Quantity (RQ): CERCLA, 5,000 lb (2,270 kg)

Description: Dipropylamine is a colorless, water-white liquid with a strong ammonia-like odor.

Potential Exposure: Used as a chemical intermediate and in the manufacture of herbicides. Boiling point = 110°C. Freezing/Melting point = -63°C. Flash point = 17°C. Autoignition temperature = 299°C. Hazard Identification (based on NFPA-704 M Rating System): Health 3, Flammability 3, Reactivity 0. Insoluble in water.

Incompatibilities: Forms explosive mixture with air. Incompatible with acids, organic anhydrides, isocyanates, vinyl acetate, acrylates, substituted allyls, alkylene oxides, epichlorohydrin, ketones, aldehydes, alcohols, glycols, mercury, phenols, cresols, caprolactum solution, strong oxidizers. Attacks aluminum, copper, lead, tin, zinc and their alloys.

Permissible Exposure Limits in Air: No OELs established.

Routes of Entry: Inhalation.

Harmful Effects and Symptoms

Short Term Exposure: Contact with the eyes and skin can cause severe irritation and burns. Inhaling the aerosol can irritate the respiratory tract and cause coughing, wheezing and/or shortness of breath; headache, nausea, dizziness, anxiety, unconsciousness. Higher exposures can cause pulmonary edema, a medical emergency that can be delayed for several hours. This can cause death.

Long Term Exposure: Repeated exposure may cause bronchitis with cough, phlegm and/or shortness of breath.

Points of Attack: Lungs.

Medical Surveillance: Lung function tests. Consider chest x-ray following acute overexposure.

First Aid: If this chemical gets into the eyes, remove any contact lenses at once and irrigate immediately for at least 15 minutes, occasionally lifting upper and lower lids. Seek medical attention immediately. If this chemical contacts the skin, remove contaminated clothing and wash immediately with soap and water. Seek medical attention immediately. If this chemical has been inhaled, remove from exposure, begin rescue breathing (using universal precautions) if breathing has stopped and CPR if heart action has stopped. Transfer promptly to a medical facility. When this chemical has been swallowed, get medical attention. If victim is conscious, administer water or milk. Do not induce vomiting. Medical observation is recommended for 24 – 48 hours after breathing overexposure, as pulmonary edema may be delayed. As first aid for pulmonary edema, a doctor or authorized paramedic may consider administering a corticosteroid spray.

Personal Protective Methods: Wear protective gloves and clothing to prevent any reasonable probability of skin contact. Safety equipment suppliers/manufacturers can provide recommendations on the most protective glove/clothing material for your operation. Teflon and Polycarbonate are among the recommended protective materials. All protective clothing (suits, gloves, footwear, headgear) should be clean, available each day, and put on before work. Contact lenses should not be worn when working with this chemical. Wear splash-proof chemical goggles and face shield unless full facepiece respiratory protection is worn. Employees should wash immediately with soap when skin is wet or contaminated. Provide emergency showers and eyewash.

Respirator Selection: Where the potential for exposure to this chemical, use a MSHA/NIOSH approved supplied-air respirator with a full facepiece operated in the positive pressure mode or with a full facepiece, hood, or helmet in the continuous flow mode, or use a MSHA/NIOSH approved self-contained breathing apparatus with a full facepiece operated in pressure-demand or other positive pressure mode.

Storage: Prior to working with dipropylamine you should be trained on its proper handling and storage. Before entering confined space where this chemical may be present, check to make sure that an explosive concentration does not exist. Store in tightly closed containers in a cool, well ventilated area. Metal containers involving the transfer of this chemical should be grounded and bonded. Where possible, automatically pump liquid from drums or other storage containers to process containers. Drums must be equipped with self-closing valves, pressure vacuum bungs, and flame arresters. Use only non-sparking tools and equipment, especially when opening and closing containers of this chemical. Sources of ignition such as smoking and open flames, are prohibited where this chemical is used, handled, or stored in a manner that could create a potential fire or explosion hazard. Wherever this chemical is used, handled, manufactured, or stored, use explosion-proof electrical equipment and fittings.

Shipping: Label "Flammable Liquid, Corrosive." Hazard Class 3. Packing Group II.

Spill Handling: Evacuate and restrict persons not wearing protective equipment from area of spill or leak until cleanup is complete. Remove all ignition sources. Ventilate area of spill or leak. Absorb liquids in vermiculite, dry sand, earth, peat, carbon, or a similar material and deposit in sealed containers. Keep this chemical out of a confined space, such as a sewer, because of the possibility of an explosion, unless the sewer is designed to prevent the build-up of explosive concentrations. It may be necessary to contain and dispose of this chemical as a hazardous waste. If material or contaminated runoff enters waterways, notify downstream users of potentially contaminated waters. Contact your Department of Environmental Protection or your regional office of the federal EPA for specific recommendations. If employees are required to clean-up spills, they must be properly trained and equipped. OSHA 1910.120(q) may be applicable.

Fire Extinguishing: This chemical is a flammable liquid. Poisonous gases including carbon monoxide and nitrogen oxide are produced in fire. Use dry chemical, carbon dioxide, or alcohol or polymer foam extinguishers. Vapors are heavier than air and will collect in low areas. Vapors may travel long distances to ignition sources and flashback. Vapors in confined areas may explode when exposed to fire. Containers may explode in fire. Storage containers and parts of containers may rocket great distances, in many directions. If material or contaminated runoff enters waterways, notify downstream users of potentially contaminated waters. Notify local health and fire officials and pollution control agencies. From a secure, explosion-proof location, use water spray to cool exposed containers. If cooling streams are ineffective (venting sound increases in volume and pitch, tank discolors, or shows any signs of deforming), withdraw immediately to a secure position. If employees are expected to fight fires, they must be trained and equipped in OSHA 1910.156.

Disposal Method Suggested: Consult with environmental regulatory agencies for guidance on acceptable disposal practices. Generators of waste containing this contaminant (≥100 kg/mo) must conform with EPA regulations governing storage, transportation, treatment, and waste disposal.

References

New Jersey Department of Health and Senior Services, "Hazardous Substance Fact Sheet, Dipropylamine." Trenton, NJ (March 1999)

Dipropylene Glycol Methyl Ether

Molecular Formula: $C_7H_{16}O_3$

Common Formula: $CH_3OCH_2CH(CH_3)OCH_2CH(CH_3)OH$

Synonyms: Arcosolv; 1,4-Dimethyl-3,6-dioxa-1heptanol; Dipropylene glycol monomethyl ether; Dowanol 50B®; Dowanol DPM®; DPGME; 2-Methoxymethylethoxypropanol

CAS Registry Number: 34590-94-8

RTECS Number: JV1575000

Regulatory Authority

- Air Pollutant Standard Set (ACGIH)[1] (Australia) (DFG)[3] (Israel) (Mexico) (OSHA)[58] (Several States)[60] (Several Canadian Provinces)
- Canada, WHMIS, Ingredients Disclosure List

Cited in U.S. State Regulations: Alaska (G), California (A, G), Connecticut (A), Illinois (G), Maine (G), Massachusetts (G), Nevada (A), New Hampshire (G), New Jersey (G), North Dakota (A), Pennsylvania (G), Rhode Island (G), Virginia (A), West Virginia (G).

Description: DPGME, $CH_3OCH_2CH(CH_3)$ OCH_2CH (CH_3) OH is a colorless liquid with a weak odor. Boiling point = 190°C. Freezing/Melting point = -80°C. Flash point = 85°C. Autoignition temperature = 270°C. Explosive limits: LEL = 1.3%; UEL = 10.4. Hazard Identification (based on NFPA-704 M Rating System): Health 0, Flammability 2, Reactivity 0. Highly soluble in water.

Potential Exposure: This material is used as a solvent for nitrocellulose and other synthetic resins.

Incompatibilities: Strong oxidizers. The substance can presumably form explosive peroxides in contact with air. Reacts violently with strong oxidants.

Permissible Exposure Limits in Air: The Federal Limit (OSHA PEL)[58] and the ACGIH recommended value is 100 ppm (600 mg/m^3). The ACGIH STEL value is 150 ppm (900 mg/m^3). The NIOSH IDLH = 600 ppm. The DFG MAK[3] is 50 ppm (300 mg/m^3). Several states have set guidelines or standards for ambient air ranging from 6.0-9.0 mg/m^3 (North Dakota) to 10 mg/m^3 (Virginia) to 12.0 mg/m^3 (Connecticut) to 14.3 mg/m^3 (Nevada).

Determination in Air: Charcoal adsorption, workup with CS_2, analysis by gas chromatography. See NIOSH Method #S-69.

Permissible Concentration in Water: No criteria set.

Routes of Entry: Inhalation, skin absorption, ingestion, skin and/or eye contact.

Harmful Effects and Symptoms

Short Term Exposure: DPGME can be absorbed through the skin, thereby increasing exposure. Exposure causes irritation of eyes and nose. Extremely high levels can cause dizziness, lightheadedness, headaches, unconsciousness.

Long Term Exposure: Repeated exposure to very high levels may affect the liver.

Points of Attack: Eyes, respiratory system, central nervous system.

Medical Surveillance: If symptom develop or overexposure is suspected, the following may be useful: Liver function tests.

First Aid: If this chemical gets into the eyes, remove any contact lenses at once and irrigate immediately for at least 15 minutes, occasionally lifting upper and lower lids. Seek medical attention immediately. If this chemical contacts the skin, remove contaminated clothing and wash immediately with soap and water. Seek medical attention immediately. If this chemical has been inhaled, remove from exposure, begin rescue breathing (using universal precautions) if breathing has stopped and CPR if heart action has stopped. Transfer promptly to a medical facility. When this chemical has been swallowed, get medical attention. Give large quantities of water and induce vomiting. Do not make an unconscious person vomit.

Personal Protective Methods: Wear protective gloves and clothing to prevent any reasonable probability of skin contact. Safety equipment suppliers/manufacturers can provide

recommendations on the most protective glove/clothing material for your operation.. All protective clothing (suits, gloves, footwear, headgear) should be clean, available each day, and put on before work. Contact lenses should not be worn when working with this chemical. Wear splash-proof chemical goggles and face shield unless full facepiece respiratory protection is worn. Employees should wash immediately with soap when skin is wet or contaminated. Provide emergency showers and eyewash.

Respirator Selection: NIOSH/OSHA: *Up to 600 ppm:* SA (any supplied-air respirator); or SCBAF (any self-contained breathing apparatus with a full facepiece). *Emergency or planned entry into unknown concentrations or IDLH conditions:* SCBAF:PD,PP (any self-contained breathing apparatus that has a full facepiece and is operated in a pressure-demand or other positive-pressure mode); or SAF:PD,PP:ASCBA (any supplied-air respirator that has a full facepiece and is operated in a pressure-demand or other positive-pressure mode in combination with an auxiliary self-contained breathing apparatus operated in a pressure-demand or other positive pressure mode). *Escape:* GMFOVHiE [any air-purifying, full-facepiece respirator (gas mask) with a chin-style, front- or back-mounted organic vapor canister having a high-efficiency particulate filter]; or SCBAE (any appropriate escape-type, self-contained breathing apparatus).

Storage: Store in tightly closed containers in a cool, well-ventilated area away from strong oxidizers (such as chlorine, bromine and fluorine). Before entering confined space where this chemical may be present, check to make sure that an explosive concentration does not exist. Store in tightly closed containers in a cool, well-ventilated area away from heat and incompatible materials. Sources of ignition such as smoking and open flames are prohibited where this chemical is used, handled, or stored in a manner that could create a potential fire or explosion hazard. Metal containers involving the transfer of 5 gallons or more of this chemical should be grounded and bonded. Drums must be equipped with self-closing valves, pressure vacuum bungs, and flame arresters. Use only nonsparking tools and equipment, especially when opening and closing containers of this chemical.

Shipping: Requires a label of "Flammable Liquid." It fall in UN/DOT Hazard Class 3.

Spill Handling: Evacuate and restrict persons not wearing protective equipment from area of spill or leak until cleanup is complete. Remove all ignition sources. Establish forced ventilation to keep levels below explosive limit. Absorb liquids in vermiculite, dry sand, earth, peat, carbon, or a similar material and deposit in sealed containers. Keep this chemical out of a confined space, such as a sewer, because of the possibility of an explosion, unless the sewer is designed to prevent the build-up of explosive concentrations. It may be necessary to contain and dispose of this chemical as a hazardous waste. If material or contaminated runoff enters waterways, notify downstream users of potentially contaminated waters. Contact your Department of Environmental Protection or your regional office of the federal EPA for specific recommendations. If employees are required to clean-up spills, they must be properly trained and equipped. OSHA 1910.120(q) may be applicable.

Fire Extinguishing: This chemical is a combustible liquid. Poisonous gases are produced in fire including carbon monoxide. Use dry chemical, carbon dioxide, or foam extinguishers. Vapors are heavier than air and will collect in low areas. Vapors may travel long distances to ignition sources and flashback. Vapors in confined areas may explode when exposed to fire. Containers may explode in fire. Storage containers and parts of containers may rocket great distances, in many directions. If material or contaminated runoff enters waterways, notify downstream users of potentially contaminated waters. Notify local health and fire officials and pollution control agencies. From a secure, explosion-proof location, use water spray to cool exposed containers. If cooling streams are ineffective (venting sound increases in volume and pitch, tank discolors, or shows any signs of deforming), withdraw immediately to a secure position. If employees are expected to fight fires, they must be trained and equipped in OSHA 1910.156.

Disposal Method Suggested: Concentrated waste containing no peroxides-discharge liquid at a controlled rate near a pilot flame. Concentrated waste containing peroxides-perforation of a container of the waste from a safe distance followed by open burning.

References

New Jersey Department of Health, "Hazardous Substance Fact Sheet: Dipropylene Glycol, Methyl Ether," Trenton, NJ (December 1998)

Dipropyl Ketone

Molecular Formula: $C_7H_{14}O$

Common Formula: $C_3H_7COC_3H_7$

Synonyms: Butyrone; Dipropil cetona (Spanish); DPK; GBL; 4-Heptanone; Heptan-4-one; Propyl ketone

CAS Registry Number: 123-19-3

RTECS Number: MJ5600000

DOT ID: UN 2710

EEC Number: 606-027-00-X

EINECS Number: 204-608-9

Regulatory Authority

- Air Pollutant Standard Set (ACGIH)[1]
- Canada, WHMIS, Ingredients Disclosure List

Cited in U.S. State Regulations: Alaska (G), California (A, G), Connecticut (A), Florida (G), Illinois (G), Maine (G), Massachusetts (G), Nevada (A), New Hampshire (G), New Jersey (G), Pennsylvania (G), Rhode Island (G), West Virginia (G).

Description: Dipropyl Ketone, $C_3H_7COC_3H_7$, is a colorless liquid with a pleasant odor. Boiling point = 143°C. Flash point =

49°C. Hazard Identification (based on NFPA-704 M Rating System): Health 2, Flammability 2, Reactivity 0. Insoluble in water.

Potential Exposure: This compound is used as a solvent for nitrocellulose and many other natural and synthetic resins. It is used in lacquer formulations and in food flavorings.

Incompatibilities: Forms explosive mixture with air. Contact with oxidizers may cause a fire and explosion hazard. Incompatible with strong bases and strong reducing agents.

Permissible Exposure Limits in Air: There is no OSHA PEL.[58] NIOSH (10-hr. workshift) and ACGIH (8-hr. workshift) recommend a TLV of 50 ppm (235 mg/m^3) TWA. States which have set guidelines or standards for dipropyl ketone in ambient air[60] include Connecticut at 4.7 mg/m^3 and Nevada at 5.595 mg/m^3.

Determination in Air: Charcoal tube; CS_2; Gas chromatography/Flame ionization detection; OSHA Adapt Method #7.

Permissible Concentration in Water: No criteria set.

Routes of Entry: Inhalation, ingestion, skin and/or eye contact.

Harmful Effects and Symptoms

Short Term Exposure: Dipropyl ketone can affect you when breathed in and by passing through your skin. Contact can irritate the eyes and skin. Dipropyl ketone is a central nervous system depressant; can cause depression, dizziness, somnolence (sleepiness, unnatural drowsiness), decreased breathing, unconsciousness.

Long Term Exposure: High or repeated exposures may damage the liver and kidney. Repeated contact can dry out the skin, causing irritation and rash.

Points of Attack: Eyes, skin, central nervous system, liver.

Medical Surveillance: If symptoms develop or overexposure is suspected, the following may be useful: Liver and kidney function tests. Interview for brain effects, including memory, mood, concentration, headaches, malaise, and altered sleep patterns. Consider cerebellar, autonomic and peripheral nervous system evaluation. Positive and borderline individuals should be referred for neuropsychological testing.

First Aid: If this chemical gets into the eyes, remove any contact lenses at once and irrigate immediately for at least 15 minutes, occasionally lifting upper and lower lids. Seek medical attention immediately. If this chemical contacts the skin, remove contaminated clothing and wash immediately with soap and water. Seek medical attention immediately. If this chemical has been inhaled, remove from exposure, begin rescue breathing (using universal precautions) if breathing has stopped and CPR if heart action has stopped. Transfer promptly to a medical facility. When this chemical has been swallowed, get medical attention. Give large quantities of water and induce vomiting. Do not make an unconscious person vomit.

Personal Protective Methods: Wear protective gloves and clothing to prevent any reasonable probability of skin contact. Safety equipment suppliers/manufacturers can provide recommendations on the most protective glove/clothing material for your operation.. All protective clothing (suits, gloves, footwear, headgear) should be clean, available each day, and put on before work. Contact lenses should not be worn when working with this chemical. Wear splash-proof chemical goggles and face shield unless full facepiece respiratory protection is worn. Employees should wash immediately with soap when skin is wet or contaminated. Provide emergency showers and eyewash.

Respirator Selection: Where the potential exists for exposures over 50 ppm, use an MSHA/NIOSH approved supplied-air respirator with a full facepiece operated in the positive pressure mode or with a full facepiece, hood, or helmet in the continuous flow mode, or use an MSHA/NIOSH approved self-contained breathing apparatus with a full facepiece operated in pressure-demand or other positive pressure mode.

Storage: Prior to working with dipropyl ketone you should be trained on its proper handling and storage. Store in tightly closed containers in a cool, well ventilated area away from oxidizers. Metal containers involving the transfer of this chemical should be grounded and bonded. Where possible, automatically pump liquid from drums or other storage containers to process containers. Drums must be equipped with self-closing valves, pressure vacuum bungs, and flame arresters. Use only non-sparking tools and equipment, especially when opening and closing containers of this chemical. Sources of ignition such as smoking and open flames, are prohibited where this chemical is used, handled, or stored in a manner that could create a potential fire or explosion hazard. Wherever this chemical is used, handled, manufactured, or stored, use explosion-proof electrical equipment and fittings.

Shipping: This compound requires a shipping label of: "Flammable Liquid." It falls in DOT Hazard Class 3 and Packing Group III.

Spill Handling: Evacuate and restrict persons not wearing protective equipment from area of spill or leak until cleanup is complete. Remove all ignition sources. Ventilate area of spill or leak. Absorb liquids in vermiculite, dry sand, earth, peat, carbon, or a similar material and deposit in sealed containers. Keep this chemical out of a confined space, such as a sewer, because of the possibility of an explosion, unless the sewer is designed to prevent the build-up of explosive concentrations. It may be necessary to contain and dispose of this chemical as a hazardous waste. If material or contaminated runoff enters waterways, notify downstream users of potentially contaminated waters. Contact your Department of Environmental Protection or your regional office of the federal EPA for specific recommendations. If employees are required to clean-up spills, they must be properly trained and equipped. OSHA 1910.120(q) may be applicable.

Fire Extinguishing: This chemical is a flammable liquid. Poisonous gases are produced in fire. Use dry chemical, carbon dioxide, or alcohol foam extinguishers. Vapors are heavier than air and will collect in low areas. Vapors may travel long distances to ignition sources and flashback. Vapors in confined areas may explode when exposed to fire. Containers may explode in fire. Storage containers and parts of containers may rocket great distances, in many directions. If material or contaminated runoff enters waterways, notify downstream users of potentially contaminated waters. Notify local health and fire officials and pollution control agencies. From a secure, explosion-proof location, use water spray to cool exposed containers. If cooling streams are ineffective (venting sound increases in volume and pitch, tank discolors, or shows any signs of deforming), withdraw immediately to a secure position. If employees are expected to fight fires, they must be trained and equipped in OSHA 1910.156.

Disposal Method Suggested: Incineration.

References

New Jersey Department of Health, "Hazardous Substance Fact Sheet: Dipropyl Ketone," Trenton, NJ (February 1989)

Diquat

Molecular Formula: $C_{12}H_{12}N_2Br_2$

Synonyms: Aquacide; Cleansweep; Deiquat; Dextrone; 9,10-Dihydro-8a,10,-diazoniaphenanthrene dibromide; 9,10-Dihydro-8a,10a-diazoniaphenanthrene(1,1'-ethylene-2,2'-bipyridylium)dibromide; 5,6-Dihydro-dipyrido (1,2a,2,1c)pyrazinium dibromide; 5,6-Dihydro-dipyrido (1,2-a:2,1'-c)pyrazinium dibromide; 6,7-Dihydropyridol(1,2-a:2',1'-c)pyrazinedium dibromide; 6,7-Dihydropyrido(1,2-a:2',1'-c)pyrazinedium dibromide; Dipyrido(1,2-a:2',1'-c) pyrazinediium, 6,7-dihydro-, dibromide; *o* Diquat; Diquat dibromide; 1,1'-Ethylene-2,2'-bipyridyliumdibromide; 1,1-Ethylene 2,2-dipyridylium dibromide; 1,1'-Ethylene-2,2'-dipyridylium dibromide; Ethylene dipyridylium dibromide; Farmon PDQ; FB/2; Feglox; Groundhog soltair; Pathclear; Preeglone; Reglon; Reglone; Reglox; Weedol (ICI); Weedtrine-D

CAS Registry Number: 85-00-7

RTECS Number: JM5690000

DOT ID: UN 2781 (solid); UN 2782 (liquid)

Regulatory Authority

- Air Pollutant Standard Set (ACGIH)[1] (HSE)[33] (former USSR)[35] (OSHA)[58] (Several States)[60]
- Clean Water Act: Section 311 Hazardous Substances/RQ 40CFR117.3 (same as CERCLA, see below)
- Safe Drinking Water Act: MCL, 0.02 mg/l; MCLG, 0.02 mg/l; Regulated chemical (47FR9352)
- Superfund/EPCRA 40CFR302.4 Reportable Quantity (RQ): CERCLA, 1,000 lb (454 kg)

Cited in U.S. State Regulations: Alaska (G), California (A, G), Connecticut (A), Florida (G), Illinois (G), Maine (G), Massachusetts (G), Minnesota (A), Nevada (A), New Hampshire (G), New Jersey (G), North Dakota (A), Pennsylvania (G), Rhode Island (G), Virginia (A), West Virginia (G).

Description: Diquat, $C_{12}H_{12}N_2Br_2$, forms a monohydrate which consists of colorless to yellow crystals. Dibromide salt (herbicide) is yellow crystals. The commercial product may be found in a liquid concentrate or a solution. Boiling point = decomposes. Freezing/Melting point = 335°C. Hazard Identification (based on NFPA-704 M Rating System): Health 2, Flammability 0, Reactivity 0. Soluble in water.

Potential Exposure: Those involved in the manufacture, formulation and application of this herbicide.

Incompatibilities: Alkalis, UV light, basic solutions. Concentrated diquat solutions attacks aluminum.

Permissible Exposure Limits in Air: There is no OSHA PEL.[58] NIOSH (10-hr.workshift) and ACGIH (8-hr. workshift) recommend a TWA value of 0.5 mg/m^3. The HSE uses the same TWA and has set a STEL of 1.0 mg/m^3. The former USSR[35] has set a MAC in workplace air of 0.2 mg/m^3. Several states have set guidelines or standards for diquat in ambient air[60] ranging from 5 $\mu g/m^3$ (North Dakota) to 8.0 $\mu g/m^3$ (Virginia) to 10.0 $\mu g/m^3$ (Connecticut) to 12.0 $\mu g/m^3$ (Nevada).

Determination in Air: No method available.

Permissible Concentration in Water: No criteria set.

Routes of Entry: Inhalation, skin absorption, ingestion, skin and/or eye contact.

Harmful Effects and Symptoms

Short Term Exposure: Diquat can affect you when breathed in and by passing through your skin. Skin contact can cause burns. High exposure can cause nausea, diarrhea, lung, liver, and kidney damage. Higher exposures can cause pulmonary edema, a medical emergency that can be delayed for several hours.

Long Term Exposure: Long-term or repeated exposure may cause cataracts. Repeated contact causes dry, cracked skin and nail damage. Exposure can cause nosebleeds. Diquat may cause mutations. Handle with extreme caution. Diquat may damage the developing fetus. Lung damage may occur.

Points of Attack: Eyes, skin, respiratory system, kidneys, liver, central nervous system.

Medical Surveillance: If symptoms develop or overexposure has occurred, the following may be useful: Lung function tests. Examination of the eyes. Kidney function tests.

First Aid: If this chemical gets into the eyes, remove any contact lenses at once and irrigate immediately for at least 15 minutes, occasionally lifting upper and lower lids. Seek medical attention immediately. If this chemical contacts the skin, remove contaminated clothing and wash immediately with soap and water. Seek medical attention immediately. If

this chemical has been inhaled, remove from exposure, begin rescue breathing (using universal precautions) if breathing has stopped and CPR if heart action has stopped. Transfer promptly to a medical facility. When this chemical has been swallowed, get medical attention. Give large quantities of water and induce vomiting. Do not make an unconscious person vomit.

Personal Protective Methods: Wear protective gloves and clothing to prevent any reasonable probability of skin contact. Safety equipment suppliers/manufacturers can provide recommendations on the most protective glove/clothing material for your operation.. All protective clothing (suits, gloves, footwear, headgear) should be clean, available each day, and put on before work. Contact lenses should not be worn when working with this chemical. Wear dust-proof chemical goggles and face shield unless full facepiece respiratory protection is worn. Employees should wash immediately with soap when skin is wet or contaminated. Provide emergency showers and eyewash.

Respirator Selection: Where the potential exists for exposures over 0.5 mg/m^3, use a MSHA/NIOSH approved full facepiece respirator with a pesticide cartridge. Increased protection is obtained from full facepiece air purifying respirators. If while wearing filter, cartridge or canister respirator, you can smell, taste, or otherwise detect diquat, or in the case of a full facepiece respirator you experience eye irritation, leave the area immediately. Check to make sure the respirator to face seal is still good. If it is, replace the filter, cartridge, or canister. If the seal is no longer good, you may need a new respirator. Where the potential for high exposures exists, use a MSHA/NIOSH approved supplied-air respirator with a full facepiece operated in the positive pressure mode or with a full facepiece, hood, or helmet in the continuous flow mode, or use a MSHA/NIODH approved self-contained breathing apparatus with a full facepiece operated in pressure-demand or other positive pressure mode.

Storage: Prior to working with diquat you should be trained on its proper handling and storage. Store in tightly closed containers in a cool, well ventilated area away from oxidizers and reducing agents. Where possible, automatically pump liquid from drums or other storage containers to process containers. Drums must be equipped with self-closing valves, pressure vacuum bungs, and flame arresters. Use only non-sparking tools and equipment, especially when opening and closing containers of this chemical. Sources of ignition such as smoking and open flames, are prohibited where this chemical is used, handled, or stored in a manner that could create a potential fire or explosion hazard. Wherever this chemical is used, handled, manufactured, or stored, use explosion-proof electrical equipment and fittings.

Shipping: Diquat falls in the category of Bipyridilium pesticides, solid, toxic, n.o.s. and in Packing Group III of Hazard Class 6.1. This imposes a requirement for a "Keep Away From Food" label.

Spill Handling: *Dry material:* Evacuate persons not wearing protective equipment from area of spill or leak until clean-up is complete. Remove all ignition sources. Collect powdered material in the most convenient and safe manner and deposit in sealed containers. Ventilate area after clean-up is complete. It may be necessary to contain and dispose of this chemical as a hazardous waste. If material or contaminated runoff enters waterways, notify downstream users of potentially contaminated waters. Contact your Department of Environmental Protection or your regional office of the federal EPA for specific recommendations. If employees are required to clean-up spills, they must be properly trained and equipped. OSHA 1910.120(q) may be applicable.

Liquid: Evacuate and restrict persons not wearing protective equipment from area of spill or leak until cleanup is complete. Remove all ignition sources. Ventilate area of spill or leak. Absorb liquids in vermiculite, dry sand, earth, peat, carbon, or a similar material and deposit in sealed containers. Oil-skimming equipment may be used to remove slicks from water. Keep this chemical out of a confined space, such as a sewer, because of the possibility of an explosion, unless the sewer is designed to prevent the build-up of explosive concentrations. It may be necessary to contain and dispose of this chemical as a hazardous waste. If material or contaminated runoff enters waterways, notify downstream users of potentially contaminated waters. Contact your Department of Environmental Protection or your regional office of the federal EPA for specific recommendations. If employees are required to clean-up spills, they must be properly trained and equipped. OSHA 1910.120(q) may be applicable.

Fire Extinguishing: Diquat is a combustible solid but does not readily ignite and burns with difficulty. Use dry chemical, carbon dioxide, water spray, or foam extinguishers. Poisonous gases are produced in fire. If material or contaminated runoff enters waterways, notify downstream users of potentially contaminated waters. Notify local health and fire officials and pollution control agencies. From a secure, explosion-proof location, use water spray to cool exposed containers. If cooling streams are ineffective (venting sound increases in volume and pitch, tank discolors, or shows any signs of deforming), withdraw immediately to a secure position. If employees are expected to fight fires, they must be trained and equipped in OSHA 1910.156.

Disposal Method Suggested: Diquat is inactivated by inert clay or by anionic surfactants. Therefore, an effective and environmentally safe disposal method would be to mix the product with ordinary household detergent and bury the mixture in clay soil.[22] In accordance with 40CFR165 recommendations for the disposal of pesticides and pesticide containers. Must be disposed properly by following package label directions or by contacting your state pesticide or environmental control agency or by contacting your regional EPA office.

References

New Jersey Department of Health, "Hazardous Substance Fact Sheet: Diquat," Trenton, NJ (June 1986)

Direct Black 38

Molecular Formula: $C_{34}H_{25}N_9O_7S_2 \cdot 2Na$

Synonyms: AHCO direct black GX; Airedale black ED; Aizen direct deep black EH; Aizen direct deep black GH; Aizen direct deep black RH; Amanil black GL; Amanil black WD; Apomine black GX; Atlantic black BD; Atlantic black C; Atlantic black E; Atlantic black EA; Atlantic black GAC; Atlantic black GG; Atlantic black GXCW; Atlantic black GXOO; Atlantic black SD; Atul black E; Azine deep black EW; Azine direct black E; Azocard black EW; Azomine black EWO; Belamine black GX; Bencidal black E; Benzanil black E; Benzo deep black E; Benzoform black BCN-CF; Benzo leather black E; Black 2EMBL; Black 4EMBL; Brasilamina black GN; Brilliant chrome leather black H; Calcomine black; Calcomine black EXL; Carbide black E; Chloramine black C; Chloramine black EC; Chloramine black ERT; Chloramine black EX; Chloramine black EXR; Chloramine black XO; Chloramine carbon black S; Chloramine carbon black SJ; Chloramine carbon black SN; Chlorazol black EA; Chlorazol black EN; Chlorazol Burl black E; Chlorazol leather black E; Chlorazol leather black EC; Chlorazol leather black EM; Chlorazol leather black ENP; Chlorazol silk black G; Chrome leather black E; Chrome leather black EC; Chrome leather black EM; Chrome leather black G; Chrome leather brilliant black ER; C.I. 30235; C.I. direct black 38, disodium salt; Coir deep black F; Diacotton deep black; Diacotton deep black RX; Diamine deep black EC; Diamine direct black E; Diaphtamine black V; Diazine black E; Diazine direct black G; Diazol black 2V; Diphenyl deep black G; Direct black 38; Direct black A; Direct black BRN; Direct black CX; Direct black CXR; Direct black E; Direct black EW; Direct black EX; Direct black FR; Direct black GAC; Direct black GW; Direct black GX; Direct black GXR; Direct black jet; Direct black meta; Direct black methyl; Direct black N; Direct black RX; Direct black SD; Direct black WS; Direct black Z; Direct black ZSH; Direct deep black E; Direct deep black EAC; Direct deep black EA-CF; Direct deep black E-EX; Direct deep black E extra; Direct deep black EW; Direct deep black EX; Direct deep black WX; Erie black BF; Erie black GXOO; Erie black jet; Erie black NUG; Erie black RXOO; Erie brilliant black S; Erie fibre black VP; Fenamin black VF; Fixanol black E; Formaline black C; Formic black C; Formic black CW; Formic black EF; Formic black MTG; Formic black TG; Hispamin black EF; Interchem direct black Z; Kayaku direct deep black EX; Kayaku direct deep black GX; Kayaku direct deep black S; Kayaku direct special black AAX; Lurazol black BA; Meta black; Mitsui direct black EX; Mitsui direct black GX; 2,7-Naphthalenedisulfonic acid,4-amino-3-[{4'-[(2,4-diaminophenyl)azo](1,1'-biphenyl)-4-yl}azo]-5-hydroxy-6-(phenylazo)-, disodium salt; Nippon deep black; Nippon deep black GX; Paper black BA; Paper black T; Paper deep black C; Paramine black B; Paramine black E; Peermine black E; Peermine black GXOO; Phenamine black BCN-CF; Phenamine black clphenamine black E 200; Phenamine black E; Phenamine black EP; Pheno black EP; Pheno black SGN; Pontamine black E; Pontamine black EBN; Sandopel black EX; Seristan black B; Telon fast black E; Tertrodirect black EFD; Tetrazo deep black G; Union black EM; Vondacel black N

CAS Registry Number: 1937-37-7

RTECS Number: QJ6160000

DOT ID: UN 3077

Regulatory Authority

- Carcinogen (Animal Positive) (IARC) (NCI) (NPT)[9]
- In 1998 EPA set an MCL for TTHM (total trihalomethane) at MCLs to 0.80 mg/l (down from 0.100 mg/l set in 1976), and Maximum Residual Disinfectant level Goals (MRDG) for chloramines was set at 4 mg/l
- Air Pollutant Standard Set (North Dakota)[60]
- EPCRA Section 313 Form R *de minimis* concentration reporting level: 0.1%

Cited in U.S. State Regulations: California (G), Florida (G), Illinois (G, W), Maryland (G), Massachusetts (G), New Jersey (G), North Dakota (A), Pennsylvania (G), Rhode Island (G), West Virginia (G).

Description: Direct Black 38, $C_{34}H_{25}N_9O_7S_2 \cdot 2Na$, is a black powder. Freezing/Melting point ≥400°C.Hazard Identification (based on NFPA-704 M Rating System): Health 1, Flammability 1, Reactivity 0.

Potential Exposure: Direct Black 38 is possibly being used to dye fabric, leather, cotton, cellulosic materials, and paper; these are then used in consumer products. The chemical may be used by artists (CPSC, EPA). The FDA has indicated that although Direct Black 38 is identified in the literature as a hair-dye component, it is currently not used by the cosmetic industry. In view of a health hazard alert issued in December 1980 by OSHA, which cautioned workers and employers of the carcinogenic effect of benzidine-derived Direct Black 38, new nonbenzidine Direct Black dyes have been developed and used successfully in commercial applications by the paper and leather industry. These nonbenzidine dyes were developed with the hope of replacing benzidine-based dyes throughout industry.

Permissible Exposure Limits in Air: North Dakota[60] has set a guideline for Direct Black 38 in ambient air of zero.

Permissible Concentration in Water: In 1998 EPA set an MCL for TTHM (total trihalomethane) at MCLs to 0.80 mg/l (down from 0.100 mg/l set in 1976), and Maximum Residual Disinfectant level Goals (MRDG) for chloramines was set at 4 mg/l.

Routes of Entry: Human exposure to Black 38 may occur through inhalation, skin absorption, and unintentional ingestion. Consumer exposure to Direct Black 38 depends upon the ability of the dye to migrate out of consumer products and either penetrate the skin or degrade prior to penetrating the skin. No additional data quantifying the rate of migration or degradation of this dye are currently available.

Harmful Effects and Symptoms

Short Term Exposure: Causes eye irritation. Toxic if inhaled or ingested.

Long Term Exposure: There is sufficient evidence that commercial Direct Black 38 is carcinogenic to experimental animals. In an occupational hazard review, it was concluded that all benzidine-based dyes, including Direct Black 38, regardless of their physical state or proportion in the mixture, should be recognized as potential human carcinogens. Mutation data reported.

First Aid: *Skin Contact:*[52] Flood all areas of body that have contacted the substance with water. Don't wait to remove contaminated clothing; do it under the water stream. Use soap to help assure removal. Isolate contaminated clothing when removed to prevent contact by others.

Eye Contact: Remove any contact lenses at once. Flush eyes well with copious quantities of water or normal saline for at least 20 – 30 minutes. Seek medical attention.

Inhalation: Leave contaminated area immediately; breathe fresh air. Proper respiratory protection must be supplied to any rescuers. If coughing, difficult breathing or any other symptoms develop, seek medical attention at once, even if symptoms develop many hours after exposure.

Ingestion: If convulsions are not present, give a glass or two of water or milk to dilute the substance. Assure that the person's airway is unobstructed and contact a hospital or poison center immediately for advice on whether or not to induce vomiting.

Personal Protective Methods: Wear protective gloves and clothing to prevent any reasonable probability of skin contact. Safety equipment suppliers/manufacturers can provide recommendations on the most protective glove/clothing material for your operation.. All protective clothing (suits, gloves, footwear, headgear) should be clean, available each day, and put on before work. Contact lenses should not be worn when working with this chemical. Wear dust-proof chemical goggles and face shield unless full facepiece respiratory protection is worn. Employees should wash immediately with soap when skin is wet or contaminated. Provide emergency showers and eyewash.

Respirator Selection: Where the potential exists for exposure to direct black 38, use a MSHA/NIOSH approved supplied-air respirator with a full facepiece operated in the positive pressure mode or with a full facepiece, hood, or helmet in the continuous flow mode, or use a MSHA/NIOSH approved self-contained breathing apparatus with a full facepiece operated in pressure-demand or other positive pressure mode.

Storage: Store in a cool, dry place. A regulated, marked area should be established where this chemical is handled, used, or stored in compliance with OSHA standard 1910.1045.

Shipping: Direct Black 38 may be considered as a Hazardous Substance, solid n.o.s.[52]

Spill Handling: Dry material: Evacuate persons not wearing protective equipment from area of spill or leak until clean-up is complete. Remove all ignition sources. Dampen dry material with water. Collect powdered material in the most convenient and safe manner and deposit in sealed containers. Ventilate area after clean-up is complete. It may be necessary to contain and dispose of this chemical as a hazardous waste. If material or contaminated runoff enters waterways, notify downstream users of potentially contaminated waters. Contact your Department of Environmental Protection or your regional office of the federal EPA for specific recommendations. If employees are required to clean-up spills, they must be properly trained and equipped. OSHA 1910.120(q) may be applicable.

Fire Extinguishing: This chemical is a combustible solid. Use dry chemical, carbon dioxide, water spray, or alcohol foam extinguishers. Poisonous gases are produced in fire including nitrogen and sulfur oxides. If material or contaminated runoff enters waterways, notify downstream users of potentially contaminated waters. Notify local health and fire officials and pollution control agencies. From a secure, explosion-proof location, use water spray to cool exposed containers. If cooling streams are ineffective (venting sound increases in volume and pitch, tank discolors, or shows any signs of deforming), withdraw immediately to a secure position. If employees are expected to fight fires, they must be trained and equipped in OSHA 1910.156.

References

National Institute for Occupational Safety and Health, Special Occupational Hazard Review for Benzidine-Based Dyes, Washington, DC: U.S. Government Printing Office (1980)

Direct Blue 6

Molecular Formula: $C_{32}H_{24}N_6O_{14}S_4 \cdot 4Na$

Synonyms: Airedale blue 2BD; Aizen direct blue 2BH; Amanil blue 2BX; Atlantic blue 2B; Atul direct blue; Azocard blue 2B; Azomine blue 2B; Belamine blue 2B; Bencidal blue 2B; Benzanil blue 2B; Benzo blue GS; Blue 2B; Brasilamina blue 2B; Calcomine blue 2B; Chloramine blue 2B; Chlorazol blue B; Chrome leather blue 2B; C.I. direct blue 6; C.I. direct blue 6, tetrasodium salt; Cresotine blue 2B; Diacotton blue B; Diamine blue 2B; Diaphtamine blue BB; Diazine blue 2B; Diazol blue 2B; Diphenyl blue 2B; Enianil blue 2B; Fixanol blue 2B; Hispamin blue 2B; Indigo blue 2B; Kayaku direct;

Mitsui direct blue 2BN; Naphtamine blue 2B; 2,7-Naphthalenedisulfonic acid, 3,3'-[(4,4'-biphenylene)-biphenylene)bis-(azo)]bis(5-amino-4-hydroxy-), tetrasodium salt; NB2B; NCI-C54579; Nfenamin blue 2B; Niagara blue 2; Nippon blue BB; Paramine blue 2B; Phenamine blue BB; Pheno blue 2B; Pontamine blue BB; Sodium diphenyl-4,4'-bis-azo-2"-8"-amino-1"-naphthol-3",6 " disulphonate; Tertrodirect blue 2B; Vondacel blue 2B

CAS Registry Number: 2602-46-2

RTECS Number: QJ6400000

Regulatory Authority

- Carcinogen (Animal Positive) (IARC) (NCI) (NTP)[9]
- In 1998 EPA set an MCL for TTHM (total trihalomethane) at MCLs to 0.80 mg/l (down from 0.100 mg/l set in 1976), and Maximum Residual Disinfectant level Goals (MRDG) for chloramines was set at 4 mg/l
- Air Pollutant Standard Set (North Dakota)[60]
- EPCRA Section 313 Form R *de minimis* concentration reporting level: 0.1%

Cited in U.S. State Regulations: California (G), Florida (G), Illinois (G), Maryland (G), Massachusetts (G), New Jersey (G), North Dakota (A), Pennsylvania (G), Rhode Island (G), West Virginia (G).

Description: Direct blue 6, $C_{32}H_{24}N_6O_{14}S_4 \cdot 4Na$, is a benzidine-based dyestuff somewhat analogous to Direct Black 38 (which see). It is a dark blue powder. Freezing/Melting point ≥400°C. Hazard Identification (based on NFPA-704 M Rating System): Health 1, Flammability 1, Reactivity 0.

Potential Exposure: Direct Blue 6 may be used by artists. It is potentially used to dye fabric, leather, cotton, cellulosic materials, and paper; these are then used in consumer products (CPSC, EPA). The FDA has indicated that although Direct Blue 6 has been identified in the literature as a hair dye component, it is not presently used by the cosmetic industry. The primary source for exposure to Direct Blue 6 is at the production site. The initial production step is in a closed system. However, other production operations, such as filter press, drying, and blending, may be performed in the open and, therefore, may afford a greater potential for worker exposure. The general population may be exposed to Direct Blue 6 through the use of retail packaged dyes containing this benzidine-based dye.

Permissible Exposure Limits in Air: The state of North Dakota[60] has set a guideline for Direct Blue 6 in ambient air of zero.

Permissible Concentration in Water: In 1998 EPA set an MCL for TTHM (total trihalomethane) at MCLs to 0.80 mg/l (down from 0.100 mg/l set in 1976), and Maximum Residual Disinfectant level Goals (MRDG) for chloramines was set at 4 mg/l.

Routes of Entry: Human exposure to Direct Blue 6 may occur through inhalation, skin absorption, and to a lesser extent, unintentional ingestion, when the dye is in the press cake or dry powder form.

Harmful Effects and Symptoms

Long Term Exposure: There is sufficient evidence for the carcinogenicity of Direct Blue 6 (technical grade) in experimental animals. In an occupational hazard review, it was concluded that all benzidine-based dyes, including Direct Blue 6, regardless of their physical state or proportion in the mixture should be recognized as potential human carcinogens. Mutation data reported.

First Aid: If this chemical gets into the eyes, remove any contact lenses at once and irrigate immediately for at least 15 minutes, occasionally lifting upper and lower lids. Seek medical attention immediately. If this chemical contacts the skin, remove contaminated clothing and wash immediately with soap and water. Seek medical attention immediately. If this chemical has been inhaled, remove from exposure, begin rescue breathing (using universal precautions) if breathing has stopped and CPR if heart action has stopped. Transfer promptly to a medical facility. When this chemical has been swallowed, get medical attention. Give large quantities of water and induce vomiting. Do not make an unconscious person vomit.

Personal Protective Methods: Wear protective gloves and clothing to prevent any reasonable probability of skin contact. Safety equipment suppliers/manufacturers can provide recommendations on the most protective glove/clothing material for your operation.. All protective clothing (suits, gloves, footwear, headgear) should be clean, available each day, and put on before work. Contact lenses should not be worn when working with this chemical. Wear dust-proof chemical goggles and face shield unless full facepiece respiratory protection is worn. Employees should wash immediately with soap when skin is wet or contaminated. Provide emergency showers and eyewash.

Respirator Selection: Where the potential exists for exposure to direct blue 6, use a MSHA/NIOSH approved supplied-air respirator with a full facepiece operated in the positive pressure mode or with a full facepiece, hood, or helmet in the continuous flow mode, or use a MSHA/NIOSH approved self-contained breathing apparatus with a full facepiece operated in pressure-demand or other positive pressure mode.

Storage: Prior to working with direct blue 6 you should be trained on its proper handling and storage. Store in tightly closed containers in a cool, well ventilated area. A regulated, marked area should be established where this chemical is handled, used, or stored in compliance with OSHA standard 1910.1045.

Shipping: Direct Blue 6 may be considered as a Hazardous Substance, solid n.o.s.[52]

Spill Handling: Evacuate persons not wearing protective equipment from area of spill or leak until clean-up is complete. Remove all ignition sources. Collect powdered material in the most convenient and safe manner and deposit in sealed containers. Ventilate area after clean-up is complete. It may be necessary to contain and dispose of this chemical as a hazardous waste. If material or contaminated runoff enters waterways, notify downstream users of potentially contaminated waters. Contact your Department of Environmental Protection or your regional office of the federal EPA for specific recommendations. If employees are required to clean-up spills, they must be properly trained and equipped. OSHA 1910.120(q) may be applicable.

Fire Extinguishing: Use dry chemical, carbon dioxide, water spray, or alcohol foam extinguishers. Poisonous gases are produced in fire. If material or contaminated runoff enters waterways, notify downstream users of potentially contaminated waters. Notify local health and fire officials and pollution control agencies. From a secure, explosion-proof location, use water spray to cool exposed containers. If cooling streams are ineffective (venting sound increases in volume and pitch, tank discolors, or shows any signs of deforming), withdraw immediately to a secure position. If employees are expected to fight fires, they must be trained and equipped in OSHA 1910.156.

References

National Institute for Occupational Safety and Health, Special Occupational Hazard Review for Benzidine-Based Dyes, Washington, DC: U.S. Government, Printing Office (1980)

Disulfiram

Molecular Formula: $C_{10}H_{20}N_2S_4$

Synonyms: Abstensil; Abstinyl; Alcophobin; Alk-Aubs; Antabus®; Antabuse®; Antadix; Antaenyl; Antaethan; Antaethyl; Antaetil; Antalcol; Antetan; Antethyl; Antetil; Anteyl; Antiaethan; Antietanol; Anti-ethyl; Antietil; Antikol; Antivitium; Aversan; Averzan; [Bis(diethylamino)thioxomethyl] disulphide; Bis(*N,N*-diethylthiocarbamoyl) disulfide; Bis(diethylthiocarbamoyl) disulfide; Bis(*N,N*-diethylthiocarbamoyl) disulphide; Bonibal; Contralin; Contrapot; Cronetal®; Dicupral; Disetil; Disulfan; Disulfuram; Disulphuram; 1,1'-Dithiobis(*N,N*-diethylthioformamide); Ekagom teds; Ephorran; Espenal; Esperal (France); Etabus; Ethyldithiourame; Ethyldithiurame; Ethyl thiram; Ethyl thiudad; Ethyl thiurad; Ethyl tuads; Ethyl tuex; Exhoran; Exhorran; HOCA; Krotenal; NCI-C02959; Nocbin; Noxal; Refusal (Netherlands); Ro-Sulfiram®; Stopaethyl; Stopethyl; Stopetyl; TATD; Tenurid; Tenutex; TETD; Tetidis; Tetradin; Tetradine; Tetraethylthioperoxydicarbonic diamide; Tetraethylthiram disulphide; Tetraethylthiuram; Tetraethylthiuram disulfide; *N,N,N',N'*-Tetraethylthiuram disulphide; Tetraethylthiuram disulphide; Tetraetil; Teturam; Teturamin; Thiosan; Thioscabin; Thireranide; Thiuram E; Thiuranide; Tillram; Tiuram; TTD; TTS

CAS Registry Number: 97-77-8

RTECS Number: JO1225000

DOT ID: UN 2588

Regulatory Authority

- Air Pollutant Standard Set (ACGIH)[1] (DFG)[3]
- Canada, WHMIS, Ingredients Disclosure List

Cited in U.S. State Regulations: Alaska (G), California (G), Florida (G), Illinois (G), Maine (G), Massachusetts (G), Minnesota (G), New Hampshire (G), New Jersey (G), Pennsylvania (G), Rhode Island (G), West Virginia (G).

Description: Disulfiram, is a white to off-white or light gray powder with a slight odor. Boiling point = 117°C at 17 mm Hg. Freezing/Melting point = 70°C. Very slightly soluble in water.

Potential Exposure: Disulfiram is used as a rubber accelerator and vulcanizer. It is used as a seed disinfectant and fungicide. It is used in therapy as an alcohol deterrent. Used in adhesives.

Incompatibilities: Oxidizers.

Permissible Exposure Limits in Air: There is no OSHA PEL.[58] NIOSH recommended REL, DFG MAK,[3] and the ACGIH recommended TLV is 2 mg/m^3 TWA. NIOSH warns that precautions should be taken to avoid concurrent exposure to ethylene dibromide.

Determination in Air: No method available.

Permissible Concentration in Water: Zero, according to USSR-UNEP/IRPTC project.[43]

Routes of Entry: Inhalation, ingestion, skin and/or eye contact.

Harmful Effects and Symptoms

Short Term Exposure: Irritates the eyes, skin, and respiratory system. Eye contact can lead to damage. Disulfiram can affect you when breathed in and by passing through your skin. Exposure to Disulfiram and alcohol within 1 – 2 days to each other can cause a reaction with flushing of the face and neck, rapid heart beat and vomiting. This could be fatal. If working with Disulfiram you should never be exposed to ethylene dibromide because of possible severe reaction. Symptoms of exposure include lassitude (weakness, exhaustion), fatigue, tremor, restlessness, headache, dizziness; metallic taste; vomiting, peripheral neuropathy.

Long Term Exposure: May cause liver and kidney damage. It may damage the developing fetus. Damage to vision, nervous system with numbness, "pins and needles," weakness and poor coordination can result from repeated exposure. May cause personality changes of depression, anxiety or irritability. Can cause skin sensitization dermatitis. Enlarged thyroid and skin rash can also occur.

Points of Attack: Eyes, skin, respiratory system, central nervous system, peripheral nervous system, liver.

Medical Surveillance: If symptoms develop or overexposure is suspected, the following may be useful: Liver, kidney, and thyroid function tests. Skin testing with dilute Disulfiram may help diagnose allergy, if done by a qualified allergist.

Exam of the nervous system, eyes, and vision. Evaluate for brain effects. Alcohol use may increase liver damage.

First Aid: If this chemical gets into the eyes, remove any contact lenses at once and irrigate immediately for at least 15 minutes, occasionally lifting upper and lower lids. Seek medical attention immediately. If this chemical contacts the skin, remove contaminated clothing and wash immediately with soap and water. Seek medical attention immediately. If this chemical has been inhaled, remove from exposure, begin rescue breathing (using universal precautions) if breathing has stopped and CPR if heart action has stopped. Transfer promptly to a medical facility. When this chemical has been swallowed, get medical attention. Give large quantities of water and induce vomiting. Do not make an unconscious person vomit.

Note: For alcohol/disulfiram or ethylene dibromide/disulfiram reaction, remove the person from exposure. Begin rescue breathing if breathing has stopped and CPR if heart action has stopped. Transfer promptly to a medical facility.

Personal Protective Methods: Wear protective gloves and clothing to prevent any reasonable probability of skin contact. Safety equipment suppliers/manufacturers can provide recommendations on the most protective glove/clothing material for your operation.. All protective clothing (suits, gloves, footwear, headgear) should be clean, available each day, and put on before work. Contact lenses should not be worn when working with this chemical. Wear dust-proof chemical goggles and face shield unless full facepiece respiratory protection is worn. Employees should wash immediately with soap when skin is wet or contaminated. Provide emergency showers and eyewash.

Respirator Selection: Where the potential exists for exposures over 2 mg/m^3, use an MSHA/NIOSH approved full facepiece respirator with a pesticide cartridge. Increased protection is obtained from full facepiece air purifying respirators. If while wearing a filter, cartridge or canister respirator, you can smell, taste, or otherwise detect Disulfiram, or in the case of a full facepiece respirator you experience eye irritation, leave the area immediately. Check to make sure the respirator-to-face seal is still good. If it is, replace the filter, cartridge, or canister. If the seal is no longer good, you may need a new respirator. Where the potential for high exposures exists, use an MSHA/NIOSH approved supplied-air respirator with a full facepiece operated in the positive pressure mode or with a full facepiece, hood, or helmet in the continuous flow mode, or use an MSHA/NIOSH approved self-contained breathing apparatus with a full facepiece operated in the pressure-demand or other positive pressure mode.

Storage: Prior to working with this chemical you should be trained on its proper handling and storage. Store in tightly closed containers in a cool, well-ventilated area away from oxidizers.

Shipping: Disulfiram is not cited in DOT's Performance-Oriented Packaging Standards.[19]

Spill Handling: Evacuate persons not wearing protective equipment from area of spill or leak until clean-up is complete. Remove all ignition sources. Collect powdered material in the most convenient and safe manner and deposit in sealed containers. Ventilate area after clean-up is complete. It may be necessary to contain and dispose of this chemical as a hazardous waste. If material or contaminated runoff enters waterways, notify downstream users of potentially contaminated waters. Contact your Department of Environmental Protection or your regional office of the federal EPA for specific recommendations. If employees are required to clean-up spills, they must be properly trained and equipped. OSHA 1910.120(q) may be applicable.

Fire Extinguishing: Extinguish fire using an agent suitable for type of surrounding fire. Disulfiram itself does not burn readily. Poisonous gases are produced in fire. If material or contaminated runoff enters waterways, notify downstream users of potentially contaminated waters. Notify local health and fire officials and pollution control agencies. From a secure, explosion-proof location, use water spray to cool exposed containers. If cooling streams are ineffective (venting sound increases in volume and pitch, tank discolors, or shows any signs of deforming), withdraw immediately to a secure position. If employees are expected to fight fires, they must be trained and equipped in OSHA 1910.156.

Disposal Method Suggested: Disulfiram can be dissolved in alcohol or other flammable solvent and burned in an incinerator equipped with afterburner and scrubber. In accordance with 40CFR165 recommendations for the disposal of pesticides and pesticide containers. Must be disposed properly by following package label directions or by contacting your state pesticide or environmental control agency or by contacting your regional EPA office.

References

Sax N. I., Ed., "Dangerous Properties of Industrial Materials Report," 1, No. 5, 40 (1981)

New Jersey Department of Health, "Hazardous Substance Fact Sheet: Disulfiram," Trenton, NJ (December 1998)

Disulfoton

Molecular Formula: $C_8H_{19}O_2PS_3$

Synonyms: Bay 19639; Bayer 19639; *O,O*-Diaethyl-*S*-(2-aethylthio-aethyl)-dithiophosphat (German); *O,O*-Diaethyl-*S*-(3-thia-pentyl)-dithiophosphat (German); *O,O*-Diethyl *S*-(2-eththioethyl) phosphorodithioate; *O,O*-Diethyl *S*-(2-eththioethyl) thiothionophosphate; *O,O*-Diethyl *S*-(2-ethylmercaptoethyl)dithiophosphate; *O,O*-Diethyl-*S*-(2-ethylthio-ethyl)-dithiofosfaat (Dutch); *O,O*-Diethyl 2-ethylthioethylphosphorodithioate; *O,O*-Diethyl *S*-2-(ethylthio)ethyl phosphorodithioate; *O,O*-Dietil-S-(2-etiltio-etil)-ditiofosfato (Italian); Dimaz; Disulfaton; Di-Syston; Disystox; Dithiodemeton; Dithiophosphate de *O,O*-diethyle ETDE *S*-(2-ethylthio-ethyle) (French); Dithiosystox; ENT 23,437; *O,O*-

Ethyl *S*-2(ethylthio)ethyl phosphorodithioate; Ethylthiodemeton; *S*-2-(Ethylthio)ethyl *O,O*-diethyl ester of phosphorodithioic acid; Frumin-Al®; Frumin G®; M-74; Phosphorodithionic acid, *O,O*-diethyl *S*-2-[(ethylthio)ethyl] ester; Phosphorodithionic acid, *S*-(2-(ethylthio)ethyl *O,O*-diethyl ester; S 276; Solvirex®; Thiodemeton®; Thiodemetron®

CAS Registry Number: 298-04-4

RTECS Number: TD9275000

DOT ID: UN 2783

Regulatory Authority

- Banned or Severely Restricted (Various Countries) (UN)[13]
- Very Toxic Substance (World Bank)[15]
- Air Pollutant Standard Set (ACGIH)[1] (HSE)[33] (Several States)[60]
- Clean Water Act: Section 311 Hazardous Substances/RQ 40CFR117.3 (same as CERCLA, see below)
- EPA Hazardous Waste Number (RCRA No.): P039
- RCRA, 40CFR261, Appendix 8 Hazardous Constituents
- RCRA 40CFR268.48; 61FR15654, Universal Treatment Standards: Wastewater (mg/l), 0.017; Nonwastewater (mg/kg), 6.2
- RCRA 40CFR264, Appendix 9; TSD Facilities Ground Water Monitoring List. Suggested test method(s) (PQL μg/l): 8140 (2)
- Superfund/EPCRA 40CFR355, Appendix B Extremely Hazardous Substances: TPQ = 500 lb (227 kg)
- Superfund/EPCRA 40CFR302.4 Reportable Quantity (RQ): CERCLA, 1 lb (0.454 kg)
- U.S. DOT Regulated Marine Pollutant (49CFR172.101, Appendix B)
- Canada, WHMIS, Ingredients Disclosure List

Cited in U.S. State Regulations: Alaska (G), California (A, G), Connecticut (A), Florida (G), Illinois (G), Kansas (G), Louisiana (G), Maine (G), Massachusetts (G), Michigan (G), Nevada (A), New Hampshire (G, W), New Jersey (G), North Dakota (A), Oklahoma (G), Pennsylvania (G), Rhode Island (G), Vermont (G), Virginia (G, A), Washington (G), West Virginia (G), Wisconsin (G).

Description: Disulfoton, is a combustible, colorless to yellowish oil with a characteristic odor. Technical product is a brown liquid. Boiling point = 132 – 133°C @ 1.5 mm pressure. Freezing/Melting point ≥- 25°C. Flash point ≥82°C. Hazard Identification (based on NFPA-704 M Rating System): Health 4, Flammability 1, Reactivity 0. Practically insoluble in water.

Potential Exposure: Those involved in the manufacture, formulation and application of this insecticide and acaricide.

Incompatibilities: Alkalies, strong oxidizers.

Permissible Exposure Limits in Air: There is no OSHA PEL.[58] HSE[33] and the recommended ACGIH[1] TWA value is 0.1 mg/m^3 and the HSE[33] has set a STEL of 0.3 mg/m^3. Several states have set guidelines or standards for disulfoton in ambient air[60] ranging from 1.0 μg/m^3 (North Dakota) to 1.6 μg/m^3 (Virginia) to 2.0 μg/m^3 (Connecticut and Nevada).

Determination in Air: OSHA versatile sampler-2; Toluene/Acetone; Gas chromatography/Flame photometric detection for sulfur, nitrogen, or phosphorus; NIOSH Method IV, Method #5600, Organophosphorus Pesticides.

Permissible Concentration in Water: A long term health advisory for an adult has been calculated by EPA as 0.009 mg/l (9 μg/l) and a lifetime health advisory of 0.3 μg/l.

Determination in Water: Extraction with Methylene chloride followed by measurement by gas chromatography using a nitrogen-phosphorus detector.

Routes of Entry: Inhalation, skin absorption, ingestion, skin and/or eye contact.

Harmful Effects and Symptoms

Short Term Exposure: Contact may cause burns to skin and eyes. Symptoms include pinpoint pupils, blurred vision, headache, dizziness, muscle spasms, and profound weakness. Vomiting, diarrhea, abdominal pain, seizures, and coma may also occur. The heart rate may decrease following oral exposure or increase following dermal exposure. Hypotension (low blood pressure) and chest pain may be noted. Hypertension (high blood pressure) is not uncommon. Respiratory symptoms include dyspnea (shortness of breath), respiratory depression, and respiratory paralysis. Psychosis may occur.

It is classified as super toxic. Probable oral lethal dose in humans is less that 5 mg/kg or a taste (less than 7 drops) for a 70 kg (150 lb) person. It is poisonous and may be fatal if inhaled, swallowed, or absorbed through the skin.

Long Term Exposure: Cholinesterase inhibitor; cumulative effect is possible. This chemical may damage the nervous system with repeated exposure, resulting in convulsions, respiratory failure. May cause liver damage.

Points of Attack: Respiratory system, lungs, central nervous system, cardiovascular system, skin, eyes, plasma and red blood cell cholinesterase.

Medical Surveillance: Before employment and at regular times after that, the following are recommended: Plasma and red blood cell cholinesterase levels (tests for the enzyme poisoned by this chemical). If exposure stops, plasma levels return to normal in 1 – 2 weeks while red blood cell levels may be reduced for 1 – 3 months.

When cholinesterase enzyme levels are reduced by 25% or more below preemployment levels, risk of poisoning is increased, even if results are in lower ranges of "normal." Reassignment to work not involving organophosphate or carbamate pesticides is recommended until enzyme levels recover. If symptoms develop or overexposure occurs, repeat the above tests as soon as possible and get an exam of the nervous system. Also consider complete blood count. Consider chest x-ray following acute overexposure. Do not drink any alcoholic beverages before or during use. Alcohol promotes absorption of organic phosphates.

First Aid: If this chemical gets into the eyes, remove any contact lenses at once and irrigate immediately for at least 15 minutes, occasionally lifting upper and lower lids. Seek medical attention immediately. If this chemical contacts the skin, remove contaminated clothing and wash immediately with soap and water. Speed in removing material from skin is of extreme importance. Shampoo hair promptly if contaminated. Seek medical attention immediately. If this chemical has been inhaled, remove from exposure, begin rescue breathing (using universal precautions) if breathing has stopped and CPR if heart action has stopped. Transfer promptly to a medical facility. When this chemical has been swallowed, get medical attention. Give large quantities of water and induce vomiting. Do not make an unconscious person vomit. Medical observation is recommended following acute overexposure.

Personal Protective Methods: Wear protective gloves and clothing to prevent any reasonable probability of skin contact. Safety equipment suppliers/manufacturers can provide recommendations on the most protective glove/clothing material for your operation.. All protective clothing (suits, gloves, footwear, headgear) should be clean, available each day, and put on before work. Contact lenses should not be worn when working with this chemical. Wear splash-proof chemical goggles and face shield unless full facepiece respiratory protection is worn. Employees should wash immediately with soap when skin is wet or contaminated. Provide emergency showers and eyewash.

Respirator Selection: Where the potential exists for exposures over 0.1 mg/m^3, use a MSHA/NIOSH approved supplied-air respirator with a full facepiece operated in the positive pressure mode or with a full facepiece, hood, or helmet in the continuous flow mode, or use a MSHA/NIOSH approved self-contained breathing apparatus with a full facepiece operated in pressure-demand or other positive pressure mode.

Storage: Prior to working with disulfoton you should be trained on its proper handling and storage. Store in tightly closed containers in a cool, well ventilated area away from alkalies. Where possible, automatically pump liquid from drums or other storage containers to process containers.

Shipping: Organophosphorus pesticides, liquid, toxic, n.o.s. fall in DOT Hazard Class 6.1 and disulfoton in Packing Group I. The label required is "Poison."

Spill Handling: Evacuate and restrict persons not wearing protective equipment from area of spill or leak until cleanup is complete. Remove all ignition sources. Ventilate area of spill or leak. Do not touch spill material. Exposure by skin contact is likely to be more significant than inhalation. *Small spills:* take up with sand or other noncombustible absorbent materials and place into containers for later disposal. *Large spills:* dike far ahead of spill for later disposal. It may be necessary to contain and dispose of this chemical as a hazardous waste. If material or contaminated runoff enters waterways, notify downstream users of potentially contaminated waters. Contact your Department of Environmental Protection or your regional office of the federal EPA for specific recommendations. If employees are required to clean-up spills, they must be properly trained and equipped. OSHA 1910.120(q) may be applicable.

Fire Extinguishing: A combustible liquid, but will not ignite easily. Poisonous gases may be generated from the fire including sulfur oxides, and phosphorus oxides. Extinguish with dry chemical, carbon dioxide, water spray, fog, or foam. Fight fire from maximum distance. Dike fire control water for later disposal; do not scatter the material. Wear positive pressure breathing apparatus and special protective clothing. Containers may explode in fire. Storage containers and parts of containers may rocket great distances, in many directions. If material or contaminated runoff enters waterways, notify downstream users of potentially contaminated waters. Notify local health and fire officials and pollution control agencies. From a secure, explosion-proof location, use water spray to cool exposed containers. If cooling streams are ineffective (venting sound increases in volume and pitch, tank discolors, or shows any signs of deforming), withdraw immediately to a secure position. If employees are expected to fight fires, they must be trained and equipped in OSHA 1910.156.

Disposal Method Suggested: Incineration with added flammable solvent in a furnace with alkali scrubber. Acid or alkaline hydrolysis may also be used.[22] In accordance with 40CFR165 recommendations for the disposal of pesticides and pesticide containers. Must be disposed properly by following package label directions or by contacting your state pesticide or environmental control agency or by contacting your regional EPA office. Consult with environmental regulatory agencies for guidance on acceptable disposal practices. Generators of waste containing this contaminant ($\geq$100 kg/mo) must conform with EPA regulations governing storage, transportation, treatment, and waste disposal.

References

U.S. Environmental Protection Agency, Disulfoton, Health and Environmental Effects Profile No. 97, Washington, DC, Office of Solid Waste (April 30, 1980)

U.S. Environmental Protection Agency, "Chemical Profile: Disulfoton," Washington, DC, Chemical Emergency Preparedness Program (November 30, 1987)

U.S. Environmental Protection Agency, "Health Advisory: Disulfoton," Washington, DC, Office of Drinking Water (August 1987)

New Jersey Department of Health, "Hazardous Substance Fact Sheet: Disulfoton," Trenton, NJ (January 1999)

Sax N. I., Ed., "Dangerous Properties of Industrial Materials Report," 8, No. 5, 74–85 (1988)

Dithiazanine Iodide

Molecular Formula: $C_{23}H_{24}IN_2S_2$

Common Formula: $C_{23}H_{24}N_2S_2I$

Synonyms: Abminthic; Anelmid; Anguifugan; Compound 01748; DEJO; Delvex; 3,3'-Diethylthiadicarbocyanine iodide; Diethylthiadicarbocyanine iodide; Dilombrin; Dithiazanine iodide; Dithiazanin iodide; Dithiazinine; Eastman 7663; 3-Ethyl-2-[5-(3-ethyl-2-benzothiazolinylidene)-1,3-pentadienyl]benzothiazolium iodide; L-01748; Netocyd; NK 136; Omnipassin; Partel; Telmicid; Telmid; Telmide; Vercidon

CAS Registry Number: 514-73-8

RTECS Number: DL7060000

Regulatory Authority

- Banned or Severely Restricted (Several Countries) (UN)[13]
- Superfund/EPCRA 40CFR355, Appendix B Extremely Hazardous Substances: TPQ = 500/10,000 lb (227/4,540 kg)
- Superfund/EPCRA 40CFR302.4 Reportable Quantity (RQ): EHS, 1 lb (0.454 kg)

Cited in U.S. State Regulations: California (G), Massachusetts (G), New Jersey (G), Pennsylvania (G).

Description: Dithiazanine iodide, $C_{23}H_{24}N_2S_2I$, is a green needle-like crystalline solid which decomposes at 248°C.

Potential Exposure: This material is used as a veterinary anthelmintic, sensitizer for photographic emulsions and for insecticides. Not registered as a pesticide in the U.S.

Harmful Effects and Symptoms

Short Term Exposure: Poisonous if swallowed, or if dust is inhaled. LD_{50} oral (mouse) 20 mg/kg (highly toxic).

First Aid: If this chemical gets into the eyes, remove any contact lenses at once and irrigate immediately for at least 15 minutes, occasionally lifting upper and lower lids. Seek medical attention immediately. If this chemical contacts the skin, remove contaminated clothing and wash immediately with soap and water. Seek medical attention immediately. If this chemical has been inhaled, remove from exposure, begin rescue breathing (using universal precautions) if breathing has stopped and CPR if heart action has stopped. Transfer promptly to a medical facility. When this chemical has been swallowed, get medical attention. Give large quantities of water and induce vomiting. Do not make an unconscious person vomit.

Personal Protective Methods: Wear protective gloves and clothing to prevent any reasonable probability of skin contact. For emergency situations, wear a positive pressure, pressure-demand, full facepiece self-contained breathing apparatus (SCBA) or pressure-demand supplied air respirator with escape SCBA and a fully-encapsulating, chemical resistant suit. Safety equipment suppliers/manufacturers can provide recommendations on the most protective glove/clothing material for your operation.. All protective clothing (suits, gloves, footwear, headgear) should be clean, available each day, and put on before work. Contact lenses should not be worn when working with this chemical. Wear dust-proof chemical goggles and face shield unless full facepiece respiratory protection is worn. Employees should wash immediately with soap when skin is wet or contaminated. Provide emergency showers and eyewash.

Respirator Selection: Where the potential for exposure to this chemical, use a MSHA/NIOSH approved supplied-air respirator with a full facepiece operated in the positive pressure mode or with a full facepiece, hood, or helmet in the continuous flow mode, or use a MSHA/NIOSH approved self-contained breathing apparatus with a full facepiece operated in pressure-demand or other positive pressure mode.

Storage: Prior to working with this chemical you should be trained on its proper handling and storage. Store in tightly closed containers in a cool, well ventilated area.

Shipping: Pesticides, solid, toxic, n.o.s. fall in Hazard Class 6.1 and this material in Packing Group II. The label required is "Poison."

Spill Handling: Evacuate persons not wearing protective equipment from area of spill or leak until clean-up is complete. Remove all ignition sources. Do not touch spilled material; stop leak if you can do so without risk. Small spills: absorb with sand or other non-combustible absorbent material and place into containers for later disposal. Small dry spills: with clean shovel place material into clean, dry container and cover; move containers from spill area. Large spills: dike far ahead of spill for later disposal. Ventilate area after clean-up is complete. It may be necessary to contain and dispose of this chemical as a hazardous waste. If material or contaminated runoff enters waterways, notify downstream users of potentially contaminated waters. Contact your Department of Environmental Protection or your regional office of the federal EPA for specific recommendations. If employees are required to clean-up spills, they must be properly trained and equipped. OSHA 1910.120(q) may be applicable.

Fire Extinguishing: This material may burn but does not ignite readily. Poisonous gases are produced in fire. Small fires: dry chemical, carbon dioxide, water spray or foam. Large fires; water spray, fog, or foam. Move container from fire area if you can do it without risk. Keep unnecessary people away; isolate hazard area and deny entry. Stay upwind; keep out of low areas. Wear self-contained, positive pressure if available, breathing apparatus and full protective clothing. If material or contaminated runoff enters waterways, notify downstream users of potentially contaminated waters. Notify local health and fire officials and pollution control agencies. From a secure, explosion-proof location, use water spray to cool exposed containers. If cooling streams are ineffective (venting sound increases in volume and pitch, tank discolors, or shows any signs of deforming), withdraw immediately to a secure position. If employees are expected to fight fires, they must be trained and equipped in OSHA 1910.156.

Disposal Method Suggested: In accordance with 40CFR165 recommendations for the disposal of pesticides and pesticide containers. Must be disposed properly by following package label directions or by contacting your state pesticide or environmental control agency or by contacting your regional EPA office.

References

U.S. Environmental Protection Agency, "Chemical Profile: Dithiazanine Iodine," Washington, DC, Chemical Emergency Preparedness Program (November 30, 1987)

Dithiobiuret

Molecular Formula: $C_2H_5N_3S_2$

Common Formula: $H_2NCSNHCSNH_2$

Synonyms: AI3-14762; Biuret, 2,4-dithio-; Biuret, dithio-; 2,4-Dithiobiuret; Dithiobiuret; 2,4-Ditiobiuret (Spanish); DTB; Imidodicarbonimidothioic diamide; Imidodicarbonodithioic diamide; Thioimidodicarbonic diamide; Thio-1-(thiocarbamoyl)urea; Urea, 2-thio-1-(thiocarbamoyl)-

CAS Registry Number: 541-53-7

RTECS Number: EC1575000

DOT ID: UN 2771

Regulatory Authority

- EPA Hazardous Waste Number (RCRA No.): P049
- RCRA, 40CFR261, Appendix 8 Hazardous Constituents
- Superfund/EPCRA 40CFR355, Appendix B Extremely Hazardous Substances: TPQ = 100/10,000 lb (45.4/4,540 kg)
- Superfund/EPCRA 40CFR302.4 Reportable Quantity (RQ): CERCLA, 100 lb (45.4 kg)
- EPCRA Section 313 Form R *de minimis* concentration reporting level: 1.0%

Cited in U.S. State Regulations: California (G), Kansas (G), Louisiana (G), Massachusetts (G), New Jersey (G), Pennsylvania (G), Vermont (G), Virginia (G), Washington (G), Wisconsin (G).

Description: Dithiobiuret, $C_2H_5N_3S_2$, $H_2NCSNHCSNH_2$, is a colorless, crystalline solid. Freezing/Melting point = (decomposition) 181°C. Hazard Identification (based on NFPA-704 M Rating System): Health 3, Flammability 1, Reactivity 0. Soluble in boiling water.

Potential Exposure: This material is used as plasticizers, as a rubber accelerator, and as an intermediate in manufacturing of pesticides.

Permissible Exposure Limits in Air: No standards set.

Permissible Concentration in Water: No criteria set.

Routes of Entry: Inhalation.

Harmful Effects and Symptoms

Short Term Exposure: Irritates the eyes on contact. May cause muscular weakness and/or paralysis leading to difficulty in breathing. This may be fatal. Can cause nausea, watery diarrhea, dehydration and weight loss. The material is highly toxic. It may cause respiratory failure. Symptoms include respiratory paralysis. LD_{50} oral (rat) 5 mg/kg (extremely toxic).

Long Term Exposure: May affect the nervous system.

Points of Attack: Nervous system, eyes.

Medical Surveillance: Examination of the nervous system.

First Aid: If this chemical gets into the eyes, remove any contact lenses at once and irrigate immediately for at least 15 minutes, occasionally lifting upper and lower lids. Seek medical attention immediately. If this chemical contacts the skin, remove contaminated clothing and wash immediately with soap and water. Seek medical attention immediately. If this chemical has been inhaled, remove from exposure, begin rescue breathing (using universal precautions) if breathing has stopped and CPR if heart action has stopped. Transfer promptly to a medical facility. When this chemical has been swallowed, get medical attention. Give large quantities of water and induce vomiting. Do not make an unconscious person vomit. Keep victim quiet and maintain normal body temperature. Effects may be delayed; keep victim under observation.

Personal Protective Methods: Wear protective gloves and clothing to prevent any reasonable probability of skin contact. Safety equipment suppliers/manufacturers can provide recommendations on the most protective glove/clothing material for your operation.. All protective clothing (suits, gloves, footwear, headgear) should be clean, available each day, and put on before work. Contact lenses should not be worn when working with this chemical. Wear dust-proof chemical goggles and face shield unless full facepiece respiratory protection is worn. Employees should wash immediately with soap when skin is wet or contaminated. Provide emergency showers and eyewash.

Respirator Selection: Where the potential for exposure to this chemical, use a MSHA/NIOSH approved supplied-air respirator with a full facepiece operated in the positive pressure mode or with a full facepiece, hood, or helmet in the continuous flow mode, or use a MSHA/NIOSH approved self-contained breathing apparatus with a full facepiece operated in pressure-demand or other positive pressure mode.

Storage: Prior to working with Dithiobiuret you should be trained on its proper handling and storage. Store in tightly closed containers in a cool, well ventilated area away from oxidizers and reducing agents. Where possible, automatically pump liquid from drums or other storage containers to process containers.

Shipping: Dithiocarbamate pesticides, solid, toxic, n.o.s., fall in Hazard Class 6.1 and Dithiobiuret in Packing Group I. This requires a "Poison" label.

Spill Handling: Evacuate persons not wearing protective equipment from area of spill or leak until clean-up is complete.

Keep sparks, flames and other sources of ignition away. Keep material out of water sources and sewers. Avoid breathing dusts and fumes from burning material. Keep upwind; avoid bodily contact with the material. Do not handle broken packages without protective equipment. Wash away any material which may have contacted the body with copious amounts of water or soap and water. Use a vacuum or wet method to reduce dust during cleanup. *DO NOT* dry sweep. Collect powdered material in the most convenient and safe manner and deposit in sealed containers. Ventilate area after clean-up is complete. It may be necessary to contain and dispose of this chemical as a hazardous waste. If material or contaminated runoff enters waterways, notify downstream users of potentially contaminated waters. Contact your Department of Environmental Protection or your regional office of the federal EPA for specific recommendations. If employees are required to clean-up spills, they must be properly trained and equipped. OSHA 1910.120(q) may be applicable.

Fire Extinguishing: Extinguish fire using agent suitable for type of surrounding fire. (Material itself does not burn or burns with difficulty). Use water in flooding quantities as fog. Use alcohol foam, carbon dioxide or dry chemical. Poisonous gases are produced in fire including sulfur oxides and nitrogen oxides. If material or contaminated runoff enters waterways, notify downstream users of potentially contaminated waters. Notify local health and fire officials and pollution control agencies. From a secure, explosion-proof location, use water spray to cool exposed containers. If cooling streams are ineffective (venting sound increases in volume and pitch, tank discolors, or shows any signs of deforming), withdraw immediately to a secure position. If employees are expected to fight fires, they must be trained and equipped in OSHA 1910.156.

Disposal Method Suggested: In accordance with 40CFR165 recommendations for the disposal of pesticides and pesticide containers. Must be disposed properly by following package label directions or by contacting your state pesticide or environmental control agency or by contacting your regional EPA office. Consult with environmental regulatory agencies for guidance on acceptable disposal practices. Generators of waste containing this contaminant (≥100 kg/mo) must conform with EPA regulations governing storage, transportation, treatment, and waste disposal.

References

U.S. Environmental Protection Agency, "Chemical Profile: Dithiobiuret," Washington, DC, Chemical Emergency Preparedness Program (November 30, 1987)

Diuron

Molecular Formula: $C_9H_{10}Cl_2N_2O$

Common Formula: $Cl_2C_6H_3NHCON(CH_3)_2$

Synonyms: 330541; AF 101; AI3-61438; Caswell No. 410; Cekiuron; Crisuron; Dailon; DCMU (in Japan); Diater; 3-(3,4-Dichloor-fenyl)-1,1-dimethylureum (Dutch); Dichlorfenidim; 3-(3,4-Dichlorophenol)-1,1-dimethylurea; 3-(3,4-Dichlorophenyl)-1,1-demethylurea; *N'*-(3,4-Dichlorophenyl)-*N,N*-dimethylurea; *N*-(3,4-Dichlorophenyl)-*N',N'*-dimethylurea; 1-(3,4-Dichlorophenyl)-3,3-dimethylurea; 1(3,4-Dichlorophenyl)-3,3-dimethyluree (French); 3-(3,4-Dichlorphenyl)-1,1-dimethylharnstoff (German); 3-(3,4-Diclorofenil)-1,1-dimetilurea (Spanish); 3-(3,4-Dicloro-fenyl)-1,1-dimetilurea (Italian); 1,1-Dimethyl-3-(3,4-dichlorophenyl)urea; Di-On; Diurex; Diurol; Diuron 4L; DMU; Drexel; Drexel diuron 4L; Duran; Dynex; EPA pesticide chemical code 035505; Farmco diuron; HW 920; Karmex; Karmex diuron herbicide; Karmex DW; Marmer; RTECS no. YS8925000; STCC 4962622; Sup'r flo; Telvar; Telvar diuron weed killer; Tigrex; Urea, *N'*-(3,4-dichlorophenyl)-*N,N*-dimethyl-; Urea, 3-(3,4-dichlorophenyl)-1,1-dimethyl-; Urox D; Vonduron

CAS Registry Number: 330-54-1

RTECS Number: YS8925000

DOT ID: UN 2767 (solid); UN 3002 (liquid)

Regulatory Authority

- Air Pollutant Standard Set (ACGIH)[1] (Argentina) (HSE)[33] (former USSR)[35][43] (Several States)[60]
- Clean Water Act: Section 311 Hazardous Substances/RQ 40CFR117.3 (same as CERCLA, see below)
- Superfund/EPCRA 40CFR302.4 Reportable Quantity (RQ): CERCLA, 100 lb (45.4 kg)
- EPCRA Section 313 Form R *de minimis* concentration reporting level: 1.0%
- Canada, WHMIS, Ingredients Disclosure List

Cited in U.S. State Regulations: Alaska (G), California (A, G), Connecticut (A), Florida (G), Illinois (G), Maine (G), Massachusetts (G), Minnesota (G), Nevada (A), New Hampshire (G), New Jersey (G), North Dakota (A), Pennsylvania (G), Rhode Island (G), Virginia (A), West Virginia (G).

Description: Diuron, $Cl_2C_6H_3NHCON(CH_3)_2$, is a white, odorless crystalline solid. Boiling point = (decomposes) 180°C. Freezing/Melting point = 158 – 159°C. Hazard Identification (based on NFPA-704 M Rating System): Health 1, Flammability 0, Reactivity 0. Slightly soluble in water.

Potential Exposure: Those involved in the manufacture, formulation and application of the herbicide.

Incompatibilities: Strong acids.

Permissible Exposure Limits in Air: There is no OSHA PEL.[58] The HSE[33] and recommended ACGIH TLV is 10 mg/m^3 TWA. Argentina[35] has adopted 10 mg/m^3 as a TWA and set 20 mg/m^3 as a STEL. The former USSR has set a MAC in ambient air of residential areas of 0.5 mg/m^3 either on a momentary or a daily average basis.[35][43] Several states have set guidelines or standards for diuron in ambient air[60] ranging from 100 μg/m^3 (North Dakota) to 160 μg/m^3 (Virginia) to 200 μg/m^3 (Connecticut) to 238 μg/m^3 (Nevada).

Determination in Air: OSHA versatile sampler-2; Reagent; High-pressure liquid chromatography/Ultraviolet detection; NIOSH IV, Method #5601.

Permissible Concentration in Water: The former USSR[35][43] has set a MAC in water bodies used for domestic purposes of 1.0 mg/l and a MAC in water bodies used for fishery purposes of 1.5 μg/l. A long-term health advisory of 0.875 mg/l has been calculated by EPA and a lifetime health advisory of 0.014 mg/l (14 μg/l).

Determination in Water: High performance liquid chromatography may be used after extraction with Methylene chloride. Measurement is made using an ultraviolet detector.

Routes of Entry: Inhalation, ingestion, skin and/or eye contact.

Harmful Effects and Symptoms

Short Term Exposure: Exposure may irritate the skin, eyes, and throat.

Long Term Exposure: May damage the developing fetus. In animals: anemia, methemoglobinemia

Points of Attack: Eyes, skin, respiratory system, blood

Medical Surveillance: If symptoms develop or overexposure is suspected, the following may be useful: Complete blood count.

First Aid: If this chemical gets into the eyes, remove any contact lenses at once and irrigate immediately for at least 15 minutes, occasionally lifting upper and lower lids. Seek medical attention immediately. If this chemical contacts the skin, remove contaminated clothing and wash immediately with soap and water. Seek medical attention immediately. If this chemical has been inhaled, remove from exposure, begin rescue breathing (using universal precautions) if breathing has stopped and CPR if heart action has stopped. Transfer promptly to a medical facility. When this chemical has been swallowed, get medical attention. Give large quantities of water and induce vomiting. Do not make an unconscious person vomit.

Note to Physician: Treat for methemoglobinemia. Spectrophotometry may be required for precise determination of levels of methemoglobinemia in urine.

Personal Protective Methods: Wear protective gloves and clothing to prevent any reasonable probability of skin contact. Safety equipment suppliers/manufacturers can provide recommendations on the most protective glove/clothing material for your operation.. All protective clothing (suits, gloves, footwear, headgear) should be clean, available each day, and put on before work. Contact lenses should not be worn when working with this chemical. Wear dust-proof chemical goggles and face shield unless full facepiece respiratory protection is worn. Employees should wash immediately with soap when skin is wet or contaminated. Provide emergency showers and eyewash.

Respirator Selection: Where the potential exists for exposures over 10 mg/m^3, use an MSHA/NIOSH approved respirator with a pesticide cartridge. More protection is provided by a full facepiece respirator than by a half-mask respirator, and even greater protection is provided by a powered-air purifying respirator. If while wearing a filter, cartridge or canister respirator, you can detect diuron, or in the case of a full facepiece respirator you experience eye irritation, leave the area immediately. Check to make sure the respirator-to-face seal is still good. If it is, replace the filter, cartridge, or canister. If the seal is no longer good, you may need a new respirator. Where the potential for high exposures exists, use an MSHA/NIOSH approved supplied-air respirator with a full facepiece operated in the positive pressure mode or with a full facepiece, hood, or helmet in the continuous flow mode, or use a MSHA/ NIOSH approved self-contained breathing apparatus with a full facepiece operated in pressure-demand or other positive pressure mode.

Storage: Store in tightly closed containers in a cool, well-ventilated area.

Shipping: Phenyl urea pesticides, solid, toxic, n.o.s. fall in Hazard Class 6.1 and diuron in Packing Group III. This requires a "Keep Away From Food" label

Spill Handling: Evacuate persons not wearing protective equipment from area of spill or leak until clean-up is complete. Remove all ignition sources. Collect powdered material in the most convenient and safe manner and deposit in sealed containers. Ventilate area after clean-up is complete. It may be necessary to contain and dispose of this chemical as a hazardous waste. If material or contaminated runoff enters waterways, notify downstream users of potentially contaminated waters. Contact your Department of Environmental Protection or your regional office of the federal EPA for specific recommendations. If employees are required to clean-up spills, they must be properly trained and equipped. OSHA 1910.120(q) may be applicable.

Fire Extinguishing: Diuron may burn, but does not readily ignite. Use dry chemical, CO_2, water spray, or foam extinguishers. Poisonous gas is produced in fire: toxic gases dimethylamine and 3,4-dichlorophenyl isocyanate are produced at temperatures exceeding 180°C/355°F. If material or contaminated runoff enters waterways, notify downstream users of potentially contaminated waters. Notify local health and fire officials and pollution control agencies. From a secure, explosion-proof location, use water spray to cool exposed containers. If cooling streams are ineffective (venting sound increases in volume and pitch, tank discolors, or shows any signs of deforming), withdraw immediately to a secure position. If employees are expected to fight fires, they must be trained and equipped in OSHA 1910.156.

Disposal Method Suggested: Diuron, stable under normal conditions, decomposes on heating to 180 – 190°C giving dimethylamine and 3,4-dichlorophenyl isocyanate. Treatment at elevated temperatures by acid or base yields dimethylamine and 3,4-dichloroaniline. Hydrolysis is not recommended as a disposal procedure because of the generation of the toxic products, 3,4-dichloroaniline and dimethylamine.[22] Incineration is recommended. In accordance with 40CFR165 recommendations for the disposal of pesticides and pesticide containers. Must be disposed properly by following package label directions or by contacting your state pesticide or environmental control agency or by contacting your regional EPA office.

References

Sax N. I., Ed., "Dangerous Properties of Industrial Materials Report," 7, No. 5, 49–55 (1987)

U.S. Environmental Protection Agency, "Health Advisory: Diuron," Washington, DC, Office of Drinking Water (August 1987)

New Jersey Department of Health, "Hazardous Substance Fact Sheet: Diuron," Trenton, NJ (April 1997)

1,4-Divinyl Benzene

Molecular Formula: $C_{10}H_{10}$

Common Formula: $C_6H_4(CH{=}CH_2)_2$

Synonyms: Benzene, divinyl-; Diethenylbenzene; Diethyl benzene; 1,4-Divinyl benzene; DVB; DVB-22; DVB-27; DVB-55; Vinylstyrene

CAS Registry Number: 108-57-6 (m-isomer); 1321-74-0 (mixed isomers)

RTECS Number: CZ9400000 (m-isomer); CZ9370000 (mixed isomers)

DOT ID: UN 2049

EINECS Number: 215-325-5

Regulatory Authority

- Air Pollutant Standard Set (ACGIH)[1] (HSE)[33] (Several States)[60]
- Canada, WHMIS, Ingredients Disclosure List

Cited in U.S. State Regulations: Alaska (G), California (A, G), Connecticut (A), Florida (G), Illinois (G), Maine (G), Massachusetts (G), Minnesota (G), Nevada (A), New Hampshire (G, W), New Jersey (G), North Dakota (A), Pennsylvania (G), Rhode Island (G), Virginia (A), West Virginia (G).

Description: DVB, $C_6H_4(CH{=}CH_2)_2$, is a pale straw colored liquid. Boiling point = 195 – 200°C. Flash point = 76°C. Explosive limits: LEL = 0.7%; UEL = 6.2%. Hazard Identification (based on NFPA-704 M Rating System): Health 1, Flammability 2, Reactivity 2. Insoluble in water. DVB exists as o-, m-, and p-isomers. The commercial product contains all 3 isomers, but m-isomer predominates. Usually contains an inhibitor to prevent polymerization. The CAS number for the mixed isomers appears on most regulatory lists.

Potential Exposure: This compound is used as a monomer for the preparation of special synthetic rubbers, drying oils, ion-exchange resins and casting resins, and in polyester resin manufacture.

Incompatibilities: Forms explosive mixture with air. Violent reaction with strong oxidizers, metallic salts. Able to polymerize; add inhibitor and monitor to insure effective levels are maintained at all times. May accumulate static electrical charges, and may cause ignition of its vapors.

Permissible Exposure Limits in Air: There is no OSHA PEL.[58] NIOSH (10 hr. workshift) and ACGIH (8-hr. workshift) recommends TWA off 10 ppm (50 mg/m^3). HSE set the same TWA. Several states have set guidelines or standards for divinyl benzene in ambient air[60] ranging from 500 μg/m^3 (North Dakota) to 800 μg/m^3 (Virginia) to 1,000 μg/m^3 (Connecticut) to 1,190 μg/m^3 (Nevada).

Determination in Air: Charcoal tube (special); Toluene; Gas chromatography/Flame ionization detection; OSHA Method #89.

Permissible Concentration in Water: No criteria set.

Routes of Entry: Inhalation, ingestion, skin and/or eye contact.

Harmful Effects and Symptoms

Short Term Exposure: Skin, eye and respiratory system irritation. Divinyl benzene can affect you when breathed in and by passing through your skin. Irritates the eyes, skin and respiratory tract. Prolonged skin contact can cause burns and rash. Exposure can irritate the nose and throat. A central nervous system depressant, higher exposure can cause dizziness, drowsiness and passing out.

Long Term Exposure: Repeated or prolonged contact may cause skin dryness, burns, and rash.

Points of Attack: Eyes, skin, respiratory system, central nervous system.

Medical Surveillance: There is no special test for this chemical. However, if illness occurs or overexposure is suspected, medical attention is recommended.

First Aid: If this chemical gets into the eyes, remove any contact lenses at once and irrigate immediately for at least 15 minutes, occasionally lifting upper and lower lids. Seek medical attention immediately. If this chemical contacts the skin, remove contaminated clothing and wash immediately with soap and water. Seek medical attention immediately. If this chemical has been inhaled, remove from exposure, begin rescue breathing (using universal precautions) if breathing has stopped and CPR if heart action has stopped. Transfer promptly to a medical facility. When this chemical has been swallowed, get medical attention. Give large quantities of water and induce vomiting. Do not make an unconscious person vomit.

Personal Protective Methods: Wear protective gloves and clothing to prevent any reasonable probability of skin contact. Safety equipment suppliers/manufacturers can provide recommendations on the most protective glove/clothing material for your operation. Butyl Rubber, Viton, Silvershield,

Polyvinyl Alcohol and Nitrile are among the recommended protective materials. All protective clothing (suits, gloves, footwear, headgear) should be clean, available each day, and put on before work. Contact lenses should not be worn when working with this chemical. Wear splash-proof chemical goggles and face shield unless full facepiece respiratory protection is worn. Employees should wash immediately with soap when skin is wet or contaminated. Provide emergency showers and eyewash.

Respirator Selection: Where the potential exists for exposures over 10 ppm, use a MSHA/NIOSH approved supplied-air respirator with a full facepiece operated in the positive pressure mode or with a full facepiece, hood, or helmet in the continuous flow mode, or use a MSHA/NIOSH approved self-contained breathing apparatus with a full facepiece operated in pressure-demand or other positive pressure mode.

Storage: Prior to working with DVB you should be trained on its proper handling and storage. Before entering confined space where this chemical may be present, check to make sure that an explosive concentration does not exist. Store in tightly closed containers in a cool, well-ventilated area away from metallic salts (such as ferric and aluminum chlorides) and oxidizers (such as perchlorates, peroxides, permanganates, chlorates, and nitrates). Store in tightly closed containers in a cool, well-ventilated area away from heat. Sources of ignition such as smoking and open flames are prohibited where this chemical is used, handled, or stored in a manner that could create a potential fire or explosion hazard. Metal containers involving the transfer of 5 gallons or more of this chemical should be grounded and bonded. Drums must be equipped with self-closing valves, pressure vacuum bungs, and flame arresters. Use only nonsparking tools and equipment, especially when opening and closing containers of this chemical.

Shipping: Dietylbenzene requires a label of "Flammable Liquid." It fall in UN/DOT Hazard Class 3 and Packing Group III.

Spill Handling: Evacuate and restrict persons not wearing protective equipment from area of spill or leak until cleanup is complete. Remove all ignition sources. Establish forced ventilation to keep levels below explosive limit. Absorb liquids in vermiculite, dry sand, earth, peat, carbon, or a similar material and deposit in sealed containers. Keep this chemical out of a confined space, such as a sewer, because of the possibility of an explosion, unless the sewer is designed to prevent the build-up of explosive concentrations. It may be necessary to contain and dispose of this chemical as a hazardous waste. If material or contaminated runoff enters waterways, notify downstream users of potentially contaminated waters. Contact your Department of Environmental Protection or your regional office of the federal EPA for specific recommendations. If employees are required to clean-up spills, they must be properly trained and equipped. OSHA 1910.120(q) may be applicable.

Fire Extinguishing: This chemical is a combustible liquid. Poisonous gases are produced in fire. Use dry chemical, carbon dioxide, or foam extinguishers. Vapors are heavier than air and will collect in low areas. Vapors may travel long distances to ignition sources and flashback. Vapors in confined areas may explode when exposed to fire. Containers may explode in fire. Storage containers and parts of containers may rocket great distances, in many directions. If material or contaminated runoff enters waterways, notify downstream users of potentially contaminated waters. Notify local health and fire officials and pollution control agencies. From a secure, explosion-proof location, use water spray to cool exposed containers. If cooling streams are ineffective (venting sound increases in volume and pitch, tank discolors, or shows any signs of deforming), withdraw immediately to a secure position. If employees are expected to fight fires, they must be trained and equipped in OSHA 1910.156.

Disposal Method Suggested: Incineration in a furnace equipped with afterburner and scrubber.

References

New Jersey Department of Health, "Hazardous Substance Fact Sheet: Divinyl Benzene," Trenton, NJ (April 986)

Dodecylbenzenesulfonic Acid

Molecular Formula: $C_{18}H_{30}O_3S$

Synonyms: Acido dodecilbencenosulfonico (Spanish); Benzenesulfonic acid, dodecyl-; Benzene sulfonic acid, dodecyl ester; Benzenesulphonic acid, dodecyl-; Benzene sulphonic acid, dodecyl ester; Calsoft LAS 99; DDBSA; Dodanic acid 83; Dodecyl benzenesulfonate; *n*-Dodecyl benzenesulfonic acid; Dodecyl benzenesulphonate; *n*-Dodecyl benzenesulphonic acid; Dodecylbenzenesulphonic acid; E 7256; Elfan WA sulphonic acid; Lauryl benzenesulfonate; Laurylbenzenesulfonic acid; Lauryl benzenesulphonate; Laurylbenzenesulphonic acid; Nacconol 98 SA; Nansa SSA; Pentine acid 5431; Rhodacal ABSA; Richonic acid; Sulframin acid 1298

CAS Registry Number: 27176-87-0

DOT ID: NA 2584

Regulatory Authority

- Clean Water Act: Section 311 Hazardous Substances/RQ 40CFR117.3 (same as CERCLA, see below)
- Superfund/EPCRA 40CFR302.4 Reportable Quantity (RQ): CERCLA, 1,000 lb (454 kg)

Cited in U.S. State Regulations: California (G), Massachusetts (G), New Jersey (G), Pennsylvania (G).

Description: Dodecylbenzenesulfonic Acid is a light yellow to brown liquid with a slight SO_2 odor. Boiling point = 315°C. Freezing/Melting point = 10°C. Hazard Identification (based on NFPA-704 M Rating System): Health 0, Flammability 1, Reactivity 1. Soluble in water.

Potential Exposure: Used as a laboratory chemical, to make detergents, to electronically cleaning, pickling baths.

Incompatibilities: May attack metals, forming flammable hydrogen gas. Keep away from combustible materials.

Permissible Exposure Limits in Air: No OELs established.

Routes of Entry: Inhalation.

Harmful Effects and Symptoms

Short Term Exposure: A corrosive. Contact with the eyes and skin can cause severe irritation and burns. Inhalation can irritate the respiratory tract.

Long Term Exposure: Repeated skin contact may cause dermatitis. Corrosive materials may affect the lungs or cause bronchitis with coughing, phlegm and/or shortness of breath.

Points of Attack: Lungs.

Medical Surveillance: Lung function tests. Consider chest x-ray following acute overexposure.

First Aid: If this chemical gets into the eyes, remove any contact lenses at once and irrigate immediately for at least 15 minutes, occasionally lifting upper and lower lids. Seek medical attention immediately. If this chemical contacts the skin, remove contaminated clothing and wash immediately with soap and water. Seek medical attention immediately. If this chemical has been inhaled, remove from exposure, begin rescue breathing (using universal precautions) if breathing has stopped and CPR if heart action has stopped. Transfer promptly to a medical facility. When this chemical has been swallowed, get medical attention. If victim is conscious, administer water or milk. Do not induce vomiting.

Personal Protective Methods: Wear protective gloves and clothing to prevent any reasonable probability of skin contact. Safety equipment suppliers/manufacturers can provide recommendations on the most protective glove/clothing material for your operation.. All protective clothing (suits, gloves, footwear, headgear) should be clean, available each day, and put on before work. Contact lenses should not be worn when working with this chemical. Wear splash-proof chemical goggles and face shield unless full facepiece respiratory protection is worn. Employees should wash immediately with soap when skin is wet or contaminated. Provide emergency showers and eyewash.

Respirator Selection: Where the potential for exposure to this chemical, use a MSHA/NIOSH approved supplied-air respirator with a full facepiece operated in the positive pressure mode or with a full facepiece, hood, or helmet in the continuous flow mode, or use a MSHA/NIOSH approved self-contained breathing apparatus with a full facepiece operated in pressure-demand or other positive pressure mode.

Storage: Prior to working with DDBSA you should be trained on its proper handling and storage. Store in tightly closed containers in a cool, well ventilated area away from metals. Where possible, automatically pump liquid from drums or other storage containers to process containers.

Shipping: Shipping label "Corrosive." Hazard Class 8. Packing Group II.

Spill Handling: Evacuate and restrict persons not wearing protective equipment from area of spill or leak until cleanup is complete. Remove all ignition sources. Ventilate area of spill or leak. Absorb liquids in vermiculite, dry sand, earth, peat, carbon, or a similar material and deposit in sealed containers. Keep this chemical out of a confined space, such as a sewer, because of the possibility of an explosion, unless the sewer is designed to prevent the build-up of explosive concentrations. It may be necessary to contain and dispose of this chemical as a hazardous waste. If material or contaminated runoff enters waterways, notify downstream users of potentially contaminated waters. Contact your Department of Environmental Protection or your regional office of the federal EPA for specific recommendations. If employees are required to clean-up spills, they must be properly trained and equipped. OSHA 1910.120(q) may be applicable.

Fire Extinguishing: This chemical may burn but does not easily ignite. Poisonous gases including sulfur oxides and hydrogen sulfide are produced in fire. Use dry chemical, carbon dioxide, or alcohol foam extinguishers. Vapors are heavier than air and will collect in low areas. Vapors may travel long distances to ignition sources and flashback. Vapors in confined areas may explode when exposed to fire. Containers may explode in fire. Storage containers and parts of containers may rocket great distances, in many directions. If material or contaminated runoff enters waterways, notify downstream users of potentially contaminated waters. Notify local health and fire officials and pollution control agencies. From a secure, explosion-proof location, use water spray to cool exposed containers. If cooling streams are ineffective (venting sound increases in volume and pitch, tank discolors, or shows any signs of deforming), withdraw immediately to a secure position. If employees are expected to fight fires, they must be trained and equipped in OSHA 1910.156.

References

New Jersey Department of Health and Senior Services, "Hazardous Substance Fact Sheet, Dodecylbenzenesulfonic Acid," Trenton NJ (October 1996)

Dodecyl Trichlorosilane

Molecular Formula: $C_{12}H_{25}Cl_3Si$

Synonyms: Dodeciltriclorosilano (Spanish); Dodecyl trichlorosilane; Silane, dodecyltrichloro-; Silane, trichlorododecyl-; Trichlorododecylsilane

CAS Registry Number: 4484-72-4

RTECS Number: VV3940000

DOT ID: UN 1771

EINECS Number: 224-769-9

Cited in U.S. State Regulations: New Hampshire (G), New Jersey (G), Oklahoma (G).

Description: Dodecyl Trichlorosilane is a colorless to yellow liquid with a pungent odor. Boiling point ≥149°C. Flash point ≥65°C. Hazard Identification (based on NFPA-704 M Rating System): Health 3, Flammability 2, Reactivity 2. Water reactive.

Potential Exposure: This material is used in silicone polymer manufacture.

Incompatibilities: Forms explosive mixture with air. Reacts with strong bases and oxidizers. Hydrolyzes with moist air, water or steam, forming corrosive and toxic chloride gases, including hydrogen chloride. Corrosive to common metals in the presence of moisture.

Permissible Exposure Limits in Air: No standards set. However, this chemical is highly corrosive.

Permissible Concentration in Water: No criteria set.

Routes of Entry: Inhalation.

Harmful Effects and Symptoms

Dodecyl Trichlorosilane can affect you when breathed in. This substance can cause sever eye burns leading to permanent damage. Dodecyl Trichlorosilane is a corrosive chemical and contact can cause severe skin burns. Breathing Dodecyl Trichlorosilane can irritate the lungs causing coughing and/or shortness of breath. Higher exposures can cause a build-up of fluid in the lungs (pulmonary edema). This can cause death. Exposure can irritate the eyes, nose and throat.

Short Term Exposure: Dodecyl Trichlorosilane can affect you when breathed in. This substance can cause sever eye burns leading to permanent damage. Dodecyl Trichlorosilane is a corrosive chemical and contact can cause severe skin burns. Breathing Dodecyl Trichlorosilane can irritate the lungs causing coughing and/or shortness of breath. Higher exposures can cause a build-up of fluid in the lungs (pulmonary edema), a medical emergency that can be delayed for several hours. This can cause death.

Long Term Exposure: May cause bronchitis to develop with cough, phlegm, and/or shortness of breath.

Points of Attack: Lungs.

Medical Surveillance: Before beginning employment and at regular times after that, for those with frequent or potentially high exposures, the following is recommended: Lung function tests. If symptoms develop or overexposure is suspected, the following may be useful: Consider chest x-ray after acute overexposure.

First Aid: If this chemical gets into the eyes, remove any contact lenses at once and irrigate immediately for at least 15 minutes, occasionally lifting upper and lower lids. Seek medical attention immediately. If this chemical contacts the skin, remove contaminated clothing and wash immediately with soap and water. Seek medical attention immediately. If this chemical has been inhaled, remove from exposure, begin rescue breathing (using universal precautions) if breathing has stopped and CPR if heart action has stopped. Transfer promptly to a medical facility. When this chemical has been swallowed, get medical attention. If victim is conscious, administer water or milk. Do not induce vomiting. Medical observation is recommended for 24 – 48 hours after breathing overexposure, as pulmonary edema may be delayed. As first aid for pulmonary edema, a doctor or authorized paramedic may consider administering a corticosteroid spray.

Personal Protective Methods: Wear protective gloves and clothing to prevent any reasonable probability of skin contact. Safety equipment suppliers/manufacturers can provide recommendations on the most protective glove/clothing material for your operation.. All protective clothing (suits, gloves, footwear, headgear) should be clean, available each day, and put on before work. Contact lenses should not be worn when working with this chemical. Wear splash-proof chemical goggles and face shield unless full facepiece respiratory protection is worn. Employees should wash immediately with soap when skin is wet or contaminated. Provide emergency showers and eyewash.

Respirator Selection: Where the potential exists for exposure to Dodecyl Trichlorosilane, use a MSHA/NIOSH approved supplied/air respirator with a full facepiece operated in the positive pressure mode or with a full facepiece, hood, or helmet in the continuous flow mode, or use a MSHA/NIOSH approved self-contained breathing apparatus with a full facepiece operated in pressure-demand or other positive pressure mode.

Storage: Prior to working with this chemical you should be trained on its proper handling and storage. Store in tightly closed containers in a cool, well-ventilated area away from water, steam and moisture, because toxic and corrosive chloride gases including hydrogen chloride can be produced. Where possible, automatically pump liquid from drums or other storage containers to process containers. Sources of ignition, such as smoking and open flames, are prohibited where Dodecyl Trichlorosilane is used, handled, or stored in a manner that could create a potential fire or explosion hazard.

Shipping: This compound requires a shipping label of: "Corrosive." It falls in DOT Hazard Class 8 and Packing Group II.

Spill Handling: Evacuate and restrict persons not wearing protective equipment from area of spill or leak until cleanup is complete. Remove all ignition sources. Ventilate area of spill or leak. Absorb liquids in vermiculite, dry sand, crushed limestone, soda ash, earth, peat, carbon, or a similar material and deposit in sealed containers. Keep this chemical out of a confined space, such as a sewer, because of the possibility of an explosion, unless the sewer is designed to prevent the build-up of explosive concentrations. It may be necessary to contain and dispose of this chemical as a hazardous waste. If material or contaminated runoff enters waterways, notify downstream users of potentially contaminated waters. Contact your Department of Environmental Protection or your regional office of the federal EPA for specific recommendations. If employees are required to clean-up spills, they must be properly trained and equipped. OSHA 1910.120(q) may be applicable.

Fire Extinguishing: This chemical is a combustible liquid, but does not easily ignite. Poisonous gases including hydrogen chloride and phosgene are produced in fire. Use dry chemical, carbon dioxide, or alcohol resistant foam extinguishers. DO NOT use water; may react violently. Fire may restart after it has been extinguished. Vapors are heavier than air and will collect in low areas. Vapors may travel long distances to ignition sources and flashback. Vapors in confined areas may explode when exposed to fire. Containers may explode in fire. Storage containers and parts of containers may rocket great distances, in many directions. If material or contaminated runoff enters waterways, notify downstream users of potentially contaminated waters. Notify local health and fire officials and pollution control agencies. From a secure, explosion-proof location, use water spray to cool exposed containers. If cooling streams are ineffective (venting sound increases in volume and pitch, tank discolors, or shows any signs of deforming), withdraw immediately to a secure position. If employees are expected to fight fires, they must be trained and equipped in OSHA 1910.156.

References

New Jersey Department of Health, "Hazardous Substance Fact Sheet: Dodecyl Trichlorosilane," Trenton, NJ (April 1997)

Emetine Dihydrochloride

Molecular Formula: $C_{29}H_{42}Cl_2N_2O_4$

Common Formula: $C_{29}H_{40}N_2O_4 \cdot 2HCl$

Synonyms: Amebicide; (-)Emetine, dihydrochloride; 1-Emetine, dihydrochloride; Emetine, hydrochloride; NSC-33669

CAS Registry Number: 316-42-7

RTECS Number: JY5250000

DOT ID: UN 2811 (poisonous solid, organic, n.o.s.)

Regulatory Authority

- Banned or Severely Restricted (Mauritius) (UN)[13]
- Superfund/EPCRA 40CFR355, Appendix B Extremely Hazardous Substances: TPQ = 1/10,000 lb (0.454/4,540 kg)
- Superfund/EPCRA 40CFR302.4 Reportable Quantity (RQ): EHS, 1 lb (0.454 kg)

Cited in U.S. State Regulations: California (A, G), Massachusetts (G), New Jersey (G), Pennsylvania (G).

Description: Emetine dihydrochloride, $C_{29}H_{40}N_2O_4 \cdot 2HCl$, is a colorless powder which turns yellow on exposure to heat or light. Freezing/Melting point = 235 – 255°C (decomposes).

Potential Exposure: Emetine dihydrochloride is an injectable form of emetine. It is an antiamebic. Emetine is the active ingredient of Ipecac.

Permissible Exposure Limits in Air: No standards set.

Permissible Concentration in Water: No criteria set.

Routes of Entry: Inhalation, ingestion, eye and/or skin contact.

Harmful Effects and Symptoms

Symptoms include nausea, vomiting, diarrhea, muscle weakness, pain, tenderness, hypotension, precordial pain, and rapid heartbeat. This material is highly toxic orally. It is an eye irritant. Probable oral lethal dose for humans if 5 – 50 mg/kg, or between 7 drops and 1 teaspoon for a 150-lb person. The LD_{50} oral-rat is 0.012 mg/kg (12 μg/kg) (extremely toxic).

Short Term Exposure: Irritates the eyes. Symptoms include nausea, vomiting, diarrhea, muscle weakness, pain, tenderness, hypotension, precordial pain, and rapid heartbeat, diarrhea, dyspnea, hallucinations, nausea or vomiting. This material is highly toxic orally. Probable oral lethal dose for humans if 5 – 50 mg/kg, or between 7 drops and 1 teaspoon for a 150-lb person. The LD_{50} oral-rat is 0.012 mg/kg (12 μg/kg) (extremely toxic).

First Aid: If this chemical gets into the eyes, remove any contact lenses at once and irrigate immediately for at least 15 minutes, occasionally lifting upper and lower lids. Seek medical attention immediately. If this chemical contacts the skin, remove contaminated clothing and wash immediately with soap and water. Seek medical attention immediately. If this chemical has been inhaled, remove from exposure, begin rescue breathing (using universal precautions) if breathing has stopped and CPR if heart action has stopped. Transfer promptly to a medical facility. When this chemical has been swallowed, get medical attention. Give large quantities of water and induce vomiting. Do not make an unconscious person vomit.

Emetine is an alkaloid. Procedures for alkaloid salts are as follows. Move victim to fresh air; call emergency medical care. If not breathing, give artificial respiration. If breathing is difficult, give oxygen. In case of contact with material, immediately flush skin or eyes with running water for at least 15 minutes. Speed in removing material from skin is of extreme importance. Remove and isolate contaminated clothing and shoes at the site. Keep victim quiet and maintain normal body temperature. Effects may be delayed; keep victim under observation.

Personal Protective Methods: Wear protective gloves and clothing to prevent any reasonable probability of skin contact. Safety equipment suppliers/manufacturers can provide recommendations on the most protective glove/clothing material for your operation.. All protective clothing (suits, gloves, footwear, headgear) should be clean, available each day and put on before work. Contact lenses should not be worn when working with this chemical. Wear dust-proof chemical goggles and face shield unless full facepiece respiratory protection is worn. Employees should wash immediately with soap when skin is wet or contaminated. Provide emergency showers and eyewash. For emergency situations, wear a positive pressure, pressure-demand, full facepiece self-contained breathing apparatus (SCBA) or pressure-demand supplied air respirator with escape SCBA and a fully-encapsulating, chemical resistant suit.

Respirator Selection: Where the potential for exposure to this chemical, use a MSHA/NIOSH approved supplied-air respirator with a full facepiece operated in the positive pressure mode or with a full facepiece, hood, or helmet in the continuous flow mode, or use a MSHA/NIOSH approved self-contained breathing apparatus with a full facepiece operated in pressure-demand or other positive pressure mode.

Storage: Prior to working with this chemical you should be trained on its proper handling and storage. Store in a refrigerator or a cool, dry place and protect from light.

Shipping: Emetine may be classified as a poisonous solid, n.o.s. and falls in Hazard Class 6.1 and Packing Group I. It requires a "Poison" label.

Spill Handling: The following procedures should be used for alkaloid salts. Evacuate persons not wearing protective equipment from area of spill or leak until clean-up is complete. Keep unnecessary people away; isolate hazard area and deny entry. Stay upwind; keep out of low areas. Ventilate closed spaces before entering them. Remove and isolate contaminated clothing at the site. Do not touch spilled material; stop leak if you can do it without risk. Use water spray to reduce vapors. Small spills: take up with sand or other noncombustible absorbent material and place into containers for later disposal. Small dry spills: with clean shovel place material into clean, dry container and cover; move containers from spill area. Large spills: dike far ahead of spill for later disposal. It may be necessary to contain and dispose of this chemical as a hazardous waste. If material or contaminated runoff enters waterways, notify downstream users of potentially contaminated waters. Contact your Department of Environmental Protection or your regional office of the federal EPA for specific recommendations. If employees are required to clean-up spills, they must be properly trained and equipped. OSHA 1910.120(q) may be applicable.

Fire Extinguishing: Extinguishing methods for alkaloid salts are as follows. Small fires: dry chemical, carbon dioxide, water spray, or foam. Large fires: water spray, fog, or foam. Procedures for alkaloid salts include the following. Move container from fire area if you can dot so without risk. Fight fire from maximum distance. Dike fire control water for later disposal; do not scatter the material. Keep unnecessary people away; isolate hazard area and deny entry. Stay upwind; keep out of low areas. Ventilate closed spaces before entering them. Wear positive pressure breathing apparatus and special protective clothing. Poisonous gases are produced in fire including chlorine and nitrogen oxides. If material or contaminated runoff enters waterways, notify downstream users of potentially contaminated waters. Notify local health and fire officials and pollution control agencies. From a secure, explosion-proof location, use water spray to cool exposed containers. If cooling streams are ineffective (venting sound increases in volume and pitch, tank discolors, or shows any signs of deforming), withdraw immediately to a secure position. If employees are expected to fight fires, they must be trained and equipped in OSHA 1910.156.

References

U.S. Environmental Protection Agency, "Chemical Profile: Emetine Dihydrochloride," Washington, DC, Chemical Emergency Preparedness Program (November 30, 1987)

Endosulfan

Molecular Formula: $C_9H_6Cl_6O_3S$

Synonyms: Benzoepin (in Japan); Beosit; Bio 5,462; Chlorthiepin; Crisufan; Cyclodan; Devisulphan; Endocel; Endosol; Endosulfan chlorinated hydrocarbon insecticide; Endosulphan; Endox; Ensodulfan (Spanish); Ensure; ENT 23,979; FMC5462; α,β-1,2,3,4,7,7-Hexachlorobiclo(2,2,1)-hepten-5,6-bioxymethylenesulfite; 1,2,3,4,7,7-Hexachlorobiclo(2,2,1)hepten-5,6-bioxymethylenesulfite; Hexachlorohexahydromethano 2,4,3-benzodioxathiepin-3-oxide; C,C'-(1,4,5,6,7,7-Hexachloro-8,9,10-trinorborn-5-en-2,3-ylene)(dimethylsulphite)6,7,8,9,10,10-hexachloro-1,5,5a,6,9,9a-hexahydro-6,9-methano-2,4,3-benzodioxathiepin 3-oxide; 6,7,8,9,10,10-Hezachloro-1,5,5a,6,9,9a-hexahydro-6,9-methano-2,4,3-benzodioxathiepin-3-oxide; 1,4,5,6,7,7-Hezachloro-5-norborene-2,3-dimethanol cyclic sulfite; Hildan, HOE 2671; Insecto; Insectophene; Kop-Thiodan; Malix; Maux; 6,9-Methano-2,4,3-benzodioxathiepin, 6,7,8,9,10,10-hexachloro-1,5,5a,6,9,9a-hexahydro-, 3-oxide, NCI-C00566; MOS-570; NIA 5462; Niagra 5,462; OMS570; Rasayansulfan; Sulfurous acid cyclic ester with 1,4,5,6,7,7-hexachloro-5-norborene-2,3-dimethanol; Thidan; Thifor; Thimul; Thiodan (in Russia); Thiodan®; Thiodan 35; Thiofor; Thiomul; Thionex; Thiosulfan; Thiosulfan Thionel; Tiovel

CAS Registry Number: 115-29-7; 959-98-8 (alpha); 33213-65-9 (beta)

RTECS Number: RB9275000; RB9875200 (beta)

DOT ID: UN 2761

EEC Number: 602-052-00-5

Regulatory Authority

- Banned or Severely Restricted (Many Countries) (UN))[13]
- Air Pollutant Standard Set (ACGIH)[1] (HSE)[33] (Argentina)[35] (former USSR)[35][43] (OSHA)[58] (Several States)[60]
- Clean Water Act: Section 311 Hazardous Substances/RQ 40CFR117.3 (same as CERCLA, see below); Toxic Pollutant (Section 401.15)
- EPA Hazardous Waste Number (RCRA No.): P050
- RCRA, 40CFR261, Appendix 8 Hazardous Constituents.
- Superfund/EPCRA 40CFR355, Appendix B Extremely Hazardous Substances: TPQ = 10/10,000 lb (4.54/4,540 kg)

- Superfund/EPCRA 40CFR302.4 Reportable Quantity (RQ): CERCLA, 1 lb (0.454 kg)
- U.S. Dot Regulated Marine Pollutant (49CFR172.101, Appendix B). Severe pollutant

alpha:

- Clean Water Act: 40CFR423, Appendix A, Priority Pollutants; Toxic Pollutant (Section 401.15)
- RCRA 40CFR268.48; 61FR15654, Universal Treatment Standards: Wastewater (mg/l), 0.023; Nonwastewater (mg/kg), 0.066
- RCRA 40CFR264, Appendix 9; TSD Facilities Ground Water Monitoring List. Suggested test method(s) (PQL μg/l): 8080 (0.1); 8250 (10)
- Superfund/EPCRA 40CFR302.4 Reportable Quantity (RQ): CERCLA, 1 lb (0.454 kg)
- U.S. Dot Regulated Marine Pollutant (49CFR172.101, Appendix B). Severe pollutant; as endosulfan

beta:

- Clean Water Act: 40CFR423, Appendix A, Priority Pollutants; Toxic Pollutant (Section 401.15)
- RCRA 40CFR268.48; 61FR15654, Universal Treatment Standards: Wastewater (mg/l), 0.029; Nonwastewater (mg/kg), 0.13
- RCRA 40CFR264, Appendix 9; TSD Facilities Ground Water Monitoring List. Suggested test method(s) (PQL μg/l): 8080 (0.05)
- Superfund/EPCRA 40CFR302.4 Reportable Quantity (RQ): CERCLA, 1 lb (0.454 kg)
- U.S. Dot Regulated Marine Pollutant (49CFR172.101, Appendix B). Severe pollutant; as endosulfan

Cited in U.S. State Regulations: Alaska (G), California (A, G), Connecticut (A), Florida (G), Illinois (G), Kansas (G, A, W), Louisiana (G), Maine (G, W), Massachusetts (G), Michigan (G), Nevada (A), New Hampshire (G, W), New Jersey (G), New York (G), North Dakota (A), Oklahoma (G), Pennsylvania (G, A), Rhode Island (G, A, W), South Carolina (A), South Dakota (A, W), Tennessee (W), Utah (W), Vermont (G), Virginia (G, A), Washington (G), West Virginia (G), Wisconsin (G).

Description: Endosulfan, $C_9Cl_6H_6O_3S$, is a chlorinated cyclodiene insecticide. The pure product is a colorless crystalline solid. The technical product is a light to dark brown waxy solid a rotten egg or sulfur odor. Freezing/Melting point = 70 – 100°C (technical); 106°C (pure); 106°C (α); 212°C (β). Hazard Identification (based on NFPA-704 M Rating System): Health 4, Flammability 1, Reactivity 0. Insoluble in water.

Potential Exposure: Those engaged in the manufacture, formulation, and application of this material.

Incompatibilities: Strong acids, strong bases. Hydrolysed by acids. Contact with alkalis forms toxic sulfur dioxide fumes. Corrosive to iron in the presence of moisture.

Permissible Exposure Limits in Air: There is no OSHA PEL. The ACGIH, NIOSH, and HSE[33] have set a TWA of 0.1 mg/m³. HSE[33] adds an STEL of 0.3 mg/m³. The notation "skin" is added to indicate the possibility of cutaneous absorption. Argentina[35] has the same limits. The Former USSR[35][43] has set a MAC of 0.1 mg/m³ in workplace air as well as a MAC in ambient air of residential areas of 0.005 mg/m³ on a momentary basis and 0.001 mg/m³ on a daily average basis. Several states have set guidelines or standards for endosulfan in ambient air[60] ranging from 0.238 μg/m³ (Kansas) to 1.0 μg/m³ (North Dakota) to 1.6 μg/m³ (Virginia) to 2.0 μg/m³ (Connecticut and Nevada) to 2.4 μg/m³ (Pennsylvania).

Determination in Air: No test available.

Permissible Concentration in Water: To protect freshwater aquatic life – 0.056 μg/l as a 24 hr average, never to exceed 0.22 μg/l. To protect saltwater aquatic life – 0.0087 μg/l as a 24 hr average, never to exceed 0.034 μg/l. To protect human health – 74.0 μg/l.[6] Mexico[35] has set a MAC of 2 μg/l in estuaries and 0.2 μg/l in coastal waters. Kansas[61] has set a guideline of 74.0 μg/l for endosulfan in drinking water.

Determination in Water: Methylene chloride extraction followed by gas chromatography with electron capture or halogen specific detection (EPA Method 608) or gas chromatography plus mass spectrometry (EPA Method 625).

Routes of Entry: Inhalation, ingestion, eye and/or skin contact. Can be absorbed through the skin.

Harmful Effects and Symptoms

Short Term Exposure: Endosulfan may affect the central nervous system, blood, resulting in irritability, convulsions, and renal failure. High level exposure at may result in death. The effects may be delayed. Ingestion of endosulfan may result in nausea, vomiting, and diarrhea. Dizziness, agitation, nervousness, tremor, incoordination, and convulsions may also occur. Central nervous system depression may terminate in respiratory failure. Contact with endosulfan may irritate or burn the skin, eyes, and mucous membranes. The probable oral lethal dose is 50 – 500 mg/kg, or 1 teaspoonful – 1 ounce for a 150 lb person. The LD_{50} oral-rat is 18 mg/kg (highly toxic). Death has occurred within 2 hours of heavy dust exposure during bagging operations.

Long Term Exposure: Repeated exposure may cause brain damage, causing convulsions, loss of coordination, and memory loss. May cause liver and kidney damage. May damage the testes.

Points of Attack: Respiratory system, lungs, central nervous system, cardiovascular system, skin, eyes, plasma, and red blood cell cholinesterase.

Medical Surveillance: Before employment and at regular times after that, the following are recommended: Plasma and red blood cell cholinesterase levels (tests for the enzyme poisoned by this chemical). If exposure stops, plasma levels return to normal in 1 – 2 weeks while red blood cell levels may be reduced for 1 – 3 months.

When cholinesterase enzyme levels are reduced by 25% or more below preemployment levels, risk of poisoning is increased, even if results are in lower ranges of "normal." Reassignment to work not involving organophosphate or carbamate pesticides is recommended until enzyme levels recover. If symptoms develop or overexposure occurs, repeat the above tests as soon as possible and get an exam of the nervous system. Also consider complete blood count. Consider chest x-ray following acute overexposure. Do not drink any alcoholic beverages before or during use. Alcohol promotes absorption of organic phosphates. Liver and kidney function tests. Examination of the nervous system. EEG.

First Aid: If this chemical gets into the eyes, remove any contact lenses at once and irrigate immediately for at least 15 minutes, occasionally lifting upper and lower lids. Seek medical attention immediately. If this chemical contacts the skin, remove contaminated clothing and wash immediately with soap and water. Speed in removing material from skin is of extreme importance. Shampoo hair promptly if contaminated. Seek medical attention immediately. If this chemical has been inhaled, remove from exposure, begin rescue breathing (using universal precautions) if breathing has stopped and CPR if heart action has stopped. Transfer promptly to a medical facility. When this chemical has been swallowed, get medical attention. Consult hospital or poison control center on use of antidotes. Transport to health care facility.

Personal Protective Methods: Wear protective gloves and clothing to prevent any reasonable probability of skin contact. Safety equipment suppliers/manufacturers can provide recommendations on the most protective glove/clothing material for your operation. Viton or neoprene materials are recommended. All protective clothing (suits, gloves, footwear, headgear) should be clean, available each day and put on before work. Contact lenses should not be worn when working with this chemical. Wear dust-proof chemical goggles and face shield unless full facepiece respiratory protection is worn. Employees should wash immediately with soap when skin is wet or contaminated. Provide emergency showers and eyewash. If used out of doors adequate emergency water should be available.

Respirator Selection: Where the potential for exposure to this chemical, use a MSHA/NIOSH approved supplied-air respirator with a full facepiece operated in the positive pressure mode or with a full facepiece, hood, or helmet in the continuous flow mode, or use a MSHA/NIOSH approved self-contained breathing apparatus with a full facepiece operated in pressure-demand or other positive pressure mode. A dust respirator approved for pesticide use may be worn. An organic vapor respirator may be necessary if exposure to the solvent is significant.

Storage: Prior to working with this chemical you should be trained on its proper handling and storage. Store in a cool, dry, well-ventilated area, free of alkalis, acids, and acid fumes. Where possible, automatically pump liquid from drums or other storage containers to process containers.

Shipping: Organochlorine pesticides, solid, toxic, n.o.s. fall in Hazard Class 6.1 and endosulfan in Packing Group II. The label required is "Poison."

Spill Handling: Evacuate persons not wearing protective equipment from area of spill or leak until clean-up is complete. Remove all ignition sources. In case of spills, stay upwind; stay out of low areas. Use water spray to reduce vapors. Do not dry sweep. Do not touch spilled material; stop leak if you can do it without risk. Small wet spills: take up with sand or other noncombustible absorbent material and place into containers for later disposal. Small dry spills: with clean shovel place material into clean, dry container and cover; move containers from spill area. Large spills: dike far ahead of spill for later disposal. Ventilate area after clean-up is complete. It may be necessary to contain and dispose of this chemical as a hazardous waste. If material or contaminated runoff enters waterways, notify downstream users of potentially contaminated waters. Contact your Department of Environmental Protection or your regional office of the federal EPA for specific recommendations. If employees are required to clean-up spills, they must be properly trained and equipped. OSHA 1910.120(q) may be applicable.

Fire Extinguishing: This chemical is a noncombustible solid. Use any extinguishing agent suitable for surrounding fire. Poisonous gases are produced in fire including sulfur oxides and chlorine. If material or contaminated runoff enters waterways, notify downstream users of potentially contaminated waters. Notify local health and fire officials and pollution control agencies. From a secure, explosion-proof location, use water spray to cool exposed containers. If cooling streams are ineffective (venting sound increases in volume and pitch, tank discolors, or shows any signs of deforming), withdraw immediately to a secure position. If employees are expected to fight fires, they must be trained and equipped in OSHA 1910.156.

Disposal Method Suggested: A recommended method for disposal is burial 18 inches deep in noncropland, away from water supplies, but bags can be burned.[22] Large quantities should be incinerated at high temperature in a unit with effluent gas scrubbing. Consult with environmental regulatory agencies for guidance on acceptable disposal practices. Generators of waste containing this contaminant (≥100 kg/mo) must conform with EPA regulations governing storage, transportation, treatment, and waste disposal. In accordance with 40CFR165 recommendations for the disposal of pesticides and pesticide containers. Must be disposed properly by following package label directions or by contacting your state pesticide or environmental control agency or by contacting your regional EPA office.

References

U.S. Environmental Protection Agency, Endosulfan: Ambient Water Quality Criteria, Washington, DC (1980)

U.S. Environmental Protection Agency, Endosulfan, Health and Environmental Effects Profile No. 98, Washington, DC, Office of Solid Waste (April 30, 1980)

U.S. Environmental Protection Agency, "Chemical Profile: Endosulfan," Washington, DC, Chemical Emergency Preparedness Program (November 30, 1987)

New York State Department of Health, "Chemical Fact Sheet: Endosulfan," Albany, NY, Bureau of Toxic Substance Assessment (April 1986)

New Jersey Department of Health and Senior Services, "Hazardous Substance Fact Sheet, Endosulfan," Trenton NJ (May 1999)

Endothall

Molecular Formula: $C_8H_{10}NO_5PS$

Synonyms: Accelerate; Aquathol; Des-I-Cate; 1,2-Dicarboxy 3,6-endoxocyclohexane; 3,6-Endo-epoxy-1,2-cyclohexanedicarboxylic acid; 3,6-Endooxohexahydrophthalic acid; Endothal chlorophenoxy herbicide; Endothall technical; 3,6-Endoxohexahydrophthalic acid; 3,6-Epoxycyclohexane-1,2-dicarboxylic acid; Hexahydro-3,6-endooxyphthalic acid; Hydout; Hydrothal-47; Hydrothol; 7-Oxabicyclo(2.2.1)heptane-2,3-dicarboxylic acid; +Tri-endothal+

CAS Registry Number: 145-73-3; 129-67-9 (disodium salt)

RTECS Number: RN7875000; RN8225000 (disodium salt)

Regulatory Authority

- Banned or Severely Restricted (Several Countries) (UN)[13]
- EPA Hazardous Waste Number (RCRA No.): P088
- RCRA, 40CFR261, Appendix 8 Hazardous Constituents.
- Safe Drinking Water Act: MCL, 0.1 mg/l; MCLG, 0.1 mg/l; Regulated chemical (47 FR 9352)
- Superfund/EPCRA 40CFR302.4 Reportable Quantity (RQ): CERCLA, 1,000 lb (454 kg)

Cited in U.S. State Regulations: California (G), Kansas (G), Louisiana (G), Oklahoma (G), Pennsylvania (G), Vermont (G), Virginia (G), Washington (G), Wisconsin (G).

Description: Endothall, $C_8H_{10}O_5NSP$, when pure is a white crystalline solid. The technical grade is a light brown liquid. Freezing/Melting point = 144°C with conversion to the anhydride. Soluble in water.

Potential Exposure: Endothall is used as a defoliant and a herbicide on both terrestrial and aquatic weeds.

Permissible Exposure Limits in Air: No standard set.

Permissible Concentration in Water: A no-observed-adverse effects-level (NOAEL) of 2 mg/kg/day has been determined by EPA. This gives a reference dose (or acceptable daily intake) of 0.02 mg/kg/day on the basis of which a lifetime health advisory of 0.14 mg/l (140 μg/l) was calculated.

Routes of Entry: Inhalation, ingestion, eye and/or skin contact.

Harmful Effects and Symptoms

Short Term Exposure: Irritates eyes, skin, and respiratory tract. Poisonous: approximate lethal dose (human) is about 2.5 teaspoonful. Little information was found in the available literature on the health effects of endothall in humans except for one case history of a young male suicide victim who ingested an estimated 7 – 8 g of disodium Endothall in solution (approximately 100 mg Endothall ion/kg) Repeated vomiting was evident. Autopsy revealed focal hemorrhages and edema in the lungs and gross hemorrhage of the gastrointestinal (GI) tract.

Long Term Exposure: May be mutagenic.

First Aid: If this chemical gets into the eyes, remove any contact lenses at once and irrigate immediately for at least 15 minutes, occasionally lifting upper and lower lids. Seek medical attention immediately. If this chemical contacts the skin, remove contaminated clothing and wash immediately with soap and water. Seek medical attention immediately. If this chemical has been inhaled, remove from exposure, begin rescue breathing (using universal precautions) if breathing has stopped and CPR if heart action has stopped. Transfer promptly to a medical facility. When this chemical has been swallowed, get medical attention. Give large quantities of water and induce vomiting. Do not make an unconscious person vomit.

Personal Protective Methods: Wear rubber gloves for all handling. Wear protective gloves and clothing to prevent any reasonable probability of skin contact. Safety equipment suppliers/manufacturers can provide recommendations on the most protective glove/clothing material for your operation.. All protective clothing (suits, gloves, footwear, headgear) should be clean, available each day, and put on before work. Contact lenses should not be worn when working with this chemical. Wear dust-proof chemical goggles and face shield unless full facepiece respiratory protection is worn. Employees should wash immediately with soap when skin is wet or contaminated. Provide emergency showers and eyewash. Wear hats, protective suits, and boots for all handling.

Respirator Selection: Where the potential for exposure to this chemical, use a MSHA/NIOSH approved supplied-air respirator with a full facepiece operated in the positive pressure mode or with a full facepiece, hood, or helmet in the continuous flow mode, or use a MSHA/NIOSH approved self-contained breathing apparatus with a full facepiece operated in pressure-demand or other positive pressure mode.

Storage: Prior to working with endothall you should be trained on its proper handling and storage. Store in tightly closed containers in a cool, well ventilated area.

Shipping: Pesticides, solid, toxic, n.o.s. fall in Hazard Class 6.1 and endothall in Packing Group II. This requires a "Poison" label.

Spill Handling: Evacuate persons not wearing protective equipment from area of spill or leak until clean-up is complete.

Remove all ignition sources. Do not dry sweep. Use industrial vacuum. Collect powdered material in the most convenient and safe manner and deposit in sealed containers. Ventilate area after clean-up is complete. It may be necessary to contain and dispose of this chemical as a hazardous waste. If material or contaminated runoff enters waterways, notify downstream users of potentially contaminated waters. Contact your Department of Environmental Protection or your regional office of the federal EPA for specific recommendations. If employees are required to clean-up spills, they must be properly trained and equipped. OSHA 1910.120(q) may be applicable.

Fire Extinguishing: This chemical is a combustible solid. Use dry chemical, carbon dioxide, water spray, or alcohol foam extinguishers. Poisonous gases are produced in fire. If material or contaminated runoff enters waterways, notify downstream users of potentially contaminated waters. Notify local health and fire officials and pollution control agencies. From a secure, explosion-proof location, use water spray to cool exposed containers. If cooling streams are ineffective (venting sound increases in volume and pitch, tank discolors, or shows any signs of deforming), withdraw immediately to a secure position. If employees are expected to fight fires, they must be trained and equipped in OSHA 1910.156.

Disposal Method Suggested: Small quantities may be disposed of by burial in soil which is rich in organic matter. Large quantities are best disposed of by incineration[22]. Consult with environmental regulatory agencies for guidance on acceptable disposal practices. Generators of waste containing this contaminant (=100 kg/mo) must conform with EPA regulations governing storage, transportation, treatment, and waste disposal. In accordance with 40CFR165 recommendations for the disposal of pesticides and pesticide containers. Must be disposed properly by following package label directions or by contacting your state pesticide or environmental control agency or by contacting your regional EPA office.

References

Sax, N. I., Ed., "Dangerous Properties of Industrial Materials Report," 8, No. 6, 51–56 (1988)

U.S. Environmental Protection Agency, "Health Advisory: Endothall," Washington, DC, Office of Drinking Water (August 1987)

Endothion

Molecular Formula: $C_9H_{13}O_6PS$

Synonyms: AC-18,737; *O,O*-Dimethyl *S*-(5-methoxy-4-oxo-4H-pyran-2-yl) phosphorothioate; *O,O*-Dimethyl-S-[(5-methoxy-pyron-2-yl)-methyl]-thiolphosphat (German); *O,O*-Dimethyl *S*-(5-methoxypyronyl-2-methyl) thiolphosphate; Endocid; Endocide; Endotiona (Spanish); ENT 24,653; Exothion; 5-Methoxy-2-(dimethoxyphosphinylthiomethyl)pyrone-4; *S*-5-Methoxy-4-oxopyran-2-ylmethyl dimethyl phosphorothioate; *S*-[(5-Methoxy-4H-pyron-2-yl)-methyl]-*o,o*-dimethyl-monothiofosfaat (Dutch); *S*-[(5-Methoxy-4H-pyron-2-yl)-methyl]-*O,O*-dimethyl-monothiophosphat (German); *S*-(5-Methoxy-4-pyron-2-ylmethyl) dimethyl phosphorothiolate; NIA-5767; Niagra 5767; Phosphate 100; Phosphopyron; Phosphopyrone; Phosphorothioate; Thiophosphate de *O,O*-dimethyle et de *S*-[(5-methoxy-4-pyronyl)-methyle] (French)

CAS Registry Number: 2778-04-3

RTECS Number: IF8225000

DOT ID: UN 2783 (organic phosphorus compound, solid)

Regulatory Authority

- Superfund/EPCRA 40CFR355, Appendix B Extremely Hazardous Substances: TPQ = 500/10,000 lb (227/4,540 kg)
- Superfund/EPCRA 40CFR302.4 Reportable Quantity (RQ): EHS, 1 lb (0.454 kg)

Cited in U.S. State Regulations: Massachusetts (G), Oklahoma (G), New Jersey, Pennsylvania (G).

Description: Endothion, $C_9H_{13}O_6PS$, is a white crystalline solid with a slight odor. Freezing/Melting point = 96°C. Hazard Identification (based on NFPA-704 M Rating System): Health 4, Flammability 1, Reactivity 0. Soluble in water.

Potential Exposure: This material is a systemic insecticide. It is not sold in the U.S. or Canada.

Incompatibilities: Strong oxidizers.

Permissible Exposure Limits in Air: No standards set.

Permissible Concentration in Water: No criteria set.

Routes of Entry: Inhalation, ingestion, skin contact.

Short Term Exposure: Organic phosphorus insecticides are absorbed by the skin, as well as by the respiratory and gastrointestinal tracts. They are cholinesterase inhibitors. Symptoms of exposure include headache, giddiness, blurred vision, nervousness, weakness, nausea, cramps, diarrhea, and discomfort in the chest. Signs include sweating, tearing, salivation, vomiting, cyanosis, convulsions, coma, loss of reflexes and loss of sphincter control. Exposure may cause psychotic behavior, loss of coordination, unconsciousness, and rarely, convulsions. This material is poisonous to humans. Its toxic effects are most likely related to action on the nervous system, The LD_{50} oral-rat is 23 mg/kg (highly toxic).

Long Term Exposure: Cholinesterase inhibitor; cumulative effect is possible. Endothion may damage the nervous system with repeated exposure, resulting in convulsions, respiratory failure. May cause liver damage.

Points of Attack: Respiratory system, lungs, central nervous system, cardiovascular system, skin, eyes, plasma, and red blood cell cholinesterase.

Medical Surveillance: Before employment and at regular times after that, the following are recommended: Plasma and red blood cell cholinesterase levels (tests for the enzyme poisoned by this chemical). If exposure stops, plasma levels return to normal in 1 – 2 weeks while red blood cell levels may be reduced for 1 – 3 months.

When cholinesterase enzyme levels are reduced by 25% or more below preemployment levels, risk of poisoning is increased, even if results are in lower ranges of "normal." Reassignment to work not involving organophosphate or carbamate pesticides is recommended until enzyme levels recover. If symptoms develop or overexposure occurs, repeat the above tests as soon as possible and get an exam of the nervous system. Also consider complete blood count. Consider chest x-ray following acute overexposure. Do not drink any alcoholic beverages before or during use. Alcohol promotes absorption of organic phosphates.

First Aid: If this chemical gets into the eyes, remove any contact lenses at once and irrigate immediately for at least 15 minutes, occasionally lifting upper and lower lids. Seek medical attention immediately. If this chemical contacts the skin, remove contaminated clothing and wash immediately with soap and water. Speed in removing material from skin is of extreme importance. Shampoo hair promptly if contaminated. Speed in removing material from skin is of extreme importance. Seek medical attention immediately. If this chemical has been inhaled, remove from exposure, begin rescue breathing (using universal precautions) if breathing has stopped and CPR if heart action has stopped. Transfer promptly to a medical facility. When this chemical has been swallowed, get medical attention. Give large quantities of water and induce vomiting. Do not make an unconscious person vomit. Keep victim quiet and maintain normal body temperature. Effects may be delayed; keep victim under observation.

Personal Protective Methods: Wear protective gloves and clothing to prevent any reasonable probability of skin contact. Safety equipment suppliers/manufacturers can provide recommendations on the most protective glove/clothing material for your operation.. All protective clothing (suits, gloves, footwear, headgear) should be clean, available each day, and put on before work. Contact lenses should not be worn when working with this chemical. Wear dust-proof chemical goggles and face shield unless full facepiece respiratory protection is worn. Employees should wash immediately with soap when skin is wet or contaminated. Provide emergency showers and eyewash.

Respirator Selection: Where the potential for exposure to this chemical, use a MSHA/NIOSH approved supplied-air respirator with a full facepiece operated in the positive pressure mode or with a full facepiece, hood, or helmet in the continuous flow mode, or use a MSHA/NIOSH approved self-contained breathing apparatus with a full facepiece operated in pressure-demand or other positive pressure mode.

Storage: Prior to working with this chemical you should be trained on its proper handling and storage. Store in tightly closed containers in a cool, well ventilated area away from strong oxidizers.

Shipping: Organophosphorus pesticides, solid, toxic, n.o.s. fall in Hazard Class 6.1 and Endothion in Packing Group II. The label required is "Poison."

Spill Handling: Stay upwind; keep out of low areas. Ventilate closed spaces before entering them. Do not touch spilled material; stop leak if you can do so without risk. Use water spray to reduce vapors. Small spills: absorb with sand or other noncombustible absorbent material and place into containers for later disposal. Small dry spills: with clean shovel place material into clean, dry containers and cover; move containers from spill area. Large spills: dike far ahead of spill for later disposal. It may be necessary to contain and dispose of this chemical as a hazardous waste. If material or contaminated runoff enters waterways, notify downstream users of potentially contaminated waters. Contact your Department of Environmental Protection or your regional office of the federal EPA for specific recommendations. If employees are required to clean-up spills, they must be properly trained and equipped. OSHA 1910.120(q) may be applicable.

Fire Extinguishing: This material may burn, but does not ignite readily. For small fires, use dry chemical, carbon dioxide, water spray, or foam. For large fires, use water spray, fog, or foam. Stay upwind; keep out of low areas. Move containers from fire area if you can do it without risk. Fight fire from maximum distance. Dike fire control water for later disposal; do not scatter the material. Ear positive pressure breathing apparatus and special protective clothing. Poisonous gases are produced in fire including phosphorus oxides and sulfur oxides. If material or contaminated runoff enters waterways, notify downstream users of potentially contaminated waters. Notify local health and fire officials and pollution control agencies. From a secure, explosion-proof location, use water spray to cool exposed containers. If cooling streams are ineffective (venting sound increases in volume and pitch, tank discolors, or shows any signs of deforming), withdraw immediately to a secure position. If employees are expected to fight fires, they must be trained and equipped in OSHA 1910.156.

Disposal Method Suggested: In accordance with 40CFR 165 recommendations for the disposal of pesticides and pesticide containers. Must be disposed properly by following package label directions or by contacting your state pesticide or environmental control agency or by contacting your regional EPA office.

References

U.S. Environmental Protection Agency, "Chemical Profile: Endothion," Washington, DC, Chemical Emergency Preparedness Program (November 30, 1987)

Endoxan

Molecular Formula: $C_7H_{15}C_{12}N_2O_2P$

Synonyms: *N,N*-Bis-(β-chloraethyl)-*N'*,*O*-propylen-phosphorsaeure-ester-diamid (German); 2-[Bis(2-chloroethyl)amino]-1-oxa-3-aza-2-phosphocyclohexane 2-oxide monohydrate; 2-[Bis(2-chloroethyl)amino)-2H-1,3, 2-oxazaphosphorine 2-oxide; 1-Bis(2-chloroethyl)amino-1-oxo-2-aza-5-oxaphosphoridine monohydrate; [Bis(chloro-

2-ethyl)amino]-2-tetrahydro-3,4,5,6-oxazaphosphorine-1,3, 2-oxide-2 hydrate; 2-[Bis(2-chloroethyl)amino]tetrahydro(2H)-1,3,2-oxazaphosphorine 2-oxide monohydrate; *N,N*-Bis(2-chloroethyl)-*N'*-(3-hydroxypropyl)phosphorodiamidic acid intramol ester hydrate; Bis(2-chloroethyl)phosphoramide-cyclic propanolamide ester; Bis(2-chloroethyl) phosphoramide cyclic propanolamide ester monohydrate; *N,N*-Bis(β-chloroethyl)-*N',O*-propylenephosphoric acid ester; *N,N*-Bis(2-chloroethyl)-*N',O*-propylenephosphoric acid ester diamide; Cyclophosphamide; Cyclophosphamidum; Cyclophosphan; Cyclostin; Cytophosphan; Cytoxan; *N,N*-Di(2-chloroethyl)-*N,O*-propylenephosphoric acid ester diamide; Endoxanal; Genoxal; Hexadrin; Mitoxan; NCI-C04900; Neosar; NSC-26271; 2-H-1,3,2-Oxazaphosphorinane; Procytox; Semdoxan; Senduxan; SK 20501; Zyklophosphamid (German)

CAS Registry Number: 50-18-0

RTECS Number: RP5950000

DOT ID: UN 1851

Regulatory Authority

- CARCINOGEN: (human positive) (IARC), (animal positive)(NTP)
- EPA Hazardous Waste Number (RCRA No.): U058
- RCRA, 40CFR261, Appendix 8 Hazardous Constituents
- Superfund/EPCRA 40CFR302.4 Reportable Quantity (RQ): CERCLA, 10 lb (4.54 kg)

Cited in U.S. State Regulations: California (G), Massachusetts (G), New Jersey (G), Pennsylvania (G).

Description: Endoxan is a white crystalline powder. It may be used or shipped in solution. Freezing/Melting point = 41 – 43°C. Soluble in water.

Potential Exposure: It is a synthetic antineoplastic drug used for treating leukemia.

Incompatibilities: Should be protected from temperatures above 86°F/30°C.

Permissible Exposure Limits in Air: No OELs set.

Routes of Entry: Inhalation, passing through the skin.

Harmful Effects and Symptoms

Short Term Exposure: When used as a medical drug it can cause nausea, vomiting, and may interfere with the body's manufacture of blood cells (anemia). It is not known if this also occurs in workplace exposures. Contact may cause eye damage.

Long Term Exposure: This chemical is a carcinogen in humans. It can cause leukemia, Hodgkin's disease, gastrointestinal, and bladder cancer. It is s probable teratogen in humans and may cause sterility in males and females. Chronic exposure may cause hair loss, darkening of the skin, and blood in the urine. May lead to scarring of the lungs. May cause kidney damage.

Points of Attack: Blood, lungs, bladder.

Medical Surveillance: Complete blood count. Lung function tests. Kidney function tests. GI tests. Tests for blood in urine.

First Aid: If this chemical gets into the eyes, remove any contact lenses at once and irrigate immediately for at least 15 minutes, occasionally lifting upper and lower lids. Seek medical attention immediately. If this chemical contacts the skin, remove contaminated clothing and wash immediately with soap and water. Seek medical attention immediately. If this chemical has been inhaled, remove from exposure, begin rescue breathing (using universal precautions) if breathing has stopped and CPR if heart action has stopped. Transfer promptly to a medical facility. When this chemical has been swallowed, get medical attention. Give large quantities of water and induce vomiting. Do not make an unconscious person vomit.

Personal Protective Methods: Wear protective gloves and clothing to prevent any reasonable probability of skin contact. Safety equipment suppliers/manufacturers can provide recommendations on the most protective glove/clothing material for your operation.. All protective clothing (suits, gloves, footwear, headgear) should be clean, available each day, and put on before work. Contact lenses should not be worn when working with this chemical. Wear dust-proof chemical goggles and face shield unless full facepiece respiratory protection is worn. Employees should wash immediately with soap when skin is wet or contaminated. Provide emergency showers and eyewash. If you are required to work in a "sterile" environment you require special training.

Respirator Selection: Where the potential for exposure to this chemical, use a MSHA/NIOSH approved supplied-air respirator with a full facepiece operated in the positive pressure mode or with a full facepiece, hood, or helmet in the continuous flow mode, or use a MSHA/NIOSH approved self-contained breathing apparatus with a full facepiece operated in pressure-demand or other positive pressure mode.

Storage: Prior to working with this chemical you should be trained on its proper handling and storage. Store in tightly closed containers in a cool, well ventilated area. Store at below 25°C. A regulated, marked area should be established where this chemical is handled, used, or stored in compliance with OSHA standard 1910.1045.

Shipping: This compound requires a shipping label of: "Poison." It falls in DOT Hazard Class 6.1 and Packing Group II.

Spill Handling: Evacuate persons not wearing protective equipment from area of spill or leak until clean-up is complete. Remove all ignition sources. Collect powdered material in the most convenient and safe manner and deposit in sealed containers. Ventilate area after clean-up is complete. It may be necessary to contain and dispose of this chemical as a hazardous waste. If material or contaminated runoff enters waterways, notify downstream users of potentially contaminated waters. Contact your Department of Environmental Protection or your regional office of the federal EPA for specific recommendations. If employees are required to clean-up spills, they must be properly trained and equipped. OSHA 1910.120(q) may be applicable.

Fire Extinguishing: Use dry chemical, carbon dioxide, water spray, or foam extinguishers. Poisonous gases are produced in fire including chlorine, nitrogen oxides, phosphorus oxides. If material or contaminated runoff enters waterways, notify downstream users of potentially contaminated waters. Notify local health and fire officials and pollution control agencies. From a secure, explosion-proof location, use water spray to cool exposed containers. If cooling streams are ineffective (venting sound increases in volume and pitch, tank discolors, or shows any signs of deforming), withdraw immediately to a secure position. If employees are expected to fight fires, they must be trained and equipped in OSHA 1910.156.

Disposal Method Suggested: Consult with environmental regulatory agencies for guidance on acceptable disposal practices. Generators of waste containing this contaminant (≥100 kg/mo) must conform with EPA regulations governing storage, transportation, treatment, and waste disposal.

References

New Jersey Department of Health and Senior Services, "Hazardous Substance Fact Sheet, Cyclophosphamide," Trenton NJ (1998)

Endrin

Molecular Formula: $C_{12}H_8Cl_6O$

Synonyms: Compound 269; 2,7:3,5-Dimethanonaphth(2,3-b)oxirene, 3,4,5,6,9,9-hexachloro-1a,2,2a,3,6,6a,7,7a-octahydro-, (aα,2.β,2aβ,2aβ,3α,6α,6aβ,7β,7aα)-; Endrex; Endrina (Spanish); Endrin chlorinated hydrocarbon insecticide; Endrine (French); ENT 17,251; Hexachloroepoxy-octahydro-endo, endo-dimethanonapthalene; (1r, 4s, 4as, ss, 7r, 8r, 8ar)-1,2,3,4,10-Hexachloro-1,4,4a,5,6,7,8,8a-octahydro-6,7-epoxy-1,4:5,8-dimethano naphthalene; Hexadrin; 1,2,3,4,10,10-Hezachloro-6,7-epoxy-1,4,4a,5,6, 7,8,8a-octahydro-1,4-endo-endo-1,4,5,8-dimethano-naphthalene; Mendrin; NCI-C00157; Nendrin

CAS Registry Number: 72-20-8

RTECS Number: IO1575000

DOT ID: NA 2761

Regulatory Authority

- Banned or Severely Restricted (Many Countries) (UN)[13]
- Air Pollutant Standard Set (ACGIH)[1] (Australia) (HSE)[33] (DFG)[3] (Israel) (Mexico) (OSHA)[58] (Several States)[60] (Several Canadian Provinces)
- Clean Water Act: Section 311 Hazardous Substances/RQ 40CFR117.3 (same as CERCLA, see below); 40CFR423, Appendix A, Priority Pollutants.
- EPA Hazardous Waste Number (RCRA No.): P051; D012
- RCRA Toxicity Characteristic (Section 261.24), Maximum Concentration of Contaminants, regulatory level, 0.02 mg/l
- RCRA, 40CFR261, Appendix 8 Hazardous Constituents.
- RCRA 40CFR268.48; 61FR15654, Universal Treatment Standards: Wastewater (mg/l), 0.0028; Nonwastewater (mg/kg), 0.13
- RCRA 40CFR264, Appendix 9; TSD Facilities Ground Water Monitoring List. Suggested test method(s) (PQL μg/l): 8080 (0.1); 8250 (10)
- Safe Drinking Water Act: MCL, 0.002 mg/l; MCLG, 0.002 mg/l; Regulated chemical (47 FR 9352)
- Superfund/EPCRA 40CFR355, Appendix B Extremely Hazardous Substances: TPQ = 500/10,000 lb (227/4,540 kg)
- Superfund/EPCRA 40CFR302.4 Reportable Quantity (RQ): CERCLA, 1 lb (0.454 kg)
- U.S. DOT Regulated Marine Pollutant (49CFR172.101, Appendix B)
- Mexico, Drinking Water Criteria, 0.07 mg/l.

Cited in U.S. State Regulations: Alaska (G), California (A, G), Connecticut (A), Florida (G), Illinois (G), Kansas (G, A), Louisiana (G), Maine (G, W), Massachusetts (G, W), Michigan (G), Nevada (A), New Hampshire (G), New Jersey (G), New York (G), North Dakota (A), Oklahoma (G), Pennsylvania (G, A), Rhode Island (G), Vermont (G), Virginia (G, A), Washington (G), West Virginia (G), Wisconsin (G).

Description: Endrin is the common name of one member of the cyclodiene group of pesticides. It is a cyclic hydrocarbon having a chlorine-substituted methano-bridge structure. Endrin is a white, crystalline solid. Freezing/Melting point = 230°C (decomposes). *Mixture in xylene:* Flash point = 27°C. Explosive limits: LEL = 1.1%; UEL = 7.0%. Hazard Identification (based on NFPA-704 M Rating System): Health 4, Flammability 1, Reactivity 0. Insoluble in water.

Potential Exposure: Those involved in manufacture, formulation and field application of this pesticide.

Incompatibilities: Parathion, strong acids (forms explosive vapors), strong oxidizers. Slightly corrosive to metal.

Permissible Exposure Limits in Air: The OSHA PEL[58], the HSE TWA[33] and the DFG MAC[3] and ACGIH value is 0.1 mg/m^3 TWA. The notation "skin" indicates the possibility of cutaneous absorption. Australia, Israel, Mexico, and the Canadian provinces of Alberta, BC, Ontario, and Quebec use the same TWA as OSHA and Alberta, BC, HSE[33] and Mexico set a STEL of 0.3 mg/m^3 and that set in Germany is 1.0 mg/m^3.[35] The NIOSH IDLH level is 2 mg/m^3. Several states have set guidelines or standards for Endrin in ambient air[60] ranging from 0.07 μg/m^3 (Pennsylvania) to 0.238 μg/m^3 (Kansas) to 1.0 μg/l (North Dakota) to 1.6 μg/l (Virginia) to 2.0 μg/l (Connecticut and Nevada).

Determination in Air: Collection by filter/chromosorb tube-102; workup with toluene; analysis by gas chromatography/electrochemical detection; NIOSH IV, Method #5519.

Permissible Concentration in Water: To protect freshwater aquatic life – 0.0023 μg/l as a 24 hr average, never to exceed 0.18 μg/l. To protect saltwater aquatic life – 0.0023 μg/l as a 24 hr average, never to exceed 0.037 μg/l. To protect human health – 1.0 μg/l[6]. According to a UN publication[35], the limit on Endrin in drinking water delivered at the tap is 0.2 μg/l. Mexico has imposed limits on Endrin in water[35] as follows: 2 μg/l in estuaries; 1 μg/l in receiving waters used for drinking water supply and 0.2 μg/l in coastal waters. The EPA has derived a no-observed-adverse effects-level (NOAE) of 0.045 mg/kg/day on the basis of which they have arrived at a long-term health advisory of 16 μg/l and a lifetime health advisory of 0.32 μg/l. Massachusetts has set a standards of 0.2 μg/l and Maine a guideline of 0.2 μg/l for Endrin in drinking water.[61] Mexico's drinking water criteria is 0.07 mg/l.

Determination in Water: Methylene chloride extraction followed by gas chromatography with electron capture or halogen specific detection (EPA Method 608) or gas chromatography plus mass spectrometry (EPA Method 625).

Routes of Entry: Inhalation, skin absorption, ingestion, skin and/or eye contact. Quickly passes through the skin.

Harmful Effects and Symptoms

Short Term Exposure: Contact can irritate the skin and eyes and may affect vision. Inhalation can cause irritation of the respiratory tract. Exposure can cause headache, nausea, vomiting, diarrhea, loss of appetite, sweating and weakness, lightheadedness, dizziness, convulsions and unconciousness. Lower exposure can affect concentration, memory and muscle coordination. Endrin can cause death by respiratory arrest. Symptoms include headache, nausea, vomiting, dizziness, tremors, convulsions, loss of consciousness, rise in blood pressure, fever, frothing of the mouth, deafness, coma, and death. This material is extremely toxic. It is rapidly absorbed through the skin. Symptoms appear between 20 minutes and 12 hours after exposure. Doses of 1 mg/kg can cause symptoms. Also, it is a central nervous system depressant and hepatotoxin. Inhalation may cause irritation to nose and throat, and sudden convulsions, which may occur from 30 minutes to 10 hours after exposure. Recovery is usually rapid, but headache, dizziness, lethargy, weakness, and weight loss may persist to 2 – 4 weeks. Prolonged breathing or ingestion can result in an onset of symptoms in 3 hours at a dose of 1 mg per kg of body weight. Ingestion of 12 grams has caused death. Pregnant women are considered to be at special risk. A rebuttable presumption notice against pesticide registration was issued on July 27, 1976 by EPA on the basis of oncogenicity, teratogenicity, and reductions in endangered species and non-target species.

Long Term Exposure: May damage the developing fetus. May damage the nervous system causing numbness and weakness in the extremities. Repeated exposure may cause personality changes of depression, anxiety and/or irritability. May cause anorexia. High or repeated exposure may cause liver damage.

Points of Attack: Respiratory system, lungs, central nervous system, cardiovascular system, skin, eyes, plasma and red blood cell cholinesterase.

Medical Surveillance: Before employment and at regular times after that, the following are recommended: Plasma and red blood cell cholinesterase levels (tests for the enzyme poisoned by this chemical). If exposure stops, plasma levels return to normal in 1 – 2 weeks while red blood cell levels may be reduced for 1 – 3 months.

When cholinesterase enzyme levels are reduced by 25% or more below preemployment levels, risk of poisoning is increased, even if results are in lower ranges of "normal." Reassignment to work not involving organophosphate or carbamate pesticides is recommended until enzyme levels recover. If symptoms develop or overexposure occurs, repeat the above tests as soon as possible and get an exam of the nervous system. Also consider complete blood count. Consider chest x-ray following acute overexposure. Do not drink any alcoholic beverages before or during use. Alcohol promotes absorption of organic phosphates. Examination of the nervous system. Electroencephalogram (a test for abnormal seizure activity). Blood Endrin level. Liver and kidney function tests.

First Aid: If this chemical gets into the eyes, remove any contact lenses at once and irrigate immediately for at least 15 minutes, occasionally lifting upper and lower lids. Seek medical attention immediately. If this chemical contacts the skin, remove contaminated clothing and wash immediately with soap and water. Speed in removing material from skin is of extreme importance. Shampoo hair promptly if contaminated. Seek medical attention immediately. If this chemical has been inhaled, remove from exposure, begin rescue breathing (using universal precautions) if breathing has stopped and CPR if heart action has stopped. Transfer promptly to a medical facility. When this chemical has been swallowed, get medical attention. Give large quantities of water and induce vomiting. Do not make an unconscious person vomit.

Personal Protective Methods: Wear protective gloves and clothing to prevent any reasonable probability of skin contact. Safety equipment suppliers/manufacturers can provide recommendations on the most protective glove/clothing material for your operation.. All protective clothing (suits, gloves, footwear, headgear) should be clean, available each day, and put on before work. Contact lenses should not be worn when working with this chemical. Wear dust-proof chemical goggles and face shield unless full facepiece respiratory protection is worn. Employees should wash immediately with soap when skin is wet or contaminated. Provide emergency showers and eyewash.

Respirator Selection: NIOSH/OSHA: 1 mg/m^3: CCROVDMFu [any chemical cartridge respirator with organic vapor cartridge(s) in combination with a dust, mist, and fume filter]; or SA (any supplied-air respirator); 2 mg/m^3: SA:CF (any supplied-air respirator operated in a continuous-flow mode); PAPROVDMFu [any powered, air purifying respirator with organic vapor cartridge(s) in combination with a dust, mist and, fume filter]; or CCRFOVHiE [any chemical cartridge respirator with a full facepiece and organic vapor cartridge(s) in combination with a high efficiency particulate filter]; or

GMFOVHiE [any air-purifying, full-facepiece respirator (gas mask) with a chin-style, front- or back-mounted organic vapor canister having a high-efficiency particulate filter]; or SCBAF (any self-contained breathing apparatus with full facepiece); or SAF (any supplied-air respirator with a full facepiece). *Emergency or planned entry into unknown concentrations or IDLH conditions:* SCBAF:PD,PP (any self-contained breathing apparatus that has a full facepiece and is operated in a pressure-demand or other positive-pressure mode); or SAF:PD,PP:ASCBA (any supplied-air respirator that has a full facepiece and is operated in a pressure-demand or other positive-pressure mode in combination with an auxiliary self-contained breathing apparatus operated in a pressure-demand or other positive pressure mode). *Escape:* GMFOVHiE [any air-purifying, full-facepiece respirator (gas mask) with a chin-style, front-or back-mounted canister having a high efficiency particulate filter]; or SCBAE (any appropriate escape-type, self-contained breathing apparatus).

Storage: Prior to working with Endrin you should be trained on its proper handling and storage. Before entering confined space where this chemical may be present, check to make sure that an explosive concentration does not exist. Store in tightly closed containers in a cool, well ventilated area away from oxidizers and heat. Where possible, automatically pump liquid from drums or other storage containers to process containers.

Shipping: Organophosphorus pesticides, solid, toxic, n.o.s. fall in Hazard Class 6.1 and Endrin in Packing Group I.

Spill Handling: For leaks or spills, use water spray to disperse vapor and to flush spills. Liquid containing this material should be absorbed in vermiculite, dry sand, earth. Do not touch spilled material; stop leak if you can do it without risk. Establish forced ventilation to keep levels below explosive limit. Small dry spills: collect powdered material in the most convenient and safe manner and deposit in sealed containers; move containers from spill area. Large spills: dike far ahead of spill for later disposal. Keep unnecessary people away; isolate hazard area and deny entry. It may be necessary to contain and dispose of this chemical as a hazardous waste. If material or contaminated runoff enters waterways, notify downstream users of potentially contaminated waters. Contact your Department of Environmental Protection or your regional office of the federal EPA for specific recommendations. If employees are required to clean-up spills, they must be properly trained and equipped. OSHA 1910.120(q) may be applicable.

Fire Extinguishing: Extinguish fire using an agent suitable for type of surrounding fire. Endrin itself does not burn but it may be dissolved in a flammable liquid. Poisonous gases are produced in fire including phosgene and hydrogen chloride. Use dry chemical, foam, carbon dioxide, water spray for solution. Small fires: dry chemical, carbon dioxide, water spray, or foam. Large fires: water spray, fog, or foam. Use water to keep fire-exposed containers cool. Keep unnecessary people away; isolate hazard area and deny entry. Stay upwind: keep out of low areas. Establish forced ventilation to keep levels below explosive limit. Wear positive pressure breathing apparatus and special protective clothing. Remove and isolate contaminated clothing at the site. Move container from fire area if you can do it without risk. Fight fire from maximum distance. Dike fire control water for later disposal; do not scatter the material. If material or contaminated runoff enters waterways, notify downstream users of potentially contaminated waters. Notify local health and fire officials and pollution control agencies. From a secure, explosion-proof location, use water spray to cool exposed containers. If cooling streams are ineffective (venting sound increases in volume and pitch, tank discolors, or shows any signs of deforming), withdraw immediately to a secure position. If employees are expected to fight fires, they must be trained and equipped in OSHA 1910.156.

Disposal Method Suggested: A disposal procedure recommended by the manufacturer consists of absorption, if necessary, and burial at least 18 inches deep, preferably in sandy soil in a flat or depressed location away from wells, livestock, children, wildlife, etc. Incineration is the recommended method.[22] Consult with environmental regulatory agencies for guidance on acceptable disposal practices. Generators of waste containing this contaminant (≥100 kg/mo) must conform with EPA regulations governing storage, transportation, treatment, and waste disposal. In accordance with 40CFR165 recommendations for the disposal of pesticides and pesticide containers. Must be disposed properly by following package label directions or by contacting your state pesticide or environmental control agency or by contacting your regional EPA office.

References

U.S. Environmental Protection Agency, Endrin: Ambient Water Quality Criteria, Washington, DC (1980)

U.S. Environmental Protection Agency, Reviews of the Environmental Effects of Pollutants: XIII, Endrin, Report EPA-600/1-79-005, Cincinnati, OH (1979)

U.S. Environmental Protection Agency, Endrin, Health and Environmental Effects Profile No. 99, Washington, DC, Office of Solid Waste (April 30, 1980)

Sax, N. I., Ed., "Dangerous Properties of Industrial Materials Report," 1, No. 5, 55–57 (1981)

U.S. Environmental Protection Agency, "Chemical Profile: Endrin," Washington, DC, Chemical Emergency Preparedness Program (November 30, 1987)

New Jersey Department of Health and Senior Services, "Hazardous Substance Fact Sheet: Endrin," Trenton, NJ (December, 1998)

New York State Department of Health, "Chemical Fact Sheet: Endrin," Albany, NY, Bureau of Toxic Substance Assessment (May 1986)

Enflurane

Molecular Formula: $C_3H_2ClF_5O$

Common Formula: CHF_2OCF_2CHFCl

Synonyms: 2-Chloro-1-(difluoromethoxy)-1,1,2-trifluoroethane; 2-Chloro-1,1,2-trifluoroethyl difluoromethyl ether; Ethrane®; Ethrane methylflurether

CAS Registry Number: 13838-16-9

RTECS Number: KN6800000

DOT ID: No citation.

Regulatory Authority

- Air Pollutant Standard Set (ACGIH)[1]

Cited in U.S. State Regulations: Alaska (G), California (G), Massachusetts (G), New Hampshire (G), New Jersey (G), Pennsylvania (G), West Virginia (G).

Description: Enflurane, CHF_2OCF_2CHFCl, is a clear, colorless liquid with a mild, sweet odor, that easily turns into a nonflammable gas. Boiling point = 56 – 57°C. Very slightly soluble in water.

Potential Exposure: This compound is used as an anesthetic.

Incompatibilities: May be able to form unstable peroxides. Decomposes on heating, forming toxic and corrosive fumes of hydrogen chloride, hydrogen fluoride, phosgene.

Permissible Exposure Limits in Air: There is no OSHA PEL but NIOSH recommends a ceiling of 2 ppm averaged over 60 minutes and not to be exceeded at any time. ACGIH recommends an OEL of 75 ppm TWA for an 8-hour workshift.

Determination in Air: Collection by Anasorb tube; workup with CS_2; analysis by gas chromatography/flame ionization detection; OSHA Method #103.

Permissible Concentration in Water: No standards set.

Routes of Entry: Inhalation of vapors.

Harmful Effects and Symptoms

Short Term Exposure: Enflurane can affect you when breathed in. There is an association between exposure to anesthetic vapors and increased cancers, miscarriages, and birth defects. Enflurane's role in these increased risks in unclear. Medical patients receiving enflurane as surgical anesthesia have had rare cases of seizures (fits) and liver damage. It is not known whether this happens with workplace exposures. Exposure can cause you to become dizzy, lightheaded and to pass out. The LD_{50} oral-rat is 5,450 mg/kg (insignificantly toxic).

Long Term Exposure: There is limited evidence that enflurane causes lung and liver cancer in animals. It may damage the developing fetus. Exposure to high levels may cause seizures and affect the liver.

Points of Attack: Eyes, central nervous system, liver.

Medical Surveillance: Liver function tests.

First Aid: If this chemical gets into the eyes, remove any contact lenses at once and irrigate immediately for at least 15 minutes, occasionally lifting upper and lower lids. Seek medical attention immediately. If this chemical contacts the skin, remove contaminated clothing and wash immediately with soap and water. Seek medical attention immediately. If this chemical has been inhaled, remove from exposure, begin rescue breathing (using universal precautions) if breathing has stopped and CPR if heart action has stopped. Transfer promptly to a medical facility. When this chemical has been swallowed, get medical attention. Give large quantities of water and induce vomiting. Do not make an unconscious person vomit.

Personal Protective Methods: Wear protective gloves and clothing to prevent any reasonable probability of skin contact. Safety equipment suppliers/manufacturers can provide recommendations on the most protective glove/clothing material for your operation.. All protective clothing (suits, gloves, footwear, headgear) should be clean, available each day, and put on before work. Contact lenses should not be worn when working with this chemical. Wear splash-proof chemical goggles and face shield unless full facepiece respiratory protection is worn. Employees should wash immediately with soap when skin is wet or contaminated. Provide emergency showers and eyewash.

Respirator Selection: Where the potential exists for exposures over 2 ppm, use an MSHA/NIOSH approved supplied-air respirator with a full facepiece operated in the positive pressure mode or with a full facepiece, hood, or helmet in the continuous flow mode, or use an MSHA/NIOSH approved self-contained breathing apparatus with a full facepiece operated in pressure-demand or other positive pressure mode.

Storage: Prior to working with enflurane you should be trained on its proper handling and storage. Store in tightly closed containers in a cool, well ventilated area. Where possible, automatically pump liquid from drums or other storage containers to process containers. Procedures for the handling, use and storage of cylinders should be in compliance with OSHA 1910.101 and 1910.169 as with the recommendations of the Compressed Gas Association.

Shipping: Enflurane is not cited by DOT in its Performance-Oriented-Packaging Standards.[19]

Spill Handling: Evacuate persons not wearing protective equipment from area of spill or leak until clean-up is complete. Remove all ignition sources. If the gas is leaking from a vaporizer or other equipment Stop the flow of gas if it can be done safely. Ventilate area of spill or leak to disperse gas. Absorb liquids in vermiculite, dry sand, earth, peat, carbon, or a similar material and deposit in sealed containers. It may be necessary to contain and dispose of this chemical as a hazardous waste. If material or contaminated runoff enters waterways, notify downstream users of potentially contaminated waters. Contact your Department of Environmental Protection or your regional office of the federal EPA for specific recommendations. If employees are required to clean-up spills, they must be properly trained and equipped. OSHA 1910.120(q) may be applicable.

Fire Extinguishing: This chemical is a noncombustible liquid or gas. Poisonous gases including hydrogen chloride, hydrogen fluoride, phosgene are produced in fire. Use any extinguisher suitable for surrounding fires. Containers may explode in fire. Storage containers and parts of containers may rocket great distances, in many directions. If material or contaminated runoff enters waterways, notify downstream users of potentially contaminated waters. Notify local health and fire officials and pollution control agencies. From a secure, explosion-proof location, use water spray to cool exposed containers. If cooling streams are ineffective (venting sound increases in volume and pitch, tank discolors, or shows any signs of deforming), withdraw immediately to a secure position. If employees are expected to fight fires, they must be trained and equipped in OSHA 1910.156.

References

National Institute for Occupational Safety and Health, Criteria for a Recommended Standard: Occupational Exposure to Waste Anesthetic Gases and Vapors, NIOSH Doc. No. 77–140 (1977)

New Jersey Department of Health and Senior Services, "Hazardous Substance Fact Sheet: Enflurane," Trenton, NJ (April, 1997)

Epichlorohydrin

Molecular Formula: C_3H_5ClO

Synonyms: 1-Chlor-2,3-epoxy-propan (German); 1-Chloro-2,3-epossipropano (Italian); 1-Chloro-2,3-epoxy-propane; 3-Chloro-1,2-epoxypropane; (Chloromethyl)ethylene oxide; (Chloromethyl)oxirane; 2-(Chloromethyl)oxirane; Chloromethyloxirane; 3-Chloropropene-1,2-oxide; γ-Chloropropylene oxide; 3-Chloropropylene oxide; -Chloro-1,2-propylene oxide; Chloropropylene oxide; 1-Cloor-2,3-epoxy-propaan (Dutch); ECH; Epichloorhydrine (Dutch); Epichlorhydrin (German); Epichlorhydrine (French); α-Epichlorohydrin; (dl)-α-Epichlorohydrin; EPI-chlorohydrin; Epichlorohydryna (Polish); Epiclorhidrina (Spanish); Epicloridrina (Italian); 1,2-Epoxy-3-chloropropane; 2,3-Epoxypropyl chloride; Epoxy resin component; Glycerol epichlorohydrin; Glycidyl chloride; Oxirane, (chloromethyl)-; Phenoxy resin component; Propane, 1-chloro-2,3-epoxy-; Skekhg

CAS Registry Number: 106-89-8

RTECS Number: TX4900000

DOT ID: UN 2023

EEC Number: 603-026-00-6

Regulatory Authority

- Carcinogen (Animal Positive) (IARC)[9] (DFG)[3]
- Air Pollutant Standard Set (ACGIH)[1] (Australia) (HSE)[33] (Sweden, Former USSR)[35] (Israel) (Mexico) (OSHA)[58] (Several States)[60] (Several Canadian Provinces)
- Clean Air Act: Hazardous Air Pollutants (Title I, Part A, Section 112); Accidental Release Prevention/Flammable substances, (Section 112®, Table 3), TQ = 20,000 lb (9,080 kg)
- Clean Water Act: Section 311 Hazardous Substances/RQ 40CFR117.3 (same as CERCLA, see below); Section 313 Water Priority Chemicals (57FR41331, 9/9/92).
- EPA Hazardous Waste Number (RCRA No.): U041
- RCRA, 40CFR261, Appendix 8 Hazardous Constituents.
- Safe Drinking Water Act: Regulated chemical (47 FR 9352)
- Superfund/EPCRA 40CFR355, Appendix B Extremely Hazardous Substances: TPQ = 1,000 lb (454 kg)
- Superfund/EPCRA 40CFR302.4 Reportable Quantity (RQ): CERCLA, 100 lb (45.4 kg)
- EPCRA Section 313 Form R *de minimis* concentration reporting level: 0.1%
- Canada, WHMIS, Ingredients Disclosure List

Cited in U.S. State Regulations: Alaska (G), Arizona (W), California (A, G), Connecticut (A), Florida (G), Illinois (G), Indiana (A), Massachusetts (G, A), Michigan (G), Minnesota (W), Nevada (A), New Hampshire (G), New Jersey (G), New York (G, A), North Carolina (A), North Dakota (A), Oklahoma (G), Pennsylvania (G, A), Rhode Island (G, A), South Carolina (A), Vermont (G), Virginia (G, A), Washington (G), West Virginia (G), Wisconsin (G).

Description: Epichlorohydrin, is a colorless liquid with a slightly-irritating, chloroform-like odor. Boiling point = 115°C. Freezing/Melting point = -48°C. Flash point = 31°C. Autoignition temperature = 411°C. Explosive limits: LEL = 3.8%; UEL = 21.0%. Hazard Identification (based on NFPA-704 M Rating System): Health 3, Flammability 3, Reactivity 2. Soluble in water.

Potential Exposure: Epichlorohydrin is used in the manufacture of many glycerol and glycidol derivatives and epoxy resins, as a stabilizer in chlorine-containing materials, as an intermediate in the preparation of cellulose esters and ethers, paints, varnishes, nail enamels, and lacquers, and as a cement for celluloid. It is an intermediate in the manufacture of a number of drugs.

Incompatibilities: Forms explosive mixture with air. Slowly decomposes on contact with water. Heat or strong acids, alkalies, metallic halides, or contaminants can cause explosive polymerization. Violent reaction with strong oxidizers, aliphatic amines, alkanolamines, amines (especially aniline), alkaline earths, chemically active metals (chlorides of aluminum, iron, zinc), powdered metals (aluminum, zinc), alcohols, phenols, organic acids causing fire and explosion hazard. Will pit steel in the presence of water. Decomposition forms highly toxic phosgene gas. May accumulate static electrical charges, and may cause ignition of its vapors.

Permissible Exposure Limits in Air: The OSHA PEL TWA value is 5 ppm (19 mg/m^3). NIOSH recommends that occupational exposure be minimized. The HSE and ACGIH have set a TWA value of 2 ppm (7.6 mg/m^3). The notation "skin" is added to indicate the possibility of cutaneous absorption. The NIOSH IDLH = Ca [75 ppm]. HSE[33] has also set an STEL of 5 ppm (20 mg/m^3). Sweden[35] has set a TWA in workplace air of 0.5 ppm (2.0 mg/m^3) and an STEL of 1.0 ppm (4 mg/m^3).

The former USSR[35] has set a MAC in workplace air of 10 mg/m^3 and MAC values in the ambient air in residential areas[35][43] of 0.2 mg/m^3 (200 $\mu g/m^3$) on either a momentary or a daily average basis. Several states have set guidelines or standards for Epichlorohydrin in ambient air[60] ranging from 2.7 μg/m^3 (Massachusetts) to 8.3 μg/m^3 (North Carolina) to 12.0 μg/m^3 (Pennsylvania) to 20.0 μg/m^3 (Connecticut) to 33.3 μg/m^3 (New York) to 50.0 μg/m^3 (Indiana, South Carolina) to 0 – 100.0 μg/m^3 (North Dakota) to 0.8 – 200.0 μg/m^3 (Rhode Island) to 160.0 μg/m^3 (Virginia) to 238.0 μg/m^3 (Nevada).

Determination in Air: Charcoal tube adsorption; CS_2; Gas chromatography/Flame ionization detection; NIOSH IV, Method #1010.[18]

Permissible Concentration in Water: EPA has suggested a permissible ambient goal of 276 μg/l based on health effects. The EPA more recently calculated[48] a no-observed-adverse-effects-level (NOAEL) of 2 mg/kg/day on the basis of which they have calculated a lifetime health advisory of 70 μg/l. The Former USSR-UNEP/IRPTC project[43] has set a MAC in water bodies used for drinking purposes of 10 μg/l. Some states have set guidelines for Epichlorohydrin in drinking water[61] ranging from 3.5 μg/l (Arizona) to 35.4 μg/l (Minnesota) to 150 μg/l (Maine).

Determination in Water: A purge-and-trap gas chromatographic/mass spectrometric procedure for the determination of volatile organic compounds may be used for Epichlorohydrin according to EPA.[48]

Routes of Entry: Inhalation, skin absorption, ingestion, skin and/or eye contact.

Harmful Effects and Symptoms

Short Term Exposure: Epichlorohydrin is corrosive to the eyes, skin, and respiratory tract. Inhalation can cause pulmonary edema, a medical emergency that can be delayed for several hours. This can cause death. May affect the nervous system. Exposure may result in unconsciousness. Acute exposure to Epichlorohydrin may result in nausea, vomiting, and abdominal pain. Liver and kidney effects may be observed. The respiratory tract may become irritated, dyspnea (shortness of breath) may occur, and in acute cases, respiratory paralysis has been observed. Central nervous system and respiratory depression have been noted.

Inhalation: Levels below 0.05 ppm have not caused adverse effects. Levels above 20 ppm may cause fatigue, stomach pains, nausea, vomiting, slowed breathing, loss of muscle strength and blue coloration of the skin.

Skin: May cause severe blistering, burns and severe pain which may be delayed. Skin absorption may cause or increase the severity of symptoms listed above.

Eyes: Levels above 5 ppm may cause severe irritation or burns. Direct contact of the liquid may cause clouding of the cornea and tissue death.

Ingestion: May cause nausea and stomach pain. Animal studies indicate that the probable lethal dose for an adult is 14 grams (1/2 ounce).

Long Term Exposure: Repeated or prolonged contact to epichlorohydrin may cause skin allergy to develop. Repeated or prolonged inhalation exposure may cause asthma. May decrease fertility in males. Repeated overexposure may cause changes in the eyes and lungs, kidney and liver damage and chronic asthmatic bronchitis. Epichlorohydrin causes cancer and changes in the genetic material of laboratory animals and has been linked to cancers and change in the genetic material of occupationally exposed humans. There is some evidence of lung cancer in humans.

Points of Attack: Eyes, skin, respiratory system, kidneys, liver, reproductive system. *Cancer site* (in animals): nasal cancer.

Medical Surveillance: Before beginning employment and at regular times after that, the following are recommended: Lung function tests. If symptoms develop or overexposure is suspected, the following may be useful: Liver and kidney function tests. Consider chest x-ray after acute overexposure. Evaluation by a qualified allergist, including careful exposure history and special testing, may help diagnose skin allergy.

First Aid: If this chemical gets into the eyes, remove any contact lenses at once and irrigate immediately for at least 15 minutes, occasionally lifting upper and lower lids. Seek medical attention immediately. If this chemical contacts the skin, remove contaminated clothing and wash immediately with soap and water. Seek medical attention immediately. If this chemical has been inhaled, remove from exposure, begin rescue breathing (using universal precautions) if breathing has stopped and CPR if heart action has stopped. Transfer promptly to a medical facility. When this chemical has been swallowed, get medical attention. Give large quantities of water and induce vomiting. Do not make an unconscious person vomit. Medical observation is recommended for 24 – 48 hours after breathing overexposure, as pulmonary edema may be delayed. As first aid for pulmonary edema, a doctor or authorized paramedic may consider administering a corticosteroid spray.

Personal Protective Methods: Wear protective gloves and clothing to prevent any reasonable probability of skin contact. Safety equipment suppliers/manufacturers can provide recommendations on the most protective glove/clothing material for your operation. Epichlorohydrin slowly penetrates rubber, so all contaminated clothing should be thoroughly washed. Teflon is among the recommended protective materials. All protective clothing (suits, gloves, footwear, headgear) should be clean, available each day, and put on before work. Contact lenses should not be worn when working with this chemical. Wear splash-proof chemical goggles and face shield unless full facepiece respiratory protection is worn. Employees should wash immediately with soap when skin is wet or contaminated. Provide emergency showers and eyewash.

Respirator Selection: NIOSH *At any detectable concentration:* SCBAF:PD,PP (any MSHA/NIOSH approved self-contained breathing apparatus that has a full facepiece and is operated in a pressure-demand or other positive-pressure mode);

or SAF:PD,PP:ASCBA (any supplied-air respirator that has a full facepiece and is operated in a pressure-demand or other positive-pressure mode in combination with an auxiliary, self-contained breathing apparatus operated in a pressure-demand or other positive pressure mode). *Escape:* GMFOVAG [any air-purifying, full-facepiece respirator (gas mask) with a chin-style, front- or back-mounted organic vapor and acid gas canister]; or SCBAE (any appropriate escape-type, self-contained breathing apparatus).

Storage: Prior to working with this chemical you should be trained on its proper handling and storage. Before entering confined space where this chemical may be present, check to make sure that an explosive concentration does not exist. Epichlorohydrin must be stored to avoid contact with strong oxidizers (such as chlorine, bromine and fluorine), strong acids (such as hydrochloric, sulfuric, and nitric) and chemically active metals (such as aluminum, caustics, chlorides of iron and aluminum and zinc) since violent reactions occur. Sources of ignition, such as smoking and open flames, are prohibited where Epichlorohydrin is handled, used, or stored. Metal containers involving the transfer of 5 gallons or more of Epichlorohydrin should be grounded and bonded. Drums must be equipped with self-closing valves, pressure vacuum bungs, and flame arresters. Where possible, automatically pump liquid from drums or other storage containers to process containers. Use only non-sparking tools and equipment, especially when opening and closing containers of Epichlorohydrin. Wherever Epichlorohydrin is used, handled, manufactured, or stored, use explosion proof electrical equipment and fittings. A regulated, marked area should be established where this chemical is handled, used, or stored in compliance with OSHA standard 1910.1045.

Shipping: This compound requires a shipping label of: "Poison." It falls in DOT Hazard Class 6.1 and Packing Group II.

Spill Handling: Evacuate persons not wearing protective equipment from area of spill or leak until clean-up is complete. Remove all ignition sources. Collect powdered material in the most convenient and safe manner and deposit in sealed containers. Establish forced ventilation to keep levels below explosive limit. Keep this chemical out of a confined space, such as a sewer, because of the possibility of an explosion, unless the sewer is designed to prevent the build-up off explosive concentrations. It may be necessary to contain and dispose of this chemical as a hazardous waste. If material or contaminated runoff enters waterways, notify downstream users of potentially contaminated waters. Contact your Department of Environmental Protection or your regional office of the federal EPA for specific recommendations. If employees are required to clean-up spills, they must be properly trained and equipped. OSHA 1910.120(q) may be applicable.

Fire Extinguishing: This chemical is a flammable liquid. Use water spray, dry chemical, foam or carbon dioxide. Water spray may be used to dilute spills to non-flammable mixtures. If leak or spill has not ignited, use water spray to disperse the vapors. Evacuate for a radius of 1,500 feet. Isolate for one-half mile in all directions if tank car or truck is involved in fire. Epichlorohydrin may react violently with water. Poisonous gases including hydrogen chloride and carbon monoxide are produced in fire. Vapors are heavier than air and will collect in low areas. Vapors may travel long distances to ignition sources and flashback. Vapors in confined areas may explode when exposed to fire. Containers may explode in fire. Storage containers and parts of containers may rocket great distances, in many directions. If material or contaminated runoff enters waterways, notify downstream users of potentially contaminated waters. Notify local health and fire officials and pollution control agencies. From a secure, explosion-proof location, use water spray to cool exposed containers. If cooling streams are ineffective (venting sound increases in volume and pitch, tank discolors, or shows any signs of deforming), withdraw immediately to a secure position. If employees are expected to fight fires, they must be trained and equipped in OSHA 1910.156.

Disposal Method Suggested: Incineration, preferably after mixing with another combustible fuel. Care must be exercised to assure complete combustion to prevent the formation of phosgene. An acid scrubber is necessary to remove the halo acids produced.[22] Consult with environmental regulatory agencies for guidance on acceptable disposal practices. Generators of waste containing this contaminant (≥100 kg/mo) must conform with EPA regulations governing storage, transportation, treatment, and waste disposal.

References

National Institute for Occupational Safety and Health, Criteria for a Recommended Standards: Occupational Exposure to Epichlorohydrin, NIOSH Doc. No. 76–206, Washington, DC (1976)

U.S. Environmental Protection Agency, Epichlorohydrin, Health and Environmental Effects Profile No. 100, Washington, DC, Office of Solid Waste (April 30, 1980)

Sax, N. I., Ed., "Dangerous Properties of Industrial Materials Report," 1, No. 4, 57–59 (1981) and 3, 68–71 (1983) and 6, No. 5, 50–51 (1986)

U.S. Environmental Protection Agency, "Chemical Profile: Epichlorohydrin," Washington, DC, Chemical Emergency Preparedness Program (November 30, 1987)

New Jersey Department of Health and Senior Services, "Hazardous Substance Fact Sheet: Epichlorohydrin," Trenton, NJ (September 8, 1987)

New York State Department of Health, "Chemical Fact Sheet: Epichlorohydrin," Albany, NY, Bureau of Toxic Substance Assessment (March 1986 and Version 2)

EPN

Molecular Formula: $C_{14}H_{14}NO_4PS$

Synonyms: *O*-Aethyl-*O*-*N*(4-nitrophenyl)-phenylmonothiophosphonat (German); ENT 17,798; *O*-Ester of *p*-nitrophenol with *O*-ethylphenyl phosphonothioate; Ethoxy-4-nitrophenoxyphenylphosphine sulfide; *O*-Ethyl-*O*-[(4-

nitrofenyl)-fenyl]monothiofosfonaat (Dutch); Ethyl *p*-nitrophenyl benzenethionophosphate; *O*-Ethyl *O*-(4-nitrophenyl) benzenethionophosphonate; Ethyl *p*-nitrophenyl benzenethionophosphonate; Ethyl *p*-nitrophenyl benzenethiophosphonate; *O*-Ethyl *O*-(4-nitrophenyl) phenyl phosphonothioate; *O*-Ethyl *O,p*-nitrophenyl phenyl phosphonothioate; Ethyl *p*-nitrophenyl phenylphosphonothioate; Ethyl *p*-nitrophenyl thionobenzenephosphate; *O*-Ethyl phenyl-*p*-nitrophenyl thiophosphonate; Pin; Santox; Thionobenzenephosphonic acid ethyl *p*-nitrophenyl ester

CAS Registry Number: 2104-64-5

RTECS Number: TB1925000

DOT ID: UN 2783

EEC Number: 015-036-00-2

Regulatory Authority

- Banned or Severely Restricted (Several Countries) (UN)[13]
- Very Toxic Substance (World Bank)[15]
- Air Pollutant Standard Set (ACGIH)[1] (DFG)[3] (Argentina)[35] (Several States)[60]
- Water Pollution Standard Proposed (Japan)[35]
- Superfund/EPCRA 40CFR355, Appendix B Extremely Hazardous Substances: TPQ = 100/10,000 lb (45.4/4,540 kg)
- Superfund/EPCRA 40CFR302.4 Reportable Quantity (RQ): EHS, 1 lb (0.454 kg)

Cited in U.S. State Regulations: Alaska (G), Connecticut (A), Florida (G), Illinois (G), Maine (G), Massachusetts (G), Michigan (G), Nevada (A), New Hampshire (G), New Jersey (G), North Dakota (A), Oklahoma (G), Pennsylvania (G), Rhode Island (G), West Virginia (G).

Description: EPN is a light yellow crystalline solid with an aromatic odor (as a pesticide) or a brown liquid above 36°C. Freezing/Melting point = 36°C. Hazard Identification (based on NFPA-704 M Rating System): Health 4, Flammability 1, Reactivity 0. Insoluble in water.

Potential Exposure: Those involved in the manufacture, formulation and field application of this pesticide.

Incompatibilities: Reaction with oxidizers. Contact with alkalies causes decomposition (hydrolysis) producing p-nitrophenol.

Permissible Exposure Limits in Air: The OSHA PEL, the DFG[3] an Argentina[35] TWA value[1] is 0.5 mg/m^3. The ACGIH recommends a TWA of 0.1 mg/m^3. The notation "skin" indicates the possibility of cutaneous absorption. The NIOSH IDLH level is 5 mg/m^3. Various STEL values have been set in various countries: 1.5 mg/m^3 in Argentina, 2.0 mg/m^3 in the US and 5.0 in Germany.[35] Some states have set guidelines or standards for EPN in ambient air[60] ranging from 5.0 μg/m^3 (North Dakota) to 10.0 μg/m^3 (Connecticut) to 12.0 μg/m^3 (Nevada).

Determination in Air: OSHA versatile sampler-2; Toluene/Acetone; Gas chromatography/Flame photometric detection for sulfur, nitrogen, or phosphorus; NIOSH Method IV Method #5600, Organophosphorus Pesticides. Collection on a filter, workup with isooctane, Gas chromatography/Flame photometric detection for sulfur, nitrogen, or phosphorus; NIOSH IV, Method #5012.

Permissible Concentration in Water: Japan[35] has set an effluent maximum of 1 mg/l and an environmental water quality standard of zero.

Routes of Entry: Inhalation, ingestion, skin absorption, skin and/or eye contact. Passes through the skin.

Harmful Effects and Symptoms

Short Term Exposure: A cholinesterase inhibitor. EPN can affect the nervous system, causing convulsions and possible respiratory failure. Exposure may result in unconsciousness or death. The effects may be delayed. Medical observation is indicated. This material may be fatal is swallowed. It is poisonous if inhaled and extremely hazardous by skin contact. Repeated exposure may, without symptoms, be increasingly hazardous. The estimated fatal oral dose is 0.3 grams for a 150 lb (70 kg) person. Acute exposure to EPN may produce the following signs and symptoms: pinpoint pupils, blurred vision, headache, dizziness, muscle spasms, and profound weakness. Vomiting, diarrhea, abdominal pain, seizures, and coma may also occur. The heart rate may decrease following oral exposure or increase following dermal exposure. Hypertension (high blood pressure) is not uncommon. Respiratory symptoms include dyspnea (shortness of breath), respiratory depression and respiratory paralysis. Giddiness, slurred speech, confusion, and psychosis may also be observed. A rebuttable presumption against pesticide registration was issued for EPN on September 19, 1979 by EPA on the basis of neurotoxicity.

Long Term Exposure: Cholinesterase inhibitor; cumulative effect is possible. EPN may damage the nervous system with repeated exposure, resulting in convulsions, respiratory failure. May cause liver damage.

Points of Attack: Respiratory system, lungs, cardiovascular system, central nervous system, eyes, skin, blood cholinesterase.

Medical Surveillance: Before employment and at regular times after that, the following are recommended: Plasma and red blood cell cholinesterase levels (tests for the enzyme poisoned by this chemical). If exposure stops, plasma levels return to normal in 1 – 2 weeks while red blood cell levels may be reduced for 1 – 3 months.

When cholinesterase enzyme levels are reduced by 25% or more below preemployment levels, risk of poisoning is increased, even if results are in lower ranges of "normal." Reassignment to work not involving organophosphate or carbamate pesticides is recommended until enzyme levels recover. If symptoms develop or overexposure occurs, repeat the above tests as soon as possible and get an exam of the nervous system. Also consider complete blood count. Consider chest x-ray following acute overexposure. Do not drink any alcoholic beverages before or during use. Alcohol promotes absorption of organic phosphates.

First Aid: If this chemical gets into the eyes, remove any contact lenses at once and irrigate immediately for at least 15 minutes, occasionally lifting upper and lower lids. Seek medical attention immediately. If this chemical contacts the skin, remove contaminated clothing and wash immediately with soap and water. Speed in removing material from skin is of extreme importance. Shampoo hair promptly if contaminated. Seek medical attention immediately. If this chemical has been inhaled, remove from exposure, begin rescue breathing (using universal precautions) if breathing has stopped and CPR if heart action has stopped. Transfer promptly to a medical facility. When this chemical has been swallowed, get medical attention. Give large quantities of water and induce vomiting. Do not make an unconscious person vomit.

Personal Protective Methods: Wear protective gloves and clothing to prevent any reasonable probability of skin contact. Safety equipment suppliers/manufacturers can provide recommendations on the most protective glove/clothing material for your operation.. All protective clothing (suits, gloves, footwear, headgear) should be clean, available each day, and put on before work. Contact lenses should not be worn when working with this chemical. Wear dust-proof chemical goggles and face shield unless full facepiece respiratory protection is worn. Employees should wash immediately with soap when skin is wet or contaminated. Provide emergency showers and eyewash.

Respirator Selection: NIOSH/OSHA: up to 5 mg/m^3 SA (any supplied-air respirator); or SCBA (any self-contained breathing apparatus with a full facepiece). *Emergency or planned entry into unknown concentrations or IDLH conditions:* SCBAF:PD,PP (any self-contained breathing apparatus that has a full facepiece and is operated in a pressure-demand or other positive-pressure mode); or SAF: PD,PP:ASCBA (any supplied-air respirator that has a full facepiece and is operated in a pressure-demand or other positive-pressure mode in combination with an auxiliary self-contained breathing apparatus operated in a pressure-demand or other positive pressure mode). *Escape:* GMFOVHiE [any air-purifying, full-facepiece respirator (gas mask) with a chin-style, front- or back-mounted organic vapor canister having a high-efficiency particulate filter]; or SCBAE (any appropriate escape-type, self-contained breathing apparatus).

Storage: Prior to working with EPN you should be trained on its proper handling and storage. Store in tightly closed containers in a cool, well ventilated area away from oxidizers.

Shipping: Organophosphorus pesticides, solid, toxic, n.o.s. fall in Hazard Class 6.1 and EPN in Packing Group I. The label required is "Poison."

Spill Handling: Evacuate persons not wearing protective equipment from area of spill or leak until clean-up is complete. Remove all ignition sources. Cover with soda ash, mix, and spray with water. Place in container of water and allow to stand for 2 days, then neutralize with 6 molar hydrochloric acid. Collect powdered material in the most convenient and safe manner and deposit in sealed containers. Ventilate area after clean-up is complete. It may be necessary to contain and dispose of this chemical as a hazardous waste. If material or contaminated runoff enters waterways, notify downstream users of potentially contaminated waters. Contact your Department of Environmental Protection or your regional office of the federal EPA for specific recommendations. If employees are required to clean-up spills, they must be properly trained and equipped. OSHA 1910.120(q) may be applicable.

Fire Extinguishing: This material may burn but does not ignite readily. For small fires, use dry chemicals, carbon dioxide, water spray, or foam. For large fires, use water spray, fog, or foam. Stay upwind; keep out of low areas. Move containers from fire area if you can do it without risk. Dike fire control water for later disposal; do not scatter the material. Wear positive pressure breathing apparatus and special protective clothing. Poisonous gases are produced in fire including carbon monoxide, phosphine, phosphoric acid, nitrogen oxides, phosphorous oxides, sulfur oxides. If material or contaminated runoff enters waterways, notify downstream users of potentially contaminated waters. Notify local health and fire officials and pollution control agencies. Fight fire from maximum distance. Containers may explode in fire. From a secure, explosion-proof location, use water spray to cool exposed containers. If cooling streams are ineffective (venting sound increases in volume and pitch, tank discolors, or shows any signs of deforming), withdraw immediately to a secure position. If employees are expected to fight fires, they must be trained and equipped in OSHA 1910.156.

Disposal Method Suggested: EPN plant wastes are treated by preaeration, activated sludge treatment, recycle, chlorination and final polishing where additional natural biological stabilization occurs. EPN is also relatively rapidly hydrolyzed in alkaline solution to benzene thiophosphoric acid, alcohol and p-nitrophenol and soil burial with alkali may be used. For large quantities, however, incineration is recommended.[22] In accordance with 40CFR165 recommendations for the disposal of pesticides and pesticide containers. Must be disposed properly by following package label directions or by contacting your state pesticide or environmental control agency or by contacting your regional EPA office.

References

U.S. Environmental Protection Agency, "Chemical Profile: EPN," Washington, DC, Chemical Emergency Preparedness Program (November 30, 1987)

New Jersey Department of Health and Senior Services, "Hazardous Substance Fact Sheet, EPN," Trenton NJ (May 1999)

Epoxy Ethyloxy Propane

Molecular Formula: $C_5H_{10}O_2$

Synonyms: 1,2-Epoxy-3-ethoxy-propane; (Ethoxymethyl)oxirane; Ethyl glycidyl ether; Oxirane, (ethoxymethyl); Ppropane, 1,2-epoxy-3-ethoxy-

CAS Registry Number: 4016-11-9

RTECS Number: TZ3200000

DOT ID: UN 2752

Cited in U.S. State Regulations: New Hampshire (G), New Jersey (G).

Description: Epoxy Ethyloxy Propane is a colorless, flammable liquid.

Potential Exposure: Used in organic synthesis. It has limited commercial use.

Incompatibilities: Strong oxidizers.

Permissible Exposure Limits in Air: No standards set.

Permissible Concentration in Water: No criteria set.

Routes of Entry: Epoxy ethyloxy propane can affect you when breathed in and by passing through your skin.

Harmful Effects and Symptoms

Short Term Exposure: Contact can irritate or burn the eyes and skin with possible permanent damage. Exposure can irritate the eyes, nose and throat. Higher levels can irritate the lungs, causing coughing and/or shortness of breath. Very high levels could cause you to feel dizzy, lightheaded, and pass out.

Long Term Exposure: Epoxy ethyloxy propane can cause a skin allergy to develop. Similar very irritating substances can affect the lungs.

Points of Attack: Skin, lungs.

Medical Surveillance: For those with frequent or potentially high exposure, the following are recommended before beginning work and at regular times after that: Lung function tests. If symptoms develop or overexposure is suspected, the following may be useful: Evaluation by a qualified allergist, including careful exposure history and special testing, may help diagnose skin allergy. Lung function tests.

First Aid: If this chemical gets into the eyes, remove any contact lenses at once and irrigate immediately for at least 15 minutes, occasionally lifting upper and lower lids. Seek medical attention immediately. If this chemical contacts the skin, remove contaminated clothing and wash immediately with soap and water. Speed in removing material from skin is of extreme importance. Seek medical attention immediately. If this chemical has been inhaled, remove from exposure, begin rescue breathing (using universal precautions) if breathing has stopped and CPR if heart action has stopped. Transfer promptly to a medical facility. When this chemical has been swallowed, get medical attention. Give large quantities of water and induce vomiting. Do not make an unconscious person vomit.

Personal Protective Methods: Wear protective gloves and clothing to prevent any reasonable probability of skin contact. Safety equipment suppliers/manufacturers can provide recommendations on the most protective glove/clothing material for your operation.. All protective clothing (suits, gloves, footwear, headgear) should be clean, available each day, and put on before work. Contact lenses should not be worn when working with this chemical. Wear splash-proof chemical goggles and face shield when working with liquid, unless full facepiece respiratory protection is worn. Wear dust-proof goggles and face shield when working with powders or dust, unless full facepiece respiratory protection is worn. Employees should wash immediately with soap when skin is wet or contaminated. Provide emergency showers and eyewash.

Respirator Selection: Where the potential for exposures to epoxy ethyloxy propane exists, use a MSHA/NIOSH approved supplied-air respirator with a full facepiece operated in the positive pressure mode or with a full facepiece, hood, or helmet in the continuous flow mode, or use a MSHA/NIOSH approved self-contained breathing apparatus with a full facepiece operated in pressure-demand or other positive pressure mode.

Storage: Prior to working with this chemical you should be trained on its proper handling and storage. Store in tightly closed containers in a cool, well-ventilated area. Sources of ignition, such as smoking and open flames, are prohibited where epoxy ethyloxy propane is used, handled, or stored in a manner that could create a potential fire or explosion hazard. Where possible, automatically pump liquid from drums or other storage containers to process containers.

Shipping: This compound requires a shipping label of: "Flammable Liquid." It falls in DOT Hazard Class 3 and Packing Group III.

Spill Handling: Restrict persons not wearing protective equipment from area of spill or leak until clean-up is complete. Remove all ignition sources. Ventilate area of spill or leak. Absorb liquids in vermiculite, dry sand, earth, or a similar material and deposit in sealed containers. Collect powdered material in the most convenient and safe manner and deposit in sealed containers. Keep epoxy ethyloxy propane out of a confined space, such as a sewer, because of the possibility of an explosion, unless the sewer is designed to prevent the build-up of explosive concentrations. It may be necessary to contain and dispose of this chemical as a hazardous waste. If material or contaminated runoff enters waterways, notify downstream users of potentially contaminated waters. Contact your Department of Environmental Protection or your regional office of the federal EPA for specific recommendations. If employees are required to clean-up spills, they must be properly trained and equipped. OSHA 1910.120(q) may be applicable.

Fire Extinguishing: This chemical is a combustible liquid. Poisonous gases are produced in fire. Use dry chemical, carbon dioxide, or alcohol foam extinguishers. Vapors are heavier than air and will collect in low areas. Vapors may travel long distances to ignition sources and flashback. Vapors in confined areas may explode when exposed to fire. Containers may explode in fire. Storage containers and parts of containers may rocket great distances, in many directions. If material or contaminated runoff enters waterways, notify downstream users of potentially contaminated waters. Notify local health and fire officials and pollution control agencies. From a secure, explosion-proof location, use water spray to cool exposed containers. If cooling streams are ineffective (venting sound increases in volume and pitch, tank discolors, or shows any signs of deforming), withdraw immediately to a secure position. If employees are expected to fight fires, they must be trained and equipped in OSHA 1910.156.

Disposal Method Suggested: Incineration.

References

New Jersey Department of Health and Senior Services, "Hazardous Substance Fact Sheet: Epoxy Ethyloxy Propane," Trenton, NJ (February 1987)

Ergocalciferol

Molecular Formula: $C_{28}H_{44}O$

Synonyms: Activated ergosterol; D-Arthin; Calciferol; Calciferon; Candacaps; Condocaps; Condol; Crtron; Crystallina; Dacitin; Daral; Davitamon D; Decaps; Dee-Osterol; Dee-Ron; Dee-Ronal; Dee-Roual; Deltalin; Deratol; Detalup; Diactol; Ergorone; Ergosterol; Ergosterol, Activated; Ertron; 1,2-Ethylidene dichloride; Fortodyl; Geltabs; Hi-Deratol; Infron; Irradiated; Irradiated ergosta-5,7,22, -trien-3-β-ol; Metadee; Mulsiferol; Mykostin; Oleovitamin D; Ostelin; Radiostol; Radsterin; 9,10,Secoergosta-5,7,10(19), 22-tetraen-3-β-ol; Shock-Ferol; Sterogly; Vigantol; Viosterol; Vitamin D2; Vitavel-D

CAS Registry Number: 50-14-6

RTECS Number: KE1050000

Regulatory Authority

- Superfund/EPCRA 40CFR355, Appendix B Extremely Hazardous Substances: TPQ = 1,000/10,000 lb (454/4,540 kg)
- Superfund/EPCRA 40CFR302.4 Reportable Quantity (RQ): EHS, 1 lb (0.454 kg)

Cited in U.S. State Regulations: California (G), Massachusetts (G), New Jersey (G), Pennsylvania (G).

Description: Ergocalciferol, $C_{28}H_{44}O$, is an odorless white crystalline solid. Freezing/Melting point = 115 – 118°C. Insoluble in water.

Potential Exposure: Used as a nutrient and/or dietary supplement food additive.

Permissible Exposure Limits in Air: No standards set.

Permissible Concentration in Water: No criteria set.

Routes of Entry: Ingestion.

Harmful Effects and Symptoms

Short Term Exposure: Ergocalciferol in a single acute ingestion presents no toxic hazard. Daily ingestion in excess of 5,000 units/day in children or 7,500 units/day in adults will produce toxic symptoms associated with hypervitaminosis D. Initial symptoms of Ergocalciferol poisoning include anorexia, nausea and vomiting. It often mimics hyperparathyroidism with thirst, muscular weakness, nervousness, kidney impairment, hypertension and excessive urination.

Long Term Exposure: Ergocalciferol poisoning disturbs calcium metabolism and causes kidney damage.

Medical Surveillance: Kidney function tests.

First Aid: If this chemical gets into the eyes, remove any contact lenses at once and irrigate immediately for at least 15 minutes, occasionally lifting upper and lower lids. Seek medical attention immediately. If this chemical contacts the skin, remove contaminated clothing and wash immediately with soap and water. Seek medical attention immediately. If this chemical has been inhaled, remove from exposure, begin rescue breathing (using universal precautions) if breathing has stopped and CPR if heart action has stopped. Transfer promptly to a medical facility. When this chemical has been swallowed, get medical attention. Give large quantities of water and induce vomiting. Do not make an unconscious person vomit.

Personal Protective Methods: Wear protective gloves and clothing to prevent any reasonable probability of skin contact. Safety equipment suppliers/manufacturers can provide recommendations on the most protective glove/clothing material for your operation.. All protective clothing (suits, gloves, footwear, headgear) should be clean, available each day, and put on before work. Contact lenses should not be worn when working with this chemical. Wear splash-proof chemical goggles and face shield unless full facepiece respiratory protection is worn. Employees should wash immediately with soap when skin is wet or contaminated. Provide emergency showers and eyewash.

Respirator Selection: Where the potential for exposure to this chemical, use a MSHA/NIOSH approved supplied-air respirator with a full facepiece operated in the positive pressure mode or with a full facepiece, hood, or helmet in the continuous flow mode, or use a MSHA/NIOSH approved self-contained breathing apparatus with a full facepiece operated in pressure-demand or other positive pressure mode.

Storage: Prior to working with this chemical you should be trained on its proper handling and storage. Store in tightly closed containers in a cool, well ventilated area.

Shipping: Medicines, poisonous, solid n.o.s. fall in Hazard Class 6.1 and Ergocalciferol in Packing Group I. The label required is "Poison."

Spill Handling: Evacuate and restrict persons not wearing protective equipment from area of spill or leak until cleanup is complete. Remove all ignition sources. Ventilate area of spill or leak. Absorb liquids in vermiculite, dry sand, earth, peat, carbon, or a similar material and deposit in sealed containers. Oil-skimming equipment may be used to remove slicks from water. It may be necessary to contain and dispose of this chemical as a hazardous waste. If material or contaminated runoff enters waterways, notify downstream users of potentially contaminated waters. Contact your Department of Environmental Protection or your regional office of the federal EPA for specific recommendations. If employees are required to clean-up spills, they must be properly trained and equipped. OSHA 1910.120(q) may be applicable.

Fire Extinguishing: Use dry chemical, carbon dioxide, water spray, or foam extinguishers. If material or contaminated runoff enters waterways, notify downstream users of potentially contaminated waters. Notify local health and fire officials and pollution control agencies. From a secure, explosion-proof location, use water spray to cool exposed containers. If cooling streams are ineffective (venting sound increases in volume and pitch, tank discolors, or shows any signs of deforming), withdraw immediately to a secure position. If employees are expected to fight fires, they must be trained and equipped in OSHA 1910.156.

References

U.S. Environmental Protection Agency, "Chemical Profile: Ergocalciferol," Washington, DC, Chemical Emergency Preparedness Program (November 30, 1987)

Ergotamine Tartrate

Molecular Formula: $C_{70}H_{76}N_{10}O_{16}$

Common Formula: $C_{66}H_{70}N_{10}O_{10} \cdot C_4H_6O_6$

Synonyms: Ercal; Ergam; Ergate; Ergomar; Ergostat; Ergotamine bitartrate; Ergotartrate; Etin; Exmigra; Femergin; Gotamine tartrate; Gynergen; Lingraine; Lingran; Medihaler ergotamine; Migraine Dolviran; Neo-Ergotin; Rigetamin; Secagyn; Secupan; Tartrato de ergosterol (Spanish)

CAS Registry Number: 379-79-3

RTECS Number: KE8225000

DOT ID: UN 1544

Regulatory Authority

- Superfund/EPCRA 40CFR355, Appendix B Extremely Hazardous Substances: TPQ = 500/10,000 lb (227/4,540 kg)
- Superfund/EPCRA 40CFR302.4 Reportable Quantity (RQ): EHS, 1 lb (0.454 kg)

Cited in U.S. State Regulations: Massachusetts (G).

Description: Ergotamine tartrate, $C_{66}H_{70}N_{10}O_{10} \cdot C_4H_6O_6$ is a crystalline solid. Freezing/Melting point = 203°C (decomposes).

Potential Exposure: The major uses off the ergot alkaloids fall into two categories: applications in obstetrics and treatment of migraine headaches.

May be used in the production of "street" drugs.

Incompatibilities: Light and heat.

Permissible Exposure Limits in Air: No standards set.

Permissible Concentration in Water: No criteria set.

Routes of Entry: Inhalation, ingestion.

Harmful Effects and Symptoms

Short Term Exposure: Has high oral toxicity and acts as a convulsant in humans. People with liver damage are at a greater risk. Nausea and vomiting occur in some patients after oral administration. Weakness in the legs is common and muscle pains in the extremities may occur. Numbness and tingling of the fingers and toes may also occur.

Long Term Exposure: May cause liver damage.

Points of Attack: Liver, nervous system.

Medical Surveillance: Liver function tests.

First Aid: If this chemical gets into the eyes, remove any contact lenses at once and irrigate immediately for at least 15 minutes, occasionally lifting upper and lower lids. Seek medical attention immediately. If this chemical contacts the skin, remove contaminated clothing and wash immediately with soap and water. Speed in removing material from skin is of extreme importance. Seek medical attention immediately. If this chemical has been inhaled, remove from exposure, begin rescue breathing (using universal precautions) if breathing has stopped and CPR if heart action has stopped. Transfer promptly to a medical facility. When this chemical has been swallowed, get medical attention. Give large quantities of water and induce vomiting. Do not make an unconscious person vomit. Keep victim quiet and maintain normal body temperature. Effects may be delayed; keep victim under observation.

Personal Protective Methods: Wear protective gloves and clothing to prevent any reasonable probability of skin contact. Safety equipment suppliers/manufacturers can provide recommendations on the most protective glove/clothing material for your operation.. All protective clothing (suits, gloves, footwear, headgear) should be clean, available each day, and put on before work. Contact lenses should not be worn when working with this chemical. Wear dust-proof chemical goggles and face shield unless full facepiece respiratory protection is worn. Employees should wash immediately with soap when skin is wet or contaminated. Provide emergency showers and eyewash.

Respirator Selection: Where the potential for exposure to this chemical, use a MSHA/NIOSH approved supplied-air respirator with a full facepiece operated in the positive pressure mode or with a full facepiece, hood, or helmet in the continuous flow mode, or use a MSHA/NIOSH approved self-contained breathing apparatus with a full facepiece operated in pressure-demand or other positive pressure mode.

Storage: Prior to working with this chemical you should be trained on its proper handling and storage. Store in tightly closed containers in a cool, well ventilated area.

Shipping: Alkaloid salts, n.o.s., poisonous solid falls in Hazard Class 6.1 and Ergotamine tartrate in Packing Group I. The label required is "Poison."

Spill Handling: Evacuate persons not wearing protective equipment from area of spill or leak until clean-up is complete. Remove all ignition sources. Stay upwind; keep out of low areas. Do not touch spilled material; stop leak if you can do so without risk. Use water vapor to reduce vapors. Absorb spills with sand or other noncombustible absorbent material. Small dry spills: with clean shovel place material into clean, dry container and cover; move containers from spill area. For large spills, dike far ahead of spill for later disposal. Ventilate area after clean-up is complete. It may be necessary to contain and dispose of this chemical as a hazardous waste. If material or contaminated runoff enters waterways, notify downstream users of potentially contaminated waters. Contact your Department of Environmental Protection or your regional office of the federal EPA for specific recommendations. If employees are required to clean-up spills, they must be properly trained and equipped. OSHA 1910.120(q) may be applicable.

Fire Extinguishing: Use dry chemical, carbon dioxide, water spray, or alcohol foam extinguishers. Poisonous nitrogen oxide gases are produced in fire. If material or contaminated runoff enters waterways, notify downstream users of potentially contaminated waters. Notify local health and fire officials and pollution control agencies. From a secure, explosion-proof location, use water spray to cool exposed containers. If cooling streams are ineffective (venting sound increases in volume and pitch, tank discolors, or shows any signs of deforming), withdraw immediately to a secure position. If employees are expected to fight fires, they must be trained and equipped in OSHA 1910.156.

References

U.S. Environmental Protection Agency, "Chemical Profile: Ergotamine Tartrate," Washington, DC, Chemical Emergency Preparedness Program (November 30, 1987)

Sax, N. I., Ed., "Dangerous Properties of Industrial Materials Report," 1, No. 3, 64–65 (1981)

Estradiol 17β

Molecular Formula: $C_{18}H_{24}O_2$

Synonyms: Altrad; Bardiol; Dihydroxyestrin; Dihydroxy follicular hormone; Oestra-1,3,5(10)triene-3,17-β-diol; Syndiol

CAS Registry Number: 50-28-2

RTECS Number: KG2975000

Regulatory Authority

- Carcinogen (Animal Positive) (IARC)[9]

Cited in U.S. State Regulations: California (G), Massachusetts (G), New Hampshire (G), Pennsylvania (G).

Description: Estradiol, 17-beta, $C_{18}H_{24}O_2$, is an odorless white to yellow crystalline substance. Boiling point = (decomposes). Freezing/Melting point = 173 – 179°C. Hazard Identification (based on NFPA-704 M Rating System): Health 1, Flammability 1, Reactivity 0.

Potential Exposure: The working environment may be contaminated during sex hormone manufacture, especially during the extraction and purification of natural steroid hormones, grinding of raw materials, handling of powdered products and recrystallization. Airborne particles of sex hormones may be absorbed through the skin, ingested or inhaled. Enteric absorption results in quick inactivation of sex hormones in the liver. The rate of inactivation is decreased for the oral, alkylated steroid hormones (methyl testosterone, anabolic steroids, etc.) Sex hormones may accumulate and reach relatively high levels even if their absorption is intermittent. Consequently, repeated absorption of small amounts may be detrimental to health. Intoxication by sex hormones may occur in almost all the exposed workers if preventive measures are not taken. The effect in the industrial sector is more successful than the agricultural one (chemical caponizing of cockerels by stilbestrol implants and incorporation of oestrogens in feed for body weight gain promotion in beef cattle), where measures taken are summary and the number of cases of intoxication is consequently bigger.

Permissible Exposure Limits in Air: No standards set but an industrial hygiene guideline of 0.05 μg airborne oestrogen dust per cubic meter of air, for an 8-hr workday (time weighted average) was adopted at one manufacturing company.

Permissible Concentration in Water: No criteria set.

Routes of Entry: Ingestion, inhalation.

Harmful Effects and Symptoms

Adverse Effects of Oestrogens in Men: Anorrhexia, nausea, vomiting, edema, a feminization syndrome characterized by gynaecomastia (uni- or bi-lateral), increased pigmentation of the areollae, tenderness of the nipples, with or without secretion, slight loss of libido with difficulty in erection, with or without involution of the secondary sex organs and sterility (by inhibition of spermatogenesis), may occur. Urinary oestrogens are increased. A differential diagnosis of breast tumors in men occupationally exposed to oestrogen is needed. Gynaecomastia may occur in cases of functional insufficiency of organs involved in the metabolism of oestrogens (chronic liver or kidney diseases) or of increased endogenous oestrogen synthesis by neoplastic growths such as pituitary, adrenal and testicular tumors or ectopic secretion of gonadotropins or prolactin by lung tumors. Endocrinopathies may be accompanied by gynaecomastia. The conversion of androgen into oestrogen may explain the occurrence of gynaecomastia in the case of gonadotropin or androgen excess in men. Some drugs (spironolactone, digitalis, etc.) may induce gynaecomastia. Physiological gynaecomastia at birth, puberty and old age, as well as the possibility of the occurrence of breast cancer in men, should be also considered.

Adverse Effects of Oestrogen in Women: Prolonged oestrogen therapy has caused malignant endometrial changes in predisposed persons. Endometrial hyperplasia and endometrial carcinoma occurred also after exposure to diethylstilbestrol. Clear cell adenocarcinoma of the vagina and cervix uteri occurred in young women exposed prenatally to diethylstilbestrol or other non-steroidal oestrogens. Women treated with oestrogen may complain of menstrual disorders, nausea, headaches, etc. Similar effects were observed during occupational exposure to natural or synthetic oestrogens and oral contraceptives. Menstrual disorders were frequent. Metrorrhagia was observed in women who started to work. Menorrhagia occurred in women workers with endometrial hyperplasia. In menopausal women diethylstilbestrol may produce abnormal uterine bleeding which often leads to suspicion of cancer of the uterine body. Manifestations like metrorrhagia, nausea and headache usually disappear after discontinuance of exposure.

Excess of progesterone may be responsible for weight gain, acne, mastalgia and breast enlargement and recurrent monilial virginities. Toxicity may also include headache, nausea, chloasma, breakthrough bleeding, weight gain, loss of libido, cholestatic liver damage; sterility may also occur.

Medical Surveillance: Persons directly involved in the hazardous sectors of manufacture should be examined every 2 weeks. Persons engaged in the pharmaceutical stage of manufacture should be examined monthly. The frequency of the physical examination of the employees should increase with the age of the worker. Feminization in men, masculinisation or menstrual disorders in women, changes in certain metabolic parameters and other symptoms detrimental to health are indications for changing the workplace.

Before admission of a new employee a very careful health examination will serve to exclude persons at risk, namely women of childbearing age, epileptics (steroid hormones may increase the frequency of seizures by changes in fluid retention) and persons with hepatic insufficiency, for example.

First Aid: If this chemical gets into the eyes, remove any contact lenses at once and irrigate immediately for at least 15 minutes, occasionally lifting upper and lower lids. Seek medical attention immediately. If this chemical contacts the skin, remove contaminated clothing and wash immediately with soap and water. Seek medical attention immediately. If this chemical has been inhaled, remove from exposure, begin rescue breathing (using universal precautions) if breathing has stopped and CPR if heart action has stopped. Transfer promptly to a medical facility. When this chemical has been swallowed, get medical attention. Give large quantities of water and induce vomiting. Do not make an unconscious person vomit.

Personal Protective Methods: Adequate preventive measures, taken in the pharmaceutical industry, have succeeded in eliminating the occurrence of intoxications almost completely. The airpolluting processes are isolated in areas having adequate exhaust facilities and hermetically sealed machines. The isolated areas are entered only by workers wearing special clothing including underwear, socks, longsleeved overalls (with no pockets) buttoned to the neck and with ties at the bottom of the trousers for tying over the boots, rubber gloves, head cover and dust respirators. Air-supplied vinyl suits may be used by the groups at highest risk. When workers leave the polluted area they should undress, take a shower, wash their hair, clean their nails and put on their own clothes if the working day is over. Workers indirectly exposed to hormonal dust (i.e. mechanics who change the filters in the ventilation system) must also be provided with adequate protective equipment. The contaminated work clothing should be thoroughly cleaned. Disposable paper garments can be burned. Gloves should be rinsed with acetone or methanol, then washed and dried. Respiratory protection equipment should also be cleaned before re-use. Workrooms must be kept very clean. An alkaline detergent has been found to be most suitable for washing clothes and wiping surfaces. Mixers, stirring rods, spatulas, glassware, dishes, etc., should be rinsed after use with acetone or methanol in a hood or near a vacuum, and then washed in the conventional manner. Methanol should be used on plastic items.

It is important to inform workers of the risk represented by the workplace and to win their cooperation in lowering of occupational exposure; for example, not putting on rubber gloves after having worked barehanded, not rubbing the face or nose with contaminated gloves, not contaminating the inside of respirators by leaving them exposed to processing dust, not using outside the work area items which have been inside it (cigarettes, pipes, handkerchiefs) and being alert to detect deficiencies in preventive measures. In some pharmaceutical companies the workers alternate 1 week in the polluted areas with 2 weeks in another place of work. In other companies rotation is carried out only when signs of intoxication occur. If women are employed the alternating work environment should respect their cyclic hormonal pattern. Workers who develop symptoms of chronic intoxication in the presence of a low concentration of hormones in the air must be forbidden to return to the polluted area.

The effectiveness of preventive measures should be checked by analyses of the amount of hormonal compounds in the air of the working environment and in the plasma of the employees, and by clinical examinations. The fact that cases of chronic intoxication may occur in spite of considerable efforts to minimize occupational exposure to sex hormones suggests the need for a closed system of the entire production process.

Respirator Selection: Where the potential for exposure to this chemical, use a MSHA/NIOSH approved supplied-air respirator with a full facepiece operated in the positive pressure mode or with a full facepiece, hood, or helmet in the continuous flow mode, or use a MSHA/NIOSH approved self-contained breathing apparatus with a full facepiece operated in pressure-demand or other positive pressure mode.

Storage: Prior to working with this chemical you should be trained on its proper handling and storage. Store in tightly closed containers in a cool, well ventilated area. A regulated, marked area should be established where this chemical is handled, used, or stored in compliance with OSHA standard 1910.1045.

Shipping: Medicines, poisonous, solid, n.o.s. fall in Hazard Class 6.1 and estradiol in Packing Group II. The label required is "Poison."

Spill Handling: Evacuate persons not wearing protective equipment from area of spill or leak until clean-up is complete. Remove all ignition sources. Collect powdered material in the most convenient and safe manner and deposit in sealed containers. Ventilate area after clean-up is complete. It may be necessary to contain and dispose of this chemical as a hazardous waste. If material or contaminated runoff enters waterways, notify downstream users of potentially contaminated waters. Contact your Department of Environmental Protection or your regional office of the federal EPA for specific recommendations. If employees are required to clean-up spills, they must be properly trained and equipped. OSHA 1910.120(q) may be applicable.

Fire Extinguishing: Use dry chemical, carbon dioxide, water spray, or alcohol foam extinguishers. If material or contaminated runoff enters waterways, notify downstream users of potentially contaminated waters. Notify local health and fire officials and pollution control agencies. From a secure, explosion-proof location, use water spray to cool exposed containers. If cooling streams are ineffective (venting sound increases in volume and pitch, tank discolors, or shows any signs of deforming), withdraw immediately to a secure position. If employees are expected to fight fires, they must be trained and equipped in OSHA 1910.156.

References

Sax, N. I., Ed., "Dangerous Properties of Industrial Materials Report," 1, No. 4, 59–60 (1981)

Parmeggiani, L., Ed., Encyclopedia of Occupational Health & Safety, Third Edition, Vol. 1, pp 1049–1052, Geneva, International Labour Office (1983)

Estrone

Molecular Formula: $C_{18}H_{22}O_2$

Synonyms: Follicular hormone; 3-Hydroxyestra-1,3, 5(10)-trien-17-one; 3-Hydroxy-17-keto-estra-1,3,5-triene; 3-Hydroxy-17-ketoestra-1,3,5-triene; 3-Hydroxy-1,3 ,5(10)-oestratrien-17-one; 3-Hydroxy-oestra-1,3,5(10)-trien-17-one

CAS Registry Number: 53-16-7

RTECS Number: KG8575000

Regulatory Authority

- Carcinogen (Animal Positive) (IARC)[9]

Cited in U.S. State Regulations: California (G), Illinois (G), Massachusetts (G), New Hampshire (G), Pennsylvania (G).

Description: Estrone, $C_{18}H_{22}O_2$ is an odorless white crystalline powder. Freezing/Melting point = 258 – 262°C. Hazard Identification (based on NFPA-704 M Rating System): Health 1, Flammability 1, Reactivity 0. Insoluble in water.

Potential Exposure: Synthesized from ergosterol. See corresponding section in entry on estradiol 17β.

Permissible Exposure Limits in Air: No standards set but an industrial guideline of 0.05 μg airborne oestrogen dust per cubic meter of air, for an 8-h workday (time weighted average) was adopted at one manufacturing company.

Permissible Concentration in Water: No criteria set.

Routes of Entry: Ingestion, inhalation.

Harmful Effects and Symptoms

Adverse effects of oestrogens in men: Anorrhexia, nausea, vomiting, edema, a feminization syndrome characterized by gynaecomastia (uni- or bi-lateral), increased pigmentation of the areollae, tenderness of the nipples, with or without secretion, slight loss of libido with difficulty in erection, with or without involution of the secondary sex organs and sterility (by inhibition of spermatogenesis), may occur. Urinary oestrogens are increased. A differential diagnosis of breast tumors in men occupationally exposed to oestrogen is needed. Gynaecomastia may occur in cases of functional insufficiency of organs involved in the metabolism of oestrogens (chronic liver or kidney diseases) or of increased endogenous oestrogen synthesis by neoplastic growths such as pituitary, adrenal and testicular tumors or ectopic secretion of gonadotropins or prolactin by lung tumors. Endocrinopathies may be accompanied by gynaecomastia. The conversion of androgen into oestrogen may explain the occurrence of gynaecomastia in the case of gonadotropin or androgen excess in men. Some drugs (spironolactone, digitalis, etc.) may induce gynaecomastia. Physiological gynaecomastia at birth, puberty and old age, as well as the possibility of the occurrence of breast cancer in men, should be also considered.

Adverse effects of oestrogen in women: Prolonged oestrogen therapy has caused malignant endometrial changes in predisposed persons. Endometrial hyperplasia and endometrial carcinoma occurred also after exposure to diethylstilbestrol. Clear cell adenocarcinoma of the vagina and cervix uteri occurred in young women exposed prenatally to diethylstilbestrol or other non-steroidal oestrogens. Women treated with oestrogen may complain of menstrual disorders, nausea, headaches, etc. Similar effects were observed during occupational exposure to natural or synthetic oestrogens and oral contraceptives. Menstrual disorders were frequent. Metrorrhagia was observed in women who started to work. Menorrhagia occurred in women workers with endometrial hyperplasia. In menopausal women diethylstilbestrol may produce abnormal uterine bleeding which often leads to suspicion of cancer of the uterine body. Manifestations like metrorrhagia, nausea and headache usually disappear after discontinuance of exposure.

Excess of progesterone may be responsible for weight gain, acne, mastalgia and breast enlargement and recurrent monilial virginities. Toxicity may also include headache, nausea, chloasma, breakthrough bleeding, weight gain, loss of libido, cholestatic liver damage; sterility may also occur.

Long Term Exposure: Confirmed carcinogen.

Medical Surveillance: Persons directly involved in the hazardous sectors of manufacture should be examined every 2 weeks. Persons engaged in the pharmaceutical stage of manufacture should be examined monthly. The frequency of the physical examination of the employees should increase with the age of the worker. Feminization in men, masculinisation or menstrual disorders in women, changes in certain metabolic parameters and other symptoms detrimental to health are indications for changing the workplace.

Before admission of a new employee a very careful health examination will serve to exclude persons at risk, namely women of childbearing age, epileptics (steroid hormones may increase the frequency of seizures by changes in fluid retention) and persons with hepatic insufficiency, for example.

First Aid: If this chemical gets into the eyes, remove any contact lenses at once and irrigate immediately for at least 15 minutes, occasionally lifting upper and lower lids. Seek medical attention immediately. If this chemical contacts the skin, remove contaminated clothing and wash immediately with soap and water. Seek medical attention immediately. If this chemical has been inhaled, remove from exposure, begin rescue breathing (using universal precautions) if breathing has stopped and CPR if heart action has stopped. Transfer promptly to a medical facility. When this chemical has been swallowed, get medical attention. Give large quantities of water and induce vomiting. Do not make an unconscious person vomit.

Personal Protective Methods: Adequate preventive measures, taken in the pharmaceutical industry, have succeeded in eliminating the occurrence of intoxications almost completely. The airpolluting processes are isolated in areas having adequate exhaust facilities and hermetically sealed machines. The isolated areas are entered only by workers wearing special clothing including underwear, socks, lonsleeved overalls (with no pockets) buttoned to the neck and with ties at the bottom of the trousers for tying over the boots, rubber gloves, head cover and dust respirators. Air-supplied vinyl suits may be used by the groups at highest risk. When workers leave the polluted area they should undress, take a shower, wash their hair, clean their nails and put on their own clothes if the working day is over. Workers indirectly exposed to hormonal dust (i.e. mechanics who change the filters in the ventilation system) must also be provided with adequate protective equipment. The contaminated work clothing should be thoroughly cleaned. Disposable paper garments can be burned. Gloves should be rinsed with acetone or methanol, then washed and dried. Respiratory protection equipment should also be cleaned before re-use. Workrooms must be kept very clean. An alkaline detergent has been found to be most suitable for washing clothes and wiping surfaces. Mixers, stirring rods, spatulas, glassware, dishes, etc., should be rinsed after use with acetone or methanol in a hood or near a vacuum, and then washed in the conventional manner. Methanol should be used on plastic items.

It is important to inform workers of the risk represented by the workplace and to win their cooperation in lowering of occupational exposure; for example, not putting on rubber gloves after having worked barehanded, not rubbing the face or nose with contaminated gloves, not contaminating the inside of respirators by leaving them exposed to processing dust, not using outside the work area items which have been inside it (cigarettes, pipes, handkerchiefs) and being alert to detect deficiencies in preventive measures. In some pharmaceutical companies the workers alternate 1 week in the polluted areas with 2 weeks in another place of work. In other companies rotation is carried out only when signs of intoxication occur. If women are employed the alternating work environment should respect their cyclic hormonal pattern. Workers who develop symptoms of chronic intoxication in the presence of a low concentration of hormones in the air must be forbidden to return to the polluted area.

The effectiveness of preventive measures should be checked by analyses of the amount of hormonal compounds in the air of the working environment and in the plasma of the employees, and by clinical examinations. The fact that cases of chronic intoxication may occur in spite of considerable efforts to minimize occupational exposure to sex hormones suggests the need for a closed system of the entire production process.

Respirator Selection: Where the potential for exposure to this chemical, use a MSHA/NIOSH approved supplied-air respirator with a full facepiece operated in the positive pressure mode or with a full facepiece, hood, or helmet in the continuous flow mode, or use a MSHA/NIOSH approved self-contained breathing apparatus with a full facepiece operated in pressure-demand or other positive pressure mode.

Storage: Prior to working with this chemical you should be trained on its proper handling and storage. Store in tightly closed containers in a cool, well ventilated. A regulated, marked area should be established where this chemical is handled, used, or stored in compliance with OSHA standard 1910.1045.

Shipping: Medicines, poisonous, solid, n.o.s. fall in Hazard Class 6.1 and estrone in Packing Group II. The label required is "Poison."

Spill Handling: Evacuate persons not wearing protective equipment from area of spill or leak until clean-up is complete. Remove all ignition sources. Collect powdered material in the most convenient and safe manner and deposit in sealed containers. Ventilate area after clean-up is complete. It may be necessary to contain and dispose of this chemical as a hazardous waste. If material or contaminated runoff enters waterways, notify downstream users of potentially contaminated waters. Contact your Department of Environmental Protection or your regional office of the federal EPA for specific recommendations. If employees are required to clean-up spills, they must be properly trained and equipped. OSHA 1910.120(q) may be applicable.

Fire Extinguishing: This chemical is a combustible solid. Use dry chemical, carbon dioxide, water spray, or alcohol foam extinguishers. Poisonous gases are produced in fire. If material or contaminated runoff enters waterways, notify downstream users of potentially contaminated waters. Notify local health and fire officials and pollution control agencies. From a secure, explosion-proof location, use water spray to cool exposed containers. If cooling streams are ineffective (venting sound increases in volume and pitch, tank discolors, or shows any signs of deforming), withdraw immediately to a secure position. If employees are expected to fight fires, they must be trained and equipped in OSHA 1910.156.

References

Sax, N. I., Ed., "Dangerous Properties of Industrial Materials Report," 1, No. 4, 63–64 (1981)

Parmeggiani, L., Ed., Encyclopedia of Occupational Health & Safety, Third Edition, Vol. 1, pp 1049–1052, Geneva, International Labour Office (1983)

Ethane

Molecular Formula: C_2H_6

Synonyms: Bimethyl; Dimethyl; Etano (Spanish); Ethyl hydride; Methylmethane

CAS Registry Number: 74-84-0

RTECS Number: KH3800000

DOT ID: UN 1035 (compressed gas); UN 1961 (refrigerated liquid)

EEC Number: 601-002-00-X

Regulatory Authority

- Clean Air Act: Accidental Release Prevention/Flammable substances, (Section 112[r], Table 3), TQ = 10,000 lb (4,540 kg)
- Canada, WHMIS, Ingredients Disclosure List

Cited in U.S. State Regulations: Alaska (G), California (A, G), Illinois (G), Maine (G), Massachusetts (G), New Hampshire (G), New Jersey (G), Pennsylvania (G), Rhode Island (G).

Description: Ethane, C_2H_6, is a compressed, liquefied, colorless gas with a mild, gasoline-like odor. Odorless when pure. Odor threshold = 899 ppm. Boiling point = -89°C. Freezing/ Melting point = -183°C. Flash point = flammable gas. Autoignition temperature = 472°C. Explosive limits: LEL = 3.0%; UEL = 12.5%. Hazard Identification (based on NFPA-704 M Rating System): Health 1, Flammability 4, Reactivity 0. Insoluble in water.

Potential Exposure: Ethane is used as a fuel, in making chemicals or as a freezing agent. The health effects caused by ethane exposure are much less serious than the fire and explosion risk posed by this chemical.

Incompatibilities: Flammable gas; forms explosive mixture with air. Strong oxidizers may cause fire and explosions. May accumulate static electrical charges, and may cause ignition of its vapors.

Permissible Exposure Limits in Air: No OELs have been established for ethane. Ethane is listed by ACGIH[1] as a simple asphyxiant, also by HSE.[33]

Permissible Concentration in Water: No criteria set.

Routes of Entry: Inhalation.

Harmful Effects and Symptoms

Short Term Exposure: Exposure can cause headache, dizziness and make you feel lightheaded. Very high levels can cause suffocation from lack of oxygen. Contact with liquid ethane can cause frostbite. Ethane is highly flammable. It is a dangerous fire hazard.

Long Term Exposure: Chronic health effects are unknown at this time.

First Aid: Remove the person from exposure. Begin rescue breathing if breathing has stopped and CPR if heart action has stopped. Transfer promptly to a medical facility. If frostbite has occurred, seek medical attention immediately; do *NOT* rub the affected areas or flush them with water. In order to prevent further tissue damage, do *NOT* attempt to remove frozen clothing from frostbitten areas. If frostbite has *NOT* occurred, immediately and thoroughly wash contaminated skin with soap and water.

Personal Protective Methods: Wear protective gloves and clothing to prevent any reasonable probability of skin contact. Safety equipment suppliers/manufacturers can provide recommendations on the most protective glove/clothing material for your operation. Polyethylene is among the recommended protective materials. All protective clothing (suits, gloves, footwear, headgear) should be clean, available each day, and put on before work. Contact lenses should not be worn when working with this chemical. Wear splash-proof chemical goggles and face shield unless full facepiece respiratory protection is worn. Wear gas-proof goggles, unless full facepiece respiratory protection is worn. Employees should wash immediately with soap when skin is wet or contaminated. Provide emergency showers and eyewash. Where exposure to the liquefied compressed gas may occur, employees should be provided with special clothing designed to prevent frostbite.

Respirator Selection: Chemical cartridge respirators should not be used where ethane exposure occurs. For high exposures use air-supplied respirators. Exposure to ethane is dangerous because it can replace oxygen and lead to suffocation. Only MSHA/NIOSH approved self-contained breathing apparatus with a full facepiece operated in the positive pressure mode should be used in oxygen deficient environments.

Storage: Prior to working with this chemical you should be trained on its proper handling and storage. Before entering confined space where this chemical may be present, check to make sure that an explosive concentration does not exist. Ethane must be stored to avoid contact with oxidizers (such as perchlorates, peroxides, permanganates, chlorates, and nitrates) since violent reactions occur. Sources of ignition such as smoking and open flames are prohibited where ethane is handled, used, or stored. Use only non-sparking tools and equipment, especially when opening and closing containers of ethane. Wherever ethane is used, handled, manufactured, or stored, use explosion-proof electrical equipment and fittings. Piping should be electrically bonded and grounded. Procedures for the handling, use and storage of cylinders should be in compliance with OSHA 1910.101 and 1910.169 as with the recommendations of the Compressed Gas Association.

Shipping: Ethane falls in DOT Hazard Class 2.1. The label required is "Flammable Gas." In the case of compressed ethane, shipment by passenger aircraft or railcar is forbidden; shipment by cargo aircraft is limited. In the case of liquid ethane passenger aircraft or railcar and even cargo aircraft shipment is forbidden.

Spill Handling: If in a building, evacuate building and confine vapors by closing doors and shutting down HVAC systems. Restrict persons not wearing protective equipment from area of spill or leak until cleanup is complete. Remove all ignition sources. Establish forced ventilation to keep levels below explosive limit and to disperse the gas. Wear chemical protective suit with self-contained breathing apparatus to combat spills. Stay upwind and use water spray to "knock down" vapor; contain runoff. Stop the flow of gas, if it can be done safely from a distance. If source is a cylinder and the leak cannot be stopped in place, remove the leaking cylinder to a safe place, and repair leak or allow cylinder to empty. Keep this chemical out of confined spaces, such as a sewer, because of the possibility of explosion, unless the sewer is designed to prevent the buildup of explosive concentrations. If employees are required to clean-up spills, they must be properly trained and equipped. OSHA 1910.120(q) may be applicable.

Fire Extinguishing: Ethane is a flammable gas. Use dry chemical, carbon dioxide, water spray, or foam extinguishers. Evacuate and restrict persons not wearing protective equipment from area of spill or leak until cleanup is complete. Remove all ignition sources. Ventilate area of leak to disperse the gas. Stop the flow of gas if it can be done safely. If source of leak is a cylinder and the leak cannot be stopped in place, remove leaking cylinder to a safe place in the open air, and repair leak or allow cylinder to empty. Keep this chemical out of confined space, such as sewer because of the possibility of explosion, unless the sewer is designed to prevent the buildup of explosive concentrations. It may be necessary to contain and dispose of this chemical as a hazardous waste. Contact your Department of Environmental Protection or your regional office of the federal EPA for specific recommendations. If employees are required to clean-up spills, they must be properly trained and equipped. OSHA 1910.120(q) may be applicable.

Disposal Method Suggested: Incineration.

References

New Jersey Department of Health and Senior Services, "Hazardous Substance Fact Sheet: Ethane," Trenton, NJ (January 1997)

Ethanolamines

Molecular Formula: C_2H_7NO; $C_4H_{11}NO$; $C_6H_{15}NO_3$

Common Formula: $H_2NCH_2CH_2OH$; $HN(CH_2CH_2OH)_2$; $N(CH_2CH_2OH)_3$

Synonyms: *diethanolamine: N,N*-Bis(2-hydroxyethyl)amine; Bis(2-hydroxyethyl)amine; Bis(hydroxyethyl)amine; DEA; Diaethanolamin (German); Dietanolamina (Spanish); Diethanolamin (Czech); *N,N*-Diethanolamine; Diethylolamine; 2,2'-Dihydroxydiethylamine; Di(2-hydroxyethyl)amine; Diolamine; Ethanol, 2,2'-iminobis-; Ethanol, 2,2'-iminodi-; 2-[(2-Hydroxyethyl)amino]ethanol; 2,2'-Iminobis(ethanol); 2,2'-Iminodi-1-ethanol; 2,2'-Iminodiethanol; Iminodiethanol; NCI C55174

monoethanolamine: 2-Aminoethanol; β-Aminoethyl alcohol; Colamine; Ethylolamine; 2-Hydroxyethylamine; MEA; Monoethanolamine

triethanolamine: Daltogen; 2,2',2''-Nitrilo-triethanol; TEA; Thiofaco T-35; Triethylolamine; Tri(2-hydroxyethyl)amine; Tri(hydroxytriethyl)amine; Tris(hydroxyethyl)amine; Trolamine

CAS Registry Number: 141-43-5 (mono-); 111-42-2 (di-); 102-71-6 (tri-)

RTECS Number: KJ5775000 (mono-); KL2975000 (di-); KL9275000 (tri-)

DOT ID: UN 2491

EEC Number: 603-030-00-8 (mono-); 603-071-00-1 (di-)

Regulatory Authority

Diethanolamine:

- Air Pollutant Standard Set (ACGIH)[1] (DFG)[3] (HSE)[33] (OSHA)[58] (Several States)[60]
- Clean Air Act: Hazardous Air Pollutants (Title I, Part A, Section 112)
- Superfund/EPCRA 40CFR302.4 Reportable Quantity (RQ): CERCLA, 1 lb (0.454 kg)
- EPCRA Section 313 Form R *de minimis* concentration reporting level: 1.0%
- Canada, WHMIS, Ingredients Disclosure List

Cited in U.S. State Regulations: Alaska (G), California (A, G), Connecticut (A), Florida (G, A), Illinois (G), Kansas (A), Maine (G), Maryland (G), Massachusetts (G), Nevada (A), New Hampshire (G), New Jersey (G), New York (G, A), North Dakota (A), Pennsylvania (G), Rhode Island (G), South Carolina (A), Virginia (A), West Virginia (G).

Description: $H_2NCH_2CH_2OH$ – monoethanolamine, $HN(CH_2CH_2OH)_2$ – diethanolamine, $N(CH_2CH_2OH)_3$ – triethanolamine. All three compounds are water soluble liquids. Ethanolamines can be detected by odor as low as 2 – 3 ppm. Some physical properties are:

	Melting Point, °C	Boiling Point, °C	Flash Point, °C
Mono-	11	172	85 (cc)
Di	28	268	172 (cc)
Tri-	21	343	179 (cc)

Monoethanolamine is a colorless, viscous liquid or solid (below 111°C) with an unpleasant, ammonia-like odor. Autoignition temperature = 410°C. Explosive limits: LEL = 3.0% (@140°C); UEL =23.5%. Hazard Identification (based on NFPA-704 M Rating System): Health 3, Flammability 2, Reactivity 0. Soluble in water.

Diethanolamine is colorless crystals or a syrupy, white liquid (above 28°C) with a mild, ammonia-like odor. Autoignition temperature = 662°C. Hazard Identification (based on NFPA-704 M Rating System): Health 1, Flammability 1, Reactivity 0.

Triethanolamine has only a faint non-ammonia odor. Hazard Identification (based on NFPA-704 M Rating System): Health 2, Flammability 1, Reactivity 1. Soluble in water.

Potential Exposure: Monoethanolamine is widely used in industry to remove carbon dioxide and hydrogen from natural gas, to remove hydrogen sulfide and carbonyl sulfide, as an alkaline conditioning agent, and as an intermediate for soaps, detergents, dyes, and textile agents. Diethanolamine is an absorbent for gases, a solubilizer for 2,4-dichlorophenoxyacetic acid (2,4-D), and a softener and emulsifier intermediate for detergents It also finds use in the dye and textile industry.

Triethanolamine is used as a plasticizers, neutralizer for alkaline dispersions, lubricant additive, corrosion inhibitor, and in the manufacture of soaps, detergents, shampoos, shaving preparations, face and hand creams, cements, cutting oils, insecticides, surface active agents, waxes, polishes, and herbicides.

Incompatibilities: *mono-:* This chemical is a medium-strong base. Reacts violently with strong oxidizers, acetic acid, acetic anhydride, acrolein, acrylic acid, acrylonitrile, cellulose nitrate, chlorosulfonic acid, epichlorohydrin, hydrochloric acid, hydrogen fluoride, mesityl oxide, nitric acid, oleum, sulfuric acid, b-propiolactone, vinyl acetate. Reacts with iron. May attack copper, aluminum and their alloys, and rubber. *di-*: Oxidizers, strong acids, acid anhydrides, halides. Reacts with CO_2 in the air. Hygroscopic (i.e., absorbs moisture from the air). Corrosive to copper, zinc, and galvanized iron (di-). The aqueous solution is a medium strong base. Reacts violently with oxidizers, strong acids and anhydrides. Attacks copper, zinc, aluminum, and their alloys.

Permissible Exposure Limits in Air: The OSHA PEL, ACGIH[1], HSE[33] and DFG[3] values for monoethanolamine is 3 ppm (8 mg/m^3) TWA and the STEL for monoethanolamine, set by ACGIH[1] and HSE,[33] is 6 ppm (15 mg/m^3). The NIOSH IDLH = (mono-) 30 ppm. The DFG MAK [2-aminoethanol] is 2 ppm (5.1 mg/m^3). Several states have set guidelines or standards for diethanolamine in ambient air[60] ranging from 150 μg/m^3 (North Dakota) to 250 μg/m^3 (Virginia) to 300 μg/m^3 (Connecticut) to 357 μg/m^3 (Nevada). A large number of states have set guidelines or standards for monoethanolamine in ambient air[60] ranging from 19.048 μg/m^3 (Kansas) to 26.7 μg/m^3 (New York) to 80.0 μg/m^3 (Florida) to 80.0 – 150.0 μg/m^3 (North Dakota) to 120 μg/m^3 (Connecticut) to 130.0 μg/m^3 (Virginia) to 190.0 μg/m^3 (Nevada) to 200.0 μg/m^3 (South Carolina).

Determination in Air: *mono-:* Adsorption on silica gel; Methanol/Water; Gas chromatography/Flame ionization detection; NIOSH IV Method #2007, Aminoethanol Compounds. *di-:* Impinger; Reagent; Ion chromatography; NIOSH IV, Method #3509, Aminoethanol Compounds II.

Permissible Concentration in Water: EPA[32] has suggested permissible ambient limits of 83 μg/l. The Former USSR-UNEP/IRPTC project[43] has set MAC values in water bodies used for domestic purposes of 0.8 mg/l for diethanolamine and 1.4 mg/l for triethanolamine but no MAC for monoethanolamine.

Routes of Entry: Inhalation of vapor, percutaneous absorption, ingestion and skin and/or eye contact.

Harmful Effects and Symptoms

The LD_{50} oral-rat for the isomers is as follows:

mono-	2,050 mg/kg (slightly toxic)
di-	710 mg/kg (slightly toxic)
tri-	8,000 mg/kg (insignificantly toxic)

All three compounds cause irritation to nose, throat and lungs upon inhalation. Skin contact causes irritation, stinging and burns. Eye contact can cause severe irritation. Ingestion can cause nausea and intestinal irritation. Fatal doses are estimated by NY State to be 6 g of monethanolamine for a 150-pound adjustment and 1 pint to 1 quart of di- or triethanolamine.

Short Term Exposure: *mono-:* Corrosive to the eyes and irritates the skin and respiratory tract. Inhalation may cause asthmatic reactions. May affect central nervous system and may cause unconsciousness. *di-:* Corrosive to the eyes. Irritates the eyes, skin, respiratory tract.

Long Term Exposure: *mono-:* Repeated or prolonged contact with skin may cause dermatitis and ulceration and lungs may be affected. May affect the central nervous system, kidneys, liver, and blood, causing asthenia and hematological changes and tissue lesions. *di-:* May affect the kidneys and liver.

Points of Attack: Eyes, skin, respiratory system, central nervous system

Medical Surveillance: With monoethanolamine, if symptoms develop or overexposure is suspected, the following may be useful: Liver and kidney function tests. Evaluation by a qualified allergist, including careful exposure history and special testing, may help diagnose skin allergy.

With diethanolamine: Before beginning employment and at regular times after that, for those with frequent or potentially high exposures, the following is recommended: Lung function tests. If symptoms develop or overexposure is suspected the following may be useful: Skin testing with dilute diethanolamine may help diagnose allergy if done by a qualified allergist.

First Aid: If this chemical gets into the eyes, remove any contact lenses at once and irrigate immediately for at least 30 minutes, occasionally lifting upper and lower lids. Seek medical attention immediately. If this chemical contacts the skin, remove contaminated clothing and wash immediately with soap and water. Seek medical attention immediately. If this chemical has been inhaled, remove from exposure, begin rescue breathing (using universal precautions) if breathing has stopped and CPR if heart action has stopped. Transfer promptly to a medical facility. When this chemical has been swallowed, get medical attention. If victim is conscious, administer water or milk. Do not induce vomiting.

Personal Protective Methods: Wear protective gloves and clothing to prevent any reasonable probability of skin contact. Safety equipment suppliers/manufacturers can provide recommendations on the most protective glove/clothing material for your operation.. All protective clothing (suits, gloves, footwear, headgear) should be clean, available each day, and put on before work. Butyl rubber, neoprene, nitrile, nitrile + PVC, Polyethylene, Polyvinyl Alcohol, or polyvinyl chloride are among the recommended protective materials. Contact lenses should not be worn when working with this chemical. Wear splash-proof chemical goggles and face shield unless full facepiece respiratory protection is worn. Employees should wash immediately with soap when skin is wet or contaminated. Provide emergency showers and eyewash.

Respirator Selection: *Monoethanolamine:* NIOSH/OSHA: *Up to 30 ppm:* CCRS [any chemical cartridge respirator with cartridge(s) providing protection against the compound of concern];* or GMFS [any air-purifying, full-facepiece respirator (gas mask) with a chin-style, front- or back-mounted canister providing protection against the compound of concern]; or PAPRS [any powered, air-purifying respirator with cartridge(s) providing protection against the compound of concern];* or SA (any supplied-air respirator);* or SCBAF (any self-contained breathing apparatus with a full facepiece). *Emergency or planned entry into unknown concentrations or IDLH conditions:* SCBAF:PD,PP (any self-contained breathing apparatus that has a full facepiece and is operated in a pressure-demand or other positive-pressure mode); or SAF:PD,PP:ASCBA (any supplied-air respirator that has a full facepiece and is operated in a pressure-demand or other positive-pressure mode in combination with an auxiliary self-contained positive-pressure breathing apparatus). *Escape:* GMFS [any air-purifying, full-facepiece respirator (gas mask) with a chin-style, front- or back-mounted canister providing protection against the compound of concern]; or SCBAE (any appropriate escape-type, self-contained breathing apparatus).

* Substance reported to cause eye irritation or damage; may require eye protection.

Storage: Prior to working with this chemical you should be trained on its proper handling and storage. Before entering confined space where this chemical may be present, check to make sure that an explosive concentration does not exist. Ethanolamine must be stored to avoid contact with strong oxidizers (such as chlorine and bromine) and strong acids, (such as hydrochloric, sulfuric and nitric acids), because violent reactions occur. Store in tightly closed containers in a cool, well-ventilated area away from heat. Sources of ignition such as smoking and open flames are prohibited where ethanolamine is used, handled, or stored in a manner that could crate a potential fire or explosion hazard.

Shipping: This compound requires a shipping label of: "Corrosive." It falls in DOT Hazard Class 8 and Packing Group III.

Spill Handling: Evacuate and restrict persons not wearing protective equipment from area of spill or leak until cleanup is complete. Remove all ignition sources. Establish forced ventilation to keep levels below explosive limit. Absorb liquids in vermiculite, dry sand, earth, peat, carbon, or a similar material and deposit in sealed containers. It may be necessary to contain and dispose of this chemical as a hazardous waste. If material or contaminated runoff enters waterways, notify downstream users of potentially contaminated waters. Contact your Department of Environmental Protection or your regional office of the federal EPA for specific recommendations. If employees are required to clean-up spills, they must be properly trained and equipped. OSHA 1910.120(q) may be applicable.

Fire Extinguishing: This chemical is a combustible liquid. Poisonous gases including nitrogen oxides are produced in fire. Use dry chemical, carbon dioxide, or foam extinguishers. Vapors are heavier than air and will collect in low areas. Vapors may travel long distances to ignition sources and flashback. Vapors in confined areas may explode when exposed to fire. Containers may explode in fire. Storage containers and parts of containers may rocket great distances, in many directions. If material or contaminated runoff enters waterways, notify downstream users of potentially contaminated waters. Notify local health and fire officials and pollution control agencies. From a secure, explosion-proof location, use water spray to cool exposed containers. If cooling streams are ineffective (venting sound increases in volume and pitch, tank discolors, or shows any signs of deforming), withdraw immediately to a secure position. If employees are expected to fight fires, they must be trained and equipped in OSHA 1910.156.

Disposal Method Suggested: Controlled incineration; incinerator is equipped with a scrubber or thermal unit to reduce No_x emissions.

References

U.S. Environmental Protection Agency, "Chemical Hazard Information Profile: Ethanolamines," Washington, DC (April 14, 1978)

Sax, N. I., Ed., "Dangerous Properties of Industrial Materials Report," 4, No. 1, 66-69 (1984) (Mono-)

New Jersey Department of Health and Senior Services, "Hazardous Substance Fact Sheet: Ethanolamine," Trenton, NJ (August 1985)

New Jersey Department of Health and Senior Services, "Hazardous Substance Fact Sheet: Diethanolamine," Trenton, NJ (June 1996)

New York State Department of Health, "Chemical Fact Sheet: Ethanolamine," Albany, NY, Bureau of Toxic Substance Assessment (March 1986)

New York State Department of Health, "Chemical Fact Sheet: Diethanolamine," Albany, NY, Bureau of Toxic Substance Assessment (March 1986)

New York State Department of Health, "Chemical Fact Sheet: Triethanolamine," Albany, NY, Bureau of Toxic Substance Assessment (April 1986)

Ethinylestradiol

Molecular Formula: $C_{20}H_{24}O_2$

Synonyms: 3,17-β-Dihydroxy-17-α-ethynyl-1,3,5(10)-estratriene; 3,17-β-Dihydroxy-17-α-ethynyl-1,3,5(10)-oestratriene; Estrogen; 17-α-Ethinyl-3,17-dihydroxy-d1,3,5-estratriene; 17-α-Ethinyl-3,17-dihydroxy-d1,3,5-oestratriene; 17-α-Ethinyl-17-β-estradiol; 17-α-Ethinylestradiol; 17-Ethinyl-3,17-estradiol; 17-Ethinylestradiol; 17-α-Ethinylestra-1,3,5(10)-triene-3,17-β-diol; Ethinylestriol; 17-Ethinyl-3,17-oestradiol; Ethinyloestradiol; Ethinyl-oestranol; 17-α-Ethinyl-d1,3,5(10)oestratriene-3,17-β-diol; 17-α-Ethinyloestra-1,3,5(10)-triene-3,17-β-diol; Ethinyloestriol; 17-Ethynyl-3,17-dihydroxy-1,3,5-oestratriene; 17-α-Ethynylestradiol; 17-α-Ethynylestradiol-17-β; Ethynylestradiol; 17-α-Ethynyl-1,3,5(10)-estratriene-3,17-β-diol; 17-α-Ethynylestra-1,3,5(10)-triene-3,17-β-diol; 17-α-Ethynyl-17-β-oestradiol; 17-α-Ethynyloestradiol; 17-α-Ethynyloestradiol-17-β; 17-Ethynyloestradiol; Ethynyloestradiol; 17-α-Ethynyl-1,3,5(10)-oestratriene-3,17-β-diol; 17-α-Ethynyl-1,3,5-oestratriene-3,17-β-diol; 17-α-Ethynyloestra-1,3,5(10)-triene-3,17-β-diol; 17-Ethynyloestra-1,3,5(10)-triene-3,17-β-diol; (17-α)-19-Norpregna-1,3,5(10)-trien-20-yne-3,17, diol; 19-Nor-17-α-pregna-1,3,5(10)-trien-2-yne-3,17-diol

CAS Registry Number: 57-63-6

RTECS Number: RC8925000

Regulatory Authority

- Carcinogen (Animal Positive) (IARC)[9]

Cited in U.S. State Regulations: California (G), Florida (G), Illinois (G), Massachusetts (G), New Hampshire (G), New Jersey (G), Pennsylvania (G).

Description: Ethinylestradiol, $C_{20}H_{24}O_2$, is a white to creamy-white odorless powder. Freezing/Melting point = 142 – 146°C. Hazard Identification (based on NFPA-704 M Rating System): Health 1, Flammability 1, Reactivity 0.

Potential Exposure: Used in medicine as an estrogenic hormone and as an oral contraceptive.

Permissible Exposure Limits in Air: No standards set.

Permissible Concentration in Water: No criteria set.

Harmful Effects and Symptoms

Short Term Exposure: Nausea, vomiting, abdominal cramps, bloating, cholasma, cholestatic jaundice, skin eruptions, loss of scalp hair, hirsutism, changes in menstrual flow, headache, migraine, dizziness, mental depression, increase or decrease in weight, edema, changes in libido, and breast enlargement.

Long Term Exposure: A confirmed carcinogen. May have teratogenic effects and reproductive effects. Human mutation data reported.

First Aid: *Skin Contact:*[52] Flood all areas of body that have contacted the substance with water. Don't wait to remove contaminated clothing; do it under the water stream. Use soap to help assure removal. Isolate contaminated clothing when removed to prevent contact by others.

Eye Contact: Remove any contact lenses at once. Immediately flush eyes well with copious quantities of water or normal saline for at least 20 – 30 minutes. Seek medical attention.

Inhalation: Leave contaminated area immediately; breathe fresh air. Proper respiratory protection must be supplied to any rescuers. If coughing, difficult breathing or any other symptoms develop, seek medical attention at once, even if symptoms develop many hours after exposure.

Ingestion: Contact a physician, hospital or poison center at once. If the victim is unconscious or convulsing, do not induce vomiting or give anything by mouth. Assure that his airway is open and lay him on his side with his head lower than his body and transport immediately to a medical facility. If conscious and not convulsion, give a glass of water to dilute the substance. Vomiting should not be induced without a physician's advice.

Personal Protective Methods: Wear protective gloves and clothing to prevent any reasonable probability of skin contact. Safety equipment suppliers/manufacturers can provide recommendations on the most protective glove/clothing material for your operation.. All protective clothing (suits, gloves, footwear, headgear) should be clean, available each day, and put on before work. Contact lenses should not be worn when working with this chemical. Wear dust-proof chemical goggles and face shield unless full facepiece respiratory protection is worn. Employees should wash immediately with soap when skin is wet or contaminated. Provide emergency showers and eyewash.

Respirator Selection: Where the potential for exposure to this chemical, use a MSHA/NIOSH approved supplied-air respirator with a full facepiece operated in the positive pressure mode or with a full facepiece, hood, or helmet in the continuous flow mode, or use a MSHA/NIOSH approved self-contained breathing apparatus with a full facepiece operated in pressure-demand or other positive pressure mode.

Storage: Prior to working with this chemical you should be trained on its proper handling and storage. Store in a refrigerator under an inert atmosphere and protect from exposure to light.

Shipping: Hazardous substance, solid, n.o.s. fall in Hazard Class ORM-E and Packing Group III. This imposes no label requirement or any maximum allowable limits on shipping weights.

Spill Handling: Evacuate persons not wearing protective equipment from area of spill or leak until clean-up is complete. Remove all ignition sources and dampen spilled material with 60 – 70% ethanol. Collect powdered material in the most convenient and safe manner and deposit in sealed containers. Ventilate area after clean-up is complete. It may be necessary to contain and dispose of this chemical as a hazardous waste. If material or contaminated runoff enters waterways, notify downstream users of potentially contaminated waters. Contact your Department of Environmental Protection or your regional office of the federal EPA for specific recommendations. If employees are required to clean-up spills, they must be properly trained and equipped. OSHA 1910.120(q) may be applicable.

Fire Extinguishing: Use dry chemical, carbon dioxide, water spray, or alcohol foam extinguishers. If material or contaminated runoff enters waterways, notify downstream users of potentially contaminated waters. Notify local health and fire officials and pollution control agencies. From a secure, explosion-proof location, use water spray to cool exposed containers. If cooling streams are ineffective (venting sound increases in volume and pitch, tank discolors, or shows any signs of deforming), withdraw immediately to a secure position. If employees are expected to fight fires, they must be trained and equipped in OSHA 1910.156.

Ethion

Molecular Formula: $C_9H_{22}O_4P_2S_4$

Synonyms: AC 3422; Bis[S-(diethoxyphosphinothioyl)mercapto]methane; Bis (dithiophosphatede *O,O*-diethyle) de *S,S*'-methylene (French); Bladan; Diethion; Embathion; ENT 24,105; Ethanox; Ethiol; Ethodan; Ethyl methylene phosphorodithioate; Etion (Spanish); FMC-1240; Fosfono 50; Hylemox; Itopaz; Kwit; Methanedithiol, *S,S*-diester with *O,O*-diethyl phosphorodithioate acid; Methyleen-*S,S*'-bis(*O,O*-diethyl-dithiofosfaat) (Dutch); Methylene-*S,S*'-bis(*O,O*-diaethyl-dithiophosphat) (German); a,*S*'-Methylene *O,O,O',O'*-tetraethyl ester phosphorodithioic acid; *S,S*'-Methylene *O,O,O',O'*-tetraethyl phosphorodithioate; Metilen-*S,S*'-bis(*O,O*-dietil-ditiofosfato) (Italian); NIA 1240; Niagara 1240; Nialate; Phosphorodithioic acid, *O,O*-diethyl ester, *S,S*-diester with methanedithiol; Phosphotox E; Rhodiacide; Rhodocide; Rodocid; RP 8167; RTECS No.TE4550000; Soprathion; STCC 4921565; *O,O,O',O'*-Tetraaethyl-bis(dithiophosphat) (German); *O,O,O',O'*-Tetraethyl *S,S*'-methylenebis(dithiophosphate); *O,O,O',O'*-Tetraethyl *S,S*'-methylenebis-phosphordithioate; Tetraethyl *S,S*'-methylene bis(phosphorothiolothionate); *O,O,O',O'*-Tetraethyl *S,S*'-methylene di(phosphorodithioate); Vegfru Fosmite; Vegfrufosmite

CAS Registry Number: 563-12-2

RTECS Number: TE4550000

DOT ID: UN 2783

EEC Number: 015-047-00-2

Regulatory Authority

- Very Toxic Substance (World Bank)[15]
- Air Pollutant Standard Set (ACGIH)[1] (Several States)[60]
- Clean Water Act: Section 311 Hazardous Substances/RQ 40CFR117.3 (same as CERCLA, see below)
- Superfund/EPCRA 40CFR355, Appendix B Extremely Hazardous Substances: TPQ = 1,000 lb (454 kg)
- Superfund/EPCRA 40CFR302.4 Reportable Quantity (RQ): CERCLA, 10 lb (4.54 kg)
- Canada, WHMIS, Ingredients Disclosure List

Cited in U.S. State Regulations: Alaska (G), California (W), Connecticut (A), Florida (A), Illinois (G), Maine (G), Massachusetts (G), Michigan (G), Nevada (A), New Hampshire (G), New Jersey (G), North Dakota (A), Oklahoma (G), Pennsylvania (G), Rhode Island (G), Virginia (A), West Virginia (G).

Description: Ethion, is a colorless to amber-colored, odorless liquid. The technical product has a very disagreeable odor. Freezing/Melting point = -12 – -13°C. Boiling point = 164°C. Flash point = 176°C. Hazard Identification (based on NFPA-704 M Rating System): Health 3, Flammability 1, Reactivity 0. Soluble in water.

Potential Exposure: Those involved in the manufacture, formulation and application of this insecticide and acaricide. Ethion is a preharvest topical insecticide used primarily on citrus fruits, deciduous fruits, nuts and cotton. It is also used as a cattle dip for ticks and as a treatment for buffalo flies.

Incompatibilities: Incompatible with alkaline formulations and strong acids. Decomposes violently when heated above 150°C. Mixtures with magnesium may be explosive.

Permissible Exposure Limits in Air: There is no OSHA PEL. NIOSH and ACGIH recommend a TWA value of 0.4 mg/m^3. The notation "skin" is added to the TWA indicating the possibility of cutaneous absorption. Argentina[35] has set a TWA of 0.4 mg/m^3 also and has added an STEL at the same level. Several states have set guidelines or standards for Ethion in ambient air[60] ranging from 4.0 μg/m^3 (North Dakota) to 6.0 μg/m^3 (Connecticut) to 9.0 μg/m^3 (Nevada).

Determination in Air: OSHA versatile sampler-2; Toluene/Acetone; Gas chromatography/Flame photometric detection for sulfur, nitrogen, or phosphorus; NIOSH IV, Method #5600, Organophosphorus Pesticide.

Permissible Concentration in Water: California[61] has set a guideline for Ethion in drinking water of 35 μg/l.

Harmful Effects and Symptoms

Symptoms may include nausea, vomiting, abdominal cramps, diarrhea, excessive salivation, headache, giddiness, weakness, muscle twitching, difficult breathing, blurring or dimness of vision, and loss of muscle coordination. Death may occur from failure of the respiratory center, paralysis of the respiratory muscles, intense bronchoconstriction, or all three. This material is very toxic; the probable oral lethal dose for humans is 50 – 500 mg/kg, which is between 1 teaspoonful and 1 ounce for a 150 lb person.

Short Term Exposure: Organic phosphorus insecticides are absorbed by the skin, as well as by the respiratory and gastrointestinal tracts. They are cholinesterase inhibitors. Symptoms of exposure include headache, giddiness, blurred vision, nervousness, weakness, nausea, cramps, diarrhea, and discomfort in the chest. Signs include sweating, tearing, salivation, vomiting, cyanosis, convulsions, coma, loss of reflexes and loss of sphincter control. Exposure may cause unconsciousness and death. The effects may be delayed and medical observation is recommended.

Long Term Exposure: Cholinesterase inhibitor; cumulative effect is possible. Ethion may damage the nervous system with repeated exposure, resulting in convulsions, respiratory failure. May cause liver damage.

Points of Attack: Respiratory system, lungs, central nervous system, cardiovascular system, skin, eyes, plasma and red blood cell cholinesterase.

Medical Surveillance: Before employment and at regular times after that, the following are recommended: Plasma and red blood cell cholinesterase levels (tests for the enzyme poisoned by this chemical). If exposure stops, plasma levels return to normal in 1 – 2 weeks while red blood cell levels may be reduced for 1 – 3 months.

When cholinesterase enzyme levels are reduced by 25% or more below preemployment levels, risk of poisoning is increased, even if results are in lower ranges of "normal." Reassignment to work not involving organophosphate or carbamate pesticides is recommended until enzyme levels recover. If symptoms develop or overexposure occurs, repeat the above tests as soon as possible and get an exam of the nervous system. Also consider complete blood count. Consider chest x-ray following acute overexposure. Do not drink any alcoholic beverages before or during use. Alcohol promotes absorption of organic phosphates.

First Aid: If this chemical gets into the eyes, remove any contact lenses at once and irrigate immediately for at least 15 minutes, occasionally lifting upper and lower lids. Seek medical attention immediately. If this chemical contacts the skin, remove contaminated clothing and wash immediately with soap and water. Speed in removing material from skin is of extreme importance. Shampoo hair promptly if contaminated. Seek medical attention immediately. If this chemical has been inhaled, remove from exposure, begin rescue breathing (using universal precautions) if breathing has stopped and CPR if heart action has stopped. Transfer promptly to a medical facility. When this chemical has been swallowed, get medical attention. Give large quantities of water and induce vomiting. Do not make an unconscious person vomit. Effects may be delayed. Keep victim under observation.

Personal Protective Methods: Wear protective gloves and clothing to prevent any reasonable probability of skin contact. Safety equipment suppliers/manufacturers can provide recommendations on the most protective glove/clothing material for your operation.. All protective clothing (suits, gloves, footwear, headgear) should be clean, available each day, and put on before work. Contact lenses should not be worn when working with this chemical. Wear splash-proof chemical goggles and face shield when working with liquid or wear dust-proof goggles when working with powders or dusts, unless full facepiece respiratory protection is worn. Employees should wash immediately with soap when skin is wet or contaminated. Provide emergency showers and eyewash.

Respirator Selection: Where the potential exists for exposures over 0.4 mg/m^3, use an MSHA/NIOSH approved supplied-air respirator with a full facepiece operated in the positive pressure mode or with a full facepiece, hood, or helmet in the continuous flow mode, or use an MSHA/NIOSH approved self-contained breathing apparatus with a full facepiece operated in pressure-demand or other positive pressure mode.

Storage: Prior to working with this chemical you should be trained on its proper handling and storage. Store in tightly closed containers in a cool, well ventilated area away from alkaline material, strong acids and other incompatible materials listed above. Where possible, automatically pump liquid from drums or other storage containers to process containers.

Shipping: Ethion falls in DOT Hazard Class 6.1 and ethion in Packing Group I. This calls for a "Poison" label.

Spill Handling: Evacuate and restrict persons not wearing protective equipment from area of spill or leak until cleanup is complete. Remove all ignition sources. Ventilate area of spill or leak. Absorb liquids in vermiculite, dry sand, earth, peat, carbon, or a similar material and deposit in sealed containers. It may be necessary to contain and dispose of this chemical as a hazardous waste. If material or contaminated runoff enters waterways, notify downstream users of potentially contaminated waters. Contact your Department of Environmental Protection or your regional office of the federal

EPA for specific recommendations. If employees are required to clean-up spills, they must be properly trained and equipped. OSHA 1910.120(q) may be applicable.

Fire Extinguishing: Use dry chemical, carbon dioxide, water spray, or foam extinguishers. Poisonous gases are produced in fire including sulfur oxides. If material or contaminated runoff enters waterways, notify downstream users of potentially contaminated waters. Notify local health and fire officials and pollution control agencies. Containers may explode in fire. From a secure, explosion-proof location, use water spray to cool exposed containers. If cooling streams are ineffective (venting sound increases in volume and pitch, tank discolors, or shows any signs of deforming), withdraw immediately to a secure position. If employees are expected to fight fires, they must be trained and equipped in OSHA 1910.156.

Disposal Method Suggested: Small amounts can be burned with alkali.[22] For larger amounts, the suggested method is incineration with added solvent in furnace equipped with afterburner and alkali scrubber. In accordance with 40CFR165 recommendations for the disposal of pesticides and pesticide containers. Must be disposed properly by following package label directions or by contacting your state pesticide or environmental control agency or by contacting your regional EPA office.

References

U.S. Environmental Protection Agency, S,S'-Methylene-O,O,O',O'-Tetraethyl Phosphorodithioate, Health and Environmental Effects Profile No. 127, Washington, DC, Office of Solid Waste (April 30, 1980)

Sax, N. I., Ed., "Dangerous Properties of Industrial Materials Report," 4, No. 1, 69–74 (1984) and 7, No. 1, 9–37 (1987)

U.S. Environmental Protection Agency, "Chemical Profile: Ethion," Washington, DC, Chemical Emergency Preparedness Program (November 30, 1987)

New Jersey Department of Health and Senior Services, "Hazardous Substance Fact Sheet: Ethion," Trenton, NJ (December 1998)

Ethoprophos

Molecular Formula: $C_8H_{19}O_2PS_2$

Synonyms: AI3-27318; Caswell No.434C; ENT 27,318; EPA pesticide chemical code 041101; Ethoprophos; *O*-Ethyl *S,S*-dipropyl dithiophosphate; *O*-Ethyl *S,S*-dipropyl phosphorodithioate; Jolt; Mobil V-C 9-104; Mocap; Mocap 10G; Phosethoprop; Phosphorodithioic acid, *O*-ethyl *S,S*-dipropyl ester; V-C 9-104; V-C Chemical V-C 9-104; Virginia-Carolina VC 9-104

CAS Registry Number: 13194-48-4

RTECS Number: TE4025000

Regulatory Authority

- Banned or Severely Restricted (E. Germany, Malaysia, Philippines) (UN)[13]
- Superfund/EPCRA 40CFR355, Appendix B Extremely Hazardous Substances: TPQ = 1,000 lb (454 kg)
- Superfund/EPCRA 40CFR302.4 Reportable Quantity (RQ): EHS, 1 lb (0.454 kg)
- EPCRA Section 313 Form R *de minimis* concentration reporting level: 1.0%
- U.S. DOT Regulated Marine Pollutant (49CFR172.101, Appendix B)

Cited in U.S. State Regulations: California (G), Massachusetts (G), New Jersey (G), Pennsylvania (G).

Description: Ethoprophos is a pale yellow liquid. Boiling point = 86 – 91°C @ 0.2 mm Hg. Hazard Identification (based on NFPA-704 M Rating System): Health 3, Flammability 1, Reactivity 0. Slightly soluble in water.

Potential Exposure: Those involved in the manufacture, formulation and application of this nematocide and soil insecticide.

Permissible Exposure Limits in Air: No standards set.

Determination in Air: OSHA versatile sampler-2; Toluene/Acetone; Gas chromatography/Flame photometric detection for sulfur, nitrogen, or phosphorus; NIOSH Method IV Method #5600, Organophosphorus Pesticides.

Permissible Concentration in Water: No criteria set.

Routes of Entry: Inhalation, ingestion, eye and/or contact.

Harmful Effects and Symptoms

Short Term Exposure: Symptoms are similar to parathion and may include nausea, vomiting, abdominal cramps, diarrhea, excessive salivation, headache, giddiness, weakness, muscle twitching, difficult breathing, blurring or dimness of vision, and loss of muscle coordination. Death may occur from failure of the respiratory center, paralysis of the respiratory muscles, intense bronchoconstriction, or all three. This material is extremely toxic; the probable oral lethal dose for humans is 5 – 50 mg/kg, or between 7 drops and 1 teaspoonful for a 150 lb person. It is a cholinesterase inhibitor which affects the nervous system.

Long Term Exposure: Cholinesterase inhibitor; cumulative effect is possible. This chemical may damage the nervous system with repeated exposure, resulting in convulsions, respiratory failure. May cause liver damage.

Points of Attack: Respiratory system, lungs, central nervous system, cardiovascular system, skin, eyes, plasma and red blood cell cholinesterase.

Medical Surveillance: Before employment and at regular times after that, the following are recommended: Plasma and red blood cell cholinesterase levels (tests for the enzyme poisoned by this chemical). If exposure stops, plasma levels return to normal in 1 – 2 weeks while red blood cell levels may be reduced for 1 – 3 months.

When cholinesterase enzyme levels are reduced by 25% or more below preemployment levels, risk of poisoning is increased, even if results are in lower ranges of "normal." Reassignment to work not involving organophosphate or carbamate pesticides is recommended until enzyme levels recover. If symptoms develop or overexposure occurs, repeat the above tests as soon as possible and get an exam of the nervous system. Also consider complete blood count. Consider chest x-ray following acute overexposure. Do not drink any alcoholic beverages before or during use. Alcohol promotes absorption of organic phosphates.

First Aid: If this chemical gets into the eyes, remove any contact lenses at once and irrigate immediately for at least 15 minutes, occasionally lifting upper and lower lids. Seek medical attention immediately. If this chemical contacts the skin, remove contaminated clothing and wash immediately with soap and water. Speed in removing material from skin is of extreme importance. Shampoo hair promptly if contaminated. Seek medical attention immediately. If this chemical has been inhaled, remove from exposure, begin rescue breathing (using universal precautions) if breathing has stopped and CPR if heart action has stopped. Transfer promptly to a medical facility. When this chemical has been swallowed, get medical attention. Give large quantities of water and induce vomiting. Do not make an unconscious person vomit. Remove and isolate contaminated clothing and shoes at the site. Keep victim quiet and maintain normal body temperature. Effects may be delayed; keep victim under observation.

Personal Protective Methods: Wear protective gloves and clothing to prevent any reasonable probability of skin contact. Safety equipment suppliers/manufacturers can provide recommendations on the most protective glove/clothing material for your operation.. All protective clothing (suits, gloves, footwear, headgear) should be clean, available each day, and put on before work. Contact lenses should not be worn when working with this chemical. Wear splash-proof chemical goggles and face shield unless full facepiece respiratory protection is worn. Employees should wash immediately with soap when skin is wet or contaminated. Provide emergency showers and eyewash.

Respirator Selection: Where the potential for exposure to this chemical, use a MSHA/NIOSH approved supplied-air respirator with a full facepiece operated in the positive pressure mode or with a full facepiece, hood, or helmet in the continuous flow mode, or use a MSHA/NIOSH approved self-contained breathing apparatus with a full facepiece operated in pressure-demand or other positive pressure mode.

Storage: Prior to working with this chemical you should be trained on its proper handling and storage. Store in tightly closed containers in a cool, well ventilated area. Where possible, automatically pump liquid from drums or other storage containers to process containers.

Shipping: Organophosphorus pesticides, liquid, toxic, n.o.s. fall in DOT Hazard Class 6.1 and ethoprophos in Packing Group I. The label required is "Poison."

Spill Handling: Evacuate persons not wearing protective equipment from area of spill or leak until clean-up is complete. Remove all ignition sources. Collect powdered material in the most convenient and safe manner and deposit in sealed containers. Ventilate area after clean-up is complete. It may be necessary to contain and dispose of this chemical as a hazardous waste. If material or contaminated runoff enters waterways, notify downstream users of potentially contaminated waters. Contact your Department of Environmental Protection or your regional office of the federal EPA for specific recommendations. If employees are required to clean-up spills, they must be properly trained and equipped. OSHA 1910.120(q) may be applicable.

Fire Extinguishing: This material may burn but does not ignite readily. For small fires, use dry chemical, carbon dioxide, water spray, or foam. For large fires, use water spray, fog, or foam. Stay upwind; keep out of low areas. Move containers from fire area if you can do it without risk. Fight fire from maximum distance. Dike fire control water for later disposal; do not scatter the material. Wear positive pressure breathing apparatus and special protective clothing. Poisonous gases are produced in fire including phosphorus oxides and sulfur oxides. If material or contaminated runoff enters waterways, notify downstream users of potentially contaminated waters. Notify local health and fire officials and pollution control agencies. From a secure, explosion-proof location, use water spray to cool exposed containers. If cooling streams are ineffective (venting sound increases in volume and pitch, tank discolors, or shows any signs of deforming), withdraw immediately to a secure position. If employees are expected to fight fires, they must be trained and equipped in OSHA 1910.156.

Disposal Method Suggested: In accordance with 40CFR 165 recommendations for the disposal of pesticides and pesticide containers. Must be disposed properly by following package label directions or by contacting your state pesticide or environmental control agency or by contacting your regional EPA office.

References

Sax, N. I., Ed., "Dangerous Properties of Industrial Materials Report," 2, No. 4, 85–88 (1982)

U.S. Environmental Protection Agency, "Chemical Profile: Ethoprophos," Washington, DC, Chemical Emergency Preparedness Program (November 30, 1987)

2-Ethoxyethanol

Molecular Formula: $C_4H_{10}O_2$

Common Formula: $C_2H_5OCH_2CH_2OH$

Synonyms: Athylenglykol-monoathylather (German); Cellosolve; Cellosolve solvent; DAG 154; Developer 1002; Dowanol E; Dowanol EE; Dynasolve MP-500; Dynasolve MP aluminium grade; 2EE; Ektasolve EE; Emkanol; Ethanol, 2-ethoxy-; Ether monoethylique de l'ethylene glycol (French); β-Ethoxyethanol; 2-Ethoxyethyl alcohol; Ethyl cellosolve; Ethylene glycol ethyl ether; Ethylene glycol monoethyl ether; Etoksyetylowy alkohol (Polish); 2-Etoxietanol (Spanish); Glycol ethyl ether; Glycol monoethyl ether; Hydroxy ether; Jeffersol EE; Justrite thinner and cleaner; NCI-C54853; Oxitol; Poly-Solv E; Poly-Solv EE; Pyralin PI 2563; Ultramac 55

CAS Registry Number: 110-80-5

RTECS Number: KK8050000

DOT ID: UN 1171

EEC Number: 603-012-00-X

Regulatory Authority

- Air Pollutant Standard Set (ACGIH)[1] (DGF)[3] (HSE)[33] (Others)[35] (Former USSR)[43] (Several States)[60]
- EPA Hazardous Waste Number (RCRA No.): U359
- RCRA, 40CFR261, Appendix 8 Hazardous Constituents
- Superfund/EPCRA 40CFR302.4 Reportable Quantity (RQ): CERCLA, 1,000 lb (454 kg)
- EPCRA Section 313 Form R *de minimis* concentration reporting level: 1.0%
- California's Proposition 65: Reproductive toxin (male)
- Canada, WHMIS, Ingredients Disclosure List

Cited in U.S. State Regulations: Alaska (G), California (A, G), Connecticut (A), Florida (G, A), Illinois (G), Kansas (G), Louisiana (G), Maine (G), Maryland (G), Massachusetts (G), Nevada (A), New Hampshire (G), New Jersey (G), New York (A), North Carolina (A), North Dakota (A), Pennsylvania (G), Rhode Island (G), Vermont (G), Virginia (G, A), Washington (G), West Virginia (G), Wisconsin (G).

Description: 2-Ethoxyethanol, $C_2H_5OCH_2CH_2OH$, is a colorless, viscous liquid with a sweetish odor. Odor threshold = 2.7 ppm. Boiling point = 135°C. Flash point = 43°C. Autoignition temperature = 235°C. Explosive limits: LEL = 1.7% @ 93°C; UEL = 15.6% @ 93°C. Hazard Identification (based on NFPA-704 M Rating System): Health 2, Flammability 2, Reactivity 0. Soluble in water.

Potential Exposure: This material is used as a solvent for nitrocellulose and alkyd resins in lacquers. It is used in dyeing leathers and textiles and in the formulation of cleaners and varnish removers. It is also used as an anti-icing additive in brake fluids and auto and aviation fuels.

Incompatibilities: Forms explosive mixture with air. Strong oxidizers may cause fire and explosions. Attacks some plastics, rubber and coatings. Able to form peroxides. Incompatible with strong acids, aluminum and its alloys.

Permissible Exposure Limits in Air: The OSHA PEL[58] is 200 ppm (740 mg/m^3). The ACGIH has set a TWA value of 5 ppm (19 mg/m^3). The notation "skin" indicates the possibility of cutaneous absorption. The NIOSH IDLH level is 500 ppm. Other TWA limits have been variously set including those by HSE[33] of 10 ppm (37 mg/m^3) by DFG[3] of 20 ppm (75 mg/m^3) and the same by Sweden.[35] Argentina[35] has set 100 ppm (370 mg/m^3). STEL values vary from 40 ppm (150 mg/m^3) in Germany[35] to 50 ppm (190 mg/m^3) in Sweden to 150 ppm (560 mg/m^3) in Argentina.[35] Finally, the Former USSR has set a MAC in ambient air in residential areas of 0.7 mg/m^3 on a once daily basis.[35] Several states have set guidelines or standards for ethoxyethanol in ambient air[60] ranging from 120 – 190 μg/m^3 (North Carolina) to 180 μg/m^3 (Florida and New York) to 190 μg/m^3 (North Dakota) to 320 μg/m^3 (Virginia) to 380 μg/m^3 (Connecticut) to 452 μg/m^3 (Nevada).

Determination in Air: Collection by charcoal tube, Methanol/CH_2C_{12}; analysis by gas chromatography/flame ionization detection; NIOSH IV Method #1403, Alcohols IV.

Permissible Concentration in Water: The Former USSR[35] has set a MAC in surface water of 1.0 mg/l.

Routes of Entry: Inhalation, skin absorption, ingestion, skin and/or eye contact.

Harmful Effects and Symptoms

Short Term Exposure: 2-Ethoxyethanol can affect you when breathed in and by passing through your skin. Irritates the eyes, skin and respiratory tract. High levels may cause headache, drowsiness, dizziness, lightheadedness and even passing out. Exposure may cause central nervous system depression and liver and kidney damage. The oral LD_{50} rat is 3,000 mg/kg (slightly toxic). Medical observation is recommended.

Long Term Exposure: 2-Ethoxyethanol should be handled as a teratogen-with extreme caution. It may damage the testes, resulting in decreased fertility. Exposure may affect blood cells causing a low blood count (anemia) and lesions of blood cells. Prolonged or repeated contact defats the skin. May cause liver and kidney damage.

Points of Attack: Eyes, respiratory system, blood, kidneys, liver, reproductive system, hematopoietic system

Medical Surveillance: For those with frequent or potentially high exposure (half the TLV or greater, or significant skin contact) the following are recommended before beginning work and at regular times after that: Complete blood count. If symptoms develop or overexposure is suspected, the following may be useful: Kidney function tests. Liver function tests.

First Aid: If this chemical gets into the eyes, remove any contact lenses at once and irrigate immediately for at least 15 minutes, occasionally lifting upper and lower lids. Seek medical attention immediately. If this chemical contacts the skin, remove contaminated clothing and wash immediately with soap and water. Seek medical attention immediately. If this chemical has been inhaled, remove from exposure, begin rescue breathing (using universal precautions) if breathing has stopped and CPR if heart action has stopped. Transfer promptly to a medical facility. When this chemical has been swallowed, get medical attention. Give large quantities of water and induce vomiting. Do not make an unconscious person vomit. Keep victim under medical observation.

Personal Protective Methods: Wear protective gloves and clothing to prevent any reasonable probability of skin contact. Safety equipment suppliers/manufacturers can provide recommendations on the most protective glove/clothing material for your operation.. All protective clothing (suits, gloves, footwear, headgear) should be clean, available each day, and put on before work. Contact lenses should not be worn when working with this chemical. Wear splash-proof chemical goggles and face shield unless full facepiece respiratory protection is worn. Employees should wash immediately with soap when skin is wet or contaminated. Provide emergency showers and eyewash.

Respirator Selection: NIOSH: *up to 5 ppm:* SA (any supplied-air respirator). *Up to 12.5 ppm:* SA:CF (any supplied-air respirator operated in a continuous-flow mode). *Up to 25 ppm:* SCBAF (any self-contained breathing apparatus with a full facepiece); or SAF (any supplied-air respirator with a full facepiece). *Up to 500 ppm:* SA:PD,PP (any supplied-air respirator operated in a pressure-demand or other positive-pressure mode). *Emergency or planned entry into unknown concentrations or IDLH conditions:* SCBAF:PD,PP (any self-contained breathing apparatus that has a full facepiece and is operated in a pressure-demand or other positive-pressure mode); or SAF:PD,PP:ASCBA (any supplied-air respirator that has a full facepiece and is operated in a pressure-demand or other positive-pressure mode in combination with an auxiliary self-contained breathing apparatus operated in a pressure-demand or other positive pressure mode). *Escape:* GMFOV [any air-purifying, full-facepiece respirator (gas mask) with a chin-style, front-or back-, mounted organic vapor canister]; or SCBAE (any appropriate escape-type, self-contained breathing apparatus).

Note: Substance reported to cause eye irritation or damage; may require eye protection.

Storage: Prior to working with this chemical you should be trained on its proper handling and storage. Before entering confined space where this chemical may be present, check to make sure that an explosive concentration does not exist. 2-Ethoxyethanol must be stored to avoid contact with strong oxidizers, such as nitrates, permanganates, chlorine, bromine, or chlorine dioxide, since violent reaction occur. Store in tightly closed containers in a dark, cool, well-ventilated area away from heat. Sources of ignition such as smoking and open flames are prohibited where 2-Ethoxyethanol is used, handled, or stored in a manner that could create a potential fire or explosion hazard. Keep in dark because of possible formation of explosive peroxides.

Shipping: This compound requires a shipping label of: "Flammable Liquid." It falls in DOT Hazard Class 3 and Packing Group III.

Spill Handling: Evacuate and restrict persons not wearing protective equipment from area of spill or leak until cleanup is complete. Remove all ignition sources. Establish forced ventilation to keep levels below explosive limit. Absorb liquids in vermiculite, dry sand, earth, peat, carbon, or a similar material and deposit in sealed containers. It may be necessary to contain and dispose of this chemical as a hazardous waste. If material or contaminated runoff enters waterways, notify downstream users of potentially contaminated waters. Contact your Department of Environmental Protection or your regional office of the federal EPA for specific recommendations. If employees are required to clean-up spills, they must be properly trained and equipped. OSHA 1910.120(q) may be applicable.

Fire Extinguishing: This chemical is a combustible liquid. Poisonous gases are produced in fire. Use dry chemical, carbon dioxide, or alcohol foam extinguishers. Vapors are heavier than air and will collect in low areas. Vapors may travel long distances to ignition sources and flashback. Vapors in confined areas may explode when exposed to fire. Containers may explode in fire. Storage containers and parts of containers may rocket great distances, in many directions. If material or contaminated runoff enters waterways, notify downstream users of potentially contaminated waters. Notify local health and fire officials and pollution control agencies. From a secure, explosion-proof location, use water spray to cool exposed containers. If cooling streams are ineffective (venting sound increases in volume and pitch, tank discolors, or shows any signs of deforming), withdraw immediately to a secure position. If employees are expected to fight fires, they must be trained and equipped in OSHA 1910.156.

Disposal Method Suggested: Incineration.[22] Consult with environmental regulatory agencies for guidance on acceptable disposal practices. Generators of waste containing this contaminant (≥100 kg/mo) must conform with EPA regulations governing storage, transportation, treatment, and waste disposal.

References

National Institute for Occupational Safety and Health, Glycol Ethers: 2-Methoxyethanol and 2-Ethoxyethanol, Current Intelligence Bulletin No. 39, Cincinnati, Ohio (May 2, 1983)

New Jersey Department of Health and Senior Services, "Hazardous Substance Fact Sheet: 2-Ethoxyethanol," Trenton, NJ (April 1996)

Sax, N. I., Ed., "Dangerous Properties of Industrial Materials Report," 4, No. 2, 61–64 (1984)

2-Ethoxyethyl Acetate

Molecular Formula: $C_6H_{12}O_3$

Common Formula: $C_2H_5OCH_2CH_2OCOCH_3$

Synonyms: Acetato de 2-etoxietilo (Spanish); Acetic acid, 2-ethoxyethyl ester; 1-Acetoxy-2-ethoxyethane; Aristoline(+); AZ 1310-SF(+); AZ 1312-SFD(+); AZ 1318-SFD(+); AZ 1350J(+); AZ 1370(+); AZ 1370-SF(+); AZ 1375(+); AZ 1470(+); AZ 4140(+); AZ 4210(+); AZ 4330(+); AZ 4620(+); AZ protective coating; AZ thinner; Cellosolve acetate; EE acetate; Egeea; Egmea; Ethanol, 2-ethoxy-, acetate; Ethoxyethanol acetate; β-Ethoxyethyl acetate; Ethoxyethyl acetate; Ethyl cellosolve acetate; Ethylene glycol acetate monoetthy ether; Ethylene glycol ethyl ether acetate; Ethylene glycol monoethyl ether acetate; Ethylene glycol monoethyl ether monoacetate; Ethyloxitol acetate; Glycol monoethyl ether acetate; H.M.D.S. III; Hydroxy ether; Kodak MX-936; Kodak photoresist developer; KTI 1300 thinner; KTI 1350J(+); KTI 1370; KTI 1375(+); KTI 1470(+); KTI 820(+); KTI 820J (+); KTI 9000; KTI 9000K; KTI 9010(+); KTI II; Liquid alkaline Strip 7463; Markem thinner XF; Microposit 111S(+); Microposit 119S(+); Microposit 119 thinner; Microposit 1375(+); Microposit 1400-33(+); Microposit 1400S(+); Microposit 1450J(+); Microposit 1470(+); Microposit Sal 601-ER7(+); Microposit XP-6009(+); Microposit XP-6012(+); MS-470 Urethane coating; OFPR-800 AR-15(+); Poly-Solv EE acetate; PR-21 Resist; PR-55 Resist; Selectilux P-15(+); Sensolve EEA; Thinner E; TSMR 8800(+); Ultamac PR-68 Resin; Ultramac PR-1024 MB-628 Resin; Waxivation compound; Waycoat 204(+); Waycoat 207(+); Waycoat HPR 205; Waycoat RX 507(+); Xanthochrome(+)

CAS Registry Number: 111-15-9

RTECS Number: KK8225000

DOT ID: UN 1172

EEC Number: 607-037-00-7

Regulatory Authority

- Air Pollutant Standard Set (ACGIH)[1] (DFG)[3] (HSE)[33] (OSHA)[58] (Several States)[60]
- Canada, WHMIS, Ingredients Disclosure List

Cited in U.S. State Regulations: Alaska (G), California (A, G), Connecticut (A), Illinois (G), Maine (G), Massachusetts (G), Nevada (A), New Hampshire (G), New Jersey (G), North Dakota (A), Pennsylvania (G), Rhode Island (G), South Dakota (A), Virginia (A), West Virginia (G).

Description: 2-Ethoxyethyl acetate, $C_2H_5OCH_2CH_2OCOCH_3$, is a colorless liquid with a mild, nonresidual odor. The odor threshold is 0.056 ppm in air. Boiling point = 156°C. Flash point = 47°C. Autoignition temperature = 380°C. Explosive limits: LEL = 1.7%; UEL = 12.7%. The UEL is also reported as 14%. Hazard Identification (based on NFPA-704 M Rating System): Health 2, Flammability 2, Reactivity 0. Soluble in water.

Potential Exposure: This material is used as a solvent for many different purposes including for nitrocellulose and other resins. Used in automobile lacquers to retard evaporation and impart a high gloss.

Incompatibilities: Forms explosive mixture with air. Incompatible with strong acids, strong alkalies, nitrates. Violent reaction with oxidizers. May form unstable peroxides. Softens many plastics. Attacks some plastics, rubber and coatings.

Permissible Exposure Limits in Air: The OSHA PEL is 100 ppm (540 mg/m^3) TWA. The NIOSH REL is 0.5 ppm TWA. The ACGIH recommends a TWA of 5 ppm (27 mg/m^3). The notation "skin" is added indicating the possibility of cutaneous absorption. The NIOSH IDLH level is 500 ppm The HSE[33] has set a TWA of 10 ppm (54 mg/m^3) and the DFG[3] and MAK of 20 ppm (100 mg/m^3). Several states have set guidelines or standards for ethoxyethyl acetate in ambient air[60] ranging from 270 μg/m^3 (North Dakota) to 450 μg/m^3 (Virginia) to 540 μg/m^3 (Connecticut and South Dakota) to 643 μg/m^3 (Nevada).

Determination in Air: Charcoal adsorption, workup with CS_2, analysis by gas chromatography /flame ionization. See NIOSH Method #1450 for esters[18].

Permissible Concentration in Water: No criteria set.

Routes of Entry: Inhalation, ingestion, eyes and skin contact.

Harmful Effects and Symptoms

Short Term Exposure: 2-Ethoxyethylacetate can affect you when breathed in and by passing through your skin. Exposure can irritate the eyes, nose and throat. High levels could cause you to become dizzy, lightheaded, and to pass out. Very high exposures could cause kidney damage and even death. May affect the blood and central nervous system.

Long Term Exposure: 2-Ethoxyehylacetate may damage the developing fetus. It may damage the testes (male reproductive glands), resulting in decreased fertility. It may affect the blood, liver and kidneys. Many similar petroleum-based solvents have been shown to cause brain and nerve damage.

Points of Attack: Inhalation, skin absorption, ingestion, skin and/or eye contact

Medical Surveillance: If symptoms develop or overexposure is suspected, the following may be useful: Kidney function tests. Complete blood count. Evaluate for brain effects and consider evaluation of the nervous systems.

First Aid: If this chemical gets into the eyes, remove any contact lenses at once and irrigate immediately for at least 15 minutes, occasionally lifting upper and lower lids. Seek medical attention immediately. If this chemical contacts the skin, remove contaminated clothing and wash immediately with soap and water. Seek medical attention immediately. If this chemical has been inhaled, remove from exposure, begin rescue breathing (using universal precautions) if breathing has stopped and CPR if heart action has stopped. Transfer promptly to a medical facility. When this chemical has been swallowed, get medical attention. Give large quantities of water and induce vomiting. Do not make an unconscious person vomit.

Personal Protective Methods: Wear protective gloves and clothing to prevent any reasonable probability of skin contact. Safety equipment suppliers/manufacturers can provide recommendations on the most protective glove/clothing material for your operation. Neoprene is among the recommended protective materials. All protective clothing (suits, gloves, footwear, headgear) should be clean, available each day, and put on before work. Contact lenses should not be worn when working with this chemical. Wear splash-proof chemical goggles and face shield unless full facepiece respiratory protection is worn. Employees should wash immediately with soap when skin is wet or contaminated. Provide emergency showers and eyewash.

Respirator Selection: NIOSH: *5 ppm:* CCROV [any chemical cartridge respirator with organic vapor cartridge(s)]; SA (any supplied-air respirator). *12.5 ppm:* SA:CF (any supplied-air respirator operated in a continuous-flow mode); PAPROV [any powered, air-purifying respirator with organic vapor cartridge(s)]. *25 ppm:* CCROV [any chemical cartridge respirator with organic vapor cartridge(s)]; GMFOV [any air-purifying, full-facepiece respirator (gas mask) with a chin-style, front- or back-mounted acid gas canister]; PAPRTOV [any powered, air-purifying respirator with a tight-fitting facepiece and organic vapor cartridge(s)]; or SCBAF (any self-contained breathing apparatus with a full facepiece); or SAF (any supplied-air respirator with a full facepiece). *500 ppm:* SA:PD,PP (any supplied-air respirator operated in a pressure-demand or other positive-pressure mode). *Emergency or planned entry into unknown concentrations or IDLH conditions:* SCBAF: PD,PP (any self-contained breathing apparatus that has a full facepiece and is operated in a pressure-demand or other positive-pressure mode); or SAF:PD,PP:ASCBA (any supplied-air respirator that has a full facepiece and is operated in a pressure-demand or other positive-pressure mode in combination with an auxiliary, self-contained breathing apparatus operated in a pressure-demand or other positive-pressure mode). *Escape:* GMFOV [any air-purifying, full-facepiece respirator (gas mask) with a chin-style, front- or back-mounted organic vapor canister]; or SCBAE (any appropriate escape-type, self-contained breathing apparatus).

Note: Substance reported to cause eye irritation or damage; may require eye protection.

Storage: Prior to working with this chemical you should be trained on its proper handling and storage. Before entering confined space where this chemical may be present, check to make sure that an explosive concentration does not exist. 2-Ethoxyethylacetate must be stored to avoid contact with strong oxidizers, such as nitrates, permanganates, bromine, chlorine, and chlorine dioxide; strong alkalis, such as sodium hydroxide and potassium hydroxide; and strong acids, such as nitric, hydrochloric, and sulfuric acids, since violent reactions occur. Store in tightly closed containers in a cool, well-ventilated area away from heat. Sources of ignition such as smoking and open flames are prohibited where 2-ethoxyethylacetate is used, handled, or stored in a manner that could create a potential fire or explosion hazard.

Shipping: This compound requires a shipping label of: "Flammable Liquid." It falls in DOT Hazard Class 3 and Packing Group III.

Spill Handling: Evacuate and restrict persons not wearing protective equipment from area of spill or leak until cleanup is complete. Remove all ignition sources. Establish forced ventilation to keep levels below explosive limit. Absorb liquids in vermiculite, dry sand, earth, peat, carbon, or a similar material and deposit in sealed containers. It may be necessary to contain and dispose of this chemical as a hazardous waste. If material or contaminated runoff enters waterways, notify downstream users of potentially contaminated waters. Contact your Department of Environmental Protection or your regional office of the federal EPA for specific recommendations. If employees are required to clean-up spills, they must be properly trained and equipped. OSHA 1910.120(q) may be applicable.

Fire Extinguishing: This chemical is a combustible liquid. Poisonous and irritating gases are produced in fire. Use dry chemical, carbon dioxide, or alcohol foam extinguishers. Vapors are heavier than air and will collect in low areas. Vapors may travel long distances to ignition sources and flashback. Vapors in confined areas may explode when exposed to fire. Containers may explode in fire. Storage containers and parts of containers may rocket great distances, in many directions. If material or contaminated runoff enters waterways, notify downstream users of potentially contaminated waters. Notify local health and fire officials and pollution control agencies. From a secure, explosion-proof location, use water spray to cool exposed containers. If cooling streams are ineffective (venting sound increases in volume and pitch, tank discolors, or shows any signs of deforming), withdraw immediately to a secure position. If employees are expected to fight fires, they must be trained and equipped in OSHA 1910.156.

Disposal Method Suggested: Incineration.

References

New Jersey Department of Health and Senior Services, "Hazardous Substance Fact Sheet: 2-Ethoxyethyl Acetate," Trenton, NJ (June 1996)

Sax, N. I., Ed., "Dangerous Properties of Industrial Materials Report," 4, No. 2, 64–67 (1984)

Ethyl Acetate

Molecular Formula: C_4H_8O2

Common Formula: $CH_3COOC_2H_5$

Synonyms: Acetato de etilo (Spanish); Acetic acid, ethyl ester; Acetic ester; Acetic ether; Acetidin; Acetoxyethane; Aethylacetat (German); Arsenosilica film 0308; AS-1; AS 1400; AS 18CZ10A; AS 18CZ5E; AS 18CZ6E; AS 1CE; AS 5CE; B446; Essigester (German); Ethylacetaat (Dutch); Ethyl acetic ester; Ethyle (acetate d') (French); Ethyl ester of acetic acid; Ethyl ethanoate; Etile (acetato di) (Italian); KTI 1470(+); Octan etylu (Polish); Vinegar naphtha

CAS Registry Number: 141-78-6

RTECS Number: AH5425000

DOT ID: UN 1173

EEC Number: 607-022-00-5

EINECS Number: 205-500-4

Regulatory Authority

- Air Pollutant Standard Set (ACGIH)[1] (DFG)[3] (HSE)[33] (Others)[35] (Former USSR)[43] (OSHA)[58] (Several States)[60]
- EPA Hazardous Waste Number (RCRA No.): U112
- RCRA, 40CFR261, Appendix 8 Hazardous Constituents
- RCRA 40CFR268.48; 61FR15654, Universal Treatment Standards: Wastewater (mg/l), 0.34; Nonwastewater (mg/kg), 33
- Superfund/EPCRA 40CFR302.4 Reportable Quantity (RQ): CERCLA, 5,000 lb (2,270 kg)
- Canada, WHMIS, Ingredients Disclosure List

Cited in U.S. State Regulations: Alaska (G), California (A, G), Connecticut (A), Florida (G, A), Illinois (G), Kansas (G), Louisiana (G), Maine (G), Massachusetts (G, A), Nevada (A), New Hampshire (G), New Jersey (G), New York (A), North Carolina (A), North Dakota (A), Oklahoma (G), Pennsylvania (G), Rhode Island (G), South Dakota (A), Vermont (G), Virginia (G, A), Washington (G), West Virginia (G), Wisconsin (G).

Description: Ethyl acetate, $CH_3COOC_2H_5$, is a colorless liquid with a pleasant, fruity odor (odor threshold = 3.9 ppm in air). Boiling point = 77°C. Flash point = -4°C. Autoignition temperature = 426°C. Explosive limits: LEL = 2.0%; UEL = 11.5%. Hazard Identification (based on NFPA-704 M Rating System): Health 1, Flammability 3, Reactivity 0. Slightly soluble in water.

Potential Exposure: This material is used as a lacquer solvent. It is also used in making dyes, flavoring and perfumery and in smokeless powder manufacture.

Incompatibilities: Forms explosive mixture with air. Heating may cause violent combustion or explosion. Incompatible with strong acids, strong alkalies, nitrates, strong oxidizers, chlorosulfonic acid, lithium aluminum hydride, oleum will hydrolyze on standing forming acetic acid and ethyl alcohol. This reaction is greatly accelerated by alkalies. Decomposes under influence of UV light, bases, acids. Attacks aluminum and plastics.

Permissible Exposure Limits in Air: The OSHA PEL and the ACGIH TWA value is 400 ppm (1,400 mg/m^3). This is also the DFG MAK value[3] and the HSE TWA value.[33] It is also the TWA value in Argentina and Japan.[35] However, lower limits have been set in Sweden[35] at 200 ppm (700 mg/m^3), in Czechoslovakia[35] at 400 mg/m^3 and in the Former USSR[35][43] at 200 mg/m^3. Further the Former USSR[35][43] has set a MAC value in the ambient air of residential areas of 0.1 mg/m^3 on either a momentary or a daily average basis. There is no tentative STEL value set. The NIOSH IDLH level is 2,000 ppm.

Several States have set guidelines or standards for ethyl acetate in ambient air[60] ranging from 2.0 mg/m^3 (Massachusetts) to 14.0 mg/m^3 (North Dakota) to 23.0 mg/m^3 (Virginia) to 28.0 mg/m^3 (Connecticut, Florida, New York and South Dakota) to 33.333 mg/m^3 (Nevada) to 140.0 mg/m^3 (North Carolina).

Determination in Air: Charcoal adsorption, workup with CS_2, analysis by gas chromatography/flame ionization. See NIOSH Method #1450 for esters.

Permissible Concentration in Water: The Former USSR[35] has set a MAC in surface water of 0.2 mg/l.

Routes of Entry: Inhalation, ingestion, eye and/or skin contact.

Harmful Effects and Symptoms

Short Term Exposure: Ethyl acetate can affect you when breathed in and by passing through your skin. Exposure to high levels can cause you to feel dizzy and lightheaded. Very high levels could cause you to pass out. Repeated contact can cause drying and cracking of the skin. The vapor can irritate the eyes and respiratory tract. Ethyl acetate is a flammable liquid and a fire hazard. May affect the central nervous system. Very high exposure may result in death.

Long Term Exposure: May decrease the fertility in males. Repeated contact can cause drying and cracking of the skin. Many similar petroleum-based chemicals can cause brain and nerve damage.

Points of Attack: Eyes, skin, and respiratory system.

Medical Surveillance: Consider the points of attack in preplacement and periodic physical examinations. Evaluate for brain and nervous system damage.

First Aid: If this chemical gets into the eyes, remove any contact lenses at once and irrigate immediately for at least 15 minutes, occasionally lifting upper and lower lids. Seek medical attention immediately. If this chemical contacts the skin, remove contaminated clothing and wash immediately with soap and water. Seek medical attention immediately. If this chemical has been inhaled, remove from exposure, begin rescue breathing (using universal precautions) if breathing has stopped and CPR if heart action has stopped. Transfer promptly to a medical facility. When this chemical has been swallowed, get medical attention. Give large quantities of water and induce vomiting. Do not make an unconscious person vomit.

Personal Protective Methods: Wear protective gloves and clothing to prevent any reasonable probability of skin contact. Safety equipment suppliers/manufacturers can provide recommendations on the most protective glove/clothing material for your operation. Silvershield and Butyl Rubber are among the recommended protective materials. All protective clothing (suits, gloves, footwear, headgear) should be clean, available each day, and put on before work. Contact lenses should not be worn when working with this chemical. Wear splash-proof chemical goggles and face shield unless full facepiece respiratory protection is worn. Employees should wash immediately with soap when skin is wet or contaminated. Provide emergency showers and eyewash.

Respirator Selection: NIOSH/OSHA: *2,000 ppm:* SA:CF (any supplied-air respirator operated in a continuous-flow mode); or PAPROV [any powered, air-purifying respirator with organic vapor cartridge(s)]; or CCRFOV [any chemical cartridge respirator with a full facepiece and organic vapor cartridge(s)]; or GMFOV [any air-purifying, full-facepiece respirator (gas mask) with a chin-style, front- or back-mounted acid gas canister]; or SCBAF (any self-contained breathing apparatus with a full facepiece); or SAF (any supplied-air respirator with a full facepiece). *Emergency or planned entry into unknown concentrations or IDLH conditions:* SCBAF: PD,PP (any self-contained breathing apparatus that has a full facepiece and is operated in a pressure-demand or other positive-pressure mode); or SAF:PD,PP:ASCBA (any supplied-air respirator that has a full facepiece and is operated in a pressure-demand or other positive-pressure mode in combination with an auxiliary self-contained breathing apparatus operated in a pressure-demand or other positive pressure mode). *Escape:* GMFOV [any air-purifying, full-facepiece respirator (gas mask) with a chin-style, front-or back-, mounted organic vapor canister]; or SCBAE (any appropriate escape-type, self-contained breathing apparatus).

Note: Substance causes eye irritation or damage; eye protection needed.

Storage: Prior to working with this chemical you should be trained on its proper handling and storage. Before entering confined space where this chemical may be present, check to make sure that an explosive concentration does not exist. Ethyl acetate must be stored to avoid contact with nitrates; strong oxidizers, such as chlorine, bromine, chlorine dioxide, nitrates, and permanganates; strong alkalis, such as sodium hydroxide and potassium hydroxide; or strong acids, such as sulfuric acid, hydrochloric acid, and nitric acid, since violent reactions occur. Store in tightly closed containers in a cool, well-ventilated area away from heat. Sources of ignition such as smoking and open flames are prohibited where ethyl acetate is used, handled, or stored in a manner that could create a potential fire or explosion hazard. Metal containers involving the transfer of 5 gallons or more of ethyl acetate should be grounded and bonded. Drums must be equipped with self-closing valves, pressure vacuum bungs, and flame arresters. Use only non-sparking tools and equipment, especially when opening and closing containers of ethyl acetate. Store in containers that are properly labeled with health hazard information and safe handling procedures. Wherever ethyl acetate is used, handled, manufactured, or stored, use explosion-proof electrical equipment and fittings.

Shipping: This compound requires a shipping label of: "Flammable Liquid." It falls in DOT Hazard Class 3 and Packing Group II.

Spill Handling: Evacuate and restrict persons not wearing protective equipment from area of spill or leak until cleanup is complete. Remove all ignition sources. Establish forced ventilation to keep levels below explosive limit. Absorb liquids in vermiculite, dry sand, earth, peat, carbon, or a similar material and deposit in sealed containers. It may be necessary to contain and dispose of this chemical as a hazardous waste. If material or contaminated runoff enters waterways, notify downstream users of potentially contaminated waters. Contact your Department of Environmental Protection or your regional office of the federal EPA for specific recommendations. If employees are required to clean-up spills, they must be properly trained and equipped. OSHA 1910.120(q) may be applicable.

Fire Extinguishing: This chemical is a flammable liquid. Poisonous gases are produced in fire. Use dry chemical, carbon dioxide, or alcohol foam extinguishers. Vapors are heavier than air and will collect in low areas. Vapors may travel long distances to ignition sources and flashback. Vapors in confined areas may explode when exposed to fire. Containers may explode in fire. Storage containers and parts of containers may rocket great distances, in many directions. If material or contaminated runoff enters waterways, notify downstream users of potentially contaminated waters. Notify local health and fire officials and pollution control agencies. From a secure, explosion-proof location, use water spray to cool exposed containers. If cooling streams are ineffective (venting sound increases in volume and pitch, tank discolors, or shows any signs of deforming), withdraw immediately to a secure position. If employees are expected to fight fires, they must be trained and equipped in OSHA 1910.156.

Disposal Method Suggested: Incineration.[22] Consult with environmental regulatory agencies for guidance on acceptable disposal practices. Generators of waste containing this contaminant (≥100 kg/mo) must conform with EPA regulations governing storage, transportation, treatment, and waste disposal.

References

Sax, N. I., Ed., "Dangerous Properties of Industrial Materials Report," 4, No. 1, 75–78 (1984)

New Jersey Department of Health and Senior Services, "Hazardous Substance Fact Sheet: Ethyl Acetate," Trenton, NJ (June 1996)

Ethyl Acetylene

Molecular Formula: C_4H_6

Common Formula: $C_2H_5C{\equiv}CH$

Synonyms: 1-Butino (Spanish); 1-Butyne; Ethyl ethyne

CAS Registry Number: 107-00-6

RTECS Number: ER9553000

DOT ID: UN 2452

Regulatory Authority

- Clean Air Act: Accidental Release Prevention/Flammable substances, (Section 112[r], Table 3), TQ = 10,000 lb (4,540 kg)

Cited in U.S. State Regulations: California (A), New Hampshire (G), New Jersey (G), Pennsylvania (G).

Description: Ethyl acetylene, $C_2H_5C\equiv CH$, is a colorless, compressed gas or liquid. Boiling point = 8.1°C. Freezing/Melting point = -130°C. Flash point ≤-7°C. Hazard Identification (based on NFPA-704 M Rating System): Health 1, Flammability 4, Reactivity 2. Insoluble in water.

Potential Exposure: Ethyl acetylene is used as a fuel and as a chemical intermediate.

Incompatibilities: Forms explosive mixture with air. Strong oxidizers may cause fire and explosion danger.

Permissible Exposure Limits in Air: No standards set.

Permissible Concentration in Water: No criteria set.

Routes of Entry: Inhalation.

Harmful Effects and Symptoms

Short Term Exposure: Ethyl acetylene can affect you when breathed in. Exposure can cause you to feel dizzy, lightheaded and to pass out. Exposure to very high levels can cause suffocation and death due to lack of oxygen.

Long Term Exposure: Unknown at this time.

First Aid: If this chemical gets into the eyes, remove any contact lenses at once and irrigate immediately for at least 15 minutes, occasionally lifting upper and lower lids. Seek medical attention immediately. If this chemical contacts the skin, remove contaminated clothing and wash immediately with soap and water. Seek medical attention immediately. If this chemical has been inhaled, remove from exposure, begin rescue breathing (using universal precautions) if breathing has stopped and CPR if heart action has stopped. Transfer promptly to a medical facility. When this chemical has been swallowed, get medical attention. Give large quantities of water and induce vomiting. Do not make an unconscious person vomit. If frostbite has occurred, seek medical attention immediately; do *NOT* rub the affected areas or flush them with water. In order to prevent further tissue damage, do *NOT* attempt to remove frozen clothing from frostbitten areas. If frostbite has *NOT* occurred, immediately and thoroughly wash contaminated skin with soap and water.

Personal Protective Methods: Wear protective gloves and clothing to prevent any reasonable probability of skin contact. Safety equipment suppliers/manufacturers can provide recommendations on the most protective glove/clothing material for your operation.. All protective clothing (suits, gloves, footwear, headgear) should be clean, available each day, and put on before work. Contact lenses should not be worn when working with this chemical. Wear splash-proof chemical goggles and face shield when working with liquid unless full facepiece respiratory protection is worn. Wear gas-proof goggles and face shield, unless full facepiece respiratory protection is worn. Employees should wash immediately with soap when skin is wet or contaminated. Provide emergency showers and eyewash.

Respirator Selection: Exposure to ethyl acetylene is dangerous because it can replace oxygen and lead to suffocation. Only MSHA/NIOSH approved self-contained breathing apparatus with a full facepiece operated in positive pressure mode should be used in oxygen deficient environments.

Storage: Prior to working with this chemical you should be trained on its proper handling and storage. Ethyl acetylene must be stored to avoid contact with strong oxidizers (such as chlorine, bromine and fluorine) since violent reactions occur. Protect cylinders from physical damage and store away from potential heat sources. Sources of ignition, such as smoking and open flames, are prohibited where ethyl acetylene is handled, used, or stored. Wherever ethyl acetylene is used, handled, manufactured, or stored, use explosion-proof electrical equipment and fittings. Procedures for the handling, use and storage of cylinders should be in compliance with OSHA 1910.101 and 1910.169 as with the recommendations of the Compressed Gas Association.

Shipping: This compound requires a shipping label of: "Flammable Gas." It falls in DOT Hazard Class 2.1.

Spill Handling: Restrict persons not wearing protective equipment from area of leak until clean-up is complete. Remove all ignition sources. Ventilate area of leak to disperse the gas. Stop flow of gas. If source of leak is a cylinder and the leak cannot be stopped in place, remove the leaking cylinder to a safe place in the open air, and repair leak or allow cylinder to empty. Water spray may be used to reduce vapors. Keep ethyl acetylene out of a confined space, such as a sewer, because of the possibility of an explosion, unless the sewer is designed to prevent the build-up of explosive concentrations.

Fire Extinguishing: This chemical is a flammable gas. Poisonous gases are produced in fire. Do not extinguish the fire unless the flow of gas can be stopped and any remaining gas is out of the line. Specially trained personnel may use fog lines to cool exposures and let the fire burn itself out. Vapors are heavier than air and will collect in low areas. Vapors may travel long distances to ignition sources and flashback. Vapors in confined areas may explode when exposed to fire. Containers may explode in fire. Storage containers and parts of containers may rocket great distances, in many directions. If material or contaminated runoff enters waterways, notify downstream users of potentially contaminated waters. Notify local health and fire officials and pollution control agencies. From a secure, explosion-proof location, use water spray to cool exposed containers. If cooling streams are ineffective (venting sound increases in volume and pitch, tank discolors, or shows any signs of deforming), withdraw immediately to a secure position. If cylinders are exposed to excessive heat from fire or flame contact, withdraw immediately to a secure location. If employees are expected to fight fires, they must be trained and equipped in OSHA 1910.156.

References

New Jersey Department of Health and Senior Services, "Hazardous Substance Fact Sheet: Ethyl Acetylene," Trenton, NJ (February 1987)

Ethyl Acrylate

Molecular Formula: $C_5H_8O_2$

Common Formula: CH_2=$CHCOOC_2H_5$

Synonyms: Acrilato de etilo (Spanish); Acrylate d'ethyle (French); Acrylic acid, ethyl ester; Acrylsaeuraethylester (German); Aethylacrylat (German); Ethoxy carbonyl ethylene; Ethylacrylaat (Dutch); Ethylakrylat (Czech); Ethyl 2-propenoate; Ethyl propenoate; Etil acrilato (Italian); Etilacrilatului (Rumanian); NCI-C50384; 2-Propenoic acid, ethyl ester

CAS Registry Number: 140-88-5

RTECS Number: AT0700000

DOT ID: UN 1917 (inhibited)

EEC Number: 607-032-00-X

- ***Regulatory Authority:*** Carcinogen (Animal Positive) (IARC) (NTP)[9] (ACGIH)[1]
- Air Pollutant Standard Set (ACGIH)[1] (DFG)[3] (HSE)[33] (Other Countries)[35] (OSHA)[58] (Several States)[60]
- Clean Air Act: Hazardous Air Pollutants (Title I, Part A, Section 112)
- EPA Hazardous Waste Number (RCRA No.): U113
- RCRA, 40CFR261, Appendix 8 Hazardous Constituents
- Superfund/EPCRA 40CFR302.4 Reportable Quantity (RQ): CERCLA, 1,000 lb (454 kg)
- EPCRA Section 313 Form R *de minimis* concentration reporting level: 0.1%
- Marine Pollutant (49CFR, Subchapter 172.101, Appendix B)
- Canada, WHMIS, Ingredients Disclosure List

Cited in U.S. State Regulations: Alaska (G), California (A, G), Connecticut (A), Florida (G), Illinois (G), Kansas (G), Louisiana (G), Maine (G), Maryland (G), Massachusetts (G, A), Nevada (A), New Hampshire (G), New Jersey (G), North Dakota (A), Pennsylvania (G), Rhode Island (G), Vermont (G), Virginia (G, A), Washington (G), West Virginia (G), Wisconsin (G).

Description: Ethyl acrylate, CH_2=$CHCOOC_2H_5$, is a colorless liquid with a sharp, acrid odor. The odor threshold is 0.00024 ppm. Boiling point = 100°C. Flash point = 10°C (oc). Autoignition temperature = 372°C. Hazard Identification (based on NFPA-704 M Rating System): Health 2, Flammability 3, Reactivity 2. Explosive limits: LEL = 1.4%; UEL = 14%. Slightly soluble in water.

Potential Exposure: This material is used as a monomer in the manufacture of homopolymer and copolymer resins for the production of paints and plastic films.

Incompatibilities: Forms explosive mixture with air. Atmospheric moisture and strong alkalies may cause fire and explosions. Unless properly inhibited (*Note:* Inert gas blanket not recommended), heat, light or peroxides can cause polymerization. Incompatible with oxidizers (may be violent), peroxides, polymerizers, strong alkalis, moisture, chlorosulfonic acid, strong acids, amines. May accumulate static electrical charges, and may cause ignition of its vapors. Polymerizes readily unless an inhibitor such as hydroquinone is added. Uninhibited vapors may plug vents by the formation of polymers.

Permissible Exposure Limits in Air: The OSHA PEL 25 ppm TWA. The ACGIH recommended TWA is 5 ppm (20 mg/m³) and the STEL is 25 ppm (100 mg/m³). The NIOSH IDLH level is Ca[300ppm]. HSE[33] TWA is the same as ACGIH and the HSE STEL is 15 ppm (60 mg/m³). DFG[3] has the same TWA value as ACGIH. Sweden[35] has set a TWA of 10 ppm (40 mg/m³). Argentina[35] has set a TWA of 25 ppm (100 mg/m³) and the same values for STEL limits. Czechoslovakia[35] and Former USSR have set MAC values in ambient air of residential areas at 0.05 mg/m³ (50 μg/m³). OSHA and HSE add the notation "skin" indicating the possibility of cutaneous absorption. Several states have set guidelines or standards for ethyl acrylate in ambient air[60] ranging from 0.28 μg/m³ (Massachusetts) to 350.0 μg/m³ (Virginia) to 400 μg/m³ (Connecticut) to 200 – 1,000 μg/m³ (North Dakota) to 476.0 μg/m³ (Nevada).

Determination in Air: Charcoal adsorption, workup with CS_2, analysis by gas chromatography/flame ionization. See NIOSH Method #1450 for esters[18].

Permissible Concentration in Water: The former USSR[35][43] has set a MAC in water bodies used for domestic purposes of 0.005 mg/l (5 μg/l).

Routes of Entry: Inhalation, ingestion, eye and/or skin contact. Passes through the skin.

Harmful Effects and Symptoms

Short Term Exposure: Ethyl acrylate can affect you when breathed in and by passing through your skin. Ethyl acrylate is corrosive and can severely irritate and burn the eyes and skin. Inhalation can cause severe irritation and pulmonary edema, a medical emergency that can be delayed for several hours. This can cause death. Breathing very high levels of ethyl acrylate can cause dizziness, difficulty breathing and even death. Contact may cause a skin allergy.

Long Term Exposure: Repeated or prolonged contact with skin may cause dermatitis and skin allergy. May affect the liver and kidneys. This chemical may be a human carcinogen; it has been show to cause stomach cancer in animals. May damage the developing fetus. Highly irritating substances may cause lung damage.

Points of Attack: Eyes, skin, respiratory system Cancer Site in animals: tumors of the forestomach.

Medical Surveillance: Before beginning employment and at regular times after that, for those with frequent or potentially high exposures, the following are recommended: Lung function tests. Liver function tests. Kidney function tests. If symptoms develop or overexposure is suspected, the following may be useful: Consider chest x-ray after acute overexposure.

Evaluation by a qualified allergist, including careful exposure history and special testing, may help diagnose skin allergy.

First Aid: If this chemical gets into the eyes, remove any contact lenses at once and irrigate immediately for at least 15 minutes, occasionally lifting upper and lower lids. Seek medical attention immediately. If this chemical contacts the skin, remove contaminated clothing and wash immediately with soap and water. Seek medical attention immediately. If this chemical has been inhaled, remove from exposure, begin rescue breathing (using universal precautions) if breathing has stopped and CPR if heart action has stopped. Transfer promptly to a medical facility. When this chemical has been swallowed, get medical attention. Give large quantities of water and induce vomiting. Do not make an unconscious person vomit. Medical observation is recommended for 24 – 48 hours after breathing overexposure, as pulmonary edema may be delayed. As first aid for pulmonary edema, a doctor or authorized paramedic may consider administering a corticosteroid spray.

Personal Protective Methods: Wear protective gloves and clothing to prevent any reasonable probability of skin contact. Safety equipment suppliers/manufacturers can provide recommendations on the most protective glove/clothing material for your operation. Teflon, Saranex, Polyvinyl Alcohol, and Butyl/Neoprene are among the recommended protective materials. All protective clothing (suits, gloves, footwear, headgear) should be clean, available each day, and put on before work. Contact lenses should not be worn when working with this chemical. Wear splash-proof chemical goggles and face shield unless full facepiece respiratory protection is worn. Remove clothing immediately if wet or contaminated to avoid flammability hazard. Employees should wash immediately with soap when skin is wet or contaminated. Provide emergency showers and eyewash.

Respirator Selection: NIOSH: *At any concentrations above the NIOSH REL, or where there is no REL, at any detectable concentration:* SCBAF:PD,PP (any self-contained breathing apparatus that has a full facepiece and is operated in a pressure-demand or other positive-pressure mode); or SAF:PD,PP:ASCBA (any supplied-air respirator that has a full facepiece and is operated in a pressure-demand or other positive-pressure mode in combination with an auxiliary self-contained breathing apparatus operated in a pressure-demand or other positive pressure mode). *Escape:* GMFOV [any air-purifying, full-facepiece respirator (gas mask) with a chin-style, front-or back-, mounted organic vapor canister]; or SCBAE (any appropriate escape-type, self-contained breathing apparatus).

Storage: Ethyl acrylate must be stored to avoid contact with oxidizers (such as peroxides, perchlorates, chlorates, nitrates and permanganates), strong alkalis (such as sodium hydroxide and potassium hydroxide) and moisture, since violent reactions occur. Store in tightly closed containers in a cool, well-ventilated area away from heat. Heat can cause ethyl acrylate to react by itself. If this takes place in a closed container, an explosion could occur. Ethyl acrylate usually contains as inhibitor such as hydroquinone or its methyl ether to prevent a self-reaction. If it does not contain an inhibitor, the reaction may occur without the application of heat. Sources of ignition, such as smoking and open flames, are prohibited where ethyl acrylate is handled, used, or stored.

Metal containers involving the transfer of 5 gallons or more of ethyl acrylate should be grounded and bonded. Drums must be equipped with self-closing valves, pressure vacuum bungs and flame arresters. Use only non-sparking tools and equipment, especially when opening and closing containers of ethyl acrylate. Wherever ethyl acrylate is used, handled, manufactured, or stored, use explosion-proof electrical equipment and fittings.

Shipping: This compound requires a shipping label of: "Flammable Liquid." It falls in DOT Hazard Class 3 and Packing Group II.

Spill Handling: Evacuate and restrict persons not wearing protective equipment from area of spill or leak until cleanup is complete. Remove all ignition sources. Ventilate area of spill or leak. Absorb liquids in vermiculite, dry sand, earth, peat, carbon, or a similar material and deposit in sealed containers. It may be necessary to contain and dispose of this chemical as a hazardous waste. If material or contaminated runoff enters waterways, notify downstream users of potentially contaminated waters. Contact your Department of Environmental Protection or your regional office of the federal EPA for specific recommendations. If employees are required to clean-up spills, they must be properly trained and equipped. OSHA 1910.120(q) may be applicable.

Fire Extinguishing: This chemical is a flammable liquid. Poisonous gases or irritating fumes are produced in fire. Use dry chemical, carbon dioxide, or foam extinguishers. Vapors are heavier than air and will collect in low areas. Vapors may travel long distances to ignition sources and flashback. Vapors in confined areas may explode when exposed to fire. Containers may explode in fire. Storage containers and parts of containers may rocket great distances, in many directions. If material or contaminated runoff enters waterways, notify downstream users of potentially contaminated waters. Notify local health and fire officials and pollution control agencies. From a secure, explosion-proof location, use water spray to cool exposed containers. If cooling streams are ineffective (venting sound increases in volume and pitch, tank discolors, or shows any signs of deforming), withdraw immediately to a secure position. If employees are expected to fight fires, they must be trained and equipped in OSHA 1910.156.

Disposal Method Suggested: Incineration[22] or by absorption and landfill disposal.[22]

References

Sax, N. I., Ed., "Dangerous Properties of Industrial Materials Report," 1, No. 2, 35–37 (1980)

New Jersey Department of Health and Senior Services, "Hazardous Substance Fact Sheet: Ethyl Acrylate," Trenton, NJ (January 1996)

Ethyl Alcohol

Molecular Formula: C_2H_6O

Common Formula: CH_3CH_2OH

Synonyms: Absolute ethanol; Aethanol (German); Aethyldlkohol (German); Alcohol; Alcohol, anhydrous; Alcohol C-2; Alcohol, dehydrated; Alcool ethylique (French); Alcool etilico (Italian); Algrain; Alkohol (German); Alkoholu etylowego (Polish); Anhydrol; Cologne spirit; Cologne spirits; Etaholo (Italian); Ethanol; Ethanol 200 proof; Ethylalcohol (Dutch); Ethyl alcohol anhydro-S; Ethyl hydrate; Ethyl hydroxide; Etylowy alkohol (Polish); Fermentation alcohol; Grain alcohol; Jaysol S; Methyl carbinol; Molasses alcohol; NCI-CO3134; Potato alcohol; Pure grain alcohol; SD alcohol 23-hydrogen; Spirit; Spirits of wine; Tescol

CAS Registry Number: 64-17-5

RTECS Number: KQ6300000

DOT ID: UN 1170

EEC Number: 603-002-00-5

- ***Regulatory Authority:*** Air Pollutant Standard Set (ACGIH)[1] (DFG)[3] (HSE)[33] (Former USSR)[43] (OSHA)[58] (Several States)[60]
- Canada, WHMIS, Ingredients Disclosure List

Cited in U.S. State Regulations: Alaska (G), California (A, G), Connecticut (A), Florida (G), Illinois (G), Maine (G), Massachusetts (G, A), Nevada (A), New Hampshire (G), New Jersey (G), New York (G), North Dakota (A), Oklahoma (G), Pennsylvania (G), Rhode Island (G), South Dakota (A), Virginia (A), West Virginia (G).

Description: Ethyl alcohol, CH_3CH_2OH, is a colorless, volatile, flammable liquid with a sweet, fruity odor. The odor threshold is 5 – 10 ppm. Boiling point = 78 – 79°C. Flash point = 12.8°C (96%). Explosive limits: LEL = 3.3%; UEL = 19.0%. Autoignition temperature = 363°C. Hazard Identification (based on NFPA-704 M Rating System): Health 0, Flammability 3, Reactivity 0. Soluble in water.

Potential Exposure: Ethyl alcohol is used in the chemical synthesis of a wide variety of compounds such as acetaldehyde, ethyl ether, ethyl chloride, and butadiene. It is a solvent or processing agent in the manufacture of pharmaceuticals, plastics, lacquers, polishes, plasticizers, perfumes, cosmetics, rubber accelerators, explosives, synthetic resins, nitrocellulose, adhesives, inks, and preservatives. It is also used as an antifreeze and as a fuel. It is an intermediate in the manufacture of many drugs and pesticides.

Incompatibilities: Forms explosive mixture with air. May accumulate static electrical charges, and may cause ignition of its vapors. Reactions may be violent with oleum, sulfuric acid, nitric acid, bases, aliphatic amines, isocyanates, strong oxidizers Also incompatible with potassium dioxide, bromine pentafluoride, acetyl bromide, acetyl chloride, platinum, sodium.

Permissible Exposure Limits in Air: The OSHA PEL, the DFG value,[3] the HSE value[33] and the recommended ACGIH TWA value is 1,000 ppm (1,900 mg/m³). There is not tentative STEL value. The former USSR-UNEP/IRPTC project[43] has set a MAC in workplace air of 1,000 mg/m³ and a MAC in ambient air of residential areas of 5.0 mg/m³ on either a momentary or a daily average basis. Several states have set guidelines or standards for ethanol in ambient air[60] ranging from 0.26 mg/m³ (Massachusetts) to 16.0 mg/m³ (Virginia) to 19.0 mg/m³ (North Dakota) to 38.0 mg/m³ (Connecticut and South Dakota) to 45.238 mg/m³ (Nevada).

Determination in Air: Collection by charcoal tube, 2-Butanol/CS_2; analysis by gas chromatography/flame ionization detection; NIOSH IV Method #1400, Alcohols I

Permissible Concentration in Water: No criteria set, but EPA[32] has suggested a permissible ambient goal of 26,000 μg/l based on health effects.

Routes of Entry: Inhalation of vapor and percutaneous absorption, ingestion, skin and/or eye contact.

Harmful Effects and Symptoms

Short Term Exposure: *Inhalation:* Levels of 5,000 – 10,000 ppm may result in irritation of mouth, nose and throat and coughing, leading to sleep and stupor.

Skin: Pure ethyl alcohol may cause drying, redness and irritation. May be absorbed through damaged skin.

Eyes: Irritation and tearing may result at 5,000 ppm of vapor. Contact with liquid may cause severe irritation.

Ingestion: 1 ounce of pure ethyl alcohol may cause reddening of face and neck and an exaggerated feeling of well-being. 3 ounces of pure ethyl alcohol may cause an initial burst of excitement and activity followed by increasing loss of coordination, slurred speech, nausea and drowsiness. This may proceed to stupor, coma and death. Lethal dose of ethyl alcohol ranges from 2 – 5 ounces, depending on age and size of the individual. *Note:* Many denatured alcohols contain additives which are extremely poisonous and cannot be removed by normal methods. Ingestion of denatured alcohol will produce much more serious poisoning.

Long Term Exposure: Prolonged inhalation of concentrations above 5,000 ppm may produce symptoms listed under inhalation and the additional symptoms of headache, dizziness, tremor and fatigue. Additives in denatured alcohol may result in other more severe symptoms. Alcohol has been linked to birth defects in humans. Ethyl alcohol may cause mutations. Repeated exposure (including alcoholic beverages) may cause spontaneous abortions, as well as birth defects and other developmental problems, including "fetal alcohol syndrome." Chronic use of ethanol may cause cirrhosis of the liver.

Points of Attack: Eyes, skin, respiratory system, central nervous system, liver, blood, reproductive system.

Medical Surveillance: For those with frequent or potentially high exposure (half the TLV or greater, or significant skin contact), the following are recommended before beginning work and at regular times after that: Liver function tests. Ethyl alcohol can be measured in the blood, urine and exhaled breath.

First Aid: If this chemical gets into the eyes, remove any contact lenses at once and irrigate immediately for at least 15 minutes, occasionally lifting upper and lower lids. Seek medical attention immediately. If this chemical contacts the skin, remove contaminated clothing and wash immediately with soap and water. Seek medical attention immediately. If this chemical has been inhaled, remove from exposure, begin rescue breathing (using universal precautions) if breathing has stopped and CPR if heart action has stopped. Transfer promptly to a medical facility. When this chemical has been swallowed, get medical attention. Give large quantities of water and induce vomiting. Do not make an unconscious person vomit.

Personal Protective Methods: Wear protective gloves and clothing to prevent any reasonable probability of skin contact. Safety equipment suppliers/manufacturers can provide recommendations on the most protective glove/clothing material for your operation. Nitrile, Polyvinyl Alcohol, Teflon, Silvershield and Butyl Rubber are among the recommended protective materials. All protective clothing (suits, gloves, footwear, headgear) should be clean, available each day, and put on before work. Contact lenses should not be worn when working with this chemical. Wear splash-proof chemical goggles and face shield unless full facepiece respiratory protection is worn. Employees should wash immediately with soap when skin is wet or contaminated. Provide emergency showers and eyewash.

Respirator Selection: NIOSH/OSHA: *3,300 ppm:* SA (any supplied-air respirator); or SCBAF (any self-contained breathing apparatus with a full facepiece). *Emergency or planned entry into unknown concentrations or IDLH conditions:* SCBAF:PD,PP (any self-contained breathing apparatus that has a full facepiece and is operated in a pressure-demand or other positive-pressure mode); or SAF:PD,PP:ASCBA (any supplied-air respirator that has a full facepiece and is operated in a pressure-demand or other positive-pressure mode in combination with an auxiliary self-contained breathing apparatus operated in a pressure-demand or other positive pressure mode). *Escape:* SCBAE (any appropriate escape-type, self-contained breathing apparatus).

Storage: Prior to working with this chemical you should be trained on its proper handling and storage. Before entering confined space where this chemical may be present, check to make sure that an explosive concentration does not exist. Ethyl alcohol must be stored to avoid contact with oxidizers such as perchlorates, peroxides, chlorates, nitrates, and permanganates, because violent reactions occur. Store in tightly closed containers in a cool, well-ventilated area away from heat or flame. Sources of ignition such as smoking and open flames are prohibited where ethyl alcohol is used, handled, or stored in a manner that could create a potential fire or explosion hazard. Metal containers involving the transfer of 5 gallons or more of ethyl alcohol should be grounded and bonded. Drums must be equipped with self-closing valves, pressure vacuum bungs, and flame arresters. Use only non-sparking tools and equipment, especially when opening and closing containers of ethyl alcohol.

Shipping: This compound requires a shipping label of: "Flammable Liquid." It falls in DOT Hazard Class 3 and Packing Group II.

Spill Handling: Evacuate and restrict persons not wearing protective equipment from area of spill or leak until cleanup is complete. Remove all ignition sources. Ventilate area of spill or leak. Absorb liquids in vermiculite, dry sand, earth, peat, carbon, or a similar material and deposit in sealed containers. It may be necessary to contain and dispose of this chemical as a hazardous waste. If material or contaminated runoff enters waterways, notify downstream users of potentially contaminated waters. Contact your Department of Environmental Protection or your regional office of the federal EPA for specific recommendations. If employees are required to clean-up spills, they must be properly trained and equipped. OSHA 1910.120(q) may be applicable.

Fire Extinguishing: This chemical is a flammable liquid. Poisonous gases are produced in fire. Use dry chemical, carbon dioxide, or foam extinguishers. Vapors are heavier than air and will collect in low areas. Vapors may travel long distances to ignition sources and flashback. Vapors in confined areas may explode when exposed to fire. Containers may explode in fire. Storage containers and parts of containers may rocket great distances, in many directions. If material or contaminated runoff enters waterways, notify downstream users of potentially contaminated waters. Notify local health and fire officials and pollution control agencies. From a secure, explosion-proof location, use water spray to cool exposed containers. If cooling streams are ineffective (venting sound increases in volume and pitch, tank discolors, or shows any signs of deforming), withdraw immediately to a secure position. If employees are expected to fight fires, they must be trained and equipped in OSHA 1910.156.

Disposal Method Suggested: Incineration. Consult with environmental regulatory agencies for guidance on acceptable disposal practices. Generators of waste containing this contaminant (≥100 kg/mo) must conform with EPA regulations governing storage, transportation, treatment, and waste disposal.

References

Sax, N. I., Ed., "Dangerous Properties of Industrial Materials Report," 1, No. 7, 55–57 (1981)

New Jersey Department of Health and Senior Services, "Hazardous Substance Fact Sheet: Ethyl Alcohol," Trenton, NJ (January 1996)

New York State Department of Health, "Chemical Fact Sheet: Ethyl Alcohol," Albany, NY, Bureau of Toxic Substance Assessment (March 1986)

Ethylamine

Molecular Formula: C_2H_7N

Common Formula: $C_2H_5NH_2$

Synonyms: Aethylamine (German); 1-Aminoethane; Aminoethane; EA; Etanamina (Spanish); Ethanamine; Etilamina (Italian, Spanish); Etyloamnia (Polish); MEA; Monoethylamine

CAS Registry Number: 75-04-7

RTECS Number: KH2100000

DOT ID: UN 1036

Regulatory Authority

- Air Pollutant Standard Set (ACGIH)[1] (DFG)[3] (HSE)[33] (OSHA)[58] (Other Countries)[35] (Several States)[60]
- OSHA 29CFR1910.119, Appendix A. Process Safety List of Highly Hazardous Chemicals, TQ = 7,500 lb
- Clean Air Act: Accidental Release Prevention/Flammable substances, (Section 112[r], Table 3), TQ = 10,000 lb (4,540 kg)
- Clean Water Act: Section 311 Hazardous Substances/RQ 40CFR117.3 (same as CERCLA, see below)
- Superfund/EPCRA 40CFR302.4 Reportable Quantity (RQ): CERCLA, 100 lb (45.4 kg)
- Canada, WHMIS, Ingredients Disclosure List

Cited in U.S. State Regulations: Alaska (G), California (A, G), Connecticut (A), Florida (G), Illinois (G), Maine (G), Massachusetts (G), Nevada (A), New Hampshire (G), New Jersey (G), North Dakota (A), Oklahoma (G), Pennsylvania (G), Rhode Island (G), Virginia (A), West Virginia (G).

Description: Ethylamine, $C_2H_5NH_2$, is a colorless gas or water-white liquid (below 17°C) with a strong, ammonia-like odor. Shipped as a liquefied compressed gas. Boiling point = 15 – 17°C. Freezing/Melting point = -81°C. Flash point ≤-18°C. Explosive limits: LEL = 3.5%; UEL = 14.0%. Autoignition temperature = 385°C. Hazard Identification (based on NFPA-704 M Rating System): Health 3, Flammability 4, Reactivity 0.

Potential Exposure: Monoethylamine (MEA) is used as an intermediate in the manufacture off the following chemicals: triazine herbicides, 1,3-diethylthiourea (a corrosion inhibitor), ethylamino-ethanol, 4-ethylmorpholine (urethane foam catalyst), ethyl isocyanate, and dimethylolethyltriazone (agent used in wash-and-wear fabrics). The cuprous chloride salts of MEA are used in the refining of petroleum and vegetable oil.

Incompatibilities: The aqueous solution a strong base. Forms explosive mixture with air. Reacts violently with strong acids; strong oxidizers, cellulose nitrate, and organic compounds causing fire and explosion hazard. Also incompatible with organic anhydrides, isocyanates, vinyl acetate, acrylates, substituted allyls, alkylene oxides, epichlorohydrin, ketones, aldehydes, alcohols, glycols, phenols, cresols, caprolactum solution. Attacks nonferrous metals: aluminum, copper, lead, tin, zinc and alloys, some plastics, rubber and coatings.

Permissible Exposure Limits in Air: The OSHA PEL is 10 ppm TWA. The DFG MAK[3] and the HSE TWA[33] and the ACGIH TWA value is 5 ppm (9 mg/m^3). The NIOSH IDLH level is 600 ppm. Japan and Sweden[35] have also set 10 ppm as a TWA value. Sweden adds an STEL of 30 mg/m^3. Brazil[35] has set a TWA of 8 ppm (14 mg/m^3) in workplace air. The Former USSR has set a MAC in ambient air in residential areas of 0.01 mg/m^3 on both a momentary and a daily average basis. Several states have set guidelines or standards for monoethylamine in ambient air[60] ranging from 180 μg/m^3 (North Dakota) to 300 μg/m^3 (Virginia) to 360 μg/m^3 (Connecticut) to 429 μg/m^3 (Nevada).

Determination in Air: Adsorption on silica, workup with H_2SO_4; Gas chromatography/Flame ionization detection; NIOSH II(3) Method #S144.

Permissible Concentration in Water: EPA[32] has suggested an ambient environmental goal of 248 μg/l on a health basis. The Former USSR[35] has set a MAC in water bodies used for domestic purposes of 0.5 mg/l.

Routes of Entry: Inhalation, skin absorption (liquid), ingestion (liquid), skin and/or eye contact (liquid).

Harmful Effects and Symptoms

Short Term Exposure: Ethylamine can affect you when breathed in and by passing through your skin. Contact can severely burn the eyes and skin. Inhalation can severely irritate the eyes, nose, throat and lungs causing cough, wheezing, and/or shortness of breath. Repeated exposure may damage the lungs, kidneys and heart.

Long Term Exposure: Repeated exposure can cause damage to the kidneys, liver and heart. Can affect the eyes causing blurred vision and/or cause the victim to see halos around lights, and result in permanent damage. Repeated exposure can affect the lungs, causing bronchitis, and tissue lesions.

Points of Attack: Eyes, skin, respiratory system, liver, kidneys.

Medical Surveillance: Before beginning employment and at regular times after that, the following are recommended: Lung function tests. Exam of the eyes and vision. If symptoms develop or overexposure is suspected, the following may be useful: Exam of the heart. Kidney function tests. Liver function tests.

First Aid: If this chemical gets into the eyes, remove any contact lenses at once and irrigate immediately for at least 15 minutes, occasionally lifting upper and lower lids. Seek medical attention immediately. If this chemical contacts the skin, remove contaminated clothing and wash immediately with soap and water. Seek medical attention immediately. If this chemical has been inhaled, remove from exposure, begin rescue breathing (using universal precautions) if breathing has stopped and CPR if heart action has stopped. Transfer promptly to a medical facility. When this chemical has been swallowed, get medical attention. Give large quantities of water and induce vomiting. Do not make an unconscious person vomit.

Personal Protective Methods: Wear protective gloves and clothing to prevent any reasonable probability of skin contact. Safety equipment suppliers/manufacturers can provide recommendations on the most protective glove/clothing material for your operation. Nitrile Teflon, and Butyl Rubber are among the recommended protective materials. All protective clothing (suits, gloves, footwear, headgear) should be clean, available each day, and put on before work. Contact lenses should not be worn when working with this chemical. Wear splash-proof chemical goggles and face shield unless full facepiece respiratory protection is worn. Employees should wash immediately with soap when skin is wet or contaminated. Provide emergency showers and eyewash.

Respirator Selection: NIOSH/OSHA: *250 ppm:* SA:CF (any supplied-air respirator operated in a continuous-flow mode); or PAPRS [any powered, air-purifying respirator with cartridge(s) providing protection against the compound of concern]. *500 ppm:* CCRFS [any chemical cartridge respirator with a full facepiece and cartridge(s) providing protection against the compound of concern] organic vapor and acid gas cartridge(s); or GMFS [any air-purifying, full-facepiece respirator (gas mask) with a chin-style, front- or back-mounted canister providing protection against the compound of concern]; or SCBAF (any self-contained breathing apparatus with a full facepiece); or SAF (any supplied-air respirator with a full facepiece). *600 ppm:* SAF:PD,PP (any supplied-air respirator that has a full facepiece and is operated in a pressure-demand or other positive-pressure mode). *Emergency or planned entry into unknown concentrations or IDLH conditions:* SCBAF:PD,PP (any self-contained breathing apparatus that has a full facepiece and is operated in a pressure-demand or other positive-pressure mode); or SAF:PD,PP: ASCBA (any supplied-air respirator that has a full facepiece and is operated in a pressure-demand or other positive-pressure mode in combination with an auxiliary self-contained breathing apparatus operated in a pressure-demand or other positive pressure mode). *Escape:* GMFS [any air-purifying, full-facepiece respirator (gas mask) with a chin-style, front- or back-mounted canister providing protection against the compound of concern]; or SCBAE (any appropriate escape-type, self-contained breathing apparatus).

Note: Substance causes eye irritation or damage; eye protection needed.

Storage: Ethylamine must be stored to avoid contact with strong acids (such as hydrochloric, sulfuric and nitric), or strong oxidizers (such as chlorine and bromine) because violent reactions occur. Before entering confined space where this chemical may be present, check to make sure that an explosive concentration does not exist. Store in tightly closed containers in a cool, well-ventilated area away from heat. Sources of ignition, such as smoking and open flames, are prohibited where ethylamine is handled, used, or stored. Metal containers used in the transfer of 5 gallons or more of ethylamine should be grounded and bonded. Drums must be equipped with self-closing valves, pressure vacuum bungs, and flame arresters. Use only non-sparking tools and equipment, especially when opening and closing containers of ethylamine. Procedures for the handling, use and storage of cylinders should be in compliance with OSHA 1910.101 and 1910.169 as with the recommendations of the Compressed Gas Association.

Shipping: This compound requires a shipping label of: "Flammable Gas." It falls in DOT Hazard Class 2.1. Passenger aircraft or railcar shipment is forbidden as is cargo aircraft shipment.

Spill Handling: Evacuate and restrict persons not wearing protective equipment from area of spill or leak until cleanup is complete. Remove all ignition sources. Ventilate area of spill or leak. Absorb liquids in vermiculite, dry sand, earth, peat, carbon, or a similar material and deposit in sealed containers. It may be necessary to contain and dispose of this chemical as a hazardous waste. If material or contaminated runoff enters waterways, notify downstream users of potentially contaminated waters. Contact your Department of Environmental Protection or your regional office of the federal EPA for specific recommendations. If employees are required to clean-up spills, they must be properly trained and equipped. OSHA 1910.120(q) may be applicable.

Fire Extinguishing: *Gas:* This chemical is flammable. Poisonous gases are produced in fire including nitrogen oxides. Do not extinguish the fire unless the flow of gas can be stopped and any remaining gas is out of the line. Specially trained personnel may use fog lines to cool exposures and let the fire burn itself out. Vapors are heavier than air and will collect in low areas. Vapors may travel long distances to ignition sources and flashback. Vapors in confined areas may explode when exposed to fire. Containers may explode in fire. Storage containers and parts of containers may rocket great distances, in many directions. If material or contaminated runoff enters waterways, notify downstream users of potentially contaminated waters. Notify local health and fire officials and pollution control agencies. From a secure, explosion-proof location, use water spray to cool exposed containers. If cooling streams are ineffective (venting sound increases in volume and pitch, tank discolors, or shows any signs of deforming), withdraw immediately to a secure position. If cylinders are exposed to excessive heat from fire or flame contact, withdraw immediately to a secure location. If employees are expected to fight fires, they must be trained and equipped in OSHA 1910.156.

Liquid: This chemical is flammable. Poisonous gases including nitrogen oxides are produced in fire. Shut off supply. If not possible, and no risk to surroundings, let the fire burn itself out; in other cases use dry chemical, carbon dioxide. Vapors are heavier than air and will collect in low areas. Vapors may travel long distances to ignition sources and flashback. Vapors in confined areas may explode when exposed to fire. Containers may explode in fire. Storage containers and parts of containers may rocket great distances, in many directions. If material or contaminated runoff enters waterways, notify downstream users of potentially contaminated waters. Notify local health and fire officials and pollution control agencies. From a secure, explosion-proof location, use water spray to cool exposed containers. If cooling streams are ineffective (venting sound increases in volume and pitch, tank discolors, or shows any signs of deforming), withdraw immediately to a secure position. If employees are expected to fight fires, they must be trained and equipped in OSHA 1910.156.

Disposal Method Suggested: Controlled incineration; incinerator is equipped with a scrubber or thermal unit to reduce No_x emissions.[22]

References

U.S. Environmental Protection Agency, "Chemical Hazard Information Profile: Ethylamines," Washington, DC (April 1, 1978)

New Jersey Department of Health and Senior Services, "Hazardous Substance Fact Sheet: Ethylamine," Trenton, NJ (December 1996)

Ethyl Amyl Ketone

Molecular Formula: $C_8H_{16}O$

Common Formula: $CH_3CH_2COCH_2CH(CH_3)CH_2CH_3$

Synonyms: Amyl ethyl ketone; EAK; Ethyl *sec*-amyl ketone; 5-Methyl-3-heptanone (OSHA); 5-Metilheptano-3-ona (Spanish); 3-Octanone

CAS Registry Number: 541-85-5

RTECS Number: MJ7350000

DOT ID: UN 2271

EEC Number: 606-020-00-1

EINECS Number: 208-793-7

Regulatory Authority

- Air Pollutant Standard Set (ACGIH)[1] (HSE)[33] (Several States)[60]
- Canada, WHMIS, Ingredients Disclosure List

Cited in U.S. State Regulations: Alaska (G), California (A), Connecticut (A), Florida (G), Illinois (G), Maine (G), Massachusetts (G), Nevada (A), New Hampshire (G), New Jersey (G), North Dakota (A), Pennsylvania (G), Rhode Island (G), Virginia (A), West Virginia (G).

Description: Ethyl amyl ketone, $CH_3CH_2COCH_2CH(CH_3)CH_2CH_3$, is a colorless liquid with a mild, fruity odor. Boiling point = 173 – 174°C. Flash point = 57°C. Hazard Identification (based on NFPA-704 M Rating System): Health 0, Flammability 2, Reactivity 0. Insoluble in water.

Potential Exposure: Ethyl amyl ketone is used as a solvent for resins, in the manufacture of perfume, and as an organic intermediate.

Incompatibilities: Forms explosive mixture with air. Contact with oxidizers may cause fire and explosions. Incompatible with strong bases, reducing agents, aldehydes, nitric acid, aliphatic amines.

Permissible Exposure Limits in Air: The HSE TWA[33] and the ACGIH TWA value is 25 ppm (130 mg/m^3). The NIOSH IDLH level is 100 ppm. Some states have set guidelines or standards for ethyl amyl ketone in ambient air[60] ranging from 1.3 mg/m^3 (North Dakota) to 2.2 mg/m^3 (Virginia) to 2.6 mg/m^3 (Connecticut) to 3.095 mg/m^3 (Nevada).

Determination in Air: Charcoal tube; Methanol/CS2; Gas chromatography/Flame ionization detection; NIOSH IV, Method #1301, Ketones II.

Permissible Concentration in Water: No criteria set.

Routes of Entry: Inhalation, ingestion, eye and/or skin contact. Passes through the skin.

Harmful Effects and Symptoms

Short Term Exposure: Contact can irritate the eyes and skin. Prolonged contact can cause skin rash. Inhalation can irritate the respiratory tract with coughing and shortness of breath. Exposure can cause headache, nausea, dizziness, lightheadedness, and unconsciousness.

Long Term Exposure: Can cause skin rash and drying and cracking of skin. Similar petroleum-based chemicals cause brain and nerve damage.

Points of Attack: Eyes, skin, respiratory system, central nervous system.

Medical Surveillance: Consider the points of attack in preplacement and periodic physical examinations. Evaluate for brain and nervous system damage.

First Aid: If this chemical gets into the eyes, remove any contact lenses at once and irrigate immediately for at least 15 minutes, occasionally lifting upper and lower lids. Seek medical attention immediately. If this chemical contacts the skin, remove contaminated clothing and wash immediately with soap and water. Seek medical attention immediately. If this chemical has been inhaled, remove from exposure, begin rescue breathing (using universal precautions) if breathing has stopped and CPR if heart action has stopped. Transfer promptly to a medical facility. When this chemical has been swallowed, get medical attention. Give large quantities of water and induce vomiting. Do not make an unconscious person vomit.

Personal Protective Methods: Wear protective gloves and clothing to prevent any reasonable probability of skin contact. Safety equipment suppliers/manufacturers can provide recommendations on the most protective glove/clothing material for your operation.. All protective clothing (suits, gloves, footwear, headgear) should be clean, available each day, and put on before work. Contact lenses should not be worn when working with this chemical. Wear splash-proof chemical goggles and face shield unless full facepiece respiratory protection is worn. Employees should wash immediately with soap when skin is wet or contaminated. Provide emergency showers and eyewash. Wear eye protection to prevent any reasonable probability of eye contact. Employees should wash promptly when skin is wet or contaminated. Remove nonimpervious clothing promptly if wet or contaminated.

Respirator Selection: NIOSH/OSHA: *100 ppm:* CCROV [any chemical cartridge respirator with organic vapor cartridge(s)]; or PAPROV [any powered, air-purifying respirator with organic vapor cartridge(s)]; or GMFOV[any air-purifying, full-facepiece respirator (gas mask) with a chin-style, front- or back-mounted acid gas canister]; or SA (any supplied-air respirator); or SCBAF (any self-contained breathing apparatus with a full facepiece). *Emergency or planned entry into unknown concentrations or IDLH conditions:* SCBAF:PD,PP (any self-contained breathing apparatus that has a full facepiece and is operated in a pressure-demand or other positive-pressure mode); or SAF:PD,PP:ASCBA (any supplied-air respirator that has a full facepiece and is operated in a pressure-demand or other positive-pressure mode in combination with an auxiliary, self-contained breathing apparatus operated in a pressure-demand or other positive-pressure mode). *Escape:* GMFOV [any air-purifying, full-facepiece respirator (gas mask) with a chin-style, front- or back-mounted organic vapor canister]; or SCBAE (any appropriate escape-type, self-contained breathing apparatus).

Storage: Prior to working with this chemical you should be trained on its proper handling and storage. Ethyl amyl ketone must be stored to avoid contact with oxidizers (such as perchlorates, peroxides, chlorates, nitrates, and permanganates) since violent reactions occur. Store in tightly closed containers in a cool, well-ventilated area. Use only non-sparking tools and equipment, especially when opening and closing containers of ethyl amyl ketone. Wherever ethyl amyl ketone is used, handled, manufactured, or stored, use explosion-proof electrical equipment and fittings.

Shipping: This compound requires a shipping label of: "Flammable Liquid." It falls in DOT Hazard Class 3 and Packing Group III.

Spill Handling: Evacuate and restrict persons not wearing protective equipment from area of spill or leak until cleanup is complete. Remove all ignition sources. Ventilate area of spill or leak. Absorb liquids in vermiculite, dry sand, earth, peat, carbon, or a similar material and deposit in sealed containers. Keep EAK out of a confined space, such as a sewer, because of the possibility of an explosion, unless the sewer is designed to prevent the build-up of explosive concentrations. It may be necessary to contain and dispose of this chemical as a hazardous waste. If material or contaminated runoff enters waterways, notify downstream users of potentially contaminated waters. Contact your Department of Environmental Protection or your regional office of the federal EPA for specific recommendations. If employees are required to clean-up spills, they must be properly trained and equipped. OSHA 1910.120(q) may be applicable.

Fire Extinguishing: This chemical is a combustible liquid. Poisonous gases including carbon monoxide are produced in fire. Use dry chemical, carbon dioxide, or foam extinguishers. Vapors are heavier than air and will collect in low areas. Vapors may travel long distances to ignition sources and flashback. Vapors in confined areas may explode when exposed to fire. Containers may explode in fire. Storage containers and parts of containers may rocket great distances, in many directions. If material or contaminated runoff enters waterways, notify downstream users of potentially contaminated waters. Notify local health and fire officials and pollution control agencies. From a secure, explosion-proof location, use water spray to cool exposed containers. If cooling streams are ineffective (venting sound increases in volume and pitch, tank discolors, or shows any signs of deforming), withdraw immediately to a secure position. If employees are expected to fight fires, they must be trained and equipped in OSHA 1910.156.

Disposal Method Suggested: Incineration.

References

New Jersey Department of Health and Senior Services, "Hazardous Substance Fact Sheet: Ethyl Amyl Ketone," Trenton, NJ (January 1999)

2-Ethylaniline

Molecular Formula: $C_8H_{11}N$

Synonyms: 2-Aethylanilin (German); *o*-Aminoethylbenzene; Aniline, *o*-ethyl-; Benzenamine, 2-ethyl-; 2-Ethyl aniline; 2-Ethylbenzenamine; 2-Ethylbenzenamino; 2-Etilanilina (Spanish)

CAS Registry Number: 578-54-1

RTECS Number: BX9800000

DOT ID: UN 2273

Cited in U.S. State Regulations: California (A, G), Florida (G), Massachusetts (G), New Hampshire (G), New Jersey (G), Pennsylvania (G), Rhode Island (G).

Description: 2-Ethylaniline, is a yellow liquid that turns brown on standing in air. Boiling point = 215°C. Freezing/Melting point = -64°C. Also reported at -44°C. Flash point = 85°C. Insoluble in water.

Potential Exposure: This material is used in making drugs, dyes and pesticides.

Incompatibilities: Decomposes on contact with light or air. Violent reaction on contact with strong oxidizers, strong, acid anhydrides, and chloroformates.

Permissible Exposure Limits in Air: No standards set.

Permissible Concentration in Water: No criteria set.

Routes of Entry: Inhalation, passing through the skin.

Harmful Effects and Symptoms

Short Term Exposure: N-ethylaniline can affect you when breathed in and by passing through your skin. High exposure can reduce the blood's ability to supply oxygen to the body (methemoglobinemia) and can cause the skin and lips to turn a blue color. Headache, weakness and passing out may occur.

Death could result from high skin or breathing exposures. Contact can irritate or burn skin and eyes.

Long Term Exposure: Exposure can cause skin allergy to develop, with rash and itching. Once allergy is present, even low exposures may trigger symptoms. May affect the nervous system causing headache, drowsiness, irritability and confusion. May cause liver damage.

Points of Attack: Skin, blood, nervous system.

Medical Surveillance: If symptoms develop or overexposure is suspected, the following may be useful: Blood test for methemoglobin (the blood change caused by ethylaniline). Evaluation by a qualified allergist, including careful exposure history and special testing, may help diagnose skin allergy. Liver function tests.

First Aid: *Eye Contact:* Immediately remove any contact lenses and flush with large amounts of water for at least 15 minutes, occasionally lifting upper and lower lids. Seek medical attention if any symptoms are present.

Skin Contact: Quickly remove contaminated clothing. Immediately wash area with large amounts of soap, promptly seek medical attention.

Breathing: Remove the person from exposure. Begin rescue breathing if breathing has stopped and CPR if heart action has stopped. Transfer promptly to a medical facility.

Note to Physician: Treat for methemoglobinemia. Spectrophotometry may be required for precise determination of levels of methemoglobinemia in urine.

Personal Protective Methods: Wear protective gloves and clothing to prevent any reasonable probability of skin contact. Safety equipment suppliers/manufacturers can provide recommendations on the most protective glove/clothing material for your operation.. All protective clothing (suits, gloves, footwear, headgear) should be clean, available each day, and put on before work. Contact lenses should not be worn when working with this chemical. Wear splash-proof chemical goggles and face shield unless full facepiece respiratory protection is worn. Employees should wash immediately with soap when skin is wet or contaminated. Provide emergency showers and eyewash.

Respirator Selection: Where the potential exists for exposure to ethylaniline use a MSHA/NIOSH approved supplied-air respirator with a full facepiece operated in the positive pressure mode or with a full facepiece, hood, or helmet in the continuous flow mode, or use a MSHA/NIOSH approved self-contained breathing apparatus with a full facepiece operated in pressure-demand or other positive pressure mode.

Storage: Prior to working with this chemical you should be trained on its proper handling and storage. 2-Ethylaniline must be stored to avoid contact with strong acids (such as hydrochloric, sulfuric and nitric), strong oxidizers, acid anhydrides (such as maleic anhydride) and chloroformates since violent reactions occur. Store in tightly closed containers in a cool, well-ventilated area. Sources of ignition, such as smoking and open flames, are prohibited where 2-ethylaniline is used, handled, or stored in a manner that could crate a potential fire or explosion hazard.

Shipping: This compound requires a shipping label of: "Keep Away From Food." It falls in DOT Hazard Class 6.1 and Packing Group III.

Spill Handling: Evacuate and restrict persons not wearing protective equipment from area of spill or leak until cleanup is complete. Remove all ignition sources. Ventilate area of spill or leak. Absorb liquids in vermiculite, dry sand, earth, peat, carbon, or a similar material and deposit in sealed containers. Keep 2-ethylaniline out of a confined space, such as a sewer, because of the possibility of an explosion, unless the sewer is designed to prevent the build-up off explosive concentrations. It may be necessary to contain and dispose of this chemical as a hazardous waste. If material or contaminated runoff enters waterways, notify downstream users of potentially contaminated waters. Contact your Department of Environmental Protection or your regional office of the federal EPA for specific recommendations. If employees are required to clean-up spills, they must be properly trained and equipped. OSHA 1910.120(q) may be applicable.

Fire Extinguishing: This chemical is a combustible liquid. Poisonous gases including aniline and nitrogen oxides are produced in fire. Use dry chemical, carbon dioxide, or alcohol foam extinguishers. Vapors are heavier than air and will collect in low areas. Vapors may travel long distances to ignition sources and flashback. Vapors in confined areas may explode when exposed to fire. Containers may explode in fire. Storage containers and parts of containers may rocket great distances, in many directions. If material or contaminated runoff enters waterways, notify downstream users of potentially contaminated waters. Notify local health and fire officials and pollution control agencies. From a secure, explosion-proof location, use water spray to cool exposed containers. If cooling streams are ineffective (venting sound increases in volume and pitch, tank discolors, or shows any signs of deforming), withdraw immediately to a secure position. If employees are expected to fight fires, they must be trained and equipped in OSHA 1910.156.

References

New Jersey Department of Health and Senior Services, "Hazardous Substance Fact Sheet: 2-Ethylaniline," Trenton, NJ (September 1996)

N-Ethylaniline

Molecular Formula: $C_8H_{11}N$

Common Formula: $C_6H_5NHC_2H_5$

Synonyms: Aethylanilin (German); Anilinoethane; Benzenamine, *N*-ethyl-; *N*-Ethylaminobenzene; Ethylaniline; *N*-Ethylbenzenamine; *N*-Ethylbenzenamino; Ethylphenylamine; *N*-Etilanilina (Spanish)

CAS Registry Number: 103-69-5

RTECS Number: BX9780000

DOT ID: UN 2272

EINECS Number: 203-135-5

Cited in U.S. State Regulations: California (A, G), Florida (G), Massachusetts (G), New Hampshire (G), New Jersey (G), Pennsylvania (G), Rhode Island (G).

Description: N-ethylaniline, $C_6H_5NHC_2H_5$, is yellow-brown oil with a weak fishy odor. Boiling point = 205°C. Freezing/Melting point = -64°C. Flash point = 85°C (oc). Flammable Limits: LEL 1.6%; UEL 9.5%. Hazard Identification (based on NFPA-704 M Rating System): Health 3, Flammability 2, Reactivity 0. Insoluble in water.

Potential Exposure: This material is used in organic synthesis.

Incompatibilities: Forms explosive mixture with air. Decomposes on contact with light or air. Reacts with many materials. Contact with strong oxidizers, strong acids can cause fire, explosions with formation of toxic vapors of aniline and oxides of nitrogen.

Permissible Exposure Limits in Air: No standards set.

Permissible Concentration in Water: No criteria set.

Routes of Entry: Inhalation, passing through the skin.

Harmful Effects and Symptoms

Short Term Exposure: N-ethylaniline can affect you when breathed in and by passing through your skin. High exposure can reduce the blood's ability to supply oxygen to the body (methemoglobinemia) and can cause the skin and lips to turn a blue color. Headache, weakness and passing out may occur. Death could result from high skin or breathing exposures. Contact can irritate or burn skin and eyes.

Long Term Exposure: Exposure can cause skin allergy to develop, with rash and itching. Once allergy is present, even low exposures may trigger symptoms. May affect the nervous system causing headache, drowsiness, irritability and confusion.

Points of Attack: Skin, blood, nervous system.

Medical Surveillance: If symptoms develop or overexposure is suspected, the following may be useful: Blood test for methemoglobinemia. Evaluation by a qualified allergist, including careful exposure history and special testing, may help diagnose skin allergy.

First Aid: *Eye Contact:* Immediately remove any contact lenses and flush with large amounts of water for at least 15 minutes, occasionally lifting upper and lower lids. Seek medical attention if any symptoms are present.

Skin Contact: Quickly remove contaminated clothing. Immediately wash area with large amounts of soap, promptly seek medical attention.

Breathing: Remove the person from exposure. Begin rescue breathing if breathing has stopped and CPR if heart action has stopped. Transfer promptly to a medical facility.

Note to Physician: Treat for methemoglobinemia. Spectrophotometry may be required for precise determination of levels of methemoglobinemia in urine.

Personal Protective Methods: Wear protective gloves and clothing to prevent any reasonable probability of skin contact. Safety equipment suppliers/manufacturers can provide recommendations on the most protective glove/clothing material for your operation.. All protective clothing (suits, gloves, footwear, headgear) should be clean, available each day, and put on before work. Contact lenses should not be worn when working with this chemical. Wear splash-proof chemical goggles and face shield unless full facepiece respiratory protection is worn. Employees should wash immediately with soap when skin is wet or contaminated. Provide emergency showers and eyewash.

Respirator Selection: Where the potential exists for exposure to ethylaniline use a MSHA/NIOSH approved supplied-air respirator with a full facepiece operated in the positive pressure mode or with a full facepiece, hood, or helmet in the continuous flow mode, or use a MSHA/NIOSH approved self-contained breathing apparatus with a full facepiece operated in pressure-demand or other positive pressure mode.

Storage: Prior to working with this chemical you should be trained on its proper handling and storage. Ethylaniline must be stored to avoid contact with strong acids (such as hydrochloric, sulfuric and nitric), and strong oxidizers since violent reactions occur. Store in tightly closed containers in a cool, well-ventilated area. Sources of ignition, such as smoking and open flames, are prohibited where ethylaniline is used, handled, or stored in a manner that could create a potential fire or explosion hazard.

Shipping: This compound requires a shipping label of: "Keep Away From Food." It falls in DOT Hazard Class 6.1 and Packing Group III.

Spill Handling: Evacuate and restrict persons not wearing protective equipment from area of spill or leak until cleanup is complete. Remove all ignition sources. Ventilate area of spill or leak. Absorb liquids in vermiculite, dry sand, earth, peat, carbon, or a similar material and deposit in sealed containers. Keep ethylaniline out of a confined space, such as a sewer, because of the possibility of an explosion, unless the sewer is designed to prevent the build-up of explosive concentrations. It may be necessary to contain and dispose of this chemical as a hazardous waste. If material or contaminated runoff enters waterways, notify downstream users of potentially contaminated waters. Contact your Department of Environmental Protection or your regional office of the federal EPA for specific recommendations. If employees are required to clean-up spills, they must be properly trained and equipped. OSHA 1910.120(q) may be applicable.

Fire Extinguishing: This chemical is a combustible liquid. Poisonous gases including aniline and nitrogen oxides are produced in fire. Use dry chemical, carbon dioxide, or alcohol foam extinguishers. Vapors are heavier than air and will collect in low areas. Vapors may travel long distances to ignition sources and flashback. Vapors in confined areas may explode when exposed to fire. Containers may explode in fire. Storage containers and parts of containers may rocket great distances, in many directions. If material or contaminated runoff enters waterways, notify downstream users of potentially contaminated waters. Notify local health and fire officials and pollution control agencies. From a secure, explosion-proof location, use water spray to cool exposed containers. If cooling streams are ineffective (venting sound increases in volume and pitch, tank discolors, or shows any signs of deforming), withdraw immediately to a secure position. If employees are expected to fight fires, they must be trained and equipped in OSHA 1910.156.

References

New Jersey Department of Health and Senior Services, "Hazardous Substance Fact Sheet: Ethylaniline," Trenton, NJ (December 1996)

Ethylbenzene

Molecular Formula: C_8H_{10}

Common Formula: $C_6H_5CH_2CH_3$

Synonyms: Aethylbenzol (German); Aristoline(+); AZ 1470(+); AZ 4210(+); Benzene, ethyl-; CEM 388; EB; Ethylbenzeen (Dutch); Ethylbenzol; Etilbenceno (Spanish); Etilbenzene (Italian); Etylobenzen (Polish); KTI 1350J(+); KTI photoresist standard (-); α-Methyltoluene; NCI-C56393; Phenylethane

CAS Registry Number: 100-41-4

RTECS Number: DA0700000

DOT ID: UN 1175

EEC Number: 601-023-00-4

Regulatory Authority:

- Air Pollutant Standard Set (ACGIH)[1] (DFG)[3] (HSE)[33] (Other Countries)[35] (OSHA)[58] (Several States)[60] (Several Canadian Provinces)
- Clean Air Act: Hazardous Air Pollutants (Title I, Part A, Section 112)
- Clean Water Act: Section 311 Hazardous Substances/RQ 40CFR117.3 (same as CERCLA, see below); 40CFR423, Appendix A, Priority Pollutants; Section 313 Water Priority Chemicals (57FR41331, 9/9/92); 40CFR401.15 Section 307 Toxic Pollutants
- RCRA 40CFR268.48; 61FR15654, Universal Treatment Standards: Wastewater (mg/l), 0.057; Nonwastewater (mg/kg), 10
- RCRA 40CFR264, Appendix 9; TSD Facilities Ground Water Monitoring List. Suggested test method(s) (PQL μg/l): 8020 (2); 8240 (5)
- Safe Drinking Water Act: MCL, 0.7 mg/l; MCLG, 0.7 mg/l
- Superfund/EPCRA 40CFR302.4 Reportable Quantity (RQ): CERCLA, 1,000 lb (454 kg)
- EPCRA Section 313 Form R *de minimis* concentration reporting level: 1.0%
- Canada, WHMIS, Ingredients Disclosure List

Cited in U.S. State Regulations: Alaska (G), Arizona (A), California (A), Connecticut (A), Florida (G, A), Illinois (G, W), Kansas (W), Maine (G), Maryland (G), Massachusetts (G, A), Minnesota (W), Nevada (A), New Hampshire (G), New Jersey (G), New Mexico (W), New York (G, A), North Dakota (A), Oklahoma (G), Pennsylvania (G), Rhode Island (G), South Carolina (A), Vermont (W), Virginia (A), West Virginia (G), Wisconsin (W).

Description: Ethylbenzene, $C_6H_5CH_2CH_3$, is a colorless liquid with a pungent aromatic odor. The odor threshold is 0.092 – 0.60 ppm. Boiling point = 136°C. Freezing/Melting point = -95°C. Flash point = 21°C. Autoignition temperature = 432°C. Explosive limits: LEL = 1.0%; UEL = 6.7%. Hazard Identification (based on NFPA-704 M Rating System): Health 2, Flammability 3, Reactivity 0. Insoluble in water.

Potential Exposure: Ethyl benzene is used in the manufacture of cellulose acetate, styrene, and synthetic rubber. It is also used as a solvent or diluent and as a component of automotive and aviation gasoline. Significant quantities of EB are present in mixed xylenes. These are used as dilatants in the paint industry, in agricultural sprays for insecticides and in gasoline blends (which may contain as much as 20% EB). In light of the large quantities of EB produced and the diversity of products in which it is found, there may exist environmental sources for ethylbenzene, e.g., vaporization during solvent use, pyrolysis of gasoline and emitted vapors at filling stations.

Groups of individuals who are exposed to EB to the greatest extent and could represent potential pools for the expression of EB toxicity include: (1) individuals in commercial situations where petroleum products or by-products are manufactured (e.g., rubber or plastics industry); (2) individuals residing in areas with high atmospheric smog generated by motor vehicle emissions.

Incompatibilities: Forms explosive mixture with air. Incompatible with strong oxidizers, nitric acid. Attacks plastics and rubber. May accumulate static electrical charges, and may cause ignition of its vapors.

Permissible Exposure Limits in Air: The OSHA PEL is 100 ppm (435 mg/m^3); there is no STEL. The DFG MAK and the HSE TWA[33] and the ACGIH TWA value[1] is 100 ppm (435 mg/m^3) and the STEL is 125 ppm (545 mg/m^3). The NIOSH IDLH level is 800 ppm [10%LEL]. In Brazil[35] the TWA is 78 ppm (340 mg/m^3); in Sweden the TWA is 80 ppm (350 mg/m^3). In Japan the 100 ppm value set in the US and UK is used. In Czechoslovakia the TWA is only 200 mg/m^3. The Former USSR[35][43] has set a MAC in the ambient air in residential areas of 0.02 mg/m^3 on either a momentary or a daily average basis. Several states have set guidelines or standards for ethylbenzene in ambient air[60] ranging from 0.12 mg/m^3 (Massachusetts) to 1.45 mg/m^3 (New York) to 4.35 mg/m^3 (Florida, South Carolina) to 4.35 – 5.45 mg/m^3 (North Dakota) to 7.25 mg/m^3 (Virginia) to 8.7 mg/m^3 (Connecticut) to 10.357 mg/m^3 (Nevada).

Determination in Air: Charcoal adsorption, workup with CS_2, analysis by gas chromatography/flame ionization. NIOSH Method #1501 for aromatic hydrocarbons.

Permissible Concentration in Water: To protect freshwater aquatic life – 32,000 µg/l, on an acute toxicity basis. To protect saltwater aquatic life – 430 µg/l, on an acute toxicity basis. For the protection of human health – 1.4 mg/l.[6] The EPA[48] has developed a lifetime health advisory for ethylbenzene of 0.68 mg/l, rounded to 0.7.[61] The Former USSR[35][43] has set a MAC in water bodies used for domestic purposes of 0.01 mg/l and in water bodies used for fishery purposes of 0.011 mg/l.

Several states have set guidelines for ethylbenzene in drinking water[61] ranging from 1.0 µg/l (Illinois) to 680 µg/l (Arizona, California, Kansas, Minnesota) to 750 µg/l (New Mexico) to 1,400 µg/l (Vermont and Wisconsin).

Determination in Water: Inert gas purge followed by gas chromatography and photoionization detection (EPA Method 602) or gas chromatography plus mass spectrometry (EPA Method 624).

Routes of Entry: Inhalation, ingestion, eye and/or skin contact.

Harmful Effects and Symptoms

Short Term Exposure: Ethyl benzene irritates the eyes, skin, and respiratory tract. Exposure to high concentrations can cause dizziness, lightheadedness and unconsciousness. Very high exposures (above the OEL) can cause difficult breathing, narcosis, coma, and even death. Swallowing the liquid may cause aspiration into the lungs, resulting in chemical pneumonitis. May affect the central nervous system. Concentration of 200 ppm can cause irritation.

Long Term Exposure: Repeated or prolonged exposure to the skin may cause drying, scaling and blistering. May cause kidney disease, liver disease, chronic respiratory disease, skin disease, as follows: EB is not nephrotoxic. Concern is expressed because the kidney is the primary route of excretion of EB and its metabolites. EB is not hepatotoxic. Since EB is metabolized by the liver, concern is expressed for these tissues. Exacerbation of pulmonary pathology might occur following exposure to EB. Individuals with impaired pulmonary function might be at risk. EB is a defating agent and may cause dermatitis following prolonged exposure. Individuals with preexisting skin problems may be more sensitive to EB. There is limited evidence that EB may damage the developing fetus, and may cause mutations.

Points of Attack: Eyes, skin, respiratory system, central nervous system.

Medical Surveillance: Consider the points of attack in preplacement and periodic physical examinations. Liver function tests. Kidney function tests. Evaluate for brain and nervous system effects.

First Aid: If this chemical gets into the eyes, remove any contact lenses at once and irrigate immediately for at least 15 minutes, occasionally lifting upper and lower lids. Seek medical attention immediately. If this chemical contacts the skin, remove contaminated clothing and wash immediately with soap and water. Seek medical attention immediately. If this chemical has been inhaled, remove from exposure, begin rescue breathing (using universal precautions) if breathing has stopped and CPR if heart action has stopped. Transfer promptly to a medical facility. When this chemical has been swallowed, get medical attention. Give large quantities of water and induce vomiting. Do not make an unconscious person vomit.

Personal Protective Methods: Wear protective gloves and clothing to prevent any reasonable probability of skin contact. Safety equipment suppliers/manufacturers can provide recommendations on the most protective glove/clothing material for your operation. Teflon is among the recommended protective materials. All protective clothing (suits, gloves, footwear, headgear) should be clean, available each day, and put on before work. Contact lenses should not be worn when working with this chemical. Wear splash-proof chemical goggles and face shield unless full facepiece respiratory protection is worn. Employees should wash immediately with soap when skin is wet or contaminated. Provide emergency showers and eyewash.

Respirator Selection: NIOSH/OSHA: *800 ppm:* CCROV [any chemical cartridge respirator with organic vapor cartridge(s)]; or GMFOV [any air-purifying, full-facepiece respirator (gas mask) with a chin-style, front- or back-mounted acid gas canister]; or PAPROV [any powered, air-purifying respirator with organic vapor cartridge(s)]; or SA (any supplied-air respirator); or SCBAF (any self-contained breathing apparatus with a full facepiece). *Emergency or planned entry into unknown concentrations or IDLH conditions:* SCBAF:PD,PP (any self-contained breathing apparatus that has a full facepiece and is operated in a pressure-demand or other positive-pressure mode); or SAF:PD, PP:ASCBA (any supplied-air respirator that has a full facepiece and is operated in a pressure-demand or other positive-pressure mode in combination with an auxiliary, self-contained breathing apparatus operated in a pressure-demand or other positive-pressure mode). *Escape:* GMFOV [any air-purifying, full-facepiece respirator (gas mask) with a chin-style, front- or back-mounted organic vapor canister]; or SCBAE (any appropriate escape-type, self-contained breathing apparatus).

Note: Substance reported to cause eye irritation or damage; may require eye protection.

Storage: Prior to working with this chemical you should be trained on its proper handling and storage. Before entering confined space where this chemical may be present, check to make sure that an explosive concentration does not exist. Protect against physical damage. Outside or detached storage is preferable. Inside storage should be in a standard flammable liquids storage room or cabinet. Isolate from acute fire hazards and oxidizing agents. Store in tightly closed containers in a cool, well-ventilated area away from heat. Sources of ignition such as smoking and open flames are prohibited where this chemical is used, handled, or stored in a manner that could create a potential fire or explosion hazard. Metal containers involving the transfer of 5 gallons or more of this chemical should be grounded and bonded. Drums must be equipped with self-closing valves, pressure vacuum bungs, and flame arresters. Use only nonsparking tools and equipment, especially when opening and closing containers of this chemical.

Shipping: This compound requires a shipping label of: "Flammable Liquid." It falls in DOT Hazard Class 3 and Packing Group II.

Spill Handling: Evacuate and restrict persons not wearing protective equipment from area of spill or leak until cleanup is complete. Remove all ignition sources. Ventilate area of spill or leak. Absorb liquids in vermiculite, dry sand, earth, peat, carbon, or a similar material and deposit in sealed containers. Keep EB out of a confined space, such as a sewer, because of the possibility of an explosion, unless the sewer is designed to prevent the build-up off explosive concentrations. It may be necessary to contain and dispose of this chemical as a hazardous waste. If material or contaminated runoff enters waterways, notify downstream users of potentially contaminated waters. Contact your Department of Environmental Protection or your regional office of the federal EPA for specific recommendations. If employees are required to clean-up spills, they must be properly trained and equipped. OSHA 1910.120(q) may be applicable.

Fire Extinguishing: This chemical is a combustible liquid. Poisonous gases are produced in fire. Use dry chemical, carbon dioxide, or foam extinguishers. Vapors are heavier than air and will collect in low areas. Vapors may travel long distances to ignition sources and flashback. Vapors in confined areas may explode when exposed to fire. Containers may explode in fire. Storage containers and parts of containers may rocket great distances, in many directions. If material or contaminated runoff enters waterways, notify downstream users of potentially contaminated waters. Notify local health and fire officials and pollution control agencies. From a secure, explosion-proof location, use water spray to cool exposed containers. If cooling streams are ineffective (venting sound increases in volume and pitch, tank discolors, or shows any signs of deforming), withdraw immediately to a secure position. If employees are expected to fight fires, they must be trained and equipped in OSHA 1910.156.

Disposal Method Suggested: Incineration.[22]

References

U.S. Environmental Protection Agency, Ethylbenzene: ambient Water Quality Criteria, Washington, DC (1980)

Sax, N. I., Ed., "Dangerous Properties of Industrial Materials Report," 2, No. 6, 57–60 (1982) and 7, No. 2, 13–35 (1987)

New York State Department of Health, "Chemical Fact Sheet: Ethyl Benzene," Albany, NY, Bureau of Toxic Substance Assessment (March 1986)

N-Ethylbenzyltoluidine

Molecular Formula: $C_{16}H_{19}N$

Common Formula: $H_3CC_6H_4N(C_2H_5)(CH_2C_6H_5)$

Synonyms: Benzenemethanamine; Ethylbenzyltoluidine; *N*-Ethyl-*N*-(3-methylphenyl)toluidine

CAS Registry Number: 119-94-8

RTECS Number: XU3676000

DOT ID: UN 2753

Cited in U.S. State Regulations: New Jersey (G).

Description: Ethylbenzyltoluidine, $H_3CC_6H_4N(C_2H_5)(CH_2C_6H_5)$ is a yellowish to light brown liquid with an unpleasant odor. Boiling point = 230°C. Flash point = 167°C. Insoluble in water.

Potential Exposure: This compound is used in dye synthesis.

Incompatibilities: Strong oxidizers.

Permissible Exposure Limits in Air: No standards set.

Permissible Concentration in Water: No criteria set.

Routes of Entry: Inhalation, passing through the skin.

Harmful Effects and Symptoms

Short Term Exposure: Exposure can interfere with the ability of the blood to carry oxygen (methemoglobinemia), causing headaches, dizziness, weakness, a bluish skin color and even death. This substance can damage the kidneys and bladder, causing painful, bloody urine. Eye contact causes irritation and can lead to permanent damage. Skin contact can cause a rash and burning feeling.

Points of Attack: Blood, skin

Medical Surveillance: If symptoms develop or overexposure has occurred, the following may be useful: Blood methemoglobin level. Kidney function tests. Urine tests for blood, and for N-acetyl p-aminophenol.

First Aid: *Eye Contact:* Immediately remove any contact lenses and flush with large amounts of water for at least 15 minutes, occasionally lifting upper and lower lids. Seek medical attention immediately.

Skin Contact: Quickly remove contaminated clothing. Immediately wash area with large amounts of soap and water. Seek medical attention.

If symptoms develop remove the person from exposure. Begin rescue breathing if breathing has stopped and CPR if heart action has stopped. Transfer promptly to a medical facility.

Note to Physician: Treat for methemoglobinemia. Spectrophotometry may be required for precise determination of levels of methemoglobinemia in urine.

Personal Protective Methods: Wear protective gloves and clothing to prevent any reasonable probability of skin contact. Safety equipment suppliers/manufacturers can provide recommendations on the most protective glove/clothing material for your operation.. All protective clothing (suits, gloves, footwear, headgear) should be clean, available each day, and put on before work. Contact lenses should not be worn when working with this chemical. Wear splash-proof chemical goggles and face shield unless full facepiece respiratory protection is worn. Employees should wash immediately with soap when skin is wet or contaminated. Provide emergency showers and eyewash.

Respirator Selection: Where the potential for exposures to Ethylbenzyltoluidine exists, use a MSHA/NIOSH approved supplied-air respirator with a full facepiece operated in the positive pressure mode or with a full facepiece, hood, or helmet in the continuous flow mode, or use a MSHA/NIOSH approved self-contained breathing apparatus with a full facepiece operated in pressure-demand or other positive pressure mode.

Storage: Prior to working with this chemical you should be trained on its proper handling and storage. Store in tightly closed containers in a cool, well ventilated area away from oxidizers and light. Where possible, automatically pump liquid from drums or other storage containers to process containers.

Shipping: This compound requires a shipping label of: "Keep Away From Food." It falls in DOT Hazard Class 6.1 and Packing Group III.

Spill Handling: Evacuate and restrict persons not wearing protective equipment from area of spill or leak until cleanup is complete. Remove all ignition sources. Ventilate area of spill or leak. Absorb liquids in vermiculite, dry sand, earth, peat, carbon, or a similar material and deposit in sealed containers. It may be necessary to contain and dispose of this chemical as a hazardous waste. If material or contaminated runoff enters waterways, notify downstream users of potentially contaminated waters. Contact your Department of Environmental Protection or your regional office of the federal EPA for specific recommendations. If employees are required to clean-up spills, they must be properly trained and equipped. OSHA 1910.120(q) may be applicable.

Fire Extinguishing: Ethylbenzyltoluidine may burn, but does not readily ignite. Poisonous gases are produced in fire. Use dry chemical, carbon dioxide, or alcohol foam extinguishers. Vapors in confined areas may explode when exposed to fire. Containers may explode in fire. Storage containers and parts of containers may rocket great distances, in many directions. If material or contaminated runoff enters waterways, notify downstream users of potentially contaminated waters. Notify local health and fire officials and pollution control agencies. From a secure, explosion-proof location, use water spray to cool exposed containers. If cooling streams are ineffective (venting sound increases in volume and pitch, tank discolors, or shows any signs of deforming), withdraw immediately to a secure position. If employees are expected to fight fires, they must be trained and equipped in OSHA 1910.156.

References

New Jersey Department of Health and Senior Services, "Hazardous Substance Fact Sheet: Ethylbenzyltoluidine," Trenton, NJ (September 1998)

Ethylbis(2-Chloroethyl)Amine

Molecular Formula: $C_6H_{13}Cl_2N$

Common Formula: $C_2H_5N(CH_2CH_2Cl)_2$

Synonyms: 2,2'-Dichlorotriethylamine; Ethyl-S; Ethylbis(β-chloroethyl)amine; HNI; TL 1149; TL329

CAS Registry Number: 538-07-8

RTECS Number: YE1225000

Regulatory Authority

- Superfund/EPCRA 40CFR355, Appendix B Extremely Hazardous Substances: TPQ = 500 lb (227 kg)
- Superfund/EPCRA 40CFR302.4 Reportable Quantity (RQ): EHS, 1 lb (0.454 kg)

Cited in U.S. State Regulations: Massachusetts (G), New Jersey (G), Pennsylvania (G).

Description: Ethylbis(2-chloroethyl)amine, $C_2H_5N(CH_2CH_2Cl)_2$, is a clear liquid with a faint, fishy amine odor. Boiling point = 85°C @ 12 mm Hg.

Potential Exposure: This is a delayed-action military casualty agent.

Permissible Exposure Limits in Air: No standards set.

Permissible Concentration in Water: No criteria set.

Harmful Effects and Symptoms

Short Term Exposure: Poisonous by all routes. This compound is a nitrogen mustard. It is highly irritating to skin, eyes, and mucous membranes. Nitrogen mustards have preferential toxicity for rapidly dividing cells. Workers exposed briefly to estimated concentrations of 10 – 100 ppm by inhalation became severely ill. The median lethal dosage is 1,500 mg-minute/m^3. Irritate the eyes in quantities which do not significantly damage the skin or respiratory tract, insofar as single exposures are concerned. After mild vapor exposure, there may be no skin lesions. After severe vapor exposures, or after exposure to the liquid, erythema may appear. Irritation and itching may occur. Later, blisters may appear in the erythematous areas. Effects on the respiratory tract include irritation of the nose and throat, hoarseness progressing to loss of voice, and a persistent cough. Fever, labored respiration, and moist riles develop. Bronchial pneumonia may appear after the first 24 hours. Following ingestion or systemic absorption, material causes inhibition of cell mitosis, resulting in depression of the blood-forming mechanism and injury to other tissues. Severe diarrhea, which may be hemorrhagic, occurs. Lesions are most marked in the small intestine and consist of degenerative changes and narcosis in the mucous membranes. Ingestion of 2 – 6 milligrams causes nausea and vomiting.

First Aid: Stop exposure and treat symptomatically. Move victim to fresh air; call emergency medical care. If not breathing, give artificial respiration. If breathing is difficult, give oxygen. In case of contact with material, immediately flush skin or eyes with running water for at least 15 minutes. Speed in removing material from skin is of extreme importance. Remove and isolate contaminated clothing and shoes at the site. Keep victim quiet and maintain normal body temperature. Effects may be delayed; keep victim under observation.

Personal Protective Methods: Wear protective gloves and clothing to prevent any reasonable probability of skin contact. Safety equipment suppliers/manufacturers can provide recommendations on the most protective glove/clothing material for your operation. All protective clothing (suits, gloves, footwear, headgear) should be clean, available each day, and put on before work. Contact lenses should not be worn when working with this chemical. Wear splash-proof chemical goggles and face shield unless full facepiece respiratory protection is worn. Employees should wash immediately with soap when skin is wet or contaminated. Provide emergency showers and eyewash. For emergency situations, wear a pressure, pressure-demand, full facepiece self-contained breathing apparatus (SCBA) or pressure-demand supplied air respirator with escape SCBA and a full-encapsulating, chemical resistant suit.

Respirator Selection: Where the potential for exposure to this chemical, use a MSHA/NIOSH approved supplied-air respirator with a full facepiece operated in the positive pressure mode or with a full facepiece, hood, or helmet in the continuous flow mode, or use a MSHA/NIOSH approved self-contained breathing apparatus with a full facepiece operated in pressure-demand or other positive pressure mode.

Storage: Prior to working with this chemical you should be trained on its proper handling and storage. Store in tightly closed containers in a cool, well ventilated area. Where possible, automatically pump liquid from drums or other storage containers to process containers.

Shipping: This material may be classified as a poisonous liquid, n.o.s. which falls in Hazard Class 6.1 and the specific material is in Packing Group I. This requires a "Poison" label.

Spill Handling: Evacuate and restrict persons not wearing protective equipment from area of spill or leak until cleanup is complete. Remove all ignition sources. Ventilate area of spill or leak. Avoid inhalation and skin contact. Do not touch spilled material; stop leak if you can do so without risk. Use water spray to reduce vapors. Small spills: absorb with sand or other noncombustible absorbent material and place into containers for later disposal. Large spills: dike far ahead of spill for later disposal. Absorb liquids in vermiculite, dry sand, earth, peat, carbon, or a similar material and deposit in sealed containers. It may be necessary to contain and dispose of this chemical as a hazardous waste. If material or contaminated runoff enters waterways, notify downstream users of potentially contaminated waters. Contact your Department of Environmental Protection or your regional office of the federal EPA for specific recommendations. If employees are required to clean-up spills, they must be properly trained and equipped. OSHA 1910.120(q) may be applicable.

Fire Extinguishing: Use dry chemical, carbon dioxide, water spray, or alcohol foam extinguishers. Poisonous gases are produced in fire. If material or contaminated runoff enters waterways, notify downstream users of potentially contaminated waters. Notify local health and fire officials and pollution control agencies. From a secure, explosion-proof location, use water spray to cool exposed containers. If cooling streams are ineffective (venting sound increases in volume and pitch, tank discolors, or shows any signs of deforming), withdraw immediately to a secure position. If employees are expected to fight fires, they must be trained and equipped in OSHA 1910.156.

References

U.S. Environmental Protection Agency, "Chemical Profile: Ethylbis(2-Chloroehtyl)Amine," Washington, DC, Chemical Emergency Preparedness Program (November 30, 1987)

Ethyl Bromide

Molecular Formula: C_2H_5Br

Synonyms: Bromic ether; Bromoethane; Bromure d'ethyle (French); Etylu bromek (Polish); Halon 2001; Hydrobromic ether; Monobromoethane; NCI-C55481

CAS Registry Number: 74-96-4

RTECS Number: KH6475000

DOT ID: UN 1891

Regulatory Authority

- Air Pollutant Standard Set (ACGIH)[1] (DFG)[3] (HSE)[33] (Former USSR)[43] (OSHA)[58] (Several States)[60]
- Canada, WHMIS, Ingredients Disclosure List

Cited in U.S. State Regulations: Alaska (G), California (A, G), Connecticut (A), Florida (G), Illinois (G), Maine (G), Massachusetts (G), Nevada (A), New Hampshire (G, W), New Jersey (G), North Dakota (A), Pennsylvania (G), Rhode Island (G), Virginia (A), West Virginia (G).

Description: Ethyl Bromide, C_2H_5Br, is a colorless liquid (turns yellow on contact with air) with an ether-like odor and a burning taste. A gas above the boiling point. Odor threshold = 3.1 ppm. Boiling point = 37 to 38°C. It has no flash point.[17] Autoignition temperature = 511°C. Explosive limits: LEL = 6.8%; UEL = 8.0%. Hazard Identification (based on NFPA-704 M Rating System): Health 2, Flammability 1, Reactivity 0. Slightly soluble in water.

Potential Exposure: This chemical is used as an ethylating agent in organic synthesis and gasoline, as a refrigerant, and as an extraction solvent. It has limited use as a local anesthetic.

Incompatibilities: Forms explosive mixture with air. Hydrolyzes in water forming hydrogen bromide. Oxidizers may cause fire or explosions. Fire and explosions may be caused by contact with chemically active metals: aluminum, magnesium or zinc powders; lithium, potassium, sodium. Attacks some plastic, rubber and coatings.

Permissible Exposure Limits in Air: The Federal limit (OSHA PEL)[58] TWA value is 200 ppm (890 mg/m^3). The recommended ACGIH TWA value is 5 ppm. The NIOSH IDLH level is 2,000 ppm. The former USSR-UNEP/IRPTC joint project[43] has set a MAC in workplace air of 5 mg/m^3. Several states have set guidelines for ethyl bromide in ambient air[60] ranging from 8.9 – 11.1 mg/m^3 (North Dakota) to 14.8 mg/m^3 (Virginia) to 17.8 mg/m^3 (Connecticut) to 21.19 mg/m^3 (Nevada).

Determination in Air: Charcoal adsorption, workup with isopropanol, analysis by gas chromatography/flame ionization. See NIOSH Method #1011.

Permissible Concentration in Water: No criteria set.

Routes of Entry: Inhalation, ingestion, skin and/or eye contact. Passes through the skin.

Harmful Effects and Symptoms

Short Term Exposure: Irritates the skin, eyes, and respiratory tract. Inhalation can cause lung irritation with coughing and/or shortness of breath. Higher exposures can cause pulmonary edema, a medical emergency that can be delayed for several hours. This can cause death. Exposure can cause central nervous system depression, headache, nausea, dizziness, loss of balance, slurred speech, numbness, unconsciousness and death. Exposure to high concentrations can cause cardiac arrhythmia. This can be fatal.

Points of Attack: Skin, liver, kidneys, respiratory system, lungs, cardiovascular system, central nervous system.

Medical Surveillance: Liver and kidney function tests. EKG. Consider chest x-ray following acute overexposure.

First Aid: If this chemical gets into the eyes, remove any contact lenses at once and irrigate immediately for at least 15 minutes, occasionally lifting upper and lower lids. Seek medical attention immediately. If this chemical contacts the skin, remove contaminated clothing and wash immediately with soap and water. Seek medical attention immediately. If this chemical has been inhaled, remove from exposure, begin rescue breathing (using universal precautions) if breathing has stopped and CPR if heart action has stopped. Transfer promptly to a medical facility. When this chemical has been swallowed, get medical attention. Give large quantities of water and induce vomiting. Do not make an unconscious person vomit.

Personal Protective Methods: Wear protective gloves and clothing to prevent any reasonable probability of skin contact. Safety equipment suppliers/manufacturers can provide recommendations on the most protective glove/clothing material for your operation. Polyvinyl Alcohol is among the recommended protective materials. All protective clothing (suits, gloves, footwear, headgear) should be clean, available each day, and put on before work. Contact lenses should not be worn when working with this chemical. Wear splash-proof chemical goggles and face shield unless full facepiece respiratory protection is worn. Employees should wash immediately with soap when skin is wet or contaminated. Provide emergency showers and eyewash.

Respirator Selection: OSHA: *up to 2,000 ppm:* SA (any supplied-air respirator); or SCBA (any self-contained breathing apparatus with a full facepiece). *Emergency or planned entry into unknown concentrations or IDLH conditions*: SCBAF:PD,PP (any self-contained breathing apparatus that has a full facepiece and is operated in a pressure-demand or other positive-pressure mode); or SAF:PD,PP:ASCBA (any supplied-air respirator that has a full facepiece and is operated in a pressure-demand or other positive-pressure mode in combination with an auxiliary self-contained breathing apparatus operated in a pressure-demand or other positive-pressure mode). *Escape:* GMFOV [any air-purifying, full-facepiece respirator (gas mask) with a chin-style, front-or back-mounted organic vapor canister]; or SCBAE (any appropriate escape-type, self-contained breathing apparatus).

Storage: Prior to working with this chemical you should be trained on its proper handling and storage. Before entering confined space where this chemical may be present, check to make sure that an explosive concentration does not exist. Store in an explosion-proof refrigerator under an inert atmosphere and protect from light. Keep away from incompatible materials listed above. Where possible, automatically pump liquid from drums or other storage containers to process containers.

Shipping: This compound requires a shipping label of: "Poison." It falls in DOT Hazard Class 6.1 and Packing Group II.

Spill Handling: Evacuate and restrict persons not wearing protective equipment from area of spill or leak until cleanup is complete. Remove all ignition sources. Ventilate area of spill or leak. Absorb liquids in vermiculite, dry sand, earth, peat, carbon, or a similar material and deposit in sealed containers. Keep ethyl bromide out of a confined space, such as a sewer, because of the possibility of an explosion, unless the sewer is designed to prevent the build-up of explosive concentrations. It may be necessary to contain and dispose of this chemical as a hazardous waste. If material or contaminated runoff enters waterways, notify downstream users of potentially contaminated waters. Contact your Department of Environmental Protection or your regional office of the federal EPA for specific recommendations. If employees are required to clean-up spills, they must be properly trained and equipped. OSHA 1910.120(q) may be applicable.

Fire Extinguishing: This chemical is a flammable liquid. Poisonous gases including carbon monoxide and hydrogen bromide are produced in fire. Use dry chemical, carbon dioxide, or alcohol foam extinguishers. Vapors are heavier than air and will collect in low areas. Vapors may travel long distances to ignition sources and flashback. Vapors in confined areas may explode when exposed to fire. Containers may explode in fire. Storage containers and parts of containers may rocket great distances, in many directions. If material or contaminated runoff enters waterways, notify downstream users of potentially contaminated waters. Notify local health and fire officials and pollution control agencies. From a secure, explosion-proof location, use water spray to cool exposed containers. If cooling streams are ineffective (venting sound increases in volume and pitch, tank discolors, or shows any signs of deforming), withdraw immediately to a secure position. If employees are expected to fight fires, they must be trained and equipped in OSHA 1910.156.

Disposal Method Suggested: Controlled incineration with adequate scrubbing and ash disposal facilities.

Ethyl Bromoacetate

Molecular Formula: $C_4H_7BrO_2$

Synonyms: Acetic acid, bromo-, ethyl ester; Antol; Bromoacetic acid, ethyl ester; Ethoxycarbonylmethyl bromide; Ethyl α-bromoacetate; Ethyl bromoacetate; Ethyl monobromoacetate

CAS Registry Number: 105-36-2

RTECS Number: AF6000000

DOT ID: UN 1603

Cited in U.S. State Regulations: California (G), New Jersey (G), Pennsylvania (G).

Description: Ethyl bromoacetate is a clear, colorless to light-yellow liquid. Boiling point = 159°C. Freezing point ≤-20°C. Flash point = 48°C. Hazard Identification (based on NFPA-704 M Rating System): Health 2, Flammability 2, Reactivity 0. Insoluble in water.

Potential Exposure: Used for making pharmaceuticals, as a warning gas in poisonous, odorless gasses, as a tear gas.

Incompatibilities: Forms explosive mixture with air. Oxidizers, strong acids, strong bases, and reducing agents.

Permissible Exposure Limits in Air: No OELs have been established. However this chemical can be absorbed through the skin, thereby increasing exposure.

Routes of Entry: Inhalation, absorbed through the skin.

Harmful Effects and Symptoms

Short Term Exposure: Inhalation can irritate the respiratory tract and cause headache, nausea, and vomiting. Higher exposures can cause pulmonary edema, a medical emergency that can be delayed for several hours. This can cause death. Can irritate the skin causing rash or a burning sensation on contact. Can cause severe eye burns leading to permanent damage.

Long Term Exposure: May cause skin allergy and irritate the lungs causing bronchitis to develop with cough, phlegm, and/or shortness of breath.

Points of Attack: Lungs, skin.

Medical Surveillance: Lung function tests. Evaluation by a qualified allergist. Consider chest x-ray following acute overexposure.

First Aid: If this chemical gets into the eyes, remove any contact lenses at once and irrigate immediately for at least 15 minutes, occasionally lifting upper and lower lids. Seek medical attention immediately. If this chemical contacts the skin, remove contaminated clothing and wash immediately with soap and water. Seek medical attention immediately. If this chemical has been inhaled, remove from exposure, begin rescue breathing (using universal precautions) if breathing has stopped and CPR if heart action has stopped. Transfer promptly to a medical facility. When this chemical has been swallowed, get medical attention. Give large quantities of water and induce vomiting. Do not make an unconscious person vomit. Medical observation is recommended for 24 – 48 hours after breathing overexposure, as pulmonary edema may be delayed. As first aid for pulmonary edema, a doctor or authorized paramedic may consider administering a corticosteroid spray.

Personal Protective Methods: Wear protective gloves and clothing to prevent any reasonable probability of skin contact. Safety equipment suppliers/manufacturers can provide recommendations on the most protective glove/clothing material for your operation.. All protective clothing (suits, gloves, footwear, headgear) should be clean, available each day, and put on before work. Contact lenses should not be worn when working with this chemical. Wear splash-proof chemical goggles and face shield unless full facepiece respiratory protection is worn. Employees should wash immediately with soap when skin is wet or contaminated. Provide emergency showers and eyewash.

Respirator Selection: Where the potential for exposure to this chemical, use a MSHA/NIOSH approved supplied-air respirator with a full facepiece operated in the positive pressure mode or with a full facepiece, hood, or helmet in the continuous flow mode, or use a MSHA/NIOSH approved self-contained breathing apparatus with a full facepiece operated in pressure-demand or other positive pressure mode.

Storage: Prior to working with this chemical you should be trained on its proper handling and storage. Store in tightly closed containers in a cool, well ventilated area away from oxidizers, strong acids, strong bases, reducing agents, heat, and sources of ignition. Where possible, automatically pump liquid from drums or other storage containers to process containers

Shipping: DOT label requirement of "Poison." Ethyl bromoacetate falls in UN/DOT Hazard Class 6.1 and Packing Group II.[19][20]

Spill Handling: Evacuate and restrict persons not wearing protective equipment from area of spill or leak until cleanup is complete. Remove all ignition sources. Ventilate area of spill or leak. Absorb liquids in vermiculite, dry sand, earth, peat, carbon, or a similar material and deposit in sealed containers. It may be necessary to contain and dispose of this chemical as a hazardous waste. If material or contaminated runoff enters waterways, notify downstream users of potentially contaminated waters. Contact your Department of Environmental Protection or your regional office of the federal EPA for specific recommendations. If employees are required to clean-up spills, they must be properly trained and equipped. OSHA 1910.120(q) may be applicable.

Fire Extinguishing: This chemical is a flammable liquid. Poisonous gases including carbon monoxide and hydrogen bromide are produced in fire. Use dry chemical, carbon dioxide, or alcohol or polymer foam extinguishers. Vapors are heavier than air and will collect in low areas. Vapors may travel long distances to ignition sources and flashback. Vapors in confined areas may explode when exposed to fire. Containers may explode in fire. Storage containers and parts of containers may rocket great distances, in many directions. If material or contaminated runoff enters waterways, notify downstream users of potentially contaminated waters. Notify local health and fire officials and pollution control agencies. From a secure, explosion-proof location, use water spray to cool exposed containers. If cooling streams are ineffective (venting sound increases in volume and pitch, tank discolors, or shows any signs of deforming), withdraw immediately to a secure position. If employees are expected to fight fires, they must be trained and equipped in OSHA 1910.156.

References

New Jersey Department of Health and Senior Services, "Hazardous Substance Fact Sheet, Ethyl Bromoacetate," Trenton, NJ (April 1999)

Ethylbutanol

Molecular Formula: $C_6H_{14}O$

Synonyms: 2-Ethyl-1-butanol; 2-Ethylbutanol-1; 2-Ethylbutyl alcohol; *sec*-Hexanol; *sec*-Hexyl alcohol; 3-Methylolpentane; *sec*-Pentylcarbinol; 3-Pentylcarbinol; Pseudohexyl alcohol

CAS Registry Number: 97-95-0

RTECS Number: EL3850000

DOT ID: UN 2275

Cited in U.S. State Regulations: California (G), New Jersey (G), Pennsylvania (G).

Description: Ethylbutanol is a colorless liquid with a mild, alcoholic odor. Boiling point = 149°C. Freezing point = -114°C. Flash point = 58°C (oc). Hazard Identification (based on NFPA-704 M Rating System): Health 1, Flammability 2, Reactivity 0. Slightly soluble in water.

Potential Exposure: Used as a solvent, for making dyes, perfumes, flavorings, and drugs.

Incompatibilities: Forms explosive mixture with air. Incompatible with oxidizers, strong acids, caustics, isocyanates, amines, isocyanates.

Permissible Exposure Limits in Air: No OELs have been established.

Routes of Entry: Inhalation, eyes, through the skin.

Harmful Effects and Symptoms

Short Term Exposure: Contact can cause skin irritation. Eye contact can cause severe irritation and burns; possible permanent damage. Inhalation can cause irritation of the respiratory tract causing coughing and wheezing. Exposure can cause headache, dizziness, nausea, and vomiting.

Long Term Exposure: Unknown at this time.

First Aid: If this chemical gets into the eyes, remove any contact lenses at once and irrigate immediately for at least 15 minutes, occasionally lifting upper and lower lids. Seek medical attention immediately. If this chemical contacts the skin, remove contaminated clothing and wash immediately with soap and water. Seek medical attention immediately. If this chemical has been inhaled, remove from exposure, begin rescue breathing (using universal precautions) if breathing has stopped and CPR if heart action has stopped. Transfer promptly to a medical facility. When this chemical has been swallowed, get medical attention. Give large quantities of water and induce vomiting. Do not make an unconscious person vomit.

Personal Protective Methods: Wear protective gloves and clothing to prevent any reasonable probability of skin contact. Safety equipment suppliers/manufacturers can provide recommendations on the most protective glove/clothing material for your operation.. All protective clothing (suits, gloves, footwear, headgear) should be clean, available each day, and put on before work. Contact lenses should not be worn when working with this chemical. Wear splash-proof chemical goggles and face shield unless full facepiece respiratory protection is worn. Employees should wash immediately with soap when skin is wet or contaminated. Provide emergency showers and eyewash.

Respirator Selection: Where the potential for exposure to this chemical, use a MSHA/NIOSH approved supplied-air respirator with a full facepiece operated in the positive pressure mode or with a full facepiece, hood, or helmet in the continuous flow mode, or use a MSHA/NIOSH approved self-contained breathing apparatus with a full facepiece operated in pressure-demand or other positive pressure mode.

Storage: Prior to working with this chemical you should be trained on its proper handling and storage. Store in tightly closed containers in a cool, well ventilated area away from oxidizers, strong acids, strong bases, reducing agents, heat, and sources of ignition. Where possible, automatically pump liquid from drums or other storage containers to process containers

Shipping: Label required is "Flammable Liquid". Ethylbutanol is in DOT/UN Hazard Class 3 and Packing Group III.[19][20]

Spill Handling: Evacuate and restrict persons not wearing protective equipment from area of spill or leak until cleanup is complete. Remove all ignition sources. Ventilate area of spill or leak. Absorb liquids in vermiculite, dry sand, earth, peat, carbon, or a similar material and deposit in sealed containers. Oil-skimming equipment may be used to remove slicks from water. Keep this chemical out of a confined space, such as a sewer, because of the possibility of an explosion, unless the sewer is designed to prevent the build-up of explosive concentrations. It may be necessary to contain and dispose of this chemical as a hazardous waste. If material or contaminated runoff enters waterways, notify downstream users of potentially contaminated waters. Contact your Department of Environmental Protection or your regional office of the federal EPA for specific recommendations. If employees are required to clean-up spills, they must be properly trained and equipped. OSHA 1910.120(q) may be applicable.

Fire Extinguishing: This chemical is a flammable liquid. Poisonous gases are produced in fire. Use dry chemical, carbon dioxide, or alcohol or polymer foam extinguishers. Vapors are heavier than air and will collect in low areas. Vapors may travel long distances to ignition sources and flashback. Vapors in confined areas may explode when exposed to fire. Containers may explode in fire. Storage containers and parts of containers may rocket great distances, in many directions. If material or contaminated runoff enters waterways, notify downstream users of potentially contaminated waters. Notify local health and fire officials and pollution control agencies. From a secure, explosion-proof location, use water spray to cool exposed containers. If cooling streams are ineffective (venting sound increases in volume and pitch, tank discolors, or shows any signs of deforming), withdraw immediately to a secure position. If employees are expected to fight fires, they must be trained and equipped in OSHA 1910.156.

Disposal Method Suggested: Incineration by spraying or in paper packaging. Flammable solvent may be added.

References

New Jersey Department of Health and Senior Services, "Hazardous Substance Fact Sheet, Ethylbutanol," Trenton, NJ (March 1999)

Ethyl Butyl Ether

Molecular Formula: $C_6H_{14}O$

Synonyms: Butane, 1-ethoxy-; Butyl ethyl ether; Ether ethylbutylique (French); Ethyl-*n*-butyl ether

CAS Registry Number: 628-81-9

RTECS Number: KN4725000

DOT ID: UN 1179

Cited in U.S. State Regulations: California (G), New Jersey (G), Pennsylvania (G).

Description: Ethyl butyl ether is a colorless liquid. Boiling point = 92°C. Freezing/Melting point = -124°C. Flash point = 4°C. Hazard Identification (based on NFPA-704 M Rating System): Health 2, Flammability 3, Reactivity 0. Slightly soluble in water.

Potential Exposure: Used as a solvent for extraction and in making other chemicals

Incompatibilities: Forms explosive mixture with air. Heat or prolonged storage may cause the formation of unstable peroxides. Violent reaction with strong oxidizers. Attacks some plastics, rubber and coatings. May accumulate static electrical charges, and may cause ignition of its vapors.

Permissible Exposure Limits in Air: No OELs established.

Routes of Entry: Inhalation, ingestion, eyes, through the skin.

Harmful Effects and Symptoms

Short Term Exposure: Irritates the eyes, skin and respiratory tract with coughing and wheezing. High levels can cause dizziness and loss of consciousness.

Long Term Exposure: Repeated or high exposures may affect the nervous system. Repeated exposure may remove the oils from the skin causing dryness, rash or cracking.

Points of Attack: Skin, nervous system.

Medical Surveillance: Test the nervous system.

First Aid: If this chemical gets into the eyes, remove any contact lenses at once and irrigate immediately for at least 15 minutes, occasionally lifting upper and lower lids. Seek medical attention immediately. If this chemical contacts the skin, remove contaminated clothing and wash immediately with soap and water. Seek medical attention immediately. If this chemical has been inhaled, remove from exposure, begin rescue breathing (using universal precautions) if breathing has stopped and CPR if heart action has stopped. Transfer promptly to a medical facility. When this chemical has been swallowed, get medical attention. Give large quantities of water and induce vomiting. Do not make an unconscious person vomit.

Personal Protective Methods: Wear solvent-resistant gloves and clothing to prevent any reasonable probability of skin contact. Safety equipment suppliers/manufacturers can provide recommendations on the most protective glove/

clothing material for your operation.. All protective clothing (suits, gloves, footwear, headgear) should be clean, available each day, and put on before work. Contact lenses should not be worn when working with this chemical. Wear splash-proof chemical goggles and face shield unless full facepiece respiratory protection is worn. Employees should wash immediately with soap when skin is wet or contaminated. Provide emergency showers and eyewash.

Respirator Selection: Where the potential for exposure to this chemical, use a MSHA/NIOSH approved supplied-air respirator with a full facepiece operated in the positive pressure mode or with a full facepiece, hood, or helmet in the continuous flow mode, or use a MSHA/NIOSH approved self-contained breathing apparatus with a full facepiece operated in pressure-demand or other positive pressure mode.

Storage: Prior to working with this chemical you should be trained on its proper handling and storage. Store in tightly closed containers in a cool, well ventilated area away from oxidizers, heat and sources of ignition. Where possible, automatically pump liquid from drums or other storage containers to process containers. Drums must be equipped with self-closing valves, pressure vacuum bungs, and flame arresters. Use only non-sparking tools and equipment, especially when opening and closing containers of this chemical. Sources of ignition such as smoking and open flames, are prohibited where this chemical is used, handled, or stored in a manner that could create a potential fire or explosion hazard. Wherever this chemical is used, handled, manufactured, or stored, use explosion-proof electrical equipment and fittings.

Shipping: Label required is "Flammable Liquid". Ethyl butyl ether is in DOT/UN Hazard Class 3 and Packing Group II.[19][20]

Spill Handling: Evacuate and restrict persons not wearing protective equipment from area of spill or leak until cleanup is complete. Remove all ignition sources. Ventilate area of spill or leak. Absorb liquids in vermiculite, dry sand, earth, peat, carbon, or a similar material and deposit in sealed containers. It may be necessary to contain and dispose of this chemical as a hazardous waste. If material or contaminated runoff enters waterways, notify downstream users of potentially contaminated waters. Contact your Department of Environmental Protection or your regional office of the federal EPA for specific recommendations. If employees are required to clean-up spills, they must be properly trained and equipped. OSHA 1910.120(q) may be applicable.

Fire Extinguishing: This chemical is a flammable liquid. Poisonous gases including carbon monoxide are produced in fire. Use dry chemical, carbon dioxide, or alcohol foam extinguishers. Vapors are heavier than air and will collect in low areas. Vapors may travel long distances to ignition sources and flashback. Vapors in confined areas may explode when exposed to fire. Containers may explode in fire. Storage containers and parts of containers may rocket great distances, in many directions. If material or contaminated runoff enters waterways, notify downstream users of potentially contaminated waters. Notify local health and fire officials and pollution control agencies. From a secure, explosion-proof location, use water spray to cool exposed containers. If cooling streams are ineffective (venting sound increases in volume and pitch, tank discolors, or shows any signs of deforming), withdraw immediately to a secure position. If employees are expected to fight fires, they must be trained and equipped in OSHA 1910.156.

Disposal Method Suggested: Incineration.

References

New Jersey Department of Health and Senior Services, "Hazardous Substance Fact Sheet, Ethylbutyl Ether," Trenton, NJ (March, 1999)

Ethyl Butyl Ketone

Molecular Formula: $C_7H_{14}O$

Common Formula: $C_2H_5COC_4H_9$

Synonyms: *n*-Butyl ethyl ketone; Butyl ethyl ketone; 3-Heptanone; Heptan-3-one

CAS Registry Number: 106-35-4

RTECS Number: MJ5250000

DOT ID: UN 1224

EEC Number: 606-003-00-9

Regulatory Authority

- Air Pollutant Standard Set (ACGIH)[1] (HSE)[33] (OPSHA)[58] (Several States)[60]
- Canada, WHMIS, Ingredients Disclosure List

Cited in U.S. State Regulations: Alaska (G), California (A, G), Connecticut (A), Florida (G), Illinois (G), Maine (G), Massachusetts (G), Nevada (A), New Hampshire (G, W), New Jersey (G), North Dakota (A), Pennsylvania (G), Rhode Island (G), Virginia (A), West Virginia (G).

Description: Ethyl butyl ketone, $C_2H_5COC_4H_9$, is a colorless liquid with a powerful, fruity odor. Boiling point = 147 – 148°C. Freezing/Melting point = -39°C. Flash point = 46°C. Hazard Identification (based on NFPA-704 M Rating System): Health 1, Flammability 2, Reactivity 0. Insoluble in water.

Potential Exposure: Ethyl butyl ketone is used as a solvent and as an intermediate in organic synthesis. It is a solvent for vinyl and nitrocellulose resins. It is used in food flavoring.

Incompatibilities: Forms explosive mixture with air Violent reaction with strong oxidizers, acetaldehyde, perchloric acid. Attacks some plastics, rubber and coatings.

Permissible Exposure Limits in Air: The OSHA TWA,[58] HSE value, and the ACGIH TWA value is 50 pm (230 mg/m^3). The HSE STEL is 75 ppm (345 mg/m^3).[33] The NIOSH IDLH level is 1,000 ppm. Several states have set guidelines or standards for ethyl butyl ketone in ambient air[60] ranging from

2.3 – 3.45 mg/m³ (North Dakota) to 3.8 mg/m³ (Virginia) to 4.6 mg/m³ (Connecticut) to 5.476 mg/m³ (Nevada).

Determination in Air: Charcoal adsorption, workup with CS_2, analysis by gas chromatography/flame ionization. See NIOSH IV, Method #1301, Ketones II.[18]

Permissible Concentration in Water: No criteria set.

Routes of Entry: Inhalation, ingestion, skin and/or eye contact.

Harmful Effects and Symptoms

Short Term Exposure: Ethyl butyl ketone irritates the eyes, skin, and respiratory tract. Exposure to high concentrations can affect the central nervous system, cause dizziness, lightheadedness, and to lose consciousness.

Long Term Exposure: Repeated or prolonged contact with skin may cause dryness and cracking.

Points of Attack: Eyes, skin, respiratory system, central nervous system.

Medical Surveillance: Consider the points of attack in preplacement and periodic physical examinations.

First Aid: If this chemical gets into the eyes, remove any contact lenses at once and irrigate immediately for at least 15 minutes, occasionally lifting upper and lower lids. Seek medical attention immediately. If this chemical contacts the skin, remove contaminated clothing and wash immediately with soap and water. Seek medical attention immediately. If this chemical has been inhaled, remove from exposure, begin rescue breathing (using universal precautions) if breathing has stopped and CPR if heart action has stopped. Transfer promptly to a medical facility. When this chemical has been swallowed, get medical attention. Give large quantities of water and induce vomiting. Do not make an unconscious person vomit.

Personal Protective Methods: Wear protective gloves and clothing to prevent any reasonable probability of skin contact. Safety equipment suppliers/manufacturers can provide recommendations on the most protective glove/clothing material for your operation.. All protective clothing (suits, gloves, footwear, headgear) should be clean, available each day, and put on before work. Contact lenses should not be worn when working with this chemical. Wear splash-proof chemical goggles and face shield unless full facepiece respiratory protection is worn. Employees should wash immediately with soap when skin is wet or contaminated. Provide emergency showers and eyewash. Repeated or prolonged skin contact. Wear eye protection to prevent any reasonable probability of eye contact. Employees should wash promptly when skin is wet or contaminated. Remove nonimpervious clothing promptly if wet or contaminated.

Respirator Selection: NIOSH/OSHA: *500 ppm:* CCROV [any chemical cartridge respirator with organic vapor cartridge(s)]; or SA (any supplied-air respirator). *1,000 ppm:* SA:CF (any supplied-air respirator operated in a continuous-flow mode); or PAPROV [any powered, air-purifying respirator with organic vapor cartridge(s)]; or CCRFOV [any air-purifying, full-facepiece respirator (gas mask) with a chin-style, front- or back-mounted acid gas canister]; or GMFOV [any air-purifying, full-facepiece respirator (gas mask) with a chin-style, front- or back-mounted acid gas canister]; or SCBAF (any self-contained breathing apparatus with a full facepiece); or SAF (any supplied-air respirator with a full facepiece). *Emergency or planned entry into unknown concentrations or IDLH conditions:* SCBAF:PD,PP (any self-contained breathing apparatus that has a full facepiece and is operated in a pressure-demand or other positive-pressure mode); or SAF:PD,PP:ASCBA (any supplied-air respirator that has a full facepiece and is operated in a pressure-demand or other positive-pressure mode in combination with an auxiliary self-contained breathing apparatus operated in a pressure-demand or other positive pressure mode). *Escape:* GMFOV [any air-purifying, full-facepiece respirator (gas mask) with a chin-style, front-or back-, mounted organic vapor canister]; or SCBAE (any appropriate escape-type, self-contained breathing apparatus).

Note: Substance reported to cause eye irritation or damage; may require eye protection.

Storage: Prior to working with this chemical you should be trained on its proper handling and storage. Ethyl butyl ketone must be stored to avoid contact with oxidizers such as peroxides, chlorates, perchlorates, permanganates, and nitrates, because violent reactions occur. Store in tightly closed containers in a cool, well-ventilated area away from heat. Sources of ignition such as smoking and open flames are prohibited where ethyl butyl ketone is used, handled, or stored in a manner that could create a potential fire or explosion hazard.

Shipping: Ketones, liquid, n.o.s. fall in Hazard Class 3 and ethyl butyl ketone in Packing Group II. The label required is "Flammable Liquid."

Spill Handling: Evacuate and restrict persons not wearing protective equipment from area of spill or leak until cleanup is complete. Remove all ignition sources. Ventilate area of spill or leak. Absorb liquids in vermiculite, dry sand, earth, peat, carbon, or a similar material and deposit in sealed containers. Keep ethyl butyl ketone out of a confined space, such as a sewer, because of the possibility of an explosion, unless the sewer is designed to prevent the build-up off explosive concentrations. It may be necessary to contain and dispose of this chemical as a hazardous waste. If material or contaminated runoff enters waterways, notify downstream users of potentially contaminated waters. Contact your Department of Environmental Protection or your regional office of the federal EPA for specific recommendations. If employees are required to clean-up spills, they must be properly trained and equipped. OSHA 1910.120(q) may be applicable.

Fire Extinguishing: This chemical is a combustible liquid. Poisonous gases are produced in fire. Use dry chemical, carbon dioxide, or foam extinguishers. Vapors are heavier than air and will collect in low areas. Vapors may travel long distances to ignition sources and flashback. Vapors in confined areas may explode when exposed to fire. Containers may explode in fire. Storage containers and parts of containers may rocket great distances, in many directions. If material or contaminated runoff enters waterways, notify downstream users of potentially contaminated waters. Notify local health and fire officials and pollution control agencies. From a secure, explosion-proof location, use water spray to cool exposed containers. If cooling streams are ineffective (venting sound increases in volume and pitch, tank discolors, or shows any signs of deforming), withdraw immediately to a secure position. If employees are expected to fight fires, they must be trained and equipped in OSHA 1910.156.

Disposal Method Suggested: Incineration.

References

National Institute for Occupational Safety and Health, Criteria for a Recommended Standard: Occupational Exposure to Ketones, NIOSH Doc. No. 78–173, Washington, DC (1978)

New Jersey Department of Health and Senior Services, "Hazardous Substance Fact Sheet: Ethyl Butyl Ketone," Trenton, NJ (February 1996)

Ethyl Butyraldehyde

Molecular Formula: $C_6H_{12}O$

Common Formula: $(C_2H_5)_2CHCHO$

Synonyms: Aldehyde-2-ethylbutyrique (French); Butyraldehyde, 2-ethyl-; Diethyl acetaldehyde; 2-Ethylbutanal; α-Ethylbutyraldehyde; 2-Ethylbutyraldehyde; Ethylbutyraldehyde; 2-Ethylbutyric aldehyde

CAS Registry Number: 97-96-1

RTECS Number: ES2625000

DOT ID: UN 1178

Cited in U.S. State Regulations: Maine (G), Massachusetts (G), New Jersey (G), Pennsylvania (G).

Description: Ethyl butyraldehyde, $(C_2H_5)_2CHCHO$, is a colorless liquid. Boiling point = 116.8°C. Flash point = 21°C (oc). Explosive limits: LEL = 1.2%; UEL = 7.7%. Hazard Identification (based on NFPA-704 M Rating System): Health 2, Flammability 3, Reactivity 1. Insoluble in water.

Potential Exposure: Involved in use in organic synthesis of pharmaceuticals, rubber chemicals.

Incompatibilities: Strong oxidizers, strong bases, reducing agents.

Permissible Exposure Limits in Air: No standards set.

Permissible Concentration in Water: No criteria set.

Routes of Entry: Inhalation.

Harmful Effects and Symptoms

Short Term Exposure: Irritates the skin and eyes on contact. Inhalation irritates the respiratory tract causing coughing and wheezing.

Long Term Exposure: Unknown at this time.

First Aid: If this chemical gets into the eyes, remove any contact lenses at once and irrigate immediately for at least 15 minutes, occasionally lifting upper and lower lids. Seek medical attention immediately. If this chemical contacts the skin, remove contaminated clothing and wash immediately with soap and water. Seek medical attention immediately. If this chemical has been inhaled, remove from exposure, begin rescue breathing (using universal precautions) if breathing has stopped and CPR if heart action has stopped. Transfer promptly to a medical facility. When this chemical has been swallowed, get medical attention. Give large quantities of water and induce vomiting. Do not make an unconscious person vomit.

Personal Protective Methods: Wear protective gloves and clothing to prevent any reasonable probability of skin contact. Safety equipment suppliers/manufacturers can provide recommendations on the most protective glove/clothing material for your operation.. All protective clothing (suits, gloves, footwear, headgear) should be clean, available each day, and put on before work. Contact lenses should not be worn when working with this chemical. Wear splash-proof chemical goggles and face shield unless full facepiece respiratory protection is worn. Employees should wash immediately with soap when skin is wet or contaminated. Provide emergency showers and eyewash.

Respirator Selection: Where the potential for exposure to this chemical, use a MSHA/NIOSH approved supplied-air respirator with a full facepiece operated in the positive pressure mode or with a full facepiece, hood, or helmet in the continuous flow mode, or use a MSHA/NIOSH approved self-contained breathing apparatus with a full facepiece operated in pressure-demand or other positive pressure mode.

Storage: Prior to working with this chemical you should be trained on its proper handling and storage. Before entering confined space where this chemical may be present, check to make sure that an explosive concentration does not exist. Store in tightly closed containers in a cool, well ventilated area away from oxidizers. Where possible, automatically pump liquid from drums or other storage containers to process containers.

Shipping: This compound requires a shipping label of: "Flammable Liquid." It falls in DOT Hazard Class 3 and Packing Group II.

Spill Handling: Evacuate and restrict persons not wearing protective equipment from area of spill or leak until cleanup is complete. Remove all ignition sources. Ventilate area of spill or leak. Absorb liquids in vermiculite, dry sand, earth, peat, carbon, polyurethane foams, or a similar material and deposit in sealed containers. Keep ethyl butyraldehyde out of a confined space, such as a sewer, because of the possibility of an explosion, unless the sewer is designed to prevent the build-up off explosive concentrations. It may be necessary to contain and dispose of this chemical as a hazardous waste. If material or contaminated runoff enters waterways, notify downstream users of potentially contaminated waters. Contact your Department of Environmental Protection or your regional office of the federal EPA for specific recommendations. If employees are required to clean-up spills, they must be properly trained and equipped. OSHA 1910.120(q) may be applicable.

Fire Extinguishing: This chemical is a flammable liquid. Poisonous gases are produced in fire. Use dry chemical, carbon dioxide, or alcohol foam extinguishers. Water may be ineffective. Vapors are heavier than air and will collect in low areas. Vapors may travel long distances to ignition sources and flashback. Vapors in confined areas may explode when exposed to fire. Containers may explode in fire. Storage containers and parts of containers may rocket great distances, in many directions. If material or contaminated runoff enters waterways, notify downstream users of potentially contaminated waters. Notify local health and fire officials and pollution control agencies. From a secure, explosion-proof location, use water spray to cool exposed containers. If cooling streams are ineffective (venting sound increases in volume and pitch, tank discolors, or shows any signs of deforming), withdraw immediately to a secure position. If employees are expected to fight fires, they must be trained and equipped in OSHA 1910.156.

Disposal Method Suggested: Incineration.

References

Sax, N. I., Ed., "Dangerous Properties of Industrial Materials Report," 1, No. 8, 69–71 (1981) and 3, No 2, 85–87 (1983)

New Jersey Department of Health and Senior Services, "Hazardous Substance Fact Sheet, Ethylbutyraldehyde," Trenton, NJ (March 1999)

Ethyl Butyrate

Molecular Formula: $C_6H_{12}O_2$

Common Formula: $C_3H_7COOC_2H_5$

Synonyms: Butonic acid ethyl ester; Butyric acid, ethyl ester; Butyric ether; Ethyl butanoate; Ethyl *n*-butyrate

CAS Registry Number: 105-54-4

RTECS Number: ET1660000

DOT ID: UN 1180

Cited in U.S. State Regulations: Florida (G), Maine (G), Massachusetts (G), New Hampshire (G), New Jersey (G), Oklahoma (G), Pennsylvania (G).

Description: Ethyl butyrate, $C_3H_7COOC_2H_5$, is a colorless liquid with a pineapple-like odor. The odor threshold is 0.015 ppm[41]. Boiling point = 120°C. Flash point = 24°C. Autoignition temperature =463°C. Hazard Identification (based on NFPA-704 M Rating System): Health 0, Flammability 3, Reactivity 0. Insoluble in water.

Potential Exposure: Ethyl butyrate is used in flavorings, extracts, perfumery, and as a solvent.

Incompatibilities: Forms explosive mixture with air. Incompatible with strong oxidizers, strong acids, strong bases, heat.

Permissible Exposure Limits in Air: No standards set.

Permissible Concentration in Water: No criteria set.

Routes of Entry: Inhalation. Passes through the skin.

Harmful Effects and Symptoms

Short Term Exposure: Ethyl butyrate can affect you when breathed in and by passing through your skin. Contact can irritate the skin and eyes. The vapor irritates the eyes, nose, throat and lungs. Symptoms may include nosebleeds, sore throat, cough with phlegm, and/or difficulty breathing. Overexposure may cause headaches and make you feel dizzy and lightheaded. Higher levels can make you pass out.

Long Term Exposure: Unknown at this time.

Points of Attack: Lungs.

Medical Surveillance: If symptoms develop or overexposure is suspected, the following may be useful: Consider lung function tests, especially with irritation symptoms. Consider live function tests after suspected overexposure.

First Aid: If this chemical gets into the eyes, remove any contact lenses at once and irrigate immediately for at least 15 minutes, occasionally lifting upper and lower lids. Seek medical attention immediately. If this chemical contacts the skin, remove contaminated clothing and wash immediately with soap and water. Seek medical attention immediately. If this chemical has been inhaled, remove from exposure, begin rescue breathing (using universal precautions) if breathing has stopped and CPR if heart action has stopped. Transfer promptly to a medical facility. When this chemical has been swallowed, get medical attention. Give large quantities of water and induce vomiting. Do not make an unconscious person vomit.

Personal Protective Methods: Wear protective gloves and clothing to prevent any reasonable probability of skin contact. Safety equipment suppliers/manufacturers can provide recommendations on the most protective glove/clothing material for your operation.. All protective clothing (suits, gloves, footwear, headgear) should be clean, available each day, and put on before work. Contact lenses should not be worn when working with this chemical. Wear splash-proof chemical goggles and face shield unless full facepiece respiratory protection is worn. Employees should wash immediately with soap when skin is wet or contaminated. Provide emergency showers and eyewash.

Respirator Selection: Where the potential for exposure to ethyl butyrate exists, use a MSHA/NIOSH approved supplied-air respirator with a full facepiece operated in the positive pressure mode or with a full facepiece, hood, or helmet in the continuous flow mode, or use a MSHA/NIOSH approved self-contained breathing apparatus with a full facepiece operated in pressure-demand or other positive pressure mode.

Storage: Prior to working with this chemical you should be trained on its proper handling and storage. Ethyl butyrate is incompatible with oxidizers (such as perchlorates, peroxides, permanganates, chlorates and nitrates), strong acids (such as hydrochloric, sulfuric and nitric), bases, and heat. Store in tightly closed containers in a cool, well-ventilated area. Sources of ignition, such as smoking and open flames, are prohibited where ethyl butyrate is handled, used, or stored. Metal containers involving the transfer of 5 gallons or more of ethyl butyrate should be grounded and bonded. Drums must be equipped with self-closing valves, pressure vacuum bungs, and flame arresters. Use only non-sparking tools and equipment, especially when opening and closing containers of ethyl butyrate.

Shipping: This compound requires a shipping label of: "Flammable Liquid." It falls in DOT Hazard Class 3 and Packing Group II.

Spill Handling: Evacuate and restrict persons not wearing protective equipment from area of spill or leak until cleanup is complete. Remove all ignition sources. Ventilate area of spill. Cover spill with activated carbon adsorbent, take up and deposit in sealed containers. Keep ethyl butyrate out of a confined space, such as a sewer, because of the possibility of an explosion, unless the sewer is designed to prevent the build-up off explosive concentrations. It may be necessary to contain and dispose of this chemical as a hazardous waste. If material or contaminated runoff enters waterways, notify downstream users of potentially contaminated waters. Contact your Department of Environmental Protection or your regional office of the federal EPA for specific recommendations. If employees are required to clean-up spills, they must be properly trained and equipped. OSHA 1910.120(q) may be applicable.

Fire Extinguishing: This chemical is a flammable liquid. Poisonous gases including carbon monoxide are produced in fire. Use dry chemical, carbon dioxide, or alcohol foam extinguishers. Vapors are heavier than air and will collect in low areas. Vapors may travel long distances to ignition sources and flashback. Vapors in confined areas may explode when exposed to fire. Containers may explode in fire. Storage containers and parts of containers may rocket great distances, in many directions. If material or contaminated runoff enters waterways, notify downstream users of potentially contaminated waters. Notify local health and fire officials and pollution control agencies. From a secure, explosion-proof location, use water spray to cool exposed containers. If cooling streams are ineffective (venting sound increases in volume and pitch, tank discolors, or shows any signs of deforming), withdraw immediately to a secure position. If employees are expected to fight fires, they must be trained and equipped in OSHA 1910.156.

Disposal Method Suggested: Incineration.

References

New Jersey Department of Health and Senior Services, "Hazardous Substance Fact Sheet: Ethyl Butyrate," Trenton, NJ (February 9, 1988)

Ethyl Chloride

Molecular Formula: C_2H_5Cl

Common Formula: CH_3CH_2Cl

Synonyms: Aethylchlorid (German); Aethylis; Aethylis chloridum; Anesthetic chloryl; Anodynon; Chelen; Chloroethaan (Dutch); Chlorene; Chlorethyl; Chloridum; Chlorure d'ethyle (French); Chloryl; Chloryl anesthetic; Cloretilo; Cloroetano (Italian); Cloroetano (Spanish); Clorudo di etile (Italian); Dublofix; Ethane, chloro-; Ether chloratus; Ether hydrochloric; Ether muriatic; Etylu chlorek (Polish); Hydrochloric ether; Kelene; Monochlorethane; Monochloroethane; Muriatic ether; Narcotile; NCI-C06224

CAS Registry Number: 75-00-3

RTECS Number: KH7525000

DOT ID: UN 1037

EEC Number: 602-009-00-0

Regulatory Authority

- Air Pollutant Standard Set (ACGIH)[1] (DFG)[3] (HSE)[33] (Other Countries)[35] (OSHA)[58] (Several States)[60] (Several Canadian Provinces)
- Clean Air Act: Hazardous Air Pollutants (Title I, Part A, Section 112); Accidental Release Prevention/Flammable substances, (Section 112[r], Table 3), TQ = 10,000 lb (4,540 kg)
- Clean Water Act: 40CFR423, Appendix A, Priority Pollutants; Section 313 Water Priority Chemicals (57FR41331, 9/9/92); 40CFR401.15 Section 307 Toxic Pollutants as chlorinated ethanes
- RCRA 40CFR268.48; 61FR15654, Universal Treatment Standards: Wastewater (mg/l), 0.27; Nonwastewater (mg/kg), 6.0
- RCRA 40CFR264, Appendix 9; TSD Facilities Ground Water Monitoring List. Suggested test method(s) (PQL μg/l): 8010 (5); 8240 (10)
- Safe Drinking Water Act: Priority List (55 FR 1470)
- Superfund/EPCRA 40CFR302.4 Reportable Quantity (RQ): CERCLA, 100 lb (45.4 kg)
- EPCRA Section 313 Form R *de minimis* concentration reporting level: 1.0%
- Canada, WHMIS, Ingredients Disclosure List

Cited in U.S. State Regulations: Alaska (G), California (A, G), Connecticut (A), Florida (G, A), Illinois (G), Maine (G), Massachusetts (G, A), Nevada (A), New Hampshire (G), New Jersey (G), New York (A), North Dakota (A), Oklahoma (G), Pennsylvania (G), Rhode Island (G), Virginia (A), West Virginia (G).

Description: Ethyl chloride, CH_3CH_2Cl, is a colorless gas or liquid (below 12°C) with a pungent, ether-like odor and a burning taste. Shipped as a liquefied compressed gas. Boiling point = 12 – 13°C. Flash point = -50°C. Autoignition temperature = 519°C. Explosive limits: LEL = 3.8%; UEL = 15.4%. Hazard Identification (based on NFPA-704 M Rating System): Health 1, Flammability 4, Reactivity 0. Slightly soluble in water.

Potential Exposure: Ethyl chloride is used as an ethylating agent in the manufacture of tetraethyl lead, dyes, drugs, and ethyl cellulose. It can be used as a refrigerant and as a local anesthetic (freezing).

Incompatibilities: Flammable gas. Slow reaction with water; forms hydrogen chloride gas. Contact with moisture (water, steam) forms hydrochloric acid and/or fumes of hydrogen chloride. May accumulate static electrical charges, and may cause ignition of its vapors. Forms explosive mixture with air. Contact with chemically active metals: aluminum, lithium, magnesium, sodium, potassium, zinc may cause fire and explosions. Attacks some plastics and rubber.

Permissible Exposure Limits in Air: The Federal limit [OSHA PEL],[58] the DFG MAK,[3] the HSE TWA[33] and the ACGIH TWA value is 1,000 ppm (2,600 mg/m^3). The HSE STEL value is 1,250 ppm (3,250 mg/m^3). The NIOSH IDLH level is 3,800 ppm [10%LEL]. Japan[35] has also set 1,000 ppm as the TWA; Sweden, however has set 500 mg/m^3 (1,300 mg/m^3) as the TWA and 1,900 mg/m^3 as the STEL. Brazil[35] has set 780 ppm as the TWA (2,030 mg/m^3). The former USSR[35][43] has set a MAC in workplace air of 50 mg/m^3.

Limits on ambient air in residential areas have been set by the former USSR[35] at 0.2 mg/m^3 on a daily average basis and by Czechoslovakia at 0.1 mg/m^3 on a daily average basis. Several states have set guidelines or standards for ethyl chloride in ambient air[60] ranging from zero (Virginia) to 0.36 mg/m^3 (Massachusetts) to 26.0 mg/m^3 (North Dakota) to 52.0 mg/m^3 (Connecticut, Florida, New York) to 61.88 mg/m^3 (Nevada).

Determination in Air: Adsorption on charcoal, workup with CS_2, gas chromatographic analysis/flame ionization. See NIOSH Method 2519.[18]

Permissible Concentration in Water: No criteria set due to volatility and low specific gravity.

Routes of Entry: Inhalation of gas, slight percutaneous absorption, ingestion, skin and/or eye contact.

Harmful Effects and Symptoms

Short Term Exposure: Ethyl chloride irritates the eyes, skin, and respiratory tract. Inhalation of vapor may have a narcotic effect. Skin contact with the liquid may cause frostbite. Ethyl chloride exposure may produce headache, dizziness, lack of coordination, stomach cramps, and eventual loss of consciousness. In high concentrations, it is a respiratory tract irritant, and death due to cardiac arrest has been recorded. Renal damage has been reported in animals.

Long Term Exposure: There is limited evidence the ethyl chloride causes skin or uterine cancer in animals. May affect the nervous system, liver, and kidneys.

Points of Attack: Liver, kidneys, respiratory system, cardiovascular system.

Medical Surveillance: If symptoms develop or overexposure I suspected, the following may be useful: Special 24-hour EKG (Holter monitor) to look for irregular heart beat. Tests of liver and kidney function.

First Aid: If this chemical gets into the eyes, remove any contact lenses at once and irrigate immediately for at least 15 minutes, occasionally lifting upper and lower lids. Seek medical attention immediately. If this chemical contacts the skin, remove contaminated clothing and wash immediately with soap and water. Seek medical attention immediately. If this chemical has been inhaled, remove from exposure, begin rescue breathing (using universal precautions) if breathing has stopped and CPR if heart action has stopped. Transfer promptly to a medical facility. When this chemical has been swallowed, get medical attention. Give large quantities of water and induce vomiting. Do not make an unconscious person vomit.

Personal Protective Methods: Wear protective gloves and clothing to prevent any reasonable probability of skin contact. Safety equipment suppliers/manufacturers can provide recommendations on the most protective glove/clothing material for your operation.. All protective clothing (suits, gloves, footwear, headgear) should be clean, available each day, and put on before work. Contact lenses should not be worn when working with this chemical. Wear splash-proof chemical goggles and face shield unless full facepiece respiratory protection is worn. Employees should wash immediately with soap when skin is wet or contaminated. Provide emergency showers and eyewash.

Respirator Selection: OSHA: *up to 3,800 ppm:* SA (any supplied-air respirator);* or SCBAF (any self-contained breathing apparatus with a full facepiece). *Emergency or planned entry into unknown concentrations or IDLH conditions*: SCBAF:PD,PP (any self-contained breathing apparatus that has a full facepiece and is operated in a pressure-demand or other positive-pressure mode); or SAF:PD,PP:ASCBA (any supplied-air respirator that has a full facepiece and is operated in a pressure-demand or other positive-pressure mode in combination with an auxiliary self-contained breathing apparatus operated in a pressure-demand or other positive-pressure mode). *Escape:* GMFOV [any air-purifying, full-facepiece respirator (gas mask) with a chin-style, front-or back-mounted organic vapor canister]; or SCBAE (any appropriate escape-type, self-contained breathing apparatus).

* Substance reported to cause eye irritation or damage; may require eye protection.

Storage: Prior to working with this chemical you should be trained on its proper handling and storage. Before entering confined space where this chemical may be present, check to make sure that an explosive concentration does not exist. Ethyl chloride must be stored to avoid contact with oxidizers (such as peroxides, chlorates, perchlorates, nitrates, and permanganates) or chemically active metals (such as sodium, potassium, calcium, powdered aluminum, zinc, and magnesium) because violent reactions occur. Store in tightly closed containers in a cool, well-ventilated area away from heat. Sources of ignition such as smoking and open flames are prohibited where ethyl chloride is handled, used, or stored. Metal containers used in the transfer of 5 gallons or more of ethyl chloride should be grounded and bonded. Drums must be equipped with self-closing valves, pressure vacuum bungs, and flame arresters. Use only non-sparking tools and equipment, especially when opening and closing containers of ethyl chloride. Procedures for the handling, use and storage of cylinders should be in compliance with OSHA 1910.101 and 1910.169 as with the recommendations of the Compressed Gas Association.

Shipping: This compound requires a shipping label of: "Flammable Gas." It falls in DOT Hazard Class 2.1 and Packing Group I.

Spill Handling: Evacuate and restrict persons not wearing protective equipment from area of spill or leak until cleanup is complete. Remove all ignition sources. Establish forced ventilation to keep levels below explosive limit. Absorb liquids in vermiculite, dry sand, earth, peat, carbon, or a similar material and deposit in sealed containers. *If ethyl chloride gas is leaked,* take the following steps: Restrict persons not wearing protective equipment from area of leak until cleanup is complete. Remove all ignition sources. Ventilate area of leak to disperse the gas. Stop flow of gas. If source of leak is a cylinder and the leak cannot be stopped in place, remove the leaking cylinder to a safe place in the open air, and repair leak or allow cylinder to empty. Keep ethyl chloride out of a confined space, such as a sewer, because of the possibility of an explosion, unless the sewer is designed to prevent the build-up of explosive concentrations. It may be necessary to contain and dispose of this chemical as a hazardous waste. If material or contaminated runoff enters waterways, notify downstream users of potentially contaminated waters. Contact your Department of Environmental Protection or your regional office of the federal EPA for specific recommendations. If employees are required to cleanup spills, they must be properly trained and equipped. OSHA 1910.120(q) may be applicable.

Fire Extinguishing: This chemical is a flammable liquid or gas. Poisonous gases including phosgene and hydrogen chloride are produced in fire.

Liquid: Establish forced ventilation to keep levels below explosive limit. Use dry chemical, carbon dioxide, or foam extinguishers. Vapors are heavier than air and will collect in low areas. Vapors may travel long distances to ignition sources and flashback. Vapors in confined areas may explode when exposed to fire. Containers may explode in fire. Storage containers and parts of containers may rocket great distances, in many directions. If material or contaminated runoff enters waterways, notify downstream users of potentially contaminated waters. Notify local health and fire officials and pollution control agencies. From a secure, explosion-proof location, use water spray to cool exposed containers. If cooling streams are ineffective (venting sound increases in volume and pitch, tank discolors, or shows any signs of deforming), withdraw immediately to a secure position. If employees are expected to fight fires, they must be trained and equipped in OSHA 1910.156.

Gas: Establish forced ventilation to keep levels below explosive limit. Do not extinguish the fire unless the flow of gas can be stopped and any remaining gas is out of the line. Specially trained personnel may use fog lines to cool exposures and let the fire burn itself out. Vapors are heavier than air and will collect in low areas. Vapors may travel long distances to ignition sources and flashback. Vapors in confined areas may explode when exposed to fire. Containers may explode in fire. Storage containers and parts of containers may rocket great distances, in many directions. If material or contaminated runoff enters waterways, notify downstream users of potentially contaminated waters. Notify local health and fire officials and pollution control agencies. From a secure, explosion-proof location, use water spray to cool exposed containers. If cooling streams are ineffective (venting sound increases in volume and pitch, tank discolors, or shows any signs of deforming), withdraw immediately to a secure position. If cylinders are exposed to excessive heat from fire or flame contact, withdraw immediately to a secure location. If employees are expected to fight fires, they must be trained and equipped in OSHA 1910.156.

Disposal Method Suggested: Incineration, preferably after mixing with another combustible fuel. Care must be exercised to assure complete combustion to prevent the formation of phosgene. An acid scrubber is necessary to remove the halo acids produced.[22]

References

U.S. Environmental Protection Agency, "Chemical Hazard Information Profile: Chloroethane," Washington, DC (1979)

U.S. Environmental Protection Agency, Chloroethanes: ambient Water Quality Criteria, Washington, DC (1980)

U.S. Environmental Protection Agency, Chloroethane, Health and Environmental Effects Profile No. 44, Washington, DC, Office of Solid Waste (April 30, 1980)

Sax, N. I., Ed., "Dangerous Properties of Industrial Materials Report," 1, No. 4, 64–66 (1981)

U.S. Public Health Service, "Toxicological Profile for Chloroethane," Atlanta, Georgia, Agency for Toxic Substances and Disease Registry (December 1988)

New Jersey Department of Health and Senior Services, "Hazardous Substance Fact Sheet: Ethyl Chloride," Trenton, NJ (June 1996)

Ethyl Chloroacetate

Molecular Formula: $C_4H_7ClO_2$

Common Formula: $CH_3CH_2OCO{\cdot}CH_2Cl$

Synonyms: Chloroacetic acid, ethyl ester; Ethyl α-chloroacetate; Ethyl chloroacetate; Ethyl chloroethanoate; Ethyl monochloracetate; Ethyl monochloroacetate

CAS Registry Number: 105-39-5

RTECS Number: AF9110000

DOT ID: UN 1181

Cited in U.S. State Regulations: California (G), New Jersey (G), Pennsylvania (G).

Description: Ethyl chloroacetate is a water-white liquid with a pungent, fruity odor. Boiling point = 145°C. Freezing point = -27°C. Flash point = 64°C. Hazard Identification (based on NFPA-704 M Rating System): Health –, Flammability 3, Reactivity 0. Insoluble in water.

Potential Exposure: Used to make rodenticides, dyes, and other chemicals. Also used as a military poison.

Incompatibilities: Forms explosive mixture with air. Incompatible with strong bases, strong acids, reducing agents. Moisture, water, steam contact forms toxic and corrosive fumes. Violent reaction with oxidizers, alkaline earth metals (barium, calcium, magnesium, strontium, etc.), alkaline metals, sodium cyanide. Attacks metals in the presence of moisture.

Permissible Exposure Limits in Air: No OELs established. However, this chemical can be absorbed through the skin, thereby increasing exposure.

Routes of Entry: Inhalation, eyes, through the skin.

Harmful Effects and Symptoms

Short Term Exposure: Contact can severely irritate and burn eyes and skin. Inhalation can irritate the respiratory tract with coughing and wheezing. Higher exposures can cause pulmonary edema, a medical emergency that can be delayed for several hours. This can cause death. Exposure can cause headache, nausea, and vomiting.

Long Term Exposure: Very irritating substances may cause lung damage.

Points of Attack: Lungs.

Medical Surveillance: Lung function tests. Consider chest x-ray following acute overexposure.

First Aid: If this chemical gets into the eyes, remove any contact lenses at once and irrigate immediately for at least 15 minutes, occasionally lifting upper and lower lids. Seek medical attention immediately. If this chemical contacts the skin, remove contaminated clothing and wash immediately with soap and water. Seek medical attention immediately. If this chemical has been inhaled, remove from exposure, begin rescue breathing (using universal precautions) if breathing has stopped and CPR if heart action has stopped. Transfer promptly to a medical facility. When this chemical has been swallowed, get medical attention. Give large quantities of water and induce vomiting. Do not make an unconscious person vomit. Medical observation is recommended for 24 – 48 hours after breathing overexposure, as pulmonary edema may be delayed. As first aid for pulmonary edema, a doctor or authorized paramedic may consider administering a corticosteroid spray.

Personal Protective Methods: Wear protective gloves and clothing to prevent any reasonable probability of skin contact. Safety equipment suppliers/manufacturers can provide recommendations on the most protective glove/clothing material for your operation.. All protective clothing (suits, gloves, footwear, headgear) should be clean, available each day, and put on before work. Contact lenses should not be worn when working with this chemical. Wear splash-proof chemical goggles and face shield unless full facepiece respiratory protection is worn. Employees should wash immediately with soap when skin is wet or contaminated. Provide emergency showers and eyewash.

Respirator Selection: Where the potential for exposure to this chemical exists, use a MSHA/NIOSH approved supplied-air respirator with a full facepiece operated in the positive pressure mode or with a full facepiece, hood, or helmet in the continuous flow mode, or use a MSHA/NIOSH approved self-contained breathing apparatus with a full facepiece operated in pressure-demand or other positive pressure mode.

Storage: Prior to working with this chemical you should be trained on its proper handling and storage. Store in tightly closed containers in a cool, well ventilated area away from oxidizers, strong acids, strong bases, reducing agents, heat and sources of ignition. Where possible, automatically pump liquid from drums or other storage containers to process containers. Drums must be equipped with self-closing valves, pressure vacuum bungs, and flame arresters. Use only non-sparking tools and equipment, especially when opening and closing containers of this chemical. Sources of ignition such as smoking and open flames, are prohibited where this chemical is used, handled, or stored in a manner that could create a potential fire or explosion hazard. Wherever this chemical is used, handled, manufactured, or stored, use explosion-proof electrical equipment and fittings.

Shipping: Label required is "Poison". Is in DOT/UN Hazard Class 3 and Packing Group II.[19][20]

Spill Handling: Evacuate and restrict persons not wearing protective equipment from area of spill or leak until cleanup is complete. Remove all ignition sources. Ventilate area of spill or leak. Absorb liquids in vermiculite, dry sand, earth, peat, carbon, or a similar material and deposit in sealed containers. It may be necessary to contain and dispose of this chemical as a hazardous waste. If material or contaminated runoff enters waterways, notify downstream users of potentially contaminated waters. Contact your Department of Environmental Protection or your regional office of the federal EPA for specific recommendations. If employees are required to clean-up spills, they must be properly trained and equipped. OSHA 1910.120(q) may be applicable.

Fire Extinguishing: This chemical is a flammable liquid. Poisonous gases including hydrogen chloride and carbon monoxide are produced in fire. Use dry chemical, carbon dioxide, or alcohol foam extinguishers. Vapors are heavier than air and will collect in low areas. Vapors may travel long distances to ignition sources and flashback. Vapors in confined areas may explode when exposed to fire. Containers may explode in fire. Storage containers and parts of containers may rocket great distances, in many directions. If material or contaminated runoff enters waterways, notify downstream users of potentially contaminated waters. Notify local health and fire officials and pollution control agencies. From a secure, explosion-proof location, use water spray to cool exposed containers. If cooling streams are ineffective (venting sound increases in volume and pitch, tank discolors, or shows any signs of deforming), withdraw immediately to a secure position. If employees are expected to fight fires, they must be trained and equipped in OSHA 1910.156.

Disposal Method Suggested: Incineration.

References

New Jersey Department of Health and Senior Services, "Hazardous Substance Fact Sheet, Ethyl Chloroacetate," Trenton, NJ (March 1999)

Ethyl 2-Chloropropionate

Molecular Formula: $C_5H_9ClO_2$

Common Formula: $CH_3CHClCOOC_2H_5$

Synonyms: Propanoic acid, 2-chloro-, ethyl ester

CAS Registry Number: 535-13-7

RTECS Number: UE8870000

DOT ID: UN 2935

Cited in U.S. State Regulations: New Jersey (G).

Description: Ethyl-2-chloropropionate, $CH_3CHClCOOC_2H_5$, is a liquid with a pleasant odor. Flash point = 38°C.

Potential Exposure: This material is used in organic synthesis.

Incompatibilities: Forms explosive mixture with air. Acids, bases, oxidizers, reducing agents.

Permissible Exposure Limits in Air: No standards set.

Permissible Concentration in Water: No criteria set.

Routes of Entry: Inhalation.

Harmful Effects and Symptoms

Short Term Exposure: Ethyl-2-chloropropionate can affect you when breathed in. Eye or skin contact may irritate the nose, throat and bronchial tubes. High exposures might cause a dangerous fluid build-up in the lungs (pulmonary edema), a medical emergency that can be delayed for several hours. This can cause death.

Long Term Exposure: Some related chemicals may cause skin allergy. Very irritating substances may cause lung problems.

Points of Attack: Respiratory tract.

Medical Surveillance: If symptoms develop or overexposure is suspected, the following may be useful: Consider lung function tests, especially if lung symptoms are present. Evaluation by a qualified allergist, including careful exposure history and special testing, may help diagnose skin allergy. Consider chest x-ray after acute overexposure.

First Aid: If this chemical gets into the eyes, remove any contact lenses at once and irrigate immediately for at least 15 minutes, occasionally lifting upper and lower lids. Seek medical attention immediately. If this chemical contacts the skin, remove contaminated clothing and wash immediately with soap and water. Seek medical attention immediately. If this chemical has been inhaled, remove from exposure, begin rescue breathing (using universal precautions) if breathing has stopped and CPR if heart action has stopped. Transfer promptly to a medical facility. When this chemical has been swallowed, get medical attention. Give large quantities of water and induce vomiting. Do not make an unconscious person vomit. Medical observation is recommended for 24 – 48 hours after breathing overexposure, as pulmonary edema may be delayed. As first aid for pulmonary edema, a doctor or authorized paramedic may consider administering a corticosteroid spray.

Personal Protective Methods: Wear protective gloves and clothing to prevent any reasonable probability of skin contact. Safety equipment suppliers/manufacturers can provide recommendations on the most protective glove/clothing material for your operation. All protective clothing (suits, gloves, footwear, headgear) should be clean, available each day, and put on before work. Contact lenses should not be worn when working with this chemical. Wear splash-proof chemical goggles and face shield unless full facepiece respiratory protection is worn. Employees should wash immediately with soap when skin is wet or contaminated. Provide emergency showers and eyewash.

Respirator Selection: Where the potential exists for exposure to ethyl-2-chloropropionate, use a MSHA/NIOSH approved supplied-air respirator with a full facepiece operated in the positive pressure mode or with a full facepiece, hood, or helmet in the continuous flow mode, or use a MSHA/NIOSH approved self-contained breathing apparatus with a full facepiece operated in pressure-demand or other positive pressure mode.

Storage: Prior to working with this chemical you should be trained on its proper handling and storage. Store in tightly closed containers in a cool, well-ventilated area away from acids, bases, oxidizing and reducing agents. Sources of ignition, such as smoking and open flames, are prohibited where ethyl-2-chloropropionate is used, handled, or stored in a manner that could create a potential fire or explosion hazard.

Shipping: This compound requires a shipping label of: "Flammable Liquid." It falls in DOT Hazard Class 3 and Packing Group III.

Spill Handling: Evacuate and restrict persons not wearing protective equipment from area of spill or leak until cleanup is complete. Remove all ignition sources. Ventilate area of spill or leak. Absorb liquids in vermiculite, dry sand, earth, peat, carbon, or a similar material and deposit in sealed containers. Keep this chemical out of a confined space, such as a sewer, because of the possibility of an explosion, unless the sewer is designed to prevent the build-up off explosive concentrations. It may be necessary to contain and dispose of this chemical as a hazardous waste. If material or contaminated runoff enters waterways, notify downstream users of potentially contaminated waters. Contact your Department of Environmental Protection or your regional office of the federal EPA for specific recommendations. If employees are required to clean-up spills, they must be properly trained and equipped. OSHA 1910.120(q) may be applicable.

Fire Extinguishing: This chemical is a flammable liquid. Poisonous gases including chlorides are produced in fire. Use dry chemical, carbon dioxide, or foam extinguishers. Vapors are heavier than air and will collect in low areas. Vapors may travel long distances to ignition sources and flashback. Vapors in confined areas may explode when exposed to fire. Containers may explode in fire. Storage containers and parts of containers may rocket great distances, in many directions. If material or contaminated runoff enters waterways, notify downstream users of potentially contaminated waters. Notify local health and fire officials and pollution control agencies. From a secure, explosion-proof location, use water spray to cool exposed containers. If cooling streams are ineffective (venting sound increases in volume and pitch, tank discolors, or shows any signs of deforming), withdraw immediately to a secure position. If employees are expected to fight fires, they must be trained and equipped in OSHA 1910.156.

References

New Jersey Department of Health and Senior Services, "Hazardous Substance Fact Sheet: Ethyl-2-Chloropropionate," Trenton, NJ (January 1, 1988)

Ethyl Cyanoacetate

Molecular Formula: $C_5H_7NO_2$

Synonyms: Acetic acid, cyano-, ethyl ester; Cyanacetate ethyle (German); Cyanoacetic acid ethyl ester; Cyanoacetic ester; Estere cianoacetico; Ethyl cyanoacetate; Ethyl cyanoethanoate; Ethylester kyseliny kyanoctove; Malonic acid, ethyl ester nitrile

CAS Registry Number: 105-56-6

RTECS Number: AG4110000

DOT ID: UN 2666

Regulatory Authority

- Clean Air Act: Hazardous Air Pollutants (Title I, Part A, Section 112)
- Clean Water Act: 40CFR423, Appendix A, Priority Pollutants as cyanide, total
- EPA Hazardous Waste Number (RCRA No.): P030 as cyanides soluble salts and complexes, n.o.s.
- RCRA, 40CFR261, Appendix 8 Hazardous Constituents. As cyanides, soluble salts and complexes, n.o.s.
- EPCRA (Section 313): X+CN- where X = H+ or any other group where a formal dissociation may occur. For example, KCN or $Ca(CN)_2$. Form R *de minimis* concentration reporting level: 1.0%
- U.S. DOT Regulated Marine Pollutant (49CFR172.101, Appendix B) as cyanide mixtures, cyanide solutions or cyanides, inorganic, n.o.s.

Cited in U.S. State Regulations: California (G), New Jersey (G), Pennsylvania (G).

Description: Ethyl cyanoacetate is a colorless to straw colored liquid with a mild pleasant odor. Boiling point = 207°C. Freezing/Melting point = -22.5°C. Flash point = 110°C. Insoluble in water.

Potential Exposure: Used to manufacture dyes, pharmaceuticals, and other chemicals.

Incompatibilities: Oxidizers, strong acids, strong bases, and reducing agents. Reacts with moisture, water, steam, forming toxic fumes.

Permissible Exposure Limits in Air: The NIOSH REL for nitriles is a ceiling limit of 6 mg/m^3, not to be exceeded in any 15-minute work period.

Determination in Air: See NIOSH Criteria Document 78-212 NITRILES.

Routes of Entry: Inhalation. May be absorbed through the skin.

Harmful Effects and Symptoms

Short Term Exposure: An organic cyanide compound. Irritates the eyes, skin, and respiratory tract. Exposure can cause headache, nausea, and vomiting. Poisonous if ingested.

Long Term Exposure: Unknown at this time.

First Aid: If this chemical gets into the eyes, remove any contact lenses at once and irrigate immediately for at least 15 minutes, occasionally lifting upper and lower lids. Seek medical attention immediately. If this chemical contacts the skin, remove contaminated clothing and wash immediately with soap and water. Seek medical attention immediately. If this chemical has been inhaled, remove from exposure, begin rescue breathing (using universal precautions) if breathing has stopped and CPR if heart action has stopped. Transfer promptly to a medical facility. When this chemical has been swallowed, get medical attention. Give large quantities of water and induce vomiting. Do not make an unconscious person vomit.

Personal Protective Methods: Wear protective gloves and clothing to prevent any reasonable probability of skin contact. Safety equipment suppliers/manufacturers can provide recommendations on the most protective glove/clothing material for your operation.. All protective clothing (suits, gloves, footwear, headgear) should be clean, available each day, and put on before work. Contact lenses should not be worn when working with this chemical. Wear splash-proof chemical goggles and face shield unless full facepiece respiratory protection is worn. Employees should wash immediately with soap when skin is wet or contaminated. Provide emergency showers and eyewash. See NIOSH Criteria Document 78-212 NITRILES.

Respirator Selection: Where the potential for exposure to this chemical, use a MSHA/NIOSH approved supplied-air respirator with a full facepiece operated in the positive pressure mode or with a full facepiece, hood, or helmet in the continuous flow mode, or use a MSHA/NIOSH approved self-contained breathing apparatus with a full facepiece operated in pressure-demand or other positive pressure mode.

Storage: Prior to working with this chemical you should be trained on its proper handling and storage. Store in tightly closed containers in a cool, well ventilated area away from oxidizers, strong bases, strong acids, reducing agents, moisture and sources of ignition. Where possible, automatically pump liquid from drums or other storage containers to process containers.

Shipping: DOT label requirement of "Keep away from Food" Ethyl cyanoacetate falls in UN/DOT Hazard Class 6.1 and Packing Group III.

Spill Handling: Evacuate and restrict persons not wearing protective equipment from area of spill or leak until cleanup is complete. Remove all ignition sources. Ventilate area of spill or leak. Absorb liquids in vermiculite, dry sand, earth, peat, carbon, or a similar material and deposit in sealed containers. It may be necessary to contain and dispose of this chemical as a hazardous waste. If material or contaminated runoff enters waterways, notify downstream users of potentially contaminated waters. Contact your Department of Environmental Protection or your regional office of the federal EPA for specific recommendations. If employees are required to clean-up spills, they must be properly trained and equipped. OSHA 1910.120(q) may be applicable.

Fire Extinguishing: This chemical is a combustible liquid. Poisonous gases including carbon monoxide, nitrogen oxides, and cyanides are produced in fire. Use dry chemical, carbon dioxide, water spray, or alcohol foam extinguishers. Vapors are heavier than air and will collect in low areas. Vapors may travel long distances to ignition sources and flashback. Vapors in confined areas may explode when exposed to fire. Containers may explode in fire. Storage containers and parts of containers may rocket great distances, in many directions. If material or contaminated runoff enters waterways, notify downstream users of potentially contaminated waters. Notify local health and fire officials and pollution control agencies. From a secure, explosion-proof location, use water spray to cool exposed containers. If cooling streams are ineffective (venting sound increases in volume and pitch, tank discolors, or shows any signs of deforming), withdraw immediately to a secure position. If employees are expected to fight fires, they must be trained and equipped in OSHA 1910.156.

Disposal Method Suggested: Consult with environmental regulatory agencies for guidance on acceptable disposal practices. Generators of waste containing this contaminant (≥100 kg/mo) must conform with EPA regulations governing storage, transportation, treatment, and waste disposal.

References

New Jersey Department of Health and Senior Services, "Hazardous Substance Fact Sheet, Ethyl CYANOACETATE," Trenton, NJ (March 1999)

Ethyl-4,4'-Dichlorobenzilate

Molecular Formula: $C_{16}H_{14}C_{12}O_3$

Synonyms: Acar; Acaraben; Acaraben 4E; Akar 338; Benzeneacetic acid, 4-chloro-α-(4-chlorophenyl)-α-hydroxy-, ethyl ester; Benzilan; Benzilic acid, 4,4'-dichloro-, ethyl ester; Benzilic acid,4,4'-dichloro, ethyl ester; Benz-o-chlor; Chlorbenzalate; Chlorbenzilat; 4-Chloro-α-(4-chlorophenyl)-α-hydroxybenzeneacetic acid ethyl ether; 4,4'-Cichlorbenzilsaeureaethylester (German); Compound 338; 4,4'-Dichlorobenzilate; 4,4'-Dichlorobenzilic acid ethyl ester; ECB; ENT 18,596; Ethyl 4-chloro-α-(4-chlorophenyl)-α-hydroxybenzene acetate; Ethyl *p,p'*-dichlorobenzilate; Ethyl 4,4'-dichlorobenzilate; Ethyl-4,4'-dichlorodiphenyl glycollate; Ethyl-4,4'-dichlorophenyl glycollate; Ethyl ester of 4,4'-dichlorobenzilic acid; Ethyl 2-hydroxy-2,2-bis(4-chlorophenyl)acetate; Folbex; Folbex smoke strips; G 23992; G 338; Geigy 338; Kop Mite; NCI-C00408; NCI-C60413

CAS Registry Number: 510-15-6

RTECS Number: DD2275000

DOT ID: UN 2996 (liquid)

Regulatory Authority

- CARCINOGEN: (animal positive) (IARC) (NCI)
- Clean Air Act: Hazardous Air Pollutants (Title I, Part A, Section 112)
- EPA Hazardous Waste Number (RCRA No.): U038
- RCRA, 40CFR261, Appendix 8 Hazardous Constituents
- RCRA 40CFR268.48; 61FR15654, Universal Treatment Standards: Wastewater (mg/l), 0.10; Nonwastewater (mg/kg), N/A
- RCRA 40CFR264, Appendix 9; TSD Facilities Ground Water Monitoring List. Suggested test method(s) (PQL µg/l): 8270 (10)
- Superfund/EPCRA 40CFR302.4 Reportable Quantity (RQ): CERCLA, 10 lb (4.54 kg)
- EPCRA Section 313 Form R *de minimis* concentration reporting level: 1.0%
- U.S. DOT Regulated Marine Pollutant (49CFR172.101, Appendix B)

Cited in U.S. State Regulations: California (G), New Jersey (G), Pennsylvania (G).

Description: Ethyl 4,4'-dichlorobenzilate is a colorless solid when pure. The technical product is a yellow or brown liquid. Boiling point = 146 – 148°C. Freezing/Melting point = 36 – 37.3°C. Slightly soluble in water.

Potential Exposure: It is used to kill mites, ticks, and other insects, and as a synergist for DDT.

Incompatibilities: Strong acids, strong bases, lime.

Permissible Exposure Limits in Air: No OELs have been established.

Routes of Entry: Inhalation, ingestion, passing through the skin.

Harmful Effects and Symptoms

Short Term Exposure: Contact irritates the skin causing rash or burning sensation. Exposure can cause headache, loss of appetite, nausea, vomiting, and diarrhea.

Long Term Exposure: This chemical may damage the male reproductive glands. May cause liver and kidney damage. Repeated or prolonged exposure may affect the nervous system, causing a loss of coordination, muscle weakness, tremors, convulsions, dizziness, and possible coma. This chemical may be a human carcinogen.

Points of Attack: Liver, kidneys, nervous system.

Medical Surveillance: Examination of the nervous system. Liver and kidney function tests. More than light alcohol consumption may exacerbate liver damage.

First Aid: If this chemical gets into the eyes, remove any contact lenses at once and irrigate immediately for at least 15 minutes, occasionally lifting upper and lower lids. Seek medical attention immediately. If this chemical contacts the skin, remove contaminated clothing and wash immediately with soap and water. Seek medical attention immediately. If this chemical has been inhaled, remove from exposure, begin rescue breathing (using universal precautions) if breathing has stopped and CPR if heart action has stopped. Transfer promptly to a medical facility. When this chemical has been swallowed, get medical attention. Give large quantities of water and induce vomiting. Do not make an unconscious person vomit.

Personal Protective Methods: Wear protective gloves and clothing to prevent any reasonable probability of skin contact. Safety equipment suppliers/manufacturers can provide recommendations on the most protective glove/clothing material for your operation.. All protective clothing (suits, gloves, footwear, headgear) should be clean, available each day, and put on before work. Contact lenses should not be worn when working with this chemical. Wear splash-proof chemical goggles and face shield unless full facepiece respiratory protection is worn. Employees should wash immediately with soap when skin is wet or contaminated. Provide emergency showers and eyewash.

Respirator Selection: Where the potential for exposure to this chemical, use a MSHA/NIOSH approved supplied-air respirator with a full facepiece operated in the positive pressure mode or with a full facepiece, hood, or helmet in the continuous flow mode, or use a MSHA/NIOSH approved self-contained breathing apparatus with a full facepiece operated in pressure-demand or other positive pressure mode.

Storage: Prior to working with this chemical you should be trained on its proper handling and storage. Store in tightly closed containers in a cool, well ventilated area away from strong acids, strong bases. Where possible, automatically pump liquid from drums or other storage containers to process containers. Sources of ignition are prohibited where this chemical is used, handled or stored. A regulated, marked area should be established where this chemical is handled, used, or stored in compliance with OSHA standard 1910.1045.

Shipping: Organochlorine pesticide, liquid, toxic. DOT label requirement of "Poison." This chemical falls in UN/DOT Hazard Class 6.1 and Packing Group II.[19][20]

Spill Handling: Evacuate and restrict persons not wearing protective equipment from area of spill or leak until cleanup is complete. Remove all ignition sources. Ventilate area of spill or leak. Absorb liquids in vermiculite, dry sand, earth, peat, carbon, or a similar material and deposit in sealed containers. It may be necessary to contain and dispose of this chemical as a hazardous waste. If material or contaminated runoff enters waterways, notify downstream users of potentially contaminated waters. Contact your Department of Environmental Protection or your regional office of the federal EPA for specific recommendations. If employees are required to clean-up spills, they must be properly trained and equipped. OSHA 1910.120(q) may be applicable.

Fire Extinguishing: This chemical may burn but does not easily ignite. Poisonous gases including chlorine are produced in fire. Use dry chemical, carbon dioxide, or alcohol foam extinguishers. Vapors are heavier than air and will collect in low areas. Vapors may travel long distances to ignition sources and flashback. Vapors in confined areas may explode when exposed to fire. Containers may explode in fire. Storage containers and parts of containers may rocket great distances, in many directions. If material or contaminated runoff enters waterways, notify downstream users of potentially contaminated waters. Notify local health and fire officials and pollution control agencies. From a secure, explosion-proof location, use water spray to cool exposed containers. If cooling streams are ineffective (venting sound increases in volume and pitch, tank discolors, or shows any signs of deforming), withdraw immediately to a secure position. If employees are expected to fight fires, they must be trained and equipped in OSHA 1910.156.

Disposal Method Suggested: Consult with environmental regulatory agencies for guidance on acceptable disposal practices. Generators of waste containing this contaminant (≥100 kg/mo) must conform with EPA regulations governing storage, transportation, treatment, and waste disposal. In accordance with 40CFR165 recommendations for the disposal of pesticides and pesticide containers. Must be disposed properly by following package label directions or by contacting your state pesticide or environmental control agency or by contacting your regional EPA office.

References

New Jersey Department of Health and Senior Services, "Hazardous Substance Fact Sheet, Ethyl-4,4'-Dichlorobenzilate," Trenton NJ (May 1999)

Ethyl Dichlorosilane

Molecular Formula: $C_2H_6Cl_2Si$

Common Formula: $Cl_2SiHC_2H_5$

Synonyms: Dichloroethylsilane; Ethyldichlorosilane; Monoethyldichlorosilane; Silane, dichloroethyl-

CAS Registry Number: 1789-58-8

RTECS Number: VV3230000

DOT ID: UN 1183

Cited in U.S. State Regulations: Florida (G), Maine (G), Massachusetts (G), New Hampshire (G), New Jersey (G), Pennsylvania (G).

Description: Ethyl dichlorosilane, $Cl_2SiHC_2H_5$, is a colorless liquid with a sharp, irritating odor. Boiling point = 75.5°C. Flash point = -1°C. Explosive Limits: LEL: 2.9%; UEL: NA. Reacts with water.

Potential Exposure: This material is used in silicone polymer manufacture.

Incompatibilities: Forms explosive mixture with air. Reacts vigorously with water, steam, or surface moisture generating hydrogen chloride. Incompatible with strong oxidizers, strong bases. Attacks most common metals in the presence of moisture.

Permissible Exposure Limits in Air: No standards set.

Permissible Concentration in Water: No criteria set.

Routes of Entry: Inhalation.

Harmful Effects and Symptoms

Short Term Exposure: Ethyl dichlorosilane is corrosive. It can affect you when breathed in. Exposure can irritate the lungs causing coughing and/or shortness of breath. Higher exposures can cause a build-up of fluid in the lungs (pulmonary edema), a medical emergency that can be delayed for several hours. This can cause death. Ethyl dichlorosilane is a corrosive chemical and contact can cause severe skin and eye burns leading to permanent eye damage.

Long Term Exposure: This chemical can irritate the lungs. Repeated exposure may cause bronchitis with cough, phlegm, and/or shortness of breath.

Medical Surveillance: For those with frequent or potentially high exposure, the following are recommended before beginning work and at regular times after that: Lung function tests. If symptoms develop or overexposure is suspected, the following may be useful: Consider chest x-ray after acute overexposure.

First Aid: If this chemical gets into the eyes, remove any contact lenses at once and irrigate immediately for at least 30 minutes, occasionally lifting upper and lower lids. Seek medical attention immediately. If this chemical contacts the skin, remove contaminated clothing and wash immediately with soap and water. Seek medical attention immediately. If this chemical has been inhaled, remove from exposure, begin rescue breathing (using universal precautions) if breathing has stopped and CPR if heart action has stopped. Transfer promptly to a medical facility. When this chemical has been swallowed, get medical attention. If victim is conscious, administer water or milk. Do not induce vomiting. Medical observation is recommended for 24 – 48 hours after breathing overexposure, as pulmonary edema may be delayed.

Personal Protective Methods: Wear protective gloves and clothing to prevent any reasonable probability of skin contact. Safety equipment suppliers/manufacturers can provide recommendations on the most protective glove/clothing material for your operation.. All protective clothing (suits, gloves, footwear, headgear) should be clean, available each day, and put on before work. Contact lenses should not be worn when working with this chemical. Wear splash-proof chemical goggles and face shield unless full facepiece respiratory protection is worn. Employees should wash immediately with soap when skin is wet or contaminated. Provide emergency showers and eyewash.

Respirator Selection: Where the potential for exposure to ethyl dichlorosilane exists, use a MSHA/NIOSH approved supplied-air respirator with a full facepiece operated in the positive pressure mode or with a full facepiece, hood, or helmet in the continuous flow mode, or use a MSHA/NIOSH approved self-contained breathing apparatus with a full facepiece operated in pressure-demand or other positive pressure mode.

Storage: Prior to working with this chemical you should be trained on its proper handling and storage. Before entering confined space where this chemical may be present, check to make sure that an explosive concentration does not exist. Ethyl dichlorosilane must be stored to avoid contact with moisture and oxidizers (such as perchlorates, peroxides, permanganates, chlorates and nitrates) since violent reactions occur. Store in tightly closed containers in a cool, well-ventilated area away from water, steam and moisture because toxic and corrosive chloride gases including, hydrogen chloride can be produced. Sources of ignition, such as smoking and open flames, are prohibited where ethyl dichlorosilane is handled, used or stored. Metal containers involving the transfer of 5 gallons or more of ethyl dichlorosilane should be grounded and bonded. Drums must be equipped with self-closing valves, pressure vacuum bungs, and flame arresters. Use only non-sparking tools and equipment, especially when opening and closing containers of ethyl dichlorosilane.

Shipping: This compound requires a shipping label of "Dangerous When Wet, Corrosive, Flammable Liquid." It falls in DOT Hazard Class 4.3 and Packing Group I. Passenger aircraft or railcar shipment is forbidden.

Spill Handling: Evacuate and restrict persons not wearing protective equipment from area of spill or leak until cleanup

is complete. Remove all ignition sources. Establish forced ventilation to keep levels below explosive limit. Absorb liquids in vermiculite, dry sand, earth, peat, carbon, or a similar material and deposit in sealed containers. Keep this chemical out of a confined space, such as a sewer, because of the possibility of an explosion, unless the sewer is designed to prevent the build-up off explosive concentrations. It may be necessary to contain and dispose of this chemical as a hazardous waste. If material or contaminated runoff enters waterways, notify downstream users of potentially contaminated waters. Contact your Department of Environmental Protection or your regional office of the federal EPA for specific recommendations. If employees are required to clean-up spills, they must be properly trained and equipped. OSHA 1910.120(q) may be applicable.

Fire Extinguishing: This chemical is a flammable liquid. Poisonous gases including phosgene and hydrogen chloride are produced in fire. Use dry chemical. Do not use water (forms hydrogen chloride gas). Fire may restart after it has been extinguished. Vapors are heavier than air and will collect in low areas. Vapors may travel long distances to ignition sources and flashback. Vapors in confined areas may explode when exposed to fire. Containers may explode in fire. Storage containers and parts of containers may rocket great distances, in many directions. If material or contaminated runoff enters waterways, notify downstream users of potentially contaminated waters. Notify local health and fire officials and pollution control agencies. From a secure, explosion-proof location, use water spray to cool exposed containers. If cooling streams are ineffective (venting sound increases in volume and pitch, tank discolors, or shows any signs of deforming), withdraw immediately to a secure position. If employees are expected to fight fires, they must be trained and equipped in OSHA 1910.156.

References

New Jersey Department of Health and Senior Services, "Hazardous Substance Fact Sheet: Ethyl Dichlorosilane," Trenton, NJ (December 1998)

Ethylene

Molecular Formula: C_2H_4

Synonyms: Acetene; Athylen (German); Bicarburretted hydrogen; Dicarburetted hydrogen; Elayl; Eteno (Spanish); Ethene; Etherin; Heavy carburetted hydrogen; Olefiant gas

CAS Registry Number: 74-85-1

RTECS Number: KU5340000

DOT ID: UN 1038 (liquid); UN 1962 (compressed)

Regulatory Authority

- Air Pollutant Standard Set (Former USSR)[43] (Virginia)[60]
- Clean Air Act: Accidental Release Prevention/Flammable substances, (Section 112[r], Table 3), TQ = 10,000 lb (4,540 kg)
- EPCRA Section 313 Form R *de minimis* concentration reporting level: 1.0%

Cited in U.S. State Regulations: Alaska (G), California (A), Florida (G), Illinois (G), Maine (G), Maryland (G), Massachusetts (G), New Hampshire (G), New Jersey (G), New York (G), Pennsylvania (G), Rhode Island (G).

Description: Ethylene, C_2H_4, is a colorless gas (at room temperature) with a sweet odor. The minimum detectable by odor is 260 ppm. Boiling point = -104°C. Autoignition temperature = 450°C. Explosive limits: LEL = 2.7%; UEL = 36.0%.[17] Hazard Identification (based on NFPA-704 M Rating System): Health 1, Flammability 4, Reactivity 2. Soluble in water.

Potential Exposure: Ethylene is used to manufacture ethylene oxide, polyethylene for plastics, alcohol, mustard gas and other organics. It is used to accelerate ripening of fruit, as an anesthetic and for oxyethylene welding and cutting of metals.

Incompatibilities: A highly flammable gas at room temperature. Contact with oxidizers may cause explosive polymerization and fire. May be spontaneously explosive in sunlight or ultraviolet light when mixed with chlorine. Reacts violently with mixtures of carbon tetrachloride and benzoyl peroxide, bromotrichloromethane, aluminum chloride and ozone. Incompatible with acids, halogens, nitrogen oxides, hydrogen bromide, aluminum chloride, chlorine dioxide, nitrogen dioxide. May accumulate static electrical charges, and may cause ignition of its vapors.

Permissible Exposure Limits in Air: ACGIH[1] and HSE[33] classify ethylene as an asphyxiant and specify no TWA values. The former USSR-UNEP/IRPTC project cites a MAC for ambient air of residential areas of 3 mg/m^3 on either a momentary or a daily average basis. Virginia[60] has set a guideline for ethylene in ambient air of 3.0 μg/m^3.

Permissible Concentration in Water: The Former USSR-UNEP/IRPTC project[43] gives a MAC in water bodies used for domestic purposes of 0.5 mg/l.

Routes of Entry: Inhalation.

Harmful Effects and Symptoms

Short Term Exposure: *Inhalation:* Increasingly severe exposures may cause faintness, incoordination, excitement, stupor, unconsciousness, convulsions, stopped breathing, paralysis, and heart, liver and kidney damage. 20 – 25% (200,000 – 250,000 ppm) gas has caused loss of sense of pain. 80 – 90% (800,000 – 900,000 ppm) has caused anesthesia.

Skin: Contact with liquid can cause a "freezing burn."

Eyes: Same as skin.

Ingestion: No information available.

Long Term Exposure: Inhalation may cause loss of appetite and weight, irritability, insomnia, increase in red blood cell count, and inflammation of the kidneys.

Points of Attack: Kidneys

Medical Surveillance: Kidney function tests.

First Aid: If this chemical gets into the eyes, remove any contact lenses at once and irrigate immediately for at least 15 minutes, occasionally lifting upper and lower lids. Seek medical attention immediately. If this chemical contacts the skin, remove contaminated clothing and wash immediately with soap and water. Seek medical attention immediately. If this chemical has been inhaled, remove from exposure, begin rescue breathing (using universal precautions) if breathing has stopped and CPR if heart action has stopped. Transfer promptly to a medical facility. When this chemical has been swallowed, get medical attention. Give large quantities of water and induce vomiting. Do not make an unconscious person vomit. If frostbite has occurred, seek medical attention immediately; do *NOT* rub the affected areas or flush them with water. In order to prevent further tissue damage, do *NOT* attempt to remove frozen clothing from frostbitten areas. If frostbite has *NOT* occurred, immediately and thoroughly wash contaminated skin with soap and water.

Personal Protective Methods: Wear protective gloves and clothing to prevent any reasonable probability of skin contact. Safety equipment suppliers/manufacturers can provide recommendations on the most protective glove/clothing material for your operation.. All protective clothing (suits, gloves, footwear, headgear) should be clean, available each day, and put on before work. Contact lenses should not be worn when working with this chemical. Wear splash-proof chemical goggles and face shield unless full facepiece respiratory protection is worn. Employees should wash immediately with soap when skin is wet or contaminated. Provide emergency showers and eyewash. Where exposure to the liquefied compressed gas may occur, employees should be provided with special clothing designed to prevent frostbite.

Respirator Selection: Exposure to ethylene gas is dangerous because it can replace oxygen and lead to suffocation. Only MSHA/NIOSH approved self-contained breathing apparatus with a full facepiece operated in positive pressure mode should be used in oxygen deficient environments.

Storage: Prior to working with this chemical you should be trained on its proper handling and storage. Before entering confined space where this chemical may be present, check to make sure that an explosive concentration does not exist. Store in tightly closed containers in a cool, well-ventilated area away from chlorine compounds, oxidizing agents, and combustible materials. Sources of ignition such as smoking and open flames are prohibited where ethylene is handled, used, or stored, Use only non-sparking tools and equipment, especially when opening and closing containers of ethylene. Wherever ethylene is used, handled, manufactured, or stored, use explosion-proof electrical equipment and fittings. Piping should be electrically bonded and grounded. Procedures for the handling, use and storage of cylinders should be in compliance with OSHA 1910.101 and 1910.169 as with the recommendations of the Compressed Gas Association.

Shipping: Compressed ethylene requires a shipping label of: "Flammable Gas." It falls in DOT Hazard Class 2.1.

Spill Handling: Restrict persons not wearing protective equipment from area of leak until clean-up is complete. Remove all ignition sources. Establish forced ventilation to keep levels below explosive limit and to disperse the gas. Stop flow of gas if it can be done safely. If source of leak is a cylinder and the leak cannot be stopped in place, remove the leaking cylinder to a safe place in the open air, and repair leak or allow cylinder to empty. Ventilate area of spill or leak. Absorb liquids in vermiculite, dry sand, earth, peat, carbon, or a similar material and deposit in sealed containers. It may be necessary to contain and dispose of this chemical as a hazardous waste. If material or contaminated runoff enters waterways, notify downstream users of potentially contaminated waters. Contact your Department of Environmental Protection or your regional office of the federal EPA for specific recommendations. If employees are required to clean-up spills, they must be properly trained and equipped. OSHA 1910.120(q) may be applicable.

Fire Extinguishing: This chemical is a combustible gas or liquid. Poisonous gases are produced in fire.

Liquid: Use dry chemical, carbon dioxide, or water spray in large amounts. Vapors are heavier than air and will collect in low areas. Vapors may travel long distances to ignition sources and flashback. Vapors in confined areas may explode when exposed to fire. Containers may explode in fire. Storage containers and parts of containers may rocket great distances, in many directions. If material or contaminated runoff enters waterways, notify downstream users of potentially contaminated waters. Notify local health and fire officials and pollution control agencies. From a secure, explosion-proof location, use water spray to cool exposed containers. If cooling streams are ineffective (venting sound increases in volume and pitch, tank discolors, or shows any signs of deforming), withdraw immediately to a secure position. If employees are expected to fight fires, they must be trained and equipped in OSHA 1910.156.

Gas: Ventilate area of leak to disperse gas. Do not extinguish the fire unless the flow of gas can be stopped and any remaining gas is out of the line. Specially trained personnel may use fog lines to cool exposures and let the fire burn itself out. Vapors are heavier than air and will collect in low areas. Vapors may travel long distances to ignition sources and flashback. Vapors in confined areas may explode when exposed to fire. Containers may explode in fire. Storage containers and parts of containers may rocket great distances, in many directions. If material or contaminated runoff enters waterways, notify downstream users of potentially contaminated waters. Notify local health and fire officials and pollution control agencies. From a secure, explosion-proof location, use water spray to cool exposed containers. If cooling streams are ineffective (venting sound increases in volume and pitch, tank discolors, or shows any signs of deforming), withdraw immediately to a secure position. If cylinders are exposed to excessive heat from fire or flame contact, withdraw immediately to a secure location. If employees are expected to fight fires, they must be trained and equipped in OSHA 1910.156.

References

Sax, N. I., Ed., "Dangerous Properties of Industrial Materials Report," 4, No. 1, 79–81 (1984)

New Jersey Department of Health and Senior Services, "Hazardous Substance Fact Sheet: Ethylene," Trenton, NJ (June 1996)

New York State Department of Health, "Chemical Fact Sheet: Ethylene," Albany, NY, Bureau of Toxic Substance Assessment (March 1986)

Ethylene Chlorohydrin

Molecular Formula: C_2H_5ClO

Common Formula: CH_2ClCH_2OH

Synonyms: Aethylenechlorhydrin (German); 2-Chloorethanol (Dutch); 2-Chlorethanol (German); β-Chlorethyl alcohol; β-Chloroethanol; 2-Chloroethanol; Chloroethanol; 2-Chloroethyl alcohol; Chloroethylowy alkohol (Polish); 2-Cloroetanol (Spanish); Cloroetanol (Spanish); delta-Chloroethanol; Ethanol, 2-chloro-; Ethylene chlorohydrine; Glycol chlorohydrin; Glycol monochlorohydrin; Monochlorhydrine du glycol (French); 2-Monochloroethanol; NCI-C50135

CAS Registry Number: 107-07-3

RTECS Number: KK0875000

DOT ID: UN 1135

EEC Number: 603-028-00-7

- ***Regulatory Authority:*** Air Pollutant Standard Set (ACGIH)[1] (DFG)[3] (HSE)[33] (Former USSR)[43] (OSHA)[58] (Several States)[60]
- Superfund/EPCRA 40CFR355, Appendix B Extremely Hazardous Substances: TPQ = 500 lb (227 kg)
- Superfund/EPCRA 40CFR302.4 Reportable Quantity (RQ): EHS, 1 lb (0.454 kg)
- Canada, WHMIS, Ingredients Disclosure List

Cited in U.S. State Regulations: Alaska (G), California (A, G), Connecticut (A), Florida (G), Illinois (G), Maine (G), Massachusetts (G), Nevada (A), New Hampshire (G), New Jersey (G), New York (G), North Dakota (A), Pennsylvania (G), Rhode Island (G), West Virginia (G).

Description: Ethylene chlorohydrin, CH_2ClCH_2OH, is a colorless liquid with a faint, ether-like odor. Boiling point = 130°C. Freezing/Melting point = -67°C. Flash point = 60°C. Autoignition temperature = 425°C. Explosive limits: LEL = 4.9%; UEL = 15.9%. Hazard Identification (based on NFPA-704 M Rating System): Health 4, Flammability 2, Reactivity 0. Soluble in water.

Potential Exposure: Ethylene chlorohydrin is used in the synthesis of ethylene glycol, ethylene oxide, amines, carbitols, indigo, malonic acid, novocaine, and in other reactions where the hydroxyethyl group is introduced into organic compounds, for the separation of butadiene from hydrocarbon mixtures, in dewaxing and removing cycloalkanes from mineral oil, in the refining of rosin, in the manufacture off certain pesticides and in the extraction of pine lignin. In the lacquer industry, it is used as a solvent for cellulose acetate, cellulose esters, resins and waxes, and in the dyeing and cleaning industry, it is used to remove tar spots, as a cleaning agent for machines, and as a solvent in fabric dyeing. It has also found use in agriculture in speeding up sprouting of potatoes and in treating seeds to inhibit biological activity. Making chemical warfare agents.

Incompatibilities: Forms explosive mixture with air. Strong oxidizers may cause fire and explosions. Incompatible with strong caustics (formation of ethylene gas), strong acids, alkaline metals, aliphatic amines, isocyanates. Violent reaction with ethylene diamine, chlorosulfonic acid. Attacks some plastics, rubber and coatings. Reacts with water or steam producing toxic and corrosive fumes.

Permissible Exposure Limits in Air: The OSHA PEL is 5 ppm. DFG,[3] HSE[33] and ACGIH have set 1 ppm (3 mg/m^3) as a ceiling value(not to be exceeded at any time) with the notation "skin" indicating the possibility of cutaneous absorption. The NIOSH IDLH level is 7 ppm. The Former USSR-UNEP/IRPTC project[43] has set a MAC in workplace air of 0.5 mg/m^3. Some states have set guidelines or standards for 2-chloroethanol in ambient air[60] ranging from 30 μg/m^3 (North Dakota) to 71 μg/m^3 (Nevada) to 320 μg/m^3 (Connecticut).

Determination in Air: Adsorption with charcoal (petroleum-based), workup with 2-propanol/CS_2, analysis by gas chromatography/flame ionization. See NIOSH IV, Method #2513.[18]

Permissible Concentration in Water: The former USSR-UNEP/IRPTC project[43] has set a MAC in water bodies used for domestic purposes of 0.7 mg/l.

Routes of Entry: Inhalation of vapor, percutaneous absorption of liquid, ingestion, skin and/or eye contact.

Harmful Effects and Symptoms

Short Term Exposure: Irritates the eyes, skin and respiratory tract. Inhalation can cause severe irritation, with cough and/or shortness of breath. Higher exposures can cause pulmonary edema, a medical emergency that can be delayed for several hours. This can cause death. Exposure to 18 ppm has resulted in vomiting and shock. Additional symptoms may include irritation of the nose, throat and lungs. Poor coordination, numbness, visual disturbance, headache, difficult breathing and unconsciousness, may also occur. Death has resulted from inhalation of 300 ppm for 2½ hours. Signs and symptoms of acute exposure to this chemical may include weakness, dizziness, confusion, visual disturbances, shock, seizures, and coma. Weak pulse, Hypotension (low blood pressure), and cyanosis (blue tint to the skin and mucous membranes) may be observed. Nausea, vomiting, and hematuria (bloody urine) may be seen after exposure. Chloroethanol affect the central nervous system, cardiovascular system, liver and kidneys. This chemical readily absorbed through the skin even though no irritation occurs and may cause symptoms listed above. Death has occurred from skin absorption of 1 teaspoon of liquid. Ingestion can cause irritation to the throat. May cause symptoms listed under inhalation. About 1/5 ounce may be lethal to a 150 pound adult.

Long Term Exposure: Can cause damage to the kidneys, liver, cardiovascular system, and nervous system. May cause reproductive damage. Can irritate the lungs. Repeated exposure may cause bronchitis. High or repeated exposure can cause nerve damage with loss of coordination in arms and legs.

Points of Attack: Respiratory system, liver, kidneys, central nervous system, cardiovascular system, eyes.

Medical Surveillance: Preplacement examination, including a complete history and physical should be performed. Examination of the respiratory system, liver, kidneys, and central nervous system should be stressed. The skin should be examined. A chest x-ray should be taken and pulmonary function tests performed (FVC-FEV). The above procedures should be repeated on an annual basis, except that the x-ray is needed only when indicated by pulmonary function testing.

First Aid: If this chemical gets into the eyes, remove any contact lenses at once and irrigate immediately for at least 15 minutes, occasionally lifting upper and lower lids. Seek medical attention immediately. If this chemical contacts the skin, remove contaminated clothing and wash immediately with soap and water. Seek medical attention immediately. If this chemical has been inhaled, remove from exposure, begin rescue breathing (using universal precautions) if breathing has stopped and CPR if heart action has stopped. Transfer promptly to a medical facility. When this chemical has been swallowed, get medical attention. Give large quantities of water and induce vomiting. Do not make an unconscious person vomit. Medical observation is recommended for 24 – 48 hours after breathing overexposure, as pulmonary edema may be delayed. As first aid for pulmonary edema, a doctor or authorized paramedic may consider administering a corticosteroid spray.

Personal Protective Methods: Wear protective gloves and clothing to prevent any reasonable probability of skin contact. Safety equipment suppliers/manufacturers can provide recommendations on the most protective glove/clothing material for your operation. Butyl rubber, neoprene, poly vinyl alcohol and Viton offer good resistance to this chemical. *Beware:* the liquid penetrates rubber. All protective clothing (suits, gloves, footwear, headgear) should be clean, available each day, and put on before work. Contact lenses should not be worn when working with this chemical. Wear splash-proof chemical goggles and face shield unless full facepiece respiratory protection is worn. Employees should wash immediately with soap when skin is wet or contaminated. Provide emergency showers and eyewash. Wear eye protection to prevent any possibility of eye contact. Remove nonimpervious clothing immediately if wet or contaminated. Provide emergency showers and eyewash.

Respirator Selection: NIOSH/OSHA: *7 ppm:* SA (any supplied-air respirator); or SCBAF (any self-contained breathing apparatus with a full facepiece). *Emergency or planned entry into unknown concentrations or IDLH conditions:* SCBAF: PD,PP (any self-contained breathing apparatus that has a full facepiece and is operated in a pressure-demand or other positive-pressure mode); or SAF:PD,PP:ASCBA (any supplied-air respirator that has a full facepiece and is operated in a pressure-demand or other positive-pressure mode in combination with an auxiliary self-contained breathing apparatus operated in a pressure-demand or other positive pressure mode). *Escape:* GMFOV [any air-purifying, full-facepiece respirator (gas mask) with a chin-style, front-or back-, mounted organic vapor canister]; or SCBAE (any appropriate escape-type, self-contained breathing apparatus).

Note: Substance reported to cause eye irritation or damage; may require eye protection.

Storage: Prior to working with this chemical you should be trained on its proper handling and storage. Before entering confined space where this chemical may be present, check to make sure that an explosive concentration does not exist. Keep in a cool, well-ventilated area, protected from physical damage, ignition sources and separated from strong oxidizers or caustics. Sources of ignition such as smoking and open flames are prohibited where this chemical is used, handled, or stored in a manner that could create a potential fire or explosion hazard. Metal containers involving the transfer of 5 gallons or more of this chemical should be grounded and bonded. Drums must be equipped with self-closing valves, pressure vacuum bungs, and flame arresters. Use only nonsparking tools and equipment, especially when opening and closing containers of this chemical.

Shipping: This compound requires a shipping label of: "Poison." It falls in DOT Hazard Class 6.1 and Packing Group I. Passenger aircraft or railcar shipment is forbidden and cargo aircraft shipment is also forbidden.

Spill Handling: Evacuate and restrict persons not wearing protective equipment from area of spill or leak until cleanup is complete. Remove all ignition sources. Establish forced ventilation to keep levels below explosive limit. Absorb liquids in vermiculite, dry sand, earth, peat, carbon, or a similar material and deposit in sealed containers. It may be necessary to contain and dispose of this chemical as a hazardous waste. If material or contaminated runoff enters waterways, notify downstream users of potentially contaminated waters. Contact your Department of Environmental Protection or your regional office of the federal EPA for specific recommendations. If employees are required to clean-up spills, they must be properly trained and equipped. OSHA 1910.120(q) may be applicable.

Fire Extinguishing: This chemical is a combustible liquid. Poisonous gases including phosgene and hydrogen chloride are produced in fire. Use dry chemical, carbon dioxide, alcohol foam, or polymer foam extinguishers. Vapors are heavier than air and will collect in low areas. Vapors may travel long distances to ignition sources and flashback. Vapors in confined areas may explode when exposed to fire. Containers may explode in fire. Storage containers and parts of containers may rocket great distances, in many directions. If material or contaminated runoff enters waterways, notify downstream users of potentially contaminated waters. Notify local health and fire officials and pollution control agencies. From a secure, explosion-proof location, use water spray to cool exposed containers. If cooling streams are ineffective (venting sound increases in volume and pitch, tank discolors, or shows any signs of deforming), withdraw immediately to a secure position. If employees are expected to fight fires, they must be trained and equipped in OSHA 1910.156.

Disposal Method Suggested: Incineration, preferably after mixing with another combustible fuel. Care must be exercised to assure complete combustion to prevent the formation of phosgene. An acid scrubber is necessary to remove the halo acids produced.

References

U.S. Environmental Protection Agency, "Chemical Profile: Chloroethanol," Washington, DC, Chemical Emergency Preparedness Program (November 30, 1987)

New York State Department of Health, "Chemical Fact Sheet: Ethylene Chlorohydrin," Albany, NY, Bureau of Toxic Substance Assessment (April 1999)

Ethylenediamine

Molecular Formula: $C_2H_8N_2$

Common Formula: $H_2NCH_2CH_2NH_2$

Synonyms: Aethaldiamin (German); Aethylenediamin (German); β-Aminoethylamine; 1,2-Diaminoaethan (German); 1,2-Diamino-ethaan (Dutch); 1,2-Diaminoethane, anhydrous; Dimethylenediamine; 1,2-Ethanediamine; 1,2-Ethylenediamine; Etilendiamina (Spanish); NCI-C60402; Sel-Rex Circuitprep SC replinisher/makeup; Sel-Rex XR-170A pretreatment

CAS Registry Number: 107-15-3

RTECS Number: KH8575000

DOT ID: UN 1604

EEC Number: 612-006-00-6

EINECS Number: 203-468-6

Regulatory Authority

- Air Pollutant Standard Set (ACGIH)[1] (DFG)[3] (HSE)[33] (Former USSR)[43] (OSHA)[58] (Several States)[60] (Several Canadian Provinces)
- Clean Air Act: Accidental Release Prevention/Flammable substances, (Section 112[r], Table 3), TQ = 20,000 lb (9,080 kg)
- Clean Water Act: Section 311 Hazardous Substances/RQ 40CFR117.3 (same as CERCLA, see below)
- Superfund/EPCRA 40CFR355, Appendix B Extremely Hazardous Substances: TPQ = 10,000 lb (4,540 kg)
- Superfund/EPCRA 40CFR302.4 Reportable Quantity (RQ): CERCLA, 5,000 lb (2,270 kg)
- Canada, WHMIS, Ingredients Disclosure List

Cited in U.S. State Regulations: Alaska (G), California (A, G), Connecticut (A), Florida (G), Illinois (G), Maine (G), Massachusetts (G), Nevada (A), New Hampshire (G), New Jersey (G), North Carolina (A), North Dakota (A), Pennsylvania (G), Rhode Island (G), Virginia (A), West Virginia (G).

Description: Ethylenediamine, $H_2NCH_2CH_2NH_2$, is a strongly alkaline, colorless, clear, thick liquid with an ammonia odor. A solid below 8.5°C. The odor threshold is 1.0 ppm. Boiling point = 116 – 117°C. Freezing/Melting point = 8.5°C. Flash point = 40°C. Autoignition temperature = 385°C. Explosive limits: LEL = 4.2%; UEL = 14.4%. Hazard Identification (based on NFPA-704 M Rating System): Health 3, Flammability 2, Reactivity 0. Soluble in water.

Potential Exposure: Ethylenediamine is used as a solvent, an emulsifier for casein and shellac solutions, a stabilizer in rubber latex, a chemical intermediate in the manufacture of dyes, corrosion inhibitors, synthetic waxes, fungicides, resins, insecticides, asphalt wetting agents, and pharmaceuticals. Ethylenediamine is a degradation product of the agricultural fungicide Maneb.

Incompatibilities: Forms explosive mixture with air. Ethylenediamine is a medium strong base. Violent reaction with strong acids, strong oxidizers, chlorinated organic compounds, acetic acid, acetic anhydride, acrolein, acrylic acid, acrylonitrile, allyl chloride, carbon disulfide, chlorosulfonic acid, epichlorohydrin, ethylene chlorohydrin, oleum, methyl oxide, vinyl acetate. Also incompatible with silver perchlorate, 3-propiolactone, mesityl oxide, ethylene dichloride, organic anhydrides, isocyanates, acrylates, substituted allyls, alkylene oxides, ketones, aldehydes, alcohols, glycols, phenols, cresols, caprolactum solution. Attacks aluminum, copper, lead, tin, zinc and alloys, some plastics, rubber and coatings.

Permissible Exposure Limits in Air: The OSHA TWA,[58] the DFG MAK,[3] the HSE TWA[33] and the ACGIH TWA value is 10 ppm (25 mg/m^3). There are no STEL values set. The NIOSH IDLH level is 1,000 ppm. The Former USSR-UNEP/IRPTC project[43] has set a MAC in workplace air of 2 mg/m^3. Several states have set guidelines or standards for ethylenediamine in ambient air[60] ranging from 250 μg/m^3 (North Dakota) to 300 – 2,500 μg/m^3 (North Carolina) to 400 μg/m^3 (Virginia) to 500 μg/m^3 (Connecticut) to 595 μg/m^3 (Nevada).

Determination in Air: XAD-2® (tube); DMF (any dust and mist respirator with a full facepiece); High-pressure liquid chromatography/Ultraviolet detection; NIOSH IV, Method #2540.

Permissible Concentration in Water: The former USSR-UNEP/IRPTC project[43] has set a MAC of 0.7 mg/l in water bodies used for domestic purposes.

Routes of Entry: Inhalation, skin absorption, ingestion, skin and/or eye contact.

Harmful Effects and Symptoms

Short Term Exposure: Ethylenediamine is corrosive to the eyes, skin, and respiratory tract. Skin contact can cause blistering. Eye contact can cause pain, serious injury and permanent damage. Inhalation can cause pulmonary edema, a medical emergency that can be delayed for several hours. This can cause death. Acute exposure to ethylenediamine may result in cough, difficulty in breathing, irritation of the lungs, and pneumonia. Nausea, vomiting, and diarrhea are often seen. Contact with ethylenediamine may result in redness, pain, irritation, and burns. Vapor inhalations at a concentration of 200 ppm for 5 – 10 minutes will lead to nasal irritation and produce a tingling sensation. Inhalation at concentrations of 400 ppm or greater leads to severe nasal irritation. Respiratory irritation may result. Many individuals are hypersensitive to ethylenediamine exposure; therefore, safe threshold limits are difficult to set.

Long Term Exposure: Repeated or prolonged contact with skin may cause skin allergy. Repeated or prolonged exposure can cause an asthma-like allergy. Repeated high exposure may cause liver, kidney, and lung damage.

Points of Attack: Respiratory system, liver, kidneys, skin.

Medical Surveillance: For those with frequent or potentially high exposure (half the TLV or greater), the following are recommended before beginning work and at regular times after that: Lung function tests. These may be normal if the person is not having an attack at the time of the test. If symptoms develop or overexposure is suspected, the following may be useful: Kidney function tests. Liver function tests. Evaluation by a qualified allergist, including careful exposure history and special testing, may help diagnose skin allergy. Consider chest x-ray after acute overexposure.

First Aid: If this chemical gets into the eyes, remove any contact lenses at once and irrigate immediately for at least 15 minutes, occasionally lifting upper and lower lids. Seek medical attention immediately. If this chemical contacts the skin, remove contaminated clothing and wash immediately with soap and water. Seek medical attention immediately. If this chemical has been inhaled, remove from exposure, begin rescue breathing (using universal precautions) if breathing has stopped and CPR if heart action has stopped. Transfer promptly to a medical facility. When this chemical has been swallowed, get medical attention. If victim is conscious, administer water or milk. Do not induce vomiting.

Personal Protective Methods: Wear protective gloves and clothing to prevent any reasonable probability of skin contact. Safety equipment suppliers/manufacturers can provide recommendations on the most protective glove/clothing material for your operation. Teflon, Saranex, Butyl rubber, and Neoprene are among the recommended protective materials. All protective clothing (suits, gloves, footwear, headgear) should be clean, available each day, and put on before work. Contact lenses should not be worn when working with this chemical. Wear splash-proof chemical goggles and face shield unless full facepiece respiratory protection is worn. Employees should wash immediately with soap when skin is wet or contaminated. Provide emergency showers and eyewash if liquids containing >5% contaminants are involved.

Respirator Selection: NIOSH/OSHA: *250 ppm:* SA:CF (any supplied-air respirator operated in a continuous-flow mode); or PAPRS [any powered, air-purifying respirator with cartridge(s) providing protection against the compound of concern]. *500 ppm:* CCRFS [any chemical cartridge respirator with a full facepiece and cartridge(s) providing protection against the compound of concern] organic vapor and acid gas cartridge(s); or GMFS [any air-purifying, full-facepiece respirator (gas mask) with a chin-style, front- or back-mounted canister providing protection against the compound of concern]; or PAPRTS [any powered, air-purifying respirator with a tight-fitting facepiece and cartridge(s) providing protection against the compound of concern]; or SCBAF (any self-contained breathing apparatus with a full facepiece); or SAF (any supplied-air respirator with a full facepiece). *1,000 ppm:* SAF:PD,PP (any supplied-air respirator that has a full facepiece and is operated in a pressure-demand or other positive-pressure mode). *Emergency or planned entry into unknown concentrations or IDLH conditions:* SCBAF:PD,PP (any self-contained breathing apparatus that has a full facepiece and is operated in a pressure-demand or other positive-pressure mode); or SAF:PD,PP:ASCBA (any supplied-air respirator that has a full facepiece and is operated in a pressure-demand or other positive-pressure mode in combination with an auxiliary self-contained breathing apparatus operated in a pressure-demand or other positive pressure mode). *Escape:* GMFS [any air-purifying, full-facepiece respirator (gas mask) with a chin-style, front- or back-mounted canister providing protection against the compound of concern]; or SCBAE (any appropriate escape-type, self-contained breathing apparatus).

Note: Substance reported to cause eye irritation or damage; may require eye protection.

Storage: Prior to working with this chemical you should be trained on its proper handling and storage. Before entering confined space where this chemical may be present, check to make sure that an explosive concentration does not exist. Ethylenediamine must be stored to avoid contact with acetic acid, acetic anhydride, acrolein, acrylic acid, acrylonitrile, allyl chloride, carbon disulfide, chlorosulfonic acid, epichlorhydrin, ethylene chlorohydrin oleum methyl oxide, vinyl acetate, hydrogen chloride and sulfuric acid since violent reactions occur. Store in tightly closed containers in a cool, well-ventilated area away from oxidizers (such as perchlorates, peroxides, permanganates, chlorates and nitrates). Detached outdoor storage is preferred. Sources of ignition, such as smoking and open flames, are prohibited where ethylenediamine is handled, used, or stored. Metal containers involving the transfer of 5 gallons or more of ethylenediamine should be grounded and bonded. Drums must be equipped with self-closing valves, pressure vacuum bungs, and flame arresters. Use only non-sparking tools and equipment, especially when opening and closing containers of ethylenediamine.

Shipping: This compound requires a shipping label of: "Corrosive, Flammable Liquid." It falls in DOT Hazard Class 8 and Packing Group II.

Spill Handling: Evacuate and restrict persons not wearing protective equipment from area of spill or leak until cleanup is complete. Stay upwind and keep out of low areas. Isolate area for 1/2 mile in all directions if tank car or truck is involved in fire. Remove all ignition sources. Establish forced ventilation to keep levels below explosive limit. Absorb liquids in vermiculite, dry sand, earth, peat, carbon, or a similar material and deposit in sealed containers. It may be necessary to contain and dispose of this chemical as a hazardous waste. If material or contaminated runoff enters waterways, notify downstream users of potentially contaminated waters. Contact your Department of Environmental Protection or your regional office of the federal EPA for specific recommendations. If employees are required to clean-up spills, they must be properly trained and equipped. OSHA 1910.120(q) may be applicable.

Fire Extinguishing: This chemical is a flammable liquid. Poisonous gases including nitrogen oxides are produced in fire. Use water spray, dry chemical, alcohol foam, or carbon dioxide. Wear full protective clothing including gloves and boots. If necessary to enter closed area, wear full-faced gas masks with self-contained breathing apparatus. Do not use water in case of drum or tank fires. If a leak or spill has not ignited, use water spray to reduce the vapors and dilute spills to nonflammable mixtures. Vapors are heavier than air and will collect in low areas. Vapors may travel long distances to ignition sources and flashback. Vapors in confined areas may explode when exposed to fire. Containers may explode in fire. Storage containers and parts of containers may rocket great distances, in many directions. If material or contaminated runoff enters waterways, notify downstream users of potentially contaminated waters. Notify local health and fire officials and pollution control agencies. From a secure, explosion-proof location, use water spray to cool exposed containers. If cooling streams are ineffective (venting sound increases in volume and pitch, tank discolors, or shows any signs of deforming), withdraw immediately to a secure position. If employees are expected to fight fires, they must be trained and equipped in OSHA 1910.156.

Disposal Method Suggested: Controlled incineration (oxides of nitrogen are removed from the effluent gas by scrubbers and/or thermal devices).

References

U.S. Environmental Protection Agency, "Chemical Hazard Information Profile: Ethylenediamine," Washington, DC (May 9, 1978)

Sax, N. I., Ed., "Dangerous Properties of Industrial Materials Report," 4, No. 2, 54–57 (1984)

U.S. Environmental Protection Agency, "Chemical Profile: Ethylenediamine," Washington, DC, Chemical Emergency Preparedness Program (November 30, 1987)

New Jersey Department of Health and Senior Services, "Hazardous Substance Fact Sheet: Ethylene Diamine," Trenton, NJ (May 1986)

Ethylenediamine Tetraacetic Acid (EDTA)

Molecular Formula: $C_{10}H_{16}N_2O$

Synonyms: Acetic acid (ethylenedinitrilo)tetra-; Acide ethylenediaminetetracetique (French); Acido etilendiaminotetraacetico (Spanish); Aroquest 75; Celon A; Celon ATH; Cheelox; Chemcolox 340; Complexon II; 3,6-Diazaoctanedioic acid, 3,6-bis(carboxymethyl)-; Edathamil; Edetic; Edetic acid; EDTA; EDTA acid; Endrate; Ethylenediamine tetraacetate; Ethylenediamine-*N,N,N',N'*-tetraacetic acid; Ethylenediaminetetraacetic acid; Ethylenedinitrilotetraacetic acid; Glycine, *N,N'*-1,2-ethanediylbis[*N*-(carboxymethyl)-9CI]; Hamp-ENE acid; Havidote; Kalex acids; Metaquest A; Nervanaid B acid; Nullapon B acid; Nullapon BF acid; Perma Kleer 50 acid; Questric acid 5286; SEQ-100; Sequestrene AA; Sequestric acid; Sequestrol; Tetrine acid; Titriplex; Tricon BW; Trilon B; Trilon BS; Trilon BW; UN 9117; Versene; Versene acid; Warkeelate acid

CAS Registry Number: 60-00-4

RTECS Number: AH4025000

DOT ID: NA 9117

EINECS Number: 200-449-4

Regulatory Authority

- Clean Water Act: Section 311 Hazardous Substances/RQ 40CFR117.3 (same as CERCLA, see below)
- Superfund/EPCRA 40CFR302.4 Reportable Quantity (RQ): CERCLA, 5,000 lb (2,270 kg)

Cited in U.S. State Regulations: California (G), New Jersey (G), Pennsylvania (G).

Description: EDTA is a white, odorless, crystalline material or white powder. Boiling point = 150°C (decomposes). Hazard Identification (based on NFPA-704 M Rating System): Health 1, Flammability 0, Reactivity 0. Slightly soluble in water.

Potential Exposure: EDTA is used as a drug, food additive, in cosmetics, household and textile articles, in pharmaceutical products and in biochemicals. It is also used as a laboratory chemical in research.

Incompatibilities: Reacts with strong oxidants, strong bases, copper, copper alloys and nickel.

Permissible Exposure Limits in Air: No OELs have been established.

Determination in Air: None available.

Routes of Entry: Inhalation.

Harmful Effects and Symptoms

Short Term Exposure: EDTA irritates the eyes, skin, and respiratory tract. Skin contact may cause a burning sensation and rash. May affect the kidneys.

Long Term Exposure: May damage the developing fetus. May cause kidney damage.

Points of Attack: Kidneys.

Medical Surveillance: Kidney function tests.

First Aid: If this chemical gets into the eyes, remove any contact lenses at once and irrigate immediately for at least 15 minutes, occasionally lifting upper and lower lids. Seek medical attention immediately. If this chemical contacts the skin, remove contaminated clothing and wash immediately with soap and water. Seek medical attention immediately. If this chemical has been inhaled, remove from exposure, begin rescue breathing (using universal precautions) if breathing has stopped and CPR if heart action has stopped. Transfer promptly to a medical facility. When this chemical has been swallowed, get medical attention. Give large quantities of water and induce vomiting. Do not make an unconscious person vomit.

Personal Protective Methods: Wear protective gloves and clothing to prevent any reasonable probability of skin contact. Safety equipment suppliers/manufacturers can provide recommendations on the most protective glove/clothing material for your operation.. All protective clothing (suits, gloves, footwear, headgear) should be clean, available each day, and put on before work. Contact lenses should not be worn when working with this chemical. Wear dust-proof chemical goggles and face shield unless full facepiece respiratory protection is worn. Employees should wash immediately with soap when skin is wet or contaminated. Provide emergency showers and eyewash.

Respirator Selection: Where the potential for exposure to this chemical, use a MSHA/NIOSH approved supplied-air respirator with a full facepiece operated in the positive pressure mode or with a full facepiece, hood, or helmet in the continuous flow mode, or use a MSHA/NIOSH approved self-contained breathing apparatus with a full facepiece operated in pressure-demand or other positive pressure mode.

Storage: Prior to working with this chemical you should be trained on its proper handling and storage. Store in tightly closed containers in a cool, well ventilated area away from oxidizers. Where possible, automatically pump liquid from drums or other storage containers to process containers.

Spill Handling: Evacuate persons not wearing protective equipment from area of spill or leak until clean-up is complete. Remove all ignition sources. Collect powdered material in the most convenient and safe manner and deposit in sealed containers. Ventilate area after clean-up is complete. It may be necessary to contain and dispose of this chemical as a hazardous waste. If material or contaminated runoff enters waterways, notify downstream users of potentially contaminated waters. Contact your Department of Environmental Protection or your regional office of the federal EPA for specific recommendations. If employees are required to clean-up spills, they must be properly trained and equipped. OSHA 1910.120(q) may be applicable.

Fire Extinguishing: This chemical is a combustible solid. It does not easily ignite. Use dry chemical, carbon dioxide, water spray, or alcohol or polymer foam extinguishers. Poisonous gases are produced in fire including nitrogen oxides and carbon monoxide. If material or contaminated runoff enters waterways, notify downstream users of potentially contaminated waters. Notify local health and fire officials and pollution control agencies. From a secure, explosion-proof location, use water spray to cool exposed containers. If cooling streams are ineffective (venting sound increases in volume and pitch, tank discolors, or shows any signs of deforming), withdraw immediately to a secure position. If employees are expected to fight fires, they must be trained and equipped in OSHA 1910.156.

References

New Jersey Department of Health and Senior Services, "Hazardous Substance Fact Sheet, Ethylenediamine Tetraacetic Acid," Trenton, NJ (January 1999)

Ethylene Dibromide

Molecular Formula: $C_2H_4Br_2$

Common Formula: $BrCH_2CH_2Br$

Synonyms: Aadibroom; Aethylenbromid (German); Bromofume; Bromuro di etile (Italian); Celmide; DBE; 1,2-Dibromaethan (German); 1,2-Dibromoetano (Italian, Spanish); α,β-Dibromoethane; *sym*-Dibromoethane; 1,2-Dibromoethane; Dibromoethane; Dibromudo de etileno (Spanish); Dibromure d'ethylene (French); 1,2-Dibromomethaan (Dutch); Dowfume 40; Dowfume EDB; Dowfume W-8; Dowfume W-85; Dwubromoetan (Polish); EDB; EDB-85; E-D-BEE; ENT 15,349; Ethane, 1,2-dibromo-; Ethylene bromide; 1,2-Ethylene dibromide; Fumo-gas; Glycol bromide; Glycol dibromide; Iscobrome D; Kopfume; NCI-C00522; Nefis; Nephis; Pestmaster; Pestmaster EDB-85; Sanhyuum; Soilbrom; Soilbrom-40; Soilbrom-85; Soilbrom-90EC; Soilbrome-85; Soilfume; Unifume

CAS Registry Number: 106-93-4

RTECS Number: KH9275000

DOT ID: UN 1605

EEC Number: 602-010-00-6

Regulatory Authority

- Carcinogen (Animal Positive) (IARC) (NCI) (NTP)[9] (ACGIH)[1] (DFG)[3] (Several Canadian Provinces)
- Banned or Severely Restricted (Many Countries) (UN)[3][35]
- Toxic Substance (World Bank)[15]

- Air Pollutant Standard Set (ACGIH)[1] (Other Countries)[35] (OSHA)[58] (Several States)[60]
- Clean Air Act: Hazardous Air Pollutants (Title I, Part A, Section 112); Accidental Release Prevention/Flammable substances, (Section 112[r], Table 3), TQ = 20,000 lb (9,080 kg)
- Clean Water Act: Section 311 Hazardous Substances/RQ 40CFR117.3 (same as CERCLA, see below); Section 313 Water Priority Chemicals (57FR41331, 9/9/92)
- EPA Hazardous Waste Number (RCRA No.): U067
- RCRA, 40CFR261, Appendix 8 Hazardous Constituents
- RCRA 40CFR268.48; 61FR15654, Universal Treatment Standards: Wastewater (mg/l), 0.028; Nonwastewater (mg/kg), 15
- RCRA 40CFR264, Appendix 9; TSD Facilities Ground Water Monitoring List. Suggested test method(s) (PQL μg/l): 8010 (10); 8240 (5)
- Safe Drinking Water Act: MCL, 0.00005 mg/l; MCGL, zero
- Superfund/EPCRA 40CFR355, Appendix B Extremely Hazardous Substances: TPQ = 10,000 lb (4,540 kg)
- Superfund/EPCRA 40CFR302.4 Reportable Quantity (RQ): CERCLA, 1 lb (0.454 kg)
- EPCRA Section 313 Form R *de minimis* concentration reporting level: 0.1%
- Canada, WHMIS, Ingredients Disclosure List

Cited in U.S. State Regulations: Alaska (G), Arizona (W), California (A, G, W), Connecticut (A, W), Florida (G), Illinois (G), Indiana (A), Kansas (G, W), Louisiana (G), Maine (G, W), Massachusetts (G, W), Michigan (G), Minnesota (W), New Hampshire (G), New Jersey (G), New Mexico (W), New York (G, A), North Carolina (A), North Dakota (A), Pennsylvania (G, A), Rhode Island (G), South Carolina (A), Vermont (G), Virginia (G, A), Washington (G, W), West Virginia (G), Wisconsin (G, W).

Description: Ethylene dibromide, $BrCH_2CH_2Br$, is a colorless nonflammable liquid or solid (below 10°C) with a sweet, chloroform-like odor. The minimum concentration detectable by odor is 10 ppm. Boiling point = 131°C. Freezing/Melting point = 10°C. Soluble in water.

Potential Exposure: Ethylene dibromide is used principally as a fumigant for ground pest control and as a constituent of ethyl gasoline. It is also used in fire extinguishers, gauge fluids, and waterproofing preparations; and it is used as a solvent for celluloid, fats, oils, and waxes.

Incompatibilities: Reacts vigorously with chemically active metals, liquid ammonia, strong bases, strong oxidizers, causing fire and explosion hazard. Light, heat, and moisture can cause slow decomposition, forming hydrogen bromide. Attacks fats, rubber, some plastics and coatings.

Permissible Exposure Limits in Air: ACGIH, Germany and Sweden have set no numerical limits, stating only that the substance is a carcinogen and that exposure should be avoided. OSHA[58] has set a TWA of 20 ppm, a ceiling of 30 ppm and a maximum peak above the ceiling of 50 ppm for 5 minutes. Argentina[35] has set a TWA of 20 ppm (140 mg/m^3) and an STEL of 30 ppm (220 mg/m^3). Czechoslovakia[35] has set a TWA of 10 mg/m^3 with a ceiling value of 20 mg/m^3. NIOSH IDLH = Ca [100 ppm]. Several states have set guidelines or standards for ethylene dibromide in ambient air[60] ranging from zero (North Dakota and New York) to 0.045 μg/m^3 (North Carolina) to 2.47 μg/m^3 (Pennsylvania) to 720 μg/m^3 (Indiana) to 770 μg/m^3 (South Carolina) to 1,500 μg/m^3 (Virginia) to 1,550 μg/m^3 (Connecticut).

Determination in Air: Charcoal tube absorption; workup with Benzene/Methanol; Gas chromatography/Electrochemical detection; NIOSH IV, Method #1008.

Permissible Concentration in Water: Several states have set guidelines or standards for ethylene dibromide in drinking water[61] ranging from 0.005 μg/l (Kansas) to 0.008 μg/l (Minnesota) to 0.01 μg/l (Arizona) to 0.02 μg/l (California and Washington) to 0.04 μg/l (Massachusetts) to 0.10 μg/l (Connecticut and New Mexico) to 0.50 μg/l (Wisconsin) to 1.0 μg/l (Maine). EPA[62] has proposed a maximum concentration level in drinking water of 0.05 μg/l.

Routes of Entry: Inhalation of the vapor, absorption through the skin, ingestion and skin and/or eye contact.

Harmful Effects and Symptoms

Short Term Exposure: Contact can cause severe skin and eye burns, with permanent eye damage. Exposure to the vapor may also damage the eyes. Inhalation may irritate and damage the lungs. Higher exposures can cause pulmonary edema, a medical emergency that can be delayed for several hours. This can cause death. High exposure can cause dizziness, drowsiness, vomiting, unconsciousness, and death. High exposure can damage the liver or kidneys enough to cause death.

Inhalation: Levels of 75 ppm may cause irritation of the nose, throat and lungs. 100 – 200 ppm for 1 hour may cause diarrhea, abdominal pain and vomiting. Other symptoms may include headache, loss of appetite, swollen glands, pale skin coloring, insomnia, dizziness and depression. Accidental high exposure has caused symptoms as listed above, internal bleeding and death.

Skin: Contact with as little as 1 gram (1/28 ounce) may cause itching, swelling, redness, burning and blistering. May be absorbed through the skin and cause symptoms as listed under inhalation.

Eyes: May cause irritation of eyes and eyelids.

Ingestion: May cause vomiting, diarrhea, abdominal pain, nausea and damage to the liver and kidneys. As little as 4.5 ml (about 1 teaspoon) has caused death.

Long Term Exposure: May cause irritation of the throat, headaches, loss of appetite, swollen glands, paleness, insomnia, vomiting, diarrhea, abdominal pain and damage to the liver and kidneys. Ethylene dibromide is a probable carcinogen in humans. It causes birth defects and changes in the genetic material of laboratory animals. It may damage the reproductive system, causing abnormal sperm in males and decreased fertility in females. Handle with extreme caution.

Points of Attack: Eyes, skin, respiratory system, liver, kidneys, reproductive system. Cancer Site in animals: skin and lung tumors.

Medical Surveillance: Preemployment and periodic examinations should evaluate the skin and eyes, respiratory tract, and liver and kidney functions.

First Aid: If this chemical gets into the eyes, remove any contact lenses at once and irrigate immediately for at least 15 minutes, occasionally lifting upper and lower lids. Seek medical attention immediately. If this chemical contacts the skin, remove contaminated clothing and wash immediately with soap and water. Seek medical attention immediately. If this chemical has been inhaled, remove from exposure, begin rescue breathing (using universal precautions) if breathing has stopped and CPR if heart action has stopped. Transfer promptly to a medical facility. When this chemical has been swallowed, get medical attention. Give large quantities of water and induce vomiting. Do not make an unconscious person vomit. Medical observation is recommended for 24 – 48 hours after breathing overexposure, as pulmonary edema may be delayed. As first aid for pulmonary edema, a doctor or authorized paramedic may consider administering a corticosteroid spray.

Personal Protective Methods: Wear protective gloves and clothing to prevent any reasonable probability of skin contact. Safety equipment suppliers/manufacturers can provide recommendations on the most protective glove/clothing material for your operation. Polyvinyl Alcohol, Viton/Neoprene, and Teflon are among the recommended protective materials. All protective clothing (suits, gloves, footwear, headgear) should be clean, available each day, and put on before work. Contact lenses should not be worn when working with this chemical. Wear splash-proof chemical goggles and face shield unless full facepiece respiratory protection is worn. Employees should wash immediately with soap when skin is wet or contaminated. Provide emergency showers and eyewash.

Respirator Selection: NIOSH: *At any detectable concentration:* SCBAF:PD,PP (any MSHA/NIOSH approved self-contained breathing apparatus that has a full facepiece and is operated in a pressure-demand or other positive-pressure mode); or SAF:PD,PP:ASCBA (any supplied-air respirator that has a full facepiece and is operated in a pressure-demand or other positive-pressure mode in combination with an auxiliary, self-contained breathing apparatus operated in a pressure-demand or other positive pressure mode). *Escape:* GMFOV [any air-purifying, full-facepiece respirator (gas mask) with a chin-style, front-or back-mounted organic vapor canister]; or SCBAE (any appropriate escape-type, self-contained breathing apparatus).

Storage: Store in a cool, dry place that is well-ventilated. Keep in tightly sealed containers and away from light, heat, active metals and liquid ammonia. A regulated, marked area should be established where this chemical is handled, used, or stored in compliance with OSHA standard 1910.1045.

Shipping: This compound requires a shipping label of: "Poison." It falls in DOT Hazard Class 6.1 and Packing Group I. Passenger aircraft or railcar shipment is forbidden as is cargo aircraft shipment.

Spill Handling: Evacuate and restrict persons not wearing protective equipment from area of spill or leak until cleanup is complete. Remove all ignition sources. Ventilate area of spill or leak. Absorb liquids in vermiculite, dry sand, earth, peat, carbon, or a similar material and deposit in sealed containers. It may be necessary to contain and dispose of this chemical as a hazardous waste. If material or contaminated runoff enters waterways, notify downstream users of potentially contaminated waters. Contact your Department of Environmental Protection or your regional office of the federal EPA for specific recommendations. If employees are required to clean-up spills, they must be properly trained and equipped. OSHA 1910.120(q) may be applicable.

Fire Extinguishing: EDB is a non-combustible liquid. Use an extinguishing agent suited to a surrounding fire. Poisonous gases are produced in fire including bromine and hydrogen bromide. If material or contaminated runoff enters waterways, notify downstream users of potentially contaminated waters. Notify local health and fire officials and pollution control agencies. From a secure, explosion-proof location, use water spray to cool exposed containers. If cooling streams are ineffective (venting sound increases in volume and pitch, tank discolors, or shows any signs of deforming), withdraw immediately to a secure position. If employees are expected to fight fires, they must be trained and equipped in OSHA 1910.156.

Disposal Method Suggested: Controlled incineration with adequate scrubbing and ash disposal facilities.[22]

References

Environmental Protection Agency, Sampling and Analysis of Selected Toxic Substances, Task II-Ethylene Dibromide, Final Report. Office of Toxic Substances, EPA, Washington, DC (September 1975)

Occupational Health and Safety Administration, Criteria for a Recommended Standard: Occupational Exposure to ethylene Dibromide, NIOSH Doc. No. 77–221 (1977)

National Institute for Occupational Safety and Health, Current Intelligence Bulletin No. 3: Ethylene Dibromide, Rockville, MD (July 7, 1975), and Current Intelligence Bulletin No. 37, Ethylene Dibromide, Cincinnati, Ohio (October 26, 1981)

Sax, N. I., Ed., "Dangerous Properties of Industrial Materials Report," 1, No. 5, 58–60 (1981)

New York State Department of Health, "Chemical Fact Sheet: Ethylene Dibromide," Albany, NY, Bureau of Toxic Substance Assessment (March 1986 and Version 2)

Ethylene Dichloride

Molecular Formula: $C_2H_4Cl_2$

Common Formula: $ClCH_2CH_2Cl$

Synonyms: Aethylenchlorid (German); 1,2-Bichloroethane; Bichlorure d'ethylene (French); Borer Sol; Brocide; Chlorure d'ethylene (French); Cloruro di ethene (Italian); Destruxol Borer-Sol; 1,2-Dichloorethaan (Dutch); 1,2-Dichlor-aethan (German); Dichloremulsion; Di-chlor-mulsion; α,β-Dichloroethane; *sym*-Dichloroethane; 1,2-Dichloroethane; Dichloro-1,2-ethane (French); Dichloroethylene; 1,2-Dicloroetano (Italian, Spanish); Dutch liquid; Dutch oil; EDC; ENT 1,656; Ethane dichloride; Ethane, 1,2-dichloro-; Ethyleendichloride (Dutch); Ethylene chloride; 1,2-Ethylene dichloride; Ethylene dichloride; Freson 150; Glycol dichloride; NCI-C00511

CAS Registry Number: 107-06-2

RTECS Number: KI0525000

DOT ID: UN 1184

EEC Number: 602-012-00-7

Regulatory Authority

- Carcinogen (Animal Positive) (IARC) (NCI)[9] (ACGIH)[1]
- Banned or Severely Restricted (E. Germany and Saudi Arabia) (UN)[13]
- Air Pollutant Standard Set (ACGIH)[1] (DFG)[3] (HSE)[33] (OSHA)[58] (Other Countries)[35] (Several States)[60]
- Clean Air Act: Hazardous Air Pollutants (Title I, Part A, Section 112)
- Clean Water Act: Section 311 Hazardous Substances/RQ 40CFR117.3 (same as CERCLA, see below); 40CFR423, Appendix A, Priority Pollutants; Section 313 Water Priority Chemicals (57FR41331, 9/9/92); Toxic Pollutant (Section 401.15)
- EPA Hazardous Waste Number (RCRA No.): U077, D028
- RCRA, 40CFR261, Appendix 8 Hazardous Constituents
- RCRA Toxicity Characteristic (Section 261.24), Maximum Concentration of Contaminants, regulatory level, 0.5 mg/l
- RCRA 40CFR268.48; 61FR15654, Universal Treatment Standards: Wastewater (mg/l), 0.21; Nonwastewater (mg/kg), 6.0
- RCRA 40CFR264, Appendix 9; TSD Facilities Ground Water Monitoring List. Suggested test method(s) (PQL μg/l): 8010 (0.5); 8240 (5)
- Safe Drinking Water Act: MCL, 0.005 mg/l; MCLG, zero; Regulated chemical (47 FR 9352)
- Superfund/EPCRA 40CFR302.4 Reportable Quantity (RQ): CERCLA, 100 lb (45.4 kg)
- U.S. DOT Regulated Marine Pollutant (49CFR172.101, Appendix B)
- Canada, WHMIS, Ingredients Disclosure List

Cited in U.S. State Regulations: Alaska (G), California (G, W), Connecticut (A, W), Florida (G, W), Illinois (G), Indiana (A), Kansas (G), Louisiana (G), Maine (G, W), Massachusetts (G, A), Minnesota (W), Nevada (A), New Hampshire (G, W), New Jersey (G, W), New Mexico (W), New York (G, A), North Carolina (A), North Dakota) (A), Oklahoma (G), Pennsylvania (G, A), Rhode Island (G, A), South Carolina (A), Vermont (G), Virginia (G, A), Washington (G), West Virginia (G), Wisconsin (G).

Description: 1,2-Dichloroethane, $ClCH_2CH_2Cl$, is a colorless, flammable liquid which has a pleasant, chloroform-like odor, sweetish taste. Decomposes slowly: turns dark and acidic on contact with air, moisture, and light. The odor threshold is 100 ppm. Boiling point = 84°C. Flash point = 13°C. Autoignition temperature = 413°C. Explosive limits: LEL = 6.2%; UEL = 16.0%. Hazard Identification (based on NFPA-704 M Rating System): Health 2, Flammability 3, Reactivity 0. Insoluble in water.

Potential Exposure: In recent years, 1,2-dichloroethane has found wide use in the manufacture of ethylene glycol, diaminoethylene, polyvinyl chloride, nylon, viscose rayon, styrenebutadiene rubber, and various plastics. It is a solvent for resins, asphalt, bitumen, rubber, cellulose acetate, cellulose ester, and paint; a degreaser in the engineering, textile and petroleum industries; and an extracting agent for soybean oil and caffeine. It is also used as an antiknock agent in gasoline, a pickling agent, a fumigant, and a dry-cleaning agent. It has found use in photography, xerography, water softening, and also in the production of adhesives, cosmetics, pharmaceuticals, and varnishes.

Incompatibilities: Forms explosive mixture with air. Reacts violently with strong oxidizers and caustics; chemically-active metals such as magnesium or aluminum powder, sodium and potassium, alkali metals, alkali amides; liquid ammonia Decomposes to vinyl chloride and HCl above 1,112°F. Attacks plastics, rubber, coatings. Attacks many metals in presence of water.

Permissible Exposure Limits in Air: The OSHA PEL is 50 ppm TWA and ceiling of 100 ppm, 200 ppm 5-minute maximum peak in any 3 hours. ACGIH recommends a limit of 10 ppm (40 mg/m^3). The NIOSH IDLH level is potential occupational carcinogen [50 ppm]. The Former USSR-UNEP/IRPTC project[43] has set MAC values in ambient air in residential areas of 3 mg/m^3 on a momentary basis and 1 mg/m^3 on a daily average basis. Several states have set guidelines or standards for ethylene dichloride in ambient air[60] ranging from 0 – 400 μg/m^3 (North Dakota) to 0.038 μg/m^3 (North Carolina) to 0.04 μg/m^3 (Rhode Island) to 0.2 μg/m^3 (New York) to 0.39 μg/m^3 (Massachusetts) to 20.0 μg/m^3 (Connecticut) to 148.0 μg/m^3 (Pennsylvania) to 200.0 μg/m^3 (South Carolina) to 650.0 μg/m^3 (Nevada) to 1,000.0 μg/m^3 (Indiana).

Determination in Air: Charcoal adsorption, workup with CS_2, analysis by gas chromatography/flame ionization. See NIOSH IV, Method #1003 for halogenated hydrocarbons.[18]

Permissible Concentration in Water: To protect freshwater aquatic life – 118,000 μg/l on an acute toxicity basis and 20,000 μg/l on a chronic basis. To protect saltwater aquatic life – 113,000 μg/l on an acute toxicity basis. To protect human health – preferably zero. An additional lifetime cancer risk of 1 in 100,000 occurs at a concentration of 9.4 μg/l.[6]

The WHO has set a limit for drinking water of 0.01 mg/l (10 μg/l).[35] The Former USSR has set a limit in water bodies used for domestic purposes variously quoted at 2.0 mg/l[43] and 0.02 mg/l.[35] EPA[48] has set a longer term health advisory of 2.6 mg/l for an adult.

Several states have set guidelines or standards for ethylene dichloride in drinking water.[61] The standards set range from 2.0 μg/l (New Jersey) to 3.0 μg/l (Florida) to 10.0 μg/l (New Mexico). The guidelines range from 0.38 μg/l (New Hampshire) to 1.0 μg/l (California and Connecticut) to 3.8 (Minnesota) to 5.0 μg/l (Maine).

Determination in Water: Inert gas purge followed by chromatography with halide specific detection (EPA Method 601) or gas chromatography plus mass spectrometry (EPA Method 624).

Routes of Entry: Inhalation of vapor, skin absorption of liquid, ingestion, skin and/or eye contact.

Harmful Effects and Symptoms

Short Term Exposure: Irritates the eyes, skin, and respiratory tract. Inhalation of the vapors can cause pulmonary edema, a medical emergency that can be delayed for several hours. This can cause death. Exposure can cause nausea, vomiting, headaches, drowsiness, and loss of consciousness. Overexposure to ethylene dichloride may damage the central nervous system, kidneys, liver.

Inhalation: Levels of 10 – 30 ppm may cause dizziness, nausea, and vomiting. Levels up to 50 ppm may cause weakness, trembling, headaches, abdominal cramps, liver and kidney damage, and fluid build up in lungs. May cause coma and death at high levels.

Skin: Contact may cause irritation and skin rash, and irritates the eyes.

Eyes: May cause redness, pain, and blurred vision. Vapor can damage the cornea.

Ingestion: Ingestion of 2 ounces has resulted in nausea, vomiting, faintness, drowsiness, difficulty breathing, pale skin, internal bleeding, kidney damage, and death due to respiratory failure. Other possible symptoms may include abdominal spasms, severe headache, lethargy, lowered blood pressure, diarrhea, shock, physical collapse, and coma.

Long Term Exposure: Repeated or prolonged contact can chronically irritate the skin causing dryness, redness and a rash. Prolonged or repeated exposure may cause eye, nose and throat irritation, nerve damage, liver and kidney damage. This substance has been determined to cause cancer of the lung, stomach, breast and other sites in laboratory animals, and may be a human carcinogen. Can irritate the lungs and bronchitis may develop. Repeated or prolonged exposure can cause loss of appetite, nausea and vomiting, trembling and low blood sugar.

Points of Attack: Eyes, skin, kidneys, liver, central nervous system, cardiovascular system. Cancer Site in animals: forestomach, mammary gland and circulatory system cancer.

Medical Surveillance: Before beginning employment and at regular times after that, the following are recommended: Lung function tests. Liver and kidney function tests. If symptoms develop or overexposure is suspected, the following may be useful: Consider chest x-ray after acute overexposure.

First Aid: If this chemical gets into the eyes, remove any contact lenses at once and irrigate immediately for at least 15 minutes, occasionally lifting upper and lower lids. Seek medical attention immediately. If this chemical contacts the skin, remove contaminated clothing and wash immediately with soap and water. Seek medical attention immediately. If this chemical has been inhaled, remove from exposure, begin rescue breathing (using universal precautions) if breathing has stopped and CPR if heart action has stopped. Transfer promptly to a medical facility. When this chemical has been swallowed, get medical attention. Give large quantities of water and induce vomiting. Do not make an unconscious person vomit. Medical observation is recommended for 24 – 48 hours after breathing overexposure, as pulmonary edema may be delayed. As first aid for pulmonary edema, a doctor or authorized paramedic may consider administering a corticosteroid spray.

Personal Protective Methods: Wear protective gloves and clothing to prevent any reasonable probability of skin contact. Safety equipment suppliers/manufacturers can provide recommendations on the most protective glove/clothing material for your operation. Silvershield and Viton are among the recommended protective materials. All protective clothing (suits, gloves, footwear, headgear) should be clean, available each day, and put on before work. Contact lenses should not be worn when working with this chemical. Wear splash-proof chemical goggles and face shield unless full facepiece respiratory protection is worn. Employees should wash immediately with soap when skin is wet or contaminated. Provide emergency showers and eyewash.

Respirator Selection: NIOSH: *At any detectable concentration:* SCBAF:PD,PP (any MSHA/NIOSH approved self-contained breathing apparatus that has a full facepiece and is operated in a pressure-demand or other positive-pressure mode); or SAF:PD,PP:ASCBA (any supplied-air respirator that has a full facepiece and is operated in a pressure-demand or other positive-pressure mode in combination with an auxiliary, self-contained breathing apparatus operated in a pressure-demand or other positive pressure mode). *Escape:* GMFOV [any air-purifying, full-facepiece respirator (gas mask) with a chin-style, front-or back-mounted organic vapor canister]; or SCBAE (any appropriate escape-type, self-contained breathing apparatus).

Storage: Prior to working with this chemical you should be trained on its proper handling and storage. Before entering confined space where this chemical may be present, check to make sure that an explosive concentration does not exist. Store in tightly closed containers in a cool, well-ventilated area away from oxidizers (such as perchlorates, peroxides, permanganates, chlorates and nitrates), strong acids (such as hydrochloric, sulfuric and nitric), chemically active metals (such as potassium, sodium, magnesium and zinc), strong caustics (such as sodium hydroxide) and dimethylaminopropylamine since violent reactions occur. A regulated, marked area should be established where this chemical is handled, used, or stored in compliance with OSHA standard 1910.1045.

Shipping: This compound requires a shipping label of: "Flammable Liquid, Poison." It falls in DOT Hazard Class 3 and Packing Group II.

Spill Handling: Evacuate and restrict persons not wearing protective equipment from area of spill or leak until cleanup is complete. Remove all ignition sources. Establish forced ventilation to keep levels below explosive limit. Absorb liquids in vermiculite, dry sand, earth, peat, carbon, or a similar material and deposit in sealed containers. Keep ethylene dichloride out of a confined space, such as a sewer, because of the possibility of an explosion, unless the sewer is designed to prevent the build-up of explosive concentrations. It may be necessary to contain and dispose of this chemical as a hazardous waste. If material or contaminated runoff enters waterways, notify downstream users of potentially contaminated waters. Contact your Department of Environmental Protection or your regional office of the federal EPA for specific recommendations. If employees are required to clean-up spills, they must be properly trained and equipped. OSHA 1910.120(q) may be applicable.

Fire Extinguishing: This chemical is a flammable liquid. Poisonous gases including phosgene and hydrogen chloride are produced in fire. Use dry chemical, carbon dioxide, or alcohol foam extinguishers. Vapors are heavier than air and will collect in low areas. Vapors may travel long distances to ignition sources and flashback. Vapors in confined areas may explode when exposed to fire. Containers may explode in fire. Storage containers and parts of containers may rocket great distances, in many directions. If material or contaminated runoff enters waterways, notify downstream users of potentially contaminated waters. Notify local health and fire officials and pollution control agencies. From a secure, explosion-proof location, use water spray to cool exposed containers. If cooling streams are ineffective (venting sound increases in volume and pitch, tank discolors, or shows any signs of deforming), withdraw immediately to a secure position. If employees are expected to fight fires, they must be trained and equipped in OSHA 1910.156.

Disposal Method Suggested: Incineration, preferably after mixing with another combustible fuel. Care must be exercised to assure complete combustion to prevent the formation of phosgene. An acid scrubber is necessary to remove the halo acids produced.[22]

References

National Institute for Occupational Safety and Health, Criteria for a Recommended Standard: Occupational Exposure to Ethylene Dichloride, NIOSH Doc. No. 76–139 (1976)

National Institute for Occupational Safety and Health, Ethylene Dichloride, NIOSH Current Intelligence Bulletin No. 25, Washington, DC (April 19, 1978)

National Institute for Occupational Safety and Health, Chloroethanes: Review of Toxicity, Current Intelligence Bulletin No. 27, Washington, DC (August 21, 1978)

U.S. Environmental Protection Agency, "Chemical Hazard Information Profile: 1,2-Dichloroethane," Washington, DC (September 1, 1977)

U.S. Environmental Protection Agency, Chlorinated Ethanes: Ambient Water Quality Criteria, Washington, DC (1980)

U.S. Environmental Protection Agency, 1,2-Dichloroethane, Health and Environmental Effects Profile No. 70, Washington, DC, Office of Solid Waste (April 31, 1980)

Sax, N. I., Ed., "Dangerous Properties of Industrial Materials Report," 1, No. 4, 50–52 (1981)

U.S. Environmental Protection Agency, "Health Advisory: 1,2-Dichloroethane," Washington, DC, Office of Drinking Water (March 31, 1987)

U.S. Public Health Service, "Toxicological Profile for 1,2-Dichloroethane," Atlanta, Georgia, Agency for Toxic Substance & Disease Registry (December 1988)

New Jersey Department of Health and Senior Services, "Hazardous Substance Fact Sheet: 1,2-Dichloroethane," Trenton, NJ (July 1986)

New York State Department of Health, "Chemical Fact Sheet: 1,2-Dichloroethane," Albany, NY, Bureau of Toxic Substance Assessment (Version 2)

Ethylene Fluorohydrin

Molecular Formula: C_2H_5FO

Common Formula: $HOCH_2CH_2F$

Synonyms: 2-Fluoetanol (Spanish); β-Fluoroethanol; 2-Fluroethanol; TL 741

CAS Registry Number: 371-62-0

RTECS Number: KL1575000

- ***Regulatory Authority:*** OSHA 29CFR1910.119, Appendix A, Process Safety List of Highly Hazardous Chemicals, TQ = 100 lb (45 kg)
- Superfund/EPCRA 40CFR355, Appendix B Extremely Hazardous Substances: TPQ = 10 lb (4.54 kg)
- Superfund/EPCRA 40CFR302.4 Reportable Quantity (RQ): EHS, 1 lb (0.454 kg)

Cited in U.S. State Regulations: Massachusetts (G), New Jersey (G), Pennsylvania (G).

Description: Ethylene fluorohydrin, $HOCH_2CH_2F$, is a colorless liquid. Boiling point = 103.5°C. Flash point = 31°C. Soluble in water.

Potential Exposure: Ethylene fluorohydrin is used as a rodenticide, insecticide, and acaricide. Not registered as a pesticide in the U.S.

Incompatibilities: Strong oxidizers.

Permissible Exposure Limits in Air: No standards set.

Permissible Concentration in Water: No criteria set.

Harmful Effects and Symptoms

Short Term Exposure: Symptoms include tremors, severe muscular weakness, nausea, headache, and slight swelling of the liver. Delayed convulsant. Toxicity rating is the same as for fluoroacetate, super toxic. The probable oral lethal dose in humans is a taste (less than 7 drops) for a 70 kg (150 lb) person. The chemical is highly toxic when inhaled or absorbed through the skin. Toxicity depends on its oxidation to fluoroacetate by tissue alcohol dehydrogenase.

First Aid: Acute poisoning should be treated like poisoning by fluoroacetate. Ethylene fluorohydrin (2-fluoroethanol) is listed among the organic fluorine derivatives of fluoroacetic acid. The emergency procedures for fluoroacetic acid are: move victim to fresh air; call emergency medical care. If not breathing, give artificial respiration. If breathing is difficult, give oxygen. In case of contact with material, immediately flush skin or eyes with running water for at least 15 minutes. Remove and isolate contaminated clothing and shoes at the site. Keep victim quiet and maintain normal body temperature. Effects may be delayed; keep victim under observation.

Personal Protective Methods: For emergency situations, wear a positive pressure, pressure-demand, full facepiece self-contained breathing apparatus (SCBA) or pressure-demand supplied air respirator with escape SCBA and a fully-encapsulating, chemical resistant suit. Wear protective gloves and clothing to prevent any reasonable probability of skin contact. Safety equipment suppliers/manufacturers can provide recommendations on the most protective glove/clothing material for your operation.. All protective clothing (suits, gloves, footwear, headgear) should be clean, available each day, and put on before work. Contact lenses should not be worn when working with this chemical. Employees should wash immediately with soap when skin is wet or contaminated. Provide emergency showers and eyewash.

Respirator Selection: Where the potential for exposure to this chemical, use a MSHA/NIOSH approved supplied-air respirator with a full facepiece operated in the positive pressure mode or with a full facepiece, hood, or helmet in the continuous flow mode, or use a MSHA/NIOSH approved self-contained breathing apparatus with a full facepiece operated in pressure-demand or other positive pressure mode.

Storage: Prior to working with you should be trained on its proper handling and storage. Store in tightly closed containers in a cool, well ventilated area away from oxidizers and reducing agents. Where possible, automatically pump liquid from drums or other storage containers to process containers.

Shipping: The regulations for fluoroacetic acid may be applicable. This compound requires a shipping label of: "Poison." It falls in DOT Hazard Class 6.1 and Packing Group I.

Spill Handling: Do not touch spilled material; stop leak if you can do so without risk. Use water spray to reduce vapors. *Small spills*: absorb liquids in vermiculite, dry sand, earth, peat, carbon, or a similar material and deposit in sealed containers. *Large spills*: dike far ahead of spill for later disposal. Evacuate and restrict persons not wearing protective equipment from area of spill or leak until cleanup is complete. Remove all ignition sources. Ventilate area of spill or leak. It may be necessary to contain and dispose of this chemical as a hazardous waste. If material or contaminated runoff enters waterways, notify downstream users of potentially contaminated waters. Contact your Department of Environmental Protection or your regional office of the federal EPA for specific recommendations. If employees are required to clean-up spills, they must be properly trained and equipped. OSHA 1910.120(q) may be applicable.

Fire Extinguishing: This chemical is a combustible liquid. Poisonous gases including toxic fluoride are produced in fire. Use dry chemical, carbon dioxide, or alcohol foam extinguishers. Vapors are heavier than air and will collect in low areas. Vapors may travel long distances to ignition sources and flashback. Vapors in confined areas may explode when exposed to fire. Containers may explode in fire. Storage containers and parts of containers may rocket great distances, in many directions. If material or contaminated runoff enters waterways, notify downstream users of potentially contaminated waters. Notify local health and fire officials and pollution control agencies. From a secure, explosion-proof location, use water spray to cool exposed containers. If cooling streams are ineffective (venting sound increases in volume and pitch, tank discolors, or shows any signs of deforming), withdraw immediately to a secure position. If employees are expected to fight fires, they must be trained and equipped in OSHA 1910.156.

Disposal Method Suggested: In accordance with 40CFR 165 recommendations for the disposal of pesticides and pesticide containers. Must be disposed properly by following package label directions or by contacting your state pesticide or environmental control agency or by contacting your regional EPA office.

References

U.S. Environmental Protection Agency, "Chemical Profile: Ethylene Fluorohydrin," Washington, DC, Chemical Emergency Preparedness Program (November 30, 1987)

Ethylene Glycol

Molecular Formula: $C_2H_6O_2$

Common Formula: $HOCH_2CH_2OH$

Synonyms: Athylenglykol (German); 1,2-Dihydroxyethane; Dowtherm SR 1; EG; ETG; 1,2-Ethanediol; Ethylene alcohol; Ethylene dihydrate; Etilenglicol (Spanish); Fridex; Glycol;

Glycol alcohol; 2-Hydroxyethanol; Ilexan E; Lutrol-9; Macrogol 400; Macrogol 400 BPC; MEG; Monoethylene glycol; NCI-C00920; Norkool; Ramp; Tescol; UCAR 17; Zerex

CAS Registry Number: 107-21-1

RTECS Number: KW2795000

DOT ID: UN 3082

EEC Number: 603-027-00-1

EINECS Number: 203-473-3

Regulatory Authority

- Air Pollutant Standard Set (ACGIH)[1] (HSE)[33] (Several States)[60] (Several Canadian Provinces)
- Clean Air Act: Hazardous Air Pollutants (Title I, Part A, Section 112)
- Superfund/EPCRA 40CFR302.4 Reportable Quantity (RQ): CERCLA, 1 lb (0.454 kg)
- EPCRA Section 313 Form R *de minimis* concentration reporting level: 1.0%
- Canada, WHMIS, Ingredients Disclosure List

Cited in U.S. State Regulations: Alaska (G), Arizona (W), California (A, G), Connecticut (W), Florida (G), Illinois (G), Maine (G, W), Maryland (G), Massachusetts (G, A, W), Nevada (A), New Hampshire (G, W), New Jersey (G, W), New York (G), North Dakota (A), Pennsylvania (G), Rhode Island (G), Virginia (A), West Virginia (G).

Description: Ethylene glycol, $HOCH_2CH_2OH$, is as colorless, odorless, viscous, hydroscopic liquid with a sweetish taste. The odor threshold in air is 25 ppm. Boiling point = 197°C. Flash point = 111°C. Autoignition temperature = 398°C. Explosive limit: LEL = 3.2%. Hazard Identification (based on NFPA-704 M Rating System): Health 1, Flammability 1, Reactivity 0. Soluble in water.

Potential Exposure: Because of ethylene glycol's physical properties, it is used in antifreeze, hydraulic fluids, electrolytic condensers, and heat exchangers. It is also used as a solvent and as a chemical intermediate for ethylene glycol dinitrate, glycol esters, resins, and for pharmaceuticals.

Incompatibilities: Sulfuric acid, oleum, chlorosulfonic acid, strong oxidizing agents, strong bases, chromium trioxide, potassium permanganate, sodium peroxide. Hygroscopic (i.e., absorbs moisture from the air).

Permissible Exposure Limits in Air: There is no OSHA PEL. ACGIH[1] and HSE[33] have recommended a ceiling value of 50 ppm (125 mg/m^3) for the vapor. The DFG MAK is 10 ppm (26 mg/m^3). There was no STEL set for the vapor. Several states have set guidelines or standards for ethylene glycol in ambient air[60] ranging from 0.17 mg/m^3 (Massachusetts) to 1.0 mg/m^3 (Virginia) to 1.25 mg/m^3 (North Dakota) to 2.976 mg/m^3 (Nevada).

Determination in Air: Collection on OSHA versatile sampler-7; Methanol; Gas chromatography/Flame ionization detection; NIOSH IV, Method #5523.

Permissible Concentration in Water: No criteria set, but EPA[32] has suggested a permissible ambient goal of 140 μg/l based on health effects and also a lifetime health advisory of 7,000 μg/l.[48] Several states have set guidelines for ethylene glycol in drinking water[61] ranging from 100 μg/l (Connecticut) to 290 μg/l (New Jersey)[59] to 5,500 μg/l (Arizona, Massachusetts and Maine).

Routes of Entry: Inhalation of particulate or vapor. Percutaneous absorption may also contribute to intoxication.

Harmful Effects and Symptoms

Short Term Exposure: Ethylene glycol irritates the eyes, skin, and respiratory tract.

Inhalation: Mild throat irritation resulted from exposure of 28 mg/m^3. Levels above 140 mg/m^3 resulted in more marked irritation, with levels of more than 250 mg/m^3 being unbreathable. These levels are only reached at elevated temperature.

Skin: May cause mild irritation if not promptly removed.

Eyes: Accidental eye contact with concentrated ethylene glycol resulted in extreme swelling of the eyes, cloudy vision and slow response to light. These symptoms lasted a month.

Ingestion: May cause symptoms in the nervous system, heart, lungs and kidneys. Earliest effects are usually felt in the nervous system between 1/2 – 12 hours after ingestion. Symptoms from 1 liquid ounce may include nausea, vomiting, dizziness, loss of coordination and abdominal pain. Large amounts may cause stupor, coma, convulsions and death. Survival of this stage may lead to development of rapid heart beat, enlarged heart and fluid in the lungs which, too, can lead to death usually after 1 – 3 days. Some individuals who drank 3 – 4 fluid ounces who survived both the above stages because of prompt medical treatment, later (3 – 17 days) died of kidney failure.

Long Term Exposure: Occupational exposure to heated ethylene glycol has caused involuntary eye movement that may indicate nerve damage. Some individuals also reported attacks of unconsciousness lasting 5 – 10 minutes which went away when they stopped working with ethylene glycol. Ethylene glycol may affect the central nervous system and eyes. Has been shown to be a teratogen in animals. Ethylene glycol may damage the developing fetus. May cause kidney and brain damage.

Points of Attack: Eyes, skin, respiratory system, central nervous system

Medical Surveillance: If symptoms develop or overexposure is suspected, the following may be useful: Kidney function test. Urine oxalate level. Exam of the nervous system. Evaluation by a qualified allergist, including careful exposure history and special testing, may help diagnose skin allergy.

First Aid: If this chemical gets into the eyes, remove any contact lenses at once and irrigate immediately for at least 15 minutes, occasionally lifting upper and lower lids. Seek medical attention immediately. If this chemical contacts the

skin, remove contaminated clothing and wash immediately with soap and water. Seek medical attention immediately. If this chemical has been inhaled, remove from exposure, begin rescue breathing (using universal precautions) if breathing has stopped and CPR if heart action has stopped. Transfer promptly to a medical facility. When this chemical has been swallowed, get medical attention. Give large quantities of water and induce vomiting. Do not make an unconscious person vomit.

Personal Protective Methods: Wear protective gloves and clothing to prevent any reasonable probability of skin contact. Safety equipment suppliers/manufacturers can provide recommendations on the most protective glove/clothing material for your operation. Teflon Neoprene+Natural rubber, Nitrile+ PVC, Polyethylene, Natural Rubber, Neoprene and Butyl Rubber are among the recommended protective materials. All protective clothing (suits, gloves, footwear, headgear) should be clean, available each day, and put on before work. Contact lenses should not be worn when working with this chemical. Wear splash-proof chemical goggles and face shield unless full facepiece respiratory protection is worn. Employees should wash immediately with soap when skin is wet or contaminated. Provide emergency showers and eyewash.

Respirator Selection: Where the potential exists for exposures over 50 ppm, use an MSHA/NIOSH approved supplied-air respirator with a full facepiece operated in the positive pressure mode or with a full facepiece, hood, or helmet in the continuous flow mode, or use an MSHA/NIOSH approved self-contained breathing apparatus with a full facepiece operated in pressure-demand or other positive pressure mode.

Storage: Prior to working with this chemical you should be trained on its proper handling and storage. Before entering confined space where this chemical may be present, check to make sure that an explosive concentration does not exist. Ethylene glycol must be stored to avoid contact with sulfuric acid since violent reactions occur. Store in tightly closed containers in a cool, well-ventilated area away from oxidizing agents (such as perchlorates, peroxides, permanganates, chlorates, and nitrates).

Shipping: Environmentally hazardous liquid, n.o.s. Hazard Class: 9. Label: "Class 9."

Spill Handling: Evacuate and restrict persons not wearing protective equipment from area of spill or leak until cleanup is complete. Remove all ignition sources. Establish forced ventilation to keep levels below explosive limit. Absorb liquids in vermiculite, dry sand, earth, peat, carbon, or a similar material and deposit in sealed containers. Keep ethylene glycol out of a confined space, such as a sewer, because of the possibility of an explosion, unless the sewer is designed to prevent the build-up of explosive concentrations. It may be necessary to contain and dispose of this chemical as a hazardous waste. If material or contaminated runoff enters waterways, notify downstream users of potentially contaminated waters. Contact your Department of Environmental Protection or your regional office of the federal EPA for specific recommendations. If employees are required to clean-up spills, they must be properly trained and equipped. OSHA 1910.120(q) may be applicable.

Fire Extinguishing: This chemical is a combustible liquid. Poisonous gases are produced in fire. Use dry chemical, carbon dioxide, or alcohol foam extinguishers. Vapors are heavier than air and will collect in low areas. Vapors may travel long distances to ignition sources and flashback. Vapors in confined areas may explode when exposed to fire. Containers may explode in fire. Storage containers and parts of containers may rocket great distances, in many directions. If material or contaminated runoff enters waterways, notify downstream users of potentially contaminated waters. Notify local health and fire officials and pollution control agencies. From a secure, explosion-proof location, use water spray to cool exposed containers. If cooling streams are ineffective (venting sound increases in volume and pitch, tank discolors, or shows any signs of deforming), withdraw immediately to a secure position. If employees are expected to fight fires, they must be trained and equipped in OSHA 1910.156.

Disposal Method Suggested: Incineration. Alternatively, ethylene glycol can be recovered from polyester plant wastes.

References

Sax, N. I., Ed., "Dangerous Properties of Industrial Materials Report," 4, No. 3, 70–74 (1984)

New Jersey Department of Health and Senior Services, "Hazardous Substance Fact Sheet: Ethylene Glycol," Trenton, NJ (June 1996)

New York State Department of Health, "Chemical Fact Sheet: Ethylene Glycol," Albany, NY, Bureau of Toxic Substance Assessment (January 1986)

Ethylene Glycol Diethyl Ether

Molecular Formula: $C_6H_{14}O_2$

Common Formula: $C_2H_5OCH_2CH_2OC_2H_5$

Synonyms: 1,2-Diethoxyethane; Diethoxyethane; Diethyl cellosolve; Ethyl glyme

CAS Registry Number: 629-14-1

RTECS Number: KI1225000

DOT ID: UN 1153

Regulatory Authority

As glycol ethers:

- Clean Air Act: Hazardous Air Pollutants (Title I, Part A, Section 112) includes mono- and di- ethers of ethylene glycol, diethyl glycol, and triethylene glycol R-$(OCH_2CH_2)_n$-OR' where n = 1,2, or 3; R = alkyl or aryl groups; R' = R, H, or groups which when removed, yield glycol ethers with the structure: R-$(OCH_2CH)_n$-OH. Polymers are excluded from the glycol category.
- EPCRA Section 313: Certain glycol ethers are covered. R-$(OCH_2CH_2)_n$-OR'; Where n = 1,2 or 3; R = alkyl C_7 or less; or R = phenyl or alkyl substituted phenyl; R' + H, or

alkyl C_7 or less; or OR' consisting of carboxylic ester, sulfate, phosphate, nitrate or sulfonate.

- Form R *de minimis* concentration reporting level: 1.0%

Cited in U.S. State Regulations: California (A, G), Florida (G), Massachusetts (G), New Hampshire (G), New Jersey (G), Pennsylvania (G).

Description: Ethylene glycol diethyl ether, $C_2H_5OCH_2CH_2OC_2H_5$, is a colorless liquid. Boiling point = 122°C. Freezing/Melting point = -74°C. Flash point = 35°C. Autoignition temperature = 406°C. Hazard Identification (based on NFPA-704 M Rating System): Health 1, Flammability 3, Reactivity 0. Slightly soluble in water.

Potential Exposure: Ethylene glycol diethyl ether is used in chemical manufacturing, as a solvent for detergents, and in other cleaning products.

Incompatibilities: Forms explosive mixture with air. Strong oxidizers may cause fire and explosions. Attacks some plastics, rubber and coatings. Able to form peroxides. Also incompatible with strong acids, aluminum and its alloys.

Permissible Exposure Limits in Air: No standards set.

Permissible Concentration in Water: No criteria set.

Harmful Effects and Symptoms

Short Term Exposure: Ethylene glycol diethyl ether can affect you when breathed in and by passing through your skin. Exposure can irritate the eyes, nose and throat. Contact can irritate the skin and eyes. High levels can cause you to feel drowsy and dizzy. Very high levels could cause you to pass out. Repeated high or single very high exposures may damage the kidneys.

Long Term Exposure: Repeated or high exposures may damage the kidneys. There is limited evidence that ethylene glycol diethyl ether may damage the developing fetus.

Points of Attack: Kidneys.

Medical Surveillance: If symptoms develop or overexposure is suspected, the following may be useful: Complete blood count (CBC). Kidney function tests.

First Aid: If this chemical gets into the eyes, remove any contact lenses at once and irrigate immediately for at least 15 minutes, occasionally lifting upper and lower lids. Seek medical attention immediately. If this chemical contacts the skin, remove contaminated clothing and wash immediately with soap and water. Seek medical attention immediately. If this chemical has been inhaled, remove from exposure, begin rescue breathing (using universal precautions) if breathing has stopped and CPR if heart action has stopped. Transfer promptly to a medical facility. When this chemical has been swallowed, get medical attention. Give large quantities of water and induce vomiting. Do not make an unconscious person vomit.

Personal Protective Methods: Wear protective gloves and clothing to prevent any reasonable probability of skin contact. Safety equipment suppliers/manufacturers can provide recommendations on the most protective glove/clothing material for your operation. Glove manufacturers have recommended gloves of neoprene or nitrile butyl rubber construction for protection against liquid ethylene glycol diethyl ether. All protective clothing (suits, gloves, footwear, headgear) should be clean, available each day, and put on before work. Contact lenses should not be worn when working with this chemical. Wear splash-proof chemical goggles and face shield unless full facepiece respiratory protection is worn. Employees should wash immediately with soap when skin is wet or contaminated. Provide emergency showers and eyewash.

Respirator Selection: Where the potential exists for exposure to ethylene glycol diethyl ether, use a MSHA/NIOSH approved supplied-air respirator with a full facepiece operated in the positive pressure mode or with a full facepiece, hood, or helmet in the continuous flow mode, or use a MSHA/NIOSH approved self-contained breathing apparatus with a full facepiece operated in pressure-demand or other positive pressure mode.

Storage: Prior to working with this chemical you should be trained on its proper handling and storage. Ethylene glycol diethyl ether must be stored to avoid contact with strong oxidizers (such as chlorine, bromine and fluorine) and strong acids (such as hydrochloric, sulfuric and nitric) since violent reactions occur. Sources of ignition, such as smoking and open flames, are prohibited where ethylene glycol diethyl ether is handled, used, or stored. Metal containers involving the transfer of 5 gallons or more of ethylene glycol diethyl ether should be grounded and bonded. Drums must be equipped with self-closing valves, pressure vacuum bungs, and flame arresters.

Use only non-sparking tools and equipment, especially when opening and closing containers of ethylene glycol diethyl ether. Wherever ethylene glycol diethyl ether is used, handled, manufactured, or stored, use explosion-proof electrical equipment and fittings.

Shipping: This compound requires a shipping label of: "Flammable Liquid." It falls in DOT Hazard Class 3 and Packing Group III.

Spill Handling: Evacuate persons not wearing protective equipment from area of spill or leak until clean-up is complete. Remove all ignition sources. Collect powdered material in the most convenient and safe manner and deposit in sealed containers. Ventilate area after clean-up is complete. Keep ethylene glycol diethyl ether out of a confined space, such as a sewer, because of the possibility of an explosion, unless the sewer is designed to prevent the build-up of explosive concentrations. It may be necessary to contain and dispose of this chemical as a hazardous waste. If material or contaminated runoff enters waterways, notify downstream users of potentially contaminated waters. Contact your Department of Environmental Protection or your regional office of the federal EPA for specific recommendations. If employees are required to clean-up spills, they must be properly trained and equipped. OSHA 1910.120(q) may be applicable.

Fire Extinguishing: This chemical is a flammable liquid. Poisonous gases including carbon monoxide are produced in fire. Use dry chemical, carbon dioxide, or foam extinguishers. Vapors are heavier than air and will collect in low areas. Vapors may travel long distances to ignition sources and flashback. Vapors in confined areas may explode when exposed to fire. Containers may explode in fire. Storage containers and parts of containers may rocket great distances, in many directions. If material or contaminated runoff enters waterways, notify downstream users of potentially contaminated waters. Notify local health and fire officials and pollution control agencies. From a secure, explosion-proof location, use water spray to cool exposed containers. If cooling streams are ineffective (venting sound increases in volume and pitch, tank discolors, or shows any signs of deforming), withdraw immediately to a secure position. If employees are expected to fight fires, they must be trained and equipped in OSHA 1910.156.

References

New Jersey Department of Health and Senior Services, "Hazardous Substance Fact Sheet: Ethylene Glycol Diethyl Ether," Trenton, NJ (May 1998)

Ethylene Glycol Dinitrate

Molecular Formula: $C_2H_4N_2O_6$

Common Formula: $O_2NOCH_2CH_2ONO_2$

Synonyms: EGDN; 1,2-Ethanediol dinitrate; Ethanediol dinitrate; Ethylene dinitrate; Ethylene nitrate; Ethylenglykoldinitrat (Czech); Glycoldinitraat (Dutch); Glycol (dinitrate de) (French); Glycol dinitrate; Glykoldinitrat (German); Nitroglycol; Nitroglykol (Czech)

CAS Registry Number: 628-96-6

RTECS Number: KW5600000

DOT ID: Forbidden to be transported

Regulatory Authority

- Explosive Substance (World Bank)[15]
- Air Pollutant Standard Set (ACGIH)[1] (DFG)[3] (HSE)[33] (OSHA)[58] (Several States)[60]
- Canada, WHMIS, Ingredients Disclosure List

Cited in U.S. State Regulations: Alaska (G), California (A, G), Connecticut (A), Florida (G), Illinois (G), Maine (G), Massachusetts (G), Nevada (A), New Hampshire (G, W), New Jersey (G), North Dakota (A), Pennsylvania (G), Rhode Island (G), West Virginia (G).

Description: Ethylene glycol dinitrate, $O_2NOCH_2CH_2ONO_2$, is a colorless to yellow, oily, odorless liquid. An explosive ingredient (60 – 80%) in dynamite along with nitroglycerine (40 – 20%). It may be detonated by mechanical shock, heat, or spontaneous chemical reaction. Boiling point = 114°C (explodes). Freezing/Melting point = -22°C. Flash point = 215°C. Very slightly water soluble.

Potential Exposure: Although ethylene glycol dinitrate is an explosive in itself, it is primarily used to lower the Freezing/Melting point of nitroglycerin; together these compounds are the major constituents of commercial dynamite, cordite, and blastine gelatin. Occupational exposure generally involves a mixture of the two compounds. Ethylene glycol dinitrate is 160 times more volatile than nitroglycerin.

Incompatibilities: Highly explosive. Heating may cause violent combustion or explosion producing toxic fumes (nitrogen oxides). May explosively detonate violently from mechanical shock, friction, impact or concussion. Not compatible with strong acids and alkalies.

Permissible Exposure Limits in Air: The OSHA PEL is 0.2 ppm (1.0 mg/m^3) (ceiling), not to be exceeded at any time. ACGIH[1] recommends a TWA of 0.05 ppm (0.31 mg/m^3). Both carry the notation "skin" indicating the possibility of cutaneous absorption. The DFG[3] has set the same value as ACGIH for their MAK. The HSE[33] has set a TWA of 0.2 ppm (1.2 mg/m^3) and an STEL of the same level with the "skin" notation. The NIOSH IDLH = 75 mg/m^3. Some states have set guidelines or standards for EGDN in ambient air[60] ranging from 3.0 μg/m^3 (North Dakota) to 6.0 μg/m^3 (Connecticut) to 7.0 μg/m^3 (Nevada).

Determination in Air: Adsorption on Tenax, workup with ethanol; Gas chromatography/Electrochemical detection; NIOSH IV, Method #2507.

Permissible Concentration in Water: No criteria set.

Routes of Entry: Inhalation of dust or vapor; ingestion of dust; percutaneous absorption, skin and/or eye contact.

Harmful Effects and Symptoms

Short Term Exposure: EGDN can cause headache, dizziness, nausea, vomiting, abdominal pain and may affect the cardiovascular system, causing a fall in blood pressure. High levels can interfere with the blood's ability to carry oxygen (methemoglobinemia). Exposure may result in death. The effects may be delayed. Skin contact can cause a rash or burning feeling on contact. Exposure to small amounts of ethylene glycol dinitrate and/or nitroglycerin by skin exposure, inhalation, or swallowing may cause severe throbbing headaches. With large exposure, nausea, vomiting, cyanosis, palpitations of the heart, coma, cessation of breathing, and death may occur. A temporary tolerance to the headache may develop, but this is lost after a few days without exposure. On some occasions a worker may have anginal pains a few days after discontinuing repeated daily exposure.

Long Term Exposure: EGDN can damage the heart causing pain in the chest and/or increased heart rate or cause arrhythmia (irregular heartbeat). This can be fatal. High exposure may affect the nervous system. May damage the red blood cells leading to anemia.

Points of Attack: Skin, cardiovascular system, blood, liver, kidneys.

Medical Surveillance: Placement and periodic examinations should be concerned with central nervous system, blood, glaucoma, and especially history of alcoholism. Urinary and blood ethylene glycol dinitrate may be determined by gas chromatography. Complete blood count (CBC). EKG (immediately, if any chest discomfort is felt). Blood methemoglobin level. Examination of the nervous system.

First Aid: If this chemical gets into the eyes, remove any contact lenses at once and irrigate immediately for at least 15 minutes, occasionally lifting upper and lower lids. Seek medical attention immediately. If this chemical contacts the skin, remove contaminated clothing and wash immediately with soap and water. Seek medical attention immediately. If this chemical has been inhaled, remove from exposure, begin rescue breathing (using universal precautions) if breathing has stopped and CPR if heart action has stopped. Transfer promptly to a medical facility. When this chemical has been swallowed, get medical attention. Give large quantities of water and induce vomiting. Do not make an unconscious person vomit.

Note to Physician: Treat for methemoglobinemia. Spectrophotometry may be required for precise determination of levels of methemoglobinemia in urine.

Personal Protective Methods: Both EGDN and nitroglycerin are readily absorbed through the skin, lungs, and mucous membranes. It is, therefore, essential that adequate skin protection by provided for each worker: impervious clothing where liquids are likely to contaminate and full body clothing where dust creates the problem. All clothing should be discarded at the end of the shift and prior to changing to street clothing. In case of spill or splash that contaminates work clothing, the clothes should be changed at once and the skin area washed thoroughly. Masks of the dust type or organic vapor canister type may be necessary in areas of concentration of dust or vapors. Employees should wash immediately with soap when skin is wet or contaminated. Provide emergency showers and eyewash.

Respirator Selection: NIOSH: Up to 1 mg/m^3: SA (any supplied-air respirator).* *Up to 2.5* mg/m^3*:* SA:CF (any supplied-air respirator operated in a continuous-flow mode).* *Up to 5* mg/m^3: SAT:CF (any supplied-air respirator that has a tight-fitting facepiece and is operated in a continuous-flow mode);* or SCBAF (any self-contained breathing apparatus with a full facepiece); or SAF (any supplied-air respirator with a full facepiece). *Up to 75* mg/m^3*:* SAF:PD/PP (any supplied-air respirator that has a full facepiece and is operated in a pressure-demand or other positive-pressure mode). *Emergency or planned entry in unknown concentration or IDLH conditions:* SCBAF:PD,PP (any self-contained breathing apparatus that has a full facepiece and is operated in a pressure-demand or other positive-pressure mode); or SAF:PD,PP: ASCBA (any supplied-air respirator that has a full facepiece and is operated in a pressure-demand or other positive-pressure mode in combination with an auxiliary, self-contained breathing apparatus operated in a pressure-demand or other positive-pressure mode). *Escape:* GMFOVHiE [any air-purifying, full-facepiece respirator (gas mask) with a chin-style, front- or back-mounted organic vapor canister having a high-efficiency particulate filter]; or SCBAE (any appropriate escape-type, self-contained breathing apparatus).

* Substance reported to cause eye irritation or damage; may require eye protection.

Storage: Prior to working with this chemical you should be trained on its proper handling and storage. Before entering confined space where EGDN is present, check to make certain that explosive concentrations do not exist. EGDN is highly explosive and can detonate violently upon heating or impact. Store in tightly closed containers in a cool, well ventilated area away from strong acids and alkalies. All sources of ignition are prohibited. Use non-sparking tools and equipment especially when opening or closing containers. Use explosion-proof electrical equipment and fittings in all areas of handling, use or manufacture. Where possible, automatically pump liquid from drums or other storage containers to process containers.

Shipping: DOT[19] simply lists this material as forbidden to transport.

Spill Handling: Spread sodium bisulfate over the spill area and sprinkle with water. Then flush to sewer with water.[24] Evacuate and restrict persons not wearing protective equipment from area of spill or leak until cleanup is complete. Remove all ignition sources. Ventilate area of spill or leak. Absorb liquids in vermiculite, dry sand, earth, peat, carbon, or a similar material and deposit in sealed containers. Keep EGDN out of a confined space, such as a sewer, because of the possibility of an explosion, unless the sewer is designed to prevent the build-up of explosive concentrations. It may be necessary to contain and dispose of this chemical as a hazardous waste. If material or contaminated runoff enters waterways, notify downstream users of potentially contaminated waters. Contact your Department of Environmental Protection or your regional office of the federal EPA for specific recommendations. If employees are required to clean-up spills, they must be properly trained and equipped. OSHA 1910.120(q) may be applicable.

Fire Extinguishing: Evacuate area of fire. Heat may cause violent combustion (explodes at 114°C) or explosion. Fight fires only from a secure, explosion-resistant position. This chemical is a combustible liquid. Poisonous gases including nitrogen oxides are produced in fire. Use dry chemical, carbon dioxide, water spray, or foam extinguishers. Vapors are heavier than air and will collect in low areas. Vapors may travel long distances to ignition sources and flashback. Vapors in confined areas may explode when exposed to fire. Containers may explode in fire. Storage containers and parts of containers may rocket great distances, in many directions. If material or contaminated runoff enters waterways, notify downstream users of potentially contaminated waters. Notify local health and fire officials and pollution control agencies. From a secure, explosion-proof location, use water spray to cool exposed containers. If cooling streams are ineffective (venting sound increases in volume and pitch, tank discolors, or shows any signs of deforming), withdraw immediately to a secure position. If employees are expected to fight fires, they must be trained and equipped in OSHA 1910.156.

References

New Jersey Department of Health and Senior Services, "Hazardous Substance Fact Sheet, Ethylene Glycol Dinitrate," Trenton NJ (March 1999)

Ethylene Glycol Monomethyl Ether

Molecular Formula: $C_3H_8O_2$

Common Formula: $CH_3OCH_2CH_2OH$

Synonyms: Aethylengykol-monomethylaether (German); Dowanol EM; EGM; EGME; Ether monomethylique de l'ethylene-glycil (French); Ethylene glycol methyl ether; Glycol ether EM; Glycol methyl ether; Glycol monomethyl ether; Jeffersol EM; MECS; 2-Methoxyethanol; Methoxyhydroxyethane; Methyl Cellosolve® (NIOSH); Methyl ethoxol; Methyl glycol; Methyl oxitol; Metil cellosolve (Italian); Metoksyetylowy alcohol (Polish); Poly-Solv EM; Prist

CAS Registry Number: 109-86-4

RTECS Number: KL5775000

DOT ID: UN 1188

EEC Number: 603-011-00-4

- ***Regulatory Authority:*** Air Pollutant Standard Set (ACGIH) (DFG)[3] (HSE)[33] (OSHA)[58] (Other Countries)[35] (Several States)[60]
- EPCRA Section 313 Form R *de minimis* concentration reporting level: 1.0%

As glycol ethers:

- Clean Air Act: Hazardous Air Pollutants (Title I, Part A, Section 112) includes mono- and di- ethers of ethylene glycol, diethyl glycol, and triethylene glycol R-$(OCH_2CH_2)_n$-OR' where n = 1,2, or 3; R = alkyl or aryl groups; R' = R, H, or groups which when removed, yield glycol ethers with the structure: R-$(OCH_2CH)_n$-OH. Polymers are excluded from the glycol category.
- Canada, WHMIS, Ingredients Disclosure List

Cited in U.S. State Regulations: Alaska (G), California (A, G), Connecticut (A), Florida (G, A), Illinois (G), Maine (G), Maryland (G), Massachusetts (G, A), Nevada (A), New Hampshire (G), New Jersey (G), New York (G, A), North Dakota (A), Pennsylvania (G), Rhode Island (G, A), Virginia (A), West Virginia (G).

Description: Ethylene glycol monomethyl ether, CH_3OCH_2 CH_2OH, is a colorless liquid with a slight ether-like odor. The odor threshold is 0.9 – 2.3 ppm. Boiling point = 124°C. Flash point = 39°C. Freezing/Melting point = -85°C. Autoignition temperature = 285°C. Explosive limits: LEL = 1.8%; UEL = 14%, both @ STP. Hazard Identification (based on NFPA-704 M Rating System): Health 2, Flammability 2, Reactivity 0. Soluble in water.

Potential Exposure: Ethylene glycol ethers are used as solvents for resins used in the electronics industry, lacquers, paints, varnishes, gum, perfume, dyes and inks, and as a constituent of painting pastes, cleaning compounds, liquid soaps, cosmetics, nitrocellulose, and hydraulic fluids.

Incompatibilities: Above flash point, explosive vapor/air mixtures may be formed. Heat or oxidizers may cause the formation of unstable peroxides. Attacks many metals. Strong oxidizers cause fire and explosions. Strong bases cause decomposition and the formation of toxic gas. Attacks some plastics, rubber and coatings. May accumulate static electrical charges, and may cause ignition of its vapors.

Permissible Exposure Limits in Air: OSHA[58] has set a TWA of 25 ppm (80 mg/m^3). The NIOSH REL is 0.1 ppm (0.3 mg/m^3). ACGIH,[1] DFG,[3] HSE[33], and Japan[35] have set a TWA of 5 ppm (16 mg/m^3) with the notation "skin" indicating the possibility of cutaneous absorption. Sweden[35] has a TWA of 10 ppm (30 mg/m^3) and an STEL of 25 ppm (80 mg/m^3). The NIOSH IDLH = 200 ppm.

The former USSR[35] has set a MAC for ambient air in residential areas of 0.3 mg/m^3 (on a once daily basis). Several states have set guidelines or standards for 2-methoxyethanol in ambient air[60] ranging from 21.0 μg/m^3 (Massachusetts) to 53.3 μg/m^3 (New York) to 100 μg/m^3 (Rhode Island) to 160.0 μg/m^3 (Florida and North Dakota) to 270.0 μg/m^3 (Virginia) to 320.0 μg/m^3 (Connecticut) to 381.0 μg/m^3 (Nevada).

Determination in Air: Adsorption with charcoal tube; workup with methanol/methylene chloride; Gas chromatography/Flame ionization detection; NIOSH IV, Method #1403, Alcohols IV.

Permissible Concentration in Water: No criteria set.

Routes of Entry: Inhalation of vapor, ingestion, eye and/ or skin contact, plus cutaneous absorption of liquid.

Harmful Effects and Symptoms

Short Term Exposure: EGME irritates the eyes and the respiratory tract. More severe exposures may cause damage to brain, nervous system, liver and kidneys, and death.

Inhalation: Exposure to levels above 25 ppm may cause headache, drowsiness, lethargy, weakness and tremors. Higher exposures can cause pulmonary edema, a medical emergency that can be delayed for several hours. This can cause death.

Skin: Readily absorbed through the skin. May cause stinging and reddening upon contact and cause or contribute to symptoms in the lungs.

Eyes: Contact with liquid or high vapor concentrations may cause burning irritation and clouded vision.

Ingestion: 3 liquid ounces (100 ml) has been reported to cause confusion, nausea, weakness and rapid breathing. Symptoms were delayed 8 – 18 hours.

Long Term Exposure: Prolonged exposure can cause anemia and lesions of blood cells. May cause symptoms listed under inhalation. Repeated exposure can cause headaches, weakness, drowsiness, personality changes, loss of weight, upset stomach and tremors. May cause kidney damage. Extended exposures of 25 – 75 ppm has caused anemia and nervous system effects such as fatigue, weakness, loss of muscle control and mental changes. Ethylene glycol monomethyl ether has been shown to cause reproductive problems, damage to the testes and cause birth defects in animal studies.

Points of Attack: Eyes, respiratory system, central nervous system, blood, kidneys, reproductive system, hematopoietic system.

Medical Surveillance: For those with frequent or potentially high exposure (half the TLV or greater, or significant skin contact), the following are recommended before beginning work and at regular times after that: Lung function tests. Complete blood count. Kidney function tests. If symptoms develop or overexposure is suspected, the following may be useful: Consider chest x-ray after acute overexposure. Exam of the nervous system.

First Aid: If this chemical gets into the eyes, remove any contact lenses at once and irrigate immediately for at least 15 minutes, occasionally lifting upper and lower lids. Seek medical attention immediately. If this chemical contacts the skin, remove contaminated clothing and wash immediately with soap and water. Seek medical attention immediately. If this chemical has been inhaled, remove from exposure, begin rescue breathing (using universal precautions) if breathing has stopped and CPR if heart action has stopped. Transfer promptly to a medical facility. When this chemical has been swallowed, get medical attention. Give large quantities of water and induce vomiting. Do not make an unconscious person vomit. Medical observation is recommended for 24 – 48 hours after breathing overexposure, as pulmonary edema may be delayed. As first aid for pulmonary edema, a doctor or authorized paramedic may consider administering a corticosteroid spray.

Personal Protective Methods: Wear protective gloves and clothing to prevent any reasonable probability of skin contact. Safety equipment suppliers/manufacturers can provide recommendations on the most protective glove/clothing material for your operation.. All protective clothing (suits, gloves, footwear, headgear) should be clean, available each day, and put on before work. Contact lenses should not be worn when working with this chemical. Wear splash-proof chemical goggles and face shield unless full facepiece respiratory protection is worn. Employees should wash immediately with soap when skin is wet or contaminated. Provide emergency showers and eyewash.

Respirator Selection: NIOSH: *up to 1 ppm:* SA (any supplied-air respirator).* *Up to 2.5 ppm:* SA:CF (any supplied-air respirator operated in a continuous-flow mode).* *Up to 5 ppm:* SCBAF (any self-contained breathing apparatus with a full facepiece); or SAF (any supplied-air respirator with a full facepiece). *Up to 100 ppm:* SA:PD,PP (any supplied-air respirator operated in a pressure-demand or other positive-pressure mode).* *Up to 200 ppm*: SAF:PD,PP (any supplied-air respirator that has a full facepiece and is operated in a pressure-demand or other positive-pressure mode). *Emergency or planned entry into unknown concentrations or IDLH conditions:* SCBAF:PD,PP (any MSHA/NIOSH approved self-contained breathing apparatus that has a full facepiece and is operated in a pressure-demand or other positive-pressure mode); or SAF:PD,PP:ASCBA (any supplied-air respirator that has a full facepiece and is operated in a pressure-demand or other positive-pressure mode in combination with an auxiliary, self-contained breathing apparatus operated in a pressure-demand or other positive pressure mode). *Escape:* GMFOV [any air-purifying, full-facepiece respirator (gas mask) with a chin-style, front-or back-mounted organic vapor canister]; or SCBAE (any appropriate escape-type, self-contained breathing apparatus).

* Substance reported to cause eye irritation or damage; may require eye protection.

Storage: Prior to working with this chemical you should be trained on its proper handling and storage. Before entering confined space where this chemical may be present, check to make sure that an explosive concentration does not exist. 2-Methoxyethanol must be stored to avoid contact with oxidizers (such as perchlorates, peroxides, permanganates, chlorates, and nitrates) and strong caustics since violent reactions occur. Sources of ignition such as smoking and open flames are prohibited where 2-methoxyethanol is used, handled, or stored in a manner that could create a potential fire or explosion hazard. Wherever 2-methoxyethanol is used, handled, manufactured, or stored, use explosion-proof electrical equipment and fittings.

Shipping: This compound requires a shipping label of: "Flammable Liquid." It falls in DOT Hazard Class 3 and Packing Group III.

Spill Handling: Evacuate and restrict persons not wearing protective equipment from area of spill or leak until cleanup is complete. Remove all ignition sources. Establish forced ventilation to keep levels below explosive limit. Use water spray to reduce vapors. Absorb liquids in vermiculite, dry sand, earth, peat, carbon, or a similar material and deposit in sealed containers. Keep EGME out of a confined space, such as a sewer, because of the possibility of an explosion, unless the sewer is designed to prevent the build-up off explosive concentrations. It may be necessary to contain and dispose of this chemical as a hazardous waste. If material or contaminated runoff enters waterways, notify downstream users of potentially contaminated waters. Contact your Department of Environmental Protection or your regional office of the federal EPA for specific recommendations. If employees are required to clean-up spills, they must be properly trained and equipped. OSHA 1910.120(q) may be applicable.

Fire Extinguishing: This chemical is a combustible liquid. Poisonous gases are produced in fire. Use dry chemical, carbon dioxide, or alcohol foam extinguishers. Vapors are heavier than air and will collect in low areas. Vapors may travel long distances to ignition sources and flashback. Vapors in confined areas may explode when exposed to fire. Containers may explode in fire. Storage containers and parts of containers may rocket great distances, in many directions. If material or contaminated runoff enters waterways, notify downstream users of potentially contaminated waters. Notify local health and fire officials and pollution control agencies. From a secure, explosion-proof location, use water spray to cool exposed containers. If cooling streams are ineffective (venting sound increases in volume and pitch, tank discolors, or shows any signs of deforming), withdraw immediately to a secure position. If employees are expected to fight fires, they must be trained and equipped in OSHA 1910.156.

Disposal Method Suggested: Concentrated waste containing no peroxides: discharge liquid at a controlled rate near a pilot flame. Concentrated waste containing peroxides: perforation of a container of the waste from a safe distance followed by open burning.[24]

References

National Institute for Occupational Safety and Health, Profiles on Occupational Hazards for Criteria Document Priorities: Glycol Ethers, pp 110–115, Report PB 274,073, Cincinnati, OH (1977)

National Institute for Occupational Safety and Health, Glycol Ethers: 2-Methoxyethanol and 2-Ethoxyethanol, Current Intelligence Bulletin No. 39, Cincinnati, Ohio (May 2, 1983)

Sax, N. I., Ed., "Dangerous Properties of Industrial Materials Report," 4, No. 2, 67–70 (1984)

New York State Department of Health, "Chemical Fact Sheet: Ethylene Glycol Monomethyl Ether," Albany, NY, Bureau of Toxic Substance Assessment (January 1986 and Version 2)

New Jersey Department of Health and Senior Services, "Hazardous Substance Fact Sheet: 2-Methoxyethanol," Trenton, NJ (April 1996)

Ethyleneimine

Molecular Formula: C_2H_5N

Synonyms: Aethylenimin (German); Aminoethylene; Azacyclopropane; Azirane; Aziridin (German); Aziridina (Spanish); Aziridine; Azirine; 1H-Azirine,dihydro-; Dihydro-1H-azirine; Dihydroazirine; Dimethyleneimine; Dimethylenimine; E-1; ENT 50,324; Ethyleenimine (Dutch); Ethyleneimine; Ethylimine; Etilenimina (Italian); TL 337

CAS Registry Number: 151-56-4

RTECS Number: KX5075000

DOT ID: UN 1185

EEC Number: 613-001-00-1

Regulatory Authority

- Carcinogen (Animal Positive) (IARC)[9] (ACGIH)[1]
- OSHA, 29CFR1910 Specifically Regulated Chemicals (See CFR 1910.1012)
- Banned or Severely Restricted (Belgium, Sweden) (UN)[13]
- Toxic Substance (World Bank)[15]
- Air Pollutant Standard Set (ACGIH)[1] (DFG)[3] (HSE)[33] (former USSR)[43] (Several States)[60] (Several Canadian Provinces)
- Clean Air Act: Hazardous Air Pollutants (Title I, Part A, Section 112); Section 112(r), Accidental Release Prevention/Flammable substances (40CFR/68.130; 59 FR 4497), TQ = 10,000 lb (4,550 kg)
- Reportable Quantity (RQ): CERCLA, 1 lb(0.454 kg)
- EPA Hazardous Waste Number (RCRA No.): P054
- RCRA, 40CFR261, Appendix 8 Hazardous Constituents
- Superfund/EPCRA 40CFR355, Appendix B, Extremely Hazardous Substances: TPQ = 500 lb(228 kg).
- EPCRA Section 313 Form R *de minimis* concentration reporting level: 0.1%
- Canada, WHMIS, Ingredients Disclosure List

Cited in U.S. State Regulations: Alaska (G), California (G), Connecticut (A), Florida (G, A), Illinois (G), Kansas (G), Louisiana (G), Maine (G), Maryland (G), Massachusetts (G), Michigan (G), Nevada (A), New Hampshire (G), New Jersey (G), New York (A), North Carolina (A), North Dakota (A), Oklahoma (G), Pennsylvania (G), Rhode Island (G), South Carolina (A), Vermont (G), Virginia (G), Washington (G), West Virginia (G), Wisconsin (G).

Description: Ethyleneimine, is a colorless volatile liquid with an ammoniacal odor. Boiling point = 56 – 57°C. Freezing/Melting point = -74°C. Flash point = -11°C. Autoignition temperature = 322°C. Explosive limits: LEL = 3.3%; UEL = 54.8%. Hazard Identification (based on NFPA-704 M Rating System): Health 4, Flammability 3, Reactivity 3. Highly soluble in water.

Potential Exposure: Ethyleneimine is a highly reactive compound and is used in many organic syntheses. Used as an intermediate and monomer for fuel oil and lubricating refining. The polymerization products, polyethyleneimines, are used as auxiliaries in the paper industry and as flocculation aids in the clarification of effluents. It is also used in the textile industry for increasing wet strength, flameproofing, shrinkproofing, stiffening, and waterproofing.

Incompatibilities: Forms explosive mixture with air. Ethyleneimine is a medium strong base. Contact with acids, aqueous acid conditions, oxidizers, aluminum, carbon dioxide or silver may cause explosive polymerization. Self reactive with heat or atmospheric carbon dioxide. May accumulate static electrical charges, and may cause ignition of its vapors. Attacks rubber, coatings, plastics, and chemically active metals.

Permissible Exposure Limits in Air: There is no OSHA PEL. Ethyleneimine is included in the Federal standard for carcinogens; all contact with it should be avoided. DFG[3] takes the same stand and sets no numerical limits. ACGIH and HSE[33] set a TWA of 0.5 ppm (1.0 mg/m^3) with the notation "skin" indicating the possibility of cutaneous absorption. The NIOSH IDLH = Ca [100 ppm]. The former USSR-UNEP/IRPTC joint project[43] have set a MAC in workplace air of 0.02 mg/m^3 and a MAC in ambient air of residential areas of 0.001 mg/m^3 on either a momentary or a daily average basis. Several states have set guidelines or standards for ethyleneimine in ambient air[60] ranging from zero (North Carolina) to 3.3 μg/m^3 (New York) to 5.0 μg/m^3 (South Carolina) to 10.0 μg/m^3 (Florida and North Dakota) to 20.0 μg/m^3 (Connecticut) to 24.0 μg/m^3 (Nevada).

Determination in Air: Bubbler; $CHCl_3$; High-pressure liquid chromatography/Ultraviolet detection; NIOSH IV, Method #3514.

Permissible Concentration in Water: No criteria set, but EPA[32] has suggested a permissible ambient goal of 14 μg/l based on health effects.

Routes of Entry: Inhalation and percutaneous absorption, ingestion, skin and/or eye contact.

Harmful Effects and Symptoms

Short Term Exposure: Ethyleneimine is corrosive to the eyes, skin, and respiratory tract. Eye contact can cause severe illness or death. Inhalation of the vapors can cause pulmonary edema, a medical emergency that can be delayed for several hours. This can cause death. Symptoms include tearing and burning of the eyes, sore throat, nausea, vomiting, coughing (may persist for weeks or months) and a slow healing dermatitis due to severe blistering. Ethyleneimine is classified as extremely toxic with a probable oral lethal dose of 5 – 50 mg/kg which is approximately 7 drops to 1 teaspoonful for a 75 kg (150 lb) person. Ethyleneimine gives inadequate warning when over-exposure is by inhalation or skin absorption. It is a severe blistering agent, causing third degree chemical burns of the skin. Also, it has a corrosive effect on mucous membranes and may cause scarring of the esophagus. It is corrosive to eye tissue and may cause permanent corneal opacity and conjunctival scarring. Severe exposure may result in overwhelming pulmonary edema. Renal damage has been described. Hemorrhagic congestion of all internal organs has been observed. May cause affect the central nervous system, kidneys and liver.

Long Term Exposure: May be a carcinogen in humans. May damage the developing fetus. May damage the testes. Repeated or prolonged contact with skin may skin allergy. May damage the liver and kidneys.

Points of Attack: Eyes, skin, respiratory system, liver, kidneys. Cancer Site in animals: lung and liver tumors.

Medical Surveillance: Before beginning employment and at regular times after that, the following are recommended: Lung function tests. If symptoms develop or overexposure is suspected, the following may be useful: Consider chest x-ray after acute overexposure. Evaluation by a qualified allergist, including careful exposure history and special testing, may help diagnose skin allergy. Liver and kidney function tests.

First Aid: If this chemical gets into the eyes, remove any contact lenses at once and irrigate immediately for at least 45 minutes, occasionally lifting upper and lower lids. Seek medical attention immediately since eye splashes can cause severe illness or death. If this chemical contacts the skin, remove contaminated clothing and wash immediately with soap and water. Seek medical attention immediately. If this chemical has been inhaled, remove from exposure, begin rescue breathing (using universal precautions) if breathing has stopped and CPR if heart action has stopped. Transfer promptly to a medical facility. When this chemical has been swallowed, get medical attention. Give large quantities of water and induce vomiting. Do not make an unconscious person vomit. Medical observation is recommended for 24 – 48 hours after breathing overexposure, as pulmonary edema may be delayed. As first aid for pulmonary edema, a doctor or authorized paramedic may consider administering a corticosteroid spray.

Personal Protective Methods: Wear protective gloves and clothing to prevent any reasonable probability of skin contact. Safety equipment suppliers/manufacturers can provide recommendations on the most protective glove/clothing material for your operation.. All protective clothing (suits, gloves, footwear, headgear) should be clean, available each day, and put on before work. Contact lenses should not be worn when working with this chemical. Wear splash-proof chemical goggles and face shield unless full facepiece respiratory protection is worn. Employees should wash immediately with soap when skin is wet or contaminated. Provide emergency showers and eyewash.

Respirator Selection: NIOSH: *At any detectable concentration:* SCBAF:PD,PP (any MSHA/NIOSH approved self-contained breathing apparatus that has a full facepiece and is operated in a pressure-demand or other positive-pressure mode); or SAF:PD,PP:ASCBA (any supplied-air respirator that has a full facepiece and is operated in a pressure-demand or other positive-pressure mode in combination with an auxiliary, self-contained breathing apparatus operated in a pressure-demand or other positive pressure mode). *Escape:* GMFOV [any air-purifying, full-facepiece respirator (gas mask) with a chin-style, front-or back-mounted organic vapor canister]; or SCBAE (any appropriate escape-type, self-contained breathing apparatus).

Storage: Prior to working with this chemical you should be trained on its proper handling and storage. Before entering confined space where this chemical may be present, check to make sure that an explosive concentration does not exist. Ethyleneimine must be inhibited and stored to avoid contact with strong acids (such as hydrochloric, sulfuric and nitric) and oxidizers (such as perchlorates, peroxides, permanganates, chlorates and nitrates) since violent reactions occur. Store in tightly closed containers in a cool, well-ventilated area. Sources of ignition, such as smoking and open flames, are prohibited where ethyleneimine is handled, used, or stored. Metal containers involving the transfer of 5 gallons or more of ethyleneimine should be grounded and bonded. Drums must be equipped with self-closing valves, pressure vacuum bungs, and flame arresters. Use only non-sparking tools and equipment, especially when opening and closing containers of ethyleneimine. Wherever ethyleneimine is used, handled, manufactured, or stored, use explosion-proof electrical equipment and fittings. A regulated, marked area should be established where this chemical is handled, used, or stored in compliance with OSHA standard 1910.1045.

Shipping: This compound requires a shipping label of: "Poison, Flammable Liquid." It falls in DOT Hazard Class 6.1 and Packing Group I. Passenger aircraft or railcar shipment is forbidden as is any cargo aircraft shipment.

Spill Handling: Evacuate and restrict persons not wearing protective equipment from area of spill or leak until cleanup is complete. Avoid breathing vapors. Stay upwind, keep out of low areas. Avoid bodily contact with the material. DO NOT handle broken packages without protective equipment. Wash away any material which may have contacted the body with copious amounts of water or soap and water. Remove all ignition sources. Establish forced ventilation to keep levels below explosive limit. Absorb liquids in vermiculite, dry sand, earth, peat, carbon, or a similar material and deposit in sealed containers. Keep this chemical out of a confined space, such as a sewer, because of the possibility of an explosion, unless the sewer is designed to prevent the build-up off explosive concentrations. It may be necessary to contain and dispose of this chemical as a hazardous waste. If material or contaminated runoff enters waterways, notify downstream users of potentially contaminated waters. Contact your Department of Environmental Protection or your regional office of the federal EPA for specific recommendations. If employees are required to clean-up spills, they must be properly trained and equipped. OSHA 1910.120(q) may be applicable.

Initial isolation and protective action distances

Distances shown are likely to be affected during the first 30 minutes after materials are spilled and could increase with time. If more than one tank car, cargo tank, portable tank, or large cylinder is involved in the incident is leaking, the protective action distance may need to be increased. You may need to seek emergency information from CHEMTREC at (800) 424-9300 or seek professional environmental engineering assistance from the U.S. EPA Environmental Response Team at (908) 548-8730 (24-hour response line).

Small spills (From a small package or a small leak from a large package)

First: Isolate in all directions (feet)..... 300

Then: Protect persons downwind (miles)

Day 0.2

Night 0.8

Large spills (From a large package or from many small packages)

First: Isolate in all directions (feet) 800

Then: Protect persons downwind (miles)

Day 0.7

Night 3.0

Fire Extinguishing: This chemical is a flammable liquid. Poisonous gases including nitrogen oxides are produced in fire. Dry chemical, alcohol foam, or carbon dioxide are useful for small fires. For large fires: water spray, fog or foam. Do not extinguish fire unless flow can be stopped; use water in flooding quantities as a fog. Solid streams of water may be ineffective. Apply water from as far a distance as possible. If tank car or truck is involved in fire, isolate the surrounding area as shown above. May polymerize in fires with evolution of heat and container rupture. Runoff to sewer may create fire or explosion hazard. Ethyleneimine vapors are not inhibited and may form polymers in vents or flames arresters, resulting in stopping of the vents. Vapors are heavier than air and will collect in low areas. Vapors in confined areas may explode when exposed to fire. Containers may explode in fire. Storage containers and parts of containers may rocket great distances, in many directions. If material or contaminated runoff enters waterways, notify downstream users of potentially contaminated waters. Notify local health and fire officials and pollution control agencies. From a secure, explosion-proof location, use water spray to cool exposed containers. If cooling streams are ineffective (venting sound increases in volume and pitch, tank discolors, or shows any signs of deforming), withdraw immediately to a secure position. If employees are expected to fight fires, they must be trained and equipped in OSHA 1910.156.

Disposal Method Suggested: Controlled incineration; incinerator is equipped with a scrubber or thermal unit to reduce No_x emissions.

References

Dermer, O. C. and Ham, G. E., Ethyleneimine and Other Aziridines, Academic Press, NY (1969)

Sax, N. I., Ed., "Dangerous Properties of Industrial Materials Report," 1, No. 2, 37–38 (1980)

U.S. Environmental Protection Agency, "Chemical Profile: Ethyleneimine," Washington, DC, Chemical Emergency Preparedness Program (November 30, 1987)

New Jersey Department of Health and Senior Services, "Hazardous Substance Fact Sheet: Ethylene Imine," Trenton, NJ (April 1997)

Ethylene Oxide

Molecular Formula: C_2H_4O

Synonyms: Aethylenoxid (German); Amprolene; Anprolene; Anproline; Dihydrooxirene; Dimethylene oxide; ENT 26,263; E.O.; 1,2-Epoxyaethan (German); 1,2-Epoxyethane; Epoxyethane; Ethene oxide; Ethyleenoxide (Dutch); Ethylene (oxide d') (French); Etilene(ossido di) (Italian); ETO; Etylenu tlenek (Polish); Merpol; NCI-C50088; Odido de etileno (Spanish); Oxacyclopropane; Oxane; α,β-Oxidoethane; Oxidoethane; Oxiraan (Dutch); Oxirane; Oxirene, dihydro-; Oxyfume; Oxyfume 12; Sterilizing gas ethylene oxide 100%; T-gas; UN 1040

CAS Registry Number: 75-21-8

RTECS Number: KX2450000

DOT ID: UN 1040

EEC Number: 603-023-00-X

EINECS Number: 200-849-9

Regulatory Authority

- Carcinogen (Animal Positive) (IARC)[9] (DFG)[3] (ACGIH)[1]
- Banned or Severely Restricted (In Agriculture) (Germany)[13]
- Highly Reactive Substance and Explosive (World Bank)[15]
- OSHA, 29CFR1910 Specifically Regulated Chemicals (See CFR 1910.1047)
- Air Pollutant Standard Set (ACGIH)[1] (HSE)[33] (OSHA)[63] (Other Countries)[35] (Several States)[60]
- OSHA 29CFR1910.119, Appendix A. Process Safety List of Highly Hazardous Chemicals, TQ = 5,000 lb (2,270 kg)
- Clean Air Act: Hazardous Air Pollutants (Title I, Part A, Section 112); Accidental Release Prevention/Flammable substances, (Section 112[r], Table 3), TQ = 10,000 lb (45,40 kg)
- EPA Hazardous Waste Number (RCRA No.): U115
- RCRA, 40CFR261, Appendix 8 Hazardous Constituents.
- RCRA 40CFR268.48; 61FR15654, Universal Treatment Standards: Wastewater (mg/l), 0.12; Nonwastewater (mg/kg), N/A
- Superfund/EPCRA 40CFR355, Appendix B Extremely Hazardous Substances: TPQ = 1,000 lb (454 kg)
- Superfund/EPCRA 40CFR302.4 Reportable Quantity (RQ): CERCLA, 10 lb (4.54 kg)
- EPCRA Section 313 Form R *de minimis* concentration reporting level: 0.1%
- FDA tolerance limit, 50 ppm in ground spices
- Canada, WHMIS, Ingredients Disclosure List

Cited in U.S. State Regulations: Alaska (G), California (A, G), Connecticut (A), Florida (G), Illinois (G), Indiana (A), Kansas (G), Louisiana (G), Maine(G), Maryland (G), Massachusetts (G), Michigan (G), Nevada (A), New Hampshire (G), New Jersey (G), New York (G, A), North Carolina (A), North Dakota (A), Oklahoma (G), Pennsylvania (G, A), Rhode Island (G, A), South Carolina (A), South Dakota (A), Vermont (G), Virginia (G, A), Washington (G), West Virginia (G), Wisconsin (G).

Description: Ethylene oxide, is a colorless, compressed, liquefied gas or liquid (below 11°C) with a sweetish odor. The odor threshold is 50 ppm. Boiling point = 11°C. Freezing/Melting point = -112°C. Flash point = -20°C. Autoignition temperature = 1,058°C. Explosive limits: LEL = 3.0%; UEL = 100%. Hazard Identification (based on NFPA-704 M Rating System): Health 3, Flammability 4, Reactivity 3. Easily dissolved in water.

Potential Exposure: Ethylene oxide is a man-made chemical used as an intermediate in organic synthesis for ethylene glycol, polyglycols, glycol ethers, esters, ethanolamines, acrylonitrile, plastics, and surface-active agents. It is also used as a fumigant for foodstuffs and textiles, an agricultural fungicide, and for sterilization, especially for surgical instruments. It is used in drug synthesis and as a pesticide intermediate.

Incompatibilities: Forms explosive mixture with air. Dangerously reactive; may rearrange chemically and/or polymerize violently with evolution of heat, when in contact with highly active catalytic surfaces such as anhydrous chlorides of iron, tin and aluminum, pure oxides of iron and aluminum, and alkali metal hydroxides. Even small amounts of strong acids, alkalis, oxidizers can cause a reaction. Avoid contact with copper. Protect container from physical damage, sun and heat. Attacks some plastics, rubber or coatings.

Permissible Exposure Limits in Air: The OSHA PEL and the recommended ACGIH[1] TLV is 1 ppm (2 mg/m^3) TWA. The HSE[33] and Sweden[35] have set a TWA of 5 ppm (10 mg/m^3); Sweden adds an STEL of 10 ppm (18 mg/m^3), Japan[35] has set a TWA of 50 ppm (90 mg/m^3). At the other extreme, the former USSR[35][43] has set a MAC in workplace air of 1 mg/m^3; Czechoslovakia has adopted this level also. NIOSH IDLH = Ca [800 ppm]. The former USSR[35][43] ha also set a MAC for ambient air in residential areas of 0.3 mg/m^3 on a momentary basis and 0.03 mg/m^3 on a daily average basis. The NIOSH IDLH level is 800 ppm. Several states have set guidelines or standards for ethylene oxide in ambient air[60] ranging from zero (North Dakota) to 0.01 mg/m^3 (Rhode Island) to 0.1 μg/m^3 (North Carolina) to 4.87 μg/m^3 (Pennsylvania) to 6.67 μg/m^3 (New York) to 10.0 μg/m^3 (South Carolina) to 20.0 μg/m^3 (Connecticut, South Dakota and Virginia) to 48.0 μg/m^3 (Nevada) to 450.0 μg/m^3 (Indiana).

Determination in Air: Collection by charcoal tube (petroleum-based); DMF Any dust and mist respirator with a full facepiece; analysis by gas chromatography/electrochemical detection; NIOSH IV, Method #1614.

Permissible Concentration in Water: No criteria set.

Routes of Entry: Inhalation of gas, ingestion, eye and/or skin contact.

Harmful Effects and Symptoms

Short Term Exposure: Signs and symptoms of acute exposure to ethylene oxide may be severe, and include dyspnea (shortness of breath), cough, pulmonary edema, pneumonia, and respiratory failure. Lethargy, headache, dizziness, twitching, convulsions, paralysis, and coma may be observed. Cardiac arrhythmias and cardiovascular collapse may also occur. Gastrointestinal effects of acute exposure may include nausea, vomiting, and abdominal pain. Ethylene oxide irritates the eyes, skin, and respiratory tract. Very high exposures can cause pulmonary edema, a medical emergency that can be delayed for several hours.

Inhalation: Exposure to 500 – 700 ppm for 2 – 3 minutes, resulted in nausea, vomiting, headache, disorientation, fluid in the lungs, followed by seizures. Human volunteers breathing a concentration of about 2,500 ppm experienced slight irritation of the respiratory tract; breathing in 12,500 ppm showed definite respiratory tract irritation within 10 seconds. Symptoms may not occur for hours after exposure. Other symptoms reported at unknown concentrations include headache, nausea, coughing, vomiting, difficult breathing, respiratory tract irritation, weakness, incoordination, seizures and fluid in the lungs.

Skin: The pure liquid may cause frostbite. A 1% water solution can cause irritation and redness. A 40 – 80% water solution may cause extensive blister formation. Ethylene oxide may severely irritate or burn mucous membranes and moist skin.

Eyes: May cause irritation and severe burns. May affect the eyes, causing delayed development of cataract. Eye contact may result in conjunctivitis (red, inflamed eyes) and erosion of the cornea.

Ingestion: May cause gastric irritation and liver injury.

Long Term Exposure: Repeated or prolonged contact with skin may cause skin allergy. ETO may affect the nervous system, kidneys, adrenal glands, skeletal muscles and cause reproductive effects. It may be carcinogenic to humans. Exposure to low concentrations of gaseous ETO may cause irritation to the respiratory tract and eyes; loss of sense of smell; and nausea and vomiting. Repeated skin exposure can cause scaling, cracking and redness. Exposure to 5 – 10 ppm for 11 years or to 1 ppm for 15 years has caused blood changes. ETO has caused cancer in several species of laboratory animals. It has also caused changes in genetic material and reproductive problems in laboratory animals. May damage the developing fetus. It may cause inheritable genetic damage in humans. There is an increased incidence of gynecological disorders and spontaneous abortions among workers in ethylene oxide production. Its role in this increase is unclear at this time. (NJ) Increased incidence of leukemia and stomach cancer have been reported; however the evidence is not considered conclusive. Leukemia, brain tumors, lung tumors, and other cancers have been observed in laboratory animals. (DHHS).

Points of Attack: Eyes, skin, respiratory system, liver, central nervous system, blood, kidneys, reproductive system. *Cancer Site:* peritoneal cancer, leukemia.

Medical Surveillance: For those with frequent or potentially high exposure (half the TLV or greater), the following are recommended before beginning work and at regular times after that: Lung function tests. If symptoms develop or overexposure is suspected, the following may be useful: Consider chest x-ray after acute overexposure. Evaluation by a qualified allergist, including careful exposure history and special testing, may help diagnose skin allergy. Liver and kidney function tests.

First Aid: If this chemical gets into the eyes, remove any contact lenses at once and irrigate immediately for at least 15 minutes, occasionally lifting upper and lower lids. Seek medical attention immediately. If this chemical contacts the skin, remove contaminated clothing and wash immediately with soap and water. Seek medical attention immediately. If this chemical has been inhaled, remove from exposure, begin rescue breathing (using universal precautions) if breathing has stopped and CPR if heart action has stopped. Transfer promptly to a medical facility. When this chemical has been swallowed, get medical attention. Give large quantities of water and induce vomiting. Do not make an unconscious person vomit. If frostbite has occurred, seek medical attention immediately; do *NOT* rub the affected areas or flush them with water. In order to prevent further tissue damage, do *NOT* attempt to remove frozen clothing from frostbitten areas. If frostbite has *NOT* occurred, immediately and thoroughly wash contaminated skin with soap and water. Medical observation is recommended for 24 – 48 hours after breathing overexposure, as pulmonary edema may be delayed. As first aid for pulmonary edema, a doctor or authorized paramedic may consider administering a corticosteroid spray.

Personal Protective Methods: Wear protective gloves and clothing to prevent any reasonable probability of skin contact. Safety equipment suppliers/manufacturers can provide recommendations on the most protective glove/clothing material for your operation.. All protective clothing (suits, gloves, footwear, headgear) should be clean, available each day, and put on before work. Contact lenses should not be worn when working with this chemical. Wear gas and splash-proof chemical goggles and face shield unless full facepiece respiratory protection is worn. Employees should wash immediately with soap when skin is wet or contaminated. Provide emergency showers and eyewash.

Respirator Selection: NIOSH: *5 ppm:* GMFS^{+} [any air-purifying, full-facepiece respirator (gas mask) with a chin-style, front- or back-mounted canister providing protection against the compound of concern]; or SCBAF (any self-contained breathing apparatus with a full facepiece); or SAF (any supplied-air respirator with a full facepiece). *Emergency or planned entry into unknown concentrations* *or IDLH conditions:* SCBAF:PD,PP (any self-contained breathing apparatus that has a full facepiece and is operated in a pressure-demand or other positive-pressure mode); or SAF:PD,PP: ASCBA (any supplied-air respirator that has a full facepiece and is operated in a pressure-demand or other positive-pressure mode in combination with an auxiliary self-contained breathing apparatus operated in a pressure-demand or other positive pressure mode). *Escape:* GMFS+ [any air-purifying, full-facepiece respirator (gas mask) with a chin-style, front- or back-mounted canister providing protection against the compound of concern]; or SCBAE (any appropriate escape-type, self-contained breathing apparatus).

+ End of service life indicator (ESLI) required.

Storage: Prior to working with this chemical you should be trained on its proper handling and storage. Before entering confined space where this chemical may be present, check to make sure that an explosive concentration does not exist. It must be stored to avoid contact with even small amounts of acids (such as nitric or sulfuric acids); alkalis (such as sodium hydroxide or potassium hydroxide); catalytic anhydrous chlorides of iron, aluminum or tin; iron or aluminum oxide; or metallic potassium, since it may react by itself, liberating much heat and causing a possible explosion. Ethylene oxide should not contact oxidizers (such as perchlorates, peroxides, permanganates, chlorates, and nitrates) since an explosion could occur. Store in tightly closed containers in a cool, well-ventilated area away from heat, sparks, or sunlight. Sources of ignition such as smoking and open flames are prohibited where ethylene oxide is handled, used, or stored. Metal containers involving the transfer of 5 gallons or more of ethylene oxide should be grounded and bonded. Drums must be equipped with self-closing valves, pressure vacuum bungs, and flame arresters. Use only non-sparking tools and equipment, especially when opening and closing containers of ethylene oxide. Wherever ethylene oxide is used, handled, manufactured, or stored, use explosion-proof electrical equipment and fittings. Procedures for the handling, use and storage of cylinders should be in compliance with OSHA 1910.101 and 1910.169 as with the recommendations of the Compressed Gas Association. A regulated, marked area should be established where this chemical is handled, used, or stored in compliance with OSHA standard 1910.1045.

Shipping: This compound requires a shipping label of: "Poison, Gas, Flammable Gas." It falls in DOT Hazard Class 2.3 and Packing Group II. Passenger aircraft or railcar shipment is forbidden as is any cargo aircraft shipment.

Spill Handling: *Gas:* Evacuate and restrict persons not wearing protective equipment from area of spill or leak until cleanup is complete. Remove all ignition sources. Establish forced ventilation to keep levels below explosive limit. Stop the flow of gas if it can be done safely. If source of leak is a cylinder and the leak cannot be stopped in place, remove leaking cylinder to a safe place in the open air, and repair leak or allow cylinder to empty. Keep this chemical out of confined space, such as sewer because of the possibility of explosion, unless the sewer is designed to prevent the buildup of explosive concentrations.

Liquid: For small spills flush area with flooding amounts of water. For large spills dike spill for later disposal. Absorb liquids in vermiculite, dry sand, earth, peat, carbon, or a similar material and deposit in sealed containers. It may be necessary to contain and dispose of this chemical as a hazardous waste. If material or contaminated runoff enters waterways, notify downstream users of potentially contaminated waters. Contact your Department of Environmental Protection or your regional office of the federal EPA for specific recommendations. If employees are required to clean-up spills, they must be properly trained and equipped. OSHA 1910.120(q) may be applicable.

Fire Extinguishing: *Gas:* Poisonous gases are produced in fire. Do not extinguish the fire unless the flow of gas can be stopped and any remaining gas is out of the line. Specially trained personnel may use fog lines to cool exposures and let the fire burn itself out. Vapors are heavier than air and will collect in low areas. Vapors may travel long distances to ignition sources and flashback. Vapors in confined areas may explode when exposed to fire. Containers may explode in fire. Storage containers and parts of containers may rocket great distances, in many directions. If material or contaminated runoff enters waterways, notify downstream users of potentially contaminated waters. Notify local health and fire officials and pollution control agencies. From a secure, explosion-proof location, use water spray to cool exposed containers. If cooling streams are ineffective (venting sound increases in volume and pitch, tank discolors, or shows any signs of deforming), withdraw immediately to a secure position. If cylinders are exposed to excessive heat from fire or flame contact, withdraw immediately to a secure location. If employees are expected to fight fires, they must be trained and equipped in OSHA 1910.156.

Liquid: Poisonous gases are produced in fire. Use dry chemical, carbon dioxide, or foam extinguishers. Although soluble in water, solutions will continue to burn until diluted to approximately 22 volumes of water to one volume of ethylene oxide. Vapors are heavier than air and will collect in low areas. Vapors may travel long distances to ignition sources and flashback. Vapors in confined areas may explode when exposed to fire. Containers may explode in fire. Storage containers and parts of containers may rocket great distances, in many directions. If material or contaminated runoff enters waterways, notify downstream users of potentially contaminated waters. Notify local health and fire officials and pollution control agencies. From a secure, explosion-proof location, use water spray to cool exposed containers. If cooling streams are ineffective (venting sound increases in volume and pitch, tank discolors, or shows any signs of deforming), withdraw immediately to a secure position. If employees are expected to fight fires, they must be trained and equipped in OSHA 1910.156.

Disposal Method Suggested: Concentrated waste containing no peroxides-discharge liquid at a controlled rate near a pilot flame. Concentrated waste containing peroxides-perforation of a container of the waste from a safe distance followed by open burning.[22]

References

Bogyo, D. A., Lande, S. S., Meylan, W. M., Howard, P. H. and Santodonate, J., (Syracuse Research Corp. Center for Chemical Hazard Assessment), Investigation of Selected Potential Environmental Contaminants: Epoxides, Report 560/11-80-005, Washington, DC, U.S. Environmental Protection Agency, (March 1980)

National Institute for Occupational Safety and Health, Ethylene Oxide, Current Intelligence Bulletin N. 35, DHHS (NIOSH) Publication No. 81–130, Cincinnati, Ohio (May 22, 1981)

Sax, N. I., Ed., "Dangerous Properties of Industrial Materials Report," 4, No. 2, 70–73 (1984)

U.S. Environmental Protection Agency, "Chemical Profile: Ethylene Oxide," Washington, DC, Chemical Emergency Preparedness Program (November 30, 1987)

New Jersey Department of Health and Senior Services, "Hazardous Substance Fact Sheet: Ethylene Oxide," Trenton, NJ (November 3, 1986)

ATSDR, ToxFAQs, Ethylene Oxide, Atlanta GA (July 1999)

Ethylene Thiourea

Molecular Formula: $C_3H_6N_2S$

Synonyms: Accel 22; Akrochem ETU-22; 4,5-Dihydroimidazole-2(3H)-thione; 4,5-Dihydro-2-mercaptoimidazole; *N,N'*-Ethylenethiourea; 1,3-Ethylenethiourea; Etilentiourea (Spanish); ETU; 2-Imidazolidinethione; Imidazolidinethione; 2-Imidazoline-2-thiol; Imidazoline-2-thiol; Imidazoline-2(3H)-thione; 2-Mercapto-2-imidazoline; 2-Mercaptoimidazoline; Mercaptoimidazoline; Mercazin I; 2-Merkaptoimidazolin (Czech); NA 22; NCI-C03372; Nocceler 22; Pennac CRA; Rhenogran ETU; Rhodanin S-62 (Czech); Sodium-22 neoprene accelerator; Soxinol 22; Tetrahydro-2H-imidazole-2-thione; 2-Thioimidazolidine; 2-Thiol-dihydroglyoxaline; 2-Thionomidazolidine; Thiourea, *N,N'*-(1,2-ethanediyl)-; USAF EL-62; Vulkacit NPV/C2; Warecure C

CAS Registry Number: 96-45-7

RTECS Number: NI9625000

DOT ID: UN 3077

EEC Number: 613-039-00-9

EINECS Number: 202-506-9

Regulatory Authority

- Carcinogen (Animal Positive) (IARC)[9]
- Banned or Severely Restricted (Sweden)[13]
- Air Pollutant Standard Set (North Dakota, Pennsylvania)[6]
- Clean Air Act: Hazardous Air Pollutants (Title I, Part A, Section 112)
- EPA Hazardous Waste Number (RCRA No.): U116
- RCRA, 40CFR261, Appendix 8 Hazardous Constituents
- Safe Drinking Water Act: Priority List (55 FR 1470)
- Superfund/EPCRA 40CFR302.4 Reportable Quantity (RQ): CERCLA, 10 lb (4.54 kg)
- EPCRA Section 313 Form R *de minimis* concentration reporting level: 0.1%

Cited in U.S. State Regulations: California (G), Florida (G), Illinois (G), Kansas (G), Louisiana (G), Maine (G, W), Maryland (G), Massachusetts (G), Michigan (G), New Hampshire (G), New Jersey (G), North Dakota (A), Pennsylvania (G, A), Rhode Island (G), Vermont (G), Virginia (G), Washington (G), West Virginia (G), Wisconsin (G).

Description: Ethylene thiourea, is a white to light green, needle-like crystalline solid with a faint amine odor. Boiling point = 230 – 313°C. Freezing/Melting point = 203 – 204°C. Flash point = 252°C. Hazard Identification (based on NFPA-704 M Rating System): Health 2, Flammability 1, Reactivity 0. Slightly soluble in cold water; highly soluble in hot. Commercial ethylene thiourea is available as a solid powder, as a dispersion in oil (which retards the formation of fine dust dispersions in workplace air), and "encapsulated" in a matrix of compatible elastomers. In this latter form, ethylene thiourea may be least likely to escape into the work-place air.

Potential Exposure: Ethylene thiourea is used extensively as an accelerator in the curing of polychloroprene (Neoprene) and other elastomers. In addition, exposure to ethylene thiourea also results from the very widely used ethylene bisdithiocarbamate fungicides. Ethylene thiourea may be present as a contaminant in the ethylene bisdithiocarbamate fungicides and can also be formed when food containing the fungicides is cooked.

Incompatibilities: Strong oxidizers, acids, acid anhydrides, acrolein.

Permissible Exposure Limits in Air: There is no current Occupational Safety and Health Administration (OSHA) exposure standard for ethylene thiourea. The ACGIH has set no limits either. NIOSH IDLH = not determined but contains potential occupational carcinogen notation. States which have set guidelines or standards for ethylene thiourea in ambient air[60] include North Dakota (zero level) and Pennsylvania (0.7 $\mu g/m^3$).

Determination in Air: Filter collection, extraction with water, complexation with pentacyanoamineferrate and spectrophotometric measurement. See NIOSH IV, Method #5011.

Permissible Concentration in Water: Maine[61] has set a guideline for drinking water of 4.4 μg/l. The USEPA in a health advisory (see reference below) has developed a no-observed-adverse-effect level (NOAEL) of 1.25 mg/kg/day based on absence of thyroid effects in male rats exposed to ETU in the diet for up to 12 months. This results in a longer term health advisory for an adult of 0.44 mg/l.

Routes of Entry: Inhalation, ingestion, skin and/or eye contact.

Harmful Effects and Symptoms

Short Term Exposure: Inhalation can cause irritation of the respiratory tract with soreness, hoarseness, cough and phlegm. High exposure can cause sweating, thirst, nausea, an increase in the heart rate and blood pressure that can last for hours or days. Higher exposures can cause pulmonary edema, a medical emergency that can be delayed for several hours. This can cause death. Contact can cause irritation of the skin and eyes, and may cause eye burns. A related chemical, *ziram*, can cause brain swelling and hemorrhage with muscle weakness and liver and kidney effects.

Long Term Exposure: Ethylene thiourea has been shown to be carcinogenic and teratogenic (causing malformation in offspring) in laboratory animals. In addition, ethylene thiourea can cause myxedema (the drying and thickening of skin, together with a slowing down of physical and mental activity), goiter, and other effects related to decreased output of thyroid hormone. *Maneb*, a related fungicide, can cause nerve damage.

Points of Attack: Eyes, skin, thyroid, reproductive system. Cancer Site in animals: liver, thyroid and lymphatic system tumors.

Medical Surveillance: Initial and routine employee exposure surveys should be made by competent industrial hygiene and engineering personnel. These surveys are necessary to determine the extent of employee exposure and to ensure that controls are effective. The *NIOSH Occupational Exposure Sampling Strategy Manual,* NIOSH Publication #77-173, may be helpful in developing efficient programs to monitor employee exposures to ethylene thiourea. The manual discusses determination of the need for exposure measurements, selection of appropriate employees for exposure evaluation, and selection of sampling times.

Employee exposure measurements should consist of 8-hour TWA (time-weighted average) exposure estimates calculated from personal or breathing zone samples (air that would most nearly represent that inhaled by the employees). Area and source measurements may be useful to determine problem areas, processes, and operations. Thyroid function tests. Examination of the nervous system. Consider chest x-ray following acute overexposure.

First Aid: If this chemical gets into the eyes, remove any contact lenses at once and irrigate immediately for at least 15 minutes, occasionally lifting upper and lower lids. Seek medical attention immediately. If this chemical contacts the skin, remove contaminated clothing and wash immediately with soap and water. Seek medical attention immediately. If this chemical has been inhaled, remove from exposure, begin rescue breathing (using universal precautions) if breathing has stopped and CPR if heart action has stopped. Transfer promptly to a medical facility. When this chemical has been swallowed, get medical attention. Give large quantities of water and induce vomiting. Do not make an unconscious person vomit. Medical observation is recommended for 24 – 48 hours after breathing overexposure, as pulmonary edema may be delayed. As first aid for pulmonary edema, a doctor or authorized paramedic may consider administering a corticosteroid spray.

Personal Protective Methods: There are four basic methods of limiting employee exposure to ethylene thiourea. None of these is a simple industrial hygiene or management decision and careful planning and thought should be used prior to implementation of any of these.

Product Substitution: The substitution of an alternative material with a lower potential health and safety risk in one method. However, extreme care must be used when selecting possible substitutes. Alternatives to ethylene thiourea should be fully evaluated with regard to possible human effects. Unless the toxic effects of the alternative have been thoroughly evaluated, a seemingly safe replacement, possibly only after years of use, may be found to induce serious health effects.

Contaminant Control: The most effective control of ethylene thiourea, where feasible, is at the source of contamination by enclosure of the operation and/or local exhaust ventilation. If feasible, the process or operation should be enclosed with a slight vacuum so that any leakage will result in the flow of air into the enclosure. The next most effective means of control would be a well designed local exhaust ventilation system that physically encloses the process as much as possible, with sufficient capture velocity to keep the contaminant from entering the work atmosphere.

To ensure that ventilation equipment is working properly, effectiveness (e.g., air velocity, static pressure, or air volume) should be checked at least every 3 months. System effectiveness should be checked soon after any change in production, process or control which might result in significant increases in airborne exposures to ethylene thiourea.

Employee Isolation: A third alternative is the isolation of employees. It frequently involves the use of automated equipment operated by personnel observing from a closed control booth or room. The control room is maintained at a greater air pressure than that surrounding the process equipment so that air flow is out of, rather than into, the room. This type of control will not protect those employees that must do process checks, adjustments, maintenance, and related operations.

Personal Protective Equipment: The least preferred method is the use of personal protective equipment. This equipment, which may include respirators, goggles, gloves, and related items, should not be used as the only means to prevent or minimize exposure during routine operations.

Exposure to ethylene thiourea should not be controlled with the use of respirators except: During the time necessary to install or implement engineering or work practice controls; or, in work situation in which engineering and work practice controls are technically not feasible; or, for maintenance; or, for operations which require entry into tanks or closed vessels; or, in emergencies.

Respirator Selection: *At any detectable concentration:* SCBAF:PD,PP (any MSHA/NIOSH approved self-contained breathing apparatus that has a full facepiece and is operated in a pressure-demand or other positive-pressure mode); or SAF: PD,PP:ASCBA (any supplied-air respirator that has a full facepiece and is operated in a pressure-demand or other positive-pressure mode in combination with an auxiliary, self-contained breathing apparatus operated in a pressure-demand or other positive pressure mode). *Escape:* GMFOVHiE [any air-purifying, full-facepiece respirator (gas mask) with a chin-style, front- or back-mounted organic vapor canister having a high-efficiency particulate filter]; or SCBAE (any appropriate escape-type, self-contained breathing apparatus).

Storage: Prior to working with this chemical you should be trained on its proper handling and storage. Store in tightly closed containers in a cool, well ventilated area away from strong oxidizers, acids, acid anhydrides, acrolein. Store in a refrigerator or a cool, dry place. A regulated, marked area should be established where this chemical is handled, used, or stored in compliance with OSHA standard 1910.1045.

Shipping: Environmentally hazardous solid, n.o.s. Hazard Class: 9. Label: "Class 9."

Spill Handling: Evacuate persons not wearing protective equipment from area of spill or leak until clean-up is complete. Remove all ignition sources. Dampen spilled material with water to avoid dust. Collect powdered material in the most convenient and safe manner and deposit in sealed containers. Ventilate area after clean-up is complete. It may be necessary to contain and dispose of this chemical as a hazardous waste. If material or contaminated runoff enters waterways, notify downstream users of potentially contaminated waters. Contact your Department of Environmental Protection or your regional office of the federal EPA for specific recommendations. If employees are required to clean-up spills, they must be properly trained and equipped. OSHA 1910.120(q) may be applicable.

Fire Extinguishing: Use dry chemical, carbon dioxide, water spray, or foam extinguishers. Poisonous gases are produced in fire. If material or contaminated runoff enters waterways, notify downstream users of potentially contaminated waters. Notify local health and fire officials and pollution control agencies. From a secure, explosion-proof location, use water spray to cool exposed containers. If cooling streams are ineffective (venting sound increases in volume and pitch, tank discolors, or shows any signs of deforming), withdraw immediately to a secure position. If employees are expected to fight fires, they must be trained and equipped in OSHA 1910.156.

Disposal Method Suggested: Incineration in a furnace equipped with afterburner and scrubber.[22]

References

National Institute for Occupational Safety and Health, Ethylene Thiourea, Current Intelligence Bulletin 22, Washington, DC (April 11, 1978)

Sax, N. I., Ed., "Dangerous Properties of Industrial Materials Report," 1, No. 2, 38–39 (1980); 7 No. 3, 106–111 (1987)

U.S. Environmental Protection Agency, "Health Advisory: Ethylene Thiourea," Washington, DC, Office of Drinking Water (August 1987)

Ethyl Ether

Molecular Formula: $C_4H_{10}O$

Common Formula: $CH_3CH_2OCH_2CH_3$

Synonyms: Aether; Anaesthetic ether; Anesthesia ether; Anesthetic ether; Diaethylaether (German); Diethyl ether; Diethyl oxide; Dwuetylowyeter (Polish); Etere etilico (Italian); Eter etilico (Spanish); Ethane, 1,1'-oxybis-; Ether; Ether, ethyl; Ether ethylique (French); Ethoxyethane; Oxyde d'ethyle (French); Solvent ether; Sulfuric ether

CAS Registry Number: 60-29-7

RTECS Number: KI5775000

DOT ID: UN 1155

EEC Number: 603-022-00-4

EINECS Number: 200-467-2

Regulatory Authority

- Air Pollutant Standard Set (ACGIH)[1] (DFG)[3] (HSE)[33] (Other Countries)[35] (OSHA)[58] (Several States)[60] (Several Canadian Provinces)
- Clean Air Act: Accidental Release Prevention/Flammable substances, (Section 112[r], Table 3), TQ = 10,000 lb (4,540 kg)
- EPA Hazardous Waste Number (RCRA No.): U117
- RCRA, 40CFR261, Appendix 8 Hazardous Constituents
- Superfund/EPCRA 40CFR302.4 Reportable Quantity (RQ): CERCLA, 100 lb (45.4 kg)
- RCRA 40CFR268.48; 61FR15654, Universal Treatment Standards: Wastewater (mg/l), 0.12; Nonwastewater (mg/kg), 160
- Canada, WHMIS, Ingredients Disclosure List

Cited in U.S. State Regulations: Alaska (G),California (A, G), Connecticut (A), Florida (G, A), Illinois (G), Kansas (G), Louisiana (G), Maine (G), Massachusetts (G, A), Nevada (A), New Hampshire (G), New Jersey (G), New York (A), North Dakota (A), Pennsylvania (G), Rhode Island (G), Vermont (G), Virginia (G, A), Washington (G), West Virginia (G), Wisconsin (G).

Description: Ethyl ether, $CH_3CH_2OCH_2CH_3$, is a colorless, mobile, highly flammable, volatile liquid with a characteristic pungent odor. The odor threshold is 0.63 ppm.[41] Boiling point = 35°C. Flash point = -45°C. Autoignition temperature = 180°C. Explosive limits: LEL = 1.9%; UEL = 36.0%.[17] Hazard Identification (based on NFPA-704 M Rating System): Health 1, Flammability 4, Reactivity 0. Slightly soluble in water.

Potential Exposure: Ethyl ether is used as a solvent for waxes, fats, oils, perfumes, alkaloids, dyes, gums, resins, nitrocellulose, hydrocarbons, raw rubber, and smokeless powder. It is also used as an inhalation anesthetic, a refrigerant, in diesel fuels, in dry cleaning, as an extractant, and as a chemical reagent for various organic reactions.

Incompatibilities: Forms explosive mixture with air. Incompatible with strong acids, strong oxidizers halogens, sulfur, sulfur compounds causing fire and explosion hazard. Can form peroxides from air, heat, sunlight; may explode when container is unstoppered or otherwise opened. Attacks some plastics, rubber and coatings. Being a non-conductor, chemical may accumulate static electric charges that may result in ignition of vapor.

Permissible Exposure Limits in Air: The OSHA TWA,[58] the HSE TWA,[33] the DFG MAK[3] and the Swedish TWA[35] and the ACGIH value is 400 ppm (1,200 mg/m^3) and the ACGIH STEL value is 500 ppm (1,500 mg/m^3). The NIOSH IDLH level is 1,900 ppm. The former USSR[35][43] have set a ceiling value in workplace air of 300 mg/m^3. Czechoslovakia[35] has set a TWA in 500 mg/m^3 and an STEL of 1,500 mg/m^3. The former USSR[35] set a MAC in the ambient air of residential areas of 1.0 mg/m^3 on a once-a-day basis and 0.6 mg/m^3 on a daily average basis. Several states have set guidelines or standards for ethyl ether in ambient air[60] ranging from 0.16 mg/m^3 (Massachusetts) to 12.0 – 15.0 mg/m^3 (North Dakota) to 20.0 mg/m^3 (Virginia) to 24.0 mg/m^3 (Connecticut, Florida, New York) to 28.571 mg/m^3 (Nevada).

Determination in Air: Charcoal adsorption, workup with ethyl acetate, analysis by gas-liquid chromatography/flame ionization. See NIOSH Method #1610.[18]

Permissible Concentration in Water: The former USSR[35][43] has set a MAC in water bodies used for domestic purposes of 0.3 mg/l.

Routes of Entry: Inhalation of vapor, ingestion, skin and/or eye contact.

Harmful Effects and Symptoms

Local: Ethyl ether vapor is mildly irritating to the eyes, nose, and throat. Contact with liquid may produce a dry, scaly, fissured dermatitis.

Systemic: Ethyl ether has predominantly narcotic properties. Overexposed individuals may experience drowsiness, vomiting, and unconsciousness. Death may result from severe overexposure. Chronic exposure results in some persons in anorexia, exhaustion, headache, drowsiness, dizziness, excitation, and psychic disturbances. Albuminuria has been reported. Chronic exposure may cause an increased susceptibility to alcohol. There is an association between exposure to anesthetic vapors and increased miscarriages and birth defects. Diethyl ether's role in these increased risks in unclear.

Short Term Exposure: Ethyl ether irritates the eyes and respiratory tract. Inhalation can cause drowsiness, excitement, dizziness, vomiting, irregular breathing and increased saliva. Swallowing the liquid may cause chemical pneumonitis. High exposure can affect the central nervous system, causing unconsciousness and even death.

Long Term Exposure: Repeated or prolonged contact can cause skin cracking, scaling and extreme drying. Repeated exposure may cause an addiction.

Points of Attack: Central nervous system, skin, respiratory system, eyes.

Medical Surveillance: Preplacement or periodic examinations should evaluate the skin and respiratory tract, liver, and kidney function. Persons with a past history of alcoholism may be at some increased risk due to possibility of ethyl ether addiction (known as "ether habit"). Tests for exposure may include expired breath for unmetabolized ethyl ether and blood for ethyl ether content by oxidation with chromate solution or by gas chromatographic methods.

First Aid: If this chemical gets into the eyes, remove any contact lenses at once and irrigate immediately for at least 15 minutes, occasionally lifting upper and lower lids. Seek medical attention immediately. If this chemical contacts the skin, remove contaminated clothing and wash immediately with soap and water. Seek medical attention immediately. If this chemical has been inhaled, remove from exposure, begin rescue breathing (using universal precautions) if breathing has stopped and CPR if heart action has stopped. Transfer promptly to a medical facility. When this chemical has been swallowed, get medical attention. Give large quantities of salt water and induce vomiting. Do not make an unconscious person vomit.

Personal Protective Methods: Wear protective gloves and clothing to prevent any reasonable probability of skin contact. Safety equipment suppliers/manufacturers can provide recommendations on the most protective glove/clothing material for your operation. Teflon, Silvershield, and Polyvinyl Alcohol are among the recommended protective materials. All protective clothing (suits, gloves, footwear, headgear) should be clean, available each day, and put on before work. Contact lenses should not be worn when working with this chemical. Wear splash-proof chemical goggles and face shield unless full facepiece respiratory protection is worn. Employees should wash immediately with soap when skin is wet or contaminated. Provide emergency showers and eyewash.

Respirator Selection: OSHA: *1,900 ppm:* CCROV [any chemical cartridge respirator with organic vapor cartridge(s)]; or GMFOV [any air-purifying, full-facepiece respirator (gas mask) with a chin-style, front- or back-mounted acid gas canister]; or PAPROV [any powered, air-purifying respirator with organic vapor cartridge(s)]; or SA (any supplied-air respirator); or SCBAF (any self-contained breathing apparatus with a full facepiece). *Emergency or planned entry into unknown concentrations or IDLH conditions:* SCBAF:PD,PP (any self-contained breathing apparatus that has a full facepiece and is operated in a pressure-demand or other positive-pressure mode); or SAF:PD,PP:ASCBA (any supplied-air respirator that has a full facepiece and is operated in a pressure-demand or other positive-pressure mode in combination with an auxiliary self-contained breathing apparatus operated in a pressure-demand or other positive pressure mode). *Escape:* GMFOV [any air-purifying, full-facepiece respirator (gas mask) with a chin-style, front-or back-, mounted organic vapor canister]; or SCBAE (any appropriate escape-type, self-contained breathing apparatus).

Note: Substance reported to cause eye irritation or damage; may require eye protection.

Storage: Prior to working with this chemical you should be trained on its proper handling and storage. Before entering confined space where this chemical may be present, check to make sure that an explosive concentration does not exist. Ethyl ether must be stored to avoid contact with strong oxidizers (such as bromine, chlorine, chlorine dioxide, and nitrates) since violent reactions occur. Store in tightly closed containers in a cool, well-ventilated area away from heat and sunlight. Unstable peroxides may form if diethyl ether is exposed for a long time to air or sunlight, causing explosions. Sources of ignition such as smoking and open flames are prohibited where ethyl ether is handled, used, or stored. Metal containers involving the transfer of 5 gallons or more of ethyl ether should be grounded and bonded. Drums must be equipped with self-closing valves, pressure vacuum bungs, and flame arresters. Use only non-sparking tools and equipment, especially when opening and closing containers of ethyl ether. Wherever ethyl ether is used, handled, manufactured, or stored, use explosion-proof electrical equipment and fittings.

Shipping: This compound requires a shipping label of: "Flammable Liquid." It falls in DOT Hazard Class 3 and Packing Group I.

Spill Handling: Evacuate and restrict persons not wearing protective equipment from area of spill or leak until cleanup is complete. Remove all ignition sources. Establish forced ventilation to keep levels below explosive limit. Absorb liquids in vermiculite, dry sand, earth, peat, carbon, or a similar material and deposit in sealed containers. Keep diethyl ether out of a confined space, such as a sewer, because of the possibility of an explosion, unless the sewer is designed to prevent the build-up of explosive concentrations. It may be necessary to contain and dispose of this chemical as a hazardous waste. If material or contaminated runoff enters waterways, notify downstream users of potentially contaminated waters. Contact your Department of Environmental Protection or your regional office of the federal EPA for specific recommendations. If employees are required to clean-up spills, they must be properly trained and equipped. OSHA 1910.120(q) may be applicable.

Fire Extinguishing: This chemical is an extremely flammable liquid. Poisonous gases are produced in fire. Use dry chemical, carbon dioxide, or alcohol foam extinguishers. Vapors are heavier than air and will collect in low areas. Vapors may travel long distances to ignition sources and flashback. Vapors in confined areas may explode when exposed to fire. Containers may explode in fire. Storage containers and parts of containers may rocket great distances, in many directions. If material or contaminated runoff enters waterways, notify downstream users of potentially contaminated waters. Notify local health and fire officials and pollution control agencies. From a secure, explosion-proof location, use water spray to cool exposed containers. If cooling streams are ineffective (venting sound increases in volume and pitch, tank discolors, or shows any signs of deforming), withdraw immediately to a secure position. If employees are expected to fight fires, they must be trained and equipped in OSHA 1910.156.

Disposal Method Suggested: Concentrated waste containing no peroxides-discharge liquid at a controlled rate near a pilot flame. Concentrated waste containing peroxides-perforation of a container of the waste from a safe distance followed by open burning.[22] Consult with environmental regulatory agencies for guidance on acceptable disposal practices. Generators of waste containing this contaminant (≥100 kg/mo) must conform with EPA regulations governing storage, transportation, treatment, and waste disposal.

References

Sax, N. I., Ed., "Dangerous Properties of Industrial Materials Report," 4, No. 1, 81–84 (1984)

New Jersey Department of Health and Senior Services, "Hazardous Substance Fact Sheet: Diethyl Ether," Trenton, NJ (January 1986)

Ethyl Formate

Molecular Formula: $C_3H_6O_2$

Common Formula: $HCOOC_2H_5$

Synonyms: Aethylformiat (German); Areginal; Ethyle (formiate d') (French); Ethylformiaat (Dutch); Ethyl formic ester; Ethyl methanoate; Etile (formiato di) (Italian); Formic acid, ethyl ester; Formic ether; Mrowczan etylu (Polish)

CAS Registry Number: 109-94-4

RTECS Number: LQ8400000

DOT ID: UN 1190

EEC Number: 607-015-00-7

Regulatory Authority

- Banned or Severely Restricted (In Agriculture) (UN)[13]
- Air Pollutant Standard Set (ACGIH)[1] (DFG)[3] (HSE)[33] (OSHA)[58] (Several States)[60] (Several Canadian Provinces)
- Canada, WHMIS, Ingredients Disclosure List

Cited in U.S. State Regulations: Alaska (G), California (A, G), Connecticut (A), Florida (G), Illinois (G), Maine (G), Massachusetts (G), Nevada (A), New Hampshire (G), New Jersey (G), North Dakota (A), Oklahoma (G), Pennsylvania (G), Rhode Island (G), Virginia (A), West Virginia (G).

Description: Ethyl formate, $HCOOC_2H_5$, is a colorless liquid with a fruity odor. Boiling point = 54 – 55°C. Flash point = -20°C. Autoignition temperature = 440°C. Explosive limits: LEL = 2.8%; UEL = 16.0%. Hazard Identification (based on NFPA-704 M Rating System): Health 2, Flammability 3, Reactivity 0. Insoluble in water.

Potential Exposure: This material is used as a solvent for cellulose nitrate and acetate; it is used as a fumigant and in the production of synthetic flavors. It is also a raw material in pharmaceutical manufacture.

Incompatibilities: Forms explosive mixture with air. Reacts violently with nitrates, strong oxidizers, strong alkalis, and strong acids. Decomposes slowly in water to form ethyl alcohol and formic acid. May accumulate static electrical charges, and may cause ignition of its vapors.

Permissible Exposure Limits in Air: The Federal limit [OSHA PEL],[58] the DFG MAK,[3] the ACGIH TWA,[1] and the HSE TWA[33] is 100 ppm (300 mg/m^3). The STEL set by HSE is 150 ppm (450 mg/m^3). The NIOSH IDLH level is 1,500 ppm. Several states have set guidelines or standards for ethyl formate in ambient air[60] ranging from 3.0 – 4.5 mg/m^3 (North Dakota) to 5.0 mg/m^3 (Virginia) to 6.0 mg/m^3 (Connecticut) to 7.143 mg/m^3 (Nevada).

Determination in Air: Charcoal (tube) adsorption, workup with CS_2; analysis by gas chromatography/flame ionization detection; NIOSH IV, Method #1452.

Permissible Concentration in Water: No criteria set.

Routes of Entry: Inhalation, ingestion, eye and/or skin contact.

Harmful Effects and Symptoms

Short Term Exposure: Either contact or the vapor can cause skin and eye irritation. Inhalation irritates the respiratory tract. Higher exposures can cause pulmonary edema, a medical emergency that can be delayed for several hours. This can cause death. Ethyl formate may affect the central nervous system. Exposure can cause headache, nausea, and vomiting.

Long Term Exposure: Prolonged or repeated contact can cause skin dryness and cracking. May affect the nervous system.

Points of Attack: Eyes, respiratory system, central nervous system.

Medical Surveillance: Consider the points of attack in preplacement and periodic physical examinations. Consider chest x-ray following acute overexposure. Nervous system tests.

First Aid: If this chemical gets into the eyes, remove any contact lenses at once and irrigate immediately for at least 15 minutes, occasionally lifting upper and lower lids. Seek medical attention immediately. If this chemical contacts the skin, remove contaminated clothing and wash immediately with soap and water. Seek medical attention immediately. If this chemical has been inhaled, remove from exposure, begin rescue breathing (using universal precautions) if breathing has stopped and CPR if heart action has stopped. Transfer promptly to a medical facility. When this chemical has been swallowed, get medical attention. Give large quantities of water and induce vomiting. Do not make an unconscious person vomit. Medical observation is recommended for 24 – 48 hours after breathing overexposure, as pulmonary edema may be delayed. As first aid for pulmonary edema, a doctor or authorized paramedic may consider administering a corticosteroid spray.

Personal Protective Methods: Wear protective gloves and clothing to prevent any reasonable probability of skin contact. Safety equipment suppliers/manufacturers can provide recommendations on the most protective glove/clothing material for your operation. Neoprene, Nitrile and Styrene-Butadiene Rubber are among the recommended protective materials. All protective clothing (suits, gloves, footwear, headgear) should be clean, available each day, and put on before work. Contact lenses should not be worn when working with this chemical. Wear splash-proof chemical goggles and face shield unless full facepiece respiratory protection is worn. Employees should wash immediately with soap when skin is wet or contaminated. Provide emergency showers and eyewash.

Respirator Selection: NIOSH/OSHA: *up to 1,500 ppm:* SA:CF (any supplied-air respirator operated in a continuous-flow mode); or PAPROV [any powered, air-purifying respirator with organic vapor cartridge(s)]; or CCROV [any chemical cartridge respirator with a full facepiece and organic vapor cartridge(s)]; or GMFOV [any air-purifying, full-facepiece respirator (gas mask) with a chin-style, front- or back-mounted organic vapor canister]; or SCBAF (any self-contained breathing apparatus with a full facepiece); SAF (any supplied-air respirator with a full facepiece). *Emergency or planned entry into unknown concentrations or IDLH conditions:* SCBAF: PD,PP (any MSHA/NIOSH approved self-contained breathing apparatus that has a full facepiece and is operated in a pressure-demand or other positive-pressure mode); or SAF: PD,PP:ASCBA (any supplied-air respirator that has a full facepiece and is operated in a pressure-demand or other positive-pressure mode in combination with an auxiliary, self-contained breathing apparatus operated in a pressure-demand or other positive pressure mode). *Escape:* GMFOV [any air-purifying, full-facepiece respirator (gas mask) with a chin-style, front-or back-mounted organic vapor canister]; or SCBAE (any appropriate escape-type, self-contained breathing apparatus).

Storage: Prior to working with this chemical you should be trained on its proper handling and storage. Before entering confined space where this chemical may be present, check to make sure that an explosive concentration does not exist. Store in tightly closed containers in a cool, well ventilated area away from oxidizers, strong bases, moisture and heat. Where possible, automatically pump liquid from drums or other storage containers to process containers. Metal containers involving the transfer of this chemical should be grounded and bonded.

Where possible, automatically pump liquid from drums or other storage containers to process containers. Drums must be equipped with self-closing valves, pressure vacuum bungs, and flame arresters. Use only non-sparking tools and equipment, especially when opening and closing containers of this chemical. Sources of ignition such as smoking and open flames, are prohibited where this chemical is used, handled, or stored in a manner that could create a potential fire or explosion hazard. Wherever this chemical is used, handled, manufactured, or stored, use explosion-proof electrical equipment and fittings.

Shipping: This compound requires a shipping label of: "Flammable Liquid." It falls in DOT Hazard Class 3 and Packing Group II.

Spill Handling: Evacuate and restrict persons not wearing protective equipment from area of spill or leak until cleanup is complete. Remove all ignition sources. Establish forced ventilation to keep levels below explosive limit. Absorb liquids in vermiculite, dry sand, earth, peat, carbon, or a similar material and deposit in sealed containers. It may be necessary to contain and dispose of this chemical as a hazardous waste. If material or contaminated runoff enters waterways, notify downstream users of potentially contaminated waters. Contact your Department of Environmental Protection or your regional office of the federal EPA for specific recommendations. If employees are required to clean-up spills, they must be properly trained and equipped. OSHA 1910.120(q) may be applicable.

Fire Extinguishing: This chemical is a flammable liquid. Poisonous gases including carbon monoxide are produced in fire. Use dry chemical, carbon dioxide, or alcohol foam extinguishers. Vapors are heavier than air and will collect in low areas. Vapors may travel long distances to ignition sources and flashback. Vapors in confined areas may explode when exposed to fire. Containers may explode in fire. Storage containers and parts of containers may rocket great distances, in many directions. If material or contaminated runoff enters waterways, notify downstream users of potentially contaminated waters. Notify local health and fire officials and pollution control agencies. From a secure, explosion-proof location, use water spray to cool exposed containers. If cooling streams are ineffective (venting sound increases in volume and pitch, tank discolors, or shows any signs of deforming), withdraw immediately to a secure position. If employees are expected to fight fires, they must be trained and equipped in OSHA 1910.156.

Disposal Method Suggested: Spray into a furnace in admixture with a flammable solvent.[24]

References

New Jersey Department of Health and Senior Services, "Hazardous Substance Fact Sheet, Ethyl Formate," Trenton NJ (March 1999)

2-Ethyl Hexaldehyde

Molecular Formula: $C_8H_{16}O$

Common Formula: $(C_2H_5)(C_4H_9)CHCHO$

Synonyms: Butyl ethyl acetaldehyde; 2-Ethylcaproaldehyde; Ethylhexanal; 3-Formylheptane; Hexanal, 2-ethyl-; Octyl aldehyde; β-Propyl-α-ethylacrolein

CAS Registry Number: 123-05-7

RTECS Number: MN7525000

DOT ID: UN 1191

Cited in U.S. State Regulations: Florida (G), Massachusetts (G), New Hampshire (G), New Jersey (G), Pennsylvania (G).

Description: Ethyl hexaldehyde, $(C_2H_5)(C_4H_9)CHCHO$, is a colorless liquid with a mild, pleasant odor. Boiling point = 163°C. Flash point = 44.4°C. Autoignition temperature = 190°C. Explosive limits: LEL = 0.85% @ 93°C; UEL = 7.2% @ 135°C. Hazard Identification (based on NFPA-704 M Rating System): Health 2, Flammability 2, Reactivity 1. Slightly soluble in water.

Potential Exposure: Uses include organic synthesis, perfume formulation, disinfectant.

Incompatibilities: Forms explosive mixture with air. Violent reaction with oxidizers. Incompatible with strong acids, caustics, ammonia, amines. May ignite spontaneously when spilled on clothing or other absorbent materials. May form unstable peroxides on contact with air; under certain conditions ignites spontaneously with air.

Permissible Exposure Limits in Air: No standards set.

Permissible Concentration in Water: No criteria set.

Routes of Entry: Skin absorption, inhalation, ingestion.

Harmful Effects and Symptoms

Short Term Exposure: Skin or eye contact can cause severe irritation or burns. The vapors can irritate the eyes, throat and bronchial tubes, with coughing and difficulty breathing. Headaches and nausea may occur. Higher exposures can cause pulmonary edema, a medical emergency that can be delayed for several hours. This can cause death.

Long Term Exposure: Repeated exposure may cause bronchitis with cough, phlegm and shortness of breath.

Medical Surveillance: Before beginning employment and at regular times after that, for those with frequent or potentially high exposures, the following are recommended: Lung function tests. If symptoms develop or overexposure is suspected, the following may be useful: Consider chest x-ray after acute overexposure. Evaluation by a qualified allergist, including careful exposure history and special testing may help diagnose skin allergy.

First Aid: If this chemical gets into the eyes, remove any contact lenses at once and irrigate immediately for at least 15 minutes, occasionally lifting upper and lower lids. Seek medical attention immediately. If this chemical contacts the skin, remove contaminated clothing and wash immediately with soap and water. Seek medical attention immediately. If this chemical has been inhaled, remove from exposure, begin rescue breathing (using universal precautions) if breathing has stopped and CPR if heart action has stopped. Transfer promptly to a medical facility. When this chemical has been swallowed, get medical attention. Give large quantities of water and induce vomiting. Do not make an unconscious person vomit. Medical observation is recommended for 24 – 48 hours after breathing overexposure, as pulmonary edema may be delayed.

Personal Protective Methods: Wear solvent-resistant gloves and clothing to prevent any reasonable probability of skin contact. Safety equipment suppliers/manufacturers can provide recommendations on the most protective glove/clothing material for your operation.. All protective clothing (suits, gloves, footwear, headgear) should be clean, available each day, and put on before work. Contact lenses should not be worn when working with this chemical. Wear splash-proof chemical goggles and face shield unless full facepiece respiratory protection is worn. Employees should wash immediately with soap when skin is wet or contaminated. Provide emergency showers and eyewash.

Respirator Selection: Where the potential for exposures to ethyl hexaldehyde exists, use a MSHA/NIOSH approved supplied-air respirator with a full facepiece operated in the positive pressure mode or with a full facepiece, hood, or helmet in the continuous flow mode, or use a MSHA/NIOSH approved self-contained breathing apparatus with a full facepiece operated in pressure-demand or other positive pressure mode.

Storage: Prior to working with ethyl hexaldehyde you should be trained on its proper handling and storage. Before entering confined space where this chemical may be present, check to make sure that an explosive concentration does not exist. Store in tightly closed containers in a cool, well ventilated area away from oxidizers, strong bases and combustible materials. Metal containers involving the transfer of this chemical should be grounded and bonded. Where possible, automatically pump liquid from drums or other storage containers to process containers. Drums must be equipped with self-closing valves, pressure vacuum bungs, and flame arresters. Use only non-sparking tools and equipment, especially when opening and closing containers of this chemical. Sources of ignition such as smoking and open flames, are prohibited where this chemical is used, handled, or stored in a manner that could create a potential fire or explosion hazard. Wherever this chemical is used, handled, manufactured, or stored, use explosion-proof electrical equipment and fittings.

Shipping: This falls under "Octyl Aldehydes, Flammable," in DOT regulations. This class of compounds requires a shipping label of "Flammable Liquid." It falls in DOT Hazard Class 3 and Packing Group III.

Spill Handling: Evacuate and restrict persons not wearing protective equipment from area of spill or leak until cleanup is complete. Remove all ignition sources. Establish forced ventilation to keep levels below explosive limit. Absorb liquids in vermiculite, dry sand, earth, peat, carbon, or a similar material and deposit in sealed containers. Keep ethyl hexaldehyde out of a confined space, such as a sewer, because of the possibility of an explosion, unless the sewer is designed to prevent the build-up of explosive concentrations. It may be necessary to contain and dispose of this chemical as a hazardous waste. If material or contaminated runoff enters waterways, notify downstream users of potentially contaminated waters. Contact your Department of Environmental Protection or your regional office of the federal EPA for specific recommendations. If employees are required to clean-up spills, they must be properly trained and equipped. OSHA 1910.120(q) may be applicable.

Fire Extinguishing: This chemical is a combustible liquid. Poisonous gases including carbon monoxide are produced in fire. Use dry chemical, carbon dioxide, or foam extinguishers. Vapors are heavier than air and will collect in low areas. Vapors may travel long distances to ignition sources and flashback. Vapors in confined areas may explode when exposed to fire. Containers may explode in fire. Storage containers and parts of containers may rocket great distances, in many directions. If material or contaminated runoff enters waterways, notify downstream users of potentially contaminated waters. Notify local health and fire officials and pollution control agencies. From a secure, explosion-proof location, use water spray to cool exposed containers. If cooling streams are ineffective (venting sound increases in volume and pitch, tank discolors, or shows any signs of deforming), withdraw immediately to a secure position. If employees are expected to fight fires, they must be trained and equipped in OSHA 1910.156.

Disposal Method Suggested: Incineration.

References

Sax, N. I., Ed., "Dangerous Properties of Industrial Materials Report," 1, No. 8, 71–72 (1981) and 3, No. 2, 47–48 (1983)

New Jersey Department of Health and Senior Services, "Hazardous Substance Fact Sheet: Ethyl Hexaldehyde," Trenton, NJ (January 1999)

2-Ethylhexyl Acrylate

Molecular Formula: $C_{11}H_{20}O_2$

Common Formula: $CH_2{=}CHCOOCH_2CH(C_2H_5)(C_4H_9)$

Synonyms: Acrilato de 2-etilhexilo (Spanish); Acrylic acid, 2-ethylhexyl ester; 2-Ethylhexyl 2-propenoate; Octyl acrylate; 2-Propenoic acid 2-ethylhexyl ester

CAS Registry Number: 103-11-7

RTECS Number: AT0855000

EEC Number: 607-107-00-7

EINECS Number: 203-080-7

Cited in U.S. State Regulations: Florida (G), Massachusetts (G), New Hampshire (G), Pennsylvania (G), Rhode Island (G).

Description: 2-Ethylhexyl Acrylate, CH_2=$CHCOOCH_2CH$ $(C_2H_5)(C_4H_9)$, is a colorless liquid with a pleasant odor. Boiling point = 214°C. Also listed at 130°C @ 50 mm. Freezing/Melting point = -90°C. Flash point = 82°C. Autoignition temperature = 252°C. Hazard Identification (based on NFPA-704 M Rating System): Health 2, Flammability 2, Reactivity 2. Insoluble in water.

Potential Exposure: Those involved in the use of this monomer in plastics manufacture, protective coatings, paper treatment and water-based paints.

Incompatibilities: Unless inhibited, sunlight, heat, contaminants or peroxides can cause polymerization. Violent reaction with strong oxidizers, with risk of fire and explosions. Forms explosive mixture with air. Incompatible with strong acids, aliphatic amines, alkanolamines. Vapors are uninhibited and may polymerize, blocking vents.

Permissible Exposure Limits in Air: None established.

Permissible Concentration in Water: No criteria set.

Routes of Entry: Skin absorption.

Harmful Effects and Symptoms

Short Term Exposure: Severely irritates the eyes and skin. Irritates the respiratory tract. High exposure may cause respiratory difficulty and collapse. CNS stimulation following ingestion.

Long Term Exposure: Repeated or prolonged contact may cause skin allergy.

Points of Attack: Skin.

Medical Surveillance: Evaluation by a qualified allergist.

First Aid: *Skin Contact:*[52] Flood all areas of body that have contacted the substance with water. Don't wait to remove contaminated clothing; do it under the water stream. Use soap to help assure removal. Isolate contaminated clothing when removed to prevent contact by others.

Eye Contact: Remove any contact lenses at once. Flush eyes well with copious quantities of water or normal saline for a least 20 – 30 minutes. Seek medical attention.

Inhalation: If convulsions are not present, give a glass or two of water or milk to dilute the substance. Assure that the person's airway is unobstructed and contact a hospital or poison center immediately for advice on whether or not to induce vomiting.

Ingestion: If convulsions are not present, give a glass or two of water or milk to dilute the substance. Assure that the person's airway is unobstructed and contact a hospital or poison center immediately for advice on whether or not to induce vomiting.

Personal Protective Methods: Wear protective gloves and clothing to prevent any reasonable probability of skin contact. Safety equipment suppliers/manufacturers can provide recommendations on the most protective glove/clothing material for your operation.. All protective clothing (suits, gloves, footwear, headgear) should be clean, available each day, and put on before work. Contact lenses should not be worn when working with this chemical. Wear splash-proof chemical goggles and face shield unless full facepiece respiratory protection is worn. Employees should wash immediately with soap when skin is wet or contaminated. Provide emergency showers and eyewash.

Respirator Selection: Where the potential for exposure to this chemical, use a MSHA/NIOSH approved supplied-air respirator with a full facepiece operated in the positive pressure mode or with a full facepiece, hood, or helmet in the continuous flow mode, or use a MSHA/NIOSH approved self-contained breathing apparatus with a full facepiece operated in pressure-demand or other positive pressure mode.

Storage: Prior to working with this chemical you should be trained on its proper handling and storage. Store in tightly closed containers in a cool, well ventilated or refrigerator away from oxidizers and light. Where possible, automatically pump liquid from drums or other storage containers to process containers.

Shipping: May be classified as a combustible liquid n.o.s. This class of compounds requires no particular shipping label. It falls in DOT Hazard Class 3 and Packing Group III.

Spill Handling: Evacuate and restrict persons not wearing protective equipment from area of spill or leak until cleanup is complete. Remove all ignition sources. Ventilate area of spill or leak. Absorb liquids in vermiculite, dry sand, earth, peat, carbon, or a similar material and deposit in sealed containers. It may be necessary to contain and dispose of this chemical as a hazardous waste. If material or contaminated runoff enters waterways, notify downstream users of potentially contaminated waters. Contact your Department of Environmental Protection or your regional office of the federal EPA for specific recommendations. If employees are required to clean-up spills, they must be properly trained and equipped. OSHA 1910.120(q) may be applicable.

Fire Extinguishing: This chemical is a combustible liquid. Poisonous and irritating gases produced in fire. Use dry chemical or carbon dioxide. Water or foam may cause frothing.[41] Vapors are uninhibited and may polymerize, blocking vents. Vapors are heavier than air and will collect in low areas. Vapors may travel long distances to ignition sources and flashback. Vapors in confined areas may explode when exposed to fire. Containers may explode in fire. Storage containers and parts of containers may rocket great distances, in many directions. If material or contaminated runoff enters waterways, notify downstream users of potentially contaminated waters. Notify local health and fire officials and pollution control agencies. From a secure, explosion-proof location, use water spray to cool exposed containers. If cooling streams are ineffective (venting sound increases in volume and pitch, tank discolors, or shows any signs of deforming), withdraw immediately to a secure position. If employees are expected to fight fires, they must be trained and equipped in OSHA 1910.156.

Disposal Method Suggested: Spray into incinerator with added flammable solvent.

References

Sax, N. I., Ed., "Dangerous Properties of Industrial Materials Report," 1, No. 7, 57–59 (1981) and 3, No. 2, 83–85 (1983)

Ethylidene Norbornene

Molecular Formula: C_9H_{12}

Synonyms: ENB; 5-Ethylidenebicyclo(2,21)hept-2-ene; 5-Ethylidene-2-norbornene (stabilized)

CAS Registry Number: 16219-75-3

RTECS Number: RB9450000

DOT ID: UN 1993

Regulatory Authority

- Air Pollutant Standard Set (ACGIH)[1] (Several States)[60]
- Canada, WHMIS, Ingredients Disclosure List

Cited in U.S. State Regulations: Alaska (G), California (A, G), Florida (G), Illinois (G), Maine (G), Massachusetts (G), Nevada (A), New Hampshire (G), New Jersey (G), North Dakota (A), Pennsylvania (G), Rhode Island (G), Virginia (A), West Virginia (G).

Description: 5-Ethylidene-2-norbornene, C_9H_{12}, is a colorless liquid with a turpentine-like odor. The odor threshold is 0.007 ppm.[41] Boiling point = 148°C. Freezing/Melting point = -80°C Flash point = 38°C. Hazard Identification (based on NFPA-704 M Rating System): Health 2, Flammability 2, Reactivity 0. Insoluble in water.

Potential Exposure: Those engaged in the synthesis of pharmaceuticals or pesticides or in the preparation of specialty resins.

Incompatibilities: Forms explosive mixture with air. Reacts violently with strong oxidants. Violent reaction with oxygen and strong oxidizers. May accumulate static electrical charges, and may cause ignition of its vapors. The substance may polymerize. Inhibit peroxide formation with tert-butyl catechol. ENB should be stored in a nitrogen atmosphere since it reacts with oxygen.

Permissible Exposure Limits in Air: There is no OSHA PEL. NIOSH and ACGIH recommended ceiling value is 5 ppm (25 mg/m^3). Some states have set guidelines on standards for ethylidene norbornene in ambient air[60] ranging from 200.0 μg/m^3 (Virginia) to 250.0 μg/m^3 (North Dakota) to 595.0 μg/m^3 (Nevada).

Determination in Air: No test available.

Permissible Concentration in Water: No criteria set.

Routes of Entry: Inhalation, skin absorption, ingestion, skin and/or eye contact.

Harmful Effects and Symptoms

Short Term Exposure: Exposure can irritate the eyes, nose, and respiratory tract. Contact can irritate and may burn, the skin and eyes. Exposure can cause headache, confusion, nausea and vomiting. Ingestion of the liquid may cause chemical pneumonitis. Exposure to high concentrations may cause unconsciousness and cause death.

Long Term Exposure: Repeated or high exposure may damage the liver, lungs and kidneys. Ethylidene norbornene may damage the testes (male reproductive glands).

Points of Attack: Eyes, skin, respiratory system, central nervous system, liver, kidneys, urogenital system, bone marrow

Medical Surveillance: If symptoms develop or overexposure is suspected, the following may be useful: Liver and kidney function tests. Lung function tests. Reproductive history and possibly semen analysis with sperm count.

First Aid: If this chemical gets into the eyes, remove any contact lenses at once and irrigate immediately for at least 15 minutes, occasionally lifting upper and lower lids. Seek medical attention immediately. If this chemical contacts the skin, remove contaminated clothing and wash immediately with soap and water. Seek medical attention immediately. If this chemical has been inhaled, remove from exposure, begin rescue breathing (using universal precautions) if breathing has stopped and CPR if heart action has stopped. Transfer promptly to a medical facility. When this chemical has been swallowed, get medical attention. Give large quantities of water and induce vomiting. Do not make an unconscious person vomit.

Personal Protective Methods: Wear protective gloves and clothing to prevent any reasonable probability of skin contact. Safety equipment suppliers/manufacturers can provide recommendations on the most protective glove/clothing material for your operation.. All protective clothing (suits, gloves, footwear, headgear) should be clean, available each day, and put on before work. Contact lenses should not be worn when working with this chemical. Wear splash-proof chemical goggles and face shield unless full facepiece respiratory protection is worn. Employees should wash immediately with soap when skin is wet or contaminated. Provide emergency showers and eyewash.

Respirator Selection: Where the potential exists for exposures over 5 ppm, use an MSHA/NIOSH approved full facepiece respirator with an organic vapor cartridge/canister. Increased protection is obtained from full facepiece powered air purifying respirators. Where the potential for high exposures exists, use an MSHA/NIOSH approved supplied-air respirator with a full facepiece operated in the positive pressure mode or with a full facepiece, hood, or helmet in the continuous flow mode, or use an MSHA/NIOSH approved self-contained breathing apparatus with a full facepiece operated in pressure-demand or other positive pressure mode.

Storage: Prior to working with this chemical you should be trained on its proper handling and storage. Ethylidene norbornene should be stored in a nitrogen atmosphere. Do not allow it to come in contact with oxygen because violent reactions can occur.

Shipping: This material may be classified as a flammable liquid n.o.s. This class requires shipping label of: "Flammable Liquid." It falls in DOT Hazard Class 3 and Packing Group III.

Spill Handling: Evacuate and restrict persons not wearing protective equipment from area of spill or leak until cleanup is complete. Remove all ignition sources. Ventilate area of spill or leak. Absorb liquids in vermiculite, dry sand, earth, peat, carbon, or a similar material and deposit in sealed containers. It may be necessary to contain and dispose of this chemical as a hazardous waste. If material or contaminated runoff enters waterways, notify downstream users of potentially contaminated waters. Contact your Department of Environmental Protection or your regional office of the federal EPA for specific recommendations. If employees are required to clean-up spills, they must be properly trained and equipped. OSHA 1910.120(q) may be applicable.

Fire Extinguishing: This chemical is a combustible liquid. Poisonous gases including carbon monoxide are produced in fire. Use dry chemical, carbon dioxide, or polymer foam extinguishers. Vapors are heavier than air and will collect in low areas. Vapors may travel long distances to ignition sources and flashback. Vapors in confined areas may explode when exposed to fire. Containers may explode in fire. Storage containers and parts of containers may rocket great distances, in many directions. If material or contaminated runoff enters waterways, notify downstream users of potentially contaminated waters. Notify local health and fire officials and pollution control agencies. From a secure, explosion-proof location, use water spray to cool exposed containers. If cooling streams are ineffective (venting sound increases in volume and pitch, tank discolors, or shows any signs of deforming), withdraw immediately to a secure position. If employees are expected to fight fires, they must be trained and equipped in OSHA 1910.156.

Disposal Method Suggested: Incineration.

References

New Jersey Department of Health and Senior Services, "Hazardous Substance Fact Sheet: Ethylidene Norbornene," Trenton, NJ (January 1999)

Ethyl Isocyanate

Molecular Formula: C_3H_5NO

Common Formula: C_2H_5NCO

Synonyms: Isocyanatoethane; Isocyanic acid, ethyl ester

CAS Registry Number: 109-90-0

RTECS Number: NQ8825000

DOT ID: UN 2481

Cited in U.S. State Regulations: California (A, G), New Hampshire (G), New Jersey (G), Pennsylvania (G).

Description: Ethyl Isocyanate, C_2H_5NCO, is a colorless liquid with a pungent odor. Boiling point = 60°C. Flash point ≤ 23°C.

Potential Exposure: Ethyl isocyanate is used to make pharmaceuticals (drug products) and pesticides.

Incompatibilities: Forms explosive mixture with air. Violent reaction with water and strong oxidizers. Incompatible with acids, bases, ammonia, amines, amides, alcohols, glycols, caprolactum solution. May accumulate static electrical charges, and may cause ignition of its vapors.

Permissible Exposure Limits in Air: No standards set.

Permissible Concentration in Water: No criteria set.

Routes of Entry: Inhalation. Passes through the skin.

Harmful Effects and Symptoms

Short Term Exposure: Ethyl isocyanate can irritate the eyes, skin and respiratory tract. Skin contact can cause a rash. Exposure can irritate the nose and throat. Higher exposures can cause pulmonary edema, a medical emergency that can be delayed for several hours. This can cause death.

Long Term Exposure: Similar chemicals cause lung and skin allergies. It is not known if ethyl isocyanate does this.

Points of Attack: Skin, lungs.

Medical Surveillance: For those with frequent or potentially high exposure, the following are recommended before beginning work and at regular times after that: Lung function tests. These may be normal if the person is not having an attack at the time of the test. If symptoms develop or overexposure is suspected, the following may be useful: Consider chest x-ray after acute overexposure. Evaluation by a qualified allergist, including careful exposure history and special testing, may help diagnose skin allergy.

First Aid: If this chemical gets into the eyes, remove any contact lenses at once and irrigate immediately for at least 15 minutes, occasionally lifting upper and lower lids. Seek medical attention immediately. If this chemical contacts the skin, remove contaminated clothing and wash immediately with soap and water. Seek medical attention immediately. If this chemical has been inhaled, remove from exposure, begin rescue breathing (using universal precautions) if breathing has stopped and CPR if heart action has stopped. Transfer promptly to a medical facility. When this chemical has been swallowed, get medical attention. Give large quantities of water and induce vomiting. Do not make an unconscious person vomit. Medical observation is recommended for 24 – 48 hours after breathing overexposure, as pulmonary edema may be delayed. As first aid for pulmonary edema, a doctor or authorized paramedic may consider administering a corticosteroid spray.

Personal Protective Methods: Wear protective gloves and clothing to prevent any reasonable probability of skin contact. Safety equipment suppliers/manufacturers can provide recommendations on the most protective glove/clothing material for

your operation.. All protective clothing (suits, gloves, footwear, headgear) should be clean, available each day, and put on before work. Contact lenses should not be worn when working with this chemical. Wear splash-proof chemical goggles and face shield unless full facepiece respiratory protection is worn. Employees should wash immediately with soap when skin is wet or contaminated. Provide emergency showers and eyewash.

Respirator Selection: Where the potential for exposures to ethyl isocyanate exists, use a MSHA/NIOSH approved supplied-air respirator with a full facepiece operated in the positive pressure mode or with a full facepiece, hood, or helmet in the continuous flow mode, or use a MSHA/NIOSH approved self-contained breathing apparatus with a full facepiece operated in pressure-demand or other positive pressure mode.

Storage: Prior to working with this chemical you should be trained on its proper handling and storage. Store in tightly closed containers in a cool, well-ventilated area. Sources of ignition such as smoking and open flames are prohibited where ethyl isocyanate is used, handled, or stored in a manner that could create a potential fire or exposition hazard. Use only non-sparking tools and equipment, especially when opening and closing containers of ethyl isocyanate.

Shipping: This compound requires a shipping label of: "Flammable Liquid, Poison." It falls in DOT Hazard Class 3 and Packing Group I. Passenger aircraft or railcar shipment is forbidden as is any cargo aircraft shipment.

Spill Handling: Evacuate and restrict persons not wearing protective equipment from area of spill or leak until cleanup is complete. Remove all ignition sources. Ventilate area of spill or leak. Absorb liquids in vermiculite, dry sand, earth, peat, carbon, or a similar material and deposit in sealed containers. Keep this chemical out of a confined space, such as a sewer, because of the possibility of an explosion, unless the sewer is designed to prevent the build-up of explosive concentrations. It may be necessary to contain and dispose of this chemical as a hazardous waste. If material or contaminated runoff enters waterways, notify downstream users of potentially contaminated waters. Contact your Department of Environmental Protection or your regional office of the federal EPA for specific recommendations. If employees are required to clean-up spills, they must be properly trained and equipped. OSHA 1910.120(q) may be applicable.

Fire Extinguishing: This chemical is a flammable liquid. Poisonous gases including nitrogen oxides are produced in fire. Use dry chemical, carbon dioxide, or alcohol foam extinguishers. Vapors are heavier than air and will collect in low areas. Vapors may travel long distances to ignition sources and flashback. Vapors in confined areas may explode when exposed to fire. Containers may explode in fire. Storage containers and parts of containers may rocket great distances, in many directions. If material or contaminated runoff enters waterways, notify downstream users of potentially contaminated waters. Notify local health and fire officials and pollution control agencies. From a secure, explosion-proof location, use water spray to cool exposed containers. If cooling streams are ineffective (venting sound increases in volume and pitch, tank discolors, or shows any signs of deforming), withdraw immediately to a secure position. If employees are expected to fight fires, they must be trained and equipped in OSHA 1910.156.

References

New Jersey Department of Health and Senior Services, "Hazardous Substance Fact Sheet: Ethyl Isocyanate," Trenton, NJ (April 1986)

Ethyl Mercaptan

Molecular Formula: C_2H_6S

Common Formula: C_2H_5SH

Synonyms: Ethanethiol; Ethyl hydrosulfide; Ethyl sulfhydrate; Ethyl thioalcohol; LPG ethyl mercaptan 1010; Mercaptoethane; Thioethanol; Thioethyl alcohol

CAS Registry Number: 75-08-1

RTECS Number: KI9625000

DOT ID: UN 2363

EEC Number: 016-022-00-9

- ***Regulatory Authority:*** Air Pollutant Standard Set (ACGIH)[1] (DFG)[33] (HSE)[33] (Former USSR)[35][43] (OSHA)[58] (Several States)[60]

Cited in U.S. State Regulations: Alaska (G), California (A, G), Connecticut (A), Florida (G, A), Illinois (G), Maine (G), Massachusetts (G), Nevada (A), New Hampshire (G), New Jersey (G), New York (A), North Carolina (A), North Dakota (A), Oklahoma (G), Pennsylvania (G), Rhode Island (G), South Carolina (A), Virginia (G, A), West Virginia (G).

Description: Ethyl Mercaptan, C_2H_5SH, is a yellowish liquid (or a colorless gas above the BP) with a strong garlic or skunk-like odor. Boiling point = 35°C. Flash point ≤-18°C. Autoignition temperature = 300°C. Explosive limits: LEL = 2.8%; UEL = 18.0%. Hazard Identification (based on NFPA-704 M Rating System): Health 2, Flammability 4, Reactivity 0. Insoluble in water.

Potential Exposure: This material is used as a warning odorant for liquefied petroleum gases. It is used as an intermediate in the manufacture of many pesticides and other organic chemicals.

Incompatibilities: Forms explosive mixture with air. This material is a weak acid. Reacts with oxidizers causing fire and explosion hazard. Reacts with strong acids evolving toxic and flammable hydrogen sulfide. May accumulate static electrical charges, and may cause ignition of its vapors. Attacks some forms of plastics, coatings and rubber.

Permissible Exposure Limits in Air: The OSHA PEL is 20 ppm (ceiling), not to be exceeded at any time. NIOSH recommends an REL of 0.5 ppm [ceiling], not to be exceeded at any time. The ACGIH recommended TWA value is 10 ppm. The HSE[33] has the same 8-hour TWA but adds an STEL of 2 ppm (3 mg/m^3). The DFG MAK[3] is 0.5 ppm (1.3 mg/m^3) TWA. The NIOSH IDLH level is 500 ppm.

Several states have set guidelines or standards for ethyl mercaptan in ambient air[60] ranging from 3.3 μg/m^3 (New York) to 10.0 μg/m^3 (Florida, North Dakota, South Carolina) to 16.0 μg/m^3 (Virginia) to 20.0 μg/m^3 (Connecticut) to 24.0 μg/m^3 (Nevada) to 100.0 μg/m^3 (North Carolina).

Determination in Air: Collection filter (special); workup with hydrochloric acid/1,2-dichloroethane; analysis with gas chromatography/flame photometric detection for sulfur, nitrogen, or phosphorus; NIOSH IV, Method #2542.

Permissible Concentration in Water: No criteria set but EPA[32] has suggested a permissible ambient goal or 13.8 μg/l based on health effects.

Routes of Entry: Inhalation, ingestion, eye and/or skin contact.

Harmful Effects and Symptoms

Short Term Exposure: Contact can cause skin and eye irritation. Inhalation can irritate the respiratory tract and pulmonary edema, a medical emergency that can be delayed for several hours. This can cause death. Exposure can cause headaches, nausea, vomiting, diarrhea, muscle weakness, and tiredness. May affect the central nervous system, causing convulsions and respiratory failure. High levels may cause dizziness, lightheadedness, coma and death.

Long Term Exposure: Repeated exposure can cause lung irritation and bronchitis. May cause liver and kidney damage. Repeated or long term exposure may damage the red blood cells causing anemia.

Points of Attack: Eyes, respiratory system, liver, kidneys, blood.

Medical Surveillance: Consider the points of attack in preplacement and periodic physical examinations. Complete blood cell count (CBC). Liver and kidney function tests. Consider chest x-ray following acute overexposure.

First Aid: If this chemical gets into the eyes, remove any contact lenses at once and irrigate immediately for at least 15 minutes, occasionally lifting upper and lower lids. Seek medical attention immediately. If this chemical contacts the skin, remove contaminated clothing and wash immediately with soap and water. Seek medical attention immediately. If this chemical has been inhaled, remove from exposure, begin rescue breathing (using universal precautions) if breathing has stopped and CPR if heart action has stopped. Transfer promptly to a medical facility. When this chemical has been swallowed, get medical attention. Give large quantities of salt water and induce vomiting. Do not make an unconscious person vomit. Medical observation is recommended for 24 – 48 hours after breathing overexposure, as pulmonary edema may be delayed. As first aid for pulmonary edema, a doctor or authorized paramedic may consider administering a corticosteroid spray.

Personal Protective Methods: Wear protective gloves and clothing to prevent any reasonable probability of skin contact. Safety equipment suppliers/manufacturers can provide recommendations on the most protective glove/clothing material for your operation.. All protective clothing (suits, gloves, footwear, headgear) should be clean, available each day, and put on before work. Contact lenses should not be worn when working with this chemical. Wear splash-proof chemical goggles and face shield unless full facepiece respiratory protection is worn. Employees should wash immediately with soap when skin is wet or contaminated. Provide emergency showers and eyewash.

Respirator Selection: NIOSH/OSHA: *5 ppm:* CCRFOV [any chemical cartridge respirator with a full facepiece and organic vapor cartridge(s)]; or SA (any supplied-air respirator). *12.5 ppm:* SA:CF (any supplied-air respirator operated in a continuous-flow mode); or PAPROV [any powered, air-purifying respirator with organic vapor cartridge(s)]. *25 ppm:* CCRFOV [any chemical cartridge respirator with a full facepiece and organic vapor cartridge(s)]; or GMFOV [any air-purifying, full-facepiece respirator (gas mask) with a chin-style, front- or back-mounted acid gas canister]; or SAT:CF (any supplied-air respirator that has a tight-fitting facepiece and is operated in a continuous-flow mode); or PAPRTOV [any powered, air-purifying respirator with a tight-fitting facepiece and organic vapor cartridge(s)]; or SCBAF (any self-contained breathing apparatus with a full facepiece); or SAF (any supplied-air respirator with a full facepiece). *500 ppm:* SA:PD:PP (any supplied-air respirator operated in a pressure-demand or other positive-pressure mode). *Emergency or planned entry into unknown concentrations or IDLH conditions:* SCBAF: PD,PP (any self-contained breathing apparatus that has a full facepiece and is operated in a pressure-demand or other positive-pressure mode); or SAF:PD,PP:ASCBA (any supplied-air respirator that has a full facepiece and is operated in a pressure-demand or other positive-pressure mode in combination with an auxiliary self-contained breathing apparatus operated in a pressure-demand or other positive pressure mode). *Escape:* GMFOV [any air-purifying, full-facepiece respirator (gas mask) with a chin-style, front-or back-, mounted organic vapor canister]; or SCBAE (any appropriate escape-type, self-contained breathing apparatus).

Storage: Prior to working with ethyl mercaptan you should be trained on its proper handling and storage. Before entering confined space where this chemical may be present, check to make sure that an explosive concentration does not exist. Store in tightly closed containers in a cool, well ventilated area away from oxidizers and acids. Where possible, automatically pump liquid from drums or other storage containers to process containers.

Shipping: This compound requires a shipping label of: "Flammable Liquid." It falls in DOT Hazard Class 3 and Packing Group I. Passenger aircraft or railcar shipment is forbidden.

Spill Handling: Evacuate and restrict persons not wearing protective equipment from area of spill or leak until cleanup is complete. Use self-contained breathing apparatus during cleanup.[57] Remove all ignition sources. Establish forced ventilation to keep levels below explosive limit. Absorb liquids in vermiculite, dry sand, earth, peat, carbon, or a similar material and deposit in sealed containers. It may be necessary to contain and dispose of this chemical as a hazardous waste. If material or contaminated runoff enters waterways, notify downstream users of potentially contaminated waters. Contact your Department of Environmental Protection or your regional office of the federal EPA for specific recommendations. If employees are required to clean-up spills, they must be properly trained and equipped. OSHA 1910.120(q) may be applicable.

Fire Extinguishing: This chemical is a flammable gas or liquid. Poisonous gases including sulfur oxides and carbon monoxide are produced in fire.

Liquid: Use dry chemical, carbon dioxide, halon, or polymer foam extinguishers. Vapors are heavier than air and will collect in low areas. Vapors may travel long distances to ignition sources and flashback. Vapors in confined areas may explode when exposed to fire. Containers may explode in fire. Storage containers and parts of containers may rocket great distances, in many directions. If material or contaminated runoff enters waterways, notify downstream users of potentially contaminated waters. Notify local health and fire officials and pollution control agencies. From a secure, explosion-proof location, use water spray to cool exposed containers. If cooling streams are ineffective (venting sound increases in volume and pitch, tank discolors, or shows any signs of deforming), withdraw immediately to a secure position. If employees are expected to fight fires, they must be trained and equipped in OSHA 1910.156.

Gas: Do not extinguish the fire unless the flow of gas can be stopped and any remaining gas is out of the line. Specially trained personnel may use fog lines to cool exposures and let the fire burn itself out. Vapors are heavier than air and will collect in low areas. Vapors may travel long distances to ignition sources and flashback. Vapors in confined areas may explode when exposed to fire. Containers may explode in fire. Storage containers and parts of containers may rocket great distances, in many directions. If material or contaminated runoff enters waterways, notify downstream users of potentially contaminated waters. Notify local health and fire officials and pollution control agencies. From a secure, explosion-proof location, use water spray to cool exposed containers. If cooling streams are ineffective (venting sound increases in volume and pitch, tank discolors, or shows any signs of deforming), withdraw immediately to a secure position. If cylinders are exposed to excessive heat from fire or flame contact, withdraw immediately to a secure location. If employees are expected to fight fires, they must be trained and equipped in OSHA 1910.156.

Disposal Method Suggested: Incineration (2,000°F) followed by scrubbing with a caustic solution.[22]

References

New Jersey Department of Health and Senior Services, "Hazardous Substance Fact Sheet, Ethyl Mercaptan," Trenton, NJ (March 1999)

Ethyl Mercuric Chloride

Molecular Formula: C_2H_5ClHg

Common Formula: C_2H_5HgCl

Synonyms: Ceresan; Chloroethyl mercury; EMC; Ethylmercuric chloride; Ethylmercury chloride; Granosan®; Granozan

CAS Registry Number: 107-27-7

RTECS Number: OV9800000

DOT ID: UN 2777

Regulatory Authority

- Air Pollutant Standard Set (ACGIH)[1] (Former USSR)[43] (OSHA)[58]
- Water Pollution Standard Proposed (Former USSR)[43]
- Canada, WHMIS, Ingredients Disclosure List

Cited in U.S. State Regulations: New Jersey (G).

Description: Ethyl mercuric chloride, C_2H_5HgCl, is silvery white, forming leaf-like crystals. Freezing/Melting point = 192°C. crystals. Insoluble in water.

Potential Exposure: It is used as an organic fungicide for seed treatment.

Incompatibilities: Oxidizers. Do not use water in.

Permissible Exposure Limits in Air: The OSHA PEL is 0.01 mg/m^3 TWA, averaged over an 8-hour workshift and 0.04 mg/m^3 as a ceiling, not to be exceeded at any time. The NIOSH IDLH = 2 mg/m^3 (as Hg) The ACGIH[1] recommended TWA is the same as OSHA and a STEL of 0.04 mg/m^3. The former USSR-UNEP/IRPTC project[43] has set a MAC in workplace air of 0.005 mg/m^3 and a MAC for ambient air in residential areas of 0.0009 mg/m^3 on a momentary basis and 0.0001 mg/m^3 on an average daily basis.

Determination in Air: No test available.

Permissible Concentration in Water: A MAC in water bodies used for domestic purposes has been set by the former USSR-UNEP/IRPTC project[43] at 0.0001 mg/l.

Routes of Entry: Inhalation, skin absorption, ingestion, skin and/or eye contact.

Harmful Effects and Symptoms

Short Term Exposure: Ethyl mercuric chloride is an extremely toxic chemical that can cause permanent brain damage weeks after exposure with little or no warning during exposure. Severe poisoning can cause death. It enters the body through the lungs, skin and contaminated hands. Poisoning causes a "pins and needles" feeling, becoming clumsy and weak, hearing loss, abnormal walking, tremors, personality changes and other brain damage. Eye contact may cause severe irritation.

Long Term Exposure: Ethyl mercuric chloride should be handled as a teratogen-with extreme caution. It may cause mutations. Handle with extreme caution. Mercury accumulates in the body. May cause kidney damage. It can take months or years for the body to get rid of excess mercury.

Points of Attack: Eyes, skin. central nervous system, peripheral nervous system, kidneys.

Medical Surveillance: Before first exposure and every 6 – 12 months after, a complete medical history and exam is strongly recommended, with: Exam of the nervous system including handwriting. Visual exam, including "visual field" exam. Hearing tests. Test for mercury in hair and blood. After suspected illness or overexposure, repeat these tests promptly, again in 4 – 6 weeks and then as recommended by your doctor. Kidney function tests.

First Aid: If this chemical gets into the eyes, remove any contact lenses at once and irrigate immediately for at least 15 minutes, occasionally lifting upper and lower lids. Seek medical attention immediately. If this chemical contacts the skin, remove contaminated clothing and wash immediately with soap and water. Seek medical attention immediately. If this chemical has been inhaled, remove from exposure, begin rescue breathing (using universal precautions) if breathing has stopped and CPR if heart action has stopped. Transfer promptly to a medical facility. When this chemical has been swallowed, get medical attention. Give large quantities of water and induce vomiting. Do not make an unconscious person vomit.

Personal Protective Methods: Wear protective gloves and clothing to prevent any reasonable probability of skin contact. Safety equipment suppliers/manufacturers can provide recommendations on the most protective glove/clothing material for your operation.. All protective clothing (suits, gloves, footwear, headgear) should be clean, available each day, and put on before work. Contact lenses should not be worn when working with this chemical. Wear dust-proof chemical goggles and face shield unless full facepiece respiratory protection is worn. Employees should wash immediately with soap when skin is wet or contaminated. Provide emergency showers and eyewash.

Respirator Selection: NIOSH/OSHA: *up to 0.1 mg/m³*: SA (any supplied-air respirator). *Up to 0.25 mg/m³:* SA:CF (any supplied-air respirator operated in a continuous-flow mode). *Up to 0.5 mg/m³*: SAT:CF (any supplied-air respirator that has a tight-fitting facepiece and is operated in a continuous-flow mode); or SCBAF (any self-contained breathing apparatus with a full facepiece); or SAF (any supplied-air respirator with a full facepiece). *Up to 2 mg/m³:* SA:PD,PP (any supplied-air respirator operated in a pressure-demand or other positive-pressure mode). *Emergency or planned entry into unknown concentrations or IDLH conditions:* SCBAF:PD,PP (any self-contained breathing apparatus that has a full facepiece and is operated in a pressure-demand or other positive-pressure mode); or SAF:PD,PP:ASCUBA (any supplied-air respirator that has a full facepiece and is operated in a pressure-demand or other positive-pressure mode in combination with an auxiliary self-contained positive-pressure breathing apparatus). *Escape:* SCBAE (any appropriate escape-type, self-contained breathing apparatus).

Storage: Prior to working with this chemical you should be trained on its proper handling and storage. Store in tightly closed containers in a cool, well-ventilated area away from oxidizers (such as perchlorates, peroxides, permanganates, chlorates and nitrates).

Shipping: This material falls in the class of mercury-based pesticides, solid, toxic n.o.s. and in Packing Group II. This compound requires a shipping label of: "Poison."

Spill Handling: Evacuate persons not wearing protective equipment from area of spill or leak until clean-up is complete. Remove all ignition sources. Spills should be collected with special mercury vapor suppressant or special vacuums. Kits specific for clean-up of mercury spills are available. Collect powdered material in the most convenient and safe manner and deposit in sealed containers. Ventilate area after clean-up is complete. It may be necessary to contain and dispose of this chemical as a hazardous waste. If material or contaminated runoff enters waterways, notify downstream users of potentially contaminated waters. Contact your Department of Environmental Protection or your regional office of the federal EPA for specific recommendations. If employees are required to clean-up spills, they must be properly trained and equipped. OSHA 1910.120(q) may be applicable.

Fire Extinguishing: Use dry chemicals appropriate for extinguishing metal fires. Do not use water. Poisonous gases are produced in fire. If material or contaminated runoff enters waterways, notify downstream users of potentially contaminated waters. Notify local health and fire officials and pollution control agencies. From a secure, explosion-proof location, use water spray to cool exposed containers. If cooling streams are ineffective (venting sound increases in volume and pitch, tank discolors, or shows any signs of deforming), withdraw immediately to a secure position. If employees are expected to fight fires, they must be trained and equipped in OSHA 1910.156.

Disposal Method Suggested: In accordance with 40CFR 165 recommendations for the disposal of pesticides and pesticide containers. Must be disposed properly by following package label directions or by contacting your state pesticide or environmental control agency or by contacting your regional EPA office.

References

New Jersey Department of Health and Senior Services, "Hazardous Substance Fact Sheet: Ethyl Mercuric Chloride," Trenton, NJ (January 11, 1988)

For a related compound which is not regulated but which was on the USEPA Extremely Hazardous Chemical List in 1985 (dropped in 1988), see the following: U.S. Environmental Protection

Agency, "Chemical Profile: Ethylmercuric Phosphate," Washington, DC, Chemical Emergency Preparedness Program (October 31, 1985)

Ethyl Methacrylate

Molecular Formula: $C_6H_{10}O_2$

Common Formula: $CH_2=C(CH_3)COOC_2H_5$

Synonyms: Ethyl 1-2-methacrylate; Ethyl α-methylacrylate; Ethyl 2-methyl-2-propenoate; Metacrilato de etilo (Spanish); 1-2-Methacrylic acid, ethyl ester; 2-Methyle-2-propenoic acid, ethyl ester; 2-Propenoic acid, 1-methyl-, ethyl ester; Rhoplex; Rhoplex AC-33 (Rohm & Haas)

CAS Registry Number: 97-63-2

RTECS Number: OZ4550000

DOT ID: UN 2277

EEC Number: 607-071-00-2

EINECS Number: 202-597-5

Regulatory Authority

- Air Pollutant Standard Set (Former USSR)[43]
- EPA Hazardous Waste Number (RCRA No.): U118
- RCRA, 40CFR261, Appendix 8 Hazardous Constituents
- RCRA 40CFR268.48; 61FR15654, Universal Treatment Standards: Wastewater (mg/l), 0.14; Nonwastewater (mg/kg), 160
- RCRA 40CFR264, Appendix 9; TSD Facilities Ground Water Monitoring List. Suggested test method(s) (PQL μg/l): 8015 (10); 8240 (5)
- Superfund/EPCRA 40CFR302.4 Reportable Quantity (RQ): CERCLA, 1,000 lb (454 kg)

Cited in U.S. State Regulations: California (A, G), Florida (G), Massachusetts (G), New Hampshire (G), New Jersey (G), Pennsylvania (G).

Description: Ethyl Methacrylate, $CH_2=C(CH_3)COOC_2H_5$, is a colorless liquid. Boiling point = 115 – 120°C. Freezing/Melting point ≤-75°C. Flash point = 20°C. Autoignition temperature = 393°C. Explosive limits: LEL = 1.8%; UEL = unknown. Hazard Identification (based on NFPA-704 M Rating System): Health 2, Flammability 3, Reactivity 0. Slightly soluble in water.

Potential Exposure: Widely known as "Plexiglass" (in the polymer form), ethyl methacrylate is used to make polymers, which in turn are used for building, automotive, aerospace, and furniture industries. It is also used by dentists as dental plates, artificial teeth, and orthopedic cement.

Incompatibilities: Forms explosive mixture with air. Incompatible with strong acids, amines, oxidizers. Corrodes some metals. Unless inhibited, violent polymerization can occur from heat, sunlight, and contact with strong oxidizers.

Permissible Exposure Limits in Air: The former USSR-UNEP/IRPTC project[43] has set a MAC in workplace air of 50 mg/m^3.

Permissible Concentration in Water: No criteria set.

Routes of Entry: Inhalation.

Harmful Effects and Symptoms

Short Term Exposure: A lacrimator. Exposure can irritate the eyes, skin, and respiratory tract. Very high vapor levels could cause you to feel dizzy, lightheaded and even to pass out. Contact can irritate the eyes and skin.

Long Term Exposure: Repeated or prolonged contact with skin may cause allergy with rash and itching. May damage the nervous system. There is limited evidence that ethyl methacrylate may damage the developing fetus.

Points of Attack: Skin, central nervous system.

Medical Surveillance: If symptoms develop or overexposure is suspected, the following may be useful: Evaluation by a qualified allergist, including careful exposure history and special testing, may help diagnose skin allergy. Exam of the nervous system.

First Aid: If this chemical gets into the eyes, remove any contact lenses at once and irrigate immediately for at least 15 minutes, occasionally lifting upper and lower lids. Seek medical attention immediately. If this chemical contacts the skin, remove contaminated clothing and wash immediately with soap and water. Seek medical attention immediately. If this chemical has been inhaled, remove from exposure, begin rescue breathing (using universal precautions) if breathing has stopped and CPR if heart action has stopped. Transfer promptly to a medical facility. When this chemical has been swallowed, get medical attention. Give large quantities of water and induce vomiting. Do not make an unconscious person vomit.

Personal Protective Methods: Wear protective gloves and clothing to prevent any reasonable probability of skin contact. Safety equipment suppliers/manufacturers can provide recommendations on the most protective glove/clothing material for your operation.. There is evidence that fluorinated ethylene propylene materials offer protection from ethyl methacrylate for up to 3 hours. Polyvinyl Alcohol is also among the recommended protective materials. All protective clothing (suits, gloves, footwear, headgear) should be clean, available each day, and put on before work. Contact lenses should not be worn when working with this chemical. Wear splash-proof chemical goggles and face shield unless full facepiece respiratory protection is worn. Employees should wash immediately with soap when skin is wet or contaminated. Provide emergency showers and eyewash.

Respirator Selection: Where the potential for exposures to ethyl methacrylate exists, use a MSHA/NIOSH approved supplied-air respirator with a full facepiece operated in the positive pressure mode or with a full facepiece, hood, or helmet in the continuous flow mode, or use a MSHA/NIOSH approved self-contained breathing apparatus with a full facepiece operated in pressure-demand or other positive pressure mode.

Storage: Prior to working with this chemical you should be trained on its proper handling and storage. Before entering confined space where this chemical may be present, check to make sure that an explosive concentration does not exist. Store in tightly closed containers in a cool, well-ventilated area away from oxidizers, (such as perchlorates, peroxides, permanganates, chlorates and nitrates). Sources of ignition, such as smoking and open flames, are prohibited where ethyl methacrylate is handled, used, or stored. Metal containers involving the transfer of 5 gallons or more of ethyl methacrylate should be grounded and bonded. Drums must be equipped with self-closing valves, pressure vacuum bungs, and flame arresters. Use only non-sparking tools and equipment, especially when opening and closing containers of ethyl methacrylate.

Shipping: This compound (inhibited) requires a shipping label of: "Flammable Liquid." It falls in DOT Hazard Class 3 and Packing Group II.

Spill Handling: Evacuate and restrict persons not wearing protective equipment from area of spill or leak until cleanup is complete. Remove all ignition sources. Establish forced ventilation to keep levels below explosive limit. Absorb liquids in vermiculite, dry sand, earth, peat, carbon, or a similar material and deposit in sealed containers. Keep ethyl methacrylate out of a confined space, such as a sewer, because of the possibility of an explosion, unless the sewer is designed to prevent the build-up of explosive concentrations. It may be necessary to contain and dispose of this chemical as a hazardous waste. If material or contaminated runoff enters waterways, notify downstream users of potentially contaminated waters. Contact your Department of Environmental Protection or your regional office of the federal EPA for specific recommendations. If employees are required to clean-up spills, they must be properly trained and equipped. OSHA 1910.120(q) may be applicable.

Fire Extinguishing: This chemical is a flammable liquid. Poisonous gases are produced in fire. Use dry chemical, carbon dioxide, or alcohol foam extinguishers. Vapors are heavier than air and will collect in low areas. Vapors may travel long distances to ignition sources and flashback. Vapors in confined areas may explode when exposed to fire. Containers may explode in fire. Storage containers and parts of containers may rocket great distances, in many directions. If material or contaminated runoff enters waterways, notify downstream users of potentially contaminated waters. Notify local health and fire officials and pollution control agencies. From a secure, explosion-proof location, use water spray to cool exposed containers. If cooling streams are ineffective (venting sound increases in volume and pitch, tank discolors, or shows any signs of deforming), withdraw immediately to a secure position. If employees are expected to fight fires, they must be trained and equipped in OSHA 1910.156.

Disposal Method Suggested: Consult with environmental regulatory agencies for guidance on acceptable disposal practices. Generators of waste containing this contaminant (≥100 kg/mo) must conform with EPA regulations governing storage, transportation, treatment, and waste disposal.

References

U.S. Environmental Protection Agency, Ethyl Methacrylate, Health and Environmental Effects Profile No. 101, Washington, DC, Office of Solid Waste (April 30, 1980)

New Jersey Department of Health and Senior Services, "Hazardous Substance Fact Sheet: Methyl Methacrylate," Trenton, NJ (April 1996)

Ethyl Methane Sulfonate

Molecular Formula: $C_3H_8O_3S$

Common Formula: $C_2H_5OSO_2CH_3$

Synonyms: EMS; ENT 26,396; Ethyl ester of methanesulfonic acid; Ethyl ester of methylsulfonic acid; Ethyl ester of methylsulphonic acid; Ethyl mesylate; Ethyl methanesulphonate; Ethyl methansulfonate; Ethyl methansulphonate; Ethyl methyl sulfonate; Half-myderan; Methanesulphonic acid ethyl ester; Methylsulfonic acid, ethyl ester; NSC 26805

CAS Registry Number: 62-50-0

RTECS Number: PB2100000

DOT ID: NA 1693

Regulatory Authority

- Carcinogen (Animal Positive) (IARC)[9]
- EPA Hazardous Waste Number (RCRA No.): U119
- RCRA, 40CFR261, Appendix 8 Hazardous Constituents
- RCRA 40CFR264, Appendix 9; TSD Facilities Ground Water Monitoring List. Suggested test method(s) (PQL μg/l): 8270 (10)
- Superfund/EPCRA 40CFR302.4 Reportable Quantity (RQ): CERCLA, 1 lb (0.454 kg)

Cited in U.S. State Regulations: California (G), Florida (G), Kansas (G), Louisiana (G), Massachusetts (G), Michigan (G), New Hampshire (G), New Jersey (G), Pennsylvania (G), Vermont (G), Virginia (G), Washington (G), Wisconsin (G).

Description: Ethyl methanesulfonate, $C_2H_5OSO_2CH_3$, is a clear liquid. Boiling point = 85°C. Hazard Identification (based on NFPA-704 M Rating System): Health 2, Flammability 1, Reactivity 0.

Potential Exposure: Was considered as a possible human male contraceptive. Also considered as a reversible male hemosterilant for insects and mammalian pests. Used as a research tool for mutagenesis and carcinogenesis studies.

Incompatibilities: Acids, acid fumes, nitrates.

Permissible Exposure Limits in Air: No standards set.

Permissible Concentration in Water: No criteria set.

Harmful Effects and Symptoms

Irritation of skin, eyes and mucous membranes, sores and burns.

Long Term Exposure: A carcinogen in animals. A teratogen.

First Aid: *Skin Contact:*[52] Flood all areas of body that have contacted the substance with water. Don't wait to remove contaminated clothing; do it under the water stream. Use soap to help assure removal. Isolate contaminated clothing when removed to prevent contact by others.

Eye Contact: Remove any contact lenses at once. Flush eyes well with copious quantities of water or normal saline for at least 20 – 30 minutes. Seek medical attention.

Inhalation: Leave contaminated area immediately; breathe fresh air. Proper respiratory protection must be supplied to any rescuers. If coughing, difficult breathing or any other symptoms develop, seek medical attention at once, even if symptoms develop many hours after exposure.

Ingestion: If convulsions are not present, give a glass or two of water or milk to dilute the substance. Assure that the person's airway is unobstructed and contact a hospital or poison center immediately for advice on whether or not to induce vomiting.

Personal Protective Methods: Wear protective gloves and clothing to prevent any reasonable probability of skin contact. Safety equipment suppliers/manufacturers can provide recommendations on the most protective glove/clothing material for your operation.. All protective clothing (suits, gloves, footwear, headgear) should be clean, available each day, and put on before work. Contact lenses should not be worn when working with this chemical. Wear splash-proof chemical goggles and face shield unless full facepiece respiratory protection is worn. Employees should wash immediately with soap when skin is wet or contaminated. Provide emergency showers and eyewash.

Respirator Selection: Where the potential for exposure to this chemical, use a MSHA/NIOSH approved supplied-air respirator with a full facepiece operated in the positive pressure mode or with a full facepiece, hood, or helmet in the continuous flow mode, or use a MSHA/NIOSH approved self-contained breathing apparatus with a full facepiece operated in pressure-demand or other positive pressure mode.

Storage: Prior to working with this chemical you should be trained on its proper handling and storage. Store in a cool, dry place or a refrigerator away from acids and acid fumes. A regulated, marked area should be established where this chemical is handled, used, or stored in compliance with OSHA standard 1910.1045.

Shipping: This compound may be classified as an irritating agent and requires a shipping label of: "Poison." Passenger aircraft or railcar shipment is forbidden.

Spill Handling: Evacuate and restrict persons not wearing protective equipment from area of spill or leak until cleanup is complete. Remove all ignition sources. Ventilate area of spill or leak. *Land spill:* Build dikes to contain the flow of spilled compound and fire-fighting water, if any. Absorb the compound on bentonite, sawdust, clay or carbon. Alternatively, add a 5% solution of sodium hydroxide. Collect the waste in barrels for later disposal or treatment. *Spill on water body:* If the spill occurs on a large flowing water body, add a 5% solution of sodium hydroxide. Collect the waste in barrels for later disposal or treatment. In case of spill on nonflowing small water body, add 5% caustic followed by carbon adsorption or biological treatment depending on the end uses of the water. It may be necessary to contain and dispose of this chemical as a hazardous waste. If material or contaminated runoff enters waterways, notify downstream users of potentially contaminated waters. Contact your Department of Environmental Protection or your regional office of the federal EPA for specific recommendations. If employees are required to clean-up spills, they must be properly trained and equipped. OSHA 1910.120(q) may be applicable.

Fire Extinguishing: Poisonous gases including sulfur oxides are produced in fire. Use dry chemical, carbon dioxide, or alcohol foam extinguishers. Vapors are heavier than air and will collect in low areas. Vapors may travel long distances to ignition sources and flashback. Vapors in confined areas may explode when exposed to fire. Containers may explode in fire. Storage containers and parts of containers may rocket great distances, in many directions. If material or contaminated runoff enters waterways, notify downstream users of potentially contaminated waters. Notify local health and fire officials and pollution control agencies. From a secure, explosion-proof location, use water spray to cool exposed containers. If cooling streams are ineffective (venting sound increases in volume and pitch, tank discolors, or shows any signs of deforming), withdraw immediately to a secure position. If employees are expected to fight fires, they must be trained and equipped in OSHA 1910.156.

Disposal Method Suggested: Nonaqueous wastes may be treated in an incinerator, boiler, or cement kiln where temp and residence time are in excess of 1,000°C and 25 seconds, respectively. Alternatively, the nonaqueous wastes may be disposed of in an RCRA-approved landfill. The detoxified aqueous waste may be treated using biological wastewater methods or disposed of through deep-well injection. Consult with environmental regulatory agencies for guidance on acceptable disposal practices. Generators of waste containing this contaminant (≥100 kg/mo) must conform with EPA regulations governing storage, transportation, treatment, and waste disposal.

References

Sax, N. I., Ed., "Dangerous Properties of Industrial Materials Report," 7, No. 2, 67–74 (1987)

N-Ethylmorpholine

Molecular Formula: $C_6H_{13}NO$

Synonyms: 4-Ethylmorpholine; NEM

CAS Registry Number: 100-74-3

RTECS Number: QE4025000

Regulatory Authority

- Air Pollutant Standard Set (ACGIH)[1] (HSE)[33] (Former USSR)[43] (OSHA)[58] (Several States)[60]
- Canada, WHMIS, Ingredients Disclosure List

Cited in U.S. State Regulations: Alaska (G), California (A, G), Connecticut (A), Florida (G), Illinois (G), Maine (G), Massachusetts (G), Nevada (A), New Hampshire (G), New Jersey (G), North Dakota (A), Pennsylvania (G), Rhode Island (G), Virginia (A), West Virginia (G).

Description: 4-Ethylmorpholine is a colorless liquid with an ammonia-like odor. Boiling point = 138°C. Flash point = 32°C. Explosive limits: LEL = 1.0%; UEL = 9.8%. Hazard Identification (based on NFPA-704 M Rating System): Health 2, Flammability 3, Reactivity 0. Soluble in water.

Potential Exposure: This material is used as a catalyst in polyurethane foam production. It is a solvent for dyes and resins. It is used as an intermediate in surfactant, dye, pharmaceutical, and rubber chemical manufacture.

Incompatibilities: Forms explosive mixture with air. Incompatible with strong acids, strong oxidizers. Attacks some plastics, rubber and coatings.

Permissible Exposure Limits in Air: The OSHA PEL is 20 ppm TWA. The ACGIH[1] and HSE[33] have set a TWA of 5 ppm (23 mg/m^3). HSE[33] adds an STEL of 20 ppm (95 mg/m^3). The notation "skin" is added to indicate the possibility of cutaneous absorption. The NIOSH IDLH level is 100 ppm. The former USSR-UNEP/IRPTC project[43] has set a MAC in workplace air of 5 mg/m^3. Several states have set guidelines or standards for N-ethylmorpholine in ambient air[60] ranging from 230 μg/m^3 (North Dakota) to 400 μg/m^3 (Virginia) to 460 μg/m^3 (Connecticut) to 548 μg/m^3 (Nevada).

Determination in Air: Adsorption on Si gel, workup with H_2SO_4 using ultrasonics; analysis by gas chromatography/ flame ionization. See NIOSH Method #S-146.

Permissible Concentration in Water: No criteria set.

Routes of Entry: Inhalation, skin absorption, ingestion, skin and/or eye contact.

Harmful Effects and Symptoms

Short Term Exposure: Can irritate and burn the skin and eyes. Exposure to high concentration of the vapor can cause drowsiness, foggy vision, visual disturbance: corneal edema, blue-gray vision, seeing colored halos around light.

Long Term Exposure: Unknown at this time.

Points of Attack: Respiratory system, eyes, skin.

Medical Surveillance: Consider the points of attack in preplacement and periodic physical examinations.

First Aid: If this chemical gets into the eyes, remove any contact lenses at once and irrigate immediately for at least 15 minutes, occasionally lifting upper and lower lids. Seek medical attention immediately. If this chemical contacts the skin, remove contaminated clothing and wash immediately with soap and water. Seek medical attention immediately. If this chemical has been inhaled, remove from exposure, begin rescue breathing (using universal precautions) if breathing has stopped and CPR if heart action has stopped. Transfer promptly to a medical facility. When this chemical has been swallowed, get medical attention. Give large quantities of water and induce vomiting. Do not make an unconscious person vomit.

Personal Protective Methods: Wear protective gloves and clothing to prevent any reasonable probability of skin contact. Safety equipment suppliers/manufacturers can provide recommendations on the most protective glove/clothing material for your operation.. All protective clothing (suits, gloves, footwear, headgear) should be clean, available each day, and put on before work. Contact lenses should not be worn when working with this chemical. Wear splash-proof chemical goggles and face shield unless full facepiece respiratory protection is worn. Employees should wash immediately with soap when skin is wet or contaminated. Provide emergency showers and eyewash.

Respirator Selection: NIOSH: *Up to 50 ppm:* CCROV [any chemical cartridge respirator with organic vapor cartridge(s)];* or SA (any supplied-air respirator).* *Up to 100 ppm:* SA:CF (any supplied-air respirator operated in a continuous-flow mode);* or PAPROV [any powered, air-purifying respirator with organic vapor cartridge(s)];* or CCRFOV [any chemical cartridge respirator with a full facepiece and organic vapor cartridge(s)]; or GMFOV [any air-purifying, full-facepiece respirator (gas mask) with a chin-style, front- or back-mounted organic vapor canister]; or SCBAF (any self-contained breathing apparatus with a full facepiece); or SAF (any supplied-air respirator with a full facepiece). *Emergency or planned entry into unknown concentrations or IDLH conditions*: SCBAF: PD,PP (any self-contained breathing apparatus that has a full facepiece and is operated in a pressure-demand or other positive-pressure mode); or SAF:PD, PP:ASCBA (any supplied-air respirator that has a full facepiece and is operated in a pressure-demand or other positive-pressure mode in combination with an auxiliary self-contained breathing apparatus operated in a pressure-demand or other positive-pressure mode). *Escape:* GMFOV [any air-purifying, full-facepiece respirator (gas mask) with a chin-style, front-or back-mounted organic vapor canister]; or SCBAE (any appropriate escape-type, self-contained breathing apparatus).

* Substance reported to cause eye irritation or damage; may require eye protection.

Storage: Prior to working with this chemical you should be trained on its proper handling and storage. Before entering confined space where this chemical may be present, check to make sure that an explosive concentration does not exist. Store in tightly closed containers in a cool, well ventilated area away from strong acids, strong oxidizers. Where possible, automatically pump liquid from drums or other storage containers to process containers Store in a refrigerator under an inert atmosphere for prolonged storage.

Shipping: This material may be classed as a flammable liquid, n.o.s. This class requires a shipping label of: "Flammable Liquid." It falls in DOT Hazard Class 3 and Packing Group III.

Spill Handling: Evacuate and restrict persons not wearing protective equipment from area of spill or leak until cleanup is complete. Remove all ignition sources. Ventilate area of spill or leak. Absorb liquids in vermiculite, dry sand, earth, peat, carbon, or a similar material and deposit in sealed containers. Keep this chemical out of a confined space, such as a sewer, because of the possibility of an explosion, unless the sewer is designed to prevent the build-up of explosive concentrations. It may be necessary to contain and dispose of this chemical as a hazardous waste. If material or contaminated runoff enters waterways, notify downstream users of potentially contaminated waters. Contact your Department of Environmental Protection or your regional office of the federal EPA for specific recommendations. If employees are required to clean-up spills, they must be properly trained and equipped. OSHA 1910.120(q) may be applicable.

Fire Extinguishing: This chemical is a flammable liquid. Poisonous gases including carbon monoxide and nitrogen oxides are produced in fire. Water may be ineffective. Alcohol foam is recommended.[17] Vapors are heavier than air and will collect in low areas. Vapors may travel long distances to ignition sources and flashback. Vapors in confined areas may explode when exposed to fire. Containers may explode in fire. Storage containers and parts of containers may rocket great distances, in many directions. If material or contaminated runoff enters waterways, notify downstream users of potentially contaminated waters. Notify local health and fire officials and pollution control agencies. From a secure, explosion-proof location, use water spray to cool exposed containers. If cooling streams are ineffective (venting sound increases in volume and pitch, tank discolors, or shows any signs of deforming), withdraw immediately to a secure position. If employees are expected to fight fires, they must be trained and equipped in OSHA 1910.156.

Disposal Method Suggested: Controlled incineration (oxides of nitrogen are removed from the effluent gas by scrubbers and/or thermal devices).

References

New Jersey Department of Health and Senior Services, "Hazardous Substance Fact Sheet, N-Ethyl Morpholine," Trenton, NJ (March 1999)

Ethyl Phenyl Dichlorosilane

Molecular Formula: $C_8H_{10}Cl_2Si$

Common Formula: $C_2H_5(C_6H_5)SiCl_2$

Synonyms: Dichloroethyl phenylsilane

CAS Registry Number: 1125-27-5

RTECS Number: VV3270000

DOT ID: UN 2435

Cited in U.S. State Regulations: Maine (G), New Hampshire (G), New Jersey (G), Oklahoma (G).

Description: Ethyl phenyl dichlorosilane, $C_2H_5(C_6H_5)SiCl_2$, is a colorless liquid. Fumes in humid air. Boiling point = 149°C. Flash point ≥65°C. Reactive with water.

Potential Exposure: Used in the manufacture of silicone polymers.

Incompatibilities: Forms explosive mixture with air. Contact with oxidizers, combustible materials, strong bases can cause a fire or explosion hazard. Contact with water, steam, moisture forms corrosive chloride gases including hydrogen chloride. Attacks most metals. Attacks some plastics, rubber and coatings.

Permissible Exposure Limits in Air: No standards set.

Permissible Concentration in Water: No criteria set.

Routes of Entry: Inhalation.

Harmful Effects and Symptoms

Short Term Exposure: Exposure can irritate the eyes, nose and throat. It is a corrosive chemical and contact can cause severe skin and eye burns leading to permanent eye damage. High levels can irritate the lungs causing coughing and/or shortness of breath. Higher exposures can cause pulmonary edema, a medical emergency that can be delayed for several hours. This can cause death.

Long Term Exposure: Very irritating substances may affect the lungs.

Points of Attack: Lungs.

Medical Surveillance: For those with frequent or potentially high exposure, the following are recommended before beginning work and at regular times after that: Lung function tests. If symptoms develop or overexposure is suspected, the following may be useful: Consider chest x-ray after acute overexposure.

First Aid: If this chemical gets into the eyes, remove any contact lenses at once and irrigate immediately for at least 30 minutes, occasionally lifting upper and lower lids. Seek medical attention immediately. If this chemical contacts the skin, remove contaminated clothing and wash immediately with soap and water. Seek medical attention immediately. If this chemical has been inhaled, remove from exposure, begin rescue breathing (using universal precautions) if breathing has stopped and CPR if heart action has stopped. Transfer promptly to a medical facility. When this chemical has been swallowed, get medical attention. Give large quantities of water and do not induce vomiting. Do not make an unconscious person vomit. Medical observation is recommended for 24 – 48 hours after breathing overexposure, as pulmonary edema may be delayed.

Personal Protective Methods: Wear protective gloves and clothing to prevent any reasonable probability of skin contact. Safety equipment suppliers/manufacturers can provide recommendations on the most protective glove/clothing material for your operation.. All protective clothing (suits, gloves, footwear, headgear) should be clean, available each day, and put on before work. Contact lenses should not be worn when working with this chemical. Wear splash-proof chemical goggles and face shield unless full facepiece respiratory protection is worn. Employees should wash immediately with soap when skin is wet or contaminated. Provide emergency showers and eyewash.

Respirator Selection: Where the potential for exposure to ethyl phenyl dichlorosilane exists, use a MSHA/NIOSH approved supplied-air respirator with a full facepiece operated in the positive pressure mode or with a full facepiece, hood, or helmet in the continuous flow mode, or use a MSHA/ NIOSH approved self-contained breathing apparatus with a full facepiece operate in pressure-demand or other positive pressure mode.

Storage: Ethyl phenyl dichlorosilane must be stored to avoid contact with water, steam, oxidizers (such as perchlorates, peroxides, permanganates, chlorates, and nitrates), and combustible materials (such as wood, paper and oil) since violent reactions occur. Store in tightly closed containers in a cool, well-ventilated area away from water, steam and moisture because toxic and corrosive chloride gases including hydrogen chloride can be produced. Sources of ignition such as smoking and open flames are prohibited where ethyl phenyl dichlorosilane is used, handled, or stored in a manner that could create a potential fire or explosion hazard.

Shipping: This compound requires a shipping label of: "Corrosive." It falls in DOT Hazard Class 8 and Packing Group II. Passenger aircraft or railcar shipment is forbidden.

Spill Handling: Evacuate and restrict persons not wearing protective equipment from area of spill or leak until cleanup is complete. Remove all ignition sources. Ventilate area of spill or leak. Absorb liquids in vermiculite, dry sand, earth, peat, carbon, or a similar material and deposit in sealed containers. It may be necessary to contain and dispose of this chemical as a hazardous waste. If material or contaminated runoff enters waterways, notify downstream users of potentially contaminated waters. Contact your Department of Environmental Protection or your regional office of the federal EPA for specific recommendations. If employees are required to clean-up spills, they must be properly trained and equipped. OSHA 1910.120(q) may be applicable.

Fire Extinguishing: This chemical is a combustible liquid. Poisonous gases including phenol and chlorine are produced in fire. Use dry chemical or carbon dioxide. Do not use water or foam extinguishers. Fire may restart after it has been extinguished. Vapors are heavier than air and will collect in low areas. Vapors may travel long distances to ignition sources and flashback. Vapors in confined areas may explode when exposed to fire. Containers may explode in fire. Storage containers and parts of containers may rocket great distances, in many directions. If material or contaminated runoff enters waterways, notify downstream users of potentially contaminated waters. Notify local health and fire officials and pollution control agencies. From a secure, explosion-proof location, use water spray to cool exposed containers. If cooling streams are ineffective (venting sound increases in volume and pitch, tank discolors, or shows any signs of deforming), withdraw immediately to a secure position. If employees are expected to fight fires, they must be trained and equipped in OSHA 1910.156.

References

New Jersey Department of Health and Senior Services, "Hazardous Substance Fact Sheet: Ethyl Phenyl Dichlorosilane," Trenton, NJ (April 1986)

1-Ethyl Piperidine

Molecular Formula: $C_7H_{15}N$

Common Formula: $C_2H_5(C_5H_9NH)$

Synonyms: *N*-Aethylpiperidin (German); Piperidine, 1-ethyl

CAS Registry Number: 766-09-6

RTECS Number: IN0250000

DOT ID: UN 2386

Cited in U.S. State Regulations: New Jersey (G).

Description: Ethyl piperidine, $C_2H_5(C_5H_9NH)$, is a flammable liquid. Boiling point = 128°C. Flash point = 19°C.

Potential Exposure: Used in organic synthesis.

Incompatibilities: Forms explosive mixture with air. Strong oxidizers, acids can cause fire and explosion hazard. Incompatible with carbon dioxide.

Permissible Exposure Limits in Air

No standards set.

Permissible Concentration in Water: No criteria set.

Routes of Entry: Inhalation. Passing through the skin.

Harmful Effects and Symptoms

Short Term Exposure: 1-Ethyl piperidine can affect you when breathed in and by passing through your skin. Contact can cause severe eye burns, leading to permanent damage and can irritate the skin. Exposure can irritate the nose and throat. High exposure can cause increased blood pressure and heart rate, weakness and shortness of breath. Higher levels may cause seizures or death.

Long Term Exposure: Can cause lung irritation. Bronchitis may develop with cough, phlegm and/or shortness of breath.

Points of Attack: Lungs.

Medical Surveillance: Before beginning employment and at regular times after that, for those with frequent or potentially high exposures, the following are recommended: Lung function tests. Consider chest x-ray following acute overexposure.

First Aid: If this chemical gets into the eyes, remove any contact lenses at once and irrigate immediately for at least 30 minutes, occasionally lifting upper and lower lids. Seek

medical attention immediately. If this chemical contacts the skin, remove contaminated clothing and wash immediately with soap and water. Seek medical attention immediately. If this chemical has been inhaled, remove from exposure, begin rescue breathing (using universal precautions) if breathing has stopped and CPR if heart action has stopped. Transfer promptly to a medical facility. When this chemical has been swallowed, get medical attention. Give large quantities of water and induce vomiting. Do not make an unconscious person vomit.

Personal Protective Methods: Wear protective gloves and clothing to prevent any reasonable probability of skin contact. Safety equipment suppliers/manufacturers can provide recommendations on the most protective glove/clothing material for your operation.. All protective clothing (suits, gloves, footwear, headgear) should be clean, available each day, and put on before work. Contact lenses should not be worn when working with this chemical. Wear splash-proof chemical goggles and face shield unless full facepiece respiratory protection is worn. Employees should wash immediately with soap when skin is wet or contaminated. Provide emergency showers and eyewash.

Respirator Selection: Where the potential for exposure to 1-ethyl piperidine exists, use a MSHA/NIOSH approved supplied-air respirator with a full facepiece operated in the positive pressure mode or with a full facepiece, hood, or helmet in the continuous flow mode, or use a MSHA/NIOS approved self-contained breathing apparatus with a full facepiece operated in pressure-demand or other positive pressure mode.

Storage: Prior to working with this chemical you should be trained on its proper handling and storage. Store in tightly closed containers in a cool, well-ventilated area away from strong oxidizers (such as chlorine, bromine an fluorine), acids (such as hydrochloric, sulfuric and nitric) and carbon dioxide. Where possible, automatically pump liquid from drums or other storage containers to process containers.

Shipping: This compound requires a shipping label of: "Flammable Liquid." It falls in DOT Hazard Class 3 and Packing Group II.

Spill Handling: Evacuate and restrict persons not wearing protective equipment from area of spill or leak until cleanup is complete. Remove all ignition sources. Ventilate area of spill or leak. Absorb liquids in vermiculite, dry sand, earth, peat, carbon, or a similar material and deposit in sealed containers. Water may be ineffective. Alcohol foam is recommended.[17] Keep 1-ethyl piperidine out of a confined space, such as a sewer, because off the possibility of an explosion, unless the sewer is designed to prevent the buildup of explosive concentrations. It may be necessary to contain and dispose of this chemical as a hazardous waste. If material or contaminated runoff enters waterways, notify downstream users of potentially contaminated waters. Contact your Department of Environmental Protection or your regional office of the federal EPA for specific recommendations. If employees are required to clean-up spills, they must be properly trained and equipped. OSHA 1910.120(q) may be applicable.

Fire Extinguishing: This chemical is a flammable liquid. Poisonous gases including nitrogen oxides, carbon dioxide, carbon monoxide, are produced in fire. Use dry chemical, water spray, or foam extinguishers. Vapors are heavier than air and will collect in low areas. Vapors may travel long distances to ignition sources and flashback. Vapors in confined areas may explode when exposed to fire. Containers may explode in fire. Storage containers and parts of containers may rocket great distances, in many directions. If material or contaminated runoff enters waterways, notify downstream users of potentially contaminated waters. Notify local health and fire officials and pollution control agencies. From a secure, explosion-proof location, use water spray to cool exposed containers. If cooling streams are ineffective (venting sound increases in volume and pitch, tank discolors, or shows any signs of deforming), withdraw immediately to a secure position. If employees are expected to fight fires, they must be trained and equipped in OSHA 1910.156.

References

New Jersey Department of Health and Senior Services, "Hazardous Substance Fact Sheet: 1-Ethyl Piperidine," Trenton, NJ (April 1997)

2-Ethyl-3-Propyl Acrolein

Molecular Formula: $C_8H_{14}O$

Synonyms: 2-Ethyl-2-hexenal; 2-Ethylhexenal; α-Ethyl-β-*n*-propylacrolein; 2-Ethyl-3-propyl acrolein

CAS Registry Number: 645-62-5

RTECS Number: MP6300000

- ***Regulatory Authority:*** Air Pollutant Standard Set (former USSR)[43]

Cited in U.S. State Regulations: Massachusetts (G), New Hampshire (G), Pennsylvania (G).

Description: 2-Ethyl-3-Propyl Acrolein is a colorless or yellowish liquid with a sharp, powerful, irritating odor. Boiling point = 175°C. Freezing/Melting point = 100°C. Flash point = 68°C. Hazard Identification (based on NFPA-704 M Rating System): Health 2, Flammability 2, Reactivity 1. Slightly soluble in water.

Potential Exposure: Those involved in organic synthesis operations and use of this material as a warning agent.

Incompatibilities: Strong oxidizers, acids, caustics, ammonia, amines.

Permissible Exposure Limits in Air: The former USSR-UNEP/IRPTC project[43] has set a MAC in workplace air of 3 mg/m^3.

Permissible Concentration in Water: No criteria set.

Harmful Effects and Symptoms

Short Term Exposure: Is an irritant. Is a moderate inhalative and ingestive toxin. Is slightly toxic (LD_{50} for rats is 3,000 mg/kg) Irritates eyes, skin and mucous membranes. Causes smarting of the skin and first degree burns on short exposure. May cause second degree burns on long exposure.

Long Term Exposure: Very irritating substances may cause lung irritation and bronchitis may develop.

Points of Attack: Lungs.

Medical Surveillance: Lung function tests. Consider chest x-ray following acute overexposure.

First Aid: *Skin Contact:*[52] Flood all areas of body that have contacted the substance with water. Don't wait to remove contaminated clothing; do it under the water stream. Use soap to help assure removal. Isolate contaminated clothing when removed to prevent contact by others.

Eye Contact: Remove any contact lenses at once. Flush eyes well with copious quantities of water or normal saline for at least 20 – 30 minutes. Seek medical attention.

Inhalation: Leave contaminated area immediately; breathe fresh air. Proper respiratory protection must be supplied to any rescuers. If coughing, difficult breathing or any other symptoms develop, seek medical attention at once, even if symptoms develop many hours after exposure.

Ingestion: If convulsions are not present, give a glass or two of water or milk to dilute the substance. Assure that the person's airway is unobstructed and contact a hospital or poison center immediately for advice on whether or not to induce vomiting.

Personal Protective Methods: Wear protective gloves and clothing to prevent any reasonable probability of skin contact. Safety equipment suppliers/manufacturers can provide recommendations on the most protective glove/clothing material for your operation.. All protective clothing (suits, gloves, footwear, headgear) should be clean, available each day, and put on before work. Contact lenses should not be worn when working with this chemical. Wear splash-proof chemical goggles and face shield unless full facepiece respiratory protection is worn. Employees should wash immediately with soap when skin is wet or contaminated. Provide emergency showers and eyewash.

Respirator Selection: Where the potential for exposure to this chemical, use a MSHA/NIOSH approved supplied-air respirator with a full facepiece operated in the positive pressure mode or with a full facepiece, hood, or helmet in the continuous flow mode, or use a MSHA/NIOSH approved self-contained breathing apparatus with a full facepiece operated in pressure-demand or other positive pressure mode.

Storage: Prior to working with this chemical you should be trained on its proper handling and storage. Store in cool, dry place and protect from air.

Shipping: DOT[19] has no specific citation for the material.

Spill Handling: Evacuate and restrict persons not wearing protective equipment from area of spill or leak until cleanup is complete. Remove all ignition sources. Ventilate area of spill or leak. Absorb liquids in vermiculite, dry sand, earth, peat, carbon, or a similar material and deposit in sealed containers. It may be necessary to contain and dispose of this chemical as a hazardous waste. If material or contaminated runoff enters waterways, notify downstream users of potentially contaminated waters. Contact your Department of Environmental Protection or your regional office of the federal EPA for specific recommendations. If employees are required to clean-up spills, they must be properly trained and equipped. OSHA 1910.120(q) may be applicable.

Fire Extinguishing: This chemical is a combustible liquid. Poisonous gases are produced in fire. Use dry chemical, carbon dioxide, or alcohol foam (recommended) extinguishers. Vapors are heavier than air and will collect in low areas. Vapors may travel long distances to ignition sources and flashback. Vapors in confined areas may explode when exposed to fire. Containers may explode in fire. Storage containers and parts of containers may rocket great distances, in many directions. If material or contaminated runoff enters waterways, notify downstream users of potentially contaminated waters. Notify local health and fire officials and pollution control agencies. From a secure, explosion-proof location, use water spray to cool exposed containers. If cooling streams are ineffective (venting sound increases in volume and pitch, tank discolors, or shows any signs of deforming), withdraw immediately to a secure position. If employees are expected to fight fires, they must be trained and equipped in OSHA 1910.156.

Disposal Method Suggested: Incineration, or dissolve in flammable solvent and spray into incinerator containing afterburner.

References

Sax, N. I., Ed., "Dangerous Properties of Industrial Materials Report," 1, No. 8, 72–73 (1981) and 3, No. 2, 48–50 (1983)

Ethyl Silicate

Molecular Formula: $C_8H_{20}O_4Si$

Common Formula: $(C_2H_5O)_4Si$

Synonyms: Ethyl orthosilicate; Ethyl silicate, condensed; Ethyl silicate 40; Etylu krzemian (Polish); Extrema; Silibone; Silicate d'ethyle (French); Silicic acid tetraethyl ester; TEOS; Tetraethoxysilane; Tetraethyl orthosilicate; Tetraethyl silicate

CAS Registry Number: 78-10-4

RTECS Number: VV9450000

DOT ID: UN 1292

EEC Number: 014-005-00-0

Regulatory Authority

- Air Pollutant Standard Set (ACGH)[1] (DFG)[3] (HSE)[33] (former USSR)[43] (OSHA)[58]

- Canada, WHMIS, Ingredients Disclosure List

Cited in U.S. State Regulations: Alaska (G), California (A, G), Florida (G), Illinois (G), Maine (G), Massachusetts (G), New Hampshire (G), New Jersey (G), Pennsylvania (G), Rhode Island (G), West Virginia (G).

Description: Ethyl silicate, $(C_2H_5O)_4Si$, is a colorless, flammable liquid with a sharp odor detectable at 85 ppm. Boiling point = 169°C. Freezing/Melting point = -77°C. Flash point = 52°C (oc); 37°C (cc). Explosive limits: LEL = 1.3% - UEL = 23%. Hazard Identification (based on NFPA-704 M Rating System): Health 2, Flammability 2, Reactivity 0. Slowly decomposes in water.

Potential Exposure: Ethyl silicate is used in production of cases and molds for investment casting of metals. The next largest application is in corrosion-resistant coatings, primarily as a binder for zinc dust paints. Miscellaneous uses include the protection of white-light bulbs, the preparation of soluble silicas, catalyst preparation and regeneration, and as a crosslinker and intermediate in the production of silicones.

Incompatibilities: Forms explosive mixture with air. Strong oxidizers, strong acids, water.

Permissible Exposure Limits in Air: The OSHA PEL is 100 ppm. The ACGIH[1] has recommended a TWA of 10 ppm (85 mg/m^3) as has NIOSH.[58] HSE[33] has set an 8-hour TWA of the same value and added an STEL of 30 ppm (255 mg/m^3). The DFG[3] has set a higher allowable MAK of 100 ppm (850 mg/m^3) and the former USSR[43] a lower MAC of 20 mg/m^3. The NIOSH IDLH level is 1,000 ppm.

Determination in Air: XAD-2® (tube); adsorption on resin, workup with CS_2, analysis by gas chromatography/flame ionization. NIOSH II,[3] Method S-264.

Permissible Concentration in Water: No criteria set.

Routes of Entry: Inhalation of vapor, ingestion, skin and/or eye contact.

Harmful Effects and Symptoms

Short Term Exposure: Skin or eye contact can cause severe irritation or burns. The vapor can irritate the nose, eyes, throat and bronchial tubes, causing cough and difficulty breathing Higher exposures can cause pulmonary edema, a medical emergency that can be delayed for several hours. This can cause death.

Long Term Exposure: Repeated skin contact can cause drying and cracking. Exposure can affect the nervous system causing tremors, weakness, dizziness and unconsciousness. High or repeated exposure may damage the liver, kidneys, lungs and red blood cells.

Points of Attack: Eyes, respiratory system, liver, kidneys, blood, skin

Medical Surveillance: For those with frequent or potentially high exposure (half the TLV or greater), the following are recommended before beginning work and at regular times after that: Lung function tests. Kidney and liver function test. If symptoms develop or overexposure is suspected, the following may be useful: Consider chest x-ray after acute overexposure. Liver function tests. Complete blood count (CBC).

First Aid: If this chemical gets into the eyes, remove any contact lenses at once and irrigate immediately for at least 15 minutes, occasionally lifting upper and lower lids. Seek medical attention immediately. If this chemical contacts the skin, remove contaminated clothing and wash immediately with soap and water. Seek medical attention immediately. If this chemical has been inhaled, remove from exposure, begin rescue breathing (using universal precautions) if breathing has stopped and CPR if heart action has stopped. Transfer promptly to a medical facility. When this chemical has been swallowed, get medical attention. Give large quantities of water and induce vomiting. Do not make an unconscious person vomit. Medical observation is recommended for 24 – 48 hours after breathing overexposure, as pulmonary edema may be delayed. As first aid for pulmonary edema, a doctor or authorized paramedic may consider administering a corticosteroid spray.

Personal Protective Methods: Wear protective gloves and clothing to prevent any reasonable probability of skin contact. Safety equipment suppliers/manufacturers can provide recommendations on the most protective glove/clothing material for your operation. Remove clothing immediately if wet or contaminated to avoid flammability hazard. All protective clothing (suits, gloves, footwear, headgear) should be clean, available each day, and put on before work. Contact lenses should not be worn when working with this chemical. Wear splash-proof chemical goggles and face shield unless full facepiece respiratory protection is worn. Employees should wash immediately with soap when skin is wet or contaminated. Provide emergency showers and eyewash.

Respirator Selection: NIOSH/OSHA: *100 ppm:* SA (any supplied-air respirator). *250 ppm:* SA:CF (any supplied-air respirator operated in a continuous-flow mode). *500 ppm:* SCBAF (any self-contained breathing apparatus with full facepiece); or SAF (any supplied-air respirator with a full facepiece). *700 ppm:* SAF:PD,PP (any supplied-air respirator that has a full facepiece and is operated in a pressure-demand or other positive-pressure mode). *Emergency or planned entry into unknown concentrations or IDLH conditions:* SCBAF:PD,PP (any self-contained breathing apparatus that has a full facepiece and is operated in a pressure-demand or other positive-pressure mode); or SAF:PD, PP:ASCBA (any supplied-air respirator that has a full facepiece and is operated in a pressure-demand or other positive-pressure mode in combination with an auxiliary self-contained breathing apparatus operated in a pressure-demand or other positive pressure mode). *Escape:* GMFOV [any air-purifying, full-facepiece respirator (gas mask) with a chin-style, front-or back-, mounted organic vapor canister]; or SCBAE (any appropriate escape-type, self-contained breathing apparatus).

Note: Substance reported to cause eye irritation or damage; may require eye protection.

Storage: Ethyl silicate must be stored to avoid contact with strong oxidizers (such as chlorine, bromine and fluorine) since violent reactions occur. Before entering confined space where this chemical may be present, check to make sure that an explosive concentration does not exist. Store in tightly closed containers in a cool, well-ventilated area away from water. Sources of ignition, such as smoking and open flames, are prohibited where ethyl silicate is handled, used, or stored. Metal containers involving the transfer of 5 gallons or more of ethyl silicate should be grounded and bonded. Drums must be equipped with self-closing valves, pressure vacuum bungs, and flame arresters. Use only non-sparking tools and equipment, especially when opening and closing containers of ethyl silicate.

Shipping: This compound requires a shipping label of: "Flammable Liquid." It falls in DOT Hazard Class 3 and Packing Group III.

Spill Handling: Evacuate and restrict persons not wearing protective equipment from area of spill or leak until cleanup is complete. Remove all ignition sources. Establish forced ventilation to keep levels below explosive limit. Absorb liquids in vermiculite, dry sand, earth, peat, carbon, or a similar material and deposit in sealed containers. Keep ethyl silicate out of a confined space, such as a sewer, because of the possibility of an explosion, unless the sewer is designed to prevent the build-up of explosive concentrations. It may be necessary to contain and dispose of this chemical as a hazardous waste. If material or contaminated runoff enters waterways, notify downstream users of potentially contaminated waters. Contact your Department of Environmental Protection or your regional office of the federal EPA for specific recommendations. If employees are required to clean-up spills, they must be properly trained and equipped. OSHA 1910.120(q) may be applicable.

Fire Extinguishing: This chemical is a flammable liquid. Poisonous gases including carbon monoxide and silicon oxide are produced in fire. Use dry chemical, carbon dioxide, or foam extinguishers. Vapors are heavier than air and will collect in low areas. Vapors may travel long distances to ignition sources and flashback. Vapors in confined areas may explode when exposed to fire. Containers may explode in fire. Storage containers and parts of containers may rocket great distances, in many directions. If material or contaminated runoff enters waterways, notify downstream users of potentially contaminated waters. Notify local health and fire officials and pollution control agencies. From a secure, explosion-proof location, use water spray to cool exposed containers. If cooling streams are ineffective (venting sound increases in volume and pitch, tank discolors, or shows any signs of deforming), withdraw immediately to a secure position. If employees are expected to fight fires, they must be trained and equipped in OSHA 1910.156.

Disposal Method Suggested: Incineration in admixture with a more flammable solvent.

References

National Institute for Occupational Safety and Health, Information Profiles on Potential Occupational Hazards-Single Chemicals: Ethyl Silicate, pp 98–105, Report No. RT79-607, Rockville, MD (December 1979)

New Jersey Department of Health and Senior Services, "Hazardous Substance Fact Sheet: Ethyl Silicate," Trenton, NJ (January 1999)

Ethylthiocyanate

Molecular Formula: C_3H_5NS

Common Formula: C_2H_5SCN

Synonyms: Aethylrhodanid (German); Ethyl rhodanate; Ethyl sulfocyanate; Thiocyanatoethane; Thiocyanic acid, ethyl ester

CAS Registry Number: 542-90-5

RTECS Number: XK9900000

Regulatory Authority

- Superfund/EPCRA 40CFR355, Appendix B Extremely Hazardous Substances: TPQ = 10,000 lb (4,540 kg)
- Superfund/EPCRA 40CFR302.4 Reportable Quantity (RQ): EHS, 1 lb (0.454 kg)

Cited in U.S. State Regulations: Massachusetts (G), New Jersey (G), Pennsylvania (G).

Description: Ethyl Thiocyanate, C_2H_5SCN, is a liquid. Boiling point = 146°C. Freezing/Melting point = -86°C. Insoluble in water.

Potential Exposure: This material is used as an agricultural insecticide.

Incompatibilities: Contact with chlorates, nitrates, nitric acid, organic peroxides, peroxides may cause a violent reaction.

Permissible Exposure Limits in Air: No standards set.

Permissible Concentration in Water: No criteria set.

Routes of Entry: Inhalation, skin contact.

Harmful Effects and Symptoms

Short Term Exposure: The ingestion of a concentrated solution may lead to vomiting. The principal systemic reaction is probably one of central nervous depression, interrupted by periods of restlessness, abnormally fast and deep respiratory movements and convulsions. Death is usually due to respiratory arrest from paralysis of the medullary centers. In nonfatal cases injures to the liver and kidneys may appear.

Long Term Exposure: Prolonged absorption may produce various skin eruptions, runny nose, and occasionally dizziness, cramps, nausea, vomiting and mild or severe disturbances of the nervous system.

Points of Attack: Nervous system, liver, kidneys.

Medical Surveillance: Liver function tests. Kidney function tests. Examination of the nervous system.

First Aid: Treatment is as for aliphatic thiocyanates. Because cyanide is probably largely responsible for poisonings, antidotal measures against cyanide should be instituted promptly for cyanide includes moving the victim to fresh air. Call emergency medical care. If not breathing, give artificial respiration. IF breathing is difficult, give oxygen. In case of contact with material, immediately flush skin or eyes with running water for at least 15 minutes. Speed in removing material from skin is of extreme importance. Remove and isolate contaminated clothing and shoes at the site. Keep victim quiet and maintain normal body temperature. Effects may be delayed, keep victim under observation.

Personal Protective Methods: Wear protective gloves and clothing to prevent any reasonable probability of skin contact. Safety equipment suppliers/manufacturers can provide recommendations on the most protective glove/clothing material for your operation.. All protective clothing (suits, gloves, footwear, headgear) should be clean, available each day, and put on before work. Contact lenses should not be worn when working with this chemical. Wear splash-proof chemical goggles and face shield unless full facepiece respiratory protection is worn. Employees should wash immediately with soap when skin is wet or contaminated. Provide emergency showers and eyewash. For emergency situations, wear a positive pressure, pressure-demand, full facepiece self-contained breathing apparatus (SCBA) or pressure-demand supplied air respirator with escape SCBA and a fully-encapsulating, chemical resistant suit.

Respirator Selection: Where the potential for exposure to this chemical, use a MSHA/NIOSH approved supplied-air respirator with a full facepiece operated in the positive pressure mode or with a full facepiece, hood, or helmet in the continuous flow mode, or use a MSHA/NIOSH approved self-contained breathing apparatus with a full facepiece operated in pressure-demand or other positive pressure mode.

Storage: Prior to working with this chemical you should be trained on its proper handling and storage. Store in tightly closed containers in a cool, well ventilated area away from oxidizers, chlorates, nitrates, nitric acid. Where possible, automatically pump liquid from drums or other storage containers to process containers.

Shipping: This material may be classed as a pesticide, liquid, n.o.s. This class requires a shipping label of: "Poison." It falls in DOT Hazard Class 6.1 and Packing Group I.

Spill Handling: Evacuate and restrict persons not wearing protective equipment from area of spill or leak until cleanup is complete. Remove all ignition sources. Ventilate area of spill or leak. Absorb liquids in vermiculite, dry sand, earth, peat, carbon, or a similar material and deposit in sealed containers. For large spills, dike far ahead of spill for later disposal. It may be necessary to contain and dispose of this chemical as a hazardous waste. If material or contaminated runoff enters waterways, notify downstream users of potentially contaminated waters. Contact your Department of Environmental Protection or your regional office of the federal EPA for specific recommendations. If employees are required to cleanup spills, they must be properly trained and equipped. OSHA 1910.120(q) may be applicable.

Fire Extinguishing: This chemical is a combustible liquid. Poisonous gases including nitrogen and sulfur oxides are produced in fire. Small fires: dry chemical, carbon dioxide, water spray, or foam. Large fires: water spray, fog, or foam. Keep unnecessary people away; isolate hazard area and deny entry. Stay upwind; keep out of low areas. Ventilate closed spaces before entering them. Wear positive pressure breathing apparatus and special protective clothing. Move container from fire area if you can do it without risk. Fight fire from maximum distance. Dike fire control water for later disposal; do not scatter the material. Vapors are heavier than air and will collect in low areas. Vapors may travel long distances to ignition sources and flashback. Vapors in confined areas may explode when exposed to fire. Containers may explode in fire. Storage containers and parts of containers may rocket great distances, in many directions. If material or contaminated runoff enters waterways, notify downstream users of potentially contaminated waters. Notify local health and fire officials and pollution control agencies. From a secure, explosion-proof location, use water spray to cool exposed containers. If cooling streams are ineffective (venting sound increases in volume and pitch, tank discolors, or shows any signs of deforming), withdraw immediately to a secure position. If employees are expected to fight fires, they must be trained and equipped in OSHA 1910.156.

Disposal Method Suggested: In accordance with 40CFR 165 recommendations for the disposal of pesticides and pesticide containers. Must be disposed properly by following package label directions or by contacting your state pesticide or environmental control agency or by contacting your regional EPA office.

References

U.S. Environmental Protection Agency, "Chemical Profile: Ethylthiocyanate," Washington, DC, Chemical Emergency Preparedness Program (November 30, 1987)

Ethyl Trichlorosilane

Molecular Formula: $C_2H_5Cl_3Si$

Synonyms: Ethyl silicon trichloride; Silane, trichloroethyl-; Silicane, trichloroethyl-; Trichloroethylsilane; Trichloroethylsilicane

CAS Registry Number: 115-21-9

RTECS Number: VV4200000

DOT ID: UN 1196

Cited in U.S. State Regulations: Florida (G), Maine (G), Massachusetts (G), New Hampshire (G), New Jersey (G), Oklahoma (G), Pennsylvania (G), Rhode Island (G).

Description: Ethyl trichlorosilane, $C_2H_5SiCl_3$, is a colorless liquid with an irritating odor. Boiling point = 98°C. Freezing/Melting point = -106°C. Flash point = 22°C. Hazard Identification (based on NFPA-704 M Rating System): Health 3, Flammability 3, Reactivity 2. Water reactive.

Potential Exposure: Used in the manufacture of silicone polymers.

Incompatibilities: Forms explosive mixture with air. Contact with water, steam, moisture forms toxic and corrosive hydrogen chloride. Contact with strong oxidizers, strong bases may cause a violent reaction. Corrodes most common metals.

Permissible Exposure Limits in Air: No standards set.

Permissible Concentration in Water: No criteria set.

Routes of Entry: Inhalation.

Harmful Effects and Symptoms

Short Term Exposure: Ethyl trichlorosilane can affect you when breathed in. Ethyl trichlorosilane is a corrosive chemical and contact can cause skin and eye burns. Exposure can irritate the respiratory tract. Exposure can irritate the lungs causing coughing and/or shortness of breath. Higher exposures can cause pulmonary edema, a medical emergency that can be delayed for several hours. This can cause death.

Long Term Exposure: May cause kidney damage. Highly irritating substances can cause lung irritation and bronchitis.

Points of Attack: Lungs, kidneys.

Medical Surveillance: For those with frequent or potentially high exposure, the following are recommended before beginning work and at regular times after that: Lung function tests. Kidney function tests. If symptom develop or overexposure is suspected, the following may be useful: Consider chest x-ray after acute overexposure.

First Aid: If this chemical gets into the eyes, remove any contact lenses at once and irrigate immediately for at least 15 minutes, occasionally lifting upper and lower lids. Seek medical attention immediately. If this chemical contacts the skin, remove contaminated clothing and wash immediately with soap and water. Seek medical attention immediately. If this chemical has been inhaled, remove from exposure, begin rescue breathing (using universal precautions) if breathing has stopped and CPR if heart action has stopped. Transfer promptly to a medical facility. When this chemical has been swallowed, get medical attention. If victim is conscious, administer water or milk. Do not induce vomiting. Medical observation is recommended for 24 – 48 hours after breathing overexposure, as pulmonary edema may be delayed. As first aid for pulmonary edema, a doctor or authorized paramedic may consider administering a corticosteroid spray.

Personal Protective Methods: Wear protective gloves and clothing to prevent any reasonable probability of skin contact. Safety equipment suppliers/manufacturers can provide recommendations on the most protective glove/clothing material for your operation.. All protective clothing (suits, gloves, footwear, headgear) should be clean, available each day, and put on before work. Contact lenses should not be worn when working with this chemical. Wear splash-proof chemical goggles and face shield unless full facepiece respiratory protection is worn. Employees should wash immediately with soap when skin is wet or contaminated. Provide emergency showers and eyewash.

Respirator Selection: Where the potential exists for exposure to ethyl trichlorosilane, use a MSHA/NIOSH approved supplied-air respirator with a full facepiece operated in the positive pressure mode or with a full facepiece, hood, or helmet in the continuous flow mode, or use a MSHA/NIOSH approved self-contained breathing apparatus with a full facepiece operated in pressure-demand or other positive pressure mode.

Storage: Ethyl trichlorosilane must be stored to avoid contact with oxidizers (such as perchlorates, peroxides, permanganates, chlorates and nitrates) since violent reactions occur. Store in tightly closed containers in a cool, well-ventilated area away from water, steam and moisture because toxic and corrosive chloride gases including hydrogen chloride can be produced. Sources of ignition, such as smoking and open flames, are prohibited where ethyl trichlorosilane is handled, used, or stored. Metal containers involving the transfer of 5 gallons or more of ethyl trichlorosilane should be grounded and bonded. Drums must be equipped with self-closing valves, pressure vacuum bungs, and flame arresters. Use only nonsparking tools and equipment, especially when opening and closing containers of ethyl trichlorosilane. Protect containers from physical damage.

Shipping: This compound requires a shipping label of: "Flammable Liquid, Corrosive." It falls in DOT Hazard Class 3 and Packing Group I. Passenger aircraft or railcar shipment is forbidden.

Spill Handling: Evacuate and restrict persons not wearing protective equipment from area of spill or leak until cleanup is complete. Remove all ignition sources. Ventilate area of spill or leak. Absorb liquids in vermiculite, dry sand, earth, peat, carbon, or a similar material and deposit in sealed containers. Keep this chemical out of a confined space, such as a sewer, because of the possibility of an explosion, unless the sewer is designed to prevent the build-up of explosive concentrations. It may be necessary to contain and dispose of this chemical as a hazardous waste. If material or contaminated runoff enters waterways, notify downstream users of potentially contaminated waters. Contact your Department of Environmental Protection or your regional office of the federal EPA for specific recommendations. If employees are required to clean-up spills, they must be properly trained and equipped. OSHA 1910.120(q) may be applicable.

Fire Extinguishing: This chemical is a flammable liquid. Poisonous gases including phosgene are produced in fire. Use dry chemical or carbon dioxide extinguishers. Do not use water or foam. Fire may restart after it has been extinguished. Vapors are heavier than air and will collect in low areas. Vapors may travel long distances to ignition sources and flashback. Vapors in confined areas may explode when exposed to fire. Containers may explode in fire. Storage containers and parts of containers may rocket great distances, in many directions. If material or contaminated runoff enters waterways, notify downstream users of potentially contaminated waters. Notify local health and fire officials and pollution control agencies. From a secure, explosion-proof location, use water spray to cool exposed containers. If cooling streams are ineffective (venting sound increases in volume and pitch, tank discolors, or shows any signs of deforming), withdraw immediately to a secure position. If employees are expected to fight fires, they must be trained and equipped in OSHA 1910.156.

References

New Jersey Department of Health and Senior Services, "Hazardous Substance Fact Sheet: Ethyl Trichlorosilane," Trenton, NJ (March 1999)

F

Fenamiphos

Molecular Formula: $C_{13}H_{22}NO_3PS$

Common Formula: $CH_3SC_6H_3(CH_3)OPO(OC_2H_5)NHCH(CH_3)_2$

Synonyms: *O*-Aethyl-*O*-(3-methyl-4-methylthiophenyl)-isopropylamido-phosphorsaeure ester (German); BAY 68138; Bayer 68138; ENT 27,572; Ethyl 3-methyl-4-(methylthio)phenyl(1-methylethyl)phosphoramidate; Ethyl 4-(methylthio)-*m*-tolylisopropylphosphoramidate; Fenaminphos; Fenamiphos nematicide; Isopropylamino-*O*-ethyl (4-methylmercapto-3-methylphenyl) phosphate; Isopropylphosphoramidic acid ethyl 4-(methylthio)-*m*-toyl ester; 1-(Methylethyl)-ethyl 3-methyl-4-(methylthio)phenylphosphoramidate; (1-Methylethyl) phosphoramidic acid ethyl 3-methyl-4-(methylthio)phenyl ester; Nemacur; Nemacurp; NSC-195106; Phenamiphos; Phosphoramidic acid, isopropyl-, ethyl 4-(methylthio)-*m*-tolyl ethyl ester; Phosphoramidic acid, (1-methylethyl)-, ethyl [3-methyl-4-(methylethylthio)phenyl] ester; Phosphoramidic acid, (1-methylethyl)-, ethyl 3-methyl-4-(methylthio)phenyl ester

CAS Registry Number: 22224-92-6

RTECS Number: TB3675000

DOT ID: UN 2783

EEC Number: 015-123-00-5

Regulatory Authority:

- Air Pollutant Standard Set (ACGIH)[1] (Several States)[60] (Several Canadian Provinces)
- Superfund/EPCRA 40CFR355, Appendix B Extremely Hazardous Substances: TPQ = 10/10,000 lb (4.54/4,540 kg)
- Superfund/EPCRA 40CFR302.4 Reportable Quantity (RQ): EHS, 1 lb (0.454 kg)
- Marine Pollutant (49CFR, Subchapter 172.101, Appendix B
- Canada, WHMIS, Ingredients Disclosure List

Cited in U.S. State Regulations: California (A, G), Connecticut (A), Illinois (G), Massachusetts (G), New Jersey (G), North Dakota (A), Pennsylvania (G), Rhode Island (G), Virginia (A), West Virginia (G).

Description: Fenamiphos, $CH_3SC_6H_3(CH_3)OPO(OC_2H_5)NHCH(CH_3)_2$, is an off-white to tan, waxy solid. Found commercially as a granular ingredient (5 – 15%) or in an emulsifiable concentrate (400 g/l). Freezing/Melting point = 40°C (technical grade) and 49.2°C (pure compound). Hazard Identification (based on NFPA-704 M Rating System): Health 3, Flammability 1, Reactivity 0. Slightly soluble in water.

Potential Exposure: Those involved in the manufacture, formulation or application of this nematocide.

Incompatibilities: May hydrolyze under alkaline conditions. Keep away from moisture.

Permissible Exposure Limits in Air: There is no OSHA PEL. ACGIH and NIOSH recommend a TWA of 0.1 mg/m³, with the notation that skin absorption is possible. Several states have set guidelines or standards for fenamiphos in ambient air[60] ranging from 1.0 μg/m³ (North Dakota) to 1.6 μg/m³ (Virginia) to 2.0 μg/m³ (Connecticut).

Determination in Air: OSHA versatile sampler-2; Toluene/Acetone; Gas chromatography/Flame ionization detection; NIOSH IV, Method #5600, Organophosphorus Pesticides.

Permissible Concentration in Water: A long term health advisory set by EPA is 18 μg/l and a lifetime health advisory is 9 μg/l.

Routes of Entry: Inhalation, skin absorption, ingestion, skin and/or eye contact.

Harmful Effects and Symptoms

Short Term Exposure: This is a highly toxic chemical (the LD_{50} for rats is 8 mg/kg). It is a cholinesterase inhibitor with effects typical of such compounds. Acute exposure to fenamiphos may produce the following sings and symptoms: pinpoint pupils, blurred vision, headaches, dizziness, muscle spasm, and profound weakness. Vomiting, diarrhea, abdominal pain, seizures, and coma may also occur. The heart rate may increase following oral exposure or decrease following dermal exposure. Hypotension (low blood pressure) may occur although hypertension (high blood pressure) is not uncommon. Chest pain may be noted. Respiratory symptoms include dyspnea (shortness of breath), respiratory depression, and respiratory paralysis. Psychosis may occur. This material is highly toxic orally, by inhalation, and by absorption through the skin. Death may occur from respiratory failure.

Long Term Exposure: Cholinesterase inhibitor; cumulative effect is possible. Organophosphates may damage the nervous system with repeated exposure, resulting in convulsions, respiratory failure. May cause liver damage.

Points of Attack: Respiratory system, central nervous system, cardiovascular system, blood cholinesterase.

Medical Surveillance: Before employment and at regular times after that, the following are recommended: Plasma and red blood cell cholinesterase levels (tests for the enzyme poisoned by this chemical). If exposure stops, plasma levels return to normal in 1 – 2 weeks while red blood cell levels may be reduced for 1 – 3 months.

When cholinesterase enzyme levels are reduced by 25% or more below preemployment levels, risk of poisoning is increased, even if results are in lower ranges of "normal." Reassignment to work not involving organophosphate or carbamate pesticides is recommended until enzyme levels recover. If symptoms develop or overexposure occurs, repeat the above tests as soon as possible and get an exam of the nervous system. Also consider complete blood count. Consider chest x-ray following acute overexposure. Do not drink any alcoholic beverages before or during use. Alcohol promotes absorption of organic phosphates.

First Aid: If this chemical gets into the eyes, remove any contact lenses at once and irrigate immediately for at least 15 minutes, occasionally lifting upper and lower lids. Seek medical attention immediately. If this chemical contacts the skin, remove contaminated clothing and wash immediately with soap and water. Speed in removing material from skin is of extreme importance. Shampoo hair promptly if contaminated. Seek medical attention immediately. If this chemical has been inhaled, remove from exposure, begin rescue breathing (using universal precautions) if breathing has stopped and CPR if heart action has stopped. Transfer promptly to a medical facility. When this chemical has been swallowed, get medical attention. Give large quantities of water and induce vomiting. Do not make an unconscious person vomit.

Personal Protective Methods: Wear solvent resistant gloves and clothing to prevent any reasonable probability of skin contact. Safety equipment suppliers/manufacturers can provide recommendations on the most protective glove/clothing material for your operation. All protective clothing (suits, gloves, footwear, headgear) should be clean, available each day, and put on before work. Contact lenses should not be worn when working with this chemical. Wear splash-proof chemical goggles and face shield unless full facepiece respiratory protection is worn. Wear air tight gas-proof goggles and face shield, unless full facepiece respiratory protection is worn. Employees should wash immediately with soap when skin is wet or contaminated. Provide emergency showers and eyewash.

Respirator Selection: Where the potential exists for exposures over 0.1 mg/m^3, use an MSHA/NIOSH approved supplied-air respirator with a full facepiece operated in the positive pressure mode or with a full facepiece, hood, or helmet in the continuous flow mode, or use an MSHA/NIOSH approved self-contained breathing apparatus with a full facepiece operated in pressure-demand or other positive pressure mode.

Storage: Store in tightly closed containers in a cool, well-ventilated area.

Shipping: This material falls under organophosphorus pesticides, solid, toxic, n.o.s. This compound requires a shipping label of: "Poison." It falls in DOT Hazard Class 6.1 and Packing Group I.

Spill Handling: Evacuate and restrict persons not wearing protective equipment from area of spill or leak until cleanup is complete. Remove all ignition sources. Ventilate area of spill or leak. Stay upwind; keep out of low areas. Ventilate closed spaces before entering them. Remove an isolate contaminated clothing at the site. Do not touch spilled material. Use water spray to reduce vapors. Absorb liquids in vermiculite, dry sand, earth, peat, carbon, or a similar material and deposit in sealed containers. It may be necessary to contain and dispose of this chemical as a hazardous waste. If material or contaminated runoff enters waterways, notify downstream users of potentially contaminated waters. Contact your Department of Environmental Protection or your regional office of the federal EPA for specific recommendations. If employees are required to clean-up spills, they must be properly trained and equipped. OSHA 1910.120(q) may be applicable.

Fire Extinguishing: Poisonous gases including oxides of nitrogen, phosphorus, and sulfur are produced in fire. This material may burn, but does not ignite readily. For small fires, use dry chemical, carbon dioxide, water spray, or foam. For large fires, use water spray, fog, or foam. Stay upwind; keep out of low areas. Move container from fire area if you can do it without risk. Fight fire from maximum distance. Dike fire control water for later disposal; do not scatter the material. Wear positive pressure breathing apparatus and special protective clothing. Vapors in confined areas may explode when exposed to fire. Containers may explode in fire. Storage containers and parts of containers may rocket great distances, in many directions. If material or contaminated runoff enters waterways, notify downstream users of potentially contaminated waters. Notify local health and fire officials and pollution control agencies. From a secure, explosion-proof location, use water spray to cool exposed containers. If cooling streams are ineffective (venting sound increases in volume and pitch, tank discolors, or shows any signs of deforming), withdraw immediately to a secure position. If employees are expected to fight fires, they must be trained and equipped in OSHA 1910.156.

Disposal Method Suggested: Incineration. In accordance with 40CFR165 recommendations for the disposal of pesticides and pesticide containers. Must be disposed properly by following package label directions or by contacting your state pesticide or environmental control agency or by contacting your regional EPA office.

References

Sax, N. I., Ed., "Dangerous Properties of Industrial Materials Report," 3, No. 1, 52–56, New York, Van Nostrand Reinhold Co. (1983)

U.S. Environmental Protection Agency, "Chemical Profile: Fenamiphos," Washington, DC, Chemical Emergency Preparedness Program (November 30, 1987)

U.S. Environmental Protection Agency, "Health Advisory: Fenamiphos," Washington, DC, Office of Drinking Water (August 1987)

New Jersey Department of Health and Senior Services, "Hazardous Substance Fact Sheet: Fenamiphos," Trenton, NJ (February 1999)

Fenitrothion

Molecular Formula: $C_9H_{12}NO_5PS$

Common Formula: $(CH_3O)_2PSO\text{-}C_6H_3(NO_2)(CH_3)$

Synonyms: 8057HC; Accothion; Aceothion; Agria 1050; Agriya 1050; Agrothion; American Cyanamid CL-47,300; Arbogal; BAY 41831; Bayer 41831; Bayer S 5660; Cekutrothion; CL 47300; CP47114; Cyfen; Cytel; Cyten; Dicathion; Dicofen; *O,O*-Dimethyl-*O*-(3-methyl-4-nitrofenyl)-monothiofosfaat (Dutch); *O,O*-Dimethyl-*O*-(3-methyl-4-nitrophenyl)-monothiophosphat (German); *O,O*-Dimethyl *O*-(3-methyl-4-nitrophenyl) phosphorothioate; *O,O*-Dimethyl *O*-(3-methyl-4-nitrophenyl) thiophosphate; *O,O*-Dimethyl *O*-(4-nitro-3-methylphenyl) phosphorothioate; *O,O*-Dimethyl *O*-(4-nitro-3-methylphenyl) thiophosphate; *O,O*-Dimethyl-*O*-4-nitro-*m*-toyl phosphorothioate; Dybar; EI 47300; ENT 25,715; Falithion; Fenitex; Fenitox; Fenitrothion; Fenitrotion (Hungarian); Folethion; Folithion; H-35-F 87 (BVM); Keen Superkill Ant and Roach exterminator; Killgerm tetracide insecticidal spray; Kotion; MEP (Pesticide); Metathion; Metathione; Metation; Methylnitrophos; Micromite; Monsanto CP 47114; Nitrophos; Novathion; Nuvand; Nuvanol; Oleosumifene; OMS 43; Ovadofos; Pennwalt C-4852; Phenitrothion; Phosphorothioic acid, *O,O*-dimethyl *O*-(3-methyl-4-nitrophenyl) ester; Phosphorothioic acid, *O,O*-dimethyl *O*-(4-nitro-*m*-tolyl) ester; S 112A; S 5660; SMT; Sumithion; Thiophosphate de *O,O*-dimethyle et de *O*-(3-methyl-4-nitrophenyle) (French); Turbair grain storage insecticide; Verthion

CAS Registry Number: 122-14-5

RTECS Number: TG0350000

DOT ID: UN 3018

EEC Number: 015-054-00-0

- ***Regulatory Authority:*** Air Pollutant Standard Set (former USSR)[43] (Japan)[35]
- Superfund/EPCRA 40CFR355, Appendix B Extremely Hazardous Substances: TPQ = 500 lb (227 kg)
- Superfund/EPCRA 40CFR302.4 Reportable Quantity (RQ): EHS, 1 lb (0.454 kg)

Cited in U.S. State Regulations: California (G), Massachusetts (G), New Jersey (G), Pennsylvania (G).

Description: Fenitrothion, $(CH_3O)_2PSO\text{-}C_6H_3(NO_2)(CH_3)$, is a volatile brownish-yellow oil. Boiling point = 118°C @ 0.05 mm Hg. (also found in the literature at 140 – 145°C). Decomposes below the BP. Freezing/Melting point = 0.3°C. Flash point = 157°C. Hazard Identification (based on NFPA-704 M Rating System): Health 3, Flammability 1, Reactivity 0. Insoluble in water.

Potential Exposure: Those involved in the manufacture, formulation and application off this insecticide. It is a selective acaricide and a contact and stomach insecticide. Used to control chewing and sucking insects on rice, orchard fruits, vegetables, cereals, cotton and forest. Also protects against flies, mosquitoes, and cockroaches.

Incompatibilities: Strong oxidizers, strong bases.

Permissible Exposure Limits in Air: The former USSR-UNEP/IRPTC project[43] has set a MAC in workplace air of 0.1 mg/m^3 and a MAC in ambient air in residential areas of 0.008 mg/m^3 on a momentary basis and 0.001 mg/m^3 on a daily average basis. Japan[35] has set a MAC in workplace air of 1.0 mg/m^3 in workplace air.

Determination in Air: OSHA versatile sampler-2; Toluene/Acetone; Gas chromatography/Flame photometric detection for sulfur, nitrogen, or phosphorus; NIOSH Method IV Method #5600, Organophosphorus Pesticides.

Permissible Concentration in Water: The former USSR-UNEP/IRPTC project[43] has set a MAC in water bodies for domestic purposes of 0.25 mg/l; in water bodies used for fishery purposes of zero.

Routes of Entry: Inhalation, through the skin, ingestion.

Harmful Effects and Symptoms

Short Term Exposure: Irritates the eyes and skin. Nausea is often the first symptom, followed by vomiting; abdominal cramps; diarrhea; excessive salivation; headache; giddiness; dizziness; weakness; tightness in the chest; loss of muscle coordination; slurring of speech, muscle twitching (particularly the tongue and eyelid); respiratory difficulty; blurring or dimness of vision; pinpoint pupils; profound weakness; mental confusion; disorientation and drowsiness. This compound is an organophosphate insecticide. It is a highly toxic cholinesterase inhibitor that acts on the nervous system. Does not cause delayed neurotoxicity and contact produces little irritation. The effects may be delayed. Keep exposed victim under medical observation.

Long Term Exposure: Cholinesterase inhibitor; cumulative effect is possible. This chemical may damage the nervous system with repeated exposure, resulting in convulsions, respiratory failure. May cause liver damage.

Points of Attack: Respiratory system, lungs, central nervous system, cardiovascular system, skin, eyes, plasma and red blood cell cholinesterase.

Medical Surveillance: Before employment and at regular times after that, the following are recommended: Plasma and red blood cell cholinesterase levels (tests for the enzyme poisoned by this chemical). If exposure stops, plasma levels return to normal in 1 – 2 weeks while red blood cell levels may be reduced for 1 – 3 months.

When cholinesterase enzyme levels are reduced by 25% or more below preemployment levels, risk of poisoning is increased, even if results are in lower ranges of "normal." Reassignment to work not involving organophosphate or carbamate pesticides is recommended until enzyme levels recover. If symptoms develop or overexposure occurs, repeat the above tests as soon as possible and get an exam of the nervous system. Also consider complete blood count. Consider chest x-ray following acute overexposure. Do not drink any alcoholic beverages before or during use. Alcohol promotes absorption of organic phosphates. Also consider complete blood count. Consider chest x-ray following acute overexposure.

First Aid: If this chemical gets into the eyes, remove any contact lenses at once and irrigate immediately for at least 15 minutes, occasionally lifting upper and lower lids. Seek medical attention immediately. If this chemical contacts the skin, remove contaminated clothing and wash immediately with soap and water. Speed in removing material from skin is of extreme importance. Shampoo hair promptly if contaminated. Seek medical attention immediately. If this chemical has been inhaled, remove from exposure, begin rescue breathing (using universal precautions) if breathing has stopped and CPR if heart action has stopped. Transfer promptly to a medical facility. When this chemical has been swallowed, get medical attention. Give large quantities of water and induce vomiting. Do not make an unconscious person vomit. Effects may be delayed. Keep victim under medical observation.

Personal Protective Methods: Wear protective gloves and clothing to prevent any reasonable probability of skin contact. Safety equipment suppliers/manufacturers can provide recommendations on the most protective glove/clothing material for your operation. All protective clothing (suits, gloves, footwear, headgear) should be clean, available each day, and put on before work. Contact lenses should not be worn when working with this chemical. Wear splash-proof chemical goggles and face shield unless full facepiece respiratory protection is worn. Employees should wash immediately with soap when skin is wet or contaminated. Provide emergency showers and eyewash.

Respirator Selection: Where the potential for exposure to this chemical, use a MSHA/NIOSH approved supplied-air respirator with a full facepiece operated in the positive pressure mode or with a full facepiece, hood, or helmet in the continuous flow mode, or use a MSHA/NIOSH approved self-contained breathing apparatus with a full facepiece operated in pressure-demand or other positive pressure mode.

Storage: Prior to working with this chemical you should be trained on its proper handling and storage. Prior to working with this chemical you should be trained on its proper handling and storage. Store in tightly closed containers in a cool, well ventilated area away from oxidizers. Where possible, automatically pump liquid from drums or other storage containers to process containers.

Shipping: This material falls in the class of organophosphorus pesticides, liquid, toxic, n.o.s. This compound requires a shipping label of: "Keep Away From Food." It falls in DOT Hazard Class 6.1 and Packing Group III.

Spill Handling: Evacuate and restrict persons not wearing protective equipment from area of spill or leak until cleanup is complete. Remove all ignition sources. Ventilate area of spill or leak. Stay upwind; keep out of low areas. Ventilate closed spaces before entering them. Remove and isolate contaminated clothing at the site. Do not touch spilled material; stop leak if you can do so without risk. Use water spray to reduce vapors. Small spills; absorb with sand or other noncombustible absorbent material and place into containers for later disposal. Large spills: dike far ahead of spill for later disposal. It may be necessary to contain and dispose of this chemical as a hazardous waste. If material or contaminated runoff enters waterways, notify downstream users of potentially contaminated waters. Contact your Department of Environmental Protection or your regional office of the federal EPA for specific recommendations. If employees are required to clean-up spills, they must be properly trained and equipped. OSHA 1910.120(q) may be applicable.

Fire Extinguishing: This chemical is a combustible liquid. Poisonous gases including oxides of sulfur and phosphorus are produced in fire. Small fires: dry chemical, carbon dioxide, water spray, or foam. Large fires: water spray, fog or foam. Move containers from fire area if you can do so without risk. Fight fire from maximum distance. Dike fire control water for later disposal; do not scatter the material. Wear positive pressure breathing apparatus and special protective clothing. Vapors in confined areas may explode when exposed to fire. Containers may explode in fire. Storage containers and parts of containers may rocket great distances, in many directions. If material or contaminated runoff enters waterways, notify downstream users of potentially contaminated waters. Notify local health and fire officials and pollution control agencies. From a secure, explosion-proof location, use water spray to cool exposed containers. If cooling streams are ineffective (venting sound increases in volume and pitch, tank discolors, or shows any signs of deforming), withdraw immediately to a secure position. If employees are expected to fight fires, they must be trained and equipped in OSHA 1910.156.

Disposal Method Suggested: Incineration (for large amounts); alkaline hydrolysis and landfill (for small amounts).[22] In accordance with 40CFR165 recommendations for the disposal of pesticides and pesticide containers. Must be disposed properly by following package label directions or by contacting your state pesticide or environmental control agency or by contacting your regional EPA office.

References

Sax, N. I., Ed., "Dangerous Properties of Industrial Materials Report," 2, No. 4, 88–92 (1982)

U.S. Environmental Protection Agency, "Chemical Profile: Fenitrothion," Washington, DC, Chemical Emergency Preparedness Program (November 30, 1987)

Fensulfothion

Molecular Formula: $C_{11}H_{17}O_4PS_2$

Synonyms: BAY 25141; Bayer 25141; Bayer S767; Chemagro 25141; Dasanit; *O,O*-Diaethyl-*O*-4-methyl-sulfinyl-phenyl-monothiophosphat (German); *O,O*-Diethyl *O*-[*p*-(methylsulfinyl)phenyl] phosphorothioate; *O,O*-Diethyl *O*-[*p*-(methylsulfinyl)phenyl] thiophosphate; DMSP; ENT 24,945; Entphosphorothioate; Ester; Fensulfotiona (Spanish); Phosphorothioic acid, *O,O*-diethyl *O*-[*p*-(methylsulfinyl)phenyl]; S 767; Terracur P

CAS Registry Number: 115-90-2

RTECS Number: TF3850000

DOT ID: UN 3018

Regulatory Authority

- Very Toxic Substance (World Bank)[15]
- Air Pollutant Standard Set (ACGIH)[1] (Several States)[60]
- Superfund/EPCRA 40CFR355, Appendix B Extremely Hazardous Substances: TPQ = 500 lb (227 kg)
- Superfund/EPCRA 40CFR302.4 Reportable Quantity (RQ): EHS, 1 lb (0.454 kg)
- U.S. DOT Regulated Marine Pollutant (49CFR172.101, Appendix B)

Cited in U.S. State Regulations: Alaska (G), California (G), Connecticut (A), Florida (G), Illinois (G), Maine (G), Massachusetts (G), Michigan (G), Nevada (A), New Hampshire (G), New Jersey (G), North Dakota (A), Oklahoma (G), Pennsylvania (G), Rhode Island (G), Virginia (A), West Virginia (G).

Description: Fensulfothion, O,O-diethyl O-4-(methylsulfinyl)phenyl phosphorothioate, is a yellow oil. Boiling point = 138 – 141°C at 0.01 mm Hg. Hazard Identification (based on NFPA-704 M Rating System): Health 4, Flammability 1, Reactivity 0. Slightly soluble in water.

Potential Exposure: Those involved in the manufacture, formulation or application of this insecticide and nematocide.

Incompatibilities: Strong bases.

Permissible Exposure Limits in Air: There is no OSHA PEL. NIOSH and ACGIH recommend a TWA value of 0.1 mg/m³. Several states have set guidelines or standards for fensulfothion in ambient air[60] ranging from 1.0 μg/m³ (North Dakota) to 1.6 μg/m³ (Virginia) to 2.0 μg/m³ (Connecticut and Nevada).

Determination in Air: OSHA versatile sampler-2; Toluene/Acetone; Gas chromatography/Flame photometric detection for sulfur, nitrogen, or phosphorus; NIOSH Method IV Method #5600, Organophosphorus Pesticides.

Permissible Concentration in Water: No criteria set.

Routes of Entry: Inhalation, through the skin, ingestion, eyes.

Harmful Effects and Symptoms

Short Term Exposure: Contact can irritate the skin and eyes. Exposure can cause rapid, fatal organophosphate poisoning. The acute oral LD_{50} value for rats is 5 – 10 mg/kg which is highly to extremely toxic. It is a cholinesterase inhibitor. This material may cause nausea, vomiting, abdominal cramps, diarrhea, headache, giddiness, vertigo, weakness, lack of muscle control, tearing, slurring of speech, difficult breathing, convulsions, excessive salivation, tightness in chest, and death from respiratory arrest. Death results primarily from respiratory arrest stemming from failure of the respiratory center, paralysis of respiratory muscles and intensive bronchoconstriction.

Long Term Exposure: Cholinesterase inhibitor; cumulative effect is possible. This chemical may damage the nervous system with repeated exposure, resulting in convulsions, respiratory failure. May cause liver damage.

Points of Attack: Respiratory system, lungs, central nervous system, cardiovascular system, skin, eyes, plasma and red blood cell cholinesterase.

Medical Surveillance: Before employment and at regular times after that, the following are recommended: plasma and red blood cell cholinesterase levels (tests for the enzyme poisoned by this chemical). If exposure stops, plasma levels return to normal in 1 – 2 weeks while red blood cell levels may be reduced for 1 – 3 months.

When cholinesterase enzyme levels are reduced by 25% or more below preemployment levels, risk of poisoning is increased, even if results are in lower ranges of "normal." Reassignment to work not involving organophosphate or carbamate pesticides is recommended until enzyme levels recover. If symptoms develop or overexposure occurs, repeat the above tests as soon as possible and get an exam of the nervous system. Also consider complete blood count. Consider chest x-ray following acute overexposure. Do not drink any alcoholic beverages before or during use. Alcohol promotes absorption of organic phosphates.

First Aid: If this chemical gets into the eyes, remove any contact lenses at once and irrigate immediately for at least 15 minutes, occasionally lifting upper and lower lids. Seek medical attention immediately. If this chemical contacts the

skin, remove contaminated clothing and wash immediately with soap and water. Speed in removing material from skin is of extreme importance. Shampoo hair promptly if contaminated. Seek medical attention immediately. If this chemical has been inhaled, remove from exposure, begin rescue breathing (using universal precautions) if breathing has stopped and CPR if heart action has stopped. Transfer promptly to a medical facility. When this chemical has been swallowed, get medical attention. Give large quantities of water and induce vomiting. Do not make an unconscious person vomit. Keep victim quiet and maintain normal body temperature. Effects may be delayed; keep victim under observation.

Personal Protective Methods: Wear protective gloves and clothing to prevent any reasonable probability of skin contact. Safety equipment suppliers/manufacturers can provide recommendations on the most protective glove/clothing material for your operation. All protective clothing (suits, gloves, footwear, headgear) should be clean, available each day, and put on before work. Contact lenses should not be worn when working with this chemical. Wear splash-proof chemical goggles and face shield unless full facepiece respiratory protection is worn. Employees should wash immediately with soap when skin is wet or contaminated. Provide emergency showers and eyewash.

Respirator Selection: Where the potential exists for exposures over 0.1 mg/m^3, use a MSHA/NIOSH approved supplied-air respirator with a full facepiece operated in the positive pressure mode or with a full facepiece, hood, or helmet in the continuous flow mode, or use a MSHA/NIOSH approved self-contained breathing apparatus with a full facepiece operated in pressure-demand or other positive pressure mode.

Storage: Prior to working with this chemical you should be trained on its proper handling and storage. Store in tightly closed containers in a cool, well-ventilated area away from strong bases, strong oxidizers. Where possible, automatically pump liquid from drums or other storage containers to process containers.

Shipping: This material is classed as an organophosphorus pesticide, liquid, toxic, n.o.s. This compound requires a shipping label of: "Poison." It falls in DOT Hazard Class 6.1 and Packing Group I.

Spill Handling: Evacuate and restrict persons not wearing protective equipment from area of spill or leak until cleanup is complete. Remove all ignition sources. Ventilate area of spill or leak. Absorb liquids in vermiculite, dry sand, earth, peat, carbon, or a similar material and deposit in sealed containers. It may be necessary to contain and dispose of this chemical as a hazardous waste. If material or contaminated runoff enters waterways, notify downstream users of potentially contaminated waters. Contact your Department of Environmental Protection or your regional office of the federal EPA for specific recommendations. If employees are required to clean-up spills, they must be properly trained and equipped. OSHA 1910.120(q) may be applicable.

Fire Extinguishing: This chemical is a combustible liquid. Poisonous gases including oxides of sulfur and phosphorus are produced in fire. Use dry chemical, carbon dioxide, or alcohol foam extinguishers. Vapors are heavier than air and will collect in low areas. Vapors may travel long distances to ignition sources and flashback. Vapors in confined areas may explode when exposed to fire. Containers may explode in fire. Storage containers and parts of containers may rocket great distances, in many directions. If material or contaminated runoff enters waterways, notify downstream users of potentially contaminated waters. Notify local health and fire officials and pollution control agencies. From a secure, explosion-proof location, use water spray to cool exposed containers. If cooling streams are ineffective (venting sound increases in volume and pitch, tank discolors, or shows any signs of deforming), withdraw immediately to a secure position. If employees are expected to fight fires, they must be trained and equipped in OSHA 1910.156.

Disposal Method Suggested: Alkaline hydrolysis.[22] In accordance with 40CFR165 recommendations for the disposal of pesticides and pesticide containers. Must be disposed properly by following package label directions or by contacting your state pesticide or environmental control agency or by contacting your regional EPA office.

References

U.S. Environmental Protection Agency, "Chemical Profile: Fensulfothion," Washington, DC, Chemical Emergency Preparedness Program (November 30, 1987)

New Jersey Department of Health and Senior Services, "Hazardous Substance Fact Sheet: Fensulfothion," Trenton, NJ (May 1998)

Fenthion

Molecular Formula: $C_{10}H_{15}O_3PS_2$

Synonyms: AI3-25540; BAY 29493; Baycid; Bayer 29493; Bayer 9007; Bayer S-1752; Baytex; Caswell No.456F; *m*-Cresol, 4-(methylthio)-, *O*-ester with *O,O*-dimethyl phosphorothioate; DALF; *O,O*-Dimethyl *O*-4-(methylmercapto)-3-methylphenyl phosphorothioate; *O,O*-Dimethyl *O*-4-(methylmercapto)-3-methylphenyl thiophosphate; *O,O*-Dimethyl *O*-(3-methyl-4-methylmercaptophenyl) phosphorothioate; *O,O*-Dimethyl *O*-[3-methyl-4-(methylthio) phenyl] ester, phosphorothioic acid; *O,O*-Dimethyl *O*-[3-methyl-4-(methylthio)phenyl] phosphorothioate; *O,O*-Dimethyl *O*-(4-methylthio-3-methylphenyl) phosphorothioate; *O,O*-Dimethyl *O*-(4-methylthio-3-methylphenyl) thiophosphate; *O,O*-Dimethyl *O*-[4-(methylthio)-*m*-tolyl] phosphorothioate; DMTP; ENT 25,540; Entex; EPA pesticide chemical code 053301; Fenthion 4E; Fenthionon; Fentiona (Spanish); Lebaycid; Mercaptophos; 4-Methylmercapto-3 -methylphenyldimethylthiophosphate; MPP; MPP (in Japan);

NCI-C08651; OMS 2; Phenthion; Phosphorothioic acid, *O,O*-dimethyl *O*-[3-methyl-4-(methylthio)phenyl] ester; Phosphorothioic acid, *O,O*-dimethyl *O*-[4-(methylthio)-*m*-tolyl] ester; Queletox; S 1752; Spotton; Thiophosphate de *O,O*-dimethyle et de *O*-(3-methyl-4-methylthiophenyle) (French); Tiguvon; Tolodex

CAS Registry Number: 55-38-9

RTECS Number: TF9625000

DOT ID

UN 3018

EEC Number: 015-048-00-8

Regulatory Authority

- Carcinogen (Animal Positive) (NCI)[9]
- Air Pollutant Standard Set (ACGIH)[1] (DFG)[3] (former USSR)[35] (Several States)[60]
- EPCRA Section 313 Form R *de minimis* concentration reporting level: 1.0%
- U.S. DOT Regulated Marine Pollutant (49CFR172.101, Appendix B), severe pollutant
- Canada, WHMIS, Ingredients Disclosure List

Cited in U.S. State Regulations: Alaska (G), California (A, G), Connecticut (A), Florida (G), Illinois (G), Maine (G), Massachusetts (G), Michigan (G), Nevada (A), New Jersey (G), North Dakota (A), Pennsylvania (G), Rhode Island (G), Virginia (A), West Virginia (G).

Description: Fenthion, $C_{10}H_{15}O_3PS_2$, is a colorless liquid when pure with a weak, garlic-like odor. The technical grade is a yellow to brown oil. Boiling point = 87°C at 0.01 mm Hg. Decomposes below the BP. Freezing/Melting point = 6°C. Very slightly soluble in water.

Potential Exposure: Those involved in the manufacture, formulation or application of this insecticide.

Incompatibilities: Oxidizers, bases, alkaline insecticides.

Permissible Exposure Limits in Air: There is no OSHA PEL. DFG MAK value[3] as well as the recommended ACGIH TWA value is 0.2 mg/m^3. They have added the notation "skin" indicating the possibility of cutaneous absorption. The former USSR[35] has set a MAC in workplace air of 0.3 mg/m^3. Several states have set guidelines or standards for fenthion in ambient air[60] ranging from 2.0 μg/m^3 (North Dakota) to 3.2 μg/m^3 (Virginia) to 4.0 μg/m^3 (Connecticut) to 5.0 μg/m^3 (Nevada).

Determination in Air: OSHA versatile sampler-2; Toluene/Acetone; Gas chromatography/Flame photometric detection for sulfur, nitrogen, or phosphorus; NIOSH Method IV Method #5600, Organophosphorus Pesticides.

Permissible Concentration in Water: Mexico[35] has set a MAC in coastal waters of 0.03 μg/l and 0.0003 mg/l (0.3 μg/l) in estuaries. The former USSR[35] has set a MAC in water bodies used for domestic purposes of 1.0 μg/l and in water used for fishery purposes of zero.

Determination in Water: No method available.

Routes of Entry: Inhalation, skin absorption, ingestion, skin and/or eye contact.

Harmful Effects and Symptoms

Short Term Exposure: Fenthion can affect you when breathed in and quickly enters the body by passing through the skin. Fatal poisoning can occur from skin contact. It is highly toxic. It may damage the developing fetus. Exposure can cause rapid fatal organophosphate poisoning with headaches, sweating, nausea and vomiting, muscle twitching, coma and death. Sometimes effects are delayed for 1 – 2 days. Poisoning can happen from skin contact, even if no irritation is felt.

Long Term Exposure: Cholinesterase inhibitor; cumulative effect is possible. Fenthion may damage the nervous system with repeated exposure, resulting in convulsions, respiratory failure. May cause liver damage.

Points of Attack: Respiratory system, lungs, central nervous system, cardiovascular system, skin, eyes, plasma and red blood cell cholinesterase.

Medical Surveillance: Before employment and at regular times after that, the following are recommended: Plasma and red blood cell cholinesterase levels (tests for the enzyme poisoned by this chemical). If exposure stops, plasma levels return to normal in 1 – 2 weeks while red blood cell levels may be reduced for 1 – 3 months.

When cholinesterase enzyme levels are reduced by 25% or more below preemployment levels, risk of poisoning is increased, even if results are in lower ranges of "normal." Reassignment to work not involving organophosphate or carbamate pesticides is recommended until enzyme levels recover. If symptoms develop or overexposure occurs, repeat the above tests as soon as possible and get an exam of the nervous system. Also consider complete blood count. Consider chest x-ray following acute overexposure. Do not drink any alcoholic beverages before or during use. Alcohol promotes absorption of organic phosphates.

First Aid: If this chemical gets into the eyes, remove any contact lenses at once and irrigate immediately for at least 15 minutes, occasionally lifting upper and lower lids. Seek medical attention immediately. If this chemical contacts the skin, remove contaminated clothing and wash immediately with soap and water. Speed in removing material from skin is of extreme importance. Shampoo hair promptly if contaminated. Seek medical attention immediately. If this chemical has been inhaled, remove from exposure, begin rescue breathing (using universal precautions) if breathing has stopped and CPR if heart action has stopped. Transfer promptly to a medical facility. When this chemical has been swallowed, get medical attention. Give large quantities of water and induce vomiting. Do not make an unconscious person vomit. If exposure and/or symptoms have occurred, the person should be under medical observation for several days, as some symptoms may be delayed.

Personal Protective Methods: Wear protective gloves and clothing to prevent any reasonable probability of skin contact. Safety equipment suppliers/manufacturers can provide recommendations on the most protective glove/clothing material for your operation. All protective clothing (suits, gloves, footwear, headgear) should be clean, available each day, and put on before work. Contact lenses should not be worn when working with this chemical. Wear splash-proof chemical goggles and face shield unless full facepiece respiratory protection is worn. Employees should wash immediately with soap when skin is wet or contaminated. Provide emergency showers and eyewash.

Respirator Selection: Where the potential exists for exposures over 0.2 mg/m^3, use a MSHA/NIOSH approved supplied-air respirator with a full facepiece operated in the positive pressure mode or with a full facepiece, hood, or helmet in the continuous flow mode, or use a MSHA/NIOSH approved self-contained breathing apparatus with a full facepiece operated in pressure-demand or other positive pressure mode.

Storage: Prior to working with this chemical you should be trained on its proper handling and storage. Store in tightly closed containers in a cool, well-ventilated area away from oxidizers, alkalies and alkaline pesticides. A regulated, marked area should be established where this chemical is handled, used, or stored in compliance with OSHA standard 1910.1045.

Shipping: This material may be classed under organophosphorus pesticides, liquid, toxic, n.o.s. This compound requires a shipping label of: "Keep Away From Food." It falls in DOT Hazard Class 6.1 and Pacing Group III.

Spill Handling: Evacuate and restrict persons not wearing protective equipment from area of spill or leak until cleanup is complete. Remove all ignition sources. Ventilate area of spill or leak. Absorb liquids in vermiculite, dry sand, earth, peat, carbon, or a similar material and deposit in sealed containers. It may be necessary to contain and dispose of this chemical as a hazardous waste. If material or contaminated runoff enters waterways, notify downstream users of potentially contaminated waters. Contact your Department of Environmental Protection or your regional office of the federal EPA for specific recommendations. If employees are required to clean-up spills, they must be properly trained and equipped. OSHA 1910.120(q) may be applicable.

Fire Extinguishing: Extinguish fire using an agent suitable for type of surrounding fire. Fenthion itself does not burn readily. Poisonous gases are produced in fire, including oxides of phosphorus and sulfur. Use dry chemical, carbon dioxide, or alcohol foam extinguishers. Vapors are heavier than air and will collect in low areas. Vapors may travel long distances to ignition sources and flashback. Vapors in confined areas may explode when exposed to fire. Containers may explode in fire. Storage containers and parts of containers may rocket great distances, in many directions. If material or contaminated runoff enters waterways, notify downstream users of potentially contaminated waters. Notify local health and fire officials and pollution control agencies. From a secure, explosion-proof location, use water spray to cool exposed containers. If cooling streams are ineffective (venting sound increases in volume and pitch, tank discolors, or shows any signs of deforming), withdraw immediately to a secure position. If employees are expected to fight fires, they must be trained and equipped in OSHA 1910.156.

Disposal Method Suggested: Hydrolysis and landfill for small quantities; incineration with flue gas scrubbing for large amounts.[22] In accordance with 40CFR165 recommendations for the disposal of pesticides and pesticide containers. Must be disposed properly by following package label directions or by contacting your state pesticide or environmental control agency or by contacting your regional EPA office.

References

Sax, N. I., Ed., "Dangerous Properties of Industrial Materials Report," 3, No. 1, 56–61 (1983) New York, Van Nostrand Reinhold Co. (1983)

New Jersey Department of Health and Senior Services, "Hazardous Substance Fact Sheet: Fenthion," Trenton, NJ (May 1986)

Ferbam

Molecular Formula: $C_9H_{18}FeN_3S_6$

Common Formula: $Fe[(CH_3)_2NCS_2]_3Fe$

Synonyms: Aafertis; AI3-14689; Bercema Fertam 50; Carbamic acid, dimethyldithio-, iron salt; Caswell No.458; Dimethylcarbamo dithioic acid, iron complex; Dimethylcarbamodithioic acid, iron(3+) salt; Dimethyldithiocarbamic acid, iron salt; Dimethyldithiocarbamic acid, iron(3+) salt; Eisendimethyldithiocarbamat (German); Eisen(III)-tris(*N,N*-dimethyldithiocarbamat) (German); ENT 14,689; EPA pesticide chemical code 034801; Ferbam 50; Ferbam, iron salt; Ferbeck; Fermate ferbam fungicide; Fermocide; Ferradour; Ferradow; Ferric dimethyl dithiocarbamate; Fuklasin ultra; Hexaferb; Hokmate; Iron dimethyldithiocarbamate; Iron(III) dimethyldithiocarbamate; Iron, tris(dimethylcarbamodithioato-*S,S*'-)-; Iron, tris(dimethylcarbamodithioato-*S,S*')-, (OC-6-11)-; Iron tris(dimethyldithiocarbamate); Iron, tris(dimethyldithiocarbamato)-; Karbam black; Karbam carbamate; Knockmate; Niacide; Stauffer ferbam; Sup'r-Flo Ferbam flowable; Trifungol; (OC-6-11)-Tris(dimethylcarbamodithioato-*S,S*')iron; Tris(dimethylcarbamodithioato-*S,S*')iron; Tris(*N,N*-dimethyldithiocarbamato)iron(III); Tris(dimethyldithiocarbamato)iron; Vancide FE95

CAS Registry Number: 14484-64-1

RTECS Number: NO8750000

DOT ID: UN 2771

EEC Number: 006-051-00-5

Regulatory Authority

- Air Pollutant Standard Set (ACGIH)[1] (DFG)[3] (HSE)[33] (Argentina)[35] (OSHA)[58] (Several States)[60] (Several Canadian Provinces)
- EPA Hazardous Waste Number (RCRA No.): U396
- RCRA, 40CFR261, Appendix 8 Hazardous Constituents
- EPCRA Section 313 Form R *de minimis* concentration reporting level: 1.0%
- Canada, WHMIS, Ingredients Disclosure List

Cited in U.S. State Regulations: Alaska (G), California (A, G), Connecticut (A), Florida (G), Illinois (G), Maine (G), Massachusetts (G), Nevada (A), New Hampshire (G), New Jersey (G), North Dakota (A), Pennsylvania (G), Rhode Island (G), West Virginia (G).

Description: Ferbam, $[(CH_3)_2NCSS]_3Fe$, is a combustible, odorless dark brown to black powder or granular solid. Freezing/Melting point = 180°C (decomposes). Insoluble in water.

Potential Exposure: Those involved in the production, formulation and application of this dithiocarbamate fungicide.

Incompatibilities: Strong oxidizers, strong bases. Heat alkalies (lime), moisture can cause decomposition. Decomposes on prolonged storage.

Permissible Exposure Limits in Air: The Federal standard (OSHA PEL) is 15 ppm (total dust) TWA.[58] NIOSH, the HSE,[33] Argentina,[35] and ACGIH have adopted a TWA value of 10 mg/m^3. The DFG[3] has no MAK at present. The HSE has set an STEL value of 20 mg/m^3. The NIOSH IDLH = 800 mg/m^3. Several states have set guidelines or standards for ferbam in ambient air[60] ranging from 100 μg/m^3 (North Dakota) to 200 μg/m^3 (Connecticut) to 238 μg/m^3 (Nevada).

Determination in Air: Collection by filter; Gravimetric; NIOSH IV, Method #0500, Particulates NOR (total).

Permissible Concentration in Water: No criteria set by degradation produces ethylene thiourea.

Routes of Entry: Inhalation, ingestion, eye and/or skin contact.

Harmful Effects and Symptoms

Short Term Exposure: Ferbam can affect you when breathed in. Breathing ferbam can irritate the nose and throat. Ferbam can cause skin and eye irritation. High exposure to ferbam may affect the nervous system and thyroid; dizziness, confusion, loss of coordination, seizures, paralysis, and coma.

Long Term Exposure: Repeated or prolonged contact with skin may cause allergy with skin rash and itching. Exposure to ferbam may damage the kidneys and liver. Ferbam may damage the developing fetus.

Points of Attack: Eyes, skin, respiratory system, gastrointestinal tract.

Medical Surveillance: For those with frequent or potentially high exposure (half the TLV or greater), the following are recommended before beginning work and at regular times after that: Kidney function tests. Liver function tests. Evaluation by a qualified allergist, including careful exposure history and special testing, may help diagnose skin allergy.

First Aid: If this chemical gets into the eyes, remove any contact lenses at once and irrigate immediately for at least 15 minutes, occasionally lifting upper and lower lids. Seek medical attention immediately. If this chemical contacts the skin, remove contaminated clothing and wash immediately with soap and water. Seek medical attention immediately. If this chemical has been inhaled, remove from exposure, begin rescue breathing (using universal precautions) if breathing has stopped and CPR if heart action has stopped. Transfer promptly to a medical facility. When this chemical has been swallowed, get medical attention. Give large quantities of water and induce vomiting. Do not make an unconscious person vomit.

Personal Protective Methods: Wear protective gloves and clothing to prevent any reasonable probability of skin contact. Safety equipment suppliers/manufacturers can provide recommendations on the most protective glove/clothing material for your operation. All protective clothing (suits, gloves, footwear, headgear) should be clean, available each day, and put on before work. Contact lenses should not be worn when working with this chemical. Wear dust-proof chemical goggles and face shield unless full facepiece respiratory protection is worn. Employees should wash immediately with soap when skin is wet or contaminated. Provide emergency showers and eyewash.

Respirator Selection: NIOSH: *up to 50 mg/m^3:* D (any dust respirator). *Up to 100 mg/^{3}:* DXSQ (any dust respirator except single-use and quarter-mask respirators);* or HiE (any air-purifying respirator with a high-efficiency particulate filter);* or SA (any supplied-air respirator).* *Up to 250 mg/m^3:* SA:CF (any supplied-air respirator operated in a continuous-flow mode);* or PAPRD (any powered, air-purifying respirator with a dust filter).* *Up to 500 mg/m^3*: HiEF (any air-purifying, full-facepiece respirator with a high-efficiency particulate filter); or SAT:CF (any supplied-air respirator that has a tight-fitting facepiece and is operated in a continuous-flow mode);* or PAPRTiE (any powered, air-purifying respirator with a tight-fitting facepiece and a high-efficiency particulate filter);* or SCBAF (any self-contained breathing apparatus with a full facepiece); or SAF (any supplied-air respirator with a full facepiece). *Up to 800 mg/m^3:* SAF:PD:PP (any supplied-air respirator that has a full facepiece and is operated in a pressure-demand or other positive-pressure mode). *Emergency or planned entry in unknown concentration or IDLH conditions:* SCBAF:PD,PP (any MSHA/NIOSH approved self-contained breathing apparatus that has a full facepiece and is operated in a pressure-demand or other positive-pressure mode); or SAF:PD,PP:ASCBA (any supplied-air respirator that has a full facepiece and is operated in a pressure-demand or other positive-pressure mode in combination with an auxiliary, self-contained breathing apparatus operated in a pressure-demand or other positive pressure mode). *Escape:* HiEF [any air-purifying, full-facepiece respirator (gas mask) with a chin-style, front- or back-mounted organic vapor canister having a high-efficiency particulate filter]; or SCBAE (any appropriate escape-type, self-contained breathing apparatus).

* Substance reported to cause eye irritation or damage; may require eye protection.

Storage: Prior to working with this chemical you should be trained on its proper handling and storage. Ferbam is incompatible with strong oxidizers (such as chlorine, bromine and fluorine). Store in tightly closed containers in a cool, well-ventilated area away from oxidizers, strong bases, heat and moisture. Ferbam can decompose upon long-term storage.

Shipping: This can fall into the category of Dithiocarbamate pesticides, solid, toxic, n.o.s. This compound requires a shipping label of: "Keep Away From Food." It falls in DOT Hazard Class 6.1 and Packing Group III.

Spill Handling: Evacuate persons not wearing protective equipment from area of spill or leak until clean-up is complete. Remove all ignition sources. Collect powdered material in the most convenient and safe manner and deposit in sealed containers. Ventilate area after clean-up is complete. It may be necessary to contain and dispose of this chemical as a hazardous waste. If material or contaminated runoff enters waterways, notify downstream users of potentially contaminated waters. Contact your Department of Environmental Protection or your regional office of the federal EPA for specific recommendations. If employees are required to clean-up spills, they must be properly trained and equipped. OSHA 1910.120(q) may be applicable.

Fire Extinguishing: Ferbam may burn, but does not readily ignite. Use dry chemical, CO_2, water spray, or foam extinguishers. Use dry chemical, carbon dioxide, water spray, or alcohol foam extinguishers. Poisonous gases are produced in fire including nitrogen oxides and sulfur oxides. If material or contaminated runoff enters waterways, notify downstream users of potentially contaminated waters. Notify local health and fire officials and pollution control agencies. From a secure, explosion-proof location, use water spray to cool exposed containers. If cooling streams are ineffective (venting sound increases in volume and pitch, tank discolors, or shows any signs of deforming), withdraw immediately to a secure position. If employees are expected to fight fires, they must be trained and equipped in OSHA 1910.156.

Disposal Method Suggested: Ferbam is hydrolyzed by alkali and is unstable to moisture, lime and heat. Ferbam can be incinerated.[22] In accordance with 40CFR165 recommendations for the disposal of pesticides and pesticide containers. Must be disposed properly by following package label directions or by contacting your state pesticide or environmental control agency or by contacting your regional EPA office. Consult with environmental regulatory agencies for guidance on acceptable disposal practices. Generators of waste containing this contaminant (≥100 kg/mo) must conform with EPA regulations governing storage, transportation, treatment, and waste disposal.

References

New Jersey Department of Health and Senior Services, "Hazardous Substance Fact Sheet: Ferbam," Trenton, NJ (April 1999)

Sax, N. I., Ed., "Dangerous Properties of Industrial Materials Report," 1, No. 6, 56–58 (1981) and 8, No. 6, 57–63 (1988)

Ferric Ammonium Citrate

Molecular Formula: $C_{12}H_{18}FeN_2O_{14}$

Common Formula: $FeC_6H_5O_7{\cdot}(NH_4)_2C_6H_5O_7$

Synonyms: Ammonium ferric citrate; Ammonium iron(III) citrate; Citrato ferrico amonico (Spanish); FAC; Ferric ammonium citrate, brown; Ferric ammonium citrate, green; Iron(III) ammonium citrate; Soluble ferric citrate

CAS Registry Number: 1185-57-5

RTECS Number: GE7540000

DOT ID: NA 9118

Regulatory Authority

- Air Pollutant Standard Set (ACGIH)[1] (HSE)[33] (former USSR)[43]
- Clean Water Act: Section 311 Hazardous Substances/RQ 40CFR117.3 (same as CERCLA, see below).
- Superfund/EPCRA 40CFR302.4 Reportable Quantity (RQ): CERCLA, 1,000 lb (454 kg)
- Canada, WHMIS, Ingredients Disclosure List

Cited in U.S. State Regulations: Alaska (G), California (A, G), Maine (G), New Hampshire (G), New Jersey (G), Pennsylvania (G).

Description: Ferric Ammonium Citrate, $FeC_6H_5O_7{\cdot}(NH_4)_2C_6H_5O_7$, forms reddish brown flakes or grains or a brownish-yellow powder. It has a slight ammonia odor. There is also a green form that is odorless. Highly soluble in water.

Potential Exposure: Ferric ammonium citrate is used in blueprinting, photography, medical treatment, and as an animal food additive.

Permissible Exposure Limits in Air: The recommended airborne exposure limit is 1 mg/m^3 (as iron) averaged over an 8-hour workshift. (This exposure limit is recommended for all soluble iron salts.) This is the value set by ACGIH, NIOSH, and by HSE.[33] HSE adds an STEL of 2 mg/m^3. Other countries including Australia, Belgium, Finland, Italy, Netherlands, and Switzerland set the same TWA.

Determination in Air: See OSHA Method ID121.[58] Filter; Acid; Inductively coupled plasma; NIOSH IV, Method #7300, Elements.

Permissible Concentration in Water: The Former USSR-UNEP/IRPTC project[43] has set a MAC for iron in water bodies used for domestic purposes of 0.5 mg/l and a MAC in sea water bodies used for fishery purposes of 0.05 mg/l.

Routes of Entry: Inhalation, ingestion, skin and/or eye contact.

Harmful Effects and Symptoms

Short Term Exposure: Ferric ammonium citrate can affect you when breathed in. Irritates the eyes, skin and respiratory tract. Ingestion can cause abdominal pain, diarrhea and vomiting. Inhaling iron oxide fumes may cause a pneumoconiosis in the lungs. Iron oxide fumes can cause "metal fume fever."

Long Term Exposure: Prolonged or repeated high exposure may cause liver damage Prolonged eye contact can cause a brownish discoloration of the eye.

Points of Attack: Eyes, skin, respiratory system, liver, gastrointestinal tract.

Medical Surveillance: Liver function tests. For those exposed to this chemical, taking dietary supplements or vitamins containing iron is not recommended without medical advice.

First Aid: If this chemical gets into the eyes, remove any contact lenses at once and irrigate immediately for at least 15 minutes, occasionally lifting upper and lower lids. Seek medical attention immediately. If this chemical contacts the skin, remove contaminated clothing and wash immediately with soap and water. Seek medical attention immediately. If this chemical has been inhaled, remove from exposure, begin rescue breathing (using universal precautions) if breathing has stopped and CPR if heart action has stopped. Transfer promptly to a medical facility. When this chemical has been swallowed, get medical attention. Give large quantities of water and induce vomiting. Do not make an unconscious person vomit. The symptoms of metal fume fever may be delayed for 4 – 12 hours following exposure: it may last less than 36 hours.

Note to Physician: In case of fume inhalation, treat pneumonitis. Give prednisone or other corticosteroid orally to reduce tissue response to fume. Positive pressure ventilation may be necessary. Treat metal fume fever with bed rest, analgesics and antipyretics.

Personal Protective Methods: Wear protective gloves and clothing to prevent any reasonable probability of skin contact. Safety equipment suppliers/manufacturers can provide recommendations on the most protective glove/clothing material for your operation. All protective clothing (suits, gloves, footwear, headgear) should be clean, available each day, and put on before work. Contact lenses should not be worn when working with this chemical. Wear dust-proof chemical goggles and face shield unless full facepiece respiratory protection is worn. Employees should wash immediately with soap when skin is wet or contaminated. Provide emergency showers and eyewash.

Respirator Selection: Where the potential exists for exposures over 1 mg/m^3, use a MSHA/NIOSH approved respirator equipped with particulate (dust/fume/mist) filters. Particulate filters must be checked every day before work for physical damage, such as rips or tears, and replaced as needed. Where the potential for high exposures exists, use a MSHA/NIOSH approved supplied-air respirator with a full facepiece operated in the positive pressure mode or with a full facepiece, hood, or helmet in the continuous flow mode, or use a MSHA/NIOSH approved self-contained breathing apparatus with a full facepiece operated in pressure-demand or other positive pressure mode.

Storage: Prior to working with this chemical you should be trained on its proper handling and storage. Store in tightly closed containers in a cool, well-ventilated area away from light. Sources of ignition such as smoking and open flames are prohibited where ferric ammonium citrate is used, handled, or stored in a manner that could create a potential fire or explosion hazard.

Shipping: DOT[19] has no specific citation for iron ammonium citrate in its performance-oriented packaging standards.

Spill Handling: Evacuate persons not wearing protective equipment from area of spill or leak until clean-up is complete. Remove all ignition sources. Absorb liquids containing ferric ammonium citrate in vermiculite, dry sand, earth, or a similar material and deposit in sealed containers. Collect powdered material in the most convenient manner and deposit in sealed containers. Ventilate area after clean-up is complete. It may be necessary to contain and dispose of this chemical as a hazardous waste. If material or contaminated runoff enters waterways, notify downstream users of potentially contaminated waters. Contact your Department of Environmental Protection or your regional office of the federal EPA for specific recommendations. If employees are required to clean-up spills, they must be properly trained and equipped. OSHA 1910.120(q) may be applicable.

Fire Extinguishing: Ferric ammonium citrate may burn, but does not readily ignite. Use dry chemical, CO_2, water spray, or foam extinguishers. Poisonous gases, including ammonia and nitrogen oxides, are produced in fire. If material or contaminated runoff enters waterways, notify downstream users of potentially contaminated waters. Notify local health and fire officials and pollution control agencies. From a secure, explosion-proof location, use water spray to cool exposed containers. If cooling streams are ineffective (venting sound increases in volume and pitch, tank discolors, or shows any signs of deforming), withdraw immediately to a secure position. If employees are expected to fight fires, they must be trained and equipped in OSHA 1910.156.

References

New Jersey Department of Health and Senior Services, "Hazardous Substance Fact Sheet: Ferric Ammonium Citrate," Trenton, NJ (January 1986)

Ferric Ammonium Oxalate

Molecular Formula: $C_6H_{18}FeN_3O_{15}$

Common Formula: $Fe(NH_4)_3(C_2O_4)_3{\cdot}3H_2O$

Synonyms: Ammonium ferric oxalate trihydrate; Ammonium ferrioxalate; Ammonium trioxalatoferrate(3+); Ammonium trioxalatoferrate(III); Ethanedioic acid, ammonium iron(3+) salt; Ethanedioic acid, ammonium iron(III) salt; Oxalato ferrico amonico (Spanish); Oxalic acid, ammonium iron(3+) salt (3:3:1); Oxalic acid, ammonium iron(III) salt (3:3:1); Triammonium tris-(ethanedioato(2-)-*o,o'*)ferrate(3-1)

CAS Registry Number: 14221-47-7; 2944-67-4

RTECS Number: LI8932000

DOT ID: NA 9119

Regulatory Authority

- Air Pollutant Standard Set (ACGIH)[1] (HSE)[33] (Former USSR)[43] (OSHA)[58] (Several Canadian Provinces)
- Clean Water Act: Section 311 Hazardous Substances/RQ 40CFR117.3 (same as CERCLA, see below)
- Superfund/EPCRA 40CFR302.4 Reportable Quantity (RQ): CERCLA, 1,000 lb (454 kg)
- Canada, WHMIS, Ingredients Disclosure List

Cited in U.S. State Regulations: Alaska (G), California (A, G), Maine (G), New Hampshire (G), New Jersey (G); Pennsylvania (G).

Description: Ferric Ammonium Oxalate, $Fe(NH_4)_3(C_2O_4)_3{\cdot}3H_2O$, is a bright green solid with a granular or salt-like appearance. Soluble in water.

Potential Exposure: Ferric ammonium oxalate is used in photography and making blueprints.

Incompatibilities: Strong oxidizers.

Permissible Exposure Limits in Air: The recommended airborne exposure limit is 1 mg/m^3 (as iron) averaged over an 8-hour workshift. (This exposure limit is recommended for all soluble iron salts.) This is the value set by ACGIH, NIOSH, and by HSE.[33] HSE adds an STEL of 2 mg/m^3. Other countries including Australia, Belgium, Finland, Italy, Netherlands, and Switzerland set the same TWA.

Determination in Air: Filter; Acid; Inductively coupled plasma; NIOSH IV, Method #7300, Elements.

Permissible Concentration in Water: The Former USSR-UNEP/IRPTC project[43] has set a MAC for iron in water bodies used for domestic purposes of 0.5 mg/l and a MAC in sea water bodies used for fishery purposes of 0.05 mg/l.

Routes of Entry: Inhalation of dust, ingestion, skin and/or eye contact.

Harmful Effects and Symptoms

Short Term Exposure: Ferric ammonium oxalate can affect you when breathed in. Irritates the eyes, skin and respiratory tract. Ingestion can cause abdominal pain, diarrhea and vomiting. Inhaling iron oxide fumes may cause a pneumoconiosis in the lungs. Iron oxide fumes can cause "metal fume fever;" irritation eyes, skin, mucous membrane; abdominal pain, diarrhea, vomiting.

Long Term Exposure: Prolonged or repeated high exposure may cause liver damage. Prolonged eye contact can cause a brownish discoloration of the eye. May cause fibrosis of the lungs. Repeated overexposure may cause kidney stones.

Points of Attack: Eyes, skin, respiratory system, liver, kidneys, lungs, gastrointestinal tract.

Medical Surveillance: Liver function tests. Kidney function tests. Lung function tests. For those exposed to this chemical, taking dietary supplements or vitamins containing iron is not recommended without medical advice.

First Aid: If this chemical gets into the eyes, remove any contact lenses at once and irrigate immediately for at least 15 minutes, occasionally lifting upper and lower lids. Seek medical attention immediately. If this chemical contacts the skin, remove contaminated clothing and wash immediately with soap and water. Seek medical attention immediately. If this chemical has been inhaled, remove from exposure, begin rescue breathing (using universal precautions) if breathing has stopped and CPR if heart action has stopped. Transfer promptly to a medical facility. When this chemical has been swallowed, get medical attention. Give large quantities of water and induce vomiting. Do not make an unconscious person vomit. The symptoms of metal fume fever may be delayed for 4 – 12 hours following exposure: it may last less than 36 hours.

Personal Protective Methods: Wear protective gloves and clothing to prevent any reasonable probability of skin contact. Safety equipment suppliers/manufacturers can provide recommendations on the most protective glove/clothing material for your operation. All protective clothing (suits, gloves, footwear, headgear) should be clean, available each day, and put on before work. Contact lenses should not be worn when working with this chemical. Wear dust-proof chemical goggles and face shield unless full facepiece respiratory protection is worn. Employees should wash immediately with soap when skin is wet or contaminated. Provide emergency showers and eyewash.

Respirator Selection: Where the potential exists for exposures over 1 mg/m^3, use a MSHA/NIOSH approved full facepiece respirator equipped with particulate (dust/fume/mist) filters. Particulate filters must be checked every day before work for physical damage, such as rips or tears, and replaced as needed. Where the potential for high exposures exists, use a MSHA/NIOSH approved supplied-air respirator with a full facepiece operated in the positive pressure mode or with a full facepiece, hood, or helmet in the continuous flow mode, or use a MSHA/NIOSH approved self-contained breathing apparatus with a full facepiece operated in pressure-demand or other positive pressure mode.

Storage: Prior to working with this chemical you should be trained on its proper handling and storage. Store in tightly closed containers in a cool, well-ventilated area away from light.

Shipping: This material is not specifically cited in DOT's Performance-Oriented Packaging Standards.[19]

Spill Handling: Evacuate persons not wearing protective equipment from area of spill or leak until clean-up is complete. Remove all ignition sources. Collect powdered material in the most convenient and safe manner and deposit in sealed containers. Ventilate area after clean-up is complete. It may be necessary to contain and dispose of this chemical as a hazardous waste. If material or contaminated runoff enters waterways, notify downstream users of potentially contaminated waters. Contact your Department of Environmental Protection or your regional office of the federal EPA for specific recommendations. If employees are required to clean-up spills, they must be properly trained and equipped. OSHA 1910.120(q) may be applicable.

Fire Extinguishing: Ferric ammonium oxalate may burn, but does not readily ignite. Use dry chemical, CO_2, water spray, or foam extinguishers. Poisonous gases are produced in fire. nitrogen oxides and ammonia. If material or contaminated runoff enters waterways, notify downstream users of potentially contaminated waters. Notify local health and fire officials and pollution control agencies. From a secure, explosion-proof location, use water spray to cool exposed containers. If cooling streams are ineffective (venting sound increases in volume and pitch, tank discolors, or shows any signs of deforming), withdraw immediately to a secure position. If employees are expected to fight fires, they must be trained and equipped in OSHA 1910.156.

References

New Jersey Department of Health and Senior Services, "Hazardous Substance Fact Sheet: Ferric Ammonium Oxalate," Trenton, NJ (August 1986)

Ferric Chloride

Molecular Formula: Cl_3Fe

Common Formula: $FeCl_3$

Synonyms: Anhydrous ferric chloride; Chlorure perrique (French); Cloruro ferrico anhidro (Spanish); Flores Martis; Iron chloride; Iron(3+) chloride; Iron(III) chloride; Iron sesquichloride; Iron trichloride; Perchlorure de fer (French); PF etchant

CAS Registry Number: 7705-08-0

RTECS Number: LJ9100000

DOT ID: UN 1773 (anhydrous); UN 2582 (solution)

EINECS Number: 231-729-4

Regulatory Authority

- Air Pollutant Standard Set (ACGIH)[1] (HSE)[33] (former USSR)[35] (OSHA)[58]
- Clean Water Act: Section 311 Hazardous Substances/RQ 40CFR117.3 (same as CERCLA, see below).
- Superfund/EPCRA 40CFR302.4 Reportable Quantity (RQ): CERCLA, 1,000 lb (454 kg)
- Canada, WHMIS, Ingredients Disclosure List

Cited in U.S. State Regulations: Alaska (G), California (A, G), Maine (G), New Hampshire (G), New Jersey (G), New York (G), Pennsylvania (G).

Description: Ferric Chloride, $FeCl_3$, is a black-brown, dark-green, or black crystalline solid. Freezing/Melting point = 306°C. Decomposes at Boiling point = 315°C. Hazard Identification (based on NFPA-704 M Rating System): Health 1, Flammability 0, Reactivity 0. Soluble in water.

Potential Exposure: Iron chloride is used to treat sewage and industrial waste. It is also used in engraving, textiles, photography, as a disinfectant and as a feed additive.

Incompatibilities: Aqueous solutions are a strong acid. Violent reaction with bases, allyl chloride, sulfuric acid, water. Shock- and friction-sensitive explosive material formed with potassium, sodium and other active metals. Attacks metals when wet.

Permissible Exposure Limits in Air: The recommended airborne exposure limit is 1 mg/m^3 (as iron) averaged over an 8-hour workshift. (This exposure limit is recommended for all soluble iron salts.) This is the value set by ACGIH, NIOSH, and by HSE.[33] HSE adds an STEL of 2.0 mg/m^3. Other countries including Australia, Belgium, Finland, Italy, Netherlands, and Switzerland set the same TWA.

Determination in Air: Filter; Acid; Inductively coupled plasma; NIOSH IV, Method #7300, Elements.

Permissible Concentration in Water: The former USSR-UNEP/IRPTC project[43] has set a MAC for iron in water bodies used for domestic purposes of 0.5 mg/l and a MAC in sea water bodies used for fishery purposes of 0.05 mg/l.

Routes of Entry: Inhalation of dust, ingestion, skin and/or eye contact.

Harmful Effects and Symptoms

Short Term Exposure: Ferric chloride is corrosive to the eyes, skin and respiratory tract. Eye contact may cause permanent damage. Inhaling iron oxide fumes may cause a pneumoconiosis in the lungs. Iron oxide fumes can cause "metal fume fever; " irritation eyes, skin, mucous membrane; abdominal pain, diarrhea, vomiting; Ingestion may cause severe irritation to mouth and throat, weak and rapid pulse, low blood-pressure, nausea, bloody vomiting, violent diarrhea, shock, dark purple skin discoloration and coma. Animal studies suggest that death may result from 2½ ounce for a 150 pound person.

Long Term Exposure: Prolonged or repeated high exposure may cause liver damage Prolonged eye contact can cause a brownish discoloration of the eye. Repeated overexposure may cause kidney stones. Excessive intake of iron compounds may result in increased accumulation of iron in body, especially the liver, spleen and lymphatic system. Inhalation of iron dusts may cause mottling of the lung. Iron chloride may reduce the fertility of both males and females.

Points of Attack: Eyes, skin, respiratory system, liver, kidneys, lungs, gastrointestinal tract.

Medical Surveillance: Liver function tests. Kidney function tests. Lung function tests. For those exposed to this chemical, taking dietary supplements or vitamins containing iron is not recommended without medical advice.

First Aid: If this chemical gets into the eyes, remove any contact lenses at once and irrigate immediately for at least 15 minutes, occasionally lifting upper and lower lids. Seek medical attention immediately. If this chemical contacts the skin, remove contaminated clothing and wash immediately with soap and water. Seek medical attention immediately. If this chemical has been inhaled, remove from exposure, begin rescue breathing (using universal precautions) if breathing has stopped and CPR if heart action has stopped. Transfer promptly to a medical facility. When this chemical has been swallowed, give milk and get medical attention. Do not induce vomiting. The symptoms of metal fume fever may be delayed for 4 – 12 hours following exposure: it may last less than 36 hours.

Note to Physician: Gastric lavage should be performed followed by saline catharsis and anodyne. Administer deferoxamine, I.V. Watch for late stricture. Serum, plasma, or urinary iron levels may be employed to estimate amount ingested.

Personal Protective Methods: Wear protective gloves and clothing to prevent any reasonable probability of skin contact. Safety equipment suppliers/manufacturers can provide recommendations on the most protective glove/clothing material for your operation. Neoprene and PVC are among the recommended protective materials. All protective clothing (suits, gloves, footwear, headgear) should be clean, available each day, and put on before work. Contact lenses should not be worn when working with this chemical. Wear dust-proof goggles and face shield when working with powders or dust, unless full facepiece respiratory protection is worn. Wear splash-proof chemical goggles and face shield when working with liquid, unless full facepiece respiratory protection is worn. Employees should wash immediately with soap when skin is wet or contaminated. Provide emergency showers and eyewash.

Respirator Selection: Where the potential exists for exposures over 1 mg/m^3, use a MSHA/NIOSH approved full facepiece respirator with a high efficiency particulate filter. Greater protection is provided by a powdered-air purifying respirator. Where the potential for high exposures exists, use a MSHA/NIOSH approved supplier-air respirator with a full facepiece operated in the positive pressure mode or with a full facepiece, hood, or helmet in the continuous flow mode, or use a MSHA/NIOSH approved self-contained breathing apparatus with a full facepiece operated in pressure-demand or other positive pressure mode.

Storage: Prior to working with this chemical you should be trained on its proper handling and storage. Iron chloride must be stored in tightly closed containers to avoid contact with sulfuric acid, sodium, potassium, allyl chloride and water since violent reactions occur and toxic vapors may be produced.

Shipping: This compound requires a shipping label of: "Corrosive." It falls in DOT Hazard Class 8 and Packing Group III.

Spill Handling: *Liquid:* Evacuate and restrict persons not wearing protective equipment from area of spill or leak until cleanup is complete. Remove all ignition sources. Ventilate area of spill or leak. Absorb liquids in vermiculite, dry sand, earth, peat, carbon, or a similar material and deposit in sealed containers. It may be necessary to contain and dispose of this chemical as a hazardous waste. If material or contaminated runoff enters waterways, notify downstream users of potentially contaminated waters. Contact your Department of Environmental Protection or your regional office of the federal EPA for specific recommendations. If employees are required to clean-up spills, they must be properly trained and equipped. OSHA 1910.120(q) may be applicable.

Powder: Evacuate persons not wearing protective equipment from area of spill or leak until clean-up is complete. Remove all ignition sources. Collect powdered material in the most convenient and safe manner and deposit in sealed containers. Ventilate area after clean-up is complete. It may be necessary to contain and dispose of this chemical as a hazardous waste. If material or contaminated runoff enters waterways, notify downstream users of potentially contaminated waters. Contact your Department of Environmental Protection or your regional office of the federal EPA for specific recommendations. If employees are required to clean-up spills, they must be properly trained and equipped. OSHA 1910.120(q) may be applicable.

Fire Extinguishing: This chemical is a noncombustible solid. Use any extinguishing agents suitable for surrounding fire. Poisonous gases are produced in fire including hydrogen chloride. If material or contaminated runoff enters waterways, notify downstream users of potentially contaminated waters. Notify local health and fire officials and pollution control agencies. From a secure, explosion-proof location, use water spray to cool exposed containers. If cooling streams are ineffective (venting sound increases in volume and pitch, tank discolors, or shows any signs of deforming), withdraw immediately to a secure position. If employees are expected to fight fires, they must be trained and equipped in OSHA 1910.156.

Disposal Method Suggested: Neutralize with lime or soda ash and bury in an approved landfill.[22]

References

Sax, N. I., Ed., "Dangerous Properties of Industrial Materials Report," 3, No. 4, 42–45 (1983)

New Jersey Department of Health and Senior Services, "Hazardous Substance Fact Sheet: Iron Chloride," Trenton, NJ (April, 1997)

New York State Department of Health, "Chemical Fact Sheet: Iron (III) Chloride," Albany, NY, Bureau of Toxic Substance Assessment (February 1986)

Ferric Nitrate

Molecular Formula: C_3FeO_9

Common Formula: $Fe(CO_3)_3$

Synonyms: Ferric(3+) nitrate; Ferric(III) nitrate; Ferric nitrate, Nonhydrate; Iron nitrate; Iron(3+) nitrate, anhydrous; Iron(III) nitrate, anhydrous; Iron trinitrate; Nitrato ferrico (Spanish); Nitric acid, iron(3+) salt; Nitric acid, iron(III) salt

CAS Registry Number: 10421-48-4

RTECS Number: QU8915000

DOT ID: UN 1466

Regulatory Authority

- Air Pollutant Standard Set (ACGIH)[1] (HSE)[33]
- Clean Water Act: Section 311 Hazardous Substances/RQ 40CFR117.3 (same as CERCLA, see below)
- Superfund/EPCRA 40CFR302.4 Reportable Quantity (RQ): CERCLA, 1,000 lb (454 kg)
- U.S. DOT Regulated Marine Pollutant (49CFR172.101, Appendix B)
- Canada, WHMIS, Ingredients Disclosure List

Cited in U.S. State Regulations: Alaska (G), California (A, G), Maine (G), New Hampshire (G), New Jersey (G), Pennsylvania (G).

Description: Ferric nitrate, $Fe(NO_3)_3$, is a pale violet, green, or grayish-white, odorless solid in lumpy crystals (like salt). Freezing/Melting point = 35°C; 47°C (nonhydrate). Highly soluble in water.

Potential Exposure: It is used in textile dyeing, tanning, and weighting silk.

Incompatibilities: An oxidizer. Oxidizable materials and combustibles including aluminum powder, sulfur and organic materials. Aqueous solution is corrosive to metals.

Permissible Exposure Limits in Air: The recommended airborne exposure limit is 1 mg/m^3 (as iron) averaged over an 8-hour workshift. (This exposure limit is recommended for all soluble iron salts.) This is the value set by ACGIH, NIOSH, and by HSE.[33] HSE adds an STEL of 2.0 mg/m^3. Other countries including Australia, Belgium, Finland, Italy, Netherlands, and Switzerland set the same TWA.

Determination in Air: Filter; Acid; Inductively coupled plasma; NIOSH IV, Method #7300, Elements.

Permissible Concentration in Water: The former USSR-UNEP/IRPTC project[43] has set a MAC in water bodies used for domestic purposes of 0.5 mg/l and in seawater for fishery purposes of 0.05 mg/l (as iron).

Routes of Entry: Inhalation, ingestion, skin and/or eye contact.

Harmful Effects and Symptoms

Short Term Exposure: Ferric nitrate can affect you when breathed in. Exposure can irritate the eyes, nose, throat, and skin. Large amounts of iron in the body can cause nausea, stomach pain, constipation, and black bowel movements. Ingestion can cause abdominal pain, diarrhea and vomiting. Inhaling iron oxide fumes may cause a pneumoconiosis in the lungs. Iron oxide fumes can cause "metal fume fever." irritation eyes, skin, mucous membrane; abdominal pain, diarrhea, vomiting.

Long Term Exposure: Prolonged or repeated high exposure may cause liver damage Prolonged eye contact can cause a brownish discoloration of the eye. May cause lung fibrosis. Repeated overexposure may cause kidney stones.

Points of Attack: Eyes, skin, respiratory system, liver, lungs, gastrointestinal tract.

Medical Surveillance: If illness occurs or overexposure is suspected, medical attention is recommended. Blood test for serum iron can detect excess body iron. If increased, consider liver function tests. Examination of the eyes. Lung function tests.

First Aid: If this chemical gets into the eyes, remove any contact lenses at once and irrigate immediately for at least 15 minutes, occasionally lifting upper and lower lids. Seek medical attention immediately. If this chemical contacts the skin, remove contaminated clothing and wash immediately with soap and water. Seek medical attention immediately. If this chemical has been inhaled, remove from exposure, begin rescue breathing (using universal precautions) if breathing has stopped and CPR if heart action has stopped. Transfer promptly to a medical facility. When this chemical has been swallowed, get medical attention. Give large quantities of water and induce vomiting. Do not make an unconscious person vomit. The symptoms of metal fume fever may be delayed for 4 – 12 hours following exposure: it may last less than 36 hours.

Personal Protective Methods: Wear protective gloves and clothing to prevent any reasonable probability of skin contact. Safety equipment suppliers/manufacturers can provide recommendations on the most protective glove/clothing material for your operation. All protective clothing (suits, gloves, footwear, headgear) should be clean, available each day, and put on before work. Contact lenses should not be worn when working with this chemical. Wear dust-proof chemical goggles and face shield unless full facepiece respiratory protection is worn. Employees should wash immediately with soap when skin is wet or contaminated. Provide emergency showers and eyewash.

Respirator Selection: Where the potential exists for exposures over 1 mg/m^3, use an MSHA/NIOSH approved respirator equipped with particulate (dust/fume/mist) filters. Particulate filters must be checked every day before work for physical damage, such as rips or tears, and replaced as needed. Where the potential for high exposures exists, use an MSHA/NIOSH approved supplied-air respirator with a full facepiece operated in the positive pressure mode or with a full facepiece, hood, or helmet in the continuous flow mode, or use an MSHA/NIOSH approved self-contained breathing apparatus with a full facepiece operated in pressure-demand or other positive pressure mode.

Storage: Prior to working with this chemical you should be trained on its proper handling and storage. Store in tightly closed containers in a cool, well-ventilated area away from organic materials and other combustible materials or aluminum powder. See OSHA standard 1910.104 and NFPA 43A *Code for the Storage of Liquid and Solid Oxidizers* for detailed handling and storage regulations.

Shipping: This compound requires a shipping label of: "Oxidizer." It falls in DOT Hazard Class 5.1 and Packing Group III.

Spill Handling: Evacuate persons not wearing protective equipment from area of spill or leak until clean-up is complete. Remove all ignition sources. Absorb liquids in vermiculite, dry sand, earth, or a similar material and deposit in sealed containers. Collect powdered material in the most convenient manner and deposit in sealed containers. It may be necessary to contain and dispose of this chemical as a hazardous waste. If material or contaminated runoff enters waterways, notify downstream users of potentially contaminated waters. Contact your Department of Environmental Protection or your regional office of the federal EPA for specific recommendations. If employees are required to clean-up spills, they must be properly trained and equipped. OSHA 1910.120(q) may be applicable.

Fire Extinguishing: This chemical is a noncombustible solid. Use extinguishers suitable for surrounding fire. Poisonous gases are produced in fire including nitrogen oxides, nitric acid. If material or contaminated runoff enters waterways, notify downstream users of potentially contaminated waters. Notify local health and fire officials and pollution control agencies. From a secure, explosion-proof location, use water spray to cool exposed containers. If cooling streams are ineffective (venting sound increases in volume and pitch, tank discolors, or shows any signs of deforming), withdraw immediately to a secure position. If employees are expected to fight fires, they must be trained and equipped in OSHA 1910.156.

References

New Jersey Department of Health and Senior Services, "Hazardous Substance Fact Sheet: Ferric Nitrate," Trenton, NJ (August 1985)

Ferric Sulfate

Molecular Formula: $Fe_2O_{12}S_3$

Common Formula: $Fe_2(SO_4)_3$

Synonyms: Diiron trisulfate; Greenmaster autumn; Iron persulfate; Iron sesquisulfate; Iron sulfate (2:3); Iron(3+) sulfate; Iron(III) sulfate; Iron tersulfate; Maxicrop Moss Killer; Sulfato ferrico (Spanish); Sulfuric acid, iron(3+) salt (3:2); Sulfuric acid, iron(III) salt (3:2); Vitax micro gran; Vitax Turf tonic

CAS Registry Number: 10028-22-5

RTECS Number: NO8505000

DOT ID: NA 9121

Regulatory Authority

- Air Pollutant Standard Set (ACGIH)[1] (HSE)[33]
- Clean Water Act: Section 311 Hazardous Substances/RQ 40CFR117.3 (same as CERCLA, see below)
- Superfund/EPCRA 40CFR302.4 Reportable Quantity (RQ): CERCLA, 1,000 lb (454 kg)
- Canada, WHMIS, Ingredients Disclosure List

Cited in U.S. State Regulations: Alaska (G), California (A, G), Maine (G), New Hampshire (G), New Jersey (G), New York (G), Pennsylvania (G).

Description: Ferric Sulfate, $Fe_2(SO_4)_3$, is a grayish-white powder or yellow lumpy crystals. Freezing/Melting point = 480°C (decomposes). Hazard Identification (based on NFPA-704 M Rating System): Health 1, Flammability 0, Reactivity 0. Slightly soluble in water.

Potential Exposure: This material is used in pigments, textile dyeing, water treatment, and metal pickling.

Incompatibilities: Hydrolyzed slowly in aqueous solution. Incompatible with magnesium, aluminum. Corrosive to copper and its alloys, mild and galvanized steel. Light sensitive.

Permissible Exposure Limits in Air: The recommended airborne exposure limit is 1 mg/m^3 (as iron) averaged over an 8-hour workshift (This exposure limit is recommended for all soluble iron salts). This is the value set by ACGIH, NIOSH, and by HSE.[33] HSE adds an STEL of 2.0 mg/m^3. Other countries including Australia, Belgium, Finland, Italy, Netherlands, and Switzerland set the same TWA.

Determination in Air: Filter; Acid; Inductively coupled plasma; NIOSH IV, Method #7300, Elements.

Permissible Concentration in Water: The former USSR-UNEP/IRPTC project[43] has set a MAC for iron in water bodies used for domestic purposes of 0.5 mg/l and a MAC in sea water bodies used for fishery purposes of 0.05 mg/l.

Routes of Entry: Inhalation, ingestion, skin and/or eye contact.

Harmful Effects and Symptoms

Short Term Exposure: *Inhalation:* May cause irritation of nose and throat, coughing and difficulty in breathing. 0.075 mg/m^3 for 2 hours did not cause any change in breathing functions. Inhaling iron oxide fumes may cause a pneumoconiosis in the lungs. Iron oxide fumes can cause "metal fume fever." irritation eyes, skin, mucous membrane; abdominal pain, diarrhea, vomiting.

Skin: Contact causes irritation. Remove promptly.

Eyes: Contact causes irritation.

Ingestion: May cause irritation of mouth and stomach, nausea, vomiting, diarrhea, drowsiness, liver damage, coma and death. The estimated lethal dose is 30 kg (1 ounce).

Long Term Exposure: Excessive intake of iron compounds may result in increased accumulations of iron in body, especially the liver, spleen and lymphatic system. May cause nausea, vomiting, stomach pain, constipation, and black bowel movements. Inhalation of iron dusts may cause mottling of the lung. Prolonged or repeated high exposure may cause liver damage. Prolonged eye contact can cause a brownish discoloration of the eye. Repeated overexposure may cause kidney stones.

Points of Attack: Eyes, skin, respiratory system, liver, lungs, gastrointestinal tract.

Medical Surveillance: If symptoms develop or overexposure is suspected, the following may be useful: Blood test for iron level (serum iron). Liver function tests. Lung function tests. For those exposed to this chemical, taking dietary supplements or vitamins containing iron is not recommended without medical advice.

First Aid: If this chemical gets into the eyes, remove any contact lenses at once and irrigate immediately for at least 15 minutes, occasionally lifting upper and lower lids. Seek medical attention immediately. If this chemical contacts the skin, remove contaminated clothing and wash immediately with soap and water. Seek medical attention immediately. If this chemical has been inhaled, remove from exposure, begin rescue breathing (using universal precautions) if breathing has stopped and CPR if heart action has stopped. Transfer promptly to a medical facility. When this chemical has been swallowed, get medical attention. Give large quantities of water and induce vomiting. Do not make an unconscious person vomit. The symptoms of metal fume fever may be delayed for 4 – 12 hours following exposure: it may last less than 36 hours.

Personal Protective Methods: Wear protective gloves and clothing to prevent any reasonable probability of skin contact. Safety equipment suppliers/manufacturers can provide recommendations on the most protective glove/clothing material for your operation. All protective clothing (suits, gloves, footwear, headgear) should be clean, available each day, and put on before work. Contact lenses should not be worn when working with this chemical. Wear dust-proof chemical goggles and face shield unless full facepiece respiratory protection is worn. Employees should wash immediately with soap when skin is wet or contaminated. Provide emergency showers and eyewash.

Respirator Selection: Where the potential exists for exposures over 1 mg/m^3 of iron, use an MSHA/NIOSH approved respirator equipped with particulate (dust/fume/mist) filters. Where the potential for high exposures exists, use an MSHA/NIOSH approved supplied-air respirator with a full facepiece operated in the positive pressure mode or with a full facepiece, hood, or helmet in the continuous flow mode, or use an MSHA/NIOSH approved self-contained breathing apparatus with a full facepiece operated in pressure-demand or other positive pressure mode.

Storage: Prior to working with this chemical you should be trained on its proper handling and storage. Store in tightly closed containers in a cool, well-ventilated area away from light, moisture, aluminum, magnesium, copper and its alloys, zinc, galvanized and mild steels.

Shipping: Ferric sulfate is not specifically cited by DOT[19] in its Performance-Oriented Packaging standards.

Spill Handling: Evacuate and restrict persons not wearing protective equipment from area of spill or leak until cleanup is complete. Remove all ignition sources. Ventilate area of spill or leak. Absorb liquids containing ferric sulfate in vermiculite, dry sand, earth, or a similar material and deposit in sealed containers. Collect powdered material in the most convenient manner and deposit in sealed containers. It may be necessary to contain and dispose of this chemical as a hazardous waste. If material or contaminated runoff enters waterways, notify downstream users of potentially contaminated waters. Contact your Department of Environmental Protection or your regional office of the federal EPA for specific recommendations. If employees are required to clean-up spills, they must be properly trained and equipped. OSHA 1910.120(q) may be applicable.

Fire Extinguishing: This chemical is a noncombustible solid. Use extinguishers suitable for surrounding fire. Poisonous gases are produced in fire including sulfur oxide. If material or contaminated runoff enters waterways, notify downstream users of potentially contaminated waters. Notify local health and fire officials and pollution control agencies. From a secure, explosion-proof location, use water spray to cool exposed containers. If cooling streams are ineffective (venting sound increases in volume and pitch, tank discolors, or shows any signs of deforming), withdraw immediately to a secure position. If employees are expected to fight fires, they must be trained and equipped in OSHA 1910.156.

Disposal Method Suggested: Treat with soda ash or dilute NaOH. Separate any precipitate and landfill. Flush solution to sewer.[22]

References

Sax, N. I., Ed., "Dangerous Properties of Industrial Materials Report," 3, No. 4, 45–47 (1983) and 7, No. 2, 75–79 (1987)

New Jersey Department of Health and Senior Services, "Hazardous Substance Fact Sheet: Ferric Sulfate," Trenton, NJ (March 1999)

New York State Department of Health, "Chemical Fact Sheet: Iron (III) Sulfate," Albany, NY, Bureau of Toxic Substance Assessment (March 1986)

Ferrocene

Molecular Formula: $C_{10}H_{10}Fe$

Synonyms: Biscyclopentadienyliron; Di-2,4-cyclopentadien-1-yl iron; Dicyclopentadienyl iron; Iron bis(cyclopentadiene); Iron dicyclopentadienyl

CAS Registry Number: 102-54-5

RTECS Number: LK0700000

DOT ID: No citation.

Cited in U.S. State Regulations: California (G), New Jersey (G), Pennsylvania (G).

Description: Ferrocene is a bright orange salt-like crystals from alcohol with a camphor odor. Boiling point = 249°C. (sublimes). Freezing/Melting point = 173°C. Insoluble in water.

Potential Exposure: An additive in fuel oil, antiknock agent in gasoline fuel, used in making rubber, silicone resins and high temperature polymers high-temperature lubricant, intermediate for high-temperature polymers.

Incompatibilities: Ammonium perchlorate.

Permissible Exposure Limits in Air: ACGIH recommends a TWA of 10 mg/m^3.

Determination in Air: Filter; Acid; Inductively coupled plasma; NIOSH IV, Method #7300, Elements. Ferrocene is approximately 30% iron.

Routes of Entry: Inhalation.

Harmful Effects and Symptoms

Short Term Exposure: Eye contact can cause irritation. Dust may irritate the air passages, with possible cough, phlegm and tightness in the chest.

Long Term Exposure: There is limited evidence may damage the testes. Repeated exposure to iron containing compounds can damage the liver and other body organs. Heavy or repeated exposure may cause mood changes such as irritability. Repeated exposure may cause lung damage.

Points of Attack: Lungs.

Medical Surveillance: Lung function tests. Serum iron level.

First Aid: If this chemical gets into the eyes, remove any contact lenses at once and irrigate immediately for at least 15 minutes, occasionally lifting upper and lower lids. Seek medical attention immediately. If this chemical contacts the skin, remove contaminated clothing and wash immediately with soap and water. Seek medical attention immediately. If this chemical has been inhaled, remove from exposure, begin rescue breathing (using universal precautions) if breathing has stopped and CPR if heart action has stopped. Transfer promptly to a medical facility. When this chemical has been swallowed, get medical attention. Give large quantities of water and induce vomiting. Do not make an unconscious person vomit.

Personal Protective Methods: Wear protective gloves and clothing to prevent any reasonable probability of skin contact. Safety equipment suppliers/manufacturers can provide recommendations on the most protective glove/clothing material for your operation. All protective clothing (suits, gloves, footwear, headgear) should be clean, available each day, and put on before work. Contact lenses should not be worn when working with this chemical. Wear dust-proof chemical goggles and face shield unless full facepiece respiratory protection is worn. Employees should wash immediately with soap when skin is wet or contaminated. Provide emergency showers and eyewash.

Respirator Selection: Where the potential for exposure to this chemical, use a MSHA/NIOSH approved supplied-air respirator with a full facepiece operated in the positive pressure mode or with a full facepiece, hood, or helmet in the continuous flow mode, or use a MSHA/NIOSH approved self-contained breathing apparatus with a full facepiece operated in pressure-demand or other positive pressure mode.

Storage: Prior to working with this chemical you should be trained on its proper handling and storage. Store in tightly closed containers in a cool, well ventilated area away from ammonium perchlorate.

Spill Handling: Evacuate persons not wearing protective equipment from area of spill or leak until clean-up is complete. Remove all ignition sources. Collect powdered material in the most convenient and safe manner and deposit in sealed containers. Ventilate area after clean-up is complete. It may be necessary to contain and dispose of this chemical as a hazardous waste. If material or contaminated runoff enters waterways, notify downstream users of potentially contaminated waters. Contact your Department of Environmental Protection or your regional office of the federal EPA for specific recommendations. If employees are required to clean-up spills, they must be properly trained and equipped. OSHA 1910.120(q) may be applicable.

Fire Extinguishing: This chemical is a combustible solid. Use dry chemical, carbon dioxide, water spray, or alcohol foam extinguishers. Poisonous gases are produced in fire. If material or contaminated runoff enters waterways, notify downstream users of potentially contaminated waters. Notify local health and fire officials and pollution control agencies. From a secure, explosion-proof location, use water spray to cool exposed containers. If cooling streams are ineffective (venting sound increases in volume and pitch, tank discolors, or shows any signs of deforming), withdraw immediately to a secure position. If employees are expected to fight fires, they must be trained and equipped in OSHA 1910.156.

References

New Jersey Department of Health and Senior Services, "Hazardous Substance Fact Sheet, Dicyclopentadienyl Iron," Trenton, NJ (April 1986)

Ferrous Ammonium Sulfate

Molecular Formula: $Fe_2H_8N_2O_4S$

Common Formula: $Fe_2(NH_4)_2SO_4$; $Fe(NH_4)_2(SO_4)_2{\cdot}6H_2O$; $O_8S_2{\cdot}H_8N_2{\cdot}Fe$

Synonyms: Ammonium iron sulfate; Ammonium iron sulphate; Iron ammonium sulfate; Iron ammonium sulphate; Mohr's salt; Sulfato ferroso amonico (Spanish); Sulfuric acid, ammonium iron(2+), salt (2:2:1); Sulphate ammonium sulfate hexahydrate; Sulphate ammonium sulphate; Sulphate ammonium sulphate hexahydrate

CAS Registry Number: 10045-89-3

RTECS Number: WS5850000

DOT ID: NA 9122

Regulatory Authority

- Air Pollutant Standard Set (ACGIH)[1] (HSE)[33] (OSHA)[58]
- Clean Water Act: Section 311 Hazardous Substances/RQ 40CFR117.3 (same as CERCLA, see below)
- Superfund/EPCRA 40CFR302.4 Reportable Quantity (RQ): CERCLA, 1,000 lb (454 kg)
- Canada, WHMIS, Ingredients Disclosure List

Cited in U.S. State Regulations: Alaska (G), California (G), Maine (G), New Hampshire (G), New Jersey (G), Pennsylvania (G).

Description: Ferrous ammonium sulfate, $Fe_2(NH_4)_2SO_4$, is a pale green or blue-green solid (powder or lumpy crystals). Soluble in water.

Potential Exposure: This substance is used in photography, analytical chemistry and in dosimeters.

Permissible Exposure Limits in Air: The recommended airborne exposure limit is 1 mg/m^3 (as iron) averaged over an 8-hour workshift. (This exposure limit is recommended for all soluble iron salts.) This is the value set by ACGIH, NIOSH, and by HSE.[33] HSE adds an STEL of 2.0 mg/m^3. Other countries including Australia, Belgium, Finland, Italy, Netherlands, and Switzerland set the same TWA.

Determination in Air: Filter; Acid; Inductively coupled plasma; NIOSH IV, Method #7300, Elements.

Permissible Concentration in Water: The former USSR-UNEP/IRPTC project[43] has set a MAC for iron in water bodies used for domestic purposes of 0.5 mg/l and a MAC in sea water bodies used for fishery purposes of 0.05 mg/l.

Routes of Entry: Inhalation, ingestion, skin and/or eye contact.

Harmful Effects and Symptoms

Short Term Exposure: Ferrous ammonium sulfate can affect you when breathed in. Exposure can irritate the eyes, nose and throat. Large amounts of iron in the body can cause nausea and stomach pain, constipation, and black bowel movements. This is more common with swallowed iron. Repeated high exposures may damage the liver. Exposure may irritate the lungs.

Long Term Exposure: Prolonged or repeated high exposure may cause liver damage. Prolonged eye contact can cause a brownish discoloration of the eye. Repeated overexposure may cause kidney stones.

Points of Attack: Eyes, skin, respiratory system, liver, gastrointestinal tract.

Medical Surveillance: For those with frequent or potentially high exposure (half the TLV or grater), the following are recommended before beginning work and at regular times after that: Lung function tests. If symptoms develop or overexposure has occurred, the following may be useful: Blood tests for iron level (serum iron). Liver function tests. For those exposed to this chemical, taking dietary supplements or vitamins containing iron is not recommended without medical advice.

First Aid: If this chemical gets into the eyes, remove any contact lenses at once and irrigate immediately for at least 15 minutes, occasionally lifting upper and lower lids. Seek medical attention immediately. If this chemical contacts the skin, remove contaminated clothing and wash immediately with soap and water. Seek medical attention immediately. If this chemical has been inhaled, remove from exposure, begin rescue breathing (using universal precautions) if breathing has stopped and CPR if heart action has stopped. Transfer promptly to a medical facility. When this chemical has been swallowed, get medical attention. Give large quantities of water and induce vomiting. Do not make an unconscious person vomit.

Personal Protective Methods: Wear protective gloves and clothing to prevent any reasonable probability of skin contact. Safety equipment suppliers/manufacturers can provide recommendations on the most protective glove/clothing material for your operation. All protective clothing (suits, gloves, footwear, headgear) should be clean, available each day, and put on before work. Contact lenses should not be worn when working with this chemical. Wear dust-proof chemical goggles and face shield unless full facepiece respiratory protection is worn. Employees should wash immediately with soap when skin is wet or contaminated. Provide emergency showers and eyewash.

Respirator Selection: Where the potential exists for exposures over 1 mg/m^3, use a MSHA/NIOSH approved respirator equipped with particulate (dust/fume/mist) filters. Particulate filters must be checked every day before work for physical damage, such as rips or tears, and replaced as needed. Where the potential for high exposures exists, use a MSHA/NIOSH approved supplied-air respirator with a full facepiece operated in the positive pressure mode o with a full facepiece, hood, or helmet in the continuous flow mode, or use a MSHA/NIOSH approved self-contained breathing apparatus with a full facepiece operated in pressure-demand or other positive pressure mode.

Storage: Prior to working with this chemical you should be trained on its proper handling and storage. Store in tightly closed containers in a cool, well-ventilated area away from light.

Shipping: This compound is not specifically cited in DOT's Performance-Oriented Packaging Standards.[19]

Spill Handling: Evacuate and restrict persons not wearing protective equipment from area of spill or leak until cleanup is complete. Remove all ignition sources. Ventilate area of spill or leak. Absorb liquids containing ferrous ammonium sulfate in vermiculite, dry sand, earth, or a similar material and deposit in sealed containers. Collect powdered material in the most convenient manner and deposit in sealed containers. It may be necessary to contain and dispose of this chemical as a hazardous waste. If material or contaminated runoff enters waterways, notify downstream users of potentially contaminated waters. Contact your Department of Environmental Protection or your regional office of the federal EPA for specific recommendations. If employees are required to clean-up spills, they must be properly trained and equipped. OSHA 1910.120(q) may be applicable.

Fire Extinguishing: This chemical is a noncombustible solid. Use extinguishers suitable for surrounding fire. Poisonous gases are produced in fire. If material or contaminated runoff enters waterways, notify downstream users of potentially contaminated waters. Notify local health and fire officials and pollution control agencies. From a secure, explosion-proof location, use water spray to cool exposed containers. If cooling streams are ineffective (venting sound increases in volume and pitch, tank discolors, or shows any signs of deforming), withdraw immediately to a secure position. If employees are expected to fight fires, they must be trained and equipped in OSHA 1910.156.

References

New Jersey Department of Health and Senior Services, "Hazardous Substance Fact Sheet: Ferrous Ammonium Sulfate," Trenton, NJ (January 1986)

Ferrous Chloride

Molecular Formula: Cl_2Fe

Common Formula: $FeCl_2$

Synonyms: Iron(II) chloride (1:2); Iron dichloride; Iron protochloride; Lawrencite; Sulphate chloride tetrahydrate

CAS Registry Number: 7758-94-3

RTECS Number: NO5400000

DOT ID: NA 1759 (solid); NA 1760 (solution)

Regulatory Authority

- Air Pollutant Standard Set (ACGIH)[1] (HSE)[33]
- Clean Water Act: Section 311 Hazardous Substances/RQ 40CFR117.3 (same as CERCLA, see below)
- Superfund/EPCRA 40CFR302.4 Reportable Quantity (RQ): CERCLA, 100 lb (45.4 kg)
- Canada, WHMIS, Ingredients Disclosure List

Cited in U.S. State Regulations: Alaska (G), California (A, G), Maine (G), New Hampshire (G), New Jersey (G), Pennsylvania (G).

Description: Ferrous chloride, $FeCl_2$, is a pale greenish salt-like crystal or power. Boiling point = 1,012°C. Freezing/Melting point = 676°C. Hazard Identification (based on NFPA-704 M Rating System): Health 2, Flammability 0, Reactivity 0. Soluble in water.

Potential Exposure: It is used in textile dyeing, metallurgy, the pharmaceutical industry and sewage treatment.

Incompatibilities: Solution attacks metals. Contact with ethylene oxide may initiate polymerization. Contact with potassium or sodium forms an impact-sensitive material.

Permissible Exposure Limits in Air: The recommended airborne exposure limit is 1 mg/m^3 (as iron) averaged over an 8-hour workshift. (This exposure limit is recommended for all soluble iron salts.) This is the value set by ACGIH, NIOSH, and by HSE.[33] HSE adds an STEL of 2.0 mg/m^3. Other countries including Australia, Belgium, Finland, Italy, Netherlands, and Switzerland set the same TWA.

Determination in Air: Filter; Acid; Inductively coupled plasma; NIOSH IV, Method #7300, Elements.

Permissible Concentration in Water: The former USSR-UNEP/IRPTC project[43] has set a MAC for iron in water bodies used for domestic purposes of 0.5 mg/l and a MAC in sea water bodies used for fishery purposes of 0.05 mg/l.

Routes of Entry: Inhalation, ingestion, skin and/or eye contact.

Harmful Effects and Symptoms

Short Term Exposure: Ferrous chloride can affect you when breathed in. Repeated or high level exposures may lead to too much iron in the body and possible liver damage. Exposure can irritate the nose and throat. Ferrous chloride is a corrosive chemical and contact can irritate and may burn the eyes and skin.

Long Term Exposure: Prolonged or repeated high exposure may cause liver damage Prolonged eye contact can cause a brownish discoloration of the eye. Repeated overexposure may cause kidney stones.

Points of Attack: Eyes, skin, respiratory system, liver, gastrointestinal tract.

Medical Surveillance: If symptoms develop or overexposure is suspected, the following may be useful: Serum iron test. Liver function test. For those exposed to this chemical, taking dietary supplements or vitamins containing iron is not recommended without medical advice.

First Aid: If this chemical gets into the eyes, remove any contact lenses at once and irrigate immediately for at least 15 minutes, occasionally lifting upper and lower lids. Seek medical attention immediately. If this chemical contacts the skin, remove contaminated clothing and wash immediately with soap and water. Seek medical attention immediately. If this chemical has been inhaled, remove from exposure, begin rescue breathing (using universal precautions) if breathing has stopped and CPR if heart action has stopped. Transfer promptly to a medical facility. When this chemical has been swallowed, get medical attention. If victim is conscious, administer water or milk. Do not induce vomiting.

Personal Protective Methods: Wear protective gloves and clothing to prevent any reasonable probability of skin contact. Safety equipment suppliers/manufacturers can provide recommendations on the most protective glove/clothing material for your operation. All protective clothing (suits, gloves, footwear, headgear) should be clean, available each day, and put on before work. Contact lenses should not be worn when working with this chemical. Wear dust-proof chemical goggles and face shield unless full facepiece respiratory protection is worn. Employees should wash immediately with soap when skin is wet or contaminated. Provide emergency showers and eyewash.

Respirator Selection: Where the potential exists for exposures over 1 mg/m^3, use a MSHA/NIOSH approved full facepiece respirator equipped with particulate (dust/fume/mist) filters. Particulate filters must be checked every day before work for physical damage, such as rips or tears, and replaced as needed. Where the potential for high exposures exists, use a MSHA/NIOSH approved supplied-air respirator with a full facepiece operated in the positive pressure mode or with a full facepiece, hood, or helmet in the continuous flow mode, or use a MSHA/NIOSH approved self-contained breathing apparatus with a full facepiece operated in pressure-demand or other positive pressure mode.

Storage: Prior to working with this chemical you should be trained on its proper handling and storage. Store in tightly closed containers in a cool, well-ventilated area away from potassium, sodium metals, or ethylene oxide.

Shipping: This compound requires a shipping label of: "Corrosive." It falls in DOT Hazard Class 8 and packing Group II.

Spill Handling: Evacuate persons not wearing protective equipment from area of spill or leak until clean-up is complete. Remove all ignition sources. Collect powdered material in the most convenient and safe manner and deposit in sealed containers. Ventilate area after clean-up is complete. It may be necessary to contain and dispose of this chemical as a hazardous waste. If material or contaminated runoff enters waterways, notify downstream users of potentially contaminated waters. Contact your Department of Environmental Protection or your regional office of the federal EPA for specific recommendations. If employees are required to clean-up spills, they must be properly trained and equipped. OSHA 1910.120(q) may be applicable.

Fire Extinguishing: This chemical is a noncombustible solid. Use extinguishers suitable for surrounding fire. Poisonous gases are produced in fire including hydrogen chloride and chlorine. If material or contaminated runoff enters waterways, notify downstream users of potentially contaminated waters. Notify local health and fire officials and pollution control agencies. From a secure, explosion-proof location, use water spray to cool exposed containers. If cooling streams are ineffective (venting sound increases in volume and pitch, tank discolors, or shows any signs of deforming), withdraw immediately to a secure position. If employees are expected to fight fires, they must be trained and equipped in OSHA 1910.156.

References

New Jersey Department of Health and Senior Services, "Hazardous Substance Fact Sheet: Ferrous Chloride," Trenton, NJ (March 1998)

Ferrous Sulfate

Molecular Formula: FeO_4S

Common Formula: $FeSO_4$

Synonyms: Copperas; Duretter; Duroferon; Exsiccated sulphate sulfate; Exsiccated sulphate sulphate; Feosol; Feospan; Fer-In-Sol; Ferralyn; Ferro-gradumet; Ferrosulfat (German); Ferrosulfate; Ferrosulphate; Ferro-Theron; Fersolate; Green Vitriol iron monosulfate; Iron protosulfate; Iron sulfate (1:1); Iron(2+) sulfate; Iron(2+) sulfate (1:1); Iron(II) sulfate; Iron Vitriol; Irospan; Irosul; Slow-Fe; Sulfato ferroso (Spanish); Sulfuric acid iron salt (1:1); Sulfuric acid, iron(2+) salt (1:1); Sulfuric acid, iron(II) salt (1:1); Sulphate sulphate (1:1); Sulsulphate

CAS Registry Number: 7720-78-7

RTECS Number: NO8500000

DOT ID: NA 9125

Regulatory Authority

- Air Pollutant Standard Set (ACGIH)[1] (HSE)[33] (OSHA)[58]
- Clean Water Act: Section 311 Hazardous Substances/RQ 40CFR117.3 (same as CERCLA, see below)
- Superfund/EPCRA 40CFR302.4 Reportable Quantity (RQ): CERCLA, 1,000 lb (454 kg)
- Canada, WHMIS, Ingredients Disclosure List

Cited in U.S. State Regulations: Alaska (G), California (A, G), Maine (G), New Hampshire (G), New Jersey (G), New York (G), Pennsylvania (G).

Description: Ferrous sulfate, $FeSO_4$, is a greenish or yellowish, solid in fine or lumpy crystals. Hazard Identification (based on NFPA-704 M Rating System): Health 2, Flammability 0, Reactivity 0. Slowly soluble in water.

Potential Exposure: It is used as a fertilizer, food or feed additive, and in herbicides, process engraving, dyeing, and water treatment.

Incompatibilities: Aqueous solution is acidic. Contact with alkalies form iron. Keep away from alkalies, soluble carbonates, gold and silver salts, lead acetate, lime water, potassium iodide, potassium and sodium tartrate, sodium borate, tannin.

Permissible Exposure Limits in Air: The recommended airborne exposure limit is 1 mg/m^3 (as iron) averaged over an 8-hour workshift. (This exposure limit is recommended for all soluble iron salts). This is the value set by ACGIH, NIOSH, and by HSE.[33] HSE adds an STEL of 2.0 mg/m^3. Other countries including Australia, Belgium, Finland, Italy, Netherlands, and Switzerland set the same TWA.

Determination in Air: Filter; Acid; Inductively coupled plasma; NIOSH IV, Method #7300, Elements.

Permissible Concentration in Water: The former USSR-UNEP/IRPTC project[43] has set a MAC for iron in water bodies used for domestic purposes of 0.5 mg/l and a MAC in sea water bodies used for fishery purposes of 0.05 mg/l.

Routes of Entry: Inhalation, ingestion, skin and/or eye contact.

Harmful Effects and Symptoms

Short Term Exposure: Ferrous sulfate can affect you when breathed in. Irritates the eyes, skin and respiratory tract. Ingestion: Less than 5 grams (1/6 ounce) can cause drowsiness, irritability, weakness, abdominal pain, nausea, vomiting and black, bloody stools. Delayed symptoms include fluid in the lungs, liver abnormalities, shock, coma, intestinal blockage and breakdown of the stomach and intestinal lining. Death has resulted from ingestion off less than an ounce.

Long Term Exposure: Excessive intake of iron compounds may result in increased accumulation of iron in body, especially the liver, spleen and lymphatic system. Inhalation of iron dusts may cause mottling of the lung. Prolonged eye contact can cause a brownish discoloration of the eye. Repeated overexposure may cause kidney stones.

Points of Attack: Eyes, skin, respiratory system, liver, gastrointestinal tract.

Medical Surveillance: If symptoms develop or overexposure is suspected, the following may be useful: Serum iron test. Liver function test. Kidney function tests. For those exposed to this chemical, taking dietary supplements or vitamins containing iron is not recommended without medical advice.

First Aid: If this chemical gets into the eyes, remove any contact lenses at once and irrigate immediately for at least 15 minutes, occasionally lifting upper and lower lids. Seek medical attention immediately. If this chemical contacts the skin, remove contaminated clothing and wash immediately with soap and water. Seek medical attention immediately. If this chemical has been inhaled, remove from exposure, begin rescue breathing (using universal precautions) if breathing has stopped and CPR if heart action has stopped. Transfer promptly to a medical facility. When this chemical has been swallowed, get medical attention. Give large quantities of water and induce vomiting. Do not make an unconscious person vomit.

Note to Physician: Gastric lavage with large amounts of 5% sodium phosphate or water. Follow this with a large amount of 1% sodium bicarbonate over a 3 hour period.

Personal Protective Methods: Wear protective gloves and clothing to prevent any reasonable probability of skin contact. Safety equipment suppliers/manufacturers can provide recommendations on the most protective glove/clothing material for your operation. All protective clothing (suits, gloves, footwear, headgear) should be clean, available each day, and put on before work. Contact lenses should not be worn when working with this chemical. Wear dust-proof chemical goggles and face shield unless full facepiece respiratory protection is worn. Employees should wash immediately with soap when skin is wet or contaminated. Provide emergency showers and eyewash.

Respirator Selection: Where the potential exists for exposures over 1 mg/m^3, use an MSHA/NIOSH approved respirator equipped with particulate (dust/fume/mist) filters. Particulate filters must be checked every day before work for physical damage, such as rips or tears, and replaced as needed. Where the potential for high exposures exists, use an MSHA/NIOSH approved supplied-air respirator with a full facepiece operated in the positive pressure mode or with a full facepiece, hood, or helmet in the continuous flow mode, or use an MSHA/NIOSH approved self-contained breathing apparatus with a full facepiece operated in pressure-demand or other positive pressure mode.

Storage: Prior to working with this chemical you should be trained on its proper handling and storage. Store in tightly closed containers in a cool, well-ventilated area away from alkalis.

Shipping: Ferrous sulfate is not specifically cited in DOT's Performance-Oriented Packaging Standards.[19]

Spill Handling: Evacuate persons not wearing protective equipment from area of spill or leak until clean-up is complete. Remove all ignition sources. Collect powdered material in the most convenient and safe manner and deposit in sealed containers. Ventilate area after clean-up is complete. It may be necessary to contain and dispose of this chemical as a hazardous waste. If material or contaminated runoff enters waterways, notify downstream users of potentially contaminated waters. Contact your Department of Environmental Protection or your regional office of the federal EPA for specific recommendations. If employees are required to clean-up spills, they must be properly trained and equipped. OSHA 1910.120(q) may be applicable.

Fire Extinguishing: This chemical is a noncombustible solid. Use extinguishers suitable for surrounding fire. Poisonous gases are produced in fire including sulfur oxides. If material or contaminated runoff enters waterways, notify downstream users of potentially contaminated waters. Notify local health and fire officials and pollution control agencies. From a secure, explosion-proof location, use water spray to cool exposed containers. If cooling streams are ineffective (venting sound increases in volume and pitch, tank discolors, or shows any signs of deforming), withdraw immediately to a secure position. If employees are expected to fight fires, they must be trained and equipped in OSHA 1910.156.

References

Sax, N. I., Ed., "Dangerous Properties of Industrial Materials Report," 3, No. 4, 45–47 (1983) and 7, No. 1, 55–60 (1987)

New Jersey Department of Health and Senior Services, "Hazardous Substance Fact Sheet, Ferrous Sulfate," Trenton, NJ (February 1999)

Ferrovanadium Dust

Molecular Formula: FeV

Synonyms: Ferro "V"

CAS Registry Number: 12604-58-9

RTECS Number: LK2900000

DOT ID: No citation.

Regulatory Authority

- Air Pollutant Standard Set (ACGIH)[1] (DFG)[3] (OSHA)[58] (Several States)[60]
- Canada, WHMIS, Ingredients Disclosure List

Cited in U.S. State Regulations: Alaska (G), California (A), Connecticut (A), Florida (G), Illinois (G), Maine (G), Massachusetts (G), Nevada (A), New Hampshire (G), New Jersey (G), North Dakota (A), Pennsylvania (G), Rhode Island (G), Virginia (A), West Virginia (G).

Description: This material consists off dark, odorless solid particles in air. An alloy having the composition 50 – 80% vanadium with iron, carbon, and trace silicon, manganese, chromium, nickel, etc. Freezing/Melting point = 1,482 – 1,521°C. Insoluble in water.

Potential Exposure: This material is added to steel to produce fineness of grain, toughness, torsion properties, and resistance to high temperatures.

Incompatibilities: Oxidizers.

Permissible Exposure Limits in Air: The OSHA PEL is 1 mg/m^3. The DFG MAK[3] and ACGIH TWA is 1 mg/m^3 and a STEL value of 3 mg/m^3. The NIOSH IDLH level is 500 mg/m^3. Several states have set guidelines or standards for ferrovanadium dust in ambient air[60] ranging from 10.0 – 30.0 μg/m^3 (North Dakota) to 16.0 μg/m^3 (Virginia) 20.0 μg/m^3 (Connecticut) to 238.0 μg/m^3 (Nevada).

Determination in Air: Collection by filter; Acid; analysis by flame atomic absorption spectrometry; OSHA Methods #ID121, #ID125G.

Permissible Concentration in Water: No criteria set.

Routes of Entry: Inhalation, skin and/or eye contact.

Harmful Effects and Symptoms

Short Term Exposure: Irritation of eyes and respiratory system. Inhalation can cause irritation with coughing and wheezing.

Long Term Exposure: Repeated exposure can cause bronchitis, pneumonitis, with cough, phlegm and/or shortness of breath.

Points of Attack: Respiratory system, eyes.

Medical Surveillance: Consider the points of attack in preplacement and periodic physical examination. Lung function tests.

First Aid: If this chemical gets into the eyes, remove any contact lenses at once and irrigate immediately for at least 15 minutes, occasionally lifting upper and lower lids. Seek medical attention immediately. If this chemical contacts the skin, remove contaminated clothing and wash immediately with soap and water. Seek medical attention immediately. If this chemical has been inhaled, remove from exposure, begin rescue breathing (using universal precautions) if breathing has stopped and CPR if heart action has stopped. Transfer promptly to a medical facility. When this chemical has been swallowed, get medical attention. Give large quantities of water and induce vomiting. Do not make an unconscious person vomit.

Personal Protective Methods: Wear protective gloves and clothing to prevent any reasonable probability of skin contact. Safety equipment suppliers/manufacturers can provide recommendations on the most protective glove/clothing material for your operation. All protective clothing (suits, gloves, footwear, headgear) should be clean, available each day, and put on before work. Contact lenses should not be worn when working with this chemical. Wear dust-proof chemical goggles and face shield unless full facepiece respiratory protection is worn. Employees should wash immediately with soap when skin is wet or contaminated. Provide emergency showers and eyewash.

Respirator Selection: NIOSH/OSHA: *up to 5 mg/m^3*: DM (any dust and mist respirator).* *Up to 10 mg/m^3:* DMXSQ (any dust and mist respirator except single-use and quarter-mask respirators);* or SA (any supplied-air respirator).* *Up to 25 mg/m^3*: SA:CF (any supplied-air respirator operated in a continuous-flow mode);* or PAPRDM (any powered, air-purifying respirator with a dust and mist filter).* *Up to 50 mg/m^3*: HiEF (any air-purifying, full-facepiece respirator with a high-efficiency particulate filter); or SAT:CF (any supplied-air respirator that has a tight-fitting facepiece and is operated in a continuous-flow mode);* or PAPRTHiE (any powered, air-purifying respirator with a tight-fitting facepiece and a high-efficiency particulate filter);* or SCBAF (any self-contained breathing apparatus with a full facepiece); or SAF (any supplied-air respirator with a full facepiece). *Up to 500 mg/m³*: SAF:PD,PP (any supplied-air respirator that has a full facepiece and is operated in a pressure-demand or other positive-pressure mode). *Emergency or planned entry in unknown concentration or IDLH conditions:* SCBAF:PD, PP (any MSHA/NIOSH approved self-contained breathing apparatus that has a full facepiece and is operated in a pressure-demand or other positive-pressure mode); or SAF:PD, PP:ASCBA (any supplied-air respirator that has a full facepiece and is operated in a pressure-demand or other positive-pressure mode in combination with an auxiliary, self-contained breathing apparatus operated in a pressure-demand or other positive pressure mode). *Escape:* HiEF [any air-purifying, full-facepiece respirator (gas mask) with a chin-style, front- or back-mounted organic vapor canister having a high-efficiency particulate filter]; or SCBAE (any appropriate escape-type, self-contained breathing apparatus).

* Substance reported to cause eye irritation or damage; may require eye protection.

Storage: Prior to working with this chemical you should be trained on its proper handling and storage. Store in tightly closed containers in a cool, well ventilated area away from oxidizers.

Spill Handling: Evacuate persons not wearing protective equipment from area of spill or leak until clean-up is complete. Remove all ignition sources. Collect powdered material in the most convenient and safe manner and deposit in sealed containers. Ventilate area after clean-up is complete. It may be necessary to contain and dispose of this chemical as a hazardous waste. If material or contaminated runoff enters waterways, notify downstream users of potentially contaminated waters. Contact your Department of Environmental Protection or your regional office of the federal EPA for specific recommendations. If employees are required to clean-up spills, they must be properly trained and equipped. OSHA 1910.120(q) may be applicable.

Fire Extinguishing: Noncombustible Solid, but dust may be an explosion hazard. Use extinguishers suitable for metal fires. If material or contaminated runoff enters waterways, notify downstream users of potentially contaminated waters. Notify local health and fire officials and pollution control agencies. From a secure, explosion-proof location, use water spray to cool exposed containers. If cooling streams are ineffective (venting sound increases in volume and pitch, tank discolors, or shows any signs of deforming), withdraw immediately to a secure position. If employees are expected to fight fires, they must be trained and equipped in OSHA 1910.156.

Disposal Method Suggested: Disposal in a sanitary landfill.

References

National Institute for Occupational Safety and Health, Criteria for a Recommended Standard: Occupational Exposure to Vanadium, NIOSH Document No. 77–222, Washington, DC (1977)

New Jersey Department of Health and Senior Services, "Hazardous Substance Fact Sheet, Ferrovanadium," Trenton, NJ (April 1999)

Fibrous Glass

Synonyms: Fiber Glas®; Fiberglass; Glass fibers; Glass wool

RTECS Number: LK3651000

DOT ID: No citation.

Regulatory Authority

- Air Pollutant Standard Set (ACGIH)[1] (OSHA)[45] (North Dakota)[60]

Cited in U.S. State Regulations: Alaska (G), New Hampshire (G), New Jersey (G), North Dakota (A), Rhode Island (G), West Virginia (G).

Description: Fibrous glass is the name for a manufactured fiber in which the fiber-forming substance is glass. Glasses are a class of materials made from silicon dioxide with oxides of various metals and other elements, that solidify from the molten state without crystallization. A fiber is considered to be a particle with a length-to-diameter ratio of 3 – 1 or greater. Most fibrous glass that is manufactured consists of fibers with diameters 3.5 μm or larger. The volume of small diameter fiber production has not been determined. Fibers with diameters less than 1 μm are estimated to comprise less than 1% of the fibrous glass market. Insoluble in water.

Potential Exposure: The major uses of fibrous glass are in thermal, electrical, and acoustical insulation, weatherproofing, plastic reinforcement, filtration media, and in structural and textile materials.

Permissible Exposure Limits in Air: NIOSH recommends that occupational exposure to fibrous glass be controlled so that no worker is exposed at an airborne concentration greater than 3,000,000 fibers/m³ of air (3 fibers/cc of air) having a diameter equal to or less than 3.5 μm and a length equal to or greater than 10 μm determined as time weighted average (TWA) concentration for up to a 10-hour work shift in a 40-hour work week; airborne concentrations determined as total fibrous glass shall be limited to a TWA concentration of 5 mg/m³ off air. This differs from the present federal standard which classifieds fibrous glass as an inert or nuisance dust with the limits off exposure being 15 million particles per cubic foot or 5 mg/m³ for the respirable fraction ACGIH recommends a TWA of 10 mg/m³ for glass in fibrous or dust form. North Dakota[60] has set a guideline for fibrous glass dust in ambient air of 0.10 mg/m³.

Determination in Air: Collection on a filter; none; phase contrast microscopy; NIOSH IV Method #7400, Asbestos and Other Fibers.

Permissible Concentration in Water: No criteria set.

Routes of Entry: Inhalation, skin and/or eye contact.

Harmful Effects and Symptoms

Different dimensions of fibrous glass will produce different biologic effects. Large diameter (greater than 3.5 μm) glass fibers have been found to cause skin, eye, and upper respiratory tract irritation; a relatively low frequency of fibrotic changes; and a very slight indication of fan excess mortality due to nonmalignant respiratory disease. Smaller diameter (less than 3.5 μm) fibrous glass has not been conclusively related to health effects in humans but glass fibers of this dimension have only been regularly produced since the 1960s.

Smaller diameter fibers have the ability to penetrate to the alveoli. This potential is cause for concern and the primary reason that fibers 3.5 μm or smaller are subject to special controls. Experimental studies in animals have demonstrated carcinogenic effects with the long (greater than 10 μm) and thin fibers (usually less than 1 μm in diameter). However, these studies were performed by implanting fibrous glass in the pleural or peritoneal cavities.

The data from studies with these routes of exposure cannot be directly extrapolated to conditions of human exposure. On the basis of available information, NIOSH does not consider fibrous glass to be a substance that produces cancer as a result of occupational exposure. The data on which to base this conclusion are limit. Fibrous glass does not appear to possess the same potential as asbest or for causing health hazard. Glass fibers are not usually of the fine submicron diameters as are asbestos fibrils and the concentrations of glass fibers in workplace air are generally orders of magnitude less than for asbestos. In one study, glass fibers were found to be cleared from the lungs more readily than asbestos.

Short Term Exposure: Irritates eyes, skin, nose, throat. Inhalation can cause dyspnea (breathing difficulty)

Long Term Exposure: A suspected carcinogen, but there is no consistent evidence of chronic health effects in exposed workers.

Points of Attack: Eyes, skin, respiratory system

Medical Surveillance: NIOSH recommends that workers subject to fibrous glass exposure have comprehensive preplacement medical examinations with emphasis on skin susceptibility and prior exposure in dusty trades. Subsequent annual examinations should give attention to the skin and respiratory system with attention to pulmonary function.

First Aid: If this chemical gets into the eyes, remove any contact lenses at once and irrigate immediately for at least 15 minutes, occasionally lifting upper and lower lids. Seek medical attention immediately. If this chemical contacts the skin, remove contaminated clothing and wash immediately with soap and water. If this chemical has been inhaled, remove from exposure, and move to fresh air.

Personal Protective Methods: Protective clothing shall be worn to prevent fibrous glass contact with skin especially hands, arms, neck, and underarms. Safety goggles or face shields and goggles shall be worn during tear-out or blowing operations or when applying fibrous glass materials overhead. They should be used in all areas where there is a likelihood that airborne glass fibers may contact the eyes. Engineering controls should be used wherever feasible to maintain fibrous glass concentrations at or below the prescribed limits. Respirators should only be used when engineering controls are not feasible; for example, in certain nonstationary operations where permanent controls are not feasible.

Respirator Selection: NIOSH: Up to 5 × REL: D (any dust respirator). *Up to 10 × REL:* DXSQ (any dust respirator except single-use and quarter-mask respirators); or HiE (any air-purifying respirator with a high-efficiency particulate filter); or SA (any supplied-air respirator). *Up to 25 × REL:* SA:CF (any supplied-air respirator operated in a continuous-flow mode); or PAPRDM (any powered, air-purifying respirator with a dust filter). *Up to 50x REL:* HiEF (any air-purifying, full-facepiece respirator with a high-efficiency particulate filter); or PAPRTHiE (any powered, air-purifying respirator with a tight-fitting facepiece and a high-efficiency particulate filter); or SCBAF (any self-contained breathing apparatus with a full facepiece); or SAF (any supplied-air respirator with a full facepiece). *Up to 1,000 × REL:* SAF:PD,PP (any supplied-air respirator that has a full facepiece and is operated in a pressure-demand or other positive-pressure mode). *Emergency or planned entry into unknown concentrations or IDLH conditions*: SCBAF:PD,PP: SCBAF:PD, PP (any MSHA/NIOSH approved self-contained breathing apparatus that has a full facepiece and is operated in a pressure-demand or other positive-pressure mode); or SAF:PD, PP:ASCBA (any supplied-air respirator that has a full facepiece and is operated in a pressure-demand or other positive-pressure mode in combination with an auxiliary, self-contained breathing apparatus operated in a pressure-demand or other positive pressure mode). Escape*:* HiEF [any air-purifying, full-facepiece respirator (gas mask) with a chin-style, front- or back-mounted organic vapor canister having a high-efficiency particulate filter]; or SCBAE (any appropriate escape-type, self-contained breathing apparatus).

Storage: Prior to working with this chemical you should be trained on its proper handling and storage.

Spill Handling: Evacuate persons not wearing protective equipment from area of spill or leak until clean-up is complete. Collect material in the most convenient and safe manner and deposit in sealed containers. Ventilate area after clean-up is complete.

Fire Extinguishing: Noncombustible fibers. Use extinguisher suitable for surrounding fire. If employees are

expected to fight fires, they must be trained and equipped in OSHA 1910.156.

Disposal Method Suggested: Fibrous glass waste and scrap should be collected and disposed of in a manner which will minimize its dispersal into the atmosphere. Emphasis should be placed on covering waste containers, proper storage of materials, and collection of fibrous glass dust. Clean-up of fibrous glass dust should be performed using vacuum cleaners or wet cleaning methods. Dry sweeping should not be performed.

References

National Institute for Occupational Safety and Health, Occupational Exposure to Fibrous Glass: A Symposium, NIOSH Doc. No. 76–151 (1976)

National Institute for Occupational Safety and Health, Criteria for a Recommended Standard: Occupational Exposure to Fibrous Glass, NIOSH Doc. No. 77–152 (1977)

National Institute for Occupational Safety and Health, Criteria for a Recommended Standard: Occupational Exposure to Crystalline Silica, NIOSH Doc. No. 75–120, Washington, DC (1975)

New Jersey Department of Health and Senior Services, "Hazardous Substance Fact Sheet: Fibrous Glass," Trenton, NJ (November 1985)

Fluenetil

Molecular Formula: $C_{16}H_{15}FO_2$

Synonyms: (1,1'-Biphenyl)-4-acetic acid, 2-fluoroethyl ester; 4-Biphenylacetic acid, 2-fluoroethyl ester; Fluenyl; β-Fluorethyl 4-biphenylacetate; Lambrol

CAS Registry Number: 4301-50-2

RTECS Number: DV8335000

DOT ID: UN 3077

Regulatory Authority

- Very Toxic Substance (World Bank)[15]
- Superfund/EPCRA 40CFR355, Appendix B Extremely Hazardous Substances: TPQ = 100/10,000 lb (45.4/4,540 kg)
- Superfund/EPCRA 40CFR302.4 Reportable Quantity (RQ): EHS, 1 lb (0.454 kg)

Cited in U.S. State Regulations: California (G), Massachusetts (G), New Jersey (G), Pennsylvania (G).

Description: $C_{16}H_{15}FO_2$ is a crystalline solid. Hazard Identification (based on NFPA-704 M Rating System): Health 3, Flammability 1, Reactivity 0.

Potential Exposure: This material has been used as an acaricide and insecticide. Its main use was as a dormant spray for orchard fruit. It is no longer made. Not registered as a pesticide in the U.S.

Incompatibilities: Nitrates. Moisture may cause material to hydrolyze.

Permissible Exposure Limits in Air: No standards set.

Permissible Concentration in Water: No criteria set.

Harmful Effects and Symptoms

Short Term Exposure: Fluenetil is highly toxic. The LD_{50} oral-rat is 6 mg/kg.

First Aid: If this chemical gets into the eyes, remove any contact lenses at once and irrigate immediately for at least 15 minutes, occasionally lifting upper and lower lids. Seek medical attention immediately. If this chemical contacts the skin, remove contaminated clothing and wash immediately with soap and water. Speed in removing material from skin is of extreme importance. Seek medical attention immediately. If this chemical has been inhaled, remove from exposure, begin rescue breathing (using universal precautions) if breathing has stopped and CPR if heart action has stopped. Transfer promptly to a medical facility. When this chemical has been swallowed, get medical attention. Give large quantities of water and induce vomiting. Do not make an unconscious person vomit. Effects may be delayed; keep victim under observation.

Personal Protective Methods: Wear protective gloves and clothing to prevent any reasonable probability of skin contact. Safety equipment suppliers/manufacturers can provide recommendations on the most protective glove/clothing material for your operation. All protective clothing (suits, gloves, footwear, headgear) should be clean, available each day, and put on before work. Contact lenses should not be worn when working with this chemical. Wear dust-proof chemical goggles and face shield unless full facepiece respiratory protection is worn. Employees should wash immediately with soap when skin is wet or contaminated. Provide emergency showers and eyewash.

Respirator Selection: Where the potential for exposure to this chemical, use a MSHA/NIOSH approved supplied-air respirator with a full facepiece operated in the positive pressure mode or with a full facepiece, hood, or helmet in the continuous flow mode, or use a MSHA/NIOSH approved self-contained breathing apparatus with a full facepiece operated in pressure-demand or other positive pressure mode.

Storage: Prior to working with this chemical you should be trained on its proper handling and storage. Store in tightly closed containers in a cool, well ventilated area away from nitrates and moisture.

Shipping: Environmentally hazardous solid, n.o.s. Hazard Class: 9. Label: "Class 9."

Spill Handling: Evacuate persons not wearing protective equipment from area of spill or leak until clean-up is complete. Remove all ignition sources. Do not touch spilled material; stop leak if you can do so without risk. Use water spray to reduce vapors. Small spills: absorb with sand or other noncombustible absorbent material and place into containers for later disposal. Small dry spills: with clean shovel place material into clean, dry container and cover; move containers from spill area. Large spills: dike far ahead of spill for later disposal. Ventilate area after clean-up is complete. It may be necessary to contain and dispose of this chemical as a hazardous waste. If material or contaminated runoff enters waterways, notify downstream users of potentially contaminated waters. Contact your Department of Environmental Protection or your regional office of the federal EPA for specific recommendations. If employees are required to clean-up spills, they must be properly trained and equipped. OSHA 1910.120(q) may be applicable.

Fire Extinguishing: This material is a combustible solid. Poisonous gases including fluorine are produced in fire. Small fires: dry chemical, carbon dioxide, water spray, or foam. Large fires: water spray, fog, or foam. Move container from fire area if you can do so without risk. Fight fire from maximum distance. Dike fire control water for later disposal; do not scatter the material. If material or contaminated runoff enters waterways, notify downstream users of potentially contaminated waters. Notify local health and fire officials and pollution control agencies. From a secure, explosion-proof location, use water spray to cool exposed containers. If cooling streams are ineffective (venting sound increases in volume and pitch, tank discolors, or shows any signs of deforming), withdraw immediately to a secure position. If employees are expected to fight fires, they must be trained and equipped in OSHA 1910.156.

Disposal Method Suggested: In accordance with 40CFR 165 recommendations for the disposal of pesticides and pesticide containers. Must be disposed properly by following package label directions or by contacting your state pesticide or environmental control agency or by contacting your regional EPA office.

References

U.S. Environmental Protection Agency, "Chemical Profile: Fluenetil," Washington, DC, Chemical Emergency Preparedness Program (November 30, 1987)

Fluoboric Acid

Molecular Formula: HBF_4

Synonyms: Acide fluoroborique (French); Acido fluoborico (Spanish); Borate(1-), tetrafluoro-, hydrogen; Borofluoric acid; Fluboric acid; Fluoroboorzuur (Dutch); Fluoro-boric acid; Fluoroborsaeure (German); Hydrofluoboric acid; Hydrogen tetrafluoroborate; Kester 5569 Solder-NU; Prepared bath 2137; Starter 2000; Starter 2137; Tetrofluoroboric acid; Tetrofluoro hydrogen borate

CAS Registry Number: 16872-11-0

RTECS Number: ED2685000

DOT ID: UN 1775

EEC Number: 009-010-00-X

Regulatory Authority

- Safe Drinking Water Act (as F): Regulated chemical (47 FR 9352); MCL, 4.0 mg/l; MCLG, 4.0 mg/l; SMCL, 2.0 mg/l.

Cited in U.S. State Regulations: New Jersey (G), Pennsylvania (G).

Description: Fluoboric Acid is a colorless liquid which does not exist as a free, pure substance. Used as an aqueous solution. Boiling point = 130°C (decomposes). Hazard Identification (based on NFPA-704 M Rating System): Health 3, Flammability 0, Reactivity 0. Soluble in water.

Potential Exposure: Used as a catalyst for acetal synthesis and cellulose esters, a metal surface cleaning agent, an aluminum electrolytic finishing agent, a stripping solution for the removal of solder and plated metals, and in intermediate in making fluoroborate salt.

Incompatibilities: A strong acid. Reacts violently with chemically active metals, strong bases releasing flammable hydrogen gas.

Permissible Exposure Limits in Air: The Federal limit (OSHA PEL)[58] and the DFG MAK[3] is 0.1 ppm (0.2 mg/m^3). The ACGIH has adopted a TWA of 1.0 ppm (2.0 mg/m^3) and an STEL value of 2.0 ppm (4.0 mg/m^3) as has HSE.[33] The NIOSH IDLH level is 25 ppm. Several states have set guidelines or standards for fluorine in ambient air[60] ranging from zero (North Carolina) to 4.0 μg/m^3 (Connecticut) to 6.7 μg/m^3 (New York) to 20.0 μg/m^3 (Florida) to 30.0 μg/m^3 (Virginia) to 20.0 – 40.0 μg/m^3 (North Dakota) to 48.0 μg/m^3 (Nevada).

Determination in Air: No tests available.

Permissible Concentration in Water: The former USSR-UNEP/IRPTC project[43] has set a MAC in water bodies used for domestic purposes of 1.5 mg/l. See also the entry on fluorides.

Routes of Entry: Inhalation, eye and/or skin contact. Absorbed through the skin.

Harmful Effects and Symptoms

Short Term Exposure: Highly corrosive. Contact can cause severe skin burns and eye irritation and burns with possible eye damage. Inhalation can irritate the lungs causing coughing and shortness of breath. High exposure can cause headache, weakness, convulsions, collapse and death. Higher exposures can cause pulmonary edema, a medical emergency that can be delayed for several hours. This can cause death.

Long Term Exposure: Repeated exposure can cause nausea, vomiting, diarrhea, loss of appetite and weight, hair loss, skin rash, and bone and teeth changes (fluorosis). May cause kidney damage, anemia, and lung irritation with the possible development of bronchitis with cough, phlegm, and/or shortness of breath.

Points of Attack: Eyes, skin, respiratory system, kidneys, blood.

Medical Surveillance: Complete blood count (CBC). Kidney function tests. Consider chest x-ray following acute overexposure.

First Aid: If this chemical gets into the eyes, remove any contact lenses at once and irrigate immediately for at least 15 minutes, occasionally lifting upper and lower lids. Seek medical attention immediately. If this chemical contacts the skin, remove contaminated clothing and wash immediately with soap and water. Seek medical attention immediately. If this chemical has been inhaled, remove from exposure, begin rescue breathing (using universal precautions) if breathing has stopped and CPR if heart action has stopped. Transfer promptly to a medical facility. When this chemical has been swallowed, get medical attention. If victim is conscious, administer water or milk. Do not induce vomiting. Medical observation is recommended for 24 – 48 hours after breathing overexposure, as pulmonary edema may be delayed. As first aid for pulmonary edema, a doctor or authorized paramedic may consider administering a corticosteroid spray.

Personal Protective Methods: Wear protective gloves and clothing to prevent any reasonable probability of skin contact. Safety equipment suppliers/manufacturers can provide recommendations on the most protective glove/clothing material for your operation. All protective clothing (suits, gloves, footwear, headgear) should be clean, available each day, and put on before work. Contact lenses should not be worn when working with this chemical. Wear splash-proof chemical goggles and face shield unless full facepiece respiratory protection is worn. Employees should wash immediately with soap when skin is wet or contaminated. Provide emergency showers and eyewash.

Respirator Selection: NIOSH/OSHA as fluorides *12.5 mg/m³*: DM (any dust and mist respirator). *25 mg/m³*: DMXSQ* (any dust and mist respirator except single-use and quarter mask respirators); or SA* (any supplied-air respirator). *62.5 mg/m³*: SA:CF* (any supplied-air respirator operated in a continuous-flow mode); PAPRDM*+ (if not present as a fume) (any powered, air-purifying respirator with a dust and mist filter). *125 mg/m³:* HiEF+ (any air-purifying, full-facepiece respirator with a high-efficiency particulate filter); or SCBAF (any self-contained breathing apparatus with a full facepiece); or SAF (any supplied-air respirator with a full facepiece). *250 mg/m³*: SA:PD,PP (any supplied-air respirator operated in a pressure-demand or other positive-pressure mode). *Emergency or planned entry into unknown concentrations or IDLH conditions:* SCBAF:PD,PP (any self-contained breathing apparatus that has a full faceplate and is operated in a pressure-demand or other positive-pressure mode); or SAF:PD,PP:ASCBA (any supplied-air respirator that has a full facepiece and is operated in a pressure-demand or other positive-pressure mode in combination with an auxiliary, self-contained breathing apparatus operated in a pressure-demand or other positive- pressure mode). *Escape:* HiEF+ (any air-purifying, full-facepiece respirator with a high-efficiency particulate filter); or SCBAE (any appropriate escape-type, self-contained breathing apparatus).

* Substance reported to cause eye irritation or damage; may require eye protection.

\+ May need acid gas sorbent.

Storage: Prior to working with this chemical you should be trained on its proper handling and storage. Store in tightly closed containers in a cool, well ventilated area away from chemically active metals, strong bases. Where possible, automatically pump liquid from drums or other storage containers to process containers

Shipping: Label required is "Corrosive". DOT/UN Hazard Class 8. Packing Group II.

Spill Handling: Evacuate and restrict persons not wearing protective equipment from area of spill or leak until cleanup is complete. Remove all ignition sources. Ventilate area of spill or leak. Absorb liquids in vermiculite, dry sand, earth, peat, carbon, or a similar material and deposit in sealed containers. It may be necessary to contain and dispose of this chemical as a hazardous waste. If material or contaminated runoff enters waterways, notify downstream users of potentially contaminated waters. Contact your Department of Environmental Protection or your regional office of the federal EPA for specific recommendations. If employees are required to clean-up spills, they must be properly trained and equipped. OSHA 1910.120(q) may be applicable.

Fire Extinguishing: This chemical may burn but does not easily ignite. Use dry chemical powder extinguishers. Poisonous gases are produced in fire including hydrogen fluoride, fluorine, and boron oxides. If material or contaminated runoff enters waterways, notify downstream users of potentially contaminated waters. Notify local health and fire officials and pollution control agencies. From a secure, explosion-proof location, use water spray to cool exposed containers. If cooling streams are ineffective (venting sound increases in volume and pitch, tank discolors, or shows any signs of deforming), withdraw immediately to a secure position. If employees are expected to fight fires, they must be trained and equipped in OSHA 1910.156.

References

New Jersey Department of Health and Senior Services, "Hazardous Substance Fact Sheet, Fluoboric acid," Trenton, NJ (April 1999)

Fluometuron

Molecular Formula: $C_{10}H_{11}F_3N_2O$

Common Formula: $C_6H_4(CF_3)NHCON(CH_3)_2$

Synonyms: C 2059; CIBA 2059; Cotoran; Cotoran Multi 50WP; Cottonex; *N,N*-Dimethyl-*N*'-[3-(trifluoromethyl) phenyl]urea; 1,1-Dimethyl-3-(3-trifluoromethylphenyl)urea; Fluometuron; Herbicide C-2059; Lanex; Meturone; NCI-C08695; Pakhtaran; 3-(5-Trifluormethylphenyl)-, dimethylharnstoff (German); *N*-(*m*-Trifluoromethylphenyl)-*N'*,*N'*-dimethylurea; *N*-(3-Trifluoromethylphenyl)-*N'*,*N'*-dimethylurea; 3-(*m*-Trifluoromethylphenyl)-1,1-dimethylurea; 3-(3-Trifluoromethylphenyl)-1,1-dimethylurea; Urea, *N,N*-Dimethyl-*N*'-[3-(trifluoromethyl)phenyl]-; Urea, 1,1-dimethyl-3-(α,α,α-trifluoro-*m*-tolyl)-

CAS Registry Number: 2164-17-2

RTECS Number: YT1575000

DOT ID: UN 3077

Regulatory Authority

- EPCRA Section 313 Form R *de minimis* concentration reporting level: 1.0%

Cited in U.S. State Regulations: New Jersey (G), Pennsylvania (G).

Description: Fluometuron, is a white crystalline solid often used in liquid solution that may be flammable. Freezing/Melting point = 163 – 165°C. Hazard Identification (based on NFPA-704 M Rating System): Health 1, Flammability 1, Reactivity 0. Slightly soluble in water.

Potential Exposure: This material is used as a herbicide.

Incompatibilities: Liquid solutions are incompatible with oxidizers.

Permissible Exposure Limits in Air: No standards set.

Permissible Concentration in Water: A no-observed-adverse effect level (NOAEL) of 0.0125 mg/kg/day has been calculated by EPA. On this basis a long-term health advisory of 5.3 mg/l and a lifetime health advisory of 0.09 mg/l have been calculated.

Routes of Entry: Inhalation, through the skin.

Harmful Effects and Symptoms

Short Term Exposure: Contact can cause eye and skin irritation. Inhalation can irritate the respiratory tract. Symptoms of exposure include increased leukocyte content in circulation blood. The material is a mild cholinesterase inhibitor. The LD_{50} oral-rat is 6,400 mg/kg (insignificantly toxic).

Long Term Exposure: May cause skin allergy. Mild cholinesterase inhibitor; cumulative effect is possible. Repeated exposure may cause in the red blood cell count. May cause liver damage.

Points of Attack: Respiratory system, lungs, central nervous system, skin, eyes, plasma and red blood cell cholinesterase.

Medical Surveillance: Liver function tests. Complete blood count (CBC). Evaluation by a qualified allergist.

First Aid: *Skin Contact:*[52] Flood all areas of body that have contacted the substance with water. Don't wait to remove contaminated clothing; do it under the water stream. Use soap to help assure removal. Isolate contaminated clothing when removed to prevent contact by others.

Eye Contact: Remove any contact lenses at once. Flush eyes well with copious quantities of water or normal saline for at least 20 – 30 minutes. Seek medical attention.

Inhalation: Leave contaminated area immediately; breathe fresh air. Proper respiratory protection must be supplied to any rescuers. If coughing, difficult breathing or any other symptoms develop, seek medical attention at once, even if symptoms develop many hours after exposure.

Ingestion: Consult a physician, hospital or poison center at once. If the victim is unconscious or convulsing, do not induce vomiting or give anything by mouth. Assure that the airway is open, lay on side and keep head lower than body and transport immediately to medical facility. If conscious and not convulsing, give a glass of water to dilute the substance. Do not induce vomiting without a physician's advice.

Personal Protective Methods: Safety equipment suppliers/manufacturers can provide recommendations on the most protective glove/clothing material for your operation. All protective clothing (suits, gloves, footwear, headgear) should be clean, available each day, and put on before work. Contact lenses should not be worn when working with this chemical. Wear dust-proof chemical goggles and face shield unless full facepiece respiratory protection is worn. Employees should wash immediately with soap when skin is wet or contaminated. Provide emergency showers and eyewash.

Respirator Selection: Where the potential for exposure to this chemical, use a MSHA/NIOSH approved supplied-air respirator with a full facepiece operated in the positive pressure mode or with a full facepiece, hood, or helmet in the continuous flow mode, or use a MSHA/NIOSH approved self-contained breathing apparatus with a full facepiece operated in pressure-demand or other positive pressure mode.

Storage: Prior to working with this chemical you should be trained on its proper handling and storage. Store in a refrigerator or a cool, dry place and protect from exposure to ultraviolet light. Keep liquid away from oxidizers. Where possible, automatically pump liquid from drums or other storage containers to process containers.

Shipping: Environmentally hazardous solid, n.o.s. Hazard Class: 9. Label: "Class 9."

Spill Handling: *Dry material:* Evacuate persons not wearing protective equipment from area of spill or leak until clean-up is complete. Remove all ignition sources. Dampen spilled material with 60 – 70% acetone and avoid airborne dust. Collect powdered material in the most convenient and safe manner and deposit in sealed containers. Ventilate area after clean-up is complete. It may be necessary to contain and dispose of this chemical as a hazardous waste. If material or contaminated runoff enters waterways, notify downstream users of potentially contaminated waters. Contact your Department of Environmental Protection or your regional office of the federal EPA for specific recommendations. If employees are required to clean-up spills, they must be properly trained and equipped. OSHA 1910.120(q) may be applicable.

Liquid: Evacuate and restrict persons not wearing protective equipment from area of spill or leak until cleanup is complete. Remove all ignition sources. Ventilate area of spill or leak. Absorb liquids in vermiculite, dry sand, earth, peat, carbon, or a similar material and deposit in sealed containers. Oil-skimming equipment may be used to remove slicks from water. Keep this chemical out of a confined space, such as a sewer, because of the possibility of an explosion, unless the sewer is designed to prevent the build-up of explosive concentrations. It may be necessary to contain and dispose of this chemical as a hazardous waste. If material or contaminated runoff enters waterways, notify downstream users of potentially contaminated waters. Contact your Department of Environmental Protection or your regional office of the federal EPA for specific recommendations. If employees are required to clean-up spills, they must be properly trained and equipped. OSHA 1910.120(q) may be applicable.

Fire Extinguishing: *Dry material:* Use dry chemical, carbon dioxide, water spray, or alcohol foam extinguishers. Poisonous gases are produced in fire including hydrogen fluoride and nitrogen oxides. If material or contaminated runoff enters waterways, notify downstream users of potentially contaminated waters. Notify local health and fire officials and pollution control agencies. From a secure, explosion-proof location, use water spray to cool exposed containers. If cooling streams are ineffective (venting sound increases in volume and pitch, tank discolors, or shows any signs of deforming), withdraw immediately to a secure position. If employees are expected to fight fires, they must be trained and equipped in OSHA 1910.156.

Combustible solution: Poisonous gases including nitrogen oxides and fluorine are produced in fire. Use dry chemical, carbon dioxide, or alcohol foam extinguishers. Vapors are heavier than air and will collect in low areas. Vapors may travel long distances to ignition sources and flashback. Vapors in confined areas may explode when exposed to fire. Containers may explode in fire. Storage containers and parts of containers may rocket great distances, in many directions. If material or contaminated runoff enters waterways, notify downstream users of potentially contaminated waters. Notify local health and fire officials and pollution control agencies. From a secure, explosion-proof location, use water spray to cool exposed containers. If cooling streams are ineffective (venting sound increases in volume and pitch, tank discolors, or shows any signs of deforming), withdraw immediately to a secure position. If employees are expected to fight fires, they must be trained and equipped in OSHA 1910.156.

References

U.S. Environmental Protection Agency, "Health Advisory: Fluometuron," Washington, DC, Office of Drinking Water (August 1987)

New Jersey Department of Health and Senior Services, "Hazardous Substance Fact Sheet, FLUMETURON," Trenton, NJ (March 1999)

Fluoranthene

Molecular Formula: $C_{16}H_{10}$

Synonyms: 1,2-Benzacenaphthene; Benzo(jk)Fluorene; Fluoranteno (Spanish); Idryl; 1,2-(1,8-Naphthalenediyl)Benzene; 1,2-(1,8-Naphthylene)Benzene

CAS Registry Number: 206-44-0

RTECS Number: LL4025000

DOT ID: UN 3077

Regulatory Authority

- OSHA, 29CFR1910 Specifically Regulated Chemicals (See CFR 1910.1002) as coal tar pitch volatiles
- Clean Water Act: 40CFR401.15 Section 307 Toxic Pollutants; 40CFR423, Appendix A, Priority Pollutants
- EPA Hazardous Waste Number (RCRA No.): U120
- RCRA, 40CFR261, Appendix 8 Hazardous Constituents
- RCRA 40CFR268.48; 61FR15654, Universal Treatment Standards: Wastewater (mg/l), 0.068; Nonwastewater (mg/kg), 3.4
- RCRA 40CFR264, Appendix 9; TSD Facilities Ground Water Monitoring List. Suggested test method(s) (PQL μg/l): 8100 (200); 8270 (10)
- Superfund/EPCRA 40CFR302.4 Reportable Quantity (RQ): CERCLA, 100 lb (45.4 kg)

Cited in U.S. State Regulations: California (G), Illinois (G), Kansas (G), Louisiana (G), New Hampshire (G), New Jersey (G), Pennsylvania (G), Vermont (G), Virginia (G), Washington (G), Wisconsin (G).

Description: Fluoranthene, $C_{16}H_{10}$, is a polycyclic hydrocarbon and a colorless crystalline solid. Freezing/Melting point = 111°C. Boiling point = about 375°C. Hazard Identification (based on NFPA-704 M Rating System): Health 0, Flammability 1, Reactivity 0. Virtually insoluble in water.

Potential Exposure: Fluoranthene, a polynuclear aromatic hydrocarbon (PAH), is produced from the pyrolytic processing of organic raw materials such as coal and petroleum at high temperatures. It is also known to occur naturally as a product of plant biosynthesis. Fluoranthene is ubiquitous in the environment and has been detected in U.S. air, in foreign and domestic drinking waters and in food-stuffs. It is also contained in cigarette smoke.

Individuals living in areas which are heavily industrialized, and in which large amounts of fossil fuels are burned, would be expected to have greatest exposure from ambient sources of fluoranthene. In addition, certain occupations (e.g., coke oven workers, steelworkers, roofers, automobile mechanics) would also be expected to have elevated levels of exposure relative to the general population. Exposure to fluoranthene will be considerably increased among tobacco smokers or those who are exposed to smokers in closed environments (i.e., indoors).

Incompatibilities: Strong oxidizers.

Permissible Exposure Limits in Air: NIOSH recommends this chemical be regulated as an occupational carcinogen (coal tar pitch volatiles); ACGIH TLV as asphalt (petroleum) fumes 5 mg/m^3 TWA.

Determination in Air: Collection on a filter, extraction with benzene, chromatographic separation, spectrophotometric analysis. See NIOSH Methods #5506 (HPLC) and #5515 (GC) for polynuclear aromatic hydrocarbons.

Permissible Concentration in Water: The only existing standard which takes fluoranthene into consideration is a drinking water standard for PAHs. The 1970 World Health Organization European Standards for Drinking Water recommends a concentration of PAHs not exceeding 0.2 µg/l. This recommended standard is based upon the analysis of six PAHs in drinking water as follows: fluoranthene; Benzo(a)pyrene; Benzo(ghi)perylene; benzo(b) fluoranthene; benzo(k)fluoranthene; and indeno(1,2,3-cd) pyrene.

More recently EPA has established[6] ambient water criteria as follows: To protect freshwater aquatic life: 3,980 µg/l based on acute toxicity. To protect saltwater aquatic life: 40 µg/l based on acute toxicity and 16 µg/l based on chronic toxicity. To protect human health: 42 µg/l. Kansas[61] has set a guideline of 0.029 µg/l for fluoranthene in drinking water.

Determination in Water: Methylene chloride extraction followed by high pressure liquid chromatography with fluorescence as UV detection; or gas chromatography (EPA Method 610), or gas chromatography plus mass spectrometry (EPA Method 625).

Routes of Entry: Inhalation.

Harmful Effects and Symptoms

Long Term Exposure: There is concern about the toxicity of fluoranthene because it is widespread in the human environment and belongs to a class of compounds (polynuclear aromatic hydrocarbons) that contain numerous potent carcinogens. Experimentally, fluoranthene does not exhibit properties of a mutagen or primary carcinogen but it is a potent cocarciongen. In the laboratory, fluoranthene has also demonstrated toxicity to various freshwater and marine organisms. This finding, coupled with the cocarcinogenic properties of the compound, points out the need to protect humans and aquatic organisms from the potential hazards associated with fluoranthene in water.

First Aid: If this chemical gets into the eyes, remove any contact lenses at once and irrigate immediately for at least 15 minutes, occasionally lifting upper and lower lids. Seek medical attention immediately. If this chemical contacts the skin, remove contaminated clothing and wash immediately with soap and water. Seek medical attention immediately. If this chemical has been inhaled, remove from exposure, begin rescue breathing (using universal precautions) if breathing has stopped and CPR if heart action has stopped. Transfer promptly to a medical facility. When this chemical has been swallowed, get medical attention. Give large quantities of water and induce vomiting. Do not make an unconscious person vomit.

Personal Protective Methods: Wear protective gloves and clothing to prevent any reasonable probability of skin contact. Safety equipment suppliers/manufacturers can provide recommendations on the most protective glove/clothing material for your operation. All protective clothing (suits, gloves, footwear, headgear) should be clean, available each day, and put on before work. Contact lenses should not be worn when working with this chemical. Wear dust-proof chemical goggles and face shield unless full facepiece respiratory protection is worn. Employees should wash immediately with soap when skin is wet or contaminated. Provide emergency showers and eyewash.

Storage: Prior to working with this chemical you should be trained on its proper handling and storage. Store in tightly closed containers in a cool, well ventilated area away from oxidizers. Where possible, automatically transfer material from other storage containers to process containers.

Shipping: Environmentally hazardous solid, n.o.s. Hazard Class: 9. Label: "Class 9."

Spill Handling: Evacuate persons not wearing protective equipment from area of spill or leak until clean-up is complete. Remove all ignition sources. Collect powdered material in the most convenient and safe manner and deposit in sealed containers. Ventilate area after clean-up is complete. Contain and isolate spill to limit spread. Construct clay/bentonite swale to divert uncontaminated portion of watershed. It may be necessary to contain and dispose of this chemical as a hazardous waste. If material or contaminated runoff enters waterways, notify downstream users of potentially contaminated waters. Contact your Department of Environmental Protection or your regional office of the federal EPA for specific recommendations. If employees are required to clean-up spills, they must be properly trained and equipped. OSHA 1910.120(q) may be applicable.

Fire Extinguishing: This chemical is a combustible solid. Use dry chemical, carbon dioxide, water spray, or alcohol foam extinguishers. Poisonous gases are produced in fire. If material or contaminated runoff enters waterways, notify downstream users of potentially contaminated waters. Notify local health and fire officials and pollution control agencies. From a secure, explosion-proof location, use water spray to cool exposed containers. If cooling streams are ineffective (venting sound increases in volume and pitch, tank discolors, or shows any signs of deforming), withdraw immediately to a secure position. If employees are expected to fight fires, they must be trained and equipped in OSHA 1910.156.

Disposal Method Suggested: Incineration.[22] Consult with environmental regulatory agencies for guidance on acceptable disposal practices. Generators of waste containing this contaminant (≥100 kg/mo) must conform with

EPA regulations governing storage, transportation, treatment, and waste disposal.

References

U.S. Environmental Protection Agency, Fluoranthene: Ambient Water Quality Criteria, Washington, DC (1980)

U.S. Environmental Protection Agency, Fluoranthene, Health and Environmental Effects Profile No. 103, Office of Solid Waste, Washington, DC (April 30, 1980)

Sax, N. I., Ed., "Dangerous Properties of Industrial Materials Report," 7, No. 2, 80–84 (1987)

Fluorene

Molecular Formula: $C_{13}H_{10}$

Synonyms: *o*-Biphenylenemethane; *o*-Biphenylmethane; α-Diphenylenemethane; Diphenylenemethane; 9H-Fluorene; Fluoreno (Spanish); 2,2'-Methylenebiphenyl

CAS Registry Number: 86-73-7

RTECS Number: LL5670000

DOT ID: UN 3077

Regulatory Authority

- OSHA, 29CFR1910 Specifically Regulated Chemicals (See CFR 1910.1002) as coal tar pitch volatiles
- Clean Water Act: 40CFR423, Appendix A, Priority Pollutants
- RCRA 40CFR268.48; 61FR15654, Universal Treatment Standards: Wastewater (mg/l), 0.059; Nonwastewater (mg/kg), 3.4
- RCRA 40CFR264, Appendix 9; TSD Facilities Ground Water Monitoring List. Suggested test method(s) (PQL μg/l): 8100 (200); 8270 (10)
- Superfund/EPCRA 40CFR302.4 Reportable Quantity (RQ): CERCLA, 5,000 lb (2,270 kg)

Cited in U.S. State Regulations: California (G), New Jersey (G), Pennsylvania (G).

Description: Fluorene, when pure, is found as dazzling-white flakes or small, crystalline plates. Fluorescent when impure. Boiling Point = 293°C (decomposes). Freezing/Melting point = 116 – 117°C. Flash point = 151°C. Hazard Identification (based on NFPA-704 M Rating System): Health 1, Flammability 1, Reactivity 0. Insoluble in water.

Potential Exposure: Fluorene is used in resins, dyes, and is a chemical intermediate. A polycyclic aromatic hydrocarbon (PAH).

Incompatibilities: Oxidizers.

Permissible Exposure Limits in Air: As coal tar pitch volatiles, the OSHA PEL is 0.2 mg/m^3 TWA (benzene-soluble fraction). NIOSH recommends 0.1 mg/m^3 TWA, and ACGIH recommends 0.2 mg/m^3 TWA. The NIOSH IDLH = (Ca) potential occupational carcinogen 80 mg/m^3. Fluorene has been tested for cancer and the results are inconclusive.

Determination in Air: See NIOSH analytical methods for polynuclear hydrocarbons #5506 (GLC) and # 5515 (G.C). See also OSHA Method #58.

Permissible Concentration in Water: In view of the carcinogenicity of polynuclear aromatic, the concentration in water is preferably zero as noted by EPA.[6]

Routes of Entry: Inhalation, skin and/or eyes.

Harmful Effects and Symptoms

Short Term Exposure: Fluorene can irritate and burn the eyes and skin.

Long Term Exposure: No chronic effects are known at this time.

First Aid: If this chemical gets into the eyes, remove any contact lenses at once and irrigate immediately for at least 15 minutes, occasionally lifting upper and lower lids. Seek medical attention immediately. If this chemical contacts the skin, remove contaminated clothing and wash immediately with soap and water. Seek medical attention immediately. If this chemical has been inhaled, remove from exposure, begin rescue breathing (using universal precautions) if breathing has stopped and CPR if heart action has stopped. Transfer promptly to a medical facility. When this chemical has been swallowed, get medical attention. Give large quantities of water and induce vomiting. Do not make an unconscious person vomit.

Personal Protective Methods: Wear protective gloves and clothing to prevent any reasonable probability of skin contact. Safety equipment suppliers/manufacturers can provide recommendations on the most protective glove/clothing material for your operation. All protective clothing (suits, gloves, footwear, headgear) should be clean, available each day, and put on before work. Contact lenses should not be worn when working with this chemical. Wear dust-proof chemical goggles and face shield unless full facepiece respiratory protection is worn. Employees should wash immediately with soap when skin is wet or contaminated. Provide emergency showers and eyewash.

Respirator Selection: Where the potential for exposure to this chemical, use a MSHA/NIOSH approved supplied-air respirator with a full facepiece operated in the positive pressure mode or with a full facepiece, hood, or helmet in the continuous flow mode, or use a MSHA/NIOSH approved self-contained breathing apparatus with a full facepiece operated in pressure-demand or other positive pressure mode.

Storage: Prior to working with this chemical you should be trained on its proper handling and storage. Store in tightly closed containers in a cool, well ventilated area away from oxidizers and source of ignition.

Shipping: Environmentally hazardous solid, n.o.s. Hazard Class: 9. Label: "Class 9." Packing Group: III.

Spill Handling: Evacuate persons not wearing protective equipment from area of spill or leak until clean-up is complete. Remove all ignition sources. Use vacuum or wet method to reduce dust during cleanup. Do not dry sweep. Collect powdered material in the most convenient and safe manner and deposit in sealed containers. Ventilate area after clean-up is complete. It may be necessary to contain and dispose of this chemical as a hazardous waste. If material or contaminated runoff enters waterways, notify downstream users of potentially contaminated waters. Contact your Department of Environmental Protection or your regional office of the federal EPA for specific recommendations. If employees are required to clean-up spills, they must be properly trained and equipped. OSHA 1910.120(q) may be applicable.

Fire Extinguishing: This chemical may burn but does not easily ignite. Use dry chemical, carbon dioxide, water spray, or alcohol or polymer foam extinguishers. Poisonous gases are produced in fire. If material or contaminated runoff enters waterways, notify downstream users of potentially contaminated waters. Notify local health and fire officials and pollution control agencies. From a secure, explosion-proof location, use water spray to cool exposed containers. If cooling streams are ineffective (venting sound increases in volume and pitch, tank discolors, or shows any signs of deforming), withdraw immediately to a secure position. If employees are expected to fight fires, they must be trained and equipped in OSHA 1910.156.

References

New Jersey Department of Health and Senior Services, "Hazardous Substance Fact Sheet, Fluorene," Trenton, NJ (May 1999)

Fluorides

Molecular Formula: F_yM_x

Synonyms: Fluoride(1-); Fluoride ion; Fluoride ion(1-); Perfluoride

CAS Registry Number: 16984-48-8 (fluoride)

RTECS Number: LM6290000

Regulatory Authority

- Air Pollutant Standard Set (ACGIH)[1] (DFG)[3] (HSE)[33] (former USSR)[43] (OSHA)[58] (Several States)[60]
- Safe Drinking Water Act: Regulated chemical (47 FR 9352); MCL, 4.0 mg/l; MCLG, 4.0 mg/l; SMCL, 2.0 mg/l.
- Canada, WHMIS, Ingredients Disclosure List

Note: Several specific fluoride compounds are regulated by USEPA, OSHA and International regulatory bodies.

Cited in U.S. State Regulations: Alaska (G), Arizona (W), Illinois (G), Iowa (A), Maine (G, W), Massachusetts (A), New Hampshire (G), New Jersey (G), North Dakota (A), Oklahoma (G), Pennsylvania (G), Rhode Island (G), Virginia (A), West Virginia (G).

Description: Of the general formula M_xF_y, appearance, odor and properties vary with specific compounds.

Potential Exposure: Fluorides are used as an electrolyte in aluminum manufacture, a flux in smelting nickel, copper, gold, and silver, as a catalyst for organic reactions, a wood preservative, fluoridation agent for drinking water, a bleaching agent for cane seats, in pesticides, rodenticides, and as a fermentation inhibitor. They are utilized in the manufacture of steel, iron, glass, ceramics, pottery, enamels, in the coagulation of latex, in coatings for welding rods, and in cleaning graphite, metals, windows, and glassware. Exposure to fluorides may also occur during preparation of fertilizer from phosphate rock by addition of sulfuric acid. Air pollution by fluoride dusts and gases has done substantial damage to vegetation and to animals in the vicinity of industrial fluoride sources. However, the contribution of ambient air to human fluoride intake is only a few hundredths of a milligram per day, an amount that is insignificant in comparison with other sources o fluoride. Operations that introduce fluoride dusts and gases into the atmosphere include: Grinding, drying, and calcining of fluoride-containing minerals; acidulation of the minerals; smelting; electrochemical reduction of metals with fluoride fluxes or melts as in the aluminum and steel industry; kiln firing of brick and other clay products and the combustion of coal.

Incompatibilities: Fluorides form explosive gases on contact with strong acids or acid fumes.

Permissible Exposure Limits in Air: The OSHA TWA[58] is 2.5 mg/m^3 which is also the value set by ACGIH,[1] DFG[3] and HSE.[33] The former USSR-UNEP/IRPTC project[43] has set MAC values for ambient air in residential areas as follows (values all in mg/m^3):

Form	Momentary MAC	Average Daily MAC
Gaseous (HF, SiF_4)	0.02	0.005
Readily soluble (NaF)	0.03	0.01
Poorly soluble (CaF_2)	0.20	0.03

A number of states have set guidelines or standards for fluorides in ambient air[60] ranging from 2.85 μg/m^3 (Iowa) to 25.0 μg/m^3 (North Dakota) to 34.0 μg/m^3 (Massachusetts) to 40.0 μg/m^3 (Virginia).

Determination in Air: Gaseous fluorides collected by impinger using caustic; particulates by filter. Analysis is by ion-specific electrode per NIOSH Method 7902.[18]

Permissible Concentration in Water: The EPA has set a standard of 4 mg/l for fluoride[61] and the State of Maine has set 2.4 mg/l as a guideline for drinking water. Arizona[61] has set 1.8 mg/l as a standard for drinking water.

Routes of Entry: Inhalation, ingestion, eye and/or skin contact.

Harmful Effects and Symptoms

Short Term Exposure: Fluorides can affect you when breathed in. Fluorides can irritate and may damage the eyes. Skin contact can cause irritation, rash or burning sensation. High repeated exposure can cause nausea, vomiting, loss of appetite, and bone and teeth changes. Extremely high levels could be fatal. Breathing can irritate the nose and throat, and cause nausea, headaches and nosebleeds. Higher exposures can cause pulmonary edema, a medical emergency that can be delayed for several hours. This can cause death. Very high exposure can cause fluoride poisoning with stomach pain, weakness, convulsions, collapse and death. These effects do not occur at the level of fluorides used in water for preventing cavities in teeth.

Long Term Exposure: Repeated high exposures may affect kidneys. Repeated high exposures can cause deposits of fluorides in the bones (fluorosis) that may cause pain, disability and mottling of the teeth. Repeated exposure can cause nausea, vomiting, loss of appetite, diarrhea or constipation. Nosebleeds and sinus problems can also occur.

Points of Attack: Eyes, respiratory system, central nervous system, skeleton, kidneys, skin.

Medical Surveillance: For those with frequent or potentially high exposure (half the TLV or greater), the following are recommended before beginning work and at regular times after that: Lung function tests. Fluoride level in urine. (for fluoride in Urine use NIOSH #8308). Levels higher than 4 mg/l may indicate overexposure. If symptoms develop or overexposure is suspected, the following may be useful: Consider chest x-ray after acute overexposure. Kidney function tests. Consider chest x-ray following acute overexposure.

First Aid: If this chemical gets into the eyes, remove any contact lenses at once and irrigate immediately for at least 15 minutes, occasionally lifting upper and lower lids. Seek medical attention immediately. If this chemical contacts the skin, remove contaminated clothing and wash immediately with large amounts of soap and water. Seek medical attention immediately. If this chemical has been inhaled, remove from exposure, begin rescue breathing (using universal precautions) if breathing has stopped and CPR if heart action has stopped. Transfer promptly to a medical facility. When this chemical has been swallowed, get medical attention. Give large quantities of water and induce vomiting. Do not make an unconscious person vomit. Medical observation is recommended for 24 – 48 hours after breathing overexposure, as pulmonary edema may be delayed. As first aid for pulmonary edema, a doctor or authorized paramedic may consider administering a corticosteroid spray.

Personal Protective Methods: In areas with excessive gas or dust levels for any type of fluorine, worker protection should be provided. Respiratory protection by dust masks or gas masks with an appropriate canister or supplied air respirator should be provided. Goggles or fullface masks should be used. In areas where there is a likelihood of splash or spill, acid resistant clothing including gloves, gauntlets, aprons, boots, and goggles or face shield should be provided to the worker.

Personal hygiene should be encouraged, with showering following each shift and before changing to street clothes. Work clothes should be changed following each shift, especially in dusty areas. Attention should be given promptly to any burns from fluorine compounds due to absorption of the fluorine at the burn site and the possibility of developing systemic symptoms from absorption from burn sites.

Respirator Selection: NIOSH: *fluorides: 12.5 mg/m³:* DM (any dust and mist respirator). *25 mg/m³:* DMXSQ* (any dust and mist respirator except single-use and quarter mask respirators); or SA* (any supplied-air respirator). *62.5 mg/m³:* SA:CF* (any supplied-air respirator operated in a continuous-flow mode); or PAPRDM*+ *If not present as a fume* (any powered, air-purifying respirator with a dust and mist filter). *125 mg/m³:* HiEF+ (any air-purifying, full-facepiece respirator with a high-efficiency particulate filter); or SCBAF (any self-contained breathing apparatus with a full facepiece); or SAF (any supplied-air respirator with a full facepiece). *250 mg/m³:* SA:PD,PP (any supplied-air respirator operated in a pressure-demand or other positive-pressure mode). *Emergency or planned entry into unknown concentrations or IDLH conditions*: SCBAF:PD,PP (any self-contained breathing apparatus that has a full faceplate and is operated in a pressure-demand or other positive-pressure mode); or SAF:PD,PP:ASCBA (any supplied-air respirator that has a full facepiece and is operated in a pressure-demand or other positive-pressure mode in combination with an auxiliary, self-contained breathing apparatus operated in a pressure-demand or other positive-pressure mode). *Escape:* HiEF+ (any air-purifying, full-facepiece respirator with a high-efficiency particulate filter); or SCBAE (any appropriate escape-type, self-contained breathing apparatus).

* Substance reported to cause eye irritation or damage; may require eye protection.

\+ May need acid gas sorbent.

Storage: Prior to working with this chemical you should be trained on its proper handling and storage. Fluorides must be stored to avoid contact with strong acids (such as hydrochloric, sulfuric, and nitric) since violent reactions occur. Fluorides form explosive gases on contact with nitric acid. Store in tightly closed containers in a cool, well-ventilated area away from water.

Spill Handling: *Liquid:* Evacuate and restrict persons not wearing protective equipment from area of spill or leak until cleanup is complete. Remove all ignition sources. Ventilate area of spill or leak. Absorb liquids in vermiculite, dry sand, earth, peat, carbon, or a similar material and deposit in sealed containers. It may be necessary to contain and dispose of this chemical as a hazardous waste. If material or contaminated runoff enters waterways, notify downstream users of potentially contaminated waters. Contact your Department of Environmental Protection or your regional office of the federal EPA for specific recommendations. If employees are required to clean-up spills, they must be properly trained and equipped. OSHA 1910.120(q) may be applicable.

Solid material: Evacuate persons not wearing protective equipment from area of spill or leak until clean-up is complete. Remove all ignition sources. Collect powdered material in the most convenient and safe manner and deposit in sealed containers. Ventilate area after clean-up is complete. It may be necessary to contain and dispose of this chemical as a hazardous waste. If material or contaminated runoff enters waterways, notify downstream users of potentially contaminated waters. Contact your Department of Environmental Protection or your regional office of the federal EPA for specific recommendations. If employees are required to clean-up spills, they must be properly trained and equipped. OSHA 1910.120(q) may be applicable.

Fire Extinguishing: Do not use water. Poisonous gases are produced in fire. If material or contaminated runoff enters waterways, notify downstream users of potentially contaminated waters. Notify local health and fire officials and pollution control agencies. Containers may explode in fire. From a secure, explosion-proof location, use water spray to cool exposed containers. If cooling streams are ineffective (venting sound increases in volume and pitch, tank discolors, or shows any signs of deforming), withdraw immediately to a secure position. If employees are expected to fight fires, they must be trained and equipped in OSHA 1910.156.

Disposal Method Suggested: Reaction of aqueous waste with an excess of lime, followed by lagooning, and either recovery or land disposal of the separated calcium fluoride.

References

National Institute for Occupational Safety and Health, Criteria for a Recommended Standard: Occupational Exposure to Inorganic Fluoride, NIOSH Doc. No. 76–103 (1976)

National Academy of Sciences, Medical and Biologic Effect of Environmental Pollutants: Fluoride, Washington, DC (1971)

New Jersey Department of Health and Senior Services, "Hazardous Substance Fact Sheet: Fluoride," Trenton, NJ (January 1986)

Fluorine

Molecular Formula: F_2

Synonyms: Bifluoriden (Dutch); Fluor (Dutch, French, German, Polish, Spanish); Fluorine-19; Fluoro (Italian); Fluorures acide (French); Fluoruri acidi (Italian); Saeure fluoride (German)

CAS Registry Number: 7782-41-4

RTECS Number: LM6475000

DOT ID: UN 1045 (gas); UN 9192 (liquid)

EEC Number: 009-001-00-0

Regulatory Authority

- Air Pollutant Standard Set (ACGIH)[1] (DFG)[3] (HSE)[33] (OSHA)[58] (Several States)[60] (Several Canadian Provinces)
- Clean Air Act: Accidental Release Prevention/Flammable substances, (Section 112[r], Table 3), TQ = 1,000 lb (454 kg)
- EPA Hazardous Waste Number (RCRA No.): P056
- RCRA, 40CFR261, Appendix 8 Hazardous Constituents
- EPCRA Section 313 Form R *de minimis* concentration reporting level: 1.0%.
- Superfund/EPCRA 40CFR355, Appendix B Extremely Hazardous Substances: TPQ = 500 lb (227 kg)
- Superfund/EPCRA 40CFR302.4 Reportable Quantity (RQ): CERCLA, 10 lb (4.54 kg)
- Canada, WHMIS, Ingredients Disclosure List

Cited in U.S. State Regulations: Alaska (G), Connecticut (A), Florida (G, A), Illinois (G), Kansas (G), Louisiana (G), Maine (G), Massachusetts (G), Nevada (A), New Hampshire (G), New Jersey (G), New York (G, A), North Carolina (A), North Dakota (A), Oklahoma (G), Pennsylvania (G), Rhode Island (G), Vermont (G), Virginia (G, A), Washington (G), West Virginia (G), Wisconsin (G).

Description: Fluorine, F_2, is a yellow compressed, gas with a characteristic pungent odor. The odor threshold is 0.035 ppm.[41] Boiling point = -188°C. Freezing/Melting point = -219°C. Reaction with water.

Potential Exposure: Elemental fluorine is used in the conversion of uranium tetrafluoride to uranium hexafluoride, in the synthesis of organic and inorganic fluorine compounds, and as an oxidizer in rocket fuel.

Incompatibilities: Water, nitric acid, oxidizers, organic compounds. Reacts violently with reducing agents, ammonia, all combustible materials, metals [except the metal containers in which it is shipped]. Reacts violently with H_2O to form hydrofluoric acid and ozone

Permissible Exposure Limits in Air: The Federal limit (OSHA PEL)[58] and the DFG MAK[3] is 0.1 ppm (0.2 mg/m^3). The ACGIH has adopted a TWA of 1.0 ppm (2.0 mg/m^3) and an STEL value of 2.0 ppm (4.0 mg/m^3) as has HSE.[33] The NIOSH IDLH level is 25 ppm. Several states have set guidelines or standards for fluorine in ambient air[60] ranging from zero (North Carolina) to 4.0 μg/m^3 (Connecticut) to 6.7 μg/m^3 (New York) to 20.0 μg/m^3 (Florida) to 30.0 μg/m^3 (Virginia) to 20.0 – 40.0 μg/m^3 (North Dakota) to 48.0 μg/m^3 (Nevada).

Determination in Air: No test available.

Permissible Concentration in Water: The former USSR-UNEP/IRPTC project[43] has set a MAC in water bodies used for domestic purposes of 1.5 mg/l. See also the entry on fluorides.

Routes of Entry: Inhalation, eye and/or skin contact.

Harmful Effects and Symptoms

Short Term Exposure: Small amounts of gas in air can have a strong caustic effect on the cornea, eyelids, nose.

Inhalation: Corrosive. Inhalation of fluorine causes coughing, choking, and chills. Higher exposures can cause pulmonary edema, a medical emergency that can be delayed for several hours. This can cause death. A symptomatic period of 1 – 2 days followed by fever, cough, tightness in chest, and cyanosis indicate pulmonary edema. Inhalation of extremely high levels may cause suffocation. It is also reported that volunteers exposed to levels of 10 ppm for up to 15 minutes reported no irritation. Levels of 25 – 75 ppm caused increasing irritation. Exposure to levels of 100 ppm for 1 minute caused strong respiratory irritation.

Skin: Corrosive to the skin. It is also reported that exposure to levels between 90 – 240 ppm produced slight irritation of the skin and a "sticky" feeling of face. Contact with liquid fluorine may cause chemical burns and frostbite.

Eyes: Corrosive to the eyes; can lead to permanent eye damage. Contact with the liquid may cause frostbite. It is also reported that mild irritation was reported at exposure to levels of 25 ppm for 5 minutes. Levels of 100 ppm for 1 minute produced marked irritation.

Ingestion: No reported exposures by this route.

Long Term Exposure: Prolonged exposure can cause fluorine to concentrate in the bones causing osteosclerosis which may be disabling. The teeth can become mottled. Repeated exposure can cause nosebleeds, nausea, vomiting, loss of appetite, diarrhea and kidneys. These effects do not occur when fluorine is used to treat drinking water to prevent cavities. It is also reported that levels of 0.1 ppm intermittently, over a prolonged period produced no ill effects.

Points of Attack: Eyes, skin, respiratory system, liver, kidneys.

Medical Surveillance: Before beginning employment and at regular times after that, the following is recommended: Urinary fluoride level. Consider chest x-ray following acute overexposure. Liver and kidney function tests. DEXA bone density scan (Dual Energy X-Ray Absorptiometry).

First Aid: If this chemical gets into the eyes, remove any contact lenses at once and irrigate immediately for at least 15 minutes, occasionally lifting upper and lower lids. Seek medical attention immediately. If this chemical contacts the skin, remove contaminated clothing and wash immediately with soap and water. Seek medical attention immediately. If this chemical has been inhaled, remove from exposure, begin rescue breathing (using universal precautions) if breathing has stopped and CPR if heart action has stopped. Transfer promptly to a medical facility. When this chemical has been swallowed, get medical attention. Give large quantities of water and induce vomiting. Do not make an unconscious person vomit. If frostbite has occurred, seek medical attention immediately; do *NOT* rub the affected areas or flush them with water. In order to prevent further tissue damage, do *NOT* attempt to remove frozen clothing from frostbitten areas. If frostbite has *NOT* occurred, immediately and thoroughly wash contaminated skin with soap and water.

Personal Protective Methods: Wear protective gloves and clothing to prevent any reasonable probability of skin contact. Safety equipment suppliers/manufacturers can provide recommendations on the most protective glove/clothing material for your operation. All protective clothing (suits, gloves, footwear, headgear) should be clean, available each day, and put on before work. Contact lenses should not be worn when working with this chemical. Wear nonvented, impact resistant goggles when working with fumes, gases, or vapors. When working with liquids, wear splash-proof chemical goggles and face shield unless full facepiece respiratory protection is worn. Employees should wash immediately with soap when skin is wet or contaminated. Provide emergency showers and eyewash. Where exposure to the liquefied compressed gas may occur, employees should be provided with special clothing designed to prevent frostbite.

Respirator Selection: NIOSH/OSHA as fluorides *12.5 mg/m³:* DM (any dust and mist respirator). *25 mg/m³:* DMXSQ* (any dust and mist respirator except single-use and quarter mask respirators); or SA* (any supplied-air respirator). *62.5 mg/m³:* SA:CF* (any supplied-air respirator operated in a continuous-flow mode); PAPRDM*+ (if not present as a fume) (any powered, air-purifying respirator with a dust and mist filter). *125 mg/m³:* HiEF+ (any air-purifying, full-facepiece respirator with a high-efficiency particulate filter); or SCBAF (any self-contained breathing apparatus with a full facepiece); or SAF (any supplied-air respirator with a full facepiece). *250 mg/m³:* SA:PD,PP (any supplied-air respirator operated in a pressure-demand or other positive-pressure mode). *Emergency or planned entry into unknown concentrations or IDLH conditions:* SCBAF:PD,PP (any self-contained breathing apparatus that has a full faceplate and is operated in a pressure-demand or other positive-pressure mode); or SAF:PD,PP:ASCBA (any supplied-air respirator that has a full facepiece and is operated in a pressure-demand or other positive-pressure mode in combination with an auxiliary, self-contained breathing apparatus operated in a pressure-demand or other positive- pressure mode). *Escape:* HiEF+ (any air-purifying, full-facepiece respirator with a high-efficiency particulate filter); or SCBAE (any appropriate escape-type, self-contained breathing apparatus).

* Substance reported to cause eye irritation or damage; may require eye protection.

\+ May need acid gas sorbent.

Storage: Prior to working with this chemical you should be trained on its proper handling and storage. Fluorine must be stored to avoid contact with most oxidizable materials because it frequently causes them to start on fire. It will frequently ignite bromine; iodine; sulfur; alkaline metals (such as sodium and potassium) and a number of organic chemicals (such as benzene and ethyl alcohol). It should not contact nitric acid, because an explosive gas will be produced. Contact with hydrogen or amorphous silicon dioxide will cause an explosion. Store in containers such as cylinders, in a cool, well-ventilated area away from heat, water and steam. Heat can cause cylinders, to burst. Contact with water or steam can produce heat and corrosive and poisonous gases.

Shipping: Compressed fluorine requires a shipping label of: "Poison Gas, Oxidizer." It falls in DOT Hazard Class 2.3 and Packing Group I. Passenger aircraft or railcar shipment is forbidden and cargo aircraft shipment is forbidden also.

Spill Handling: Restrict persons not wearing protective equipment from area of leak until clean-up is complete. Ventilate area of leak to disperse the gas. Stop flow of gas. If source of leak is a cylinder and the leak cannot be stopped in place, remove the leaking cylinder to a safe place in the open air, and repair leak or allow cylinder to empty. If liquid fluorine is spilled, clear the area and allow it to evaporate. Vapors are heavier than air and will collect in low areas. Do not use water or wet method. It may be necessary to contain and dispose of this chemical as a hazardous waste. Contact your Department of Environmental Protection or your regional office of the federal EPA for specific recommendations. If employees are required to clean-up spills, they must be properly trained and equipped. OSHA 1910.120(q) may be applicable.

Fire Extinguishing: Non-flammable, but reacts chemically with any material capable of burning. Poisonous gases are produced in fire. Fluorine can ignite combustible materials and may increase the severity of an ongoing fire. Use extinguisher appropriate to the burning material. Do not spray water directly on leaking fluorine as poisonous gases are produced. Keep sealed fire-exposed cylinder cool by spraying with water. For small fire, used dry chemical or carbon dioxide. For large fire, use water spray, fog, or foam. For massive fire in cargo area, use unmanned hose holder or monitor nozzles. A few whiffs of the gas or vapor could cause death. Gas, vapor of liquid could be fatal on penetrating the firefighters' normal full protective clothing. Only special protective clothing designed to protect against fluorine should be used; the normal full protective clothing available to the average fire department will not provide adequate protection. Do not direct water onto fluorine leaks as the fire may be intensified. Do *NOT* spray water on leaking cylinder (to prevent corrosion of cylinder). Turn leaking cylinder with the leak up to prevent escape of gas in liquid state.

Disposal Method Suggested: Fluorine can be combusted by means of a fluorine-hydrocarbon air burner followed by a caustic scrubber and stack. Consult with environmental regulatory agencies for guidance on acceptable disposal practices. Generators of waste containing this contaminant (≥100 kg/mo) must conform with EPA regulations governing storage, transportation, treatment, and waste disposal.

References

Sax, N. I., Ed., "Dangerous Properties of Industrial Materials Report," 1, No. 4, 68–70 (1981) and 3, No. 4, 50–53 (1983)

U.S. Environmental Protection Agency, "Chemical Profile: Fluorine," Washington, DC, Chemical Emergency Preparedness Program (November 30, 1987)

New Jersey Department of Health and Senior Services, "Hazardous Substance Fact Sheet: Fluorine," Trenton, NJ (March 1999)

New York State Department of Health, "Chemical Fact Sheet: Fluorine," Albany, NY, Bureau of Toxic Substance Assessment (May 1986)

Fluoroacetamide

Molecular Formula: C_2H_4FNO

Common Formula: CH_2FCONH_2

Synonyms: AFL 1081; Compound 1081; FAA; Fluorakil 100; 2-Fluoroacetamide; Fluoroacetic acid amide; Fussol; Megatox; Monofluoroacetamide; Navron; Rodex; Yanock

CAS Registry Number: 640-19-7

RTECS Number: AC1225000

Regulatory Authority

- Banned or Severely Restricted (In Agriculture) (Several Countries) (UN)[13]
- EPA Hazardous Waste Number (RCRA No.): P057
- RCRA, 40CFR261, Appendix 8 Hazardous Constituents.
- Superfund/EPCRA 40CFR355, Appendix B Extremely Hazardous Substances: TPQ = 100/10,000 lb (45.4/4,540 kg)
- Superfund/EPCRA 40CFR302.4 Reportable Quantity (RQ): CERCLA, 100 lb (45.4 kg)

Cited in U.S. State Regulations: California (G), Kansas (G), Louisiana (G), Massachusetts (G), New Hampshire (G), New Jersey (G), Pennsylvania (G), Vermont (G), Virginia (G), Washington (G), Wisconsin (G).

Description: Fluoroacetamide, CH_2FCONH_2, is a colorless crystalline solid. Freezing/Melting point = 107 – 109°C. Hazard Identification (based on NFPA-704 M Rating System): Health 4, Flammability 1, Reactivity 0. Soluble in water.

Potential Exposure: This material is a rodenticide; insecticide proposed mainly for use on fruits to combat scale insects, aphids, and mites. Use is largely restricted to licensed pest control operators.

Incompatibilities: Strong oxidizers.

Permissible Exposure Limits in Air: No standards set.

Permissible Concentration in Water: No criteria set.

Routes of Entry: Inhalation, skin and/or eyes.

Harmful Effects and Symptoms

Short Term Exposure: Signs and symptoms may be extremely severe and range from nausea, vomiting, and diarrhea to convulsions, coma, and heart failure. Other symptoms include hyperactivity, respiratory depression or arrest, cyanosis (blue tint to the skin and mucous membranes), and ventricular fibrillation.

This material is super toxic; probable oral lethal dose in humans is less than 5 mg/kg, or a taste (less than 7 drops) for a 150-lb person. Chemically inhibits oxygen metabolism by cells with critical damage occurring to the heart, brain, and lungs resulting in heart failure, respiratory arrest, convulsions, and death.

Warning: Effects usually appear within 30 minutes of exposure but may be delayed as long as 20 hours. Caution is advised. Vital signs should be monitored closely.

Long Term Exposure: See Fluorides.

Medical Surveillance: See Fluorides.

First Aid: Acute exposure to fluoroacetamide may require decontamination and life support for the victim. Emergency personnel should wear protective clothing appropriate to the type and degree of contamination. Air-purifying or supplied-air respiratory equipment should also be worn, as necessary. Rescue vehicles should carry supplies such as plastic sheeting and disposable plastic bags to assist in preventing spread of contamination.

Inhalation: Move victim to fresh air. Evaluate vital signs. If no pulse is detected, provide CPR. If not breathing, provide artificial respiration. If breathing is labored, administer oxygen. Rush to a health care facility.

Eye Exposure: Remove any contact lenses at once and flush eyes with lukewarm water for 15 minutes.

Skin Exposure: Follow steps under inhalation above. Wash exposed skin areas 3 times with soap and water. Rush to health care facility.

Ingestion: Evaluate vital signs including pulse and respiratory rate, and note any trauma. If no pulse is detected, provide CPR. If not breathing, provide artificial respiration. If breathing is labored, administer oxygen or other respiratory support. Rush to health care facility. Obtain authorization and/or further instructions from the local hospital for performance of other invasive procedures.

Warning: Effects usually appear within 30 minutes of exposure but may be delayed as long as 20 hours. Caution is advised. Vital signs should be monitored closely.

Personal Protective Methods: For emergency situations, wear a positive pressure, pressure-demand, full facepiece self-contained breathing apparatus (SCBA) or pressure-demand supplied air respirator with escape SCBA and a fully-encapsulating, chemical resistant suit. Wear protective gloves and clothing to prevent any reasonable probability of skin contact. Safety equipment suppliers/manufacturers can provide recommendations on the most protective glove/clothing material for your operation. All protective clothing (suits, gloves, footwear, headgear) should be clean, available each day, and put on before work. Contact lenses should not be worn when working with this chemical. Wear dust-proof chemical goggles and face shield unless full facepiece respiratory protection is worn. Employees should wash immediately with soap when skin is wet or contaminated. Provide emergency showers and eyewash.

Respirator Selection: Where the potential for exposure to this chemical, use a MSHA/NIOSH approved supplied-air respirator with a full facepiece operated in the positive pressure mode or with a full facepiece, hood, or helmet in the continuous flow mode, or use a MSHA/NIOSH approved self-contained breathing apparatus with a full facepiece operated in pressure-demand or other positive pressure mode.

Storage: Prior to working with this chemical you should be trained on its proper handling and storage. Store in tightly closed containers in a cool, well ventilated area.

Shipping: The restrictions for sodium fluoroacetate may be applicable. This compound requires a shipping label of: "Poison." It falls in DOT Hazard Class 6.1 and Packing Group I.

Spill Handling: Evacuate persons not wearing protective equipment from area of spill or leak until clean-up is complete. Remove all ignition sources. Do not touch spilled material; stop leak if you can do so without risk. Small spills: absorb with sand or other non-combustible absorbent material and place into containers for later disposal. Large spills: dike spill for later disposal. Keep unnecessary people away; isolate hazard area and deny entry. Stay upwind; keep out of low areas. Collect powdered material in the most convenient and safe manner and deposit in sealed containers. Ventilate area after clean-up is complete. It may be necessary to contain and dispose of this chemical as a hazardous waste. If material or contaminated runoff enters waterways, notify downstream users of potentially contaminated waters. Contact your Department of Environmental Protection or your regional office of the federal EPA for specific recommendations. If employees are required to clean-up spills, they must be properly trained and equipped. OSHA 1910.120(q) may be applicable.

Fire Extinguishing: This chemical is a combustible solid. Use dry chemical, carbon dioxide, water spray, or alcohol foam extinguishers. Poisonous gases are produced in fire including fluorine and nitrogen oxides. If material or contaminated runoff enters waterways, notify downstream users of potentially contaminated waters. Notify local health and fire officials and pollution control agencies. From a secure, explosion-proof location, use water spray to cool exposed containers. If cooling streams are ineffective (venting sound increases in volume and pitch, tank discolors, or shows any signs of deforming), withdraw immediately to a secure position. If employees are expected to fight fires, they must be trained and equipped in OSHA 1910.156.

Disposal Method Suggested: Consult with environmental regulatory agencies for guidance on acceptable disposal practices. Generators of waste containing this contaminant ($\geq$100 kg/mo) must conform with EPA regulations governing storage, transportation, treatment, and waste disposal.

References

U.S. Environmental Protection Agency, "Chemical Profile: Fluoroacetamide," Washington, DC, Chemical Emergency Preparedness Program (November 30, 1987)

Fluoroacetic Acid

Molecular Formula: $C_2H_3FO_2$

Common Formula: FCH_2COOH

Synonyms: Acide monofluoracetique (French); Acido fluoroacetico (Spanish); Acido monofluoroacetio (Italian); Cymonic acid; FAA; Fluoroacetate; 2-Fluoroacetic acid; Fluoroethanoic acid; Gifblaar poison; HFA; MFA; Monofluorazijnzuur (Dutch); Monofluoressigsaeure (German); Monofluoroacetate; Monofluoroacetic acid

CAS Registry Number: 144-49-0

RTECS Number: AH5950000

DOT ID: UN 2642

EEC Number: 607-081-00-7

Regulatory Authority

- Very Toxic Substance (World Bank)[13]
- Superfund/EPCRA 40CFR355, Appendix B Extremely Hazardous Substances: TPQ = 10/10,000 lb (4.54/4,540 kg)
- Superfund/EPCRA 40CFR302.4 Reportable Quantity (RQ): EHS, 1 lb (0.454 kg)

Cited in U.S. State Regulations: California (G), Massachusetts (G), New Hampshire (G), New Jersey (G), Oklahoma (G), Pennsylvania (G).

Description: Fluoroacetic acid, FCH_2COOH, is a colorless crystalline solid. Boiling point = 165°C. Freezing/Melting point = 35°C. Soluble in water.

Potential Exposure: This material is used as a rodenticide.

Incompatibilities: Strong oxidizers.

Permissible Exposure Limits in Air: The sodium salt has a TWA of 0.05 mg/m^3 as set by ACGIH,[1] DFG[3] and HSE[33] as well as OSHA[58] and STEL of 0.15 mg/m^3 with the notation "skin" indication the possibility of cutaneous absorption.

Permissible Concentration in Water: No standards set for the acid but see the entry on "Sodium Fluoroacetate."

Harmful Effects and Symptoms

Short Term Exposure: Corrosive to the eyes, skin, and respiratory tract. The major symptoms of fluoroacetic acid poisoning include severe epileptiform convulsions alternating with coma and depression; death may result from asphyxiation during convulsion or from respiratory failure. Cardiac irregularities, such as ventricular fibrillation and sudden cardiac arrest, nausea, vomiting, excessive salivation, numbness, tingling sensations, epigastric pain, mental apprehension, muscular twitching, low blood pressure, and blurred vision may also occur. This material is very toxic; the LD_{50} oral-rat is 4.7 mg/kg (extremely toxic), and may affect the cardiovascular system, central nervous system, and kidneys and may cause cardiac and renal failure. This may cause death.

Long Term Exposure: See information for short-term exposure.

Points of Attack: Central nervous system, heart, kidneys, lungs.

Medical Surveillance: Kidney function tests. EKG. Lung function tests. Examination of the nervous system. Consider chest x-ray following acute exposure.

First Aid: If this chemical gets into the eyes, remove any contact lenses at once and irrigate immediately for at least 15 minutes, occasionally lifting upper and lower lids. Seek medical attention immediately. If this chemical contacts the skin, remove contaminated clothing and wash immediately with soap and water. Seek medical attention immediately. If this chemical has been inhaled, remove from exposure, begin rescue breathing (using universal precautions) if breathing has stopped and CPR if heart action has stopped. Transfer promptly to a medical facility. When this chemical has been swallowed, get medical attention. Give large quantities of water and induce vomiting. Do not make an unconscious person vomit. The symptoms of central nervous system, cardiac, and renal failure do not become manifest until a few hours have passed. Specific treatment is necessary in case of poisoning with this substance; the appropriate means with instructions must be available.

Personal Protective Methods: For emergency situations, wear a positive pressure, pressure-demand, full facepiece self-contained breathing apparatus (SCBA) or pressure-demand supplied air respirator with escape SCBA and a fully-encapsulating, chemical resistant suit. Wear protective gloves and clothing to prevent any reasonable probability of skin contact. Safety equipment suppliers/manufacturers can provide recommendations on the most protective glove/clothing material for your operation. All protective clothing (suits, gloves, footwear, headgear) should be clean, available each day, and put on before work. Contact lenses should not be worn when working with this chemical. Wear dust-proof chemical goggles and face shield unless full facepiece respiratory protection is worn. Employees should wash immediately with soap when skin is wet or contaminated. Provide emergency showers and eyewash.

Respirator Selection: Where the potential for exposure to this chemical, use a MSHA/NIOSH approved supplied-air respirator with a full facepiece operated in the positive pressure mode or with a full facepiece, hood, or helmet in the continuous flow mode, or use a MSHA/NIOSH approved self-contained breathing apparatus with a full facepiece operated in pressure-demand or other positive pressure mode.

Storage: Prior to working with this chemical you should be trained on its proper handling and storage. Store in tightly closed containers in a cool, well ventilated area away from oxidizers.

Shipping: This compound requires a shipping label of: "Poison." It falls in DOT Hazard Class 6.1 and Packing Group I.

Spill Handling: Evacuate persons not wearing protective equipment from area of spill or leak until clean-up is complete. Remove all ignition sources. Do not touch spilled material. Stop leak if you can do so without risk. Stay upwind; keep out of low areas. Use water spray to reduce vapors. For small spills, take up with sand or other noncombustible absorbent material and place into containers for later disposal. For large spills, dike spill for later disposal. Ventilate area after clean-up is complete. It may be necessary to contain and dispose of this chemical as a hazardous waste. If material or contaminated runoff enters waterways, notify downstream users of potentially contaminated waters. Contact your Department of Environmental Protection or your regional office of the federal EPA for specific recommendations. If employees are required to clean-up spills, they must be properly trained and equipped. OSHA 1910.120(q) may be applicable.

Fire Extinguishing: Small fires: use dry chemical, carbon dioxide, water spray, or foam. For large fires, use water spray, fog, or foam. Stay upwind; keep out of low areas. Wear self-contained, positive pressure breathing apparatus and full protective clothing. Move container from fire area. Cool containers that are exposed to flames with water from the side until well after fire is out. Use dry chemical, carbon dioxide, water spray, or alcohol foam extinguishers. Poisonous gases are produced in fire including nitrogen oxides. If material or contaminated runoff enters waterways, notify downstream users of potentially contaminated waters. Notify local health and fire officials and pollution control agencies. From a secure, explosion-proof location, use water spray to cool exposed containers. If cooling streams are ineffective (venting sound increases in volume and pitch, tank discolors, or shows any signs of deforming), withdraw immediately to a secure position. If employees are expected to fight fires, they must be trained and equipped in OSHA 1910.156.

Disposal Method Suggested: Dissolve in a combustible solvent. Spray the solution into a furnace equipped with an afterburner.[24]

References

U.S. Environmental Protection Agency, "Chemical Profile: Fluoroacetic Acid," Washington, DC, Chemical Emergency Preparedness Program (November 30, 1987)

Fluoroacetyl Chloride

Molecular Formula: C_2H_2ClFO

Common Formula: FCH_2COCl

Synonyms: Acetyl chloride, fluoro-; TL 670

CAS Registry Number: 359-06-8

RTECS Number: AO6825000

Regulatory Authority

- Superfund/EPCRA 40CFR355, Appendix B Extremely Hazardous Substances: TPQ = 10 lb (4.54 kg)

Cited in U.S. State Regulations: Massachusetts (G), New Jersey (G), Pennsylvania (G).

Description: Fluoroacetyl chloride, FCH_2COCl, is a liquid. Hazard Identification (based on NFPA-704 M Rating System): Health 4, Flammability 1, Reactivity 2. Reacts with water.

Potential Exposure: May be used in organic synthesis.

Incompatibilities: Violent reaction with water.

Permissible Exposure Limits in Air: No standards set.

Permissible Concentration in Water: No criteria set.

Harmful Effects and Symptoms

Short Term Exposure: Highly toxic by inhalation. Corrosive to skin and irritating to eyes.

First Aid: If this chemical gets into the eyes, remove any contact lenses at once and irrigate immediately for at least 15 minutes, occasionally lifting upper and lower lids. Seek medical attention immediately. If this chemical contacts the skin, remove contaminated clothing and wash immediately with soap and water. Seek medical attention immediately. If this chemical has been inhaled, remove from exposure, begin rescue breathing (using universal precautions) if breathing has stopped and CPR if heart action has stopped. Transfer promptly to a medical facility. When this chemical has been swallowed, get medical attention. Give large quantities of water and induce vomiting. Do not make an unconscious person vomit. Effects may be delayed; keep victim under observation.

Personal Protective Methods: For emergency situations, wear a positive pressure, pressure-demand, full facepiece self-contained breathing apparatus (SCBA) or pressure-demand supplied air respirator with escape SCBA and a fully-encapsulating, chemical resistant suit. Wear protective gloves and clothing to prevent any reasonable probability of skin contact. Safety equipment suppliers/manufacturers can provide recommendations on the most protective glove/clothing material for your operation. All protective clothing (suits, gloves, footwear, headgear) should be clean, available each day, and put on before work. Contact lenses should not be worn when working with this chemical. Wear splash-proof chemical goggles and face shield unless full facepiece respiratory protection is worn. Employees should wash immediately with soap when skin is wet or contaminated. Provide emergency showers and eyewash.

Respirator Selection: Where the potential for exposure to this chemical, use a MSHA/NIOSH approved supplied-air respirator with a full facepiece operated in the positive pressure mode or with a full facepiece, hood, or helmet in the continuous flow mode, or use a MSHA/NIOSH approved self-contained breathing apparatus with a full facepiece operated in pressure-demand or other positive pressure mode.

Storage: Prior to working with this chemical you should be trained on its proper handling and storage. Store in tightly closed containers in a cool, well ventilated area.

Spill Handling: Evacuate and restrict persons not wearing protective equipment from area of spill or leak until cleanup

is complete. Remove all ignition sources. Ventilate area of spill or leak. Stay upwind; keep out of low areas. Do not touch spilled material; stop leak if you can do so without risk. Use water spray to reduce vapors (may react violently with water). Small spills: absorb with sand or other non-combustible absorbent material and place into containers for later disposal. Large spills: dike far ahead of spill for later disposal. It may be necessary to contain and dispose of this chemical as a hazardous waste. If material or contaminated runoff enters waterways, notify downstream users of potentially contaminated waters. Contact your Department of Environmental Protection or your regional office of the federal EPA for specific recommendations. If employees are required to clean-up spills, they must be properly trained and equipped. OSHA 1910.120(q) may be applicable.

Fire Extinguishing: This material is a combustible solid and water reactive. Extinguish with dry chemical, carbon dioxide. Spray cooling water on unopened containers that are exposed to flames until well after fire is out. Move container from fire area if you can do so without risk. Do not get water in container, as material may react violently with water. Poisonous gases are produced in fire including phosgene. If material or contaminated runoff enters waterways, notify downstream users of potentially contaminated waters. Notify local health and fire officials and pollution control agencies. From a secure, explosion-proof location, use water spray to cool exposed containers. If cooling streams are ineffective (venting sound increases in volume and pitch, tank discolors, or shows any signs of deforming), withdraw immediately to a secure position. If employees are expected to fight fires, they must be trained and equipped in OSHA 1910.156.

References

U.S. Environmental Protection Agency, "Chemical Profile: Fluoroacetyl Chloride," Washington, DC, Chemical Emergency Preparedness Program (November 30, 1987)

Fluorobenzene

Molecular Formula: C_6H_5F

Synonyms: Benzene fluoride; Benzene, fluoro-; MFB; Monofluorobenzene; Phenyl fluoride

CAS Registry Number: 462-06-6

RTECS Number: DA0800000

DOT ID: UN 2387

Cited in U.S. State Regulations: Florida (G), Massachusetts (G), New Hampshire (G), New Jersey (G), Pennsylvania (G).

Description: Fluorobenzene is a colorless liquid. Boiling point = 85°C. Flash point = -15°C. Insoluble in water.

Potential Exposure: Fluorobenzene is used as an insecticide and as a reagent for plastic or resin polymers.

Incompatibilities: Oxidizers.

Permissible Exposure Limits in Air: No standards set.

Permissible Concentration in Water: No criteria set.

Routes of Entry: Inhalation, passing through the skin.

Harmful Effects and Symptoms

Short Term Exposure: Fluorobenzene can irritate the eyes, nose, throat and lungs. Higher exposures can cause pulmonary edema, a medical emergency that can be delayed for several hours. This can cause death. A closely related chemical, chlorobenzene, can damage the liver and kidneys with high or repeated exposure. It is unknown if fluorobenzene causes these effects. Overexposure could cause headache, nausea and make you dizzy.

Long Term Exposure: May cause liver and kidney damage. Repeated exposure may damage the lungs and affect the nervous system.

Points of Attack: Lungs, liver, kidney, nervous system.

Medical Surveillance: Before beginning employment and at regular times after that, for those with frequent or potentially high exposures, the following are recommended: Periodic lung function tests. If symptoms develop or overexposure is suspected, the following may also be useful: Tests for kidney and liver function. Examination of the nervous system. Consider chest x-ray after acute overexposure.

First Aid: If this chemical gets into the eyes, remove any contact lenses at once and irrigate immediately for at least 15 minutes, occasionally lifting upper and lower lids. Seek medical attention immediately. If this chemical contacts the skin, remove contaminated clothing and wash immediately with soap and water. Seek medical attention immediately. If this chemical has been inhaled, remove from exposure, begin rescue breathing (using universal precautions) if breathing has stopped and CPR if heart action has stopped. Transfer promptly to a medical facility. When this chemical has been swallowed, get medical attention. Give large quantities of water and induce vomiting. Do not make an unconscious person vomit. Medical observation is recommended for 24 – 48 hours after breathing overexposure, as pulmonary edema may be delayed.

Personal Protective Methods: Wear protective gloves and clothing to prevent any reasonable probability of skin contact. Safety equipment suppliers/manufacturers can provide recommendations on the most protective glove/clothing material for your operation. All protective clothing (suits, gloves, footwear, headgear) should be clean, available each day, and put on before work. Contact lenses should not be worn when working with this chemical. Wear splash-proof chemical goggles and face shield unless full facepiece respiratory protection is worn. Employees should wash immediately with soap when skin is wet or contaminated. Provide emergency showers and eyewash.

Respirator Selection: Prior to working with this chemical you should be trained on its proper handling and storage.

Where the potential for exposures to fluorobenzene exists, use a MSHA/NIOSH approved supplied-air respirator with a full facepiece operated in the positive pressure mode or with a full facepiece, hood, or helmet in the continuous flow mode, or use a MSHA/NIOSH approved self-contained breathing apparatus with a full facepiece operated in pressure-demand or other positive pressure mode.

Storage: Fluorobenzene must be stored to avoid contact with oxidizers since violent reactions occur. Store in tightly closed containers in a cool, well-ventilated area. Sources of ignition, such as smoking and open flames, are prohibited where fluorobenzene is handled, used, or stored. Metal containers involving the transfer of 5 gallons or more of fluorobenzene should be grounded and bonded. Drums must be equipped with self-closing valves, pressure vacuum bungs, and flame arresters. Use only non-sparking tools and equipment, especially when opening and closing containers of fluorobenzene.

Shipping: This compound requires a shipping label of: "Flammable Liquid." It falls in DOT Hazard Class 3 and Packing Group II.

Spill Handling: Evacuate and restrict persons not wearing protective equipment from area of spill or leak until cleanup is complete. Remove all ignition sources. Ventilate area of spill or leak. Absorb liquids in vermiculite, dry sand, earth, peat, carbon, or a similar material and deposit in sealed containers. It may be necessary to contain and dispose of this chemical as a hazardous waste. If material or contaminated runoff enters waterways, notify downstream users of potentially contaminated waters. Contact your Department of Environmental Protection or your regional office of the federal EPA for specific recommendations. If employees are required to clean-up spills, they must be properly trained and equipped. OSHA 1910.120(q) may be applicable.

Fire Extinguishing: This chemical is a flammable liquid. Poisonous gases including carbon monoxide and fluorine are produced in fire. Use dry chemical, carbon dioxide, or foam extinguishers. Vapors are heavier than air and will collect in low areas. Vapors may travel long distances to ignition sources and flashback. Vapors in confined areas may explode when exposed to fire. Containers may explode in fire. Storage containers and parts of containers may rocket great distances, in many directions. If material or contaminated runoff enters waterways, notify downstream users of potentially contaminated waters. Notify local health and fire officials and pollution control agencies. From a secure, explosion-proof location, use water spray to cool exposed containers. If cooling streams are ineffective (venting sound increases in volume and pitch, tank discolors, or shows any signs of deforming), withdraw immediately to a secure position. If employees are expected to fight fires, they must be trained and equipped in OSHA 1910.156.

References

New Jersey Department of Health and Senior Services, "Hazardous Substance Fact Sheet: Fluorobenzene," Trenton, NJ (March 1999)

Fluorotrichloromethane

Molecular Formula: CCl_3F

Synonyms: Algofrene type 1; Arctron 9; CFC-11; Electro-CF 11; Eskimon 11; F 11; FC 11; Fluorocarbon 11; Fluorochloroform; Fluorotrojchlorometan (Polish); Freon 11; Freon HE; Freon MF; Frigen 11; Genetron 11; Halocarbon 11; Isceon 131; Isotron 11; Ledon 11; Methane, fluorotrichloro; Methane, trichlorofluoro-; Monoflurotrichloromethane; NCI-C04637; Propellant 11; R 11; Refrigerant 11; Trichlorofluoromethane; Trichloromonofluoromethane; Ucon fluorocarbon 11; Ucon refrigerant 11

CAS Registry Number: 75-69-4

RTECS Number: PB6125000

EINECS Number: 200-892-3

Regulatory Authority

- Air Pollutant Standard Set (ACGIH)[1] (DFG)[3] (HSE)[33] (OSHA)[58] (Other Countries)[35] (Several States)[60] (Several Canadian Provinces)
- Clean Air Act: Stratospheric ozone protection (Title VI, Subpart A, Appendix A), Class I, Ozone Depletion Potential = 1.0.
- EPA Hazardous Waste Number (RCRA No.): U121.
- RCRA, 40CFR261, Appendix 8 Hazardous Constituents
- RCRA 40CFR268.48; 61FR15654, Universal Treatment Standards: Wastewater (mg/l), 0.020; Nonwastewater (mg/kg), 30
- RCRA 40CFR264, Appendix 9; TSD Facilities Ground Water Monitoring List. Suggested test method(s) (PQL μg/l): 8010 (10); 8240 (5)
- Superfund/EPCRA 40CFR302.4 Reportable Quantity (RQ): CERCLA, 5,000 lb (2,270 kg)
- EPCRA Section 313 Form R *de minimis* concentration reporting level: 1.0%
- Canada, WHMIS, Ingredients Disclosure List

Cited in U.S. State Regulations: Alaska (G), Arizona (W), California (A,W), Connecticut (A), Florida (G), Illinois (G), Kansas (G, W), Louisiana (G), Maine (G, W), Maryland (W), Massachusetts (G), Nevada (A), New Hampshire (G), New Jersey (G), North Carolina (A), North Dakota (A), Pennsylvania (G), Rhode Island (G), Vermont (G), Virginia (G, A), Washington (G), West Virginia (G), Wisconsin (G, W).

Description: Fluorotrichloromethane, CCl_3F, is a colorless liquid or gas with a chlorinated solvent odor. The odor threshold is 5.0 ppm. Boiling point = 24°C. Freezing/Melting point = -111°C. Soluble in water.

Potential Exposure: This material is used as a refrigerant, aerosol propellant and foaming agent.

Incompatibilities: Chemically active and powdered metals: aluminum, barium, sodium, potassium, calcium, powdered aluminum, zinc, magnesium.

Permissible Exposure Limits in Air: The OSHA PEL is 1,000 ppm (5,600 mg/m^3) TWA. The recommended ACGIH ceiling is 1,000 ppm. HSE[33] has set this same value as an 8-hour TWA and added 1,250 ppm (7,000 mg/m^3) as an STEL value. DFG[3] has set a MAC of 1,000 ppm (5,600 mg/m^3). Sweden[35] has set a TWA of 500 ppm (3,000 mg/m^3) and an STEL of 750 ppm (4,500 mg/m^3). The former USSR has set a MAC in workplace air of 180 ppm (1,000 mg/m^3) and a MAC in ambient air in residential areas of 100 mg/m^3 on a once-daily basis and 10 mg/m^3 on a daily average basis.

Several states have set guidelines or standards for F-11 in ambient air[60] ranging from 13.0 mg/m^3 (Virginia) to 56.0 mg/m^3 (North Dakota) to 112.0 mg/m^3 (Connecticut) to 133.33 mg/m^3 (Nevada) to 560 mg/m^3 (North Carolina). Beyond this, general concern about destruction of the ozone layer above the earth has prompted restrictions on chlorofluorocarbon use and venting to the atmosphere. Substitute materials are being developed to reduce emissions drastically. The NIOSH IDLH level is 10,000 ppm.

Determination in Air: Charcoal adsorption, workup with CS_2, analysis by gas chromatography/flame ionization. See NIOSH IV, Method #1006.[18]

Permissible Concentration in Water: For the protection of human health: preferably zero. An additional lifetime cancer risk of 1 in 100,000 results at a level of 1.9 μg/l. In January 1981 EPA (46FR2266) removed F-11 from the priority toxic pollutant list. A number of states have developed guidelines for F-11 in drinking water[61] ranging from 1.0 μg/l (Arizona) to 2,300 μg/l (Maine) to 3,400 μg/l (California) to 3,500 μg/l (Wisconsin) to 8,000 μg/l (Kansas) to 8,750 μg/l (Maryland).

Determination in Water: Inert gas purge followed by gas chromatography with halide specific detection (EPA Method 601) or gas chromatography plus mass spectrometry (EPA Method 624).

Routes of Entry: Inhalation, ingestion, skin and/or eye contact. Absorbed through the skin.

Harmful Effects and Symptoms

Short Term Exposure: Irritates the skin and eyes. Overexposure can cause lightheadedness, dizziness, incoherence, tremors, cardiac arrythmia, asphyxiation., cardiac arrest. This can occur without other warning symptoms. The liquid may cause frostbite.

Long Term Exposure: Repeated or prolonged contact with skin may cause dryness and cracking. Can irritate the lungs causing coughing and/or shortness of breath.

Points of Attack: Skin, cardiovascular system.

Medical Surveillance: For those with frequent or potentially high exposure (half the TLV or grater), the following are recommended before beginning work and at regular times after that: Lung function tests. If symptoms develop or overexposure is suspected, the following may be useful: Consider Holter monitor (a special 24 hour EKG to look for irregular heart beat).

First Aid: If this chemical gets into the eyes, remove any contact lenses at once and irrigate immediately for at least 15 minutes, occasionally lifting upper and lower lids. Seek medical attention immediately. If this chemical contacts the skin, remove contaminated clothing and wash immediately with soap and water. Seek medical attention immediately. If this chemical has been inhaled, remove from exposure, begin rescue breathing (using universal precautions) if breathing has stopped and CPR if heart action has stopped. Transfer promptly to a medical facility. When this chemical has been swallowed, get medical attention. Give large quantities of water and induce vomiting. Do not make an unconscious person vomit.

Personal Protective Methods: Wear protective gloves and clothing to prevent any reasonable probability of skin contact. Safety equipment suppliers/manufacturers can provide recommendations on the most protective glove/clothing material for your operation. All protective clothing (suits, gloves, footwear, headgear) should be clean, available each day, and put on before work. Contact lenses should not be worn when working with this chemical. Wear eye protection and face shield unless full facepiece respiratory protection is worn. Employees should wash immediately with soap when skin is wet or contaminated. Provide emergency showers and eyewash.

Respirator Selection: NIOSH/OSHA: *2,000 ppm:* SA (any supplied-air respirator); or SCBAF (any self-contained breathing apparatus with a full facepiece). *Emergency or planned entry into unknown concentrations or IDLH conditions:* SCBAF:PD,PP (any self-contained breathing apparatus that has a full facepiece and is operated in a pressure-demand or other positive-pressure mode); or SAF:PD,PP:ASCBA (any supplied-air respirator that has a full facepiece and is operated in a pressure-demand or other positive-pressure mode in combination with an auxiliary self-contained breathing apparatus operated in a pressure-demand or other positive pressure mode). *Escape:* GMFOV [any air-purifying, full-facepiece respirator (gas mask) with a chin-style, front-or back-, mounted organic vapor canister]; or SCBAE (any appropriate escape-type, self-contained breathing apparatus).

Storage: Prior to working with this chemical you should be trained on its proper handling and storage. Trichlorofluoromethane must be stored to avoid contact with chemically active metals such as aluminum or lithium since violent reactions occur. Store in tightly closed containers in a cool, well-ventilated area away from sources of heat.

Shipping: This material is not specifically cited by DOT. It may be considered as a hazardous substance, liquid, n.o.s. which imposes no label requirements or shipping weight limits.

Spill Handling: Evacuate and restrict persons not wearing protective equipment from area of spill or leak until cleanup is complete. Remove all ignition sources. Ventilate area of spill or leak. Absorb liquids in vermiculite, dry sand, earth, peat, carbon, or a similar material and deposit in sealed containers. It may be necessary to contain and dispose of this chemical as a hazardous waste. If material or contaminated runoff enters waterways, notify downstream users of potentially contaminated waters. Contact your Department of Environmental Protection or your regional office of the federal EPA for specific recommendations. If employees are required to clean-up spills, they must be properly trained and equipped. OSHA 1910.120(q) may be applicable.

Fire Extinguishing: This chemical is a noncombustible liquid. Poisonous gases including hydrogen fluoride, hydrogen chloride and phosgene are produced in fire. Use extinguishing agents suitable for surrounding fire. Vapors are heavier than air and will collect in low areas. Vapors may travel long distances to ignition sources and flashback. Vapors in confined areas may explode when exposed to fire. Containers may explode in fire. Storage containers and parts of containers may rocket great distances, in many directions. If material or contaminated runoff enters waterways, notify downstream users of potentially contaminated waters. Notify local health and fire officials and pollution control agencies. From a secure, explosion-proof location, use water spray to cool exposed containers. If cooling streams are ineffective (venting sound increases in volume and pitch, tank discolors, or shows any signs of deforming), withdraw immediately to a secure position. If employees are expected to fight fires, they must be trained and equipped in OSHA 1910.156.

Disposal Method Suggested: Incineration, preferably after mixing with another combustible fuel. Care must be exercised to assure complete combustion to prevent the formation of phosgene. An acid scrubber is necessary to remove the halo acids produced. Consult with environmental regulatory agencies for guidance on acceptable disposal practices. Generators of waste containing this contaminant (≥100 kg/mo) must conform with EPA regulations governing storage, transportation, treatment, and waste disposal.

References

U.S. Environmental Protection Agency, Halomethanes: Ambient Water Quality Criteria, Washington, DC (1980)

U.S. Environmental Protection Agency, Trichlorofluoromethane and Dichlorodifluoromethane, Health and Environmental Effects Profile No. 167, Office of Solid Waste, Washington, DC (April 30, 1980)

Sax, N. I., Ed., "Dangerous Properties of Industrial Materials Report," 5, No. 6, 92–95 (1985)

New Jersey Department of Health and Senior Services, "Hazardous Substance Fact Sheet: Trichlorofluoromethane," Trenton, NJ (June 1998)

Fluorouracil

Molecular Formula: $C_4H_3FN_2O_2$

Synonyms: Adrucil; AI3-25297; Arumel; 2,4-Dioxo-5-fluoropyrimidine; Effluderm (free base); Efudex; Efudix; Efurix; Fluoroblastin; Fluoroplex; 5-Fluoro-2,4(1H,3H)-pyrimidinedione; 5-Fluoropyrimidine-2,4-dione; 5-Fluoro-2,4-pyriminedione; 5-Fluorouracil; Fluorouracile; Fluorouracilo; Fluorouracilum; 5-Fluoruracil (German); Fluracilum; Fluri; Fluril; Fluro Uracil; FT-207; 5-FU; FU; Kecimeton; NSC 19893; 2,4(1H,3H)-Pyrimidinedione, 5-fluoro-; Queroplex; RO 2-9757; Timazin; U-8953; Ulup; Uracil, 5-fluoro-

CAS Registry Number: 51-21-8

RTECS Number: YR0350000

Regulatory Authority

- Superfund/EPCRA 40CFR355, Appendix B Extremely Hazardous Substances: TPQ = 500/10,000 lb (227/4,540 kg)
- Superfund/EPCRA 40CFR302.4 Reportable Quantity (RQ): EHS, 1 lb (0.454 kg)
- EPCRA Section 313 Form R *de minimis* concentration reporting level: 1.0%
- California's Proposition 65: Reproductive toxin

Cited in U.S. State Regulations: California (G), Massachusetts (G), New Jersey (G), Pennsylvania (G).

Description: Fluorouracil, $C_4H_3FN_2O_2$, is a white crystalline solid which is practically odorless. Freezing/Melting point = 282 – 283°C (decomposes). Soluble in water.

Potential Exposure: This material is used as an antineoplastic drug, for cancer treatment, and as a chemosterilant for insects.

Incompatibilities: Oxidizers, strong bases, heat.

Permissible Exposure Limits in Air: No standards set.

Permissible Concentration in Water: No criteria set.

Routes of Entry: Ingestion, skin contact. Passes through the skin.

Harmful Effects and Symptoms

Short Term Exposure: Contact can irritate and burn the eyes and skin. Can cause headache, fatigue, dizziness, and mental confusion; nausea, vomiting, diarrhea, and abdominal pain. Minimum toxic dose in humans is approximately 450 mg/kg (total dose) over 30 days for the ingested drug. Intravenous minimum toxic dose in humans is a total dose of 6 mg/kg over 3 days. Depression of white blood cells occurred after intravenous administrative of a total dose of 480 mg/kg over 32 days. Occasional neuropathy and cardiac toxicity have been reported. Do not use during pregnancy. Patients with impaired hepatic or renal function, with a history of high-dose pelvic irradiation or previous use of alkylating agents should be treated with extreme caution. Patients with nutritional deficiencies and protein depletion have a reduced tolerance to fluorouracil.

Long Term Exposure: May decrease fertility in males and females. May cause skin allergy. Very high exposure may affect the heart. Loss of appetite and nausea are earliest symptoms, with other symptoms of diarrhea, inflammation or sores in the mouth, gastric burning, and intestinal discomfort. More serious symptoms are due to the suppression of bone marrow, with decrease of white cell count and blood platelets, and anemia. Hair loss, nail changes, dermatitis, and pigmentation and atrophy of skin also occur. Sunlight can exacerbate these effects.

Points of Attack: Blood, heart, skin.

Medical Surveillance: Complete blood count (CBC), EKG, evaluation by a qualified allergist.

First Aid: If this chemical gets into the eyes, remove any contact lenses at once and irrigate immediately for at least 15 minutes, occasionally lifting upper and lower lids. Seek medical attention immediately. If this chemical contacts the skin, remove contaminated clothing and wash immediately with soap and water. Seek medical attention immediately. If this chemical has been inhaled, remove from exposure, begin rescue breathing (using universal precautions) if breathing has stopped and CPR if heart action has stopped. Transfer promptly to a medical facility. When this chemical has been swallowed, get medical attention. Give large quantities of water and induce vomiting. Do not make an unconscious person vomit. Keep victim quiet and maintain normal body temperature.

Personal Protective Methods: For emergency situations, wear a positive pressure, pressure-demand, full facepiece self-contained breathing apparatus (SCBA) or pressure-demand supplied air respirator with escape SCBA and a fully-encapsulating, chemical resistant suit. Wear protective gloves and clothing to prevent any reasonable probability of skin contact. Safety equipment suppliers/manufacturers can provide recommendations on the most protective glove/clothing material for your operation. All protective clothing (suits, gloves, footwear, headgear) should be clean, available each day, and put on before work. Contact lenses should not be worn when working with this chemical. Wear dust-proof chemical goggles and face shield unless full facepiece respiratory protection is worn. Employees should wash immediately with soap when skin is wet or contaminated. Provide emergency showers and eyewash.

Respirator Selection: Where the potential for exposure to this chemical, use a MSHA/NIOSH approved supplied-air respirator with a full facepiece operated in the positive pressure mode or with a full facepiece, hood, or helmet in the continuous flow mode, or use a MSHA/NIOSH approved self-contained breathing apparatus with a full facepiece operated in pressure-demand or other positive pressure mode.

Storage: Prior to working with this chemical you should be trained on its proper handling and storage. Store in tightly closed containers in a cool place, possibly a refrigerator. Store at 59 – 86°F/15 – 30°C. Keep away from oxidizers, strong bases, and heat.

Shipping: This material is not specifically cited by DOT. It may be considered as a Hazardous Substance, solid, n.o.s. which imposes no label requirements or restrictions on shipping weight.

Spill Handling: Evacuate persons not wearing protective equipment from area of spill or leak until clean-up is complete. Remove all ignition sources. Stay upwind; keep out of low areas. If water pollution occurs, notify appropriate authorities. Spill or leak; shut off ignition sources; no flares, smoking or flames in hazard area. Keep combustibles (wood, paper, oil etc.) away from spilled material. Do not touch spilled material. Small spills: Collect powdered material in the most convenient and safe manner and deposit in sealed containers. Large spills: dike far ahead of spill for later disposal. Ventilate area after clean-up is complete. It may be necessary to contain and dispose of this chemical as a hazardous waste. If material or contaminated runoff enters waterways, notify downstream users of potentially contaminated waters. Contact your Department of Environmental Protection or your regional office of the federal EPA for specific recommendations. If employees are required to clean-up spills, they must be properly trained and equipped. OSHA 1910.120(q) may be applicable.

Fire Extinguishing: Use dry chemical, carbon dioxide, or alcohol or polymer foam extinguishers. Poisonous gases, including nitrogen oxides, hydrogen fluoride, and carbon monoxide, are produced in fire. If material or contaminated runoff enters waterways, notify downstream users of potentially contaminated waters. Notify local health and fire officials and pollution control agencies. From a secure, explosion-proof location, use water spray to cool exposed containers. If cooling streams are ineffective (venting sound increases in volume and pitch, tank discolors, or shows any signs of deforming), withdraw immediately to a secure position. If employees are expected to fight fires, they must be trained and equipped in OSHA 1910.156.

References

Sax, N. I., Ed., "Dangerous Properties of Industrial Materials Report," 8, No. 6, 64–73 (1988)

U.S. Environmental Protection Agency, "Chemical Profile: Fluorouracil," Washington, DC, Chemical Emergency Preparedness Program (November 30, 1987)

New Jersey Department of Health and Senior Services, "Hazardous Substance Fact Sheet, 5-Fluorouracil," Trenton, NJ (June 1999)

Fluosilicic Acid

Molecular Formula: F_6H_2Si

Common Formula: H_2SiF_6

Synonyms: Dihyrogen hydrofluorosilicate; Hexafluorosilicate(2-) dihydrogen; Hexafluorosilicic acid; Hexafluosilicic acid; Hydrofluosilicic acid; Hydrogen hexafluorosilicate; Hydrosilicofluoric acid; Kiezelfluorwaterstofzuur (Dutch); Sand acid; Silicofluoric acid

CAS Registry Number: 16961-83-4

RTECS Number: VV8225000

DOT ID: UN 1778

EEC Number: 009-011-00-5

Regulatory Authority

Air Pollutant Standard Set (ACGIH)[1] (DFG)[3] (HSE)[33] (OSHA)[58] (UNEP)[43]

Cited in U.S. State Regulations: Maine (G), New Hampshire (G), New Jersey (G), New York (G), Pennsylvania (G).

Description: Fluorosilicic acid, H_2SiF_6, is a transparent, colorless fuming liquid. Boiling point = about 100°C (decomposes). Freezing/Melting point ≤-20°C. It is not flammable. Soluble in water.

Potential Exposure: A solution of fluorosilicic acid is used for sterilization in the brewing and bottling industry, electrolytic refining of lead, electroplating, hardening cement, removing mold and others.

Incompatibilities: The aqueous solution is a strong acid reacts with water or steam to produce toxic and corrosive fumes of hydrogen fluoride. Incompatible, and may react violently with bases, aliphatic amines, alkanolamines, alkylene oxides, aromatic amines, amides, ammonia, ammonium hydroxide, calcium oxide, epichlorohydrin, isocyanates, oleum, organic anhydrides, sulfuric acid, strong oxidizers, vinyl acetate, water. Attacks glass, concrete, and ceramics. The anhydrous form dissociates almost instantly into silicon tetrafluoride and hydrogen fluoride.

Permissible Exposure Limits in Air: The OSHA TWA[58] is 2.5 mg/m^3 which is also the value set by ACGIH,[1] DFG[3] and HSE.[33] The former USSR-UNEP/IRPTC project[43] has set MAC values for ambient air in residential areas as 0.03 mg/m^3 on a momentary basis and 0.01 mg/m^3 on a daily average basis.[43]

Permissible Concentration in Water: No criteria set.

Routes of Entry: Inhalation, ingestion.

Harmful Effects and Symptoms

Short Term Exposure: Corrosive to the eyes, skin, and respiratory tract. Higher exposures can cause pulmonary edema, a medical emergency that can be delayed for several hours. This can cause death. Medical observation is recommended.

Inhalation: May cause difficult breathing and burning of the mouth, throat and nose which may result in bleeding. These may be felt at 7.5 mg/m^3. Nausea, vomiting, profuse sweating and excess thirst may occur at higher levels.

Ingestion: Corrosive. Most reported instances of fluoride toxicity are due to accidental ingestion and it is difficult to associate symptoms with dose. 5 – 40 mg may cause nausea, diarrhea, and vomiting. More severe symptoms of burning and painful abdomen, sores in mouth, throat and digestive tract, tremors, convulsions and shock will occur around a dose of 1 gram. Death may result in ingestion of 2 – 5 grams.

Long Term Exposure: Fluoride may increase bone density, stimulate new bone growth or cause calcium deposits in ligaments. This may become a problem at levels of 20 – 50 mg/m^3 or higher. May cause mottling of the bones teeth at this level, resulting in fluorosis. May cause lung damage.

Points of Attack: Bones, lungs.

Medical Surveillance: Lung function tests. Consider chest x-ray following acute overexposure. DEXA bone densitometry scan.

First Aid: If this chemical gets into the eyes, remove any contact lenses at once and irrigate immediately for at least 15 minutes, occasionally lifting upper and lower lids. Seek medical attention immediately. If this chemical contacts the skin, remove contaminated clothing and wash immediately with soap and water. Seek medical attention immediately. If this chemical has been inhaled, remove from exposure, begin rescue breathing (using universal precautions) if breathing has stopped and CPR if heart action has stopped. Transfer promptly to a medical facility. When this chemical has been swallowed, get medical attention. If victim is conscious, administer water or milk. Do not induce vomiting. Medical observation is recommended for 24 – 48 hours after breathing overexposure, as pulmonary edema may be delayed. As first aid for pulmonary edema, a doctor or authorized paramedic may consider administering a corticosteroid spray.

Note to Physician: Inject intravenously 10 ml of 10% calcium gluconate solution. Gastric lavage with lime water of 1% calcium chloride.

Personal Protective Methods: Use only with an effective and properly maintained exhaust ventilation or with a fully enclosed process. Wear protective gloves and clothing to prevent any reasonable probability of skin contact. Safety equipment suppliers/manufacturers can provide recommendations on the most protective glove/clothing material for your operation. All protective clothing (suits, gloves, footwear, headgear) should be clean, available each day, and put on before work. Contact lenses should not be worn when working with this chemical. Wear splash-proof chemical goggles and face shield unless full facepiece respiratory protection is worn. Remove any clothing that you think may have become chemically soiled and wash before reuse. Employees should wash immediately with soap when skin is wet or contaminated. Provide emergency showers and eyewash.

Respirator Selection: NIOSH/OSHA as fluorides *12.5 mg/m^3*: DM (any dust and mist respirator). *25 mg/m^3*: DMXSQ* (any dust and mist respirator except single-use and quarter mask respirators); or SA* (any supplied-air respirator). *62.5 mg/m^3*: SA:CF* (any supplied-air respirator operated in a continuous-flow mode); PAPRDM*+ (if not present as a fume) (any powered, air-purifying respirator with a dust and mist filter). *125 mg/m^3*: HiEF+ (any air-purifying, full-facepiece respirator with a high-efficiency particulate filter); or SCBAF (any self-contained breathing apparatus with a full facepiece); or SAF (any supplied-air respirator with a full facepiece). *250 mg/m^3*: SA:PD,PP (any supplied-air respirator operated in a pressure-demand or other positive-pressure mode). *Emergency or planned entry into unknown concentrations or IDLH conditions:* SCBAF:PD,PP (any self-contained breathing apparatus that has a full faceplate and is operated in a pressure-demand or other positive-pressure mode); or SAF: PD,PP:ASCBA (any supplied-air respirator that has a full facepiece and is operated in a pressure-demand or other positive-pressure mode in combination with an auxiliary, self-contained breathing apparatus operated in a pressure-demand or other positive- pressure mode). *Escape:* HiEF+ (any air-purifying, full-facepiece respirator with a high-efficiency particulate filter); or SCBAE (any appropriate escape-type, self-contained breathing apparatus).

* Substance reported to cause eye irritation or damage; may require eye protection.

\+ May need acid gas sorbent.

Storage: Prior to working with this chemical you should be trained on its proper handling and storage. Store in a cool, dry area that is well-ventilated. Protect from damage. Avoid acids. Concentrated solution can be stored in glass, but lead is preferred.

Shipping: This compound requires a shipping label of: "Corrosive." It falls in DOT Hazard Class 8 and Packing Group II.

Spill Handling: Evacuate and restrict persons not wearing protective equipment from area of spill or leak until cleanup is complete. Remove all ignition sources. Ventilate area of spill or leak. Enter only with protective clothing and devices. Treat with soda ash or slaked lime. Dilute with water. Use an industrial vacuum cleaner to remove the spill. It may be necessary to contain and dispose of this chemical as a hazardous waste. If material or contaminated runoff enters waterways, notify downstream users of potentially contaminated waters. Contact your Department of Environmental Protection or your regional office of the federal EPA for specific recommendations. If employees are required to cleanup spills, they must be properly trained and equipped. OSHA 1910.120(q) may be applicable.

Fire Extinguishing: Material itself is not flammable. Do not use water. Use CO_2 or dry chemicals on surrounding fire. Containers may explode in fire. Storage containers and parts of containers may rocket great distances, in many directions. If material or contaminated runoff enters waterways, notify downstream users of potentially contaminated waters. Notify local health and fire officials and pollution control agencies. From a secure, explosion-proof location, use water spray to cool exposed containers. If cooling streams are ineffective (venting sound increases in volume and pitch, tank discolors, or shows any signs of deforming), withdraw immediately to a secure position. If employees are expected to fight fires, they must be trained and equipped in OSHA 1910.156.

Disposal Method Suggested: Add slowly to a large amount of soda ash in solution.[24] Discharge to sewer with large volumes of water.

References

New York State Department of Health, "Chemical Fact Sheet: Fluosilicic Acid," Albany, NY, Bureau of Toxic Substance Assessment (March 1986)

Flurazepam

Molecular Formula: $C_{21}H_{23}ClFN_3O$

Synonyms: 7-Chloro-1-[2-(diethylamino)ethyl]-5-(2-fluorophenyl)-1H-1,4-benzodiazepin-2(3H)-one; Dalmane®; Felmane; Noctosom; Ro-5-6901/3; Stauroderm

CAS Registry Number: 17617-23-1

RTECS Number: DF2368050

Cited in U.S. State Regulations: New Jersey (G).

Description: Flurazepam, $C_{21}H_{23}ClFn_3O$, is a pale yellow crystalline solid. Freezing/Melting point = 80°C.

Potential Exposure: Flurazepam is used as a sedative in capsules or liquid form.

Permissible Exposure Limits in Air: No standards set.

Permissible Concentration in Water: No criteria set.

Routes of Entry: Inhalation, ingestion.

Harmful Effects and Symptoms

Short Term Exposure: Flurazepam can affect you when breathed in. Flurazepam is used as a medical drug. When taken in that way, it can cause drowsiness and difficulty with coordination, concentration and balance. It can also cause irritability, anxiety, weakness, headache, upset stomach and joint pains. Effects may last for 1 – 3 days. It is not known for certain if these effects occur from occupational exposure. Drinking alcohol after exposure may worsen the symptoms caused by flurazepam. Person taking lithium (a medication) could have a serious reaction with flurazepam exposure.

Long Term Exposure: Exposure to flurazepam may cause jaundice, skin rash, and a low white blood cell count. Similar compounds are known teratogens. Suddenly discontinuing exposure after high exposure for at 3 months may cause shakiness, irritability, and convulsions.

Points of Attack: Blood, liver, skin.

Medical Surveillance: If symptoms develop or overexposure is suspected, the following may be useful: Blood flurazepam level. Complete blood Count (CBC). Liver function tests. Evaluation by a qualified allergist.

First Aid: *Eye Contact:* Immediately remove any contact lenses and flush with large amounts of water for at least 15 minutes, occasionally lifting upper and lower lids.

Skin Contact: Remove contaminated clothing. Wash contaminated skin with water.

Personal Protective Methods: Wear protective gloves and clothing to prevent any reasonable probability of skin contact. Safety equipment suppliers/manufacturers can provide

recommendations on the most protective glove/clothing material for your operation. All protective clothing (suits, gloves, footwear, headgear) should be clean, available each day, and put on before work. Contact lenses should not be worn when working with this chemical. Wear dust-proof chemical goggles and face shield unless full facepiece respiratory protection is worn. Employees should wash immediately with soap when skin is wet or contaminated. Provide emergency showers and eyewash.

Respirator Selection: Where the potential for high exposures exists, use a MSHA/NIOSH approved supplied-air respirator with a full facepiece operated in the positive pressure mode or with a full facepiece, hood, or helmet in the continuous flow mode, or use a MSHA/NIOSH approved self-contained breathing apparatus with a full facepiece operated in pressure-demand or other positive pressure mode.

Storage: Store in tightly closed containers in a cool, well-ventilated area away from sources of heat. If you are required to work in a "sterile" environment you require special training.

Shipping: This material falls into the DOT category of medicines, poisonous, solid, n.o.s. This category requires a shipping label of: "Keep Away From Food." It falls in DOT Hazard Class 6.1 and Packing Group III.

Spill Handling: Evacuate and restrict persons not wearing protective equipment from area of spill or leak until cleanup is complete. Remove all ignition sources. Ventilate area of spill or leak. Absorb liquids in vermiculite, dry sand, earth, peat, carbon, or a similar material and deposit in sealed containers. Collect powdered material in the most convenient and safe manner and deposit in sealed containers. It may be necessary to contain and dispose of this chemical as a hazardous waste. If material or contaminated runoff enters waterways, notify downstream users of potentially contaminated waters. Contact your Department of Environmental Protection or your regional office of the federal EPA for specific recommendations. If employees are required to clean-up spills, they must be properly trained and equipped. OSHA 1910.120(q) may be applicable.

Fire Extinguishing: Use extinguishing agents suitable for surrounding fire. Poisonous gases are produced in fire including chlorine, fluorine and nitrogen oxides. If material or contaminated runoff enters waterways, notify downstream users of potentially contaminated waters. Notify local health and fire officials and pollution control agencies. From a secure, explosion-proof location, use water spray to cool exposed containers. If cooling streams are ineffective (venting sound increases in volume and pitch, tank discolors, or shows any signs of deforming), withdraw immediately to a secure position. If employees are expected to fight fires, they must be trained and equipped in OSHA 1910.156.

References

New Jersey Department of Health and Senior Services, "Hazardous Substance Fact Sheet: Flurazepam," Trenton, NJ (January 1986)

Fonofos

Molecular Formula: $C_{10}H_{15}OPS_2$

Common Formula: $C_6H_5SPS(OC_2H_5)C_2H_5$

Synonyms: *O*-Aethyl-*S*-phenyl-aethyl-dithiophosphonat (German); Capfos; Cudgel; Difonate; Double down; Dyfonate; Dyphonate; ENT 25,796; *O*-Ethyl *S*-phenyl ethyldithiophosphonate; *O*-Ethyl *S*-phenyl (RS)-ethylphosphonodithioate; *O*-Ethyl *S*-phenyl ethylphosphonodithioate; Fonophos; N-2790; Phosphonodithioic acid, ethyl-*O*-ethyl, *S*-phenyl ester

CAS Registry Number: 944-22-9

RTECS Number: TA5950000

DOT ID: UN 2783

EEC Number: 015-091-00-2

Regulatory Authority

- Banned or Severely Restricted (In Agriculture) (Malaysia, E. Germany) (UN)[13]
- Air Pollutant Standard Set (ACGIH)[1] (Several States)[60]
- Superfund/EPCRA 40CFR355, Appendix B Extremely Hazardous Substances: TPQ = 500 lb (227 kg)
- Superfund/EPCRA 40CFR302.4 Reportable Quantity (RQ): EHS, 1 lb (0.454 kg)
- U.S. DOT Regulated Marine Pollutant (49CFR172.101, Appendix B)

Note: Classified for restricted use for direct supervision of a certified applicator.

Cited in U.S. State Regulations: Alaska (G), California (G), Connecticut (A), Florida (G), Illinois (G), Maine (G), Massachusetts (G), Nevada (A), New Hampshire (G), New Jersey (G), North Dakota (A), Pennsylvania (G), Rhode Island (G), Virginia (A), West Virginia (G).

Description: Fonofos, $C_{10}H_{15}OS_2P$, $C_6H_5SPS(OC_2H_5)C_2H_5$, is a pale yellow liquid with a pungent, mercaptan-like odor. Boiling point = 130°C at 0.1 mm. Freezing/Melting point = 30°C. Flash point = 94°C (cc). Hazard Identification (based on NFPA-704 M Rating System): Health 4, Flammability 1, Reactivity 0. Insoluble in water.

Potential Exposure: Those involved in the manufacture, formulation and application off this soil insecticide.

Incompatibilities: Strong acids, alkalies.

Permissible Exposure Limits in Air: There is no OSHA PEL. NIOSH as well as the ACGIH recommends a TWA value of 0.1 mg/m^3 with the notation "skin" indicating the possibility of skin absorption. Several states have developed guidelines or standards for fonofos in ambient air[60] ranging from 1.0 μg/m^3 (North Dakota) to 2.0 μg/m^3 (Connecticut, Nevada and Virginia).

Determination in Air: OSHA versatile sampler-2; Toluene/Acetone; Gas chromatography/Flame photometric detection for sulfur, nitrogen, or phosphorus; NIOSH IV Method #5600, Organophosphorus Pesticides.

Permissible Concentration in Water: The EPA has developed health advisories for fonofos as follows; long term health advisory is 70 μg/l and lifetime health advisory is 14 μg/l (See Reference Below).

Routes of Entry: Inhalation, skin absorption, ingestion, skin and/or eye contact.

Harmful Effects and Symptoms

Short Term Exposure: Symptoms include nausea, vomiting, abdominal cramps, diarrhea, excessive salivation, headache, giddiness, vertigo, sensation of tightness in chest, blurring of vision, ocular pain, loss of muscle coordination, slurring in speech, muscle twithching, drowsiness, excessive secretion of respiratory tract mucous, and convulsions. This material is cholinesterase inhibitor. It can cause severe symptoms and death from respiratory arrest. The LD_{50} oral-rat is 3 mg/kg (extremely toxic). Exposure above the airborne exposure limit may result in death. The effects may be delayed. Medical observation is recommended.

Long Term Exposure: Cholinesterase inhibitor; cumulative effect is possible. Fonofos may damage the nervous system with repeated exposure, resulting in convulsions, respiratory failure. May cause liver damage.

Points of Attack: Respiratory system, central nervous system, cardiovascular system, blood cholinesterase.

Medical Surveillance: Before employment and at regular times after that, the following are recommended: Plasma and red blood cell cholinesterase levels (tests for the enzyme poisoned by this chemical). If exposure stops, plasma levels return to normal in 1 – 2 weeks while red blood cell levels may be reduced for 1 – 3 months.

When cholinesterase enzyme levels are reduced by 25% or more below preemployment levels, risk of poisoning is increased, even if results are in lower ranges of "normal." Reassignment to work not involving organophosphate or carbamate pesticides is recommended until enzyme levels recover. If symptoms develop or overexposure occurs, repeat the above tests as soon as possible and get an exam of the nervous system. Also consider complete blood count. Consider chest x-ray following acute overexposure. Do not drink any alcoholic beverages before or during use. Alcohol promotes absorption of organic phosphates.

First Aid: If this chemical gets into the eyes, remove any contact lenses at once and irrigate immediately for at least 15 minutes, occasionally lifting upper and lower lids. Seek medical attention immediately. If this chemical contacts the skin, remove contaminated clothing and wash immediately with soap and water. Speed in removing material from skin is of extreme importance. Shampoo hair promptly if contaminated. Seek medical attention immediately. If this chemical has been inhaled, remove from exposure, begin rescue breathing (using universal precautions) if breathing has stopped and CPR if heart action has stopped. Transfer promptly to a medical facility. When this chemical has been swallowed, get medical attention. Give large quantities of water and induce vomiting. Do not make an unconscious person vomit. Keep victim quiet and maintain normal body temperature. Effects may be delayed; keep victim under observation.

Personal Protective Methods: Wear protective gloves and clothing to prevent any reasonable probability of skin contact. Safety equipment suppliers/manufacturers can provide recommendations on the most protective glove/clothing material for your operation. All protective clothing (suits, gloves, footwear, headgear) should be clean, available each day, and put on before work. Contact lenses should not be worn when working with this chemical. Wear splash-proof chemical goggles and face shield unless full facepiece respiratory protection is worn. Employees should wash immediately with soap when skin is wet or contaminated. Provide emergency showers and eyewash.

Respirator Selection: Where the potential exists for exposures over 0.1 mg/m^3, use an MSHA/NIOSH approved supplied-air respirator with a full facepiece operated in the positive pressure mode or with a full facepiece, hood, or helmet in the continuous flow mode, or use an MSHA/NIOSH approved self-contained breathing apparatus with a full facepiece operated in pressure-demand or other positive pressure mode.

Storage: Prior to working with this chemical you should be trained on its proper handling and storage. Store in tightly closed containers in a cool area.

Shipping: Fonofos is an organophosphorus pesticide, liquid, toxic, n.o.s. This compound requires a shipping label of: "Poison." It falls in DOT Hazard Class 6.1 and Packing Group I.

Spill Handling: Evacuate and restrict persons not wearing protective equipment from area of spill or leak until cleanup is complete. Remove all ignition sources. Ventilate area of spill or leak. Absorb liquids in vermiculite, dry sand, earth, peat, carbon, or a similar material and deposit in sealed containers. It may be necessary to contain and dispose of this chemical as a hazardous waste. If material or contaminated runoff enters waterways, notify downstream users of potentially contaminated waters. Contact your Department of Environmental Protection or your regional office of the federal EPA for specific recommendations. If employees are required to clean-up spills, they must be properly trained and equipped. OSHA 1910.120(q) may be applicable.

Fire Extinguishing: This chemical is a combustible liquid. Poisonous gases including sulfur and phosphorus oxides are produced in fire. Use dry chemical, carbon dioxide, water spray, fog, or foam extinguishers. Vapors are heavier than air and will collect in low areas. Vapors may travel long distances to ignition sources and flashback. Vapors in confined areas may explode when exposed to fire. Containers may explode in fire. Storage containers and parts of containers may rocket great distances, in many directions. If material or contaminated runoff enters waterways, notify downstream users of potentially contaminated waters. Notify local health and fire officials and pollution control agencies. From a secure, explosion-proof location, use water spray to cool exposed containers. If cooling streams are ineffective (venting sound increases in volume and pitch, tank discolors, or shows any signs of deforming), withdraw immediately to a secure position. If employees are expected to fight fires, they must be trained and equipped in OSHA 1910.156.

Disposal Method Suggested: This phosphono compound is reported to be satisfactorily decomposed by hypochlorite. In accordance with 40CFR165 recommendations for the disposal of pesticides and pesticide containers. Must be disposed properly by following package label directions or by contacting your state pesticide or environmental control agency or by contacting your regional EPA office.

References

U.S. Environmental Protection Agency, "Chemical Profile: Fonofos," Washington, DC, Chemical Emergency Preparedness Program (November 30, 1987)

U.S. Environmental Protection Agency, "Health Advisory: Fonofos," Washington, DC, Office of Drinking Water (August 1987)

New Jersey Department of Health and Senior Services, "Hazardous Substance Fact Sheet: Fonofos," Trenton, NJ (August 1985)

Formaldehyde

Molecular Formula: CH_2O

Common Formula: HCHO

Synonyms: Aldehyde formique (French); Aldeide formica (Italian); BFV; Dynoform; FA; Fannoform; Formaldehido (Spanish); Formaldehyd (Czech); Formaldehyd (Polish); Formalin; Formalin 40; Formalina (Italian, Spanish); Formaline (German); Formalin-loesungen (German); Formalith; Formic aldehyde; Formol; Fyde; Hercules 37M6-8; Hoch; Ivalon; Karsan; Low dye-fast dry ink; Lysoform; Magnifloc 156C flocculant; Meethanal; Methyl aldehyde; Methylene glycol; Methylene oxide; Morbicid; NCI-C02799; Oplossingen (Dutch); Oxomethane; Oxymethylene; Polyoxymethylene glycols; Steriform; Superlysoform; Tetraoxymethylene; Trioxane

CAS Registry Number: 50-00-0

RTECS Number: LP8925000

DOT ID: UN 2209 (or 1198)

EEC Number: 605-001-00-5

EINECS Number: 200-001-8

Regulatory Authority

- Carcinogen (Animal Positive) (IARC)[9] (ACGIH)[1] (DFG)[3]
- Banned or Severely Restricted (Several Countries) (UN)[13]
- Toxic Substance (World Bank)[15]
- OSHA, 29CFR1910 Specifically Regulated Chemicals (See CFR 1910.1048)
- Air Pollutant Standard Set (ACGIH)[1] (OSHA)[58] (DFG)[3] (HSE)[33] (Other Countries)[35] (Several States)[60] (Several Canadian Provinces)
- Clean Air Act: Hazardous Air Pollutants (Title I, Part A, Section 112); Accidental Release Prevention/Flammable substances, (Section 112[r], Table 3), TQ = 15,000 lb (6,810 kg)
- OSHA 29CFR1910.119, Appendix A. Process Safety List of Highly Hazardous Chemicals, TQ = 1,000 lb (450 kg)
- Clean Water Act: Section 311 Hazardous Substances/RQ 40CFR117.3 (same as CERCLA, see below); Section 313 Water Priority Chemicals (57FR41331, 9/9/92)
- EPA Hazardous Waste Number (RCRA No.): U122
- RCRA, 40CFR261, Appendix 8 Hazardous Constituents
- Superfund/EPCRA 40CFR355, Appendix B Extremely Hazardous Substances: TPQ = 500 lb (227 kg)
- Superfund/EPCRA 40CFR302.4 Reportable Quantity (RQ): CERCLA, 100 lb (45.4 kg)
- EPCRA Section 313 Form R *de minimis* concentration reporting level: 0.1%
- Canada, WHMIS, Ingredients Disclosure List

Cited in U.S. State Regulations: Alaska (G), California (G, W), Connecticut (A), Florida (G), Illinois (G), Indiana (A), Kansas (G), Louisiana (G), Maine (G, W), Maryland (G, W), Massachusetts (G, A), Michigan (G), Nevada (A), New Hampshire (G), New Jersey (G, W), New York (G, A), North Carolina (A), North Dakota (A), Oklahoma (G), Pennsylvania (G, A), Rhode Island (G), South Carolina (A), South Dakota (A), Vermont (G), Virginia (G, A), Washington (G, A), West Virginia (G), Wisconsin (G).

Description: Formaldehyde, HCHO, is a colorless, pungent gas. The odor threshold is 0.8 ppm.[41] Formalin (as formaldehyde) is sold as an aqueous solution containing 30 – 50% formaldehyde and 6 to 15% methanol, which is added to prevent polymerization. Boiling point (gas) = -19.5°C. The 37% commercial solution BP = 101°C. Flash point = 50°C (commercial 37% solution, 15% methanol). Explosive limits (gas): LEL = 7.0%; UEL = 73.0%. NFPA 704 M Hazard Identification (gas): Health 3, Flammability 4, Reactivity 0. NFPA 704 M Hazard Identification (37% solution, 15% methanol): Health 3, Flammability 20, Reactivity 0.

Potential Exposure: Formaldehyde has found wide industrial usage as a fungicide, germicide, and in disinfectants and embalming fluids. It is also used in the manufacture of artificial silk and textiles, latex, phenol, urea, thiourea and melamine resins, dyes, and inks, cellulose esters and other organic molecules, mirrors, and explosives. It is also used in the paper, photographic, and furniture industries. It is an intermediate in drug manufacture and is a pesticide intermediate.

Incompatibilities: Pure formaldehyde may polymerize unless properly inhibited (usually with methanol). Forms explosive mixture with air Incompatible with strong acids, amines, strong oxidizers, alkaline materials, nitrogen dioxide, performic acid, phenols, urea. Reaction with hydrochloric acid forms bis-chloromethyl ether, a carcinogen. Formalin is incompatible with strong oxidizers, alkalis, acids, phenols, urea, oxides, isocyanates, caustics, anhydrides.

Permissible Exposure Limits in Air: The OSHA TWA[58] is 3 ppm determined as a TWA. The ceiling concentration is 5 ppm which shall not be exceeded at any time, and 10 ppm which shall not be exceeded in any 30 minute period. ACGIH has recommended a TWA of 1.0 ppm (1.5 mg/m^3) and STEL of 2.0 ppm (3 mg/m^3) with the notation that formaldehyde is "an industrial substance suspect of carcinogenic potential for man." There are a number of values set in other countries for TWA values (some using different conversion factors from ppm to mg/m^3). The NIOSH IDLH = Ca (20 ppm).

Some off these are as follows:[35]

Country	TWA	
	ppm	mg/m^3
Former USSR	—	0.5
Czechoslovakia	—	0.5
W. Germany	0.5	0.6
Sweden	0.8	1.0
Brazil	1.6	2.3
Japan	2.0	2.5
U.K.[33]	2.0	3.0

The Former USSR has also set MAC values for ambient air in residential areas of 0.035 mg/m^3 on a momentary basis and 0.012 mg/m^3 on a daily average basis,[43] also cited as 0.003 mg/m^3.[35] A number of states have set guidelines or standards for formaldehyde in ambient air[60] ranging from zero (North Carolina and North Dakota) to 0.77 μg/m^3 (Massachusetts) to 5.0 μg/m^3 (New York) to 7.2 μg/m^3 (Pennsylvania) to 7.5 μg/m^3 (South Carolina) to 12.0 μg/m^3 (Connecticut, South Dakota and Virginia) to 18.0 μg/m^3 (Indiana) to 71.0 μg/m^3 (Nevada) to 75.0 μg/m^3 (Washington).

Determination in Air: Collection with Si gel cartridge coated with DNPH; Workup with acetonitrile; analysis with high-pressure liquid chromatography/ultraviolet detection; NIOSH IV, Method #2016. See also Method(s) #2541, #3500.

Permissible Concentration in Water: EPA[32] has suggested a permissible ambient goal of 41.4 μg/l based on health effects. The former USSR has set a MAC in water bodies used for domestic purposes of 0.01 mg/l,[43] also quoted as 0.05 mg/l.[35] Further, they have set a MAC in water bodies used for fishery purposes of 0.25 mg/l. Several states have set guidelines for formaldehyde in drinking water[61] ranging from 0.7 μg/l (New Jersey)[59] to 10.0 μg/l (Maryland) to 30.0 μg/l (California and Maine).

Routes of Entry: Inhalation, ingestion, skin and/or eye contact.

Harmful Effects and Symptoms

Short Term Exposure: Corrosive to the eyes, skin, and respiratory tract. Acute exposure to formaldehyde may result in burns to the skin, eyes, and mucous membranes; lacrimation (tearing); nausea; vomiting (may be bloody); abdominal pain; and diarrhea. Difficulty in breathing, cough, pneumonia, and pulmonary edema may occur. Sensitized people may experience asthmatic reactions, even when exposed briefly. Hypotension (low blood pressure) and hypothermia (reduced body temperature) may precede cardiovascular collapse. Lethargy, dizziness, convulsions, and coma may be noted. Nephritis (inflammation of the kidneys), hematuria (bloody urine), and liver toxicity have been reported. Exposure at concentrations well above the PEL may cause death. The effects may be delayed.

Note: There is considerable individual variation in sensitivity to formaldehyde.

Inhalation: Irritation of the nose and throat can occur after an exposure of 0.25 – 0.45 ppm. Levels between 0.4 ppm and 0.8 ppm can give rise to coughing and wheezing, tightness of the chest and shortness of breath. Sudden exposures to concentrations of 4 ppm may lead to irritation of lung and throat severe enough to give rise to bronchitis and laryngitis. Breathing may be impaired at levels above 100 ppm and serious lung damage may occur at 50 ppm.

Skin: Direct contact with the liquid can lead to irritation, itching, burning and drying. It is also possible to develop an allergic reaction to the compound following exposure by any route.

Eyes: Exposure to airborne levels of formaldehyde of 0.4 ppm have brought on tearing and irritation. Small amounts of liquid splashed in the eye can cause damage to the cornea. Eye irritation was reported at levels between 0.05 – 2.0 ppm.

Ingestion: As little as 1 liquid ounce has resulted in death to humans. Smaller amounts can damage the throat, stomach and intestine resulting in nausea, vomiting, abdominal pain and diarrhea. Accidental exposure may also cause loss of consciousness, lowered blood pressure, kidney damage and, if the person is pregnant, the possibility of the fetus being aborted.

Long Term Exposure: Inhalation can result in respiratory congestion with associated coughing and shortness of breath. Repeated skin contact can lead to drying and scaling. Some individuals may experience allergic reactions after initial contact with the chemical. Subsequent contact may cause skin rashes and asthma and reactions may become more severe if exposure persists. Long tem inhalation of high levels of formaldehyde vapor (14 ppm) in rats resulted in an elevated incidence of cancer of the nose. Genetic damage from exposure has been shown in bacteria and some insects. Whether it causes effects in humans is uncertain. May be carcinogenic to humans.

Points of Attack: Eyes, respiratory system. Cancer Site: nasal cancer.

Medical Surveillance: For those with frequent or potentially high exposure (half the TLV or greater) the following are recommended before beginning work and at regular times after that: Lung function tests. If symptoms develop or overexposure is suspected, the following may be useful: Evaluation by a qualified allergist, including careful exposure history and special testing, may help diagnose skin allergy. Consider chest x-ray after acute overexposure.

First Aid: If this chemical gets into the eyes, remove any contact lenses at once and irrigate immediately for at least 30 minutes, occasionally lifting upper and lower lids. Seek medical attention immediately. If this chemical contacts the skin, remove contaminated clothing and wash immediately with soap and water. Seek medical attention immediately. If this chemical has been inhaled, remove from exposure, begin rescue breathing (using universal precautions) if breathing has stopped and CPR if heart action has stopped. Transfer promptly to a medical facility. When this chemical has been swallowed, get medical attention. Give large quantities of water and induce vomiting. Do not make an unconscious person vomit. Medical observation is recommended for 24 – 48 hours after breathing overexposure, as pulmonary edema may be delayed. As first aid for pulmonary edema, a doctor or authorized paramedic may consider administering a corticosteroid spray.

Personal Protective Methods: Prevention of intoxication may be easily accomplished by supplying adequate ventilation and protective clothing. Barrier creams may also be helpful. Wear protective gloves and clothing to prevent any reasonable probability of skin contact. Safety equipment suppliers/manufacturers can provide recommendations on the most protective glove/clothing material for your operation Teflon, Viton, Butyl Rubber, Polyethylene, and Styrene-Butadiene Rubber are among the recommended protective materials. All protective clothing (suits, gloves, footwear, headgear) should be clean, available each day, and put on before work. Contact lenses should not be worn when working with this chemical. Wear splash-proof chemical goggles and face shield unless full facepiece respiratory protection is worn. Employees should wash immediately with soap when skin is wet or contaminated. Provide emergency showers and eyewash.

Respirator Selection: NIOSH: *At any concentrations above the NIOSH REL, or where there is no REL, at any detectable concentration:* SCBAF:PD,PP (any self-contained breathing apparatus that has a full facepiece and is operated in a pressure-demand or other positive-pressure mode); or SAF:PD,PP:ASCBA (any supplied-air respirator that has a full facepiece and is operated in a pressure-demand or other positive-pressure mode in combination with an auxiliary self-contained breathing apparatus operated in a pressure-demand or other positive pressure mode). *Escape:* GMFS [any air-purifying, full-facepiece respirator (gas mask) with a chin-style, front- or back-mounted canister providing protection against the compound of concern]; or SCBAE (any appropriate escape-type, self-contained breathing apparatus).

Storage: Prior to working with this chemical you should be trained on its proper handling and storage. Before entering confined space where this chemical may be present, check to make sure that an explosive concentration does not exist. Formaldehyde must be stored to avoid contact with oxidizers (such as permanganates, nitrates, peroxides, chlorates and perchlorates), alkaline materials, since violent reactions occur. Store in tightly closed containers in a cool, well-ventilated area away from heat, sparks or flames. Sources of ignition such as smoking and open flames are prohibited where formaldehyde is used, handled, or stored in a manner that could create a potential fire or explosion hazard. Where possible, automatically pump liquid from drums or other storage containers to process containers. Procedures for the handling, use and storage of cylinders should be in compliance with OSHA 1910.101 and 1910.169 as with the recommendations of the Compressed Gas Association. A regulated, marked area should be established where this chemical is handled, used, or stored in compliance with OSHA standard 1910.1045.

Shipping: The regulations apply to commercial formaldehyde solutions, which have a flash point of 85°C. This type of compound requires a shipping label of: "Class 9. It falls in DOT Hazard Class 9 and Packing Group III.

Spill Handling: Shut off ignition sources; no flares, smoking or flames in hazard area. Do not touch spilled material; stop leak if you can do so without risk. Use water spray to reduce vapors; do not get water inside container. *Small spills:* absorb with sand or other noncombustible absorbent material and place into containers for later disposal. *Large spills:* dike far ahead of spill for later disposal. Use fluorocarbon water spray, cellosize, and Hycar to diminish vapors. Use sodium carbonate, ammonium hydroxide or sodium sulfite to neutralize spill. Use universal gel, fly ash, universal sorbent material, or cement powder to absorb the spill. Keep formaldehyde out of a confined space, such as a sewer, because of the possibility of an explosion, unless the sewer is designed to prevent the build-up off explosive concentrations. It may be necessary to contain and dispose of this chemical as a hazardous waste. If material or contaminated runoff enters waterways, notify downstream users of potentially contaminated waters. Contact your Department of Environmental Protection or your regional office of the federal EPA for specific recommendations. If employees are required to clean-up spills, they must be properly trained and equipped. OSHA 1910.120(q) may be applicable.

Fire Extinguishing: This chemical is a flammable gas or a combustible liquid. Poisonous gases are produced in fire. Small fires: dry chemical, carbon dioxide, water spray of foam. Large fires: water spray, fog or foam. Move container from fire area if you can do so without risk. Do not get water inside container. Spray cooling water on containers exposed to flames until well after fire is out. Withdraw immediately in case of rising sound from venting safety device or any discoloration of tank due to fire. Vapors are heavier than air and will collect in low areas. Vapors may travel long distances to ignition sources and flashback. Vapors in confined areas may explode when exposed to fire. Containers may explode in fire. Storage containers and parts of containers may rocket great distances, in many directions. If material or contaminated runoff enters waterways, notify downstream users of potentially contaminated waters. Notify local health and fire officials and pollution control agencies. From a secure, explosion-proof location, use water spray to cool exposed containers. If cooling streams are ineffective (venting sound increases in volume and pitch, tank discolors, or shows any signs of deforming), withdraw immediately to a secure position. If employees are expected to fight fires, they must be trained and equipped in OSHA 1910.156.

Disposal Method Suggested: Incineration in solution in combustible solvent.[22] Consult with environmental regulatory agencies for guidance on acceptable disposal practices. Generators of waste containing this contaminant (≥100 kg/mo) must conform with EPA regulations governing storage, transportation, treatment, and waste disposal.

References

Environmental Protection Agency, Investigation of Selected Potential Environmental Contaminants-Formaldehyde, Final Report, Office of Toxic Substances, Environmental Protection Agency, August, 1976

National Institute for Occupational Safety and Health, Criteria for a Recommended Standard: Occupational Exposure to Formaldehyde, NIOSH Doc. No. 77–126 (1977)

U.S. Environmental Protection Agency, "Chemical Hazard Information Profile: Formaldehyde," Washington, DC (1979)

U.S. Environmental Protection Agency, Formaldehyde, Health and Environmental Effects Profile No. 104, Office of Solid Waste, Washington, DC (April 30, 1980)

Sax, N. I., Ed., "Dangerous Properties of Industrial Materials Report," 1, No. 4, 70–72 (1981) and 3, No. 3, 71–76 (1983)

National Institute for Occupational Safety and Health, Formaldehyde: Evidence of Carcinogenicity, Current Intelligence Bulletin No. 34, DHHS (NIOSH) Publication No. 81–111, Cincinnati, Ohio (April 15, 1981)

Clary, J. J., Gibson, J. E. and Waritz, R. S., Formaldehyde Toxicology, Epidemiology, Mechanisms, New York, Marcel Dekker, Inc. (1983)

U.S. Environmental Protection Agency, "Chemical Profile: Formaldehyde," Washington, DC, Chemical Emergency Preparedness Program (November 30, 1987)

New Jersey Department of Health and Senior Services, "Hazardous Substance Fact Sheet: Formaldehyde," Trenton, NJ (February 1989)

New York State Department of Health, "Chemical Fact Sheet: Formaldehyde," Albany, NY, Bureau of Toxic Substance Assessment (March 1986)

Formaldehyde Cyanohydrin

Molecular Formula: C_2H_3NO

Common Formula: $HOCH_2CN$

Synonyms: Cyanomethanol; Formaldehido cianhidrina (Spanish); Glycolonitrile glyconitrile; Gyycolic nitrile; α-Hydroxyacetonitrile; 2-Hydroxyacetonitrile; Hydroxyacetonitrile; a-Hydroxymethylcyanide; Hydroxymethylinitrile; Methylene cyanohydrine

CAS Registry Number: 107-16-4

RTECS Number: AM0350000

DOT ID: UN 3226 (nitriles, toxic, n.o.s.)

Regulatory Authority

- Air Pollutant Standard Set (NIOSH)[9]
- Superfund/EPCRA 40CFR355, Appendix B Extremely Hazardous Substances: TPQ = 1,000 lb (454 kg)
- Superfund/EPCRA 40CFR302.4 Reportable Quantity (RQ): EHS, 1 lb (0.454 kg)
- Canada, WHMIS, Ingredients Disclosure List

Cited in U.S. State Regulations: California (A, G), Massachusetts (G), New Jersey (G), Pennsylvania (G).

Description: Formaldehyde cyanohydrin, $NOCH_2CN$, is a colorless, odorless oily liquid with a sweetish taste. Boiling point = 186°C (slight decomposition). Freezing/Melting point ≤-72°C. Flash point ≥ 93°C. Soluble in water.

Potential Exposure: This material is used in the manufacture of intermediates in pharmaceutical production and as a component of synthetic resins as a chemical intermediate for organic compounds, and as a solvent.

Incompatibilities: Alkalis and exposure to heat. Traces of alkalis promote violent polymerization.

Permissible Exposure Limits in Air: The NIOSH REL for nitriles is a ceiling limit of 6 mg/m^3, not to be exceeded in any 15-minute work period.

Determination in Air: No test available.

Permissible Concentration in Water: No criteria set.

Routes of Entry: Inhalation, skin absorption, ingestion, skin and/or eye contact.

Harmful Effects and Symptoms

Short Term Exposure: Irritates eyes, skin, respiratory system. The symptoms are similar to cyanide poisoning. Odor of bitter almonds on patient's breath may or may not be present. Headache, dizziness, weakness, giddiness, confusion, convulsions, vertigo, dyspnea (breathing difficulty), abdominal pain, nausea, vomiting may follow exposure. Respiration may initially be rapid, then slow and labored, followed by coma and convulsions. This material is extremely toxic. Exposure by any route should be avoided; may have fatal consequences; death from asphyxiation may occur similar to that resulting from hydrogen cyanide.

Long Term Exposure: Repeated exposure may cause personality changes of depression, anxiety, or irritability. Prolonged or repeated exposure may damage the nervous system; affect respiratory system, cardiovascular system.

Points of Attack: Eyes, skin, respiratory system, central nervous system, cardiovascular system.

Medical Surveillance: Blood test for cyanide. Examination of the nervous system.

First Aid: If this chemical gets into the eyes, remove any contact lenses at once and irrigate immediately for at least 15 minutes, occasionally lifting upper and lower lids. Seek medical attention immediately. If this chemical contacts the skin, remove contaminated clothing and wash immediately with soap and water. *Speed in removing material from skin is of extreme importance.* Seek medical attention immediately. If this chemical has been inhaled, remove from exposure, begin rescue breathing (using universal precautions) if breathing has stopped and CPR if heart action has stopped. Transfer promptly to a medical facility. When this chemical has been swallowed, get medical attention. Give large quantities of water and induce vomiting. Do not make an unconscious person vomit. Keep victim quiet and maintain normal body temperature. Effects may be delayed; keep victim under observation.

Use amyl nitrate capsules if symptoms develop. All area employees should be trained regularly in emergency measures for cyanide poisoning and in CPR. A cyanide antidote kit should be kept in the immediate work area and must be rapidly available. Kit ingredients should be replaced every 1 – 2 years to ensure freshness. Persons trained in the use of this kit, oxygen use, and CPR must be quickly available.

Personal Protective Methods: Wear protective gloves and clothing to prevent any reasonable probability of skin contact. Safety equipment suppliers/manufacturers can provide recommendations on the most protective glove/clothing material for your operation. All protective clothing (suits, gloves, footwear, headgear) should be clean, available each day, and put on before work. Contact lenses should not be worn when working with this chemical. Wear splash-proof chemical goggles and face shield unless full facepiece respiratory protection is worn. Employees should wash immediately with soap when skin is wet or contaminated. Provide emergency showers and eyewash. See NIOSH Criteria Document 78-212 NITRILES.

Respirator Selection: NIOSH: *up to 20 ppm:* SA (any supplied-air respirator). *Up to 50 ppm:* SA:CF (any supplied-air respirator operated in a continuous-flow mode). *Up to 100 ppm:* SCBAF (any self-contained breathing apparatus with a full facepiece); or SAF (any supplied-air respirator with a full facepiece). *Up to 250 ppm:* SAF:PD,PP (any supplied-air respirator that has a full facepiece and is operated in a pressure-demand or other positive-pressure mode). *Emergency or planned entry into unknown concentrations or IDLH conditions*: SCBAF:PD,PP (any self-contained breathing apparatus that has a full facepiece and is operated in a pressure-demand or other positive-pressure mode); or SAF:PD,PP:ASCBA (any supplied-air respirator that has a full facepiece and is operated in a pressure-demand or other positive-pressure mode in combination with an auxiliary self-contained breathing apparatus operated in a pressure-demand or other positive-pressure mode). *Escape:* GMFOV [any air-purifying, full-facepiece respirator (gas mask) with a chin-style, front-or back-mounted organic vapor canister] or SCBAE (any appropriate escape-type, self-contained breathing apparatus).

Storage: Prior to working with this chemical you should be trained on its proper handling and storage. Store in a refrigerator[52] under an inert atmosphere and away from all alkaline materials. Where possible, automatically pump liquid from drums or other storage containers to process containers.

Shipping: This compound requires a shipping label of: "Poison" It falls in DOT Hazard Class 6.1.

Spill Handling: Evacuate and restrict persons not wearing protective equipment from area of spill or leak until cleanup is complete. Remove all ignition sources. Ventilate area of spill or leak. Build dikes to contain flow as necessary. Use water spray to disperse vapors and dilute standing pools of liquid. Avoid breathing vapors. Keep upwind. Avoid bodily contact with the material. Absorb liquids in vermiculite, dry sand, earth, peat, carbon, or a similar material and deposit in sealed containers. It may be necessary to contain and dispose of this chemical as a hazardous waste. If material or contaminated runoff enters waterways, notify downstream users of potentially contaminated waters. Contact your Department of Environmental Protection or your regional office of the federal EPA for specific recommendations. If employees are required to clean-up spills, they must be properly trained and equipped. OSHA 1910.120(q) may be applicable.

Initial isolation and protective action distances

Distances shown are likely to be affected during the first 30 minutes after materials are spilled and could increase with time. If more than one tank car, cargo tank, portable tank, or large cylinder is involved in the incident is leaking, the protective action distance may need to be increased. You may need to seek emergency information from CHEMTREC at (800) 424-9300 or seek professional environmental engineering assistance from the U.S. EPA Environmental Response Team at (908) 548-8730 (24-hour response line).

Small spills (From a small package or a small leak from a large package)

First: Isolate in all directions (feet) 700

Then: Protect persons downwind (miles)

Day 1.2

Night 5.5

Large spills (From a large package or from many small packages)

First: Isolate in all directions (feet) 2,000

Then: Protect persons downwind (miles)

Day 4.6

Night .... 7.0+

Fire Extinguishing: This chemical is a combustible liquid. Poisonous gases including cyanides and nitrogen oxides are produced in fire. Use dry chemical, carbon dioxide, or halon extinguishers. Vapors are heavier than air and will collect in low areas. Vapors may travel long distances to ignition sources and flashback. Vapors in confined areas may explode when exposed to fire. Containers may explode in fire. Storage containers and parts of containers may rocket great distances, in many directions. If material or contaminated runoff enters waterways, notify downstream users of potentially contaminated waters. Notify local health and fire officials and pollution control agencies. From a secure, explosion-proof location, use water spray to cool exposed containers. If cooling streams are ineffective (venting sound increases in volume and pitch, tank discolors, or shows any signs of deforming), withdraw immediately to a secure position. If employees are expected to fight fires, they must be trained and equipped in OSHA 1910.156.

References

U.S. Environmental Protection Agency, "Chemical Profile: Formaldehyde Cyanohydrin," Washington, DC, Chemical Emergency Preparedness Program (November 30, 1987)

New Jersey Department of Health and Senior Services, "Hazardous Substance Fact Sheet, Glycolonitrile," Trenton, NJ (April 1999)

Formamide

Molecular Formula: CH_3NO

Common Formula: $HCONH_2$

Synonyms: Carbamaldehyde; Formic acid amide; Methanamide

CAS Registry Number: 75-12-7

RTECS Number: LQ0525000

Regulatory Authority

- Air Pollutant Standard Set (ACGIH)[1] (HSE)[33] (OSHA)[58] (Several States)[60]
- Water Pollution Standard Proposed (EPA)[32]
- Canada, WHMIS, Ingredients Disclosure List

Cited in U.S. State Regulations: Alaska (G), California (A, G), Connecticut (A), Florida (G, A), Illinois (G), Maine (G), Massachusetts (G), Nevada (A), New Hampshire (G), New Jersey (G), New York (A), North Dakota (A), Pennsylvania (G), Rhode Island (G), South Carolina (A), Virginia (A), West Virginia (G).

Description: Formamide, $HCONH_2$, is a colorless, viscous liquid with a faint ammonia odor. Boiling point = 210°C (decomposes). Freezing/Melting point = 2.5°C. Flash point = 154°C.[17] Autoignition temperature ≥ 500°C. Hazard Identification (based on NFPA-704 M Rating System): Health 2, Flammability 1, Reactivity - Soluble in water.

Potential Exposure: Formamide is a powerful solvent. It is also used as an intermediate in pharmaceutical manufacture. It may be pyrolyzed to give HCN.

Incompatibilities: Forms hydrocyanic acid with water solutions. Hygroscopic (absorbs moisture from air). Incompatible with non-oxidizing mineral acids, strong acids, ammonia, cresols, iodine, isocyanates, oleum, phenols, pyridine, sulfur trioxide, oxidizers, iodine, pyridine. Formamide decomposes on heating at 180°C forming ammonia, water, carbon monoxide and hydrogen cyanide. Attacks metals such as aluminum, iron, copper, brass, lead, and natural rubber.

Permissible Exposure Limits in Air: There is no OSHA PEL. The HSE[33] have adopted a TWA value of 20 ppm (30 mg/m³) and set an STEL of 30 ppm (45 mg/m³). A former USSR limit is 3 mg/m³ for workplace air.[35][43] NIOSH and the ACGIH recommend a TWA of 10 ppm (15 mg/m³) with the notation "skin" indicating the possibility of cutaneous absorption.

Several states have set guidelines or standards for formamide in ambient air[60] ranging from 100 μg/m³ (New York) to 300 μg/m³ (Florida) to 300 – 450 μg/m³ (North Dakota) to 480 μg/m³ (Virginia) to 600 μg/m³ (Connecticut) to 714 μg/m³ (Nevada) to 750 μg/m³ (South Carolina).

Determination in Air: No test available.

Permissible Concentration in Water: No criteria set, but EPA[32] has suggested a permissible ambient goal of 414 μg/l based on health effects.

Routes of Entry: Inhalation, ingestion, skin and/or eye contact.

Harmful Effects and Symptoms

Short Term Exposure: Formamide irritates the skin, eyes, and mucous membranes. Contact may cause eye burns. Exposure can cause skin irritation and a rash. May affect the central nervous system.

Long Term Exposure: Formamide may also damage the testes (male reproductive glands). There is limited evidence that formamide is a teratogen in animals and may have a toxic effects upon human reproduction.

Points of Attack: Eyes, skin, respiratory system, central nervous system, reproductive system.

Medical Surveillance: There is no special test for this chemical. However if illness occurs or overexposure is suspected, medical attention is recommended.

First Aid: If this chemical gets into the eyes, remove any contact lenses at once and irrigate immediately for at least 15 minutes, occasionally lifting upper and lower lids. Seek medical attention immediately. If this chemical contacts the skin, remove contaminated clothing and wash immediately with soap and water. Seek medical attention immediately. If this chemical has been inhaled, remove from exposure, begin rescue breathing (using universal precautions) if breathing has stopped and CPR if heart action has stopped. Transfer promptly to a medical facility. When this chemical has been swallowed, get medical attention. Give large quantities of water and induce vomiting. Do not make an unconscious person vomit.

Personal Protective Methods: Wear protective gloves and clothing to prevent any reasonable probability of skin contact. Safety equipment suppliers/manufacturers can provide recommendations on the most protective glove/clothing material for your operation. All protective clothing (suits, gloves, footwear, headgear) should be clean, available each day, and put on before work. Contact lenses should not be worn when working with this chemical. Wear splash-proof chemical goggles and face shield unless full facepiece respiratory protection is worn. Employees should wash immediately with soap when skin is wet or contaminated. Provide emergency showers and eyewash.

Respirator Selection: Where the potential exists for exposures over 20 ppm, use a MSHA/NIOSH approved supplied-air respirator with a full facepiece operated in the positive pressure mode or with a full facepiece, hood, or helmet in the continuous flow mode, or use a MSHA/NIOSH approved self-contained breathing apparatus with a full facepiece operated in pressure-demand or other positive pressure mode.

Storage: Prior to working with this chemical you should be trained on its proper handling and storage. Store in tightly closed containers in a cool, well-ventilated area away from moisture and other incompatible materials listed above. Where possible, automatically pump liquid from drums or other storage containers to process containers.

Shipping: Formamide is not specifically cited by DOT[19] and may be classed as a hazardous substance, liquid, n.o.s. This imposes neither label requirements or maximum limits on air or rail shipments.

Spill Handling: Evacuate and restrict persons not wearing protective equipment from area of spill or leak until cleanup is complete. Remove all ignition sources. Ventilate area of spill or leak. Absorb liquids in vermiculite, dry sand, earth, peat, carbon, or a similar material and deposit in sealed containers. Keep formamide out of a confined space, such as a sewer, because of the possibility of an explosion, unless the sewer is designed to prevent the build-up off explosive concentrations. It may be necessary to contain and dispose of this chemical as a hazardous waste. If material or contaminated runoff enters waterways, notify downstream users of potentially contaminated waters. Contact your Department of Environmental Protection or your regional office of the federal EPA for specific recommendations. If employees are required to clean-up spills, they must be properly trained and equipped. OSHA 1910.120(q) may be applicable.

Fire Extinguishing: This chemical is a combustible liquid. Poisonous gases including nitrogen oxides are produced in fire. Use dry chemical, carbon dioxide, water spray, or alcohol-resistant foam extinguishers. Vapors in confined areas may explode when exposed to fire. Containers may explode in fire. Storage containers and parts of containers may rocket great distances, in many directions. If material or contaminated runoff enters waterways, notify downstream users of potentially contaminated waters. Notify local health and fire officials and pollution control agencies. From a secure, explosion-proof location, use water spray to cool exposed containers. If cooling streams are ineffective (venting sound increases in volume and pitch, tank discolors, or shows any signs of deforming), withdraw immediately to a secure position. If employees are expected to fight fires, they must be trained and equipped in OSHA 1910.156.

Disposal Method Suggested: Dissolve in a combustible solvent and dispose by burning in a furnace equipped with an alkali scrubber for the exit gases.[22]

References

Sax, N. I., Ed., "Dangerous Properties of Industrial Materials Report," 1, No. 1, 44–45 (1980)

New Jersey Department of Health and Senior Services, "Hazardous Substance Fact Sheet: Formamide," Trenton, NJ (December 1996)

Formetanate Hydrochloride

Molecular Formula: $C_{11}H_{16}ClN_3O_2$

Common Formula: $C_{11}H_{15}N_3O_2{\cdot}HCl$

Synonyms: Carzol; Carzol SP; Dicarzol; *m*-([(Dimethylamino)methylene]amino)phenylcarbamate, hydrochloride; 3-Dimethylaminomethyleneaminophenyl-*N*-methylcarbamate, hydrochloride; *N*,*N*-Dimethyl-*N*'-([(methylamino)carbonyl]oxy)phenylmethanimidamide monohydrochloride; ENT 27,566; EP-332; Formetanate hydrochloride; Morton EP332; NOR-AM EP 332; Schering 36056; SN 36056

CAS Registry Number: 23422-53-9

RTECS Number: FC2514000

Regulatory Authority

- EPA Hazardous Waste Number (RCRA No.): P198
- RCRA, 40CFR261, Appendix 8 Hazardous Constituents
- RCRA 40CFR268.48; 61FR15654, Universal Treatment Standards: Wastewater (mg/l), 0.056; Nonwastewater (mg/kg), 1.4
- Superfund/EPCRA 40CFR355, Appendix B Extremely Hazardous Substances: TPQ = 500/10,000 lb (227/4,540 kg)
- Superfund/EPCRA 40CFR302.4 Reportable Quantity (RQ): EHS, 1 lb (0.454 kg)
- U.S. DOT Regulated Marine Pollutant (49CFR172.101, Appendix B), severe pollutant as formetanate

Cited in U.S. State Regulations: California (G), Massachusetts (G), New Jersey (G), Pennsylvania (G).

Description: Formetanate hydrochloride, $C_{11}H_{15}N_3O_2{\cdot}HCl$, is a white or yellowish, crystalline solid or powder with a faint odor. Freezing/Melting point = 200 – 202°C (decomposes). Hazard Identification (based on NFPA-704 M Rating System): Health 3, Flammability 1, Reactivity 0. Highly soluble in water.

Potential Exposure: Those involved in the manufacture, formulation and application of this plant insecticide, acaricide, and miticide.

Permissible Exposure Limits in Air: No standards set.

Permissible Concentration in Water: No criteria set.

Routes of Entry: Inhalation, ingestion.

Harmful Effects and Symptoms

Short Term Exposure: Diarrhea, nausea, vomiting, excessive salivation, headache, pinpoint pupils and uncoordinated muscle movements are all common symptoms. Extremely toxic to humans. Not absorbed through contact with skin. Inhalation or ingestion may cause poisoning. Inhibits cholinesterase activity so effects are in relation to action on nervous system and can result in death.

First Aid: If this chemical gets into the eyes, remove any contact lenses at once and irrigate immediately for at least 15 minutes, occasionally lifting upper and lower lids. Seek medical attention immediately. If this chemical contacts the skin, remove contaminated clothing and wash immediately with soap and water. Seek medical attention immediately. If this chemical has been inhaled, remove from exposure, begin rescue breathing (using universal precautions) if breathing has stopped and CPR if heart action has stopped. Transfer promptly to a medical facility. When this chemical has been swallowed, get medical attention. Give large quantities of water and induce vomiting. Do not make an unconscious person vomit.

Personal Protective Methods: Wear protective gloves and clothing to prevent any reasonable probability of skin contact. Safety equipment suppliers/manufacturers can provide recommendations on the most protective glove/clothing material for your operation. All protective clothing (suits, gloves, footwear, headgear) should be clean, available each day, and put on before work. Contact lenses should not be worn when working with this chemical. Wear dust-proof chemical goggles and face shield unless full facepiece respiratory protection is worn. Employees should wash immediately with soap when skin is wet or contaminated. Provide emergency showers and eyewash.

Respirator Selection: Where the potential for exposure to this chemical, use a MSHA/NIOSH approved supplied-air respirator with a full facepiece operated in the positive pressure mode or with a full facepiece, hood, or helmet in the continuous flow mode, or use a MSHA/NIOSH approved self-contained breathing apparatus with a full facepiece operated in pressure-demand or other positive pressure mode.

Storage: Prior to working with this chemical you should be trained on its proper handling and storage. Store in tightly closed containers in a cool, well ventilated area.

Shipping: This material falls under carbamate pesticides, solid, n.o.s. This compound requires a shipping label of: "Poison." It falls in DOT Hazard Class 6.1 and Packing Group II.

Spill Handling: Evacuate persons not wearing protective equipment from area of spill or leak until clean-up is complete. Remove all ignition sources. Stay upwind and keep out of low areas. Do not touch spilled material or breathe the dust, vapors, or fumes from burning materials. Do not handle broken packages without protective equipment. Wash away any material that may have contacted the body with soap and water. Collect powdered material in the most convenient and safe manner and deposit in sealed containers. Ventilate area after clean-up is complete. It may be necessary to contain and dispose of this chemical as a hazardous waste. If material or contaminated runoff enters waterways, notify downstream users of potentially contaminated waters. Contact your Department of Environmental Protection or your regional office of the federal EPA for specific recommendations. If employees are required to clean-up spills, they must be properly trained and equipped. OSHA 1910.120(q) may be applicable.

Fire Extinguishing: Extinguish fire using agent suitable for type of surrounding fire, as the material itself does not burn or burns with difficulty. Use water in flooding quantities as a fog. Use alcohol foam, carbon dioxide, or dry chemical. Poisonous gases are produced in fire. If material or contaminated runoff enters waterways, notify downstream users of potentially contaminated waters. Notify local health and fire officials and pollution control agencies. From a secure, explosion-proof location, use water spray to cool exposed containers. If cooling streams are ineffective (venting sound increases in volume and pitch, tank discolors, or shows any signs of deforming), withdraw immediately to a secure position. If employees are expected to fight fires, they must be trained and equipped in OSHA 1910.156.

Disposal Method Suggested: In accordance with 40CFR 165 recommendations for the disposal of pesticides and pesticide containers. Must be disposed properly by following package label directions or by contacting your state pesticide or environmental control agency or by contacting your regional EPA office. Consult with environmental regulatory agencies for guidance on acceptable disposal practices. Generators of waste containing this contaminant (≥100 kg/mo) must conform with EPA regulations governing storage, transportation, treatment, and waste disposal.

References

U.S. Environmental Protection Agency, "Chemical Profile: Formetanate HCl," Washington, DC, Chemical Emergency Preparedness Program (November 30, 1987)

Formic Acid

Molecular Formula: CH_2O_2

Common Formula: HCOOH

Synonyms: Acide formique (French); Acido formico (Italian); Acido formico (Spanish); ADD-F; AI3-24237; Amasil; Ameisensaeure (German); Aminic acid; Bilorin; Collo-Bueglatt; Collo-Didax; Formisoton; Formylic acid; Hydrogen carboxylic acid; Methanoic acid; Mierenzur (Dutch); Myrmicyl

CAS Registry Number: 64-18-6

RTECS Number: LQ4900000

DOT ID: UN 1779

EEC Number: 607-001-00-0

Regulatory Authority

- Air Pollutant Standard Set (ACGIH)[1] (DFG)[3] (HSE)[33] (OSHA)[58] (Other Countries)[35] (Several States)[60] (Several Canadian Provinces)
- Clean Water Act: Section 311 Hazardous Substances/RQ 40CFR117.3 (same as CERCLA, see below)
- Superfund/EPCRA 40CFR302.4 Reportable Quantity (RQ): CERCLA, 5,000 lb (2,270 kg)
- EPA Hazardous Waste Number (RCRA No.): U123
- RCRA, 40CFR261, Appendix 8 Hazardous Constituents.
- EPCRA Section 313 Form R *de minimis* concentration reporting level: 1.0%
- Canada, WHMIS, Ingredients Disclosure List

Cited in U.S. State Regulations: Alaska (G), California (A, G), Connecticut (A), Florida (G, A), Illinois (G), Kansas (G), Louisiana (G), Maine (G), Massachusetts (G), Nevada (A), New Hampshire (G), New Jersey (G), New York (G, A), North Dakota (A), Oklahoma (G), Pennsylvania (G), Rhode Island (G), South Carolina (A), Vermont (G), Washington (G), West Virginia (G), Wisconsin (G).

Description: Formic acid, HCOOH, is a colorless, flammable, fuming liquid, with a pungent odor. Boiling point = 101°C. Flash point = 69°C; (90% solution) 50°C. Autoignition temperature = 434°C. Explosive limits: LEL = 18%; UEL = 57%. Hazard Identification (based on NFPA-704 M Rating System): Health 3, Flammability 2, Reactivity 0. Soluble in water.

Potential Exposure: Formic acid is a strong reducing agent and is used as a decalcifier. It is used in dyeing color fast wool, electroplating, coagulating latex rubber, regeneration old rubber, and dehairing, plumping, and tanning leather. It is also used in the manufacture of acetic acid, airplane dope, allyl alcohol, cellulose formate, phenolic resins, and oxalate; and it is used in the laundry, textile, insecticide, refrigeration, and paper industries, as well as in drug manufacture.

Incompatibilities: Forms explosive mixture with air. A medium strong acid and a strong reducing agent. Violent reaction with oxidizers, furfuryl alcohol, hydrogen peroxide, nitromethane. Incompatible with strong acids, bases, ammonia, aliphatic amines, alkanolamines, isocyanates, alkylene oxides, epichlorohydrin. Decomposes on heating and on contact with strong acids forming carbon monoxide. Attacks metals: aluminum, cast iron and steel, many plastics, rubber and coatings.

Permissible Exposure Limits in Air: The OSHA TWA[58] and the recommended ACGIH TWA value[1] is 5 ppm (9 mg/m^3). The NIOSH IDLH level is 30 ppm. The ACGIH has set an STEL of 10 ppm (18 mg/m^3). The 5 ppm TWA is endorsed by Argentina,[35] Germany,[3] the U.K.[33] and Japan.[35] The Former USSR[35][43] has set a MAC in workplace air of 1.0 mg/m^3. The NIOSH IDLH = 30 ppm. Several states have set guidelines or standards for formic acid in ambient air[60] ranging from 30 μg/m^3 (New York) to 90 μg/m^3 (Florida and North Dakota) to 150 μg/m^3 (Virginia) to 180 μg/m^3 (Connecticut) to 214 μg/m^3 (Nevada) to 225 μg/m^3 (South Carolina).

Determination in Air: Collection using Si gel (special); workup with water; analysis by ion chromatography; NIOSH IV, Method #2011. OSHA: #ID-112.

Permissible Concentration in Water: No criteria set, but EPA[32] has suggested a permissible ambient goal of 124 μg/l based on health effects.

Routes of Entry: Inhalation of vapor, percutaneous absorption, ingestion, eye and/or skin contact.

Harmful Effects and Symptoms

Short Term Exposure: Formic acid is very corrosive to the eyes, skin, and respiratory tract.

Inhalation: Workers exposed to 15 ppm experience nausea. Other symptoms include irritation of the nose, throat and lungs; coughing, runny nose and tearing eyes. Higher exposures can cause pulmonary edema, a medical emergency that can be delayed for several hours. This can cause death.

Skin: Concentrated solutions may cause severe irritation, burning and blistering. Accidental exposure has resulted in death.

Eyes: May cause irritation and tearing. Concentrated solutions may cause severe chemical burns.

Ingestion: Corrosive. May affect the energy metabolism, causing acidosis. May cause salivation, vomiting, burning sensation in the mouth, vomiting of blood, diarrhea and pain. In severe cases, person may go into shock and develop difficulty in breathing. Death may result. Animal data suggest that ingestion of about 3 ounces may be fatal to a 150 pound individual.

Long Term Exposure: Prolonged or repeated exposure to formic acid may cause skin irritation and allergy with rash and itching. May affect the kidneys May cause genetic changes in living cells.

Points of Attack: Respiratory system, lungs, skin, kidneys, liver, eyes.

Medical Surveillance: Consideration should be given to possible irritant effects on the skin, eyes, and lungs in any placement or periodic examinations. Lung function tests. Kidney function tests checking for blood and urine. Consider chest x-ray following acute overexposure. Evaluation by a qualified allergist.

First Aid: If this chemical gets into the eyes, remove any contact lenses at once and irrigate immediately for at least 15 minutes, occasionally lifting upper and lower lids. Seek medical attention immediately. If this chemical contacts the skin, remove contaminated clothing and wash immediately with soap and water. Seek medical attention immediately. If this chemical has been inhaled, remove from exposure, begin rescue breathing (using universal precautions) if breathing has stopped and CPR if heart action has stopped. Transfer promptly to a medical facility. When this chemical has been swallowed, get medical attention. If victim is conscious, administer water or milk. Do not induce vomiting.

Personal Protective Methods: Wear protective gloves and clothing to prevent any reasonable probability of skin contact. Safety equipment suppliers/manufacturers can provide recommendations on the most protective glove/clothing material for your operation Butyl Rubber, Natural Rubber, Neoprene, Nitrile+PVC, Polyurethane, Styrene-Butadiene, and PVC are among the recommended protective materials. All protective clothing (suits, gloves, footwear, headgear) should be clean, available each day, and put on before work. Contact lenses should not be worn when working with this chemical. Wear splash-proof chemical goggles and face shield unless full facepiece respiratory protection is worn. Employees should wash immediately with soap when skin is wet or contaminated. Provide emergency showers and eyewash.

Respirator Selection: NIOSH/OSHA: *30 ppm:* SA (any supplied-air respirator); or; or SCBAF (any self-contained breathing apparatus with a full facepiece). *Emergency or planned entry into unknown concentrations or IDLH conditions:* SCBAF:PD,PP (any self-contained breathing apparatus that has a full facepiece and is operated in a pressure-demand or other positive-pressure mode); or SAF:PD,PP: ASCBA (any supplied-air respirator that has a full facepiece and is operated in a pressure-demand or other positive-pressure mode in combination with an auxiliary self-contained breathing apparatus operated in a pressure-demand or other positive pressure mode). *Escape:* GMFOVHiE [any air-purifying, full-facepiece respirator (gas mask) with a chin-style, front- or back-mounted organic vapor canister having a high-efficiency particulate filter]; or SCBAE (any appropriate escape-type, self-contained breathing apparatus).

Note: Substance reported to cause eye irritation or damage; may require eye protection.

Storage: Prior to working with this chemical you should be trained on its proper handling and storage. Before entering confined space where this chemical may be present, check to make sure that an explosive concentration does not exist. Keep in sealed containers in well-ventilated area. Protect from heat or flame and materials listed above under "Incompatibilities." Where possible, automatically pump liquid from drums or other storage containers to process containers.

Shipping: This compound requires a shipping label of: "Corrosive." It falls in DOT Hazard Class 8 and Packing Group II.

Spill Handling: Evacuate and restrict persons not wearing protective equipment from area of spill or leak until cleanup is complete. Remove all ignition sources. Establish forced ventilation to keep levels below explosive limit. Absorb liquids in vermiculite, dry sand, earth, peat, carbon, or a similar material and deposit in sealed containers. It may be necessary to contain and dispose of this chemical as a hazardous waste. If material or contaminated runoff enters waterways, notify downstream users of potentially contaminated waters. Contact your Department of Environmental Protection or your regional office of the federal EPA for specific recommendations. If employees are required to clean-up spills, they must be properly trained and equipped. OSHA 1910.120(q) may be applicable.

Fire Extinguishing: This chemical is a combustible liquid. Poisonous gases including carbon monoxide are produced in fire. Use dry chemical, carbon dioxide, or alcohol foam extinguishers. Vapors are heavier than air and will collect in low areas. Vapors may travel long distances to ignition sources and flashback. Vapors in confined areas may explode when exposed to fire. Containers may explode in fire. Storage containers and parts of containers may rocket great distances, in many directions. If material or contaminated runoff enters waterways, notify downstream users of potentially contaminated waters. Notify local health and fire officials and pollution control agencies. From a secure, explosion-proof location, use water spray to cool exposed containers. If cooling streams are ineffective (venting sound increases in volume and pitch, tank discolors, or shows any signs of deforming), withdraw immediately to a secure position. If employees are expected to fight fires, they must be trained and equipped in OSHA 1910.156.

Disposal Method Suggested: Incineration with added solvent.[22] Consult with environmental regulatory agencies for guidance on acceptable disposal practices. Generators of waste containing this contaminant (≥100 kg/mo) must conform with EPA regulations governing storage, transportation, treatment, and waste disposal.

References

U.S. Environmental Protection Agency, Formic Acid, Health and Environmental Effects Profile No. 105, Office of Solid Waste, Washington, DC (April 30, 1980)

Sax, N. I., Ed., "Dangerous Properties of Industrial Materials Report," 1, No. 2, 39-41 (1980) and 3, No. 4, 53-56 (1983)

New York State Department of Health, "Chemical Fact Sheet: Formic Acid," Albany, NY, Bureau of Toxic Substance Assessment (March 1986 and Version 2)

New Jersey Department of Health and Senior Services, "Hazardous Substance Fact Sheet, Formic acid," Trenton, NJ (January 1996)

Formothion

Molecular Formula: $C_6H_{12}NO_4PS_2$

Common Formula: $(CH_3O)_2PSSCH_2CON(CH_3)CHO$

Synonyms: Aflix; Anthio; Antio; Carbamoylmethyl phosphorodithioate; CP 53926; *O,O*-Dimethyldithiophosphorylacetic acid *N*-methyl-*N*-formylamide; *O,O*-Dimethyl *S*-(*N*-formyl-*N*-methylcarbamoylmethyl) phosphorodithioate; *O,O*-Dimethyl-*S*-(3-methyl-2,4-dioxo-3-azabutyl)-dithiofosfaat (Dutch); *O,O*-Dimethyl-*S*-(3-methyl-2,4-dioxo-3-aza-butyl)-dithiophosphat (German); *O,O*-Dimethyl-*S*-(*N*-methyl-*N*-formyl-carbamoylmethyl)-dithiophosphat (German); *O,O*-Dimethyl *S*-(*N*-methyl-*N*-formylcarbamoylmethyl) dithiophosphate; *O,O*-Dimethyl *S*-(*N*-methyl-*N*-formylcarbamoylmethyl) phosphorodithioate; *O,O*-Dimethyl phosphorodithioate *N*-formyl-2-mercapto-*N*-methylacetamide *S*-ester; ENT 27,257; Formotion (Spanish); *S*-[2-(Formylmethylamino)-2-oxoethyl] *O,O*-dimethyl phosphorodithioate; *N*-Formyl-*N*-methylcarbamoylmethyl *O,O*-dimethyl phosphorodithioate; *S*-(*N*-Formyl-*N*-methylcarbamoylmethyl) *O,O*-dimethyl phosphorodithioate; *S*-(*N*-Formyl-*N*-methylcarbamoylmethyl) dimethyl phosphorodithiolothionate; S 6900; SAN 244 I; SAN 6913 I; SAN 71071; Spencer S-6900; VEL 4284

CAS Registry Number: 2540-82-1

RTECS Number: TE1050000

Regulatory Authority

- Air Pollutant Standard Set (former USSR)[35][43]
- Superfund/EPCRA 40CFR355, Appendix B Extremely Hazardous Substances: TPQ = 100 lb (45.4 kg)
- Superfund/EPCRA 40CFR302.4 Reportable Quantity (RQ): EHS, 1 lb (0.454 kg)

Cited in U.S. State Regulations: California (G), Massachusetts (G), New Jersey (G), Pennsylvania (G).

Description: Formothion, $(CH_3O)_2PSSCH_2CON(CH_3)$ CHO, is an odorless, yellowish viscous oil or crystalline mass. Freezing/Melting point = 25°C. Hazard Identification (based on NFPA-704 M Rating System): Health 2, Flammability 1, Reactivity 0. Slightly soluble in water.

Potential Exposure: An insecticide and acaricide on crops and ornamentals. It is not presently produced commercially in the U.S.

Incompatibilities: Alkaline materials.

Permissible Exposure Limits in Air: The former USSR[35][43] has set a MAC in workplace air of 0.5 mg/m^3 and a MAC in ambient air in residential areas of 0.01 mg/m^3 on a momentary basis and 0.006 mg/m^3 on a daily average basis.

Determination in Air: OSHA versatile sampler-2; Toluene/Acetone; Gas chromatography/Flame photometric detection for sulfur, nitrogen, or phosphorus; NIOSH Method IV Method #5600, Organophosphorus Pesticides.

Permissible Concentration in Water: The former USSR[35][43] has set a MAC in water bodies used for domestic purposes of 0.004 mg/l (4 μg/l).

Routes of Entry: Inhalation, absorbed by the skin.

Harmful Effects and Symptoms

Short Term Exposure: Early symptoms of poisoning include: headache, dizziness, weakness, perspiring, nausea, vomiting, and sensation of tightness in chest. Chronic low doses may produce symptoms similar to influenza. Formothion is one of the least toxic systemic organophosphates. Formothion is a compound of low to moderate toxicity. It causes the depression of cholinesterase leading to accumulation of acetylcholine in the nervous system, which is believed to be responsible for the symptoms. Organic phosphorus insecticides are absorbed by the skin, as well as by the respiratory and gastrointestinal tracts. They are cholinesterase inhibitors. Symptoms of exposure include headache, giddiness, blurred vision, nervousness, profound weakness, nausea, cramps, diarrhea, and discomfort in the chest. Signs include sweating, tearing, salivation, vomiting, cyanosis, convulsions, coma, loss of reflexes and loss of sphincter control.

Long Term Exposure: Cholinesterase inhibitor; cumulative effect is possible. This chemical may damage the nervous system with repeated exposure, resulting in convulsions, respiratory failure. May cause liver damage.

Points of Attack: Respiratory system, lungs, central nervous system, cardiovascular system, skin, eyes, plasma and red blood cell cholinesterase.

Medical Surveillance: Before employment and at regular times after that, the following are recommended: Plasma and red blood cell cholinesterase levels (tests for the enzyme poisoned by this chemical). If exposure stops, plasma levels return to normal in 1 – 2 weeks while red blood cell levels may be reduced for 1 – 3 months.

When cholinesterase enzyme levels are reduced by 25% or more below preemployment levels, risk of poisoning is increased, even if results are in lower ranges of "normal." Reassignment to work not involving organophosphate or carbamate pesticides is recommended until enzyme levels recover. If symptoms develop or overexposure occurs, repeat the above tests as soon as possible and get an exam of the nervous system. Also consider complete blood count. Consider chest x-ray following acute overexposure. Do not drink any alcoholic beverages before or during use. Alcohol promotes absorption of organic phosphates.

First Aid: If this chemical gets into the eyes, remove any contact lenses at once and irrigate immediately for at least 15 minutes, occasionally lifting upper and lower lids. Seek medical attention immediately. If this chemical contacts the skin, remove contaminated clothing and wash immediately with soap and water. Speed in removing material from skin is of extreme importance. Shampoo hair promptly if contaminated.

Speed in removing the material form the skin is of extreme importance. Seek medical attention immediately. If this chemical has been inhaled, remove from exposure, begin rescue breathing (using universal precautions) if breathing has stopped and CPR if heart action has stopped. Transfer promptly to a medical facility. When this chemical has been swallowed, get medical attention. Give large quantities of water and induce vomiting. Do not make an unconscious person vomit. Keep victim quiet and maintain normal body temperature. Effects may be delayed; keep victim under observation.

Personal Protective Methods: Wear protective gloves and clothing to prevent any reasonable probability of skin contact. Safety equipment suppliers/manufacturers can provide recommendations on the most protective glove/clothing material for your operation. All protective clothing (suits, gloves, footwear, headgear) should be clean, available each day, and put on before work. Contact lenses should not be worn when working with this chemical. Wear splash-proof chemical goggles and face shield unless full facepiece respiratory protection is worn. Employees should wash immediately with soap when skin is wet or contaminated. Provide emergency showers and eyewash.

Respirator Selection: Where the potential for exposure to this chemical, use a MSHA/NIOSH approved supplied-air respirator with a full facepiece operated in the positive pressure mode or with a full facepiece, hood, or helmet in the continuous flow mode, or use a MSHA/NIOSH approved self-contained breathing apparatus with a full facepiece operated in pressure-demand or other positive pressure mode.

Storage: Prior to working with this chemical you should be trained on its proper handling and storage. Store in tightly closed containers in a cool, well ventilated area away from. alkaline materials. Where possible, automatically pump liquid from drums or other storage containers to process containers.

Shipping: This falls under organophosphorus pesticides, liquid, toxic, n.o.s. This compound requires a shipping label of: "Keep Away From Food." It falls in DOT Hazard Class 6.1 and Packing Group III.

Spill Handling: Evacuate and restrict persons not wearing protective equipment from area of spill or leak until cleanup is complete. Remove all ignition sources. Stay upwind; keep out of low areas. Ventilate closed spaces before entering them. Do not touch spilled material; keep leak if you can do so without risk. Use water spray to reduce vapor. Small spills: absorb with sand or other noncombustible absorbent material and place into containers for later disposal. Small dry spills: with clean shovel, place material into clean, dry containers and cover; move containers from spill area. Large spills: dike far ahead of spill for later disposal. It may be necessary to contain and dispose of this chemical as a hazardous waste. If material or contaminated runoff enters waterways, notify downstream users of potentially contaminated waters. Contact your Department of Environmental Protection or your regional office of the federal EPA for specific recommendations. If employees are required to clean-up spills, they must be properly trained and equipped. OSHA 1910.120(q) may be applicable.

Fire Extinguishing: Small fires: dry chemical, carbon dioxide, water spray, or foam. Large fires: water spray, fog, or foam. Move container from fire area if you can do so without risk. Fight fire from maximum distance. Dike fire control water for later disposal; do not scatter material. When heated to decomposition, it emits very toxic fumes off nitrogen oxides, phosphorus oxides and sulfur oxides. This compound is an organophosphorus insecticide. Some of these materials may burn but none of them ignite readily. Fire and runoff from fire control water may produce irritating or poisonous gases. Vapors are heavier than air and will collect in low areas. Containers may explode in fire. Storage containers and parts of containers may rocket great distances, in many directions. If material or contaminated runoff enters waterways, notify downstream users of potentially contaminated waters. Notify local health and fire officials and pollution control agencies. From a secure, explosion-proof location, use water spray to cool exposed containers. If cooling streams are ineffective (venting sound increases in volume and pitch, tank discolors, or shows any signs of deforming), withdraw immediately to a secure position. If employees are expected to fight fires, they must be trained and equipped in OSHA 1910.156.

Disposal Method Suggested: In accordance with 40CFR 165 recommendations for the disposal of pesticides and pesticide containers. Must be disposed properly by following package label directions or by contacting your state pesticide or environmental control agency or by contacting your regional EPA office.

References

U.S. Environmental Protection Agency, "Chemical Profile: Formothion," Washington, DC, Chemical Emergency Preparedness Program (November 30, 1987)

Fosthietan

Molecular Formula: $C_6H_{12}NO_3PS_2$

Synonyms: AC 64475; Acconem; CL64475; (Diethoxyphosphinylimino)-1,3-dithietane; Diethoxyphosphinylimino-2-dithietanne-1,3 (French); 1,3-Dithietan-2-ylidenephosphoramidic acid diethyl ester; Geofos; NEM-A-TAK

CAS Registry Number: 21548-32-3

RTECS Number: NJ6490000

Regulatory Authority

- Superfund/EPCRA 40CFR355, Appendix B Extremely Hazardous Substances: TPQ = 500 lb (227 kg)
- Superfund/EPCRA 40CFR302.4 Reportable Quantity (RQ): EHS, 1 lb (0.454 kg)

Cited in U.S. State Regulations: California (G), Massachusetts (G), New Jersey (G), Pennsylvania (G).

Description: $C_6H_{12}NO_3PS_2$, Fosthietan, is a pale yellow oil with a mercaptan-like odor. Moderately soluble in water.

Potential Exposure: This material is used as a nematocide and insecticide. Not registered as a pesticide in the U.S.

Incompatibilities: Alkaline material.

Permissible Exposure Limits in Air: No standards set.

Determination in Air: OSHA versatile sampler-2; Toluene/Acetone; Gas chromatography/Flame photometric detection for sulfur, nitrogen, or phosphorus; NIOSH Method IV Method #5600, Organophosphorus Pesticides.

Permissible Concentration in Water: No criteria set.

Routes of Entry: Inhalation, ingestion, absorbed through the skin.

Harmful Effects and Symptoms

Short Term Exposure: This compound is a liquid organophosphorus insecticide. Organic phosphorus insecticides are absorbed by the skin, as well as by the respiratory and gastrointestinal tracts. Organic phosphorus insecticides are absorbed by the skin, as well as by the respiratory and gastrointestinal tracts. They are cholinesterase inhibitors. Symptoms of exposure include headache, giddiness, blurred vision, nervousness, weakness, nausea, cramps, diarrhea, and discomfort in the chest. Signs include sweating, tearing, salivation, vomiting, cyanosis, convulsions, coma, loss of reflexes and loss of sphincter control.

Long Term Exposure: Cholinesterase inhibitor; cumulative effect is possible. This chemical may damage the nervous system with repeated exposure, resulting in convulsions, respiratory failure. May cause liver damage.

Points of Attack: Respiratory system, lungs, central nervous system, cardiovascular system, skin, eyes, plasma and red blood cell cholinesterase.

Medical Surveillance: Before employment and at regular times after that, the following are recommended: Plasma and red blood cell cholinesterase levels (tests for the enzyme poisoned by this chemical). If exposure stops, plasma levels return to normal in 1 – 2 weeks while red blood cell levels may be reduced for 1 – 3 months.

When cholinesterase enzyme levels are reduced by 25% or more below preemployment levels, risk of poisoning is increased, even if results are in lower ranges of "normal." Reassignment to work not involving organophosphate or carbamate pesticides is recommended until enzyme levels recover. If symptoms develop or overexposure occurs, repeat the above tests as soon as possible and get an exam of the nervous system. Also consider complete blood count. Consider chest x-ray following acute overexposure. Do not drink any alcoholic beverages before or during use. Alcohol promotes absorption of organic phosphates.

First Aid: If this chemical gets into the eyes, remove any contact lenses at once and irrigate immediately for at least 15 minutes, occasionally lifting upper and lower lids. Seek medical attention immediately. If this chemical contacts the skin, remove contaminated clothing and wash immediately with soap and water. Speed in removing material from skin is of extreme importance. Shampoo hair promptly if contaminated. Seek medical attention immediately. If this chemical has been inhaled, remove from exposure, begin rescue breathing (using universal precautions) if breathing has stopped and CPR if heart action has stopped. Transfer promptly to a medical facility. When this chemical has been swallowed, get medical attention. Give large quantities of water and induce vomiting. Do not make an unconscious person vomit. Remove and isolate contaminated clothing and shoes at the site. Keep victim quiet and maintain normal body temperature. Effects may be delayed; keep victim under observation.

Personal Protective Methods: Wear protective gloves and clothing to prevent any reasonable probability of skin contact. Safety equipment suppliers/manufacturers can provide recommendations on the most protective glove/clothing material for your operation. All protective clothing (suits, gloves, footwear, headgear) should be clean, available each day, and put on before work. Contact lenses should not be worn when working with this chemical. Wear splash-proof chemical goggles and face shield unless full facepiece respiratory protection is worn. Employees should wash immediately with soap when skin is wet or contaminated. Provide emergency showers and eyewash.

Respirator Selection: Where the potential for exposure to this chemical, use a MSHA/NIOSH approved supplied-air respirator with a full facepiece operated in the positive pressure mode or with a full facepiece, hood, or helmet in the continuous flow mode, or use a MSHA/NIOSH approved self-contained breathing apparatus with a full facepiece operated in pressure-demand or other positive pressure mode.

Storage: Prior to working with this chemical you should be trained on its proper handling and storage. Store in tightly closed containers in a cool, well ventilated area away from alkaline material. Where possible, automatically pump liquid from drums or other storage containers to process containers.

Shipping: This material falls under organophosphorus pesticides, liquid, toxic, n.o.s. This compound requires a shipping label of: "Poison." It falls in DOT Hazard Class 6.1 and Packing Group I.

Spill Handling: Evacuate and restrict persons not wearing protective equipment from area of spill or leak until cleanup is complete. Remove all ignition sources. Stay upwind; keep out of low areas. Ventilate closed spaces before entering them. Do not touch spilled material; stop leak if you can do so without risk. Use water spray to reduce vapors. *Small spills:* absorb with sand or other non-combustible absorbent material and place into containers for later disposal. *Large spills:* dike far ahead of spill for later disposal.

Fire Extinguishing: This chemical may burn but does not ignite readily. Poisonous gases including nitrogen, phosphorus and sulfure oxides are produced in fire. Use dry chemical, carbon dioxide, or alcohol foam extinguishers. Vapors are heavier than air and will collect in low areas. Vapors may travel long distances to ignition sources and flashback. Vapors in confined areas may explode when exposed to fire. Containers may explode in fire. Storage containers and parts of containers may rocket great distances, in many directions. If material or contaminated runoff enters waterways, notify downstream users of potentially contaminated waters. Notify local health and fire officials and pollution control agencies. From a secure, explosion-proof location, use water spray to cool exposed containers. If cooling streams are ineffective (venting sound increases in volume and pitch, tank discolors, or shows any signs of deforming), withdraw immediately to a secure position. If employees are expected to fight fires, they must be trained and equipped in OSHA 1910.156.

Disposal Method Suggested: In accordance with 40CFR 165 recommendations for the disposal of pesticides and pesticide containers. Must be disposed properly by following package label directions or by contacting your state pesticide or environmental control agency or by contacting your regional EPA office.

References

U.S. Environmental Protection Agency, "Chemical Profile: Fosthietan," Washington, DC, Chemical Emergency Preparedness Program (November 30, 1987)

Fuberidazole

Molecular Formula: $C_{11}H_8N_2O$

Synonyms: Baycor; Bayer 33172; Baytan; Bitertanol, fuberidazole; Fuberidatol; Fuberisazol; Fubridazole; 2-(2-Furanyl)-1H-benzimidazole; Furidazol; Furidazole; 2-(2'-Furyl)-benzimidazole; 2-(2-Furyl)benzimidazole; ICI Baytan; Neovoronit; RTECS No.DD9010000; Sibutol; Sibutrol; Voronit; Voronite; W VII/117

CAS Registry Number: 3878-19-1

RTECS Number: DD9010000

Regulatory Authority

- Superfund/EPCRA 40CFR355, Appendix B Extremely Hazardous Substances: TPQ = 100/10,000 lb (45.4/4,540 kg)
- Superfund/EPCRA 40CFR302.4 Reportable Quantity (RQ): EHS, 1 lb (0.454 kg)

Cited in U.S. State Regulations: California (G), Massachusetts (G), New Jersey (G), Pennsylvania (G).

Description: Fuberidazole, $C_{11}H_8N_2O$, is a crystalline solid. Melting point = 280°C (decomposition). Hazard Identification (based on NFPA-704 M Rating System): Health 1, Flammability 1, Reactivity 0. Slightly soluble in water.

Potential Exposure: Uses include cereal seed dressing; and fungicidal non-mercurial seed dressing with special action against fusarium. Not registered as a pesticide in the U.S.A.

Permissible Exposure Limits in Air: No standards set.

Permissible Concentration in Water: No criteria set.

Routes of Entry: Inhalation, ingestion, skin contact.

Harmful Effects and Symptoms

Short Term Exposure: Fuberidazole is classified as moderately toxic. Its probable oral lethal dose in humans is 0.5 – 5 g/kg or between 1 ounce and 1 pint for a 70 kg (150 lb) person. The oral LD_{50} rat is 1,100 mg/kg.[9]

First Aid: If this chemical gets into the eyes, remove any contact lenses at once and irrigate immediately for at least 15 minutes, occasionally lifting upper and lower lids. Seek medical attention immediately. If this chemical contacts the skin, remove contaminated clothing and wash immediately with soap and water. Seek medical attention immediately. If this chemical has been inhaled, remove from exposure, begin rescue breathing (using universal precautions) if breathing has stopped and CPR if heart action has stopped. Transfer promptly to a medical facility. When this chemical has been swallowed, get medical attention. Give large quantities of water and induce vomiting. Do not make an unconscious person vomit.

Personal Protective Methods: Wear protective gloves and clothing to prevent any reasonable probability of skin contact. Safety equipment suppliers/manufacturers can provide recommendations on the most protective glove/clothing material for your operation. All protective clothing (suits, gloves, footwear, headgear) should be clean, available each day, and put on before work. Contact lenses should not be worn when working with this chemical. Wear dust-proof chemical goggles and face shield unless full facepiece respiratory protection is worn. Employees should wash immediately with soap when skin is wet or contaminated. Provide emergency showers and eyewash. For emergency situations, wear a positive pressure, pressure-demand, full facepiece self-contained breathing apparatus (SCBA) or pressure-demand supplied air respirator with escape SCBA and a fully-encapsulating, chemical resistant suit.

Respirator Selection: Where the potential for exposure to this chemical, use a MSHA/NIOSH approved supplied-air respirator with a full facepiece operated in the positive pressure mode or with a full facepiece, hood, or helmet in the continuous flow mode, or use a MSHA/NIOSH approved self-contained breathing apparatus with a full facepiece operated in pressure-demand or other positive pressure mode.

Storage: Prior to working with this chemical you should be trained on its proper handling and storage. Store in tightly closed containers in a cool, well ventilated area.

Shipping: This compound is not cited by DOT in its performance-oriented packaging standards.

Spill Handling: Evacuate persons not wearing protective equipment from area of spill or leak until clean-up is complete. Remove all ignition sources. Do not touch spilled material; stop leak if you can do so without risk. Use water spray to reduce vapors. Small spills: absorb with sand or other noncombustible absorbent material and place into containers for later disposal. Small dry spills: with clean shovel place material into clean, dry container and cover; move containers from spill area. Large spills: dike far ahead of spill for later disposal. Ventilate area after clean-up is complete. It may be necessary to contain and dispose of this chemical as a hazardous waste. If material or contaminated runoff enters waterways, notify downstream users of potentially contaminated waters. Contact your Department of Environmental Protection or your regional office of the federal EPA for specific recommendations. If employees are required to clean-up spills, they must be properly trained and equipped. OSHA 1910.120(q) may be applicable.

Fire Extinguishing: This material is a combustible solid. Move container from fire area if you can do so without risk. Fight fire from maximum distance. Dike fire control water for later disposal; do not scatter the material. Keep unnecessary people away; isolate hazard area and deny entry. Stay upwind; keep out of low areas. Ventilate closed spaces before entering them. Wear positive pressure breathing apparatus and special protective clothing. Use dry chemical, carbon dioxide, water spray, or alcohol foam extinguishers. Poisonous gases are produced in fire including nitrogen oxides. If material or contaminated runoff enters waterways, notify downstream users of potentially contaminated waters. Notify local health and fire officials and pollution control agencies. From a secure, explosion-proof location, use water spray to cool exposed containers. If cooling streams are ineffective (venting sound increases in volume and pitch, tank discolors, or shows any signs of deforming), withdraw immediately to a secure position. If employees are expected to fight fires, they must be trained and equipped in OSHA 1910.156.

Disposal Method Suggested: In accordance with 40CFR 165 recommendations for the disposal of pesticides and pesticide containers. Must be disposed properly by following package label directions or by contacting your state pesticide or environmental control agency or by contacting your regional EPA office.

References

U.S. Environmental Protection Agency, "Chemical Profile: Fuberidazole," Washington, DC, Chemical Emergency Preparedness Program (November 30, 1987)

Fumaric Acid

Molecular Formula: $C_4H_4O_4$

Synonyms: Allomaleic acid; Boletic acid; (E)-Butenedioic acid; *trans*-Butenedioic acid; 2-Butenedioic acid (E); Butenedioic acid, (E)-; 1,2-Ethenedicarboxylic acid, *trans*-; *trans*-1,2-Ethylenedicarboxylic acid; 1,2-Ethylenedicarboxylic acid, (E); Kyselina fumarova (Czech); Lichenic acid; NSC-2752; U-1149

CAS Registry Number: 110-17-8

RTECS Number: LS9625000

DOT ID: NA 9126

EEC Number: 607-146-00-X

Regulatory Authority

- Clean Water Act: Section 311 Hazardous Substances/RQ 40CFR117.3 (same as CERCLA, see below)
- Superfund/EPCRA 40CFR302.4 Reportable Quantity (RQ): CERCLA, 5,000 lb (2,270 kg)

Cited in U.S. State Regulations: California (G), New Jersey (G), Pennsylvania (G).

Description: Fumaric acid is a colorless to white, odorless crystalline powder. Fruity-acidic taste. Sublimation temperature = 200°C. Freezing/Melting point = 287°C. Flash point = 230°C. Also listed as >156°F (>69°C). Autoignition temperature (dust cloud) = 740°C. Hazard Ranking (based on NFPA 704 M Hazard Identification): Health 0, Flammability 1, Reactivity 0. Slightly soluble in water.

Potential Exposure: Fumaric acid is used as a food additive, as an antioxidant in resins, and to make dyes. Decomposes above 350°C forming irritating fumes of maleic anhydride.

Incompatibilities: Dust cloud from powder or granular form mixed with air can explode. Reacts with strong oxidizers. Incompatible with sulfuric acid, caustics, ammonia, amines, isocyanates, alkylene oxides, epichlorohydrin.

Permissible Exposure Limits in Air: No OELs established.

Determination in Air: No test available.

Routes of Entry: Inhalation. Absorbed through the skin.

Harmful Effects and Symptoms

Short Term Exposure: Irritates the eyes, skin, and respiratory tract.

Long Term Exposure: Repeated exposure may cause liver damage.

Points of Attack: Liver.

Medical Surveillance: Liver function tests.

First Aid: If this chemical gets into the eyes, remove any contact lenses at once and irrigate immediately for at least 15 minutes, occasionally lifting upper and lower lids. Seek medical attention immediately. If this chemical contacts the skin, remove contaminated clothing and wash immediately with soap and water. Seek medical attention immediately. If this chemical has been inhaled, remove from exposure, begin rescue breathing (using universal precautions) if breathing has stopped and CPR if heart action has stopped. Transfer promptly to a medical facility. When this chemical has

been swallowed, get medical attention. Give large quantities of water and induce vomiting. Do not make an unconscious person vomit.

Personal Protective Methods: Wear protective gloves and clothing to prevent any reasonable probability of skin contact. Safety equipment suppliers/manufacturers can provide recommendations on the most protective glove/clothing material for your operation. All protective clothing (suits, gloves, footwear, headgear) should be clean, available each day, and put on before work. Contact lenses should not be worn when working with this chemical. Wear dust-proof chemical goggles and face shield unless full facepiece respiratory protection is worn. Employees should wash immediately with soap when skin is wet or contaminated. Provide emergency showers and eyewash.

Respirator Selection: Where the potential for exposure to this chemical, use a MSHA/NIOSH approved supplied-air respirator with a full facepiece operated in the positive pressure mode or with a full facepiece, hood, or helmet in the continuous flow mode, or use a MSHA/NIOSH approved self-contained breathing apparatus with a full facepiece operated in pressure-demand or other positive pressure mode.

Storage: Prior to working with this chemical you should be trained on its proper handling and storage. Store in tightly closed containers in a cool, well ventilated area away from heat and incompatible materials listed above.

Spill Handling: Evacuate persons not wearing protective equipment from area of spill or leak until clean-up is complete. Remove all ignition sources. Collect powdered material in the most convenient and safe manner and deposit in sealed containers. Ventilate area after clean-up is complete. It may be necessary to contain and dispose of this chemical as a hazardous waste. If material or contaminated runoff enters waterways, notify downstream users of potentially contaminated waters. Contact your Department of Environmental Protection or your regional office of the federal EPA for specific recommendations. If employees are required to clean-up spills, they must be properly trained and equipped. OSHA 1910.120(q) may be applicable.

Fire Extinguishing: This chemical is a combustible solid. Use dry chemical, carbon dioxide, water spray, or foam extinguishers. Poisonous gases are produced in fire including maleic anhydride. If material or contaminated runoff enters waterways, notify downstream users of potentially contaminated waters. Notify local health and fire officials and pollution control agencies. Containers may explode in fire. From a secure, explosion-proof location, use water spray to cool exposed containers. If cooling streams are ineffective (venting sound increases in volume and pitch, tank discolors, or shows any signs of deforming), withdraw immediately to a secure position. If employees are expected to fight fires, they must be trained and equipped in OSHA 1910.156.

References

New Jersey Department of Health and Senior Services, Hazardous Substance Fact Sheet, Trenton, NJ (1998)

Furan

Molecular Formula: C_4H_4O

Synonyms: Divinylene oxide; Furano (Spanish); Furfuran; NCI-C56202; Oxacyclopentadiene; Oxole; Tetrole

CAS Registry Number: 110-00-9

RTECS Number: LT8524000

DOT ID: UN 2389

Regulatory Authority

- Air Pollutant Standard Set (former USSR)[43]
- Clean Air Act: Accidental Release Prevention/Flammable substances, (Section 112[r], Table 3), TQ = 5,000 lb (2,270 kg)
- OSHA 29CFR1910.119, Appendix A. Process Safety List of Highly Hazardous Chemicals, TQ = 500 lb (227 kg)
- EPA Hazardous Waste Number (RCRA No.): U124
- RCRA, 40CFR261, Appendix 8 Hazardous Constituents
- Superfund/EPCRA 40CFR355, Appendix B Extremely Hazardous Substances: TPQ = 500 lb (227 kg)
- Superfund/EPCRA 40CFR302.4 Reportable Quantity (RQ): CERCLA, 100 lb (45.4 kg)

Cited in U.S. State Regulations: California (G), Florida (G), Kansas (G), Louisiana (G), Massachusetts (G), New Hampshire (G), New Jersey (G), Oklahoma (G), Pennsylvania (G), Vermont (G), Virginia (G), Washington (G), Wisconsin (G).

Description: Furan, C_4H_4O, is a colorless liquid. Boiling point = 32°C. Freezing/Melting point = -85.65°C. Flash point = -36°C. Explosive limits: LEL = 2.3%; UEL = 14.3%.[17] Hazard Identification (based on NFPA-704 M Rating System): Health 1, Flammability 4, Reactivity 1. Insoluble in water.

Potential Exposure: Furan is used as a chemical intermediate for tetrahydrofuran in formation of lacquers, as a solvent for resins in organic synthesis, especially for pyrrole, thiophene.

Incompatibilities: Forms explosive mixture with air. Violent reaction with acids, oxidizers. Unless stabilized with an inhibitor, air exposure forms unstable peroxides.

Permissible Exposure Limits in Air: The Former USSR-UNEP/IRPTC project[43] has set a MAC in workplace air of 0.5 mg/m^3.

Permissible Concentration in Water: The Former USSR-UNEP/IRPTC project[43] has set a MAC in water bodies used for domestic purposes of 0.2 mg/l.

Routes of Entry: Inhalation, ingestion, skin contact.

Harmful Effects and Symptoms

Short Term Exposure: Contact can irritate and burn the skin and eyes. Vapors can irritate the respiratory tract and are a central nervous system depressant. Higher exposures can cause pulmonary edema, a medical emergency that can be delayed for several hours. This can cause death. Exposure can cause headache, dizziness, shortness of breath, unconsciousness, and suffocation are among the symptoms. The vapors are narcotic. Acute exposure to furan by inhalation may involve both reversible and irreversible changes. Acute exposure by ingestion or skin absorption, as well as chronic exposure, are associated with high toxicity.

Long Term Exposure: Furan may be a carcinogen since it has been shown to cause cancer of the liver and white blood cells in animals. May cause skin allergy. May damage the liver and kidneys.

Points of Attack: Skin, liver, kidneys.

Medical Surveillance: Liver and kidney function tests. Consider chest x-ray following acute overexposure. Evaluation by a qualified allergist.

First Aid: If this chemical gets into the eyes, remove any contact lenses at once and irrigate immediately for at least 15 minutes, occasionally lifting upper and lower lids. Seek medical attention immediately. If this chemical contacts the skin, remove contaminated clothing and wash immediately with soap and water. Seek medical attention immediately. If this chemical has been inhaled, remove from exposure, begin rescue breathing (using universal precautions) if breathing has stopped and CPR if heart action has stopped. Transfer promptly to a medical facility. When this chemical has been swallowed, get medical attention. Give large quantities of water and induce vomiting. Do not make an unconscious person vomit. Medical observation is recommended for 24 – 48 hours after breathing overexposure, as pulmonary edema may be delayed. As first aid for pulmonary edema, a doctor or authorized paramedic may consider administering a corticosteroid spray.

Personal Protective Methods: Wear protective gloves and clothing to prevent any reasonable probability of skin contact. Safety equipment suppliers/manufacturers can provide recommendations on the most protective glove/clothing material for your operation Polyvinyl Alcohol is among the recommended protective materials. All protective clothing (suits, gloves, footwear, headgear) should be clean, available each day, and put on before work. Contact lenses should not be worn when working with this chemical. Wear splash-proof chemical goggles and face shield unless full facepiece respiratory protection is worn. Employees should wash immediately with soap when skin is wet or contaminated. Provide emergency showers and eyewash. For emergency situations, wear a positive pressure, pressure-demand, full facepiece self-contained breathing apparatus (SCBA) or pressure-demand supplied air respirator with escape SCBA and a fully-encapsulating, chemical resistant suit.

Respirator Selection: Where the potential for exposure to this chemical, use a MSHA/NIOSH approved supplied-air respirator with a full facepiece operated in the positive pressure mode or with a full facepiece, hood, or helmet in the continuous flow mode, or use a MSHA/NIOSH approved self-contained breathing apparatus with a full facepiece operated in pressure-demand or other positive pressure mode.

Storage: Prior to working with this chemical you should be trained on its proper handling and storage. Before entering confined space where this chemical may be present, check to make sure that an explosive concentration does not exist. Store in an explosion-proof refrigerator.[52] Keep in a tightly closed container under an inert atmosphere and protect from light for long-term storage. A regulated, marked area should be established where this chemical is handled, used, or stored in compliance with OSHA standard 1910.1045.

Shipping: This compound requires a shipping label of: "Flammable Liquid." It falls in DOT Hazard Class 3 and Packing Group I.

Spill Handling: Evacuate and restrict persons not wearing protective equipment from area of spill or leak until cleanup is complete. Avoid breathing vapors. Keep upwind. Do not handle broken packages without protective equipment. Wash away any material which may have contacted the body with copious amounts of water or soap and water. Shut off ignition sources; no flares, smoking, or flames in hazard area. Stop leak if you can do so without risk. Use water spray to reduce vapors. Small spills: absorb with sand or other noncombustible absorbent material and place into containers for later disposal. Large spills: dike far ahead of spill for later disposal. The exposure concentration limit of 10 ppm together with the low boiling point of furan requires that adequate ventilation be provided in areas handling this chemical. Establish forced ventilation to keep levels below explosive limit. Contact with liquid must be avoided since this chemical can be absorbed through the skin. Keep furan out of a confined space, such as a sewer, because of the possibility of an explosion, unless the sewer is designed to prevent the build-up off explosive concentrations. Thorough washing with soap and water followed by prolonged rinsing should be done immediately after accidental contact. It may be necessary to contain and dispose of this chemical as a hazardous waste. If material or contaminated runoff enters waterways, notify downstream users of potentially contaminated waters. Contact your Department of Environmental Protection or your regional office of the federal EPA for specific recommendations. If employees are required to clean-up spills, they must be properly trained and equipped. OSHA 1910.120(q) may be applicable.

Fire Extinguishing: This chemical is a flammable liquid. Water may be ineffective. Small fires: use dry chemical, carbon dioxide, water spray, or alcohol foam. Large fires: water spray, fog, or alcohol foam. Move container from fire area if this can be accomplished without risk. Spray cooling water on containers that are exposed to flames until well after fire is out. For massive fires in cargo area, use unmanned hose holder or monitor nozzles; if this is impossible withdraw from area and let fire burn. Withdraw immediately in case off rising sound from venting safety device or any discoloration of tank due to fire. Isolate for 1/2 mile in all directions if a tank car or truck is involved. Vapors are heavier than air and will collect in low areas. Vapors may travel long distances to ignition sources and flashback. Vapors in confined areas may explode when exposed to fire. Containers may explode in fire. Storage containers and parts of containers may rocket great distances, in many directions. If material or contaminated runoff enters waterways, notify downstream users of potentially contaminated waters. Notify local health and fire officials and pollution control agencies. From a secure, explosion-proof location, use water spray to cool exposed containers. If cooling streams are ineffective (venting sound increases in volume and pitch, tank discolors, or shows any signs of deforming), withdraw immediately to a secure position. If employees are expected to fight fires, they must be trained and equipped in OSHA 1910.156.

Disposal Method Suggested: Consult with environmental regulatory agencies for guidance on acceptable disposal practices. Generators of waste containing this contaminant (≥100 kg/mo) must conform with EPA regulations governing storage, transportation, treatment, and waste disposal.

References

Sax, N. I., Ed., "Dangerous Properties of Industrial Materials Report," 7, No. 3, 93-95 (1987)

U.S. Environmental Protection Agency, "Chemical Profile: Furan," Washington, DC, Chemical Emergency Preparedness Program (November 30, 1987)

New Jersey Department of Health and Senior Services, "Hazardous Substance Fact Sheet, Furan," Trenton, NJ (May 1999)

Furfural

Molecular Formula: $C_5H_4O_2$

Synonyms: Artificial Ant oil; Fural; 2-Furaldehyde; Furale; 2-Furanaldehyde; 2-Furancarbonal; 2-Furan carboxaldehyde; Furfuraldehyde; α-Furole; Furole; 2-Furyl-methanal; NCI-C56177; Oil of Ants, Artificial; Pyromucic aldehyde

CAS Registry Number: 98-01-1

RTECS Number: LT7000000

DOT ID: UN 1199

EEC Number: 605-010-00-4

EINECS Number: 202-627-7

Regulatory Authority

- Air Pollutant Standard Set (ACGIH)[1] (DFG)[3] (HSE)[33] (OSHA)[58] (Other Countries)[35] (Several States)[60] (Several Canadian Provinces)
- Clean Water Act: Section 311 Hazardous Substances/RQ 40CFR117.3 (same as CERCLA, see below).
- EPA Hazardous Waste Number (RCRA No.): U125
- RCRA, 40CFR261, Appendix 8 Hazardous Constituents
- Superfund/EPCRA 40CFR302.4 Reportable Quantity (RQ): CERCLA, 5,000 lb (2,270 kg)
- Canada, WHMIS, Ingredients Disclosure List

Cited in U.S. State Regulations: Alaska (G), California (A, G), Connecticut (A), Florida (G, A), Illinois (G), Kansas (G), Louisiana (G), Maine (G), Massachusetts (G), Nevada (A), New Hampshire (G), New Jersey (G), New York (A), North Dakota (A), Pennsylvania (G), Rhode Island (G), South Carolina (A), Vermont (G), Virginia (G), Washington (G), West Virginia (G), Wisconsin (G).

Description: Furfural, $C_5H_4O_2$, is a colorless to yellow aromatic heterocyclic aldehyde with an almond-like odor. Turns amber on exposure to light and air. Boiling point = 161 – 162°C. Flash point = 60°C. Autoignition temperature = 316°C. Explosive limits: LEL = 2.1%; UEL = 19.3%. Hazard Identification (based on NFPA-704 M Rating System): Health 3, Flammability 2, Reactivity 0. Soluble in water.

Potential Exposure: Furfural is used as a solvent for wood resin, nitrated cotton, cellulose acetate, and gums. It is used in the production of phenolic plastics, thermosetting resins, refined petroleum oils, dyes, and varnishes. It is also utilized in the manufacture of pyromucic acid, vulcanized rubber, insecticides, fungicides, herbicides, germicides, furan derivatives, polymers, and other organic chemicals.

Incompatibilities: Forms explosive mixture with air. Acids and bases can cause polymerization causing fire or explosion hazard. Reacts violently with oxidants. Incompatible with strong acids, caustics, ammonia, aliphatic amines, alkanolamines, aromatic amines, oxidizers. Attacks many plastics and Attacks many plastics.

Permissible Exposure Limits in Air: The OSHA TWA[58] as well as ACGIH has adopted a TWA value of 2 ppm (8 mg/m^3). HSE[33] adds an STEL value of 10 ppm (40 mg/m^3). The notation "skin" is added to indicate the possibility of cutaneous absorption. The NIOSH IDLH level is 100 ppm. The DFG[3] and Sweden[35] have set a TWA of 5 ppm (20 mg/m^3). Japan and Czechoslovakia and the Former USSR have set MAC values of 2.5 ppm (10 mg/m^3) in workplace air. The Former USSR[35][43] adds a MAC in ambient air of residential areas of 0.05 mg/m^3 both on a momentary and on a daily average basis. Several states have set guidelines or standards for furfural in ambient air[60] ranging from 26.7 μg/m^3 (New York) to 80.0 μg/m^3 (Florida) to 160.0 μg/m^3 (Connecticut) to 190 μg/m^3 (Nevada) to 200.0 μg/m^3 (South Carolina) to 80.0 – 400.0 μg/m^3 (North Dakota).

Determination in Air: Collection by XAD-2® (special tube); workup with toluene; analysis by gas chromatography/flame ionization detection; NIOSH IV, Method #2529.

Permissible Concentration in Water: The former USSR[35][43] has set a MAC in water bodies used for domestic purposes of 1.0 mg/l.

Routes of Entry: Inhalation of vapor, percutaneous absorption, ingestion, skin and/or eye contact.

Harmful Effects and Symptoms

Short Term Exposure: Contact may cause skin irritation, causing rash and a burning sensation. Liquid and concentrated vapor are irritating to the eyes, skin, upper respiratory tract. Higher exposures can cause pulmonary edema, a medical emergency that can be delayed for several hours. This can cause death.

Workers chronically exposed to the vapor have had complaints of headache, fatigue, itching of the throat, lacrimation, loss of the sense of taste, numbness of the tongue, and tremor. Occupational overexposure is relatively rare due to the liquid's low vapor pressure, and symptoms usually disappear rapidly after removal from exposure.

Long Term Exposure: Eczematous dermatitis as well as skin sensitization, resulting in allergic contact dermatitis and photosensitivity, may develop following repeated exposure. The substance may have effects on the liver. Furan causes mutations. Repeated exposure may cause loss of taste, numbness of the tongue, and may cause headaches, tiredness, tremors, itchy throat, watery eyes. Long term exposure may cause skin to sunburn more easily.

Points of Attack: Eyes, respiratory system, skin.

Medical Surveillance: For those with frequent or potentially high exposure (half the TLV or greater) the following are recommended before beginning work and at regular times after that: Lung function tests. If symptoms develop or overexposure is suspected, the following may be useful: Liver function tests. Consider chest x-ray after acute overexposure. Evaluation by a qualified allergist, including careful exposure history and special testing, may help diagnose skin allergy.

First Aid: If this chemical gets into the eyes, remove any contact lenses at once and irrigate immediately for at least 15 minutes, occasionally lifting upper and lower lids. Seek medical attention immediately. If this chemical contacts the skin, remove contaminated clothing and wash immediately with soap and water. Seek medical attention immediately. If this chemical has been inhaled, remove from exposure, begin rescue breathing (using universal precautions) if breathing has stopped and CPR if heart action has stopped. Transfer promptly to a medical facility. When this chemical has been swallowed, get medical attention. Give large quantities of water and induce vomiting. Do not make an unconscious person vomit.

Personal Protective Methods: Wear protective gloves and clothing to prevent any reasonable probability of skin contact. Safety equipment suppliers/manufacturers can provide recommendations on the most protective glove/clothing material for your operation. Teflon, Silvershield, Viton, Polyvinyl Alcohol, and Butyl Rubber are among the recommended protective materials. All protective clothing (suits, gloves, footwear, headgear) should be clean, available each day, and put on before work. Contact lenses should not be worn when working with this chemical. Wear splash-proof chemical goggles and face shield unless full facepiece respiratory protection is worn. Employees should wash immediately with soap when skin is wet or contaminated. Provide emergency showers and eyewash.

Respirator Selection: OSHA: *50 ppm:* CCROV [any chemical cartridge respirator with organic vapor cartridge(s)]; or SA (any supplied-air respirator). *100 ppm:* SA:CF (any supplied-air respirator operated in a continuous-flow mode); or CCRFOV [any chemical cartridge respirator with a full facepiece and organic vapor cartridge(s)]; or PAPROV [any powered, air-purifying respirator with organic vapor cartridge(s)]; or GMFOV [any air-purifying, full-facepiece respirator (gas mask) with a chin-style, front- or back-mounted acid gas canister]; or SCBAF (any self-contained breathing apparatus with a full facepiece); or SAF (any supplied-air respirator with a full facepiece). *Emergency or planned entry into unknown concentrations or IDLH conditions:* SCBAF:PD,PP (any self-contained breathing apparatus that has a full facepiece and is operated in a pressure-demand or other positive-pressure mode); or SAF:PD,PP:ASCBA (any supplied-air respirator that has a full facepiece and is operated in a pressure-demand or other positive-pressure mode in combination with an auxiliary, self-contained breathing apparatus operated in a pressure-demand or other positive pressure mode). *Escape:* GMFOV [any air-purifying, full-facepiece respirator (gas mask) with a chin-style, front- or back-mounted organic vapor canister]; or SCBAE (any appropriate escape-type, self-contained breathing apparatus).

Note: Substance reported to cause eye irritation or damage; may require eye protection.

Storage: Furfural must be stored to avoid contact with oxidizing materials (such as perchlorates, peroxides, chlorates, nitrates, and permanganates) and strong acids because violent reactions occur. Store in tightly closed containers in a cool, well-ventilated area away from heat. Sources of ignition such as smoking and open flames are prohibited where furfural is used, handled, or stored in a manner that could create a potential fire or explosion hazard. Before entering confined space where this chemical may be present, check to make sure that an explosive concentration does not exist.

Shipping: This compound requires a shipping label of: "Flammable Liquid." It falls in DOT Hazard Class 3 and Packing Group III.

Spill Handling: Evacuate and restrict persons not wearing protective equipment from area of spill or leak until cleanup is complete. Remove all ignition sources. Establish forced ventilation to keep levels below explosive limit. Absorb liquids in vermiculite, dry sand, earth, peat, carbon, or a similar material and deposit in sealed containers. It may be necessary to contain and dispose of this chemical as a hazardous waste. Keep furfural out of a confined space, such as a sewer, because of the possibility of an explosion, unless the sewer is designed to prevent the build-up off explosive concentrations. If material or contaminated runoff enters waterways, notify downstream users of potentially contaminated waters. Contact your Department of Environmental Protection or your regional office of the federal EPA for specific recommendations. If employees are required to clean-up spills, they must be properly trained and equipped. OSHA 1910.120(q) may be applicable.

Fire Extinguishing: This chemical is a combustible liquid. Poisonous gases are produced in fire. Use dry chemical, carbon dioxide, or alcohol foam extinguishers. Vapors are heavier than air and will collect in low areas. Vapors may travel long distances to ignition sources and flashback. Vapors in confined areas may explode when exposed to fire. Containers may explode in fire. Storage containers and parts of containers may rocket great distances, in many directions. If material or contaminated runoff enters waterways, notify downstream users of potentially contaminated waters. Notify local health and fire officials and pollution control agencies. From a secure, explosion-proof location, use water spray to cool exposed containers. If cooling streams are ineffective (venting sound increases in volume and pitch, tank discolors, or shows any signs of deforming), withdraw immediately to a secure position. If employees are expected to fight fires, they must be trained and equipped in OSHA 1910.156.

Disposal Method Suggested: Incineration. Consult with environmental regulatory agencies for guidance on acceptable disposal practices. Generators of waste containing this contaminant (≥100 kg/mo) must conform with EPA regulations governing storage, transportation, treatment, and waste disposal.

References

Sax, N. I., Ed., "Dangerous Properties of Industrial Materials Report," 1, No. 2, 41–42 (1980), and 7, No. 3, 96–102 (1987)

New Jersey Department of Health and Senior Services, "Hazardous Substance Fact Sheet: Furfural," Trenton, NJ (February 1989)

Furfuryl Alcohol

Molecular Formula: $C_5H_6O_2$

Common Formula: $C_4H_3OCH_2OH$

Synonyms: Alcohol furfurilico (Spanish); Alcool furfurylique (French); 2-Furancarbinol; 2-Furanmethanol; Furfural alcohol; Furfuralcohol; Furfurylalkohol (German); Furylalcohol; 2-Furylcarbinol; 2-Hydroxymethylfuran; Microposit remover 1112A; NCI-C56224

CAS Registry Number: 98-00-0

RTECS Number: LU9100000

DOT ID: UN 2874

EEC Number: 603-018-00-2

EINECS Number: 202-626-1

Regulatory Authority

- Air Pollutant Standard Set (ACGIH)[1] (DFG)[33] (HSE)[33] (former USSR)[35] (OSHA)[58] (Several States)[60] (Several Canadian Provinces)
- Canada, WHMIS, Ingredients Disclosure List

Cited in U.S. State Regulations: Alaska (G), California (A, G), Connecticut (A), Florida (G, A), Illinois (G), Maine (G), Massachusetts (G), Nevada (A), New Hampshire (G), New Jersey (G), New York (A), North Dakota (A), Pennsylvania (G), Rhode Island (G), South Carolina (A), Virginia (A), West Virginia (G).

Medical Surveillance: Consider the points of attack in preplacement and periodic physical examinations. Lung function tests. Nervous system.

First Aid: If this chemical gets into the eyes, remove any contact lenses at once and irrigate immediately for at least 15 minutes, occasionally lifting upper and lower lids. Seek medical attention immediately. If this chemical contacts the skin, remove contaminated clothing and wash immediately with soap and water. Seek medical attention immediately. If this chemical has been inhaled, remove from exposure, begin rescue breathing (using universal precautions) if breathing has stopped and CPR if heart action has stopped. Transfer promptly to a medical facility. When this chemical has been swallowed, get medical attention. Give large quantities of water and induce vomiting. Do not make an unconscious person vomit.

Personal Protective Methods: Wear protective gloves and clothing to prevent any reasonable probability of skin contact. Safety equipment suppliers/manufacturers can provide recommendations on the most protective glove/clothing material for your operation. All protective clothing (suits, gloves, footwear, headgear) should be clean, available each day, and put on before work. Contact lenses should not be worn when working with this chemical. Wear splash-proof chemical goggles and face shield unless full facepiece respiratory protection is worn. Employees should wash immediately with soap when skin is wet or contaminated. Provide emergency showers and eyewash.

Respirator Selection: NIOSH/OSHA: *75 ppm:* CCROV [any chemical cartridge respirator with organic vapor cartridge(s)]; or GMFOV [any air-purifying, full-facepiece respirator (gas mask) with a chin-style, front- or back-mounted acid gas canister]; or PAPROV [any powered, air-purifying respirator with organic vapor cartridge(s)]; or SA (any supplied-air respirator); or SCBAF (any self-contained breathing apparatus with a full facepiece). *Emergency or planned entry into unknown concentrations or IDLH conditions:* SCBAF: PD,PP (any self-contained breathing apparatus that has a full facepiece and is operated in a pressure-demand or other positive-pressure mode); or SAF:PD,PP: ASCBA (any supplied-air respirator that has a full facepiece and is operated in a pressure-demand or other positive-pressure mode in combination with an auxiliary, self-contained breathing apparatus operated in a pressure-demand or other positive-pressure mode). *Escape:* GMFOV [any air-purifying, full-facepiece respirator (gas mask) with a chin-style, front- or back-mounted organic vapor canister]; or SCBAE (any appropriate escape-type, self-contained breathing apparatus).

Note: Substance reported to cause eye irritation or damage; may require eye protection.

Storage: Prior to working with this chemical you should be trained on its proper handling and storage. Metal containers involving the transfer of 5 gallons or more of ethyl acetate should be grounded and bonded. Drums must be equipped with self-closing valves, pressure vacuum bungs, and flame arresters. Use only non-sparking tools and equipment, especially when opening and closing containers of ethyl acetate. Store in containers that are properly labeled with health hazard information and safe handling procedures. Wherever ethyl acetate is used, handled, manufactured, or stored, use explosion-proof electrical equipment and fittings. Furfuryl alcohol must be stored to avoid contact with strong oxidizers (such as chlorine, bromine, and fluorine) and any acid, since violent reactions occur. Store in tightly closed containers in a cool, well-ventilated area away from heat. Sources of ignition such as smoking and open flames are prohibited where furfuryl alcohol is used, handled, or stored in a manner that could create a potential fire or explosion hazard. Wherever Furfuryl alcohol is used, handled, manufactured, or stored, use explosion-proof electrical equipment and fittings.

Shipping: This compound requires a shipping label of: "Keep Away From Food." It falls in DOT Hazard Class 6.1 and Packing Group III.

Spill Handling: Evacuate and restrict persons not wearing protective equipment from area of spill or leak until cleanup is complete. Remove all ignition sources. Establish forced ventilation to keep levels below explosive limit. Absorb liquids in vermiculite, dry sand, earth, peat, carbon, or a similar material and deposit in sealed containers. It may be necessary to contain and dispose of this chemical as a hazardous waste. If material or contaminated runoff enters waterways, notify downstream users of potentially contaminated waters. Contact your Department of Environmental Protection or your regional office of the federal EPA for specific recommendations. If employees are required to clean-up spills, they must be properly trained and equipped. OSHA 1910.120(q) may be applicable.

Fire Extinguishing: This chemical is a combustible liquid. Poisonous gases are produced in fire. Use dry chemical, carbon dioxide, or foam extinguishers. Vapors are heavier than air and will collect in low areas. Vapors may travel long distances to ignition sources and flashback. Vapors in confined areas may explode when exposed to fire. Containers may explode in fire. Storage containers and parts of containers may rocket great distances, in many directions. If material or contaminated runoff enters waterways, notify downstream users of potentially contaminated waters. Notify local health and fire officials and pollution control agencies. From a secure, explosion-proof location, use water spray to cool exposed containers. If cooling streams are ineffective (venting sound increases in volume and pitch, tank discolors, or shows any signs of deforming), withdraw immediately to a secure position. If employees are expected to fight fires, they must be trained and equipped in OSHA 1910.156.

Disposal Method Suggested: Incineration in admixture with a more flammable solvent.

References

National Institute for Occupational Safety and Health, Information Profiles on Potential Occupational Hazards: Furfuryl Alcohol, Report PB-276,678, pp. 12–15, Rockville, Md. (1977)

National Institute for Occupational Safety and Health, Criteria for a Recommended Standard: Occupational Exposure to Furfuryl Alcohol, NIIOSH Document No. 79–133 (1979)

New Jersey Department of Health and Senior Services, "Hazardous Substance Fact Sheet: Furfuryl Alcohol," Trenton, NJ (May 1998)

Sax, N. I., Ed., "Dangerous Properties of Industrial Materials Report," 7, No. 6, 56–60 (1987)

G

Gallium

Molecular Formula: Ga

Synonyms: Elemental gallium

CAS Registry Number: 7440-55-3

RTECS Number: LW8600000

DOT ID: UN 2803

Cited in U.S. State regulations: New Jersey (G), Pennsylvania (G).

Description: Gallium is a lustrous, silvery liquid or metal or gray solid. Boiling point = 2,403°C. Freezing/Melting point = 30°C. Hazard Identification (based on NFPA-704 M Rating System): (metal) Health 2, Flammability 0, Reactivity 0. Insoluble in water.

Potential Exposure: Those involved in preparing such semiconductor compounds as gallium arsenide. Used in light-emitting diodes, batteries, microwave equipment.

Incompatibilities: Violent reaction with acids, halogens.

Permissible Exposure Limits in Air: No OELs have been established.

Permissible Concentration in Water: No criteria set.

Routes of Entry: Inhalation, ingestion, skin and/or eyes.

Harmful Effects and Symptoms

Short Term Exposure: Gallium can affect you when breathed in. Gallium is a corrosive chemical and exposure can irritate or burn the eyes, nose and throat. Inhalation can cause pulmonary edema, a medical emergency that can be delayed for several hours. This can cause death. Repeated or high exposures can cause metallic taste, nausea, vomiting, skin rash and may damage the kidneys.

Long Term Exposure: May cause kidney damage. Some gallium compounds affect the nervous system. Highly irritating substances, such as gallium, may affect the lungs.

Points of Attack: Kidneys.

Medical Surveillance: If symptoms develop or overexposure is suspected, the following may be useful: Kidney function tests.

First Aid: *Eye Contact:* Immediately remove any contact lenses and flush with large amounts of water for at least 15 minutes, occasionally lifting upper and lower lids. *Skin Contact:* Remove contaminated clothing. Wash contaminated skin with water. *Breathing:* Remove the person from exposure. Begin rescue breathing if breathing has stopped and CPR if heart action has stopped. Transfer promptly to a medical facility. If swallowed, do not induce vomiting. Medical observation is recommended for 24 – 48 hours after breathing overexposure, as pulmonary edema may be delayed.

Personal Protective Methods: Wear protective gloves and clothing to prevent any reasonable probability of skin contact. Safety equipment suppliers/manufacturers can provide recommendations on the most protective glove/clothing material for your operation.. All protective clothing (suits, gloves, footwear, headgear) should be clean, available each day, and put on before work. Contact lenses should not be worn when working with this chemical. Wear splash-proof chemical goggles and face shield when working with liquid, unless full facepiece respiratory protection is worn. Wear dust-proof goggles and face shield when working with powders or dust, unless full facepiece respiratory protection is worn. Employees should wash immediately with soap when skin is wet or contaminated. Provide emergency showers and eyewash.

Respirator Selection: Where the potential exists for exposure to solid gallium, use a MSHA/NIOSH approved full facepiece respirator with a high efficiency particulate filter. Greater protection is provided by a powered-air purifying respirator. Where the potential for exposure to liquid gallium or high exposures exists, use a MSHA/NIOSH approved supplied-air respirator with a full facepiece operated in the positive pressure mode or with a full facepiece, hood, or helmet in the continuous flow mode, or use a MSHA/NIOSH approved self-contained breathing apparatus with a full facepiece operated in the pressure-demand or other positive pressure mode.

Storage: Prior to working with this chemical you should be trained on its proper handling and storage. Gallium must be stored to avoid contact with acids and halogens since violent reactions occur. Store in tightly closed containers in a cool, well-ventilated area.

Shipping: This compound requires a shipping label of: "Corrosive." It falls in DOT Hazard Class 8 and Packing Group I.

Spill Handling: Evacuate persons not wearing protective equipment from area of spill or leak until clean-up is complete. Remove all ignition sources. Collect powdered material in the most convenient and safe manner and deposit in sealed containers. Ventilate area after clean-up is complete. It may be necessary to contain and dispose of this chemical as a hazardous waste. Absorb liquids in vermiculite, dry sand, earth, peat, carbon, or a similar material and deposit in sealed containers. If material or contaminated runoff enters waterways, notify downstream users of potentially contaminated waters. Contact your Department of Environmental Protection or your regional office of the federal EPA for specific recommendations. If employees are required to clean-up spills, they must be properly trained and equipped. OSHA 1910.120(q) may be applicable.

Fire Extinguishing: Use dry chemical, carbon dioxide, water spray, or foam extinguishers. Poisonous gases are produced in fire. If material or contaminated runoff enters waterways, notify downstream users of potentially contaminated waters. Notify local health and fire officials and pollution control agencies. From a secure, explosion-proof location, use water spray to cool exposed containers. If cooling streams are ineffective (venting sound increases in volume and pitch, tank discolors, or shows any signs of deforming), withdraw immediately to a secure position. If employees are expected to fight fires, they must be trained and equipped in OSHA 1910.156.

References

New Jersey Department of Health and Senior Services, "Hazardous Substance Fact Sheet: Gallium," Trenton, NJ (February, 1988)

Gallium Trichloride

Molecular Formula: Cl_3Ga

Common Formula: $GaCl_3$

Synonyms: Gallium chloride; Gallium(3+) chloride; Gallium(III) chloride; Tricloruro de galio (Spanish)

CAS Registry Number: 13450-90-3

RTECS Number: LW9100000

Regulatory Authority

- SUPERFUND/EPCRA 40CFR355, Appendix B Extremely Hazardous Substances: TPQ = 500/10,000 lb (227/4,540 kg)
- SUPERFUND/EPCRA 40CFR302.4 Reportable Quantity (RQ): EHS, 1 lb (0.454 kg)

Cited in U.S. State regulations: California (G), Massachusetts (G), New Jersey (G), Pennsylvania (G).

Description: Gallium Trichloride, $GaCl_3$, is a colorless solid which forms needle-like crystals. Boiling point = 201°C. Freezing/Melting point = 78°C. Hazard Identification (based on NFPA-704 M Rating System): Health 3, Flammability 0, Reactivity 1.

Potential Exposure: Used as a raw material in the production of metallic gallium; and in the processing of monocrystal semi-conductor compounds.

Permissible Exposure Limits in Air: No standards set.

Permissible Concentration in Water: No criteria set.

Routes of Entry: Inhalation, skin, eyes.

Harmful Effects and Symptoms

Short Term Exposure: May act on skin or mucous membranes. May cause rash and neuritis. Can cause respiratory center paralysis and death in animals.

Points of Attack: Skin.

Medical Surveillance: In view of the toxicity of gallium and its compounds, as shown by experiments, all persons involved in work with these substances should undergo periodic medical examinations, during which special attention should be paid to the condition of the liver, respiratory organs, and skin.

First Aid: If this chemical gets into the eyes, remove any contact lenses at once and irrigate immediately for at least 15 minutes, occasionally lifting upper and lower lids. Seek medical attention immediately. If this chemical contacts the skin, remove contaminated clothing and wash immediately with soap and water. Seek medical attention immediately. If this chemical has been inhaled, remove from exposure, begin rescue breathing (using universal precautions) if breathing has stopped and CPR if heart action has stopped. Transfer promptly to a medical facility. When this chemical has been swallowed, get medical attention. Give large quantities of water and induce vomiting. Do not make an unconscious person vomit.

Personal Protective Methods: Wear protective gloves and clothing to prevent any reasonable probability of skin contact. Safety equipment suppliers/manufacturers can provide recommendations on the most protective glove/clothing material for your operation.. All protective clothing (suits, gloves, footwear, headgear) should be clean, available each day, and put on before work. Contact lenses should not be worn when working with this chemical. Wear dust-proof chemical goggles and face shield unless full facepiece respiratory protection is worn. Employees should wash immediately with soap when skin is wet or contaminated. Provide emergency showers and eyewash.

Respirator Selection: Where the potential for exposure to this chemical, use a MSHA/NIOSH approved supplied-air respirator with a full facepiece operated in the positive pressure mode or with a full facepiece, hood, or helmet in the continuous flow mode, or use a MSHA/NIOSH approved self-contained breathing apparatus with a full facepiece operated in pressure-demand or other positive pressure mode.

Storage: Prior to working with this chemical you should be trained on its proper handling and storage. Store in tightly closed containers in a cool, well ventilated area.

Spill Handling: Evacuate and restrict persons not wearing protective equipment from area of spill or leak until cleanup is complete. Remove all ignition sources. Ventilate area of spill or leak. Absorb liquids in vermiculite, dry sand, earth, peat, carbon, or a similar material and deposit in sealed containers. Keep this chemical out of a confined space, such as a sewer, because of the possibility of an explosion, unless the sewer is designed to prevent the build-up of explosive concentrations. It may be necessary to contain and dispose of this chemical as a hazardous waste. If material or contaminated runoff enters waterways, notify downstream users of potentially contaminated waters. Contact your Department of Environmental Protection or your regional office of the federal EPA for specific recommendations. If employees are required to clean-up spills, they must be properly trained and equipped. OSHA 1910.120(q) may be applicable.

Fire Extinguishing: This chemical is a combustible solid. Use dry chemical, carbon dioxide, water spray, or alcohol foam extinguishers. Poisonous gases are produced in fire. If material or contaminated runoff enters waterways, notify downstream users of potentially contaminated waters. Notify local health and fire officials and pollution control agencies. From a secure, explosion-proof location, use water spray to cool exposed containers. If cooling streams are ineffective (venting sound increases in volume and pitch, tank discolors, or shows any signs of deforming), withdraw immediately to a secure position. If employees are expected to fight fires, they must be trained and equipped in OSHA 1910.156.

References

U.S. Environmental Protection Agency, "Chemical Profile: Gallium Trichloride," Washington, DC, Chemical Emergency Preparedness Program (Nov. 30, 1987)

Gasoline

Molecular Formula: C_5H_{12} to C_9H_{20}

Synonyms: Benzine; Essance (French); Gasolina (Spanish); Motor fuel; Motor spirits; Natural gasoline; Petrol

CAS Registry Number: 8006-61-9

RTECS Number: LX3300000

DOT ID: UN 1203

Regulatory Authority

- Air Pollutant Standard Set (ACGIH)[1] (Several States)[60]
- Water Pollution Standard Proposed (Several States)[61]

Cited in U.S. State regulations: Alaska (G), California (G), Connecticut (A), Florida (G), Illinois (G), Maine (G, W), Massachusetts (G), Nevada (A), New Hampshire (G), New Jersey (G), New York (G), Rhode Island (G, W), Virginia (A), West Virginia (G).

Description: Gasoline is a highly flammable, mobile liquid with a characteristic odor. A complex mixture of volatile hydrocarbons (paraffins, cycloparaffins and aromatics. The odor threshold is 0.25 ppm.[41] Boiling point = 38 – 204°C. Flash point = -38 – -45°C (depending on octane). Explosive limits: LEL=1.4 %; UE=7.6%. Hazard Identification (based on NFPA-704 M Rating System): Health 1, Flammability 3, Reactivity 0. Insoluble in water. Physical property values may vary depending on grade.

Potential Exposure: Gasoline is used as a fuel, diluent, and solvent throughout industry.

Incompatibilities: Forms explosive mixture with air. Strong oxidizers may cause fire and explosions. Incompatible with nitric acid. May accumulate static electrical charges, and may cause ignition of its vapors.

Permissible Exposure Limits in Air: Presently, the composition of gasoline is so varied that a single standard for all types of gasoline is not applicable.[3] It is recommended, however, that atmospheric concentrations should be limited by the aromatic hydrocarbon content. There is no OSHA PEL. ACGIH[1] recommends a TWA of 300 ppm (900 mg/m^3) and a STEL value of 500 ppm (1.500 mg/m^3). Several states have set guidelines or standards for gasoline in ambient air[60] ranging from 15.0 mg/m^3 (Virginia) to 18.0 mg/m^3 (Connecticut) to 21.429 mg/m^3 (Nevada).

Determination in Air: No test available.

Permissible Concentration in Water: States which have set guidelines for gasoline in drinking water include Rhode Island (zero concentration) and Maine (100 μg/l).

Routes of Entry: Inhalation, percutaneous absorption.

Harmful Effects and Symptoms

Short Term Exposure: Contact can cause eye and skin irritation. Inhalation can cause irritation of the respiratory tract. High levels can cause headache, nausea, dizziness, irregular heartbeat, loss of coordination, seizures, coma, and possible death. *Inhalation:* Nose and throat irritation have been reported after exposure to 900 ppm for 1 hour. Drowsiness, dizziness, nausea and numbness may occur at 1,000 ppm after 15 minutes exposure. In animal studies, death occurred after 30,000 ppm for five minutes. *Skin:* May cause itching and burning of the skin and after a longer exposure, redness and blistering. *Eyes:* Moderate irritation of the eye has been reported after one hour exposure to 500 ppm. Mild irritation has been reported after an 8 hour exposure to 140 ppm. *Ingestion:* Gasoline causes a burning sensation in the mouth, throat and stomach. Vomiting, diarrhea, drowsiness and intoxication may follow. As little as 3 – 4 ounces may be fatal. Inhalation of liquid gasoline into the lungs following ingestion or vomiting may result in pulmonary edema (an accumulation of fluid in the lungs and a medical emergency that can be delayed for several hours), rapid breathing or death.

Long Term Exposure: There is evidence that gasoline can cause kidney cancer in animals. Repeated exposure can cause permanent eye damage. Prolonged contact can cause drying of the skin with cracking and rash. Repeated high exposure may damage the lungs. and/or cause brain damage. May cause kidney damage. Gasolines often contain hexane, benzene and lead. Hexane, a component of gasoline, can produce nerve damage result in tremors, numbness of hands and feet and loss of muscle control. Benzene, has been linked to blood disorders in man, including leukemia. Lead additives can produce nausea, cramps, loss of appetite, sleep problems, headaches and agitation. See also entries for *tetraethyl lead, benzene, hexane, and ethylene dibromide.*

Points of Attack: Kidneys, skin, eyes, nervous system, brain.

Medical Surveillance: If symptoms develop or overexposure has occurred, the following may be useful: urinary lead level. Evaluation by a qualified allergist, including careful exposure history and special testing, may help diagnose skin allergy. Liver function tests.

First Aid: If this chemical gets into the eyes, remove any contact lenses at once and irrigate immediately for at least 15 minutes, occasionally lifting upper and lower lids. Seek medical attention immediately. If this chemical contacts the skin, remove contaminated clothing and wash immediately with soap and water. Seek medical attention immediately. If this chemical has been inhaled, remove from exposure, begin rescue breathing (using universal precautions) if breathing has stopped and CPR if heart action has stopped. Transfer promptly to a medical facility. If swallowed, use gastric lavage (stomach wash) followed by saline catharsis. Get medical attention. Medical observation is recommended for 24 – 48 hours after ingestion, as pulmonary edema may be delayed.

Personal Protective Methods: Barrier creams and impervious gloves. Wear protective gloves and clothing to prevent any reasonable probability of skin contact. Safety equipment suppliers/manufacturers can provide recommendations on the most protective glove/clothing material for your operation.. Neoprene, VITON/Neoprene, Nitrile, PV Alcohol, and Polyurethane are recommended. All protective clothing (suits, gloves, footwear, headgear) should be clean, available each day, and put on before work. Contact lenses should not be worn when working with this chemical. Wear splash-proof chemical goggles and face shield unless full facepiece respiratory protection is worn. Employees should wash immediately with soap when skin is wet or contaminated. Provide emergency showers and eyewash, protective clothing, masks in heavy exposure to vapors.

Respirator Selection: Where the potential exists for exposure over 300 ppm, use a MSHA/NIOSH approved respirator with an organic vapor cartridge/canister. More protection is provided by a powered-air purifying respirator. Where the potential for high exposures exists, use a MSHA/NIOSH approved supplied-air respirator with a full facepiece operated in the positive pressure mode or with a full facepiece, hood, or helmet in the continuous flow mode, or use a MSHA/NIOSH approved self-contained breathing apparatus with a full facepiece operated in pressure-demand or other positive pressure mode.

Storage: Prior to working with this chemical you should be trained on its proper handling and storage. Before entering confined space where this chemical may be present, check to make sure that an explosive concentration does not exist. Sources of ignition such as smoking and open flames are prohibited where gasoline is handled, used, or stored. Metal containers involving the transfer of 1 gallons or more of gasoline should be grounded and bonded. Drums must be equipped with self-closing valves, pressure vacuum bungs, and flame arresters. Use only non-sparking tools and equipment, especially when opening and closing containers of gasoline. Wherever gasoline is used, handled, manufactured, or stared, use explosion-proof electrical equipment and fittings.

Shipping: This compound requires a shipping label of: "Flammable Liquid." It falls in DOT Hazard Class 3 and Packing Group II.

Spill Handling: Evacuate and restrict persons not wearing protective equipment from area of spill or leak until cleanup is complete. Remove all ignition sources. Establish forced ventilation to keep levels below explosive limit. Absorb liquids in vermiculite, dry sand, earth, peat, carbon, or a similar material and deposit in sealed containers. Keep this chemical out of a confined space, such as a sewer, because of the possibility of an explosion, unless the sewer is designed to prevent the build-up of explosive concentrations. It may be necessary to contain and dispose of this chemical as a hazardous waste. If material or contaminated runoff enters waterways, notify downstream users of potentially contaminated waters. Contact your Department of Environmental Protection or your regional office of the federal EPA for specific recommendations. If employees are required to clean-up spills, they must be properly trained and equipped. OSHA 1910.120(q) may be applicable.

Fire Extinguishing: This chemical is a flammable liquid. Poisonous gases are produced in fire. Use dry chemical, carbon dioxide, or foam extinguishers. Vapors are heavier than air and will collect in low areas. Vapors may travel long distances to ignition sources and flashback. Vapors in confined areas may explode when exposed to fire. Containers may explode in fire. Storage containers and parts of containers may rocket great distances, in many directions. If material or contaminated runoff enters waterways, notify downstream users of potentially contaminated waters. Notify local health and fire officials and pollution control agencies. From a secure, explosion-proof location, use water spray to cool exposed containers. If cooling streams are ineffective (venting sound increases in volume and pitch, tank discolors, or shows any signs of deforming), withdraw immediately to a secure position. If employees are expected to fight fires, they must be trained and equipped in OSHA 1910.156.

Disposal Method Suggested: Incineration. Alternatively, gasoline vapors may be recovered from fuel transfer operations by various techniques.

References

New Jersey Department of Health and Senior Services, "Hazardous Substance Fact Sheet: Gasoline," Trenton, NJ (December 1996)

New York State Department of Health, "Chemical Fact Sheet: Gasoline," Albany, NY, Bureau of Toxic Substance Assessment (Mart 1986)

Sax, N. I., Ed., "Dangerous Properties of Industrial Materials Report" 1, No. 8, 75–76 (1981)

Germanium

Molecular Formula: Ge

Synonyms: Elemental germanium; Germanium element

CAS Registry Number: 7440-56-4

Regulatory Authority

- Air Pollutant Standard Set (former USSR)[43]

Description: Germanium, Ge, is a grayish-white, lustrous, brittle metalloid. It is never found free and occurs most commonly in ergyrodite and germanite. It is generally recovered as a by-product in zinc production, coal processing, or other sources. Boiling point = 2,700°C. Freezing/Melting point= 937°C. Insoluble in water.

Potential Exposure: Because of its semiconductor properties, germanium is widely used in the electronic industry in rectifiers, diodes, and transistors. It is alloyed with aluminum, aluminum-magnesium, antimony, bronze, and tin to increase strength, hardness, or corrosion resistance. In the process of alloying germanium and arsenic, arsine may be released; stibine is released from the alloying of germanium and antimony. Germanium is also used in the manufacture of optical glass for infrared applications, red-fluorescing phosphors, and cathodes for electronic valves, and in electroplating, in the hydrogenation of coal, and as a catalyst, particularly at low temperatures. Certain compounds are used medically. Industrial exposures to the dust and fumes of the metal or oxide generally occur during separation and purification of germanium, welding, multiple-zone melting operations, or cutting and grinding of crystals. Germanium tetrahydride (germanium hydride, germane, monogermane) and other hydrides are produced by the action of a reducing acid on a germanium alloy.

Incompatibilities: Finely divided metal is incompatible with ammonia, bromine, oxidizers, aqua regia; concentrated sulfuric acid, carbonates, halogens, nitrates. Explosive reaction or ignition with potassium chlorate, potassium nitrate, chlorine, bromine, oxygen, potassium hydroxide, in the presence of heat. Violent reaction with nitric acid.

Permissible Exposure Limits in Air: The former USSR-UNEP/IRPTC project[43] has set a MAC in workplace air of 2 mg/m^3.

Permissible Concentration in Water: No criteria set, but EPA[32] has suggested a permissible ambient goal of 8 μg/l based on health effects.

Determination in Water: Germanium may be determined by atomic absorption spectroscopy, emission spectrography and spectrophotometry with phenylfluorone.

Routes of Entry: Inhalation of gas, vapor, fume, or dust.

Harmful Effects and Symptoms

Short Term Exposure: Symptoms of germanium exposure[24] include low temperature, languor, diarrhea, cyanosis, extreme depression of heart and breath, edema, lung hemorrhage, hemorrhage from the small intestinal wall, peritoneal extravasation. The dust of germanium dioxide is irritating to the eyes. Germanium tetrachloride causes irritation of the skin. Germanium tetrachloride is an upper respiratory irritant and may cause bronchitis and pneumonitis.

Long Term Exposure: Prolonged exposure to high level concentrations may result in damage to the liver, kidney, and other organs.

Medical Surveillance: Consider respiratory, liver, and kidney disease in any placement or periodic examinations.

First Aid: If this chemical gets into the eyes, remove any contact lenses at once and irrigate immediately for at least 15 minutes, occasionally lifting upper and lower lids. Seek medical attention immediately. If this chemical contacts the skin, remove contaminated clothing and wash immediately with soap and water. Seek medical attention immediately. If this chemical has been inhaled, remove from exposure, begin rescue breathing (using universal precautions) if breathing has stopped and CPR if heart action has stopped. Transfer promptly to a medical facility. When this chemical has been swallowed, get medical attention. Give large quantities of water and induce vomiting. Do not make an unconscious person vomit.

Personal Protective Methods: Wear protective gloves and clothing to prevent any reasonable probability of skin contact. Safety equipment suppliers/manufacturers can provide recommendations on the most protective glove/clothing material for your operation.. All protective clothing (suits, gloves, footwear, headgear) should be clean, available each day, and put on before work. Contact lenses should not be worn when working with this chemical. Wear dust-proof chemical goggles and face shield unless full facepiece respiratory protection is worn. Employees should wash immediately with soap when skin is wet or contaminated. Provide emergency showers and eyewash.

Respirator Selection: Where the potential for exposure to this chemical, use a MSHA/NIOSH approved supplied-air respirator with a full facepiece operated in the positive pressure mode or with a full facepiece, hood, or helmet in the continuous flow mode, or use a MSHA/NIOSH approved self-contained breathing apparatus with a full facepiece operated in pressure-demand or other positive pressure mode.

Storage: Prior to working with this chemical you should be trained on its proper handling and storage. Store in tightly closed containers in a cool, well ventilated area away from oxidizers and other incompatibles listed above.

Spill Handling: Evacuate persons not wearing protective equipment from area of spill or leak until clean-up is complete. Remove all ignition sources. Collect powdered material in the most convenient and safe manner and deposit in sealed containers. Ventilate area after clean-up is complete. It may be necessary to contain and dispose of this chemical as a hazardous waste. If material or contaminated runoff enters waterways, notify downstream users of potentially contaminated waters. Contact your Department of Environmental Protection or your regional office of the federal EPA for specific recommendations. If employees are required to clean-up spills, they must be properly trained and equipped. OSHA 1910.120(q) may be applicable.

Fire Extinguishing: Poisonous gases are produced in fire. If material or contaminated runoff enters waterways, notify downstream users of potentially contaminated waters. Notify local health and fire officials and pollution control agencies. From a secure, explosion-proof location, use water spray to cool exposed containers. If cooling streams are ineffective (venting sound increases in volume and pitch, tank discolors, or shows any signs of deforming), withdraw immediately to a secure position. If employees are expected to fight fires, they must be trained and equipped in OSHA 1910.156.

Disposal Method Suggested: Recovery and return to suppliers for reprocessing is preferable.

References

U.S. Environmental Protection Agency, Toxicology of Metal, Vol. II: Germanium, pp 222–223, Report EPA-600/1-77-022, Research Triangle Park, NC (May 1977)

Germanium Tetrahydride (Germane)

Molecular Formula: GeH_4

Synonyms: Germane; Germanium hydride; Monogermane

CAS Registry Number: 7782-65-2

RTECS Number: LY4900000

DOT ID: UN 2192

Regulatory Authority

- Air Pollutant Standard Set (ACGIH)[1] (HSE)[33] (former USSR)[43] (OSHA)[58] (Several States)[60]
- Canada, WHMIS, Ingredients Disclosure List

Cited in U.S. State regulations: Alaska (G), California (A, G), Connecticut (A), Florida (G), Illinois (G), Maine (G), Massachusetts (G), Nevada (A), New Hampshire (G), North Dakota (A), Pennsylvania (G), Rhode Island (G), Virginia (A), West Virginia (G).

Description: Germane, GeH_4, is a colorless gas. Boiling point = -88.5°C. Freezing/Melting point = -165°C. Decomposes at 350°C. Insoluble in water.

Potential Exposure: This material is used as a doping agent in solid state electronic component manufacture.

Incompatibilities: May ignite spontaneously in air. Reacts explosively with bromine.

Permissible Exposure Limits in Air: There is no OSHA PEL. NIOSH and ACGIH recommend a TWA of 0.2 ppm (0.6 mg/m^3). HSE[33] also sets the TWA value but adds an STEL of 0.6 ppm (1.8 mg/m^3). The former USSR-UNEP/IRPTC project[43] has set a MAC in workplace of 1 mg/m^3. Several states have set guidelines or standards for germane in ambient air[60] ranging from 6.0 $\mu g/m^3$ (North Dakota) to 10.0 $\mu g/m^3$ (Virginia) to 12.0 $\mu g/m^3$ (Connecticut) to 14.0 $\mu g/m^3$ (Nevada).

Determination in Air: No methods available.

Permissible Concentration in Water: No criteria set.

Routes of Entry: Inhalation.

Harmful Effects and Symptoms

Short Term Exposure: Inhalation can cause headache, giddiness, fainting, nausea, vomiting. May cause Data on toxicity are limited but ACGIH reports that it is between ten hydride (stannone) and arcine in toxicity. Germanium tetrahydride is a toxic hemolytic gas capable of production kidney damage. This effect is similar to that of arsine and stibine (antimony hydride). One hour exposure tests on animals yielded the following results: animal rabbit, concentration 100 ppm, effect survived; animal mouse, concentration 150 ppm, effect fatal; animal mouse, concentration 185 ppm, effect fatal; animal guinea pig, concentration 150 ppm, effect sickened; animal guinea pig, concentration 185 ppm, effect fatal. The TWA value was set on the rather arbitrary basis that germane can be considered half as toxic as stibine.[53]

Long Term Exposure: May cause kidney injury. Hemolytic effects.

Points of Attack: Central nervous system, kidneys, blood.

Medical Surveillance: Workers exposed to germane should undergo periodic physical examinations.[30] Kidney function tests. Complete blood count (CBC). Examination of the nervous system.

First Aid: If this chemical gets into the eyes, remove any contact lenses at once and irrigate immediately for at least 15 minutes, occasionally lifting upper and lower lids. If this chemical has been inhaled, remove from exposure, begin rescue breathing (using universal precautions) if breathing has stopped and CPR if heart action has stopped. Transfer promptly to a medical facility.

Personal Protective Methods: Wear protective gloves and clothing to prevent any reasonable probability of skin contact. Safety equipment suppliers/manufacturers can provide recommendations on the most protective glove/clothing material for your operation.. All protective clothing (suits, gloves, footwear, headgear) should be clean, available each day, and put on before work. Contact lenses should not be worn when working with this chemical. Wear gas-proof chemical goggles and face shield unless full facepiece respiratory protection is worn. Employees should wash immediately with soap when skin is wet or contaminated. Provide emergency showers and eyewash.

Respirator Selection: Where the potential for exposure to this chemical, use a MSHA/NIOSH approved supplied-air respirator with a full facepiece operated in the positive pressure mode or with a full facepiece, hood, or helmet in the continuous flow mode, or use a MSHA/NIOSH approved self-contained breathing apparatus with a full facepiece operated in pressure-demand or other positive pressure mode.

Storage: Prior to working with this chemical you should be trained on its proper handling and storage. Store in tightly closed containers in a cool, well ventilated area away from bromine. Procedures for the handling, use and storage of cylinders should be in compliance with OSHA 1910.101 and 1910.169 as with the recommendations of the Compressed Gas Association.

Shipping: This compound requires a shipping label of: "Poison Gas, Flammable Gas." It falls in DOT Hazard Class 2.3 and Packing Group I. Passenger aircraft or railcar shipment is forbidden as is any cargo aircraft shipment.

Spill Handling: Evacuate and restrict persons not wearing protective equipment from area of spill or leak until cleanup is complete. Remove all ignition sources. Ventilate area of leak to disperse the gas. Stop the flow of gas if it can be done safely. If source of leak is a cylinder and the leak cannot be stopped in place, remove leaking cylinder to a safe place in the open air, and repair leak or allow cylinder to empty. Keep this chemical out of confined space, such as sewer because of the possibility of explosion, unless the sewer is designed to prevent the buildup of explosive concentrations. It may be necessary to contain and dispose of this chemical as a hazardous waste. Contact your Department of Environmental Protection or your regional office of the federal EPA for specific recommendations. If employees are required to clean-up spills, they must be properly trained and equipped. OSHA 1910.120(q) may be applicable.

Initial isolation and protective action distances

Distances shown are likely to be affected during the first 30 minutes after materials are spilled and could increase with time. If more than one tank car, cargo tank, portable tank, or large cylinder is involved in the incident is leaking, the protective action distance may need to be increased. You may need to seek emergency information from CHEMTREC at (800) 424-9300 or seek professional environmental engineering assistance from the U.S. EPA Environmental Response Team at (908) 548-8730 (24-hour response line).

Small spills (From a small package or a small leak from a large package)

First: Isolate in all directions (feet)..... 200

Then: Protect persons downwind (miles)

Day 0.1

Night 0.4

Large spills (From a large package or from many small packages)

First: Isolate in all directions (feet)..... 600

Then: Protect persons downwind (miles)

Day 0.4

Night 1.4

Fire Extinguishing: This chemical is a flammable gas. Poisonous gases are produced in fire. Do not extinguish the fire unless the flow of gas can be stopped and any remaining gas is out of the line. Specially trained personnel may use fog lines to cool exposures and let the fire burn itself out. Vapors are heavier than air and will collect in low areas. Vapors may travel long distances to ignition sources and flashback. Vapors in confined areas may explode when exposed to fire. Containers may explode in fire. Storage containers and parts of containers may rocket great distances, in many directions. If material or contaminated runoff enters waterways, notify downstream users of potentially contaminated waters. Notify local health and fire officials and pollution control agencies. From a secure, explosion-proof location, use water spray to cool exposed containers. If cooling streams are ineffective (venting sound increases in volume and pitch, tank discolors, or shows any signs of deforming), withdraw immediately to a secure position. If cylinders are exposed to excessive heat from fire or flame contact, withdraw immediately to a secure location. If employees are expected to fight fires, they must be trained and equipped in OSHA 1910.156.

Glucose Oxidase

Synonyms: Deoxin-1; E.C. 1.1.3.4; Glucose aerodehydrogenase; β-D-Glucose oxidase; Microcide; Notatin; Orylophyline; Oxidase glucose; Penatin

CAS Registry Number: 9001-37-0

RTECS Number: RQ8452000

Cited in U.S. State regulations: New Jersey (G).

Description: Glucose Oxidase takes the form of amorphous powder or crystals. Soluble in water.

Potential Exposure: Glucose Oxidase is used as a food preservative, as a stabilizer for vitamins and in laboratories.

Permissible Exposure Limits in Air: No standards set.

Permissible Concentration in Water: No criteria set.

Harmful Effects and Symptoms

Short Term Exposure: Glucose Oxidase can affect you when breathed in.

Long Term Exposure: Exposure can cause an asthma-like lung allergy. Once allergy develops, even very small future exposures can cause asthma attacks with shortness of breath, wheezing, cough and chest tightness. Allergy symptoms can also resemble hay fever, with itching eyes, sneezing and watery or stuffy nose.

Medical Surveillance: Before beginning employment and at regular times after that, for those with frequent or potentially high exposures, the following are recommended: Lung function tests. These may be normal if the person is not having an attack at the time of the test.

First Aid: If this chemical gets into the eyes, remove any contact lenses at once and irrigate immediately for at least 15 minutes, occasionally lifting upper and lower lids. Seek medical attention immediately. If this chemical contacts the skin, remove contaminated clothing and wash immediately with soap and water. Seek medical attention immediately. If this chemical has been inhaled, remove from exposure, begin rescue breathing (using universal precautions) if breathing has stopped and CPR if heart action has stopped. Transfer promptly to a medical facility. When this chemical has been swallowed, get medical attention. Give large quantities of water and induce vomiting. Do not make an unconscious person vomit.

Personal Protective Methods: Wear protective gloves and clothing to prevent any reasonable probability of skin contact. Safety equipment suppliers/manufacturers can provide recommendations on the most protective glove/clothing material for your operation.. All protective clothing (suits, gloves, footwear, headgear) should be clean, available each day, and put on before work. Contact lenses should not be worn when working with this chemical. Wear dust-proof chemical goggles and face shield unless full facepiece respiratory protection is worn. Employees should wash immediately with soap when skin is wet or contaminated. Provide emergency showers and eyewash.

Respirator Selection: Where the potential exists for exposures to Glucose Oxidase, use a MSHA/NIOSH approved respirator with a high efficiency paniculate filter. More protection is provided by a full facepiece respirator than by a half-mask respirator, and even greater protection is provided by a powered-air purifying respirator. Paniculate filters must be checked every day before work for physical damage, such as rips or tears, and replaced as needed. Where the potential for high exposures exists, use a MSHA/NIOSH approved supplied-air respirator with a full facepiece operated in the positive pressure mode or with a full facepiece, hood, or helmet in the continuous flow mode, or use a MSHA/NIOSH approved self-contained breathing apparatus with a full facepiece operated in pressure-demand or other positive pressure mode.

Storage: Glucose oxidase should be stored in a freezer.

Spill Handling: Evacuate persons not wearing protective equipment from area of spill or leak until clean-up is complete. Remove all ignition sources. Collect powdered material in the most convenient and safe manner and deposit in sealed containers. Ventilate area after clean-up is complete. It may be necessary to contain and dispose of this chemical as a hazardous waste. If material or contaminated runoff enters waterways, notify downstream users of potentially contaminated waters. Contact your Department of Environmental Protection or your regional office of the federal EPA for specific recommendations. If employees are required to clean-up spills, they must be properly trained and equipped. OSHA 1910.120(q) may be applicable.

Fire Extinguishing: Use extinguishing agent suitable for surrounding fire. If employees are expected to fight fires, they must be trained and equipped in OSHA 1910.156.

References

New Jersey Department of Health and Senior Services, "Hazardous Substance Fact Sheet: Glucose Oxidase," Trenton, NJ (April 1986)

Glutaraldehyde

Molecular Formula: $C_5H_8O_2$

Common Formula: $HCO(CH_2)_3CHO$

Synonyms: Cidex; Cudex; 1,3-Diformal propane; Glutamic dialdehyde; Glutaral; Glutaraldehyd (Czech); Glutard dialdehyde; Glutaric acid dialdehyse; Glutaric dialdehyde; NCI-C55425; 1,5-Pentanedial; Pentanedial; 1,5-Pentanedione; Potentiated acid glutaraldehyde; Sonacide

CAS Registry Number: 111-30-8

RTECS Number: MA2450000

Regulatory Authority

- Air Pollutant Standard Set (ACGIH)[1] (DFG)[3] (HSE)[33] (NIOSH) (Several States)[60]
- Canada, WHMIS, Ingredients Disclosure List

Cited in U.S. State regulations: Alaska (G), California (A, G), Connecticut (A), Florida (G), Illinois (G), Maine (G), Massachusetts (G), Nevada (A), New Hampshire (G), New Jersey (G), New York (G), North Dakota (A), Oklahoma (G), Pennsylvania (G), Rhode Island (G), Virginia (A), West Virginia (G).

Description: Glutaraldehyde, $HCO(CH_2)_3CHO$, is a colorless liquid with a pungent odor, which readily changes to a glossy polymer. The odor threshold is 0.04 ppm (NY) and 0.2 ppm (NJ). Boiling point = 187 – 189°C (decomposes). Freezing/Melting point = -14°C. Hazard Identification (based on NFPA-704 M Rating System): Health 2, Flammability 0, Reactivity 0. Soluble in water.

Potential Exposure: Glutaraldehyde is used as a cross-linking agent for protein and polyhydroxy materials. It has been used in tanning and as a fixative for tissues. It is also used as an inter-mediate. Buffered solutions are used as antimicrobial agents in hospitals.

Incompatibilities: Water contact forms a polymer solution. A strong reducing agent. Incompatible with strong acids, caustics, ammonia, amines, strong oxidizers. *Note:* Alkaline solutions of glutaraldehyde (i.e., activated glutaraldehyde) react with alcohol, ketones, amines, hydrazines and proteins.

Permissible Exposure Limits in Air: There is no OSHA PEL. NIOSH and ACGIH recommend a ceiling value of 0.2 ppm (0.8 mg/m^3) but proposed no STEL. DFG[3] and

HSE[33] have adopted the same value as an 8-hour TWA and a MAK. The HSE[33] uses the same values as an STEL. Several states have set guidelines or standards for glutaraldehyde in ambient air[60] ranging from 6.0 μg/m^3 (Virginia) to 7.0 μg/m^3 (North Dakota) to 14.0 μg/m^3 (Connecticut) to 17.0 μg/m^3 (Nevada).

Determination in Air: Si gel; Acetonitrile; High-pressure liquid chromatography/Ultraviolet; NIOSH IV, Method #2532.

Permissible Concentration in Water: No criteria set.

Routes of Entry: Inhalation, skin absorption, ingestion, skin and/or eye contact. Can be absorbed through the skin.

Harmful Effects and Symptoms

Short Term Exposure: Irritates the eyes, skin, and respiratory tract. *Inhalation:* 0.3 ppm can cause nose and throat irritation. 0.4 ppm has caused headaches. 0.5 ppm has been described as intolerably irritating. *Skin:* Can cause irritation. Contact with a 5% solution can sensitize the skin and cause an allergic response to subsequent contact of much lower concentrations. *Eyes:* Vapors of a 2% solution (0.4 ppm) have produced irritation. *Ingestion:* Can cause irritation of the mouth and stomach. The LD_{50} oral rat is 134 mg/kg (moderately toxic).

Long Term Exposure: Repeated or prolonged contact with skin may cause chemical sensitization, skin allergy and asthma. Exposure may cause liver and nervous system damage. Glutaraldehyde may cause mutations, handle with extreme caution.

Points of Attack: Eyes, skin, respiratory system, liver, nervous system.

Medical Surveillance: If symptoms develop or overexposure has occurred, the following may be useful: liver function tests. Evaluation by a qualified allergist, including careful exposure history and special testing, may help diagnose skin allergy.

Note: Testing by NIOSH has not been completed to determine the carcinogenicity of glutaraldehyde and related low-molecular-weight-aldehydes. However, the limited studies to date indicate that these substances have chemical reactivity and mutagenicity similar to acetaldehyde and malonaldehyde. NIOSH recommends that acetaldehyde and malonaldehyde be considered potential occupational carcinogens in conformance with the OSHA carcinogen policy. Therefore, NIOSH recommends that careful consideration should be given to reducing exposures to related aldehydes such as glutaraldehyde. Exposure to acetaldehyde has produced nasal tumors in rats and laryngeal tumors in hamsters, and exposure to malonaldehyde has produced thyroid gland and pancreatic islet cell tumors in rats. Further information can be found in the "NIOSH Current Intelligence Bulletin 55: Carcinogenicity of Acetaldehyde and Malonaldehyde, and Mutagenicity of Related Low-Molecular-Weight Aldehydes" (NIOSH Publication No. 91–112).

First Aid: If this chemical gets into the eyes, remove any contact lenses at once and irrigate immediately for at least 30 minutes, occasionally lifting upper and lower lids. Seek medical attention immediately. If this chemical contacts the skin, remove contaminated clothing and wash immediately with soap and water. Seek medical attention immediately. If this chemical has been inhaled, remove from exposure, begin rescue breathing (using universal precautions) if breathing has stopped and CPR if heart action has stopped. Transfer promptly to a medical facility. When this chemical has been swallowed, get medical attention. Give large quantities of water and induce vomiting. Do not make an unconscious person vomit.

Personal Protective Methods: Wear protective gloves and clothing to prevent any reasonable probability of skin contact. Safety equipment suppliers/manufacturers can provide recommendations on the most protective glove/clothing material for your operation.. Butyl Rubber, Neoprene, Viton, and Polyvinyl Chloride are among the recommended protective materials. All protective clothing (suits, gloves, footwear, headgear) should be clean, available each day, and put on before work. Contact lenses should not be worn when working with this chemical. Wear splash-proof chemical goggles and face shield unless full facepiece respiratory protection is worn. Employees should wash immediately with soap when skin is wet or contaminated. Provide emergency showers and eyewash.

Respirator Selection: Where the potential exists for exposures over 0.2 ppm, use an MSHA/NIOSH approved respirator with an organic vapor cartridge/canister and a dust/mist/fume prefilter. More protection is provided by a full facepiece respirator than by a half-mask respirator, and even greater protection is provided by a powered-air purifying respirator. Where the potential for high exposures exists, use an MSHA/NIOSH approved supplied-air respirator with a full facepiece operated in the positive pressure mode or with a full facepiece, hood, or helmet in the continuous flow mode, or use an MSHA/NIOSH approved self-contained breathing apparatus with a full facepiece operated in pressure-demand or other positive pressure mode.

Storage: Prior to working with this chemical you should be trained on its proper handling and storage. Store in tightly closed containers in a cool, well ventilated area away from strong acids, caustics, ammonia, amines, oxidizers.

Shipping: The DOT[19] has no specific requirements for Glutaraldehyde in their Performance-Oriented Packaging Standards.

Spill Handling: Evacuate and restrict persons not wearing protective equipment from area of spill or leak until cleanup is complete. Remove all ignition sources. Ventilate area of spill or leak. Absorb liquids in vermiculite, dry sand, earth, peat, carbon, or a similar material and deposit in sealed containers. Keep this chemical out of a confined space, such as a sewer, because of the possibility of an explosion, unless the sewer is designed to prevent the build-up of explosive concentrations. It may be necessary to contain and dispose of this chemical as a hazardous waste. If material or contaminated runoff enters waterways, notify downstream users of potentially contaminated waters. Contact your Department of Environmental Protection or your regional office of the federal EPA for specific recommendations. If employees are required to clean-up spills, they must be properly trained and equipped. OSHA 1910.120(q) may be applicable.

Fire Extinguishing: Extinguish fire using an agent suitable for type of surrounding fire. Glutaraldehyde itself does not burn. Poisonous gases are produced in fire. Use dry chemical, carbon dioxide, or alcohol foam extinguishers. Vapors are heavier than air and will collect in low areas. Vapors may travel long distances to ignition sources and flashback. Vapors in confined areas may explode when exposed to fire. Containers may explode in fire. Storage containers and parts of containers may rocket great distances, in many directions. If material or contaminated runoff enters waterways, notify downstream users of potentially contaminated waters. Notify local health and fire officials and pollution control agencies. From a secure, explosion-proof location, use water spray to cool exposed containers. If cooling streams are ineffective (venting sound increases in volume and pitch, tank discolors, or shows any signs of deforming), withdraw immediately to a secure position. If employees are expected to fight fires, they must be trained and equipped in OSHA 1910.156.

Disposal Method Suggested: Incineration.

References

New York State Department of Health, "Chemical Fact Sheet; Glutaraldehyde," Albany, NY, Bureau of Toxic Substance Assessment (April 1986)

New Jersey Department of Health and Senior Services, "Hazardous Substance Fact Sheet: Glutaraldehyde," Trenton, NJ (February, 1989)

Sax, N. I., Ed., "Dangerous Properties of Industrial Materials Report," 1, No. 7, 2–4 (1981)

Glycerin (Mist)

Molecular Formula: $C_3H_8O_3$

Common Formula: $HOCH_2CHOHCH_2OH$

Synonyms: 90 Technical glycerin; Glycerin, anhydrous; Glycerin, synthetic; Glyceritol; Glycerol; Glycyl alcohol; Grocolene; Moon; 1,2,3-Propanetriol; Synthetic glycerin; 1,2,3-Trihydroxypropane; Trihydroxypropane

CAS Registry Number: 56-81-5

RTECS Number: MA8050000

DOT ID: No citation.

Regulatory Authority

- Air Pollutant Standard Set (ACGIH)[1] (HSE)[33] (OSHA)[58] (Several States)[60]
- Canada, WHMIS, Ingredients Disclosure List

Cited in U.S. State regulations: Alaska (G), California (A, G), Florida (A), Illinois (G), Maine (G), Nevada (A), New Hampshire (G), New Jersey (G), New York (A), Pennsylvania (G), Rhode Island (G), Virginia (A), West Virginia (G).

Description: Glycerol, $HOCH_2CHOHCH_2OH$, is a viscous colorless or pale yellow, odorless, syrupy liquid. Boiling point = 171°C (decomposes). Freezing/Melting point = 18.2°C. Flash point = 199°C. Autoignition temperature = 370°C. Hazard Identification (based on NFPA-704 M Rating System): Health 1, Flammability 1, Reactivity 0. Soluble in water.

Potential Exposure: Glycerol is used as a humectant in tobacco; it is used in cosmetics, antifreezes and inks. It is used as a fiber lubricant. It is used as a raw material for alkyd resins and in explosives manufacture.

Incompatibilities: Able to polymerize above 300°F (150°C). Incompatible with acetic anhydrides (especially in the presence of a catalyst), strong acids, caustics, aliphatic amines, isocyanates. Strong oxidizers (e.g., chromium trioxide, potassium chlorate, potassium permanganate) can cause fire and explosion hazard. Hygroscopic (i.e., absorbs moisture from the air). Decomposes when heated producing corrosive gas of acrolein.

Permissible Exposure Limits in Air: OSHA[58] sets the TWA at 10 mg/m^3 (total) and 5 mg/m^3 (respirable fraction). ACGIH classifies glycerin mist as a nuisance particulate with a TLV of 10 mg/m^3. HSE[33] sets the TWA at 10 mg/m^3 and adds an STEL of 20 mg/m^3. Several states have set guidelines or standards for glycerol in ambient air[60] ranging from 3.0 μg/m^3 (Virginia) to 200.0 μg/m^3 (Florida and New York) to 238.0 μg/m^3 (Nevada).

Determination in Air: Determined gravimetrically after collection on a filter. See NIOSH Analytical Methods #0500 for total nuisance dust and #0600 for respirable nuisance dust.

Permissible Concentration in Water: No criteria set.

Harmful Effects and Symptoms

Short Term Exposure: Glycerin can be irritating to the eyes, skin, and respiratory tract. When swallowed, it can cause insomnia, nausea, vomiting, diarrhea, fever, hemoglobinuria, convulsions and paralysis. Toxic in high concentrations; it is somewhat dehydrating and irritating to exposed tissues. Symptoms include headache, dizziness, insomnia, nausea, vomiting, diarrhea, fever, elevated blood sugar and diabetic coma; very large doses may cause irritation and dehydration of tissues, hemolysis, renal failure, hemoglobinuria, convulsions, and paralysis.

Long Term Exposure: May cause kidney damage.

Points of Attack: Eyes, skin, respiratory system, kidneys.

Medical Surveillance: Kidney function tests.

First Aid: *Skin Contact:* Flood all areas of body that have contacted the substance with water. Don't wait to remove contaminated clothing; do it under the water stream. Use soap to help assure removal. Isolate contaminated clothing when removed to prevent contact by others. *Eye Contact:* Remove any contact lenses at once. Immediately flush eyes well with copious quantities of water or normal saline for at least 20 – 30 minutes. Seek medical attention. *Inhalation:* Leave contaminated area immediately; breathe fresh air. Proper respiratory protection must be supplied to any rescuers. If coughing, difficult breathing or any other symptoms develop, seek medical attention at once, even if symptoms develop many hours after exposure. *Ingestion:* Contact a physician, hospital or poison center at once. If the victim is unconscious or convulsing, do not induce vomiting or give anything by mouth. Assure that his airway is open and lay him on his side with his head lower than his body and transport immediately to a medical facility. If conscious and not convulsing, give a glass of water to dilute the substance. Vomiting should not be induced without a physician's advice.

Personal Protective Methods: Wear protective gloves and clothing to prevent any reasonable probability of skin contact. Safety equipment suppliers/manufacturers can provide recommendations on the most protective glove/clothing material for your operation.. All protective clothing (suits, gloves, footwear, headgear) should be clean, available each day, and put on before work. Contact lenses should not be worn when working with this chemical. Wear splash-proof chemical goggles and face shield unless full facepiece respiratory protection is worn. Employees should wash immediately with soap when skin is wet or contaminated. Provide emergency showers and eyewash.

Respirator Selection: Where the potential for exposure to this chemical, use a MSHA/NIOSH approved supplied-air respirator with a full facepiece operated in the positive pressure mode or with a full facepiece, hood, or helmet in the continuous flow mode, or use a MSHA/NIOSH approved self-contained breathing apparatus with a full facepiece operated in pressure-demand or other positive pressure mode.

Storage: Prior to working with this chemical you should be trained on its proper handling and storage. Store in tightly closed containers in a cool, well ventilated area away from oxidizers. Where possible, automatically pump liquid from drums or other storage containers to process containers.

Shipping: The DOT imposes no specific requirements for glycerol in its Performance-Oriented Packaging Standards.[19]

Spill Handling: Evacuate and restrict persons not wearing protective equipment from area of spill or leak until cleanup is complete. Remove all ignition sources. Ventilate area of spill or leak. Absorb liquids in vermiculite, dry sand, earth, peat, carbon, or a similar material and deposit in sealed containers. Keep this chemical out of a confined space, such as a sewer, because of the possibility of an explosion, unless the sewer is designed to prevent the build-up of explosive concentrations. It may be necessary to contain and dispose of this chemical as a hazardous waste. If material or contaminated runoff enters waterways, notify downstream users of potentially contaminated waters. Contact your Department of Environmental Protection or your regional office of the federal EPA for specific recommendations. If employees are required to clean-up spills, they must be properly trained and equipped. OSHA 1910.120(q) may be applicable.

Fire Extinguishing: This chemical is a combustible liquid. Poisonous and corrosive gases including acrolein are produced in fire. Use dry chemical, carbon dioxide, water fog, or alcohol foam extinguishers. Vapors are heavier than air and will collect in low areas. Vapors may travel long distances to ignition sources and flashback. Vapors in confined areas may explode when exposed to fire. Containers may explode in fire. Storage containers and parts of containers may rocket great distances, in many directions. If material or contaminated runoff enters waterways, notify downstream users of potentially contaminated waters. Notify local health and fire officials and pollution control agencies. From a secure, explosion-proof location, use water spray to cool exposed containers. If cooling streams are ineffective (venting sound increases in volume and pitch, tank discolors, or shows any signs of deforming), withdraw immediately to a secure position. If employees are expected to fight fires, they must be trained and equipped in OSHA 1910.156.

Disposal Method Suggested: Mixture with a more flammable solvent followed by incineration.

References

Sax, N. I., Ed., Dangerous Properties of Industrial Materials Report, 1, No. 5, 61–63 (1981) and 3, No. 4, 58–60 (1983)

Glycidol

Molecular Formula: $C_3H_6O_2$

Common Formula: $HOCH_2CH(O)CH_2$

Synonyms: Epihydrin alcohol; 2,3-Epoxy-1-propanol; 2,3-Epoxypropanol; Glycide; Glycidyl alcohol; 3-Hydroxy-1,2-epoxypropane; 3-Hydroxypropylene oxide; Methanol, oxiranyl-; NCI-C55549; Oxiranemethanol

CAS Registry Number: 556-52-1

RTECS Number: UB4375000

Regulatory Authority

- Air Pollutant Standard Set (ACGIH)[1] (DFG)[3] (former USSR)[43] (OSHA)[58] (Several States)[60]
- Canada, WHMIS, Ingredients Disclosure List

Cited in U.S. State regulations: Alaska (G), California (A, G), Connecticut (A), Florida (G), Illinois (G), Maine (G), Massachusetts (G), New Hampshire (G), New Jersey (G), North Dakota (A), Pennsylvania (G), Rhode Island (G), Virginia (A), West Virginia (G).

Description: Glycidol, $C_3H_6O_2$, $HOCH_2CH(O)CH_2$, is a colorless liquid. Boiling point = 166°C. Freezing/Melting point = -45°C. Flash point = 72°C. Autoignition temperature = 415°C. Soluble in water.

Potential Exposure: Glycidol is used as an intermediate in the synthesis of glycerol, glycidyl ethers, esters and amines.

Incompatibilities: Forms explosive mixture with air. Violent reaction with strong oxidizers, nitrates. Decomposes on contact (especially in the presence of heat) with strong acids, strong bases, water, metal salts (e.g., aluminum chloride, ferric chloride, tin chloride) or metals (copper, zinc), causing fire and explosion hazard. Contact with barium, lithium, sodium, magnesium, titanium may cause polymerization. Attacks some plastics, rubber and coatings.

Permissible Exposure Limits in Air: The OSHA PEL and DFG MAK value[3] is 50 ppm (150 mg/m^3). NIOSH and ACGIH recommend a TWA of 25 ppm (75 mg/m^3). The NIOSH IDLH is 150 ppm. The former USSR-UNEP/IRPTC project[43] has set a MAC in workplace air of 5 mg/m^3. Several states have set guidelines or standards for glycidol in ambient air[6] ranging from 750 – 3,000 μg/m^3 (North Dakota) to 1,300 μg/m^3 (Virginia) to 1,500 μg/m^3 (Connecticut).

Determination in Air: Adsorption on charcoal workup with tetrahydrofuran, analysis by gas chromatography/flame ionization. See NIOSH IV, Method #1608.[18]

Permissible Concentration in Water: No criteria set.

Routes of Entry: Inhalation, ingestion, skin and/or eye contact. Passes through the skin.

Harmful Effects and Symptoms

Short Term Exposure: Glycidol affect you when breathed in and by passing through your skin. Irritates the eyes, skin, and respiratory tract. Vapor exposure can damage vision. Exposure can irritate the eyes, nose, throat and lungs. Higher exposures can cause pulmonary edema, a medical emergency that can be delayed for several hours. This can cause death. High levels can cause you to feel dizzy, lightheaded, confused, excited, to pass out and even die. May affect the central nervous system.

Long Term Exposure: Repeated or prolonged contact may cause skin sensitization. Can irritate the lungs; bronchitis may develop. May cause personality changes; depression, anxiety, or irritability. Possibly carcinogenic to humans. Because this is a mutagen, handle it as a possible carcinogen, with extreme caution. May cause sterility in males.

Points of Attack: Eyes, skin, respiratory system, central nervous system. Cancer site in animals: stomach, brain, and breast.

Medical Surveillance: Before beginning employment and at regular times after that, the following are recommended: Lung function tests. Examination of the nervous system. If symptoms develop or overexposure is suspected, the following may be useful: Evaluation by a qualified allergist, including careful exposure history and special testing, may help diagnose skin allergy. Consider chest x-ray after acute overexposure. Exam of the eyes and vision.

First Aid: If this chemical gets into the eyes, remove any contact lenses at once and irrigate immediately for at least 15 minutes, occasionally lifting upper and lower lids. Seek medical attention immediately. If this chemical contacts the skin, remove contaminated clothing and wash immediately with soap and water. Seek medical attention immediately. If this chemical has been inhaled, remove from exposure, begin rescue breathing (using universal precautions) if breathing has stopped and CPR if heart action has stopped. Transfer promptly to a medical facility. When this chemical has been swallowed, get medical attention. Give large quantities of water and induce vomiting. Do not make an unconscious person vomit. Medical observation is recommended for 24 – 48 hours after breathing overexposure, as pulmonary edema may be delayed. As first aid for pulmonary edema, a doctor or authorized paramedic may consider administering a corticosteroid spray.

Personal Protective Methods: Wear protective gloves and clothing to prevent any reasonable probability of skin contact. Safety equipment suppliers/manufacturers can provide recommendations on the most protective glove/clothing material for your operation.. All protective clothing (suits, gloves, footwear, headgear) should be clean, available each day, and put on before work. Contact lenses should not be worn when working with this chemical. Wear splash-proof chemical goggles and face shield unless full facepiece respiratory protection is worn. Employees should wash immediately with soap when skin is wet or contaminated. Provide emergency showers and eyewash.

Respirator Selection: NIOSH: *Up to 150 ppm:* SA (any supplied-air respirator);* or SCBAF (any self-contained breathing apparatus with a full facepiece). *Emergency or planned entry into unknown concentrations or IDLH conditions:* SCBAF:PD,PP (any self-contained breathing apparatus that has a full facepiece and is operated in a pressure-demand or other positive-pressure mode); or SAF:PD,PP:ASCBA (any supplied-air respirator that has a full facepiece and is operated in a pressure-demand or other positive-pressure mode in combination with an auxiliary, self-contained breathing apparatus operated in a pressure-demand or other positive-pressure mode). *Escape:* GMFOVHiE [any air-purifying, full-facepiece respirator (gas mask) with a chin-style, front- or back-mounted organic vapor canister having a high-efficiency particulate filter]; or SCBAE (any appropriate escape-type, self-contained breathing apparatus).

* Substance reported to cause eye irritation or damage; may require eye protection.

Storage: Prior to working with this chemical you should be trained on its proper handling and storage. Glycidol must be stored to avoid contact with strong oxidizers (such as perchlorates, peroxides, permanganates, chlorates and nitrates). Store in tightly closed containers in a cool, well-ventilated area away from heat. Metal containers involving the transfer of this chemical should be grounded and bonded. Where possible, automatically pump liquid from drums or other storage containers to process containers. Drums must be equipped with self-closing valves, pressure vacuum bungs, and flame arresters. Use only non-sparking tools and equipment, especially when opening and closing containers of this chemical. Sources of ignition such as smoking and open flames, are prohibited where this chemical is used, handled, or stored in a manner that could create a potential fire or explosion hazard. Wherever this chemical is used, handled, manufactured, or stored, use explosion-proof electrical equipment and fittings.

Shipping: This material is not specifically cited by DOT but may be considered as a combustible liquid n.o.s. This type of compound requires no specific shipping label. It falls in DOT Hazard Class 3 and Packing Group III.

Spill Handling: Evacuate and restrict persons not wearing protective equipment from area of spill or leak until cleanup is complete. Remove all ignition sources. Ventilate area of spill or leak. Absorb liquids in vermiculite, dry sand, earth, peat, carbon, or a similar material and deposit in sealed containers. Keep this chemical out of a confined space, such as a sewer, because of the possibility of an explosion, unless the sewer is designed to prevent the build-up of explosive concentrations. It may be necessary to contain and dispose of this chemical as a hazardous waste. If material or contaminated runoff enters waterways, notify downstream users of potentially contaminated waters. Contact your Department of Environmental Protection or your regional office of the federal EPA for specific recommendations. If employees are required to clean-up spills, they must be properly trained and equipped. OSHA 1910.120(q) may be applicable.

Fire Extinguishing: This chemical is a combustible liquid. Acrid smoke and poisonous gases including carbon monoxide are produced in fire. Use dry chemical, carbon dioxide, or alcohol or polymer foam extinguishers. Vapors are heavier than air and will collect in low areas. Vapors may travel long distances to ignition sources and flashback. Vapors in confined areas may explode when exposed to fire. Containers may explode in fire. Storage containers and parts of containers may rocket great distances, in many directions. If material or contaminated runoff enters waterways, notify downstream users of potentially contaminated waters. Notify local health and fire officials and pollution control agencies. From a secure, explosion-proof location, use water spray to cool exposed containers. If cooling streams are ineffective (venting sound increases in volume and pitch, tank discolors, or shows any signs of deforming), withdraw immediately to a secure position. If employees are expected to fight fires, they must be trained and equipped in OSHA 1910.156.

Disposal Method Suggested: *Concentrated waste containing no peroxides:* discharge liquid at a controlled rate near a pilot flame. *Concentrated waste containing peroxides:* perforation of a container of the waste from a safe distance followed by open burning.

References

New Jersey Department of Health and Senior Services, "Hazardous Substance Fact Sheet: 2, 3-Epoxy-1-Propanol," Trenton, NJ (December 1998)

Glycidyl Aldehyde

Molecular Formula: $C_3H_4O_2$

Common Formula: CH_2–(O)–CH–CHO

Synonyms: Epihydrinaldehyde; Epihydrine aldehyde; 2,3-Epoxy-1-propanal; 2,3-Epoxypropanal; 2,3-Epoxypropionaldehyde; Glycidal; Glycidaldehyde; Oxiranecarboxaldehyde

CAS Registry Number: 765-34-4

RTECS Number: MB3150000

DOT ID: UN 2622

Regulatory Authority

- Air Pollutant Standard Set (Several States)[60]
- Carcinogen (Animal Positive) (IARC)[9]
- EPA HAZARDOUS WASTE NUMBER (RCRA No.): U126
- RCRA, 40CFR261, Appendix 8 Hazardous Constituents
- SUPERFUND/EPCRA 40CFR302.4 Reportable Quantity (RQ): CERCLA, 10 lb (4.54 kg)

Cited in U.S. State regulations: California (A, G), Florida (G), Illinois (G), Kansas (G), Louisiana (G), Massachusetts (G), Nevada (A), New Hampshire (G), New Jersey (G), New York (A), Pennsylvania (G), South Carolina (A), Vermont (G), Virginia (G), Washington (G), Wisconsin (G).

Description: Glycidyl aldehyde, CH_2–(O)–CH–CHO, $C_3H_4O_2$, is a colorless liquid with a pungent, aldehyde-like odor. Boiling point = 112 – 113°C. Freezing/Melting point = -62°C. Flash point = 31°C. Hazard Identification (based on NFPA-704 M Rating System): Health 2, Flammability 3, Reactivity 0. Soluble in water.

Potential Exposure: This material has been used in the finishing of wool and the tanning of leather and surgical sutures in the U.K. It has been tested as a disinfectant.

Incompatibilities: Oxidizers cause a fire and explosion hazard.

Permissible Exposure Limits in Air: No OELs have been established, but this chemical may be a carcinogen in humans, and can be absorbed through the skin.

Permissible Exposure Limits in Air: Several states have set guidelines or standards for glycidyl aldehyde in ambient air[60] ranging from zero (Nevada) to 0.03 $\mu g/m^3$ (New York) to 75.0 $\mu g/m^3$ (South Carolina).

Permissible Concentration in Water: No criteria set.

Routes of Entry: Inhalation, ingestion, skin and/or eye contact.

Harmful Effects and Symptoms

Short Term Exposure: May be fatal if inhaled, ingested, or absorbed through the skin. Contact can cause severe skin and eye irritation and burns with possible permanent eye damage. Skin burns can be slow healing, followed by pigmentation.

Inhalation can cause respiratory tract irritation. Higher exposures can cause pulmonary edema, a medical emergency that can be delayed for several hours. This can cause death.

Long Term Exposure: May be a carcinogen in humans. May cause skin sensitization and allergy. May cause liver and kidney damage. May affect the nervous system. Can cause lung irritation and bronchitis.

Points of Attack: Eyes, skin, liver, kidneys, nervous system, lungs, mucous membranes.

Medical Surveillance: Liver and kidney function tests. examination of the nervous system. Evaluation by a qualified allergist. Consider chest x-ray following acute overexposure.

First Aid: If this chemical gets into the eyes, remove any contact lenses at once and irrigate immediately for at least 15 minutes, occasionally lifting upper and lower lids. Seek medical attention immediately. If this chemical contacts the skin, remove contaminated clothing and wash immediately with soap and water. Seek medical attention immediately. If this chemical has been inhaled, remove from exposure, begin rescue breathing (using universal precautions) if breathing has stopped and CPR if heart action has stopped. Transfer promptly to a medical facility. When this chemical has been swallowed, get medical attention. Give large quantities of water and induce vomiting. Do not make an unconscious person vomit. Medical observation is recommended for 24 – 48 hours after breathing overexposure, as pulmonary edema may be delayed. As first aid for pulmonary edema, a doctor or authorized paramedic may consider administering a corticosteroid spray.

Personal Protective Methods: Wear protective gloves and clothing to prevent any reasonable probability of skin contact. Safety equipment suppliers/manufacturers can provide recommendations on the most protective glove/clothing material for your operation.. All protective clothing (suits, gloves, footwear, headgear) should be clean, available each day, and put on before work. Contact lenses should not be worn when working with this chemical. Wear splash-proof chemical goggles and face shield unless full facepiece respiratory protection is worn. Employees should wash immediately with soap when skin is wet or contaminated. Provide emergency showers and eyewash.

Respirator Selection: Where the potential for exposure to this chemical, use a MSHA/NIOSH approved supplied-air respirator with a full facepiece operated in the positive pressure mode or with a full facepiece, hood, or helmet in the continuous flow mode, or use a MSHA/NIOSH approved self-contained breathing apparatus with a full facepiece operated in pressure-demand or other positive pressure mode.

Storage: Prior to working with this chemical you should be trained on its proper handling and storage. Store in tightly closed containers in a cool, well ventilated area away from oxidizers, nitrates, heat, and flames. Metal containers involving the transfer of this chemical should be grounded and bonded. Where possible, automatically pump liquid from drums or other storage containers to process containers. Drums must be equipped with self-closing valves, pressure vacuum bungs, and flame arresters. Use only non-sparking tools and equipment, especially when opening and closing containers of this chemical. Sources of ignition such as smoking and open flames, are prohibited where this chemical is used, handled, or stored in a manner that could create a potential fire or explosion hazard. Wherever this chemical is used, handled, manufactured, or stored, use explosion-proof electrical equipment and fittings. A regulated, marked area should be established where this chemical is handled, used, or stored in compliance with OSHA standard 1910.1045.

Shipping: This compound requires a shipping label of: "Flammable Liquid, Poison." It falls in DOT Hazard Class 3 and Packing Group II.

Spill Handling: Evacuate and restrict persons not wearing protective equipment from area of spill or leak until cleanup is complete. Remove all ignition sources. Use water spray to reduce vapors. Ventilate area of spill or leak. Absorb liquids in vermiculite, dry sand, earth, peat, carbon, or a similar material and deposit in sealed containers. Keep this chemical out of a confined space, such as a sewer, because of the possibility of an explosion, unless the sewer is designed to prevent the build-up of explosive concentrations. It may be necessary to contain and dispose of this chemical as a hazardous waste. If material or contaminated runoff enters waterways, notify downstream users of potentially contaminated waters. Contact your Department of Environmental Protection or your regional office of the federal EPA for specific recommendations. If employees are required to clean-up spills, they must be properly trained and equipped. OSHA 1910.120(q) may be applicable.

Fire Extinguishing: This chemical is a flammable liquid. Poisonous gases are produced in fire. Use dry chemical, carbon dioxide, water spray, fog, or polymer foam extinguishers. Vapors are heavier than air and will collect in low areas. Vapors may travel long distances to ignition sources and flashback. Vapors in confined areas may explode when exposed to fire. Containers may explode in fire. Storage containers and parts of containers may rocket great distances, in many directions. If material or contaminated runoff enters waterways, notify downstream users of potentially contaminated waters. Notify local health and fire officials and pollution control agencies. From a secure, explosion-proof location, use water spray to cool exposed containers. If cooling streams are ineffective (venting sound increases in volume and pitch, tank discolors, or shows any signs of deforming), withdraw immediately to a secure position. If employees are expected to fight fires, they must be trained and equipped in OSHA 1910.156.

Disposal Method Suggested: Consult with environmental regulatory agencies for guidance on acceptable disposal practices. Generators of waste containing this contaminant

(≥100 kg/mo) must conform with EPA regulations governing storage, transportation, treatment, and waste disposal.

References

Sax, N. I., Ed., "Dangerous Properties of Industrial Materials Report" 7, No. 3, 103–105 (1987)

New Jersey Department of Health and Senior Services, "Hazardous Substance Fact Sheet, Glycidaldehyde," Trenton, NJ (May 1999)

Glyphosate

Molecular Formula: $C_3H_8NO_5P$

Common Formula: $HOCOCH_2NHCH_2PO(OH)_2$

Synonyms: Glycine, *N*-(phosphonomethyl)-; MON0573; MON 2139; *N*-(Phosphonomethyl)-glycine; Phosphonomethyliminoacetic acid; Rodeo®; Roundup®

CAS Registry Number: 1071-83-6

RTECS Number: MC1075000

DOT ID: UN 2783

Regulatory Authority

- Air Pollutant Standard Set (former USSR)[35][43]

Cited in U.S. State regulations: California (G, W), New Jersey (G), Pennsylvania (G).

Description: Glyphosate, $HOCOCH_2NHCH_2PO(OH)_2$ is a colorless crystalline powder. Freezing/Melting point = 230°C. Soluble in water. Often used as a liquid in a carrier solvent which may change physical and toxicological properties. It is a organophosphorus chemical.

Potential Exposure: Those involved in the manufacture, formulation and application of this non-selective, non-residual pre-emergence herbicide. Has wide residential use in the United States for the control of weeds.

Incompatibilities: Solutions are corrosive to iron, unlined steel, and galvanized steel forming a highly combustible or explosive gas mixture. *Do not* store Glyphosate in containers made from these materials.

Permissible Exposure Limits in Air: No OELs have been established in the U.S. for this chemical. The former USSR has set a ceiling value in workplace air of 1.5 mg/m^3.[35][43]

Determination in Air: OSHA versatile sampler-2; Toluene/Acetone; Gas chromatography/Flame photometric detection for sulfur, nitrogen, or phosphorus; NIOSH Method IV, Method #5600, Organophosphorus Pesticides.

Permissible Concentration in Water: The U.S.E.P.A. has developed data on Glyphosate including a no-observed-adverse effects level (NOAEL) of 10 mg/kg/day. This corresponds to a drinking water equivalent level of 3.5 mg/l from which a lifetime health advisory of 0.7 mg/l was derived. California[61] has set a guideline of 0.5 mg/l for drinking water.

Determination in Water: Analysis of Glyphosate is by a high-performance liquid chromatographic (HPLC) method.

Routes of Entry: Inhalation, ingestion, through the skin.

Harmful Effects and Symptoms

Short Term Exposure: Irritates the eyes, skin, and respiratory tract. Exposure to high levels can cause nausea, vomiting, diarrhea, decreased blood pressure, and convulsions. High exposures can cause arrhythmia and possible death. The acute LD_{50} oral for rats is 5,600 mg/kg (insignificantly toxic).

Long Term Exposure: May cause liver and kidney damage. It does not seem to exhibit reproductive effects, mutagenicity or carcinogenicity in animal studies.

Points of Attack: Respiratory system, lungs, central nervous system, cardiovascular system, skin, eyes, plasma and red blood cell cholinesterase, liver, kidney, heart.

Medical Surveillance: Before employment and at regular times after that, the following are recommended: Plasma and red blood cell cholinesterase levels (tests for the enzyme poisoned by this chemical). If exposure stops, plasma levels return to normal in 1 – 2 weeks while red blood cell levels may be reduced for 1 – 3 months.

When cholinesterase enzyme levels are reduced by 25% or more below preemployment levels, risk of poisoning is increased, even if results are in lower ranges of "normal." Reassignment to work not involving organophosphate or carbamate pesticides is recommended until enzyme levels recover. If symptoms develop or overexposure occurs, repeat the above tests as soon as possible and get an exam of the nervous system. Also consider complete blood count. Consider chest x-ray following acute overexposure. Do not drink any alcoholic beverages before or during use. Alcohol promotes absorption of organic phosphates. Liver and kidney function tests. Special 24-hour EKG (Holter monitor) for irregular heartbeat.

First Aid: If this chemical gets into the eyes, remove any contact lenses at once and irrigate immediately for at least 15 minutes, occasionally lifting upper and lower lids. Seek medical attention immediately. If this chemical contacts the skin, remove contaminated clothing and wash immediately with soap and water. Speed in removing material from skin is of extreme importance. Shampoo hair promptly if contaminated. Seek medical attention immediately. If this chemical has been inhaled, remove from exposure, begin rescue breathing (using universal precautions) if breathing has stopped and CPR if heart action has stopped. Transfer promptly to a medical facility. When this chemical has been swallowed, get medical attention. Give large quantities of water and induce vomiting. Do not make an unconscious person vomit.

Personal Protective Methods: Wear protective gloves and clothing to prevent any reasonable probability of skin contact. Safety equipment suppliers/manufacturers can provide recommendations on the most protective glove/clothing material for your operation.. All protective clothing (suits, gloves, footwear, headgear) should be clean, available each day, and put on before work. Contact lenses should not be worn when working with this chemical. Wear splash-proof chemical

goggles and face shield unless full facepiece respiratory protection is worn. Employees should wash immediately with soap when skin is wet or contaminated. Provide emergency showers and eyewash.

Respirator Selection: Where the potential for exposure to this chemical, use a MSHA/NIOSH approved supplied-air respirator with a full facepiece operated in the positive pressure mode or with a full facepiece, hood, or helmet in the continuous flow mode, or use a MSHA/NIOSH approved self-contained breathing apparatus with a full facepiece operated in pressure-demand or other positive pressure mode.

Storage: Prior to working with this chemical you should be trained on its proper handling and storage. Store in tightly closed containers in a cool, well ventilated area away from oxidizers. Where possible, automatically pump liquid from drums or other storage containers to process containers

Shipping: This is an organic phosphorus compound, solid, toxic requiring a "Poison" label. It is in Hazard Class 6.1.

Spill Handling: Evacuate and restrict persons not wearing protective equipment from area of spill or leak until cleanup is complete. Remove all ignition sources. Ventilate area of spill or leak. Absorb liquids in vermiculite, dry sand, earth, peat, carbon, or a similar material and deposit in sealed containers. Collect powdered material in the most convenient and safe manner and deposit in sealed containers for later disposal. It may be necessary to contain and dispose of this chemical as a hazardous waste. If material or contaminated runoff enters waterways, notify downstream users of potentially contaminated waters. Contact your Department of Environmental Protection or your regional office of the federal EPA for specific recommendations. If employees are required to clean-up spills, they must be properly trained and equipped. OSHA 1910.120(q) may be applicable.

Fire Extinguishing: Solid Glyphosate may burn, but does not readily ignite. Use dry chemical, carbon dioxide, water spray, or alcohol foam extinguishers. Poisonous gases are produced in fire including nitrogen oxides and phosphorus oxides. If material or contaminated runoff enters waterways, notify downstream users of potentially contaminated waters. Notify local health and fire officials and pollution control agencies. From a secure, explosion-proof location, use water spray to cool exposed containers. If cooling streams are ineffective (venting sound increases in volume and pitch, tank discolors, or shows any signs of deforming), withdraw immediately to a secure position. If employees are expected to fight fires, they must be trained and equipped in OSHA 1910.156.

References

U.S. Environmental Protection Agency, "Health Advisory: Glyphosate," Washington, DC, Office of Drinking Water (August 1987)

New Jersey Department of Health and Senior Services, "Hazardous Substance Fact Sheet, Glyphosate," Trenton, NJ (June 1999)

Grain Dust

Molecular Formula: None.

Synonyms: None.

CAS Registry Number: None.

RTECS Number: MD7900000

Regulatory Authority

- Air Pollutant Standard Set (ACGIH)[1] (OSHA)[58]

Cited in U.S. State regulations: Alaska (G), Illinois (G), New Hampshire (G), New Jersey (G), Pennsylvania (G), West Virginia (G).

Description: Oats, barley or wheat dust containing microbial flora and fauna.

Potential Exposure: Grain elevator workers, grain harvesters.

Permissible Exposure Limits in Air: The OSHA PEL is 10 mg/m^3 TWA. NIOSH and ACGIH recommend a TWA of 4 mg/m^3 (total dust).

Determination in Air: By gravimetric means after filter separation. NIOSH IV, Method #0500, Particulates NOR (total).

Permissible Concentration in Water: No criteria set.

Routes of Entry: Inhalation.

Harmful Effects and Symptoms

Short Term Exposure: Irritates the eyes, skin, upper respiratory system. Can cause coughing, wheezing.

Long Term Exposure: Impaired lung function, chronic bronchitis, both immediate and delayed asthmatic reactions, grain fever upon exposure to dust concentrations in excess of 15 mg/m^3.[53]

Points of Attack: Respiratory system.

Medical Surveillance: Lung function tests.

First Aid: If this material gets into the eyes, remove any contact lenses at once and irrigate immediately for at least 15 minutes, occasionally lifting upper and lower lids. Seek medical attention immediately. If this chemical contacts the skin, remove contaminated clothing and wash immediately with soap and water. If this chemical has been inhaled, remove from exposure, begin rescue breathing (using universal precautions) if breathing has stopped and CPR if heart action has stopped. Transfer promptly to a medical facility.

Personal Protective Methods: Safety equipment suppliers/manufacturers can provide recommendations on the most protective glove/clothing material for your operation.. All protective clothing (suits, gloves, footwear, headgear) should be clean, available each day, and put on before work. Wear dust-proof chemical goggles and face shield unless full facepiece respiratory protection is worn. Employees should wash immediately with soap when skin is wet or contaminated. Provide emergency showers and eyewash.

Respirator Selection: Dust respirator required.

Spill Handling: Evacuate persons not wearing protective equipment from area of spill or leak until clean-up is complete. Remove all ignition sources. Collect powdered material in the most convenient and safe manner and deposit in sealed containers. Ventilate area after clean-up is complete. If employees are required to clean-up spills, they must be properly trained and equipped. OSHA 1910.120(q) may be applicable.

Fire Extinguishing: This material may be combustible solid depending on specific components of the grain dust including fertilizers, pesticides and microorganisms. Use dry chemical, carbon dioxide, water spray, or foam extinguishers. Poisonous gases may be produced in fire. If employees are expected to fight fires, they must be trained and equipped in OSHA 1910.156.

Graphite

Molecular Formula: C

Synonyms: Black lead; Mineral carbon; Plumbago; Silver graphite; Stove black

CAS Registry Number: 7782-42-5

RTECS Number: MD9659600

DOT ID: UN 1362 (carbon, activated)

EINECS Number: 231-955-3

Regulatory Authority

- Air Pollutant Standard Set (ACGIH)[1] (HSE)[33] (OSHA)[58]
- Canada, WHMIS, Ingredients Disclosure List

Cited in U.S. State regulations: Alaska (G), California (G), Florida (G), Illinois (G), Maine (G), Massachusetts (G), New Hampshire (G), New Jersey (G), Pennsylvania (G), Rhode Island (G), West Virginia (G).

Description: Graphite is crystallized carbon and usually appears as soft, black scales. There are two types of graphite, natural and artificial (activated). Sublimation point = 3,652°C. Insoluble in water.

Potential Exposure: Natural graphite is used in foundry facings, steel making, lubricants, refractories, crucibles, pencil "lead," paints, pigments, and stove polish. Artificial graphite may be substituted for these uses with the exception of clay crucibles; other types of crucibles may be produced from artificial graphite. Additionally, it may be used as a high temperature lubricant or for electrodes. It is utilized in the electrical industry in electrodes, brushes, contacts, and electronic tube rectifier elements; as a constituent in lubricating oils and greases; to treat friction elements, such as brake linings; to prevent molds from sticking together; and in moderators in nuclear reactors. In addition, concerns have been expressed about synthetic graphite in fibrous form. Those exposed are involved in production of graphite fibers from pitch or acrylonitrile fibers and the manufacture and use of composites of plastics, metals or ceramics reinforced with graphite fibers.

Incompatibilities: Graphite is a strong reducing agent and reacts violently with oxidizers, such as fluorine; chlorine trifluoride; potassium peroxide.

Forms an explosive mixture with air. May be spontaneously combustible in air.

Permissible Exposure Limits in Air: The OSHA PEL is 15 mppcf. The ACGIH recommends a TWA for natural graphite of 2.0 mg/m^3 for respirable dust and a TWA for synthetic graphite of 10 mg/m^3. OSHA seta a dual limit for synthetic graphite of 10 mg/m^3 of total dust and 5 mg/m^3 of a respirable fraction. HSE[33] sets these same dual limits for graphite containing less than 1% quartz.

Determination in Air: By filter separation and gravity measurement.

Permissible Concentration in Water: No criteria set.

Routes of Entry: Inhalation of dust, eye and skin contact.

Harmful Effects and Symptoms

Short Term Exposure: Causes upper respiratory irritation.

Long Term Exposure: Lungs may be affected by repeated or prolonged exposure to dusts, resulting in graphite pneumoconiosis. Exposure to natural graphite may produce a progressive and disabling pneumoconiosis similar to anthracosilicosis. Symptoms include headache, coughing, depression, decreased appetite, dyspnea, and the production of black sputum. Some individuals may be asymptomatic for many years then suddenly become disabled. It has not yet been determined whether the free crystalline silica in graphite is solely responsible for development of the disease. There is evidence that artificial graphite may be capable of producing a pneumoconiosis.

Points of Attack: Respiratory system; lungs; cardiovascular system.

Medical Surveillance: Preemployment and periodic examinations should be directed toward detecting significant respiratory disease, through chest x-rays and pulmonary function tests.

First Aid: If this chemical gets into the eyes, remove any contact lenses at once and irrigate immediately for at least 15 minutes, occasionally lifting upper and lower lids. Seek medical attention immediately. If this chemical contacts the skin, remove contaminated clothing and wash immediately with soap and water. Seek medical attention immediately. If this chemical has been inhaled, remove from exposure, begin rescue breathing (using universal precautions) if breathing has stopped and CPR if heart action has stopped. Transfer promptly to a medical facility. When this chemical has been swallowed, get medical attention. Give large quantities of water and induce vomiting. Do not make an unconscious person vomit.

Personal Protective Methods: Wear protective gloves and clothing to prevent any reasonable probability of skin contact. Safety equipment suppliers/manufacturers can provide recommendations on the most protective glove/clothing material for your operation.. All protective clothing (suits, gloves, footwear, headgear) should be clean, available each day, and put on before work. Contact lenses should not be worn when working with this chemical. Wear dust-proof chemical goggles and face shield unless full facepiece respiratory protection is worn. Employees should wash immediately with soap when skin is wet or contaminated. Provide emergency showers and eyewash

Respirator Selection: NIOSH: (natural) *12.5 mg/m³*: D (any dust respirator). *25 mg/m³*: DXSQ (any dust respirator except single-use and quarter-mask respirators); or SA (any supplied respirator). *62.5 mg/m³*: PAPRD (any powered, air-purifying respirator with a dust filter); or SA:CF (any supplied-air respirator operated in a continuous-flow mode). *125 mg/m³*: HiEF (any air-purifying, full-facepiece respirator with a high-efficiency particulate filter); or PAPRTHiE (any powered, air-purifying respirator with a tight-fitting facepiece and a high-efficiency particulate filter); or SAT:CF (any supplied-air respirator that has a tight-fitting facepiece and is operated in a continuous-flow mode); SCBAF (any self-contained breathing apparatus with a full facepiece). *1,250* mg/m³: SAF:PD,PP (any supplied-air respirator that has a full facepiece and is operated in a pressure-demand or other positive-pressure mode). *Emergency or planned entry into unknown concentrations or IDLH conditions:* SCBAF:PD,PP (any self-contained breathing apparatus that has a full faceplate and is operated in a pressure-demand or other positive-pressure mode); or SAF:PD,PP:ASCBA (any supplied-air respirator that has a full facepiece and is operated in a pressure-demand or other positive-pressure mode in combination with an auxiliary, self-contained breathing apparatus operated in a pressure-demand or other positive-pressure mode). *Escape:* HiEF (any air-purifying, full-facepiece respirator with a high-efficiency particulate filter); or SCBAE (any appropriate escape-type, self-contained breathing apparatus).

Storage: Prior to working with this chemical you should be trained on its proper handling and storage. Store in tightly closed containers in a cool, well ventilated area

Shipping: Carbon, activated (synthetic) is UN 1362. Label "SPONTANEOUSLY COMBUSTIBLE." in Hazard Class 4.2, Packing III or II.

Spill Handling: Evacuate persons not wearing protective equipment from area of spill or leak until clean-up is complete. Remove all ignition sources. Collect powdered material in the most convenient and safe manner and deposit in sealed containers. Ventilate area after clean-up is complete. It may be necessary to contain and dispose of this chemical as a hazardous waste. If material or contaminated runoff enters waterways, notify downstream users of potentially contaminated waters. Contact your Department of Environmental Protection or your regional office of the federal EPA for specific recommendations. If employees are required to clean-up spills, they must be properly trained and equipped. OSHA 1910.120(q) may be applicable.

Fire Extinguishing: This chemical is a combustible solid. Use dry chemical, carbon dioxide, water spray, or alcohol foam extinguishers. Poisonous gases are produced in fire. If material or contaminated runoff enters waterways, notify downstream users of potentially contaminated waters. Notify local health and fire officials and pollution control agencies. From a secure, explosion-proof location, use water spray to cool exposed containers. If cooling streams are ineffective (venting sound increases in volume and pitch, tank discolors, or shows any signs of deforming), withdraw immediately to a secure position. If employees are expected to fight fires, they must be trained and equipped in OSHA 1910.156.

Disposal Method Suggested: Carbon (graphite) fibers are difficult to dispose of by incineration. Waste fibers should be packaged and disposed of in a landfill authorized for the disposal of special wastes of this nature, or as otherwise may be required by law. Do not incinerate.

Gum Arabic

Molecular Formula: None listed.

Synonyms: Acacia; Acacia dealbata gum; Acacia gum; Acacia Senegal; Acacia syrup; Arabic gum; Australian gum; Gum acacia; Gum ovaline; Gum Senegal; Indian gum; NCI-C50748; Senegal gum; Starsol No. 1; Wattle gum

CAS Registry Number: 9000-01-1

RTECS Number: CE5945000

DOT ID: No citation.

Cited in U.S. State regulations: New Jersey (G), Pennsylvania (G).

Description: Gum Arabic is a white or yellowish solid. It has a molecular weight of about 240,000 but no clearly defined structure.[52] It is an exudate from acacia trees. A combustible solid when present as a dust.

Potential Exposure: Gum Arabic is used in granular or powder form in foods, pharmaceuticals, cosmetics, adhesives, inks and textile printing.

Incompatibilities: Oxidizers.

Permissible Exposure Limits in Air: No standards set.

Determination in Air: No OELs have been established.

Permissible Concentration in Water: No criteria set.

Routes of Entry: Inhalation.

Harmful Effects and Symptoms

Short Term Exposure: Unknown at this time.

Long Term Exposure: Exposure can cause an asthma-like lung allergy to develop. Once allergy develops, even very small future exposures can cause attacks with wheezing, coughing, chest tightness and shortness of breath. Gum Arabic can also cause allergy-like symptoms of chronic runny nose and skin allergy with rash.

Points of Attack: Lungs, skin.

Medical Surveillance: Before beginning employment and at regular times after that, for those with frequent or potentially high exposures, the following are recommended: Lung function tests. These may be normal if person is not having an attack at the time of the test. If symptoms develop or overexposure is suspected, the following may be useful: Evaluation by a qualified allergist, including careful exposure history and special testing, may help diagnose skin allergy.

First Aid: If this chemical gets into the eyes, remove any contact lenses at once and irrigate immediately for at least 15 minutes, occasionally lifting upper and lower lids. Seek medical attention immediately. If this chemical contacts the skin, remove contaminated clothing and wash immediately with soap and water. Seek medical attention immediately. If this chemical has been inhaled, remove from exposure, begin rescue breathing (using universal precautions) if breathing has stopped and CPR if heart action has stopped. Transfer promptly to a medical facility.

Personal Protective Methods: Wear protective gloves and clothing to prevent any reasonable probability of skin contact. Safety equipment suppliers/manufacturers can provide recommendations on the most protective glove/clothing material for your operation.. All protective clothing (suits, gloves, footwear, headgear) should be clean, available each day, and put on before work. Contact lenses should not be worn when working with this chemical. Wear dust-proof chemical goggles and face shield unless full facepiece respiratory protection is worn. Employees should wash immediately with soap when skin is wet or contaminated. Provide emergency showers and eyewash.

Respirator Selection: Engineering controls should be established to eliminate hazardous exposures to Gum Arabic. If, however, you experience irritation, discomfort, or sensitization, or where the potential for high exposures exists, use a MSHA/NIOSH approved high efficiency particulate filter. More protection is provided by a full facepiece respirator than by a half-mask respirator, and even greater protection is provided by a powered-air purifying respirator. Particulate filters must be checked every day before work for physical damage, such as rips or tears, and replaced as needed. Where the potential for high exposures exists, use a MSHA/NIOSH approved supplied-air respirator with a full facepiece operated in the positive pressure mode or with a full facepiece, hood, or helmet in the continuous flow mode, or use a MSHA/NIOSH approved self-contained breathing apparatus with a full facepiece operated in pressure-demand or other positive pressure mode.

Storage: Store in tightly closed containers in a cool, well-ventilated area away from oxidizers (such as perchlorates, peroxides, permanganates, chlorates, and nitrates). Dust is combustible. Sources of ignition such as smoking and open flames are prohibited where Gum Arabic is used, handled, or stored in a manner that could create a potential fire or explosion hazard.

Shipping: While not specifically cited by DOT,[17] Gum Arabic may be classed as a hazardous substance, solid, n.o.s.[52] This, however, imposes no label requirement or shipping weight maximum.

Spill Handling: Evacuate persons not wearing protective equipment from area of spill or leak until clean-up is complete. Remove all ignition sources. Collect powdered material in the most convenient and safe manner and deposit in sealed containers. Ventilate area after clean-up is complete. It may be necessary to contain and dispose of this chemical as a hazardous waste. If material or contaminated runoff enters waterways, notify downstream users of potentially contaminated waters. Contact your Department of Environmental Protection or your regional office of the federal EPA for specific recommendations. If employees are required to clean-up spills, they must be properly trained and equipped. OSHA 1910.120(q) may be applicable.

Fire Extinguishing: This chemical is a combustible solid. Use dry chemical, carbon dioxide, water spray, or foam extinguishers. Poisonous gases are produced in fire. If material or contaminated runoff enters waterways, notify downstream users of potentially contaminated waters. Notify local health and fire officials and pollution control agencies. From a secure, explosion-proof location, use water spray to cool exposed containers. If cooling streams are ineffective (venting sound increases in volume and pitch, tank discolors, or shows any signs of deforming), withdraw immediately to a secure position. If employees are expected to fight fires, they must be trained and equipped in OSHA 1910.156.

References

New Jersey Department of Health and Senior Services, "Hazardous Substance Fact Sheet: Gum Arabic," Trenton, NJ (January, 1986)

H

Hafnium

Molecular Formula: Hf

Synonyms: Celtium; Elemental hafnium; Hafnium metal

CAS Registry Number: 7440-58-6

RTECS Number: MG4600000

DOT ID: UN 1326 (powder, wet); UN 2545 (powder, dry)

Regulatory Authority

- Air Pollutant Standard Set (ACGIH)[1] (DFG)[3] (HSE)[33] (OSHA)[58] (Several States)[60]
- Canada, WHMIS, Ingredients Disclosure List

Cited in U.S. State Regulations: Alaska (G), California (A, G), Connecticut (A), Florida (G), Illinois (G), Maine (G, W), Massachusetts (G), Nevada (A), New Hampshire (G), New Jersey (G), North Dakota (A), Oklahoma (G), Pennsylvania (G), Rhode Island (G), Virginia (A), West Virginia (G).

Description: Hafnium is a refractory metal which occurs in nature in zirconium minerals. Boiling point = 5,197°C. Freezing/Melting point = 2,230°C. Insoluble in water.

Potential Exposure: Hafnium metal has been used as a control rod material in nuclear reactors. Thus, those engaged in fabrication and machining of such rods may be exposed.

Incompatibilities: Moisture may cause self ignition or explosions. The powder may spontaneously explode. Powder or dust may spontaneously ignite on contact with air, and at higher temperatures with nitrogen, phosphorous, oxygen, halogens, and sulfur. A powerful oxidizer. Being an oxidizer, substance may have a violent reaction with many substances. Contact with air, hot nitric acid, heat, shock, friction, strong oxidizers, or ignition sources may cause explosions of powdered form.

Permissible Exposure Limits in Air: The OSHA PEL and the ACGIH TLV is 0.5 mg/m^3 TWA. There is no DFG MAK at present. The NIOSH IDLH level is 50 mg/m^3. Several states have set guidelines or standards for hafnium in ambient air[60] ranging from 5.0 μg/m^3 (North Dakota) to 8.0 μg/m^3 (Virginia) to 10.0 μg/m^3 (Connecticut) to 12.0 μg/m^3 (Nevada).

Determination in Air: Collection on a filter; workup with acid; analysis by plasma emission spectroscopy; NIOSH II(5), Method #S194.

Permissible Concentration in Water: No criteria set.

Routes of Entry: Inhalation, ingestion, eye and/or skin contact.

Harmful Effects and Symptoms

Short Term Exposure: Irritation of eyes, skin and mucous membranes.

Long Term Exposure: Lungs may be affected by repeated or prolonged exposure. May cause liver damage.

Points of Attack: Eyes, skin, lungs, liver, and mucous membranes.

Medical Surveillance: Consider the points of attack in preplacement and periodic physical examinations. Liver function tests. Lung function tests.

First Aid: If this chemical gets into the eyes, remove any contact lenses at once and irrigate immediately for at least 15 minutes, occasionally lifting upper and lower lids. Seek medical attention immediately. If this chemical contacts the skin, remove contaminated clothing and wash immediately with soap and water. Seek medical attention immediately. If this chemical has been inhaled, remove from exposure, begin rescue breathing (using universal precautions) if breathing has stopped and CPR if heart action has stopped. Transfer promptly to a medical facility. When this chemical has been swallowed, get medical attention. Give large quantities of water and induce vomiting. Do not make an unconscious person vomit.

Personal Protective Methods: Wear protective gloves and clothing to prevent any reasonable probability of skin contact. Safety equipment suppliers/manufacturers can provide recommendations on the most protective glove/clothing material for your operation. All protective clothing (suits, gloves, footwear, headgear) should be clean, available each day, and put on before work. Contact lenses should not be worn when working with this chemical. Wear dust-proof chemical goggles and face shield unless full facepiece respiratory protection is worn. Employees should wash immediately with soap when skin is wet or contaminated. Provide emergency showers and eyewash.

Respirator Selection: NIOSH/OSHA: *Up to 2.5 mg/m^3:* DM (any dust and mist respirator). *Up to 5 mg/m^3:* DMXSQ (any dust and mist respirator except single-use and quarter-mask respirators); or SA (any supplied-air respirator). *Up to 12.5 mg/m³:* (any supplied-air respirator operated in a continuous-flow mode);* or SA:CF (any powered, air-purifying respirator with a dust and mist filter).* *Up to 25 mg/m³:* HiEF (any air-purifying, full-facepiece respirator with a high-efficiency particulate filter); or SAT:CF (any supplied-air respirator that has a tight-fitting facepiece and is operated in a continuous-flow mode);* or PAPRTHiE (any powered, air-purifying respirator with a tight-fitting facepiece and a high-efficiency particulate filter);* or SCBAF (any self-contained breathing apparatus with a full facepiece); or SAF (any supplied-air respirator with a full facepiece). *Up to 50 mg/m³:* SAF:PD,PP (any supplied-air respirator that has a full facepiece and is operated in a pressure-demand or other positive-pressure mode). *Emergency or planned entry into unknown concentrations or IDLH conditions:* SAF:PD,PP (any self-contained breathing apparatus that has a full facepiece and is operated in a pressure-demand or other positive-pressure mode); or SAF:PD,PP:ASCBA (any supplied-air respirator that has a full facepiece and is operated in a pressure-demand or other positive-pressure mode in combination with an auxiliary self-contained positive-pressure breathing apparatus). *Escape:* HiEF (any air-purifying, full-facepiece respirator with a high-efficiency particulate filter); or SCBAE (any appropriate escape-type, self-contained breathing apparatus).

* Substance reported to cause eye irritation or damage; may require eye protection.

Storage: Prior to working with this chemical you should be trained on its proper handling and storage. Store in tightly closed containers in a cool, well ventilated area.

Shipping: Dry hafnium powder requires a shipping label of: "Spontaneously Combustible." It falls in DOT Hazard Class 4.2 and Packing Group II. W.et with 25% or more of water requires a shipping label of: "Flammable Solid." It falls in DOT Hazard Class 4.1 and Packing Group II.

Spill Handling: Evacuate persons not wearing protective equipment from area of spill or leak until clean-up is complete. Remove all ignition sources. Collect powdered material in the most convenient and safe manner and deposit in sealed containers. Ventilate area after clean-up is complete. It may be necessary to contain and dispose of this chemical as a hazardous waste. If material or contaminated runoff enters waterways, notify downstream users of potentially contaminated waters. Contact your Department of Environmental Protection or your regional office of the federal EPA for specific recommendations. If employees are required to clean-up spills, they must be properly trained and equipped. OSHA 1910.120(q) may be applicable.

Fire Extinguishing: This chemical is a combustible solid. Use dry chemical, carbon dioxide, water spray, or alcohol foam extinguishers. Poisonous gases are produced in fire. If material or contaminated runoff enters waterways, notify downstream users of potentially contaminated waters. Notify local health and fire officials and pollution control agencies. From a secure, explosion-proof location, use water spray to cool exposed containers. If cooling streams are ineffective (venting sound increases in volume and pitch, tank discolors, or shows any signs of deforming), withdraw immediately to a secure position. If employees are expected to fight fires, they must be trained and equipped in OSHA 1910.156.

Disposal Method Suggested: Recovery.

Hafnium and Compounds

Molecular Formula: Hf

CAS Registry Number: 7440-58-6

RTECS Number: MG4600000

DOT ID: UN 1326 (wet); UN 2545 (dry)

Regulatory Authority

- Air Pollutant Standard Set (ACGIH)[1] (DFG)[3] (HSE)[33] (OSHA)[58] (Several States)[60]
- Canada, WHMIS, Ingredients Disclosure List

Cited in U.S. State Regulations: Alaska (G), California (A, G), Connecticut (A), Florida (G), Illinois (G), Maine (G, W), Massachusetts (G), Nevada (A), New Hampshire (G), New Jersey (G), North Dakota (A), Oklahoma (G), Pennsylvania (G), Rhode Island (G), Virginia (A), West Virginia (G).

Description: Hafnium is a refractory metal which occurs in nature in zirconium minerals.

Potential Exposure: Hafnium metal has been used as a control rod material in nuclear reactors. Thus, those engaged in fabrication and machining of such rods may be exposed.

Incompatibilities: Moisture may cause self ignition or explosions. The powder may spontaneously explode. Powder or dust may spontaneously ignite on contact with air, and at higher temperatures with nitrogen, phosphorous, oxygen, halogens, and sulfur. A powerful oxidizer. Being an oxidizer, substance may have a violent reaction with many substances. Contact with air, hot nitric acid, heat, shock, friction, strong oxidizers, or ignition sources may cause explosions of powdered form.

Permissible Exposure Limits in Air: The OSHA PEL, the DFG MAK[3] and the ACGIH recommended value[1] is 0.5 mg/m³. The STEL value set by HSE[33] is 1.5 mg/m³. The NIOSH IDLH level is 250 mg/m³. Several states have set guidelines or standards for hafnium in ambient air[60] ranging from 5.0 μg/m³ (North Dakota) to 8.0 μg/m³ (Virginia) to 10.0 μg/m³ (Connecticut) to 12.0 μg/m³ (Nevada).

Determination in Air: Collection on a filter; workup with acid; analysis by plasma emission spectroscopy; NIOSH II,[5] Method #S194.

Permissible Concentration in Water: No criteria set.

Routes of Entry: Inhalation, ingestion, eye and/or skin contact.

Harmful Effects and Symptoms

Short Term Exposure: Irritation of eyes, skin and mucous membranes.

Long Term Exposure: May cause lung and liver damage.

Points of Attack: Eyes, skin, liver, lungs.

Medical Surveillance: Consider the points of attack in preplacement and periodic physical examinations. Lung function tests. Liver function tests.

First Aid: If this chemical gets into the eyes, remove any contact lenses at once and irrigate immediately for at least 15 minutes, occasionally lifting upper and lower lids. Seek medical attention immediately. If this chemical contacts the skin, remove contaminated clothing and wash immediately with soap and water. Seek medical attention immediately. If this chemical has been inhaled, remove from exposure, begin rescue breathing (using universal precautions) if breathing has stopped and CPR if heart action has stopped. Transfer promptly to a medical facility. When this chemical has been swallowed, get medical attention. Give large quantities of water and induce vomiting. Do not make an unconscious person vomit.

Personal Protective Methods: Wear protective gloves and clothing to prevent any reasonable probability of skin contact. Safety equipment suppliers/manufacturers can provide recommendations on the most protective glove/clothing material for your operation. All protective clothing (suits, gloves, footwear, headgear) should be clean, available each day, and put on before work. Contact lenses should not be worn when working with this chemical. Wear dust-proof chemical goggles and face shield unless full facepiece respiratory protection is worn. Employees should wash immediately with soap when skin is wet or contaminated. Provide emergency showers and eyewash.

Respirator Selection: NIOSH/OSHA: *Up to 2.5 mg/m³:* DM (any dust and mist respirator). *Up to 5 mg/m³:* DMXSQ (any dust and mist respirator except single-use and quarter-mask respirators); or SA (any supplied-air respirator). *Up to 12.5 mg/m³:* (any supplied-air respirator operated in a continuous-flow mode);* or SA:CF (any powered, air-purifying respirator with a dust and mist filter).* *Up to 25 mg/m³:* HiEF (any air-purifying, full-facepiece respirator with a high-efficiency particulate filter); or SAT:CF (any supplied-air respirator that has a tight-fitting facepiece and is operated in a continuous-flow mode);* or PAPRTHiE (any powered, air-purifying respirator with a tight-fitting facepiece and a high-efficiency particulate filter);* or SCBAF (any self-contained breathing apparatus with a full facepiece); or SAF (any supplied-air respirator with a full facepiece). *Up to 50 mg/m³:* SAF:PD,PP (any supplied-air respirator that has a full facepiece and is operated in a pressure-demand or other positive-pressure mode). *Emergency or planned entry into unknown concentrations or IDLH conditions:* SAF:PD,PP (any self-contained breathing apparatus that has a full facepiece and is operated in a pressure-demand or other positive-pressure mode); or SAF:PD,PP:ASCBA (any supplied-air respirator that has a full facepiece and is operated in a pressure-demand or other positive-pressure mode in combination with an auxiliary self-contained positive-pressure breathing apparatus). *Escape:* HiEF (any air-purifying, full-facepiece respirator with a high-efficiency particulate filter); or SCBAE (any appropriate escape-type, self-contained breathing apparatus).

* Substance reported to cause eye irritation or damage; may require eye protection.

Storage: Prior to working with this chemical you should be trained on its proper handling and storage. Store in tightly closed containers in a cool, well ventilated area away from strong oxidizers and other incompatible materials listed above.

Shipping: Dry hafnium powder requires a shipping label of: "Spontaneously Combustible." It falls in DOT Hazard Class 4.2 and Packing Group II. wet with 25% or more of water requires a shipping label of: "Flammable Solid." It falls in DOT Hazard Class 4.1 and Packing Group II.

Spill Handling: Evacuate persons not wearing protective equipment from area of spill or leak until clean-up is complete. Remove all ignition sources. Collect powdered material in the most convenient and safe manner and deposit in sealed containers. Ventilate area after clean-up is complete. It may be necessary to contain and dispose of this chemical as a hazardous waste. If material or contaminated runoff enters waterways, notify downstream users of potentially contaminated waters. Contact your Department of Environmental Protection or your regional office of the federal EPA for specific recommendations. If employees are required to clean-up spills, they must be properly trained and equipped. OSHA 1910.120(q) may be applicable.

Fire Extinguishing: This chemical is a combustible solid. Use dry chemical, carbon dioxide, water spray, or alcohol foam extinguishers. Poisonous gases are produced in fire. If material or contaminated runoff enters waterways, notify downstream users of potentially contaminated waters. Notify local health and fire officials and pollution control agencies. From a secure, explosion-proof location, use water spray to cool exposed containers. If cooling streams are ineffective (venting sound increases in volume and pitch, tank discolors, or shows any signs of deforming), withdraw immediately to a secure position. If employees are expected to fight fires, they must be trained and equipped in OSHA 1910.156.

Disposal Method Suggested: Recovery.

Halothane

Molecular Formula: $C_2HClBrF_3$

Common Formula: $CF_3CHBrCl$

Synonyms: Anestan; 2-Bromo-2-chloro-1,1,1-trifluoro-; 2-Bromo-2-chloro-1,1,1-trifluoroethane; Chalothane; Ethane, 2-bromo-2-chloro-1,1,1-thrifluoro-; Fluotane; Fluothane; Halotan; Halsan; Narcotane; Narcotann Ne-Spofa (Russian); 1,1,1-Trifluoro-2-bromo-2-chloroethane;

1,1,1-Trifluoro-2-chloro-2-bromoethane; 2,2,2-Trifluoro-1-chloro-1-bromoethane

CAS Registry Number: 151-67-1

RTECS Number: KH6475000

DOT ID: UN 1610 (halogenated, irritating liquids, n.o.s.)

Regulatory Authority

- Air Pollutant Standard Set (ACGIH)[1] (NIOSH) (DFG)[3]
- Canada, WHMIS, Ingredients Disclosure List

Cited in U.S. State Regulations: Alaska (G), New Hampshire (G), New Jersey (G), New York (G), Pennsylvania (G), West Virginia (G).

Description: Halothane, $CF_3CHBrCl$, is a highly volatile, colorless liquid with a sweetish odor. Odor threshold = 33 ppm. Boiling point = 50°C. Freezing/Melting point = -118°C. Practically insoluble in water. It is non-flammable.

Potential Exposure: Halothane is used as an inhalation anesthetic. It has been estimated that halothane accounts for two-thirds of all anesthesias.

Incompatibilities: May attack rubber and some plastics; sensitive to light. Light causes decomposition. May be stabilized with 0.01% thymol.

Permissible Exposure Limits in Air: There is no OSHA PEL. NIOSH recommends a (ceiling) 2 ppm (16.2 mg/m^3) (60-minute). *Note:* The NIOSH REL for exposure to waste anesthetic gas. ACGIH recommends a TWA value of 50 ppm (404 mg/m^3) but no STEL value. The DFG[3] has set an MAK value of 5 ppm (40 mg/m^3).

Determination in Air: Collection with charcoal tube;[2] workup with CS_2; Analysis by gas chromatography/flame ionization detection; OSHA Method #29.

Permissible Concentration in Water: No criteria set.

Routes of Entry: Inhalation, skin absorption, ingestion, skin and/or eye contact Passes through the skin.

Harmful Effects and Symptoms

Short Term Exposure: Contact can irritate the eyes and skin. Inhalation can irritate the respiratory tract causing coughing and wheezing. May affect the cardiovascular system with low blood pressure and abnormal heartbeat, and central nervous system. High levels of exposure may cause dizziness, lightheadedness, nausea, vomiting, and very high levels can cause unconsciousness. Medical observation is indicated. *Inhalation:* Exposure to 4,000 ppm can cause amnesia and impairment of manual dexterity. Exposure to 10,000 ppm can cause anesthesia. Liver impairment has occurred from use as an anesthetic. *Skin:* Repeated or prolonged application can destroy the skin. Halothane gives rise to only a very low incidence of postoperative nausea and is generally safe which accounts for its wide-spread use.

Long Term Exposure: May affect the liver and kidneys. Reported to have caused irregular menstrual period, headache, fatigue, and unconsciousness. Halothane has also been shown to cause birth defects in rats. May cause reproductive toxicity in humans. Experiments in laboratory animals have shown that continuous exposures of 15 – 50 ppm may cause liver damage.

Points of Attack: Liver, kidneys, skin, respiratory system, cardiovascular system, central nervous system, reproductive system.

Medical Surveillance: For those with frequent or potentially high exposure (half the TLV or greater) the following are recommended before beginning work and at regular times after that: Liver and kidney function tests. EKG. If symptoms develop or overexposure has occurred, repeat these tests. More than light alcohol consumption may exacerbate liver damage.

First Aid: If this chemical gets into the eyes, remove any contact lenses at once and irrigate immediately for at least 15 minutes, occasionally lifting upper and lower lids. Seek medical attention immediately. If this chemical contacts the skin, remove contaminated clothing and wash immediately with soap and water. Seek medical attention immediately. If this chemical has been inhaled, remove from exposure, begin rescue breathing (using universal precautions) if breathing has stopped and CPR if heart action has stopped. Transfer promptly to a medical facility. When this chemical has been swallowed, get medical attention. Give large quantities of water and induce vomiting. Do not make an unconscious person vomit.

Personal Protective Methods: Wear protective gloves and clothing to prevent any reasonable probability of skin contact. Safety equipment suppliers/manufacturers can provide recommendations on the most protective glove/clothing material for your operation. Polyvinyl Chloride is among the recommended protective materials. All protective clothing (suits, gloves, footwear, headgear) should be clean, available each day, and put on before work. Contact lenses should not be worn when working with this chemical. Wear splash-proof chemical goggles and face shield unless full facepiece respiratory protection is worn. Employees should wash immediately with soap when skin is wet or contaminated. Provide emergency showers and eyewash.

Respirator Selection: Where the potential exists for exposures over 2 ppm, use an MSHA/NIOSH approved supplied-air respirator with a full facepiece operated in the positive pressure mode or with a full facepiece, hood, or helmet in the continuous flow mode, or use an MSHA/NIOSH approved self-contained breathing apparatus with a full facepiece operated in pressure demand or other positive pressure mode.

Storage: Prior to working with this chemical you should be trained on its proper handling and storage. Store in tightly closed containers in a dark, cool, well ventilated area, away from direct sunlight, oxidizers.

Shipping: DOT label requirement of "Poison." Halothane is a halogenated, irritating liquid, n.o.s. and falls in UN/DOT Hazard Class 6.1 and Packing Group I.

Spill Handling: Evacuate and restrict persons not wearing protective equipment from area of spill or leak until cleanup is complete. Remove all ignition sources. Ventilate area of spill or leak. Absorb liquids in vermiculite, dry sand, earth, peat, carbon, or a similar material and deposit in sealed containers. Keep this chemical out of a confined space, such as a sewer, because of the possibility of an explosion, unless the sewer is designed to prevent the build-up of explosive concentrations. It may be necessary to contain and dispose of this chemical as a hazardous waste. If material or contaminated runoff enters waterways, notify downstream users of potentially contaminated waters. Contact your Department of Environmental Protection or your regional office of the federal EPA for specific recommendations. If employees are required to clean-up spills, they must be properly trained and equipped. OSHA 1910.120(q) may be applicable.

Fire Extinguishing: This chemical is a nonflammable liquid. Poisonous gases including hydrogen chloride, hydrogen fluoride, and hydrogen bromide are produced in fire. Use dry chemical, carbon dioxide, or foam extinguishers. If material or contaminated runoff enters waterways, notify downstream users of potentially contaminated waters. Notify local health and fire officials and pollution control agencies. From a secure, explosion-proof location, use water spray to cool exposed containers. If cooling streams are ineffective (venting sound increases in volume and pitch, tank discolors, or shows any signs of deforming), withdraw immediately to a secure position. If employees are expected to fight fires, they must be trained and equipped in OSHA 1910.156.

References

Sax, N. I., Ed., Dangerous Properties of Industrial Materials Report, 1, No. 5, 63 (1981)

New Jersey Department of Health and Senior Services, "Hazardous Substance Fact Sheet: Halothane," Trenton, NJ (June 1999)

New York State Department of Health, "Chemical Fact Sheet: Halothane," Albany, NY, Bureau of Toxic Substance Assessment (May 1986)

Helium

Molecular Formula: He

Synonyms: Helium, elemental

CAS Registry Number: 7440-59-1

RTECS Number: MH6520000

DOT ID: UN 1046 (compressed gas); UN 1963 (refrigerated liquid)

Cited in U.S. State Regulations: Alaska (G), California (A), Florida (G), Illinois (G), Maine (G), Massachusetts (G), New Hampshire (G), New Jersey (G), Pennsylvania (G), Rhode Island (G).

Description: Helium, He, is a colorless, odorless and tasteless gas. It is non-flammable. Boiling point = -269°C. Freezing/Melting point = -272°C. Soluble in water.

Potential Exposure: It is used in weather balloons and in welding gases. Liquid helium is used as a closed system cooling agent.

Permissible Exposure Limits in Air: No occupational exposure limits have been established for this substance. Large amounts of helium will, however, decrease the amount of available oxygen. Oxygen content should never be below 19%.

Permissible Concentration in Water: No criteria set.

Routes of Entry: Inhalation.

Harmful Effects and Symptoms

Short Term Exposure: Helium can affect you when breathed in. Exposure to high levels can cause you to feel dizzy and lightheaded. Very high levels can cause you to pass out and even die due to suffocation from lack of oxygen. Contact with liquid helium can cause frostbite.

Long Term Exposure: Unknown at this time.

First Aid: If frostbite has occurred, seek medical attention immediately; *do not* rub the affected areas or flush them with water. In order to prevent further tissue damage, *do not* attempt to remove frozen clothing from frostbitten areas. If frostbite has NOT occurred, immediately and thoroughly wash contaminated skin with soap and water. Seek medical attention promptly. *Breathing:* Remove the person from exposure. Begin rescue breathing if breathing has stopped and CPR if heart action has stopped. Transfer promptly to a medical facility.

Personal Protective Methods: *Clothing:* Where exposure to cold equipment, vapors, or liquid may occur, employees should be provided with special clothing designed to prevent the freezing of body tissues. All protective clothing (suits, gloves, footwear, headgear) should be clean, available each day, and put on before work. *Eye Protection:* Wear splash-proof chemical goggles and face shield when working with liquid, unless full facepiece respiratory protection is worn.

Respirator Selection: Exposure to helium gas is dangerous because it can replace oxygen and lead to suffocation. Only MSHA/NIOSH approved self-contained breathing apparatus with a full facepiece operated in the positive pressure mode should be used in oxygen deficient environments.

Storage: Prior to working with this chemical you should be trained on its proper handling and storage. Store in tightly closed containers in a cool, well-ventilated area. Liquid helium should be stored and transferred under positive pressure to prevent infiltration of air and other gasses. Procedures for the handling, use and storage of cylinders should be in compliance with OSHA 1910.101 and 1910.169 as with the recommendations of the Compressed Gas Association.

Shipping: Compressed helium requires a shipping label of: "Nonflammable Gas." It falls in DOT Hazard Class 2.2. Liquid (cryogenic) helium requires a shipping label of: "Nonflammable Gas" also. It falls in DOT Hazard Class 2.2.

Spill Handling: If liquid helium is spilled or leaked, take the following steps: Restrict persons not wearing protective equipment from area of spill or leak until cleanup is complete. Stop the leak or move container to a safe area and allow the liquid to evaporate. If helium gas is leaked, take the following steps: Restrict persons not wearing protective equipment from area of teak until clean-up is complete. Ventilate area of leak to disperse the gas. Stop flow of gas. If source of leak is a cylinder and the leak cannot be stopped in place, remove the leaking cylinder to a safe place in the open air, and repair leak or allow cylinder to empty.

Fire Extinguishing: Helium gas may burn but does not readily ignite. Stop flow of gas. Use extinguisher suitable for surrounding fire. Containers may explode in fire. Storage containers and parts of containers may rocket great distances, in many directions. From a secure, explosion-proof location, use water spray to cool exposed containers. If cooling streams are ineffective (venting sound increases in volume and pitch, tank discolors, or shows any signs of deforming), withdraw immediately to a secure position. If employees are expected to fight fires, they must be trained and equipped in OSHA 1910.156.

References

New Jersey Department of Health and Senior Services, "Hazardous Substance Fact Sheet: Helium," Trenton, NJ (August 1998)

Hematite (Underground Mining)

Molecular Formula: Fe_2O_3

Synonyms: Bloodstone; Haematite; Hematite, red; Iron ore; Iron oxide; Red iron ore

CAS Registry Number: 1317-60-8; 1309-37-1 (ferric oxide, Fe_2O_3)

RTECS Number: MH7600000

Regulatory Authority

- Carcinogen (NTP)[10] (OSHA)
- Air Pollutant Standard Set (ACGIH)[1] (DFG)[3] (HSE)[33] (former USSR)[43] (OSHA)[58]

Cited in U.S. State Regulations: California (G), Illinois (G), Maine (G), New Jersey (G), Pennsylvania (G).

Description: Hematite is a noncombustible, black to black-red or brick-red mineral (iron ore) composed mainly of ferric oxide, Fe_2O_3. See also *Iron Oxide* entry.

Potential Exposure: It, as an iron ore composed mainly of ferric oxide, is a major source of iron and is used as a pigment for rubber, paints, paper, linoleum, ceramics, dental restoratives, and as a polishing agent for glass and precious metals. It is also used in electrical resistors, semiconductors, magnets, and as a catalyst. Human exposure to hematite from underground hematite mining is principally through inhalation and/or ingestion of dusts. No estimates are available concerning the number of underground miners exposed.

Incompatibilities: Contact with hydrogen peroxide, ethylene oxide, calcium hypochlorite will cause explosion. Violent reaction with powdered aluminum, hydrazine, hydrogen trisulfide.

Permissible Exposure Limits in Air: ACGIH recommends a TWA of 2 ppm (5 mg/m^3) for iron oxide fume. OSHA[58] has set a TWA of 10 mg/m^3. HSE[33] has set an 8-hour TWA of 5 mg/m^3 and an STEL of 10 mg/m^3. DFG[3] has set an MAK of 6 mg/m^3 for iron oxide fine dust. The former USSR-UN EP/IRPTC project[43] has set a MAC in workplace air of 4 mg/m^3.

Permissible Concentration in Water: No criteria set.

Routes of Entry: Dust inhalation.

Harmful Effects and Symptoms

Long Term Exposure: There is sufficient evidence for the carcinogenicity in humans of underground hematite mining (with exposure to radon). Underground hematite miners have a high incidence of lung cancer, whereas surface hematite miners do not. It is not known whether this excess risk may be due to hematite; to radon (a known lung carcinogen); to inhalation of ferric oxide or silica; or to a combination of these or other factors. Some studies of metal workers exposed to ferric oxide dusts have shown an increased incidence of lung cancer, while other studies have not. The influence of factors in the workplace, other than ferric oxide, cannot be eliminated.

First Aid: If this chemical gets into the eyes, remove any contact lenses at once and irrigate immediately for at least 15 minutes, occasionally lifting upper and lower lids. Seek medical attention immediately. If this chemical contacts the skin, remove contaminated clothing and wash immediately with soap and water. Seek medical attention immediately. If this chemical has been inhaled, remove from exposure, begin rescue breathing (using universal precautions) if breathing has stopped and CPR if heart action has stopped. Transfer promptly to a medical facility. When this chemical has been swallowed, get medical attention. Give large quantities of water and induce vomiting. Do not make an unconscious person vomit.

Personal Protective Methods: Wear protective gloves and clothing to prevent any reasonable probability of skin contact. Safety equipment suppliers/manufacturers can provide recommendations on the most protective glove/clothing material for your operation. All protective clothing (suits, gloves, footwear, headgear) should be clean, available each day, and put on before work. Contact lenses should not be worn when working with this chemical. Wear dust-proof chemical goggles and face shield unless full facepiece respiratory protection is worn. Employees should wash immediately with soap when skin is wet or contaminated. Provide emergency showers and eyewash.

Respirator Selection: Where the potential for exposure to this chemical, use a MSHA/NIOSH approved supplied-air respirator with a full facepiece operated in the positive pressure mode or with a full facepiece, hood, or helmet in the continuous flow mode, or use a MSHA/NIOSH approved self-contained breathing apparatus with a full facepiece operated in pressure-demand or other positive pressure mode.

Storage: Prior to working with this chemical you should be trained on its proper handling and storage. Store in tightly closed containers in a cool, well ventilated area. Where possible, automatically transfer material from other storage containers to process containers A regulated, marked area should be established where this chemical is handled, used, or stored in compliance with OSHA standard 1910.1045.

Spill Handling: Evacuate persons not wearing protective equipment from area of spill or leak until clean-up is complete. Remove all ignition sources. Collect powdered material in the most convenient and safe manner and deposit in sealed containers. Ventilate area after clean-up is complete. It may be necessary to contain and dispose of this chemical as a hazardous waste. If material or contaminated runoff enters waterways, notify downstream users of potentially contaminated waters. Contact your Department of Environmental Protection or your regional office of the federal EPA for specific recommendations. If employees are required to clean-up spills, they must be properly trained and equipped. OSHA 1910.120(q) may be applicable.

Fire Extinguishing: This chemical is a noncombustible solid. Use extinguishing agents suitable for surrounding fires. Poisonous gases are produced in fire. If material or contaminated runoff enters waterways, notify downstream users of potentially contaminated waters. Notify local health and fire officials and pollution control agencies. If employees are expected to fight fires, they must be trained and equipped in OSHA 1910.156.

Heptachlor

Molecular Formula: $C_{10}H_5Cl_7$

Synonyms: Aahepta; Agroceres; Arbinex 30TN; 3-Chlorochlordene; Drinox; E 3314; ENT15,152; Eptacloro (Italian); 1,4,5,6,7,8,8-Eptacloro-3a, 4,7,7a-tetraidro-4,7-endo-metano-indene (Italian); GPKH; Hepachloor-3a,4,7,7a-tetrahydro-4,7-endo-methano-indeen (Dutch); Hepta; Heptachlorane; Heptachlore; Heptachlore (French); 3,4,5,6,7,8,8a-Heptachlorodicyclopentadiene; 3,4,5,6,7,8,8-Heptachlorodicyclopentadiene; 1,4,5,6,7,10,10-Heptachloro-4,7,8,9-tetrahydro-4,7-endomethyleneindene; 1(3a),4,5,6,7,8,8-Heptachloro-3a(1),4,7,7a-tetrahydro-4,7-methanoindene; 1,4,5,6,7,8,8a-Heptachloro-3a,4,7,7a-tetrahydro-4,7-methanoindene; 1,4,5,6,7,8,8-Heptachloro-3a,4,7,7a-tetrahydro-4,7-methano-1H-indene; 1,4,5,6,7,8,8-Heptachloro-3a,4,7,7a-tetrahydro-4,7-methanoindene; 1,4,5,6,7,8,8-Heptachloro-3a,4,7,7a-tetrahydro-4,7-methanol-1H-indene; 1,4,5,6,7,8,8-Heptachloro-3a,4,7,7,7a-tetrahydro-4,7-methelene indene; 1,4,5,6,7,8,8-Heptachlor-3a,4,7,7,7a-tetrahydro-4,7-endo-methano inden (German); Heptacloro (Spanish); Heptagran; Heptamul; Heptox; Indene; 4,7-Methanoindene, 1,4,5,6,7,8,8-Heptachloro-3a,4,7,7a-tetrahydro-; NCI-C00180; Rhodiachlor; Velsicol 104

CAS Registry Number: 76-44-8

RTECS Number: PC0700000

DOT ID: UN 2761

EEC Number: 602-046-00-2

Regulatory Authority

- Air Pollutant Standard Set (ACGIH)[1] (DFG)[3] (HSE)[33] (OSHA)[58] (Argentina)[35] (former USSR)[43] (Several States)[60]
- Banned or Severely Restricted (Many, Many Countries) (UN)[13]
- Carcinogen (Animal Positive) (IARC) (NCI)[9]
- CLEAN AIR ACT: Hazardous Air Pollutants (Title I, Part A, Section 112)
- CLEAN WATER ACT: Section 311 Hazardous Substances/ RQ 40CFR117.3 (same as CERCLA, see below); 40CFR 423, Appendix A, Priority Pollutants; Section 313 Water Priority Chemicals (57FR41331, 9/9/92); 40CFR401.15 Section 307 Toxic Pollutants
- EPA HAZARDOUS WASTE NUMBER (RCRA No.): P059; D031
- RCRA, 40CFR261, Appendix 8 Hazardous Constituents
- RCRA Toxicity Characteristic (Section 261.24), Maximum Concentration of Contaminants, regulatory level, 0.008 mg/l
- RCRA 40CFR268.48; 61FR15654, Universal Treatment Standards: Wastewater (mg/l), 0.0012; Nonwastewater (mg/kg), 0.066
- RCRA 40CFR264, Appendix 9; TSD Facilities Ground Water Monitoring List. Suggested test method(s) (PQL μg/l): 8080 (0.05); 8270 (10)
- SAFE DRINKING WATER ACT: MCL, 0.0004 mg/l; MCLG, zero
- SUPERFUND/EPCRA 40CFR302.4 Reportable Quantity (RQ): CERCLA, 1 lb (0.454 kg)
- EPCRA Section 313 Form R *de minimis* concentration reporting level: 1.0%
- U.S. DOT Regulated Marine Pollutant (49CFR172.101, Appendix B), severe pollutant
- Canada, WHMIS, Ingredients Disclosure List

Cited in U.S. State Regulations: Alaska (G), Arizona (W), California (W), Connecticut (A), Florida (G, A), Illinois (G, W), Kansas (G, A, W), Louisiana (G), Maine (G, W), Massachusetts (G, A), Michigan (G), Minnesota (W), Nevada (A), New Hampshire (G), New Jersey (G, Dept. of Envir. Prot., Not Dept. of Health), New York (A), North Dakota (A), Oklahoma (G), Pennsylvania (G, A), Rhode Island (G), South Carolina (A), Vermont (G), Virginia (G, A), Washington (G), West Virginia (G), Wisconsin (G).

Description: Heptachlor, $C_{10}H_5Cl_7$, is a white, sand-like solid with a camphor-like odor. Boiling point = 135 – 145°C. Freezing/Melting point = 95 – 96°C (pure) and 46 – 74°C (technical product). Hazard Identification (based on NFPA-704 M Rating System): Health 3, Flammability 0, Reactivity 0. Insoluble in water.

Potential Exposure: Those involved in the manufacture, formulation and application of this insecticide. Registration of heptachlor-containing pesticides has been canceled by the U.S. EPA with the exception of its use for termite control outside of dwellings by in-ground (subsurface) insertion. Infants have been exposed to heptachlor and heptachlor epoxide through mothers' milk, cows' milk, and commercially prepared baby foods. It appears that infants raised on mothers' milk run a greater risk of ingesting heptachlor epoxide than if they were fed cows' milk and/or commercially prepared baby food. Persons living and working in or near heptachlor treated areas have a particularly high inhalation exposure potential.

Incompatibilities: Reacts with strong oxidizers. Attacks metal. Forms hydrogen chloride gas with iron and rust above 74°C.

Permissible Exposure Limits in Air: The OSHA PEL is 0.5 mg/m^3 TWA, with the notation "Skin" indicating the possibility of cutaneous absorption. NIOSH and ACGIH recommend the same airborne limit as OSHA. This same TWA has been set by the Argentine[35] by Germany[3] and the U.K.,[33] but with different STEL's in each case: 1.5 mg/m^3 in the Argentine, 2.0 in the U.K. and 5.0 in Germany. The NIOSH IDLH level is (Ca) 35 mg/m^3. The former USSR has set[35][43] a much lower limit of 0.01 mg/m^3 in workplace air and values in ambient air of residential areas of 0.001 mg/m^3 on a once daily basis and 0.0002 mg/m^3 on a daily average basis. A number of states have set guidelines of standards for heptachlor in ambient air[60] ranging from 0.0068 μg/m^3 (Massachusetts) to 0.18 μg/m^3 (Pennsylvania) to 1.19 μg/m^3 (Kansas) to 1.7 μg/m^3 (New York) to 2.5 μg/m^3 (Connecticut and South Carolina) to 5.0 μg/rn^3 (Florida and North Dakota) to 8.0 μg/m^3 (Virginia) to 12.0 μg/m^3 (Nevada).

Determination in Air: Collection by Chromosorb tube – 102; Toluene; Gas chromatography/Electrochemical detection; NIOSH II(5), Method #S287.

Permissible Concentration in Water: To protect freshwater aquatic life: 0.0038 μg/l as a 24 hour average, never to exceed 0.52 μg/l. To protect saltwater aquatic life: 0.0036 μg/l as a 24 hour average, never to exceed 0.053 μg/l. To protect human health: preferably zero. An additional lifetime cancer risk of 1 in 100,000 is imposed by a concentration of 2.78 ng/l (0.00278 μg/l).[6] The USEPA has set health advisories[47] for heptachlor and heptachlor epoxide. The lifetime health advisory is 17.5 μg/l for heptachlor and 0.4 μg/l for heptachlor epoxide. Mexico[35] has set limits of 0.018 mg/l (18 μg/l) for both heptachlor and heptachlor epoxide in drinking water, 0.2 μg/l for heptachlor in coastal watersand 2.0 μg/l for heptachlor in estuaries. The former USSR has set a limit of 50 μg/l of heptachlor in water bodies used for domestic purposes.[35][43] WHO[35] has set a limit of 0.1 μg/l in drinking water for heptachlor. Several states have set guidelines and standards for heptachlor and heptachlor epoxide in drinking water.[61] Illinois has set a standard of 0.1 μg/l for both heptachlor and heptachlor epoxide. Guidelines have been set for heptachlor ranging from 0.02 μg/l (California) to 0.1 μg/l (Minnesota) to 0.104 μg/l (Kansas) to 0.23 μg/l (Maine) to 0.50 μg/l (Arizona). Guidelines have been set for heptachlor epoxide in drinking water[61] ranging from 0.006 μg/l (Kansas and Minnesota) to 0.10 μg/l (California). The EPA has recently[62] proposed drinking water maximum contaminant levels for heptachlor at 0.4 μg/l and heptachlor epoxide at 0.2 μg/l.

Determination in Water: Methylene chloride extraction followed by gas chromatography with electron capture or halogen specific detection (EPA Method 608) or gas chromatography plus mass spectrometry (EPA Method 625).

Routes of Entry: Inhalation, skin absorption, ingestion, eye and/or skin contact.

Harmful Effects and Symptoms

Short Term Exposure: Heptachlor can cause a feeling of anxiety, headache, dizziness, weakness, a sensation of "pins and needles" on the skin, and muscle twitching. Heptachlor has been demonstrated to be highly toxic to aquatic life, to persist for prolonged periods in the environment, to bioconcentrate in organisms at various trophic levels, and to exhibit carcinogenic activity in mice. Exposure symptoms in animals include tremors, convulsions and liver damage. The principal metabolite of heptachlor, heptachlor epoxide is more acutely toxic than heptachlor.

Long Term Exposure: High or repeated exposure may cause brain damage with personality changes, decreased memory, difficult coordination and concentration. Higher levels can cause tremor, seizures, unconsciousness and death. This substance is possibly carcinogenic to humans. There is limited evidence that heptachlor may damage the developing fetus. May cause liver and kidney damage.

Points of Attack: Central nervous system, liver. Cancer site in animals: liver cancer.

Medical Surveillance: Consider the points of attack in preplacement and periodic physical examinations. Liver and kidney function tests. Evaluation for brain effects.

First Aid: If this chemical gets into the eyes, remove any contact lenses at once and irrigate immediately for at least 15 minutes, occasionally lifting upper and lower lids. Seek medical attention immediately. If this chemical contacts the skin, remove contaminated clothing and wash immediately with soap and water. Seek medical attention immediately. If this chemical has been inhaled, remove from exposure, begin rescue breathing (using universal precautions) if breathing has stopped and CPR if heart action has stopped. Transfer promptly to a medical facility. When this chemical has been swallowed, get medical attention. Give large quantities of water and induce vomiting. Do not make an unconscious person vomit.

Personal Protective Methods: Wear protective gloves and clothing to prevent any reasonable probability of skin contact. Safety equipment suppliers/manufacturers can provide recommendations on the most protective glove/clothing material for your operation. All protective clothing (suits, gloves, footwear, headgear) should be clean, available each day, and put on before work. Contact lenses should not be worn when working with this chemical. Wear dust-proof chemical goggles and face shield unless full facepiece respiratory protection is worn. Employees should wash immediately with soap when skin is wet or contaminated. Provide emergency showers and eyewash.

Respirator Selection: NIOSH: *At any concentrations above the NIOSH REL, or where there is no REL, at any detectable concentration:* SCBAF:PD,PP (any self-contained breathing apparatus that has a full facepiece and is operated in a pressure-demand or other positive-pressure mode); or SAF:PD,PP: ASCBA (any supplied-air respirator that has a full facepiece and is operated in a pressure-demand or other positive-pressure mode in combination with an auxiliary self-contained breathing apparatus operated in a pressure-demand or other positive pressure mode). *Escape:* GMFOVHiE [any air-purifying, full-facepiece respirator (gas mask) with a chin-style, front- or back-mounted organic vapor canister having a high-efficiency particulate filter]; or SCBAE (any appropriate escape-type, self-contained breathing apparatus).

Storage: Prior to working with this chemical you should be trained on its proper handling and storage. Store in tightly closed containers in a cool, well ventilated area. Protect storage containers from physical damage. Where possible, automatically pump liquid from drums or other storage containers to process containers. A regulated, marked area should be established where this chemical is handled, used, or stored in compliance with OSHA standard 1910.1045.

Shipping: While not specifically cited by DOT,[19] heptachlor falls under organochloride pesticides, solid, toxic, n.o.s. This compound requires a shipping label of: "Poison." It falls in DOT Hazard Class 6.1 and Packing Group II.

Spill Handling: Evacuate persons not wearing protective equipment from area of spill or leak until clean-up is complete. Remove all ignition sources. Collect powdered material in the most convenient and safe manner and deposit in sealed containers. Ventilate area after clean-up is complete. It may be necessary to contain and dispose of this chemical as a hazardous waste. If material or contaminated runoff enters waterways, notify downstream users of potentially contaminated waters. Contact your Department of Environmental Protection or your regional office of the federal EPA for specific recommendations. If employees are required to clean-up spills, they must be properly trained and equipped. OSHA 1910.120(q) may be applicable.

Fire Extinguishing: Use dry chemical, carbon dioxide, water spray, or foam extinguishers. Poisonous gases including chlorine are produced in fire. If material or contaminated runoff enters waterways, notify downstream users of potentially contaminated waters. Notify local health and fire officials and pollution control agencies. From a secure, explosion-proof location, use water spray to cool exposed containers. If cooling streams are ineffective (venting sound increases in volume and pitch, tank discolors, or shows any signs of deforming), withdraw immediately to a secure position. If employees are expected to fight fires, they must be trained and equipped in OSHA 1910.156.

Disposal Method Suggested: Incineration (1,500°F, 0.5 sec minimum for primary combustion; 3,200°F, 1.0 sec for secondary combustion) with adequate scrubbing and ash disposal facilities. Consult with environmental regulatory agencies for guidance on acceptable disposal practices. Generators of waste containing this contaminant (≥100 kg/mo) must conform with EPA regulations governing storage, transportation, treatment, and waste disposal. In accordance with 40CFR165 recommendations for the disposal of pesticides and pesticide containers. Must be disposed properly by following package label directions or by contacting your state pesticide or environmental control agency or by contacting your regional EPA office.

References

U.S. Environmental Protection Agency, Heptachlor: Ambient Water Quality Criteria, Washington, DC (1980)

U.S. Environmental Protection Agency, Heptachlor, Health and Environmental Effects Profile No. 108, Office of Solid Waste, Washington, DC (April 30, 1980)

U.S. Environmental Protection Agency, Heptachlor Epoxide, Health and Environmental Effects Profile No. 109, Office of Solid Waste, Washington, DC (April 30, 1980)

Sax, N. I., Ed., Dangerous Properties of Industrial Materials Report, 1, No. 8, 76–78 (1981) and 6, No. 5, 16–49 (1986)

U.S. Public Health Service, "Toxicological Profile for Heptachlor/Heptachlor Epoxide," Atlanta, Georgia, Agency for Toxic Substances and Disease Registry (Oct. 1987)

New Jersey Department of Health and Senior Services, "Hazardous Substance Fact Sheet, Heptachlor," Trenton, NJ (March, 1998)

Heptachlor Epoxide

Molecular Formula: $C_{10}H_5Cl_7O$

Synonyms: ENT 25,584; Epoxyheptachlor; HCE; 1,4,5,6,7,8,8-Heptachloro-2,3-epoxy-2,3,3a,4,7,7a-hexahydro-4,7-methanoindene; 1,4,5,6,7,8,8-Heptachloro-2,3-epoxy-3a,4,7,7a-tetrahydro-4,7-methanoindan; 2,3,5,6,7,7-Heptachloro-1a,1b,5,5a,6,6a-hexahydro-2,5-methano-2H-indeno (1,2-b)oxirene; Heptaclorepoxido (Spanish); 4,7-Methanoindan, 1,4,5,6,7,8,8-heptachloro-2,3-epoxy-3a, 4,7,7a-tetrahydro; Velsicol 53-CS-17

CAS Registry Number: 1024-57-3

RTECS Number: PB9450000

Regulatory Authority

- Air Pollutant Standard Set (ACGIH)[1] (DFG)[3] (HSE)[33] (OSHA)[58] (Argentina)[35] (former USSR)[43] (Several States)[60]

- Banned or Severely Restricted (Many Countries) (UN)[13]
- Carcinogen (Animal Positive) (IARC) (NCI)[9]
- CLEAN WATER ACT: 40CFR423, Appendix A, Priority Pollutants; 40CFR401.15 Section 307 Toxic Pollutants as hexachlorocyclohexane
- EPA HAZARDOUS WASTE NUMBER (RCRA No.): D031
- RCRA, 40CFR261, Appendix 8 Hazardous Constituents., waste number not listed
- RCRA Toxicity Characteristic (Section 261.24), Maximum Concentration of Contaminants, regulatory level, 0.008 mg/l
- RCRA 40CFR268.48; 61FR15654, Universal Treatment Standards: Wastewater (mg/l), 0.016; Nonwastewater (mg/kg), 0.066
- RCRA 40CFR264, Appendix 9; TSD Facilities Ground Water Monitoring List. Suggested test method(s) (PQL μg/l): 8080 (1); 8270 (10)
- SAFE DRINKING WATER ACT: MCL, 0.0002 mg/l; MCLG, zero
- SUPERFUND/EPCRA 40CFR302.4 Reportable Quantity (RQ): CERCLA, 1 lb (0.454 kg)

Cited in U.S. State Regulations: Alaska (G), Arizona (W), California (W), Connecticut (A), Florida (G, A), Illinois (G, W), Kansas (G, A, W), Louisiana (G), Maine (G, W), Massachusetts (G, A), Michigan (G), Minnesota (W), Nevada (A), New Hampshire (G), New Jersey (G, Dept. of Envir. Prot., Not Dept. of Health), New York (A), North Dakota (A), Oklahoma (G), Pennsylvania (G, A), Rhode Island (G), South Carolina (A), Vermont (G), Virginia (G, A), Washington (G), West Virginia (G), Wisconsin (G).

Description: Heptachlor epoxide, $C_{10}H_5Cl_7O$, is a solid. Freezing/Melting point = 160 – 162°C. Hazard Identification (based on NFPA-704 M Rating System): Health 3, Flammability 1, Reactivity 0. Soluble in water. It is an oxidation product of heptachlor formed by plants and animals, including humans, after exposure to heptachlor. It is also present as a contaminant in heptachlor.

Potential Exposure: Those involved in the manufacture, formulation and application of this insecticide. Infants have been exposed to heptachlor and heptachlor epoxide through mothers' milk, cows' milk, and commercially prepared baby foods. It appears that infants raised on mothers' milk run a greater risk of ingesting heptachlor epoxide than if they were fed cows' milk and/or commercially prepared baby food. Persons living and working in or near heptachlor treated areas have a particularly high inhalation exposure potential.

Incompatibilities: Melted heptachlor with iron and rust.

Permissible Exposure Limits in Air: The ACGIH[1] and OSHA[58] have set a TWA of 0.5 mg/m^3 with the notation "Skin" indicating the possibility of cutaneous absorption. This same TWA has been set by the Argentine[35] by Germany[3] and the U.K.,[33] but with different STEL's in each case: 1.5 mg/m^3 in the Argentine, 2.0 in the U.K. and 5.0 in Germany. The NIOSH IDLH level is 100 mg/m^3. The former USSR has set[35][43] a much lower limit of 0.01 mg/m^3 in workplace air and values in ambient air of residential areas of 0.001 mg/m^3 on a once daily basis and 0.0002 mg/m^3 on a daily average basis. A number of states have set guidelines of standards for heptachlor in ambient air[60] ranging from 0.0068 μg/m^3 (Massachusetts) to 0.18 μg/m^3 (Pennsylvania) to 1.19 μg/m^3 (Kansas) to 1.7 μg/m^3 (New York) to 2.5 μg/m^3 (Connecticut and South Carolina) to 5.0 μg/rn^3 (Florida and North Dakota) to 8.0 μg/m^3 (Virginia) to 12.0 μg/m^3 (Nevada).

Determination in Air: Collection by Chromosorb tube – 102; Toluene; Gas chromatography/Electrochemical detection; NIOSH II(5), Method #S287.

Permissible Concentration in Water: To protect freshwater aquatic life: 0.0038 μg/l as a 24 hour average, never to exceed 0.52 μg/l. To protect saltwater aquatic life: 0.0036 μg/l as a 24 hour average, never to exceed 0.053 μg/l. To protect human health: preferably zero. An additional lifetime cancer risk of 1 in 100,000 is imposed by a concentration of 2.78 ng/l (0.00278 μg/l).[6] The USEPA has set health advisories[47] for heptachlor and heptachlor epoxide. The lifetime health advisory is 17.5 μg/l for heptachlor and 0.4 μg/l for heptachlor epoxide. Mexico[35] has set limits of 0.018 mg/l (18 μg/l) for both heptachlor and heptachlor epoxide in drinking water, 0.2 μg/l for heptachlor in coastal watersand 2.0 μg/l for heptachlor in estuaries. The former USSR has set a limit of 50 μg/l of heptachlor in water bodies used for domestic purposes.[35][43] WHO[35] has set a limit of 0.1 μg/l in drinking water for heptachlor. Several states have set guidelines and standards for heptachlor and heptachlor epoxide in drinking water.[61] Illinois has set a standard of 0.1 μg/l for both heptachlor and heptachlor epoxide. Guidelines have been set for heptachlor ranging from 0.02 μg/l (California) to 0.1 μg/l (Minnesota) to 0.104 μg/l (Kansas) to 0.23 μg/l (Maine) to 0.50 μg/l (Arizona). Guidelines have been set for heptachlor epoxide in drinking water[61] ranging from 0.006 μg/l (Kansas and Minnesota) to 0.10 μg/l (California). The EPA has recently[62] proposed drinking water maximum contaminant levels for heptachlor at 0.4 μg/l and heptachlor epoxide at 0.2 μg/l.

Routes of Entry: Inhalation, skin absorption, ingestion, eye and/or skin contact.

Harmful Effects and Symptoms

Short Term Exposure: Heptachlor can cause a feeling of anxiety, headache, dizziness, weakness, a sensation of "pins and needles" on the skin, and muscle twitching. Heptachlor has been demonstrated to be highly toxic to aquatic life, to persist for prolonged periods in the environment, to bioconcentrate in organisms at various trophic levels, and to exhibit carcinogenic activity in mice. Exposure symptoms in animals include tremors, convulsions and liver damage. The principal metabolite of heptachlor, heptachlor epoxide is more acutely toxic than heptachlor.

Long Term Exposure: High or repeated exposure may cause brain damage with personality changes, decreased memory, difficult coordination and concentration. Higher levels can cause tremor, seizures, unconsciousness and death. This substance is possibly carcinogenic to humans. There is limited evidence that heptachlor may damage the developing fetus. May cause liver and kidney damage.

Points of Attack: Central nervous system, liver. Cancer site in animals: liver cancer.

Medical Surveillance: Consider the points of attack in preplacement and periodic physical examinations. Liver and kidney function tests. Evaluation for brain effects.

First Aid: If this chemical gets into the eyes, remove any contact lenses at once and irrigate immediately for at least 15 minutes, occasionally lifting upper and lower lids. Seek medical attention immediately. If this chemical contacts the skin, remove contaminated clothing and wash immediately with soap and water. Speed in removing material from skin is of extreme importance. Shampoo hair promptly if contaminated. Seek medical attention immediately. If this chemical has been inhaled, remove from exposure, begin rescue breathing (using universal precautions) if breathing has stopped and CPR if heart action has stopped. Transfer promptly to a medical facility. When this chemical has been swallowed, get medical attention. Give large quantities of water and induce vomiting. Do not make an unconscious person vomit.

Personal Protective Methods: Wear protective gloves and clothing to prevent any reasonable probability of skin contact. Safety equipment suppliers/manufacturers can provide recommendations on the most protective glove/clothing material for your operation. All protective clothing (suits, gloves, footwear, headgear) should be clean, available each day, and put on before work. Contact lenses should not be worn when working with this chemical. Wear dust-proof chemical goggles and face shield unless full facepiece respiratory protection is worn. Employees should wash immediately with soap when skin is wet or contaminated. Provide emergency showers and eyewash.

Respirator Selection: NIOSH: *At any concentrations above the NIOSH REL, or where there is no REL, at any detectable concentration:* SCBAF:PD,PP (any self-contained breathing apparatus that has a full facepiece and is operated in a pressure-demand or other positive-pressure mode); or SAF:PD,PP:ASCBA (any supplied-air respirator that has a full facepiece and is operated in a pressure-demand or other positive-pressure mode in combination with an auxiliary self-contained breathing apparatus operated in a pressure-demand or other positive pressure mode). *Escape:* GMFOVHiE [any air-purifying, full-facepiece respirator (gas mask) with a chin-style, front- or back-mounted organic vapor canister having a high-efficiency particulate filter]; or SCBAE (any appropriate escape-type, self-contained breathing apparatus).

Storage: Prior to working with this chemical you should be trained on its proper handling and storage. Store in tightly closed containers in a cool, well ventilated area away from ferrous metals. A regulated, marked area should be established where this chemical is handled, used, or stored in compliance with OSHA standard 1910.1045.

Shipping: While not specifically cited by DOT,[19] heptachlor falls under organochloride pesticides, solid, toxic, n.o.s. This compound requires a shipping label of: "Poison." It falls in DOT Hazard Class 6.1 and Packing Group II.

Spill Handling: Evacuate persons not wearing protective equipment from area of spill or leak until clean-up is complete. Remove all ignition sources. Collect powdered material in the most convenient and safe manner and deposit in sealed containers. Ventilate area after clean-up is complete. It may be necessary to contain and dispose of this chemical as a hazardous waste. If material or contaminated runoff enters waterways, notify downstream users of potentially contaminated waters. Contact your Department of Environmental Protection or your regional office of the federal EPA for specific recommendations. If employees are required to clean-up spills, they must be properly trained and equipped. OSHA 1910.120(q) may be applicable.

Fire Extinguishing: This chemical is a combustible solid. Use dry chemical, carbon dioxide, water spray, or alcohol foam extinguishers. Poisonous gases are produced in fire. If material or contaminated runoff enters waterways, notify downstream users of potentially contaminated waters. Notify local health and fire officials and pollution control agencies. From a secure, explosion-proof location, use water spray to cool exposed containers. If cooling streams are ineffective (venting sound increases in volume and pitch, tank discolors, or shows any signs of deforming), withdraw immediately to a secure position. If employees are expected to fight fires, they must be trained and equipped in OSHA 1910.156.

Disposal Method Suggested: Incineration (1,500°F, 0.5 sec minimum for primary combustion; 3,200°F, 1.0 sec for secondary combustion) with adequate scrubbing and ash disposal facilities. Consult with environmental regulatory agencies for guidance on acceptable disposal practices. Generators of waste containing this contaminant (≥100 kg/mo) must conform with EPA regulations governing storage, transportation, treatment, and waste disposal. In accordance with 40CFR165 recommendations for the disposal of pesticides and pesticide containers. Must be disposed properly by following package label directions or by contacting your state pesticide or environmental control agency or by contacting your regional EPA office.

References

U.S. Environmental Protection Agency, Heptachlor: Ambient Water Quality Criteria, Washington, DC (1980)

U.S. Environmental Protection Agency, Heptachlor, Health and Environmental Effects Profile No. 108, Office of Solid Waste, Washington, DC (April 30, 1980)

U.S. Environmental Protection Agency, Heptachlor Epoxide, Health and Environmental Effects Profile No. 109, Office of Solid Waste, Washington, DC (April 30, 1980)

Sax, N. I., Ed., Dangerous Properties of Industrial Materials Report, 1, No. 8, 76–78 (1981) and fi. No. 5, 16–49 (1986)

U.S. Public Health Service, "Toxicological Profile for Heptachlor/Heptachlor Epoxide," Atlanta, Georgia, Agency for Toxic Substances and Disease Registry (Oct. 1987)

n-Heptane

Molecular Formula: C_7H_{16}

Common Formula: $CH_3(CH_2)_5CH_3$

Synonyms: Dipropal methane; Dipropyl methane; Dipropylmethane; Eptani (Italian); Heptan (Polish); *n*-Heptane; Heptanen (Dutch); Heptyl hydride; Normal heptane; Skelly-Solve C

CAS Registry Number: 142-82-1

RTECS Number: MI7700000

DOT ID: UN 1206

EEC Number: 601-008-00-2

Regulatory Authority

- Air Pollutant Standard Set (ACGIH)[1] (DFG)[3] (HSE)[33] (OSHA)[58] (Several States)[60]
- Canada, WHMIS, Ingredients Disclosure List

Cited in U.S. State Regulations: Alaska (G), California (A, G), Connecticut (A), Florida (G, A), Illinois (G), Maine (G), Massachusetts (G), Nevada (A), New Hampshire (G), New Jersey (G), New York (A), North Dakota (A), Oklahoma (G), Pennsylvania (G), Rhode Island (G), South Dakota (A), Virginia (A), West Virginia (G).

Description: n-Heptane, $CH_3(CH_2)_5CH_3$, is a clear liquid which is highly flammable and volatile with a mild, gasoline-like odor. The odor threshold is 40 – 547 ppm; also reported at 230 ppm. Boiling point = 98°C. Freezing/Melting point = -91°C. Flash point = -4°C. Autoignition temperature = 204°C. Explosive limits: LEL = 1.05%; UEL = 6.7%. Hazard Identification (based on NFPA-704 M Rating System): Health 1, Flammability 3, Reactivity 0. Insoluble in water.

Potential Exposure: n-Heptane is used as an industrial solvent and in petroleum refining process; as a standard in testing knock of gasoline engines.

Incompatibilities: Forms explosive mixture with air. Strong oxidizers may cause fire and explosions. Attacks some plastics, rubber and coatings. May accumulate static electric charges that can ignite its vapors.

Permissible Exposure Limits in Air: The OSHA PEL is TWA of 500 ppm (2,000 mg/m^3) NIOSH recommends a TWA of 85 ppm and STEL of 44 ppm. The HSE[33] and the ACGIH has set a TWA of 400 ppm (1,600 mg/m^3) and an STEL of 500 ppm (2,000 mg/m^3). The NIOSH IDLH level is 750 ppm. The DFG has set an MAK of 500 ppm. Several states have set guidelines or standards for heptane in ambient air[60] ranging from 7.0 mg/m^3 (Connecticut and South Dakota) to 16.0 – 20.0 mg/m^3 (North Dakota) to 24.5 mg/m^3 (Virginia) to 32.0 mg/m^3 (Florida and New York) to 38.095 mg/m^3 (Nevada).

Determination in Air: Charcoal adsorption, workup with CS_2; Analysis by gas chromatography/flame ionization detection; NIOSH IV Method #1500, Hydrocarbons.

Permissible Concentration in Water: No criteria set.

Routes of Entry: Inhalation of the vapor, ingestion, skin and/or eye contact.

Harmful Effects and Symptoms

Short Term Exposure: n-Heptane irritates the eyes, skin, and respiratory tract. A narcotic at high concentrations. n-Heptane can cause dermatitis and mucous membrane irritation. Aspiration of the liquid may result in chemical pneumonitis, pulmonary edema, and hemorrhage. Systemic effects may arise without complaints of mucous membrane irritation. Exposure to high concentrations causes narcosis producing vertigo, incoordination, intoxication characterized by hilarity, slight nausea, loss of appetite, and a persisting gasoline taste in the mouth. These effects may be first noticed on entering a contaminated area. n-Heptane may cause low order sensitization of the myocardium to epinephrine. Swallowing the liquid may cause chemical pneumonitis.

Long Term Exposure: The liquid defeats the skin causing dryness and irritation. May affect the central nervous system, liver. Many petroleum solvents similar to heptane can cause brain damage that can affect memory, concentration, mood, sleep patters.

Points of Attack: Skin, respiratory system, lungs, peripheral nervous system.

Medical Surveillance: Preplacement examinations should evaluate the skin and general health, including respiratory, liver, and kidney function. Interview for brain effects. Consider cerebellar, autonomic and peripheral nervous system evaluation. Refer positive and borderline individuals for neuropsychological testing.

First Aid: If this chemical gets into the eyes, remove any contact lenses at once and irrigate immediately for at least 15 minutes, occasionally lifting upper and lower lids. Seek medical attention immediately. If this chemical contacts the skin, remove contaminated clothing and wash immediately with soap and water. Seek medical attention immediately. If this chemical has been inhaled, remove from exposure, begin rescue breathing (using universal precautions) if breathing has stopped and CPR if heart action has stopped. Transfer promptly to a medical facility. When this chemical has been swallowed, get medical attention. Give large quantities of water and induce vomiting. Do not make an unconscious person vomit.

Personal Protective Methods: Wear protective gloves and clothing to prevent any reasonable probability of skin contact. Safety equipment suppliers/manufacturers can provide

recommendations on the most protective glove/clothing material for your operation. Nitrile, Viton, Polyethylene, and Nitrile+PVC are among the recommended protective materials. All protective clothing (suits, gloves, footwear, headgear) should be clean, available each day, and put on before work. Contact lenses should not be worn when working with this chemical. Wear splash-proof chemical goggles and face shield unless full facepiece respiratory protection is worn. Employees should wash immediately with soap when skin is wet or contaminated. Provide emergency showers and eyewash.

Respirator Selection: NIOSH/OSHA: *750 ppm:* CCROV [any chemical cartridge respirator with organic vapor cartridge(s)]; or GMFOV [any air-purifying, full-facepiece respirator (gas mask) with a chin-style, front- or back-mounted acid gas canister]; or PAPROV [any powered, air-purifying respirator with organic vapor cartridge(s)]; or SA (any supplied-air respirator); or SCBAF (any self-contained breathing apparatus with a full facepiece). *Emergency or planned entry into unknown concentrations or IDLH conditions:* SCBAF:PD,PP (any self-contained breathing apparatus that has a full facepiece and is operated in a pressure-demand or other positive-pressure mode); or SAF:PD,PP:ASCBA (any supplied-air respirator that has a full facepiece and is operated in a pressure-demand or other positive-pressure mode in combination with an auxiliary self-contained breathing apparatus operated in a pressure-demand or other positive pressure mode). *Escape:* GMFOV [any air-purifying, full-facepiece respirator (gas mask) with a chin-style, front-or back-mounted organic vapor canister]; or SCBAE (any appropriate escape-type, self-contained breathing apparatus).

Storage: Prior to working with heptane you should be trained on its proper handling and storage. Before entering confined space where heptane may be present, check to make sure that an explosive concentration does not exist. Store in tightly closed containers in a cool, well ventilated area. Metal containers involving the transfer of this chemical should be grounded and bonded. Where possible, automatically pump liquid from drums or other storage containers to process containers. Drums must be equipped with self-closing valves, pressure vacuum bungs, and flame arresters. Use only non-sparking tools and equipment, especially when opening and closing containers of this chemical. Sources of ignition such as smoking and open flames, are prohibited where this chemical is used, handled, or stored in a manner that could create a potential fire or explosion hazard. Wherever this chemical is used, handled, manufactured, or stored, use explosion-proof electrical equipment and fittings.

Shipping: This compound requires a shipping label of: "Flammable Liquid." It falls in DOT Hazard Class 3 and Packing Group II.

Spill Handling: Evacuate and restrict persons not wearing protective equipment from area of spill or leak until cleanup is complete. Remove all ignition sources. Stay upwind and use water spray to "knock down" vapor. Establish forced ventilation to keep levels below explosive limit. Absorb liquids in vermiculite, dry sand, earth, peat, carbon, or a similar material and deposit in sealed containers. Keep this chemical out of a confined space, such as a sewer, because of the possibility of an explosion, unless the sewer is designed to prevent the build-up of explosive concentrations. It may be necessary to contain and dispose of this chemical as a hazardous waste. If material or contaminated runoff enters waterways, notify downstream users of potentially contaminated waters. Contact your Department of Environmental Protection or your regional office of the federal EPA for specific recommendations. If employees are required to clean-up spills, they must be properly trained and equipped. OSHA 1910.120(q) may be applicable.

Fire Extinguishing: This chemical is a flammable liquid. Poisonous gases are produced in fire. Use dry chemical, carbon dioxide, water spray, or foam extinguishers. Vapors are heavier than air and will collect in low areas. Vapors may travel long distances to ignition sources and flashback. Vapors in confined areas may explode when exposed to fire. Containers may explode in fire. Storage containers and parts of containers may rocket great distances, in many directions. If material or contaminated runoff enters waterways, notify downstream users of potentially contaminated waters. Notify local health and fire officials and pollution control agencies. From a secure, explosion-proof location, use water spray to cool exposed containers. If cooling streams are ineffective (venting sound increases in volume and pitch, tank discolors, or shows any signs of deforming), withdraw immediately to a secure position. If employees are expected to fight fires, they must be trained and equipped in OSHA 1910.156.

Disposal Method Suggested: Incineration.

References

National Institute for Occupational Safety and Health, Criteria for a Recommended Standard: Occupational Exposure to Alkanes, NIOSH Document No. 77–151, Washington, DC (1977)

New Jersey Department of Health and Senior Services, "Hazardous Substance Fact Sheet: Heptane," Trenton, NJ (April, 1997)

Sax, N. I., Ed., "Dangerous Properties of Industrial Materials Report," 1. No. 6, 58–59 (1981)

1-Heptanethiol

Molecular Formula: $C_7H_{16}S$

Synonyms: *n*-Heptylmercaptan; Heptyl mercaptan

CAS Registry Number: 1639-09-4

RTECS Number: MJ1400000

DOT ID: UN 1228

Regulatory Authority

- Air Pollutant Standard Set (NIOSH)

Cited in U.S. State Regulations: New Jersey (G), Pennsylvania (G).

Description: 1-Heptanethiol is a flammable, colorless liquid with a strong odor. Boiling point = 176°C. Freezing/Melting point = –43°C. flash point = 46°C.

Potential Exposure: Used as a chemical intermediate for fuels, dyes, pharmaceuticals, and to make other chemicals.

Incompatibilities: Oxidizers, strong acids, strong bases, alkali metals and reducing agents.

Permissible Exposure Limits in Air: The NIOSH recommended airborne exposure limit is (ceiling) 0.5 ppm (2.7 mg/m^3) (15 minute).

Determination in Air: No method available.

Routes of Entry: Inhalation, ingestion, skin and/or eye contact.

Harmful Effects and Symptoms

Short Term Exposure: Irritates the eyes, skin and respiratory tract. Exposure can cause headache, dizziness, nausea, and vomiting. High concentrations of mercaptans may cause cold extremities, rapid pulse, increased respiration, drowsiness, cyanosis, and unconsciousness.

Long Term Exposure: Unknown at this time.

Points of Attack: Eyes, skin, respiratory system, central nervous system, blood.

First Aid: If this chemical gets into the eyes, remove any contact lenses at once and irrigate immediately for at least 15 minutes, occasionally lifting upper and lower lids. Seek medical attention immediately. If this chemical contacts the skin, remove contaminated clothing and wash immediately with soap and water. Seek medical attention immediately. If this chemical has been inhaled, remove from exposure, begin rescue breathing (using universal precautions) if breathing has stopped and CPR if heart action has stopped. Transfer promptly to a medical facility. When this chemical has been swallowed, get medical attention. Give large quantities of water and induce vomiting. Do not make an unconscious person vomit.

Personal Protective Methods: Wear protective gloves and clothing to prevent any reasonable probability of skin contact. Safety equipment suppliers/manufacturers can provide recommendations on the most protective glove/clothing material for your operation. All protective clothing (suits, gloves, footwear, headgear) should be clean, available each day, and put on before work. Contact lenses should not be worn when working with this chemical. Wear splash-proof chemical goggles and face shield unless full facepiece respiratory protection is worn. Employees should wash immediately with soap when skin is wet or contaminated. Provide emergency showers and eyewash.

Respirator Selection: NIOSH: *Up to 5 ppm*: CCROV [any chemical cartridge respirator with organic vapor cartridge(s)]; or SA (any supplied-air respirator). *Up to 12.5 ppm*: SA:CF (any supplied-air respirator operated in a continuous-flow mode); or PAPROV [any powered, air-purifying respirator with organic vapor cartridge(s)]. *Up to 25 ppm:* CCRFOV [any chemical cartridge respirator with a full facepiece and organic vapor cartridge(s)]; or GMFOV [any air-purifying, full-facepiece respirator (gas mask) with a chin-style, front- or back-mounted organic vapor canister]; or PAPRTOV [any powered, air-purifying respirator with a tight-fitting facepiece and organic vapor cartridge(s)]; or SCBAF (any self-contained breathing apparatus with a full facepiece); or SAF (any supplied-air respirator with a full facepiece). *Emergency or planned entry into unknown concentrations or IDLH conditions*: SCBAF:PD,PP (any self-contained breathing apparatus that has a full facepiece and is operated in a pressure-demand or other positive-pressure mode); or SAF:PD, PP:ASCBA (any supplied-air respirator that has a full facepiece and is operated in a pressure-demand or other positive-pressure mode in combination with an auxiliary self-contained breathing apparatus operated in a pressure-demand or other positive-pressure mode). *Escape:* GMFOV [any air-purifying, full-facepiece respirator (gas mask) with a chin-style, front- or back-mounted organic vapor canister] or SCBAE (any appropriate escape-type, self-contained breathing apparatus).

Storage: Prior to working with this chemical you should be trained on its proper handling and storage. Store in tightly closed containers in a cool, well-ventilated area Metal containers involving the transfer of this chemical should be grounded and bonded. Where possible, automatically pump liquid from drums or other storage containers to process containers. Drums must be equipped with self-closing valves, pressure vacuum bungs, and flame arresters. Use only non-sparking tools and equipment, especially when opening and closing containers of this chemical. Sources of ignition such as smoking and open flames, are prohibited where this chemical is used, handled, or stored in a manner that could create a potential fire or explosion hazard. Wherever this chemical is used, handled, manufactured, or stored, use explosion-proof electrical equipment and fittings.

Shipping: Label: Flammable Liquid, Poison; Hazard Class: 3.

Spill Handling: Evacuate and restrict persons not wearing protective equipment from area of spill or leak until cleanup is complete. Remove all ignition sources. Ventilate area of spill or leak. Absorb liquids in vermiculite, dry sand, earth, peat, carbon, or a similar material and deposit in sealed containers. Keep this chemical out of a confined space, such as a sewer, because of the possibility of an explosion, unless the sewer is designed to prevent the build-up of explosive concentrations. It may be necessary to contain and dispose of this chemical as a hazardous waste. If material or contaminated runoff enters waterways, notify downstream users of potentially contaminated waters. Contact your Department of Environmental Protection or your regional office of the federal EPA for specific recommendations. If employees are required to clean-up spills, they must be properly trained and equipped. OSHA 1910.120(q) may be applicable.

Fire Extinguishing: This chemical is a flammable liquid. Poisonous gases including sulfur oxides, carbon monoxide, and hydrogen sulfide are produced in fire. Use dry chemical, carbon dioxide, or alcohol or polymer foam extinguishers. Vapors are heavier than air and will collect in low areas. Vapors may travel long distances to ignition sources and flashback. Vapors in confined areas may explode when exposed to fire. Containers may explode in fire. Storage containers and parts of containers may rocket great distances, in many directions. If material or contaminated runoff enters waterways, notify downstream users of potentially contaminated waters. Notify local health and fire officials and pollution control agencies. From a secure, explosion-proof location, use water spray to cool exposed containers. If cooling streams are ineffective (venting sound increases in volume and pitch, tank discolors, or shows any signs of deforming), withdraw immediately to a secure position. If employees are expected to fight fires, they must be trained and equipped in OSHA 1910.156.

Disposal Method Suggested: Incineration.

References

New Jersey Department of Health and Senior Services, "Hazardous Substance Fact Sheet, 1-Heptanethiol," Trenton NJ (June 1999)

Heptene

Molecular Formula: C_7H_{14}

Common Formula: $C_3H_7CH{=}CHC_2H_5$

Synonyms: 1-Ethyl-2-propyl ethylene; *n*-Heptene; 1-Heptene; 1-Heptylene; Heptylene

CAS Registry Number: 592-76-7

RTECS Number: MU8815000

DOT ID: UN 2278

Cited in U.S. State Regulations: Massachusetts (G), New Hampshire (G).

Description: Heptene, $C_3H_7CH{=}CHC_2H_5$, is a colorless liquid with a mild, gasoline-like odor. Boiling point = 94°C. Flash point ≤ 0°C. Autoignition temperature = 260°C. Explosive limits: LEL = 1.0%; UEL – unknown. Hazard Identification (based on NFPA-704 M Rating System): Health 0, Flammability 3, Reactivity 0. Insoluble in water.

Potential Exposure: Those involved in use as a plant growth retardant or in organic synthesis.

Incompatibilities: Strong oxidizers can cause a fire and explosion hazard.

Permissible Exposure Limits in Air: No standards set.

Permissible Concentration in Water: No criteria set.

Routes of Entry: Inhalation, skin contact.

Harmful Effects and Symptoms

Short Term Exposure: Irritates the eyes, skin and respiratory tract. Narcotic at higher concentrations; may as a slight anesthetic. Inhalation can cause coughing, wheezing, and/or shortness of breath. May also act as simple asphyxiant.

First Aid: If this chemical gets into the eyes, remove any contact lenses at once and irrigate immediately for at least 15 minutes, occasionally lifting upper and lower lids. Seek medical attention immediately. If this chemical contacts the skin, remove contaminated clothing and wash immediately with soap and water. Seek medical attention immediately. If this chemical has been inhaled, remove from exposure, begin rescue breathing (using universal precautions) if breathing has stopped and CPR if heart action has stopped. Transfer promptly to a medical facility. When this chemical has been swallowed, get medical attention. Give large quantities of water and induce vomiting. Do not make an unconscious person vomit.

Personal Protective Methods: Wear protective gloves and clothing to prevent skin contact. Safety equipment suppliers/manufacturers can provide recommendations on the most protective glove/clothing material for your operation. All protective clothing (suits, gloves, footwear, headgear) should be clean, available each day, and put on before work. Contact lenses should not be worn when working with this chemical. Wear splash-proof chemical goggles and face shield unless full facepiece respiratory protection is worn. Employees should wash immediately with soap when skin is wet or contaminated. Provide emergency showers and eyewash.

Respirator Selection: Where the potential for exposure to this chemical, use a MSHA/NIOSH approved supplied-air respirator with a full facepiece operated in the positive pressure mode or with a full facepiece, hood, or helmet in the continuous flow mode, or use a MSHA/NIOSH approved self-contained breathing apparatus with a full facepiece operated in pressure-demand or other positive pressure mode.

Storage: Prior to working with heptene you should be trained on its proper handling and storage. Before entering confined space where heptene may be present, check to make sure that an explosive concentration does not exist. Store in tightly closed containers in a cool, well ventilated area away from oxidizers. Metal containers involving the transfer of this chemical should be grounded and bonded. Where possible, automatically pump liquid from drums or other storage containers to process containers. Drums must be equipped with self-closing valves, pressure vacuum bungs, and flame arresters. Use only non-sparking tools and equipment, especially when opening and closing containers of this chemical. Sources of ignition such as smoking and open flames, are prohibited where this chemical is used, handled, or stored in a manner that could create a potential fire or explosion hazard. Wherever this chemical is used, handled, manufactured, or stored, use explosion-proof electrical equipment and fittings.

Shipping: This compound requires a shipping label of: "Flammable Liquid." It falls in DOT Hazard Class 3 and Packing Group II.

Spill Handling: Evacuate and restrict persons not wearing protective equipment from area of spill or leak until cleanup is complete. Remove all ignition sources. Establish forced ventilation to keep levels below explosive limit. Absorb liquids in vermiculite, dry sand, earth, peat, carbon, or a similar material and deposit in sealed containers. Keep this chemical out of a confined space, such as a sewer, because of the possibility of an explosion, unless the sewer is designed to prevent the build-up of explosive concentrations. Spills on water may be handled with oil skimming equipment and sorbent (polyurethane) foams. It may be necessary to contain and dispose of this chemical as a hazardous waste. If material or contaminated runoff enters waterways, notify downstream users of potentially contaminated waters. Contact your Department of Environmental Protection or your regional office of the federal EPA for specific recommendations. If employees are required to clean-up spills, they must be properly trained and equipped. OSHA 1910.120(q) may be applicable.

Fire Extinguishing: This chemical is a flammable liquid. Poisonous gases are produced in fire. Use dry chemical, carbon dioxide, or alcohol or polymer foam extinguishers. Vapors are heavier than air and will collect in low areas. Vapors may travel long distances to ignition sources and flashback. Vapors in confined areas may explode when exposed to fire. Containers may explode in fire. Storage containers and parts of containers may rocket great distances, in many directions. If material or contaminated runoff enters waterways, notify downstream users of potentially contaminated waters. Notify local health and fire officials and pollution control agencies. From a secure, explosion-proof location, use water spray to cool exposed containers. If cooling streams are ineffective (venting sound increases in volume and pitch, tank discolors, or shows any signs of deforming), withdraw immediately to a secure position. If employees are expected to fight fires, they must be trained and equipped in OSHA 1910.156.

Use foam, CO_2, or dry chemical.

Disposal Method Suggested: Incineration.

References

Sax, N. I., Ed., Dangerous Properties of Industrial Materials Report, 2, No. 2, 29–30 (1982)

New Jersey Department of Health and Senior Services, "Hazardous Substance Fact Sheet, Heptene," Trenton NJ (June 1999)

Hexachlorobenzene

Molecular Formula: C_6Cl_6

Synonyms: Amatin; Anticarie; Benzene, hexachloro-; Bunt-Cure; Bunt-No-More; Ceku C.B.; Co-Op Hexa; Esachlorobenzene (Italian); Granox NM; HCB; Hexa C.B.; Hexachlorbenzol (German); Hexaclorobenceno (Spanish); Julin's carbon chloride; No Bunt; No Bunt 40; No Bunt 80; No Bunt liquid; Pentachlorophenyl chloride; Perchlorobenzene; Saatbenizfungizid (German); Sanocid; Sanocide; Smut-Go; Snieciotox; Zaprawa nasienna sneciotox

CAS Registry Number: 118-74-1

RTECS Number: DA2975000

DOT ID: UN 2729

EEC Number: 602-065-00-6

Regulatory Authority

- Air Pollutant Standard Set (ACGIH) (former USSR)[35][43] (Czechoslovakia)[35] (Several States)[60]
- Banned or Severely Restricted (Many Countries) (UN)[13]
- Carcinogen (Animal Positive) (IARC)[9]
- CLEAN AIR ACT: Hazardous Air Pollutants (Title I, Part A, Section 112)
- CLEAN WATER ACT: Section 313 Water Priority Chemicals (57FR41331, 9/9/92)
- EPA HAZARDOUS WASTE NUMBER (RCRA No.): U127; D032
- RCRA, 40CFR261, Appendix 8 Hazardous Constituents
- RCRA Toxicity Characteristic (Section 261.24), Maximum Concentration of Contaminants, regulatory level, 0.13 mg/l
- RCRA 40CFR268.48; 61FR15654, Universal Treatment Standards: Wastewater (mg/l), 0.055; Nonwastewater (mg/kg), 10
- RCRA 40CFR264, Appendix 9; TSD Facilities Ground Water Monitoring List. Suggested test method(s) (PQL μg/l): 8120 (0.05); 8270 (10)
- SAFE DRINKING WATER ACT: MCL, 0.001 mg/l; MCLG, zero
- SUPERFUND/EPCRA 40CFR302.4 Reportable Quantity (RQ): CERCLA, 10 lb (4.54 kg)
- EPCRA Section 313 Form R *de minimis* concentration reporting level: 0.1%
- Canada, WHMIS, Ingredients Disclosure List

Cited in U.S. State Regulations: Arizona (W), California (A, G), Florida (G), Illinois (G), Kansas (G, W), Louisiana (G), Maine (G, W), Maryland (G), Massachusetts (G), Michigan (G), Minnesota (W), New Hampshire (G), New Jersey (G), New York (G, A), North Dakota (A), Pennsylvania (G, A), Vermont (G), Virginia (G), Washington (G), West Virginia (G), Wisconsin (G).

Description: Hexachlorobenzene, C_6Cl_6, is a solid, crystallizing in needles. Boiling point = 323 – 326°C. Freezing/Melting point =231°C. Flash point = 242°C. Hazard Identification (based on NFPA-704 M Rating System): Health 1, Flammability 1, Reactivity 0. Slightly soluble in water.

Potential Exposure: Hexachlorobenzene was used widely as a pesticide to protect seeds of onions and sorghum, wheat, and other grains against fungus until 1965. This material was used to make fireworks, ammunition for military uses, synthetic rubber, as a porosity controller in the manufacture of electrodes, as an intermediate in dye manufacture, in organic synthesis, and as a wood preservative. It is formed as a by-product of making other chemicals, in the waste streams of chloralkali and wood-preserving plants, and when burning municipal waste. Currently, there are no commercial uses of hexachlorobenzene in the United States.

Incompatibilities: Reacts violently with oxidizers, dimethyl formamide above 65°C.

Permissible Exposure Limits in Air: There is not OSHA PEL. ACGIH recommends a TLV of 0.002 TWA (skin); Animal Carcinogen. The former USSR has set a MAC in workplace air of 0.9 mg/m^3.[35][43] They have also set[35] a MAC in ambient air of residential areas of 0.013 mg/m^3. Czechoslovakia[35] has set a TWA in workplace air of 1.0 mg/m^3 and an STEL of 2.0 mg/m^3. Several states have set guidelines or standards for hexachlorobenzene in ambient air[60] ranging from zero in North Dakota to 0.48 ppb (Pennsylvania) to 0.03 $\mu g/m^3$ (New York).

Determination in Air: Use NIOSH IV, Method # 1003, Halogenated hydrocarbons.

Permissible Concentration in Water: The US EPA recommended that drinking water should not contain more than 0.05 milligrams of hexachlorobenzene per liter of water (0.05 mg/l) in water that children drink and should not contain more than 0.2 mg/l in water that adults drink for longer periods (about 7 years. The EPA has set a maximum contamination level (MCL) of 0.001 mg/l in drinking water. The former USSR has set a MAC of 0.05 mg/l in water bodies used for domestic purposes.[35][43] Several states have set guidelines for hexachlorobenzene in drinking water[61] ranging from 0.02 μg/l (Arizona) to 0.20 μg/l (Kansas) to 0.21 μg/l (Minnesota) to 5.4 μg/l (Maine). The World Health Organization (WHO)[35] has set a limit in drinking water of 0.01 μg/l.

Determination in Water: Methylene chloride extraction followed by concentration and gas chromatography with electron capture detection (EPA Method 612) or gas chromatography plus mass spectrometry (EPA Method 625).

Routes of Entry: Inhalation, ingestion, eye and skin contact.

Harmful Effects and Symptoms

Short Term Exposure: Irritates the eyes, skin, and respiratory tract. *Inhalation:* Coughing, shortness of breath and labored breathing have been reported from large, unmeasured doses or by decomposition to chlorine. *Skin:* Can cause irritation. Exposure to sunlight with (or soon after) exposure can increase effects. Following this reaction, changes in skin pigment and blistering may follow. Red or dark urine may be noticed. High doses may cause redness, pain and serious burns. *Eyes:* May cause irritation. Higher doses may cause redness, pain and blurred vision. *Ingestion:* Headache, dizziness, nausea, vomiting, numbness of hands and arms, apprehension, excitement, tremors, partial paralysis of arms and legs, loss of muscle control, loss of sensory perception, convulsions and coma may result from high doses.

Long Term Exposure: May affect the lungs, liver, skin, and nervous system. This substance causes cancer in laboratory animals, and may be carcinogenic to humans. May damage the developing fetus. May cause liver, thyroid, kidney and immune system damage. High, prolonged or repeated exposure may affect the nervous system. Repeated skin exposure can lead to permanent skin changes and increased hair growth. Animal tests show that this substance possibly causes toxic effects upon human reproduction Ingestion of contaminated grain, estimated at doses of 0.05 – 0.2 grams/day, resulted in porphyria cutanea tarda (PCT) in Turkey which is characterized by red-colored urine, skin sores, change in skin color, arthritis, and problems of the liver, nervous system, and stomach. The following symptoms were also reported: enlarged livers, porphyria in the blood, loss or appetite, weight loss and wasting of skeletal muscles. Severe and long-standing poisoning caused abnormal hair growth, loss of vision, wasting of hands, black discoloration, and skin sores which became ulcerated, healing with pigmented scars. Breast-fed children developed "pink-sore," a condition which was 95% fatal. Toxic effects on blood and active symptoms persisted up to 20 years. Studies in animals show that ingestion of this chemical can damage the liver, thyroid, nervous system, bones, kidneys, blood, and immune and endocrine system.

Points of Attack: Liver, skin, thyroid.

Medical Surveillance: Liver function tests. Thyroid function tests. Evaluation by a qualified allergist and/or dermatologist. Iron as a dietary supplement could increase liver damage. consult a physician before taking supplements. Guard against sunlight exposure to contaminated skin.

First Aid: If this chemical gets into the eyes, remove any contact lenses at once and irrigate immediately for at least 15 minutes, occasionally lifting upper and lower lids. Seek medical attention immediately. If this chemical contacts the skin, remove contaminated clothing and wash immediately with soap and water. Seek medical attention immediately. If this chemical has been inhaled, remove from exposure, begin rescue breathing (using universal precautions) if breathing has stopped and CPR if heart action has stopped. Transfer promptly to a medical facility. When this chemical has been swallowed, get medical attention. Give large quantities of water and induce vomiting. Do not make an unconscious person vomit.

Note to Physician: For ingestions of less than 10 mg/kg body weight occurring less than an hour before treatment, induce emesis. For ingestions of more than 10 mg/kg body weight occurring less than an hour before treatment, use gastric lavage. For ingestion occurring more than an hour before treatment, use activated charcoal. There is no specific antidote, and supervision for at least 72 hours is recommended.

Personal Protective Methods: Wear protective gloves and clothing to prevent any reasonable probability of skin contact. Safety equipment suppliers/manufacturers can provide recommendations on the most protective glove/clothing material for your operation. All protective clothing (suits, gloves, footwear, headgear) should be clean, available each day, and put on before work. Contact lenses should not be worn when working with this chemical. Wear dust-proof chemical goggles and face shield unless full facepiece respiratory protection is worn. Employees should wash immediately with soap when skin is wet or contaminated. Provide emergency showers and eyewash.

Respirator Selection: *At any detectable concentration:* SCBAF:PD,PP (any MSHA/NIOSH approved self-contained breathing apparatus that has a full facepiece and is operated in a pressure-demand or other positive-pressure mode); or SAF:PD,PP:ASCBA (any supplied-air respirator that has a full facepiece and is operated in a pressure-demand or other positive-pressure mode in combination with an auxiliary, self-contained breathing apparatus operated in a pressure-demand or other positive pressure mode). *Escape:* GMFOV [any air-purifying, full-facepiece respirator (gas mask) with a chin-style, front-or back-mounted organic vapor canister]; or SCBAE (any appropriate escape-type, self-contained breathing apparatus).

Storage: Prior to working with this chemical you should be trained on its proper handling and storage. Store in tightly closed containers in a cool, well ventilated area away from oxidizers, dimethyl formamide and heat. Where possible, automatically pump liquid from drums or other storage containers to process containers. A regulated, marked area should be established where this chemical is handled, used, or stored in compliance with OSHA standard 1910.1045.

Shipping: This compound requires a shipping label of: "Keep Away From Food." It falls in DOT Hazard Class 6.1 and Packing Group III.

Spill Handling: Evacuate persons not wearing protective equipment from area of spill or leak until clean-up is complete. Remove all ignition sources. Collect powdered material in the most convenient and safe manner and deposit in sealed containers. Ventilate area after clean-up is complete. It may be necessary to contain and dispose of this chemical as a hazardous waste. If material or contaminated runoff enters waterways, notify downstream users of potentially contaminated waters. Contact your Department of Environmental Protection or your regional office of the federal EPA for specific recommendations. If employees are required to clean-up spills, they must be properly trained and equipped. OSHA 1910.120(q) may be applicable.

Fire Extinguishing: This chemical may burn but does not easily ignite. Use dry chemical, carbon dioxide, water spray, or foam extinguishers. Poisonous gases including chlorine are produced in fire. If material or contaminated runoff enters waterways, notify downstream users of potentially contaminated waters. Notify local health and fire officials and pollution control agencies. Containers may explode in fire. From a secure, explosion-proof location, use water spray to cool exposed containers. If cooling streams are ineffective (venting sound increases in volume and pitch, tank discolors, or shows any signs of deforming), withdraw immediately to a secure position. If employees are expected to fight fires, they must be trained and equipped in OSHA 1910.156.

Disposal Method Suggested: Incineration is most effective at 1,300°C and 0.25 sec. Consult with environmental regulatory agencies for guidance on acceptable disposal practices. Generators of waste containing this contaminant (≥100 kg/mo) must conform with EPA regulations governing storage, transportation, treatment, and waste disposal.

References

U.S. Environmental Protection Agency, Chlorinated Benzenes: Ambient Water Quality Criteria. Washington, DC (1980)

U.S. Environmental Protection Agency, Status Assessment of Toxic Chemicals: Hexachlorobenzene, Report EPA-600/2-79-210g, Cincinnati, Ohio (December 1979)

U.S. Environmental Protection Agency, Hexachlorobenzene, Health and Environmental Effects Profile No. 110, Office of Solid Waste, Washington, DC (April 30, 1980)

Sax, N. I., Ed., Dangerous Properties of Industrial Materials Report, 4, No. 1, 88–92 (1984)

New York State Department of Health, "Chemical Fact Sheet: Hexachlorobenzene (HCB)," Albany, NY, Bureau of Toxic Substance Assessment (May 1986)

New Jersey Department of Health and Senior Services, "Hazardous Substance Fact Sheet, Hexachlorobenzene," Trenton, NJ (November 1988)

U.S. Department of Health and Human Services, "ATSDR ToxFAQs, Hexachlorobenzene," (Atlanta, GA, September 1997)

Hexachlorobutadiene

Molecular Formula: C_4Cl_6

Common Formula: CCl_2=CCI-CCI=CCl_2

Synonyms: 1,3-Butadiene, 1,1,2,3,4,4-hexachloro-; Butadiene, hexachloro-; C46; Dolen-Pur; GP-40-66:120; HCBD; Hexachlor-1,3-butadien (Czech); 1,1,2,3,4,4-Hexachloro-1,3-butadiene; Hexachloro-1,3-butadiene; Hexachlorobutadiene; Hexaclorobutadieno (Spanish); Perchloro-1,3-butadiene; Perchlorobutadiene

CAS Registry Number: 87-68-3

RTECS Number: EJ0700000

DOT ID: UN 2279

Regulatory Authority

- Air Pollutant Standard Set (ACGIH)[1] (NIOSH) (former USSR)[35][43] (Several States)[60]
- Carcinogen (Suspected Human) (ACGIH)[1] (DFG)[3]
- CLEAN AIR ACT: Hazardous Air Pollutants (Title I, Part A, Section 112)
- CLEAN WATER ACT: Section 313 Water Priority Chemicals (57FR41331, 9/9/92); 40CFR401.15 Section 307 Toxic Pollutants
- EPA HAZARDOUS WASTE NUMBER (RCRA No.): U128; D033
- RCRA, 40CFR261, Appendix 8 Hazardous Constituents
- RCRA Toxicity Characteristic (Section 261.24), Maximum Concentration of Contaminants, regulatory level, 0.5 mg/l

- RCRA 40CFR268.48; 61FR15654, Universal Treatment Standards: Wastewater (mg/l), 0.055; Nonwastewater (mg/kg), 5.6
- RCRA 40CFR264, Appendix 9; TSD Facilities Ground Water Monitoring List. Suggested test method(s) (PQL µg/l): 8120 (5); 8270 (10)
- SAFE DRINKING WATER ACT: Priority List (55 FR 1470)
- SUPERFUND/EPCRA 40CFR302.4 Reportable Quantity (RQ): CERCLA, 1 lb (0.454 kg)
- EPCRA Section 313 Form R *de minimis* concentration reporting level: 1.0%
- Canada, WHMIS, Ingredients Disclosure List

Cited in U.S. State Regulations: Alaska (G), California (G, A), Connecticut (A), Florida (G, A), Illinois (G), Kansas (G, W), Louisiana (G), Maine (G), Maryland (G), Massachusetts (G), Michigan (G), Minnesota (W), Nevada (A), New Hampshire (G), New Jersey (G), New York (A), North Dakota (A), Pennsylvania (G, A), Rhode Island (G), South Carolina (A), Vermont (G), Virginia (G, A), Washington (G), West Virginia (G), Wisconsin (G).

Description: Hexachlorobutadiene, CCl_2=CCl-CCl=CCl_2, is a clear, colorless liquid with a faint, turpentine-like odor. Boiling point = 212°C. Freezing/Melting point = -21°C. Flash point = 90°C. Autoignition temperature = 610°C. Hazard Identification (based on NFPA-704 M Rating System): Health 2, Flammability 1, Reactivity 1. Insoluble in water.

Potential Exposure: Hexachlorobutadiene is used as a solvent for elastomers, a heat-transfer fluid, a transformer and hydraulic fluid, and a wash liquor for removing higher hydrocarbons.

Incompatibilities: Strong reaction with oxidizers, aluminum powder. Attacks aluminum, some plastics, rubber and coatings

Permissible Exposure Limits in Air: There is no OSHA PEL. NIOSH and ACGIH recommend a TWA of 0.02 ppm (0.24 mg/m^3). The former USSR[35][43] has set MAC values for ambient air in residential areas of 0.001 mg/m^3 on a momentary basis and 0.0002 mg on a daily average basis. Several states have set guidelines or standards for hexachlorobutadiene in ambient air[60] ranging from zero (North Dakota) to 0.72 µg/m^3 (Pennsylvania) to 0.80 µg/m^3 (New York) to 1.2 µg/m^3 (South Carolina) to 2.4 µg/m^3 (Connecticut, Florida and Virginia) to 6.0 µg/m^3 (Nevada).

Determination in Air: Collection by XAD-2® (tube); workup with hexane; analysis by gas chromatography/electrochemical detection; NIOSH IV, Method #2543.

Permissible Concentration in Water: To protect freshwater aquatic life – 90 µg/l on an acute toxicity basis and 9.3 µg/l on a chronic basis. To protect saltwater aquatic life – 32 µg/l on an acute toxicity basis. For the protection of human health-preferably zero. An additional lifetime cancer risk of 1 in 100,000 is imposed by a concentration of 4.47 µg/l.[6] The former USSR[35][43] has set a MAC in water bodies used for domestic purposes of 0.01 mg/l. Kansas and Minnesota have set guidelines for hexachlorobutadiene in drinking water of 4.5 µg/l.[61]

Determination in Water: Methylene chloride extraction followed by concentration, gas chromatography with electron capture detection (EPA Method 612) or gas chromatography plus mass spectrometry (EPA Method 625).

Routes of Entry: Inhalation, skin absorption, ingestion, skin and/or eye contact.

Harmful Effects and Symptoms

Short Term Exposure: Hexachlorobutadiene can affect you when breathed in and by passing through your skin. The liquid is corrosive; contact can irritate and burn the skin and eyes. Vapors can irritate the eyes, skin, and respiratory tract.

Long Term Exposure: Hexachlorobutadiene should be handled as a carcinogen — with extreme caution. It may damage the developing fetus. Exposure can cause severe kidney damage. Hexachlorobutadiene may damage the liver. Repeated or prolonged contact may cause skin sensitization and allergy.

Points of Attack: Eyes, skin, respiratory system, kidneys. Cancer site in animals: kidney tumors.

Medical Surveillance: Before beginning employment and at regular times after that, the following are recommended: Kidney function tests. Liver function tests. evaluation by a qualified allergist.

First Aid: If this chemical gets into the eyes, remove any contact lenses at once and irrigate immediately for at least 15 minutes, occasionally lifting upper and lower lids. Seek medical attention immediately. If this chemical contacts the skin, remove contaminated clothing and wash immediately with soap and water. Seek medical attention immediately. If this chemical has been inhaled, remove from exposure, begin rescue breathing (using universal precautions) if breathing has stopped and CPR if heart action has stopped. Transfer promptly to a medical facility. When this chemical has been swallowed, get medical attention. Give large quantities of water and induce vomiting. Do not make an unconscious person vomit.

Personal Protective Methods: Wear protective gloves and clothing to prevent any reasonable probability of skin contact. Safety equipment suppliers/manufacturers can provide recommendations on the most protective glove/clothing material for your operation. All protective clothing (suits, gloves, footwear, headgear) should be clean, available each day, and put on before work. Contact lenses should not be worn when working with this chemical. Wear splash-proof chemical goggles and face shield unless full facepiece respiratory protection is worn. Employees should wash immediately with soap when skin is wet or contaminated. Provide emergency showers and eyewash.

Respirator Selection: Where the potential exists for exposures over 0.02 ppm, use a MSHA/ NIOSH approved supplied-air respirator with a full facepiece operated in the positive pressure mode or with a full facepiece, hood, or helmet in the continuous flow mode, or use a MSHA/NIOSH approved self-contained breathing apparatus with a full facepiece operated in pressure-demand or other positive pressure mode.

Storage: Prior to working with this chemical you should be trained on its proper handling and storage. Store in tightly closed containers in a cool, well ventilated area away from oxidizers. Where possible, automatically pump liquid from drums or other storage containers to process containers A regulated, marked area should be established where hexachlorobutadiene is handled, used, or stored. A regulated, marked area should be established where this chemical is handled, used, or stored in compliance with OSHA standard 1910.1045.

Shipping: This compound requires a shipping label of: "Keep Away From Food." It falls in DOT Hazard Class 6.1 and Packing Group III.

Spill Handling: Evacuate and restrict persons not wearing protective equipment from area of spill or leak until cleanup is complete. Remove all ignition sources. Ventilate area of spill or leak. Absorb liquids in vermiculite, dry sand, earth, peat, carbon, or a similar material and deposit in sealed containers. Keep this chemical out of a confined space, such as a sewer, because of the possibility of an explosion, unless the sewer is designed to prevent the build-up of explosive concentrations. It may be necessary to contain and dispose of this chemical as a hazardous waste. If material or contaminated runoff enters waterways, notify downstream users of potentially contaminated waters. Contact your Department of Environmental Protection or your regional office of the federal EPA for specific recommendations. If employees are required to clean-up spills, they must be properly trained and equipped. OSHA 1910.120(q) may be applicable.

Fire Extinguishing: This material is not flammable. Use extinguishing agents suitable for surrounding fire. Poisonous gases are produced in fire. Vapors are heavier than air and will collect in low areas. Vapors may travel long distances to ignition sources and flashback. Vapors in confined areas may explode when exposed to fire. Containers may explode in fire. Storage containers and parts of containers may rocket great distances, in many directions. If material or contaminated runoff enters waterways, notify downstream users of potentially contaminated waters. Notify local health and fire officials and pollution control agencies. From a secure, explosion-proof location, use water spray to cool exposed containers. If cooling streams are ineffective (venting sound increases in volume and pitch, tank discolors, or shows any signs of deforming), withdraw immediately to a secure position. If employees are expected to fight fires, they must be trained and equipped in OSHA 1910.156.

Disposal Method Suggested: High temperature incineration with flue gas scrubbing.[22] Consult with environmental regulatory agencies for guidance on acceptable disposal practices. Generators of waste containing this contaminant (≥100 kg/mo) must conform with EPA regulations governing storage, transportation, treatment, and waste disposal.

References

U.S. Environmental Protection Agency, Hexachlorobutadiene: Ambient Water Quality Criteria, Washington, DC (1980)

U.S. Environmental Protection agency. Sampling and Analysis of Selected Toxic Substances. Task 1B — Hexachlorobutadiene, EPA Rep. No. 560/6-76-015, Off. Toxic Subst., Washington, DC (1976)

U.S. Environmental Protection Agency, Hexachlorobutadiene, Health and Environmental Effects Profile No. 111, Office of Solid Waste, Washington, DC (April 30, 1980)

Sax, N. I., Ed. Dangerous Properties of Industrial Materials Report, 2, No. 5, 71–75 (1982)

New Jersey Department of Health and Senior Services, "Hazardous Substance Fact Sheet: Hexachlorobutadiene," Trenton, NJ (March, 1998)

Hexachlorocyclohexanes

Molecular Formula: $C_6H_6Cl_6$

Synonyms: *technical grade containing 68.7% α-BHC, 6.5% β-BHC, 13.5% γ-BHC:* BHC; Compound 666; DBH; ENT 8,601; Gammexane; HCCH; Hexa; Hexachlorocyclohexane; Hexachlorocyclohexane (mixed isomers); Hexachlorocyclohexane isomers; Hexacloran (in Russia); Hexaclorociclohexano (Spanish); Hexaklon (in Sweden); Hexhexane; Hexylan; Jacutin; Latka-666 HCH

α-isomer: A13-09232; α-Benzenehexachloride; Benzene hexachloride-α-isomer; Benzene-*trans*-hexachloride; α-BHC; Cyclohexane 1,2,3,4,5,6-hexachloro-; Cyclohexane 1,2,3,4,5,6-hexachloro-α; Cyclohexane 1,2,3,4,5,6-hexachloro-α isomer; Cyclohexane 1,2,3,4,5,6-hexachloro-(α, dl); Cyclohexane 1,2,3,4,5,6-hexachloro-(1α,2α,3β,4α,5β,6β)-; Cyclohexane, α-1,2,3,4,5,6-hexachloro-; ENT9,232; Forlin; Gamaphex; α-HCH; α-Hexachloran; α-Hexachlorane; Hexachlorcyclohexan (German); α-1,2,3,4,5,6-Hexachlorcyclohexane; Hexachlorocyclohexan (German); α-Hexachlorocyclohexane; 1-α,2α,3β,4α,5β,6β-Hexachlorocyclohexane; 1,2,3,4,5,6-Hexachlorocyclohexane; 1a,2a,3b,4a,5b,6b-Hexachlorocyclohexane; Hexachlorocyclohexane; 1,2,3,4,5,6-Hexaclorociclohexano (Spanish); Isotox; Lindagam; α-Lindane; Silvano

β-isomer: β-Benzenehexachloride; β-BHC; ENT 9,233; β-HCH; β-1,2,3,4,5,6-Hexachlorocyclohexane; β-Hexachlorocyclohexane; 1-α,2-β,3-α,4-β,5-α,6-β-Hexachlorocyclohexane; β-Lindane

γ-isomer: see Lindane

δ-isomer: δ-Benzenehexachloride; δ-BHC; ENT9,234; δ-HCH; HCH, δ-; HCH-delta; δ-1,2,3,4,5,6-Hexachlorocyclohexane; δ-Hexachlorocyclohexane; 1-α,2-α,3-α,4-β,5-α,6-β-Hexachlorocyclohexane; δ-Lindane

CAS Registry Number: 608-73-1 (technical grade); 319-84-6 (α-isomer); 319-85-7 (β-isomer); 58-89-9 (γ-isomer) see LINDANE; 319-86-8 (δ-isomer)

RTECS Number: GV3150000

DOT ID: UN 2761

Regulatory Authority

- Air Pollutant Standard Set (DFG)[3] (former USSR)[43]
- Banned or Severely Restricted (Many Countries) (UN)[13]
- Carcinogen (Animal Positive) (IARC)[9]

all isomers:

- CLEAN WATER ACT: 40CFR401.15 Section 307 Toxic Pollutants
- U.S. DOT Regulated Marine Pollutant (49CFR172.101, Appendix B), severe pollutant

α-isomer:

- CLEAN WATER ACT: 40CFR423, Appendix A, Priority Pollutants; 40CFR401.15 Section 307 Toxic Pollutants as hexachlorocyclohexane
- RCRA 40CFR268.48; 61FR15654, Universal Treatment Standards: Wastewater (mg/l), 0.00014; Nonwastewater (mg/kg), 0.066
- RCRA 40CFR264, Appendix 9; TSD Facilities Ground Water Monitoring List. Suggested test method(s) (PQL μg/l): 8080 (0.05); 8250 (10)
- SUPERFUND/EPCRA 40CFR302.4 Reportable Quantity (RQ): CERCLA, 10 lb (4.54 kg)
- EPCRA Section 313 Form R *de minimis* concentration reporting level: 1.0

β-isomer:

- CLEAN WATER ACT: 40CFR423, Appendix A, Priority Pollutants; 40CFR401.15 Section 307 Toxic Pollutants as hexachlorocyclohexane
- RCRA 40CFR268.48; 61FR15654, Universal Treatment Standards: Wastewater (mg/l), 0.00014; Nonwastewater (mg/kg), 0.066
- RCRA 40CFR264, Appendix 9; TSD Facilities Ground Water Monitoring List. Suggested test method(s) (PQL μg/l): 8080 (0.05); 8250 (40)
- SUPERFUND/EPCRA 40CFR302.4 Reportable Quantity (RQ): CERCLA, 1 lb (0.454 kg)

γ-isomer:

- See LINDANE

δ-isomer:

- CLEAN WATER ACT: 40CFR423, Appendix A, Priority Pollutants; 40CFR401.15 Section 307 Toxic Pollutants as hexachlorocyclohexane
- RCRA 40CFR268.48; 61FR15654, Universal Treatment Standards: Wastewater (mg/l), 0.023; Nonwastewater (mg/kg), 0.066
- RCRA 40CFR264, Appendix 9; TSD Facilities Ground Water Monitoring List. Suggested test method(s) (PQL μg/l): 8080 (0.1); 8250 (30)
- SUPERFUND/EPCRA 40CFR302.4 Reportable Quantity (RQ): CERCLA, 1 lb (0.454 kg)
- Canada, WHMIS, Ingredients Disclosure List

Cited in U.S. State Regulations: California (G), Florida (G), Illinois (G), Kansas (G), Louisiana (G), Massachusetts (G), Michigan (G), New Hampshire (G), New Jersey (G), Pennsylvania (G), Rhode Island (G), Virginia (G), Washington (G), Wisconsin (G), Wisconsin (G).

Description: HCH, $C_6H_6Cl_6$, is a white-to-brownish crystalline solid with a phosgene-like odor. Freezing/Melting point = 65°C. Hazard Identification (based on NFPA-704 M Rating System): Health 2, Flammability 1, Reactivity 0. It consists of eight stereoisomers of which the gamma (γ) isomer is most insecticidally active and hence most important. See also "Lindane."

Potential Exposure: The major commercial usage of HCH is based upon its insecticidal properties. The 7-isomer has the highest acutetoxicity, but the other isomers are not without activity. It is generally advantageous to purify the 7-isomer from the less active isomers. The γ-isomer acts on the nervous system of insects, principally at the level of the nerve ganglia. As a result, lindane has been used against insects in a wide range of applications including treatment of animals, buildings, man for ectoparasites, clothes, water for mosquitoes, living plants, seeds and soils. Some applications have been abandoned due to excessive residues, e.g., stored foodstuffs. By voluntary action, the principal domestic producer of technical grade BHC requested cancellations of its BHC registrations on September 1, 1976. As of July 21, 1978, all registrants of pesticide products containing BHC voluntarily canceled their registrations or switched their former BHC products to lindane formulations.

Incompatibilities: Decomposes on contact with powdered iron, aluminum, and zinc, and on contact with strong bases producing trichlorobenzene.

Permissible Exposure Limits in Air: ACGIH recommends a TLV of 0.5 mg/m^3 TWA (skin). The DFG[3] has set an MAK of 0.5 mg/m^3 for mixed HCH isomers. The former USSR-UNEP/IRPTC project[43] has set a MAC in workplace air of 0.1 mg/m^3 and MAC value for ambient air in residential areas of 0.03 mg/m^3 both on a momentary and on a daily average basis.

Determination in Air: Collection by filter/bubbler; workup with isooctane; analysis by gas chromatography/electrolytic conductivity detection; NIOSH IV, Method #5502 (recommended for Lindane).

Permissible Concentration in Water: The EPA has set a limit in drinking water of 0.2 ppb. There are no criteria for the protection of freshwater or saltwater aquatic life from technical BHC (mixed isomers) due to insufficient data. To protect human health-preferably zero for technical product. An additional cancer risk of 1 in 100,000 is imposed by a concentration of 0.123 μg/l.[6] The former USSR-UNEP/IRPTC project[43] has set a MAC in water bodies used for domestic purposes of 0.02 mg/l and zero in water bodies used for fishery purposes.

Determination in Water: Methylene chloride extraction followed by gas chromatography with electron capture or halogen specific detection (EPA Method 608) or gas chromatography plus mass spectrometry (EPA Method 625).

Routes of Entry: Inhalation, skin absorption, ingestion, skin and/or eye contact.

Harmful Effects and Symptoms

Short Term Exposure: Irritates the eyes and respiratory tract. May affect the central nervous system, causing convulsions, respiratory failure, and collapse. Effects may be delayed. Exposure may result in death. See also below.

Long Term Exposure: Repeated or prolonged skin contact may cause irritation, redness. The effects of lindane and/or the α-, β-, and δ-isomers of HCH observed in humans are lung irritation, heart disorders, blood disorders, headache, convulsions, and changes in the levels of sex hormones. These effects have occurred in workers exposed to HCH vapors during pesticide formulation and/or in individuals exposed accidentally or intentionally to large amounts of HCH. Exposure to excessive amounts of HCH can also result in death in humans and animals. Convulsions and kidney disease have been reported in animals fed lindane or β-HCH. Liver disease has been reported in animals fed lindane and α-, β-, or technical grade HCH. Longer exposure to lindane and α-, β-, or technical-grade HCH has been reported to result in liver cancer. Reduced ability to fight infection was reported in animals fed lindane and injury to the ovaries and testes was reported in animals exposed to lindane or β-HCH. In animals, there is evidence that oral exposure to lindane during pregnancy results in an increased incidence of fetuses with extra ribs. HCH is processed by the body into other chemical products, some of which probably are responsible for the harmful effects. The Department of Health and Human Services has determined that HCH may reasonably be anticipated to be carcinogenic. Liver cancer has been seen in laboratory rodents that ate HCH for long periods of time.

Points of Attack: Eyes, skin, respiratory system, central nervous system, blood, liver, kidneys

Medical Surveillance: NIOSH and OSHA recommend tests of whole blood (chemical/metabolite). See "Occupational Health Guidelines for Chemical Hazards." NIOSH Pub Nos. 81-123; 88-118, Suppls. I-IV. 1981-1995. Blood Serum. Complete Blood count.

First Aid: If this chemical gets into the eyes, remove any contact lenses at once and irrigate immediately for at least 30 minutes, occasionally lifting upper and lower lids. Seek medical attention immediately. If this chemical contacts the skin, remove contaminated clothing and wash immediately with soap and water. Speed in removing material from skin is of extreme importance. Shampoo hair promptly if contaminated. Seek medical attention immediately. If this chemical has been inhaled, remove from exposure, begin rescue breathing (using universal precautions) if breathing has stopped and CPR if heart action has stopped. Transfer promptly to a medical facility. When this chemical has been swallowed, get medical attention. Give large quantities of water and induce vomiting. Do not make an unconscious person vomit.

Personal Protective Methods: Wear protective gloves and clothing to prevent any reasonable probability of skin contact. Safety equipment suppliers/manufacturers can provide recommendations on the most protective glove/clothing material for your operation. All protective clothing (suits, gloves, footwear, headgear) should be clean, available each day, and put on before work. Contact lenses should not be worn when working with this chemical. Wear dust-proof chemical goggles and face shield unless full facepiece respiratory protection is worn. Employees should wash immediately with soap when skin is wet or contaminated. Provide emergency showers and eyewash.

Respirator Selection: NIOSH/OSHA: *Up to 5 mg/m³:* CCROVDMFu [any chemical cartridge respirator with organic vapor cartridge(s) in combination with a dust, mist, and fume filter]; or SA (any supplied-air respirator). *Up to 12.5 mg/m³:* SA:CF (any supplied-air respirator operated in a continuous-flow mode);* or PAPROVDMFu [any powered, air-purifying respirator with organic vapor cartridge(s) in combination with a dust, mist, and fume filter].* *Up to 25 mg/m³:* CCRFOVHiE [any chemical cartridge respirator with a full facepiece and organic vapor cartridge(s) in combination with a high-efficiency particulate filter]; or GMFOVHiE [any air-purifying, full-facepiece respirator (gas mask) with a chin-style, front- or back-mounted organic vapor canister having a high-efficiency particulate filter]; or PAPRTOVHiE [any powered, air-purifying respirator with a tight-fitting facepiece and organic vapor cartridge(s) in combination with a high-efficiency particulate filter];* or SCBAF (any self-contained breathing apparatus with a full facepiece); or SAF (any supplied-air respirator with a full facepiece). *Up to 50 mg/m³:* SAF:PD,PP (any supplied-air respirator that has a full facepiece and is operated in a pressure-demand or other positive-pressure mode). *Emergency or planned entry into unknown concentrations or IDLH conditions:* SCBAF:PD,PP (any self-contained breathing apparatus that has a full facepiece and is operated in a pressure-demand or other positive-pressure mode); or SAF:PD,PP: ASCBA (any supplied-air respirator that has a full facepiece and is operated in a pressure-demand or other positive-pressure mode in combination with an auxiliary self-contained positive-pressure breathing apparatus). *Escape:* GMFOVHiE [any air-purifying, full-facepiece respirator (gas mask) with a chin-style, front- or back-mounted organic vapor canister having a high-efficiency particulate filter]; or SCBAE (any appropriate escape-type, self-contained breathing apparatus).

* Substance reported to cause eye irritation or damage; may require eye protection.

Storage: Prior to working with this chemical you should be trained on its proper handling and storage. Store in tightly closed containers in a cool, well ventilated area away from alkalis, powdered iron, aluminum, and zinc. Where possible, automatically pump liquid from drums or other storage containers to process containers. A regulated, marked area should be established where this chemical is handled, used, or stored in compliance with OSHA standard 1910.1045.

Shipping: This compound falls under the category of organochlorine pesticides, solid, toxic, n.o.s. This compound requires a shipping label of: "Keep Away From Food." It falls in DOT Hazard Class 6.1 and Packing Group III.

Spill Handling: Evacuate persons not wearing protective equipment from area of spill or leak until clean-up is complete. Remove all ignition sources. Collect powdered material in the most convenient and safe manner and deposit in sealed containers. Ventilate area after clean-up is complete. It may be necessary to contain and dispose of this chemical as a hazardous waste. If material or contaminated runoff enters waterways, notify downstream users of potentially contaminated waters. Contact your Department of Environmental Protection or your regional office of the federal EPA for specific recommendations. If employees are required to clean-up spills, they must be properly trained and equipped. OSHA 1910.120(q) may be applicable.

Fire Extinguishing: This chemical may be dissolved in flammable liquids. Use dry chemical, carbon dioxide, water spray, or foam extinguishers. Poisonous gases are produced in fire including phosgene and hydrogen chloride. If material or contaminated runoff enters waterways, notify downstream users of potentially contaminated waters. Notify local health and fire officials and pollution control agencies. From a secure, explosion-proof location, use water spray to cool exposed containers. If cooling streams are ineffective (venting sound increases in volume and pitch, tank discolors, or shows any signs of deforming), withdraw immediately to a secure position. If employees are expected to fight fires, they must be trained and equipped in OSHA 1910.156.

Disposal Method Suggested: A process has been developed for the destructive pyrolysis of benzene hexachloride at 400 – 500°C with a catalyst mixture which contains 5 – 10% of either cupric chloride, ferric chloride, zinc chloride, or aluminum chloride on activated carbon.

References

U.S. Environmental Protection Agency, Hexachtorocyclohexane: Ambient Water Quality Criteria. Washington, DC (1980)

U.S. Environmental Protection Agency, Hexachlorocyclohexane. Health and Environmental Effects Profile No. 112, Office of Solid Waste, Washington, DC (April 30, 1980)

Sax, N. I., Ed., Dangerous Properties of Industrial Materials Report, 7, No. 4, 26–38 (1987) New York, Van Nostrand Reinhold Co. (1983)

U.S. Department of Health and Human Services, "ATSDR ToxFAQs, Hexachlorocyclohexanes," Atlanta, GA (June 1999)

Note: See also Lindane

Hexachlorocyclopentadiene

Molecular Formula: C_5Cl_6

Synonyms: C-56®; 1,3-Cyclopentadiene, 1,2,3,4,5,5-hexachloro-; Graphlox; HCCPD; Hex; Hexachlorcyklopentadien (Czech); 1,2,3,4,5,5-Hexachloro-1,3-cyclopentadiene; Hexachloro-1,3-cyclopentadiene; Hexaclorociclopentadieno (Spanish); HRS1655; NCI-C55607; PCL; Perchlorocyclopentadiene

CAS Registry Number: 77-47-4

RTECS Number: GY1225000

DOT ID: UN 2646

Regulatory Authority

- Air Pollutant Standard Set (ACGIH)[1] (former USSR)[35][43] (Several States)[60]
- CLEAN AIR ACT: Hazardous Air Pollutants (Title I, Part A, Section 112)
- CLEAN WATER ACT: Section 311 Hazardous Substances/ RQ 40CFR117.3 (same as CERCLA, see below); Section 313 Water Priority Chemicals (57FR41331, 9/9/92); 40CFR401.15 Section 307 Toxic Pollutants
- EPA HAZARDOUS WASTE NUMBER (RCRA No.): U130
- RCRA, 40CFR261, Appendix 8 Hazardous Constituents
- RCRA 40CFR268.48; 61FR15654, Universal Treatment Standards: Wastewater (mg/l), 0.057; Nonwastewater (mg/kg), 2.4
- RCRA 40CFR264, Appendix 9; TSD Facilities Ground Water Monitoring List. Suggested test method(s) (PQL μg/l): 8120 (5); 8270 (10)
- SAFE DRINKING WATER ACT: MCL, 0.05 mg/l; MCLG, 0.05 mg/l
- SUPERFUND/EPCRA 40CFR355, Appendix B Extremely Hazardous Substances: TPQ = 100 lb (45.4 kg)
- SUPERFUND/EPCRA 40CFR302.4 Reportable Quantity (RQ): CERCLA, 10 lb (4.54 kg)
- EPCRA Section 313 Form R *de minimis* concentration reporting level: 1.0%
- Canada, WHMIS, Ingredients Disclosure List

Cited in U.S. State Regulations: Alaska (G), California (A, G), Connecticut (A), Florida (G, A), Illinois (G), Indiana (A), Kansas (G, W), Louisiana (G), Maine (G), Maryland (G), Massachusetts (G, A), Michigan (G), Nevada (A), New Hampshire (G), New Jersey (G), New York (G, A), North Carolina (A), North Dakota (A), Pennsylvania (G), Rhode Island (G), South Carolina (A), Vermont (G), Virginia (G, A), Washington (G), West Virginia (G), Wisconsin (G), Wisconsin (G).

Description: Hexachlorocyclopentadiene, C_5Cl_6, is a pale-yellow to amber-colored, oily liquid with a pungent, unpleasant odor. The odor threshold is 0.15 – 0.33 ppm. Boiling point = 239°C. Freezing/Melting point = -8°C.

Hazard Identification (based on NFPA-704 M Rating System): Health 2, Flammability 1, Reactivity 0. Slightly soluble in water (reactive).

Potential Exposure: Hexachlorocyclopentadiene is used to produce the flame retardant chlorendic anhydride, which has applications in polyesters, and to produce chlorendic acid which is used as a flame retardant in resins. Hexachlorocyclopentadiene is also used as an intermediate in the production of pesticides, such as aldrin, dieldrin, and endosulfan.

Incompatibilities: Reacts slowly with water to form hydrochloric acid; will corrode iron and most metals in presence of moisture. Explosive hydrogen gas may collect in enclosed spaces in the presence of moisture. Contact with sodium may be explosive.

Permissible Exposure Limits in Air: There is no OSHA PEL. NIOSH and ACGIH recommend a TWA of 0.01 ppm (0.1 mg/m^3). The former USSR[35][43] has set a MAC in workplace air of 0.01 mg/m^3. A number of states have set guidelines or standards for hexachtorocyclopentadiene in ambient air[60] ranging from 0.015 μg/m^3 (Massachusetts) to 0.33 μg/m^3 (New York) to 0.50 μg/m^3 (Indiana and South Carolina) to 0.6 – 10.0 μg/m^3 (North Carolina) to 1.0 μg/m^3 (Florida and North Dakota) to 2.0 μg/m^3 (Connecticut, Nevada and Virginia).

Determination in Air: Sample collection on Porapak® tube,[2] desorption with hexane and analysis by gas chromatography with electron capture detection. See NIOSH IV, Method # 2518.[18]

Permissible Concentration in Water: The US EPA regulates HCCPD in water. The MCL is 50 ppb. EPA recommends that exposure in children should not exceed 2 ppm in water for 10-day periods or no more then 0.7 ppm for up to 7 years. EPA requires that spills or accidental releases of 10 pounds or more of HCCPD be reported to the EPA. The former USSR[35][43] has set a MAC in water bodies used for domestic purposes of 0.001 mg/l (1 μg/l). Kansas has set a guideline for drinking water of 206 μg/l.[61]

Determination in Water: Methylene chloride extraction followed by concentration, gas chromatography with electron capture detection (EPA Method 612) or gas chromatography plus mass spectrometry (EPA Method 625).

Routes of Entry: Inhalation, skin absorption, ingestion, skin and/or eye contact.

Harmful Effects and Symptoms

Short Term Exposure: This compound is very toxic and may be fatal if inhaled, swallowed, or absorbed through the skin. Eye contact may result in severe irritation. Contact of liquid with the skin may cause blistering and burning. Inhalation of mist is highly irritating to mucous membranes, causing tearing, sneezing, and salivation. Higher exposures can cause pulmonary edema, a medical emergency that can be delayed for several hours. This can cause death. Headaches and throat irritation have also been reported as a result of exposure to this compound. The probable human lethal dose is 50 – 500 mg/kg, or between 1 teaspoon and 1 ounce for a 150 lb (70 kg) person. Severe exposure induces pulmonary hyperemia and edema, degenerative and necrotic changes in brain, heart and adrenal glands and necrosis of liver and kidney tubules. Pulmonary edema is a medical emergency that can be delayed for several hours. This can cause death.

Long Term Exposure: May damage the liver, kidneys, nervous system, and heart. Can irritate the lungs. Repeated exposure may cause bronchitis with cough, phlegm and/or shortness of breath.

Points of Attack: Eyes, skin, respiratory system, liver, kidneys

Medical Surveillance: Before beginning employment and at regular times after that, the following are recommended: Liver and kidney function tests. Lung function tests. If symptoms develop or overexposure is suspected, the following may be useful: lung, liver and kidney function tests. Exam of the nervous system. Consider chest x-ray after acute overexposure.

First Aid: If this chemical gets into the eyes, remove any contact lenses at once and irrigate immediately for at least 15 minutes, occasionally lifting upper and lower lids. Seek medical attention immediately. If this chemical contacts the skin, remove contaminated clothing and wash immediately with soap and water. Speed in removing material from the skin is of extreme importance. Seek medical attention immediately. If this chemical has been inhaled, remove from exposure, begin rescue breathing (using universal precautions) if breathing has stopped and CPR if heart action has stopped. Transfer promptly to a medical facility. When this chemical has been swallowed, get medical attention. Give large quantities of water and induce vomiting. Do not make an unconscious person vomit. Keep victim quiet and maintain normal body temperature. Medical observation is recommended for 24 – 48 hours after breathing overexposure, as pulmonary edema may be delayed. As first aid for pulmonary edema, a doctor or authorized paramedic may consider administering a corticosteroid spray.

Personal Protective Methods: Wear protective gloves and clothing to prevent any reasonable probability of skin contact. Safety equipment suppliers/manufacturers can provide recommendations on the most protective glove/clothing material for your operation. Butyl Rubber/Neoprene and PVC are among the recommended protective materials. All protective clothing (suits, gloves, footwear, headgear) should be clean, available each day, and put on before work. Contact lenses should not be worn when working with this chemical. Wear splash-proof chemical goggles and face shield unless full facepiece respiratory protection is worn. Employees should wash immediately with soap when skin is wet or contaminated. Provide emergency showers and eyewash.

Respirator Selection: Where the potential exists for exposures over 0.01 ppm, use a MSHA/NIOSH approved supplied-air respirator with a full facepiece operated in the positive pressure mode or with a full facepiece, hood, or helmet in the continuous flow mode, or use a MSHA/NIOSH approved self-contained breathing apparatus with a full facepiece operated in pressure-demand or other positive pressure mode.

Storage: Prior to working with this chemical you should be trained on its proper handling and storage. Store in tightly closed containers in a cool, well-ventilated area away from water. Protect containers from physical damage. In the presence of moisture, hexachlorocyclopentadiene will corrode iron and other metals.

Shipping: This compound requires a shipping label of: "Poison." It falls in DOT Hazard Class 6.1 and Packing Group I. Passenger aircraft or railcar shipment is forbidden and cargo aircraft shipment is also forbidden.

Spill Handling: Evacuate and restrict persons not wearing protective equipment from area of spill or leak until cleanup is complete. Remove all ignition sources. Absorb liquids in vermiculite, dry sand, earth, peat, carbon, or a similar material and deposit in sealed containers. Do not use water or wet method. Ventilate area of spill or leak after clean-up is complete. Keep this chemical out of a confined space, such as a sewer, because of the possibility of an explosion, unless the sewer is designed to prevent the build-up of explosive concentrations. It may be necessary to contain and dispose of this chemical as a hazardous waste. If material or contaminated runoff enters waterways, notify downstream users of potentially contaminated waters. Contact your Department of Environmental Protection or your regional office of the federal EPA for specific recommendations. If employees are required to clean-up spills, they must be properly trained and equipped. OSHA 1910.120(q) may be applicable.

Initial isolation and protective action distances

Distances shown are likely to be affected during the first 30 minutes after materials are spilled and could increase with time. If more than one tank car, cargo tank, portable tank, or large cylinder is involved in the incident is leaking, the protective action distance may need to be increased. You may need to seek emergency information from CHEMTREC at (800) 424-9300 or seek professional environmental engineering assistance from the U.S. EPA Environmental Response Team at (908) 548-8730 (24-hour response line).

Small spills (From a small package or a small leak from a large package)

First: Isolate in all directions (feet)..... 500

Then: Protect persons downwind (miles)

Day 0.6

Night 2.6

Large spills (From a large package or from many small packages)

First: Isolate in all directions (feet)..... 1,400

Then: Protect persons downwind (miles)

Day 2.1

Night .... 7.0 +

Fire Extinguishing: Extinguish fire using an agent suitable for type of surrounding fire. Hexachlorocyclopentadiene itself does not burn. Poisonous gases including chlorine, hydrogen chloride, phosgene, and carbon monoxide are produced in fire. Vapors are heavier than air and will collect in low areas. Containers may explode in fire. Storage containers and parts of containers may rocket great distances, in many directions. If material or contaminated runoff enters waterways, notify downstream users of potentially contaminated waters. Notify local health and fire officials and pollution control agencies. From a secure, explosion-proof location, use water spray to cool exposed containers. If cooling streams are ineffective (venting sound increases in volume and pitch, tank discolors, or shows any signs of deforming), withdraw immediately to a secure position. If employees are expected to fight fires, they must be trained and equipped in OSHA 1910.156.

Disposal Method Suggested: Incineration after mixing with another combustible fuel. Care must be exercised to assure complete combustion to prevent the formation of phosgene. An acid scrubber is necessary to remove the halo acids produced. Consult with environmental regulatory agencies for guidance on acceptable disposal practices. Generators of waste containing this contaminant (≥100 kg/mo) must conform with EPA regulations governing storage, transportation, treatment, and waste disposal.

References

U.S. Environmental Protection Agency, Hexachlorocyclopentadiene: Ambient Water Quality Criteria, Washington, DC (1980)

U.S. Environmental Protection Agency, Chemical Hazard Information Profile: Hexachloropentadiene, Washington, DC (March 15, 1977)

National Institute for Occupational Safety and Health, Information Profiles on Potential Occupational Hazards: Hexachlorocyclopentadiene, pp 16–18, Report PB-276, 678, Rockville, Maryland (October 1977)

National Academy of Sciences, Kepone, Mirex, Hexachlorocyclopentadiene: An Environmental Assessment, Washington, DC (1978)

U.S. Environmental Protection Agency, Reviews of the Environmental Effects of Pollutants: XII. Hexachlorocyclopentadiene. Report No. EPA-600/1-78-047, Cincinnati, Ohio (1978)

U.S. Environmental Protection Agency, Hexachlorocyclopentadiene, Health and Environmental Effects Profile No. 114, Office of Solid Waste, Washington, DC (April 30, 1980)

Sax, N. I., Ed., Dangerous Properties of Industrial Materials Report, 4, No. 2, 76–79 (1984)

U.S. Environmental Protection Agency, "Chemical Profile: Hexachlorocyclopentadiene," Washington, DC, Chemical Emergency Preparedness Program (Nov. 30, 1987)

New Jersey Department of Health and Senior Services, "Hazardous Substance Fact Sheet: Hexachlorocyclopentadiene," Trenton, NJ (April, 1999)

New York State Department of Health, "Chemical Fact Sheet: Hexachlorocyclopentadiene," Albany, NY, Bureau of Toxic Substance Assessment (May 1986)

U.S. Department of Health and Human Services, "ATSDR ToxFAQs, Hexachlorocyclopentadiene (HCCPD)" Atlanta, GA (June 1999)

Hexachloroethane

Molecular Formula: C_2Cl_6

Common Formula: CCl_3CCl_3

Synonyms: Avlothane; Carbon hexachloride; Distokal; Distopan; Distopin; Egitol; Ethane hexachloride; Ethane, hexachloro-; Ethylene hexachloride; Falkitol; Fasciolin; HCE; Hexachloraethan (German); Hexachlorethane; 1,1,1,2,2,2-Hexachloroethane; Hexachloroethylene; Hexacloroetano (Spanish); Mottenhexe; NCI-C04604; Perchloroethane; Phenohep

CAS Registry Number: 67-72-1

RTECS Number: K14025000

DOT ID: UN 9037

Regulatory Authority

- Air Pollutant Standard Set (ACGIH)[1] (DFG)[3] (HSE)[33] (OSHA)[58] (Several States)[60]
- Carcinogen (Animal Positive) (IARC)[9]
- CLEAN AIR ACT: Hazardous Air Pollutants (Title I, Part A, Section 112)
- CLEAN WATER ACT: Section 313 Water Priority Chemicals (57FR41331, 9/9/92); Toxic Pollutant (Section 401.15) as chlorinated ethanes
- EPA HAZARDOUS WASTE NUMBER (RCRA No.): U131; D034
- RCRA, 40CFR261, Appendix 8 Hazardous Constituents
- RCRA Toxicity Characteristic (Section 261.24), Maximum Concentration of Contaminants, regulatory level, 3.0 mg/l
- RCRA 40CFR268.48; 61FR15654, Universal Treatment Standards: Wastewater (mg/l), 0.055; Nonwastewater (mg/kg), 30
- RCRA 40CFR264, Appendix 9; TSD Facilities Ground Water Monitoring List. Suggested test method(s) (PQL μg/l): 8120 (0.5); 8270 (10)
- SAFE DRINKING WATER ACT: Priority List (55 FR 1470)
- SUPERFUND/EPCRA 40CFR355, Appendix B Extremely Hazardous Substances: TPQ = 100 lb (45.4 kg)
- SUPERFUND/EPCRA 40CFR302.4 Reportable Quantity (RQ): CERCLA, 100 lb (45.4 kg)
- EPCRA Section 313 Form R *de minimis* concentration reporting level: 1.0%
- Canada, WHMIS, Ingredients Disclosure List

Cited in U.S. State Regulations: Alaska (G), California (A, G), Connecticut (A), Florida (G), Illinois (G), Kansas (G, A, W), Louisiana (G), Maine (G), Massachusetts (G, A), Michigan (G), Minnesota (W), Nevada (A), New Hampshire (G), New Jersey (G), North Dakota (A), Pennsylvania (G), Rhode Island (G), Vermont (G), Virginia (G, A), Washington (G), West Virginia (G), Wisconsin (G).

Description: Hexachloroethane, CCl_3CCl_3, is a colorless solid with a camphor-like odor. It gradually evaporates when it is exposed to air. Boiling point = (sublimes) @ 183 – 187°C. Freezing/Melting point = (sublimes) 187°C. Insoluble in water.

Potential Exposure: In the US, about half the HCE is used by the military for smoke-producing devices. It is also used to remove air bubbles in melted aluminum. It may be present as an ingredient in some fungicides, insecticides, lubricants, and plastics. It is no longer made in the United States, but it is formed as a by-product in the production of some chemicals. Can be formed by incinerators when materials containing chlorinated hydrocarbons are burned. Some HCE can also be formed when chlorine reacts with carbon compounds in drinking water. As a medicinal, HCE is used as an anthelmintic to treat fascioliasis in sheep and cattle. It is also added to the feed of ruminants, preventing methanogenesis and increasing feed efficiency. HCE is used in metal and alloy production, mainly in refining aluminum alloys. It is also used for removing impurities from molten metals, recovering metals from ores or smelting products and improving the quality of various metals and alloys. HCE is contained in pyrotechnics. It inhibits the explosiveness of methane and the combustion of ammonium perchlorate. Smoke containing HCE is used to extinguish fires. HCE has various applications as a polymer additive. It has flameproofing qualities, increases sensitivity to radiation crosslinking, and is used as a vulcanizing agent. Added to polymer fibers, HCE acts as a swelling agent and increases affinity for dyes.

Incompatibilities: Incompatible with metals such as aluminum, cadmium, hot iron, mercury, or zinc. Alkalis forms spontaneously explosive chloroacetylene. Attacks some plastics, rubber and coatings.

Permissible Exposure Limits in Air: The Federal standard[58] is 1.0 ppm (10 mg/m^3). NIOSH and ACGIH recommends the same TWA. The notation "skin" is added to indicate the possibility of cutaneous absorption. The NIOSH IDLH level is (Ca) 300 ppm. The DFG[3] has set an MAK value of 1.0 ppm (10 mg/m^3) but HSE[33] has more complex limits; they are 5 ppm (50 mg/m^3) for vapor; 10 mg/m^3 for total inhalable dust and 5 mg/m^3 for respirable dust. Several states have set guidelines or standards for hexachloroethane in ambient air[60] ranging from 0.18 μg/m^3 (Massachusetts) to 50.0 μg/m^3 (Connecticut) to 238.095 μg/m^3 (Kansas) to 1000 (μg/m^3 (North Dakota) to 1,600 μg/m^3 (Virginia) to 2,381 μg/m^3 (Nevada).

Determination in Air: Charcoal adsorption, workup with CS_2, analysis by gas chromatography/flame ionization. See NIOSH Method 1003 for halogenated hydrocarbons.

Permissible Concentration in Water: The US EPA suggests (1997)that water consumed over a lifetime contain no more than 1 part HCE per billion parts of water. (1 ppb). In an older citation, the EPA suggested the following: To protect freshwater aquatic life – 118,000 μg/l based on acute toxicity and 20,000 μg/l based on chronic toxicity. To protect saltwater aquatic life – 113,000 μg/l based on acute toxicity. To protect human health – 9.4 μg/l to keep lifetime cancer risk below 5 – 10 μg/l.[6] The former USSR[35][43] has set a MAC in water bodies used for domestic purposes of 0.01 mg/l (10 μg/l). States which have set guidelines for hexachloroethane in drinking water[61] include Kansas (1.9 μg/l) and Minnesota (24.6 μg/l).

Determination in Water: Methylene chloride extraction followed by concentration, gas chromatography with electron capture detection (EPA Method 612) or gas chromatography plus mass spectrometry (EPA Method 625).

Routes of Entry: Inhalation, skin absorption, ingestion, eye and/or skin contact.

Harmful Effects and Symptoms

Short Term Exposure: Contact can irritate and burn the eyes and skin. Exposure can irritate the respiratory tract. High levels of exposure can cause dizziness, lightheadedness, and unconsciousness. Irritation occurs when there is an excessive amount of hexachloroethane dust in the air or when it is heated and vapors are formed. Hexachloroethane acts primarily as a central nervous system depressant, and in high concentrations it causes tremors, narcosis. It should be noted that the low vapor pressure of this compound as well as its solid state minimize its inhalation hazards.

Long Term Exposure: A potential occupational carcinogen. May cause kidney and liver damage.

Points of Attack: Eyes, skin, respiratory system, kidneys. Cancer site in animals: liver cancer.

Medical Surveillance: For those with frequent or potentially high exposure (half the TLV or greater) the following are recommended before beginning work and at regular times after that: Liver and kidney function tests. More than light alcohol consumption can exacerbate liver damage. Sample of blood, urine, or feces can be tested for exposure to HCE. These tests are useful only if exposure occurred 24 – 48 hours prior to testing.

First Aid: If this chemical gets into the eyes, remove any contact lenses at once and irrigate immediately for at least 15 minutes, occasionally lifting upper and lower lids. Seek medical attention immediately. If this chemical contacts the skin, remove contaminated clothing and wash immediately with soap and water. Seek medical attention immediately. If this chemical has been inhaled, remove from exposure, begin rescue breathing (using universal precautions) if breathing has stopped and CPR if heart action has stopped. Transfer promptly to a medical facility. When this chemical has been swallowed, get medical attention. Give large quantities of water and induce vomiting. Do not make an unconscious person vomit.

Personal Protective Methods: Wear protective gloves and clothing to prevent any reasonable probability of skin contact. Safety equipment suppliers/manufacturers can provide recommendations on the most protective glove/clothing material for your operation. All protective clothing (suits, gloves, footwear, headgear) should be clean, available each day, and put on before work. Contact lenses should not be worn when working with this chemical. Wear dust-proof chemical goggles and face shield unless full facepiece respiratory protection is worn. Employees should wash immediately with soap when skin is wet or contaminated. Provide emergency showers and eyewash.

Respirator Selection: NIOSH: *At any concentrations above the NIOSH REL, or where there is no REL, at any detectable concentration:* SCBAF:PD,PP (any MSHA/NIOSH approved self-contained breathing apparatus that has a full facepiece and is operated in a pressure-demand or other positive-pressure mode) or SAF:PD,PP:ASCBA (any supplied-air respirator that has a full facepiece and is operated in a pressure-demand or other positive-pressure mode in combination with an auxiliary self-contained breathing apparatus operated in a pressure-demand or other positive pressure mode). *Escape:* GMFOV [any air-purifying, full-facepiece respirator (gas mask) with a chin-style, front-or back-mounted organic vapor canister]; or SCBAE (any appropriate escape-type, self-contained breathing apparatus).

Storage: Prior to working with this chemical you should be trained on its proper handling and storage. Hexachloroethane must be stored to avoid contact with hot iron, zinc, aluminum, and alkalis, since violent reactions occur. Store in tightly closed containers in a cool, well-ventilated area away from heat. A regulated, marked area should be established where this chemical is handled, used, or stored in compliance with OSHA standard 1910.1045.

Shipping: Hexachloroethane has no specific label or shipping weight regulations.[52]

Spill Handling: Evacuate persons not wearing protective equipment from area of spill or leak until clean-up is complete. Remove all ignition sources. Collect powdered material in the most convenient and safe manner and deposit in sealed containers. Ventilate area after clean-up is complete. It may be necessary to contain and dispose of this chemical as a hazardous waste. If material or contaminated runoff enters waterways, notify downstream users of potentially contaminated waters. Contact your Department of Environmental Protection or your regional office of the federal EPA for specific recommendations. If employees are required to clean-up spills, they must be properly trained and equipped. OSHA 1910.120(q) may be applicable.

Fire Extinguishing: Extinguish fire using an agent suitable for type of surrounding fire. Hexachloroethane itself does not burn. Poisonous gases including phosgene and hydrogen chloride are produced in fire. If material or contaminated runoff enters waterways, notify downstream users of potentially contaminated waters. Notify local health and fire officials and pollution control agencies. From a secure, explosion-proof location, use water spray to cool exposed containers. If cooling streams are ineffective (venting sound increases in volume and pitch, tank discolors, or shows any signs of deforming), withdraw immediately to a secure position. If employees are expected to fight fires, they must be trained and equipped in OSHA 1910.156.

Disposal Method Suggested: Incineration after mixing with another combustible fuel. Care must be exercised to assure complete combustion to prevent the formation of phosgene. An acid scrubber is necessary to remove the halo acids produced. Consult with environmental regulatory agencies for guidance on acceptable disposal practices. Generators of waste containing this contaminant (≥100 kg/mo) must conform with EPA regulations governing storage, transportation, treatment, and waste disposal.

References

U.S. Environmental Protection Agency, Chlorinated Ethanes: Ambient Water Quality Criteria, Washington, DC (1980)

U.S. Environmental Protection Agency, Chemical Hazard Information Profile: Hexachloroethane, Washington, DC (1979)

U.S. Environmental Protection agency, Hexachloroethane, Health and Environmental Effects Profile No. 116, Office of Solid Waste, Washington, DC (April 30, 1980)

New Jersey Department of Health and Senior Services, "Hazardous Substance Fact Sheet: Hexachloroethane," Trenton, NJ (May 1998)

Sax, N. I., Ed., "Dangerous Properties of Industrial Materials Report" 2, No. 6, 75–78 (1982) and 6, No. 4, 70–83 (1986)

U.S. Department of Health and Human Services, "ATSDR ToxFAQs, Hexachloroethane," Atlanta, GA (September 1997)

Hexachlorophene

Molecular Formula: $C_{13}H_6Cl_6O_2$

Common Formula: $C_6H(OH)Cl_3CH_2C_6H(OH)Cl_3$

Synonyms: Acigena; AI3-02372; Almederm; AT-17; AT-7; B 32; B&B Flea kontroller for dogs only; Bilevon; Bis(2-hydroxy-3,5, 6-trichlorophenyl)methane; Bis-2,3,5-trichlor-6-hydroxyfenylmethan (Czech); Bis(3,5,6-trichlor *o*-2-hydroxyphenyl)methane; Blockade anti bacterial finish; Brevity blue liquid bacteriostatic scouring cream; Brevity blue liquid sanitizing scouring cream; Compound G-11; Cotofilm; Dermadex; 2,2'-Dihydroxy-3,3',5,5',6,6'-hexachlorodiphenylmethane; 2,2'-Dihydroxy-3,5,6,3',5',6'-hexachlorodiphenylmethane; Distodin; Enditch pet shampoo; EN-Viron D concentrated phenolic disinfectant; Esaclorofene; Exofene; Fesia-Sin; Fomac; Fostril; G-11; Gamophen; Gamophene; Germa-Medica; HCP; Hexabalm; 2,2',3,3',5,5'-Hexachloro-6,6'-dihydroxydiphenylmethane; Hexachlorofen (Czech); Hexachlorophane; Hexachlorophen; Hexachlorophene; Hexaclorofeno (Spanish); Hexafen; Hexaphene-LV; Hexide; Hexophene; Hexosan; Hilo Cat Flea powder; Hilo Flea powder; Hilo Flea powder with Rotenone and Dichlorophrene; Isobac; Isobac 20; Methane, bis(2,3,5-trichloro-6-hydroxyphenyl); 2,2'-Methylenebis(3,4,6-trichlorophenol); 2,2'-Methylenebis(3,5,6-trichlorophenol); Nabac; Nabac 25 EC; NCI-C02653; Neosept V; NSC4911; Pedigree dog shampoo bar; Phenol, 2,2'-methylenebis(3,4,6-trichloro)-; Phenol, 2,2'-methylenebis(3,5,6-trichloro-); Phisodan; Phisohex; Ritosept; Septisol; Septofen; Staphene O; Steral; Steraskin; Surgi-Cen; Surgi-Cin; Surofene; Tersaseptic; Trichlorophene; Turgex

CAS Registry Number: 70-30-4

RTECS Number: SM0700000

DOT ID: UN 2875

EEC Number: 604-015-00-9

Regulatory Authority

- Air Pollutant Standard Set (Massachusetts)[60]
- Banned or Severely Restricted (In Pharmaceuticals) (UN)[13]
- SUPERFUND/EPCRA 40CFR302.4 Reportable Quantity (RQ): CERCLA, 100 lb (45.4 kg)
- EPA HAZARDOUS WASTE NUMBER (RCRA No.): U132
- RCRA 40CFR264, Appendix 9; TSD Facilities Ground Water Monitoring List. Suggested test method(s) (PQL μg/l): 8270 (10)
- EPCRA Section 313 Form R *de minimis* concentration reporting level: 1.0%

Cited in U.S. State Regulations: California (G), Kansas (G), Louisiana (G), Massachusetts (A), Maine (W), New Jersey (G), Pennsylvania (G), Vermont (G), Virginia (G), Washington (G), Wisconsin (G).

Description: Hexachlorophene, $C_6H(OH)Cl_3CH_2C_6H(OH)Cl_3$, is a crystalline compound. Freezing/Melting point = 165°C. Insoluble in water.

Potential Exposure: HCP has been used as an antibacterial agent in a wide variety of consumer products, including soaps and deodorants. It has also been used as an antifungal agent to treat various citrus fruits and vegetables.

Incompatibilities: Oxidizers.

Permissible Exposure Limits in Air: No OELs have been established. Massachusetts[61] has set a guideline for ambient air of zero.

Permissible Concentration in Water: A no-adverse-effect level in drinking water has been calculated by NAS/NRC as 0.008 mg/l. An ADI was calculated on the basis of the available chronic toxicity data to be 0.0012 mg/kg/day. The former USSR[35] has set a MAC in water bodies used for domestic purposes of 0.03 mg/l. Maine has set a guideline for drinking water of 2.0 μg/l.[61]

Routes of Entry: Inhalation, passing through the skin.

Harmful Effects and Symptoms

Short Term Exposure: Hexachlorophene may irritate the eyes and skin, and cause a skin allergy to develop. May cause permanent eye damage. Inhaling can irritate the respiratory tract.

May affect the central nervous system, causing dizziness, weakness, convulsions (fits), coma, or death. Exposure can cause loss of appetite, nausea, vomiting, cramps and diarrhea.

Long Term Exposure: There is an association between exposure to pregnant women to hexachlorophene and birth defects. There is limited evidence that hexachlorophene is a teratogen in animals. May cause an asthma-like allergy. May cause liver damage. Repeated exposure may cause brain damage leading to paralysis and blindness.

Points of Attack: Nervous system, skin, liver, reproductive system.

Medical Surveillance: If symptoms develop or overexposure is suspected, the following may be useful: examination of the nervous system. Eye exam. Evaluation by a qualified allergist, including careful exposure history and special testing, may help diagnose skin allergy. More than light alcohol consumption may exacerbate liver damage.

First Aid: If this chemical gets into the eyes, remove any contact lenses at once and irrigate immediately for at least 15 minutes, occasionally lifting upper and lower lids. Seek medical attention immediately. If this chemical contacts the skin, remove contaminated clothing and wash immediately with soap and water. Seek medical attention immediately. If this chemical has been inhaled, remove from exposure, begin rescue breathing (using universal precautions) if breathing has stopped and CPR if heart action has stopped. Transfer promptly to a medical facility. When this chemical has been swallowed, get medical attention. Give large quantities of water and induce vomiting. Do not make an unconscious person vomit.

Personal Protective Methods: Wear protective gloves and clothing to prevent any reasonable probability of skin contact. Safety equipment suppliers/manufacturers can provide recommendations on the most protective glove/clothing material for your operation. All protective clothing (suits, gloves, footwear, headgear) should be clean, available each day, and put on before work. Contact lenses should not be worn when working with this chemical. Wear dust-proof chemical goggles and face shield unless full facepiece respiratory protection is worn. Employees should wash immediately with soap when skin is wet or contaminated. Provide emergency showers and eyewash.

Respirator Selection: Where the potential exists for exposures to hexachlorophene, use a MSHA/NIOSH approved full facepiece respirator with a high efficiency paniculate filter. Greater protection is provided by a powered-air purifying respirator. Particulate filters must be checked every day before work for physical damage, such as rips or tears, and replaced as needed.

Storage: Prior to working with this chemical you should be trained on its proper handling and storage. Store in tightly closed containers in a cool, well-ventilated area away from oxidizers.

Shipping: This compound requires a shipping label of: "Keep Away From Food." It falls in DOT Hazard Class 6.1 and Packing Group III.

Spill Handling: Evacuate persons not wearing protective equipment from area of spill or leak until clean-up is complete. Remove all ignition sources. Collect powdered material in the most convenient and safe manner and deposit in sealed containers. Ventilate area after clean-up is complete. It may be necessary to contain and dispose of this chemical as a hazardous waste. If material or contaminated runoff enters waterways, notify downstream users of potentially contaminated waters. Contact your Department of Environmental Protection or your regional office of the federal EPA for specific recommendations. If employees are required to clean-up spills, they must be properly trained and equipped. OSHA 1910.120(q) may be applicable.

Fire Extinguishing: Hexachlorophene itself does not burn. Extinguish fire using an agent suitable for surrounding fire. Poisonous gases including chlorine are produced in fire. If material or contaminated runoff enters waterways, notify downstream users of potentially contaminated waters. Notify local health and fire officials and pollution control agencies. From a secure, explosion-proof location, use water spray to cool exposed containers. If cooling streams are ineffective (venting sound increases in volume and pitch, tank discolors, or shows any signs of deforming), withdraw immediately to a secure position. If employees are expected to fight fires, they must be trained and equipped in OSHA 1910.156.

Disposal Method Suggested: Incineration, preferably after mixing with another combustible fuel. Care must be exercised to assure complete combustion to prevent the formation of phosgene. An acid scrubber is necessary to remove the halo acids produced.[22] Consult with environmental regulatory agencies for guidance on acceptable disposal practices. Generators of waste containing this contaminant (≥100 kg/mo) must conform with EPA regulations governing storage, transportation, treatment, and waste disposal.

References

U.S. Environmental Protection Agency, Hexachlorophene, Health and Environmental Effects Profile No. 116, Office of Solid Waste, Washington, DC (April 30, 1980)

Sax, N. I., Ed., "Dangerous Properties of Industrial Materials Report" 6, No 2, 62–66 (1986)

New Jersey Department of Health and Senior Services, "Hazardous Substance Fact Sheet: Hexachlorophene," Trenton, NJ (April 1999)

Hexafluoroacetone

Molecular Formula: C_3F_6O

Common Formula: CF_3COCF_3

Synonyms: 6FK; Acetone, hexafluoro-; HFA; NCI-C56440; 2-Propanone, 1,1,1,2,2,2-hexafluoro-

CAS Registry Number: 684-16-2

RTECS Number: UC2450000

DOT ID: UN 2420

Regulatory Authority

- Air Pollutant Standard Set (ACGIH)[1] (OSHA)[58] (Several States)[60]
- Canada, WHMIS, Ingredients Disclosure List

Cited in U.S. State Regulations: Alaska (G), California (G), Connecticut (A), Florida (G), Illinois (G), Maine (G), Massachusetts (G), Nevada (A), New Hampshire (G), New Jersey (G), North Dakota (A), Pennsylvania (G), Rhode Island (G), Virginia (A), West Virginia (G).

Description: Hexafluoroacetone, CF_3COCF_3, is a colorless, nonflammable gas, with a musty odor. Shipped as a liquefied compressed gas. Boiling point = -26 – -28°C. Freezing/Melting point = -122°C. Hazard Identification (based on NFPA-704 M Rating System): Health 3, Flammability 0, Reactivity 0. Reacts with water.

Potential Exposure: Hexafluoroacetone is used as a chemical intermediate. A gas at room temperature, it forms various hydrates with water which are used as solvents for resins and polymers. Other derivatives are used to make water repellent coatings for textiles and also to produce polymers.

Incompatibilities: Reacts with water, oxidizers, strong acids. Hygroscopic (i.e., absorbs moisture from the air); reacts with moisture to form a highly acidic sesquihydrate and considerable heat.

Permissible Exposure Limits in Air: There is no OSHA PEL. NIOSH and ACGIH recommend a TWA of 0.1 ppm (0.7 mg/m^3). They both add the notation "Skin" indicating the possibility of cutaneous absorption. Several states have set guidelines or standards for hexafluoroacetone in ambient air[60] ranging from 7.0 μg/m^3 (North Dakota) to 11.0 μg/m^3 (Virginia) to 14.0 μg/m^3 (Connecticut) to 17.0 μg/m^3 (Nevada).

Determination in Air: No method available.

Permissible Concentration in Water: No criteria set.

Routes of Entry: Inhalation, passing through the skin.

Harmful Effects and Symptoms

Short Term Exposure: Exposure can severely irritate the eyes, nose, throat, and skin. Higher exposures can cause pulmonary edema, a medical emergency that can be delayed for several hours. This can cause death. Exposure can cause headache, nausea, vomiting, dizziness, and lightheadedness. Skin or eye contact with the liquid can cause frostbite.

Long Term Exposure: There is a limited evidence that this chemical is a teratogen in animals. It may damage the testes (male reproductive glands) and affect sperm production. Hexafluoroacetone can damage the liver, kidneys and lungs. Repeated exposure may cause bronchitis. Prolonged exposure may affect blood cells.

Points of Attack: Eyes, skin, respiratory system, kidneys, reproductive system.

Medical Surveillance: For those with frequent or potentially high exposure (half the TLV or greater), the following are recommended before beginning work and at regular times after that: lung function tests, reproductive history for men. If symptoms develop or overexposure is suspected, the following may be useful: liver and kidney function tests. Consider chest x-ray after acute overexposure. Complete blood count (CBC).

First Aid: If this chemical gets into the eyes, remove any contact lenses at once and irrigate immediately for at least 15 minutes, occasionally lifting upper and lower lids. Seek medical attention immediately. If this chemical contacts the skin, remove contaminated clothing and wash immediately with soap and water. Seek medical attention immediately. If this chemical has been inhaled, remove from exposure, begin rescue breathing (using universal precautions) if breathing has stopped and CPR if heart action has stopped. Transfer promptly to a medical facility. When this chemical has been swallowed, get medical attention. Give large quantities of water and induce vomiting. Do not make an unconscious person vomit. Medical observation is recommended for 24 – 48 hours after breathing overexposure, as pulmonary edema may be delayed. As first aid for pulmonary edema, a doctor or authorized paramedic may consider administering a corticosteroid spray. If frostbite has occurred, seek medical attention immediately; do *NOT* rub the affected areas or flush them with water. In order to prevent further tissue damage, do *NOT* attempt to remove frozen clothing from frostbitten areas. If frostbite has *NOT* occurred, immediately and thoroughly wash contaminated skin with soap and water.

Personal Protective Methods: Wear solvent resistant gloves and clothing to prevent any reasonable probability of skin contact. Safety equipment suppliers/manufacturers can provide recommendations on the most protective glove/clothing material for your operation. All protective clothing (suits, gloves, footwear, headgear) should be clean, available each day, and put on before work. Contact lenses should not be worn when working with this chemical. Wear splash-proof chemical goggles and face shield when working with the liquid unless full facepiece respiratory protection is worn. Employees should wash immediately with soap when skin is wet or contaminated. Provide emergency showers and eyewash. Where exposure to the liquefied compressed gas may occur, employees should be provided with special clothing designed to prevent frostbite.

Respirator Selection: Where the potential exists for exposures over 0.1 ppm, use a MSHA/NIOSH approved supplied-air respirator with a full facepiece operated in the positive pressure mode or with a full facepiece, hood, or helmet in the continuous flow mode, or use a MSHA/NIOSH approved self-contained breathing apparatus with a full facepiece operated in pressure-demand or other positive pressure mode.

Storage: Prior to working with this chemical you should be trained on its proper handling and storage. Store in tightly closed containers in a cool, well-ventilated area away from direct sunlight, water, heat, reducing agents, nitrates, and nitric acid. Procedures for the handling, use and storage of cylinders should be in compliance with OSHA 1910.101 and 1910.169 as with the recommendations of the Compressed Gas Association.

Shipping: This compound requires a shipping label of: "Poison Gas." It falls in DOT Hazard Class 2.3 and Packing Group I. Passenger aircraft or railcar shipment is forbidden and cargo aircraft shipment is forbidden also.

Spill Handling: Evacuate and restrict persons not wearing protective equipment from area of spill or leak until cleanup is complete. Remove all ignition sources. Ventilate area of spill or leak. Absorb liquids in vermiculite, dry sand, earth, peat, carbon, or a similar material and deposit in sealed containers. If hexafluoroacetone gas is leaked, take the following steps: Restrict persons not wearing protective equipment from area of leak until clean-up is complete. Ventilate area of leak to disperse the gas. Stop flow of gas. If source of leak is a cylinder and the leak cannot be stopped in place, remove the leaking cylinder to a safe place in the open air, and repair leak or allow cylinder to empty. Keep this chemical out of a confined space, such as a sewer, because of the possibility of an explosion, unless the sewer is designed to prevent the build-up of explosive concentrations. It may be necessary to contain and dispose of this chemical as a hazardous waste. If material or contaminated runoff enters waterways, notify downstream users of potentially contaminated waters. Contact your Department of Environmental Protection or your regional office of the federal EPA for specific recommendations. If employees are required to clean-up spills, they must be properly trained and equipped. OSHA 1910.120(q) may be applicable.

Initial isolation and protective action distances

Distances shown are likely to be affected during the first 30 minutes after materials are spilled and could increase with time. If more than one tank car, cargo tank, portable tank, or large cylinder is involved in the incident is leaking, the protective action distance may need to be increased. You may need to seek emergency information from CHEMTREC at (800) 424-9300 or seek professional environmental engineering assistance from the U.S. EPA Environmental Response Team at (908) 548-8730 (24-hour response line).

Small spills (From a small package or a small leak from a large package)

First: Isolate in all directions (feet) 200

Then: Protect persons downwind (miles)

Day 0.2

Night 0.6

Large spills (From a large package or from many small packages)

First: Isolate in all directions (feet) 700

Then: Protect persons downwind (miles)

Day 0.5

Night 2.2

Fire Extinguishing: HFA is non-flammable reactive gas. Poisonous gases including carbon monoxide and hydrogen fluoride are produced in fire. Use dry chemical, carbon dioxide, or alcohol foam extinguishers. Vapors are heavier than air and will collect in low areas. Containers may explode in fire. Storage containers and parts of containers may rocket great distances, in many directions. If material or contaminated runoff enters waterways, notify downstream users of potentially contaminated waters. Notify local health and fire officials and pollution control agencies. From a secure, explosion-proof location, use water spray to cool exposed containers. If cooling streams are ineffective (venting sound increases in volume and pitch, tank discolors, or shows any signs of deforming), withdraw immediately to a secure position. If employees are expected to fight fires, they must be trained and equipped in OSHA 1910.156.

References

U.S. Environmental Protection Agency, Chemical Hazard Information Profile: Hexafluoroacetone, Washington, DC (1979)

Sax, N. I., Ed., Dangerous Properties of Industrial Materials Report, 1, No. 4, 75–76 (1981)

New Jersey Department of Health and Senior Services, "Hazardous Substance Fact Sheet: Hexafluoroacetone," Trenton, NJ (Feb. 1986)

Hexafluoroethane

Molecular Formula: C_2F_6

Synonyms: Ethane, hexafluoro-; F116; Freon 116; Perfluoroethane; R-116

CAS Registry Number: 76-16-4

RTECS Number: KI4110000

DOT ID: UN 2193

Cited in U.S. State Regulations: New Hampshire (G), New Jersey (G), Pennsylvania (G).

Description: Hexafluoroethane is a colorless and odorless gas, or liquid under pressure. Boiling point = -78°C. Hazard Identification (based on NFPA-704 M Rating System): Health 1, Flammability 0, Reactivity 0. Insoluble in water.

Potential Exposure: It is used as a coolant, in dielectric fluids and as a propellant and refrigerant.

Incompatibilities: Active metals. Keep away from heat and sunlight.

Permissible Exposure Limits in Air: No standards set.

Permissible Concentration in Water: No criteria set.

Routes of Entry: Inhalation.

Harmful Effects and Symptoms

Short Term Exposure: Hexafluoroethane can affect you when breathed in. Contact with the liquefied gas could cause frostbite. High levels in the air can cause you to feel dizzy, lightheaded, and to pass out. Very high levels could cause suffocation from lack of oxygen and death. Overexposure may also cause abnormal heart rhythms and even cause the heart to stop.

Points of Attack: Heart.

Medical Surveillance: Special 24-hour EKG (Holter monitor). Consider chest x-ray following acute overexposure.

First Aid: If this chemical gets into the eyes, remove any contact lenses at once and irrigate immediately for at least 15 minutes, occasionally lifting upper and lower lids. Seek medical attention immediately. If this chemical contacts the skin, remove contaminated clothing and wash immediately with soap and water. Seek medical attention immediately. If this chemical has been inhaled, remove from exposure, begin rescue breathing (using universal precautions) if breathing has stopped and CPR if heart action has stopped. Transfer promptly to a medical facility. When this chemical has been swallowed, get medical attention. Give large quantities of water and induce vomiting. Do not make an unconscious person vomit. If frostbite has occurred, seek medical attention immediately; do *NOT* rub the affected areas or flush them with water. In order to prevent further tissue damage, do *NOT* attempt to remove frozen clothing from frostbitten areas. If frostbite has *NOT* occurred, immediately and thoroughly wash contaminated skin with soap and water.

Personal Protective Methods: Wear protective gloves and clothing to prevent any reasonable probability of skin contact. Safety equipment suppliers/manufacturers can provide recommendations on the most protective glove/clothing material for your operation. All protective clothing (suits, gloves, footwear, headgear) should be clean, available each day, and put on before work. Contact lenses should not be worn when working with this chemical. Wear gas-proof goggles, unless full facepiece respiratory protection is worn. Wear splash chemical goggles and face shield when working with the liquid unless full facepiece respiratory protection is worn. Employees should wash immediately with soap when skin is wet or contaminated. Provide emergency showers and eyewash. Where exposure to cold equipment, vapors, or liquid may occur employees should be provided with special clothing designed to prevent the freezing of body tissues.

Respirator Selection: Where the potential for exposures to hexafluoroethane exists, use a MSHA/NIOSH approved supplied-air respirator with a full facepiece operated in the positive pressure mode or with a full facepiece, hood, or helmet in the continuous flow mode, or use a MSHA/NIOSH approved self-contained breathing apparatus with a full facepiece operated in pressure-demand or other positive pressure mode.

Storage: Prior to working with this chemical you should be trained on its proper handling and storage. Store in tightly closed containers in a cool, well-ventilated area away from metals, including aluminum, zinc and beryllium, and from open flames or temperatures above 52°C. Procedures for the handling, use and storage of cylinders should be in compliance with OSHA 1910.101 and 1910.169 as with the recommendations of the Compressed Gas Association.

Shipping: This compound requires a shipping label of: "Nonflammable Gas." It falls in DOT Hazard Class 2.2.

Spill Handling: Evacuate and restrict persons not wearing protective equipment from area of spill or leak until cleanup is complete. Remove all ignition sources. Ventilate area of spill or leak. Ventilate area of leak to disperse the gas. Stop flow of gas. If source of leak is a cylinder and the leak cannot be stopped in place, remove the leaking cylinder to a safe place in the open air, and repair leak or allow cylinder to empty. Absorb liquids in vermiculite, dry sand, earth, or a similar material and deposit in sealed containers. Keep this chemical out of a confined space, such as a sewer, because of the possibility of an explosion, unless the sewer is designed to prevent the build-up of explosive concentrations. It may be necessary to contain and dispose of this chemical as a hazardous waste. If material or contaminated runoff enters waterways, notify downstream users of potentially contaminated waters. Contact your Department of Environmental Protection or your regional office of the federal EPA for specific recommendations. If employees are required to clean-up spills, they must be properly trained and equipped. OSHA 1910.120(q) may be applicable.

Fire Extinguishing: Hexafluoroethane is a nonflammable liquid and gas. Use extinguishing agents suitable for type of surrounding fire. Poisonous gases including hydrofluoric acid and other fluoride gases are produced in fire. Vapors are heavier than air and will collect in low areas. Containers may explode in fire. Storage containers and parts of containers may rocket great distances, in many directions. If material or contaminated runoff enters waterways, notify downstream users of potentially contaminated waters. Notify local health and fire officials and pollution control agencies. From a secure, explosion-proof location, use water spray to cool exposed containers. If cooling streams are ineffective (venting sound increases in volume and pitch, tank discolors, or shows any signs of deforming), withdraw immediately to a secure position. If employees are expected to fight fires, they must be trained and equipped in OSHA 1910.156.

References

New Jersey Department of Health and Senior Services, "Hazardous Substance Fact Sheet: Hexafluoroethane," Trenton, NJ (March, 1999)

Hexamethylenediamine

Molecular Formula: $C_6H_{16}N_2$

Common Formula: $H_2N(CH_2)_6NH_2$

Synonyms: 1,6-Diaminohexane; 1,6-Hexamethylenediamine; Hexamethylenediamine; 1,6-Hexanediamine; HMDA; NCI-C61405

CAS Registry Number: 124-09-4

RTECS Number: MO1180000

DOT ID: UN 2280 (solid); UN 1783 (solution)

Regulatory Authority

- Air Pollutant Standard Set (former USSR)[43]

Cited in U.S. State Regulations: Maine (G), New Hampshire (G), New Jersey (G), Oklahoma (G), Pennsylvania (G).

Description: HMDA, $C_6H_{16}N_2$, is a hygroscopic, colorless solid (pellets or flakes). The odor threshold is 0.004 mg/rn^3. Boiling point = 199°C.

Freezing/Melting point = 39 – 42°C. Flash point = 71°C; 81°C (solution). Autoignition temperature= 310°C. Explosive limits: LEL = 0.7%; UEL = 6.3%. Soluble in water.

Potential Exposure: HMDA is used as a raw material for nylon fiber and plastics; in the manufacture of oil-modified and moisture-area types of urethane coatings; in the manufacture of polyamides for printing inks, dimer acids, and textiles; and as an oil and lubricant additive (probably as a corrosion inhibitor); also used in paints and as a curing agent for epoxy resins.

Incompatibilities: Forms explosive mixture with air. The aqueous solution is a strong base. A strong reducing agent. Reacts violently with oxidizers, acids, acid chlorides, acid anhydrides, carbon dioxide. Reacts with ethylene dichloride, organic anhydrides, isocyanates, vinyl acetate, acrylates, substituted allyls, alkylene oxides, epichlorohydrin, ketones, aldehydes, alcohols, glycols, phenols, cresols, caprolactum solution. Attacks aluminum, copper, lead, tin, zinc and alloys.

Permissible Exposure Limits in Air: The former USSR set a workplace MAC (PDK) of 0.1 mg/m^3, and values in the ambient air of residential areas of 0.001 mg/m^3 both on a momentary and a daily average basis.

Permissible Concentration in Water: The former USSR[43] has set a MAC in water bodies used for domestic purposes of 0.01 mg/l.

Harmful Effects and Symptoms

Short Term Exposure: A corrosive substance. Severely irritates the eyes, skin, and respiratory tract. Can cause permanent eye damage. Inhalation can cause nosebleeds, sore throat, hoarseness, cough, phlegm and/or difficult breathing. Higher exposures can cause pulmonary edema, a medical emergency that can be delayed for several hours. This can cause death. The oral LD_{50} rat is 750 mg/kg (slightly toxic).

Long Term Exposure: Repeated or prolonged exposure can cause dermatitis, eczema, and liver damage including hepatitis. There is limited evidence that this chemical can damage the developing fetus. Repeated exposure may cause bronchitis.

Points of Attack: Lungs, skin, liver.

Medical Surveillance: Lung function tests, liver function tests. Consider chest x-ray following acute overexposure.

First Aid: If this chemical gets into the eyes, remove any contact lenses at once and irrigate immediately for at least 15 minutes, occasionally lifting upper and lower lids. Seek medical attention immediately. If this chemical contacts the skin, remove contaminated clothing and wash immediately with soap and water. Seek medical attention immediately. If this chemical has been inhaled, remove from exposure, begin rescue breathing (using universal precautions) if breathing has stopped and CPR if heart action has stopped. Transfer promptly to a medical facility. When this chemical has been swallowed, get medical attention. If convulsions are not present, give a glass or two of water or milk to dilute the substance. Do not induce vomiting. Assure that the person's airway is unobstructed and contact a hospital or poison center immediately for advice on whether or not to induce vomiting. Medical observation is recommended for 24 – 48 hours after breathing overexposure, as pulmonary edema may be delayed. As first aid for pulmonary edema, a doctor or authorized paramedic may consider administering a corticosteroid spray.

Personal Protective Methods: Wear protective gloves and clothing to prevent any reasonable probability of skin contact. Safety equipment suppliers/manufacturers can provide recommendations on the most protective glove/clothing material for your operation. All protective clothing (suits, gloves, footwear, headgear) should be clean, available each day, and put on before work. Contact lenses should not be worn when working with this chemical. Wear splash-proof chemical goggles and face shield unless full facepiece respiratory protection is worn. Employees should wash immediately with soap when skin is wet or contaminated. Provide emergency showers and eyewash.

Respirator Selection: Where the potential for exposure to this chemical, use a MSHA/NIOSH approved supplied-air respirator with a full facepiece operated in the positive pressure mode or with a full facepiece, hood, or helmet in the continuous flow mode, or use a MSHA/NIOSH approved self-contained breathing apparatus with a full facepiece operated in pressure-demand or other positive pressure mode.

Storage: Prior to working with this chemical you should be trained on its proper handling and storage. Before entering confined space where this chemical may be present, check to make sure that an explosive concentration does not exist. Store in a cool, dry place away from oxidizers, strong acids, acid chlorides, acid anhydrides, carbon dioxide, and metals. Preferably store under an inert atmosphere in a tightly closed container.

Shipping: Solid HMDA requires a shipping label of "Corrosive." It falls in DOT Hazard Class 8 and Packing Group III. A solution of HMDA requires a shipping label of: "Corrosive Poison." It falls in DOT Hazard Class 8 and Packing Group II.

Spill Handling: *Solid:* Evacuate persons not wearing protective equipment from area of spill or leak until clean-up is complete. Remove all ignition sources. Collect powdered material in the most convenient and safe manner and deposit in sealed containers. Establish forced ventilation to keep levels below explosive limit. It may be necessary to contain and dispose of this chemical as a hazardous waste. If material or contaminated runoff enters waterways, notify downstream users of potentially contaminated waters. Contact your Department of Environmental Protection or your regional office of the federal EPA for specific recommendations. If employees are required to clean-up spills, they must be properly trained and equipped. OSHA 1910.120(q) may be applicable.

Liquid: Evacuate and restrict persons not wearing protective equipment from area of spill or leak until cleanup is complete. Remove all ignition sources. Establish forced ventilation to keep levels below explosive limit. Absorb liquids in vermiculite, dry sand, earth, peat, carbon, or a similar material and deposit in sealed containers. Keep this chemical out of a confined space, such as a sewer, because of the possibility of an explosion, unless the sewer is designed to prevent the build-up of explosive concentrations. It may be necessary to contain and dispose of this chemical as a hazardous waste. If material or contaminated runoff enters waterways, notify downstream users of potentially contaminated waters. Contact your Department of Environmental Protection or your regional office of the federal EPA for specific recommendations. If employees are required to clean-up spills, they must be properly trained and equipped. OSHA 1910.120(q) may be applicable.

Fire Extinguishing: *Solid:* This chemical as a solid may burn but does not easily ignite. Use dry chemical, carbon dioxide, water spray, or foam extinguishers. Poisonous gases including nitrogen oxides are produced in fire. If material or contaminated runoff enters waterways, notify downstream users of potentially contaminated waters. Notify local health and fire officials and pollution control agencies. From a secure, explosion-proof location, use water spray to cool exposed containers. If cooling streams are ineffective (venting sound increases in volume and pitch, tank discolors, or shows any signs of deforming), withdraw immediately to a secure position. If employees are expected to fight fires, they must be trained and equipped in OSHA 1910.156.

Liquid: This chemical in solution is a combustible liquid. Poisonous gases including nitrogen oxides are produced in fire. Use dry chemical, carbon dioxide, or foam extinguishers. Vapors in confined areas may explode when exposed to fire. Containers may explode in fire. Storage containers and parts of containers may rocket great distances, in many directions. If material or contaminated runoff enters waterways, notify downstream users of potentially contaminated waters. Notify local health and fire officials and pollution control agencies. From a secure, explosion-proof location, use water spray to cool exposed containers. If cooling streams are ineffective (venting sound increases in volume and pitch, tank discolors, or shows any signs of deforming), withdraw immediately to a secure position. If employees are expected to fight fires, they must be trained and equipped in OSHA 1910.156.

Disposal Method Suggested: Incineration; incinerator is equipped with a scrubber or thermal unit to reduce NO_x emissions.

References

U.S. Environmental Protection Agency, Chemical Hazard Information Profile: 1,6-Diaminohexane. Washington, DC (June 6, 1978)

Sax, N. I., Ed., Dangerous Properties of Industrial Materials Report, 2, No. 1. 30–31 (1982) and 8, No. 1, 46–50 (1988)

Hexamethylene Diisocyanate

Molecular Formula: $C_8H_{12}N_2O_2$

Common Formula: $OCN(CH_2)_6NCO$

Synonyms: AI3-28285; Desmodur H; Desmodur N; Diisocianto de hexametileno (Spanish); 1,6-Diisocyanatohexane; HDI; Hexamethyl 1,6-diisocyanate; 1,6-Hexamethylene diisocyanate; Hexamethylene 1,6-diisocyanate; 1,6-Hexane diisocyanate; Hexane, 1,6-diisocyanato-; 1,6-Hexanediol diisocyanate; 1,6-Hexylene diisocyanate; HMDI; Isocyanic acid, diester with 1,6-hexanediol; Isocyanic acid, hexamethylene ester; Metyleno-bis-fenyloizocyjanian; NSC 11687; Szesciometylenodwuizocyjanian (Polish); TL78

CAS Registry Number: 822-06-0

RTECS Number: MO1740000

DOT ID: UN 2281

EEC Number: 615-011-00-1

Regulatory Authority

- Air Pollutant Standard Set (ACGIH)[1] (DFG)[3] (former USSR)[43] (Connecticut)[60]
- CLEAN AIR ACT: Hazardous Air Pollutants (Title I, Part A, Section 112)
- SUPERFUND/EPCRA 40CFR302.4 Reportable Quantity (RQ): CERCLA, 1 lb (0.454 kg)
- EPCRA Section 313 Form R *de minimis* concentration reporting level: 1.0%
- Canada, WHMIS, Ingredients Disclosure List

Cited in U.S. State Regulations: California (G), Connecticut (A), Illinois (G), New Hampshire (G), New Jersey (G), Pennsylvania (G).

Description: Hexamethylene diisocyanate, $OCN(CH_2)_6$ NCO, is a colorless liquid with a sharp, irritating odor. Boiling point = 255°C. Freezing/Melting point = -67°C. Flash point = 140°C. Autoignition temperature = 454°C. Explosive limits: LEL = 0.9%; UEL = 9.5. Hazard Identification (based on NFPA-704 M Rating System): Health 3, Flammability 1, Reactivity 1. Reacts with water.

Potential Exposure: Used to make other chemicals, coatings, and polyurethane. It is also used as a hardener in automobile and airplane paints.

Incompatibilities: Reacts violently with alcohols, amines, oxidizers, strong bases, carboxylic acids, organotin catalysts Reacts slowly with water to form carbon dioxide, amine and polyureas. Temperatures above 200°C can cause polymerization (also reported as 93°C). Attacks copper.

Permissible Exposure Limits in Air: There is no OSHA PEL NIOSH recommends a TWA of 0.005 ppm (0.035 mg/m^3) and ceiling of 0.020 ppm (0.140 mg/m^3) (10-minute). ACGIH recommends a TWA of 0.005 ppm (0.035 mg/m^3) with no STEL. The DFG[3] has set an MAK of 0.01 ppm (0.07 mg/m^3). The former USSR-UNEP/IRPTC project[43] has set a MAC in workplace air of 0.05 mg/m^3. The State of Connecticut has set a guideline for hexamethylene diisocyanate in ambient air of 0.7 μg/m^3.[60]

Determination in Air: Collection by impinger; Reagent; High-pressure liquid chromatography/Fluorescence/Electrochemical detection; NIOSH IV, Method #5522, Isocyanates. See also Method #5521.

Permissible Concentration in Water: No criteria set.

Routes of Entry: Inhalation, ingestion, skin and/or eye contact.

Harmful Effects and Symptoms

Short Term Exposure: Hexamethylene diisocyanate can affect you when breathed in and by passing through your skin. Contact can irritate and may burn the eyes and skin. Skin contact may cause skin blisters. Eye contact may cause corneal damage. Exposure can cause headache, nausea, vomiting and irritability. Just a few breaths of high levels of hydrogen sulfide in air can cause death. High levels can irritate the lungs causing coughing and/or shortness of breath. Higher exposures can cause pulmonary edema, a medical emergency that can be delayed for several hours. This can cause death. Respiratory sensitization may result from high levels of exposure.

Long Term Exposure: Repeated or prolonged contact may cause skin sensitization and allergy with itching and skin rash. Repeated or prolonged inhalation exposure may cause a lung allergy (asthma) to develop. Once allergy develops, future exposure can cause cough, wheezing, and shortness of breath.

Points of Attack: Eyes, skin, respiratory system.

Medical Surveillance: Before beginning employment and at regular times after that, the following are recommended: Lung function tests. These may be normal if person is not having an attack at the time of the test. If symptoms develop or overexposure is suspected, the following may be useful: evaluation by a qualified allergist, including careful exposure history and special testing, may help diagnose skin allergy. Consider chest x-ray after acute overexposure.

First Aid: If this chemical gets into the eyes, remove any contact lenses at once and irrigate immediately for at least 15 minutes, occasionally lifting upper and lower lids. Seek medical attention immediately. If this chemical contacts the skin, remove contaminated clothing and wash immediately with soap and water. Seek medical attention immediately. If this chemical has been inhaled, remove from exposure, begin rescue breathing (using universal precautions) if breathing has stopped and CPR if heart action has stopped. Transfer promptly to a medical facility. When this chemical has been swallowed, get medical attention. Give large quantities of water and induce vomiting. Do not make an unconscious person vomit. Medical observation is recommended for 24 – 48 hours after breathing overexposure, as pulmonary edema may be delayed. As first aid for pulmonary edema, a doctor or authorized paramedic may consider administering a corticosteroid spray.

Personal Protective Methods: Wear protective gloves and clothing to prevent any reasonable probability of skin contact. Safety equipment suppliers/manufacturers can provide recommendations on the most protective glove/clothing material for your operation. All protective clothing (suits, gloves, footwear, headgear) should be clean, available each day, and put on before work. Contact lenses should not be worn when working with this chemical. Wear splash-proof chemical goggles and face shield unless full facepiece respiratory protection is worn. Employees should wash immediately with soap when skin is wet or contaminated. Provide emergency showers and eyewash.

Respirator Selection: NIOSH: *Up to 0.05 ppm:* SA (any supplied-air respirator).* *Up to 0.125 ppm:* SA:CF (any supplied-air respirator operated in a continuous-flow mode).* *Up to 0.25 ppm:* SCBAF (any self-contained breathing apparatus with a full facepiece); or SAF (any supplied-air respirator with a full facepiece). *Up to 1 ppm:* SAF:PD,PP (any supplied-air respirator that has a full facepiece and is operated in a pressure-demand or other positive-pressure mode). *Emergency or planned entry into unknown concentrations or IDLH conditions:* SCBAF:PD,PP (any MSHA/NIOSH approved self-contained breathing apparatus that has a full facepiece and is operated in a pressure-demand or other positive-pressure mode) or SAF:PD,PP:ASCBA (any supplied-air respirator that has a full facepiece and is operated in a pressure-demand or other positive-pressure mode in combination with an auxiliary self-contained breathing apparatus operated in a pressure-demand or other positive pressure mode). *Escape:* GMFOV [any air-purifying, full-facepiece respirator (gas mask) with a chin-style, front-or back-mounted organic vapor canister]; or SCBAE (any appropriate escape-type, self-contained breathing apparatus).

* Substance reported to cause eye irritation or damage; may require eye protection.

Storage: Prior to working with this chemical you should be trained on its proper handling and storage. Before entering confined space where this chemical may be present, check to make sure that an explosive concentration does not exist. Hexamethylene diisocyanate should be stored away from moisture or water. This contact will cause it to polymerize and explode its container. Hexamethylene diisocyanate must be stored to avoid contact with amines, carboxylic acids, strong bases and alcohols since violent reactions occur. Store in tightly closed containers in a cool, well-ventilated area at temperatures below 93°C/200°F.

Shipping: This compound requires a shipping label of: "Poison." It falls in DOT Hazard Class 6.1 and Packing Group II.

Spill Handling: Evacuate and restrict persons not wearing protective equipment from area of spill or leak until cleanup is complete. Remove all ignition sources. Establish forced ventilation to keep levels below explosive limit. Absorb liquids in vermiculite, dry sand, earth, peat, carbon, or a similar material and deposit in sealed containers. Keep this chemical out of a confined space, such as a sewer, because of the possibility of an explosion, unless the sewer is designed to prevent the build-up of explosive concentrations. It may be necessary to contain and dispose of this chemical as a hazardous waste. If material or contaminated runoff enters waterways, notify downstream users of potentially contaminated waters. Contact your Department of Environmental Protection or your regional office of the federal EPA for specific recommendations. If employees are required to clean-up spills, they must be properly trained and equipped. OSHA 1910.120(q) may be applicable.

Fire Extinguishing: This chemical is a combustible liquid. Poisonous gases including carbon monoxide, nitrogen oxides and hydrogen cyanide are produced in fire. Use dry chemical, carbon dioxide, or foam extinguishers. Vapors are heavier than air and will collect in low areas. Vapors may travel long distances to ignition sources and flashback. Vapors in confined areas may explode when exposed to fire. Containers may explode in fire. Storage containers and parts of containers may rocket great distances, in many directions. If material or contaminated runoff enters waterways, notify downstream users of potentially contaminated waters. Notify local health and fire officials and pollution control agencies. From a secure, explosion-proof location, use water spray to cool exposed containers. If cooling streams are ineffective (venting sound increases in volume and pitch, tank discolors, or shows any signs of deforming), withdraw immediately to a secure position. If employees are expected to fight fires, they must be trained and equipped in OSHA 1910.156.

References

New Jersey Department of Health and Senior Services, "Hazardous Substance Fact Sheet: Hexamethylene Diisocyanate," Trenton, NJ (April 1999)

U.S. Department of Health and Human Services, "ATSDR ToxFAQs, Hexamethylene Diisocyanate," Atlanta, GA (August 1999)

Hexamethylphosphoric Triamide

Molecular Formula: $C_6H_{18}N_3OP$

Common Formula: $[(CH_3)_2N]_3PO$

Synonyms: Eastman inhibitor HPT; ENT 50,882; Hempa; Hexametapol; Hexamethylorthophosphoric triamide; Hexamethylphosphoramide; Hexamethylphosphoric acid triamide; *N,N,N,N,N,N*-Hexamethylphosphoric triamide; Hexamethylphosphoric triamide; Hexamethylphosphorotriamide; Hexamethylphosphotriamide; HMPA; HMPT; HMPTA; HPT; Memta; Phosphoric acid hexamethyltriamide; Phosphoric hexamethyltriamide; Phosphoric triamide, hexamethyl-; Phosphoric tris(dimethylamide); Phosphoryl hexamethyltriamide; Triamida hexametilfosforica (Spanish); Tris(dimethyamino)phosphorus oxide; Tris(dimethylamino) phosphine oxide

CAS Registry Number: 680-31-9

RTECS Number: TD0875000

DOT ID: UN 3082

EEC Number: 015-106-00-2

Regulatory Authority

- Very Toxic Substance (World Bank)[15]
- Air Pollutant Standard Set (Several States)[60]
- Banned or Severely Restricted (Sweden)[13]
- Carcinogen (Animal Positive) (IARC)[9] (ACGIH)[1] (DFG)[3]
- CLEAN AIR ACT: Hazardous Air Pollutants (Title I, Part A, Section 112)
- SUPERFUND/EPCRA 40CFR302.4 Reportable Quantity (RQ): CERCLA, 1 lb (0.454 kg)
- EPCRA Section 313 Form R *de minimis* concentration reporting level: 0.1%

Cited in U.S. State Regulations: Alaska (G), California (G), Florida (G), Illinois (G), Maine (G), Maryland (G), Massachusetts (G), Michigan (G), New Hampshire (G), New Jersey (G), New York (A), North Dakota (A), Pennsylvania (G, A), Rhode Island (G), South Carolina (A), Virginia (A), West Virginia (G).

Description: Hexamethylphosphoric triamide, $[(CH_3)_2N]_3PO$, is a colorless liquid with a spicy odor. Boiling point = 232°C. Freezing/Melting point = 5 – 7°C. Flash point = 105°C. Soluble in water.

Potential Exposure: Hexamethylphosphoric triamide is a material possessing unique solvent properties and is widely used as a solvent, in small quantities, in organic and organometallic reactions in laboratories. This is the major source of occupational exposure to HMPA in the United States. It is also used as a processing solvent in the manufacture of aramid fibers. HMPA has been evaluated for use as an ultraviolet light inhibitor in polyvinylchloride formulations, as an additive for antistatic effects, as a flame retardant, and as a deicing additive for jet fuels. Hexamethylphosphoric triamide has also been extensively investigated as an insect chemosterilant.

Incompatibilities: Violent reactions with oxidizers, strong acids, chemically active metals.

Permissible Exposure Limits in Air: There is no current OSHA standard for hexamethylphosphoric triamide exposure. ACGIH classifies HMPA as an "industrial substance suspect of carcinogenic potential for man" with no suggested threshold limit value, but with the notation "skin" indicating the possibility of cutaneous absorption. DFG[3] follows this same policy. Several states have set guidelines or standards for HMPA in ambient air[60] ranging from zero (North Dakota) to 0.0024 ppb (Pennsylvania) to 0.03 μg/m³ (New York) to 3.0 μg/m³ (Virginia) to 14.5 μg/m³ (South Carolina).

Determination in Air: No method available.

Permissible Concentration in Water: No criteria set.

Routes of Entry: Inhalation, skin absorption, ingestion, skin and/or eye contact. Through the skin.

Harmful Effects and Symptoms

Short Term Exposure: Hexamethyl phosphoramide can affect you when breathed in and by passing through your skin. Exposure may irritate and damage the nose, throat and lungs causing a nasal discharge and lung changes. Very high levels may cause kidney and lung damage.

Long Term Exposure: Potential occupational carcinogen. May cause cancer of the nose. Hexamethyl phosphoramide may cause mutations. Handle with extreme caution. There is limited evidence that HEMPA may damage the testes of males and affect sperm production. Repeated exposure may severely damage the kidneys and lungs. May damage the nose, causing chronic nasal discharge.

Points of Attack: Eyes, skin, respiratory system, central nervous system, gastrointestinal tract. Cancer Site in animals: nasal cavity.

Medical Surveillance: For those with frequent or potentially high exposure (or significant skin contact), the following are recommended before beginning work and at regular times after that: Exam of the nose. Lung function tests. Kidney function tests.

First Aid: If this chemical gets into the eyes, remove any contact lenses at once and irrigate immediately for at least 15 minutes, occasionally lifting upper and lower lids. Seek medical attention immediately. If this chemical contacts the skin, remove contaminated clothing and wash immediately with soap and water. Seek medical attention immediately. If this chemical has been inhaled, remove from exposure, begin rescue breathing (using universal precautions) if breathing has stopped and CPR if heart action has stopped. Transfer promptly to a medical facility. When this chemical has been swallowed, get medical attention. Give large quantities of water and induce vomiting. Do not make an unconscious person vomit.

Personal Protective Methods: Wear protective gloves and clothing to prevent any reasonable probability of skin contact. Safety equipment suppliers/manufacturers can provide recommendations on the most protective glove/clothing material for your operation. Butyl Rubber is among the recommended protective materials. All protective clothing (suits, gloves, footwear, headgear) should be clean, available each day, and put on before work. Contact lenses should not be worn when working with this chemical. Wear splash-proof chemical goggles and face shield unless full facepiece respiratory protection is worn. Employees should wash immediately with soap when skin is wet or contaminated. Provide emergency showers and eyewash.

Respirator Selection: *At any concentrations above the NIOSH REL, or where there is no REL, at any detectable concentration:* SCBAF:PD,PP (any MSHA/NIOSH approved self-contained breathing apparatus that has a full facepiece and is operated in a pressure-demand or other positive-pressure mode); or SAF:PD,PP:ASCBA (any supplied-air respirator that has a full facepiece and is operated in a pressure-demand or other positive-pressure mode in combination with an auxiliary self-contained breathing apparatus operated in a pressure-demand or other positive pressure mode). *Escape:* GMFOV [any air-purifying, full-facepiece respirator (gas mask) with a chin-style, front-or back-mounted organic vapor canister]; or SCBAE (any appropriate escape-type, self-contained breathing apparatus).

Storage: Prior to working with this chemical you should be trained on its proper handling and storage. Hexamethyl phosphoramide must be stored to avoid contact with oxidizers (such as perchlorates, peroxides, permanganates, chlorates, and nitrates); strong acids (such as hydrochloric, sulfuric, and nitric) and chemically active metals (such as potassium, sodium, magnesium, and zinc) since violent reactions occur. Store in tightly closed containers in a cool, well-ventilated area away from area. Sources of ignition such as smoking and open flames are prohibited where hexamethyl phosphoramide is handled, used, or stored. A regulated, marked area should be established where this chemical is handled, used, or stored in compliance with OSHA standard 1910.1045.

Shipping: Environmentally hazardous liquid, n.o.s. Hazard Class: 9. Label: "Class 9."

Spill Handling: Evacuate and restrict persons not wearing protective equipment from area of spill or leak until cleanup is complete. Remove all ignition sources. Ventilate area of spill or leak. Absorb liquids in vermiculite, dry sand, earth, peat, carbon, or a similar material and deposit in sealed containers. Keep this chemical out of a confined space, such as a sewer, because of the possibility of an explosion, unless the sewer is designed to prevent the build-up of explosive concentrations. It may be necessary to contain and dispose of this chemical as a hazardous waste. If material or contaminated runoff enters waterways, notify downstream users of potentially contaminated waters. Contact your Department of Environmental Protection or your regional office of the federal EPA for specific recommendations. If employees are required to clean-up spills, they must be properly trained and equipped. OSHA 1910.120(q) may be applicable.

Fire Extinguishing: This chemical is a combustible liquid. Poisonous gases including phosphine, nitrogen oxides, phosphorus oxides are produced in fire. Use dry chemical or carbon dioxide extinguishers. Vapors are heavier than air and will collect in low areas. Containers may explode in fire. Storage containers and parts of containers may rocket great distances, in many directions. If material or contaminated runoff enters waterways, notify downstream users of potentially contaminated waters. Notify local health and fire officials and pollution control agencies. From a secure, explosion-proof location, use water spray to cool exposed containers. If cooling streams are ineffective (venting sound increases in volume and pitch, tank discolors, or shows any signs of deforming), withdraw immediately to a secure position. If employees are

expected to fight fires, they must be trained and equipped in OSHA 1910.156.

References

National Institute for Occupational Safety and Health, Current Intelligence Bulletin No. 6: Hexamethylphosphoric Triamide (HMPAI), Rockville, Maryland (October 24, 1975)

U.S. Environmental Protection Agency, Chemical Hazard Information Profile: Hexamethylphosphoramide, Washington, DC (August, 1976)

National Institute for Occupational Safety and Health, Information Profiles on Potential Occupational Hazards-Single Chemicals: Hexamethyl Phosphoramide, pp. 106–113, Rockville, Maryland (December, 1979)

New Jersey Department of Health and Senior Services, "Hazardous Substance Fact Sheet: Hexamethyl Phosphoramide," Trenton, NJ (November 1986)

n-Hexane

Molecular Formula: C_6H_{14}

Common Formula: $CH_3(CH_2)_4CH_3$

Synonyms: Exxsol hexane; Genesolv 404 azeotrope; Gettysolve-B; Hexane; *n*-Hexano (Spanish); Hexano (Spanish); Hexyl hydride; NCI-C60571; NSC68472; Skellysolve B

CAS Registry Number: 110-54-3

RTECS Number: MN9275000

DOT ID: UN 1208

EEC Number: 601-037-00-0

Regulatory Authority

- Air Pollutant Standard Set (ACGIH)[1] (DFG)[3] (HSE)[33] (former USSR)[35][43] (OSHA)[58] (Several States)
- CLEAN AIR ACT: Hazardous Air Pollutants (Title I, Part A, Section 112)
- SUPERFUND/EPCRA 40CFR302.4 Reportable Quantity (RQ): CERCLA, 1 lb (0.454 kg)
- EPCRA Section 313 Form R *de minimis* concentration reporting level: 1.0%
- Canada, WHMIS, Ingredients Disclosure List

Cited in U.S. State Regulations: Alaska (G), Arizona (W), California (A, G), Connecticut (A), Florida (G), Illinois (G), Maine (G, W), Massachusetts (G), Nevada (A), New Hampshire (G), New Jersey (G, W), New York (G), North Carolina (A), North Dakota (A), Oklahoma (G), Pennsylvania (G), Rhode Island (G), Virginia (A), West Virginia (G).

Description: Hexane, $CH_3(CH_2)_4CH_3$, is a highly flammable, colorless, volatile liquid with a gasoline-like odor. The water/odor threshold is 0.0064 mg/l and the air/odor threshold is 230 – 875 mg/m^3. Boiling point = 69°C. Flash point = -22°C. Autoignition temperature = 225°C. Explosive limits: LEL = 1.1%, UEL = 7.5%. Hazard Identification (based on NFPA-704 M Rating System): Health 1, Flammability 3, Reactivity 0. Practically insoluble in water.

Potential Exposure: n-Hexane is used as a solvent, particularly in the extraction of edible fats and oils, as a laboratory reagent, and as the liquid in low temperature thermometers. Technical and commercial grades consist of 45 – 85% hexane, as well as cyclopentanes, isohexane, and 1 – 6% benzene.

Incompatibilities: Forms explosive mixture with air. Contact with strong oxidizers may cause fire and explosions. Contact with dinitrogen tetraoxide may explode at 28°C. Attacks some plastics, rubber and coatings. May accumulate static electrical charges, and may cause ignition of its vapors.

Permissible Exposure Limits in Air: The OSHA PEL is 500 ppm (1,800 mg/m^3) TWA. The recommended NIOSH REL, as well as Sweden[35] and the DFG[3] and the ACGIH has proposed a TWA of 50 ppm (180 mg/m^3). The NIOSH recommendation for other hexane isomers is 100 ppm TWA and STEL of 510 ppm. The NIOSH IDLH level is 1,100 ppm (10% LEL). HSE[33] has set an 8-hour TWA of 100 ppm (360 mg/m^3) and an STEL of 125 ppm (450 mg/m^3) for -hexane. They have also set an 8-hour TWA of 500 ppm (1,800 mg/m^3) for all isomers except the normal isomer and an STEL of 1,000 ppm (3,600 mg/m^3) for these other isomers, as has ACGIH.[1] Japan[35] has set a TWA of 40 ppm (140 mg/m^3). The former USSR[35][43] has set a MAC in workplace air of 300 mg/m^3 as well as a limit for ambient air in residential areas of 60 mg/m^3 on a momentary basis. Several states have set guidelines or standards for -hexane in ambient air[60] ranging from 1.1 mg/m^3 (North Carolina) to 1.8 mg/m^3 (North Dakota) to 4.28 mg/m^3 (Nevada) to 3.6 – 36.0 mg/m^3 (Connecticut) to 30.0 mg/m^3 (Virginia).

Determination in Air: Charcoal adsorption, workup with CS_2, analysis by gas chromatography/flame ionization. See NIOSH Method 1500.[18]

Permissible Concentration in Water: The EPA[48] has set a health advisory for n-hexane involving the calculation of a LOAEL of 570 mg/kg/day which results in a long-term health advisory for an adult of 14.3 mg/l. A health-based maximum contaminant level of 33 μg/l has been derived by the State of New Jersey[59] based on neurotoxic effects observed in rats exposed to n-Hexane by inhalation. Arizona and Maine[61] have set guidelines for n-Hexane in drinking water of 4,000 μg/l.

Routes of Entry: Inhalation of vapor, ingestion, eye and skin contact. Passes through the skin.

Harmful Effects and Symptoms

Short Term Exposure: Irritates the eyes, nose, and respiratory tract. Exposure can cause lightheadedness, giddiness, headaches, and nausea. High levels can lead to unconsciousness and death. *Inhalation:* Exposure to levels above 500 ppm may cause headaches, abdominal cramps, a burning feeling of the face, numbness and weakness of the fingers and toes. Levels above 1,300 ppm may cause the above plus nausea and irritation of the nose and throat. Levels above 1,500 ppm may cause the above plus blurred vision, loss of appetite and loss of weight. Most symptoms disappear within a few months if exposure ceases. Breathing liquid into the lungs may cause a chemical pneumonia. *Skin:* Contact may cause irritation, redness, swelling, blisters and pain. Skin exposure may contribute to symptoms listed under inhalation. *Eyes:* Levels over 880 ppm may cause irritation. *Ingestion:* May contribute to symptoms listed under inhalation. Estimated lethal dose is one ounce to one pint.

Long Term Exposure: High or repeated exposure can damage the nervous system, causing numbness, tingling, and/or muscle weakness in the hands, feet, arms and legs. Repeated skin contact can cause irritation, dryness and cracking and can lead to rash. May cause symptoms listed under inhalation. Exposure to levels above 650 ppm for two to four months can result in weakness and numbness of the arms and legs. Symptoms go away within a few months if exposure stops. Use by older children in the US and Europe who have "sniffed" household chemicals containing n-hexane in an attempt to get "high" has caused paralysis of the arms and legs. In laboratory studies, animals exposed to high levels of n-hexane had signs of nerve damage, lung damage and damage to the sperm-forming cells.

Points of Attack: Eyes, skin, respiratory system, central nervous system, peripheral nervous system.

Medical Surveillance: Consider the skin, respiratory system, central and peripheral nervous system, and general health in preplacement and periodic examinations.

First Aid: If this chemical gets into the eyes, remove any contact lenses at once and irrigate immediately for at least 15 minutes, occasionally lifting upper and lower lids. Seek medical attention immediately. If this chemical contacts the skin, remove contaminated clothing and wash immediately with soap and water. Seek medical attention immediately. If this chemical has been inhaled, remove from exposure, begin rescue breathing (using universal precautions) if breathing has stopped and CPR if heart action has stopped. Transfer promptly to a medical facility. When this chemical has been swallowed, get medical attention. Do not induce vomiting.

Personal Protective Methods: Wear protective gloves and clothing to prevent any reasonable probability of skin contact. Safety equipment suppliers/manufacturers can provide recommendations on the most protective glove/clothing material for your operation. Polyvinyl Alcohol, Viton, Polyurethane, Teflon, Viton/Chlorobutyl Rubber, Silvershield, and Chlorinated Polyethylene are among the recommended protective materials. All protective clothing (suits, gloves, footwear, headgear) should be clean, available each day, and put on before work. Contact lenses should not be worn when working with this chemical. Wear splash-proof chemical goggles and face shield unless full facepiece respiratory protection is worn. Employees should wash immediately with soap when skin is wet or contaminated. Provide emergency showers and eyewash.

Respirator Selection: NIOSH/OSHA: *500 ppm:* SA (any supplied-air respirator). *1,100 ppm:* SA:CF (any supplied-air respirator operated in a continuous-flow mode); or SCBAF (any self-contained breathing apparatus with a full facepiece); or SAF (any supplied-air respirator with a full facepiece). *Emergency or planned entry into unknown concentrations or IDLH conditions:* SCBAF:PD,PP (any self-contained breathing apparatus that has a full facepiece and is operated in a pressure-demand or other positive-pressure mode); or SAF: PD,PP:ASCBA (any supplied-air respirator that has a full facepiece and is operated in a pressure-demand or other positive-pressure mode in combination with an auxiliary self-contained breathing apparatus operated in a pressure-demand or other positive pressure mode). *Escape:* GMFOV [any air-purifying, full-facepiece respirator (gas mask) with a chin-style, front-or back-mounted organic vapor canister]; or SCBAE (any appropriate escape-type, self-contained breathing apparatus).

Note: Substance reported to cause eye irritation or damage; may require eye protection.

Storage: n-Hexane must be stored to avoid contact with strong oxidizers (such as chlorine, bromine, and fluorine), because violent reactions occur. Before entering confined space where this chemical may be present, check to make sure that an explosive concentration does not exist. Store in tightly closed containers in a cool, well-ventilated area away from heat. Sources of ignition such as smoking and open flames are prohibited where n-Hexane is used, handled, or stored in a manner that could create a potential fire or explosion hazard. Metal containers used in the transfer of 5 gallons or more of n-Hexane should be grounded and bonded. Drums must be equipped with self-closing valves, pressure vacuum bungs, and flame arresters. Use only non-sparking tools and equipment, especially when opening and closing containers of n-Hexane.

Shipping: This compound requires a shipping label of: "Flammable Liquid." It falls in DOT Hazard Class 3 and Packing Group II.

Spill Handling: Evacuate and restrict persons not wearing protective equipment from area of spill or leak until cleanup is complete. Remove all ignition sources. Establish forced ventilation to keep levels below explosive limit. Absorb liquids in vermiculite, dry sand, earth, peat, carbon, or a similar material and deposit in sealed containers. Keep this chemical out of a confined space, such as a sewer, because of the possibility of an explosion, unless the sewer is designed to prevent the build-up of explosive concentrations. It may be necessary to contain and dispose of this chemical as a hazardous waste. If material or contaminated runoff enters waterways, notify downstream users of potentially contaminated waters. Contact your Department of Environmental Protection or your regional office of the federal EPA for specific recommendations. If employees are required to clean-up spills, they must be properly trained and equipped. OSHA 1910.120(q) may be applicable.

Fire Extinguishing: This chemical is a flammable liquid. Poisonous gases are produced in fire. Use dry chemical, carbon dioxide, or alcohol foam extinguishers. Vapors are heavier than air and will collect in low areas. Vapors may travel long distances to ignition sources and flashback. Vapors in confined areas may explode when exposed to fire. Containers may explode in fire. Storage containers and parts of containers may rocket great distances, in many directions. If material or contaminated runoff enters waterways, notify downstream users of potentially contaminated waters. Notify local health and fire officials and pollution control agencies. From a secure, explosion-proof location, use water spray to cool exposed containers. If cooling streams are ineffective (venting sound increases in volume and pitch, tank discolors, or shows any signs of deforming), withdraw immediately to a secure position. If employees are expected to fight fires, they must be trained and equipped in OSHA 1910.156.

Disposal Method Suggested: Incineration.

References

U.S. Environmental Protection Agency, Clinical Hazard Information Profile: -Hexane, Washington, DC (May 13, 1977)

National Institute for Occupational Safety and Health, Criteria for a Recommended Standard: Occupational Exposure to Alkanes, NIOSH Document No. 77–151, Washington, DC (1977)

New Jersey Department of Health and Senior Services, "Hazardous Substance Fact Sheet: n-Hexane," Trenton, NJ (April, 1997)

New York State Department of Health, "Chemical Pact Sheet: Hexane," Albany, NY, Bureau of Toxic Substance Assessment (April 1986)

Sax, N. I., Ed., "Dangerous Properties of Industrial Materials Report" 1, No. 6, 59–61 (1981)

Hexanol

Molecular Formula: $C_6H_{14}O$

Common Formula: $CH_3(CH_2)_4CH_2OH$

Synonyms: Alcohol C-6; *n*-Amyl carbinol; Amylcarbinol; Caproyl alcohol; Epal-6; *n*-Hexanol; Hexanol; *n*-Hexyl alcohol; Hexyl alcohol; 1-Hydroxyhexane; Pentyl carbinol

CAS Registry Number: 111-27-3

RTECS Number: MQ4025000

DOT ID: UN 2282

EEC Number: 603-059-00-6

Regulatory Authority

- Air Pollutant Standard Set (former USSR)[43]

Cited in U.S. State Regulations: California (G), New Hampshire (G), New Jersey (G), Pennsylvania (G), Rhode Island (G).

Description: Hexanol, $CH_3(CH_2)_4CH_2OH$, is a flammable, colorless liquid. Boiling point = 157°C. Flash point = 63°C. Explosive limits: LEL = 1.2%; UEL = 7.7%.[41] Hazard Identification (based on NFPA-704 M Rating System): Health 1, Flammability 2, Reactivity 0. Slightly soluble in water.

Potential Exposure: To those using hexanol as a solvent or in the synthesis of pharmaceuticals, plasticizers and textile chemicals.

Incompatibilities: Forms explosive mixture with air. Incompatible with strong acids, caustics, aliphatic amines, isocyanates, strong oxidizers.

Permissible Exposure Limits in Air: The former USSR has set a TLV of 2.4 ppm (10 mg/m^3).[43]

Determination in Air: Charcoal tube; CS2; Gas chromatography/Flame ionization detection; IV (#1500, Hydrocarbons)

Permissible Concentration in Water: A value of 0.03 mg/l is the maximum allowable in drinking water.[111]

Routes of Entry: Skin contact, inhalation, ingestion.

Harmful Effects and Symptoms

Short Term Exposure: Vapor irritates the eyes, skin and respiratory tract. Contact causes smarting of the skin and first-degree burns on short exposure; may cause second-degree burns on long exposure. Believed to be moderately toxic upon ingestion. Exposure can cause headache, dizziness, confusion, muscle weakness, nausea, vomiting and diarrhea.

Long Term Exposure: Repeated exposure may damage the nervous system.

Points of Attack: Skin, eyes, nervous system.

Medical Surveillance: Examination of the nervous system.

First Aid: If this chemical gets into the eyes, remove any contact lenses at once and irrigate immediately for at least 15 minutes, occasionally lifting upper and lower lids. Seek medical attention immediately. If this chemical contacts the skin, remove contaminated clothing and wash immediately with soap and water. Seek medical attention immediately. If this chemical has been inhaled, remove from exposure, begin rescue breathing (using universal precautions) if breathing has stopped and CPR if heart action has stopped. Transfer promptly to a medical facility. When this chemical has been swallowed, get medical attention. Give large quantities of water and induce vomiting. Do not make an unconscious person vomit.

Personal Protective Methods: Wear protective gloves and clothing to prevent any reasonable probability of skin contact. Safety equipment suppliers/manufacturers can provide recommendations on the most protective glove/clothing material for your operation. All protective clothing (suits, gloves, footwear, headgear) should be clean, available each day, and put on before work. Contact lenses should not be worn when working with this chemical. Wear splash-proof chemical goggles and face shield unless full facepiece respiratory protection is worn. Employees should wash immediately with soap when skin is wet or contaminated. Provide emergency showers and eyewash.

Respirator Selection: Where the potential for exposure to this chemical, use a MSHA/NIOSH approved supplied-air respirator with a full facepiece operated in the positive pressure mode or with a full facepiece, hood, or helmet in the continuous flow mode, or use a MSHA/NIOSH approved self-contained breathing apparatus with a full facepiece operated in pressure-demand or other positive pressure mode.

Storage: Prior to working with hexanol you should be trained on its proper handling and storage. Before entering confined space where hexanol may be present, check to make sure that an explosive concentration does not exist. Store in tightly closed containers in a cool, well ventilated area away from oxidizers, strong acids. Metal containers involving the transfer of this chemical should be grounded and bonded. Where possible, automatically pump liquid from drums or other storage containers to process containers. Drums must be equipped with self-closing valves, pressure vacuum bungs, and flame arresters. Use only non-sparking tools and equipment, especially when opening and closing containers of this chemical. Sources of ignition such as smoking and open flames, are prohibited where this chemical is used, handled, or stored in a manner that could create a potential fire or explosion hazard. Wherever this chemical is used, handled, manufactured, or stored, use explosion-proof electrical equipment and fittings.

Shipping: This compound requires a shipping label of: "Flammable Liquid." It falls in DOT Hazard Class 3 and Packing Group III.

Spill Handling: Evacuate and restrict persons not wearing protective equipment from area of spill or leak until cleanup is complete. Remove all ignition sources. Establish forced ventilation to keep levels below explosive limit. Absorb liquids in vermiculite, dry sand, earth, peat, carbon, or a similar material and deposit in sealed containers. Keep this chemical out of a confined space, such as a sewer, because of the possibility of an explosion, unless the sewer is designed to prevent the build-up of explosive concentrations. It may be necessary to contain and dispose of this chemical as a hazardous waste. If material or contaminated runoff enters waterways, notify downstream users of potentially contaminated waters. Contact your Department of Environmental Protection or your regional office of the federal EPA for specific recommendations. If employees are required to clean-up spills, they must be properly trained and equipped. OSHA 1910.120(q) may be applicable.

Fire Extinguishing: This chemical is a flammable liquid. Poisonous gases including carbon monoxide are produced in fire. Use dry chemical, carbon dioxide, or foam extinguishers. Water may be ineffective. Vapors are heavier than air and will collect in low areas. Vapors may travel long distances to ignition sources and flashback. Vapors in confined areas may explode when exposed to fire. Containers may explode in fire. Storage containers and parts of containers may rocket great distances, in many directions. If material or contaminated runoff enters waterways, notify downstream users of potentially contaminated waters. Notify local health and fire officials and pollution control agencies. From a secure, explosion-proof location, use water spray to cool exposed containers. If cooling streams are ineffective (venting sound increases in volume and pitch, tank discolors, or shows any signs of deforming), withdraw immediately to a secure position. If employees are expected to fight fires, they must be trained and equipped in OSHA 1910.156.

Use alcohol foam to fight fire.

Disposal Method Suggested: Incineration.

References

Sax, N. I., Ed., Dangerous Properties of Industrial Materials Report, 2, No. 2, 32–33 (1982) and 7, No. 6, 65–67 (1987)

New Jersey Department of Health and Senior Services, "Hazardous Substance Fact Sheet, n-HEXANOL," Trenton, NJ (May, 1999)

Hexazinone

Molecular Formula: $C_{11}H_{20}O_2N_3$

Synonyms: Brushkiller; 3-Cyclohexyl-6-dimethylamino-1-methyl-1,2,3,4-tetrahydro-1,3,5-triazine-2-,4-dione; 3-Cyclohexyl-6-(dimethylamino)-1-methyl-*s*-triazine-2,4(1H,3H)-dione; 3-Cyclohexyl-6-(dimethylamino)-1-methyl-1,3,5-triazine-2,4(1H,3H)-dione; 3-Cyclohexyl-1-methyl-6-(dimethylamino)-*s*-trazine-2,4(1H,3H)-dione; DPX 3674; *s*-Triazine-2,4(1H,3H)-dione, 3-cyclohexyl-6-(dimethylamino)-1-methyl-; 1,3,5-Triazine-2,4(1H,3H)-dione, 3-cyclohexyl-6-(dimethylamino)-1-methyl-; Velpar; Velpar weed killer

CAS Registry Number: 51235-04-2

RTECS Number: XY7850000

DOT ID: UN 2763

Regulatory Authority

- EPCRA Section 313 Form R *de minimis* concentration reporting level: 1.0%

Cited in U.S. State Regulations: New Jersey (G), Pennsylvania (G)

Description: Hexazinone, $C_{11}H_{20}O_2N_3$, is a white crystalline solid that is practically odorless. Freezing/Melting point = 115 – 117°C. Soluble in water.

Potential Exposure: Those involved in the manufacture, formulation and application of this contact and residual herbicide. It is a broad spectrum herbicide is used in industrial and government right-of-way weed control for pipelines, drainage ditches, etc.[23]

Permissible Exposure Limits in Air: No standards set.

Permissible Concentration in Water: The EPA has analyzed data on hexazinone and developed a no observed adverse effect level (NOAEL) of 25 mg/kg/day based on studies of dogs which resulted in a long-term health advisory of 8.75 mg/l. A NOAEL of 10 mg/kg/day was developed based on studies of rats which yielded a lifetime health advisory of 0.21 mg/l.

Determination in Water: Solvent extraction with methylene chloride followed by analysis by gas chromatography with a thermionic bead detector.

Routes of Entry: Skin, inhalation, ingestion.

Harmful Effects and Symptoms

Short Term Exposure: May cause eye and skin irritation. The acute oral LD_{50} for rats is 1,690 mg/kg (slightly toxic). In experience with humans, only one report was available on hexazinone. It involved a 26 year-old woman who inhaled hexazinone dust. Vomiting occurred within 24 hours.

First Aid: If this chemical gets into the eyes, remove any contact lenses at once and irrigate immediately for at least 15 minutes, occasionally lifting upper and lower lids. Seek medical attention immediately. If this chemical contacts the skin, remove contaminated clothing and wash immediately with soap and water. Seek medical attention immediately. If this chemical has been inhaled, remove from exposure, begin rescue breathing (using universal precautions) if breathing has stopped and CPR if heart action has stopped. Transfer promptly to a medical facility. When this chemical has been swallowed, get medical attention. Give large quantities of water and induce vomiting. Do not make an unconscious person vomit.

Personal Protective Methods: Wear protective gloves and clothing to prevent any reasonable probability of skin contact. Safety equipment suppliers/manufacturers can provide recommendations on the most protective glove/clothing material for your operation. All protective clothing (suits, gloves, footwear, headgear) should be clean, available each day, and put on before work. Contact lenses should not be worn when working with this chemical. Wear dust-proof chemical goggles and face shield unless full facepiece respiratory protection is worn. Employees should wash immediately with soap when skin is wet or contaminated. Provide emergency showers and eyewash.

Respirator Selection: Where the potential for exposure to this chemical, use a MSHA/NIOSH approved supplied-air respirator with a full facepiece operated in the positive pressure mode or with a full facepiece, hood, or helmet in the continuous flow mode, or use a MSHA/NIOSH approved self-contained breathing apparatus with a full facepiece operated in pressure-demand or other positive pressure mode.

Storage: Prior to working with this chemical you should be trained on its proper handling and storage. Store in tightly closed containers in a cool, well ventilated area above 32°C.

Shipping: Not specifically cited by DOT, hexazinone could be considered under triazine pesticides, solid, toxic n.o.s. This compound requires a shipping label of: "Keep Away From Food." It falls in DOT Hazard Class 6.1 and Packing Group III.

Spill Handling: Evacuate persons not wearing protective equipment from area of spill or leak until clean-up is complete. Remove all ignition sources. Collect powdered material in the most convenient and safe manner and deposit in sealed containers. Ventilate area after clean-up is complete. It may be necessary to contain and dispose of this chemical as a hazardous waste. If material or contaminated runoff enters waterways, notify downstream users of potentially contaminated waters. Contact your Department of Environmental Protection or your regional office of the federal EPA for specific recommendations. If employees are required to clean-up spills, they must be properly trained and equipped. OSHA 1910.120(q) may be applicable.

Fire Extinguishing: Use dry chemical, carbon dioxide, water spray, or foam extinguishers. Poisonous gases including nitrogen oxides are produced in fire. If material or contaminated runoff enters waterways, notify downstream users of potentially contaminated waters. Notify local health and fire officials and pollution control agencies. From a secure, explosion-proof location, use water spray to cool exposed containers. If cooling streams are ineffective (venting sound increases in volume and pitch, tank discolors, or shows any signs of deforming), withdraw immediately to a secure position. If employees are expected to fight fires, they must be trained and equipped in OSHA 1910.156.

References

U.S. Environmental Protection Agency, "Health Advisory: Hexazinone," Washington, DC, Office of Drinking Water (August 1987)

1-Hexene

Molecular Formula: C_6H_{12}

Common Formula: $CH_3(CH_2)_3CH{=}CH_2$

Synonyms: Butyl ethylene; Butylethylene; 1-*n*-Hexene; 1-Hexene; Hexylene

CAS Registry Number: 592-41-6

RTECS Number: MP6601000

DOT ID: UN 2370

Cited in U.S. State Regulations: Florida (G), Massachusetts (G), New Hampshire (G), New Jersey (G), Oklahoma (G), Pennsylvania (G).

Description: Hexene, $CH_3(CH_2)_3CH{=}CH_2$, is a colorless liquid. Boiling point = 63 – 65°C. Freezing/Melting point = -140°C. Flash point = -7°C. Autoignition temperature = 253°C. Explosive Limits: LEL = 1.2%; UEL = 6.9%. Hazard Identification (based on NFPA-704 M Rating System): Health 1, Flammability 3, Reactivity 0. Insoluble in water.

Potential Exposure: Those involved in its use in organic synthesis. Used in fuels, and to make flavors, perfumes, dyes, and plastic resins.

Incompatibilities: Forms explosive mixture with air. Violent reaction with oxidizers, strong acids.

Permissible Exposure Limits in Air: No standards set.

Permissible Concentration in Water: No criteria set.

Routes of Entry: Inhalation.

Harmful Effects and Symptoms

Short Term Exposure: Irritating to skin, eyes and respiratory tract causing coughing and wheezing. Death may occur. Ingestion may cause chemical pneumonitis. Exposure may affect the central nervous system. Exposure can cause headache, nausea, dizziness and unconsciousness.

Long Term Exposure: Removes the skin's natural oils causing dryness and cracking.

First Aid: If this chemical gets into the eyes, remove any contact lenses at once and irrigate immediately for at least 15 minutes, occasionally lifting upper and lower lids. Seek medical attention immediately. If this chemical contacts the skin, remove contaminated clothing and wash immediately with soap and water. Seek medical attention immediately. If this chemical has been inhaled, remove from exposure, begin rescue breathing (using universal precautions) if breathing has stopped and CPR if heart action has stopped. Transfer promptly to a medical facility. When this chemical has been swallowed, get medical attention. Give large quantities of water and induce vomiting. Do not make an unconscious person vomit.

Personal Protective Methods: Use organic vapor respirator or air line Wear protective gloves and clothing to prevent any reasonable probability of skin contact. Safety equipment suppliers/manufacturers can provide recommendations on the most protective glove/clothing material for your operation. All protective clothing (suits, gloves, footwear, headgear) should be clean, available each day, and put on before work. Contact lenses should not be worn when working with this chemical. Wear splash-proof chemical goggles and face shield unless full facepiece respiratory protection is worn. Employees should wash immediately with soap when skin is wet or contaminated. Provide emergency showers and eyewash. mask. Wear protective goggles or face shield.

Respirator Selection: Where the potential for exposure to this chemical, use a MSHA/NIOSH approved supplied-air respirator with a full facepiece operated in the positive pressure mode or with a full facepiece, hood, or helmet in the continuous flow mode, or use a MSHA/NIOSH approved self-contained breathing apparatus with a full facepiece operated in pressure-demand or other positive pressure mode.

Storage: Prior to working with hexene you should be trained on its proper handling and storage. Before entering confined space where hexene may be present, check to make sure that an explosive concentration does not exist. Store in tightly closed containers in a cool, well ventilated area away from oxidizers and strong acids. Metal containers involving the transfer of this chemical should be grounded and bonded. Where possible, automatically pump liquid from drums or other storage containers to process containers. Drums must be equipped with self-closing valves, pressure vacuum bungs, and flame arresters. Use only non-sparking tools and equipment, especially when opening and closing containers of this chemical. Sources of ignition such as smoking and open flames, are prohibited where this chemical is used, handled, or stored in a manner that could create a potential fire or explosion hazard. Wherever this chemical is used, handled, manufactured, or stored, use explosion-proof electrical equipment and fittings.

Shipping: This compound requires a shipping label of: "Flammable Liquid." It falls in DOT Hazard Class 3 and Packing Group II.

Spill Handling: Evacuate and restrict persons not wearing protective equipment from area of spill or leak until cleanup is complete. Avoid contact with liquid or vapor. Stay upwind and use water spray to "knock down" vapor. Remove all ignition sources. Establish forced ventilation to keep levels below explosive limit. Absorb liquids in vermiculite, dry sand, earth, peat, carbon, or a similar material and deposit in sealed containers. Oil skimming equipment and sorbent (urethane) foams may be used for spills on water. Keep this chemical out of a confined space, such as a sewer, because of the possibility of an explosion, unless the sewer is designed to prevent the build-up of explosive concentrations. It may be necessary to contain and dispose of this chemical as a hazardous waste. If material or contaminated runoff enters waterways, notify downstream users of potentially contaminated waters. Contact your Department of Environmental Protection or your regional office of the federal EPA for specific recommendations. If employees are required to clean-up spills, they must be properly trained and equipped. OSHA 1910.120(q) may be applicable.

Fire Extinguishing: This chemical is a flammable liquid. Poisonous gases are produced in fire. Water may be ineffective. Use alcohol or polymer foam, dry chemical, or carbon dioxide. Vapors are heavier than air and will collect in low areas. Vapors may travel long distances to ignition sources and flashback. Vapors in confined areas may explode when exposed to fire. Containers may explode in fire. Storage containers and parts of containers may rocket great distances, in many directions. If material or contaminated runoff enters waterways, notify downstream users of potentially contaminated waters. Notify local health and fire officials and pollution control agencies. From a secure, explosion-proof location, use water spray to cool exposed containers. If cooling streams are ineffective (venting sound increases in volume and pitch, tank discolors, or shows any signs of deforming), withdraw immediately to a secure position. If employees are expected to fight fires, they must be trained and equipped in OSHA 1910.156.

Disposal Method Suggested: Incineration.

References

Sax, N. I., Ed., Dangerous Properties of Industrial Materials Report. 1, No. 8, 78–79 (1981) and 3, No. 2, 50–51 (1983)

sec-Hexyl Acetate

Molecular Formula: $C_8H_{16}O_2$

Common Formula: $CH_3COOCH(CH_3)CH_2CH(CH_3)CH_3$

Synonyms: 1,3-Dimethylbutyl acetate; Hexyl acetate; Methylamyl acetate; Methylisoamyl acetate; Methylisobutyl carbinol acetate

CAS Registry Number: 108-84-9

RTECS Number: A10875000

DOT ID: UN 1233

Regulatory Authority

- Air Pollutant Standard Set (ACGIH)[1] (DFG)[3] (OSHA)[58] (Several States)[60]

Cited in U.S. State Regulations: Alaska (G), Connecticut (A), Florida (G), Illinois (G), Maine (G), Nevada (A), New Hampshire (G), North Dakota (A), Rhode Island (G).

Description: sec-Hexyl acetate, $CH_3COOCH(CH_3)CH_2CH(CH_3)CH_3$, is a colorless liquid with a mild, pleasant, fruity odor. Boiling point = 147 – 148°C. Freezing/Melting point = -64°C. Flash point = 45°C. Autoignition temperature = 266°C. Explosive limits: LEL = 0.9%; UEL = 5.0%. Hazard Identification (based on NFPA-704 M Rating System): Health 1, Flammability 2, Reactivity 0. Insoluble in water.

Potential Exposure: This material is used as a solvent in the spray lacquer industry. It is a good solvent for cellulose esters and other resins.

Incompatibilities: Forms explosive mixture with air. Incompatible with strong acids, strong alkalis, nitrates, strong oxidizers.

Permissible Exposure Limits in Air: The Federal standard,[58] the DFG MAK value[3] and the ACGIH recommended TWA value is 50 ppm (300 mg/m^3). The NIOSH IDLH level is 500 ppm. Some states have set guidelines and standards for sec-hexyl acetate in ambient air[60] ranging from 3.0 mg/m^3 (North Dakota) to 6.0 mg/m^3 (Connecticut) to 7.143 mg/m^3 (Nevada).

Determination in Air: Charcoal adsorption, workup with CS_2, gas chromatography/flame ionization. See NIOSH Method 1450 for Esters.[18]

Permissible Concentration in Water: No criteria set.

Routes of Entry: Inhalation, ingestion, eye and skin contact.

Harmful Effects and Symptoms

Short Term Exposure: Irritates eyes and respiratory tract. May affect the central nervous system causing headache, dizziness, nausea, narcosis.

Long Term Exposure: May cause skin drying and cracking.

Points of Attack: Eyes, central nervous system.

Medical Surveillance: Consider the points of attack in preplacement and periodic examinations.

First Aid: If this chemical gets into the eyes, remove any contact lenses at once and irrigate immediately for at least 15 minutes, occasionally lifting upper and lower lids. Seek medical attention immediately. If this chemical contacts the skin, remove contaminated clothing and wash immediately with soap and water. Seek medical attention immediately. If this chemical has been inhaled, remove from exposure, begin rescue breathing (using universal precautions) if breathing has stopped and CPR if heart action has stopped. Transfer promptly to a medical facility. When this chemical has been swallowed, get medical attention. Give large quantities of water and induce vomiting. Do not make an unconscious person vomit.

Personal Protective Methods: Wear protective gloves and clothing to prevent any reasonable probability of skin contact. Safety equipment suppliers/manufacturers can provide recommendations on the most protective glove/clothing material for your operation. All protective clothing (suits, gloves, footwear, headgear) should be clean, available each day, and put on before work. Contact lenses should not be worn when working with this chemical. Wear splash-proof chemical goggles and face shield unless full facepiece respiratory protection is worn. Employees should wash immediately with soap when skin is wet or contaminated. Provide emergency showers and eyewash. Wear eye protection to prevent any reasonable probability of eye contact. Employees should wash promptly when skin is wet or contaminated. Remove nonimpervious clothing promptly if wet or contaminated.

Respirator Selection: NIOSH/OSHA: *Up to 500 ppm:* CCROV [any chemical cartridge respirator with organic vapor cartridge(s)];* or GMFOV [any air-purifying, full-facepiece respirator (gas mask) with a chin-style, front- or back-mounted organic vapor canister]; or PAPROV [any powered, air-purifying respirator with organic vapor cartridge(s)];* or SA (any supplied-air respirator);* SCBAF (any self-contained breathing apparatus with a full facepiece). *Emergency or planned entry into unknown concentrations or IDLH conditions:* SCBAF:PD,PP (any MSHA/NIOSH approved self-contained breathing apparatus that has a full facepiece and is operated in a pressure-demand or other positive-pressure mode); or SAF:PD,PP:ASCBA (any supplied-air respirator that has a full facepiece and is operated in a pressure-demand or other positive-pressure mode in combination with an auxiliary self-contained breathing apparatus operated in a pressure-demand or other positive pressure mode). *Escape:* GMFOV [any air-purifying, full-facepiece respirator (gas mask) with a chin-style, front-or back-mounted organic vapor canister]; or SCBAE (any appropriate escape-type, self-contained breathing apparatus).

* Substance reported to cause eye irritation or damage; may require eye protection.

Storage: Prior to working with hexyl acetate you should be trained on its proper handling and storage. Before entering confined space where this chemical may be present, check to make sure that an explosive concentration does not exist. Store in tightly closed containers in a cool, well ventilated area. Metal containers involving the transfer of this chemical should be grounded and bonded. Where possible, automatically pump liquid from drums or other storage containers to process containers. Drums must be equipped with self-closing valves, pressure vacuum bungs, and flame arresters. Use only non-sparking tools and equipment, especially when opening and closing containers of this chemical. Sources of ignition such as smoking and open flames, are prohibited where this chemical is used, handled, or stored in a manner that could create a potential fire or explosion hazard. Wherever this chemical is used, handled, manufactured, or stored, use explosion-proof electrical equipment and fittings.

Shipping: This compound requires a shipping label of: "Flammable Liquid." It falls in DOT Hazard Class 3 and Packing Group III.

Spill Handling: Evacuate and restrict persons not wearing protective equipment from area of spill or leak until cleanup is complete. Remove all ignition sources. Establish forced ventilation to keep levels below explosive limit. Absorb liquids in vermiculite, dry sand, earth, peat, carbon, or a similar material and deposit in sealed containers. Keep this chemical out of a confined space, such as a sewer, because of the possibility of an explosion, unless the sewer is designed to prevent the build-up of explosive concentrations. Oil-skimming equipment may be used to remove slicks from water. It may be necessary to contain and dispose of this chemical as a hazardous waste. If material or contaminated runoff enters waterways, notify downstream users of potentially contaminated waters. Contact your Department of Environmental Protection or your regional office of the federal EPA for specific recommendations. If employees are required to clean-up spills, they must be properly trained and equipped. OSHA 1910.120(q) may be applicable.

Fire Extinguishing: This chemical is a combustible liquid. Poisonous gases are produced in fire. Water may be ineffective. Use alcohol foam, CO_2 or dry chemical extinguishers. Vapors are heavier than air and will collect in low areas. Vapors may travel long distances to ignition sources and flashback. Vapors in confined areas may explode when exposed to fire. Containers may explode in fire. Storage containers and parts of containers may rocket great distances, in many directions. If material or contaminated runoff enters waterways, notify downstream users of potentially contaminated waters. Notify local health and fire officials and pollution control agencies. From a secure, explosion-proof location, use water spray to cool exposed containers. If cooling streams are ineffective (venting sound increases in volume and pitch, tank discolors, or shows any signs of deforming), withdraw immediately to a secure position. If employees are expected to fight fires, they must be trained and equipped in OSHA 1910.156.

Disposal Method Suggested: Incineration.

Hexylene Glycol

Molecular Formula: $C_6H_{14}O_2$

Common Formula: $(CH_3)_2C(OH)CH_2CHOHCH_3$

Synonyms: 2,4-Dihydroxy-2-methylpentane; Diolane; 1,2-Hexanediol; Isol; 2- Methylpentane-2,4-diol; 2-Methyl-2,4-pentanediol; 4-Methyl-2,4-pentanediol; 2,4-Pentanediol, 2-methyl-; Pinakon; α,α,α'-trimethylene glycol

CAS Registry Number: 107-41-1

RTECS Number: SA0810000

DOT ID: No citation.

EEC Number: 603-053-00-3

Regulatory Authority

- Air Pollutant Standard Set (ACGIH)[1] (HSE)[33] (North Dakota, Nevada)[60]

Cited in U.S. State Regulations: Alaska (G), California (A, G), Florida (G), Illinois (G), Massachusetts (G), Michigan (G), Minnesota (W), Nevada (A), New Hampshire (G), New Jersey (G), North Dakota (A), Pennsylvania (G), Rhode Island (G), West Virginia (G).

Description: Hexylene glycol, $(CH_3)_2C(OH)CH_2CHOHCH_3$, is a colorless liquid with a mild, sweet odor. The odor threshold is 50 ppm. Boiling point = 196 – 198°C. Freezing/Melting point = -50°C. Flash point = 102°C. Autoignition temperature = 260°C. Explosive limits: LEL = 1.2%; UEL = 8.1%.[41] Hazard Identification (based on NFPA-704 M Rating System): Health 1, Flammability 1, Reactivity 0. Soluble in water.

Potential Exposure: Hexylene glycol is used in the formulation of hydraulic brake fluids and printing inks. It is used as a fuel and lubricant additive, as an emulsifying agent and as a cement additive.

Incompatibilities: Incompatible with strong acids, caustics, aliphatic amines, isocyanates, strong oxidizers. Hygroscopic (i.e., absorbs moisture from the air).

Permissible Exposure Limits in Air: There is no OSHA PEL. NIOSH and ACGIH recommend a ceiling value of 25 ppm (125 mg/m^3). HSE[33] set the same value as an STEL. States which have set guidelines or standards for hexylene glycol in ambient air[60] include 1.25 mg/m^3 (North Dakota) and 2.976 mg/m^3 (Nevada).

Determination in Air: No test available.

Permissible Concentration in Water: No criteria set.

Routes of Entry: Inhalation, ingestion, skin and eye contact.

Harmful Effects and Symptoms

Short Term Exposure: Irritates the eyes, skin, and respiratory tract. Contact can irritate and may burn the eyes and skin. Exposure may effect the central nervous system. Eye and throat irritation and respiratory discomfort were slight upon exposure to 100 ppm but more pronounced at 1,000 ppm.

Ingestion produces central nervous system depression. High exposure can cause dizziness, loss of coordination, and unconsciousness. Extremely high exposures can cause coma and kidney damage.

Long Term Exposure: Repeated exposure may cause dry skin, rash, sensitization and allergy. May damage the kidney and liver and may affect the nervous system. Many similar petroleum-based solvents have been shown to cause brain and nerve damage.

Points of Attack: Eyes, skin, respiratory system, central nervous system.

Medical Surveillance: If symptoms develop or overexposure is suspected, the following may be useful: liver and kidney function tests. Evaluation by a qualified allergist. Evaluate for brain effects and refer positive and borderline individuals for neuropsychological testing.

First Aid: If this chemical gets into the eyes, remove any contact lenses at once and irrigate immediately for at least 15 minutes, occasionally lifting upper and lower lids. Seek medical attention immediately. If this chemical contacts the skin, remove contaminated clothing and wash immediately with soap and water. Seek medical attention immediately. If this chemical has been inhaled, remove from exposure, begin rescue breathing (using universal precautions) if breathing has stopped and CPR if heart action has stopped. Transfer promptly to a medical facility. When this chemical has been swallowed, get medical attention. Give large quantities of water and induce vomiting. Do not make an unconscious person vomit.

Personal Protective Methods: Wear protective gloves and clothing to prevent any reasonable probability of skin contact. Safety equipment suppliers/manufacturers can provide recommendations on the most protective glove/clothing material for your operation. All protective clothing (suits, gloves, footwear, headgear) should be clean, available each day, and put on before work. Contact lenses should not be worn when working with this chemical. Wear splash-proof chemical goggles and face shield unless full facepiece respiratory protection is worn. Employees should wash immediately with soap when skin is wet or contaminated. Provide emergency showers and eyewash.

Respirator Selection: Where the potential exists for exposures over 25 ppm, use an MSHA/NIOSH approved supplied-air respirator with a full facepiece operated in the positive pressure mode or with a full facepiece, hood, or helmet in the continuous flow mode, or use an MSHA/NIOSH approved self-contained breathing apparatus with a full facepiece operated in pressure demand or other positive pressure mode.

Storage: Prior to working with this chemical you should be trained on its proper handling and storage. Before entering confined space where hexanol may be present, check to make sure that an explosive concentration does not exist. Store in tightly closed containers in a cool, well ventilated area away from oxidizers (such as peroxides, perchlorates, chlorates, permanganates, and nitrates). Where possible, automatically pump liquid from drums or other storage containers to process containers.

Shipping: Hexylene glycol is not cited in DOT'S Performance-Oriented Packaging Standards.[19]

Spill Handling: Evacuate and restrict persons not wearing protective equipment from area of spill or leak until cleanup is complete. Remove all ignition sources. Establish forced ventilation to keep levels below explosive limit. Absorb liquids in vermiculite, dry sand, earth, peat, carbon, or a similar material and deposit in sealed containers. Keep this chemical out of a confined space, such as a sewer, because of the possibility of an explosion, unless the sewer is designed to prevent the build-up of explosive concentrations. It may be necessary to contain and dispose of this chemical as a hazardous waste. If material or contaminated runoff enters waterways, notify downstream users of potentially contaminated waters. Contact your Department of Environmental Protection or your regional office of the federal EPA for specific recommendations. If employees are required to clean-up spills, they must be properly trained and equipped. OSHA 1910.120(q) may be applicable.

Fire Extinguishing: This chemical is a combustible liquid. Poisonous gases are produced in fire. Use dry chemical, carbon dioxide, or alcohol foam extinguishers. Vapors are heavier than air and will collect in low areas. Vapors in confined areas may explode when exposed to fire. Containers may explode in fire. Storage containers and parts of containers may rocket great distances, in many directions. If material or contaminated runoff enters waterways, notify downstream users of potentially contaminated waters. Notify local health and fire officials and pollution control agencies. From a secure, explosion-proof location, use water spray to cool exposed containers. If cooling streams are ineffective (venting sound increases in volume and pitch, tank discolors, or shows any signs of deforming), withdraw immediately to a secure position. If employees are expected to fight fires, they must be trained and equipped in OSHA 1910.156.

Disposal Method Suggested: Incineration.

References

Sax, N. I., Ed., Dangerous Properties of Industrial Materials Report, 2, No. 2, 33–35 (1982)

New Jersey Department of Health and Senior Services, "Hazardous Substance Fact Sheet: Hexylene Glycol," Trenton, NJ (April, 1997)

Hexyl Trichlorosilane

Molecular Formula: $C_6H_{13}Cl_3Si$

Common Formula: $C_6H_{13}SiCl_3$

Synonyms: Silane, trichlorohexyl-; Trichlorohexylsilane

CAS Registry Number: 928-65-4

RTECS Number: VV4320000

DOT ID: UN 1784

Cited in U.S. State Regulations: Maine (G), New Hampshire (G), New Jersey (G), Oklahoma (G), Pennsylvania (G).

Description: Hexyl trichlorosilane, $C_6H_{13}SiCl_3$, is a colorless liquid which fumes in moist air. Boiling point = 191 – 192°C. Reacts with water.

Potential Exposure: Used in the manufacture of other silicon chemicals.

Incompatibilities: Reacts violently with strong oxidizers. Reacts violently with water, moisture, and steam producing chlorine and hydrogen chloride. Attacks active metals (e.g. aluminum and magnesium). Chlorosilanes on contact with ammonia, forms a self-igniting product.

Permissible Exposure Limits in Air: No standards set.

Permissible Concentration in Water: No criteria set.

Routes of Entry: Inhalation.

Harmful Effects and Symptoms

Short Term Exposure: Hexyl trichlorosilane can affect you when breathed in. Hexyl trichlorosilane is a corrosive chemical and can cause severe skin and eye burns leading to permanent eye damage. Exposure can severely irritate the respiratory tract causing coughing and/or shortness of breath. Higher exposures can cause pulmonary edema, a medical emergency that can be delayed for several hours. This can cause death.

Long Term Exposure: Highly irritating substances can cause lung irritation and bronchitis with cough, phlegm, and/or shortness of breath.

Points of Attack: Lungs.

Medical Surveillance: Before beginning employment and at regular times after that, for those with frequent or potentially high exposures, the following are recommended: Lung function tests. If symptoms develop or overexposure is suspected, the following may be useful: consider chest x-ray after acute overexposure.

First Aid: If this chemical gets into the eyes, remove any contact lenses at once and irrigate immediately for at least 15 minutes, occasionally lifting upper and lower lids. Seek medical attention immediately. If this chemical contacts the skin, remove contaminated clothing and wash immediately with soap and water. Seek medical attention immediately. If this chemical has been inhaled, remove from exposure, begin rescue breathing (using universal precautions) if breathing has stopped and CPR if heart action has stopped. Transfer promptly to a medical facility. When this chemical has been swallowed, get medical attention. If victim is conscious, administer water or milk. Do not induce vomiting. Do not make an unconscious person vomit. Medical observation is recommended for 24 – 48 hours after breathing overexposure, as pulmonary edema may be delayed. As first aid for pulmonary edema, a doctor or authorized paramedic may consider administering a corticosteroid spray.

Personal Protective Methods: Wear protective gloves and clothing to prevent any reasonable probability of skin contact. Safety equipment suppliers/manufacturers can provide recommendations on the most protective glove/clothing material for your operation. All protective clothing (suits, gloves, footwear, headgear) should be clean, available each day, and put on before work. Contact lenses should not be worn when working with this chemical. Wear splash-proof chemical goggles and face shield unless full facepiece respiratory protection is worn. Employees should wash immediately with soap when skin is wet or contaminated. Provide emergency showers and eyewash.

Respirator Selection: Where the potential for exposure to hexyl trichlorosilane exists, use a MSHA/NIOSH approved supplied-air respirator with a full facepiece operated in the positive pressure mode or with a full facepiece, hood, or helmet in the continuous flow mode, or use a MSHA/NIOSH approved self-contained breathing apparatus with a full facepiece operated in pressure-demand or other positive pressure mode.

Storage: Prior to working with this chemical you should be trained on its proper handling and storage. Store in tightly closed containers in a cool, well-ventilated area away from water and steam. Hexyl trichlorosilane can give off corrosive hydrogen chloride gas on contact with water, steam or moisture. Where possible, automatically pump liquid from drums or other storage containers to process containers.

Shipping: This compound requires a shipping label of: "Corrosive." It falls in DOT Hazard Class 8 and Packing Group II.

Spill Handling: Evacuate and restrict persons not wearing protective equipment from area of spill or leak until cleanup is complete. Remove all ignition sources. Ventilate area of spill or leak. Absorb liquids in vermiculite, dry sand, earth, peat, carbon, or a similar material and deposit in sealed containers. Keep this chemical out of a confined space, such as a sewer, because of the possibility of an explosion, unless the sewer is designed to prevent the build-up of explosive concentrations. It may be necessary to contain and dispose of this chemical as a hazardous waste. If material or contaminated runoff enters waterways, notify downstream users of potentially contaminated waters. Contact your Department of Environmental Protection or your regional office of the federal EPA for specific recommendations. If employees are required to clean-up spills, they must be properly trained and equipped. OSHA 1910.120(q) may be applicable.

Fire Extinguishing: This chemical is a combustible liquid. Poisonous gases including chlorine and hydrogen chloride are produced in fire. Do not use water as poisonous and corrosive gases will form Use dry chemical, carbon dioxide, or foam extinguishers. Vapors are heavier than air and will collect in low areas. Vapors may travel long distances to ignition sources and flashback. Vapors in confined areas may explode when exposed to fire. Containers may explode in fire. Storage containers and parts of containers may rocket great distances, in many directions. If material or contaminated runoff enters waterways, notify downstream users of potentially contaminated waters. Notify local health and fire officials and pollution control agencies. From a secure, explosion-proof location, use water spray to cool exposed containers. If cooling streams are ineffective (venting sound increases in volume and pitch, tank discolors, or shows any signs of deforming), withdraw immediately to a secure position. If employees are expected to fight fires, they must be trained and equipped in OSHA 1910.156.

References

New Jersey Department of Health and Senior Services, "Hazardous Substance Fact Sheet: Hexyl Trichlorosilane," Trenton, NJ (April 1999)

Hydrazine

Molecular Formula: H_4N_2

Common Formula: ($H_2N{-}NH_2$)

Synonyms: Amerzine; Diamide; Diamine; Diamine, hydrazine base; Hidrazina (Spanish); Hydrazine base; Hydrazyna (Polish); Levoxine; Mannitol mustard; Oxytreat 35; SCAV-OX; SCAV-OX 35%; SCAV-OX II; Ultra Pure

CAS Registry Number: 302-01-2

RTECS Number: MU7175000

DOT ID: UN 2029 (anhydrous); UN 3293 (<37% solution); UN 2030 (37 – 64% solution); UN 2029 (>64% solution)

EEC Number: 007-008-00-3

Regulatory Authority

- Air Pollutant Standard Set (ACGIH)[1] (HSE)[33] (OSHA)[58] (Other Countries)[35] (Several States)[60]
- Banned or Severely Restricted (Belgium, Denmark) (UN)[13]
- Carcinogen (Animal Positive) (IARC) (ACGIH)[1] (DFG)[3]
- CLEAN AIR ACT: Hazardous Air Pollutants (Title I, Part A, Section 112); Accidental Release Prevention/Flammable substances, (Section 112[r], Table 3), TQ = 15,000 lb (6,810 kg)
- EPA HAZARDOUS WASTE NUMBER (RCRA No.): U133
- RCRA, 40CFR261, Appendix 8 Hazardous Constituents
- SUPERFUND/EPCRA 40CFR355, Appendix B Extremely Hazardous Substances: TPQ = 1,000 lb (454 kg)
- SUPERFUND/EPCRA 40CFR302.4 Reportable Quantity (RQ): CERCLA, 1 lb (0.454 kg)
- EPCRA Section 313 Form R *de minimis* concentration reporting level: 0.1%
- Canada, WHMIS, Ingredients Disclosure List

Cited in U.S. State Regulations: Alaska (G), California (A, G), Connecticut (A), Florida (G, A), Illinois (G), Kansas (G), Louisiana (G), Maine (G), Maryland (G), Massachusetts (G, A), Michigan (G), Nevada (A), New Hampshire (G), New Jersey (G), New York (A), North Carolina (A), North Dakota (A), Oklahoma (G), Pennsylvania (G, A), Rhode Island (G, A), South Carolina (A), Vermont (G), Virginia (G, A), Washington (G), West Virginia (G), Wisconsin (G).

Description: Hydrazine, ($H_2N{-}NH_2$), is a colorless, oily liquid with an ammoniacal odor. The odor threshold is 3.7 ppm. Boiling point = 113°C. Freezing/Melting point = 2°C. Flash point = 38°C. Autoignition temperature = from 24°C (on a rusty iron surface) to 270°C (on glass surfaces). Explosive limits: LEL = 2.9%; UEL = 98%.[17] Hazard Identification (based on NFPA-704 M Rating System): Health 3, Flammability 3, Reactivity 0. Soluble in water.

Potential Exposure: Because of its strong reducing capabilities, hydrazine is used as an intermediate in chemical synthesis and in photography and metallurgy. It is also used in the preparation of anticorrosives, textile agents, and pesticides, and as a scavenging agent for oxygen in boiler water. Hydrazine is widely used in pharmaceutical synthesis. It is also used as a rocket fuel.

Incompatibilities: Forms explosive mixture with air. A highly reactive reducing agent and a medium strong base. Can ignite SPONTANEOUSLY on contact with oxidizers or porous materials such as earth, wood and cloth. Air or oxygen is not required for decomposition. Oxidizers, hydrogen peroxide, nitric acid, metallic oxides, acids, halogens can cause fire and explosions. Attacks cork, glass, some plastics, rubber and coatings.

Permissible Exposure Limits in Air: The Federal standard[58] and the HSE TWA[33] is 1.0 ppm (1.0 mg/m^3). ACGIH recommended a TWA of 0.1 ppm (0.1 mg/m^3) with the notation that hydrazine is a substance suspect of carcinogenic potential for man. The notation "skin" also indicates the possibility of cutaneous absorption. The NIOSH IDLH = Ca (50 ppm). No tentative STEL value is given except by Sweden[35] which gives 0.3 ppm (0.4 mg/m^3). Czechoslovakia[35] sets 0.05 mg/m^3 as a MAC in workplace air and the former USSR[35][43] set 0.1 mg/m^3 as a MAC. A number of states have set guidelines or standards for hydrazine in ambient air[60] ranging from zero (North Carolina and North Dakota) to 0.003 μg/m^3 (Rhode Island) to 0.018 mg/m^3 (Massachusetts) to 0.24 μg/m^3 (Pennsylvania) to 0.33 μg/m^3 (New York) to 0.5 μg/m^3 (South Carolina) to 1.0 μg/m^3 (Connecticut, Florida and Virginia) to 2.0 μg/m^3 (Nevada).

Determination in Air: Collection in a bubbler with HCl, reaction with -dimethylaminobenzaldehyde, colorimetric measurement. See NIOSH Method #3503 for hydrazine[18] and OSHA Method 20.[58]

Permissible Concentration in Water: EPA[32] has suggested a permissible ambient goal of 18 μg/l based on health effects.

The former USSR[35][43] has set a MAC in water bodies used for domestic purposes of 0.01 mg/l (10 μg/l).

Routes of Entry: Inhalation, skin absorption, ingestion, eye and/or skin contact.

Harmful Effects and Symptoms

Short Term Exposure: Hydrazine is corrosive to the eyes, skin, and respiratory tract. Inhalation of the vapors can cause pulmonary edema, a medical emergency that can be delayed for several hours. This can cause death. May affect the liver, kidneys, and central nervous system. Exposure may result in death. The effects may be delayed. Signs and symptoms of acute exposure to hydrazine may include severe eye irritation, facial numbness, facial swelling, and increased salivation. Hydrazine vapor may immediately irritate the nose and throat. Headache, twitching, seizures, convulsions, and coma may also occur. Gastrointestinal signs and symptoms include anorexia, nausea, and vomiting. Pulmonary edema and hypotension (low blood pressure) are common. Hydrazine is toxic to the liver, ruptures red blood cells, and may cause kidney damage. Dermal contact may result in irritation or severe burns. Target organs affected include central nervous system; respiratory system; skin and eyes. Chronic exposure in humans may cause pneumonia, liver and kidney damage. Liver damage may be more severe than kidney damage. It is a suspected human carcinogen.

Long Term Exposure: Repeated or prolonged contact may cause sensitization and skin allergy. Hydrazine may affect the liver, kidneys, and central nervous system. A probable carcinogen in humans. Can irritate the lungs and may cause bronchitis. Can damage the nervous system causing weakness, shaking, and loss balance and coordination. Exposure can cause blood damage and may cause anemia.

Points of Attack: Eyes, skin, respiratory system, central nervous system, liver, kidneys. Cancer site in animals: tumors of the lungs, liver, blood vessels and intestine.

Medical Surveillance: For those with frequent or potentially high exposure (half the TLV or greater, or significant skin contact), the following are recommended before beginning work and at regular times after that: Complete blood count. Liver and kidney function tests. Lung function tests. Exam of the nervous system. If symptoms develop or overexposure is suspected, the following may also be useful: consider chest x-ray after acute overexposure. Evaluation by a qualified allergist, including careful exposure history and special testing, may help diagnose skin allergy.

First Aid: If this chemical gets into the eyes, remove any contact lenses at once and irrigate immediately for at least 15 minutes, occasionally lifting upper and lower lids. Seek medical attention immediately. If this chemical contacts the skin, remove contaminated clothing and wash immediately with soap and water. Seek medical attention immediately. If this chemical has been inhaled, remove from exposure, begin rescue breathing (using universal precautions) if breathing has stopped and CPR if heart action has stopped. Transfer promptly to a medical facility. When this chemical has been swallowed, get medical attention. Give large quantities of water and induce vomiting. Do not make an unconscious person vomit. Medical observation is recommended for 24 – 48 hours after breathing overexposure, as pulmonary edema may be delayed. As first aid for pulmonary edema, a doctor or authorized paramedic may consider administering a corticosteroid spray.

Note to Physician: Consider Pyridoxine (25 mg/kg) which has been shown to be an effective anticonvulsant for hydrazine poisoning.

Personal Protective Methods: Wear protective gloves and clothing to prevent any reasonable probability of skin contact. Safety equipment suppliers/manufacturers can provide recommendations on the most protective glove/clothing material for your operation. Nitrile, Silvershield, PVC, Neoprene, and Butyl Rubber are among the recommended protective material. All protective clothing (suits, gloves, footwear, headgear) should be clean, available each day, and put on before work. Contact lenses should not be worn when working with this chemical. Wear splash-proof chemical goggles and face shield unless full facepiece respiratory protection is worn. Employees should wash immediately with soap when skin is wet or contaminated. Provide emergency showers and eyewash.

Respirator Selection: NIOSH: *At any concentrations above the NIOSH REL, or where there is no REL, at any detectable concentration:* SCBAF:PD,PP (any self-contained breathing apparatus that has a full facepiece and is operated in a pressure-demand or other positive-pressure mode); or SAF:PD,PP: ASCBA (any supplied-air respirator that has a full facepiece and is operated in a pressure-demand or other positive-pressure mode in combination with an auxiliary self-contained breathing apparatus operated in a pressure-demand or other positive pressure mode). *Escape:* SCBAE (any appropriate escape-type, self-contained breathing apparatus).

Storage: Prior to working with this chemical you should be trained on its proper handling and storage. Before entering confined space where hexanol may be present, check to make sure that an explosive concentration does not exist. Hydrazine must be stored to avoid contact with oxidizers (such as perchlorates, peroxides, permanganates, chlorates and nitrates), strong acids (such as hydrochloric, sulfuric and nitric), hydrogen peroxide and metal oxides since violent reactions occur. Store in tightly closed containers in a cool, well-ventilated area away from heat. Sources of ignition, such as smoking and open flames, are prohibited where hydrazine is used, handled, or stored in a manner that could create a potential fire or explosion hazard. Wherever hydrazine is used, handled, manufactured, or stored, use explosion-proof electrical equipment and fittings. A regulated, marked area should be established where this chemical is handled, used, or stored in compliance with OSHA standard 1910.1045.

Shipping: This compound requires a shipping label of: "Flammable Liquid, Poison." It falls in DOT Hazard Class 3 and Packing Group I. Passenger aircraft or railcar shipment is forbidden and cargo aircraft shipment is limited.

Spill Handling: Evacuate and restrict persons not wearing protective equipment from area of spill or leak until cleanup is complete. Remove all ignition sources. Stay up wind; keep out of low areas. In case of contact with material, immediately flush skin or eyes with running water for at least 15 minutes. Establish forced ventilation to keep levels below explosive limit. Absorb liquids in vermiculite, dry sand, earth, peat, carbon, or a similar material and deposit in sealed containers. Keep this chemical out of a confined space, such as a sewer, because of the possibility of an explosion, unless the sewer is designed to prevent the build-up of explosive concentrations. It may be necessary to contain and dispose of this chemical as a hazardous waste. If material or contaminated runoff enters waterways, notify downstream users of potentially contaminated waters. Contact your Department of Environmental Protection or your regional office of the federal EPA for specific recommendations. If employees are required to clean-up spills, they must be properly trained and equipped. OSHA 1910.120(q) may be applicable.

Fire Extinguishing: This chemical is a flammable liquid. Poisonous gases including ammonia fumes, hydrogen and nitrogen oxides are produced in fire. Small fires: dry chemical, carbon dioxide, water spray or foam. Large fires: water spray, fog, or foam. Stay upwind; keep out of low areas. Wear positive pressure breathing apparatus and protective clothing. Isolate for one-half mile in all directions if tank car or truck is involved in fire. Move container from fire area if you can do so without risk. Dike fire control water for later disposal, do not scatter material. Spray cooling water on containers that are exposed to flames until well after fire is out. It is a flammable/combustible material and may be ignited by heat, sparks, or flames. Vapors are heavier than air and will collect in low areas. Vapors may travel long distances to ignition sources and flashback. Vapors in confined areas may explode when exposed to fire. Containers may explode in fire. Storage containers and parts of containers may rocket great distances, in many directions. If material or contaminated runoff enters waterways, notify downstream users of potentially contaminated waters. Notify local health and fire officials and pollution control agencies. From a secure, explosion-proof location, use water spray to cool exposed containers. If cooling streams are ineffective (venting sound increases in volume and pitch, tank discolors, or shows any signs of deforming), withdraw immediately to a secure position. If employees are expected to fight fires, they must be trained and equipped in OSHA 1910.156.

Disposal Method Suggested: Controlled incineration with facilities for effluent scrubbing to abate any nitrogen compounds formed in the combustion process.[22] Consult with environmental regulatory agencies for guidance on acceptable disposal practices. Generators of waste containing this contaminant (≥100 kg/mo) must conform with EPA regulations governing storage, transportation, treatment, and waste disposal.

References

National Institute for Occupational Safety and Health, Criteria for a Recommended Standard: Occupational Exposure to Hydrazines, NIOSH Document No. 78–172, Washington, DC (1978)

Sax, N. I., Ed., Dangerous Properties of Industrial Materials Report, 1, No. 1, 45–46 (1981) and 3. No. 4, 65–68 (1983)

U.S. Environmental Protection Agency, "Chemical Profile: Hydrazine," Washington, DC, Chemical Emergency Preparedness Program (Nov. 30, 1987)

New Jersey Department of Health and Senior Services, "Hazardous Substance Fact Sheet: Hydrazine," Trenton, NJ (September 1986)

Hydrazine Sulfate

Molecular Formula: $H_4N_2{\cdot}H_2O_4S$

Synonyms: HS; Hydrazine hydrogen; Hydrazine monosulfate; Hydrazine sulphate; Hydrazinium sulfate; Hydrazonium sulfate; Idrazina solfato (Italian); NSC-150014; Siran hydrazinu (Czech)

CAS Registry Number: 10034-93-2

RTECS Number: MV9625000

DOT ID: UN 3077

Regulatory Authority

- Carcinogen: (NTP) sufficient animal evidence (IARC) (OSHA)
- Air Pollutant Standard Set (NIOSH)
- EPCRA Section 313 Form R *de minimis* concentration reporting level: 0.1%.

Cited in U.S. State Regulations: California (G), New Jersey (G), Pennsylvania (G).

Description: Hydrazine sulfate is a white or colorless, crystalline powder. Freezing/Melting point = 254°C (decomposition). Soluble in water.

Potential Exposure: Used in analysis and refining of minerals, rare metals, determination of arsenic in metals, as a catalyst and antioxidant, and in fungicides, germicides and blood tests. Used as a catalyst for making acetate fibers.

Incompatibilities: A strong reducing agent. Reacts with oxidizers, bases.

Permissible Exposure Limits in Air: The NIOSH airborne exposure limit for hydrazines is (ceiling) 0.03 ppm (0.04 mg/m^3) (2-hour).

Determination in Air: Use OSHA Method #20 or NIOSH Method #3503, Hydrazine.

Routes of Entry: Inhalation, skin and/or eye contact. Absorbed through the skin.

Harmful Effects and Symptoms

Short Term Exposure: Irritates the eyes and respiratory tract. Exposure can affect the brain and nervous system, causing dizziness and lightheadedness at first, followed by trembling and convulsions.

Long Term Exposure: Hydrazine sulfate has been shown to cause liver and lung cancers in animals. Exposure can damage the liver and kidneys. Repeated exposure can damage blood cells causing a low blood count (anemia). It can also cause methemoglobinemia with fatigue, shortness of breath, and even a bluish color to the nose, finger tips and lips. May cause skin allergy to develop.

Points of Attack: Liver, kidneys, blood, central nervous system, skin.

Medical Surveillance: Liver and kidney function tests. Complete blood count (CBC). Examination of the nervous system. Blood methemoglobin level. Evaluation by a qualified allergist.

First Aid: If this chemical gets into the eyes, remove any contact lenses at once and irrigate immediately for at least 15 minutes, occasionally lifting upper and lower lids. Seek medical attention immediately. If this chemical contacts the skin, remove contaminated clothing and wash immediately with soap and water. Seek medical attention immediately. If this chemical has been inhaled, remove from exposure, begin rescue breathing (using universal precautions) if breathing has stopped and CPR if heart action has stopped. Transfer promptly to a medical facility. When this chemical has been swallowed, get medical attention. Give large quantities of water and induce vomiting. Do not make an unconscious person vomit.

Note to Physician: Treat for methemoglobinemia. Spectrophotometry may be required for precise determination of levels of methemoglobinemia in urine. Pyridoxine (25 mg/kg) is an effective anticonvulsant for hydrazine poisoning.

Personal Protective Methods: Wear protective gloves and clothing to prevent any reasonable probability of skin contact. Safety equipment suppliers/manufacturers can provide recommendations on the most protective glove/clothing material for your operation. All protective clothing (suits, gloves, footwear, headgear) should be clean, available each day, and put on before work. Contact lenses should not be worn when working with this chemical. Wear dust-proof chemical goggles and face shield unless full facepiece respiratory protection is worn. Employees should wash immediately with soap when skin is wet or contaminated. Provide emergency showers and eyewash.

Respirator Selection: Where the potential for exposure to this chemical, use a MSHA/NIOSH approved supplied-air respirator with a full facepiece operated in the positive pressure mode or with a full facepiece, hood, or helmet in the continuous flow mode, or use a MSHA/NIOSH approved self-contained breathing apparatus with a full facepiece operated in pressure-demand or other positive pressure mode.

Storage: Prior to working with this chemical you should be trained on its proper handling and storage. A regulated, marked area should be established where this chemical is handled, used, or stored in compliance with OSHA standard 1910.1045. Store in tightly closed containers in a cool, well-ventilated area away from oxidizers, bases.

Shipping: Environmentally hazardous solid, n.o.s. Hazard Class: 9. Label: "Class 9." Packing Group: III.

Spill Handling: Evacuate persons not wearing protective equipment from area of spill or leak until clean-up is complete. Remove all ignition sources. Collect powdered material in the most convenient and safe manner and deposit in sealed containers. Ventilate area after clean-up is complete. It may be necessary to contain and dispose of this chemical as a hazardous waste. If material or contaminated runoff enters waterways, notify downstream users of potentially contaminated waters. Contact your Department of Environmental Protection or your regional office of the federal EPA for specific recommendations. If employees are required to clean-up spills, they must be properly trained and equipped. OSHA 1910.120(q) may be applicable.

Fire Extinguishing: Use dry chemical, carbon dioxide, water spray, or alcohol foam extinguishers. Poisonous gases including nitrogen oxides and sulfur oxides are produced in fire. If material or contaminated runoff enters waterways, notify downstream users of potentially contaminated waters. Notify local health and fire officials and pollution control agencies. From a secure, explosion-proof location, use water spray to cool exposed containers. If cooling streams are ineffective (venting sound increases in volume and pitch, tank discolors, or shows any signs of deforming), withdraw immediately to a secure position. If employees are expected to fight fires, they must be trained and equipped in OSHA 1910.156.

Hydrazoic Acid

Molecular Formula: HN_3

Synonyms: Azoimide; Diazoimide; Hydrogen azide; Hydronitric acid; Stickstoffwasserstoffsaeure (German); Triazoic acid

CAS Registry Number: 7782-79-8

RTECS Number: MW2800000

Regulatory Authority

- Air Pollutant Standard Set (DFG)[3]

Cited in U.S. State Regulations: Oklahoma (G).

Description: Hydrazoic acid, HN_3, is a colorless liquid with an intolerable, pungent odor. Boiling point = 37°C. Freezing/Melting point = -80°C. Soluble in water.

Potential Exposure: May be used in organic synthesis. Used in making heavy metal azide detonators for explosives.

Incompatibilities: Forms unstable heavy metal azides with heavy metals. Forms explosive salts with carbon disulfide. Violent reaction with cadmium, copper, nickel, nitric acid, fluorine, heat and shock. A highly sensitive explosive hazard when subject to shock or exposed to heat.

Permissible Exposure Limits in Air: The MAK value set by DFG[3] is 0.1 ppm (0.27 mg/m^3).

Permissible Concentration in Water: No criteria set.

Routes of Entry: Inhalation, ingestion, skin contact.

Harmful Effects and Symptoms

Short Term Exposure: Irritating to skin, eyes and respiratory tract. Continued inhalation causes, cough, headache, dizziness, weakness, fall in blood pressure, chills and fever and collapse. Prolonged exposure to high concentration can cause fatal convulsions and death. Highly toxic.

Long Term Exposure: Chronic exposure can affect the central nervous system, hypotension, palpitation, ataxia and weakness.

Points of Attack: Central nervous system.

First Aid: If this chemical gets into the eyes, remove any contact lenses at once and irrigate immediately for at least 15 minutes, occasionally lifting upper and lower lids. Seek medical attention immediately. If this chemical contacts the skin, remove contaminated clothing and wash immediately with soap and water. Seek medical attention immediately. If this chemical has been inhaled, remove from exposure, begin rescue breathing (using universal precautions) if breathing has stopped and CPR if heart action has stopped. Transfer promptly to a medical facility. When this chemical has been swallowed, get medical attention. Give large quantities of water and induce vomiting. Do not make an unconscious person vomit.

Personal Protective Methods: Wear protective gloves and clothing to prevent any reasonable probability of skin contact. Safety equipment suppliers/manufacturers can provide recommendations on the most protective glove/clothing material for your operation. All protective clothing (suits, gloves, footwear, headgear) should be clean, available each day, and put on before work. Contact lenses should not be worn when working with this chemical. Wear splash-proof chemical goggles and face shield unless full facepiece respiratory protection is worn. Employees should wash immediately with soap when skin is wet or contaminated. Provide emergency showers and eyewash.

Respirator Selection: Where the potential for exposure to this chemical, use a MSHA/NIOSH approved supplied-air respirator with a full facepiece operated in the positive pressure mode or with a full facepiece, hood, or helmet in the continuous flow mode, or use a MSHA/NIOSH approved self-contained breathing apparatus with a full facepiece operated in pressure-demand or other positive pressure mode.

Storage: Prior to working with this chemical you should be trained on its proper handling and storage. Store in tightly closed containers in a cool, well ventilated area away from incompatible materials listed above. Protect from heat or shock.

Spill Handling: Evacuate and restrict persons not wearing protective equipment from area of spill or leak until cleanup is complete. Remove all ignition sources. Ventilate area of spill or leak. Absorb liquids in vermiculite, dry sand, earth, peat, carbon, or a similar material and deposit in sealed containers. Keep this chemical out of a confined space, such as a sewer, because of the possibility of an explosion, unless the sewer is designed to prevent the build-up of explosive concentrations. It may be necessary to contain and dispose of this chemical as a hazardous waste. If material or contaminated runoff enters waterways, notify downstream users of potentially contaminated waters. Contact your Department of Environmental Protection or your regional office of the federal EPA for specific recommendations. If employees are required to clean-up spills, they must be properly trained and equipped. OSHA 1910.120(q) may be applicable.

Fire Extinguishing: This chemical is dangerously explosive material Poisonous gases including nitrogen oxides are produced in fire. Containers may explode in fire. Storage containers and parts of containers may rocket great distances, in many directions. If material or contaminated runoff enters waterways, notify downstream users of potentially contaminated waters. Notify local health and fire officials and pollution control agencies. From a secure, explosion-proof location, use water spray to cool exposed containers. If cooling streams are ineffective (venting sound increases in volume and pitch, tank discolors, or shows any signs of deforming), withdraw immediately to a secure position. If employees are expected to fight fires, they must be trained and equipped in OSHA 1910.156.

Disposal Method Suggested: May be destroyed by converting hydrazoic acid to sodium azide and the reaction mixture decomposed with nitrous acid (National Research Council, 1983).

Hydriodic Acid

Molecular Formula: HI

Synonyms: Hydriodic acid solution; Hydrogen iodide

CAS Registry Number: 10034-85-2

RTECS Number: MW3760000

DOT ID: UN 1787 (solution); UN 2197 (anhydrous)

Cited in U.S. State Regulations: Florida (G), Maine (G), Massachusetts (G), New Hampshire (G), New Jersey (G), Oklahoma (G), Pennsylvania (G), Rhode Island (G).

Description: Hydriodic acid, HI, is colorless when freshly made but rapidly turns yellowish or brown on exposure to light or air. Aqueous solution of hydrogen iodide which is a gas (anhydrous) at room temperature. Boiling point = -35°C. Freezing/Melting point = -51°C. Soluble in water.

Potential Exposure: Hydriodic acid is used as a disinfectant, analytical reagent, raw material for pharmaceuticals and to make iodine salts.

Incompatibilities: Contact with water forms toxic and corrosive fumes. A strong reducing agent and reducing agent. Violent actions with strong acids, chemically active metals, magnesium, phosphorus, perchloric acid, strong oxidizers.

Explodes on contact with ethyl hydroperoxide. Protect from moisture, heat and shock.

Permissible Exposure Limits in Air: No standards set.

Permissible Concentration in Water: No criteria set.

Routes of Entry: Inhalation, ingestion, skin contact.

Harmful Effects and Symptoms

Short Term Exposure: Corrosive to skin, eyes and mucous membranes. Skin or eye contact can cause severe burns and permanent damage. Inhalation causes, cough, headache, fall in blood pressure, chills and fever and collapse; irritation of the nose and throat and can cause spasms of the windpipe which can be fatal. High concentration can cause fatal convulsions. A Japanese source[24] states that HI is painful at 0.15 – 0.2 ppm; intolerable at 0.3 ppm.

Long Term Exposure: Chronic exposure can cause injury to kidneys and spleen, hypotension, palpitation, ataxia and weakness. Very irritating substances may cause lung damage. Prolonged absorption of iodides can cause skin rash, headache, irritation of mucous membranes with running nose.

Points of Attack: Lungs, kidneys.

Medical Surveillance: Before beginning employment and at regular times after that, for those with frequent or potentially high exposures, the following are recommended: lung function tests, chest x-ray following acute overexposure, kidney function tests.

First Aid: If this chemical gets into the eyes, remove any contact lenses at once and irrigate immediately for at least 15 minutes, occasionally lifting upper and lower lids. Seek medical attention immediately. If this chemical contacts the skin, remove contaminated clothing and wash immediately with soap and water. Seek medical attention immediately. If this chemical has been inhaled, remove from exposure, begin rescue breathing (using universal precautions) if breathing has stopped and CPR if heart action has stopped. Transfer promptly to a medical facility. When this chemical has been swallowed, get medical attention. If victim is conscious, administer water or milk. Do not induce vomiting.

Personal Protective Methods: Wear protective gloves and clothing to prevent any reasonable probability of skin contact. Safety equipment suppliers/manufacturers can provide recommendations on the most protective glove/clothing material for your operation. All protective clothing (suits, gloves, footwear, headgear) should be clean, available each day, and put on before work. Contact lenses should not be worn when working with this chemical. Wear splash-proof chemical goggles and face shield unless full facepiece respiratory protection is worn. Employees should wash immediately with soap when skin is wet or contaminated. Provide emergency showers and eyewash.

Respirator Selection: Where the potential for exposure to hydriodic acid exists, use a MSHA/NIOSH approved supplied-air respirator with a full facepiece operated in the positive pressure mode or with a full facepiece, hood, or helmet in the continuous flow mode, or use a MSHA/NIOSH approved self-contained breathing apparatus with a full facepiece operated in pressure-demand or other positive pressure mode.

Storage: Prior to working with this chemical you should be trained on its proper handling and storage. Hydriodic acid must be stored to avoid contact with strong acids (such as hydrochloric, sulfuric and nitric), chemically active metals (such as potassium, sodium, magnesium and zinc), and strong oxidizers (such as chlorine, bromine and fluorine) since violent reactions occur. Store in tightly closed containers in a cool, well-ventilated area away from heat and moisture. Protect storage containers from physical damage. Procedures for the handling, use and storage of cylinders should be in compliance with OSHA 1910.101 and 1910.169 as with the recommendations of the Compressed Gas Association.

Shipping: Hydriodic acid solution requires a shipping label of: "Corrosive." It falls in DOT Hazard Class 8 and Packing Group II. Hydrogen iodide, anhydrous requires a shipping label of: "Nonflammable Gas, Corrosive." It falls in DOT Hazard Class 2.2. Passenger aircraft or railcar shipment is forbidden and cargo aircraft shipment is forbidden as well.

Spill Handling: Evacuate and restrict persons not wearing protective equipment from area of spill or leak until cleanup is complete. Remove all ignition sources. Ventilate area of spill or leak. Neutralize with chemically basic substances such as sodium bicarbonate, soda ash, or slaked lime. Absorb liquids in vermiculite, dry sand, earth, peat, carbon, or a similar material and deposit in sealed containers. Keep this chemical out of a confined space, such as a sewer, because of the possibility of an explosion, unless the sewer is designed to prevent the build-up of explosive concentrations. It may be necessary to contain and dispose of this chemical as a hazardous waste. If material or contaminated runoff enters waterways, notify downstream users of potentially contaminated waters. Contact your Department of Environmental Protection or your regional office of the federal EPA for specific recommendations. If employees are required to clean-up spills, they must be properly trained and equipped. OSHA 1910.120(q) may be applicable.

Fire Extinguishing: Poisonous gases including iodine are produced in fire. Use water on fires in which hydriodic acid is involved. Containers may explode in fire. Storage containers and parts of containers may rocket great distances, in many directions. If material or contaminated runoff enters waterways, notify downstream users of potentially contaminated waters. Notify local health and fire officials and pollution control agencies. From a secure, explosion-proof location, use water spray to cool exposed containers. If cooling streams are ineffective (venting sound increases in volume and pitch, tank discolors, or shows any signs of deforming), withdraw immediately to a secure position. If employees are expected to fight fires, they must be trained and equipped in OSHA 1910.156.

Disposal Method Suggested: Add slowly to a large amount of soda ash and slaked lime in solution with stirring. Flush resulting solution to a sewer.[24]

References

New Jersey Department of Health and Senior Services, "Hazardous Substance Fact Sheet: Hydriodic Acid," Trenton, NJ (September 1986)

Hydrogen

Molecular Formula: H_2

Synonyms: Hidrogeno (Spanish); Para Hydrogen; Hydrogen, compressed; Hydrogen, refrigerated liquid; Liquid hydrogen

CAS Registry Number: 1333-74-0

RTECS Number: MW8900000

DOT ID: UN 1049 (gas); UN 1966 (liquid)

EEC Number: 001-001-00-9

Regulatory Authority

- Highly Reactive Substance and Explosive (World Bank)[15]
- CLEAN AIR ACT: Accidental Release Prevention/Flammable substances, (Section 112[r], Table 3), TQ = 10,000 lb (4,540 kg)

Cited in U.S. State Regulations: Alaska (G), California (A, G), Florida (G), Illinois (G), Maine (G), Massachusetts (G), New Hampshire (G), New Jersey (G), Pennsylvania (G), Rhode Island (G).

Description: Hydrogen, H_2, is a highly flammable, colorless liquid or gas. Boiling point = -252°C. Autoignition temperature: 500 – 571°C. Explosive limits: LEL = 4.0%; UEL = 75%. Hazard Identification (based on NFPA-704 M Rating System): Health 0, Flammability 4, Reactivity 0. Slightly soluble in water.

Potential Exposure: Hydrogen is used as a fuel in weeding, as a raw material for ammonia manufacture and in organic hydrogenation reactions.

Incompatibilities: Vapors form explosive or combustible mixture with air over a wide range of concentrations. Heating may cause violent reaction, combustion or explosion. Ignites easily with oxygen. Violent reaction with strong oxidizers, halogens, acetylene, bromine, chlorine, fluorine, nitrous oxide and other gases; metal catalysts (e.g. platinum and nickel) greatly enhance these reactions. Mild steel and most iron alloys become brittle at liquid hydrogen temperatures.

Permissible Exposure Limits in Air: There are no numerical limits but hydrogen is classed as a simple asphyxiant.[1][33] Large amounts of hydrogen will decrease the amount of available oxygen. Oxygen content should be tested to ensure that it is at least 19% by volume in confined spaces. The health effects caused by exposure to hydrogen are much less serious than its fire and explosion risk.

Permissible Concentration in Water: No criteria set.

Routes of Entry: Inhalation.

Harmful Effects and Symptoms

Short Term Exposure: Exposure to high levels can cause suffocation from lack of oxygen. Contact with liquid hydrogen can cause severe burns and frostbite.

Medical Surveillance: There are not special tests.

First Aid: *Contact With Liquid Hydrogen:* Put affected part into warm water. Seek medical attention. *Breathing:* Remove the person from exposure. Begin rescue breathing if breathing has stopped and CPR if heart action has stopped. Transfer promptly to a medical facility.

Personal Protective Methods: *Clothing:* Where exposure to cold equipment, vapors, or liquid may occur, employees should be provided with special clothing designed to prevent the freezing of body tissues. All protective clothing (suits, gloves, footwear, headgear) should be clean, available each day, and put on before work. *Eye Protection:* Wear splash-proof chemical goggles and face shield when working with liquid, unless full face piece respiratory protection is worn.

Respirator Selection: Exposure to hydrogen is dangerous because it can replace oxygen and lead to suffocation. Only MSHA/NIOSH approved self-contained breathing apparatus with a full face piece operated in positive pressure mode should be used in oxygen deficient environments.

Storage: Prior to working with this chemical you should be trained on its proper handling and storage. Before entering confined space where hexanol may be present, check to make sure that an explosive concentration does not exist. Hydrogen must be stored to avoid contact with heat, flames, sparks, and oxygen. Sources of ignition such as smoking and open flames are prohibited where hydrogen is used, handled, or stored. Metal containers involving the transfer of 5 gallons or more of hydrogen should be grounded and bonded. Drums must be equipped with self-closing valves, pressure vacuum bungs, and flame arresters. Use only non-sparking tools and equipment, especially when opening and closing containers of hydrogen. Wherever hydrogen is used, handled, manufactured or stored, use explosion-proof electrical equipment and fittings. Piping should be electrically bonded and grounded. Procedures for the handling, use and storage of cylinders should be in compliance with OSHA 1910.101 and 1910.169 as with the recommendations of the Compressed Gas Association.

Shipping: Compressed hydrogen requires a shipping label of: "Flammable Gas." It falls in DOT Hazard Class 2.1. Refrigerated liquid hydrogen requires a shipping label of "Flammable Gas." It falls in DOT Hazard Class 2.1. Passenger aircraft or railcar shipment is forbidden and the cargo aircraft shipment is forbidden as well.

Spill Handling: If hydrogen gas is leaked, take the following steps: Restrict persons not wearing protective equipment from area of leak until clean-up is complete. Remove all ignition sources. Establish forced ventilation to keep levels below explosive limit. Stop flow of gas. If source of leak is a cylinder and the leak cannot be stopped in place, remove the leaking cylinder to a safe place in the open air, and repair leak or allow cylinder to empty. If liquid hydrogen is spilled or leaked, take the following steps: Restrict persons not wearing protective equipment from area of spill or leak until clean-up is complete. Remove all ignition sources. Stop the leak or move the container to a safe area and allow the liquid to evaporate. Keep hydrogen out of a confined space, such as a sewer, because of the possibility of an explosion, unless the sewer is designed to prevent the build-up of explosive concentrations.

Fire Extinguishing: If spill has not ignited, use water spray to direct flammable gas-air mixtures away from sources of ignition. If it is desirable to evaporate a spill quickly, water spray may be used to increase the rate of evaporation, if the increased vapor evolution can be controlled. Do not discharge solid streams into liquid. Because of danger of reignition, hydrogen fires normally should not be extinguished until the supply of hydrogen has been shut off. If liquid hydrogen has ignited, use water to keep fire-exposed containers cool and to protect men stopping the source of a spill. If it is necessary to extinguish small hydrogen fires, use dry chemical, carbon dioxide, or halogenated extinguishing agent. Vapors may travel long distances to ignition sources and flashback. Vapors in confined areas may explode when exposed to fire. Containers may explode in fire. Storage containers and parts of containers may rocket great distances, in many directions. From a secure, explosion-proof location, use water spray to cool exposed containers. If cooling streams are ineffective (venting sound increases in volume and pitch, tank discolors, or shows any signs of deforming), withdraw immediately to a secure position. If employees are expected to fight fires, they must be trained and equipped in OSHA 1910.156.

Disposal Method Suggested: Combustion.

References

New Jersey Department of Health and Senior Services, "Hazardous Substance Fact Sheet: Hydrogen," Trenton, NJ (June, 1996)

Hydrogenated Terphenyls

Molecular Formula: $(C_6H_n)_3$

Common Formula: C_6H_{11}-C_6H_4-C_6H_5

Synonyms: Hydrogenated diphenylbenzenes; Hydrogenated phenylbiphenyls; Hydrogenated triphenyls

CAS Registry Number: 61788-32-7

Regulatory Authority

- Air Pollutant Standard Set (ACGIH)[1] (OSHA)[58] (Several States)[60]

Cited in U.S. State Regulations: Alaska (G), Connecticut (A), Illinois (G), Maine (G), Massachusetts (G), Nevada (A), New Hampshire (G), Rhode Island (G), Virginia (A).

Description: Hydrogenated terphenyls, $(C_6Hn)_3$, are complex mixture of terphenyl isomers that are partially hydrogenated, and clear, oily, pale-yellow liquids with a faint odor. Boiling point = 340°C (40% hydrogenated). Freezing/Melting point = 148°C (40% hydrogenated). Flash point = 157°C. Autoignition temperature = 374°C. Insoluble in water.

Potential Exposure: These materials are used as high temperature heat transfer media and as plasticizers.

Incompatibilities: Strong oxidizers.

Permissible Exposure Limits in Air: There is no OSHA PEL. NIOSH and ACGIH[1] recommend a TWA value of 0.5 ppm (5 mg/m^3). States which have set guidelines or standards for hydrogenated terphenyls in ambient air include Virginia at 80.0 μg/m^3, Connecticut at 100.0 μg/m^3, and Nevada at 119.0 μg/m^3.[60]

Determination in Air: No method available.

Permissible Concentration in Water: No criteria set.

Routes of Entry: Inhalation, ingestion, skin and/or eye contact.

Harmful Effects and Symptoms

Short Term Exposure: Irritates eyes, skin, respiratory system. Potential acute hazards consist of damage to the lungs and damage to the skin and eyes from burns from the hot coolant.

Long Term Exposure: Potential chronic hazards comprise damage to liver, kidney and blood-forming organs with the possibility of induction of metabolic disorders and cancer.[53]

Points of Attack: Eyes, skin, respiratory system, liver, kidneys, hemato system.

Medical Surveillance: Liver and kidney function tests. Complete blood count (CBC). Lung function tests.

First Aid: If this chemical gets into the eyes, remove any contact lenses at once and irrigate immediately for at least 15 minutes, occasionally lifting upper and lower lids. Seek medical attention immediately. If this chemical contacts the skin, remove contaminated clothing and wash immediately with soap and water. Seek medical attention immediately. If this chemical has been inhaled, remove from exposure, begin rescue breathing (using universal precautions) if breathing has stopped and CPR if heart action has stopped. Transfer promptly to a medical facility. When this chemical has been swallowed, get medical attention. Give large quantities of water and induce vomiting. Do not make an unconscious person vomit.

Personal Protective Methods: Wear protective gloves and clothing to prevent any reasonable probability of skin contact. Safety equipment suppliers/manufacturers can provide recommendations on the most protective glove/clothing material for your operation. All protective clothing (suits, gloves, footwear, headgear) should be clean, available each day, and put on before work. Contact lenses should not be worn when working with this chemical. Wear splash-proof chemical goggles and face shield unless full facepiece respiratory

protection is worn. Employees should wash immediately with soap when skin is wet or contaminated. Provide emergency showers and eyewash.

Respirator Selection: Where the potential for exposure to this chemical, use a MSHA/NIOSH approved supplied-air respirator with a full facepiece operated in the positive pressure mode or with a full facepiece, hood, or helmet in the continuous flow mode, or use a MSHA/NIOSH approved self-contained breathing apparatus with a full facepiece operated in pressure-demand or other positive pressure mode.

Storage: Prior to working with this chemical you should be trained on its proper handling and storage. Store in tightly closed containers in a cool, well ventilated area away from oxidizers. Where possible, automatically pump liquid from drums or other storage containers to process containers.

Spill Handling: Evacuate and restrict persons not wearing protective equipment from area of spill or leak until cleanup is complete. Remove all ignition sources. Ventilate area of spill or leak. Absorb liquids in vermiculite, dry sand, earth, peat, carbon, or a similar material and deposit in sealed containers. Keep this chemical out of a confined space, such as a sewer, because of the possibility of an explosion, unless the sewer is designed to prevent the build-up of explosive concentrations. It may be necessary to contain and dispose of this chemical as a hazardous waste. If material or contaminated runoff enters waterways, notify downstream users of potentially contaminated waters. Contact your Department of Environmental Protection or your regional office of the federal EPA for specific recommendations. If employees are required to clean-up spills, they must be properly trained and equipped. OSHA 1910.120(q) may be applicable.

Fire Extinguishing: This chemical is a combustible liquid. Irritating vapors are produced in fire. Use dry chemical, carbon dioxide, or foam extinguishers. Vapors are heavier than air and will collect in low areas. Vapors in confined areas may explode when exposed to fire. Containers may explode in fire. Storage containers and parts of containers may rocket great distances, in many directions. If material or contaminated run-off enters waterways, notify downstream users of potentially contaminated waters. Notify local health and fire officials and pollution control agencies. From a secure, explosion-proof location, use water spray to cool exposed containers. If cooling streams are ineffective (venting sound increases in volume and pitch, tank discolors, or shows any signs of deforming), withdraw immediately to a secure position. If employees are expected to fight fires, they must be trained and equipped in OSHA 1910.156.

Hydrogen Bromide

Molecular Formula: HBr

Synonyms: Acide bromhydrique (French); Acido bromidrico (Italian); Anhydrous hydrobromic acid; Bromowodor (Polish); Bromwasserstoff (German); Broomwaterstof (Dutch); HBr; Hydrobromic acid; Hydrobromic acid, anhydrous; Hydrogen bromide, anhydrous

CAS Registry Number: 10035-10-6

RTECS Number: MW3850000

DOT ID: UN 1048 (anhydrous); UN 1788 (hydrobromic acid solution)

EEC Number: 035-002-00-0

Regulatory Authority

- Air Pollutant Standard Set (ACGIH)[1] (DFG)[3] (HSE)[33] (former USSR)[43] (OSHA)[58] (Several States)[60]
- OSHA 29CFR1910.119, Appendix A. Process Safety List of Highly Hazardous Chemicals, TQ = 5,000 lb (2,270 kg)
- Canada, WHMIS, Ingredients Disclosure List

Cited in U.S. State Regulations: Alaska (G), California (A, G), Connecticut (A), Florida (G), Illinois (G), Maine (G), Massachusetts (G), Nevada (A), New Hampshire (G), New Jersey (G), New York (A), North Dakota (A), Oklahoma (G), Pennsylvania (G), Rhode Island (G), Virginia (A), West Virginia (G).

Description: Hydrogen bromide, HBr, is a corrosive colorless gas with a sharp, irritating odor. The odor threshold is 2 – 6.6 ppm. Boiling point = -67°C (anhydrous). Freezing/Melting point = -87°C (anhydrous). Soluble in water. A constant boiling dihydrate melts at -11°C and boils at 126°C. Shipped as a liquefied compressed gas. Often used in an aqueous solution.

Potential Exposure: Hydrogen bromide and its aqueous solutions are used in the manufacture of organic and inorganic bromides, as a reducing agent and catalyst in controlled oxidations, in the alkylation of aromatic compounds, and in the isomerization of conjugated diolefins. It is used in the production of many drugs.

Incompatibilities: The aqueous solution is a strong acid. Violent reaction with strong oxidizers, strong caustics, and many organic compounds causing fire and explosion hazard. Reacts with water forming toxic hydrobromic acid. Incompatible with aliphatic amines, alkanolamines, alkylene oxides, aromatic amines, amides, ammonia, ammonium hydroxide, calcium oxide, epichlorohydrin, fluorine, isocyanates, oleum, organic anhydrides, sulfuric acid, sodium tetrahydroborate, vinyl acetate. Hydrobromic acid is highly corrosive to most metals forming flammable hydrogen.

Permissible Exposure Limits in Air: The OSHA PEL is 3.0 pm (10 mg/m^3) TWA. The NIOSH and ACGIH recommended ceiling[1] value for hydrogen bromide is 3.0 pm (10 mg/m^3). The NIOSH IDLH level is 30 ppm. The HSE[33] has set the 3 ppm (10 mg/m^3) as an 8-hour TWA value. DFG has set a 5 pm (17 mg/m^3) as an MAK value.[3] The former USSR-UNEP/IRPTC project[43] has set a MAC in workplace air of 2 mg/m^3. Several states have set guidelines or standards for hydrogen bromide in ambient air[60] ranging from 80.0 μg/m^3 (Virginia) to 100.0 μg/m^3 (North Dakota) to 200.0 μg/m^3 (Connecticut and New York) to 238.0 μg/m^3 (Nevada).

Determination in Air: Collection using Si gel; workup with $NaHCO_3$/Na_2CO_3; Analysis by ion chromatography; NIOSH IV, Method #7903, Inorganic Acids.

Routes of Entry: Inhalation, ingestion, eye and/or skin contact.

Harmful Effects and Symptoms

Short Term Exposure: Hydrogen bromide is a corrosive chemical and contact can severely burn the eyes, with permanent damage. It can cause severe burns of the skin. Exposure can irritate the eyes, nose, throat and lungs. Higher exposures can cause pulmonary edema, a medical emergency that can be delayed for several hours. Contact with liquid can cause frostbite.

Long Term Exposure: Long-term exposure can irritate the lungs and cause a chronic discharge. Bronchitis may develop with cough, phlegm, and/or shortness of breath. It may damage the sense of smell. Long-term exposure can also cause chronic indigestion and may damage the nervous system. Repeated skin contact can cause an acne-like rash to develop.

Points of Attack: Eyes, skin, respiratory system.

Medical Surveillance: Before beginning employment and at regular times after that, the following are recommended: Lung function tests. If symptoms develop or overexposure is suspected, the following may be useful: consider chest x-ray after acute overexposure. Serum Bromine level. Exam of the nervous system.

First Aid: If this chemical gets into the eyes, remove any contact lenses at once and irrigate immediately for at least 15 minutes, occasionally lifting upper and lower lids. Seek medical attention immediately. If this chemical contacts the skin, remove contaminated clothing and wash immediately with soap and water. Seek medical attention immediately. If this chemical has been inhaled, remove from exposure, begin rescue breathing (using universal precautions) if breathing has stopped and CPR if heart action has stopped. Transfer promptly to a medical facility. When this chemical has been swallowed, get medical attention. If victim is conscious, administer water or milk. Do not induce vomiting. Medical observation is recommended for 24 – 48 hours after breathing overexposure, as pulmonary edema may be delayed. As first aid for pulmonary edema, a doctor or authorized paramedic may consider administering a corticosteroid spray. If frostbite has occurred, seek medical attention immediately; do *NOT* rub the affected areas or flush them with water. In order to prevent further tissue damage, do *NOT* attempt to remove frozen clothing from frostbitten areas. If frostbite has *NOT* occurred, immediately and thoroughly wash contaminated skin with soap and water.

Personal Protective Methods: Wear protective gloves and clothing to prevent any reasonable probability of skin contact. Safety equipment suppliers/manufacturers can provide recommendations on the most protective glove/clothing material for your operation. All protective clothing (suits, gloves, footwear, headgear) should be clean, available each day, and put on before work. Contact lenses should not be worn when working with this chemical. Wear eye protection to prevent any possibility of eye contact. Wear splash-proof chemical goggles and face shield when working with the liquid unless full facepiece respiratory protection is worn. Employees should wash immediately with soap when skin is wet or contaminated. Provide emergency showers and eyewash.

Respirator Selection: NIOSH/OSHA: *30 ppm:* SA:CF (any supplied-air respirator operated in a continuous-flow mode); or PAPRAG [any powered, air-purifying respirator with acid gas cartridge(s)]; or GMFAG [any air-purifying, full-facepiece respirator (gas mask) with a chin-style, front-or back-mounted organic vapor canister]; or SCBAF (any self-contained breathing apparatus with a full facepiece); or SAF (any supplied-air respirator with a full facepiece). *Emergency or planned entry into unknown concentrations or IDLH conditions:* SCBAF:PD,PP (any self-contained breathing apparatus that has a full facepiece and is operated in a pressure-demand or other positive-pressure mode); or SAF:PD,PP:ASCBA (any supplied-air respirator that has a full facepiece and is operated in a pressure-demand or other positive-pressure mode in combination with an auxiliary self-contained breathing apparatus operated in a pressure-demand or other positive pressure mode). *Escape:* GMFAG [any air-purifying, full-facepiece respirator (gas mask) with a chin-style, front-or back-mounted organic vapor canister]; or SCBAE (any appropriate escape-type, self-contained breathing apparatus).

Note: Substance causes eye irritation or damage; eye protection needed.

Storage: Prior to working with this chemical you should be trained on its proper handling and storage. Hydrogen bromide must be stored to avoid contact with strong oxidizers, caustics, metals and moisture, because violent reactions occur. Store in tightly closed containers in a cool, well-ventilated area.

Shipping: Anhydrous HBr requires a shipping label of: "Poison Gas, Corrosive." It falls in DOT Hazard Class 2.3 and Packing Group II. Passenger aircraft or railcar shipment is forbidden and cargo aircraft shipment is forbidden as well. Hydrobromic acid solution requires a shipping label of: "Corrosive." It falls in DOT Hazard Class 8 and Packing Group II.

Spill Handling: Evacuate and restrict persons not wearing protective equipment from area of spill or leak until cleanup is complete. Remove all ignition sources. Ventilate area of spill or leak. Ventilate area of leak to disperse the gas. Stop flow of gas. If source of leak is a cylinder and the leak cannot be stopped in place, remove the leaking cylinder to a safe place in the open air, and repair leak or allow cylinder to empty. Absorb liquids in vermiculite, dry sand, earth, peat, carbon, or a similar material and deposit in sealed containers. Keep this chemical out of a confined space, such as a sewer, because of the possibility of an explosion, unless the sewer is designed to prevent the build-up of explosive concentrations. It may be necessary to contain and dispose of this chemical as a hazardous waste. If material or contaminated runoff enters waterways, notify downstream users of potentially contaminated waters. Contact your Department of Environmental Protection or your regional office of the federal EPA for specific recommendations. If employees are required to clean-up spills, they must be properly trained and equipped. OSHA 1910.120(q) may be applicable.

Initial isolation and protective action distances

Distances shown are likely to be affected during the first 30 minutes after materials are spilled and could increase with time. If more than one tank car, cargo tank, portable tank, or large cylinder is involved in the incident is leaking, the protective action distance may need to be increased. You may need to seek emergency information from CHEMTREC at (800) 424-9300 or seek professional environmental engineering assistance from the U.S. EPA Environmental Response Team at (908) 548-8730 (24-hour response line).

UN 1048:

Small spills (From a small package or a small leak from a large package)

First: Isolate in all directions (feet)..... 200

Then: Protect persons downwind (miles)

Day 0.1

Night 0.2

Large spills (From a large package or from many small packages)

First: Isolate in all directions (feet)..... 400

Then: Protect persons downwind (miles)

Day 0.2

Night 0.7

Fire Extinguishing: Hydrogen bromide is non-combustible. Reacts with water to produce toxic hydrobromic acid. Fight surrounding fire with an agent appropriate for surrounding fire. Poisonous gases including bromine and hydrogen bromide are produced in fire. Vapors are heavier than air and will collect in low areas. Vapors may travel long distances to ignition sources and flashback. Vapors in confined areas may explode when exposed to fire. Containers may explode in fire. Storage containers and parts of containers may rocket great distances, in many directions. If material or contaminated runoff enters waterways, notify downstream users of potentially contaminated waters. Notify local health and fire officials and pollution control agencies. From a secure, explosion-proof location, use water spray to cool exposed containers. If cooling streams are ineffective (venting sound increases in volume and pitch, tank discolors, or shows any signs of deforming), withdraw immediately to a secure position. If employees are expected to fight fires, they must be trained and equipped in OSHA 1910.156.

Disposal Method Suggested: Soda ash/slaked lime is added to give a neutral bromide solution which is discharged to sewers or streams with water dilution.

References

U.S. Environmental Protection Agency, Chemical Hazard Information Profile: Bromine and Bromine Compounds, Washington, DC (November 1, 1976)

New Jersey Department of Health and Senior Services, "Hazardous Substance Fact Sheet: Hydrogen Bromide," Trenton, NJ (April 1999)

Hydrogen Chloride

Molecular Formula: HCl

Synonyms: Acide chlorhydrique (French); Acido cloridrico (Italian); Anhydrous hydrochloric acid; Aqueous hydrogen chloride; Chloorwaterstof (Dutch); Chlorohydric acid; Chlorowodor (Polish); Chlorwasserstoff (German); HCl; Hydrochloric acid; Hydrochloric acid, anhydrous; Hydrochloride; Hydrogen chloride; Muriatic acid; Spirits of salt

CAS Registry Number: 7647-01-0

RTECS Number: MW9610000; MW4025000

DOT ID: UN 1050 (anhydrous); UN 1789 (solution); UN 2186 (refrigerated liquid)

EEC Number: 017-002-00-2

Regulatory Authority

- Air Pollutant Standard Set (ACGIH)[1] (DFG)[3] (HSE)[33] (Argentina)[35] (former USSR)[43] (OSHA)[58] (Several States)[60]
- Toxic Substance (World Bank)[13]
- OSHA 29CFR1910.119, Appendix A. Process Safety List of Highly Hazardous Chemicals, TQ = 5,000 lb (2,270 kg)
- CLEAN AIR ACT: Hazardous Air Pollutants (Title I, Part A, Section 112); Accidental Release Prevention/Flammable substances, (Section 112[r], Table 3), TQ = 5,000 lb (2,270 kg)
- CLEAN WATER ACT: Section 311 Hazardous Substances/ RQ 40CFR117.3 (same as CERCLA, see below); Section 313 Water Priority Chemicals (57FR41331, 9/9/92)
- SUPERFUND/EPCRA 40CFR302.4 Reportable Quantity (RQ): CERCLA, 5,000 lb (2,270 kg)
- EPCRA Section 313 Form R *de minimis* concentration reporting level: 1.0%
- Note: Non-aerosol forms of hydrochloric acid have been deleted from EPCRA/SARA 313 reporting, 7/29/96 (FR vol. 61, No. 146, p. 39356–39357)
- Canada, WHMIS, Ingredients Disclosure List

Cited in U.S. State Regulations: Alaska (G), California (A, G), Florida (G), Illinois (G), Maine (G), Maryland (G),

Massachusetts (G, A), Nevada (A), New Hampshire (G), New Jersey (G), New York (G, A), North Carolina (A), North Dakota (A), Oklahoma (G), Pennsylvania (G), Rhode Island (G, A), South Carolina (A), South Dakota (A), Virginia (A), West Virginia (G).

Description: Hydrogen chloride, HCl, is a colorless to slightly yellow gas or fuming liquid with a pungent, irritating odor. May be shipped and stored as a cryogenic liquid. The odor threshold is 0.77 ppm. Boiling point = -85°C. Freezing/Melting point = -114°C. Soluble in water. The aqueous solution is known as hydrochloric acid or muriatic acid and may contain as much as 38% HCl. Shipped as a liquefied compressed gas.

Potential Exposure: Hydrogen chloride itself is used in the manufacture of pharmaceutical hydrochlorides, chlorine, vinyl chloride from acetylene, alkyl chlorides from olefins, arsenic trichloride from arsenic trioxide; in the chlorination of rubber; as a gaseous flux for babbitting operations; and in organic synthesis involving isomerization, polymerization, alkylation, and nitration reactions. The acid is used in the production of fertilizers, dyes, dyestuffs, artificial silk, and paint pigments; in refining edible oils and fats; in electroplating, leather tanning, ore refining, soap refining, petroleum extraction, pickling of metals, and in the photographic, textile, and rubber industries.

Incompatibilities: The aqueous solution is a strong acid. Corrosive fumes emitted on contact with air. Reacts violently with bases, oxidizers forming toxic chlorine gas. Reacts, often violently, with acetic anhydride, active metals, aliphatic amines, alkanolamines, alkylene oxides, aromatic amines, amides, 2-aminoethanol, ammonia, ammonium hydroxide, calcium phosphide, chlorosulfonic acid, ethylene diamine, ethyleneimine, epichlorohydrin, isocyanates, metal acetylides, oleum, organic anhydrides, perchloric acid, 3-propiolactone, uranium phosphide, sulfuric acid, vinyl acetate, vinylidene fluoride. Highly corrosive to most metals, forming flammable hydrogen gas. Attacks some plastics, rubber and coatings.

Permissible Exposure Limits in Air: The OSHA PEL, NIOSH REL and ACGIH ceiling value for hydrogen chloride is 5 ppm (7 mg/m^3), not to be exceeded at any time. There is no STEL value. The NIOSH IDLH level is 50 ppm. The HSE[33] has set the same 5 ppm value as an 8-hour TWA and also as an STEL, as has Argentina.[35] The DFG[3] has also set the 5 ppm as an MAK value. The former USSR-UNEP/IRPTC project[43] has set a MAC in workplace air of 5 mg/m^3 and a MAC in ambient air in residential areas of 0.2 mg/m^3 both on a momentary and a daily average basis. Several states have set guidelines or standards for HCl in ambient air[60] ranging from 10 μg/m^3 (Massachusetts) to 70 μg/m^3 (North Dakota) to 120 μg/m^3 (Virginia) to 140 μg/m^3 (New York and South Dakota) to 167 μg/m^3 (Nevada) to 175 μg/m^3 (South Carolina) to 600 – 2,000 μg/m^3 (Rhode Island) to 700 μg/m^3 (North Carolina).

Determination in Air: Si gel; $NaHCO_3/Na_2CO_3$; Ion chromatography; NIOSH IV, Method #7903, Inorganic Acids.

Permissible Concentration in Water: No criteria set.

Routes of Entry: Inhalation of gas or mist, ingestion, eye and/or skin contact.

Harmful Effects and Symptoms

Signs and symptoms of acute ingestion of hydrogen chloride may be severe and include salivation, intense thirst, difficulty in swallowing, chills, pain, and shock. Oral, esophageal, and stomach burns are common. Vomitus generally has a coffee-ground appearance. The potential for circulatory collapse is high following ingestion of hydrogen chloride. Acute inhalation exposure of hydrogen chloride may result in sneezing, hoarseness, choking, laryngitis, and respiratory tract irritation. Bleeding of nose and gums, ulceration of the nasal and oral mucosa, bronchitis, pneumonia, dyspnea (shortness of breath), chest pain, and pulmonary edema may also occur. If the eyes have come in contact with hydrogen chloride, irritation, pain, swelling, corneal erosion, and blindness may result. Dermal exposure may result in dermatitis (red, inflamed skin), severe burns, and pain.

Short Term Exposure: HCl is corrosive to the eyes, skin, and respiratory tract. Inhalation of high concentrations of vapors can cause pulmonary edema, a medical emergency that can be delayed for several hours. This can cause death. May cause inflammation and destruction of the nasal passages, dental erosion, loss of voice, coughing, pneumonia, headaches and rapid throbbing of the heart. May cause death in the range of 1,000 – 2,000 ppm. Skin contact can cause irritation or burns of the skin. Eye contact may cause irritation and severe damage to the surface of the eye, severe burns and loss of sight. Contact with the liquid may cause frostbite. Ingestion may cause irritation of mouth, throat and stomach; salivation, nausea, vomiting, chills and fever; holes in the intestinal tract; inflammation of the kidneys; and shock.

Long Term Exposure: Irritates the lungs, causing chronic bronchitis. May cause irritation and skin rash. Long term exposure to low levels (greater than 5 ppm) of hydrogen chloride can cause some dental erosion. Aside from such dental erosions, no significant abnormalities have been associated with long term low level exposures. There is limited evidence that workers involved in the manufacturing of hydrogen chloride have an increase of respiratory cancers.

Points of Attack: Eyes, skin, respiratory system.

Medical Surveillance: For those with frequent or potentially high exposure (half the TLV or greater) the following are recommended before beginning work and at regular times after that: lung function tests. If symptoms develop or overexposure is suspected, the following may be useful: consider chest x-ray after acute overexposure.

First Aid: If this chemical gets into the eyes, remove any contact lenses at once and irrigate immediately for at least

15 minutes, occasionally lifting upper and lower lids. Seek medical attention immediately. If this chemical contacts the skin, remove contaminated clothing and wash immediately with soap and water. Seek medical attention immediately. If this chemical has been inhaled, remove from exposure, begin rescue breathing (using universal precautions) if breathing has stopped and CPR if heart action has stopped. Transfer promptly to a medical facility. When this chemical has been swallowed, get medical attention. If victim is conscious, administer water or milk. Do not induce vomiting. Medical observation is recommended for 24 – 48 hours after breathing overexposure, as pulmonary edema may be delayed. As first aid for pulmonary edema, a doctor or authorized paramedic may consider administering a corticosteroid spray. If frostbite has occurred, seek medical attention immediately; do *NOT* rub the affected areas or flush them with water. In order to prevent further tissue damage, do *NOT* attempt to remove frozen clothing from frostbitten areas. If frostbite has *NOT* occurred, immediately and thoroughly wash contaminated skin with soap and water.

Personal Protective Methods: Wear protective gloves and clothing to prevent any reasonable probability of skin contact. Safety equipment suppliers/manufacturers can provide recommendations on the most protective glove/clothing material for your operation. All protective clothing (suits, gloves, footwear, headgear) should be clean, available each day, and put on before work. Contact lenses should not be worn when working with this chemical. Wear eye protection to prevent any possibility of eye contact. Employees should wash immediately with soap when skin is wet or contaminated. Provide emergency showers and eyewash.

Respirator Selection: NIOSH/OSHA: *50 ppm:* CCRS [any chemical cartridge respirator with cartridge(s) providing protection against the compound of concern]; or GMFS [any air-purifying, full-facepiece respirator (gas mask) with a chin-style, front- or back-mounted canister providing protection against the compound of concern]; or PAPRS [any powered, air-purifying respirator with cartridge(s) providing protection against the compound of concern]; or SA (any supplied-air respirator); or SCBAF (any self-contained breathing apparatus with a full facepiece). *Emergency or planned entry into unknown concentrations or IDLH conditions:* SCBAF:PD,PP (any self-contained breathing apparatus that has a full facepiece and is operated in a pressure-demand or other positive-pressure mode); or SAF:PD,PP:ASCBA (any supplied-air respirator that has a full facepiece and is operated in a pressure-demand or other positive-pressure mode in combination with an auxiliary self-contained breathing apparatus operated in a pressure-demand or other positive pressure mode). *Escape:* GMFAG [any air-purifying, full-facepiece respirator (gas mask) with a chin-style, front-or back-mounted organic vapor canister]; or SCBAE (any appropriate escape-type, self-contained breathing apparatus).

Note: Substance reported to cause eye irritation or damage; may require eye protection.

Storage: Hydrogen chloride must be stored to avoid contact with any alkali or active metals (such as potassium, sodium, and zinc), because violent reactions occur. Store in tightly closed containers in cool, well-ventilated area away from heat.

Shipping: Anhydrous HCl requires a shipping label of: "Poison Gas, Corrosive." It falls in DOT Hazard Class 2.3 and Packing Group II. Passenger aircraft or railcar shipment is forbidden and cargo aircraft shipment is forbidden also. HCl solution requires a shipping label of: "Corrosive." It falls in DOT Hazard Class 8 and Packing Group II.

Spill Handling: *Gas:* If hydrogen chloride gas is leaked, take the following steps: Restrict persons not wearing protective equipment from area of leak until clean-up is complete. Ventilate area of leak to disperse the gas. Stop flow of gas. If source of leak is a cylinder and the leak cannot be stopped in place, remove the leaking cylinder to a safe place in the open air, and repair leak or allow cylinder to empty. If hydrogen chloride solution is spilled or leaked, take the following steps: Restrict persons not wearing protective equipment from area of spill or leak until cleanup is complete. Collect material in a convenient manner and deposit in sealed containers. If necessary, dilute and/or neutralize the material before collection. It may be necessary to contain and dispose of hydrogen chloride as a hazardous waste. Contact the NJ Department of Environmental Protection (DEP) or your regional office of the federal Environmental Protection Agency (EPA) for specific recommendations.

Liquid: Evacuate and restrict persons not wearing protective equipment from area of spill or leak until cleanup is complete. Remove all ignition sources. Ventilate area of spill or leak. Absorb liquids in vermiculite, dry sand, earth, peat, carbon, or a similar material and deposit in sealed containers. Keep this chemical out of a confined space, such as a sewer, because of the possibility of an explosion, unless the sewer is designed to prevent the build-up of explosive concentrations. It may be necessary to contain and dispose of this chemical as a hazardous waste. If material or contaminated runoff enters waterways, notify downstream users of potentially contaminated waters. Contact your Department of Environmental Protection or your regional office of the federal EPA for specific recommendations. If employees are required to clean-up spills, they must be properly trained and equipped. OSHA 1910.120(q) may be applicable.

Initial isolation and protective action distances

Distances shown are likely to be affected during the first 30 minutes after materials are spilled and could increase with time. If more than one tank car, cargo tank, portable tank, or large cylinder is involved in the incident is leaking, the protective action distance may need to be increased. You may need to seek emergency information from CHEMTREC at (800) 424-9300 or seek professional environmental engineering assistance from the U.S. EPA Environmental Response Team at (908) 548-8730 (24-hour response line).

UN 1048:

Small spills (From a small package or a small leak from a large package)

First: Isolate in all directions (feet)..... 200

Then: Protect persons downwind (miles)

Day 0.1

Night 0.2

Large spills (From a large package or from many small packages)

First: Isolate in all directions (feet)..... 400

Then: Protect persons downwind (miles)

Day 0.2

Night 0.7

UN 1050, UN 2186:

Small spills (From a small package or a small leak from a large package)

First: Isolate in all directions (feet)..... 200

Then: Protect persons downwind (miles)

Day 0.1

Night 0.3

Large spills (From a large package or from many small packages)

First: Isolate in all directions (feet)..... 500

Then: Protect persons downwind (miles)

Day 0.3

Night 1.1

Fire Extinguishing: Extinguish fire using an agent suitable for type of surrounding fire. (The material itself does not burn, but contact with metals produces hydrogen gas which will increase the chance of explosion). Poisonous gases including chlorine and hydrogen chloride are produced in fire. Use extinguishers suitable for surrounding fires other than water. Contact with water will produce heat and hydrochloric acid. If water is used, it must be used in flooding quantities. Vapors are heavier than air and will collect in low areas. Containers may explode in fire. Storage containers and parts of containers may rocket great distances, in many directions. If material or contaminated runoff enters waterways, notify downstream users of potentially contaminated waters. Notify local health and fire officials and pollution control agencies. From a secure, explosion-proof location, use water spray to cool exposed containers. Do not apply water to cryogenic liquid containers. If cryogenic liquid containers are exposed to direct flame or elevated temperatures for prolonged periods, with draw to a secure location. If cooling streams are ineffective (venting sound increases in volume and pitch, tank discolors, or shows any signs of deforming), withdraw immediately to a secure position. If employees are expected to fight fires, they must be trained and equipped in OSHA 1910.156.

Disposal Method Suggested: Soda ash-slaked lime is added to form the neutral solution of chloride of sodium and calcium. This solution can be discharged after dilution with water.[22] Alternatively, hydrogen chloride can be recovered from a variety of process waste streams.

References

National Institute for Occupational Safety and Health, Information Profiles on Potential Occupational Hazards: Hydrogen Chloride (Gas), pp 23–28, Report PB-276, 678, Rockville, Maryland (1977)

Sax, N. I., Ed., Dangerous Properties of Industrial Materials Report, 1, No. 7, 62–65 (1981)

U.S. Environmental Protection Agency, "Chemical Profile: Hydrogen Chloride," Washington, DC, Chemical Emergency Preparedness Program (Nov. 30, 1987)

New Jersey Department of Health and Senior Services, "Hazardous Substance Fact Sheet: Hydrogen Chloride," Trenton, NJ (February 1989)

New York State Department of Health, "Chemical fact Sheet: Hydrogen Chloride," Albany, NY, Bureau of Toxic Substance Assessment (March 1986)

Hydrogen Cyanide

Molecular Formula: CHN

Common Formula: HCN

Synonyms: Acide cyanhydrique (French); Acido cianidrico (Italian); Aero liquid HCN; Blausaeure (German); Blauwzuur (Dutch); Cyaanwaterstof (Dutch); Cyanwasserstoff (German); Cyclon; CycloneB; Cyjanowodor (Polish); Formonitrile; HCN; Hydrocyanic acid; Prussic acid; Zaclon Discoids

CAS Registry Number: 74-90-8

RTECS Number: MW6825000

DOT ID: UN 1051 (anhydrous, stabilized); UN 1613 (solution); UN 1614 (absorbed)

EEC Number: 006-006-00-X

Regulatory Authority

- Toxic Substance (World Bank)[15]
- Air Pollutant Standard Set (ACGIH)[1] (DFG)[3] (HSE)[33] (Other Countries)[35] (OSHA)[58] (Several States)[60]
- OSHA 29CFR1910.119, Appendix A, Process Safety List of Highly Hazardous Chemicals, TQ = 1,000 lb (454 kg)
- Banned of Severely Restricted (Belgium, E. Germany, Philippines) (UN)[13]
- CLEAN AIR ACT: Accidental Release Prevention/Flammable substances, (Section 112[r], Table 3), TQ = 2,500 lb (1,135 kg)
- CLEAN WATER ACT: Section 311 Hazardous Substances/RQ 40CFR117.3 (same as CERCLA, see below); Section 313 Water Priority Chemicals (57FR41331, 9/9/92)
- EPA HAZARDOUS WASTE NUMBER (RCRA No.): P063
- U.S. DOT Regulated Marine Pollutant (49CFR172.101, Appendix B)

- SUPERFUND/EPCRA 40CFR355, Appendix B Extremely Hazardous Substances: TPQ = 100 lb (45.4 kg)
- SUPERFUND/EPCRA 40CFR302.4 Reportable Quantity (RQ): CERCLA, 10 lb (4.54 kg)
- EPCRA Section 313 Form R *de minimis* concentration reporting level: 1.0%
- U.S. DOT Regulated Marine Pollutant (49CFR172.101, Appendix B)
- Canada, WHMIS, Ingredients Disclosure List

Cited in U.S. State Regulations: Alaska (G), California (A, G), Connecticut (A), Florida (G), Illinois (G), Kansas (G), Louisiana (G), Maine (G), Maryland (G), Massachusetts (G), Nevada (A), New Hampshire (G), New Jersey (G), New York (G, A), North Carolina (A), North Dakota (A), Oklahoma (G), Pennsylvania (G), Virginia (G, A), Washington (G), West Virginia (G), Wisconsin (G).

Description: Hydrogen cyanide, is a volatile, colorless or pale-blue liquid or colorless gas with a bitter, almond-like odor. The odor threshold is 0.58 ppm. Odor is *not* a reliable indicator of toxic amounts of vapor. Often used as a 96% solution in water. It is intensely poisonous, highly flammable, explosive, and is a weak acid. Boiling point = 26°C. Freezing/Melting point = -13°C. Flash point = -18°C. Autoignition temperature = 538°C. Explosive limits: LEL = 5.6%; UEL = 40.0%.[17] Hazard Identification (based on NFPA-704 M Rating System): Health 4, Flammability 4, Reactivity 2. Soluble in water.

Potential Exposure: Hydrogen cyanide is used as a fumigant, in electroplating, and in chemical synthesis of acrylates and nitriles, particularly acrylonitrile. It may be generated in blast furnaces, gas works, and coke ovens. Cyanide salts have a wide variety of uses, including electroplating, steel hardening, fumigating, gold and silver extraction from ores, and chemical synthesis.

Incompatibilities: Unless stabilized and maintained, samples stored more than 90 days are hazardous. The gas can form an explosive mixture with air. Material containing more than 2 – 5% water are less stable than dry material and can be self-reactive, forming an explosive mixture with air. Heat above 50 – 60°C or contact with amines or strong bases can cause polymerization. The aqueous solution is a weak acid. Violent reaction with oxidizers, acetaldehyde, hydrogen chloride in alcoholic mixtures, causing fire and explosion hazard. Incompatible with amines, strong acids, sodium hydroxide, calcium hydroxide, sodium carbonate, water, ammonia. Attacks some plastics, rubber and coatings.

Permissible Exposure Limits in Air: The OSHA PEL is 10 ppm (11 mg/m^3) STEL, not to be exceeded during any 15 minute work period. The NIOSH REL is 4.7 ppm (5 mg/m^3) STEL, not to be exceeded during any 15 minute work period. ACGIH recommends a limit of 4.7 ppm Ceiling, not to be exceeded at any time. Sweden[35] has also set 5 mg/m^3 as a ceiling value. Brazil[35] has set 8 ppm (9 mg/m^3) as a TWA value and a number of countries (Argentina, Brazil, Germany, Japan) have set 10 ppm (11 mg/m^3) as a TWA value. The UK[33] has set no TWA, only an STEL of 10 ppm. ACGIH[1] has set 10 ppm (10 mg/m^3) as a ceiling value. Czechoslovakia and the former USSR have set 0.3 mg/m^3 as a MAC in workplace air. The notation "skin" is added to indicate the possibility of cutaneous absorption. The NIOSH IDLH level is 50 ppm. In ambient air in residential areas, the former USSR has set a MAC of 0.01 mg/m on a daily average basis. Czechoslovakia[35] has set 0.008 mg/m^3 on both a momentary and daily average basis. A number of states have set guidelines or standards for hydrogen cyanide in ambient air[60] raging from 33 μg/m^3 (New York) to 80 μg/m^3 (Virginia) to 100 μg/m^3 (North Dakota) to 220 μg/m^3 (Connecticut) to 238 μg/m^3 (Nevada) to 250 μg/m^3 (South Carolina) to 120 – 1,000 μg/m^3 (North Carolina).

Determination in Air: Soda lime; Water; Visible spectrophotometry; NIOSH IV, Method #6010.

Permissible Concentration in Water: A USPHS drinking water criterion for alternate source selection is 100 μg/l.[32]

Routes of Entry: Inhalation of vapor, percutaneous absorption of liquid and concentrated vapor, ingestion and eye and skin contact.

Harmful Effects and Symptoms

Short Term Exposure: Hydrogen cyanide can irritate the and burn the skin and eyes. Inhalation can irritate the respiratory tract. Lacrimation (tearing) and a burning sensation of the mouth and throat are common. Can cause dizziness, headache, weakness, anxiety, confusion, pounding heart, difficult breathing and nausea. These can rapidly lead to convulsions and death unless exposure is immediately stopped and proper first aid applied. High exposure can cause sudden death. Signs and symptoms of acute exposure to hydrocyanic acid may include hypertension (high blood pressure) and tachycardia (rapid heart rate), followed by hypotension (low blood pressure) and bradycardia (slow heart rate). Cherry red mucous membranes and blood may be noted. Cardiac arrhythmias and other cardiac abnormalities are common. Cyanosis (blue tint to the skin and mucous membranes) may be observed. Weakness, headache, vertigo (dizziness), agitation, giddiness, salivation, nausea, and vomiting, may be followed by combative behavior, convulsions, paralysis, protruding eyeballs, dilated and unreactive pupils, and coma. Tachypnea (rapid, shallow respirations) or hyperpnea (rapid, deep respirations) may be followed by respiratory depression. Lung hemorrhage and pulmonary edema may also occur.

Inhalation: At less than 20 ppm, exposure to hydrogen cyanide may produce headache, dizziness, nausea and vomiting. Concentrations greater than 50 ppm may cause difficulty in breathing, rapid throbbing of the heart, paralysis, unconsciousness, respiratory arrest or death. 30 minutes exposure to 135 ppm may cause death. 270 ppm has caused immediate death. *Skin:* Hydrogen cyanide is readily absorbed through the skin. Symptoms are similar to above. *Eyes:* Hydrogen cyanide is irritating to the eye and rapidly absorbed.

Ingestion: Symptoms are similar to above. Death has resulted from ingestion of 570 mg/kg of 1.4 oz for a 150 pound person.

Long Term Exposure: Repeated exposure can interfere with thyroid function and can cause goiter. Itching scarlet rash, red bumps, severe nose itch leading to bleeding, and possibly holes in the nose, may result from long term exposure to hydrogen cyanide. Headache, nausea, vomiting, weakness and enlarged thyroid gland have also been reported at exposures from 4 to 12 ppm. May damage the nervous system.

Points of Attack: Central nervous system, cardiovascular system, thyroid, blood.

Medical Surveillance: Preplacement and periodic examinations should include the cardiovascular and central nervous systems, liver and kidney function, blood, history of fainting or dizzy spells. Blood cyanide test. Evaluation of thyroid function. Examination of the nervous system. Urinary thiocyanate levels have been used, but are nonspecific and are elevated in smokers.

First Aid: If this chemical gets into the eyes, remove any contact lenses at once and irrigate immediately for at least 15 minutes, occasionally lifting upper and lower lids. Seek medical attention immediately. If this chemical contacts the skin, remove contaminated clothing and wash immediately with soap and water. Seek medical attention immediately. If this chemical has been inhaled, remove from exposure, begin rescue breathing (using universal precautions) if breathing has stopped and CPR if heart action has stopped. Transfer promptly to a medical facility. When this chemical has been swallowed, get medical attention. Give large quantities of water and induce vomiting. Do not make an unconscious person vomit. Use amyl nitrate capsules if symptoms develop. All area employees should be trained regularly in emergency measures for cyanide poisoning and in CPR. A cyanide antidote kit should be kept in the immediate work area and must be rapidly available. Kit ingredients should be replaced every 1 – 2 years to ensure freshness. Persons trained in the use of this kit, oxygen use, and CPR must be quickly available.

Personal Protective Methods: Wear protective gloves and clothing to prevent any reasonable probability of skin contact. Safety equipment suppliers/manufacturers can provide recommendations on the most protective glove/clothing material for your operation. All protective clothing (suits, gloves, footwear, headgear) should be clean, available each day, and put on before work. Contact lenses should not be worn when working with this chemical. Wear eye protection to prevent any possibility of eye contact. Employees should wash immediately with soap when skin is wet or contaminated. Provide emergency showers and eyewash. See NIOSH Criteria Document 78–212 NITRILES.

Respirator Selection: NIOSH/OSHA: *47 ppm:* SA (any supplied-air respirator). *50 ppm:* SA:CF (any supplied-air respirator operated in a continuous-flow mode); or SCBAF (any self-contained breathing apparatus with a full facepiece); or SAF (any supplied-air respirator with a full facepiece). *Emergency or planned entry into unknown concentrations or IDLH conditions:* SCBAF:PD,PP (any self-contained breathing apparatus that has a full facepiece and is operated in a pressure-demand or other positive-pressure mode); or SAF:PD,PP: ASCBA (any supplied-air respirator that has a full facepiece and is operated in a pressure-demand or other positive-pressure mode in combination with an auxiliary self-contained breathing apparatus operated in a pressure-demand or other positive pressure mode). *Escape:* GMFS [any air-purifying, full-facepiece respirator (gas mask) with a chin-style, front- or back-mounted canister providing protection against the compound of concern]; or SCBAE (any appropriate escape-type, self-contained breathing apparatus).

Storage: Prior to working with this chemical you should be trained on its proper handling and storage. Before entering confined space where hexanol may be present, check to make sure that an explosive concentration does not exist. Before entering confined space where this chemical may be present, check to make sure that an explosive concentration does not exist. Store in tightly closed containers in a cool, well ventilated area away from incompatible materials and conditions (see above). Store outdoors, if possible, or indoors in standard combustible liquid storage room or cabinet, away from sources of ignition. Protect containers against physical damage. Metal containers involving the transfer of this chemical should be grounded and bonded. Where possible, automatically pump liquid from drums or other storage containers to process containers. Drums must be equipped with self-closing valves, pressure vacuum bungs, and flame arresters. Use only non-sparking tools and equipment, especially when opening and closing containers of this chemical. Sources of ignition such as smoking and open flames, are prohibited where this chemical is used, handled, or stored in a manner that could create a potential fire or explosion hazard. Wherever this chemical is used, handled, manufactured, or stored, use explosion-proof electrical equipment and fittings.

Shipping: Hydrocyanic acid aqueous solution (<5% HCN) requires a shipping label of: "Poison." It falls in DOT Hazard Class6.1 and Packing Group II. Passenger aircraft or railcar is forbidden and the limit on cargo aircraft shipment is limited. Shipment of solutions from 5 – 20% is forbidden. Stabilized, anhydrous HCN requires a shipping label of: "Poison, Flammable Liquid." It forbidden and cargo aircraft shipment is forbidden as well.

Spill Handling: Evacuate and restrict persons not wearing protective equipment from area of spill or leak until cleanup is complete. Remove all ignition sources. Establish forced ventilation to keep levels below explosive limit. Stay upwind; keep out of low areas. Ventilate closed spaces before entering. Use water spray to reduce vapors. Do not touch spilled material; stop leak if you can do it without risk. Shut off ignition; no flares, smoking, or flames in hazard area. Isolate area until gas dispersed. Absorb liquids in vermiculite, dry sand, earth, peat, carbon, or a similar material and deposit in sealed containers. Keep this chemical out of a confined space, such as a sewer, because of the possibility of an explosion, unless the sewer is designed to prevent the build-up of explosive concentrations. It may be necessary to contain and dispose of this chemical as a hazardous waste. If material or contaminated runoff enters waterways, notify downstream users of potentially contaminated waters. Contact your Department of Environmental Protection or your regional office of the federal EPA for specific recommendations. If employees are required to clean-up spills, they must be properly trained and equipped. OSHA 1910.120(q) may be applicable.

Initial isolation and protective action distances

Distances shown are likely to be affected during the first 30 minutes after materials are spilled and could increase with time. If more than one tank car, cargo tank, portable tank, or large cylinder is involved in the incident is leaking, the protective action distance may need to be increased. You may need to seek emergency information from CHEMTREC at (800) 424-9300 or seek professional environmental engineering assistance from the U.S. EPA Environmental Response Team at (908) 548-8730 (24-hour response line).

UN 1613, UN 1051:

Small spills (From a small package or a small leak from a large package)

First: Isolate in all directions (feet)..... 200

Then: Protect persons downwind (miles)

Day 0.1

Night 0.5

Large spills (From a large package or from many small packages)

First: Isolate in all directions (feet)..... 600

Then: Protect persons downwind (miles)

Day 0.4

Night 1.7

Fire Extinguishing: Poisonous gases including cyanide are produced in fire. If material is on fire and conditions permit, do not extinguish; combustion products are less toxic than the material itself. Cool exposures using unattended monitors. Do not use water directly on fire as cyanide gas may form. Use water to control vapors only. May react with itself without warning with explosive violence.

Gas: This chemical is a flammable gas. Keep unnecessary people away; isolate hazard area and deny entry. Stay upwind; keep out of low areas. Ventilate closed spaces before entering them. Wear positive pressure breathing apparatus and special protective clothing. Evacuate area endangered by gas. See isolation distance above if tank car or truck is involved in fire. Use dry chemicals, alcohol resistant (AFFF) foam, or carbon dioxide. Small fires; let burn unless leak can be stopped immediately. Large fires: water spray, fog or foam. Move container from fire area if you can do it without risk. Stay away from ends of tanks. Withdraw immediately in case of rising sound from venting safety device or any discoloration of tank due to fire. Cool container with water using unmanned device until well after fire is out. Isolate area until gas has dispersed. Firefighting should be done from a safe distance. A few whiffs of gas, or liquid penetrating firefighter's protective clothing, could be fatal. Only special protective clothing should be worn.

Liquid: This chemical is a flammable liquid. Use dry chemical, carbon dioxide, or alcohol resistant (AFFF) foam extinguishers. Vapors are heavier than air and will collect in low areas. Vapors may travel long distances to ignition sources and flashback. Vapors in confined areas may explode when exposed to fire. Containers may explode in fire. Storage containers and parts of containers may rocket great distances, in many directions.

If material or contaminated runoff enters waterways, notify downstream users of potentially contaminated waters. Notify local health and fire officials and pollution control agencies. From a secure, explosion-proof location, use water spray to cool exposed containers. Under favorable conditions, experienced crews can use coordinated fog streams to sweep the flames off the surface of the burning liquid. Do not direct straight streams into the liquid. If cooling streams are ineffective (venting sound increases in volume and pitch, tank discolors, or shows any signs of deforming), withdraw immediately to a secure position. If employees are expected to fight fires, they must be trained and equipped in OSHA 1910.156.

Disposal Method Suggested: Chemical conversion to ammonia and carbon dioxide using chlorine or hypochlorite in a basic media. Controlled incineration is also adequate to totally destroy cyanide.[22] Alternatively, HCN can be recovered, from ammonoxidation process waste streams for example. Consult with environmental regulatory agencies for guidance on acceptable disposal practices. Generators of waste containing this contaminant (≥100 kg/mo) must conform with EPA regulations governing storage, transportation, treatment, and waste disposal.

References

National Institute for Occupational Safety and Health, Criteria for a Recommended Standard: Occupational Exposure to Hydrogen Cyanide, NIOSH Doc. No. 77–108 (1977)

U.S. Environmental Protection Agency, "Chemical Profile: Hydrocyanic Acid," Washington, DC, Chemical Emergency Preparedness Program (Nov. 30, 1987)

New York State Department of Health, "Chemical Fact Sheet: Hydrogen Cyanide," Albany, NY, Bureau of Toxic Substance Assessment (Feb. 1986 and Version 3)

Sax, N. I., Ed., "Dangerous Properties of Industrial Materials Report," 1, No. 6, 61–64 (1981)

New Jersey Department of Health and Senior Services, "Hazardous Substance Fact Sheet, Hydrogen Cyanide," Trenton, NJ (June, 1998)

Hydrogen Fluoride

Molecular Formula: HF

Synonyms: Acido fluorhidrico (Spanish); Anhydrous hydrofluoric acid; Antisal 2B; C-P 8 solution; Doped poly etch; Fluorhydric acid; Fluoric acid; Fluoruro de hidrogeno (Spanish); Freckle etch; Hydrofluoric acid; Hydrofluoric acid gas; Hydrogen fluoride, anhydrous; Implanter fumer; KTI buffered oxide etch 50:1; KTI buffered oxide etch 6:1; KTI oxide etch 10:1; KTI oxide etch 5:1; KTI oxide etch 50:1; Mae etchants; Mixed acid etch; Poly etch 95%; Rubigine; Silicon etch solution

CAS Registry Number: 7664-39-3

RTECS Number: MW7875000

DOT ID: UN 1052 (anhydrous); UN 1790 (solution)

EEC Number: 009-002-00-6

Regulatory Authority

- Air Pollutant Standard Set (ACGIH)[1] (DFG)[3] (HSE)[33] (Other Countries)[35] (OSHA)[58] (Several States)[60]
- Toxic Substance (World Bank)[15]
- OSHA 29CFR1910.119, Appendix A, Process Safety List of Highly Hazardous Chemicals, TQ = 1,000 lb (454 kg)

Hydrogen fluoride:

- CLEAN AIR ACT: Hazardous Air Pollutants (Title I, Part A, Section 112)
- CLEAN WATER ACT: Section 311 Hazardous Substances/RQ 40CFR117.3 (same as CERCLA, see below); Section 313 Water Priority Chemicals (57FR41331, 9/9/92)
- EPA HAZARDOUS WASTE NUMBER (RCRA No.): U134
- SUPERFUND/EPCRA 40CFR355, Appendix B Extremely Hazardous Substances: TPQ = 100 lb (45.4 kg)
- SUPERFUND/EPCRA 40CFR302.4 Reportable Quantity (RQ): CERCLA, 100 lb (45.4 kg)
- EPCRA Section 313 Form R *de minimis* concentration reporting level: 1.0%

Hydrofluoric acid (conc. 50% or greater)

- CLEAN AIR ACT: Hazardous Air Pollutants (Title I, Part A, Section 112); (conc. 50% or greater, or anhydrous) Accidental Release Prevention/Flammable substances, (Section 112[r], Table 3), TQ = 1,000 lb (454 kg)
- CLEAN WATER ACT: Section 311 Hazardous Substances/RQ 40CFR117.3 (same as CERCLA, see below); Section 313 Water Priority Chemicals (57FR41331, 9/9/92)
- EPA HAZARDOUS WASTE NUMBER (RCRA No.): U134
- SUPERFUND/EPCRA 40CFR355, Appendix B Extremely Hazardous Substances: TPQ = 100 lb (45.4 kg)
- SUPERFUND/EPCRA 40CFR302.4 Reportable Quantity (RQ): CERCLA, 100 lb (45.4 kg)
- EPCRA Section 313 Form R *de minimis* concentration reporting level: 1.0%
- Canada, WHMIS, Ingredients Disclosure List

Cited in U.S. State Regulations: Alaska (G), California (A, G), Connecticut (A), Florida (G), Illinois (G), Kansas (G), Kentucky (A), Louisiana (G), Maine (G), Maryland (G), Massachusetts (G, A), Nevada (A), New Hampshire (G), New Jersey (G), New York (A), North Carolina (A), North Dakota (A), Oklahoma (G), Pennsylvania (G), Rhode Island (G, A), South Carolina (A), South Dakota (A), Vermont (G), Virginia (G, A), Washington (G), West Virginia (G), Wisconsin (G).

Description: Hydrogen fluoride, HF, is colorless, fuming liquid or gas with a strong, irritating odor. Odor threshold = 0.03 mg/m^3. Boiling point =19 – 20°C. Freezing/Melting point = -83°C. Highly soluble in water. It is not flammable.

Potential Exposure: Hydrogen fluoride, its aqueous solution hydrofluoric acid, and its salts are used in production of organic and inorganic fluorine compounds such as fluorides and plastics; as a catalyst, particularly in paraffin alkylation in the petroleum industry; as an insecticide; and to arrest the fermentation in brewing. It is utilized in the fluorination processes, especially in the aluminum industry, in separating uranium isotopes, in cleaning cast iron, copper, and brass, in removing efflorescence from brick and stone, in removing sand from metallic castings, in frosting and etching glass and enamel, in polishing crystal, in decomposing cellulose, in enameling and galvanizing iron, in working silk, in dye and analytical chemistry, and to increase the porosity of ceramics.

Incompatibilities: A super-strong acid; aqueous solutions are less strong. Reacts violently with bases. Reacts, possibly with violence, with many compounds including acetic anhydride, aliphatic amines, alcohols, alkanolamines, alkylene oxides, aromatic amines, amides, 2-aminoethanol, ammonia, ammonium hydroxide, arsenic trioxide, bismuthic acid, calcium oxide, ethylene diamine, ethyleneimine, epichlorohydrin, isocyanates, metal acetylides, nitrogen trifluoride, oleum, organic anhydrides, oxygen difluoride, phosphorous pentoxide, sulfuric acid, strong oxidizers, vinyl acetate, vinylidene fluoride. Attacks glass, concrete, ceramics, and other silicon-containing compounds. Attacks metals, some plastics, rubber and coatings.

Permissible Exposure Limits in Air: The OSHA PEL and recommended ACGIH for hydrogen fluoride (measured as fluoride) is 3 ppm TWA. The NIOSH recommended REL is 3 ppm (2.5 mg/m^3) and a ceiling of 6 ppm (5 mg/m^3) (15 minute). DFG[3] and the HSE[33] have set TWA values of 3 ppm and STEL values of 6 ppm. The NIOSH IDLH level is 30 ppm. Argentina, Japan and Sweden have set TWA value of 3 ppm.[35] The former USSR[35][43] has set a MAC in workplace air of 0.05 mg/m^3 and Czechoslovakia has set 1.0 mg/m^3. Limits in ambient air in residential areas have been set by the former USSR[35] at 0.02 mg/m on a momentary basis and 0.005 mg/m^3 on a daily average basis. A number of states have set guidelines or standards for hydrogen fluoride in ambient air[60] ranging from 3.4 μg/m^3 (Massachusetts) to 8.3 μg/m^3 (New York) to 20.0 μg/m^3 (Virginia) to 25.0 μg/m³ (North Dakota and South Carolina) to 30.0 μg/m³ (Rhode Island) to 50.0 μg/m³ (Connecticut and South Dakota) to 60.0 μg/m³ (Nevada) to 25.0 – 250.0 μg/m³ (North Carolina) to 830.0 μg/m³ (Kentucky).

Determination in Air: Si gel; $NaHCO_3/Na_2CO_3$; Ion chromatography; NIOSH IV, Method #7903, Inorganic Acids. See also #7902, #7906.

Permissible Concentration in Water: No criteria set.

Routes of Entry: Inhalation, skin absorption, ingestion, eye and/or skin contact.

Harmful Effects and Symptoms

Short Term Exposure: Hydrogen fluoride is corrosive to the eyes, skin, and the respiratory tract. Eye burns may not be immediately painful. Inhalation of this gas can cause pulmonary edema, a medical emergency that can be delayed for several hours. This can cause death. High levels of exposure can cause death. Acute exposure to hydrogen fluoride will result in irritation, burns, ulcerous lesions, and necrosis of the eyes, skin, and mucous membranes. Total destruction of the eyes is possible. Other effects include nausea, vomiting, diarrhea, pneumonitis (inflammation of the lungs), and circulatory collapse. Ingestion of an estimated 1.5 grams produced sudden death without gross pathological damage. Repeated ingestion of small amounts resulted in moderately advanced hardening of the bones. Contact of skin with anhydrous liquid produces severe burns. Inhalation of anhydrous hydrogen fluoride or hydrogen fluoride mist or vapors can cause severe respiratory tract irritation that may be fatal. Hydrogen fluoride may induce hypocalcemia, causing cardiac and renal failure.

Long Term Exposure: The substance may cause fluorosis. Can irritate the lungs and may cause bronchitis. Long-term exposure may damage the liver and kidneys.

Points of Attack: Eyes, skin, respiratory system, bones.

Medical Surveillance: Before beginning employment and at regular times after that, the following is recommended. Lung function tests. If symptoms develop or overexposure has occurred, the following may be useful: liver, and kidney function tests. Consider chest x-ray after acute overexposure.

First Aid: If this chemical gets into the eyes, remove any contact lenses at once (contact lenses *should not be worn* when working with HF) and irrigate immediately for at least 30 minutes, occasionally lifting upper and lower lids. Seek medical attention immediately. If this chemical contacts the skin, remove contaminated clothing and flush immediately with large amounts of water. Immerse exposed skin area in iced 70% ethyl alcohol. Seek medical attention immediately. If this chemical has been inhaled, remove from exposure, begin rescue breathing (using universal precautions) if breathing has stopped and CPR if heart action has stopped. Transfer promptly to a medical facility. When this chemical has been swallowed, get medical attention. If victim is conscious, administer water or milk. Do not induce vomiting. Medical observation is recommended for 24 – 48 hours after breathing overexposure, as pulmonary edema may be delayed. As first aid for pulmonary edema, a doctor or authorized paramedic may consider administering a corticosteroid spray.

Personal Protective Methods: Wear protective gloves and clothing to prevent any reasonable probability of skin contact. Safety equipment suppliers/manufacturers can provide recommendations on the most protective glove/clothing material for your operation. All protective clothing (suits, gloves, footwear, headgear) should be clean, available each day, and put on before work. Contact lenses should *not* be worn when working with this chemical. Wear eye protection to prevent any possibility of eye contact. Employees should wash immediately flush with soap when skin is wet or contaminated. Provide emergency showers and eyewash.

Respirator Selection: NIOSH/OSHA: *30 ppm:* CCRS [any chemical cartridge respirator with cartridge(s) providing protection against the compound of concern]; or PAPRS [any powered, air-purifying respirator with cartridge(s) providing protection against the compound of concern]; or GMFS [any air-purifying, full-facepiece respirator (gas mask) with a chin-style, front- or back-mounted canister providing protection against the compound of concern]; or SA (any supplied-air respirator); or SCBA (any self-contained breathing apparatus). *Emergency or planned entry into unknown concentrations or IDLH conditions:* SCBAF:PD,PP (any self-contained breathing apparatus that has a full facepiece and is operated in a pressure-demand or other positive-pressure mode); or SAF:PD,PP:ASCBA (any supplied-air respirator that has a full facepiece and is operated in a pressure-demand or other positive-pressure mode in combination with an auxiliary self-contained breathing apparatus operated in a pressure-demand or other positive pressure mode). *Escape:* GMFS [any air-purifying, full-facepiece respirator (gas mask) with a chin-style, front- or back-mounted canister providing protection against the compound of concern]; or SCBAE (any appropriate escape-type, self-contained breathing apparatus).

Note: Substance reported to cause eye irritation or damage; may require eye protection.

Storage: Prior to working with this chemical you should be trained on its proper handling and storage. Hydrogen fluoride must be stored to avoid contact with metals, concrete, glass and ceramics, because it can severely corrode these materials. See also incompatible listed above. Contact with metals may form a flammable gas. Keep away from heat. Where possible, automatically pump liquid from drums or other storage containers to process containers.

Shipping: Hydrogen fluoride, anhydrous requires a shipping label of: "Corrosive, Poison." It falls in DOT Hazard Class 8 and Packing Group I. Passenger aircraft or railcar shipment is forbidden and cargo aircraft shipment is forbidden as well. Hydrofluoric acid solution (below 60%) requires a shipping label of: "Corrosive, Poison." It falls in DOT Hazard

Class 8 and Packing Group II. Hydrofluoric acid solution (stronger than 60%) requires a shipping label of: "Corrosive, Poison." It falls in DOT Hazard Class 8 and Packing Group I.

Spill Handling: Evacuate and restrict persons not wearing protective equipment from area of spill or leak until cleanup is complete. Remove all ignition sources. Ventilate area of spill or leak. Restrict persons not wearing protective equipment from area of leak until clean-up is complete. Ventilate area of spill or leak. If a gas, stop flow of gas. If source of leak is a cylinder and the leak cannot be stopped in place, remove the leaking cylinder to a safe place in the open air, and repair leak or allow cylinder to empty. If in liquid form, allow to vaporize and disperse the gas, or cover with sodium carbonate or an equal mixture of soda ash and slaked lime. After mixing, add water if necessary to form a slurry. Absorb liquids in vermiculite, dry sand, earth, peat, carbon, or a similar material and deposit in sealed containers. Keep this chemical out of a confined space, such as a sewer, because of the possibility of an explosion, unless the sewer is designed to prevent the build-up of explosive concentrations. It may be necessary to contain and dispose of this chemical as a hazardous waste. If material or contaminated runoff enters waterways, notify downstream users of potentially contaminated waters. Contact your Department of Environmental Protection or your regional office of the federal EPA for specific recommendations. If employees are required to clean-up spills, they must be properly trained and equipped. OSHA 1910.120(q) may be applicable.

Initial isolation and protective action distances

Distances shown are likely to be affected during the first 30 minutes after materials are spilled and could increase with time. If more than one tank car, cargo tank, portable tank, or large cylinder is involved in the incident is leaking, the protective action distance may need to be increased. You may need to seek emergency information from CHEMTREC at (800) 424-9300 or seek professional environmental engineering assistance from the U.S. EPA Environmental Response Team at (908) 548-8730 (24-hour response line).

UN 1052:

Small spills (From a small package or a small leak from a large package)

First: Isolate in all directions (feet)..... 200

Then: Protect persons downwind (miles)

Day 0.1

Night 0.4

Large spills (From a large package or from many small packages)

First: Isolate in all directions (feet)..... 500

Then: Protect persons downwind (miles)

Day 0.3

Night 1.4

Fire Extinguishing: Hydrogen fluoride is a non-combustible liquid or gas. Contact with metals may form flammable hydrogen gas which can cause fire and explosion. Use chemical extinguishers. Poisonous gases including fluorine are produced in fire. Vapors are slightly lighter than air. Firefighting gear (including SCBA) does not provide adequate protection. If exposure occurs, remove and isolate gear immediately and thoroughly decontaminate personnel. If material or contaminated runoff enters waterways, notify downstream users of potentially contaminated waters. Notify local health and fire officials and pollution control agencies. Do not extinguish the fire unless the flow of the gas can be stopped and any remaining gas is out of the line. Specially trained personnel may use fog lines to cool exposures and let the fire burn itself out. Use dry chemicals, foam, carbon dioxide. Containers may explode in fire. Storage containers and parts of containers may rocket great distances, in many directions. From a secure, explosion-proof location, use water spray to cool exposed containers. If cooling streams are ineffective (venting sound increases in volume and pitch, tank discolors, or shows any signs of deforming), withdraw immediately to a secure position. If employees are expected to fight fires, they must be trained and equipped in OSHA 1910.156.

Disposal Method Suggested: Reaction with excess lime followed by lagooning and either recovery or landfill disposal of the separated calcium fluoride. The supernatant liquid from this process is diluted and discharged to the sewer.[22] Alternatively, hydrogen can be recovered and recycled in many cases. In accordance with 40CFR165 recommendations for the disposal of pesticides and pesticide containers. Must be disposed properly by following package label directions or by contacting your state pesticide or environmental control agency or by contacting your regional EPA office. Consult with environmental regulatory agencies for guidance on acceptable disposal practices. Generators of waste containing this contaminant (≥100 kg/mo) must conform with EPA regulations governing storage, transportation, treatment, and waste disposal.

References

National Institute for Occupational Safety and Health, Criteria for a Recommended Standard: Occupational Exposure to Hydrogen Fluoride, NIOSH Doc. No. 76–143 (1976)

U.S. Environmental Protection Agency, Hydrofluoric Acid, Health and Environmental Effects Profile No. 117, Office of Solid Waste, Washington, DC (April 30, 1980)

Sax, N. I., Ed., "Dangerous Properties of Industrial Materials Report," 1, No. 6, 64–66 (1981) and 5, No. 6, 52–56 (1985)

U.S. Environmental Protection Agency, "Chemical Profile: Hydrogen Fluoride," Washington, DC, Chemical Emergency Preparedness Program (Now. 30, 1987)

New Jersey Department of Health and Senior Services, "Hazardous Substance Fact Sheet: Hydrogen Fluoride," Trenton, NJ (August 1998)

Hydrogen Peroxide

Molecular Formula: H_2O_2

Synonyms: Albone; Carro's acid; Dihydrogen dioxide; Hydrogen dioxide; Hydroperoxide; Hyoxyl; Inhibine; Lea Ronal NP-A/NP-B solder stripper; Nanostrip; Oxydol; Patclin 948 solder stripper; Perhydrol; Perone; Perossido di idrogeno (Italian); Peroxan; Peroxide; Peroxido de hidrogeno (Spanish); Peroxyde d'hydrogene (French); Piranha etch; RCA Clean (steps 1 and 2); Superoxol

CAS Registry Number: 7722-84-1

RTECS Number: MX0900000

DOT ID: UN 2014 (>20%<60%, solution); UN 2015 (>60%, stabilized); UN 2984 (8 – 19%, solution)

EEC Number: 008-003-00-9

EINECS Number: 231-765-0

Regulatory Authority

- Air Pollutant Standard Set (ACGIH)[1] (DFG)[3] (HSE)[33] (Argentina)[35] (OSHA)[58] (Several States)[60]
- Carcinogen (Animal Suspected) (IARC)[9]

Conc. >52%:

- OSHA 29CFR1910.119, Appendix A, Process Safety List of Highly Hazardous Chemicals, TQ = 7,500 lb
- SUPERFUND/EPCRA 40CFR355, Appendix B Extremely Hazardous Substances: TPQ = 1,000 lb (454 kg)
- SUPERFUND/EPCRA 40CFR302.4 Reportable Quantity (RQ): EHS, 1 lb (0.454 kg)
- Canada, WHMIS, Ingredients Disclosure List

Cited in U.S. State Regulations: Alaska (G), California (A, G), Connecticut (A), Florida (G), Illinois (G), Maine (G), Massachusetts (G), Nevada (A), New Hampshire (G), New Jersey (G), New York (G), North Dakota (A), Oklahoma (G), Pennsylvania (G), Rhode Island (G), Virginia (A), West Virginia G).

Description: Hydrogen peroxide, H_2O_2, is a colorless liquid with a sharp odor and bitter taste. Boiling point =125°C (70%); 106°C (30%); 141°C (90%). Freezing/Melting point = -39°C (70%); -11°C (90%). Hazard Identification (based on NFPA-704 M Rating System): (35% solution) Health 1, Flammability 0, Reactivity 1. Hydrogen peroxide is completely miscible with water and is commercially sold in concentrations of 3, 35, 50, 70, and 90% solutions. It is not flammable.

Potential Exposure: Hydrogen peroxide is used in the manufacture of acetone, antichlor, antiseptics, benzoyl peroxide, buttons, disinfectants, pharmaceuticals, felt hats, plastic foam, rocket fuel, sponge rubber and pesticides. It is also used in bleaching bone, feathers, flour, fruit, fur, gelatin, glue, hair, ivory, silk, soap, straw, textiles, wax, and wood pulp, and as an oxygen source in respiratory protective equipment. Other specific occupations with potential exposure include liquor and wine agers, dyers, electroplaters, fat refiners, photographic film developers, wool printers, veterinarians, and water treaters.

Incompatibilities: Contact with combustible material may result in SPONTANEOUS combustion. A powerful oxidizer; attacks many substances. Attacks many organic substances such as wood, textile, and paper. Contact with most organic, readily oxidizable materials, reducing agents and combustibles cause fire and explosions particularly in the presence of metals. Contact with iron, copper, brass, bronze, chromium, zinc, lead, manganese, silver and catalytic metals (and their salts), especially in a basic (pH 7 or above) environment, cause rapid decomposition with evolution of oxygen gas, which increases fire hazard. Attacks, and may ignite, some plastics, rubber and coatings. Decomposes slowly at ordinary temperatures and builds up pressure in a closed container. The rate of decomposition doubles for each 50°F/10°C rise (1.5 times 10°C rise) in temperature and becomes self-sustaining at 285°F/141°C.

Permissible Exposure Limits in Air: The OSHA PEL and the NIOSH and ACGIH recommended value is 1 ppm (1.4 mg/m^3). The NIOSH IDLH level is 75 ppm. The DFG[3] has set an MAK value of 1 ppm and HSE[33] has set an 8-hour TWA of 1 ppm (1.5 mg/m^3) and an STEL of 2 ppm (3 mg/m^3). Argentina[35] has these same limits. Several states have set guidelines or standards for hydrogen peroxide in ambient air[60] ranging from 15 μg/m^3 (North Dakota) to 25 μg/m^3 (Virginia) to 28 μg/m^3 (Connecticut) to 36 μg/m^3 (Nevada).

Determination in Air: Bubbler; $TiOSO_4$; Visible spectrophotometry; OSHA Method #ID126SG.

Permissible Concentration in Water: No criteria set.

Routes of Entry: Inhalation of vapor or mist, ingestion, eye and/or skin contact.

Harmful Effects and Symptoms

Short Term Exposure: Hydrogen peroxide is corrosive to the eyes, skin, and respiratory tract. Higher exposures can cause pulmonary edema, a medical emergency that can be delayed for several hours. This can cause death. Signs and symptoms of acute exposure to hydrogen peroxide may be severe and include irritation or burns to the skin, eyes, respiratory tract, mouth, esophagus, stomach, and intestines. Distension or rupture of the stomach and other hollow viscera may occur; vomiting is common. Corneal ulceration may develop.

Long Term Exposure: Because this is a mutagen, handle it as a possible cancer-causing substance-with extreme caution. Can irritate the lungs. Repeated exposure may cause bronchitis. Repeated skin contact can cause a rash with redness and blisters.

Points of Attack: Eyes, skin, respiratory system.

Medical Surveillance: For those with frequent or potentially high exposure (half the TLV or greater) the following are recommended before beginning work and at regular times after that: lung function tests. If symptoms develop or overexposure is suspected: consider chest x-ray if pulmonary edema is suspected.

First Aid: If this chemical gets into the eyes, remove any contact lenses at once and irrigate immediately for at least 15 minutes, occasionally lifting upper and lower lids. Seek medical attention immediately. If this chemical contacts the skin, remove contaminated clothing and wash immediately with soap and water. Seek medical attention immediately. If this chemical has been inhaled, remove from exposure, begin rescue breathing (using universal precautions) if breathing has stopped and CPR if heart action has stopped. Transfer promptly to a medical facility. When this chemical has been swallowed, get medical attention. If victim is conscious, administer water or milk. Do not induce vomiting. Medical observation is recommended for 24 – 48 hours after breathing overexposure, as pulmonary edema may be delayed. As first aid for pulmonary edema, a doctor or authorized paramedic may consider administering a corticosteroid spray.

Personal Protective Methods: Wear protective gloves and clothing to prevent any reasonable probability of skin contact. Safety equipment suppliers/manufacturers can provide recommendations on the most protective glove/clothing material for your operation. For hydrogen peroxide (30 – 70%), Neoprene+Natural Rubber and Neoprene/Natural Rubber are among the recommended protective materials. All protective clothing (suits, gloves, footwear, headgear) should be clean, available each day, and put on before work. Contact lenses should not be worn when working with this chemical. Wear splash-proof chemical goggles and face shield unless full face-piece respiratory protection is worn. Where fumes or vapor are excessive, workers should be provided with gas masks with full face pieces and proper canisters or supplied air respirators. Additional health hazards may occur from the decomposition of hydrogen peroxide. Oxygen, possibly at high pressure, may form, which may create an explosion hazard. Hydrogen peroxide is generally handled in a closed system to prevent contamination. Employees should wash immediately with soap when skin is wet or contaminated. Provide emergency showers and eyewash.

Respirator Selection: NIOSH/OSHA: *10 ppm:* SA (any supplied-air respirator). *25 ppm:* SA:CF (any supplied-air respirator operated in a continuous-flow mode). *50 ppm:* SCBAF (any self-contained breathing apparatus with a full facepiece); or SAF (any supplied-air respirator with a full facepiece). *75 ppm:* SAF:PD,PP (any supplied-air respirator that has a full facepiece and is operated in a pressure-demand or other positive-pressure mode). *Emergency or planned entry into unknown concentrations or IDLH conditions:* SCBAF:PD, PP (any self-contained breathing apparatus that has a full facepiece and is operated in a pressure-demand or other positive-pressure mode); or SAF:PD,PP:ASCBA (any supplied-air respirator that has a full facepiece and is operated in a pressure-demand or other positive-pressure mode in combination with an auxiliary self-contained breathing apparatus operated in a pressure-demand or other positive pressure mode). *Escape:* GMFS [any air-purifying, full-facepiece respirator (gas mask) with a chin-style, front- or back-mounted canister providing protection against the compound of concern]; or SCBAE (any appropriate escape-type, self-contained breathing apparatus).

Note: Substance reported to cause eye irritation or damage; may require eye protection.

Storage: Prior to working with this chemical you should be trained on its proper handling and storage. Hydrogen peroxide must be stored to avoid contact with iron, copper, chromium, brass, bronze, lead, silver, manganese, and their salts since violent reactions occur. Store in tightly closed containers in a cool, well-ventilated area away from alcohols, glycerol. Organic materials and radiant heat (sunlight). Containers should be protected from physical and mechanical disturbances. Sources of ignition such as smoking and open flames are prohibited where hydrogen peroxide is used, handled, or stored in manner that could create a potential fire or explosion hazard. Wherever hydrogen peroxide is used, handled, manufactured, or stored, use explosion-proof electrical equipment and fittings. See OSHA standard 1910.104 and NFPA 43A *Code for the Storage of Liquid and Solid Oxidizers* for detailed handling and storage regulations. A regulated, marked area should be established where this chemical is handled, used, or stored in compliance with OSHA standard 1910.1045.

Shipping: Stabilized H_2O_2or solutions >40% require a shipping label of: "Oxidizer, Corrosive." It falls in DOT Hazard Class 5.1 and Packing Group I. Passenger aircraft or railcar shipment is forbidden and cargo aircraft shipment is forbidden as well. An H_2O_2 solution (>8%<20%) requires a shipping label of: "Oxidizer." It falls in DOT Hazard Class 5.1 and Packing Group III. An H_2O_2 solution, stabilized if necessary (>20%<40%) requires a shipping label of: "Oxidizer, Corrosive." It falls in DOT Hazard Class 5.1 and Packing Group II. An H_2O_2 solution, stabilized (>40%<60%) requires a shipping label of: "Oxidizer, Corrosive." It falls in DOT Hazard Class 5.1 and Packing Group II. An H_2O_2 solution, stabilized (>60%) requires a shipping label of: "Oxidizer, Corrosive." It falls in DOT Hazard Class 5.1 and Packing Group I.

Spill Handling: Evacuate and restrict persons not wearing protective equipment from area of spill or leak until cleanup is complete. Remove all ignition sources. Ventilate area of spill or leak. Use a large quantity of water to wash down spills and reduce the flammable vapors. Absorb liquids in vermiculite, dry sand, earth, peat, carbon, or a similar material and deposit in sealed containers. Keep this chemical out of a confined space, such as a sewer, because of the possibility of an explosion, unless the sewer is designed to prevent the build-up of explosive concentrations. It may be necessary to contain and dispose of this chemical as a hazardous waste. If material or contaminated runoff enters waterways, notify downstream users of potentially contaminated waters. Contact your Department of Environmental Protection or your regional office of the federal EPA for specific recommendations. If employees are required to clean-up spills, they must be properly trained and equipped. OSHA 1910.120(q) may be applicable.

Fire Extinguishing: Noncombustible, but highly reactive and can increase the intensity of fire. Fires should be fought with water only. Do not use dry chemical, carbon dioxide, or foams. Large fires: flood fire area with water. Flammable vapors may accumulate in storage areas and containers. Containers may explode in fire. Storage containers and parts of containers may rocket great distances, in many directions. If material or contaminated runoff enters waterways, notify downstream users of potentially contaminated waters. Notify local health and fire officials and pollution control agencies. From a secure, explosion-proof location, use water spray to cool exposed containers. If cooling streams are ineffective (venting sound increases in volume and pitch, tank discolors, or shows any signs of deforming), withdraw immediately to a secure position. If employees are expected to fight fires, they must be trained and equipped in OSHA 1910.156.

Disposal Method Suggested: Dilution with water to release the oxygen. After decomposition, the waste stream may be discharged safely.[22]

References

U.S. Environmental Protection Agency, "Chemical Profile: Hydrogen Peroxide (>52%)" Washington, DC, Chemical Emergency Preparedness Program (Nov. 30, 1987)

New Jersey Department of Health and Senior Services, "Hazardous Substance Fact Sheet: Hydrogen Peroxide," Trenton, NJ (October 1998)

New York State Department of Health, "Chemical Fact Sheet: Hydrogen Peroxide," Albany, NY, Bureau of Toxic Substance Assessment (March 1986)

Sax, N. I., Ed., "Dangerous Properties of Industrial Materials Report," 1, No. 6, 66–68 (1981)

Hydrogen Selenide

Molecular Formula: H_2Se

Synonyms: Anhydrous hydrogen selenide; Electronic E-2; Selane; Selenium dihydride; Selenium hydride; Seleniuro de hidrogeno (Spanish)

CAS Registry Number: 7783-07-5

RTECS Number: MX1050000

DOT ID: UN 2202 (anhydrous)

EEC Number: 034-002-00-8

Regulatory Authority

- Air Pollutant Standard Set (ACGIH)[1] (DFG)[3] (HSE)[33] (OSHA)[58] (Several States)[60]
- OSHA 29CFR1910.119, Appendix A. Process Safety List of Highly Hazardous Chemicals, TQ = 150 lb
- CLEAN AIR ACT: Accidental Release Prevention/Flammable substances, (Section 112[r], Table 3), TQ = 500 lb (227 kg)
- SUPERFUND/EPCRA 40CFR355, Appendix B Extremely Hazardous Substances: TPQ = 10 lb (4.54 kg)
- SUPERFUND/EPCRA 40CFR302.4 Reportable Quantity (RQ): EHS, 1 lb (0.454 kg)
- Canada, WHMIS, Ingredients Disclosure List

Cited in U.S. State Regulations: Alaska (G), California (A, G), Connecticut (A), Florida (G), Illinois (G), Maine (G), Massachusetts (G), Nevada (A), New Hampshire (G), New Jersey (G), New York (G), North Dakota (A), Oklahoma (G), Pennsylvania (G), Rhode Island (G), Virginia (A), West Virginia (G).

Description: Hydrogen selenide, H_2Se, is a colorless, flammable gas with a very offensive odor resembling decayed horseradish. The odor threshold is 0.3 ppm; many people rapidly lose the ability to detect the odor of hydrogen selenide so it is not a reliable warning of exposure. Boiling point = -41°C. Freezing/Melting point = -64 – 66°C. Soluble in water.

Potential Exposure: Hydrogen selenide is used in semiconductor manufacture. Also, it may be produced by the reaction of acids or water and metal selenides or hydrogen and soluble selenium compounds.

Incompatibilities: Forms an explosive mixture with air. Contact with air causes the emissions of toxic and corrosive fumes of selenium dioxide A strong reducing agent. Reacts violently with oxidizers causing fire and explosion hazard. Incompatible with strong acids, water, halogenated hydrocarbons. Decomposes above 100°C forming toxic and flammable products including selenium and hydrogen.

Permissible Exposure Limits in Air: The OSHA PEL, NIOSH REL and the ACGIH TWA value is 0.05 ppm (0.2 mg/m^3). The NIOSH IDLH level is 1 ppm. The DFG[3] and the HSE[33] have also set 0.05 ppm (0.2 mg/m^3) as an allowable limit. Several states have set guidelines or standards for H_2Se in ambient air[60] ranging from 2 g/m^3 (North Dakota) to 3 μg/m^3 (Virginia) to 4 μg/m^3 (Connecticut) to 5 μg/m^3 (Nevada).

Determination in Air: No method available.

Permissible Concentration in Water: No criteria set, but EPA[32] suggests a permissible ambient goal of 10 μg/l (the same as for selenium) based on health effects.

Routes of Entry: Inhalation, eye and/or skin contact. Can be absorbed through the skin.

Harmful Effects and Symptoms

Short Term Exposure: Irritates the eyes, skin, and respiratory tract. Exposure can cause dizziness, fatigue, nausea, vomiting and diarrhea. *Inhalation:* The odor of hydrogen selenide in small concentrations (e.g. below 1 ppm) disappears rapidly because of olfactory fatigue. The odor and irritating effects are not a reliable warning to gradually increasing concentrations. Low levels have caused coughing, sneezing and difficulty in breathing. Levels of 0.2 ppm may cause nausea, vomiting, a metallic taste in the mouth and garlic breath. Levels of 1.5 ppm may cause intolerable irritation of mouth and nose. Inhalation of the gas may cause pneumonitis.

Higher exposures can cause pulmonary edema, a medical emergency that can be delayed for several hours. This can cause death. *Skin:* May cause irritation and red coloration of nails. Contact can cause a burning sensation on contact and rash. Contact with liquid may cause freezing burns. *Eyes:* Levels of 1.5 ppm are described as intolerable.

Long Term Exposure: Repeated exposure can cause garlic odor on breath, dizziness, nausea, vomiting, labored breathing, bluing of skin, pulmonary edema, metallic taste, coughing, nasal secretion, pain in the chest, difficulty in breathing, irritation of the eyes, irritation of the respiratory tract resulting in bronchitis and conjunctivitis. It may also cause anemia. Animal studies suggest that liver damage and lung impairment are possible.

Points of Attack: Eyes, respiratory system, liver, blood.

Medical Surveillance: Consider the points of attack in replacement and periodic physical examinations. Liver function tests. Complete blood count (CBC). Urine test for selenium (normal is less than 100 micrograms per liter of urine).

First Aid: If this chemical gets into the eyes, remove any contact lenses at once and irrigate immediately for at least 15 minutes, occasionally lifting upper and lower lids. Seek medical attention immediately. If this chemical contacts the skin, remove contaminated clothing and wash immediately with soap and water. Seek medical attention immediately. If this chemical has been inhaled, remove from exposure, begin rescue breathing (using universal precautions) if breathing has stopped and CPR if heart action has stopped. Transfer promptly to a medical facility. When this chemical has been swallowed, get medical attention. Give large quantities of water and induce vomiting. Medical observation is recommended for 24 – 48 hours after breathing overexposure, as pulmonary edema may be delayed. As first aid for pulmonary edema, a doctor or authorized paramedic may consider administering a corticosteroid spray. Do not make an unconscious person vomit. If frostbite has occurred, seek medical attention immediately; do *NOT* rub the affected areas or flush them with water. In order to prevent further tissue damage, do *NOT* attempt to remove frozen clothing from frostbitten areas. If frostbite has *NOT* occurred, immediately and thoroughly wash contaminated skin with soap and water.

Personal Protective Methods: Wear protective gloves and clothing to prevent any reasonable probability of skin contact. Safety equipment suppliers/manufacturers can provide recommendations on the most protective glove/clothing material for your operation. All protective clothing (suits, gloves, footwear, headgear) should be clean, available each day, and put on before work. Contact lenses should not be worn when working with this chemical. Wear eye protection to prevent any possibility of eye contact. Wear splash-proof chemical goggles and face shield when working with the liquid unless full facepiece respiratory protection is worn. Employees should wash immediately with soap when skin is wet or contaminated. Provide emergency showers and eyewash.

Respirator Selection: NIOSH/OSHA: *Up to 0.5 ppm:* SA (any supplied-air respirator). *Up to 1 ppm:* SA:CF (any supplied-air respirator operated in a continuous-flow mode);* or SCBAF (any self-contained breathing apparatus with a full facepiece); or SAF (any supplied-air respirator with a full facepiece). *Emergency or planned entry into unknown concentrations or IDLH conditions:* SCBAF:PD,PP (any self-contained breathing apparatus that has a full facepiece and is operated in a pressure-demand or other positive-pressure mode); or SAF:PD,PP:ASCBA (any supplied-air respirator that has a full facepiece and is operated in a pressure-demand or other positive-pressure mode in combination with an auxiliary self-contained breathing apparatus operated in a pressure-demand or other positive pressure mode). *Escape:* GMFS [any air-purifying, full-facepiece respirator (gas mask) with a chin-style, front-or back-mounted canister providing protection against the compound of concern]; or SCBAE (any appropriate escape-type, self-contained breathing apparatus).

* Substance reported to cause eye irritation or damage; may require eye protection.

Storage: Prior to working with hydrogen selenide you should be trained on its proper handling and storage. Before entering confined space where this chemical may be present, check to make sure that an explosive concentration does not exist. Store at room temperature away from contact with oxidizers, acids, water, and halogenated hydrocarbons. Metal containers involving the transfer of this chemical should be grounded and bonded. Where possible, automatically pump liquid from drums or other storage containers to process containers. Drums must be equipped with self-closing valves, pressure vacuum bungs, and flame arresters. Use only non-sparking tools and equipment, especially when opening and closing containers of this chemical. Sources of ignition such as smoking and open flames, are prohibited where this chemical is used, handled, or stored in a manner that could create a potential fire or explosion hazard. Wherever this chemical is used, handled, manufactured, or stored, use explosion-proof electrical equipment and fittings. Procedures for the handling, use and storage of cylinders should be in compliance with OSHA 1910.101 and 1910.169 as with the recommendations of the Compressed Gas Association.

Shipping: This compound requires a shipping label of: "Poison Gas, Flammable Gas." It falls in DOT Hazard Class 2.3 and Packing Group I. Passenger aircraft or railcar shipment is forbidden and cargo aircraft shipment is forbidden as well.

Spill Handling: Evacuate and restrict persons not wearing protective equipment from area of spill or leak until cleanup is complete. Remove all ignition sources. Ventilate area of spill or leak. Warn other workers of leak. Evacuate area. Put on proper protective clothing and equipment. Stop flow of gas. Ventilate area. Remove sources of ignition. If leak cannot be stopped, move leaking cylinder to safe place out of doors. Absorb liquids in vermiculite, dry sand, earth, peat, carbon, or a similar material and deposit in sealed containers.

Keep this chemical out of a confined space, such as a sewer, because of the possibility of an explosion, unless the sewer is designed to prevent the build-up of explosive concentrations. It may be necessary to contain and dispose of this chemical as a hazardous waste. If material or contaminated runoff enters waterways, notify downstream users of potentially contaminated waters. Contact your Department of Environmental Protection or your regional office of the federal EPA for specific recommendations. If employees are required to clean-up spills, they must be properly trained and equipped. OSHA 1910.120(q) may be applicable.

Initial isolation and protective action distances

Distances shown are likely to be affected during the first 30 minutes after materials are spilled and could increase with time. If more than one tank car, cargo tank, portable tank, or large cylinder is involved in the incident is leaking, the protective action distance may need to be increased. You may need to seek emergency information from CHEMTREC at (800) 424-9300 or seek professional environmental engineering assistance from the U.S. EPA Environmental Response Team at (908) 548-8730 (24-hour response line).

Small spills (From a small package or a small leak from a large package)

First: Isolate in all directions (feet)..... 500

Then: Protect persons downwind (miles)

Day 0.8

Night 3.6

Large spills (From a large package or from many small packages)

First: Isolate in all directions (feet)..... 1,600

Then: Protect persons downwind (miles)

Day 2.9

Night .... 7.0+

Fire Extinguishing: This chemical is a flammable gas. Do not extinguish fire unless flow of gas can be stopped. Small fires: dry chemical or carbon dioxide. Stay upwind and uphill; keep out of low areas. Wear self-contained breathing apparatus and full protective clothing. Use water in flooding quantities as fog. Cool containers that are exposed to flames with water until well after the fire is out. Decomposes above 100°C forming toxic and flammable selenium and hydrogen. Gas is heavier than air and will collect in low areas. Vapors may travel long distances to ignition sources and flashback. Vapors in confined areas may explode when exposed to fire. Containers may explode in fire. Storage containers and parts of containers may rocket great distances, in many directions. If material or contaminated runoff enters waterways, notify downstream users of potentially contaminated waters. Notify local health and fire officials and pollution control agencies. From a secure, explosion-proof location, use water spray to cool exposed containers. If cooling streams are ineffective (venting sound increases in volume and pitch, tank discolors, or shows any signs of deforming), withdraw immediately to a secure position. If employees are expected to fight fires, they must be trained and equipped in OSHA 1910.156.

References

U.S. Environmental Protection Agency, "Chemical Profile: Hydrogen Selenide," Washington, DC, Chemical Emergency Preparedness Program (Nov. 30, 1987)

New York State Department of Health, "Chemical Fact Sheet: Hydrogen Selenide," Albany, NY, Bureau of Toxic Substance Assessment (April 1986)

New Jersey Department of Health and Senior Services, "Hazardous Substance Fact Sheet, Hydrogen Selenide," Trenton, NJ (May 1999)

Hydrogen Sulfide

Molecular Formula: H_2S

Synonyms: Acide sulhydrique (French); Dihydrogen monosulfide; Dihydrogen sulfide; Hydrogene sulfure (French); Hydrogen sulfuric acid; Hydrogen sulphide; Hydrosulfuric acid; Idrogeno solforato (Italian); Sewer gas; Shwefelwasserstoff (German); Stink Damp; Sulfureted hydrogen; Sulfur hydride; Sulfuro de hidrogeno (Spanish); Zwavelwaterstof (Dutch)

CAS Registry Number: 7783-06-4

RTECS Number: MX1225000

DOT ID: UN 1053

EEC Number: 016-001-00-4

Regulatory Authority

- Air Pollutant Standard Set (ACGIH)[1] (DFG)[3] (HSE)[33] (OSHA)[58] (former USSR)[43] (Several States)[60]
- OSHA 29CFR1910.119, Appendix A. Process Safety List of Highly Hazardous Chemicals, TQ = 1,500 lb
- Extremely Hazardous Substance (EPA-SARA) (TPQ = 500)[7]
- CLEAN AIR ACT: Accidental Release Prevention/Flammable substances, (Section 112[r], Table 3), TQ = 10,000 lb (4,540 kg)
- CLEAN WATER ACT: Section 311 Hazardous Substances/ RQ 40CFR117.3 (same as CERCLA, see below)
- EPA HAZARDOUS WASTE NUMBER (RCRA No.): U135
- RCRA, 40CFR261, Appendix 8 Hazardous Constituents
- SUPERFUND/EPCRA 40CFR355, Appendix B Extremely Hazardous Substances: TPQ = 500 lb (227 kg)
- SUPERFUND/EPCRA 40CFR302.4 Reportable Quantity (RQ): CERCLA, 100 lb (45.4 kg)
- EPCRA Section 313; reporting required
- Canada, WHMIS, Ingredients Disclosure List

Cited in U.S. State Regulations: Alaska (G), California (A, G), Connecticut (A), Florida (G), Illinois (G), Kansas (G), Kentucky (A), Louisiana (G), Maine (G), Massachusetts (G, A),

Michigan (G), Montana (A), Nevada (A), New Hampshire (G), New Jersey (G), New York (G), Oklahoma (G), Pennsylvania (G), Rhode Island (G), South Carolina (A), Vermont (G), Virginia (G, A), Washington (G), West Virginia (G), Wisconsin (G).

Description: Hydrogen sulfide, H_2S, is a flammable, colorless gas with a characteristic rotten egg odor. The odor threshold is 0.008 ppm. Sense of smell becomes rapidly fatigued and can NOT be relied upon to warn of the continuous presence of H_2S. Shipped as a liquefied compressed gas. Boiling point = -60°C. Freezing/Melting point = -86°C. Autoignition temperature = 260°C. Explosive limits: LEL = 4.0%; UEL = 44.0%. Hazard Identification (based on NFPA-704 M Rating System): Health 4, Flammability 4, Reactivity 0. Soluble in water.

Potential Exposure: Hydrogen sulfide is used in the synthesis of inorganic sulfides, sulfuric acid, and organic sulfur compounds, as an analytical reagent, as a disinfectant in agriculture, and in metallurgy. It is generated in many industrial processes as a by-product and also during the decomposition of sulfur-containing organic matter, so potential for exposure exists in a variety of situations. Hydrogen sulfide is found in natural gas, volcanic gas, and in certain natural spring waters. Its may also be encountered in the manufacture of barium carbonate, barium salt, cellophane, depilatories, dyes, and pigments, felt, fertilizer, adhesives, viscose rayon, lithopone, synthetic petroleum products; in the processing of sugar beets; in mining, particularly where sulfide ores are present; in sewers and sewage treatment plants; during excavation of swampy or filled ground for tunnels, wells, and caissons; during drilling of oil and gas wells; in purification of hydrochloric acid and phosphates; during the low temperature carbonization of coal; in tanneries, breweries, slaughterhouses; in fat rendering; and in lithography and photoengraving.

Incompatibilities: A highly flammable and reactive gas; heating may cause violent combustion or explosion. Forms explosive mixture with air. Incompatible with acetaldehyde, barium pentafluoride, chlorine monoxide, chlorine trifluoride, chromic anhydride, copper, lead dioxide, nitric acid, nitrogen iodide, nitrogen trichloride, nitrogen trifluoride, oxygen difluoride, oxidizers, phenyl diazonium chloride, sodium, sodium peroxide. Reacts with alkali metals. Attacks some plastics.

Permissible Exposure Limits in Air: The OSHA PEL is ceiling 20 ppm; 50 ppm (10-minute maximum peak). NIOSH recommends a REL of ceiling 10 ppm (15 mg/m^3) (10-minute). NIOSH IDLH =100 ppm. A TWA of 10 ppm (14 mg/m^3) and the STEL of 15 ppm (31 mg/m^3) is recommended by ACGIH,[1] DFG,[3] and HSE.[33] The former USSR-UNEP/IRPTC joint project[43] has set an MAK of 10 mg/m^3 in workplace air; they have also set a MAC in ambient air of residential areas of 0.008 mg/m^3 on both a momentary and a daily average basis. Several states have set guidelines or standards for hydrogen sulfide in ambient air[60] ranging from 14.0 μg/m^3 (South Carolina) to 230.0 μg/m^3 (Virginia) to 280.0 μg/m^3 (Connecticut) to 333.0 μg/m^3

Determination in Air: Collection by charcoal tube; workup with NH_4OH/H_2O_2; analysis by ion chromatography; NIOSH IV, Method #6013.

Permissible Concentration in Water: EPA[32] suggests a permissible ambient goal of 207 μg/l based on health effects. EPA has established that hydrogen sulfide is a regulated toxic substance and is a hazardous substance defined under the Federal Water Pollution Control Act.

Routes of Entry: Inhalation of gas, ingestion, eye and/or skin contact.

Harmful Effects and Symptoms

Signs and symptoms of acute exposure to hydrogen sulfide may include tachycardia (rapid heart rate) or bradycardia (slow heart rate), hypertension (low blood pressure), cyanosis (blue tint to skin and mucous membrane), cardiac palpitations, and cardiac arrhythmias. Dyspnea (shortness of breath), tachypnea (rapid respiratory rate), bronchitis, pulmonary edema, respiratory depression, and respiratory paralysis may occur. Neurological effects include giddiness, irritability, drowsiness, weakness, confusion, delirium, amnesia, headache, sweating, and dizziness. Muscle cramping, tremor, excessive salivation, cough, convulsions, and coma may be noted. Nausea, vomiting, and diarrhea are commonly seen. Exposure to hydrogen sulfide gas may result in skin irritation, lacrimation (tearing), inability to detect odors, photophobia (heightened sensitivity to light), and blurred vision.

Short Term Exposure: Irritates the eyes, skin, and respiratory tract. May affect the central nervous system. Inhalation can cause pulmonary edema, a medical emergency that can be delayed for several hours. This can cause death. Levels of 20 ppm may cause headache, loss of appetite and dizziness. 50 ppm may cause muscle fatigue. 300 ppm may cause muscle cramps, low blood pressure, and unconsciousness after 20 minutes. Levels of 500 ppm can cause immediate loss of consciousness, slowed respiration and death in 30 – 60 minutes. At levels of 700 ppm and above respiratory paralysis and death can occur in seconds. Non-fatal cases may recover fully or may experience abnormal reflexes, dizziness, sleep disturbances and low of appetite that last for months or years. *Skin:* Readily absorbed. May cause irritation, reddening and swelling. Contact with liquid can cause breezing burns. *Eyes:* Irritation may be felt at levels as low as 0.1 ppm. Levels of 10 ppm and above can cause irritation, pain, tearing, and increased light sensitivity. Liquid may cause freezing burns.

Long Term Exposure: Long term exposure to low levels can cause pain and redness of the eyes with blurred vision. Repeated exposure can cause fatigue, loss of appetite, headaches, irritability, poor memory, dizziness, troubled sleeping, and nausea. Can cause irritation of the lungs and bronchitis with cough, phlegm, and/or shortness of breath. Animals studies showed that pigs that ate food containing hydrogen sulfide had diarrhea after a few days and weight loss after about 105 days.

Points of Attack: Eyes, lungs.

Medical Surveillance: For those with frequent or potentially high exposure (half the TLV or greater), the following are recommended before beginning work and at regular times after that: Lung function tests. If symptoms develop or overexposure is suspected, the following may be useful: blood sulfide level (normal is less than 0.05 mg/l). Consider chest x-ray after acute overexposure.

First Aid: If this chemical gets into the eyes, remove any contact lenses at once and irrigate immediately for at least 15 minutes, occasionally lifting upper and lower lids. Seek medical attention immediately. If frostbite has occurred, seek medical attention immediately; do *NOT* rub the affected areas or flush them with water. In order to prevent further tissue damage, do *NOT* attempt to remove frozen clothing from frostbitten areas. If frostbite has *NOT* occurred, immediately and thoroughly wash contaminated skin with soap and water. If this chemical has been inhaled, remove from exposure, begin rescue breathing (using universal precautions) if breathing has stopped and CPR if heart action has stopped. Transfer promptly to a medical facility. Medical observation is recommended for 24 – 48 hours after breathing overexposure, as pulmonary edema may be delayed. As first aid for pulmonary edema, a doctor or authorized paramedic may consider administering a corticosteroid spray.

Personal Protective Methods: Wear protective gloves and clothing to prevent any reasonable probability of skin contact. Safety equipment suppliers/manufacturers can provide recommendations on the most protective glove/clothing material for your operation. Neoprene and Polyvinyl Chloride are among the recommended protective materials. All protective clothing (suits, gloves, footwear, headgear) should be clean, available each day, and put on before work. Contact lenses should not be worn when working with this chemical. Because of poor warning signs, it may cause olfactory paralysis, and some persons are congenitally unable to smell H_2S. Accidental exposure may occur when workers enter sewage tanks and other confined areas in which hydrogen sulfide is formed by decomposition. In a number of cases workers enter unsuspectingly and collapse almost immediately. Workers, therefore, should not enter enclosed spaces without proper precautions. All Federal standard and other safety precautions must be observed when tanks or other confined spaces are to be entered. In areas where the exposure to hydrogen sulfide exceeds the standards, workers should be provided with fullface canister gas masks or preferable supplied air respirators. When liquid H_2S is involved, wear clothing to prevent skin freezing. Wear eye protection to prevent any reasonable probability of eye contact. Remove clothing immediately if wet or contaminated to avoid flammability hazard. Employees should wash immediately with soap when skin is wet or contaminated. Provide emergency showers and eyewash.

Respirator Selection: NIOSH/OSHA: *Up to 100 ppm:* PAPRS [any powered, air-purifying respirator with cartridge(s) providing protection against the compound of concern]; or GMFS [any air-purifying, full-facepiece respirator (gas mask) with a chin-style, front- or back-mounted canister providing protection against the compound of concern]; or SA (any supplied-air respirator); or SCBAF (any self-contained breathing apparatus with a full facepiece). *Emergency or planned entry into unknown concentrations or IDLH conditions:* SCBAF: PD,PP (any self-contained breathing apparatus that has a full facepiece and is operated in a pressure-demand or other positive-pressure mode); or SAF:PD,PP:ASCBA (any supplied-air respirator that has a full facepiece and is operated in a pressure-demand or other positive-pressure mode in combination with an auxiliary self-contained breathing apparatus operated in a pressure-demand or other positive pressure mode). *Escape:* GMFS [any air-purifying, full-facepiece respirator (gas mask) with a chin-style, front- or back-mounted canister providing protection against the compound of concern]; or SCBAE (any appropriate escape-type, self-contained breathing apparatus).

Note: Substance reported to cause eye irritation or damage; may require eye protection.

Storage: Prior to working with hydrogen sulfide you should be trained on its proper handling and storage. Before entering confined space where this chemical may be present, check to make sure that an explosive concentration does not exist. Hydrogen sulfide must be stored to avoid contact with strong oxidizers (such as chlorine, bromine and fluorine) and nitric acid since violent reactions occur. Outdoor or detached storage is preferred. Indoors, store in a cool, well-ventilated area. Sources of ignitions, such as smoking and open flames, are prohibited where hydrogen sulfide is handled, used, or stored. Use only non-sparking tools and equipment, especially when opening and closing containers of hydrogen sulfide. Wherever hydrogen sulfide is used, handled, manufactured, or stored, use explosion-proof electrical equipment and fittings. Procedures for the handling, use and storage of cylinders should be in compliance with OSHA 1910.101 and 1910.169 as with the recommendations of the Compressed Gas Association.

Shipping: This compound requires a shipping label of: "Poison Gas, Flammable Gas." It falls in DOT Hazard Class 2.3 and Packing Group I. The limit on passenger aircraft or rail-car shipment is forbidden and the limit on cargo aircraft shipment is forbidden as well.

Spill Handling: Evacuate and restrict persons not wearing protective equipment from area of spill or leak until cleanup is complete. If in a building, evacuate building and confine vapors by closing doors and shutting down HVAC systems. Restrict persons not wearing protective equipment from area of spill or leak until cleanup is complete. Remove all ignition sources. Establish forced ventilation to keep levels below explosive limit. Wear chemical protective suit with self-contained breathing apparatus to combat spills. Stay upwind and use water spray to "knock down" vapor; contain runoff. Stop the flow of gas, if it can be done safely from a distance. If source is a cylinder and the leak cannot be stopped in place, remove the leaking cylinder to a safe place, and repair leak or allow cylinder to empty. Keep this chemical out of confined spaces, such as a sewer, because of the possibility of explosion, unless the sewer is designed to prevent the buildup of explosive concentrations. If employees are required to clean-up spills, they must be properly trained and equipped. OSHA 1910.120(q) may be applicable.

Initial isolation and protective action distances

Distances shown are likely to be affected during the first 30 minutes after materials are spilled and could increase with time. If more than one tank car, cargo tank, portable tank, or large cylinder is involved in the incident is leaking, the protective action distance may need to be increased. You may need to seek emergency information from CHEMTREC at (800) 424-9300 or seek professional environmental engineering assistance from the U.S. EPA Environmental Response Team at (908) 548-8730 (24-hour response line).

Small spills (From a small package or a small leak from a large package)

First: Isolate in all directions (feet)..... 200

Then: Protect persons downwind (miles)

Day 0.1

Night 0.3

Large spills (From a large package or from many small packages)

First: Isolate in all directions (feet)..... 400

Then: Protect persons downwind (miles)

Day 0.2

Night 0.9

Fire Extinguishing: A very flammable gas. This chemical is a flammable gas. Poisonous gases including sulfur dioxide are produced in fire. Do not extinguish the fire unless the flow of gas can be stopped and any remaining gas is out of the line. Specially trained personnel may use fog lines to cool exposures and let the fire burn itself out. Vapors are heavier than air and will collect in low areas. Vapors may travel long distances to ignition sources and flashback. Vapors in confined areas may explode when exposed to fire. Containers may explode in fire. Storage containers and parts of containers may rocket great distances, in many directions. If material or contaminated runoff enters waterways, notify downstream users of potentially contaminated waters. Notify local health and fire officials and pollution control agencies. From a secure, explosion-proof location, use water spray to cool exposed containers. If cooling streams are ineffective (venting sound increases in volume and pitch, tank discolors, or shows any signs of deforming), withdraw immediately to a secure position. If cylinders are exposed to excessive heat from fire or flame contact, withdraw immediately to a secure location. If employees are expected to fight fires, they must be trained and equipped in OSHA 1910.156.

Disposal Method Suggested: Hydrogen sulfide can be recovered as such or converted to elemental sulfur or sulfuric acid. Consult with environmental regulatory agencies for guidance on acceptable disposal practices. Generators of waste containing this contaminant ($\geq$100 kg/mo) must conform with EPA regulations governing storage, transportation, treatment, and waste disposal.

References

National Institute for Occupational Safety and Health, Criteria for a Recommended Standard: Occupational Exposure to Hydrogen Sulfide, NIOSH Doc. No. 77–158 (1977)

U.S. Environmental Protection Agency, Hydrogen Sulfide, Health and Environmental Effects Profile No. 118, Office of Solid Waste, Washington, DC (April 30, 1980)

Sax, N. I., Ed., Dangerous Properties of Industrial Materials Report, 3, No. 4, 68–73 (1983)

U.S. Environmental Protection Agency, "Chemical Profile: Hydrogen Sulfide," Washington, DC, Chemical Emergency Preparedness Program (Nov. 30, 1987)

New Jersey Department of Health and Senior Services, "Hazardous Substance Fact Sheet: Hydrogen Sulfide," Trenton, NJ (May 1986)

New York State Department of Health, "Chemical Fact Sheet: Hydrogen Sulfide," Albany, NY, Bureau of Toxic Substance Assessment (March 1986 and Version 2)

U.S. Department of Health and Human Services, "ATSDR ToxFAQs, Hydrogen Sulfide," Atlanta, GA (June 1999)

Hydroquinone

Molecular Formula: $C_6H_6O_2$

Common Formula: $C_6H_4(OH)_2$

Synonyms: Arctuvin; Benzene, *p*-dihydroxy-; *p*-Benzenediol; 1,4-Benzenediol; Benzohydroquinone; Benzoquinol; Black and white bleaching cream; Boydes PTS developer; Cronaflex PDC developer; DIAK5; Dihydroquinone; 1,4-Dihydroxybenzen (Czech); *p*-Dihydroxybenzene; 1,4-Dihydroxybenzene; Dihydroxybenzene; 1,4-Dihydroxy-benzol (German); 1,4-Dihyroxy-benzeen (Dutch); 1,4-Diidrobenzene (Italian); *p*-Dioxobenzene; *p*-Dioxybenzene; Eldopaque; Eldoquin; HE5; Hidroquinona (Spanish); Hydrochinon (Czech, Polish); Hydroquinol; α-Hydroquinone; *p*-Hydroquinone; *p*-Hydroxyphenol; 4-Hydroxyphenol; Idrochinone (Italian); Kodagraph liquid developer; Kodak 55/66 developer; NCI-C55834; PD-86 developer; Phiaquin; Pyrogentisic acid; β-Quinol; Quinol; SR-201; Tecquinol; Tenox HQ; Tequinol

CAS Registry Number: 123-31-9

RTECS Number: MX3500000

DOT ID: UN 2662

EEC Number: 604-005-00-4

EINECS Number: 204-617-8

Regulatory Authority

- Air Pollutant Standard Set (ACGIH)[1] (DFG)[3] (HSE)[33] (former USSR)[35] (OSHA)[58] (Several States)[60]

- CLEAN AIR ACT: Hazardous Air Pollutants (Title I, Part A, Section 112)
- SUPERFUND/EPCRA 40CFR355, Appendix B Extremely Hazardous Substances: TPQ = 500/10,000 lb (227/4,540 kg)
- SUPERFUND/EPCRA 40CFR302.4 Reportable Quantity (RQ): CERCLA, 1 lb (0.454 kg)
- EPCRA Section 313 Form R *de minimis* concentration reporting level: 1.0%
- Canada, WHMIS, Ingredients Disclosure List

Cited in U.S. State Regulations: Alaska (G), California (A, G), Connecticut (A), Florida (G, A), Illinois (G), Maine (G), Maryland (W), Massachusetts (G), Michigan (G), Nevada (A), New Hampshire (G), New Jersey (G), New York (A), North Dakota (A), Rhode Island (G), South Carolina (A), Virginia (A), West Virginia (G).

Description: Hydroquinone, $C_6H_4(OH)_2$, is alight-tan, light-gray, or colorless crystals. Boiling point =286°C. Freezing/ Melting point = 172 – 174°C. Flash point =165°C. Autoignition temperature = 516°C. Hazard Identification (based on NFPA-704 M Rating System): Health 2, Flammability 1, Reactivity 0. Insoluble in water.

Potential Exposure: Hydroquinone is a reducing agent and is used as a photographic developer and as an antioxidant or stabilizer for certain materials which polymerize in the presence of oxidizing agents. Many of its derivatives are used as bacteriostatic agents, and others, particularly 2,5-bis(ethyleneimino) hydroquinone, gave been reported to be good antibiotic and tumor-inhibiting agents.

Incompatibilities: Hydroquinone is a reducing agent. Dust forms an explosive mixture with air. May explode on contact with oxygen. Incompatible with strong oxidizers, caustics; reacts violently with sodium hydroxide. May be oxidized to quinone at room temperatures in the presence of moisture.

Permissible Exposure Limits in Air: The Federal standard,[58] the DFG MAK[3] and the HSE TWA[33] and the recommended ACGIH TWA value is 2 mg/m^3. The STEL value set by HSE[33] is 4 mg/m^3. The NIOSH IDLH level is 50 mg/m^3. The TWA set in Sweden[35] is 0.5 mg/m^3 and the STEL is 1.5 mg/m^3. The former USSR has set a MAC in ambient air in residential areas of 0.02 mg/m^3 on a once-daily basis.[35] Several states have set guidelines or standards for hydroquinone in ambient air[60] ranging from 6.67 μg/m^3 (New York) to 20.0 μg/m^3 (Florida and South Carolina) to 20.0 – 40.0 μg/m^3 (North Dakota) to 35.0 μg/m^3 (Virginia) to 40.0 μg/m^3 (Connecticut) to 48.0 μg/m^3 (Nevada).

Determination in Air: Collection on a filter; workup with CH_3COOH; analysis by high-pressure liquid chromatography/ ultraviolet detection; NIOSH IV, Method #5004.

Permissible Concentration in Water: The former USSR[35] has set a MAC in water bodies used for domestic purposes of 0.5 mg/l.

Routes of Entry: Inhalation of dust, ingestion, eye and/or skin contact.

Harmful Effects and Symptoms

Short Term Exposure: Irritates the eyes, skin, and respiratory tract. Can cause conjunctivitis; keratitis (inflammation of the cornea). Can affect the central nervous system and cause excitement. Signs and symptoms of acute exposure to hydroquinone may be severe and include dyspnea (shortness of breath), a sense of suffocation, increased respiratory rate, and respiratory failure. Pallor (paleness of skin), cyanosis (blue tint to skin and mucous membranes), and cardiovascular collapse may occur. Neurologic effects include headache, tinnitus (ringing in the ears), dizziness, delirium, muscle twitching, tremor, and convulsions. Nausea, vomiting, and the production of green to brown-green urine may also occur. Ingestion may cause respiratory failure. This material is very toxic; the probable oral lethal dose for humans is 50 – 500 mg/kg, or between 1 teaspoon and 1 ounce for a 150 lb person. Fatal human doses have ranged from 5 – 12 grams, but 300 – 500 mg have been ingested daily for 3 – 5 months without ill effects. Death is apparently initiated by respiratory failure or anoxia.

Long Term Exposure: Repeated or prolonged contact may cause skin sensitization. Hydroquinone cause changes in color of the conjunctiva, cornea, and skin. Over the years this can cause clouding of the eyes and permanent vision damage. May cause genetic damage, mutations in humans. Such chemicals have a cancer risk.

Points of Attack: Eyes, skin, respiratory system, central nervous system.

Medical Surveillance: Careful examination of the eyes, including visual acuity and slit lamp examinations, should be carried out in preplacement and periodic examinations. Also the skin should be examined. Hydroquinone is excreted in the urine as a sulfate ester, although this has not been helpful in following worker exposure to dust.

First Aid: If this chemical gets into the eyes, remove any contact lenses at once and irrigate immediately for at least 15 minutes, occasionally lifting upper and lower lids. Seek medical attention immediately. If this chemical contacts the skin, remove contaminated clothing and wash immediately with soap and water. Seek medical attention immediately. If this chemical has been inhaled, remove from exposure, begin rescue breathing (using universal precautions) if breathing has stopped and CPR if heart action has stopped. Transfer promptly to a medical facility. When this chemical has been swallowed, get medical attention. Give large quantities of water and induce vomiting. Do not make an unconscious person vomit.

Personal Protective Methods: Wear protective gloves and clothing to prevent any reasonable probability of skin contact. Safety equipment suppliers/manufacturers can provide recommendations on the most protective glove/clothing material for your operation. Neoprene, Natural Rubber, and Viton are among the recommended protective materials. All protective clothing (suits, gloves, footwear, headgear) should be clean, available each day, and put on before work. Contact lenses should not be worn when working with this chemical. Wear dust-proof chemical goggles and face shield unless full face-piece respiratory protection is worn. Employees should wash immediately with soap when skin is wet or contaminated. Provide emergency showers and eyewash. Specific engineering controls are recommended for this chemical by NIOSH. See NIOSH Criteria Document #78–155.

Respirator Selection: NIOSH/OSHA: *Up to 50 mg/m³:* PAPRD (any powered, air-purifying respirator with a dust filter); or HiEF (any air-purifying, full-facepiece respirator with a high-efficiency particulate filter); or SAT:CF (any supplied-air respirator that has a tight-fitting facepiece and is operated in a continuous-flow mode); or SCBAF (any self-contained breathing apparatus with a full facepiece); or SAF (any supplied-air respirator with a full facepiece). *Emergency or planned entry into unknown concentrations or IDLH conditions:* SCBAF:PD,PP (any self-contained breathing apparatus that has a full faceplate and is operated in a pressure-demand or other positive-pressure mode); or SAF:PD,PP: ASCBA (any supplied-air respirator that has a full facepiece and is operated in a pressure-demand or other positive-pressure mode in combination with an auxiliary self-contained breathing apparatus operated in a pressure-demand or other positive pressure mode). *Escape:* HiEF (any air-purifying, full-facepiece respirator with a high-efficiency particulate filter); or SCBAE (any appropriate escape-type, self-contained breathing apparatus).

Note: Substance causes eye irritation or damage; eye protection needed at 50 and 100 mg/m³.

Storage: Prior to working with this chemical you should be trained on its proper handling and storage. Hydroquinone must be stored to avoid contact with sodium hydroxide since violent reactions occur. Store in tightly closed containers in a cool, well-ventilated area away from oxidizing materials. Where possible, automatically pump liquid from drums or other storage containers to process containers.

Shipping: This compound requires a shipping label of: "Keep Away From Food." It falls in DOT Hazard Class 6.1 and Packing Group III.

Spill Handling: Evacuate persons not wearing protective equipment from area of spill or leak until clean-up is complete. Stay upwind; keep out of low areas. Remove all ignition sources. Do not touch spilled material; stop leak if you can do it without risk. Small spills: take up with sand or other non-combustible absorbent material and place into containers for later disposal. Small dry spills: with clean shovel place material into clean, dry container and cover; move containers from spill area. Large spills: dike far ahead of spill for later disposal. Ventilate area after clean-up is complete. It may be necessary to contain and dispose of this chemical as a hazardous waste. If material or contaminated runoff enters waterways, notify downstream users of potentially contaminated waters. Contact your Department of Environmental Protection or your regional office of the federal EPA for specific recommendations. If employees are required to clean-up spills, they must be properly trained and equipped. OSHA 1910.120(q) may be applicable.

Fire Extinguishing: Hydroquinone may burn but does not readily ignite. For small fires use dry chemical, carbon dioxide, water spray or foam. Move container from fire area if you can do so without risk. Poisonous gases are produced in fire. If material or contaminated runoff enters waterways, notify downstream users of potentially contaminated waters. Notify local health and fire officials and pollution control agencies. Containers may explode in fire. From a secure, explosion-proof location, use water spray to cool exposed containers. If cooling streams are ineffective (venting sound increases in volume and pitch, tank discolors, or shows any signs of deforming), withdraw immediately to a secure position. If employees are expected to fight fires, they must be trained and equipped in OSHA 1910.156.

Disposal Method Suggested: Incineration (1,800°F, 2.0 sec minimum), then scrub to remove harmful combustion products.[22]

References

National Institute for Occupational Safety and Health, Criteria for a Recommended Standard: Occupational Exposure to Hydroquinone, NIOSH Doc. No. 78–155 (1978)

Sax, N. I., Ed., Dangerous Properties of Industrial Materials Report, 2, No. 2, 35–37 (1982) and 8, No. 1, 51–60 (1988)

New Jersey Department of Health and Senior Services, "Hazardous Substance Fact Sheet: Hydroquinone," Trenton, NJ (July 1996)

Hydroxylamine

Molecular Formula: H_3NO

Common Formula: $HONH_2$

Synonyms: Oxammonium

CAS Registry Number: 7803-49-8

RTECS Number: NC2975000

DOT ID: No citation.

EEC Number: 612-122-00-7

Regulatory Authority

OSHA 29CFR1910.119, Appendix A, Process Safety List of Highly Hazardous Chemicals, TQ = 2,500 lb (1,135 kg)

Cited in U.S. State Regulations: California (G), Florida (G), Massachusetts (G), New Hampshire (G), New Jersey (G), Pennsylvania (G), Rhode Island (G).

Description: Hydroxylamine, $HONH_2$ is a white crystalline substance. Boiling point =70°C (decomposes below BP). Freezing/Melting point = 32°C; it is very hygroscopic and unstable. Flash point = 129°C (explosive). Autoignition temperature = 265°C. Soluble in water.

Potential Exposure: Those involved in chemical synthesis or use of hydroxylamine as a reducing agent.

Incompatibilities: Self reactive. Contaminants, temperatures above 65°C, or open flame can cause explosive decomposition, especially in presence of moisture and carbon dioxide. Incompatible with strong acids, organic anhydrides, isocyanates, aldehydes, sodium, finely divided zinc, some metal oxides. Aqueous solution is a weak base. Contact with strong oxidizers may cause a fire and explosion hazard. Attacks some metals. Contact with calcium or zinc forms a heat-sensitive explosive [bis(hydroxylamide)] (Sax).

Permissible Exposure Limits in Air: No standards set.

Permissible Concentration in Water: No criteria set.

Routes of Entry: Inhalation and ingestion of dust.

Harmful Effects and Symptoms

Short Term Exposure: Corrosive to skin, eyes, and mucous membranes. May cause methemoglobinemia. The effects may be delayed. Exposure can cause headache; vertigo (dizziness); tinnitus (ringing of ear); dyspnea (difficult breathing); nausea and vomiting, cyanosis; proteinuria and hematuria; jaundice; restlessness and convulsion; yellowish brown deposit on conjunctiva and cornea; astigmatism; reddening of hair and exposed skin.

Long Term Exposure: Repeated or prolonged contact may cause skin sensitization and eczema. May affect the nervous system. May cause anemia. May cause liver damage.

Points of Attack: Skin, blood, liver, nervous system.

Medical Surveillance: Annual physical exams including renal and hepatic. Examination by a qualified allergist. Complete blood count (CBC). Examination of the nervous system.

First Aid: If this chemical gets into the eyes, remove any contact lenses at once and irrigate immediately for at least 15 minutes, occasionally lifting upper and lower lids. Seek medical attention immediately. If this chemical contacts the skin, remove contaminated clothing and wash immediately with soap and water. Seek medical attention immediately. If this chemical has been inhaled, remove from exposure, begin rescue breathing (using universal precautions) if breathing has stopped and CPR if heart action has stopped. Transfer promptly to a medical facility. When this chemical has been swallowed, get medical attention. If victim is conscious, administer water or milk. Do not induce vomiting.

Note to Physician: Treat for methemoglobinemia. Spectrophotometry may be required for precise determination of levels of methemoglobinemia in urine.

Personal Protective Methods: Wear protective gloves and clothing to prevent any reasonable probability of skin contact. Safety equipment suppliers/manufacturers can provide recommendations on the most protective glove/clothing material for your operation. All protective clothing (suits, gloves, footwear, headgear) should be clean, available each day, and put on before work. Contact lenses should not be worn when working with this chemical. Wear dust-proof chemical goggles and face shield unless full facepiece respiratory protection is worn. Employees should wash immediately with soap when skin is wet or contaminated. Provide emergency showers and eyewash.

Respirator Selection: Where the potential for exposure to this chemical, use a MSHA/NIOSH approved supplied-air respirator with a full facepiece operated in the positive pressure mode or with a full facepiece, hood, or helmet in the continuous flow mode, or use a MSHA/NIOSH approved self-contained breathing apparatus with a full facepiece operated in pressure-demand or other positive pressure mode.

Storage: Prior to working with this chemical you should be trained on its proper handling and storage. Protect against physical damage. Store in cool, noncombustible buildings and separate from oxidizing materials. Open airtight containers occasionally to relieve pressure from decomposition products.

Shipping: Hydroxylamine itself is not cited by DOT but hydroxylamine sulfate requires a shipping label of: "Corrosive." It falls in DOT Hazard Class 8 and Packing Group III.

Spill Handling: Evacuate persons not wearing protective equipment from area of spill or leak until clean-up is complete. Remove all ignition sources. Cover spill with sodium bisulfite and sprinkle with water. Collect material in the most convenient and safe manner and deposit in sealed containers. Ventilate area after clean-up is complete. It may be necessary to contain and dispose of this chemical as a hazardous waste. If material or contaminated runoff enters waterways, notify downstream users of potentially contaminated waters. Contact your Department of Environmental Protection or your regional office of the federal EPA for specific recommendations. If employees are required to clean-up spills, they must be properly trained and equipped. OSHA 1910.120(q) may be applicable.

Fire Extinguishing: Use extreme caution in approaching a fire because the material may explode when exposed to heat or flame. No attempt should be made to fight fires except for remotely activated fire extinguishing equipment. Evacuate the surrounding area. Water or foam may cause frothing. Use dry chemical, carbon dioxide, or alcohol foam extinguishers. Poisonous gases including nitrogen oxides and sulfur oxides are produced in fire. If material or contaminated runoff enters waterways, notify downstream users of potentially contaminated waters. Notify local health and fire officials and pollution control agencies. From a secure, explosion-proof location, use water spray to cool exposed containers. If cooling streams are ineffective (venting sound increases in volume and pitch, tank discolors, or shows any signs of deforming), withdraw immediately to a secure position. If employees are expected to fight fires, they must be trained and equipped in OSHA 1910.156.

Disposal Method Suggested: Add sodium bisulfite solution and flush to sewer; or incinerate.

References

Sax, N. I., E., Dangerous Properties of Industrial Materials Report, 2, No. 2, 37–39 (1982) and 8, No. 4, 34–39 (1988)

Hydroxypropyl Acrylate

Molecular Formula: $C_6H_{10}O_3$

Common Formula: CH_2=$CHCOOCH_2CHOHCH_3$

Synonyms: Acrylic acid 2-hydroxypropyl ester; HPA; β-Hydroxypropyl acrylate; 1,2-Propanediol 1-acrylate; Propylene glycol monoacrylate

CAS Registry Number: 999-61-1

RTECS Number: AT1925000

DOT ID: UN 1760

EEC Number: 607-108-00-2

Regulatory Authority

- Air Pollutant Standard Set (ACGIH)[1] (HSE)[33] (Several States)[60]

Cited in U.S. State Regulations: Alaska (G), California (A, G), Connecticut (A), Florida (G), Illinois (G), Maine (G), Massachusetts (G), Nevada (A), New Hampshire (G), New Jersey (G), North Dakota (A), Pennsylvania (G), Rhode Island (G), Virginia (A), West Virginia (G).

Description: 2-Hydroxypropyl acrylate, CH_2=CHCOO $CH_2CHOHCH_3$, is a clear to light-yellow liquid with a sweetish, solvent odor. Boiling point = 225°C. Flash point = 65°C. Explosive limits: LEL= 1.8%; UEL = unknown. Hazard Identification (based on NFPA-704 M Rating System): Health 3, Flammability 1, Reactivity 0. Soluble in water.

Potential Exposure: This material is used as a bifunctional monomer for acrylic resins; it is used as a binder in nonwoven fabrics and may be used in the production of detergent lube oil additives.

Incompatibilities: The substance can polymerize due to heating, initiators, UV light. Can become unstable at high temperatures and pressures or may react with water with some release of energy, but not violently. Reacts with oxidizers, strong acids, nitrates.

Permissible Exposure Limits in Air: There is no OSHA PEL. The NIOSH REL is 0.5 ppm (3 mg/m^3)TWA with "skin" notation. The HSE[33] and the ACGIH recommend a TWA value of 0.5 ppm (3 mg/m^3) with the notation "skin" indicating the possibility of cutaneous absorption. There is no STEL. Several states have set guidelines or standards for hydroxypropyl acrylate in ambient air[60] ranging from 30 $\mu g/m^3$ (North Dakota) to 50 $\mu g/m^3$ (Virginia) to 60 $\mu g/m^3$ (Connecticut) to 71 $\mu g/m^3$ (Nevada).

Determination in Air: No method available.

Permissible Concentration in Water: No criteria set.

Routes of Entry: Inhalation, skin contact.

Harmful Effects and Symptoms

Short Term Exposure: The substance is corrosive to the eyes, the skin, and the respiratory tract. Eye contact can cause burns and permanent damage. Inhalation of the vapors can cause pulmonary edema, a medical emergency that can be delayed for several hours. This can cause death. The oral LD_{50} rat is 250 gm/kg (moderately toxic).

Long Term Exposure: Repeated or prolonged contact with skin may cause dermatitis and eczema; sensitization.

Points of Attack: Skin, lungs.

Medical Surveillance: Lung function tests. Evaluation by a qualified allergist.

First Aid: If this chemical gets into the eyes, remove any contact lenses at once and irrigate immediately for at least 15 minutes, occasionally lifting upper and lower lids. Seek medical attention immediately. If this chemical contacts the skin, remove contaminated clothing and wash immediately with soap and water. Seek medical attention immediately. If this chemical has been inhaled, remove from exposure, begin rescue breathing (using universal precautions) if breathing has stopped and CPR if heart action has stopped. Transfer promptly to a medical facility. When this chemical has been swallowed, get medical attention. Give large quantities of water and do not induce vomiting. Do not make an unconscious person vomit. Medical observation is recommended for 24 – 48 hours after breathing overexposure, as pulmonary edema may be delayed. As first aid for pulmonary edema, a doctor or authorized paramedic may consider administering a corticosteroid spray.

Personal Protective Methods: Wear protective gloves and clothing to prevent any reasonable probability of skin contact. Safety equipment suppliers/manufacturers can provide recommendations on the most protective glove/clothing material for your operation. All protective clothing (suits, gloves, footwear, headgear) should be clean, available each day, and put on before work. Contact lenses should not be worn when working with this chemical. Wear splash-proof chemical goggles and face shield unless full facepiece respiratory protection is worn. Employees should wash immediately with soap when skin is wet or contaminated. Provide emergency showers and eyewash.

Respirator Selection: Where the potential for exposure to this chemical, use a MSHA/NIOSH approved supplied-air respirator with a full facepiece operated in the positive pressure mode or with a full facepiece, hood, or helmet in the continuous flow mode, or use a MSHA/NIOSH approved self-contained breathing apparatus with a full facepiece operated in pressure-demand or other positive pressure mode.

Storage: Prior to working with this chemical you should be trained on its proper handling and storage. Before entering confined space where hexanol may be present, check to make sure that an explosive concentration does not exist. Store in tightly closed containers in a cool, well ventilated area away from oxidizers and reducing agents. Where possible, automatically pump liquid from drums or other storage containers to process containers

Shipping: Label "Corrosive." Hazard Class 8.

Spill Handling: Evacuate and restrict persons not wearing protective equipment from area of spill or leak until cleanup is complete. Remove all ignition sources. Establish forced ventilation to keep levels below explosive limit. Absorb liquids in vermiculite, dry sand, earth, peat, carbon, or a similar material and deposit in sealed containers. Keep this chemical out of a confined space, such as a sewer, because of the possibility of an explosion, unless the sewer is designed to prevent the build-up of explosive concentrations. It may be necessary to contain and dispose of this chemical as a hazardous waste. If material or contaminated runoff enters waterways, notify downstream users of potentially contaminated waters. Contact your Department of Environmental Protection or your regional office of the federal EPA for specific recommendations. If employees are required to clean-up spills, they must be properly trained and equipped. OSHA 1910.120(q) may be applicable.

Fire Extinguishing: This chemical is a combustible liquid. Poisonous and acid fumes including acrylic acid, acrolein are produced in fire. Water may be ineffective. Use dry chemical, alcohol foam or carbon dioxide. Vapors are heavier than air and will collect in low areas. Vapors may travel long distances to ignition sources and flashback. Vapors in confined areas may explode when exposed to fire. Containers may explode in fire. Storage containers and parts of containers may rocket great distances, in many directions. If material or contaminated runoff enters waterways, notify downstream users of potentially contaminated waters. Notify local health and fire officials and pollution control agencies. From a secure, explosion-proof location, use water spray to cool exposed containers. If cooling streams are ineffective (venting sound increases in volume and pitch, tank discolors, or shows any signs of deforming), withdraw immediately to a secure position. If employees are expected to fight fires, they must be trained and equipped in OSHA 1910.156.